THE FOCAL ENCYCLOPEDIA OF PHOTOGRAPHY

THE FOCAL ENCYCLOPEDIA OF PHOTOGRAPHY

Desk Edition

McGRAW-HILL BOOK COMPANY
New York San Francisco Toronto

Library of Congress Catalog Card Number 69-11420
Code No. 21410

First Published	October 1956
Reprint	August 1957
Reprint	December 1958
The Caxton Edition	June 1959
Revised Reprint	June 1960
Caxton Reprint	September 1960
Desk Edition	September 1960
Desk Edition Reprint	April 1961
Reprint	June 1962
Fully Revised Edition	October 1965
Revised Desk Edition	January 1969
Desk Edition Reprint	June 1971
Desk Edition Reprint	August 1972
Desk Edition Reprint	October 1973

Printed in Great Britain by
Fletcher & Son Ltd, Norwich
and bound by
Richard Clay (The Chaucer Press) Ltd, Bungay, Suffolk

PREFACE TO THE REVISED EDITION

I

THE world of photography has changed more and faster during the last ten years than in any other decade of its history.

As it has become cheaper and easier to make better pictures, the use of the camera—one of the symbols of contemporary social evolution—has spread to new masses.

Professional photography has equally grown in dimension and scope serving commerce and industry. News photography is now an ever-present witness to events everywhere.

Photographic materials provide the most efficient means of recording and documentation in nearly all fields of business and professional activity. Contemporary reproduction and printing are unthinkable without them. New methods of education and tools of perception are based on their use.

Photography helps in pioneering research far beyond the limitations of sight. Cameras observe, isolate and interpret; they hold on to passing time, vary its pace and look into the unknown.

Photography, which used to mean just "taking pictures", now provides the most penetrating and all-embracing medium of communication and understanding by images.

New aims call for new means as in turn new means will point to new ends.

Sensitized materials which serve photography are now undergoing fundamental changes. They are no longer based solely on silver halide systems. Alternative mechanisms provide and evolve altogether new technologies and economic requirements of their own.

Not surprisingly, the dimensions, structure, and geography of the industries producing the new equipment and materials are appreciably affected by all that.

2

THE FOCAL ENCYCLOPEDIA OF PHOTOGRAPHY was first published October 1956; it was reissued with minor corrections eleven times within nine years.

To keep step with the developments in photography, not only was a major revision in depth and breadth inevitable—but it had to be envisaged on almost the same scale as the original production of the book. This work commenced early in 1962 and continued into July 1965.

Every effort has been made to ensure as lasting an expectation of usefulness for the Revised Edition as the First Edition proved to possess. The Editors were as much concerned with the question of what to include as with the balancing problem of what to reject. They were careful to account for progress as distinct from mere novelty. Similarly they were concerned with preserving information which may appear to be obsolescent but is basic to understanding.

3

Any revision of a reference work must be guided by the present needs of its users.

The changes that have come over photography pointed first of all to the need for paying particular attention to the new technologies, the new subjects to which they are applied and the new terminology which has thus emerged. By far the largest part of the new text that has gone into the Revised Edition had to be devoted to them.

A rising level of education also called for providing in depth factual material of a type which readers so often look for in reference books. A fair amount of additional historical, statistical and geographical information has been included for this purpose.

Thus whilst the spirit of the work has not changed, with the spread of the contents into new areas some shift of balance became inevitable as the text had grown from 1,300,000 to 1,700,000 words, and the number of illustrations from 1,500 to 1,750. The Revised Edition is one-third larger than the First Edition was.

4

The initiative and control of the revisions was largely in the hands of the Principal Technical Editor; the organization of the formidable amount of work rested with the Administrative Editor. Their efforts were aided by several new consultants, notably in the fields of colour, economics, education, science and techniques.

All contributors to the First Edition were invited to revise their original contributions and, whenever they were not in a position to do so, alternative experts were coopted. Their signatures were added (in brackets) to those of the contributors of the original articles to denote collective responsibility.

Articles devoted to the national aspects of photographic activities have been extended to cover recent developments and achievements. Their number has grown to twenty-eight.

Nearly one hundred new biographies have also been added.

The number of contributors—including authors of new articles and the revising experts—in the Revised Edition has grown from 197 to 276.

EDITORIAL BOARD

KEY TO SIGNATURES

A.C.	Arthur Cox
A.C.K.	A. C. Kerger
A.E.S.	Arnold E. Stanley
A.F.A.	A. Finnis Attwell
A.F.B.	A. F. Bucknell
A.F.C.B.	A. F. C. Brown
A.G.	Alison Gernsheim
A.G.B.	Anthony G. Bennett
A.H.	Arthur Hamer
A.Ja.	Alan Jackson
A.Jo.	Adam Johann
A.J.G.	A. J. Garratt
A.J.W.	Anthony John Hunting Weber
A.K.-K.	A. Kraszna-Krausz
A.L.M.S.	A. L. M. Sowerby
A.M.	Adolf Morath
A.N.	Alfred Neumann
A.O.	Alfredo Ornano
A.Pg.	André Page
A.Pr.	Arthur Pastor
A.P.J.	A. Philip Jenkins
A.R.A.	Albert R. Andrews
A.R.P.	A. R. Pippard
A.S.C.	A. S. Cross
A.S.H.	Arthur S. Hughes
A.S.W.	A. S. Wilson
A.W.G.	A. W. Gorrill
A.W.S.	A. W. Smith
B.A.	Bernard Alfieri
B.B.	Berthold Beiler
B.D.	Branibor Debeljković
B.H.G.	B. H. Grimes
B.K.	B. Kelley
B.M.	Betti Mautner
B.N.	Beaumont Newhall
B.T.	Barnard Thornton
B.U.	Bodo Ulrich
B.W.C.	Brian W. Coe
C.D.	Charles Duncan
C.D.M.	C. Douglas Milner
C.E.B.	Charles E. Brown
C.F.	Cecil Farthing
C.H.G.	Charles H. Giles
C.I.J.	C. I. Jacobson
C.J.V.R.	C. J. V. Roberts
C.M.	Corrado Marin
C.S.	C. Staff
C.T.H.	Cecil Thomas Holmes
C.W.G.W.	C. W. G. Walmsley
C.W.L.	C. W. Long
C.W.P.	C. Wilson Peck
D.A.	Donald Allen
D.A.H.	D. A. Harrington
D.A.S.	D. A. Spencer
D.A.W.	David A. Wilson
D.B.	Dick Boer
D.Br.	Dennis Brownbill
D.G.W.	D. G. Wright
D.H.O.J.	D. H. O. John
D.J.F.	Douglas James Fry
D.L.	Douglas Liversidge
D.L.H.	Dennis L. Hooker
D.M.	David Mindline
D.M.N.	D. M. Neale
D.P.C.	Dennis P. Craft
D.P.L.	Donald Patrick Leggatt
D.W.D.	David W. Dewhirst
E.A.M.	E. A. Miskin
E.B.	Ernest Boesiger
E.Bo.	E. Bovey
E.C.	Eric Coop
E.E.	Ezra Eliovson
E.E.L.	E. E. Loening
E.F.T.	E. F. Teal
E.H.	E. Heimann
E.M.	Eric Moore
E.M.H.	Eugene M. Hanson
E.V.	Ernö Vadas
F.A.M.	F. A. Mills
F.B.	Fred Barlow
F.G.W.	F. G. Wallis
F.H.S.	Frank H. Smith

KEY TO SIGNATURES

F.J.B.W.	F. J. B. Wall
F.P.	Frederick Purves
F.S.J.	F. S. Judd
F.W.B.	F. W. Beken
F.W.G.	Frederick William Goodman
G.B.H.	G. B. Harrison
G.B.M.	G. B. Macalpine
G.Cl.	G. Campbell
G.Cg.	Gordon Catling
G.C.B.	G. C. Brock
G.C.C.	G. C. Collins
G.C.F.	Geoffrey Carlton Farnell
G. de H.	Geoffrey de Havilland
G.E.L.	Grace Eleanor Lockie
G.G.T.	G. G. Twidle
G.H.C.	G. H. Cook
G.H.L.	George H. Lunn
G.H.S.	George H. Sewell
G.I.B.	Gerald I. Bearcroft
G.I.P.L.	G. I. P. Levenson
G.M.	Gérard Morisset
G.R.H.	G. R. Hall
G.S.D.	Gordon S. Drinkall
G.T.S.	G. T. Schwartz
G.W.	George Wells
G.W.A.	George W. Ashton
G.W.P.	G. W. Payne
H.	Viscount Hanworth
H.A.S.	Hans A. Schreiner
H.B.	Howard Byrne
H.B.J.C.	H. B. J. Cramer
H.Co.	H. Cooper
H.C.	H. Craeybeckx
H.E.D.	Horace E. Dobbs
H.E.E.	Harold E. Edgerton
H.G.	Helmut Gernsheim
H.G.C.	Hans G. Casparius
H.H.	Herbert Holmes
H.J.H.	Henry J. Howard
H.J.Wa.	H. J. Walls
H.J.Wi.	Herbert J. Williams
H.M.W.	H. M. Wagner
H.P.	H. Phillips
H.P.R.	H. P. Rooksby
H.R.	Herbert Rittlinger
H.R.V.	H. R. Verry
H.Sc.	H. Scott
H.S.	Harry Shorland
H.v.W.	Hugo van Wadenoyen
H.W.	H. Wolff
H.W.G.	H. W. Greenwood
H.W.H.	Howard W. Hoadley
H.Y.F.	Hector Y. Faita
I.A.	Ivor Ashmore
I.H.	Iván Hevesy
I.V.S.	Ippolit Vasilevich Sokolov
J.A.H.	J. A. Harrison
J.A.R.	J. A. Radley
J.Bl.	John Blaxland
J.Bo.	John N. Borwick
J.Br.	John Brown
J.B.W.	John Barras Walker
J.C.	Julien Caunter
J.D.C.	John D. Chittock
J.D.F.	J. Denis Forman
J.E.	John Eggert
J.E.S.	J. E. Sparshott
J.F.C.M.	John F. C. Morris
J.H.	John Hall
J.H.-D.	J. Home-Dickson
J.L.A.H.	J. L. A. Hunt
J.Mi.	James Mitchell
J.Mo.	J. Morgan
J.M.C.	Jose-Maria Casademont
J.M.W.	J. M. Whitaker
J.N.U.	J. N. Unwalla
J.N.W.	J. N. Whyte
J.S.F.	Joseph S. Friedman
J.V.	John Vickers
J.W.G.	J. W. Gates
K.B.	Keast Burke
K.C.M.S.	K. C. M. Symons
K.G.B.	K. G. Birch
K.J.H.	K. J. Habell
K.P.	K. Pfister
K.S.	Karl Sandels
K.S.M.	K. S. Meakin
L.A.M.	L. A. Mannheim
L.B.C.C.	Leonard Bruce Charles Cook
L.C.M.	L. C. Mason
L.E.H.	L. E. Hallett
L.F.G.	L. Fritz Gruber
L.G.F.	L. G. Frost
L.G.S.	Leslie G. Sandys
L.H.-F.	L. Herbert-Felton
L.H.N.	Leo. H. Narodny
L.M.F.	Leslie M. Froude
L.S.	Leonard Smith
L.V.	Lancelot Vining
M.A.	Maurice Ambler
M.B.C.	M. B. Cookson
M.C.-B.	Manuel Closa-Bosser
M.C.K.	M. C. Kennard
M.D.	Maurice Déribéré
M.F.	Max Factor
M.F.H.	Margaret F. Harker
M.G.F.	M. G. Fisher
M.H.T.	M. H. Tester
M.K.K.	Maurice K. Kidd
M.L.V.G.	Marie L. V. Gayler
N.L.	N. Lord
N.W.	Norah Wilson
O.B.	Oswell Blakeston
O.G.P.	Oliver G. Pike
O.H.	Othmar Helwich
O.R.C.	O. R. Croy

P.A.M.	Peter A. Merigold
P.B.	Pal Bence
P.B.C.	P. B. Clapham
P.B.M.	P. B. Meadway
P.C.P.	P. C. Poynter
P.G.	Pierre Gassmann
P.G.L.	P. G. Law
P.H.	Peter Hansell
P.J.	Philip Johnson
P.La.	Peter Laurie
P.R.-W.	P. Ransome-Wallis
P.S.	P. Sonthonnax
P.W.	P. Warlow
P.W.H.	Percy W. Harris
P.Z.A.	P. Z. Adelstein
R.A.	Rudolf Arnheim
R.A.H.	Rodney Angus Hoggard
R.B.F.	R. B. Flint
R.B.M.	R. B. Morris
R. de L.G.	Roland de L. Garlick
R.E.	Ronald Elvey
R.E.W.	R. E. Withrington
R.F.	Richard Farrand
R.G.	R. Groves
R.H.C.	R. Howard Cricks
R.H.H.	R. H. Herz
R.J.L.	Raymond John Lunnon
R.J.N.	R. J. North
R.K.	R. Knapmann
R.K.N.	Raisen K. Narusawa
R.L.T.	R. L. Taylor
R.May.	Roger Mayne
R.Md.	Ray Mackland
R.Mo.	Rolf Mortensen
R.M.F.	R. M. Fanstone
R.M.H.N.	R. M. H. Noble
R.McV.W.	R. McV. Weston
R.N.W.	Roger Noel Watson
R.R.	Robert Rose
R.S.	Ronald Spillman
R.Sh.	R. Shaw
R.W.B.T.	Ronald W. B. Truscott
R.W.D.	R. W. Dear
R.W.J.U.	R. W. J. Uffen
S.B.	S. Beaufoy
S.F.	Shinnasuke Fukushima
S.-h.S.	Shao-hua Shih
S.J.H.	Sydney J. Hargrave
S.K.	Shigene Kanamaru
S.M.	Steve McCutcheon
S.M.N.	Sidney M. Newhall
S.U.	Sadao Umemoto
T.A.L.	T. A. Longmore
T.H.J.	T. Herbert Jones
T.J.K.	T. J. Kucera
T.P.F.	Terence Pitts Fenby
T.R.	Tony Rose
U.V.B.	U. V. Bogaerde
V.A.Y.-G.	V. A. Yashtold-Govorko
V.C.W.	V. C. Wilson
V.G.	Vernon Gorter
V.J.	Václav Jirů
W.A.B.	Wolfgang A. Bajohr
W.C.	Walter Clark
W.D.E.	W. D. Emanuel
W.F.B.	W. F. Berg
W.H.D.	W. H. Dimsdale
W.L.W.	W. L. Wilcock
W.M.H.	W. M. Hampton
W.S.	W. Suschitzky
W.W.	Wilhelm Wohlfahrt
W.Z.H.	W. Z. Herz

A

A. Prefix used in scientific terminology to convey negation—e.g., aspherical, asymmetrical.

ABAXIAL. Term used in optics to denote points off the axis of an optical system. Rays near the axis are termed paraxial.

ABBE, ERNST, 1840–1905. German physicist. Professor at University of Jena. Joined the optical works of Carl Zeiss in 1866 and became partner in 1875. Encouraged technical glass laboratory of Schott and Genossen in Jena in 1884 from which came many notable optical glasses. Biography by F. Auerbach (1918). Collected works: Jena 1904–6, 2 vols.

ABBE NUMBER. Number expressing the extent to which the shorter and longer wavelengths of light are separated by refraction through a particular glass—i.e., the amount the glass disperses the various colours.

Abbe numbers, formulated by E. Abbe (1840–1905), in practice vary from about 25·5 for Double Extra Dense Flint glass to 60·3 for Hard Crown glass. The greater the number, the smaller the dispersion—i.e., the less the separation of the colours after refraction.

ABBREVIATIONS. The following abbreviations are commonly used in photographic literature.

A	Angstrom; (on cameras) advance (film), or open (camera back)
a	ampere
A.C.	alternating current
A.C./D.C.	alternating current and direct current—i.e., universal
A.I.I.P.	Associate of the Institute of Incorporated Photographers
A.P.S.A.	Associate of the Photographic Society of America
A.R.P.S.	Associate of the Royal Photographic Society
A.S.A.	American Standards Association
A.S.C.	American Society of Cinematographers
A.S.C.C.	American small centre cap (lamp fitting)
B	bulb or brief
B.C.	bayonet cap (lamp fitting)
B.F.I.	British Film Institute
B.K.S.	British Kinematograph Society
b. & w.	black and white
B.P.	British Pharmacopoeia
B.P.M.A.	British Photographic Manufacturers' Association Ltd.
B.S.I.	British Standards Institution
C	Centigrade
c.cm.	cubic centimetre
C.I.E.	Commission Internationale d'Eclairage (International Committee of Illumination)
c.f.	cut film
c.f.h.	cut film holder
cm.	centimetre
c.p.	candle power
C.U.	close-up (cine)
d.a.	delayed action
D. and P.	developing and printing
d.b.	double bayonet (fitting)
D.C.	direct current
D.G.Ph.	Deutsche Gesellschaft für Photographie (German Photographic Society)
DIN	Deutsche Industrie-Norm (German standards specification used for emulsion speed system)
d.v.f.	direct vision finder
d.w.	double weight
e.r.c.	ever-ready case
E.S.	Edison screw (lamp fitting)
F	Fahrenheit; fast (applied to flash synchronization)
f	focal length
f	aperture
F.B.K.S.	Fellow of the British Kinematograph Society

F.I.A.P. Fédération Internationale de l'Art Photographique (International Federation of Photographic Art)
F.I.I.P. Fellow of the Institute of Incorporated Photographers
fl. dr. fluid drachm
fl. oz. fluid ounce
f.p. focal plane
f.p.a. film-pack adapter
f.p.m. feet per minute
f.p.s. frames per second (cine); feet per second; focal plane shutter
F.P.S.A. Fellow of the Photographic Society of America
F.R.A.S. Fellow of the Royal Astronomical Society
F.R.M.S. Fellow of the Royal Microscopical Society
F.R.P.S. Fellow of the Royal Photographic Society
F.R.S. Fellow of the Royal Society
F.R.S.A. Fellow of the Royal Society of Arts
F.Z.S. Fellow of the Zoological Society (Great Britain)
γ gamma
g. gram (Continental abbreviation sometimes misapplied to grain)
G.D.L. Gesellschaft Deutscher Lichtbildner (Society of German Photographers)
G.E.S. Giant Edison Screw (lamp fitting)
gm. gram
gr. grain
H. & D. Hurter and Driffield (emulsion speed system)
I instantaneous
I.C.I. International Committee of Illumination
I.I.P. Institute of Incorporated Photographers
inf. (or ∞) infinity
i.p.s. inches per second (magnetic tape)
I.R. infra-red
K Kelvin degrees
kg. kilogram
km. kilometre
l. litre
l.c. leather case
lm. lumen
log. logarithm
l.p.m. lines per millimetre
L.S. long shot (cine); lumen second
μ micron
m. metre
M medium, (applied to flash synchronization)
M.B.K.S. Member of the British Kinematograph Society
M.E.S. midget Edison screw (lamp fitting)
mg. milligram
min. minim
ml. millilitre
mm. millimetre
mμ millimicron
M.P.S. Master Photographer's Association
MQ metol-hydroquinone
M.S. medium shot (cinematography)
MV filter monochromatic vision filter
n.m. nanometer (= 1/1,000,000 mm.)
N.P.L. National Physical Laboratory
ortho orthochromatic
oz. ounce
P.A. Photographic Alliance (of Great Britain)
pan panchromatic
P.D.A. Photographic Dealers' Association (Great Britain)
P.E. cell photo-electric cell
pH symbol for hydrogen ion concentration—a measure of acidity or alkalinity of a solution
P.O.P. print(ing) out paper
pref. prefocus
P.S.A. Photographic Society of America
pt. pint; part
PV filter panchromatic vision filter
qt. quart
RF rangefinder; rollfilm
r.p.m. revolutions per minute
r.p.s. revolutions per second
R.P.S. Royal Photographic Society (of Great Britain)
R.R. rapid rectilinear (lens)
S slow (applied to flash synchronization)
S.B.C. small bayonet cap (lamp fitting)
S.C.C. small centre cap (lamp fitting)
Sch. Scheiner (emulsion speed system)
sec. second
s.f. sheet film
S.F.T.A. Society of Film and Television Arts, Ltd.
SLR single lens reflex
SM speed midget (flash bulb); single metal (plate holder)
S.M.P.T.E. Society of Motion Picture and Television Engineers (U.S.A.)
sp. gr. specific gravity
S.P.S.E. Society of Photographic Scientists and Engineers (U.S.A.)
s/s same size
s.w. single weight
T time; transmission (stops)
T. & B. time and brief (or bulb)
TLR twin lens reflex
U.S. uniform system (diaphragm aperture numbering)
U.V. ultra-violet
V volt (electricity); (on German cameras), Vorlauf = delayed action (shutter)
V.P. vest pocket (film size = $1\frac{5}{8} \times 2\frac{1}{2}$ ins.)
w watt
w.a. wide angle (lens)
X instantaneous (applied to flash synchronization)
Z (on German shutters), Zeit = time

ABERRATIONS OF LENSES

A perfect lens would show the image of a point as a point, and a straight line as a straight line. But in practice, lenses are never perfect; they reproduce a point as a patch, and a straight line as a more or less curved band. Most of the trouble is caused by faults, or aberrations, inherent in the lens construction.

The designer controls most of the aberrations by combining a number of single lenses in such a way that the aberrations of one lens tend to be cancelled out by opposing aberrations in the others.

Aberrations which affect an image point on the axis of the lens are classified as axial aberrations. The principal axial aberrations are chromatic and spherical.

Chromatic Aberration. A prism deviates blue light more than red light. Simple lenses do the same thing with the result that blue light comes to a focus nearer to the lens than red light. This defect is called the chromatic aberration of the lens. The human eye is most sensitive to the light towards the middle of the spectrum, but some photographic emulsions are most sensitive to light at the blue end. So while a lens suffering from chromatic aberration may form an image that looks sharp, it gives a blurred, out-of-focus negative.

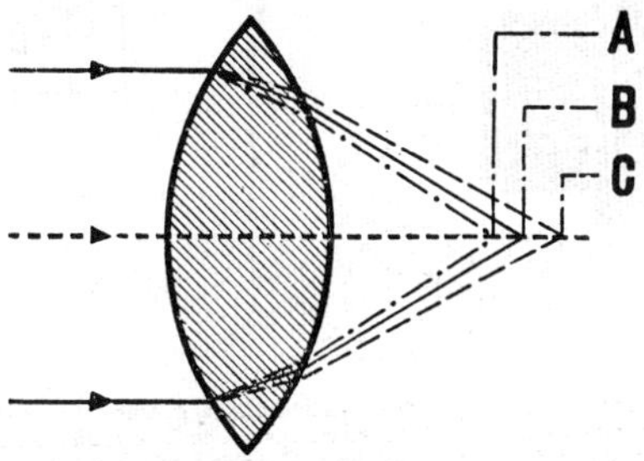

CHROMATIC ABERRATION. A simple lens refracts blue rays A more strongly than green rays B or red rays C The latter therefore come to a focus farther behind the lens, and the image shows reddish-bluish colour fringes.

To obtain any correction at all it is necessary to utilize different types of optical glass in one lens system. Unfortunately the characteristics of the optical glass available today are such that complete correction for all colours is impossible and the situation is aggravated by the fact that the balance of other aberrations, particularly spherical aberration, varies with the colour of the light. In many instances the best chromatic correction of a narrow bundle of rays close to the lens axis has to be sacrificed in order to obtain a better compromise correction over the whole lens aperture and this compromise correction may become unbalanced when the aperture is reduced by several stops.

Spherical Aberration. Rays coming from an object on the axis and going through the centre of the lens come to a focus at a certain point on the axis of the lens. Rays from an axial object going through the lens near the edges should come to a focus at the same point, but in practice, because of spherical aberration, they tend to come to a different point of focus. The difference between these focal points is the spherical aberration of the lens.

When the lens is more complex in construction it is possible to correct this aberration for

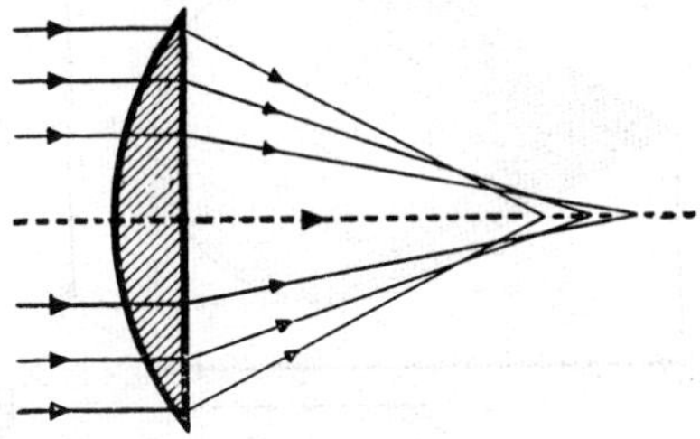

SPHERICAL ABERRATION. With a simple converging lens the rays farthest from the lens axis converge more strongly, and come to a focus nearer the lens than the central rays close to the lens axis. The image is never fully sharp.

rays passing through one selected annular zone of the lens leaving rays passing through other zones to cross the axis at positions slightly different from the position where the corrected rays and a narrow central bundle of rays come to a focus. The small residual errors are called zonal spherical aberration and, in the case of a corrected lens, performance in the centre of the negative will depend on their magnitude.

Spherical aberration increases with the lens aperture.

The spherical and chromatic aberrations affect the whole of the field covered by the lens, but there are others which affect only the off-axis parts of the field. The off-axis aberrations are: coma, astigmatism, curvature of field and distortion.

Coma. This aberration is restricted to off-axis image points. In some respects it can be

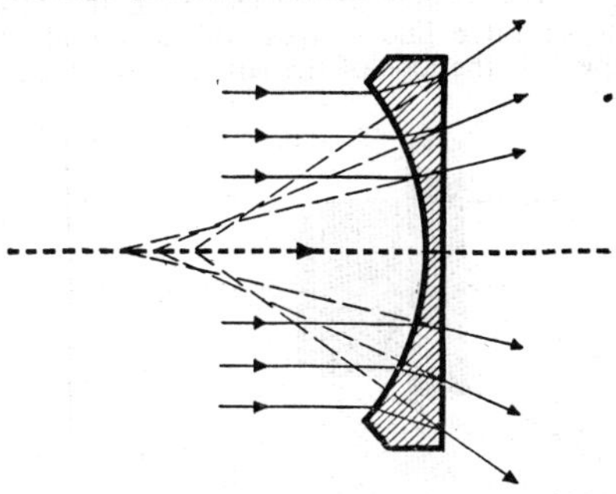

SPHERICAL ABERRATION. With a simple diverging lens marginal rays appear to come from points nearer the lens.

considered as spherical aberration of an oblique bundle of rays. Since the lens has no symmetry about a line passing through the centre of the lens and an off-axis point in the image, the effects of coma are complex and unsymmetrical. Complete lack of comatic correction results in an image of a bright point object having an unsymmetrical light patch flaring away to one side like the tail of a comet—hence the name coma. In a lens corrected for coma, residual comatic errors result in complicated light

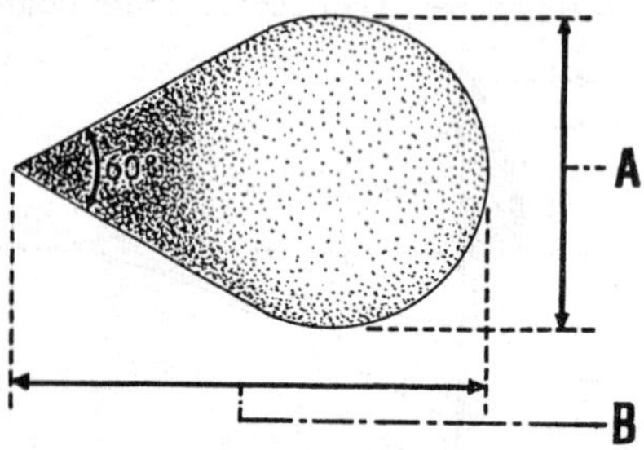

COMA. An uncorrected lens tends to reproduce off-axis image points as unsymmetrical patches flaring away to one side. A first order coma patch is still regular in shape, the length B being three times the width A. Higher order coma patches are more complicated and may have multiple tails.

patches corresponding to bright points in the object and these shapes can vary at different points in the field of view. This aberration is very difficult to eliminate in lenses covering wide angular fields at wide relative apertures and because of its complexity no hard and fast rules can be laid down.

The size of the coma flare in an uncorrected lens is directly proportional to the distance of the image point from the axis, and proportional to the square of the aperture of the lens.

Astigmatism. In a correctly centred lens system astigmatism cannot be present at an image point lying on the lens axis—it only affects definition away from the centre of the field. If astigmatism is the only aberration present, an object point will be imaged as two short straight lines occupying different positions in space, one nearer to the lens than the other. One of these line images will be radial with respect to the axis of the lens and the other at

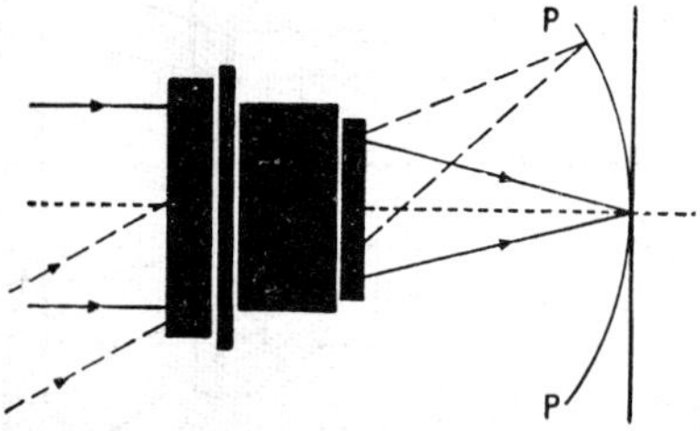

CURVATURE OF FIELD. When a lens shows curvature of field, all image points lie on a curved surface P instead of a flat plane.

right angles to it. The distance between them is a measure of the astigmatism that is present at that point in the field and the position of best focus is halfway between them.

The length of the lines and thus the deterioration of the image will increase with increased astigmatism and with wider relative aperture of the lens.

If the radial line is in the plane of the emulsion it will blurr edges of images which are tangential with respect to the centre of the field. If the other tangential line is in focus it will blurr radial image edges. In a corrected lens system astigmatism can be eliminated at points in the image corresponding to one semi-angular field of view and at other points there will be small residual errors of the same nature which are called zonal astigmatism.

As the field is increased beyond the corrected point astigmatic errors increase very rapidly and this is one of the reasons why lenses should always be used over the field for which they were designed.

Curvature of Field. Instead of a sharp image being formed on a flat plane, it is sometimes found that the sharpest image is formed on a curved surface. This is a most undesirable

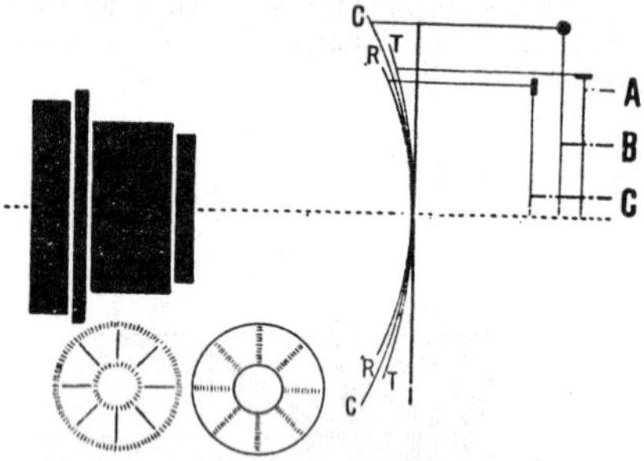

ASTIGMATISM. Top: With astigmatism present in a lens, transverse lines A are focused on a surface T, circular discs B on a surface C, and radial lines C on a surface R. Bottom: An object containing both transverse and radial lines can never show both sharp at the same time.

defect in photographic lenses since photographic plates are always flat and even films cannot easily be bent over a spherical surface. Curvature of field is very closely connected with the astigmatism of the lens and can be reduced by adjusting the astigmatism. In many simple lenses astigmatism is deliberately introduced as a compromise method of flattening curvatures which otherwise would not be acceptable. When both curvature and astigmatism are eliminated at a point well out in the field of view the lens can be termed an anastigmat.

Field curvature increases rapidly with the distance from the lens axis; it cannot be reduced by using a smaller lens aperture.

Transverse Chromatic Aberration. This is sometimes known as colour magnification error and only occurs away from the centre of the field. It

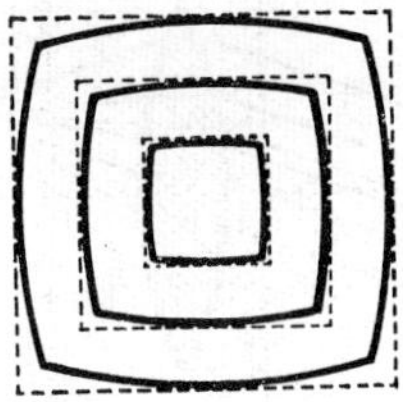

BARREL DISTORTION. *This results when the image magnification decreases as the distance of the image point from the axis increases. The off-axis image points are reproduced nearer to the axis than they should be.*

PIN-CUSHION DISTORTION. *This results when the image magnification increases as the distance of the image points from the axis increases. The off-axis image points are reproduced further from the axis than they should be.*

is independent of axial chromatic aberration and can exist when all the other aberrations are eliminated. It can be considered as a variation of the focal length of the lens for light of different colours which causes the scale of the image to vary with colour. When it is the only error present, an image of a bright object point out in the field appears as a series of points of different colours. This gives a coloured edge to the objects being photographed which affects definition when panchromatic or colour emulsions are used. Since this colour displacement is radial with respect to the lens axis it will be most apparent on image edges which are tangential with respect to the centre of the field.

Distortion. Distortion does not affect the sharpness of the image, it only alters the shape. It occurs because the off-axis part of the image is not reproduced on the same scale as the axial portion. If a test object consisting of a grid of straight lines is photographed, the lines in the photograph will be straight near the centre of the field, but they will curve inwards or outwards towards the edges. According to their shape, distortions are described as "barrel" or "pin-cushion."

Distortion is not affected by altering the size of the lens aperture.

Flare. In addition to aberrations, lenses sometimes suffer from flare. This is caused by light multiple reflections at the surfaces of the lens elements.

At each glass-air surface approximately 5 per cent of incident light is reflected and 95 per cent is transmitted. The reflected light will also have 5 per cent reflections in turn at each glass-air surface.

Any light with two reflections may reach the image area and reduce the contrast of the original image. Sometimes a bright light (or the sun) even if outside the field of the camera will form a patch in the image area. This form of flare is most objectionable. However, as the coating of lenses reduces surface reflection from 5 to 1 per cent there is not often serious flare with coated lenses.

A further source of flare light is stray reflections from the metal mounts of lenses. These reflections are minimised by blackening the interior of lens mounts.

Effect of Aperture. The position of the aperture is of importance, because it is one of the methods for controlling lens aberrations. The best position for the aperture depends upon the type of lens.

The usual type of anastigmatic lens has the aperture diaphragm between the lens components. Simple meniscus lenses usually have the diaphragm behind.

Spherical aberration is the only axial aberration that is appreciably affected by a change of aperture. But spherical aberration does not normally affect the definition of off—axis points as much as the other aberrations; so it

VARIATION OF ABERRATIONS IN UNCORRECTED LENSES

Aberration	*Variation with Image Size (h)*	*Variation with Aperture (A)*
Spherical aberration	Independent	A^2
Coma	h	A^3
Astigmatism (difference in focus)	h^2	Independent
Astigmatism (length of focal lines)	h^2	A
Curvature of field	h^2	Independent (but depth of field is improved by stopping down)*
Distortion	h^3	Independent
Chromatic Aberration	Independent	Independent

*The curvature of field is not affected by the size of the aperture, but making the aperture smaller does in fact lessen the effect simply because it increases the depth of focus.

may not seriously limit the maximum aperture of the lens. With an appreciable field, the coma and astigmatism are much more important, and these—in particular, the astigmatism—increase with aperture. In practice, the maximum useful aperture of a lens is primarily controlled by astigmatism and coma.

A maximum aperture of *f* 8 used to be considered good, but to-day an aperture of *f* 2·8 is regarded as average, and many good fast lenses work at *f* 1·5.

Off-axis aberrations increase rapidly with the angle of field and the normal photographic lens is only satisfactory for fields up to 50°. Special wide-angle lenses can cover fields of over 90° but they must work at comparatively small apertures.

Aberrations with Different Types of Lenses. The aberrations in a lens depend upon:

(*a*) the type of lens
(*b*) the skill of the designer
(*c*) the reliability and accuracy of the manufacture of the lens.

A single meniscus lens has large aberrations which are acceptable for some uses. It is easy to design and make.

A triplet lens is capable of much better performance at a larger aperture than a single meniscus lens, but design is more difficult and much greater accuracy is required during manufacture.

A lens consisting of four separated lens elements has significantly smaller aberrations than a triplet lens, but design and manufacture are more difficult.

A designer can reduce aberrations by using more complex designs, but a point is reached where the increased difficulties and errors of manufacture prevent any improvement in the reduction of aberrations.

Zoom lenses are complex mechanical—optical systems which are intended to:

(*a*) give a range of focal lengths
(*b*) have good correction for aberrations
(*c*) maintain accurate focus while zooming the lens.

These requirements are very difficult to combine and normally zoom lenses have larger aberrations than lenses of fixed focal length.

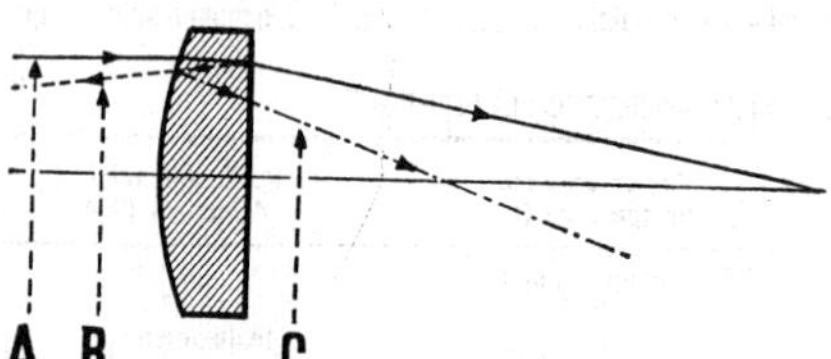

FLARE. Part of the light reaching any glass air surface in a lens is reflected back along its path. Successive reflections may direct some stray light to the film in places where it has no business to go and may even form ghost images. This is flare. A, incident light beam, producing the main image. B, light ray after one partial reflection. C, light ray after two partial reflections, forming flair.

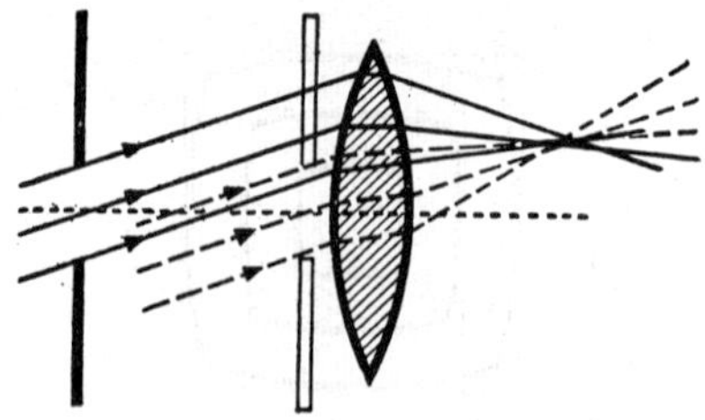

POSITION OF LENS STOP. In a simple lens aberrations can be reduced to some extent by careful choice of the position of the stop. This determines which part of the lens forms image points due to oblique rays (i.e., inclined to the axis).

Simple Lenses. A simple lens made of a single piece of glass, generally suffers from all the aberrations described above. Even so, lenses of this type are used in the cheaper cameras and give good results. The field is made reasonably flat by giving the lens a meniscus shape and by carefully choosing the position of the stop. The effects of spherical aberration, coma, astigmatism, and field curvature are reduced by keeping the aperture down to about *f* 13.

No attempt is made to correct the chromatic aberration because chromatic aberration is not a serious fault where the stop is so small and the negative is intended only for contact printing. For instance, a 4 ins. lens may have as much as 0·08 in. chromatic aberration, but this error corresponds to a circle of confusion of only 0·006-in. at *f* 13, which is quite sharp enough for contact prints.

Compound Lenses. Lenses intended for making negatives for enlargment or for colour photography have to be very carefully corrected. In consequence, they are complex and more expensive than the simpler lenses in use in the early days of photography. To produce a flat field, the designer may have to accept some astigmatism and possibly a little spherical aberration—he cannot afford much or a change in aperture of the lens will shift the focus. Coma and distortion can normally be fairly well corrected; chromatic aberration can be practically eliminated for the two important colours, yellow and blue. (There remains a small degree of chromatic aberration, described as residual.) The other colours may come to slightly different foci, but the error is negligible for ordinary photography. The infra-red focus, however, is quite different from the yellow-blue focus, and when a photographic lens is to be used for infra-red photography, the mount is specially marked with the infra-red focusing position.

G.H.C. & D.P.C.

See also: *Colour Correction of lenses; Lens history; Lens performance; Lens testing; Optical design.*

Books: *Photographic Optics* and *A System of Lens Design* both by Arthur Cox (London).

ABNEY, SIR WILLIAM, 1843–1920. English photo-chemist. Gave practical directions for manufacturing emulsions. Invented copper bromide-silver nitrate intensification, 1877. Introduced hydroquinone as a developer, 1880. Invented gelatin chloride printing-out paper (P.O.P.) 1882. Worked on photo-chemistry, sensitometry, densitometry, solarization, photometry, colorimetry, spectral analysis. Wrote numerous books and articles. President of the Photographic Society of Great Britain (which became "Royal" in 1894) for 5 years (1892–94, 1896, 1903–5). Received twice the Progress Medal of the R.P.S. (1878, 1890). Knighted in 1900. Was adviser in Science to the Board of Education.

ABRASION MARKS. Marks on a processed photographic material due to the effect of abrasion or milder stress either before or after exposure. Also known as pressure marks or stress marks.

Abrasion which actually tears the emulsion itself will usually result in black marks, due to the easier penetration of the developer and to the freeing of grains from the protective effect of the surrounding gelatin. Mild pressure which leaves the emulsion surface undisturbed will, if effected before exposure, desensitize the underlying grains. This effect cannot occur after exposure, so that post-exposure pressure will almost invariably result in a plus-density mark.

Marks of the type described are inseparable from the image and consequently cannot be selectively removed by after-treatment. Mild abrasion marks produced on a processed negative, i.e. simple surface scratches, may often be removed by careful rubbing with cotton wool soaked in methylated spirit. Such marks may often be avoided in the final print by the use of diffuse illumination in enlarging.

Alternatively the negative surface may be treated with a fluid which has a similar refractive index to that of gelatin, so that in effect the scratches are filled in and are not visible on projection. D.Br.

See also: *Faults.*

ABSOLUTE. Term used in special cases in chemistry to denote a pure product, e.g., absolute alcohol.

ABSOLUTE TEMPERATURE. Temperature measured from the Absolute Zero, i.e., the temperature at which virtually all molecular movement ceases (−273° C. or −459° F.). The degrees of the Absolute scale are spaced the same as Centigrade degrees; to turn the latter into °Absolute, add 273. Their only significance to photographers is that all colour temperatures are expressed in terms of degrees on the Absolute scale (also known as the Kelvin scale).

See also: *Colour temperature.*

ABSORPTION. All surfaces absorb some of the light that falls on them. Some, like lampblack, absorb almost all the light; others like white paper, absorb very little. The light energy absorbed is usually converted into heat, but it may be transformed into light of another colour and sent out again or it may produce a chemical change, as in a photographic emulsion.

Colour is the result of selective absorption of light rays of certain wavelengths. If white light falls on a surface that absorbs the red and green rays, the surface appears blue to the eye because that is the only visible light reflected by the surface.

ABSORPTION CURVE. Graphic representation of proportion of light absorbed by a medium, plotted against the wavelength of the light. Generally used to give a scientific indication of the characteristics of colour filters and dyes in colour films.

See also: *Filters.*

ABSTRACT PHOTOGRAPHY

The camera is basically very like the eye. It records the relations of light and shade and draws the relative positions and sizes of various parts of the subject in scientifically correct perspective. So long as the whole photographic process is carried out correctly it gives what is generally accepted to be a true transcription of the scene as it appeared from a given viewpoint. The results are entirely realistic.

This objective realism is very different from the subjective versions of the world produced by the painters, many of whom nowadays hardly ever work direct from nature, but draw their images from impressions absorbed, stored and synthesized in the mind. They commonly express themselves in distorted representations of real things, or even in completely unrecognizable abstract forms.

In the hands of an artist, the camera too, can be made to produce either representational or abstract pictures of aesthetic worth.

There is no rigid dividing line between the two types; whether the picture appears to be representational or abstract will, in some cases, depend on the viewer. Thus specialized photographs made by the scientist may appear entirely abstract when seen by the layman to whom the subject is unfamiliar.

A photograph, though intended to be purely representational, may nevertheless have a strong abstract element in it. This is especially noticeable when an object, possibly of unusual form and texture, is photographed from an unfamiliar angle, or so close up that it can hardly be recognized, or in strange lighting conditions, or with a lack of local colour, or by the use of some special technique, such as infra-red.

Early Examples. Many techniques known in early days produced straight photographs with marked abstract qualities. Geological photographs, for example, revealed the abstract beauty of rock formations and strata. This form of abstraction was used from about 1930 with great effect by Edward Weston in his pictures of rocks, sand dunes and close-ups of pebbles and tree forms. Fifteen years earlier, Paul Strand had revealed the beauty of abstract forms.

Studies of motion produced another peculiarly photographic form of abstraction. As early as 1881 Eakins, Muybridge's collaborator, took superimposed action shots which foreshadowed the multiple electronic flash pictures taken later by Professor Edgerton of such subjects as a golf swing and a drop of milk splashing into a bowl. Many photographs of this type, though made for purely scientific purposes, possess fascinating aesthetic qualities.

The first who deliberately made purely abstract photographs was Alvin Langdon Coburn. In 1917 he made abstracts by taking small objects as seen in the middle of three mirrors, clamped together as in a kaleidoscope: he called them vortographs.

In 1918 Christian Schad made schadographs by placing flat objects and cut-outs on the sensitive emulsion and then exposing it to light. These pictures resembled the cubist collages of the period. This and similar methods of producing images without the use of a camera, developed in 1922 by Man Ray, Moholy-Nagy and others, are known as rayographs and photograms. Man Ray and Moholy Nagy were pioneers also in more strictly photographic methods of making abstracts, such as the use of negative prints, multiple exposures on either plate or paper, and the process known as solarization.

Recent Examples. There has been a considerable development of abstract photography since the war. This is probably a result of the increased acceptance of the more abstract forms of painting, for photographers are very prone to emulate contemporary trends in older arts.

Methods. Among the special techniques that create abstract images of rare beauty are many that can only be applied by scientists with laboratory equipment. While these are taken primarily for scientific reasons, pictures made by such methods have often been included in books of photographic art and are accepted at some advanced exhibitions.

Among these scientific techniques are: X-ray photography, X-ray diffraction photography, the use of the principle of diffraction to photograph air-waves made by a bullet, variations in temperature round a heated object, the stress in a rod, etc., photomicrography, the Lichtenberg figure made directly on film by an electric discharge in gas, astronomical photography, electron micography, high-speed electronic flash, and time-motion studies in which the movements of a worker are recorded by attaching lights to appropriate parts of the body.

But abstract photographs may be made by less ingenious means, without involved specialized techniques and equipment. For many of them only the simplest of cameras is necessary. But the photographer must possess an experimental turn of mind, aesthetic sensibility and imagination.

A camera with focusing screen and with ample extension is almost an essential for some abstraction techniques. It encourages the beginner to study the photographic image on the screen rather than the object being photographed. The camera should, however, be mobile so that the subject may be tackled from any promising angle. A fixed lens reflex camera is excellent in many ways, but has obvious limitations. Most of these are overcome by the single lens reflex camera, which takes fully interchangeable lenses. Other roll film and miniature cameras can of course be used, but as the optical finder does not give the typically photographic image, this has to be visualized in the imagination. The small size may be a drawback in cases where it is planned to apply special after-technique—e.g., solarization.

Generally the urge to abstraction starts with the desire to imitate the rather obvious type of pattern picture that appears nowadays in all the salons—photographs that pick out in close-up or semi-close-up clearly marked lines of rounded kerb, or intersecting curves of plates or other staged still-life objects. These subjects may be treated in a realistic way with full tonal range or may be in high-key.

Both the simplification of subject matter (primary control) and the modification of tones (secondary control) are forms of abstraction which, carried further, can produce surprising and fascinating results. These provide the beginnings of the technique for abstraction.

Primary Controls. Among the effects to be studied are scale, viewpoint, sharpness, perspective, and lighting.

Scale—i.e., the size in relation to the frame—should be studied to observe how the subject's importance is increased by coming closer; how detail can be stressed, how the perspective becomes more violent until the shape of the object is lost and only local surface features remain in almost unrecognizable form.

The viewpoint or angle of approach also changes the appearance of the subject: both in perspective and in the distribution of light and dark masses. A low viewpoint will seem to increase height and importance and looking down may produce a striking, unusual pattern.

Sharpness, blur and combinations of both may be created by suitably focusing the lens. Parts of the subject may be shown bitingly sharp, while more distant ones are left nebulous. This effect is common in extreme close-ups seen through a fairly large aperture and applied to small objects—e.g., fragments of broken glass or glittering Christmas tree decorations—produces fascinating evocative images. Extreme all-over sharpness, especially in very close-up shots concentrating on the surface pattern and texture of such subjects as wood grain, rock erosion and crumbling plaster, can also give many strange abstractions.

Perspective and proportion as controlled by the use of camera movements and optical devices can give further twists to the subject—e.g., by using (or misusing) the tilting of back or front, or both; changing the shapes of things by taking them as reflections in a distorting mirror or bent metal glazing plate; or through a distorting filter of wobbly glass, or through a crude, uncorrected lens.

Lighting effects also play their part in transforming the subject or in stressing certain aspects of it. Surface undulations can be emphasized by oblique side lighting, form can be simplified by strong contrasts in light and shade, or by depriving a part of the subject of light altogether and reducing it to a silhouette.

In dealing with small objects a good method is to use a "light box"—i.e., a box sealed except for holes through which spotlights can be directed and an opening for the camera lens. A variety of objects can be placed inside or suspended. In this way the subject and its shadows cast on to the walls can be made to form abstract patterns.

Out of doors lighting effects can be modified only by varying the direction of approach or waiting for the sun to move, but subjects can be found in unlimited number by looking at things through the camera from all sorts of angles. Any material can be used once the necessary photographic imagination has been cultivated.

Secondary Controls. In addition to the effects that can be directly seen on the focusing screen, the experienced abstract photographer can look at the image and visualize how he can change it in the final print by one or more of a number of after-processes—e.g., after it has been enormously enlarged, modified by a filter, or photographed on an infra-red emulsion. If the subject is moving or has moving parts, he will try to visualize it as wholly or partly blurred. He may even visualize it with another image superimposed on it.

The following are some of the methods that may be used. Many of them are processes of simplification. The aim may be to cut out half-tones and reduce the image to a simple statement in black and white, or to eliminate local colour in order to stress form, or to modify the colour rendering so as to set off one part of the subject against another.

Exposure and development are the first and most obvious controls. Over-exposure with full development applied to a flatly-lighted subject can give an interesting effect: printed on a contrast paper the lighter tones will be almost invisible, but the edges of objects and other darker areas will stand out boldly. The effect can be made stronger by applying Farmer's reducer to the negative. Under-exposure is useful for simplifying shadows and stressing the planes and ups and downs of the subject. If in addition the negative is greatly over-developed, there will be very fierce contrasts of black and white, especially if there was also contrasty lighting.

In flatly-lighted subjects contrast can be obtained by the use of blue-sensitive or special contrast emulsions, combined if necessary with prolonged development. This technique is useful for stressing local detail and texture; such things as the grain in wood or the surface texture of stone. Local colour patterns can be emphasized in various ways by the many filters available. Infra-red plates give great scope for creating weirdly unreal landscapes, and pictures of plant forms. They can also give surprising results when applied to other subjects: the results are not always predictable, but experiments are well worth while.

The negative can be modified either during or after development. Solarization is one of the most promising ways of making radical changes in the normal negative image. The process is, however, somewhat uncertain and it is always advisable not to use a valuable negative, but to experiment with duplicates. Enlargements of solarized negatives have a peculiar beauty of their own, abrupt contrasts are absent and contours appear as if outlined with a light crayon. Solarization gives the best effects with subjects having strong contrast and good outlines. The method can be applied equally well to positive transparencies, when the outline will print black.

The abstract qualities of the negative itself may be exploited by making a positive transparency from it to give a negative instead of a positive print. Or by placing negative and positive in contact, slightly out of register, and printing them together on a contrasty paper, the contours of the subject will be outlined in black or white. By altering the position of the positive on the negative, the width of line may be varied, and by varying the degree of hardness of the plates and printing paper, the print may be made in either strong black and white or grey tones.

Greatly enlarging a coarse-grained negative, or better still, deliberately inducing reticulation may destroy fine details and local textural qualities, thus simplifying the image and giving it a monumental quality. Grain and textural effects can also be introduced by printing through a texture screen, but the method of using a grainy negative or positive is more interesting, because the graininess forms an organic part of the image. A granulated negative will give a white network in the print, the positive a black one. Further effects may be created by combining negative and positive, one or both granulated.

Another technique that has been occasionally used with effect consists of distorting the image by heating a wet negative until the gelatin just melts and begins to run. The whole or part of the negative may be dipped in hot water, or heat may be applied locally by other means.

In enlarging, the negative or the paper, or both, can be tilted and the picture can be elongated or compressed. Or the paper can be curved or distorted in various ways. Opaque and semi-transparent masks suitably shaped make it possible to hold back or change the definition of certain parts.

Multiple Images. In addition to superimposing negative and positive versions of the same image, entirely different images can be combined in negative or positive form or both. They can be either enlarged together, or one printed and developed first, the print rinsed, but not fixed, and replaced on the enlarger for a further exposure from another negative. For further control, parts of the images may be held back in the printing or wholly or partially bleached away.

Superimposition can, of course, also be done direct in the camera, and every photographer will know that even accidental double exposures can sometimes be very attractive. When they are planned, an effort should be made to memorize the arrangement of masses in the first picture so that the second image falls in the best place to achieve the desired contrasts, balance and rhythm. Subjects with large, simple masses of shadow generally give the best results. Interesting effects may also be created by making multiple exposures on the same subject, and slightly moving—or better still, revolving—the camera between each exposure.

Odd combinations and strange contrasts can be obtained also by photo-montage done on a fairly large scale and then copying the mounted composite print. This method, however, unless done very skilfully, tends to look unconvincing and rather unphotographic, because the blending of the images is not really organic.

The recording by the camera of movement, moving objects and lights, has been extensively exploited for obtaining abstract images, either as blurs or clearly-defined patterns. And multiple exposures of arrested movement have also produced interesting abstractions. Complex linear patterns have been made in large number by recording the diminishing travel of a small light attached to a pendulum. Often referred to as physiographs, the idea can be further elaborated by using several lights fixed to a whole system of pendulums, or by making overlapping patterns by means of multiple exposures. Interesting linear effects can be created in other ways—e.g., by recording the lights of moving traffic. H.v.W. & R.May.

See also: *Photograms; Tricks and effects.*
Books: *All the Photo Tricks*, by Edwin Smith (London); *The Complete Art of Printing and Enlarging* and *Design by Photography*, both by O. R. Croy (London).

ACCELERATOR. Another name for the alkali in a developer. That constituent of a developer which increases its speed of action.

ACCESSORY SHOE. Metal shoe fitted to the top of the camera to allow supplementary items such as a rangefinder, a flash unit, or a viewfinder to be fixed to it. Most manufacturers of cameras and accessories follow the dimensions adopted by the leading Continental miniature cameras, but some in the past have not, so the point should be checked when buying an accessory for this type of fitting.

Some cameras have an insulated flash synchronising contact built into the accessory shoe. This automatically connects to a corresponding contact in the base of the maker's own flash attachment and there is no need for an external plug-in synchronising lead.

ACCIDENTS. Photographs of road accidents may be taken for their news value or for use as evidence in a court of law. For the latter, special requirements may exist.

See also: *Evidence by photographs; Traffic photography.*

ACCUMULATOR. Secondary electric storage cell which produces a current after being charged from an outside current supply. When the current stored in an accumulator is used up, it can be recharged. The best known type is the lead acid accumulator.

See also: *Batteries.*

ACETATE FILM. Also tri-acetate film. So-called safety film. The term refers to the base of the film which consists of cellulose acetate instead of the inflammable nitrate, which it has almost completely replaced in all film types.

See also: *Supports for emulsions.*

ACETIC ACID. Chemical used as acid in stop baths, acid hardening fixers and certain bleachers and toners.

Formula and molecular weight: CH_3COOH; 60.

Characteristics: Colourless, corrosive liquid with strong irritating smell of vinegar. The pure acid (glacial acetic acid) solidifies below 62° F. (16·5° C.).

Solubility: Mixes with water in all proportions.

ACETONE. Dimethyl ketone. Solvent in certain processing solutions which contain substances not easily soluble in water. Also solvent in film cements, preparations for varnishing negatives, etc.

Formula and molecular weight: $(CH_3)_2CO$; 58.

Characteristics: Colourless, pleasant smelling liquid. Highly inflammable.

Solubility: Mixes with water in all proportions.

ACHROMATIC. Term applied to a lens which has been corrected for chromatic aberration so that it brings light rays of two colours in the visible spectrum to the same focus.

The term is also used in colour physics to describe colours which have no hue or saturation and possess only lightness. Thus white, greys and black are examples of achromatic colours.

See also: *Aberration of lenses; Colour measurement.*

ACID. Hydrogen compound in which the hydrogen can be replaced by a metal (or by a group of molecules with similar characteristics) to form a compound known as a salt, e.g., the metal sodium will replace the hydrogen atom in hydrochloric acid to form the salt, sodium chloride.

The strengths of the various acids are compared on the basis of the pH values which depend both on the characteristics of the acid and on its concentration. Strong acids are generally corrosive. Dilute acids have a sharp taste and most will turn blue litmus red. Any solution with a pH value below 7 is acidic.

See also: *Indicator chemicals*

ACID FIXER. Fixing bath for negatives or prints which contains an acid component in addition to the fixing agent, sodium thiosulphate (hypo).

The acid neutralizes the developer carried over on the sensitized material and stops its action. This prevents prints from going darker in the fixing bath and lessens the risk of staining and uneven development when a number of prints are in contact in the bath.

ACID RINSE. Another name for stop bath used to stop development before transferring material to the fixer.

ACRES, BIRT. Dates unknown. British photographer and manufacturer. Pioneer of cinematography and of the film industry who obtained, independently of his French contemporaries, patents on a chronophotograph (1893) and a cinematograph (1895). Collaborated with R. W. Paul. Also designed the first narrow-gauge (half of 35 mm.) camera (Birtac) and made, in 1895, the first British film.

ACTINIC. Term referring to the ability of light to change the nature of materials exposed to it. The rays at the blue-violet end of the spectrum are generally regarded as being the more highly actinic, since it is these rays that most strongly affect the photographic emulsion, can cause chemical action, and also cause sunburn.

ACTINOMETER. From the mid-1850s there were exposure meters, e.g. Louis Bing's (1866) and Walter Woodbury's (1879). In these meters a piece of sensitized paper was exposed to the light until it darkened to a standard tint. The number of second or minutes it took to reach this tint was measured by stop watch, and a calculator converted this time into the correct exposure for any type of sensitive material.

If the sensitive surface were held at an angle to the main lighting the exposure would be incorrect. Actinometers were apt to give unreliable readings in hazy or cloudy weather. At such times, because there is then no major source of light, the general level of illumination is lower, and shadows where they exist are much lighter than they are in clear sunshine. Hence actinometers were liable to give overexposed negatives when the weather was cloudy.

By choosing a suitable stop, the photographer could use this type of meter to give him the length of the actual exposure. He simply opened the camera shutter and exposed the sensitized paper at the same time; when the paper darkened to the standard tint, he closed the shutter again.

Meters of this sort were usually about the size and shape of a pocket watch. The sensitized paper was supplied as circular discs which were loaded into the back of the instrument. Small areas of the paper could be exposed in succession through a hole in the dial. A standard tint printed next to the hole told the user when the paper had darkened to the right degree. The face of the dial carried the necessary calculator for converting "time to darken" into "exposure." These meters were very suitable for the slow sensitive materials of the earlier days, but are now inconveniently slow.

The modern integrating meters proper are large instruments which not only measure the amount of light necessary for the correct exposure of the negative, but operate a shutter, switch off lamps, or do anything else that may be required at the end of the exposure. They usually have a vacuum-type photo-electric cell, the output of which is amplified by a valve amplifier and used to operate a counter geared to count exposure units. A clock is pre-set to the required number of exposure units and as the exposure units are received by the instrument, a hand travels across the face of the clock. When the counter hand reaches the pre-set hand on the clock face a relay operates to work any apparatus connected to the meter.

A variation on this idea was used when exposing slow printing daylight printing papers for carbon, gum bichromate and other processes in which no visible image appears on the paper during exposure. The actinometer consisted of a continuous tone step wedge carrying a series of opaque numbers. A strip of P.O.P. paper was exposed under this step wedge in a separate printing frame alongside the main printing frame. The figures on the step wedge disappeared one by one as the paper under them darkened to the same tint. Exposure of the print was complete when the darkening of the wedge just reached a particular figure found by previous trial and error. The method was independent of variations in the strength of the light during exposure.

See also: *Exposure meter.*

ACTION. Photography of moving subjects demands special techniques, according to the nature of the subject, its direction of movement, speed and distance from the camera. Exposure times are dependent on these considerations, on the degree of enlargement required from the resultant negative and on the pictorial effect desired.

See also: *Chronophotography; Flash (electronic); High speed cinematography; High speed photography; Movement; Sport; Stroboscopic flash.*

ACUTANCE. Physical measure of image sharpness. The concept of acutance arose from the need for measuring objectively the ability of a photographic process to produce a sharp picture. Whereas sharpness is a subjective measure and necessarily based on a visual comparison of at least two photographs, acutance is based on physical measurements intended to correlate with the visual assessment of sharpness. According to Higgins and Jones acutance is determined by printing a sharp knife edge on the material under test, using a point source of light or a well collimated beam. Sideways scatter of the exposing light in the emulsion layer causes the developed image of the knife edge to be smeared out, producing an S-shaped density distribution of maximum and minimum densities D_1 and D_2 and width w. This width w is measured between two points at which the slope of the S-curve (of density plotted against distance) has a given minimum value. If the slope of the curve at any one point is G_n and is measured at N points within w, acutance is defined as

$$A = \frac{1}{N}\sum_{n=1}^{n} (G_n)^2 \times (D_1 - D_2)$$

i.e., as the average of the squared slopes times the density range across the knife edge.

See also: *Lens performance; Negative materials; Resolving power.*

ADAM-SALOMON, ANTOINE SAMUEL, 1811–81. French sculptor in Paris, part-time photographer from 1856. His portraits in the style of old masters were considered by his contemporaries more artistic than those of Nadar and Carjat; today, however, they seem artificial in pose.

ADAMSON, ROBERT, 1821–48. Scottish photographer, brother of Dr. John Adamson, professor of chemistry at St. Andrews University, who taught him the Calotype process. Robert Adamson opened a portrait studio at Rock House, Edinburgh, 1842 or early 1843, and was joined there by D. O. Hill, summer 1843. Adamson was mainly concerned with the technical aspect of photography, Hill with artistic posing. Their collaboration lasted until Adamson's failing health forced him to retire in autumn 1847.

ADDITIVE SYNTHESIS. The formation of a colour by mixing light of two or more other colours. As a rule the three additive primary colours red, green and blue are used since they can, when mixed, produce a full range of colours including white. A limited range of colours can be produced with two primaries. The mixing of the light occurs at the retina of the eye. The light beams can enter the eye simultaneously or successively, or else form on the retina a mosaic which is too small to be resolved.

See also: *Colour synthesis; Subtractive synthesis.*

ADHESIVE. Binding agent to join surfaces of similar or different materials. Adhesives can be divided into traditional types (mostly animal or fish glues) which are available in the small quantities normally required by the photographer for odd repairs, and the newer types of synthetic resin adhesives. The latter may require special equipment and techniques. They are used extensively in industry but are not generally sold in small quantities across the counter.

For simple repair jobs, the traditional types of adhesive, especially glue, are quite satisfactory. Final choice depends on the materials to be joined.

Wood. For making or repairing wooden articles which are not likely to be exposed to continuous wetting there is nothing better than a liquid glue. Both halves of the joint are coated fairly thinly with the glue, brought together, and kept under pressure for about twenty-four hours.

If the part is going to be continuously wet, the repair should be given two good coats of varnish, or alternatively joined with a waterproof glue (below).

Paper, Cardboard. For paper, any of the proprietary brands of gum, paste, or dextrine adhesives are satisfactory. If the pieces may have to be stripped apart later—e.g., prints mounted on a temporary support—a rubber solution adhesive is best.

For cardboard, a liquid glue (above) or a waterproof glue (below) makes the strongest joint.

Leather, Fabric. A liquid glue or rubber solution will be satisfactory.

Bakelite. No glue or cement normally on sale to the public will make a satisfactory job of sticking bakelite. Parts that appear to have been joined are held together only by the thin line of glue that has dried out along the outer edges of the joint. Since bakelite is non-porous, it seals the adhesive between the surfaces of the joint as effectively as the container in which the glue is normally kept. Unless the material is sufficiently porous to allow the solvents to evaporate, the glue stays liquid.

Glass, Porcelain. For amateur use there are no satisfactory adhesives for glass or similar materials with a non-porous structure. Broken glass surfaces can sometimes be fused together, but the job calls for very special skill and heating facilities. There are, however, synthetic resins which will bond together glass surfaces.

Lens components and glass and gelatin filter assemblies are cemented together with Canada Balsam or more modern optical cements.

Plastics. Plastics vary widely in composition. Some kinds can be joined satisfactorily with a cement made up of a quantity of the plastic itself dissolved in a suitable solvent. Perspex may be joined by a solution of Perspex in chloroform. The fumes are dangerous and the job should be done out of doors.

Wet Surfaces. Surfaces which are to be constantly wet should be joined with a waterproof glue consisting of celluloid or something similar dissolved in a volatile liquid. The surfaces to be joined must be sufficiently porous to allow the solvent to evaporate or the glue will never set.

Contact Adhesives. Also called impact adhesives, these are liquid glues which are applied to both surfaces to be stuck together and then allowed to dry out completely.

If the two surfaces are then pressed into contact they adhere strongly and make a permanent and generally waterproof joint. This type of adhesive is suitable for non-porous materials—plastics, metals and china, as well as for flexible, porous materials like cardboard, leather and cloth, but the joint is no stronger than the adhesive itself. For really strong joints in plastics and other rigid materials, the two-solution synthetic resin glues which set much harder are better. F.P.

Synthetic Resins. The development of these adhesives has enabled many materials in common use to be bonded permanently, the joints obtained being largely unaffected by heat, moisture, micro-biological or chemical attack. Nevertheless, there is no single all-purpose cement capable of adhering strongly to all types of material and selection of the most suitable adhesive for each purpose depends upon the nature of the components to be joined and upon the conditions in use.

Synthetic resin adhesives are virtually plastics in the liquid state, though they are often supplied in other forms, e.g., as powder, to prolong their storage life. The thermosetting types can be made to set into infusible, insoluble solids by the addition of an acid-forming compound known as a "hardener" or "accelerator". They offer outstanding adhesion to many materials with a degree of durability under severe conditions that could not have been obtained until recent years. The glueing of most materials in common use falls within the scope of synthetic resin adhesives.

Asbestos. A resorcinol-formaldehyde resin provides a strong bond between asbestos and asbestos or between asbestos and cork, fabric, leather, wood, rubber or laminated plastics. The adhesive is generally marketed as a reddish-brown liquid which must be mixed with a white powder hardener before use. No heating is required and the resultant bond withstands even boiling water.

Cork. The same resin can be used as for asbestos, above.

Glass. Epoxy resin adhesives give unequalled adhesion to glass and ceramics generally. Cold-setting forms are available, but higher joint strength and greater durability are obtainable with the hot-setting epoxies. The strength of these resins is such that, in tests to destruction, the glass fractures before the adhesive fails.

Leather. The same resin can be used as for asbestos, above.

Metals. Very strong joints can be obtained between metal surfaces or between metals and other materials such as ceramics by using epoxy resin adhesives. When bonding metals it is particularly important to start with clean surfaces, free even from contact with the hands.

Plastics. Laminated plastics (e.g., as used for table tops on kitchen furniture), can be successfully joined together or to rubber, wood and other non-porous materials with resorcinol-formaldehyde adhesives as for asbestos. Thermo-plastic plastics such as Perspex, P.V.C., polythene are usually best bonded by solvents for these materials.

Porcelain. Both resorcinol-formaldehyde and epoxy resins adhere strongly to this material, but for glazed porcelain surfaces the epoxies only should be used.

Rubber. Great durability is obtainable with resorcinol-formaldehyde resins if the rubber can first be treated by acid or chlorine water as instructed by the makers.

Wood. Urea-formaldehyde and resorcinol-formaldehyde resin adhesives offer great durability with wood joins under exposed conditions. Animal, vegetable and fish glues are inexpensive and simple to use but have poor water-resistance, while casein, with improved water-resistance, is very susceptible to mould and microbiological attack. K.S.M.

Optical Cements. Optical cements are usually classified as either thermo-setting or thermo-plastic materials. Canada balsam, which has been used for two centuries or longer, is still widely employed. It is representative of the thermo-plastic group. For many optical requirements, however, other substances must take its place. This is especially the case when considerations of resistance to high and low temperatures, moisture-proof qualities, and excellent adhesion are very important. The more modern cements which have these improved properties include cellulose caprate, n-butyl methacrylate, polyester resins, vinyl acetate, and certain formulations based on epoxy resins. In general these new types are thermo-setting since they become rigid on heating; the chemical process involved is known as polymerization. Cellulose caprate is a notable exception for it is a solid which softens in the region 95–100°C. If the final cement is too brittle it may be improved by the addition of a plasticizer which is either a high-boiling liquid or a low-melting solid.

When using optical cements it is important to ensure the utmost care in preserving a clean material. This is especially so if the cemented film appears in the focal plane of the instrument as with graticules and the field lenses of binoculars. Light transmission of a high order and an appropriate refractive index are other valuable features of a satisfactory optical cement.

In cementing lenses it is usual first to apply the material to the glass surfaces as a viscous liquid taking care to avoid the introduction of air bubbles. Further treatment varies according to the type of cement. For a cold-setting formulation, say, of a polyester resin, the cemented units are allowed to remain in a suitable jig until the polymerization process is complete. Some cements require heat treatment either to remove a solvent as for Canada balsam or to complete the polymerization process as with n-butyl methacrylate. The temperature of baking may be in the range 40°C. to 100°C. and the time from a few hours to as many days, the conditions selected being determined by a number of factors including the nature of the cement, the degree of permissible strain in the joint and subsequent likely exposure to thermal and mechanical shock. E.Bo.

See also: *Adhesive tape.*

ADHESIVE TAPE. Many of the wide variety of types of self-adhesive tape are used in photography in place of other agents like gum, glue, drawing pins and so on. Here are the principal types with photographic applications.

Transparent Cellulose Tape. This is the popular sticky tape used for repairing torn papers, books and photographs. It makes a quick, foolproof fastener for anchoring the end of bulk film to the film spool or for rejoining a cut length of roll film to the paper backing. It can be used to protect bottle labels from contact with water and dark room chemicals and it will hold papers flat for copying.

P.V.C. Tape. This is a tough, flexible, waterproof tape used for electrical insulation, for putting a waterproof seal around film containers, for temporary repairs to processing dishes and tanks, camera bellows and so on.

Cloth-based Plasticized Tape. This is a thicker and stronger type of self-adhesive tape which is suitable for book and album binding, for repairing camera cases and hold-alls and for covering polished and plated parts, e.g. control knobs, levers and handles, to make them easier to grip.

Double-sided Tape. This is a tape that is sticky on both sides. When bought in a roll, one side is covered with a protective layer of paper which is stripped off before use. Double-sided tape can be used for holding the printing paper down on the enlarging easel and is specially useful for big enlargements that have to be stuck to a vertical wall surface. It is also used for holding printed material for copying.

Light-reflecting Tape. Various kinds of light-reflecting tape—UV-sensitive, fluorescent, glass-beaded and metallized tape are used in commercial and industrial darkrooms where panchromatic materials are manufactured or processed by very dim safelight.

Splicing Tapes. Self-adhesive cine film splicing tape is made up in lengths equal to 4 frames of film with a peel-off protective paper over the sticky side. Self-adhesive tape for splicing magnetic recording tape is supplied in rolls and is coated with a specially hard adhesive. In this case, the splicing tape is applied only to the back—i.e. the shiny side—of the tape because nothing must come between the magnetic coating and the face of the recording head.

See also: *Adhesive; Splicing.*

AERIAL FOG. Type of chemical fog affecting negative materials if they are excessively exposed to air when wet with developer.

See also: *Faults; Fogging.*

AERIAL PERSPECTIVE. General effect of depth created by haze in a photograph of a landscape. There is usually a certain amount of haze—water vapour, smoke, or dust—in the air and the amount between the lens and the subject increases with the distance of the subject. So distant objects have softer outlines and lighter tones than those closer to the camera. The eye accepts this increase of haziness with distance as an indication of depth and spaciousness. Haziness can be controlled by using a yellow, orange, or red filter with a panchromatic film or plate. In telephotography of distant views where aerial fog is a drawback it can often be cut out altogether by using an infra-red film or plate and filter in combination.

AERIAL PHOTOGRAPHY. Photography of terrain on the ground and objects in the air from cameras mounted in aircraft. Civilian and military applications range from aerial survey to reconnaissance, strike recording, missile tracking and satellite photography.

Air-to-ground photography comprises all pictures taken of features on the ground from aircraft. The type of camera and technique employed depends on the purpose for which pictures are required. For aerial survey the ground is photographed in a systematic manner to cover the whole of a specified area. This provides the necessary data for map-making and other survey applications. Reconnaissance photography, especially in wartime, aims at recording individual ground features.

Air-to-air photography may include the recording of other aircraft in flight, but the widest field of application is the photography (usually with motion picture cameras) of bullets and missiles aimed at target aircraft. The recording cameras may either be mounted alongside the guns, or in the target aircraft itself to record strikes by guided or automatic missiles. Different types of equipment are required for these two applications.

Ground-to-air photography is concerned with tracking flying objects, from missiles to satellites, during flight. Cameras for this purpose are usually very large and have long focus lenses and complicated movable mountings to track a flying object.

See also: *Aerial survey; Aircraft camera; Air Force photography; Air photographs; Air-to-air photography.*

AERIAL SURVEY

Photographic recording of the topography of the ground by airborne cameras. The value of photographic surveys made from the air was realized as early as 1858 when, on 23rd October, Nadar took out a French patent for "an entirely new system of photography, whereby photographic bird's-eye views of portions of the earth's surface may be taken". The camera was fixed either vertically in the bottom of the balloon car or in the side of the car for angle shots. After many failures, Nadar succeeded in obtaining a distinct, though faint, image from about 80 metres above the village of Petite-Bicêtre near Paris in winter 1858. In the 1860s. Nadar took some good views of Paris from a captive balloon from heights of up to 500 metres, but balloon photography with collodion wet plates was not reliable enough ever to carry out his plan that the whole of France should be photographed from the air for the purpose of ordnance survey. A United States patent was taken out to cover the making of maps from a balloon in 1873. Since that time the techniques have been extensively developed, and today complete national surveys are made by air, e.g. Great Britain and Canada.

In general, aerial survey falls into two categories: qualitative or quantitative, and photogrammetric. The first is mainly concerned with size, shape, tone and texture, some overall distortion being acceptable in many cases: its applications lie in agriculture, archaeology, military reconnaissance, etc. In photogrammetric survey, position and direction, and therefore undistorted reproduction, is usually the most important factor. This type is used in map making and it accounts for the majority of aerial photographic effort in peacetime. Both categories demand the best possible definition. Both also involve three phases: planning; photography; and interpretation or analysis..

Equipment and Materials. In contrast to military reconnaissance, cost is a major factor in commercial survey. While the use of long-focus lenses from great heights may seem preferable, costs as well as atmospheric conditions usually dictate otherwise. By covering more ground from lower altitudes, wide-angle lenses entail among other things lower fuel consumption, less mileage flown and less film used. There is also the important advantage of having larger areas of overlap for stereoscopy, with greater distances between viewpoints, giving exaggerated and more measurable perspective displacement. The distortion and errors involved can be more economically corrected in specialized printing and analyzing equipment. Since accuracy is generally more important than time, slower, more suitable and more economical aircraft can be used, often with relatively simple equipment.

The type of work determines the size and complexity of equipment used. It ranges from light, fast-operating hand-held cameras using film up to about 5½ ins. wide, to large, automatic, remote-controlled fixed instruments taking up to 18-ins. wide film or even plates. All are characterized by fast recycling operation, highly corrected lenses, usually interchangeable, and special precautions to ensure film flatness. In many, the latter is achieved by pressing the film against a register glass which may be engraved with a grid or fiducial marks to assist measurement; in others, the film is flattened by suction against a back platen.

Shutters may be of the focal plane type but inter-lens shutters are preferable since they introduce no distortion. These have to be large and may be electrically or spring driven, giving exposures up to about 1/1,000 second. Fast shutter speeds to overcome blurring due to the aircraft's movement involve wide-aperture lenses and fast film. An alternative is Image Motion Compensation (IMC), in which the film is moved during exposure in step with the image movement, either by moving the platen or register glass or by advancing the film itself. IMC permits slower shutter speeds and therefore the use of slower, more highly corrected lenses and fine-grain film. The amount of film movement necessary is computed from the aircraft's height and speed and the lens focal length, and is set on a special control.

An extension of this technique is the continuous strip camera which makes an unbroken record of the ground beneath, along the whole length of the film. It employs a fixed focal plane slit, similar to a focal plane shutter but motionless, past which the film is drawn at a speed matched to the image movement. The slit width is adjusted to give the correct exposure. A stereo version has two lenses, slightly offset from each other in the direction of flight, each covering half the width of the film. The camera is useful in applications normally covered by line overlaps.

Other specialized techniques include the photography of airborne radarscopes, and panoramic cameras. Ordinary air cameras may also be mounted in split pairs or in fans of three or more to give wider coverage. However these methods are generally more suitable for reconnaissance than for survey.

Colour or black-and-white emulsions may be employed. Special colour films with one primary colour omitted are sometimes used where difficult hues must be separated. Black-and-white emulsions are more common, generally fast panchromatic, but again, where tonal separation is important, infra-red emulsions may be better. The emulsions are coated on a dimensionally stable "topographic" base, capable of withstanding extremes of temperature, and fast high-temperature processing. Glass remains the most stable support however, and some makes of survey camera use plates, fed from automatic magazines.

Photography. Survey photographs may be verticals or obliques. Verticals are taken with a camera mounted in the aircraft and pointing vertically downwards. They provide a plan view of the ground, the scale depending on aircraft altitude and on lens focal length. Obliques are usually taken from a relatively low altitude,

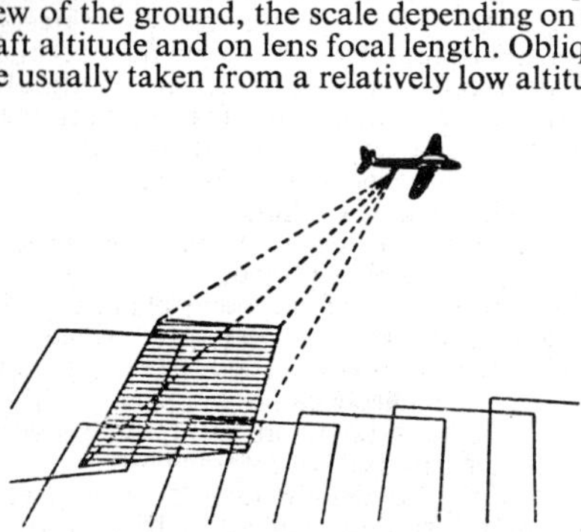

OBLIQUE AERIAL SURVEY. The camera is pointed at an angle to the ground. Special photogrammetric methods are required for quantitative information from the pictures.

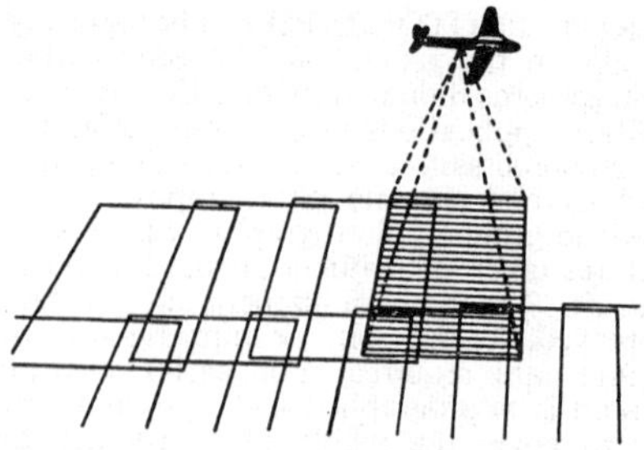

VERTICAL AERIAL SURVEY. The camera points vertically downwards, making exposures at regular intervals as the aircraft flies along the lines of a plotted grid.

either with a hand-held or fixed camera slanted obliquely at the ground. They give a perspective view with a better impression of land forms, etc. than verticals and, although useful alone, are often used to complement the latter.

Depending on the subject, the pictures may be pin-points, line over-laps or mosaics. Pin-points are one or more vertical or oblique photographs of a single feature. Line overlaps are generally used for long narrow features such as roads, beaches, railways, etc. They are a series of vertical or oblique photographs taken at intervals such that each overlaps the one before by at least 60 per cent. This ensures that nothing is missed and provides overlapping coverage for stereo viewing.

Mosaics consist of a series of vertical line overlaps flown on parallel tracks, overlapping each other by at least 30 per cent. They form a photographic map covering a broad area. The prints are matched for scale and stuck down in jig-saw fashion on a card base. Where possible, they are cut along a feature such as a road, river or hedge to hide the join.

Such careful presentation is not always necessary. Photogrammetric survey requires only accurate positional and directional data plus adequate overlap on uncut prints for stereoscopic analysis. This also demands a high standard of flying and navigation, nowadays generally assisted by radio beacons and electronic aids. Where large tracts of relatively featureless country must be covered, as in the Canadian prairie, widely spaced line overlaps only are flown, which form a photographic reference grid for ground survey. On the other hand vast areas of African jungle have been completely covered by taking overlapping strips in concentric circles about a mobile ground radio beacon.

Errors include aircraft drift due to side winds, and tilt, which may be eliminated by rotating and gyro-stabilized camera mounts; while height variations can be corrected by re-scaling in the enlarger. For added accuracy, many cameras record such data as the aircraft's height, attitude, heading and position, and the time at the moment of exposure, on the film rebate. Difference in scale between the edges and centre of the picture is unavoidable but can be reduced by using long-focus lenses. However, economics or low cloud may prevent this. Haze, ever-present at any altitude, can usually be overcome by use of appropriate filters and emulsions.

Stereoscopic Analysis. Although simple measurements from single prints can provide much information, in most cases stereoscopic examination of overlapping pictures is more informative. Here the 60 per cent minimum overlap is important. Apart from facilitating the identification of objects and giving a more realistic view, stereoscopy can provide an accurate measure of ground height.

Because the viewpoints are much farther apart than the normal spacing of the eyes, the stereoscopic effect is much more vivid than when the scene is viewed normally. In fact, when viewed from the height of the aircraft, the scene does not appear stereoscopically to the unaided human eye. Aerial photographs—particularly those taken during war-time reconnaissance—are easier to interpret and yield much more information when they are made for stereoscopic viewing.

Measurements can be taken by hand, but are usually done by electronic scanning or on complex plotting machines. These employ twin projectors over a large plotting table, coupled to a small movable table or "floating mark". The results may either be converted directly to contours (analogue method) or stored for subsequent processing in a computer (analytical method). In either case, the end product is usually a contoured map or series of elevations, even perhaps a three-dimensional terrain model, complete with miniature buildings and trees.

Applications. The commonest application of aerial survey is in the geographical and geological fields. Besides eliminating a vast amount of tedious field work, it is sometimes the only way of mapping and analyzing large and inaccessible areas for mineral search projects. However, it has not yet completely supplanted ground survey, where this is possible. Apart from this, repetitive surveys of inhabited regions are valuable in showing the growth of land development, communications, coast erosion, etc., particularly where buildings hamper ground work.

In the same way, it is of use in agriculture and forestry, to show the growth of crops and the condition of land. In ecology and botany it has proved its value in showing the distribution of animal and vegetable life and in the treatment of pests such as locusts and cotton bollweevil. Because the surface of the ground, once disturbed, is never the same again, the aerial camera can be used in archaeology. Here it can show the tonal difference in soil or vegetation over buried ruins that would be indistinguishable by ground observation. P.B.M.

See also: *Aircraft camera; Air Force photography; Archaeology; Balloons and kites; Ecological photography; Moving image photography; Photogrammetry.*

Books: *Aerial Mapping and Photogrammetry*, by L. G. Thorey (Cambridge); *Aerial Photographs and their Applications*, by H. T. U. Smith (New York).

AFOCAL LENS. Attachment for changing the focal length of a camera lens. An afocal lens is, in fact, a Galilean telescope with variable separation between the front and rear components. In use, the camera lens is set on infinity and the image is correctly focused by adjusting the separation of the lenses in the attachment. The result is an increase in image size as with a long-focus lens; but no extra camera extension is necessary.

A similar attachment may be used to reduce the focal length of the camera lens, thus producing a smaller image but wider angle of view.

See also: *Supplementary lens.*

AFTERTREATMENT. Faults in negatives and prints caused by errors in exposure and development can be rectified to a certain extent by aftertreatment. While the cure is never as effective as prevention, it can at times convert an unprintable result into a passable one.

Negatives. All forms of aftertreatment of negatives involve some risk; they may even spoil the negative instead of improving it, so if the negative is particularly valuable, it is as well to make a duplicate first, just in case anything goes wrong.

The methods of aftertreatment are:

(1) Chemical. These methods include the various wet processes of reduction and intensification, contrast adjustment by redevelopment, and scratchproofing.

(2) Physical. These are mainly methods of working on the dry negative, e.g., abrasive reduction, spotting, retouching, blocking out, and varnishing.

(3) Optical. In optical aftertreatment the whole negative is copied and its contrast or density adjusted during the copying process.

Prints. Methods of print aftertreatment fall into two broad groups:

(1) Physical aftertreatment. The surface of the print may be glazed, the image can be hand coloured with dye or water colour and blemishes or weak tone areas can be retouched.

(2) Chemical aftertreatment. A print may be treated chemically so that the character of the image is changed. This may be done in order to intensify or reduce the image, or to tone it to a different colour.

See also: *Colouring prints; Doping prints; Glazing; Hardening bath; Intensification; Oil reinforcement; Optical aftertreatment; Redevelopment; Reducing; Retouching; Scratch-proofing; Toning; Varnishing negatives; Waxing prints.*

AGAINST THE LIGHT

The very expression "against the light" (or "contre-jour") implies something out of the way and difficult. This is a pity as it makes many photographers keep to the safe rule of having the light source behind the camera when they could get more attractive pictures with the light source in front. There is nothing particularly difficult about the technique: the main necessity is to keep sunlight or direct light from any other source from reaching the lens, where it would cause fog and flare on the film.

There are many good reasons for taking pictures against the light. One is to deal with messy and fussy backgrounds or even ugly plain ones—e.g., brick walls. Against the light shots show less detail in any case, and the background is likely to be in the shade and therefore under-exposed and dark.

It is the unusual patterns of light and shade that make the majority of against the light photographs. By careful positioning it is possible to include all the principal shadows cast by the subject in the picture frame. The best effects are produced in the morning and evening, when the shadows are long. And if the camera looks down from a high viewpoint the effect is even better as the shadows will not be foreshortened. Long shadows have another valuable effect apart from forming interesting patterns: by projecting towards the camera they give the picture more depth and three-dimensional reality than can be achieved by other normal means.

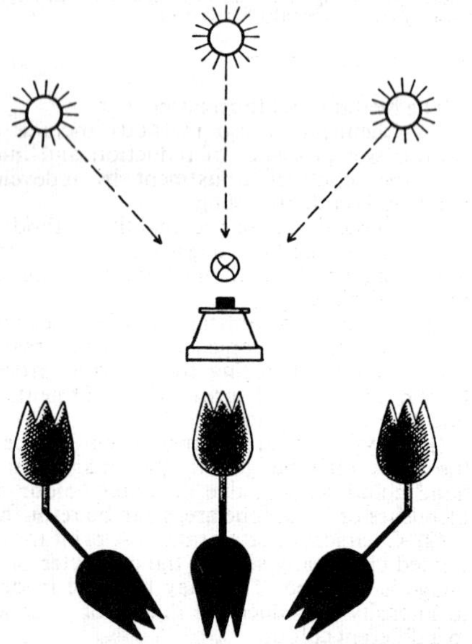

DIRECTION OF THE LIGHT. The sun directly behind the subject casts the shadow straight towards the camera. Both edges of the object may be outlined with a bright light rim. When the sun comes slightly from one side, the shadow is cast in the opposite direction and only one side is rim-lit.

The sun also produces a bright rim of light round the edges of objects and this rim-lighting is a further effect that helps to make the subject stand out from its background. In addition, the rim of light makes the subject appear attractively luminous. This luminosity is the greatest charm of against-the-light pictures. It can be increased still more by using a soft focus lens or fitting a diffusing disc before the camera lens. When the image is diffused in this way, the luminosity pervades even the shadow areas.

Shielding the Lens. A really efficient lens hood is always advisable but against the light it becomes an absolute necessity. When lens hoods leave the factory they are usually dull black inside, but if after long use or by accident they become bright inside they must be given a coat of dead black paint or they will cause the very trouble they are intended to prevent.

The lens hood should be as long and narrow as possible without cutting off the corners of the negative; the safest policy is to buy the one made for the particular camera and lens.

When shooting so much into the sun that even the lens hood cannot shade the lens completely, it is often possible to get over the difficulty by choosing a camera position in the shade of a doorway, building, tree, or even a telegraph pole, or a friend can hold a hat or wallet to shield the camera if there is no natural shade. Of course, he must take care not to intrude into the field of view. If the camera is on a tripod, there is no difficulty in shielding the lens in this way with the free hand, and even without a tripod it is still possible to hold the camera with one hand at shutter speeds of 1/100 second or shorter.

Suitable Subjects. People, boats, and in fact most snapshot subjects and all pictures of activity come out well against the light. Pictures of snow and most subjects that owe their appeal to texture or sparkle are always improved by being taken against the light. With foliage, flowers, icicles and anything thin enough to let the light shine through, there is a striking translucent effect which is most attractive. Sun shining on steam or mist can be successfully exploited in against the light shots.

Landscape photographs when normally lighted are often difficult because they so frequently suffer from a lack of tone contrast. Against the light, they come to life because the shadows separate the receding planes and give them depth and atmosphere. This technique cannot be used if detail must appear in the far distance because everything near the horizon tends to disappear in haze.

When the sun hides behind clouds or shows through a thin cloud or through mist it can safely be included in the picture. Even the dullest landscape can be made into a successful

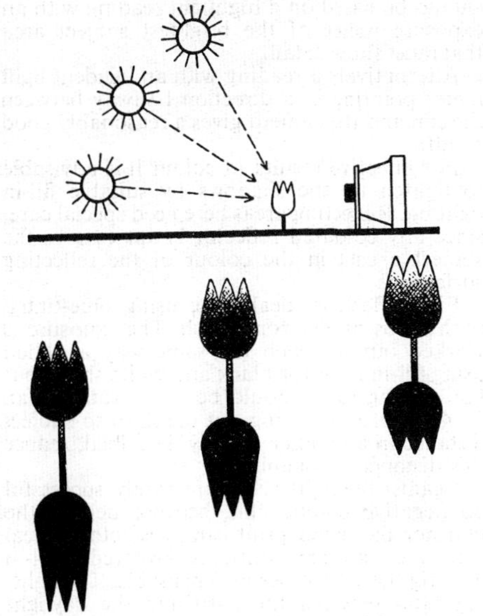

HEIGHT OF THE SUN. *The higher the sun, the shorter the shadow of the subject. Low sun provides the best lighting for against-the-light subjects, as it creates more pictorial shadow pattern and the shadows tend to be mellower.*

picture by putting the horizon to the bottom of the picture and making the sky the main theme with the sun painting dazzling rims round the clouds. This technique is particularly effective with sunsets.

Artificial light sources—e.g., street lamps—can often be usefully included in the picture by taking the photograph at dusk when the sky is still quite light but not bright enough to cancel them out. After dark, weak and diffused lamps can still be included but the camera should be kept at least thirty feet away from them. Strong lights, like motor car headlamps, must not shine straight into the lens but can be photographed if the actual bulb is shaded by its own reflector—i.e., if the picture is taken from the side. Street scenes at night are vastly improved when the pavement is wet because then the picture is enriched with reflections of the light sources.

Sunbeams shining through windows into a room make fine pictures and it can be worth while to shake a mat or rug to raise some extra dust to give the beams more brilliance.

Balancing Contrast. When taking close-ups against the light there may be too much contrast between light and shadow areas for the film to cope with both extremes. In such cases it is an advantage if the shadows can be brightened by the use of a reflector or fill-in flash. Out of doors, the greatest contrast exists when the sun shines from a cloudless sky; it is always much reduced by reflections from white clouds and still more by any light haze or mist.

Light-coloured surroundings such as a sandy beach or snow reflect so much light into the shadows that they provide the ideal location for against the light portraits or animal close-ups. When there is no natural reflection it may be possible to place the subject near a light coloured wall, or to use an artificial reflector—e.g., a piece of white or light mounting board or even a newspaper.

Flash can also be employed to lighten the shadows and has the advantage of being much more flexible and independent of the other prevailing light. The flash can be mounted on the camera, and either flash bulbs or electronic units are suitable.

With fill-in flash, the exposure for the daylight and for the flash must be balanced to avoid overlighting the shadows and so destroying the natural against-the-light effect. The best way of ensuring this is to keep the lighting level from the flash at about one-quarter the light intensity from the daylight. In other words, the flash-lit shadow areas should receive about one-quarter the exposure they would need for full illumination.

A handy method of doing that is to work out the daylight exposure by pointing the exposure meter with incident light attachment at the light source, and ascertaining the aperture at a shutter speed at which the flash can be synchronized (with bulbs and simple cameras this would probably be 1/30 second). This aperture is then divided into double the guide number normally used for the flash, and the result used as the flash distance for the fill-in lighting. If this is different from the camera distance, the flash may be used off the camera (with an extension flash cable) to bring it nearer to or further away from the shadow areas. Alternatively, if the flash is too strong, it may be possible to reduce the light by one or more thicknesses of a normal handkerchief in front of the flash reflector.

Exposure. The exposure depends on the type of picture aimed at. This applies particularly when the sky is used as a background and there is no possibility of illuminating the foreground (or the subject in a close-up). If these parts of the picture must show detail there is nothing for it but to over-expose the sky. A great improvement can be effected by renouncing the close-up—which always demands extra exposure anyway—and taking the picture from a greater distance. By finding the half-way exposure between highlights and shadows it is then possible to get detail in both. If it is not possible to take the picture from a greater distance there remains the possibility, especially when the background is a fine cloudscape, of making the foreground object into a silhouette by exposing for the clouds.

Open landscapes taken against the light need a fairly short exposure. But the strong and dazzling reflections of sun on water are apt to mislead the photographer into giving too short an exposure. If the exposure is based on these highlights the result will be a picture with a night effect.

Sensitized Material. The fastest films are the softest and therefore best able to manage the greater contrasts of against the light photographs. Similarly, soft developers help to prevent the highlights from blocking up and becoming impossible to print. With correct exposure on a good film containing all kinds of subjects there is no need to alter normal development technique, but there must never be any tendency towards over-development. It may be necessary to print on a softer grade of bromide paper, and luminous against the light shots often print better on soft paper.

Colour Shots Against the Light. A colour transparency can reproduce particularly effectively the great brightness range of against-the-light scenes and still preserve the luminosity of the shadows. The transparency can thus give an enhanced impression of depth when a darkish foreground object is lit from behind against the bright background.

Since colour films have a limited exposure latitude, the exposure for against-the-light shots is particularly critical. Generally it is impossible to avoid some colour distortion in the darker parts. Exposures with transparency film should however be based on the highlights, since colour errors there are most noticeable. Only extremely bright catch lights and haloes of light are acceptable overexposed. Generally the exposure should be based on a highlight reading with an exposure meter of the brightest subject area that must show detail.

Alternatively a reading with an incident light meter pointing in a direction halfway between the sun and the camera gives a reasonably good result.

For effective results in colour it is advisable to lighten up the shadows by suitable fill-in lighting. Reflecting areas here need special care, since any coloured reflector is apt to give the subject a cast in the colour of the reflecting surface.

Fill-in flash is ideal here, using blue-tinted flash bulbs or electronic flash. The exposure is worked out in much the same way as when using fill-in flash for black and white shots, but the lighting ratio should be 1 : 3, rather than 1 : 4. So after working out the flash to subject distance in a manner already described, reduce this distance by one-fifth.

Against-the-light shots are rarely successful on negative colour film, because neither the film nor the paper print can satisfactorily deal with the extreme contrasts involved. Fill-in lighting there becomes indispensable. The lighting ratio between the flash and the daylight should be higher still—around 1 : 2. In practice this can be allowed for by assuming for fill-in purposes a guide number 50 per cent higher than the normal guide number of the flash.

H.W.

See also: *Contrast control; Fill-in light; lighting the subject.*

AGEING OF MATERIALS. Sensitized materials that are stored for long periods gradually lose their sensitivity, and may also change other characteristics. Troubles such as fogging and mottle can also arise. The extent to which these effects occur depends a great deal on the conditions of storage.

See also: *Expiry date; Faults; Keeping qualities of materials.*

AGENCIES. Almost invariably, the photographer who has produced a saleable picture or picture sequence stands to gain most by letting an agency market his work for him. Taking saleable pictures and selling them require two entirely different kinds of experience.

The agent is in constant communication with editors and others who buy photographs, so he is highly informed about current market requirements and can place saleable work before the most likely buyer with minimum delay and maximum economy. The agent is also well informed about the relative prices paid for mediocre, good or excellent pictures in various markets, so he knows what a particular photograph is worth and how much to ask for it.

In return for his services the agent takes a commission on sales.

News Agencies. A news agency is a big organization. It employs a staff of photographers who cover the main daily news events, and their pictures are circulated to a large number of journals.

The news agency also accepts work on a percentage basis from free-lances or amateurs who happen to be on the scene when something of news value occurs.

The usual procedure for a free-lance who has a live news picture that he wants to place through an agent is first to telephone the agency, explain the happening and state what pictures were obtained. Then if the agency is close at hand, he should take or send the undeveloped film to the agency. If he is in another town, he should pack the film, clearly marked "Press matter. Urgent. To be called for", and address it with the agency's name and the terminus station and send it by rail or, where appropriate, by air.

Inside the pack, he should enclose a note marked "Undeveloped film". In addition, he must include caption material based on the six imperatives of news reporting: who, what, why, when, where and how? Many a first-class picture fails to get published through lack of full information.

The film will be processed by the agency and prints made and circulated (often within half an hour of getting the film) to a number of news editors on a non-exclusive basis. (Top-line agencies now have branches with wire machines in all large towns.) No photographer working alone could hope to do this. The news agency service often covers foreign markets, too, and it is not unusual for a single news picture of merit to be sold fifty times.

The news agency will take 50 or 60 per cent of sales revenue. This is a fair return in view of the printing costs, staff hours and the highly organized distribution that makes so many sales possible.

Feature Agencies. The feature agency deals with pictures of general interest which are neither as topical nor as urgent as those handled by a news agency. A feature picture deals with a particular aspect of the news—e.g., photographs of a race meeting are news pictures, whereas photographs of the faces of the spectators, showing their various reactions, are feature pictures. The best feature agencies probably supply editors with the greater part of the pictures they use on personalities, beauty, interesting occurrences, ways of doing things, people with unusual hobbies or queer pets, new developments in the social sciences and amenities, and anything of a curious or humorous nature—in fact, whatever is of general interest, but not direct news.

The free-lance who tries to cover the news interest of an event at which staff photographers are present is often wasting his time. His best plan is to look for a feature that has escaped their notice. Good pictures in this class can be sold days after the event, when straight news pictures are dead.

Feature Agency Terms. The photographer who makes use of the best type of feature agency can expect the following conditions.

The photographer will supply negatives and contact prints. From these, the agency editor will order from his own printer enough single pictures or sets to supply all interested markets at home and abroad. The agent deducts a fixed percentage share. As terms do vary somewhat with different agencies the photographer should get exact terms from his agent in writing.

Advice on sales possibilities of a proposed picture or picture-story will be given before it is undertaken. The agent is always ready to discuss the photographer's ideas for pictures or series and his wider experience often enables him to add just the necessary angle, or twist, that will make a story saleable. There should be no charge for this or any other extra service.

Current information on editorial trends and requirements should be passed on by the agent to the free-lances who deal with him. He should point out in particular when there is a possibility of a suitable feature in the photographer's own district.

Assignments are received by some agents from editors, mainly of overseas magazines, who wish to run a story on an event or personality in this country. This well-paid work is handed out by the agents to free-lances, and preference is naturally given to those who bring in original material of their own.

The agent may finance a story of the photographer's choosing, if he feels that the chance of success is worth the risk. This enables the free-lance to undertake a worth-while story which he could not afford to proceed with on his own.

It sometimes takes a month or more for the agent to receive payment from an editor. If the agent feels that a photographer is able to produce regular, saleable work, he may advance him a weekly sum of money until payment for sales catches up.

Pictures can lie in a photographer's files for years without being sold. If they are kept in an agent's files, they will constantly be turned over by book illustrators, calendar printers, advertisers, artists needing a picture for reference, and magazine representatives looking for pictures of a particular subject. Landscapes, happy children, pretty girls, pets and beach scenes, are among the file pictures which may sell five weeks or five years after they were taken.

For National Press, and most other uses, a fee of two guineas upwards is normally paid in the United Kingdom, depending on the purpose for which the picture is required. With the provincial press payment may be lower, around £1 7s. 6d. In certain cases a calendar printer may pay fifteen guineas. Twenty would be quite ordinary for a national magazine cover, and a picture chosen for a national advertising campaign might well fetch more.

There is a ready market for colour material for editorial, advertising, calendar, etc., use and the leading agencies maintain comprehensive libraries of colour photographs of many subjects. Fees paid vary between ten guineas and 100 guineas, and may in special cases even exceed the latter. 35 mm. transparencies are extremely difficult to market in the United Kingdom. The minimum size normally accepted by an agency is 2¼ ins. square, but the standard material for the colour market is 4 × 5 ins. and larger. Only the highest quality colour pictures are generally considered by an agency.

The commission varies from agent to agent. One agent will charge a commission of 50 per cent on home and overseas sales, and pay all of the costs of printing. Another may charge 40 per cent and pay half the printing costs. Probably the lowest commission asked by an agent is 35 per cent on home sales, 50 per cent on overseas sales (which have to be shared with the overseas agent) and a share of the printing cost.

Exclusive Rights. These are the most important words in picture selling and in the agent-photographer relationship. News agencies normally offer pictures to news editors on a non-exclusive basis. The editor pays a standard rate, and knows that the picture may be bought by his competitors as well. On the other hand, a feature editor will pay five or ten times as much for a feature picture on an exclusive basis, knowing that it will enhance his paper's reputation for original pictures. He will seldom buy a good feature picture offered on a non-exclusive basis, because he knows it may also be used by a competitor.

The good agent obtains the highest fees possible by selling "first publication rights" to a single paper or magazine. The photographer must of course tell his agent if a picture supplied by him has already been sold or he may be responsible for a serious infringement of copyright or exclusive rights as a result of which both parties might have to pay heavily in cash and reputation.

Specialist Agencies. Apart from the general agencies, there are a number of small specialist agencies, each concerned with supplying pictures to a limited market, such as art or technical journals, or those concerned with the theatre, fashion or travel. Each of these is well known to editors, dealers, the trade, and publicity firms concerned with that particular market. These agents know what pictures are in demand in their own spheres and so know exactly where to place free-lance pictures that meet their requirements.

Having found a good agent, the photographer will do well to stay with him. Changing about and sending sets to various agents in order to find out which makes most money, serves no useful purpose. A picture can only be offered by one agent at a time, so direct comparison is not possible.

Buying from an Agency. Anyone requiring a photograph for reproduction or display should telephone an agency that specializes in the particular subject or is likely to have a selection of pictures dealing with it. This is better than making a personal call in the first instance, as it will give the agency's librarian time to look out suitable material in advance. Some agencies charge a search fee for this.

Agencies do not like sending out selections of file copies except to regular clients. The trouble involved in sorting and sending, and the incidental costs of recovering unreturned prints and reprinting damaged ones, is not covered by the fees paid for those sold. R.S.

See also: *Print libraries; Reproduction fees; Selling photographs.*

Books: *How to Take Photographs that Editors will Buy*, by R. Spillman (London).

AGITATION. Term applied to the various methods of keeping the solution or sensitive material in motion during processing. The movement is intended to bring fresh chemicals into contact with the emulsion surface.

The usual method of agitation is to move the film either intermittently or continuously during processing. Daylight processing tanks for single film lengths incorporate a rod which engages with the film bobbin so that it can be rotated; some types also allow vertical agitation by pushing downwards on the rod. A more recent type of daylight tank is designed for agitation by inversion—i.e. the top of the tank is stoppered, and the tank is completely inverted and returned to its right-way-up position at the specified intervals. The older method of agitation by means of a rod is probably the more satisfactory, and cleaner. Plate tanks do not usually have agitators and it is necessary to rock the whole tank body.

Mechanical devices for automatic agitation have been tried, but motor driven agitation is liable to cause uneven processing due to the regular intermittent effect of the power with all but the smoothest motors.

The most widely used current method in colour processing is nitrogen burst agitation. This is suitable only for larger tanks, as used for sheet films, and involves blowing a short intermittent burst of a neutral gas (usually nitrogen) into the developing solution (ordinary air would oxidize the developer too quickly). The gas is derived from cylinders, and burst intensity controlled by a pressure valve. In most cases a timing system opens the valve automatically at pre-determined intervals, to provide a regular and repeatable degree of agitation. This system needs additional equipment (gas supply, valve gear and tubes to lead the nitrogen gas into the tanks) but once installed requires little or no attention. Outfits are available in various sizes, according to the capacity of the tank system.

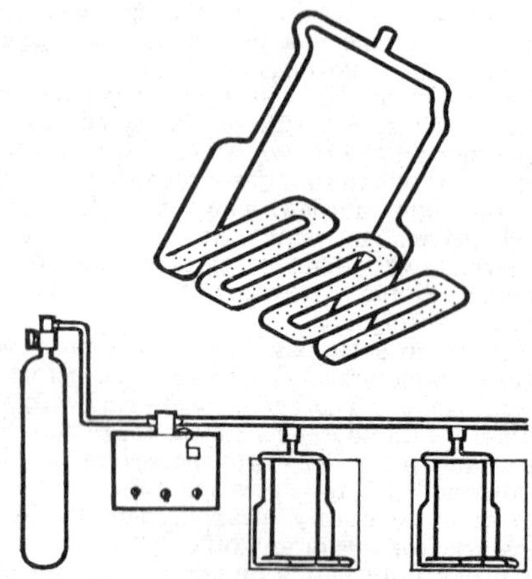

NITROGEN BURST EQUIPMENT. Gas burst agitation relies on the insertion of a hollow perforated grid-like gas distributor (top) into the developing tank. Usually every tank has such a distributor. These are fed from a gas cylinder via a timing valve which releases bursts of nitrogen gas at pre-set intervals to stir up the liquid.

Other agitation methods that have been tried include ultrasonic vibrations, produced by a quartz crystal or other generator. These vibrations break up the local concentrations of chemicals formed during the processing reactions and thus provide a uniform degree of mixing throughout the solution, without requiring the bodily movement of large solution quantities.

For highly standardized processing conditions (as specified for the development of film samples tested for speed) continuous agitation is nearly always essential. This may take the form of constant inversion of the vessel containing the developer, or of sweeping it across the surface of the sensitive material mechanically.

Continuous processing machines for roll films or roll papers as well as motion picture films produce the necessary agitation by the constant movement of the sensitive material through the solution. This may be supplemented by other methods such as gas burst agitation. Where films or papers are fixed to drums dipping into the developing solution constant movement is essential to keep the surface of the material wet all the time.

When processing large sheets of material (for example for photo murals) the processing solutions are usually applied by a spray (frequently connected to a recirculating pump system) or by swabbing over the surface. Both correspond to continuous agitation. L.A.M.

AIR BELLS. Bubbles of air which adhere to the surface of the emulsion on plates, films and printing papers. Unless removed by brushing the surface with the finger or by agitating the solution, these bubbles prevent the developer from acting on the emulsion and leave clear spots on the finished negative. The use of a wetting agent does much to stop air bells from forming.

See also: *Faults.*

AIR BRUSH

The air brush has been in existence for well over fifty years and during that time has become an essential tool for every commercial art studio. It brings within the scope of the artist effects that are difficult, if not impossible, to obtain with the ordinary brush. Colour can be much more quickly applied than with any type of brush with the added quality of a perfectly even coat and the ability to achieve an infinite variation in tones.

In this way it becomes possible to blend in tone and outline corrections with the original image in such a way that the handwork becomes almost invisible. This is especially valuable when two or more images have to be combined—the airbrush work here hides joins and abrupt changes of line and tone.

The air brush projects liquid colours on to the paper or other surface in the form of a cone-shaped spray pattern of extremely finely atomized colour by means of compressed air.

The air brush is used by photographers for colouring photographs, working up positives and negatives, introducing cloud effects, removing unwanted backgrounds and general touching-up. It is also very widely used for process retouching, working up litho plates and even for applying gum to photographs.

Other airbrush effects include the emphasis or suppression of selected picture areas, and the introduction of lettering or shadows.

Construction. The air brush is an instrument about the size and shape of a fountain pen. In place of a nib, there is a pointed nozzle with a needle, through which the spray of colour is projected. Immediately behind the nozzle there is a paint reservoir, and behind that there is a control button placed so that the tip of the forefinger rests naturally on it when the brush is held pencil-wise in the hand. Below the control button is a connexion for the flexible compressed air pipe.

The control button can be moved in two directions; it can be pressed down to allow compressed air to enter from the pipe and pass through the air cap, and it can be pressed back

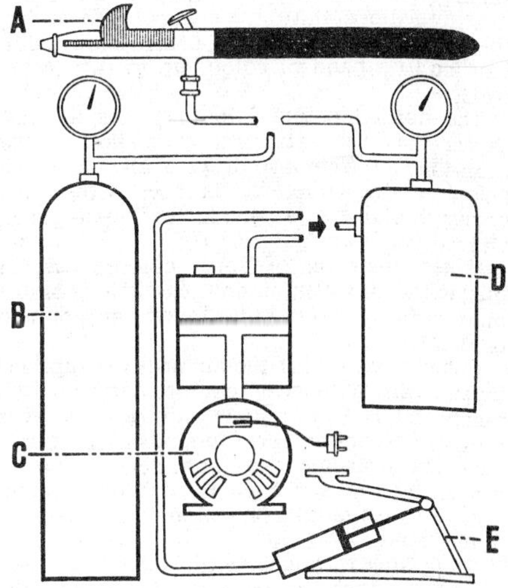

AIR BRUSH COMPONENTS. The essential parts of an air brush are the brush itself A, and the air supply. The latter may be derived from either a cylinder of compressed air B, or an air reservoir D. Such a reservoir would be replenished by a pressure pump C or a foot pump E. In every case a pressure gauge and control valve are desirable to check and regulate pressure of the supply and so ensure even spraying.

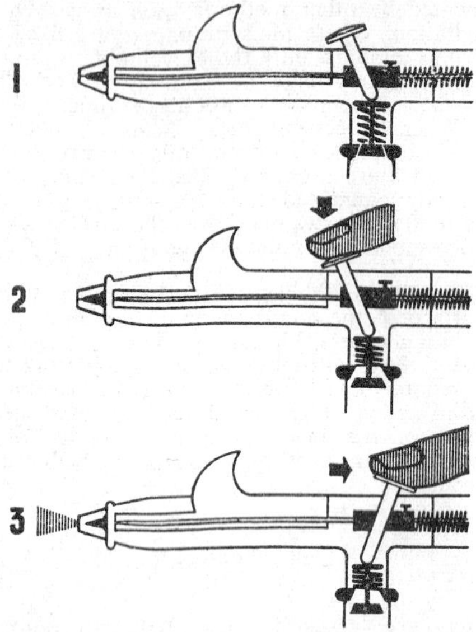

AIR BRUSH OPERATION. 1. Normally both the air valve from the compressed air supply and the needle valve of the paint reservoir are closed. 2. Pressure on the control button opens the air valve. 3. Pulling back the button opens the needle valve; the compressed air stream carries the paint away from the needle tip in a stream of tiny droplets.

to withdraw the needle controlling the paint supply to the nozzle. By suitable manipulation of the finger button, it is possible to control the covering power of the brush from a fine hairline to a band of colour up to two inches wide.

The normal colour container has a small capacity as the artist rarely wishes to use more than a dozen or so drops of any particular colour at one time. Air brushes with larger colour containers are made for small poster and repetition work.

Water, light oil or spirit colours can be used in the air brush, but all pigmented colours must be finely ground and free from sediment or lumps.

Compressed air for the air brush is supplied by an electrically-driven compressor or a foot pump and is stored in an air receiver at a pressure of about forty pounds per square inch. The instrument is connected to the air receiver by means of a length of flexible rubber hose with suitable connexions to make an airtight joint.

On passing through the air valve, the compressed air goes through air passages to the air cap at the front of the instrument. This air cap entirely surrounds the fluid nozzle, the ...of which lies centrally in the hole at the ...f the air cap. When the compressed air ...t through the air cap it drives forward ...ding cone, at the same time creating a slight vacuum within the fluid tip and sucking colour from the cup along the needle. On mixing with the air stream the colour is atomized into very finely divided particles.

Use. The instrument is held between the index finger and the thumb exactly as a pencil is held, but with the tip of the forefinger resting lightly on the top of the finger button. The nozzle of the air brush should be pointing at about forty-five degrees to the surface to be sprayed whilst the air hose lies naturally across the back of the hand.

To operate the instrument the finger button is first pressed downwards when compressed air immediately passes through the air cap, but no colour emerges. The finger button is now drawn backwards slowly, but still keeping the downward pressure, and the instrument will start to spray colour. The farther the finger button is drawn back, the greater will be the amount of colour sprayed. When ceasing to spray, the finger button must be allowed to go right forward before releasing the downward pressure on it. If this is not done, a particle of unatomized colour will remain in the nozzle and blow on to the paper in a blob the next time the button is depressed.

A certain amount of practice is necessary to enable the user to master precise control of the flow of colour, but once he gets the "feel" of the finger-button action proficiency will quickly follow.

To draw a very fine line, the instrument is held with the nozzle practically touching the paper, but at an angle to allow the released air to escape. The finger button must be drawn back only very slightly, as otherwise far too much colour will be released, and the air brush must be kept moving fairly quickly along the line or the colour will build up to form a blob.

For filling in areas of solid colour the air brush should be from three to four inches from the surface and the area to be covered filled in with steady, even strokes, holding the instrument at the same distance from the surface throughout the operation. Each stroke should overlap half the width of the previous stroke

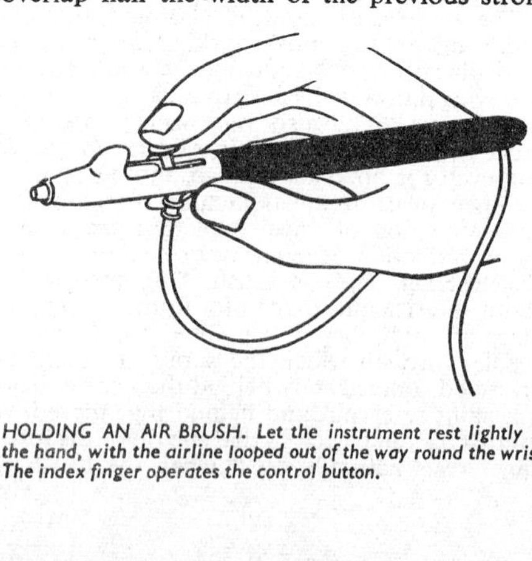

HOLDING AN AIR BRUSH. Let the instrument rest lightly in the hand, with the airline looped out of the way round the wrist. The index finger operates the control button.

23

to obtain an even coverage. As the movement of the air brush ceases at the end of the stroke, the finger button is allowed to go forward to cut off the flow of paint and it is not drawn back again until the brush is starting the next stroke.

Graduated shading is produced simply by controlling the distance the finger button is drawn back in relation to the distance of the brush from the paper. For instance, if the button is drawn back very slightly as when drawing a fine line, but with the instrument some distance from the surface, an almost imperceptible haze is applied which can be graduated right up to solid colour by gradually drawing back the finger button.

It is possible to impart a stippled finish with an air brush simply by reducing the air pressure. This process is particularly suitable for some litho work where a comparatively coarse grain is essential.

Stencils and Masks. For photographic retouching, particularly of engineering subjects where sharp images and good contrasts are required, it is a simple matter to confine the effect of the brush to a sharply defined area by cutting a stencil to mask off the rest of the photograph. These stencils are cut from a special transparent varnish paper which will adhere lightly but firmly to the surface of the photograph. A sheet of the varnish paper is smoothed on to the print and the artist cuts around the outline of the part to be sprayed

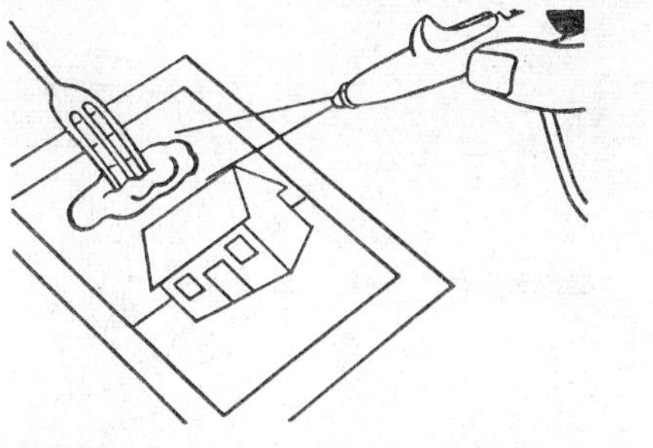

SPRAYING CLOUDS. Fluff out a piece of cotton wool of suitable size, hold in position on print with a fork, and spray over surrounding portion. Add modelling after removing cotton wool.

with a very sharp-pointed stencil knife, applying just sufficient pressure to cut through the varnish paper only. The cut-out portion is then lifted out, leaving the required part of the picture exposed.

Some artists prefer to lay a piece of draughtsman's tracing paper over the photograph and draw around the area with a fine pencil. The tracing paper is then laid on a sheet of cardboard and the mask cut out by going over the pencil line with a sharp penknife. This method has the drawback that the mask does not adhere to the surface of the print and, no matter how carefully it is pressed down, there is always a risk that the air pressure will blow colour under the edge and blur the outline of the picture.

Masks used in close contact with the paper give a sharp outline which is suitable for technical subjects, but is generally too hard for normal subjects—e.g., portraits or landscapes. In such cases the mask is traced on to stiff paper or card which can be held away from the surface of the print. The farther the mask is held from the surface when the spray is applied, the softer the outline. This technique would be suitable for darkening the lips in a portrait where a mask in contact would produce an obvious edge.

Compressed Air. The air should be maintained at a regular pressure to ensure first-class work, and it should also be perfectly clean and free from oil and condensed moisture. It is always best to use a compressor designed specially for studio use and to follow instructions given by the manufacturer to drain accumulated water from the air receiver at regular intervals.

Before connecting up the air brush, the air should always be allowed to blow through the hose for a short time to remove any dust particles that may have lodged there. To avoid unnecessary wastage of compressed air, it is essential for the air hose connexions to be screwed firmly on to the nipples provided on the air receiver and air brush. This must always be done with the fingers only; it should never be necessary to use a tool.

Where the air brush is used for a variety of

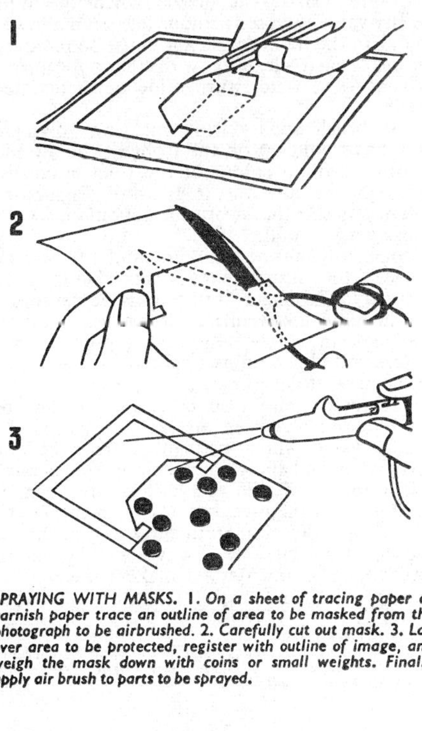

SPRAYING WITH MASKS. 1. On a sheet of tracing paper or varnish paper trace an outline of area to be masked from the photograph to be airbrushed. 2. Carefully cut out mask. 3. Lay over area to be protected, register with outline of image, and weigh the mask down with coins or small weights. Finally apply air brush to parts to be sprayed.

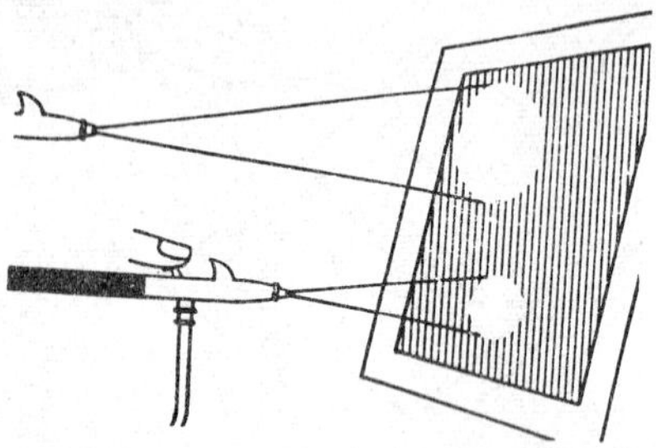

SPRAYING DISTANCE. The farther the air brush is held from the surface to be sprayed, the larger the area covered by the spray. For fine lines and detail hold the brush really close.

processes, it is advisable to have a supplementary air-pressure regulating tank which enables the pressure to be varied from high to low and back again without having to waste time waiting for the air compressor to build up the air pressure.

Air Pressure. Whilst each air brush can be relied upon to do the work claimed for it by the manufacturers, it will be found that each instrument has a certain amount of individuality and no two air brushes are exactly alike in their touch and performance, particularly after they have been in use for some time—one instrument will work best with an air pressure of 28 pounds per square inch, whilst another will need 32 pounds. That is why even in a big studio an artist generally prefers to work with his own air brush.

Cleaning. It is most important to keep the air brush scrupulously clean. If colour is allowed to harden in the fluid nozzle, it will most probably be damaged beyond repair. Immediately after use and whilst the colour is still in a moist condition any colour remaining in the colour cup should be emptied out and replaced with clean water for water colours, turpentine for oil colours or methylated spirit for spirit colours. Any colour adhering to the cup is then removed with a small, stiff brush.

The colour cup is then emptied and refilled with the appropriate solvent and a little is sprayed through the instrument. Next a fingertip is pressed lightly over the front of the air cap and the finger button is depressed and pulled back so that the compressed air is blown back through the front of the fluid nozzle into the colour cup. By agitating the solvent in this way it cleans the fluid passage thoroughly from the cup to the nozzle. It may be necessary to empty the cup and repeat the cleaning process until all traces of colour have been removed from the instrument. Great care should be taken not to spill any liquid into the slot through which the finger button lever works.

the air cap becomes clogged with dried it is unscrewed from the instrument and refully inside and outside with a stiff brush and the appropriate solvent. The hole in the air cap can be cleaned with a splint of soft wood, but on no account should a metal implement be used or it may injure the cap and produce an imperfect spray pattern. When the air cap has been removed, the exterior of the fluid nozzle can be carefully cleaned with a soft rag immersed in cleaning fluid. Fluffy material should not be used for this purpose. When the air cap is replaced, it should be tightened up with the fingers only.

Common Troubles. The following notes indicate the probable causes of the commoner air brush troubles.

Air brush sprays broken line: the trouble may be caused by using too coarse a pigment. It can be prevented by ensuring that the colour is finely ground and filtered through a fine mesh muslin to eliminate any large particles. Thoroughly clean the fluid nozzle, needle and air cap.

Coarse texture line and spray pattern: the air pressure is probably too low. The normal working pressure is about thirty pound per square inch, but it may have to be increased to as high as forty-five pounds when heavy bodied colour is used. This defect may also be caused by holding the air brush too close to the surface of the paper.

Uneven line or one-sided spray pattern: there is probably an obstruction in the air cap which should be removed and cleaned out.

Colour leaking at nozzle when the fluid needle valve is closed: colour has been allowed to dry in the fluid nozzle and must be removed by prolonged cleaning with the appropriate solvent—i.e., water, turpentine or methylated spirit.

Air brush spits during spraying: pigment may have built up on the front of the air cap through using a colour of too thick a consistency or too low an air pressure. Pigmented colour suitable for air brushing should have the consistency of milk.

Air brush spits at beginning and end of each stroke: the instrument is not being used correctly. The finger button must be depressed all the time the needle is drawn back, otherwise colour will be left on the point of the nozzle and blown off as a spot at the beginning of the next stroke.

This defect may also be caused by dirt or solid colour particles on the seating of the needle valve which prevent it from closing completely and allow liquid colour to flow past.

The faults referred to are those that can be dealt with by the user, but any of them may be caused by a defect in an integral part of the air brush, in which case the wisest plan is to return the instrument to the maker for examination.

J.E.S.

See also: *Retouching*.

Books: *Airbrush Technique of Photographic Retouching*, by W. F. King and A. L. Slade (New York); *Retouching*, by O. R. Croy (London).

AIRCRAFT CAMERA

Air photography proper is generally taken to mean the photography of the ground from the air. It has two main purposes: to provide maps and economic information in peacetime and reconnaissance in wartime. There is some overlapping because air survey maps are also made during a war.

Survey cameras tend to resemble one another and show little departure from basic essentials, while reconnaissance cameras take varied forms according to their rôle. Special cameras are used in missile recording.

Common Features. All air cameras must be very robust in construction to stand relatively rough handling and they are generally much larger and heavier than ground cameras. While some of the smaller ones are hand-held, the majority are designed for fixed mounting in aircraft.

In principle the air camera is a rigid fixed-focus high precision instrument designed to take roll films. Operation may be manual but is usually electrically powered, with fast recycling rates and varying degrees of automation.

Formats range from $2\frac{1}{4} \times 2\frac{1}{4}$ ins. to 18×18 ins., the commonest being 9×9 ins. and, on the Continent, 7×7 ins. (18×18 cm.). Films, both perforated and unperforated, from 70 mm. to $18\frac{1}{2}$ ins. wide are used. These are carried in interchangeable magazines holding up to 1,000 feet.

Fast panchromatic emulsions are most common, used in conjunction with a minus-blue filter, but slower types and infra-red are employed for special purposes. Colour is also proving increasingly valuable. These emulsions are usually coated on dimensionally stable "topographic" bases (usually polyester film).

Optical System. Air photographs must provide the best possible definition, so highly corrected lenses are necessary. In particular, the lens must as far as possible be free from distortion. Since the subject is always at infinity, no focusing is required. The lens is fixed permanently at one end of the cone or optical unit; the register glass which defines the focal plane being likewise fixed permanently at the other. The survey camera is required to measure angles between objects on the ground with the precision of a theodolite, and for this to be possible, the optical unit must be rigid with no possibility of relative movement between lens and focal plane once assembled.

The register glass against which the film is located by the pressure plate is a feature of British survey cameras, but American and Continental designers favour a different system in which the film is held flat by vacuum against a rigid flat metal plate while pressed down against a metal frame defining the focal plane. The film should be held flat to 0·0005 in. or better because small departures from flatness which have a negligible effect on definition cause appreciable errors in the map.

The cone is an alloy casting, the locating faces for lens mount and focal plane being machined flat and parallel within fine limits. Diaphragms are often cast on the inside of the cone to keep flare light to a minimum.

In cameras fitted with a register glass this is often engraved with an accurately calibrated grid. By reference to the grid pattern on each picture it is possible to correct for small residual distortions which occur even with survey quality film base.

With a distortion-free lens, rays entering at any angle with the optical axis would emerge on the image side at the same angle, thus the angles subtended at the lens by image points would be exactly the same as those subtended by corresponding objects on the ground. Since no lens is perfect, the camera is calibrated before use by determining the departures from this exact angular relationship at every point in the field and drawing a distortion curve. The lens is set at the focus position which gives best over-all definition, and a value of calibrated focal length is adopted which gives the lowest average distortion over the negative area.

Modern survey lenses have strikingly low distortion figures; in the best examples no image point is displaced from its theoretical position by more than 0·01 mm.

Survey cameras are very commonly used at heights of 15,000–20,000 feet, a 10 foot object then being reproduced by a 6 ins. lens as a patch 0·004 in. diameter. Thus the need for good resolution can be appreciated. Formerly the lens resolution was the limiting factor, even with the somewhat grainy high-speed film normally used. With the best modern lenses this is no longer true except in the outer zones of the field.

Shutters and Image Movement. The exposure time used depends on the speed of the aircraft and the scale of the photograph. Focal plane or diaphragm shutters may be fitted, the former speeded up to 1/2,000 second. Focal plane shutters introduce distortion however, due to the aircraft's forward movement during the shutter blind transit.

Diaphragm shutters are therefore preferable. They are basically similar to those on ground cameras but much larger. The stresses involved in fast recycling over long periods make efficiency and reliability difficult to attain, especially at high speeds and large diameters. Exposure times range between 1/250 and 1/800 second. Certain types of electrically driven disc shutters overcome these difficulties and also give shorter exposures.

Fast shutter speeds freeze image movement on the film. But where this is not feasible, the alternative is to move the film in step with the image. This is the principle of Image Motion Compensation (IMC), embodied in some air cameras or separately in the magazines. Th film is moved during exposure, usually by the transport mechanism or by moving the film aperture or pressure platen. Speed of movement is calculated from the aircraft's height and speed and the lens focal length. The image thus stays stationary on the film and longer exposures are possible.

Instrumentation. An increasing amount of information is today recorded on the film rebate. It may include the time, the aircraft's attitude, even its geographical position. It is piped optically to the film from instruments on the camera, and recorded with each exposure. Alternatively, a separate camera, synchronized with the air cameras, may be focused on the pilot's instrument panel.

This information facilitates analysis of the photographs, since it shows exactly what the aircraft was doing when each picture was taken.

Mounting and Installation. The mounting serves to locate the camera positively in the aircraft, to insulate it from vibration and landing shocks, and to provide drift adjustment. The latter is required because the axis of the aircraft does not point along the line of flight, when flying on a given course with any wind coming from the side. In other words, the aircraft moves simultaneously forward (due to its engines) and drifts sideways (due to the wind). So the camera must be rotated about a vertical axis through a corresponding angle to ensure that the edges of the photographs are parallel to the line of flight. Drift adjustment up to 30° may be provided and also some vertical adjustment. The anti-vibration elements are so designed that negligible angular movement is transmitted to the camera during the time of exposure.

Heating is provided to counteract the low temperatures at altitude, by enclosing the camera and photographing through an optically flat window while maintaining a stream of warm air around the installation.

Operation. Air photographs are taken in a series of overlapping exposures. The time interval between them is determined by the aircraft speed and height, format size and focal length of lens. It varies between 1/8 second for fast low-level photography to 2–60 seconds used in high altitude work.

Control may be entirely manual—a release button and stopwatch—ranging to fully automatic from a remote panel. The former is often employed in surveying, the latter in military aircraft where the cameras are often inaccessible.

Operation of the camera throughout the complete cycle is automatic and repeats at intervals determined by setting the control unit. This unit provides electrical pulses at regular intervals, each pulse triggering the shutter, causing the film to wind on and re-ming the shutter ready for the next exposure. control unit incorporates a computing ometer that triggers the camera electrically. It may also have lens iris, shutter, drift setting and IMC controls, functioning indicator lamps and frame counters. In sophisticated installations many of these settings are monitored electronically and adjusted automatically.

Hand-Held Cameras. Hand-held cameras have few if any of these refinements. Used mainly for taking oblique "pin-points" of single features, they are designed as compactly as possible, protectively cowled and with double hand grips and trigger release for maximum steadiness. Other characteristics are fast lever-wind operation and open frame sights. Maximum film size is about 7 ins., the commonest being 70 mm., usually held in magazines.

Some fixed type air cameras are convertible to hand operation simply by fitting a special mount. Others may be rigid, fixed-focus versions of standard ground cameras and may even take plates.

Most precision ground cameras can still be used for this type of photography, provided they are protected against slipstream and rough handling. Miniature cameras are less suitable, owing to the inherent low information capacity of 35 mm. film.

Survey Cameras. Commercial survey demands the maximum accuracy at the minimum cost. Speed, while important, is not essential, since there is nothing but cost to prevent several attempts being made on one objective. So the most suitable aircraft can be chosen, relatively slow and large, to allow the operator to sit with the cameras and supervise them manually. The cameras themselves are therefore relatively simple.

Since they are required to measure angles between objects on the ground with the accuracy of a theodolite, the cameras must also be of the highest precision. The film must be held flat to 0·0005 in. or better because small departures from flatness that barely affect definition, cause appreciable distortion errors. An accurate grid engraved on the register glass and transferred to each picture helps to correct small residual distortions of the film base. Glass is more stable in this respect, and some survey cameras use plates in a quick-change magazine. Focal plane shutters introduce distortion, and all true survey instruments are characterized by inter-lens shutters.

Modern survey lenses have strikingly low distortion figures, although many are wide-angle objectives. In the best examples, no image point is displaced from its theoretical position by more than 0·01 mm.

Survey cameras are commonly used at heights of 15,000–20,000 feet, a 10-foot object then being reproduced as a patch 0·004 in. in diameter. So resolution is important. Typical values for modern lenses, for low contrast detail on aero type film are 23 lines per mm. axially and 15 lines at the limits of the field.

Apertures do not normally exceed *f*4 for normal and *f* 5·6 for wide-angle objectives.

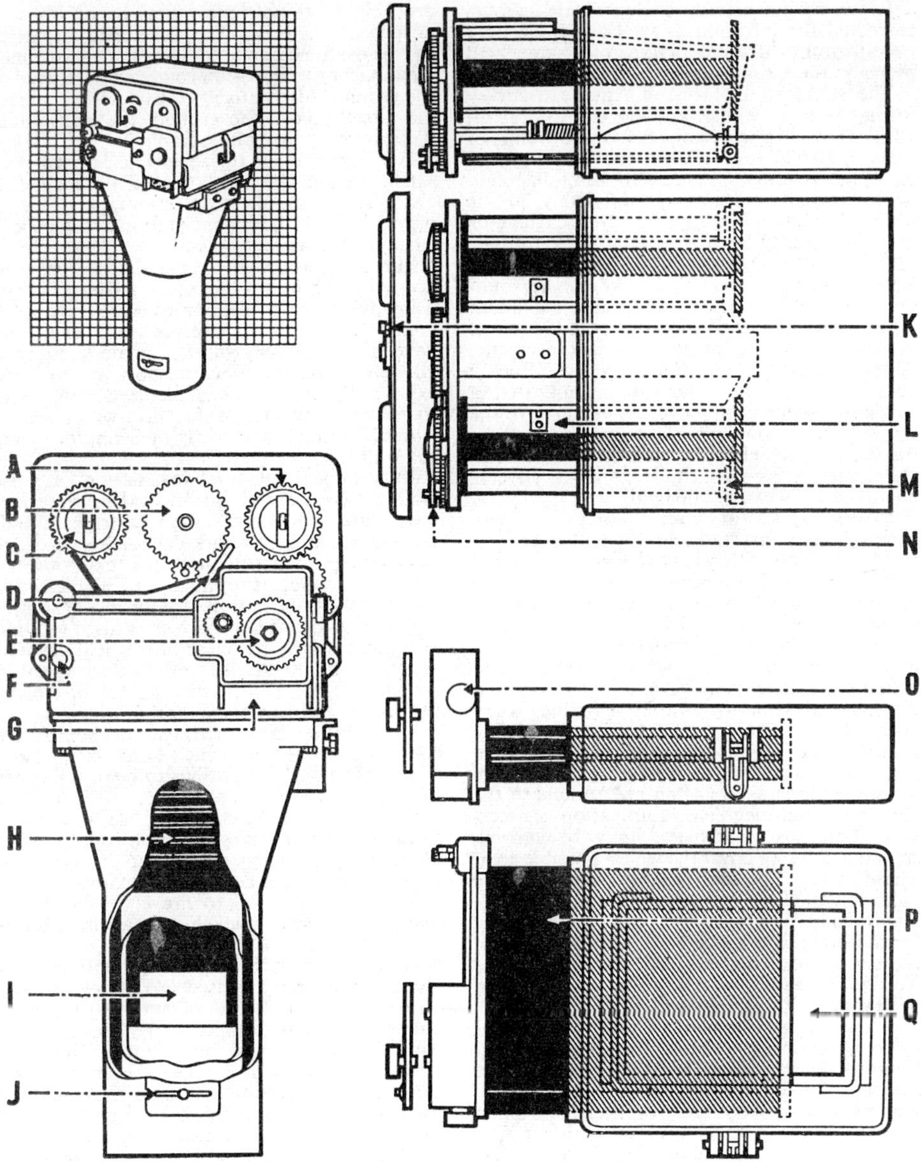

AIRCRAFT CAMERA. The camera consists essentially of three sections: the cone, the shutter assembly, and the film magazine and transport unit.

Top Left: *Complete camera unit against 1-in. grid to show scale.*

Bottom left: *The main features of the camera (shown without the magazine cover) are* A, *take-up spool friction;* B, *counting unit;* C, *feed spool friction;* D, *counter arm adjuster;* E, *gear box assembly of shutter;* F, *shutter speed adjustment;* G, *plug block for electrical connexions;* H, *baffles;* I, *lens;* J, *iris diaphragm adjustment.*

Top right: *The main parts of the film magazine are:* K, *magazine cover;* L, *pressure plate;* M, *driving sprocket;* N, *take-up spool drive.*

Bottom right: *The shutter unit includes:* O, *hand release button;* P, *shutter blind;* Q, *register glass to keep film flat.*

The Camera takes roll film yielding 500 *exposures* $7\frac{1}{2} \times 9$ *ins. and takes interchangeable lenses in detachable cones.*

Some special air survey lenses are colour-corrected for infra-red as well as visible light rays, forming infra-red images in the same focal plane as the visual image.

The standard British and American survey format is 9 × 9 ins., and on the Continent, 18 × 18 cm. In wide-angle survey for maps of scale 1:20,000 to 1:60,000, the lens angle of view is commonly 45° across the semi-diagonal, while for larger scales approximately 30°, is usual. Corresponding approximate focal lengths are 6 ins. and 12 ins. for the 9 × 9 ins. format; 11·5 cm. and 21 cm. for the 18 × 18 cm. image.

Reconnaissance Cameras. Unlike survey, military reconnaissance demands the maximum information in the shortest possible time. Instead of precise angles and distances, the pictures must show what is where. Enemy action may deny any other opportunity than a single high-speed run, covering as much of the target area as possible. This may have to be made at any height from near ground level to the highest possible altitude, often in adverse conditions of weather and light.

Diverse types of aircraft may be used, generally very fast and seldom ideal as camera platforms. At best they may be specially adapted versions of combat aircraft; at worst, standard aircraft modified to carry a camera on occasional photographic sorties.

All these factors profoundly affect camera design. The emphasis is on high output and definition rather than precise operation and dimensional accuracy. Rapid recycling, wide-aperture lenses up to *f* 1·5 and fast reliable shutters, diaphragm or focal plane, are called for, some distortion errors being acceptable. Since the cameras are often inaccessible to the aircrew, a high degree of automation is necessary. Batteries of cameras may be carried, spread in a fan to cover the widest possible area. These must operate reliably under severe vibration and high "g" (gravitational pulls).

Unorthodox Techniques. Although it is desirable for a reconnaissance camera to do many jobs, the search for greater speed and coverage as well as increased specialization, has led to a wide variety of types and techniques.

A logical development of IMC is the continuous strip camera. This moves the film continuously past a fixed focal plane slit at the calculated image motion compensation speed, making an unbroken record of the ground along the film. Both shutter and intermittent film movement are thus eliminated but a fairly complex control system is needed. Another development is the panoramic camera, in which a rotating prism scans from horizon to horizon across the flight path and records on film moving past a fixed slit in step with it. The camera replaces bulky multiple-fan installations.

To save time, high speed in-flight processing is sometimes used. The film is carried in special magazines in which it is developed and stabilized immediately after exposure. The chemicals are held in thixotropic pastes or on rolls of absorbent material. The developed negatives can be examined on landing or may be scanned by an airborne television camera and relayed to the ground. Alternatively, the television camera may itself record the ground, the transmitted pictures being then photographed.

An extension of this principle is used with satellite cameras, which record on magnetic, electrostatic or thermoplastic tape. This is played back to the ground station at high speed when within range.

Night Cameras. Night reconnaissance cameras normally work on the open flash principle, using pyrotechnic flash bombs or cartridges, for high or low level work respectively. They may have simple capping shutters or no shutters at all, relying on the film transport to move the exposed frames out of the picture aperture and so prevent further exposure. Ultra-wide aperture lenses are fitted, some loss of definition being acceptable.

In a typical low-level installation, a pulse from the control unit advances the film, opens the shutter and fires a flash cartridge simultaneously. After a short delay, the cartridge bursts clear of the picture area, exposing the film and the cycle is repeated. In a high altitude version, the control unit may open the shutter and release a flash bomb which is fused to burst at a given height and clear of the field of view. When the bomb bursts, a photo-electric cell in the aircraft triggers the mechanism for closing the shutter and advancing the film.

Usually the flash cartridges or bombs are contained in ejectors which hold six or more flashes. The flash output has to be of the order of millions of lumen-seconds.

The release of the cartridges, as well as the shutter operation, can still be controlled to take place automatically at pre-set intervals to provide continuous coverage of an area.

Should a flash fail to fire, a further automatic unit closes the shutter again after a specified time.

Missile Recording. Special 16 and 35 mm cine and still cameras are used in radio-controlled target aircraft to record guided missile attacks and measure miss-distances. They are carried in wing-tip "pods" and have ultra wide-angle lenses, sometimes up to 180°, so that several cameras give all-round coverage. The cine types work at framing rates up to 200 f.p.s.; the still cameras sometimes have shutters actuated by small explosive charges. Operation is by radio or radar control.

Such cameras have to be extremely tough, to be recoverable if the target aircraft crashes. Special steps are taken to protect the film and this may take the form of a recoverable or perhaps jettisonable cassette, in which the exposed film is sealed in an armoured watertight casing. G.C.B. & P.B.M.

See also: *Aerial survey; Air Force photography; Photogrammetry.*

AIR FORCE PHOTOGRAPHY

The primary role of military air photography is reconnaissance; since air forces are the only means available in wartime, another important task is survey. The two rôles often overlap sufficiently to be treated as one. Other uses of photography include the recording of weapon attacks, and a wide range of ground work.

Unlike most civilian air photography, the requirement is normally for quantity of information and speed of delivery rather than photogrammetric accuracy. As much information as possible may have to be gathered in a single sortie, which may be made at any height between ground level and 60,000 feet or over, perhaps in the face of enemy action, in adverse weather or at night.

The aircraft are usually combat types, not designed primarily for photography, although special pilotless "drones" have been developed. The aircraft may be specially adapted for photographic reconnaissance or perhaps only slightly modified for an occasional sortie. They are likely to be very fast and are often unarmed, relying on speed and height for protection. The choice is influenced by the fact that the photographic reconnaissance aircraft must return to base with its films whereas a combat aircraft's mission is completed after delivering its attack.

More than one camera is normally carried, mounted internally or in streamlined "bulges" or pods and photographing through optically flat windows. As they are not always accessible, remote control and automation are needed. Efficient heating is required to keep the film flexible and the lenses and windows clear of condensation, particularly after a fast dive from the sub-zero temperatures at high altitude. Special periscopic sights are fitted to let the pilot aim at the ground below.

Reconnaissance. A characteristic of all air force photography is the amount of logistic support required, both in materials and in trained personnel. This is particularly true of reconnaissance, which may be broadly divided into planning, the photography itself, processing and interpretation. To give some idea of the scale of organization required, a single bomber-type aircraft may carry eight cameras, each with 1,000 feet of film and producing a total of some 10,000 negatives.

The type of operation ranges from long-range strategic reconnaissance of factories and installations deep in enemy territory, to impromptu low-level tactical missions in support of troops. The requirement may be anything from full-scale vertical and oblique coverage of large industrial areas to a few close-range oblique pin-points of enemy positions. This wide range of tasks demands a variety of equipment and techniques. The cameras must be robust high-definition instruments with fast shutters and image motion compensation, and wide-aperture lenses for operation in poor light. Since stereoscopic examination helps interpretation of the pictures, each print must overlap the last by at least 60 per cent; so that fast automatic recycling is needed, particularly at low levels.

A single-seater aircraft normally carries at least two cameras: one vertical and one sideways- or forward-facing oblique. In large aircraft a battery of cameras may be spread in a fan, covering the ground from horizon to horizon. In some cases the cameras are "packaged" by mounting them in interchangeable crates, each designed for a specific task, e.g. high-level, low-level or beach reconnaissance, or survey.

Special requirements lead to special techniques. For long narrow features like roads, beaches, railways, etc., continuous strip cameras may be used. A single panoramic camera can replace the multiple fan, or a small version may be used with a fan to identify the pictures.

Various methods may be employed to ensure the safe and rapid receipt of the pictures. High-speed in-flight processing of the film as it leaves the camera produces negatives ready to print on landing. They may also be scanned in the air by a television camera and immediately transmitted to base. Alternatively, the whole operation may be done by television.

This leads logically to direct recording on video tape, but so far the only reported use of this is in satellites.

Where such methods are not used, the films have to be processed on the ground. Large quantities are handled by rapid continuous processing and printing machines like those used in cinematograph laboratories. Manual processing of smaller numbers can be hastened by printing from wet negatives and by spirit drying, and it is possible for the finished prints to be ready to be air-dropped at a forward position within 15 minutes of landing. All this may be carried out in the field, in specially equipped tents and caravans.

Interpretation extracts the maximum information from the pictures. It is necessary because the identity of objects viewed from the air, particularly from vertically above, is not always apparent. Thus a road bridge may be invisible until viewed stereoscopically or identified by the shape of its shadow, while a vague blur in a single picture assumes shape and size under stereo examination. Other factors are size—found by scaling and measurement—shape and tone (and colour when used). This type of interpretation is done on vertical pictures, backed up where possible by oblique views to add realism and perspective. In cases of urgency, the negatives may be interpreted without printing.

Night Reconnaissance. Air photography in the Royal Air Force is not confined to clear sunny days. Height and speed are normally the only safety precautions that photo reconnaissance aircraft can employ against enemy opposition and, unlike the bomber, its mission cannot succeed unless it lands back safely in its country of origin. Under these circumstances and also because of the need for the immediate assessment of the damage caused by raids and other military operations, night photography plays a most important rôle.

For certain types of subject it is possible to employ an electronic flash produced by releasing through a special tube a condenser charge built up by an engine-driven generator in the course of the outward flight of the aircraft. But so far the pyrotechnic flash holds its own as the best type of illuminant and there is no great difficulty in making flash bombs big enough to cover any target area from any altitude. Unfortunately, as the size of the pyrotechnic charge increases, so does the duration of the flash. Open flash technique is essential, since the entire light output must be utilized, and 1/10 second is about the minimum that can be expected with this type of flash. Again, therefore, the requirements for high-level daylight photography — moving film, anti-vibration mountings, and vertical stabilization—are also essential in successful work at night.

A number of infra-red applications have also evolved from the necessity of photographing at night, in low light and through haze and cloud. The recording of airborne radar is an alternative method, useful in marine reconnaissance and in making radar maps for subsequent blind navigation above cloud.

Survey. Air force survey techniques differ little from those of commercial mapping, except where they are affected by wartime conditions, and in the use of faster aircraft. In fact air forces often assist with peacetime surveys, as with the post war 6 inch-to-1 mile survey of Great Britain. Planning and execution are generally the same as for reconnaissance.

The limits set by aircraft design and performance are just as decisive in low altitude oblique reconnaissance. Aircraft considerably smaller than the type used for high-level reconnaissance are in the main employed for this type of work. These machines are capable of speeds faster than sound and in the design of any photographic equipment carried aerodynamic considerations must be given the first priority. Odd protuberances are out of the question on the surface of a modern fighter aircraft and it may even be necessary to incorporate the outer component of the lens as part of the fuselage of the aircraft. Cameras must be kept as small as possible and since at such high speeds it is impossible for the pilot to gauge the exact instant of exposure, it is desirable to use either a cinematograph camera or at least a camera that will automatically make a series of exposures to "bracket" the target area.

Obviously wartime air survey can seldom be done in conjunction with ground work and it may also have applications not normally met in commercial survey. For instance, it is possible to estimate the depth of water off an invasion beach from air photographs but this would normally be done by surface soundings in peacetime.

The greatest difference lies in the end product. In commercial survey this is usually a map. Wartime requirements are more often for mosaics—photographic maps composed of overlapping vertical pictures, covering a wide area of ground. Mosaics may comprise merely a stack of numbered photographs for the user to fit together or examine separately as necessary, or it may be a large composite picture, each print being cut, fitted together and stuck down on a mount, together with scales, compass directions, etc.

The final refinement, used in briefing for major operations, etc., is the three-dimensional terrain model. It is built up on a solid contoured base derived from stereo measurements, coloured and equipped with model trees, houses and other features.

Strike Recording. An important photographic task in any air force is recording weapon attacks, both "live" in action and simulated in practices.

The best-known application is the cine camera gun, usually a 16 mm. camera, aligned with the aircraft's guns and operated by the firing button. In war, the films help to substantiate claims of damage; in training, they take the place of live ammunition and permit assessment of results. The same cameras or larger 35 mm. or 70 mm. versions are used to record dive-bombing and rocket attacks and practices.

The cameras run at speeds from 8 to 120 frames per second and are designed to overrun for a few seconds after release of the firing button, to allow time for the last round to reach the target. Since the pilot starts violent evasive action during this period, wide-angle lenses are needed to keep the target within the field of view.

Strike cameras may also be mounted in radio-controlled target aircraft (drones). Such cameras are usually cine instruments with wide-angle lenses, mounted in specially robust pods to protect the camera if the target craft crashes. Sometimes the film magazines are armoured to preserve the film for processing and evaluation even when the rest of the aircraft (and camera) breaks up completely on impact.

High-level bombing is recorded with standard vertical air cameras. Where cloud, smoke or debris obscures the target, a separate reconnaissance must be mounted subsequently.

Again, this branch of military photography is supported by a closely-integrated ground organization for maintaining and reloading the cameras, processing films and assessing the results as quickly as possible.

Ground Photography. Apart from aerial work, much of the air force photographer's time is taken up with ordinary ground photography, covering many aspects of commercial and industrial work. It includes portraits for identity cards, instrumentation and oscilloscope recording, press and publicity material, and the production of full-scale training films, often involving animation and slow-motion photography.

Photographs are required of all damaged aircraft parts and of the scenes of accidents, even of routine evolutions. In the Royal Navy for example, every individual landing on an aircraft carrier is recorded on 16 mm. film, often supplemented by still cameras. The results have proved invaluable in assessing the causes of accidents. P.B.M. & R.W.B.T.

See also: *Aerial survey; Aircraft camera; Air-to-air photography; Missile photography; Naval photography.*

AIRGRAPH. Letter transmitted as a microfilm copy from which the receiver was given an enlarged print; chiefly used during World War II.

See also: *Microcopying.*

AIR PHOTOGRAPHS. Photographs can easily be taken from aircraft using equipment normally employed for outdoor work with little or no modification.

Amateur Pictures. The view from the windows of airliners is not always good, but, with luck in getting a seat and clear weather, quite reasonable records of a flight can be obtained with small hand cameras. If the photographer is fortunate enough to fly in a private aircraft, the view will be better and the scope for composition correspondingly greater.

Photography should only be attempted when close to the ground, say 1,000 feet or lower, unless another aeroplane or a striking group of clouds or mountains forms the motif of the picture. While the choice of apparatus is not critical, a direct vision finder is to be preferred, panchromatic film and, under good atmospheric conditions, a pale yellow filter are usually best. Shutter speeds of 1/100 second or slower should not be used. Exposure may be calculated as for distant landscapes, but a meter is preferable. Development should be for high contrast. Colour work can be done quite successfully, but an ultra-violet absorbing filter or haze filter is almost essential.

Yellow, orange, or red filters may be needed for black-and-white photography to eliminate haze.

Vibration may be a problem, and the camera should not be rested on window frames, etc. Airliners are not normally troubled by bumpiness, but the smaller aircraft will present problems in turbulent weather, and the proportion of successful shots may be small. The larger airliners usually experience the greatest bumpiness when at low altitudes—soon after take off and just before landing—which are usually also the best times for air to ground photography; this kind of turbulence is, however, seldom sufficiently serious to prevent the production of sharp photographs with the use of a fast shutter speed. Photography from open cockpit aircraft calls for considerable skill and some thought should be given to protection of bellows cameras against the force of the wind.

Before attempting photography from airliners, the amateur should verify that this is not forbidden by the regulations of the country over which the aircraft is flying.

Press Pictures. By logical extension the same approach would serve for occasional professional work of a pictorial, press, or record kind, yielding pictures with a broadly pictorial motif or news interest such as appear regularly in the press. These are taken for general interest and are not essentially different in technique or outlook from ground photography, nor do they account for more than a small fraction of all the film exposed in aircraft. Hand-held cameras are manufactured especially for this class of work. They have some of the features of the aircraft camera proper, such as rigid construction, absence of focusing, shutter for short exposures only, but are usually smaller and adapted for use in the hand.

An essential characteristic of all these pictures is that they show the scene in normal perspective, as it would be seen from the side window of the aircraft. Features appear in much the same spatial relationship as seen from the ground; the horizon may be visible; at least we can feel where it should be. The camera axis, in fact, has been at some small angle with the horizontal; in technical terms the photographs are obliques. The more important class of air photographs, however, is that taken with the camera axis pointing vertically downwards —known as verticals.

For certain purposes oblique photography also may be used in mapping and for military reconnaissance, but as there the scale varies across the picture verticals are more usual.

Vertical Photographs. To the uninitiated, the vertical air photograph is often meaningless because it shows the landscape in an unfamiliar aspect—the map view. Perspective is almost entirely absent because the objects are all clustered near to a mean plane, the earth's surface, which is usually at a great distance from the camera. Yet such photographs are of immense value in peace-time and military value in war, for map making and reconnaissance, because the scale is uniform across the picture, and because they contain a complete catalogue of all visible details on the piece of ground shown and indicate the exact location of each with respect to the others.

The positioning is not confined to two dimensions; if two shots taken immediately after each other are viewed in a stereoscopic viewer, the pictures leap to life in a startling manner, with every house and tree standing up realistically and every contour of the ground clearly displayed. By measurements taken from an accurate stereoscopic record, it is possible to draw a map in both contour and plan. G.C.B.

See also: *Aerial survey; Air-to-air photography; Balloons and kites; Infra red photography; Photogrammetry.*
Books: *Aerial Photographs and their Applications*, by H. T. U. Smith (New York).

AIR-TO-AIR PHOTOGRAPHY. Photographing an aeroplane in flight from another aircraft, or air-to-air photography, is highly specialized work calling for close co-ordination between the two pilots and the photographer. Air-to-air photographs were formerly taken from the passenger seat of an open two-seater, the camera sometimes being held in the slipstream. Nowadays the photographer can work in comfort through the open window of a cabin aircraft, but frequently he is handicapped by the great disparity of speed between such aircraft and the modern high-speed jet-propelled types, which he may be called upon to photograph. The high altitudes at which these sometimes operate necessitate a pressurized cabin, the windows of which may consist of a double sheet of Perspex, destroying definition in a photographic negative. Thus, though the speed of sensitized materials has kept pace with that of aircraft, the design of the latter no longer provides a vantage point for a photographer, who is therefore still obliged to use the slower machine. This greatly limits his freedom of action and choice of viewpoints.

Camera. Owing to the limited demand, there are only a few special "air-to-air" cameras on the market, although other cameras are, in limited quantities, adapted to the appropriate specifications. Cameras constructed for air-to-ground work are not suitable. For casual air-to-air work the normal direct vision type of press camera, taking $2\frac{1}{4} \times 3\frac{1}{4}$ or 4×5 ins. negatives, is quite suitable, as is also the single lens reflex type of camera, preferably in the roll-film size. When flying in an open aircraft, any bellows on a camera should be replaced by a suitable solid box. Air-to-air cameras usually are of rigid design with hooding cones to protect most of the camera controls against wind pressure. Interchangeable lenses may be provided. Some models take magazines of 70 mm. film for 50 or so exposures, obviating the need for frequent film changes.

Photographer's Aircraft. The aircraft from which the photographer is to work should have a speed comparable to that of the target aircraft and afford good vision. Speeds vary so greatly that the target aircraft may be stalling while flying faster than the maximum speed of the photographic aircraft. Under these conditions there is little time to compose the picture, and success depends on a combination of experience and luck.

In airliners having the engines, whether jet or piston, mounted in the wings in the conventional manner, the best camera position is in front of the wings. At any point behind the wing the heat trail of the exhaust from the engines will render the image unsharp on the film. If the only available seats on an aircraft are rear ones, it is best to take one as far back as possible, to shoot toward the tailplane, rather than in a forward direction.

The most suitable of all photographic aircraft is the now obsolete four-engined bomber with tail turret. Seated in the turret, the photographer has an angle of view of almost 180° in both horizontal and vertical planes.

The Subject. Air-to-air photography is a combination of portrait, landscape and high-speed photography, with the target aircraft as the model and the background constantly changing. The background should be chosen to suit the aircraft—e.g., clouds for a bomber, coastline for an airliner and sea or a ship for a naval machine.

Technique. In air-to-air photography at least three people are involved: the pilot of the photographer's aircraft, the pilot of the target aircraft, and the photographer. They must work as a team, and it is desirable for the pilots to have previous experience of this form of flying. It is also a help if the pilot of the photographic aircraft has an eye for a picture.

Work on a photographic sortie begins with briefing before take-off. A complete plan of the intention and order of execution should be arranged. If this is not done failure may result and flying time be wasted. This is important with large modern aircraft, which may cost hundreds of pounds for every hour they are in the air.

Once the team is airborne, inter-communication between pilot and photographer is essential, and radio communication between the pilots almost as important.

When the target aircraft is in close formation the lighting should be carefully studied before an exposure is made, and in case of need the pilot asked to alter course to effect an improvement. He can also pass messages to the pilot of the target aircraft, requesting alterations of relative position; or this can sometimes be done by hand signals. While a skilled pilot can reduce speed of a target aircraft flying singly for the benefit of the photographer, this is impossible in formation flying, and the photographer will have to take quick decisions when a picture presents itself.

Camera Manipulation. Fast panchromatic plates or films are used, and with a 2× or 3× yellow filter a lens aperture of *f* 8 will provide well-exposed negatives on most occasions even at high shutter speeds. Except for special effects, over-correction of skies with orange or red filters is not recommended. In very intense light above clouds it is advisable to allow sufficient exposure so as to permit soft development, otherwise the negatives will be harsh.

When working with the camera at an open cabin window, it must on no account be held out into the slipstream, as it will be wrenched away by the force of the air blast. The photographer must never allow the camera itself, or his elbow, to touch the aircraft during an exposure or the vibration will spoil the definition of the negative. C.E.B.

See also: *Air Force photography; Air photographs.*

AIRY DISC. Central area of the image of a point source of light focused by a lens free from aberrations. The Airy disc is surrounded by concentric rings of light of feebler intensity which can in practice be ignored. The effect is an optical phenomenon connected with the wave nature of light and is called after its discoverer, G. B. Airy, Astronomer Royal in 1830.

ALBADA FINDER. Direct vision optical viewfinder devised by Van Albada in which the picture area is indicated by a white outline apparently suspended in space. This viewfinder is useful in sports photography. Present bright-line finders work on a similar principle.

See also: *Viewfinder.*

ALBERT, JOSEPH, 1825–86. German photographer. Made collotype commercially practicable in 1868, introduced as Albertype.

ALBERT EFFECT. Particular type of image reversal produced by treating an exposed plate with dilute chromic acid and then exposing it to white light. Subsequent development brings up a positive image formed on the grains which were not affected by the initial exposure and bleaching process.

ALBUMEN. Complex organic compound at one time much used in photography for making albumenized paper—an early type of daylight printing paper. For this purpose it was prepared from the white of egg, which is almost pure albumen.

ALBUMEN-ON-GLASS PROCESS. The first practical photographic process on glass, made public in 1848 by Abel Niépce de Saint-Victor. The glass plate was coated with white of egg containing potassium iodide. When dry, the plate was sensitized with an acid solution of silver nitrate and, after exposure, developed with gallic acid. Very fine detail was obtained, making the process excellent for architecture and art reproductions, but the exposure—5 to 15 minutes, according to circumstances—was far too long for portraiture. Albumen positives were widely used for lantern slides and stereoscopic pictures until about 1857.

See also: *Hyalotype.*

ALBUMEN PAPER. Glossy positive printing paper introduced by L. D. Blanquart-Evrard in 1850, in general use until *c.* 1900. The paper was coated with a solution of albumen and an ammonia salt in rectified spirit and then sensitized with silver nitrate. It was usually toned with chloride of gold to improve the colour and permanence of the picture. With the exception of the inventor, who developed his prints, it was used as a printing-out paper.

See also: *Sensitized materials history.*

ALBUMS. An album is the most convenient place for keeping a collection of photographs so that they can be looked at without being handled. It displays the prints while keeping them clean and free from creases.

The simplest form of album is just a book with stiff covers and blank pages. Prints are stuck in position with adhesive on each page. There are just enough pages to take up all the space between the covers when the album is full.

Loose Leaf Types. Loose-leaf albums allow the removal or insertion of pages. One of the most convenient types has the pages attached to the spine of the cover by split metal rings. This type lies flat when open.

Albums are available with both vertical and horizontal formats and in a wide range of sizes. The covers may be finished in materials like decorative cardboard, fabric or leather. For special subjects—e.g., collections dealing with a single event or a particular period—covers can also be bought ready embossed in silver or gilt with suitable titles such as "Our Wedding," "Baby" and "Summer Holidays." In addition to the title the subject may be indicated by mounting a representative photograph or a cut-out on the cover.

There is usually a choice of black, grey or white pages, but it is advisable to buy only

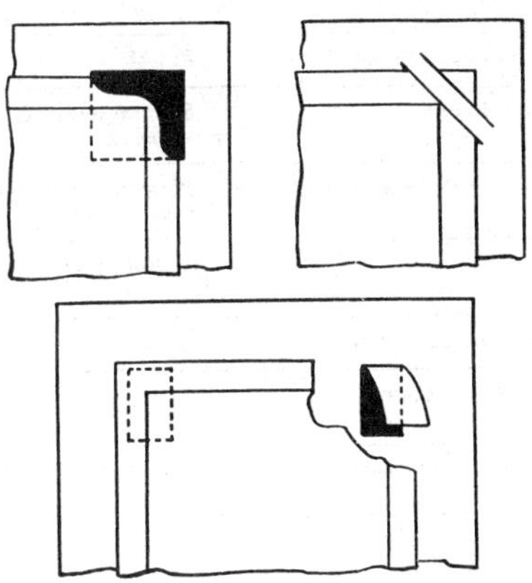

MOUNTING IN ALBUMS. Top left: *Mounting a print with gummed corners.* Top right: *Slits cut into the album page itself hold the print in a fixed position.* Bottom: *Double gummed corner pieces are a further alternative.*

those made specially for photographs. In ordinary paper there may be impurities which in time will stain or bleach the print.

Doubled-over celluloid or plastic pages are also obtainable to hold slip-in prints of the same size.

Mounting the Prints. There are several ways of mounting the photographs. Some albums have diagonal slits in each page, four to a print, so placed that the corners of the print can be slipped into the appropriate slits. With this system prints can be mounted on one side of the page only, and only prints of standard sizes can be used unless fresh slits are made. This system is now little used.

A more popular method is to use small adhesive corner pockets. These can be stuck to the page to accept the corners of any sized print; they are cheap to buy and are sold gummed ready to use. Another idea is to use paper corner pieces gummed on both sides so that they can be first stuck to the print, and then to the page.

When large prints are to be fixed to a page permanently, the use of dry mounting tissue is best. This would be appropriate in the case of weddings, where each print occupies a page.

Titling. It is usual to add captions or titles to each picture. Special white ink is made for titles on black or grey paper; ordinary Indian ink is suitable for white pages. In general, it is better to keep the captions brief and factual, giving only the essential information of who, where and when. Clever or literary captions are best avoided, as is also fancy lettering. J.D.C.

ALCOHOL. Used as a solvent in varnishes, for rapidly drying negatives, also used in certain processing solutions which contain substances not easily soluble in water.

Formula and molecular weight: C_2H_5OH; 46.

Characteristics: Colourless liquid with faint pleasant smell.

Solubility: Mixes with water in all proportions.

Ethyl alcohol (above formula) is available in several pure forms (absolute alcohol, rectified spirit) and also in denatured versions (i.e., with additions to render it unfit for drinking and so exempt it from excise duty), such as industrial methylated alcohol (colourless), and the common purple methylated spirits.

In addition there are a large number of other alcohols of which methyl alcohol (CH_3OH, wood spirit) is sometimes used in photography.

All these kinds of alcohol mentioned above (except for the purple methylated spirit) can be used in photography.

ALDEHYDE. Term in chemistry denoting a compound containing the group CHO, e.g., formaldehyde, the aldehyde of formic acid. Formaldehyde hardens gelatin.

ALKALI. In chemistry, the term alkali is applied to a number of soluble metallic

ALKALIS COMPARED

Alkali	*Other Names*	*Formula*	*Solubility at Normal Temperature*	*Alkalinity*
Sodium carbonate	Soda ash	Na_2CO_3	Freely soluble	Medium
Potassium carbonate	Potash	K_2CO_3	Highly soluble	Medium
Sodium hydroxide	Sodium hydrate, caustic soda	NaOH	Highly soluble	Very high
Potassium hydroxide	Potassium hydrate, caustic potash	KOH	Highly soluble	Very high
Borax	Sodium tetraborate	$Na_2B_4O_7$ $10H_2O$	Slightly soluble	Low
Borax plus boric acid	—	—	—	Very low
Sodium metaborate	—	$NaBO_2$ $.4\,H_2O$	Fairly soluble	Medium
Trisodium phosphate	Sodiumtriphosphate	Na_3PO_4 $.12H_2O$	Fairly soluble	High
Trisodium phosphate plus disodium phosphate	—	—	—	Low
Sodium sulphite	—	Na_2SO_3	Fairly soluble	Low

*Alkaline solutions with a high buffering action change little in pH with appreciable variations of concentration.

hydroxides, with certain definite characteristics, e.g., they turn red litmus blue, they feel soapy, and they react with acids to form salts. Metals which form such alkalis are known as the alkali metals, while salt solutions with similar characteristics are termed alkaline.

The degree of alkalinity of a solution is measured by its pH value; all values above 7 are alkaline and the higher the figure, the stronger the alkali.

The term has a special application in photography where it refers to an important constituent of the developer. Most developers must contain an alkali because they will only work in alkaline solutions. The alkalinity of the developer solution greatly influences its activity and development speed. The more strongly alkaline (i.e., the higher the pH value) the more energetic the developer.

There are, however, some developing agents which can work in nearly neutral or even slightly acid solutions, i.e. with a pH below 7.

How alkaline a solution is, depends mainly on the concentration, and to some extent on the nature, of the alkali.

Generally, the more alkali a solution contains, the higher its pH value. Some alkalis, like sodium or potassium hydroxide solutions, change greatly in pH when the alkali concentration is slightly varied. Developers containing these alkalis rapidly lose their activity in use and are soon exhausted, since the alkali concentration and hence pH value rapidly drops as the alkali is used up.

Other alkalis, like sodium carbonate or trisodium phosphate solutions, change their pH much more gradually in certain ranges as their concentration varies. They have some alkalinity in reserve which is brought out when the concentration decreases. The reason is that these alkalis form a very small amount of sodium hydroxide when they are dissolved in water. As this hydroxide is used up, more is formed all the time. They thus behave like very weak solutions of sodium hydroxide which remain more or less constant during use.

Substances like sodium carbonate are strictly speaking salts of a weak acid—that is why they behave as if they were proper alkalis.

In effect not all of the alkali in the solution comes into action simultaneously. Hence the pH value—the real hydroxide (actually hydroxyl ion) concentration remains nearly the same as long as there is enough alkalinity reserve left to trickle through all the time.

This property is particularly marked with certain mixtures of alkalis (strictly speaking alkaline salts in equilibrium with an acid form)—the so-called buffer solutions. Such developers lose their activity much more slowly.

The alkalis sodium or potassium hydroxide are commonly used for energetic developers, and sodium carbonate for normal formulae.

Sodium carbonate is used in three forms: anhydrous, monohydrate (chiefly in the U.S.A.) and crystalline (decahydrate). These (like the two forms of sodium sulphite) con-

ALKALIS COMPARED

*Buffering Action**	*Use* *High Concentration*	*Low Concentration*	*Remarks*
Very high	Most normal negative and positive developers	Some fine grain formulae	Exists in several forms; anhydrous, crystals, monohydrate
Very high	A few normal developers	—	Very hygroscopic
Nil	High contrast and high energy developers	—	Deliquesces; must be kept sealed in bottles
Nil	High contrast and high energy developers	—	Deliquesces; must be kept sealed in bottles
Moderate	—	Fine grain developers	—
Very high	—	Fine grain developers	—
High	Normal developers	Fine grain developers	*Kodalk* is a proprietary metaborate alkali
High	—	Fine grain developers	—
Very high	—	Fine grain and low energy developers	Typical buffer mixture
Moderate	Fine-grain and low-energy developers	Colour developers	Also acts as preservative

The pH of solutions with low buffering actions changes greatly with small variations of concentration.

tain varying amounts of water of crystallization. They are freely interchangeable.

EQUIVALENTS OF SODIUM CARBONATE

Decahydrate Crystals $Na_2CO_3.10H_2O$	Monohydrate $Na_2CO_3.H_2O$	Anhydrous Na_2CO_3
100 parts	43½ parts	37 parts
230 parts	100 parts	86 parts
270 parts	117 parts	100 parts

One disadvantage of sodium carbonate as an alkali is that it gives off carbon dioxide gas when it is neutralized (e.g., in an acid stop bath). As the carbon dioxide is formed within the emulsion layer, it can cause blisters both on negatives and on prints.

Milder alkalis, like weak solutions of borax, or sodium metaborate, are used in low energy fine grain developers. Even sodium sulphite by itself acts as an alkali.

Substances like triethanolamine, and combinations of sodium sulphite and formaldehyde or acetone have also been used. L.A.M.

See also: *Chemical calculations; Chemicals; Developers.*

ALPHA PAPER. Obsolete name for gelatine chloro-bromide paper introduced by J. M. Eder, January 1883, and produced by Marion & Co. of London under this name, 1884.

ALPHA PARTICLE. Positively charged particle—in fact a helium atom nucleus—with a velocity of about 1/12 to 1/20 that of light, emitted by radioactive substances. The path of the particle can be recorded in specially made photographic emulsions—a technique which has important applications in nuclear research.

See also: *Nuclear physics.*

ALPINE PHOTOGRAPHY. Another name for photography of mountains and mountain scenery that has probably come into use because of the popularity of Switzerland as a holiday resort. But the photography of Alpine scenery is no different from that practised in any other mountainous country.

See also: *Mountains.*

ALTERNATING CURRENT (A.C.). Electric current produced by a potential which is constantly and regularly varying between zero and a maximum, alternately positive and negative.

Electricity generated by an alternator. Most of the public electricity supply in Great Britain is alternating current.

ALUM, AMMONIUM. Ammonium aluminium sulphate. Sometimes replaces potash alum as hardener in fixing baths.

Formula and molecular weight: $NH_4Al(SO_4)_2.12H_2O$; 453.

Characteristics: White crystalline salt.

Solubility: Fairly soluble in water at room temperature.

ALUM, CHROME. Potassium chromium sulphate. Hardener in stop baths and fixing baths.

Formula and molecular weight: $KCr(SO_4)_2.12H_2O$; 499. Ammonium chrome alum $NH_4Cr(SO_4)_2.12H_2O$ is similar.

Characteristics: Violet crystals forming a purplish solution. When heated, or in the presence of sulphite, the solution gradually turns greenish and loses its hardening properties.

Solubility: Fairly soluble in water at room temperature.

ALUMINIUM AMMONIUM SULPHATE. Chemical sometimes used as a hardener in fixing baths.

See also: *Alum, Ammonium.*

ALUMINIUM CHLORIDE. Often used in concentrated liquid hardener solutions.

Formula and Molecular weight: $AlCl_3$; 133.

Characteristics: white crystalline powder usually caked into lumps because of strong deliquescence. Fumes of hydrogen chloride.

Solubility: very soluble in water and is usually handled as a concentrated liquor of 25–30 per cent AlCl3.

ALUMINIUM POTASSIUM SULPHATE. Chemical commonly used as a hardener in fixing baths.

See also: *Alum, Potash.*

ALUMINIUM SULPHATE. Possible substitute for the alum hardeners.

Formula and molecular weight: $Al_2(SO_4)_3.18H_2O$; 666.

Characteristics: White crystals with sweet, astringent taste.

Solubility: Highly soluble in water.

ALUM, IRON. Ferric alum; ferric ammonium sulphate. Chemical used in certain toners and reducers.

Formula and molecular weight: $NH_4Fe(SO_4)_2.12H_2O$; 482. Ferric potassium sulphate $KFe(SO_4)_2.12H_2O$ is similar.

Characteristics: Purplish crystals, give a yellowish solution.

Solubility: Highly soluble in water at room temperature.

ALUM, POTASH. Potassium aluminium sulphate; potassium alum. This is the common alum. Used as a hardener in fixing baths.

Formula and molecular weight: $KAl(SO_4)_2.12H_2O$; 474.

Characteristics: Colourless crystals or crystalline powder.

Solubility: Fairly soluble in water at room temperature.

AMATEUR PHOTOGRAPHY

Potentially creative, amateur photography gets closer to being the folk-art of the twentieth century than other contemporary media, e.g., radio, cinema, television. Its popularity is international; the interest and diversity of its applications almost unlimited.

Not only does this variety of application defy all attempts at a common definition but it also leads to frequent protests from some amateurs against being classed with some others. One amateur may very much differ from another in respect of the degree to which he cultivates his hobby, the subject matter, mode of expression and techniques he is attracted by and his relative freedom from financial limitations.

There are people who expose only a few rolls of films during their holidays and leave the processing to a dealer. There are others who very regularly devote a great deal of time to photography and so become experts at many aspects of it. Between these two extremes there are numerous groups, intermediate in experience, enthusiasm and purpose. The flag-bearers of amateur photography will be found organized in clubs, societies and associations. They are the people most consistently concerned with pictorial quality and technical progress. They read periodicals, frequent lectures, participate in exhibitions and foregather at congresses. Many brilliant amateurs, of course, remain lone workers and will not join in any of these activities. In fact the membership of Great Britain's approximately 3,000 amateur organizations has been estimated to be no more than a fraction of 1 per cent of the number of people who take amateur photographs in this country. Yet, in Britain as well as in other Western countries, the organized amateur movement speaks for all amateurs, even if it may not be truly representative of them. The photographic industry, to whom the more vociferous elements in the clubs may occasionally cause a little inconvenience, nevertheless makes every effort to keep the movement happy and active because of its importance as a testing ground and listening post.

Types and Interests. Both within the organizations and outside of them there is a legion of different types following as many different fields of interest. There are amateurs producing fine pictures without preference to specific subjects and there are others who are all the time after some specific subject putting pictorial excellence second. There are amateurs who pride themselves on advanced technical skills and there are others who prefer to explore the theoretical backgrounds of the very same techniques. There are die-hard perfectionists and there are restless experimenters. There are disciplined followers of "schools" and there are eternal controversialists. There are ardent collectors of all types of equipment and there are untiring inventors of home produced gadgets (which once made may never be used). There are plenty of photographic enthusiasts who never succeed in making a reasonable picture and some who would not even attempt to take one. There are many Royal or otherwise distinguished amateurs and masses of very ordinary ones. There are youngsters accustomed to handle a camera even before they have learned to read or write and aged people still taking pictures with their eyesight almost gone.

The economic background to amateur photography is also enlivened by varied patterns and peculiarly contrasting elements. Probably dominant is the fact that people are not sensible about their hobbies, and their habits of spending money on them are governed to a greater degree by their enthusiasms than their circumstances. At least at periods of economic crisis the average man is comparatively slower in cutting down on his hobbies than on daily necessities. Of course people have been known to give up photography altogether but more often than not it grows on them. This is particularly true between the ages of 20 and 30 when courting, young family life, travel and sport may provide the initial stimulus. In Great Britain, within this age group, sixty-seven people in every hundred took photographs in 1951. Once a person, often to his own surprise, has succeeded in taking a few genuinely good photographs the chances are that he will acquire a taste for the sheer creative satisfaction gained in the process. As his enthusiasm grows so will his expenses and he may try to make his hobby pay for itself. So it is by no means uncommon for amateurs methodically to follow cash prize competitions or even to freelance, employing the winnings and profits to acquire better equipment and more material.

Vice versa, there are well-known professional photographers who enjoy using their cameras to occupy their leisure in an amateur capacity. These participate in predominantly amateur events and take an active part in the life of amateur organizations. In fact, many leading figures of the amateur movement are professionals. Editors of photographic periodicals, authors, lecturers, scientists, technicians and even business executives of the photographic industry have, with a varying degree of justification, claimed to be "only amateurs". It is hard to know where to draw the line and, in any case, probably quite unnecessary to do so.

Contributions to Progress. There always has been a busy two-way traffic between amateurs and the business of photography. A number of basic inventions originated with amateurs. Richard L. Maddox, M.D., who made the first silver bromide emulsion and the dry plate and yet never received any financial benefit, is a classic example. Yet isolated names like his illuminate as little the importance of amateurs to photography as does the equally marginal fact that amateurs were the first photographic book illustrators, that they pioneered in geographical and architectural exploration with the camera, that new applications of photography to scientific research are still more likely to be initiated by amateurs than by professionals.

A more typical impact of amateurs on photography, of an altogether different order, became first apparent on the introduction for sale of ready-made dry plates. Professor Herman W. Vogel, who produced the orthochromatic emulsion, tried to quell the alarm of professional photographers by saying, "Not everybody who got hold of a violin is as yet a violinist and not everybody who got hold of a dry plate is as yet a photographer." But it is much easier to take a picture than to play the fiddle—as events were soon to prove. These events followed and later repeated a familiar cycle; first, simplification of technique spread photography among amateurs; secondly, masses of amateurs applying the new techniques in new ways upset the prevailing patterns of professional photography. This is exactly what happened after 1880 when dry plates in the hands of amateurs marked the beginning of pictorial photography; again after 1900 when film and light-weight hand cameras brought with the "snapshot" the fashion of informal photography; and more recently, since the 1930's, when precision-made cameras, high-speed lenses and fast panchromatic materials turned first amateur and then professional photography into an ever present and candid tool for recording life as it is. In the 1960s trends towards a contempt of "technique" and in particular a deliberately careless play with colour often originated with amateurs or at least with very young people with a camera who have not as yet settled into patterns of established professionalism.

The importance of the part played by amateurs in all these developments is not so much due to the merits of particularly gifted individuals or even the spirit of freedom permeating any activity unhampered by commercial consideration as simply to numbers. Manufacturing equipment and materials for hundreds of thousands of amateur consumers is a very different proposition from producing tools for a few hundred professionals. The competitive element, tempted by an ever expanding market, stimulates ideas, accelerates research and makes it a practica proposition to tool up to high quality standards. Thus the amateur's greatest gift to photographic progress is, in fact, mass consumption.

Purchasing Power. According to the United States Department of Commerce the value of photographic materials supplied at manufacturers' prices in 1961, exceeded 1,269 million dollars. The estimated figure for 1962 exceeded 1,362 million dollars. Thirty-four per cent of this output went to the amateur market (as compared, according to the U.S. National Association of Photographic Manufacturers, with 28 per cent to professional and industrial users, 13 per cent to the motion picture industry, 14 per cent to blockmakers and printers and 11 per cent to medical institutions). Over and above this figure, amateurs fired off in 1961 39 million dollars worth of flash bulbs, in 1962 an estimated 43·7 million dollars worth (at manufacturers' prices) as well as paying for 390 million dollars worth of developing and printing services in 1961 and an estimated 385 million dollars worth in 1962 (at retailers' prices).

In the United States and in Great Britain at least every fourth person has a camera, in Germany nearly every fourth. Camera ownership in Japan is—at least in the bigger cities—also reaching similar proportions. In the countries without a photographic equipment industry of comparable output, however, fewer people tend to own cameras. At the beginning of the 1960s large previously untapped markets were opened up by the introduction of increasingly automated low-price cameras.

In the United States at present nearly 670 million flash bulbs are consumed per year, in Great Britain probably less than 30 million. But the price of a flash bulb in the United States is less than that of a Sunday paper, in Great Britain it is equal to that of two Sunday papers.

American amateurs in 1962 made 2,300 million pictures, 43 per cent of them in colour; British amateurs took some 500 million pictures in 1963, about 30 per cent in colour.

The uneven distribution of the photographic industry coupled with restrictive international trading practices inevitably leads to uneven progress of amateur photography in various countries. While this is regrettable and perhaps unfair, the steady advance of amateur photography in those countries where it is not handicapped is an encouraging signpost on the road towards a brighter future everywhere in the world. A.K-K.

See also: *Clubs and associations.*

AMBROTYPE. Under-exposed collodion on glass negative which appears as a positive when backed with black varnish, black paper, etc. A variant of the collodion process; devised by Frederick Scott Archer in collaboration with Peter W. Fry in 1851. Mainly used for portraiture as a cheap substitute for the daguerreotype. Collodion positives on glass are called Ambrotype in England and the United States, and Melainotype on the Continent of Europe.

AMERICAN SOCIETY OF MAGAZINE PHOTOGRAPHERS. Society established in 1945 (and now having world-wide membership), "to safeguard and promote the interests of magazine photographers, to maintain and promote high professional standards and to cultivate friendship and mutual understanding among professional magazine photographers."

The ASMP is incorporated as a labour union to enable it to work more effectively for its members. Since 1950 the activities of the Society have broadened to include all forms of photography for publication, including advertising. The ASMP acts as a clearing house for information on rates, practices, market conditions. Its *Code of Minimum Standards*, established in 1951, has become accepted by the industry as a minimum scale for editorial work.

Monthly meetings are held in New York, Los Angeles and San Francisco. From time to time the Society sponsors exhibitions, conferences and symposia.

The Society publishes a directory of its members twice yearly and also an illustrated monthly magazine, *Infinity*, sold by subscription and on a single issue basis. A confidential bulletin for members only contains technical, market and other information of interest. The ASMP has also published three distinguished photographic annuals, *Photographic* 1949, *Picture* 1957 and *Picture* 1960.

AMIDOL. 2 : 4-diaminophenol hydrochloride. A developing agent.

Formula and molecular weight: $C_6H_3OH(NH_2HCl)_2$; 197.

Characteristics: Whitish grey powder; keeps well in air-tight bottles, but slowly turns brown in the air. Solutions do not keep, therefore dissolve amidol just before use.

Solubility: Freely soluble in water at room temperature.

AMMONIA. Used in solution in the form of ammonium hydroxide; also in gaseous form for dry development of certain types of dyeline materials.

Formula and molecular weight: NH_3.17.

Characteristics: Colourless gas with a pungent smell.

Solubility: A saturated solution in water contains approximately 308 grams of gas per litre and has a specific gravity of 0·880.

AMMONIUM ALUM. Chemical sometimes used as a hardener in fixing baths in place of potash alum.

See also: *Alum, Ammonium.*

AMMONIUM BICHROMATE. Ammonium dichromate. Used in sensitizers for the carbon, carbro, and gum bichromate processes.

Formula and molecular weight: $(NH_4)_2Cr_2O_7$; 252.

Characteristics: Orange crystals.

Solubility: Freely soluble in water at room temperature.

AMMONIUM BROMIDE. Restrainer in special warm tone developers, also in certain bromide paper emulsions.

Formula and molecular weight: NH_4Br; 98.

Characteristics: White crystalline solid; hygroscopic.

Solubility: Highly soluble in water at room temperature.

AMMONIUM CARBONATE. Alkali in special warm tone developers.

Formula and molecular weight: $(NH_4)_2CO_3$; 96. The substance sold as ammonium carbonate is often a mixture of ammonium bicarbonate and carbamate, which acts very much like the carbonate.

Characteristics: White waxy cubes or white crystalline powder. Smells strongly of ammonia.

Solubility: Freely soluble in water at room temperature.

AMMONIUM CHLORIDE. Sal-ammoniac. Chemical used in some toners, bleachers, sensitizers, etc.

Formula and molecular weight: NH_4Cl; 53·5.

Characteristics: White crystalline powder.

Solubility: Freely soluble in water at room temperature.

AMMONIUM DICHROMATE. Another name commonly used for ammonium bichromate.

AMMONIUM HYDROXIDE. Ammonia solution. Alkali used in certain developers, intensifiers and toners. Also in hypersensitizing.

Formula and molecular weight: NH_4OH; 35.

Characteristics: Colourless liquid with pungent smell. It is a solution of ammonia gas in water. The concentrated solution of sp. gr. 0·880, known as 0·880 ammonia, contains about 30 per cent by weight of ammonia gas.

AMMONIUM PERSULPHATE. Chemical used in some reducers.

Formula and molecular weight: $(NH_4)_2S_2O_8$; 228.

Characteristics: White crystalline powder. The solution does not keep.

Solubility: Highly soluble in water at room temperature.

AMMONIUM SULPHIDE. Used as a toner in the sulphide toning process and as a darkener for the bleached image in mercuric chloride intensification.

Formula and molecular weight: indefinite.

Characteristics: A yellow-brown solution with a most unpleasant smell.

AMMONIUM SULPHOCYANIDE. Ammonium rhodanate; ammonium thiocyanate. Used in gold toners and certain dye toners.

Formula and molecular weight: NH_4CNS; 76.

Characteristics: Very hygroscopic white crystals. Keep in stoppered bottles.

Solubility: Highly soluble in water at room temperature.

AMMONIUM THIOCYANATE. Another name for ammonium sulphocyanide. Used in gold toners.

AMMONIUM THIOSULPHATE. Used as a fixing salt in high-speed fixers.

Formula and molecular weight: $(NH_4)_2S_2O_3$; 148.

Characteristics: colourless crystalline salt.

Solubility: Freely soluble in water.

AMP. More commonly used abbreviation of ampere, which is the unit of volume of current or flow used in electricity.

AMPHITYPE. Name given by Sir John Herschel to his process (1844) by which either positive or negative pictures could be produced, according to manipulation. Fox Talbot gave the same name to an albumen-on-glass process which he patented in 1851. The pictures appeared either positive or negative, depending on the light in which they were viewed.

AMPLIFIER, IMAGE. Electronic system using an image converter tube to amplify or intensify a very faint image falling on a photocathode. The amplified image appears on a fluorescent screen or can be recorded on a photographic film.

See also: *Image amplifier.*

AMYL ACETATE. Solvent in certain film cements and negative varnishes.

Formula and molecular weight: $CH_3CO_2C_5H_{11}$; 130.

Characteristics: Colourless liquid with pear-drop-like smell. Very inflammable.

Solubility: Does not mix with water. Mixes with alcohol, acetone, etc., in all proportions.

ANAGLYPH. A method of stereoscopy invented by Louis Ducos de Hauron, 1891, and re-introduced as a novelty under different names several times since.

Two positives, forming a stereoscopic pair, are printed on one support, each in a different colour—generally blue-green and red. The observer looks at the picture through coloured glasses so that each eye looks through a glass of the complementary colour of the image it is to see.

If the right eye looks through a blue-green glass at the red image, that image will look black and the other will be invisible. Similarly the left eye will see the blue-green image and not see the red. In this way each eye sees a single image and the two impressions fuse to produce a stereoscopic sensation.

ANAMORPHOSCOPE. Cylindrical convex viewing mirror used for looking at distorted pictures of objects taken by photographing their reflections in a similar type of mirror. Seen normally, the pictures are meaningless; through the viewer they appear correct.

ANAMORPHOTIC LENS. Optical system in which image formation occurs on different scales in the vertical and horizontal directions. The system may consist either of cylindrical lenses or of prisms.

History. The first proposal for deliberately producing a regularly deformed optical image came from Sir David Brewster over a hundred years ago. He found it was possible to obtain achromatic deviation from two prisms made of the same material. In 1889 J. Anderton proposed the use of such a pair of prisms in front of a camera lens and in 1898 Dr. Rudolph proposed to replace one of Brewster's prisms by a double prism to achromatise the system. Of more recent years similar effects have been demonstrated by the use of either cylindrical lenses or mirrors. It is, however, only in the last fifteen years that anamorphotic systems have become something more than a novelty effect and have found a wide use in the motion picture field.

Taking and Projection. With the normal type of optical system the image produced is of the same geometry as the object viewed. With the anamorphotic lens system the geometry in one plane is "squeezed" or "unsqueezed" relative to the other plane. The usual arrangement is to

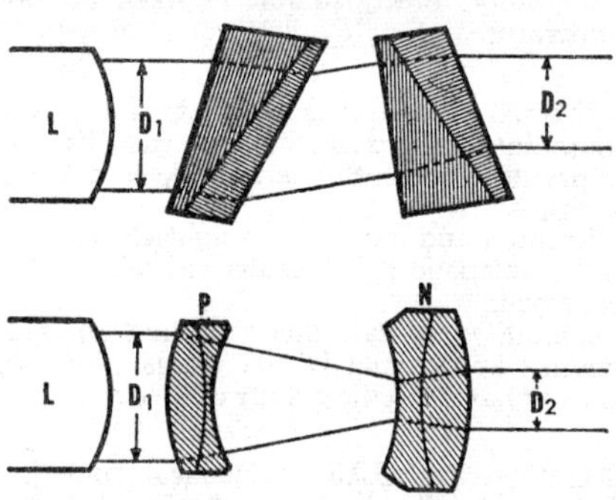

ANAMORPHIC SYSTEMS. A prism anamorphic attachment depends for its operation on the achromatized prisms which successively compress a parallel bundle of light rays (top). A cylindrical lens system (bottom) combines a positive cylindrical lens P and a negative one N to compress the light beam. The compression factor is the ratio D_1/D_2, in other words the width of the original beam divided by the width of the compressed beam. L is the taking or projection lens in front of which the anamorphic attachment is mounted.

restore the geometry of the image either at the subsequent printing or projection stages.

Anamorphotic systems have not in themselves produced any startling change in the appearance of motion picture screen pictures. They are, however, a means of providing wide screen pictures of the panoramic type, whilst avoiding some of the practical difficulties which prevent conventional systems from providing an adequate standard of performance.

These difficulties are (1) that the extra screen width has to be provided by short focus projection lenses, leading to exaggerated image grain due to the greater optical magnification; and (2) that the projected picture has to be cropped top and bottom, with less light passing through this smaller aperture to illuminate the larger screen.

Anamorphotic projection systems largely overcome both problems. For example the largest area that can be projected by a standard projector has an aspect ratio of about $1\frac{1}{3} : 1$. If all this area is to be projected on to a screen having the Cinemascope aspect ratio of about $2\frac{1}{2}$ to 1 the optical projection magnification must be nearly twice as great horizontally as it is vertically. To maintain correct geometrical reproduction of the original scene, the image on the positive film must be unnaturally compressed so that its scale horizontally is half its scale vertically. Thus if the original object seen by the camera was a circle then the image on the positive film, achieved through an anamorphotic taking lens, must be an ellipse. This ellipse when projected on the screen via an anamorphotic projection attachment, will then appear once more as a circle.

Copying. Ancillary to the straightforward taking and projection through anamorphotic attachments, is the use of anamorphotic copying systems. This type of system can restore the normal geometry of a "squeezed" negative print on to the positive film stock for use in theatres having conventional 4 : 3 screen proportions. This sometimes entails both a reduction and an "unsqueezing" operation from the same anamorphotic copying lens. Such copying lenses employ a basic spherical system on either side of which are disposed cylindrical members. In one plane the cylindrical members have no power and the spherical member operates at a particular magnification whilst in the other plane the cylindrical members combine with the spherical member to achieve a different magnification. A development of this type of process has led to a picture quality improvement.

One wide-screen process utilizes the fact that positive film stock has less grain than negative film. The negative is obtained by normal taking lenses on larger than normal film stock (e.g. 65 mm. wide). At an anamorphotic printing stage the negative is reduced and a compression ratio introduced to produce a standard "squeezed" 35 mm. positive projection print.

In this process the use of normal lenses avoids

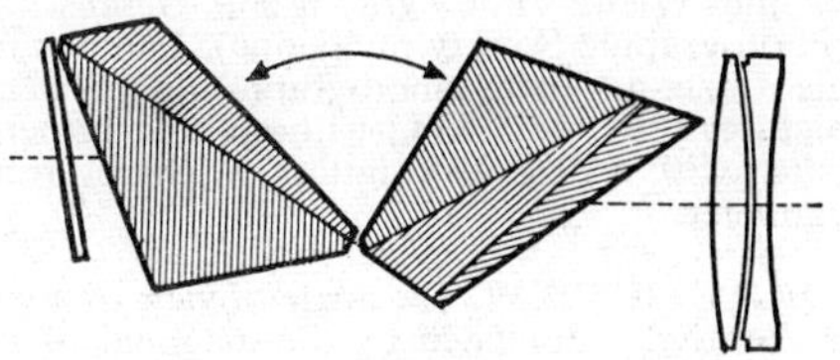

VARIABLE EXPANSION ATTACHMENT. The two prisms of an anamorphic system can be movable to give a variable compression or expansion. Swinging the prisms apart increases the compression, swinging them together decreases it.

light transmission and definition losses arising from bulky anamorphotic attachments. The reduction of the larger negative areas also takes a greater advantage of the finer grain of positive film stock. The angular fields covered are small at the printing stage and the copying system need only work at a fixed conjugate distance hence definition losses are reduced. Finally, one negative only need be used for the production of release prints compressed in any ratio to suit any screen aspect ratio or, alternatively for the production of normal release prints. P.A.M.

See also: *Cine history; Lens history; Projection practice.*

ANASTIGMAT. A lens in which all optical defects, including astigmatism, have been corrected in its design. The first of this kind, the Protar lens, designed by Paul Rudolph, was introduced by Zeiss in 1889. All high class modern lenses are anastigmats.

ANDERSON, JAMES (born Isaac Atkinson), 1813–77. English photographer, originally a watercolour artist. Settled in Rome, 1838, and took the name of James Anderson. 1849, began taking photographs of antique sculpture and views of Rome. Later the firm of Anderson became known chiefly for reproductions of art in Italian museums; it continued until *c.* 1955.

ANDREE, SALOMON AUGUST, 1854–97. Swedish engineer and aeronaut. Perished in an attempt at crossing the North Pole by balloon in 1897. His photographic films were found in 1930 and developed to acceptable negatives, documenting the fate of the expedition.

ANDRESEN, MOMME, 1857–1951. German photographic scientist, later director of Agfa. Investigated and introducted new developing agents: para-phenylene-diamine in 1888; Eikonogen in 1889; para-aminophenol in 1891. Worked also on diazo printing. Wrote articles and books.

ANGERER, LUDWIG, 1827–79. Austrian professional photographer. Leading member of the Wiener Photographische Gesellschaft (Photographic Society of Vienna). Famous for his Carte-de-Visite photographs and for his pictures (80 × 100 cm., without enlargement) taken with a Petzval-Voigtländer lens of 8 ins. diameter.

ANGLE OF VIEW. The angle of view of a lens is the angle subtended by the diagonal of the plate at the node of emergence of the lens. This is not necessarily the same as the angle of the lens, i.e., the angle that indicates the maximum covering capacity of the lens. In cameras fitted with one or more movements the lens must be able to cover more than the angle of view corresponding to the plate size.

See also: *Covering power.*

ANGLE SHOT. The orthodox way of photographing people or things is to point the camera at the subject from a viewpoint at eye level, with the back of the camera vertical and parallel to the principal plane of the subject—e.g., the front of a building. This approach avoids distortion of perspective and makes it easier to focus the important part of the subject.

When the camera is pointed at an angle to the subject, particularly from a close viewpoint, proportions tend to look distorted, parallel lines begin to converge more and more steeply, and it becomes more difficult to bring the whole depth of the subject into focus.

Sometimes distortion of perspective is introduced into photographs intentionally to achieve certain effects, but it is generally desirable to avoid it. This can be done either by holding the camera parallel to the plane which is to remain undistorted, or by using camera movements, as on technical cameras.

Pictorially, the extremes of angle shots are high and low angle shots.

A high angle shot is one taken from such a height that the subject is seen against the background of the ground itself or in which the horizon runs close to the top of the picture. Such a view tends to dwarf the subject and make it look top-heavy. This is because the vertical lines are foreshortened and the top of the subject, being much nearer to the camera than the bottom, looks so much bigger in proportion. In addition, the subject tends to become lost in a mass of surrounding detail of the same tone. High angle shots are only resorted to where the resulting distortion is sought deliberately—as in a portrait of a sitter with a heavy jaw which must be played down. They should not be used if the tones of the subject are much the same as those of its surroundings.

A low angle shot is one taken with the camera held as low as possible—generally with the photographer kneeling or even lying prone on the ground. Shooting from this angle has the advantage of showing the subject against a background either of the sky, or of objects so far away that they do not compete in tone with the subject itself and do not surround it with fussy detail. So a low angle simplifies the composition and isolates the subject.

Just as the high angle shot makes the subject look top heavy, the low angle shot emphasizes the width of the base. And by giving the subject a broad base and making the verticals taper towards the top of the picture, it imparts both stability and grace to the proportions. This treatment often improves pictures of tall slender buildings and of the human figure. But it is rarely a happy choice for portraiture because it makes the lower jaw and the fleshy part of the nose bulge unpleasantly. Where the lower part of the face needs emphasis, however, a low angle may be chosen deliberately.

Angle shots belong essentially to the field of the miniature because the small camera can be used at angles which would be practically impossible with a more cumbersome instrument. And when used at an extreme angle—e.g., looking up the front of a building—the miniature can still show everything in sharp focus from the brickwork a few feet away to the television aerial on the roof. This is possible because of the camera's short focal length lens, and the depth of field this allows. With a technical camera the same overall depth is achieved by the use of the tilting camera front. Even so, absolute sharpness is not always necessary or desirable for certain effects. F.P.

See also: *Camera movements.*

ANGSTROM (A). Unit that is now generally used to express the wavelength of light. There are 10 million Angstrom units to a millimetre. The visible spectrum extends from 7,000 Angstrom units (red light) to 4,000 Angstrom units (violet light).

See also: *Light.*

ANGULAR FIELD. Angle subtended at the lens by the largest object which the lens can image on the plate with acceptable sharpness.

See also: *Covering power.*

ANHYDROUS. Term indicating that a substance is free from water. In photography it distinguishes a chemical sold as a powder without water of crystallization from the same chemical in the hydrated form. When making up solutions it is important to use the anhydrous form of the chemical if that is specified since, weight for weight, it contains more of the active substance than its hydrates.

See also: *Chemical calculations.*

ANILINE PROCESS. Early line-printing process invented in 1865 in which a sheet of printing paper sensitized with ammonium-bichromate and phosphoric acid was developed by exposing it to the vapour of a suitable aniline dye.

ANIMALS. Photographing native wild animals in their haunts is very much more difficult than, for instance, photographing birds at their nests. Birds have their eggs or young to form an irresistible bait, but animals as a rule keep their young well out of sight. And whereas most birds rely on their eyes and ears to warn them of the presence of the photographer, animals have keen noses which can often detect a watcher a hundred yards away.

Many of the very small creatures, like mice, voles, and shrews, remain hidden in the thick foliage that conceals their haunts. These are almost impossible to photograph in the wild state and must be captured and taken indoors.

Then again, many animals that do come out into the open do so only at night. These have to be photographed by flashlight.

There are, however, other animals that show themselves outside their lairs in daylight, particularly in the early morning and just before sunset. These may be photographed from a specially constructed hide erected down wind from the animals so that the photographer's scent is carried away.

Deer may pick up human scent at a hundred yards, so it is very difficult to get near enough to them to get a reasonable picture.

Bait. Foxes, badgers and rabbits can usually be induced to leave their lairs in daylight by tempting them with a suitable bait. This should be laid several days in advance near the entrance to the lair, and renewed regularly. At the same time as the first bait is laid, the hide should be set up so that the animals will become accustomed to it.

Foxes will come out of their holes for a bait of meat, or bread and milk; badgers like honey, jam or anything sweet; rabbits can be tempted with chopped parsley, and stoats and weasels will come out for fresh meat, particularly when they are feeding a litter.

The otter is a creature of regular habits, and it is possible to pick up his accustomed track and set the camera to cover it. Otters prefer to kill their food for themselves so they cannot be attracted by a bait, but as they show themselves on the bank and in the water by daylight, they can be photographed by waiting.

Equipment. The camera, tripod, and hide are similar to those used for bird photography. But because animals are more difficult to approach, the hide is generally farther away than for bird photography, and a long focus, or telephoto lens, should be used. Photographs can, however, be made with a normal angle lens if the camera is concealed near the lair and operated by some form of remote control.

Remote Exposure Control. Some photographers prefer to use remote control shutter release for dealing with wild subjects. This technique dispenses with the hide and calls for only a small amount of camouflage for the camera and its support.

Once the camera has been set up and focused on the bait, nest, or the entrance to the lair, the photographer settles down some distance away to wait for the subject to make its appearance. If necessary he may use field glasses to tell him when the best picture offers itself. When that time comes, he simply fires the shutter by remote electrical or mechanical control.

The disadvantage of this method is that the photographer has to visit the camera after every exposure and also whenever changes in the light call for adjustment of the camera controls. Some cameras are designed to wind on the film and tension the shutter automatically after each exposure, but the noise of the camera mechanism is inclined to frighten the subjects, necessitating a waiting period, until they return to the scene. In a hide the photographer need not disburb the subject.

Automatic Exposure. There are several ways of making the subject take its own photograph while the photographer is away. One method of automatic shutter release is to get the animal to complete an electric circuit either by standing on a particular spot, breaking a thread stretched across its path, or seizing a bait.

This technique is uncertain in its results because it carries no guarantee that the subject will be in a suitable pose when the shutter clicks. And, as with remote exposure, the camera must be visited after every exposure.

But automatic exposure coupled with flash is very useful for securing pictures of nocturnal creatures which could not in any case be observed from a hide. In this way many striking photographs have been taken of subjects ranging from big game at a water hole to a fox robbing a hen run. The photographer can always be sure of getting a correctly exposed negative with the standardized light of the flash, but the success of exposures in daylight photographs depends upon the state of the lighting when the shutter clicks.

The most successful method of getting flying birds, flying mammals and insects, such as bats and butterflies, to "take their own photographs"

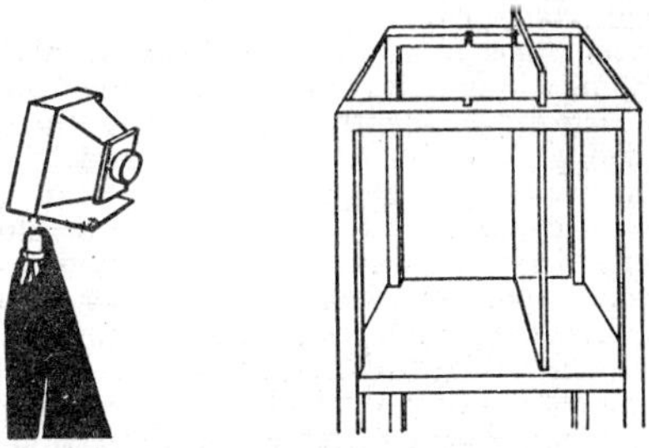

GLASS BOX SET-UP Small animals can be photographed in a glass-sided box containing suitable materials (earth, twigs, etc.) corresponding to natural surroundings. Removable glass partitions confine the animal to a limited distance zone from the camera and thus ensure overall sharpness.

is to fix up an electric photo-cell trip in conjunction with electronic flash. This fires the flash and releases the shutter of the camera at the moment the subject passes through the pre-arranged beam. This cuts out the human element that makes hand release of the shutter so uncertain. Such electrical equipment should be used in the open with the greatest care under the advice of an experienced electrician, for the high voltages obtained can be fatal if carelessly used.

Animals in Captivity. Large animals in captivity have to be photographed in their existing surroundings.

The smaller mammals are easy to keep under control and photograph in a glass-sided box. This should measure about 18 ins. wide by 12 ins. high and 6 ins. deep from front to back. The front must be made of good polished plate glass to give an undistorted picture, but the top and sides may be of ordinary glass. Normally the back of the box is painted a neutral grey or brown and suitable scenery is built up in front of it. Sometimes the case has a glass back to it and the scenery is continued behind. However, because the camera is only a few feet away and the depth of field is small, the scenery behind the case is well out of focus.

If the camera is set up in front of the glass case, its reflection is almost certain to appear in the finished photograph. The best method is to set the camera slightly higher than the case and pointing down on it. A sheet of dull black paper is then fixed to the tripod, opposite the glass front and reaching up to the bottom of the lens. If the size or distance of the animal permits, an alternative way of avoiding reflections is to bring the lens against the glass.

Any form of visual focusing camera may be used for this type of subject—even an ordinary hand camera, so long as its focusing scale extends down to 4 feet. 35 mm. miniature camera users will find a lens of about 90 mm. very convenient at 4 feet from the subject.

This small studio set-up may be used out doors in a good diffused light, but never in direct sunlight. Indoors, it should be stood on a table close to a north window with a flood-light of about 500 watts to light up the shadow side, fixed about 2 feet above the level of the camera. With this lighting and a fast pan film, the exposure should be around 1/60th to 1/125th second at *f* 8. O.G.P.

Illustrations: Plate Section III.

See also: *Big game; Birds; Cats and kittens; Dogs and puppies; Fish; Insects; Pets; Reptiles; Underwater photography; Zoo.*

Books: *Camera on Safari*, by John Sherman (London).

ANIMATION

Animation is a method of producing the illusion of movement from a successive presentation of still images. In that sense cinematography itself relies on animation. In a more restricted sense animation is taken to cover the creation of movement of inanimate objects such as drawings, models, puppets, etc. The best-known application of this is the cartoon film by animation of drawings and going back in principle to the zoetrope invented even before photography itself.

While animated cartoons and drawings have widespread uses in the cinema besides pure entertainment, similar principles can produce more primitive forms of animated displays in still photographs and projected slides.

Cartoon films. Cartoons are made by photographing a long series of drawings each of which has been drawn slightly different from the one preceding. The differences correspond to the successive stages of the movements of the cartoon characters. These stages are carefully planned in such a way that on projection at the standard speed of 24 or 16 frames per second the succession of images conveys the illusion of movement. It does this in the same way in which the succession of individual frames recording a natural movement recreate the impression of this movement on the screen.

Cartooning involves the technique of producing the individual drawings and of photographing them.

The photography is the simplest part. It needs a cine camera which permits the exposure of single frames one by one. It must also be able to focus on a sufficiently close distance to cover just a field of view corresponding to the size of the drawings. For this purpose the focal length of the lens is immaterial. A short focus lens, however, has to be set to a much closer distance to cover a given drawing size than a long focus lens.

The latter thus gives more elbow room for arranging the drawings, illumination, etc., on a cartoon bench. (Perspective effects in the cartoon are obtained by appropriate drawing and not by camera technique.)

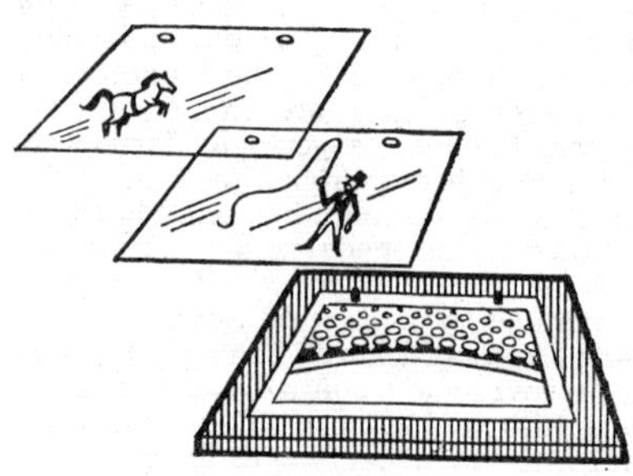

CELS IN CARTOONING. Different parts of the action are animated on separate cels (celluloid sheets) which are superimposed in their correct position over the background.

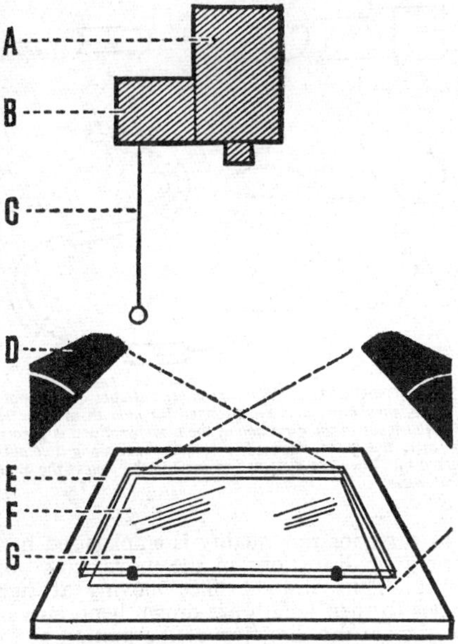

CARTOON BENCH SET-UP. *The main parts are:* A, *camera;* B, *single-frame unit;* C, *release cord;* D, *lamp units;* E, *baseboard;* F, *glass sheet to hold drawings flat;* G, *register pins.*

Also required is a cartoon bench which is in effect a glorified copying stand. It consists of a baseboard to hold the drawings, a column which carries the camera pointing vertically downwards on the baseboard (and permits raising and lowering the camera) plus the lighting units to illuminate the drawings. The arrangement of the camera on the stand as well as the lamps requires preliminary tests to establish the exact area taken in by the camera (this will be the frame within which all the cartooning action takes place) and to ensure that the lamps illuminate the drawings absolutely evenly.

In addition to single frame operation, the camera should also permit back winding for double exposure and super-imposition effects. For tracking effects it is further useful to be able to adjust the distance between the easel of the cartooning stand and the camera. Alternatively a zoom lens provides a convenient way. For maximum image steadiness professional cartooning cameras have gate register pins which hold each film frame accurately in the same position in the camera gate. A frame counter is a valuable aid.

The animation board—the easel of this camera set-up—carries register pins just outside the picture area to locate the drawings accurately. These register pins are spaced and located in the same way relative to the picture area as similar register pins in the animation desk used for preparing the drawings. The actual drawings photographed for a cartoon are often assembled from individual drawings or part drawings on a number of sheets of transparent plastic or celluloid (the so-called cels). A glass plate is needed to hold these cels absolutely flat.

Planning the animation. Any movement to be animated must be analyzed first. This involves determining the time—and hence the number of frames on the film—which the movement is to occupy. Then the action is split up into phases within this period and key drawings made for the beginning and the end of each phase. The intermediate frames can then be filled in with additional drawings showing the intermediate stages of the movement.

Making complete new drawings for every frame exposed in the camera is a long process, but various ways of saving work are possible.

Firstly, many movements are repetitive, for example the steps of a character walking across the scene. So a series of movement drawings can be repeated (for example by tracing) to simplify matters. For animated drawings of a mechanism—for instance an internal combustion engine—the same sequence of drawings can be filmed several times over to demonstrate the working cycle.

Secondly, a movement sequence of drawings nearly always consists of changing as well as unchanging items. To save effort it is usual to split up the drawing—for example in the simplest case into a figure and its background. The background then remains the same for the sequence and only the figure is animated, drawn on transparent plastic cels. For further simplification the body of the figure may remain the same throughout a series and only the movements of arms or legs animated. In this case the figure is split up into further cels of the body, the successive stages of the arm movement and the successive stages of the leg movement. In

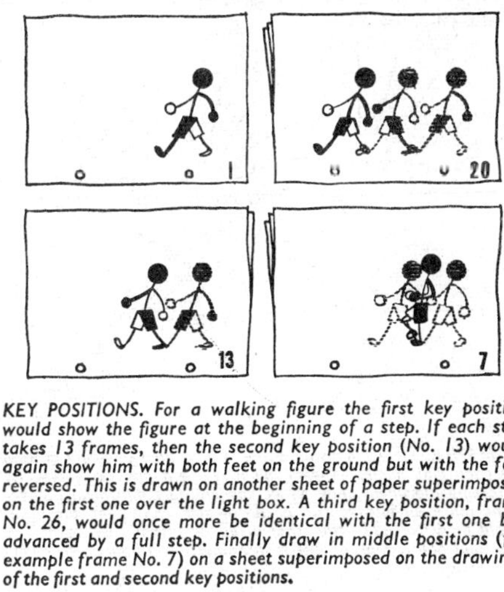

KEY POSITIONS. *For a walking figure the first key position would show the figure at the beginning of a step. If each step takes 13 frames, then the second key position (No. 13) would again show him with both feet on the ground but with the feet reversed. This is drawn on another sheet of paper superimposed on the first one over the light box. A third key position, frame No. 26, would once more be identical with the first one but advanced by a full step. Finally draw in middle positions (for example frame No. 7) on a sheet superimposed on the drawings of the first and second key positions.*

photographing the series, only the cels for the arms and for the legs would then be changed.

Thirdly, certain elements of a drawing may move as a whole, without being separately animated. Thus in a scene of a car travelling across the countryside, the car might be animated with a periodic movement sequence while the background slides past it behind (underneath on the cartoon bench). The background drawing in this case is made on a long strip which is moved across the field of view of the camera bit by bit between each successive frame. Such movement needs accurate calibration, and the cartoon bench may carry calibrated scales to help the cartoonist to shift a drawing through pre-determined distances.

Where animated drawings are split up into several layers of cels, the different parts of a drawing must of course register and match up exactly. The register pins on a cartoon desk (where the drawings are made) and on the baseboard of the camera ensure this. In more complex scenes several layers of cels may be animated independently of each other. Background and other features which move in their entirety are either not mounted on register pins, or are mounted on a second set of pins set in a block which slides in a groove along the cartoon bench.

Drawing Technique. For natural animation of diagrams, etc., it is sufficient to draw the various movement stages in a straightforward manner and photograph them in sequence. For cartooning, such movements tend, however, to appear jerky on the screen. This is partly because they lack the smoothness created by the slight blurring of images in a real motion picture, and partly because cartoons rely on exaggeration as well as simplification for their visual appeal. A strictly natural movement of a cartoon character would therefore appear subdued against the cartoon quality of the figures and other objects in the picture.

ANIMATING LINES. Top: *If a line is animated to swing around the radius of a circle* (left) *the movement appears jerky. By distorting the lines* (centre) *we can animate an upward movement more smoothly; for a downward movement the lines would be distorted in the other direction* (right). Bottom: *The practical application of this is in animating a figure waving its arm up and down.*

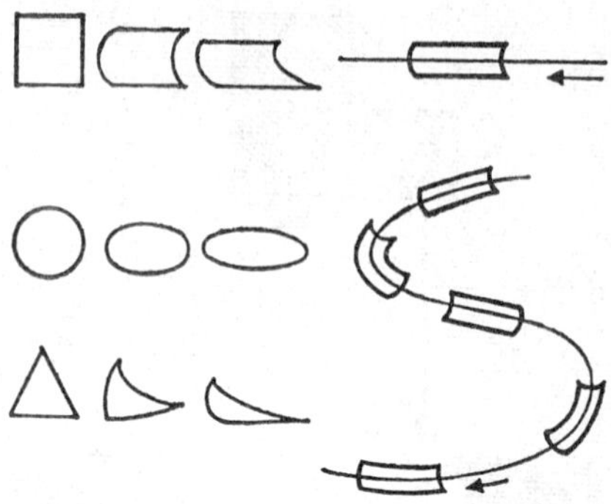

DISTORTION OF AREAS. *When geometric shapes are animated to move, they must also be distorted for smooth motion. The distortion goes in the direction of the movement and is greater, the faster the speed.* Right: *If a shape moves along a complex path, all its lines are distorted to assume the course the object is following.*

In cartoons this quality is emphasized by a variety of distortions in the movements portrayed. Thus straight lines moving at right angles to their length are drawn bent, curving away from the direction of the motion at the beginning of the movement. They bend in the opposite direction at the end. This conveys an exaggerated impression of inertia and so makes this inertia visible graphically on the film.

The same principle is applied to solid bodies. As a result such bodies appear squashed when they start to move, then unsquash and squash once more as they come to rest. The effect is that all objects seem to be endowed with an exaggerated elasticity. The degree of exaggeration is used to convey weight, speed of movement and inherent elasticity itself. Thus while in real life a cricket ball hit by a bat will hardly distort at all, a tennis ball (observable when filmed with a high speed cine camera) is noticeably squashed. In a cartoon the cricket ball is drawn visibly squashed, while a tennis ball is literally flattened at each impact.

Puppet Animation. While straightforward cartoons only involve the copying of a series of two-dimensional drawings, puppet films use three-dimensional sets in miniature. Some of the principles involved are the same, especially those relating to planning and timing the animation. Making a puppet film is, however, in many respects much more tricky since the illusion of perspective as well as movement must be conveyed in a tabletop set-up.

One practical problem is that of depth of field. The greater the scale of reproduction of any three-dimensional scene, the more restricted the depth of field becomes. For example a studio set of a room 15 feet wide may be recorded on a 16 mm. frame with a 500-fold reduction. A corresponding table-top scene 18 ins. wide only undergoes a 50-fold reduction. To obtain the same differential depth effect or depth of field would require a lens aperture with ten times as high an *f* number.

Secondly, at close range the perspective effects are quite different. In particular it becomes necessary to create artificial perspectives by drawing them into the background and by making objects nearer the camera deliberately larger than similar objects farther away.

Thirdly, the animation itself calls for considerable skill since the limbs and other parts of the puppets must be moved by just the right amount between each frame. And while it is comparatively easy for someone who can draw to make a cartoon figure do almost anything, a puppet is limited by its physical capabilities such as joints, etc. Specialists in making professional puppet films sometimes employ highly complex and specially designed puppets with a very wide scope of movement. The head only of a puppet may be made in several dozen different versions to indicate varying facial expressions, these heads being interchanged on the body as required.

Animated Titles. Animation by single frame shooting can also produce a wide variety of trick title effects. These should still be planned frame by frame, but there is of course almost unlimited latitude since the timing does not have to conform to any nature-bound rules such as the rate of a human walk.

To make an animated title of for instance letters gradually falling into place, the individual letters are pushed one by one gradually into their required positions while the camera photographs frame after frame. Plastic or felt letters which can be rearranged at will indefinitely are most suitable for this purpose.

Alternatively a title may be drawn or painted step by step on its background, the camera exposing frame after frame as the title grows. Arrows on a map can be animated in the same way to show the progress of a journey.

If the map is required again subsequently, the arrow may be drawn or painted on a transparent plastic sheet and this placed over the map. It can then be extended exposure by exposure, so making the arrow appear to grow.

Animation of Still Pictures. With still transparencies the possibilities of animation are much more limited, since the successive presentation of different or partly different images is more complicated.

With slide projectors simple animation effects involving two transparencies are possible if two projectors are available. Each projector projects an image of its transparency on the screen, so that the image outlines that are not animated appear exactly superimposed. For the movement the light from the one projector is then faded out while the second projector fades in. This change-over should be fairly rapid and may be repeated to and fro to show simple periodic movements.

Another method, suitable for displaying transparencies, uses polarizing sheets. The stationary parts of the image to be animated are drawn or painted on an ordinary plastic sheet, and the alternative positions of portions to be animated laid on in the polarizing sheet material. This is orientated with its plane of polarization at right angles for the two positions. A further polarizing filter is placed in front of the light source and rotated. This makes the two polarizing areas on the image go light and dark alternately, producing the illusion of movement.

A similar system uses a bi-refringent plastic which is stuck over areas to be animated. On rotation of the polarizing filter in front of the light source these areas appear in fluctuating brightness. This method is suitable for creating the illusion of liquid flows, etc.

Yet another method of animating still pictures is image dissection. Here the subject (for example a diagram) is photographed through a lenticular screen which breaks up the image into strips with blank intervals between them. For the second image the lenticular screen is moved slightly at right angles to its line structure. Then the second image is recorded in the spaces between the lines belonging to the first image. For viewing, a similar lenticular screen is laid over the transparency (or even print) and slowly moved at right angles to the line pattern. This alternately brings the one and the other image into view. The line orientation should in this case be horizontal, otherwise an observer may see one image with one eye and the other image with the other eye, as in a parallax stereogram. By suitable choice of the lenticular screen it may be possible to get several different or partly different images on one picture. L.A.M.

See also: *Cine technique; Table-top photography; Titling.*

Books: *How to Cartoon*, by J. Halas and B. Privett (London); *The Technique of Film Animation*, by J. Halas and R. Manvell (London); *How to Animate Cut-outs*, by C. H. Barton (London).

ANNAN, JAMES CRAIG, 1864–1946. Scottish photographer. Son of Thomas Annan, whose portrait studio in Glasgow he took over. Introduced the photogravure process of Karl Klič into Britain, 1883. Leading pictorialist and member of the Linked Ring; first president of the International Society of Pictorial Photographers. Propagated the work of D. O. Hill and R. Adamson *c.* 1900 in exhibitions and by issuing a small selection of photogravures.

ANNAN, THOMAS, 1829–87. Scottish portrait photographer. Father of James Craig Annan, friend of D. O. Hill and of Sir J. W. Swan. Made (1879–81) carbon prints of Hill's calotypes, and introduced Swan's carbon process in Scotland. He photographed Hill's painting of the signing of the Deed of Demission at Edinburgh with a specially built camera (14 × 32 ins.). Made carbon prints as large as 21¼ × 48 ins.

ANSCHÜTZ, OTTOMAR, 1846–1907. German photographer. 1882 began taking instantaneous serial pictures of animals in motion using a hand camera fitted with a focal plane shutter. Devised electrical rapid viewing instrument for stroboscopic synthesis of these pictures (Electrotachyscope) 1887—one of the forerunners of cinematography.

ANTHONY, EDWARD, 1818–88. American engineer. Became one of the early daguerreotypists and made a photographic survey of parts of the U.S./Canadian boundary. Set up in business as a photographer in 1842, later became one of the largest photographic dealers. He founded *Anthony's Bulletin of Photography*. The firm of E. & H. T. Anthony & Co. combined in 1901 with Scovill & Adams Co. to form the Anthony & Scovill Co. which in 1907 became Ansco Co. and in 1928 Agfa Ansco Co. Biography by American Museum of Photography (New York, 1942).

ANTI-FOGGING AGENT. Chemical included in a developer to retard the formation of development fog. The oldest known anti-fogging agent is potassium bromide; nowadays a variety of organic substances have been found more effective and more selective.

See also: *Developer improver.*

ANTI-HALATION BACKING. Most modern negative materials have some form of anti-halation backing to absorb the light that passes through the emulsion. If the light is not absorbed, some of it is reflected and falls on the under side of the emulsion layer, causing halation.

There are several forms of halation backing: a coat of dye or pigment painted on dyed gelatin; an intermediate layer of dyed gelatin between the emulsion and the support; and a dye that is incorporated in the base of the material itself.

Dyes or pigments may be applied directly to the back of the support. Red or purple dyes are used for orthochromatic, and blue-green for panchromatic, materials. The dye in each case absorbs those wavelengths of light to which the emulsion is most sensitive.

This type of anti-halation backing is commonly used for glass plates. The backing dye or pigment is generally mixed with gum, which make it adhere when dry, but allows it to dissolve away during processing.

A backing layer is a gelatin layer incorporating the dye and is applied to the back of the support. This type of anti-halation backing is almost always used on roll and sheet films, because the gelatin layer counteracts the tendency of the emulsion layer to make the film curl. The backing layer remains on the film all the time, but the dyes bleach out during processing.

Reversal materials and colour films frequently have an intermediate anti-halo layer, applied between the emulsion (or emulsions) and their support. This layer contains colloidal silver and is the most effective single anti-halation protection. This is because any light reflected from the back of the support has to pass through the layer twice: once before, and once after reflection.

Since reversal processing and colour processing procedures involve a silver bleaching stage, the colloidal silver automatically disappears during processing, leaving the layer clear. For this reason black-and-white reversal films are not suitable for processing to a negative. An intermediate layer of this kind does not by itself provide any counteraction to the curl of the film.

The film base of many 35 mm. films is itself dyed with a blue-grey or neutral-grey dye which acts in the same way as an intermediate layer, i.e., any light that passes through the emulsion is filtered both before and after it is reflected from the back of the support.

The dye used is not removed in processing, but it does not interfere with printing.

Several modern 35 mm. films carry an anti-halo backing, similar to roll films, in addition to the dyed film base.

ANTINOUS RELEASE. Originally a proprietary name, now commonly applied to all flexible wire or cable types of shutter release.

ANTIQUES. The photography of museum and collector's pieces needs considerable technical skill. The aim is usually to produce a factual presentation of the object, from which the connoisseur could get much of the information he would expect to obtain by a visual examination of the object itself. It must be to scale, free from distortion, indicate surface texture and show all decipherable detail.

Equipment. To do this sort of work properly a rear-focusing field camera is essential. It must have an extension more than twice the focal length of the normal camera lens, the length of this being at least equal to the diagonal of the film or plate it has to cover. focal length of the lens employed and the focal length of the lens should be at least equal to the diagonal of the film or plate it has to cover. Back focusing is necessary for working to scale, as otherwise any movement of the lens backwards or forwards in focusing changes the distance between the lens and the object and thus alters the size of the image.

The usual practice when photographing a number of articles to the same scale (snuff boxes, coins, miniatures and the like) is to set the camera by focusing on a foot rule and measuring the image on the focusing screen. It is then only necessary to move the camera

backwards and forwards until the object is sharp on the focusing screen.

A double extension camera is necessary because so often in this work the object must be reproduced same-size and to do this the camera must be extended to twice the focal length of the lens in use.

A swing back is absolutely essential as converging verticals are no more permissible in a photograph of, say, a cabinet, than they are in a photograph of a building. This statement applies equally to converging horizontals, which should also be corrected by use of the relevant camera movements.

The equipment should include at least three lenses—normal, long-focus, and wide angle—to meet all possible needs. These lenses should of course be of the highest possible quality.

Lighting. No hard and fast rules can be laid down on the lighting of antiques, as such objects are extremely varied. Some require soft diffused light; with others even spotlights might be suitably brought into play.

As exposure time is seldom important, it is always a good plan to use a smaller stop than would normally give the depth of field necessary to cover the subject. Extreme stopping down can, however, lead to an ultimate loss of quality, caused by diffraction. The correct use of tilt of the lens panel, on a camera where such movement is possible, can achieve overall sharpness without extensive stopping down.

In this class of work it is essential to expose for the shadows; modern slow emulsions have enough latitude to look after the highlight detail if development is not carried too far.

The background needs to be carefully chosen to give a clear contrast with the outline of the subject on both lighted and shadow sides. It should be coloured or lighted to come out on the print as a neutral grey; it should never be white, as the strong reflected light tends to be spread by halation and destroy the crisp definition around the edges of the subject.

Sensitized Material. Many experienced technical photographers prefer to use sheet film or possibly plates, as these can be processed individually. When using roll-film it is important, on any one roll of film, to shoot only similar subjects under similar lighting conditions, so that processing can be suited to the entire roll. Any work such as blocking out or titling that may have to be carried out on the negative is easier on a large sheet film or plate than on a comparatively small piece of roll-film. Whether panchromatic or ordinary colour-blind emulsions are used will depend on the subject.

Ordinary colour-blind emulsions are preferable when statuary or the like has to be reproduced in its natural grey tones, but, for subjects containing colour, a panchromatic emulsion with a suitable filter is essential. The slower emulsions, whether panchromatic or ordinary, are preferable unless the subject has to be photographed *in situ* in poor light.

Nineteen out of twenty subjects will be suitably corrected on pan plates by using a 3× yellow filter but, very occasionally in antiques, one gets rich colours in, say, rosewood or marquetry or the tortoiseshell inlay on Buhl furniture which call for an orange filter. Focusing should always be done with the filter in position because the filter may cause a slight shift in the focus of the rays that form the photographic image.

Processing. In many branches of photography, processing which gives approximately correct tone values is all that is required. This standard is not good enough for a negative which may at any moment be the only record of a unique piece which has met with disaster. For that reason the normal practice of taking a spool of mixed subjects at varying exposures and developing it by time and temperature in a standard developer will not do. Such important records must be developed separately to a gradation and contrast scale which is complementary to that of the subject and yet may vary from one subject to the next.

Photographers who specialize in this critical work often develop by inspection. The developer employed is almost invariably a standard pyro-soda formula. With this it is possible to pre-determine the character of a negative by varying the pyro content; a solution rich in pyro will tend to produce a contrasty negative, whereas by reducing the concentration the same exposed material can be made to yield the softest image it is possible to print.

Photographs of antiques are mostly used for records and for catalogue reproduction, and for that reason they should preferably be printed on glossy or smooth paper which gives the maximum resolution and rendering of detail. C.D.(V.G.)

See also: *Ceramics; Coins; Glassware; Jewellery; Lighting the subject; Manuscripts and old documents; Paintings and drawings; Sculpture; Silverware.*

ANTI-REFLECTION COATING. Very thin coating of a metallic fluoride applied to a glass surface—e.g., of a lens—to reduce light lost by reflection.

See also: *Coated lens.*

ANTI-SCREEN PLATES. Commercially used term for plates whose blue sensitivity has been reduced by dyes. Such plates may be used without a colour filter to produce effects normally obtained with an orthochromatic plate and a yellow filter.

APERTURE. Size of the lens opening through which the light passes. There are several definitions of aperture, according to the exact parameter referred to by the term. Some of these are specified in the British Standard BS. 1019: 1963.

Limiting Aperture. This is the physical limit—for example the iris diaphragm or the lens mount itself—to a light beam from an object point at the lens axis. If the lens has an iris diaphragm, the limiting aperture at any setting is the actual diameter of the diaphragm opening.

Effective Aperture. This is the diameter of the light beam reaching the lens from a distant object point along its axis, which passes through the edge of the limiting aperture. The object here is assumed to be at an infinite distance, so that the light beam can be assumed to be parallel. The physical stop diameter (limiting aperture) can be larger or smaller than the effective aperture, depending on the lens construction. It is however the effective aperture that determines the light passing power of the lens.

The effective aperture is a diameter; if the limiting aperture is not a circle, the size of the effective aperture is specified by the diameter of a circle of equal area.

Relative Aperture. This is the ratio of the focal length of the lens to the diameter of the effective aperture. For example if a lens has a focal length of 100 mm. and the effective aperture has a diameter of 25 mm., the relative aperture is 100/25 = 4. This is also known as the *f*-number, and is expressed as f/4, *f*4, 1 : 4, etc.

The light passing power of the lens, and hence the exposure required in a photograph, depends on the relative aperture. Lenses of different focal length and different effective apertures still form an image of the same brightness, other things being equal, if the relative aperture or *f*-number is the same. (In practice light losses in the lens by absorption and reflection complicate the situation slightly.) In common photographic parlance the term aperture is taken to mean the relative aperture.

The light transmitted through the lens is proportional to the area of the effective aperture, and hence inversely proportional to the square of the relative aperture of *f*-number. In other words a lens set to *f* 4 passes four times the light of a lens set to *f* 8.

Aperture Ratio. When the light beam going through the lens comes from a nearer object, and hence is no longer parallel, the light passing power depends on the aperture ratio. This is the value of $2n \sin \alpha$, where n is the refractive index of the medium in the image space (practically 1 with air) and α is the angle measured in the same medium, between the axis and a ray from the axial object point which passes through the edge of the limiting aperture.

When the object is at infinity, the aperture ratio with a medium of a refractive index of 1 is equal to the relative aperture. With objects at nearer distances the aperture ratio can be taken as the image distance divided by the effective aperture at the infinity setting. For example when a lens of relative aperture *f* 4 is used for same-size copying, the lens-to-image distance is twice the focal length. The aperture ratio then becomes 1 : 8.

The exposure for a photograph taken under these conditions must be increased in proportion to the square of the aperture ratio divided by the relative aperture—in other words $(8/4)^2 =$ 4 times the exposure with the lens set to infinity. In practice it is inconvenient to work out aperture ratios for different distance settings, and a simple value for the exposure factor is $(M + 1)^2$, where M is the magnification (image size/object size). Where the magnification is less than 0·1, the exposure factor is small enough to be negligible in practice.

Numerical Aperture. This is a parameter used to express the light gathering power as well as the angle of view of microscope objectives. The numerical aperture (N.A.) is the sine of half the angle of view of the objective, multiplied by the refractive index of the medium in which the objective is used (air, oil, etc.).

See also: *Diaphragm; Depth of field; Exposure; Exposure automation; Lens; Optical calculations; Photomicrography.*

APEX SYSTEM. Abbreviated term for *A*dditive system of *P*hotographic *E*xposure. This is based on a logarithmic way of expressing the camera exposure equation:

$$A_V + T_V = B_V + S_V = E_V$$

where $A_V = \text{Log}_2 f^2$ (*f* being the lens aperture); $T_V = \text{Log}_2\ 1/T$ (T being the exposure time in seconds); $B_V = \text{Log}_2\ B/6$ (B being the lighting level in foot-candles); $S_V = \text{Log}_2\ kS$ (S being the arithmetic film speed and k a constant); E_V = exposure value.

This form of expression of the camera exposure equation is of interest in connection with the exposure value scale marked on many modern camera shutters, and also in connection with the ASA and BSI logarithmic speed figures adopted in the corresponding standards since 1960. The logarithmic speeds correspond directly to values for S_V.

The terms in the above equation all carry whole-number integers, and are thus easy to manipulate mathematically. These terms are also known as unit values, and some cameras even carry such unit values for their aperture and shutter speed settings. Adding together the unit value for the lens aperture and the unit value for the shutter speed thus gives the exposure value, and there is a wide range of possible combinations of aperture unit value and time unit value to correspond to the same exposure value. (This disregards for the moment the effect of reciprocity failure.)

Further, if the unit value for the lighting (B_V in the above equation) is known or can be determined from an instrument, it is only necessary to add it to the logarithmic film speed value to get the required exposure setting on the camera. The use of the logarithmic scale to the base 2 means that every increase of one step in any of the unit values corresponds to a halving of the exposure, in the same way as with the exposure value scale.

UNIT VALUES COMPARED

Unit Value	Corresponds to Aperture	Exposure Time (sec.)	Lighting Level (ft. candles*)	Film Speed A.S.A. Log	Film Speed A.S.A. and B.S.I. Arith.
0	f 1	1	12 (0·35)	0°	3
1	f 1·4	1/2	25 (0·70)	1°	6
2	f 2	1/4	50 (1·41)	2°	12
3	f 2·8	1/8	100 (2·82)	3°	25
4	f 4	1/15	200 (5·63)	4°	50
5	f 5·6	1/30	400 (11·3)	5°	100
6	f 8	1/60	800 (22·5)	6°	200
7	f 11	1/125	1,600 (45·1)	7°	400
8	f 16	1/250	3,200 (90·1)	8°	800
9	f 22	1/500	6,400 (180)	9°	1,600
10	f 32	1/1000	13,000 (360)	10°	3,200
11	f 45	1/2000	26,000 (721)	11°	6,400

* Candle as per square foot given in brackets.

Owing to the suitability of the ASA log (to the base 2) speed values, the latter are often—though incorrectly—referred to as APEX speeds.

APLANAT. Lens corrected to give an image as free as possible from chromatic and spherical aberration. The first true aplanat was introduced by Karl August von Steinheil in 1866.

APOCHROMAT. Lens (usually microscope object glass or special process lens), which has been corrected for the three primary spectral colours as opposed to achromatic lenses which are corrected for two colours only.

APRON. Long celluloid or plastic band with dimpled edges in contact with which the film is wrapped around the reel of one type of 35 mm. developing tank. Used in some roll film tanks.

An apron is also the endless fabric band used on rotary glazers.

See also: *Glazers and glazing machines; Tanks.*

AQUARIA. Both black-and-white and colour photographs may be taken of fish and other aquatic life in tanks in private or public aquariums. Successful pictures can be made both by natural or artificial lighting.

See also: *Fish.*

AQUATINT. Another name for the gum bichromate process following a fine art term for a form of etching.

ARABIC GUM. Form of gum used in the Arabin or gum bichromate process. It is more usually called gum arabic.

See also: *Gum arabic.*

ARABIN PROCESS. Another name for the gum-bichromate process perfected by Nelson K. Cherril who published an account of it in June, 1909.

ARAGO, FRANÇOIS, 1786–1853. French astronomer, physicist and politician. Used his influence in 1839 to obtain official recognition for the daguerreotype, and Government pensions for Daguerre and Isidore Niépce. Made the first official announcement of the daguerreotype, 7th January, 1839, and made the process public, 19th August, 1839, at a joint session of the French Academies of the Sciences and Fine Arts. Biographies by Lunel (1853), and Audigamme: *Arago, son genie et son influence* (1869).

ARC. Electric discharge across a gap in a circuit. The arc may be struck between carbons, as in the arc lamp, when it provides an intense concentrated source of light; or between electrodes in a gas-filled bulb or tube, as in the discharge tube, and electronic flash tubes. In the form of an intermittent spark the arc may provide illumination for high-speed cinematography.

The term is also used to indicate sparking on faulty contacts.

ARCHAEOLOGY

As scientific methods are now applied to archaeological work, photography has come to be a vital part of every excavation. It falls roughly into three categories: aerial photography, field work and indoor work.

What is needed in every case is a clear, sharp picture. The photographer cannot choose his composition; his job is to present the truth, and this requirement governs his choice of equipment, preparation and lighting. And he must be prepared to tackle a great variety of subjects. Archaeological photographs may be required to reveal the details of walls, floors, foundations, hearths, post holes, doorways, mosaic pavements, or small finds *in situ*, or they may have to record changes of colour or texture of the soil in a vertical of horizontal plane, denoting layers of occupation, foundation, trenches, etc. The newcomer to archaeological photography will find it helpful to study the illustrations in books and periodicals devoted to this exacting branch of photography.

AERIAL PHOTOGRAPHY

Aerial photographs have been used for a long time by archaeologists to help them in their study of the remains of ancient civilizations. The camera from the air can show the layout of a camp, a village, or a city, in a picture that would otherwise have called for a tedious and expensive ground survey.

The more general study of the face of the earth, known as geomorphology, also makes frequent use of aerial photography.

More recently observers have noticed that from time to time patterns appeared in aerial photographs that could not at first be seen on the ground. Research on the site confirmed that the markings indicated the presence of the buried ruins of ancient streets and houses. From then on, it showed that many otherwise invisible sites could be rendered clearly on photographs taken from about 5,000 feet in the very early morning or late evening. The extremely oblique rays of the sun at such times throw up shallow relief that could not otherwise be seen.

Even where there is no relief at all, vanished sites are often revealed by variations in the colour of crops growing on the land. These differences are more marked at some growing seasons than others. Colour differences can sometimes be traced to the stronger initial growth of crops over silted-up ditches which in dry seasons retain more moisture than the surrounding earth.

It is an interesting fact that no matter how often or for how many years the land has been ploughed and re-ploughed over such sites, it is impossible to obliterate all traces. Even when the visible surface signs have disappeared, aerial photography, if necessary with special—e.g., infra-red—sensitive materials and filters, can make them appear again.

Method. There are two types of photography:

(1) The vertical pin-point of a selected site. This has the overwhelming advantage of giving a definite scale to the photograph when the height from ground, focal length of lens, and width of negative are known. A number of such vertical shots may also be taken in the form of a vertical line overlap or even a mosaic.

(2) The oblique photograph taken from the side of the aircraft (either port or starboard). This is more of a picture and is taken at a lower level than the vertical type. It is easier to photograph this way, but the working out of a true scale is a highly complicated mathematical procedure. For a vertical photograph the fore and aft line of the aircraft and wing levels must be truly horizontal to the ground, and the camera must be suspended in the aircraft in such a manner as to avoid or lessen vibration. When taking oblique shots, the operator and his camera must not be in contact with the side of the aircraft at the moment of exposure or camera shake may be produced.

The work calls for close co-operation between pilot and photographer. Before the flight, contours and other information about the area are studied with the aid of maps and the timing and lines of flight are decided on.

There is always one type of lighting that gives the most information according to the relief of the subject and the state of the vegetation.

Oblique shots are generally best taken from a height of not more than 500 feet, in order not to lose the height of the subjects like cairns or burial mounds and to get the subject as large as possible on the negative.

Vertical pin-points require extreme care to get the subject in the centre of the negative; where possible it is better to make a series of pictures, each with a 60 per cent forward overlap. This method also allows any two of the pictures to be observed stereoscopically (a process known as interpretation). At the magnification given by the stereoscope, all the contours and details in the photograph are revealed.

Subjects in very low relief call for rising or setting sun and good results will often be obtained, particularly in the Middle or Far East, by flying into the sun.

Spring and autumn are very good seasons to photograph and show individual characteristics, because immediately over buried stone and brick walls the earth provides less moisture and the crops do not grow quite so thickly as in the earth surrounding the walls. The ripening and yellowing of the two occur at different times and the difference in colour can be distinguished by a panchromatic emulsion. The reverse happens with ditches or hut circles, because the ditch or hut marking, no matter how often ploughed, always provides a greater depth of good soil; this holds more water than the adjacent ground and provides a more lush growth of crop.

While the work is better carried out by two people with good ground work first, it can be done in a small light aircraft or even helicopter by a good pilot, using remote camera control.

Equipment. Costly equipment is not necessary for a survey. A good quarter plate or $2\frac{1}{2} \times 3\frac{1}{2}$ ins. film camera, with a first-class anastigmat lens and speeded shutter, fitted with a pale yellow or a pale green filter, and loaded with a fine grain material, is satisfactory for normal work. The shots must be made through the open window with the bellows of the camera protected against the airstream. If the windows cannot be opened, there must be no reflections from lights inside the cabin.

Where possible, the photographs are taken from an aeroplane, but archaeological expeditions also use balloons and kites. These are suitable for comparatively small areas and for taking photographs up to about 150 feet. The balloon or kite is flown over the site and the camera shutter is released by a simple time-switch.

Both oblique and vertical photographs are used, each type having special advantages.

The sensitive material normally used in this and other kinds of aerial photography is the ordinary aero film. This is simply a standard panchromatic material with high contrast, developing with normal technique to about gamma 1·5.

Prints from aerial negatives should be on the soft side to retain the maximum shadow detail.

In addition to black-and-white pictures, much useful work is nowadays done on colour film. Often details which give no visible contrast on black-and-white material show up clearly as a colour difference on colour film. Infra-red also offers possibilities. G.T.S. & M.B.C.

GROUND PHOTOGRAPHY

Before commencing work in the field or indoors, much time must be spent preparing the ground, whether the photographs are to accompany a report or merely for the records. In order to bring out the details—by which some historical point may later be illustrated—every subject needs careful preparation. In the case of trenches, the edges must be sharp and straight, the dump kept at least two feet from the edge and the intervening space swept with a hard brush. Floors must be swept to remove all loose earth and until the surface looks clean and hard; stones must be thoroughly brushed, and mosaic pavements should be washed. The cleaning is generally the joint responsibility of the photographer and the site supervisor.

It may be desirable to obtain a comprehensive view of a complicated site from a high staging or an extension ladder. Under such conditions the photographer may have to work under conditions of some difficulty being restricted by the size of the platform available.

Cameras. The choice of a camera may be governed by personal taste, or the finances of the excavation, but since the object is sharpness and detail in the photographs, the equipment should if possible include a field camera, with focusing screen, or a twin-lens reflex. The photographer must be able to see exactly what he is getting into his picture and this cannot be done adequately with a viewfinder.

A whole-plate or half-plate field camera has an enormous advantage over the miniature and folding cameras in having all the camera movements so necessary in this kind of work; rising front, swing back, cross front and double or triple extension. For work in a restricted space, such as a narrow trench or a tomb, a 9 × 12 cm. or quarter-plate camera is ideal; it has most of the movements of the larger makes and the negatives are big enough to do away with the need for enlarging, except for reproduction in publications.

Full recording involves taking a great number of photographs; this is specially true of foreign excavations, when many, if not all, of the finds have to be left in the country, and will not be available at a later date. Only a selection of these photographs will be used for publication in the final report, but the rest are an essential part of the records. They must provide a detailed record of every stage of the excavation. As all 35 mm. negatives have to be enlarged, while from the larger camera a contact print is all that is necessary, the larger camera actually saves much time and expense.

Where running costs have to be kept down, the miniature and twin-lens reflex cameras cost less for sensitive materials than the field camera, and the darkroom requirements will be correspondingly modest. But even if a miniature camera must be chosen, it is advisable to buy one with a focusing screen and interchangeable lenses—i.e., a miniature reflex. A camera of this type is very useful on any excavation both for photographing small finds and for making colour transparencies. Transparencies are used very sparingly, if at all, for illustrating reports because of the cost of colour reproduction, but they are particularly useful for lantern lectures.

Lens. It is not necessary for the lens to work at a wide aperture; the archaeological photographer, whose subjects do not move, seldom requires anything larger than *f* 8. However, he does need at least two lenses, one of medium focal length, and one of short focal length (wide angle), with, if possible, a third of long focal length. The wide-angle lens is essential when there are architectural remains on the site. A telephoto lens can be very useful on occasions, but it is an expensive extra and by no means necessary in normal archaeological work.

Tripod. The tripod should be very rigid, have a tilting head for vertical work, and be adjustable in height to at least six feet. A small tripod with a ball joint head is a necessity for work in restricted areas. It should be as simple to adjust as possible, especially for work in tombs, where there is little freedom of movement.

Exposure Meter. A good photo-electric exposure meter is a necessary part of the equipment, but it will not always be possible to get an accurate reading for the whole subject. When photographing a deep trench and in similar instances where the light strikes unevenly, the reading for the one part will be necessarily much higher than for the other. In such cases the reading should be taken for the darkest shadow; the over-exposed highlight areas of the negative can be corrected in the printing.

Scales. An archaeological photograph which does not include some sort of scale is quite meaningless. Surveying rods are generally used and when they are shown vertically they should always be parallel with the sides of the focusing screen or viewfinder. When used horizontally it is frequently better to lay them along the bottom of a wall or structure of some sort, than to have them parallel to the base of the picture.

For less extensive views a smaller scale should be used and in every case the scales should be clearly marked and kept repainted. They should not be so large and clumsy that they take away attention from the subject; they should be carefully placed where they will give scale to the subject, but not where, by their size and position, they form a major distrac-

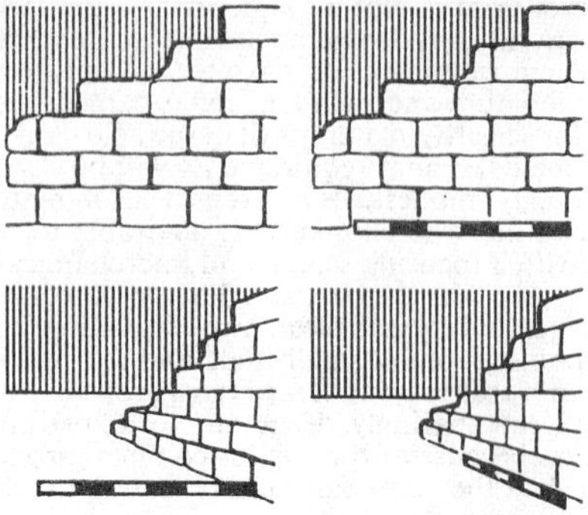

SCALE: Top left: *A record shot by itself gives little idea of size.* Top right: *A surveyor's 6 feet scale rod against the base of the object will provide a comparison.* Bottom left: *In oblique views the scale rod may be parallel to the base of the pictures.* Bottom right: *If parallel to the base of the object the divisions help to indicate actual depth of the subject.*

tion. For small finds, groups of pottery, etc., cardboard scales of varying sizes can be cut and painted to show alternate black and white divisions.

Filters. Filters are useful in archaeological work for emphasizing changes of colour in stratification and bringing out the colour and texture of pottery and the colour of mosaics. A set of tri-colour red, tri-colour green, light yellow or light green is all that will be needed. For special filters, the necessary increase of exposure is always given on a printed slip included with the filters as supplied by the makers.

Additional Equipment. Other items of necessary equipment are a small spirit level, a plumb line and a lens hood. The photographer will always find it useful to have with him one or two small decorator's paint brushes for last-minute brushing, drawing pins, plasticine and a set of small tools.

Negative Materials. Panchromatic materials are the most satisfactory for general use, though there are occasions when it is useful to have a supply of ortho as well. The speed and make of film is a matter of choice, but it is better to sacrifice speed for fine grain qualities where the negatives will have to be enlarged later. Developers made up in packets are preferable because they can be readily mixed in even the most primitive conditions in the field. It is also advisable to use tanks for processing because they save time, keep out the dust and obviate finger marks and other evils.

If miniature or roll film cameras are to be used, the films should be processed in daylight developing tanks. A changing bag can be used for loading them into the tank, thus doing away with the necessity for a darkroom. The printing is best done when the expedition returns to base.

Lighting. With archaeological subjects, the right lighting is of the utmost importance and from the beginning of the excavation the photographer must study the conditions on the site at all hours of the day and make notes for future reference. For most work in eastern countries at least it is advisable to avoid full sunlight. A photograph of a long trench is not improved by having the upper part in strong sunlight and the lower levels in deep shadow, and a collection of large storage jars photographed *in situ* in brilliant sunshine is confusing and unsatisfactory. For such photographs, sun obscured by light cloud is ideal. As this cannot often be arranged, the best times are early morning and just after sunset.

The viewpoint will usually have been chosen by the director of the excavation, and the photographer and site supervisor must co-operate in seeing that everything is ready and no time is lost in last-minute tidying up and arranging the gear. The position of the camera and ranging rods should always be tried out beforehand. The light at the foot of a deep trench or in a dark corner can often be improved by the use of large sheets of white cardboard or mirrors as reflectors.

Subjects such as inscriptions, objects with a raised pattern and decorated masonry are better not taken in the middle of the day, but when the sun strikes them obliquely; some are best taken with the sun directly behind them with the lens carefully shaded against the direct rays.

In photographing small finds and potsherds vertically, if the subject rests on a square of ground glass supported across two boxes, with

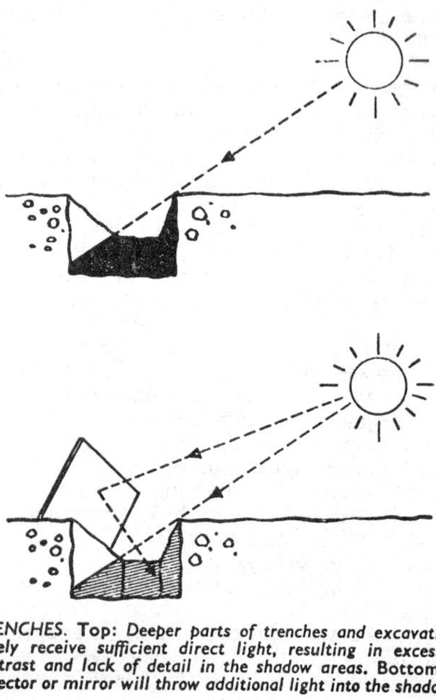

TRENCHES. Top: *Deeper parts of trenches and excavations rarely receive sufficient direct light, resulting in excessive contrast and lack of detail in the shadow areas.* Bottom: *A reflector or mirror will throw additional light into the shadows.*

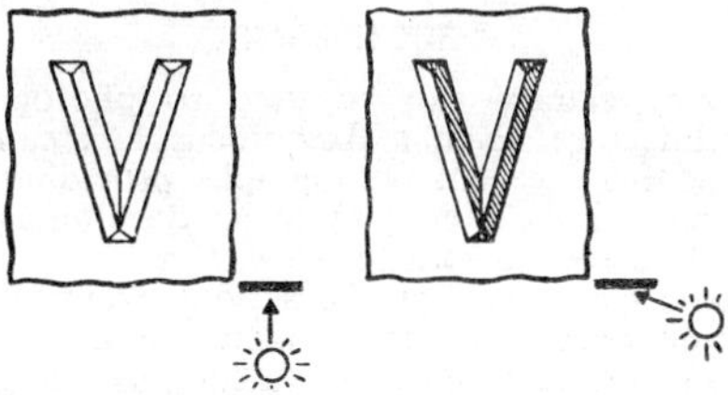

LIGHTING RELIEFS. Right: Engraved inscriptions etc., need side lighting to make them stand out. Left: Front light illuminates the surface evenly without casting shadows. It therefore fails to bring out any relief in the subject.

a sheet of white paper underneath, there will be no confusing or featureless shadows.

Photographing tombs and confined spaces where there is little or no light presents many difficulties; flash is quick and convenient, but tends to give hard shadows. If a portable generator is available, a very satisfactory type of diffused light is obtained by throwing the beam on to the ceiling or walls. (This can also be achieved by "bouncing" flash off the same surfaces.) However, in many cases there is so little room to work that there might not be space for the necessary electrical apparatus. A makeshift can be improvised with car head lamps at the end of a long flex, but the ideal is a proper adjustable lighting stand.

Indoor Work. In photographing specimens indoors, the technique does not differ basically from ordinary studio work. In the field, however, the light may not be under the control of the photographer, but good results can be obtained by natural daylight if sheets of white cardboard are used to light the side of a subject farthest from the light coming through a window or door. Large sheets of drawing paper will serve as a background and these must be continually renewed as they become creased or soiled. The pottery is normally washed before it reaches the photographer; it may, however, need a final brush and in some cases a pattern can be emphasized if the pottery is damped with a wet brush. Plasticine lightly rubbed over troublesome highlights on glazed or shiny objects will help to tone them down.

Filing System. A reliable system of filing is essential so that the director, even after a lapse of years, will be able to find a particular photograph without searching through hundreds. The method usually adopted is to number the negatives on the clear edge with Indian ink, prefacing the number with the name of the site. The number is written on the outside of the negative envelope, and a record made of the date, viewpoint, time of day, lighting, make and speed of film and exposure. These details are entered in a negative register and copied on to the back of the record print. N.L.

See also: *Aerial survey; Antiques.*

Book: *Photography for Archaeologists*, by M. B. Cookson (London).

ARCHER, FREDERICK SCOTT, 1813–57. English professional photographer and sculptor. Inventor of the wet collodion process on glass plates, which he presented as a free gift to the world, March 1851. This first practicable photographic process not patented in Britain superseded the daguerreotype and Calotype and was used by photographers for thirty years. False claimants tried to rob Archer of the honour as inventor. After Talbot's claim that the process was covered by his Calotype patent was refuted by law in December 1851, photography for the first time became widespread and fortunes were made, yet Archer died in extreme poverty. He also devised, in collaboration with Peter W. Fry, the Ambrotype process, or collodion positive on glass, and invented the stripping of collodion films, since applied in much photomechanical work. Archer was also author of the first manual on collodion photography.

ARC (HIGH INTENSITY). Type of arc used in some arc lamps. With low intensity A.C. arcs, certain drawbacks prevent their use in cinematography where a steady point-source light is required. By employing carbons run at a sufficiently high current density, a relatively flickerless and compact source of illumination is obtained which is suitable for projection.

ARCHITECTURE

An architectural photograph might show an entire building, a section of a building, either interior or exterior, or some detail of a building in close-up. Architectural photography is used for scientific purposes, for reproduction either in the form of picture postcards or in magazines, for advertising, or for the records of builders and architects. For the latter purpose photographs are normally taken at regular intervals from specific points of view, following through the stages of development to the completion of a project. Architectural photography can, of course, be applied in numerous other ways.

Records. Accurate records, e.g., of historic buildings such as those made for the National Buildings Record, are best taken under strictly controlled conditions, among them being that the camera should have a rigid body, be set absolutely square and level and record the whole building in a series of views, each of which shows two sides and an indication of one of the other camera stations. Preferably each picture should include a measuring rod in contact with one of the main walls. Such work is outside the scope of the ordinary photographer unless he works in co-operation with the National Buildings Record.

Pictorial Studies. Buildings are excellent subjects for the pictorial photographer in so far as they exhibit beauty of proportion and texture, and in the way they lend themselves to bold effects of light and shade. The pictorial worker might approach the subject as a means of personal expression; he may deliberately introduce distortion with angle shots, or suppress detail by using soft focus, or give extra emphasis to certain features when composing.

In creating a pictorial study of a piece of architecture, the photographer may include one or more of the surrounding features, or more or less of the background. He can control the importance of such items by varying his viewpoint and distance to alter the angle or perspective. Foliage forms one of the most useful aids of this kind; often the inclusion of adjacent trees introduces a balancing element to the composition or provides a useful contrast for the hard outlines of the building. A wide angle lens and a close viewpoint can achieve deliberately exaggerated perspective effects.

Normal Photographs. Outside these specialized fields, architectural photography is simply concerned with producing a reasonably good likeness of the building which will convey a truthful impression of its shape and proportions. As a rule, the surroundings, too, are important, contributing to the distinctive character of the building.

Architectural Styles. It is useful for the photographer of architecture to have some knowledge of the various architectural styles, even if his interest goes no further than the surface. Each period has its own pictorial features and the photographer who can recognize them knows what to expect from the guide book classification of any particular building. So he can tell in advance which places will yield the type of picture that interests him most.

The principal guide book classifications of English ecclesiastical and domestic architecture are Norman (1050–1180), Early English Gothic (1180–1280), Decorated Gothic (1280–1370), Perpendicular Gothic (1370–1500), Early Tudor (1500–1550), Late Tudor (1550–1600), Palladian Classic (1600–1660), Wren Classic (1660–1700), Georgian (1700–1800), Regency (1800–1837), Victorian (1837–1900), and Modern. Photographically the appeal of each is distinct—from the simple stone masses of the Norman period to the sophisticated elegance of the Regency and the functional excellence of the best Modern styles. Many guide books contain a brief introduction to the various styles.

EQUIPMENT

Any camera can be used to photograph architecture. Even a fixed focus box camera will give excellent photographs providing the photographer is content to let the limitations of the camera dictate the view.

Stand Cameras. For the serious architectural photographer, a stand camera equipped with interchangeable lenses and a complete range of camera movements is essential. Those who wish to take up this form of photography professionally might invest in a whole-plate, but much useful work can be done with a smaller size, e.g., 4 × 5 ins. To increase the extent of the lens movements, the lenses might be mounted off the centre of the lens panel. But lens movements must not be excessive or they will cause "cutting-off" at the extreme corners of the negative.

The falling front is most valuable in photographing church architecture because it allows the lens to be dropped below the base of the plate, to photograph the top of the object, i.e., the top of a font and its base at the same time, to show what the surface plan is like.

Hand Cameras. Ordinary hand cameras are unsuitable for serious architectural photography. The main drawback is the absence of camera movements—in particular, the rising front.

As practically every architectural subject lies mainly above eye level, a camera without an adequate amount of rising front must be tilted upwards at a steep angle and this produces distortion of verticals which is impossible to correct accurately.

Alternatively, either a wide angle lens can be used (if the camera takes interchangeable lenses) or the viewpoint must be moved sufficiently far away to eliminate the tilt. But this means that the negative has to be greatly enlarged (with an inevitable loss of quality).

Miniature Cameras. Although the miniature camera has its limitations in the field of architecture, it has advantages in some directions over the hand camera.

The principal advantage of the miniature for architectural photography is its great depth of field even at wide lens apertures. This enables sharp pictures to be obtained even when a fast shutter speed must be used to arrest the movement of people around the building.

Again, the miniature worker can carry a comprehensive range of lenses which can be quickly interchanged and focused. This allows him to make a complete record of the building in a set of photographs including long shots of distant details and wide angle shots in confined spaces. He can do this in a fraction of the time that would be taken with the conventional equipment.

With a camera taking an oblong negative, e.g., 24 × 36 mm., it is even possible to provide the equivalent of rising front movement by keeping the back of the camera vertical and photographing the subject on the bottom half of the negative. And even when there are uncorrected verticals on the negative, they can be restored to some extent by tilting the negative carrier and easel of the enlarger.

Lenses. A suitable set of lenses for architectural work would be one lens of normal focal length, a wide angle lens and one of longer focal length than the normal. If at all possible, supplementary attachments should be avoided, since the best quality is not achieved with such combinations. Many of the older lenses of quite simple construction can be used to achieve really good results in this field.

Lens Hood. A lens hood should be used always and adjusted afresh for each focal length of lens. The best type for this work has a rectangular shape which just permits the negative area to be covered without cutting off within the field of view. It should be wide enough for the plate when in the horizontal position and tall enough to allow the maximum rise of front without cutting off. Loose pieces of thin black card are useful when photographing detail under brightly lighted windows. A suitable piece is rested on top of the existing hood to cut off any strong beams of light. Exceptionally, it is advisable not to use a lens hood with a wide angle lens and considerable camera movements. The danger of cutting off the negative corners is considerable under such conditions.

Tripod. The most suitable form of tripod has a maximum height of about 5 feet 10 ins. and a minimum of 2 feet with all the legs sliding into one another so that the camera can be adjusted to any height between these two extremes.

Tripods for architectural work should be fitted with a very strong universal ball-joint at the top or a pan and tilt head, providing independent forward backward and side to side tilts of the camera platform.

To prevent the tripod legs from slipping on polished floors there are special rubber feet that fit over the ends of the legs. The same purpose can be served by using three rubber heels, fastened to pieces of wood about 2 ins. square, drilled with holes to take the tripod legs. A mat will also serve the same purpose.

A useful accessory is a small spirit level, to allow the back of the camera to be set accurately in the vertical and horizontal planes. The single bubble type used by engineers is preferable to the circular or T-type sometimes offered for photography, and the casing should be cut away at the sides so that the bubble can be seen at eye level. Some of the better cameras have spirit levels built into the camera body.

Negative Materials. Medium speed, backed panchromatic sheet film, or plates, are the most suitable negative material and in a roll-film camera, a medium speed panchromatic film. These materials have the necessary latitude to deal with the extreme tone range of architectural subjects, and the fine grain for recording detail and texture.

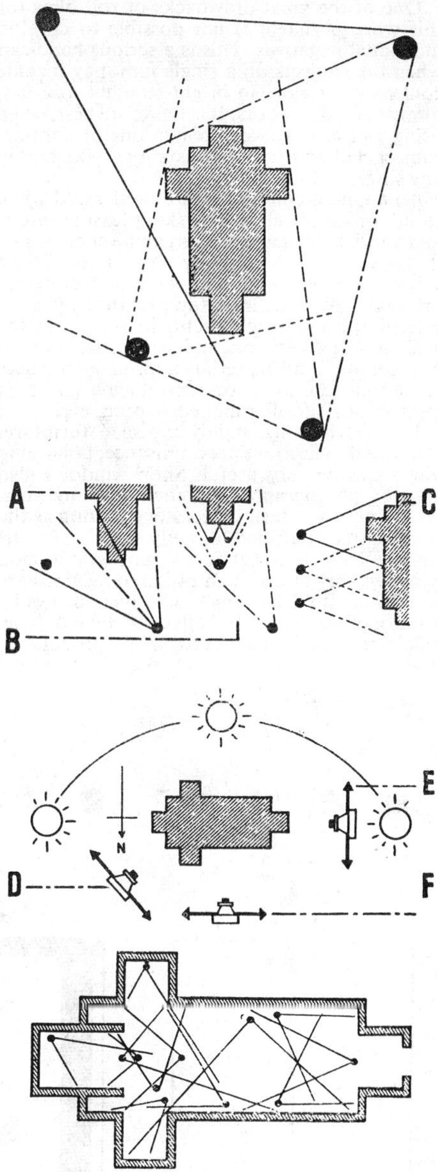

CHURCHES. Top: *For general coverage take overlapping views of the whole building, including the previous camera position in each shot.* Upper Centre: *If necessary take a second overlapping shot A to ensure covering the earlier camera position. General views can be followed by secondary views B and detail shots. To record the whole of a long wall C take overlapping shots with the camera positions moved parallel to the direction of the wall.* Lower Centre: *The eastern end of a church is best photographed from position D between 5 and 7 a.m., the west side from E between 4 and 8 p.m., and the north face from F between 6 and 9 p.m.* Bottom: *A complete interior record requires a large number of shots from pre-planned positions.*

One of the great drawbacks of roll films for this work is that it is not possible to develop individual negatives. This is a serious handicap when the subjects on a single film may include contrasty exteriors in bright sunlight and flat, dimly lighted interiors. It is therefore best, when using roll film, to keep conditions of lighting, subject matter, etc., as constant as possible on any single roll.

Filters. Under conditions of blue sky and white cloud, or simply clear blue sky, pleasing results and varying dramatic effects can be achieved by the use of filters—ranging from red to light yellow. Architectural photographs are greatly improved by a definite tone in the sky area. Filters are also often useful in recording the grain in woodwork, or emphasizing the contrast between different materials and, in such places as church interiors, for reproducing coloured inscriptions and illuminated writing, etc.

For shop fronts, highly polished furniture, etc., where reflections are a nuisance, polarizing filters can be very useful. Shop windows also may be photographed by their own lighting, helped out by a flash if necessary so long as the flash is not reflected directly back into the camera lens. It is better to photograph shop windows very late at night, when reflections are at a minimum. The camera can then always be positioned to lessen the effect of unavoidable reflections from street lamps, shop signs, etc.

TECHNIQUE

Before setting up the camera, viewpoint and lighting must be considered.

Viewpoint. Once the viewpoint is chosen, the subject is adjusted to fill the picture space by using a lens of a suitable focal length—not by moving the viewpoint closer to or farther away from the subject (because that would change the perspective).

Lighting. The broad principles of the lighting of outdoor subjects described under daylight apply here. Generally, the best lighting for architectural subjects is sunlight softened by thin clouds and with plenty of reflection from large white cloud masses. Direct sunlight from a cloudless sky gives a hard picture with inky shadows, while the shadowless lighting of an overcast sky loses all the relief and texture of the subject. These faults can be corrected to some extent in development.

The direction of the lighting calls for special attention. Frontal lighting is generally unsuitable because it casts no shadows that can be seen from the camera position and thus does not show in relief the features of the building which stand out towards the camera.

Back lighting again gives no impression of relief to the side of the building facing the camera, and it introduces violent and dramatic contrasts that are out of place in this context.

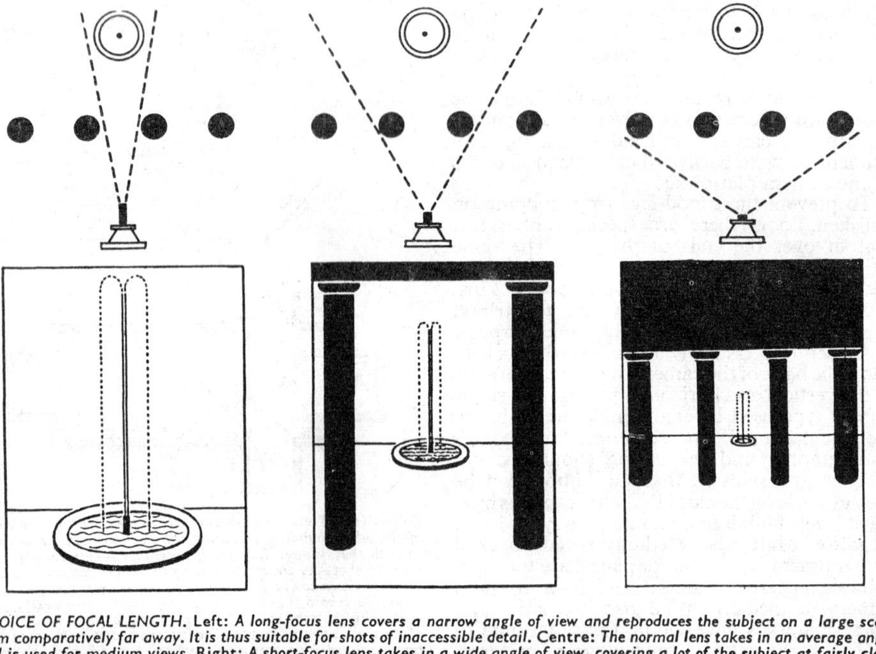

CHOICE OF FOCAL LENGTH. Left: *A long-focus lens covers a narrow angle of view and reproduces the subject on a large scale from comparatively far away. It is thus suitable for shots of inaccessible detail.* Centre: *The normal lens takes in an average angle and is used for medium views.* Right: *A short-focus lens takes in a wide angle of view, covering a lot of the subject at fairly close range. It is ideal for photographs in confined spaces where it is impossible to back sufficiently from the subject. Extreme wide-angle lenses may, however, produce some image distortion (e.g., the widening of the outside columns in this case).*

As a rule the most suitable lighting is from the side and front at a fairly low angle, e.g., 45 degrees. To emphasize texture at the expense of everything else, it should come from the side and simply skim the surface.

The only way in which the photographer can get the light from the right direction is to choose the right time of day. So for the best results each aspect of the building must be photographed at a different time of day, and a little planning with this in mind will save time and wasted journeys.

Many important buildings including churches and cathedrals are floodlit on special occasions. Pictures of buildings by this form of lighting can be most effective, and provide a valuable supplement to a set taken in daylight.

Perspective. When the lighting conditions are right, the camera is set up at the chosen viewpoint with the baseboard horizontal and the back vertical. The image is roughly focused and examined on the focusing screen. If the top of the building (or the spire or tower of the church) does not come into the picture, it is brought in if possible by raising the camera front. If there is not enough movement available, then front and back swings are used and the baseboard tilted until the subject is all on the screen.

The above procedure takes care of the vertical perspective. Next the horizontal perspective is corrected if necessary, by swinging the back of the camera to produce the most pleasing perspective of the horizontal lines.

If the back is swung to bring it parallel to one face of the building, parallel horizontal lines on that face will reproduce as parallel lines in the print. At the same time, the perspective along the second side of the building may become too steep and it may be necessary to reduce the lateral back swing to an intermediate position giving reasonably good perspective on each face.

At this stage it may be found that the lens has been moved so far off centre that it no longer covers the plate. This is corrected by using front swing until the whole of the picture once more lies inside the area covered and circle of illumination. By this time the focusing screen should show a picture of the building in which all vertical lines are parallel with the sides of the frame and in which the whole subject is included.

Where other features are included in the picture—e.g., trees, or a gateway in the foreground, or hills in the background—their size in relation to the building is controlled by adjusting the distance of the camera from the subject. Moving the camera closer, for instance, would make the objects in the foreground appear larger and the hills in the background smaller in relation to the building. Moving the camera farther away would make the foreground objects appear smaller and the hills larger. The image size must then be adjusted to fill the negative by changing the lens for one

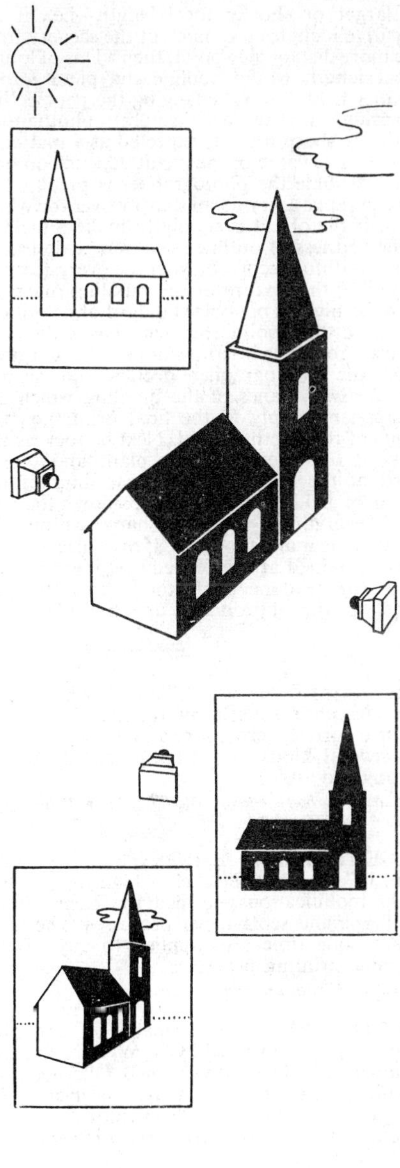

DIRECTION OF THE SUN. To display the structure of a building to its best advantage, the lighting must emphasize its shape. This is governed by the direction of the light, which can only be controlled by waiting for the sun to move to the best position, and then moving the camera position to the best viewpoint. Top: When the sun is directly behind the camera, everything is evenly lit, and the result is flat and lifeless, without modelling. Lower centre: With the sun behind the subject, the shadow side faces the camera and the image consists of nearly detailless silhouettes. Bottom: The light from the side throws the subject into relief by casting shadows yet retains highlight and half-tone detail.

of larger or shorter focal length—i.e., if the negative includes too much of the scenery from the more distant viewpoint, then a lens of longer focal length would confine the picture to a smaller field without changing the perspective.

Focusing. In an architectural photograph, critically sharp focus is expected as a matter of course. The value of the result depends on how much trouble the photographer is prepared to take to focus every detail sharply over the whole area. If the object is simply to make a picture, some softness of outline may even be desirable, but a faithful record must show every feature, including the texture of the building material, with the highest possible standard of definition.

Up to this point, the lens has only been focused roughly; it is now adjusted accurately. First the photographer decides the nearest and farthest planes of the building which are to appear sharply in the final print, i.e., the limits of the depth of field. Next he focuses the lens (at full aperture) on a plane about one-third of the way from the near limit of the depth of field. Finally he stops down the lens until everything appears sharp within the required near and far limits. If overall sharpness is not attained at the desired lens aperture, this may still be done by further tilting the lens panel in the appropriate direction. He then replaces the focusing screen with a loaded plate holder, and makes the exposure normally.

Exposure. Exposure for architectural exteriors can be measured by any of the usual methods and the values are estimated as for other outdoor subjects—e.g., landscapes—making due allowance for such factors as the tone of the subject, the contrast of the scene (including the setting) and the need for reproducing detail in the deepest shadows.

Normally the exposure will be made with a very small stop to achieve the greatest possible depth of field.

Development. Specialists in this field mostly use negative material of a fairly large format and develop by inspection. But the amateur is well advised to stick to his usual developer and adjust his technique to give a clean, fine-grain, not over-contrasty negative. This type of negative will bring out the shadow and highlight detail that is an essential feature of any good architectural record. In extreme cases of halation, however, special developers may be needed.

L.H.-F.(V.G.)

See also: *Camera movements; Interiors; National Building Record; Photogrammetry.*

Books: *Focus on Architecture and Sculpture*, by H. Gernsheim (London); *Photography and Architecture* by Eric de Maré (London).

ARC LAMP. Type of lamps which uses incandescence caused by the passage of an electric current across a gap in a circuit. There are several kinds of arc lamp used in photography and projection.

See also: *Carbon arc lamp; Discharge lamp; Point source lamp; Zirconium lamp.*

ARGENTOTYPE. Early type of positive print invented by Herschel in 1842 which, with slight modifications, formed the basis of the Kallitype and sepia paper processes. The term was at one time also applied to any of the bromide printing processes.

See also: *Printing processes (obsolete).*

ARISTOTYPE. A print made on collodio-chloride paper invented by G. W. Simpson and commercially introduced by J. B. Obernetter of Munich, 1868. The same name was given to Sir W. de W. Abney's gelatin-chloride paper, commercially introduced by Liesegang in Düsseldorf, 1886.

ARMAT, THOMAS, 1866–1919. American inventor (originally an estate agent in Washington, D.C.). Tried to convert the Edison Kinetoscope for projection on to a screen, modified the Demeny Chronophotographe, developed with Jenkins the Phantascope (1895) and exhibited motion pictures with his Vitascope in New York, April 1896 (several months after Lumière showed his Cinematograph in Paris.

ARTIFICIAL LIGHT. Natural light is the light of the sun, moon or stars; all other kinds of lighting are classed as artificial. Electricity provides the principal source of artificial lighting used in photography.

Photographic materials are generally less sensitive to artificial light than to daylight, although in practice the difference is very slight and is dependent on the nature of the light source. However, where accuracy is important, this speed difference must be taken into account, so films often have two speed ratings: one for daylight and one for artificial light by normal filament bulbs.

Exposure for subjects may be either calculated from the type, power and distance of the light, or it may be measured.

There are various types of calculator and tables available for the first method, but there are so many variables that, unless the photographer has some previous experience with the lighting set-up, the results may be no more than 50 per cent satisfactory.

Measurement with an exposure meter is much more reliable—and quicker. At the same time the readings must be taken in such a way that they give a true picture of the illumination reflected in the direction of the camera, and the results must be interpreted intelligently.

Exposures for flash illumination can be calculated fairly satisfactorily from the flash factor of the bulb or electronic outfit in use.

See also: *Lighting the subject; Light sources.*

ARTIFICIAL LIGHT HISTORY

Oxy-hydrogen Light (Limelight). The first artificial source of illumination applied to photography in order to shorten the exposure was the limelight invented by Thomas Drummond in 1826.

A jet of oxygen projected through a flame of hydrogen gas impinged on a disk or cylinder of lime, which became brilliantly incandescent. During experiments with this light in the winter and spring of 1839–40, L. L. B. Ibbetson of the Royal Polytechnic Institution, London, succeeded in making in less than 5 minutes a daguerreotype of a piece of coral magnified 12½ times in a microscope—a subject which he stated would have required 25 minutes by sunlight. J. B. Dancer, a well-known microscopist, in July 1840 publicly demonstrated the daguerreotype process at the Liverpool Mechanics' Institution, including taking a daguerreotype of a flea magnified by a gas microscope to fill a 7 × 5 ins. plate.

After chemical acceleration had speeded up the daguerreotype plate, J. F. Goddard photographed a white bust by oxy-hydrogen light with an exposure of 3 minutes in March 1841. Two months later Antoine Claudet took the first portraits from life by artificial light in 15 seconds. (This and other artificial light sources "employed to take portraits at night" were patented by Claudet on 18 December 1841).

These specimens were exhibited at the Royal Institution as a curiosity proving the relative speeds of Goddard's and Claudet's acceleration methods, rather than suggesting the application of limelight to portraiture. Claudet admitted that the light was too harsh, resulting in chalk-white faces.

Limelight was the usual illuminant for optical projection lanterns until after World War I. Since then it has been replaced by electric light.

Pyrotechnic Light. In France, Gaudin and Delamarre patented in 1854 a method of photographing with a kind of Bengal light, with which an exposure of 2 to 3 seconds was sufficient for stereoscopic portraits. A similar pyrotechnic compound of sulphur, potassium nitrate and antimony was used in John Moule's Photogen patented in February 1857 and advertised as "The Rival of the Sun". The object was to make the portrait photographer independent of the weather, and able to work in the evening. The compound when burned in a hexagonal lantern with glass sides produced a brilliant bluish-white light. One side of the lamp had a blue glass screen to protect the sitter from the dazzling light and soften the effect, and it was claimed that better portraits were obtained than with oxy-hydrogen light, though a critic called them "hideous portraits, ghastly and grave-like". The time of combustion was about 15 seconds, during which the exposure was completed, the fumes being carried out of the window by a stove-pipe.

There was a revival of pyrotechnic compound with the Luxograph of Alder and Clarke patented in April 1878. The apparatus had a 4 foot reflector lined with silvered glass, and being fitted with wheels, could easily be moved about. Semi-transparent screens interposed between sitter and light-source softened the effect.

The compound was ignited by a gas jet, and the cost per exposure was stated to be about 2d.—the same as with Moule's Photogen.

Magnesium Light. In 1859 Professor Robert Wilhelm Bunsen (1811–99) of Heidelberg University in collaboration with (Sir) Henry Roscoe reported on the extreme brightness and actinic qualities of burning magnesium wire, and its value for photography. Bunsen devised an apparatus with a spool of magnesium wire unwound by clockwork to bring the end of the wire regularly into contact with the flame of an ordinary spirit-lamp. The first portrait by magnesium light was taken on 22 February 1864 by Alfred Brothers of Manchester, the sitter being Roscoe. Magnesium light was on several occasions used for photography in coal-mines, in the Great Pyramid by C. Piazzi Smyth (1865) and in the Mammoth Caves, Kentucky, by Charles Waldack (1866). Various forms of lamps and reflectors were invented, but almost insuperable objections were the expense of magnesium, as well as the smoke, and magnesium was not much used until the introduction of flashlight powder in 1886. Powdered magnesium had been first proposed in 1865 by W. White, and the idea was revived nearly twenty years later by G. A. Kenyon, suggesting a mixture of powdered magnesium and potassium chlorate. Subsequently this mixture became the most commonly used compound—though some photographers mixed gunpowder with magnesium powder and accidents were not rare! With skill, good results could be obtained, the length of the flash varying according to the compound used from 1/25th to ½ second, using fast gelatin dry plates. The price of magnesium had greatly fallen, but smoke formation remained a great drawback, for the so-called "smokeless" flash powders were only smokeless before ignition.

Attempts were made to catch the smoke in a bag.

Louis Boutan, a specialist in underwater photography, used an electrically ignited magnesium flash lamp designed in 1893 by an electrical engineer named Chauffour. It consisted of coiled magnesium wire in a watertight glass globe containing oxygen, and connected by a fine platinum wire to a battery. When the current was released the platinum wire became red-hot and ignited the magnesium, producing a vivid flash which was intensified by a reflector fixed to the globe. Chauffour's system was a forerunner of the modern smokeless flashbulb

Louis Boutan's electrically-ignited magnesium flash lamp for underwater photography (1893).

with electric ignition, but it had the disadvantage that as a result of the high temperature the oxygen expanded and burst the globe. The innovation patented by Dr. Paul Vierkoetter in 1925 consisted in avoiding excessive expansion by burning the magnesium wire in a near-vacuum. This improvement eliminated the shattering of the glass bulb. The modern safety flashbulb was introduced, and has been in general use since 1929, when Johannes Ostermeier devised a refinement substituting aluminium wire or foil in oxygen, for the magnesium wire, ribbon, and other shapes specified by Vierkoetter.

Electric Light. Probably the earliest use of electric light for photography was made in 1844 when some of the micro-daguerreotypes of medical specimens taken by Léon Foucault to illustrate Alfred Donné's *Cours de Microscopie* were illuminated by a carbon arc lamp.

In June 1851 Fox Talbot successfully photographed by the flash of a Leyden battery a page of *The Times* fastened to a revolving wheel—a forerunner of stroboscopic light photography.

Three years later Gaudin and Delamarre patented in France a parabolic reflector of silver-plated copper for portrait photography, the light-source being either the pyrotechnic compound already mentioned, or a carbon arc lamp. The eyes of the sitter were protected from the glare by a small concave mirror in front of the carbon points, which threw the light back into the reflector, and by a blue glass screen placed between the light and the sitter. This patent fell into the public domain because the annual payments were not made, which implies that it was not worth keeping on. In fact, the first electric light portrait is claimed to have been taken in 1857 by the St. Petersburg photographer Levitzky with an arc lamp powered by a Bunsen battery consisting of no fewer than 800 elements.

About 1860 Nadar installed a Bunsen battery of 50 elements in his Paris studio, and by 1863 had achieved good results in portraiture with exposures of 60 to 85 seconds, on wet collodion plates, and using white reflectors to light up the shadows. His electric light pictures of the sewers and catacombs of Paris caused a sensation at the International Exhibition in London, 1862.

Adolf Ost of Vienna improved upon Nadar's system by using, instead of white reflecting screens, a smaller lamp powered by 40 Bunsen elements and placed further from the sitter, the main one being powered by 80 elements. Blue glass screens were interposed between the lights and the sitter. Ost arrived at exposures of 7 seconds (using wet collodion), but the running costs of the installation were exceedingly high, so its use was restricted to dark days. Ost showed portraits taken by electric light, up to a size of 13 × 16 ins., at the first photographic exhibition in Vienna in May and June 1864.

Big photomechanical printing firms, and some ordinary photographic establishments, turned to electric light to allow longer working hours. The Woodbury Permanent Photographing Printing Co., for instance, installed about

1870 a powerful dynamo to provide the necessary illumination for making gelatin reliefs. Compared with 10 minutes' exposure by sunlight, the 1,200 candle-power electric light required 4 hours to give the same result. The limelight formerly used at this establishment necessitated an exposure of double that time.

The first studio in the world relying exclusively on electric light was opened by Henry Van der Weyde in Regent Street, London, in September 1877. Van der Weyde installed a powerful gas-driven Siemens dynamo in the basement, which supplied current to a carbon arc light of 6,000 candle-power in a 4 foot diameter parabolic reflector fitted with a Fresnel lighthouse lens. To shade the sitter's eyes, a 4 in. concave mirror (similar to Gaudin's and Delamarre's) was fixed in the centre of the reflector, the positive carbon projecting through it. Exposures were stated to be 8 to 10 seconds for "Promenade" portraits ($7\frac{1}{2} \times 3\frac{3}{4}$ ins.) and 2 to 3 seconds for cartes. The Prince of Wales (later Edward VII) set the fashion for having one's portrait taken in evening dress at Van der Weyde's studio, and within a few years similar electric light studios were opened in other European capitals, as well as rival studios in London.

J. E. Mayall's establishment in Bond Street had 12,000 candle-power.

After the introduction of the incandescent filament bulb in 1880, its inventor (Sir) Joseph Wilson Swan expected that electric light for photography would become comparatively inexpensive. However, only a few electricity supply undertakings were formed and public utility electricity did not become general until the present century.

Photoflood lamps of 500 and 1,000 watt power with aluminium reflectors were introduced in the 1920s and have remained an essential light-source in photographic studios and for the photography of architectural interiors and sculpture. A more concentrated beam of light is obtained by using photoflood lamps fitted in spotlights.

Gas Light. As early as 1854 there are references to gas light being used both for taking and printing photographs. Coal-gas is a feeble source of illumination, and being yellowish, weak in actinic rays. Each Bunsen burner with a mantle was equivalent to only about 60 candle-power, and a limit was set to the number of burners by the great heat generated. Nevertheless, a few provincial studios used gas lighting augmented by reflectors, in imitation of the London electric light studios. One of the earliest was that of P. Maitland Laws in Newcastle-on-Tyne, about 1880. He used a Wigham burner with 68 jets yielding 1250 candle-power, a movable reflector lined with silvered glass, and a blue glass screen. Exposures were several times as long as with electric light and Laws regarded the installation only as an auxiliary for the winter season, but his clients were so infected with the craze for artificial light photography that they insisted on gas light portraits even in the middle of the summer.

Gas light was extensively used between about 1885 and 1920 for contact printing on so-called gaslight paper, i.e. chloride and chlorobromide papers. H. & A.G.

Books: *The March of Photography*, by Erich Stenger; *A Source Book of Photographic History*, by Wolfgang Baier (both London).

ARTIGUE PROCESS. Variation of carbon printing, introduced by M. Artigue. In the process no transfer took place and the print was not reversed. The special materials are no longer obtainable.

See also: *Printing processes (obsolete)*.

ART PHOTOGRAPHS. Euphemism used in commercial advertising for describing nude or semi-nude studies posed for sex appeal.

A.S.A. SPEED. System of rating the speed of sensitized materials, laid down by the American Standards Association in the A.S.A. standards PH2.5–1960 for black-and-white negative materials and PH2.21–1961 for colour reversal materials. The A.S.A. speed for black-and-white negative emulsions is based on measurement of a prescribed point of the characteristic curve which is at a minimum density level above base + fog density, and satisfies certain conditions of slope. The A.S.A. speed can be expressed either as an arithmetical speed—e.g. A.S.A. 100—or in a logarithmic form where each successive step represents a change in sensitivity by a factor of 2—e.g. A.S.A. 5°. It has become widespread practice to quote both, e.g. A.S.A. 100/5°. This form distinguishes the A.S.A. speed from the older numerically similar exposure index figure.

With reversal colour films the speed is based on a mean exposure derived from two reference points on the characteristic curve, but is expressed in the same way.

Numerically A.S.A. speeds are identical with B.S. speeds as specified by B.S. 1380 : 1962, and related to the speeds specified in the German standard DIN 4512 of 1962.

See also: *Speed of sensitized materials.*

ASPECT RATIO. Shape of the screen image of the projected cine film, expressed in terms of the proportion of the width to the height. Normally this is 4:3 (1·33:1) but is greater with wide-screen motion picture systems—for example 1·8 :1, 2·5:1, or even 3:1.

See also: *Projection practice.*

ASPHALT. Another name for bitumen, important in photography as being the substance used by Niépce for making photographs. Asphalt is rendered insoluble in turpentine by the action of light.

ASPHERIC LENS. Term generally applied to optical systems in which one or more lens (or mirror) elements have non-spherical surfaces. An aspheric surface can be described as the deformation of a spherical surface of equal vertex radius. The effect of such deformation, on either a reflecting or refracting surface, is to advance or retard the incident wave-front relative to the action of the spherical surface by itself. Simple examples of such surfaces of

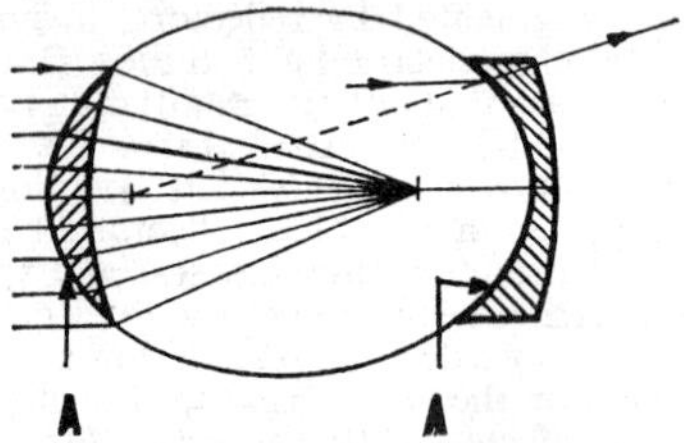

SIMPLE ASPHERIC LENS SYSTEM. The non-spherical surfaces (A) are here elliptic, being the surfaces of an ellipsoid.

revolution are the ellipse, parabola and the hyperbola derived from a conic section (i.e., slicing off the top of a round cone at an angle). Each has particular properties when used as reflecting or refracting surfaces.

The use of an aspheric or non-spherical surface can reduce optical aberrations, especially spherical aberration. A related idea is a bispheric lens, which has a spherical surface but with different curvatures in the centre and at the

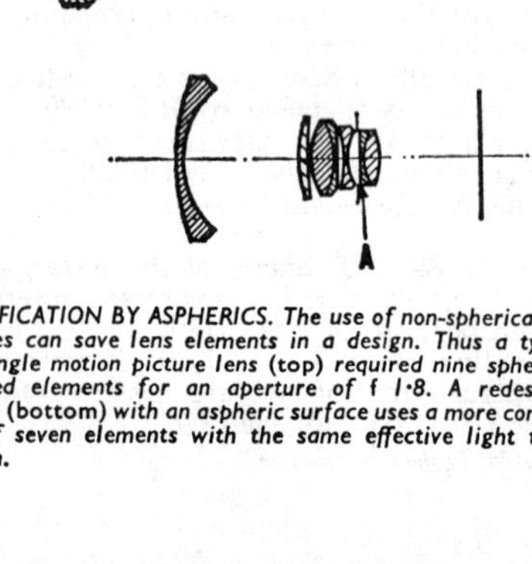

SIMPLIFICATION BY ASPHERICS. The use of non-spherical lens surfaces can save lens elements in a design. Thus a typical wide-angle motion picture lens (top) required nine spherical-surfaced elements for an aperture of f 1·8. A redesigned version (bottom) with an aspheric surface uses a more compact lens of seven elements with the same effective light transmission.

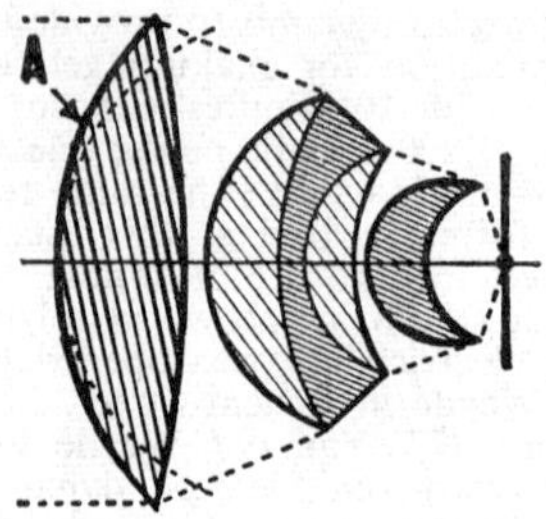

ULTRA-SPEED DESIGN. An experimental triplet achieved an aperture of f 0·519 with only five elements, one having an aspheric surface (A). The dotted line indicates the spherical curve.

edges. This again helps to bring axial and marginal rays more closely to the same focus. Aspherical surfaces are a powerful tool to the lens designer, and in an established system can lead to a new level of performance. The extra degree of freedom it offers can also lead to a more simplified optical construction resulting in fewer glass elements to achieve a particular result. The "modern" aspheric surface is often a much more complex surface of revolution than any of the conic sections and its use is not restricted to correcting axial aberration only.

Economically, aspheric surfaces are more expensive to produce than spherical surfaces, since they must be manufactured singly. However, a particular system need not cost very much more if in using such a surface, several spherical surfaces are eliminated.

One of the more well known applications of an aspheric surface is the corrector plate of a Schmidt mirror system. In the last few years a small number of photographic lenses and lenses for cine and television purposes have utilized aspheric components. Condenser lenses in projectors frequently include one or more components with an aspheric surface. As such lenses are often moulded (or pre-moulded) rather than ground, and the required standard of precision lower, this manufacture is not appreciably more expensive than for normal condenser lenses. P.A.M.

See also: *Aberrations of lenses; Lens history; Optical design.*

ASSOCIATION OF CINEMATOGRAPH, TELEVISION AND ALLIED TECHNICIANS. The Association of Cine and Allied Technicians was founded in 1933 and registered as a trade union in the same year. Its aims and activities correspond to those of other trades unions, and are devoted to the interests of those employed in the motion picture industry and in television. Membership is normally open to film technicians engaged in film production and film processing, including directors, cameramen, sound recording engineers, art directors, editors and assisting grades. Similar grades employed in television are also eligible for membership. The Association has negotiated agreements on pay and working conditions with employers in features, newsreel and short films, and also in film processing.

It is affiliated to the Trades Union Congress and the Labour Party.

The Association is governed by a General Council representing all grades, members of the Council being elected by ballot at the annual conference where policy is also decided. Membership of the Association in 1963 was 11,637. Entry to most grades is dependent upon employment conditions in the industry.

The official publication of the Association, the *Film and TV Technician*, is issued monthly.

ASTIGMATISM. One of the off-axis aberrations of lenses which makes the lens reproduce a point light source as two lines at right angles lying in different focal planes.

See also: *Aberrations of lenses; Lens performance; Lens testing.*

ASTRONOMY

Astronomical photography is concerned with the photography of everything that lies outside the earth's atmosphere. From the beginnings of photography, photographic science and astronomy have been closely linked (the word *photography* was coined by the astronomer John Herschel). The first successful daguerrotype of the sun's disc was made in 1845, but it was the introduction of the dry emulsion plate about 1875–80 that revolutionized astronomy. Since then, much of our vast increase in knowledge of the Universe has come by the application of photography. Although some measurements of the brightnesses of stars and other objects have recently been made more accurately by photoelectric methods, photography is likely to remain a major technique in astronomy for many years. In return, the exacting demands of the astronomer for materials sensitive to weak light fluxes and to particular wavelengths have stimulated many improvements in the manufacture of emulsions that have had applications in other fields.

Photography has various advantages to the astronomer; the following are the more important.

(1) The sensitive emulsion can add up the light received over a period and thus give an image of a celestial object that does not emit enough light to make it visible to the eye even through the most powerful telescope.

(2) Sensitive materials can be manufactured with a wider range of colour response than that of the human eye (e.g. ultra-violet and infra-red).

(3) The camera gives a permanent record of transient events.

Modern Apparatus. Four main forms of camera (or telescope adapted for photography) are used for different purposes:

(1) Refracting telescope. With an achromatic doublet lens working at about *f* 12 to *f* 15: the largest (Yerkes Observatory, Wisconsin) has a 40 in. diameter lens. It is used for photography of the moon and planets, and the very accurate measurement of the positions of stars to determine their distance by the parallax method.

(2) Reflecting telescope. The image is formed at the focus of a parabolic mirror, typically of about *f* 5. The telescope is perfectly achromatic, and the mirror can be made very large to gather much light and so enable faint objects to be studied. The useful field is however severely limited by coma to about ½°, even if a correcting lens is placed just in front of the photographic plate. Reflecting telescopes are the most generally used equipment today for the study of individual stars and the more distant galaxies. The largest (Hale telescope, Palomar Mountain Observatory) has a mirror 200 ins. in diameter working at *f* 3·3.

(3) Astrographic camera. The lens is relatively small (typically 5 to 10 ins. diameter) and of three or four components, to give a working field several degrees in diameter at about *f* 8. It is used to measure accurately the relative positions of stars and their motions with respect to one another over periods of years.

(4) Schmidt camera. This ingenious design of camera was announced by Bernhard Schmidt in 1932. It is primarily a reflecting telescope, with all its advantages. The mirror is however spherical, and a thin corrector plate (a weak aspherical lens) at the centre of curvature of the mirror removes the spherical aberration and coma without introducing significant chromatic aberration. It can be made in large sizes to work at about *f* 3 and give essentially perfect images of stars over a field of several degrees Used for survey work it can reach much fainter stars than the astrographic camera. The largest camera has a corrector plate a little over 48 ins. in diameter and works at *f* 2·5 (Palomar Schmidt).

Other instruments are used for specialized purposes, e.g. lenses and mirrors of great focal length (100 feet or more) for solar work, and fast super-Schmidt designs (20 ins. aperture, *f* 1·5) for recording meteors and artificial satellites.

Since celestial objects move across the sky because of the rotation of the earth, most large telescopes are mounted on equatorial mountings that follow the objects automatically during the long exposure. The polar axis of these mountings is parallel to the axis of rotation of the earth, and is driven in the contrary direction by an accurate clock.

The precision required of astronomical equipment, in both optics and mounting, is the most exacting of any form of photography. Stars are effectively perfect points of light at infinity, and must be recorded as small symmetrical images alike at the centre and at the edge of the photographic plate. Most lenses designed for pictorial work of the highest quality, or even for photogrammetry, show intolerable defects when attempts are made to apply them to astronomy. Moreover the quantitative interpretation of the astronomical negatives, by such methods as densitometry to give the brightnesses of stars, etc., requires the most meticulous attention to cleanliness in working, uniformity of development, and reproducibility of processing techniques. Most practical astronomers are necessarily skilled technical photographers.

Spectrographs. The study of the composition of the light from celestial objects gives information about the temperature, constitution, and chemical composition of the source. In addition the measured wavelengths of features in the spectrum reveal the velocity of the object. A line in the spectrum is displaced in wavelength by an amount directly proportional to the velocity, to the red end of the spectrum if the object is receding. Spectrography is therefore a most important branch of astrophysics. Usually a reflecting telescope is used to image the star or other object on to the slit of a spectrograph.

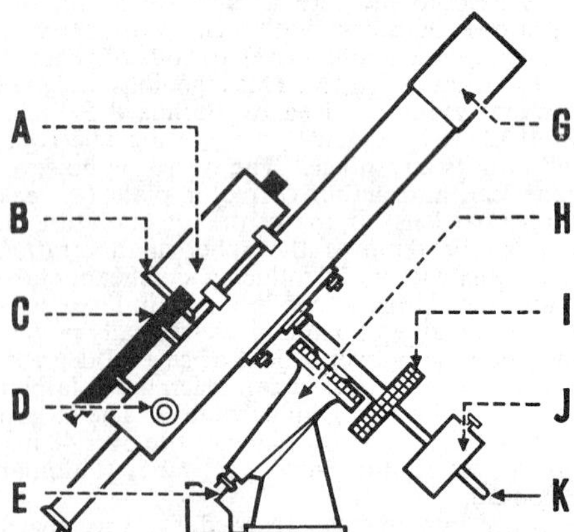

EQUATORIAL MOUNTING, carrying a telescope and small camera. The rotation of the polar axis once a day compensates for the rotation of the earth and makes the telescope follow the apparent motion of the stars. A, camera; B, plateholder; C, finder telescope; D, eyepiece focusing knob; E, polar axis (angle of elevation = latitude of observatory); F, telescope eyepiece; G, objective lens and dew cap; H, polar axis with right ascension circle; I, declination circle; J, counter weight; K, declination axis; L, driving clock.

In modern work reflecting diffraction gratings are found more efficient than prisms. The spectrum so formed is photographed with cameras: all-mirror optics are used for work in the ultra-violet region, and small fast Schmidt cameras for faint light sources.

Emulsions. The choice of emulsion for any particular photograph depends on such factors as the brightness of the object, the wavelength region to be studied, the steadiness of the earth's atmosphere, the aperture and the focal length of the telescope, and the purpose of the photograph. The resolving power and dispersion of the spectrograph are also relevant. In the past empirical experience played a large part in the choice; more recently progress has been made in choosing materials for their informational sensitivity characteristics, by the application of information theory to the imaging and photographic processes.

Although astronomical objects cover an enormous brightness range the indisposable factor is usually the extreme faintness of the objects to be studied. Even with the largest telescopes and the most sensitive emulsions exposures may run to several hours. For such work all other considerations must be sacrificed to that of speed, even though the emulsion for any other purpose would have an intolerably coarse grain. Emulsions considered fast in commercial and pictorial work are useless because of the failure of the reciprocity law for long exposures. Special emulsions with low reciprocity failure are used. The speed of some emulsions can also be increased by such techniques as pre-flashing or baking (heating at about 60°C for some hours) before the exposure. Colour sensitized (dyed) emulsions are used in limited wavelength regions.

Although spectacular colour photographs of a few objects have been taken, the low speed and small exposure range of colour reversal and similar processes have limited the application of direct colour photography in astronomy.

BRIGHTNESS OF CELESTIAL OBJECTS

Object	*Magnitude*	*Relative Brightness*
Sun	-26	10^{10}
Moon	-11	10^{4}
Stars:		
Brightest	-1	1
Faintest visible with naked eye	$+6$	$1{\cdot}6 \times 10^{-3}$
Faintest photographed	$+22$	6×10^{-10}

The magnitude scale is logarithmic. An interval of 5 magnitudes corresponds to a factor of 100 in brightness; 1 magnitude = 2·512 in brightness.

The Sun. The sun is the only star near enough for its surface to be investigated in detail.

Direct photography of the sun is of two types:

(1) photography in white light to record sunspots and the finer structure, called granulation, on the disc;

(2) photographs in very narrow wavelength regions centred on such Fraunhofer lines as the red *C* line of hydrogen at 6563 A, which show the distribution of that element in the solar atmosphere and such features as prominences and flares.

The earth's atmosphere is normally turbulent because it is not of uniform temperature, and small irregularities of refractive index blur the finest detail: exposures on the disc are therefore kept short (0.001 second) to "freeze" the image, and there is sufficient light to use fine grain plates of high contrast.

Photographs in the light of a single wavelength (spectroheliograms) are taken with scanning spectrographs or through narrow band filters of the Lyot type, using a suitable colour-sensitized emulsion. They show the prominences and flares that are features of the chromosphere, the lowest layer in the solar atmosphere.

Above this is the still fainter corona which constitutes the outer atmosphere of the sun. This could originally be studied only during total eclipses. A coronagraph is a telescope designed to minimize scattered light and produce an artificial eclipse; used at high altitude observatories, it enables the brighter inner corona to be photographed hour by hour in good sky conditions. The first coronagraph was designed by Bernard Lyot and was installed at the Pic-du-Midi Observatory, in the Pyrenees, in 1935. Time lapse films of prominence and coronal activity have been made.

Some investigations, however, must still be made during the brief few minutes of a total eclipse of the sun. Suitable eclipses occur every few years and require an expedition to the narrow band on the earth's surface where the eclipse is total. Exacting photographic requirements have often to be met under tropical or other extreme conditions, with poor water supply, dust, etc.

It is not possible to record all the various luminous zones around the sun on one photographic plate because the range of brightness is too great. The prominences and more intense features near the surface of the sun can be photographed in a few seconds but the long streamers that reach out for great distances into space need much longer exposures.

The Moon. J. W. Draper first photographed the moon in March 1840, shortly after the invention of photography. Detailed photographs of the surface of the moon are usually taken near the terminator (the boundary between the illuminated and dark parts). On these parts of the moon's surface the sun is just rising or setting, and hence mountains, craters, etc., are thrown into relief. Turbulence in the earth's atmosphere at night still produces blurring of the image, though it is less serious than in the daytime. Typical exposures are one or two seconds. The smallest features detectable have dimensions of a few hundred yards.

A most striking recent feat has been the photography of the coarser features of that area of the moon's surface that is invisible from the earth. This was achieved by the Russian Automatic Interplanetary Station (AIS) which passed behind the moon in October 1959. The AIS carried two cameras with 200 mm. (*f* 5·6) and 500 mm. (*f* 9·5) lenses respectively, which were automatically directed at the moon. The photographs were processed and dried in the AIS, and were then transmitted to the earth by radio, for reconstruction by methods analagous to those used in the formation of a picture on a television screen.

The Planets. Of the principal planets of the solar system only the brighter (Venus, Mars, Jupiter and Saturn) have discs large enough to show appreciable detail, and these have been extensively studied. Mercury, Uranus and Neptune are difficult subjects and Pluto is too small to show a photographable disc.

The first photographs to show planetary surface markings, of Mars and Jupiter, were taken independently by B. A. Gould and A. A. Common in 1879.

The method and exposures are generally similar to those used for the moon. Mars shows some features similar to the earth, with an atmosphere, clouds, and markings on the solid surface. Photographs in blue light show clouds and atmospheric features, red and infra-red rays penetrate the atmosphere and show surface markings. The cloud layers on Venus are too dense for even infra-red rays to penetrate to the surface. Spectrography gives indications of the gaseous constituents of the atmospheres of the planets.

The minor planets, orbiting between Mars and Jupiter, are too small to show a disc and appear star-like: they are detected by their motion against the background of the stars on photographs taken with astrographs or Schmidt cameras. The equatorially mounted camera, driven to follow the stars, records the stars as point images, but the minor planets give elongated trails because they move during the long exposure.

Astrometry. Although moving through space with velocities up to a few hundred kilometres a second, the stars are so distant that their relative motion in the sky is very difficult to detect. A typical value is 0·1 second of arc in ten years, and even with cameras of long focus the displacement of the star images may be only a few thousandths of a millimetre from one negative to another taken several years later. Great care must be taken to maintain the exact adjustment of the lenses of the astrographic camera during the interval. Careful processing and drying are necessary to avoid distorting the emulsion when it is still wet and soft, and plates must also be stored in suitable conditions since they may be needed for measurement decades later.

Sky Surveys. The light gathering power and wide field of the Schmidt cameras have enabled surveys of the sky to be made to very faint magnitude limits: they are thus surveys of the Universe to great distances. The most comprehensive is the National Geographic Society—Palomar Survey, covering the whole sky visible to the 48 inch Schmidt camera. It records tens of millions of stars and galaxies down to 20^m, the more distant galaxies being at least 3×10^9 light years away and receding at velocities of one-third the velocity of light. Contact print copies of the Survey are in the world's principal observatories. Individual objects of interest can be selected from these surveys for detailed study with reflectors and spectrographs.

Artificial Satellites. Most of the work of tracking rockets and satellites in the first stages of flight after launching is similar to the techniques of missile photography and uses special long focus camera systems.

Once the satellite is above the atmosphere and in free flight the precise orbit must be obtained, and small changes in the orbit determined from time to time. Photographs are therefore required which show the track and velocity of the satellite, rather than details of the object itself. When in orbit most of the artificial satellites are faint objects, and to photograph them cameras of large linear aperture are required: the field must also be large since the satellites are moving relatively quickly.

Schmidt cameras are evidently suitable, and one of the cameras developed for this work is a modified Schmidt design with a 31 ins. mirror working at *f* 1·0. A crystal clock photographed on to the same film frame can be read to 0·1 millisecond, and the expected accuracy of the observation is 2 seconds of arc at 1 millisecond. A 6 ins. satellite has been recorded in this way at a distance of 2,500 miles.

One method of determining the path and velocity is to mount the camera on the usual equatorial mounting and drive it to keep the star images stationary on the film. The moving satellite then appears as a line of light against the star background. The exposure is interrupted at regular and accurately determined times so that the image of the satellite appears as a dashed line on the film. The length and midpoint of each dash gives the instantaneous velocity and position of the satellite.

To be of scientific value amateur photographs of satellites must satisfy certain minimum conditions, specified by the Society of Photographic Scientists and Engineers as follows:

The camera should take negatives of at least 4 × 5 ins. and have a lens of a focal length of at least 5 ins. The sensitive material should have an emulsion speed of at least 200 ASA, and the satellite should be photographed against a star background (preferably a well known constellation for easy recognition), with two gaps in the track timed to 0·1 second.

Photography from Balloons and Rockets. Turbulence in the atmosphere near the earth produces blurring of the images of celestial objects, and prevents the photography of the finest detail. The highest accessible mountain sites do not remove this difficulty. Since 1957 a group of Princeton University astronomers have made successful observations with a 12 ins. aperture telescope which is carried to heights above 80,000 feet by an unmanned balloon. The balloons are huge envelopes of plastic film which, containing a small amount of helium gas at launching, expand to their full size at altitude. The telescope is automatically guided on the sun and operated by remote radio control. At the end of each flight the telescope is returned to earth by parachute, or by releasing gas from the balloon.

A similar telescope of 36 ins. aperture for photography of stars and planets is now used.

The earth's atmosphere is also opaque to ultra-violet radiation below about 3,300 A, but this interesting region of the spectrum is becoming accessible to cameras and spectrographs mounted in rockets and satellites. Most of the negatives so obtained, e.g. of the ultra-violet spectrum of the sun, have been exposed in rockets above the atmosphere and processed after recovery from the fallen rocket. The practical difficulties of research in space are immense, and space research will not make terrestrial telescopes obsolete in the foreseeable future. The application of image intensifier devices to terrestrial telescopes may yield results as important as space research. These devices are interposed between the primary image, formed by the telescope, and the photographic plate; they amplify the incident light signal by photoelectric methods. Various types are undergoing development and useful results have been obtained. This seems a more promising approach than attempts to make even larger reflecting telescopes. D.W.D. & G.T.S.

See also: *Information in the Image; Missile photography; Reciprocity law; Schmidt camera; Spectrography.*

Books: *Astronomical Photography*, by G. de Vaucouleurs (London), historical and general; *Stars and Stellar Systems: Vol. I, Techniques*, edited by W. A. Hiltner (Chicago), advanced text; *Astronomical Photography for the Amateur*, by T. Rackham (London).

ATGET, JEAN EUGÈNE AUGUSTE, 1857–1927. French actor and photographer. Specialized in picturing Paris from 1898 onwards; the artistic and documentary value of his photographs was recognized only after his death. Biographies: *Atget, Photographe de Paris*, by Pierre Mac-Orlan (Paris 1930); *World of Atget* by B. Abbott (New York 1964).

ATMOSPHERE. That quality in a photograph which evokes the natural sensations associated with particular scenes. The atmosphere of a March landscape might be conveyed by a cold, clear sky, pale sunshine, bare, wind-blown branches and scurrying clouds. A cathedral interior would possess atmosphere if, by subdued tones and distant, highlighted windows, it suggested the characteristic echoing twilight of such places.

ATMOSPHERIC ACTION. Action of the air on photographic materials which generally causes slow deterioration. In addition to active oxygen which is particularly damaging to developers, the air may contain varying amounts of harmful impurities such as sulphur, suspended moisture and dust, and many kinds of industrial fume and smoke. For this reason all sensitive materials should be kept out of contact with free air as much as possible.

See also: *Fogging*.

ATOMIC THEORY. Supposition that all matter consists of basic and chemically indivisible units, the atoms. During chemical reactions they combine with each other in proportions that are either simple or a multiple of the number of atoms involved.

AUDIO-VISUAL AIDS. Equipment and systems for projecting pictures with or without sound accompaniment, used as an aid to teaching, instruction and training. "Audio-visual" implies the use of both sound and image.

Aids of this nature cover cine film in 8 and 16 mm., especially with optical or magnetic sound tracks; film strips and slides and projectors for them, as well as tape recorders used to provide a sound commentary and/or control the picture presentation; finally closed circuit television systems used for education and instruction.

See also: *Visual Aids*.

AUSTRALIA. Australia as a nation can be said to have grown up with photography. Records show that an experiment in daguerrotype photography (a street scene) was made by a visitor from France as early as May 1841 while in the following year (9th December to be exact) the first professional daguerrotypist, G. B. Goodman, had set himself up in business in a studio on the roof of the Royal Hotel, Sydney. Goodman subsequently travelled Australia visiting all the principal centres and taking the new portraits which attracted much public attention. Technical developments in photography reached Australia fairly quickly and by the early fifties a number of permanent businesses had been established in all the towns which are now the state capitals. In Sydney the photographic work of such men as Oswald Allen, W. Bradley, W. Hetzer, Thomas B. Glaister, Edward Dalton, Wheeler and W. C. W. Freeman was acclaimed as being the equal of any in the world. In Melbourne the new art was associated with the names of Townsend Duryea, P. M. Batchelder, T. A. Hill, John Noon, W. Asquith, W. W. Pentland, Antoine Fauchery, Davis & Co., and Burman.

Early Exhibition. Amateurs were soon active and by the year 1858, interest had grown sufficiently for a display of photographs to be held by the Sydney Philosophical Society; on this occasion, in addition to many photographs by the above-mentioned Sydney professionals, examples were displayed by a number of amateurs—these including Professor J. Smith, Stanley Jevons and R. Hunt, the latter two being officers of the then newly-established Sydney Royal Mint. Large collections of either negatives or prints by these three workers have survived and may well represent the world's most important collections of early amateur photography. Active, although not an exhibitor on this occasion, was Sir William MacArthur of Camden, believed to be the first Australian-born photographer.

Historical Collections. A comparatively large cross-section of early photography of all types has been preserved in Australia, and most state libraries, as well as the Australian National Library at Canberra, show increasing interest in this subject. Two of the most important collections are perhaps that of Professor John Smith, who came from Aberdeen (*c.* 1857) to be first Professor of Chemistry at Sydney University and who recorded photographically the growth of his University, and the Holterman Collection, which was brought to public notice by the Australasian Photo-Review in 1953. The latter collection comprises several thousand original negatives of the period 1872–83. Under the sponsorship of the gold miner, Bernard Otto Holtermann (famous for his discovery of the world's greatest specimen of reef gold) photographers Beaufoy Merlin and Charles Bayliss covered all the principal towns of the south-east of the continent, besides producing many hundreds of negatives representing a complete portrayal of life on the Hill End and Gulgong goldfields in 1872. So complete are these two last-named series that it has been possible to effect perfect reconstructions of the two towns. The collection now in The Mitchell Library (Sydney), includes the largest negatives ever made by the wet plate process.

All historical aspects are now being actively attended to by the Historical Committee of the Australian Photographic Society and by a large number of state libraries and by the National Library at Canberra, which has already acquired important collections representing the work of most of our pioneer pictorialists.

Technical and Manufacturing Progress. Australia can claim at least four "firsts" in photography. These include the first practical photolithography (by James W. Osborne, Victorian Government Photographer, in September, 1859, and also by John Degotardi of Sydney); the first high-quality method of half-tone photo-reproduction (the Woodburytype, by Walter Woodbury, 1864) the full length feature film (by Ensign Perry of the Salvation

Army, Melbourne, 1900, being Commandant Herbert Booth's *Soldiers of the Cross*) and the first film for children (*Kidstakes*, *c*. 1930). *Soldiers of the Cross* also featured colour, which was introduced by means of hand-coloured lantern slides during the changing of the reels.

Commercial dry plates began to reach Australia in the early eighties while the first Australian-made plates appeared *c*. 1883–4 when Thomas Baker opened a small factory for the coating of Austral Dry Plates. A little later he entered into partnership with J. J. Rouse and established shops in several centres. By the turn of the century, the partnership was representing the Eastman Kodak Co. in Australia and about 1907 became a Kodak subsidiary as did, about twenty years later, the only other large distributing organization, Harringtons Ltd., which had ceased manufacturing during World War I. Today at Coburg there is a modern forty acre plant in which is coated a wide variety of sensitized materials. The production programme also covers a range of accessories, and chemical preparations for amateur, professional and specialized uses, as well as colour film processing. Various firms make photographic and other lenses and ranges of modern slide projectors. Local packaging is undertaken by leading importers and there are a large number of modern colour processing laboratories catering for the increasing public interest.

Applications. In every department of life, Australia may be regarded as a most photographically-minded country. Possibly the most characteristically Australian aspect of scientific photography lies in the fields of cosmic rays and radio astronomy. In the clinical photography of operative surgery, the photographic technicians of the leading hospitals are recognized as being amongst the world's foremost, considerable progress having been made with many special techniques, including the production of stereoscopic colour films. Governmental agricultural, forestry and soil conservation services use photography extensively. There has been great added scope for advanced technical photography through such projects as the Snowy Mountain Hydro-Electric Scheme and the Long Range Weapons Establishment at Woomera, S.A.

Recent research work has led to improvement in the electrophotographic process—in particular, to the invention of liquid development of the latent electrostatic image (thereby enabling the process to be extended to full colour printing) and to the establishment of new coating techniques based upon a chargeless system. These processes have now been internationally patented.

A forward step in the field of projection has resulted in a well-defined and exciting art form in which symbolical and exciting abstract slides dissolve into each other to the accompaniment of electronic music. A further recent development involves the use of revolving polaroid filters.

Photographers and Organizations. Pictorial photography has been followed in Australia for almost one hundred years. Its greatest period was in the mid-twenties when the salons of the world were distinguished by the contributions of Dr. Julian Smith, Harold Cazneaux, John B. Eaton, Norman Deck, and Henri Mallard.

The Australian Photographic Society was established in 1961 and rapidly achieved full status as the co-ordinating body for Australian Photography both nationally and internationally. Its forerunner, the Australian Photographic Federation continues as a division linking the governing bodies of the camera club movement in every state; it compromises the Photographic Society of Queensland, Federation of Camera Clubs (N.S.W.), Canberra Photographic Society, Victorian Association of Photographic Societies, Tasmanian Photographic Federation, South Australian Photographic Federation, Western Australian Photographic Federation together with Australian Portfolio Photographic Society and the Stereoscopic Society, Australian section. Through direct membership of the A.P.S. and its 350 affiliated camera clubs, more than 6,000 photographers are now linked together and thereby associated with the F.I.A.P.

A large number of national and international exhibitions are regularly held. The A.P.S. will eventually issue its own honours but, in the meantime, a start is being made with awards of medals and plaques of striking national design.

Increased interest in exhibitions has resulted in greater support by Australians to overseas salons. Our most notable successes are those associated with natural history photography and colour photography. Australia continues to supply an increasing proportion of the entries in Asian Exhibitions and also receives important support from those countries for its own salons. In addition there are amateur cine societies, trade and professional associations.

Publications. Portfolios apart, there have been few purely photographic volumes published in Australia. Outstanding is Jack Cato's monumental *The Story of the Camera in Australia*—an authoritative volume of over 200 pages of letterpress and seventy-six pages of illustrations, published in December 1955.

The longest established journal, *The Australasian Photo-Review*, ceased publication in 1956, its 63rd year. Photographic magazines now being published are the monthlies *Australian Photography* and *Camera World*, and the quarterly *Industrial Photography*. There are two trade monthlies, while the journal *Professional Photography* represents the official organ for the professional photographer.

K.B. & A.R.A.

AUSTRIA

Austria has made a valuable contribution to the progress of photography, notably through the activities of certain photographic organizations in Austria—some of which are almost as old as photography itself. Among the individuals who are prominent for their contribution to basic research, the names of Petzval, Eder, Valenta and Schinzel are particularly worthy of mention.

The history of photography in Austria began in the year 1839, when von Ettingshausen, professor of Physics at Vienna University, returned from Paris where he had learned Daguerre's process from the inventor himself. He passed his knowledge on to his friends, who formed the nucleus from which the whole Austrian photographic movement evolved.

Inventions and Discoveries. In 1840 the Viennese mathematician, Josef M. Petzval, computed his portrait lens. It has since become world famous, and even today its aperture of *f* 3·6 is noteworthy. In the same year the Viennese, Franz Kratochwila, succeeded in increasing the sensitivity of photographic plates five times by exposing them to bromide and chloride vapours, as well as the iodine vapour used by Daguerre. The combined effect of these improvements was to reduce the long exposure times of 15 minutes or more to a matter of seconds, thus greatly facilitating portrait photography. Petzval's lens was manufactured in Vienna by Peter Friedrich Voigtländer in 1841.

In 1876 Josef Maria Eder (1855–1944), introduced important improvements in the collodion process. The daguerrotype and Fox Talbot's calotype were superseded by Scott Archer's collodion process. Eder now discovered that the keeping properties and sensitivity of collodion plates could both be considerably enhanced by means of the new cadmium double salts which he had discovered. The discovery was awarded a bronze medal at the 1878 World Exhibition in Paris.

In collaboration with von Tóth, Eder evolved a new method of intensifying collodion negatives with lead salts and potassium ferricyanide, which was published under the name of lead intensification. The two men also performed pioneering work on toning processes, in the course of which they demonstrated that by converting the silver image to silver ferrocyanide, it was then possible to dye it in various colours by the use of metallic chlorides. Their work laid the foundations of all later toning processes.

In 1880 Eder and G. Pizzighelli introduced an important improvement in the method of preparing the dry plate by their invention of the silver oxide ammonia method, which is still in use today. This method combines the advantages of rapid working and low ripening temperature, and produces vigorous, clear negatives.

In the same year Eder and von Tóth discovered pyrocatechin as a developing agent for silver bromide gelatin emulsions. Together with the discovery of the developing properties of hydroquinone by the Englishman, Abney, this achievement headed the long list of organic developers.

In 1881, Eder and Pizzighelli invented a method of making positive plates and gaslight papers of low sensitivity with pure silver chloride emulsion requiring alkaline development. Hitherto only printing-out papers, using excess silver nitrate, had been known. Twelve years later this discovery was popularized throughout the world by the Americans in the form of Velox silver chloride paper. Eder later also introduced chlorobromide paper, which was first manufactured commercially in England and is today the standard emulsion for the warm-tone papers popularly used by pictorial photographers.

In 1884 Eder discovered that erythrosin among the eosin group of coal-tar dyes was a particularly good green and yellow sensitizer. He thereby succeeded in making the first orthochromatic emulsion, the term orthochromatic itself being coined in Vienna at that time. Shortly after this discovery the manufacture of orthochromatic erythrosin plates began throughout the world.

In 1889 Professor Eduard Valenta found that glycin red could act as a sensitizing dye for the red-orange spectral region. Six years later, only one year after Rœntgen's discovery of X-rays, Valenta in collaboration with Eder produced a series of unique works on photographic science entitled *Röntgenphotographie*, also an *Atlas Typischer Spektren* (Atlas of Typical Spectra) and, in 1904 *Beiträge zur Photochemie und Spektralanalyse* (Contributions to photographic chemistry and spectral analysis).

In 1898 Eder read a paper to the third International Congress on Applied Chemistry in Vienna on the principles of a standard sensitometer. His paper brought about the use of Scheiner's Sensitometer for determining sensitometric standards.

In 1906 Theodor Scheimpflug invented a device for eliminating distortion in aerial photographs which he called a photoperspectograph. His invention was published in a paper entitled *Die Herstellung von Karten und Plänen auf photographischem Wege* (Map and Plan Making by Photography) which still stands as an exposition of the essentials of aerial photogrammetry.

In 1919 Eder and Hecht produced a new grey scale wedge sensitometer for determining emulsion speeds. This had become necessary owing to the constant increase in the speed of photographic materials during the first World War.

Among Eder's considerable literary output mention must be made of his *Ausführliches Handbuch der Photographie* (Comprehensive Handbook of Photography), which was first published in 1884. Many accounts of his extensive work on photographic science were published in the *Photographische Korrespondenz*, the organ of the Photographische Gesellschaft (Photographic Society).

In 1922 the Viennese, Otto Schuloff, discovered pinaflavol as a green sensitizer, which led to still further improvement in orthochromatic plates. At the same time he discovered the desensitizing properties of the pinacryptol group of dyes.

In the field of colour photography Dr. Karl Schinzel (1886–1951) in 1905 suggested a method of colour photography involving three superimposed emulsions sensitized respectively for the primary colours, blue, green and red and dyed in the complementary colours. Schinzel's idea is the basis of integral tripack colour film which was to sweep the world thirty years later. In 1936 Schinzel published a series of articles on colour photography involving tripack materials and colour development, which examined the whole question from the chemical point of view.

Josef Rheden (died 1946), an astronomer at the Wiener Universitätssternwarte (Vienna University Observatory) made a number of contributions to photography. Over a period of 25 years he took well over 2,000 photographs of small planets, star clusters, nebulae, comets and the Milky Way, which aroused much interest among specialists. Rheden was one of the first to use a miniature camera for photographing nebulae and star clusters. His purpose was to obtain a reduced, point-size image from which to measure the brightness of the star by photographic means. He also worked on sensitometry, and by means of latensification succeeded for the first time in photographing a number of very weak stars, without increase of exposure. Rheden's name became widely known through his exposure tables which still retain their value today and, together with the accompanying instructions, these tables can legitimately be described as unique. In the form of a small pocket book, the tables give full data for every half-hour of the day, for all latitudes, from sunrise to sunset throughout the year, and for a multitude of different subjects.

Organizations and Institutions. Die Photographische Gesellschaft (The Photographic Society) which was founded in 1861 in the Wiener Akademie der Wissenschaften (Vienna Academy of Science) has played a leading part in guiding the progress of photography from the beginning. Its first president was Anton Martin, author of the first German language manual of photography (apart from Daguerre manuals), the *Repertorium der Photographie*, which appeared as early as 1846. The Society has always been active in promoting study of both the theoretical and the practical aspects of photography. In 1864 it organized in Vienna the first exhibition of photographs to be held in German-speaking Europe, and innumerable exhibitions have followed up to the present time. Members of the Society have won the highest awards in foreign exhibitions of pictorial photography; and a series of grants enabled the Society to offer awards for distinguished services and achievements in photography and other closely related fields. It is noteworthy that the greater part of these awards and grants have been made to foreigners.

The Society worked to obtain legal backing for protection of copyright in photographic work, and its efforts were successful at the turn of the last century. It also influenced foreign copyright legislation.

In 1906 the Society obtained a charter from Emperor Francis Joseph I with the right to call itself the Kaiserlich-königliche Photographische Gesellschaft (Imperial and Royal Photographic Society) and to use the Austrian coat of arms. The Republic later confirmed the Society's claim to the latter privilege.

The Graphische Lehr- und Versuchsanstalt (Graphic Arts Institute) had its origin in the photographic department of the then existing Höhere Staatsgewerbeschule (Higher State Training School), which, however, had more the character of a scientific research department. Eder therefore put forward plans in 1885 for an establishment which would both pursue research and offer training in the profession of photography. The Institute was duly opened on 1st March, 1888, on the premises of a former school under the leadership of Eder, who had meanwhile been appointed Professor of Chemistry at the Wiener Technische Hochschule (Vienna Technical College). Its main activities at the outset, which it has maintained to this day, are professional training, scientific research, and the attainment of high technical and artistic standards. In the succeeding years a number of additional departments were formed, the valuable collections were increased, and the library is today recognized as one of the most complete collections of technical reference books, particularly in the field of photography.

The journal *Photographische Korrespondenz* (Photographic Review), founded by Ludwig Schrank in 1864, is the official organ both of the Photographische Gesellschaft and of the Graphische Lehr- und Versuchsanstalt.

The Austrian Stereoskopische Gesellschaft (Stereoscopic Society) which meets in the Graphic Arts Institute, was formed in 1928 to encourage and promote stereoscopy, stereo photography and stereo projection. It is noteworthy that polarized light has been used for projection at the Society's meetings since about 1940.

Research Laboratories. Austria has two research establishments for specialized branches of photography. The Mikrobiologische Station der Stadt Linz (Linz Microbiological Station) was founded in 1953 to continue the work of the Vienna Forschungslaboratorium für Wissenschaftliche and Angewandte Mikroskopie, Mikrophotographie und Mikrokinematographie (Research Laboratory for Scientific and Applied Microscopy, Photomicrography and Kinemicrography) founded 1921. The Linz establishment is mainly concerned with photographic and cinematographic recording of micro-organisms and their life cycle, in particular by means of Schild's method of dish microscopy.

The Versuchs- und Forschungslaboratorium für Infrarot-Photographie (Experimental and Research Laboratory for Infra-Red Photography) founded in Vienna in 1933 is active in the natural sciences and technology, and is mainly concerned with new applications in these fields.

Industry. The principal Austrian products are photographic papers (one firm), plates (one firm) and chemicals. In addition, well-known sub-standard cine cameras and projectors are produced by one firm. Production of still cameras and accessories is on a smaller scale. Austria depends to a large extent on imports of apparatus and sensitive materials, particularly films, of which there is no domestic production.

Professional Photography. In the early days in Austria, as in other countries, photographs were taken by persons of the most varied professions, so that there was no distinction between professionals and amateurs. Similarly many portrait painters took up photography as a profession, and exercised a noticeable influence on the style of photographic portraits and even of landscape photographs.

In the first few years studios were few in number, but the introduction of the negative process led to a rapid increase. Methods were improved and in 1860 the technique of negative retouching was introduced by the photographer Rabending and became widely practised. Amongst prominent professionals in the period up to the turn of the century the following are worthy of mention: Ludwig Angerer, Viktor Angerer, Hermann Vogel, Fritz Luckhardt and Perlmutter.

In 1905 the Photographengenossenschaft (Photographers' Association) was formed in Vienna, the Reichsverband der Photographengenossenschaften (National Union of Photographers' Associations) followed in 1906, and in 1908 a school for photographic apprentices was opened in Vienna by the Photographengenossenschaft.

Today all Austrian professional photographers must belong to the Photographeninnung (Photographers' Guild) of the federal province where they practise their profession. All provincial guilds are affiliated to the central Bundesinnung (Federal Guild). These guilds support the economic and professional interests of their members and are also responsible for professional training.

Amateur Photography. The first organization of amateur photographers to be formed in German-speaking territory was the Club der Wiener Amateurphotographen (Vienna Amateur Photographic Club) which dates from 1887. It later changed its name to Kamera Klub. The Wiener Photo Club (Vienna Photo Club) was formed in 1897, the Club der Amateurphotographen (Amateur Photographic Club) in 1903. In 1911 the Wiener Lichtbildner-Club (Vienna Photographic Club) was formed from the earlier Gesellschaft für Photographische Kunst (Society for Photographic Art). Many more clubs came into being, in particular a large number of photographic sections of the Naturfreunde (Naturalists) touring club, which popularized photography among hikers and mountain-climbers. Today the majority of Austrian amateur photographic clubs are affiliated to the Verband Österreichischer Amateur-Photographen-Vereine (Association of Austrian Amateur Photographic Clubs). In 1948 Austria became a member of the Fédération Internationale de l'Art Photographique (F.I.A.P.)

Among the workers who have had an important influence on the progress of photography in Austria the following are particularly worthy of mention: Dr. Hugo Henneberg, Dr. Friedrich Spitzer, Hans Watzek, Heinrich Kühn, and the late Rudolf Koppitz, of the Graphische Lehr- und Versuchsanstalt. Kühn was responsible for the special Imagon lens, which was manufactured by Rodenstock to his specification.

Over the years many different trends have of course manifested themselves in Austrian photography. Today's workers may be said to be pursuing a middle course; they are not exclusively wedded to modernism, nor have they remained rooted in the methods of the past. They try instead to maintain a balance between the traditional and the modern approach.

Periodicals. Apart from the *Photographische Korrespondenz* (Photographic Review) already mentioned, there are the *Allgemeine Photographische Zeitung* (official publication of the Photographers' Guild), and *Photo-Wirtschaft* (trade publication).

Amateurs are served by *Galerie* (formerly *Photo-Digest*), the official publication of the Verband Österreichischer Amateur-Photographen-Vereine, mentioned above. There is also a photographic art magazine which has been published since 1950 under the title *Österreichische Photo-Zeitung* (Austrian Photo Journal). O.H.

AUTOCHROME PROCESS. The first practical system of colour photography, invented 1903 by Auguste and Louis Lumière, who introduced, in 1907, Autochrome plates, manufactured by their firm at Lyons. A colour "screen" was formed by coating a mixture of starch grains dyed in the primary colours on to a glass plate. A panchromatic emulsion layer was exposed and, after reversal development, viewed through the same "screen".

See also: *Colour synthesis.*

AUTOCORRELATION FUNCTION. Mathematical function which provides a complete statistical measure of the physical (as opposed to visual) effects of granularity. By so correlating the fluctuations in density due to granularity at any point of an image with the fluctuations at all other points, both the size of the fluctuations and their distribution over the image surface can be statistically represented by a single function.

See also: *Information in the image.*

AUTOMATIC EXPOSURE CONTROL. System of exposure setting in cameras and other photographic gear, where the action of light on a photo cell system directly controls the amount of exposure of the sensitive material without manual interference. In still cameras automatic exposure control systems may consist of an iris diaphragm or a shutter setting mechanism directly linked to a built-in photoelectric meter. Sometimes the current of the meter circuit directly controls the period during which the shutter remains open. In enlargers an automatic control system generally uses a photo cell to control the enlarger illumination.

See also: *Enlarging; Exposure automation.*

AUTOMATIC FOCUSING. Term generally applied to focusing devices incorporated in cameras by which the act of sighting the subject through a rangefinder automatically brings the lens into focus on the subject. On enlargers it means that the lens is automatically kept focused on the easel while the head of the enlarger is raised or lowered. True automatic focusing is the means by which an independent mechanism keeps the lens focused on a given subject.

See also: *Focusing (automatic).*

AUTORADIOGRAPHY. Photographic method for recording the distribution and location of radioactive material within a specimen. The basis of this method is the placing of a sample (such as a plant leaf or a histological section containing radioactive material) in contact with a suitable photographic emulsion. After exposure and processing, the photographic layer reveals the location of the radioactive material within the sample. The image thus obtained is called an autoradiograph.

The techniques may conveniently be divided into those applied to macro- and those applied to micro-autoradiography. The former concerns the recording of the distribution of radioactivity for example in the skin of a fish or in a sheet of metal. Microautoradiography is used in studying the location of radioactive material within a minute biological cell, where very high resolution is required.

Macro-autoradiography. A simple application of this method is the study of the distribution of radioactive material in a plant leaf. If a plant was nourished for some time with a solution containing radioactive phosphorus (P^{32}) this will be taken up by the plant stem and leaves. The plant is then placed in contact with a fast X-ray film for a few hours exposure, preferably in a vacuum cassette, in order to achieve intimate contact between the specimen and the film. The radioactive phosphorus decays with the emission of high-speed electrons, thus exposing the photographic layer and rendering the exposed areas developable. The result indicates the location and relative concentration of radioactive phosphorus within the plant as a whole or within the structure of a leaf.

Micro-autoradiography. Probably the most widely used method in micro-autoradiography is the stripping technique. Special autoradiographic stripping emulsions are now available, using a nuclear track type emulsion. The advantage of a stripped emulsion layer is that it can be brought in very intimate contact with the specimen, thus permitting high resolution, which is aided by the close packing of grains in a nuclear track type emulsion. Since a stripped layer of about 5 microns thickness is easily damaged and difficult to handle, it is reinforced e.g. by a 10-micron thick gelatine layer carried on a temporary glass support. The two layers can then be stripped off together. An area of the emulsion layer is cut out sufficient to cover a histological section on a microscope slide, with a margin at least ¼ in. all round. The slide bearing the specimen containing the radioactive material is then placed on the bottom of a glass dish filled with distilled water. The emulsion and its gelatine support are stripped from the glass plate and placed on the surface of the water with the emulsion side facing the specimen. As it swells for about 3–4 minutes, the emulsion-gelatine composite first crumples and then stretches out flat. It should then be lifted from the water by raising the slide underneath it. After drying the specimen with the superimposed emulsion layer, it is placed in a light-tight box for exposure, which may take a few days to several weeks, depending on the activity of the radioactive isotope and its half-life.

Processing is carried out with the emulsion layer in permanent contact with the specimen. The autoradiograph is then viewed under a microscope when superimposed on the specimen. This technique permits the accurate alignment of the autoradiographic image with the microscopic details of the specimen. A resolution of the order of 1–2 microns can be achieved under most favourable geometrical conditions. The basic requirement that a thin sample should be brought into intimate contact with a thin emulsion layer, can also be met in other ways. These include mounting the sections directly on the emulsion layer, coating mounted sections with liquid emulsion, or dipping in the emulsion.

Applications. In biology, autoradiography has proved to be of great value in studies of the metabolism of plants and animals, also in investigations of the events occurring during mitosis and in the distribution of labelled atoms in chromosomes and in many other fields. More recently a new stimulus was given by the synthesis of a variety of organic compounds with radioactive hydrogen (H^3) which can be used in cellular investigations.

In engineering, autoradiography may be used as a tool for studies of lubrication problems, where the transfer of radio-active material from one surface to another can be examined. The degree of segregation of small amounts of foreign materials containing tracers, e.g. in steel may be studied or the uniformity of distribution of certain labelled atoms in a metal may be checked. Because of the cumulative action of the photographic layer, autoradiography is often preferable to the use of Geiger or scintillation counters, particularly if high resolution is required in the simultaneous recording of the distribution of radioactive material within several specimens. R.H.H.

See also: *Industrial radiography; Medical radiography; Nuclear physics; X-ray film.*

AUTOTRANSFORMER. Transformer in which the primary and secondary windings are formed by tapping one continuous winding at suitable points. The transformation ratio is the inverse of the ratio between the turns of wire around the whole winding and the turns on the tapped section.

See also: *Transformer.*

AUXILIARY LENS. Simple lens placed in front of the camera lens to change its focal length. Usually called a supplementary lens.

AVAILABLE LIGHT. Loosely applied term intended to describe photography in poor lighting conditions without supplementary illumination.

AWARDS. Most national photographic societies and associated organizations make awards for merit in their various fields of activity—in addition to the traditional medals for exhibition successes. These awards may take the form of medals, cash, or kind and are provided either out of the Society's funds or from the interest on money bequeathed for the purpose. The list of awards below is a selection of most of the more important ones.

America. The following are the principal photographic awards made in America by the leading societies.

Principal awards made by the Photographic Society of America include:

Progress Medal: bronze medal awarded annually, if a suitable nominee is forthcoming, for an outstanding contribution to photography or an allied subject.

P.S.A. Service Medal: awarded for the encouragement of work for the Society and photography.

Journal Award: awarded for the best scientific or technical paper published originally in an official publication of the P.S.A.

Awards made by the Society of Motion Picture and Television Engineers include:

Progress Medal: gold medal awarded annually, if a suitable nominee is forthcoming, in recognition of a significant contribution to the advance and development of motion picture technology.

Samuel L. Warner Memorial Award: made for outstanding contributions in the field of recorded sound on motion picture films.

Herbert T. Kalmus Award: gold medal for major contributions to the scientific progress of colour in motion pictures.

David Sarnoff Medal: gold medal awarded for an outstanding technical contribution in the field of television.

Journal Award: awarded for the most outstanding paper originally published in the Journal of the Society the preceding year.

The Society of Photographic Engineers makes the following award:

Progress Award: silver medal awarded for achievements in the advancement of the science of photography.

The National Press Photographers Association makes the following awards:

Joseph A. Sprague Memorial Award: bronze plaque and scroll and up to three gold rings presented in the course of a year to outstanding press photographers.

The Photographers' Association of America makes the following award:

George Harris Award: sculptured medal presented annually to the professional photographer who has achieved most for his profession during the preceding year.

Specialized awards are also made by the American Society of Photogrammetry and the Biological Photographic Association, while a number of other organizations also grant honours or degrees.

Australia. Awards of medals and plaques are made by the Australian Photographic Society.

Austria. The Wiener Photographische Gesellschaft (Photographic Society of Vienna) makes the following awards:

Society Medals: gold, silver and bronze medals awarded to suitable persons of any nationality for special services to, or achievements in, scientific and applied photography.

Voigtländer Medals: gold, silver and bronze medals awarded as above.

Society Diploma: awarded for photographic achievements and services irrespective of membership or of nationality.

Britain. In Great Britain the principal awards are made by the Royal Photographic Society:

Progress Medal: silver medal awarded annually, if a suitable candidate can be found, for some important contribution to the advancement of photography. It is one of the oldest and best-known photographic awards in the world.

Henderson Award: bronze medal and approximately £4 10s. in cash awarded annually for the best paper read or published on a photochemical subject.

Hood Medal: bronze plaque awarded annually for meritorious performance in some branch of photography.

Newman Memorial Award: bronze plaque awarded in conjunction with the British Kinematograph Society. Four plaques are awarded for outstanding work in cinematography.

Memorial Lecture Awards: medals, together with an honorarium, in each case awarded for lectures commissioned specially by the Lectures Committee. There are three: the Renwick (annually), Hurter and Driffield (bi-annually), and Trail Taylor (bi-annually).

Olaf Bloch Memorial Award: prize of books to the value of £10 awarded annually in conjunction with the Institute of British Photographers for the best essay on a set subject.

Williamson Research Award: grant in cash or kind awarded annually for research directed to the solution of one or more technical problems connected with photography.

France. The following awards are made by the Société Française de Photographie (Photographic Society of France):

Davanne Medal: bronze medal awarded annually for any notable improvement in photography.

Janssen Medal: bronze medal awarded every other year for some outstanding service to photography.

Peligot Medal: bronze medal awarded as with the Janssen Medal every other year.

Germany. The following award is made by the Gesellschaft Deutscher Lichtbildner (Society of German Photographers):

David Octavius Hill Medal: gold medal awarded from time to time to outstanding personalities of any nationality who have contributed notably to progress in photographic research, manufacture or practice.

The Deutsche Gesellschaft für Photographie (German Photographic Society), have awarded an annual Kulturpreis (cultural award) since 1959 consisting of DM 10,000 and a lens symbol mounted in gold.

The West German Government has awarded, since 1961, a unique annual Youth Photo Prize, to the value of D.M. 10,000.

The Deutsche Kinotechnische Gesellschaft (German Cine Technical Society) make one important cine award:

OskarMesster Medal: bronze medal awarded for outstanding contributions to cinematography.

Holland. The Dutch Society of Press Photographers awards cash prizes for the best exhibits at its World Press Contest.

Sweden. Scholarships are awarded by the Sweden-America Foundation for study in the United States; by the Swedish Photographers' Association; by the Stockholm municipal authorities, and by many publishing houses. In addition there is the following:

Svenska Dagbladet Award: gold medal and money prize for the year's best photography. It is presented annually by, and named after, Sweden's leading newspaper.

B

BABIES. To the parents, the pictures of a baby provide a permanent record of the child's development. In the first few months babies are changing almost daily, and the camera is the only way of capturing the fleeting phases.

Technique. From the photographer's point of view, the main problem of baby photography is the smallness of the subject. The bodies of very young babies are unattractive, so the camera must confine itself to the facial features and expression. This means either getting close, or using a long focus lens, according to the type of camera. If the camera cannot be focused for close-ups, the difficulty can be overcome by using a supplementary lens.

Working close to the subject means that the photographer must avoid parallax errors unless he is using a single lens reflex camera or one with automatic parallax compensation. Then the closer the subject, the shallower the depth of field, so that it becomes more difficult—and more important—to focus accurately. A miniature camera will give the required depth of field at quite wide apertures, but with the popular roll film types it will be necessary to stop down to as little as *f* 11. With such small apertures, the correct exposure is often too long to arrest movement unless the fastest available panchromatic film is used. The minimum exposure for small babies, unless they are asleep, is 1/50th second, and shorter still for bigger—and livelier—children.

Perspective distortion is also apt to be a nuisance at short distances, and the baby's arms and legs must not be allowed to point towards the camera. The baby's eyes are always the best points to focus on—whatever else is blurred, the eyes must be sharp.

Subject. As a baby cannot understand what the photographer wants, it cannot be expected to co-operate, nor can it be threatened or forced. The only answer is to be patient until the right moment comes naturally. In any case the baby's mother or someone responsible must always be present. Even if the photographer is on the best of terms with his subject, he will be much too occupied in taking the photograph to act if the baby is in danger of rolling off the table or hurting itself in any other way.

Babies live according to a very strict routine and if it is upset, it makes them unhappy. Also, small babies tire quickly and sessions must be very brief. So the work must go quickly and smoothly. Everything possible must be prepared before the baby is brought into the room. Lights and background should be set up, camera loaded, exposure measured. This leaves the photographer free to watch for the right expression. When it comes, it never lasts more than a fleeting second, and a last-minute change in lighting arrangements or exposure may mean missing a brilliant picture.

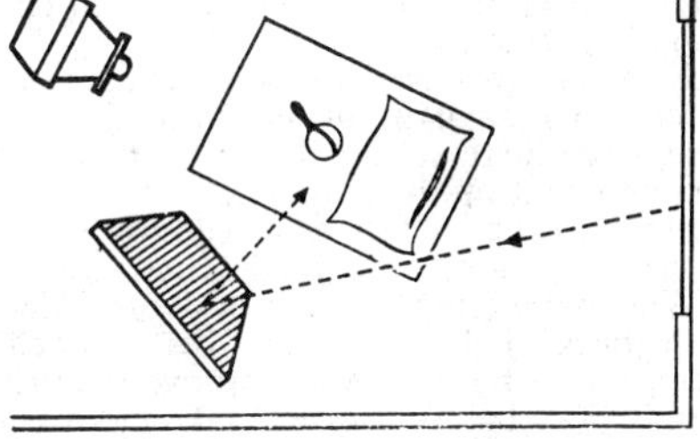

BABIES INDOORS. *For indoor daylight shots, the window should be to one side of or even somewhat behind, the baby. Use a reflector to light the shadow side.*

Lighting. For indoor pictures, two Photofloods on the subject are sufficient for most occasions, and as they give off a good deal of heat, the baby must be dressed lightly or not at all. It is a good plan to light the background separately, choosing a plain wall or other neutral surface for the purpose.

Flashlight can be a most convenient means of photographing baby. Even the small bulbs, or the smaller electronic sets, are powerful enough to permit the use of a small stop. This will give the necessary depth of field at close

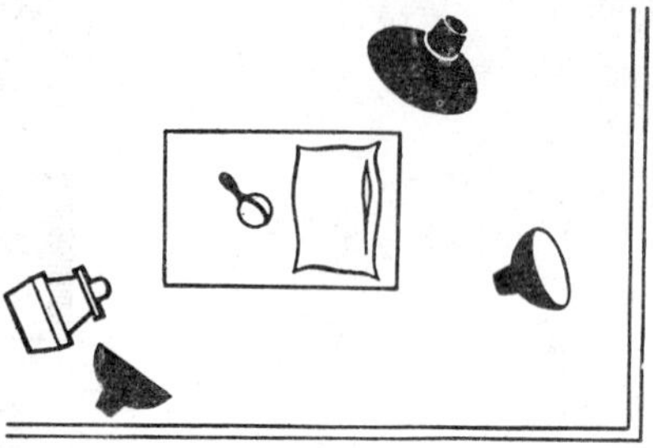

BABIES BY ARTIFICIAL LIGHT. The main light shines down on the baby, a fill-in lamp near the camera lights up the shadows, and a third lamp illuminates the background.

distances. The general principles of lighting apply to flash too, i.e., to obtain any modelling in the subject, do not use the light source close to the camera. As the camera and flash will be rather close to the baby it will often be sufficient to put the flashlight at a distance of one foot from the lens. This distance should be increased when taking the picture from farther away.

With one bulb only, put the baby near the background so that this will also get some of the light. A second flash put behind the baby and directed on the background can be very effective.

Flash bulbs are usually coated with a safety lacquer, but there is a very slight chance that they might explode on firing. Always have a safety screen in front of the bulb when photographing babies.

Good pictures may also be taken near or in a window by daylight or in sunshine. In this case, since there are no clouds to act as reflectors and light up the shadows, a white sheet or card should be held up as a reflector just clear of the picture area to soften the contrast.

When taking pictures out of doors, side—or even back—light gives the best modelling. Here, too, a reflector will improve the picture greatly. The baby's eyes will take more kindly to the light from the reflector than to direct sunlight and will look larger and more expressive in consequence.

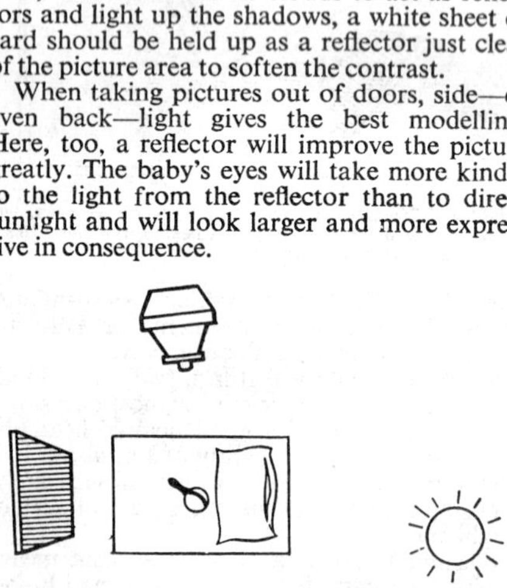

BABIES OUTDOORS. Keep the sun to one side or behind the baby to prevent the infant from screwing up its eyes. Use a reflector to provide fill-in light in the shadows.

The First Weeks. Most of the routine of the baby's day will make good pictures. The first few weeks are mostly spent in sleeping and feeding, and the baby can be photographed asleep, or in the arms of its mother. The periods of wakefulness at this age are short, so the photographer must work fast. The baby's body does not photograph well in the early weeks, and bathing pictures are best left until later, or at least taken only for record purposes.

The Third Month. In the third month the first really lively pictures become possible. The baby uses its hands much more; it discovers things by touch, plays with its fingers or the blanket, and smiles when talked to. Best of all for the photographer, it begins to raise its head when laid face down. This pose must not be overdone, however, and care must be taken to see that the baby's arms are tucked under its chest to support it.

The Fourth Month. As the baby gets older the possibilities for pictures multiply. At four months it can hold up its head without difficulty when lying face down; it reaches for toys, and reacts to faint sounds. By this time it looks attractive without its clothes, and since it loves playing in the bath, that is the place to photograph it at its best. A high viewpoint—e.g., standing on a chair by the bath—will give the photographer the best shots, but he must work quickly and have everything ready as the time is strictly limited.

The Fifth Month. At this age the baby begins to hold and chew things—a rusk or a bit of banana—and it will hold other objects in its hand. It can now be propped against cushions and photographed, and there is ample scope for similar pictures on the floor, and in the pram or playpen.

Later. From eight months the baby begins to crawl and must be kept still by giving it a toy to occupy its attention, or have its movements restricted in some way—e.g., by sitting it in a high chair. After eleven months it begins to feed itself and may be photographed holding a cup or spoon. There are many possibilities here, but messy foods like chocolate or spinach should be reserved for special effects.

From this date its development makes rapid progress, and the opportunities for making pictures increase as it learns first to kneel, then to stand, and finally to toddle. There are endless pictures to be made of its playful moments, its moods, and the regular events of its daily life, and the more automatic the photographer can make his technique, the more of the really significant instants he will be free to observe and capture. W.S.

See also: *Children; Flash technique; Portraiture at home; Portraiture outdoors.*

BACK FOCUS. Actual distance that separates the vertex of the back surface of the lens and the film when the lens is focused on infinity. May be greater or less than the focal length.

BACKGROUND

The background of a photograph should be selected and handled with the same care as the subject. Some photographs have no background—e.g., a landscape or part of a subject taken close up—but a great number consist of a single object or a group of objects seen against a background—e.g., portraits, figure studies and articles of all sorts. In all such cases the background must be thought about; it is easy to ignore or forget about it when making the exposure but it is impossible to overlook it in the print.

Generally speaking, the background should be unobtrusive—i.e., even in tone, without any sharp contrasts and softly defined or even blurred. It should be either lighter or darker than the parts of the subject against which it is viewed. The background of the finished print may consist of: the normal setting of the subject (as with most fixed subjects), a chosen natural artificial background. F.P.

History. The idea of using backgrounds of painted landscapes and interiors to enliven daguerreotype portraits was patented by Antoine Claudet in December 1841, but they were little used in England. After carte-de-visite portraits became fashionable in 1859, the photographer's studio became a stage full of properties and pictorial backgrounds. The classical column and curtain drawn back to reveal a landscape, which had served portrait painters, were pressed into service. Camille Silvy frequently designed painted backgrounds specially to indicate the status of a well-known sitter. In general, the humbler the sitter's home surroundings, the greater his desire for splendour; therefore the grander the studio, the more business the photographer could expect. Marion & Co. of London, Smedley & Co. of Blackburn and Seavey of New York manufactured backgrounds and accessories to suit the clientèle of every photographer; e.g., military props for a garrison town, a ship's deck with mast and guns for a naval base. Alexander Bassano in London had a large variety of indoor and outdoor scenes suitable for all seasons painted on a roll 80 feet long. William Notman of Montreal arranged trees, rocks, tents and camp-fires in the studio, with stuffed deer and bears, so that the armed trappers waiting for their kill seemed genuinely on the trail. Salt made an excellent substitute for snow.

Each decade in the Carte and later Cabinet period had a specially characteristic accessory. In the 'sixties the balustrade, column and curtain; in the 'seventies the rustic bridge and stile; in the 'eighties the hammock, swing, and railway carriage; in the 'nineties palm-trees, cockatoos, and bicycles; and in the early twentieth century, a motorcar for snobs.

Architectural and garden backgrounds have remained a favourite with society photographers, but in general most leading portrait photographers have always used plain backgrounds, which do not distract attention from the sitter's personality. H. & A.G.

The Normal Setting. The background of a fixed subject—like a tree, a statue, or a building—is largely out of the photographer's control, but much can be done to change the way the background finally appears.

(1) Suitably placed artificial lighting can be used to give the subject a darker or lighter background indoors (and even sometimes out of doors with flash). Out of doors the same thing may be possible by choosing a time of day when the sun is on either the subject or the background according to the tone contrast sought.

(2) By using a wide aperture and taking the camera up to the subject, the background can be thrown out of focus so that the subject stands out sharply by contrast.

(3) By shooting from a higher or lower viewpoint it is often possible to show the subject against a less "busy" background than by shooting at eye level.

(4) The background of a moving subject may be blurred while the subject remains sharp by swinging the camera as the exposure is made.

(5) Where the subject and background are of different colours, the tones can be separated by

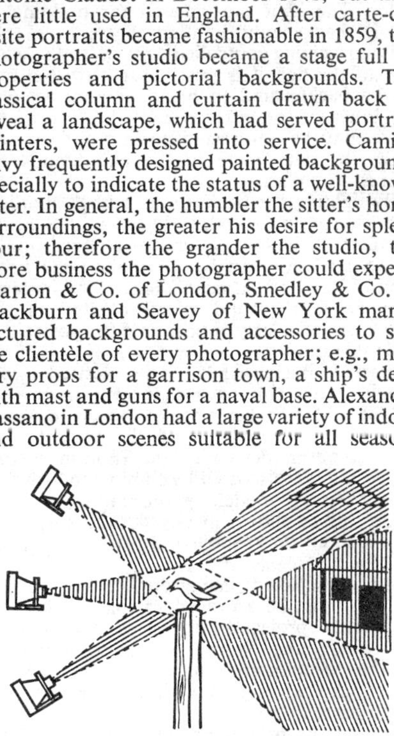

BACKGROUND AND VIEWPOINT. A high camera angle emphasizes the background at the expense of the subject. A low-angle makes the subject stand out (often against the sky).

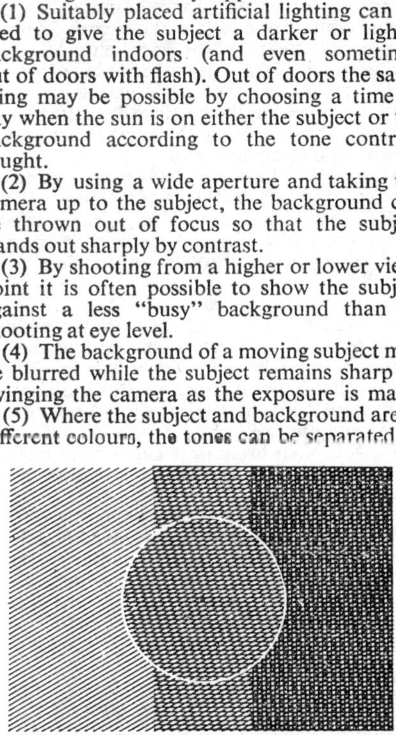

BACKGROUND TONE. A subject will stand out against a lighter or darker background, but tends to merge into surroundings of the same or reasonably similar tone.

BACKGROUND SHARPNESS. The subject tends to get lost in a sharp background, but will be set off by an unsharp (e.g., out-of-focus) one. This may be achieved by differential focusing.

using a suitable contrast filter which deepens the tone of one and lightens the other.

(6) Where it is not practicable to apply any of the above expedients, an unsuitable background can always be dealt with by blocking out the background on the negative and substituting a new one by combination printing or retouching.

Natural Backgrounds. The sky, the sea, grassy banks, and stretches of sand are good natural backgrounds for outdoor pictures of people, animals, birds, and isolated landscape features, e.g., rocks, trees, motor cars, etc. All these backgrounds have the advantage of being fairly uniform in tone and texture, and the subject can generally be placed against them either by posing or, with fixed objects, by a suitable choice of viewpoint.

Masses of foliage and hedges are apt to be covered with distracting white spots in the print. This is because glossy leaf surfaces reflect spots of light, and the sun or the sky light often filters through in patches from behind.

Brick walls make suitable backgrounds for figure studies so long as they are far enough away from the subject to be thrown out of focus. The bricks take on a vague general texture that is not distracting. But the lighting on the wall and the subject should not be the same, and the wall should have no odd features. A broken brick or a patch of discoloration will pull the observer's attention away from any subject.

When the sky is used as a background for a portrait, the relative tones need careful handling. If the subject's face is sunburned, then the yellow filter that darkens the blue of the sky will lighten the sun tan. On the other hand, if the exposure is made without a filter, the tanned complexion will look unnaturally dark against a too-pale sky.

The solution varies with the relative tones of skin and sky and with the photographer's intention. Generally a green filter used with a pan film will give a natural rendering of tone values so long as the exposure is correct.

Artificial Backgrounds. The simplest artificial background is a length of cloth—a sheet, a roller blind, a curtain or even a carpet—hung from a rod or picture rail. As long as the material is stretched taut, this type of background will serve fairly well, but it is not ideal. Creases and folds can never be completely smoothed out, and they usually reproduce with an undesirable clarity.

For regular use it is better to use a rigid background of wall-board or plywood fastened to a frame, or failing that, a sheet of fabric stretched tightly across a rigid frame. A background of this type is known as a flat and is a standard item in the equipment of the commercial studio.

Backgrounds of this type can be made big enough for portraiture, but in larger sizes they are cumbersome. The most convenient large background is either a plain distempered wall, or a sized and painted sheet nailed on to two battens and rolled up when not in use.

An advantage of the rigid background is that it can be turned or tilted so that it appears darker or lighter than the subject.

Background Lighting Effects. Where possible, artificial backgrounds are coloured white or silver on one side and a neutral tint on the other. The light side is used for building up special background lighting effects.

By projecting controlled light on to the white side of the flat, it is possible to vary the tone of the background to blend or contrast with the tones of the subject, e.g., the face of the sitter may be contrasted against a dark area while the dark mass of the hair stands out against a light area.

By lighting the background with a spotlight and interposing suitable shapes in the beam, an abstract shadow pattern, or a suggestion of an actual scene can be created on the background to lend atmosphere to the subject. Artificial background effects created in this way are particularly appropriate for portraiture and nude studies.

Projected Backgrounds. Any desired scene which is available as a slide or transparency can be added by back projection on to a projection screen hung behind the subject. This is a favourite device of the motion picture studio but it is also used by commercial studios. It enables a model, photographed in the studio, to be shown in any desired surroundings without the trouble and expense of a trip to the actual scene.

With projected backgrounds, the subject lighting must be arranged so that it does not fall on the projection screen or it obliterates the background picture. It is also important for the direction and angle of the light on the subject to correspond with that in the projected scene.

The relative brightness of the projected background and of the foreground objects must also be matched to ensure that both are correctly exposed.

Another method of obtaining a projected background is by front projection. This is more compact, since it eliminates the need for an extensive projection set-up behind the background screen. To avoid the projected image from interfering with the foreground objects, a special background screen and projection set-up is required. The special screen is a highly reflective and highly directional beaded screen which reflects virtually all light falling on it along the original light path. The projected image is therefore visible only within about 5° of the projection axis, but in this region is far more intense than it would be with a normal beaded screen. Any image projected on foreground objects is also invisible, since the scattering power of normal surfaces—dress fabrics, people and other objects—reduces the projected image brightness far below the lighting level of the foreground objects. For the same reason any spill light falling on the background from outside the projection axis has no effect on the brightness of the projected image.

To combine the projected background with the foreground object on the film, the camera must also photograph along the projection axis. This is achieved with a beam splitting arrangement using a semi-reflecting mirror or prism. The background is projected into the camera axis, and is reflected back along this axis, fusing perfectly with the foreground objects. Accordingly, no shadows are visible from the viewpoint of the lens. The camera, beam splitter and background projector are mounted together on a platform.

Background projectors for front projection can take black-and-white or colour slides of varying sizes, and the illumination is provided by either a high intensity projection lamp or by electronic flash. In one commercial outfit of this kind flash tubes up to 5,000 watt seconds are used, synchronized with the camera shutter and also with electronic flash units illuminating the foreground scene. Background image brightnesses obtained in this way can be adequate for exposures at *f*16 to *f*22 with a medium speed colour film, with a projected

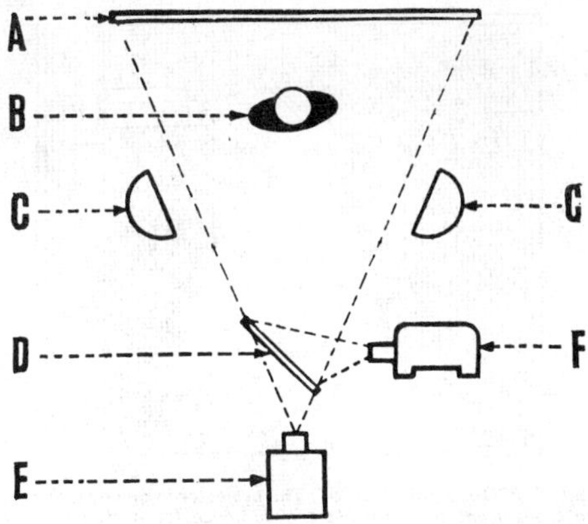

BACKGROUND BY FRONT PROJECTION. A, special highly directional reflecting screen. B, foreground subject. C, lamps illuminating the foreground. D, semi-silvered reflector. E camera. F, projector.

background size of 10 × 13 feet. The foreground scene is then lit to a lighting level to obtain the same exposures.

In addition to photographic images, transparencies of art work, drawn designs, and other graphic material can be used; this also provides a method for optical combination of foreground and background images in different scales (for example showing a model against an outsize perfume bottle for cosmetic advertisements, and so on). Subsidiary tungsten lamps in the projector and in the studio lighting set-up (modelling lights) permit preliminary setting up of the combined foreground and background composition.

Backgrounds are apt to call for more attention from the miniature camera worker than from users of larger types. This is because the greater depth of field of the miniature lens brings the background into sharp focus when the larger camera would blur it out of recognition. F.P.

See also: *Back projection.*

BACKING. Special coating of dye, pigment, or paper applied to the back of the support of a plate or film to reduce halation.

See also: *Anti-halation backing.*

BACKLIGHTING. Lighting projected on to the subject from behind to produce a light fringe or halo so that the subject stands out distinctly from a darker background. It is also used to add sheen to hair and accent its soft texture.

As a lighting effect, the results can be very pictorial. But extra care in estimating the correct exposure is necessary.

See also: *Against the light; Lighting the subject.*

BACK PROJECTION. Projection of a picture on to the back of a translucent screen instead of on to the front of a reflecting one. This has the advantage that the projection apparatus may be behind the screen and therefore completely hidden and out of the way of the audience.

The method has been used for many years for showing slides and films in shop window displays, on exhibition stands, etc.; but it is finding increasing favour for home projection. It is also used to provide realistic background scenes for plays in television studios, as well as in commercial photography, where the model may be photographed in the required setting to save going on location.

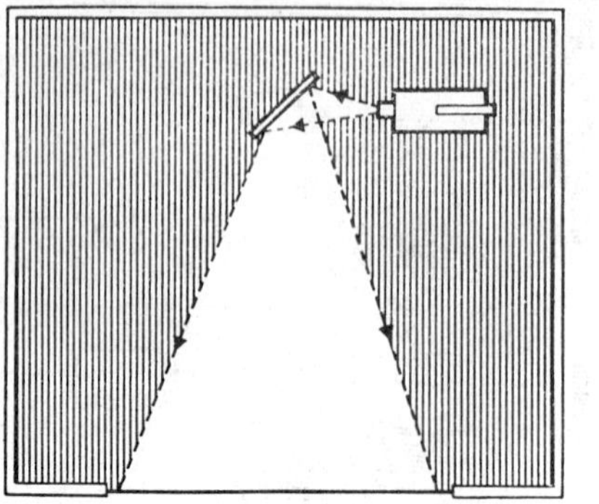

BACK PROJECTION SET-UP. The projector is arranged behind a translucent screen while the audience (or a studio set-up requiring a projected background) is in front. If space is limited, the projection beam can be reflected on to the screen with one or more mirrors arranged at suitable angles.

To economize on space behind the screen a short focus projector lens is normally used to keep the screen-to-projector distance to a minimum. A 45° mirror is necessary to reverse the picture from left to right, so that this introduces further economy since the projector is set up parallel to the screen; the beam is deflected on to the screen by the mirror and the space behind the screen requires to be only deep enough to accommodate this. Further mirrors can be used to zig-zag the beam and save even more space.

The screen may be thin fabric or matt surfaced glass or plastic. The modern black nylon screen is the most efficient since the picture brilliance is not affected by light falling upon it. This is essential in television and photographic studios where the bright floodlights would otherwise kill the background picture, and in the home it enables slides and films to be enjoyed in full room lighting. Daylight screens of this type, used for back projection, are also employed in visual aid education. H.H.

See also: *Background; Projection principles; Projection practice.*

BACON, ROGER, 1214–94, English alchemist, philosopher and friar. A successful pioneer in experimental physics, was familiar with convex lenses and invented the magnifying glass. He knew the principle of the camera obscura from Arabian writings, but did not invent it. Explained the rainbow, perspective, and tides.

BAEKELAND, LEO HENDRIK, 1863–1944. American chemist (of Belgian descent). Invented the Bakelite plastic. Investigated silver chloride developable emulsions and produced the unwashed emulsion that gave the Velox paper and revolutionized the production of printing papers (1893). It is said that this discovery was anticipated by Eder and Pizzighelli. Biography by A. R. Mathis (Brussels, 1948).

BAIRD, JOHN LOGIE, 1888–1946, Scottish inventor of the first television system. Gave first demonstration, to members of the Royal Institution, January 1926, by reflecting light from an image on to a photo-electric cell. 1929, B.B.C. began experimental transmissions. April, 1932, Baird demonstrated ultra-short wave TV. 1937, B.B.C. decided to use the Marconi system.

BAIT. Pieces of food of a suitable nature placed to lure wild creatures into range of the camera; often fastened to a mechanical or electrical shutter release or synchronized flash so that the animal takes its own photograph.

BALANCE. The disposition of the various units making up a picture. When the size, lighting and position of the parts of the subject on one side of the picture are in a pleasing relation to those on the other, the picture is said to possess balance.

See also: *Composition*

BALANCES AND SCALES. Balances and scales in photography are used for weighing out chemicals when making up processing solutions.

Two types of balance are commonly used in photography: the spring balance and the chemical balance.

A spring balance consists of a weighing pan suspended from a spring associated with a scale and pointer so that a weight placed in the pan moves the pointer along the scale. The pointer indicates the weight directly on the scale. Such balances usually incorporate a zero adjustment. They are not very accurate for small quantities but are good enough for weighing down to about ½ oz. A convenient size for the photographer would weigh up to 1 lb.

Modern rough balances weighing down to 1 gram use the displacement of a weighted arm, similar to old-fashioned letterweights.

A chemical balance has two weighing pans suspended from a beam balanced on agate or knife edges. The whole balance either hangs from a hook, or is supported on a pillar. When weighing, the beam should swing to and fro during observation; when the pointer attached to it swings evenly in both directions, the two sides are accurately balanced. The accuracy of this type of balance is very high, especially with the laboratory type which is enclosed in a glass case to shut out air currents. Such balances can weigh accurately to 0·00001 gram. For photographic work, however, an accuracy of 0·01 gram is perfectly adequate; even an accuracy of 0·1 gram is often sufficient.

A set of scales has two weighing pans, mounted above a beam supported on knife edges. The material to be weighed is placed in one pan (generally a scoop or tray which can be lifted off), and has to be balanced by standard weights which are placed in the other pan. Often a small movable weight is mounted on the beam itself and slides along a graduated scale; this serves for weighing very small quantities. Scales can be made to measure small quantities accurately, but as the principle lends itself to a sturdy construction they are more often used for weights up to 14 lbs. with an accuracy of $\frac{1}{2}$ to 1 oz.

Weighing with a Balance. For accurate weighing the pans and weights must be kept scrupulously clean. Chemicals must always be weighed on watch-glasses or at least on pieces of paper and not allowed to come into contact with the pans. The weight of the watch glass or paper is counterbalanced by a similar weight in the second pan.

Chemicals are always added or removed with a small spoon or something similar; they are never poured straight from the bottle or tin because a sudden rush of fine powder—e.g., amidol or metol—out of the bottle could swamp the pan and be difficult to clean up. For the same reason the pan should never be filled to capacity; it is safer to weight out large quantities in several lots.

Weights are usually housed in a small wooden box, with a pair of tweezers for handling. The smaller weights are best handled with the tweezers, so as not to soil them and so upset the accuracy.

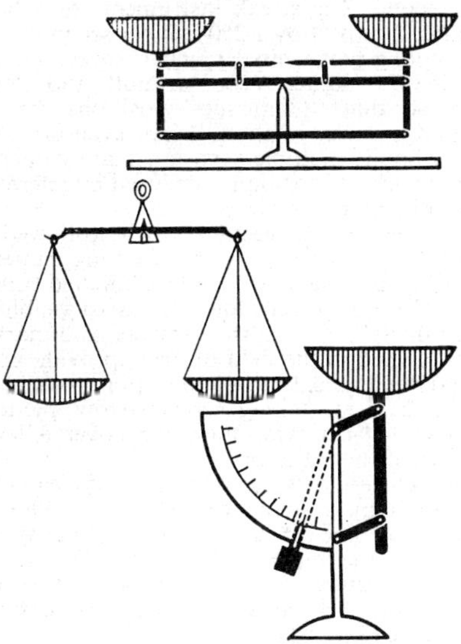

TYPES OF BALANCE. Top: Normal pair of scales mounted above a beam supported on knife edges. The parallelogram arrangement keeps the pans vertical. Centre: Hanging scales, with pans below the beam. The whole balance may be hung from a nail, or in the case of a chemical precision balance, fixed to a vertical pillar. In both instances the requisite weights go into one pan, the matter to be weighed out into the other. Bottom: Counterweight balance reading directly on a calibrated scale. The counter weight may be adjustable.

Systems of Weights. Two systems of weights are in common use in Britain and the U.S.A.: avoirdupois and metric. Most photographic formulae give quantities in both systems, but the metric is more convenient, as it is easier to scale quantities down or up in the metric system than in avoirdupois.

A set of avoirdupois weights may consist of the following units: 4 ounces, two weights of 2 ounces, 1 ounce, $\frac{1}{2}$ ounce, $\frac{1}{4}$ ounce, 60 grains, 40 grains, 30 grains, two of 20 grains, 10 grains, 5 grains, two of 2 grains, and 1 grain. This range covers anything from 1 grain to 10 ounces.

The most practical metric set may comprise: 100 grams, 50 grams, two of 20 grams, 10 grams, 5 grams, two of 2 grams, 1 gram, 0·5 gram, two of 0·2 gram, 0·1 gram, and 0·05 gram. This will weigh anything from 0·05 to over 200 grams. Specially accurate weights will go down to 0·005 gram, but these are hardly necessary in photography.

Coins as Weights. If no proper weights are available, British coins can be used as weights as follows:

Five shillingsworth of silver coins of any value	= 1	oz.
Half-crown	= 218	grains
Two-shilling piece	= 175	grains
Shilling	= 88	grains
Sixpence	= 44	grains
Silver threepence	= 22	grains
Three pennies or five half-pennies	= 1	oz.
Penny	= 145	grains
Halfpenny	= 88	grains

In some cases the coins of other countries are similarly related. L.A.M.

See also: *Chemicals; Solutions; Weights and measures.*

BALDUS, EDOUARD, 1820– (after) 1869. French architectural photographer born in Germany. Recorded historic monuments in Burgundy, the Dauphiné and Fontainebleau for the French Government, 1851. Founder member of Société Héliographique. Changed from the Calotype to the collodion process in 1854 and made for the Government a complete documentation of the new wing of the Louvre in over 1,500 detail photographs. Also noted for views, especially of mountains. Invented a photo-engraving process, 1854.

BALL AND SOCKET HEAD. Universal type of camera mounting made to screw on the top of a tripod.

See also: *Camera supports.*

BALLISTIC PHOTOGRAPHY. Photographic study of projectiles. Where these move at speeds in the region of 15,000 m.p.h., photography may provide the only means of investigating their characteristics in flight. These include velocity, aerodynamics—shape, behaviour and shock wave generation—and behaviour on impact.

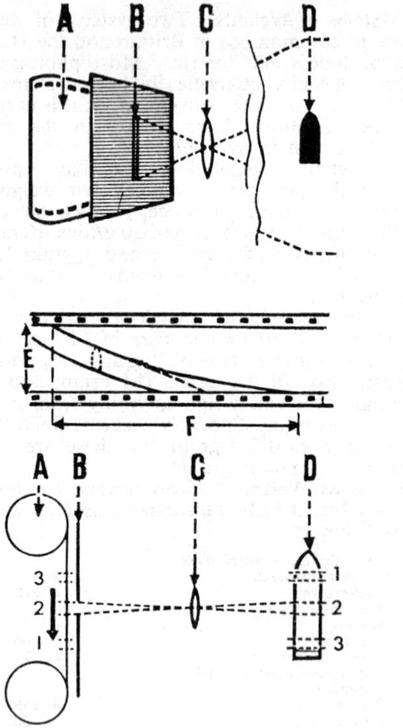

PHOTOGRAPHING SHELLS AND BULLETS. Top: *To measure the speed of a missile, a moving film camera is set up so that the image moves along a slit in front of the film while the latter travels across the slit. A, film. B, slit. C, lens. D, shell.* Centre: *The resultant picture is a curved smear F, equivalent to the time of travel through the distance E. The mean slope of this smear is also equivalent to the speed.* Bottom: *For a sharp picture of the shell in flight, the image must move across the slit in front of the film, while the latter advances in step with the image. 1, 2, and 3 are projections of the shell progressively recorded in the same relative positions on the film, building up a continuous and sharp picture.*

The first ballistic photographs (E. Mach, 1885/6) recorded moving bullets and shells in silhouette against the light of an electric spark.

The result was not a photograph in the generally accepted sense of the word (i.e., formed by a lens) but a record of the shadow of the projectile on the sensitized material. The short duration of the spark and its relatively great distance from the projectile and the sensitized material, combined to give a sharp image.

A similar technique is used today for the same purpose except that an electronic flash tube takes the place of the old open spark between metal electrodes. Photographs made in this way show the projectile as a sharp-edged black silhouette and may also show the pressure waves set up in the air by the passage of the projectile—(like the bow wave of a ship). This phenomenon, known as the schlieren (German for streaks) effect, is caused by the change in the refractive index of the air along the pressure wave and the consequent distortion of the rays of light before they reach the sensitized material. The number, shape and disposition of the waves provide an indication of the air resistance and stability of the projectile and so are important in designing for range and accuracy.

A high speed camera can record the projectile throughout its travel across the field of view. If the distance travelled across the field is known, this is represented by the width of the film frame; while the time of travel is indicated by the length of film on which the projectile's image appears, providing camera running speed has been accurately measured. So projectile velocity is equivalent to frame width divided by length of film. If ultra-short framing exposures are used, they will "freeze" the projectile's image, showing any erratic behaviour in flight.

The main requirement is therefore for good temporal and spatial resolution rather than pictoral accuracy, and for frame-by-frame analysis instead of "slow-motion" moving pictures. Still cameras, when used, generally have special ultra-fast shuttering devices or simple capping shutters for open flash work with short-duration and stroboscopic flash. High speed framing cameras and streak cameras are used more often, the former operating at rates between a few thousand and several million frames per second. The streak instruments may have writing speeds of over 250 feet per second.

For long-range missiles, widely spaced groups of cine theodolites record azimuth and elevation, so that the missiles' positions can be plotted by triangulation. More accurate still, are Schmidt multiple-exposure plate cameras, from which the position is deduced by reference to star images.

Conventional projectiles can be followed in much the same way by fast framing cameras covering fixed fields across the estimated trajectory. Flare or tracer indicators assist visibility, and fiducial marks in the cameras, and marker boards sited in the field of view, provide positional references. In this application, the ribbon frame camera, taking long narrow pictures across a 5½-ins. wide film, can cover a large section of the trajectory at once.

For closer work, velocity may typically be measured by a streak camera. This is aimed across the line of fire, with the optical slit parallel to the trajectory. The projectile is therefore recorded as a streak slanting across the film. Again, distance travelled is measured across the film and time along its length.

The streak camera can also give a pictorial image for aerodynamic study. For this it is mounted with the slit perpendicular to the trajectory and with the film moving in the same direction as the projectile's image. Having found the velocity, the film speed past the slit can be adjusted to match the calculated image speed, resulting in a still picture of the projectile by virtually orthodox moving film photography.

The same principle is applied in a multiple shadowgraph system, in which the projectile is fired through a tube having vertical slits at measured intervals along its length. Collimated light is projected on each slit through ports on the opposite side of the tube. From the slits, optical relays of equal length lead to a single streak camera, recording side-by-side images from the slits as the projectile passes each one. The distance apart between any two images along the film represents the time of flight between the corresponding slits, and hence the velocity.

A shell's behaviour on impact—say, with a sheet of armour plate—would normally be studied with a framing camera, photographing the sheet edge-on. Since the entire event might take less than a millisecond, electronic flash is the usual illuminant. It can be triggered by the shell piercing a metal foil or gauze screen in front of the armour plate and making an electrical contact between them.

Other forms of illumination for ballistic photography, apart from conventional sources, include spark gaps, argon bombs and rare gas discharge tubes. Complex circuitry is needed to ensure a square-wave output of constant intensity and, where necessary, rapid stroboscopic repetition. Where the flash duration determines exposure time, sparks can give exposures down to 1/1,000,000 second, while 1/100,000,000 second has been achieved with discharges through rare gases.

Synchronization, both of the camera with the projectile and of the projectile with the flash must be exact. Cameras need time to accelerate to high speeds, and electronic timing and frame-counting circuits may be required to programme starting the camera, opening the shutter, firing the gun and triggering the flash. The projectile itself can also initiate events, as already described, by connecting foil screens, cutting light beams or breaking wire circuits.

Since time is measured in distance along the film, it is usual to flash timing marks on the rebate, through an optical relay from an electronic timer. This compensates for any irregular running of the camera mechanism. P.B.M.

See also: *High speed cinematography; High speed photography; Missile photography; Schlieren photography; Spark photography; Stroboscopic flash.*

BALLOONS AND KITES. Most photographs that have to be taken from the air are best taken from an aeroplane, but there are occasions when it is more convenient to take them from a kite or captive balloon. The first really satisfactory aerial photograph, from a tethered balloon, was taken by J. W. Black in October 1860 from a height of 1,200 feet over Boston, U.S.A.

Nadar was the earliest photographer to work from a balloon in 1858, but most of his pictures were taken from a moving free balloon. Since the course of the balloon cannot be predicted accurately, and it is not possible to repeat a run over the picture area, there is little point in such photography today. (The exception is the automatic photography of the earth from a stratosphere balloon from altitudes which cannot be reached by an aeroplane.) But a captive balloon or kite offers a vantage point that can be controlled within reasonable limits both in position and altitude.

Kites. The pioneer of this field of aerial photography was a Belgian, A. Batut, who in 1887 took photographs of the ground with a camera attached to a kite. The exposure was timed by lighting a slow-burning fuse which released a stretched rubber band attached to the shutter control after the kite had reached the desired height.

At the beginning of the century, American meteorological observatories were using clockwork operated cameras, suspended below kites, for cloud photography.

The most stable kite for this type of photography is the Cody kite—a box-form kite divided into cells and made of reinforced silk, nylon or cotton panels stretched between bamboo spars. Any weight of equipment can be lifted by simply adding kites, one below the other on the same cable.

The photographic equipment is never attached directly to the first kite. This kite always acts as a pilot. It is launched first and when it is flying well clear of the ground—at least 150 to 200 feet—the next kite is attached. If the pilot kite will give all the lift required, the camera cradle is attached to the cable. The camera cradle is a small waterproof box with a window in the bottom through which the photographs are taken. The box is fitted with air stabilizing fins and a small drag parachute to prevent it from yawing and to keep it standing clear of the cable.

The camera with its clockwork automatic release and rewind mechanism is not loaded into the cradle until there is plenty of steady tension to ensure that it will be carried clear of the ground as soon as the cable is paid out.

Equipment of this type must be flown from a mobile power-operated winch—generally mounted on the back of a truck—because the cable tension may be several hundredweights. An amateur set-up is not difficult to contrive with two ordinary box kites flown one above the other to give stability, and a miniature camera.

The timed release is easy to arrange with either an alarm clock mechanism or the stretched rubber band and fuse mentioned above. In this case only one exposure can be made per flight.

But kites are not an ideal form of suspension for delicate photographic equipment. They are apt to yaw violently from side to side, or fall to the ground when the wind drops.

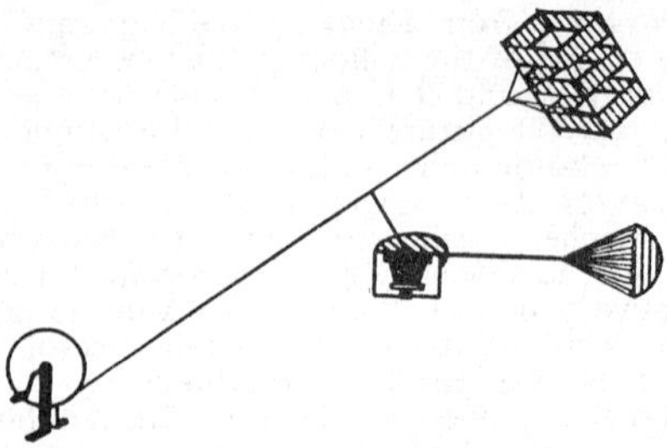

CAMERA KITE. The camera unit is supported from the main cable holding the kite from the winch, and steadied by stabilizing fins and a small drag parachute.

Balloons. A much more satisfactory method is to fly the camera from a balloon on a still day. This was the method adopted in the American Civil War, the South African War and in World War I. In these wars the photographs were taken by observers from captive observation balloons. This function has now been taken over by the highly developed air force reconnaissance units.

Where photographs are to be taken from a balloon, it is no longer necessary to use a man-carrying type. The cost can be cut to a fraction by using a balloon just big enough to raise the camera and the required length of cable. If the pictures have to be taken on days when the wind is more than about 5 m.p.h. spherical balloons are useless and a properly stabilized balloon flown from a winch must be used.

By operating only when the wind is light it is possible to get good pictures using one or more spherical rubber balloons like the type used in meteorological offices. If the balloon is filled with ordinary coal gas from the domestic supply, every 1,000 cubic feet will lift roughly 50 pounds of gear, including the weight of the envelope and flying wire or cord.

The camera with its release gear is housed in a box as for kite photography (above) and the box is fitted with an aerodynamic stabilizer and a small drag parachute or drogue. This unit is attached to a suspension wire or cord long enough to let the balloon be flown well clear of the ground before it is added.

When the unit leaves the ground, the slight wind drag on the drogue swings it away from the flying wire so that the wire does not appear in the picture.

The layer of atmosphere near the earth prevents photography of optimum quality of heavenly bodies from the earth. Scientists are constantly working to find ways of getting astronomical telescopes and cameras into the clearer upper layers of the atmosphere. One of the most successful methods to date is to sling the apparatus below a high-altitude free balloon. By using a small amount of hydrogen or helium in a giant envelope of extremely light plastic sheet, photographic gear and telescopes up to 36 ins. have been raised to over 100,000 feet to take photographs of the planets. When the balloon reaches its ceiling, the gas inside fills out the plastic envelope. In time the pressure builds up and the balloon bursts. The envelope is designed so that after bursting it forms a parachute which lowers the apparatus gently to the ground. Photographs taken in this way show details which were previously beyond the scope of the finest telescopes. Both polythene and foil material balloons are used today for such high altitude work.

Camera. A camera with a normal angle lens will cover a square of ground approximately 70 × 70 feet when looking down from an altitude of 100 feet. The area covered at any other height is given by

Side of Square Covered
= Height of Camera from Ground × ·706

In aerial photography from kites or balloons, the camera should be at least 50 feet from the ground to keep it out of the worst of the ground wind eddies. The air is always turbulent in the lee of trees and tall buildings.

The shutter speed for such photographs should be at least 1/100 second to allow for the constant movement of the camera.

When taking single shots, it saves time to make the shutter release gear operate a visible indicator, e.g., make it unroll an orange-coloured streamer. This provides a warning that the exposure has been made and that it is time to haul down and reset the mechanism.

From all points of view, short of using a specially designed camera, a 35 mm. precision miniature is the best camera for this work; it is small, light, and easy to couple to an automatic exposure mechanism.

Negative Material. Medium speed panchromatic plates or films generally give the best results; they allow for enlargement of sections of the negative without unpleasant graininess. A pale yellow filter counteracts any ground haze and brightens up the tones of grass and foliage.

Uses. This photographic technique—with balloons especially—is particularly suitable for such enterprises as archaeological expeditions, site excavation, and prospecting; the equipment is cheap and self-contained and it is capable of yielding detailed plan views of relatively small areas of terrain without the difficulty or expense that would be involved in commissioning an aircraft. F.P.

See also: *Aerial survey; Astronomy.*

BANQUETS. The work of the professional photographer at banquets, dinners and similar functions falls into two main headings; he may be called upon to take formal groups showing all the assembled diners, as well as a series of informal, intimate close-ups of individual parties seated at their tables.

The formal group, properly taken, can make a very impressive picture. It is not a difficult photograph to take provided that the correct technique is adopted and full use made of the normal room lighting. The exposure is always based on a brief time exposure of a duration of 1–5 seconds, according to the intensity of the lighting. Many of the large banqueting halls have their lighting installation overhauled and cleaned during the summer months, in time for the season. As the installations collect a certain amount of dust, the regular banquet photographer will find it necessary to increase his exposures slightly, as the season progresses. The basic exposure. is usually supplemented by exposing a flash bulb of medium power to add brilliance to the foreground.

If the photographer ignored the natural lighting and made use only of a high-intensity flash he would produce a negative that would be almost unprintable. The picture would be grossly over-exposed in the foreground, and the far distance would be just as seriously under-exposed.

Professional photographers normally use half-plate or whole-plate cameras for this work in conjunction with short-focus or even wide-angle lenses. There are obvious disadvantages in making use of such lenses for there is always some distortion at the edges of the picture, but conditions rarely permit the photographer to use a lens of normal focal length.

Technique. A tall, rigid tripod is an essential. Its weight will depend on the camera. Firms who specialize in such work use a 6-foot step-ladder modified to accommodate the camera on the top platform. Failing such aids, it is usual to work from the stage, the band platform, or (best of all) from a balcony of medium height. The camera must be able to look down on the assembly if everyone is to be seen clearly in the finished photograph.

It is standard practice to work obliquely across the lines of tables. When the camera looks directly down the length of a table it is usually impossible to see more than a few of the guests near the front.

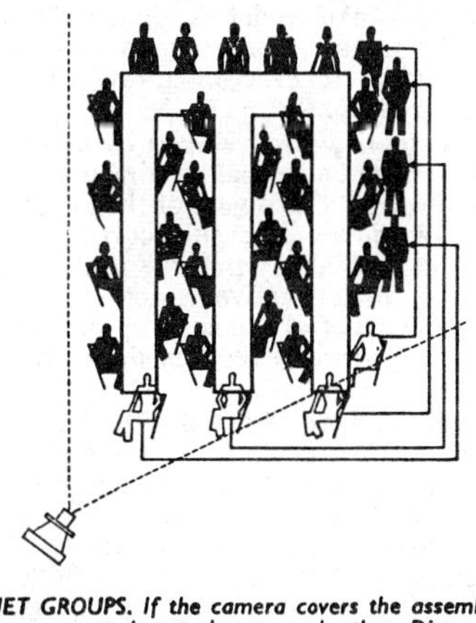

BANQUET GROUPS. If the camera covers the assembly from one corner, guests do not obscure each other. Diners at ends of tables are moved to stand at edge of group.

As the principal guests are normally seated at the top table it is usual to ask them to stand throughout the exposure. It is also customary to take the photograph as soon as the guests have taken their seats after the saying of Grace. Later in the meal the tables become disarranged and the scene is never free from the presence of hurrying waiters. Furthermore, trouble is likely to be experienced with haze from cigarette smoke, particularly if the room is very long and if many guests have taken the opportunity to smoke.

The necessary over-all sharpness is achieved by focusing the camera on a point about a third of the way down the tables and then stopping down to *f* 16. Electronic flash, being fast, easy to use and portable, is popular for photographing dancing, floor shows, etc. Using fastest panchromatic material in a normally well-lit room at this aperture, a time exposure of 4 seconds should be ample to produce a negative of fairly even over-all density. The slight falling off in illumination in the foreground is rectified by firing a medium power flash bulb just before closing the shutter.

Subject. The great secret of success in this field of photography is a knowledge of "audience-control." The photographer must make certain that everyone present knows that a photograph is about to be taken and that they realize that they have to keep absolutely still throughout the exposure. This information is conveyed either by the photographer (who must speak slowly, clearly and compellingly) or, at larger functions, by the toastmaster to whom the photographer has previously explained what is wanted.

Examination of the subsequent prints shows unmistakably how far the operator has succeeded in controlling his audience.

The intimate type of picture can be taken with the aid of small flash bulbs or more often nowadays with an electronic flash. It is usual to pose these small groups to be sure of getting pleasing expressions and to avoid catching someone while eating. If poor lighting conditions should make a small aperture impracticable, good overall sharpness can be achieved at a wider aperture by using the front-tilt on the camera.

The Business Side. The business side of banquet photography is normally conducted on "high-pressure" lines. If possible, the photographer circulates proofs of all pictures among the guests before they leave so that he can take orders on the spot. Successful exponents of such methods make a point of adding sufficient to their charges to enable the finished prints to be posted to customers with secure wrapping to protect them. G. Cg.

See also: *Groups; Weddings.*

Books: *Group Photography*, by Gordon Catling (London).

BARIUM SULPHATE. Used to prepare baryta paper on which most photographic printing papers are coated. As barium sulphate is opaque to X-rays, it is largely used in radiography when it is administered to patients either orally or by injection. This enables the operator to "see" the shape and position of organs which, without this aid, would not be recorded.

Formula and molecular weight: $BaSO_4$; 233.
Characteristics: A heavy white powder.
Solubility: Insoluble in water and alcohol.

BARNACK, OSKAR, 1879–1936. German camera designer. Joined the firm of Ernst Leitz in 1911 as mechanic in the experimental workshop. Made a small roll-film camera for trial exposures of cine film employing double standard 35 mm. frame size (24 × 36 mm.) and so in 1914 produced the Leica prototype. The first series of six Leicas appeared in 1924.

BARN DOORS. Colloquial term for the flaps placed in front of a spotlight to control the spread of the light. The flaps are hinged so that they can be adjusted as required. Barn doors are sometimes necessary when only part of a subject is to be illuminated.

BARREL DISTORTION. Type of distortion of the image formed by an insufficiently corrected lens, where magnification decreases with increasing obliqueness of the image-forming rays. Lines parallel with the picture edge appear to bend inwards near the corners crowding in more subject area on a reduced scale.

See also: *Aberrations of lenses.*

BARYTA COATING. Base of the photographic image in most types of printing paper. It consists of an emulsion of barium sulphate in gelatin, spread in a thin film over the paper backing and subsequently coated with the sensitive emulsion. Its main purpose is to provide a smooth white chemically inert base.

See also: *Supports for emulsions.*

BASE. In photography, the support on which a sensitized emulsion is coated—e.g., the glass of a negative, or the paper of a print—is called the base.

In chemistry, a base is a metallic or metalloid (e.g., ammonium) oxide or hydroxide which combines with acids to form salts; a base that is soluble in water is an alkali.

See also: *Supports for emulsions.*

BASKETT'S REDUCER. Physical reducer for negatives, which works by removing the silver deposit by abrasion. The original Baskett's formula consisted of equal parts of terebene, salad oil and metal polish. Any similar mixture of an abrasive metal polish and oil or vaseline will do.

See also: *Reducing.*

BAS-RELIEFS. Carvings, sculptures or other forms of images in low relief. In photographing them strong oblique lighting is usually required to obtain sufficient modelling.

Double Image Method. Prints resembling photographs of bas-reliefs can be made from negatives of normal subjects. The method is to make a transparency from the negative by contact printing on to a lantern plate or similar material. The negative and positive are then placed together in the enlarger slightly out of register, and a print made in the normal way. The result is a picture with the main outlines indicated by dark or light contours, and with somewhat distorted over-all tonal gradation. Both the negative and positive should be rather soft for the best results.

This technique has also been used to obtain strong pictures of actual bas-reliefs having very shallow modelling, such as old and worn coins. In this case only the slightest displacement of negative and positive is called for—sufficient to accentuate the characters of an inscription, for instance, without giving a double image.

The direction of displacement should coincide with the direction of the lighting on the subject so as to accentuate the natural shadows and highlights.

Swelled Gelatin Method. If a light sensitive layer of bichromated gelatin is exposed under a negative, the emulsion under the shadow areas is hardened by the actinic rays but that under the highlights remains unaffected.

The unaffected gelatin absorbs relatively large quantities of water and swells up while the hardened areas absorb little water and so do not swell as much. The result is an image in relief in which the highlights stand out and the shadows are depressed. This image may be cast in plaster while still wet, or if it is gently heated to just below melting point it will persist in the dried emulsion. Either way it is possible to obtain relief facsimiles from it in plaster or metal. Correctly carried out the method yields relief comparable to that of coins or medallions.

This method was developed by Carlo Baer, who constructed a special lighting arrangement for the purpose. The object of the arrangement was to illuminate the subject (invariably a portrait) in such a way that the resulting densities in the final negative varied according to the physical relief of the surface photographed and not the normal effects of light, shade and colour.

If the method is attempted without some such arrangement it is essential to treat the subject to remove colour contrasts—i.e., by powdering the hair and face—and to light it evenly from each side to give symmetrical modelling. Finally, the negative may need considerable local intensification and working up to give the necessary strong contrasts.

The plate on which the negative is printed can be prepared by coating glass with the following mixture:—

Gelatin	2 ounces	50 grams
Water	7 ounces	175 c.cm.

Leave the gelatin to soak for 4 hours and then heat on a water bath until dissolved; stir in 120 minims (6·5 c.cm.) of glycerin, strain, and coat on to glass plates while still warm. The above quantity is enough to coat 40 square cm. (6·5 square ins.).

At any time after drying, the plates may be sensitized by bathing for 15 minutes in a 6 per cent solution of ammonium bichromate and drying in darkness.

The plate is contact printed by daylight to give a good strong image, and then soaked in water until the gelatin swells to form the relief.

The swelled gelatin method is only suitable for producing reliefs of well defined objects and for portraits; it is useless for landscapes. As the amount of relief is small, the finished bas-relief should not be more than 3 ins. or so high, or the effect will be lost. F.P.

See also: *Photosculpture ,Tricks and effects.*

BATCH NUMBERS. Serial numbers stamped by the manufacturer on all packages of sensitized materials coated from a particular batch of emulsion. While variations between one batch of emulsion and the next are never very great, they may be serious enough to affect exposure and contrast, particularly where large quantities of the material are being exposed and processed to a rigid technique—e.g., in photo agencies, photofinishing and picture postcard manufacture.

Where variations in speed from one batch to another might be costly, the user keeps a check on the batch numbers of the materials as they come from the supplier. When the batch number changes, a test run is made with the new material to see whether any change is necessary in the technique standardized for the previous batch. This is particularly important with colour materials, which have a more limited exposure latitude. Colour render- to batch, and where constant quality is required in a large number of exposures, all the material should come from the same manufacturer's batch.

See also: *Keeping qualities of materials.*

BATTERIES

In recent years there have been spectacular improvements in portable batteries, and many items of photographic equipment can now be run off small built-in batteries. These can power film transport motors in cine and still cameras, solenoid-operated remote shutter controls, motor-driven focusing mechanisms and portable tape recorders. Batteries are also used for firing expendable flash bulbs and operating electronic flash units, for the comparison lamp in certain kinds of exposure meter, and for the projector lamp in some low voltage slide projectors.

Though there are many battery types, a photographer with battery-operated equipment can usually be sure of getting an identical size and type of replacement when a battery runs down. However. there may be emergencies when he may have to use or adapt an alternative type. In addition, the prices of the different types vary widely for what often appears to be the same specification.

Practical Aspects. Strictly speaking a distinction must be made between electric cells and batteries. A cell is the basic current-generating unit. A battery is an arrangement of two or more cells connected in series to provide increased voltage, or in parallel to provide increased current. The cells may be sold already connected as a battery in a common case, or they may be sold as single units which can be made up into a battery of any required multiple of the voltage of the basic cell.

The important properties of a cell are:

Shelf Life. How long will it stay good in store?

Discharge Characteristics. How does it hold its voltage under loads of all kinds: normal, light heavy, continuous, intermittent and at temperatures above and below normal?

Capacity. How much energy will it deliver before the voltage drops below the lowest acceptable level?

Economics. How much does it cost both in actual cash and in relation to the service it gives?

In addition to the electrical characteristics, there are the physical characteristics to be thought of—size, weight and the risk of damage to equipment by corrosive liquids or gases given off by the cell in use.

The accompanying table compares these practical aspects over the range of cells suitable for photographic equipment. Generally it is safest to follow the makers' recommendation or, in their absence, to renew cells with others of the exact type. However, there may be exceptions to this rule as, for instance, when a piece of equipment is used under abnormal conditions.

What Kinds Are There? Cells can be divided into two broad categories: primary and secondary. Primary cells cannot be regenerated and have to be discarded when their voltage drops below the minimum working value. Secondary cells can be re-charged from an external power supply when they run down.

PRIMARY CELLS

The elementary form of primary cell consists of two electrodes immersed in a liquid electrolyte. The electrodes are most commonly carbon and zinc and the electrolyte may be an acid, an alkali or a metal salt. Under these conditions there is a difference in electrical potential between the two electrodes and if they are joined by a conductor a current will flow from one to the other.

The voltage of the cell is always the same for electrodes of the same materials; increasing the size or area of the electrodes does not increase the voltage, but only the capacity of the cell. When a simple cell is under load a layer of ionized gas tends to build up on one electrode, raising the internal resistance and cutting down the output of the cell. For this reason the cell always contains a depolarizing chemical which disperses or absorbs the gas and helps to maintain performance.

There are five kinds of primary cell used in photographic equipment: simple carbon-zinc, leakproof carbon-zinc, layer type carbon-zinc, high energy alkaline manganese, and mercury.

Simple Carbon-Zinc Cells. These are the same small cylindrical cells that are used in ordinary torches. They are a development of the Le Clanché wet cell. The positive pole is a carbon rod surrounded by a mixture of powdered carbon and manganese dioxide. The negative pole is the cylindrical zinc case which is generally protected by a cardboard sleeve. The contents of the cell are impregnated with a jelly of ammonium chloride and sealed off with a layer of pitch. Cells of this simple type are cheap and reliable but they tend to swell near the end of their life—particularly if they are kept in a damp place—and may jam tight in the equipment. They may even burst and exude electrolyte which quickly corrodes and ruins metal parts. So they should always be looked at every month or so and never left in the equipment once they are run down.

Leakproof Carbon-Zinc Cells. These are a big improvement on the ordinary type and are well worth the small extra cost. The cell is enclosed in a sealed steel jacket and an internal air space above the electrolyte makes the normal gas vent unnecessary. This type of cell is safe to fit inside photographic equipment where leakage could cause costly damage. The voltage and other characteristics are the same as that of the ordinary carbon-zinc cell.

Both the ordinary and leakproof types are suitable for running small lamps, preferably on intermittent duties, and for operating capacitor flash units.

Layer Type Batteries. Compact high voltage batteries of carbon-zinc cells are made by assembling suitable numbers of thin-section layer-type cells. These consist of a flat sandwich of carbon and zinc with an electrolyte-impregnated card in between. When one cell in a high voltage assembly fails the entire battery has to be scrapped.

Layer-type batteries are frequently used in capacitor flash guns. They have also been employed to power electronic flash equipment but nowadays the high operating voltage is more conveniently obtained from a converter, which converts the low voltage direct current from an ordinary battery into direct or alternating current of the correct voltage for the equipment.

Alkaline Manganese Primary Cells. These look like leakproof carbon-zinc cells but use manganese dioxide in contact with the steel container as the positive pole and an alkaline electrolyte—usually potassium hydroxide. They have many advantages—low internal resistance, ability to supply heavy currents for long periods, good performance at low temperatures and a useful capacity of as much as ten times that of the comparable carbon-zinc cell.

Alkaline primary cells are excellent for driving the small motors of cine cameras and sound recording equipment, for electronic flash units and for running slide projectors. They give the same voltage as the ordinary carbon-zinc cell. Whether they cost more depends upon the type of use they get. Broadly speaking the alkaline cell works out cheaper on continuous heavy current duties and the carbon-zinc wins on long-term intermittent use at low currents.

Mercury Cells. Mercury cells use electrodes of zinc and mercuric oxide with potassium hydroxide as the electrolyte. In operation the mercuric oxide liberates mercury to form the positive electrode. It also acts as a self-depolarizer—i.e. it dissipates the film of ionized gas which tends to form on the positive electrode and slow down the action of the cell. So this type of cell can be discharged continuously for extremely long periods without its voltage falling off.

Weight for weight the mercury cell has six to seven times the capacity of an ordinary carbon-zinc cell, it is leakproof and its storage life is longer than that of any other primary cell. It costs more than other cells but may in certain circumstances be cheaper in operation.

Mercury cells are ideal for miniaturization. In the form of "button" cells they are fitted in photographic equipment where small size and weight are an advantage—e.g. built-in flash equipment, automatic camera mechanisms, and cadmium sulphide exposure meters.

SECONDARY CELLS

The secondary cell is rechargeable and is based on a reversible electro-chemical action. As in the primary cell, there are two electrodes immersed in an electrolyte. If an electric current is passed through the cell from an external source it produces a chemical change in the electrodes and charges the cell. If the electrodes are then connected to a load, a current will flow in the reverse direction from the charging current and the chemical state of the cell will go back to what it was before charging. This means that the cell can be used as a source of electricity that can be replenished when it runs down.

As it stores up electricity, a secondary cell

battery is also called an accumulator. In certain types of photographic equipment—e.g. where currents of the order of one or more amperes are required, a secondary battery, in spite of its higher cost, is a better proposition than a primary battery.

There are five kinds of secondary cell used in photographic equipment: lead-acid, nickel-iron alkaline, nickel-cadmium, silver-zinc and silver-cadmium.

Lead-Acid Accumulator. The commonest type of storage battery employs electrodes of lead and lead oxide in a dilute sulphuric acid electrolyte. During charging the cells give off gas, so they cannot be hermetically sealed. When lead-acid accumulators are fitted in portable photographic equipment they must be of the non-spill type which have filler plugs designed so that while they will allow gas to escape they retain the acid—even if the equipment is turned upside down.

Lead-acid batteries provide an efficient source of heavy currents. However, if they are left discharged the surfaces of the plates become coated with insoluble lead sulphate and the battery is ruined. Miniature lead-acid batteries are used in some electronic flash units and large, heavy duty types are used to drive camera motors and sound equipment used by professional film units working on location away from the mains.

The lead-acid accumulator requires regular maintenance. In addition to making sure that the accumulator is kept charged (the charge is gradually lost even during storage), this involves topping up the electrolyte with distilled water (never acid) which is used up in the course of the charge/discharge cycle.

Nickel-Iron Alkaline Batteries. This type of rechargeable battery employs electrodes of nickel and iron in an electrolyte of sodium hydroxide. Usually, the case of the cell is made of nickel-plated steel and forms one of the electrodes.

Nickel-iron batteries are extremely robust; they will stand heavy over-charging and short circuits and can also be left discharged for years without deteriorating. They need thus no maintenance. However, they have a lower voltage than the lead acid type and, for an equal amount of energy stored, they are bulkier, heavier and more expensive. For this reason they are not used in portable photographic equipment although they form a reliable power supply for outdoor filming operations.

Nickel-Cadmium Batteries. This type of battery

BATTERIES COMPARED

Cell	*Type**	*Working Voltage*	*Shelf Life*	*Special Characteristics*	*Best Suited for:*
Zinc-carbon	N	1·5	Fair	Low cost	Short term loads with intermittent use. Bulb flash equipment, hand torches, etc.
Leakproof zinc-carbon ...	N	1·5	Good	No risk of leakage of corrosive electrolyte	Photographic uses generally, bulb and electronic flash, small motors
Layer type zinc-carbon ...	N	1.5	Poor	Compact high-tension source	Capacitor flash guns; some types of electronic flash
High-energy alkaline ...	N	1·5	Good	High energy and current capacity	Continuous, heavy drain—e.g. for motor drives in movie cameras, tape recorders and record players.
Mercury	N	1·35	Excellent	High capacity and constant voltage	Portable transistorized equipment. Constant voltage applications. Cadmium sulphide photo cells
Nickel-iron	R	1·20	Excellent	Robust and foolproof	Heavy-duty power supplies, driving professional cine cameras
Nickel-cadmium	R	1·5	Excellent	Not affected by over-charging or being left discharged for long periods. Sealed-in electrolyte	Portable, high-consumption equipment electronic flash, tape recorders.
Silver-zinc	R	1.5	Good	Highest capacity for bulk of any secondary cell	All applications where bulk and weight must be kept to minimum
Silver-cadmium	R	1·5	Excellent	Similar to above with emphasis on long life and constant voltage	As above but where robustness, long life and constant voltage justify more weight and bulk
Lead-acid	R	2·0	Good	Low cost	Heavy duty where weight is not a disadvantage.

* N = Non-rechargeable; R = Rechargeable.

has been used on heavy duty equipment for over half a century but in recent years small, hermetically-sealed versions have been developed and these have a wide application in photographic work.

The basic nickel-cadmium cell consists of nickel hydroxide (positive) and cadmium (negative) electrodes immersed in a potassium hydroxide electrolyte. Normally, when charging is complete, the cell liberates free hydrogen at the negative and free oxygen at the positive electrode.

However, by designing the cell in such a way that there is an excess of cadmium, the nickel hydroxide electrode reaches full charge first and liberates oxygen which at once reacts with the cadmium and does not exist as a free gas. No hydrogen forms on the cadmium, since it is still short of full charge, so there is no free gas to disperse. As a result the cell can be completely sealed and used in the most vulnerable equipment.

Miniature nickel-cadmium accumulators are sold in a packaged unit complete with their own charging system consisting of a mains transformer, a half-wave rectifier, voltage dropping resistance and a 2-pin mains plug. The whole unit is simply plugged into a mains outlet for charging whenever necessary. As a rule charging is completed in less than 24 hours, so that the equipment can be charged overnight ready for the next day.

This is the form in which the accumulator is contained in portable electronic flash units or tape recorders.

Nickel-cadmium accumulators will stand more abuse than any other type of storage battery, and—like the nickel-iron type—need no maintenance. The sealed types are being increasingly used in photographic equipment, particularly for small motorized systems—even of cameras.

Silver-Zinc Accumulators. Silver-zinc accumulators are approximately one-third the weight and volume of other types of storage battery of the same capacity. At low discharge rates they maintain a practically constant voltage and they can be discharged at a high rate without loss of capacity.

The silver (positive) and zinc (negative) plates are immersed in an alkaline (caustic soda) electrolyte and encased in a plastic case under lateral pressure which prevents shedding of the active material.

This type of storage battery is more expensive than lead-acid or nickel-iron alkaline batteries, but its small size and high output make it very suitable for powering portable photographic equipment and apparatus. Silver-zinc units completely sealed in resin are manufactured to drive automatic camera mechanisms in rocket-borne space equipment.

Silver-Cadmium Accumulators. This battery has many of the advantages of the silver-zinc type. Capacity-weight ratio, while not as good as for a silver-zinc battery, is still 2½ times greater than with other rechargeable batteries. Its two great virtues are its long life and its resistance to knocks and vibration. In these respects it is superior to the silver-zinc accumulator. F.P.

BAYARD, HIPPOLYTE, 1801–87. French amateur photographer and Civil Servant. Discovered an original silver iodide negative paper process early in 1839 and exhibited direct positive prints produced by another process in June, 1839 (i.e., before Daguerre's invention was published) at a charity bazaar in Paris. Founder member of the Société Heliographique, 1851, and was for 15 years (1866–81) the Honorary Secretary of the Société Française de Photographie (French Society of Photography). Biography by Lo Duca (Paris 1943).

BAYLEY, ROGER CHILD, 1869–1934. English writer of textbooks and guides to photography. Editor of *Photography* and later of *Focus* and *Amateur Photographer*.

BEAM SPLITTER. Arrangement of mirrors or prisms to divide a beam of light into two or more parts.

There are two principal forms: one which divides the beam transversely and deflects each half in different directions—for instance, a beam directed at the apex of a prism is divided at the apex and each part reflected in different directions by each of the oblique faces.

The other form is when the light energy within the whole beam is divided along two or more channels; for instance, a beam falling on the surface of a transparent medium is partially reflected and partially transmitted. The surface of the mirror or prism is usually semi-metalized to increase the proportion of the light reflected.

When reflected and transmitted beams are required to be of equal intensity, metallized glass beam splitters have the disadvantage of considerable light absorption. By using films (on glass) of bismuth oxide or zinc sulphide, working on an interference principle, almost zero absorption is achieved.

An example of the first type of beam splitter is that used to project stereoscopic films or slides in which the pairs of pictures are arranged side by side, the projector having a single optical system. The stereo attachment, as it is called, is fitted to the projector lens. The beam is divided by a 90° prism, the light from one picture going to the left and that from the other going to the right.

These two beams are then reflected forward again by two further prisms set at the necessary angles to super-impose the two pictures on the screen.

A similar device is used on the camera to take the stereo pictures. Used in reverse in this way it is strictly no longer a beam splitter, although it is sometimes referred to as such.

The type based on semi-metallized reflectors used to be used in one-shot colour cameras. The requirement was to triplicate the image produced by a single lens so as to obtain three colour-separation negatives. Two semi-metallized mirrors of glass or collodion (pellicles) were arranged at suitable angles within the camera to intercept the light from the lens and direct part to one side and part to the other. At the same time they allowed part of the light to continue to the third focal plane. Thus three matched images were produced. A comparable system is still used in present day television cameras.

If two of the films or plates were used together as a bipack, only two images were required and therefore one reflector.

Certain automatic exposure control systems in enlargers and special cameras use beam splitters to direct some of the light to an exposure meter or photo-multiplier cell. The amount of light diverted in this way is usually around 5–10 per cent of the image forming light.

Prismatic beam splitters are normally used for small format cameras, such as professional 35 mm. cine equipment. They also find application in various optical instruments, such as binocular microscopes.

Amateur cine cameras frequently use beam splitters to provide a through-the-lens reflex finder. Here two prisms are cemented together with a semi-reflecting interface. This diverts a small proportion of the light from the optical path inside the lens to another optical system belonging to the finder. A further beam splitter may also be used to divert part of this light to an exposure meter cell for through-the-lens exposure measurement.

Beam splitters need not divide the light simultaneously along two channels, but can also do so alternately. An example is the reflex shutter of cine cameras which allows the light to pass through to the film during the open phase of the shutter, and reflects it into a viewfinder during the closed phase.

See also: *Cine camera; Colour camera; Sterescope; Stereoscopic camera.*

BEARD, RICHARD. Dates unknown. British daguerreotypist. Formerly a London coal merchant and patent speculator, he bought Daguerre's British patent for the daguerreotype process in 1841 and held a monopoly until 1853.

Introduced into Britain Wolcott's mirror camera and use of blue glass for glazing the studio. These, coupled with his employee J. F. Goddard's acceleration of the plate with bromine, enabled Beard to open the first photographic portrait studio in Britain—probably in Europe—at the Royal Polytechnic Institution, London, on 23 March 1841. In 1842 he patented colouring daguerreotypes, using stencils and powder colours. Beard quickly made a fortune from his four London studios and many licencees but lost it again in protracted lawsuits against infringers of the patent, and became bankrupt in 1850.

BECK EFFECT. Increase in the light intensity and other characteristics of a D.C. carbon arc when the positive carbon is loaded with certain metallic salts and the arc is worked at a high current density. This is the basic principle of the high intensity arc.

BECQUEREL, EDMOND, 1820–91. French physicist. Investigated the photographic process and demonstrated in 1840 that an underexposed daguerreotype plate could be intensified by diffused supplementary exposure under red glass (the Becquerel effect). Observed the direct coloration of light-sensitive silver chloride about 1847. Discovered the photogalvanic effect, 1839.

BECQUEREL EFFECT. Intensification of a latent image by exposure to light to which the emulsion is otherwise insensitive. An image on print out paper can be intensified if it is exposed to green light before fixing, and if a latent X-ray image is exposed to white light a silver image eventually builds up on the areas that have already received an exposure. Apart from this, silver bromide emulsions do not exhibit the effect.

BEDFORD, FRANCIS, 1816–94. English landscape and architectural photographer (originally a lithographer); became well-known in the mid-nineteenth century. Was commanded by Queen Victoria to accompany the Prince of Wales (Edward VII) on his educational tour of the Middle East in 1862. One hundred and seventy-two of Bedford's several hundred photographs were published the following year.

BEEF-CAKE. (Slang.) Photograph of a man posed to display his physique; masculine equivalent of "cheese-cake". The subjects are usually professional or amateur "strong men", pugilists and the like.

BEESWAX. Hard yellow wax, soluble in hot turpentine, obtained from the honeycomb of the honey bee. It is the basis of dopes for giving depth and lustre to exhibition prints; it is sometimes made into a polish for glass surfaces used in glazing prints and it may be used on the temporary support surface in the double transfer carbon process.

See also: *Doping prints; Waxing prints.*

BELGIUM

Belgium has played an important part in the development of photography and cinematography.

History. In a thesis written for his doctor's degree in 1829 the physicist, J. A. F. Plateau (1801–83), formulated the laws of the persistence of vision on which cinematography is based. He calculated that the duration of a visual impression from the moment of maximum intensity to the moment when it is only just perceptible is about one-third second.

Basing his design on Faraday's description of the stroboscopic illusion as published in the Journal of the Royal Institution of Great Britain in 1831, Plateau constructed the phenakistoscope, the first known cinematograph apparatus. The instrument, which appeared in 1833, consisted of a cardboard disc with a number of holes punched near the circumference and a series of figures painted on one side. The disc was spun round on its axis with the drawings facing a mirror and when the operator observed the images in the mirror through the holes, an illusion of movement was produced.

The first production model of Plateau's phenakistocope was sold in Paris and London. Meanwhile an improved instrument made by Stampfer of Vienna as well as the Zoetrope of English manufacture had appeared. In collaboration with C. W. Ackermann, Plateau soon produced an improved apparatus under the name of fantascope. This achieved great success, but though it already embodied the essential principles of cinematography, another sixty years were to pass before the first film appeared on the screen.

The modern reflex method of document copying was invented by the German, Albrecht Breyer (1812–70), when studying at Liége University. It was described in a letter to the Brussels Academy of Sciences, dated 14th August, 1839, five days before the meeting of the Paris Academy at which D. J. F. Arago gave the full description of Daguerre's process. Albrecht Breyer used a silver chloride paper of his own manufacture which he called "heliographic" and succeeded in making direct copies of line documents, without the need for optical apparatus, within seven minutes. As his letter to the Brussels Academy indicates, he was fully aware of the theoretical principles of his process, which have remained basically unchanged to this day, but did not himself exploit the process further. It was revived by P. Yvon in 1891 and J. H. Player in 1896, and is often now referred to as Player-type in consequence.

The first Belgian manufacturer of sensitized materials on a large scale was D. C. E. van Monckhoven (1834–82). In 1852 van Monckhoven published a manual of photography which was translated into four languages. In 1879 he discovered that the ripening of silver bromide is notably accelerated by the addition of ammonia. This discovery had a decisive influence on the manufacture of sensitive materials, in particular, the dry plate.

In his laboratory at Ghent, van Monckhoven prepared emulsions for sale both direct to photographers and to other manufacturers who coated them on to glass plates. One such manufacturer produced 1,300 plates per day. In a prospectus van Monckhoven stated that his dry gelatin-bromide emulsion contained exactly half its weight of silver bromide and was ideal for photographers wishing to prepare large-size glass plates. The Ghent plant also produced carbon paper in various types for single and double transfer.

After van Monckhoven's death the manufacture of sensitized materials was carried on by smaller firms, one of which was owned by L. A. H. Baekeland (1863–1944). Baekeland settled in the U.S.A. in 1889, where he formed a company which manufactured the first silver chloride developing paper. The company was later taken over by Eastman Kodak and Baekeland devoted himself to electro-chemistry and research into plastics. In 1909 he invented a synthetic substance which was called "bakelite" after him and has made his name famous. Baekeland became a professor at Columbia University and held a number of honorary degrees from other seats of learning.

Amongst other photographic research workers in Belgium the physicist André Callier is noteworthy for his discovery in 1909 of the absorption and scatter of light by the silver image grains in a photographic negative. The result of this selective scattering of the transmitted light is to increase the contrast of the projected image and emphasize scratches, etc., in enlargements.

At the time of Baekeland's departure for the U.S.A. Lieven Gevaert (1868–1935) opened a small laboratory for the manufacture of photographic paper. A company bearing the name Société L. Gevaert & Cie. was formed in 1894; within a few years it had grown considerably.

Science. There are adequate facilities for study of the theory of the photographic process at the universities of Ghent, Liége and Louvain. Typical subjects of study are: the nature of the centres of sensitivity formed on silver halide crystals during the preparation of the emulsion; action of light on light-sensitive crystals and the formation of the latent image; the nature of the latent image and the changes it undergoes under the influence of light and the action of the developer. For many years extensive research has been devoted to evolve scientifically founded methods for measuring emulsion characteristics. Remarkable results have been achieved, e.g., in connection with resolving power.

At Louvain University important research is proceeding with a view to finding macromolecular products to serve as protective colloids in the preparation of emulsions, to replace gelatin and as a support in place of the cellulose derivatives used at present.

Important research is also being carried out in the laboratories of Belgian factories.

Organizations and Periodicals. The principal organizations in Belgium are la Fédération des Unions Professionnelles Photographiques de Belgique (Federation of Professional Photographic Unions of Belgium), which is the body of professional photographers and photographic dealers, and the Chambre Syndicale des Négociants en Articles Photographiques (Photographic Retailers Association).

Photographic periodicals published in Belgium are called *Photo-tribune*, *Video* and *Spotlight*.

Amateur Photography. At the turn of the century the most notable successes at international photographic exhibitions were obtained by the Belgian pictorialists Alexandre, F. Leys, G. Marisiaux, Léonard Misonne, L. Sneyers, Vanderkindere and Willems, all of them except the first-named being amateurs. Misonne's masterly landscapes are still an inspiration to the young workers of today. In the portrayal of the everyday utensils of modern life and the use of close-ups the Frenchman Dubreuil, who lived in Belgium at the beginning of the century, anticipated the modern approach to photography.

There is now a large and well organized amateur photographic movement in Belgium. Clubs are active in all towns and large centres of population, the majority being affiliated to the F.B.C.P. (Belgian Federation of Photographic Clubs). The Federation was founded in 1947 and at the end of 1964 comprised 160 clubs with a total membership of about 7,500. It arranges national and international competitions.

There is also a strong movement of amateur cinematographers, Facineb, uniting 60 clubs, total membership about 1,500. H.C.

Industry. The manufacture of photographic goods in Belgium is confined almost entirely to sensitized materials and chemicals, the industry being dominated by the major company Gevaert-Agfa N.V., established at Antwerp. This organization undertakes the production of all types of photographic materials, in particular X-ray film and sensitized paper, and has shown steady growth in size and the value of output over the past five years. The Company currently employs 9,300 workpeople nationally, and merged with Agfa in July 1964.

As a result of the industry's structure, no production statistics are published. It is possible, however, to estimate the annual value of output by relating the total exports of sensitized goods each year to information contained in the Gevaert published Report, and the figures resulting from this are shown in the table.

Imports. The value of imported photographic equipment and materials has shown considerable expansion since 1959, reaching a total of Fr. 655·9 million in 1963.

As might be expected, a large part of this is due to apparatus, among which still cameras hold the leading position. In 1963, imports of these attained 239,000 units with a value of Fr. 99·2 million, the principal sources of supply being West Germany (74,600 cameras worth Fr. 53·4 million) and Japan (19,800 cameras worth Fr. 17·1 million).

Cine camera imports have followed a rising trend during the past five years to reach 14,100 units valued at Fr. 41·7 million in 1963, the majority of which are 8 mm. models. Japan is the leading supplier of this gauge, with 5,300 cameras worth Fr. 9·3 million, whilst West Germany supplied 4,500 units with a value of Fr. 13·5 million in the year. Imports of cine projectors have also risen during the period totalling 11,000 units valued at Fr. 31·4 million in 1963, among which 8 mm. gauge types also predominated. The bulk of the latter came from Austria (4,800 projectors worth 9·8 million), whilst West Germany supplied 1,200 units worth Fr. 3·4 million.

Still projectors were supplied chiefly by West Germany in 1963, whose share amounted to 12,800 units with a value of Fr. 18·2 million out of the total of 19,200 projectors worth Fr. 30·4 million.

Imports of sensitized materials and chemicals have nearly doubled in value in the period 1959 to 1963, reaching a total of Fr. 281·6 million in the latter year. Within the overall figure, sensitized paper is the most important product with a value of Fr. 118·8 million, the principal sources of which are (in Fr. million), France (49·4), West Germany (14·4), and the U.S.A. (8·4).

Among other leading items in the field is still film, the 1963 total value of which being composed of roll film worth Fr. 27·6 million and miniature format film worth Fr. 16·7 million. The main suppliers of roll film during the year were West Germany (Fr. 9·9 million) and the U.S.A. (Fr. 9·7 million), whilst the U.S.A. also provided Fr. 12·9 million's worth of the miniature type.

Cine film imports have continued to follow a rising trend over the past five years to a total of 6·1 million metres worth Fr. 33·1 million in 1963. Of this, 16 mm. and smaller gauge stock accounted for 5·3 million metres valued at Fr. 26·7 million, the principal source being the U.S.A. with 2·7 million metres worth Fr. 18·2 million.

Exports. The value of exported photographic goods has recorded a steady increase in the past five years to a total of Fr. 3,505 million in 1963. By far the greater part of this is composed of sensitized materials, and the relatively small amount of equipment consists almost entirely of re-exported items manufactured in other countries.

All sensitized products have shared in the general growth during the period, the two principal materials being photographic paper and X-ray film and plates. Exports of the former reached Fr. 888·7 million in 1963, of which the U.S.A. received Fr. 227·2 million, West Germany Fr. 142·5 million, and the United Kingdom Fr. 119·6 million. The total of Fr. 1067·2 million for X-ray film and plates during the year was composed mainly of sheet film (Fr. 958·6 million), whilst the main markets for the group as a whole were again the U.S.A. (Fr. 263·1 million) and West Germany (Fr. 163·4 million).

Cine film exports constitute a substantial part of the total, being composed in 1963 of 194·2 million metres worth Fr. 443·9 million of 35 mm. stock and 107·5 million metres worth Fr. 208·3 million of 16 mm. and smaller gauge film. Major purchasers of both types during the year were the U.S.A. (96·0 million metres valued at Fr. 145·7 million) and West Germany (29·0 million metres with a value of 72·7 million).

Exports of still film in 1963 consisted of roll film worth Fr. 173·1 million (Fr. 135·1 million being black-and-white stock), together with miniature format type valued at Fr. 64·8 million (of which Fr. 47·3 million was colour film). Leading markets for these products are the United States and West Germany, receiving totals of Fr. 76·6 million and Fr. 25·7 million respectively. G.R.H. & A.G.B.

BELGIAN PRODUCTION OF SENSITISED MATERIAL

	1959	1960	1961	1962	1963
Fr. million	2,270	2,500	3,170	3,530	3,625
(£ million)	16·3	17·9	22·7	25·3	26·0

BELGIAN IMPORTS OF PHOTOGRAPHIC GOODS

Million Francs c.i.f.

	1959	1960	1961	1962	1963
Still cameras	68·8	65·8	72·6	79·4	99·2
Cine cameras }	63·2	30·6	34·8	35·6	41·7
Cine projectors }		29·9	28·5	37·0	31·4
Still projectors	7·6	19·3	21·0	24·6	30·4
Other equipment and parts	n/a	78·0	123·3	135·4	171·6
Total apparatus	n/a	223·6	280·2	312·0	374·3
Sensitized paper	42·2	45·7	47·4	77·8	118·8
X-ray film and plates	25·0	27·5	27·0	29·2	33·0
Roll film	17·5	24·8	29·4	33·5	44·3
Cine film	26·0	21·8	29·1	52·0	33·1
Plates and sheet film	24·4	26·7	43·9	48·5	33·1
Chemicals	12·0	13·6	16·1	15·8	19·3
Total sensitized material and chemicals	145·1	160·1	192·9	236·8	281·6
Total imports	n/a	383·7	473·1	548·8	655·9
(£ million)	—	2·8	3·4	3·9	4·7

BELGIAN EXPORTS OF PHOTOGRAPHIC GOODS

Million Francs f.o.b.

	1959	1960	1961	1962	1963
Still cameras	4·0	4·4	4·1	3·0	4·4
Cine cameras	1·2	1·5	2·3	2·7	2·3
Still projectors	5·8	2·5	6·0	18·5	19·4
Cine projectors	3·5	0·9	1·0	1·5	1·6
Other equipment and parts	n/a	17·2	25·3	33·9	31·0
Total apparatus	n/a	26·5	38·7	59·6	58·7
Sensitized paper	537·8	644·8	763·3	930·7	888·7
X-ray film and plates	527·1	650·3	837·4	927·2	1,067·2
Roll film	127·7	178·8	229·9	231·3	237·9
Cine film	436·7	474·9	604·8	673·4	652·2
Plates and sheet film	351·9	350·2	395·1	483·6	485·5
Chemicals	41·8	65·5	93·9	109·2	114·4
Total sensitized materials	2,023·0	2,364·5	2,924·4	3,355·4	3,445·9
Total exports	n/a	2,391·0	2,964·3	3,415·0	3,504·6
(£ million)	—	17·1	21·3	24·5	25·1

BELGIAN EXPORT MARKETS FOR SENSITIZED MATERIALS 1963

Million Francs f.o.b.

Country	*Sensitized Paper*	*X-ray Film and Plates*	*Still Film*	*Cine Film*
European Common Market:				
France	38·6	64·7	67·8	53·7
West Germany	142·5	163·4	93·4	72·9
Italy	34·3	26·9	53·1	37·4
Netherlands	53·7	56·1	58·6	17·1
E.C.M. Total	269·1	311·1	272·9	181·1
European Free Trade Area:				
Austria	14·5	7·8	11·6	10·6
Denmark	14·6	9·5	11·3	4·3
Finland	7·2	7·6	8·2	3·0
Norway	16·6	5·1	4·8	2·3
Portugal	8·4	8·2	3·0	2·8
Sweden	71·4	54·7	31·4	9·3
Switzerland	34·5	8·7	14·3	7·2
U.K.	119·6	80·5	42·4	90·6
E.F.T.A. Total	286·8	182·1	127·0	130·1
Australia	3·5	3·1	2·8	8·6
Brazil	3·5	23·3	9·9	6·1
Canada	1·0	48·4	6·4	12·0
Hong Kong	2·0	—	0·5	2·7
India	4·0	12·8	9·6	19·1
Mexoci	6·5	23·0	7·2	26·9
New Zealand	1·7	0·9	1·8	0·8
Union of South Africa	5·6	15·7	4·8	5·7
U.S.A.	227·2	263·1	178·5	148·2
All other countries	77·8	183·7	102·0	110·9
Total exports	888·7	1,067·2	723·4	652·2

BELITSKI'S REDUCER. Chemical reducer for negatives, consisting of a solution of ferric potassium citrate or oxalate in an acid-fixing bath.

See also: *Reducing.*

BELLOWS. Collapsible and light-tight sleeve joining the lens to the back of the camera in all but a number of small hand cameras, box cameras, and miniature cameras. Nicéphore Niépce had a square bellows camera constructed, probably by Charles Chevalier, sometime between 1826 and Niépce's death in 1833. (This camera is preserved at the Musée Denon, Chalon-sur-Saône.) Camera bellows were at one time made of either leather or rubberized fabric. Nowadays bellows of plastic or plasticized fabric are being increasingly used.

It is not difficult to make bellows either in these materials or in stout opaque paper to meet temporary requirements. For parallel bellows, cut a rectangle of the material with the width equal to one and one-third times the maximum bellows extension required plus an allowance for attachment. The length should be the average of the distances round the outside and inside of the bellows when compressed, plus an allowance for the overlap in joining the two ends together. The material is then folded and joined up to complete the bellows.

The method of repairing bellows is dictated

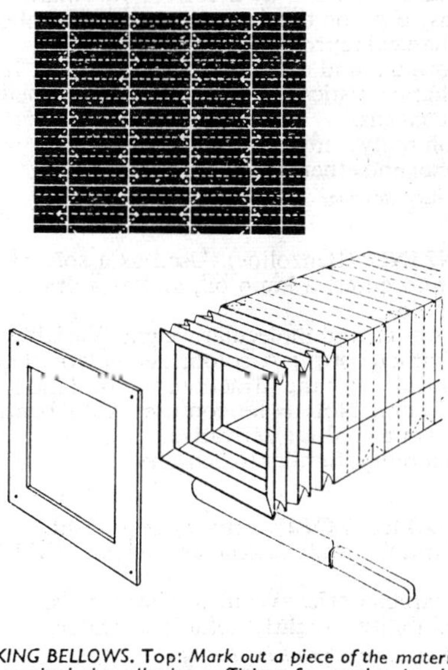

MAKING BELLOWS. Top: Mark out a piece of the material in the required size allowing sufficient for overlap in glueing together. Score the folds as shown. Bottom: Bend up the tube and glue together. Then fold along the scored lines: in one direction on the solid lines, and in the opposite direction along the dotted ones. A knife will help in completing each fold. Crease the folds to make them permanent. Finally stick the ends of the bellows to suitable panels.

by the particular nature of the damage. The usual faults that develop are pinholes at the corners of the folds. If they are extensive, as they may be if the material is perishing, repairs are not likely to be satisfactory. But small holes due to abrasions or accident can be repaired by applying a patch of black paper or other thin opaque material. Another effective method is to seal the hole with rubber solution or cellulose cement into which a little lamp black has been worked.

See also: *Home-made equipment; Repairs to cameras.*

BELLOWS EXTENSION. Adjustable extension unit, consisting of bellows connecting a movable front panel with a fixed rear section, intended for fitting on cameras with interchangeable lenses. The rear plate attaches to the front of the camera body, while the lens goes into the front panel. Such a unit provides a wide range of focusing distances, especially for close-ups and macro-photography.

See also: *Extension bellows.*

BENNETT, CHARLES, 18??–1927. English amateur photographer. In 1878 published his method of increasing the photographic sensitivity of gelatine emulsions by prolonged heating at a temperature of about 90° F.

BENZENE. Used as a solvent for waxes and resins, e.g., of the bitumen in certain photomechanical reproduction processes.

Formula and molecular weight: C_6H_6; 78.

Characteristics: A colourless liquid smelling like coal gas.

Solubility: mixes in all proportions with alcohol and ether; insoluble in water.

See also: *Benzine.*

BENZINE. (Benzoline). Used as a solvent for fats and gums, a lamp oil, and as a dry cleaning fluid.

Formula and molecular weight: Variable.

Characteristics: A colourless, volatile liquid obtained by the fractional distillation of petroleum. Not to be confused with benzene which is a coal-tar derivative.

Solubility: Similar to benzene.

BENZTRIAZOLE· Anti-fogging agent.

Formula and molecular weight: $C_6H_5N_3$; 119.

Characteristics: White needle crystals.

Solubility: Slightly soluble in water.

See also: *Developer improver.*

BERTALL (Vicomte d'Arnoux), 1820–82. French caricaturist and book illustrator, also active as a portrait photographer of celebrities in the 1860s and 1870s.

BERTHON, RODOLPHE. Dates unknown. French engineer and astronomical optician. Developed the lenticular screen additive process which was originally patented by Keller-Dorian in 1908, since made commercially available as Kodacolor (1928) and Agfacolor (1931) for amateur cinematography, also used (1936) in commercial cinemas.

BERTILLON, ALPHONSE, 1854–1914. French anthropologist. Introduced the identification of criminals by anthropometry which he described in his *Photographie Judiciaire* (1890).

BERTSCH, ADOLPHE, Dates unknown. French microscopist and photographer. Introduced, in July 1860, the chambre noire automatique, a 4 ins. square metal box with frame viewfinder, spirit level, and fixed focus (automatic) lens of 3 ins. focal length. The 2¼ ins. square negatives could be enlarged 10 × diameter with Bertsch's megascope or solar enlarger. This was a forerunner of the modern system of a small camera producing negatives for enlargement at a period when enlarging was seldom practised.

BETWEEN-LENS SHUTTER. Shutter that operates between the back and front lens components. This type of shutter includes some of the finest made and is the most popular type for hand cameras with non-interchangeable lenses. Sometimes this term is applied to all shutters of the same basic construction although they may in fact be placed behind the lens.

BICHROMATED ALBUMEN PROCESS. Normal photomechanical reproduction process for making line blocks. A metal plate is coated with albumen sensitized with ammonium bichromate. When it is exposed behind the negative the coating is rendered insoluble wherever the light acts. Washing the plate leaves behind an image in undissolved albumen. The line block is made by treating this image with powdered bitumen and etching away the unprotected metal. This process is rapidly becoming obsolete.

BICHROMATE POISONING. Painful rash or outbreak of small ulcers on the hands sometimes suffered by workers who handle potassium bichromate.

See also: *Skin affections.*

BICHROMATE PROCESSES. Bichromate is the usual trade term for chromium salts of potassium, ammonium, or sodium of the Cr_2O_7 group, which are more popularly called dichromates. Their light hardening effect in association with organic colloids such as gum, glue, albumen and gelatin is exploited in a number of photographic and photomechanical processes. Among the many such bichromate processes are most of the control processes and a number of photomechanical reproduction processes.

See also: *Control processes; Dusting-on process; Glass pctures; Pigment processes.*

BI-CONCAVE. Term applied to a lens indicating that each of its faces curves inwards towards the centre.

BI-CONVEX. Term indicating that each of the outer faces of a lens curves outwards away from the centre.

BIG GAME. The authorities in the world's big-game areas are becoming more interested in preserving wild animal life than in allowing it to be indiscriminately killed by the hunter. This is having a two-fold effect: travellers are being forced to do their shooting with the camera instead of the rifle, and the animals are becoming tamer. In many of the big game preserves of Africa and North America, it is possible to photograph more wild animals in the course of a day's motor ride than a hunter could hope to see in a month.

The Subject. As in every other branch of photography the man who knows the most about his subject gets the best pictures. Wild animals vary in their habits and the technique that works best with one species is useless with another.

Elephants have very keen scent but relatively poor vision and hearing. So long as there is a breeze, however slight, the photographer can work upwind to within fairly close range. Even so, there is more hope of getting a picture with a telephoto than with a normal angle lens.

Lions, surprisingly, are probably the easiest of all the larger animals to photograph. They pay very little attention to the photographer as long as he is in a car and keeps more than about four yards away. In addition to having light-coloured skins, lions frequent open scrub country where the light is in any case good for photography.

Photographers who specialize in big game photography agree that the danger involved in photographing lions is overrated. Lions are actually known to warn off photographers who approach too close. They do this by making a rush at the photographer, but once he retreats make no attempt to take the matter further.

Rhinoceros are both short-sighted and short-tempered. The rhino has very keen hearing and if he is being watched from a hide is apt to shy away at the slightest unusual sound. In common with most other animals in game reserves, the rhinoceros has become accustomed to seeing motor cars and generally takes little notice of them. For this reason it is easier to photograph this animal from a motor car than from a hide, and safer than on foot.

Buffalo are the most dangerous and difficult of all African wild animals to photograph. They are nocturnal in their habits, and tend to stay in the thick bush. The African buffalo is a formidable creature and one of the very few animals that will attack a man on sight. There are places, however, like the Belgian Congo National Park and the Queen Elizabeth National Park in Uganda where buffalo can be approached near enough to give good results with a telephoto lens. These particular beasts are relatively tame because they have been allowed to go unmolested for some generations. Another way to photograph buffalo is to arrange an automatic shutter release device at a water hole that the animal is known to use.

Zebra, antelopes and other swift-footed creatures can be photographed early or late in the day from a hide set up near their water hole. In game parks and reserves where they are unharassed by man, they can be approached by car or lorry close enough to be photographed through a telephoto lens. A number of striking action shots have, however, been taken of them from low-flying aircraft but the practice is frowned on by the game wardens.

Giraffes belong to the relatively tame species that can be photographed from a car and are quite safe to approach on foot. Once they become aware of the presence of the photographer, however, their long legs rapidly carry them out of range of even a telephoto lens.

Hippopotamuses, again, are animals not so much dangerous as shy, and they must be stalked cautiously from the down-wind side. The hippo is usually found in herds and it spends most of the day in the water, coming out mostly at night and only rarely in the daytime. So the most characteristic photographs are those which show the animal partly submerged in its natural element. The hippopotamus has capacious lungs; once scared it dives and can remain submerged long enough to exhaust the patience of the most ardent photographer.

Technique. There are three ways of going after big game with a camera: on foot, by car and from a hide. In practice, stalking the game on foot is the least effective method and where the terrain is suitable, photographing from a car is the easiest and safest way of getting pictures. A hide has its own advantages, but it ties the photographer down to one place with no choice of viewpoint and lighting.

Photographing from a Car. When working from a car, the photographer generally stands on the front seat and shoots from the open sunshine roof. In this way he can get the driver to take him around the animal to find the best viewpoint and lighting. It is a fact that most animals are less disturbed by a car than by a man on foot. Fortunately, too, the surface of the ground in many game reserves is reasonably level and there is no need for the car to stick to the road.

Most drivers quite naturally prefer to keep the engine of the car running when close to potentially dangerous animals. For this reason it is never wise to rest the camera on the edge of the sunshine roof or vibration is almost certain to spoil the picture.

Photographing from a Hide. Hides are always constructed well above the ground in the neighbourhood of a water hole or salt lick known to be frequented by the game. Some hides are luxurious affairs equipped with beds and other amenities intended to make the hours of waiting as comfortable as possible. The photographer must be prepared to enter the hide many hours before the game shows itself—that is, during the early morning and late evening—and generally he will have to stay there until the following day.

The photographer who works from a hide can always find plenty of places to steady his camera during the exposure and can, if the light is poor, use a tripod and give quite long time exposures. It is always better if the subject can be photographed with a telephoto lens some distance away from the hide. This helps to offset the high viewpoint; if animals are photographed at close range with an ordinary lens the camera looks down and gives an uninteresting plan view.

The most serious disadvantage of a hide is that it imposes a fixed viewpoint and direction of lighting. Once in the hide the photographer has no way of shifting his viewpoint.

Lighting. The light in the real bush or jungle is always too weak for instantaneous exposures even at midday. Fortunately, however, most big game roams in the more open scrub country where the light is good. The hours around midday are practically useless for photography because then the sun is more or less vertically overhead, giving hard uninteresting lighting. Early morning and late afternoon in the dry and sunny season are the best times.

Sensitized Material. The colouring of most animals and their surroundings is of a sort that comes out well on fast panchromatic film. This usually gives excellent rendering of the tones of dark green foliage, red earth and reddish-brown skin and hide. As the sky seldom if ever comes into the picture, there is rarely any need for a filter, all the tones of the subject being at one end of the spectrum.

Exposure. As a tripod is generally out of the question and the camera must be held, exposures are never longer than 1/100th second. Even at this speed it is not easy to avoid camera shake when using a telephoto lens. However, using fast panchromatic film in the average morning and late afternoon light it is generally possible to use this shutter speed or even 1/300th second with a lens aperture of *f* 8. This aperture is usually just small enough to give the necessary depth of field for the average big game photograph. Animals in motion can be taken at much higher shutter speeds and bigger lens apertures provided that they are far enough away from the camera.

Equipment. The lens is the most important item of equipment in big game photography. Because it is rarely possible or advisable to go closer than about 40 yards from the subject, a good telephoto lens of 3× to 4× magnifying power is absolutely essential. This in its turn dictates the use of either a reflex camera or a miniature with coupled focusing.

In practice both of these types of camera are used. Those who favour the miniature often have it mounted on a gun stock with triggers for firing the shutter and winding on the film. The camera is then aimed as though it were a rifle; this method of mounting helps to overcome the added risk of camera shake with the long focus lens.

The single lens reflex, taking, say, a 2¼ ins. square film, has the advantage of taking a bigger negative that requires less enlargement and consequently gives crisper rendering of hair and other essential fine detail. G. de H.

See also: *Animals; Zoo.*
Book: *Camera on Safari*, by John Sherman (London).

BINOCULAR VISION. Visual perception of objects with two eyes.

The fundamental difference between seeing with one and with two eyes is in the appreciation of the depth of the subject. When an object is looked at through two eyes, each eye sees a little more than the other of its own side of the object. In addition each eye sees the object against a different part of the background. The first difference increases as the object comes closer to the eye, the second increases as it gets farther away from the background. When the separate pictures seen by the two eyes are fused into a single impression in the brain, these effects combine to give an impression of the depth of the subject and its position in space between the observer and the background—i.e., a stereoscopic picture.

The stereoscopic effect of binocular vision grows less as the subjects get farther away—i.e., as the separation of the eyes becomes less and less in proportion to the distance of the subjects. After a certain point is reached, both eyes see substantially the same picture and the sensation of depth is no longer apparent. For a scene viewed at a given distance, the stereoscopic effect is greater the more the scene extends in depth.

See also: *Eye; Stereoscopic photography; Vision.*

BIOLOGY. The study of the various phenomena of living plant and animal matter, most of which are capable of being recorded by photographic methods, is known as biology. The subjects studied may range from microscopic spores to complete botanical and zoological specimens. So biological photography makes use of normal, macro- and micro-photographic techniques. Some work is also done nowadays with cinematography, and both still and moving pictures may be in black-and-white or colour.

Biological photography is simply an application of the equipment and techniques of botanical, medical and zoological photography. It may be carried out in the field or the laboratory; some projects can be covered by a single worker with simple portable equipment, others call for a staff of scientists and assistants with highly specialized and costly equipment.

Elementary Organisms. The biologist's interest starts with the study of living cells. In this branch, while specimens may be collected in the field, all the photographic work takes place in the studio and is generally a matter of straightforward photomicrography. The most serious problem of photographing living micro-organisms used to be the difficulty of finding a light bright enough to enable instantaneous exposures to be given without at the same time drying out or killing the specimen. The advent of electronic flash, however, has made both still and cine high speed photomicrography of living specimens very much easier. As the light is practically cold, there is no danger to the specimen. Still exposures can be made by simply operating the flash while the normal room illumination is switched off, the duration of the flash being short enough to arrest the movement of the liveliest specimen. Cine photography can be carried out by using a stroboscopic flash synchronized with the camera shutter.

Primitive Life. In the study of primitive forms of life it is necessary to photograph fossil remains, both in the field and the laboratory. Field pictures are necessary to show the relation of the fossil to the material in which it is embedded. The laboratory pictures show the fossil from several angles against a plain background and include a graduated rule to show the actual size. Such photographs alongside those of the present day forms of related organisms show how the subject has changed in the course of centuries of evolution.

Growth. The biologist uses photography to record the development of the embryo. This involves a series of pictures of a number of similar embryonic specimens, taken at intervals to give the effect of a single embryo photographed at different stages of development. The successive specimens must therefore be taken from the same angle and under identical lighting conditions. Such a photographic series may also be supplemented by a series of X-ray photographs of a single developing specimen. The X-ray exposures must be made from the same angle as the photographs so that the two can be compared. The subsequent growth of the organism or creature is recorded by photographing it at regular intervals alongside a scale or preferably against a background ruled off in squares. Photographs made in this way are invaluable for studying the effect on animal growth of different methods of feeding. This branch of biology may call successively for micro, macro, normal photographic and X-ray equipment.

Structure. The internal and external form of the subject is one of its most important characteristics and one that lends itself to photographic methods of recording. Here the camera is first called upon to make accurate records of the outward appearance, showing coloration, texture and modelling. Further records are then made at various stages of dissection to show, for instance, the muscular and nervous systems, the bodily organs, and the skeletal structure. All this work is carried out in the laboratory; it has no relation to the subject in its natural surroundings.

Form. Morphology, or the study of organic form and how it develops, is an important branch of biology. Here also photographic records are used extensively. Typical specimens are photographed under controlled conditions—if possible in the laboratory—to reveal characteristic features of the species or class.

Most forms of life, while conforming to definite characteristics, exhibit certain typical variations—e.g., dwarfs, giants and "sports". These deviations from the standard form can be studied by comparative photographs showing the normal and the variant. Such photographs must be taken under comparable conditions, and the biologist must be certain that his standard is in fact typical of the species.

Heredity. Biologists are also interested in the way in which characteristics are handed on from one generation to the next, and in this field the camera is employed to record the characteristics of successive generations of a wide variety of plants, insects, fish and animals. The points of interest may range from the cell structure to the whole organism and involve extensive series of pictures made in both field and laboratory.

Behaviour. The behaviour of the subject under normal and abnormal conditions provides an unlimited choice of material for photography. Studies of this type include the courtship, mating and breeding activities of insects, birds, fish and animals, life cycles of parasites, response to seasonal climatic and temperature changes and the like. Although much of the bread-and-butter work in this field is carried out with still cameras of all types, an increasing amount is being done with cine cameras, and the results often have an educational and even an entertainment value far beyond the immediate intention of the makers of the film. Their early success, in fact, has inspired the production of a type of nature film in which, by careful production, cutting, and "dubbing" of sound, factual studies of insect and animal behaviour turn a biological study into a popular cinema feature.

Ethnology. Ethnographic surveys—i.e., the recording of race varieties and evolutionary changes—form an important part of the science of biology. The scope of such surveys extends to the various race types scattered over the earth's surface and backwards in time to the earliest traces of human life.

Much has been done in attempting to establish racial characteristics by making "average" portraits, consisting of dozens of superimposed photographs of separate individuals. The biologist, however, is more interested in producing sharp pictures of typical individuals which show clearly the details of skin and hair character, bone formation, stature and proportions peculiar to the race as a whole.

Equipment. The equipment for biological photography is simply the most suitable selection of normal photographic equipment for the particular project being followed; there is no special biological camera, since the work may call for anything from a microscope camera to a large plate or reflex camera. Where the funds available will not cover the cost of special cameras and equipment for each branch of the work, a 35 mm. miniature system—including camera, short, normal and long focus lenses, close-up focusing device, portable electronic flash unit, microscope attachment and possibly a reflex attachment—is the best compromise. The availability of attachments may depend on the camera make. F.P.

See also: *Animals; Autoradiography; Big game; Birds; Botany; Fish; Insects, Macrophotography; Medical photography; Photomicrography; Reptiles; Zoo.*

Books: *Amateur Photomicrography*, by A. Jackson; *Medical Photography*, by T. A. Longmore; *Practical Photomicrography*, by J. Bergner, W. Mehliss and E. Gelbkte (all London).

BIOT, JEAN BAPTISTE, 1774–1864. French astronomer, chemist, mathematician, and member of the French Academy. He was one of the scientists appointed to report on Daguerre's discovery. Biot had known of Daguerre's work and so was able to support M. Arago and Alexander von Humbold in the report made to the Academy on 7th January 1839. This caused a sensation and in July 1839, a bill was passed which gave to Daguerre a pension of 6,000 francs, and to Isidore Niépce a pension of 4,000 francs on condition that Daguerre described his process in full.

BIOW, HERMANN, 1810–50. German daguerreotypist. Opened a studio at Altona near Hamburg 15th September 1841. With C. F. Stelzner, his partner for seven months, took earliest news photographs of great fire at Hamburg, May 1842. 1846–50, took portraits of many royalties and celebrities in Dresden, Berlin and Frankfurt, for his projected National Gallery of Photographic Portraits, 126 of which were published as lithographs in 1849.

BI-PACK. Combination of two films placed emulsion to emulsion and exposed as one.

The two films are differently sensitized to yield colour separation negatives. The two negatives can be used for two-colour photography.

Alternatively the bi-pack can be used in one focal plane of a special single-reflector one-shot colour camera, with a single film in the other focal plane to give the third separation negative. This arrangement used to be employed extensively in colour cinematography.

See also: *Colour camera.*

BIRDS

The photography of birds is concerned with three main types of picture, each of which calls for its own technique and equipment. There are the pictures of nests with eggs in them; then there are the pictures of parent birds at the nest, either sitting on the eggs or feeding the young; and finally there is the most difficult branch of all—pictures of birds in the open country, on the tidal flats by the seashore or actually in flight.

NEST PHOTOGRAPHS

Photographs of just the nest are much easier to take than those which include birds or are of birds in the open.

Equipment. Nowadays, for photographs of nests and eggs, experienced bird photographers tend to use a hand-held miniature camera—preferably equipped with a long focus lens—in conjunction with a flash mounted on the camera (where the nest is in shadow).

This arrangement makes it possible to take excellent pictures of nests even when they are tucked away in a hedge or a hole. The powerful illumination of the flash allows the lens to be stopped right down to give sharp definition over the whole subject and at the same time to work at a shutter speed fast enough to avoid camera shake.

With this type of equipment it is possible to approach the nest without a lot of fuss and noise, take the photograph through any suitable break in the foliage, and then get away quickly and quietly. The whole operation is unlikely to worry even the shyest bird; some birds in fact will sit tight and if the eggs are the important thing, it may even be necessary to scare the bird off them.

Where possible, eggs in a nest should be photographed in colour, and with fast miniature lenses and high speed film this is no difficulty. However, it is important to get the colours exact; it isn't enough to produce a pleasing picture. Often the only way to tell the eggs of one species from another is by slight differences

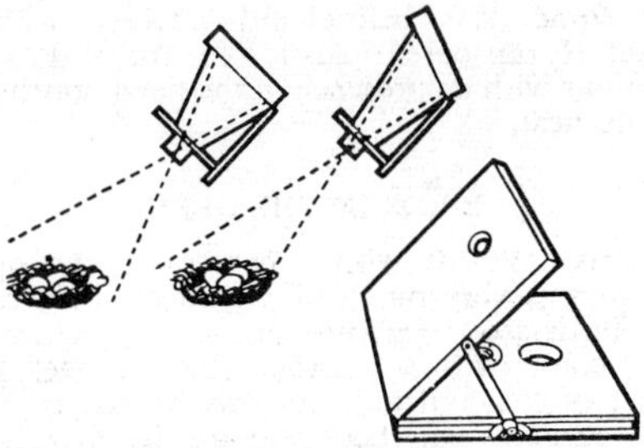

SWING BACK AND TILTING BASE. Left and Centre: *A swing back on the camera will extend the effective depth of field available.* Right: *A tilting base allows easy yet rigid adjustment of camera angle. Useful with heavy cameras.*

in the tint of the shell or the colour of the markings on it. So the recommendations of the manufacturer of the film about exposure and filter technique should be followed to the letter, especially when using bulb or electronic flash (below).

It is as well to remember that, for the record, the colour and nature of the nest materials—grass, moss, leaves, mud—and the lining—wool, fur, feathers—are as important as the details of the eggs themselves. So the lighting should be directed—by a white card reflector if necessary—into the nest to light up the interior.

Again, the viewpoint must not be so close that the nest fills the whole frame. To provide a complete record the picture should include enough of the surroundings to illustrate the general character of the nesting site favoured by that particular bird—rocky ledge, grassy hollow, undergrowth or open branches.

Made in this way, pictures of nests with eggs take the place of the old fashioned birds egg collections. They convey more information and do no harm to the birds or their young, and they are so much more convienent to store, catalogue and exhibit.

In the past, the standard equipment for photographing nests and eggs was a plate camera mounted on a tripod. By comparison with modern miniature camera practice, this method has a number of disadvantages. When focusing close, the depth of field of the lens required to cover a 5 × 4 or half-plate is extremely limited. The only way to get a sharp image is to use the smallest possible stop in conjunction with the swing back. In addition, with lens panel focusing, every movement of the lens in or out alters the scale of magnification and calls for a corresponding movement of the camera support. So a focusing back is essential. Finally, the long exposures necessary with this set-up demand a tripod—a piece of equipment that can make photography in awkward places a difficult and frustrating business. However, many excellent nest photographs are still being produced by patient specialists who put up with all the disadvantages for the sake of the slightly better quality that comes from working with a bigger format.

Lighting. Subdued daylight is quite the best type of lighting as it shows up the markings on the eggs and the modelling of the shell to the best advantage. On bright days the direct rays of the sun must be prevented from shining on the subject and the photograph made by the diffused light from the sky.

However, flash illumination has the twin advantages of being independent of weather conditions and dispensing with the tripod. Bulb or electronic flash are equally suitable so long as the film maker's instructions about exposure and filter are followed. With this form of lighting it is important to ensure a clear field for the flash beam as well as the camera lens.

BIRDS AT THEIR NESTS

Birds are easiest to photograph when they are busy at their nests, either sitting on their eggs or feeding their young. Once the nest has been located, the photographer can be sure of being able to find the birds when and where he wants them. This is quite the easiest type of bird photography; once the nest has been located, the rest is just a matter of patience and straightforward photography. However, it is almost always necessary to work from cover, the form it takes depending on how far the camera is away from the subject. With a miniature camera equipped with a telephoto lens it is possible to work far enough away not to worry the bird, but close enough for a good picture. With a lens that dictates a closer position, the photographer and camera must be concealed completely in a specially constructed hide.

The Hide. The hide can always be improvised from branches or other suitable material gathered on the spot, but the serious bird photographer almost always uses a portable erection made up of light wooden struts and covered with cloth or fabric, so designed that it can be put together or dismantled in a few minutes. The hide is always big enough for the photographer to be able to work in comfort in view of the fact that he may have to spend hours waiting for his opportunity. Every sign of movement must be avoided, so if the sides

INSIDE THE HIDE. *Have method in the hide, know where all equipment is placed so as to get at it instantly.*

and front of the hide are made of cloth they must be fixed firmly to prevent them from flapping in the wind.

The hide should be made as comfortable as possible because it may be in use for long periods of waiting. The camera should be set up and roughly focused before erecting the hide, otherwise it will be very difficult to erect the tripod and arrange the camera in the small space available.

Technique. The lens should point through a small hole in the front of the hide, and there should be a small peep-hole in a convenient place or a periscope in the top of the hide to give the photographer a clear view of the surroundings of the site focused.

The subject is generally very nervous on its first approach to the nest, and it is wise to allow the bird to feed its young three or four times before starting to make exposures. If the parent hears the slight click made by the shutter on its first visit, it may be scared off for a long time. But when once it discovers that there is no danger from the strange erection that has appeared close to its nest, it will take very little notice of unusual noises.

Equipment. The best camera for shooting from available cover is a miniature equipped with a telephoto lens—200 mm. to 400 mm. on a 35 mm. film is ideal. With such a long focal length a firm tripod is essential to avoid camera shake. Rangefinder focusing is preferable since it allows the lens to be focused at its working aperture, whereas an eye level reflex may have to be stopped down after focusing and the bird may move away in the meantime.

For working from a hide, a tripod-held miniature eye level reflex camera with a long focus lens—135 mm. on a 35 mm. film—is the best arrangement since it requires only one hole in the front of the hide.

The conventional large-format reflex camera is also suitable since its extra bulk is unimportant. With a reflex camera taking quarter-plates or 6 × 6 cm. films, a lens of about 9 ins. focal length serves for most subjects. A telephoto lens of 14 or 17 ins. focal length is better for photographs of waterfowl like ducks, geese and grebes on the open water, and waders feeding on the shore at low tide. As nests are usually built in shady places the lens should have a maximum aperture of at least *f* 4·5.

A shutter speed as low as 1/25th second can be used if the right instant is chosen for firing the shutter. At that speed, and with fast panchromatic material, the lens can be stopped down far enough to bring the nest and its surroundings into sharp focus.

In shady locations—or for walking at night—synchronised flash may have to be used. This should preferably be the flash bulb type because the hum from the charging circuit of an electronic flash is enough to keep many birds away And since the reflector will have to show outside the hide, it should be left in position for some days beforehand—covered with a sheet of transparent plastic—for the bird to get familiar with its presence in the neighbourhood of its nest.

BIRDS IN THE OPEN

Stalking birds with a camera in the open country is very much of a gamble; as a rule, the birds see the photographer long before he can come within photographing range. But most of the difficulties of this branch of bird photography can be overcome by attracting the birds with a suitable bait, particularly in winter when other food is scarce.

Garden birds can be tempted to come to a bait of suitable food placed on a bird table or wherever is the most convenient place for photography. Robins come to mealworms or cake crumbs, blackbirds to soaked bread or raisins; the tit family will cling to a hanging feeder containing nuts. The thrush species—field-fares, redwings, and mistle—thrushes—can be tempted with wild berries. The berries should be collected in the autumn and saved until the birds have stripped the hedgerows.

Large birds of prey like kites, buzzards, ravens and crows can be attracted by carrion such as a dead rabbit, pegged down to prevent it from being carried away.

All these forms of baiting ensure that the bird will remain in one place long enough to be photographed. It is only necessary to set the camera up in a hide and wait for the subject to appear. The hide should be erected several days in advance to give the birds time to get used to it.

Equipment. Any camera that will take a long-focus lens is suitable, but the miniature has many advantages if the bird has to be stalked. Where a bait is used to attract the bird, one very successful technique is to set up the camera on a tripod, focus on the bait, and then operate the shutter from a distance by remote control. With this technique any sort of camera will do. A suitable type of control that can be bought quite cheaply consists of a length of fine PVC tubing with the shutter release plunger at one

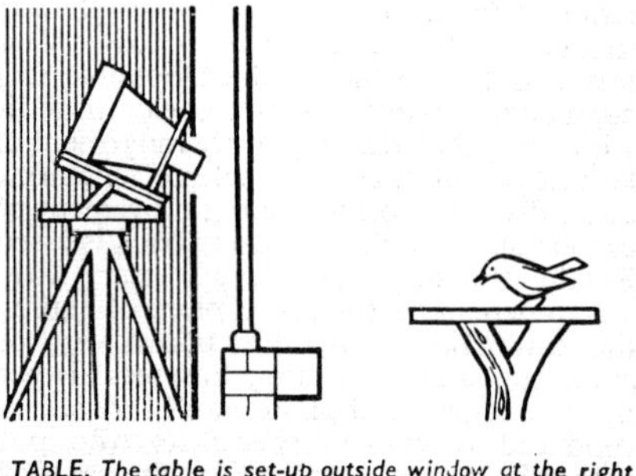

BIRD TABLE. The table is set-up outside window at the right height. Crumbs attract the birds for feeding. The camera is hidden behind a curtain with a hole for the lens.

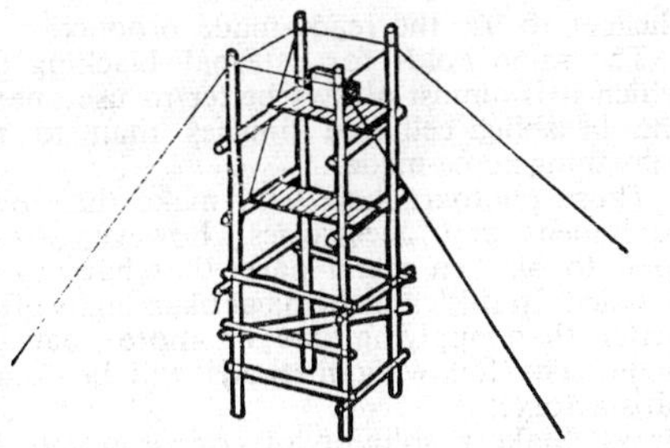

PYLON HIDE. *To photograph birds in their nests in trees, pylon hides may have to be used. These may be as high as 60 feet. Strong cables keep the structure steady.*

end and a press bulb at the other. If the camera is to work close to the bird, the spot should be baited for some days beforehand and a wooden box with a lens hole in the front should be left at the camera position. When the birds get accustomed to feeding in front of the box—which they will do in a day or so—the camera can be set up inside and shooting can start.

Exposure. The shutter speed to be used for birds in flight varies according to the direction of movement. A bird flying across the field of view and less than 20 feet away may need an exposure of 1/1000 second, but the same bird flying towards the camera could be photographed with a much lower shutter speed.

At the same time, the size of the bird has a bearing on the permissible shutter speed. As a rule, the smaller the bird, the faster it beats its wings and the higher the shutter speed needed.

The high speed of electronic flash is useful to arrest the wing action of birds. Use an aperture and flash intensity which allows a certain amount of daylight to be recorded in the background. When working from a hide it is not easy to judge the moment of exposure as the bird flashes past the small space covered by the lens; a fraction of a second either way makes all the difference between success and failure.

Radio Remote Control. An elaborate, but extremely successful method of photographing shy birds in close-up makes use of both radio and photo-electric equipment. The camera is set up near the bird's nest or a bait and either camouflaged or left in position long enough for the birds to ignore it. A photo-cell-operated shutter release and recycling device are coupled to the camera in such a way that when the bird crosses the point focused by the camera lens it releases the shutter and—if conditions require—fires a synchronized electronic flash. The equipment is switched to ready or switched off by a radio relay triggered by a small transmitter at the operating point. This is to economize in the local battery power while waiting for the bird to arrive. The photographer can be stationed as much as a quarter of a mile away—far enough to avoid disturbing the bird. He watches the scene through binoculars or a telescope and switches the circuit on only when he sees a chance of a picture. O.G.P.(F.P.)

See also: *Pets; Zoo.*
Books: *The Technique of Bird Photography*, by John Warham (London).

BISCHOF, WERNER, 1916–54. Swiss photographer. After World War II, travelled for leading illustrated papers throughout the world. Killed in car crash in the Andes 16th May 1954 while working on a picture story on Peru. Several books of his photographs were published.

BISSON, AUGUSTE ROSALIE, 1826–??. French daguerreotype portrait photographer in Paris from 1841, when he and his brother opened a studio. During 1850s they became leading architectural photographers and equally noted for their Alpine views. July 1861 Auguste Bisson was the first to photograph from the summit of Mont Blanc. The expedition lasted three days and twenty-five porters were needed to carry the equipment for the wet collodion process. Gave up photography after the introduction of carte-de-visite views which reduced the demand for large prints.

BISSON, LOUIS AUGUSTE, 1814–??. French daguerreotype portrait photographer in Paris from 1841, when he and his brother opened a studio. During 1850s they became leading architectural photographers and equally noted for their Alpine views. They gave up photography after the introduction of carte-de-visite views which reduced the demand for large prints.

BITUMEN. General name of a group of mineral hydrocarbons. Only one form is of interest to photographers. This is the solid bitumen known as asphalt.

BITUMEN GRAIN PROCESS. Early printing process from which present-day photogravure was developed. The copper plate on which the image is etched is first dusted over with finely separated grains of bitumen. The plate is then heated to make the grains adhere during the etching process which follows.

A sheet of carbon tissue, exposed behind a positive transparency of the subject, is transferred to the surface of the copper plate and "developed" with hot water as in the carbon process. The plate is then etched with ferric chloride beginning with strong and then successively weaker baths which attack the copper in proportion to the thickness of the developed gelatin print. After etching, the plate bears an impression of the image, deep in the shadows, and shallow in the highlights. This impression is covered with minute heterogeneous points of unetched copper where the surface has been protected by the grains of bitumen.

When the plate is inked and wiped, these points prevent the wiping action from removing the ink from the hollows forming the image. In this state, the inked image is printed by pressing it into contact with damped paper.

The bitumen grain process lends itself to a number of methods of control in the hands of the artist or craftsman. For this reason it is still used for individual work and for limited editions of fine art reproduction. F.H.S.

See also: *Photomechanical reproduction.*

BLACK BODY. Theoretically perfect source of radiant energy.

The conception of such a hypothetical reference source is found useful in stating the spectral composition of light, or, in photography, its colour temperature.

When a solid substance is heated, it glows first dull red, then brighter red, and finally white. Discounting any light reflected from the surface, the energy distribution of the radiation emitted by the substance depends on the temperature to which it is heated. This is completely true for a black body—i.e., one which reflects no light falling on it.

In experimental physics such a black body is obtained by taking a hollow metal sphere and cutting a circular opening in one side. Owing to the shape of the body, all light entering the opening is absorbed inside the sphere and virtually none emerges again. This aperture is therefore black—i.e., the "black body" and on heating the sphere the light emerging is solely composed of radiant energy due to the heating.

The spectral composition of any incandescent light source can be stated in terms of the absolute temperature at which a black body emits rays of the same wavelength distribution. This temperature is referred to as the colour temperature of the light.

Apart from the above mentioned sphere there is no such thing as a perfectly black body; all substances reflect a certain amount of light which mixes with that emitted on heating. So the spectral composition of the light they emit is not absolutely proportional to their temperature. In many cases, however, the discrepancy is not serious.

See also: *Colour Temperature.*

BLACKING. Many formulae have been published in the past for making up paints and dopes for blacking the inside of cameras, lens hoods and lens tubes, dark slides, and enlarger bellows. Nowadays, however, there are preparations available for a shilling or so that dry in a few minutes and leave behind a dull black anti-reflection coating. It is always easier and cheaper to use the ready-made product.

The same holds for external blacking for which it is almost always better to use one of the brushing cellulose finishes than to try something home-made.

Those photographers who make their own equipment and accessories, however, often need to blacken metal parts that have to be handled. In this case chemical blacking is often better than applying one or more coats of paint. The following methods will be found satisfactory:

Brass. Make a solution of copper nitrate by dissolving 1 ounce of copper foil or turnings in 2 ounces of nitric acid (in open air). Clean the brass with emery cloth and cover it with the solution for 2 minutes. Now lift the part out with a wire hook and hold it over a blue gas flame. In a few moments a fine black coating will cover the surface. Repeat the process several times to build up a thick enough coating.

Aluminium. (1) The easiest way to blacken aluminium parts is to rub them bright with emery cloth, dip them in linseed or olive oil, and then heat in a gas flame to turn the oil into a film of black carbon. This type of finish tends to leave glazed patches and it does not stand up very well to handling.

(2) A more robust finish can be given like this: dissolve 1 part of arsenious oxide and 1 part of ferrous sulphate in 12 parts of concentrated hydrochloric acid. Add the solution to an equal volume of water (do not add the water to the acid). Clean the part with emery and immerse in the solution until it turns black. Wash and dry.

The finish can be given a gloss by rubbing lightly with linseed oil.

Iron and Steel. Ferrous metals can be blackened in the same way as aluminium (2) above.

BLACK LIGHT. Black light is the name for ultra-violet illumination when it is used to make objects visible in the dark. The objects are treated with a dye or pigment which fluoresces under the ultra-violet radiation.

The source consists of a mercury arc or other suitable light screened by a filter (e.g., covered with a lacquer) which cuts out all the visible rays but allows the ultra-violet through.

It is used in theatrical transformation scenes, and for publicity and advertising.

See also: *Ultra-violet and fluorescence photography.*

BLANCHARD BRUSH. Simple spreader for applying liquid to a flat surface. It was used for coating plates in the collodion process.

See also: *Brushes.*

BLANQUART-EVRARD, LOUIS DÉSIRÉE, 1802–72. French amateur photographer and inventor. Introduced an important modification of the Calotype process 1847, albumen paper 1850. Founded at Lille the first photographic printing and publishing firm 1851. Blanquart-Evrard's technique of developing prints, instead of the slow business of printing them out, speeded up the mass production of prints for sale, and the publication of a large number of books and albums illustrated with original photographs.

BLEACHING. Converting a silver image (negative or positive) into a more or less colourless silver compound, such as silver chloride, bromide, iodide, or complex salts containing chromates, ferrocyanides, etc., of other metals as well. Bleaching forms the initial stage of a large number of toning processes for transparencies and prints (where the additional metal compounds often produce the colour of the toned image); as well as of many intensification methods.

Other applications of bleaching are in the formation of gelatin relief images and matrices for various control processes, and reversal processing of black and white or colour reversal materials.

BLEACH-OUT PROCESS. This is a method for making drawings by going over a photographic image in ink or pencil and afterwards removing the image by chemical bleaching. The photograph should preferably be a fairly pale print of a clearly outlined subject. The artist first draws over the outlines of the subject in either pencil or water-proof ink.

Two baths are used in the bleach-out process. The first is the bleacher; it consists of flake iodine dissolved in a solution of potassium iodide (about ¼ ounce of each to 5 ounces of water). The second is a normal fixing bath for prints; it may be acid or plain hypo.

The print is first immersed in the bleacher until the image disappears. It is then rinsed in running water and transferred to the fixing bath. After some time, the slight discoloration clears and leaves the drawing on a white background. The paper is then washed and dried in the usual way.

Line drawings for reproduction can be made easily and quickly by drawing over a blue print made from the photographic negative. When the composite image is photographed in the process camera, the blue does not reproduce.

See also: *Sketch photograph.*

BLISTER. Blemish on negatives and prints caused by bubbles of gas or liquid trapped under the surface of the emulsion. The bubbles may be formed either in the emulsion layer or between the emulsion and the support. Blistering occurs on films, plates and paper prints, and the cause can be diagnosed according to whether the blisters appear after development, fixing or washing.

Blistering During Development. When plates are developed at a high temperature the gelatin may swell and leave the support. Where this happens around the edges, it causes frilling of the emulsion; where it happens inside the margins it causes blistering. This type of blister can be avoided by hardening the emulsion before development—e.g., with an alum hardening bath or a 10 per cent solution of formalin. Once formed the blisters may disappear if the material is immersed in methylated spirit after washing and then dried off normally.

Blistering after Development. This is generally the result of transferring the sensitized material from a developer containing carbonate to a strongly acid stop bath or fixing bath. The resulting chemical action generates carbon dioxide which may collect in bubbles in the emulsion. Absence of hardener in the fixer or the use of a fixing bath much colder than the developer makes matters worse.

The remedies are: to rinse the material thoroughly after development; to use a hardening and fixing bath; and to keep the temperatures of all baths within two or three degrees of each other.

Blistering after Fixing. Water blisters may form on the surface of the material (more especially on prints) during washing immediately after being fixed in a strong fixing bath. The washing water penetrates the gelatin faster than the fixing solution can escape and detaches the emulsion from the support. Once the emulsion has been loosened in this way it is very difficult to get it to adhere after the water has dried out. Local damage to the emulsion —e.g., by creasing or handling with warm fingers—increases the likelihood of blisters.

Paper that is carefully handled and fixed in a hardening fixer of the correct strength very rarely suffers from this cause.

Blistering after Washing. Very cold washing water containing an excess of dissolved air (as most high pressure main supplies are likely to do) is another cause of blistering. In this case the trouble appears after washing and results from air bubbles liberated from the water as it warms up to the temperature of the atmosphere.

The remedy here is to allow the water to stand for some time and then to wash in successive changes.

Modern sensitized materials will stand a surprising amount of ill-treatment without blistering, and if they are handled with normal care and processed according to the maker's recommendations the risk of trouble on this score is negligible. F.P.

See also: *Faults.*

BLOCH, OLAF, 1872–1944. English emulsion chemist. Worked in the Research Laboratory of Ilford Ltd. Responsible for producing many of the new types of plates for modern scientific work, including spectrum photography, the recording of nuclear tracks, cosmic radiation and spectroscopy in high vacua. Received the Progress Medal of the Royal Photographic Society in 1929, and was President of the R.P.S. in 1931–3. The Olaf Bloch Memorial Award of the R.P.S. and the Institute of British Photographers was established in his memory.

BLOCKING OUT. Painting out an unwanted background on a negative before printing. Mostly used for commercial photographs.
See also: *Retouching.*

BLOCK. Name for mounted line and half-tone plates, and stereos and electros used for letterpress printing.
See also: *Photomechanical reproduction.*

BLOOMED LENS. Lens which has had one or more of its air/glass surfaces coated with a tin film of metallic fluoride to reduce the amount of light lost by reflection.
See also: *Coated lens.*

BLOW UP. (Slang.) Expression sometimes used to describe the enlarging of a negative; can also refer to the enlargement itself.

BLUE PRINT. Print made by the ferro-prussiate process. So named because it gives a white image on a blue ground. Invented by Sir John Herschel, 1842, and named by him Cyanotype.
See also: *Document copying.*

BLUR. Lack of definition in a negative or positive, caused by misfocusing or by movement of the image during exposure.
See also: *Faults.*

BODY RELEASE. Shutter release button or lever mounted on the body of a camera instead of on the actual shutter assembly.

BOGISCH, A. 1857–1929. German chemist. Worked at the Hauff factory at Feuerbach, near Stuttgart. Discovered in 1893 the new developers metol, amidol, glycin, ortol.

BOISSONNAS, FRED, 1858–1947. Swiss professional photographer in Geneva; took over father's portrait studio 1888. Also made pictorial photographs for exhibition. Obtained first successful photographs of Mont Blanc from Geneva with telephoto lens, 1891, on orthochromatic plates manufactured by himself and his brother Edmond. Boissonnas took for Lord Napier thousands of views in Greece from 1902 onward, and was commissioned by King Fuad to do similar work on Egypt.

BOLTON, WILLIAM BLANCHARD, 1848–99. English amateur photographer in Liverpool; editor of *The British Journal of Photography* 1879–85. Associated with B. J. Sayce in the introduction of collodio-bromide of silver emulsion, 1864, but admitted, 1873, that Sayce alone was responsible for the invention. 1874, published an important improvement—a washed collodion emulsion process.

BOOKS ON PHOTOGRAPHY

Books on photography are as old as photography itself; Daguerre's own handbook on his technique was published in the same year as his invention.

To-day the literature of photography would fill large libraries. The library of the Royal Photographic Society contains 6,000 volumes and it does not claim to be complete even in respect of publications in the English language.

Only a very small proportion of these books consists of original works in the sense of containing information not available from alternative sources. Just as in literature on other subjects there is a large amount of duplication, of mutual borrowing and of pretension to knowledge.

The ideal book on photography should have as its author a person of sound technological training, practical experience, methodical mind, writing ability and some visual imagination. Such authors are rare. Many books are written by people who lack these attributes and are either too young or too old for the task. The reason for this is that experts of exceptional ability at the height of their career can seldom spare the time to pursue literary sidelines.

There is, however, an effectively compensating factor provided by the selective influence and the editorial resources of experienced publishers. It is by no means accidental that the best books on photography are published, in most countries, by a few specialist publishers, whose imprints serve as a reasonably reliable guide to the reading public.

Elementary Textbooks. The average amateur rarely needs an extensive library, but any purposeful and satisfying photographic activity warrants the study of some elementary introduction to the subject to avoid the disappointment of haphazard and accidental results. There are scores of elementary books on the market and the safest way of choosing among them is to be guided by the degree of their proven popularity. In view of its reprint record since its first publication W. D. Emanuel's *All-in-One Camera-Book* appears at present to be the most widely read independent "first book" on photography in the English language.

Manufacturers' introductions to photography vary a great deal in size, scope and purpose. Their presentation is often more lavish than independent publishers' products. Technically they are very reliable and authoritative within the limits defined by the individual manufacturer's interests and policies.

Comprehensive Works. Books on photography suited to training professionals and serving other serious students are less numerous than they used to be. The older generation of authors who could command and transmit encyclopedic surveys of photography as a whole is no longer with us and keeping their works up to date has proved an almost insuperable task.

Attempts at replacing individual works of this order by comprehensive symposia assembled from the work of a number of experts suffer as a rule from serious editorial difficulties in co-ordinating the material across varying levels of outlook and methods of approach.

The British *Manual of Photo Technique* avoids the difficulties inherent in multi-authorship by splitting up the subject into a series of smaller but self-contained volumes of which, so far, twelve have been published: H. J. Walls' *Camera-Technique*, Arthur Cox's *Optics*, R. H. Cricks's *Illumination*, W. F. Berg's *Exposure*, C. I. Jacobson's *Developing* and *Enlarging*, O. R. Croy's *Retouching*, J. H. Coote's *Colour Prints*, C. L. Thomson's *Colour Films*, P. Jenkins' *Colour Separation Negatives*, K. C. M. Symons' *Stereo Photography*, L. Lobel's and M. Dubois' *Sensitometry*.

Special Techniques. The application to definite subjects like portraiture, scenery, natural history, etc., is covered by many useful books. In these fields of specialization the reader has a reasonable chance of finding the book he wants by basing his choice on the author's reputation or on his skill as a photographer as evidenced in his illustrations.

When selecting a book on special techniques —e.g. lighting, photomicrography, medical photography, etc.—the frequency of its reissue is perhaps the most truthful guide to its reliability and completeness. No sizable work of this kind can hope to be entirely free of misprints or omissions at its first publication and so if there is a comparable choice between a book that is well established and one where the merit appears to lie in novelty, the established work is likely to be preferable provided the date of its last re-editing is reasonably recent.

Books which purport to pass on the secrets of photography as an Art form are now less frequently published than in the past. They have, in fact, never been particularly successful; the technically minded reader may be eternally thirsting for further knowledge but people with predominantly artistic interests are apt to be too individual to wish to follow someone else's approach, particularly from books.

The advent of several high performance cameras during the last few decades has also led to the appearance of some fairly ambitious volumes which serve as complete textbooks based on the techniques of the particular camera and its accessories. The genuine usefulness of such works is beyond doubt when they are devoted to sufficiently mature instruments and their authors are in the position to pass on specific and independent information. This is, however, by no means always the case. Some manufacturers are quick to encourage any further addition to the literary background of their merchandise and so the potential reader is often faced with a bewildering assortment of competent technical advice, diluted journalistic hand-outs and glossy pictorial puffs. But here, as always, the buyer can safely let his choice be guided by the standing of the publisher, the reputation of the author, and the number of recent reprints of the work. As the growing pressure of competition propels manufacturers into continually overtaking each other with new models of their products (just as the makers of automobiles do) books on cameras have to be very recent indeed to be fully useful to the owner of the very latest model of that particular camera.

Scientific Books. The tendency to specialize within a wide subject has become most marked in the scientific field. Almost the only modern book to offer a wide and yet penetrating view of the subject as a whole is *The Theory of the Photographic Process* by C. E. Kenneth Mees. This work is the product of a brilliant creative mind and mature experience furnished with the resources available only to the research laboratories of a giant organization for manufacturing photographic materials. Scientific publications emanating from such centres are among the most authoritative and valuable contributions to the photographic libraries.

We have today a young and very populous generation of scientists who record the results of their research in short papers on some narrow sector of specific subjects rather than in self-contained monographs. The best all-over view of their work was presented periodically in the volumes of *Progress in Photography*, an international survey and summary which is now replaced by the quarterly *Perspective*.

Originating at the same source *The Focal Library* now serves as a central pool of information on the science and technology of photography, cinematography, image and sound recording. It covers research, manufacturing processes and special applications of both by teaching texts, symposia and monographs. The principle language of the library is English including translations from other languages, notably Russian and German; symposia are multilingual. The Library is governed by an Editorial Board of leading experts in high academic and other scientific positions from ten countries.

General Reference Books. Among general reference books offering a wide variety of photographic data the American *Photo-Lab Index* is the most ambitious and correspondingly elaborate publication. Leading manufacturers of photographic materials in Great Britain, U.S.A., Germany, Belgium, etc., publish their own data books which offer the advantage of the most authentic and up-to-date information on their own products.

Amongst annual publications of reference the *British Journal Photographic Almanac* (first published for the year 1860) holds undisputed seniority and is known all over the world, chiefly among professionals. *The Photo-Amateur's Pocketbook* is for more popular use.

Trade directories are provided in Great Britain by *The Photographic Dealer's Pocketbook* and in the U.S.A. by *The Photo-Dealer's Directory*.

Pictorial Collections. Pictorial annuals are found in every country but only few of them stay. They are expensive to produce and so have to rely on the fickle support of advertisers. Among the best known are the British *Photography Year Book* and the *U.S. Camera*, both representing a revitalized pictorialism following patterns of contemporary illustrative and commercial photography.

Books by photographers are not necessarily on photography and their success or otherwise is no indication of their importance to readers on photographic matters. Cecil Beaton's semi-autobiographical books, for example, have never failed to attract a public, but this has little to do with the fact that their author happens to be a photographer, among other things.

Fine volumes devoted to the work of individual photographers are a publisher's nightmare. They are gladly looked at but rarely bought—unless devoted to a specific subject which commands attention in its own right.

Inexpensive Publications. Photographic publishing has its "paper backs" and their impact is equally impressive. For a few pennies these little books provide straightforward answers to practically any problem "the man in the street" using a camera may encounter. Their influence in recent years has been responsible for turning millions of snapshotters into enthusiastic and reasonably successful amateur photographers. For these millions the *Focal Photo Guides* and *Viewfinder Books* and other series fashioned on their example act as an unfailing source of commonsense information and assistance.

The *Camera Guides* are a closely related type of publication. These books have improved on the function of the manufacturers' instruction literature about specific cameras by being more informative, sometimes more clearly written and certainly more impartial when it comes to dealing with the limitations as well as the scope of the camera in question. They are constantly kept up-to-date. A.K.-K.

See also: *Literature on photography; Periodicals; Training and education in photography.*

BOOM LIGHT. Light source attached to the end of an adjustable boom to give it a wider range of horizontal movement than is possible with a simple telescopic stand. A boom light enables a subject to be lit from above or behind, without getting the light-stand in the picture.

See also: *Light supports.*

BORAX. Sodium tetraborate, weak alkali in fine-grain developers.

Formula and molecular weight: $Na_2B_4O_7 \cdot 10H_2O$; 381.

Characteristics: White crystals or crystalline powder.

Solubility: Slightly soluble in water at room temperature; fairly soluble in hot water.

BORDER. Margin of the print, which may be left white, printed black, or treated in a number of other ways to suit the subject. A white border is the most common, although prints trimmed down to the picture area without leaving a border are also popular.

See also: *Masks; Paper holder.*

BORIC ACID. Boracic acid. Used in hardening fixers containing potash alum—extends the hardening life of such fixers; also in certain buffered fine-grain developers.

Formula and molecular weight: H_3BO_3; 62.

Characteristics: White crystalline powder.

Solubility: Slightly soluble in water at room temperature.

BOTANY

Botanical photography is essentially concerned with making accurate records of plant life. In every branch of botanical study and experiment the camera supplements the notebook of the naturalist and scientist and it has therefore a very wide range of applications. The principal fields are: life histories of individual species, comparative studies, plant variation in relation to environment, plant structure. Subjects range from high plants to microscopic grains of pollen, fungus spores and the like.

In general there are three ways of photographing plants: growing in their natural surroundings, complete in the laboratory, and dissected in the laboratory. Each class of photograph poses different problems and to some extent calls for different equipment. The first two classes require normal photographic equipment, the third class usually involves photomacro- and micro-graphic apparatus.

Equipment. Accurate detail in the final print is essential in such scientific records, and very often the negative will be printed as a lantern slide which will be greatly magnified in projection. So in general the largest convenient plate camera is preferred—e.g., a quarter-plate or at least a $2\frac{1}{4} \times 3\frac{1}{4}$ ins. stand type. There must in any case be provision for focusing by screen, and the camera must have a double extension for taking close-ups of small plants. A reflex camera is suitable so long as it has some means —e.g., extension tubes—of providing the necessary extension. And since focusing extremely close-up subjects is critical, a focusing magnifier is essential. There is very little need for a wide aperture lens in botanical photography because the normal practice is to stop well down—to *f* 11 or *f* 16—to get the maximum depth of field. Occasionally, however, it is convenient to be able to use differential focusing to separate the subject from the background. There are times, too, in a wind, or if photographing an uprooted specimen that is wilting, when it is necessary to use a wide aperture and an instantaneous exposure. So the lens should have a maximum aperture of about *f* 4·5. Colour materials are being increasingly used for botanical records, making it necessary for lenses to be fully colour corrected.

Ideally there should be a selection of lenses to cope with the wide variety of subjects and conditions—e.g., a normal lens for routine exposures; a long focus lens to give a big image with good perspective of small subjects *in situ*; a short focus lens for macro-scale photography of small thin objects where perspective is unimportant, such as for sections; a wide angle lens for confined situations, and a telephoto lens for distant shots of inaccessible specimens. If only a single alternative is permissible, it should be a long focus lens giving about one and a half times the image size of the normal lens.

Since time exposures are normal, a really rigid tripod is necessary. The tubular telescopic type used without the legs extended is generally satisfactory. In addition the tripod must be equipped with a tilting head which allows the camera to be pointed down at any angle including the vertical. A suitable tilting platform can be made from two pieces of $\frac{1}{2}$ in. thick plywood hinged at one end to open and shut like a book. A length of slotted brass attached to one half engaging with a thumbscrew on the other enables it to be locked at any angle. One of the hinged pieces is attached to the top of the tripod and the camera is fastened to the other with a bolt screwed into the tripod bush.

As far as possible, the tone values of the subject should be natural since the photograph is intended to give an accurate impression of the colours or their monochrome equivalent. Normally only a pale green correction filter should be used with medium speed panchromatic material. Where a contrast filter is used to emphasize certain features, the fact should always be noted on the print or included in the caption if the picture is published.

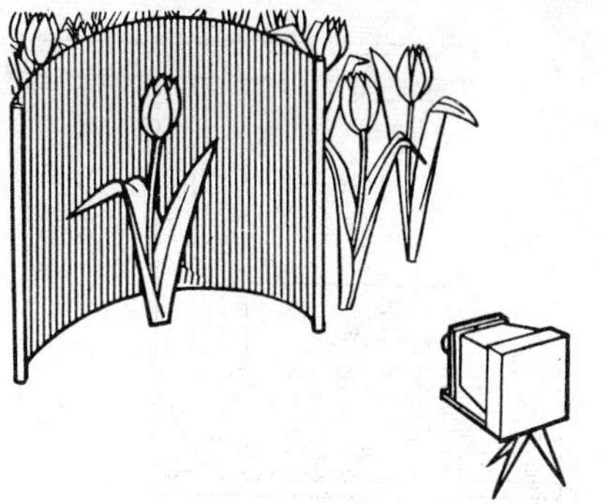

SPECIMENS IN THE FIELD. For a detail close-up of a plant set up the camera at the same level as the subject, and arrange a background of plain light grey paper behind the plant to be photographed. This also screens the plant against air movement.

The most suitable form of sensitized material for exposures in the field is a medium speed or soft gradation panchromatic plate. In the laboratory—and in particular for macro and micro work—it is sometimes better to use a slow panchromatic plate for the sake of the crisper contrasts.

Lighting. Where the subject is out in the open, daylight is the obvious form of lighting, but much botanical photography must be done in woods and forests or under hedges and banks—where the light is feeble and exposures would be unduly long. Under these circumstances it is better to rely on a lightweight portable electronic flash. This need not exceed 30 to 50 joules because the subject will usually be small enough to be covered by a flash two or three feet away. Flash has the added advantages that it arrests any movement due to wind or wilting and can be directed from any angle to conform with the camera viewpoint. (If the sun shines from the wrong angle, the photographer may have to wait a long time for it to move around.)

Where the weight of an electronic flash is a disadvantage, and only one or two exposures are contemplated, ordinary bulbflash may be used.

Field Work. Some time is always spent in finding a specimen that is typical—not necessarily the largest.

The main object of photographing a specimen in the field is to show it in its natural surroundings, so there is no point in choosing a subject that happens to have taken root in uncharacteristic terrain. For the same reason the subject must not take up the whole of the picture space to the exclusion of the setting. Under these circumstances it is not always easy to emphasize the subject—e.g., by differential focusing or lighting—without giving an un-

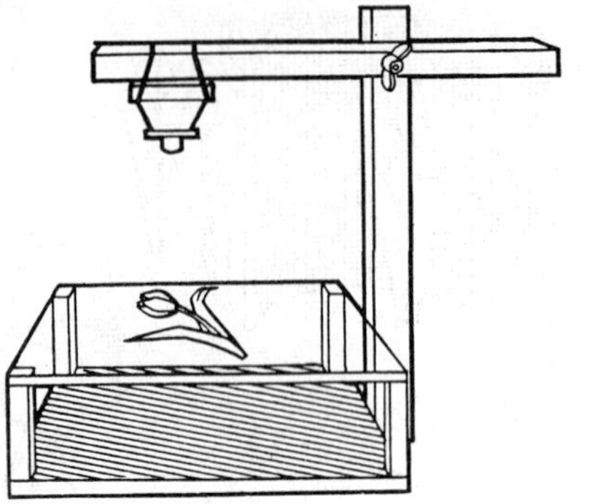

CAMERA ROSTRUM. For laboratory shots with simple apparatus the specimen is laid out on a glass plate above a plain background. The camera is mounted squarely over the glass.

natural impression. Often the best compromise is to take two photographs: one a straight record of the plant in its setting, showing both sharp and in their normal tone values, and the other a close-up of the plant to show its details. Plants more than 6 ins. or so high are generally best shown close up against a plain background of light grey paper.

Plants should always be photographed with the camera down on their level, and not looking down from normal eye level. White paper reflectors or a fill-in flash are generally necessary to light the lower surfaces of the plant because these receive little or no natural reflected light and are apt to come out solid black.

In pictures of fungi it is important to show the underside of the cap, so one good specimen is always pulled up and laid alongside the others to show the gills or tubes.

Laboratory Work. When a specimen is photographed in the laboratory it is with the object of showing its shape, colouring, root development and general proportions. It is most conveniently photographed spread flat out with the camera looking down on it. In this case the specimen chosen should be a single complete growth, freshly dug up and with the earth washed out of the roots. The plant is then laid out on a clean glass plate supported horizontally above a plain sheet of neutral coloured paper. Both the plant and the background are lighted to give the required contrast—i.e., showing the specimen either darker or lighter than the background.

Some specimens — delicate flowers and orchids in particular—start to wilt rapidly as soon as they are prepared for photographing, and the heat from normal electric lights makes matters worse. The result is that the specimen looks tired and, during a time exposure, leaves and petals may even show some movement. In such cases, use an electronic flash with a good reflector to fill in the shadows and, if possible, a second flash directed on to the background.

The subject may also be photographed in daylight near a window with the background paper sloped to catch the light and a hand mirror used to reflect light into the shadows.

Small Subjects. Much of the laboratory work is concerned with making records of dissected plants, seeds, pollens and such-like small subjects. Some plant sections are large enough to be taken by normal close-up photography, but cross-sections of flower stems, individual blooms, seed-pods and leaves call for macrophotographic techniques. Such pictures can, however, still be taken with the normal camera and lens equipment by adding extra extension tubes or bellows and using a short focus lens in place of the normal camera lens.

Further down the scale of sizes there are the seeds, pollens and spores that call for photomicrographic equipment. The technique here is simply normal photomicrography and differs in no other basic way.

The photographer who has no special equipment for making highly magnified photographs can, within limits, make do with his enlarger. It is a simple matter to substitute a mounted specimen for the negative in the enlarger and project it in darkness on to a plate laid emulsion side up on a black card on the enlarger baseboard. In this way enlargements can be made up to 20 ×. The negative made in this way can be further enlarged, although the normal camera or enlarger lens cannot be expected to give results equal to those of a properly corrected microscope objective designed specially for this work. H.J.H.

See also: *Biology; Close-ups; Flowers; Gardens; Macrophotography; Photomicrography.*

Books: *Camera Close-up*, by O. R. Croy (London).

BOTTLES. Photographic solutions are most conveniently stored in glass bottles with rubber stoppers. Glass stoppers tend to stick unless they are greased regularly, and in the darkroom, grease on anything that has to be handled is a nuisance. Cork stoppers are unsafe because they are attacked by many photographic solutions. Concentrated acids that are usually stored in glass-stoppered bottles are not normally used in photography.

Ribbed bottles are used for poisons, amber or brown for solutions that deteriorate under the action of the actinic rays, and clear glass for most other liquids. Concentrated developers are always sold in brown glass bottles.

The commonest sizes are 10 ounce (284 cc.) or half-pint; 20 ounce (568 cc.) or pint; 40 ounce (1,136 cc.), or quart; and 80 ounce (2,272 cc.) or Winchester quart.

The labels on bottles should be painted over

with clear cellulose lacquer or covered with transparent tape to prevent them from being discoloured by the solution or washed off when the bottle is cleaned.

Polythene bottles are made in all the sizes normally used in photography. They are unbreakable, and unaffected by any of the chemicals used in ordinary photographic processing. They have a further useful property: being flexible they can be squeezed to exclude any air, if the liquid contents do not quite fill the bottle. This makes them particularly suitable for storing liquids which are oxidized easily—e.g. most photographic developers.

See also: *Stoppers*.

BOURNE, SAMUEL, 1834–1912. English photographer of Indian views. Took up photography in the mid-1850s and *c.* 1860 became a partner in the firm founded by Shepherd in Simla 1843 (later also in Calcutta). Bourne became famous for his photographs of the Himalayas taken on expeditions lasting several months and requiring up to sixty porters to carry the wet collodion equipment and other baggage. 1868, photographed at an altitude of 18,600 feet—a record held until 1880.

BOX CAMERA. The box camera is the simplest and cheapest type of camera. It is mainly intended for people who want to take snapshots in good light and who do not know —or want to know—a lot about photography. In its traditional form, it is simply an oblong or square box fitted with a very elementary lens, shutter, and viewfinder, and some means of registering and winding a paper-backed roll film.

Modern box cameras are usually more streamlined, resembling miniature cameras in appearance. In recent years designers have made extensive use of the possibilities of styling offered by the simple design and by the utilization of moulded plastic body components.

Traditionally box cameras used to take $2\frac{1}{4} \times 3\frac{1}{4}$ inch (6 × 9 cm.) or $2\frac{1}{4} \times 2\frac{1}{4}$ inch (6 × 6 cm.) pictures on size 120 roll film. A more modern trend is the box camera taking 16 exposures $1\frac{5}{8} \times 1\frac{5}{8}$ inches (4 × 4 cm), again on size 120 roll film, yielding greater film economy (even though part of the film area is wasted). This has become popular with reversal colour films, especially as the $1\frac{5}{8} \times 1\frac{5}{8}$ inch format is suitable for mounting in 2 × 2 inch super slide mounts.

An alternative and still popular film size is 127, yielding usually 12 exposures $1\frac{5}{8} \times 1\frac{5}{8}$ inches (4 × 4 cm.). Other sizes also exist (for example 828 for 28 × 40 mm. images on unperforated 35 mm. film).

The Body. The box itself is generally made of plastic or of sheet metal, covered with leather or plastic. It is made to open so that the film can be inserted. There is one chamber to hold the full spool of film and another for the empty spool. Between the two chambers the film is led behind a metal aperture that frames the picture area. A spring-loaded pressure plate fixed inside the back of the box keeps the film pressed flat against the picture aperture.

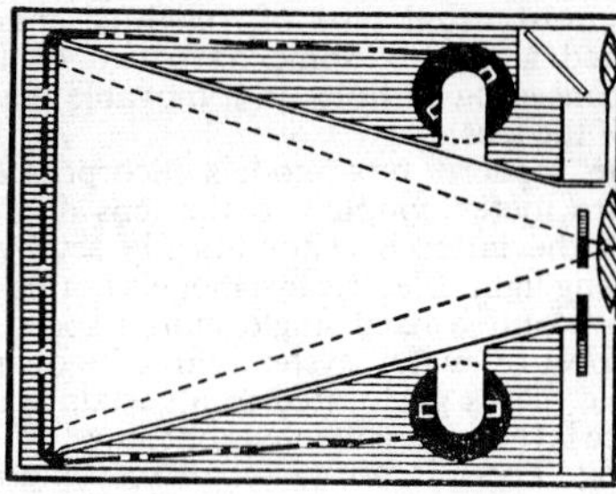

BOX CAMERA. Feed and take-up spools are on each side of a cone, pulling the film across the rear opening. The front contains the lens, shutter, and a viewfinder.

With some designs the film plane is curved, so that the film is deliberately not allowed to lie flat. This to some extent compensates for the curvature of field of the cheap lenses employed.

The film is transported by turning a key or knob on the outside of the box. There is a window—and sometimes two—located in the back of the camera through which the frame numbers on the paper backing of the film can be read when winding on from one frame to the next. This window has a red or green filter in it to stop active light from getting to the film.

Lens. The lens in the cheapest box cameras is made of a single meniscus-shaped piece of glass. It is not corrected for any of the aberrations and has a single fixed stop that confines the light rays to the centre of the lens and gives it an aperture between *f* 11 and *f* 16. Some of the better types have doublet lenses with a maximum aperture of about *f* 8.

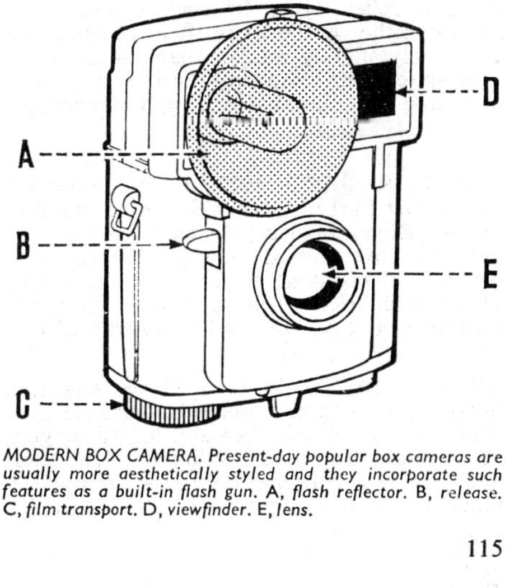

MODERN BOX CAMERA. Present-day popular box cameras are usually more aesthetically styled and they incorporate such features as a built-in flash gun. A, flash reflector. B, release. C, film transport. D, viewfinder. E, lens.

The most usual type of lens fitted has only one fixed stop, but some have two or three stops consisting of holes in a movable plate in front of the lens.

Some modern box models incorporate an exposure meter coupled to the lens aperture, so that the latter is automatically set by the prevailing light. Electric eye models of this kind still have only a fixed single shutter speed, but the exposure meter system directly operates movable plates perforated in a certain way to produce larger or smaller lens openings.

Focusing. There is no means of focusing on a typical box camera; the lens is set at a fixed distance from the film to give reasonably sharp negatives of everything beyond about 7 feet from the camera. More expensive types focus down to 3 feet. To do this the lens is made to slide or screw outwards in its mount.

Shutter. Box camera shutters are very simple affairs; they admit light to the film through a hole in a metal disc rotated by pressure on the shutter release lever.

Nowadays the shutter incorporates a set of contacts to synchronize the firing of flash bulbs with the full open position of the shutter. In some cases special flash guns are available which plug directly into the camera, without the need for a connecting cable between the flash and the shutter.

Some box cameras have the flash gun built in, with a miniature reflector in the front of the housing—above or to one side of the lens—taking miniature flash bulbs. A special compartment of the camera then takes the batteries for firing the flash. As an alternative the reflector may rise up from the top of the camera (and be collapsible) taking the smallest types of flash bulb. Box cameras with a built-in electronic flash unit have also been proposed.

Viewfinder. The viewfinder on the smallest box cameras is usually of the direct-vision type, either frame or optical and on larger sizes, a reflex type. In such cameras the picture in the viewfinder is often the same size as or only slightly smaller than the negative.

In spite of its simplicity the box is capable of taking photographs of a high standard.

See also: *Flash camera.*

BOYLE, ROBERT, 1627–91. Irish scientist, son of first Earl of Cork. One of the founders of the Royal Society; became President in 1680. Discoverer of the law (governing gases) which bears his name. Described the darkening of silver chloride when exposed to light, but attributed it to air. Devised a camera obscura.

BRADY, MATTHEW B., 1823–96. American portrait photographer; opened a daguerreotype studio in New York 1844 and in Washington 1847. Photographed many famous Americans and formed a National Photographic Portrait Gallery. In 1861 he organized the photographic documentation of the Civil War by teams of photographers in his employment, the most important of whom was Alexander Gardner. The undertaking impoverished Brady and he was obliged to sell his collection to the U.S. Government in 1875. Biography by James Horan (New York, 1955) and by Roy Meredith (New York, 1946).

BRAGG, SIR WILLIAM HENRY, 1862–1942. British physicist. With his son, Sir William Laurence Bragg (born 1890), received the Nobel Prize in 1915 for fundamental work in X-ray crystallography.

BRAUN, ADOLPHE, 1811–77. Alsatian textile designer, photographer and founder of the firm *Braun et Cie* in Dornach and Paris. Started as a daguerreotypist *c.* 1842 and became widely known in the 1850s for his photographs of flowers, and in the 1860s with his instantaneous street views of Paris and views of Switzerland. Introduced into France Swan's carbon process, particularly for the reproduction of old master paintings.

BREW. (Slang.) Developer, usually homemade, of an unorthodox type, or more jocularly of uncertain age.

BREWSTER, SIR DAVID, 1781–1868. Scottish scientist. Friend of Fox Talbot, introduced D. O. Hill to R. Adamson, 1843. Perfected the refracting stereoscope, 1849, which was commercially introduced two years later. With it began stereoscopic photography, in which at first the pictures were taken by moving the camera, or by two separate cameras. Brewster suggested the twin-lens stereo camera, 1849, the prototype of which was not constructed until 1853.

BREYER, ALBRECHT, 1812–76. German physician, born in Berlin but living in Belgium. Invented in 1839 reflectography (Breyertypes), i.e., reflex copying by exposing a silver chloride paper through the back of the printed page, without a camera. His method was later perfected in the Playertype and Manul processes.

BRIEF EXPOSURE (B). Term for a time exposure in which the shutter remains open only while the shutter release is held down. Sometimes also called a bulb exposure.

BRIGHT-LINE FINDER. Optical viewfinder which shows a white frame within the field of view, outlining the actual view taken in by the camera lens. The frame may incorporate parallax correction marks or may even be coupled with the focusing movement to compensate automatically for parallax. On cameras with interchangeable lenses the finder sometimes shows several bright frames alternatively or simultaneously to indicate the fields of view of different focal lengths.

See also: *Viewfinder.*

BRIGHTNESS RANGE. Term indicating the ratio of the brilliance of the extreme highlights to the darkest shadows of a scene or image.

See also: *Contrast; Contrast control; Exposure; Gamma.*

BRILLIANCE. Another name for luminosity, i.e., the intensity of light reflected from a surface (= incident light intensity × reflectivity of surface).

BRILLIANT FINDER. Reflex viewfinder fitted to many cheap cameras. It has a plano-convex lens instead of a ground glass viewing screen.

See also: *Viewfinder.*

BRITAIN

Photography as we know it today represents a technological triumph whose foundations were laid by many men working in many lands, all driven by the same compelling purpose, with minds actively searching for ways and means of achieving that common aim and each quick to seize upon new ideas and new discoveries that might bear upon the problem. This is the British part of the story.

Sensitized Materials. In 1802 there was presented at the Royal Institution of Great Britain a paper entitled, *An Account of a Method of Copying Paintings upon Glass and of Making Profiles by the Agency of Light upon Nitrate of Silver.* Its authors were Thomas Wedgwood, son of the famous Staffordshire potter, and Sir Humphry Davy, and in the work described they were following up the earlier experiments of J. H. Schulze who, 75 years earlier, had produced simple patterns on chalk, made light-sensitive with silver nitrate, and contained in a bottle. As far as we know, Wedgwood was the first to try to make pictures on a light-sensitive surface. The process had very limited utility because of the impermanence of the record—no means were known by which the image could be fixed—but it was the acorn from which the oak tree grew.

Wedgwood had little opportunity to take his experiments further; he died in 1805. Davy turned his attention to other matters and for a time no further progress was made in this country.

In France and elsewhere the daguerreotype did an enormous amount to stimulate public interest in these matters, but the next step in the straight line development of photographic techniques was again made by an Englishman, William Henry Fox Talbot, of Lacock Abbey in Wiltshire. Much has been written about the various dates on which Daguerre and Fox Talbot made their discoveries and published them, but the significant fact is that the daguerreotype, which made all the running at the beginning, was a branch development which by its nature could not fail to be superseded and eventually entirely eclipsed.

Fox Talbot was a country gentleman with exceptionally wide interests and a mathematician and botanist of great ability. While on holiday on Lake Como he attempted to make drawings with the help of the camera lucida and began to reconsider the use of the camera obscura which he had previously used for a like purpose. He wondered how he could make camera images permanent. On his return to England he began to experiment with paper treated with silver nitrate and found that paper treated first with sodium chloride and then bathed in silver nitrate was more sensitive to light. To render his results less fugitive, he "fixed" them in a strong solution of common salt or potassium iodide. It was not a completely satisfactory method but it enabled him to make records which could be kept for examination at a later date. At first, like Wedgwood and Davy, he was dealing with shadowgraphs in which lights and darks were reversed. But from such a print, waxed to make the paper transparent, he was able to produce another with lights and darks reversed again, which reproduced the original subject exactly. So the negative-positive system of photography came into being.

Wedgwood had tried without success to make camera pictures. Fox Talbot began his research in 1834 and in 1835 he succeeded in making a picture in the camera obscura—the famous *Latticed Window* of Lacock. In 1839 he read a paper to the Royal Society in London entitled *Some Account of the Art of Photogenic Drawing or the Process by which Natural Objects may be made to Delineate themselves without the Aid of the Artist's Pencil.* This paper appeared in the *Proceedings of the Royal Society* 1839.

In 1841 Fox Talbot patented the calotype process which employed development of a latent image based on the developing action of gallic acid, and he followed up with another patent covering fixation with sodium thiosulphate. But there were other inventors in the field at the same time. Gallic acid had been used by the Rev. J. B. Reade, a minister of the Church of England, as an accelerator before and during the exposure and he claimed also to have been the first (in 1837) to use sodium thiosulphate to fix the image. The fact that it was a good solvent of silver salts was discovered by Sir John Herschel in 1819 and he is generally given the credit for the introduction of hypo as a fixing agent. He also suggested the use of the terms negative and positive in connexion with the new process.

In 1844 Fox Talbot began to publish *The Pencil of Nature*, a work in six parts containing twenty-four calotypes. This was the first book to be illustrated with actual photographs. Here, then, was the practical process employing a camera exposure of reasonably short duration, with development and fixation to give a stable negative image from which any number of positive copies could be made.

It needed only the close attention of other minds for improvement to follow improvement and for photography to become the practical process which we know today.

Paper is not a very good support for the negative material. It was clear that a transparent base such as glass would be better, but there had to be a medium in which the silver salts could be held and applied. Albumen had been used by Niepce de St. Victor, and collodion had been suggested by Gustave le Gray in Paris, but the practical realization of the collodion process must be credited to another Englishman, Frederick Scott Archer of Bishops Stortford, who published his method in 1851. Here was a process capable of excellent results, but cumbersome in that the sensitization had to be done on the spot with development following immediately after exposure. This was known as the wet collodion process and it survives in a limited fashion to this day. Collodion dry plates were manufactured and sold at home and abroad by Dr. Hill Norris of Birmingham in 1856, but they were very slow in speed.

Dr. R. L. Maddox made the next great contribution. He used gelatin as the medium for the sensitive salts and described his experiments in a letter to the *British Journal of Photography* in 1871. Following improvement by others the gelatin dry plates were faster than the wet plates and after exposure there was no need for immediate development. The latent image was reasonably stable. From that date progress has been rapid and continuous.

Silver salts are naturally sensitive mainly to radiations at the blue end of the spectrum, so the early photographic emulsions were less sensitive to yellow and red light. Then, in 1873, Herman Vogel, Professor from Berlin, discovered that a box of plates manufactured in England and containing a dyestuff to suppress halation had a band of sensitivity wider than the normal. It was some time before his observation was confirmed by E. Becquerel in 1874 and J. Waterhouse in 1875 and even longer before sensitizers were found which did not adversely affect the emulsion. Pioneer work in England was done by C. E. K. Mees and then Sir William Pope and Professor W. H. Mills in conjunction with F. F. Renwick during the first Word War when it was necessary to make good the loss of German supplies; later work by other British investigators resulted in the description of the merocyanines and in the general elucidation of the essential structure of all colour sensitizers. This was followed by new work on fog restrainers, stabilizers and developers—work which has recently culminated in the discovery of Phenidone—phenylpyrazolidone.

In the field of colour, Britain can claim the fundamental conception of the three-colour theory formulated by Clerk Maxwell, a Scot, and the first demonstration of successful additive projection by him at the Royal Institution in 1861 when he reproduced, even though somewhat imperfectly, the colours of a piece of tartan ribbon.

A great deal of work was done in Britain on the screen plate process by J. Joly, C. L. Finlay and others, leading to the use of the integral screen process for still and cine work.

In subtractive colour processes, three-colour carbro was developed by the Autotype Company and as a practical factory process under complete control in England by H. D. Murray, P. C. Bull and D. A. Spencer. One of the integral tripack processes now available originated and was developed in Great Britain.

The chemistry and physics of the processes of photographic emulsion-making are complicated and the principles have been elucidated only very slowly. In the same way the laws relating amount of exposure to the depth of image developed had to be discovered by patient experiment. The popular idea was that by varying the development, or by intensifying or reducing the image, tone relationships could be varied more or less at will. This conception did not satisfy Dr. Ferdinand Hurter, a Swiss chemist working in England. He had been induced to take up photography by his friend, Vero C. Driffield, an English engineer. They began to examine the problem of photographic sensitivity together, and the work became the classic example of brilliant research in the photographic field. On the 31st May, 1890, they read their paper, *Photochemical Investigations and a New Method of Determination of the Sensitiveness of Photographic Plates*, before the local section of the Society of Chemical Industry at Liverpool. After preliminary criticism this work was accepted and the H. & D. curve established as the guide to the behaviour of the photographic emulsion.

The Support. Glass was a fairly adequate support, so long as the photographic material was required in flat sheets, and its rigidity makes it still the most suitable base for a number of important purposes today. Where flexibility was called for, it failed and was supplanted by cellulose nitrate. This material was first produced in Birmingham by Parkes, and it later made practicable the provision of sensitive material in roll form—although Eastman's first roll film was in fact a gelatin emulsion on a paper base which was afterwards stripped.

In the years 1888–90 William Friese-Greene, a fashionable London studio photographer, was using emulsion coated on strips of celluloid in his moving picture camera for which he took out the famous master patent in 1889; today, emulsion on film support has ousted glass from most fields other than graphic arts.

Celluloid held the field for many years as the universal support for flexible film, but its inflammability led to continuous work on slow-burning or "safety" supports and by now it has been almost entirely displaced.

For many years Britain lagged behind in the manufacture of film base, but just before the second World War a start was made by two British firms working in conjunction, which led to the production in this country of first-class film base in quantities commensurate with the requirements of the home market.

Special Processes. Side by side with the main photographic process there developed other methods of making pictures by light, many of them being evolved or improved in Britain.

In 1839 Mungo Ponton of Edinburgh made a light-sensitive bichromate paper. Fox Talbot knew of the work and realized that it was the size in the paper which was affected by the bichromate. By 1852 he was using bichromated gelatin and albumen in important pioneer work in photo-engraving.

In 1842 Sir John Herschel introduced the blue print or ferro-prussiate process which depended on the reduction of ferric salts to the ferrous state.

In 1864 Sir Joseph Wilson Swan patented the first practical method of pigment printing. He coated glass with collodion and then with pigmented gelatin. After drying he stripped off the film from the glass. Later he coated his pigmented layer direct on paper. After sensitization with potassium bichromate and exposure the surface was coated with indiarubber solution and then the unhardened gelatin and the original paper support were washed away. Another transfer was then employed to bring the image the right way round. Improvements were made by J. R. Johnson in 1868 and J. R. Sawyer in 1874.

Carbro, in which the hardening is brought about by means of a chemical reaction between the pigment paper and a bromide print, was a development by H. F. Farmer of a technique introduced by T. Manly in 1905, following up his Ozotype paper of 1899. The development of the three-colour carbro process has already been mentioned.

In 1873 William Willis invented platinotype, a rapid printing-out material containing salts of iron and platinum in which ferric iron was reduced to the ferrous state by the action of light. It was developed by immersion in potassium oxalate when the ferrous salt brought about the reduction of the platinum salt to metal. In 1917 palladiotype, a variant, was introduced by the same company.

In 1891 C. F. Cross, E. J. Bevan and A. Green introduced the primuline process, the first of the diazo-dye printing processes.

Eadweard Muybridge, whose series of photographs of people and animals in motion constituted one of the first steps in the development of cinematography, worked in America but he himself was English. Other notable British pioneers in this field were Friese-Greene and R. W. Paul.

Lenses. The achromatic lens was invented by an Englishman, W. Chester Moor Hall, in 1733, who had a few telescopes made for his own pleasure but did not publish his discovery. John Dollond demonstrated an achromatic prism to the Royal Society in 1758 and applied it to the construction of the telescope. W. H. Wollaston's meniscus lens, devised in 1812, was used by Fox Talbot and by Daguerre. One of the most famous lens-making firms in England was founded in 1830 by Andrew Ross, an instrument maker skilled in optics and the making of microscopes and telescopes. His son, Thomas, in 1841 made a doublet of two cemented achromatic lenses, and Thomas Davidson of Edinburgh made a symmetrical doublet the same year, but the intrinsic merits of these lenses were not appreciated at the time. Andrew Ross claimed to be the first to make achromatic single photographic lenses of two glasses only and he made for John Kibble of Glasgow a monster lens which had a clear aperture of 13 ins. and a focal length of 6 feet. The camera to which it was attached was mounted on wheels and drawn by a horse and it took a picture 36 × 44 ins.

In 1857 Thomas Grubb patented an achromatic meniscus lens, and in 1858 Thomas Sutton a symmetrical separated triplet.

John Henry Dallmeyer had married the daughter of Andrew Ross and set up on his own account in 1839. In 1858 he made lenses for Thomas Skaife's Pistolgraph camera, and in 1860 produced a rapid lens of the Petzval type which became known as the "baby" lens because it was so suitable for subjects in that class. In 1866 Dallmeyer patented an aplanat. H. L. Aldis came to him as a mathematician, so these three companies, Ross, Dallmeyer and Aldis, have long established associations.

After the introduction of the new barium crown glasses of the Schott glass factory at Jena the firm of Ross went on to make a variety of famous lenses including the Ross-Goerz double anastigmat, the Homocentric, the Teleros and the Xpres. In 1891, T. R. Dallmeyer had taken out the original patent for a telephoto lens, and in 1899 the Adon lens was introduced. In 1919 they introduced the Dallon, and in 1920 the Pentac.

The firm of R. & J. Beck had produced an iris diaphragm for photographic lenses in 1882 and were actively engaged in photographic lens manufacture from then onwards. Beck also introduced the idea of using magnifiers in front of the lens with their Frena camera in 1894.

The biggest single English contribution to lens manufacture was undoubtedly the introduction by W. H. Dennis Taylor of the Cooke lens under patents taken out in 1893, 1895 and 1898 while working for Cooke of York. It was made and marketed by Taylor, Taylor & Hobson.

Cameras. In 1851 Richard Willats exhibited a portable collapsible camera at the Great Exhibition. It took pictures $8\frac{1}{2} \times 10\frac{1}{2}$ ins. and collapsed to a depth of 4 ins. Up to 1875 there was little enlarging, so plate sizes were large. Pockets must have been larger too, because in 1855 the *Photographic Journal* carried an advertisement for a pocket camera taking a plate $5\frac{1}{2} \times 7$ ins. and weighing 2 pounds 6 ounces.

Kinnear of Edinburgh introduced the bellows in 1857 and "Kinnear pattern" became a popular description.

Thomas Sutton introduced the first single lens reflex in 1861 and box cameras containing magazines of plates were advertised in England from 1880 onwards.

Melhuish and Spencer took out a patent for the first roll holder in 1854 and in 1888 Morgan and Kidds' roller slide in its largest form took twenty-four exposures 12×15 ins. and weighed 5 pounds 12 ounces.

The Frena camera of R. & J. Beck (1892) took up to forty cut films and in the last years of the century we had the various detective cameras including the Walking Stick and the Demon.

With the popularization of the celluloid roll film, development of simple cameras went rapidly ahead, Blair's "folding Kamaret" having been probably the first of the folding cameras which have been such a marked feature of the present century.

In recent years Britain has been noted mainly for the production of the simpler types of camera, in which it has a very large export trade, and for specialized types for the motion picture industry, for the services and for survey work.

British lenses have always been held in high repute, but there is still no large-scale manufacture of better quality shutters in this country. In this connexion it is worth recording that the focal plane shutter was invented by William England in 1861 and reintroduced by B. J. Edwards in 1879.

Industry and Research. In 1873 Chapman of Manchester and Burgess of Peckham started small-scale production of gelatin dry plates. By 1876 plates for the trade were being produced by the Liverpool Dry Plate Company and they were followed a few years later by Wratten & Wainwright and the Britannia Works Company, a private business founded by A. H. Harman who began to make plates in Ilford in 1879. In 1898 a public company was floated, known as the Britannia Works Company (1898) Limited, to take over the Harman enterprise and in 1900 the name of this company became Ilford Limited. It is interesting to note that in 1903 a proposal for the amalgamation of this company with Kodak was put forward and only narrowly defeated. The objectors held that it was important to maintain a native photographic manufacturing industry with control residing in the country and the plan was dropped. In 1912 Eastman absorbed Wratten & Wainwright. The work of Sir Ivor Philipps brought the amalgamation of many of the smaller manufacturing companies with Ilford to completion in 1930.

In the last thirty years Britain has changed from being the biggest importer of photographic materials to being one of the greatest exporters in the world. This success has been due first to continuity of effort. In Britain today there are companies with family names which were associated with the industry at the beginning. The name of Houghton goes back to 1834 when Mr. George Houghton entered into partnership with Antoine Claudet to sell glass in High Holborn. Claudet knew Daguerre, and Houghton and Claudet became for two years the sole licensees for the daguerreotype.

Secondly, the industry has had to face intense competition from abroad and that is stimulating. Thirdly, it has put an enormous amount of effort into photographic research and plant development.

In England, we tend to favour the amateur approach in scientific investigation, as in sport. Certainly in photography many important advances were made by men for whom photography was a pursuit of leisure hours. Fox Talbot himself was led to make his experiments because he wanted to draw and had no skill in draughtsmanship. Maddox was a doctor and J. B. Reade a minister. Sir John Herschel was a man of outstanding scientific ability with a flair for investigational work. Hurter and Driffield were technical men who interested themselves in photography as a hobby. Sir W. Abney was an instructor in chemistry to the Royal Engineers at Chatham.

In the manufacturing companies, research was at first very much an *ad hoc* affair and ingredients were added to emulsions simply to see what would happen. But when men like Sheppard, Mees and Renwick came along the position began to change. S. E. Sheppard and C. E. K. Mees were students at University College in 1900 under Sir William Ramsay, who suggested that they should repeat the work of Hurter and Driffield using improved apparatus. In 1903 the work gained for them the degree of B.Sc. by research. In 1903 they published joint papers on the theory of development in which the concept of gamma infinity first appeared. This collaboration went on until 1906 when Sheppard was awarded an 1851 Exhibition Scholarship and went to Marburg to work with Karl Schaum on sensitizing dyes. The association was resumed in 1913 when he rejoined Mees in the new Eastman Research Laboratory at Rochester, U.S.A., where he worked at first on gelatin. Later he turned his attention to the structure and properties of silver halide emulsions in gelatin and went on to demonstrate that minute amounts of a labile sulphur compound in gelatin are responsible for its sensitising properties.

Mees, who had joined Wratten and Wainwright of Croydon in 1906, was invited by Eastman in 1912 to go to Rochester to direct the new laboratory. To enable him to go, Eastman acquired the Wratten and Wainwright concern and the subsequent story of his career is the story of the Eastman Kodak Research Laboratory. In 1913 he was joined by Capstaff, another research man of English birth.

Frank Forster Renwick was active in the affairs of the Royal Photographic Society, being President (1927–9), and very largely responsible for the emergence of the Scientific and Technical Group in 1919 and for the birth of the Photographic Alliance. He was particularly interested in the work of the British Photographic Research Association, formed after the first World War and sponsored by the manufacturers and the Department of Scientific and Industrial Research to promote the idea of co-operative research into fundamental problems.

This Association did much useful work but came to an end after the amalgamation of a number of British companies to form the Ilford group. The Research Association Staff were largely absorbed into this group.

Renwick was deeply interested in the work of Hurter and Driffield and made many contributions to the problem of tone reproduction. An early piece of work was the development with Ferguson and Benson of a special photometer for the measurement of photographic densities. He collaborated with Sir William Pope at Cambridge on sensitizers during the first World War. In 1922 he went to America, but by 1925 he was back with his old company, where he began building up the Ilford Laboratories with B. V. Storr and Olaf Bloch.

The Kodak Laboratories at Harrow, backed by the resources of the Eastman organization, also played a worthy part on this side of the Atlantic in the great advance in fundamental knowledge and in its practical application to new products. In Britain especially the photographic process has played a great part in nuclear physics since this science was born in 1896, when it was observed that radiation from a uranium salt fogged a photographic plate. Photography was used by Rutherford for much of his classical research on the alpha particle, and by Moseley in 1913 for his work which established the concept of the atomic number.

For Aston's research on isotopes Ilford Limited produced special plates in which the silver halide crystals protrude from the gelatin medium. These are still employed to record radiation which has minimal power to penetrate matter.

British scientists and manufacturers have led the world in the applications of special photographic emulsions to research on fundamental nuclear particles. For his work on special emulsions Dr. Waller of Ilford Limited has received the Duddell medal of the Physical Society and the Longstreth medal of the Franklin Institute of Pennsylvania. Professor C. F. Powell of the University of Bristol received the Nobel Prize for the discovery of the π-meson in cosmic radiation using photographic emulsions exposed on mountains. Balloon flights carrying emulsion more than 20 miles above the earth resulted in the discovery of many other transitory particles produced in atomic collisions at great energies, and this work has been extended to research with the machines that produce particles with energies up to 25,000 million electron volts.

Photographic Societies. An important factor in the development of photography in Britain is the interest in photographic clubs and societies, with the Royal Photographic Society at their head. Founded in 1853 at a meeting of the Royal Society of Arts as the Photographic Society of Great Britain, its first President was Sir Charles Locke Eastlake, who was also President of the Royal Academy and a painter of very considerable ability. Its founder and first secretary was Roger Fenton, famous for his pictures of the Crimea. In 1894, when Abney was President, it became the Royal Photographic Society of Great Britain, and ever since it has presented to the world a balanced conception of photography in all its aspects—aesthetic, scientific and learned.

There are about 1,000 photographic Societies in Great Britain, and these together comprise the Photographic Alliance.

Publications. The English photographic press has been another big influence. The senior member is The Royal Photographic Society's *Photographic Journal*, dating from 1853. Then follows *The British Journal of Photography*, the lineal descendant of *The Liverpool Photographic Journal*, first published in 1854 and edited by a committee of that society. Three years afterwards it was acquired by Henry Greenwood and became *The Liverpool and Manchester Journal*. In 1859 it became *The Photographic Journal* and it acquired its present title in 1860. The R.P.S. also publishes the *Journal of Photographic Abstracts*. Other contemporary photographic magazines in Britain include the following: *Photography*, *Photoguide Magazine*, *Amateur Photographer*, the Photographic Dealers' Association's *Photographic Retailer*, The Institute of British Photographers' *Record*, *Good Photography*, and the *Amateur Cine World*.

Since the second World War, British books on photography have advanced into a leading position all over the world. They are read in larger numbers than books of any other origin and have been translated into more foreign languages than any other. British books have set patterns of approach to photographic subjects on almost every level which have stimulated foreign publishers far and wide.

British Photographers. The work of the early British photographers is still held in high regard the world over. The first of these pioneers was a Scot, David Octavius Hill, born at Perth in 1802. He studied painting at Edinburgh under Andrew Wilson. In 1843 he set out to portray on one canvas the 470 personages present at the General Assembly at which the Free Church of Scotland was founded, and at the suggestion of Sir David Brewster he made use of the calotype for his portrait studies. His portraits are still recognized as masterpieces, simple in treatment and straightforward in pose, full of quiet and noble dignity. It is widely believed that the technical control was in the hands of Robert Adamson of Burnside, Scotland, who was taught the calotype process by his brother, a colleague of Sir David Brewster, a friend of Fox Talbot.

The next was Roger Fenton, born in 1819, a founder and first secretary of what is now the Royal Photographic Society. He is mainly remembered for his work in photographing the Crimean War in 1855, some of his pictures being reproduced as wood engravings in the *Illustrated London News*. He was thus the first war photographer, although he was closely followed by others, notably Alexander Gardner of Paisley, Scotland, who went to America in 1856 with details of the wet plate process and became photographer to the army of the Potomac 1861–5.

The next landmark was the emergence of the composite picture which dominated the field for fourteen years. The development came about because of the impossibility with the collodion process of getting straight landscape negatives from which one could make prints with clouds showing in the skies. Then it came to be used as a means of making elaborate compositions with highly dramatic content. In England the phase began with the work of W. Lake Price and of O. G. Rejlander, a Swede living in this country. For one celebrated picture entitled *The Two Ways of Life* in the maner of a Renaissance painting, a composite picture 40 ins. in length which was exhibited at the Art Treasurer's Exhibition in Manchester in 1857, he used more than thirty negatives.

He was closely followed by Henry Peach Robinson, an artist who, like Fenton and Rejlander, had distinguished himself by having a painting hung in the Royal Academy. In 1858 he produced *Fading Away*, a composite picture made from five negatives portraying the death of a young girl. The technical quality of the picture was unchallenged, but its good taste was questioned and the controversy no doubt helped to make the name of the artist who became a dominating figure in photography and retained this position for over forty years. During this time he issued annually for subscribers a series of large prints made for the most part by combination printing, the publication of which is said to have marked the high point of the photographic year.

Mrs. Julia Margaret Cameron took up photography in the Isle of Wight in 1863, when she was 48. She knew many famous people and made them all sit for her. Her work is notable for the broad effects which gave character and greatness to her portraits.

With the coming of the dry plate the number of photographers greatly increased. Representative of this period is the work of Payne Jennings, B. Gay Wilkinson, J. B. B. Wellington and F. M. Sutcliffe.

In 1889 appeared a book, *Naturalistic Photography* by Dr. P. H. Emerson, which had a tremendous influence on the photographers of the day; it rejected handwork on prints and suggested that rules for composition were valueless.

The new approach to pictorialism was first apparent in the work of George Davison, whose picture *The Onion Field* has been described as the inspiration of the impressionist school of pictorial photography. Then came A. Horsley Hinton and Alexander Keighley who brought the technique of combination printing to perfection, perhaps with a little more artistry than was shown by some of their predecessors. Hinton's influence was enormous, not only because of his pictures but also because of his prolific writing on photographic matters. Over a period of thirty years Keighley exhibited hundreds of pictures of exceptionally high standard and in a romantic rather than a naturalistic vein. His early work was of the English scene, but he is remembered especially for pictures made in France and Italy.

With the formation in 1892 of the Linked Ring, pictorial photography in Great Britain arrived at its renaissance. Keighley, Horsley Hinton, H. P. Robinson and Davison were all concerned in its foundation, which came about because of the lack of encouragement which they found within the Photographic Society.

Around about this time too the work of Craig Annan began to be noticed. A townsman, he loved the country and had a strong sympathy for the Dutch school of painting. It was his practice to make straight negatives and to make his reproductions by photogravure. Also prominent at this time was Charles Job.

The next development concerned the introduction of the various printing processes involving control—e.g., gum-bichromate. Originating in France, these methods were taken up by Maskell, Mummery, Holcroft and Page Croft in this country. The platinum and gum processes followed and the multiple gum prints of Batkin of Birmingham became notable features of exhibitions from 1903 to 1911.

In 1910 one of the most important events was the formation of the London Salon of Photography, in consequence of the decline of the "Linked Ring", which was first founded in 1892. The London Salon of Photography annual International Exhibition has continued without interruption ever since, and today it is universally recognized as one of the most important displays in the world, attracting thousands of entries every year.

Developments in America were closely followed in Britain. Stieglitz and Holland Day between them had a great deal to do with the founding of American pictorial photography and the work of both was well known in this country. Holland Day actually organized in London an exhibition of the new American School and on this occasion introduced his cousin, Alvin Langdon Coburn, who later worked and exhibited in Britain.

The period of the gum print came to an end about 1911 although some workers, notably Harold Leighton, continued to exhibit in this medium.

The oil and bromoil methods came next and in this period mention must be made of the names of Bertram Cox, J. H. Gear and F. J. Mortimer, whose seascapes are still unsurpassed. Like Hinton, Mortimer's influence went beyond his pictures. He was Secretary of the London Salon, editor of *Photograms* (suspended in 1962) and of the *Amateur Photographer*. He was killed in the second World War. J.Mi.

Industry. The United Kingdom photographic industry has shown a gradual overall expansion in recent years, both in terms of numbers employed and the value of output, although the number of manufacturing establishments has followed a declining trend. This has been due almost entirely to a reduction among firms and their employees producing apparatus, whilst sensitized goods manufacturers have remained almost constant in number and increased their workpeople during the period.

The most recent figures available are those of the 1958 Census of Production, which recorded 47 photographic manufacturing establishments with more than 25 employees each, together providing employment for 15,718 workpeople in that year. These were divided into 19 establishments with 3,356 employees producing photographic equipment, and 28 establishments employing 12,362 workpeople engaged in the manufacture of sensitized materials.

The industry is centred in the London area, where the two leading companies, Kodak and Ilford, are located and which together account for the major part of production and employment.

The larger organization, Kodak Ltd., is a subsidiary of the Eastman-Kodak Company of the United States, and undertakes the manufacture of all types of sensitized film, plates, paper and photographic chemicals, in addition to still and cine cameras, projectors and accessories, and photocopy equipment. It operates factories at Harrow, Middlesex (sensitized goods), Stevenage, Herts (equipment), Kirkby, Lancs. (chemicals) and a film processing plant at Hemel Hempstead, Herts., employing in 1963 a total of 10,700 persons.

Ilford Ltd., now partly owned by Imperial Chemical Industries, is a major producer of sensitized materials, in particular X-ray, black-and-white and colour film, in addition to marketing cameras and projectors made under its name by the West German firm of Dacora and the Japanese Elmo company. It also markets photocopy equipment. The company has its headquarters at Ilford, Essex (sensitized goods, especially plates), a film production plant at Brentwood, Essex, a factory at Mobberly, Cheshire (silver sensitized papers), and a plant at Basildon, Essex (chemicals and photocopy papers in addition to the processing of film).

The total number of persons employed in 1963 was 4,500.

The third largest photographic company is Johnsons of Hendon, Ltd., manufacturers of chemicals, dark room equipment and accessories, and importing agents for several leading makes of cameras and projectors.

Among other manufacturers of photographic goods in the U.K. are Coronet Ltd. at Birmingham, producing the cheaper range of still cameras, Specto Ltd. at Windsor, Berks., active in the field of still and cine projectors, Gnome Ltd. at Cardiff, Wales, well known for enlarging and projection equipment, and Wray Optical Ltd. at Bromley, Kent, manufacturing still projectors, enlargers and lenses. Important producers of 16 mm. and 35 mm. cine projection equipment are Rank Precision Industries and Associated Electrical Industries.

The British Photographic Manufacturers' Association is the official trade body representing the common interests of the industry.

It is supported by the majority of producers and derives its funds from their annual subscriptions.

Production. The value of photographic goods production has increased by slightly more than one-quarter over the five years from 1959 to reach a total of £57·2 million in 1963. Of this, about 15 per cent was accounted for by equipment, among which still cameras formed an important product with 1,240,000 units worth £2·1 million during the year, the majority of which were box-type models.

Output of both cine cameras and projectors has recorded a fall in value since 1959 to respective totals of £0·5 million and £2·0 million in 1963, the larger part in each case being attributable to 16 mm. and smaller gauge equipment.

The remaining four-fifths of the total value of production is due to sensitized materials, which have shown substantial growth during the past five years to a total of £47·9 million in 1963.

The principal item is formed by the group comprising still film and plates, which includes X-ray film, and this has risen by 45 per cent in value in the period to £23·6 million in 1963. Sensitized paper and cloth together with cine film hold the shares of other sensitized goods produced, the former being worth £14·5 million and the latter £9·8 million during 1963, although the output value of paper and cloth has grown more rapidly since 1959 than that of cine film.

Imports. The value of apparatus and sensitized materials imported by the United Kingdom has shown an impressive increase during the past five years. In 1963, just over two-thirds of the £20·1 million total was composed of apparatus, among which still cameras formed an important item. Imports of these reached 937,000 units with a value of £3·6 million, of which 333,000 cameras worth £2·5 million were classified as 35 mm. or smaller format types.

The principal sources of still cameras in 1963 were West Germany, with 145,000 units valued at £1·2 million and Japan with 96,000 units worth £879,000; whilst Hong Kong supplied 564,000 cameras with a value of £328,000, the majority of which were box-type models.

Imports of cine cameras have followed a sharply rising trend since 1959 reaching 151,000 units worth £2·6 million in 1963. Nearly all of these are accounted for by 8 and 16 mm. models, supplied chiefly by Japan which has now gained the leading position among the sources with 66,600 units valued at £693,800, whilst West Germany was responsible for 18,300 cameras worth £327,000. Major items among imports of other equipment are still and cine projection apparatus, which together totalled 221,400 units with a value of £3·3 million during the year.

The value of sensitized materials imported is fairly evenly divided among the various types of product. The largest single item in 1963 was photographic paper and cloth, totalling £2·6 million, of which two-thirds in value was supplied by Belgium. Imports of cine film are

UNITED KINGDOM PRODUCTION OF PHOTOGRAPHIC GOODS

£000's

(Manufacturers' Net Selling Price ex-works)

	1959	1960	1961	1962	1963
Still cameras	1,540	1,969	1,905	1,632	2,078
Cine cameras	972	938	1,106	709	507
Still projectors	459	472	868	1,023	903
Cine projectors	2,110	1,973	1,703	1,773	2,015
Other equipment and parts	2,312	2,667	2,936	2,457	3,737
Total apparatus	7,393	8,019	8,518	7,594	9,240
Sensitized paper and cloth	5,330	6,197	6,413	6,384	14,503
Still film and plates	18,188	20,336	21,698	23,770	23,597
Cine film	6,729	6,659	6,704	6,497	9,840
Total sensitized material	30,247	33,192	34,815	36,651	47,940
Total production	37,640	41,211	43,333	44,245	57,180

UNITED KINGDOM IMPORTS OF PHOTOGRAPHIC GOODS

£000's c.i.f.

	1959	1960	1961	1962	1963
Still cameras	2,203	3,009	2,766	2,606	3,574
Cine cameras	1,079	1,419	1,622	2,261	2,594
Still projectors } Cine projectors }	1,068	1,401	2,120	2,423	1,261 2,068
Other equipment and parts	2,389	3,663	4,899	4,658	4,308
Total apparatus	6,739	9,492	11,407	11,948	13,805
Sensitized paper and cloth	958	1,472	1,722	1,965	2,256
Roll film and packs	381	489	598	589	894
Sheet film and plates	732	1,059	1,282	1,379	1,311
Cine film	1,018	1,443	1,884	1,519	1,567
Chemicals	123	145	180	254	267
Total sensitized material and chemicals	3,212	4,608	5,666	5,706	6,295
Total imports	9,951	14,100	17,073	17,654	20,100

UNITED KINGDOM EXPORTS OF PHOTOGRAPHIC GOODS

£000's f.o.b.

	1959	1960	1961	1962	1963
Still cameras	472	412	684	592	999
Cine cameras	882	625	945	1,034	1,062
Cine projectors	768	866	780	623	575
Other equipment and parts	2,674	3,198	4,504	7,565	12,136*
Total apparatus	4,796	5,101	6,913	9,814	14,772
Sensitized paper and cloth	1,747	2,169	2,227	2,299	2,590
X-ray film	2,862	3,277	3,167	3,538	3,817
Roll films, film packs and plates ...	2,173	2,316	2,396	2,570	2,674
Other still film	839	1,063	1,175	1,168	1,106
Cine film	2,643	3,075	3,232	3,191	3,502
Chemicals	722	859	885	1,095	1,056
Total sensitized materials and chemicals	10,986	12,759	13,082	13,861	14,745
Total exports	15,782	17,860	19,995	23,675	29,517*

* (Owing to changes in Customs classifications in 1963 these figures may not be strictly comparable with those for earlier years.)

UNITED KINGDOM EXPORT MARKETS FOR X-RAY FILM AND CINE FILM, 1963

Country	*X-ray Film* Million Sq. Feet	*X-ray Film* £000's f.o.b.	*Cine Film* Million Linear Feet	*Cine Film* £000's f.o.b.
European Common Market:				
Belgium	0·6	57·7	0·9	12·6
France	0·8	64·6	3·6	35·9
West Germany	0·1	14·8	35·8	375·4
Italy	0·5	45·9	1·7	29·2
Netherlands	0·5	56·8	5·2	65·5
E.C.M. total	2·5	239·8	47·2	518·6
European Free Trade Area:				
Austria	0·6	54·2	5·5	72·5
Denmark	2·1	238·1	21·4	304·0
Finland	2·0	196·6	3·2	31·7
Norway	1·5	153·8	8·0	123·4
Portugal	0·4	39·6	1·3	16·6
Sweden	2·6	298·5	38·1	478·8
Switzerland	0·6	71·7	3·5	36·5
E.F.T.A. total	9·8	1,052·8	81·0	1,063·1
Australia	1·8	146·5	30·5	171·6
Canada	5·4	576·8	0·5	11·0
Hong Kong	0·4	37·5	4·9	37·2
New Zealand	0·6	59·4	2·7	30·9
Union of South Africa	1·0	94·3	3·2	57·8
U.S.A.	11·7	864·5	1·9	18·4
All other countries	7·1	745·1	125·7	1,593·0
Total exports	40·3	3,816·7	297·6	3,501·6

also of importance, reaching 168·7 million linear feet worth £1·6 million in 1963, among which 35 mm. stock accounted for 127·1 million linear feet with a value of £0·9 million, supplied chiefly by Belgium (73·3 million linear feet worth £0·4 million) and the U.S.A. (33·1 million linear feet worth £0·4 million).

Exports. The composition of exports of photographic goods reflects the pattern of production, with sensitized materials accounting for about a half of the £29·5 million total in 1963.

Apparatus exported has trebled in value during the past five years, among which cine cameras and cine projectors hold a leading place, the former reaching 6,600 units with a value of £1·1 million and the latter totalling 27,300 projectors worth £575,000 in 1963. The principal destinations of cine cameras were Italy (2,350 units valued at £110,000) and France (450 units worth £188,000). Among cine projectors, 16 mm. and smaller gauge models totalled 9,650 units worth £498,000, of which Australia received 1,500 projectors valued at £85,600 and Italy took 780 units with a value of £25,000.

Exports of still cameras in the same year

reached 132,300 units with a value of £999,000 the box type accounting for 117,400 units worth £124,000. Markets for the latter were widely spread, among the most important being France (12,200 cameras worth £12,500) and the Irish Republic (12,400 units valued at £11,600).

The value of sensitized materials has expanded by over one-third since 1959 to a total of £14·7 million in 1963. The most important single item, X-ray film, reached 35·8 million square feet worth £3·8 million, the principal markets being the U.S.A. and Canada, whilst the E.F.T.A. countries were also large customers.

Cine film represents the other most important product exported, with a total of 297·6 million linear feet valued at £3·5 million in 1963, approximately half of which in both quantity and value is accounted for by 35 mm. stock. Major purchasers of this are to be found among the E.F.T.A. countries, in particular Denmark and Sweden, whilst West Germany also receives a considerable amount. G.R.H. & A.G.B.

BRITISH BOARD OF FILM CENSORS. Statutory powers of film censorship in Great Britain are vested in the local authorities, who normally rely on the British Board of Film Censors to supervise the character of the films exhibited. The Board is an independent body, set up in 1912 on the initiative of the cinema industry. It views all films intended for public exhibition except newsreels. Three certificates are issued to approved films: "U"—"suitable for general exhibition"; "A"—"more suitable for exhibition to adult audiences"; "X"—"suitable for adults only". Children under 16 are not admitted to "A" films unless accompanied by an adult. They are excluded from performances of "X" films.

BRITISH FILM INSTITUTE. The British Film Institute was founded in 1933. It is governed by a board whose members are appointed by the Chancellor of the Exchequer, and part of its revenue is derived from a government grant; its function, as defined by the Redcliffe Committee in 1948, is "to encourage the development of the art of the film, to promote its use as a record of contemporary life and manners, and to foster public appreciation and study of it from those points of view." This purpose has recently been extended to include television.

One of the Institute's most important functions is to maintain the National Film Archive, in which are preserved films selected from current production for their special artistic, historical, social or scientific importance. The National Film Archive ranks as one of the major collections in the world and the films preserved under scientifically-controlled conditions (at Aston Clinton, Buckinghamshire) are valued at over a million pounds. Copies of some Archive films are available for hire from the Institute's own film library, which specializes in films on art and film classics, especially those of the silent era.

The National Film Theatre, opened in 1952, provides an admirable exhibition space for this and other material. It was constructed as the Telekinema for the Festival of Britain, 1951, and is specially equipped to show silent films at their proper speeds, as well as large-screen television and stereoscopic films. Its programmes present a steady repertory of acknowledged film masterpieces, as well as series devoted to the work of outstanding actors and directors, or to significant trends in the cinema. Every October the London Film Festival shows the best films of the year from all over the world. Associate membership of the Institute carries the right to receive the illustrated theatre programme and buy tickets.

With the object of raising the standards of the intelligent filmgoer, the Institute organizes conferences and publishes pamphlets, and each year has university extension courses in London and a Summer School in the country. It runs a lecture agency, arranging about 1,000 lectures a year.

The Institute has also pioneered in introducing film appreciation and practical film-making into school curricula, and given support to the Film Society Movement.

For full members, the Institute provides a library of books on film and television with loan and reference sections and an information and research service. Copies of rare stills may be bought from the stills library, the second largest collection in the world. Full members receive the Institute's quarterly film journal *Sight and Sound*, and concessionary rates for *The Monthly Film Bulletin*, containing reviews of all feature films, and for *Contrast*, a quarterly television journal.

The Institute has currently approximately 23,000 associate members and 10,000 full members. The headquarters and offices are in London.

BRITISH KINEMATOGRAPH SOCIETY. The British Kinematograph Society was founded in 1931 to encourage and further the scientific and technical aspects of kinematography and its allied arts and sciences.

There are five grades of membership, namely, fellows, corporate members, associate members, associates and students.

Membership is open to those gainfully occupied in the kinematograph industry or an allied or contributory industry, to those who, with the intention of entering the industry, are undergoing an approved course or serving under an approved apprenticeship scheme, and to those who, although not gainfully occupied in the industry, can establish the satisfactory use of kinematographic processes for professional, scientific, educational or artistic purposes.

Within the Society four Divisions operate, representing the theatre, film production, industrial and specialized films and television. Each Division has an elected committee.

This organization affords means through its various channels for disseminating information on film production and exhibition, including television production as related to film and large screen reproduction.

The Society has representatives in India, Italy (both Rome and Milan), Denmark, South Africa, Australia, Holland, Sweden, Argentina, Japan and Brazil, and through these agencies promotes the interchange of technical information. The Society works in close liaison with kindred societies in the U.S.A.

The affairs of the Society are administered by a Council of which a proportion is elected annually. Five Officers, of which the President is chief, are vested with executive powers. In addition to the Divisional Committees. Theatre, Film Production, Industrial and Specialized Films and Television, business committees are appointed annually to be responsible for the work connected with Publications, Education, International Relations, Papers, Fellowship, Membership and Branches.

The Society and its Divisions hold monthly meetings in the winter season and organize part-time courses for laboratory technicians and theatre engineers. *British Kinematography*, the Society's journal, is published monthly.

BRITISH PHOTOGRAPHIC MANUFACTURERS ASSOCIATION. National association of manufacturers of photographic apparatus and materials which was incorporated in 1916 as a company limited by guarantee and not having a share capital.

The primary objects of the association are to organize, promote and protect the rights and interests of members of the association and to foster the development of the British photographic trade generally; it also facilitates the interchange by members of the association of their views on matters of common interest and promotes the consideration and discussion of all questions affecting the industry.

The association is regarded as the representative organization for British photographic manufacturers. It maintains close contact with government departments and is consulted on matters of general interest to the industry.

Contact is also maintained with kindred trade associations with a view to consultation and joint action wherever it is in their common interest.

The affairs of the association are conducted by a council, the members of which serve in an honorary capacity and are elected annually by the general body of members.

BROKEN NEGATIVE. If the glass of a negative is cracked but the emulsion film is undamaged, the trouble can be remedied by transferring the emulsion to another support.

See also: *Faults; Stripping*.

BROLOID PROCESS. One of the many variants of the basic carbro technique for building up a pigment image on a bromide print.

BROMETCHING. One method of preparing a bromide print so that the image assumes the texture of the support and looks something like an etching. The process involves over-exposing and then over-developing the print.

After washing, the print is exposed to the action of the etching solution (equal parts of 25 per cent sodium chloride solution, 5 per cent potassium permanganate solution and 4 per cent sulphuric acid solution in 20 to 30 times the volume of water). Fresh batches of solution are applied until the highlights are clear, and the print is then washed well, fixed in fresh acid fixer and then given a final wash. The result is a bold image with partcularly strong, pure blacks.

See also: *Sketch photograph*.

BROMIDE. Salt of hydrobromic acid. The principal bromides used in photography are silver bromide, ammonium bromide, potassium bromide.

BROMIDE PAPER. Most popular type of development paper for enlarging. The sensitive salt with which the paper is coated is silver bromide usually with a small silver iodide content. Invented and first produced in 1873 by Peter Mawdsley.

See also: *Papers*.

BROMIDE PENCIL. Special pencils once much used for working on bromide prints. They are made of crayon instead of graphite so that they will leave a matt finish to the work. The print must be either steamed or sprayed with artist's fixatif after applying the pencil to prevent it from rubbing off. They are manufactured in blue-black and brown-black tones. Nowadays most workers find retouching dyes easier to use.

See also: *Retouching*.

BROMOIL PROCESS. The bromoil process has largely replaced the former oil-pigment process because it can be applied equally to contact prints or enlargements. The print is made on the type of paper sold for bromoil work. It is then bleached—an operation which removes the black silver image but leaves it in a condition in which it will take up oil pig-

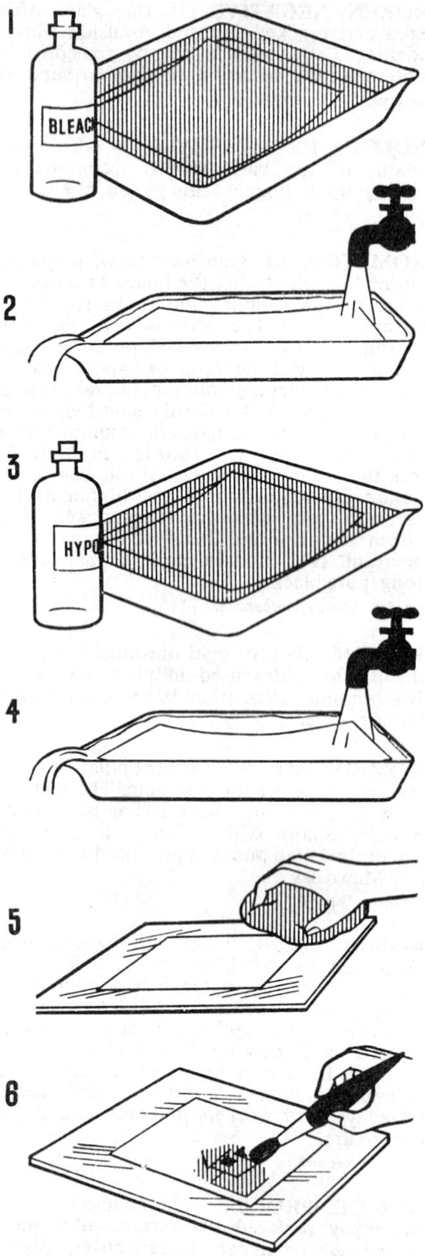

BROMOIL PROCEDURE. 1. Bleach bromide print (should be made on paper without supercoating). 2. Wash until all stain is discharged. 3. Fix in a plain fixing bath. 4. Wash thoroughly. 5. Place wet bleached print on a glass plate and blot off excess moisture with a damp cloth or cotton wool. ***6. Ink up*** *with special bromoil brush and suitable greasy ink.*

ment applied with a brush. So the final image is formed by oil pigment of any colour the photographer chooses.

Bleaching the Print. Prints for bromoil must be thoroughly fixed and washed. They should be dried and soaked in tepid water between 75° F. and 80° F., from 20 to 40 minutes according to the instructions issued by the maker of the paper.

There are several bleaching solutions. The following is satisfactory:—

Stock solution A		
Copper chloride	1½ ounces	37·5 grams
Sodium chloride	12 ounces	300 grams
Hydrochloric acid	10 minims	0·5 c.cm.
Water	40 ounces	1,000 c.cm.
Stock solution B		
Potassium bichromate	265 grains	15 grams
Water	40 ounces	1,000 c.cm.

The print is immersed in a bath consisting of equal parts of A and B and 2 parts of water. The solution is kept at a temperature of 65° F. to 75° F. When bleaching is complete, the print is washed until the yellow bichromate stain disappears, and it is then fixed in a plain hypo fixing bath (1 ounce hypo to 20 ounces water). It should be washed for about 20 minutes. Pigmenting may follow at once or the print may be dried and put to one side for pigmenting later.

If the print has been dried, it must be soaked in water at a temperature of 75° F. for half an hour before pigmenting.

Pigmenting. The wet bleached print is laid on a sheet of plate glass or similar smooth surface and the surface moisture is gently blotted off.

The print is then "inked up" with one of the special greasy inks supplied for the process. The condition of the ink is important. If it is too hard it will not adhere to the print; if too soft it will adhere equally to highlights and shadows. Special thinning medium is available for bringing the ink to the required consistency.

The ink is applied with a special flat bristle bromoil brush held vertically and dabbed gently over the surface. The art of bromoil is in the brush action—sometimes called hopping—and it must be acquired before the worker can achieve real control of the tones of the finished picture.

The ideal method of working is for the gelatin to absorb the pigment gradually so that the image is gradually built up. Some parts of the print may be made darker than the original bromide and in others, action with a dry brush will remove the pigment and lighten the tone.

Bromoil Transfer. By an extension of the process an impression of the inked image can be transferred by pressure to another base. Bromoil transfer, as it is termed, needs expert manipulation and a specially constructed press. The results have a delicate and recognizable character of their own.

The bromoil print is made as before described. Success will depend upon the choice of suitable inks and the paper upon which the work is done. The special papers made for the bromoil process should be used.

Generally a hard ink will transfer more easily and without the risk of spoiling the outlines of the image. The print for transfer should be vigorous, and strongly inked if it is to transfer without loss of quality. Any control necessary is best done on the bromoil print before transfer.

For transfer a specially designed press is used. The selected paper must have a slightly rough surface and be uncoated. The press generally consists of pressure rollers. It is important that the pressure be even and regular. If it is too heavy, the transfer is unsharp; if too light the transfer will not take place. The old type office letter press has been used, as well as the roller type domestic ringer, but it will be necessary to experiment with such improvised equipment before attempting serious work.

The transfer must take place while the print is still moist, but not too wet. The bromoil print and the paper, which should be larger than the print, are laid in contact. It is best to lower the transfer paper on to the print from one side as this avoids air being trapped between the two. Care must be taken not to move the two once they are together as the result would be a blurred image.

The two papers are then passed through the press. The bromoil is gently pulled away from the transfer paper and if all has gone well the image will have been transferred.

The image will be reversed left for right after the transfer. If this is a disadvantage it may be corrected by making the exposure through the back of the negative. This is quite easy when enlarging and should always be done when a transfer may be required. R.M.F.

See also: *Control processes; Pigment processes.*
Book: *Bromoil and Transfer*, by C. J. Symes (London).

BROOM, CHRISTINA, 1863–1939. English photographer, probably the first woman press photographer. Began free-lance work at the age of forty and published her photographs as postcards and sometimes also in newspapers. Photographed suffragette and other political meetings and royal occasions.

BROWN, GEORGE EDWARD, 1872–1934. English chemist, writer and journalist. Editor of *The British Journal of Photography* and the *B.J. Almanac* from 1906–34.

BROWNIE. Trade name of the first of the simple, mass-produced box cameras—often indiscriminately applied in describing any make of simple camera. It was first manufactured by Kodak in 1900 and helped spread of snapshot photography among the general public.

BRUGUIERE, FRANCIS, 1880–1945. American photographer born in France. Started photographic career in San Francisco, of which he published a book of his photographs in 1918. Believing that current practice of imitating painting in photography was futile, he photographed abstract constructions, using the camera to record light effects on them. In 1927, he held an exhibition of his abstract photographs at The Art Centre in New York, followed by exhibitions in London and Berlin. In 1937, produced the photo-murals for the British Pavilion at the Paris Exhibition.

BRUSH DEVELOPMENT. Development carried out by swabbing the solution on to the exposed material with a brush or mop—usually applied to prints too large to fit the available dishes.

BRUSHES. The most suitable brushes for colouring, retouching and spotting are of the type used by artists.

Apart from artist's brushes, photographers occasionally need convenient tools for spreading and coating solutions on to various supports.

Retouching Brushes. The best brushes for spotting and touching out small blemishes are those manufactured for miniature painting. These are known as "red sable" brushes and are available in a range of sizes. The lower the

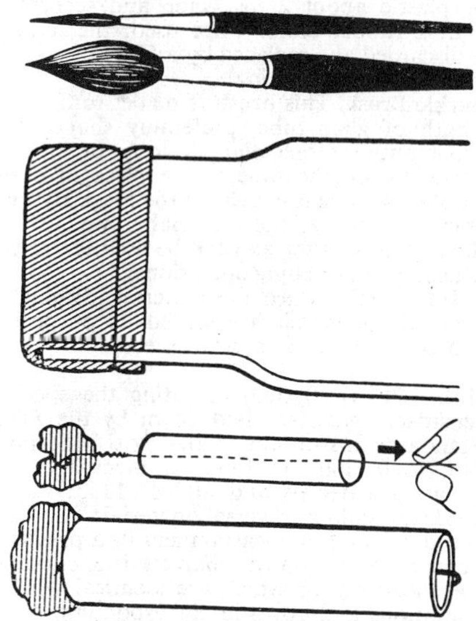

BRUSH TYPES. Top: *Retouching brushes. Usually these are very fine, but a thick brush can easily come to a point, too.* Centre: *Blanchard brush, consisting of a piece of fluffless material tied over the end of a piece of glass or plastic.* Bottom: *Buckle brush made by pulling a ball of cotton wool into the end of a glass tube. A loop of wire holds the cotton wool.*

number stamped on the shank, the finer the brush. For spotting negatives a No. 00 is satisfactory while the slightly larger No. 0 is best for enlargements. For painting out small areas and outlining on the negative before blocking out, a No. 1 water colour sable is the right size, while laying broad washes calls for a No. 3, 4 or 5.

When the point of a brush is marred by projecting hairs, the brush should be moistened, drawn to a point, and then lightly passed over the flame of a match. It is worth noting that the smallest brush does not necessarily give the finest point.

Colour Brushes. Good quality artist's water colour brushes are the best for colouring prints and slides. The No. 1 size is suitable for colouring details, No. 3 for normal work on the print and for really broad washes, a No. 5 or larger will save time and give a more even result. Where a number of colours are being used on the same print, it is more economical to buy a set of quills, one for each colour, and use them as required on the end of a common shank.

The tip of a good quality brush should spring straight again immediately after being bent by pressure on a flat surface.

Blanchard Brush. The name for a spreader used for coating surfaces with a fluid medium, e.g., for applying the sensitizer to the paper base in the gum bichromate process.

The brush is made by folding a piece of fluffless material over the end of a piece of glass or plastic about 2 ins. wide and securing it with a rubber band. Once used, the material is discarded and replaced by a fresh piece when the brush is used again.

Buckle Brush. This brush is made from a short length of glass tube, preferably shaped to a bell-mouth at one end. A loop of string is passed down the tube and a piece of cotton wool or sponge is caught in the loop. By tightening the string, the material is wedged into the end of the tube and can be used for various brushing or swabbing operations.

Here again, when the material forming the head of the brush has served its purpose, it can be replaced by a clean piece.

B.S. SPEED. System of rating the speed of sensitized materials, laid down by the British Standards Institution in the British Standard No. 1380: Part 1: 1962 for black-and-white negative materials and in No. 1380: Part 2: 1963 for colour reversal materials. The B.S. speed is based on measurement of a prescribed point or points on the characteristic curve, the specifications for which are identical with the conditions laid down in corresponding A.S.A. (American) Standards.

The B.S. speed is expressed either as an arithmetical speed—e.g., B.S. 100—or in a logarithmic form to a base 2—e.g. B.S. 5°. This logarithmic form is frequently quoted in conjunction with the arithmetic form—e.g. B.S. 100/5°.

These notations are identical with the A.S.A. system, and are based on the same criterion as the speeds specified in the German Standard DIN 4512 of 1961. The B.S. Standard No. 1380: 1962 however also retains as obsolescent a logarithm base 10 speed scale, where every doubling of the speed corresponds to an increase of 3° in the logarithmic figure. Hence B.S. 100 is equivalent to B.S. Log 31°. This latter figure is never quoted in conjunction with the arithmetic speed (unlike the log base 2 speed), and is intended to be discontinued eventually. Its transitional retention is based on the fairly widespread use of logarithm base 10 speed figures in Britain before the publication of B.S. 1380: Part 1: 1962.

See also: *A.S.A. speed; Speed of sensitized materials.*

BUBBLE CHAMBER PHOTOGRAPHY. Technique of photographing nuclear particles by means of the bubbles they release in a super-heated liquid. Has in recent years become a specialized tool in nuclear research, quite distinct from the recording of tracks in nuclear emulsions, and the photography of the events in cloud chambers, although the latter problem is very similar to that of bubble chamber photography.

Basically a bubble chamber is a means of studying the interaction of charged particles (e.g., protons, electrons and other elementary parts of matter) with the atomic nuclei of the medium in the chamber. The particles are released into the chamber with extremely high energy, usually after acceleration by an electrostatic field in a proton-synchrotron. The particles can enter the bubble chamber directly, or can be made to strike a thin metal target, so producing particles other than protons. As an example of the energy levels attained, the proton-synchrotron at C.E.R.N., Geneva, supplies energies of 30 GeV, and the particles are travelling at speeds so close to that of light that the relativistic mass increase is about 30 times.

Types of Bubble Chamber. A bubble chamber operates on the principle that a charged particle passing through a super-heated liquid causes boiling to occur along its path. The analogous situation in a cloud chamber is that of an ionizing particle leaving a trail of droplets in a supercooled vapour. The advantage of a bubble chamber lies in the greater density of the medium, giving a much higher probability of a particle being stopped within the chamber, and ensuring a greater number of collisions with the atomic nuclei of the medium.

The most favoured liquids for use in a chamber are hydrogen, helium, or a simple hydrocarbon such as propane. Hydrogen is considered particularly useful because the nucleus is very simple, consisting of one proton only. Propane, however, is more easy to work

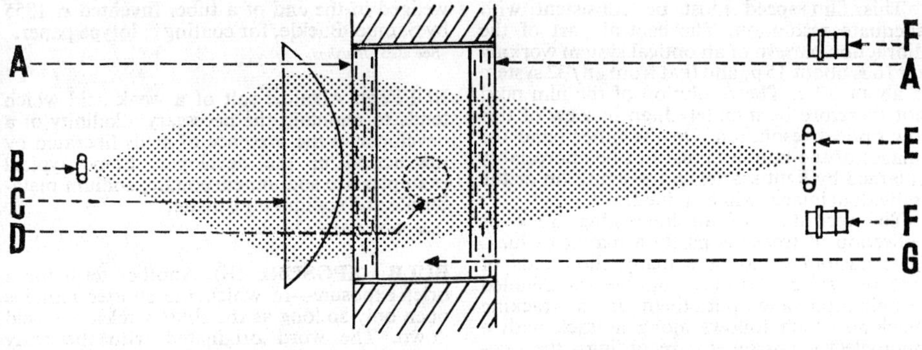

BUBBLE CHAMBER SET-UP. The light beam from the flash is projected through the bubble chamber to a point in between three cameras spaced 120 degrees apart. The cameras thus do not record the flash itself, but photographs any bubble tracks formed by nuclear particles. The bubbles scatter the light beam and so show up as bright against the dark background. The use of multiple cameras permits more accurate location in depth as well as separation of several tracks. A, glass windows of bubble chamber, usually about 50 cm. in diameter. B, flash source. C, condenser. D, beam aperture. E, image of flash between cameras. F, cameras.

with, and presents less severe optical problems although having a more complicated nucleus.

For photographing particle tracks in a hydrogen bubble chamber, a condenser behind the chamber focuses an image of an electronic flash close to the camera lenses. Usually at least three cameras are used to obtain at least two good views over the whole of the chamber. This provides dark-ground illumination with no direct light reaching the camera lenses. A bubble is visible on account of the small amount of light scattered sideways from the main beam. Hydrogen has a very low refractive index (about 1·09) so that at 10° from the main beam the scattered light has about $\frac{1}{3}$ the intensity of that passing straight through the bubble, and at 20° it has dropped to 1/10. The image of the flash tube must therefore be very close to the camera lenses. Only the blue light from the flash is used (with a non-colour sensitized film), partly to eliminate partial light spill into the camera lens by the chromatic aberration of the condenser, and partly to obtain maximum image resolution.

Because the liquid in the chamber is under high pressure, the transparent faces of the chamber must be of very thick glass of high optical quality to avoid distortion of the bubble tracks. The glass faces are coated to reduce the reflectance in the part of the spectrum used. (The inner faces of the glass are not optically eliminated by being in contact with the liquid, as is usually the case, because of the very low refractive index of hydrogen.)

With propane, the light scattered out of the main beam is so much higher that the flash tubes can be placed at one side of the chamber, with the light refracted through approximately 90°. Thus no rear window or condenser system is required, and a wider band of the spectrum can be used for recording. However, these advantages in the optical system are considered by many workers not to outweigh the advantages of the simple nucleus of the hydrogen atom.

For operation of the chamber, the pressure of the liquid is normally kept above that of the vapour. Just before the entry of the particle beam, the pressure is reduced, and after a delay of about 5 milliseconds, depending on the exact conditions and the availability of light, a flash exposure of about 100 to 200 microseconds duration is made of the bubbles, and the pressure immediately restored.

Photographic Problems. Bubble chamber photography needs a fast enough film, as the camera lenses are stopped down to f16 or even f32 for adequate depth of field through the chamber. With a bubble size of less than 0·5 mm. and a 15 times reduction, the film records merely the light points due to the bubbles. After formation the bubbles begin to grow, which to the camera lenses appears as an increase in brightness. With a fast film the bubbles can be photographed immediately, before currents in the liquid distort the tracks, or the liquid as a whole begins to boil.

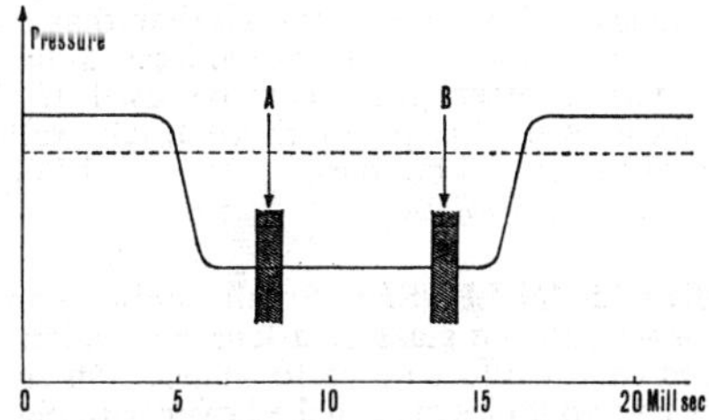

THE BUBBLE CHAMBER OPERATIONAL CYCLE. At the beginning of the cycle the liquid pressure (solid line) is lowered and then the beam allowed to pass through (A). The flash is fired and the photograph taken at point B, after which the liquid pressure is restored while the film is advanced and the system got ready for the next cycle. The dotted line indicates the vapour pressure; when the liquid pressure drops below this level the bubbles are formed when any ionizing particles pass through the chamber.

This film speed must be consistent with adequate resolution. The central part of the diffraction pattern of an optical system working at *f* 16 is about 15 μ, and that from an *f* 32 system is about 30 μ. The resolution of the film need not therefore be infinitely high in view of this low optical resolution. On the other hand the images of the bubbles must not be appreciably enlarged by light scatter in the emulsion or the individual images will be joined.

The recording of an interesting track or collection of tracks is partly a matter of luck. It is customary to take many thousands of pictures, select the ones suitable for detailed examination and put them in a tracking machine which follows along a track with a photoelectric scanner, recording the co-ordinates on punched tape, ready for insertion into a computer. The film must therefore have many of the properties normally associated with ciné film: freedom from static discharge when wound on rapidly, stability in storage both before exposure and between exposure and development, and tolerance to rapid processing—possibly at a high temperature. The positions of individual bubbles may have to be measured with an accuracy of about ± 2 μ; this is not possible over several inches of film owing to the unpredictable distortions in the base. However, by placing fiducial marks on the glass windows of the chamber so that reference points appear every centimetre or so on the film, this accuracy can be achieved on ordinary commercial films using triacetate base. The tracks must also be of sufficient density to be detected by the tracking machines. If the image is to be closely confined to the region of the original optical image, this calls for a film with as high a contrast as possible, consistent with accepting a wide enough range of exposures. Further, the granularity of the developed image must be sufficiently low to avoid errors in interpreting the positions of bubbles.

Usually the most satisfactory compromise between the various requirements of the sensitive material is a film of 40–160 ASA, developed to a considerably higher contrast than that used in normal camera work. Sometimes a faster film with coarser grain can be used if the emphasis is on the positions of tracks rather than of individual bubbles. M.G.F.

See also: *Nuclear physics.*

BUBBLES IN LENSES. Small bubbles sometimes seen in the glass of a lens but which do not normally affect its performance. On some lenses, the bubbles are touched over with blacking to minimize any light scatter.

See also: *Optical glass; Second-hand equipment.*

BUCKLE BRUSH. Swab holder for applying liquids, consisting of a ball of cotton wool wedged in the end of a tube. Invented *c.* 1855 by Samuel Buckle, for coating calotype paper.

See also: *Brushes.*

BUFFER. Alkaline salt of a weak acid which tends to maintain the necessary alkalinity of a developer in the presence of acids liberated by the action of the developer. Some typical examples are sodium carbonate, sodium metaborate and sodium phosphate.

See also: *Alkali; pH value.*

BULB EXPOSURE (B). Another term for a brief exposure—in which the shutter remains open only so long as the shutter release is held down. The word originated with the early pneumatic shutter release.

BULB FITTING. Connector cap, forming part of electric lamp, secures the lamp in the holder and connects it to the electricity supply. The shape and size of the fitting varies with the type of lamp.

See also: *Lamp caps and fittings.*

BUNSEN, ROBERT, 1811–99. German Professor of Chemistry at Heidelberg University. In collaboration with the English chemist, Henry (later Sir Henry) E. Roscoe, he discovered the Bunsen-Roscoe Law, sometimes called the law of photochemical reciprocity. With R. Kirchhoff, Bunsen laid the foundations of spectrum analysis in 1860. Bunsen invented the "grease spot" photometer and introduced magnesium (ribbon) light (1859).

BURGESS, JOHN. Dates unknown. English photographer. Improved Maddox's silver bromide gelatin process and started small-scale commercial production of dry plates in 1873. From 1880 also manufactured silver bromide paper.

BURNING-IN. Local extra exposure to a print, to darken selected picture areas which will otherwise show the image lighter than desired. When this is carried out by additional exposure to the image, detail becomes stronger. Burning-in can also be done with a fogging light (for example a thin pencil-like torch beam) and this darkens the print locally and also obliterates detail.

See also: *Shading and spot printing.*

BUSCH, EMIL, 1820–88. German optical manufacturer. Founded optical works in Rathenow which produced in 1857 the famous 6 ins. and 7 ins. diameter lenses for cameras and in 1866 a 10 ins. diameter lens giving pictures 30 × 30 ins. with a focal length of 34 ins. Busch also designed well-known combination lenses for portraits and landscapes.

C

CABINET PHOTOGRAPH. The cabinet format, introduced by G. W. Wilson for views, 1862, became general only after its application to portraits by F. R. Window four years later. The picture usually measures 4 × 5½ ins. on a mount 4¼ × 6½ ins. It was largely superseded by the postcard size around 1900, and died out during World War I.

See also: *Sizes and packings.*

CABLE RELEASE. Camera accessory used for releasing the shutter without touching the camera itself. It consists of a stiff wire inside a flexible outer casing similar to that of a Bowden cable but smaller in diameter. The casing carries a threaded nipple at one end to screw into a suitable socket on the camera shutter. The other end has a plunger which, when depressed, releases the shutter.

A modified form of cable release has a locking screw or a spring-loaded catch that can be engaged with the plunger to hold it in the depressed position. This enables the release to be used for holding the shutter open for making long exposures with the shutter control set at B.

See also: *Shutter release.*

CADMIUM LAMP. Mercury vapour discharge lamp containing cadmium. This compensates for the complete absence of red in a mercury vapour discharge and thus produces a reasonably balanced spectral emission.

CADMIUM SULPHIDE. Semi-conductor which increases in conductivity of electric currents when exposed to light. The conductivity is proportional to the logarithm of the light intensity. Used in photo-resistor cells for exposure control and exposure measurement. Unlike a selenium cell, a cadmium sulphide cell requires an external source of current. Exposure meters incorporating cadmium sulphide cells therefore measure the current flowing through the cell.

See also: *Exposure meter; Photo-electric cell.*

CALGON. Proprietary name for sodium hexametaphosphate. It is used to soften hard water.

CALLIER, ANDRÉ, 1877–1938. Belgian amateur physicist. Did valuable work on sensitometry and densitometry for which he received the Progress Medal of the Royal Photographic Society in 1915. Discovered that diffuse density differs from specular density by a factor Q (the Callier coefficient).

CALLIER EFFECT. Selective scattering of light in an enlarger or similar optical system; first investigated by André Callier in 1909.

In an enlarger the condenser is normally adjusted so that the light leaving it comes to a focus at the lens when there is no negative in the carrier. With a negative in the carrier, part of this light is scattered by the silver grains forming the image. So this part does not reach the enlarging lens. The highlights of the negative, corresponding to the heaviest deposit of silver, scatter and lose most light; the clearer parts, corresponding to the shadows, scatter least. This accentuates the difference between the highlights and shadows and gives a print of higher contrast than if printed by contact.

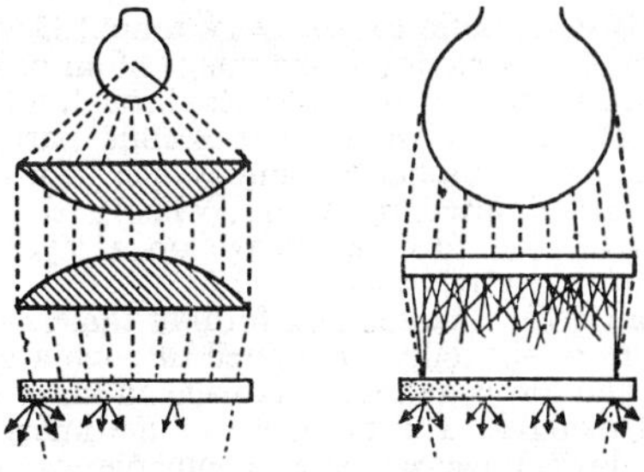

CALLIER EFFECT. Left: *In a condenser enlarger the directional light is preferentially scattered by the negative densities, increasing effective image contrast.* Right: *In a diffuser enlarger all negative areas scatter equally.*

Because of the Callier effect, the density of an even image area on a photographic negative or transparency is greater when measured by directed light (specular density) than when measured by diffused light (diffuse density). In the former case the density is due to absorption as well as scattering of light; in the latter case to absorption only. The ratio of specular to diffuse density is the Callier quotient or coefficient, and depends largely on the grain size of the image.

The effect is only noticeable when the light falling on each point of the negative is in the form of a bundle of more or less parallel rays, i.e., when a small light source and a condenser are used. When the light falling on the negative is already diffused the contrast of the negative is reproduced in the print without extra emphasis.

See also: *Density; Enlarger.*

CALOTYPE. (Talbotype). Process for making paper negatives in the camera. Not to be confused with collotype, which is a photomechanical reproduction process. Calotype was patented by Fox Talbot in 1841.

The basic process employed iodized paper—i.e., best quality letter paper brushed over with a solution of silver iodide and potassium iodide and allowed to dry. Before use the paper was treated with a solution consisting of silver nitrate, acetic acid and gallic acid crystals. The paper could then be exposed in the camera either wet or after drying, exposure being around five minutes at *f* 8 in bright sunlight. After exposure, development was carried out in a silver nitrate and gallic acid solution similar to that used for preparing the iodized paper for exposure. It was then rinsed, fixed in "hypo", washed and dried. The printing time of the resulting paper negative could be shortened by waxing the back of the paper to make it more translucent.

After the introduction of the collodion process the calotype process became obsolete.

See also: *Discovery of photography; Printing processes (obsolete).*

CAMERA. (Latin camera = a room.) Device for making a record of the image of an object formed when rays of light pass through a lens and fall on a flat surface. The essential parts of a camera are a chamber into which light can only be admitted by a lens, when a shutter is opened, to form an image on a layer of light-sensitive material.

Basic Design. The simplest form of camera has no lens at all; light is admitted through a pinhole and forms an inverted image on the sensitized material. The brilliance of the image is very low, however, and a pinhole camera calls for unduly long exposures in consequence. A lens allows more light to pass and makes it possible to use short exposures.

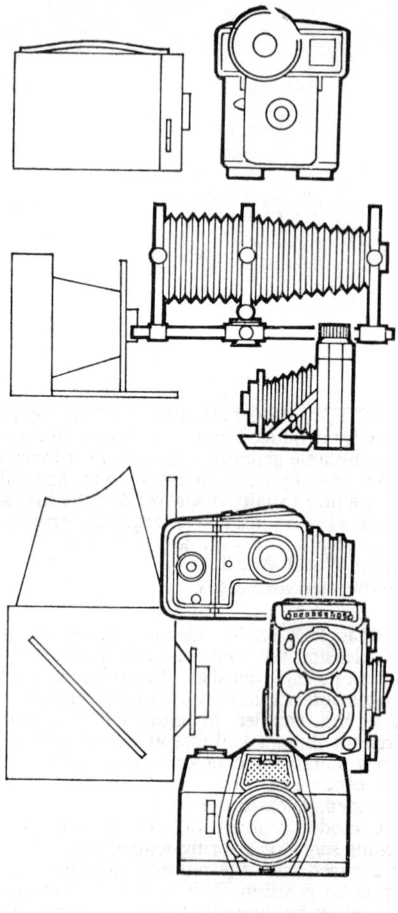

CAMERA TYPES. The photographic camera evolved along a few basic lines, from which all current types are derived. Top: Box camera; the simplest form (available for amateur use from the end of the nineteenth century) with its modern streamlined counterpart. Upper centre: Studio camera with variable bellows extension; giving rise to present-day technical, field, and press models as well as the amateur folding roll-film camera. Lower centre: Reflex camera, and modern twin-lens reflex and miniature single lens reflex with eye-level finder. Bottom: 35 mm. cine camera, with the 35 mm. precision miniature developed from an experimental camera for testing cine film, and provided with most advanced features.

The light passing through the lens is controlled in intensity by an adjustable diaphragm. A shutter is used to regulate the duration of the exposure. The diaphragm may be located in front of the lens, between its components, or just behind it. The shutter may be fitted in any of these positions or it may work immediately in front of the sensitized material. The shutter allows the light falling on the lens to pass through on to the sensitized material for periods of time as brief as 1/1000 of a second or less or for as long as required.

Many modern cameras incorporate an exposure meter which indicates the required settings of lens and aperture or even adjusts them automatically to the prevailing light.

Most cameras are made to take sensitized materials in the form of either films or plates. The camera has provision for holding these materials so that they can easily be changed or removed.

The camera must also have some means for sighting so that the image of the subject to be photographed will fall in the required position on the sensitized material when the shutter opens. For this purpose nearly all cameras have some form of viewfinder. This may be either an optical sighting device, or a focusing screen fitted so that it receives the same image as the sensitized material.

The lens position is usually adjustable to enable the camera to record sharply objects at different distances. The actual distance setting may be carried out with the aid of a distance scale or with an optical focusing device (e.g. rangefinder) or by visual observation on a focusing screen.

Basic Types. Modern cameras, although following this elementary design, usually incorporate individual features which have led to the creation of basic types, as follows: the sub-miniature, extremely small in both construction and film size; the miniature, taking perforated 35 mm. film and usually employing a range of accessories; the folding camera, medium size, folding into a compact form when not in use; the box camera, cheap and simple in construction; the reflex camera, any size, but embodying a special viewfinder system through which correct focusing can be accurately determined; the technical camera, usually for larger formats, with a wide range of adjustments of the lens and image plane.

These different types may include further refinements which make the camera particularly suitable for specialized work—e.g., scientific research, photomechanical reproduction and even applications such as gunnery practice and underwater photography.

Modern precision cameras are also available with comprehensive accessories designed to adapt the camera for specialized uses; the miniature camera is the most notable example.

See also: *Aircraft camera; Box camera; Camera history; Camera lucida; Camera manufacture; Camera obscura; Close-up camera; Colour camera; Electroplane camera; Eye camera; Field camera; Flash camera; Folding camera; Half-frame camera; Miniature camera;Panoramic camera; Photo finish camera; Pinhole camera; Polypose camera; Portrait studio camera; Press camera; Process camera; Quick-fire camera; Reflex camera; Schmidt camera; Spy camera; Stereoscopic camera; Technical camera, Ultra-miniature camera; Underwater photography.*

Books: *Cameras*, by W. D. Emanuel and A. Matheson; *Camera Technique*, by H. J. Walls; *How to Choose the Camera You Need*, by W. D. Emanuel and L. Gaunt (all London).

CAMERA GUIDES. Series of pocket books published by the Focal Press, each dealing with a well-established make of camera and its accessories. The pedigree, construction, method of use and scope of each camera is presented comprehensively, accurately and objectively, and new editions of each Guide are published to keep pace with improvements and new models.

CAMERA HISTORY

The invention of photography is generally ascribed to those experimenters who first succeeded in producing a permanent camera image. The development of the camera, however, started considerably earlier though its application to photography had to wait until chemical knowledge had advanced sufficiently.

The Camera Obscura. The photographic camera derives directly from the camera obscura. This was originally, as its name implies, literally a darkroom, with a tiny hole in the roof, wall, or window shutters through which the view outside was projected upside down on to the opposite wall, or a white screen placed opposite the hole. Its invention has been erroneously ascribed by various writers to Roger Bacon, Alberti, Leonardo da Vinci, and Giovanni Battista Porta. In fact it was described by the Arabian scholar Alhazen before 1039, i.e., about 250 years before Bacon. Since Alhazen's account does not in any way imply that he divulges a novel observation, it may be assumed that this knowledge was fairly widespread amongst Arab scholars.

In 1550, Gardano had fitted a bi-convex lens in the hole. In 1568 Daniel Barbaro recommended the addition of a diaphragm to sharpen the image, and in 1573 Danti suggested reflecting the images from a concave mirror to make them upright.

After Porta had in 1558 recommended the use of the camera obscura as an aid to artists, various forms were evolved, including tents, cubicles carried on poles, and even small portable and, later on, pocket models.

The portable reflex camera, with a mirror set at an angle of 45° to the lens, appears to be due to the mathematician Johann Sturm in

1676. A further reduction in size, replacement of the ordinary spectacle lens by a lens combination fitted in a brass tube, and substitution of a ground glass for the sheet of oiled paper, made the various camera types illustrated by Johann Zahn in 1685 the forerunners of nineteenth century box and reflex cameras.

The Photographic Camera. The evolution of the photographic camera in the nineteenth century —from the cumbersome apparatus needing a porter to the "detective" camera hidden in a walking-stick or a book—is in many ways analogous to that of the earlier drawing instrument.

Nicéphore Niépce was the first successfully to apply the camera obscura to photography (1826). One of his cameras incorporated an accordion-like square bellows and another variable iris diaphragms—prototypes of modern camera features. Also noteworthy is the fact that one of Niépce's instruments was made of zinc. In order to observe the progress of the image during the long exposure Niépce made a little spy-hole in the camera box, which he plugged up. The same device was later used by Fox Talbot.

The various lenses used by Niépce no longer exist, but from his letters it is known that W. H. Wollaston's periscopic lens (a meniscus prism) which he bought from Chevalier in 1828 gave better all-round sharpness than any previous lens he had used.

Daguerreotype Cameras. The camera which became generally known by Daguerre's name at the time the daguerreotype process was published in 1839 consisted of two boxes, the rear part with the ground-glass sliding within the front part containing the lens. Those which were made by Alphonse Giroux, a relative of Madame Daguerre's who for a few months had a monopoly in the manufacture of these cameras, bear Giroux's guarantee and seal and Daguerre's signature. They are 12¼ ins. high, 14½ ins. wide, 10½ ins. long when closed, extending to 20 ins., for use with metal plates 6½ × 8½ ins. and cost 400 francs (then £16). The camera was not patented however (except in England) and numerous similar and cheaper

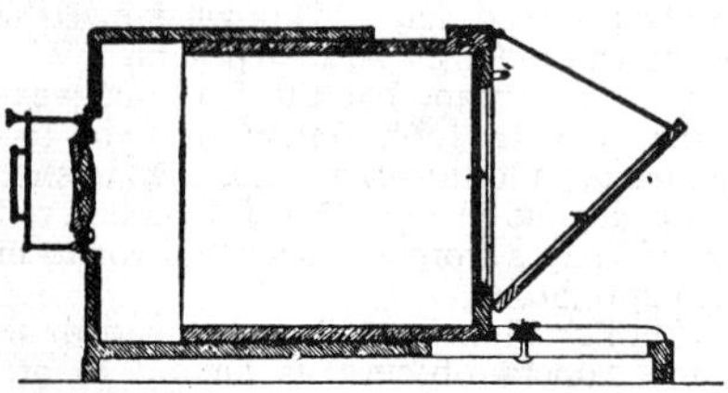

DAGUERRE'S CAMERA. Made originally by Alphonse Giroux, the camera consisted of two boxes which slid inside each other for focusing the lens on the ground-glass screen at the back. A simple brass disc in front of the lens acted as a shutter. The plate size was 6½ × 8½ ins. (i.e., the original "whole plate"). The lens was a Wollaston type achromatic lens with a focal length of 15 ins., and an effective aperture of f 14.

CHEVALIER'S COLLAPSIBLE WOODEN CAMERA. The body is hinged in the middle and folds up sideways after removal of lens panel and focusing screen.

models were soon made by other firms all over the world. Daguerre's design remained the standard type for several decades, both for studio and fieldwork.

Most of these Giroux cameras were fitted with a Wollaston type achromatic meniscus prism of 1829 by Chevalier. The lens had a focal length of 15 ins. and was contained in a brass mount with a simple brass disc acting as shutter. The effective aperture of the lens was reduced to f 14 by a 1⅛ ins. stop fixed inside the lens tube, which gave great sharpness to the image. Since Chevalier could not produce the lenses quickly enough to meet the demand, some of the Giroux cameras were fitted with lenses made by Lerebours and other Parisian opticians. These latter lenses are of the plano-convex type, with a focal length of 16 ins., and had an effective aperture of f 17.

Under these circumstances it is not surprising that for the first year or so exposures were 20 to 30 minutes in full sunshine. In order to reduce exposures to a reasonable length it was soon evident that smaller cameras, fitted with lenses of short focus and wide aperture, were called for. Landscape and architectural photographers, too, though less troubled by length of exposure, were eager to reduce the bulk of the apparatus, which in the daguerreotype process required, in addition to the camera, a plate box, iodizing box, mercury developing box, spirit lamp, and other paraphernalia, the total weight being 110 lbs.

Baron Séguier in December, 1839, introduced a cheaper and less bulky outfit in the form of a collapsible bellows camera. With the entire equipment, this could be packed inside a large box acting as carrying case. Séguier's outfit weighed 35 continental pounds and occupied only one-third of the volume of Giroux's though taking pictures the same size. Séguier also introduced the photographic tripod (previous cameras were placed on some kind of solid stand or table), the ball-and-socket head and the darkroom tent.

Chevalier's "Photographe", introduced in 1840, was a wooden folding camera in which the long sides were made collapsible (after the lens-board and ground-glass had been removed) by being divided horizontally along the middle and hinged. The following year he supplied also half-plate and quarter-plate models of the usual rigid box type, fitted with a single (and later double) achromatic plano-convex lens.

Buron, another Parisian instrument maker, introduced a quarter-plate camera with two lenses, one of 20 cm. ($7\frac{7}{8}$ ins.) focal length for landscapes, and one of 8 cm. ($3\frac{1}{8}$ ins.) focal length for portraits. In conjunction with one-eighth plates this was claimed to require only one minute exposure in full sunshine. In practice, however, exposures were considerably longer and everyone objected to their grimaces, due to the direct sunshine, as being a far cry from the promised likeness of "the human face divine".

In the attempt to find a solution, Alexander S. Wolcott, a New York manufacturer of dental supplies, took out a U.S. patent for a mirror camera in May, 1840. This wooden box had instead of the lens a large open front through which the light passed on to a 7 ins. concave mirror inside; this reflected the light and formed an image on the sensitized plate, which was fixed in the centre of the open front facing the mirror. This had the advantage of accepting more light, and the image was not reversed. The 12 ins. focal length mirror, however, limited the size of the portrait to 2 sq. ins., and the image was slightly soft. Exposures indoors were at first 3 to 5 minutes, but, after chemical acceleration of the daguerreotype plate had been invented, they were reduced to 60–90 seconds. The mirror camera

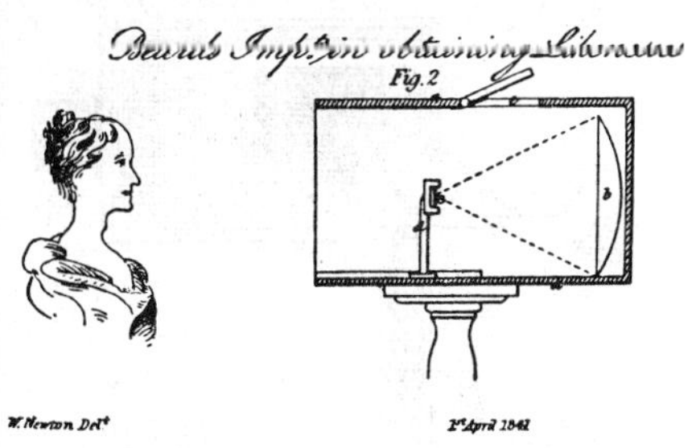

WOLCOTT'S MIRROR CAMERA. Patented in 1840, this camera had no lens, but instead formed the image by means of a large concave mirror at the rear. The image was reflected to the plate in front of the mirror. The plate holder was adjustable to allow for focusing the image. In the beginning this permitted exposures of 3–5 minutes (compared with an exposure of 20–30 minutes with Daguerre's camera).

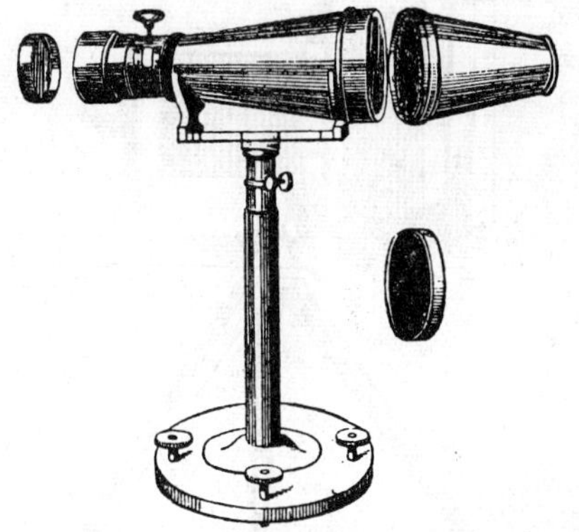

VOIGTLÄNDER'S ALL-METAL CAMERA. This was introduced in 1841, and was fitted with a Petzval lens of effective aperture f 3·6, the fastest lens of its time. It took circular pictures $3\frac{1}{2}$ ins. in diameter, with exposures of around 1 minute.

enabled Wolcott to open the world's first public portrait studio, in New York in March, 1840. Twelve months later, Richard Beard, who had acquired the exclusive rights for Wolcott's camera in Britain, opened in London the first public studio in Europe.

The general demand for a quick-acting portrait lens prompted the Viennese mathematician, Professor J. Petzval, to compute one for Voigtländer & Co., who designed a conical-shaped camera in conjunction with it. Both were introduced in January, 1841, and with them it was possible, even before methods of chemical acceleration had been devised, to take portraits in about a minute. The Voigtländer camera, which was made in two models, wood or brass, took circular pictures 9 cm. ($3\frac{1}{2}$ ins.) in diameter.

Petzval's double combination lens had a large aperture (f 3·6) and a fairly short focal length of 15 cm. (6 ins.) giving excellent definition in the picture centre without the use of a stop. It was thus thirty times faster than the Chevalier and Lerebours lenses in the Giroux cameras. The lens, which was also made in various other focal lengths for use in ordinary daguerreotype cameras, revolutionized portrait photography and remained the most generally used lens type all over the world until the introduction of Paul Rudolph's anastigmat by Zeiss in 1889. The Voigtländer and Wolcott cameras, on the other hand, were superseded as soon as the speeding up of the daguerreotype process by chemical methods had made the taking of larger portraits possible.

Cameras for the Calotype. Fox Talbot had his first cameras constructed in 1834 by a local carpenter. To arrive at exposures of reasonable length, they were made very small at first, measuring no more than $2\frac{1}{2} \times 2\frac{1}{2}$ ins., the picture area being 1 in. square. As they were fitted with a 2 ins. fixed-focus microscope lens,

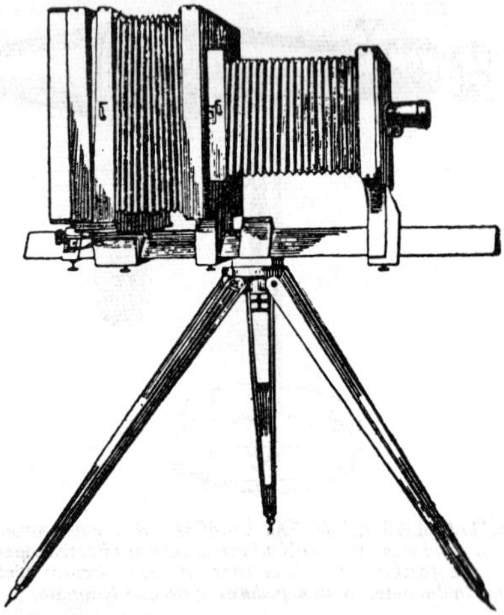

PETZVAL STUDIO CAMERA. One interesting feature is the monorail construction like an optical bench, used again a century later in modern technical cameras.

the ground-glass could be dispensed with. To see how far the image had imprinted itself on the paper, Talbot had spy-holes made in the lens-board. His final camera design, like Niépce's and Daguerre's cameras, consisted of two boxes, the rear part sliding within the front box.

Richard Willatts of London exhibited a portable camera of novel design at the Great Exhibition, 1851. Fitted with a conical expanding cloth body, it was a collapsible camera of great lightness, and though taking pictures $8\frac{1}{2} \times 10\frac{1}{2}$ ins. it was only 4 ins. deep when packed up. The back of the camera was mounted on a sliding plate which could be clamped at any distance from the lens, thus enabling the operator to use lenses of different focal length.

In 1850 Marcus Sparling, Roger Fenton's assistant during the Crimean War, designed the first magazine camera, which was extensively employed by amateurs in the British Army. Ten sheets of sensitized paper could be stored in separate holders in a kind of magazine, each sheet being dropped after exposure into a receptacle underneath the camera.

In May, 1854, A. J. Melhuish and J. B. Spencer patented the first roll-film arrangement. The sensitive material (waxed paper) was rolled up on a spool, and after each exposure the paper was moved on and wound up on the receiving spool. The roll-holder was made in several sizes, the largest holding a roll for twelve pictures 12 × 15 ins., and it could be fitted to any camera. Frank Haes used the device for photographing animals in quick succession at the London Zoo in 1855 and 1856.

About the same time, and apparently independently, Captain H. J. Barr of the Indian Service invented a system of fixing sheets of sensitized paper on a band of black calico, leaving a space about 2 ins. between each paper. The band was rolled up on to one spool and unwound on to another in the usual way. The exposed pictures were taken off the band and developed separately. This was easier than Melhuish's and Spencer's arrangement, in which the paper roll had to be marked for each picture and cut up before development.

Cameras for Collodion Plates. Whilst the work of the professional portrait photographer was considerably eased by Archer's collodion process (1851–c. 1880) which provided a much faster sensitive material, to the landscape photographer the advantages of the process were almost outweighed by the colossal amount of equipment he had to take with him on his outings. As the plates had to be exposed and developed while the collodion was still moist, he needed—in addition to camera and tripod, and a choice of several lenses—a chest full of bottles containing chemicals for coating, sensitizing, developing, and fixing the glass negatives; a good supply of glass plates; a number of dishes, scales and weights; glass measures and funnels; a pail to fetch the rinsing water (and where there was none likely to be met with, the water itself!) and above all, a portable dark-tent in which all the chemical hocus-pocus took place.

Many photographers therefore engaged a man as porter, or pushed a wheelbarrow or a small hand-cart containing all the equipment. The more successful could afford a carriage, which in some cases simply served to convey the photographer and his equipment to the scene; in others it was completely fitted up as a travelling darkroom. Roger Fenton took with him to the Crimean War a horse-drawn van rigged up as a darkroom and living and sleeping quarters.

To obviate the need for a dark-tent, William Brown adapted in 1851 a camera designed by Frederick Scott Archer. This incorporated loose sleeves of black material through which the

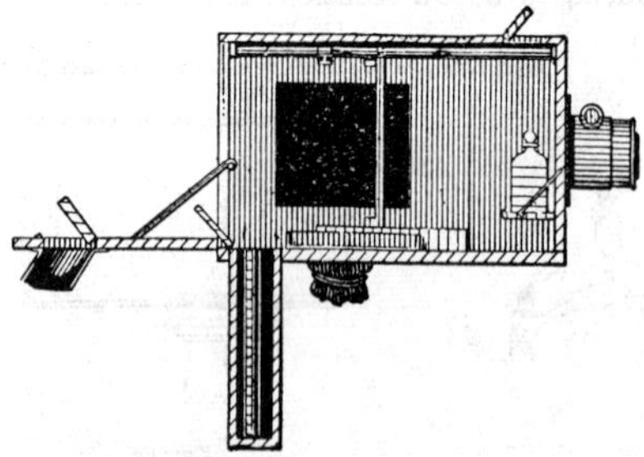

ARCHER'S CAMERA. This was designed for the wet collodion process and incorporated the equipment for sensitizing the wet plates immediately before exposure. The image was focused on a ground-glass screen held in a frame which slid forwards and backwards on rails in the top of the body. Sleeves of black fabric at the sides enabled the operator to introduce his hands inside the camera to sensitize and process the plates.

operator introduced his hands and head into the body of the camera, where he carried out the sensitizing and processing in a gutta-percha dish. During the exposure, the dish was dropped into a detachable chamber at the bottom of the camera. Light was admitted through a small yellow window in the top of the camera.

This type was modified innumerable times for newer processes, and is father of the cameras still sometimes seen in use by old-fashioned tintype photographers at seaside resorts.

In Newton's camera, introduced the following year (1852), a chamber underneath the camera was divided into four compartments for the sensitizing bath, developing solution, rinsing water and fixing bath. The collodionized plate was attached to a rod which was moved along by rack and pinion and lowered into each of these compartments in turn.

In his patent of June, 1851, Fox Talbot also suggested a camera containing a glass cell or chamber into which the chemicals were to be introduced through a funnel and drawn off at the bottom by a stopcock. The mere idea seemed so fantastic to the photographic fraternity that it was dismissed as a joke, and Talbot's camera never went beyond the specification stage.

Large Cameras. During the collodion period the camera became both larger and smaller—according to the purpose it was intended for. Realizing the possibilities of photography as an independent art medium (in the 1840's photography had been largely in the hands of professionals whose main aim was to make a living out of daguerreotype portraiture) many amateurs took up the art and competed with each other in exhibitions. Naturally, the bigger pictures were more impressive and, as the slow albumen papers ruled out enlarging, the photographer had to use large plates from which he made contact prints.

Thus 10 × 12 ins. and 12 × 16 ins. were quite ordinary plate sizes. Francis Frith took some of his Egyptian and Palestinian views on 16 × 20 ins. plates. C. Thurston Thompson, the official photographer at the South Kensington Museum, had a camera 39 ins. square by 12 feet long. It ran on rails, and was first used in 1858 for reproducing the Raphael Cartoons, then at Hampton Court Palace, on glass plates 36 ins. square.

The largest camera before 1900 was made for John Kibble, a Glasgow amateur photographer, in 1860. It was mounted on wheels and drawn by a horse and took 36 × 44 ins. glass plates, each weighing about 44 lbs. The London optician Andrew Ross constructed a special lens of 13 ins. diameter and focal length 6 feet, at a cost of £170.

Small Cameras. A contrast to this monster camera were the binocular stereoscopic cameras introduced in 1856, when a great demand arose for photographs to be viewed in Sir David Brewster's stereoscope. These models took 3 × 3¼ ins. pictures and had 5 ins. lenses, and so were the first camera type to take successful instantaneous pictures of live subjects.

Realization of the new fields suddenly opened up by the binocular camera led to the construction of small cameras for taking single pictures. The first was Thomas Skaife's "pistol" camera (June, 1856), using plates of about 1½ ins. in diameter. It had a spring shutter worked by rubber bands which were released by a trigger (hence "pistol"), and a 1 in. Dallmeyer combination portrait lens working at *f* 1·1 or *f* 1·6—one of the fastest lenses ever constructed. Similar designs followed in the next few years.

Another miniature camera was Bertsch's "chambre noire automatique" introduced in 1860. It was of the fixed-focus variety (hence "automatic"); all objects beyond 40 feet were rendered sharp. This made a ground-glass

A WET-PLATE WORKER'S OUTFIT. This was a comparatively portable outfit and included, besides the camera and tripod, a dark-tent, a stool, and all the chemicals.

unnecessary and, instead, Bertsch fixed a frame viewfinder and a spirit-level on top of his 4 ins. square metal box, which took pictures 2¼ × 2¼ ins. The camera was made for the wet collodion process, but could be adapted for albumen or dry collodion plates which were stored in a changing box.

In 1864, the French engineer Dubroni introduced a small hand camera in which the sensitizing and developing of the plate were in fact carried out inside the camera body—a ruby glass globe contained in a box with apertures for the lens and for the introduction of the plate. Through a light-tight valve in another opening at the top the solution was introduced by means of a pipette, the camera then gently rocked to make it flow evenly over the plate, and the solution removed in the same way. The camera was made in five sizes, the smallest taking pictures 2 ins. square.

Other ingenious small cameras included C. Piazzi Smyth's model for photographing the interior of the Great Pyramid in 1865 and H. Cook's opera glass camera. Smyth's camera incorporated an ebonite silver nitrate sensitizing bath in which a 1 × 3 ins. plate (an ordinary microscope slide) was actually exposed through a 1 in. square window in one side of the bath. Cook's version had one lens for focusing, with a ground-glass screen at the rear, and another for taking with an attached magazine for 50 dry collodion plates. Piazzi Smyth stressed the advantages of the small negative over the larger, and described the process of enlarging the tiny plate.

All these ingenious instruments were, however, exceptions. The great majority of photographers worked with big plate cameras, still mostly of the wooden box type. The lighter bellows camera began to establish itself only very slowly, as photographers found a big bellows extension not rigid enough for the long exposures needed.

Multiple Cameras. In 1851 Antoine Claudet exhibited a multiplying camera which enabled the operator to obtain on one negative several miniature portraits, representing the same person in a different pose, or several people grouped together. The multiplying was performed by cross sliding motions of the dark slide by means of two racks. The resulting picture gives rather the effect of cinematographic shots.

Carte-de-visite photographers worked with a camera fitted with four identical lenses of short focus (4½ ins.); the camera interior was divided into four compartments, one for each lens. By exposing first one-half of the plate and then the other by means of a sliding plate-holder, eight small portraits could be taken on a plate 8½ × 10½ ins. This facilitated the printing and multiplication of the small pictures, which were bought by the dozen by the sitter.

Apparatus for the Gelatin Process. The gelatin dry plate which came into general use in 1879–80 not only enormously simplified photographic technique, but also revolutionized camera design, reducing the photographer's equipment to approximately what it is today. The new negative material was fast enough for snapshots of moving objects, provided the cameras were small, quickly ready for action, and equipped with an instantaneous shutter. Manufacturers quickly responded, and during the 1880's and 1890's a mass of equipment of all shapes flooded the market.

Fast bromide paper had now made enlarging feasible and so quarter-plate and 4 × 5 ins. hand cameras established themselves as the most popular sizes for amateurs in the Anglo-Saxon countries, the equivalent on the Continent being the 9 × 12 cm. size. The characteristics of the new apparatus were compactness, simplicity in manipulation and light weight. In general, they can be classified into four main types: change-box cameras, magazine cameras, roll-film cameras, reflex cameras.

(1) Change-box cameras contained glass plates or cut films, usually 12 in number, in a separate change-box attached to the camera. This permitted a changeover in daylight, and the box was usually fitted with an automatic counter indicating the number of exposures made. Each plate was contained in a separate sheath, and after exposure a fresh plate was lifted into a soft leather bag attached to the top of the box, gripped with the fingers, and then manipulated into the focal plane. The prototype of this kind of change-box seems to have been that patented by H. Cook in May 1867 for use with his opera-glass camera.

(2) The magazine camera stored a number of plates or cut films (12–40) in a magazine or chamber inside the camera body, the plate being changed after each exposure by some kind of mechanism, which differs in almost every camera model.

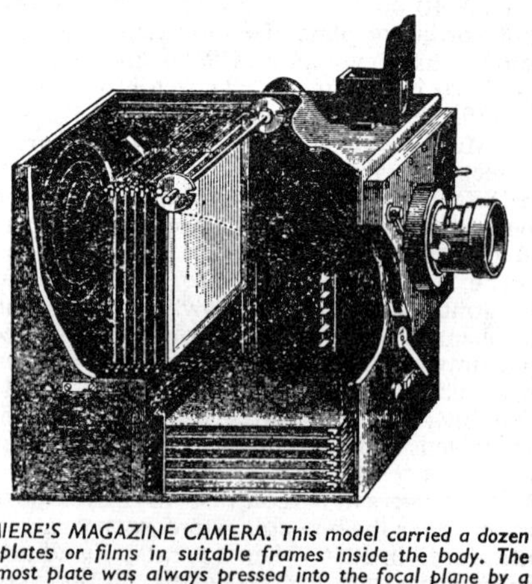

LUMIERE'S MAGAZINE CAMERA. This model carried a dozen dry plates or films in suitable frames inside the body. The foremost plate was always pressed into the focal plane by a spring at the rear, and dropped into a compartment below by a suitable mechanism after exposure.

WARNERKE'S FOLDING ROLL-FILM CAMERA. *The lens panel moved on a monorail-like sliding bracket with a stabilizing rail, and the back incorporated a draw-slide and focusing screen, as well as the film rollers and transport mechanism.*

In some magazine cameras the plates were contained in a grooved box, and dropped into position (or lifted up, as the case may be) by means of a rod with pincers through a slit in the camera. After exposure the plate was returned to its groove, and the next plate brought into position by a sliding movement of the magazine. In other magazine cameras, a dozen or more plates or cut films were stored in a chamber at the back of the apparatus. By various mechanisms, sometimes merely by shutting the camera ("shuttle magazine camera") the exposed plate or film was dropped to the bottom of the camera, and the next one pushed forward into the focal plane by a spring at the back of the magazine.

A magazine camera patented by Alfred Pumphrey in 1881, held separate pieces of coated film in frames. These were fixed by one edge to an endless band passing over two rollers, similar to the arrangement of pictures in a revolving stereoscope. By turning a knob, a film was brought into a vertical position behind the lens; another turn moved on the exposed piece and the next frame came into place. In 1882, Pumphrey introduced a camera taking 100 sheets of 3¼ × 4¼ ins. film.

The "Frena" was the first hand camera to expose a pack of celluloid cut films, and was introduced in England in 1892. It held forty 3¼ × 4¼ ins. sheets which were notched and inserted in a holder like a pack of playing cards.

(3) Roller-slides and roll-film cameras, which eventually superseded the change-box and magazine cameras, used flexible films instead of glass plates or cut films. The film was wound on two spools, the action of winding up the exposed portion unwinding a fresh portion for the next exposure. Originally the film was in a separate box, the roller-slide, which was made in many sizes for attachment to almost any camera. George Eastman popularized this idea in 1885 when he brought out his roller-slide attachment containing a paper roll coated with gelatin emulsion for 24 exposures. Three years later the roller-slide was loaded with nitrocellulose film for 48 exposures.

The Walker/Eastman roller-slide came close to Warnerke's of 1875. This was a very modern-looking folding camera which extended on a monorail.

The roll-film idea did not, however, become popular until much later when the growth of the amateur movement after 1880 created a demand and made mass production feasible, which brought down prices. Above all, the new D & P firms, which catered exclusively for the mass of amateurs unversed in photographic technique, took over the processing of these awkward long bands of negative material.

The camera which revolutionized photography in the 1890's and gave the greatest stimulus to amateur photography was the

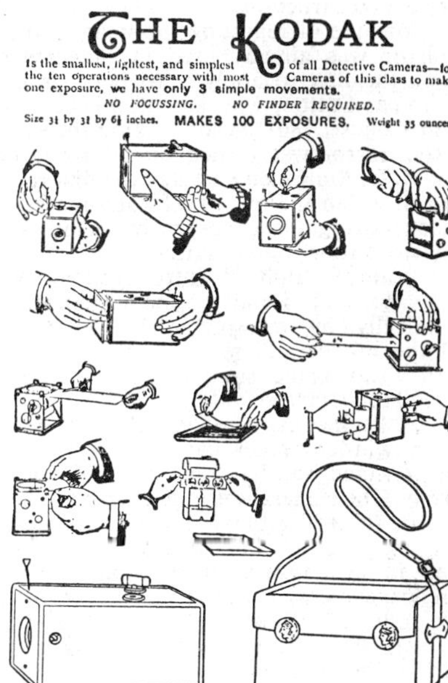

EASTMAN KODAK ROLL-FILM CAMERA. *The first amateur snapshot camera, the Kodak brought photography to the masses, very largely due to the developing and printing service that went with it. The camera took a paper roll with a stripping emulsion (later a celluloid film) and carried 100 exposures at one loading. It was also very simple to use: a fixed focus lens reproduced everything sharp from 8 feet to infinity, there was a single-speed shutter, and a single lens stop. When the film was finished, the camera could be sent back to the maker for reloading and processing of the exposed film.*

Kodak, the first roll-film camera, brought out by Eastman in August, 1888. It was a simple small hand camera measuring only 6½ × 3½ × 3½ ins. and weighing 2 lb. 3 ozs. It incorporated a roll of paper coated with a stripping emulsion on which 100 circular pictures 2½ ins. in diameter could be taken. (Nitrocellulose roll film was used after 1889.) The camera had a rectilinear fixed-focus lens giving sharp definition of everything beyond 8 feet, one speed, and fixed stop. The apparatus was thus the embodiment of simplicity, and the price of 25 dollars (then 5 guineas) was reasonable. The Kodak's appeal to the amateur unversed in photography was enhanced by Eastman's stipulation that the camera be sent back to the factory for developing and printing of the film. It was returned loaded with a new film, together with 100 mounted prints, for a service fee of 10 dollars. Thus Eastman's slogan "You press the button, we do the rest" was put into practice.

(4) The single and twin-lens reflex cameras are classed as a separate group, for although they were variously made with change-box, magazine, or roll-film attachment, they are basically of different construction.

The principle of reflecting the image with a 45° mirror was already used in the camera obscura; the first to apply it to photography was Thomas Sutton, who patented his instrument in August, 1861. When making the exposure, the mirror was turned up by a handle. In 1888 S. D. McKellen patented the first reflex camera in which the mirror was automatically displaced during the exposure by being connected with a roller blind shutter.

The "Cambier Bolton" introduced in 1898 by W. Watson, London, made exposures automatically by pressing a button which raised the mirror and released the Thornton Pickard focal plane shutter. Speeds ranged from 1/20 to 1/1000 second.

R. & J. Beck constructed in February, 1880, the first twin-lens reflex camera. Both lenses were connected so that they focused simultaneously. The camera was a quarter-plate and, except for the roller blind shutter attached to the taking lens, it does not seem to have differed in any respect from modern twin-lens reflex cameras. Large numbers of other makes followed. Ross & Co.'s "Dividend" (1891) was the most advanced in design. Two models were made: one for 48 exposure Eastman roll-film and the other with a change-box containing 12 plates.

Pocket Cameras. Many of the cameras advertised as "pocket cameras" in fact were not. The first real pocket cameras for gelatin dry plates were Marion & Co.'s metal miniature camera of 1884 and the "Tom Thumb" produced by Ford of New York about 1890. The external measurements of the latter were 3½ × 3½ × 2½ ins. and it weighed 7¾ ozs. The picture area was 2½ ins. square. Like the Kodak, it was of the fixed-focus variety (everything beyond 8 feet was sharp) but it was made for plates, each in a separate plate holder.

Eastman followed up his Kodak hand camera with a pocket model in 1895, measuring only 2¼ × 2⅞ × 3⅞ ins., and weighing 7½ ozs. Designed by Frank A. Brownell, this camera was the prototype of the "Brownie" class. It incorporated an important new feature, the daylight loading roll-film spool, taking a dozen 1½ × 2 ins. pictures. In other respects, too, this camera pointed the way to modern camera design; it was made of aluminium, and the shutter was set and released by one button. Enlargements up to 12 × 15 ins. could be made from the negatives. Three years later Eastman effected a further improvement in compactness—the Folding Pocket Kodak, measuring 1⅝ × 3½ × 6¾ ins. This was a collapsible camera with bellows; when drawn out for use the lens-board was held rigid by struts.

The "Blocknote" made by Gaumont of Paris and introduced in 1903 was a highly finished folding pocket camera fitted with either a Rapid Rectilinear, a Zeiss Protar or a Tessar lens with an aperture of *f* 9–*f* 6·8 according to lens. The 4½ × 6 ins. plates were carried in single dark slides.

Freak Cameras. In January, 1881, Thomas Bolas designed for Scotland Yard a camera looking like a wooden box and another got up in the form of a book. These were the first of the so-called detective cameras. Manufacturers were quick to take up the idea and within a few years an absolute craze started for cameras disguised in the form of parcels, picnic baskets, or a Gladstone bag—"an alligator-skin travelling handbag of innocent appearance". These were usually quarter-plate magazine cameras, but before long detective cameras degenerated into toys of little practical value, got up in the form of opera-glasses, field-glasses, revolvers, guns, books, watches, or concealed in purses, walking-sticks, hats, cravats, or beneath the waistcoat. The lenses of these cheap cameras were poor, the pictures too minute to be of any use, and, like their forerunners the eighteenth-century cameras obscuras, disguised in walking-sticks and in book form, they can hardly be regarded as proper equipment. They do, however, show a steady trend towards the miniature camera, which as a scientific precision instrument did not arrive until 1924. A number of such ultra-small cameras have kept on appearing on the market up to the present day. Some were simple in design and scope, others reached a high standard of precision.

As a contrast to the small cameras in the 1890's, Anderson of Chicago constructed the Mammoth camera in 1900. It took pictures 4½ × 8 feet—more than three times the size of Kibble's—and measured 9 feet high × 6 feet wide × 20 feet long when extended. The weight of the apparatus was 900 lbs. and, when the plate-holder was loaded, 1,400 lbs. Each glass plate, which had of course to be specially made and coated, cost £40, and the production of the print, £6 to £7. Two lenses had been specially designed for the Mammoth by Zeiss, one a wide-angle of 5½ feet focal length and the other a telescopic Rapid Rectilinear of 10 feet focal length. Obviously a camera this size needed special transport too: a railway truck 20 feet long pulled by a locomotive for long distances, and a furniture van for Chicago and district. The taking of a photograph required the services of fifteen men, and on the average an exposure of 2½ minutes. Five gallons each of developing and fixing solutions were needed. The purpose of the camera: the Chicago & Alton Railroad Co. wanted giant photographs for the Paris Exhibition of its new luxury St. Louis express to be taken in one shot instead of being made up from several views.

While the freaks in camera design—the outsize and sub-standard models—make very amusing history, the majority of amateur photographers in the period 1880 to 1930 worked with hand cameras or pocket cameras—to which were added in 1910 the vest pocket size. Practically all cameras were made of lightweight metal covered with black leather and were of the folding bellows type.

The 35 mm. Miniature Camera. The increasing volume of production of 35 mm. film for the cinema industry made its use economic for still photography. A few cameras appeared at the beginning of the twentieth century, employing this type of film. Examples were the Minnograph made by Levy-Roth of Berlin, taking 50 pictures 18 × 24 mm. on 35 mm. cine film, and an American camera designed and constructed by George P. Smith of Missouri in 1912; this latter 35 mm. camera took pictures 1 in. wide and 1½ in. long on perforated cine film.

In 1914 Oskar Barnack, a microscope designer employed by Leitz in Wetzlar, Germany, constructed the prototype of the Leica. This was at first only for his own use, in order to make test exposures on the small film he was using in a cine camera of his own construction. In 1924 Leitz put a modified form of this camera into production. The Leica of 1924 had a focal plane shutter which was automatically tensioned by the film transport mechanism, and used 5-foot lengths of 35 mm. film in daylight cassettes. Later models incorporated a coupled rangefinder for accurate focusing, and interchangeable lenses. The standard lens was a 5 cm. Elmar *f* 3·5, specially designed for the purpose by Prof. Dr. Berek. In fact, the resolving power of the Elmar was a long way in advance of the resolving power of the films of that day, and for many years the grain of the emulsion remained a serious problem to all photographers using miniature cameras. The importance of the Leica lies in the fact that it raised the miniature camera to a precision instrument. With it the era of the true miniature began, though for several years professional photographers and serious amateurs considered it a toy.

Soon other miniature cameras followed. Improvements in 35 mm. films and fine grain developers brought the quality of enlargements from miniature negatives on a par with those from larger picture sizes. In addition some of the makers of precision miniatures, from the Leica onwards, developed exclusive ranges of accessories and supplementary equipment to adapt the camera to every conceivable purpose, from amateur to scientific photography.

Candid Cameras. The 35 mm. miniature also proved popular for candid photography owing to its speed of operation. To permit even more rapid shooting, a few motorized miniatures also appeared. The most famous was first produced in 1936 and incorporated a spring motor. Pressing a button automatically exposed the picture, advanced the film and retensioned the shutter, permitting exposures at the rate of 8 per second.

At the same time wide aperture lenses made the miniature suitable for photography in poor light conditions. This, however, was not confined to the 35 mm. camera. In 1924, the Ernemann works, Dresden, brought out a 4·5 × 6 cm. and a 6 × 9 cm. plate camera, the Ernox, fitted with an *f* 2 anastigmat. This was followed in 1925 by the Ermanox, with an *f* 1·8 lens—the fastest lens ever made up to that time. This made possible snapshots of the stage, public dinners, and other indoor parties, without flashlight. Above all, since this camera was used in conjunction with fast panchromatic plates, the problem of grain which worried 35 mm. users at the time did not arise, and until that problem had been solved the Ermanox was the favourite camera for very poor lighting conditions.

Modern Reflex Cameras. The Rolleiflex, put on the market by Franke & Heidecke in 1928, has been the precursor of numerous similar twin-lens reflex cameras. It found immediate favour with all those who like to compose their picture on a focusing screen and have control over it during the exposure. There are two main sizes, 1⅝ × 1⅝ ins. (4 × 4 cm.) and 2¼ × 2¼ ins. (6 × 6 cm.), each taking 12 pictures on ordinary roll film.

As an alternative to the twin-lens reflex camera—with its limitation to a single focal length—a number of single-lens reflex cameras also appeared in the 1930's. Some of these took roll films, others 35 mm. film.

Present-day 35 mm. reflex cameras utilize pentaprism systems to permit observation of the screen image at eye level.

Camera Design Today. Fundamentally, present-day cameras differ little from corresponding types in use twenty years ago. They do, however, incorporate various refinements such as flash-synchronized shutters and double exposure lock or automatic shutter tensioning, rapid film transport levers, film indicators, better and faster lenses, and so on, to make them more foolproof and to increase their versatility. At the same time styling of cameras has become more streamlined, and pressed metal body frameworks have given place to metal or even plastic castings. Certain components were redesigned for easier mass production, rather than improved camera function, while improved testing methods were adopted to keep the standard of precision production up to the level of older hand-made cameras. **H.G.**

See also: *Camera obscura; Chronology of photographic inventions; Cine history; Colour photography; history; Development history; Discovery of photography; Historiography of photography; Leica; Lens history; Museums and collections; Rolleiflex; Sensitized materials history.*

Books: *The March of Photography*, by E. Stenger (London); *A Source Book of Photographic History* by W. Baier (London); *History of Photography*, by J. M. A. Eder (New York); *The History of Photography*, by Helmut Gernsheim (Oxford).

CAMERA LUCIDA. This is not a camera in the sense in which the word is now used. It is (or was) a four-sided reflecting prism used by draughtsmen and artists to enable them to draw outlines of objects in correct perspective. The prism generally has one angle of 90° and the opposite of 135°.

The artist holds the prism in position above a horizontal sheet of paper and sees a reflected image of the scene in front of him. He can then trace the outline by looking over the edge of the prism and referring what he has drawn to the reflected image.

The true camera lucida was invented in 1807 by Dr. William Hyde Wollaston, who also invented the meniscus lens. The camera lucida constructed by Dr. Robert Hooke in 1668 was an optical lens projection in a light room.

CAMERA MANUFACTURE

The demands by the photographer on his camera have long ceased to be confined to the optical and technical quality of his pictures. Nowadays the camera must be designed so that it is instantly ready for action, reliable under all circumstances and as foolproof as possible. The appearance of the instrument, and last but not least its price, are also of considerable importance in the design and manufacture of a camera.

Every design must to some extent be a compromise; high efficiency has to be paid for by higher retail price; the combination of a number of operations in one camera control results in bigger dimensions and greater weight of the camera which has to accommodate all the necessary mechanical linkages. So every advantage in one respect becomes a disadvantage in another. To deal with these problems, the designer requires imagination as well as technical ability.

Methods. The first cameras were craftsman-built instruments with brass-reinforced mahogany bodies and leather bellows. Even when parts were produced in quantity, they were assembled and fitted individually by skilled workers. These were drawn from the cabinet making and allied trades. Since relatively large plates and small aperture lenses were the rule, there was no need to work to close limits and as the demand was not great it could be satisfied by hand fabricated products.

Fast roll films and wide aperture lenses changed the manufacturing picture. These improvements made it possible to produce a new type of small—even pocketable—camera, and immediately increased the potential market. Small cameras had to be built to extremely fine limits and were more expensive to make, while the improved optical equipment added further to the cost. So to keep the price of the product down to an economic level the manufacturer has had to develop mass production methods to a remarkable pitch of perfection.

The design of the modern camera is naturally influenced by the manufacturing techniques available. It is an assembly of parts made of materials that lend themselves to mass production: machined or die-cast brass or light-alloy castings, sheet metal pressings or plastic injection mouldings.

Initial Stages. The specification for a particular camera is worked out in relation to the prospective demand and what the prospective buyer can be expected to pay. From rough drawings made at this stage the experimental workshop turns out a mock-up to form the basis for further practical discussions. This rough model is criticized, handled and where necessary altered until all concerned are satisfied with the principal features. Finally, an accurate prototype is built. This, apart from the fact that it is hand-made, is identical with the proposed production camera. Such prototypes call for a very high standard of craftsmanship and are the work of a special department.

Once the prototype is approved, the real work starts, and an immense amount of effort must be expended before the design can go into production. The factory building must be planned as a complete, organized whole so that everything moves steadily forward from the receipt of the raw material through the machining, assembly, inspection, packing and dispatch without unnecessary handling opera-

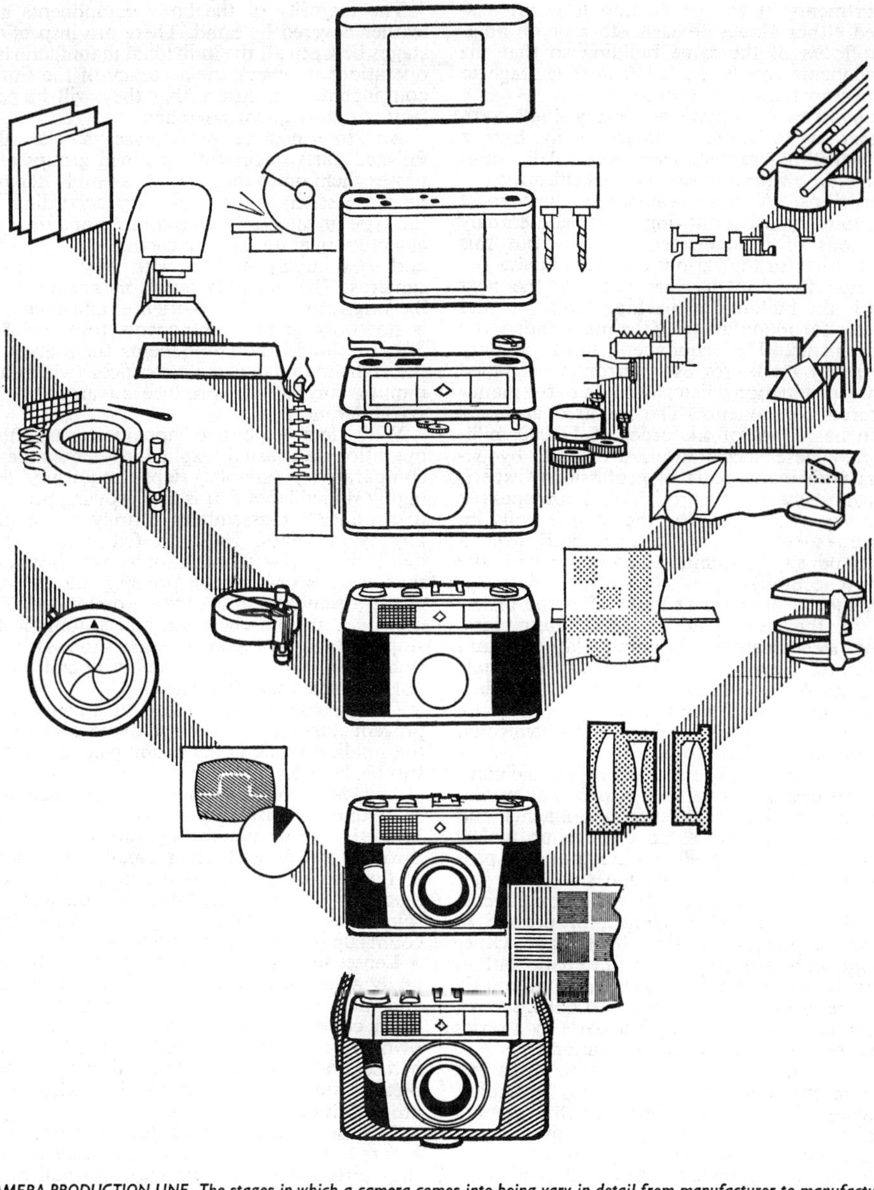

CAMERA PRODUCTION LINE. The stages in which a camera comes into being vary in detail from manufacturer to manufacturer, but the general principle, as shown in a simplified form for a typical camera, is much the same. Top: *the body casting is machined and drilled for various apertures and screw holes.* Upper left: *Certain components like rangefinder housings and the like are stamped and pressed from sheet metal on large hydraulic presses, followed by plating and polishing.* Upper right: *Screws and mechanical components like winding levers and gear wheels are mass turned on automatic lathes and finished on milling machines.* Centre left: *Exposure meters are assembled and tested, and connected to the meter cell which is built into the body.* Centre right: *Viewfinder and rangefinder elements are assembled and tested before being installed in the camera.* Lower left: *The shutter is frequently supplied by a separate manufacturer, but undergoes careful speed and other tests for accuracy, reliability and synchronization before assembly in the camera.* Lower right: *Lenses also come from outside sources and are built into the models as complete single or multiple units.* Bottom: *Finally the finished camera undergoes thorough inspection with detailed optical and mechanical tests before being passed out and packed for sale.*

tions or bottlenecks. The various successive departments in the production flow are disposed either alongside each other or on adjacent floors of the same building so that the components can be passed from one stage to the next by the shortest route.

Components. As a first stage, every single component has to be drawn up accurately. Even a simple roll-film camera needs over 200 drawings of this type with almost 3,000 specifications of dimensions. A large camera factory must maintain extensive drawing offices and employ hundreds of draughtsmen to carry out this work alone for the various camera models.

When the drawings are finished, the next step is the building of machine tools, jigs and moulds for manufacturing the many individual components. This equipment must be very accurately made, for every error is magnified during production when passed on to the manufactured components. The required accuracy often has to be of the order of 1/1000 millimetre. These special tools are made by experts with many years of professional experience, and checked by precision microscopes for accuracy. For instance, the simple roll-film camera mentioned above may need over a thousand such machine tools made with the utmost accuracy.

A special department looks after the ordering of the raw materials—e.g., aluminium, brass, steel, leather, chromium, lacquers and varnishes. Only the highest quality materials are suitable for camera manufacture, and they are prepared for use only after being specially tested in the factory, mechanically, chemically, and under the microscope.

The production of the many individual components can now go ahead. A precision miniature may need over 600 such components, and many of those have to go through up to 200 different operations of manufacture and pass through twenty or more machines.

In the machine shops the main components are either cast and machined, or pressed in hydraulic presses into their final shape. Other components are turned on automatic lathes. Here the rod of raw material is fed continuously into the machine and shaped into bolts, screws, rivets, lens mounts, and other parts by a series of automatically controlled operations.

In another department automatic drills and cutting machines cut the various apertures, notches, and screw holes into the camera bodies. Milling machines finish many different forms of notches, slits and projections.

Finishing and Assembly. After all machining operations have been carried out, the parts pass on to the finishing stage. The finishing process varies according to the part. Some components are polished in special machines with diamonds, to produce mirror smooth surfaces, some are nickel or chromium-plated in the plating shop, while others are simply protected against corrosion for example, by lacquering or dipping in chemical baths.

The majority of the body components are leather covered by hand. There are inspection stages between all the individual manufacturing operations to check the accuracy of the single components and ensure that they will fit perfectly on subsequent assembly.

An intermediate warehouse stocks the finished parts in carefully planned groups and passes them on to the various assembly departments. Assembly methods vary according to the type of camera in question. At one time all assembly operations were carried out by hand, and they largely still are with very expensive cameras. The assembly operators are qualified by long training and experience, and their job is not only to put components together, but also to choose the components themselves of the right manufacturing tolerances (which are minute, anyway) to produce an assembly of zero tolerance.

With less expensive apparatus assembly operations are usually split into a number of comparatively simple steps requiring less highly skilled labour. It is also growing practice to produce sub-assemblies entirely automatically by machines. These are fed with components by special vibrator units which ensure that every screw, wheel, pressing, etc. arrives at the assembly point correctly orientated. The machines are self-checking, and automatically stop if a faulty component reaches the assembly stage, or if any of the assembly operations is not completed satisfactorily.

During assembly great care must be taken to prevent dust from getting into the camera, as the smallest speck in the wrong place can cause trouble later on.

The testing of components and of assemblies is frequently also carried out by machine. Modern electronic testing units can pass through screws and other small components at the rate of several thousand an hour. The machine rejects items that fall outside the tolerances laid down and at the same time counts up the number of rejected pieces.

Lenses and shutters are generally bought ready assembled from specialist manufacturers and fitted into the camera as a complete unit. Even camera manufacturers who make their own lens and shutter assemblies produce them in a separate department and deliver them complete to the general assembly shop where they are built into the camera body.

In either case lenses and shutters are retested before building into the camera, to ensure that they satisfy the exact specifications laid down in the production plan. This involves optical and alignment tests for lenses and speed tests for shutters.

Testing. The manufacture of the components and their assembly is checked at every stage, but each camera has to undergo a final comprehensive test before it is released for sale. Here all individual operations are thoroughly tried out, and the alignment and definition of the lens exactly checked. If any camera fails to pass the final test, it is returned to the appropriate point of the production line where the shortcomings arose. This test department is under the direct control of the management.

The cameras of a big manufacturer are sent out to all parts of the world. They therefore have to operate perfectly under any climate conditions. Constant functional tests are made in a refrigerator room at −20° F. and in a suitable oven at 120° F. to check the adaptability of the camera to temperature variations. Selected cameras are often placed in specially designed machines which tension and release the shutter as much as a million times to establish which parts break down first under heavy use.

Once the camera is released for sale it is packed in a special padded box along with the necessary instruction leaflets and passed on to the warehouse for dispatch. W.W.

See also: *Lens manufacture; Optical glass; Shutter manufacture; Production of photographic goods.*

CAMERA MOVEMENTS

Camera movements are all the different ways in which the film plane and the lens may be turned or moved from the normal position.

In a simple camera, the plane of the negative is parallel to the plane of the lens, and the optical axis of the lens passes through the centre of the negative at right angles. This arrangement is satisfactory for the general run of photography because such a camera reproduces scenes more or less as the eye sees them (after allowing for the fact that the light-sensitive surface of the eye is curved while the camera sees everything on a flat surface).

But the large professional stand cameras and technical cameras are almost always designed so that the angle of the planes of both the negative and the lens can be varied. The axis of the lens can also be shifted up, down or sideways in relation to the centre of the film plane. Every alteration made in this way produces a characteristic change in the appearance of the image.

Rise and Cross Front. This movement is provided by mounting the lens panel in guides so that it can slide to either side of, or above and below the normal axis. Only the lens panel is mounted in this way; there is no need for similar movements of the negative since the object is simply to displace one in relation to the other.

With the lens displaced in this way, the camera can look at things from above, below, or one side without being tilted at an angle to the planes of the subject. The objection to such tilting is that it produces an unpleasant appearance of distortion because the nearer parts of the subject are reproduced on a larger scale than those farther away.

The classic example of the effect of tilting the camera is the converging of the vertical lines in a photograph taken looking up at a tall building. But the same thing occurs with the horizontal lines when the camera looks at an angle at a line of buildings (when the lines are seen to converge as they run away from the camera). And the same effect can be noticed when the camera looks down on an object like a tall box. (In this case the vertical lines converge towards the bottom of the object.)

Distortion from the same cause affects subjects where there are no obvious vertical or horizontal lines—e.g., in the human figure. Here, in the absence of straight lines for comparison, the trouble is not so obvious, but it is there just the same, and the improvement is often striking when correction is applied.

The rise or cross movements are used in all circumstances like those quoted above, and wherever the camera has to look obliquely at the subject to show parts not properly visible in the normal square-on front view.

In practice, the camera is set up at the viewpoint which covers the subject from the required angle, but instead of being pointed at the subject, it is arranged so that the plane of the negative is parallel to the plane of the subject. The image is roughly focused, and then the cross front is used to slide the lens off centre to bring the subject into place on the screen.

As a result of this manœuvre the principal plane of the subject remains parallel to the negative plane and so is reproduced without distortion—the top of the box is visible, but the sides remain parallel; the belfry of the church comes into the picture without looking as though it were about to topple backwards, and the line of houses no longer narrows in violent perspective.

Used in this way, the rise and cross movements enable the photographer to control the perspective of subject planes inclined at an angle to the line of sight of the camera.

Back Swings. To provide the vertical and lateral back movements, the back is hinged or pivoted so that it can be swung about either side of the vertical and mounted on a base support which allows it to be turned horizontally about ten degrees each side of its normal position.

These movements are used to control the perspective of subject planes which are not at right angles to the camera sight line. If the back is swung to bring it parallel to the principal plane of the subject then parallel lines lying in the subject plane considered will appear parallel.

In practice, the camera is set up to look at the subject from the required viewpoint; the image is roughly focused, and the swing back is adjusted until it lies parallel to the plane of the subject in which parallel lines are to remain

parallel. After adjusting the lens panel in such a way that it is again parallel to the film plane (or as near parallel as possible) the image is then re-focused and the lens stopped down until the depth of field includes the whole of the subject.

The result is similar to that produced by the rise and cross movement (above) since in each case the plane of the negative is made parallel to the principal plane of the subject to correct the convergence of the important parallels.

In instances such as the above, if the rise and cross movements of the camera front, or the backswing of the camera, are individually insufficient, a combination of these can be used. When correction is made by means of the swing-back, so that the film plane lies parallel to that plane of the subject in which parallel lines are to remain parallel, then overall sharpness is regained by bringing the lens panel into a position parallel to the film plane. Care should always be

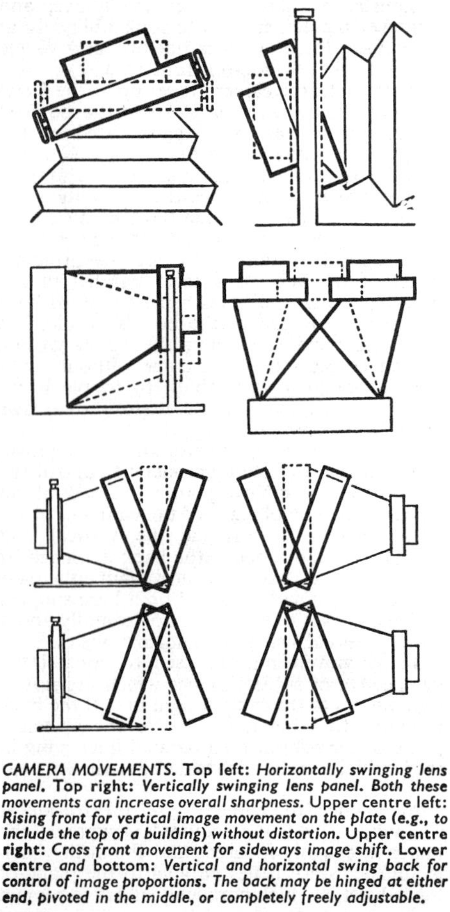

CAMERA MOVEMENTS. Top left: *Horizontally swinging lens panel.* Top right: *Vertically swinging lens panel. Both these movements can increase overall sharpness.* Upper centre left: *Rising front for vertical image movement on the plate (e.g., to include the top of a building) without distortion.* Upper centre right: *Cross front movement for sideways image shift.* Lower centre and bottom: *Vertical and horizontal swing back for control of image proportions. The back may be hinged at either end, pivoted in the middle, or completely freely adjustable.*

taken that the camera movements are not used to an extent beyond which the lens is able to cope, and the corners of the image should always be examined carefully on the ground glass screen (with the lens stopped down to the aperture which will be used, if this is practicable).

In practice, if the degree of effective rising front used on, say, a tall building is such that the top of the building is not quite sharp when film plane and lens panel are parallel to the face of the building, this can often be corrected by tilting the lens panel upwards slightly. Such movement must be only slight, however, and it calls for a little stopping down, in order to retain sharpness over the entire building.

Swing Front. The lens is mounted in a panel which can be turned about both horizontal and vertical axes; it can thus be tilted down or up or turned from side to side. Ideally, the turning axes should pass through the rear nodal point of the lens. If this condition is fulfilled, the lens may be turned (within the limits of its field) without changing the position of the image.

Front swing is used to make the image of the principal plane of the subject coincide with the plane of the film. It is not necessary when the film plane, lens, and subject planes are all parallel—e.g., for a square-on photograph of a flat-fronted building.

When the principal plane of the subject is not at right angles to the camera sight line, some parts of the subject are nearer the lens than others. If the middle of the subject is focused sharply, then the ends will be out of focus. The ends can be brought into sharp focus if the lens is stopped down far enough to include them both in the depth of field. But this means losing light and prolonging the exposure.

The other way is to move one side of the lens forward to bring the near parts of the subject into focus, and the other side back to focus the distant parts—i.e., to swing the front of the camera in the same direction as the slope of the subject plane.

There are two principal uses for front swing. It can be used simply as a focusing device for giving a sharp picture of a sloping subject without the need for stopping down, or to increase the scope of the vertical back-swing.

If the swing front is used simply to bring the focal plane to coincide with the film plane when the back of the camera has been swung parallel to a sloping subject plane, the result is exactly as though the picture had been taken by a camera with a large amount of rise or cross front movement.

The same effect can be produced in a camera with a long format by focusing the picture on one half of the plate only—e.g., by taking a picture of a tall building on the bottom half of the plate. This is equivalent to using a camera of half the negative format with a very generous amount of rising front, or the equivalent combination of front and back swing.

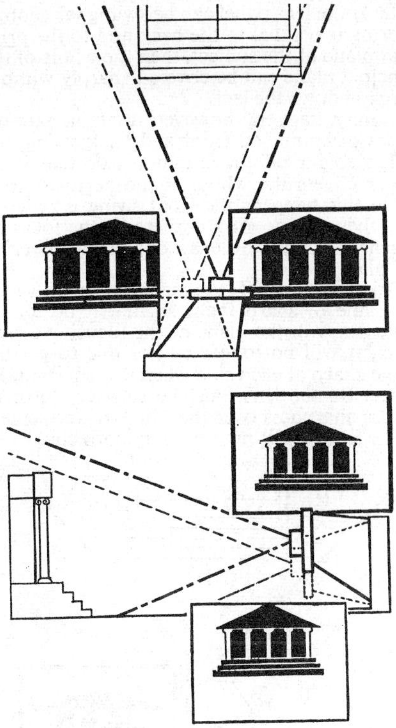

RISING AND CROSS FRONT. Top: *Sometimes the camera position cannot be altered and part of the image is cut off on the negative (thin frame and lens angle). The cross front shifts the whole image sideways, presenting a different view while avoiding distortion of perspective (heavy lines).* Bottom: *The rising front does the same thing vertically in order to include, for example, the top of a building or to reduce the foreground area without tilting the camera. Rising and cross front movements do not distort the image proportions.*

Combined Movements. Camera movements are used both independently and in conjunction.

The swing back and the rise and cross front movements may be used simultaneously to control perspective—generally to produce a more pleasing perspective although they may also be made to produce deliberate distortion.

The swing front alone has no effect on perspective; its purpose is simply to make the plane of sharp focus of the image coincide with the plane of the negative material. The object here is to avoid having to stop down the lens to bring both near and far ends of an inclined subject plane within the depth of field.

Front swing is often used alone—e.g., to bring foreground and background into focus without stopping down too much. It may always be used alone, in fact, for improving the depth of field over an inclined subject plane where there is no need to correct the perspective of the subject.

Back swing, on the other hand, is only rarely used without some degree of front swing. This is because applying back swing to compensate for camera tilt throws the plane of the negative out of register with the focal plane of the lens. Unless some front swing is added the photographer is forced to stop the lens right down and give a needlessly long exposure.

The amount of rise or cross front movement is decided by the position of the image on the focusing screen. The swing back can be set parallel with the principal plane of the subject by eye or measurement. The angle of the swing front can only be decided by careful examination of the sharpness of the image on the screen, preferably through a focusing magnifier.

Camera Movements in Practice. A photograph fully corrected by the use of camera movements is made as follows:

(1) The photographer selects the viewpoint which shows the subject to the best advantage

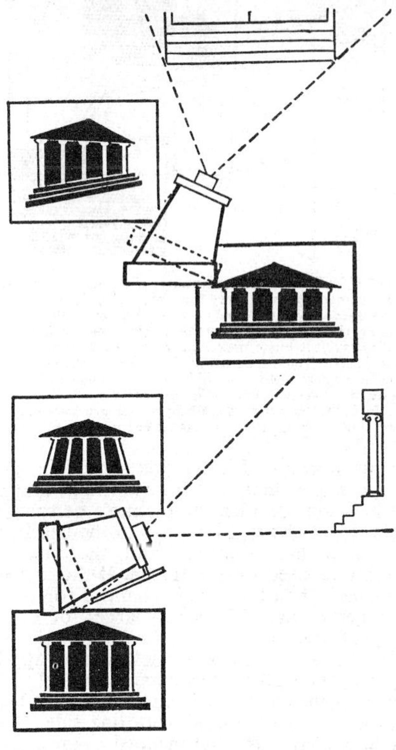

SWING BACK. Top: *When the camera points obliquely at the subject normal perspective effects appear, with near parts shown on a larger scale than distant ones. Swinging the camera back until it is parallel with the main subject direction, alters the perspective to give an isometric reproduction.* Bottom: *In the same way a vertical camera swing will straighten out converging verticals. In both cases the lens panel should be placed parallel to the film plane and the main subject direction, in order to retain overall sharpness of the image.*

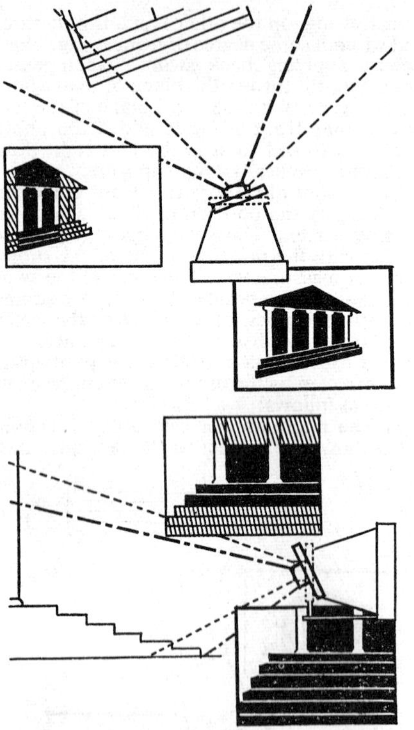

SWINGING LENS. Top: *If some parts of the subject are nearer to the camera than others, the depth of field may not be sufficient to make the image sharp all over (thin frame). Swinging the lens in the direction of (but not parallel to) the principal subject plane makes near and distant parts equally sharp (heavy frame).* Bottom: *The same principle is used when the camera looks down obliquely on a subject of appreciable depth; a vertical lens swing increases the over-all sharpness. In both cases the image proportions are somewhat distorted, and the image shifts across the negative.*

for his purpose. Ideally this should be done by eye alone since it is the viewpoint and not the lens that decides the picture's perspective.

(2) The camera is set up with the baseboard level and back vertical and the image is roughly focused on the screen. If the image is too small, then the lens is changed for one of a longer focus; if it is too large, for one of shorter focus.

(3) At this stage lateral back swing is applied to bring the back of the camera parallel to the subject plane—e.g., the front of a building being photographed obliquely from one side. If the whole camera has been inclined at an angle—e.g., to look up at a tall building—vertical swing is used to bring the back of the camera vertical again. Lateral swing is adjusted by eye but vertical swing can if necessary be checked with a plumb line.

(4) The front swing is now adjusted to give the best distribution of sharpness over the plate. If the lens panel can be swung far enough to bring it parallel to the back and to the principal plane of the subject, then the whole of the principal plane can be covered sharply without stopping down the lens.

It may happen, however, that at extreme angles of combined front and back swing, the lens is so far off the common axis that it no longer covers the whole of the negative area. When this happens the front swing is reduced until there are no dark corners on the focusing screen—i.e., until the lens is once again covering the plate.

This change, of course, will also swing the plane of sharp focus so that it no longer coincides with the plane of the focusing screen —i.e., it will no longer be possible to get the image sharp at each side when it is sharp in the centre. So the lens must be stopped down to restore sharpness over the whole of the screen.

The image will now be fully corrected—i.e.,

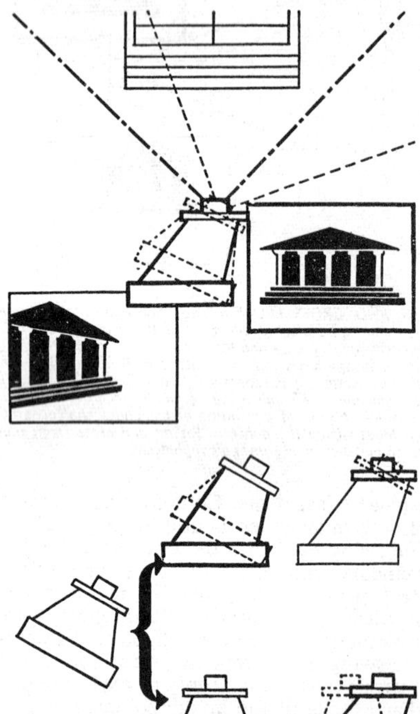

SWING BACK AND SWINGING LENS. Top: *Use of the swing back to correct the perspective rendering can be combined with the swinging lens panel to increase the image sharpness. The distortions arising from the two movements will cancel each other out if the lens panel and camera back end up parallel.* Bottom: *In that case the swing back and swinging lens act together like a cross front movement. The effect of swinging the back and then the front is the same as that of turning the camera and moving the front horizontally.*

all vertical lines will be parallel to the sides of the focusing screen; the horizontal perspective will be acceptable; the relative proportions of near and far parts of the subject will appear natural, and the image will be sharp wherever sharp definition is called for. F.P. & V.G.

See also: *Architecture; Perspective; Technical camera.*
Book: *Basic Photography*, by M. Langford (London).

CAMERA OBSCURA. Literally, a "dark chamber". The term is applied to any optical device which forms a visible image of a scene on a screen set up in a darkened tent, room, or even a large box.

Except for the shutter, the camera obscura possesses all the essential elements of the photographic camera and can be regarded as its immediate ancestor.

From the early days of our civilization men appear to have known that light passing through a very small hole in the wall of a darkened room will form an inverted picture of the scene outside on the opposite wall. The image formed in this way is neither brilliant nor sharp. Making the hole bigger increases the brightness of the image but makes the definition worse.

Both the brilliance and sharpness of the image are greatly improved when a converging lens is used instead of a simple hole. The image can then be focused sharply on the screen.

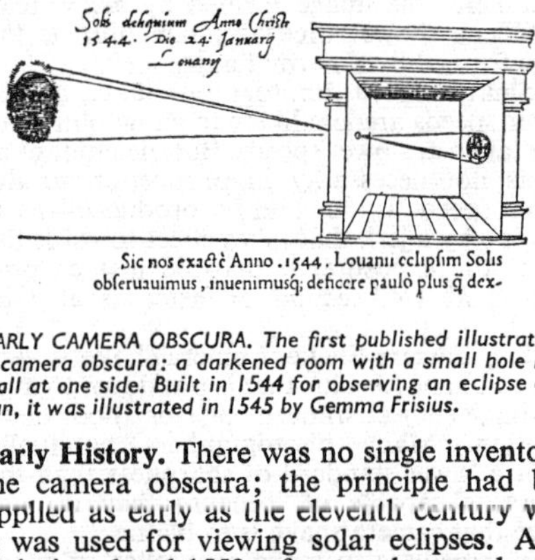

EARLY CAMERA OBSCURA. The first published illustration of a camera obscura: a darkened room with a small hole in the wall at one side. Built in 1544 for observing an eclipse of the sun, it was illustrated in 1545 by Gemma Frisius.

Early History. There was no single inventor of the camera obscura; the principle had been applied as early as the eleventh century when it was used for viewing solar eclipses. A description dated 1550 refers to a lens, and one in 1568, to a lens and diaphragm.

The first really complete description of the camera obscura was given by Giovanni Battista Della Porta in 1558. On the strength of this he is usually regarded as the inventor.

Up to the seventeenth century the camera obscura was always a room in a house. One of its commonest uses was for producing "magical" effects. Charlatans, fortune-tellers and magicians employed the principle to mystify their clients. With its aid they conjured up miniature moving pictures of people and strange animals, often making them appear inside a glass of water or wine, the real action being staged outside the house.

With all such devices the observers had to be stationed on the inside of the dark chamber. A movable version in the form of a tent was mentioned in 1620, but it was not until the middle of the century that truly portable models appeared. These were commonly large (often collapsible) boxes equipped with a single biconvex lens, a reflecting mirror, and a viewing screen of oiled paper.

In this form the camera obscura was much used by artists as a guide to outline and perspective. It was often taken along by the eighteenth century tourist making the Grand Tour because it enabled those with no draughtsmanship to make sketches to illustrate the inevitable diary of the tour. It was, in fact, the equivalent of the present day snapshot camera.

Improvements. The camera obscura remained essentially unchanged from the end of the seventeenth century until the beginning of the nineteenth when Wollaston got over the serious spherical aberration of the double convex lens by using a meniscus. Chevalier took it a step further by using a reflecting prism with curved faces in place of the lens and mirror. No significant improvements have been made since.

In the common form of camera obscura, the observer stands in a darkened room or tent and an image of the external scene is projected on a white table-top before him. Immediately above the table there is a long focus lens with its axis vertical and pointing at the centre of the table. A mirror, set at 45° to the vertical is mounted outside the chamber above the lens so that it directs rays of light from the surroundings through the lens which focuses the image on to the viewing screen. In permanent types, the mirror can be rotated by the observer to scan the whole of the view outside. Permanent installations of this type are popular at seaside resorts where, for a copper or so, visitors can enjoy the scenery at their ease.

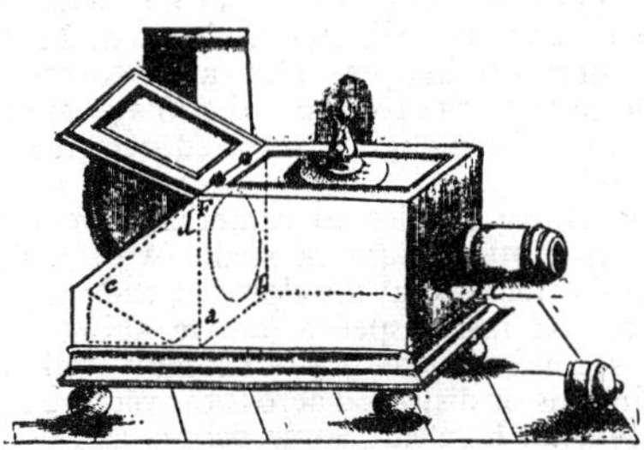

ARTIST'S CAMERA OBSCURA. Portable reflex box camera obscura with lens, first described by Johann Zahn in 1685.

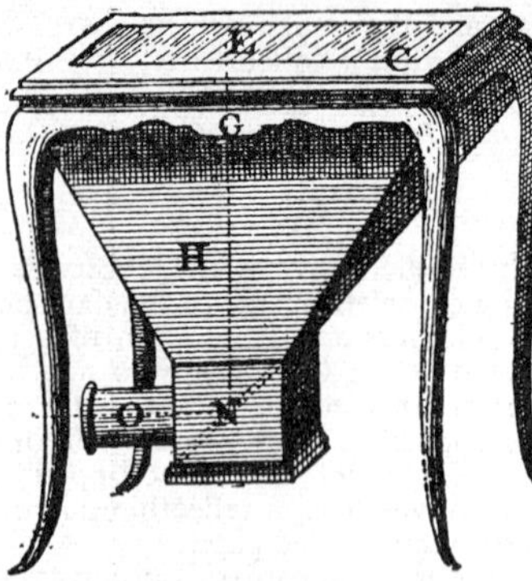

TABLE CAMERA. Guyot's table camera obscura of 1770 with lens O and reflex mirror N. The image was observed on the translucent surface E and was reversed left-to-right. The instrument made some attempt to combine semi-scientific utility with a semi-artistic finish to enable it to fit into a well-furnished drawing-room without appearing seriously out of harmony with the interior style of the period.

Artists have always made great use of the camera obscura and it was a natural step from tracing the projected image on a sheet of drawing paper to seeking for some means of fixing it with chemicals. Thomas Wedgwood about 1799 made the first photographic experiments with the camera obscura. Daguerre, who in his early days was a successful painter, made use of a camera obscura which is now in the possession of the Conservatoire des Arts et Métiers, Paris (Museum of Arts and Crafts).
F.P.

See also: *Camera supports; Holding the camera.*

CAMERA SHAKE. Blur is often caused, not by movement of the subject, but by movement of the camera at the time of exposure. This movement, generally referred to as camera shake, cannot be classified as neatly as subject movement. It depends on a number of variable factors: the human element; the camera; the focal length of lens; the shutter speed.

Easily recognizable characteristics of camera shake (distinguishing it from other types of unsharpness) are an equal degree of blurring of all parts of the picture, and frequently the appearance of two slightly displaced and somewhat unsharp images. This also covers the whole of the negative area; the image appears as if it had been slightly smeared over the film or plate.

The above applies to cameras with a diaphragm shutter. Camera shake with a focal plane shutter used at the slower speeds appears similar. At faster speeds (above about 1/30–1/60 second) image positions may look discontinuous or distorted across the picture area.

The Human Element. Some people can hold a camera steady at shutter speeds as slow as 1/10 second, others are troubled by camera shake even at a shutter speed of 1/100 second.

Part of the cause undoubtedly lies in the way the camera is held and how the shutter is released. The standard advice on holding the camera is to stand with the feet apart and steady the camera against the chest or face (depending upon the type of viewfinder). Before making the exposure some photographers find it helpful to take a breath, hold it and at the same time squeeze the shutter release steadily, rather than trying to click it erratically.

While this advice will do much to reduce serious camera shake it is always wise to avoid shake by taking advantage of any support available and using the fastest shutter speed possible.

The Camera. The design of the camera makes it more or less prone to suffer from shake. Generally, large, heavy, and square cameras give the least shake, and small, light, and streamlined cameras, the most. But the position, angle, and release pressure of the shutter control may have as much influence as the size and shape of the camera, and a camera that is free from shake when used at waist level may be impossible to hold steady when using the eye-level finder.

The Lens. Generally, the longer the focal length of the lens the greater the risk of camera shake. For the same amount of angular movement of the camera, the image formed by a 2× telephoto lens shows twice as much blur as the image formed by a normal angle lens.

The Shutter Speed. Photographs taken at fast shutter speeds are less likely to show blur than those taken at slower speeds. But the amount of blur is not necessarily in proportion to the shutter speed. If the blur is produced—as it often is—by a jerk that takes effect towards the end of the exposure, it may be just as pronounced at 1/40 second or faster as at 1/20 second.

Avoiding Camera Shake. Camera shake is very often to blame for blur that is put down to misfocusing, subject movement, or poor lens definition. Many photographers habitually accept a lower standard of sharpness than the camera is capable of, because they do not realize that camera shake is to blame.

The remedy is to make a number of exposures with the camera rigidly clamped, say, to the top of a heavy table.

This will show the standard of definition that the camera will give when there is no shake. The photographer should then experiment to find the shutter speed and methods of holding the camera and releasing the shutter that will give this standard of definition for hand-held exposures.

Trials carried out with a large number of operators show that the slowest shutter speed that the average photographer can use without undue fear of camera shake is about 1/125 second.

See also: *Holding the camera.*

CAMERA SUPPORT

No matter how good a camera may be, or how accurately it has been made, it cannot yield really acceptable results if it moves during the exposure. Much of the poor definition obtained by amateur and professional photographers is caused by camera shake during the time the shutter is open.

Tripods. At shutter speeds slower than 1/25 second some sort of camera support is always essential. A tripod is undoubtedly the most effective answer to camera shake and should be used wherever possible; certainly whenever top quality results are required.

A tripod is a three-legged stand on the top of which the camera is fixed. Except for some of the larger studio models most tripods have folding or telescopic legs. There is nothing to beat the heavy wooden tripod for strength and rigidity. It is ideal for studio and indoor use but it is rather cumbersome to carry around and for this reason it is no longer popular with modern photographers who like to travel light. The necessity for portability has led to the lighter, collapsible type of tripod. Many of these are excellent, but some of the cheaper models are unsteady and useless.

The light folding type of tripod is not suitable for a heavy plate camera. A heavy camera needs a heavy tripod; there are even people who prefer to use a heavy tripod with a small camera rather than risk camera shake.

Methods for altering the height of a tripod vary according to the type being used. With a telescopic tripod it is only necessary to slide a section of the tubing in or out until it reaches the correct height. Folding tripods can be adjusted similarly by unfolding the legs to the height required. The permanent, studio types of tripod and some of the better metal types have an adjustable centre pillar.

Stands. The simplest type of stand consists of a solid, heavy base carrying a vertical pillar with a platform on top to take the camera. The centre pillar is adjustable for height and the platform can be tilted to point the camera up or down. In the more expensive versions of such stands all the movements are controlled by wheel or handle operated screws. One handle raises or lowers the pillar, another turns it to rotate the camera in the horizontal plane and a third controls the tilt of the top. The base of the stand may be on wheels with provision for locking them or putting them out of action when required.

A stand built for heavier duty consists of two vertical pillars mounted on a heavy low trolley. The camera is carried on a counterbalanced tilting platform between the two pillars. This type of stand has the advantage of allowing the camera height to be adjusted from the top of the pillars—which may be 7 or 8 feet high—down almost to ground level.

An even more elaborate type of camera stand is used in many commercial studios. This is the "all-angle" stand. It consists of a rigid H-section alloy chassis mounted on wheels. The chassis is designed to hold a large-diameter tubular column which carries the camera platform. The column is supported in a universally-mounted housing which allows it to be swung to any angle, and extended or rotated by control handwheels. Separate controls below the camera platform allow the head to be tilted in two planes. This type of stand is also used for clinical photography in hospitals where it allows the camera to be supported over the top of a prone patient at any angle. The stand is mounted on large castors which can be raised out of contact with the floor when once the camera is in position.

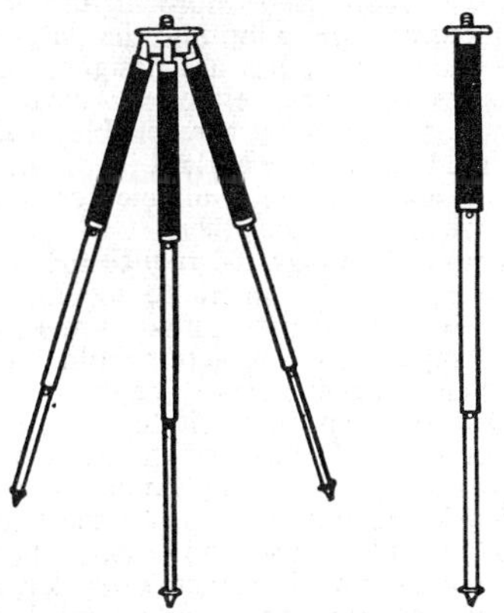

CAMERA SUPPORTS. Left: *Telescopic tripod with three extending legs. The tips may carry spikes (for rough surface) or rubber ends (for smooth floors). The tripod screw is ¼ in. (English) or ⅜ in. (Continental) Whitworth* Right: *Telescopic monopod.*

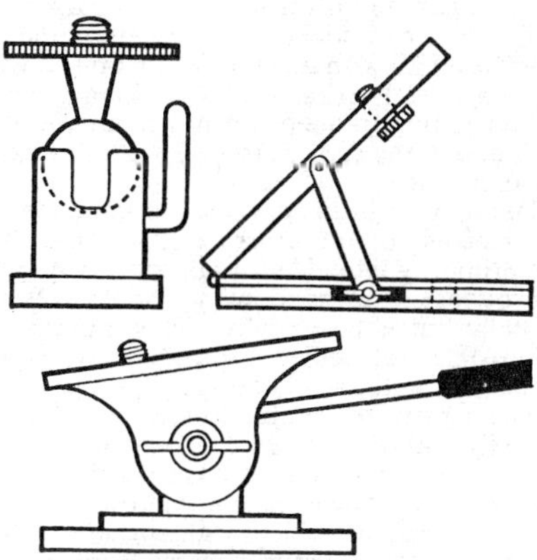

TRIPOD HEADS. Top left: *Normal ball and socket head adjustable in all directions, locked by a clamping lever.* Top right: *Tilting platform for larger studio and field cameras.* Bottom: *Pan and tilt head for a cine camera.*

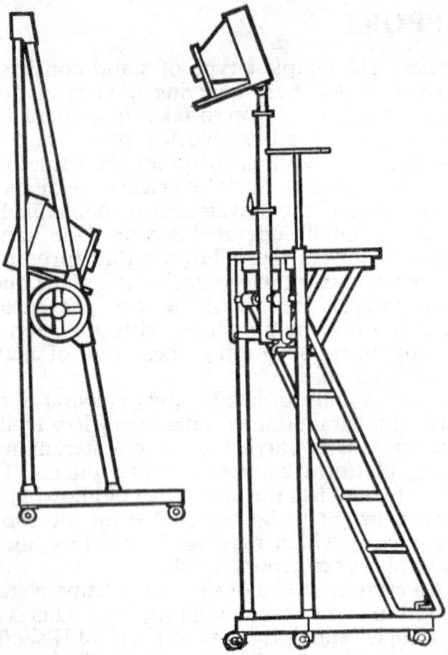

STUDIO STAND. Stands for studio cameras have to be particularly solid to support the heavy camera and provide the full range of adjustments. Left: Generally the camera is mounted on a tilting platform which moves up and down along braced vertical columns. The whole stand is mounted on castors. Elaborate professional stands may have a separate platform (with ladder) for the operator as well as for the camera.

Heads. The camera must be attached firmly to the tripod or stand, and although there is always a plate or table with an attaching screw, this is not always satisfactory as it does not allow the camera to be set at an angle to the horizontal. The two most popular tripod heads, designed to allow adjustment of the camera angle, are the "universal" ball and socket head, and the pan and tilt head.

The ball and socket head, as its name suggests, is a ball with a tripod screw on top which can turn freely in a socket. A clamp can be tightened to grip the ball firmly in any position. This allows the camera to be tilted or angled as the user wishes.

Pan and tilt heads provide the same range of movements but are rather more versatile. They are primarily intended for cinematography and can be locked so as to swing (or "pan") horizontally but not vertically. This movement is impossible to achieve satisfactorily with a simple ball and socket head but it is essential for a cine camera to be able to scan a scene horizontally without any trace of wobble.

The panoramic head is basically the same as the pan and tilt head except that the axis of the panning mechanism has a scale engraved around it. This is calibrated in degrees from 0 to 360 and either moves with the head or is fixed on the stationary part. Panning of the camera can then be measured in degrees against a suitable reference mark engraved on the head. This is invaluable for photographing a panorama in sections when, after each successive exposure, the camera is panned to a fresh position. The angle through which the camera must be moved each time is equal to the angle of field of the lens (in practice, slightly less).

Other tripod heads are made for specialized subjects. The stereo head provides two alternate positions for the camera with a separation of 2½ ins. for taking stereo pairs for three dimensional viewing. For astronomical photography there is the equatorial head which when driven by a special clock, keeps the camera following stars during time exposures. Another type of head available, particularly for twin lens reflex cameras, is designed to overcome parallax errors in sighting the camera; it allows the camera to be slid upwards, on a tube, to the exact height necessary to compensate for the parallax. This height is normally preset.

A special type of tilting head is useful for certain subjects such as still-life close-ups, copying, etc., taken with a heavy camera. The head there consists of two wooden platforms hinged book fashion at one end, and fitted with adjustable struts at the other to open the two halves to the desired extent. The bottom half is mounted on the tripod and the top carries the camera. The more the two halves are opened out, the greater the tilt; usually the adjustment possible is from 0° to 90°. Although this type of head permits tilting in only one direction, it is considerably firmer than a ball and socket or other type of tilting head.

Threads. All types of support provide a camera attachment in the form of a projecting screwed stud which screws into the camera tripod bush. The stud may be either ¼ or ⅜ in. Whitworth. Until fairly recently most British cameras were fitted with ¼ in. Whitworth tripod bushes, and Continental cameras with ⅜ in. Whitworth. Nowadays more and more British camera manufacturers are adopting the larger size. Many camera supports are designed to offer both sizes of fitting, either by means of a separate adapter or by a reversible head with a ¼ in. stud on one side and a ⅜ in. on the other. Separate adapters are available for adapting either size thread to the other.

Quick-release Fittings. To reduce the waste of time over fitting a camera to the top of the tripod by the normal screw, some camera makers supply a quick action holder for use with their own cameras. The holder has a tripod bush which allows it to be fitted onto the tripod head in place of the camera in the normal manner. With the holder in position the camera can be mounted on the tripod or removed from it in a second or two. The actual method of securing the camera body depends on the make of camera and most quick-release attachments have to be designed for a specific camera and cannot be used with other makes. However, universal quick-release fittings use an attachment on the camera (screwing into the tripod bush) which mates with a fitting on the tripod.

Gyro Supports. When the camera has to be supported on a moving base—e.g. on a vehicle moving over a rough surface, on a boat or in rocket-borne equipment—it may be mounted on a gyro-stabilized platform. This is essentially a camera support housing a spinning gyroscope driven by compressed air or electricity. The whole assembly is slung in gimbals which leave it free to turn in both horizontal and vertical planes. When the gyroscope is spinning, the support is held "straight and level" no matter how the carrier changes its altitude or direction. Simple versions use the gyroscope mounting merely to increase the camera inertia and so smooth vibrations and reduce camera shake. Alternatively, the gyro may be mounted separately and be used merely for feeding stabilizing signals to the control system of the entire vehicle in which the camera is carried.

In motion picture studios, the gyroscopic effect is applied to the camera support for another purpose. A gyroscope is mounted in the tripod panning head with its axis horizontal. When the gyroscope is spinning it opposes the panning movement of the head with a force proportional to the rate of turning—i.e. the faster the operator tries to move the head, the greater the resistance he has to overcome. The result is that the panning motion can be started, held steady or stopped smoothly and without sudden jerks or changes of speed.

A similar damping effect is obtained by incorporating an escapement gear with flywheel into the tripod head. The gear is coupled with the panning or tilting movement (or both) and is directly driven by the manual effort of moving the camera on the tripod.

Other Supports. A camera clamp can be a very useful support when it is inconvenient to carry a tripod. It is simply a pair of metal jaws which can be fixed to solid objects such as a table or a fence with a screw to hold the camera. Most of them incorporate a "universal" fitting to point the camera at any angle.

The pistol grip provides a steady single-hand hold, which is useful on both still and cine cameras yielding shake-free negatives at exposures considerably longer than are possible with the conventionally hand-held camera. A built-in cable release enables the shutter to be released by means of a trigger on the grip. The pistol grip is not, of course, suitable for time exposures.

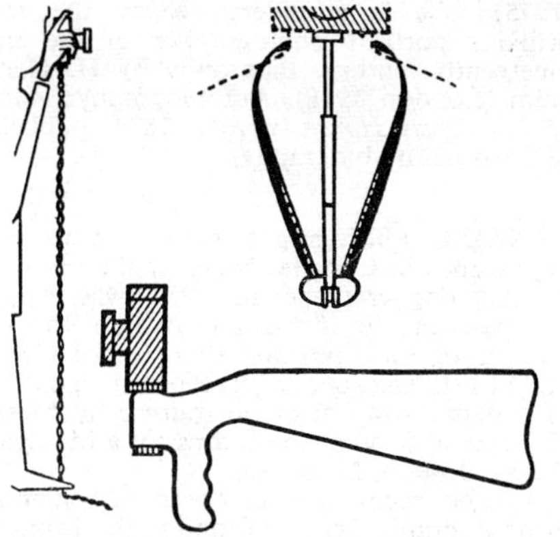

SPECIAL SUPPORTS. Left: *Steadying chain fastened to camera.* Top right: *Chestpod.* Bottom right: *Gun stock holder for long-focus sports photography with miniature cameras.*

The monopod is a support consisting of a single leg which is used for keeping the camera still during slow exposures. Usually it is a telescopic tube which extends to about 5 feet. The camera is attached to the top and the foot has a spike or a rubber ferrule to stop it from slipping. Exposures of up to 1 second are usually quite safe.

The chestpod is a shortened variation of the monopod. It consists of a metal rod about 12 ins. long with a screw for the camera at one end and a leather strap at the other. The strap is then hung round the neck so that a downward pressure on the rod tautens it, and enables the camera to be held steady at eye level.

The gunpod is another light-weight portable support. As its name implies this is shaped like a rifle stock and fits comfortably into the shoulder. Many people use a gunpod to steady their cine cameras in the absence of a tripod and find it a reasonable substitute. This type of holder is also used for wild life photography with a miniature camera and a high-powered telephoto lens.

The chainpod is an extremely simple support that works quite well in practice. A piece of chain about 5 feet in length has a screw at one end which is fixed to the camera. The other end is then dropped on the ground and the user places his foot on it. By pressing the camera upwards the chain tautens and holds the camera steady. The chainpod can be used equally well with both eye-level or waist-level cameras. G.W.P.

CAMERON, JULIA MARGARET, 1815–79. English amateur photographer. She took up portrait photography at the age of 48 and specialized in close-ups. Many notables of the Victorian age sat for her, including Browning, Tennyson (a family friend), Sir John Herschel, Carlyle, Longfellow, Darwin, and others. Many of her imaginative compositions are rather pre-Raphaelite in style, in particular her illustrations to Tennyson's *Idylles of the King*

(1875). She is considered today the most brilliant portrait photographer of the mid-nineteenth century. Biography by H. Gernsheim (London 1948). Autobiography: *Annals of My Glass House* (written 1874, published in Gernsheim's biography).

CANADA. Photography made a very early appearance in Canada. News of the invention of the daguerrotype in 1839 was eagerly reported in the Press and within fourteen months of the invention the *Quebec Gazette* (dated 13th October 1840) recorded an account of a demonstration of portraiture by Messrs. Halsey and Sadd at the address of a M. Grace, rue St. Joseph, Montreal.

Another recorded reference to the beginnings of photography is to be found in the Montreal Directory for 1842–3, in which is entered: "Prosch (G.W.), photographer, Place d'Armes, East Side". And in the *Minerve* of 28th October 1847, a reporter refers to a M. Desnoyers who proposed to devote his time to daguerrotype photography.

The demonstration of 1840 is most likely the earliest use of photography in Canada.

Commercial and Technical Progress. Perhaps the most important and influential firm in the practice of photography from its early establishment onwards in Canada was that founded by William Notman in Montreal in 1856. At the start daguerreotypes and ambrotypes were made. A notable work was a series of photographs showing the setting up of the Victoria Bridge stage by stage, which together with a series of views of Canada East and Canada West were presented to the then Prince of Wales (later King Edward VII) in bound form when he officiated at the opening of the bridge in 1860.

Then followed the use of wet plates and the popular *carte de visite* size portrait, and when dry plates came into use in 1880 W. Notman started to make them in the studio. But his output was not great enough to meet requirements, and he arranged with Stanley Bros. of Lewiston, Maine, to open a factory for their manufacture. When platinum paper was introduced in 1890 he obtained a printing machine from the Platinotype Co., London, and used this printing-out process until its discontinuation in 1914.

The London *Photographic News* (reported in the *Canadian Illustrated News* of 23rd July 1870) confirms W. Notman as one of the leading photographers of his time, and also mentions an associate of his, Frazer, as well as Inglis of Montreal, and Messrs. Sheldon and Davis of Kingston. Their work featured in exhibitions throughout the world and received much favourable comment and recognition by medals and other awards. Notman's work was forgotten until some of his portfolios were published recently by a leading Canadian magazine.

Manufacturing. When the dry plate was introduced Notman first approached British manufacturers to set up a plant in Canada, but he met with no success and this led him to make the arrangements with the American firm of Stanley Bros. referred to above. Then in 1900 the Canadian Kodak Company, Ltd., was formed in Toronto, and from its small beginnings has grown continuously. In May 1963 it was announced that a new expansion programme was being undertaken on the 45 acre site to increase facilities to 1,040,000 square feet, which would place Canadian Kodak in a position to supply all their Canadian requirements from Toronto.

Currently 90 per cent of the black-and-white film sold by Kodak in Canada is made in Toronto and there are large production figures for paper, motion picture films, X-ray materials and graphic arts film. Kodak colour- processing plants exist in Toronto and at the new Pacific Coast plant in North Vancouver.

1963 has also seen the opening of a manufacturing plant for Ansco of Canada Ltd. and the assembly in Canada for the first time of many products of Argus design.

Bell and Howell (Canada) has for some time been active in the assembly of cameras and projectors in Toronto. Since 1952 high precision optics and instruments have been produced in Midland, Ontario by Leitz (Canada) and many of the current lenses for the Leica camera have been designed and manufactured there. The company also maintains assembly lines for Leica cameras, and has been particularly active in the design and production of optical instruments to special industrial and government specifications.

Consumption of photographic materials is split evenly between amateur photographers and all professional fields. Canadian production of photographic materials approximates $30 million annually whilst the retail sales of photographic equipment and materials, including processing, approach $200 million. Photographic imports since 1953 have doubled, the 1961 value being placed at $52 million.

Associations. Most provinces have their own professional associations, and the Professional Photographers of Canada are making progress toward affiliation of the several groups. The greatest obstacle is distance.

Formed under Federal Charter in 1948, the Professional Photographers of Canada was formerly known as the Commercial and Press Photographers' Association of Canada, and its zeal in stressing education in photography both to photographers and to the general public has been exemplary.

This association has for 17 years exhibited its annual National Print Show from coast to coast in an endeavour to interest the public in fine photography. Within the ranks of its own membership it has since 1954 conducted an Annual Short Course at the University of

Western Ontario, covering a wide variety of professional techniques under the instruction of leading Canadian and non-Canadian professionals. Members travel to this course from all parts of Canada, often thousands of miles.

The Professional Photographers of Canada also organize at regular intervals a series of lectures of 8 to 10 weeks on a weekly basis on subjects allied to photography, including optics, and colour. The 1963 series was on visual communications and photography, and was held at the Doon School of Fine Arts, an affiliate of the University of Waterloo.

The Canadian Photographic Trade Association, concerned primarily with matters within the industry, was founded in 1956. It also maintains an active interest in the promotion of photography by its assistance with public exhibitions and the preparation of material for general publication.

One nation-wide association which has proved very successful is the Colour Photographic Association of Canada, which has a membership of over forty clubs. The Association is devoted solely to colour photography, and holds national conventions, regional meetings and a full complement of association activities, and provides instructional aids for the benefit of its member clubs.

The Federation of Canadian Amateur Cinematographers has been formed to serve the interest of amateur motion picture workers in Canada and embraces fifteen amateur cine clubs within Canada. It concerns itself actively with film festivals in Canada, and internationally through Union Internationale de Cinema Amateur, of which it is a member.

Activities. It has been estimated that 70 per cent of the population use cameras, and that over 250 million pictures are taken each year. Automation in cine and still photography has been a stimulus to the more casual type of amateur photographer and has led to an increase in camera club memberships.

Whilst camera clubs are active across the country, there is no central association to absorb them, although steps in the past few years to form a federation of the many clubs in the Toronto area are setting a pattern for possible extension.

The Toronto International Salon of the Toronto Camera Club, founded in 1888, is the oldest Salon of Photography on the North American continent and still continues successfully today.

Present-day Photography. Contemporary use of photography is probably as extensive in Canada as anywhere in the world, particularly with regard to commercial, industrial, and scientific applications. Uses rather special to Canada are in the aerial surveying of the country, much of which has not previously been mapped, and in the production of mail order catalogues.

A few Canadian photographers of today have established for themselves international reputations, especially in the field of portraiture.

Motion picture production in Canada has brought in a great many international awards, particularly to the National Film Borad. With the emphasis on documentary, training, educational and industrial motion pictures, and serving the most widespread TV network in the world (4,000 miles from Newfoundland to Vancouver Island), motion picture photography is a prolific recording and artistic medium in Canada.

Statistics indicate that over 44 million feet of 16 mm. film and 20 million feet of 35 mm. film were processed in Canada during 1961, and the country is well supplied with film libraries, usually under the auspices of provincial governments, education departments and the Natianal Film Board.

The vicinity of the United States, with its large photographic press, has made the publication of magazines for the amateur photographer in Canada a project of some hazard. Additionally the advertising revenue necessary to support production costs is very limited in a country of 18 million inhabitants. The monthly publication *Photo Age* serves the amateur field.

Many Canadian newspapers carry weekly photographic pages and supplements, and weekly radio programmes by well-known photographic personalities have been regular features for some years. With the opening of Toronto's first private television station in January 1961, the first of several weekly series for amateur photographers began. This was eventually followed by a similar series on the Canadian Broadcasting Corporation's television stations. These must rank with the first programmes of this type seen anywhere on a regular basis.

R. de L.G. & G.M. (J.B.W.)

CANADA BALSAM. Viscous fluid resin obtained from the Balsam Fir of North America. The liquid form of the balsam slowly changes into a hard crystalline solid with a refractive index almost exactly equal to that of glass.

It is used for cementing together pieces of optical glass—e.g., the component parts of a lens—to make an invisible joint. It is also used in some negative varnishes.

Solubility: Insoluble in water; soluble in xylol.

See also: *Adhesive.*

CANDELA. Photometric unit of luminous intensity (formerly known as candle power) and defined as 1/60 of the luminous intensity of 1 sq. cm. of a black body at the temperature of solidification of platinum (1,769° C.). Also known in common usage as the candle, and approximately equal to the former international candle. The candela is now a standard international unit.

See also: *Light units.*

CANDID PHOTOGRAPHY. Term coined by the art editor of the *Weekly Graphic*, London, on 11th January, 1930, for some revealing photographs of politicians, taken by Dr. Erich Salomon. A candid photograph is a picture of a person, or persons, engaged in some natural, believable activity. It is unposed and is often taken without the knowledge of the subject. Very early in the history of photography inventors began constructing—and sometimes marketing—cameras that the photographer could conceal about his person to enable him to take photographs without being detected. But true candid picture making only became really popular with the development of the modern miniature camera.

With the advent of the large circulation illustrated magazines throughout the world, in the 1930s, photo-journalistic work became well known. A high percentage of such work is of a candid nature.

Many photographers have achieved world fame for their candid photography. Notable names include Henri Cartier Bresson, Alfred Eisenstaedt, W. Eugene Smith, Robert Capa, and Margaret Bourke-White.

There are two ways in which the candid approach is helpful: for scenes of everyday life including or featuring people and children, engaged in their everyday activities and for procuring natural and unaffected portraits of

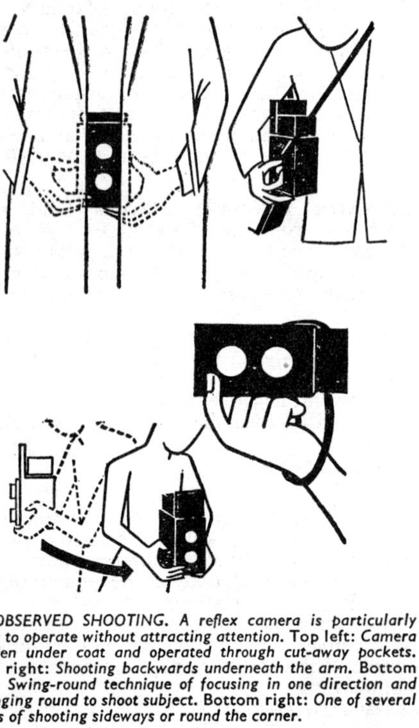

UNOBSERVED SHOOTING. A reflex camera is particularly easy to operate without attracting attention. Top left: Camera hidden under coat and operated through cut-away pockets. Top right: Shooting backwards underneath the arm. Bottom left: Swing-round technique of focusing in one direction and swinging round to shoot subject. Bottom right: One of several ways of shooting sideways or round the corner.

subjects who would be awkward and self-conscious if they knew they were being photographed.

Camera. The best camera for candid photography is the 35 mm. miniature because it is small enough to be used unobtrusively and its operation is as nearly automatic as coupling and interlocking can make it. At the same time many workers prefer the twin lens reflex because it can be used at waist level and the photographer does not need to advertise his intentions by looking at the subject. The miniature user, however, can make use of a long focus lens to let him work at a distance from his subject and thus stand a better chance of getting his pictures without being noticed.

A camera with automatic film transport is better than one that has to be checked visually at the risk of attracting the attention of the subject.

Very fast shutter speeds are not needed; probably the most useful speed is around 1/100 second although there are occasions when a short time exposure may be unavoidable.

The quieter the shutter the better; even the slight click of a high grade shutter can be embarrassingly evident. Indoors, in fact, it may be worth while using a supplementary silent shutter in front of the lens of the type made for studio portraiture and nature work.

Lens. Each type of lens has points for and against it. The long focus lens lets the photographer shoot from a safe distance, but it calls for accurate focusing and sighting because it has a narrow angle of view and a shallow depth of field. A wide angle lens can be set to give a deep zone of sharp focus and includes so much of the scene in front of the camera that it can be sighted quite casually. The virtues and shortcomings of the normal camera lens lie midway between these extremes. The photographer who can afford all three will find each one useful at different times.

On most occasions a wide aperture lens is a decided advantage because it permits a fast shutter speed to be used.

Lighting. The candid photographer must be prepared to make do with the existing lighting. About the only control he can exercise is to change his position to photograph the subject from a more favourable angle. There is one exception to this rule: when taking photographs in the dark by invisible lighting such as infra-red.

Viewfinder. A reflex viewfinder would appear to be ideal for candid photography, yet many of the top men use a miniature and a direct vision finder with complete success. The reflex finder allows the photographer to point his lens at the subject while he looks away at right angles, and the same thing can be done with the angle type of supplementary finder. These aids help to make candid photography easier for people who cannot acquire the art of direct shooting.

Technique. The candid expert rarely tries to conceal or disguise himself or his camera. He simply acquires the knack of making himself unobtrusive. This may be done in a number of ways:

(1) By wearing conventional dress with no eccentricities.

(2) By moving and working in a slow, leisurely fashion.

(3) By misdirection—i.e., by pretending to be interested in anything but the subject. This means not only looking, but turning, in another direction.

(4) By taking up a position between the subject and the strongest light source. (This makes it harder for the subject to see what he is up to.)

(5) By keeping his camera either out of sight, or pointing in another direction until the last minute.

As far as possible all the necessary adjustments are made to the camera before the photographer goes into action; he sets the shutter speed and lens aperture and focuses the camera for the approximate working distance. Then he approaches the subject casually and chooses the position that gives him the best view and lighting of the subject. If the subject is unlikely to move out of range, the photographer settles down for a while before starting to shoot to disarm suspicion. Soon his presence is accepted and ceases to arouse curiosity and he is free to take his pictures without the subject's knowledge.

The candid technique is particularly useful for getting natural pictures of children. It is not necessary for the child to be unaware that there is a camera pointing at him; the photographer simply waits until the child loses interest in the camera and turns all his attention back to his game or occupation.

For stunt picture series in magazines, elaborate ways of concealing the camera have been used—even by installing façades (such as half-silvered mirrors) in public places. F.P.

See also: *Spy camera; Ultra-miniature.*

CANDLE METRE. Unit of illumination also known as a metre candle, or a lux. It is the illumination of a surface one metre distant from a light source of one candle power.

See also: *Light units.*

CANDLE METRE SECOND. Photometric unit applied to exposure, now usually called a lux-second.

See also: *Light units.*

CANDLE POWER. Obsolescent term for luminous intensity. The present international unit is the candela. Frequently (but incorrectly) also applied to the unit of light intensity itself, e.g. by quoting a source to have *x* candle power.

See also: *Light units.*

CANVAS PRINTS. Canvas and other similar fabrics can be treated so that it is possible to print them with a photographic image. The printing may be carried out by contact printing in daylight or artificial light or by enlarging.

See also: *Fabric printing.*

CAPA, ROBERT (Robert Friedmann), 1913–54. Hungarian born war photographer and founder member of Magnum Group. Photographed five wars in eighteen years: the Spanish Civil War, the Japanese invasion of China, World War II, the Arab-Israel conflict, and the Indochina War. Killed in May 1954 by a landmine in North Vietnam while photographing French combat troops.

CAPACITOR FLASH GUN. Flash gun in which the bulb is fired by the discharge from a small condenser. The advantage of the system is that it operates even when the dry battery which charges the condenser in the first place is getting weak.

See also: *Flash equipment.*

CAPTION. Some pictures are no use without a caption, some are better if they have one, and others do not need one at all. News, sporting, and technical photographs need captions to describe the people or things they show. And photographs that aim at conveying an idea—artistic, abstract, or humorous—generally need a caption to make the idea clear. Then there are the examples of pure photography, pictures which tell the whole story without any help from the caption.

Photographers are not expected to write the captions for pictures they submit for publication—caption writing for the press is a skilled technique that is best left to the editorial department. But editors do need the material for making the captions. Unless they get the right information with the photograph they are unlikely to use it.

Essential Information. First they want to know who took the photograph. (If the sender did not take it himself, he will have to convince the editor that he owns the copyright or that he is entitled to sell on behalf of the owner.) So the sender must always attach his name and address to the photograph. It is always safer to write or stamp this on the back of the print because the covering letter might be mislaid and labels get unstuck.

Then editors want to know who or what the photograph is about, where and when it was taken, and what are its points of interest from the angle of their particular readership. This means giving complete names, dates, and a short write-up of the facts. All this can go on a typed sheet that can be stuck along one edge of the back of the print and folded over to protect the surface.

If ever the photographer is asked to write the caption himself, the rules are simple. Do not include any information in the caption that is already clear in the photograph. Tell the reader anything that is not obvious in the photograph but which he must know if he is to be interested in what does appear. Make sure that the meaning is clear, even if it involves writing a longer caption.

Whenever a photograph is taken with an eye on subsequent publication, all the relevant information should be written down at the time. If any people are featured in the picture it may be necessary to get their written permission to reproduce their portraits—or the promise of it—and their names and addresses. It may be extremely difficult or even impossible to get these facts at a later date.

Creative Captions. So much for the pictures that must have a caption if they are to mean anything. But what about the pictures that are understandable without any caption but can be improved by a good one?

These are the difficult captions to write for the simple reason that they need imagination. They are creative. They have more to do with fancy than fact. They are no longer bald descriptions; they are personal comments. This sort of caption starts before the picture is made. It starts with the idea behind the picture. That idea is the germ of the caption; if there was no idea it is unlikely that there can be an effective caption.

Nevertheless, when it comes to the literary side of writing such captions, editors often prefer to do that themselves.

Labels. But considerable thought is often given to pictures that do not need a caption at all. When the photograph tells the story itself, all the caption it needs is something to take the place of a reference number—a label. A straightforward figure study or landscape is not improved by an obscure title when all it requires is a label. Nothing is gained by trying to give a suggestion of hidden significance to a picture that ought not to want it.

A label is in fact a bare title or description, such as *Derwentwater*, or *Collie dog and pups*, without embellishments.

A human portrait should be sufficiently interesting in its own right to be able to do without such forced afterthoughts as, *Soliloquy*, *Meditation*, *Thoughts in Idleness* and similar stale clichés. A figure study that cannot be sold without a fanciful caption will not be sold with one either. The task of the caption here is simply to identify the picture, and very simple labels will suffice: *Norah*, *Dutch Girl*, *Model Resting*.

The same applies to landscapes. It is impossible to make up for lack of beauty in a picture by adding a poetic caption like *Evening's Glory*, or *The Knell of Parting Day*. And to go even further and add a complete quotation either distracts attention from a good photograph or focuses it on a bad one. Once again, the caption should be simple and unselfconscious: *Cotswold Cottages* not *Mid Rustic Charm*, and *Ice on Wire Netting*, not *Nature's Handiwork*.

For those who must have rules, the following general advice can be given. Be sincere. Be clear rather than simply brief. Be honest. (Do not use a caption to give a photograph a meaning that was not intended when the picture was taken.) Captions sell pictures, but the pictures have to be worth buying. F.P.

See also: *Lettering; Titling.*

CARBOLIC ACID. Solution of phenol. Once widely used disinfectant and preservative—e.g., of gelatin and organic mountants.

CARBON ARC LAMP. This was the earliest form of commercial photographic lighting. It is produced by passing an electric current through two carbon pencils with their points in contact and then slightly separating the points. This process is known as striking the arc; it creates a flame of burning gas between the ends of the carbons. The gas itself and the ends of the carbons glow with an intense brightness, forming a compact light source which is particularly suitable for projectors, floodlights, and other forms of controlled illumination.

Improvements. The simplest type of arc lamp is known as the open or low-intensity arc. It contains an excessive amount of red light and consumes carbons at a relatively high rate. A great improvement was made by enclosing the arc in a glass globe with only a very restricted inlet for air. The enclosed arc tends to consume the air faster than it can enter the globe with the result that the pressure in the globe falls. Under the reduced pressure the gap between the carbons can be increased to as much as 2 ins. (5 cm.), giving a long, violet flame of high actinic power. Carbons last much longer in the enclosed arc than in the open type; the burning rate may be as low as ¼ in. per hour in a lamp of average size.

Carbon arcs have been greatly improved in colour temperature and intensity by using cored carbons incorporating certain metallic salts, and by increasing the current loading of the carbons. A combination of the two methods forms the basis of the modern high intensity arc used in television studios, film studios and cinemas.

Consumption. A typical low intensity arc consumes around 2 kilowatts with an efficiency of about 15 lumens per watt, whereas a high intensity arc may consume as much as 13 or 14 kilowatts with an efficiency approaching 30 lumens per watt—i.e., almost twice as much light as the low intensity arc for a given power consumption. High intensity arcs with a working current of 200 amps. or even more are used for film studio lighting.

Electrical Characteristics of the Arc. The carbons must be brought into contact and then separated to start or strike the arc. At the instant of contact the resistance between them is practically nil, so the circuit must include some form of ballast to limit the flow of current. In D.C. arcs the ballast consists of a variable resistor; in A.C. arcs a choke, or better still, a reactive transformer, is used.

Once the carbons are separated and the arc is struck, the resistance varies according to the current passing and the type of arc. The important figure in calculations is the voltage drop across the arc which may vary from twenty or thirty to as much as 160 volts. The supply volts must be greater than the arc volts, the difference being dissipated in the resistor (unless the arc is being run on A.C., when a step-down reactive transformer may be used). Where two or more arcs can be connected in series, less ballast is needed and the efficiency of the system rises in consequence.

A.C. or D.C. When an arc is fed from a D.C. supply, the positive carbon burns away to form a hollow crater from which most of the light is emitted. This crater is treated as the light source when focusing the arc. The carbons are inclined at an angle so that the maximum amount of light is radiated in the most useful direction.

When the arc is fed from an A.C. supply, each carbon is alternatively positive and negative so that the tips of both glow with the same intensity. Sometimes both carbons are sloped to form a "scissors" in which both craters point towards the desired direction. But the twin source still remains as a defect of this type of arc.

When cored carbons are run on either D.C. or A.C. at a high current density—up to 10 amps. per square millimetre—the main light no longer comes from the positive crater; it is radiated by a ball of incandescent gas between the carbons. This ball of gas approximates to a point source and has a colour temperature almost equal to daylight. This type of arc, when produced by an alternating current, shows a reduced tendency to flicker at the frequency of the supply and it is for this reason the only

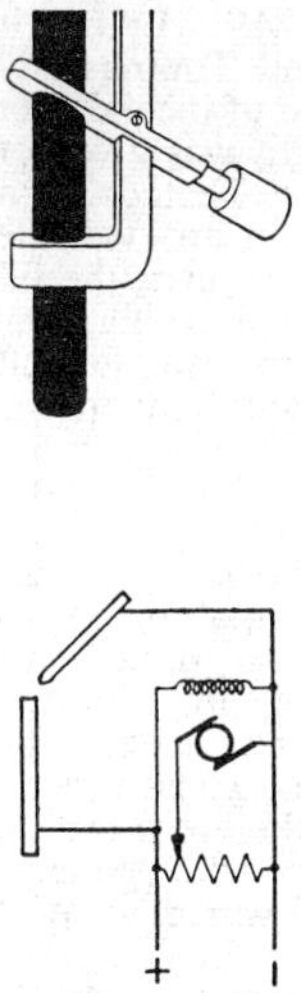

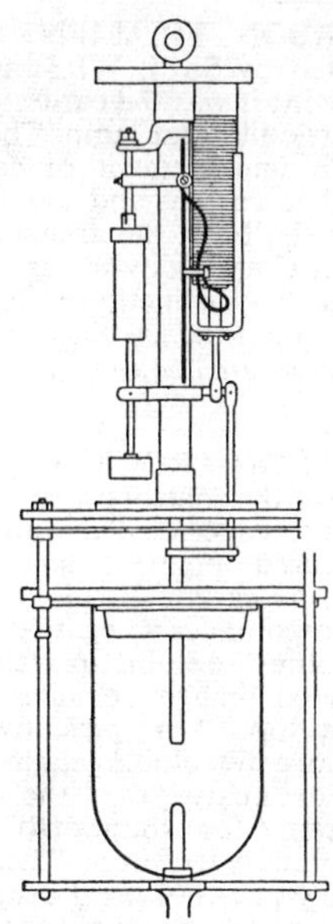

CARBON FEEDING METHODS. Top left: *Stirrup feed; lifting the counterweight allows the carbon to drop.* Bottom left: *Motor feed; the speed is adjusted by a potentiometer feeding the brushes, thus controlling the feeding rate.* Right: *Solenoid feed; as the current decreases, the upper carbon drops.*

A.C. carbon arc that is suitable for projection or film studio lighting.

Feeding the Carbons. As the carbons are burned away in the arc, they must be closed up to maintain the gap or the arc will eventually go out.

In the simple arc lamps fitted in lantern slide projectors and flood lights the carbons are individually adjusted by hand wheels on the outside of the lamp house, the operator observing the effect on the arc through a deep violet or red window.

Arcs used for overhead street and public lighting are controlled by a solenoid which automatically strikes the arc when the current is switched on and maintains the gap between the carbons.

The high intensity arcs used in film studios and cinema projectors burn at a very high rate and so have to be fed automatically by a motor connected across the arc gap. As the carbons burn, the gap widens, the volts across it increase, and the motor speeds up. As the carbons approach each other, the volts across the arc fall, and the motor slows down. By a suitable choice of reduction gearing, the carbons can be kept moving in fast enough to compensate for the amount of carbon consumed in the arc.

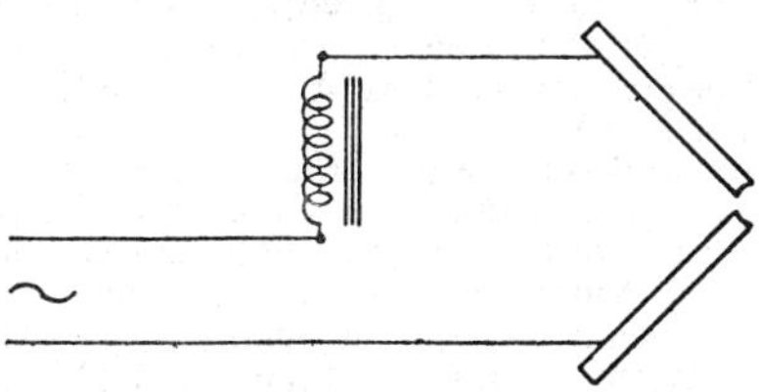

CISSORS ARC. This is an A.C. arc; craters are formed in both arbons. When used in projectors, etc., both carbons must be irected towards the condenser or other light collection system. he most efficient control is a choke. The current is increased y withdrawing the iron core from the windings.

See also: *Discharge lamp; Point source lamp; Zirconium lamp.*

CARBON FILAMENT LAMP. Lamp invented by Sir J. W. Swan and Thomas Alva Edison; it was the earliest type of incandescent electric filament lamp. The light was given out by a fine filament of carbon heated by an electric current and mounted in an evacuated glass bulb. In the absence of oxygen, the filament simply glowed instead of burning away. It gave a distinctly red light and was less efficient than the gas-filled tungsten filament type.

See also: *Artificial light history.*

CARBON PROCESS. The carbon process is one of the control processes for making positive prints. Paper coated with a layer of carbon or coloured pigment suspended in gelatin is sold as carbon tissue. This is sensitized, and exposed under a negative. The exposed gelatin becomes insoluble in water, therefore the pigmented image remains after the paper is "developed" by washing in warm water to remove the soluble gelatin.

For many years the carbon process was favoured for exhibition prints, and the best professional portraiture. The process is capable of giving prints of great beauty, and it can reproduce a wide range of tones. Prints can be made in a great variety of colours and they have the additional merit of being permanent. R.M.F.

History. In the 1850s, numerous attempts were made to devise a cure for the fading of positive prints, due chiefly to inadequate rinsing. Alphonse Louis Poitevin (1855), John Pouncy, Adolphe Fargier and others invented methods of making positive prints in permanent pigment. The principle involved in most methods was to colour with carbon—one of the most unalterable substances in nature—gelatine, gum, or similar organic substances made light-sensitive with potassium bichromate, which hardens in light. "Development" consisted in washing away the insoluble parts of the picture. These early methods were deficient in half-tone. Carbon printing was successfully introduced with (Sir) Joseph Wilson Swan's perfected process in 1866 (patented 1864). The firm of Mawson & Swan in Newcastle-on-Tyne supplied ready-made carbon tissue consisting of a film of gelatine containing finely powdered carbon, spread on paper. The tissue was sensitized by the photographer with a solution of potassium bichromate and, when dry, exposed under a negative. The face of the carbon tissue was then attached to a temporary paper support and soaked in warm water which removed the original paper backing and the soluble parts of the gelatine which had been protected from light by the more opaque parts of the negative. The insoluble gelatine image remaining on the temporary support was then dried, and as the picture was laterally reversed, it had to be transferred a second time to its final support. The picture was in relief, the gradations of the negative being represented by corresponding degrees of thickness in the coloured gelatin.

Swan's carbon process was widely practised under licence by firms producing large quantities of prints, and is still in use.

A satisfactory carbon process on a different principle was patented by Poitevin in 1860, based on the hygroscopic property given by light to a coating of perchloride of iron and tartaric acid. After exposure under a negative, powdered carbon or other permanent pigment was brushed on, and this adhered to the image in proportion to the action of light and consequent degree of stickiness. The picture was then coated with a layer of collodion and soaked in water, when it detached from the glass plate on which it had been formed and was transferred to paper. H. & A.G.

Sensitizing the Tissue. The tissue can be purchased in a range of sizes and colours. Before it is sensitized it will keep in a dry place indefinitely. The tissue can be sensitized by artificial light and will dry in a few hours.

The sensitizing solution is:—

Potassium bichromate	2 ounces	50 grams
Water	40 ounces	1,000 c.cm.
Liquid ammonia ·880	140 minims	7 c.cm.

The best working temperature is 60° F. It should not be above 65° F.

The solution is poured into a dish larger than the tissue to be sensitized and the tissue is immersed in it face down. Any air bells on the front or back of the tissue should be broken. Sensitizing takes three minutes.

At the end of this time the tissue is lifted out and laid coated side down on a sheet of glass, and the surplus liquid is driven out with a flat squeegee. The tissue is then pinned up to dry in a linen cupboard, or any other place which is warm and dark. The tissue does not become sensitive to light until it is almost dry.

When it is dry the tissue is sensitive and must be kept away from the light because the effect of light action once started is continuous. After drying, the tissue should be stored under gentle pressure to keep it flat for use.

Printing. A negative must have greater density and contrast for carbon printing than for bromide enlarging. The thin type of negative that makes a good bromide paper enlargement will not make a satisfactory carbon print. The negative must also be given a "safe edge" so that the print will have a clean white margin. This is done by masking it with lantern slide binding paper.

The print is made by contact. As the surface of the paper is coloured, exposure does not produce a visible image and the correct time has to be found by trial. The usual plan is to print a negative of similar quality side by side with the carbon print on a piece of P.O.P., or self-toning paper. When the test paper shows an image of the right depth the carbon print is fully printed.

Development. During printing, a thin layer of insoluble tissue is formed over the top surface of the tissue. This layer would prevent the action of development, so for development it is necessary to transfer the tissue to another support. If the final print is to show the image the right way round the tissue is transferred to a temporary support at this stage and from this it is transferred again to its permanent support. But if the negative image has already been reversed to allow for the transfer, the tissue is transferred straight to the final support.

Once the tissue has been exposed, the action goes on even in the dark, so the print must be developed very soon after exposure. First the tissue and a sheet of transfer paper are placed together in cold water for about three minutes. The tissue will first curl up in the water and then uncurl. The film surface of the tissue is then brought into contact with the film surface of the transfer paper. Both sheets are picked up together and laid on a sheet of glass with the tissue uppermost. They are then squeegeed firmly into contact so that the solution and any air bubbles are expelled from between the two surfaces. The two are then placed under light pressure between blotting boards for about twenty minutes.

The papers are then soaked in a dish of water at 100° F., tissue uppermost. In a few minutes small quantities of the soluble gelatin begin to emerge from the edges. When this happens, one corner of the tissue is raised and the whole sheet is lifted gently off the transfer paper. The transfer paper should slip by its own weight to the bottom of the dish and if the surfaces are at all difficult to separate, they should be allowed to soak longer, but never forced apart.

The support is now left in the warm water until "development" is complete. As the tissue is very tender at this stage it must not be touched, and it is a good plan to support the paper on a sheet of glass to avoid damaging it. Gentle rocking of the dish to make the water flow across the print helps to speed up development. Development is complete when all the soluble gelatin has gone from the print.

At this stage errors in exposure may be partially compensated for. Over-exposed prints can be reduced by raising the temperature of the water to 120° F. Under-exposed prints should be left to develop face down at a temperature of 90° F.

When the print is developed it is given a rinse in cold water and then placed in a 5 per cent alum solution to harden the gelatin and to discharge the bichromate. It is rinsed again in clean water and then dried.

Double Transfer. In ordinary single transfer carbon printing the image is reversed. It may, however, be printed the right way around by using the double transfer process.

Printing and development is the same as in single transfer printing except that a temporary support takes the place of the single transfer paper. This support may be a sheet of either opal glass or celluloid, previously waxed and polished.

The exposed tissue is transferred to this support, stripped, and developed in the usual way (see single transfer above). It is then dried, after which it is soaked in tepid water with the final support; the two are then squeegeed together and dried. After drying, the print may be stripped without difficulty.

The carbon process can also be used for making colour prints, although carbro is now the more common method.

Wet Carbon Process. When a carbon print of fair quality is needed quickly, the wet carbon process can be used. This is practically identical to the normal process except that the sensitized pigment paper is not dried before exposure.

A sheet of thin celluloid ·003 in. thick is first waxed and polished. The pigment paper is then sensitized in the normal way and then, while still wet, squeegeed into contact with the waxed celluloid. This sandwich is then placed in a printing frame (with the negative against the celluloid) and exposed.

After exposure, development of the sandwich is carried out in hot water as usual. When dry, the image on the celluloid is then transferred to a final support in the same way as in the double transfer process. R.M.F.

See also: *Control processes; Pigment processes.*

CARBON TETRACHLORIDE. Solvent for removing grease and finger marks from negatives; it is non-inflammable.

Formula and molecular weight: CCl_4; 154.

Characteristics: Colourless liquid with fairly pleasant smell. The vapour is toxic.

Solubility: Does not mix with water.

CARBRO COLOUR PRINTS

Carbro is the oldest printing process in general use to-day and many people still consider that the quality of a really good carbro colour print cannot be quite equalled by any other process. If this is true, it is partly because the three colour images which go to make up the print are in relief and give depth and sparkle to the shadows. Other advantages are that it is possible to correct the register even if the separation negatives are slightly different in size: the process takes up very little working space and it is never necessary to have more than two dishes in use at one time.

The cost of a first print is rather less than for dye transfer, but the cost of duplications is greater; the working time taken to make it is at least 2½ hours.

Carbro colour print materials are the same as for single colour carbro except that pigment papers of three colours are used and three

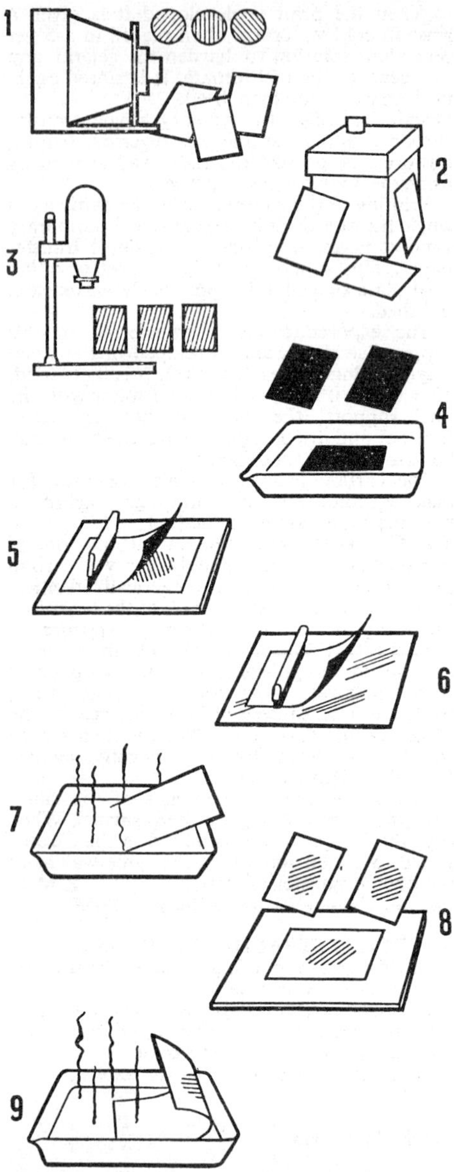

TRICHROME CARBRO PROCEDURE. 1. *Making separation negatives by exposure through red, green, and blue filters.* 2. *Processing.* 3. *Making bromide prints from the separation negatives.* 4. *Sensitizing cyan, magenta, and yellow tissues.* 5 *Squeegeeing into contact with the bromide print.* 6. *Squeegeeing tanned tissue on to temporary support.* 7. *Developing the reliefs in hot water, followed by chilling, drying, and re-soaking.* 8. *Registration and transfer.* 9. *Final transfer.*

transparent plastic supports and a paper temporary support are needed.

The starting point for the process is a good set of separation negatives which may be made either from a transparency or direct from the subject. The actual process of making the print divides itself into four stages.

(1) A set of three bromide prints is made from the separation negatives.

(2) From the bromide prints a set of three colour pigment reliefs is prepared by the standard carbro method.

(3) The three colour reliefs are superimposed in register on a temporary support.

(4) The three-colour image is transferred to its final support.

The Bromide Prints. The first step is to make three bromide prints, or enlargements, one from each separation negative. Non-supercoated paper should be used as this helps to get even chemical action between the bromide and sensitized pigment paper. The requirements for the bromide prints are that they should be:

(1) In balance, this means that any part of the subject which is white or grey should print with the same density in all three bromides. Some slight variations may have to be made to suit different batches of pigment paper.

(2) Of sufficient density in the lightest part of the print just to affect the carbro pigment.

(3) Of correct contrast for the pigment sensitizer in use.

The first requirement is a problem common to all separate colour printing processes. The most reliable way of finding the relative exposure needed to balance the prints is by trial and error with a hard grade of bromide paper. A good densitometer can save much of the work, but the measured density differences between the separation negatives will usually have to be multiplied by a factor of between 1 and 1·5 to allow for the Callier effect of the enlarger, before converting to opacities.

The density of the highlight and the contrast of the bromides is largely a question of experience and of suiting the particular pigment sensitizer in use.

If the separation negatives are of different contrast it will be found impossible to match the colourless parts of the subject both in the highlights and shadows. Adjustments for this can be made by using different grades of paper or by altering the composition of the bromide developer. However, it is usually safer to intensify one or more negatives or to vary the composition of the pigment sensitizer as between the three colour images.

The Colour Reliefs. The next stage is that of making the colour reliefs. This depends on the principle that gelatin sensitized with potassium bichromate can be made insoluble either by the action of light or, in the case of carbro, by the chromium salts liberated when a silver bromide image is bleached. Specially prepared pigment papers are supplied in three colours, cyan (blue green), magenta, and yellow. They are made by adding colouring matter to gelatin and coating it on to paper.

The pigment paper is first sensitized in a solution containing potassium bichromate and other chemicals and immediately brought into contact with the bromide print.

There are a great number of published formulae for pigment sensitizing baths. They fall into two main groups: those employing a single bath and those calling for two separate solutions.

The two bath method is sometimes recommended for those starting the process, because contrast control is simple and depends on the time of immersion in the second bath and because the manual dexterity required in bringing the bromides and tissue into contact need not be so great. The time of immersion in the second bath is, however, critical and the total time in the solution including the time taken to bring the pigment paper and bromide into contact varies from 20 to 45 seconds.

The single bath formula is in many ways more straightforward and is particularly suited to prints of whole plate size or larger. Whichever sensitizer is adopted, it is most important to keep to one formula and to find out exactly the type of bromide print which is suited to it. Nearly all published formulae can give first-class results and it is a great mistake to keep on changing formulae.

The single bath sensitizer has the characteristic that the bleaching action starts immediately the pigment paper and bromide touch. This means that the two must be brought into firm contact quickly. A simple way of doing this is to take two sheets of celluloid and to hinge them together at one end. The bromide and the pigment paper are then placed facing each other on the two sides. They can thus be held apart until the instant when contact is made between them. They are pressed together either by passing them through a mangle or by rolling with a heavy roller squeegee.

After allowing about a quarter of an hour for the chemical action between the bromide and pigment paper to take place the two are stripped apart and the pigment paper is squeegeed face downwards on to a transparent plastic support and left for 10 to 20 minutes. The type of plastic support used is rather important because the surface must be able to hold the relief image during development and also allow it to transfer later on to the support paper. In the old days celluloid was used but this needed waxing to lessen the adhesion and success depended on striking a balance between too much or too little waxing. The wax also had to be removed from the surface of the images after transfer as otherwise the next image would not adhere. Nowadays there are two or three types of plastic obtainable with the right kind of surface and which do not require any treatment.

The Temporary Support. After drying, the reliefs are transferred in turn to a temporary paper support starting with the cyan image. This, on its transparent plastic support, is brought into contact with the paper support under water. After removing the sandwich from the water the two are squeegeed together firmly and allowed to dry. The paper support with the pigment image adhering to it is then stripped off the transparent support. The magenta and yellow images are then registered in turn and transferred on top of the cyan image.

This is done in each case after removing the sandwich from the water by laying it down with the transparent support uppermost and moving the two sides in relation to each other before squeegeeing.

It is possible to obtain register between the three-colour images even if the separation negatives are not of exactly equal size. This is because the temporary paper support can be slightly stretched in relation to the plastic support of the relief. Errors in registration of up to $\frac{1}{8}$ in. on a 10 × 12 ins. print can also be corrected by bending the transparent support inwards or outwards and fixing it in this position during the time the image is being transferred.

The Final Support. After the three images have been combined it is usual to transfer the composite image to a final support. This is not absolutely essential, but the temporary paper support is designed to assist the transfer of the pigment images; also, since the finished print must have the cyan image on top, a direct transfer to the final support would mean registering the magenta on to the relatively pale yellow image, which is generally a difficult operation.

The transfer is made by bringing the temporary and final support into contact under water, squeegeeing them together, and allowing them to dry down. The sandwich is then immersed in hot water, the backing paper of the temporary support is removed and the print is hung up to dry.

With carbro it is most important to have a good working time table so that the various operations which have to be carried out for each of the three colour images can be dovetailed into one another and so that each image receives the same treatment at each stage of the process.

There are several interesting variations on the process and the colour images can be transferred to almost any surface previously coated with gelatin.

It is sometimes said that carbro is a tricky process and it is true that it does sometimes develop faults which are difficult to locate. It is quite common to find that two workers apparently using the same technique do not obtain equally successful results. The golden rule is to decide on the detailed technique to be followed and never change more than one variable at a time. Once the mechanics of bringing the bromide paper and pigment paper into contact and of making the various transfers have been mastered the process becomes straightforward and enjoyable. H.

See also: *Colour print processes; Registering images.*
Books: *Amateur Carbro Colour Prints*, by Viscount Hanworth; *Colour Photography in Practice*, by D. A. Spencer; *The Autotype Colour Printing Processes* (all London).

CARBRO PROCESS. In these days many photographers who want carbon prints usually make them by the carbro process. In this case the print is made from a bromide contact print or enlargment so there is no need for exposure by daylight and no limit to the size of the picture that can be made from the negative. As the negative can be reversed in the enlarger there is no need for double transferring to make the print appear the right way around.

The carbro process has all the advantages of the carbon process: choice of colour, permanence of the finished print, image quality.

The process was worked out by Thomas Manly about 1905. It was then called Ozotype. Manly first sensitized and made his prints upon the final support. Later he found that the silver image of a bromide print could be made to affect carbon tissue and the result was the Ovobrome process which after improvement became the carbro process as it is today.

The Bromide Print. The basis of carbro is a bromide print that has been normally exposed, developed to finality, fixed, washed, and dried. Only a good quality bromide print will give a good carbro. The print must have a white margin to act as a "safe edge" to avoid frilling.

For best results, the bromide print should be made on non-supercoated paper. Normal bromide papers have a supercoating over the emulsion to protect it from abrasion marks; this supercoating may cause chalky highlights if the print is used for carbro. For this reason, bromide papers are made specially for carbro work without the supercoating.

The bromide print should ideally be slightly soft and of very full density to obtain a good quality carbro print from it. It should also be thoroughly washed and perfectly free from surface stains or deposits.

Two-Bath Sensitizing. The following solutions are needed for the carbro process:—

Stock solution A.

Potassium bichromate	2 ounces	50 grams
Potassium ferricyanide	2 ounces	50 grams
Potassium bromide	2 ounces	50 grams
Water	40 ounces	1,000 c.cm.

One part of stock solution A is diluted with twice its bulk of water. This bath may be used repeatedly if filtered after use.

Stock solution B.

Acetic acid (glacial)	200 minims	10 c.cm.
Hydrochloric acid (pure)	200 minims	10 c.cm.
Formalin (40 per cent commercial)	9 ounces	220 c.cm.
Water	300 minims	15 c.cm.

For use 1 part of the stock solution is added to 32 parts of water. This should be used once only.

The bromide print is soaked in cold water for ten minutes. A piece of carbro tissue, ½ to 1 in. larger than the print, is sensitized by soaking it for 3 minutes in the diluted A solution. Care must be taken to brush away any air bells clinging to the front or back of the tissue. The dish should be rocked gently. Sensitizing takes 3 minutes. Just before the sensitizing time has elapsed the print is taken from the water and laid face upwards on a sheet of glass.

The tissue is taken from the A bath, allowed to drain and then placed in the diluted B solution. The time in this bath varies from 15 to 25 seconds according to the brilliancy required in the final print. The longer the period of immersion the softer the print.

The tissue is next taken from the second bath and laid face down on the bromide print, starting from one side and lowering the tissue into place to expel the air and prevent air pockets from forming. The tissue is then pressed firmly into contact with the print with a flat squeegee. Chemical action begins almost at once, so any movement between the tissue and the print must be avoided or it may cause a double image.

The print and tissue are then lifted from the glass and left for about twenty minutes between several sheets of greaseproof paper under light pressure.

During this time a piece of transfer paper is put to soak in water. Ordinary paper should have 10 minutes; thick paper rather longer. After soaking, the transfer paper is laid upon a sheet of glass ready for the carbon tissue.

The bromide print and the tissue are now carefully separated by lifting one corner of the print and pulling the two apart. The bromide print, which will be found to have bleached, is put into water to be redeveloped later, and the tissue is placed, coated side down on the transfer paper, and squeegeed into contact. The two are put under light pressure between sheets of paper for twenty to fifty minutes.

Hot Water Development. When the pigment paper and transfer paper have been in contact for the correct time, the two still in contact are placed in a dish of water at a temperature of 95° F. to 100° F. After a short time the pigmented gelatin will be seen oozing from between the tissue and the paper. The two are then separated by pulling them gently apart under water. If they do not part freely they should be left a little longer. After the two are separated it will be found that the pigmented gelatin has been transferred to the transfer paper. Development now proceeds as in the carbon process until all the free pigment dissolves away and leaves the print behind. The dish may be rocked, or water may be poured over the print to assist development. This must be done carefully because the image is very tender at this stage. For the same reason the surface should not be rubbed or fingered.

When development is complete the print is placed in a 3 per cent solution of alum to remove the yellow bichromate tint from the highlights. It is then washed for five minutes in several changes of water and pinned up to dry.

The bromide print may be redeveloped in any non-staining developer and either kept as a bromide print or used to make more carbros. (It is unwise to try to make more than five or six.) The effect of repeated bleaching is to intensify, and many bromides are in consequence better after being used for making carbros. For the same reason subsequent carbros are often better than the first.

By an extension of the carbro process, very satisfactory colour prints can be made. R.M.F.

See also: *Carbro colour prints; Control processes; Pigment processes.*
Book: *The Autotype Carbro Process* (London).

CARD NEGATIVE. Photographic negative on a card base, which is printed by reflection, as opposed to the conventional transparent negative which is printed by transmission of light.

See also: *Negative card.*

CAREERS IN PHOTOGRAPHY. Photography has much to offer as a career for both men and women. It provides interest, variety and personal satisfaction and can give high rewards for those who possess the ability, personality and, of course, necessary qualifications.

There are many branches providing careers either at a technical level or for professional photographers capable, after acquiring the necessary experience, of either running their own practices or taking control of photographic sections in industrial concerns, hospitals or research laboratories.

The best method of training to become a professional photographer is to attend a 2–3-year training course at a recognized school of photography in a technical college or college of art, studying for the Intermediate and Final Examinations of the Institute of Incorporated Photographers. At least one school of photography (at the Regent Street Polytechnic, London), conducts a further 3-year Diploma course.

The student must first pass at least five subjects (including English Language or Literature) in the G.C.E. at "O" level and some photographic schools require passes in "A" level subjects. It is possible to take up an appointment after passing the I.I.P. Intermediate Examination and to study at evening class for the Final.

In the U.S.A., certain colleges run degree courses in photography.

Portraiture. Portraiture is a branch of photography demanding determination, a pleasant personality, a love of people and strong artistic traits.

There is scope for portrait photographers of outstanding ability and, although not all can hope to reach the top, the person embarking upon portraiture as a career will have a varied and exciting life with opportunities to photograph and talk to many people of interest.

The ideal basis of this career is the training received at a photographic school, followed by entry into a studio as an assistant. It is also possible to enter a studio as a pupil without prior training, studying for the relevant qualifications at evening classes. It is essential, however, that the employer be a bona fide professional photographer of standing.

Advertising Photography. The advertising photographer is an illustrator so that, in addition to his knowledge of photography, he should know the art of pictorial composition. He should also have an inexhaustible interest in life.

A candidate must satisfy himself that in addition to his technical knowledge, he has a marked artistic potential; many photographers prefer an assistant to have had training in a recognized art school as well as the technical training received at a photographic school.

The assistant will probably remain on this level on a rising salary for four or five years.

Press Photography. There is more hard work than glamour in this demanding branch of photography.

Speed of working is important and the press photographer rarely has a second chance to get a particular picture, and little opportunity to size up the subject or wait for better conditions.

Entry to the position of press photographer on the staff of a newspaper is normally through employment as an assistant and photo-printer in the newspaper darkroom. In all but the Fleet Street newspapers and the London Agencies training is, by agreement within the newspaper industry, in the hands of the National Council for the Training of Journalists. Junior press photographers, after a probationary period of six months, are indentured for a three-year period.

To obtain the Council's certificate of proficiency, the I.I.P. Intermediate Examination should be passed (this is a necessity to proceed to the Diploma level), and the I.I.P. Final Examination is a prerequisite to the Council's Diploma. The salaries, negotiated by the National Union of Journalists, vary according to the type of newspaper and town.

Industrial Photography. The activities of the industrial photographer vary as widely as industry itself, so the industrial photographer must be versatile. He may have to produce photographs appropriate for report illustration, for brochure, for advertisement or exhibition stand, and should be able to undertake portraiture, press and scientific photography, document copying and cinematography, according to the specific needs of his firm.

In the lighter industries, there are growing opportunities for girls and already several such departments have staff which is entirely female.

The potential industrial photographer may find that, where possible, experience as a trainee in an independent commercial studio, coupled with part-time day release classes at a school of photography or technical college, offers special advantages.

The prospects of reaching executive positions commanding high salaries are good and the recognition of the value and standing of applied photography steadily increases.

Medical Photography. Medical Photography covers the fields of research, teaching, clinical application and record work. The majority of teaching hospitals, medical institutes and other institutions engaged in specialized work have photographic departments of varying size staffed by men and women. The work demands, apart from photographic qualifications, a flair for scientific work, a sympathy for the ideals which lie behind the treatment of the sick person, and a keen interest in the modern treatment of disease.

The I.I.P. Intermediate Examination is a basic qualification, followed by specialized training in a hospital photographic department or as a student at the London School of Medical Photography, St. Bartholomew's Hospital, London, E.C.1, leading to the Final Examination.

The salaries of medical photographers within the National Health Service are governed by the Whitley Council.

Cinematography. Cinematography, apart from being an entertainment medium, covers the fields of education, instruction, advertising and research. Many large organizations maintain film units to produce technical and prestige films, whilst the smaller film units in particular, require staff proficient in several aspects of production.

Training in the United Kingdom can be obtained at the Polytechnic School of Cinematography, The Polytechnic, Regent Street, London, and in other recognized photographic schools, although facilities generally are limited.

Other Employment. There are various types of employment in photographic establishments apart from the actual taking of photographs. It is usually necessary, and certainly desirable, to obtain experience in processing and printing at the beginning of a career in photography. Indeed, these and other occupations in photographic establishments provide satisfying careers in themselves.

The people who develop or process the negatives also do an important job, because unsatisfactory work at this stage may be irreparable.

Photographic printing permits a very high degree of skill. The demand for skilled printers greatly exceeds the supply, while there is generally a small surplus of camera operators.

The retoucher requires some artistic ability and again need not have photographic training, although a good general knowledge of photography is helpful. The work calls for some ability with pen and brush, a sense of judgment and a keen appreciation of light and shade, form and texture. Retouchers are mainly employed in portrait studios and in commercial and advertising work; the latter demanding the higher standard of artistic ability. There is always a shortage of good retouchers, particularly in commercial photography. Most retouchers are women.

Finishing again is usually done by women. The finisher is responsible for the final appearance of the print, spotting to remove small blemishes, and mounting. The ideal combination is a retoucher-finisher who has the skill necessary to carry out the complicated corrections so often called for in commercial subjects.

In work for reproduction where very highly-skilled craftsmanship is called for, it is usually carried out by a process finisher who must be a capable commercial artist.

The receptionist is an important member of the portrait studio staff, and in many instances the success of the business largely depends upon her skill. She does not require a photographic training, but needs to be interested in people and photography and to be able to inspire confidence; she must also be a good saleswoman and may be required to do clerical work and accounts.

Examinations. Details of the examinations mentioned in this article, including syllabuses, sessions, entry fees, exemptions, etc., may be obtained on application to the Education Officer at the Institute of Incorporated Photographers.

Training in the subjects contained in the examination syllabus can be obtained by attending photographic schools on a full or part-time basis and advice on the availability of such schools may be obtained from the Institute. With effect from the 1st September, 1965, no student is permitted to sit the Intermediate and Final Examinations of the Institute unless he or she has satisfied the educational standards required (e.g. G.C.E. "O" level passes, or equivalent, in 5 subjects, including English Language or Literature), and is a Registered Student of the Institute. J.L.A.H.

See also: *Examinations; Professional photography; Training and education in photography; Training of photographic technologists and scientists.*

CAREY-LEA, MATTHEW, 1823–97. American chemist. Carried out many photo-chemical researches: advised the use of green glass for safelights in darkrooms and for the better preservation of sight. Introduced the mordant dye pictures (1865). Investigated the latent image and the development process, especially ferrous oxalate developers (1877–80).

CARJAT, ETİENNE, 1828–1906. French caricaturist, writer and editor of *Le Boulevard* 1862–3, he also photographed celebrities 1855–*c*. 1875. His portraits are often livelier than Nadar's and have been undeservedly overshadowed by the latter's great reputation.

CARROLL, LEWIS (the Rev. Charles Lutwidge Dodgson), 1832–98. English author of *Alice's Adventures in Wonderland* and amateur photographer 1856–80. His favourite subjects were little girls, and Helmut Gernsheim's rediscovery of Carroll's photographic work proves him to have been the best photographer of children in the nineteenth century. He also photographed celebrities. Biography: *Lewis Carroll—Photographer* by Helmut Gernsheim (London, 1949).

CARTE-DE-VISITE. A photograph, usually a full-length portrait, on a mount the size of a visiting card. The carte-de-visite album containing portraits of family, friends, royalty and celebrities was a typical feature of the mid-Victorian drawing-room. The first photographer to make cartes was Dodero of Marseilles in 1851, but it was André Adolphe Disdéri of Paris who popularized them. In November, 1854 Disdéri patented a method of cheapening the production of portraits by taking ten (in fact usually eight) on one plate and cutting up the resulting print. The fashion for this format dates from May, 1859 when Napoleon III had a carte portrait taken by Disdéri. During the 1860s, at the height of the craze, cartes of celebrities were sold to the public in hundreds of thousands and fortunes were made by leading photographers. The price of a carte varied from 1s. to 1s. 6d.

CARTOON. Motion picture film produced by successively photographing series of drawings which reproduce the movement phases of cartoon characters at a realistic or imaginary rate. The first cartoon films are almost as old as cinematography itself. Today the animated cartoon is—apart from its rôle in entertainment—an important means of advertising in the cinema and on television.

See also: *Animation.*

CARTRIDGE. Term applied to a type of packing for 35 mm. film in measured lengths; nowadays generally known as a cassette.

CASE. Cameras and their accessories are fragile pieces of equipment needing protection when they are being carried about.

Block Form Cases. The old-fashioned camera case is constructed like a box with a flap covering the top and held down by press studs or a tongue and buckle. When designed to hold a folding camera, it has rounded corners and is fitted with a carrying strap which allows it to be carried slung over the shoulder. Block form cases for rigid cameras, e.g., box and reflex cameras, and for heavier types, e.g., technical and view, have a handle on top of the case and the lid is firmly held down with straps. Such cases usually have a compartment for half a dozen plate holders, alternative lenses and focusing cloth.

Ever-Ready Cases. It is usual nowadays for camera cases—and sometimes exposure meter cases—to be designed so that they can be left in position without affecting the use of the instrument. A case of this type is known as an every-ready case. The camera is generally held in its ever-ready case by a screw which fits the tripod bush. As a rule, the top and front of the case are formed by a single flap of leather that folds down uncovering the self-erecting front of the camera, or the lens and the range finder, viewfinder, etc., allowing them to be operated normally while in the case.

The flap of an ever-ready case should hang clear of the field of the lens when it is open. In some cases the material is stiff and tends to spring back and obscure the view of the lens although the intrusion is not apparent in the viewfinder. With the type of case in which the flap folds down behind the camera instead of hanging in front, the trouble cannot occur.

The screw which holds the camera in the case may itself be drilled and threaded to take the tripod screw. This enables the camera to be used on a tripod without taking it out of the case.

Water-tight Cases. One or two manufacturers produce special camera cases which close hermetically and so seal off the camera against moisture, dust or other outside influences. The cases are designed to be buoyant so that the camera and its case float even when dropped into the water. Often a cartridge of silica gel or similar water absorbing substance keeps the interior fully dry. These cases must not be confused with underwater camera cases, which permit photography under water.

Hold-all Cases. The miniature camera and its range of accessories cannot be carried about conveniently in separate cases. Specially designed hold-all cases are available to take the whole outfit. They are made either in the form of an attache case with fitted compartments or arranged as a shoulder bag. For outdoor photography it is more convenient to have a bag with a shoulder sling.

Hold-alls are made either for general use with any equipment, or tailor-made for specific makes. Some also have adjustable compartments, formed by means of press studs, so that the inside can be arranged to suit the combination of equipment in use.

Lens cases. Individual lens cases are useful for cameras equipped with a range of interchangeable lenses. The commonest type of lens case is a leather tube, closed at one end with a removable cap at the other. The case is lined with velvet or foam plastic and usually has two outside loops for threading on to the camera case shoulder strap. This type of case is practical and robust but the movement of the lens inside tends to work up dust and fluff which inevitably transfers itself to the glass surfaces.

Transparent plastic holders are cleaner and better-looking but they are bulkier, and apt to be more fragile than the lined leather type. In addition they are not suitable for fitting to the camera strap and have to be mounted in the hold-all case. Their greatest attraction is to the camera dealer since they provide dust and finger-proof display cases. This type of case consists of a circular base with a Perspex dome attached over it by a bayonet or screw fitting. The base carries a lens mount identical with that of the camera. The lens is attached to this mount and then covered with the Perspex dome.

CASSETTE. Light-trapped metal or plastic container for a length of film to enable it to be loaded into a camera in full light. It is an alternative to spooling the film with an opaque backing paper, and is used mainly for perforated 16, 35 and 70 mm. film.

In the standard type the bare film is wound on a spool in darkness or safe illumination, and then placed in a cylindrical cover with the end of the film projecting from a light-trapped slit in the side. The cassette is placed in the loading chamber of the camera and 4 ins. or so of the film drawn off to attach the end to the camera's take-up spool. On closing the camera this fogged end has to be wound off by advancing the film two or three frames.

After exposure the film usually has to be rewound into the cassette before it can be removed from the camera. The cassette spool is therefore keyed to engage with the rewinding

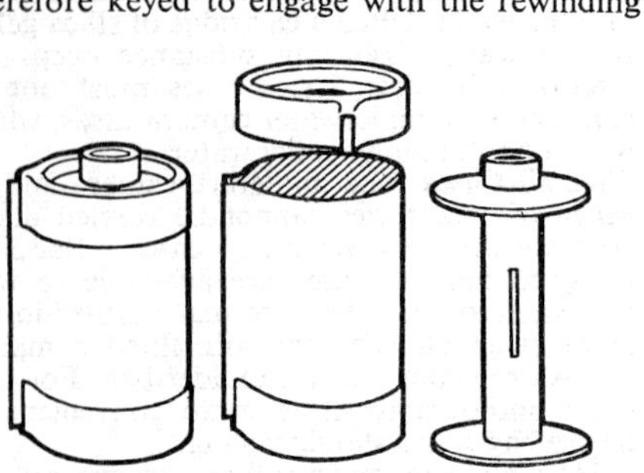

STANDARD CASSETTE. Left: *Complete cassette.* Centre: *Shell with velvet-lined slit, and lid.* Right: *Spool.*

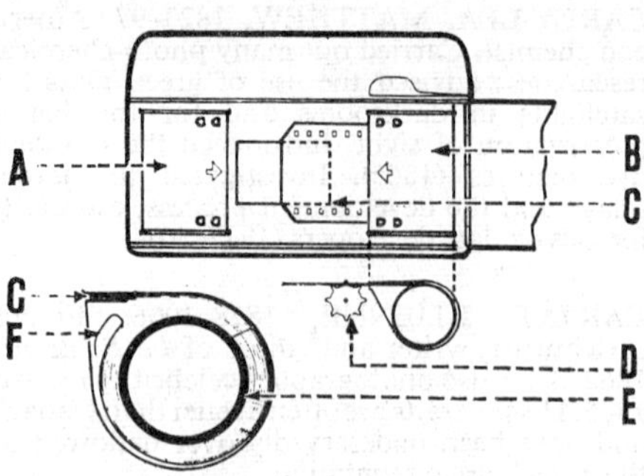

THE RAPID SYSTEM. This uses a feed cassette and a take-up cassette in the camera. On inserting the feed cassette with its protruding film, the transport sprocket inside the camera automatically pushes the film into the take-up cassette. A, take-up cassette. B, feed cassette. C, film. D, transport sprocket. Bottom: Inside the cassette the film is wound up between a loose core E and a spring F.

knob fitted to the camera. The other end of the spool has a short extension which projects from the cover so that it may be turned by hand if required. However, a few cameras are designed to use a take-up cassette to obviate rewinding.

Film Manufacturers' Cassettes. 35 mm. film is supplied in cassettes ready for loading into the camera. These are of simple construction, the slit or "mouth" of the cassette having a light trap of velvet or plush. This material is sometimes carried right round the interior of the cover, the two ends projecting from the mouth. They are not intended to be used again, but usually they may be loaded two or three times if care is taken to keep the plush clear of grit particles, which will otherwise scratch the film. The mouths of metal cassettes can be opened, after removing the end caps, to brush the plush.

The normal 35 mm. cassette holds enough film for 36 exposures 24 × 36 mm., and is suitable for most 35 mm. cameras taking other picture formats (18 × 24 mm., 24 × 24 mm., etc.).

One type of cassette holds single-perforated 35 mm. film in 12- or 20-exposure lengths for 26 × 26 mm. pictures and is designed for specific cameras only. Here the film is wound up together with a backing paper. This cassette is not reloadable and the film can only be removed by breaking up the cassette. The film is pre-threaded with a feed and a take-up chamber built into the cassette, permitting instant loading into the camera without separate threading. The rapid system also uses a quick-loading cassette, but holding 12-exposure lengths of standard 35 mm. film.

Subminiature 16 mm. film is supplied in cassettes also for loading into the camera, and of similar construction to standard 35 mm. cassettes. There are several types of 16 mm. cassette: single cassettes involve rewinding of the film when the whole film is exposed; double cassettes (usually with two chambers joined by a connecting piece) need no rewinding.

Reloadable Cassettes. These are substantial camera accessories. They are accurately made, usually in brass with a black anodized finish, and may be used only in the make of camera for which they are designed. There are, however, certain universal cassettes that can be used with many 35 mm. cameras for which reloadable cassettes are not otherwise provided.

In addition to being reloadable and more substantial, these cassettes are designed to dispense with the plush light trap; instead of the film being drawn through plush, it leaves the cassette through a wide opening to eliminate the possibility of cassette scratches.

The body consists of two accurately made sleeves, one sliding within the other. In each there is a longitudinal opening. When the spool of film is inserted and the cassette assembled, the inner sleeve is rotated so that the openings are diametrically opposite, and the cassette is therefore light-tight. The sleeves are machined to permit the end of the film to project. When the cassette is loaded into the camera, the mouth is opened automatically on closing the camera by turning the inner sleeve until the two openings coincide—hence the term "self-opening" that is often applied to these cassettes. Alternatively, the film may pass through a light-trapping channel, requiring no opening movement. Similar cassettes are available for 70 mm. film, and are loaded from bulk lengths. Here the cassette again has a light-trapped slit, and is essentially a scaled-up version of the standard 35 mm. cassette.

Loading Refills. 35 mm. film refills for reloadable cassettes are normally provided in lengths for 36 exposures 24 × 36 mm. The standard over-all length is 1·6 metres (5 ft. 3 in.) which allows for about 9 ins. of free film to provide a long leading end and a shorter spool end. The leading end is trimmed to half its width for 4 to 5 ins., which is necessary for the film to lead on to the top sprocket of a number of miniature camera models. The other end is trimmed to a "V" for attachment to the cassette spool.

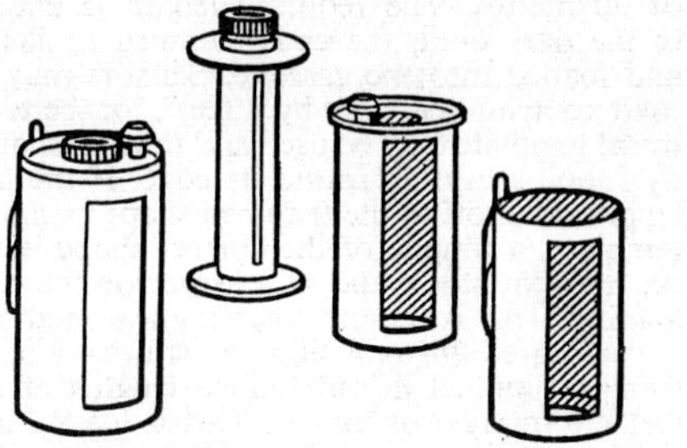

SELF-OPENING 35 mm CASSETTE. Left: *Complete cassette.* Centre and Right: *Spool, inner shell, outer shell. The cassette is opened inside the camera by turning the inner shell (usually by the camera locking key) so that the slots coincide.*

The film is loaded into the cassette in the darkroom. The end is attached to the spool by means of the slot or clip, or by means of adhesive tape, and the film wound on to it by turning the spool with the right hand whilst feeding the film from the left hand, checking the edges with the thumb and forefinger to apply light pressure. Mechanical film winders can speed up this operation.

Refills may also be supplied for daylight loading. The film is ready wound on a spool and is protected by a black paper leader attached to its end. The spool is inserted in the cassette with the paper leader projecting from the light-trapped slit. The leader is then drawn off to bring the end of the film through. Daylight refills are normally available only for 35 mm. cassettes.

Loading Bulk Film. 35 mm. film for still cameras is supplied in untrimmed lengths up to 100 feet

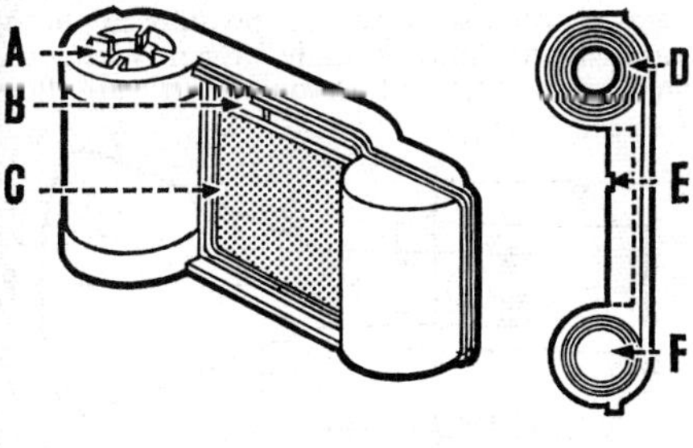

SELF-CONTAINED EASY-LOADING CASSETTE. The whole cassette goes into the camera without any film threading or other manipulation. Inside, the film runs from a feed roll on to a take-up spool but is not re-wound. The cassette must be broken open to extract the film for processing. A, take-up spool shaft to engage film transport in camera. B, perforation aperture for transport control. C, picture opening. D, take-up spool. E, notch for film speed setting in camera. F, feed roll.

FILM LENGTHS FOR VARIOUS NUMBERS OF EXPOSURES

Number of Exposures	*Length of film* in.	cm.	*Number of Exposures*	*Length of film* in.	cm.
1	10½	27	20	39	99
2	12	30	21	40½	103
3	13½	34	22	42	107
4	15	38	23	43½	110
5	16½	42	24	45	114
6	18	46	25	46½	118
7	19½	50	26	48	122
8	21	53	27	49½	126
9	22½	57	28	51	130
10	24	61	29	52½	133
11	25½	65	30	54	137
12	27	69	31	55½	141
13	28½	72	32	57	145
14	30	76	33	58½	148
15	31½	80	34	60	152
16	33	84	35	61½	156
17	34½	88	36	63	160
18	36	91	37	64½	164
19	37½	95	38	66	168

or 30 metres. The required length is cut off in the darkroom, the ends trimmed to shape, and loaded into the cassette. Scissors may be used to trim the film by "feel", or better a metal template may be used and the film shaped by running a knife round its edge. If the film length is cut off to the trimmed shape using the template, a tongue of the correct shape is left on the film stock and will serve for the next loading. This is the most economical method.

Perforated 70 mm. film for still cameras is usually supplied in untrimmed lengths of 100 feet (30 metres) or longer. This is loaded into the cassette in more or less the same way as 35 mm. film, though a leading tongue is not usually required. H.H. (L.A.M.)

See also: *Charger for cassette.*

CATCHLIGHT. In a portrait, tiny highlight on the eye created by the reflection of a bright light source.

If there are no catchlights in a portrait the eyes tend to look lifeless and the eyeball lacks roundness. If the lighting used for the portrait is such that no natural catch-lights are produced, then the addition of a weak directional light—not sufficiently powerful to affect the general lighting of the portrait—placed not far from the camera-viewpoint, will create these.

It is common practice to add catch-lights by means of retouching and knifing, on either print or negative.

These highlights should not be positioned centrally on the eye, but to one side, and if the lighting used has produced central catch-lights, these should be removed and replaced, by means of retouching, slightly to the side.

CATECHOL. Developing agent; pyrocatechin. Used in tanning developers and has been employed in some fine grain formulae.

CATHODE RAY TUBE TRACES

The cathode ray tube is a modern scientific tool with a wide and ever increasing field of applications (television is the most popular of these).

The tube is an evacuated glass bulb in which a narrow stream of electrons emitted by the cathode is made to pass between pairs of plates. When a voltage is applied to these plates it deflects the stream of electrons from its straight path. This arrangement forms a delicate and weightless recording instrument which will follow very rapid changes in the magnitude of the deflecting voltage. So any physical change (which need not necessarily be electrical) that can be made to modify the deflecting voltage can be reproduced as a visible trace on the end of the tube.

Types of Trace. The movement of the electron beam can be recorded either directly on a photographic plate, or indirectly by photographing a fluorescent screen which glows where the beam strikes it. Since the cathode ray itself leaves no lasting trace, scientists naturally want to be able to make permanent photographic records of any traces that interest them. The photographic method used depends on the type of trace to be recorded. There are four principal types: traces which record the same change repeated at regular intervals (repetitive); traces of single transient events—i.e., events that happen once only; traces that are continuously changing without ever repeating themselves (slow, non-cyclic); and traces that repeat at regular time intervals but may be different each time (cyclic, non-repetitive).

Methods. Transient and repetitive traces may be photographed on a stationary film or plate. A transient trace occurs once only, so it is more difficult to photograph than the trace of a regular cyclic event to which the plate can be left exposed until it has built up an image of the required density.

Non-repetitive and non-cyclic events may be photographed on a film wrapped around a revolving drum and moving at right angles to the path of the trace. In this case a time base may be added to the photograph by applying a regular voltage pulse to a second pair of control plates set across the beam.

Slowly changing cyclic events may also be recorded by a cinematograph camera so long as the time interval during which the shutter remains open is long enough to include more than two complete cycles.

Photographic records of traces obtained by such methods have two great advantages; they are permanent, and they can be enlarged for more detailed study. The recorded traces can also be compared directly and stored for reference. When traces are over so quickly that the eye cannot see them it is possible to photograph them automatically. Even when the phenomena occur at irregular intervals there

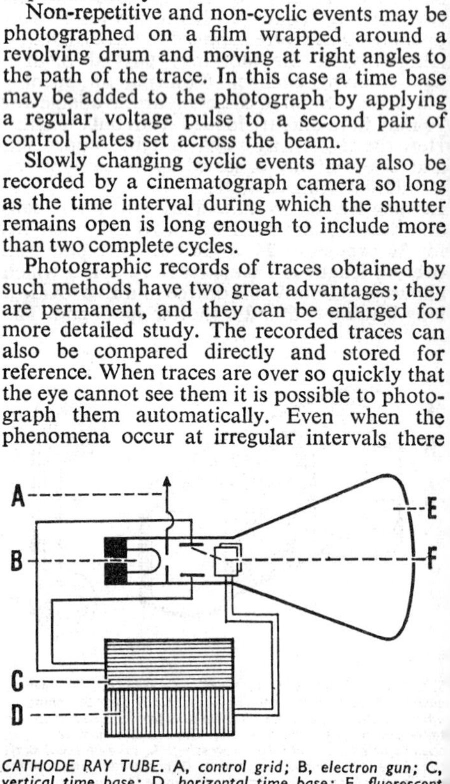

CATHODE RAY TUBE. A, control grid; B, electron gun; C, vertical time base; D, horizontal time base; E, fluorescent screen coated on inside of tube; F, deflector plates.

is no need for a constant watch to be kept on the apparatus.

Continuously Evacuated Tubes. When the trace is recorded by allowing the beam to impinge directly on the photographic material, the film or plate is actually put into the tube. Such work calls for a continuously evacuated tube. This is a tube which can be opened up to allow the sensitive material to be inserted, and then hermetically sealed. It is then evacuated by a vacuum pump.

The film or plate inside is kept covered with a metal shutter which can be opened at the time of exposure.

Exposures as short as 10^{-9} seconds can be made with this type of apparatus, but the tubes are bulky and expensive. They also take a long time to reload because the air has to be pumped out again each time the tube is opened to change the sensitive material. Tubes of this type can be made to accommodate both still and drum type cameras.

Permanently Evacuated Tubes. Permanently evacuated and sealed tubes are more generally used than the type just described. With these tubes traces may be recorded by the direct action of the electrons on the photographic material. In this method, the tube is equipped with a window (generally of thin metal foil) which is thick enough to maintain the vacuum but thin enough to let the electrons pass through. Suitable windows are made of aluminium, glass, or Cellon, about 0·001 cm. thick.

The sensitive material is exposed to the rays after they pass through the window so that all the photographic equipment remains outside the tube and at atmospheric pressure.

Fluorescent Screens. Traces can also be photographed by fitting a screen of transparent fluorescent material into the end of the tube. When the electron stream strikes the screen it creates a visible fluorescent image which persists long enough to be photographed. The fluorescent image may be recorded by normal photography, as in mass miniature radiography or by exposing sensitive paper or film in contact with the screen.

With the conventional type of cathode ray tube, prints made by contact are always slightly unsharp because of the thickness of the glass wall of the tube. This leads to a certain degree of diffusion. The grain of the screen also limits the sharpness of the images recorded by normal photography.

Tubes with the front wall in the form of a fibre window give better contact prints. Here an array of short glass fibres at right angles to the tube surface are fused together to form the window. Each fibre acts as a light guide, transmitting the luminous spot of light from the phosphor coating on the inside of the tube to the paper or film on the outside. These light points are perfectly sharp, but the resolution of the image depends on the fibre diameter.

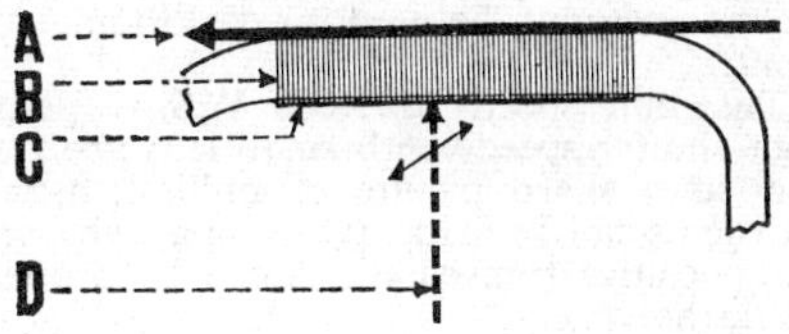

FIBRE WINDOWS. A fibre window can be used with a cathode ray tube to record an electron beam scan direct on a moving film. The beam (dotted) here scans at right angles to the plane of the diagram. A, moving film. B, cathode ray tube envelope with its face plate consisting of a fibre bundle with a fibre axis at right angles to the surface. C, phosphor layer. D, electron beam.

Cathode ray tubes with such fibre windows are widely used in computer print-out devices to record letters and figures on the sensitive material in contact with the tube.

For isolated high-speed transients slow decaying phosphors may be used for the fluorescent layer. The wave form or other image to be recorded there persists sufficiently long to make a developable impression on the photographic emulsion, even though the actual phenomenon causing the trace may have lasted only a microsecond or less.

Contact printing must, of course, be carried out in darkness or by a suitable safelight, but when a camera is used it can be connected to the tube by a light-tight extension.

Cameras for this type of fluorography need fast, coated lenses of the order of $f\,1{\cdot}9$ to $f\,1{\cdot}0$, and the shutter may have to be synchronized with the spot so that the film will not be affected by the afterglow on the screen which may last up to five seconds. The type of film used, whether ortho, or pan, depends upon the colour of the fluorescence. This varies according to the coating material. As a rule development is prolonged to give a high contrast image, and in addition, it is often necessary to intensify the image, especially when dealing with high-speed writing made at anything up to 5,000 km./sec. (one-sixtieth of the speed of light).

Drum cameras are used for recording noncyclic traces. The film is wound around a drum which is made to rotate at a suitable speed. The drum may be up to 50 cm. diameter and the speed of rotation may reach 6,000 r.p.m. It is enclosed in a light-tight housing with a slit behind which the moving film is exposed. A shutter uncovers the slit at the instant of exposure.

With this type of apparatus, very brief traces can be photographed at film speeds as high as 50 metres per second. For longer traces it is necessary to use a camera in which a length of film is wound off a storage spool and on to a take-up spool after exposure.

The Polaroid Land camera can now be attached to many oscillographs. This permits the production of a permanent record in seconds without the need for darkroom equipment.

The film speed of 3,000 A.S.A. permits high shutter speeds with small lens apertures, and so a sharp picture of millimicrosecond transients can be made. By using a transparent film, positive transparencies can be produced inside the camera.

Where the record to be made is complex, like a television picture or a radar screen, a continuous record may be made with the normal type of cinematograph camera using a wide aperture lens and a high-speed film. G.T.S.

Applications. The recording of oscilloscope wave forms, both cyclic and transient, is the most obvious and most frequent use of photographic recording in this field. Such records—where visual observation of the phenomenon is not sufficient—provide a permanent picture of machine and other test procedures using the oscilloscope, can be used for the evaluation of the traces recorded (especially in the case of infrequent transients), etc.

Photography of the television screen is another application of cathode ray trace recording. In this case the trace is a more complex one, since it has to build up a complete picture while scanning the whole tube surface in a specific pattern.

The exposure of such a photograph must therefore be long enough to cover at least one complete scanning cycle and preferably to take in exactly a whole number of such cycles. Otherwise only a partial picture is obtained in the photographic record, or a portion of the picture may be brighter than the rest.

Photographs of a radar screen are required less frequently, especially as isolated pictures. In the course of aircraft tests or special flights a series of such photographs, taken at regular intervals, provide, however, a valuable addition to the flight log. Since it is inconvenient to bring a special camera into position above the normal navigational radar screen every time an exposure is required, special sequence cameras may be trained on a radar screen set aside exclusively for photography and fed the same signals as the screen used by the navigator. The most advanced radar cameras of this kind include remote controlled operation so that they can be set up out of the way of navigating and other crew.

The exposures may be made either when required on pressing a button, or automatically at pre-set intervals. In the latter case the camera must have a sufficient film capacity to cover a reasonably long exposure sequence.

Where the cathode ray tube is used for an alpha numerical display (in other words letters and figures) of a computer output, this almost invariably needs recording. Here a modified contact printing method is mostly used, in conjunction with a cathode ray tube fitted with a fibre window. To record output data displayed at high speed, continuously moving paper is used. This travels past the cathode ray tube face while the electron beam traces out the display.

As an alternative, electro-photographic processes are also employed. Here the electron beam directly charges a suitable surface (for example a selenium-coated drum) to form an electrostatic image which is then converted into a visible image by the usual procedure of a process like xerography.

This permits particularly high recording speeds, and yields better definition and resolution than procedures involving photography via a fluorescent screen layer.

Cathode ray tubes with fibre windows are also used for certain high-speed picture transmission systems. Here the electron beam is modulated in the same way as in a television receiver tube, but scans back and forth more or less along one line instead of over the full area of a screen. The sensitive material is again pulled past the screen in contact with the fibre window at a constant speed, the scanning beam thus builds up the picture line by line as the recording material moves past.

This method can, without undue difficulty, be adapted for very high transmitting speeds; the exposure efficiency is high for no light is lost through rephotographing a displayed image via an optical system.

Two-dimensional computer output displays on a screen can also be used for making synthetic motion pictures. Here the display may represent a graphic arrangement of information fed into the computer according to a particular programme—for example the disposition of cars in particular traffic situations. The computer can further display on the tube successive such arrangements corresponding to changes at specific time intervals. A motion picture camera then photographs these displays one by one on successive frames of the film. On projection at an appropriate speed the film reproduces the imaginary movements worked out by the computer according to the programmed information.

This technique has been used for traffic and road planning, and is also suitable for making animated diagram and other films with a computer.

Apart from being a photographable image in its own right, the electron beam scan of a fluorescent screen can also serve as a light source for printing and enlarging. The fluorescent screen there illuminates the negative point by point as the electron beam scans over it. In electronic printers the light coming through the negative is continuously monitored and this information used to modulate the electron beam for automatic contrast and exposure correction. L.A.M.

See also: *Electronic printing; Instrument recording; Oscillograph recording; Television screen photography.*
Book: *The Photography of Cathode Ray Tube Traces*, by Ilford Ltd. (London).

CATS AND KITTENS. Guile, persuasion, and even tempting tit-bits are wasted on the cat that does not want to co-operate. Unlike the dog, it will not obey commands, and unless it is in the right mood, it is a waste of time to attempt to photograph it; it should be left until another day. Many cats fear strangers, and most seem to sense whether a person is genuinely fond of their species or not, and react accordingly. Cats and kittens are small animals. Only cameras which permit the use of close-up lenses, or which have sufficient focusing extension will give really satisfactory pictures. If the camera is not suitable for going close enough to fill most of the negative with the animal, it is better to try for pictures of somebody holding a cat or the cat as a part of a more general scene.

Going close with a camera means that the viewfinder needs to be free of parallax, and that focusing must be most accurate.

Out-of-Doors. A sky background is perhaps even more suitable for a cat outdoors than a dog, since the cat is by nature a climber. The photograph should always make the most of the animal's good points. For example, if it has a strikingly-marked face, it will be worth while trying to get a close-up of the head. This might be tackled by putting the cat on somebody's lap who can calm the animal by stroking it, with the hand just out of picture. If one tries to hold on to a cat by force it will only struggle to get free, and will never yield a good picture. The animal must be at ease. In the final picture only the head of the cat would be shown. The beauty of close-up shots of this sort depends largely on getting them pin sharp.

With the cat on somebody's lap and the camera close enough to photograph the head of the cat only, the background will be out of focus even at the smallest stop. A small stop is necessary because of the lack of depth of field with such a close picture. Sunshine will therefore be best, and the more effective pictures will be those taken with sidelight or against the sun.

The best place to photograph a cat is where one finds it. In other words it is usually better to take the camera to the cat and not the cat to the camera. If it becomes necessary to place the animal, it should be put in a spot from where it cannot escape so easily, like the top of a garden wall, a small table, or a window ledge.

Generally speaking, at least some pictures should be taken from the level of the cat. This means holding the camera down on the ground, or putting the cat higher up. But any angles may give a good picture.

Added Interest. Cats are so varied in their expressions and habits that there is no need to be satisfied with stereotyped studies. After the cat has had a saucer of milk, there is generally an interval for licking and cleaning. This is a good time for photographs. There are pictures to be made from a yawn or miaow, sharpening claws or stretching. In every case the secret of getting an interesting shot lies in foreseeing the picture in time to photograph it.

These and similar pictures call for a shutter speed of 1/100 second or faster.

The colour of the animal will have to be taken into consideration when determining the exposure. It is easy to see that a white cat needs less exposure than a dark one, if the picture is to show detail and texture in the animal's coat. Panchromatic film will be advisable for most cats.

White cats are every bit as difficult to photograph as black ones. Often a white cat will make a better picture in the shade. Siamese cats are also tricky; not only are they temperamental but their dark faces and light bodies present an exposure problem.

The most important features of a cat are its eyes and its whiskers. Both should be rendered perfectly sharp; the general rule is to focus on the eyes of the animal. Strong sidelight can sometimes throw shadows on part of the eyes or make one eye appear much darker than the other.

In extreme close-ups the cat may squint as it watches the camera. A most natural thing to do, but it will spoil the picture for most cat fanciers.

It is never difficult to attract the attention of the subject. Cats are keenly interested in all strange movements and noises. Flapping of a handkerchief, hissing, or calling the cat by name is enough to get it to look in the desired direction.

Food and drink are useful bait, and even when they have been disposed of, a full and contented cat is less likely to be temperamental than a hungry one.

Indoors. Left to themselves, most cats will select a sunny spot to lie down in. This weakness for warmth is a great help. Indoors the cat will usually bask happily in direct sunlight on a window-ledge, which is both a characteristic setting and an ideal spot for a photograph. Also, the warmth of the fireside and a soft cushion will hold it for some time, and allow ample opportunity to set up tripod, lights and reflectors (so long as no ends of flex are allowed to dangle temptingly during the preparation).

It is best to prepare all lights and the camera before attempting to make the cat settle down. The warmth of your lamps may help to settle the animal, but if the job takes too long, and if the lamps are near, the cat may go to sleep.

Photofloods—as opposed to flash—have the advantage of rendering the cat's pupils as slits and not as large round openings which look most un-catlike. When using flash it is advisable to have either daylight or one other lamp on at the same time in order to make the cat's pupils contract.

Kittens. The natural curiosity of kittens lends itself to picture-making. A cotton reel or an empty box never fails to inspire them to a bout of hilarious clowning. Here also the table-top technique is the most convenient. Once the subject is confined within the limits of the table-top (and the depth of field of the lens), it is easy to attract its attention in any desired direction by dangling a string or making a sudden noise.

The more convincingly small and helpless a kitten can be made to appear in a photograph, the better. This can be done by including some object in the picture area to contrast with the kitten. Hard, angular box shapes will emphasize its fluffiness, a shopping basket accentuates its relative smallness, and so on. But the animal should never be decorated with bows of ribbon and frills. In all other particulars photographing kittens does not differ from photographing cats. P.J.

See also *Animals; Pets.*

CAUSTIC OF LENS. If a simple converging lens is considered as consisting of a series of concentric rings, the focal length of these rings gets progressively shorter as the diameter of the rings increases. So when parallel rays of light fall on the lens, the rays from the edges converge more steeply than those at the centre. This effect is known as spherical aberration.

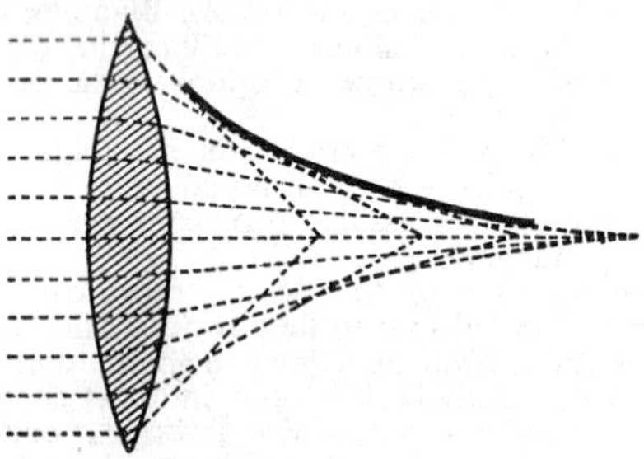

CAUSTIC CURVE. Rays from the edges of a simple positive lens converge more strongly than central rays. In crossing each other the sets of rays form a brilliant curve of light.

This means that each ray crosses the path of its next outside neighbour. Where the rays cross they produce a point of high intensity. All these points of high intensity lie on a curve called the caustic of the lens.

A curve of this type can be seen on the top of a cup of tea when the sunlight is reflected off the inside of the rim.

See also: *Aberrations of lenses; Lens performance; Lens testing.*

CAUSTIC POTASH. Corrosive alkali occasionally used in developers, especially high contrast types.

See also: *Potassium hydroxide.*

CAUSTIC SODA. Poisonous and corrosive alkali. Used in some developers, especially high contrast types.

See also: *Sodium hydroxide.*

CAVES. There are caves in practically every country in the world, formed by actual movement of parts of the earth's crust, by the dissolving action of subterranean water or by the scouring action of waves along the coast. Almost always, the interiors are of interesting and often beautiful conformation, and as such are well worth photographing.

Caves and pot-holes abound in the limestone hills of England and Wales. They lie in four main groups or districts: the north-west border of Yorkshire, the Peak District of Derbyshire, the Mendip Hills in Somerset, and the upper reaches of the Neath and Swansea valleys of South Wales. And they are all easily reached from the principal industrial centres.

The charm of cave photography lies in the novelty of the subject and in the fact that the "scenery" owes nothing to the hand of man, being entirely the result of natural action. Mountain streams boring down through the limestone have carved out a maze of subterranean passages decorated by a surprising variety of stalactite and stalagmite formations. (The stalactites hang from the roof.)

The cave photographer has his own peculiar difficulties, for he must possess endurance to reach his objective and patience in working in often damp and awkward conditions. On the other hand, he has the advantage of having everything under control. Exposure is controlled by lighting, and the subject (apart from flowing water) is static.

Equipment. Photography underground owes a lot to the 35 mm. camera, colour film, and the development of small high-output flash bulbs. This does not rule out the use of larger cameras, but it must depend upon the purpose for which the photograph is intended and whether the conditions will permit a more bulky load.

Apart from the 35 mm. camera's obvious advantage of compactness, a 3·5 cms. wide-angle lens (the most versatile lens underground) can be used at full aperture and still retain sufficient depth of field for most subjects. In large chambers where the foreground is not intrusive, maximum aperture will make the most of your flash. Synchronized flash is not essential with pictorial or record shots, but it can prevent accidental light traces from someone else's light intruding in the picture; it will certainly make more candid action shots practicable.

Lighting. Open flash can be used with flash-bulbs, and must be used with flash-powder, which serves admirably in any large chamber which cannot be adequately covered by flash bulbs. As little as a quarter of an ounce of flash-powder will illuminate a vast chamber at an aperture of f 3·5. Guidance on exposure is given by the manufacturers of both bulbs and flash powder, but it is advisable to assume the worst conditions of those given in the tables unless the subject consists almost entirely of white calcite formations. Whichever flash is used, it is advisable to keep it above ground level to avoid a burnt out foreground.

All the normal rules of photography apply to the placing of the flash, and the entire lighting is controllable. Flash illumination may be placed within the picture area, provided the light source is shielded by some object or part of the cave, so that it does not shine directly into the lens. Lighting from one side of the camera will always give the more dramatic effect and bring out texture, and a stretch of passage may be illuminated with considerable effect by back lighting from round a corner. Here the critical factor for exposure is the distance of the flash from the subject, which should always be appropriate to the lens aperture in use, whatever the position of the camera.

Electronic flash may be used if available, although there is much to be said for flash bulbs in conditions of high humidity, and where hard knocks are likely, or space is limited. The type of flash must be related to the type of film, especially when colour film is used.

Film. It is difficult to convey a true impression of a cave in monochrome to someone unfamiliar with the subject. Appropriate colours are substituted subconsciously on viewing monochrome photographs of more familiar scenes. This might be said to add to the challenge, and monochrome slides and prints of excellent quality can be produced and are of the greatest value for reproduction purposes, where costs rule out colour. But where the record is most important and transparencies are required for projection, colour film is the answer. There are some beautiful colours, ranging from bright red to the more subtle shades of the surrounding rock, the blue of still pools and running streams, and the wonderful cave paintings of France and Spain which demand the use of colour film. The emphasis must be on the accuracy of the rendering of the colouring.

The make of film is a matter for personal preference and it may be of the daylight type used with blue bulbs or electronic flash, or alternatively an artificial light film used with clear bulbs and the appropriate correction filter. An exceptionally fast film is not essential, unless a small aperture must be used, or full use must be made of the flash bulb's output.

Flash-powder can be used with artificial light film quite satisfactorily, although the colour balance may vary slightly with larger quantities of powder, or when the powder has been kept a long time after mixing.

Flash-powder used with open flash is a simple process of securing the camera on its tripod, judging the subject through the viewfinder by painting it in with the light of a torch, opening the shutter and then igniting the fuse for the flash powder. This will allow the photographer to appear in the photograph since the fuse acts as a delayed action device. Celluloid strips 9 ins. long make excellent fuses, and also illuminate the scene.

A figure in an underground photograph, in an unobtrusive position, helps to indicate scale, and adds interest. Subjects such as calcite cascades, curtains and stalactites have beauty and interest, but may still require something to give scale

Cine. Cinematography offers an even greater challenge. The intense light given in a fraction of a second by a flash bulb, which is adequate for still photography, cannot be maintained continuously whilst the cine camera is operating.

The main problem (away from a public electricity supply) is lighting. Lighting can be provided by battery, pressure lamps and types of portable gas lamps. Of these, the most satisfactory are the highly compact battery units in conjunction with high output lamps designed for low voltages. Output is in any case limited, so it is desirable to use the fastest possible film with acceptable grain for the purpose. Under extremely poor lighting conditions it might be necessary to use less than the normal number of frames per second, in order to give sufficient exposure to each frame.

Precautions. There are very many caves which the photographer can explore without danger if proper precautions are taken. But even the easiest cave can become a dangerous place to be in if you sprain an ankle or suffer any other simple mishap which prevents you moving freely.

So when going caving do not go alone; always carry spare lights—even if only 2–3 candles and water-proofed matches; find out all you can about your proposed objective; see that you are properly equipped; and tell someone responsible of your trip.

B.T. (L.B.C.C.)

See also: *Geology; Mountains.*

C.C. FILTER. Colour correction filter used in colour photography and for control in making colour prints. C.C. filters have a selective colour absorption in one or two of the three main spectrol bands. They exist in a range of densities in the three complementary and the three primary colours (cyan, magenta, yellow; red, blue, green). Their designation frequently indicates both their colour and their density: for example a C.C.20M filter is a magenta colour correcting filter with a density of 0·2 (to green light). C.C. filters can be used to eliminate colour casts, but do not have the absorption characteristics required to change the effective colour temperature of any given type of light.

See also: *Colour printing; Colour technique; Filters.*

CELL. Enclosed compartment with transparent or opaque walls containing one or more liquids or solids intended to react with each other chemically under controlled conditions, or to affect light or other energy passing through it. There are accordingly a number of completely different kinds of cell. For photography the following are relevant.

Electric Cell. This is a container with two conducting electrodes of different materials (usually different metals) dipping into a suitable solution. When the electrodes are connected together externally, a current flows from one to the other. At the same time the solution (the electrolyte) reacts with the electrodes. This is the reaction which provides the electric energy. So-called dry cells contain the electrolyte in the form of a jelly or absorbed in a suitable material. The construction of a cell may be quite complex to improve the efficiency of the current generation. A number of connected cells form a battery.

Filter Cells. A small transparent glass or plastic container, filled with a dye solution or a solution of suitable coloured salts can be used as a filter in photography and photomicrography. Since dyes and coloured chemicals in solution have accurately known spectral characteristics, it is in this way possible to prepare filters of precisely standardized specification.

Microscope Cell. Very small transparent containers are used to hold liquids carrying small organisms for investigation under the microscope. In its simplest form such a microscope cell may consist of nothing more than a glass slide with a recess to take the liquid, and a thin cover glass.

Photo-Electric Cell. Certain metal layers in contact with other metals are able to generate electricity when light falls on them, or may vary in their conductivity of an electric current when exposed to light. Such assemblies are used as light-sensitive cells.

See also: *Batteries; Filters; Photo-electric cell; Photomicrography.*

CELLULOID. Transparent and flexible solid, principally pyroxylene with some naptha, amyl acetate, oil and camphor. Invented in 1861 and much used as a base for photographic films in the 1880s.

See also: *Supports for emulsions.*

CELLULOSE. Main basis of the acetate and nitrate bases used in film manufacture. Cotton wool is pure cellulose.

CELSIUS SCALE. Continental temperature scale having 0° as freezing point and 100° as boiling point. Commonly known in England as Centigrade scale.

CEMENT. Adhesive, usually consisting of solvent-based or thermo-setting or thermoplastic synthetic resin, used for a variety of joints in photographic apparatus. Special optical cements are employed for joining lens components, prisms and other glass elements.

See also: *Adhesive.*

CENTI-. Prefix implying a hundredth part of. Principally used in metric system—e.g., centimetre.

See also: *Weights and measures.*

CENTIGRADE SCALE. Internationally termed the Celsius scale. Thermometrical scale on which zero represents the freezing point of water (and melting point of ice) and 100 degrees represents the boiling point of water at sea level. The Centigrade scale is now replacing the Fahrenheit scale in this country, and it is usual to include after a Fahrenheit reading its Centigrade equivalent.

See also: *Thermometer; Weights and measures.*

CENTRE OF INTEREST. That part of the picture to which the eye of the observer automatically travels; it is decided by the arrangement and lighting of the subject. Normally there should be only a single centre of interest; if there are more, the result generally tends to confuse the eye.

See also: *Composition.*

CERAMICS. The two biggest problems to be overcome when photographing ceramics are reflections and distortion.

Eliminating Reflections. Nearly all reflections can be eliminated by use of a suitable lighting technique. Lighting is normally from a single source (preferably a spotlight) to give crispness and modelling to the subject. Shadows are softened by a white reflector, which may be lighted up by another spotlight if the contrast of the subject is too great. The subject is lighted from the side or between side and back, roughly between 80° and 180° from the camera. Once the photographer has found a position giving satisfactory modelling, he adjusts the spotlight vertically until any highlights disappear. With tall objects such as jugs or vases this will occur when the spotlight is in a high position. With flat objects such as plates or shallow bowls, where highlights are not so troublesome, the spotlight is placed much lower to give effective modelling. This method of lighting gets rid of all reflections from flat objects and practically all from taller ones.

Dealing with Distortion. The aim of this type of photograph is to give as accurate a representation of the subject as possible. Distortion, from whatever cause, cannot be tolerated. There are two principal causes of distortion: viewpoint too low and tilting the camera.

Too low a viewpoint when photographing on

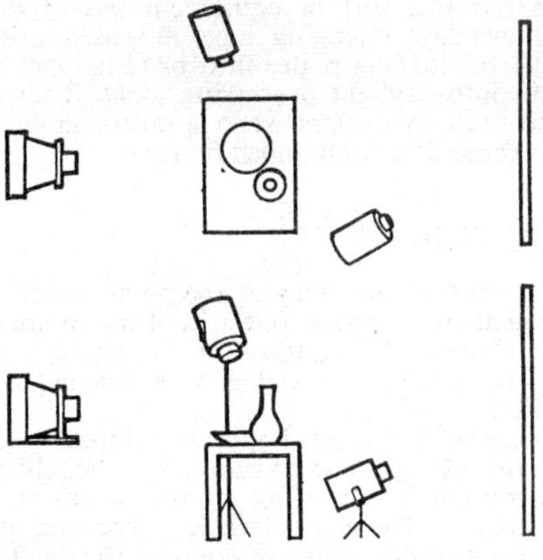

LIGHTING POTTERY. A spotlight from near the side of the subject gives crispness and modelling. It should be at a sufficiently high angle to make catchlights disappear. A white reflector (not shown) is usually necessary to light the shadows, while another spot can be used for background effects.

to a table-top causes round flat objects to appear oval, a defect intensely disliked by pottery manufacturers. So, for subjects like plates or dishes, a very high viewpoint is used so that the camera is looking almost vertically down. Where there are both flat and taller objects combined, as when photographing a tea set, the problem is best solved by standing up the plates behind the taller objects such as cups and jugs. This arrangement gives a pleasing and undistorted grouping. In this case, of course, the viewpoint is lowered so that the camera angle is between 0° and 20° above the horizontal.

Tilting the camera so that the back is no longer vertical—e.g., when photographing down on to the subject—causes perpendicular lines to run inwards to the bottom so that objects on the edges of the photograph appear to be falling over outwards. This distortion is completely rectified by swinging the back of the camera into a vertical position.

A further form of distortion arises with wide-angle lenses when the angle of view appreciably exceeds 60°. Solid objects near the edges of the field appear elongated in a direction away from the lens axis; e.g., spherical shapes become oval.

The Background. Backgrounds for ceramics are always kept fairly plain so that they do not conflict with the pattern or shape of the subject. Wood grained wallpapers offer an excellent variety of background varying from the light tone of pine to a dark walnut if required, and textured paper or fabric is also used with pleasing effect as background material.

As a rule when pottery is photographed from a low viewpoint, the subject is kept well in front of the background which then remains slightly out of focus. To increase the feeling of depth in the photograph it is common practice to shine a spotlight on the background on the shadow side of the subject. This means that the dark side of the subject stands out strongly against the lighted background.

The Camera. The photographing of ceramics is best done with a professional bellows type of camera with full camera movements to allow for the correction of distortion and the achievement of the maximum depth of field. A stop of f45 or f64 is normally used to ensure over-all sharpness.

It is customary to use a plate of fairly soft gradation, and to take care to produce the required balance of light and shade in the image on the ground-glass screen. Under these circumstances, normal development in a M.Q. developer gives a completely satisfactory negative.

Photographs of pottery or china are usually considered to look their best on a glossy printing paper. D.L.H.

See also: *Antiques; Close-ups; Glassware; Lighting the subject; Silverware.*

CHANGING BAG. Bag made of opaque fabric which allows sensitized materials to be handled and loaded into the camera in daylight. There are sleeves at each side of the bag through which the hands and the lower part of the arm are inserted. There is elastic in the open ends of the sleeves to stop light from leaking past.

The camera and film or box of plates is first pushed through one of the sleeves—or through a separate light-trapped flap—and then the sleeves are pulled well over the arms. From that point, the loading or unloading can proceed without any risk of accidentally exposing the film or plate.

Changing bags originally had a window of ruby fabric or celluloid which admitted safe-light into the bag and a pair of eye-holes through which the user could see as well as feel what was going on. And a box often took the place of the body of the bag, so that it was

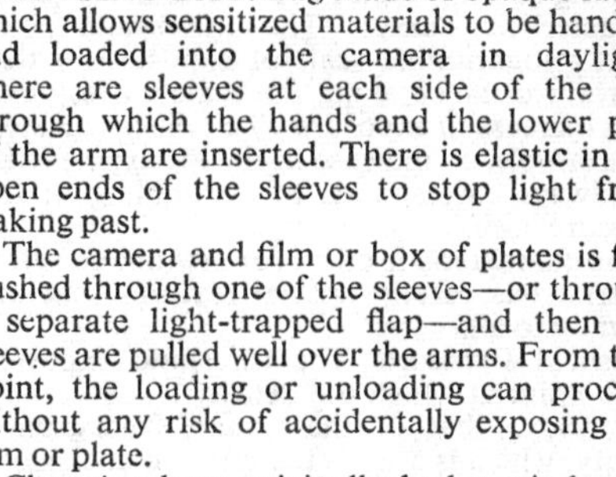

CHANGING BAG. The bag usually consists of a light-proof black fabric with sleeves bound by elastic at the ends. A zip fastener serves for placing the equipment in the bag.

actually possible to develop plates by inspection. These elaborations could only be used with slow, "ordinary" plates which could stand a fair amount of red light. The greater sensitivity of orthochromatic materials and the increased use of panchromatic materials has put this sort of equipment out of date.

Nowadays changing bags are used principally for loading plates into plate holders and films into daylight processing tanks. They are also used by dealers when a customer brings in a camera with a jammed film.

CHARACTERISTIC CURVE

This shows how light acts on a photographic material and produces a density on development. The image formed in the camera consists of patches of light of varying intensity which are all allowed to act on the photographic material for a constant length of time. So the various exposures which make up the image are exposures of constant time and varying intensity. All these different exposures reproduce as a range of densities on development. In a characteristic curve these densities are plotted against the logarithm of the exposures which produced them.

Thus from a characteristic curve for a particular material we can read off the density value which results from a given exposure.

It is not essential to understand the exact meaning of a logarithm in order to make use of characteristic curves. The important point about a logarithmic scale of exposures is that a certain distance along the scale corresponds to an increase in exposure by a constant factor. For example, an increase of the log exposure by 0·3 always corresponds to a doubling of the exposure itself.

The characteristic curve for any one sensitive material has certain special features that are common to all.

Threshold Exposure. The threshold exposure is the point at which the density is just noticeably above that of the general background density, which is always there in the unexposed portion and which is defined as the fog level. As exposure is increased from the threshold value, the density builds up increasingly rapidly. This portion of the curve is defined as the toe. On increasing the exposure still further we reach a portion of the curve where the density increases at constant rate. This is the straight-line portion of the curve. For even greater exposures the curve begins to turn into the horizontal, this portion of the curve being described as the shoulder. The height or level of the horizontal portion defines the maximum density often denoted as D_{max}.

With some materials the density may even fall again on increasing exposure. This effect is known as solarization.

Slope. The slope of the curve at any point has a special importance. The slope is the inclination of the straight line drawn through the point and tangential to the characteristic curve. If a perpendicular line is drawn down from the point, the value of the slope is the ratio of the length of this line to the length of the exposure axis between the line and the point where the tangent cuts the axis. In terms of trigonometry, the slope is the tangent of the angle which the tangential line makes with the exposure axis.

One of the most important slopes is that of the straight-line portion of the characteristic curve. The slope of this is called the gamma of the characteristic curve and is a measure of the degree of contrast to which the emulsion has been developed.

The gradient of the curve between any two points is the slope of a line drawn between the two points. It may be defined between two exposure values or two density values. The gradient is also designated as average slope and is, for many applications, a useful measure of the contrast of reproduction.

Speed. The position of the characteristic curve with respect to the exposure scale is decided by the sensitivity or speed of the photographic material. The further to the left—i.e., towards shorter exposures—the characteristic curve is situated, the less the exposure required to produce any one density on the curve and the faster the emulsion.

The actual speed figure is based on the exposure necessary to reach a certain point on the characteristic curve. The point chosen is called the speed criterion; it differs according to the purpose for which the material is to be used and for which the speed has to be defined. For pictorial photography, for example, the speed criterion is always a point near the lower density end of the characteristic curve, since it is this part of the curve which is used in practice. For process or X-ray work a much higher density would be chosen.

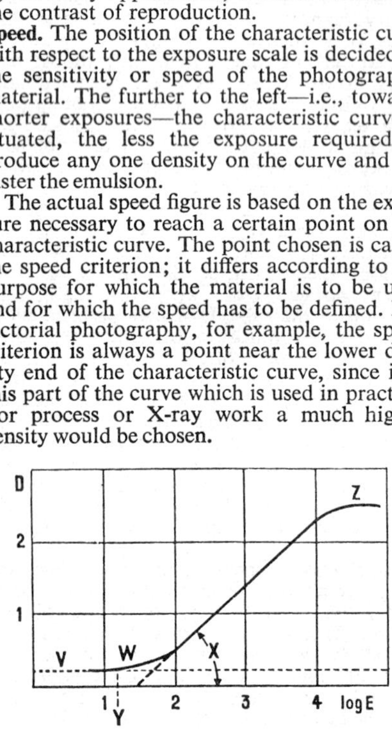

CHARACTERISTIC CURVE. Density D is plotted against the logarithm of the exposure log E. Important features: V, fog level; W, toe; X, slope of straight-line portion (gamma = tan X); Y, threshold exposure; Z, shoulder.

The speed figure is derived by means of the speed formula from the exposure necessary to reach the speed point. The speed scale is as a rule so arranged that the speed increases with decreasing exposure, and the speed formula is therefore based on the reciprocal of the exposure, i.e., on 1/E.

Tone Reproduction. The shape of the characteristic curve is of great practical importance, since upon it depends the final reproduction of the tones in the original. According to the requirements of the process, materials of varying response characteristics are available. The difference between the various materials lies as a rule less in the shape of the curve than in the value of the contrast, which is most conveniently expressed by the gamma figure, or alternatively by the average slope. A material of high contrast produces a big difference in density for a small difference in log exposure, and vice versa.

TYPICAL GAMMA VALUES

Material	*Gamma*
Portrait films and plates	0·5 to 0·8
Press materials	0·5 to 0·8
General purpose films and plates, roll films ...	0·8 to 1·2
Positive cine film	up to 1·8
Negative cine film	0·65
General purpose process materials	2·0
Line materials of high contrast	4·0
"Lith" type materials	8–12
X-ray films	3·0

Controlling Contrast. With a given material, slope and gamma values are affected by several factors. The most important is development: over a certain range gamma increases with the time and/or temperature of development and is most conveniently controlled in this way.

This control method has its limits; for example, it is impracticable to shorten development too much, since the material would develop unevenly. On the other hand, extending the development time will only increase the contrast to a given value, beyond which no further increase is possible; this value is known as gamma infinity. Therefore it is essential to choose a material of approximately the correct inherent gamma.

The slope of the characteristic curve also depends on the colour of the exposing light; the gamma is low for ultra-violet, a good deal higher for blue, and often higher still for red light. For colour separation work, where the same material is exposed through blue, green, and red filters, the difference in gamma has to be compensated for by using different times of development.

Gamma varies with time of exposure; it may be considerably lower for electronic flash than for normal snapshot exposures. To compensate for this, the time of development for films taken by electronic flash should be increased by 50 per cent.

Colour Materials. For integral-tripack materials

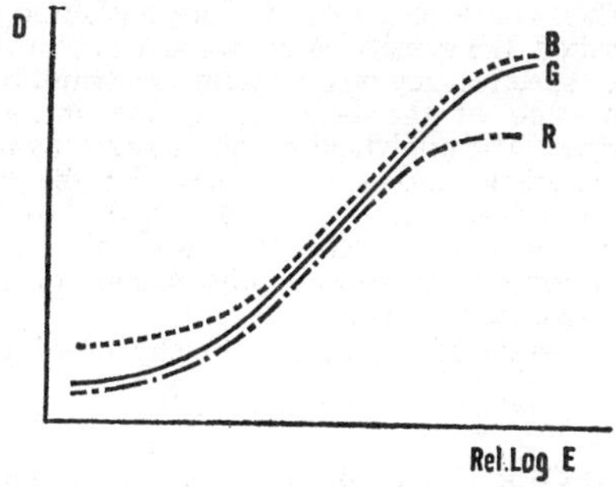

COLOUR FILM CURVES. A colour film uses three curves, one for each layer—blue, green and red sensitive. Ideally the curves should coincide; here the blue curve has a shallower toe (too much yellow in the shadows of the colour negative or too much blue in the print) while the red sensitive emulsion curve flattens out before the others in the highlights, leaving too much red there in the print.

the behaviour of each of the three constituent layers has to be described by a characteristic curve. The material may be developed to a black-and-white image, as is always done with reversal materials, when the image does not differ from that in any multi-coated material. When the layers are developed in colour, however, the value of the densities read on a densitometer depends on the filter through which the readings are taken.

For a positive colour material, the spectral sensitivity of the densitometer for the three spectral regions should have curves which correspond to the tri-stimulus curves of colorimetry or are linear transformations of these. If these conditions are fulfilled, a grey scale will be reproduced by the material as grey, provided the three characteristic curves coincide. The material will have a uniform colour cast, if the curves are identical in shape and merely displaced parallel to one another along the log exposure axis. This cast can be corrected by interposing a suitable filter during exposure. If the slopes of the curves differ, no filter can prevent the lights and the shadows having different casts. No correction is possible if the values for fog or for maximum density differ, except by viewing the picture through a correction filter.

Meaningful densitometry of colour negative materials is more difficult, since the result can be judged only on the final positive, the layers of which have spectral sensitivities which most probably differ from the tri-stimulus curves. It is therefore necessary to make arbitrary extensions or compressions on the density scales for the negative, until nominally equal densities for the three layers, will produce grey in the final picture. When this has been achieved, the same requirements as for the positive hold for the three characteristic curves of the negative material, except that uniform fog densities may not matter if the curve shapes are otherwise identical (e.g. integral masks).

When characteristic curves are measured as described, the density values for any one of the three spectral regions contain contributions from each of the layers: they are integral densities. The analytical densities and thus the characteristic curves which describe the behaviour of each layer, can be obtained from the curves for integral densities by a simple transformation, if the spectral absorptions of the three dyes are known.

The value of the density in any one layer depends not only on the exposure given to that layer but also on the exposures of—and hence the densities in—the neighbouring layers. This dependance is known as the inter-layer effect, which is studied by the use of analytical densities. The inter-layer effects may produce useful masking and hence an improvement in colour saturation. W.F.B.

See also: *Colour sensitometry; Gamma; Sensitometry; Speed of sensitized materials.*

Books: *General Sensitometry*, by Y. N. Gorokhovskii and T. M. Levenberg, *Exposure*, by W. F. Berg (both London).

CHARGER. Type of magazine for narrow-gauge cine cameras. Makers of certain 9·5 mm. cameras frequently refer to their film loads as chargers.

In electronic flash units and portable tape recorders chargers are built in or separate units for recharging accumulators or other secondary cells used for the current suppy.

See also: *Cine camera; Flash (electronic); Magazine.*

CHARGER FOR CASSETTE. Apparatus to enable reloadable 35 mm. film cassettes to be quickly and conveniently loaded. There are two types of charger—darkroom and daylight.

Film Winder. The darkroom type is simply a film winder which grips the film spool between centres so that it can be turned by a handle. It clamps to the edge of the bench. The spool is taken out of the cassette and mounted in the winder. The darkroom light is then switched off and the end of a measured length of film or a standard refill is clipped to the spool. The handle is then turned until the whole length of film—except for two or three inches to act as a leader—has been wound on. The spool of film is then removed from the winder and fitted back into the cassette. The handle end of the winder is forked to engage with the driving pin in the spool, and the handle is prevented by a ratchet from being turned backwards.

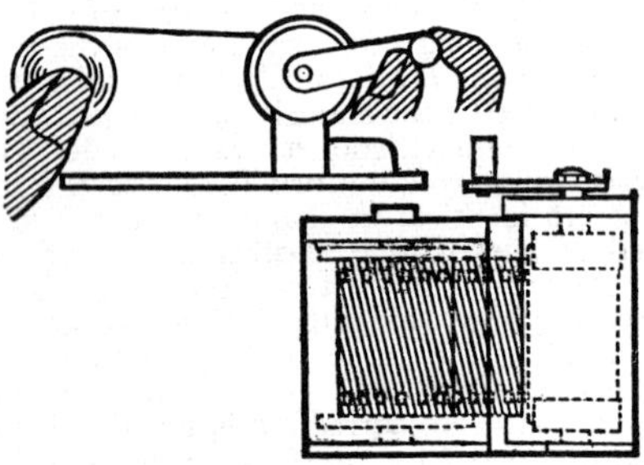

CASSETTE LOADING. Left: A mechanical film winder simplifies spooling up 35 mm. reloads. Right: A light-tight charger permits the whole job to be done in daylight.

Daylight Charger. The daylight charger is in two parts; one is a storage container that holds a bulk length of film, the other is a light-tight box which takes the complete film cassette and has a handle mounted in the end and forked to engage the spool driving pin. Each part has a light-trapped slit through which the film runs, and they can be clamped together so that film can be fed from one to the other without being exposed to the light.

To charge a cassette, the spool is removed and the end of the film from the storage box is attached to it. The cassette is then slipped over the spool and the cassette holder is fitted over it. The two parts are then clamped together; turning the handle now draws film from the storage chamber, through the light trapped slit and on to the cassette spool.

A table supplied with the charger indicates how many turns of the handle will wind on any desired length of film.

CHEESE-CAKE. (Slang.) Photograph of a young woman posed to show obvious sex appeal.

See also: *Beef-cake.*

CHEMICAL CALCULATIONS

When dealing with chemicals in photography, calculations may be required for the following purposes:—

(1) To find the amount of one chemical which can be used to take the place of a given quantity of another.

(2) To find the equivalent amounts of anhydrous and hydrated crystalline versions of a chemical.

(3) To establish the amounts of two or more chemicals which are needed for a reaction.

(4) To work out quantities required for making up solutions of given concentration of a chemical.

(5) To work out dilutions of stock and other solutions.

Substitution of Chemicals. In many processing solutions certain chemicals can be replaced by others without affecting the properties of the solution. The amounts needed are in proportion to the molecular weights of the chemicals concerned. These are worked out from tables of atomic weights.

For example, potassium bichromate can be replaced in bleaching baths by ammonium bichromate. The chemical formula of potassium bichromate is $K_2Cr_2O_7$, of ammonium bichromate $(NH_4)_2Cr_2O_7$, and the respective molecular weights 294 (2 × 39 for K; 2 × 52 for Cr; 7 × 16 for O; and 252 (2 × 14 for N; 8 × 1 for H; 2 × 52 for Cr; 7 × 16 for O). Therefore 294 parts by weight of potassium bichromate will replace or be replaced by 252 parts of ammonium bichromate.

Thus 100 parts potassium bichromate are equivalent to 100 × 252/294 = approx. 86 parts ammonium bichromate—i.e., about 7 parts of the one to 6 of the other.

Conversely, 100 parts ammonium bichromate are equivalent to 100 × 292/252 = approx. 117 parts potassium bichromate.

For any other quantity Q of ammonium bichromate therefore use Q × 117/100 or Q × 294/252 parts of potassium bichromate.

Similarly, sodium bichromate has the formula $Na_2Cr_2O_7.2H_2O$, and a molecular weight of 298. This is practically the same as for potassium bichromate, so the two can be interchanged in equal quantities.

ATOMIC WEIGHTS OF ELEMENTS IN PHOTOGRAPHIC CHEMICALS

Element	*Symbol*	*Atomic Weight*
Aluminium	Al	27
Antimony	Sb	122
Barium	Ba	137·5
Boron	B	11
Bromine	Br	80
Cadmium	Cd	112·5
Calcium	Ca	40
Carbon	C	12
Chlorine	Cl	35·5
Chromium	Cr	52
Cobalt	Co	59
Copper	Cu	63·5
Fluorine	F	19
Gold	Au	197
Hydrogen	H	1
Iodine	I	127
Iron	Fe	56
Lead	Pb	207
Magnesium	Mg	24
Manganese	Mn	55
Mercury	Hg	200
Nickel	Ni	59
Nitrogen	N	14
Oxygen	O	16
Palladium	Pd	107
Phosphorus	P	31
Platinum	Pt	195
Potassium	K	39
Selenium	Se	79
Silicon	Si	28
Silver	Ag	108
Sodium	Na	23
Sulphur	S	32
Tin	Sn	119
Titanium	Ti	48
Tungsten	W	184
Uranium	U	238
Vanadium	V	51
Zinc	Zn	65·5

The weights are rounded off to the nearest 0·5.

When substituting chemicals, the substituted chemical must not affect the action of the bath in any material way. In the above example of ammonium and potassium bichromate that holds true as long as the bleacher in which they are used is acid; if it becomes alkaline, it will liberate ammonia which may give rise to quite a different action. On the other hand, sodium and potassium bichromate are not likely to differ in their action at all.

Substitution by molecular or equivalent weights also does not apply to alkalis (other than strong ones like sodium and potassium hydroxide) in developers. The reason is that the activity of a developer depends on its pH value, and equivalent solutions of most of the alkalis used in developers rarely have the same pH value.

Anhydrous and Hydrated Crystalline Chemicals. Many chemicals exist in several forms with and without water of crystallization. The various forms are, of course, absolutely identical in their equivalent solutions. Calculation of the respective amounts follows the same principle as for the substitution of chemicals.

For example, anhydrous sodium carbonate Na_2CO_3, molecular weight 106, is identical, when dissolved, with crystalline sodium carbonate $Na_2CO_3.10H_2O$, molecular weight 286, or the monohydrate $Na_2CO_3.H_2O$, molecular weight 124. Thus 100 parts of crystals can be replaced by 100 × 106/286 = 37 parts of the anhydrous salt, or by 100 × 124/286 = approx. 43 parts of the monohydrate.

Chemical Reactions. Sometimes a chemical has to be made in solution by the interaction of two or more other chemicals. In that case the required amounts can be calculated from the equation of the reaction.

For example, sodium metaborate can be prepared by mixing solutions of sodium hydroxide and borax. In practice that means that the latter two chemicals can be used in a developer specifying sodium metaborate. The reaction is:—

$$2NaOH + Na_2B_4O_7.10H_2O + 5H_2O = 4(NaBO_2.4H_2O)$$

(The water of crystallization has been included in the equation to simplify calculation of equivalent amounts of the crystalline chemicals, but it does not take part in the reaction.)

The weights reacting are:—

$$2 \times 40 + 382 \quad (+ 90) = 4 \times 138$$

Note that both sides of the equation must balance quantitatively—i.e., in this case, both sides add up to 552.

Thus 552 parts of sodium metaborate are obtained by mixing 80 parts of sodium hydroxide (in solution) with 382 parts of borax. Therefore 100 parts of metaborate require 100 × 80/552 = approx. 14·6 parts sodium hydroxide plus 100 × 382/552 = 69·5 parts borax. In this particular case it is better

SUBSTITUTION OF CHEMICALS

100 *Parts of*	*Equivalent Parts*	*Substitute*
Ammonia, sp. gr. 0·880	155	Ammonia, sp. gr. 0·920
Ammonia, sp. gr. 0·920	65	Ammonia, sp. gr. 0·880
Ammonium bichromate	117	Potassium or sodium bichromate
Ammonium bromide	122	Potassium bromide
	142	Sodium bromide crystals
Ammonium chloride	107	Sodium chloride
	139	Potassium chloride
Ammonium persulphate	118	Potassium persulphate
Ammonium thiocyanate	110	Sodium thiocyanate
	128	Potassium thiocyanate
Chrome ammonium alum	105	Chrome potassium alum
Chrome potassium alum	96	Chrome ammonium alum
Ferric oxalate	177	Ferric ammonium oxalate
	200	Ferric potassium oxalate
Ferric ammonium oxalate	115	Ferric potassium oxalate
	56	Ferric oxalate
Ferric potassium oxalate	87	Ferric ammonium oxalate
	50	Ferric oxalate
Lead acetate	87	Lead nitrate
Lead nitrate	114	Lead acetate
Potassium bichromate	85	Ammonium bichromate
	100	Sodium bichromate
Potassium bromide	82	Ammonium bromide
	116	Sodium bromide crystals
Potassium chloride	73	Ammonium chloride
	77	Sodium chloride
Potassium citrate, neutral	93	Sodium citrate, neutral
Potassium metabisulphite	86	Sodium metabisulphite
	94	Sodium bisulphite
Potassium persulphate	84	Ammonium persulphate
Potassium thiocyanate	78	Ammonium thiocyanate
	86	Sodium thiocyanate
Sodium bichromate	85	Ammonium bichromate
	100	Potassium bichromate
Sodium bisulphite	106	Potassium metabisulphite
	91	Sodium metabisulphite
Sodium bromide crystals	71	Ammonium bromide
	86	Potassium bromide
Sodium chloride	93	Ammonium chloride
	130	Potassium chloride
Sodium citrate, neutral	108	Potassium citrate, neutral
Sodium metabisulphite	116	Potassium metabisulphite
	110	Sodium bisulphite
Sodium thiocyanate	91	Ammonium thiocyanate
	116	Potassium thiocyanate

Note.—Ammonium salts should not be used in developers or other alkaline solutions unless specifically stated.

to take a slight excess of borax rather than of hydroxide, so this ratio can be simplified to 1 part hydroxide plus 5 parts borax making 7 parts metaborate.

As another example, sodium metabisulphite can be used instead of sodium bisulphite. The latter is in fact formed on dissolving the metabisulphite:—

$$Na_2S_2O_5 + H_2O = 2NaHSO_3$$

From this, 190 parts metabisulphite yield 208 parts of bisulphite. Other quantities of sodium metabisulphite—or of potassium metabisulphite—can then be calculated as already described.

Solution Strengths. The strength of concentration of solutions can be quoted in various ways: weight per volume (W/V), weight per weight (W/W), volume per volume (V/V), or percentages (%). There are also the chemical notations of molarity and normality.

Weight per volume—e.g., grams per litre—solutions are made by dissolving the requisite number of grams in the minimum amount of water, and then adding enough water to make 1 litre. The procedure is analogous for ounces or grains per pint or any other unit. With a solution containing A grams per litre, B c.cm. are equivalent to A × B/1,000 grams; if the solution contains C grains per pint, D fluid ounces are equivalent to C × D/20 grains, and so on.

Weight per weight solutions—e.g., X grams of solute per 1,000 grams of solution—are made by dissolving X grams in 1,000-X grams of solvent. This notation is comparatively rare in photographic use. Proportionate amounts are calculated in the same way as for weight per volume solutions.

Volume per volume solutions—e.g., c.cm. per litre—are made by taking the requisite number of c.cm. and adding enough water to make 1 litre (ounces per pint or any other unit are similar). This notation is often used for solutions of liquids.

Percentage solutions may be weight per volume or per weight, or volume per volume, and in each case the figure specifies parts per 100. The most common system is that of weight per volume, as it is easiest to handle. It is, however, important to state which system is used; for instance a 10 per cent solution of sulphuric acid on a weight per volume basis is only just over half as strong as a 10 per cent volume per volume solution.

Molar solutions specify the number of gram-molecules per litre. For instance sodium sulphite (Na_2SO_3) has a molecular weight of 126; a molar solution therefore contains 126 grams per litre. (Water of crystallization is ignored in working out the active concentration of molar solutions, but has to be taken into account when weighing out the crystalline chemical.) A solution of 12·6 grams per litre is called deci-molar solution and written 0·1M. Other strengths are labelled accordingly.

Normal solutions contain the gram-equivalent weight per litre. This may be the same as the gram-molecular weight, or it may be a simple fraction of it, depending on the chemical concerned. Again there are decinormal (0·1N), twice normal (2N) etc., solutions.

Using Percentage Stock Solutions. Provided the chemicals keep well in solution, and are used at a low concentration, the ingredients of many formulae can be stored in individual solutions of a fixed percentage strength. This eliminates weighing out and dissolving the chemicals afresh each time a fresh processing bath is required. The system is specially useful where several baths with the same ingredients, but

SUBSTITUTION OF ANHYDROUS AND CRYSTALLINE SALTS

100 *Parts of*	*Equivalent Parts*	*Substitute*
Sodium carbonate, crystals	37	Sodium carbonate, anhydrous
	43	Sodium carbonate, monohydrate
Sodium carbonate, monohydrate	233	Sodium carbonate, crystals
	86	Sodium carbonate, anhydrous
Sodium carbonate, anhydrous	116	Sodium carbonate, monohydrate
	270	Sodium carbonate, crystals
Sodium phosphate, tribasic, crystals	43	Sodium phosphate, tribasic, anhydrous
	48	Sodium phosphate, tribasic, monohydrate
Sodium phosphate, tribasic, monohydrate	208	Sodium phosphate, tribasic, crystals
	90	Sodium phosphate, tribasic, anhydrous
Sodium phosphate, tribasic, anhydrous	232	Sodium phosphate, tribasic crystals
	111	Sodium phosphate, tribasic, monohydrate
Sodium sulphate, crystals	44	Sodium sulphate, anhydrous
Sodium sulphate, anhydrous	225	Sodium sulphite, crystals
Sodium sulphite, crystals	50	Sodium sulphite, anhydrous
Sodium sulphite, anhydrous	200	Sodium sulphite, crystals
Sodium thiosulphate, crystals	64	Sodium thiosulphate, anhydrous
Sodium thiosulphate, anhydrous	156	Sodium thiosulphate, crystals

in different proportions, are employed from time to time—e.g., with reducers and intensifiers of different degrees of activity.

For convenience all the solutions may be made up to 5 or 10 per cent. To make up a formula, simply multiply all the weights in grams by 20 or 10, mix the corresponding number of c.cm., and make up to the final volume with water.

Owing to the poor keeping qualities of developing agents in solution by themselves, this method is not generally suitable for developers, although separate stock solutions —e.g., developing agents plus preservative on the one hand, and alkali on the other—are practical.

Diluting Percentage Solutions. To dilute a solution of A per cent to make a solution of B per cent, take $100 \times B/A$ parts and add water to make up to 100 parts.

For example, to dilute a 40 per cent solution to 15 per cent, take $100 \times 15/40 = 37{\cdot}5$ c.cm. and make up to 100 c.cm. again with water. If the original solution contained 40 per cent of the chemical, i.e., 40 grams per 100 c.cm., 37·5 c.cm. of it will contain $40 \times 37{\cdot}5/100$ grams $= 15$ grams, which makes 15 per cent if diluted to 100 c.cm.

Dilution of Stock Solutions. Many processing solutions, notably developers, are conveniently made up in concentrated stock solutions and diluted for use. The solutions keep better that way and are easier to store.

The recommended dilution may be specified in several ways.

(1) Diluting x times. To dilute a solution a given number of times—e.g., 5 times—take 1 part stock solution and add $(x - 1)$ parts (in this example, 4 parts) of water to make a total of x—i.e., 5 parts of working solution. If a given final volume is required, the amount of stock solution used will be one-fifth.

(2) Diluting to make x parts. This is the same as diluting x times.

(3) Diluting with y parts of water. This means that the working solution consists of 1 part stock solution plus y parts of water. To obtain a given final volume, this has to be divided by $(y + 1)$ for the amount of stock solution required. For instance, for 400 c.cm. of working solution, obtained by diluting the stock solution with 4 parts of water, the required amount of stock solution is $400/(4 + 1)$, or 80 c.cm.

(4) Diluting 1 : y. This is the same as diluting with y parts of water.

Note that diluting with y parts of water (or 1 : y) is the same as diluting $(y + 1)$ times, to make $(y + 1)$ parts. For instance, diluting with 4 parts of water means diluting 5 times, i.e., to make 5 parts in all. L.A.M.

See also: *Chemicals; Solutions; Weights and measures.*

CHEMICAL FOCUS. When a beam of parallel light rays falls on a convex lens, the rays are bent towards the axis. If the lens is uncorrected for chromatic aberration, the invisible actinic rays intersect the axis at a point or points (the chemical focus) closer to the lens than the intersection of the visible rays (the visual focus).

In a fully corrected lens the chemical and visual foci coincide.

See also: *Aberrations of lenses; Lens.*

CHEMICAL FOG. Veil of metallic silver deposited all over the negative. Heavy fog is caused by prolonged or overactive development, but a certain amount of chemical fogging always arises during development.

See also: *Fogging.*

CHEMICALS. The term "photographic chemicals" is apt to suggest a range of organic and inorganic substances which are manufactured exclusively for photographic use. In fact, most of the chemicals used in photography have important application in other fields.

There are about fifty of these chemicals in common use including the developing agents, sodas and other chemicals used in black-and-white processing and the more popular reducers, intensifiers, toners, etc. This figure rises to about eighty with the chemicals used in colour processing and other more specialized work. Almost all these substances are readily available as branded chemicals in quantities to suit every requirement.

In a wider sense the term "photographic chemicals" has also come to include the many proprietary developers, fixers and other ready-made processing solutions which have become a feature of modern photographic practice.

Purity. Chemicals usually bear on their labels such terms as Commercial, Technical, Pure, Recryst., B.P. (British Pharmacopoeia), B.P.C. (British Pharmaceutical Codex), or A.R. (Analytical Reagent). To a chemist these descriptions give a more or less reliable indication of their purity, but the photographer might well be confused by the various grades of a particular chemical which he could buy.

For most purposes B.P. or B.P.C. grade chemicals made by the larger manufacturers of fine chemicals are suitable. A.R. chemicals are even purer, especially when backed by commercial trade marks for this grade, but are rarely required for photographic use and are usually very expensive.

The best policy generally is to buy the branded chemicals which are supplied specifically for photographic use. It can be assumed that the manufacturer employs analytical and photographic tests to ensure that a satisfactory and standard product is supplied at a fair price. He will also distinguish between impurities which are harmless and those which are undesirable—a job which would be quite impossible for the practical photographer. For example—a sample of anhydrous sodium sulphite containing one or two per cent of moisture or inert sodium sulphate would be considered satisfactory for photographic use, but a high percentage of carbonate, a trace of hypo or a few parts pe million of copper would be extremely undesirable.

There is always an element of risk in using unbranded chemicals, particularly commercial or technical grades, since there is no guarantee against variation from batch to batch or against the presence of undesirable impurities.

Forms. Solid chemicals are frequently classified as "crystal", "granulated" or "powder" according to their physical form. This classification can be misleading; although some chemicals naturally occur as powder or in granulated form, others consist of very small crystals or are produced by grinding or pulverizing large crystals.

In some cases, notably sodium sulphite, sodium carbonate and thiosulphate, the term "powder" has come to indicate the dried or anhydrous form of the chemical; but all powdered chemicals are not anhydrous. In the cases of borax, boric acid and potassium bromide the crystal and powder forms are identical in composition.

In general, powdered and granulated chemicals are easier to handle and they dissolve more readily than the larger crystal forms. If, however, a chemical is liable to deteriorate on exposure to air, e.g., potassium ferricyanide, potassium metabisulphite, the larger crystals are more stable.

A number of chemicals contain in their crystal form a quantity of water of crystallization. In some cases this can be partially or entirely removed by drying under carefully controlled conditions, to give free-flowing granular powders which are known as dried, desiccated or anhydrous chemicals.

Anhydrous chemicals contain more of the active ingredient weight for weight than the crystals and allowance must be made for this when making up formulae in which crystals are specified. For example—100 parts sodium carbonate cryst. = 37 parts dried; = 43 parts monohydrate. 2 parts sodium sulphite cryst. = 1 part dried. 8 parts hypo cryst. = 5 parts dried. 13 parts sodium acetate cryst. = 8 parts dried.

Anhydrous chemicals are usually more satisfactory for general photographic use. They are easier to store, weigh, and dissolve, and are less affected by high temperatures. Although their cost per pound is greater than that of the crystals, they are proportionately stronger and are often cheaper in actual practice.

One drawback of anhydrous or desiccated salts is that they may cake together in one lump if stored in slightly damp conditions. This applies as much to individual chemicals as to mixtures such as acid fixing powder. Once caked together, the chemical usually has to be broken up before dissolving, as large lumps take a long time to go into solution.

Using Individual Chemicals. The small user will generally buy developing agents in ounce bottles, and carbonate, sulphite and hypo in pound bottles or tins. The ounce and pound containers correspond to all intents and purposes, to the 25 gm. and 500 gm. containers. Even so, the cost of even a modest stock of the chemicals necessary to develop and fix will be at least 20s., not counting the cost of a balance and weights.

The main advantage of preparing solutions from individual chemicals is that any one of a wide variety of published formulae can be made up as required at comparatively low cost. This is economical for the large user, as the solutions can be made up in bulk. For the smaller user who processes an occasional film now and again, ready packed chemicals are much more convenient and handier.

Made-up Formulae. Packed chemicals made up to specific formulae are usually either supplied in powder form or as concentrated liquids. Both forms have their advantages but of recent years the convenience of liquid developers has greatly increased their popularity.

Powders, which are usually packed in cartons, foil, plastic envelopes or tins, are available in all sizes from the small individual pack costing a few pence to the largest tin making 50 gallons or more for bulk processing. Powdered chemicals are compact, easily stored and least liable to damage in transit.

Liquid preparations have the two-fold advantage of economy and convenience and many photographic solutions are now available in this form. They can usually be diluted to four or five times their volume for contact printing, to five to ten times for bromide printing and dish development of negatives and up to twenty times or more for tank development.

Storage of Solid Chemicals. A safe rule is to protect photographic chemicals from exposure to air, moisture and excessive heat and to avoid any possibility of contamination with metals or other chemicals. This implies the use of air-tight glass, plastic or stoneware containers.

Developing agents should be stored in air-tight bottles or lined tins with close-fitting lids. The chief hazard is oxidation by exposure to air, a process which is accelerated by dampness. If this is avoided, most developing agents will keep in good condition for years. Slight discoloration has no adverse effect, but a few developing agents—e.g., Amidol and paraphenylenediamine—deteriorate more rapidly and should not be stored for longer than about a year.

Anhydrous or dried chemicals absorb moisture from the air and some deteriorate in other ways if they are not stored in air-tight containers. Sodium sulphite oxidizes to the inert sulphate; sodium carbonate is slowly converted to bicarbonate. Dried hypo does not deteriorate chemically but solidifies to a rock-like mass.

Crystal chemicals containing water of crystallization (such as sodium sulphite, sodium carbonate and hypo) become powdery on exposure to air. This drying-out or efflorescence tends to increase their strength, but some, such as sulphite, also deteriorate chemically. The danger in these cases is that the chemical is no longer uniform in composition and it is impossible to be certain of the extent of the deterioration by casual examination.

Some photographic chemicals are deliquescent—i.e., they absorb moisture from the air and liquefy. These include the thiocyanates, hydroxylamine hydrochloride, sodium sulphide and the caustic alkalis. In such cases the use of completely air-tight containers is essential and it is usually more convenient to prepare stock solutions of these chemicals. Some also deteriorate quite rapidly: caustic alkalis absorb carbon dioxide and are gradually converted to carbonates; sodium sulphide is converted to a mixture of compounds including hypo which makes it unsuitable for toners.

Storage of Solutions. Nearly all photographic solutions deteriorate on exposure to air and are best stored in completely air-tight bottles. They should also be protected from strong light and extremes of temperature.

Some solutions, such as those containing Amidol and some colour developers, are unstable and should not be stored; most developers in very dilute solutions deteriorate quite rapidly, particularly after use.

Concentrated developers, including proprietary liquid developers, usually keep well in reasonably full, airtight bottles. In most cases any serious deterioration is indicated by a marked darkening in colour and it is always unwise to use developers in this condition. Once the bottle has been opened, an air space above the contents is inevitable; this will have no serious effect although concentrated developer should not be stored for a long period in a bottle which is more than, say, two-thirds empty.

Fixing baths will normally keep for several months, but eventually decompose giving a yellow deposit of sulphur.

Stoppering. Any bottle containing a photographic solution should have an efficient stopper. Corks quickly become sodden and porous and only of temporary use.

Glass stoppers are excellent for acids, solvents, etc., but are unsuitable for use with developers and alkaline solutions, owing to the risk of their becoming cemented firmly into the bottle neck.

Well-fitting soft rubber bungs, or plastic screw caps with rubber or plastic wads are the most satisfactory closures. R.L.T. & A.R.P.

See also: *Alkali; Bottles; Chemical calculations; Solubility; Solutions; Stoppers.*

Books: *Chemical Analysis in Photography*, by G. Russell; *Chemicals in Photography*, by A. Pope (London).

CHEMICAL SYMBOLS. Internationally agreed letters or groups of letters signifying individual chemical elements. Generally these symbols are the initial letters or abbreviations of the names of elements; in some cases the Latin or Greek names are used—e.g., H for hydrogen, O for oxygen, Ag (argentum) for silver, and so on.

Simple Formulae. The formula of a chemical compound is made up of such symbols and shows what elements, and how many atoms of each, are contained in a molecule of the compound. Thus the formula $Ag\ NO_3$ for silver nitrate indicates that each molecule of silver nitrate contains one atom of silver (Ag), one atom of nitrogen (N) and three atoms of oxygen (O). Similarly, the formula $Na_2S_2O_3$ for sodium thiosulphate shows two atoms of sodium (Na), two atoms sulphur (S), and three atoms oxygen (O).

The same symbols are used in formulating chemical reactions. There they show the nature and proportions of molecules taking part:—

$$CaCl_2 + 2AgNO_3 \rightarrow Ca(NO_3)_2 + 2AgCl$$

Here one molecule of calcium chloride (consisting of calcium Ca and chlorine Cl) reacts with two molecules of silver nitrate to form one molecule of calcium nitrate and two of silver chloride.

The numbers of each kind of atom on each side of the equation must be the same. The symbols in the equation thus also indicate the equivalent quantities—established by reference to atomic weight tables—of materials used up and produced. The above equation therefore shows that 2 × 170 parts by weight of silver nitrate (atomic weights Ag = 108 + N = 14 + 3 × O = 3 × 16, total 170) form 2 × 143.5 parts of silver chloride (Ag = 108 + Cl = 35·5, total 143·5).

With many reactions—especially those taking place in solution—the symbols in fact signify the quantitative reaction only; the actual mechanism of the interaction between the molecules may be more complicated.

Structural Symbols. In organic chemistry where large and complex molecules are involved, the structure of a compound may involve large numbers of similar atoms. There structural formulae are often required to show the arrangement of the elements, since large numbers of compounds may have the same empirical formula. Thus a formula $C_6H_6O_2$ might correspond to a hypothetical compound

$$CH \equiv C - CH = CH - CH_2 - COOH$$

or to

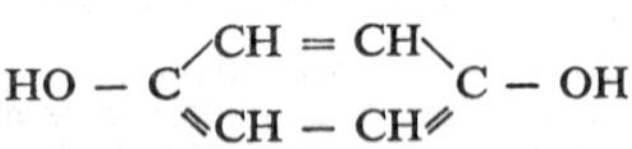

or to

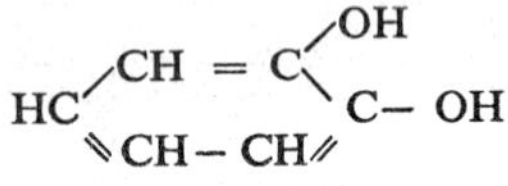

The first of these would probably be too unstable to exist, but the other two are the developing agents hydroquinone and pyrocatechin respectively.

The structure

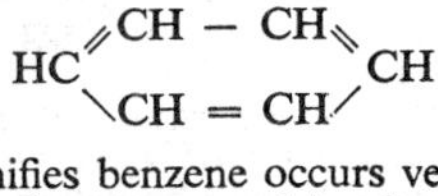

which signifies benzene occurs very frequently in organic chemistry, and is often replaced by a structural symbol, so hydroquinone may be written as

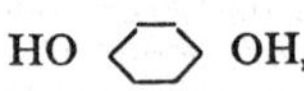

each corner of the hexagon corresponding to a carbon and a hydrogen atom (CH), except where something else (here OH) is attached, which replaces the hydrogen atom.

Other structural symbols are the lines which join the atoms together. Single, double, and triple lines indicate different types of chemical bond, and thus show quite a lot about how the compound is likely to behave. L.A.M.

See also: *Chemical calculations; Chemicals.*

CHEVALIER, CHARLES LOUIS, 1804–59. French optician and camera obscura manufacturer. Made apparatus for Niépce and gave his address to Daguerre for whom he was also working. In 1839 made lenses for the Daguerre-Giroux cameras. In 1840 introduced the Photographe, a portable daguerreotype camera. Produced photomicrographs as early as March 1840. Chevalier was one of the earliest French photographic writers and dealers. Biography by Arthur Chevalier (Paris 1862).

CHEVREUL, MICHEL EUGENE, 1786–1889. French chemist. Carried out important researches on dyes and colouring matters and particularly their behaviour under the action of moisture, light and heat. In 1849 he studied the action of light on Prussian blue and many other colouring matters and explained the part played by oxygen and moisture in bleaching. Biography by Malloizel (Paris 1887).

CHIAROSCURO (from Italian—chiaro = light + oscuro = dark). Light and shade effect. In art, the manner of indicating transition from one tone into the other or of contrast between them. The term nowadays has more generally come to mean the way things are picked out by light or submerged in shadow.

CHILDREN. A good child photograph should look as though the camera had caught the subject unawares. Seen in this way the child looks natural and makes the best photograph. This result cannot be achieved by forcing the youngster to pose, or it will look stiff and self-conscious. Children have a will of their own; they can be made to pose naturally, but it must be done by kindness and understanding.

A successful child photographer must be so practised in the technical side of his work that he can forget about it and bring all his understanding and diplomacy to managing his subject. If he is preoccupied with technicalities he is bound to miss opportunities. He must also be able to deal tactfully with the child's mother. Small children usually feel safer when the mother is near; around the age of two they tend to cling to her, and she may even have to be included in the picture. But with older children, there are times when a fussy mother can be an embarrassment to both the photographer and his subject. No child can look natural if it is being constantly tidied and told to sit up straight and smile. The photographer must persuade the mother to wait in another room if he hopes to turn out good pictures.

Being photographed is often an upsetting experience for the child. Often it has been put into clean clothes, given an extra wash, and had its hair freshly brushed before being brought along to the studio where its surroundings are completely unfamiliar. Under such circumstances it is practically impossible to get a natural likeness. At home, the child gets over all the fuss more easily and the photographer has a better chance to get it into the right mood to be photographed.

Things are much easier if the youngster can be taken when it is busy out of doors. Children playing in the open, or watching a Punch and Judy show, or the animals at the Zoo, very soon forget all about the camera. And once the child can be made to stay in one spot and forget the camera, most of the difficulties vanish. Generally the answer is to get it interested in some activity like playing with a toy, reading a book, winding a musical box, or blowing soap bubbles.

The experienced child photographer wins the co-operation of his subject by treating it as an equal. He sets out first of all to make friends with the child, even explaining the camera to it, letting it look through the finder, or help to rig up the lamps if it is old enough.

The Background. A suitable background is essential to a good child photograph. It should offer a contrast to the subject by being lighter or darker, and it should not be fussy. Indoors, a plain wall or a window will do, and it should have some light on it—black backgrounds are only for special effects. Out of doors, the sky, a hedge, or part of the house can make a good setting. In most cases the subject should be well away from the background so that it can be rendered sharp while the background is thrown out of focus.

Viewpoint. Generally the viewpoint is chosen as close as possible so that the subject fills the whole of the negative area. There are exceptions, as when the child is seen as part of the landscape, or when it is running about, but mostly the expression is the important thing, not the surroundings.

The camera is best used on the same level as the child; children always look more interesting when they are seen on their own level. The photographer can get down on to the floor where the child spends so much of its time, or he can put the child on the table or in a high chair, or even on top of a wall. Such positions also make it more difficult for the child to wander out of range.

There is no reason why the camera should look down simply because it is fitted with an eye-level viewfinder unless, for example, the child is lying in bed.

Lighting. With child photographs, the simpler the lighting, the better. Indoors, good basic lighting is given by a main light shining at about forty-five degrees and from slightly above the camera, with a second light farther away directed to soften the shadows and put highlights in the eyes. The lamps are then altered according to the position the child takes up. If there is any daylight, it may be used to advantage to replace the fill-in lamp. There may even be enough light coming through the window to make extra lights unnecessary and low sunlight coming through the window may be used with only a reflector for the shadows. Sometimes a light hanging above the child will accentuate the hair, and it is often the only way of making fair hair look really fair.

The background, however, may call for a lamp to itself.

For softly lit child studies, where only one light source (flash or photoflood) is available, the bounce light technique is ideal. This means directing the light at a suitable reflector such as a nearby wall or ceiling and using the reflected light to illuminate the subject. Such reflected light is, of course, weaker than direct light and calls for more exposure. With flash, use one and a half times to double the normal guide number for direct light. Remember also to count the distance as the path from the flash to the reflecting surface plus from the latter to the subject. For colour photography, the reflecting surfaces should be white or a neutral light grey.

Out of doors, weak, diffused sunshine is the most suitable lighting. In direct sunshine, a reflector should be used. This may be a sheet of plywood or card covered with silver paper, a white cloth, or even a newspaper. With backlighting there is no tendency for the child to screw up its eyes, and it can tolerate the soft reflected light more easily than direct sun.

Generally light coming from the direction of the camera is unsuitable; side or backlighting gives much better modelling.

Flashlight is particularly suitable for photographing children who cannot stand strong lights, and for action photographs indoors. It can also be used instead of a reflector out of doors. The general principles of lighting apply. It is better to have the flash at some distance from the camera in order to get modelling. In a medium sized room, with bright walls, there will be sufficient reflection to prevent heavy shadows.

A reflector or a second, preferably weaker flash can be used to advantage. The background should be either lit separately or it should be near enough to the child to get some light on it. You have to watch windows, mirrors and pictures which might reflect in the background. It is possible to get reasonable results with a box camera (on a tripod) and open flash. W.S.

See also: *Babies; Flash technique; Portraiture; Portraiture at home; Portraiture outdoors.*

Books: *The Art and Technique of Taking Portraits*, by H. van Wadenoyen (London); *Camera Portraiture*, by H. Williams (London).

CHINA. As early as 2,450 years ago, ancient Chinese scientists already had considerable optical knowledge and an understanding of the principles of light sensitivity.

In the fifth century B.C., the Chinese scientist and philosopher Mo Tzu expounded the relationship between the fundamental optical principles of reflection, the propagation of light in a straight line, and the formation of inverted images by light passing through a pin-hole. He also explained the relationship between an object and its image in a flat, concave or convex metal mirror.

In the fourth century B.C. a book entitled *Lord Chou's Notes on Technology* recorded the method of making fire by concentrating sunlight with a concave mirror.

A book entitled *Huei Nan Tzu* published in the Chin dynasty, 2,100 years ago, contains the following extract: "Cut a ball of ice, lift it to face the sun and put some dry grass under its shadow; fire will be created." This was a clear description of how light can be focused to a point through a convex lens.

In the North Sung dynasty (A.D. 960–1127) a scholar named Shen Kwa made a detailed and systematic explanation on the law of light in his *Pen Talk of the Dreamy Stream.* At that time the study of light already existed. It was called *Ke Shu* (The Art of Optics).

A Chinese lecturer at the University of Hong Kong claims to have found traces of china plates made sensitive to light by a chemical process 2,000 years ago. These plates, he says, might have been turned into negative pictures if used in combination with a camera obscura. He argues, therefore, that the main principles of photography were known in China 2,000 years ago.

Cameras were introduced into China around 1840. In 1845 (the twentieth year of the reign of Emperor Tao Kwang of the Chin dynasty) a scholar named Chou Shou-chang mentioned photography in his *Ssu Yi Tang Diary.*

After the May Fourth cultural movement in 1919 photography made a marked progress. During the years between 1923 and 1937 various photographic organizations were set up in Peking, Shanghai, Canton and other cities. Many books and magazines on photography were published. Photographic exhibitions were held on many occasions. During the War of resistance against the Japanese invasion, China's progressive photographers began to go amongst the people in order to record their life and struggles. This marked the beginning of a new stage in photographic progress.

After the founding of the People's Republic of China, photographic work developed rapidly with the steadily rising living standards. In the past few years, the ranks of photographers have greatly increased and photography has been widely used in industry, medicine, education and in the Press; it is also popular among the people.

Press. Chinese newspapers use large numbers of photographs in various forms. The major daily papers in China issue a full page spread of pictures every week, containing 12 to 15 photographs. In addition, they regularly devote several columns to photographic features and illustrate the news with photographs. Figures show that certain outstanding photographs often get a simultaneous circulation of more than five and a half million.

The Hsinhua News Agency supplies prompt news photographs to all newspapers and magazines in the country and periodically issues photographs for exhibition. In order to speed up the delivery of news photographs, a wide network for the radio-transmission of photographs, covering China and other countries, has been set up.

During the past few years, illustrated magazines have won wide popularity. The *China Pictorial* is published in seventeen languages—including Han, Mongolian, Tibetan, Uighur, Korean, Russian, English, French, Japanese, Indonesian and Spanish. *The Nationalities Pictorial* is published in Han, Mongolian, Tibetan, Korean and Kazakh. *The Liberation Army Pictorial* is widely read among the armed forces and its circulation has been growing every year. In addition there are provincial and regional pictorials such as the *Kwangtung Pictorial*, the *Sinkiang Pictorial*, etc.

Other magazines carry large numbers of photographs. Seventy per cent of the magazines use photographs for their front and back covers, and some of them devote special pages to photo pictures.

Photographic book publishing has made great progress, and the various publishing houses put out a great variety of works on photographic theory and techniques, as well as popular books. During the past few years a number of picture books have been published, among them being *China*, *Peking*, *Shanghai*, *Heilungkiang*, and *Inner Mongolia.* Their content is rich and varied, and the quality of photography and printing is good.

The Chinese people take a great interest in the photographic display windows in shops, schools, factory clubs, village cultural pavilions and country fairs, where they can see reflected in them their own life and work.

Lantern-slides are likewise used in large numbers. These have reached every remote corner of the country. In schools, they are used for visual education. In the villages educational slides on science and history, etc., are often projected out of doors at night. Pictorial albums, post cards and calendars are issued regularly by various publishing houses and photography is widely employed in economic construction and scientific research.

Science. Since early 1953, China has established aerial photographic survey teams to investigate and survey forest resources. These teams have photographed the forest areas in the Tahailin regions, the Greater and Lesser Khingan Mountains of North-east China, and the Altai and Tienshan Mountains. Photography was also used in the great afforestation projects of 1954, when 2,700,000 acres of forest belt were planted for crop protection, coastal defence, water conservation or timber production.

Photomicrography is used in agricultural research on pest control and crop growth, and also in medicine.

China also employs photography in industry, astronomy, geology, physics, chemistry, medicine, archaeology, and water conservancy. In the water conservancy project of the Huai River, for instance, photographs were taken for records and research through every stage of the work. All the main conservancy projects in China, including the ancient Tukiangyen irrigation system, have been photographed for research purposes. In one case, a photograph of the Shihlungkiang Reservoir in Hupeh province prompted on-the-spot research which revealed that the project had certain advantages over others.

Amateur Activities. With the steady rise in living standards, amateur photography is rapidly growing. Photographic shops with printing and developing services are plentiful in the Peking shopping centres as well as in the places of recreation and culture. With the rapid increase in the number of photography fans, the trade is experiencing a minor boom. Amateur photography clubs are also on the increase and extending their influence. The Society of Photography in Peking University has been successful in raising the students' standard of photographic technique and in fostering their interest in photography.

As a result of these developments the industry as a whole is growing. The supply of photographic materials and equipment has considerably increased in recent years. Cameras, various types of films, photographic papers, enlargers, modern optical glass, electronic flash units, glazers, printers, copying cameras and photographic chemicals have all been put into planned production, and their quality is rising steadily.

Society. The Photographic Society of China a nation-wide organization was founded in Peking in December 1956, and aims mainly at encouraging creation in photographic art and raising the level of photographic theory. It is a member of the All-China Federation of Literary and Art Circles. The Society holds an annual national exhibition and in addition regularly sponsors one-man shows and exhibitions on special themes. It publishes an annual collection of photographs and edits the bi-monthly magazine *China Photography*.

The Photographic Society of China organizes forums of various kinds, holds discussions on photographic problems and organizes visits to factories, the countryside, construction sites and famous beauty-spots to take photographs. It also arranges lecture tours for photographers and helps to provide training for young photographic workers and amateurs.

An Exhibition of Chinese Photographic Art has been seen in many countries. The Society takes part in international activities and every year Chinese photographers send their works to international exhibitions, in order to promote friendship and mutual understanding between the peoples of China and other countries. Many foreign countries have sent exhibitions to China. The Photographic Exhibition on Cuba's Revolution, held in China in the spring of 1963, attracted more than a million people.

There are 18 affiliated branches of the Society in the country's provinces, municipalities and autonomous regions.

The Future. The State is devoting increasing attention to the training of photographers. In addition to short-term photographic training classes in various organizations, courses on news photography have been instituted in noted universities, such as Peking and Futan Universities, with the aim of training news photographers. The Shanghai People's Fine Arts Publishing House has published several popular books on photography.

As photography is close to the life of the people of China and serves their interests, it holds an important position in public estimation. Although it is on the threshold of its development, the experience of the past few years shows that it has wide prospects. S.-h.S.

CHLORHYDROQUINONE. Developing agent used in warm tone developers. Similar in use to hydroquinone.

See also: *Chlorquinol.*

CHLORIDE PAPER. Development paper coated with an emulsion of silver chloride. They are known generally as contact papers and less often, nowadays, as gaslight papers.

See also: *Papers.*

CHLOROBROMIDE. Light-sensitive emulsion composed of a mixture of silver chloride and silver bromide. Emulsions of this type are coated on development papers and lantern plates. They possess the property of giving warm tones by normal development processes.

CHLORQUINOL. Adurol; chlorhydroquinone; chloro-para-dihydroxybenzene. Developing agent particularly in warm tone developers.

Formula and molecular weight: $C_6H_3(OH)_2Cl$; 144·5.

Characteristics: Fine white crystalline powder.

Solubility: Fairly soluble in water.

CHROMATIC ABERRATION. Axial aberration of simple lenses, causing light of different colours to fall in different planes of focus.

See also: *Aberrations of lenses.*

CHROMATICITY. Term for describing objectively the colour quality of a visual stimulus (e.g. a coloured light or a surface) irrespective of its luminance. Chromaticity and luminance together completely specify a colour stimulus.

In the C.I.E. system of colour measurement chromaticity is expressed mathematically by two numbers, x and y, termed chromaticity coordinates, and a diagram showing the ranges of these coordinates for all colours is termed a chromaticity diagram.

See also: *Colour measurement.*

CHROMATYPE. Several variants of photography on paper using chromium salts. Robert Hunt, who coined the name, announced his process in 1843. The paper was prepared with a solution of sulphate of copper and potassium bichromate. It produced direct positive photogenic drawings, but was not sensitive enough for use in the camera.

CHROME ALUM. Potassium chromium sulphate. Standard ingredient in some hardening baths.

See also: *Alum, Chrome.*

CHROME FILM. Alternative name for orthochromatic film; now obsolete. At one time the ending "chrome" to trade names of films used to indicate an orthochromatic material. Since such films fell into almost universal disuse for general amateur photography, a number of well known film names with the "chrome" ending became panchromatic but retained the old description.

Among colour films the ending "chrome" frequently signifies a reversal material for producing transparencies, especially in the U.S.A. (where the ending "color" signifies a negative colour material). In Europe this distinction is frequently not made.

See also: *Colour materials; Negative materials.*

CHROMIC ACID. Sensitizer used in some formulae for the carbon and carbro printing processes.

See also: *Chromium trioxide.*

CHROMIUM AMMONIUM SULPHATE. Hardener for plates and films used in some stop baths and fixing baths.

See also: *Alum, Chrome.*

CHROMIUM POTASSIUM SULPHATE. Standard ingredient in some hardening baths. Popularly called chrome alum.

See also: *Alum, Chrome.*

CHROMIUM TRIOXIDE. Anhydrous chromic acid; chromic anhydride. Sensitizer for carbon and carbro processes.

Formula and molecular weight: CrO_3; 100.

Characteristics: Very hygroscopic red crystals, forms chromic acid in solution.

Solubility: Highly soluble in water, soluble in acetic acid and ether.

CHROMOGENIC DEVELOPMENT. Process of producing a colour image simultaneously with a silver image in a photographic emulsion.

An image in colour, formed from a dyestuff, may be obtained by (*a*) the generation, or (*b*) the destruction of a dye during the process of the development of a silver image. The generation of a coloured body by development is the basis of most commercial colour processes today.

Dye destruction, which is noted here only as an interesing corollary, finds application in the formation of colour correcting masks in integral tripack materials.

The formation of a coloured image during development has been known since early times in the case of development by pyrogallol, and the fact that a dye image remained even after the silver was bleached away was demonstrated by Liesegang in 1895. Homolka (1907) carried this further by using as developers substances which would oxidize to good dyes. With indoxyl and thio-indoxyl he obtained blue and red images.

R. Fischer first obtained patents (1912) on colour development processes of this sort. He showed that an existing silver image could be bleached to silver halide and redeveloped to give a colour image and, in 1914, obtained patents covering the addition to the developer of substances that would couple with the oxidized developer to give a dye. For example, tri-chlornaphthol was to act as a coupler for oxidized paraphenylenediamine.

In the first stage of development the developing agent reacts with the silver halide, e.g. silver bromide to give oxidized developer (a quinonediamine), soluble halide and silver. The oxidized developer reacts with the coupler to give a leuco-dye. The leuco-dye reacts with further silver halide and is oxidized to the final dye. In theory, four atoms of silver are reduced in the formation of one molecule of dye.

The coupler can be added to the developer as mentioned above, but the control of the developer and of the entire processing is much simplified if the coupler is incorporated in the emulsion.

For integral, multilayer colour films it is essential that the couplers remain in their proper layers. This is achieved either by making the coupler molecule too large to permit diffusion through gelatin, or by dissolving the coupler in droplets of a fatty substance dispersed within the gelatin.

For a three-colour subtractive system, the developing agents and couplers have to be selected with care so as to provide dyes of the correct spectral absorptions. With p-amino-diethylaniline as developing agent, phenols and naphthols give cyan dyes, certain pyrazolones give magenta dyes, and esters such as ethy-lacetoacetate give yellow dyes. A wider range of couplers is available where chromogenic development is intended for toning rather than for subtractive colour reproduction. The following developing agents and couplers have been recommended:—

Developers:	paraphenylenediamine	violet
	paratoluidinediamine	blue
	dimethylparaphenylenediamine	blue-green
Couplers:	alpha naphthol	blue
	2 : 4 dichlornaphthol	blue-green
	trichlornaphthol	green-blue
	pentabromonaphtol	green
	p-nitrobenzylcyanide	yellow
	1-phenyl-3-methyl-5-pyrazolone	magenta-red

See also: *Colour film principles; Colour film processing.*
Books: *Developing* by C. I. Jacobson (London).

CHRONOLOGY OF PHOTOGRAPHIC INVENTIONS

Before 1700: Camera obscura and magic lantern.

1725 Schulze, J. H.: Experiments on light sensitivity of silver salts; contact images (from stencils) on liquid mixtures of chalk and silver nitrate in a bottle; no fixing.

1777 Scheele, C. W.: Blackening of silver chloride in the violet and the blue of the spectrum quicker than by other colours.

1800 Herschel, Sir William: Discovery of infra-red.

1800 Wedgwood T.: Contact copying of silhouettes, leaves, etc., on leather sensitized with silver nitrate; no fixing.

1801 Ritter, J. W.: Blackening of silver chloride by ultra-violet.

1816 Niépce, J. Nicéphore: Camera photographs on paper sensitized with silver chloride. Partially fixed.

1819 Herschel, Sir John F. W.: Discovery of thiosulphates and of the solution of silver halides by "hypo".

1822–5 Niépce, J. Nicéphore: Copying of engraings on glass, zinc and pewter sensitized with bitumen. Further attempts at direct photography.

1826 Niépce J. Nicéphore: First photograph from nature on pewter sensitized with bitumen (exposure 8 hours). Bellows for photographic camera; iris diaphragm.

1827 Niépce, J. Nicéphore: First prints from etched heliographic plates.

1835–7 Daguerre, L. J. M.: Daguerreotype (direct photography on silvered copper plates with a silver iodide surface); development of the latent image by mercury vapour. 1837, fixing with salt solution.

1835 Talbot, W. H. Fox: "Photogenic drawings", copied on paper sensitized with silver chloride; fixed with potassium iodide or by prolonged washing in salt water. Also tiny camera photographs.

1839 Reade, J. B.: Photomicrographs in the solar microscope on paper sensitized with silver nitrate and gallic acid (by printing out or by development with gallic acid); fixing with thiosulphate.

1839 Experimental photographic portraits in France and U.S.A.

1839 Bayard, H.: Direct positive photographs on paper (silver chloride paper blackened by light, then impregnated with potassium iodide and exposed in the camera).

1839 Daguerre, L. J. M.: Publication of working instructions for the daguerreotype by the French Government.

1839 Herschel, Sir John F. W.: First use of the word "photography"; independent invention of a photographic process on glass.

1839 Ponton, Mungo: Light sensitivity of a paper impregnated with potassium bichromate—the basis of many printing processes.

1839–41 Donné, A.; Berres, Joseph; Grove, W. R.; Fizeau, H. L.: Transformation of daguerreotypes into etched printing plates.

1840 Goddard, J. F.: Use of bromine for the acceleration of daguerreotype plates.

1840 Draper, J. W. First successful daguerreotype portrait.

1840 Petzval, J.: Portrait lens (first application of calculation for the investigation of a lens type).

1840 Fizeau, H. L.: Gold toning of daguerreotypes.

1840 Soleil, J. B. F.: Actinometer for exposure time determination.

1840 Wolcott, A.: Patents for mirror camera in U.S.A.

1840 Wolcott, A.: Opened world's first photographic portrait studio in New York.

1841 Beard, R.: Opened first photographic portrait studio in Europe.

1841 Claudet, A.: Red light for dark room.

1841 Talbot, W. H. Fox: Calotype process (negatives on silver iodide paper impregnated with silver nitrate and gallic acid; development of the latent image by gallic acid; positive prints from these paper negatives).

1842 Herschel, Sir John F. W.: Ferroprussiate paper and other processes using iron salts.

1843 Claudet, A. F. J.: Use of painted backgrounds for portraiture.
1844 Fizeau, H. L., and Foucault, L.: Discovery of reciprocity failure on daguerreotypes; photographic photometry.
1844 Hunt, R.: Ferrous oxalate process, ferrous sulphate development.
1844 Talbot, W. H. F.: Published part I of the *Pencil of Nature*, the first photographically illustrated book.
1845 Martens, F. von: Panoramic camera for daguerreotypes.
1847 Niépce de St. Victor, C.F.A.: Negatives on glass by the albumen process.
1847 Blanquart-Evrard, L. D.: Modification of Calotype process.
1847 Mathieu, P. E.: Gold toning of positive prints on paper.
1848 Becquerel, Edmond: Attempts at colour photography (heliochromy) on silver plates which were incompletely chlorinated.
1849 Brewster, Sir David: Lenticular stereoscope prototype. Introduced 1851. Suggests binocular camera, prototype 1853.
1850 Blanquart-Evrard, L. D.: Albumen paper for positive prints.
1850 Humphrey, S. D.: Publication of the first photographic journal: *The Daguerreian Journal*, New York.
1850 Le Gray, G.: Waxed paper process.
1851 Archer, F. Scott: Wet collodion process.
1851 Blanquart-Evrard, L. D.: Prints by development.
1851 Talbot, W. H. Fox: Photograph of an object in rapid movement by using an electric spark.
1851 Dodero: Portraits on visiting cards, popularized by Disdéri.
1851 Regnault, V.: Use of pyrogallol as developer (physical development).
1851 Archer, F. S., and Fry, P. W.: Ambrotypes (negative collodion-on-glass photos with black background giving positive effect).
1851 Claudet, A.; Duboscq, J.: Animated photographs (in Zoetrope and Phenakistiscope) suggested.
1852 Martin, A. A.: Positive-looking images by the wet collodion process on a blackened metal support (ferrotypes).
1852 Laussedat, A.: Photography applied to surveying.
1852 Talbot, W. H. Fox: Insolubilization of bichromated gelatin by light; first experiments with half-tone heliogravure.
1852 Lemercier, Lerebours, Barreswil, and Davanne: Photolithography with halftones on grained stone sensitized with bitumen.
1852/3 Delves, J.; Highley, S.; Shadbolt, G.: Photomicrographs by the wet collodion process.
1853 Dancer, J. B.: Prototype of stereocamera with two lenses.
1853 Dancer, J. B.: Microphotographs by the wet collodion process.
1853 Gaudin, M. A. A.: Use of potassium cyanide as fixing agent for silver images; experiments with gelatin as carrier for silver images.
1854 Melhuish, A. J. and Spencer, J. B.: Rollholders for sensitive paper negatives.
1854 Brébisson, A. de: Salted paper with starch, for positive prints.
1854 Nègre, C.: Photo engravings giving halftone with bitumen-coated steel plate.
1855 Niépce de St. Victor, C. F. A.: Heliogravure with bitumen on steel.
1855 Poitevin, A. L.: Photolithography on stone sensitized with bichromated gelatin, glue, albumen or gum—principle of collotype; first attempts to obtain "carbon prints".
1855 Lafon de Camarsac, P. M.: Photographs on enamel and porcelain (burnt-in).
1855 Taupenot, J. M.: Negatives on collodioalbumen dry plates prepared in advance and requiring only a complementary sensitization.
1855 Relandin: Roller-blind shutter mounted on the lens.
1856 Norris, R. Hill: Collodion dry plates preserved with gum arabic or gelatin put on the market.
1856 Pretsch, P.: Photogalvanography (halftone intaglio copper printing plates by galvanoplastic moulding from bichromated gelatin reliefs).
1857 Grubb, Thomas: Aplanat lens.
1857 Woodward, J. J.: Solar enlarger.
1858 Tournachon, G. F. ("Nadar"): Photography from the air in a free balloon.
1859 Bunsen, R., and Roscoe, H. E.: Use of magnesium to provide artificial light.
1859 Asser, E. J. and James, Sir H.: Photozincography.
1860 Joubert, F.: "Phototype" (collotype).
1860 Willème, F.: Photosculpture.
1861 Fargier, A.: Half-tone pigment images on bichromated gelatin with transparent support, exposed from the back.
1861 Maxwell, J. Clerk: Three-colour separation, and colour synthesis by multiple projection.
1861 Russell, C.: Dry-collodion plates preserved with tannin.
1861 England, W.: Focal plane shutter.
1862 Leahy, Thomas, and Russell, C.: Alkaline developer (pyrogallol-ammonia) for chemical development; reversal by dissolving the first negative and development of the remaining silver salt.
1864 Swan, Sir Joseph W.: First successful carbon process (with double transfer).
1864 Bolton, W. B., and Sayce, B. J.: First workable collodion emulsion.
1864 Woodbury, W. B.: Multiplication of carbon prints by moulding (Woodburytype).

1864 Ducos du Hauron, L.: Patent for an apparatus for animated photography (camera and projector) made by J. Duboscq.
1865 Simpson, G. Wharton: Positive collodio-chloride emulsion for printing-out papers, made by J. B. Obernetter (1868).
1865 White, W.: Magnesium flash powder.
1866 Sanchez, M., and Laurent, J.: Baryta coating of photographic papers.
1866 Steinheil, A.: Aplanat lens (rectilinear).
1867 Cook, H.: Opera-glass camera and plate change-box for dry collodion plates.
1868 Harrison, W. H.: Experiments with positive silver bromide gelatin emulsion for chemical development.
1868 Albert, J.: Perfected collotype (use of glass plates as supports).
1868 Ducos du Hauron, L.: Three-colour photography and the various methods of achieving it; subtractive colour synthesis.
1868 Cros, C.: Principle of three-colour separation and synthesis.
1869 Ost, A.: Citric acid stabilization of printing-out papers.
1870 Dagron, P. R. P.: Use of miniature photographs for the pigeon post during the Siege of Paris.
1871 Maddox, R. L.: First gelatin dry plates. Extremely slow. Improved by J. Burgess (1873)., R. Kennett (1874) and C. Bennett (1878).
1872 Gillot, C.: Line blocks on zinc, using the etching technique of F. Gillot (1850).
1873 Mawdsley, P.: Gelatin bromide paper for negatives and positives.
1873 Vogel, H.: Dye sensitization for the green (basis of orthochromatic plates).
1873 Willis, W.: Platinum paper for development (sold commercially in 1878 after various improvements).
1873–4 Johnston, J.; King, J.: Silver bromide gelatin emulsion (negative) prepared with excess bromide and washed before coating.
1874 Janssen, P. J.: Photographic revolver for the chronophotographic study of the Venus transit.
1875 Warnerke, L.: Stripping-film paper on reels in a roll-holder for 100 exposures.
1876 The Liverpool Dry Plate Co. began manufacture of R. Kennett's gelatin dry plates.
1877 Carey-Lea, M.: Ferrous oxalate developer.
1877–85 Muybridge, E.: Study of the movement of animals by instantaneous photography and chronophotography.
1878 Swan, J. W., and Bennett, C.: Increase in the speed of silver bromide gelatin emulsions by ripening in a neutral medium.
1878 Monckhoven, D. C. E. van: Preparation of silver bromide gelatin emulsions in the presence of ammonia.
1878 Wratten, F. C. L.: "Noodling" of silver bromide gelatin emulsions before washing.
1879 Klič, K.: Photogravure on copper by transfer of a carbon print after graining with resin.
1879 Swan, J. W., also Eastman, G.: Machine for coating plates.
1880 Abney, Sir W. de W.: Use of hydroquinone as developer.
1881 Eder, J. M., and Pizzighelli, G.: Silver chloride gelatin emulsion (positive) for chemical development.
1881 Cros, C.: Dyeing of gelatin by imbibition ("hydrotypy").
1881 Pumphrey, A.: Magazine camera (for gelatin sheet film).
1882 Berkeley, H. B.: Use of sodium sulphite in developers.
1882 Attout, P. A. and Clayton, J.: Commercial manufacture of orthochromatic silver bromide gelatin plates.
1882 Abney, Sir W. de W.: Silver chloride gelatin emulsion for printing-out papers manufactured commercially in 1886 by Liesegang.
1882 Meisenbach, G.: Commercial production of half-tone blocks (line screen turned at 90° in the middle of the exposure).
1882–96 Marey, E. J.: Chronophotography on fixed plates and on movable films (photographs in rapid succession at equal time intervals).
1883 Farmer, H. E.: Single-bath reducer (with ferricyanide and thiosulphate).
1884 Kayser, H.: Plate changing magazine.
1885 Urie, John, sen. and jun.: Patent for automatic positive printing machine.
1886 Ives, F. E.: Half-tone blocks with crossed screen and square diaphragm.
1886 Urie, John, sen. and jun.: Patent for continuous positive print processing machine.
1887 Goodwin, H.: Patent (granted only in 1898) for the manufacture of celluloid rollfilm coated with silver bromide gelatin emulsion.
1887 Pizzighelli, G.: Platinum paper for printing-out.
1887 Bausch, E.: Central shutter with blades forming also an iris diaphragm.
1888 Eastman Kodak Co.: Introduction of the first "Kodak" roll film camera (D. & P. since early '80's).
1888 Carbutt, J.: Silver bromide gelatin coated celluloid sheet film.
1888–9 Andresen, M.: Use of para-phenylenediamine and Eikonogen as developers.
1889 Edison, T. A.: Use of flexible Eastman film for Kinetoscope pictures.
1889 Namias, R.: Acid permanganate reducer.
1889 Lainer, A.: Stabilization of fixers with bisulphites.
1889 Enjalbert, T. E.: Portrait automaton for ferrotypes.

1889 Rudolph, P.: First Anastigmat lens (Protar) introduced by Zeiss.
1890 Hurter, F., and Driffield, V. C.: Scientific investigation of emulsion characteristics; creation of photographic photometry.
1890 Bonnet, G.: Bichromated glue process for block-making (so-called "enamel" process).
1891 Lippmann, G.: Colour photography by the interference method.
1891 Andresen, M.: Use of para-aminophenol as developer.
1891 Bogisch, A.: Use of metol, glycin and diaminophenol as developers.
1891 Turner, S. N.: Daylight-loading film rolls (Kodak).
1891 Edison, T. A.: Kinetoscope giving the illusion of movement by viewing images representing phases of a short scene.
1892 Ives, F. E.: One-shot colour camera.
1893 von Hoegh, E.: First double anastigmat lens (Dagor) introduced by Goerz.
1893 Schumann, V.: Emulsions with negligible gelatin content, sensitive for the extreme ultra-violet.
1893 Taylor, H. D.: Anastigmatic objectives with three uncemented lenses (Cooke lens).
1894 Rouillé-Ladêvéze, A.: Gum-bichromate prints for pictorial photography.
1894 Ives, F. E.: Patents photochromoscope, later called "Kromskop", viewer for seeing single or stereoscopic photographs in colour (invented 1892).
1895 Lumière, A. and L.: Cinematograph.
1895 Roentgen, W. C.: Discovery of X-rays and of radiography.
1895 Klič, K.: Intaglio screen photogravure and rotogravure introduced. (Invented 1890.)
1896 Méliès, G.: Directing of films; cinematographic effects and illusions.
1898 Acres, Birt: First narrow-gauge cine film (17·5 mm., Birtac camera).
1900 Lumière, A. and L.: "Photorama" for the taking and the projection of panoramic pictures of 360°.
1901 Pulfrich, C.: Stereocomparator for the accurate exploitation of stereo pairs in terrestrial photogrammetry.
1901 Gaumont, L.: Synchronization of a cinematograph with a phonograph.
1901 Eichengrün, A.: Safety films from cellulose acetate.
1902 Rudolph, P.: Tessar anastigmat lens introduced by Zeiss.
1902 Deckel, Fr.: Compound shutter.
1902 Lüppo-Cramer, H.: Metol-hydroquinone developer.
1902 Deville, E.: Principle of an apparatus for stereophotogrammetric restitution.
1902 Traube, A.: Ethyl Red used as colour sensitizer; this was the first sensitizer of the isocyanine series.
1902 Eastman Kodak Company: Gelatin backing on films to prevent curl (commercially introduced); daylight development tank.
1904 König, E., and Homolka, B.: Ortho and panchromatic sensitizers (Orthochrome, Pinachrome, Pinacyanol).
1904 Rawlins, G. E.: Fatty ink processes ("oil prints") applied to artistic photography.
1904 Korn and Glatzel: Phototelegraphy.
1906 Wratten and Wainwright Ltd.: First commercial panchromatic plates.
1906 Courtet, E. (pseudonym Cohl): Animated films.
1907 Lumière, A. and L.: Autochrome plate for colour photography introduced.
1907 Welbourne Piper, C., and Wall, E. J.: Bromoil process.
1907 Homolka, B.: Colour development with leuco dyes.
1907 Goldschmidt, R. B.: Photographic reduction of documents for library use.
1907 Smith, C. A., and Urban, C.: Kinemacolor, first commercial additive two-colour cine process.
1908 Clerc, L. P.: Theory of the photogravure screen.
1908 Keller-Dorian, A.: Patent for lenticular colour process for cinematography.
1908 Belin, E.: Phototelegraphy on telephone lines.
1910 Goldberg, E.: Moulded neutral grey wedges for sensitometry.
1910 Dufay, L.: Regular screen process of additive colour photography.
1911 Lauste, E. A.: First experiments in talking films.
1912 Deckel, Fr.: Compur shutter introduced.
1912 Gaumont, L.: Three-colour cinematography by simultaneous additive synthesis.
1912 Fischer, R., and Siegrist, H.: Colour development based on formation of indamine and indophenol dyes (later employed in three-colour photography).
1914 Eastman Kodak Co.: First Kodachrome process (two-colour subtractive).
1919 Adams, E. Q., and Haller, H. L.: Discovery of cryptocyanine, the first practical sensitizer for infra-red.
1920 Pope, Sir William J., Mills, W. H., and Hamer, Miss F. M.: Constitution of the colour sensitizers of the cyanine group.
1920 Lüppo-Cramer, H.: Desensitization.
1921 Belin, E.: Transmission of pictures by wireless.
1921 Bocage, A.: Radiographic tomography of living objects.
1921 Duclaux, J., and Jeantet, P.: Sensitization for the far ultra-violet by putting a fluorescent layer on the emulsion.
1923 Koegel, G.: Diazo prints developed with ammonia vapour.
1923 König, W.: Preparation of di- and tri-

carbocyanines, sensitizers for the extreme red and the near infra-red.

1923 Eastman Kodak Co.: Reversal processing of 16 mm. film, leading to widespread amateur cinematography.

1924 Sheppard, S. E., and Punnett, R. F.: Discovery and identification of active impurities in gelatin which led to the preparation of very rapid emulsions.

1924 Barnack, O.: Leica camera using 35 mm. film, manufactured by Leitz.

1925 Séguin, A. and L.: Practical use of a condenser discharge in a rare gas for the photography of very rapid movements.

1927 Beginning of universal use of sound films.

1928 Hamer, Miss F. M.: Polymethine sensitizers derived from benzothiazoles with alkyl substitution on the middle carbon.

1928 Franke & Heidecke: Rolleiflex twin lens reflex camera introduced.

1928 Arens, H., and Eggert, J.: Density surfaces for various types of emulsion.

1928 Commercial microfilming on 16 mm. film in rotary camera.

1929 Ostermeier, J.: Flashbulb: Combustion of aluminium wire or foils in oxygen in a sealed bulb.

1931 Schmidt, B.: Catadioptric lens with large aperture.

1935 Brooker, L. G. S., and Keyes, G. H.: Preparation of tetra- and penta-carbocyanines.

1935 Heisenberg, E.: Emulsion for direct positives and prints, without inversion of a provisional negative.

1935 Laporte, M.: Electronic flash with white light.

1935 Mannes, L. D., and Godowsky, L.: Subtractive three-colour separation and synthesis on the same film with superimposed emulsions, the colour couplers being added to the developers (Kodachrome process).

1936 Colour couplers in the emulsions of colour film (Agfacolor reversal process).

1936 Koslowski, R.: Increase in the sensitivity of emulsions through the addition of aurous thiocyanate.

1936 Strong, J.: Reflection-diminishing coatings with insoluble fluorides on lens surfaces.

1938 Mott, N. F., and Gurney, R. W.: Theory of the latent image.

1939 Agfacolor negative-positive colour printing process.

1939 James, T. H.: Investigations on the kinetics of development.

1939 Jones, L. A.: Experimental basis of the standardization of the sensitometry of negative emulsions.

1939 Edgerton, H. E.: Improvement in electronic flash lamps.

1940 Ardenne, M. von: Investigation of the structure of developed images under the electron microscope.

1941 High-resolution plates permitting 750-fold enlargement of the image.

1942 Rott, A.: Principle of reversal by the transfer of diffusion images, and practical applications.

1942 Silver halide-sensitized zinc plates for rapid photolithography.

1942 Kendall, J. D.: Patent on Phenidone, a new developing agent (1-phenyl-3-pyrazolidone).

1946 Blackner, L. L., Brown, F. M., and Kunz, C. J.: Projection of images within fifteen seconds after taking.

1947 Land, E. H.: One step photographic process (Polaroid).

1947 Magnetic recording replaces photographic sound recording during shooting in film studios.

1949 General introduction of safety base for 35 mm. motion picture films.

1950 Coloured couplers for self-masking colour correction.

R.S.S. (H. & A.G.)

See also: *Camera history; Cine history; Colour photography: history; Development history; Discovery of photography; Lens history; Sensitized materials history.*

Books: *The History of Photography*, by H. Gernsheim (Oxford); *A Source Book of Photographic History*, by W. Baier (London).

CHRONOPHOTOGRAPHY. Photographic method of analysing an action by taking a series of still pictures at regular intervals throughout its duration. This method of analysing motion is very similar to cinematography, although the apparatus used is quite different.

Muybridge's Experiments. In Paris, Professor E. J. Marey, about 1870, was investigating human and animal motion by means of various mechanical devices; he called the records chronographs. An American racehorse owner, Governor Leland Stanford of California, doubted the results of some of his investigations into the action of racehorses—in particular that there was ever an instant when a trotting horse had all its feet off the ground at the same time. E. J. Muybridge, an English photographer living in San Francisco, was commissioned to check Marey's findings.

In 1877 Muybridge succeeded in proving that Marey was right, and in doing so produced what were the first true chronophotographs. These classic pictures were made by arranging for a horse and rider to pass in front of a row of cameras with shutter releases connected to threads stretched at regular intervals across the path. As the horse came opposite each camera in turn, it broke the thread operating that particular shutter and took its own photograph. The series of photographs, when printed on one sheet, yielded a continuous chronophotograph.

In 1880 Muybridge devised a projection instrument which he called zoopraxiscope to demonstrate the movement. In 1884–5 he extended his chronophotography to the movement of a great variety of animals and human beings. In this series he used up to 36 cameras with shutters electrically operated by a clockwork device.

Marey's Single Negative. Muybridge's experiments prompted Professor Marey to adopt the photographic approach to his own researches. He constructed various cameras with this end in view, aiming at getting the whole series of chronophotographs on the one negative. His first apparatus, constructed in 1882, consisted of a plate camera equipped with a spring motor-driven rotary disc shutter having a number of regularly spaced apertures. As each aperture passed in front of the plate it recorded the position at that instant of an object moving across the field of the camera. A small object moving quickly recorded a series of separate images on the plate, the space between depending on how fast the shutter was rotated.

If the subject was large in relation to the speed of rotation of the shutter, the successive images overlapped and gave a confused picture. Marey overcame this difficulty by putting his subject into a black suit with a white stripe sewn down the arm and leg facing the camera. He then took the photographs against a black background and obtained a series of separate images of the white stripes. In this way he was able to produce records of the movements of the limbs of people walking or running past the camera.

Marey's next development was a "gun" which took a series of pictures of a moving subject and recorded them separately on a single plate. The gun carried a circular plate in front of which was a disc with twelve openings around its circumference. In front of this disc was a second disc pierced with a slit. On pressing the trigger of the gun, a clockwork mechanism rotated the discs. The disc carrying the twelve frames rotated one-twelfth of a revolution while the disc carrying the shutter slit revolved once, so that each of the twelve openings appeared in turn behind the lens and was exposed through the slit. The result was a plate carrying twelve separate photographs showing successive attitudes of the moving subject. This gun successfully recorded chronophotographs of birds in flight, taking twelve successive pictures per second with a shutter speed of 1/720 second.

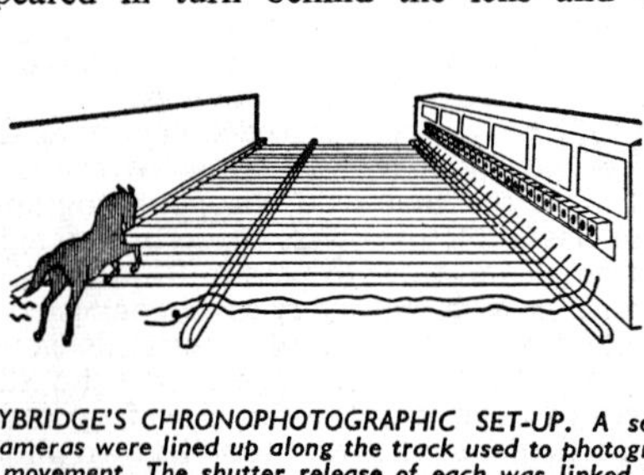

MUYBRIDGE'S CHRONOPHOTOGRAPHIC SET-UP. A series of cameras were lined up along the track used to photograph the movement. The shutter release of each was linked to a rubber band held tensioned by a trip wire. As the horse moved over the track, it broke the trip wires which released the rubber bands and exposed a picture in each camera in turn.

In 1883 the French Government established a department of physiological research in Paris where Marey carried on a chronophotographic investigation of the movements used by humans and animals in various forms of activity. Marey subsequently recorded chronophotographic sequences on long rolls of paper and, finally, roll film and in 1892 was able to project them.

All of his instruments for this suffered from an inherent disadvantage: the sensitized material had to be transported fast enough to separate the images and brought to a sudden stop at each instant of exposure. In practice this meant being satisfied with very small images to reduce the amount of shift of the material between each exposure. But this type of record is only useful if the individual pictures are large enough to be examined in detail, and the sensitized materials of the time would not permit a high degree of enlargement. For this reason a number of workers in this field chose to develop the Muybridge approach of making the successive exposures in a number of separate cameras.

Other Systems. The apparatus used by General Lebert in 1890 consisted of six full-size cameras mounted in a circle with an electric motor in the centre. The motor carried an arm which rotated and operated the shutters of each camera in turn. This arrangement was used for photographing projectiles.

A year or two after this, Albert Londe constructed a camera in which sets of lenses equipped with electro-magnetic shutters formed separate images regularly arranged to occupy a single plate. One six-lens camera used a 13 × 18 cm. plate while one with twelve lenses covered a 24 × 30 cm. plate. With these cameras there was a separate clockwork selector switch, governed by a metronome which operated each of the shutter solenoids in turn, allowing a variable time interval between the successive exposures according to the speed and range of the action to be analysed.

In the mid-eighties, O. Anschütz, working on similar lines to Muybridge, constructed and used chronophotographic arrangements of up to twenty cameras. Like Muybridge, he devised a method of reproducing the movement from his results. He mounted his pictures on a drum which could be rotated to bring each picture in turn into a viewing aperture. As each picture came into view, it was illuminated by an electric spark so that as the drum rotated the observer saw a moving picture.

Present-day Use. Various workers in other countries were also interested in synthesizing movement in this way by means of chrono-

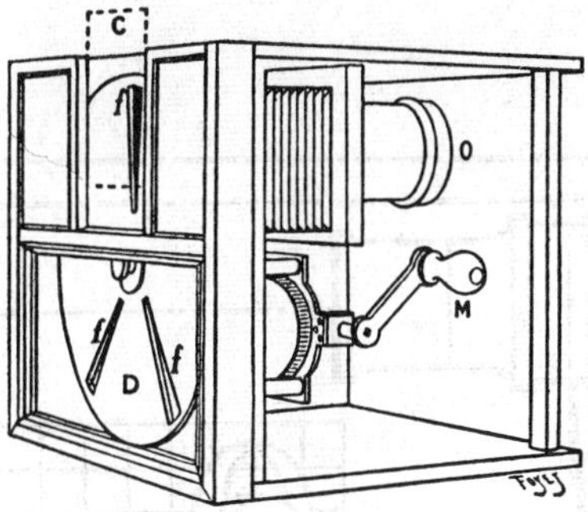

MAREY'S ORIGINAL CHRONOPHOTOGRAPHIC CAMERA. O, lens. M, crank for rotating the shutter. D, rotating shutter disc. f, shutter slit. C, plate.

photographs, but, although chronophotography was directly responsible for the development of cinematography, it pursued its own distinct course right up to the present time. There is a basic difference between the two. In both a series of still pictures is taken of a moving subject. But the object of chronophotography is to present a series of still pictures for detailed study, each picture "freezing" a particular phase of the action. In cinematography the object is to produce a moving picture of the whole event.

So the still pictures taken in chronophotography are usually of a larger format and of a higher standard of definition than cinematograph pictures.

Today the most important application of chronophotography is in time and motion study.

A special type of chronophotography which is reminiscent of Marey's early apparatus is carried out today with electronic flash. The subject moves against a dark background and is illuminated by an intermittent flash while the camera shutter is left open. This produces a succession of images on the same plate, each image being displaced both in time and space from the preceding one by an amount that depends on the frequency of the flashes and the speed of the movement.

The method has both scientific and artistic applications. Because of the high speed of the flash, it is possible to analyse extremely rapid motion, while the movements of a dancer or athlete photographed in this way can produce interesting and even beautiful patterns. F.P.

Illustrations: Plate Section XX.

See also: *Camera history; Cine history; Motion study; Stroboscopic flash.*

CHRYSOIDIN. Diaminoazobenzene. Orange dye used in some desensitizers and dye toners.

Formula and molecular weight: $C_6H_5N_2C_6H_3(NH_2)_2$; 212.

Characteristics: Orange powder, stains fabrics and mordanted gelatin.

Solubility: Slightly soluble in water at room temperature.

CHURCHES. The technique of photographing churches is a normal aspect of architectural photography. Other specific considerations which may arise involve matters of perspective and its correction by camera movements.

In addition to such technical problems, the photographer may also have to consider non-photographic aspects, e.g., permission to take photographs in the church, which must always be sought beforehand; and in the case of weddings, familiarity with the procedure of the wedding ceremony.

See also: *Architecture; Camera movements; Interiors; Permits to photograph; Perspective; Stained glass windows; Weddings.*

C.I.E. STANDARDS. Standards of illuminants and of colour perception set up by the Commission Internationale d'Éclairage (International Commission of Illumination) to provide a means of accurately describing colours. Also known as I.C.I. standards or (in Germany) I.B.K. standards.

See also: *Colour measurement.*

CINE CAMERA

The basic elements of a cine camera are similar to those of a still camera: a lens, a light-tight compartment containing the sensitive film on which the lens projects its image, and a device (shutter) for opening and closing the light path through the lens to the film while exposing the picture for a specific length of time. Unlike a still camera, a cine camera must expose a series of images, one after the other, at a steady rate on a long strip of film.

An important part of the cine camera is therefore the film transport system. This, together with the necessary motor and film supply (usually amounting to something like 4,000 exposures per load) takes up an appreciable amount of space, so that a cine camera is invariably far larger in relation to its image size than a still camera.

Nearly always the pictures on the film are below each other, so that the film runs vertically through the camera, or at least the film gate (the part where the image is projected on to the film). Suggestions have been made for running the film horizontally and recording images side by side, but—with one short-lived exception—never realized in practice for amateur cameras.

In addition to the lens, film gate, film transport system and film supply, a cine camera must

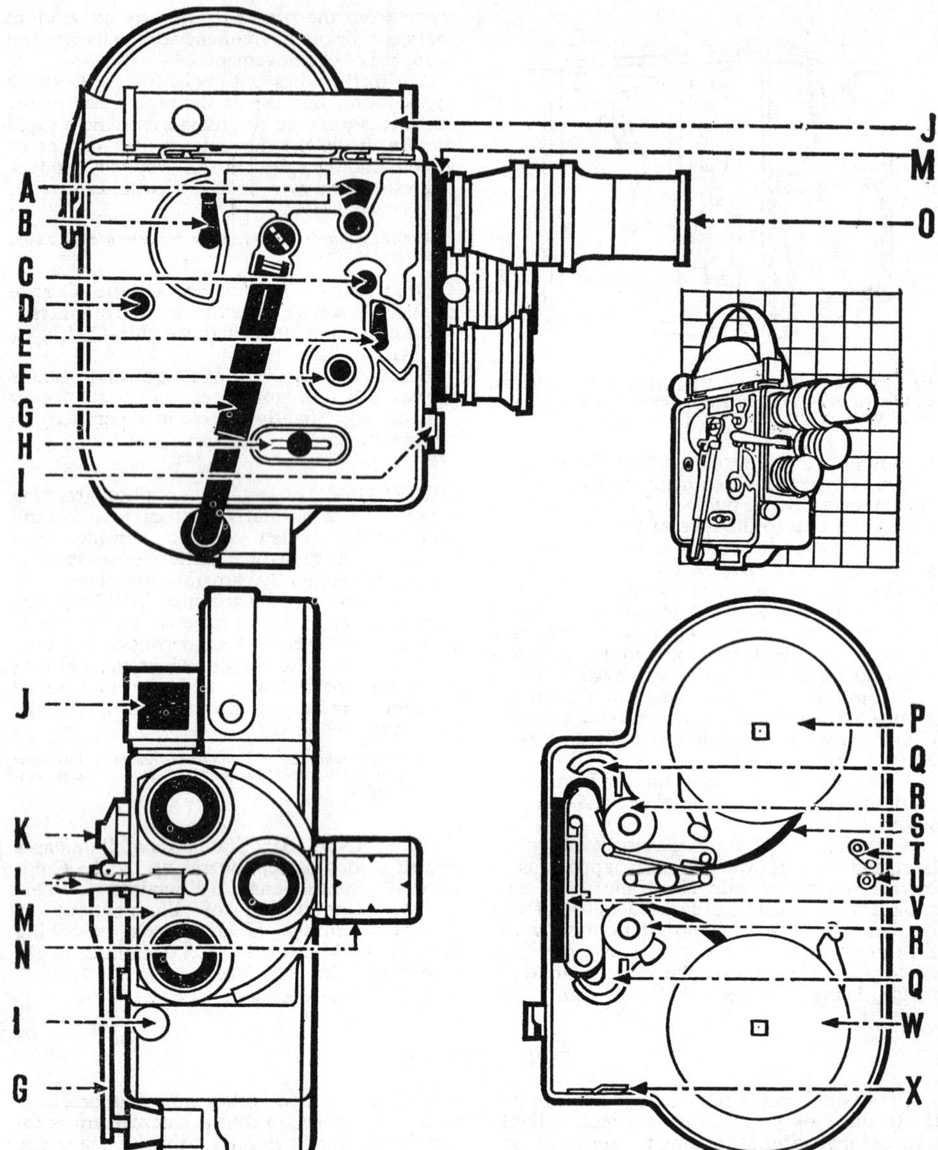

A PRECISION NARROW-GAUGE CINE CAMERA. Top left: *Side view.* Top right: *General view against 1 in. square grid.* Bottom left: *Front view.* Bottom right: *Interior with film path and transport items.*

The main components and controls of the camera are: A, *frame counter (rarely found in amateur cameras).* B, *motor clutch.* C, *crank shaft (for forward or reverse running by hand cranking).* D, *footage counter.* E, *time lever for instantaneous or time exposures when shooting single frames.* F, *speed control.* G, *folding motor winding handle.* H, *release control for continuous running or single shots.* I, *release button.* J, *through-the-lens focusing finder.* K, *motor shaft.* L, *turret handle.* M, *lens turret.* N, *adjustable multi-focus viewfinder.* O, *lens (these are mounted on the rotating turret). Often a zoom lens is used instead of several lenses on a turret.* P, *feed spool.* Q, *loop formers.* R, *sprockets.* S, *film.* T, *audible footage indicator control.* U, *footage indicator reset.* V, *film gate with pressure pad.* W, *take-up spool (driven by the camera motor through suitable gearing).* X, *film knife.*

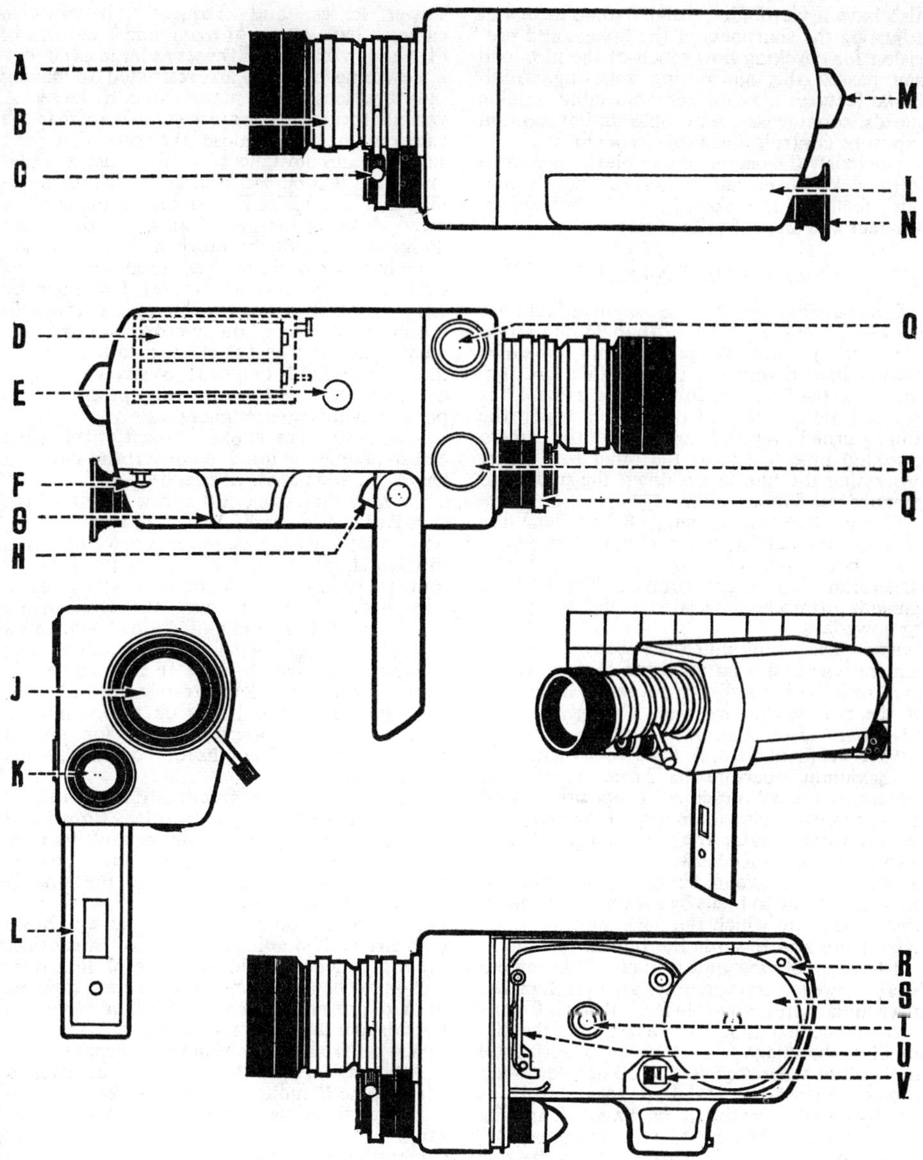

MODERN 8 mm. AMATEUR CINE CAMERA. Top: *Left-hand side view.* Upper centre: *Right-hand side view with hand grip folded open.* Lower centre left: *Front view.* Lower centre right: *General view against a 1 in. scale grid.* Bottom: *Left-hand side view with lid removed.*

The main controls and features are: A, *lens hood.* B, *focusing mount.* C, *zooming lever.* D, *motor batteries in camera.* E, *locking screw of battery compartment.* F, *reverse running switch.* G, *tripod bush.* H, *release button.* J, *zoom lens built into camera.* K, *cadmium sulphide cell of automatic exposure control.* L, *folding hand grip.* M, *head-rest.* N, *through-the-lens reflex finder (also indicates aperture set by meter and image sharpness).* O, *running speed setting.* P, *compartment for mercury battery for exposure meter.* Q, *film speed adjustment and manual control on meter cell.* R, *footage counter feeler (footage visible in finder).* S, *feed spool.* T, *shaft of take-up spool.* U, *film gate.* V, *battery power indicator for motor.*

also have a viewfinder, usually some means of adjusting the sharpness of the image, and provision for checking how much of the film load has passed through during shooting. Additional features may include adjustable running speeds, zoom lenses, a variable shutter, built-in exposure control, and other refinements.

Obviously the degree of complexity is related to the price of the camera and the more elaborate features are found on high-precision amateur and on professional cine cameras.

AMATEUR CAMERAS

Cine cameras for the amateur invariably use narrow gauge (i.e. narrower than 35 mm.) film. The most popular gauge is 8 mm. which is usually loaded into the camera in a 16 mm. width. As the film runs through the camera it is exposed along one half only, after which the film is turned over and run through the camera a second time to expose the other half. After processing the film is slit down the middle to produce two 8 mm. widths.

Charges containing a single 8 mm. wide film are used in quick loading cameras (Super-8).

A second amateur gauge is 16 mm., giving larger individual images (frames). The 9·5 mm. gauge is a third amateur size though less popular nowadays.

Lenses. The main difference between a still camera lens and a normal cine camera lens is that the latter has a shorter focal length because of the smaller picture size. This gives it an advantage over the still camera—a much greater depth of field. A fixed focus lens with a maximum aperture as large as *f*2·8 is therefore a very practical proposition with narrow gauge cine cameras. However, for serious work greater freedom is needed and, because the distance between camera and subjects will vary from a matter of inches to infinity, the lens is made to focus by means of an adjustable mount in which the lens moves bodily away from and towards the film in the gate.

The angle of view of a standard cine camera lens is, however, considerably narrower than that of a standard still camera lens. With the latter the normal focal length is taken as equal to the diagonal of the image format, giving a horizontal angle of view between 40 and 50 degrees. With a cine camera the standard focal length is usually about twice the frame diagonal, and the horizontal angle of view is 20 to 25 degrees.

Part of the reason for this is that as the cine camera records movement, it can concentrate on a smaller subject field and cover a wider field of view by swinging across it, much like the eye scans a scene it views, rather than taking it in all at once. (The eye does not take in the whole of a still picture at once either, but roams over the image area under normal viewing conditions).

The maximum aperture (effective speed) of a standard amateur cine lens is usually around *f*1·8. Only the very cheapest cameras have slower lenses, and frequently better cine cameras have lenses of maximum apertures of *f*1·4 and in exceptional cases as large as *f*0·95.

Few cine cameras are restricted to having only one lens. With many cameras, lenses of various focal lengths are available so that the cameraman may choose the one that best suits the circumstances of his shot: a wide-angle lens will be useful in cramped surroundings and a long-focus lens can secure a close view of distant action. The lens mounts are designed to make the range of lenses quickly interchangeable—in the less ambitious cameras by screw-in mounts where each lens must be unscrewed before the next can replace it; and in the more expensive equipment by an arrangement called a lens turret—a rotating fitment that usually holds two or three screw-in lenses, of which any one can be swung precisely into position with a minimum of delay.

Where the lens is fixed, afocal attachments can sometimes be fitted to convert the standard lens into a wide-angle or telephoto one.

On an 8 mm. camera the focal length of the standard lens is usually 12·5 to 13 mm. (½ inch), while wide-angle lenses range from 6·5 to 9 mm. in focal length. Long-focus lenses run from 19 mm. (1½ times magnification) upwards; usual ones are 25, 38, and 50 mm. (the latter having an image magnification of 4 times compared with the standard lens), or even longer. The standard focal length for a 16 mm. camera is 25 mm.; wide-angle lenses range from 13 to 17 mm. and long-focus lenses up to 75 mm. (3 times magnification) or longer for special purposes. The focal lengths for 9·5 mm. cameras are similar.

Lens mounts are frequently standardized with an internationally agreed lens thread and distance between the lens flange and the film. There are two such standards: the type C mount for 16 mm. cameras and the type D mount for 8 mm. models.

The type C mount for 16 mm. cameras has a diameter of 25·4 mm. (1 in.) with 32 threads per in. and a flange-to-film distance of 17·5 mm. 0·69 in.). The type D mount for 8 mm. cameras has a diameter of 15·8 mm. (0·632 in.), with 32 threads per in. and a distance of 12·3 mm. (0·484 in.) from the lens flange to the film.

A type D lens will thus fit any 8 mm. camera with a type D mount, so that the serious amateur with this type of camera has a very wide range of lenses to choose from. A number of cameras however have other lens mount standards or mounts suitable only for the lens range made by the camera manufacturer.

Lenses with a type C mount can often be fitted to cameras with a type D mount with the aid of an adaptor. The reverse is not normally possible, partly because lenses designed for 8 mm. cameras do not have a sufficient angular coverage in the larger 16 mm. field, and partly because there is no way of recessing the lens into the camera mount.

Zoom Lenses. Most modern amateur cameras, especially in the 8 mm. gauge, are fitted with a zoom lens. This has a variable focal length which can be adjusted even during filming, so altering the field of view and the image scale. It is thus possible to simulate a tracking shot—i.e. where the camera appears to approach or recede from the subject—by just working the focal length adjustment without moving the camera.

Zoom lenses are more complex in design and larger in size than units of any one of the single focal length covered by the zooming range. On the other hand the immense convenience of zooming and the fact that one lens covers a range of focal length greatly outweighs the possible drawback of bulk.

The zooming range of the lens depends largely on its price. Inexpensive zoom lenses for 8 mm. cameras have a range of 3 to 1 (for example 9 to 27 or 10 to 30 mm.), covering a medium wide-angle view to tele shots with a magnification of over 2 times. More expensive zoom lenses may have a range of 1 to 4 or even 1 to 6 (for example 8 to 48 mm.). For hand-held shooting it is generally preferable to choose a zoom lens with the widest possible angle at the near zooming limit, even at the expense of the lower tele-magnification. Where the latter is high, hand-held shots risk camera shake and hence blurring.

On 16 mm. and 9·5 mm. cameras the zooming range typically covers focal lengths from 17 to 85 mm., though extreme-range zoom lenses (for example 12–120 mm.) exist for professional 16 mm. cameras.

Zoom lenses also exist as afocal attachments to fit in front of the standard camera lens. Further, there are also afocal attachments for zoom lenses to change the focal length range. For example, one such attachment may convert an 8 to 48 mm. lens to a range of 6·25 to 37·5 mm., giving more scope at the wide-angle end of the range at the expense of the longer focal lengths.

The zooming movement is normally controlled by a lever on the lens itself. This may be radial (protruding from the lens mount and swinging up and down at the side of the lens) or longitudinal (swinging backwards and forwards below the lens). Other ways of control include a knob on the camera body itself, and power zooming arrangements. In this case a spring or electric motor increases or decreases the focal length on pressing one of two buttons on the camera or on the lens.

Finders. Cine cameras nearly always have a direct vision optical finder. On simple models the finder is built into the camera above or to one side of the lens and shows the field of view taken in by the standard lens. Additional image frames may show the fields with alternative lenses. For more advanced cameras the viewfinder may itself be adjustable, showing different fields of view either by changing the magnification or by adjusting a series of masks. A control on the finder usually shows the focal length to which it is set. For zoom lenses the focal length adjustment of the finder may be coupled to the zooming movement of the lens.

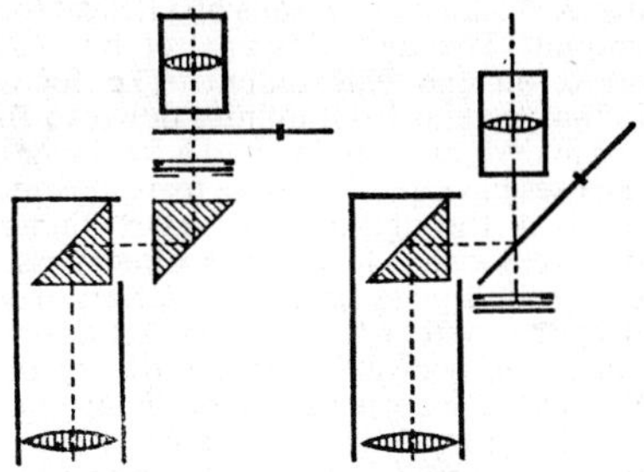

THROUGH-THE-LENS FOCUSING FINDERS. Left: *Prism system for focusing film in gate from behind.* Right: *Reflecting shutter which intermittently deflects the beam into the finder.*

Many modern cameras have a through-the-lens reflex finder. Here part of the light going through the lens to the film is diverted into the finder system. This arrangement has two advantages: first, the finder automatically shows the correct field-of-view irrespective of the focal length of the lens used (especially useful with zoom lenses); and secondly, visual focusing of the image also becomes possible.

There are two kinds of reflex finder, utilizing either a reflecting shutter or a beam splitter. In the first case the shutter which rotates in front of the film carries a mirror surface, so that the light is diverted into the finder system whenever the shutter obscures the film gate. In this way no light is lost from the exposure point of view, but the image in the finder may flicker slightly. Also, if the shutter stops in a half-open position at the end of a shot, no image is visible in the finder. (On the other hand most cameras automatically interrupt the shooting cycle with the shutter closed.)

In a beam-splitter camera the lens system contains a partially silvered reflector (usually a split prism with a semi-reflecting surface) which diverts about 10 to 20 per cent of the light from the lens into the finder system. The image in this case is always visible and always steady. If the camera is fitted with a built-in exposure meter, this must be calibrated to allow for the loss of light involved. The same applies if a separate exposure meter is used.

Generally, reflecting shutters are more common in professional cine cameras, and beam splitters in amateur models where the beam splitter frequently forms part of the zoom lens. Where the latter is not a standard fixture on the camera, it may carry its own through-the-lens reflecting finder.

Focusing. The simplest amateur cine cameras have a fixed-focus lens set for a depth of field from about 5 feet to infinity. On most models however the lens is adjustable and focuses over a range of distances by means of a helical focusing mount. The distances are set by a scale engraved on the lens mount. The focusing range may extend from infinity down to three feet, or in certain cases down to a few inches, as some lenses have an ultra-close focusing mount.

Where a through-the-lens reflex finder is fitted, direct visual focusing is often possible. Here a ground glass or grained screen may be fitted in the system to show the actual image sharpness on viewing through the eyepiece. Some cameras incorporate a split-image rangefinder of the optical wedge type, where the picture is seen split in two with the two halves displaced relatively to each other when the image is unsharp, fusing into one continuous picture when the lens is set for the correct distance.

Some cameras take a focusing attachment with a prism and magnifying lens that fit into the camera gate instead of the pressure pad. This is useful for ultra close-up work and for titling with the camera set up on a rigid support. The image is focused in the film plane with the camera open, and the attachment also shows the exact field of view taken in. Focusing takes place before shooting, as the camera can only be loaded with film afterwards.

The lenses of more advanced cine cameras carry a depth of field indicator. This may take the usual form of aperture index marks arranged symmetrically to both sides of the focusing index, or it may consist of coloured dots which become visible at different aperture settings and are coupled with the aperture adjustment. On zoom lenses the index lines of the depth-of-field indicator may be curves uncovered to varying degrees according to the focal length setting. The ends of the curves actually opposite the distance of scale at any focal length selected show the depth of field at that setting.

The Shutter. Nearly all cine cameras have a shutter which periodically covers and uncovers the film aperture. Its main purpose is to cover the film opening while the intermittent transport pulls down the film to bring the next frame into position in the gate. Usually the shutter consists of a rotating blade of semi-circular shape, with its axis at the centre of the semicircle. The shutter runs in synchronism with the film transport, so that the film is steady in the gate while it is uncovered by the shutter.

At a normal camera speed for silent pictures (16 frames per second) the shutter rotates with 16 revolutions per second; at higher speeds the shutter must be speeded up accordingly. This takes place automatically, through the gearing of the shutter and the film movement.

The shutter opening—i.e. the phase of the rotation during which the film is uncovered—is usually 180°. The shutter thus exposes the film for half the cycle of its operation, so that at 16 frames a second the effective exposure per frame is about 1/32 second. Sometimes the shutter opening may be slightly greater or slightly less than 180°.

While the angle of the shutter is usually fixed on simple cameras, more advanced models may have a variable shutter adjusted by a control on the camera. This may cover a range of movement from fully open (180°) to fully closed. When the shutter is open at 90°, the effective exposure time per frame is thus only half of what it is with the 180° opening—around 1/60 second at 16 frames per second.

Such a variable shutter has two uses. First, it permits control of the exposure of the film without adjustment of the aperture, e.g. when a limited depth of field is desired in the scene, or when the light is so bright that the film would be over-exposed even at the smallest lens opening. Secondly, it permits trick effects such as fading in and out during shooting.

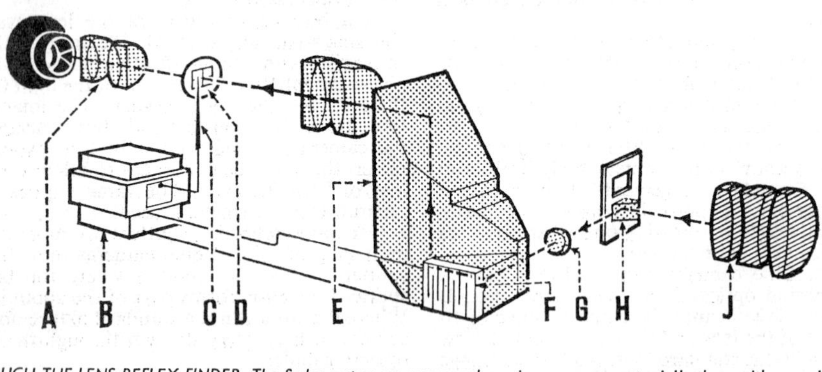

THROUGH-THE-LENS REFLEX FINDER. The finder system on many modern cine cameras—especially those with zoom lenses—sees the view through the camera lens. The light is diverted through either a beam splitter or a mirror shutter and a part of it may be further diverted to an exposure meter system. A, eyepiece lenses. B, exposure meter mechanism. C, meter needle. D, cross-hair in finder. E, prism system with semi-reflecting surface to divert light to the photo electric cell F. G, field lens. H, mirror shutter (or beam splitter). J, camera lens.

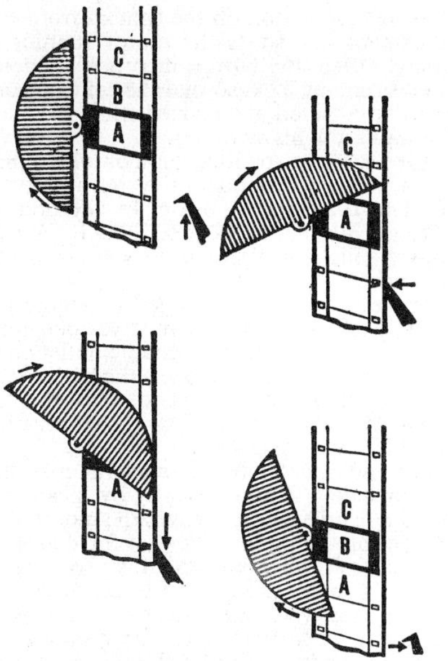

FILM TRANSPORT AND SHUTTER. Top left: *Frame A in gate, shutter open.* Top right: *Claw engages film perforation as shutter closes.* Bottom left: *Shutter covers gate as claw pulls down film.* Bottom right: *Shutter uncovers film with frame B in gate as claw withdraws from perforation.*

An alternative shutter design uses an oscillating instead of a rotating system; this may have certain constructional advantages.

The shutter can also serve other purposes. Some cine cameras have shutter blades designed as a reflecting mirror which diverts the light into a through-the-lens reflex finder during the closed phase of the shutter. Systems have also been suggested where the shutter blades carry a barrier layer of a photo-electric cell, for direct exposure measurements through the lens.

The Film Gate. The gate is the position in the camera where the film is held during the exposure period. It is constructed in two parts, one fixed and the other (the pressure plate) free to move aside to allow the film to be loaded and held between the two parts in the correct focal plane.

The pressure plate is spring-loaded and the film rides through the gate channel on raised edges which hold the picture area of the film out of contact with the gate and prevent the film surface from being scratched. Gates of the best designs open very wide or are removable from the camera to allow easy cleaning. The fixed part of the gate has an aperture in it to suit the size and shape of the picture, but with most cameras an aperture in the pressure plate is not considered necessary unless required for through-the-lens focusing.

In the camera, the film is always run through with its sensitive emulsion (dull) surface facing the lens and the subject. The only exceptions have been some types of now obsolete colour films.

As a rule the gate aperture is fixed corresponding to the frame dimensions on the film as laid down by international standards. A few cine cameras have also appeared with a variable gate aperture, where masks can reduce the image height from the full frame height to virtually nothing. This can be used either for an additional way of fading in and out (a kind of wipe-fade) and also for producing wide-screen images with the gate in an approximately half-open position. The frame area on the film then has an aspect ratio of about 1·5 to 2 and the picture yields a wide-screen effect when the film is projected on to a screen of appropriate dimensions with a wide-angle lens in the projector. This, however, requires a greater image magnification, and part of the film area is wasted. If the film gate is varied during shooting, possibly in conjunction with the zooming movement of the lens, certain dynamic screen effects can be obtained; larger views for close ups and smaller ones for more distant shots, etc.

Intermittent Feed. The film is pulled down through the gate by means of a metal pin or claw that works as near the aperture as possible. There are various devices for driving the claw to save wear on the film. Ideally the movement of the claw should describe a narrow vertical rectangle: a stroke to pull down the film, a horizontal movement to withdraw from the sprocket holes in the film, an upstroke to return the claw to the top and a horizontal movement to engage in the sprocket holes again.

While the claw is the commonest film transport mechanism, especially in amateur cine cameras, various other systems exist. Best known is the Maltese Cross where a constantly rotating shaft imparts an intermittent movement to a sprocket wheel which is permanently in contact with the film. The sprocket thus pulls down the film through the picture gate in jerks. The Maltese Cross movement is more popular in professional cine cameras, especially the 35 mm. size where the film is larger and heavier and where, therefore, the claw may more easily damage the perforations.

Simple amateur cameras, especially the 8 mm. gauge, have only a single filming speed of 16 frames per second. More advanced models may have a range of speed, e.g. 8, 12, 16, 24, 32, 48, 64, and possibly even 128 frames per second. (Most common is a range of 12 to 32 frames per second.) In addition most cameras have provision for single frame exposures, when pressing the release button exposes just one frame. This is useful for certain tricks, animation and titling.

On some models the speed can be changed during shooting, for instance from 16 to 32 frames per second for instant switching over to slow motion. Where the camera has a built-in exposure control system, this must be coupled with the change-over to keep the exposure of the film correct.

When cameras are intended for high speed work, the design of the intermittent mechanism becomes much more critical because vibration must be avoided from the rapidly oscillating parts. For speeds up to ten times normal (240 f.p.s.) the intermittently fed film is generally preferred because of the superior photographic results. But there is in any case a limit to the speeds attainable, governed by film strength.

Cameras running at higher speeds than 240 f.p.s. dispense with the intermittent motion and make use of a system of optical compensation to keep the image still in relation to the film; in effect, the film runs continuously and the image is caused to follow its movement at precisely the same speed. This principle is also used in some forms of viewer for editing.

For still higher speeds, systems have been devised of which many do not produce on the film separate images that can be projected but simply record data to be studied. Such systems are capable of an equivalent rate of exposure as high as 10 million f.p.s.

Camera Motors. The two most common drives for narrow gauge film cameras are the governor-controlled clockwork motor and the battery-powered electric motor. In either case the motor should ideally start instantly, though often the first frame or two of a scene may be over-exposed while the motor accelerated to its normal running speed.

Spring motors are wound by a key or a crank in the camera side. To prevent over-winding some cameras have a slipping clutch which comes into operation when the motor is fully wound. At one winding, the spring motor should be able to pull through at least 6 feet of film; on more modern cameras the capacity may be as much as 10 or 12 feet. The better cameras also have a device for stopping the motor automatically when it slows down near the end of the spring tension. That keeps the running speed constant. During filming, it is always advisable to rewind the motor after every scene.

Some cameras have the spring motor contained in a separate grip or handle; twisting the grip winds the motor.

Electric motors may be powered by a variety of batteries—dry cells or rechargeable miniature accumulators. Usually one set of batteries or one charge is sufficient for running about half-a-dozen films through the camera.

The advantage of the electric motor is, of course, that it eliminates rewinding and permits the shooting of longer scenes. Some cameras with a spring motor can also take an electric motor which attaches to a suitable driving shaft. Battery-powered cameras often have an indicator which shows the state of the battery to warn when it needs replacement or recharging.

The release button on the camera front or in the camera side starts the motor running for filming. Often, the button incorporates a continuous running lock, so that the camera can be set-up on a tripod and started while a cameraman walks into his own shot.

Many cameras are fitted with, or take, a pistol grip which incorporates the release button linked to the camera release. In addition, the starting button may be operated by a cable release. This can also permit remote control operations.

On electrically-driven cameras, remote control is easier, since it only involves plugging a cable with a switch into a suitable outlet on the camera body. On some models the batteries are contained in a handle or pistol grip; this may be removed and connected to the camera by a cable for remote operation.

More advanced amateur cameras may have a backwinding feature, usually by means of a separate handle which engages the motor shaft. This permits rewinding the film for double exposures—e.g. when making dissolves. If the camera is spring-driven, the backwind should only be used when the spring motor is partly run down, since it rewinds the motor at the same time as it transports the film in reverse. With electrically-driven cameras this problem does not arise and some models permit running the motor backwards over longer film lengths. For the most accurate results when making dissolves and other tricks in the camera, the rewinding should be done in conjunction with a good film measuring device—preferably a frame counter.

Footage Counter. All cameras have some means of indicating how much film has been exposed. Some of the dials work in jumps of so many feet, but the better ones move smoothly and may allow an estimate to the nearest 3 ins. of film, which is good enough for most purposes. An additional dial that is needed for the most accurate work is a frame counter by means of which any frame on the film in the camera can be found.

Some of the footage counters are self-setting to zero when the camera is loaded. A further convenient arrangement is one whereby the footage counter is visible in the viewfinder and can be watched during filming.

Exposure Control. On simple cine cameras with a fixed shutter opening and a single running speed, the exposure time per frame is also fixed and the lens aperture is the only means of controlling the exposure. The lens thus carries an adjustable diaphragm which may be in the form of an iris or—on simpler cameras—a pair of diaphragm plates with slots of varying width which reduce the effective opening as the plates move relative to each other. On some cameras the iris can close down fully and is used for fading a scene out or in at the end or beginning, respectively, of a shot.

Variable filming speeds and variable shutter

openings provide further means of exposure control. Generally the exposure time is inversely proportional to the camera speed: the exposure per frame at 32 frames per second is half as long as at 16 frames per second. While filming at different speeds the aperture must be adjusted to compensate; the variable shutter opening however is a means of deliberate exposure control when the aperture adjustment is inadequate or undesirable (e.g. for reasons of depth of field).

The correct exposure—i.e. usually the aperture setting—can be obtained from exposure tables or guides supplied with the camera or with the film. Some cameras have an exposure calculator with symbolic indications of different weather and subject conditions.

The majority of modern amateur cine cameras have, however, some form of built-in exposure meter.

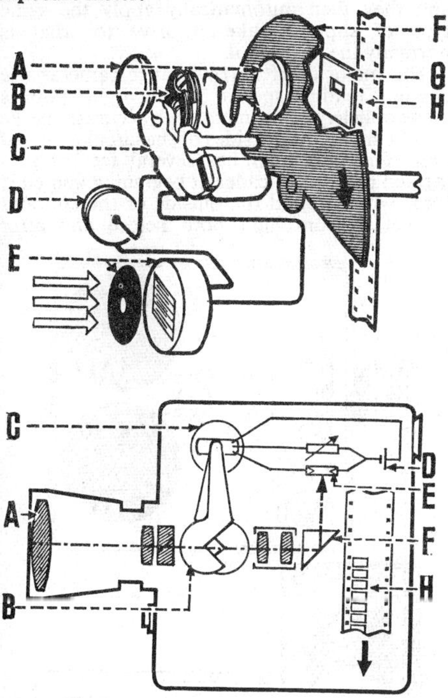

ELECTRIC EYE SYSTEMS: Top: *Conventional electric eye system. The light falling on the exposure meter cell controls the current flowing through the moving coil instrument which in turn works the diaphragm plates to set a larger or smaller aperture. A diaphragm in front of the meter cell adjusts the sensitivity to allow for different camera speeds and film speeds. Bottom: Self-compensating system. The meter cell measures the light, deflected by a mirror shutter, after it has passed through the lens diaphragm. The moving coil instrument is connected to the cell through a bridge circuit The diaphragm opens or closes to ensure that a standard light intensity reaches the meter cell. Errors in diaphragm calibration are automatically compensated. Camera speed and film speed settings are allowed for by adjustment of the bridge circuit. A, camera lens. B, diaphragm plates. C, moving coil instrument. D, battery. E, cadmium sulphide cell. F, shutter. G, film gate. H, film.*

In its simplest form this may be a straightforward exposure meter which indicates the correct aperture value at a specified film speed and camera running speed. This value is then simply transferred to the camera controls. Slightly more elaborate are coupled exposure meters, where a meter needle is visible in the finder and lined up with an index mark by adjusting the lens aperture control. The latter is thus coupled with the meter system and directly sets the correct exposure.

Electric Eye Systems. The most popular exposure control system in cine cameras is the fully automatic arrangement where the exposure meter directly sets the lens aperture. This is usually implied when a camera is specified as having electric eye exposure control. There are several types, and they may use either selenium or cadmium sulphide cells.

In the commonest system the current from the photo-electric cell passes through a pair of coils which control the movement of two slotted aperture plates in the optical path. When little or no current passes, the plates are in their rest position with the wider ends of their slots overlapping, so that the lens is effectively at its full opening. As more current passes through the coils (due to more light falling on the cell) the aperture plates move so that the overlapping slots become narrower, reducing the lens opening. A pointer attached to one of the aperture plates may show the actual aperture set in the viewfinder, or may merely appear in the finder when the light is inadequate for a correctly exposed shot. The aperture is adjusted continuously, even during shooting.

Alternatively, a servo motor may adjust the lens aperture according to the prevailing light. In one set-up the motor drives the lens iris as well as an iris diaphragm in front of the photo cell, until the cell current balances in a bridge circuit. More light means more cell current which has to be reduced by closing the iris in front of the cell and hence the iris in the lens. This system usually adjusts the lens aperture before the camera starts running, but can also work during shooting.

Some cine lenses, especially for cameras without exposure meter, have a built-in meter cell which controls the lens aperture in one of the ways described above.

Through-the-lens Measurement. For reflected-light exposure measurements the meter cell should ideally take in exactly the same view as the camera lens. This is comparatively easy if the camera has a single fixed lens, but becomes difficult with interchangeable lenses. In a number of cine cameras, therefore, the exposure meter measures the light coming through the lens. That compensates automatically for the different fields of view, for changes in the lens-to-film distance at extreme close-ups, and in one case even for variations in the calibration of the aperture. There are several systems.

The swing-in cell moves into the optical path

of the camera lens before shooting. The cell measures the average reflected light and the lens aperture can be adjusted semi-automatically or automatically. The cell then swings out of the optical path for shooting; hence no continuous monitoring of the exposure level is possible. Such a set-up also requires a certain minimum space between the lens and the film for the movement of the exposure meter.

In beam-splitter systems, part of the light diverted into a through-the-lens viewfinder is further diverted to the exposure meter cell. The latter can thus continuously adjust the lens aperture according to the changes in the prevailing light intensity. Since only a small proportion of the light coming through the lens is utilized for exposure measurement, this method requires a highly sensitive meter cell of the cadmium sulphide type.

In some cameras the meter cell measures some of the light reflected from a mirror shutter. The meter in this case needs appropriate calibration to allow for the fact that the light reaches the cell intermittently and not continuously. A number of designs also exist (mainly in the form of patent applications) where a photo-electric layer on the shutter blades themselves measures the light falling on them and adjusts the exposure.

Most advanced are automatic compensation systems. Here the photocell measures the light coming through the lens and through the aperture diaphragm, via a beam splitter or a reflecting shutter. The meter output is balanced against the control current in a bridge circuit or fed in opposition to this control current into the coil controlling the aperture plates. If the scene is bright, the light reaching the meter cell causes the aperture plates to close down until the light is reduced to a point when the cell current balances the control current in the meter circuit. If the light increases or decreases, the current through the meter cell automatically causes the diaphragm plates to open or close the aperture. As this arrangement measures the light intensity that reaches the film, it is independent of errors in the lens aperture through calibration or mechanical wear and tear.

Compensations and Override. The exposure meter of electric eye cameras also has to allow for other variables connected with exposure: the film speed, camera speed and shutter opening. On the simplest cine cameras, only the film speed is a variable and is allowed for by placing different sized masks in front of the meter cell. The range of film speed covered this way is usually limited over a 4 : 1 range.

More advanced cameras allow for the variable factors by an adjustable iris diaphragm in front of the meter cell or by adjusting a rheostat controlling the current in the meter circuit. On some cameras the compensation for different running speeds is automatic, especially where the camera permits switching over from normal to high framing speeds during a shot.

Advanced exposure meter systems may have a control to disengage the automatic coupling. This permits the cameraman to set his own aperture irrespective of the indicated light conditions. Often a meter needle indicated the correct aperture even when the automatic exposure control is disengaged; this aperture has to be set on the lens by hand.

Film Loading. Cine cameras normally have a feed spool on which the film is loaded into the camera, and a take-up spool to accumulate the film as it runs through and is exposed. The feed spool is usually free to rotate on its own spindle. Sometimes a weak brake is included to prevent the spool over-running when the mechanism is stopped. The take-up spool is driven through a clutch, belt or other form of drive that allows some slip to compensate for the changing diameter of the coil of film as it builds up on the spool. Some cameras will also run in reverse and they then automatically apply the same type of slipping take-up drive to what is normally the feed spool.

The majority of narrow gauge cameras use daylight loading spools. The daylight spool is made of heavy gauge metal with accurate cheeks that fit snugly to the sides of the circular coil of film. The film itself has several feet of perforated protective leader at beginning and end; these wrap around the outside of the coil and protect it from light both before and after

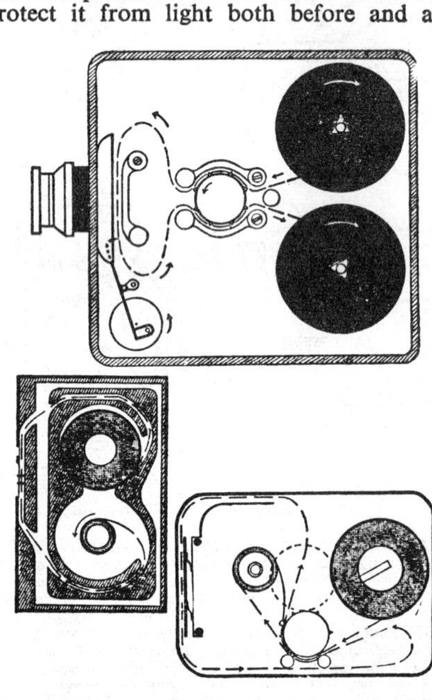

WAYS OF LOADING FILM. Top: Spools. The film is threaded through the camera mechanism and runs from one spool to the other. Bottom left: 9·5 mm. or single 8 mm. charger. The whole film is inside a light-tight container. Bottom right: 16 mm. magazine incorporating transport mechanism. This couples with the camera motor on insertion into the camera.

exposure. The light-tightness relies entirely on this arrangement, so the spool must be protected from light as much as possible when it is not in the camera. Once the film has started to run, the camera must not be opened until the entire length has been exposed—otherwise several feet of the film will be fogged in both feed and take-up spools.

As the threading of the film through the gate, sprocket and on to the take-up spool is usually a finicky business, camera designers have produced various ways of simplifying this. These are developed along two lines; automatic loading and magazines.

In automatic loading systems it is only necessary to insert the film spool in the camera and push the film into a slot. The mechanism then takes over, laces the film along the correct path and—on some cameras—even attaches it to the take-up spool.

Magazine loading is the oldest system. Here the film is preloaded into a light-tight magazine which is simply inserted in the camera. In some cases the magazine is designed as if it were the back half of the film gate; the claw is then ready to engage with the film without any further adjustment. Such magazines are used in many 9·5 mm. cameras as well as some 8 and 16 mm. models. On some 16 mm. magazines the intermittent mechanism is actually inside the magazine itself instead of in the camera. This makes

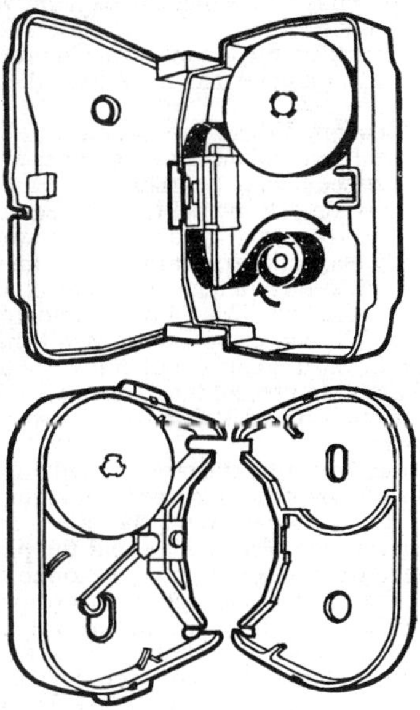

SPOOL-LOADING MAGAZINES. Certain double-8 mm. magazines can be loaded directly from standard double-8 mm. spools, and then inserted in the cine camera. This permits rapid film loading and changing over at the end of the first run. Some magazines (bottom) are used in cameras designed to take either the magazine or the standard double-8 mm. spool without magazine.

for rapid exchange of films in loading. Occasionally the loop of film may protrude from the magazine and has to be inserted in the film path of the camera.

A modern type of 8 mm. magazine consists of a plastic or metal case which is loaded with a standard 8 mm. spool. Its advantage is that film loading can be carried out at leisure, and the magazine clipped into the camera when required. This again permits rapid film changing during shooting. A few 8 mm. cameras even provide a choice of spool or magazine loading.

A special point with 8 mm. cameras is the need for reloading the double-8 mm. film at the end of the first run through the camera, for exposure of the second half of the film. In a spool loading camera this involves re-threading in the same way as the original loading procedure; the take-up spool and even the film itself may carry a marking like "exposed first half" or "half exposed" to indicate that the film must run through the camera once again.

This re-running is also simplified by a film magazine. Since the film runs through the camera in its full width, removal of the magazine in the middle of a run fogs a number of frames on both sides of the film. (With 9·5 or 16 mm. cameras the magazine permits removal of the film in the middle of a full magazine load without interfering with already exposed portions.)

Some cameras have special provision for rapid switching over at the end of the first run of double 8 mm. film. There the whole film container may swivel round, complete with film gate and intermittent movement. In one system the whole camera may be turned upside down for the second run, the lens being moved over from opposite the first film track to the second film track.

The Super-8 system uses a 50 feet length of single width 8 mm. film in a rapid loading cassette. There none of the mid-film changeover problems arise.

PROFESSIONAL AND SOUND CAMERAS

Professional motion picture cameras are mainly used for newsreels, documentary and feature film production as well as for certain special scientific purposes. They are usually 16 mm. or 35 mm. models taking film loads up to 1,000 feet with various other special provisions. Constructionally a professional camera is designed for heavy-duty work and is accordingly more robust and more bulky.

More advanced types of 16 mm. camera tend to be more complex than those for 35 mm. film, which often have less provision for special effects such as fades, dissolves, and other tricks. This is because these effects are produced by optical printing methods on 35 mm. film; such facilities are not available to most narrow-gauge users, so the only alternative for those film makers is to obtain their effects in the camera.

On the other hand, professional motion picture cameras are more versatile in a number of other respects such as the use of special accessories and attachments. Significantly, few professional cine cameras have built-in exposure automation; since the cameraman is invariably a highly trained technician, he prefers to measure his exposure by more accurate methods with a separate meter, rather than rely on electric eye systems which give an acceptable result most of the time but are unreliable in a certain percentage of cases.

Lenses. The focal lengths of 16 mm. professional camera lenses cover much the same range as for amateur models, but are more extensive. Thus long focus and tele lenses of much higher magnification are sometimes used. This is practical, since the professional cine cameraman almost invariably works with the camera mounted on a rigid support. Zoom lenses are also more versatile, and may cover a zooming range up to 10:1, e.g. with focal lengths from 12 to 120 mm.

For 35 mm. cameras, the standard focal length is 50 mm., with wide-angle units from 25 mm. upwards and extreme tele focal lengths as long as optical manufacturers can make them. (Lenses of 2,000 and even 3,000 mm. exist for 35 mm. cine use.) Zoom lenses are handy for newsreel and documentary work, but are less used in feature film production.

This is because in a scripted film it is easier to decide on exact camera positions and angles of view with lenses of fixed focal length, while tracking shots are often made by actual camera movement on a dolly.

In view of the longer focal length—and hence reduced depth of field—of lenses on 35 mm. cameras, actual focusing becomes more important. Visual focusing systems, including observation of the image in the film plane or on a ground glass screen, are frequently used. On studio film cameras the focusing adjustment is often the job of a separate operator known as the focus puller, who has to adjust the distance setting to keep the main subject in sharp focus while it moves toward or away from the camera.

Film Transport. While professional 16 mm. cameras may still employ a claw movement for intermittent film transport, this is usually more rugged than in an amateur camera. Maltese Cross transport movements are also used since they are more effective in advancing the rather heavier 35 mm. film at a constant speed.

In some 35 mm. cameras the film itself is lifted away from the gate to reduce friction as it slides down to the next position. The registration pin may hold the film accurately in place when it is stationary during exposure. This is essential for precise film location to permit accurate matching up of images for various optical printing effects and tricks.

The cameras used in motion picture studios—and often those used for making newsreels—are driven by constant speed electric motors and synchronized with the sound camera which records the sound track separately. Cameras used for location work are generally driven by "wild" motors powered from mobile batteries. Some are driven by clockwork. The speed of these cameras is not controlled as accurately as that of the studio cameras.

The shutter on professional cameras is often of the mirror type to provide a through-the-lens view of the exact field taken in on the film. Alternative systems for through-the-lens viewing may utilize prisms that swing into the optical path when the camera is not running. Beam splitters in the system itself are rare, mainly because this tends to limit the range of interchangeable lenses that can be used.

Magazines. As a professional motion picture camera often needs a considerable film reserve, the space for the film is comparatively bulky. On most 35 mm. cameras and many 16 mm. professional types the film is loaded in magazines mounted externally on the camera. The magazine reels may be driven by a belt or by gears. It is this magazine which also makes the professional cine camera so bulky.

Usual film capacities for 16 mm. cameras are 100 to 400 feet; magazines up to 1,000 feet, however, also exist. The full capacity of a 35 mm. studio camera is usually also 1,000 feet. The magazines are loaded from spools or with film on cores. The exact packing and loading method depends on the make of camera. When a film is on a spool with opaque flanges it can usually be loaded into the magazine in ordinary lighting; film on cores has to be loaded in the dark.

Sound Cameras. Where live sound is used in professional motion pictures (in other words sound recorded during the shooting of the picture) this is almost invariably recorded separately. The recording medium is nowadays commonly magnetic tape or magnetic film (perforated motion picture film stock coated with a magnetic oxide layer instead of a photographic emulsion). The cine camera itself usually has to be enclosed in a sound-proof housing—a so-called blimp—to prevent the motor noise of the camera from being picked up by the recording microphones.

The sound record is then transcribed on to an optical track by being recorded photographically on to a separate film, the latter edited together with the picture film, and the picture and sound track finally printed on to the same positive film to produce a release print carrying both the picture and the optical sound track.

In newsreel and documentary film cameras the sound is often recorded on tape in a separate tape recorder, coupled with the camera by electrical means to synchronize the speed of the two.

The sound may also be recorded directly on pre-striped film—sensitive motion picture stock carrying a magnetic stripe. In such single-system cameras the sound is recorded on the film ahead of the picture, so the editing presents special problems. Single-system sound cameras are normally used for newsreel and television filming. If the picture is shot on 16 mm. negative film, the editing may be done on a positive print and the sound track re-recorded from the stripe of the negative, similarly edited in step.

Amateur Sound Cameras. Single-system 8 mm. sound cameras record the sound on a magnetic stripe on pre-striped film. The camera contains the magnetic sound heads and may be battery or mains operated. Again the recording of the sound on the film—with picture and sound separated by the standard interval—makes editing virtually impossible, especially when a reversal film is used in the camera.

Alternative amateur live recording systems use a tape recorder powered by and coupled to the camera. Some amateur 8 mm. cameras have a special outlet for connection to a matched tape machine. The latter starts running as soon as the camera does, recording the sound effects while shooting the picture. After editing the picture and the sound, the latter may be re-corded on a strip applied to the film after processing. L.A.M.

ACCESSORIES

The accessories for normal cine use are similar in many ways to those used in still photography. One way in which they can be different is that in cinematography some of them are required to be used in movement. For example, in still photography a polarizing filter is fitted to cut out a troublesome reflection. In cine work not only can the same purpose be served, but two polarizers may be rotated against each other during the filming of a scene and produce the gradual reduction in exposure needed to make a scene fade out, a requirement that cannot apply to snap-shotting.

Another difference is due to the fact that cinematography is inclined to be a more strenuous pursuit—and the cameras are usually heavier—so there is a need for heavier duty equipment.

The Cine Tripod. A steady tripod is almost an essential accessory in motion picture making. The head of the cine tripod is designed to allow the camera to be swung horizontally (panned) and pivoted vertically (tilted); this accessory is known as a pan-and-tilt head, and is fitted with a long operating arm so that camera movements will be steady and under the control of the cameraman.

As often as not the professional uses a head that is stabilized by an internal flywheel, he also uses a tripod head that is controlled in both directions by worm gears turned by two separate handles. The type of head more often used by narrow gauge workers has a simple form of adjustable friction control for both pan and tilt movements, restrained so that they cannot turn too freely. It is important that a tripod head is steady when being rotated very slowly—any tendency to stick or jerk is enough to condemn it.

Mobile Camera Supports. The cameraman must often be able to go on filming while the camera follows behind a moving subject. Or sometimes it may be necessary to move (or track) up to or away from a static subject. So there are several types of mobile camera supports used in the professional film studio. Most of them are wheeled platforms upon which the camera on its tripod can be mounted.

The dolly is a robust three or four-wheeled platform of this type. It is pushed about the set by hand and carries the cameraman and his equipment. The dolly provides for horizontal movements only.

The velocilator is a rather more elaborate form of dolly. It is driven by hand or by an electric motor and incorporates an electrically-driven raising and lowering gear for the camera and its operator. Power for the motors is supplied through a trailing mains cable.

The camera crane is a mobile structure that supports the entire camera crew and, if necessary, the director on a platform rigidly slung from the end of the long jib. Like the velocilator the crane is completely power-driven through a trailing cable and it gives the cameraman an almost unlimited choice of camera angle and movements.

Small Accessories. Gauging exposure is a fundamental need, as with still photography. To satisfy it there are the same instruments: the exposure calculators, the optical extinction meters, the photo-cell meters. In some cases meters are also issued as cine versions which quote only the simplified shutter speeds possible with a cine camera. Sometimes attachments are sold which enable the meter to read the incident light as well as the reflected light. Most professionals use the incident light method of exposure estimation.

Filters are used extensively by professionals. Amateurs do not bother with them as much as they should. Apart from the range of normal coloured filters, use is made of the various other types, such as graduated sky filters, and ultra-violet filters. Particularly useful are neutral density filters; these are grey filters that cut down the light by a definite amount without affecting the colour values: they are used to prevent over-exposure or to make a larger lens aperture necessary so that the depth of field is reduced. Polarizing filters also have special value in cinematography; they can be used for reducing unwanted reflections, darkening sky tones, and as continuously variable neutral density filters.

Supplementary lenses, diffusion attachments, lens hoods and other normal still photography accessories also have their important uses in cinematography.

Rangefinders can be fitted to narrow-gauge cameras, but they are not vital items as the normal cine lens has such a great depth of field. Professionals do not need them because they are able to focus through the camera lens.

Reflectors are useful accessories for amateur and professional alike. They are light-surfaced boards which reflect light into the deep shadows that often occur on close-ups of people. A silvered surface will give a hard light and a matt white surface gives a soft light.

Special Accessories. In the field of special effects the accessory designer has great scope.

The most important item is the effects (or matte) box, which at its best is a comprehensive lens attachment designed to hold a wide range of accessories and ensure their accurate placing in relation to the camera and the tripod; a registration device guarantees its alignment with the optical axis of the camera. At the front of the box is a vertical plate with an accurately cut aperture that admits light from the subject area covered by the lens, and no more; because of this the effects box is an excellent lens hood.

The amateur's effects box has to be more comprehensive than the professional's because the amateur has to do most of his special effects in the camera. Into the front of the appliance can be fitted various masks to cut down the picture area so that it represents the views seen through binoculars, a telescope, a keyhole and so on; also the masks necessary for split-screen effects.

Into the body of the appliance and nearer the lens can be fitted such attachments as mirrors, prisms for reversing or inverting the scene, odd-shaped pieces of glass (from broken bottles) for distorting the image, cylindrical distortion lenses, and the combination of short-focus positive and negative lenses which when separated can put an image considerably out of focus.

There are other special accessories besides the effects box. One intended for the amateur is a reverse motion cradle, for use, if the camera cannot be run backwards. The cradle is a stoutly constructed frame fitted between the tripod and the camera and able to hold the camera firmly upside down.

A more ambitious attachment is an animation unit by means of which the camera may be operated one frame at a time at the intervals chosen to suit the particular task in hand. Such units are often used in cartoon work. One version of the mechanism is a completely automatic time-lapse unit intended for long periods of single-frame filming: it will operate not only the camera, but the lights and the cover which cuts off daylight from the subject for the exposure periods. This may be used for periods running into weeks or more.

G.H.S. & J.C.

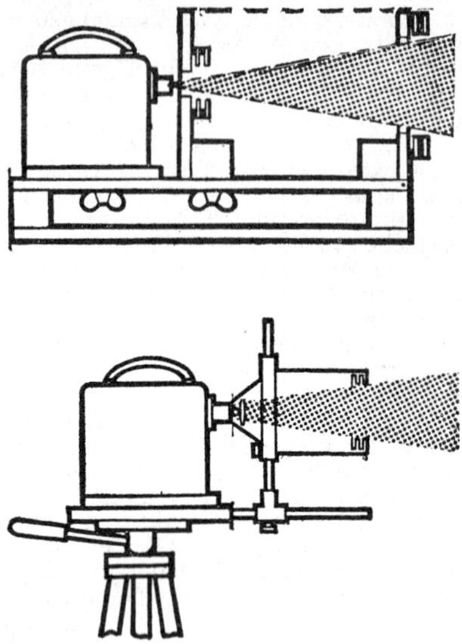

EFFECTS BOX. Top: Full-scale effects box with front and rear mask slots. These will take all kinds of cut-out masks, fading glasses, diffusion attachments, etc. At the same time the effects box also acts as a lens hood. Bottom: Simpler effects box. This is more portable and permits hand-held filming. Filters fit into the rear slot which is closest to the lens.

See also: *Animation; Cinematography; Exposure automation; High-speed cinematography; Projector* (cine); *Sound recording*.

Books: *The Complete Technique of Making Films*, by P. Monier; *How to Make* 8 *mm. Films*, by N. Bair; *How to Use 9·5 mm. Film*, by D. M. Neale; *Choosing and Using a Cine Camera* by E. Gilmour (all London).

CINE FILM

Since the cine camera exposes a long succession of pictures on a strip of film to recreate the impression of movement on projection, cine film is invariably supplied in lengths from 25 feet upwards—sometimes up to 1,000 feet. It is wound on spools or cores and carries regular perforations by which the transport mechanism advances the film in the camera. Cine film is available in a number of different widths or gauges. Standard cine film is 35 mm. wide and used almost exclusively for professional motion picture work. Narrow gauge films are 16, 9.5 or 8 mm. wide and used to be called sub-standard film to distinguish them from the standard 35 mm. gauge. This term is now largely fallen into disuse. Wider gauges than 35 mm. also exist.

NARROW GAUGE FILM

Of the three film gauges in this group, 8 mm. film is the most widely used amateur film

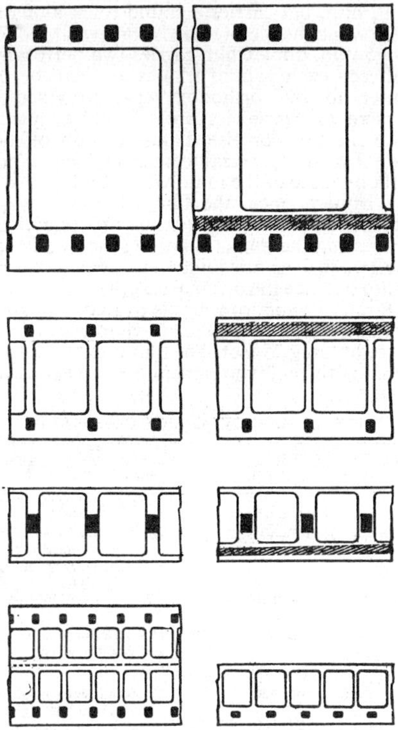

CINE FILM GAUGES. Approx. natural size. Top left: *35 mm. silent* Top right: *35 mm. sound.* Upper centre left: *16 mm. silent.* Upper centre right: *16 mm. sound.* Lower centre left: *9·5 mm. silent.* Lower centre right: *9·5 mm. sound.* Bottom left: *Double-run 8 mm.* Bottom right: *Super-8 mm.*

material. It carries 80 frames per foot of film, with an image size of 0·138 × 0·189 inch (3·51 × 4·80 mm.), the projected frame size being 0·129 × 0·172 inch (3·28 × 4·37 mm.). The film is usually supplied and exposed in the camera as double-8 film, 16 mm. wide, which is subsequently slit down the middle after processing (see below).

A second version, Super-8, is produced mainly as single-8 film in preloaded charges. The film uses smaller perforations and carries larger images—the size being 0·166 × 0·224 inch (4·22 × 5·58 mm.) recorded on the film, and 0·158 × 0·211 inch (4·01 × 5·36 mm.) as the projected image area. This film carries 72 frames per foot.

The 8 mm. film—especially with magnetic sound track—is also being used increasingly for educational and instructional purposes.

The 9·5 mm. gauge enjoys a limited popularity among amateurs, but is obsolescent. It carries 40 frames to the foot, the image size on the film being 0·276 × 0·346 ins. (7·0 × 8·8 mm.) and the projected image area 0·236 × 0·312 ins. (6·0 × 7·93 mm.).

The 16 mm. gauge originally started as an amateur film size, but is today more widespread for professional purposes. The film may carry perforations along both edges (silent film) or only along one edge (sound film). There are 40 frames to the foot and the image size on the film is 0·294 × 0·410 ins. (7·47 × 10·41 mm.), while the projected image area is 0·284 × 0·380 ins. (7·21 × 9·65 mm.).

Film Types. Narrow gauge cine films are available in several types, according to their use and the manner of processing. They are:—

(1) Negative films. These resemble normal negative film for still photography, and are processed in more or less the same manner. They serve for the production of release prints (positives) in any required number. The negative is never screened in a projector, and therefore is not liable to damage. Negative stock is used principally by professional and commercial studios and newsreel companies where several positives are required.

(2) Reversal films. These make up by far the largest proportion of amateur cine film materials, and are processed by reversal to yield a positive image directly on the actual film that was exposed in the camera. The running cost of reversal film is therefore lower than with negative film, but the reversal positive is the only one of its kind. If it should get damaged or worn by repeated projection, any duplicate made from it would show the same injury to the image. If additional copies are required, they have to be made by duplication before the original is used for projection; for few copies, they are made on reversal film.

(3) Positive film. This is used almost exclusively for making positive prints from negative films, or occasionally also for filming titles. The latter is the only use for amateur purposes, since printing negative cine films requires special apparatus.

(4) Colour films. These are almost invariably reversal films, although 16 mm. negative colour film has been made for professional filming. The characteristics of narrow gauge make. In the 8 mm. gauge a large majority of amateur cine films are nowadays shot in colour.

Non-sensitized cine film is also available for splicing purposes—e.g., as leaders. For convenience, this often has a yellow base.

Speed. Narrow-gauge cine materials fall into four general speed groups, three black-and-white and one colour.

(1) High-speed materials. These are rated between 200 and 400 A.S.A., are available as negative and as reversal emulsions, and are used for filming by poor or artificial light.

(2) Medium speed materials. These are rated between 80 and 160 A.S.A., are available in negative and in reversal emulsions, and are intended for all-round use where the extreme speed of high-speed materials is not required.

(3) Fine grain films. Intended mainly for outdoor shooting, these materials come in

speeds from 25 to 50 A.S.A. Again they are available both as reversal and as negative emulsions.

(4) Colour films. These range in speed from about 12 to 40 A.S.A.; the majority being between 16 and 25 A.S.A. Narrow-gauge colour cine films are nearly all reversal materials, and are available in daylight and artificial light type emulsions.

Special professional 16 mm. colour reversal films exist in speeds up to 160 A.S.A.

Grain and Resolving Power. As with negative materials for still photography, the grain of narrow gauge cine films tend to increase with increasing speed. In view of the considerably greater enlargement which takes place during projection, narrow gauge cine films must have a finer grain than ordinary film materials.

A point in favour of the cine film here is that as the pictures are projected at the rate of 16 or more frames per second, no image remains on the screen for long enough to make its grain structure really noticeable. In other words, the grain becomes partly (but not completely) hidden by the change of images. This effect, however, does not compensate for the loss of resolving power due to the grain of the film.

The effect of grain and resolving power is, of course, most important with the smallest narrow gauge size—i.e., 8 mm. For that reason, some manufacturers do not produce high-speed films in this size.

Colour Sensitivity. Most narrow gauge cine films—negative as well as reversal—are panchromatic—i.e., sensitive to all colours. Many, especially the high-speed types, have an increased sensitivity to red which makes them particularly suitable for artificial light work. The positive films are generally blue sensitive only, as colour sensitivity is not necessary when copying from a negative. Positive films can therefore be handled in the light of a comparatively bright safelight for processing.

Sizes and Packings. The type of packing depends on the size of the film.

16 mm. film is available on spools holding 50 or 100 feet. For certain cameras it is also available in 50 feet magazines which can be loaded in full daylight and changed at any time during the run. The film is perforated on each side, with one perforation each side per frame.

16 mm. sound film is perforated on one side only, the other side carrying the sound track. This type of film is only used for professional filming and is usually available on spools holding 100 or more feet.

9·5 mm. film is available in 30 feet chargers, or 50 feet magazines, as well as on 50 and 100 feet spools. The designs of chargers and magazines vary with the make of the camera. The film is perforated in the middle with a single perforation between each frame. The image area extends almost to the edges of the film.

9·5 mm. sound film is similar to ordinary 9·5 mm. film, but carries a sound track along one edge while the picture size is somewhat reduced.

9·5 mm. duplex film carries two perforations between each pair of frames and can be used either in the orthodox way, provided the camera and projector are suitably adapted to take the film, or else it may be run off horizontally in a special camera which exposes only one side of the film. After running through the camera once, the film is turned over to expose the other half and then processed. After processing the film is slit along the middle, and the two halves are joined end to end to yield a film of twice the original length.

8 mm. double run film is available in spools holding 25 or 50 feet or in chargers or magazines holding 25 feet. The film is exposed twice running through the camera once to expose one

TYPES OF NARROW-GAUGE CINE MATERIAL

Type	*Speed A.S.A.*	*DIN*	*Gauges (mm.)*	*Uses*
Negative Films (Panchromatic)				
High speed	200–400	24–27	16	Filming in poor lighting conditions and particularly in artificial light
Medium speed	80–160	20–23	16	All-round filming
Finegrain	25–50	15–18	16	All-round filming in good daylight
Reversal Films (Panchromatic)				
High speed	200–400	24–27		Filming in poor light and artificial light
Medium speed	80–160	20–23		All-round filming
Extra-fine grain	25–50	15–18		Filming in good daylight, yields special fine-grain and is thus ideal for 8 mm.
Positive Films (Blue-sensitive)				
Positive release	–		16	Printing films shot on negative film

half, and then again, after turning over, to expose the second half. After processing it is slit down along the middle and joined end to end to produce double the original length. 8 mm. double run is 16 mm. with twice the normal number of perforations.

Super-8 mm. film is available in single-run 50 foot chargers for Super-8 cameras. Thirty foot chargers of single-8 film also exist for few older cameras.

Magnetic sound film in all three gauges carries a magnetic oxide layer, applied as narrow strip either outside the perforations or in place of the optical sound track on 16 mm film. It may be applied before exposure or (more usually) after processing.

PROFESSIONAL FILM

Professional motion picture production was originally based exclusively on the use of

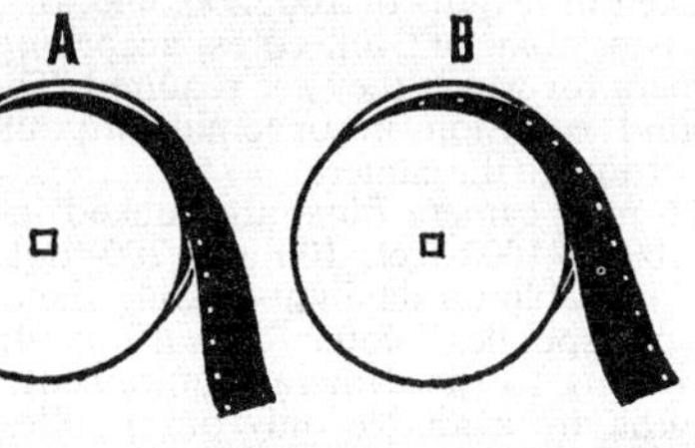

WINDING OF 16 mm. SINGLE PERFORATED FILM. There are two ways of spooling up the film, known as type A and type B winding, for camera feed spools. They differ in the disposition of the perforations; in both cases the film is wound emulsion in. Type B is more common; if this is spooled on to another spool, it becomes type A winding with respect to this second spool.

standard 35 mm. film stock. Today this is still the main film gauge, though 16 mm. film is used in documentary film production, and extra wide film—up to 75 mm.—in certain wide screen processes.

The 35 mm. gauge carries 16 frames to the foot with 4 performations per frame. The image size on the film is 0·629 × 0·866 ins. (16·0 × 22·0 mm.), the projected image area being 0·602 × 0·827 ins. (15·29 × 21·0 mm.) This applies to sound film which carries a sound track alongside the image. In order to preserve the 4 : 3 aspect ratio of the picture, there is appreciable space between individual frames.

In addition special film gauges are used for making narrow-gauge release prints. The most important is 32 mm. wide film, perforated with 16 mm. perforations. The film is used for making 16 mm. positives with or without sound track, being slit down after processing into two 16 mm. widths.

Film Types. Unlike most amateur film making, professional motion picture production mainly employs negative-positive processes. So there is a larger number of film types as well as actual emulsions available. They can be grouped as follows:

(1) Black-and-white camera films. All 35 mm. cine film for use in the camera is negative panchromatic stock with speeds between 40 and 400 ASA. The slow emulsions are intended for outdoor location shooting, background etc. Most studio shots are taken on medium speed film of 100 to 200 ASA. Materials of 400 ASA and over are mainly used for newsreel work and other occasions where filming has to take place in poor light conditions which are not under the cameraman's control.

Professional 16 mm. camera films are made in the same emulsions. In addition there are also 16 mm. reversal films, mainly (but not exclusively) for amateur use.

(2) Colour camera films. There are three main types: negative, standard reversal, and low-contrast reversal. Most 35 mm. colour film is negative stock. Usually reversal films are available only in 16 mm. widths.

Standard reversal films are used where only one copy is likely to be required—though duplication for additional copies is possible.

Low-contrast reversal colour film is intended for 16 mm. shooting where a number of release prints are needed. The film yields a low-contrast reversal positive which is easier to duplicate, but is not itself suitable for projection.

(3) Black-and-white printing films. These are intended for making the final release print, and are available in 35 mm., 16 mm., 32 mm. and special wide-screen gauges. 32 mm. film serves for simultaneous printing of two 16 mm. copies side by side the films being then slit down the middle. The printing film for printing from negatives is usually a fine grain positive film, non-colour sensitized, and of fairly high contrast.

Reversal duplicating film is a non-colour sensitized fine grain reversal material for making release positives from black-and-white reversal originals.

(4) Colour printing films. Positive colour film is used for printing the colour negative to produce release prints. It also serves for printing colour positives from internegative film. The positive material is available in 35 mm., as well as in 16 and 32 mm. widths for reduction printing from 35 mm. negatives.

Reversal colour copying film is used for making release prints from the low-contrast camera material.

(5) Intermediate printing films. These are materials used for making internegatives and intermediate duplicates from a master negative or print for subsequent production of release prints.

Duplicating positive film is a black-and-white material of extreme fine grain and fairly high contrast. It serves for making the master positive from the original negative exposed in the camera. Duplicating positive film is available in 16 mm., 32 mm., 35 mm. and wider sizes.

Duplicating negative film is a fine grain non-colour sensitized black-and-white film of low contrast to preserve the full gradation of the picture. This film produces the dupe negatives from the master positive and serves for the final stage of making release positives. The dupe negative carries the picture but not the sound track.

Duplicating reversal film is often used for making a dupe negative from the original master negative in one step. This eliminates the need for making a master positive when only a comparatively small number of release prints are needed. For large releases it is better to go via a master positive, this reduces the need for handling the master negative when further dupe negatives are required.

Interpositive colour film serves for making duplicate colour negatives from the original master colour negative. The colour negatives are printed from the interpositive onto internegative colour film.

Interregative colour film is a negative material

for making colour negatives from normal reversal colour film for subsequent production of release prints, and for dupe colour negatives from interpositive colour masters.

(6) Sound recording films. These are used for recording the sound track on film during shooting or—more usually nowadays—during transcription from magnetic tape. There are two types: low contrast for variable density sound recording, and high contrast for variable area recording. The sound recording films serve for making the optical master track, a negative derived from the master magnetic track. The optical master track is then printed together with the master negative or a dupe negative of the picture to produce the release print. This therefore contains both the optical sound track and the picture—this is the so-called married print.

(7) Special films are produced for oscillograph recording, infra-red cinematography, television recording etc. For instance a television film for photographing the display on the screen is normally orthochromatic with a maximum colour sensitivity corresponding to the emission of the fluorescent screen.

Sizes and Packings. 35 mm. camera films are packed in lengths of 100, 200, 400, or 100 feet on 1-in., 2-in. or 3-in. cores, according to the camera for which they are required. The film is wound emulsion in or emulsion out, again according to the camera.

16 mm. camera films are packed in lengths of 100 to 1000 feet, 100 and 200-foot lengths are available on daylight-loading metal spools or on spoolless cores of various diameters according to the camera requirement. Longer lengths are available only on spoolless cores. The film is wound emulsion in.

Single-perforated 16 mm. film exists in two windings, according to the side on which the perforations are located relative to the winding direction. The usual winding is the type B.

Duplicating, intermediate and printing films in 16, 32 and 35 mm. stock are supplied in 1000-foot lengths (sometimes up to 1200 feet) on spoolless cores—usually 2 ins. in diameter. Single-perforated 16 mm. film is again usually wound according to the type B system. Some 16 mm. positive films are also available in shorter lengths. L.A.M.

See also: *Cine camera; Cinematography; Perforations.*
Book: *Photo-Lab-Index*, by Henry M. Lester (New York).

CINE FILM PROCESSING

While the processing principles for cine film do not differ from those for still film, the dimensions and the length of the motion picture stock pose certain handling problems. These have been solved by the provision of specialized and often automated equipment for the purpose —particularly for professional use.

Among black-and-white cine films the material for professional use in the camera is mainly negative stock, while the preferred amateur material is reversal film. The latter provides a finished film ready for projection. Since the processing of amateur reversal films (colour as well as black-and-white) is usually included in the price of the film, user-processing is very much rarer in amateur narrow-gauge film making than it is among amateur still photographers. With black-and-white film at any rate it is, however, sometimes useful to be able to process a short test length to check exposure conditions, or even to process a number of films when results are required in a hurry or to ensure maximum consistency in the processed strips.

Professional cinematographers in the vast majority of cases use negative films—both in the standard 35 mm. and the narrow-gauge 16 mm. size. Here processing is practically never included in the price of the film and therefore has to be carried out independently. Further, the processed film is not an end product, but requires copying on to a positive film (which is also processed again) to obtain final projection copies.

PROCESSING BY THE AMATEUR

There are a number of simple processing units available which largely overcome the main objection to home processing, i.e. the difficulty of handling the long lengths of film involved. The following types of equipment may be used:

(1) Scaled-up apron or spiral reel tanks.
(2) Drum-and-trough set-ups.
(3) Motorized horizontal spool-type tanks.
(4) Small motorized continuous processing machines.

Reel Tanks. Simple processing tanks for cine

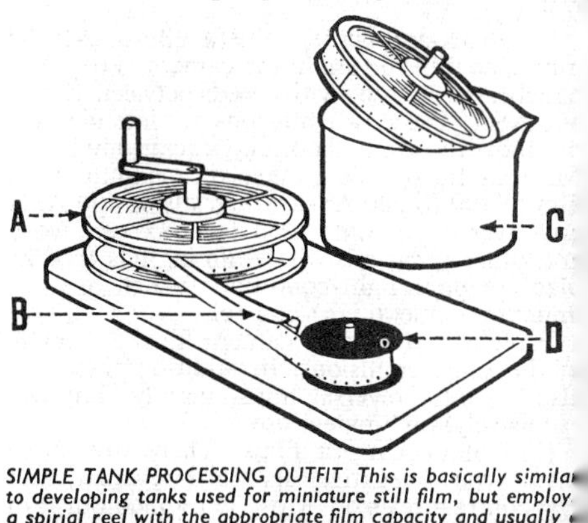

SIMPLE TANK PROCESSING OUTFIT. This is basically similar to developing tanks used for miniature still film, but employ a spiral reel with the appropriate film capacity and usually a mechanical loading aid to facilitate handling of the film length in the dark. A, reel on winding bench. B, film guide. C, tank body. D, film spool.

film are derived from normal daylight processing tanks for still 35 mm. and roll film, but with increased capacity for 25–100 feet of 16 mm. film. The 16 mm. width of course also covers double 8 mm. film which is slit down the middle after processing by special film slitters.

The tank consists of a reel of suitable size plus a light-tight plastic container with lid and openings for pouring the processing solutions.

There are two main kinds, apron tanks and spiral reel tanks..

The apron tank consists of a flexible transparent plastic apron strip with intermittent raised dimples down each edge. The film is wound on to a supporting disc interleaved with The apron strip. Lengths of apron up to 100 feet are available together with supporting discs.

The spiral reel tank works on the same principle as spiral reel type tanks for still films, with the film being carried in a groove spiralling in from the circumference to the centre.

When handling film lengths of 25 to 100 feet, loading the film into the tank by hand is obviously impractical. Loading aids provided for either type of tank usually consist of a holder for the film spool, a second one for the apron in the case of an apron tank, and a hand or motor driven spooling unit for the tank reel itself. With spiral reel tanks the film is guided into the centre of the reel, and the guide then follows the grooves as the film is laid into them while the reel turns. A similar (or the same) spooling unit unwinds the film from the reel after the complete processing sequence and spools it up on reels for projection. In the case of double 8 mm. film the film can pass through the slitting device before spooling up.

The processing reels are made of transparent plastic to facilitate a second exposure when processing reversal film.

As an alternative to processing in the tank,

FILM SLITTER. Double-8 mm. film is processed exactly like 16 mm., but then needs slitting along the middle. Top: A simple slitting set-up passes the film from the spool through a cutting roller for immediate winding up on 8 mm. projection reels. Bottom: The slitter consists of two stepped and interlocking steel rollers. As the film passes through it is sheared apart accurately along the centre.

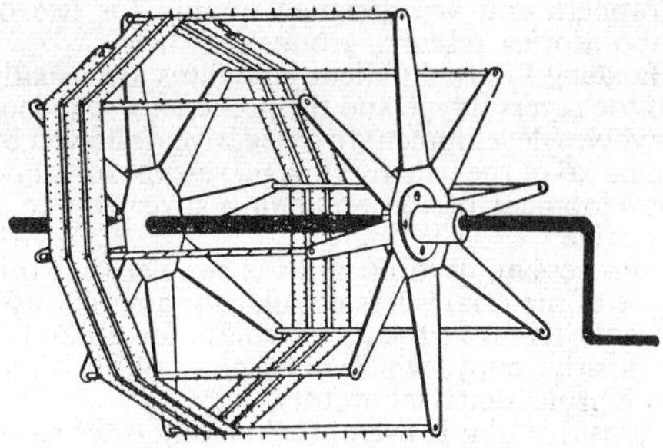

CINE FILM PROCESSING DRUM. The cine film is wound up on the outside of the framework, neighbouring turns being kept apart by suitable separators.

ordinary bowls can be used. In this case processing must take place in the darkroom.

The film stays on the reel even for drying. Special dryers available for the purpose take the complete reel and blow cold or warm air through it for about 15–20 minutes.

Drum and Trough Units. Older types of processing unit use a rack shaped like a drum on which the film runs in a spiral path round the outside. The drum is mounted on a framework dipping into a tray which contains the processing solutions which are poured in and out in turn. Lamps can be arranged above the drum for the re-exposure during reversal processing. This arrangement is economical in solution quantities, but the film is more laborious to load and processing has to take place in the dark.

Motorized Spool Tanks. This type of unit consists of two main components: a head unit with the two spools and a tank container. The head unit carries the spools on two geared shafts which may be hand or motor driven. The film is wound up directly on one spool and the end fixed to the second spool. The whole unit is then mounted on the tank container, so that both spools dip into the solution.

During processing, movement of the hand crank or the motor continuously winds the film from one spool to the other and back again to provide constant movement of the film in the solution and to prevent the turns from sticking to each other. The reversal exposure may take place either by lifting the head unit out of the tank and continuing the spooling to and fro in front of a lamp, or by including a shuttered window in the side of the tank.

Several tank units may be provided to permit the head with the spools to be lifted from one solution to the next. Alternatively the solutions may be poured in and out of the one tank body in succession. For drying the film is wound on a drying drum.

Small-scale continuous processors work on the same principle as commercial cine film processing machines, with the film running over rollers through a succession of tanks. Such units are usually too costly for amateur cinematographers and are designed mainly for use of laboratories, research groups etc.

Handling Films. Amateur cine films are usually of the reversal type and the processing sequence involves development to a negative, followed by removal of the negative image, re-exposure, and development of the remaining silver salt to a positive.

Processing negative films is simpler, but this type of material is less popular for amateur use. It only serves as the intermediate for obtaining a positive copy, which must normally be done by a professional laboratory.

Positive film is useful for making titles to be spliced into a picture on reversal film.

Black-and-White Reversal Film. The sequence of processing is as follows. In total darkness or by a deep green safelight:—

(1) Pre-soak the film for 2–3 minutes in water to ensure even development.

(2) First development to produce a negative image. A suitable formula is:—

Metol	35 grains	2 grams
Sodium sulphite, cryst. ...	7 ounces	180 grams
Hydroquinone	140 grains	8 grams
Sodium carbonate, cryst. ...	5 ounces	125 grams
Potassium thiocyanate ...	35 grains	2 grams
Water to make	40 ounces	1,000 c.cm.

Develop for 6 minutes at 65° F. (18° C.). This solution may be saved and used again for the second development.

(3) Hardening bath. This must be included in the process if the bleaching bath A, designed to prevent staining, is used. If preferred, bleach bath B may be used and the hardening bath omitted. The formula for the hardening bath is:—

Formalin (40 per cent) ...	1 ounce	25 c.cm.
Sodium hydroxide	27 grains	1·5 grams
Water to make	40 ounces	1,000 c.cm.

Harden for 5 minutes at 65° F. (18° C.).

(4) Rinse for 2 minutes in running water to remove the developer and hardener.

(5) Bleach in one of the following bleaching baths to dissolve away the negative image:—

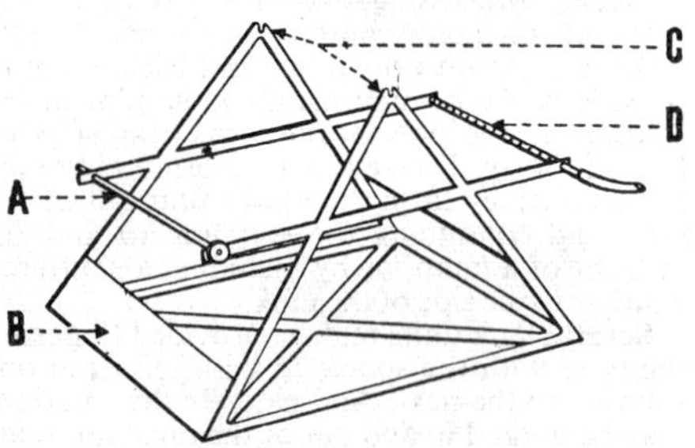

PROCESSING DRUM SUPPORT. This carries the processing drum as well as the solution tank. Accessories on the support frame include a spool bar for easy loading from the camera take-up spool, and a perforated tube for spray washing of the film between the processing stages. The solutions are replaced by changing the tank (not shown). A, spool bar. B, support for tank. C, half-bearings to support the film drum. D, spray tube with hose to water supply.

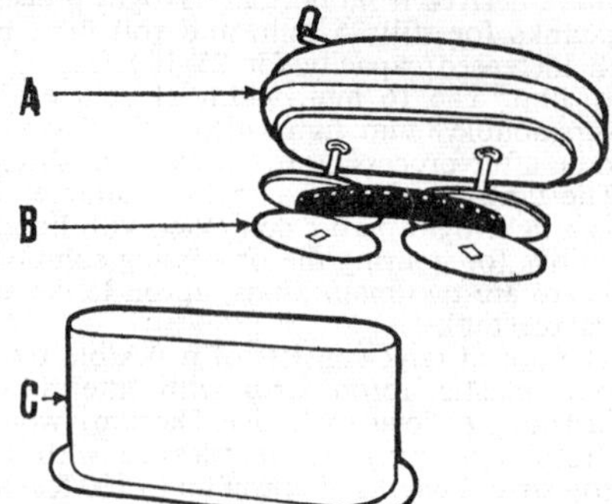

RECIPROCATING DAYLIGHT PROCESSING TANK. The film is wound continuously from one spool to the other and back again, reversing automatically each time. The spools are mounted on shafts underneath the winding head which fits over the processing tank, making a light-tight seal. The winding can be by hand crank or a motor. The film only needs winding on to one reel and attaching to the second reel in the dark. A, watertight winding head. B, film spools with film. C, tank.

Bleaching Bath A

Water	40 ounces	1,000 c.cm.
Potassium permanganate ...	35 grains	2 grams
Sulphuric acid (conc.) ...	192 minims	10 c.cm.
Silver nitrate (10 per cent)	77 minims	4 c.cm.

Bleach for 7 minutes at 65° F. (18° C.). This solution will not keep after use.

Bleaching Bath B

Water	40 ounces	1,000 c.cm.
Potassium bichromate ...	175 grains	10 grams
Sulphuric acid (conc.) ...	200 minims	10 c.cm.

Dissolve the bichromate in the water, then add the sulphuric acid slowly, and with constant stirring, to the cold solution. Bleach for 5 minutes at 65° F. (18° C.).

(6) Rinse in running water for 2 minutes.

(7) In an orange or bright green safelight: clear in the following clearing bath to remove the bleacher stains.

Sodium bisulphite	2 ounces	50 grams
Water to make	40 ounces	1,000 c.cm.

Clear for 2 minutes at 65° F. (18° C.).

(8) Rinse in running water for 1 minute.

(9) Second exposure, to make the remaining silver halide developable. The film should be exposed to the equivalent of 100 watts at 8 feet for 1 minute, whilst the drum is rotated very quickly to ensure even exposure.

(10) Second development to form the positive image. The developer used for the first development may be used again. Develop for 5 minutes at 65° F. (18° C.).

(11) Rinse in running water for 1 minute to remove the developer.

(12) Fix in this fixing bath:—

Sodium thiosulphate, cryst. ...	10 ounces	240 grams
Sodium sulphite, anhyd. ...	265 grains	15 grams
Glacial acetic acid	330 minims	17 c.cm.
Boric acid, cryst.	130 grains	7·5 grams
Potassium alum	265 grains	15 grams
Water to make	40 ounces	1,000 c.cm.

Fix for 10 minutes at 65°F. (18°C.).

In white light:

(13) Final wash. Wash in running water for 30 minutes.

(14) Drying. If a dryer is available to take the film on its developing reel, this is ideal. Otherwise a simple drying drum may be used. There are several types of drum on the market and they are effective in drying the long lengths of film without causing drying marks or impression marks.

The film must be absolutely dry before being re-spooled. While re-spooling, maintain an even tension, but never tighten the film already on the spool, as this may give rise to scratches.

Negative Film. Processing of negative film must be carried out in total darkness or by the light of a dark green safelight. The sequence is as follows:—

(1) Pre-soak for about 1 minute.

(2) Develop for a time found to be correct for the particular processing machine in use, by processing short tests on standard exposure subjects.

(3) Rinse for 1 minute.

(4) Fix for twice the time taken to clear.

The white light may now be switched on.

(5) Wash for at least 30 minutes.

(6) Dry and spool up.

Positive Film. This is non colour-sensitive, and processing may be carried out by the light of an amber safelight. The sequence is:—

(1) Pre-soak for about 2 minutes.

(2) Develop in a contrast developer for a time indicated by a test film.

(3) Rinse for 5 minutes in running water.

(4) Fix for twice the time taken to clear the transparent sections.

(5) Thoroughly wash in running water for at least 30 minutes.

(6) Dry and spool up.

Colour Film. Most reversal colour films can be processed by the user in much the same way as black-and-white reversal films. Although the processing cost by a laboratory appointed by the film manufacturer (or by the manufacturer himself) are included in the purchase price of the film, home processing again has the advantage of making test exposures quickly available for evaluation. Some manufacturers publish formulae or market processing kits for reversal colour films; independent processing formulae have also been published from time to time.

Certain types of reversal colour film can only be processed by the maker or specially equipped processing laboratories, since the procedure is rather more complex.

Negative colour films are rare in narrow-gauge sizes. Where available, processing follows largely the same procedure as for negative black-and-white film. Temperature control is much more strict, and the maker's recommended chemicals should be used. The making of colour positives from colour negatives is beyond the scope of the amateur cinematographer, so that home processing even of colour negative cine film is rarely worthwhile.

Processing Tests for Black-and-White Film. Preliminary processing tests made on a film exposed on a test chart can provide useful information about the characteristics of the film, and the efficiency of the equipment.

The chart may include black and white discs on a white and black background respectively.

The black discs on the white background are a test for streakiness in development and, should this defect occur, the images of the discs will have trailing grey tails either above or below them. This trouble is known as directional effect. It occurs when solution agitation is poor and exhausted developer remains very near to the film surface.

In addition the chart should include a grey scale of areas ranging from white through the greys to jet black. This tone range checks the reversal process generally and, particularly, shows whether the second exposure covers the whole range between highlights and shadows.

The chart must be illuminated evenly for the test exposure. Expose about 20 feet of film, and then carefully store the chart for tests.

Preliminary tests of the processing solutions employed, to establish optimum times, etc., can also be made on short lengths of test film. For this purpose an ordinary roll or miniature film tank is ideal, if it is adjustable to take 16 mm. film widths. The test itself need involve only a foot or two of film. A series of tests can for instance be made, varying the first and second development times, and so on. When making such a series, change only one variable at a time.

With this arrangement quite small quantities of processing solution are adequate; the typical amounts for an adjustable roll film tank used for 16 mm. film would be around 250–300 c.cm.

Provided the times and temperatures for the preliminary test are carefully followed in the full scale processing routine, the differences between the two results will be so slight that the full film should be almost perfect.

Professional Cine Processing. Motion picture laboratories use partly or fully automated continuous processing machines. There the films run through the machine at a constant speed, running over rollers from tank to tank. In some machines the films do not even touch the rollers, but ride over a liquid or air cushion that keep it a fraction of an inch away from the roller itself.

The processing times at each stage are determined by the length of film running through any given tank. The longer the film path, with the film moving at a constant speed, the longer obviously the time of treatment. The solutions in the tanks are thermostatically controlled and automatically agitated.

After processing sequence the film runs through a drying chamber and is then spooled up on large reels.

Films from professional cine cameras are

usually several hundred feet in length (positive films from printers up to 1,000 feet). When shorter lengths are being processed, they are frequently spliced together for going through the machine continuously.

Professional motion picture film processing machines are also extensively automated, with controlled replenishment of solutions and automatic warning signals if a film should break.

Colour film processing largely follows along the same lines, with correspondingly bigger machines to accommodate the greater number of solutions required.

Some film processing machines use special systems of solution application, such as spray units, viscous solution applicators, and so on.

A.E.S. (L.A.M.)

See also: *Developing negatives; Processing machine; Reversal process; Tanks.*

Books: *How to Process,* by L. Wheeler (London)

CINE HISTORY

The desire to produce a picture of the world around him has occupied man for many hundreds of years. Movement has played an important part in these attempts.

The earliest forms of artificially produced moving image took the form of shadows cast by a fire or lamp on to some sort of screen—the Chinese shadow plays are an example. With the invention of the magic lantern, a further advance was made with the use of articulated metal figures and painted transparencies which could be projected on a screen and given a semblance of motion. These slides, operated by levers or gears, were currently in use throughout the nineteenth century.

FUNDAMENTAL DEVELOPMENT

The motion picture is based on the phenomenon of persistence of vision—in which the eye retains the image of an object for a fraction of a second after the object is removed, an effect recognized since the days of Ptolemy. In December, 1824, P. M. Roget carried out the first experiments showing the principles behind this phenomenon. J. A. Paris introduced a toy—the Thaumatrope—in 1826 which, by the rapid rotation of a card held between two threads, enabled the eye to superimpose (by persistence of vision) the images on both faces at once—a bird and a cage, a horse and rider, and so on.

Early Animation Experiments. It was J. A. Plateau in 1829 who laid down the theory of persistence of vision. On the basis of this work, he invented in 1832 the Phenakistiscope—consisting of a circular card with a series of images drawn on its face, and a number of apertures around its circumference; when rotated on its axis and viewed through the apertures in a mirror, the drawings appeared to move.

Simultaneously, S. R. von Stampfer invented a very similar instrument, the Stroboscope, W. G. Horner in 1833 devised a cylindrical form, known by him as the Daedateum, popularized in later years as the Zoetrope. Miniature forms were made by Molteni and others for projection with the magic lantern, using glass image-bearing discs with rotating shutters on the same axis—the shutter rotating ten times for every rotation of the picture disc.

An important modification of this system was made by Molteni, and adapted by Beale in his Choreutoscope; instead of rotating continuously, the picture disc was moved intermittently by a form of Maltese cross mechanism, one picture for each revolution of the shutter; the efficiency of the system was thus greatly increased. This was the first use of an intermittent mechanism in the production of moving pictures.

In 1877 Emile Reynaud produced the Praxinoscope, a form of Zoetrope in which a cylindrical band of drawn images was viewed, not directly, but as a virtual image in a twelve-sided mirror drum in the centre of the cylinder. The results were of a very much higher quality than those of the conventional Zoetrope.

Reynaud patented in 1888 a projection Praxinoscope in which a transparent, flexible band with perforation holes was transported by a drum carrying pins engaging in the perforations; this formed the basis for his Théâtre Optique which gave public performances from 1892 until 1900, each film being drawn and coloured by Reynaud.

Photographic Experiments. Although photography had been an accomplished fact since the 1830's, it was not until after the invention of the wet collodion plate that instantaneous photography became possible. T. H. DuMont in 1861 and W. Donisthorpe in 1876 among others proposed means of exposing plates in rapid succession, the resulting negatives being printed for viewing in a Zoetrope.

Chronophotography—the use of instantaneous photographs taken at regular intervals for the study of movement—gave a tremendous impetus to the development of the motion picture. The first apparatus was made by Janssen, and called the astronomical revolver, being used to record the transit of Venus across the sun in 1874. Eadweard Muybridge, working in California, started a series of experiments in 1877 using a battery of cameras exposing in rapid succession. He thus obtained photographs of successive phases of horses in motion; these results could be viewed in a Zoetrope, or projected on a screen by his Zoopraxiscope (1880).

E. J. Marey, in France, constructed in 1882

the photographic gun, taking a circular plate on which 12 exposures were made successively by a rotating shutter, the plate being moved intermittently between exposures by a clock mechanism. It was used to describe the flight of birds. During the same period, A. Londe in France was working with a multiple lens apparatus and O. Anschütz in Germany used batteries of cameras similar to those of Muybridge.

Basic Improvements. Glass plates could be used to record only a few pictures at a time; the introduction of flexible films—at first paper rolls—overcame this difficulty. Marey in 1887 constructed the first practical cine camera which used rolls of light-sensitive paper. An intermittent mechanism carried the "film" through the camera, exposing it while stationary and covering it by a shutter while moving. Marey used the camera with great success in the study of animal movement.

By the end of 1887 the two principles of analysis and synthesis of movement had been thus demonstrated. It now remained for these principles to be exploited for commercial ends.

In the years between 1888 and 1893 a number of inventors in different countries tried to apply these principles, among them Le Prince (1888, Friese Greene and Evans (1889) and Donisthorpe (1889) in this country. None of them was able to solve successfully the mechanical problems involved in taking and projecting sequences of images. The first commercially successful apparatus was introduced in America by Thomas Edison, whose laboratory had developed, by the end of 1889, a working film camera which he called the Kinetograph. The camera had been made possible by the introduction of flexible celluloid films produced by George Eastman, and intended for use in the earliest box cameras. Edison cut these films down to 35 mm. width, and perforated the film with four pairs of rectangular perforations to each frame—essentially the same arrangement as in modern practice. Films taken with this apparatus were first shown to the public in 1893, in the form of coin-operated peepshow machines called Kinetoscopes. Edison had decided against projection, as a less lucrative form of presentation.

Public projection of movie films was first demonstrated by Louis and Auguste Lumiere, who, stimulated by the Kinetoscope, built the Cinematographe in 1895—a combined camera, printer and projector for 35 mm. film. The apparatus used a claw mechanism to move the film, and was light and portable. With it, the Lumieres gave the first of many public shows of projected moving pictures at the Grand Café in Paris, on December 28th, 1895.

Simultaneously with but independently from the Lumieres, Birt Acres and R. W. Paul in Great Britain, W. Latham, T. Armat and C. Jenkins in America, M. Skladanowsky in Germany and G. Demeny in France introduced workable apparatus, and by the end of 1896 cinematography was a recognized and profitable business.

As the industry grew, standardization became more and more essential; in 1909 an international conference decided to adopt the Edison perforation and film width as standard. It has remained so with minor modifications until the present day. The Maltese cross mechanism, operating intermittently a sprocket wheel below the gate, became employed almost universally in projector design, while a claw mechanism was used in cameras.

Detailed Improvements. There has been relatively little change in the mechanical principles of camera and projector design in recent years; in details the apparatus has been immensely improved, however. An important advance in camera design was the introduction of the registration pins which, inserted into perforations while the film is being exposed, result in a much more steady picture. The first British patent to cover the use of this device was issued to T. H. Blair in 1896.

Proceeding concurrently with the development of cameras and projectors has been the invention and elaboration of the ancillary apparatus: printers, processing machines, perforators, film stock—all of which, by their many improvements, have added to the high technical quality of modern film production.

Narrow-gauge Cinematography. The 35 mm. film width was adopted as a standard very early in the development of cinematography, but for reasons of economy or convenience a number of film widths narrower than this have been employed—principally for amateur use. One of the first to use narrow gauge film was the British pioneer, Birt Acres, who used in his Birtac camera 17·5 mm. film formed by splitting standard stock down the middle.

One of the earliest amateur cine cameras was the Biokam, 1899, which used 17·5 mm. stock with one perforation between each frame. Pathé introduced in 1912 the Pathé-KOK projector, hand operated with a built-in generator for illumination; this took 28 mm. film on a safety base, the films being reduction prints from Pathé 35 mm. productions.

Amateur cinematography can be said to have begun, however, in 1923: the Eastman Kodak Company produced the first 16 mm. reversal safety film, and Victor and Bell Howell introduced suitable cameras and projectors. In the same year Pathé introduced the 9·5 mm. gauge. Eastman Kodak introduced the 8 mm. film size in 1932; since then this gauge has overtaken 16 mm. and 9·5 mm. in popularity. Colour films have been available for some time in all three sizes; with the introduction of magnetic striping methods, 8 mm. and 16 mm. films can be easily provided with sound tracks.

Equipment for the production of stereoscopic and wide-screen films, including anamorphic lens systems, is available for amateur use.

Originally intended for amateur use, 16 mm. and 8 mm. films are used to an increasing extent by professional film producers for publicity, educational and propaganda films, while some films have been enlarged to 35 mm. for commercial distribution with great success.

SOUND AND CINEMATOGRAPHY

Sound has been associated with the motion picture from its beginning. Edison's aim was to produce motion pictures to accompany his phonograph, patented in 1877; it is claimed that the first successful experiment in his laboratories was made with a synchronized phonograph record. Many early pioneer workers included in their patents provision for use of phonographs—notably Demeny who, with his apparatus, patented in 1892, used to photograph and project facial movements as an aid to teaching the deaf.

Various methods for electrically or mechanically synchronizing phonograph records to motion pictures were patented in the years up to 1900; at the turn of the century, a number of short films of music-hall turns accompanied by phonograph records had been made—with some success. In 1911, Gaumont produced the Chronophone apparatus using synchronized discs, amplification of the records being carried out mechanically. This system was operated with great success for a number of years at the enormous Gaumont-Palace cinema in Paris.

Recording on Film. Eugene-Augustin Lauste was the first person to achieve any success with sound-on-film recording. A former associate of Edison, from 1900 he became interested in the problems of recording sound, and by 1904 had succeeded in recording sound and visual images on the same film. For the years preceding the first world war he worked at, and patented, many forms of light modulator, his success being limited only by the lack of means for amplifying the signal obtained from his film records. In principle, however, his work anticipated modern methods of optical sound recording.

The invention of the triode valve, with the electronic amplification subsequently possible, gave work on recording sound a fresh impetus. The firm Tri-Ergon, formed in Germany in 1919, produced a system employing a glow lamp which converted electrical variations from the microphone into brightness variations. In spite of an effective demonstration in 1922 in Berlin, the system was not adopted by any company.

At the same time Lee de Forest in the United States was carrying out experiments with a modulator lamp of his own design, and in collaboration with T. W. Case—who had developed a bismuth sulphide photoelectric cell—he demonstrated his Phonofilm sound system in 1923 at the Rivoli Theatre, New York. T. W. Case also devised a modulator lamp which he called the Aeo-Light; Fox acquired the right to his process in 1926 and created the Fox-Movietone system. C. Wente at the Bell Telephone Laboratories in 1922 invented an electromagnetic light valve—this became the basis of the variable density system exploited by Western Electric, who had also created a process using sound on disc known as Vitaphone.

General Adoption of Sound. In spite of these advances, no major film producer entered the field of sound films; only Warner Brothers in 1926 adopted the Vitaphone system for one and two reel vaudeville shorts and comedies.

In 1927 Fox-Case and Warner Brothers began a campaign in favour of the sound film. On the 23rd January, 1927, Warner Brothers launched *The Jazz Singer* starring Al Jolson, the first talking, singing feature film—while Fox-Case presented, in May 1927, their first sound feature *Seventh Heaven.*

A year later, R.C.A., with Westinghouse and the General Electric Company, announced their system of variable area sound recording. This was based on the Pallophotophone System, devised by C. A. Hoxie in 1920 and using the methods first demonstrated by Lauste in 1910.

Gaumont in France, in conjunction with the Danish engineers Petersen and Poulsen, adopted the G.P.P. system; this employed a separate 35 mm. sound record. They produced by this process *L'Eau de Nil* in 1928. The system was soon dropped owing to the difficulty of dealing with two film strips.

In Germany, the Tobis-Klangfilm organization was formed. There followed a severe battle with Western Electric and R.C.A. over Tri-Ergon patents, but the dispute was settled amicably in 1930.

The variable area and variable density methods soon became universally accepted; they were applied also to 16 mm., 9·5 mm. and to the now obsolete Pathé Rural 17·5 mm. substandard films. Improvements in sound quality came rapidly, in particular with the development by R.C.A. of push-pull variable area recording.

Recent Developments. A major innovation was introduced in 1941 with the stereophonic system employed for certain versions of Walt Disney's *Fantasia.* A separate 35 mm. sound recording film, run in synchronization with the visual record, carried a number of sound tracks each designed to operate individual loudspeakers in the auditorium. With the development of magnetic recording techniques, and the introduction of wide screen projection, the use of stereophonic magnetic sound tracks has now become a common practice.

The introduction of a system for applying a stripe of magnetic recording material to the edges of 35 mm. or substandard prints has resulted in an improvement in versatility and quality of sound recording. In the case of substandard films in particular, this has made it possible to record synchronized commentaries and music with less cost and difficulty than in the case of optical sound.

COLOUR FILMS

The earliest colour motion pictures were made by hand-colouring each frame. This technique was usually carried out in small factories employing a number of girls, each of whom applied one colour. The method was used, often with great effect, from the earliest days of cinematography.

As the length of films and the number of prints increased, this became impractical, and Pathé devised a method of applying dyes through stencils made from positive prints; up to six stencils were used, one stencil for each colour required. Each positive stencil was run through a printing machine, in register with the positive film and the colours were applied through the stencil by means of rollers or brushes. This method, known as Pathecolour, was in use for many years and gave, for most subjects, very pleasing results. The idea, although elegant, was costly for long films, so simpler techniques were employed to give a coloured image—the principal methods being chemical toning and dye tinting, and combinations of the two methods.

Additive Processes. The above processes were, of course, non-photographic, and it was not until G. A. Smith, in 1906, filed his patent for a two-colour additive process that colour cinematography became a practical commercial proposition. Commercialized in 1910 as Kinemacolor, the process employed a camera taking successive frames through rotating red and blue-green filters at 32 frames per second (twice the normal speed); a projector with a rotating filter disc similar in principle to the camera was used for showing the film. The results were, for many subjects, very pleasing, though they suffered from colour fringing in the projected images.

Subsequently, many processes were marketed employing successive or simultaneous admixtures of two or three colours. Examples of the more successful additive processes were the Keller-Dorian-Berthon lenticular process—marketed in 16 mm. by the Eastman Kodak Company as Kodacolor from 1928 to 1935—and Dufaycolor—a three-colour mosaic screen process introduced in 1934 in 35 mm., 16 mm. reversal film, and later as a 35 mm. negative-positive process.

Subtractive Processes. The most successful colour cine processes have been those based on the principle of subtractive colour reproduction. Among the earliest examples were Prizma Color, devised by Van Doren Kelley in 1919. From this system, a two-colour method using a double-coated film chemically toned in two subtractive colours, was derived a host of later processes.

In 1915 the Technicolor Motion Picture Organization was formed, and first produced an additive process. Then a two-colour process was devised using a beam-splitter camera and dyed relief positives, cemented back to back; this system was operated from 1922 to 1928, when the process was then changed to two-colour imbibition printing. In 1933 the first three-colour Technicolor film, Walt Disney's *Flowers and Trees*, was produced, and in 1935 the first Technicolor feature film, *Becky Sharp*, was made. The process has since then maintained immense popularity.

1935 saw the introduction of the first successful three-colour monopack reversal colour process—Kodachrome—in 16 mm. and 8 mm. gauges. Other monopack reversal processes have also appeared since then. As a result of research during and after the war a number of monopack negative-positive colour processes are now in current use.

SPECIAL TECHNIQUES

Although the fundamental pattern of cinematography has developed along straightforward lines, numerous ideas have been evolved to produce special effects or techniques as a departure from normal methods. Some of these ideas are as old as cinematography itself; others have been very short-lived.

The Animated Film. The origins of the animated film can be found in the attempts to make silhouettes and painted figures move in the lever and gear slides of the early magic lantern; also in the series of figures drawn for animation by the Phenakistiscope, Zoetrope and Praxinoscope. Emile Reynaud in the 1880's painted many hundreds of images, beautifully executed, on the strips of film used in his Théâtre Optique. The animated films produced by him before the introduction of cinematography were even accompanied by specially written music and synchronized sound effects.

The first successful cartoon films were made by Emil Cohl, working with the Gaumont Company in 1907; other pioneers at this time were J. Stuart-Blackton and Windsor McCay in the United States. An important step forward was made with J. R. Bray's 1914 American patent for the use of transparent sheets enabling figures to be easily animated on a static background. Earl Hurd, in 1915, also patented the uses of "cells" for animation, describing the methods which have become standard for cartoon making. Bray's elaboration of Hurd's patents led to the formation of the Bray-Hurd Company in 1917. In this same year Max Fleischer, the creator of "Popeye", made his first cartoons for Paramount, and "Felix" the forerunner of many cartoon cats, was conceived by Pat Sullivan. Walt Disney made his first cartoon, *Laugh-O-Grams,* in 1921; the first *Silly Symphony* was made in 1929, followed by his first colour cartoon in 1932.

Other forms of animated films have been the silhouette films of Lotte Reiniger, the puppet films of George Pal, Ptushko and Starevich, and the abstract films of Len Lye and Norman McLaren. McLaren, working with the National Film Board of Canada, made many films by drawing directly on to the 35 mm. film strip with dyes and inks; he also carried out experiments in the hand-drawing of sound tracks, and in three-dimensional abstract cartoons.

The enormous production expenses of cartoon making curtails the serious use of the animated film; production is confined mainly to comedy shorts or sponsored cartoons.

Wide Films and Wide Screens. Among the earliest cine apparatus was the 1896 Chronophotographe apparatus of Georges Demeny. The width of film employed in this camera was 60 mm., with an image 35 × 45 mm.—four times that of the 35 mm. film. This wide film, although uneconomical, was capable of giving very large pictures of high quality, and was the first of a series of apparatus intended for wide screen projection. Louis Lumière devised for an exhibition in 1900 a 75 mm. camera and projector showing pictures 65 feet wide to an audience of 25,000. R. Grimoin Samson patented in 1897 his Cineorama apparatus, which by means of 10 projectors would project a panoramic picture covering 360° on the wall of a round building. This apparatus, never used because of insufficient cooling of the small projection room containing ten 40 amp arc lamps, was the forerunner of a number of panoramic projection processes.

The first satisfactory method was devised by Abel Gance in 1927. This system employed three projectors, running from the same motor, which projected on to a screen of triple width either a triptych of three different pictures or one large panoramic scene taken with three cameras suitably arranged. Gance made a film, *Bonaparte,* using this system; it had some success, but it was not followed by further productions.

An attempt was made by Paramount and Fox in 1930 to introduce films 56 mm. and 70 mm. in width; also in the same year projection lenses of variable focal length were proposed to allow certain passages of films to be shown on a wide screen.

Wide screen systems were revived again after the war; Cinerama, a process using triple cameras and projectors to present a combined picture of great width and height was first introduced in 1952. The high cost of installation of projection equipment restricts the process to relatively few theatres; a variety of simpler methods have been introduced to overcome this difficulty.

One method employs anamorphic lenses which record a wider angle of view than normal by laterally compressing the image on the film. A similar lens fitted to the projector expands the image to its original proportions on the screen. With such lenses a picture of aspect ratio (the ratio of height to width) of 1 : 2·66 can be produced, compared with 1 : 1·33 in the conventional film. The anamorphic lens was invented by Henri Chretien in 1927, and publicly demonstrated in 1937. It was first commercially exploited in 1953 as CinemaScope. A simpler system of widescreen projection uses standard films, masked in the projector gate to ratios of up to 1 : 2, projected with short focal length lenses on to wider screens.

These wide-screen processes brought with them two disadvantages: loss of quality due to over-enlargement of the images on the film, and lack of compatability between projection apparatus for different systems. The first process effectively to overcome both disadvantages was Technirama, in which a "double-frame" horizontally arranged 35 mm. negative film can be printed to produce "squeezed" anamorphic prints or normal prints in any film size from 16 mm. to 70 mm. The use of a negative image twice the normal area gives a great improvement in clarity, a principle which has also been exploited in the use of films wider than 35 mm. The first process of this kind—Todd-AO is typical; employing a camera negative 65 mm. wide printed on to a print film 70 mm. wide, carrying six magnetic stereophonic sound tracks, it was introduced in 1955.

In recent years systems have been demonstrated in America and Russia in which a large number of 16 mm. or 35 mm. projectors are used to project a 360° panaromic picture on the walls of a cylindrical or hemispherical room—a return to the principle first demonstrated in 1897 by Grimoin Samson!

Stereoscopic Films. As in the case of colour and sound, the problem of presenting motion pictures in relief has occupied inventors from the earliest days of cinematography.

In principle, most processes of stereo cinematography have made use of a pair of stereoscopic images obtained either by a twins lens camera, two cameras, or a single lens camera with suitable beam-splitting devices. These stereo images must be viewed by methods which allow each eye to see only the appropriate image of each pair.

The earliest projection system, the anaglyph, was proposed by A. d'Almeida in 1858 and first demonstrated for motion pictures by C. Grivolas in 1897. It employed a pair of stereo images of complementary colours obtained by dyeing or projection through filters. The coloured projected images, usually red and blue-green, were viewed by filters of the appropriate colours usually mounted in the form of spectacles. Processes based on this system were in use in 1925 and 1935; they were introduced once more in 1950 but were quickly superseded by methods involving polarized light, as the anaglyph method is unsuitable for colour films.

Although many attempts had been made to employ polarizing materials in stereo projection, it was not until 1932—when a cheap form of polarizer was introduced—that it became commercially practicable.

The first public exhibition of stereoscopic motion pictures by polarized light methods took place at the New York World's Fair in 1939. The TeleKinema at the Festival of Britain in 1951 demonstrated colour stereoscopic pictures with stereophonic sound; in the following year production began in Hollywood on a number of stereoscopic feature films.

The polarizing system enjoyed a brief period of popularity; the difficulties of synchronization and registration of both images, the necessity to use two projectors simultaneously, as well as the need for the spectator to wear analysing spectacles, led to its abandonment in favour of the wide screen techniques.

It is interesting to note a British patent was issued to G. R. Wilson in February 1898 in which the principles of stereo cinematography by both the anaglyph and polarizing methods were outlined; apparatus was described, similar in principle to that of recent years, involving the use of synchronized cameras and projectors.

Another method, known as the parallax stereogram, has also been used for viewing stereoscopic motion pictures. The system relies on a screen consisting of a large number of vertical elements—wires, slats or embossed lenticles—in the form of a grid, by means of which each eye sees only the image appropriate to it. This method obviates the need for individual analysing viewers. The process, originally suggested by A. Berthier in 1896 and later developed by E. Estanave and F. E. and H. E. Ives, was used as the basis for a workable system by the Russian S. Ivanov in 1941.

Another version of this system was demonstrated by A. Matley of Paris in 1949; called the Cyclostereoscope, stereo pairs are projected on a screen, round which revolves a grid in the form of a truncated cone: as the grid is rotating rapidly, no vertical line structure is visible.

In recent years, methods for the production of stereo pairs on one film instead of the customary two have been devised. Applied to 16 mm. as well as 35 mm. films, these processes usually employ some form of beam-splitting device to produce a pair of stereo images side by side within a standard frame area.

Another development has been the production of stereoscopic projection prints employing Vectograph film. In this, two self-polarized images are actually combined on the one film.

The production of stereoscopic entertainment films has, for the time being, ceased, owing principally to the difficulties of presentation. When these problems have been resolved, we are likely to see further exploitation of the stereoscopic motion picture.

Recent Trends. In recent years, following the development of television, falling cinema audience numbers and increased costs caused a big reduction in the number of cinemas and film production companies especially in the United States. Faced with this competition, the film industry responded with an attempt to recapture lost audiences with spectacular entertainment—colour, "three-dimensional" pictures, stereophonic sound and wide screens of various kinds.

It is likely that further developments in spectacular presentation will be introduced to try and reduce still further the drift from the cinema. With this loss of interest in the cinema, however, has come a rapid increase in the non-professional applications of cinematography.

The use of colour and sound by the amateur has increased considerably. Magnetic and even optical sound projectors are available for 8 mm. film, which, with improvements in film stock and camera design, has recently reached standards of quality formerly associated with larger sizes. This has also been helped by the larger image area of Super-8 film introduced in 1965. For this reason, 8 mm. film is being increasingly employed in science, industry and education, and is likely to find many new applications in the future. B.W.C.

See also: *Anamorphotic Lens; Chronophotography; Museums and collections.*

Book: *The March of Photography*, by E. Stenger.

CINE LABORATORIES. Numerous firms specialize in the processing of cine film, operating a similar service to the popular D. & P. service in still photography.

Although some of these laboratories undertake reversal processing, this is normally done by the film manufacturers and is covered in the price paid for the film. Most laboratories deal only in negative processing and positive printing; a few specialize in colour processing and printing.

Laboratories seldom handle all gauges of cine film, most of them accepting only 35 mm. or 16 mm. film. The processing of other gauges is invariably catered for by the manufacturers of the film.

In addition to normal processing and printing, many laboratories can provide optical effects (e.g., dissolves, fades) if required. With some firms, this service extends to providing 35 mm. reduction prints to 16 mm. and other similar facilities.

See also: *Cine film processing; Wholesale photo finishing.*

CINEMA STILLS. Still photographs representing scenes or actors from cine films. These photographs, which are taken for display outside theatres and for press publicity, play a vital part in the economics of film business as a means of creating public interest. A film is an expensive product which has to be sold intensively, and photographs are the basis of the selling campaign.

The still can also assist directly in keeping down production costs, for many firms will lend expensive goods to studios in return for a still which can be used in their own advertising. For example, a manufacturer might lend a luxury model car free in return for a still of a star posing with the car. Stills also have a more permanent value as the guide of the film historian and the inspiration of serious students.

Stills are not enlargements from films. The motion picture camera is constructed to interpret motion on the screen, but its shutter speed is not fast enough to arrest motion. Examination of a long strip of cine film will show that only a few of the pictures have a sharp image. The sharp frames cannot be predetermined, and may turn out to be the least striking single pictures. So, after shooting the scene, the stills are taken separately to provide sharp photographs of well-composed dramatic incidents suitable for reproduction in the press.

Choosing the Shot. The art of the still cameraman lies in selecting the most telling moments in the film story. Because he has to concentrate on this problem, he uses simple standard equipment and employs an automatic technique. The usual camera is an 8 × 10 ins. model with swing back and front movements, which takes a negative the same size as the standard still print. The negative can therefore be easily retouched and used for bulk printing by contact. Quality and speed are all-important in turning out the large quantities of stills required. If the still man used a miniature, he could snap during the cine take and arrest action and save time; but there would be no guarantee that the action would be stopped at the right moment. An instant after the heroine is looking bewitching, she may speak a word which changes her expression disastrously.

The standard technique is modified for violent-action stills—fights, motor smashes, etc.—when dramatic impact is not spoilt if the moment caught is awkward. In this case the still man may shoot with a press-type camera during the actual scene. Outdoor shots, too, are generally action pictures—horsemen riding across moors, yachts racing, etc.—and the still man can make considerable use of his press camera on location.

Stills of an exceptional nature, that are not wanted for bulk printing, are taken with a smaller camera for the advantage of shorter exposures and increased depth of field. An example is the kind of publicity picture taken to show what goes on behind the scenes, which appeals to a limited number of editors.

Lenses. With the standard camera it is customary to use a lens of 12 ins. focal length, as the majority of film stills are medium close-ups. A long shot will not as a rule give the drama, the expression on the actors' faces, or the point of a joke. Long-shot stills are only taken to draw attention to lavish spectacles covering a lot of ground.

Lighting in the studio is controlled by the cine cameraman; but if he is working in low key, the still man, with his slower lens, may require additional light—e.g., he may want an extra flood wheeled to the side of his camera or ask for the diffusers to be removed from lamps. He never likes to increase exposure beyond 1/5 second or the actors might move. But with lighting dictated by someone else, there is not much chance for him to express his individual talent. His opportunity lies in being able to regroup the actors to give the essence of the situation with emphasis but not exaggeration.

Glamour stills of the stars are deemed so important (for the press and for sending to fans) that the still man is given his own studio in which he can pose, light and take close-ups. Here, where mobility no longer matters, he may use a lens of 20 ins. focal length to ensure that the sitter's features are portrayed without distortion. The studio is also used for taking fashion stills of the stars' dresses for press publicity.

Uses for Stills. Apart from his importance in film business, the still cameraman also has some part in film creation. He may take: research stills—e.g., of a costume in a museum for the wardrobe department to copy; location-finding stills to illustrate the photographic possibilities of a district; set stills—i.e., record photographs for use in case sets have to be rebuilt for retakes and continuity stills to record the make-up of an actor so that continuity of appearance can be maintained in scenes shot at a later date.

Other types of picture that the still photographer may be asked to take include the title still, a picture double-exposed behind the credit titles to give them pictorial appeal; the set-dressing still—e.g., a photograph of the star of the film to hang on the wall of a set; an insert still, which may be a photograph of an actor as a wanted criminal printed in a fake newspaper for a story point; or a location background still—a landscape turned into a lantern slide and projected on a translucent matt screen as a background in the Studio.

The increasing popularity of certain three-dimensional films suggest that further developments will occur in the technique of presenting films in depth. At present 3D films are advertised by conventional stills; but tomorrow may bring viewing boxes for stereoscopic samples of exciting scenes installed in the foyers. Yet another category will be added to the list of pictures which the still cameraman has to take. O.B.

CINEMATOGRAPHY

Cinematography creates the illusion of movement by presenting to the eye a rapid succession of motionless pictures. The illusion depends on a characteristic of the human eye called persistence of vision: when the eye is looking at an object which suddenly disappears, persistence of vision causes the image to linger in the brain for a fraction of a second after the object has gone. To make use of this phenomenon in the showing of a succession of still pictures, each view in the series must appear before the effect of the previous one has died away: the brain is then unable to distinguish the gaps between the separate pictures and an apparently continuous view is seen. If the pictures are progressive and occur at least 16 times per second the series is interpreted as a single picture embodying motion.

But, although persistence of vision causes each image to remain upon the retina for a short time after the stimulus is removed, decay is so rapid that at 16 pictures per second the brain would be conscious of flicker.

To obtain the impression of an evenly lit continuous picture each projected stationary image is further interrupted by means of a shutter so that the retina receives at least 48 separate impressions per second, replacement occurring more rapidly than the period of decay.

Cinematography includes both the making of the pictures and their projection on to a screen. The pictures are taken by a camera equipped to expose a succession of individual pictures or frames on a long strip of flexible film coated with a light-sensitive emulsion. By suitable photographic treatment the film is made to yield a corresponding succession of positive images which, by means of a projector, are thrown in rapid sequence on to a viewing screen.

Mechanics. There are several points of similarity between the camera and the projector. Both are equipped with mechanism that will move the strip of film in jumps of one frame at a time, allowing each frame to remain stationary in the gate behind the lens during the period of exposure or projection and then moving it rapidly into the next position. In both, the lens is covered by a rotating or reciprocating shutter during the fraction of a second that the film is being moved.

The film transport methods have much in common. Above the gate the unused film is stored on a spool or on a rotating centre or core from where it is unrolled smoothly at a steady speed and fed to the gate. There the required intermittent movements are imparted to the film by a mechanical device close to the gate. The basic difference between the intermittent mechanism of camera and projector is that in the camera it must be light and balanced to prevent vibration, while in the projector it needs to be robust to withstand heavy duty use. Below the gate the film is taken away and wound smoothly on to a spool or core on which it is stored.

Except in the simplest equipment the film is fed to and from the gate by sprocket wheels. To prevent damage to the film when it changes from smooth progression over the sprockets to intermittent movement through the gate a loop of film is left above and below the gate.

Films. To enable the rolls of film to be fed efficiently through camera and projector the manufacturers punch specially shaped holes in the film, to come beside or between the pictures, and these perforations engage with the film transport mechanisms. Any errors in the dimensions or positioning of these holes would prevent satisfactory registration and make the projected picture unsteady, so the perforations are made to high standards of accuracy and arranged according to the specifications laid down for the size of film.

Excluding certain non-standard gauges used for special purposes (e.g. 70 mm. film for certain wide-screen techniques) cinematograph films are made in four gauges.

35 mm. (roughly $1\frac{3}{8}$ ins.) is the professional gauge as used in the cinema and is known as the standard size. It is perforated along each edge

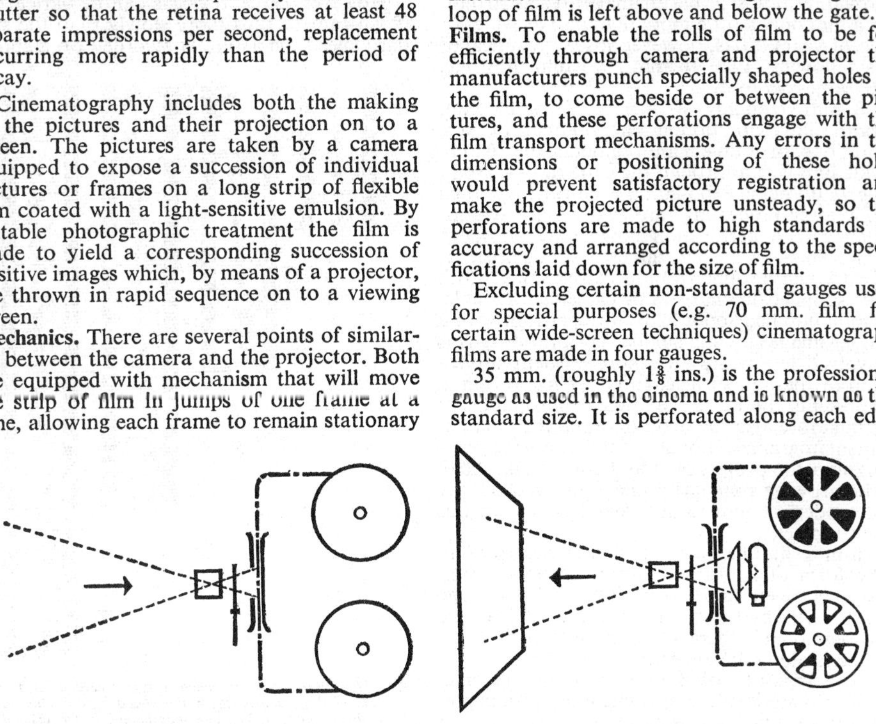

CINE CAMERA AND PROJECTOR. Left: *Cine camera with a continuous strip of film being pulled past the lens at an intermittent rate by a suitable mechanism engaging the perforation holes of the film. A shutter covers the lens while the film is actually in motion.* Right: *Projector, using the same intermittent movement principle on the processed film.*

in both silent and sound versions. Film intended for certain wide screen processes differs in some respects. 35 mm. film has 16 standard frames per foot and 4 perforations per frame along each side. The actual frame size depends on whether the film is silent or sound; sound film carries the sound track between the picture and one of the rows of perforations, so the frame size is reduced.

16 mm. (roughly ⅝ in.), the widest of the so-called narrow gauge films, has one perforation each side per frame for silent use. The sound version is different because the sound track makes use of the space that would be occupied by one row of perforations. This sound track may be optical or magnetic. Double perforated 16 mm. film with magnetic sound tracks along the outside edge of the perforations also exists. The normal 16 mm. film has 40 frames to the foot and is widely used for documentary film making, educational films and also a certain amount of amateur filming.

9·5 mm. (roughly ⅜ in.) is perforated in the centre between the frames, so there is only one per frame. Both sound and silent types are the same, the sound track being recorded down one edge of the normal film. At one time 9·5 mm. film used to be a very popular amateur gauge, providing almost as big an image area as 16 mm. in a film only a little more than half as wide. In recent years it has, however, lost a great deal of its following. Like 16 mm., 9·5 mm. film has 40 frames per foot.

8 mm. (roughly $\frac{5}{16}$ in.) is really 16 mm. width film with double the number of perforations. In double-8 mm. cameras, the 16 mm. wide film is run through the camera twice, half the width being exposed on each run; it is then split after processing into two 8 mm. widths, each with a perforation now along one edge only. This film has 80 frames per foot of single 8 mm. width. Super-8 mm. of the same width but supplied in chargers or cassettes of single-8 mm. film has about 72 frames per foot with a larger image size and smaller perforations. This involved less waste and more efficient utilization of the film surface. In either form, 8 mm. represents the most economical film gauge. It is therefore also the main material for amateur cinematography, though its use is also spreading into the professional and educational field. The growing popularity and improved quality of modern 8 mm. sound film has helped this development.

8 mm. film can also carry a sound track in the form of a magnetic stripe along the film edge between the perforations and the edge itself. Alternatively, an optical track can be accommodated there. On Super-8 mm. film, the track is along the unperforated edge.

The majority of 8 mm. films are reversal colour materials (the overwhelming proportion of 8 mm. film used is colour). There are also some black-and-white reversal materials in this gauge. 9·5 mm. films are mainly available as black-and-white or colour reversal materials, and 16 mm. films are made both as reversal stock (colour or black-and-white) for amateur use and as negative stock (mainly black-and-white) for professional cinematography. This is because the amateur movie-maker is usually satisfied with a single copy of his film, and reversal processing is the most economical and convenient way of achieving this. Professional motion pictures on the other hand require a number of release prints for projection in various places (running into thousands in the case of 35 mm. film) so there a negative film material is more suitable. The negative-positive process also provides greater scope for a wide variety of special effects obtained during printing the release positive.

For this reason also standard 35 mm. motion picture stock is almost invariably of the negative type (black-and-white or colour) for camera use, and positive film for printing.

Picture Frequencies. Projection speeds have been standardized at 16 frames per second for silent films and 24 frames per second for sound films. It is a fundamental rule that the camera must run at the same speed as the projector if action on the screen is to take place at the same speed as it was performed.

However, camera speeds can be varied, and if the camera runs more slowly than the standard speed the action is recorded on a shorter

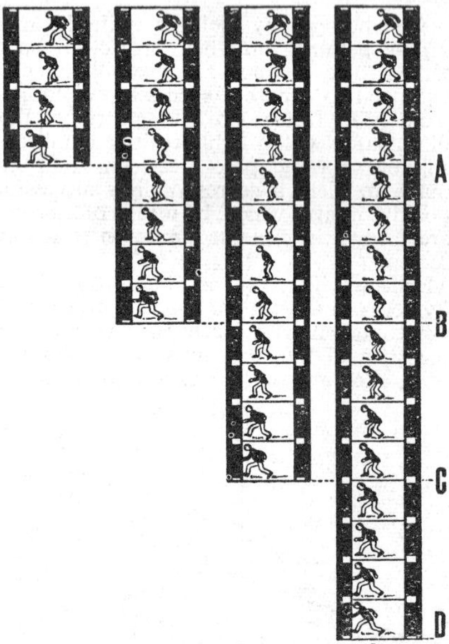

CAMERA SPEED. A given movement is recorded on different lengths of film, according to the speed of the film passing through the camera. The film lengths shown correspond to ½ second shooting time: A, fast motion of 8 f.p.s.; B, standard silent speed at 16 f.p.s.; C, standard sound speed at 24 f.p.s.; D, medium slow motion at 32 f.p.s.

length of film than normal—it takes less time to go through the projector and events on the screen happen that much faster.

Running the film more quickly through the camera spreads the action out over a longer strip of film and this will take longer to go through the projector, thus stretching the action and providing slow motion. The slow motion effect is of great value in the analysis of rapid movement, and in some cases a frequency of several thousand pictures a second is used.

At the other end of the scale, the speeding up of slow movements makes possible the production of fascinating films recording such subjects as the growth of plants and crystals. For such subjects the taking rate may be slowed down so that, instead of a few frames per second being exposed, only one per hour or more is taken; for recording astronomical changes, exposures are as few as one frame every twenty-four hours.

It is also possible to make inanimate objects appear to move by exposing individual frames and modifying the subject between exposures—thus a knife and fork can be made to dance a jig. This technique of animation extends to the making of cartoons where the subject is a long series of drawings, each drawing being slightly different from the preceding one.

The Sound Record. When characters on the screen speak it is essential that the picture and sound remain in synchronism. This means that the sound record must be carried on the positive film. But if the sound consists merely of a commentary, such close synchronism is not necessary. This distinction is important when considering methods of recording and reproducing sound.

The sound may be recorded on a track along the edge of the picture film; but far more often it is first recorded on a separate strip of material—either film or tape—which is mechanically synchronized to the picture. The separate sound record may be then later combined on to a final film with the picture.

Two methods of recording the sound are in common use: photographic (or optical) and magnetic. Today the photographic method is becoming increasingly superseded by magnetic recording because of the relative simplicity and cheapness of the latter. The magnetic method is simple because a recording can be played back immediately for checking and can be erased if unsatisfactory; it is cheap because the record can be erased and the material re-used. Photographic sound recording needs expensive equipment and is costly in use.

Cameras. All cine cameras are similar in principle in having a lens, a shutter, a gate and an intermittent film transport. Other optical and mechanical features and refinements vary according to both the film gauge employed and the purpose for which the camera is made.

Most 8 mm. cine cameras, being intended mainly for amateur use, are designed for foolproof and easy operation. Thus the exposure is usually controlled by a built-in exposure meter, modern cine cameras have a variable focus lens and one or possibly two to three running speeds. The intermittent film transport may be driven by a spring or electric motor.

The very simplest 8 mm. cameras are only equipped with a single fixed lens (usually without focusing adjustment), no exposure meter and are virtually the box cameras of movie making. At the other end of the scale, advanced 8 mm. models may have a full range of running speeds, variable shutter, interchangeable or zoom lenses, and a number of further refinements. With very few exceptions these are however still essentially amateur instruments.

The 9·5 mm. camera is also an amateur instrument, and usually on the level of the low to medium price 8 mm. model. A few semi-professional cameras in this gauge also exist.

Cine cameras for 16 mm. film are by contrast usually more versatile, being intended as much for professional as advanced amateur work. The cameras thus may have interchangeable or zoom lenses, reflex finder, variable shutter, and special refinements such as back-winding, large capacity magazines and an extensive range of accessories.

Fully professional versions are generally more robust, and may have electric motors, film capacities up to several hundred feet and provision for simultaneous sound recording.

Film Production. The making of a cine film involves a number of important stages. The actual shooting of a film is only one small part of the many operations needed to produce a complete film.

While an amateur may bring back random shots from a holiday trip and afterwards try to assemble what he has got into a coherent film, anything more ambitious needs a planning routine. This applies to all feature (story) films —amateur and professional—as well as documentary productions.

The first stage is the story outline or treatment, which is the basis for the script. The latter translates descriptions, actions, moods and impressions to be conveyed in the film into a sequence of individual shots. The script specifies as far as possible all the details of the shot, the subject, actors, etc.—and then serves as a fairly rigid shooting guide for the director as well as—later—the film editor.

Once the whole film is planned out in this way, all the sets and scenes selected, properties and other items collected and (especially in professional films) numerous other details organized, shooting itself can begin.

This is then the actual job of filming the individual scenes and actions one by one, setting up camera and lighting as required. In a professional film production the shooting is nearly always carried out by a team. This may consist of as few as two people—director and cameraman—or may include a comprehensive technical and administrative crew plus the actors and ancillary staff.

In filming the scenes, visual points and impressions may be put over not only by straightforward recording of what goes on in front of the camera, but also by a variety of camera movements, changes of angle and direction, and a wide range of special tricks.

Processing. Reversal films (black-and-white as well as colour) for amateur use are normally processed by the manufacturer, with the charges included in the price of the film. In the processing laboratory more or less automatic continuous processing machines carry the film through the various stages of development, fixing, colour development, etc. Even professional film makers get specialist laboratories to do the processing work for them, though large film studios usually have their own processing department.

Narrow-gauge amateur films can be processed by the user, though this is a somewhat laborious procedure and requires special tanks and spooling gear to handle the film lengths involved.

A reversal film yields the final image ready for projection on the film that actually went through the camera. Negative films are not projected, but serve as the intermediate stage for making a positive projection print. This invariably requires the services of a motion picture laboratory.

Editing. This is the stage of putting together the final film from the individual shots. They are not necessarily taken in the order in which they are scripted. So the editor not only has to sort out the individual shots, but also cut them to the right length and combine them to convey most effectively the story, moods and filmic effects intended by the director.

For this purpose the editor views all the film lengths on an animated viewer, to judge them technically and artistically and eventually splices them together in their final order for presentation.

In professional productions the sound is also edited at this stage and the sound tracks—usually on separate films—matched with the picture. From the edited version of both image and sound the final release print is produced. The editor also has to order the special effects required for scene transitions, printing tricks (so-called opticals since they are normally obtained by optical printing) and add the titles. These are usually filmed separately.

Projection. To get the best from a film it must be projected to give a sharp, bright and steady picture. In addition the enjoyment of the film can be considerably enhanced by various refinements in the presentation of a show. All these points are as important as the making of the film itself.

Like cameras, so projectors differ in the film gauge they accept and in the purpose for which they are made. For amateur use, 8 mm., 9·5 mm. and 16 mm. projectors are comparatively simp machines consisting of the light source an optical system to direct an intense beam of ligl through the film gate, an intermittent mov ment which transports the film frame by fram from a feed spool through the gate and on to take-up spool, and a projection lens whic projects the image on to the screen. Simp projectors run at a single or two speeds—1 frames per second for silent films and 24 fram per second for sound films. Sound projectors i addition carry the electronic circuits to pick u the signals from the optical or magnetic trac on the film and convert them into the soun reproduced through a loudspeaker.

Some 8 and 9·5 mm. projectors have pro vision for sound reproduction; nearly all 1 mm. projectors are designed for sound films.

Motion picture cinema projectors—for 1 mm. 35 mm. or even wider film gauges—a larger machines with very intense light sources t produce a sufficiently bright image at the hig degrees of magnification required for a fu scale cinema screen.

The requirements of amateurs are compara tively simpler. They may involve nothing mor than setting up the projector and arrangin seating accommodation for the audience t view the image comfortably. Projection pro visions in lecture halls are more elaborate an those in motion picture theatres are most ac vanced of all. The latter may incorporat various semi-automatic devices to let a sho almost run itself according to a pre-set pro gramme. This covers the change-over from on projector to another while showing successiv reels of a film, the control of the sound qualit and reproduction, and incidental activities suc as the switching of the illumination in th auditorium.

Special projection systems are used for film required for audio-visual presentation—fo example in the class room, training scheme etc. Specially designed projectors are also use for televising motion picture films by projectin the film into a television camera. This require appropriate synchronization between the pro jector and the framing rate of the televisio system.

Scope of Cinematography. The existence o motion picture films is nowadays taken fo granted—it is part of everyday life.

Entertainment is, of course, its main busines made up of such varied artistic offerings as th feature films, cartoons, newsreels, documen taries, and fictional films made specially fo children. The commercial world also use films for advertising; and film sequences ar made for the television services where direc television is impossible or inconvenient. Bu 35 mm. film does not have by any means monopoly of practicability.

Narrow-gauge films had been introduce solely to cater for amateurs and their persona pleasure. But 16 mm. was soon recognized a having great possibilities in serious work. It has obvious advantages in lightness of equipment and lower cost, and so it has been developed to satisfy the modern needs of many different fields.

Good colour is so easily obtained and sound is a routine facility.

Portable equipment and non-flammable film have made 16 mm. ideal for use in schools. Education films cover a range of subjects that is literally world wide. The inclusion of these visual aids at the right places in the curriculum has proved effective in teaching. The appreciation of film is so great in some schools that the children are encouraged to make their own productions which, although fictional, are considered as part of a constructive policy towards the arts.

Most 16 mm. films are nowadays made with sound—optical or magnetic—which is frequently post-synchronized in the form of commentaries, musical background, etc. The use of a magnetic sound track makes it comparatively easy to issue for instance a publicity film in a number of foreign language versions with no more trouble than the transcription of the translated commentary from a master tape to the release prints.

Improved technical quality has brought 8 mm. film into the educational and other semi-professional fields. Even with provision for sound projection, 8 mm. projectors are still more compact than 16 mm. ones and 8 mm. educational films are available, reduction printed from 16 mm. originals.

In industrial circles, not only are there the usual types of informational films for staff training but there are also research needs to be catered for—recording the progress of various processes and experiments, and investigation and fault-finding.

Science directs the eye of the camera into every possible corner. Science newsreels, for example, ably demonstrate the mixture of highly specialized activities such as the use of the electron microscope in the study of micro-organisms, with the more down-to-earth tasks of presenting, perhaps, the growth of plants as influenced by fertilizers. It is a constant urge, this study of subjects that are so often not visible to the common eye—from the amazing depths of the sea to the awe-inspiring depths of space.

One particularly important application of cinematography is in analysing rapid events.

By using high speed cinematography at speeds of well over 1,000,000 frames per second, details too fast for the eye to see can be recorded. This cine technique is widely used in science and industry and has become an important technique of its own.

In the medical world cinematography is employed for the making of training films. It is an ideal medium. Colour is of particular value here and even three-dimensional films have been attempted and found to be effective. Any surgical operations that are rare or have unusual features are photographed as a matter of course and these films are added to the valuable libraries that are being built up. By this means medical knowledge is being spread throughout the world. G.H.S. & L.A.M.

See also: *Animation; Back projection; Chronophotography; Cine camera; Cine film processing; Cine film; Cine laboratories; Cinema stills; Cine history; Cine technique; Editing; Electroplane camera; High speed cinematography; Projection practice; Projection principles; Projector (cine); Screens for projection; Sound recording; Sound synchronization; Splicing; Time lapse photography; Titling.*

Books: *How to Film*, by G. Wain; *How to Direct*, by T. Rose; *How to Use 9·5 mm. Film*, by D. M. Neale; *How to Make 8 mm. Films*, by N. Bau; *How to Use Colour*, by C. L. Thomson; *How to Choose Music*, by F. Rawlings; *How to do the Simpler Tricks*, by J. Caunter; *How to do Tricks*, by J. Caunter; *How to Cartoon*, by J. Halas and B. Privett; *How to Animate Cut-outs*, by C. H. Barton; *How to Script*, by O. Blakeston; *How to Write Film Stories*, by R. Harrison; *How to Edit*, by H. Baddeley; *How to Title*, by L. F. Minter; *How to Project*, by N. Jenkins; *How to Act*, by T. Rose and M. Benson; *How to Film Children*, by N. Natkin; *How to Make Holiday Films*, by H. Baddeley; *How to Process*, by L. Wheeler; *How to Add Sound*, by D. M. Neale; *How to Write Commentaries*, by M. Kirsch; *How to Film Indoors*, by L. F. Minter and E. J. Chard; *How to Produce Magnetic Sound*, by J. D. Chittock; *How to Plan Your Films*, by C. V. Willson; *How to Use 8 mm. at Work*, by A. Wigens; *Successful Movie Making*, by A. Wigens; *The Technique of Making Films*, by P. Monier; *The Simple Art of Making Films*, by Tony Rose; *The Technique of Documentary Film Production*, by H. Baddeley; *The Technique of Film Animation*, by J. Halas and R. Manvell; *The Technique of Film Editing*, by K. Reisz (all London).

CINEMICROGRAPHY. Cinematography through a microscope, to show the behaviour of living micro-organisms or non-living action phenomena such as crystal growth. Photographically the technique is similar to photomicrography, and the cine camera is mounted above the eyepiece of the microscope, focused on infinity.

A micro-adapter may also be used.

A camera with a through-the-lens viewfinder is useful, but observation of the specimen is usually through a beam-splitting eyepiece of the microscope.

For living micro-organisms the film is usually shot at the normal camera speed; slower or faster speeds may be necessary for other phenomena.

See also: *Photomicrography.*

CINERADIOGRAPHY. Filmed X-ray image, usually of live subjects or of moving machinery. The cine camera photographs the action on a fluorescent screen. As the light intensity of the screen is very low, large lens apertures and highly sensitive films are necessary. Alternatively the radiographic image may be intensified by being projected on the photo-cathode of an image amplifier.

CINE TECHNIQUE

By itself a length of cine film records action and movement. As part of a complete film production it also presents a story and conveys its atmosphere, mood, as well as putting over concepts and ideas in a visual form. These are realized as much by the action recorded as by the combination of different activities, scenes and images.

Making a film thus involves as much planning and preparation as shooting itself. In professional motion picture production, where the running costs of a studio, technicians and artists is very high, the planning is the most important part of production economics—besides being the key to the successful realization of a film story or idea.

The technique of making a film further involves the various aspects of photography (lighting, exposure, camera operation) and a range of effects to present movement and action in realistic or contrived form.

Scripting. Every professional motion picture film starts with an outline, generally referred to as the treatment. This treatment is written in essay or story form. It outlines the idea without worrying about the technicalities of producing it as a film. The technicalities are gone into later.

PLANNING AND PRODUCTION

The treatment, however, already visualizes the film in terms of action and images. In the case of a feature film it may have to convert the literary style and expression of an original story into a pictorial presentation.

Once the treatment has been approved the script is written. This means that the subject matter is painstakingly divided up into scenes, preferably by someone who has sufficient technical knowledge to write only what is filmable: this is especially necessary in the case of documentary and other non-fiction films made on limited budgets. The scenes are numbered consecutively and against each item is placed practical information such as location, camera position and movements, and the actors required (with their dialogue, if any).

The script analyzes every activity or sequence of activities into their component parts or shots. These shots are the actual lengths of film exposed in a continuous run. A shot may last from a few seconds to a few minutes; it is very rarely longer.

The script is also a working plan for the film editor when he puts the final film together from the individual shots. The editor may break down scenes still further by intercutting shots. For this purpose he may want alternative versions of the same scene, for example on sets involving big crowds. Such scenes may be photographed by several cameras from different viewpoints to provide the greatest scope for combination.

When the professional script is complete it will be broken down. That is, various detailed lists will be worked out from it giving such information as the number of artistes required and the scenes in which they will appear, the number of scenes to be shot in each set or location, proposed shooting dates, properties needed. There will be designs and plans for constructing and dressing the various sets. These lists and plans enable all the many specialized departments to get to work on the multifarious aspects of preparation.

The script breakdown also enables the shooting of scenes on the same or similar locations to be combined. The scenes in a film are practically never shot in their scripted order. It is rarely convenient to do so, and once a studio set is built up, it would be uneconomical to use it for one scene near the beginning of the film and leave it set up until needed again for another scene near the end. So all the shots requiring the same set are shot as one group irrespective of their sequence in the film. It is the editor's job to sort them out later.

The planning for an amateur film is usually much less detailed and comprehensive. Scripting is, however, still a valuable aid and discipline for the film maker. It helps him to shoot with a purpose and may mean a considerable saving of film which might otherwise be shot wild and eventually discarded as useless.

Preparing to Film. Much of the technique of motion picture photography is the same for both amateur and professional filming. The camera has to be adjusted for exposure, focus, speed (which does not often change from the normal) and viewfinder parallax (where such a refinement is included). The camera will be set up and the most suitable lens chosen to match the subject: if there is only one lens, as with most amateur cameras, then the subject-to-camera distance has to be altered to give the desired image size. Indoors in restricted spaces a wide-angle lens is a great asset for covering large views.

The camera is mostly used on its tripod; and even when the set-up is static a pan-tilt head will normally be used, making the levelling and centring adjustments much easier than if the tripod legs had to be shifted about instead. The pan-tilt head is essential when the scene calls for a panning movement, and setting up will then involve rehearsing the front and end set-ups of the pan. Some pan-tilt heads have a levelling adjustment so that the plane of panning can be set without moving the tripod legs.

The routine of shooting is simpler with silent films than with sound, and with narrow gauge equipment compared with the much heavier 35 mm. For making sound films the camera and the associated sound recorder may be driven by mutually synchronized motors, and some time must be allowed for the camera to run up to full speed before each shot (or "take"). For making silent films or where the sound is post-recorded, the take can be made immediately the camera has started.

There have to be enough technicians to carry out all the necessary duties involved. In professional circles there are many—a case of at least one man to each job. In newsreel and amateur crews each man generally takes on more than one set of duties. An amateur lone-worker will do everything himself, but this he can often manage because his productions are naturally less ambitious. Cine clubs are usually more concerned with keeping everyone occupied and cheerful, so they make the most of the personnel they have.

The director decides on the set-up and the camera-operator arranges the camera; the lighting cameraman deals with the lighting of the scene.

Shot Types. While the camera is running it may itself either be fixed in one spot (for a static shot) or move around for a travelling shot. In either case it may take in a big view or a small one.

According to the amount of the subject area taken in, shots are classified as long shots, medium shots and close-ups, and are described as such in the script.

A long shot takes in the full setting of a scene and may for instance introduce a sequence involving nearer shots in the same setting.

A medium shot picks out a more or less self-contained section of a setting where an action or part of the action takes place.

A close-up shows detail, for example the head and shoulders of an artiste.

There are also big close-ups—for example when just the face or even only the eyes of an actor appear in the picture—and there are occasionally other intermediates such as medium close-ups, etc.

The terms long shot, medium shot and so on tend to be relative to the sequence or group of shots to which they belong. Thus in a travel sequence a long shot might show a whole valley, while an individual house already becomes a medium shot. On the other hand in an indoor sequence a scene taking in a whole room might still be a long shot and a full-length view of a single person sitting in a chair a medium shot.

Travelling shots may be broadly grouped according to whether the camera changes its direction or its viewpoint. In the first group are the panning movement where the camera swings from side to side, and the rather rarer tilt where the camera swings up or down. Either type of swing must be completely smooth (that is why a tripod with a pan-and-tilt head is almost indispensable). If the scene has no other action in it, a pan or a tilt can be filmed at an increased camera speed (for example 32 or 48 frames per second) and carried out more quickly. On projection at a normal camera speed this makes the pan appear more smooth. At the end of a pan the camera should always hold the scene for a second or two before cutting.

If the camera travels as a whole, the movement must still be smooth. In the studio this is assured by a special rolling camera stand or dolly. A hand-held travelling shot, especially with the cameraman walking, is always too unsteady to be satisfactory—except for special effects. For certain types of travelling shot the camera may be mounted on a car, for example when following another car (or running in front of it). Here the slight unsteadiness is less serious, especially when the shot is intended to give the impression of camera movement.

The tracking shot provides a good transition from a medium shot to a close-up (or vice versa). Here the camera moves directly towards or away from the subject. During such a shot the focus setting needs constant adjustment, a job done in a professional camera team by an assistant known as the focus-puller.

With the growing use of zoom lenses, a zooming shot (from which the zoom lens derives its name) may replace the tracking shot. There the camera remains fixed and only the focal length of the lens is continuously changed to give the impression of going closer to the subject.

There is a difference in the effect between a tracking shot and a zooming shot. As the camera changes its viewpoint in a tracking shot, the perspective of the scene also changes: while the subject becomes larger in the picture frame, it becomes also more prominent against its background. In the zooming shot from a fixed viewpoint the perspective remains constant: though the subject grows larger in the picture frame the background details grow with it. In many cases this difference does not matter. Where, however, the camera is intended to move through a scene of pronounced depth, only a tracking shot should be used.

Lighting. The lighting of the subject has to satisfy two main requirements—technical and artistic. Dealing here with the technical, there must be: enough light to give correct exposure; and the difference in brilliance between the highlights and shadows of the subject must be narrowed down until it lies within the available contrast range of the film. Colour film has a more restricted contrast range and control is correspondingly more difficult. In daylight this control is attained by reflecting daylight or shining artificial light into the shadow areas and shading down the highlights. In artificial light the same effect is obtained by adjusting the number, power and position of the lamp units, or by the use of shutters on the lamps.

Professional studios tend to favour arc lamps, tungsten lamps, and compact source lamps, built into lens- or reflector-spotlights for giving hard controlled beams; also large open-faced units to give masses of soft light.

Narrow-gauge workers more often use Photofloods, giving a very white and actinic light but with a greatly curtailed burning life. These are mostly used singly or sometimes in batches of two, three or four, mounted in bowl-shaped reflectors on suitable stands. This type of lighting cannot be made to cause hard shadows like those cast by spotlights.

Quartz-iodine lamps are also useful for amateur filming, and have the advantage of giving light of a constant colour temperature suitable for type A colour film. These lamps also have a limited life (about 15 hours) but remain constant in intensity throughout their burning time.

Portable lighting units may be attached to the camera for hand-held indoor shots. The units consist of a grip fixed to a horizontal bar carrying usually two lamps. The camera is mounted in the centre. The lamps themselves may again be either photofloods or quartz-iodine lamps. Usually they are run from the normal main supply, but units also exist which run from portable accumulators and are thus independent of any house current.

The intensity of the light on the subject is controlled: by varying the distance of the soft lights from the subject; by concentrating or spreading out the beam from the spotlights; and by using shutters or suitable diffusing materials in front of the lighting units.

The shape of the light areas projected from spotlights is also controlled by various types of shield attached to the fronts of the lamps: snoots are truncated cones or tubes with the open end facing forward; barndoors consist of an arrangement of doors hinged on four sides of a square so that they can be flapped over the spotlamp beam independently.

With these devices, or with cut-out masks fitted on the lamps, brilliant light can be thrown into a shadow area of a subject without overlapping on to the adjacent highlight and making it too bright.

The camera lens is shielded from unwanted light rays by: a deep lens hood; adjusting the barndoors; and the use of niggers, which are flat black boards mounted on portable stands so that they can be set at any desired position and angle between light and camera.

Exposure. Before filming can start, the lens aperture required for obtaining correct exposure must be determined. There are two main methods of using exposure meters—the highlight system and the reflected light or average light system; the former is favoured by many cinematographers.

The highlight meter measures the strength of the illumination instead of the subject brightness. As faces reflect an almost constant proportion of the light falling on them it is possible with the meter to make face-brightness remarkably constant for all scenes. This is what most (but not all) cinematographers want, and is known as key-tone measurement. So the exposure for faces is independent of the background whether this background be in sunlight or shadow, and the amount of sky included in the picture makes no difference.

The majority of amateur cine cameras have a built-in exposure meter which measures the light reflected from the subject. In many cases the meter automatically sets the lens aperture and may do so continuously even during shooting. Such reflected light readings are reliable in about 90 per cent of all cases. But they need correction for exceptionally contrasty subjects when shooting into the light or when the scene includes a great deal of sky area.

When filming interiors most professionals use a meter that measures the intensity of the light in foot-candles. It is an incident light reading as with the highlight meter. The cameraman knows that he needs a certain number of foot-candles on the subject for him to use a chosen lens-aperture and he adjusts the various lamps until such an intensity is provided. Shadows and highlights are dealt with separately and the difference between them is kept within the contrast range of the film being used. For exteriors the meters used by professionals are usually similar to those available to amateurs.

Artistic Elements. The technical requirements of film-making are one consideration. There is also the artistic side involving various key men, not all of whom concern us closely in this largely technical survey of film production.

In the professional field the producer is the man at the top—he decides on the type and scope of the film and engages suitable technicians to carry the project through.

The script writer is the man with sufficient imagination and technical knowledge to convert the artistry of the original story into a form that suits the visual medium in use.

The art director is an artist able to design the settings that will create a realistic atmosphere suited to the theme.

In the actual filming, the technician who enhances the work of the set designer is the lighting cameraman. The artistic requirements of lighting can be divided into two phases. First there is the general key of the lighting which is adjusted to accord with the mood the director wishes to convey; this is controlled both by exposure and by the balance of light and shade within the picture area. Secondly there is the use of lighting to give artistic emphasis and hold attention to the main interest of the picture and its action.

The camera operator also will have a feeling for the artistic because he is the man who has to carry out the adjustments in each camera set-up and give the right timing to all the camera movements. Composition is a visually important factor that he will sense instinctively, although he may be influenced by any sketches made by the art director.

But the man who has the last word on these questions and who is the most important of the echnicians is the director. Ideally he is the man who understands the abilities and limitations f all the other technicians and can persuade hem to work as a unit in the production of the lm. He is a combination of the practical and he artistic. He, if anyone, will have the touch hat lifts the film into the top class. This art annot well be described, but the quality of the nd product will be influenced by the director's ecision on such matters as camera angles; the vay in which successive camera positions build p the effect of a sequence; how the camera hould be panned or tilted or tracked during a cene, or moved only slightly for a subtle effect; ow the characters will move in order to preerve continuity or heighten the flow of the ction; and in the preparation of the necessary cenes for any special effects.

In an amateur—and often even in a documentary—film unit the team is invariably much maller and most members of it may do several obs. Thus the producer and director may be ne and the same person and he may also write he script. The cameraman may look after both he lighting and the camera operation. Essenally the functions outlined remain however the ame within the scope of the film being made.

Shooting. In an amateur production the shooting stage merely involves filming the shots ccording to the script and often with appreiable improvisation on the spot. As long as the movie-maker keeps notes of the shots as he akes them, he can—since he also handles the lm himself at the editing stage—sort them out fterwards without much difficulty.

Where actors participate in a scene, they are ehearsed beforehand to establish their posiions, movements and gestures during a shot. The cameraman in the meantime rehearses the amera movements necessary, if any, to keep he main subject where he wants it in the frame luring a shot.

For more ambitious productions, and specially professional films, the preparations or shooting itself are rather more elaborate. More detailed rehearsals may be needed not only of the actors but also of the technicians. When the director is satisfied, he gives a word or the take to begin

First comes the number board. This, written on with chalk, is photographed on the front of every scene to record the name of the film, the cene number, the take number (as any scene may have to be re-taken) and several other elevant items. The information, recorded hus on the film itself and quoted in all the elative documents, will make the work of people like the film editor very much easier. f sound is being recorded, the number board has a clapper mounted on it—a hinged piece of wood that can be lifted and banged down again. This gives both a visual and an audible signal that can be identified, enabling the sound and picture to be synchronized at the editing stage.

While the number board is being held in front of the camera the director shouts "Camera" and the camera is switched on. As soon as the board has been carried out of the way the director calls "Action"—the cue for the actors to start their scene. To stop the scene the director will call "Cut".

A continuity assistant records details of everything that might change between one take and the next, such as the position of properties and details of movement. The professional continuity girl takes down shorthand notes of all dialogue spoken in each take and types it out in full on the special forms provided. The camera assistant may make technical notes about such items as lighting, exposures and the amount of film used.

In elaborate scenes the continuity notes are supplemented by still photographs. The still cameraman who does this work is also responsible for all pictures taken for publicity, production stills of the set and the camera crew, and the special publicity stills that are made in his own portrait studio.

After shooting, the exposed motion picture film is sent off to the laboratories with any special instructions that might apply, and after the processing the positive film is put into the hands of the editing department.

SPECIAL TECHNIQUES

The technique of cinematography includes the knowledge of a whole range of special effects produced not only by variations from the normal procedure but by tricks of one kind or another made possible by special apparatus. Many effects are by this time so well established that, by professionals anyway, they are almost taken for granted.

The professional is fortunate in that his special effects can be produced for him in the laboratory by means of the optical printer after his negative has been normally exposed and processed.

Transitions. The transitions are the cinematic devices by means of which one scene changes gradually to another instead of there being a sudden change, as with a cut. The three basic types of transition are the fade, the dissolve and the wipe.

With a fade the scene either gets progressively darker until it becomes black (a fadeout) or appears out of blackness (a fade-in). A density wedge, i.e., a neutral filter in which the density increases smoothly and progressively from one end to the other, is sometimes used to obtain fades in the camera: it is slid across in front of the camera lens. The more advanced cameras have a fade shutter—that is, a variable shutter that can be operated while the camera is running and which gradually reduces the exposure to zero for a fade-out, or opens up to full for a fade-in. A cruder method of fading, often used by amateurs, is to close

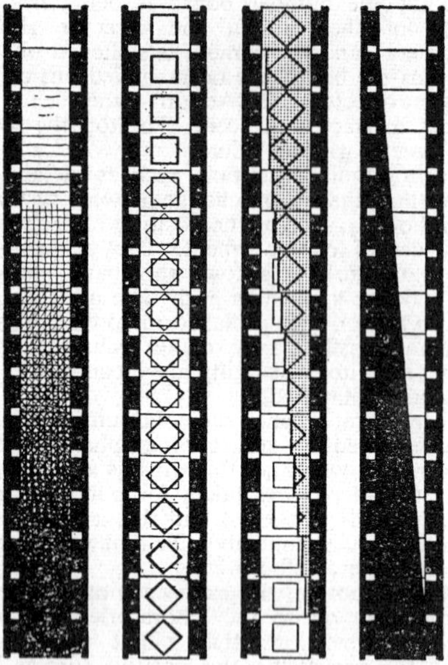

SCENE TRANSITIONS. Left: *Fade out. The image becomes progressively darker until it disappears altogether. The following scene then fades in from complete blackness.* Centre left: *Dissolve. The second scene appears weakly superimposed on the first, replacing it.* Centre right: *Wipe. The second scene displaces the first across the frame.* Right: *Wipe fade. The scene is pushed off the frame by blackness.*

down the lens aperture; but this will only work if there is a large enough range of aperture stops to produce a full black, and in any case there will be a variation in depth of field.

To be fully effective, a fade with the lens aperture needs an iris diaphragm which can close down fully. The difficult part of making a fade in this way is the timing of the closing-down stage so that the darkening of the scene is uniform and gradual. An accessory available for certain cameras closes down the aperture at the uniform rate by a motor.

A number of amateur cine cameras also have a variable shutter and may in certain cases link the control of this with a fading motor.

The dissolve is an effect whereby one scene gradually fades away at the same time as another scene gradually appears and takes its place. To produce a dissolve the first scene is faded out by one of the methods described and the camera stopped. The film is then wound back to the beginning of the fade and the next scene is shot with a fade-in that coincides with the fade-out already there.

A wipe is the type of transition in which a line or shape moves or grows on the screen, gradually wiping away one scene and disclosing the other. In the professional version both scenes are in contact along the line of the wipe and, as it is usually impossible to arrange the necessarily accurate timing with a double exposure in the camera, this transition is a product of the optical printer. The amateur compromises: he wipes away a scene by covering the field of view with a black card cut to the desired shape; there is no winding back and the next scene is uncovered from blackness with a similar shaped card. There is here an affinity with a fade, so the transition is more properly called a wipe-fade.

Various other scene transitions (such as iris fades and wipes where a scene emerges out of a rapidly growing hole in the centre of the previous scene) are almost invariably produced by optical means during printing.

A flash-pan is a sudden transition rather like a cut, but conveys the impression as if the camera had rapidly panned from one scene to the other. It is made by concluding a shot with a quick pan away from the subject. This panning movement must be sufficiently rapid to blur the subject beyond recognition. The second scene may be introduced by a similar flash-pan and the two joined in the middle of the pan. In practice, however, it is sufficient to use only the first panning movement to connect the two scenes. It is much more difficult to bring the camera to rest without any jerk and exactly aligned on the view it is to take in.

Slow Motion and Fast Motion. When the film runs at the same speed through the camera as through the projector (for example 24 frames per second for a sound film) the movement on the screen appears as fast as the original movement in front of the camera.

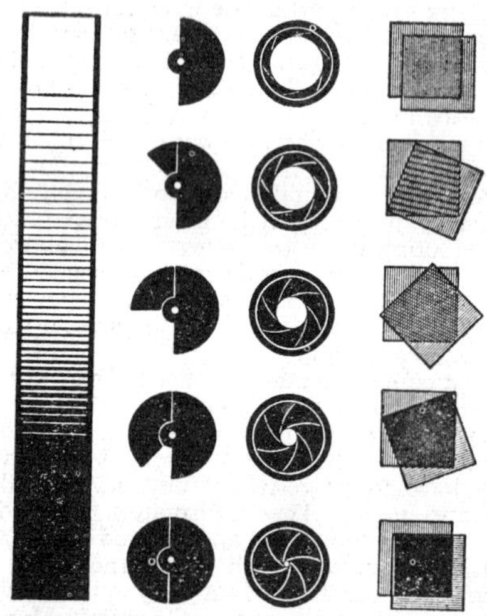

MAKING FADES. Left: *Use of a fading glass; this is progressively moved across the camera lens to reduce the amount of light reaching the film.* Centre left: *Closing the camera shutter (only suitable for cameras with variable shutters).* Centre right: *Closing down the iris. This does not, however, shut out all the light at the end of the fade, leaving an under-exposed, but still visible, image.* Right: *Use of a pair of polarizing filters in front of the lens, rotating them until they are crossed.*

For special effects the camera however may also be run at a different speed. By increasing the rate at which the film runs through the camera and projecting it at the same standard speed, the action on the screen appears to be slowed up. This so-called slow motion adds interest to sport shots as it shows the exact motions—often undetectable by the naked eye—of an athlete or a dancer. It can be used creatively to give a scene a dreamy or unrealistic effect and it is a valuable aid in industrial cinematography for analyzing rapid motion. There the scope of this technique moves into the field of high-speed cinematography.

Some projectors can also be run at a slower speed than the standard one—for example 5 or 6 frames per second. When projected at this speed, a film shot at the normal rate appears on the screen in slow motion, though with somewhat jerky movement.

As already noted, shooting at a higher camera speed can help in making a panning movement smoother. A further use is in filming models, for example a model ship in a tank in the studio. Normally the rolling and pitching of such a ship on waves made in the tank is much too rapid to be realistic. By filming the model at a higher camera speed, the movement is slowed down to a rate where it corresponds to the motion of a real full-scale ship.

By running the camera slower than the framing rate at which the film will be projected (for example at 8 or 12 frames per second for projection at 16 or 24 f.p.s.) all movement is speeded up. This is largely used for comic effects when the shots include people. However it can also serve to heighten the apparent speed of a train or car, etc.

At a slower camera speed the exposure per frame is also increased. So this is a useful way of shooting in poor light conditions, provided no familiar movements give the show away.

Time Lapse and Stop Motion. The principle of fast motion can be extended to the point where the camera exposes one frame every second, every minute or even every hour. This technique of so-called time lapse photography is useful for depicting processes which are normally far too slow to be observed live: the growth of plants, changes of weather in the sky, etc.

For time lapse films the camera operates at the single frame speed, with an automatic unit which exposes a frame at pre-set intervals. For growth studies the plant is usually set up in front of the camera in normal daylight. Immediately before every exposure the timing mechanism obscures the prevailing daylight, switches on lamps (to ensure uniform exposure for every frame). It then switches off the lamps and readmits the daylight after the exposure.

Stop motion involves stopping the camera during the run of a scene, making a change of some kind in the scene and then continuing filming. The effect is one of a sudden change. One can therefore make something or someone appear or disappear in an instant in accordance with the best tradition of magic.

Related to these techniques is the art of animation. Here inanimate objects are moved by the cameraman, a little at a time, between exposures. The objects may be three-dimensional as with puppets, or two-dimensional as with flat diagrams or drawn cartoons.

Cartoons are made by photographing a long series of drawings each of which has been drawn slightly different from the one preceding. The changes represent the progressive movement of the portion of the drawing to be animated, such as a cartoon character. To reduce the amount of labour involved in the drawing, a detailed background is usually drawn on a sheet of paper while the animated objects in the foreground are drawn on sheets of transparent plastic (cells).

Reverse Motion. All film speeds can be used with the film travelling backwards in the camera. Reverse motion is a valuable asset to a film-maker, for not only does it make fascinating tricks possible, with events happening in reverse, but a number of effects can be achieved that would be dangerous or impossible filmed in the normal way.

Narrow-gauge cameras are not usually designed to run backwards, but the cameraman can get the same result by shooting with his camera held upside down and then reversing

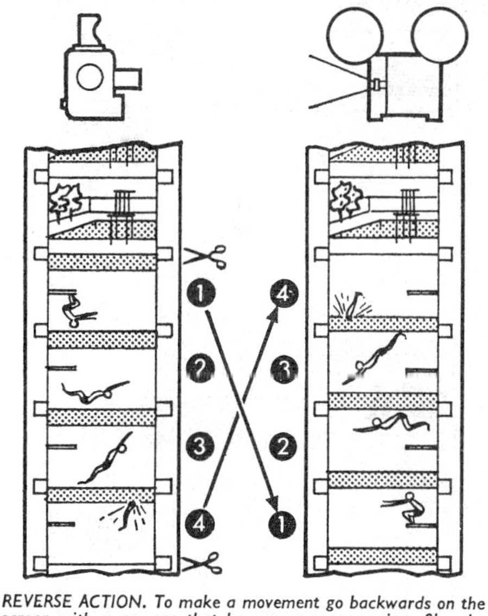

REVERSE ACTION. To make a movement go backwards on the screen with a camera that has no reverse running, film the action with the camera held upside down. Then, after the film is processed, cut out the inverted section and splice it back into the film again the right way up. On projection the action appears reversed, for example the diver floats up out of the water back on to the springboard. This technique is suitable for 16 mm. and 9·5 mm. films; with 8 mm. the film has to be reversed sideways where a section is inverted to keep the perforations on the same side. That may lead to some loss of sharpness on projection.

the film end-for-end after processing. This is not easy to do in the case of 8 mm. as this film is perforated along one side only and so will not turn over satisfactorily.

Superimposition. Two exposures can be made on the same strip of film by winding back to the start before the second exposure. This is the method used in the creation of a ghost. For this effect the actor is first photographed against a black background, and the second exposure is on the chosen empty scene where on the screen the transparent ghost will later appear to be walking.

Double exposure is also used in the familiar titling effect where white lettering is superimposed on a scene photographed in a previous exposure.

Split-screen Filming. There is another way of arranging multiple exposures. Instead of making them come on top of each other it is possible to position them side by side so that the two actions appear as though they were taking place at the same time. Only a section of the frame is exposed on each run through the camera, the rest of the frame being covered by a suitable mask (or matte) until it is needed. The masks are usually held in an effects box fitted to the front of the camera.

This method is particularly valuable in scientific work where one series of events is to be compared with a standard controlled series.

Trick Scenery. The technique of masking can be employed in a number of subtle and often more complex ways. Instead of double exposure combined with the use of black masks, only one "mask" is used and it is not black—it is a painted representation on glass of what will seem like part of the real scene, supported in front of the camera so that it will be photographed at the same time as the real scene, without double exposure.

Scale models of foreground objects are mounted relatively close in front of the lens to convey the impression of real objects at a greater distance. The expense of building the upper storeys of tall buildings is often saved by suspending elaborate scale models in front of the camera in this way.

In the Schüfftan process, a mirror set at 45° is used to reflect the image of a scale model at one side while the main subject is seen through areas scraped from the reflecting surface of the mirror.

Substitute Backgrounds. Not only may one add details to a scene to make it complete but there are devices that enable the action in a studio to be portrayed in any desired surroundings. These systems normally employ two films—one of the background, and one for the complete action.

In principle most of these devices date from the earliest days of cinematography. They involve such things as back projection, in which the action is filmed against a screen on to which a projector throws the required background from behind; travelling mattes, whereby a complicated system of masking is used during the film printing to combine separate films as if they were one; and numerous other effects based on similar methods.

Stereoscopy. Normal cinematography exposes one frame at a time (in rapid succession). Three-dimensional cinematography, which is now available to professional and amateur, exposes two frames at a time, taken from viewpoints about 2½ ins. apart.

Special projection arrangements are necessary to show these two separate records superimposed on the same screen area and yet allow the eyes of the audience to view them individually. The modern method is to project and view the two images through a pair of polarizing filters, the directions of polarization of each picture being at right angles to each other. The effect is one of looking at solid objects through the window of the screen. J.C. & L.A.M.

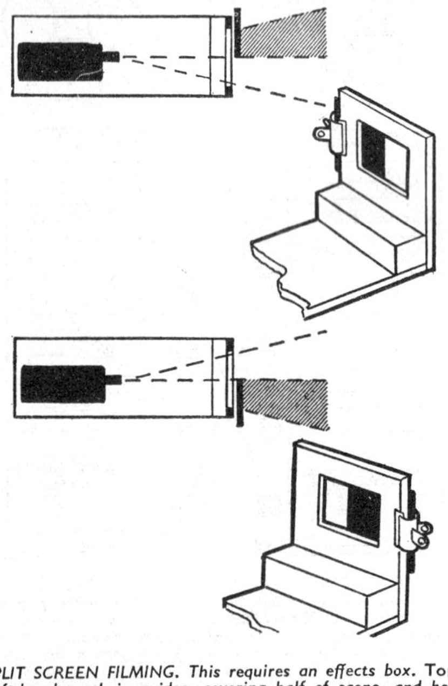

SPLIT SCREEN FILMING. This requires an effects box. Top: Left-hand mask in guides, covering half of scene, and held with a clip during filming. Bottom: Mask in right-hand side up to same centre mark as before. Left-hand mask removed, and right-hand scene filmed after rewinding film.

See also: *Animation; Background; Cinematography; Editing; Projection practice; Splicing; Titling.*

Books: *The Simple Art of Making Films*, by Tony Rose; *The Technique of Making Films*, by P. Monier; *How to Film*, by G. Wain; *How to Script*, by O. Blakeston; *How to Direct*, by Tony Rose; *How to Use Colour*, by C. L. Thomson; *How to Make Holiday Films*, by H. Baddeley; *How to Write Film Stories*, by R. Harrison; *How to Do the Simpler Tricks*, by J. Caunter; *How to Do Tricks*, by J. Caunter; *How to Film Children*, by M. Natkin; *How to Plan Your Films*, by C. V. Willson; *How to Use* 8 *mm. at Work*, by A. Wigens; *Successful Movie Making*, by A. Wigens (all London).

CINE TERMS

Accompaniment. Music and/or effects (often recorded on gramophone discs) which accompany the projection of a silent film and are not rigidly synchronized with the picture.
Action Sketch. Rough drawing indicating a stage or series of stages of an action, used in the production of animated cartoons.
Actuality. Film record of actual events, made without use of actors and without reconstructing or interfering with the action in any way.
Actual Sound. Sound effects the origin of which is visible on the screen or present in the scene (dialogue, traffic noises in a street scene)—as distinct from commentative sound.
Amplifier. Apparatus for increasing the strength of electrical impulses—i.e., to step up the signals from the play-back head of a tape recorder or the exciter lamp in a sound projector to a level where they can be reproduced in a loudspeaker. Similarly, an amplifier can amplify signals from a microphone to operate recording equipment.

The amplifier is an essential part of tape recorders, sound projectors, as well as, of course, of wireless and television receivers and transmitters. The amplifier may employ circuits with valves or with transistors.
Animated Viewer. Apparatus for the convenient inspection of film during editing; it provides a slightly enlarged moving picture and is more simple to operate than a normal projector. The speed of film travel is easily variable.
Animation. Apparent movement of inanimate objects or drawings on the cinema screen. The word is also used for the art of creating such movement.
Animation Board. Adjustable board for holding animation drawings in position while they are being photographed, e.g., in cartooning.
Answer Print. The first projection print of a newly completed film delivered by the processing laboratory. Also known as an approval print. It is submitted by the laboratory for the approval of the producer and director who may call for changes in the grading (lightness or darkness) or colour balance in the scenes as printed by the laboratory.
Auto-Couplings. Method of arranging disc recordings so that the continuation of one side of a record is made on a second disc. An automatic record changer will then play the continuations in correct sequence.
Backward Take. Recording of a scene with the camera running in reverse (or held upside down), commonly used for trick effects and in animation. Normal motion then appears in reverse on the complete film.
Balancing Stripe. Second stripe applied to the opposite margin of a striped film to ensure even spooling. A stripe is a magnetic coating applied to the margin of a film and used to record sound.
Beater (or Dog) Movement. One means of producing intermittent film motion in the camera or projector. An arm at the bottom of the gate digs at the lower film loop and lengthens it, thus pulling the film down through the gate. This arm sometimes has locating pins made in it to register in the film perforations.
Big Close-up. Very close shot, taking in only a small area such as part of the human face to draw attention to minute details.
Bin. Large receptacle into which film is allowed to fall while assembling shots or when running film through a viewer or projector. It takes the place of a take-up spool.
Bin Stick. Wooden strip with pegs or clips on which pieces of film may be hung when editing, laid across the top of a film bin.
Blimp. Case enclosing camera (or projector) to reduce mechanical noise. Mostly used for professional motion picture cameras if sound is to be recorded live during a shot. The blimp then prevents the microphone from picking up the camera noise. Projectors only need a blimp if they are in the same room as recording gear for recording a post-synchronized sound track.
Blind Shots. Shots used at the beginning or end of a sound effect in which the origin of the sound is not seen.
Bloop. Opaque, oblique line painted across a sound-track to obscure a splice or defect.
Blooping Ink. Black, opaque ink which is quick drying. It is used for blacking out areas on film and particularly for painting over splices in an optical sound track to render them inaudible.
Bracket. To move the head from side to side through a field of light. A device by which an actor can discover the position in which he is most strongly lit (felt when the light is hottest on the face.)
Breakdown. Analysis of the film script in terms of scenes with the same locations, sets, properties, etc. In this way it is possible to estimate production costs and determine the most practical order in which the scenes should be filmed.
Bridging Shot. A shot used to cover a break in continuity between successive scenes. A cutaway may be used for this purpose to avoid a jump-cut. Also, a close-up of holiday leaflets may bridge a sequence of holiday preparations with shots of the resort.
Bump In or Out. Instantaneous appearance or disappearance of a person or object during a shot. A common effect in animated cartoons.
Business. Incidental action in a film (e.g., lighting a cigarette or cleaning spectacles) which is not strictly necessary from a story-telling point of view and is often introduced during rehearsals by the director or one of the actors.
Call-Sheet. Written summons to attend a rendezvous for filming issued to actors and technicians.
Camera Operator. Person responsible for the

operation of the movie camera, as distinct from the lighting side of photography. In professional units and some amateur ones the operator works under the supervision of the lighting cameraman.

Capstan. Drum or roller pulling wire or tape through a magnetic recorder.

Cast. Any excess—generally unwanted—of a particular colour in a colour picture. Collectively also the actors in a film feature.

Cell Flash. Bright patch on film caused by reflection of light from uneven surface of a cell into the lens.

Centrifugal Governor. Device using centrifugal force to keep the motor speed of the projector constant.

Cells. Transparent sheets on which animation drawings are traced for cartooning. Cellulose acetate or colourless plastic is generally used.

Change-over. Switching smoothly from one projector to another at the end of a reel so that there is no interruption in the picture on the screen. Certain frames near the end of every reel carry a spot or other mark near the top right-hand corner—the so-called change-over cue. This warns the projectionist to start the second projector.

Clapper Board. Type of take-board for sound films. A hinged flap is banged down on the clapper board to make a synchronizing mark on the sound-track. It is operated by a junior assistant known as the clapper boy.

Close-up. Shot in which the camera is, or appears to be, very close to the subject. If the subject is a human figure a close-up would include head and shoulders only.

Claw Movement. One means of producing intermittent film motion in the camera or projector. The claw is a metal arm with teeth at one end; the other end is attached to a rotating disc which pulls the claw down while a link attached to the arm pulls the teeth into the film perforations and withdraws them as required.

Commentary. Descriptive spoken words accompanying a film, but not comprising dialogue of the action in the film.

Commentative Sound. Sound which is not directly identified with the action it accompanies in the film, such as music or commentary.

Continuity. Easy transition from one shot to the next and from one sequence to the next without awkward breaks or discrepancies.

Continuity Girl. Member of the production team who notes all the details of each take during shooting (the set, lighting, costumes, position of actors) to ensure that no discrepancies or errors of continuity occur between consecutive shots when the material is assembled.

Continuity Shot. Shot takes after a film is completed, for insertion where required to maintain continuity. It may be a bridging shot taken from the stock library, or a shot inserted in the script afterwards.

Continuity Title. Sub-title designed to tell part of the story, not shown pictorially, other than by direct quotation of speech. For example, "Fifteen years later", or the classic, "Came the dawn".

Copy. Reproduction of a reversal or colour film duplicated from the original.

Counterpoint. Use of sound to present an aspect of the story different from, but complementary to, that shown by the picture.

Credit Titles. Printed list of the cast and technical unit responsible for a film. It appears at the beginning, or less commonly at the end, of a film.

Creep. Difference between the tape speed and capstan speed in a magnetic recorder, due to the effects of tension and elasticity.

Cross-cutting. Alternation from one scene to another in editing so that two or more events are represented as taking place simultaneously.

Crystal Pick-up. Type of pick-up producing electrical impulses from a mineral crystal instead of a magnet and coil. It has an extremely high resistance.

Cut. Instantaneous transition from one shot to another. Also the act of trimming and joining film shots together.

Cut-Away. Shot that temporarily draws the spectator's attention from the main action; e.g., a close-up showing the reaction of a bystander to a street fight.

Cutter. Person engaged in film editing. Usually an assistant who carries out the mechanical part of the operation.

Cutting Copy, Cutting Print. Reversal or colour copy or a positive print used solely for editing purposes. The original or negative is cut to match it when editing is completed and is never projected, but carefully maintained as a source of subsequent prints.

Direct Sound Recording. Recording of dialogue as it is spoken during shooting.

Director. Person who is responsible for the creative side of production; he translates the script into a film and is in charge of the actual shooting.

Disc Recording. Sound recorded as a spiral, wavy groove on a disc; the ordinary gramophone record.

Dissolve. Gradual transition from one shot to another, the second being superimposed on the first at the point of transition.

Documentary. Non-fiction film, often stating a point of view on some current topic.

Dog (or Beater) Movement. One means of producing intermittent film motion.

Dolly. Wheeled platform or vehicle to carry the camera for movement while shooting. A shot taken in this way is also known as a dolly shot.

Dope Sheet. List of shots taken, made out by the cameraman or his assistant.

Double Frame. One animation drawing photographed for two frames instead of one. This either halves the speed of a movement or the number of drawings required for an action of a certain time.

Dubbing. Re-recording of sound, e.g., from

disc on to tape or film. Also a term describing the technique of replacing a sound track in one language with one in another language, whilst still retaining perfect lip synchronization.

Dupe Negative. Picture negative prepared from a positive picture original and used for the production of married prints.

Duping Print. Positive print made from a negative specifically for the purpose of making a dupe negative.

Edge Numbering. Numbers printed at every foot along the edge of a negative film to facilitate cutting of the negative to match the edited cutting print.

Editing. Process of assembling the component shots of a film into their final order and then cutting them to their final length.

Effects Box. Bellows or box-like attachment to fit in front of the camera lens. It acts as a lens hood and also holds masks (mattes) at various distances in front of the lens for special effects. Alternatively known as matte box.

Effects Track. Sound-track carrying sound other than music or speech.

Erase Head. Magnet or electro-magnet used to remove the varying magnetization representing sound recorded on magnetic tape, wire or stripe.

Establishing Shot. Shot used to present for the first time a new scene, to establish relationship of details to be shown in subsequent shots. The establishing shot is thus usually a long shot.

Exciter Lamp. Lamp used to scan the sound track in a sound projector; it excites the photo cell.

Exterior. Scene filmed out of doors.

Fade. Gradual darkening of a shot on the screen to complete blackness (fade-out) or the reverse effect (fade-in).

Fading Glass. Strip of glass transparent at one end, gradually increasing in opacity towards the other. Passed across the lens while filming, it produces a fade.

Fading Solution. Dye solution for making fades on positive, reversal or colour film.

Fan. Device driven by motor to cool the lamphouse and film in the projector.

Fast Motion. Special technique which produces abnormally fast motion by filming at a slow speed—e.g., eight frames per second, and projecting at a normal speed—e.g., sixteen frames per second.

Film Bin. Large receptacle into which film is allowed to fall while assembling shots or when running film through a viewer or projector. It takes the place of a take-up spool.

Film Horse. Frame on which separate shots are hung in numerical order during editing.

Flapover. Change in the title seen on the screen, effected by turning a card rapidly so that the reverse side is shown whilst the camera is running.

Flashback. Transition from present to past tense in film narrative, often conveyed in the form of a character's recollections. The whole of that portion of the narrative which is related in the past tense is also known as a flashback.

Flash-pan. Movement of the camera quickly away from the subject at the end of a shot so that a quick pan appears to lead straight into the next shot, no cut being apparent.

Flash Frame. First frame or frames of a shot that are over-exposed due to the time lag in the camera motor reaching the correct speed.

Flicker. Failure of persistent vision; the picture is not evenly bright all the time.

Flutter. Effect produced by very quick variations in the speed of a sound recording medium. Similar to wow, but occurring as much faster irregularities.

Footage. Length of film in feet; usually applied to the amount of film run through the camera for a shot or a sequence of shots.

Frame Glass. Sheet of glass used in the production of animated cartoons to press down upon the cells and keep them flat during filming.

Frame Line. Narrow line or area between two frames on a length of cine film.

Framing. Adjustment of the gate mask on a projector to obscure the film perforations; also the positioning of the film in the gate to avoid the frame line showing.

Full Stripe. Magnetic stripe occupying the full width of the unperforated edge of a single-perforated 16 mm. film (0·1 in. wide).

Gate. Component in camera and projector which holds each frame flat and momentarily still behind the lens.

Gate Float. Unsteadiness in the moving picture due to irregular positioning of the frame in the gate.

Gate Mask. Aperture defining the area of the film depicted on the screen.

Governed Motor. Motor which runs at a speed determined by a mechanical governor.

Grading. Estimating the printing exposure for each individual shot in a master negative of a film to produce a correctly exposed print. In a motion picture laboratory this operation is carried out by a grader. Grading also refers to the colour correction required in colour printing. Modern printing apparatus usually controls the exposure from scene to scene automatically by photo-electric measurement.

Half Track. Magnetic coating, $\frac{1}{20}$ in. wide, applied to a 16 mm. sound film to cover half the optical sound-track. Optical and magnetic sound-tracks can thus be used alternately. The term is also applied to recordings on magnetic tape which occupy only half the width of the tape; in this way the other half of the tape can be used by turning it upside down.

Hold. Photography of one animation drawing on several frames so that it appears still on the screen; also used in scripts to indicate that the camera should continue to film a scene even though the action may have ceased.

Hum. Unwanted low note produced when an

A.C. mains supply or other low parasitic frequency enters the amplifier used for recording or reproduction. It can be caused by induction in insufficiently screened leads carrying the signal from the recording or reproduction unit to the amplifier.

Humidifier. Device for introducing moisture into the cooling air in a projector. Also a moistened pad in film cans to prevent the film from becoming too brittle.

Inching. Movement of film or magnetic tape sometimes in jumps, an inch or two at a time. It is usually done by manual operation of the projector or recorder so that a particular part of the film or tape can be quickly found or precisely located.

Induction Motor. Electric motor which runs at a speed largely but not entirely determined by the mains supply frequency.

Input. Electrical impulses or voltage supplied to an apparatus.

Inserts. Shots, sometimes unscripted, such as close-ups of letters, newspaper headlines, clocks, etc., in which no actors appear. By their nature it is unnecessary for them to be taken in conjunction with other shots in the same sequence.

Intercutting. Alternation from one scene to another in editing so that two or more events are represented as taking place simultaneously.

Interior. Shot taken indoors.

Intermittent. Film transport mechanism which advances the film one frame at a time—e.g., at 16 or 24 times a second. There are several types of intermittence—claw, dog (or beater) and Maltese Cross.

Inter-Negative. Negative made for the purpose of making release prints—for example from the master negative via a duping print. Also intermediate colour negative made from a reversal master positive, for example in 16 mm. colour filming for making colour release prints. The inter-negative is also made as one stage in making opticals (mixes, fades, superimpositions, etc.).

Jitter. Uncontrolled movement in an animated cartoon. It is caused by faulty animation, tracing or camerawork.

Jump-Cut. A cut which breaks continuity of time by jumping forward from one part of an action to another obviously separated from the first by an interval of time. Also a cut between two similar views of a scene but at different distances from the camera, so that the picture appears to jump forward on the screen.

Key Line. Line of dialogue which reveals the heart of a story situation.

Keys. Animation drawings of the principal positions in a movement. Keys are made of positions where any part of the figure stops, starts or changes direction.

Keystoning. Distortion of the screen image (widening at top or bottom) when the projector axis is not at right angles to the screen. When the projector is tilted upwards, the screen image appears narrower at the top; when the projector tilts downwards from a high level, the screen image is wider at the top.

Lap-dissolve. Gradual transition from one shot to another, the second being superimposed on the first at the point of transition.

Leader. Strip of film at the beginning of a spool used for loading or threading into the camera or projector.

Level Synchronization. Placing of synchronizing marks so that the record of a sound appears immediately opposite the corresponding picture. (As opposed to printing synchronization.)

Library. Collection of stock shots which may be useful for cutting into future films. A stock shot used in a film for which it is not taken is also known as a library shot.

Light Box. Animation desk with glass drawing surface illuminated from below.

Lighting Cameraman. Person responsible for the photography of a film. In some units an assistant operates the camera.

Light Valve. Device for converting electrical impulses into corresponding fluctuations of a beam of light.

Line Test. Sequence of pencilled animation drawings photographed and projected, usually in negative, to check the quality of the animation, before proceeding with finished work.

Lip-synchronization. Synchronization of sound with film, accurate to a fraction of a second, so that speech can be reproduced in synchronism with an actor's lip movements.

Live Recording. Recording of dialogue and other sound made while a scene is actually being filmed.

Location. Natural setting, usually out of doors and away from the studio, where shots for a film are taken.

Long Shot. Shot in which the camera is, or appears to be, at a considerable distance from the subject. If the subject is a human figure it would occupy less than one-third of the height of the picture.

Loop. Short length of film joined at its ends so that it can be projected continuously or repetitively. Used by actors to rehearse the timing for post-synchronized dialogue. Sound films carrying continuous sounds such as wind, sea, crowd noises, etc. may also be made up into a loop for mixing in with other sound effects at different points of the film.

Magnetic Film. 35 mm. or 16 mm. film stock carrying a magnetic layer over its whole surface in place of a photographic emulsion. A number of magnetic tracks may be recorded on the film side by side in the course of extensive mixing operations.

Magnetic Recording. Sound recorded as variations in magnetization of a wire, or of particles in a coating on film or tape.

Main Title. Opening title bearing the name of the film.

Maltese Cross. One type of mechanism for pro-

ducing an intermittent film movement. The cross which is mounted on the same shaft as a film sprocket serves to move the sprocket round by the space of one picture at a time.

Married Print. Print of a sound film incorporating both picture and sound track; i.e., a normal print of a sound film for projection.

Master. Film from which release copies are printed and which is not itself used for projection.

Master Recording. Complete sound record, usually made on magnetic tape or film, from which the final optical or magnetic recording is later prepared.

Master Shot. Shot covering the action of a long scene which is afterwards broken up by the insertion of close-ups, etc.

Match. Checking of two shots to make sure that one will lead smoothly into the next.

Matte. Mask placed in front of the lens to blank out part of the scene. Mattes are available in various standard shapes, a common one being that which simulates the effect of looking through a pair of binoculars. For more complex effects the masked area can later be exposed, using a different subject, while the originally exposed area is masked.

Matte Box. Lens attachment in the form of bellows or a box with a frame to take masks (mattes) for split-screen and similar trick effects. Also known as effects box.

Medium Shot. Shot which in effect is about midway between a close-up and a long shot.

Microphone Howl. Noise produced when a microphone is too near the associated loudspeaker, resulting in acoustical feedback.

Mid-shot. Shot which in effect is about midway between a close-up and a long shot.

Mix. Gradual transition from one shot to another, the second being superimposed on the first at the point of transition.

Mixing. Electrical combination of sounds or recordings so that they can be reproduced simultaneously through the one reproducing system.

Modulation. Variation in an electric current induced by suitable equipment—e.g., a microphone.

Monitor Speaker. Loudspeaker (usually small) indicating the volume and balance of sound being recorded or reproduced in another room.

Montage. Process of creative editing. The term is also used to signify a series of dissolved shots, suggesting a passage of time, journey, etc.

Mood Music. Music written in various moods and styles for use in films, radio and television, but not for a specific production. Such pre-recorded music may be classified by type and filed in music libraries on disc or tape.

Mute. A picture negative or positive of a sound film without its sound-track. The term implies the separate existence of a sound-track for the film, whereas "silent" refers to a film for which no sound-track has been recorded.

Narration. Commentary spoken by one of the characters in a film, but not synchronized with the picture.

Narrow Gauge. Films and apparatus not of the standard 36 mm. type—e.g., 16 mm., 9·5 mm., 8 mm.

Neutral Density Filter. Neutral grey filter used to reduce the light passing through a lens without affecting its colour. With the relatively long shutter speeds used in cinematography, this is sometimes necessary to avoid over-exposure.

Noise Level. Unavoidable noise produced by projector mechanism or by hum and other sources in the sound equipment. Reproduced sound must have a volume level considerably higher than the noise level for satisfactory quality.

Non-sync. Recording reproduced without provision for precise synchronization with the film.

Notched Titles. Titles printed only on two or three frames and preceded by a notch in the edge of the film. On some obsolete projectors the notch trips a mechanism which arrests the film motion while the title is being read.

Number Board. Also known as take board. Board held briefly in front of the camera and photographed at the beginning of a take. It records the title of the film, the number of the scene and the number of the take, to enable the editor to identify every shot. In sound film production the number board incorporates the clapper and is then known as the clapper board.

Optical Printer. Printing machine for producing positive copies from a negative (or vice versa) by re-photographing the negative image on to the positive film through a lens system (the alternative way of printing is with a contact printer). Optical printers are used for special effects and trick work, and for making reduction prints—e.g. 16 mm. prints from a 35 mm. original.

Opticals. Dissolves, fades, etc., made by the processing laboratories on an optical printing machine after the scenes have been photographed normally.

Optical Centering. Method of centering the film in the gate by moving the optical system instead of the gate mask. In this way the projector need not be constantly re-aligned with the screen.

Optical Effects. Dissolves, fades, etc., made by the processing laboratories on an optical printing machine after the scenes have been photographed normally.

Optical Recording. Recording of sound photographically on a sound track along one margin of a film. This track modulates the light passing through it from a special lamp, and the light impulses are electrically converted into sound.

Output. Electrical impulses or voltage delivered by an apparatus.

Overlap. Double thickness of film at a splice.

Overlay. Transparent sheet on which part of

the design for a title is drawn or painted.
Pan. Movement of the camera from left to right or from right to left while taking a shot.
Pan Blur. Movement of the camera quickly away from the subject so that a quick pan appears to read straight into the next shot, no cut being apparent.
Parallel Action. Common form of film narrative in which events are represented as taking place simultaneously in different places by cutting from one to another.
Participation Shot. Shot in which the spectator associates the camera eye with the eyes of a character in the story.
Persistence of Vision. Inability of the human eye to separate individual images in rapid succession. This makes the illusion of movement possible in motion pictures.
Pick-up. Instrument using a needle or stylus to follow the waviness in the groove of a gramophone record and so generate corresponding electrical impulses.
Pilot Commentary. Shot by shot description of a film, recorded during screening as a timing guide for subsequent recording of commentary free from projector noise.
Pinch Roller. Roller holding wire or tape in firm contact with the capstan of a magnetic recorder.
Play-Back. Reproduction of a sound track in a studio during shooting to enable additional sound or action—or both—to be synchronized with it.
Post-synchronization. Recording of dialogue and other sound after production so that it matches the already existing lip movements and actions in the film.
Potentiometer. Electrical resistance with provision for drawing of a variable proportion of the voltage applied to its ends. Usually takes the form of a circular casing with a projecting spindle. Rotation of the spindle moves a contact connected to the middle one of three connexions and so selects more or less of the voltage between the other two connexions. It is commonly used for volume and tone controls.
Presence. Illusion of sound which, although coming from the screen, gives the impression of reality.
Pressure Plate/Pad. Device to enable background and cells for an animated cartoon to be photographed flat at even over-all pressure. It consists of a sheet of plate glass in a wood or metal frame hinged to the animation board. Beneath it is a pad of felt or several thicknesses of cloth, blotting paper, or sponge rubber. It is also a camera component holding the film accurately in the focal plane in the camera.
Print. Positive copy of a film, usually intended for projection.
Printing Synchronization. Placing of synchronizing marks so that the sound is in advance of the picture, as required for printing and projection. The interval between the sound and the picture is 20 frames for 35 mm. sound film, 26 frames for 16 mm. optical sound film, 28 frames for 16 mm. magnetic sound films, and 56 frames for 8 mm. magnetic sound films.
Properties. Any portable articles appearing in a film, for example furniture, flowers, fountain pens and fishing rods.
Proscenium. Surround to a stage or screen. It is often illuminated by concealed and coloured lighting.
Quick Motion. Trick of filming with the camera running abnormally slow, so that when the film is projected at normal speed, the action is quicker than it was in reality. The opposite of slow motion.
Racking. Alignment of the film in the camera or projector gate so that the frame line is correctly placed in relation to the sprocket holes or gate aperture. Incorrect racking when projecting causes the frame line to appear on the screen.
Record-Playback Head. Electro magnet which serves either to produce or detect local magnetization of the wire, tape or stripe in a magnetic recorder.
Recording Level Indicator. Instrument showing when the recording impulses approach the intensity that would produce overloading—e.g., in a magnetic recorder.
Reduction Gear. Mechanical means of reducing motor speed.
Reel. Spool of film as used on the projector. A reel is also a measure of length, corresponding to about 16 minutes at silent speed of projection or 11 minutes at sound speed (400 feet of 16 mm. or 9·5 mm., 200 feet of 8 mm.).
Registration. Correct position of animated drawings in relation to one another and to the camera. Also the accurate superimposition of the separate images used in some types of cine film printing—e.g., colour.
Register Pegs. Pegs fitted to a metal bar, designed to keep images on separate supports in correct register. The support has punched holes which correspond to the pegs.
Register Pins. Device in camera gate to ensure that each frame of film is in exactly the same position as the preceding and following frames. This ensures good picture steadiness.
Relational Editing. Editing of shots, which have no literal connexion with one another, in such a way as to suggest an association of ideas, as in a simile or metaphor.
Release Print. Positive print of a film intended for distribution—i.e. projection in cinemas, etc.
Retake. Repetition of a scene or take.
Reverse Action. Recording of a scene with the camera running in reverse (or held upside down) commonly used for trick effects and in animation. When projected, the film then shows events in reverse motion.
Reverse Shot. Shot in which the camera's viewpoint is almost exactly opposite to that o the previous shot. For example; a shot of A looking at B following a shot of B looking at A.
Rewinder. Apparatus for winding film on to its spool ready for projection.

Rostrum. Apparatus used in the production of animated cartoons, giving rigid support to the camera and animation board so that they do not alter position relative to each other in an uncontrolled way.
Rough Cut. Preliminary re-arrangement of shots in editing before their lengths are accurately adjusted.
Running Title. Long title which gradually moves up the screen so that the first lines of the title disappear from the top as others appear at the bottom.
Run Out. In animation, to cause a line or shape to grow from a point.
Rushes. Lengths of processed film which are usually viewed before editing by the director and unit immediately the film is returned from the processing laboratories.
Saturation. Condition of a magnetic recording material in which an increase in recording impulse no longer produces a proportional increase in magnetization.
Scene. Part of the action recorded on film with the camera running, or appearing to run, continuously. The term scene is sometimes used in a more general sense to describe any piece of action which appears to be continuous (i.e., is not divided by a lapse of time).
Scenario. Story of the film in its final form from which the director works during production. It usually contains a detailed description of the action, shot by shot, together with settings and camera positions. Sometimes also known as a shooting script.
Scraper. Portion of a splicer which scrapes off the emulsion leaving the film base ready for the application of the cement.
Script. Story of the film in its final form from which the director works during production. It usually contains a detailed description of the action, shot by shot, together with settings and camera positions.
Script Scene. Part of the action which, according to the script, may be recorded in a single shot although the director may often break it down into several. The word "scene" by itself is sometimes used in a more general sense to describe any piece of action which appears to be continuous (i.e., is not divided by a lapse of time).
Sequence. Phase in the development of the film story, roughly equivalent to the chapter in a novel.
Set. Artificially constructed setting used in a film.
Shooting Script. Story of the film in its final form from which the director works during production. It usually contains a detailed description of the action, shot by shot, together with settings and camera positions.
Shot. Part of the action recorded on film with the camera running, or appearing to run, continuously.
Show Print. Alternative term for release print, i.e. projection print of a finished film for distribution and showing.
Single 8. Film stock 8 mm. in width as distinct from double run 8 mm. film which, before cutting, is 16 mm. wide. Single 8 mm. film is nowadays virtually obsolete as camera stock, since practically all cameras take the film in the double 8 mm. form.
Single Frame Exposure. Exposure of one frame at a time instead of running the camera continuously; a method used in animation.
Single-system Sound Camera. Camera in which picture and sound are recorded simultaneously on a single film.
Slipping Clutch. Device which allows automatic adjustment of the relative revolving speed of spools.
Slow Motion. Trick of filming with the camera running abnormally fast so that when the film is projected at normal speed, the action appears slower than it was in reality.
Special Effect. Illusion either realistic (e.g., a snowstorm) or fantastic (e.g., a character rendered invisible) introduced into a film by means of technical trick work.
Splice. Join between two pieces of film, or the process of joining film.
Splicer. Apparatus for joining film.
Split Track Recording. Method of independently recording several sounds side by side in the width of one normal track on magnetic tape or stripe so that the recordings are reproduced simultaneously.
Spoken Title. Dialogue rendered in the form of a sub-title.
Sprocket. Toothed drum driving the film by engagement with the perforations or sprocket holes.
Sprocket Holes. Holes in cine film by which the film is driven through the camera and projector. They are placed on the edges of 35, 16 and 8 mm. film, and between the frames in the centre of 9·5 mm. film.
Sprocket Modulation. Throbbing of the reproduced sound due to the presence of sprocket holes close to the sound-track.
Sound Drum. Comparatively large roller carrying the film in a projector at the point where the sound-track is scanned.
Sound Head. Components in a sound projector or tape recorder that actually converts the recording into electrical impulses.
Sound-Track. Photographic record of sound on film.
Standard Film. Cine film 35 mm. in width.
Start Marks. Marks on a film and a disc or tape so that the picture can be started in close synchronization with the sound.
Stereophonic Sound. Reproduction of sound so that it seems to have depth and direction, even when the original source is moving—e.g., the sound of an actor's voice actually moving across the screen with the image. In principle it is very similar to stereoscopic photography.
Still. Photograph of a scene from a film or a scene in the making; a normal photograph as distinct from a cinematograph shot.

Stock. Cine film before exposure.
Stock Shot. Shot used in a film, but not taken specially for it; shot taken from a film or library source outside the unit producing the film. Alternative name for library shot.
Stop Action. Exposure of one frame at a time instead of running the camera continuously; a method used in animation.
Story Board. Board to display the action and layout sketches of an animated cartoon in sequence.
Straight Cut. Normal transition by cut as distinct from a dissolve or fade.
Stripe. Magnetic coating applied to one or both margins of a film and used for recording sound.
Stroboscopic Disc. Rotating arrangement of dark and light radial lines which appear stationary when illuminated by light flickering at a related frequency.
Sub-Standard Film. Cine film less than 35 mm. in width—e.g., 16, 9·5 and 8 mm. Nowadays usually known as narrow-gauge.
Sub-Title. Explanatory wording commonly incorporated in a silent film either to replace dialogue or to bridge a gap in the story.
Supersonic Bias. High frequency impulses mixed with the impulses to be recorded. This reduces the distortion of a magnetic recording system.
Synchronization. Accurate co-ordination between sound and its relevant picture.
Synchronizer. Usually two-way or four-way. Apparatus through which two or more lengths of film, either picture or sound track, can be drawn in step with one another for editing or track laying purposes.
Synchronizing Mark. Reference mark enabling separate picture and sound record to be brought into step.
Synchronous Motor. Electric motor which runs at a speed determined solely by the mains supply frequency.
Take. One recording of a shot. One shot is often filmed several times and the best take selected.
Take-board. Board on which scene numbers, take numbers, etc., are recorded and filmed before each take. This information is of great use to the editor when subsequently identifying and assembling the scenes. In a sound film production the take board incorporates the clapper and is then frequently known as the clapper board.
Take-up. Reel or core winding up the film after it leaves the gate of the camera or projector.
Tape. Narrow plastic or paper ribbon coated with magnetic oxide and used for magnetic recordings.
Tempo. Impression of pace in a film created by the relative length of each shot within a sequence and by the speed of the action within a shot.
Throw. Distance from the projector to the screen.
Tilt. Up or down movement of the camera whilst taking a shot.
Title. Lettering of any kind, either explanatory or in the form of dialogue in silent films.
Title Card. Card on which the lettering is arranged or inscribed.
Title Frame. Surround to ensure accurate alignment and centering of a title in filming.
Tracking Shot. Shot from a camera that is travelling forwards, sideways or backwards. Often used to follow moving subjects. A zoom shot, which is much simpler to take since the camera is stationary, is sometimes substituted for a tracking shot.
Trailer. Blank film at the end of a reel after the last shot; also short film of extracts used to advertise a forthcoming attraction.
Travelling Matte. System used to combine selectively images on separate films so that they appear to have been taken on only one film at one time and place—e.g., an actor against a jungle background. It is normally achieved by making a series of silhouette masks from the original films and interposing them during the printing stage.
Treatment. Story of a film in its preliminary form, visualized in pictorial terms but not yet worked out in technical details. The shooting script is compiled from the treatment.
Trip Gear. Apparatus which enables single frames of film or a succession of single frames to be exposed at constant speed.
Turret (Head). Rotating mount fitted to the front of a cine camera, which carries two, three or four lenses and possibly their associated viewfinders. The desired lens can be quickly brought into position by rotating the turret.
Two-core Loading. Method of loading chargers which enables the film to be attached to the core in light.
Twin-track Recorder. Magnetic tape recorder in which less than half the width of the tape is used at once so that interchange of the spools enables a second recording to be made on the same length of tape.
Two-Shot. Medium close-up taken so as to include two people only, head and shoulders.
Two-Way. Brief term for two-way synchronizer. Apparatus consisting of duplicate sprockets mounted on a single spindle for running two lengths of film (picture or sound-track) in step with one another.
Unit. Personnel engaged in making a film.
Variable Area Recording. Optical sound recording in which the track consists of clear and opaque areas, the complementary widths of which vary from point to point along the track.
Variable Density Recording. Optical sound recording in which the light absorption of a constant width track varies from point to point.
Wild Track. Sound-track recorded independently of a picture and intended to be cut into the main track later.

Wipe. Transition from shot to shot in which a line appears to pass across the screen pushing off the first shot and revealing the second shot.
Work Print. Reversal or colour copy or a positive print used solely for editing purposes. The original or negative is cut to match it when editing is completed and is never projected, but carefully maintained as a source of subsequent prints.
Wow. Slow wailing effect produced by uneven movement of the recording medium during sound recording or reproduction. Often caused by uneven transmission in the motor drive.
Zoom. Change in apparent distance between camera and subject, effected by a variable focus lens, during actual filming. T.R. (L.A.M.)

Book: *ABC of Film and TV Terms*, by O. Skilbeck (London).

CIRCLE OF CONFUSION. The average human eye cannot distinguish between a true point and a disc of a diameter smaller than 0·01 in. (0·25 mm.) looked at from a distance of 10 ins. (25 cm.). In other words, any patch of light or shade with a diameter less than 0·01 in. appears as a sharp point. This figure fixes the acceptable standard of sharpness, or "circle of confusion", for a photographic print intended to be viewed from a distance of 10 ins. That is commonly regarded as the normal viewing distance and is a suitable range at which a healthy unaided eye can look at things without fatigue.

If the print is an enlargement, the negative will be generally very much smaller and will have to conform to a stricter standard of sharpness, and therefore a smaller circle of confusion. Again, while a contact print from the negative would generally be viewed at less than 10 ins., a poster-size enlargement would be viewed at a much greater distance. In each case the circle of confusion would have a different value. But at the proportionate viewing distances, the sharpness would appear the same.

There are two ways of approaching this question: the first relates the viewing distance to a standard image perspective, the second to a standard degree of print enlargement.

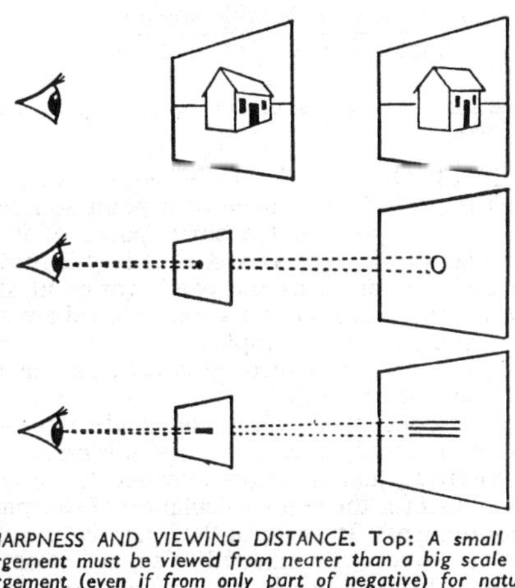

SHARPNESS AND VIEWING DISTANCE. Top: *A small enlargement must be viewed from nearer than a big scale enlargement (even if from only part of negative) for natural viewing perspective.* Centre: *The limiting blur or circle of confusion is smaller in a print viewed closely.* Bottom: *The more the negative is to be enlarged the finer must be the image resolution, since lack of detail becomes more apparent.*

Viewing Distance. The first approach assumes that to see a contact print image in its correct perspective it must be looked at from a distance equal to the focal length of the camera lens that took the picture. We should therefore hold a contact print 4 ins. away from the eye if the negative was taken with a 4 ins. lens, 8 ins. away if the negative was taken with an 8 ins. lens, or 2 ins. away if we used a 2 ins. lens.

Another way of stating the same thing is to say that a picture taken with a 4 ins. lens must be enlarged $2\frac{1}{2}$ times for correct viewing at 10 ins.

In this way the image subtends the same angle at the eye as the subject did when the photograph was taken. This is said to be the most natural way of viewing a picture.

But if a print has to be viewed from less than 10 ins. the actual diameter of the circle of confusion must be reduced because at closer range the eye can observe smaller degrees of sharpness. If the limited diameter is 0·01 in. 10 ins. away, it will be only 0·004 in. from 4 ins. away and 0·002 in. from 2 ins. away (this case would call for a magnifier to observe the image).

So a lens of 2 ins. focal length must have a circle of confusion five times smaller than a 10 ins. lens, and a 4 ins. lens must have a circle of confusion $2\frac{1}{2}$ times smaller, and so on.

In each case, however, the circle of confusion is 1/1000 of the focal length—i.e., 0·01 in. for a 10 ins. lens, 0·002 in. for a 2 ins. lens and 0·004 in. for a 4 ins. lens.

This is a satisfactory way of arriving at the circle of confusion of "normal" lenses where the focal length is approximately equal to the diagonal of the negative. The rule does not apply to close-ups where the distance between the lens and the plate is very much greater than its focal length.

Scale of Enlargement. The above method of fixing the size of the circle of confusion assumes that every photograph is looked at from the distance that gives the right perspective. But this is not true. An 8 × 10 ins. enlargement from a negative taken with a 2 ins. lens would normally be looked at from the customary viewing distance of 10 ins. The negative, in this case, would probably measure 1 × $1\frac{1}{2}$ ins. and be enlarged at least eight times. But for correct

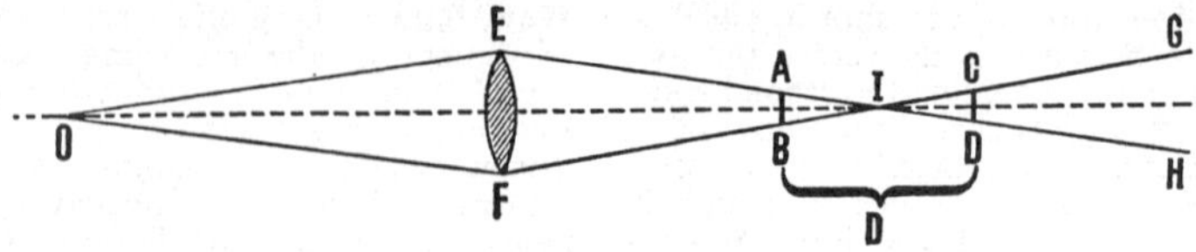

DEPTH OF FOCUS. Circle of confusion determines depth of focus. A cone of rays from the object point O meet at I after refraction by lens EF, and diverge again towards GH. Up to AB and CD (diameter of circle of confusion) the blurring of I is too small to be noticeable the limiting distance D is the depth of focus and represents the tolerance in the lens-image distance for sharp focus. Depth of field follows from analogous, though not identical, geometry.

perspective viewing at 10 ins. a negative made by a 2 ins. lens must be enlarged only five times. An 8 × 10 enlargement should, in theory, be looked at from a distance of 10 × 8/5 = 16 ins. to give the correct perspective effect.

Since people will insist on looking at big enlargements from too close a viewpoint, the "normal" circle of confusion with a diameter of 1/1000 of the focal length is not good enough. So for negatives that have to be greatly enlarged, lens makers tend to adopt circle of confusion diameters of 1/1500 or even 1/2000 of the focal length.

Long and Short Focus Lenses. Nowadays many cameras can be used with a range of lenses, some of longer and some of shorter focal length than the "normal" lens. A 35 mm. camera, for instance, with a "normal" lens of 2 ins. (5 cm.) focal length may use a telephoto lens of 6 ins. (15 cm.) focal length, or greater. According to the rules of correct perspective, the negative taken with such a lens should be enlarged 10/6 = 1·6 times.

In practice, of course, the telephoto lens is used to give a magnified image of a distant object, and the perspective is not considered at all. So the negative taken with a telephoto lens is enlarged by the same amount as a negative taken by a normal lens.

By the same reasoning, the negative taken by a wide angle lens is enlarged on the same scale. In this way each lens shows its characteristic magnified, normal, or reduced scale of reproduction for the same amount of negative enlargement. But only the normal lens will produce normal perspective; both long-focus and wide-angle lens pictures will show distortion at the normal viewing distance of 10 ins.

The final effect on the circle of confusion is simply this: all lenses intended to be used with the same negative size have the same circle of confusion, irrespective of their focal length. And the circle of confusion is in all such cases the same as the circle of confusion demanded for the "normal" lens for that negative size.

So the circle of confusion chosen for any lens depends on the size of the negative it is to cover. If a 6 ins. lens is intended for use as a long focus lens on a miniature camera, it will have a smaller circle of confusion than a 6 ins. lens intended for use as a normal lens on a 4 × 5 ins. camera.

In the same way, the circle of confusion for a wide-angle lens is larger than when used as a normal angle lens on a smaller negative.

This difference affects hyperfocal distance and depth of field values as well as the acceptable amount of blur in calculating shutter speeds to arrest movement, and tolerances in lens correction.

Each case calls for three sets of values for each focal length of lens, depending on whether it serves as a normal, long focus, or wide angle camera lens for that negative size.

For calculations it is still most convenient to relate the circle of confusion directly to the focal length of the lens. We can do that by multiplying the hyperfocal distance—for depth of field calculations—or dividing the subject distance—for shutter speed calculations—by the "power" of the long focus or telephoto lens. This is the ratio:

$$\frac{\text{focal length of long focus lens}}{\text{normal focal length for the negative size}}$$

For long focus and telephoto lenses this factor is always greater than 1.

With wide angle lenses the factor corresponding to the "power" of the lens is given by the ratio of:

$$\frac{\text{angle of view of wide angle lens}}{\text{angle of view of normal lens}}$$

L.A.M.

See also: *Circle of least confusion; Depth of field; Depth of focus.*

CIRCLE OF LEAST CONFUSION. Since no lens is perfect, the image of a point source is not a true point but a small patch of light, circular on the lens axis. As the lens is moved in and out of focus the patch varies in size. When the patch is at its minimum diameter, and the image, by implication, is in correct focus, the patch is then often described as the "circle of least confusion".

The concept of circle of confusion, and therefore circle of least confusion, is essentially theoretical, and is simply intended to be used as a basis for the ready calculation of sharpness requirements. It assumes that the image patch is an evenly illuminated disc, whereas in practical optics this is not so, and the over-all diameter is not, in itself, the sole criterion of definition.

For example, a patch consisting of a bright central area with surrounding halo (such as frequently occurs behind the image plane, and therefore records objects behind that focused on) gives better definition than a patch of the same over-all diameter, but with a bright peripheral ring (such as frequently occurs in front of the image plane, and therefore records objects in front of that focused on).

Such characteristics vary from lens to lens and determine the actual performance in practice. This covers the quality of a slightly out-of-focus image and therefore the actual depth of field, as well as the position of best focus.

See also: *Circle of confusion.*

CIRCUS. There are two principal reasons for regarding circus photography as a distinct subject and not merely as another type of theatre photography: the lighting comes mostly from directly overhead; the audience encircles the performers. These factors largely govern the choice of equipment and the best technique for taking photographs.

The lighting usually comes from above and tends to be concentrated on the centre of the ring. Spotlights are occasionally used from the side of the ring. There is rarely any change of colour or brilliance in the lighting from start to finish of the performance.

Because the audience encircles the ring, there is a wide choice of viewpoint, but only certain positions are suitable for photography. The performers always play to the higher priced seats so the photographer should take care at least not to sit on the opposite side of the ring or he will get little but back views. A ringside seat will be too close to take in much of the ring but it is useful for close-ups of individual items. Generally a seat next to the gangway and two or three rows back from the ring is the best compromise, and the photographer can stand up if necessary without spoiling the view of the people behind.

Camera. Practically any good camera with a lens of at least $f\,3{\cdot}5$ and an eye-level viewfinder can be used for circus photography. A miniature camera with coupled focusing or a reflex camera with eye-level focusing gets over the difficulty of focusing by scale in the poor lighting of the auditorium. Where the camera must be focused by scale, it is useful to remember that all circus rings are 42 feet in diameter. It is never possible to use such a small stop that the depth of field covers the whole ring.

A lens of normal focal length covers most of the requirements of circus photography as the photographer is always fairly near the subject. Even a normal lens will not cover the whole of the ring from the second or third row, so a long focus lens is only necessary when shots are wanted of just part of the ring. A deep lens hood is advisable, more especially when the camera is pointed up for shots of the trapeze artistes.

Exposure. For most shots the exposure should be made with the fastest shutter speed possible at the full aperture of the lens. Generally this will be about 1/100 second at $f\,3{\cdot}5$. Slower shutter speeds may be used for acts employing the larger animals—e.g., bears, sea-lions, elephants. It is usually safe to double the exposure meter reading with circus lighting.

There is no point in using anything but the fastest panchromatic material.

With fast moving subjects, the fastest shutter speed possible under available lighting may be still too slow to arrest movement. In such cases, the exposure should be made at the "dead point" in the movement (e.g., when a trapeze artiste is at the highest point of his swing).

Colour. The combined use of a fast colour film and a wide-aperture camera lens makes colour photography possible, even when using the available light. Whether photography is possible in any given lighting must be found by use of an exposure meter and through experience. It is essential to expose correctly for the performers in the ring, even if the entire background is thus reproduced black. A correctly exposed background serves no useful purpose, if the main point of interest is and lifeless.

The Subject. There are several kinds of subject: acts on the edge of the ring (clowns) which call for fast shutter speeds; fast moving acts in the ring proper (equestrian items) which also need fast shutter speeds; slow moving animal acts (elephants, etc.) which can be shot at 1/50 second; trapeze turns which may be shot at 1/50 second by careful timing, and the spectators'—particularly children's—faces around the edge of the ring, for which exposures as slow as 1/10 second may often be used. Anything that takes place inside a cage is not worth trying for because the bars are invariably too distracting in the finished photograph.

It is sometimes possible to get a seat at a dress rehearsal. This is a great help in establishing the most suitable viewpoint and technique for the real show. Failing this it is worth while making a preliminary visit with a notebook and no camera.

It is perhaps unnecessary to point out that the photographer should always make sure that photography is permitted (it is often encouraged) and that under no circumstances should any kind of flash photography be attempted. L.V. & F.P.

See also: *Ice rink; Theatre.*

Book: *My Way with the Miniature*, by L. Vining (London).

CITRIC ACID. Acid used in toners and clearing solutions.

Formula and molecular weight: $(CH_2CO_2H)_2COHCO_2H$; 192.

Characteristics: Colourless crystals with intensive sour taste.

Solubility: Very soluble in water at room temperature, soluble in alcohol.

CIVIALE, AIMÉ, 18??—1893. French army officer and Alpine photographer. Perfected the paper negative process and used it in topography. His equipment for large composite panorama photographs of the Pyrenees weighed 550 lbs. and required 25 men and mules for transport (1887–8).

CLAUDET, ANTOINE FRANÇOIS JEAN, 1797–1867. French daguerreotypist and scientist. Resided from 1827 in London as glass importer. Was one of the first and last to practise the daguerreotype process. Established a supply house for daguerreotype materials in partnership with G. Houghton. Claudet was also an early professional photographer (he established the second photographic studio in London, in 1841) and introduced painted backgrounds. Published in 1841 acceleration process for the daguerreotype. Introduced red light into the darkroom. He was one of the foremost photographic scientists of this early period. He suggested in 1851 the use of series of photographs for the synthesis of motion in the Phenakistiscope.

CLAYDEN EFFECT. Term for the partial reversal of a negative produced by giving the plate or film a very short exposure to an intense light before the normal exposure in the camera.

CLEANING PHOTOGRAPHIC VESSELS. Bottles and dishes that have become stained or dirty in the course of photographic processing can generally be cleaned with a solution of 1 part of strong hydrochloric acid to 3 or 4 of water. This solution removes the lime deposited from hard water. It is this deposit that holds most of the chemical discoloration; when it is removed, the other impurities go with it. The solution must not be allowed to get on to the fingers, and the vessel should be well washed in clean water to remove the acid.

Developer and fixer vessels often acquire a blackish deposit of metallic silver. This can be removed with nitric acid. Developer stains can also be removed by a strong solution of potassium permanganate or bichromate and sulphuric acid (as used in reversal processing); the brown stain is then removed with a 5–10 per cent solution of potassium metabisulphite.

Most deposits on dishes can be removed by the above methods, but stains which have penetrated the surface of porcelain dishes indicate that the glaze is faulty.

There are various ways of removing stubborn deposits from the inside of bottles, but most of them involve the use of abrasives or concentrated acids. Bottles are so cheap that it is rarely worth while using such extreme measures.

If bottles and dishes are well rinsed under the tap and dried each time after use it is a simple matter to keep them clean.

See also: *Stains.*

CLEARING BATH. Any bath included in the processing of a negative or print to remove stains or in some cases neutralize chemicals left by a previous operation.

See also: *Reversal process; Toners.*

CLEAR SPOT FOCUSING. Otherwise known as aerial focusing. A method of focusing in which the ground glass image is dispensed with and the subject focused optically in a clear spot produced by cementing a microscope cover glass to the centre of the focusing screen with Canada balsam or oiling the glass to the ground glass with a drop of cedar wood or similar oil.

See also: *Focusing.*

CLERC, LOUIS PHILIPPE, 1875–1959. French professor of chemistry at the Faculté des Sciences de Paris; also taught at the Collège Technique Estienne physics and chemistry applied to photography, and the theory of photo-engraving, Hon. Professor at the Institut d'Optique. Author of *La Technique Photographique*, *La Technique des Reproductions Photomécaniques* and other books. Founded *Science et Industries Photographiques* (1921), which under his editorship commanded world wide respect for his exceptionally lucid and correct abstracts.

CLERK MAXWELL, JAMES, 1831–79. Scottish mathematician, physicist and natural philosopher.

See also: *Maxwell, James Clerk.*

CLICHÉ-VERRE. Process in which pictures drawn or painted on glass are copied on to photographic paper. Designs are painted on a glass plate with semi-transparent varnish or oil paint, or scratched with an etching needle on an opaque coating of dark varnish or sensitized collodion darkened by exposure to light. Talbot was the first person to make "etchings" in this manner (*c.* 1835). The technique on varnish was devised, independently of Talbor, by William and John Havell and J. T. Wilmore in London, 1839. Collodion was used as the coating in 1853 by three French amateur photographers, Constant Dutilleux, L. Grandguillaume and Adalbert Cuvelier. Camille Corot, a friend of Dutilleux, made many clichés-verre; so occadid Daubigny, Millet, Théodore Rousseau, and other artists.

The cliché-verre process has been repeatedly re-invented, but has never been able to establish itself.

CLICK STOPS. Type of iris diaphragm control on camera and enlarger lenses which engages with a definite click at positions corresponding to the standard aperture settings. Lenses fitted with click stops can be stopped down by touch—e.g., in the dark—by counting the number of clicks felt or heard during movement of the setting ring from either end.

See also: *Diaphragm.*

CLIFFORD, CHARLES, ?–1863. English Court photographer to Queen Isabella II, in Madrid. Active chiefly as an architectural and landscape photographer and published several volumes of Spanish views in the 1850s.

CLINICAL PHOTOGRAPHY. Medical photography (other than radiography) of parts or the whole of the patient for diagnosis purposes of treatment, progress, etc.

See also: *Dental photography; Endoscopic photography; Medical photography; Ophthalmic photography.*

CLOSE-ACCESS PHOTOGRAPHY. Photographic or cinematographic set-up in which the finished picture is obtained within seconds of the exposure, ready for immediate viewing, evaluation or projection.

See also: *Rapid Access.*

CLOSE-UP CAMERA. Camera unit specially designed for close-up photography, frequently by untrained operators—for example in medical, dental and similar fields.

While many of the more versatile cameras, in particular miniatures, are adaptable to close-up photography with special accessories, a close-up camera usually has such accessories or features incorporated in its design. Thus the lens may be pre-focused to a certain near distance, or set to several shorter distances with the aid of supplementary lenses. The camera often includes a distance gauge or other mechanical probe to set the camera at the right distance from the object. Flash or electronic flash equipment of a specific light output may also be associated with the camera.

The emphasis is on ease of operation and restricted versatility. Hence the simplest close-up cameras are straightforward box models with a built-in flash reflector and a selection of two or three supplementary lenses. A doctor's or dentist's assistant can then take pictures of specific subjects by placing the attached distance gauge or field frame over the object area to be covered and pressing the button. The camera functions automatically look after exposure, sharpness and other technical details. More advanced types may incorporate an optical ranging device—e.g. a beam of light projected through a rangefinder system so that two beams converge into a point at the centre of the object area when the latter is at the right shooting distance.

See also: *Eye camera.*

CLOSE-UP (CINE). Film shot in which the camera appears to be sufficiently near to the subject to show or pick out the kind of detail a person would normally see at arm's length. The close-up of a person would show the head or head and shoulders.

A big close-up goes very near to show up an item of minute detail, e.g. the eyes or the mouth.

See also: *Cine technique.*

CLOSE-UPS

Photographs of subjects which are too close to the camera to be focused by the normal camera focusing mechanism. Close-ups are taken by the use of double or triple extension where the camera is so equipped, an extension tube, or a supplementary lens.

Camera Extension. The extension of the normal camera is governed by the length of the baseboard or the range of movement of the lens mount. With practically all roll film and miniatures this movement permits only a focusing range from infinity to about 3 feet.

Field and other cameras which focus by racking the whole lens board forward or backward usually have a range of focusing movement as much as two, or even three times the focal length of the camera lens. This double or triple extension is used for really close-up focusing in copying and macrophotography.

The object distance for copying to a given scale is given by the equation:—

$$u = \frac{F(M + 1)}{M}$$

where u = Distance of object from lens
F = Focal length of lens
M = Scale of reproduction

The scale of reproduction is always the ratio of image size/object size.

If the image is smaller than the object, the scale of reproduction will be a reduction. For instance, if the object is four times as large as the image, the scale of reproduction is 0·25, or 1/4, or 1 : 4. If the image is larger than the object, the scale of reproduction is greater than 1, and represents a magnification. For instance, if the image is twice as large as the object, this is expressed as magnification 2 : 1, or 2/1, or 2.

Example: the object distance for copying on a scale of 0·5 (a reduction of 1 : 2) with a 6 ins. lens is:—

$$\frac{6(0{\cdot}5 + 1)}{0{\cdot}5} = 18 \text{ ins.}$$

The extension of the lens beyond the infinity setting is given by the equation:—

$$d = FM$$

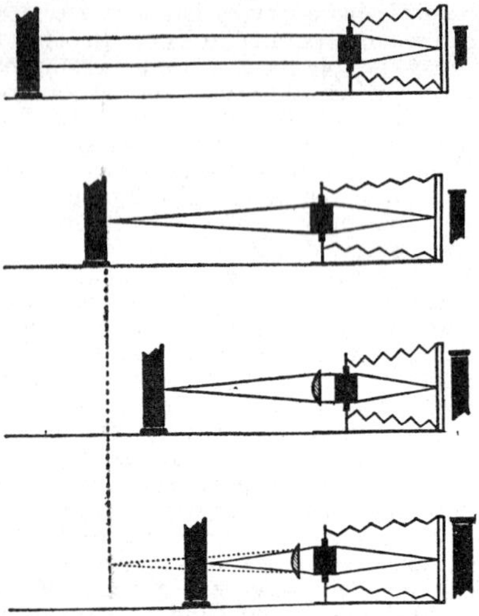

CLOSE-UP FOCUSING. Top: *The smallest image of an object is formed with the camera focused on infinity.* Upper centre: *Moving the camera lens forward focuses it at a nearer distance, giving a larger image.* Lower centre: *The image is larger still (and the subject distance shorter) with the camera focused on infinity, but a supplementary lens placed in front of it.* Bottom: *The largest image and closest subject distance is obtained by moving the lens forward to its nearest focusing limit and adding a supplementary lens.*

Where d = Focusing extension in front of infinity setting
F = Focal length of camera lens
M = Scale of reproduction

The total bellows extension needed is the focal length plus the extension beyond the infinity setting, i.e., d + F or F(M + 1).

For example, when copying on a scale of 0·5 with a 6 ins. lens the focusing extension beyond the infinity setting is:—

$$6 \times 0{\cdot}5 = 3 \text{ ins.}$$

giving a total bellows extension of 9 ins.

At such close quarters the object distance is measured from the front node of the lens. For a single lens this is approximately one-third to one-half the actual thickness of the lens behind the front surface. In practice it is not possible to guarantee sharp focus from calculation and measurement. The final adjustment is done by inspecting the image on the focusing screen of the camera.

The scale of reproduction at any object distance can be found from the equation:—

$$M = \frac{F}{u - F}$$

where M = Scale of reproduction
F = Focal length of lens
u = Object distance from lens

So for an object 24 ins. away from a 6 ins. lens the scale of reproduction is:—

$$\frac{6}{24 - 6} = 0{\cdot}33 \text{ or } 1 : 3$$

i.e., the image is one-third the size of the object.

The scale of reproduction obtainable with the camera fully extended is:—

$$M = \frac{e - F}{F}$$

where M = Maximum scale of reproduction
e = Total camera extension
F = Focal length of camera lens

For a double extension camera where e = 2F the maximum scale of reproduction is 1 : 1, or same size. For a triple extension camera where e = 3F, the maximum scale of reproduction is 2 : 1 or 2 times enlargement.

The image distance (total camera extension) at any reduction is the same as the object distance at the corresponding magnification, and vice versa. Thus the image distance which gives a scale of reproduction of 0·33 (1 : 3) is the same as the object distance which gives a scale of reproduction of 3 (3 : 1) with the same lens.

A double extension covers practically the whole range of normal close-up requirements while a triple extension brings the equipment into the realm of macrophotography. But while it is feasible, although expensive, to fit a double or triple extension to the bellows type of folding camera, it is not a practical proposition with those which have focusing mounts, front cell focusing and other systems. So with such cameras—and those which in any case are only equipped for down-to-3-feet focusing—one of the methods described below must be adopted.

Extension Tubes. These are simply distance pieces which are inserted between the lens and its mount to provide extra camera extension. Such devices can be used only where the camera lens is easily removed—e.g., by a quick screw thread or a bayonet mount. Since the diameter of the tube—especially for long separations—may approach the diagonal of the negative, the method is really only suitable for miniature cameras. With 35 mm. cameras in particular, extension tubes are an excellent means of close-up focusing.

Most of the classic miniatures list ranges of sizes of tubes made specially to suit the various lenses. The set of tubes, used in conjunction with the normal focusing movement of the lens, enables the subject to be brought progressively nearer up to as little as 2 ins. (5 cm.) giving a maximum reproduction scale of 1 : 1.

With cameras having interchangeable lenses and a focal plane shutter, the length of extension tube that can be used is limited only by considerations of vignetting due to the size of the lens mount, camera steadiness and focusing adjustment. The practical limit of such systems is normally reckoned to be reached with magnifications of around 10 ×. But this is merely because at higher magnifications photographing through a microscope becomes more convenient. Here the short focal length of the microscope objective still permits the use of reasonable dimensions in the camera set-up. With a standard 50 mm. lens on a miniature camera a magnification of say 25× would require a lens-to-film distance of some 4¼ feet!

With interchangeable camera lenses mounted in front of a diaphragm shutter, extension tubes can still be used but are much more limited. Since the shutter aperture is in effect some way along the light path between the lens and the film plane, extension tubes of any length would result in appreciable vignetting. Close-ups with such cameras are therefore frequently taken by using a combination of extension tubes and supplementary lenses.

Adjustable extension bellows can also take the place of extension tubes. Such units consist of a front and a rear panel mounted movably on a single or double rail. The distance between the two panels is continuously variable from a few millimeters to several inches. The rear panel fits to the camera body, and the front panel takes the lens or one of several lenses (often with appropriate adaptor rings or fittings). Usually such units are tailored to a particular camera model. Their main advantage is that they provide a continuous focusing movement ranging—according to the lens used—from subjects at infinity down to same size or larger reproduction. The bellows movement may include a calibrated scale marked either in scales of reproduction for a particular focal length or merely with a numerical scale to permit easy reference to a particular noted setting.

With extension tubes, bellows or mounts some means of accurate focusing, preferably with a visual image, is required. For this purpose reflex housings or focusing stages are provided for certain camera models and show the image on a ground glass screen. This is then replaced for the exposure by the camera but with the film plane in the same position as the screen.

For macrophotography the method is practicable up to magnifications of 10×.

Close-up Focusing Mounts. Long focusing extensions are normally associated with the bellows-type technical camera, but helical focusing mounts which move the whole lens forward or back can also be designed for an appreciable focusing travel to include a close-up range. There are two approaches to this.

The first, and older one, has been used with rangefinder miniature cameras with interchangeable lenses. Here the lens is removed from the camera and inserted into a supplementary focusing mount which at the same time also acts as an extension tube. Usually the focusing movement covers subject distances from the near limit of the normal focusing system (around 3½ feet) to about half this distance—i.e. a scale of reproduction of 1 : 6 to 1 : 8 with the standard camera lens.

When used with rangefinder cameras such devices frequently incorporate an optical compensator (for example a wedge) that fits in front of one of the rangefinder windows to provide additional convergence of the rangefinder beams. The rangefinder then indicates the exact focus with the close-up focusing mount as well, and can be used for hand-held close-up photography even of live subjects.

The other and more recent system is the incorporation of a helical focusing mount with extra extension into the lens itself. This may be achieved by double helical movements which in extreme cases cover a range of focusing distances down to same size reproduction. The built-in close-up mount has the advantage that it covers the full focusing range from infinity downwards. Lenses of this kind are usually intended for reflex cameras, since a corresponding rangefinder compensation is difficult to design.

Supplementary Lenses. Where the camera has only the normal focusing range, and the lens cannot be removed, or the camera is otherwise not adapted for extension tubes, close-up focusing can be carried out by adding a positive

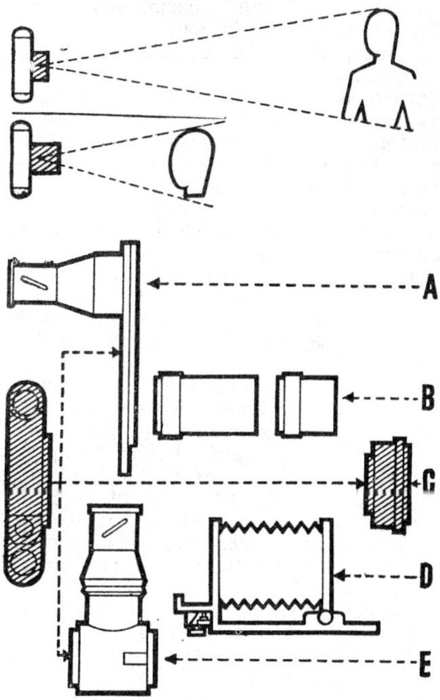

EQUIPMENT FOR CLOSE-UPS. Top: *In the usual photographic apparatus, the lens is racked out to focus near objects sharply.* Centre and bottom: *For greater-versatility, a camera with interchangeable lenses permits the use of extension tubes or bellows, and also of a focusing device to ensure maximum image sharpness.* A, *focusing stage with ground glass screen and focusing magnifier.* B, *extension tubes.* C, *camera lens.* D, *extension bellows.* E, *reflex housing with ground glass or clear glass focusing screen.*

supplementary lens in front of the camera lens. This calls for no extra extension, although the effect is greatest when the camera lens is at the forward limit of its movement.

When the camera lens is set to infinity, and a positive supplementary lens is fitted in front of it, the combination is focused on objects at a distance equal to the focal length of the supplementary lens, e.g., if a 24 ins. supplementary is added, whatever the focal length of the camera lens, the combined lenses will be sharply focused on objects 2 feet away.

The supplementary lens does not alter the marked values of the camera lens aperture, but if it is a simple, uncorrected supplementary, it degrades the performance of the camera lens. For this reason the camera lens has to be stopped well down if critical definition is required.

Certain high quality supplementary lenses of special correction also exist; these are usually intended for use with a particular camera lens. Interchangeable-lens cameras using a convertible lens system, i.e. where the front lens component is interchangeable for different focal lengths—may include in their range of alternative lens units special macro-lenses. These are suitable front components that yield high image quality at or around same size reproduction.

Focusing Aids and Gauges. Where supplementary lenses are used on a reflex camera no additional focusing aid is necessary. The reflex screen directly shows the image sharpness as well as the field of view taken in.

Various ways exist for other cameras. The simplest is a focusing table showing the exact distance (measured from the subject to either the front of the supplementary lens or to the film plane) for each setting of the camera lens with any given supplementary lens. Such a table requires accurate measurement, which is a little cumbersome in practice.

Handier are close-up rangefinders which fit into the accessory shoe of the camera. These are usually designed specifically for a given camera and set of supplementary lenses. Using them involves setting the rangefinder to the required camera lens setting and power of supplementary lens, and then approaching the subject until the double rangefinder images fuse into one. At that distance the subject will also be sharp on the film.

Distance gauges are available for a number of cameras and exist in versions for either supplementary lenses or extension tubes. They consist of a frame at the end of a rod of fixed length which is attached to the camera. Anything in the plane of the frame will be sharp on the film, while the limits of the frame also outline the subject area taken in. Alternatively, a set of four legs may be fixed to a collar mounted either between the lens and the camera or to the front of the lens mount. In this case the tips of the legs define the plane of maximum subject sharpness as well as the field of view. Frequently such distance gauges serve a set of supplementary lenses, in which case the legs or rod are extendable to several set positions. This type of distance gauge is particularly suitable for close-up copying, since the camera can simply be placed on top of the surface of the original.

Various modifications of the system exist, and work on more or less the same principle.

Field Covered by Lens. The size of field covered at any distance is given by:

$$W = \frac{N}{M}$$

where W = Width of field
N = Width of negative
M = Scale of Reproduction

For instance, photographing on a scale of 0·2 (i.e., a reduction of 1 : 5), the width of the field taken on the long side of a 4 × 5 ins. negative is 5/0·2 = 25 ins. and on the short (4 ins.) side, is 4/0·2 = 20 ins.

Since at close range the lens is considerably farther from the film than in normal photography, a given lens can cover a larger negative format in close-up photography. For instance a 25 mm. lens with a covering angle of 60° would, when focused on infinity, cover an image circle of 29 mm. in diameter. It would therefore be unable to cover the full image area of a 24 × 36 mm. negative. When used at same-size reproduction however the same covering angle corresponds to an image circle of some 58 mm. —which can comfortably accommodate a 24 × 36 mm. negative area. Hence lenses of specially short focal length are frequently used for macro-photography at large magnifications, and lead to more manageable camera set-ups.

Distortion. As the camera gets closer to the subject, the perspective of the picture it sees becomes more and more unpleasant when the print is viewed from a normal distance. There is no remedy for this; the viewpoint and the perspective have a fixed relationship.

The more the subject approaches a single plane, the less the apparent distortion becomes —e.g., there is none in a close-up of a two-dimensional drawing, very little in a sculptured relief, but a great deal in a close-up portrait where the sitter's limbs extend towards the camera. There will therefore be less in a profile portrait than in a full-face, particularly if the sitter has a long nose or jutting chin. There will also be less in a photograph of a dog lying across the picture than if its head is much closer to the camera than its tail. And a child will have more normal proportions if photographed from its own eye level than if the camera looks down on it so that its head is so much nearer—and larger in proportion—than its feet. These examples illustrate the general approach to close-up subjects to preserve natural perspective.

Depth of Field. As the camera gets closer to the

subject, the zone of sharpness shrinks closer and closer to the actual focused distance. The effect can be compensated for—up to a point—by stopping down the lens, but at close distance a very narrow zone of sharp focus must be accepted as inevitable. As the depth of field thus decreases, depth of focus increases.

Exposure. Where the camera extension is increased (by bellows or extension tubes) the marked lens aperture has to be increased in proportion to the increased lens-film distance, requiring extra exposure. In addition, a further exposure increase is necessary with near shots because the shadow areas become more important when they are seen close at hand.

When comparatively long camera extensions are used for taking close-ups, the relative lens aperture (*f*-number) must be multiplied by a correction factor to give the correct value for calculating exposures. This correction factor depends on the scale of reproduction:—

True *f*-number = nominal *f*-number × (M + 1) where M is the scale of reproduction.

When the scale is smaller than 0·1 (reduction more than 1 : 10) the error in the *f*-number is less than 10 per cent (one-fifth of a stop) and can be ignored.

Some lenses with built-in close-up mount may have the aperture control ring coupled with the focusing movement in the near focusing range. Focusing the lens on a near distance then automatically opens up the diaphragm to compensate for the required exposure increase.

The Subject. Subjects for close-up photography must be as nearly as possible in one plane if they are to be sharp because the depth of field at short distances is very small. On the other hand, the depth of focus increases and the image on the focusing screen changes only very slowly from sharp to unsharp during focusing. L.A.M.

See also: *Depth of field; Depth of focus; Extension bellows; Extension of camera; Focusing stage; Macrophotography; Optical calculations; Photomicrography; Reflex attachment; Supplementary lenses.*

Books: *Amateur Photomicrography*, by A. Jackson (London); *Camera Close-up*, and *Camera Copying and Reproduction*, by O. R. Croy (London); *Close Range Photography*, by C. H. Adams (London).

CLOTH PRINTING. Several processes have been developed for making photographic prints on woven material.

See also: *Fabric printing.*

CLOUD NEGATIVES. Cloudless views can often be improved by printing in clouds from a second cloud negative. Where the skyline is fairly straight, the procedure is quite simple. A selection of cloud negatives is useful.

Simple Shading. First the landscape negative is enlarged in the usual way, but the sky portion is shaded during exposure with a card cut to the shape of the skyline. (It is generally wise to stick scraps of white paper on the margins of the holder to indicate the position or level of the skyline.) The card is kept moving slightly during the exposure to avoid a sharp edge.

Next the cloud negative is inserted in the negative carrier and, with the orange filter over the lens, it is focused and brought into position over the same sheet of printing paper. (A rough sketch of the final effect aimed at is useful as a guide at this stage.)

Finally, the sky area of the print is exposed while the previously exposed foreground area is shaded up to the level indicated by the guide marks on the paper holder.

The cloud and landscape negatives should be alike in the direction of the lighting, viewpoint, and in the character of the daylight—i.e., season, time of day, weather, etc. The exposure times for each negative are best found by separate test strips.

Masking. Where more complicated outlines are to be shaded, fairly accurate masks and counter masks are required. They are used on a masking stage consisting of a sheet of glass supported a little way above the paper holder. It is necessary to allow some space between the mask and the paper so that the image of the mask will be slightly blurred.

The first negative is focused on the paper holder or baseboard, and a sheet of white but fairly opaque paper is placed on top of the glass above it. The outline of the image to be printed is then traced on the paper. The paper is then cut along the line.

The two parts of the paper make up the mask and counter mask. They are fitted together on the masking stage and the outside edge of each mask is taped or stuck down, so that each part can be either folded back out of the way, or brought into position without having to be re-aligned.

With the first negative correctly focused in the enlarger, the lower mask is folded out of the way, and the sky mask is folded down and held in position on the glass plate. The printing paper is inserted underneath the masking stage, and exposed.

After the first exposure, the paper is left in position and both masks are folded down on the masking stage. The orange filter is slipped over the enlarger lens and the second negative is inserted in the enlarger, and positioned so as to fit the outline of the counter mask. Finally the counter mask is folded out of the way, and the second exposure is made.

Exposure times are found beforehand by test strips.

Both negatives must be enlarged to the same degree. If there is any difference in scale, the shadow outlines of the two sides of the masks will not coincide on the printing paper. This discrepancy would produce either a white or a

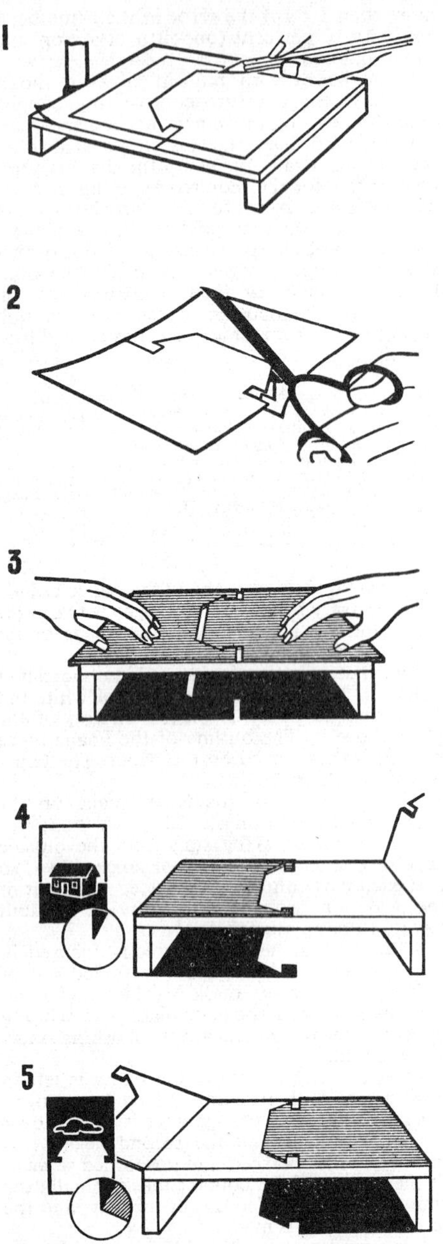

CLOUDS BY COMBINATION PRINTING. 1. *Mark out skyline of landscape on a sheet of paper supported on glass platform just above the enlarger easel.* 2. *Cut out mask and countermask.* 3. *Fit mask and countermask together on platform, tape down outside edge.* 4. *Fold mask out of way and expose paper on easel to landscape negative.* 5. *Bring back mask, remove countermask, and expose to sky negative.*

dark, uneven outline along the boundary of the two images.

For more involved masking it becomes necessary to construct special masking boxes. These are made to hold the masks on one or two glass frames and to allow the printing paper or paper holder to be removed and then replaced in exactly the same position relative to the masks. L.A.M.

See also: *Tricks and Effects.*

CLOUDS. The sky and clouds are often an important part of the scene that the photographer wants to record, but unless he knows how to deal with them, they often reproduce as blank white paper in the resulting print. There are two reasons for this:

(1) All black-and-white negative emulsions are more sensitive to light and medium blue than to any other colour, and

(2) The sky is more often than not the brightest part of any outdoor scene.

Cloud and Light. The greater the contrast between sky and cloud, the easier it is to photograph them, and the contrast varies according to the light and the type of cloud. Some clouds are much brighter than others; the brightest are the "cotton wool" cumulus clouds, and because they generally occur in a dark blue sky, they are the easiest type of cloud to photograph. Cumulus clouds are about twice as bright as the upper part of the low-level nimbus (rain) clouds which tend more to a grey tone. Half-way between those two types in brilliance, are the high, wispy cirrus clouds (mares' tails), but they usually have a very pale sky as background and therefore show very little contrast to the sky.

The darkest part of the sky is always near the zenith but slightly to the north; the blue lightens towards both the sun and the horizon. Just above the horizon the sky is almost white and clouds do not stand out from it.

The direction of the lighting also influences the brightness of clouds. Front lighting produces the greatest reflection towards the camera and therefore the greatest brightness. But what is gained in brightness is apt to be lost in modelling. The sun shining on clouds at an angle of about 45°—i.e., over the photographer's shoulder—produces both good modelling and bright clouds.

Haze. In the course of the day, over-all contrast again varies considerably. The best time for photographing clouds and skies is between one and four hours after sunrise and from four o'clock in the afternoon up to about one hour before sunset. Atmospheric haze always tends to lower the contrast of the scene, and during the middle hours of the day there is much more haze in the atmosphere than at the beginning and end of the day. And there is also more haze in summer than in spring and autumn. The finest and clearest cloud formations are often found in April and May.

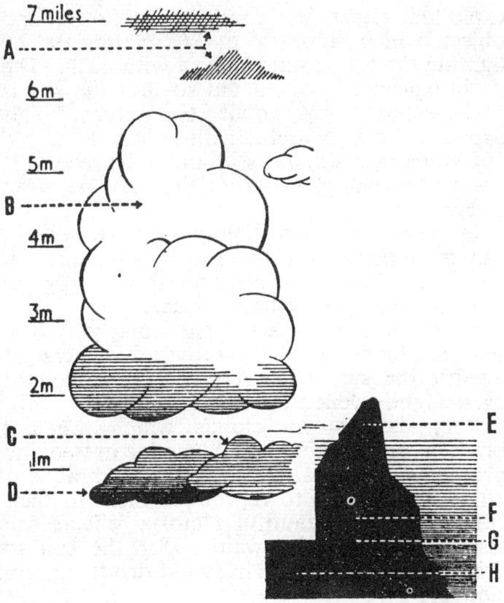

CLOUD TYPES. Left: *Cloud formations and height of occurrence. A, cirrus clouds. B, cumulus, going over into cumulonimbus on lower levels. C, nimbus. D, rain-carrying storm clouds.* Right: *Filters required to penetrate haze at different heights. E, Ultra-violet (higher mountain regions). F and G, light to medium yellow (low mountain regions). H, deep yellow to red (valleys and flat country).*

There is far more haze in the air over the British Isles than in continental climates, and more over big cities than over country districts. The amount of haze diminishes with increasing altitude and disappears altogether at about 6,000 feet in summer and 3,000 feet in winter.

Film. The brighter types of cloud can be photographed satisfactorily against the darker parts of the sky on panchromatic film without the need for correction filters.

Colour. The cloudscape for its own sake is not nearly as often attempted in colour as in black and white, except for sunset effects where colour really does score. The more delicate sunset is better than the lurid type which may come out rather exaggerated. In landscapes where the sky plays a secondary role it usually records well even without special care, and distant clouds are still separated from the sky by the film.

Filters. With black and white film medium yellow or green filters in normal conditions give a true rendering of sky tones; orange and red make them dramatically dark. At high altitudes a U.V. filter provides all the correction necessary; even a pale yellow gives too much contrast.

There are also graduated sky filters in which the tint deepens towards the upper half, and the lower half is quite clear. Such filters prevent over-exposure of the sky while leaving the landscape unaffected: they call for no increase in exposure.

With colour film, the only filters that can be used are colourless U.V. filters and neutral grey polarizing screens. These darken the blue of the sky by cutting out the polarized light rays which are reflected principally from just above the horizon. In this case the effect must be studied either on a focusing screen, or by looking through the filter and rotating it until it is in the position that gives the desired effect.

Although all makes of films and filters show as many subtle differences as there are variations in the colour of the sky, the table above will help to produce good results in most conditions.

FILTERS FOR CLOUDS

Conditions	*Desired Rendering* *Natural*	*Crisp*	*Dramatic*
Hazy sky	Medium yellow or green	Dark yellow, green or orange	Red
Clear but little contrast (pale sky)	Pale yellow or green	Dark yellow, green or orange	Red
Blue sky with good contrast	Pale yellow	Medium yellow or green	Orange
Very dark sky with brilliant contrast ...	None	Pale yellow	Medium yellow or green
The same above 6,000 feet[1]	None	U.V.	Light yellow
Sunset with blue sky ...	Pale yellow	Medium yellow	Orange or red
Sunset with red sky ...	None[2]	Pale green or blue-green	
Black storm clouds ...	None	None	None

[1]Distant cloud photography not recommended during the middle of the day.
[2]Light green filter with pan red film.
This table is intended for panchromatic film material. With ortho film double the strength of filters, but use yellow ones only.

SUN AND CLOUDS. *Side light gives best modelling, back lighting shows greatest contrast, front lighting yields flat results.*

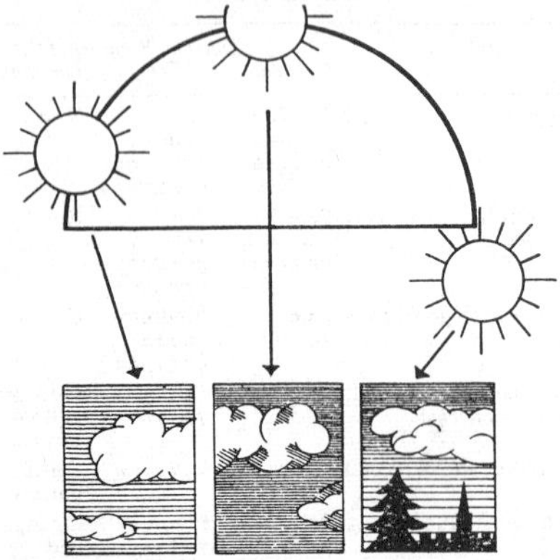

CLOUD ILLUMINATION. Left: *Early morning clouds are delicate against a light sky.* Centre: *Midday clouds show more modelling against a darker sky.* Right: *Evening sky after sunset is graded in tone with brilliant backlit clouds.*

Exposure. Correct exposure is vitally important. Over-exposure eliminates the delicate nuances of tone between sky and cloud and gross over-exposure will even cancel out the effect of filters. The danger of over-exposure is greatest when photographing against the light, when clouds are brilliantly rim-lighted, but the landscape is usually in shadow.

Under-exposure has the opposite effect and increases contrast between cloud and sky by darkening the blue tones. This effect can be exploited either by choosing a foreground object which appears in silhouette, or by lighting the foreground subject with flash. The flash exposure is calculated so that the sky is under-exposed by about two stops. This expedient is also useful with colour film.

Clouds rarely stand still and 1/10 second is about the longest exposure that can be given safely.

In the darkroom, a slightly over-exposed sky can be darkened on the print by shading; if it has been heavily over-exposed, it may be reduced with persulphate reducer.

Excellent cloud effects—particularly at sunset—can be created by photographing straight against the sun when it is partly hidden or even slightly veiled by cloud or mist.

Very dark storm clouds which are not brighter but darker than blue sky present no problems. Normal photographic technique is all that is needed to reproduce them in their correct tones. Beautiful pictorial effects can also be obtained by waiting for the sun to break through holes in wind-driven storm clouds.

Printed-in-Clouds. Combination printing can be used to print in a good sky from another negative in place of an unsuitable or over-exposed one. In this case the lighting and cloud type in the sky negative should be chosen to match the subject conditions. Clouds can also be put in or reinforced by an air brush. H.W.

See also: *Cloud negatives; Filters; Landscapes.*
Book: *Photography of Scenery* by C. D. Milner.

CLUBS AND ASSOCIATIONS

In 1852 the first photographic society in Britain was formed in Leeds. The second was the Photographic Society, now known as The Royal Photographic Society, founded on 20 January, 1853; other early societies include Liverpool Amateur Photographic Association on 9 March, 1853, and Dublin Photographic Society—now the Photographic Society of Ireland—which dates from 8 November, 1854.

Other societies in Europe which have enjoyed a continuous existence for over a century include the Société Française de Photographie (Paris, November, 1854); Manchester Photographic Society (November, 1855); Newcastle-upon-Tyne Photographic Society (December 1860); Edinburgh Photographic Society (1861); and the Photographische Gesellschaft, Vienna (1861).

The first affiliation of photographic societies took place in 1892. During the next few years there arose in different regions a desire to control their own activities more closely and in 1899 the Yorkshire Photographic Union was formed. This was followed by the formation of the present Northern Counties Photographic Federation in 1901; the Lancashire & Cheshire in 1905; the Midland Counties Photographic Federation in 1907; and the East Anglian Federation in 1910. During the complete reorganization in 1930 when the Photographic Alliance of Great Britain was formed to co-ordinate all the Federations, the old affiliation of photographic societies in areas outside the then existing federations was renamed the Central Association. It was by the transfer of clubs from this organization that the Western Counties Photographic Federation was formed in 1932; the Scottish Photographic Federation (originally founded in 1903) was consolidated in 1946; the Welsh Federation in 1951, the Northern Ireland Federation (later disbanded) in 1952, and the Overseas Federation in 1954. The Federation of Cinematographic Societies (later also disbanded) was admitted to the Photographic Alliance in 1937. The thirteenth member is the North and East Midlands Photographic Federation, admitted in 1955.

In April 1961 the right of automatic affiliation to the R.P.S., hitherto enjoyed by all clubs who joined their regional federations, was suspended and from that date all societies accepted into membership by regional federations have been affiliated to the Photographic Alliance of Great Britain. The total number of clubs in affiliation in January 1964 was 1,163, but unaffiliated clubs probably number at least 1,500 more. The registered number of clubs in the Photographic Alliance, as given above, may be compared with the following approximate numbers in other countries in 1955: Denmark, 170: Federal Germany, 300: Italy, over 50: Japan, 1,500: Norway, 50: Sweden, 173: South Africa, over 40: U.S.A., estimated 1,500. In Russia there are no separate clubs as such.

Aims and Organization. The work of each club is determined by its aims, facilities, and space available at its headquarters. In the larger societies there are darkrooms, studio space and equipment, and libraries of photographic books and journals available. The officers and committee in almost every photographic body are re-elected annually by nomination and a secret ballot. In the majority of clubs the president or chairman serves for two years.

The annual or seasonal syllabus arranged in each society is usually planned to suit the majority of members and may include travel lectures illustrated by means of monochrome or colour transparencies, film strips or motion pictures; demonstrations of photographic processes and talks on technique and apparatus. A number of clubs organize brains trust meetings, jumble sales of unwanted equipment, displays of home-made articles, and one-man shows of prints and transparencies at every weekly or fortnightly meeting. Monthly competitions are also arranged, and there are annual exhibitions open to the public. Many clubs also arrange outings and rambles, when members' families and friends are always welcome. In large clubs the weekly meeting is usually devoted to some popular aspect of photographic work, while on other evenings of the week group meetings are arranged under the leadership of a specialist.

The address of the local photographic society in any district can be obtained from the local library, the area education office, a local photographic retailer, or the local newspaper. Or a letter to one of the national photographic journals will usually yield some information. There are also clubs which specialize in certain aspects of photography, such as natural history, architectural and stereoscopic work. Circulating postal portfolios are also popular with photographers unable to attend regular meetings.

Throughout Great Britain the average annual subscription is twenty-five to thirty shillings.

Federations. The services offered to members of societies are greatly extended when the club is affiliated to its regional federation. The subscription fees payable by the local club to the district association (whose committee is elected or appointed by all the member societies in that group) depend on the costs of the services expected by all the local societies in the federation to which they belong. In the Central Association, the largest in the United Kingdom and covering an area of some 7,000 square miles with an estimated population of over ten million, the membership fee is fifty shillings a year with a half-rate for junior clubs and schools. In return, the Association offers all its member societies over 90 lecturers with a choice of over 277 subjects, as well as 84 judges; through the other federations in the Photographic Alliance local clubs may also draw on another 400 lecturers with a further 1,000 talks and demonstrations plus the services of nearly 400 other judges and critics. The Photographic Alliance also offers affiliated clubs a large programme of recorded lectures with film strips or transparencies, loan collections of prints as well as slides, and a leaflet service on club management and club services.

Central Association members may participate in an annual exhibition, free of charge, from which the best entries are selected for submission to the annual national Photographic Alliance Exhibition.

BRITISH FEDERATION MEMBERSHIP

Federation	*Number of Societies*
Central Association	287
Midland	119
Lancs. and Cheshire	120
Scottish	113
Yorkshire	96
East Anglian	84
Western	52
Overseas	127
Northern Counties	57
Welsh	46
North and East Midlands	62

The Central Association has encouraged the formation of small area organizations including Photograph Federations in Surrey; Kent; Hampshire and Wiltshire; South London and North West London, as well as the Berkshire, Buckinghamshire and Oxfordshire Association. Other groups operate in North London and West Sussex. Their activities are independent of the main body and their services include the organization of group club meetings to hear important speakers unable to visit every society in turn; inter-club competitions; regional annual displays and competitions; area conferences, and the compilation of lists giving the names of county judges and lecturers.

Founding a Club. The foundation of a new club in any district should only be contemplated when no other society exists within a reasonable distance; two or more bodies drawing on a restricted potential number of members in one community will only reduce the prestige, the strength, and the value of both organizations.

When, however, the establishment of a new society appears to be necessary, a preliminary meeting should be convened in an accessible public centre, preferably one which may eventually serve as a permanent headquarters. Town halls, church halls, libraries, local colleges or schools, professional or scientific centres, cafés, or private rooms in hotels may all be approached and considered. Notices advertising the purposes of the meeting would undoubtedly be displayed on request by chemists, photographic dealers, newsagents, libraries, and schools; an open letter to the local newspapers and the national photographic journals would also provide publicity.

After the meeting has approved the resolution that a photographic society be formed, a temporary committee should be appointed for the purpose of drafting the rules and fixing the subscription. The regulations should be as few as possible; the revenue should be as large as possible so that good facilities may be made available to members. The income must be calculated to cover rent, lighting, heating, subscriptions to the regional federation, and the purchase of equipment. Standard and miniature size projectors are most likely to be needed, as well as a screen, even if no further apparatus is required for demonstrations, darkrooms, and studios. The constant borrowing of equipment from members is to be strongly deprecated. Allowances may also be necessary for meeting lecturers' expenses, the cost of slide cabinets and print screens, an easel, blackboard and reading desk, stationery, postages, and many other incidental expenses depending entirely on local circumstances.

At the second meeting, members should be enrolled, the rules approved, a permanent committee appointed for a specified term, and a provisional programme of lectures announced. At first it may be difficult to fill a programme as many lecturers may have accepted engagements six months or more ahead. In such cases some of the trade organizations may be willing to assist by sending postal lectures which can be read by members; local dealers may also be invited to give a talk or demonstration at short notice.

If required, advice may be sought from the area federation or council, but it is not the duty of these unpaid officials and enthusiasts to nurse newly-formed clubs burdened with inexperienced or incompetent committees.

In every society the aim should be to make the legacy for future generations of photographers richer than the inheritance bequeathed from the past. Only through such a policy can photography become a force in the life of any nation. T.H.J.

See also: *Amateur Photography; American Society of Magazine Photographers; Association of Cinematograph, Television and Allied Technicians; British Film Institute; British Kinematograph Society; British Photographic Manufacturers Association; Federation Internationale de l'Art Photographique; Institute of Amateur Cinematographers; Institute of Incorporated Photographers; Photographic Alliance of Great Britain; Photographic Dealers Association; Photographic Society of America; Royal Photographic Society; Society of Film and Television Arts, Ltd.; Society of Motion Picture and Television Engineers; Society of Photographic Scientists and Engineers.*

COATED LENS. When light passes from air to glass, or from glass to air in a camera lens a small percentage is reflected back and where there are a considerable number of such surfaces the loss of light may become considerable. Perhaps more important than the loss of light is the fact that the reflected light, after possibly several internal reflections, eventually reaches the sensitive surface of the film in the camera as over-all light fog which has the effect of reducing the contrast.

Such reflection losses in a lens can be reduced by coating ("blooming") the lens with a thin, transparent film of, for example, magnesium fluoride. The coating should ideally satisfy two conditions: its refractive index must be such that light reflected from its upper (air-film) surface equals that reflected at its lower (film-glass) surface; its thickness must be such that the light reflected from its upper and lower surfaces cancels out (that is it is out of phase by half a wavelength).This gets rid of stray reflected light which normally degrades the photographic image.

The loss of light at any one surface for normal incidence is given by Fresnel as:

$$\left(\frac{N_1 - N_0}{N_1 + N_0}\right)^2$$

where N_0 is the refractive index of the first medium and N_1 of the second. Thus for a glass of refractive index 1·5 and air (refractive index 1) the loss is:—

$$\left(\frac{1{\cdot}5 - 1}{1{\cdot}5 + 1}\right)^2 = \left(\frac{{\cdot}5}{2{\cdot}5}\right)^2 = \frac{1}{25} \text{ or 4 per cent.}$$

Thus a lens system having six air-glass surfaces will transmit only about 78 per cent of incident light, 22 per cent being lost by reflection. This loss is not only undesirable because of the reduction in speed of the system, but also because much of the reflected light may reach

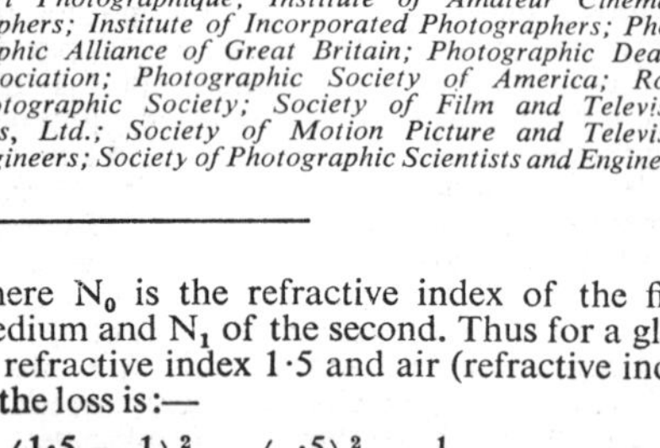

LENS COATING. Top: A normal lens reflects and scatters light at each surface. Bottom: Coating largely prevents this loss.

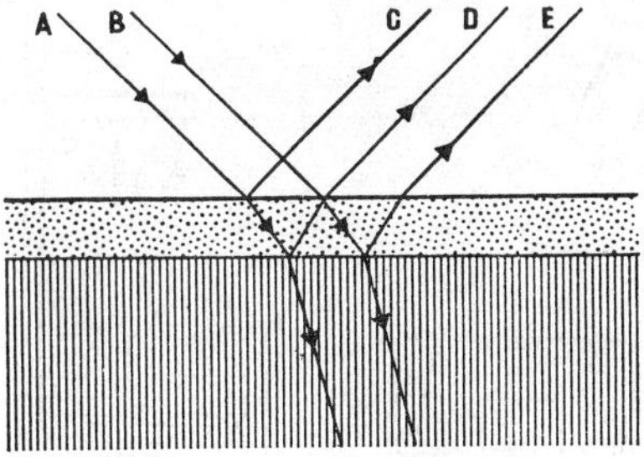

ELIMINATING REFLECTIONS. A surface film causes reflections at the air-film and film-glass interfaces. With the right film thickness, the reflected rays go out of phase and interfere (e.g., C and D for ray A, and D and E for ray B), cancelling each other out, thus eliminating the reflections.

the sensitized emulsion in the camera causing glare. However, if an air-glass surface is coated with a thin film of material whose refractive index lies between that of air and the glass, the reflection can be considerably reduced, and the transmission correspondingly increased.

For example, if a film of transparent material of refractive index N equal to 1·25 is deposited on the glass surface. Light will be reflected back at each surface of the film; these will be:—

$$\left(\frac{N_1 - N}{N_1 + N}\right)^2 + \left(\frac{N - N_0}{N + N_0}\right)^2$$

$$\left(\frac{1{\cdot}5 - 1{\cdot}25}{1{\cdot}5 + 1{\cdot}25}\right)^2 + \left(\frac{1{\cdot}25 - 1}{1{\cdot}25 + 1}\right)^2$$

$$= \left(\frac{\cdot 25}{2{\cdot}75}\right)^2 + \left(\frac{\cdot 25}{2{\cdot}25}\right)^2 = \frac{1}{121} + \frac{1}{81}$$

$$= \text{say 2 per cent.}$$

This is the saving when the thickness of the film is uncontrolled. By making the film of suitable thickness and refractive index, the two reflected rays from the upper and lower surfaces of the film can be made to interfere so that for rays of normal incidence practically no light is lost by reflection and the energy so saved increases the transmitted light.

To fulfil this condition the refractive index of the coating should be the square root of the glass's index, and the film thickness should be one-quarter of the wavelength of the light for which extinction is required. Magnesium fluoride, having a refractive index of 1.39, nearly satisfies the first condition for most glasses, and its optical and mechanical stabilities are good; a lens system having six air-glass surfaces coated with this material may lose only 10 per cent of the incident light by reflection. The thickness of the coating is usually one-quarter of the wavelength of green light, and the less complete destructive interference in the red and blue ends of the spectrum result in the characteristic magenta hue of a bloomed surface. Surfaces which are coated to give minimum reflectivity of red and blue light show blue and amber colours respectively. While the reflected colour of a bloomed surface appears intense, the spectral transmission of a lens system is only slightly affected by coating. This slight effect is nevertheless sometimes used to balance the transmission characteristics of sets of interchangeable lenses.

To increase the mechanical stability of the thin film, the lens is frequently baked to about 250°C. prior to coating, and the resulting blooming is known as "hard".

While it is impossible to obtain complete extinction of reflected light of any wavelength with a single film on most glasses, this becomes possible with a combination of two films of different refractive indices and thicknesses. Components frequently used are bismuth oxide (refractive index 2·4) and magnesium fluoride. For minimum white-light reflectivity extinction of blue-green light is necessary owing to the assymmetry of the spectral transmission curve of a two-layer system. The resulting combination usually shows an amber colour by reflection. The use of two-layer blooming is particularly valuable in optical systems having many air-glass surfaces and in systems using monochromatic light, though for most purposes it has little advantage over a single film. A.W.S. (P.B.C.)

See also: *Flare; Interference; Lens manufacture; Transmission efficiency.*

COBALT CHLORIDE. Used in certain metallic toners.

Formula and molecular weight: $CoC1_2 . 6H_2O$; 238.

Characteristics: Purple hygroscopic salt.

Solubility: About 80–90 parts in 100 parts of water at room temperature.

COHL, ÉMILE, 1857–1938. French caricaturist (real name: Emile Courtet). Produced in 1908 the first cine film consisting of animated drawings.

COINS. Collectors and numismatists often find it necessary to take photographs of coins and medals either to illustrate an article or a catalogue, or for normal records. Often an enlarged photograph of the coin, taken in suitable lighting, lends itself to examination more readily than the coin itself. Suitable photographs can be made with all the normal types of camera.

Equipment. There are three principal ways of tackling the subject depending upon the type of camera to be used. In each case, however, the most convenient arrangement is to mount the camera on a stand—e.g., in place of the enlarging head on an enlarger—so that it is looking down on the subject and can be easily set rigidly and squarely at any desired height from the easel.

Ordinary folding and box-type cameras do not extend far enough to be able to focus the subject close up. The method here is to fit a supplementary lens in front of the camera lens. A lens of 12 to 18 ins. focal length is suitable for most work in this field. The camera focusing is set at infinity and the height of the camera is adjusted so that the lens/subject distance is exactly equal to the focal length of the supplementary. In this case there is no need to make any aperture correction. The camera lens must be stopped well down to compensate for the uncorrected supplementary lens.

Double or triple extension folding cameras present no difficulties. The height of the camera is simply adjusted until the image of the coin on the focusing screen is of the required size. The aperture of the lens must be corrected to allow for the close-up viewpoint.

Cameras with removable and interchangeable lenses—e.g., miniatures and single lens reflex cameras—can have the extension increased as necessary by extension tubes. Focusing in this case is done either by measurement or by focusing stage which allows a ground-glass screen to be substituted for the camera body while focusing the image. Here, too, the aperture value must be corrected for close-up working.

A further alternative is available with one or two of the classical miniatures. This takes the form of a copying stand which supports the camera at a fixed distance above the table and allows it to be used for close-ups with either extension tubes or supplementary lenses.

Lighting. Some form of spot-light directed almost horizontally across the surface of the coin from a distance of about 3 feet will bring out the relief. Flat lighting is useless.

The low angle of illumination naturally causes a shadow on the background at the far side of the specimen. This can be blocked out in the usual way on a large negative, but it is difficult with miniature negatives and in any case it is easier to avoid it altogether when making the exposure.

This is done by photographing all coins on a sheet of glass supported about eight to ten inches above the work-table. If white paper is then placed below the glass, the result will be a shadowless picture, and the coins appear to rest on nothing at all.

If it is desired to get a really black background, the best material to use is a special paper called flock paper. It has a short pile, a little like velvet, and it reflects almost no light. It is rather difficult to buy, but a dealer in fancy papers for display purposes should be able to obtain it.

Flaws. It is often necessary for certain details on a coin to be specially clear, for example, a line-flaw or the formation of a particular letter. Then the lighting must be carefully adjusted to show up the detail, while the image is watched on the focusing screen. To show a line-flaw clearly the light must strike the flaw broadside on. If the light is turned

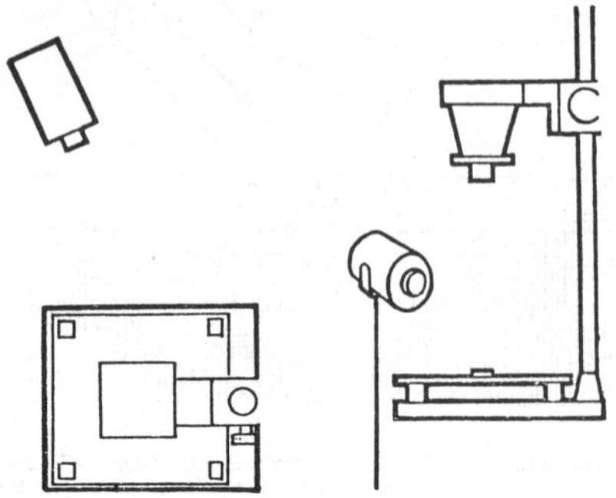

SET-UP FOR COINS. The coin lies on a glass plate supportec about 2 ins. above the enlarger baseboard. The camera is mounted on the enlarger column. A low spot lamp shining obliquely across the coin provides relief illumination to pick out the surface detail and inscriptions. At the same time it also illuminates the background.

through 90° from this position, the flaw wil be almost invisible. It might happen that twc such flaws occur at right-angles to one another. Here, if the light is directed on each at about 45°, it will make both flaws at least visible, though not so distinctly as in the case of the single flaw.

If the coin is of very bright metal which scatters the light, the image on the screen must be carefully watched while adjusting the light tc cause the minimum scatter.

A polarizing filter is the ideal solution in such cases; the surface of the coin itself must not be dulled as this may easily ruin its "tone".

Exposure. The best advice for those who contemplate much coin photography is to standardize the set-up. Use the same lighting unit at a fixed distance and always work with the same fine-grain film and developer. Then the only things which will alter the exposure will be the amount of reduction and slight variations in the tone of the coins to be photographed.

Make a few trial exposures at various reductions with the lens at a fixed aperture and after making any necessary corrections, construct a table of exposures for each ratio of reduction likely to be used.

When using an extension to the lens for close-ups the *f* values for the particular lens do not hold good. These altered values can be calculated exactly, although the trial exposure method can be used quite satisfactorily by the more experienced.

As the exposures required may run to several seconds or even minutes, it is important to bear in mind the risk of reciprocity failure, particularly when using colour film. Thus, a theoretically correct exposure of 30 seconds may actually need to be increased to 60 seconds or more.

Depth of field is not a serious problem. Sufficient depth combined with the best defini-

tion will usually be obtained with a good modern lens by stopping down about half-way. But when copying at natural size (1 : 1) or magnifying, it is best to stop right down.

Showing Both Sides. It is very desirable to show both sides of a coin side by side. Obviously two separate negatives can be taken, and the two negatives printed on the same sheet of paper. This will be the easiest method for most amateurs.

With a miniature camera taking 35 mm. film, there is a way of producing both sides of a coin on the one negative.

Choose a ratio of reduction so that the focused image can be made to occupy a little less than half the negative area and to lie close to one side of it. Make the exposure and wind on the next frame. Then turn the coin over, place it so that the new image is next to the frame previously exposed, and make the second exposure.

The film strip is afterwards cut up into normal-size frames by cutting through the middle of the frame so that the two sides of each coin appear side by side with a dividing line between them.

Using Casts. It is often necessary to reproduce both sides of several coins, all on one negative. The only successful way to do this is to use casts. These are made by taking impressions of both faces of the coins in wax from which plaster casts are subsequently made, one for each side.

Casts have so many advantages that they are used for the production of all the best photographic plates of coins, both in coin text-books and coin auction sale catalogues.

Among their advantages are: both sides of any number of coins can be taken together; as the casts are of a uniform colour the resulting photograph shows all the coins in a uniform tone; the original coins may have been unevenly toned or are badly stained, but the casts taken from them do not show these defects. C.W.P.

See also: *Antiques; Close-up.*

Books: *Camera Close-up* and *Camera Copying and Reproduction*, by O. R. Croy (London).

COLD CATHODE LAMP. Tubular fluorescent light source used principally in enlargers where its low working temperature is a great advantage. The tube is bent to form a grid which covers the whole negative area and gives such a uniformly distributed light that no condenser is required.

Cold cathode enlarger lighting is particularly suitable for plate sizes of quarter-plate and over, and the small amount of heat generated makes it possible to reduce the lamphouse to nothing more than a shallow box.

There are cold cathode attachments for fitting to the back of a plate camera to convert it into an enlarger.

One attraction of this form of enlarger illumination is that the light is principally composed of actinic rays so that the focused image and the image forming rays coincide more exactly than with a light of lower colour temperature. Another advantage exists in so far that condensers are not required to obtain even illumination, and so a range of different focal length lenses can be used in the enlarger without making adjustments or changes in the lamp housing.

The principal disadvantage of this light source is its high working voltage which calls for a special transformer.

See also: *Discharge lamp; Fluorescent lamp.*

COLD WEATHER. Weather around freezing point creates its own particular problems for the photographer. At the same time modern camera equipment is designed to stand up to a wide variety of climate and weather and generally speaking gives little cause for complaint in anything short of polar conditions.

Sluggish shutter action is the biggest problem in cold weather, but it is not likely to become noticeable until the temperature is several degrees below freezing point.

So this trouble rarely occurs in most of Britain, although it may call for attention in the north of Scotland. In parts of the Continent and in the mountainous areas everywhere, low temperatures are more common and may slow down the shutters of the best cameras. Even then the photographer will usually find that sluggishness begins only after the camera has been kept in the cold for a while.

Focal plane shutters are affected just as much as the diaphragm types. Fabric shutters are more likely to be affected than metal ones.

Some parts of the camera suffer more than others. The bellows may tend to harden at low temperatures and the only reliable way of keeping them flexible is to avoid leaving the camera extended in the cold air for longer than necessary.

Condensation. Condensation on the lens is one of the biggest nuisances in cold weather photography, especially in Britain and similarly humid climates. It is always wise to check the condition of the lens before using the camera and to wipe the lens gently if it shows any trace of moisture.

Condensation is apt to occur whenever warm, moist air comes into contact with a cold surface. It does not occur when cold air comes into contact with a warm surface. So the risk of condensation on the lens is always present when the camera is cold and is brought into contact with warm air. This means that a camera can safely be taken out into the cold air and no condensation will occur. But when it is returned to the case, slipped into the pocket or under the coat, or brought into a warm room, moisture may form on it. The next time it is taken out, the lens may be misted over and incapable of recording a sharp picture. Wiping it gently with a piece of absorbent cloth or lens tissue is all that is needed.

When the camera is finally brought back indoors after a session out in the cold, moisture will again tend to form on it. This condensation will evaporate as the camera rises to room temperature, but until it does so, it should be left open in the warm air or it may remain damp after it is put away. A useful way is to place it in a tin with activated silica gel.

Fingers are a separate cold weather problem; anything that keeps them warm and alive is apt to impede their action. It is difficult to manipulate the camera when wearing thick gloves, but a pair of cotton gloves under wool mittens will usually solve the problem. This arrangement allows the fingers their freedom and provides a reasonable degree of warmth.

Camera. If the camera is kept in a warm-lined leather case it should remain in workable condition, even when the temperatures are low. If possible, carry the camera inside the overcoat or jacket until it is required for use. This precaution will ensure that the camera will be in good working order in the coldest weather likely to arise in the temperate climates.

A camera with an all-metal body chills more rapidly in cold weather than the conventional leather or plastic types and it is also more unpleasant to handle. For this reason, a wooden tripod is preferable to a metal one.

Although no one kind of camera is technically superior to any other for cold weather photography, the most practical is probably the twin-lens reflex; it is reasonably small and can therefore be carried quite easily under clothing; it is light—an important matter when walking on ice or in snow where after a while every item of equipment seems to increase in weight at every step. Furthermore, with a reflex camera there is no worry about exposed bellows and it can be manipulated quickly—a big factor in cold weather, when exposed fingers can quickly become stiff and painful. Finally the twin-lens reflex uses roll film which is by far the most convenient form of sensitive material for cold weather work. Some photographers do, however, prefer 35 mm. miniature cameras.

Sensitized Materials. Plates are not technically inferior to roll film for cold weather use, but films are easier to handle, lighter to carry and modern fine grain films are so good that for most purposes there is nothing to choose between the quality of an enlargement from a small film and the equivalent contact print from a plate negative.

Low temperatures have no ill effect on sensitized material, in fact very cold weather tends to preserve their qualities. Panchromatic materials are generally to be preferred to orthochromatic for cold weather photography because of their greater sensitivity to light in the red region, which is usually prominent when the sun is low in the winter sky.

Loading and unloading of films presents no serious snags in cold weather—so long as it is carried out under some form of shelter if there is any risk of powdered snow or ice being blown into the open camera by the wind. Generally it is enough to shield the camera with the body or under the coat. It is also important not to breathe on the back of the lens when changing the film. Film material gets brittle, and may snap if coiled too tightly.

Filters. Yellow filters are the most useful for the cold weather photographer. In snow scenes, in particular, the deep yellow filter assists in intensifying shadows in weak winter sunlight (thus giving greater contrast between snow and shadows) and emphasizing the brilliance of the snow. And a filter helps to improve the definition of distant scenery by cutting out the atmospheric haze so often present in winter landscapes.

As a correcting filter, the pale yellow or ultra-violet filter is ideal when exposures have to be kept short, but otherwise a medium yellow will be found to be of more general service. This yields abundant correction, sharply defines clouds, and gives a very close approximation to correct tone rendering. With panchromatic films a medium yellow filter requires only a two times increase in exposure.

Exposure. An exposure meter should always be used when working among snow and ice. Light values under such conditions can be extremely deceptive and are frequently very much stronger than they appear even if there is a lot of mist about. Moreover, the light can change quite rapidly. In snow scenes, when the sun is shining, the light can be so intense that it is beyond the range of the exposure meter.

At such times, it is advisable to stop down as much as possible, and it is as well to make sure that the lens diaphragm can be set to give a very small stop. (With many lenses the smallest stop allowed for is only *f* 16 which may be too large under such conditions.)

A lens hood is at all times an advantage, but when photographing snow or ice in sunshine it is absolutely essential if the negative is not to be fogged or degraded by stray reflected light.

Some of the most beautiful effects are obtained when one shoots against the sun; in fact, texture of snow can only be recorded this way.

Often it is wise to vary the exposure on either side of the meter reading because of deceptive light. Any serious photographer should budget for such multiple shots.

Low Temperature Development. Development should only be carried out at low temperatures when there is no alternative. Low temperatures give inconveniently long development times with practically all developers.

A high speed developer should be used to keep the development time as short as possible. The Kodak SD-22 formula is specially suitable for work at low temperatures:—

Stock Solution A		
Sodium bisulphite	4 ounces	100 grams
Amidol	1½ ounces	40 grams
Pyrocatechol	1½ ounces	40 grams
Benztriazole	35 grains	2 grams
Water to make	40 ounces	1,000 c.cm.

Stock Solution B		
Sodium hydroxide	5 ounces	125 grams
Potassium bromide	350 grains	20 grams
Potassium iodide	70 grains	4 grams
Water to make	40 ounces	1,000 c.cm.

For use take 1 part each of A and B, and 2 parts water. The mixture does not keep.

At temperatures below 0° C. an anti-freezing agent such as glycol must be used to stop the solution from freezing. The glycol may replace part or all of the water used in diluting the developer. The lowest useful development temperature under these conditions is about 0° F. (– 18° C.).

Fixing and Washing. As long as the solutions are above the freezing point of water, these operations should present no difficulty, as fixation and elimination of the fixer chemicals will still take place at quite low temperatures. The chemical action is of course slower, and longer times must be allowed.

To speed up fixation at low temperatures a rapid fixer based on ammonium thiosulphate may be used. A solution of ammonium thiocyanate is also very useful at temperatures below freezing; although in normal conditions it tends to frill the film.

At temperatures below 0° C., glycol may be added to the fixer to prevent freezing. The washing water may be treated similarly, but washing is best carried out in changes of water to reduce waste of glycol.

Drying takes considerably longer, too, and there is a risk of the water freezing on the film. To counteract this and to save time, the negatives may be soaked in 80 per cent methylated spirit (the colourless industrial variety) before drying. Before using this technique it is essential to check that the spirit contains no adulterant which is precipitated by the addition of water. D.L.

See also: *Polar photography.*

Book: *Photography on Expeditions*, by D. H. O. John (London).

COLLEN, HENRY. Dates unknown. English miniature painter. Opened the first Calotype studio under licence from Talbot in Somerset Street, London, by August 1841. Used the Calotype chiefly as a base to paint on, and after a few years returned to miniature painting.

COLLODIO-BROMIDE OF SILVER EMULSION. Introduced by B. J. Sayce and W. B. Bolton, 1864, was the first workable photographic emulsion. It was a great step forward, for though somewhat slower than the wet collodion process, it had the advantage of dispensing with the inconvenience of dipping the plate in a sensitizing bath immediately before exposure. The collodion emulsion contained the more sensitive silver bromide in place of iodide, and tannin as a preservative. All the photographer had to do was to pour the sensitive emulsion on his glass plate. The commercial production of collodion emulsion dry plates was started by the Liverpool Dry Plate & Photographic Company early in 1867.

COLLODIO-CHLORIDE PAPER. Variety of printing out paper in which collodion replaces gelatin. It was popular in the middle of the last century.

See also: *Papers; Sensitized materials history.*

COLLODION. Transparent and viscous fluid used as the vehicle for the sensitive silver salts in the wet plate process. It consists of a solution of pyroxyl in equal parts of alcohol and ether. Also used for coating collodio-chloride (print-out) papers and for giving a high gloss (enamelling) to the surface of prints.

COLLODION PROCESS. The collodion (wet plate) process was invented by F. Scott Archer and published by him in 1851. It was in general use for 30 years, superseding the daguerreotype and calotype processes. Until recently, it was still used by process workers in photomechanical reproduction for making the negatives for line and half-tone blockmaking. In this particular industry, wet plates had a number of advantages: they have a very fine grain, good contrast, and they can be processed quickly and dried off by heat. They are also easy to intensify and reduce, but their sensitivity is low—only slightly faster than bromide paper.

To prepare them a clean glass plate is coated with collodion containing potassium and other iodides. It is then, in the darkroom, bathed in a solution of silver nitrate until the coating contains light sensitive silver iodide precipitate. The plate is exposed in the camera while it is still wet, and physically developed with a solution of ferrous sulphate and acetic acid, which reduces metallic silver from the silver nitrate solution on the wet plate, and deposits it on the latent image areas. Potassium or sodium cyanide is used for fixing and there are many methods of intensification and reduction by which very fine grain, thoroughly opaque and transparent line and half-tone images may be obtained—the cyanide solution can be washed out by merely rinsing under the tap for a minute; the hypo takes longer.

See also: *Physical Development; Sensitized materials history.*

COLLOID. Name for any of that class of substances—e.g., albumen, caramel, glue and starch—which when dissolved in water will not diffuse through a parchment membrane. The colloids mentioned above are used in photography in various ways.

COLLOTYPE. Photomechanical process invented by Poitevin in 1855 and perfected and commercially introduced by Joseph Albert in 1868.

It is still used for reproducing exceptionally delicate detail in monochrome and for short runs of a few thousand of colour work. It gives beautiful results with little perceptible grain structure.

A sheet of ground glass is coated with bichromated gelatin and dried in an oven so that the surface of the gelatin develops a reticulation grain of exceedingly fine wrinkles over the whole surface. After exposure to a continuous tone negative which selectively hardens the gelatin, and after a short general exposure through the back (hence the need for glass), the bichromate is washed out of the coating and the plate is treated with a solution of glycerine which renders it selectively hygroscopic in the less hardened areas.

The plate is then rolled up with a greasy lithographic ink which takes best in the areas containing the least water. This provides an ink image which is printed upon paper.

The main advantage of collotype is the minutely fine and asymmetrical nature of the reticulation grain which is excellent for reproducing delicate detail and, in particular, regular detail such as roof tiles, fabrics and the regular textures of geological specimens and steel engravings. These regular patterns tend to "clash" with a half-tone screen formation and produce an ugly interference pattern commonly known as "moiré".

See also: *Photomechanical reproduction.*

COLOUR

Most people know that white light is a combination of all the other colours. A glass prism will split up a ray of white light into a pattern of coloured bands known as the spectrum, and under the right conditions, raindrops will do the same thing; the spectrum in this case is, of course, better known as a rainbow.

COLOUR AND VISION

The Spectrum. The colours of the spectrum are divided into bands of red, orange, yellow, green, blue, indigo, violet—in that order. These colours are visual sensations brought about by electro-magnetic waves of varying lengths. The wavelength is measured in Angstrom units; one Angstrom unit is equal to one ten-millionth of a millimetre and is usually written 1A.

Our eyes can see light of wavelengths between about 4,000 A—violet—and 7,000 A—red. Above and below this visible spectrum there are so-called invisible radiations—ultra-violet which has a shorter wavelength than violet, and infra-red which has a longer wavelength than red. But it is only the visible rays that have to be considered in normal colour photography.

Things appear coloured when they have the property of absorbing light of certain wavelengths and reflecting the remainder. When the mixture of coloured light making up white falls on a red geranium, the red light is reflected back and the blue and green light are absorbed and lost, so that the flower looks red because only the red light reaches our eyes. Blue glass looks blue because it allows only light of the short wavelengths to pass through it, the light of longer wavelengths (green and red) being absorbed.

How the Eye sees Colours. The human eye can be easily compared to a camera; it has a lens to produce the image, an iris for controlling its brightness, and it can bring objects at all distances into focus on the sensitive surface at the back of the eyeball. The sensitive surface, called the retina, is equipped with cells which are light-sensitive and correspond to the photographic emulsion. These receptors are of two main sorts, "rods" and "cones" (so called from their shapes). When an image falls upon them, they send impulses along the optic nerve to the brain which then experiences the sensation of sight.

Scientists do not yet agree about how we see in colour, but the theory that seems to fit the facts best is that the receptors form three separate systems. Each system responds to about one-third of the spectrum. One set reacts to violet and blue light, one to green and yellow, and one to orange and red. These divisions overlap so that, for example, yellow light stimulates a proportion of both the green and the red systems of receptors. The brain then interprets this joint message to mean yellow.

Again, when a beam of pure green light falls on a white screen, what we see is a green patch (for white reflects all colours), and with a pure red light we see a red patch. If both lights shine on the screen at the same time it will look yellow. According to the theory this is because both the green and the red receptors are stimulated, so that the brain receives impulses from both. As this is exactly what happens when we look at a yellow object, the theory fits the facts.

Effect of Viewing Conditions on Colours. The colour we see will depend on the lighting conditions of the scene as much as the actual colours of the scene.

The quality of the light, i.e., its spectral composition, will affect the colour—obvious examples are the sodium and mercury street lamps which are in common use today. The former give almost entirely yellow, while the latter are entirely blue-green so that only those colours show. Less obvious is the distortion given by fluorescent lamps, which in many cases give light formed from a mixture of a few separate bands of wavelengths, not a continuous spectrum. Thus objects whose colour is in one of the gaps between these bands will appear darker and different in hue. There is also the difference between north sky light, sunshine, electric arc-lights and ordinary tungsten filament bulbs, which have progressively less blue and more red light respectively. This corresponds to a fall in colour temperature which alters the colours of the scene by making the yellow and red objects brighter at the expense of the blue ones and tends to make all the colours redder. The proportion of blue in sunshine can be changed by atmospheric haze or mist, which makes it more like the blue sky light, e.g., the blue haze on distant mountain scenery.

Another way of changing the light quality is by reflection, e.g., a subject against the background of a brick wall will actually be illuminated by reddish light because of reflection from the wall.

The intensity of the light will also affect the colours we see. Of the receptors, the cones only are colour sensitive and they function at normal intensities. At very low intensities, e.g., starlight, vision appears to be confined to the rods, which are colour blind. This explains the difficulty we have in discerning colours at night. It has the converse effect that a photograph in which the colours are dark and desaturated will look as if it had been taken at night, whereas one in which the colours are bright and saturated will give the effect of sunshine, although the exact opposite may be the truth.

Another factor which will modify the colour of an object is the direction of the light. If the lighting is very diffuse the colours will look desaturated and dull. This condition may be caused by an overcast sky or, even more noticeably, when this is combined with snow on the ground, or on the water. In these cases, light comes from all angles and, even though the actual intensity may be high enough to give the normal "cone" vision, the desaturation effect occurs.

On the other hand if the light is strongly directional, e.g., full sunshine, the colours will appear at their brightest, particularly if the surface is glossy and reflects well. Thus wet patches on matte objects will appear brighter than the dry parts. The actual reflections of the light source are not coloured; since the light is completely reflected from the surface, none is absorbed by the object, which therefore appears the same colour as the light source itself. This explains why the highlights, on the eyes in portraits, for instance, or on ripples on water, etc., are always white.

The eye is also capable of adapting itself to different colours of illumination. It is well known that on coming out of sunlight electric light appears yellow since the eye is adapted to the white of daylight which is relatively more blue. After a few seconds, however, the eye adapts to the new type of light and is quite satisfied that this is white. Similarly, on projection of transparencies or cine film in a darkened room, the eye will adapt to the overall colour balance of the picture. Thus pictures which are out of balance and would appear to have an objectionable colour cast when viewed by full daylight will often be quite acceptable when projected.

Colour and Contrast. Yet another factor which modifies the colours we see is the contrast of an object with its surroundings. If we take two identical patches of colour and surround one with a black border and the other with a white, the former will appear both lighter and brighter (more saturated) than the latter. The same thing happens if the borders are the same hue but a darker or lighter shade respectively. This is a well known optical illusion; it is shown very clearly in step-wedges. Where steps of increasing density adjoin each other, one edge of the step will appear lighter in contrast with the darker neighbour and the other edge will appear darker so that the density appears to vary across the step. This occurs in neutral black-and-white wedges but can be very noticeable in coloured ones.

A similar effect is that of colour contrast. A colour will always appear brighter when it is viewed against a background of its complementary colour and if subject and background are not of complementary colours, the eye will tend to see them as if they were. This can be shown more plainly if we consider green foliage against a grey wall; if the foliage is a good saturated green, the wall will appear pinky-grey because the colour balance of the eye is adapting towards green so that other objects become magenta by comparison.

Subjective and Objective Colour. In colour photography the eye tends to see what it wants to see, whereas the film sees what is actually present. Thus, in the example of the subject next to a brick wall mentioned above, the film would record the subject as much more red than it really was because of the increase of red in the light by reflection from the wall. But this might pass unnoticed by the photographer for two reasons. First, the eye will adapt itself to the new colour of the light so that the reddish light will now appear white. Secondly, even if full adaption does not take place, the brain will still reject the findings of the eye if they do not fit in with what it knows to be true. Thus we shall think we see correct flesh colours on the subject's face even if the eye is sending a message along the optic nerve that the face is red. These psychological effects are most persistent and any scene will be compared in the memory with an average of what that scene should look like and different findings by the eye are ignored. This often gives rise to big differences between photographs showing the scene as it really was and memory fitting it to agree with what the brain thinks it ought to have been.

Similarly there are associations of colours with other feelings—red colours with warmth, blue with cold, etc., and these too can operate in reverse so that a warm room will appear redder than it is, a cold scene bluer. In the same way red colours are advancing, blue colours receding tones so that distant objects may appear bluer than they really are or conversely, a blue object in a picture will look farther away. There are many of these psychological effects which change the colours we see, or think we see, but do not, of course, change the physical reality seen by the camera itself. J.Mo.

COLOUR HARMONY

When a presentation of two or more concurrent colour stimuli typically affords a pleasing impression, there is colour harmony. More strictly, colour harmony refers to the pleasingness of the combination itself, over and above any favourable effect of the components taken separately. This harmony of the combination itself seems more stable and fundamental than the liking for colours taken separately, and so the distinction is of some importance. Common usage, however, ignores the significance of the combination as a unit, and simply correlates colour harmony with over-all pleasingness.

Criterions. Everyone sees colour arrangements which he personally likes or dislikes, but colour harmony is more than a mere matter of individual taste. Though necessarily based on personal preference, colour harmony is conceived as a social phenomenon which is subject to statistical and experimental test; only if enough qualified people like an arrangement is it to be regarded as really harmonious. "Qualified people", according to the older view, comprised a relatively small group of acknowledged experts; according to the newer view, the contemporary population in general. Thus the consensus of the group provides the criterion of colour harmony.

Great difficulty has been encountered, however, in establishing the subject of colour harmony on a dependable basis. Despite the many outstanding artistic contributions through the ages, it would be possible even today for any colourist to pose with the public as a colour harmony expert and promote almost any general collection of colour samples of possibly completely haphazard relationship as a system of colour harmony.

This state of affairs is readily understandable in view of the many variables involved and the many different combinations which can afford colour effects which are at least acceptable. Moreover, aesthetic preferences themselves are so often slight, doubtful, or even lacking, that serious objections to a confident recommendation are unlikely. The general picture is complicated by variations to be expected in different times and places due to differences in personalities, colour fashions, customs, cultures, national and local traditions, and ethnic groups.

Probably the greatest difficulty is one which has only recently been removed, that is, inability to control colour stimuli closely enough; for unless various combinations can be produced and reproduced in various settings, the experimental investigator cannot verify his findings or even know what he is working with. But this unsatisfactory situation is improving significantly and may eventually be eliminated now that effective experimentation is feasible.

In the meanwhile, there are various technical groups such as artists, decorators, architects, industrial designers, lighting engineers, and photographers who are becoming increasingly concerned with the production and control of colour harmony. They have at their disposal a considerable accumulation of aesthetic rules and precautions which have gradually accumulated through the centuries and which are beginning to feel the impact of scientific corrections. Restrained use of traditional precepts, as in choosing arrangements for preliminary check viewing or getting a point of departure, can be effective.

If something exceptionally good is hit upon, a careful record of the circumstances can be made for future reference. By trial and error, an effective repertoire may be developed. Following is a summary of the more persistent older precepts, and also more recent developments.

Single Colours in Isolation. Colours which are pleasing separately tend to be pleasing in combination. Taken separately and on the average, red and blue are the most pleasing hues, while yellow and orange are the least pleasing. Within limits, increasing the lightness of a colour increases its pleasingness, and similarly for saturation. A minority group prefers relatively weak, desaturated colours. Colour space is marked by limited regions rather than sharp points of preference.

Continued contemplation of a colour sample tends to produce affective adaptation—i.e., reduction in pleasantness or unpleasantness toward indifference. In closely successive viewing, affective contrast may either increase or decrease pleasantness, depending on the direction of the contrast. Preference for particular colours is affected by age, fashion, and other factors. Colour preferences vary more between individuals than within individuals.

Combinations of Colours. In regard to binary combinations, large or small rather than moderate hue differences harmonize best. According to this, for example, a neck tie of yellow or greenish-blue (but not green) would go with a blue shirt. Experiment confirms that harmony increases with increased hue difference. Reference to Rood's hue circle is helpful in visualizing hue intervals. Complementaries are located in the hue circle about opposite each other—i.e., at maximum hue interval. They can be readily paired by reading off the colours at the opposite ends of a double-headed arrow. Exact complementaries, however, seem no more harmonious than nearby hues.

High contrast in either saturation or brightness is harsh and likely to produce dazzle or flutter; complementaries harmonize best when one or both are unsaturated. (The recommendation that triads or triple colours be equi-spaced in hue, can be followed by using a triple-armed pointer on the Rood's circle.) Complementaries may be combined with related colours and two pairs of complementaries are sometimes used together. Other "four-colour harmonies" and still more complex combinations have been endorsed. The more complex hue combinations depend greatly on effective use of saturation and brightness. In regard to aesthetic balance, strong saturation is best restricted to small areas. Large areas of weak saturation balance small areas of strong saturation; the same thing happens with lightness. Saturation and lightness have been combined in a single balance formula.

Designs and Decorations. Colour contributes to other aesthetic factors, and the more complex a design or composition the greater this contribution can be. Thus colour can be used to simplify by obscuring irregularities, as with paint; and colour can contribute to aesthetic unity by employing a dominant hue, or an extended background, or repetition. Contrasting colour can place desirable accent on regions of special interest. Lightness variation can avoid vagueness or monotony, especially in monochrome designs. Colour can contribute to mood as by subtle textural effects and modes of appearance. Dominantly warm colours in a composition create a warm impression, and dominantly cool colours create a cool impression. Dark colours are suggestive of greater weight, light colours of space.

The natural lightnesses of colours are considered safest, for instance, light for yellow and dark for purple-blue. Moreover, the pleasing colour complexes of nature may persist or even actually improve in pictorial conditions, but they are not dependable for human designs.

Over-all pleasingness often depends greatly on appropriateness or suitability in relation to function; for example, a painted industrial machine is more satisfactory if the colour contrasts are arranged to minimize fatigue, or a decorative poster is more agreeably informative if it is legible. More general environmental harmonies are achieved by colour co-ordination of larger units such as work rooms, whole interiors and exteriors, aeroplanes, trains, and housing developments.

Though many of the present available rules and precepts are helpful, they are too simple and specific to apply unreservedly to designs of complexity. There are signs that more adequate, higher-order principles of colour harmony may eventually be discovered.

Of recent years the general public have begun to take a greatly increased interest in colour. This naturally results in broader and more discriminating colour perceptions, more critical demands for adequate colour systems, and a greater sensitivity to colour harmonies and discords. These developments tend to stimulate progress in colour photography S.M.N.

See also: *Colour measurement; Colour temperature; Eye; Vision.*

Book: *Interaction of Colour*, by J. Albers (Yale); *Eye, Film and Camera in Colour Photography*, by R. M. Evans (New York).

COLOUR CAMERA. Special "one-shot" cameras used to be manufactured in which all three separation negatives were exposed at the same time. These cameras incorporated partially silvered reflectors in the camera between the lens and the focal planes. In one-shot cameras for still photography, the image forming rays, after passing through the lens, were split up into three parts by a system of mirrors or membranes called pellicles. Each part was focused through a suitable tri-colour filter on to a separate plate (or film). The filters were located immediately in front of the plates. The three plates were then developed to make the separation negatives. One-shot colour cameras have now become obsolete owing to the ease with which colour transparency and colour negative films yield direct colour prints.

Many designs have been proposed for one-shot colour cameras, but the types that have been commercially available are all of a few similar basic designs. All the cameras are of necessity somewhat bulky and cumbersome. There are three types of light-splitting systems: semi-silvered mirrors, pellicles, and prisms. Systems of the latter type are still used in colour television cameras.

Splitting Systems. Glass mirrors have the disadvantage that they refract the light passing through them so tend to distort the image. Also, they give a double reflection from the rear surface of the glass.

So almost all one-shot cameras for still photography used pellicle reflectors—very thin sheets of collodion or cellulose, less than one-thousandth of an inch thick. These were stretched across optically flat metal frames. Pellicles were so thin that they eliminated the

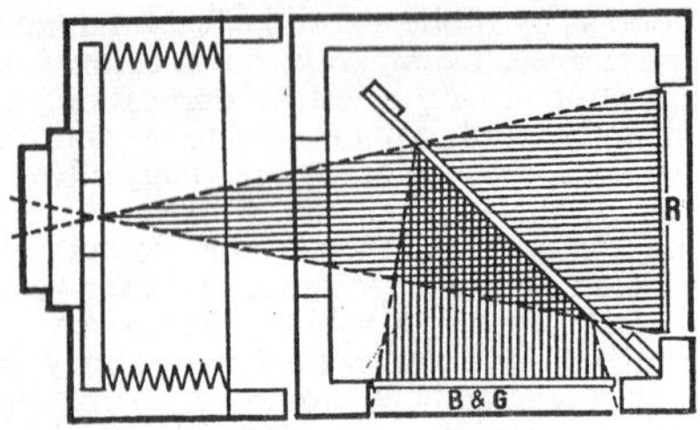

COLOUR BACK. *This single-exposure unit fits on the back of a normal camera. A single mirror deflects part of the light to a bipack* B *and* G, *the rest forms the red record* R.

risk of double reflection and refraction. The pellicles were coated with a metallic layer to obtain a suitable reflection-transmission ratio.

The principal design of still colour cameras took three separate plates and had two pellicle reflectors, each of which reflected part of the light and transmitted the rest. The rays of light from the lens fell on the first reflector which deflected roughly one-third on to the blue record plate and allowed the remaining two-thirds to pass through it to fall on the second mirror. This mirror deflected roughly half of the remaining light on to the red record plate and allowed the other half to pass through to the green record plate. Each of the plates was mounted at the correct focal distance from the lens so that sharp images of identical size were formed on all three.

Another design of one-shot camera used a single reflector and had only two focal planes. In this type of camera, one of the negatives was made on a single film in one dark slide, and the other two negatives exposed on two special films loaded in the other darkslide together face to face in the form of a bipack. For example, the front film of the bipack might give the blue filter record, and the rear film the red filter record, both being in the reflected beam in the camera. The green filter record was then given by the single film in the transmitted beam in the camera.

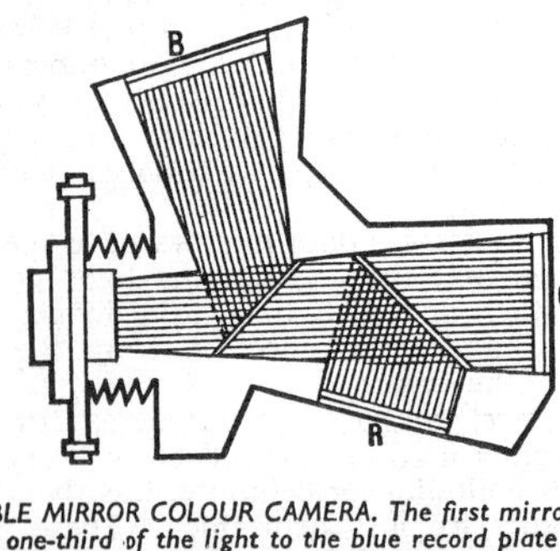

DOUBLE MIRROR COLOUR CAMERA. *The first mirror deflects about one-third of the light to the blue record plate* B *behind a blue filter, the second mirror divides the remainder between the plate behind a red filter* R *and one behind a green filter* G.

The relative simplicity of a single mirror camera led to the introduction of many models of this type for still photography until manufacture of the special bipack film was discontinued.

Prism-reflecting systems have been used in certain miniature size one-shot cameras, but the weight and expense of prisms preclude their use for larger negative sizes.

Prism type reflectors were also used in one-shot motion picture cameras, which have a single reflecting surface and two film gates, using a pair of films as a bipack in one gate, and the third film in the other.

Construction. It was usual for the camera body to be made of metal since the three separate images must stay in accurate register with each other. Any differences in the size of the images would produce colour fringes in the finished positive.

The reflectors and plateholders were almost always mounted in a rigid housing with the lens on a separate rack-out panel joined to the body by bellows. The shutter must be mounted on or near the lens so that it cuts off the light before reflection. For this reason focal plane shutters would be impractical in a colour camera.

Few camera movements were provided on one-shot colour cameras. Some designs provided for rising and cross fronts and sometimes for swinging the lens. But a swing back was never provided because of the obvious difficulty of swinging each of the plateholders exactly the same degree to maintain perfect register of the three images.

The lens for a colour camera must naturally be well corrected to give the three colour record images of equal size and sharpness. Smaller cameras used first-class anastigmat lenses, but larger cameras generally used the more highly corrected apochromatic lenses for optimum image quality.

The most popular formats were quarter-plate to 4 × 5 ins. for professionals and 2½ × 3½ ins. for amateurs and for use in the hand. Some 5 × 7 ins. cameras were made for portrait and commercial studio work.

Sensitized Material. Most one-shot colour cameras were made to take plates or sheet film. Normally panchromatic emulsions of the same type were used in all three darkslides. Some makes used one panchromatic and two orthochromatic emulsions, or one panchromatic, one orthochromatic, and one (blue non-colour sensitized) sensitive plate.

Usually the exact sensitive material employed in a colour camera depended on the filters fitted and the metallizing of the reflectors, and the camera manufacturer specified a suitable sensitive material and quoted a film speed to be used for his camera-film-filter combination as a whole. A.P.J.

See also: ***Beam splitter; Separation negatives.***

COLOUR CORRECTION OF LENSES. Elimination of lens aberrations with special reference to the reduction of faults varying with the wavelength of light. In popular photographic advertising lenses are frequently referred to as "colour corrected" to imply to their suitability for colour photography.

In fact colour correction of a lens is as important in black-and-white photography as in colour. The principal faults—several kinds of chromatic aberration—lead to a different plane of sharpness or to a different image scale (or both) for images formed by light rays of different wavelengths. In a colour photograph that may produce colour fringes and loss of definition; in a black-and-white negative made on panchromatic film the same loss of definition will occur and reduce the detail resolution and the detail contrast of the image. Lenses without colour correction would be suitable only for photography by light of a single wavelength.

Types of Colour Correction. Chromatic aberrations are corrected by combining lens elements of different dispersion so that the colour errors introduced by the individual elements cancel out. In a simple form achromatic correction makes the position and size of images coincide at two specified (and usually well-separated) wavelengths of the spectrum. This yields an acceptable compromise over the whole spectrum range, but does not necessarily correct all intermediate colours. When using certain colour filters the plane of sharpest focus may therefore shift slightly. This is one of the reasons why for maximum accuracy the image should be focused through the filter through which the exposure is made. (Another reason is that the filter may, through its actual thickness of glass, introduce slight image shifts.)

With apochromatic correction the lens is corrected for three main spectral regions. Such lenses are essential in graphic arts and colour separation work for exact registration of images corresponding to different colours.

Infra-red and Ultra-violet. Neither achromatic nor—usually—apochromatic correction covers radiations outside the visible band. Thus the image formed by ultra-violet or by infra-red rays may not be formed in the same plane as a visible image. Photographs under conditions where the prevailing light is rich in ultra-violet rays (for example alpine views) are therefore liable to show some loss of definition. An ultra-violet absorbing filter remedies this.

For pictures to be made by ultra-violet or infra-red rays exclusively, the visual and the photographic foci do not usually coincide. Hence after focusing visually the lens requires resetting to reduce its distance from the image plane (with ultra-violet radiation) or to increase this distance (for infra-red photography). Some lenses carry a second setting index for focusing with infra-red rays (since visual focusing through a nearly opaque infra-red transmitting filter is virtually impossible). Ultra-violet photography is only possible with lenses specially designed for the purpose, which have appropriate ultra-violet transmission characteristics. Such lenses may have a distance setting scale calculated for a particular wavelength of ultra-violet rays.

Colour Correction and Colour Rendering. The glass elements of the lens itself may have certain absorption characteristics which can affect the colour rendering of a picture taken through that lens on colour film. A lens which absorbs ultra-violet rays strongly (and possibly even some light of the extreme violet end of the spectrum) yields a warmer (slightly more reddish) colour rendering than a lens of less pronounced ultra-violet absorption. Hence colour pictures taken with different cameras are unlikely to show exactly the same colour balance even discounting other variables.

In practice this is disturbing only when a range of interchangeable lenses is used with a given camera. The variations in colour rendering then show up when a complete set of colour transparencies is projected in succession. Some lens manufacturers match lenses of an interchangeable series belonging to the same camera by a suitable choice of lens coatings. For a lens of maximum ultra-violet absorption the coating may be virtually non-absorbing in this spectral band, while lenses of lesser absorption may receive a coating that itself holds back some ultra-violet radiation. Appropriate balancing of these coatings then yields consistent colour rendering throughout the range of lenses. L.A.M.

See also: *Aberration of lenses; Coated lens; Infra-red photography; Optical design.*

COLOUR FILM PRINCIPLES

The basic principles of colour reproduction are described in this section. In order to illustrate these principles clearly, the discussion concentrates on subtractive reversal colour films. These are the films which produce the majority of positive transparencies for viewing in projectors.

Colour negative films and their associated colour print materials involve the same principles as reversal films. Their only difference is in the detailed processing to a final positive image, and the fact that the final image is a print instead of a transparency. Some of the more important differences between negative and reversal materials are pointed out at the end of this section.

Additive colour photography is not discussed here in any detail as this is now used only for specialized purposes, for instance, in colour television.

DYE COUPLING SYSTEMS

Integral Tripack Colour Films. Nearly all modern colour reproduction processes rely on the fact that red, green and blue light can be combined to produce a wide gamut of colours. In subtractive colour photography the proportions of these three colours in the original scene are recorded as latent images on three suitably chosen emulsions. These latent images are used to control the development of three dye images, and these dyes control the amounts of red, green and blue light projected onto the screen. The three dyes used must be superimposed, so in the film the three recording layers are situated one on top of the other on the same support (hence the name integral tripack). In principle then we need three dyes, each of which absorbs light throughout (roughly) one third (i.e. red, green or blue part) of the spectrum and transmits freely elsewhere, together with three emulsions, each sensitive to one third of the spectrum only.

All silver halide emulsions are basically sensitive to blue light. In order to make an emulsion react to green light alone the emulsion must first be made sensitive to green light by optical sensitization and then, as it still retains its blue sensitivity, it must be used behind a filter that absorbs blue light. Similarly an emulsion sensitive only to red light must be optically sensitized to red light and again used with a blue absorbing filter. The tripack colour film can then be made up, with three sensitized emulsion layers and a filter layer: the top layer is sensitive only to blue light; below this is a yellow filter layer to absorb blue light, and below this again are the green and red sensitive layers. Conventionally the red layer is at the bottom of the tripack next to the support, or base. (Other layers may be added to this basic arrangement but they are refinements which need not concern us here). Despite this stack of emulsion layers, the overall thickness of the tripack is only about 0·0005–0·001 in., and the base about 0·003–0·005 in. thick.

The choice of filter layer, optical sensitizers and emulsions requires great care. The filter layer should be capable of absorbing all the blue light to which the "green" and "red" emulsions are sensitive, but should let through all the green and red light. The green and red optical sensitizers must be chosen to give sufficient sensitivity to these colours so that all three layers are effectively the same speed, but they should not overlap in their sensitivities otherwise colour reproduction is impaired. Further, the three sensitivities should combine to cover the whole of the visible spectrum.

The three layers must not only be balanced for speed but they must also be balanced for contrast and general characteristic curve shape. The emulsions must therefore be carefully chosen to meet these and other requirements, such as balance of reciprocity failure and latent image effects, and to give good graininess and sharpness properties.

In practice not all of these requirements can be met. The filter layer lets through some blue light, which impairs colour reproduction, and absorbs some green and red light, which reduces the effective speed of the green and red layers. Due allowance of this must be made when sensitizing these layers. No set of sensitized emulsions exist which give no overlap of sensitivities for the three layers and cover the whole spectrum uniformly. The sensitivities used have to be a compromise—with some overlap and some non-uniformity of coverage.

The particular form of the sensitivities chosen will of course play a considerable part in the quality of colour reproduction of the film. These chosen sensitivities will match only one particular type of illuminant, say daylight. If the energy distribution of the source is changed, say by using tungsten light, the effective sensitivities of the three layers will no longer match each other. The film will not therefore be balanced and will reproduce colours with a strong colour cast (yellow-red in this case). Similar balance changes can occur if the amount of ultra-violet light changes. Here the camera lens is important as some lenses absorb considerably in the ultra-violet region of the spectrum. The film manufacturers try to match film sensitivities in relation to an average lens but it remains that lenses vary over a considerable range in this respect.

Exposure and Development. Just what happens during the exposure and development stages of a colour reversal film may be illustrated by considering four patches—white, green, magenta (purplish-red) and black. When the film is exposed, light from the white patch affects all three emulsion layers—that is, a latent image is formed in each layer. The light from the green patch will produce a latent image in the green layer alone; light from the magenta patch will produce latent images in the blue and red layers (magenta being a combination of red and blue), and the black patch does not emit light so it will give no latent image in the film.

The film is then processed first in a black-and-white developer to produce a negative exactly as in ordinary monochrome films. The dyes are then developed up in their appropriate layers, using the silver halide remaining after the first development. Yellow (blue-absorbing) dye is developed in the blue layer; magenta (green-absorbing) dye in the green layer, and cyan, or blue-green (red-absorbing) dye in the red layer. The film is then bleached and fixed to remove the unwanted silver image and the yellow colour of the filter layer, leaving just the three-dye image.

Let us now consider the dye images for the four colours, white, green, magenta and black. For the white patch all layers were exposed and developed in the black-and-white develo-

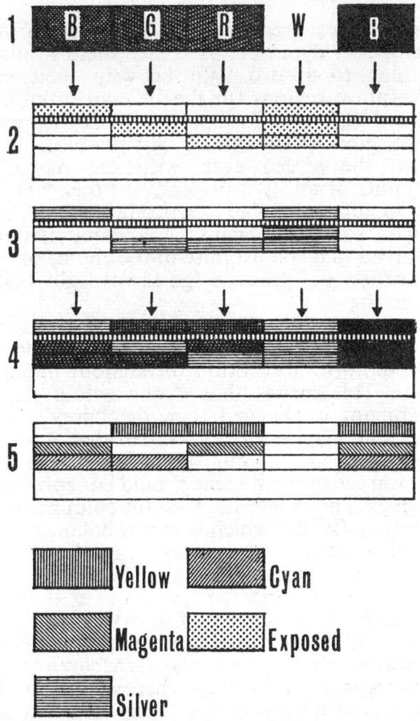

REVERSAL COLOUR FILM PROCESSING. 1. Subject consisting of blue B, green G, red R, white W, and black B. 2. Exposure produces latent images in the layers sensitive to the different colours. 3. First development yields a negative silver image. 4. After re-exposure, second development produces a positive silver image together with a positive dye image. 5. After removal of all the silver by bleaching and fixing, only the positive dye image is left.

per so that no dyes are developed up. With no dyes present all the light from the projector is transmitted onto the screen and therefore appears as a white patch.

Similarly for the green patch, yellow and cyan dyes are present so that the blue and red components of the projector light are absorbed and only the green component is transmitted onto the screen. For the magenta patch, the magenta dye absorbs green light and red and blue are transmitted. Finally, for the black patch, all three dyes are present so that all the projector light is absorbed and none is transmitted to the screen.

Substantive and Non-Substantive Process. The process by which the dyes are introduced into the film during development may take one of two forms.

In the first procedure, referred to as a substantive colour former process, the three dye colour formers are introduced into the film in their appropriate layers during manufacture of the film. In processing, the film is re-exposed, after black-and-white development, to white light to render the remaining silver halide developable and then the film is colour developed—the developer reacting with each of the three colour formers to produce the appropriate dye.

In the second system, the non-substantive colour former process, the dye formers are contained in the developer solution itself, and a different solution is used for each dye. The normal sequence in processing is similar up to the end of the black-and-white development of the negative image. Then follows a selective re-exposure of the red layer alone and development in a cyan colour developer. The blue layer is then similarly treated by selective re-exposure (from the top of the film so that the filter layer prevents exposure of the green layer) and development in a yellow colour developer. Finally the green layer is re-exposed and magenta colour developed.

Each process has advantages and disadvantages. Substantive films are complicated to manufacture but relatively simple to process (this type of film can often be home processed). The processing of non-substantive films is more complex but their manufacture is relatively straightforward, at least by colour film standards.

Safelights. Generally safelights are not recommended when processing colour film. If these are necessary on any occasion, a study of the spectral sensitivity curves shows that films are least sensitive in the blue-green and the yellow regions of the spectrum. At low illumination levels the eye is most sensitive in the blue-green region so the best safelight filter would be a narrow-cut blue-green filter matched in its particular characteristics to the film being used. The overall illumination level must still be kept very low even with this filter in use and is suitable only for checking a darkroom clock, or for finding equipment in the room, but not for judging the appearance of the developing image on the film.

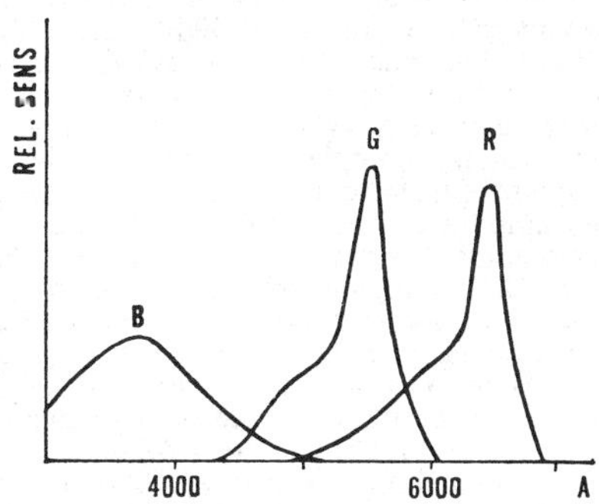

SPECTRAL SENSITIVITIES. The sensitivities shown for the blue (B), green (G), and red (R) recording layers are typical for a reversal colour film balanced for daylight use. The blue recording layer is sensitive well into the ultra-violet region, far below the limit of the eye.

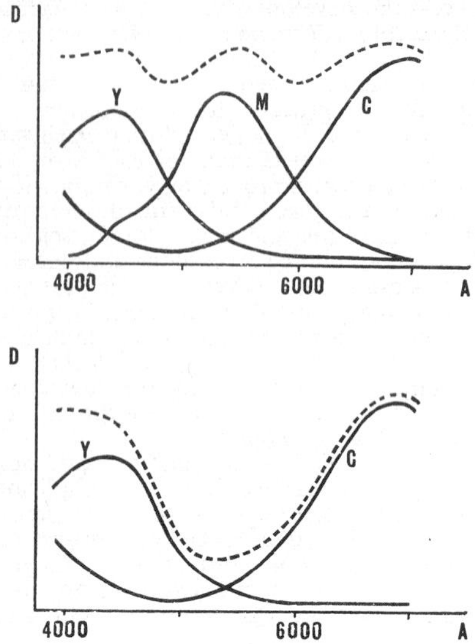

SPECTRAL ABSORPTION CURVES. These curves are typical of image dyes used in reversal colour films. Top: All three dyes together, if divided in the appropriate proportions, should yield a visual neutral or grey (dotted curve). Bottom: Addition of yellow and cyan (Y, C) produces green (dotted curve).

Image Dye Characteristics. The three dyes used to form the image are cyan (blue-green), magenta and yellow. These absorb respectively red, green and blue light. Spectral absorption curves for a typical set of these dyes show that the dyes are "imperfect" in that their absorption bands overlap and they do not absorb uniformly in their appropriate regions. (Absorption here is expressed in terms of density, D, where $D = \log {}^{1}/{}_{T}$, and T is the ratio of transmitted to incident light—the transmission). The amount of each dye that is produced to form the image depends upon the original exposure. Where the silver halide in the emulsion is completely exposed, no dye will form; where no silver halide is exposed, a maximum amount of dye will form, and there is a steady gradation between these two extremes for steadily changing exposures. The spectral absorption curves of varying amounts of dye are well defined since the density at any wavelength is proportional to the amount of dye present, e.g. doubling the amount of dye will double the density at each wavelength.

The film manufacturer tries to reduce the overlap, or unwanted absorption, of the dyes in order to obtain "clean" colour reproduction, but there is a limit to what can be done in this direction. Generally dyes which do not give significant unwanted absorption are also very "peaky"; that is, each dye only absorbs heavily over a narrow region of the spectrum. If three such peaky dyes were used then there would be nothing to absorb light between their peak absorption regions (in the blue-green and the yellow regions of the spectrum). Consequently some greenish-coloured light would always reach the screen even with the maximum amounts of all dyes present, in other words it would not be possible to reproduce good blacks and dark colours. Again then the manufacturer is forced to a compromise in the choice of dyes in order to produce good overall colour rendering.

An interesting effect is that the dyes used to produce a grey colour (visual neutral) do not give uniform absorption throughout the spectrum. This means that if the colour of the illuminant is changed then the "grey" patch may no longer look grey. One of the important criteria in all colour reproduction is that greys in the original scene should be reproduced as greys. This is referred to as the colour balance criterion. (A film which is out of balance is one where greys are not correctly reproduced and the whole picture appears to be tinted). The manufacturer aims to produce a correctly balanced transparency for its appropriate illuminant, namely projector light, and attempts to choose dyes so that the transparency also appears to be in balance when viewed in daylight (as such viewing conditions are often used with, for example, hand viewers).

Sensitometric Curves. As in black-and-white photography properties such as speed and contrast may be determined by exposing the film to a range of intensities of light and measuring the densities of the images produced. The standard exposing light is a "white" light, and a range of intensities of this light produce a grey scale. A graph of density against the logarithm of the exposure (log E) yields the characteristic curve of the material. In black-and-white photography the characteristic curve is easily obtained as the density is produced by a single silver image. In colour photography the image is made up from three dyes so it is necessary to make three density measurements. These measurements are made in the red, green and blue regions of the spectrum. The density in the red region of the spectrum will depend mainly on the amount of cyan dye present, the green density depends mainly on the magenta dye, and the blue density depends mainly on the yellow dye.

A grey scale characteristic curve for a colour film in fact consists of three curves, one for each layer. In order to simplify their assessment the densitometry is so arranged that the three densities are equal for a true grey. The relative positions of the three curves then indicate the amount by which the film is out of balance at any point. For example in the maximum density region of the curves the red density may be low and the blue high. This indicates a lack of cyan dye (therefore too much red light is trans-

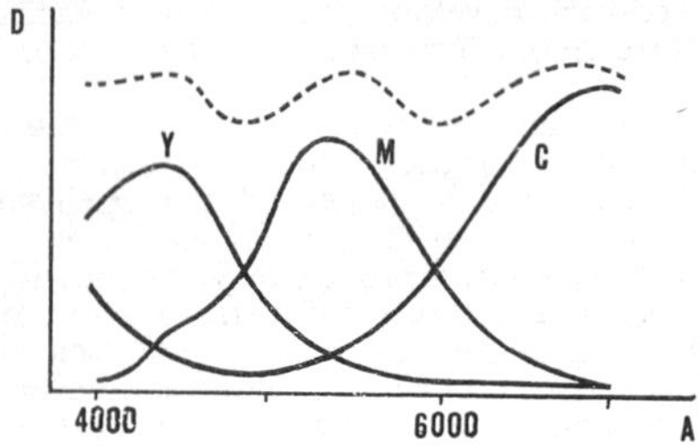

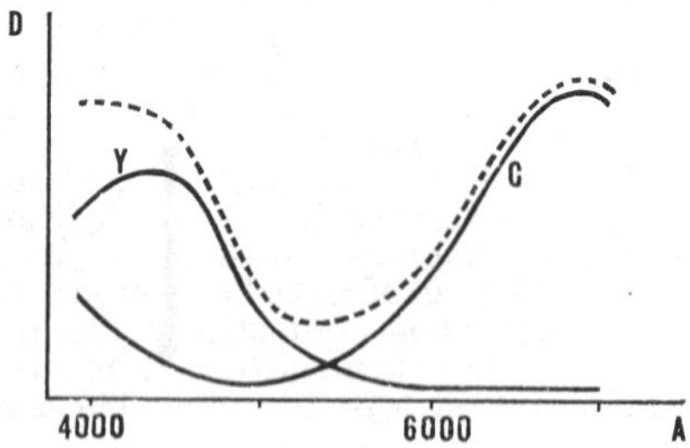

mitted) and an excess of yellow dye (too much blue light is absorbed). The maximum density will therefore appear yellowish-red.

Speed and contrast can be measured as in black-and-white photography—the speed is determined by the relative position on the log E axis and the slope of the straight-line part of the curve gives the contrast. The manufacturer can measure these various properties from the characteristic curve and from them can make any necessary adjustments to the film or the development. Similarly any processing station, or individual with the necessary facilities for measurement, can control development to a high order of accuracy.

For a full sensitometric assessment of colour films in addition to a grey scale exposure it is necessary to examine saturated and de-saturated red, green, blue, cyan, magenta and yellow scale exposures. These exposure scales are used mainly by the manufacturer. From them it is possible to detect, amongst other things, any spurious development of dyes in wrong layers and whether or not any unusual interlayer effects are occurring.

Other Colour Measurements. Density measurements of colour as just described are of relatively limited use. In order to study the general properties of colour reproduction of a film it is necessary to make much more detailed measurements.

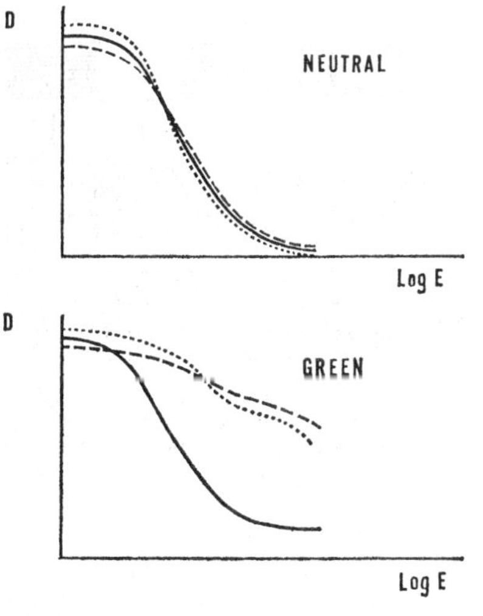

CHARACTERISTIC CURVES OF REVERSAL FILM. Top: For a visual neutral the curves should be superimposed. This composite curve is used to determine the mean contrast, speed and balance of the film. The solid curve corresponds to the green densities (i.e. magenta dye measured through a green filter), the broken curve to red densities and the dotted curve to blue densities. Bottom: Green scale characteristic curve corresponding to the reproduction of a saturated green colour at varying exposure levels. At very low exposures the film reproduces greens as near grey due to the limit of maximum density. Excessive exposure also leads to poor reproduction.

An example of the use of subjective colour measurements is in the study of the variation of saturation with luminosity. For instance in a green scale characteristic curve, the three curves converge as the exposure decreases, this indicates that the saturation of the green decreases and eventually becomes zero (a grey). An underexposed transparency will not then yield well-saturated colours. A similar effect can also occur for overexposed transparencies.

An example of the use of objective measurements is the examination of the chromaticity gamut of a set of dyes, on a C.I.E. chromaticity diagram. This gamut is very much compressed and distorted compared with the gamut of real colours. The compression can be offset to some extent in practice by using a fairly high contrast for the film, but the distortions are much more difficult to overcome. Generally the manufacturer aims to produce a pleasing reproduction—one in which the more important colours such as flesh tones, grass and sky are not obviously wrong.

Colour measurements must always be very carefully interpreted to be meaningful. In colour photography such measurements are further complicated by, for example, the fact that a scene originally viewed in daylight is viewed in reproduced form in projector light, and the fact that the "memory colour" of, say, flesh tones may not be the same as the colour observed in the original scene.

Colour Negative Processes. It is possible to make colour prints from colour transparencies but when colour prints are required it is easier and more economical to print from colour negatives.

The processing of negatives differs from transparencies in that there is only one development stage (negative materials use the substantive process). The latent image is converted directly to a dye image. This dye image is complementary in colour to the original object. When this image is printed, the process is repeated in the colour paper material so that it finally appears as the correct colour in the print.

The dyes used to form the image in the negative are chosen to give the best results in the printing stage, rather than to be suited to visual requirements as in transparency materials. It is therefore possible to use integral masks in the film. These masks are designed to reduce the adverse effects of the unwanted absorption of the dyes. For example, the blue absorption of the magenta dyes, and the green absorption of the cyan dyes are often very objectionable, and suitable masking techniques are used to overcome these faults. Such masks give rise to the typical orange appearance of colour negatives. The dyes used to form the print image are similar to those used in transparency materials as they must be matched to visual requirements. Similarly the spectral sensitivities of negative and print materials

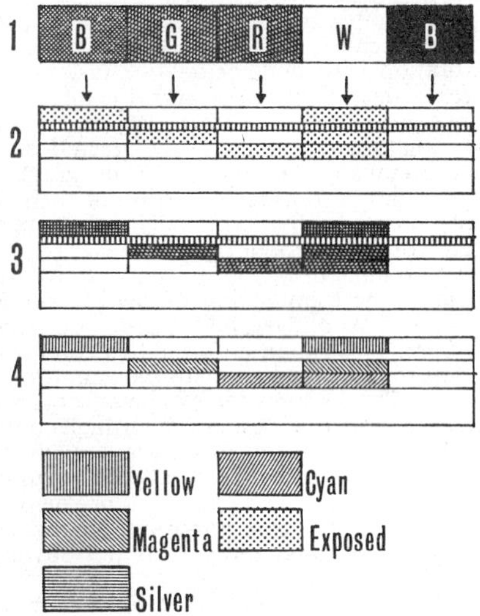

NEGATIVE COLOUR FILM PROCESSING. 1. Subject consisting of blue B, green G, red R, white W, and black B. 2. Exposure produces latent images in the layers sensitive to the colours. 3. Colour development produces a negative silver image together with negative dye images in each layer. 4. The silver is removed by bleaching and fixing, leaving only the dye images. These are negative; areas corresponding to blue in the original are yellow, green becomes magenta and red becomes cyan.

need not be the same as for transparency materials. In particular the print material has sensitivities designed to match the printing process, and as in black-and-white photography the print material sensitivity may differ considerably from that of the negative material.

Generally negative materials have a low contrast and are used with printing papers of high contrast to give an effective contrast similar to transparency materials. The advantage of a low contrast negative is that it has a reasonable latitude for exposure, and any discrepancies in exposure can be largely compensated during the printing stage. With transparency materials there is no possibility of compensating for over or under-exposure so their effective latitude is small. A further advantage of negatives is that they need not be exposed only in one type of lighting. Again the effects of exposing in, say, daylight and tungsten light can be compensated in printing so that the final print in either case is reasonably well balanced. P.W.

DYE DESTRUCTION AND DIFFUSION SYSTEMS

While the majority of colour materials are based on the chemical formation of a dye image during processing, it is equally possible to conceive the reverse system. In other words, we can start with three dye layers—cyan, magenta and yellow—and remove the dye in some way in accordance with the photographic image formed in each layer.

Two types of processes use this idea: dye destruction and dye diffusion. Both are used for subtractive colour photography; the principles of the formation of the coloured image are therefore the same, though the chemistry and certain practical aspects are significantly different.

Dye Destruction. When three differentially sensitized layers of an integral tripack film each contain a final dye, the image-forming relationships are reversed as compared with a dye-forming system. The chemical process of dye destruction involves a reaction between the dye and the by-products of development of the first (negative) silver image—or alternatively a reaction between that silver image and a subsequent processing bath (e.g. bleacher). In either case the dye is removed where the silver image is formed. The remaining dye—assuming a negative silver image—thus corresponds to a positive image. So a dye destruction system yields a reversal positive picture in one stage, without reversal processing. The complete processing procedure therefore consists only of development of the silver image and its removal (and removal of the unused silver salts).

The latent images in the three sensitive layers control the removal of dye, though indirectly. As the dyes in the layers also act as their own filters, no separate filter layer is necessary. Thus

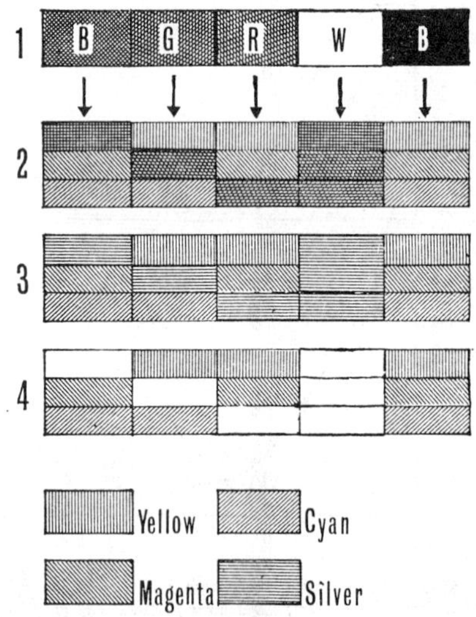

DYE DESTRUCTION PROCESS. 1. Subject (or transparency) consisting of blue B, green G, red R, white W and black B. 2. On exposure and development silver images are formed in the appropriate layers of the material, the latter already containing complementary dyes in the same layers. 3. Dye bleaching removes the dye in the image areas corresponding to the silver image. 4. Removal of the silver leaves only a positive dye image.

if the top, blue sensitive emulsion already contains a yellow dye, this automatically holds back blue light from the lower two layers. Magenta dye in the second layer holds back green light from the bottom, red sensitive emulsion. The speed balance of the three layers must, however, be different, to allow for the light lost by absorption through the dyes. In effect the bottom layer must thus be more sensitive—even to red—since the yellow and magenta dyes absorb not only blue and green light, but also some red.

Image patches corresponding to the same four colours as considered before—white, green, magenta and black—are reproduced in much the same way as with a reversal dye coupling film. Thus for the white patch all layers are exposed and during processing all dye is therefore destroyed. With no dye in the transparency, all the light during projection is transmitted on to the screen to yield a white patch. A green patch develops a negative image in the magenta dye layer, destroying the magenta dye. The yellow and cyan dyes from the other two layers are left and combine to form green. A magenta patch similarly destroys the cyan and yellow dyes in the red and blue sensitive layers, leaving only magenta in the green sensitive layer. Finally a black patch produces no image in any of the layers so that all three dyes remain intact and absorb all the light during projection.

Dye Characteristics. The colours used in dye destruction processes are still imperfect in their absorption characteristics—in other words their transmission is still not ideal. The dye deficiencies however are generally less pronounced (lower unwanted absorptions) than with dyes formed in the emulsions. This is because the choice of dyestuffs available for dye-destruction systems is considerably wider; the film manufacturer is not confined to a compromise of chemical and colourimetric dye properties. Dye destruction processes therefore can offer at least potentially better colour reproduction though the limitations of emulsion sensitivity (and their overlaps) are still present.

A second aspect of such a system is that more light-fast dyes are available for it. The comparatively restricted range of dyestuffs that can be formed as byproducts of a dye-coupling process in development is almost invariably light-sensitive to some degree; the dyes of such colour images therefore tend to fade when exposed to strong light for long periods. The dyes used in dye destruction systems are produced by different chemical processes and belong among the most stable dyestuffs as far as their sensitivity to light is concerned.

Neither of their factors guarantees perfect colour reproduction with a dye destruction process, since the sensitometry of the material is still dependent on the balance and other properties of the three individual emulsions. If the characteristic curves for the three emulsions do not coincide, or diverge over part of their range, the material will still exhibit colour casts and colour variations at different exposure levels.

Negative Dye Destruction. While a straightforward dye destruction system would be suitable for a colour transparency film, materials of this type actually available are almost exclusively intended for making colour prints. This is because the inclusion of a dye in each emulsion layer, complementary to the sensitivity of that layer, greatly reduces the effective speed. So it is not practical to produce a dye-destruction film which is sufficiently fast for a camera material.

While the system just described would produce positive prints (or transparencies) from a positive colour transparency, the process can equally be modified for negative-positive working. In contrast to a dye-forming system,

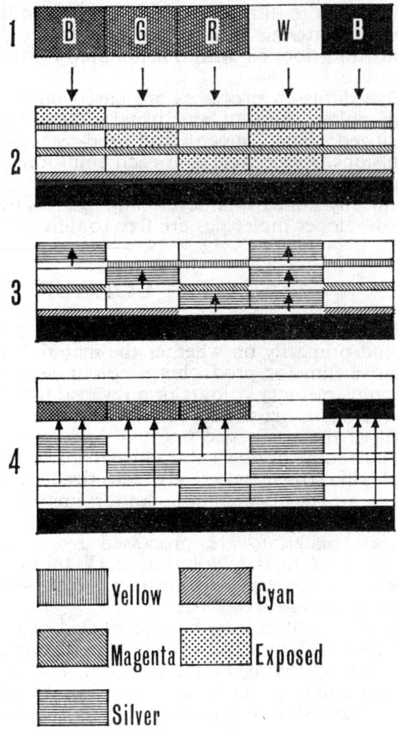

TRANSFER DIFFUSION SYSTEM. 1. Original subject consisting of blue B, green G, red R, white W, and black B. 2. On exposure, latent images are formed in the negative film layers according to their sensitivities. Adjacent layers contain dye coupled to developer. 3. At the beginning of development the dye-developer molecules from the layers diffuse into the adjacent emulsion layers to develop the silver images there and at the same time anchor the dye in place. 4. Towards the end of development the dye-developer molecules not used up in developing the negative silver image diffuse into the positive of contact with the negative, and there reconstitute the colours in the original. The whole process takes about a minute.

this becomes more complex and in fact corresponds to reversal processing. The procedure starts with ordinary reversal processing, to yield a negative silver image—via a positive one—from the colour negative. It thus involves the usual stages of first development, bleaching of a silver image, re-exposure and second development. The negative silver image is then bleached together with the dye in those image portions, leaving a positive dye image. This then is a positive of the negative dye image in the colour negative.

Dye destruction materials of this kind are again available only for making colour prints and are subject to the same problems as well as means of control as dye-forming negative-positive colour print systems.

Dye Diffusion. A further system of colour photography, which has more in common with dye-destruction than dye formation, is dye diffusion. This is at the same time the colour counterpart to diffusion transfer. It is used in a camera material, but yields positive colour prints and does so within a minute or two of exposure.

Dye diffusion processes are based on a tripack colour film of the usual differentially sensitized blue, green and red-responsive emulsions. Associated with each emulsion is a complementary dye layer containing the dyes chemically linked to a developing agent. These dye-developer molecules are free to diffuse out of their layers once the latter are permeated by an alkaline activating solution.

During exposure each sensitive layer records its corresponding component of the subject colours, forming a latent negative image. During processing, the dye developer molecules develop the image in the corresponding layers; the *unused* dye-developer (where no negative image was formed in each layer) diffuses out of the negative into the receiving layer of a positive material. The latter is in contact with the negative film during processing. The positive receiving layer thus builds up a positive dye image, complementary in tone values to the negative developed in the three film layers. This dye positive is then the final picture.

Like an integral tripack reversal colour film, the dye diffusion process provides virtually no latitude either of exposure or of colour adjustment. It thus has all the drawbacks of a reversal material, as well as the limited tone range of a print material. On the other hand it has the immense advantage of providing a finished colour picture almost instantaneously. L.A.M.

See also: *Colour film processing; Colour materials; Colour measurement; Colour photography: history; Colour printing; Colour synthesis; Colour sensitometry; Masking.*

Books: *Colour Photography in Practice*, by O. A. Spencer (London); *The Reproduction of Colour*, by R. W. G. Hunt (London); *Principles of Colour Photography*, by R. M. Evans, W. T. Hanson and W. L. Brewer (New York and London); *Colour Film* by C. L. Thomson (London).

COLOUR FILM PROCESSING

Processing procedures for colour films depend primarily on whether the material is a negative film for producing a colour negative in complementary colours or a reversal film for producing a positive colour transparency. Most of the processes, negative and reversal, are currently of the subtractive tripack type employing three selectively sensitized layers which are developed into colour images by means of chromogenic developers. Additive reversal materials are processed by similar stages to subtractive films, but black-and-white developers are used to form a black silver image in a single layer.

Some manufacturers have marketed processing kits containing all the chemicals required for user processing of the material, and in a few cases have released formulae. Suitable formulae have also been published independently.

SUBTRACTIVE NEGATIVE COLOUR FILM

The processing procedure for modern multilayer negative colour film is comparatively simple and hardly more complicated than for making black-and-white negatives. The processing of these colour films can be carried out with much the same equipment, apart from a few minor modifications and some additional items. The maker's instructions should be closely followed for each material, since different films of similar structure may react differently in processing.

Principle of Processing. The steps of producing a colour negative are:

(1) Colour development produces in each layer a dye image and a silver image, the amount of dye generated being proportional to the silver developed.

(2) An intermediate step stops development and removes developer and by-products. This can be done simply by prolonged washing but, in general, it is advisable to use a stop-bath. This solution not only interrupts the development process in a safer and more controllable way than washing, but, at the same time, facilitates the removal of certain unwanted by-products. In some processes a combined stop-fix bath is used with a hardening agent.

(3) Wash for at least 5 minutes.

(4) The next operation removes the unwanted silver image (and the unchanged silver halide if no fixing agent has been used in the stop-bath). There are various ways of carrying out this process. The traditional method is to use a potassium ferricyanide bleach bath followed by a fixing bath. This method has been largely superseded by the use of combined

bleach-fix solutions in which the ferricyanide is replaced by organic iron compounds. Bleach-fix solutions of this type have good keeping properties and less tendency to stain.

(5) The processing is finished as usual by thorough washing, in some cases preceded by a hardening bath.

Colour Development. The developing agents used universally now are salts of diethyl-para-phenylene diamine, or its derivatives. The salts may be the hydrochloride, the sulphate or the sulphite. In addition to one or other of these agents, the colour developer contains a large amount of sodium carbonate and a small quantity of sodium sulphite to improve its keeping properties by protecting it from aerial oxidation. A further ingredient is potassium bromide, while hydroxylamine hydrochloride may be included as an anti-stain agent.

The composition of the colour developer varies according to recommendations of different makers, since the characteristics of the colour materials themselves vary appreciably. Attempts have, however, been made to compound a colour developer universally suitable for an appreciable number of negative colour films. A formula suggested by H. Gordon is:

Para-diethylamino-aniline hydrochloride or sulphate (diethyl-para-phenylene diamine hydrochloride or sulphate)	52 grains	3 grams
or		
Para-hydroxyethyl-ethyl-amino aniline sulphate (hydroxyethyl-ethyl-para-phenylene diamine sulphate)	70 grains	4 grams
Disodium salt of ethylene diamine tetra-acetic acid	26 grains	1·5 grams
Potassium carbonate	3 ounces	75 grams
Sodium sulphite (anhyd.)	35 grains	2 grams
Potassium bromide	44 grains	2·5 grams
Tripotassium phosphate	175 grains	10 grams
Water to make	40 ounces	1,000 c.cm.

The development time, depending on the make of film, is in the region of 6–12 minutes at 65°F.

Stop Bath. As in black-and-white processing the stop bath is weakly acid but its pH value must be more carefully adjusted and controlled. The acid is usually acetic acid buffered with sodium acetate (which may be produced in the solution) to obtain and maintain the pH value in the best range between pH 5·0 and 6·0. A typical formula is:

Acetic acid (glacial)	420 minims	22 c.cm.
Sodium hydroxide	160 grains	9 grams
Sodium sulphite (anhyd.)	9 grains	0·5 grams
Water to make	40 ounces	1,000 c.cm.

In some cases an addition of fixing and hardening agents is recommended. A suitable stop-fixer is:

Sodium sulphite (anhyd.)	175 grains	10 grams
Sodium thiosulphate (cryst.)	8 ounces	200 grams
Ammonium chloride	350 grains	20 grams
Sodium bisulphite	265 grains	15 grams
Water ot make	40 ounce	1,000 c.cm.

For a hardening stop-fixer the sodium bisulphite might be replaced by 88 grains (5 grams) citric acid and 175 grains (10 grams) potassium alum.

The time of treatment in the stop bath is normally 5 minutes, followed by a wash of about 10 minutes in running water.

Bleach Bath. When potassium ferricyanide is used, it can either be applied as a plain solution, containing 7 per cent potassium ferricyanide and 2 per cent potassium bromide, or with the addition of pH controlling agents, such as mono- and dibasic sodium or potassium phosphates.

Bleaching normally takes 5 minutes and is followed by rinsing for 5 minutes and fixing with sodium thiosulphate. The fixing bath can be either a plain 20 per cent solution of sodium thiosulphate or hardening and pH controlling agents (alum and sodium acetate respectively) may be added.

Combined bleach-fix solutions simplify the process, but they cannot be prepared by simply mixing sodium thiosulphate with ferricyanide. This mixture, the well-known Farmer's reducer, has only very limited keeping properties.

One method of making up a bleach-fix solution is based on the use of ferric compounds in combination with a sequestering agent like ethylene-diamine-tetra-acetic acid. A formula for a bleach-fix solution of universal application, published by H. Gordon, is:

Potassium bromide	1¼ ounces	30 grams
Sodium thiosulphate (cryst.)	8 ounces	200 grams
Ferric salt of ethylene-diamine tetra-acetic acid	2¼ ounces	60 grams
Sodium carbonate (anhyd.)	88 grains	5 grams
Potassium thiocyanate	175 grains	10 grams
Water to make	40 ounces	1,000 c.cm.

The average time of treatment is 10 minutes. A thorough wash follows the bleach-fix process or the bleach formula based on potassium ferricyanide. These solutions can be used for negatives as well as prints. When no hardening agents have been used in the processing solutions, it may be advisable to apply a hardening solution containing 2 per cent formalin and 3 per cent sodium carbonate (anhyd.) for about 5 minutes, followed by a short wash. This solution also stabilizes the dyes.

Processing Equipment. When negative colour film is used on a fairly small scale, very little more equipment is needed than for black-and-white film. For negative processing, spiral tanks or cut-film tanks can be used, but if the processing is to be carried out on a large scale, deep tanks are essential. These should be similar to the normal type used for D. & P. work, although not all tanks of this type are suitable.

The developer tank has to be enclosed in a water jacket so that the development temperature can be thermostatically controlled. For this reason the developer tank should be a good conductor of heat, and a stainless steel tank is therefore advisable.

However, the bleach-fix solutions attack stainless steel and a plastic or stoneware tank must be used for this stage of the process; copper tanks may also be used with ferricyanide bleaches.

On a large scale, continuous processing is both more economical and reliable. In that case, stainless steel or hard rubber tanks are normally employed (as in black-and-white work) except for the ferricyanide bleach. The temperature is generally controlled by heat exchangers at the recirculation stage, where the solutions are pumped round from a reservoir through the tanks. Replenisher solution is added at this stage to keep the composition of the solutions constant, some of the old solution being thrown away if necessary. This replenishment must be accurately controlled by routine analysis of samples at fixed time intervals and the whole process checked by sensitometric strips being included at regular intervals.

SUBTRACTIVE REVERSAL COLOUR FILM

Reversal colour materials are of two main types—those which have the colour-forming elements, couplers, actually incorporated in the emulsion layer and those in which the couplers are added to the developers.

Non-incorporated Couplers. The processing of this type of material is complicated and only undertaken on a large scale on continuous machines, with rigorous analytical control of the solutions. These materials are integral tripacks with separate red, green and blue sensitive layers (from base to surface) and the essential steps in the processing are:

(1) Development of the latent image to a negative with a black-and-white developer of the type generally used for reversal work.

(2) Flashing to red light to fog the unused silver halides in the red sensitive layer.

(3) Development in a cyan-coupling developer to give silver plus a positive cyan dye image.

(4) Flashing to blue light and re-development to give silver plus a yellow dye image in the blue sensitive layer.

(5) Flashing to green light followed by magenta coupling development in the green sensitive layer.

(6) Bleaching all the silver, to leave the positive image composed of three subtractive dye images.

Many variations on this outline are possible, e.g., dye toning or mordanting processes instead of the more usual dye-coupling ones, fogging developers instead of exposure to light, etc., and, of course, washing, hardening and stop baths are also necessary. The processes are complicated, and in any case details are not published.

The materials are usually prepared in the standard and sub-standard cine sizes and are processed on continuous machines, identifying numbers being punched on the leaders before splicing them together. Still pictures taken on 35 mm. film are cut after processing and mounted in cardboard or plastic frames ready for viewing. These operations may also be done automatically.

Reversal materials of this type can also be used for making prints for viewing by reflection.

Incorporated Couplers. The materials which contain the colour forming couplers already in the emulsion layers are very much simpler to process, and on a large scale may be done on batch processing machines or in tanks in the same way as black-and-white materials. More baths are needed but the principles are just the same. Care must be taken that the tanks are of suitable material as with negative colour films.

For the amateur, the main points to watch are temperatures and times, and it is most important that the instructions given by the manufacturers are carefully followed.

Principle of Processing. Steps in the production of a reversal image on monopack colour film are the following:

(1) First development in a negative developer of the M.Q. type (or amidol) producing a negative silver image.

(2) Wash, or treatment in stop and hardening solutions.

(3) Re-exposure of the unaffected silver salts to a strong light source.

(4) Colour development. There the re-exposed silver salts are reduced to silver and dyes are simultaneously produced in the three layers.

(5) There may again follow some intermediate treatment in hardening and stop solutions.

(6) Bleaching to remove all the silver, leaving only the dyes and silver halide which is dissolved in the usual way with a hypo solution.

For large-scale work deep tanks with floating lids to cover the solutions and prevent oxidation are preferable. Normal stainless steel film clips and hangers may be used for holding the film during processing. Single rolls on 35 mm. film lengths can be processed in an ordinary developing tank; some types have transparent-ended reels to permit re-exposure without removing the film from the reel.

First Development. Certain features are common to all developers used at this stage.

Firstly, they all possess a very high activity and are used more concentrated than for normal black-and-white development. The alkali content is kept on the high side as well

Secondly, they contain a small concentration of thiocyanate which has the property of dissolving silver halide, and therefore helps to produce transparencies with clear whites by removing any excess of silver halide.

While most of these developers are of the

M.Q. type, amidol has also been used for some makes of colour film.

The usual range of concentrations of various published formulae is:

Metol	25–50 grains	1·5– 3 grams
Sodium sulphite (anhyd.)	1– 2 ounces	25–50 grams
Hydroquinone	70–100 grains	4– 6 grams
Sodium carbonate (anhyd.)	1–1½ ounces	25–35 grams
Sodium or potassium thiocyanate	20–100 grains	1–5 grams
Potassium bromide	20–100 grains	1–5 grams
Water	40 ounces	1,000 c.cm.

The developing time ranges between 10 and 35 minutes.

The M.Q. developers require either a stop or a combined stop-hardening bath (1 per cent acetic acid with 2 per cent sodium acetate and 3 per cent chrome alum) after development, and then a short wash. Films developed in amidol do not get any treatment in stop and hardening solutions as amidol carries on development even in an acid medium. They are thoroughly washed for about 20–30 minutes.

Re-exposure. After the first development and stop bath, the remainder of the processing can be carried out in white light, as the next step is to fog all the unused silver halide grains by exposure to light. As the speed of the wet film is very low, the exposure must be quite heavy. It is not possible to over-expose it, but the exposure must be sufficient or the full density will not be obtained from the second (colour) development.

The recommended exposure is usually about 5–10 seconds each side, 1 foot from a Photoflood lamp, or its equivalent. If any quantity of processing is to be undertaken, it is best to use an exposing box with two Photofloods, the transparency being hung between them and both sides being exposed simultaneously. Sheets of glass should be placed so as to protect the lamps from accidental splashes of cold water which might crack them.

Where roll films or 35 mm. films are being processed in a tank, they must be unrolled and hung flat on clips for re-exposure. However, if the ends of the spiral reel are made of transparent plastic, the reel may be rotated under the lamp so that the light penetrates between the coils of film and reaches every part of it.

Colour Development. The composition of the colour developer is similar in principle to the colour developers used for negative film. No universal formulae have been published, as the developer must be closely matched to the film characteristics.

The time of colour development ranges between 10 and 20 minutes at 65°F.

In all cases, it is necessary to put the films through stop and hardening baths or combined solutions and they must then be washed thoroughly.

Final Steps. The removal of the silver is carried out in bleaching solutions containing potassium ferricyanide, in most cases in mixture with potassium bromide.

These bleaching solutions convert the black metallic silver produced in the first developer as well as in the colour developer into silver bromide. The latter is then fixed out in a plain 20 per cent solution of thiosulphate. A wash of about 20 minutes ends the process.

Precautions. Accurate and careful work is essential in reversal colour processes in order to obtain the best results.

It is important that the two developers should be maintained at the correct temperature within half a degree, and the other solutions should preferably be at the same temperature, although it is not essential to maintain them quite so accurately.

Equally important is the degree of agitation which must be standardized. Cut sheet films may be raised from the solution, drained and lowered back, or a spiral tank may have the spiral rotated at regular intervals, e.g., every 30 seconds.

On a larger scale, continuous circulation of the solution, bursts of compressed nitrogen, or mechanical paddles are methods in regular use for ensuring good agitation.

With the rapid growth of colour photography, the speed of processing has been increased in order to improve plant output. Short processing times have been achieved by working at higher temperatures and reducing the number of baths, e.g. by substituting "blixes" for separate bleach and fix baths—this tendency is likely to continue.

OTHER COLOUR MATERIALS

Additive Reversal Colour Film. The processing of the additive colour materials is similar to that for black-and-white reversal materials since the colour comes from the screen or reseau and not from dyes formed by colour development.

The usual sequence is:

(1) First (negative) development.
(2) Bleaching, clearing, and hardening.
(3) Re-exposure.
(4) Second (positive) development.

Additional washes are required between each bath, and the same precautions as to accurate working should be taken as in the case of the subtractive materials.

Colour Processing in the Camera. Diffusion transfer colour film is based on a different type of tripack in which the layers contain complex dye-developer molecules which can be activated to cause migration of dye to a receiving sheet, thus forming the positive image. This special processing is very rapid, being complete within a minute, since the silver and unwanted chemicals do not have to be removed but are left in the discarded negative. J.Mo.

See also: *Colour film principles*; *Colour materials*.

Books: *Colour Photography in Practice*, by D. A. Spencer; *Colour Films*, by C. L. Thomson; *Colour Prints*, by J. Coote (all London.)

COLOUR PRINTS. Many people look on a hand-coloured photographic print as a satisfactory substitute for one made by direct colour photography. It is certainly possible to achieve very beautiful results, much depending upon the skill of the worker and his knowledge and sense of colour.

The principal types of colour used for handwork on prints are dyes, transparent oil colours, and pastel chalks. It is also possible to use chemical methods.

Most subjects are suitable for hand-colouring, but simple colour schemes are always the best.

When colours in the original are likely to render as dark tones in the print, it is advisable to use a suitable filter to lighten the tone when taking the picture.

Transparent Colours or Dyes. Dye tinting is the simplest method of colouring prints. The colours can be bought in sets either in solution or dry, coated on sheets of paper (these must be dissolved in water for use). Dyes are transparent and can be used for prints, lantern slides, or larger transparencies. The dyes can be diluted with water and they are easily applied as an even wash of colour. Dyes of different colours can be combined to give any desired tint. Being fully transparent they do not obscure any of the detail, tone and gradation of the photographic image. If the first attempt at colouring does not give a strong enough effect, a further wash or washes of colour may be applied.

The best practice is to build up to the depth of tint required by applying successive washes of colour. If the result looks exaggerated, the colours can be toned down by soaking the print in water. Dyes cannot be completely removed from the emulsion by soaking, but stubborn colours will usually yield to a weak solution of carbonate of soda.

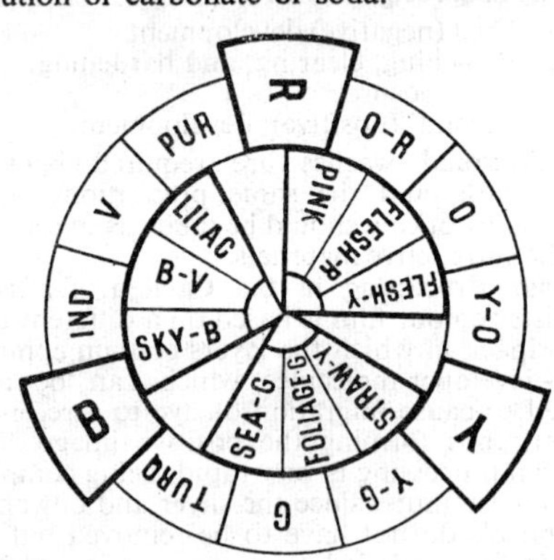

COLOUR CHART. This indicates what colours are required to mix a particular shade. The outer circle covers pure colours, the artist's primaries being red R, yellow Y, and blue B. The inner segments show the tones obtained by adding white. (These correspond respectively to the subtractive primaries cyan, magenta and yellow.)

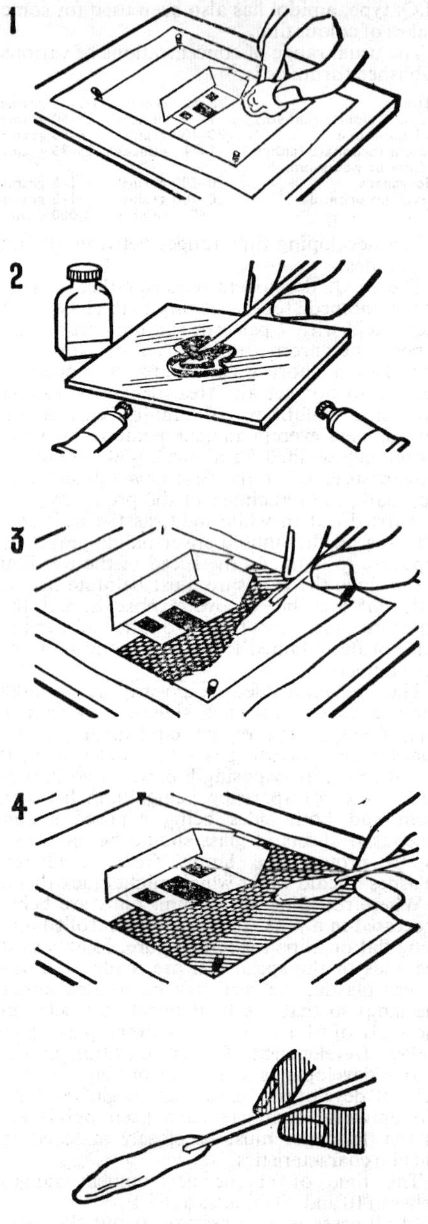

OIL COLOURING. 1. Attach print to a hard surface (e.g., board covered with blotting paper). Clean surface of print wit turpentine, and leave to dry. 2. Mix colours on a palett thinning with turpentine as necessary. 3. Apply colour to mai areas with a cotton wool rubber. 4. Grade the colour to corre shade by carefully rubbing off excess with clean cotton wo This also blends edges and hard outlines. Bottom: Cott wool rubber, consisting of a tuft of cotton wool wrapped rour the end of an orange stick. Replace cotton wool frequentl

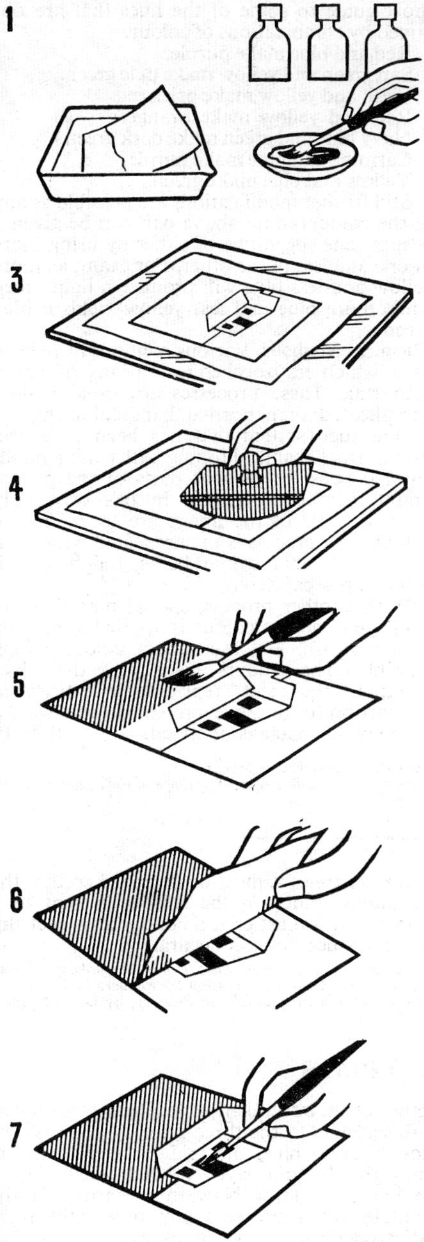

PRINT COLOURING WITH DYES. 1. Soak prints in warm water. 2. Mix dyes as required from concentrate or solid. 3. Place print on a sheet of blotting paper, and support on a glass plate. 4. Blot off surplus moisture. 5. Apply washes of diluted colour to large areas. 6. Take up excess liquid with edges or corners of clean blotting paper, to avoid sharp colour edges. 7. Use small brush to apply more concentrated colour to details.

Good quality brushes should be used, particularly for small size prints. Size O is suitable for very small details and 2 and 4 for larger areas.

Small prints can be very effective when coloured. A semi-matt surface is probably best for taking the colour. The print to be coloured is soaked for five minutes, laid upon a sheet of glass, and the surplus moisture blotted off. The print is then ready for colouring with successive faint washes. The washes should be carried right up to the edge of the outline without overlapping on to an adjacent colour. Spaces between the edges of different colour areas are then filled in with a mixture of both colours.

The depth of the tint will depend upon the subject. In general, prints for colouring should be fairly light in tone; dark prints tend to look dull and muddy.

Portraits are best kept in a high key and may be sepia-toned first.

Oil Colouring. Prints can be coloured with oil colours, but it is a difficult job and best left to the professional artist. Transparent oil colours are used; they must not be mixed with white or opaque colour. The print, preferably on semi-matt paper (sepia-toned in the case of a portrait), is prepared by applying the special medium supplied by the maker of the colours. Artists' megilp thinned with pure turpentine may be used.

The colours are applied to areas of any size on tufts of cotton wool; brushes are used for the smaller details.

The print should be fixed by its four corners to a smooth board covered with several sheets of blotting paper.

A small tuft of cotton wool is wound around the point of a small piece of wood; these are supplied in the colouring outfits. The tip of the cotton wool is worked into a fine point. This is placed in the colour, the latter being upon a palette or sheet of clean glass. The cotton wool is then applied to the area to be coloured, working it into the surface, and following the outlines carefully. The colours may be worked further into the surface of the print with another piece of clean cotton wool. Any colour found incorrect or excessively brilliant can be reduced or removed with cotton wool moistened with the thinning medium.

After colouring is completed the print should be dried in a warm, dust-free atmosphere.

Often the reds, crimson and carmine can be made brighter by the addition of a little yellow. Yellow can also be added to the greens to produce greater brilliance.

Greys are the result of equal parts of blue, yellow and crimson diluted according to the depth of the grey.

A few experiments with the actual colours along these lines will soon indicate the possibilities.

Pastel Chalks. Coloured chalks are extensively used by professional colourists, but their use calls for considerable skill; the colour is opaque and needs to be applied very lightly. The print should be light in tone and preferably of a fairly large size.

The print surface is prepared by rubbing it lightly with fine pumice powder, and the colours are then applied with small pieces of cotton wool. Or the crayon may be rubbed on a piece of cartridge paper from which it is wiped off with a piece of cotton wool and applied to the print. Excess or overlapping colour can be removed with a pointed india rubber. Fine detail can be filled in with a small piece of chamois leather twisted around a match stick. Very fine detail is best put in with water colour.

Chalk colouring needs to be fixed: the print may be held in the steam from a kettle, or sprayed with the special liquid fixative sold by the makers of the colour.

Colour Blending. All colouring should be done in soft daylight, as it is difficult to judge the effect of the colour by artificial lighting.

The successful colouring of photographs depends very largely upon an understanding of the art of colour blending.

The best results will be obtained when the colours are well diluted. Very vivid colour effects are rarely successful. For small prints the colouring should be delicate.

The beginner should study the effects of painters and especially note the relative depth of colour of near and more distant objects. The latter are much more soft in colouring than those close at hand. By this means depth can be easily conveyed.

Colours are rarely used alone. The results are most harmonious when a given colour is obtained by combinations. The following is a short guide to some of the hues that are obtained by combinations of colour.

Red and blue make purple.
Sap green and yellow make pale green.
Flesh and yellow make brick red.
Red and yellow make orange.
Navy blue and green make dark green.
Carmine and blue make purple.
Yellow and blue make green.

Still further modifications are possible as any of the colours in an above pair can be given a deeper tone suggestive of either by using more of one and less of the other. For example, more yellow and less blue will produce a light green while more blue and less yellow yields a blue green.

Chemical Methods. Various methods have been used which utilize chemical means of hand-colouring. These processes are usually more complicated than normal hand-colouring.

One such system that has been employed uses normal chemical toning baths and a hand-painted resist to prevent parts of the photograph from being toned. In this way, only selected parts of the image are toned in one colour. The process is then repeated with fresh resists and other colour baths, building each colour up separately.

With another process, special materials are employed and the colour is applied selectively with a brush, any mixture of colours being possible. The colours react with the silver image in the print, replacing it entirely in proportion to the silver present. In this way, a pure colour image is obtained. R.M.F.

See also: *Flexichrome process.*
Book: *The Complete Art of Printing and Enlarging*, by O. R. Croy (London).

COLOUR LIGHTING. The lighting needed for taking colour photographs must fulfil certain requirements.

For normal results it must be of the correct colour temperature for the colour materials being used in the camera; it must be sufficiently bright to permit a reasonably short exposure time; it must be arranged so that the brightness range of the subject is kept to a minimum; it must emit a continuous spectrum to reproduce colours accurately.

See also: *Colour materials; Colour technique; Colour temperature; Contrast; Lighting the subject; Light sources.*
Book: *Colour Photography in Practice*, by D. A. Spencer (London).

COLOUR MATERIALS

Materials for direct colour photography are generally classified according to the system of colour synthesis and type of processing employed. They fall into the following groups:

(1) Subtractive reversal materials.
(2) Subtractive negative materials.
(3) Subtractive positive materials.
(4) Additive materials.

In addition, black-and-white materials are used in various indirect processes.

Subtractive Reversal Materials. Subtractive colour materials consist of three differentially colour sensitive layers coated on top of each other as an integral tripack. These three layers are sensitive to the three primary colours: the top layer to blue, the middle layer to green, and the bottom layer to red. Generally a yellow filter layer between the top and the middle layers prevents any blue light from affecting the lower two emulsions.

The material is processed by reversal, the second development taking place in a colour developer which produces not only a silver image but also a dyed image in each layer. This dyed image is generally chosen to be complementary to the sensitivity of the layer in which it is formed. Thus the blue sensitive layer yields a yellow positive dye image, the green sensitive layer yields a positive magenta dye image, and the red sensitive layer yields a positive cyan (blue-green) dye image. The three silver images are eventually removed by chemical means, and the dye images form the composite colour picture by subtractive synthesis.

There are two main types of reversal subtractive materials, requiring distinctly different processing technique.

The original materials of this kind utilize controlled individual colour development of each layer. The colour developer in each case contains the appropriate colour forming agent.

The other type of subtractive reversal film incorporates colour couplers in each layer, so that one colour developer simultaneously develops each layer to its correct colour.

Subtractive reversal film yields positive colour transparencies on the same support which was exposed in the camera. Duplicates or prints can, however, be made by printing the transparency on to a similar reversal subtractive material.

Subtractive Negative Materials. In principle these are similar to subtractive reversal materials, but are processed to a colour negative only. The resultant image is not only reversed tonally, but also appears in colours complementary to those of the original subject. For instance blue skies are recorded brownish-yellow, greens as magenta, yellow as blue, etc.

The choice of colours with negative films is fixed by convention rather than by scientific necessity; the purpose of the colours is to enable the three layers to cover the visible spectrum in three approximately even ranges.

These negatives can be printed by contact or enlargement to produce colour prints, black-and-white prints, or colour transparencies. In each case the colour negative can be used to produce unlimited numbers of positives.

In addition, the colour negatives may be used to produce direct separation positives for various indirect colour printing processes.

Some subtractive negative materials incorporate suitable masking layers to compensate for the shortcomings of the image dyes. The absorption of these dyes often deviates appreciably from the ideal absorption required for perfect colour reproduction. In some materials of this type the colour couplers incorporated in the emulsions are themselves coloured and thus automatically modify the response of the respective emulsion layers.

Subtractive Positive Materials. In structure and general principle subtractive positive materials are similar to subtractive negative materials. They are used for printing colour negatives to yield prints or transparencies. The positive material differs from the corresponding negative material mainly in its gradation and spectral sensitivity curves. It is available on a paper or film base.

Grain and Resolving Power. Most colour emulsions are inferior in resolving power to average black-and-white emulsion. The fact that there are at least three, and often as many as five, separate layers may lead to a certain loss in sharpness. With some materials this is accentuated by a slight degree of diffusion of the colour couplers, both within each emulsion layer and from one layer to another.

Subtractive colour films are often claimed to be grainless. While the grain size in each emulsion is probably no smaller than it would be with an average black-and-white material, the graininess itself tends to get lost. This is partly because the image dyes do not form as distinct grains as the silver image, partly because the differently coloured dye grains (such as they are) are very rarely in register in the three layers, and partly because the contrast of each dye image is lower than of a corresponding black-and-white silver image.

In practice, lack of sharpness in colour images is sometimes due to inadequate colour correction of the lens used to take the picture.

Colour Balance. As all colour materials rely on individual sensitivity to the three primary colours, they will yield correct colour reproduction only if the light contains these three primaries in a specified proportion. In practice therefore colour materials are balanced for light sources of given colour temperatures by adjusting the relative emulsion sensitivities.

Reversal films are generally available as daylight type and artificial light type materials. Daylight film is balanced for average daylight (5,500 to 6,000° K) and artificial light film may be balanced either to 3,200° K (commercial studio illumination) or 3,400° K (photofloods).

If the colour emulsion is exposed by light of a different colour temperature than that for which it is balanced, the colours will be falsified. Thus a daylight type film exposed by Photoflood light will show very warm (reddish) colours, while an artificial light type film exposed by daylight will give pictures with a bluish hue. With reversal materials it is thus specially important to match the light source to the colour balance of the material, as no effective correction can be made after exposure.

Most colour materials can, however, be used with light sources for which they are not balanced, if a suitable colour correcting or colour compensating filter is fitted in front of the camera lens.

With subtractive negative film the printing stage offers an opportunity for correcting the colour balance. For that reason the matching of the film to the light source is not so important, and there are many negative colour materials which can be used equally well by daylight or by artificial light, provided light sources of different colour temperatures are not mixed.

Speed. At one time colour materials used to be comparatively slow; current emulsions do, however, reach the sensitivity of high-speed black-and-white films.

The standard method of determining the speed of reversal colour films necessarily differs from that for black-and-white materials; speeds themselves are however expressed in the same notation and when used with exposure meters or tables yield correct results. There is no standard for determining the speed of colour negative film. Figures given by manufacturers are therefore equivalent speeds and mean that the material should be exposed like a black-and-white film rated at the quoted speed.

Gradation. Most colour materials, especially reversal ones for transparencies, cover a comparatively short scale of tones. This is partly due to the fact that the maximum density obtainable with the dyes is less than with silver, and partly to the fact that the emulsion is designed to be contrasty to yield good colour saturation. Some makes provide an alternative emulsion of even greater contrast for special purposes such as photomicrography and copying in colour.

One result of the comparatively contrasty gradation of the film is a very limited exposure latitude. This is in part due to the limitations imposed by the reversal process and to the extreme thinness (and low coating weight of silver) of the emulsions. The exposure latitude is further reduced by the fact that it is difficult to match the gradation of the component emulsions of a tripack material sufficiently accurately over the whole range of tones. So incorrect exposure results not only in loss of tone at the highlight or shadow end of the scale, but also upsets the colour balance.

Subtractive negative materials usually have a softer gradation and thus an increased exposure latitude. They are, however, still limited by the lack of colour balance when the exposure deviates appreciably from its optimum value.

Special soft-gradation negative and positive materials are made for duplication from colour transparencies. The negative film, also known as inter-negative, produces a low-contrast intermediate negative from the transparency, without excessive colour distortion at the ends of the tone scale. The inter-negative is then printed on positive colour film or paper to give duplicate transparencies or colour prints.

Duplicating reversal film is a subtractive reversal material that yields low contrast direct transparencies. It is used for photography when large numbers of duplicate transparencies are likely to be required, and also serves as the master print for duplicating motion picture films by reversal. Making the original transparency on duplicating reversal material avoids excessive contrasts which would be further enhanced by the reversal printing process.

Image density. As colour films build up the image by layers of dye, the maximum density obtainable is usually rather lower than with a silver image. This is of particular significance in colour motion pictures, where the film may also have to carry a contrasty black-and-white sound track.

Positive colour materials have images containing appreciably less dye than transparency films or negatives. Since however the light has to pass through the dyes twice (to reach the white paper backing and again on reflection from the paper), the effective reflection density is of a similar order as with transparency materials.

TYPES OF COLOUR MATERIAL

Type	*Speed (ASA) D.*	*Speed (ASA) A.*	*Available as*	*Characteristics and Uses*
Subtractive negative	25-100	25-100	S, R, M	General photography; processed by maker, trade or user
Subtractive reversal I	10-64	16-40	S, R, M	General photography; emulsion incorporates no couplers; processed by maker only
Subtractive reversal II	25-64	25-64	S, R, M	General photography. Emulsion incorporates couplers; processed by trade or user
High speed subtractive reversal	100-200	80-200	S, R, M	Photography by poor light. Emulsion incorporates couplers. Processed by trade or user. Forced processing for extra speed often possible.
Positive print	—	—	Pp	Making paper prints from colour negatives
Positive transparency	—	—	S, M	Making positive transparencies from colour negatives
Reversal print	—	—	Pl, Pp	Making direct prints from colour transparencies
Duplicating reversal	—	—	S, M	Making duplicate transparencies from colour transparencies
Internegative	—	—	S, M	Negatives from transparencies for duplication

Abbreviations. D = speed of daylight type material to daylight; A = speed of artificial light type material to appropriate artificial light source. Availability: S = sheet film; R = roll film; M = 35 mm. miniature film; Pl = plastic opaque base; Pp = paper.

Supports. Most colour materials are coated on cellulose acetate safety film base. Subtractive reversal materials for print-making from transparencies are often coated on a white opaque plastic base.

Positive print materials for making prints from colour negatives may be coated either on a white plastic support or on paper.

Sizes. Subtractive colour materials (reversal and negative) are available in most roll film and sheet film sizes, and also as 35 mm. and 8–16 mm. cine film. The range offered by any one make is, however, usually limited; thus one brand may cover 35 mm. miniature and narrow-gauge cine film only, while another may be available exclusively in sheet film sizes.

Subtractive reversal and positive printing materials are available in roughly the same range of sizes as black-and-white bromide papers.

Keeping Qualities. Subtractive colour materials do not keep as well as black-and-white materials. Apart from normal factors of deterioration such as loss of speed, increase of contrast, etc., a complicating factor is the presence of three emulsions instead of one. Owing to their different sensitization, they may deteriorate at different rates, thus upsetting the colour balance. This is especially apparent with subtractive reversal films, where there is no way of compensating for inaccuracies in colour balance after the original exposure.

With many materials the colour balance shifts towards red with increasing age. For that reason some manufacturers have slightly overbalanced the sensitivity of their material to blue, on the assumption that the material would not be used until some months later when the colour balance should be about correct.

The ideal storage conditions for colour materials are on the whole similar to those for black-and-white materials. They should be kept in a cool dry place, and well away from fumes. After exposure the material should be processed as soon as possible.

Special Subtractive Colour Materials. Certain subtractive colour materials depart from the principles already described either to yield a colour material for special purposes or by using a different kind of process. Examples of the first kind are false-sensitized masking films, and of the second kind dye-destruction and transfer-diffusion colour prints.

A false sensitized film has a deliberately abnormal and non-complementary relationship between the sensitivities of the emulsion layers and the colours produced in them. The most usual purpose of such films is the production of colour masks when making separation negatives or duplicate colour prints from coloured transparencies. False-sensitified colour films are sometimes also used in certain fields of scientific photography to separate for instance radiation intensities as colour effects rather than to show actual colour values.

Dye destruction colour materials contain the final dyes in the three emulsion layers. During processing these dyes are destroyed image-wise. A dye destruction colour paper therefore yields a direct positive from a positive transparency, and requires reversal processing to yield a positive print from a colour negative.

In transfer diffusion colour material the dyes are also contained in, or associated with, three differentially sensitized emulsion layers. During processing of the exposed negative film, a corresponding positive dye image is transferred to a receiving layer; the latter forms the positive colour print.

Additive Materials. These are nowadays obsolete, having been displaced by subtractive materials. The latter offer overwhelming practical advantages over the additive processes.

Essentially an additive colour film consists of a panchromatic silver halide emulsion coated on top of a mosaic of transparent red, green and blue elements. This mosaic breaks up the light into its primary components and so produces small image units corresponding to the colour of the image at each point. These image units in combination with the colour units then yield the additive colour picture, which may be either negative or positive (by reversal processing).

The colour elements, usually in the form of a mechanically ruled pattern, act as filters and the emulsion has to be exposed through this mosaic. Alternatively, the colour mosaic can be a separate screen with a normal panchromatic plate exposed in contact with it. After processing the plate is bound up in register again with the same screen.

The limitations of additive materials, which were largely responsible for their disappearance, were low resolution (owing to the finite size of the screen elements), disturbing visibility of the mosaic in enlargements, and high image density since the image is formed by both silver and filter elements. Even with powerful projectors additive transparencies therefore tended to yield dim screen images. L.A.M.

See also: *Colour film principles; Colour film processing; Colour printing; Colour print processes; Colour synthesis.*

Books: *Colour Photography in Practice*, by D. A. Spencer; *Colour Films: The Facts*, by A. Cromwell-Clyne (both London).

COLOUR MEASUREMENT

Colour is a mental quality. It is the result of a chain of processes both physical and psychological. Radiant energy from a light source is modified by reflection from or transmission through various objects, enters the eye and is perceived, and then interpreted as a colour by the observer. The emission of energy from the source and its modification by objects are physical processes. The act of perception, involving the sensitivity of the eye, is a psychophysical process. Measurements of these processes are objective measurements. The interpretation of these by the observer is a psychological process. Classifications of colours by this process are subjective measurements.

The following is a brief survey of the psychological aspects of vision, covering colour perception and subjective colour measurements, and of the physical aspects of colour, covering objective measurements.

COLOUR PERCEPTION

The process of vision is so complex that we cannot immediately specify the exact colour of an object. Red, green and blue are obviously different colours but to try to specify precisely "which red" or "which blue" of the possible reds or blues is extremely difficult.

The colour we see depends on many factors. The object may be a surface which is either glossy or matt. It may be a translucent medium such as glass, jelly or the sky. It may be large or small, round or square. Such physical attributes can affect the perceived colour of the object. The surrounding objects may be strongly coloured, black or white. The illumination may be daylight, tungsten light or sodium light. The perceived colour will be different under all these different viewing conditions.

The observer himself is equally important. A colour blind person, by definition, does not see colours the same as does a "normal" person. Even two people with so-called normal colour vision are unlikely to see colours exactly the same. And even confining observations to just one person still admits of considerable variation. The eye is not only capable of adapting itself at least partially to the illuminant, it also changes in sensitivity to colour as the illumination level changes (notably at low illumination levels, such as moonlight) and at different points on the retina (the region outside the fovea is largely colour blind).

Yet another complication is the effect of after-images. If an observer stares first at a red patch and then at a "white" surface he will see the surface as blue-green. Such after-images occur to some extent whenever the observer looks from one coloured patch to another even slightly different patch.

With such a welter of information the task of sorting it into some logical system of measurement of colour seems at first sight doomed to failure. However steps can be taken to eliminate some of the disturbing factors mentioned above. The physical attributes of patches of colours to be matched can be fixed so that a constant size and shape are used, and measurements confined to a specific type of surface or medium. The viewing conditions can be rigidly defined by using a constant surround to the patches and using a specific illuminant. The differences in observers may also be allowed for by averaging the results of many observers. Such limitations imposed on measurements must always be borne in mind when using them. It may not be valid to change the conditions for a particular system of measurement, and comparisons between one system and another must be made with caution.

Dimensions of Colour. To the layman colour has as many dimensions as there are colours. To him orange and brown are unrelated, a red rose is somehow different from a red car.

The layman seldom attempts to abstract "colour" from the complex sensations involved in looking at objects. He does not for example analyse orange and brown and conclude that they are similar in hue and different in "lightness" (Brown may be thought of as a "dark orange"). When he does consider colour itself he uses terms such as bright green; light blue, deep purple, pale yellow, vivid red—terms which are sufficient for his purpose, but are not systematic or precise enough to be useful as measures of colour.

Colour can in fact be shown to have only three dimensions. The colours mentioned above, bright green, light blue and so on can be expressed in terms of three variables, namely: hue; luminosity and saturation.

Hue is the attribute of colour which is denoted by red, orange, yellow, green, blue, violet (i.e. the hues of the spectrum) and purple (the non-spectral hue between red and violet) and, of course, all their intermediates.

Luminosity is the attribute which denotes the amount of light which an area appears to emit. The terms lightness and brightness are also used for this attribute. (Strictly, lightness refers to diffusely reflected or transmitted light but is commonly used to refer to colours from reflected light. Brightness is still frequently used, particularly when referring to sources of light, but the term luminosity is preferable).

Saturation is the attribute which denotes the proportion of chromatic colour present. Alternatively chroma may be used for the third dimension. This denotes the amount, rather than the proportion, of chromatic colour present, irrespective of the amount of achromatic colour.

For example we may have two articles which are both green. Neither one can be classed as more blue or more yellow than the other (same hue); neither can be called lighter or darker than the other (same luminosity), but one is simply a stronger green, or "greener", than the other. They differ only in the strength of greenness, that is they differ in saturation. A colour that has zero saturation (and therefore no specific hue) is termed achromatic. These colours are black, greys and white. Other colours are termed chromatic.

Any given colour is essentially some particular combination of attributes, for instance, the colour which is typically evoked by the light from a tungsten lamp is yellow in hue, weak in saturation and high in luminosity. Similarly, a ripe tomato in the sunshine is likely to yield a slightly yellowish red hue, strong saturation and moderate luminosity. Lightning in a dark sky may elicit no hue or saturation but very high luminosity, that is, an achromatic colour. The thousands of different combinations of hue, saturation and luminosity which can occur account directly for corresponding thousands of distinguishably different colours.

Luminosity and saturation are commonly confused. A "light" blue for example implies not only high luminosity but also low saturation. A "bright" red implies high saturation as well as fairly high luminosity. This confusion arises because common terminology is related to common experience. A high luminosity blue in nature is also of low saturation. Highly saturated, high luminosity blues are nearly always "artificial" colours, produced by dyes and pigments. Similarly highly saturated, low luminosity, reds are uncommon.

Modes of Appearance. Basic descriptions of colours (in terms of hue, saturation, and luminosity) may be supplemented by reference to five principal types of stimulus situation which give rise to colours and to which they seem to belong. These modes of appearance of colour include the illuminant mode in which the colour is perceived as belonging to a source of light—e.g., the sun; the illumination mode in which colour is perceived in space—e.g., lamplight in the atmosphere; the surface mode with colour belonging to surfaces—e.g. colour prints; the volume mode—e.g., a clear mass of red jelly; and the film mode in which the colour seems soft and without definite location in depth—e.g., overcast sky.

The film mode is the simplest or basic mode in the sense that all other modes can be reduced to it by appropriate viewing through a small aperture in a screen. When a surface colour, for instance, is reduced to film colour, as by looking through an aperture screen, the surface mode is destroyed and the colour thereby becomes disassociated or unrelated. The several modes of appearance are classified as object modes (surface, volume, and illuminant) and non-object modes (illumination and film) according to whether or not they are related to particular physical objects.

COLOUR SYSTEMS

The attributes of hue, luminosity and saturation or chroma form the co-ordinates of a subjective colour system. The nearest approach to this system in practice is the Munsell colour atlas. In this the dimensions are Munsell hue, value and chroma. These co-ordinates are similar but not identical to hue, luminosity and chroma respectively. For objective measurements the C.I.E. (Commission Internationale de l'Eclairage—sometimes referred to as the I.C.I.—International Commission on Illumination) system is used frequently. In this the dimensions of dominant wavelength, luminance, and purity may be used.

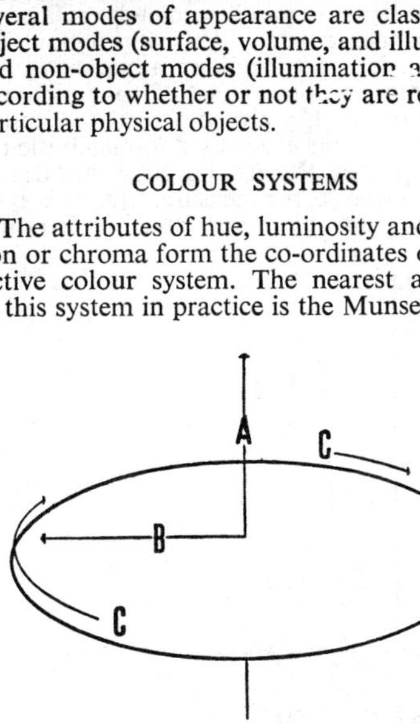

FUNDAMENTAL COLOUR SYSTEM. The colours occupy a system of cylindrical co-ordinates: A, brightness, increasing vertically; B, saturation or chroma, increasing radially from the centre; C, hue, varying with angle about central axis.

Each system has its own specific set of dimensions. Because these dimensions are in many ways similar there is a great temptation to confuse them. Hue and dominant wavelength, purity and saturation for example are often referred to as though they were simply alternative names for the same thing. This is not so. It is like saying that Magnetic North and Geographic North are the same. Each is useful for a particular purpose; it is possible to convert measurements from one system to another but they are not interchangeable at random.

Subjective Colour Systems. The most convenient way of representing a subjective colour system is in the form of a solid with cylindrical co-ordinates; with luminosity increasing vertically, saturation increasing radially, and hue varying with the angle about the luminosity axis. As an analogy we may think of lines of latitude representing luminosity, lines of longitude representing hue and the horizontal distance from the axis representing saturation.

The Munsell system of colour presents the nearest approach to this colour solid in practice. In this system the colour samples are presented in the form of about 1,000 coloured paper chips. These are suitably arrayed with respect to the dimensions of Munsell hue, value and chroma. The perceptual spacing of the colour samples is approximately uniform throughout the system.

Any principal vertical section through the Munsell system reveals varying values and chromas of two opposite constant hue planes—e.g., yellow and purple-blue—separated by the neutral (achromatic) axis. Any horizontal section, on the other hand, shows a plane of constant value with varying hue and chroma. The hue, value, and chroma scales are naturally used for describing or notating the standard colour samples. The Munsell notation Y 8/12 for instance, refers to a sample of yellow Munsell hue, value 8, and chroma 12. This particular sample normally evokes a colour of yellow hue, rather high lightness and strong chroma.

Any unknown colour sample can be identified or specified simply by matching it in daylight with a notated colour sample, if such is available; moreover, interpolation is feasible within the orderly standard arrangements of Munsell samples.

Objective Colour Measurements. If we knew enough about the process of vision we could, for simple situations, calculate the actual

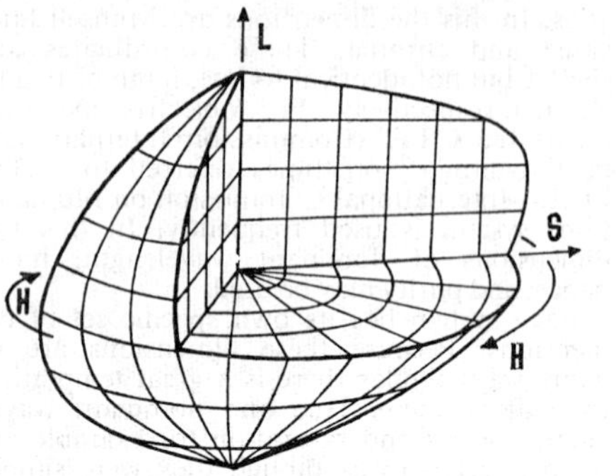

SUBJECTIVE COLOUR SOLID. This type of subjective colour system is based on cylindrical co-ordinates; luminosity (L) increases vertically, saturation (S) increases radially, and hue (H) varies with the angle about the central luminosity axis. The Munsell colour solid is of this basic type.

appearance of a colour once we knew the physical properties of the light entering the eye. As yet this cannot be done, but it is possible to calculate whether or not two lights look alike to a person with average colour vision. If they differ we cannot say what the difference will look like although we can obtain some idea of the magnitude of the difference.

Such objective measurements of colour are sufficient for many purposes. For example in colour photography we need to know whether or not the reproduction looks like the original, or by how much it differs. The following sections describe some of the physical measurements that are used for this objective description of colour.

Spectral Reflectance Data. It is well known that "white" light is composed of all colours of the spectrum. If we shine this white light on to a surface the surface will look, say, green because it reflects more of the green component of the light than of the others. If we shone a red light on to the surface it would appear black because it does not reflect red light. The surface itself has not changed in its ability to reflect light in a certain way. It is the change of illuminant that has caused the apparent change in colour.

By measuring the proportion of light that a surface can reflect at each wavelength of light throughout the spectrum, and combining this information with the spectral energy distribution of the source, we have a means of specifying objectively the colour of the surface. The reflectance (or transmission in the case of transparent media) is normally expressed as a percentage of the incident light at each wavelength, and the curve obtained by plotting out the percentage reflectance against wavelength is termed the spectral reflectance curve.

Two surfaces which have identical spectral reflectance curves will always look the same when viewed side by side. But under certain conditions it is also possible for two surfaces of different spectral reflectances to look the same. These are called metameric colours. For example a surface which reflects only yellow light will appear yellow, but so will a surface which reflects both red and green light only. Viewed in "white" light they may look the same but viewed in red light one will appear black, the other red.

Spectral reflectance measurements do not provide immediate information on the appearance of a surface colour—they are not intended to. They are of value because they specify an intrinsic property of the surface alone. In order to obtain from them some idea of the appearance of the surface, spectral reflectance curves can be combined with colour mixture data for plotting on a C.I.E. or similar colour diagram.

Colour Mixture Data. A wide range of colours can be obtained by adding red, green and blue lights in various proportions. The intensities of each light required to match a given sample provide a measure of the colour of the sample.

For highly saturated samples it is not always possible to obtain a match directly by the addition of the three lights. In such cases it is however possible to desaturate the sample by adding one of the lights to it and then matching the result with the other two lights. The light added to the sample is thought of as a negative addition of light. Expressed mathematically we can say, for example, that

Sample + Red = Green + Blue

or Sample = Green + Blue − Red

It is possible in this way to find the amounts of the three lights required to match each of the pure monochromatic lights throughout the spectrum. By plotting these amounts against wavelength we obtain a set of colour mixture curves. Strictly they are colour mixture curves for the eye, as the eye is used to match the two patches of light. We are, however, still dealing with objective colour measurement. The eye is used here purely as an instrument to determine

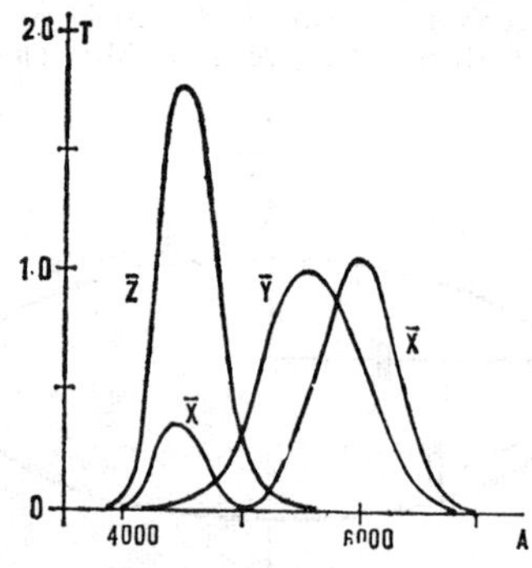

COLOUR MIXTURE CURVES: The values of x, y, z are the amounts of the C.I.E. imaginary primaries required to match a unit amount of energy for each monochromatic light throughout the spectrum. They are plotted here as tri-stimulus values against wavelengths.

a colour match, and colour mixture curves are not intended to give a subjective measure of colour. The colour mixture curves obtained depend on the energy distribution of the source of the "white" light, e.g. different curves would be obtained for tungsten light and daylight as the source.

There is nothing unique about the three lights used to obtain these curves. There is an infinite choice of lights that could be used. They need not be confined to any particular wavelengths, they need not be monochromatic. The only criterion is that it must be possible to obtain white by direct addition of all three lights in suitable proportions. The lights used to obtain a particular set of mixture curves are called the primary colours for the system.

The colour mixture curves for any set of real primaries always involve negative parts of the curves. These negative components complicate any calculations with the mixture curves. In order to simplify calculations we need a set of all positive mixture curves. These can be obtained mathematically by calculating such a set of curves from the mixture curves of a real set of primaries. The primaries that must be used for such mixture curves are imaginary colours.

The use of imaginary quantities is common in science as there are many cases where calculations are simplified by their application. An imaginary colour is no more unusual than any other imaginary quantity. To give some meaning to the concept we may think of an imaginary colour as being one with a higher saturation than any obtainable colour. The most highly saturated colour that can be obtained in practice is a monochromatic colour. We can conceive of colours more saturated than this just as we can conceive of speeds in excess of the speed of light—the highest speed obtainable in practice. (N.B. There are exceptions to these statements in both colour physics and relativity physics.)

The C.I.E. System. One set of colour mixture curves based on imaginary primaries is accepted as an international standard. This is the set adopted by the Commission Internationale de l'Eclairage (C.I.E.).

The basic light source for the C.I.E. system is an equal-energy source, that is, one that has equal output energy at all wavelengths, and the basic colour mixture curves are those required to match a unit amount of energy at all wavelengths. A number of subsidiary light sources are specified, each designed to be typical of the more commonly used sources.

It is possible to convert the basic colour mixture curves to suit these sources as required. The amounts of the primaries required to match a colour are called tristimulus values, and denoted by $\bar{x}$, $\bar{y}$, $\bar{z}$ for monochromatic light in the C.I.E. system. A feature of the C.I.E. mixture curves is that the primaries are so chosen that the $\bar{y}$ curve is identical to the luminosity curve of the eye. (The eye is that of

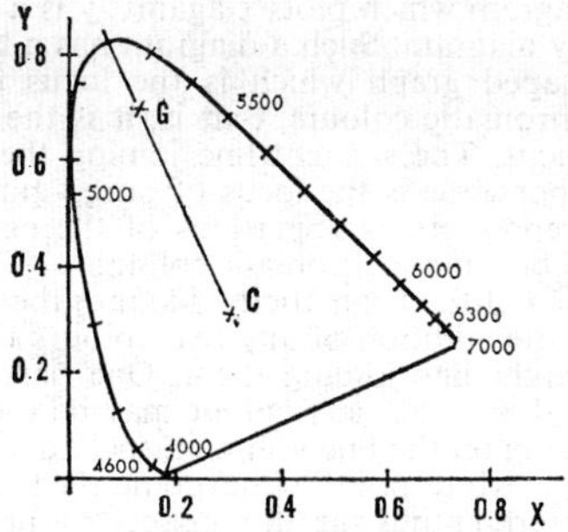

C.I.E. CHROMATICITY DIAGRAM. This shows the spectrum locus and the locus of purest purple (connecting line between the points corresponding to 4,000 and 7,000 A). All real colours lie within these loci. G is a green viewed in an illuminant whose colour is plotted at C against the white point or achromatic point. The dominant wavelength of G is 5,200 A and its excitation purity is 75 per cent. Luminance would be a dimension at right angles to the plane of this diagram, with its axis erected at C.

the C.I.E. "standard observer"—obtained by averaging the results of a number of observers.)

To find the tristimulus values for a non-monochromatic colour, the transmission curve (T) for the colour is combined with the colour mixture curves (modified for the particular illuminant being considered). This is done by multiplying the curves $\bar{x}$ with T, $\bar{y}$ with T, $\bar{z}$ with T, wavelength by wavelength and then adding the results throughout the spectrum in each case to obtain three numbers X, Y and Z. These numbers then indicate the amounts of the primaries which would be required in an additive mixture to match the colour.

These three values can be modified to a form suggestive of colour appearance. We can abstract the redness, greenness and blueness of the colour from them by taking respectively the values

$$x = \frac{X}{X + Y + Z} \text{ (approximately redness)}$$

$$y = \frac{Y}{X + Y + Z} \text{ (approximately greenness)}$$

$$z = \frac{Z}{X + Y + Z} \text{ (approximately blueness)}$$

In this way we reduce all colour to a common function since

$$x + y + z = 1$$

for all colours. It is also true from this equation that if we know x and y then z is automatically defined. For example if x is nearly equal to 1 then both y and z must be small and the colour is red: if x and y are both small z must be nearly 1 and the colour is blue. When x, y and z are all nearly equal to $\frac{1}{3}$ the colour is close to "white" since it has no predominant hue.

A diagram which plots x against y is a chromaticity diagram. Such a diagram has a horseshoe-shaped graph which is the locus of all monochromatic colours, that is, it is the spectrum locus. The straight line joining the ends of the horsehoe is the locus of purest purples, i.e. it represents combinations of the extreme red and blue monochromatic colours.

In the C.I.E. chromaticity diagram the resultant of the addition of any two colours lies on the straight line joining them. One particular case of this is that any colour may be thought of as being on the line joining a spectral colour and the "white point" (the point on the diagram representing the illuminant colour used to view the colour itself). The wavelength of the spectral colour used is called the dominant wavelength for the colour. The position of this colour on the line depends on the proportions of the spectral colour and the white needed to obtain the colour. This proportion may be expressed as the ratio of the distances of the colour and of the spectral colour from the "white point". This ratio is called the excitation purity of the colour.

The similarity to subjective colour diagrams is apparent if we think of dominant wavelength as being roughly equivalent to hue, and excitation purity as roughly equivalent to saturation.

The third dimension in the C.I.E. system is luminance. The value of luminance is in fact the value Y. This derives from the fact, mentioned previously, that the colour mixture curve for $\bar{y}$ is identical to the luminosity curve of the eye. Luminance is comparable with the subjective dimension of luminosity.

The C.I.E. diagram is particularly useful for calculations in additive systems. Its use is not confined solely to such systems, however, as it is one of the few methods available for precise objective measurements of colour. Strictly the C.I.E. system is designed for illuminant colours but is again extended to cover surface colours as well.

Other Systems. Only a few of the other valuable systems can be mentioned here.

The Ostwald surface-colour system, originally published in 1917, is now represented by a collection of over 900 mat and glossy colour samples in the *Colour Harmony Manual.*

The calibrated coloured glasses of the well-known Lovibond subtractive system continue to be useful in conjunction with the Tintometer for evaluating either reflecting or transmitting samples.

The I.S.C.C.-N.B.S. (Inter-Society Colour Council—National Bureau of Standards) system of designation consists of over 300 meaningful and carefully related colour names representing corresponding blocks of Munsell colour space which are small enough for many purposes.

Among the colour dictionaries is that of Maerz and Paul which contains 7,000 printed colour samples many of which are correlated with common colour names.

Ridgway's earlier series of over 1,000 fully-named samples was recently restandardized.

Villalobos's recent colour atlas of over 7,000 small colour samples is characterized by unusually good spacing.

There are available numerous coded sample collections for more special applications, such as the Horticultural Colour Chart of over 1,000 samples for evaluating and comparing flowers; others are for agricultural products, textiles, paints, printing inks, plastics, etc.

Various useful colorant-mixture systems have been devised. Such systems commonly provide series of colour samples together with formulae for reproducing their appearance by admixture of several standard paints or other colorants. Similarly, subtractive colour photography depends on sets of standard cyan, magenta, and yellow dyes, though these are combined in layers rather than by physical mixture.

P.W. & S.M.N.

See also: *Colour; Colour sensitometry; Spectrum; Vision.*

Books: *Colour Photography in Practice*, by D. A. Spencer (London); *The Reproduction of Colour*, by R. W. G. Hunt (London); *An Introduction to Color*, by Ralph M. Evans (New York); *Colour in Business, Science and Industry*, by Deane B. Judd (New York).

COLOUR PHOTOGRAPHY HISTORY

The historical development of colour photography could be divided into two periods, each guided by a basic philosophy entirely different and independent of the other. In one, called objective colour, an attempt was made to produce the same conditions in the image which caused the original object to be coloured. In the other, subjective colour, the attempt was made to produce conditions in the image which created the same colour sensations as did the original. The first concept held sway from the earliest days up to the close of the nineteenth century. The second, introduced in 1855, was not accepted until 1890, but from then on it became the basic philosophy for all subsequent work.

Objective Colour Photography. An object appears coloured if it preferentially absorbs some of the light rays falling upon it. Every colour is characterized by a definite pattern of absorptions for specific rays, i.e., its absorption spectrum. To match one colour with another, the matching colour must have an absorption spectrum identical with that of the colour under consideration. In colour photography this matching must be done for each image point. Processes which accomplish this form the class of objective colour processes.

Interference Processes. The starting point for such procedures lay in the disclosures of Johann Seebeck. In a letter to his friend Goethe in 1810 he reported that silver chloride, upon exposure, took on the colour of the exposing light. This observation became the basis for the researches of others, notably Edmond Becquerel and Niepce de St. Victor.

In their experiments they exposed a specially treated layer of silver chloride under a colour original. The image formed appeared to be coloured in hues approaching those in the original. Zenker gave a plausible explanation for this in 1868. He reasoned that when light was incident upon the layer of silver chloride, it passed through the layer and became reflected by the white paper backing upon which the light sensitive material was coated. This caused the formation of standing waves. At the nodes, the forward and backward beams cancelled each other, but at the anti-nodes they supplemented each other. At these points the light intensity became sufficient to cause photochemical decomposition of the silver chloride. Hence through the depth of the layer were formed lamina of silver, whose distances apart were half the wavelength of the incident light. The lamina served as gratings and interference patterns for any subsequent light incident upon them, with a resultant reproduction of the wave pattern of the original.

In 1890 Otto Wiener pointed out that it should be possible to replace the silver chloride layer with a fine-grained photographic emulsion. This would enable the operator to fix the images and thereby achieve permanency. The actual work was accomplished by Gabriel Lippmann, the following year. The process is now known as the Lippmann interference procedure.

The experimental evidence that colour was produced as a result of optical interference came in 1907, when Cajal reported upon a study he made upon the Lippmann process. Sections made through an image showed the presence of lamina. These averaged from four to eight in number, though occasionally, especially in areas of bright saturated colours, as many as thirteen were noted. Whites and achromatic colours gave diffuse images. A Lippmann slide had to be viewed normally, and by the light reflected from a mirror placed in contact with the emulsion side of the slide. This created many difficulties of operation, and served as a drawback to its further development.

Dispersion Processes. An entirely different solution was devised by Lancaster in 1895. The image of the object being photographed was broken up into a series of points by means of a screen similar to a process screen used in block-making, placed in the focal plane of the camera lens. A second lens brought each image point to a focus upon a prism with an angle of the order of 2° or 3°. The prism saw the image only as a series of point sources of light, each of which it dispersed into a spectrum. In contact with the prism, a fine-grained photographic emulsion recorded the minute spectra. The final image was a series of minute deposits of silver, each a record of the quantity and quality of the light registered at that point.

To re-create the image in colour, a positive of the record was placed between two prisms, identical to the one used in the original photograph, and positioned complementary to each other. The combination, in effect, formed a glass plate with parallel sides. On the side facing the light was placed a screen identical to the one through which the photograph was taken. On directing a parallel beam of light upon the screen, the picture area was broken up into point sources of light. Each point, upon traversing the first prism, became dispersed into a micro-spectrum. With correct registration between screen and positive, each micro-spectrum fell upon an image point in the positive, and became modified by it. The second prism reformed the modified beams. Thus the quality and relative quantity of the original light was recreated at each point of the image.

The process had to record properly the complete spectral distribution and intensities of the incident rays in minute areas, therefore only very fine grained and thus extremely slow emulsions could be used. Exposures of the order of a fraction of a second normally used in everyday photography were out of the question. For this reason objective colour systems reluctantly gave way to subjective systems.

Subjective Colour Photography. The present practice of colour photography is that of trichromatic or subjective colour reproduction. In 1802 Thomas Young promulgated the hypothesis that colour was a sensation. His main thesis was that the sensation produced by any one colour could be matched by that produced by a suitable mixture of three fundamental colours called primaries. From this point of view every colour could be conceived as being formed of some combination of the three primaries.

To duplicate a colour, it was necessary and sufficient to form a colour with the same relative intensities for the three primaries. The problem, therefore, was to determine the relative intensity of each primary in a given colour and then re-form it.

The first to suggest a method to accomplish this was James Clerk Maxwell. In 1855 he suggested the use of optical filters to isolate each primary, and the use of the photographic process to determine their intensities, point by point in a given scene. At a subsequent lecture, in 1861, he gave a demonstration of the process.

The experimental work was done by Thomas Sutton, a prominent photographic practitioner of that era and editor of the magazine *Photographic Notes*, where he published a detailed account. He made three exposures, one each through a red, green and violet filter. He then projected these images, as positive transparencies, in register with each other upon a white screen. Each was illuminated with the same coloured light to which it was exposed in the camera. The result was not exactly breath-taking. It was described by Sutton in these words, " . . . and when these different coloured images were superimposed upon a screen, a sort of photograph of the striped ribbon was produced in natural colours."

Maxwell in fact prepared a fourth plate exposed through a yellow filter; this was intended as a possible alternative to the red filter negative, but was apparently not used in the actual experiment.

Though disappointed with the result, Maxwell and Sutton had made and disclosed one of the most fundamental of all photographic discoveries, and laid the foundations of trichromatic colour photography.

Nevertheless for a hundred years Maxwell's demonstration remained something of a mystery, because at that time no photographic plates were sensitive at all to light of longer wavelength than blue, and therefore separation negatives in green and red light were impossible. The solution of the mystery was given in the centenary year, 1961, when Evans showed that the three negatives had actually been exposed to different wavebands in the blue, violet, and ultra-violet regions. Unknown to Maxwell or Sutton, the filters used happened to transmit these bands, which enabled true separation negatives to be obtained, aided by the fact that the spectrum of the red cloth which was photographed almost certainly had a secondary reflectance band in the ultra-violet.

The subject took a fresh start in 1868, when Ducos du Hauron proposed a solution based upon the same principles previously disclosed by Maxwell. He was, however, completely unaware of Maxwell's contributions. He went further than Maxwell. He outlined a number of other techniques by which colour reproduction could be achieved. Unlike Maxwell, he continued to write and lecture upon trichrome colour photography.

Slowly the popular prejudice against subjective processes wore down, but it was not until 1890 that they were finally dissipated. Further progress concerned itself mainly with the disclosure of procedures that were practical utilization of the fundamental principles disclosed by Maxwell and du Hauron.

The processes could be grouped into three classes. These were:—

(1) Processes which utilized additive colour analysis and colour synthesis;

(2) Processes which utilized additive colour analysis but subtractive colour synthesis; and

(3) Processes which utilized subtractive colour analysis and colour synthesis.

Additive Analysis and Synthesis. The very first demonstration of colour photography, that made by Maxwell before the Royal Institution in 1859, was an example of additive colour analysis and synthesis. In additive analysis each colour record or separation, to use more technical language, is formed from the whole light reflected from the subject. The primary desired is selected by the use of an optical filter which transmits but a single primary. The other two are absorbed by the filter in question, and are therefore lost or wasted.

In additive colour synthesis, the image is re-formed by projecting a positive transparency from each separation in register with the others, upon the same surface, each illuminated by the light of a single primary colour. The re-formed image consists therefore of the additive mixture of the primary coloured lights; hence the name.

A much simpler procedure to achieve the same result was described by du Hauron, in his French patent of 1868. He divided the image into a series of minute image points, by the use of a screen placed in the rear focal plane of the camera lens. The elements of the screen were primary coloured dots or lines, which completely filled the screen area. The size of the dots was such that the eye was unable to distinguish them individually, but saw only a blend. Although the dots were coloured red, green, and blue, the eye saw the blend as colourless or grey. Immediately over the screen was coated a panchromatic photographic emulsion. The exposure was made through the screen elements. Therefore the intensity of the red primary reflected from any point of the scene became recorded as a silver density immediately below a red dot at the corresponding point in the image.

On reversal processing, the silver deposit behind each coloured dot reduced the intensity of the light incident at that point to a value needed to reproduce the relative intensity of the primary in question. Thus only a single exposure was required to made a complete analysis of the scene. By the same token, only a single projector was needed to effect colour synthesis. The net result was that the beam of light reflected from each point of the object was divided into three sub-beams, and each passed through a filter which transmitted but a single primary colour.

The loss of light in this system was very considerable. The division of the beam into three meant a loss of two-thirds of the available light. The use of a filter which absorbed at least two-thirds of the light incident upon it meant the loss of another two-thirds. This was the main fault to be found with additive processes.

The first commercial additive screen plate was marketed by Lumière in 1907. The screen consisted of a mixture of dyed potato starch grains spread on to a sheet of glass coated with adhesive. The gaps between the grains were then filled with carbon black. Later materials used dyed globules of resin which were rolled to bring them into good contact without gaps.

With this type of screen thorough mixing of the dyed grains before coating was expected to yield a completely even distribution of the filter elements. Practical observation, and later on theoretical calculation, showed that a certain degree of clumping of grains of the same colour is inevitable. The reason is that in a random distribution there is an equal chance for, say, a red grain to be next to a blue, a green, or another red grain. Similiarly, there is a certain though decreasing statistical probability for three, four, five, or more red grains to come together. Microscopic examination even revealed grain clumps of 20 grains of the same colour for which again there is a very small but definite statistical probability. Visually, therefore, such a grain screen appears much coarser than suggested by the size of the grains.

A completely even arrangement of the filter elements is obtainable only in a regularly patterned screen—the exact opposite of a random distribution. The first commercially successful screen plate with a mechanically ruled screen was the Finlay plate of 1908. A few years later a modified process appeared, using a separate screen which was placed in contact with a panchromatic plate before exposure. After processing and printing as a positive transparency the plate was bound up with a similar screen of filter elements for viewing. This method enabled any number of colour positives to be produced from one exposure.

Around 1930, additive colour materials with a printed screen or reseau also appeared as roll, miniature, and cine film (Dufay). One attempt was made to develop a negative-positive additive colour process, using a negative and a positive screen film. Special printing arrangements were necessary to prevent interference between the screens of the two films.

A special type of screen process was the Kodacolor lenticular cine film, marketed in 1928 and based on patents of 1908. The back of the film support was embossed with minute lenses which, in conjunction with a striped filter in front of the camera lens, broke up the image into large numbers of coloured strips or bands. On projection of the reversal positive through the same striped filter, the colour image was synthesized on the screen.

Additive Analysis with Subtractive Synthesis. The practical needs of colour photography soon found the purely additive processes much too costly, in so far as light was concerned. This became increasingly apparent in cinematography, where it was essential that the three primary impressions be formed instantaneously and simultaneously. This was possible with the screen processes, but these were severely handicapped by the demands of definition. Other technological difficulties existed, especially the ease of accurate and satisfying duplication.

It was this last that turned attention to another of du Hauron's many disclosures, that of subtractive colour synthesis. This was accomplished by the conversion of each primary selection or colour separation into a positive formed of a dye which absorbed a single primary, absorbed it sufficiently so that the residue was the amount needed for colour duplication. Each image point, therefore, consisted of a mixture of three dye deposits, each subtracting sufficient of a single primary to leave the amount needed for duplication.

Colour analysis was still achieved by the additive methods. Special cameras were constructed in which optical devices situated behind the camera lens divided the lens beam into three sub-beams. An optical filter then filtered each beam to transmit but a single primary. In this manner three separations were made, each on a different piece of photographic material.

This created the problem whereby the separate primary records could be converted into dye positives, which could be superimposed one upon the other in exact register. A long era in the history of colour photography concerned itself with that problem. Many ingenious devices were disclosed to accomplish the desired result, but of all of these, just two have achieved widespread popularity, imbibition and carbro.

In the carbro process three separation positives are formed by selective tanning (via bromide prints) of pigmented gelatin foils. On treatment with hot water, the untanned gelatin is washed away, and pigmented yellow, magenta, and cyan images remain. These are then superimposed in register.

In the imbibition methods, first applied to colour printing in 1925 by Jos Pé, three positives are again made by selectively tanning gelatin layers followed by washing out in hot water. That leaves a set of matrices (one for each separation image) which are dyed in dye baths of the appropriate colour and printed in register on to a galetin coated paper.

This is the basic principle of the Technicolor process of cinematography.

Subtractive Analysis and Synthesis. Since it was possible to synthesize the colour image with no loss of useful light, it became desirable to do the same in the analytical part of the process. The first to suggest a method to accomplish this was again du Hauron. He suggested that it should be possible to prepare systems that were sensitive to but a single primary colour. Three such systems in the form of a pack, could be placed in the focal plane of the camera, one immediately behind the other. Each primary colour would register upon one of the three elements. Each could be processed to yield a positive image in the appropriate colour for subtractive synthesis, then joined together to form a single unit, the colour photograph of the original. In these procedures each primary was recorded and duplicated upon a separate piece of photographic material. These had to be joined together somehow, to re-form the whole.

In 1912 Siegrist and Fischer disclosed a new photographic procedure that converted the latent image directly into a dye image. The oxidation product formed when exposed silver halides were reduced to form the silver image was highly reactive and capable of union with other substances to form yellow, magenta and cyan dyes. It was possible to add such ingredients, colour formers, to each emulsion, so that the blue sensitive element could yield a final yellow dye image (blue absorbing), the green sensitive element a magenta dye image (green absorbing) and the red sensitive element a cyan dye image (red absorbing), all by the use of the same developing agent. Therefore it became possible to form a photographic material in which three light-sensitive emulsions were coated one over the other, with filters between them to insure that each element would be acted upon by the light of but a single primary. Each of the three elements contained the proper colour former so that upon processing a yellow image was formed in the blue sensitive element, a magenta in the green, and a cyan in the red. In this process both analysis and synthesis were accomplished by subtractive methods, with a minimum loss of available light.

A commercial film material based on this principle appeared as early as 1914, but utilized only two colours. Faithful reproduction of the full range of natural colours was therefore impossible.

Siegrist and Fischer's system had one major drawback: the colour formers tended to "wander" from their respective layers. The first integral subtractive tricolour taking material, Kodachrome (introduced in 1935), therefore did not incorporate colour formers in the emulsion layers at all. Instead, each layer was processed separately (although coated on the same support as the other two layers) by a complicated technique involving many carefully controlled stages.

The first commercially available colour film with couplers incorporated in the emulsion layers appeared in 1936 under the name Agfacolor. Two methods have been used to prevent the couplers from wandering from their layers. One was to adsorb the coupler substance on insoluble resin particles or oily media, which effectively anchored the coupler. Another trend has been the development of giant coupler molecules which are in themselves sufficiently immobile not to wander.

Both Kodachrome and Agfacolor were reversal processed, producing a positive colour transparency in the film originally exposed in the camera. Duplicate transparencies could be made by rephotographing the transparency, and materials of this type were introduced first with an opaque plastic base and later with a paper base, for making reversal colour prints.

In 1939 a subtractive colour film was introduced which utilized a negative-positive system. The film was developed to a colour negative in complementary colours, and that was in turn printed on to a positive material to produce a colour print. Designed originally for cinematography, this system soon became a very widespread method of producing colour photographs on paper. Its special merit is the possibility of this type of process being carried out by the skilled amateur photographer at home.

A further development of subtractive colour photography was the use of masks to improve colour rendering, followed in 1950 by the introduction of a colour negative film (Kodacolor) incorporating coloured couplers which act as integral masks for printing.

Other Dye Systems. The great majority of colour materials available today use colour development systems in which the dyes of the image are actually synthesized in the layers. There are in fact two other methods of forming colour pictures; in both of these the dyes are present in the material before exposure. During processing the dyes may be either destroyed selectively, or transferred selectively to another support.

In the first case, the conventional three emulsion layers sensitive to red, green and blue light are coated. Each, however, contains a dye of the appropriate complementary colour (e.g. the red sensitive layer is dyed cyan). After exposure and negative development in a conventional black and white developer, the silver image is used to bleach the dyes in proportion to the amount of silver present. The bleach bath is a strong acid, and may be thought of as dissolving the silver and liberating hydrogen which reduces the dyes to colourless soluble compounds removed by washing. In practice complexing agents and catalysts are also required for a practical process. Since the dye is destroyed in proportion to the amount of the negative silver image, the resulting photograph is automatically a positive. These silver-dye-bleach systems are all based on early patents by Christensen from 1918 and by Gaspar in the 1930's who first made such materials commercially available as Gasparcolor.

The other process is of much more recent introduction. The subtractive tripack is coated with the dyes in adjacent layers, and these are not the ordinary azo-dyes of the dye bleach process, but special dyes which can also function as developers. After exposure, the material is treated with an alkaline solution and under these conditions the dye developers become mobile and develop the exposed silver halide. In doing so, they are of course oxidized, but at the same time their mobility is destroyed and the dye remains fixed in the layer. Where no exposure has taken place, the dye continues to move and finally migrates to a receiving sheet which is pressed in contact with the surface of the material. Here the dye is fixed by a mordant and an acid layer stabilizes it and prevents further migration by neutralizing the alkali. Since dye migrates where there has been no exposure, a positive picture is obtained on the receiving layer. In practice the alkali is contained in a pod ruptured by pressure rollers on withdrawal from the camera. The contents are thus spread between the negative and positive sheets which are separated after an appropriate time interval for development and dye migration to occur; this is the Polacolor system.

In recent years there has been an enormous increase in the use of colour photography and a steady improvement in colour rendering, speed, and resolution of films. The transparency reversal process remains probably the most widely used type of colour photography, but the negative-positive process giving colour prints on paper is increasingly popular. Prints are exposed to light for much longer periods than are transparencies and this fact has emphasized the need for improved light fastness in the dyes used.

Progress has indeed been striking in the past fifteen years and there has been a vast increase of devotees in the field of colour photography. Whereas a generation ago the practice of colour photography required the use of a very expensive and delicate camera, and required the knowledge of very intricate and difficult processing techniques, today every camera is a colour camera, and almost every photographer a potential colour photographer.

J.S.F. (C.H.G. & J.Mo.)

See also: ***Cine history; Museums and collections; Sensitized materials history.***

Book: *A Source Book of Photographic History*, by W. Baier (London); *History of Color*, by Joseph S. Friedman (Boston).

COLOUR PRINTING TECHNIQUE

In the early days of colour photography, the only practical way to make a colour print was to photograph the original scene three times on a panchromatic film or plate, once through each of three differently-coloured filters, and prepare a black-and-white print from each of the negatives obtained. Next, each print was dyed a different colour and the three coloured images obtained were transferred in turn to a final paper support, giving the finished print. The process was lengthy and costly, for three negatives and three prints had to be prepared for each final picture, and the problems of registration (making sure that one coloured image lay exactly on top of another) were severe. It is not surprising that colour printing by this method was limited to specialists, despite the very attractive results which could be obtained.

With the advent of tripack material, however, almost all the problems were swept away. For the tripack, as its name implies, combines all three negatives on one sheet of film and all three prints on one sheet of opaque white material, usually paper. Tripack material is so simple to use that nowadays all colour photographs are taken with it, whether to produce colour negatives or transparencies, and almost all colour prints are made on it.

In essence, tripack material consists of three layers of ordinary photographic emulsion on a suitable base, sensitized in effect to blue, green and red light respectively. During processing, complementary dye images (yellow, magenta and cyan) are produced together with the black silver, which is then removed to leave only a picture in the dyes. The mechanism of image formation is thus largely the same as with tripack colour film.

The usual sequence of coating the layers is blue-sensitive on top, green-sensitive beneath and red-sensitive on the bottom. This is necessary with films to enable the blue-sensitivity of the lower layers to be neutralized by a yellow filter layer. In printing, the light used for the exposure is deficient in blue and the sensitivities of the two remaining layers can be so adjusted that their blue-sensitivity can be disregarded. Some manufacturers have therefore changed the order of the layers to secure a sharper-looking picture. So the red-sensitive layer may be on top, the green-sensitive layer in the middle and the blue-sensitive layer on the bottom. The printing and processing of the paper is not affected in any way, whatever the sequence.

Paper types. There are two kinds: positive and reversal paper. Positive paper is for printing from colour negatives which, being only a means to an end, can be made to suit the paper, regardless of the appearance of the negatives. This allows the film manufacturer considerable scope, for he can incorporate in the film emulsions dyes designed to overcome the deficiencies of photographic dyes. This accounts for the orange or yellow "masks" which characterize certain makes of film.

Reversal paper is for making prints from transparencies which allow of no such latitude, for transparencies must look correct if they are to fulfil their basic function. This precludes the use of any masking dyes, and it is more difficult to achieve good colour reproduction in prints made on reversal paper, because of the deficiencies of the dyes in both the transparency and the print.

Printing and Enlarging Equipment. To produce good colour prints from colour negatives requires no special skills and little special equipment not already available for black-and-white work. Owing to the difficulty of working with small pieces of paper in almost total darkness it is hardly ever worth while making contact prints. Colour prints are therefore implied to be enlargements.

Any enlarger will do but it should have a tungsten light source, for fluorescent light can lead to unnecessary difficulties when trying to produce a correctly-balanced print. It should also ideally have a filter drawer in which can be placed coloured gelatin foils whose function is explained below. The enlarger should be quite steady to avoid movement during the exposure, and of course a timer and a holder for the paper being printed are also necessary.

As fluctuations in mains voltage affect both the colour and intensity of the light, some form of voltage regulator should be installed. In its simplest form this consists of a voltmeter and rheostat which enables the voltage to be adjusted to a standard value immediately before making each exposure. The maximum variation of voltage permissible without seriously affecting results is about ± 2 per cent, i.e., ± 4 volts when working at 200 volts. For continuous printing, of course, some form of voltage stabilizer is essential.

The majority of good-quality enlarging anastigmats are suitable for colour work but the colour reproduction is very much better if they are coated (bloomed). An uncoated lens is likely to give considerable softening and degradation of the colours.

Stray light from the lamphouse, negative carrier, etc., must be carefully avoided and the enlarger should not be used close to a light or shiny wall which may reflect light on to the paper during exposure.

The darkroom should also be well blacked-out for colour paper is rather fast, and stray light will almost certainly fog it. As the paper is sensitive to light of all colours, a special darkroom safe light (made by most manufacturers for their colour paper) is necessary. This gives a small amount of light of a colour to which the eye is more sensitive than the paper, but the latter must still not be exposed freely to the safelight.

Filters. Ideally, one should be able to photograph a scene on a piece of colour negative film, process the film and put it into an enlarger and from this make a colour print merely by giving a simple exposure. Unfortunately, there are many reasons why this will not produce a perfectly balanced print; that is, one which is not too blue, nor too green nor with a tint or cast of some other colour. Variations in the colour of the light when the picture was taken, in the colour temperature and age of the enlarger lamp, the age of the processing solutions, and of the paper and film itself—all are liable to affect the colour balance of the print. So we need some method of controlling the colour of the light coming through the negative

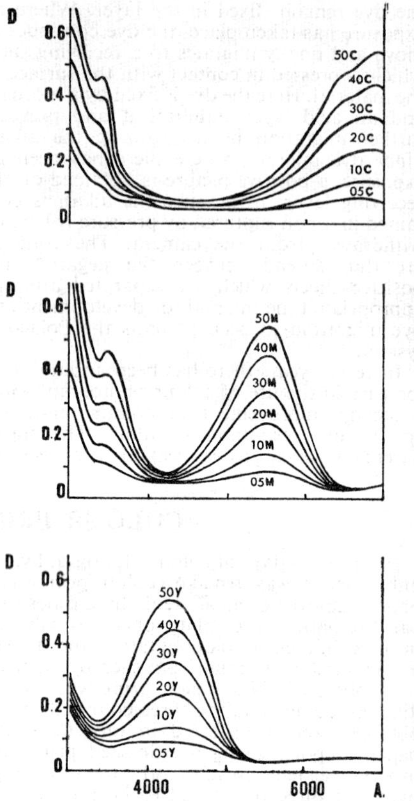

COLOUR COMPENSATING FILTERS. These curves show the densities of a typical set of cyan, magenta and yellow filters used to modify the balance of colour prints made from colour negatives by the white light exposure method.

at the printing stage. This is done with coloured gelatin filters, which are placed either above the negative in the lamphouse or immediately below the enlarger lens. In either case they modify the colour of the light reaching the paper.

They thus also modify the colour response of the enlarging paper. If a colour print is too yellow, this is due to too strong a dye image being produced in the yellow-forming (blue sensitive) layer—so this blueness must be reduced. A yellow filter, which absorbs blue, can do this. A pale yellow filter is sufficient if the print is only slightly too yellow; if the yellow cast is strong, then a deeper yellow filter is necessary. Thus a graduated series of yellow filters can correct all yellow casts. Similarly, predominant magenta tints or casts can be removed by a series of magenta filters which progressively hold back green light affecting the green-sensitive layer responsible for the magenta component of the image. Cyan filters in the same way control blue-green or cyan casts by reducing the red light reaching the red-sensitive layer of the paper.

This leads to the important rule that a colour cast is removed by a filter of its own colour. That rule also applies to casts of other colours: blue, red and green. Excess green in a colour print is due to too strong images being produced in the cyan-forming (red-sensitive) as well as the yellow-forming (blue-sensitive) layers. Hence both red and blue light must be held back and a cyan and a yellow filter are required for correction. (Cyan plus yellow make green, so the cast is again removed by a combination of filters yielding its own colour.)

In the same way a blue cast is removed by a combination of magenta and cyan filters and a red cast by a combination of magenta and yellow filters. Nor need the combination of filters be of equal strength: a cast may be yellowish-green, indicating that its yellow component is stronger than its cyan component so that the yellow filter in the combination must be denser than the cyan.

This method of colour printing, requiring a set of graduated filters each of the three colours, yellow, magenta and cyan, is referred to as subtractive or white-light printing. It calls for at least four filters of each colour, though in practice manufacturers may produce a series of up to eleven graduated filters of each colour. Prints can be made quickly by this method, once the correct filters have been determined; they are then merely put in the filter drawer and an exposure made in exactly the same way as in black-and-white enlarging.

An alternative method of printing calls for less filters. This is three-exposure or tricolour printing. Here, only three dense filters are necessary: blue, green and red. An exposure is first made through the blue filter; this affects only the blue-sensitive layer and hence is responsible for the yellow image. Without moving the enlarger, an exposure is now made through the green filter which affects the green-sensitive layer and hence produces the magenta image. Still without moving the enlarger or paper a third exposure is made through the red filter. This affects only the red sensitive layer and yields the cyan image. The print is processed and, if the exposures have been correct through each filter, a perfectly-balanced print results. If there is too much yellow, the exposure through the blue filter is reduced, if too little yellow the exposure is increased, and so on.

Where economy is the sole consideration, and colour prints are required only occasionally, the three-exposure method has the advantage that it requires the purchase of only three filters to start producing colour prints. But it is slow and tedious, for three comparatively long exposures must be given for each print. Also, the enlarger and print must remain steady between and during exposures, otherwise the colour images in the print will be out of register, with consequent general fuzziness and lack of fine detail. When a whole roll of film or more is to be printed the white-light method is to be preferred, as this requires only one short exposure for each print, and twelve prints can be made in the same time as one or two done by the tricolour method.

A further advantage of the white-light method is that the print can be dodged or shaded by the use of a sheet of black card exactly as in black-and-white printing. This is not practicable with tricolour printing as the shading has to be done three times, once while each filter is being used, and it is difficult to keep the three shaded areas in register. In practice, however, dodging or shading is rarely necessary in colour printing.

White-Light Printing. Apart from the colour balance of the colour negative, the filtering required for a correctly balanced print depends on the enlarger itself and on the colour balance of the paper. Manufacturers frequently specify the latter from batch to batch, indicating the deviation from a standard balance in terms of a basic colour correction. A first test print is therefore made, using the basic correction filter in the enlarger plus an ultra-violet filter recommended with some types of paper. Make a stepped print, giving a series of increasingly long exposures, say 2, 4, 8 and 16 seconds. Process the print and examine it when it is dry. Decide on the cast which is present in the step which appears to be about right in overall depth.

Think of the cast in terms of yellow, magenta or cyan. If it is yellowish, for example, make a further set of test prints, using yellow filters of varying strength, until a filter is found which just removes the yellow cast. This filter must now be put in the filter drawer and the final print made. It is not necessary to make a separate print for each filter and process it before proceeding; a square of print paper may be divided into four, and a small print made on each quarter of the paper, using four different filters. These filters may be held below the lens, although in this position, if they are not perfectly clean and of a high optical quality, they may affect the definition of the print. This is not important, however, at this stage as only test prints are being made. When the paper is processed it gives four differently-filtered pictures side by side, making comparison simple.

When the yellow cast has been removed, it may be necessary to use a second filter of a different colour, for there may be another, weaker, cast present. This should be removed in the same way. Note that this cast must be weaker than the first, otherwise it would have been dominant, so it will require a weaker filter to remove it.

When this second cast has been removed make the final print. In working with tripack material, only two layers can be out of balance. This means that filters of only two colours are necessary.

If a filter pack ever is found to contain three colours, remove the filter colour of the lowest density together with the same densities of the other two filter colours. For example if the pack for a balanced print consists of a cyan filter of density 30, a magenta filter of density 20 and a yellow filter of density 10 (see below for a meaning of these filter notations), the density 10 in all three filters only adds up to grey and can be removed. In other words the colour correction effect is the same with a filter pack consisting of a cyan density of 20 plus a magenta density of 10.

Filter combinations of all three colours are liable to arise when the filter pack in the enlarger already contains two filter colours as the basic correction for a particular paper batch, and a third colour is introduced to correct the cast in the first test print. From this it follows that a cast can be corrected not only by adding a filter of the same colour as the cast, but also by reducing the filter density of the other two colours. Thus to get rid of a yellow cast you can either put a yellow filter into the enlarger (as explained above), or reduce the density of the cyan and magenta filters if these are already in position. This reduces the predominance of the yellow image by increasing the strength of the other two images.

COLOUR CORRECTION FILTER ADJUSTMENTS FOR POSITIVE PRINTING

Print Shows Excess of:	*Print Shows Lack of:*	*Take out (or Reduce Density of) Following Filters:*	*Or Insert (or Increase Density of) Following Filters:*
Red	Cyan	Cyan	Yellow and magenta
Green	Magenta	Magenta	Yellow and cyan
Blue	Yellow	Yellow	Magenta and cyan
Cyan	Red	Yellow and magenta	Cyan
Magenta	Green	Yellow and cyan	Magenta
Yellow	Blue	Magenta and cyan	Yellow

Filter Notation. There are two main ways of labelling white-light correction filters. The best known is the CC notation. CC stands for colour correction and the filter notation includes a density value and a letter indicating the colour. The density values range from 05 to 50. Thus the notation CC-50M indicates a magenta correction filter of density 50. (The actual density—to green light—is 0.5, but 50 happens to be a more convenient way of putting it.) Similarly a CC-05Y is a yellow filter with a density value 05 (in fact 0.05.

The complete series in each colour may include the values 05, 10, 20, 30, 40, 50 and even higher ones. Higher filter densities may, however, be obtained by combining two lower ones, e.g. a filter value of CC-80C could be obtained by combining a cyan 50 and a cyan 30 filter. Similarly the 05 step provides intermediate combinations such as 15, 25, etc.

With this notation the complete filter combination for a particular print can be expressed in a form like CC-10M + CC-20Y.

Some manufacturers also produce a series of red, green and blue filters—ranging for example from CC-05R up to CC-50R, and the same way for the blue and green series. These filters behave like combinations of two filter colours in the cyan, magenta and yellow range of equal density. Thus a green filter CC-20G is equivalent to the combination CC-20Y + CC-20C and can be substituted for this filter pair. These red, green and blue filter series are, however, rarely essential.

The second notation dispenses with the letters altogether and uses only a series of six figures. These define individual filters or complete filter combinations. The first two figures are the yellow density, the second to the magenta density and the third to the cyan density. In this notation a filter 00-50-00 corresponds to a CC-50M; the notation 20-10-00 is equivalent to the combination CC-20Y+ CC-10M .

Exposure. Once any colour cast on the correctly exposed step of a test print is corrected, the overall density range will no longer be right. The exposure therefore needs correction as well. This can be done either by a further stepped print with the correct filter combination, or by multiplying the original exposure by the product of the filter factors. The latter vary slightly from manufacturer to manufacturer. Yellow filter factors can usually be ignored, while magenta and cyan ones may range from 1·2 to 2·5 according to filter density.

Some allowance is also desirable for the actual number of filters in the enlarger since each filter results in some light loss by scattering and reflection, quite apart from its colour filtering effect. Usually the exposure should be increased by 10 per cent for every additional filter in the pack or reduced by 10 per cent when the combination is decreased by one filter. One way to eliminate this correction is to have a pack of standard number—for example four filters—and keeping up this number with plain colourless gelatin sheets if fewer filters are required.

In practice, filtering is not so difficult as may appear from the instructions given above. One or two points should be stressed however. Logically, test prints should be made with filters of gradually increasing strength. Make a note of everything that is done, so that it is possible to determine which filter produced a particular result. The information is best written in pencil on the back of the print.

Three-Exposure Printing. This requires three filters, in effect a tricolour set, which should preferably be mounted on a rotating or sliding stage under the enlarger lens so that each can be brought in front of the lens in turn. Care must be taken not to move the paper or negative or to shake the enlarger when changing the

TRICOLOUR EXPOSURE ADJUSTMENTS FOR POSITIVE PRINTING

Print Shows Excess of:	*Print Shows Lack of:*	*Reduce Exposure through:*	*Or Increase Exposure through:*
Red	Cyan	Green and blue	Red
Green	Magenta	Red and blue	Green
Blue	Yellow	Red and green	Blue
Cyan	Red	Red	Green and blue
Magenta	Green	Green	Red and blue
Yellow	Blue	Blue	Red and green

filters between exposures, otherwise lack of registration and colour fringing may be produced in the print.

The colour balance and density of the print are adjusted entirely by regulating the duration of the three exposures (which must, of course, be determined by trial).

First make a stepped print through the red filter only, giving, say 4, 8, 16 and 32 seconds. After processing, this yields a cyan image. The exposure which gave a section showing the picture clearly yet rather faintly is the correct one.

Next put another strip of paper in the holder and give the whole strip the chosen exposure through the red filter. Without moving the paper give it additionally a stepped exposure through the green filter and process as before. There will now be an overall cyan picture superimposed by a stepped magenta picture. Note the exposure through the green filter giving an exact balance between the magenta and cyan image, that is, a blue picture.

Put a third strip of paper in the holder and give it the chosen red and green exposures and, on top of this, make the stepped picture through the blue filter. After processing, there is an overall blue picture on which is a stepped yellow image. One of these will give a balanced result, not too blue, not too yellow. The final print may now be made using the chosen exposures.

If the final picture is too light, all three exposures must be increased by the same fraction; if too dark, decreased by the same fraction. So, if the exposures given were red 8, green 10 and blue 14 seconds and this was found to be too light, make another print giving red 12, green 15 and blue 21 seconds.

Some care is necessary in the choice of the filters. Each filter should completely hold back light of a colour that might affect layers of the colour paper other than of the same sensitivity as the filter colour. Thus the blue filter must pass only blue light and no red, otherwise accurate colour control becomes impossible. An easy test is to make test strips through each filter in turn on the colour paper in use. After processing the strips should show only the pure colour images (yellow, magenta and cyan respectively) without any tinge of the second dye. If this is not the case, one or more of the printing filters may need an additional colour filter to hold back unwanted colour components.

Colour Grading Aids. The trial-and-error methods outlined above are only suited to very small-scale working where occasional prints are required. When many prints have to be produced daily from different types of negatives, a quicker method is essential. This is done by means of a densitometer, a device which measures the colour of a negative or transparency in terms of its blue, green and red densities. The negative is placed in the densitometer and a reading taken in turn through each of the three filters which are incorporated in the apparatus. By a simple piece of calculation this can be translated into the exposure which need be given through three similar filters, as in tricolour printing, or by a slightly more involved calculation it can be used to show what yellow, magenta or cyan filters are necessary when doing white-light printing.

Densitometer reading requires careful choice of the section of the negative to be read. A simplified densitometer technique measures the average light through the whole negative, in turn through a red, green and blue filter. This is considered to integrate to grey—in other words the negative is treated as if the total light coming through it were neutral and filtered to achieve such a neutral balance. This principle is based on the assumption that the overall colour of average subject in nature is neutral; this is, in fact, valid for about 80 per cent of all cases. For negatives where an integration to grey does not yield a correctly balanced print, the additional correction necessary can be estimated fairly accurately from the first test print.

Another simple aid for colour negative grading is a colour step wedge. This consists of a normal grey wedge of about 18 or 20 steps which has a red, green and blue filter strip laid side by side over all the steps. This colour step wedge is placed on top of a sheet of the colour paper on the enlarger baseboard and exposed to the light from the colour negative which has, however, passed first through a diffuser to provide the grey integration. The lightest visible step in each colour on the print indicates the relative exposures required through the blue, green and red filters for three-exposure printing, or the exposure plus the correction filters required for white-light printing.

Filter Control Aids. Some professional colour enlargers greatly simplify the mechanical selection of filter combinations. Usually they incorporate a colour head in which continuously graduated cyan, magenta and yellow filters are moved into the light path. Alternatively a single high density filter of each colour may be moved into the light to greater or lesser extent, progressively increasing the proportion of the light going through the filter as compared with the

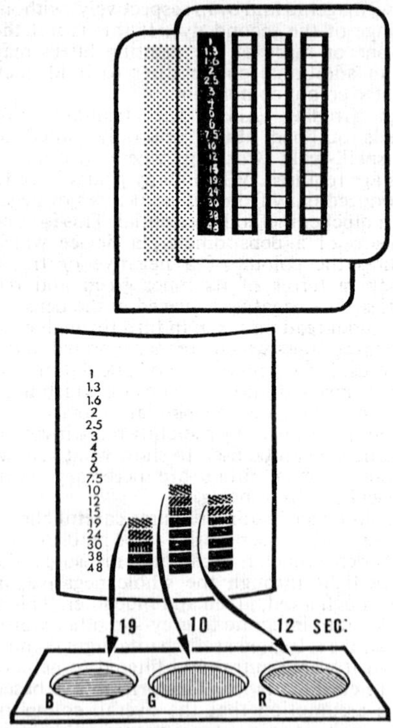

TEST STRIPS WITH COLOUR WEDGES: Place a sheet of the colour paper underneath a colour step wedge (in practice a grey step wedge overlaid with strips of red, green and blue tricolour filters. The exposure takes place to the light coming through the colour negative, but diffused by an opal screen placed below the enlarger lens. After processing, the faintest yellow, magenta and cyan steps on the test print indicate the required exposure times for three-exposure printing. This method requires preliminary calibration to a standard negative and the type of colour paper used. For single-exposure printing the difference between the lightest steps in the three colours can by suitable calibration indicate the required subtractive filter combination. This technique corresponds to integration to grey and sometimes needs additional colour correction.

proportion of unfiltered light reaching the negative. Usually the light so filtered is directed into a diffusing chamber or sphere to assure a uniformly mixed light reaching the negative. The degree to which each filter is introduced may be set and read-off on the colour head by means of calibrated wheels or dials.

Colour heads or colour enlargers for three-exposure printing may switch red, green and blue filters of high density successively into the light path. The time for which these filters are in position usually is controlled by an electronic or other timing device and selected on a separate control panel linked with the filter changing mechanism.

An alternative semi-automatic three colour printing system starts the exposure with white light and switches high-density cyan, magenta and yellow filters into the light path. These filters in effect terminate the exposure to light of their complementary colour. For example an initial white-light exposure of say 10 seconds corresponds to a simultaneous exposure of this time to red, green and blue light. If the red light exposure is intended to be 10 seconds, the cyan filter then moves into the light path to cut off the red light. For a green filter exposure of 14 seconds the magenta filter would move into position as well after a further 4 seconds, and the yellow filter would be switched in to terminate the blue exposure—and the complete exposure sequence. This arrangement is a kind of combined white-light three-exposure printing method and has the advantage of being quicker than three successive exposures through blue, green and red filters.

Processing the Print. Processing a print made from a colour negative consists of the following essential stages: development (to produce the dye images), bleaching (to remove the silver images which are an integral part of the colour development stage but which are not required in the final print), and fixing (to remove all silver salts). The bleach also removes the yellow filter layer if one is present. The complete sequence is as follows: development, an acid stop bath or a long wash in water to stop development and remove any developer chemicals, bleach, fix, and a wash to remove all chemicals from the gelatin. Sometimes the stop bath contains fixer to remove all silver salts which were not affected by the developer. This permits processing to be continued in room lighting without the risk of fogging due to residual developer affecting the silver bromide.

Developers for positive paper are similar in composition to negative developers, but in some cases a different developing agent is recommended, namely N-ethyl-ethylhydroxy-p-phenylene-diamine, sulphate, which is said to have less tendency to cause dermatitis than diethyl-para-phenylene diamine hydrochloride. It also has the advantage of a higher solubility in water it is therefore more easily removed from the paper base and so minimizes the risk of stained whites in the prints.

Processing kits, containing full instructions and chemicals for the preparation of the various solutions, are available for some of the print materials.

Processing procedure should be standardized and should be rigidly adhered to throughout, otherwise unexplained changes in colour balance and density may occur. The key points to be watched are the degree of exhaustion of the solutions, the temperature of the developer and the method of agitation of the print used. There are many methods of keeping the developer temperature constant, but the most convenient is probably the water bath. In this, the processing dishes or tanks are placed in a large volume of water which is thermostatically-controlled, thereby controlling the temperature of the solutions. All liquids tend to cool due to evaporation, so the material of which the dishes or tanks are made should be a reasonably good heat-conductor to allow for rapid re-heating.

Agitation requires a special mention. In its simplest form this is a gentle movement of the print in the solution at carefully-regulated intervals. This can be done by rocking the dish or tank or moving the print round with a pair of tongs. But this requires that an operator should be in constant attendance.

Large-scale processing plants now use nitrogen-burst agitation; in this, nitrogen gas is released into the solutions at any chosen interval by means of an electronically-controlled burst valve from a supply cylinder. Nitrogen gas is chosen as, unlike air, it does not affect the solutions. The gas is released at the bottom of the tank through a plate containing a large number of very small holes, so that the whole volume of solution is uniformly affected.

Nitrogen-burst agitation is not used with shallow dishes but only in tanks. To make economical use of the system it is necessary to process considerable numbers of prints at once and these are usually placed in a basket made of a fine plastic mesh which permits the solutions to circulate freely round the prints. Each compartment of the basket takes two prints, back to back, permitting a considerable number of prints to be processed simultaneously.

Where large-scale processing of standard-sized prints is being carried out, continuous-processing machines are used. The paper is manufactured in long rolls and these are first printed in automatic printers. When the roll has been completely exposed, it is fed into a continuous-processing machine which carries the paper through automatically until it is rolled up at the far end of the machine, processed, and sometimes glazed, ready to be cut up into individual prints.

The paper is fed through the machine at a constant rate, and so, to regulate the amount of time that it is in any one solution, a series of rollers are mounted on rods which can be raised or lowered in each tank. The paper passes under these rollers, so the deeper the rod is in a tank the longer the paper is immersed in the solution in that tank.

Glazing. The fact that colour developers are highly alkaline and that the paper has three layers of gelatin poses special problems for glazing. When glazing black-and-white material the glazer can be run very hot, but if colour print material is glazed this way there is considerable danger of the soft gelatin being displaced, ruining the picture. For this reason colour prints are often finished without a glaze, or glazed at a lower temperature, which cuts down the rate of output considerably. Another method of glazing is to dry the print first, then re-wet and glaze it. The intermediate drying reduces the danger of ruining the print.

Yet another method is to harden the print in a formalin bath after it has been washed but this is not completely satisfactory as the surface of the print becomes rather brittle, with consequent danger of cracking when it is handled.

Automatic Printing Machines. Mass-produced prints for the amateur market are made on automatic machines, as mentioned above. There are many types of these, but they all work on the same general principle—the negative is assessed by three photo-cells or groups of photo-cells covered with red, green and blue filters. The reading from each group of cells is fed into an electronic assessor which decides what filtering is necessary. This information is passed on to a series of relays which pull the appropriate filters into the light beam for the necessary amount of time and then cut off the light when sufficient exposure has been given.

Automatic machines are expensive, and where the volume of work does not merit the purchase of one a colour enlarger with photo-cell control is used. The enlarger has a lamphouse with integral colour filters which can be placed in the light path or removed by means of control knobs. The colour assessment is carried out with photo-cells which may be clustered around the enlarger lens, inside the bellows, and covered with blue, green and red filters. The enlarger is set up by moving one of the glass filters into the light beam until the output from two of the sets of photo-cells balance. A second filter is then introduced gradually until the output from the third group of photo-cells balance that from the other two. The enlarger is then ready to print, and it only remains for the exposure time to be calculated by means of another photo-cell placed on the paper-holder.

Automatic printers operate on the principle that all the colours of the average scene integrate to give grey and the purpose of the electronic assessor is to decide what filtering is necessary to achieve this. Fully-automatic printers also decide at the same time how long the exposure must be to give the print the correct density, so that it is neither too light nor too dark. The fact that the negative may have a strong orange or yellow mask does not affect the printer, as this can be allowed for when setting up the machine, as can batch-to-batch paper variations.

The great majority of negatives do integrate to grey, and printing from them poses no problem, but occasionally this is not so as, for example, when the picture is of someone in a grey suit lying on grass. This scene is obviously predominantly green, but the assessor has no way of knowing this and tries to remove the green. The result is a print which shows the suit as reddish in colour, while the grass has a washed-out look. This is known as colour failure, and needs additional correction. Experienced operators can estimate this from the colour negative.

Positive Transparencies from Negatives. Colour negatives can also be printed on positive colour film material to produce transparencies. The principles of colour grading, exposure, filter selection, are identical with those for making colour prints on paper. Judging the exposure is a little different, since the test transparencies have to be held up against a suitable light source—preferably of daylight quality (for example certain blue-filtered tungsten lamps)—to assess the required image density. Processing of positive colour films is also a little different since the emulsions on the film are much more delicate and easily damaged. Positive transparency films are usually processed in sheet film holders which go into appropriate tanks. The basic processing sequence is similar to that for colour paper though the recommended processing solutions and processing times may be different.

Printing from Transparencies. The majority of colour prints are made by printing colour negatives on to positive colour paper. Where prints are required from transparencies, there are two possibilities.

One involves making an inter-negative, which means photographing the transparency on to a special colour negative material and then printing in the normal way. Despite the apparent deviousness of the route to get the finished print, it produces very satisfying results. It is popular with processing laboratories, who prefer its reliability and simplicity to printing direct from the transparency on to reversal paper.

There are two difficulties in direct printing on to reversal paper. One is that colour transparencies have to be balanced for the best visual effect which involves compromises in colour rendering. These shortcomings are further emphasized when the transparencies are printed on to reversal paper.

Further, the contrast of a colour transparency is very much higher than the tone range which a paper print can accommodate. The reversal print made from a transparency therefore tends to lose detail in the highlights and shadows and is liable to show further colour falsifications in these tones. The most suitable transparencies for printing on reversal paper are therefore those of low contrast and with a thin image.

The actual exposing procedure is very similar to enlarging from colour negatives, and corresponding sets of cyan, magenta, and yellow correction filters are employed. The correction applied, however, is different. Excess of any one colour is compensated by reducing the filter density of that colour (or increasing the filter density of the other two colours). Similarly, lack of one colour is compensated by increasing the filter density of that colour.

With tricolour printing the sequence of operations is the same, i.e., if the print is too blue (lacking in yellow) the exposure through the blue filter must be reduced; if it is not blue enough (too yellow) the exposure through the blue filter must be increased.

As with all reversal material, there are two extra processing stages, the first development and the reversal exposure. This adds considerably to the time necessary to produce a test print and makes the process tedious when printing by trial-and-error methods.

CORRECTION FILTER ADJUSTMENTS FOR REVERSAL PRINTING

Print Shows Excess of:	*Print Shows Lack of:*	*Take out (or Reduce density of) Following Filters:*	*Or Insert (or Increase Density of) Following Filters:*
Red	Cyan	Yellow and magenta	Cyan
Green	Magenta	Yellow and cyan	Magenta
Blue	Yellow	Cyan and magenta	Yellow
Cyan	Red	Cyan	Magenta and yellow
Magenta	Green	Magenta	Yellow and cyan
Yellow	Blue	Yellow	Magenta and cyan

Dye Destruction Papers. While most types of colour printing paper form the dye images in the emulsion layers by chemical reactions while the silver image is developing, there are also systems which work in the opposite way. There the three emulsion layers already contain the dyes before the paper is exposed. Again the blue-sensitive layer contains a yellow dye, the green-sensitive layer a magenta dye and the red-sensitive layer a cyan dye. During processing these dyes are *destroyed* and not formed in proportion to the silver image produced. Straightforward processing thus yields a positive from a positive—a print from a colour transparency—while a positive colour print from a colour negative requires a sequence similar to normal reversal processing.

The filtering technique for prints from negatives and transparencies is the same as for conventional negative-positive and reversal printing procedures respectively. The actual filter combinations required for balanced prints is likely to be different from that needed with the conventional system.

The main difference is in processing. For making positives from colour transparencies, the processing steps are development (which forms the silver image), dye bleaching (which removes the dye in the areas where the silver image is formed), silver bleaching, and fixing to dissolve away the bleached silver image. This sequence is very much shorter than reversal processing with conventional colour materials.

When making a positive print from a colour negative, the processing stages are development, silver bleaching (to remove the developed silver image), re-exposure and re-development to form a new silver image in the previously unexposed areas of the emulsions, followed by dye bleaching, silver bleaching, and fixing as before.

J.M.W., J.Mo. and L.A.M.

See also: *Colour film principles; Colour materials; Colour print processes; Duplicate colour slides; Enlarger.*

Books: *Colour Photography in Practice*, by D. A. Spencer; *Colour Prints*, by Jack H. Coote; *Making Pavette Colour Prints*, by Felix Smith; *Making Colour Prints, by* J. M. Whittaker (all London).

COLOUR PRINT PROCESSES

The principle behind colour printing processes is that visible light of any colour can be resolved into a mixture of blue, green and red light in varying proportions. Further, light of any of these three colours can be absorbed by a filter of its complementary (opposite) colour. So blue light can be absorbed by a filter of its complementary colour yellow, green light by a filter of its complementary colour magenta (blue red), and red light by a filter of its complementary colour cyan (blue green). All colour printing processes are based on this rule.

It is customary to refer to the three colours blue, green and red as additive primaries (because light of all three colours added together in the right proportion gives white light) and their complementary colours yellow, magenta and cyan as subtractive primaries, because they are obtained by the subtraction of the additive primaries from white light.

Modern photographic colour printing processes are of three types. One is integral tripack printing which uses a white base coated with three successive layers of emulsion in which three dye images are formed. The second (and oldest) process involves the assembly of a colour print by the superimposition of three coloured images which have been prepared separately. The third type is a variant of the first, using a tripack of three emulsion layers which already contain dyes. The dye pictures are here formed by the image-wise destruction or removal of dye.

Owing to the ease and speed with which prints can be made on tripack material, this has almost completely supplanted the assembly method of producing colour prints photographically.

Integral Tripack Printing Papers. These printing papers work on the same principle as colour transparencies. The three image layers cannot, however, be as dense, because the contrast range is limited to the difference between the amount of light reflected by the white base and the darkest black of which the print surface is capable.

The three layers are sensitive to blue, green and red light respectively, and analyze the light coming from the negative or transparency in terms of these three colours. When the print is processed the three layers produce coloured images in their complementary colours, so that a yellow image is produced in the blue-sensitive layer, a magenta image in the green-sensitive layer and a cyan image in the red-sensitive layer. The colour balance of the print can be easily controlled by using colour filters during printing, but little or no contrast control as between the three colour images is possible.

There are two kinds of integral tripack printing papers; those intended for printing from colour negatives and those intended for printing from colour transparencies.

The colour negative is solely a means to the end because it is a negative with the colours complementary to those of the subject and hence of no value in itself. It is thus possible to include masking dyes in the negative to compensate for any deficiencies in the dyes which constitute the colour negative. Because of these deficiencies in photographic dyes, the negative-positive system of producing prints will give more accurate colour reproduction than direct prints made from a colour transparency. The latter is normally designed to be viewed direct and so, of course, cannot contain masking dyes. There is also the further advantage that neither the negative nor print require reversal; this cuts down the processing to three essential steps—colour development (in which silver images, which are not required in the finished print, are of necessity produced), bleaching (to convert the unwanted silver images back to silver bromide) and fixing (to remove all silver salts).

Reversal papers used for making prints from

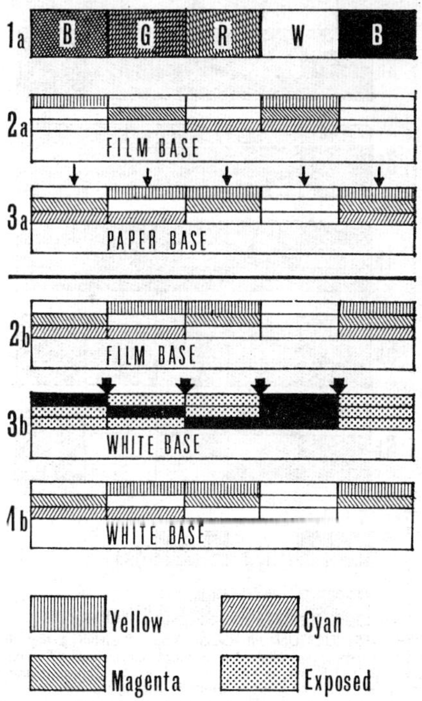

TRIPACK COLOUR PRINTING. 1. Subject consisting of Blue B, green G, red R, white W, and black B. 2A. A colour negative film records each colour only in the layer sensitive to it. The dyes formed are complementary to the original sensitivities. 3A. On printing on a positive colour material, the layers are affected in the same way, reversing the colours to form the colour print.

2B. A positive transparency consists of similar colours to a positive print. 3B. To make a reversal print, the material is exposed to the transparency and developed to a negative silver image (shown as solid black). 4B. The unexposed parts of each layer are converted into a colour image, and the silver removed.

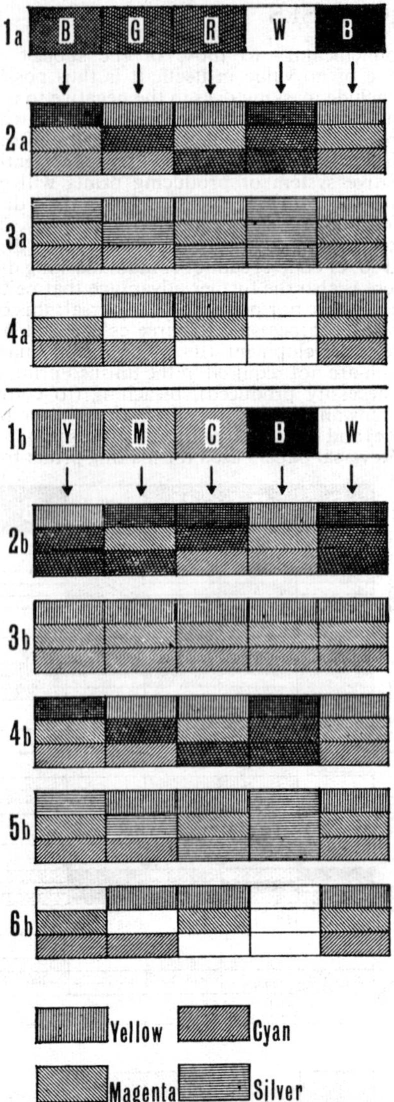

DYE DESTRUCTION PROCESS. Top: *Positive prints from positive transparencies.* 1a. *Subject consisting of blue* B, *green* G, *red* R, *white* W, *and black* B. 2a. *Exposure and development leads to a silver image in the layers sensitive to the transparency colours. These layers already contain complementary dyes.* 3a. *Dye bleaching removes the dye image in the areas of the silver image.* 4a. *The final bleach-fix also removes the silver image, leaving only the dye images.*

Bottom: *Positive printing from colour negatives.* 1b. *Colour negative consisting of yellow* Y, *magenta* M, *cyan* C, *black* B, *and white* W. 2b. *Exposure and development produces a positive silver image in the appropriate layers.* 3b. *Bleaching removes the developed silver.* 4b. *The second exposure and redevelopment produces a negative silver image from the residual silver salts.* 5b. *The dye bleach removes the dyes in the areas of the silver image.* 6b. *The final bleach-fix removes the negative silver image, leaving only a positive dye image.*

positive colour transparencies require at least five steps for processing; first development, re-exposure to light (the so-called reversal exposure), colour development, bleaching and fixing. Owing to the difficulty of obtaining good whites in the finished print this material is sometimes coated on a white opaque film base instead of on paper.

In dye-destruction tripack papers the three emulsion layers already contain the dyes. The final image is produced by bleaching the dyes in proportion to the silver image formed in each layer. Straightforward processing (development, dye bleaching, silver bleaching) produce a positive picture from a positive transparency. To obtain a positive from a colour negative requires a reversal processing procedure.

Separate Image Processes. In many commercial processes the separate images are made via three separation negatives taken direct from the original subject or from a colour transparency. An exception to this is when separation positives are printed direct from a colour negative on to panchromatic material through red, green and blue filters in succession.

Separation negatives are made direct from the subject by photographing it on to three different negatives through red, green and blue filters respectively. Unless an expensive beam-splitting camera is used, photographs taken by this method are limited to still life subjects. When a positive colour transparency is used as the original, it can be photographed three times through tricolour filters or printed or enlarged in a similar way to yield separation negatives.

Most transparencies have a very great tone range. This fact, coupled with the imperfections of the dyes used in both the transparency and in the final colour print, make it important to apply some kind of correction if the full advantages of the separate image processes are to be realised. By the use of masking—weak negatives made from the transparency and bound in register with it—it is possible to achieve improved colour rendering.

When separation negatives have been made by one of these methods, it is then necessary to produce the final print from them. Many methods were formerly used to achieve this (carbon, Carbro, Duxochrome, dye transfer, chemical toning and dye toning). Nowadays, dye transfer is the only method in common use, although the others linger on in the hands of enthusiasts.

Carbro and Carbon. These two allied colour printing processes are the oldest ones remaining in use today. They are capable of giving a gradation in the highlights and shadows which is seldom equalled and not yet surpassed by other processes.

They depend on the tanning action of chromium compounds on gelatin, and the difference between the two processes lies in the way in which this tanning action takes place.

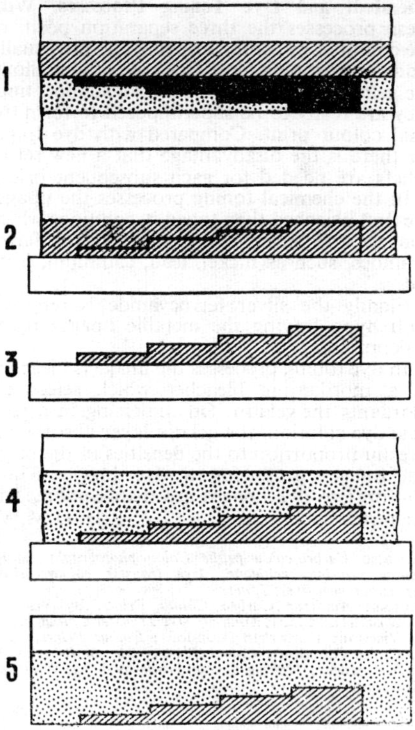

CARBRO IMAGE FORMATION. 1. Sensitized pigment paper squeegeed into contact with bromide print. This forms a tanned gelatin image. 2. Pigment paper squeegeed on to plastic temporary support. 3. Gelatin relief left after hot water treatment. 4. Relief on plastic support squeegeed on to gelatin-coated permanent support. 5. Final pigment image on permanent support after stripping off temporary support.

In the carbon process the three coloured gelatin images are made by sensitizing specially prepared sheets of pigment paper in a potassium bichromate sensitizer and then printing them directly from the separation negatives. Tanning action takes place in proportion to the exposure and after transferring the pigment papers to transparent supports the untanned gelatin can be washed away in hot water leaving three coloured relief images. These are then combined on one sheet of paper to form the final colour print.

The disadvantage of the carbon process is the very low sensitivity of the pigment paper. This means that it usually has to be printed by contact from an enlarged negative using a special light source such as an arc lamp.

Carbro overcomes this disadvantage but at the cost of an additional step. In carbro three bromide prints or enlargements are made, one from each of the separation negatives, and the tanning action takes place when the wet sensitized pigment paper and the bromide print are brought into contact.

Carbro is, on the whole, an easier process for the amateur particularly as the bromide prints give a useful visual guide for exposure and colour balance. The definition is not quite up to that of carbon because the tanning action is caused by diffusion of the chemicals between the pigment paper and the bromide print. For most practical purposes this is quite unimportant.

Duxochrome. This is a pigment process like carbro, but the image is obtained by direct development. It uses special films consisting of a gelatin layer containing both sensitive silver salts and colour pigment mixed together and coated on a celluloid base. Three separate films are provided, one for each colour. After exposure to the appropriate separation negatives the films are developed in a tanning developer and the surplus gelatin washed off in hot water. Since the Duxochrome film con-

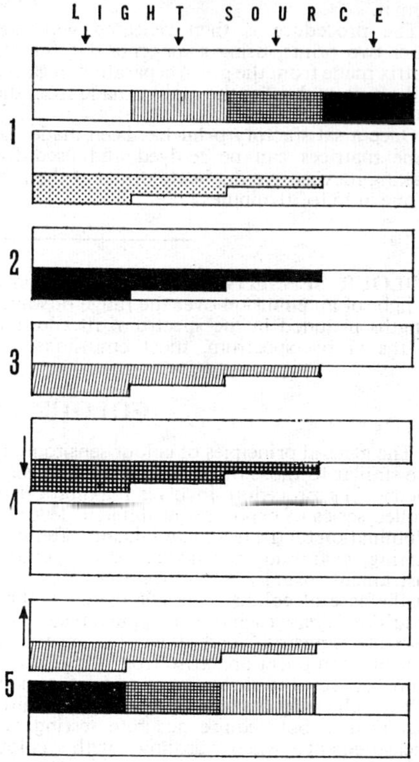

DYE TRANSFER IMAGE FORMATION. 1. Latent image formed on inside of emulsion layer of matrix film exposed through base 2. Silver image developed in tanning developer. 3. Gelatin relief left after bleaching of silver and treatment in hot water. 4. Dyed gelatin relief brought into contact with gelatin coated and mordanted paper. 5. Dye image transferred to gelatin layer of coated paper after removal of matrix.

tained coloured pigments the resulting relief images have varying depths of gelatin mixed with proportionate amounts of pigment and developed silver.

To complete the print the silver image is removed by bleaching and fixing and the reliefs are transferred and registered in turn from their celluloid supports on to a final gelatin-coated paper.

Dye Transfer. Three separation negatives are made from the original subject or transparency, using suitable blue, green and red filters. These three negatives are printed on to special matrix films which are developed in a tanning developer. The silver is bleached out and the untanned gelatin is then washed away by means of hot water.

The matrix from the red separation negative is placed in a bath containing cyan dye, which it absorbs in proportion to the amount of gelatin present in the matrix. The matrix is then washed and placed in contact with a specially-prepared (pre-modulated) paper which absorbs the dye from it.

The procedure is then repeated with the other two prints, using a magenta dye for the matrix made from the green separation negative and a yellow dye for the matrix made from the blue separation negative.

Once a satisfactory print has been made the same matrices can be re-dyed and used for making inexpensive duplicate prints at the rate of one in 15 to 30 minutes.

Chemical and Dye Toning Processes. With these processes the three separation positives are coloured by toning or dying and are usually made on stripping film. This type of film allows the gelatin images to remain on a support until they are ready to be superimposed to form the final colour print. Compared with dye transfer there is the disadvantage that a new set of reliefs are needed for each subsequent print.

In the chemical toning processes the images are first bleached then toned in solutions which convert the silver images into metal ferrocyanides, such as nickel, lead, cadmium, iron, etc.

Finally the silver ferrocyanide is removed with hypo leaving the metallic images ready for combining.

In dye toning processes the image is bleached in a mordanting bleacher which selectively mordants the gelatin. On immersing in a suitable dye solution, the gelatin layer absorbs the dyes in proportion to the densities of the original image, producing a dye image. Three appropriate dye images are then combined in the print. H. (J.M.W.)

See also: *Carbro colour prints; Colour photography history; Colour printing technique; Dye transfer colour prints; Separation negatives; Toners.*

Books: *Amateur Carbro Colour Prints*, by Viscount Hanworth (London); *Amateur Dye Transfer Colour Prints*, by Viscount Hanworth (London); *Colour Prints*, by J. H. Coote (London); *Colour Photography in Practice*, by D. A. Spencer; *Making Colour Prints* by J. M. Whittaker (London).

COLOUR SENSITIVITY. Relative response to light of an emulsion over the range of wavelengths included in the spectrum. In addition to the visible spectrum, most emulsions are partly sensitive to ultra-violet rays. Some have special responses (e.g. to infra-red).

See also: *Negative materials; Sensitizer; Spectral sensitivity; Spectrography.*

COLOUR SENSITOMETRY

The general principles of colour sensitometry are similar to those of black-and-white sensitometry. The procedure involves making a controlled series of exposures at different levels of illumination on the test material, and, after processing, examining the image with a suitable instrument, usually a densitometer. However, in the case of colour materials there are three emulsion layers, each of which must receive the correct exposure, and then be measured. Further complications arise from the relationship between the three layers and from many effects which on a single film would have little significance but require accurate testing and measurement when dealing with colour materials.

Nowadays virtually all colour films work on the subtractive principle, using three overlapping dye images that produce the appropriate colour by subtracting from white light suitable proportions of red, green or blue light with additive colour materials and separation negatives the individual images do not overlap, and therefore need not absorb only one primary colour. In fact they are usually formed of silver. However, the sensitometric considerations are much the same, and in the following sections it is assumed that the test materials are of the subtractive type.

SENSITOMETERS

A typical colour sensitometer exposes the film in contact with a density wedge to the light from a standard lamp. The shutter system consists of a sector wheel and a capping shutter. The sector wheel rotates continuously at a fixed speed, and the capping shutter remains open while the whole area of film is exposed just once. The exposure time is controlled by the speed of rotation of the wheel and the angle of the sector. Some sort of filter is usually included in the optical path to modify the colour

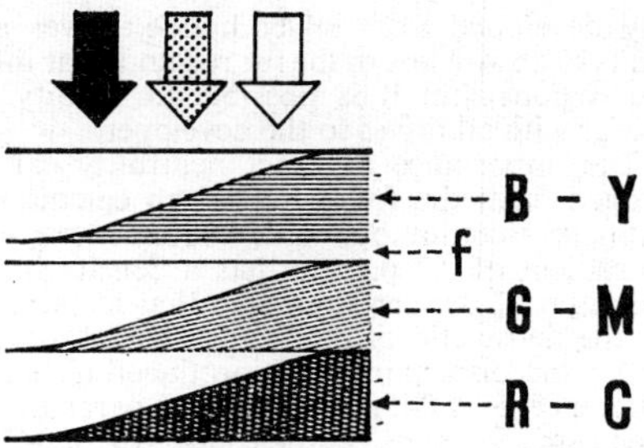

COLOUR FILM EXPOSURE. Simple representation of the build-up of the three dye layers in a colour negative film. Actually the images do not develop from the bottom of the layer upwards but are distributed more uniformly. B–Y, blue sensitive, yellow-forming layer. G–M, green sensitive magenta-forming layer. R–C, red sensitive cyan-forming layer. f, yellow filter layer.

of the light. Each component requires special attention, and these are described below.

Light Source. This must provide light similar in quality to that which the film will receive in normal use, so that the sensitometer correctly measures the relative speeds of the three emulsion layers. The most commonly used source is a tungsten projection lamp operating at a colour temperature of about 2,800 to 3,100° K. The use of tungsten lamps has several advantages: they emit a continuous spectrum and the colour temperature of the lamps can be adjusted with the applied voltage. Furthermore they are readily obtained and replaced. For simulating daylight a blue correction filter must be included with the lamp.

Although the light distribution made up from sunlight and skylight does not match that from a black body at any colour temperature, a near approximation is between 5,000 and 6,000° K., obtained with a tungsten lamp and a suitable conversion filter. Thus the effective colour temperature of the lamp can be chosen independently, and is usually made as high as possible, consistent with adequate lamp life, to ensure maximum light output.

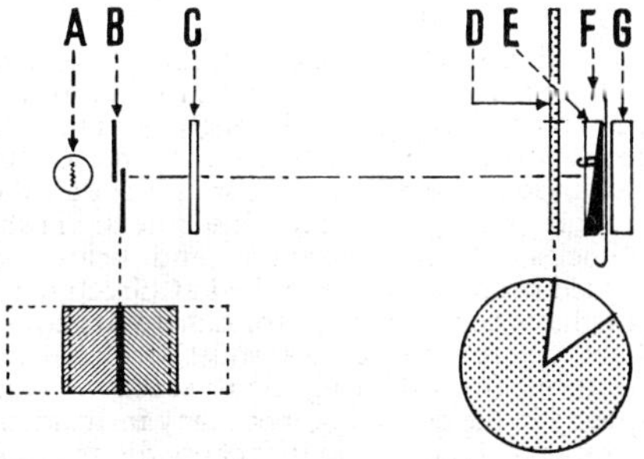

SIMPLE COLOUR SENSITOMETER. Light from a lamp passes through a shutter for a known time and reaches the film through a wedge in contact with it. The shutter consists of a rotating sector wheel and a capping shutter. A, lamp. B, capping shutter, shown below in open (dotted) and closed positions. C, filter to modify the colour of the light. D, sector wheel. E, wedge. F, film. G, pressure plate.

The original Davis and Gibson filters designed to simulate daylight exactly with a tungsten source, give a better match than glass or gelatin filters, but are inconvenient to use, and not even necessary. For the spectral composition of daylight varies considerably from place to place and with the weather conditions. It is therefore possible only to strike at an average and balance the relative speeds of the three layers of the film to suit this, knowing that the film will rarely receive light made up in just this way. Also, camera lenses all absorb a certain proportion of the ultra-violet radiation, but the exact amount depends on the camera and it is necessary to choose some average amount to adopt for the sensitometer. Even, the exposing wedge absorbs selectively (see below). It is of primary importance that the *total* amounts of light received by each layer of the film in the sensitometer are in the same proportion as in the average conditions of use, and the light distribution at each wavelength should be made as close as possible to that of the average source. Some materials, e.g. colour papers, "positive" film for cinema release prints, and artificial light films, can be exposed directly to the tungsten lamp, or through a filter converting the colour temperature to 3,200° K.

The candle power and colour temperature of the exposing source should be known, and held constant throughout the life of the lamp. This can be corrected by changing the lamp voltage, and the standard distance between the lamp and the film.

Alternative light sources suitable for sensitometry are a xenon arc, xenon flash tube or a mercury vapour lamp, if the specialized nature of the material requires that it is used with these sources.

Exposure Time. It is now generally accepted that practically all photographic materials should be tested by varying the intensity of the light while the exposure time is constant. The effects of reciprocity failure (i.e. the change of speed and contrast with exposure time) are particularly noticeable on colour films because each emulsion layer will behave differently, and a film that is in correct speed and contrast balance at one exposure time will not be so for another. Therefore the exposure time given in the sensitometer must be the same as that normally given. For daylight type colour films 1/25 or 1/50 second represents a good average, while for sheet film used in the studio one or two seconds is probably more appropriate, and an even longer time would usually apply to colour enlarging papers. On the other hand an exposure time of the order of 1/1,000 second is used to show the response of a film to an electronic flash exposure.

Exposing Wedge. Whether a wedge provides a continuous variation in density or is divided

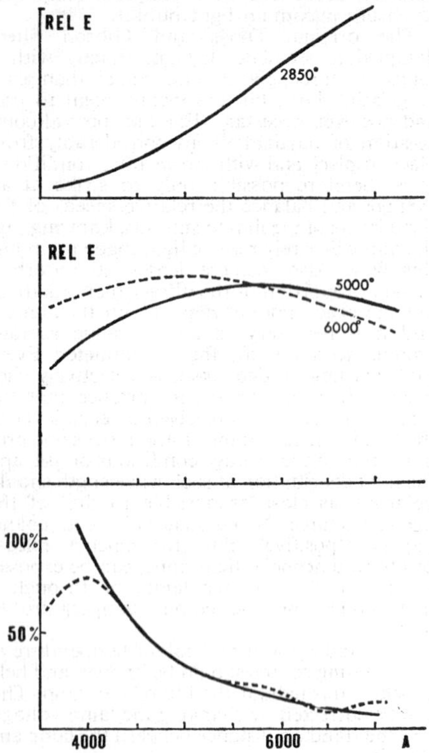

SPECTRAL ENERGY DISTRIBUTION. Top: *Tunsten lamp operating at 2850° K (arbitrary units).* Centre: *Black body emission curves for colour temperatures of 5,000° and 6,000° K.* Bottom: *Conversion filter curves in terms of transmission. It is impossible to obtain a smooth transmission curve as shown by the full line (corresponding to −154 mireds), and the dotted line shows the characteristics of a real filter.*

into steps depends on the measuring device. Automatic recording densitometers will generally work with samples that have received a gradual change of exposure with distance along the film, whereas densitometers that give a single reading from a piece of film will require the use of a step wedge. The latter is much more common, since the densitometers are less expensive, and the results from a stepped exposure are more easily assessed visually. For colour materials the density increment for a continuous wedge should not be more than 0·3 per cm. and for a step wedge the density difference between steps should not be more than 0·30 (laid down by American Standard ASA.PH. 2.21, 1961, and the British B.S. No. 1380, 1962).

Most important, a wedge should absorb equally at all wavelengths. The materials usually used as absorbing media are a suspension of carbon in gelatin, which absorbs more in the ultra-violet region and less in the infra-red than in the visible spectrum, and silver chemically developed from silver halide. Silver also tends to absorb less in the far red and near infra-red regions, but it is possible to modify the colour with additives to the developer.

The importance of exact neutrality can be gauged from the following rough calculation. Suppose a continuous wedge 15 cm. long with a gradient of 0·3 per cm. has a density in the red region 1 per cent less than that in the blue. At the dense end of the wedge, with a density of 4·5 the discrepancy has produced a density difference of 0·045, for the two parts of the spectrum. A density difference of 0·045 is equivalent to a difference in transmission of 11 per cent. Thus the quality of light reaching the film is quite different from that leaving the source and correction filter, and foils any attempts to make the source exactly resemble, say, daylight. In fact the correction filter should really be designed to produce the right result in conjunction with the central portion of the wedge, because it is here that the middle densities in the film will be produced—these are the most important for judging colour balance.

Because of the inherent difficulties in making good wedges alternative methods have been tried for varying the exposure, e.g. moving the film past an illuminated slit, while the aperture of the optical system is changed.

Standardization. As in the case of black-and-white sensitometry, careful control is essential at every stage of the testing procedure. The exposure time, the colour of the light and the level of illumination in the sensitometer must be specified. Also, the materials must be used in the correct safelight, the duration between exposure and development must be held constant, and the film storage and processing must be carried out in standard conditions. The errors likely to arise from carelessness in any of these matters are very much more complicated than in the case of black-and-white sensitometry.

DENSITOMETRY

Since it is the task of a colour film to separate the light into red, green and blue components, and then to form the final image by modifying the proportions of these colours, it is reasonable to analyse the result by measuring the densities through red, green and blue filters. Transparencies, colour cine films and prints are finished products, to be looked at directly, and it might be considered appropriate to use colour filters whose spectral transmissions resemble the three (hypothetical) light sensitive components in the eye. Although many instruments for measuring colour do in fact use filters of this type, they are not suitable for colour densitometry. Apart from reproducing colours, the first and most important task of a colour film is to reproduce correctly a neutral grey (i.e. a non-selective absorber). When formulating a colour film from its component layers we must

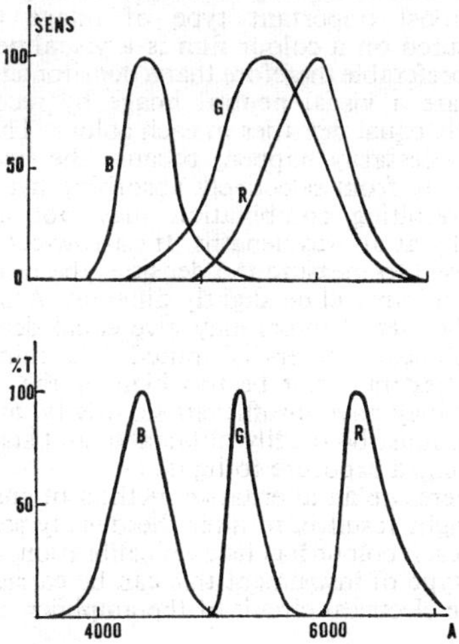

VISUAL SENSITIVITY AND FILTER TRANSMISSIONS. Top: *Variation of sensitivity with wavelengths for the components of the eye. The curves approximate to those defined for the standard observer by the C.I.E.* Bottom: *Transmission characteristics of three filters used for colour densitometry. These curves have been drawn to a peak transmission value of 100 per cent. In practice even the maximum transmission of a narrow cutting filter might be as low as 25 per cent.* B, G, R, = *blue, green, red.*

know how much of each dye is necessary to make up a visually neutral image. Therefore the densities obtained from single dye layers should in some way indicate what contribution they will make to the final image.

This is not the case if wide band filters are used because successive dye densities effectively shift the absorption maximum away from the peak transmission of the filter and the dye densities are then no longer additive. Only if measurements are made at a single wavelength would the combined density of several layers be the simple algebraic sum of their individual densities. In practice therefore, narrow-cutting filters are used, transmitting narrow bands of the spectrum centred at about 430, 530 and 630 mμ. The exact choice of filters depends on the type of material, and to some extent on the densitometer. An infra-red absorbing filter must be included in the light beam, as most colour filters and all film dyes transmit quite freely in this region, and any residual infra-red radiation in the system leads to underestimation of densities, particularly at high values. Photocells and photomultipliers vary greatly in their sensitivity to infra-red radiation. It is necessary to choose a cell with high red sensitivity, so that high densities can be measured in this colour; at the same time the high *infra*-red sensitivity associated with this calls for a strong I.R. absorbing filter.

For the measurement of colour negative films, the problem is more straightforward. The effective density of each sample depends on the colour sensitivities of the colour paper or positive film). Therefore the spectral transmissions of the filters (taken in conjunction with the emission of the lamp in the densitometer, and the sensitivity of the photocell or photomultiplier) should match the spectral sensitivity of the paper. The instrument will then record the true printing densities of the samples, and any effects due to the densities of single dyes not indicating their contribution in the presence of the other dyes is the same for the densitometer and the paper. (In fact filters matching paper sensitivities are also very narrow cutting.)

Types of Densitometer. There are two main types of densitometer. In the first the light reaches the photocell after passing through the colour filters and the sample. The signal is amplified and the reading of density is recorded on a meter, usually in such a way that the meter reading is linear with density. If a photomultiplier is used as the receiver the amplifying circuit can be designed to give a current output that is approximately linear with density. The instrument can be set to read zero when no sample is in the light beam either by adjusting the lamp brightness (this will change the colour slightly, but if the filter is very narrow cutting, it will have no effect on the result), inserting an optical wedge in the beam, or by a compensating device in the electrical circuit. Some refinements that can be made to this type of instrument are a servo mechanism to record the density on a chart (in effect to draw a characteristic curve for each dye image) and a tuned amplifier—ensuring a minimum of interference from outside sources.

In the second type of optical system the light, after passing through the filters can be directed to the photocell either via the specimen and a compensating wedge, or via a fixed or "zeroing" wedge. The principle is that the light intensity in the two light paths should be the same, and the meter is used simply as a null detector. The instrument is first set by moving the wedge in the reference beam until the light intensities are equal. When a sample is inserted the compensating wedge is moved until the combined density in the specimen beam is back to the original value.

The movement of the wedge can be recorded on a scale, which can be calibrated in terms of density once the density increment of the compensating wedge is known.

Again, refinements to this type of instrument are possible.

The light beam can, for instance, be diverted alternately through each path at a high frequency, the amplifier being tuned to this frequency, and a servo mechanism can be made to drive the wedge, attached directly to a recording pen, until the balance point is reached automatically.

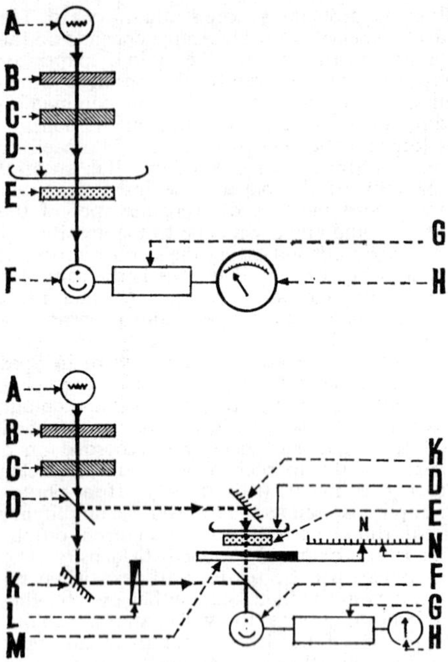

BASIC OPTICAL COMPONENTS OF DENSITOMETERS. Top: *Here the characteristics of the amplifying circuit are designed to give a reading in terms of density.* Bottom: *In this system the results are based upon the calibration of the measuring wedge. Both types have advantages and disadvantages and both are extensively used. A, lamp. B, infra-red filter. C, colour filter. D, film. E, opal diffuser. F, photo-cell. G, amplifier. H, meter. J, chopper splitting the beam from the lamp into the specimen beam (dotted) and a reference beam (dashes). K, fully reflecting mirrors. L, calibrating wedge. M, measuring wedge moved across the light beam. N, density scale. The optical systems also contain certain subsidiary lenses and mirrors.*

As a modification, the instrument can be made to read zero for each colour separately without moving the reference wedge either by inserting pale coloured filters into the reference beam, or by a suitable circuit in the amplifier.

In each optical system an opal diffuser is placed immediately below the sample to ensure that the diffuse density is measured. As in the the case of black-and-white material, all standard specifications are made on the basis of diffuse densities. (The difference between specular and diffuse densities is not very great for dye images which do not scatter light to the same extent as silver.) Measurements of reflection density are also made to the same standard as for black-and-white papers; the paper is illuminated at an angle of $45° + 5°$ to the normal and the light reflected in a cone 5° around the normal is measured (or vice versa). Both types of densitometer can be used for reflection measurements.

Correction of Density Scale. As already shown the most important type of image to be measured on a colour film is a visual neutral. It is preferable therefore that a densitometer can indicate a visual neutral image by recording exactly equal densities in each colour. This will not necessarily happen, because the image is made up from selectively absorbing dyes, and the resulting combination may not absorb equally at all wavelengths. If narrow-cut filters are used to measure the densities the results in each colour will be slightly different. Although another set of filters may give equal densities, the choice of filters is limited. The minimum densities must not be too high or the instrument may have insufficient sensitivity, and the filters must be readily obtainable and stable to prolonged exposure to light.

Therefore an alternative method of ensuring the right result is to alter the density scale so that each colour has its own calibration. In the first type of instrument this can be carried out in the electrical circuit of the amplifier, and in

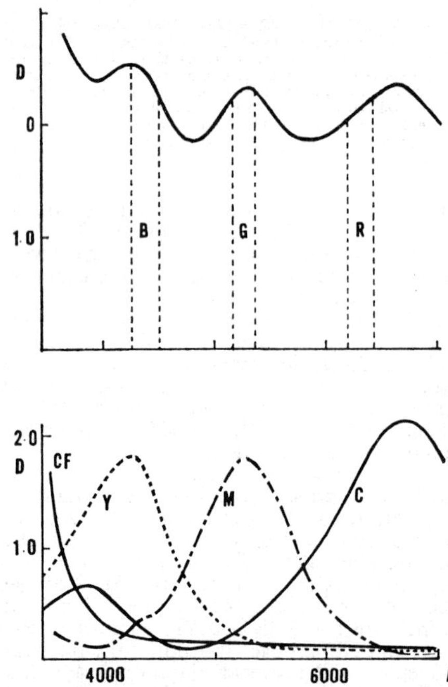

NEUTRAL IMAGE CURVES. Top: *Spectral absorption curve of a visually neutral image on reversal film, with the central regions of the transmission band of densitometric blue, green and red filters. These filters would not give equal densities in each case.* Bottom: *Absorption curves of the three dyes (yellow, magenta, cyan) that combine to form this visually neutral image. The curve CF represents uncoupled colour former and any residual image colour not due to the three basic dyes. The densities of the dyes taken singly vary greatly according to the wavelengh of light used for measurement.*

the second case the connection between the compensating wedge movement and the pen movement can be made via some sort of variable gear (which can be electrical in nature). This type of correction is a feature of only the more elaborate instruments. The original settings must be made with sample whose densities have been found by independent means. This involves making visual measurements of density and colour, using an optical system whereby the inverse square law of intensity distribution is used as the initial reference. This type of correction can also be applied to the measurement of colour negative films if the filters do not quite match the emulsion sensitivities of the paper. Again the densities of several samples must be found by experiment to supply the initial settings.

INTERPRETATION OF RESULTS

It is possible here to give only a brief indication of the various factors in the reproduction of colours by films.

Reversal Films. The significant factors in the characteristic curves of a reversal film are the shoulder speed of each layer (i.e. the point where the density first begins to decrease with increasing exposure), the contrast and shape of each curve—particularly in the lower densities, and the minimum density of each image. If the measurements are made with an instrument with corrected density scales, the three curves should be superimposed. If the maximum densities are sufficiently high (above about 2·7) they need not be equal. Typical faults are lower speed of one sensitive layer, giving a complementary bias in pictures, and a low maximum density of one of the dye images. If the contrast of the three images is not matched the transparency will have different colour casts in highlights and shadows.

Negative-Positive Material. In a negative colour film the exact speed balance between the layers is not so important, although the absolute speeds (where the density first begins to rise with increasing exposure) of course, should be sufficient for the purpose, and the minimum densities of the three images can be higher than for a reversal film. Both these factors can be taken care of at the printing stage. On the other hand the curves should all include a long straight line portion, and should all be of exactly equal contrast. For example, if the contrast of the cyan image is too low; this would give rise to cyan highlights and red shadows in the prints.

Colour paper is also a "negative" material, but since it is used with negative film the final image is positive, and must have characteristics similar to that on a reversal film. Therefore the important features of a colour paper are a very low minimum density in each colour; each layer must have the same contrast (at the required value) and the maximum densities must be

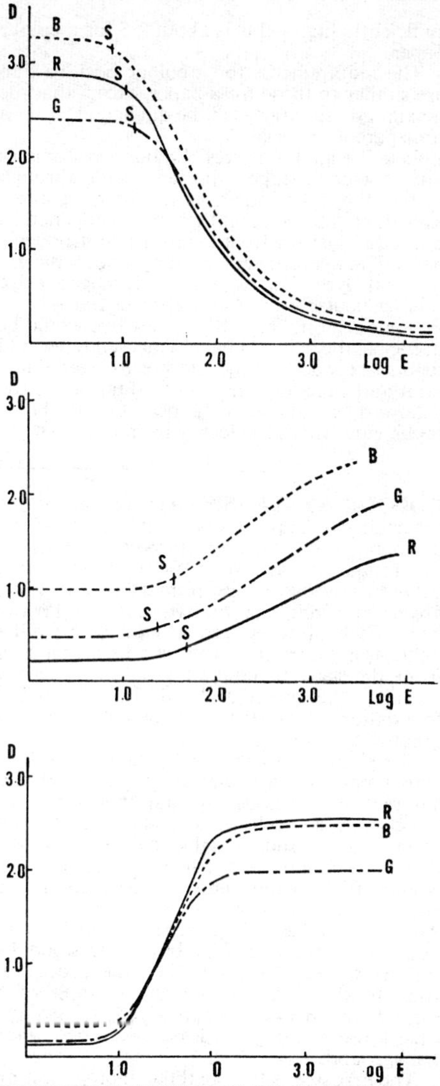

CHARACTERISTIC CURVES OF COLOUR MATERIALS. **Top:** *A reversal colour film, exposed to light of the same colour as used for camera exposures. B, R, G, indicate the three sensitive layers; S, is the shoulder speed for each layer. This curve family shows too slow a blue sensitive layer (leading to a yellow cast in the film) and too low a maximum magenta density.* **Centre:** *A colour negative film. The red curve is noticeably shallower than the other two, leading to too low a cyan contrast in the film (too much cyan in the highlights and too little—i.e. too much red—in the shadows).* **Bottom:** *A colour paper. The colour of the exposing light, though similar to that in a colour print, has been carefully adjusted with colour correction filters so that the image on the paper is close to neutral. These curves show a high fog level in the yellow-forming (blue sensitive) layer leading to yellow highlight stains. The contrast and maximum density of the magenta image (due to the green sensitive layer) are lower than for the other two images—a fault in this particular paper sample shown up by the characteristic curves.*

sufficiently high—this is about 2·5 for a colour paper.

The requirements for a colour positive film are similar to those for a paper, except that the maximum densities will be higher as it is a transparent material.

Colour Wedge Exposures. Colour sensitometry must include wedge exposures made through colour filters, in addition to white light ex-exposures. The fact that three dyes happen to be present in just the right quantities to produce a neutral image conveys nothing about the individual dyes. For example, if a magenta dye was particularly poor and absorbed heavily in the blue region, very little yellow dye would be needed in the neutral image. Thus yellows would tend to be washed out, and greens (containing cyan and yellow dyes) would contain too little yellow dye, and would appear bluish. These faults can only be detected from coloured exposures, and may need advanced colour correction.

Other examples of the type of information can be obtained from coloured exposures are: the minimum red density of a saturated red which contains the maximum quantities of magenta and yellow dyes, the change of blue density with increasing green exposure—indicating the purity of the magenta dye, or, in the case of a masked material, the masking efficiency, and the apparent speed to blue light of the green sensitive layer—indicating the efficiency of the yellow filter layer. M.G.F.

See also: *Characteristic curve; Colour film principles; Colour temperature; Exposure; Sensitometry; Speed of sensitized materials.*

Books: *Sensitometry*, by M. Dubois and L. Lobel (London); *Principles of Colour Photography*, by R. M. Evans, W. T. Hanson and W. L. Brewer (London and New York); *General Sensitometry*, by Yu. N. Gorokhovskii and T. M. Levenberg (London).

COLOUR SYNTHESIS. Any colour of the spectrum can be perceived by the agency of one or more of the systems of receptors forming the retina of the eye. Colours intermediate between any two of the primary colours blue, green, and red, act on two systems. White light affects all three systems equally. So white light, or any chosen colour, can be reproduced from the raw material of the three primary colours, blue, green, and red. This fact is the foundation of every system of colour photography in use today.

There are, however, two different ways of arriving at any particular colour: additive and subtractive synthesis. In additive synthesis, the colour, say yellow, is produced by adding together green and red light. In subtractive synthesis the colour is produced by subtracting colours from white light; in this case blue is subtracted, leaving yellow.

Additive Synthesis. The general principles of additive synthesis can be best understood by following what happens when the process is used to make a coloured photograph of a typical subject like a bunch of red, yellow, and blue flowers and green leaves against a white background.

The first step is to take three photographs on panchromatic material through three filters, red, green and blue. The filters used are called tricolour filters because each is transparent to roughly one-third of the spectrum, just as three sets of receptors in the eye are each believed to be sensitive to one-third of the spectrum.

(1) The red filter passes the red light from the red flowers, and most of the yellow light from the yellow ones. All the light passed by this filter affects the emulsion and produces a black deposit on development. But the green and blue light does not get past the filter and does not register on the plate. So the green and blue areas will remain clear on the negative when it is developed. The negative is now used to make a positive transparency on a lantern slide. The clear areas on the negative corresponding to the image of the leaves and the blue flowers reproduce as black areas on the positive.

(2) The green filter lets through the light reflected by the yellow flowers and green leaves, but stops that from the red and blue flowers. So a positive transparency from this negative will show the red and blue flowers as a blackened image while all yellow and green areas will be clear.

(3) The blue filter only lets through light from the blue flowers so its transparency will show the red and yellow flowers and the leaves as dark areas, and the blue as clear patches.

Now all three positive transparencies are projected from three lanterns on to a screen, one on top of the other, each through the filter it was taken with. The background will be clear on all slides and "add up" to white. But on two of the slides (the green and blue), the image of the red flowers consists of an opaque deposit of silver which prevents light getting to the screen. So only the light from the red lantern gets to the screen. This gives a red image of the red flowers.

The image of green leaves exists as opaque silver on the red and blue slides, so the only light that can reach the screen through the leaf image comes from the green lantern, giving a green picture of the leaves.

In the same way the blue lantern projects a picture of the blue flowers which are blocked out in metallic silver on the slides in the red and green lanterns.

The image of the yellow flowers only appears as an opaque deposit on the slide in the blue lantern, so that it allows both red and green light to shine on the screen. These two colours appear to the eye as yellow, so the whole picture is reproduced in its proper colours. The process is an additive process, because it arrives at the result by adding the three different pictures from three separate lanterns.

At the same time, however, each filter cuts out at least two-thirds of the white light available and uses only one-third at best.

Additive synthesis as here described formed the basis of the first demonstration of colour synthesis in photography. It has, however, never been used to any extent in practical photography. All additive systems of colour photography depend on splitting up the image into microscopic red, green and blue elements placed side by side. Owing to their small size, the elements are fused by the eye and produce the same effect as the three superimposed lantern slide images.

Subtractive Synthesis. The second system of synthesis works in a different way. It is concerned not with the colour of light that a filter will pass, but with the colours it holds back. From what has been said, it is clear that a filter is transparent (or nearly so) to one set of wavelengths, and nearly opaque to the others. So when a filter appears a blue-green colour, it is because it lets blue and green light through, but stops red light. It is, in fact, subtracting the red light from the white light shining on it. So a blue-green (or cyan) coloured filter can be regarded as a "minus-red" filter.

A filter that looks a magenta colour because it subtracts green rays from the white light shining on it is a "minus-green" filter.

A yellow filter—one that subtracts blue rays and passes green and red—is a "minus-blue" filter.

These "minus" filters are the key to the subtractive processes. The following illustration gives the general principles on which all such processes depend:—

As in the previous example, the subject may be a bunch of blue, red, and yellow flowers, and green leaves, against a white background. As before, three negatives are made through filters of the primary colours, red, green and blue, and these three negatives are used to make three corresponding positive transparencies.

At this point, the black areas of developed-out silver are treated with chemicals and transformed into corresponding areas of transparent colour of the minus or complementary colour of the filter through which they were taken.

To take them in order: the positive from the red filter is converted into a minus-red or cyan (blue-green) colour; the positive from the green filter is converted to a minus-green or magenta colour, and that from the blue filter into a minus-blue or yellow.

Now, instead of being projected separately, the three transparencies are bound together in register and projected as a single slide. Since there is a clear background on all three positives, the background will be white, the colour of the projection light.

(1) The red flowers are registered on the green and blue filter positives, which are now minus green and minus blue in colour respectively, that is they subtract the green and blue light from the white light in the lantern, and so leave only red, the correct colour.

(2) The blue flowers which registered on the red and green positives are now represented by minus-red and minus-green areas—i.e., areas which, correctly, allow only blue light to pass to the screen.

(3) The green leaves which registered on the red and blue positives are now represented by minus red and minus blue areas—i.e., areas which, correctly, allow only green light to pass through.

(4) The yellow flowers only leave a positive image on the blue filter positive. This image is now represented by minus blue which is, correctly, yellow.

So the final result is a picture in the natural colours of the original subject.

Subtractive Analysis. The subtractive system described above still analyses the colours by utilizing red, green, and blue taking filters. In a multi-layer subtractive system the colours are similarly analysed by having three emulsion layers, sensitive to blue, green, and red respectively, superimposed during exposure. The three layers are then processed simultaneously to produce the colour images in the complementary minus colours and yield either a colour transparency or—by interrupting the process at an intermediate stage—a colour negative.

The latter then yields a colour positive by repeating the steps of subtractive analysis and synthesis.

These systems are the basis of most present-day colour materials.

Dyes v. Pigments. In general, colours in the form of pigments are not as brilliant as those of the spectrum. This is because they do not have a very sharp break between the colours they absorb and those they reflect, but tend to absorb all wavelengths of light to some degree although reflecting some more strongly than others.

So pigments are not really ideal for photographic colour processes although they are still employed.

The modern synthetic dyes are usually more brilliant and are almost universally used in present-day colour processes. Some of the subtractive processes, in fact, depend upon the synthesis of the dye while development is proceeding.

J.Mo.

See also: *Colour film principles; Colour materials; Colour photography history; Colour print processes.*

Books: *Colour Photography in Practice*, by D. A. Spencer (London); *The Reproduction of Colour*, by R. W. G. Hunt (London).

COLOUR TECHNIQUE

In colour photography, pictorial effects depend on colour differences rather than contrast of tones as in black-and-white work. Quite small colour differences will show up much more than similar tone differences in monochrome.

The Effect of Colour. The way of using colour depends to a great extent on personal taste. Some colour combinations are, however, generally accepted as harmonizing and others as clashing. The way in which they are combined emphasizes or subdues different parts of the scene. If there is too much colour in a colour picture, all the shades will compete with each other or even clash. It is better to include only one or two brilliant patches of colour at a time. A brightly coloured costume, for instance, may show up the ability of the film to record the colours, but will distract attention from everything else in the picture. On the other hand a small red scarf may add to the character and charm of the wearer.

It is also wise to avoid many splashes of colour in the background or away from the centre of interest. Use colour to help to knit the picture together—not to pull it apart. Brightly coloured areas near the edge of the frame tend to lead the eye out of the picture.

Colours can differ in various ways. Firstly, they may be pure or impure. Pure colours are more likely to clash than impure ones. Secondly, colour schemes can be built up on similar or contrasting colours. Similar colours may blend while contrasting colours set each other off.

By association of ideas, colour may also create an atmosphere or a mood.

Colour Harmony. In harmonizing different colours, the hue as well as the saturation is important. The more saturated the colour, the more brilliant it appears. Mixing with white de-saturates it, dilution with black degrades it. Thus a mahogany brown, a bright red, and a pale pink may all be the same hue; but the brown is degraded, the red fully saturated and the pink desaturated. Colours of the same family or hue but of different saturation will easily blend and harmonize.

Colour Contrast. The contrast between different colours is greatest when they are complementary (blue and yellow, red and blue-green etc.).

Two fully saturated colours of equal area will conflict with each other and fight for attention. This disharmony is greatest when colours are not complementary; thus a pillar-box red and a magenta or a blue and a blue-green, will clash.

A bright colour against a desaturated complementary will appear brighter than against a saturated non-complementary. This use of complementary colours helps to focus the eye on the subject.

Colour and Warmth. Certain colours—e.g., yellows and reds—suggest warmth: others—e.g., blues and greens—suggest coldness. Cool colours also appear to recede and warm colours to advance. This is partly due to the fact that the eye focuses reds and yellows as if they were nearer than blues and greens. Similarly, light and desaturated tones in a picture tend to seem more distant than darker and more brilliant tones. This again is based on experience—e.g., of the way in which the planes of a landscape become paler as they recede into the distance.

These factors are able to create depth. Thus yellow and reddish colours (including brown) in the foreground make it stand out, while greens and blues in the background make it appear farther away.

Colour and Sharpness. Brilliantly coloured objects in the picture should be sharpest, shapeless blobs of saturated colours are most disturbing. Subdued and pastel shades can be less sharp; the subject should be selected so that the background colours are more subdued than those of the principal subject. When photographing a scene which takes in foreground, middle distance and background alike everything must be sharp. The depth of field requirements in colour are therefore more stringent than in black-and-white photography.

Equipment for Colour. Practically any camera can be used for colour shots provided the light is good enough, but the scope of simple

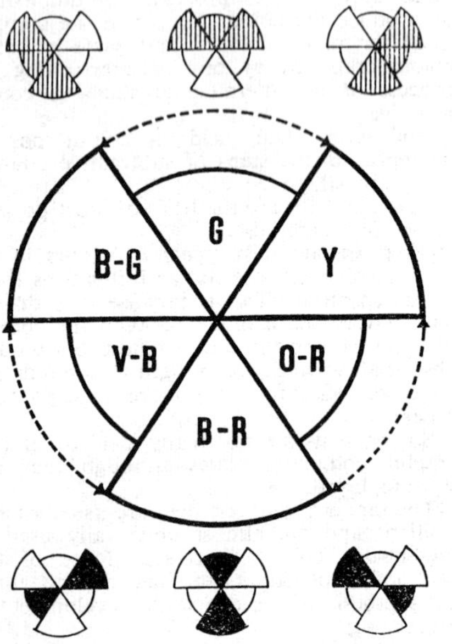

COLOUR BALANCE. Centre: *The colours of the colour wheel are G, green; Y, yellow; O-R orange-red; B-R, blue-red (magenta); V-B, violet-blue, B-G, blue-green.* Top: *Adjacent segments indicate harmonizing colours (shaded).* Bottom: *Opposing segments (solid) are contrasting colours.*

cameras is very limited because such a camera does not provide sufficient adjustment of lens aperture and shutter speed to cope with more than a very restricted range of light conditions. The shutter speed or speeds on a cheap camera are often not as reliable as they should be. An accurate shutter is a help in assuring correct exposures, as the latitude of colour film is slight. For large scale colour photography the camera lens should be well corrected for all chromatic faults It should also be coated, since stray light caused by reflections inside the lens and scattered over the film will desaturate the tones. For the same reason it is always wise to use a lens hood.

Loading of colour film into the camera follows standard practice. It is, however, advisable to cover the red film window at the back of many roll film cameras at all times, except when actually winding on the film, to avoid the risk of fogging.

It is, of course, possible to take colour photographs without using colour film. The method involves taking three black-and-white photographs of the subject, separately, through red, green and blue filters. This is nowadays used mostly for photomechanical reproduction and the negatives are known as colour separation negatives. Such negatives of moving subjects can be taken only with a one-shot colour camera taking all three pictures at once each through its correct filter. These cameras are nowadays obsolete. The separation negatives are used to make a colour print by superimposition of complementary coloured positive images.

The Colour of the Light. Since colour film records all variations in hue and tone, it will also react to the different colours of light sources by which the picture may be taken. That is why reversal colour materials are specially balanced for use in either daylight or artificial light. Within both types of lighting, however, there are variations in colour quality depending on specific conditions.

Three main factors affect the colour of daylight.

(1) Atmospheric dust filters the sun's rays, and absorbs or scatters rays of short wavelength (violets and blues) more than those of longer wave-length (reds and yellows). When the sun is low on the horizon the light becomes redder; on the other hand, just before sunrise or just after sunset the colour will sometimes be blue due to reflections from a clear blue sky. Except for special effects it is therefore best to avoid taking colour pictures within approximately one hour after sunrise and one hour before sunset.

(2) Over water, snow, or from distant views a great deal of ultra-violet radiation may reach the camera. Although this is invisible to the human eye, the colour film is sensitive to it and the resulting picture may be too blue. (In addition many lenses are not corrected for ultra-violet radiation, and such rays may therefore upset the definition of the image on the film.)

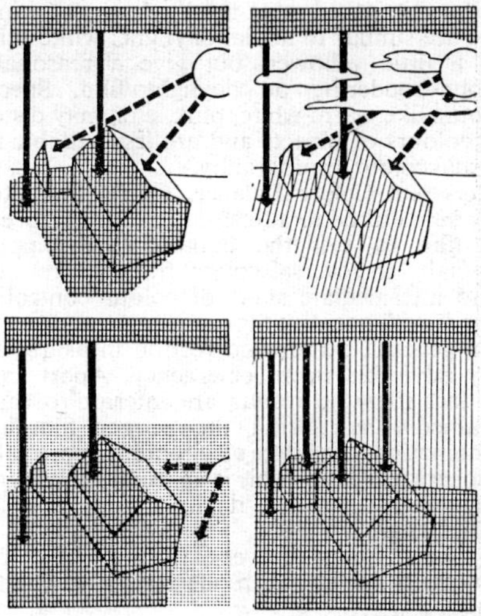

THE COLOUR OF DAYLIGHT. Top left: *Brilliant sun with a cloudless sky produces intensely blue shadows. Subjects in shadow require a correction filter.* Top right: *White clouds lighten shadows, and make them less blue.* Bottom left: *Low evening sun is yellowish, while the shadows are again blue. The yellow tinge can be corrected by a pale bluish filter.* Bottom right: *Overcast days yield a flat and cold colour rendering, requiring a pale correction filter.*

(3) Part of the light reaching the earth does not come directly from the sun but is reflected from the blue sky. When the sun is shining in a clear sky the shadow parts of the subject may be lit only by the sky and therefore tend to take on a blue tint. Colour film tends to overemphasize this. Conditions for colour photography are generally best when there is a lot of white cloud about; this lights up the shadows and also makes them less blue. In overcast weather the colour of the diffused daylight is also bluer than direct sunlight, but not so blue as the light from the clear sky.

The colour temperature of most artificial light sources (with the possible exception of electronic flash) is lower than that of daylight, and will record a yellowish or reddish over-all cast on daylight type colour film. Artificial light type films are made more sensitive to blue to compensate for the excess of red in the light.

Most amateur artificial light type films are balanced for Photoflood lamps and are ideal for colour photography at home. A few makes are balanced for high-power studio lamps of slightly lower temperature and others for clear flash bulbs. Ordinary flash is yellower than daylight. So flash bulbs for colour photography are tinted blue, to match daylight.

Fluorescent tubes of the "colour matching" type are similar to actual daylight; white tubes are a little yellower, but give a reasonable colour rendering on daylight film. Special colours like warm white, pink, etc., may distort the colours of objects and are less suitable for normal colour photography.

Filters and Colour Balance. Correction filters may be required to adjust the colour balance of the film to suit the lighting. This applies especially to reversal colour films where there is no intermediate stage of colour control as there is with negative films.

In general, extreme correction measures are only advisable in an emergency. Apart from the fact that the results are inferior to those obtained when using the proper film-light combination, the filters also require appreciable exposure increases. This is specially so with the conversion filter for daylight film used in artificial light.

Exposure. Under- or over-exposure with colour film upsets the colour rendering as well as the subject tones.

Slight over-exposure makes flesh colours pale, and slight under-exposure makes them richer. Darker colours become washed out by over-exposure, and are degraded by under-exposure. But if the subject contains few dark areas, slight under-exposure may actually improve the rendering of flesh tones.

Colour films are particularly liable to reciprocity failure—the loss of speed (and often also change in colour balance) at long or very short exposure times. Preferably adjust the lighting on the subject (or choose subject conditions) to permit exposure times not longer than about 1/10 or 1/5 second. With subjects requiring a longer exposure time (for example artificial light set-ups indoors) the exposure may have to be increased above the calculated exposure for that lighting level.

Special artificial light colour films exist which are designed for long exposure times, between 1/10 and 10 seconds.

The high sensitivity of the colour rendering to differences in exposure level is due to the fact that all three layers of a colour film may have individual reciprocity failure characteristics that can either compensate or reinforce each other. For that reason two makes of colour film of similar speed and colour rendering at normal exposure levels (1/10 second or shorter times) may diverge radically in both speed and colour rendering at long exposure times.

Exposure Charts and Tables. Most colour film packages include exposure tables issued by the manufacturers; these are reasonably accurate for normal daylight (or artificial light) conditions. But to be absolutely sure, it is usually best to take three exposures of the subject, at half a stop below and half a stop above the estimated aperture, in addition to the exposure which has been estimated as "correct".

Most charts or tables advise increased exposures for very dark subjects and decreased exposures for very light subjects. If, however,

CORRECTION FILTERS FOR COLOUR

Subject or Light Cause	*Film*	*Effect*	*Filter required for Correction*
Water, snow, mountains, distant views	Daylight	Blue cast due to ultra-violet radiation	Ultra-violet or pale straw
Distant views with haze	Daylight	Blue cast	Pale straw (haze
Outdoors with overcast sky	Daylight	Bluish, cold colours	Pale straw
Subjects in shadow under blue sky	Daylight	Strong blue cast	Medium straw
Early morning or late afternoon lit by sun	Daylight	Yellowish colours	Pale blue to give mid-day colouring; none to record effect faithfully
Normal daylight	Artificial light (Type A or B)	Strong blue cast	Yellowish to orange conversion (according to film type used). Recommended only in emergencies
Blue flash bulbs electronic flash	Daylight	Correct	None
Clear flash bulbs	Daylight	Yellowish cast	Bluish conversion
	Type A and B	Slight bluish cast	Yellowish light balancing
Indoors with Photofloods	Daylight	Yellowish cast	Blue conversion not recommended except in emergency
	Type A	Correct	None
	Type B	Very slight bluish cast	Very pale yellow light balancing
Indoors with studio lamps	Daylight	Yellow cast	Blue conversion not recommended except in emergency
	Type A	Slight yellow cast	Pale blue or mauve light balancing
	Type B	Correct	None
Indoors with household lamps	Types A and B	Yellow cast	Mauve light balancing
	Daylight	Strong orange cast	Lack of balance too great to be cured successfully by filter

Type A colour film is balanced for 3,400–3,500° Kelvin, Type B for 3,200° Kelvin.

human faces form an appreciable part of the picture, it is advisable to give the normal exposure; this may render dark clothes too dark, but it is preferable to false rendering of the skin tone.

Using Exposure Meters. A meter gives a more reliable indication of the exposure required. Average reflected light readings are sufficiently accurate with evenly lit colour subjects, but close-up individual readings are required if the surroundings are appreciably different from the subject itself in brightness. Highlight readings and incident light measurement are specially suitable for colour exposures, as they tie the exposure to the lightest tones of the scene, where faults in the colours are more noticeable. Specially light or specially dark subjects need some allowance; in the first case use half to one stop smaller, in the second case half to one stop larger.

For portraits the exposure should be based on close-up readings of the skin tone (key-tone readings).

Subject Contrast. Colour films vary in exposure range or latitude. Reversal films in particular tend to be contrasty whereas some negative films more nearly correspond to medium speed black-and-white films.

Therefore, in general, contrasty subjects should be avoided unless it is known from experience that the film can handle them. Even so high contrasts do not usually lead to a satisfying result unless a special effect is aimed at; and if strict colour accuracy is essential this can be guaranteed only in the mid-brightness range, although it depends on the particular colour and therefore the quantities of each dye required to produce it. For this reason it is often stated that the subject brightness range should not exceed 10 : 1, although the film may in fact be capable of recording a 200 : 1 range.

Lighting Outdoors. Direct sunlight produces very strong contrasts and a reflector (e.g., a white sheet, or silver paper mounted on a board, etc.) is advisable to throw light into the dark parts. Sometimes natural reflectors, such as concrete or white sand, will soften the shadows. But coloured reflectors can upset the colouring of the subject—e.g., a brick wall can reflect a rosy tint.

Fill-in flash may be used for brightening up the shadows, particularly in against-the-light shots, which have an extreme contrast range.

The ideal lighting for outdoor colour subjects, especially portraits, is slightly obscured sun, when shadows are only just visible. But colour photographs should not as a rule be made during the midday hours as the light then comes from too high an angle and makes, for instance, the eyes in a portrait appear dark and sunken.

Daylight Indoors. Owing to the strongly directional nature of daylight indoors, reflecting screens or fill-in flash are usually necessary to avoid heavy shadows.

If there is only one window, the subject should be as near to it as possible, and neutral reflecting surfaces of silver paper or white card should be used to light up the shadow side. A bay window is ideal as the light then comes from three sides instead of one and, if the room has two windows in adjacent or opposite walls, there may even be no need for reflectors.

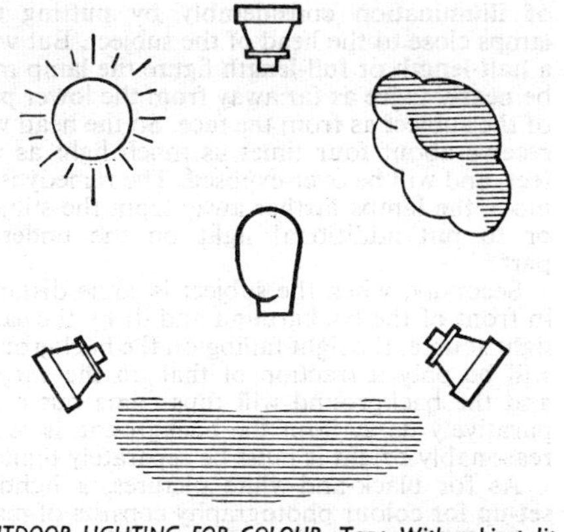

OUTDOOR LIGHTING FOR COLOUR. Top: *With subject lit by direct sun and cloud reflections, the colours will be brilliant.* Bottom left: *With direct sun and blue sky, effects are similar but shadows deeper and bluer.* Bottom right: *Front of subject lit by cloud and blue sky makes colours bluish and cold.*

Direct sunlight coming in through the window will lead to high contrast and curtains of net or similar diffusing material will have to be used to soften the lighting. This will make longer exposure times necessary.

Artificial Lighting Indoors. Artificial light gives more control of lighting effects and subject brightness range.

There is no great difference between the lighting for black-and-white and for colour, except that the heavy shadows and strong highlights sometimes used for ordinary photography are rarely satisfactory for colour.

All lamps should be fitted with reflectors, preferably painted white inside or with a matt surface to give fairly diffused lighting, free from "hot" spots. Spotlights used for special effects should have the condenser lens of white glass without any trace of green.

All lamps used in a lighting set-up must be of the same colour temperature; this is most easily achieved by having them all of the same kind. Mixing lamps of different types (e.g., Photofloods with ordinary studio lamps) will yield differences in the colour rendering over different parts of the subject. Although the colour temperature of the light itself will affect the colour rendering, an over-all change in colour is not nearly so disturbing as mixed lighting.

The inverse square law has two particular applications to colour photography.

Firstly, when taking head and shoulder portraits only it is possible to increase the level of illumination considerably by putting the lamps close to the head of the subject. But with a half-length or full-length figure the lamp may be nearly twice as far away from the lower part of the subject as from the face. So the head will receive about four times as much light as the feet, and will be over-exposed. The remedy is to move the lamps farther away from the subject or to put additional light on the under-lit part.

Secondly, when the subject is some distance in front of the background and lit by the same light source, the light falling on the background will be only a fraction of that on the subject and the background will thus come out comparatively dark. So if the background is to be reasonably bright it must be separately lighted.

As for black-and-white pictures, a lighting set-up for colour photography consists of main light, fill-in light, effect light, and background light. To start with it is best to use only one or two lamps and aim at obtaining even lighting on the subject; the effect and background lights can then be used to add highlights and contrast where necessary.

For colour subjects the fill-in illumination should only be a little weaker than the principal light and fairly diffused, to avoid creating a second set of shadows. Generally a fill-in lamp very close to the camera lens will achieve the required result. Reflecting screens will also serve for fill-in lighting, if they can be placed close enough to the subject. Yellow tinted flash bulbs can be used as fill-in lighting with Photofloods as the principal light because the colour temperatures are similar.

Flash Indoors. Synchronized flash permits instantaneous exposures, and enables moving subjects to be shot in colour—e.g., children at play, unposed and candid portraits, and people at their normal occupations.

If only one flash bulb is used it is better placed on or near the camera. With a good reflecting screen on the shadow side the flash may be also held to one side of the camera, pointing down at the subject. This increases the lighting contrast, but also improves modelling.

If several bulbs are used, lighting set-ups similar to those with Photoflood or studio lamps are possible with the added advantage of instantaneous exposures. The rules of lighting are much the same in each case.

Flash with Daylight. Blue tinted flash bulbs which match the colour temperature of daylight can be used for fill-in lighting for outdoor or indoor shots by daylight. Electronic flash has approximately the same colour temperature as daylight without any further treatment.

Flash used in this way is a great help when dealing with people who cannot avoid frowning in bright sunshine. It is also invaluable for natural portraits of children, engrossed in play with their heads in full shadow.

When used as a fill-in light the flash must be synchronized with the shutter, since the slowest suitable speeds are at least 1/30 or 1/25 second; open flash technique is rarely possible because this results in too long an exposure to the daylight and the effect of the flash is lost.

The fill-in flash must not be too strong, or its effect will be obvious by the unnatural appearance of the subject, and perhaps by double shadows.

Outdoor flash exposures are calculated in the same way as for black-and-white photography.

A special application of flash outdoors is to give a dark sky background to a portrait or figure. This is done by setting the camera aperture and shutter speed so that the background and subject would by themselves be under-exposed by about 2 stops and then using a flash bulb at the appropriate distance to yield a correctly exposed image. This technique gives a correctly exposed subject and an under-exposed, dark sky background like that produced by a deep filter in black-and-white photography. This is in fact the only way of darkening a sky in a colour picture apart from the use of a polarizing filter.

Outdoor Subjects. Most subjects are better with a simple background like the sky or a subdued area of colour. This is specially desirable with pictures of people, since violent background colours may make flesh tones appear unnaturally pale and lifeless by contrast.

Sunlight with white clouds, or slightly diffused bright daylight is ideal for outdoor colour pictures, since the lighting is then reasonably balanced, even without auxiliary shadow illumination.

Seaside shots on the sands do not usually present great problems; even in bright sunlight the sand acts as a fairly good reflector to light the shadows. If the sand is particularly yellow, it may produce a warm rendering of skin tones due to the reflected light, but this is rarely disturbing. But the tones are apt to be unpleasantly affected if the subject is coloured on one side by light reflected from the blue-sky, and on the other by yellow light reflected by sand.

When the light is as brilliant as at the seaside, it is usually best to arrange for side lighting with a reflector to light the shadow areas. Direct front lighting yields brilliant colours, but colour emulsions (especially reversal) tend to exaggerate the saturation and brilliance of the colours. With the light more to one side, or slightly diffused, the colours are less vivid and correspond more nearly to our visual impression.

Pictures on the water itself, including river and lakeside views, need a haze filter to cut out the ultra-violet radiation reflected by the water. Such views need trees or a human figure in the foreground to impart a sense of distance. The stronger contrast and more vivid colour of the foreground enhances the feeling of depth. Boats make an excellent foreground, with their brightly painted colours and the tones of ropes, nets, etc. Morning or afternoon sun will often produce more brilliant colours here than at midday, because many of the surfaces—boat sides, quay walls, etc.—are vertical and show up best when the sun is low.

Sunlight sparkling on the water is difficult to capture on colour prints but comes out well on reversal film.

Distant landscapes call for brilliant lighting which gives foreground features strong contrast and rich colours, while the distance is weaker and more subdued. The exposure should be based on the foreground objects, leaving the distance to take care of itself.

Woodland scenes with the sunlight streaming through the trees are inviting subjects but the contrast is severe unless they are taken in diffused sunlight.

If the leaves are reasonably transparent, back lighting can provide attractive effects. The exposure should be based on a highlight reading taken with a meter close up to one of the illuminated leaves.

When taking portraits in woods, parks or other green surroundings, it is necessary to watch out for green reflections on faces and clothes.

For mountain shots an ultra-violet or haze filter is almost essential. Both haze and ultra-violet radiation tend to give a blue cast without a filter.

In snow scenes the subject contrast tends to be low, but side lighting from the sun helps to create texture and modelling. The shadows in very clear weather will appear really blue. Exposure should be based on highlight readings.

Sunsets require a suitable foreground to give a base to the picture. As the colours and hues visible vary so much, almost any colour reproduction on the film, provided it is not grotesquely over-coloured, will be acceptable. Exposure readings should be taken with an incident light meter pointed straight at the sky.

Towns and cities are usually less coloured than country scenes, but this fact lends greater value to patches of brilliant colour—e.g., motor cars, summer dresses and blue sky.

In city streets the biggest problem is large areas of shadow and although diffused daylight is best for most subjects, buildings generally look better in bright sunshine. So it is preferable to wait until the sun is out and to choose a viewpoint which reduces the shadow area.

Lighted street scenes in town centres make effective night pictures. Such subjects are best taken on daylight film, as the artificial light type tends to render the lights too blue. The warmer rendering on daylight film is usually more natural and attractive. It is not possible to record the full brightness range of a night scene on the film. On the other hand, the shadows are meant to be completely black, and the amount of black is a matter of taste. So there is a surprising amount of latitude and the results are acceptable over a large range of exposures.

Indoor Subjects. The limited amount of light available indoors usually means that people—singly or in groups—have to be posed in one place. This also allows time to discover unpleasing splashes of colour such as out-of-focus flowers in the background, etc.

Groups in particular must be evenly lighted since complicated lighting schemes may result in disturbing shadows and distorted colour values.

The evenness of the lighting should be checked by taking a highlight reading on a white card at various points in the picture area. The lighting should be adjusted until the highlight readings are constant everywhere.

Walls can be used as reflectors, but only if they are neutral in colour. A flood lamp shining on a white ceiling will provide a diffused general illumination from above.

For portraits a background colour complementary to the subject will provide the best colour contrast. A background in a desaturated hue will brighten the subject colour if it is complementary to it. Flesh tones stand out well against pale blues and greens; but if the background is a warm colour such as cream or yellow, it should be appreciably lighter or darker than the skin tone.

Make-up should be natural and rather subdued, and with sleeveless or topless dresses the photographer should make sure that the skin

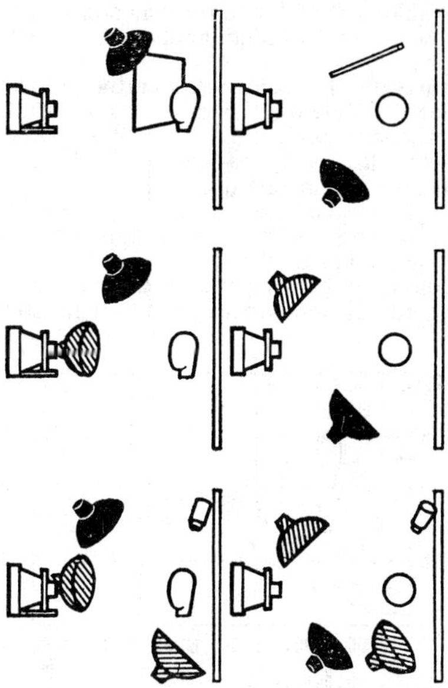

LIGHTING FOR COLOUR. Top: *A single lamp should be high and to one side, with a reflector for shadows.* Centre: *A second lamp (shaded) acts as a fill-in for the main light (solid).* Bottom: *Four lamps allow almost unlimited variations, with additional fill-in and effect lighting.*

tone of the arms and shoulders is made up to the same hue as the face. Colour film shows up variations particularly strongly.

Details of skin structure and facial blemishes show up much more plainly on colour film than in black-and-white. For glamour portraits a certain amount of disguising make-up may be necessary or the model should be moved farther away and lighted evenly.

The eyes must be well lit, especially in close-up shots, so that their colour can be seen without being degraded by shadows in the eye sockets. Where face and hands or arms appear in the picture they should be given the same amount of light; under-lighted skin tones not only change in brightness, but also in hue.

Figure studies, and particularly nudes, need carefully chosen backgrounds as well as even lighting. The large area of skin tone present can show appreciable changes in hue if unevenly lit or when near strongly coloured reflecting surfaces. Flesh tones contrast with cool background colours like blue and green and tend to merge with warm tones like yellow or orange. With interiors the greatest danger is an excessive contrast range. Windows should not appear in the picture and exposure meter readings should be made of the brightest parts and the deepest shadows to make sure that they come within the range of the film. Usually several reflecting screens will be necessary to produce more even illumination.

The background will not usually be sharp, but it should still suit the subject. For instance a pale blue-grey background will make yellow flowers stand out well, while dark wood will create a pleasing contrast to pink roses.

A soft but directional light should be used to give good modelling—e.g., a Photoflood with a large diffusing screen. A suitable form of diffused lighting is produced by shining one or more lights on to a large piece of white card and directing the reflected light on to the subject in the manner of a "bounced" flash, so that no direct light falls on the subject. Use a reflector or another diffused lamp to lighten up what shadow there is. G.W. & L.A.M.

See also: *Colour; Colour materials; Colour temperature.*
Books: *Colour Photography in Practice*, by D. A. Spencer; *Colour Transparencies*, by C. Leslie Thomson; *The Colour Book of Photography*, by L. Lorelle; *Colour in Your Camera*, by Gösta Skoglund (all London).

COLOUR TEMPERATURE. Measure of the energy distribution over the spectral range (i.e., the colour quality) of a light source with a continuous spectrum provides a fairly easily workable method for comparing and specifying the "whiteness" of illuminants in colour photography.

Kelvin Scale. The colour temperature, expressed in degrees Kelvin (°K.), is the temperature on the absolute scale (°C. + 273) to which a perfect black-body radiator would have to be heated to emit light of the same colour as the light source in question. (A perfect black-body radiator is one which does not reflect any light falling on it.)

Colour temperature measurements are applicable to all solid incandescent light sources such as flash bulbs, filament lamps, etc. (The colour temperature of a filament lamp is for all practical purposes the same as the temperature of the filament.)

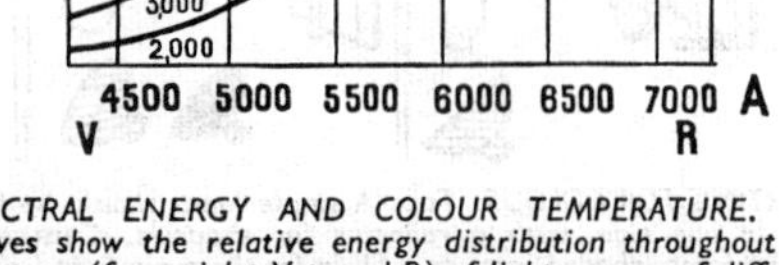

SPECTRAL ENERGY AND COLOUR TEMPERATURE. The curves show the relative energy distribution throughout the spectrum (from violet V to red R) of light sources of different colour temperatures. Tungsten lamps (about 3,000° K) are rich in red; balance shifts to blue at higher temperatures.

Precise colour temperatures cannot be stated for discharge lamps as these do not have a truly continuous spectrum. Where colour temperatures of such sources are quoted, they are only approximations.

The lower the colour temperature of a source, the richer the light is in yellow and red rays; the higher the colour temperature, the richer it is in blue rays.

Colour temperature is of special importance in colour photography as colour films are sensitized to give a true colour rendering with light of a specified colour temperature. Thus some films are balanced for a colour temperature of 6,000° K. (daylight), while others are balanced for 3,200–3,400° K. (tungsten or Photoflood light).

The colour temperature of a light source can be modified by special filters which absorb some of the blue or some of the red end of the spectrum. Filters of this type are known as photometric or light balancing filters. For overall correction, they may be used over the camera lens instead.

Mired Values. An alternative way of expressing colour temperatures is the mired (micro-reciprocal degrees) scale. The mired value of a light source is its colour temperature in degrees Kelvin divided into 1,000,000. Thus a colour temperature of 5,000° K. corresponds to 1,000,000/5,000 = 200 mireds, and 3,200° K. to about 313 mireds.

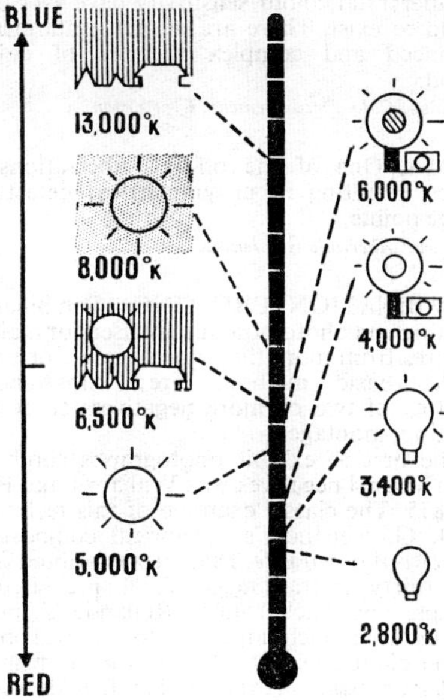

COLOUR TEMPERATURE. Left: *Typical natural light sources in order of ascending average colour temperature; mean noon sun, average daylight, overcast sky, blue sky. Right: Artificial sources are household lamp, Photoflood and other overrun photographic lamps clear flash bulb, blue flash bulb.*

COLOUR TEMPERATURES COMPARED

Type of Light Source	*Approx. Colour Temperature*
Small general service (below 100 watts) ...	2,600° K
Large general service (100 watt and above) ...	2,800° K
Sensitometric lamp (international white light SA)	2,848° K
Projector lamps	3,100° K
Over-run lamps 500 watt (100-hour life)	3,100–3,200° K
Flash bulbs, paste filling	3,300° K
Over-run lamps (275 watt, 2-hour life, and 500 watt, 6-hour life)	3,400° K
Flash bulb lamps, foil or wire filling	4,000° K
Fluorescent daylight tubes	4,800° K*
Fluorescent warm-white tubes	3,700° K*
Electronic speed flash discharge lamps	4,800–6,000° K*
English daylight (international white light SB)	4,800° K*
Mean noon sun	5,000° K*
Blue flash bulbs	6,000° K
American daylight (international white light SC)	6,500° K*
Overcast American daylight (international white light SD)	7,500° K*
North sky, blue	10,000–20,000° K*

American lamps, being run at lower voltage, operate at colour temperatures between 50° and 100° higher than the English lamps here quoted.

*These sources are not temperature sources in the strict sense, and their colour temperature therefore only approximately indicates the quality of light they emit.

The mired values are useful in assessing the effect of light-balancing filters. The degree to which such a filter raises or lowers the colour temperature depends not only on the depth of the filter itself, but also on the initial colour temperature of the source. When the colour temperature is expressed in mireds, the change in the mired value (the mired shift) is more or less independent of the initial colour temperature. Yellowish filters which lower the colour temperature raise the mired value, and have a positive mired shift. Bluish filters which raise the colour temperature lower the mired value and have a negative mired shift.

As a slightly more handy unit decamired values are also used. One decamired = 10 mireds. Light balancing filters are frequently rated in decamired values to indicate their actual correction effect. For example in the reddish series of some manufacturers an R5 filter signifies a decrease of the colour temperature corresponding to raising the mired rating by 5 decamireds.

Special meters, similar in principle to photoelectric exposure meters, are made for measuring colour temperature. L.A.M.

See also: *Colour materials; Colour technique; Colour temperature meter; Light sources.*

COLOUR TEMPERATURE METER. Instrument for measuring the colour temperature of a light source. It is based on the fact that the spectral-energy distribution of sources at different temperatures can be defined reasonably accurately by stating the relative proportions of red and blue radiation which they emit. An instrument constructed to measure this proportion of red to blue will then give a measure of the colour temperature of any light source.

A typical colour temperature meter consists of a selenium barrier layer type photo-cell connected to a micro-ammeter. Immediately in front of the photo-cell is a circular filter, half red and half blue. These filters are rotated by means of an external lever, and are spring loaded so that the red filter is normally in front of the photo-cell. By pressing the lever, the red filter is replaced by the blue. In front of the filters is a diaphragm which may be rotated to control the amount of light received by the photo-cell.

The instrument is directed towards the light source to be measured and held at eye level. The shutter is adjusted until the pointer of the meter coincides with a red index line on the scale. The filter lever is then depressed, bringing the blue filter in front of the cell in place of the red one. The cell therefore measures the amount of red light and the amount of blue in succession. If the proportion of the two is equal, the meter needle will not move, but if the blue proportion is greater or less, the blue filter reading will be higher or

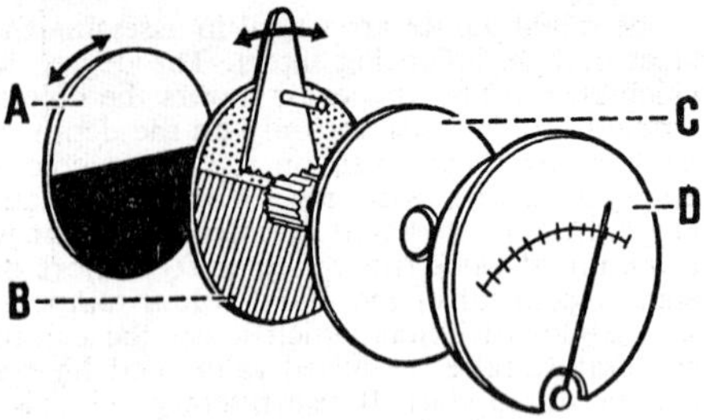

COLOUR TEMPERATURE METER. A, rotating shutter; B, blue/red filter disc; C, photo-cell; D, micro-ammeter.

lower respectively. As the zero setting to the red index also corresponds to a standard colour temperature, the meter reading directly gives the colour temperature of the light source.

Alternative systems may use a null-method with two photo-cells, balancing the red and blue content of the light on them. The drawback of this method is that the response of the two cells may not be the same at varying wavelengths, which would introduce an error.

Visual colour temperature meters rely on matching various shades of fluorescent and non-fluorescent pigments, or on matching the tint of a dichroic filter with a standard tint by means of a filter wedge. This type of meter cannot, of course, give more than a very approximate value.

When the colour temperature measurement has been made it may prove necessary to employ a colour temperature correction filter on the camera.

However, the effects of a correction filter may be assessed by using the meter to measure a light source first without and then with the filter interposed.

Colour temperature meters can also be supplied with a second scale calibrated in mireds ($= 10^6 \times 1/°K$). This scale is of particular assistance when it is desired to select a correction filter, since any true colour temperature correcting filter will give a constant change in mireds irrespective of the source employed. G.C.C.

See also: *Colour temperature*

COLOUR VISION. Ability of the eye to distinguish between radiations of different wavelength within the visible spectral range. In the eye it is the cones (one of the two types of light-sensitive cells or receptors in the retina) which are responsible for the colour sensitivity of the eye. The eye behaves as if it had three types of such receptors, each sensitive to one of the primary colours, blue, green and red. While this theory accounts for a large number of phenomena of colour perception and also provides a working basis for most processes of colour photography, no individual receptors of differential colour sensitivity have ever been found to exist. There are in fact a number of advanced and complex theories of colour vision.

See also: *Colour measurement; Eye; Vision.*

COMA. One of the off-axis aberrations of lenses, resulting in unsymmetrical defects in image points.

See also: ***Aberrations of lenses.***

COMBINATION PRINTING. Combination printing and photomontage are used for making pictures from more than one negative or print.

The basic methods are: simultaneous printing of two or more negatives; successive printing; montage.

The first to exhibit photographs combined from several negatives was William Lake Price in 1855. The classic example of this technique is O. G. Rejlander's allegorical composition *The Two Ways of Life*, 1857. It was printed from over thirty separate negatives of single figures, groups, and background. Rejlander's purely photographic technique was to mask all parts of the picture except what was being printed. H. P. Robinson, on the other hand, made scissor and pastepot photomontages which he retouched and rephotographed. His first and best-known composition was *Fading Away* (1858), from five negatives.

A widespread form of combination printing, particularly during the collodion period, was the printing in of separate cloud negatives, first suggested in 1852 by Hippolyte Bayard. Owing to the colour-blind and not very sensitive negative material, the sky became so dense that it printed white if the exposure were correct for the landscape.

Simultaneous Combination Printing. Two (or sometimes more) negatives are placed together, emulsion to emulsion, in the negative carrier of the enlarger, and printed as one. Where parts of the images overlap, they produce lighter tones in the print because the negative densities are added.

The process is often used to reproduce ghost effects and for printing genuine or faked negatives of falling snow, rain, etc., into a picture.

Two identical negatives, or a negative and a positive transparency made from it, can also be combined by simultaneous printing. If one of the two is reversed left to right, a variety of interesting image patterns and shapes is obtainable. The negative and positive used for this method must be relatively thin, because the effective printing densities are added.

Further special effects can be achieved by removing parts of one or both images by local reduction with Farmer's reducer.

The reducer is applied to the emulsion with a plug of cotton wool. If the edges of the reduced area are not to appear sharply defined, the emulsion should first be soaked in water.

If sharp outlines do not matter, the reducer can be painted on to the dry negative, or the whole negative may be immersed in the reducer after protecting it where necessary with an asphalt varnish, or similar protective lacquer.

Successive Combination Printing. Several negatives may be printed, one after the other, on the same sheet of paper. The appearance of the resulting combination print differs from that obtained by simultaneous combination in that the print tone is darker where parts of the images overlap. For this reason the separate exposures must be reduced, or the print will be too dark. The process can be used for ghost effects in which one image is seen through the other. But as a rule, the separate images are required to merge without overlapping, so a certain amount of complementary shading may be necessary during the separate exposures. This may be done with the hands, or with a suitably shaped card. For complicated combinations specially cut masks and counter masks are used.

This method is used for printing in clouds when the sky in the subject negative is blank. The usual technique is to mount a sheet of thin card on a glass plate supported just above the paper holder on the enlarger baseboard after the image has been focused. On the card the skyline of the picture is sketched in, and the sheet cut along that outline. This yields a mask and counter mask which are realigned in position on the platform. The subject portion is now printed on the enlarging paper with the sky mask in position, the subject mask replaced and sky mask removed, and the clouds exposed finally on the same paper from another negative.

Montage. Different parts of various prints can be cut out and pasted together on a base to make a composite photograph. This is a special and major technique of its own. L.A.M.

See also: *Cloud negatives; Montage; Tricks and effects.*

Books: *The Complete Art of Printing and Enlarging*, by O. R. Croy (London); *All the Photo Tricks*, by Edwin Smith (London).

COMMERCIAL PHOTOGRAPHY

Commercial photography is the branch of professional photography concerned principally with supplying the need of the sales, advertising and display side of industry, and the art departments of the press and specialized publications.

Nowadays the demand for photographs for all these purposes is so general that there is usually a commercial photographer's business in even the smallest town. The size and scope of the establishment vary widely. In the smaller towns the studio may be a one-man business turning out photographs for local industries and traders. This type of studio may even be run in conjunction with a portrait photographer's business and a photographic dealer's shop. The photographer and his staff in such cases will usually go out to cover weddings, banquets and events of local interest.

In the cities and more particularly in London, the commercial studio tends to specialize almost exclusively on work for reproduction—i.e., on photographs for catalogues, advertisements, posters and sales literature. This type of establishment employs a large staff of photographers, darkroom staff and retouchers. It may even include commercial art and block-making facilities so that the client can be supplied with the finished blocks of the illustrations he has commissioned.

While the large establishments will—and often do—undertake outside assignments they usually prefer to do all the work in the studio where they can keep all the conditions under control and repeat the shot if necessary. Modern trends have encouraged most commercial studios to handle colour; and this in turn has compelled them to adopt precision technique in place of the old rule-of-thumb methods.

The ever widening applications of photography in commerce and advertising are leading more and more studios into specialization—e.g., fashion, advertising, industry, aircraft, architecture. To-day, a studio would need to command a large organization and a team of workers to be able to handle every branch of the work. It would have to be prepared to deal with a wide range of subjects and be ready to apply scientific techniques like industrial radiography, ultra-violet light photography, or ultra-high speed photography, and cinematography. Such work is beyond the scope and equipment of any but the most highly specialized studio.

The word studio is commonly understood to include the whole organization and not merely the room where the photographs are taken.

The Small Studio. The small specialist studio generally adapts itself to the requirements of one class of work. It may, for instance, be designed to handle food pictures for a group of women's magazines. In this case the studio might consist of a room with a cooker in one corner, a typical kitchen table in another, and a plain wall against which small objects can be photographed. For lighting equipment it may manage with a minimum of a couple of flood lights, two focusing spots, and a few portable hand lamps. The camera may be anything from a half-plate to an 8 × 10. As a rule a large size is preferred since retouching can be done on the negative and the prints made by contact. A small room or large cupboard to serve as a darkroom containing processing dishes or sink, and a contact printer or an enlarger (or both) complete the set-up.

There are large numbers of such small commercial studios specializing in photographs for catalogue and advertising illustrations. They are designed to handle small objects quickly and they are often squeezed into basements or out-of-the-way corners, because they try to get as near as possible to their clients, to be able to offer a quick and efficient service. Such work is usually required for reproduction, and nearly always has to be retouched before it goes to the blockmaker. So unless the studio includes retouching in its service it usually has a working arrangement with an outside commercial retoucher.

The Larger Studio. The larger commercial studios are completely self-contained with all the facilities for carrying out a wide variety of commissions without having to call in outside help. This calls for a much more complex type of organization which can handle teams engaged on both indoor and outdoor photography.

At the same time even the largest studios tend to specialize in some degree. A studio dealing with heavy industrial subjects, for instance, must be planned in terms of concrete floors, plenty of wall space, and easy access to the street, whereas a fashion studio spends its money on carpets, a reception desk, and dressing-rooms.

The Camera. Cameras taking large-size plates or cut-film are still used in some studios, especially for reproducing fine detail and for giving an opportunity for negative retouching. But it is becoming more general to use smaller negatives and print by enlargement. This in turn calls for additional darkroom facilities. Most studios have at least one camera, fitted with the whole range of camera movements, which can be used at any angle on a heavy commercial-type camera stand. At the same time there is usually a small folding camera, and a precision 35 mm. camera available for specialized photography.

Lighting Equipment. Lighting equipment varies considerably according to the work of the studio, but it usually includes a number of movable floodlights, and a bank of floods for general illumination, stand lamps for modelling, and focusing spots for adding highlights. As most studios now handle colour, all the light sources must be of the same colour temperature or be fitted with suitable lamp filters. It is no longer practical to mix daylight and artificial light, still less fluorescent lamps and half-watt lighting. Some studios nowadays do a lot of their work with electronic-flash operated from the mains, but in general, the lighting equipment of the commercial studio consists of the normal units sold for the purpose.

Some of the larger studios which specialize in colour work have extensive electronic flash equipment of high output. Such studios can frequently be rented, by the hour or the day, by the photographer who has only occasional need for this type of lighting.

Some electronic flash outfits incorporate into each lighting unit a tungsten lamp. This system enables the photographer to arrange his lighting visually, and yet to expose by electronic flash.

In a properly run modern studio, the number of trailing cables is kept as low as possible. Lighting units are operated from convenient wall fixtures, and the cables are suspended from above instead of letting them trail along the floor. In a studio of any size, there is at least one ceiling light (or bank of lights) for general top-lighting, mounted on runners fixed to the ceiling and pulled into position from below.

If the ceiling of the studio is low, it reflects a useful amount of light, but on the other hand a high ceiling gives more scope for manipulating top lights and spots.

Accommodation. In addition to the normal studios and darkrooms, a reception or conference room is an essential part of the establishment. This is used by clients who wish to have a say in the direction of the photograph—a situation that often arises when the studio is working on an advertising photograph that has to conform to a particular layout dictated by the agent, or when making catalogue illustrations of a new technical product with special selling features that have to be emphasized to satisfy the manufacturer. These activities must be kept separate from the normal routine work of the studio, and the premises must be laid out with that condition in mind.

Negative Filing and Storage. As time goes on, negatives accumulate so a reliable negative filing system is essential. The copyright of a photograph taken to instructions may belong to the client, but the negative is the property of the photographer. Re-prints may be required long after the picture has been taken, so every print and negative is given a reference and entered in a card-index system.

Storing the negatives so that they do not deteriorate and in accordance with fire regulations, is a major problem to the large commercial studio. The difficulty is made worse by the variation in size of the negatives, which may range from 8 × 10 glass plates, to 35 mm. cine film strips. Also the effort and additional work in extracting individual negatives and deleting them from the card-index (probably with cross references) is more trouble than leaving them where they are, so that over a period of years, negative storage may account for a considerable proportion of the studio floor space.

Models. Advertising photographs often include people, so the commercial studio must be able to call on the services of professional models. This means compiling a filing index of some complexity which must record the personal information about each model, and be kept up to date. In addition to the model-register which each studio might possess, there are in the larger cities numerous model agencies, who provide models for all purposes. The agencies have photographs and details on all the male, female and child models (and even animals) on their books, and the photographer can pick the type who will best fill his particular needs. A studio must be ready to provide suitable models at short notice during the winter months for instance to stage a sunny picture advertising beach wear, or produce a cosy fireside picture in the middle of August. Such inquiries demand dressing-rooms, and plenty of floor space for the erection of simple sets. The range of effects is extended by using projected backgrounds—e.g., in the middle of winter a slide of a sunny garden with masses of flowers in full bloom may be projected to form a background for models displaying summer dresses, or enjoying a cool drink.

If the studio had to wait to take such photographs in actual sunshine, the summer would be over before the advertisement could appear.

"Props." The design and erection of studio-sets plays a major part in the make-up of the larger commercial studio. Smaller studios often work wonders with a few simple "props"—e.g., long wooden boxes that can be used horizontally as steps, or vertically as stands. A couple of "flats" are standard items in all studios for representing the corner of a room.

Backgrounds. Studios use both soft, roll-up backgrounds of painted canvas, and "flats" of three-ply or wall-board. The roll-up background has the advantage that it keeps clean when not in use; flats are stacked face-to-face for protection. Apart from a table with lamps available for photographing small objects, the studio floor is normally kept clear so that sets, corners, or short sections of wall can be built up as required with "flats" which may be either painted, distempered, or covered with wallpaper to suit the character of the interior represented. Odd flats are also used for preventing interference from the lighting of adjacent sets when more than one camera is in use in the same studio. The increasing use of colour means that any backgrounds used must be in perfect condition because it is no longer possible to touch out blemishes on the negative or print. The walls and most of the flats used in the studio are now coloured in light pastel shades which serve equally well as backgrounds for colour or black-and-white. One result of this is that nowadays the commercial studio is quite often clean and colourful. The flats, and even the walls, can be used also as reflectors. In the case of colour work it is, of course, only possible to use white surfaces for such purposes, as coloured flats would introduce a colour cast into the picture.

Retouching. Much of the work done in a commercial studio is required for reproduction, and retouching of prints and skilful use of the air-brush are part of the service. Most small commercial studios have an arrangement with an art studio where specialists will do the work; although in some of the larger studios there is an art department which also handles layout work and typography.

The air-brush is employed for ghosting effects where the important part of the picture stands out against a "misty" surround, and for reinforcing tones on large prints. When smaller prints and precision work are required, the photographs must be dealt with by an expert retoucher who is equally conversant with the use of an air-brush, knife and pencil. This form of retouching is often employed for colouring industrial pictures (colour retouching) and the fineness or boldness of the work must be in accordance with the requirements of block making and printing. Obviously the retoucher must have a working knowledge of the printer's art and be familiar with the methods of reproduction. B.A.

See also: *Professional photography; Studio; Studio photography.*
Books: *The Business of Photography*, by C. Abel (London); *Basic Photography*, by M. Langford (London).

COMMON SALT. Everyday name used for sodium choride. Household salt is not necessarily always pure sodium chloride.

COMPENSATING DEVELOPER. Obsolete term still sometimes applied to M.Q.-borax and similar types of fine grain developer of low alkalinity, because the low gamma obtained compensates to some extent for high subject contrasts. In terms of gradation, the result is identical with that obtained by development to a low gamma in any normal developer.

More correctly, compensating developers should compress the over-all subject contrast in the negative, without greatly affecting the gradation in the highlights and shadows—yielding an effect analogous to tone separation. This often happens in two-bath development, where the action of the developer absorbed in the emulsion layer keeps up or even increases the contrast of the shadows, but does not appreciably increase the highlight density.

COMPENSATING POSITIVE. Low-contrast, low-density positive image on film or glass, bound up in register with the negative during printing in order to bring the tone range of an excessively contrasty negative within the range of the printing process.

See also: *Contrast control; Masking.*

COMPETITIONS. Photographic competitions are of several types. Generally speaking the term is applied to those contests in which prizes are given.

There are competitions organized by the great national newspapers where prizes are offered, for example, for the best pictorial prints, the best sports picture, the best action picture, the best child portrait and so on. These competitions are generally run on well organized lines and the final selections are made by competent judges. Then there are commercial competitions run by manufacturers to increase sales of some particular product. In such competitions the purchase of the product is usually a condition. The competitor is asked to include a coupon or box lid or some evidence of purchase with the photograph. Many of these competitions are promoted by manufacturers of photographic products who award a prize for the best photograph taken with a particular make of camera or using a particular paper or film. Finally there are the competitions run by the photographic press in specialized subjects. These are organized to stimulate readership and to promote good photography generally.

Conditions for Success. A study of thousands of competition entries has shown quite clearly that the great majority of entrants, particularly for the national newspaper competitions, have little idea of what is required in order to win. Usually there is no limit on the size of the print submitted and for this reason the great majority of people send in contact prints of the 2¼ × 3¼ ins. size or even smaller. Experienced entrants, however, generally submit a picture measuring half plate (4¾ × 6½ ins.), whole plate (6½ × 8½ ins.) or even 8 × 10 ins. They know that the proprietors of the paper will want to reproduce the winning pictures in their journal and that it is much easier to judge a picture and to visualize it in its reproduced form when it is of a reasonable size. A certain number of very large pictures of what are generally called "exhibition size" are frequently sent in, but these stand no better chance than a whole plate print.

It should be obvious to any entrant that the picture must be clear and sharp and of a reasonable technical quality, but at least 75 per cent of the entries for the big national newspaper competitions are thrown out on the first sorting because they are thoroughly bad photographically. So before submitting a print, the sender should make sure that it is at least sharp, correctly exposed, and free from blemishes. Then the picture must make an immediate appeal to the average reader of the paper and should generally be simple in character and possess one main interest. Technical quality alone will not win a prize any more than good grammatical construction will ensure the printing of a story. A good sharp picture of a baby sitting on the sands with a spade and pail will stand no more chance than thousands of other similar pictures.

The type of photograph that wins prizes is a simple record of some particular incident with a strong sentimental appeal and the best training for success in this kind of photograph is undoubtedly to study the type of picture that is awarded the prize in similar competitions or in previous years. The judges in selecting the winning photograph always bear in mind that what is required is the picture that will have the maximum appeal to the paper's readers. If the picture has a strong appeal, slight technical faults may be overlooked.

In competitions designed to boost the sales of particular products the same general rules apply, but the picture should in addition create a desire to own the product. In fact, success in this field is largely dependent upon the photographer's sense of salesmanship.

In the general photographic competitions simplicity and direct appeal should again be aimed at and technical quality must be of the highest. The standard of a pictorial entry will be judged as keenly as for a picture entered at a salon or general exhibition.

All prints submitted to competitions should be black-and-white and on glossy paper for good reproduction. Rough paper prints, and particularly those which have been sepia toned, are almost impossible to reproduce satisfactorily. Finally, if the competitor wants his prints back, he should be sure to enclose suitable wrapping, stamped and addressed.

Copyright. The better organized competitions usually require that the winners of big prizes should make over the copyright of the picture to the promoters. This is reasonable when the prize offered is ten guineas or more, but entrants should beware of competitions, particularly for commercial products, in which one of the conditions of entry is that all entries become the copyright of the promoters. By making this condition, some firms will obtain hundreds of photographs for reproduction without having to pay any reproduction fee, simply because the pictures they use have not necessarily won prizes. The better competitions usually state that the main prizes call for the surrender of the copyright but in the case of all the other entries the copyright remains with the entrant. P.W.H.

See also: *Awards; Judging exhibition entries.*

COMPLEMENTARY COLOUR. One colour is said to be the complementary colour of another when the light of the two added together makes white light. A filter transmits light of its own colour, and absorbs the light of the complementary colour.

Negative/positive processes of colour photography produce first of all a picture made up of colours which are complementary to those of the actual scene, and then, by an extension of the same principle, give a final print in colours which are complementary to the negative and, therefore, true to the original subject.

COMPOSITION

Composition is the tool wherewith the artist handles the subject matter at his disposal in the picture space so as best to express his purpose. His purpose is not always simply to please; he may wish to instruct, entertain, or even annoy. The subject matter by itself can do these things only to a limited degree; most of the virtue lies in the way in which it is presented by the artist.

The painter has the advantage of the photographer in being able to put in or leave out bits of subject matter at will. If anything present in the actual scene does not help to make his meaning clear, he can leave it out. If the purpose of the picture would be strengthened by adding something, he can add it.

Up to a point, the photographer can include more or less of the scene by changing his viewpoint, but eventually he must accept the contents of the picture space as he finds them. When that point is reached, his only remaining way of expressing his personal feeling about his subject matter is by emphasis—playing down some objects or aspects of them, and stressing others.

The things that control emphasis in a photograph are: tone, sharpness, scale, and arrangement.

Tone. If the tone of the subject is lighter or darker than that of the background and any other objects in the picture, the subject will tend to stand out from its surroundings and the observer's attention will naturally turn to it. If the subject itself is the picture, it is sufficient to show it well lighted against a plain background of darker tone, or if it is dark in tone, against a lighter background. But if the picture must include other items all more or less equal in tone, the principal item must be emphasized in some other way.

Sharpness. The photographer can direct attention to any part of the picture by making that part sharp and detailed and leaving the rest vague and blurred. This is one of the most useful photographic tricks for achieving emphasis and it is simply a matter of focusing the lens accurately on the part to be emphasized and throwing the rest of the picture out of focus.

Scale. All other things being equal, attention is held by the biggest single unit in the picture, whether it is an object or a mass of light or shade. To give the subject importance, it is useful to make it appear larger in proportion than the rest of the subject matter. The photographer can achieve this by altering the distance between the camera and the subject.

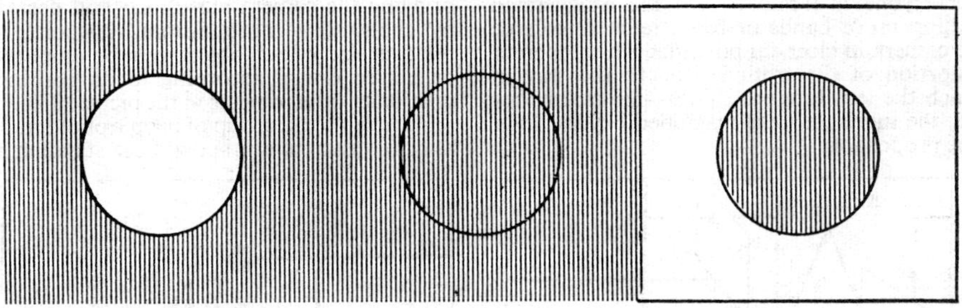

EFFECTS OF TONE. Tone differences make a subject stand out against its surroundings. **Left:** *Light subject, dark background.* **Right:** *Dark subject, light background. In both cases the separation is good and the subject contrasts effectively against the background.* **Centre:** *Subject tone and surroundings the same, resulting in poor differentiation.*

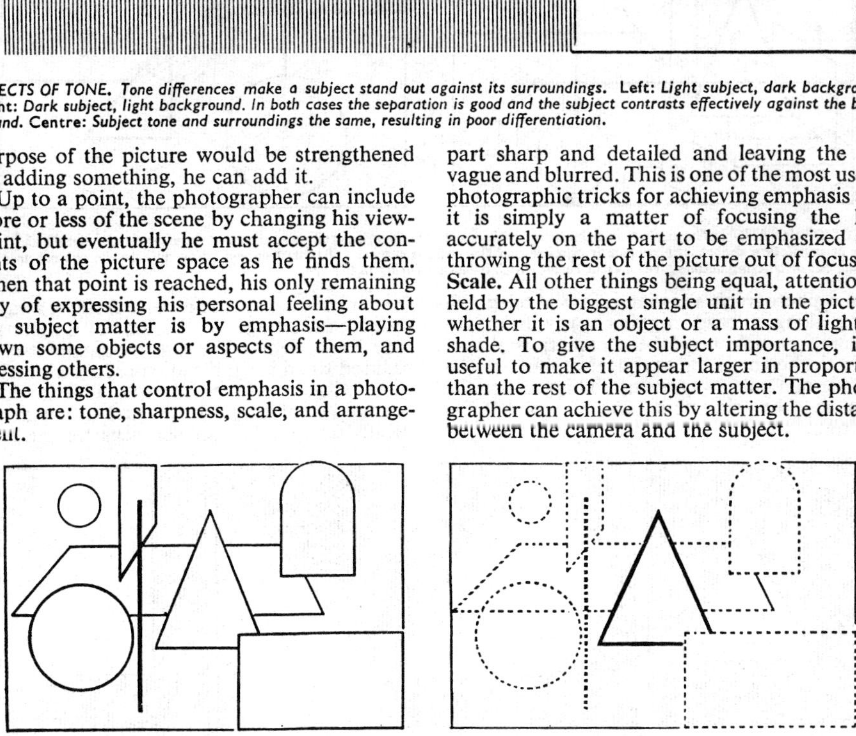

EFFECTS OF DEFINITION. **Left:** *With all parts of the sceen equally sharp, nothing stands out—the subject appears as a two-dimensional design.* **Right:** *Making the most important part only sharp (e.g., by differential focusing) isolates it, and adds an impression of three-dimensional depth. Such effects are best judged visually on a focusing screen.*

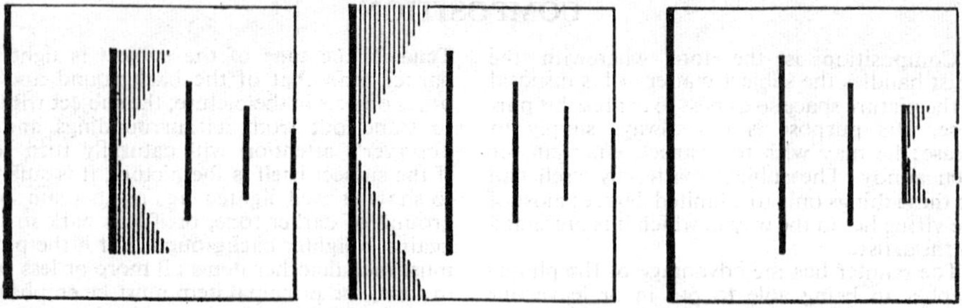

EFFECT OF SIZE. *The relative size of an object influences its importance in the composition of the picture.* **Left:** *Close camera position, and the subject dominates the picture area.* **Centre:** *Camera too close and the subject is of incidental foreground interest.* **Right:** *Distant viewpoint and the subject becomes part of the over-all picture arrangement.*

Everyone is familiar with the exaggerated proportion of hands or feet stretched towards the camera in close-up portraits. This apparent distortion of the truth is due to the way in which the lens looks at things—the closer they are, the more they are magnified beyond their true proportion.

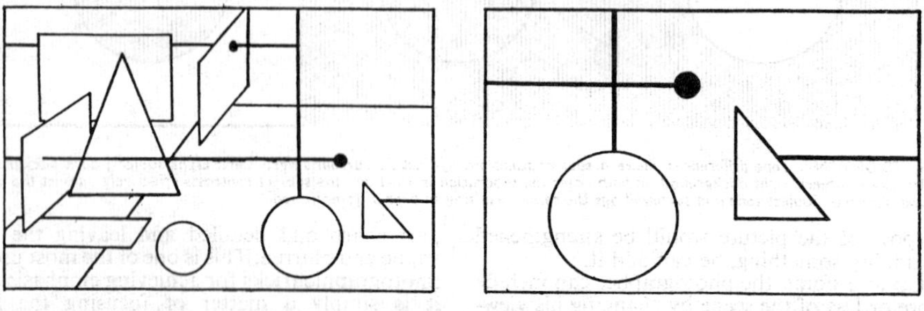

EFFECT OF ISOLATION. *The subject in a picture can be more clearly presented when isolated from irrelevant detail.* **Left:** *In a mass of conflicting shapes, the subject is obscure and the pattern confused.* **Right:** *Concentrating only on the subject area and leaving out distracting details gives a clear result which is less confusing to the eye.*

In a photograph showing a row of telegraph poles along a road running away from the camera, the pole nearest the camera is almost twice the height of the next nearest, but a pole a quarter of a mile away seems no bigger than the pole on either side of it. This shows that the apparent size of the subject is increased by bringing the camera close to it and decreased by choosing a more distant viewpoint. The relative size of an object can be emphasized or played down, according to its position in relation to other objects in the picture area.

For example, a group of people photographed at close quarters will show those standing near the camera as giants and those farthest away as dwarfs; or cottages at the foot of a mountain may look right while the mountain itself is reduced to an insignificant rise. In such cases a more distant viewpoint will give a more normal set of proportions, and although this means accepting a smaller negative image, it

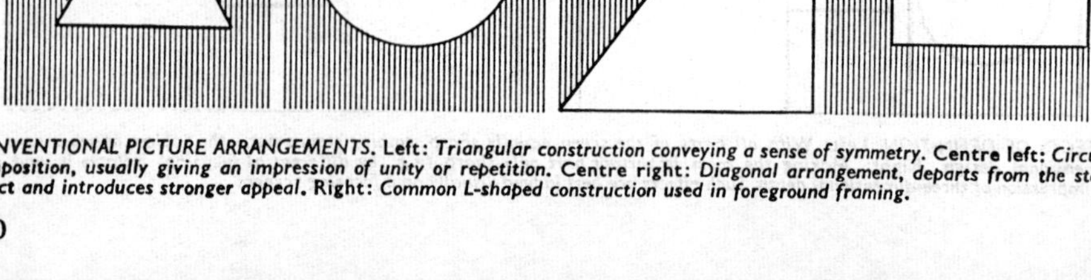

CONVENTIONAL PICTURE ARRANGEMENTS. Left: *Triangular construction conveying a sense of symmetry.* **Centre left:** *Circular composition, usually giving an impression of unity or repetition.* **Centre right:** *Diagonal arrangement, departs from the static effect and introduces stronger appeal.* **Right:** *Common L-shaped construction used in foreground framing.*

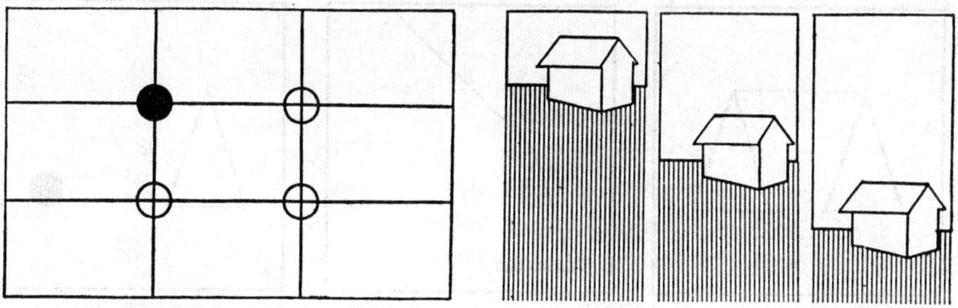

POSITIONS OF THE SUBJECT. For the most acceptable results visually, the subject may be positioned according to traditional recommendations. **Left:** *Intersections of thirds are usually strong points of interest, especially the top left intersection.* **Right:** *Horizons are best either high or low in the picture; central division tends to create two pictures.*

can always be enlarged in printing.

Arrangement. The subject receives importance from its position in the picture, that is, the place it occupies in relation to the other units making up the picture and to the actual frame of the picture. Arrangement in the picture space, like furnishing a room, is to some extent a matter of personal taste. But just as it would clearly be wrong to put a bookcase with its back to the light, or an easy-chair under a draughty window, there are rules of common-sense that can be applied in arranging the picture material and trimming the print.

When arranging the picture, the first common sense rule is to simplify the problem by cutting out all unnecessary bric-à-brac, however strong its sentimental value. Ruthless simplification makes the job of arrangement easier, and the meaning clearer and more direct.

Generally the only means of excluding unwanted material is the choice of viewpoint—a close viewpoint is often chosen because it cuts out a lot of background and surrounding objects; a low viewpoint, because it gets rid of fussy foreground and adds importance to the subject.

Having reduced the problem to its essentials, the next concern is to see that what remains helps the emphasis. All strong lines and shapes should as far as possible lead the eye towards the important part of the picture, certainly not away from it. A useful trick to give this sort of emphasis is to take advantage of the pattern into which the supporting details fall, or can be arranged, and to have the principal interest at the "strong" point of the pattern—at the top of a triangle, the centre of a circle, the inside of a spiral and so on. Such devices, however, are easier for the artist's brush than for the camera lens which must take things as it finds them.

Some thought should also be given to the balance of shapes in the picture. The principal shape should, ideally, be so positioned that it does not make the picture too heavily unsymmetrical. This is normally ensured by balancing the effect of the principal shape by a smaller, but stronger, tone in the other half of the picture.

Placing the Subject. The rules about placing the principal interest in the picture are straightforward and are as much rules of common sense as of art.

(1) Keep the subject away from the edges of the picture. If it is too close to any of the four edges, it will look uncomfortable, and there will be difficulty in filling the rest of the space without drawing the interest away.

(2) Keep the subject away from the geometrical centre of the picture. Central placing

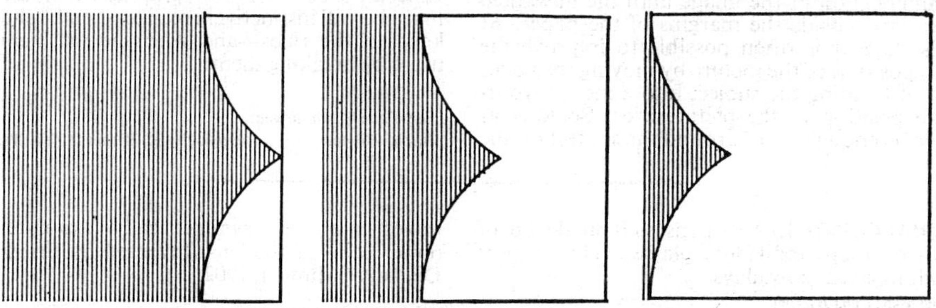

EFFECTS OF DIRECTION. **Left:** *Outward directions near the picture edge lead the eye away from the subject.* **Centre:** *Ideally there should be plenty of space in front of strong directional lines or movement to keep the interest in the picture.* **Right:** *Inward lines too near the edge lose emphasis and suggest that the important subject is outside the picture.*

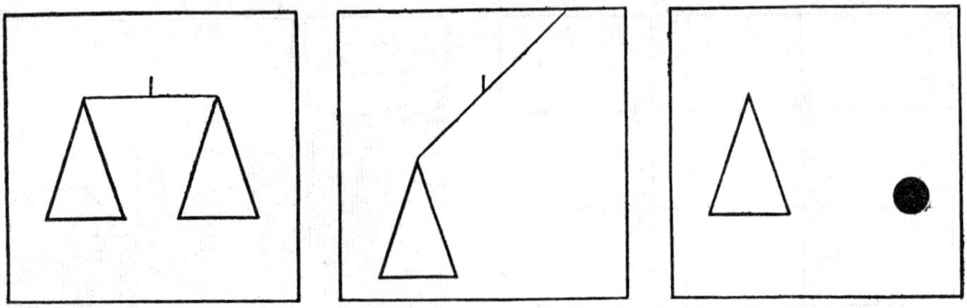

EFFECTS OF BALANCE. *Careful relative positioning of important elements may help in balancing the composition.* Left: *This symmetrical arrangement achieves balance but lacks interest.* Centre: *Concentration into a single area creates unbalance.* Right: *Large elements may be balanced by smaller ones of more dominant tone.*

is useful for strong emphasis, but it is apt to be dull and formal. This is particularly so with the horizon, which should never be allowed to cut the picture space into two equal areas. Keep it well up or well down, leaving no doubt about where the interest lies—in the sky or in the landscape. If the horizon has no business in the picture, then cut it out completely and the picture will be rid of a strong distracting force.

(3) Allow more space in front of the subject than behind it. In a portrait, there should be more space in the direction of the sitter's gaze; a moving object should be shown moving into the picture and not out of it.

As far as possible these methods of controlling emphasis are applied when taking the photograph. It is always easier to cut out an unwanted bit of the scene by a change of viewpoint or camera angle than it is to get rid of it in the print. At the same time, the enlarging process does give the photographer a valuable second chance to improve the composition.

On the Enlarger. Ideally all the composition of the picture should be arranged at the time of exposure. In practice, however, this is not always possible.

Any necessary adjustment of shape and content to improve the composition can be made at the enlarging stage by increasing the magnification of the image until the unwanted part lies outside the margins of the paper. At this stage it is often possible to improve the composition of the picture by moving the paper about to bring the subject into a more favourable position in the picture area. So long as there is enough "air" around the subject on the negative it may even be possible to choose between a horizontal and vertical picture format as an aid to composition.

Sometimes the negative may contain enough suitable subject material to form the basis of two or even more separate pictures that can be composed in this way on the enlarger. In general, however, the main constituents of the picture should be selected and arranged at the time of taking; there is no merit in shooting indiscriminately and hoping to compose a picture at the enlarging stage.

Trimming. A final opportunity to improve the composition arises at the print trimming stage. As little as possible should be done here, however, because trimming means wasted paper and a smaller final print.

Retouching. The various methods of physical and chemical after-treatment of both negative and print afford another means of improving the composition of the subject. Unwanted objects may be removed by bleaching out or subdued by reducing. Tones may be modified to adjust the emphasis of tonal arrangements, and additional images—e.g., clouds—may be added by multiple printing.

Photography, like any other medium of self-expression, is a very personal thing. General rules on composition are usually valid, and a positive aid to better work. But the more advanced the artist or photographer becomes, the more he will instinctively realize when he should keep to the rules—and when he will achieve more by breaking them. F.P.

See also: *Visual appeal.*

Book: *Composition in Pictures*, by R. Bethers (New York).

COMPOUND LENS. Lens unit made up of two or more individual glasses. The term is seldom used nowadays.

See also: *Lens History.*

COMPOUND SHUTTER. Central-opening type of shutter formed by a number of identical leaves mounted symmetrically around the optical axis of the lens. Invented by Friedrich Deckel of Munich, 1902.

COMPUR SHUTTER. Trade name of highly developed type of diaphragm shutter in which the speed of the operating mechanism is controlled by a clockwork escapement and, on the slow speeds, a gear train. The Compur shutter was devised by Friedrich Deckel of Munich in 1912, and his firm is still the sole manufacturer of Compur shutters. It was a development of the same inventor's Compound shutter introduced ten years earlier.

CONCAVE. Optical term applied to lenses and mirrors indicating that the surface is curved so that it forms a hollow.

CONCENTRATED DEVELOPER. Liquid developer in a concentrated solution that is diluted with water for use. Such solutions are more convenient to use than in powder form.

See also: *Developers; Solutions.*

CONDENSATION. When certain articles are brought from a cold atmosphere into a warm room, moisture forms on their surfaces. This is particularly evident with glass and metal parts, and is due to the condensation of moisture present as water vapour in the atmosphere.

Condensation can be very troublesome on photographic equipment, particularly lenses; even slight misting on a lens causes considerable diffusion of the image. Therefore a cold camera brought into a warm room should not be used until it has warmed up to the temperature of the room and the moisture on the lens has evaporated again. Condensation can often be removed by taking the camera out again into the cold for a short period, and is not likely to reappear when the camera is brought back into the warm room.

The same trouble occurs with enlargers kept under similar conditions.

See also: *Cold weather.*

CONDENSER. Optical system which concentrates the light from a source into a more or less defined beam, enabling the light source to be focused to any one position; e.g. into the lens on an enlarger.

A condenser consists of either a single plano-convex lens or a pair, with their plane sides facing out. It forms a very efficient light collector and can be adjusted to give suitably even lighting.

Use in Enlargers. A condenser or a condenser-diffuser type enlarger makes use of one, or more often, a pair of plano-convex lenses. Where only one such lens is used, the flat side is turned to the negative, and in fact is often used as a pressure plate to press down on the celluloid side of a film negative to hold it flat. When a pair of lenses are used, they are mounted with their convex sides facing and not quite touching. The mounting for double condensers must allow room for expansion due to heat.

Double condensers give the most even illumination and make for more compact lamphouse design. Single condensers are used only because they are cheaper, and then only for the smaller negative sizes. The double lens system is always used for the larger negative sizes because, apart from any other considerations, single lenses would be inconveniently thick at the centre and would suffer from excessive spherical aberration.

The condenser must be large enough to extend well out beyond the corners of the negative. If it does not, the corners of the print will be darkened or even completely cut off. The diameter of the condensers should be slightly greater than the length of the diagonal of the negative.

Focal Length. The focal length of the condenser system is also important, particularly in a true condenser enlarger. Even in the condenser-diffuser enlarger the correct focal length is a great aid to even lighting.

The focal lengths of the condensers and the enlarging lens are directly related when used together. If a single condenser is used, it should be capable of forming an image of the lamp filament in the enlarging lens without an inconveniently great separation between lamp and condenser. If the image is formed at any other point, the enlarging lens will be able to project another image of this condenser-image on the print. The unwanted image may not be sharply focused, but it will be sufficiently real to make lighting uneven.

When the condenser-image of the filament falls within the enlarging lens, the enlarging lens can only see the condenser as an evenly illuminated disc of light, and this is exactly what is required for even illumination of the negative. To achieve this conveniently, a single condenser should have a focal length about half to the same as that of the enlarging lens.

The same conditions must be fulfilled by a double plano-convex combination. In this case the individual components should have a focal length of approximately the same to twice that of the enlarging lens, so that the focal length of the combination will be roughly half to equal to that of the enlarging lens.

Focusing. It has been said above that the image of the light source should be focused on the enlarging lens to give the most even illumination of the negative. But the condenser-to-lens distance varies according to the degree of enlargement: the lens is moved towards the condenser to increase the scale of enlargement and away from it to reduce the scale. So in a plain condenser enlarger the condenser must be re-focused every time the magnification is changed. This is commonly done by moving the lamp towards or away from the condenser. In some horizontal enlargers it is also possible to move the condenser.

When the lighting is a combination of condenser and diffuser, refocusing is not necessary for small changes, but it is always advisable whenever large changes of magnification are made. For this reason the lamphouse should always provide for lamp adjustment.

Spotlights and Projectors. Condensers are also employed in spotlights and projectors. Spotlight condensers sometimes take the form of Fresnel lenses which have the same effect of concentrating the light in a beam. They are however much thinner than a full-sized condenser lens and are thus less liable to crack as a result of the heat of the lamp.

In projectors special cooling arrangements (e.g., a fan) are provided. G.W.A.

See also: *Enlarger; Lighting equipment; Projector (Cine); Projector (still).*

CONFERENCES

In these days of easy travel, conferences are on the increase, both on a national and international level. This also applies in the photographic field. One of the reasons is the continuing specialization in all fields of science and technology. Whereas at one time the Society Lecture was a well-attended feature, its importance has dwindled relative to that of the conference devoted to one or a few relatively narrow fields.

International Conferences. Photographic conferences are held on both national and international levels, but the differences between these tend to disappear. Starting in 1889, Photographic Congress was the organizing body of a series of nine international conferences, which set out to bring some law and order into nomenclature, copyright and manufacturing standards and gradually changed to scientific communication and discussion. The last of these was in 1935.

Since the last war, a string of informal conferences has been run in the sense that no central organizing body existed to plan and coordinate the meetings. At first devoted to all that was new in photographic science and applications, succeeding conferences became more specialized, and groups of specialists set out to organize their own meetings, such as those on nuclear-particle recording, high-speed photo-

PRE-1939 INTERNATIONAL CONGRESSES OF PHOTOGRAPHY

Congress No.	*Date*	*Place*	*Members No.*	*Communications No.*	*Highlights*
1	1889 August 6–17	Paris	249	No formal lectures	Standardization exclusive aim. Nomenclature. Copyright. Leisurely. Questions circulated before the meeting.
2	1891 August 23–29	Brussels	87	ditto	Nine sessions only. International Union of all Photographic Societies formed. Stricter regulations for lectures.
3	1900 July 23–28	Paris	492	2 lectures 19 notes*	Subjects not on standardization creep in. Lippmann interference colour photography. Levy acid blast etching for blocks. A very hot summer!
4	1905 July 16–25	Liège	150	26 notes*	First report with half-tone illustrations, art paper. Permanent commission set up. Sessions on Research and Theory and on Applications .
5	1910 August 1–5	Brussels	331	65	Large part played by L. P. Clerc, C. E. K. Mees. Interest in documentation, bibliography. No definite questions set. Reports from commissions.
6	1925 June 29–July 4	Paris	232	40 5 reports†	Changed character: communications play major part, cine. Standardization of sensitometric standard lamps still not solved.
7	1928 July 9–14	London	221	71	Preprints. Clark on U.D.C. 770. Jones speed criterion. Permanent Secretary to the Standardization Committees appointed. Standard unit for photographic intensity proposed.
8	1931 August 3–8	Dresden	411 (of whom 70 dependants)	81 + 7 reports	Name change: *International Congress of Scientific and Applied Photography.* Standard unit for photographic insensity accepted, ditto standard developer. DIN standard of speed to be tested. Safety film for all sub-standard cine film.
9	1935 July 7–13	Paris	185	89	Clerc complains of lack of discipline by authors. Kodachrome 16-mm. three-colour motion picture film. Sound just emerging. Reactions to DIN speed system. Jones and Russell on speed criteria; Reinders on theory of development.

* Notes refer to specific set questions. † Reports are by national standardizing committees.

POST-1945 INTERNATIONAL PHOTOGRAPHIC CONFERENCES

Place	*Time*	*Title*	*Partici-pants*	*Lectures*	*Publication*	*Remarks*
Liège	1949 April 15–17	Informal meetings on fundamental aspects of the photographic process and on nuclear particle recording.	48	?	*Science et Industrie Photographique*, *19*, pp. 281, 321, 361, 441 (1948). (Reports by Hautot and Cuër.)	First post-war meeting.
Zürich	1949 April 6–9	ditto	90	25	*Journees Internationales de Science Photographique*. Publ. Revue d'Optique, Paris, 1950.	Arrangement of material also published in *Science et Industrie Photographique*, *20* and *21* (1949 and 1950).
Bristol	1950 March 28–31	Conference on "Fundamental Mechanisms of Photographic Sensitivity".	officially 88 actually 100 +	48	*Fundamental Mechanisms of Photographic Sensitivity*. Ed. J. W. Mitchell. Publ. Butterworth, London, 1951.	
Paris	1951 Sept. 24–29	Colloque sur la "Sensibilité des Cristaux et des Emulsions Photographiques".	136	68	Same title. Ed. Morand and Vassy. Publ. Revue d'Optique, Paris, 1953.	
London	1953 September	R.P.S. Centenary Conference.	600–700	150	*Science and Applications of Photography*. Ed. R. S. Schultze. Publ. Royal Photographic Society, London, 1955.	The complete field o photography was covered.
Cologne	1956 Sept. 24–27	Internationale Konferenz für wissenschaftliche Photographie.	377	121	*Wissenschaftliche Photographie*. Ed. O. Helwich. Publ. Helwich, Darmstadt, 1958.	The complete field o photography was covered. Russian delegation takes part for the first time. Publication tri-lingual.
Liège	1959 Sept. 14–19	Colloque International de Photographie Scientifique.	142	23	*Scientific Photography*. Ed. H. Sauvenier. Publ. Pergamon Press, London, 1962.	Few original papers; well-organized discussion sessions on prearranged problems.
Zürich	1961 September	Kolloquium über Wissenschaftliche Photographie.	190	No formal papers	*Photographic Science*. Ed.W. F. Berg. Publ. Focal Press, London, 1963.	Meeting entirely devoted to discussions, following Presidents' introduction of original work submitted.
Turin	1963 Sept. 23–28	Colloqiuio Internzaionale Sulla Scienza Della Fotografia.	200	66		

POST-1945 INTERNATIONAL PHOTOGRAPHIC CONFERENCES—*continued*

Place	*Time*	*Title*	*Partici-pants*	*Lectures*	*Publication*	*Remarks*
New York	1964 Apr. 27– May 1	International Conference on Photographic Science and Engineering.		81 plus 3	Organized by S.P.S.E. Many Subjects, parallel sessions and 3 panel discussions.	
Budapest	1957 September	Conference on Scientific and Applied Photography.			No report.	Attended mostly by workers from the "Socialist Republics" with a sprinkling from
Budapest	1959 Sept. 2–5	Second Conference on Scientific and Applied Photography.	70	31	No report.	the West. Papers from the whole field.
Budapest	1961 Aug. 28– Sept. 1	Third Conference on Scientific and Applied Photography.	138	52	No report.	
Budapest	1963 September	Fourth Conference on Scientific and Applied Photography.			No report planned.	
Strasbourg	1957 July 1–6	Premier Colloque International de Photographie Corpusculaire.	82	47	Same title. Ed. P. Cuër. Publ. Centre National de la Recherche Scientifique, Paris, 1958.	International Committee constituted.
Montreal	1959 September	Deuxieme Colloque, etc.	48	87	*Photographie Corpusculaire II*. Ed. P. Demers. Publ. Presses Universitaires de Montréal, 1959.	
Moscow	1960	Third International Colloquium on Nuclear-particle Recording.				
München	1962 Sept. 3–8	Fourth International Colloquium on Nuclear-particle Recording.	89	82	*Korpuskularphotographie IV*. Ed. H. Frieser and G. Heimann. Publ. Inst. f. wiss. Photographie, T. H., München, 1963.	Fast publication by by photo-lithography.
Geneva (CERN)	1964 Sept. 15–18	Fifth International Conference on Nuclear Photography.				
Washington, D.C.	1952 October	International Congress on High-speed Photography.			*High-speed Photography*, Vol. V. Special issue of the JSMPTE. Publ. SMPTE, New York, 1952.	Originated from a Symposium by the SMPTE at Washington, October 1948. A permanent international committee set up. Part of their 72nd semi-annual convention.

POST-1945 INTERNATIONAL PHOTOGRAPHIC CONFERENCES—*continued*

Place	*Time*	*Title*	*Partici-pants*	*Lecturers*	*Publication*	*Remarks*
Paris	1954 Sept. 22–28	Second International Congress on High-speed Photography.		68	*Photographie et Cinematographie Ultra-Rapide*. Ed. P. Naslin and J. Vivie. Publ. Dunod, Paris, 1956.	
London	1956 Sept. 10–15	Third International Congress on High-speed Photography.	330	61	*High-speed Photography*. Ed. R. B. Collins. Publ. Butterworth, London, 1957.	
Cologne	1958 Sept. 22–23	Fourth International Congress on High-speed Photography.	350	56	*Kurzzeitphotographie*. Ed. H. Schardin and O. Helwich. Publ. Helwich, Darmstadt, 1959.	
Washington, D.C.	1960 Oct. 16–22	Fifth International Congress on High-speed Photography.		117	*High-Speed Photography*. Ed. J. S. Courtney-Pratt. Publ. SMPTE, New York, 1962.	
Scheveningen	1962 Sept. 17–22	Sixth International Congress on High-speed Photography.	400	96	*High Speed Photography* Ed. J. G. A. de Graaf and P. Tegelaar. Publ. Tjeenk Willink and Zoon, Haarlem, 1963.	
Hakone*	1953 September	Hakone Symposium on "Photographic Sensitivity".	40	19	*Photographic Sensitivity*, Vol. I. Ed. S. Fujisawa. Publ. Maruzen, Tokyo, 1956.	Attended by Prof. Seitz (U.S.A.), Prof Mott and Dr. Frank (U.K.).
Tokyo*	1957 October	Tokyo Symposium on "Photographic Sensitivity".	117	17	*Photographic Sensitivity*, Vol. II. Ed. S. Fujisawa. Publ. Maruzen, Tokyo, 1958.	Attended by Prof. Chibisov and Dr. Bogomolov (U.S.S.R.) Publication contains a report on a Second Symposium held at Tokyo, May 1954, and a Third Symposium at Tokyo in June 1955.
Düsseldorf	1960 Sept. 27–30	First International Congress on Medical Photography and Cinematography.		142	*I. Internationaler Kongress fur medizinische Photographie und Kinematographie*. Ed. H. Orbach. Publ. Thieme Stuttgart, 1962.	
Cologne	1963 Oct. 14–19	I. Internationaler Kongress für Reprographie.				

* These symposia are perhaps borderline in not being truly internationally representative.
The tables omit conferences on specialized fields such as Astronomical Photography and Radiography.

graphy and medical photography; here permanent secretariats were set up at least for the first two mentioned, thus providing a desirable degree of continuity.

The conferences on photographic science proper have undergone an interesting evolution. They first grew from an intimate colloquium to a vast assembly with parallel sessions, with all chances of fruitful discussion being drowned by the many papers presented. The tendency now is to reduce their size by limiting the number of subjects put up for discussion and to try out different methods of presentation.

An international gathering of photographic scientists at Brestenberg, November 1962, considered the following basic pattern of international conferences as desirable.

Congresses, devoted to all aspects of photography, should be held very infrequently. They might mark a celebration, and their purpose would be essentially social.

Symposia, devoted to a few chosen subjects, to be held every three years. Their purpose would be a mixture of social and scientific discussion.

Discussion meetings on very narrow subjects should be held whenever sufficient progress had been made. Attendance should be strictly limited, the purpose would be scientific discussion and, perhaps, the issue of a report.

A provisional International Committee on the Science of Photography (I.C.S.P.), with headquarters at Zürich, was set up charged with the task of pursuing these aims.

National Conferences. As national photographic science organizations grew in strength in the various countries, they began to run more ambitious conferences. The S.P.S.E. in the U.S.A., the recently created Science Committee of the R.P.S. (sprung from the Scientific and Technical Group), the Deutsche Photographische Gesellschaft, Sektion Forschung, and the Society for Scientific Photography of Japan have all been responsible for successful conferences. As they become known, they tend to attract international attendance, and the distinctions between national and international meetings are disappearing. There is certainly a need for a limited degree of co-ordination of all these endeavours. W.F.B.

CONJUGATE FOCI. (Conjugate Points). Name in optics for the associated positions of object and image in relation to the lens. For every position of the image there is a corresponding position of the object and these positions are interchangeable.

If the distance of each point is measured from the corresponding front (Df) or rear (Dr) principal focus of the lens (focal length, f) then:

$$Df \times Dr = f^2$$

See also: *Lens; Optical calculations.*

CONSTANT DENSITY RATIO LAW. According to Hurter and Driffield, relative densities measured over the different parts of a negative are decided by the exposure and not development. Any changes in the time of normal development can affect contrast only.

CONSTANT GAMMA EMULSION. Term indicating an emulsion which develops to the same contrast factor over the part of the spectrum being used, i.e., 2500–3750 A for ultraviolet spectrography, or 4000–7000 A for making separation negatives.

See also: *Gamma; Spectrography.*

CONTACT PAPER. Relatively slow development paper (usually a chloride paper) for making positives by contact printing.

See also: *Papers.*

CONTACT PRINTER

Equipment for contact printing is of three main types: frames of varying size for making single prints; boxes for quick repetition work and machines for bulk photographic printing, as in commercial photo-finishing or post-card production.

Printing Frames. The familiar printing frame is the simplest type of equipment for making contact prints from negatives. The sort used by amateur photographers is generally made from a hard, non-warping wood, like beech, although plastic materials have been used for small negative sizes.

The wooden printing frame carries a sheet of plain glass in a rebate. The back of the frame is detachable and is pressed against the glass by hinged spring arms which can be released to allow the negative and paper to be inserted.

In use, the frame is laid face downwards and the back removed. The negative is inserted, image side upwards, and a piece of printing paper laid on it, emulsion side down. If a white border is required around the print, a suitable mask is interposed between the negative and printing paper. When printing from glass negatives, the plate takes the place of the printing glass.

Exposures with printing frames are generally made under an electric lamp. A reading lamp is excellent for making occasional prints. The frame is laid face upwards on the table and the

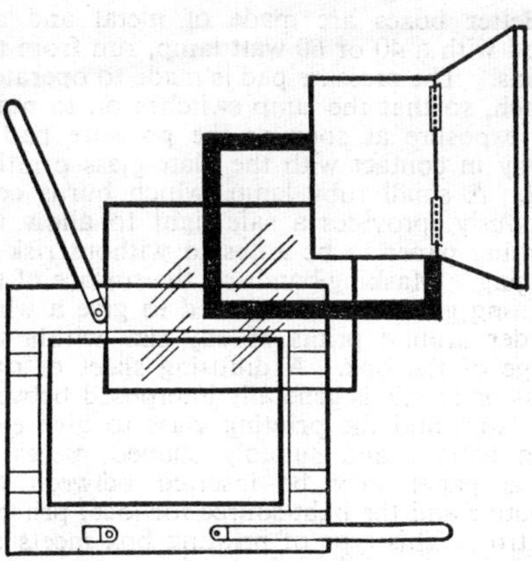

PRINTING FRAME. The parts, in order of assembly, are the frame itself, the glass (for film negatives), the border mask, and the back. In use the negative and paper are inserted between the mask and the back, and pressed into close contact by the springs which hold down the back.

lamp adjusted to throw its light down on to the front of the frame. In this way the lamp distance is kept constant and all negatives are exposed under the same conditions. Even illumination is essential. An opal or pearl lamp is satisfactory provided that it is kept at a distance of more than twice the diagonal of the negative away from the frame.

Large Frames. Printing frames for large negatives, 8 × 10 ins. or 12 × 15 ins., have to be much stronger than the amateur models. Because the negative is larger, the pressure applied must be greater to give good contact with the printing paper. Instead of wood, metal castings are often used to give the necessary strength, and plate-glass is used in the frame to stand up to the increased pressure which is applied by levers or screw-clamps.

Frames of this kind are used not so much for making continuous-tone prints from ordinary negatives (few photographers use negative sizes that large) as for a variety of graphic arts purposes. The frame there usually forms part of a printing box or other set-up with its own light source.

The limiting factor in this type of construction is the strength of the glass used in the front of the frame, and large sizes are apt to be heavy as well as expensive. For this reason it is usual to employ the vacuum principle to obtain the contact pressure in printing frames of the largest sizes, such as those used in printing from negatives of high definition or for printing-down in photomechanical work. In this, the frame carries a strong rubberized canvas pad or blanket with a heavy rubber beading. After loading the frame in the normal way, the blanket is pressed into contact with the printing glass around the edges of the negative, and the space between is evacuated by an air pump. The pump may be either hand-operated or electrically driven. The pressure of the air acts on the back of the blanket and the front of the glass, forcing the negative into close contact with the paper. Since the air pressures on the front and back of the frame are exactly counterbalanced, the printing glass and its frame can be of much lighter construction than where pressure is applied mechanically.

The contact obtained with a vacuum frame is much better than that given by a mechanical pressure frame since the air pressure is distributed evenly over the whole area. The frame can be opened instantly after admitting air to break the vacuum.

Miniature Frames. Miniature negatives on 35 mm. cine-film require special types of printing frame. One type is used for making contact prints on to strips of perforated 35 mm. paper or positive film, and the other for miniature lantern slides where the negatives are contact-printed on to 2 × 2 ins. plates.

For making diapositive films or film strips on 35 mm. film, the printing frame has a film magazine and a sprocket drive for feeding the positive film into position behind the negative in the gate of the printing box, one frame at a time. Negative and printing material (film or paper) are pressed into close contact by a pair of spring-loaded optically flat glass plates and exposure is made through the pressure glass by placing the printing frame under a lamp.

The magazine spools hold up to 10 or 12 feet of paper or film and the exposed material is cut off, removed, and developed in lengths of 5 or 6 feet in a drum-type processing outfit. Since exposures cannot be developed individually, this method of printing calls for considerable skill and judgement to produce an evenly-exposed strip of prints.

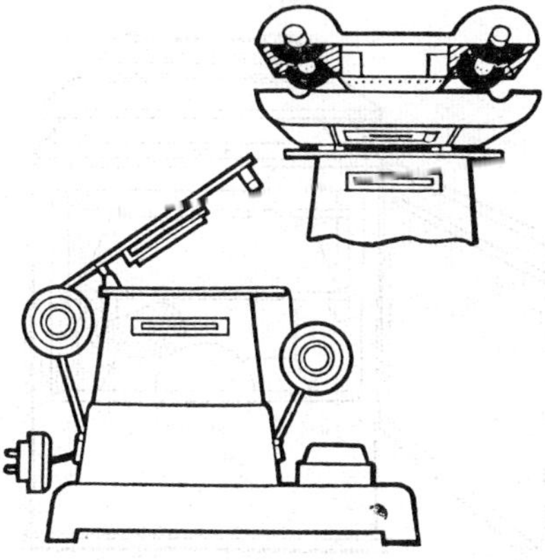

FILM STRIP PRINTERS. Left: *Printing 35 mm. negatives on to miniature lantern plates. The negative film is spooled up and held above the illuminated gate of a light box, and the plate aligned in contact above it.* Right: *Printing on to positive film uses a similar arrangement, but needs a magazine, holding the positive film, on top of the negative.*

The frame for printing lantern slides is much simpler. It is an ordinary frame for 2 × 2 ins. plates with a pressure pad and guides which keep the 35 mm. film parallel to the edges of the lantern plate.

Another special frame s used for printing stereo-negatives made in a stereoscopic camera. Such frames are sometimes known as transposing frames because the images from each pair of negatives must be interchanged before they can be viewed correctly in a stereoscope.

Luminiscent Printers. Some printing frames incorporate their own light source in the form of a luminescent panel. This yields a soft and even glow when the panel is connected to an electric supply. While the light intensity is low, the panel is in virtually direct contact with the negative so that all light is nearly 100 per cent utilized. Exposure times with such a panel accordingly are comparable with the printing exposures needed in a normal printing frame placed under a reading lamp.

Printing Boxes. The simplest printing boxes consist of little more than a printing frame with a self-contained source of illumination. Even so, they meet most of the needs of the average amateur photographer.

The cheapest type consists of a wooden box, painted white inside, with a light source provided by four flash-lamp bulbs, connected in parallel with a battery. An opening in the top of the box holds a glass plate backed up by a hinged, felt-covered pressure pad which presses the printing paper and negative tightly against the glass. A spring catch holds the pressure pad down, and when in this position, the bulbs can be switched on to give the necessary exposure to the paper. The box will take negatives up to quarter-plate size and black paper or red celluloid masks can be inserted to leave a white border on the print.

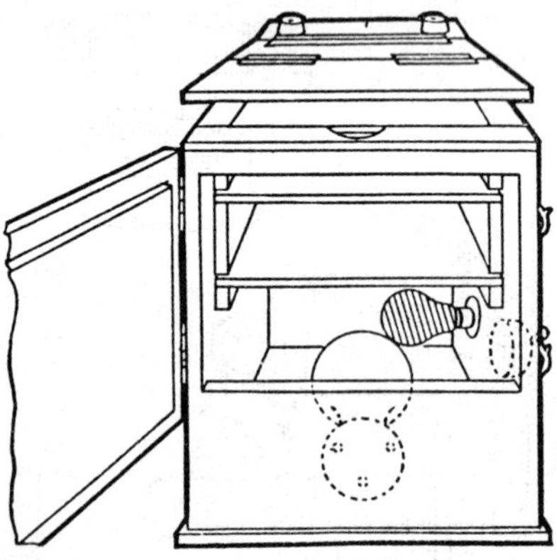

PRINTING BOX. Most designs feature a white exposing lamp fitted below a glass top with one or more diffusing sheets in between. The negative and paper are placed on the glass and pressed together by the hinged platen. An orange lamp (shaded) permits exact positioning before exposure.

Better boxes are made of metal and are fitted with a 40 or 60 watt lamp, run from the mains. The pressure pad is made to operate a switch, so that the lamp switches on to make the exposure as soon as the pressure pad is firmly in contact with the plate-glass printing bed. A small ruby lamp, which burns continuously, provides a safe light to allow the printing paper to be adjusted without risk of fogging. Masking bands on the surface of the printing glass can be adjusted to give a white border around prints of any size within the range of the box. A diffusing sheet of opal glass or plastic is generally interposed between the lamp and the printing glass to give even illumination, and suitably shaped pieces of tissue paper may be inserted between the negative and the light-source for local printing control. This type of printing box meets the requirements of the advanced amateur.

The professional printer must be able to turn out work of the very highest quality rapidly and easily and handle large negatives on both film and glass plates. For such work the illumination must be absolutely even, and strong enough for short exposures. The contact between negative and printing paper must be good enough to allow negatives of the highest definition to print without any loss of sharpness.

This type of printer is built to take negatives of any size up to 12 × 15 ins. Thin adjustable steel masking bands provide white margins on all prints up to this size. The pressure platen is made of hinged sections, faced with felt or soft rubber to distribute the pressure evenly over the area of the negative and paper. Specially selected flat plate glass is used to ensure intimate contact between negative and paper.

In general large-size printing boxes are nowadays used for special purposes only, such as printing X-ray films, making off-set litho masters from negatives or positives in photomechanical reproduction (especially at the small office offset level) and for document copying.

Illumination. There is a choice of diffused or direct illumination to suit the type of negative. Fully diffused illumination is used for negatives carrying heavy retouching or blocking-out. This type of illumination is usually provided by a bank of lamps behind diffusing sheets of ground glass or opal.

For high definition negatives or line subjects, direct illumination must be used to preserve the fine detail. Such lighting is given by a small light source placed some distance away from the negative. Even illumination is essential, so the printing light is fixed at a distance from the negative at least equal to its diagonal. The small light source may also be obtained by placing a pearl or opal lamp behind a 1 or 2 ins. diameter hole cut in an opaque baffle.

Printers sometimes have provision for varying the distance between the lamp and the printing glass to give further control of the exposure time. Other alternatives include interchangeable light sources of different intensity (from 15 watts to 1000 watts), or a variable step-by-step resistance in series with the printing light. By adjusting the strength of the light it is possible to print from a given negative on to any one of the four commonly used classes of printing paper (chloride, slow chlorobromide, fast chlorobromide, and bromide papers) and still give about the same exposure time.

Bulk Printers. Wholesale photo finishers use special printing machines for handling D. & P. and picture postcard printing. On such machines, everything possible is made automatic. The details of construction vary greatly according to the make and the output.

Reflex Printers. The reflex printer for copying documents is larger than most types of printer because it must be big enough to handle documents ranging from 8 × 10 ins. up to 30 × 40 ins. or larger.

A reflex printer is designed to give complete contact between the document and the printing paper, generally by the use of a heavy sponge-rubber pressure pad in conjunction with a rubber blanket. Large size printers use the vacuum system. A bank of lamps is generally necessary to give good, over-all illumination. In addition to the usual diffusing glass, there is a slot to hold a yellow filter when copying documents on a faded or yellow base.

Reflex printers are intended to be used by people with no knowledge of photography or printing; they are therefore simple, foolproof, and almost automatic. G.B.M.

See also: *Contact printing; Document copying.*

CONTACT PRINTING

Contact printing is any method of making a photographic print in which the sensitized material is exposed in contact with the negative.

Mostly an artificial light source is used for this purpose; special printing papers are, however, suitable for daylight printing.

ARTIFICIAL LIGHT

Contact prints are normally made on so-called contact papers (formerly also known as gaslight papers) which are developed and fixed after the exposure to yield the visible image. Contact prints are usually made in a printing frame or printing box.

The prints are the same size as the negatives. They are usually black, but there are special warm tone papers which can be made to give warm black to red-brown prints and toners that can tone contact prints to any desired colour.

Papers are available in a range of paper grades, and can deal with negatives of almost any contrast.

It is always a wise plan to collect a batch of negatives for printing and then to do the job with ample quantities of developer and fixer under reasonably convenient working conditions.

Sorting the Negatives. Negatives should first be grouped according to their suitability for printing.

Unsharp negatives, also films or plates showing deep scratches, dark or light blobs, streaks, and areas of uneven density, are useless for print-making without some preliminary preparation. Very much under-exposed negatives with practically no shadow detail should also be thrown out unless they have been deliberately under-exposed—as, for instance, in night photographs.

The remaining negatives are then grouped according to their contrast and hence the grade of paper they will require to give acceptable prints.

MATCHING NEGATIVES AND PAPERS

Type of Negative	*Correct Paper Grade*
Very contrasty, consisting almost entirely of areas of black and almost transparent film or glass	No. 0 or Extra Soft
Contrasty, with strong blacks and fairly thin shadows	No. 1 or Soft
Average, with moderate contrast and tone gradation	No. 2 or Normal
Soft, with light grey shadows and medium grey highlights	No. 3 or Hard
Very soft (and often very thin), with little tone difference between the various parts of the image	No. 4 or 5 ; Extra Hard or Ultra Hard

The over-all density of the image does not affect the grade of paper that should be used; the paper grade is decided solely by the contrast of the negative. It is thus possible to have very dense and very thin negatives in the same contrast group. For this reason it may be worth while to classify the negatives in each group according to their density. This avoids big jumps in exposure times from one negative to the next within the same group.

The density may be judged by laying the negative on the page of a book printed clearly on white paper. Using this test, a negative in each group should then be chosen as a standard. It should show good tone and detail even in the thinnest areas, and the text underneath should be just readable, even through the densest areas. The negatives in the group can then be divided into four or five degrees of density with reference to the standard negative. It is much safer to estimate densities in this way than by looking through the negative at a bright light.

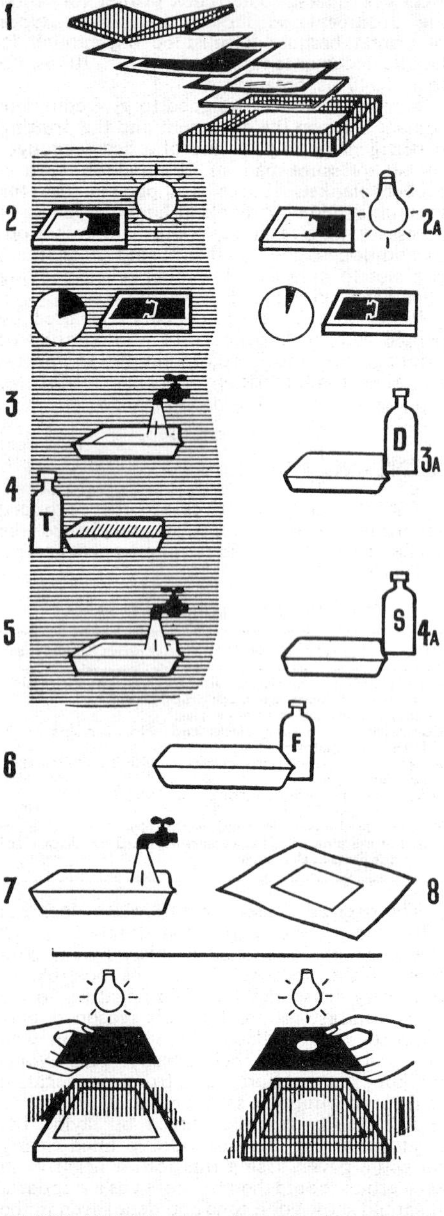

CONTACT PRINTING. Top and left: Daylight printing on P.O.P. 1. Assembling negative and paper in the printing frame. 2. Exposure to daylight. 3. Washing. 4. Toning. 5. Washing. 6. Fixing. 7. Final rinse. 8. Drying. Top and right: Printing by artificial light on chloride paper. 2A. Exposure. 3A. Development. 4A. Stop bath. Stages 6 to 8 following here as for daylight printing (the fixer used is not, however, the same). Bottom: Exposure control by shading (to hold back part of the picture) and spot printing (to get more detail into areas corresponding to exceptional negative densities).

Loading. During loading and processing the printing paper must be handled under suitable safelight illumination—generally yellow or light brown.

The printing frame is opened, the back taken out, and the negative inserted, emulsion side up. It is then covered by a sheet of printing paper, emulsion side down. The back of the frame is then replaced and clamped down, pressing the negative and paper into intimate contact.

If the print is to have a white border, a cut-out opaque paper or red celluloid printing mask is inserted in the printing frame in front of the negative. When printing from plate negatives, the glass of the printing frame is not needed.

To load a printing box the pressure plate is raised and the negative is placed, emulsion side up, on the glass window and covered by the paper. The pressure plate is then clamped down, pressing the paper and negative into close contact.

The emulsion side of the paper must be in contact with the emulsion side of the film or plate. If the negative is inserted the wrong way round, the layer of film or glass between the actual negative surface and the paper leads to a blurred print.

If the paper is inserted the wrong way round, the print will be even more blurred and it will also be very much under-exposed, since the light will have passed through the paper base. In both cases the right and left hand sides of the picture will be transposed.

Exposure. The exposure time required for a contact print depends on the paper speed, negative density, and the power and distance from the negative of the printing light. The actual exposure time is best determined by trial and error—initially with text exposures (see below).

With average contact papers suitable lamp distances are 7–8 ins. (20 cm.) with a 40-watt lamp, 12 ins. (30 cm.) with a 60-watt lamp, or 15–18 ins. (40–50 cm.) with a 100-watt lamp. Under these conditions, average negatives will call for exposure times around 5–10 seconds.

Negatives larger than $2\frac{1}{2} \times 3\frac{1}{2}$ ins. (6×9 cm.) should not be exposed nearer than 12 ins. from the printing light, and 5×7 ins. (13×18 cm.) or larger sizes should be at least 18 ins. away to avoid uneven illumination.

In a printing box the distance is in any case fixed by the design of the box.

Chlorobromide or bromide papers are very much faster than ordinary contact papers, so the printing distances have to be greater in proportion; they may be several feet.

For timing the exposures there is probably nothing better than a loud-ticking alarm clock. It is usually easier to count ticks by ear than to watch the seconds hand of a clock—particularly in the relatively dim light of the darkroom.

Alternatively, the exposure can be regulated by one of the many types of automatic timer. These switch the light off at the end of any desired interval.

Processing. After exposure the paper is removed from the frame or printer and it is then ready for development, which is basically the same as with enlarging papers. This can be carried out by subdued artificial light (no direct light must reach the print) or by a suitable safelight. The steps involved are:—

(1) Immerse the print in a suitable developer, and develop at about 65° F. for ¾–1 minute moving the print in the solution all the time.

(2) Lift the print out of the developer, drain, and immerse in an acid stop bath for 10–15 seconds.

(3) Lift out, drain, and immerse the print in the fixing bath. Keep the print moving for the first half a minute or so.

(4) After 1 minute white light can be turned on for inspection.

(5) If two fixing baths are used, transfer the print to the second fixer after about five minutes. Move the prints about from time to time in both fixers and take special care that they do not stick together.

(6) After a further 5 minutes, transfer the print to a bowl of water for washing.

(7) Wash in running water for 1 hour or in about twelve changes of water, leaving the prints for 5 minutes in each change.

(8) If a hypo eliminator is used, immerse the prints in the eliminator solution for 10 minutes, moving them about all the time.

(9) Follow this by a further wash of 10 minutes.

(10) Drain the prints and lay them face down on a clean fluffless cloth, or squeegee them on a glazing plate for glazing (glossy papers only).

(11) Leave until dry.

Certain contact papers are designed for automatic processing in hand-driven or motorized roller processors. Such papers incorporate the developing agent in the emulsion. They are passed after exposure through a processing unit which applies an activator solution to the print surface (to develop the image) followed by a stabilizing bath, instead of a fixer. The completely processed and only slightly damp print leaves the processor within about 10 to 15 seconds, requiring only drying. Such prints are reasonably permanent, though not as stable as normally processed prints. Nor must they be hot glazed. However prints processed in this way can be made fully permanent by fixing in a normal fixing bath for about five minutes, followed by washing as described above.

Test Exposures. Once the grade of paper, strength and distance of the printing light, and the method of processing have been decided, the only variable factor that remains is the actual time of exposure for each negative.

The best time is found by making small test prints. A number of these are made from an average negative, with exposures of, say, 2, 4, 8, 16 and 32 seconds. The test prints need not be larger than about 1½ × 2½ ins. (4 × 6 cm.).

If necessary the exact exposure can be pinpointed after processing by using an exposure time between the times employed for a slightly too dark and a slightly too light print. If all prints are too light or too dark, the strength or distance of the printing lamp should be adjusted.

The test prints will also show whether the paper grade is correct: if the best print is grey and dull, without proper blacks or really light tones, the paper is too soft. If the print is too black in the shadows, and at the same time white and lacking detail in all the lighter areas, the paper is too contrasty.

As the sensitivity of different paper grades is not the same, a new series of test exposures must be made to find the correct exposure time whenever the grade of paper is changed.

After some experience, both print contrast and exposure can be estimated with fair accuracy so that there is no need to make a full range of test exposures every time unless for specialized work.

Local Exposure Control. In contact printing, a limited amount of local control can be exercised in the manner of shading and spot printing enlargements.

The hand or a suitably-shaped card can be held between the printing frame and the light to control the exposure of larger areas of the picture—e.g., the foreground in landscapes—and a card with a hole in it can be used for overprinting smaller areas.

In more elaborate printing boxes, movable masks can be inserted between the lamps and the printing stage to serve a similar purpose. In some multi-lamp printers the lamps can be switched on or off individually to vary the strength of the printing light over selected areas of the negative.

DAYLIGHT

Daylight printing is the simplest way of making prints because the image can be seen growing darker as exposure progresses. By inspecting the print from time to time it is easy to see when it has had enough exposure. Once the print has been exposed, it has only to be toned and fixed.

Exposures are rather long (up to ten minutes), and the prints are sepia brown in colour. Negatives for daylight printing must be contrasty.

The procedure consists of loading the printing frame, exposing, and processing.

Since daylight paper is only affected by comparatively prolonged exposure to daylight, it can be handled safely in subdued daylight or ordinary artificial light.

The paper and negative must be free from

any traces of dampness before they are loaded into the printing frame, or the soluble silver salts in the paper emulsion may form brown or black stains on the negative.

Exposing. The loaded frame is exposed to bright daylight—e.g., light from the blue sky—or even the sun. Exposures may vary from as little as 2 minutes in direct sunlight to as long as 30 minutes in normal daylight. The light on the frame must be even and free from irregular patches of shadow. Prints made in direct sunlight are usually softer than those made by diffused daylight.

High power mercury vapour and arc lamps may also be used for exposing daylight prints.

From time to time, one half of the hinged back of the frame is opened and a corner of the paper is turned back to check the depth of the image. The inspection must take place in very weak daylight and the paper must not be allowed to move in relation to the negative or it will give a double image.

The print loses density in the fixing bath, so exposure must be continued until the image looks a little too dark.

Processing. Processing consists of: washing, toning, rinsing, fixing, washing. Toning and fixing can be combined.

Self-toning papers need no toning bath; they are simply washed, fixed, and washed again.

The purpose of the first wash is to remove all the soluble silver salts remaining in the P.O.P. emulsion. These would interfere with efficient toning and fixing.

The exposed prints are washed for about twenty minutes in running water, or in about five changes of water. Some types of P.O.P. or self-toning papers may curl violently; these are best washed in a very shallow dish in changes of water. In this way it is easier to keep them flat.

Most tap water turns milky when silver salts dissolve in it. When successive changes of such water stay quite clear, washing can be considered complete.

The simplest toning bath for P.O.P. is a mixture of equal parts of 1 per cent gold chloride and 10 per cent ammonium thiocyanate solutions, diluted fifty times. Every 100 c.cm. (4 ounces) of such a toning bath will tone 700 square cm. (120 square inches) of print area. The toning solutions should be made up at least two hours before use.

An alternative toning bath consists of one part 1 per cent gold chloride and four parts 10 per cent disodium phosphate, diluted twenty times.

Toning time and tones obtained vary as the bath becomes exhausted. If a batch of prints is to have a uniform tone, all the prints must be toned together.

When the prints arrive at the desired tone, they are washed for about ten minutes in running water, and then fixed.

P.O.P. papers are fixed in a 20 per cent sodium thiosulphate (hypo) solution. Acid fixers must not be used.

The prints should be kept moving in the fixing bath to ensure free access of the solution and complete fixation. Prints should not be left in the solution for longer than about ten minutes, or they will begin to lose density. (At the same time it is useful to know that prolonged fixing will lighten slightly over-exposed prints.)

With self-toning papers, toning takes place during fixing, and no separate toning bath is necessary. Certain self-toning papers do not even require pre-washing; they can be put in the fixing bath straight after exposure.

With most papers, 100 c.cm. of fixer will fix about 1,500 square cm. of print area—i.e., 4 ounces to 240 square inches.

P.O.P. that is normally toned and fixed in separate baths may have both operations carried out in one combined toning and fixing bath. The prints therefore only need washing before treatment.

Combined toning and fixing baths must not be over-worked. This is because they still appear to tone satisfactorily after the bath is in fact exhausted. The image in these circumstances is not permanent.

A suitable formula for a combined toning and fixing bath is:—

Sodium thiosulphate	12 ounces	300 grams
Sodium chloride	35 grains	2 grams
Lead nitrate or acetate	35 grains	2 grams
Gold chloride, 1 per cent solution	2 ounces	50 c.cm.
Water to make	40 ounces	1,000 c.cm.

The capacity of combined baths is about 1,800 square cm. of print area for every 0·1 gram of gold chloride contained in the volume of solution used—i.e., 190 square inches for every grain.

Printing-out papers should be washed for at least one hour in running water, or in twelve to fifteen changes of clean water. They are then blotted and dried in the normal way, preferably by laying them face down on a clean towel.

Glossy gelatin daylight papers can be cold-glazed. L.A.M.

See also: *Contact printer; Developing prints; Enlarging; Papers.*

CONTACT SCREEN. Half-tone screen consisting of a regular pattern of unsharp dots, and used in contact with a continuous—tone negative to produce a half-tone positive suitable for photomechanical purposes. (Can also be used for obtaining a half-tone negative from a continuous-tone positive.) An increasingly popular alternative to the ruled half-tone screen.

See also: *Half-tone screen; Photomechanical reproduction.*

CONTINUOUS-TONE IMAGE. Positive or negative print or transparency, which is composed of a range of densities from black through grey to white, and where the greys are formed by varying amount of silver, dye or pigment.

A continuous tone reproduction contrasts with a line reproduction which is composed of only two tones, black (or a colour) and white. The same applies to multi-colour line images; although there are several colours, each is present only in one depth. Nor is a photomechanical half-tone reproduction a continuous-tone image, since the illusion of tones is produced by varying areas (different sized dots) of printed ink on white paper, as in modern letterpress or photo-lithography. A half-tone reproduction simulates a continuous tone image since when viewed at the correct distance the tones appear to be the result of varying densities. On closer examination, however, it becomes obvious that the densities are in fact integrated areas of black ink and white paper.

On the other hand, photogravure does give a genuine continuous tone image. Although a cell structure is superimposed for technical reasons, the tones are the result of varying densities of ink and not varying area.

Some photomechanical methods of reproduction are often described as continuous tone printing processes, such as Collotype. In many cases this is a misnomer. The tones of a Collotype print, for example, are produced by varying areas of black ink printed on to white paper from a reticulated gelatin surface. Owing to the irregular grain structure it is however difficult to break the tones down visually into the discrete areas of black ink and white paper; thus the tones appear to be varying densities.

Confusion can be caused in the use of the term in some European countries where continuous tone images are known (in literal translation) as half-tone images, half-tone screen images being simply called "screen" images.

D.A.H.

CONTRAST

In black-and-white photography, a difference in visual brilliance between one part of the image and another. Without contrast there could be no such thing as a visible image. A line in a photograph is visible only because it is either darker or lighter in tone than its ground. Every distinguishable part of the image is the result of a contrast in tone values.

Range and Scale. The difference in brilliance between the lightest and darkest tones in the picture is called its brightness range. The brightness range of a scene, negative, or print is expressed as the ratio of the brightest to the deepest tone.

The brightness range of an image tells only one thing about its character. A line drawing in Indian ink and a photograph may have the same brightness range, from black pigment or silver to white paper. But the line drawing has only two tones: black and white, while the photograph includes intermediate grey tones as well. In the same way, two photographs may be equal in over-all brightness range or contrast while one includes a large number of intermediate grey tones between black and white, and the other only a few.

The scale of tones discernible in a photograph governs its ability to give an impression of roundness and depth. To some extent, the black and white artist can overcome the monotonal limitations of his medium by shading effects in which he builds up intermediate tone values by shading and cross-hatching. He uses lines of varying thickness, more or less widely spaced according to the tones he requires.

So the artist can show the gradual change of tone that indicates a rounded surface, even though he is restricted to two tones—black and white—in applying the medium.

A photographic print does not express intermediate tones in this way. The depth of tone in a print is decided by the density of the silver deposit. In order to reproduce a full scale of tones from black to white the density of the deposit must vary in exact proportion to the luminosity of each individual tone.

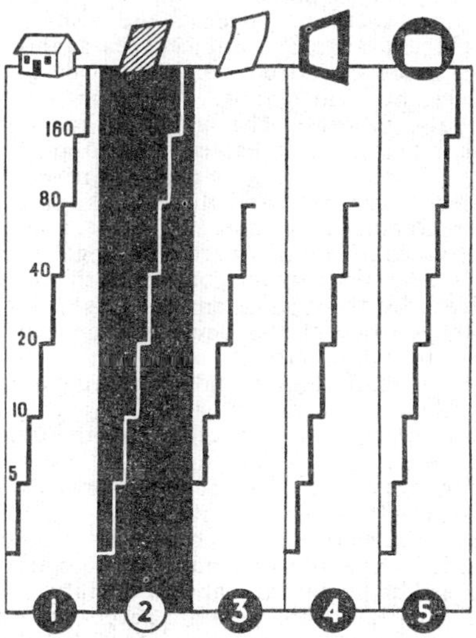

COMPARISON OF CONTRAST RANGES. 1. An average subject may consist of a house of light tone in sunshine with shadows, with a range of 160 : 1. 2. The negative records all the tones. 3. A paper print can only reproduce a portion of the range. 4. A transparency projected on screen shows increased shadow detail. 5. A transparency lit from behind is the only positive reproduction that can encompass the full original range.

In practice the tones of the picture may proceed from black to white in a small number of steep steps or a large number of shallow steps. The first results in a hard, boldly defined picture with only a few clearly recognizable tones—say black, dark grey, middle grey, pale grey, and white. This is known as a hard or contrasty print.

The second gives a print in which there are no sudden changes from dark to light tones and in which the eye can recognize a large number of tones of different depths of grey. A picture of this type can reproduce the gradual change of light over a rounded surface and the delicate differences in tone between near and more distant objects. This is known as a soft print. At the same time, since there are no strong contrasts, the picture loses brilliance and strength. It may even appear flat and dull.

For this reason the photographer generally aims at producing an intermediate type of print with enough tones to give modelling but not enough to destroy firm definition and clarity. In other words, a normal print.

Subject Contrast. The brightness range of the subject and hence the contrast of the picture it presents to the camera can vary within very wide limits. A scene where there are whitewashed walls in sunlight and, say, dark paintwork in shadow, might have a brightness range of 160 : 1, while the range of a calm seascape under an overcast sky might be only 2 : 1. A scene with a brightness range of 60 : 1 viewed in cross-lighting might not exceed 20 : 1 in frontal lighting where no shadows are visible. The brightness range of a subject seen against the light is always extreme, but it can be greatly reduced by fill-in lighting.

Negative Contrast. The brightness range of a negative can never exceed a ratio of about 200 : 1 because the clear parts are never completely transparent and the blackest parts are never completely opaque.

How contrasty any particular negative may be is determined by the contrast of the original subject, by the type of emulsion used, by the characteristics of the developer, and by the time of the development.

A contrasty subject tends to give a contrasty negative, although by suitably varying the characteristics of the developer or the time of the development it is possible to produce a contrasty negative of a soft subject or vice versa.

Print Contrast. The contrast of a glossy bromide print can never be more than about 60 : 1 and it is generally very much less because the whitest highlights are always slightly veiled, and the deepest shadows are never completely black. In practice the contrast of the picture will depend on the contrast of the negative and on the paper grade. It is also affected by the texture of the paper surface. Matt and rough textures scatter light and so give less deep shadows than a glossy or glazed surface, thus shortening the contrast range.

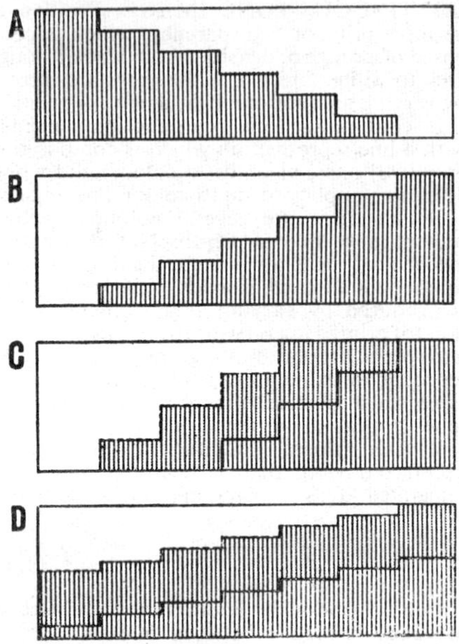

SUBJECT AND NEGATIVE CONTRAST. A, subject of normal brightness range. B, reproduction of subject tones on negative material of normal gradation. C, reproduction on contrasty negative material (either shadow or highlight gradation is lost). D, reproduction on soft gradation material (negative has no very high or low densities).

Subjective Considerations. Contrast as measured, say, by a densitometer is by no means the same thing as contrast as it is finally perceived by the human brain. For example the eye may fail to perceive a range of delicately graduated tones in a photograph which also contains a strong patch of brilliant light or deep shade. It may be much more affected by a white image on a dark ground than by an identical dark image on a white ground, although the measured contrast would be the same in each case.

Again, the ability to discriminate between tones—and therefore to perceive detail and form—varies between one observer and another. It may even vary in the same observer; if he is familiar with the subject and knows what to look for, he is more likely to be able to see significant detail in it than if he has no idea of what he is looking at.

Colour and Contrast. Finally the colours of the scene can alter the whole character of its contrasts as they appear to the eye. As the sensitized material does not respond to colours in the same way as the eye, the final positive print may present a picture of different contrast from the original scene.

If the print contrasts are to be as close a representation as possible of those of the original scene, a colour filter in front of the taking lens can be used to hold back those colours which affect the photographic emulsion more than they do the eye. The commonest example of this is the use of a yellow filter of a suitable depth of tone for correcting the over-sensitivity of the emulsion to blue. For example if the scene includes a blue sky with white clouds, the blue sky is liable to reproduce as light as the clouds, and the two will merge in the print; the filter restores the contrast.

On the other hand, where adjacent areas of the subject differ in colour but not in brightness, it is possible to make the colour contrast visible as a tone contrast. This requires a filter to distort the true tone values and make certain colours come out either darker or lighter in the black-and-white print than they would appear to the eye. The use of such contrast filters is common in medical and document photography and in the photomicrography of differentially-stained specimens.

Wherever flat frontal lighting is unavoidable—e.g. flash on the camera—there are no shadows to accentuate texture and modelling. In such cases, the only way of reproducing important detail may be to exploit the colour contrasts in the subject by the use of filters—or by taking the photograph on colour film.

Colour materials generally show the same response to the spectrum as the eye so that the colour contrasts of the print are usually a fair representation of what the eye would see in the subject.

For this reason flat lighting, while uninteresting in a black-and-white photograph, is successful in colour—and even recommended by the makers of colour materials because it keeps the brightness contrast safely within the range that the colour film can handle.

From all this it is clear that the subject of photographic contrast is complicated by a number of subjective and psychological factors, by personal idiosyncrasy and by the nature and context of the scene.

It is a matter of first importance to all photographers because it controls print quality and decides the whole character of the finished photograph. But much of its effect must be learned by observation and experience. It cannot be mastered simply by being measured with a densitometer. F.P.

See also: *Characteristic curve; Contrast control; Gamma; Papers.*

CONTRAST CONTROL. Some control of the contrast of the negative is possible during processing.

There are also various ways of controlling the contrast of the final photograph by modifying the normal technique during enlargement, or when processing the print.

Control in Negative Processing. The two main means of controlling contrast are the choice of developer and degree of development.

Energetic developers will give high contrast especially the caustic hydroquinone type of formula as used in process work. Soft-working developers, including most fine-grain types will give low contrast images.

Prolonged development also increases the image contrast, while shortened development decreases.

The maximum contrast obtainable in each case depends on the negative material employed.

Special techniques like two-bath and water-bath development also afford a measure of contrast control.

Control by Masking. The contrast of the negative can be effectively altered by making a weak positive transparency from it, and of different contrast. This is then bound in register with the negative before the print is made. The technique is known as masking, and although involving extra work, is capable of producing a wide range of results that can be strictly controlled.

In particular, masking is of special value in making colour prints or duplicates of colour transparencies.

Control in the Enlarger. The type of enlarger illumination to some extent controls the contrast of the projected image. But it is not practicable to use the enlarger illumination as a means of contrast control, because this would mean changing the lamphouse and the optical system.

One practical system of contrast control in the enlarger is to use a printing light of variable colour in conjunction with a variable contrast paper. The latter produces prints of low contrast when the printing light is rich in blue, and higher contrast as the light becomes increasingly yellow.

The colour of the light can be controlled by using a range of interchangeable filters. The filters may be balanced for high efficiency enlarger lighting, and each have the same filter factor, so that the printing exposure is unaffected by changing the filters.

Part of the exposure may be given through a filter of one colour, and the rest through a filter of the colour giving the opposite contrast effect. This method gives a continuous range of contrasts from hard to soft.

On some enlargers an automatic mixing device combines the light of two different colours, again yielding a continuous range of contrasts.

A different procedure, widely used in photo-finishing laboratories, entails the use of a hard paper grade all the time, the contrast being reduced by a controlled fogging exposure before, during, or after the printing exposure. The relative exposure times are balanced by a suitably calibrated control.

A modern printing method with automatic contrast control uses a flying spot scanner (produced by an electron beam on a cathode ray tube) as the light source. This scans the negative in the manner of a television image, while the light coming through is monitored by a photo-multiplier tube. With the aid of a suitable electronic circuit this tube controls the brightness of the scanning spot all the time, and so adjusts effective printing exposure to the negative density over each part of the image. The feed-back of the photo-multiplier tube is variable, giving any desired degree of contrast reduction or increase. The necessary apparatus is complex and comparatively expensive.

Control in Print Processing. Some local control may be exercised during development by applying concentrated developer to the print locally with a plug of cotton wool. This often gives some increase in local density and contrast. But the actual degree of control possible in this way is very small.

One way of controlling the over-all contrast of the print is to choose a developer with suitable characteristics. Otherwise, of course, there is the more normal course of using a different paper grade—or even a different make of paper.

Hydroquinone developers give slightly higher contrast than a standard metol-hydroquinone formula. By the same standard metol developers give prints of rather lower contrast. The range of control made possible by the use of pure metol, or pure hydroquinone developers, is usually equal to just under one paper grade.

Three stock developers consisting of pure hydroquinone, standard metol-hydroquinone, and metol formulae, give the whole range of control available by this method. There is nothing to be gained by using intermediate mixtures.

With a standard metol-hydroquinone formula, the strength of the solution also has some slight effect on contrast; the more concentrated the developer, the more contrasty the result.

For very close control of contrast the developer may even be changed during development. The print is made on the contrast grade estimated to be most suitable, and development is started in a metol-hydroquinone developer solution. If the image, as it begins to appear, seems too soft, the print is transferred at once to the hydroquinone developer. If it appears too hard, it is put into a metol developer, and development finished there. The sooner the print is transferred to the soft or contrasty developer, the more marked the effect.

This method calls for experience in judging the contrast of an under-developed print; this is not necessarily the same thing as the contrast of the finished picture.

The developers very quickly become contaminated in the process of changing over prints from one to the other. For this reason it is better to work with small quantities and renew the baths frequently rather than continue to use a comparatively large volume of heavily contaminated solution. L.A.M.

See also: *Electronic printing; Gamma; Hot water bath development; Masking; Optical aftertreatment; Redevelopment; Three-bath development; Tone separation; Two-bath development.*

CONTRAST FILTER. Filter that is used to lighten or darken the tone of a particular part of the subject in relation to its surroundings, as opposed to a filter employed to reproduce each tone in its normal visual brightness. Any colour filter can be employed as a contrast filter.

CONTRAST GRADES. Bromide, chlorobromide and contact (gaslight) development papers are manufactured in a range of grades of printing contrast. The number of grades varies with the maker of the paper and the type.

See also: *Papers.*

CONTRAST TRANSFER FUNCTION. Alternative term for transfer function, describing—in the case of an optical or photographic system—how the response of the system to fluctuations in exposure depends on spatial frequency (i.e. the fineness of detail to be resolved). This function—usually in the form of a graph—enables the overall performance of photographic systems to be measured and compared.

See also: *Information in the image.*

CONTRE JOUR. Term used to denote a photograph taken with the camera pointed towards the light source. Contre jour pictures are notable for the halo effect usually produced around the edges of the subject. Because photographs of this type have an extremely great brightness range, they need careful exposure.

See also: *Against the light.*

CONTROL PROCESSES. In the early days of photography the shortcomings in contrast range and colour sensitiveness of plates and film caused photographers to look for printing processes that would give them some control over the tone values of the picture. The pigment processes answered the need of the time and some of them continue to be used to-day. In the pigment processes the image is developed in light sensitive gelatin or gum, loaded with a suitable pigment. During development the pigment is removed from the highlights, revealing the white paper base. The amount of pigment removed is under the control of the photographer, and in some processes extra pigment can be added to the shadows, giving very full control over the final tones.

Many of the processes have to-day little more than historical interest, but others have been improved, and are still employed to a limited extent.

BASIC CONTROL PRINTING PROCESSES COMPARED

Name	Type	Control	Remarks
Bromoil	Bromide print image bleached out and converted to relief image: this accepts ink or pigment	Hard ink applied by brush, strengthens shadows. Soft ink softens lights and contrasts. Ink applied in different depth. Highlights picked out with soft rubber. Unsightly objects can be removed	This process needs some skill and experience before it can be used with success
Bromoil Transfer	The inked bromoil print may be used to transfer the pigment image to a sheet of plain paper by passing this through a roller press in contact with the inked print	None at this stage	Prints very attractive and are more delicate than the bromoil. Care is needed—and experience—before the process is mastered
Oil Process	Gelatin coated double transfer paper as used for carbon printing is sensitized with bichromate and exposed by daylight in contact with negative. When a brown image is seen with shadow detail, print is washed in running water until stain disappears. Then dried print soaked and ink applied with brush as for bromoil	In application of the ink by a patting action on the print. Detail will be built up and density attained	This process has largely been superseded by bromoil, which allows a small negative to be used for enlarging
Gum Bichromate	Print sensitized by coating of bichromate, gum arabic and pigment. Print exposed under negative by contact in daylight or to mercury vapour lamp. Developed with water	Application of water to the image by spray or gentle jet reduces depth. For more rapid development warmer water may be applied locally. Gentle brushing will also emphasize highlights	Prints may be made in any colour pigment. Considerable skill needed to master the process. Has largely been superseded by bromoil, as control is effected more easily

With the improvements that have been made in bromide papers and the almost universal use of small negative sizes, many printing processes that were formerly popular have gone out of use. The more important of these are mentioned in the table.

The only important pigment or control processes in use today are carbon, carbro, gum bichromate, bromoil. Of these, only gum bichromate and bromoil permit really selective control. There are numerous other processes which allow varying degrees of control; some of these are obsolete variations of a basic process, others are modern processes which, although not intended as control methods, do provide some incidental manipulation. Of all these, the most well-known processes—by name at least—are as follows: arabin, Artigue, broloid, dusting on, Flexichrome (a modern hand-colouring process), Fresson, gum platinum, ozobrome, ozotype. R.M.F.

See also: *Printing processes (obsolete); Pigment processes.*

CONVERGING LENS. Lens which is thicker in the middle than at the edge. It has at least one convex surface. Rays of light passing through a converging lens are refracted towards the optical axis. A converging lens is capable of forming a real image, i.e., that can be focused on a screen. Converging lenses are also known as positive lenses.

See also: *Lens; Supplementary lens.*

CONVERGING VERTICALS. When anyone stands in front of a tall building and looks up at it, the vertical lines appear to converge as they get farther away. At least, they converge in the image on the retina of the eye.

But we know that the sides of the building do not really converge, so the brain refuses to see them that way. Furthermore, the whole of the building is not seen at once; the eye travels up and down and the brain adds up all the impressions. So it has time to cheat itself over the discrepancy between the width at the bottom and that at the top.

But when a camera looks up at the building, it records it as it appears to the lens. And it does it with a single glance. When the brain is confronted with the true state of things on a reduced scale so that the eye can take it in as a single impression, it refuses to accept it. In some way the picture looks wrong. But if the photograph shows the sides of the building parallel (as they could never appear from an angle) the brain is satisfied and the picture looks right.

In taking the photograph, the true picture can be distorted to get one that is acceptable. This is done by first setting up the camera with the negative plane vertical and parallel with the front of the building. This ensures that the vertical lines will be parallel, but it also means that the top part of the building lies off the negative. So the rising front is brought into action, sliding the lens up until it brings the whole of the building into the picture.

If converging verticals have not been adgusted in the camera, they can be corrected during enlarging by tilting the paper easel, and the negative carrier, or tilting the paper easel and the enlarger lens. Tilting the easel alone will correct the verticals, but will prevent overall sharpness from being retained over the print.

Tilting the Paper. The vertical lines of the picture can be made to run parallel by raising the paper holder or masking frame at the edge towards which the lines of the projected image diverge. This decreases the degree of enlargement at the raised end of the paper and, within limits, it is possible to adjust the tilt until the lines run parallel to each other and to the sides of the picture. But because the surface of the paper no longer lies in the same focal plane, the image will no longer be uniformly sharp. So the image is first focused so that it is sharp in the centre of the field, and the lens is then stopped down until the depth of focus extends far enough to give sharp definition over the whole of the image. But stopping down will rarely in practice be sufficient, unless an appropriate tilt of negative carrier or lens panel is also used.

Another consequence of the tilt is that the image gets progressively brighter as the distance between the lamp and the paper decreases. So it is necessary to give the paper progressively less exposure as it comes closer to the lamp. This is done by shading the enlargement during part of the exposure.

Tilting the paper also has the effect of stretching out the image in the direction of the lines that have been made parallel. The elongation is only important in scientific photo-

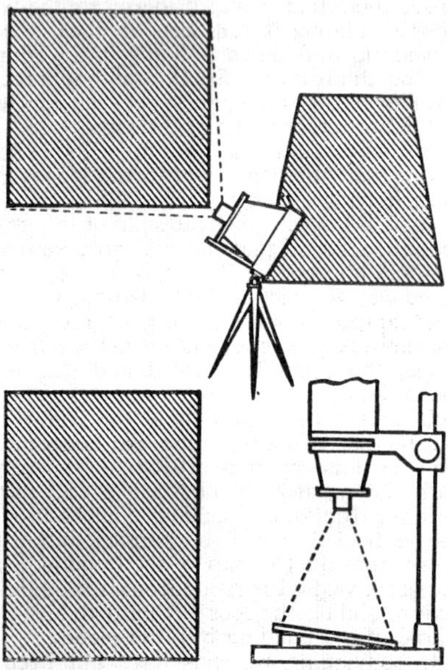

CORRECTING DISTORTION. Top: *Pointing the camera upwards at a building makes verticals converge.* Bottom: *Tilting the paper holder during enlarging makes the converging lines parallel, but usually distorts the original proportions.*

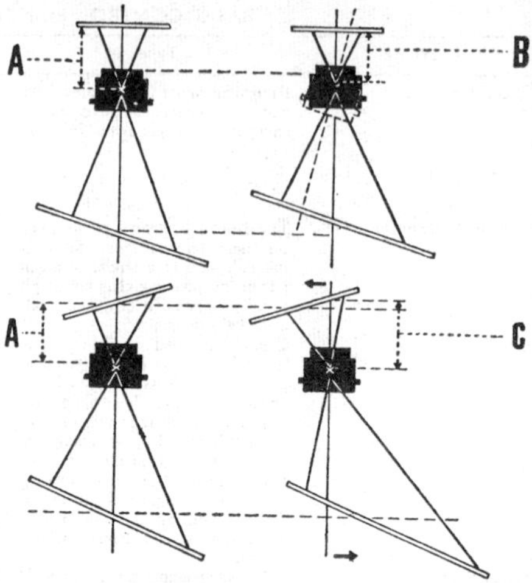

CORRECTING DISTORTION. Top left: *The degree of distortion due to straightening verticals in enlarging depends on the distance A which is normally fixed for a given enlargement.* Top right: *Tilting the lens alters the lens-negative distance B to permit approximate correction.* Bottom left: *An alternative method is to tilt both negative and baseboard in opposite directions to each other.* Bottom right: *In this case a new lens-negative distance C for more correct representation is obtained by sliding both negative and paper sideways.*

graphy; it is rarely as objectionable in pictorial work as uncorrected verticals.

Tilting Paper and Negative. The amount of possible correction can be doubled if the negative carrier also is tilted. The direction of tilt should be opposite to that of the paper, so that the side of the negative towards which the lines converge comes nearer to the lens—i.e., the negative carrier is lowered at the side on which the paper holder is raised.

This method of correction preserves the negative-lens-paper relationship, so that the image can be focused sharply over the whole area without stopping down the lens. The projected image will be perfectly sharp so long as the planes of the negative and the paper intersect in the plane of the lens.

Since the tilted paper is still farther away from the lamp at one side than the other, progressive shading is still necessary during the exposure.

The extent to which the image is elongated or compressed, depends on the lens used and on the scale of magnification.

Accurate Correction. For exact work—aerial, architectural records, etc.—no distortion of the image proportions can be tolerated. But the necessary accuracy is only possible under specific conditions of compensation; the focal length of the enlarger lens must be greater than that of the camera lens used in making the negative, and a definite degree of enlargement must be used depending on the focal lengths of camera and enlarging lens.

To correct converging verticals without image distortion, the degree of enlargement required is given by the equation:

$$M = F_e / \sqrt{(F_e)^2 - (F_c)^2}$$

where M = Degree of enlargement
F_e = Focal length of enlarger lens
F_c = Focal length of camera lens

If prints are required on any other scale of enlargement than this, a duplicate negative must be prepared at this magnification, and prints made from it in the normal way. The greater the focal length of the enlarger lens, the smaller the magnification required for the duplicate negative. L.A.M.

See also: *Camera movements; Perspective.*
Books: *Basic Photography*, by M. Langford; *The Complete Art of Printing and Enlarging*, by O. Croy (both London).

CONVERTIBLE LENS. A convertible lens is made of two lens assemblies which can be used separately or together. Such lenses are usually of the symmetrical type.

When the separate lenses have different focal lengths—as they usually have—there is a choice of three different focal lengths. To use the rear component alone, the front one is simply unscrewed, but when the front component is used alone it has to be screwed into the back of the mount in place of the rear component to bring it behind the stop.

Changing over from one arrangement to another alters the effective aperture as well as the focal length. For example, a lens may have a combined focal length of 12¼ ins. at *f* 6·8, and the component parts used alone give lenses of 26½ ins. at *f* 16 and 19 ins, at *f* 12·5.

Interchangeable Lens Systems. The principle of the convertible lens is also the basis of one kind of interchangeable lens system used by a few makers of 35 mm. cameras. In these cameras a fixed rear lens component is associated with a number of interchangeable front lens components which convert it to short, normal, or long focus as required. Where the camera has a between-lens shutter a single shutter mechanism may be incorporated in the fixed part.

Supplementary Lens Units. Corrected lens assemblies are made to combine with standard camera lenses to shorten or lengthen the focal length.—i.e., to produce wide angle or telephoto combinations. These supplementary units fit in front of the standard lens. In one case the standard lens is removed from the camera and fitted in front of the auxiliary lens. The result is an increase or decrease in the focal length of the standard lens depending on the characteristics of the auxiliary component.

See also: *Afocal lens; Lens, interchangeable; Supplementary lens.*

CONVEX. Optical term applied to a surface, e.g., of a lens or mirror, which bulges outwards at the centre.

COOKE TRIPLET. The Cooke Triplet was designed by H. D. Taylor in 1893. It was a complete breakaway from tradition, and was the basic design from which many of the leading camera lenses have been developed. In its simplest form, the Cooke Triplet uses three glasses arranged as a compound front and a single back element. The front element consists of a converging crown glass and a diverging flint glass with a small separation. The back element, widely separated from the front, is a converging crown glass.

This lens was originally manufactured in apertures up to *f* 6·3, but other lens designers have since used it as a model and improved its performance.

Zeiss replaced the single back element by a compound lens to make the now famous Tessar lenses with apertures up to *f* 2·8. A further modification of the back element gave the Ross Xpres.

Many of the well-known lenses derived from the Cooke Triplet are complicated and expensive. In its simplest form it gives a maximum aperture of *f* 6·3 and will cover a plate with a diagonal equal to the focal length of the lens. To improve on this performance, the single elements have to be replaced by more or less complex assemblies of glasses which naturally add to the cost.

See also: *Lens history.*

COOPER, JOHN THOMAS, 1790–1854. English resident chemist of the Regent Street Polytechnic. Read a paper before the Royal Society of Arts on 19th May, 1839, which was the first account of the production of sensitized material on a comparatively large scale. The title of this paper was *The Preparation of Photogenic Paper*.

COPPER CHLORIDE. Cupric chloride. Used in bleachers, some toners, intensifiers and reducers.

Formula and molecular weight: $CuCl_2.2H_2O$; 171.

Characteristics: Green hygroscopic crystals.

Solubility: Highly soluble in water.

COPPER SULPHATE. Blue vitriol; cupric sulphate. Used in bleachers, some toners, intensifiers and reducers.

Formula and molecular weight: $CuSO_4.5H_2O$; 249·7.

Characteristics: Blue crystals.

Solubility: Freely soluble in water at room temperature.

COPPER TONING. Method of toning bromide prints with copper ferricyanide which gives a wide range of warm black to purplish-red tones according to the time of immersion in the toning bath.

COPYING

Copying is the process of making photographic reproductions of graphic art, manuscript or printed matter—e.g., paintings, drawings, photographs, documents, postage stamps.

In such cases the aim may be to produce a photographic print which can be used instead of the original, either because the original is too precious or fragile to be exposed to everyday handling, or because it is too large to handle or sometimes too small—e.g., postage stamps. But the commonest reason, of course, is that any number of prints or reproductions for distribution may be made from the negative.

Photographic negatives may be copied for any of the above reasons, and also as a way of altering the tone scale, density or some other characteristic.

Approach. The normal procedure requires that the copy should be evenly illuminated, and that the camera should be perfectly square to the copy (i.e., so that the plate or film plane is parallel to the copy). The correct exposure must be then determined, and after exposure the sensitized material must be processed to suit the contrast range of the original. The only other consideration that may be necessary is that of scale: it may be desired to reproduce the copy to a pre-determined size.

Other special requirements exist when dealing with special subjects. For instance, in copying manuscripts and old documents, severe limitations may be imposed on the handling of the copy, making it impossible to adopt a normal procedure; in photographing oil paintings, it may be required to emphasize the texture of the oil paint. Such specialized subjects are separate techniques in their own right and are not considered here in this more fundamental account of copying procedure.

Equipment. It is not always necessary to use a camera to reproduce documents or line drawings. These may be printed by contact: examples are blue-prints and dye-line, widely used in drawing offices.

Reflex copying may be used for originals on thick paper, or documents printed on both sides. This is a variation of contact printing in which the sensitized paper is placed on top of the original, emulsion side down. On exposure the light passing through the sensitized paper is reflected back by the white ground of the document but not by the ink. This therefore yields a paper negative from which positive copies can be made by normal contact printing.

Special equipment and materials for these processes are produced and widely used in industry and commerce.

If the original has to be enlarged or reduced a camera must be used. Some copying machines, for example, incorporate a special camera and make copies up to 18 × 24 ins. from 36 × 48 ins. originals, using bromide paper as negative material—usually processed in the machine.

Other commercial copying systems are used to make copies of small size for filing and storage, some producing half-plate or whole-plate negatives on sheet film from drawings and large documents. Micro-copying outfits record smaller originals on 35 mm. film.

From the technical point of view the basic requirements of camera copying apply equally to the photographer making occasional copies with improvised apparatus as to commercial systems.

The essentials are that the subject must be perfectly flat, at right angles to the optical axis of the lens and parallel to the plane of the sensitive material. This condition is met conveniently by a copying stand analogous to an enlarger. The camera is mounted on either a vertical column or a pair of horizontal parallel

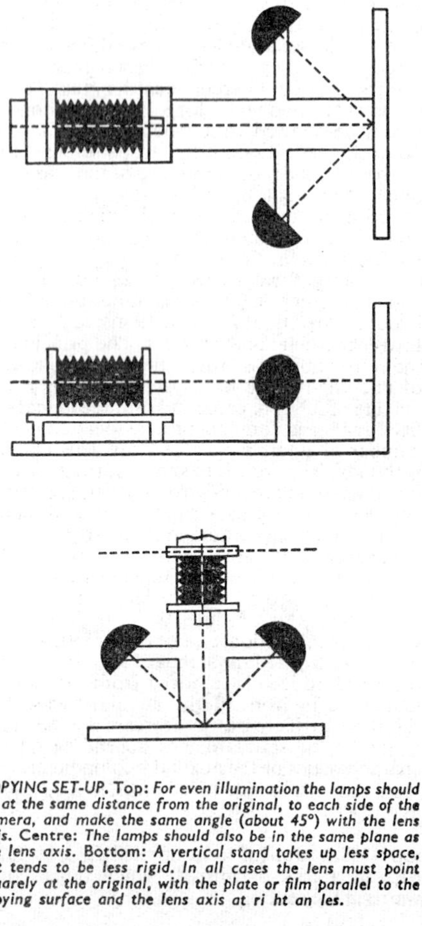

COPYING SET-UP. Top: *For even illumination the lamps should be at the same distance from the original, to each side of the camera, and make the same angle (about 45°) with the lens axis.* Centre: *The lamps should also be in the same plane as the lens axis.* Bottom: *A vertical stand takes up less space, but tends to be less rigid. In all cases the lens must point squarely at the original, with the plate or film parallel to the copying surface and the lens axis at right angles.*

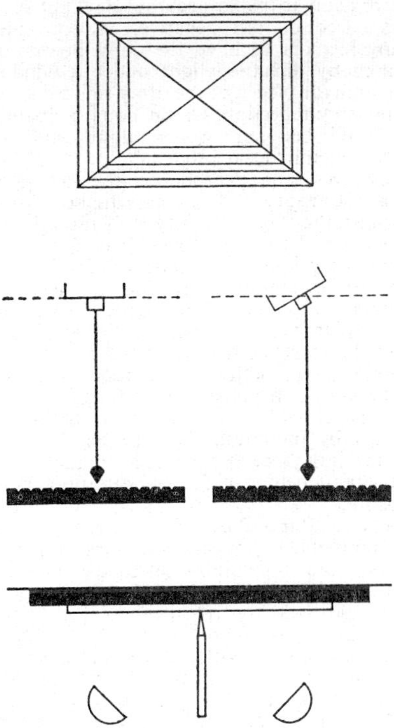

CAMERA ALIGNMENT. Top: *A baseboard marked with diagonals is a useful aid to centring the camera accurately above the original to be copied.* Centre: *Once the image is centred on the focusing screen of the camera, a plumb line will show whether the optical axis of the camera is at right angles to the baseboard.* Bottom: *Pencil test for even illumination. If both lights are equal in intensity and distance from the centre, the shadows on either side of the pencil will be equal in tone. If both lamps are at the same angle, the shadows will also be equal in length and position indicating good alignment.*

guide rails in the same way as the head of an enlarger.

Anyone who possesses an enlarger can convert it into a copying stand by simply removing the head and fixing the camera in its place. Some miniature camera systems are planned so that the camera or a focusing stage can be substituted for the enlarger head in this way: others have special stands.

The camera must be equipped for close up focusing by means of long bellows extension, extension tubes, or supplementary lenses.

For the best quality reproduction a plate camera of reasonably large negative size is normally used. An 8 × 10 ins. camera with adapters for smaller plates makes an excellent copying camera. For documents the copy can be made direct on to one of the document copying papers, which saves the cost of a plate. Negative materials for copying, too, are mainly produced in plate or sheet film form, so that a plate camera can best cope with all kinds of work. Ideally it should have a focusing back, since this is more convenient than front focusing when working close up.

Miniature cameras are widely used for copying documents and small drawings on 35 mm. micro-copying film. Rollfilm cameras are largely unsuitable since special negative materials are not available. However, adequate line and half-tone copying is often possible using the slowest films, processed for maximum contrast.

Sub-miniature cameras have the unique ability of being able to make legible copies of documents, drawings and pictures without any of the normal elaborate technical preparations. This is because the great depth of field of the subminiature lens makes accurate focusing unnecessary even at full aperture. So the camera can be held in the hand and the copy made with a short instantaneous exposure. This facility for quick, unobtrusive copying makes the subminiature popular for spying where the object is legibility and not superb photographic quality. At the same time, the "system" type of sub-miniature, when loaded with special fine grain material and used on its proper copying stand can be a surprisingly efficient copying instrument.

Illumination. The illumination must be as uniform as possible over the whole of the area to be copied.

The difficulty in achieving even illumination is due to the fact that the intensity of the light varies according to the law of inverse squares—i.e., at double the distance it is four times weaker, at three times the distance it is nine times weaker, and so on. However, in practice it is always necessary to establish the final details of the lighting set-up by experiment.

If a single lamp is used it should shine as perpendicular as possible on the original. The greater its distance, the more even the illumination. This is however suitable only for dead matt negatives.

The more usual practice is to use at least one lamp from each side. Substantially even illumination results from placing the lamps about one diagonal of the subject apart and at a distance from the subject of about three-quarters of a diagonal. This illumination is satisfactory for most subjects.

More even illumination is given by four lamps arranged over the corners of the copy board. In this case the distance between the lamps and the board may be reduced to a half diagonal of the square containing the subject. This arrangement is the one generally used for copying coloured and continuous tone subjects. It can of course be imitated by making exposures with a single lamp at each of the four corners in turn, or by making two exposures at each side with the two-lamp arrangement.

In all lighting arrangements no direct reflections from the surface should fall on the lens.

With a focusing screen there is no difficulty in checking the image for the presence of unwanted reflections.

All lamps used for illuminating the copy board should be well diffused. A pearl or opal lamp in a satin finished, matt, or white reflector makes a satisfactory source.

The most uniform lighting is given by tubular strip lighting—i.e., fluorescent lamps. Two such tubes, slightly larger than the long side of the copy and spaced a diagonal apart and at a distance rather less than a diagonal above the long sides of the subject, give a standard of illumination that can be considered uniform for all practical purposes. Fluorescent strip-lighting cannot be used when copying on colour film.

There are certain advantages in using daylight for illuminating the subject—e.g., it is uniform over the whole area. But its strength varies sometimes from moment to moment and it is difficult to make accurate daylight exposures based on test negatives. A convenient arrangement is to have the copy board illuminated by direct daylight under a window, with a mirror placed at the other side to double the light by reflection. Or the copy board may be placed between two windows and illuminated by mirrors at either side.

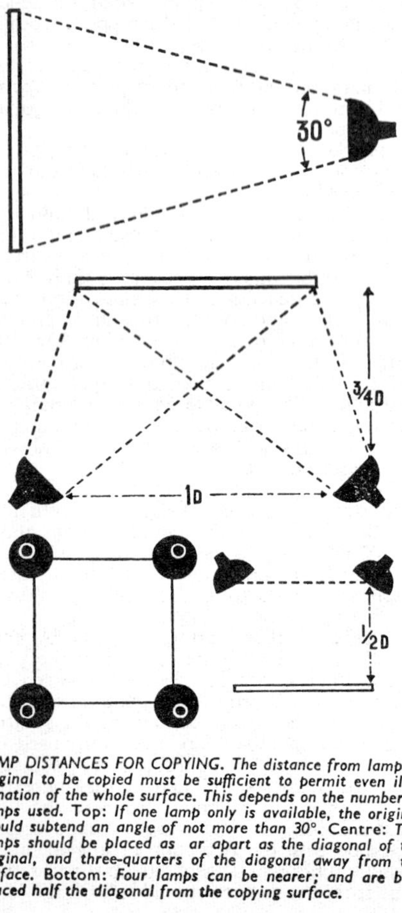

LAMP DISTANCES FOR COPYING. The distance from lamp to original to be copied must be sufficient to permit even illumination of the whole surface. This depends on the number of lamps used. Top: If one lamp only is available, the original should subtend an angle of not more than 30°. Centre: Two lamps should be placed as ar apart as the diagonal of the original, and three-quarters of the diagonal away from the surface. Bottom: Four lamps can be nearer; and are best spaced half the diagonal from the copying surface.

Negatives and transparencies and other forms of transparent documents are best copied by transmitted light. This type of illumination is also useful for softening the contrast of a hard photographic print and it may be used alone or in combination with reflected light illumination to control the degree of softening. The most convenient method of lighting here is to set the transparency up in front of a sheet of white paper uniformly illuminated by any of the above arrangements. Small subjects like lantern slides can be satisfactorily illuminated by mounting them to fill a cut-out aperture in a piece of card and moving a pearl lamp about behind it for the duration of the exposure. F.P.

Fluorescent Paper Bases. It is nowadays virtually impossible to obtain photographic printing papers suitable for continuous-tone work which do not contain artificial "brightening" agents. These additives absorb ultra-violet radiation in the illumination and re-emit the energy at longer wavelengths within the visible spectrum, so giving the illusion of extra brightness. This re-radiation is not usually distributed evenly over the whole spectrum and so a characteristic tint is given to the base-white, only just detectable by the eye; but readily distinguished by the photographic film.

Though apparently brighter than a normal white, visually, the blue-sensitive materials used so much in Graphic Arts studios will record certain of these additives as darker in tone than a normal white if the radiation is outside the blue region, or much lighter, if the emission is in the shorter wavelengths. Though an unretouched print is easy to copy, any hand-work or air-brushed areas in which process white or greys have been used to modify the photographic image of the print, appear different and may no longer match adjacent image-greys. Even with fluorographic pigments the retoucher cannot precisely judge the actinic effect under the illuminations of different studios.

Short of using printing papers (where available) on non-fluorescing bases, there is only one certain cure. The copyboard of the copying studio must be illuminated by light free from ultra-violet rays. Arcs, discharge lamps or high-efficiency tungsten light-sources need filtering by ultra-violet absorbing glass, or the first negative may be made on an ortho or panchromatic emulsion using low efficiency types of tungsten lamp. Such lamps emit only a very small proportion of ultra-violet rays. And it is the light source that must be filtered, for the fluorescence takes place before the image-forming rays reach the camera lens. J.V.

Lens. Any lens used for copying must have a flat field; the aperture is relatively unimportant. It is desirable to use a lens which has been computed for close subject distances (e.g., anenlarging lens); otherwise poor definition may result, especially with lenses of unsymmetrical design.

One method that may be successful in overcoming such a fault is to reverse the lens in its mount.

Special process lenses, usually apochromatic and of moderate maximum aperture, are used professionally. But they are very expensive and by no means essential for the average copying job for which any good anastigmat lens is quite satisfactory.

Sensitized Material. The average photographer makes copies on odd occasions and wants to see the results right away. So from this point of view alone, sheet film or, possibly, glass plates are the best form of sensitized material. Plates and sheet film are also much easier to handle singly during processing, and a further advantage is the great variety of emulsions available.

One of the most important factors in making a good copy is the correct choice of negative material. The table below gives a very general idea of the most suitable types of emulsion for the principal classes of subject, but considerable variation is possible according to the condition of the original and the type of print required—e.g., the copy may be required to give a true representation of the faded and stained appearance of an old drawing, or it may have to show as nearly as possible what the original looked like when it was new.

FILM AND SUBJECT

Original	*Type of Sensitized Material*	*Filter*
Black and white line drawings; typed or printed matter	Process; ordinary blue-sensitive	None
The same, on age-darkened, yellow or tinted ground	Orthochromatic; panchromatic if ground reddish in character.	Filter as near to ground colour as possible
Black and white photographs and half-tone illustration	Ordinary blue-sensitive; orthochromatic	None
The same on discoloured or tinted ground or where image has been toned	Orthochromatic, panchromatic if ground reddish in character.	For discoloured or tinted ground, filter of same colour
Documents to be micro-copied	35 mm. micro-film	Palest yellow that will give sufficient contrast in the print
Coloured originals	Panchromatic	Depends on colours of original; filters should be selected to give good contrast between colours of similar luminosity

The characteristics of the various sensitized materials may also be modified to suit the purpose by the use of filters as indicated.

Exposure. Correct exposure is more important in copying than in many other branches of photography. At the same time, the fact that the subject is normally quite static gives the photographer a chance to experiment.

The only certain way of arriving at the correct exposure for copying an unfamiliar type of original with a temporary lighting set-up is to expose a test negative on the same principle as a test strip is made in enlarging. The simplest arrangement is to advance the edge of a sheet of black card across the subject in successive steps. Another way of using a plate camera is to expose the plate in steps by progressively covering it with the sheath of the plate holder, but extreme care must be taken in this case to prevent the camera from being shaken or moved before the exposure is complete.

The test negatives are developed and fixed, and the best exposure is found by examining the image in white light, or, better still, by making a print. Generally, the correct exposure will be given by the strip on which the white areas of the copy register as a density that will just print out pure white while the deepest tones give the maximum black that the paper can produce. The reproduction of the intermediate tones will depend on the characteristics of the plate—in a print from a process plate, for instance, there will be no intermediate tones between intense black and pure white; a slow, fine-grain panchromatic plate will reproduce a whole range of intermediate greys if they are present in the original.

Processing. The plate must be given exactly the same development as the test negative, preferably in the developer recommended by the plate manufacturer. The experienced professional will generally develop by inspection and be guided solely by the appearance of the image. Anyone unfamiliar with the exact behaviour of the particular plate under his conditions will get on better by following the manufacturer's recommendations and developing by time and temperature.

Line subjects copied on process plates can, if necessary, be cleared by a brief immersion in a ferricyanide reducing bath. This aftertreatment leaves the lines on the negative transparent and free from any trace of fog which would prevent their reproducing as strong black lines on the print.

Printing. Negatives of continuous tone subjects are printed on the most suitable grade of bromide or contact paper in much the same way as ordinary negatives. Line subjects may be printed on the hardest grade of bromide or contact paper, or a more intensely black and white image can be produced by printing on reflex document copying paper. This is very much slower than normal printing papers and is more suitable for contact printing than enlarging. Fine pencil drawings should not, however, be regarded as normal line subjects; excessive contrast can destroy the tone recorded by soft pencil lines or shading.

Scale. The scale of reproduction of a copy is generally of interest—if not of importance—and should be recorded at the time of taking. One of the easiest ways is to mark a scale—e.g., of inches or centimetres—down the margin of the subject, or to lay a ruler alongside. This method enables the scale to be measured directly off any reproduction, either magnified or reduced, that might be made from the copy negative.

The degree of reduction or magnification is determined by the ratio of camera extension to subject distance. That is, for half scale copying, the lens-to-plate distance is half the lens-to-subject distance; for same size they are equal; and for 2× magnification the lens-to-plate distance is twice the lens-to-subject distance. The simple equation is:—

$$M = \frac{v}{u}$$

where M = magnification
v = lens-to-plate distance
u = lens-to-subject distance.

Reduction is the inverse of magnification. That is, a half scale reproduction in terms of magnification is $\frac{1}{2}$, or in terms of reduction 2:—

$$R = \frac{1}{M} = \frac{u}{v}$$

where R = reduction.

To establish the lens-to-subject distance for setting up the camera the equations are:—

$$u = F\left(1 + \frac{1}{M}\right)$$

or

$$u = F(1 + R)$$

where F = the focal length of the lens, or the combination of the camera lens and supplementary lens if used.

In practice the scaling is finally adjusted by measuring the image on the focusing screen. This is because distances should be measured from the nodes of the lens, and these are not precisely known; and further the engraved focal length of a lens is a nominal value and not necessarily an exact one. F.P.

See also: *Antiques; Close-ups; Coins; Document copying; Duplicate colour slides; Duplicate negatives; Macrophotography; Manuscripts and old documents; Micro-copying; Models (scale); Paintings and drawings; Philately; Supplementary lens.*

Books: *Camera Copying and Reproduction*, by O. R. Croy (London); *Microcopying Methods*, by H. R. Verry (London).

COPYRIGHT AND THE PHOTOGRAPHER

Copyright is a form of property in work; literary, dramatic, musical and artistic work. Ideas, however original, are not protected by the law of copyright. Copyright is an acknowledgement that a person is entitled to the fruit of his own work and the use of his brain in producing that work.

In England the law of copyright is governed by the Copyright Act, 1956. This was a major piece of legislation repealing the Copyright Act of 1911, altering the basis of definition of copyright and providing elaborate transitional provisions where the protection was altered from the position before 1956. The transitional provisions will continue to be of importance for a considerable time yet, but they are highly detailed and in a comment on the present law of copyright in relation to photography such as this, the changes and transitional provisions must be ignored. But quite shortly and oversimplified in statement the change is this: Before 1956 copyright law stressed the rights of the owner in his copyright. Since 1956 the emphasis is upon what he may prevent others doing without his authorizing, i.e. what he may restrain.

Photographs. The Copyright Act, 1956, gives copyright to every "original artistic work" and within the definition of such are photographs. Very broadly the word "artistic" means that some skill or labour has been used. "Photograph" is any product of photography or any process akin to photography (other than a part of a cinematographic film).

In relation to photography, the word "original" probably means "new", i.e., it is a fresh photograph, not a photograph of a photograph. There has been learned discussion as to what "original" means, but the leading legal textbook has narrowed the meaning in relation to photographs to that simple criterion, and the arguments are convincing.

The period of protection is from the date of making the photograph and for fifty years from the end of the calendar year on which the photograph is published. Publication in general terms means that reproductions must be issued to the public. Accordingly photographs have perpetual copyright from the date of being taken until publication, and it is only on publication that the fifty years starts to run.

To enjoy copyright the author must be a person qualified to own the right. The 1956 Act gives in Section 1 a definition of a "qualified person" as follows:—

"For the purposes of any provision of this Act which specifies the conditions under which copyright may subsist in any description of work or other subject-matter, 'qualified person' —(a) in the case of an individual, means a person who is a British subject or British protected person or a citizen of the Republic of Ireland or (not being a British subject or British protected person or a citizen of the Republic of

Ireland) is domiciled or resident in the United Kingdom or in another country to which that provision extends, and (b) in the case of a body corporate, means a body incorporated under the laws of any part of the United Kingdom or of another country to which that provision extends."

There are also the overriding requirements of common law and policy. For example, there would be no copyright for reasons of public policy in an indecent or obscene photograph. An indecent or obscene nude would probably be recognized as such without difficulty. The law has not defined these aspects. It would be easy to indulge in satire over some decisions but common sense will continue to arbitrate on them.

Ownership of Photographic Copyright. Under the Copyright Act, 1956, it is provided that the author in relation to a photograph means the person who at the time when the photograph is taken is the owner of the material on which it is taken. Prior to 1956 legal authority took the view that a photograph was made when the negative was developed, not from taking the exposure. The provision of the 1956 Act amounts to the same thing but realizes that modern photography does not necessarily involve a negative. Where a photograph is taken on commission, the person who commissioned it is the owner of the copyright with one very important exception; work in the course of employment by a newspaper proprietor (including periodicals) gives the proprietor the first right of publish but thereafter the copyright revests in the author.

The owner of the copyright can prevent any one infringing his rights and in addition claim damages where those rights have been infringed. The remedies are, of course, by way of injunction, damages and for an account.

The owner can also sell his copyright outright by assigning it for a fee. The only legal requirement is that such an assignment must be in writing. Of course he could give it away, but to perfect the gift it would have to be in writing.

Finally, the owner can sell part, i.e., he can grant a licence for a period or for a special purpose. When a copyright owner deals with his rights by assignment or licence, he himself may not reproduce what he has assigned because the assignee, on a valid assignment having been made, becomes the owner of the copyright of what has been assigned.

Dealings in Copyright. The Copyright Act, 1956, also permits an agreement to be made before a work is created whereby the copyright may be vested when it is created in someone other than the author of the work. It can also pass by will or an intestacy. Great care should be taken in framing assignments of copyright so as "not to carry rights in excess of those intended to be assigned". There can be partial assignments of copyright limited as to time or territory or as to the means of or style of reproduction.

Sometimes, of course, an outright assignment is the simplest and most direct method of carrying out the parties' intention. But as a general rule it is far wiser for the owner of copyright to grant a licence. If a publisher becomes the absolute owner of copyright he can use the work as he wishes. In the case of a licence he becomes, if the licence is clearly worded, bound by the terms of the licence. Thus a publisher under licence would not be entitled to publish a work in any different form from that authorized by the licence. Again, if an assignment is made in consideration of a royalty and the copyright is then assigned to a third party, the original owner is not a party to that dealing and could not sue for his royalties. In spite therefore of the disarmingly simple statement that an assignment or exclusive licence needs only the formality of writing any owner of copyright who wishes to dispose of his right, or part of his rights, should ask himself whether or not it might not be wise to obtain professional advice.

Freedom to Take Photographs. The photographer must not infringe another's copyright, and some special dangers outside copyright are dealt with below. But apart from this, he has practically the world and the scene of life from which to create his original artistic work. Within rather obvious limits, he can photograph whatever, wherever and whenever he pleases. He can photograph the Boat Race, or a cricket match, take a snapshot of Charlie Chaplin or the Pope (a snapshot having been described by a learned legal authority as being one of the hazards of using the highway).

Where there are restrictions, they are usually obvious. It would be clearly asking for trouble to take photographs of an atomic plant and a photographer in Russia should be circumspect. But these restrictions are so obvious that they do not need further comment. There are a number of other minor restrictions, e.g., not taking photographs in police courts, which are the subject of special laws or bye-laws, but these restrictions have nothing to do with copyright as such.

Dangers of Infringement. The greatest danger, of course, lies where there is an intention to steal another's work, and this article contains no suggestion of how that may be done with impunity. Copyright is a form of property protecting not ideas, but expression. What is protected is not original thought or information but the original expression of thought or information in some concrete form.

Photography is a relatively narrow mode of expression and limited to its genre. Thus, a photograph of a building does not amount to a breach of the architect's copyright. To infringe the copyright in an architect's building, one would have to build an infringing building. One could infringe plans by photographing them, but that would be an infringement of the plans.

A photographer may photograph sculpture or artistic craftsmanship if they are permanently situate in public places, for example in streets, squares, and national museums. Possibly statues in a theatre where the public only go by licence after payment would not be a public place.

To constitute an infringement there must be use, directly or indirectly, of the author's work. Two photographs of the same person may be almost identical but have been taken independently without reference to each other. In such a case there could be no infringement; the moral basis of copyright being the eighth commandment: "Thou shalt not steal."

Copies of Letters. Letters perhaps call for special comment, especially as photographers are more and more employed for making copies of letters for business and for legal work.

The writer of a letter is the author and owner of the copyright in the letter. By sending his letter to his addressee, he does not "publish" the letter in the sense in which publishing is used in copyright—unless of course it is a letter to a paper. The letter belongs to the recipient. He commits no offence by throwing it into the fire or selling it, but if he has it copied (and that includes photographing it) there is an infringement of copyright.

Before 1956 it was arguable if photographed correspondence were used in judicial proceedings that there might be a breach of copyright. The Copyright Act 1956, Section 9, expressly provides that there should be no infringement in such a case and the same Section sets out the general exceptions to infringing copyright in artistic work, which, of course, includes photographs. Such exceptions include private study and fair dealing, illustrating again the moral basis of copyright law.

Libellous Photographs. Happily libel is going out of fashion as a speculative and possibly remunerative form of litigation. Before the last war it was otherwise. Everyone can remember striking cases. Thus an unmarried woman whose photograph appeared next door to the picture of a married man and had an unsuitable caption underneath, claimed successfully that her chastity had had doubts cast upon it. Another (a model) claimed damages (and succeeded) because the photographed legs of another girl had been substituted for her own. She protested successfully that a girl of her position in life would not parade her legs like that and that her reputation had suffered.

Some of these cases make intriguing reading to the outsider, but they may well have been painful and distressing to the plaintiff, and one must not forget that everyone is entitled to his or her reputation, a principle guarded jealously over some hundreds of years by our Common Law.

Where models are concerned, a special danger lies. It would be unfortunate if an unmarried woman (unless she had expressly consented) were posed for a maternity scene in the rôle of a mother, or if one used a photograph of a model who was known to be a teetotaller for an advertising campaign to increase the sale of gin. Usually these dangers crop up unintentionally with no blame to anyone. The only safeguard is to insure against such a danger. The premiums are small, and well worth while, to any commercial photographer.

In one case known to the writer, an editor of a well-known publishing house, after insuring continuously for some fifteen years, discontinued his insurance. In the following year a completely innocent libel was published and although the dispute was promptly settled the costs came to much more than the fifteen years' premiums!

So far as models are concerned, it is a frequent practice to have agreements to try to guard against these dangers. The model assigns the copyright, grants unrestricted use of the photograph, and agrees that any reproduction shall be deemed to represent an imaginary person, and that no imputations shall be deemed to be made because the photograph is being used for advertisements and so on. If the model is under legal age, the parent or guardian must also sign the agreement.

Cinema Copyright. A film in the sense of a cinematograph film is, of course, a series of photographs projected so closely together as to give the illusion of movement. The Copyright Act, 1956, granted copyright in a cinematographic film independently of literary or dramatic copyright so that this right does not affect copyright in any literary, dramatic, musical or artistic work from which it is inspired. In order for a cinematographic film to enjoy copyright protection, the maker must have been qualified under the Act to enjoy protection and if it has been published, publication must have been in this country or in a country to which the 1956 Act extends. Publication has a wider meaning than for other work when used in relation to a cinematographic film including sale or hire of the film and offer for sale or hire of copies of the film. It would include, for example, a tape on which the film was recorded. The owner of the copyright in a cinematographic film is the maker, which means the person responsible for making the film. Under the Act, the owner of a copyright in a cinematographic film is also given separate broadcasting and television rights under Section 13 of the Act.

Extent of Copyright Protection. Protection extends far beyond the United Kingdom. The Berne Convention, Berlin, Rome and Brussels Conventions are all highly technical and specialized treaties and there are also reciprocal arrangements within the British Commonwealth.

The large object of these conventions and agreements was to give citizens of the countries who signed the treaties the same rights in other countries as the natives of those other countries themselves enjoyed. Thus, that a British subject be able to claim in France, or in any other country subscribing to such a treaty, national treatment to protect his copyright in France, or in that other subscribing country. Conversely, that if the owner of copyright in (say) France or some other subscribing country found his rights being infringed in England or a treaty country, that he should be able to claim protection for his copyright not only in France, but in the treaty countries as well.

The United States of America remained outside these Conventions but there were certain American Conventions, the Monte Video Convention, and others, to none of which the United Kingdom was a signatory. These conventions proceeded on the same principle as the ones referred to above, with the same fundamental objective of according national protection to works published in any of the countries who were signatories under the laws of the other contracting stages.

Roughly most of the South American countries and the United States of America were parties to these Conventions.

The latest attempt at really internationalizing copyright protection is the Universal Copyright Convention signed at Geneva in 1952. Under this the contracting states agree, broadly speaking, to give the published works of authors in the other contracting countries the same rights as they give to works first published in their own territories. The United States of America ratified the Convention in 1955. Although this Convention was an advance in internationalizing copyright the actual protection afforded by the contracting states is still a matter for the individual state. R.M.H.N.

See also: *Permits to photograph; Reproduction fees.*

Books: *Copyright*, by Copinger and Skone James (London); *Photographers and the Law*, by D. Charles (London).

CORKS. Corks are unsatisfactory as stoppers for bottles of photographic solutions because they deteriorate under the action of many of the ordinary photographic chemicals. Dipping them in hot paraffin wax helps to prevent them from becoming soft and absorbent, but it is safer to use rubber, plastic or glass stoppers.

See also: *Stoppers.*

COROT, JEAN BAPTISTE CAMILLE, 1795–1875. French painter. Executed many clichés-verre, a process "re-invented" by Adalbert Cuvelier and others in 1853, by scratching with an etching needle drawings on glass plates coated with sensitized collodion darkened by exposure to light. The prints were produced on photographic paper by contact copying.

CORRECTION FILTER. Term for any filter which alters colours to suit the colour response of the film in use. Strictly speaking, any colour filter can be regarded as a correction filter, although the term is more commonly applied to filters used in colour photography for adjusting colour temperature. Filters used in this way may be placed over the camera lens or, for indoor work, over the light source.

In professional filming, entire windows are sometimes covered with correction filters when daylight is being used with artificial light colour film; this is often more convenient than blacking-out the windows and using artificial light entirely.

CORREX TANK. Original amateur developing tank for film. The film was attached to a celluloid apron and wound around a reel before immersion in the circular, light-tight developing tank. Raised bosses along the edges of the apron kept the apron and film out of contact and allowed the solution to circulate.

For most purposes the apron tank has now been replaced by the apronless spiral groove type, but for certain special work—e.g., colour processing—in which the wet film is removed from the spool and later returned to it, the apron tank is still preferred because it makes reloading much easier.

See also: *Developing negatives; Tanks.*

CORROSIVE SUBLIMATE. Bleaching agent used in mercury intensification; another name for mercuric chloride.

COSINE LAW. Optical law which formulates the variation of illumination given across the negative by a lens.

When a perfect lens forms an image on a plane at right angles to its optical axis, the image is brightest opposite the centre of the lens. There the lens diameter subtends the greatest angle. At points nearer the edges of the field the lens aperture subtends a narrower angle, and the illuminating beam falls obliquely on the negative. The brightness of the image at such points is therefore less, and the farther away from the centre of the field the point is chosen, the fainter the illumination will be.

According to the cosine law, the illumination varies as the fourth power of the cosine of the semi-angular field measured from the lens axis. (This angle is the angle between the lens axis and an imaginary line from the centre of the lens to the image point in question.)

See also: *Covering power.*

COSTING

The cost of prints is of interest principally to the professional who sells them; his livelihood depends on being able to sell his work at such a price that it will leave him a sufficient margin of profit over his actual expenses in producing it. From the amateur's point of view, the cost of a print must be weighed against the satisfaction he gets out of making it, and that is a purely personal balance sheet.

Anyone who hopes to run a commercial establishment at a profit without knowing exactly what he can afford to charge for his work, is heading for trouble.

Before a commercial photographer knows how much to charge for a job to show a reasonable profit, he must know what it has cost him to produce in both direct and indirect expense—i.e., in wages and materials, and in overheads.

The first essential is a card or sheet for each job. Different studios will have different ideas of how this card should be made up, but the following is a typical specimen.

Job. No. 801

JOB SHEET (FRONT)

Client: Smith & Co., Regent Street, W.1.
Client's Order No. A543

Description of Work:
Photograph of girl brushing hair at dressing table.
Remarks:
Summary:

	£ s. d.	£ s. d.	£ s. d.
Studio costs:			
Wages and salary of proprietor ...	2 0 0		
Materials	10 0		
		2 10 0	
Add on cost at 71 per cent		1 15 6	
		4 5 6	
Add profit at 100 per cent		2 10 0	
			6 15 6
Outside work ...		6 6 6	
Add profit		2 2 2	
			8 8 8
Total			15 4 2
Charge			15 15 0

JOB SHEET (BACK)
STUDIO WORK
Wages

Date	Artist	Hours	Rate £ s. d.	Amount £ s. d.
May 12th, 1962 ...	Brown	2	7 0	14 0
May 13th, 1962 ...	Brown	1	7 0	7 0
May 13th, 1962 ...	Jones	2	6 6	13 0
May 13th, 1962 ...	Jones, K.	1	6 0	6 0
				2 0 0

MATERIAL

Description	£ s. d.	Amount £ s. d.
4 plates ...	2 0	8 0
6 sheets ...		2 0
		10 0

OUTSIDE WORK AND SPECIAL ITEMS

Invoice No.	Supplier	Description	Amount £ s. d.
600	Black and Co.	Hire of table	1 13 6
1	Miss Smith	Model fee	3 3 0
601	Furniture Co.	Hire of props	1 10 0
			6 6 6

The client's name and a brief description of the job are entered in the job sheet. Each job is numbered and the sheet is given the corresponding number. It follows the job through, and each individual concerned enters the time spent on taking the photograph, developing, retouching, negative printing, print finishing, etc. Details of the film, paper, flash bulbs, etc., used are also entered on it. Travelling time and waiting time, if any, and hotel and travelling expenses must be entered. If a car is used it may be charged up at so much a mile to include petrol, oil, depreciation, repairs, etc.

The cost of any outside work, hire of properties and model fees are included, in fact, all the expenses which can be definitely earmarked to the job. These expenses are what accountants term "direct expenses". The proprietor's operating wage and the employees' wages are broken down to an hourly rate in working out the cost of the labour on the job. The job sheet now shows all the direct expenses in connexion with the job.

A percentage may be added to the cost of materials to allow for waste and a percentage to the labour costs to cover lost or idle time.

To help in arriving at labour costs it may be preferred for each employee to keep a time sheet, a typical example of which is shown.

TIME SHEET

Name .... *Week ended*....................................

Day	*Description*	*Job. No.*	*Hours*	*Rate*	£ s. d.
Monday ...					
Tuesday ...					
Wednesday					
Thursday ...					
Friday ...					
Saturday ...					

A little more work will be involved in keeping and analysing time sheets, but they have the advantage that the job sheet need not follow the jobs round with the risk of getting mislaid or soiled. The information on these time sheets will be summarized weekly in the office and the result entered on a job sheet.

A Materials Used Sheet and an Expenses Sheet should also be completed for each job. Suitable printed sheets can be obtained from most stationers.

Overheads. Such items as rent, lighting, office expenses, stationery, depreciation, etc., are generally known collectively as overheads.

The figure for these is found in the annual accounts. Here is a specimen profit and loss account of a commercial photographer.

PROFIT AND LOSS ACCOUNT
Commercial Photographer

	£
Sales	8,000
Materials Consumed	700
Wages	1,500
Wages, proprietor (as photographer)	800
Outside work, including model fees, property hire, etc.	300
Travelling and hotel expenses, including car expenses	350
	3,650
Gross profit	4,350
Expenses (Overheads):	
Trade and office sundries	120
Printing and stationery	150
Post and carriage	75
Advertising	150
Office salaries	300
Rent and rates	450
Heating and lighting	100
Telephone	70
Insurance	30
Depreciation	75
Repairs and renewals	50
Accountancy	30
Bad debts	30
Proprietor's salary (as administrator) ...	500
	2,130
Net profit	2,220

This shows overheads, or on-costs, as they are frequently called, as £2,130. Now each job has got to bear its proper proportion of these overheads. The wages, including the proprietor's wages, or wages and materials, are generally a guide to the relative size of the job, so the most common method is to apply the overheads in proportion to the wages and materials. It will be noted that the proprietor's earnings fall under three headings: first as a photographer, then as an administrator and finally in the profits.

From the specimen profit and loss accounts, it will be seen that the overhead expenses at £2,130 are equivalent to 71 per cent of the £3,000 paid out for wages and materials. So to absorb the overheads into the costs there must be included by way of overheads a figure equivalent to 71 per cent of the wages and materials of each job.

Having got the amount of wages and materials which gives the prime costs; having a note of the outside work and any special expenses, and having included overheads by adding in the amount equivalent to 71 per cent of the wages and materials, the total represents the total cost of the job, including overheads.

Percentage of Profit. If a price has been quoted for the job, the difference between the price and the cost determines the profit (or loss), and the summary on the cost sheet is completed by inserting the amount to be charged and the profit which is accordingly made.

If a price is yet to be given, then the price is arrived at by adding the profit to the costs. The figure should certainly not be less than 50 per cent for an ordinary mechanical job and may run up to several hundred per cent if much creative work has been put into it, as is frequently the case in some forms of advertisement illustration.

The total cost and profit added together, make the price to be charged.

In most studios the waste of material is not sufficiently important to worry about, but if it is considered necessary to deal with such items they can best be taken into account by way of a percentage on the wages and the materials. The percentages can only be estimated by the individual photographer in the light of his experience. For example—admittedly a severe one—if an assistant can only be found employment for half the normal working hours the charge made for that part of a job constituting his services must be at double the rate he is actually paid.

Financial Account. Finally to prove if the costing has been accurate, the aggregate of the cost figures must be reconciled with the corresponding figures in the financial accounts. This is achieved by summarizing the costs and the profits on all the jobs to give yearly totals.

The totals of wages, materials, outside and special expenses, on-cost, profit and sales respectively can be compared with the corresponding items in the financial accounts. To obtain the comparison with the total of on-cost it will, of course, be necessary to total the overheads in the financial accounts that the percentage was intended to cover. Theoretically an exact match should result. In practice it seldom does.

If the comparison throws up serious discrepancies it should be possible to find out the reason. If, for example, the total of the wages taken up in the jobs costs is less than the actual year's wages bill it could be accounted for by lost time. This could be confirmed by the analysis of the wages sheets. It should be possible to check through if there is any discrepancy on outside work and special expenses. A discrepancy in the on-cost would probably be explained by the percentage worked on being too low or too high. It would, therefore, call for revision in the future. This running summary of job results also shows the profits to any given moment.

Simplified Method. There is a simplified method by which the small commercial studio, particularly the one-man business, can charge out its work or check its price list.

It is necessary first of all to arrive at the total cost of running the studio. This cost will include the salary that the photographer has fixed for himself, the wages of any assistants, all overheads such as rent, lighting, heating, telephone, advertising, printing and stationery and trade and office expenses (but eliminating materials and special expenses such as model fees, hotel and travelling expenses). These figures can, of course, be ascertained from the annual accounts. Allowing for a weekly half-day, Sundays, Bank Holidays and a fortnight's vacation, there are 271 working days. On the basis of an eight-hour day this gives 2,168 working hours in a year. This should be reduced by at least a third to allow for unproductive time, such as office work, which gives 1,445 or, say 1,400 productive hours. The total studio costs divided by 1,400 will give cost per productive hour. The number of hours spent on the job, multiplied by this hourly rate, plus the cost of material used, and any special expenses will give the cost of the job. Only the proprietor's hours should be taken into consideration. The time of the assistant should be ignored.

The rate of profit will be governed by circumstances (such circumstances have already been discussed earlier). When the profit has been added the resulting total will be the calculated charge for the job, which will enable the photographer to fix his price.

See also: *Prices of commercial photographs; Reproduction fees; Selling photographs.*
Book: *The Business of Photography*, by C. Abel (London).

COUPLER. Chemical used in colour development where it combines with certain by-products of the development process to form a dye.

Modern colour photographs are almost entirely made on subtractive materials in which the dyes that form the coloured images are synthesized in the photographic emulsion layers during development. Where exposed silver halide is being reduced to silver, the developer becomes oxidized. The oxidized developer molecules in certain cases, e.g. phenylene diamines, have the property of combining, or "coupling" with other special compounds to form dyes. Thus an image of dye is built-up at the same time as the silver image.

These special compounds are called couplers, and three different types are used to form the yellow, magenta and cyan images. For example, derivatives of phenol or naphthol are commonly used to form cyan dyes, pyrazolones for magenta, and aceto-acetanilides for yellow colours. If the couplers are of simple constitution, and fairly soluble, they can be dissolved in developers. Each film layer is then colour-developed individually with a different developer-coupler combination.

When the couplers are incorporated in the emulsion layers during manufacture, one developer produces all three dye images. However, the coupler solubility must be reduced, otherwise the couplers could wander into the wrong layer and degrade the colours. More complex molecules have been devised for this purpose; they are dispersed in the emulsion

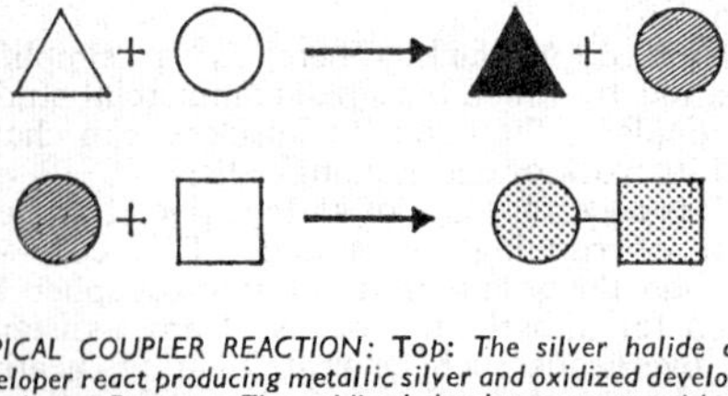

TYPICAL COUPLER REACTION: Top: The silver halide and developer react producing metallic silver and oxidized developer complexes. Bottom: The oxidized developer reacts with the coupler in the emulsion (or in the processing solution) to form dye.

layers either as colloidal suspensions, or dissolved in minute oily droplets. In either case, their water solubility is low, so that no wandering occurs. Considerable research has been required to design couplers which react at the appropriate rate and form dyes of the correct colour and stability.

Coloured Couplers. Most coupler substances are colourless until they form the dye image with the development product. But there are reasons why a coupler that is already coloured may be desirable.

The basis of colour photography is the fact that white light is made up of three broad regions of the spectrum—blue, green and red. Theoretically the three dyes that form the image in a modern colour negative material each control one of these sections. Thus the yellow dye controls the amount of blue light transmitted by the negative, magenta controls green and cyan controls the red light. If the dyes were perfect, they would do this and nothing more. Thus for example, if the magenta image could be obtained on its own and viewed through a green filter, the image would be seen. But the dye should be transparent to blue and red light, so that if we looked at it through blue or red filters, we should see no image. In fact we would see a faint image through these filters because the dye is imperfect and is absorbing a certain proportion of colours other than green. These unwanted absorptions will mean that at the next stage, printing on to the colour positive, the magenta negative will not only print on to the green sensitive layer of the positive, but on to the other two layers as well. Thus it is impossible to obtain pure colours in the positive because there is interference from these unwanted absorptions.

An elegant way of overcoming this difficulty is based on the following consideration. If we still take the magenta image on the negative as our example, and view it through a blue filter, we see a low contrast image. If the layer was coloured pale yellow and the yellow colour destroyed in exact proportion as the magenta dye is formed, we should see nothing but a uniform density. At every point on the negative the sum of the blue absorption of the magenta dye formed and the yellow dye remaining would be the same. The image would thus appear to be formed from a "perfect" dye, and would not print in the wrong places on the positive.

In practice this can be achieved by using coloured couplers. The couplers themselves are in fact dyes, and when they react with the oxidized colour developer, the dye molecules are destroyed, coupling takes place, and a new dye of a different colour is formed.

Theoretically all six unwanted absorptions—two colours for each of the three dyes—could be corrected by suitable coloured couplers. In fact it is found that yellow dyes are usually good enough without correction, magenta dyes need correction for unwanted blue absorption, while cyan dyes absorb too much of both the green and the blue. Thus magenta *couplers* that absorb blue light—i.e. are themselves yellow—are required, and the cyan couplers used absorb blue and green, i.e. are orange-red in colour. This accounts for the overall strong yellow-orange tint of many modern colour negative materials. This overall colour of course increases the amount of light required for printing, particularly the amount of blue light, and printing times or printing light output must be increased. This is a small price to pay for the considerable improvement in the saturation of the colours that result.

This type of more or less automatic colour correction corresponds to masking as used with colour separation negatives.

Similar effects can be obtained by a coupler which is colourless to start with, but where *unused* coupler left over after the formation of the dye image is turned into a masking dye at a later stage of the process. J.Mo.

See also: *Chromogenic development; Colour film principles; Masking.*

COVERING POWER. Term mainly applied to optical coverage of lenses, but can also refer to the light-obscuring ability of a silver image.

Lenses. The covering power of a lens refers to the maximum area of the focal plane over which the lens is capable of giving an evenly illuminated image of an acceptable standard of definition.

The limits of the area usefully covered are reached when either the definition or the illumination fall below the required standard. It also depends on the angle of view of the lens.

Definition. Aberrations tend to increase away from the axis of the lens so that, from one cause or another, the definition of the image falls off suddenly beyond a certain distance from the optical axis. For the same lens, the falling off occurs earlier at wide apertures—i.e., the covering power is greatest when the lens is stopped down.

Illumination. When a uniformly bright object plane is focused by the lens on the plane of the negative in a camera, the illumination of the negative is no longer uniform but falls off steadily from centre to edge.

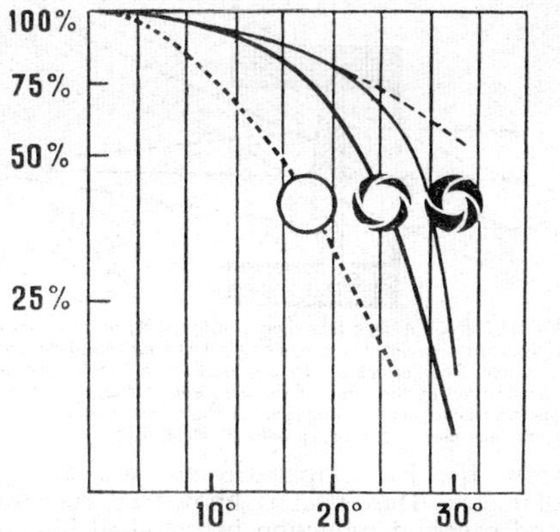

COVERING POWER. Relative image brightness (taking 100 per cent as the brightness in the centre of the negative) is here plotted against the distance in degrees from the centre of the field. The top dotted curve corresponds to the theoretical $\cos^4\theta$ *relationship, the other curves show actual falling off in brightness at a small, a medium, and full aperture respectively.*

Four effects contribute to this reduction in illumination. The first three operate whatever the lens type and whatever the relative aperture and together reduce the illumination by a factor $\cos^4\theta$, where θ is the semi-angle of view. The fourth effect is known as vignetting, and depends on the type of lens construction, generally decreasing to zero as the lens is stopped down.

(1) Marginal areas of the uniformly bright object are farther from the lens than the central area, and the lens receives less light from them in accordance with the inverse square law; the corresponding areas in the negative are therefore less brightly illuminated. The reduction in illumination is proportional to $\cos^2\theta$.

(2) For two object areas of the same size, one on the axis and one near the margin, the latter is inclined at an angle θ to the line joining it to the centre of the lens. In other words, a given cone of rays is spread over a bigger area than the central cone at right angles to the subject plane. The light reflected in the direction of the lens from such a marginal area is reduced by a factor $\cos\theta$ compared with the light reflected towards the lens from an area on the lens axis (Lambert's law).

(3) The effective aperture of the lens is reduced when viewed obliquely, appearing as an ellipse. A marginal area of the negative therefore receives less light than a central area. Again, the effective aperture and the corresponding illumination is reduced by a factor $\cos\theta$.

(4) With complicated lens types an oblique beam of light may be partly cut off by the mount. This may be caused by some obstructing metalwork, but more frequently it is due to the fact that lens elements some distance

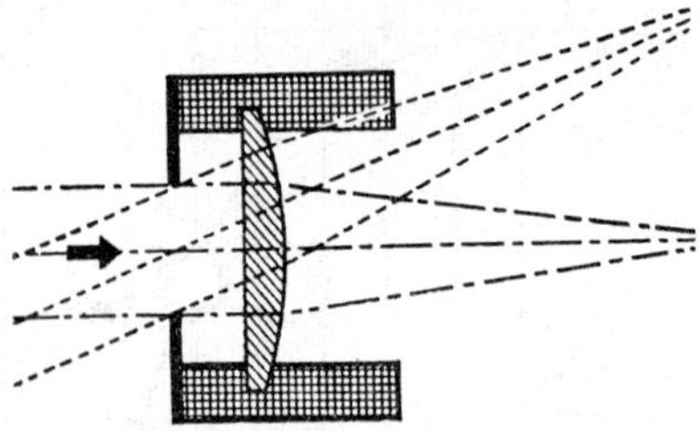

VIGNETTING: A deep lens mount may cut off marginal rays of light, thus reducing the image brightness near the corners of the negative. This effect occurs in addition to the falling off in illumination at the edge of the field due to the obliqueness of the rays reaching the negative. A shallower mount and large front lens element improve both shortcomings.

from the iris diaphragm are too small in diameter. The effect is known as vignetting, and causes a reduction in marginal illumination. This reduction in effective aperture is additional to the obliquity factor (3) above. Vignetting should not be serious with lenses covering moderate angles, but is often appreciable in wide-angle lenses. It decreases as the lens is stopped down.

Apart from vignetting, the reduction in marginal illumination due to what may be called photometric causes—the $\cos^4\theta$ relation—becomes serious only for wide-angle lenses. For lenses covering a moderate angle of 50° ($\theta = 25°$) the value of $\cos^4\theta$ is 0·68. the corresponding reduction in illumination being 32 per cent.

A reduction of about 20 per cent in the illumination of the margins of the image plane is not appreciable, so that the defect goes unnoticed in normal and narrow-angle lenses.

The effect is much more serious in wide-angle lenses, but as vignetting becomes less as the lens is stopped down it may not be usually obvious in lenses that work at a small aperture.

In extreme wide-angle lenses, however, the effect is so marked that it is necessary to reduce the central illumination with an opaque diaphragm. This is sometimes made in the form of a metal star that can be rotated during exposure by blowing on it.

Angle of View. The closer the plate is brought to the lens—i.e., the greater the distance of the object focused—the greater the angle subtended by the diagonal of the plate at the lens. This means that the maximum covering power is required when the lens is focused on infinity. As the lens is moved away from the plate to focus objects closer than infinity, the angle of view becomes smaller and the covering power required is correspondingly less.

In the same way, a long-focus lens can do with a relatively lower covering power than a normal lens for a given negative size, while a wide-angle lens needs a relatively greater covering power.

For most practical purposes, a lens should cover a negative of diagonal equal to its own focal length. Where the camera is equipped with movements, it must have a lens of greater covering power than this or the use of any of the movements will take part of the plate outside the limits of the field. The general practice is to use a lens of slightly longer focal length than the one normally specified for the negative size—i.e., for a quarter plate camera fitted with movements, a lens would be used which would comfortably cover a 4 × 5 ins. plate. K.J.H.

Covering Power of Image Silver. The weight of silver, expressed in grams per square decimeter of image area, that is required per unit of density is referred to as the photometric constant or equivalent and is a measure of the light obscuring, or covering power of the silver pigment.

The covering power changes with the size of the developed grain and to some extent with the density at which it is measured. A large grain can be regarded as stopping light in direct proportion to its projected area just as any opaque macroscopic body casts a shadow.

The grains of a developed emulsion, even if having a filamentary fine-structure, may be considered as being roughly spherical. On account of the less advantageous ratio of area to volume the larger grains will utilize silver less efficiently as a pigment. Grains of 0·5 micron diameter need about 0·01 grams of silver per square decimeter for a unit density whereas grains of 1·5 microns diameter require about three times as much.

Very small grains of diameter less than 0·05 microns behave as though their optical diameter is greater than their mechanical diameter and the covering power of images formed from these small grains can be such that only 0·002 gram of silver is required per square decimeter to give a unit density.

Physically-developed images are more compact in their fine-structure than are chemically developed ones. The above observations hold for very fine physically-developed particles, but the physically-developed images of larger particles appear grey and are of comparatively poor covering power because of the higher conductivity and reflectivity of the more massive silver.

CREDIT. Printed acknowledgment of the origin of a photograph reproduced in a publication. Photographs may be credited to the photographer and/or the agency who distributes them and/or some third party who may have acquired all the rights in them. Policy varies from publication to publication, some never giving credit, others always doing so, and yet others crediting only those photographers whose names are already well known. Many newspapers have, in recent years, begun giving credit to their own staff men.

Agencies specializing in news photographs claim as a rule exclusive credit; whilst agencies dealing with general or pictorial subjects mostly ask for photographs to be credited jointly to the photographer and themselves. If their names have a readily recognizable commercial character there is no reason to object to this practice, but if they trade under some personal name there are obvious disadvantages to the photographer.

Advertising agents and their clients, who have commissioned photographs and so acquired some rights in them, usually try to obtain a credit whenever the photographer has an additional chance of publishing the prints in question. This tendency is human but regrettable. Experienced publishers recognize these demands as requests for unpaid advertising and generally offer to provide either a credit or a reproduction fee, but not both.

The photographer cannot expect to receive a credit unless he takes the trouble to see that his name is either on or at least clearly associated with the photograph. The best method of ensuring this is to use a rubber stamp on the back of the print. Its text should preferably be confined to the photographer's name, address and telephone number. Some photographers go to great pains to include elaborate references to their rights and conditions of reproduction, but such admonitions are unnecessary.

It is wise to stamp on to one's prints the words "please credit . . .", or something similar, but the practice of permitting publication of pictures for credit instead of payment is to be strongly discouraged.

See also: *Copyright and the Photographer; Press Photography; Selling Photographs.*

CRIME PHOTOGRAPHY

Photography is used by the police to provide evidence and supplement written reports on a very extensive scale. Much of this work is of a routine character and not connected with the actual investigation of crime. The equipment and techniques employed in this branch are those of normal photography. Most of such photography is done "in the field".

Apart from such general work, however, there is the highly specialized field of photography applied to crime detection. Much of this is carried on in the laboratory with special equipment and techniques. By the use of such technical aids as infra-red and ultra-violet lighting, and radiography, in conjunction with suitable filters and sensitized materials, criminal evidence that would otherwise remain hidden can often be brought to light.

The work of the laboratory carries on from where the outdoor photographer leaves off, so there is inevitably a certain amount of overlapping in the work of the two departments.

In presenting photographs as evidence, the pictures must not be retouched—or even spotted—in any way.

Equipment. For ordinary police work the official photographer uses a $4\frac{3}{4} \times 6\frac{1}{2}$ ins. camera, with an 8 ins. lens and a wide-angle lens of about $4\frac{1}{2}$ ins. in reserve. He probably carries a comprehensive range of filters for both general and specialized effects.

For lighting subjects in and out of doors he will have a supply of enclosed flash bulbs with, possibly, several high intensity Photoflood bulbs, reflectors and flexible leads and plugs. The special circumstances of this type of photography may call for fast, medium or slow orthochromatic or panchromatic films or plates.

In the police laboratory, the photographic equipment normally includes a long extension camera which can be fitted with short focus camera lenses to make low power photomicrographs up to about ten diameters. For greater magnification than this, the photographer may use any of the normal equipment for photomicrography. Finally, in addition to the usual lighting equipment—Photofloods, flash bulbs, and normal electric lighting units—there are special lamps for generating infra-red and ultra-violet light.

The infra-red light sources are used for such work as the examination of heavily dyed textile fibres, of sections of biological material where the natural pigment is reddish or brownish in colour, and for showing up the presence of carbon on sections taken from the lung in cases of suffocation by smoke. Various inks, sealing waxes, seals, markings of stamps, etc., often yield positive evidence by the use of infra-red photomicrography. Ultra-violet lighting is extensively used for revealing forgeries and fakes. Many inks and pigments which are apparently identical by ordinary lighting fluoresce differently under ultra-violet lighting, and the difference can be photographed.

There is rarely any provision for making X-rays in the photographic department: when X-rays are required the work is done by independent specialist firms.

Routine Work. The ordinary police photographic department staffed by policemen photographers is concerned almost entirely with straightforward pictures of scenes of crime and personal identification. This makes up the overwhelming majority of photographic jobs in crime photography.

The Scene of the Crime. A typical outdoor job for the police photographer might be to take a number of exterior scenes of a house showing the point of entry of a criminal, of a field, wood or lane connected with a murder or sex offence, or a road showing the scene of an accident. Photographs are wanted here to show how the particular spot was approached, its visibility from the highway and its connexion with other points of evidence.

If the crime has been committed in a building it is usual to take several general views from different angles, using a wide angle lens in conjunction with Photoflood or flash bulbs, or using daylight, with or without artificial light of some sort, as illumination. Special attention is paid to recording the position of various objects with relation to the body (if there is one), the point of entry or exit of any intruder, etc. Care has to be taken when using the wide angle lens to avoid distortion caused by having the back of the camera out of vertical.

Close-ups of important details are generally desirable. For these and the more general shots the proprietary self-contained lighting units are very valuable. They not only ensure adequate and correct illumination, but also allow the operator the opportunity, which he does not get with flash, of examining the picture before exposure. This allows him to avoid "flares" and reflections from mirrors, polished fittings, glass, etc., in the final picture. Backed plates are normally preferable for interior work since they yield brighter and clearer negatives.

The use of available daylight brings in a number of minor complications, since its uncertain and fluctuating intensity and the often unsuitable position of windows make calculations of exposure rather rough and ready and may render it difficult to take an exposure in the direction required. These difficulties are overcome by obscuring windows with curtains and using artificial illumination only. On occasions this may be supplied by flash bulbs or powder.

In the latter case the lens is focused on an ordinary hand torch directed towards the camera from each side of the subject in turn. This also helps to define the limits of the picture.

Hans Gross, the authority on criminal investigation, lays down as an inviolable rule that one must " never alter the position of, or pick up, or even touch any object before it has been minutely described in the report". The value of the camera to make a record of the position of objects in relation to each other and to reproduce a likeness of an object, as it was when first found, thus becomes very obvious indeed.

Generally, the photographs considered essential are those showing the scene itself, any points of exit and entry, all points where traces of the crime are found or where they might have been expected but where, in fact, there are none, and the scene from spots where witnesses have seen, or could have seen, anything. A number of these points are often not apparent in the early stages of the investigation so that it is generally considered better to take too many rather than too few photographs. Views are, therefore, taken from several sides and angles, and reference points from which the photographs were taken are marked on a diagram of the scene. The diagram is also marked with measurements showing the space-relationship of various objects.

The photographs, beside being an impartial record of things present in their state when photographed, will also indicate things which should normally be there but are in fact absent. Sometimes such missing elements may lead to important deductions being made as to how, or by whom, the crime was committed.

Road Accidents. In road accident cases, photographs are taken along the lines of approach of the two vehicles at the eye height of the sitting drivers in order to demonstrate their field of vision before the accident. Photographs are also taken of the actual scene and of any marks on the cars which may have special significance. There may be skid marks on the road, or, in the case of hit-and-run incidents, tyre impressions on the victim's person. How these are photographed will depend upon the nature of the surface and the type of impression. A polaroid screen may be used over the lens to reduce reflection from the road. Or again, the photographer may actually make use of the glare caused by oblique daylight to obtain better contrast, taking vertical photographs of the surface with the camera stopped down to less than *f*11, and a process film to give extreme contrast. If it is visible, the whole tyre impression is recorded in the hope that, besides identifying the make of tyre, the particular tyre in question may show a characteristic marking or fault such as a peculiar cut, mark, etc.

Identification. Photographs for identification are taken in a studio with light-coloured walls. Profile and full face photos of one-eighth scale, and full length photos of 1/27th scale are taken on the fastest type of plate or film, with the lens at about the level of the sitter's face or chest height for the full length photograph. The focal length of the lens used is at least equal to the diameter of the picture in order to produce satisfactory perspective free from distortion of the features.

When recording diseased skin, bruises or injuries, use may be made of suitable filters and plates to obtain either faithful or deliberately exaggerated reproduction. By using an orthochromatic plate and no filter under artificial lighting, for instance, it is possible to show the extent of bruises not readily visible to the eye. They can be made even plainer by the use of colour-insensitive plates.

"Candid" motion pictures are a valuable means of identification. Pictures taken of a man "off his guard" often reveal characteristic movements of hands, feet or head which can identify him as surely as a physical peculiarity. X-ray photographs of teeth and other portions of the anatomy, preserved in practitioners' records, have, on occasions, been produced for positive identification of living people and even more particularly for identifying cadavers.

Identification after Death. It is possible to identify corpses even in such a state of advanced decomposition that the features are unrecognizable. For this the police need a photograph of the person when alive, for preference one of the full face. A life-size negative of the skull of the corpse is taken from the same angle as the existing photograph, and printed as a transparency. The existing photograph is then enlarged to life size and placed under the transparency and in register with it. The extent to which the eyes, ears and nose in the "living" photograph correspond to the sockets, the bony aperture and the nasal bones, respectively, in the transparency of the skull is noted. These observations will go a long way towards establishing whether the two photographs belong to one and the same person. If "living" photographs of the head in three different positions are available, the certainty of identification can be almost guaranteed.

Dusts. High power photomicrography of occupational dusts—i.e., fragments of material found in a place where a man works—has often proved of importance in identification, or in connecting suspects with a crime.

Coal particles can be embedded in the skin of miners and remain for a number of years; certain classes of quarrymen are found to have tiny particles of silica in their breathing tracts and lungs. The presence of shellac in the system has been used to identify the body of a french polisher. Similarly, metal particles are associated with engineers, cereal and malt debris with brewery workers, and flour and starches with adhesive makers, bakers, millers, etc. There are many similar associations.

Finger Prints. The pattern of the ridges on the finger tips is an unchanging characteristic of the individual. These ridges contain sweat glands, and pressure of the finger tip on a surface leaves an extremely fine film of grease in the characteristic pattern of the ridges. When the finger prints are dusted with an appropriate powder, the powder clings to the greasy parts and can be blown away from the rest of the area, leaving a finger print impression that is visible and can be photographed.

Suitably coloured dusting powders are used to contrast with the colour of the surfaces bearing the finger prints. Black powder is used for a white ground, white powder for a black ground. On coloured grounds, a coloured powder is used in conjunction with the plates and colour filters that give the best contrast. On multi-coloured backgrounds the photographer uses a powder which fluoresces in ultra-violet light to give a photograph in which the glowing finger prints are shown against a dull background.

The photographer's ingenuity is demonstrated at its best when finger prints occur in different places, for example, on both sides of a piece of glass. In this case the prints on one side are dusted with a light powder and photographed against a dark background, after which the prints on the reverse side are dusted with a black powder and photographed against a white background.

Prints in an inaccessible position may often be photographed by employing a mirror. In this case unbacked film loaded the wrong way round in the holder is used to counteract the reversal of the image by the mirror.

The police record finger prints by pressing the fingers first on to an inked pad and then on to a sheet of white paper, so that the ridges show as black lines on a white background—the reverse of the picture produced when white powdered or light-coloured powdered finger prints are photographed against a dark background. For this reason the tones of the photograph must be reversed, but it must not be reversed laterally—i.e., from left to right. To do this a lateral reverse negative is prepared by taking a photograph on an unbacked plate or film, loaded back to front in the camera. From this an intermediate positive transparency is prepared. The positive is then printed in the usual way to obtain prints for comparison with the actual police records, the effect being that of a finger print made with black ink on white paper. For finger prints consisting of dark coloured powder on a light background, a direct photographic method without reversal gives the required print for comparison.

There is a simpler method of correctly recording light powder finger prints on a dark ground. If the plate holder is loaded with bromide paper and the finger prints are photographed through a prism or optical mirror set up in front of the lens to correct the lateral reversal, developing and fixing in the usual way gives the desired print. But by this method only one print can be made for each exposure in the camera and any duplicate records must be made by repeating the whole process.

Another method preferred by a number of police departments for photographing light powder impressions is to expose on a thin fast film, and then develop, bleach, expose to white light, and redevelop. This gives a positive which can be printed in the normal way to give a black on white image with no lateral reversal.

For comparing finger print photographs, enlargements of about 6 diameters are used.

The skin on the finger tips of corpses is often hard and shrivelled, making straightforward finger printing impossible. In such cases, the skin is stripped from the fingers, well massaged with a barium paste, mounted between glass plates and photographed with rear illumination and the lens stopped down to at least *f* 45.

Foot Prints. Police photography also includes an appreciable amount of foot print recording either in the form of dust prints on linoleum, etc., or impressions in earth, clay, or snow.

Examination of car tyre marks also comes into this category.

Technically the photography is fairly straightforward. Oblique illumination is often used where practicable to show up impressions in relief.

Debris on the Person. Photomicrography at magnifications of the order of 200× to 750× is employed for the examination of fibres from clothing and other textiles, paper, hairs, dust and debris, which may be found in the turn-ups of trousers, in pockets, under the finger nails, and in the aural or nasal cavities.

In many instances the debris under the nails or in the turn-ups of the trousers has been obviously connected with the scene of a crime—e.g., fragments of pine needles have been found in the clothing of the murderer when the crime took place in a wood, and alum and wood fragments have been photographed on the clothing of a thief who had ripped open the fireproof lining of a safe.

Debris in Lungs. The lungs of a corpse often reveal significant traces of foreign matter that help in establishing the cause and time of death, and even of identity (when they contain characteristic occupational dusts). Thus in the examination of the lungs of a man who had died in a fire, sections of the lung tissue stained red with safranin and photographed through a red filter on panchromatic film clearly showed carbon deposits in the tissue. These deposits indicated that the man had inhaled smoke from the fire and therefore was alive before the fire had started.

The debris in the water of ponds, ditches, and streams is often characteristic, and, in the case of drowning, the water in the lungs should contain debris of a similar nature to that in the water in which it is presumed death occurred. Totally different debris present could indicate that the body had been taken away from the site of death, and put into the water in another place to divert suspicion and enquiries from the proper quarter. Debris present in the lungs, therefore, allows the real site of death to be sought and identified with accuracy.

Stains on Clothing. Bloodstains are such a vital form of evidence that many complex techniques have been developed to detect them on all types of surface and material. Photography in one form or another enters into all these methods of detection since a photograph is more convincing than a mere statement. But in many instances the evidence could not be revealed without the photographic process.

For instance, a coloured fabric may be more transparent to infra-red rays than a bloodstain of the same colour. Or the stain may have been caught in such a state that it is relatively opaque to the rays. In these circumstances the presence of otherwise invisible stains, their shape, and location, can be reproduced by infra-red photographic technique.

Blood, seminal stains, starches and occupational dusts of various types are also revealed by photomicrography which shows up the characteristic structure and colour of the material; and by ultra-violet radiation which characterizes certain substances by their fluorescence. Alternatively, infra-red illumination may be used, in particular to reveal internal structural detail normally masked by a skin or covering that is relatively opaque to visible light.

Poisons. In poisoning cases, particularly in tropical countries, the presence of certain vegetable poisons may sometimes be established by the presence of characteristic fragments of berries, leaves, stems or roots. Photomicrographs of such fragments assist the prosecution in explaining to the judge and jury reasons for identifying the material as the vegetable poison used. Various dried and ground insects, powdered glass, and finely chopped hair are other fairly common native poisons whose presence can be clearly shown in photomicrographs.

Firearms and Ammunition. Guns and bullets can usually be identified by the striations left on the bullets by the rifling of the weapon, the indentation left by the firing pin in the soft metal of the firing cap, the impression made on the base of the cartridge by being forced into contact with the gun breach by the pressure of the explosion, and the marks of the extractor and the ejector on the cartridges. Evidence of this type is photographed generally at magnifications of about 3× to 6× on slow or medium speed emulsions giving good or high contrast. The lighting is arranged to shine almost parallel to the surface so that it throws the indentation, scratches, etc., into bold relief. (This method is also used for examining markings on metal objects like coins, in forgery cases, and door locks, window hasps, etc., in "breaking and entering" cases).

Textiles may be examined for stains and powder marks from bullets, etc., under infra-red light. The opaque carbon from the explosive blown from the muzzle of the firearm is deposited around the point of impact of the bullet according to certain well-known laws.

For instance, if these marks can be seen, as on a white fabric, it is possible to calculate the distance, etc., of the muzzle from the fabric at the time of firing. With coloured fabrics, the location of this carbon deposit is not so easy to see, but when photographed by infra-red light the conformation of the carbon deposit can be clearly demonstrated, even on dark coloured textiles. Where the bullet enters the body it generally leaves a ring of metal which can be seen in an X-ray photograph where the metal, being relatively opaque to the X-rays, shows up as a distinct image.

Documents. Documents on which the writing has been obliterated by dirt or by the passage of time are regularly photographed by infra-red illumination to reveal the original writing. This is possible because infra-red light generally shows a higher penetrating power than visual light, and although it is invisible, infra-red light affects the emulsion of the special plates and produces a visible photographic image. Sometimes even charred documents can be made legible by this method provided the charring hasn't gone too far.

The same technique has also revealed alterations in passports, additions in different ink or with different pencil to documents, and it has often been successful in reproducing indecipherable drawings and sketches which have been obliterated by rubbing or other mechanical friction.

Since some inks are more opaque than others to infra-red rays, it has been possible to reproduce the hidden text in certain passages of old manuscripts which have been obliterated—e.g. by the censor of the Spanish Inquisition. More recently the method has been used to reveal the original document when dirt, inks, etc., have been deliberately applied in an endeavour to hide the underlying text.

Writing. Questions of the authorship of handwriting are tackled by comparing low power photomicrographs of specimens. Negatives of individual letters enlarged from five to ten times can be superimposed to show similarities. The same type of test is used to establish typewritten text as the work of one particular machine. No two typewriters give identical impressions; there are always slight differences in the lining up of the type bars or even microscopic differences in the shape of the letters. All these points can be demonstrated with photographically enlarged reproductions. Photography by oblique light can also reveal the impression left on a pad after the top sheet with the writing on it has been torn off.

Colour Photography. Most laboratories are making increasing use of colour photography and photomicrography to make the point of their pictures clearer. In the U.S.A. a portable back-projector is sometimes taken into court to show colour transparencies on it. J.A.R.

See also: *Evidence by photographs; Forgeries; Identity photographs; Police photography; Traffic photography.*

Books: *Photography in Crime Detection*, by J. A. Radley (London); *Photographic Evidence*, by C. C. Scott (Kansas City); *Suspect Documents*, by Wilson R. Harrison (London).

CRITICAL ANGLE. When a ray travels from a denser to a rarer medium it is refracted away from the normal: the angle of incidence at which the emergent ray just grazes the surface is the critical angle. For all greater angles of incidence most of the light will be reflected; some will be absorbed, but none will be transmitted.

See also: *Refractive index.*

CRITICISM OF PHOTOGRAPHS. Frequently photographs are submitted for appreciation, criticism and discussion at club gatherings, in postal portfolios and to the editors of amateur magazines.

Inevitably the opinions thus evoked may reveal more about the person voicing them than about the work they are supposed to evaluate. In fact, art criticism at its best is a literary form in its own right designed to inform or entertain the public or just show off the brilliance of the critic rather than to help the artist to do his job better next time.

Critics. Yet amateurs seeking criticism do so mostly in the hope of benefiting from the encouragement and advice they may receive. Anxious to be helpful, the expert is easily tempted to put forward emphatic views by making the most of any little point he happens to notice or by groping for concrete recommendations, too often based only on his own vague beliefs or established habits of thought.

Firm adherents of some school of thought, ideologists and reformers are notoriously aggravating as critics and so are, of course, active artists who have developed their own individual approach to the very type of work they are supposed to review. The critic with vigorous or rigid standards cannot be expected to possess the necessary tolerance and sympathy for work not on his own level.

And apart from the critic's own qualifications, isolated opinions based on a single piece of work are apt to be misleading. It is impossible to judge another man's work from a single print. The critic must see at least a few of his photographs to be able to detect what the photographer is after, what may be intentional, and what is accidental in his style and technique.

Scope. Certain points merit examination in any photograph.

(1) The photographer's purpose in taking it.

(2) Has his technique been wholly and successfully subordinated to that purpose?

(3) Would it be possible to produce from the negative a better print?

(4) Was the photographer's approach original or imitative?

The purpose of the photographer in making a particular picture must receive first consideration; whether his work appears to fulfil that purpose or fall short of it is, in fact, the only point that really matters. If the man behind the camera happens to be solely concerned with some rare or subtle expression in his subject's face then he should be criticized only on account of anything that may distract from that expression, no matter whether the distracting element by itself represents good or bad photography. Thus not even the degree of technical perfection is an unfailingly reliable guide to criticism.

Yet technical criticism proper is comparatively easy or, at least, harmless, even if it falls into errors. Objections like "This is over-exposed, this is unsharp, this is too contrasty" can be readily checked and so possibly refuted by anybody looking at the same picture. Often enough what may appear to be a technical fault is a deliberate mark of style.

Critical suggestions about how to trim a print, angle it differently, throw back this detail or bring out that, may serve a useful purpose, as long as all this advice does not conjure up an altogether new conception unrelated to the photographer's real intention. Criticism is reasonably safe if it is confined to discussing shifts of emphasis rather than to changes in subject matter; it may usefully suggest leaving out something from the author's work but it would be grotesque for it to demand the inclusion of anything additional. You can criticize only what is there.

Limitations. The critic ventures on to very shaky grounds indeed once he airs ideas of his own about how the picture should have been taken. He can seldom be sufficiently acquainted with all the relevant conditions to be able to pronounce really valid opinions on them. What is still more important, he is never in the position to prove his point if challenged. Judges in court aim at judgements which are both just and appear to be just; by the same token, critics must aim at opinions which are both justified and can be proved to be so.

This condition is not met by setting forth the critic's ideas in diagrammatic reconstructions which purport to represent some pictorial analysis or outline an improved version of the photograph criticized. Diagrams are very useful to picture some general statement, but in order to translate a particular photograph truthfully they would have to be drawn, not by hand, but by the same lens that took the picture and at the same spot where it was taken. Sketches not only eliminate most photographic values but may distort and evade awkward issues by using a non-photographic idiom. It is just as futile to suggest improvements to a photograph in an elegant drawing as it would be to try to correct a piece of sculpture by whistling a tune about it.

The critic is privileged to appear clever at the expense of the work criticized but it is an abuse of that privilege to make up arguments extraneous to the subject, to pronounce opinions as if they were rules and to claim ultimate authority in a realm of abstract speculation where none can possibly exist. Any honest critic realizes that as it is his job to be an expert in afterthought, he should at least try to avoid any pretence to creative talents. "Creative" criticism is a contradiction in terms. A critic should offer the product of his experience, knowledge and skill acquired from studying the work of others rather than his own creative urges, prejudices and frustrations. Looking at a print he may be able to suggest that the photographer ought to take more care of his technique or that he should change some of his tools or that he might just as well give up the game altogether. But, being another man with eyes and ideas of his own, he can never take a hand in the actual work without turning it into something that was probably never intended. A.K.-K.

See also: *Composition; Judging exhibition entries; Visual appeal.*

CROMER, G. 1873–1934. French photographer. Owner of a famous collection on the history of photography, the greater part of which is now at George Eastman House, Rochester, N.Y.

CROOKES, SIR WILLIAM, 1832–1919. British physicist. Investigated the spectral sensitivity of the wet collodion process into the ultra-violet. With John Spiller devised the first "dry" collodion process, 1854. Used light filters for the photography of coloured objects (1854–60). Founder in 1858, and for five years editor of the *Photographic News*. He was Secretary of the Royal Photographic Society in 1857.

CROPPING. American term for trimming of a print. In its wider sense it applies not only to physical cutting, but also to the elimination of unwanted picture areas in printing—i.e., trimming to improve composition.

CROS, CHARLES, 1842–88. French doctor, poet, philologist and painter. Discovered in 1881 the imbibition process, and was a notable pioneer in the theory and practice of three-colour photography (1869).

CROSS, CHARLES FREDERICK, 1855–1935. English industrial chemist. Patented in 1890, with E. J. Bevan, the Primuline process—the first successful method of diazo printing.

CROSS FRONT. Term indicating that the lens panel of a camera may be moved laterally to bring the lens to the right or left of the central optical axis.

See also: *Camera movements.*

CRYSTAL. A crystal is a piece of a homogeneous substance with a spontaneously formed regular geometrical structure characteristic of that substance. The structure may be a simple one or a derivative of a simple shape—e.g., developed from a cube, prism or pyramid. When, for instance, a saturated solution of a substance in water is further concentrated by evaporation, some of the dissolved substance may deposit in the form of crystals.

If crystals of some compounds are heated, they give off a certain amount of water (water of crystallization) and change into an anhydrous form. Such crystals are called hydrated.

Many photographic chemicals are obtainable in both hydrated crystalline and anhydrous forms. These obviously differ in the weight of active chemical present, but often they also differ in their chemical characteristics. So it is most important to know which form is intended when making up solutions from formulae.

See also: *Chemicals.*

CRYSTAL VARNISH. Clear varnish used for giving a protective finish to lantern slides and transparencies.

CRYSTOLEUM. Coloured positive made by combining a transparency and a hand-coloured image. The method is no longer popular.

CUBIC CENTIMETRE (C.CM. OR C.C.). Unit of volume in the metric system equal to 16·9 minims. In scientific work the millilitre (ml.) is largely used instead. For photographic purposes the difference between 1 c.cm. and 1 ml. is negligible.

See also: *Weights and measures.*

CURL. A number of sensitized materials—in particular, papers—tend to curl after processing. This is because of the difference between the shrinkage of the emulsion coating and its support.

The first roll films were troublesome in this way until it occurred to the manufacturers to coat both sides of the celluloid with gelatin—only the coating on the front being sensitized. This produces an even drying tension on both sides of the support so there is no tendency to curl. Modern films that employ certain plastic bases are less prone to curl.

There are several ways of flattening papers that curl after processing.

Certain photographic papers have a strong tendency to curl before use. It is best to minimize this either by keeping boxes containing the paper full, or by compressing the paper in some way within the box. Once curled, the paper should be straightened with extreme care, to prevent stress marks, and cracks in the emulsion, the latter being a particular risk with glossy surfaces.

See also: *Drying.*

CURVATURE OF FIELD. One of the off-axis aberrations of lenses, which causes the plane of focus to be curved.

See also: *Aberrations of lenses.*

CUSTOMS. Photographic equipment taken out of the United Kingdom by a photographer for his own use and brought back again by him is admitted duty free if the Customs Officer at the port of landing is satisfied that the articles are of British manufacture and, if not British made, that they had not been previously imported without payment of any duty properly due.

Purchase Tax is normally not payable if the Customs Officer is satisfied that the articles have already borne tax and that they are in substantially the same state on re-importation as when they were taken abroad—e.g., that the lens has not been exchanged for a costlier one.

The equipment must be produced to the officers on re-importation, with any evidence which can be produced in support of its title to free admission—e.g., a bill from a retailer in this country showing when and where the articles were purchased or the original customs receipt.

The Commissioners will not issue certificates to passengers leaving this country for articles contained in their baggage, nor will they keep a record of any such articles pending return.

Photographic equipment brought into the United Kingdom by tourists from abroad is usually admitted without difficulty provided the equipment is for the personal use of the owner. It must not be sold or otherwise disposed of during the owner's stay, unless duty and tax are paid first.

Customs regulations in other countries mostly follow a similar pattern; a reasonable amount of equipment for a tourist's own use is admitted (usually consisting of one still and one cine camera with accessories and 5–6 films). However, residents of the country may have to pay whatever duties and taxes are applicable.

Returning residents of the United States are granted a $100 customs exemption provided they have been outside the United States for at least 48 hours and have not claimed an exemption in the last 31 days. In addition, there are restrictions on foreign equipment bearing certain trade marks owned by firms in the U.S.A. Such trade marks must either be obliterated, or the owner of the apparatus must obtain a release from the American owners of the trade mark. Certain firms have also given a blanket release permitting a specified number of articles of their brand to be brought in without any restriction.

If in any doubt on current regulations, it is wise to contact the local office of the Customs and Excise Department for full details and also, if necessary, the Embassy or Legation of the country to be visited, who could clarify the position as related to their own regulations.

CUT FILM. Form of negative material similar to roll film but in flat sheets made to the standard plate sizes and now generally used instead of glass plates.

See also: *Films.*

CUTTING. Operation of trimming and joining film shots together; in its wider sense also the timing of such shots in length. A single instantaneous transition from one shot to another is a cut.

Cutting can be regarded as the more mechanical side of film editing, and the person who does this in a motion picture studio is the cutter—usually an assistant to the editor. He handles a print of the whole film specially made for editing purposes, this is known as a cutting copy or a cutting print. The original or negative is cut to match it when editing is completed, and then carefully stored for making final release prints (but never projected).

See also: *Editing.*

CYAN. Name applied to the blue-green colour reflected by a surface which absorbs red, for which reason it is also referred to as minus red. It is one of the three primary colours used to form the image in subtractive colour processes, the other two being minus blue (yellow) and minus green (magenta).

CYANOTYPE. (Blue print.) Printing process giving a white negative image against a blue ground.

See also: *Blue print; Document copying; Ferroprussiate process; Printing processes* (*obsolete*).

CYCLE. Complete single unit of a repetitive movement, variation or event, or sequence of such movements or events. The cycle is a convenient way of describing both the sequence and the rate at which it is repeated (in cycles per unit time).

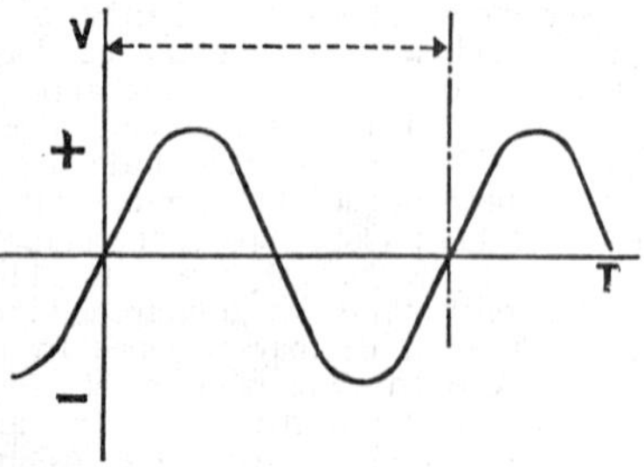

CYCLIC VARIATION. The voltage V plotted against time T rises to a maximum, drops to zero, reverses direction to a negative maximum and returns to zero. This constitutes one cycle (dotted arrow) after which the process repeats itself.

A cyclic variation always implies a change which periodically returns to its starting point. Graphically, the simplest cyclic variation is a sine wave, which can plot the change of voltage, light intensity, sound amplitude, etc. against time.

In referring to light or other radiation, the cyclic frequency is the number of waves that pass a given point per second. This depends both on the wavelength and on the speed of propagation.

Applied to an alternating current of electricity, the cycle is a single complete fluctuation in potential from zero, through the positive maximum, back to zero, the negative maximum and to zero again.

Applied to sound, the cycle is again a complete wave and the frequency in cycles per second is the number of waves passing a given point. The human ear can hear sound vibrations from about 20 to nearly 20,000 cycles per second.

See also: *Frequency, Sine wave.*

CZECHOSLOVAKIA. The discovery of photography met with an immediate response in the Czech lands. The Czechs and Slovaks in no way remained behind the other civilized nations of the world, either in the early daguerreotype or in the wet collodion process or subsequently in other fields of photography. They also made their contribution to the progress of photography.

Contributions to Photography. Among the leading figures was Dr. Josef Petzval, a native of Spišská Bělá in Slovakia, who as early as 1840 designed a lens of good quality with a high speed—for that period—of *f* 3·6. Thanks to him it was possible to cut the exposure time of daguerreotypes to several seconds.

The great reformer, philosopher and poet Jan Evangelista Purkyně (born in Libochovice in 1794) perfected a stroboscope for viewing animated pictures through a slotted disc, and in the 1860's, with his instrument known as the cinescope—i.e., motion viewer—for the first time used a series of photographic snapshots to portray motion. In this way he undoubtedly contributed to the speedier development of cinematography.

Photography also captured the attention of Jakub Husnik, 1837–1916, a professor of drawing in Prague, who in 1863, after numerous attempts, produced two-tone photography. By studying the effect of light on dichromates, Husnik made the discovery, in 1869, of a collotype process. He also discovered a process emulsion for the photography of originals yellowed with age.

Karel Václav Klíč, 1841–1926, painter, photographer, journalist, satirist, caricaturist and printer discovered a technique of heliogravure and copper plate printing which makes possible the speedy and perfect copying of half-tone photographic reproductions. In 1879 he perfected the first photogravure process. In 1895, working at Lancaster and Accrington in England with improvised lighting equipment, he etched the first intaglio copper cylinder (rotogravure).

J. Mahler perfected the method of observing stereoscopic pictures by utilizing the phenomenon of polarization in the form of vectographs—i.e., stereoscopic pairs differentially polarized and printed or projected as a single image which is subsequently split up into right and left eye pictures when viewed through suitably oriented polarizing spectacles.

Contemporary colour photography technique, utilizing colour-forming units, is based on Fischer's and Homolka's discoveries of 1912. Benno Homolka, a German chemist of Czech origin, discovered carbo-cyanine dyes for colour sensitizing panchromatic emulsions.

Photographic Education. The historic Karlova universita (Charles University) in Prague encourages the study of scientific photography. This is directed by the Institute for Scientific Photography, Photographic Chemistry and

Photographic Physics, which is part of the Physics Department at the Fakulta přírodních věd (Faculty of Natural Science). The student may study these special subjects when he has completed four terms on the basic syllabus. The University primarily encourages research work, but at the same time provides an education suitable for production work in the photographic industry.

The Department of Photography at the Průmyslová škola grafického umění (Industrial School of Graphic Arts) in Prague and the Department of Photography at the Vysoká škola umělcko průmyslová (High School of Industrial Arts) in Brno both provide training suitable for practical photographers.

At the Brno Vysoká škola umělcko průmyslová, photography is taught either as the main subject or as a supplementary subject in other fields—e.g., the graphic arts. The studies include drawing and design (drawing ability is demanded in the entrance examinations) physics, chemistry and various art subjects.

The growing demands placed on photography to-day have necessitated a considerable extension of the education of young photographers. Third year students at the Photographic Department of the Průmyslová škola grafického umění in Prague are therefore divided into special groups such as medical photography, archive and museum photography, display art photography, etc. They are given every opportunity to co-operate with scientific institutes and the press.

The highest educational institution of art and film photography in Czechoslovakia is the Filmová fakulta při Akademi musických umění (Film Faculty at the Academy of Music and Dramatic Arts) in Prague, which has a special department for film cameramen, where the techniques and aesthetics of static and film photography are taught.

Besides these institutions the Ministerstvo Kultury (Ministry of Culture) and the Ústřední ústav lidového umění (Central Institute of Folk Art) sponsor courses in photography and home movies for amateurs. Attending these courses every year are workers from factories and offices who, upon completion of the course, co-operate with the regional bodies of the National committees to develop photography and home movies either as members of advisory committees in regional and district organizations, or as consultants at local advisory centres of folk art or as instructors in photography and movie making to hobby groups attached to factories.

The magazine *Czechoslovak Photography* also deals with the ideological and specialized training of photographers. Its editorial board co-operates with representatives in all fields of Czechoslovak photography. The task of the board is to use the magazine to increase popular interest in photography and to raise the artistic level of photography. A quarterly review *Fotography*, printed also in Russian, tries to establish firm views on art and the aesthetics of applied photography. It also aims at maintaining a contact between photographers from East and West.

Photographic Organizations. Amateur photographers are members of photographic circles at the factory trade union clubs. To-day there are several tens of thousands of photographers organized in works' clubs, educational centres for adults, youth and young pioneer organizations.

For their benefit, exhibitions, courses and lectures are organized. For example, the Advisory Committee for Photography at the Dům Kultury (House of Culture) in Prague sponsors lectures on such subjects as: photographic theory and composition; colour photography; creative understanding of press photography; aims of photo art; photographing architecture, landscapes, etc.

Besides this, there are smaller groups of photographic workers in Czechoslovakia. These are, in particular, the photographic artists organized in the photographic section of the Svaz československých výtvarných umelcu (Union of Czechoslovak Artists); photoreporters, organized in the Svaz československých novinářů (Union of Czechoslovak Journalists); and professional photographers who, since 1946, have been organized in co-operative and communal enterprises.

Industry. Although the Czechoslovak precision and optical industry is long established, it has made its greatest strides in the past fifteen years. During this period several factories have been merged into a single enterprise which controls production of diverse products at various branch establishments.

The Meopta Works make enlargers, projectors both for amateurs and professionals, telescopes, laboratory instruments, lenses, microscopes, and a wide variety of accessories as well as still and movie cameras. Between 1946 and 1960 production increased by nearly 700 per cent. The goods produced are exported through the Kovo Foreign Trade Corporation to some seventy countries throughout the world. The largest market is in Great Britain, Canada, the U.S.A., West Germany and Eastern Europe.

Artistic Outlook. In artistic photography there is no desire to ape the art of painting. On the contrary it is taught that, if photography is to gain the respect it deserves as the youngest creative art, it must rid itself of the typical influences of painting. This means, first and foremost, that photographers should aim at pure photographic expression and the preservation of the character of photography.

It is life, movement, spontaneity, naturalness and truth that correspond most closely to the character of photography. The essence of photography is documentation and its choice of themes from daily life is unlimited.

Photography must not seek to remove itself from life in special forms of expression. The most modern photographic technique is an achievement of advanced science and a product of workers' hands; it permits us to capture what no other field of art can. Its mission is to reflect reality. It can fulfil this aim only if it speaks a language which is not only understandable and clear, but possesses a content and form which are rich and beautiful. In this way photography can become a significant factor in raising the creative abilities of ordinary men and women and in developing their taste and artistic understanding. V.J.

D

DAGRON, PRUDENT RENÉ PATRICÉ, 1819–1900. French photographer. Patented, 1860, microphotographs with Stanhope lenses set as jewellery and in souvenirs; they were produced on collodio-albumen dry plates. During the Siege of Paris 1870–1 Dagron worked for the official pigeon post service, by which photographically reduced messages were carried.

DAGUERRE, LOUIS JACQUES MANDÉ, 1787–1851. French artist, inventor of the diorama (1822) and of the daguerreotype. He and, independently, Niépce worked on the fixation of the image obtained in the camera obscura. The inventors were brought together by the optician Charles Chevalier, in 1826, and signed in 1829 an agreement on co-operation "for the further improvement of the invention of Niépce which was perfected by Daguerre". The decisive step in the invention of the daguerreotype was Daguerre's discovery (1835) of the possibility of developing with mercury vapour the iodized silver plates the partners had been using. The final step was to fix the picture with sodium chloride (1837). The first report of the discovery was made by the French physicist Arago, on 7th January, 1839, to the French Academy of Sciences. The publication of the process followed on 19th August, 1839. A printed handbook was published at the same time. The process was bought by the French Government and given "free to the world" on 19th August, 1839—but had been patented in England, Wales and the Colonies five days earlier. The coming of the daguerreotype aroused vivid interest over the whole of the civilized world, and for some years the process was popular. Its one grave defect, that it produced one original only, was the cause of its gradual replacement by negative-positive processes whereby copies could be produced. Biographies by Mentienne (Paris 1892), G. Potonnié (Paris 1935), H. & A. Gernsheim (London 1956).

See also: *Discovery of photography; Niépce, J. N.*

DAGUERREOTYPE. Photograph taken by the method published by Daguerre in France in 1839. It consists of a positive image formed by mercury vapour on a polished coating of silver on a copper plate.

There are five steps in the daguerreotype process:

(1) Preparing. A sheet of copper carrying a coating of metallic silver (produced by electroplating, or by the Sheffield plate process) is buffed to a very high polish.

(2) Sensitizing. The plate is exposed to iodine vapour in a light-tight "iodizing box" until it is uniformly covered with an orange coating of silver iodide. This coating is light sensitive.

(3) Exposing. The plate is exposed in a camera for anything up to twenty minutes in bright daylight. The camera lens was sometimes equipped with a reversing mirror since the image produced is a reverse (right to left) positive.

(4) Developing. The image is developed by exposing the plate to the vapour of mercury heated to 140° F. (60° C.). The mercury amalgamates with the silver and clings to the highlights.

(5) Fixing. The unused silver iodide is washed off with a weak solution of sodium hyposulphite (common salt solution was used originally).

After washing, the image is now permanent; the highlights consist of a milky deposit of mercury amalgam and the shadows of plain polished silver.

Daguerreotypes went out of fashion after the introduction of the wet collodion process.

See also: *Discovery of photography; Sensitized materials history.*

DALLASTYPE. A modification of Paul Pretsch's photo-engraving process, reintroduced in 1863 by Campbell Duncan Dallas, one-time manager of Pretsch's Photogalvanographic Company.

DALLMEYER, JOHN HENRY, 1830–83. Optician and lens manufacturer in London (of German descent), connected by employment and later by marriage to the Ross family. Produced portrait lenses based on Petzval's design; patented (in 1866) the Rectilinear lens which is very similar to Steinheil's aplanat. Made photoheliographs.

DALLMEYER, THOMAS RUDOLPH, 1859–1906. English optician, son of John Henry Dallmeyer. Introduced to England a telephoto lens based on the Porro-Steinheil design, for which he received the Progress Medal of the Royal Photographic Society in 1896. Was President of the R.P.S. from 1900 to 1903.

DANCER, JOHN BENJAMIN, 1812–87. English optician and microscopist in Liverpool, later Manchester. Made the first photomicrograph in England by the daguerreotype process, 1840, and in 1853 was the first to make microphotographs (by the collodion process). In the same year Dancer made the prototype of a binocular camera proposed by Sir David Brewster, which did not go into production until 1856.

DANCING. As a subject for the camera, dancing shares all the problems of photography of fast-moving subjects by poor light and it has certain limitations, some of which are common to theatre photography. Straightforward flash-photography may be employed for factual records but ballet and other dance-forms call for a more sensitive interpretation. The work of the professional dance photographer is mostly confined to ballet, although many of its problems apply equally to other types of dancing.

A series of ballet pictures should convey something of the atmosphere and theme of the ballet. Lighting should be planned to illuminate supporting dancers, as well as the principals, whilst favouring the latter. Good composition and lighting are very important, followed closely by careful choice of the precise moment for exposure to capture (or imply, in a posed shot) the feeling of the movement. This is of special importance in pictures taken close to the dancers. A knowledge of ballet terms, attitudes and working methods, as well as familiarity with the particular ballet being photographed, is almost indispensable. And permission to take photographs is always essential before commencing.

Choice of Negative Size. Prints of dance photographs are commonly 12 × 15 ins. and larger in size. They must be produced very quickly and be of superb quality, even when part of a negative is enlarged to cut out some supporting dancer's faulty position—a recurring nuisance. So it is not surprising that a ballet company will often supplement the pictures taken by a 35 mm. expert with other photographs taken by a specialist using a larger negative size. These pictures are posed at a special photo-call. The equipment and technique are very similar to those employed by photographers who specialize in theatre photography. Usually in fact a specialist theatre photographer is called in, and it is difficult to distinguish between the pictures of an expert in this field and the best of the 35 mm. action pictures.

Pictures from larger negatives are of superior photographic quality and it is easier to produce large numbers of prints from them quickly for display and press distribution. Some specialists use the 2¼ ins. square negative, but the leading British stage photographers tend to prefer quarter-plate or even larger sizes. The twin lens reflex, the larger single-lens reflex and the view camera are all used by different specialists.

There are three recognized ways of taking ballet photographs: by normal stage lighting, with special lighting, with synchro flash.

By Normal Lighting. Stage lighting is much more contrasty than it appears to the eye. If recorded photographically, it is apt to be over-dramatic and not always appropriate to the nature of the production. So as a rule, the general, non-directional, illumination should, if possible, be increased when the stage spots are the main source of light. Under these lighting conditions professional ballet photographers get good results with 35 mm. miniature cameras. Wide aperture lenses and the fastest films are necessary, particularly for taking photographs during a public performance when no additional light can be used. Earlier workers often employed various methods of hypersensitization and even today the fastest emulsions are only just good enough at the speeds demanded by the action. A shutter speed of 1/100 or 1/125 second at $f\,1{\cdot}5$ can be used with a fairly fast type of film. This gives very little latitude when combined with development in a low energy formula to give fine grain, but this shutter speed and aperture are frequently required for the faster movements of a dancer.

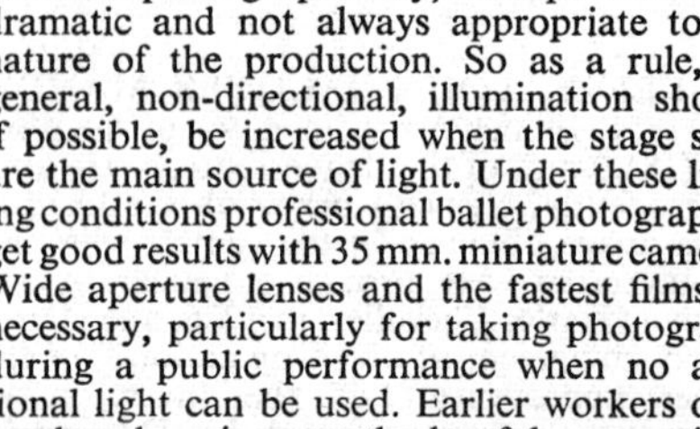

BASIC LIGHTING PLAN FOR DANCE SHOTS. Directional lights at each side of the stage provide the main illumination, while a floodlamp close to the camera fills in harsh shadows.

Special Lighting. The depth of field required demands an aperture of f 8 or f 11 at most and this means correspondingly longer exposures. For this reason certain dance movements must be omitted from a series of posed pictures. Professional stage photographers generally bring their own lighting equipment to the theatre. Two 500-watt high-efficiency spotlights, one or two 1,500-watt floodlamps and several 250-watt hand-lamps make a typical collection. With this lighting and careful choice of the "dead moment" of a climax or change of direction in the dance, exposures of $\frac{1}{4}$ or $\frac{1}{2}$ second may be made. By sensitive handling of these lamps and suitable directions to the dancers before exposure, the expert can recapture the true atmosphere and movement of the dance. With such lighting, and high-speed panchromatic plates or films, there is ample latitude for fine-grain development to a low gamma.

Theatre assignments have to be carried out in the shortest possible time as the cost per hour for rehearsals is high. Only a handful of photographers have achieved success in this sphere.

Synchro-Flash. Reliable and controllable synchro-flash equipment has made it possible to combine the advantages of snapshot technique with the pictorial and practical advantages of the larger camera. While the pressman's flash-on-camera secures only a record devoid of atmosphere, independently mounted flash units coupled remotely to the shutter contacts are capable of much more subtle effects in experienced hands. Some experiment is necessary to be able to predict the precise contribution that the combined flashes will produce; but a photographer with experience can produce in this way photographs possessing much of the atmosphere of the production itself.

The photographer must attend several rehearsals, and take notes of the most suitable moments for the pictures he will take at the photo-call when the company is his to command for an hour or so. The flash lamps are arranged before each scene, and the movements of the dancers are modified, if necessary, to suit his composition. The exposure is then made with the dancers in action, actually performing the particular part of the ballet. With ordinary flash bulbs, the shutter controls the actual exposure by opening for the required interval—e.g., 1/50 to 1/125 second at the peak illumination of the flash bulb. With the electronic flash tube the highest shutter speed—e.g., 1/500 second—may be used, because the duration of the flash is from 1/1000 to 1/5000 second. This is so short that the fastest human motion is too slow to blur the image.

Whether such speed is an advantage in dance photography is a matter of opinion. When there is no suggestion of blur some of the illusion of motion is lost, and many ballet photographers prefer expendable bulbs which, besides being compact, reliable and reasonably inexpensive, permit some control of the degree of blurring of the faster-moving parts of the subject. With properly designed circuits (usually series-capacitor-ignition) there is no limit to the number of bulbs which may be synchronized to fire simultaneously. The lampholders, etc., and associated equipment are portable and light in weight; but where large numbers of pictures must be taken, the cost of bulbs becomes prohibitive.

Electronic flash costs little in use, beyond its transportation cost, although this may be considerable because of the bulk and weight per unit of light-output. But the initial cost is high, and at least two linked units of 200 joules output each are needed to cover quite a small ballet group.

In both kinds of flash photography of the ballet the usual flash–distance–aperture factors for the particular materials employed may be used for a start; but the individual worker will have to modify the figures later to suit his own processing technique and the kind of picture he wants to make.

Other Dance Forms. It will seldom be possible actually to pose the dancers in a ballroom, even an exhibition couple at rehearsal. An opportunist technique will be more appropriate, with high-speed film and large-aperture lenses. The shutter-speed and aperture should be set to suit the lighting conditions (which will often be very poor indeed) and zonal focusing adopted, i.e., the focus is pre-set for the distance at which the best opportunities are expected to occur. This decision, and the choice of the most interesting moments in the movement, can be made early in each dance. The exposure is then made when the couple enter that zone for which the focus has been set.

Exhibition dances, by one or two couples only, will offer the best pictures, especially if the dancers are spotlighted.

Folk dancing and national dances are usually performed in the open air, so almost any hand camera may be used. Keeping the traditional nature of such dancing in mind, it is sound to include, as background, such of the surrounding location and audience as is likely to emphasize this, so long as it is subordinate to the main subject. A knowledge of the particular dance forms is valuable, enabling the photographer to anticipate the direction and movements.

Cabarets are frequently dimly lit, and large-aperture lenses with high-speed film are necessary. The characteristic atmosphere may be retained if, in a subsidiary manner, spectators at their tables are included in the picture area. The silhouettes of one or two heads framing the main subject, marginally, can be especially effective. J.V.

See also: *Theatre.*

DARK CORNERS. Dark corners on a print may indicate that the camera lens is not illuminating the whole of the plate area. This occurs when the covering power of the lens is not great enough for the plate in use. The lens is then said to vignette. The same effect happens when excessive camera movements are used. The remedy is to use a lens with a wider angle of field or to try stopping down the aperture.

Even if no vignetting occurs, the illumination at the corner of a plate is less than that at the centre, being reduced by a factor equal to the fourth power of the semi-angle of field. Naturally, this effect is accentuated in wide-angle lenses.

Dark corners may also be caused by using a lens hood which has an acceptance angle too narrow for the plate size. This cuts the corners of the negative, leaving them clear so that they print out darker than the rest of the picture. Light corners on an enlargement mean that the illumination of the negative is faulty or that the enlarging lens does not cover the negative fully. In a condenser enlarger it may be that the distance between the lamp and the condenser needs adjusting, that the condenser is the wrong focal length or that the lamphouse reflector is of the wrong shape. In a diffuser enlarger, uneven illumination or insufficient diffusion can also lead to dark corners.

See also: *Covering power; Faults.*

DARKROOM

Any room selected for use as a darkroom should have a volume of not less than 400 cubic feet; for comfortable working it should not be smaller than 8 feet × 6 feet × 8 feet 6 ins. high. It should be sited so that it keeps cool in summer, and it should not be near equipment that emits fumes, X-rays or other active radiations unless properly protected.

STRUCTURE

The design of a darkroom and the choice of building materials should be given adequate consideration.

Floor. The ideal floor should be resistant to photographic chemicals, and to staining, water-tight, durable, non-slipping when wet, resilient for foot comfort and a suitable colour. While it is impossible to obtain all these qualities in one material, a compromise can be achieved by selecting a floor that combines the most essential properties. An ordinary wooden floor is unsatisfactory because it is not water-tight and dirt and chemicals will accumulate in the cracks. Also, normal types of concrete flooring will not withstand the destructive action of photographic chemicals.

The two most serious effects on floor coverings are corrosion and staining caused by contact with photographic chemicals. The degree of staining and corrosion is usually dependent on the chemically reactive properties of the pigment of the covering material and the absorptive qualities of the floor material itself. For example, if a material is impervious to liquids, spilt developer can remain on the surface until oxidized, but the residue may still be wiped off with a damp cloth. If the floor covering material is porous enough to permit absorption of the solution, the residue will be difficult to remove and discoloration will be almost certain to result.

The darkroom flooring must permit frequent wet mopping without damaging any ceiling below. This means that, when composition tiles are used, both the adhesive material and the flooring material itself must be water-tight.

No attempt should be made to lay a waterproof floor over an existing floor unless the latter is in good condition. All boards of wooden floors should be solid and fastened securely in place. If the boards move under traffic, they will break up the waterproofing. Concrete floors must be smooth and should show no signs of disintegration.

In every darkroom there is a chance of occasional spilling of water and processing solutions. This creates a potential hazard of slipping and for this reason highly polished or waxed flooring should not be used.

When constructing any of the waterproof floors discussed here, it is general practice to lay a fabric membrane before the waterproof floor is laid. Since paper membranes crack easily, an open-mesh cotton fabric saturated with pitch or a plastic based material is recommended.

Generally asphalt is considered the best possible choice of flooring material and satisfies most of the requirements mentioned. It may be laid in one piece and coved to form a skirting, thus preventing any chemicals attacking the foundation. Any settlement or movement of the building will not cause the surface to crack. If heavy equipment is to be installed, the flooring beneath the load-bearing points should be reinforced. If necessary, asphalt flooring can be removed completely without causing damage to the foundation or sub-flooring. Asphalt is available in a variety of colours and the surface is slip-proof.

There are also many types of P.V.C.-based and non-slip tiling which give excellent results. These materials are more attractive than asphalt but also more expensive. They are resistant to photographic chemicals and are reasonably slip-proof. The main disadvantage lies in the jointing and therefore, if plastic tiles are to be used, the bonding material must be resistant to photographic chemicals.

Walls. The walls should have a hard, smooth finish of light colour; rough surfaces accumulate dust and are difficult to clean. A wall painted with cream enamel or oil paint is easily cleaned and reflects the light. Dark coloured walls have no place in a darkroom.

Tiles make a wall surface that is easily cleaned and has good reflecting properties. They should be set in acid-proof cement to prevent penetration of solutions. This is very necessary where there is risk of contamination from hypo, because hypo in contact with unprotected brickwork and concrete sets up gradual disintegration that is practically impossible to check.

Ceiling. The ceiling should be finished in matt white enamel or oil paint. It should not be lower than 8 feet 6 ins.

Light Traps. It should always be possible to enter or leave a darkroom without admitting any light. A labyrinth that turns through two right angles makes an efficient light trap. The walls should be painted matt-black to prevent reflected light from entering, and a white line, about 2 ins. wide, should be painted at about eye-level to assist entry and exit. No direct light should be allowed to shine in; if necessary there should be a baffle across the entrance.

To save space, double doors or double curtains may be used or, if space is very limited, a curtain across the door can be made to serve. When the light-lock consists entirely of curtains they should be made to open in the centre and have a generous overlap. They should be of thick limp material, well weighted at the bottom, and with a broad white band of tape stitched down each edge to mark the opening. When a curtain is placed across a door, it should be hung from a rod about two feet above the top of the door to allow easy opening and closing. Sliding doors occupy little space but are difficult to make light-proof.

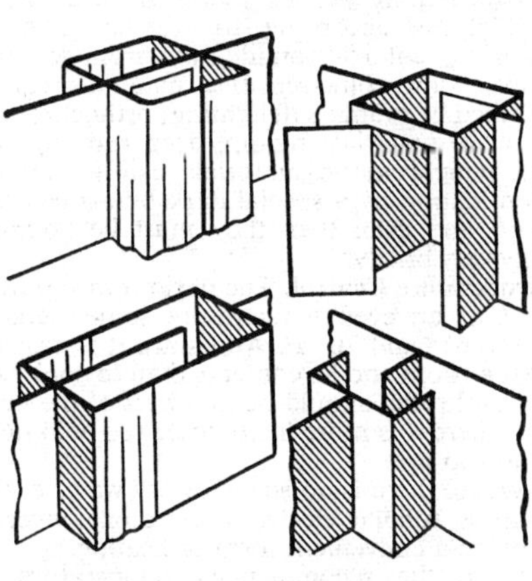

LIGHT TRAPS. Top left: *Heavily curtained door.* Top right: *Two-door chamber.* Bottom: *Doorless labyrinth, with or without curtains. Inside surfaces should be black.*

Windows. If existing windows have to be obscured to make a darkroom, they should have removable shutters so that the room can be flooded with daylight. This is a great help when the room is given a periodical clean out. Blinds are a practical alternative to solid shutters that fit closely into the window frames. They must be made of completely opaque, tough material and should run in grooves deep enough to prevent light leaking round the edges. There are two other points that must be light trapped: the housing for the roller, and the recess into which the free end of the blind fits when it is drawn.

Ventilation. A good air circulation is most important and is frequently overlooked because people are apt to shut out the air in an effort to keep out the light. At least 400 cubic feet of air should be allowed for every person using the darkroom and every effort should be made to keep it moving. In small darkrooms, this can be done by fitting a pair of light-tight ventilating units, one at the top and the other at the bottom of a wall, shutter or windows.

Another method is to construct the walls of the darkroom to within 6 ins. of ceiling height and to hang from the ceiling two black painted light baffles about 18 ins. deep, inside and outside the darkroom. Similar light-trapped vents at floor level help to promote circulation of air.

But any darkroom that is in continuous use ought to have an extraction fan. A number of fans are especially made for the darkroom. They are light-trapped and practically silent. They should always be fitted as high as possible.

All vent pipes and ventilators should be matt black inside to make them efficient light-traps.

Drainage. It is NOT safe to drain exhausted developing or fixing baths into a copper drainage system. Ordinary lead plumbing is not affected by the solutions used in the normal darkroom but they should be well diluted.

Sinks should have individual connexions to the main drain and each should have its own trap. A detachable strainer, made preferably of stainless steel, should be fitted to the outlet from each sink. Where the drainage pipes bend at a sharp angle or form a junction with another drain there should be a removable cover or plug to simplify the job of clearing any blockage. Plumbing and pipe-work should be accessible so that attention can be given to leaks and stoppages without unnecessary dismantling and removal of fixed equipment.

FITTINGS

The darkroom lighting should be as bright as is consistent with the speed of the sensitive materials being handled. This can be best done by indirect illumination, reflected off the ceilings and walls, from a properly designed safelight fitted with interchangeable screens

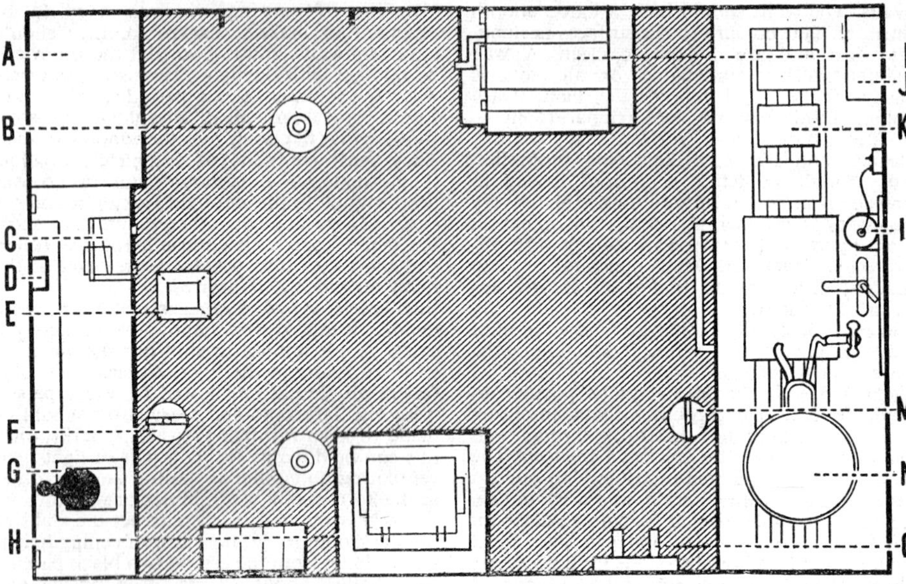

COMMERCIAL OR INDUSTRIAL DARKROOM. This type of room may be used for the photographic department of an industrial firm or of a hospital. A, film drying cabinet. B, white light. C, trimmer. D, darkroom timer. E, ceiling light for general darkroom illumination. F, darkroom lamp for local illumination of "dry" bench. G, enlarger. H, contact printer (may if necessary be replaced by a second enlarger). I, glazing machine. J, shelf for measures, bottles, etc. K, processing tanks in "wet" bench; replaced by dishes when required. L, immersion heater for tanks or dishes. M, darkroom lamp for local illumination of "wet" bench. N, rotary print washer. O, wall bracket for film hangers.

of suitable colour. Where the ceiling is broken or of poor reflecting power, the safelight should have a white-painted reflector suspended above it. The general illumination should be supplemented by direct lighting where necessary, e.g., over the processing dishes and the sink.

If two sets of safelighting are used—for example, one for negative development and one for printing—the switches for the brighter lights should be built into a wall-box with a flap cover to prevent people from switching them on by accident. The white light switches should always be boxed in separately or fixed about 18 ins. above the normal level for light-switches. Switches controlling direct safelighting are best placed under the edge of the bench near the lamps concerned.

Safelight lamps should either be hooded or hung below eye level so that they cannot shine directly into the eyes of the darkroom workers. The white light should be placed close to the surface of the ceiling and as near the centre of the room as convenient in order to give as few and as short shadows as possible.

It is useful to have a white light over the sink so that prints can be looked at immediately after fixing. A daylight lamp in a deeply hooded shade can conveniently be suspended over the fixing dish or washing tank. If controlled by a foot switch, it can be flicked on for the few seconds necessary to check print quality. In the same way a small illuminated opal, attached to the wall behind the sink, serves for scanning negatives by white light after fixing.

Safelights. When the correct safelight is chosen for the sensitized material in use, it is possible to have ample light to work by. Papers require yellow, yellow-green or orange safelighting according to whether they are chloride, chloro-bromide, or bromide emulsions. Slow, blue-sensitive materials can be handled by orange safelighting; orthochromatic negative materials require deep ruby lighting, but panchromatic materials call for indirect illumination by a special dark blue-green safelight, and even then they must be examined only very briefly.

Temperature Control. The darkroom should be kept at an even temperature somewhere between 65° and 70° F. And since it is easier to heat a room and its contents than to cool them, the darkroom should be in a part of the building where it is more likely to be too cold rather than too hot.

Where there is no steam or hot water central heating, tubular electric heaters or convectors provide a convenient form of heating.

Since the recommended temperature for most photographic processing operations is 68° F. (20° C.), the solutions used should be brought to this temperature and, if the room temperature is within this range, the rate at which their temperature changes will be very low. This is also the most comfortable working temperature.

Large dishes and tanks containing processing solutions are most conveniently heated by electric immersion heaters, thermostatically controlled to switch off at the right temperature. Tanks in continuous use are best immersed in a thermostatically-controlled water bath. This gives finer control than individual immersion heaters and it is the system generally used in colour photography, which demands a high degree of reproducibility in processing. Dishes can also be satisfactorily warmed or cooled by a water bath. There are many types of thermostatically controlled individual electric dish warming stands, but for handling a number of dishes the water bath method is cheaper and more convenient.

Benches. Darkroom work benches are either wet or dry. Processing and making up of solutions are confined to the wet bench; the dry bench is reserved for enlarging, printing and loading sensitized materials. Both benches should be kept as far apart as possible, and the wet bench should be alongside the sink.

Teak is the traditional wood used for making wet benches, but there are many other materials as good: i.e. stainless steel or plastic sheet. All these materials may be used very effectively as a covering for the top of the wet bench. Wood which is used uncovered must be made waterproof by coating it with wax. Precautions must be taken to prevent solutions spilt on the bench from running down the back at the line where the bench-top joins the wall. If lead is used to cover the bench, it should be continued up the wall 6 ins. or the joint may be sealed with a strip of beading rendered water-tight with a sealing compound, pitch or other waterproof adhesive.

Bench-tops of dry benches should be covered with thick linoleum, rubber or compressed resin-impregnated fibreboard edged with oak. The covering may be continued up the wall to prevent dust from collecting in the corner.

The average operator prefers to stand at a bench 3 feet high and sit at a bench 2 feet 6 ins. high. All working benches that are not open at ground level should have toe recesses about 3 ins. high and the same depth.

Cupboards and Drawers. The space under benches is usually the most suitable place for cupboards and drawers. The cupboard space under the wet bench should be reserved for storing bottles of solutions in frequent use and for small developing tanks. The space under the sink can be divided up into compartments to take developing dishes in various sizes, standing on end. A number of compartments should be provided over the dry bench for safelight screens. At least one of the drawers should have a light-tight sliding lid inside so that sensitive material can be left in the drawer without risk of fogging. One type of drawer has a safe lid that slides into place as the drawer is closed and remains closed the next time the drawer is pulled out.

The atmosphere in the darkroom is generally damp, so bulk stocks of sensitized materials should always be kept in another place. But one of the cupboards or drawers in the darkroom should be set aside for storing materials in actual use.

Every darkroom needs a waste bin, and it is as well to separate paper and film waste from glass to reduce the danger of accidents.

Sinks. A shallow darkroom sink can be used as a wet bench for most of the processing. It should be about nine inches deep and rather higher than usual to give as much comfort as possible when working. Hot and cold water should be available and the taps should be

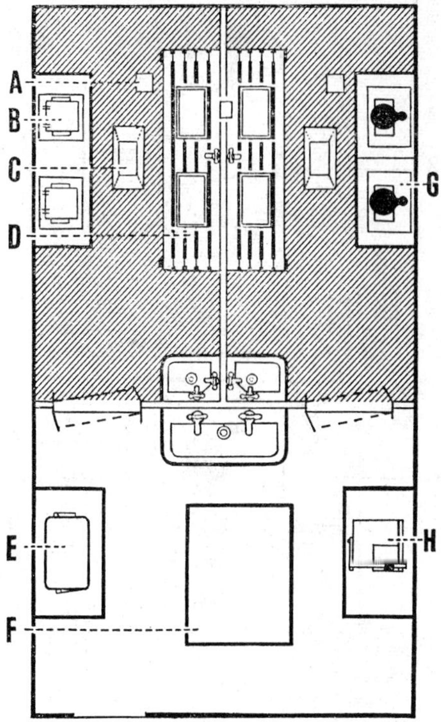

PROFESSIONAL PRINTING DARKROOM. This is divided into two sections, one for contact printing, and one for enlarging. Many professional photographers do not make contact prints at all, in which case the contact room becomes unnecessary. A third room serves as finishing room for glazing, mounting, trimming, retouching, etc. The "wet" benches carrying the developing dishes are arranged along the dividing wall together with the sinks for preliminary washing. A, one of several lamps providing local illumination. B, contact printer. C, ceiling lamp for general darkroom illumination. D, processing bench. E, electric hot glazer. F, work table for retouching and mounting. G, enlarger. H, trimmer.

high enough to allow a standard 2-gallon bucket or a Winchester bottle to stand under them (i.e., at least 15 ins).

The hot and cold taps should have different types of handle, or one of them may have a spot of luminous paint so that they can be easily distinguished in the dark.

An anti-splash device or a piece of rubber hose may be attached to the cold tap for washing out tanks and dishes or swilling down the darkroom floor (if the hot tap is left without a hose, it serves to distinguish it easily from the cold water tap in the dark).

If the water supply is contaminated by solid matter like grit or rust from the mains, a filter may have to be fitted in the water-supply line.

Suitable sinks may be made of wood (teak, cypress or maple) porcelain, chemical stoneware or stainless steel. Wooden sinks should be coated with asphalt, rubber or a sheet plastic, or lined with lead (not less than 5 pound thickness). All the seams of lead-lined sinks should be burned-in and not soldered. Bitumen-lined sinks are very practical and, although they wear in use, they can be repaired without difficulty.

Duckboards on the floor of the sink prevent breakage of dishes and glassware and stop dishes and tanks from slipping about. Draining boards alongside the sink are made of wood, or —where expense is no object—stainless steel. Some sinks have a cover which can be fitted over the top to form a useful extra bench when the sink is not in use. G.B.M.

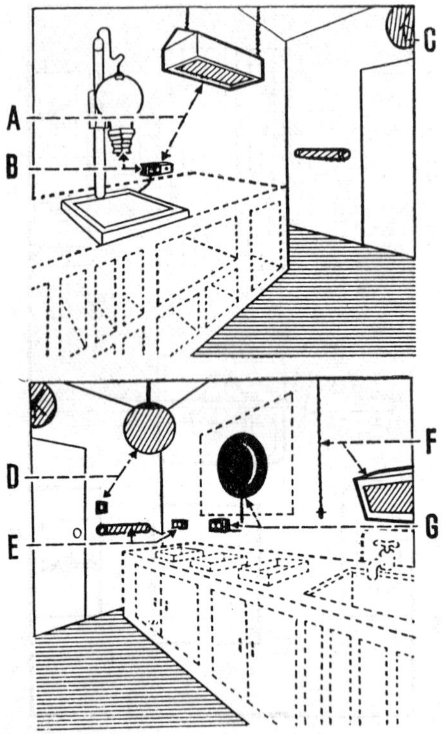

ELECTRICAL INSTALLATIONS. The various units are as far as possible fed from a number of lighting and power points, and controlled by individual switches. A, darkroom lamp and switch for general illumination. B, outlet point and switch for enlarger or printer. C, electric clock (needs no switch). D, white light and special boxed switch to prevent accidental switching on during working. E, tubular heater fed from power point. F, darkroom lamp and switch for local illumination of work bench. G, extractor fan in blacked-out window. All switches within reach of the sink and wet bench should be of the cord-operated, ceiling fitting type.

Electrical Equipment. Most of the darkroom electrical equipment is in use only intermittently and is thus usually connected into some sort of switch-plug. All such plugs should be of the three-pin type with the third pin properly earthed. The frame of all pieces of metal equipment should be connected to the third pin of the plug. This greatly reduces the risk of shock.

It is wise to provide more socket outlets than are, at the time, thought to be necessary. Darkroom equipment is frequently being added to and multi-way adaptors are no more than a dangerous expedient. It should be made a strict rule to connect only one piece of equipment to any socket outlet or point.

The average small darkroom might reasonably have the following electrical equipment:

ELECTRICAL INSTALLATIONS

Equipment	*Power Consumption—Watts*	*Total Watts*	*Total Current at 230 volts*
White ceiling light ...	60		
Safelight, general ...	40		
Safelight, local	15		
Contact printer ...	100		
Enlarger	100		
Fan	60	375	1·6 amp.
Clock	Neglible		
Dish Heater	100*		
Print Dryer and Glazer	250*		
Kettle	1,500*		
Room Heater	2,000	3,850	16 amp.

* These items would not normally be all switched on at the same time and would be adequately served by two independently fused 15 amp. circuits or as a ring main, by two 13 amp. sockets. One socket outlet would be reserved for the room heater and the other would supply any two of the other units.

These items would be supplied by two subcircuits, i.e., a 5-amp. lighting circuit and a 15-amp. heating circuit. The various items would then be conveniently installed as follows.

White ceiling light: suspended from ceiling. Controlled by switch near door (a key switch prevents accidental fogging of materials).

Safelight, general: hanging from wall over dry bench. Controlled by switch near dry bench socket outlet.

Safelight, local: hanging from wall between wet bench and sink. Controlled by cord from switch located in the ceiling.

Contact printer: fitted with 5-amp. plug to connect to socket outlet at right-hand end of dry bench. Controlled by own switch which is usually built-in.

Enlarger and fan: as for contact printer.

Clock: connected through clock fuse to 5-amp. subcircuit permanent wiring.

Room heater: on floor, with flex to reach 15-amp switch socket outlet to right of dry bench 5-amp. socket outlet, clear of dry bench top, so that heavy flex can fall straight from socket outlet to floor. Controlled by socket switch for on and off, and own switch from 1,000 to 2,000 watts.

Dish heater: with flex and 5-amp. plug to 5-amp. switch socket outlet at left hand end of dry bench. Controlled by socket switch.

Kettle: with flex and 5-amp. plug to connect to same switch socket as dish heater.

Safety Precautions. Commercial darkrooms and processing establishments come under the Factory Acts and any electrical equipment must conform to the regulations set out in the Electricity Supply Regulations, the Electricity Regulations of the Factory Acts and the Wiring Regulations of the Institution of Electrical Engineers.

The electrical wiring and equipment in amateur darkrooms must conform to the Electricity Regulations governing domestic electrical installation. Normally the electrical installation is done by a contractor who knows what the regulations require and it is inspected by a representative of the Electricity Authority before the supply is connected. However most darkroom equipment and wiring is installed and added to by the occupier and may not always be up to the required safety standards. The following safety measures are basic to any installation and are intended as a guide for amateurs without any knowledge of electricity.

Earthing.—All portable darkroom electrical equipment should be connected to a 3-pin socket outlet by a 3-core flexible cable so that all metal cases and exposed metal parts are connected to earth automatically when the equipment is plugged in. The only exceptions are lighting fittings which are out of reach and not in the way of normal darkroom operations. If anything goes astray inside properly earthed equipment and a live part of the circuit comes into contact with the chassis or case, the fuse will blow. Without an earth connexion the case would be made live and anyone touching it would get a shock which could be fatal.

Wet Bench.—No switches or socket outlets should be mounted within reach of the wet bench. It should not be possible for anyone working at the wet bench to operate a mains switch or insert a plug in a socket. Any switching should be done by a cord connected to a ceiling switch as used in bathrooms (and other places where the operator is likely to be in wet contact with water pipes or other main services connected directly to earth). F.P.

IMPROVISED DARKROOMS

For the simpler tasks in photography—e.g., changing plates or freeing a jammed film in the camera—an easy and simple substitute for the use of a darkroom is to crawl into bed head first, making sure that there are no light leaks round the body, and carry out the work in darkness under the clothes at the foot of the bed.

To load film into the developing tank, drape the bedclothes over the arms and shoulders while sitting on the floor with the legs under the bed. But by far the better way is to use the bathroom or the space under the stairs; the essential is that the place be light-tight and clean.

Under Stairs. See that the door when closed is light-tight and there is no extraneous light. Any gaps can be filled up by tacking strips of felt around the edge of the door. Take extra care that any electrical meters or mains are protected from splashes of water or accidental contact.

There must be a shelf, box or table covered with lino where sensitive materials can be handled without fear of water, solutions or dry chemicals coming into contact with them. There must also be a space for the processing dishes or tanks for negatives and prints. Washing can be carried out in the kitchen sink.

For printing in a confined space of this type the best method is to use a printing box which can either be bought ready made or contrived from a wooden box with a little ingenuity and thought. The printing box and safelight can be electrically operated from the mains or from a battery.

The walls can be painted up to 42 ins. with an acid-resisting black paint and above this and on the ceiling with a light colour such as cream or yellow. The floor should be rendered waterproof if possible with a covering of lino or, if it a cement floor, with traffic paint.

Shelves should be fitted wherever possible in odd corners to take bottles, etc.

It is not essential to have photographic dishes, for use can be made of suitable kitchen dishes and pails for washing water and waste.

Bathroom. An ordinary bathroom can be converted into a darkroom in this way: make the window light-tight by means of a well-fitting black blind running in grooves or behind slats. Then arrange a working surface over the bath.

A tray for processing with 2 ins. lip all round, made of plastic or plywood, can fit the top of the bath for half its length, while the other half is covered by a plywood bench to take the printing box or enlarger.

To prevent the light of the contact printer or enlarger from fogging the paper whilst being developed and also to prevent splashes of solution on sensitive material, a partition about 9 ins. high can be placed between the tray and the bench.

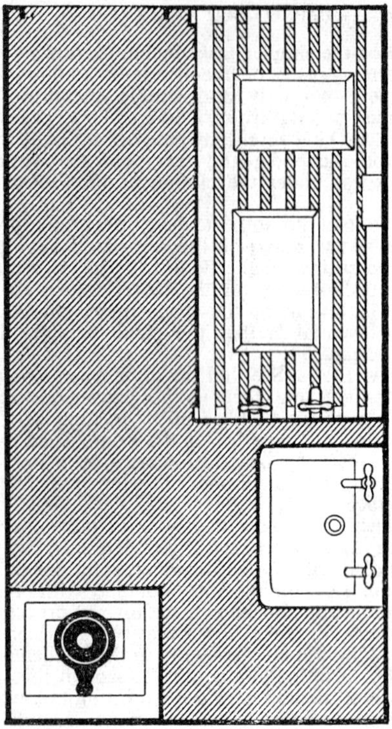

CONVERTED BATHROOM. To be practical, the conversion must be simple and quick, with little preparation before, or clearing up after, use. In this case a wooden rack fits over the bath and takes the dishes and other gear connected with processing. The enlarger stands on a suitable table or low cupboard which may house all other accessories when not in use. The only lengthy part of the conversion may be the task of blacking out awkwardly placed or shaped windows.

Illumination can be by a safelight or by fitting one of the special coloured bulbs in the ceiling light. As in all darkrooms, great care must be taken when fitting any temporary electrical switches and equipment to ensure that there is no possibility of an electric shock. A small distribution board providing fused socket outlets for the various pieces of equipment can be connected to the nearest 15-amp point by a length of three-core cable. All wire should be of a waterproof covered type and all circuits properly earth-connected.

With a little thought the tray and bench can be constructed so that when not in use they form a box to store the dishes. If there is space available in the bathroom, a small cupboard should be fixed to store solutions, chemicals and measures, etc.

Spare Room. Where there is a spare room available for conversion specially for photographic work, the layout will naturally depend on space available and the nature of work to be undertaken.

The room should be so arranged that at least one bench can be kept dry for the sensitive material while it is in the dry state. All wet operations should be carried out on a second bench (the wet bench) with a sink next to it. It must be impossible for splashes of working solutions to reach the dry bench. Similarly the benches must be made so that hypo, etc., cannot drip on to the bench or floor.

Where possible the wet and dry benches should be on opposite walls or at right angles. An ideal working bench would be 33 ins. high and up to 18 ins. wide.

The space under the dry bench can be divided in two, one half as a cupboard for storage of cameras, lenses and sensitive material. Chemicals, whether in bottles or dry packages, should never be stored in this cupboard. The other half of the space can be divided up to take a drawer at the top and a shelf with space underneath for mixing pail and stirring rods, etc.

The Darkroom Trolley. In small flats where space is at a premium, the darkroom trolley offers a way of fairly rapidly converting a normal room into a darkroom for printing and enlarging. The trolley consists of a sideboard-size cupboard on wheels and containing an enlarger, developing dishes and solutions, and storage for other incidental equipment (timer, etc.) and paper. When not in use, everything is contained inside the cupboard which can be wheeled out of the way against one wall of the room. When required for darkroom work, the top of the trolley becomes a working surface and carries the enlarger, a space for dry manipulation, and the processing dishes. A swing-up flap may increase the working area.

Setting up only requires blacking out the room. Darkroom illumination may be provided by a coloured safelight bulb, placed in, for instance, a reading lamp in place of the normal bulb. The working surfaces are preferably coated with plastic; if this has a wood pattern finish, the whole unit can look like a normal piece of furniture when not in use.

It can also have space for film developing tanks and film processing equipment; the room only needs blacking out for sufficient time to load the film into the tank, after which processing can take place in ordinary lighting without any further preparation.

Darkroom trolleys are usually designed individually, frequently by amateurs with some carpentry skill. They have on occasion been offered commercially.

A compact enlarging darkroom along similar lines is the wardrobe which contains the enlarger on a shelf, plus a folding table for the processing dishes and storage for all other equipment. A darkroom safelight may be built into the wardrobe. To get it ready for work only involves setting up the table, opening the wardrobe doors and blacking out the room. When not required all equipment is unobtrusively shut away inside.

SPECIALIZED DARKROOMS

Darkrooms for various specialized work are as numerous as the variety of work, but they still have the basic principles of a tidy, dust-free and well-ventilated, dry room and a layout suitable to the photographic task involved.

Photo-Finish Darkroom. One example is the type used in connexion with photo finish cameras, where the photograph (an 8 × 10 ins. enlargement) of the result of the race must be in the hands of the judge in the shortest possible time. This is built in a stand or a tower about thirty feet above the ground.

An average size for such darkrooms is 6 × 8 feet; the camera is off-set to one end of the 8 foot side of the darkroom.

All pieces of equipment are installed as near as possible to each other but still leaving plenty of working space.

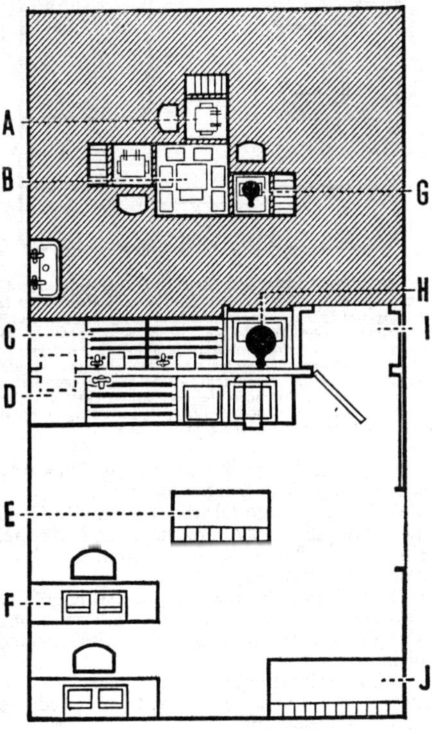

SMALL PHOTO-FINISHING LABORATORY. The darkroom contains three working places round a central bench carrying individual developer and stop bath dishes and a common fixing tank. Here two workers use contact printers or roll-head printers, the third on an enlarger. The daylight room contains the final washing, glazing, and other finishing equipment. A, contact printer or fixed-ratio roll-head printer, B, central processing bench. C, washing-tank. D, light-trapped hatch. E, trimming bench, F, sorting and dispatching bench. G, enlarger. H, special enlarger, I, light-trapped doors. J, mounting bench.

Taking the camera as the starting point of the processing operation and working clockwise, the order in which the equipment is laid out is: camera, film processing unit (with thermostat to control solution temperatures), fitted flush to the bench, safelight flush with the bench to enable the negative to be placed correctly in the negative carrier of the enlarger, fixed enlarger and paper carrier, print processing dishes, sink and dry bench.

To attain high speed of production, all the work must be carried out with the negative wet. It is therefore vitally important for the whole electrical circuit to be efficiently earthed. All switches and plugs are positioned at the most convenient place for the operator to reach without unnecessary movement. Every piece of electrical equipment is separately fused on a fuse board in the darkroom and all fuses plainly marked to enable a quick replacement to be made. The floor and benches are covered with lino with a good acid-resisting, hard-wearing surface. The walls are painted acid-resisting black up to 42 ins. and cream above.

Ventilation is effected by louvre type ventilators under the benches and an extractor fan on the wall. Heating is carried out by tubular heaters under the benches.

X-Ray Darkroom. In general, this follows the same lines of organization as any professional darkroom. In certain cases, however, it must have provision for rapid processing, e.g., when a radiograph is required during a surgical operation. This mainly involves careful temperature control and a suitable working layout of benches and equipment so as to waste as little time as possible.

Most darkrooms of this type have a light-trapped service hatch where the exposed films are handed in, and the processed radiographs passed out. Where rapid processing is used, the finished radiographs are examined wet in front of a suitable viewing box.

MOBILE DARKROOMS

Although not as necessary now as in the days of wet plates, there is still a limited requirement for portable and mobile darkroom equipment in the field.

Portable Darkrooms. Changing bags are made of a light-weight and lightproof material, in the form of a bag large enough to take camera, dark slides and plate boxes, and fitted with two sleeves with elastic openings for the arms. The whole can be folded into a very small size and packed away in the camera case.

For use on a job at a remote place there is an igloo type of tent in which a limited amount of work—e.g., developing negatives and making contact prints—can be carried out. No poles are required. This type of tent is packed in a box that when unpacked becomes the work table. The tent is made of a light-tight material;

along the four edges which come to a point at the top are sewn rubber tubes that can be inflated to form structural members, thus giving a rigid conical-shaped tent. The door is fitted with a zip fastener and the walls of the tent have flaps which tuck under the tarpaulin forming the floor, thus making all light-tight. The main use for this type of tent is to enable an operator on a long task away from the main studio—e.g., when shooting a film or taking part in an expedition—to check his results.

Full-sized darkroom tents are mainly used by the Armed Forces in the forward area and consist of a double-lined tent of about 20 × 20 feet, the inner lining being of a black light-tight material.

The interior of the tent is divided into three rooms: one room, taking half the space, forms the work room for drying, trimming, sorting, mosaic making; a second room, taking one-quarter of the space, is for negative processing and the third room is for printing and enlarging.

The door is usually zip fastened and the tent walls and floor covering overlap to prevent light leaks. All the tentage and equipment are packed in suitable cases designed so that when unpacked they become useful articles of darkroom furniture such as tables, chairs, sink, etc. The equipment includes a small, petrol-driven generator with a distribution board and waterproof leads which connect with suitable points in the tent for safelights, etc. An extractor fan is inserted in a hole in the tent wall and a small pump lifts water from a tarpaulin tank to the sink.

Lorries and Trailers. A mobile darkroom is built on a lorry and fitted out to deal with processing of plates or films, quick drying and contact printing. The principal use is for on-the-spot processing of news and sport pictures in remote districts for transmission by wire or radio.

The Royal Air Force use a darkroom trailer of this kind (T Type) in the field. These trailers are equipped to deal with aerial photographs from the 4 × 5 ins. type camera and have side awnings fitted for use as a workroom. They provide their own electric light from a small generator and may be fitted with a refrigeration plant for use in tropical climates.

Photographic Train. The extreme example of a mobile darkroom is the Royal Air Force train of four articulated vehicles more than thirteen feet high and over thirty-two feet long and weighing more than twelve tons. They are used with forward area units responsible for photographic reconnaissance. Such a train consists of the following vehicles.

(1) An office for administration, keeping records of work carried out, sorting and dispatch of prints, numbering and examination of negatives.

(2) Film processing unit capable of automatically processing and spooling up to 500 exposures of 9 × 9 ins. format film at a speed of about four feet per minute.

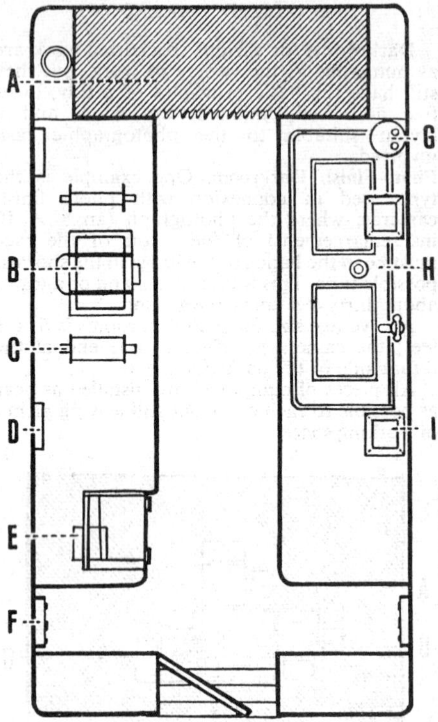

AIR FORCE PHOTOGRAPHIC TRAILER. All the equipment for processing and printing the aerial films is housed in a trailer 15 feet long and 7½ feet wide. A, film processing room, separated by curtain. B, printing box. C, spool with negatives. D, ventilation fan. E, print trimmer. F, controls for refrigeration unit under dry bench. G, water supply tank. H, sink with processing dishes. I, darkroom lamp (several are arranged around the trailer). More advanced trailers contain automatic film processing units for dealing with continuous lengths of up to 500 exposures of 9 × 9 ins. film at a time.

(3) Printing section fitted with a multiprinter which automatically produces and finishes prints from the 9 × 9 ins. film at about 1,000 prints an hour.

(4) Darkroom for manual processing of films in a series of tanks and drying on a large drum.

(5) Darkroom for manual printing and finishing.

(6) Trailer with a heavy duty generator that supplies the electric current required to operate all the equipment.

To complete the train there are several trucks (3 ton and 15 hundredweight) for film storage, stores, tentage, and all the paraphernalia necessary to make a completely self-contained unit. S.J.H.

See also: *Enlarger; Processing machine; Studio photography; Tanks; Wholesale photo-finishing.*

Books: *Developing*, by C. I. Jacobson (London); *The Business of Photography*, by C. Abel (London).

DARK SLIDE. Name sometimes used for a sheet film or plate holder. If the holder carries two sheets the term "double dark slide" is used.

DAVANNE, ALPHONSE, 1824–1912. French amateur photographer. Contributed to half-tone photolithography and wrote numerous books and articles on photography. Was one of the founders of the Société Française de Photographie (French Photographic Society) and consecutively Vice-President, President and Honorary President of its Council (1866–1912).

DAVISON, GEORGE, 1856–1930. English civil servant and amateur photographer. Follower of P. H. Emerson's Naturalistic Photography; founded the Impressionistic school of photography in 1890. Founder member of The Linked Ring, 1892. Managing Director of Kodak Ltd., 1898–1912.

DAVY, SIR HUMPHRY, 1778–1829. English chemist. Experimented in collaboration with Thomas Wedgwood before 1802 in the production of silhouettes and contact copies of leaves, etc., on white paper or leather moistened with silver nitrate and chloride solution, and under the microscope, but could not fix them. Discovered the electric arc light in 1813. Biography by J. C. Gregory (1930).

DAY, WILL, 1874(?)–1936. English cinematographer and pioneer of the cinema. Part of his large collection on the history of cinematography (and of its forerunners) is now in the Science Museum, London.

DAYLIGHT

The sun is the source of all daylight. Even when it hides behind a cloud, it provides the light that we see—and take photographs by. Direct sunshine at noon in the middle of summer has an intensity of 10,000 foot candles—equal to 400 Photoflood lamps placed 6 feet away from the subject, and almost a thousand times stronger than the artificial light illumination in an ordinary living-room.

But the strength of the sunlight is not constant; it varies throughout the year from one month to the next and throughout the day from one hour to the next.

It varies because the sun's altitude changes throughout the year and from the beginning to the end of the day. The sun is highest in the sky, and therefore its light is strongest, in June, and it is lowest, and its light is weakest, in December. In the course of the day the strength of the light varies in the same way from sunrise to noon, and from noon to sunset.

Both the yearly and daily patterns are regular—March and September lie at equal distances from the June peak and so have the same strength of sunlight; the sunlight is as strong at nine o'clock in the morning as at three o'clock in the afternoon.

The visible spectrum of uninterrupted sunlight extends from 4,000 Angstrom Units (violet) to over 7,200 Angstrom Units (red). All the coloured rays it includes in its range add up to make white light.

But sunlight is a commodity which the photographer rarely, if ever, has to deal with in its pure state. Photographs are made, not by sunlight, but by daylight, and the two are not by any means the same thing. In daylight some light always comes from the sky, and some often comes from reflections off nearby surfaces.

The direct rays of the sun are white, they travel in straight lines, and all the straight lines come directly from the sun. But daylight is more often coloured than white (generally more bluish than direct sunlight); to a large extent it is able to find its way around corners, and it comes from all directions, even from below.

Daylight is a variable mixture; part consists of white light that has arrived straight from the sun, and part of light that has been changed in direction or colour, or both.

The Direction of Daylight. The direction of the light that shines on the subject may be affected by the surroundings. Certain objects—white clouds, light painted buildings, snow, chalk cliffs and masses of sand or water, can act as reflectors. Under the right conditions they can even provide the principal lighting. When sunlight is reflected by such surfaces, it scatters in all directions and no longer casts hard shadows.

A similar scattering takes place when the sunlight has to pass through atmospheric moisture in the form of thin cloud, mist or raindrops. Denser forms of atmospheric moisture—heavy cloud and fog—completely destroy any appearance of direction in the light, and cast no shadows.

The Strength of Daylight. The strength or intensity of the daylight is governed by the altitude of the sun and the weather.

The altitude of the sun varies according to the latitude, the time of year, and the time of day. In practice, however, these factors do not affect the strength of the daylight as much as might be expected: the sun at midday in June is only about half as strong again as in January for similar weather conditions. Finally, careful research has shown that daylight quickly increases in strength once the sun is clear of the horizon. From then, thanks to the fact that it is helped by sky and cloud reflection, it remains at about the same strength for the greater part of the day. Towards evening the strength falls off again as quickly as it increased in the morning.

The weather, in fact, has a much greater effect on the strength of the daylight than any other factor. Cloud or fog can make a June afternoon as dark as December, and a sudden change in the weather at any time can raise or lower the light intensity by ten times in the space of a few minutes.

Exposures in Daylight. The correct exposure in daylight photography depends as much on the quality of the daylight as on its strength. But while the strength or intensity can be measured by an exposure meter, it is not as easy to judge the quality. The quality of daylight is a combination of contrast and colour. Both can affect the exposure profoundly.

Contrast. Lighting contrast depends upon the amount of direct sunlight present in the daylight. The strongest contrasts are seen when the sun shines out of a clear sky and there are no clouds or reflecting surfaces to illuminate the shadow areas of the subject. When photographs must be taken under such conditions, it is common practice to give a very full exposure and curtail the development time of the negative (in other words, expose for the shadows, and develop for the highlights).

This procedure flattens out the contrast of the negative and can be made to yield an acceptable picture instead of a useless example of "soot and whitewash". In strong, clear sunshine it is safe to double the normal exposure and reduce development of the negative by a third to a half.

At the other end of the scale there is the type of daylight where there are no direct rays from the sun, and the shadow areas are so illuminated with reflected light that they cease to count as shadow areas. This sort of lighting can occur when the sun is completely obscured by heavy clouds or fog and equally when it is veiled in white mist. Both of these conditions give lighting of low contrast, although one is very much more intense than the other.

In low-contrast lighting the exposure must be kept to a low value and, if possible, development should be extended. This increases the contrast of the negative and prevents the print from looking flat and lifeless.

At the same time, there is a danger of losing contrast by giving too short an exposure. If the exposure is just sufficient to record detail in the darkest shadow and there is not much difference between that and the strongest highlight, it is quite possible for the whole exposure scale to lie on the toe of the characteristic curve. When this happens, the negative will develop to a low contrast because the slope of the toe of the curve is less than that of the straight line portion.

CONTRAST OF DAYLIGHT. Top: *Direct sunlight yields high contrast with brilliant highlights and deep, sharp, shadows.* Centre: *Veiled sun gives softer light with clearly visible, but considerably more luminous shadows.* Bottom: *An overcast sky gives a lower level of even diffused illumination with virtually no discernible shadows at all.*

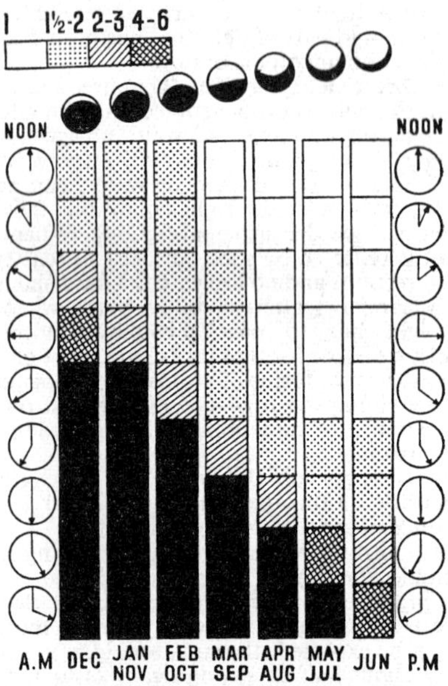

INTENSITY OF DAYLIGHT. The squares represent the relative amount of light given by the sun from a clear sky at different hours of the day and in the various months of the year for a latitude of 45–55° N. The circles show the relative altitudes of the sun at noon, and also the proportion of hours of light. In Great Britain this table applies for Greenwich Mean Time. During the months of Summer Time, one hour should be added to all the times shown.

So ideally, in this kind of daylight, the exposure should be just enough to register the deepest shadow (in which detail is to be shown) on the bottom of the straight line portion of the characteristic curve.

Between the contrasty lighting of the direct rays of the sun and the flat, diffused illumination found in mist, fog or rain, there are infinite possibilities of variation. Fortunately modern sensitive materials possess very wide latitude which looks after quite large errors in exposure. And the number of grades of printing paper is enough to look after quite large errors in judging contrast. So that it is only necessary to override the normal exposure meter reading for the more extreme contrast variations in daylight.

The Colour of Daylight. The atmosphere always contains a certain amount of water vapour. Finely divided droplets of water in this form act as a colour filter, letting yellow, orange, red, and infra-red rays through and reflecting blue, violet and ultra-violet rays. So, an observer looking at the sun through mist or fog, sees it as an orange-red disc, while if he stands with his back to the sun, the mist itself appears blue-white. Motor-car headlights seen through fog appear dull orange, but to the driver behind them they appear as a dazzling white.

So the water vapour in the air tends to colour the daylight orange and to rob it of blue rays. And the more vapour the light has to shine through, the more orange the daylight appears. Morning and evening light are both inclined to be redder than broad daylight; when the sun is low down on the horizon its rays travel obliquely through the vapour layer and so pass through more of it.

When daylight is reflected it takes up the colour of the reflector. As reflectors are usually white or almost white (clouds, snow, painted buildings, etc.) this is rarely important to the black-and-white photographer. But coloured reflecting surfaces must be handled with care in colour photography where the blue sky or reflected light from a grassy bank, or the red brick wall of a house outside the picture area may upset the colour values of the subject.

For black-and-white exposures, the photographer needs to trouble about only the extreme conditions. These are usually encountered early and late in the day, when the sun shines through relatively large amounts of water vapour. As water vapour holds back blue and violet light the daylight at such times is redder than usual. In terms of practical photography this means longer exposures, particularly with orthochromatic films and plates, because they are not very strongly sensitive to light at the red end of the spectrum.

Daylight at High Altitudes. There is generally less water vapour and dust in the atmosphere at high altitudes than at lower levels. One result of this is that the light falling on the scene is not scattered as much as usual and distant detail

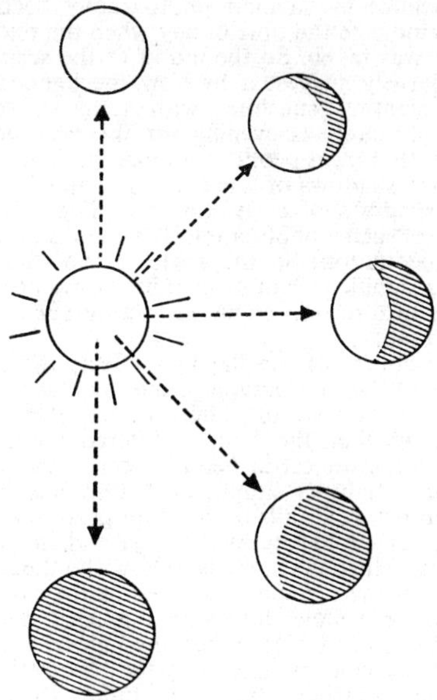

DIRECTION OF THE SUN. The direction of the light determines the relative areas of light and shade in the subject. Front lighting illuminates the whole visible area of the subject; as the sun moves round, the camera sees progressively more of the shaded side, until everything is in shadow.

is clearer. There is also less air above the scene, which means less scattered light from the sky. So the landscape, which is normally seen by the camera as being much darker than the sky, appears almost as bright.

At low altitudes the excess of blue light from the sky must be held back by a filter or the sky will come out as a blank white area on the print. But at high altitudes the balance of brightness changes and the sky tones will generally be reproduced satisfactorily without a filter. If anything more than a very pale yellow filter is used at high altitudes, the sky tones are apt to be unnaturally dark.

The absence of light-scattering water vapour also means that the light at high altitudes is rich in ultra-violet rays. The camera lens is not corrected for these rays, which thus tend to upset definition. Under such conditions, the scene can be reproduced with greater clearness by using a U.V. (ultra-violet) filter. This special filter is practically colourless, and while it allows ordinary visible light to pass through unimpeded, it is opaque to ultra-violet light.

Character. What has been said above about photography by daylight is concerned only with making an acceptable negative—i.e., one that is correctly exposed. But all the variable factors in daylight have their effect on the aesthetic character of the photograph.

The direction of the lighting has a special

significance in daylight photography because it can indicate the time of day when the photograph was taken. So the mood of the scene is very largely suggested by how the light falls. Low slanting sunshine with long shadows generally suggests evening for the very good reason that most people are more familiar with the long shadows of late evening than with the long shadows of early morning. The general mood of such a photograph is one of rest, quiet, and meditation. So the angle of the sunlight adds a significance of its own in addition to the effects of direction common to all other forms of lighting.

The colour of the daylight affects the tone values of the photograph in the same way as a filter. It makes no difference to the final result whether the light is filtered before it falls on the subject or before it enters the lens; coloured light is simply light that has been filtered before it falls on the subject.

So, early and late in the day, and in misty seasons when the light is yellowish, the tone values of the scene are changed as they would be by a yellow filter—i.e., green, yellow, orange and red objects become lighter in tone and other colours, including the blue sky, are relatively darker. At such times there is nothing to be gained by using a filter, as the light has been filtered already.

The diffusion of the light by fog, mist, clouds, and irregular reflecting surfaces, can make all the difference between a soft, well-modelled picture with a multitude of grey tones, and a hard, strongly etched study with only two or three tones between black and clear white. Generally, when the light is well scattered, it tends to destroy surface texture, although that can be an advantage in playing down wrinkles in portraiture. Such lighting gives restful effects.

Strong unbroken sunshine is suitable for vivid, dramatic treatment with a harder effect.

Here also, the character of the daylight almost always has a special significance because it indicates certain weather conditions, seasons, and even actual places—viz., the Alps, the tropics, the English Lakes, are all associated with types of daylight quality.

Supplementing Daylight. When the daylight alone is either not powerful enough or too harsh for photography, other light sources may be brought in to help. Flash, both bulb and electronic, or electric lamps may be used as a modelling light, under dull conditions, to simulate sunshine while the daylight provides the fill-in lighting. Or the artificial lighting may be used from the camera position to brighten up the deep shadows cast on a subject by direct sunlight. Such illumination can serve the further purpose of brightening a subject taken against the light.

With black-and-white material any of these light sources may be used quite safely. With colour material the colour temperature of the light has to be taken into account. The colour temperature of daylight may be anything from 4,800 K° (English daylight) to 7,500 K° (American daylight) while the colour temperature of artificial light may be anything from 2,600 K° (ordinary tungsten lamps) to 4,000 K° (flash bulbs). This means that ordinary tungsten and even photoflood lamps cannot be used. (The colour of such lighting would reproduce with an unpleasantly yellow-orange cast.) Special blue flash bulbs with a colour temperature of 6,000 K° and electronic flash (colour temperature 4,800°–6,000 K°) can be combined with daylight without any disturbing distortion of the colour values. F.P.

See also: *Portraiture outdoors; Sunrise and sunset; Sunshine.*
Book: *How to Shoot Outdoor Colour*, by G. Wells and F. Smith (London).

DAYLIGHT ENLARGER. Early type of enlarger which employed diffused daylight as a system of illumination. The image was projected into a darkened box or room.

DAYLIGHT PRINTING. Simple way of making a positive contact print from a negative, using daylight to make the exposure. At one time was a widely employed amateur procedure with printing-out papers (P.O.P.).

See also: *Contact printing; Papers.*

DEBABOV, DIMITRI GEORGIEVICH, 1901–49. Russian press photographer and artist. Originally an actor, he eventually became permanent photo-correspondent to *Izvestia*. Specialized in photographs documenting Arctic life. Died from a venomous insect bite while photographing hunters in the Taiga.

DECKLE EDGE. Name for the decorative wavy or ragged edge given to some photographs. A deckle edge may be obtained with a special print trimmer.

DEFINITION. Term in photography indicating the fineness of the detail in a negative or positive print. The standard of definition is affected by the ability of the lens and the photographic emulsion to resolve distinct lines (the resolving power) and the scale of contrast of the emulsion. A line is only clear when it is both sharply resolved and is distinctly darker or lighter than the adjacent tones.

See also: *Acutance; Circle of confusion; Grain; Lens testing; Negative materials; Resolving power.*

DELAMOTTE, PHILIP HENRY, 1820–89. English water colour artist and photographer. Professor of drawing at King's College, and organizer of the photographic department of the Manchester Art Treasures Exhibition in 1857. In 1855 published 160 photographs of the rebuilding of the Crystal Palace at Sydenham.

DELAYED ACTION. Automatic operation of the shutter at some time after the shutter release has been pressed. Many shutters incorporate a delay mechanism and there are proprietary devices which can either be attached to the normal shutter control or screwed into the flexible release socket. The delay is generally about 10–15 seconds and gives the photographer time to take up his place in the picture.

See also: *Self-portrait; Shutter release.*

DELIQUESCENCE. Property of a number of substances of absorbing so much moisture from the atmosphere that they finally dissolve in it. Many such substances—e.g., phosphorous pentoxide—are used as drying agents for desiccating the air in storage containers.

See also: *Chemicals.*

DEMACHY, ROBERT, 18??–1938. Paris banker and pictorial photographer. In 1895 he brought the gum bichromate printing process to a high degree of perfection, and introduced modern oil transfer processes for prints in 1911. From these processes the modern bromoil process was developed. Biography by F. C. Lambert (London 1904).

DEMENY, GEORGES, 1850–1917. French scientist and pioneer of moving pictures. Worked from 1882 to 1894 with Marey. Constructed Marey's second Chronographe camera (1890) which used a flexible film in magazines. Designed the Phonoscope (1891–2) for the optical synthesis of motion depicted in a series of photographs arranged round a glass disc. Made in 1893 a chronophotograph with film movement by an eccentric beater; later (1896) developed it into a projector. The apparatus was designed for, and marketed by, Gaumont. Demeny used first 60 mm., later 35 mm. film.

DENMARK. There is no record of the first photographs taken in Denmark, but interest in photography has been strong since the earliest times and today there is a small but active industry in addition to well patronized professional and amateur movements.

Industry. In Denmark there is a factory for films and photographic paper. Some small-scale photographic manufacturing is carried on as side lines by other industrial undertakings. A good deal of this work originates from the second world war, when Denmark was cut off from supplies of tripods, dishes, filters, dark-room accessories and the like from abroad. There are also several types of enlargers made, but these are mainly expensive instruments for professional use. The limited market compels the makers to go in for precision instruments of high quality and high prices rather than cheap mass production. There is an ingenious studio camera in which six or twelve plates are stored in the top of the body. A light touch of a lever makes a plate fall into position ready for exposure and another touch of the lever makes it drop to the bottom of the camera, where the back of the holder covers it and keeps it light-tight.

Professional Photography. The professional photographers are organized in about twelve local circuits, the central administration of which lies with Dansk Fotografisk Forening (League of Danish Photographers), founded in 1880. It was the first Danish photographic society and, although its membership is now exclusively professional, it originally included amateurs.

Professional photographers are trained as apprentices to other photographers for three or four years; in addition they have to take a two-months' course at the photographic school owned by the Dansk Fotografisk Forening. But in Denmark entry to the photographic trade is unrestricted and can be followed by anyone, so a good number of Danish professionals were originally amateurs who—often gradually—turned professional.

Camera Clubs. The first amateur photographic society in Denmark was probably the Københavns Fotografiske Amatørklub (Copenhagen Amateur Photographic Society) which was founded in 1895. It still has an active membership.

There are about 170 camera clubs (practically all of them admitting amateurs and professionals without distinction, but not always photographic dealers). Eighty of the clubs are affiliated to the central organization, Danske Kamera Pictorialister (Danish Camera Pictorialists), which is a member of the F.I.A.P. The D.K.P. provides clubs with programmes for their meetings: portfolios from inland and abroad, lectures in manuscript to be read at the meetings (with accompanying slides), lantern slide series, many of them with typed criticisms, and so on.

The title D.K.P. (Danish Camera Pictorialist) is awarded by the club to photographers (without distinction between amateurs and professionals) who have for a number of years shown distinguished skill and steady interest in pictorial photography, either as photographers or as critics and writers. There are at present forty holders of the title.

Exhibitions. There is an annual national exhibition of Danish pictures which is shown in turn in a number of towns. A set of slides from the best pictures in this exhibition is sent to the neighbouring countries, Sweden, Norway and Finland, with which close collaboration is maintained and which exchange corresponding sets from their own national exhibitions.

While in earlier days the subjects favoured by Danish pictorial photographers were mainly landscapes and portraiture to be rendered as true to nature as possible, the trend is now towards more sophisticated subjects. Abstract compositions play an increasing part in the exhibitions and a recent analysis showed that less than 3 per cent of exhibited pictures were landscapes (even in the widest sense of the term), and portraits were hardly more numerous. There are no awards whatever and many leading workers refuse to send pictures to exhibitions where awards are given. It is felt that such medals, prizes and the like tend to lower the dignity of photography.

Publications. The only photographic journal of general interest and circulation is *Foto-Magasinet.* It is published by an independent publisher, but as one of the two editors happens to be the president of the D.K.P. there is some connexion with the photographic societies.

There is also the monthly *Dansk fotografisk Tidsskrift*—somewhat smaller, but with a few illustrations. It is the official publication for professional photographers and is only available to members of their organization. *Smalfilmbladet* is the journal of the amateur cine societies. *Foto-Tidende* is the organ of the photographic dealers' society. Some small publications are issued more or less regularly by photographic firms and distributed free of charge for publicity purposes. H.B.J.C.

DENSITOMETRY. Technique of density measurement of silver deposits in any part of a developed photographic image. Can be carried out by transmitted light (e.g., on a negative or transparency) or by reflected light (e.g., on a paper print). The instrument used for such measurement is a densitometer; the results may be used for plotting the characteristic curve of the material concerned, or for exposure determination.

See also: *Colour sensitivity; Density; Sensitometry.*

DENSITY. Exposure followed by processing produces blackening which can be quantitatively measured as "density". A density is measured in a densitometer, an instrument in which the intensity of a beam of light can be measured before and after passing through any selected test patch on the developed image. If I_1 is the intensity of the light before it passes through, and I_2 the intensity after it passes through the test patch, the density D is defined as

$$D = \log I_1/I_2.$$

The ratio I_1/I_2 is often called the opacity, but the inverse ratio I_2/I_1 is more frequently used and called the transparency. Since I_2 is always smaller than I_1, the values for transparencies run from 0–1. If multiplied by 100, transparency values are obtained as percentages.

The density value obtained for any one density patch will depend to some extent on the design of the measuring instrument.

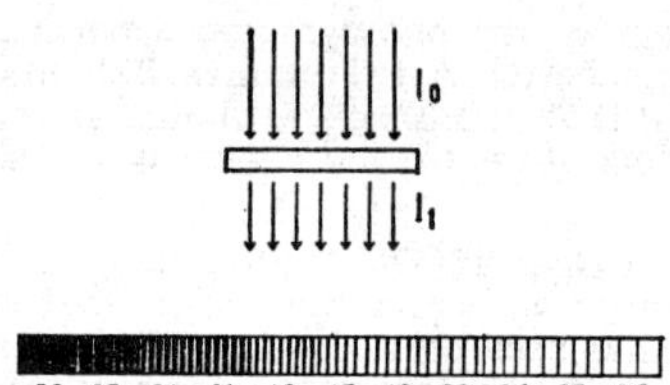

D	3.0	2.7	2.4	2.1	1.8	1.5	1.2	0.9	0.6	0.3	0.0	D
log.I_1.	0.0	0.3	0.6	0.9	1.2	1.5	1.8	2.1	2.4	2.7	3.0	
I_1	1	2	4	8	16	32	63	125	250	500	1000	

DENSITY. If a beam of intensity I_0 passes through an absorbing medium and is reduced to I_1, the opacity is I_0/I_1 and the density is log I_0/I_1. For a value of I_1 = 1,000, a range of 1,000 to 1 for I_1 corresponds to densities from 0 to 3·0.

When density values are plotted against the logarithm of the exposures which produced them, a characteristic curve of the sensitized material is obtained.

TRANSPARENCY (OR REFLECTIVITY) AND DENSITY

I_1/I_2	*Transparency or Reflectivity*	*Density*
1	100%	0
1·25	80%	0·1
1·6	62%	0·2
2	50%	0·3
2·5	40%	0·4
3·3	31%	0·5
5·6	18%	0·75
10	10%	1·0
20	5%	1·3
40	2·5%	1·6
100	1·0%	2·0
1000	0·1%	3·0
10000	0·01%	4·0

Specular Densities. The density of a test patch on a photographic negative can be measured in two ways. The first is important in projection and projection printing. It is known as the specular density and is arrived at by measuring the intensity of the transmitted pencil of light at a distance from the test patch. This method ignores the light that passes through the patch and is scattered sideways at the surface.

Diffuse Density. The second kind of density is

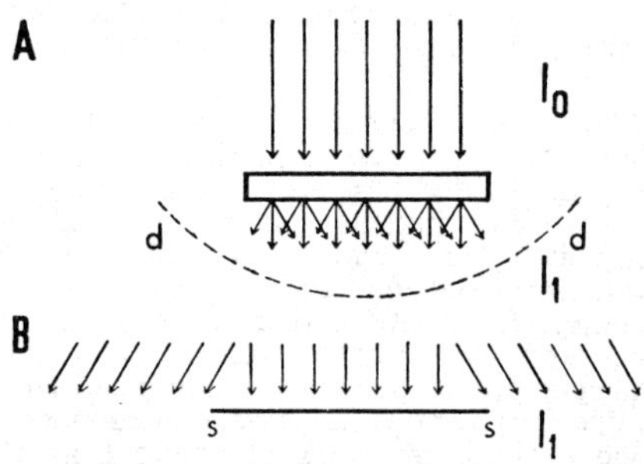

TYPES OF DENSITY. If a medium both absorbs and scatters light, the density measurement depends on the lay-out of the measuring apparatus. A: At d-d all the transmitted light is measured; the density value obtained is the diffuse density. B: At s-s a large part of the scattered light will miss the measuring surface; the density obtained is the specular density.

important in contact printing. It is known as the diffuse density and is based on a measurement of the total amount of light passing through the test patch. As the diffuse density takes into account all the light transmitted by the test patch, and the specular density measures only a part, the specular density is always greater than the diffuse density. The diffuse density is more easily measured and standardized than the specular density. Thus the British Standard Specification 1384–1947 defines the contact printing density, but makes no attempt to define the specular density. Contact-printing density is obtained by measuring the density patch in contact with a piece of translucent light-scattering material, such as pot opal glass, which is part of the densitometer.

Reflection Density. Reflection density is a measure of the depth of a tone on a surface viewed by reflected light—i.e., a photographic print. It is strictly analogous to the transmission density of an image viewed by transmitted light—i.e., a negative or positive transparency. In reflection densitometers, the measuring light as a rule falls on to the test strip at an angle of 45° and the amount of reflected light is measured at 90°, i.e., at right angles to the paper surface. The light value I_1 is that reflected from the base of the paper, and I_2 is that measured with the density patch in position. The ratio of I_2/I_1 is called the reflectivity.

Reflection density measurements may depend on the physical nature of the surface as well as the density of the image—e.g., the figure for a paper with a pronounced or regular surface texture will vary according to the direction of the incident light. W.F.B.

See also: *Callier effect; Sensitometry.*

DENTAL PHOTOGRAPHY. The camera is used in dentistry to take photographs of the teeth to show dental disease, malformations, the soft tissues, the formation of the oral cavity and frontal and side views of the face. It is here that dental photography adjoins photography in plastic surgery. Its value in orthodontics and maxillofacial surgery is undisputed. Most of these are excellent subjects for comparative photographs where the same region is photographed with visible light, infra-red and ultra-violet to include features not normally visible.

Photography is also used in dentistry in order to demonstrate the physical improvement following treatment. This often imparts a marked—and sometimes dramatic—psychological benefit. Furthermore, a photograph taken in comparative youth will often serve as a model in later life when new dentures are needed to counteract the effects of reabsorption and other changes.

Dental research organizations have carried out mass oral photography of the full upper and lower dental arches of selected groups of the community.

Education of the patient in dental hygiene by photographs is profitable if the subject matter is selected with care; the keynote being to confine the demonstrations to those things that have a direct bearing on the patient's own condition or health.

Equipment. The apparatus employed is simple and compact and it is usually kept available for instant use, generally at the chair-side. A small view camera is often used because it is versatile, permitting scaled photographs up to full size, and it gives accurate visual focusing, selective composition, and allows for the quick interchange of sensitized materials. A roll film, or even a miniature camera needs a special long focal length lens, a supplementary lens, or an extension tube, to get a sufficiently large image at the requisite working distance. Simple equipment may be devised whereby the camera is provided with a series of finder frames and corresponding supplementary lenses. This arrangement allows of very rapid working even by unskilled personnel.

More expensive equipment includes the precision miniature camera, either of the single-lens reflex type or fitted with a reflex housing. In either event, a long-focus lens and extension tubes will be found an advantage. A specialized outfit made for this type of work consists of a miniature camera with bellows focusing and carrying its own lighting close to the lens. The whole unit is mounted on an X-ray tube stand.

Dental mirrors of various shapes and sizes, together with a set of unobtrusive lip retractors complete the equipment of the dental photographer. By such means it is possible to record a full view of the occlusal and lingual aspects of the teeth.

Lighting. Axial illumination—i.e., illumination along the optical axis of the camera—is ideal for a deep cavity such as the mouth. In practice the camera and its light or lights, with or without the distance frame, are mounted together as a composite unit, an arrangement which gives nearly axial illumination. The complete assembly can be mounted on a bracket for instant use. Being close to the subject the light source need only be a small unit—such as a tubular projection lamp—in a non-reflecting

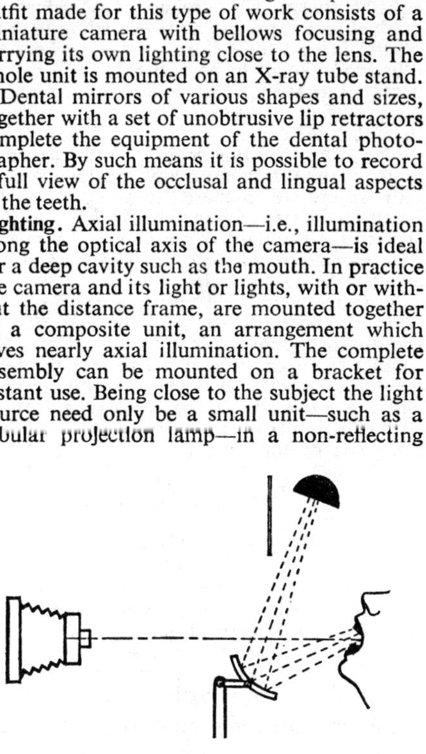

DENTAL LIGHTING. The light is concentrated on the required area of the mouth by means of a concave mirror set up just outside the field of view of the camera lens.

shield. These small sources give sharper detail and fewer troublesome reflections than larger lamps. A small flash bulb or electronic flash tube in a blackened reflector is often employed, and in some arrangements the light is focused on the subject by a concave mirror.

A circular electronic flash tube mounted round the camera lens will be found extremely useful for deep intra-oral views.

Colour. The teeth and associated structures are being increasingly reproduced in colour instead of black-and-white. For colour photographs, more powerful light sources are employed and they are balanced with appropriate filters to suit the sensitized material.

Intra-oral views have a natural background, but when facial views are taken, extraneous objects beyond the subject are excluded by setting up a conventional background—usually green.

Polished metal retractors are apt to give undesirable reflections, and painted finger nails can be even more objectionable. The general light from a large adjacent window—common in a dental surgery—will very often cancel out the advantages of the concentrated illumination of a small light source, in addition to upsetting the colour balance when using colour film.

Technique. Practically every dental photograph is only of real value when compared with another, either of the same, or a different patient. For this reason it is highly necessary to establish a standard technique by which anyone can take regular photographs for comparison without possessing any greater technical skill than that of the average amateur. T.A.L.

See also: *Medical photography; Medical radiography.*
Books: *Photographic Aids to Clinical Dental Practice*, by J. A. Dyce (London); *Longmore's Medical Photography* (London).

DEPTH OF FIELD

When the camera lens is focused to give a sharp image of a particular object, other objects closer or farther away do not appear equally sharp. The decline of sharpness is gradual and there is a zone extending in front of and behind the focused distance where the blur is too small to be noticeable and can be accepted as sharp. This zone is known as the depth of field of the lens. It is often miscalled the depth of focus.

The extent of the depth of field depends, amongst other things, on how great a degree of blur we are still prepared to accept as sharp, i.e. upon the size of the circle of confusion. The stricter the standard of sharpness, the smaller the permissible circle of confusion and the smaller the depth of field.

The depth of field also depends on the aperture and focal length of the lens, and on the distance from the lens to the subject. The smaller the aperture and the shorter the focal length of the lens, the greater the depth of field. The closer the subject, the smaller the depth of field. The limits of the depth of field of a lens are given by the equations:

$$D_n = \frac{D}{1 + Dcf/F^2}$$

$$D_f = \frac{D}{1 - Dcf/F^2}$$

where: D_n = Near limit of depth of field
D_f = Far limit of depth of field
D = Focused distance
c = Limiting circle of confusion
f = *f*-number of lens aperture
F = Focal length of lens

(All measurements must be in the same units.)

For a circle of confusion equal to 1/1000 of the focal length of the lens we have:

$$D_n = \frac{D}{1 + Df/1000F}$$

$$D_f = \frac{D}{1 - Df/1000F}$$

where: D_n = Near limit of depth of field
D_f = Far limit of depth of field
D = Focused distance
f = *f*-number of lens aperture
F = Focal length of lens

Effect of Aperture and Focal Length. With a pinhole camera the depth of field is theoretically unlimited. Everything is reproduced equally sharp, however near or far away, and irrespective of the distance between the pinhole and the film or plate. So the nearer the relative aperture of a lens approaches a pinhole the greater the depth of field at any object distance.

In fact both relative aperture and focal length govern depth of field. The depth of field depends on the ratio of *f*-number/focal length. The larger this ratio, the greater the depth of field.

For an absolute circle of confusion (i.e. independent of the focal length of the lens), the depth of field is proportional to the ratio of *f*-number/(focal length)^2.

Hyperfocal Distance. The depth of field depends also on the distance of the subject: the farther from the lens, the greater the depth of field.

For any one lens aperture, focal length, and circle of confusion, there is a point beyond which the depth of field behind the subject is infinite. The nearest point at which this happens is called the hyperfocal distance. It is also the near limit of the depth of field when the lens is set at infinity.

When a lens is focused at its hyperfocal distance, the depth of field extends from infinity to half the hyperfocal distance. At half

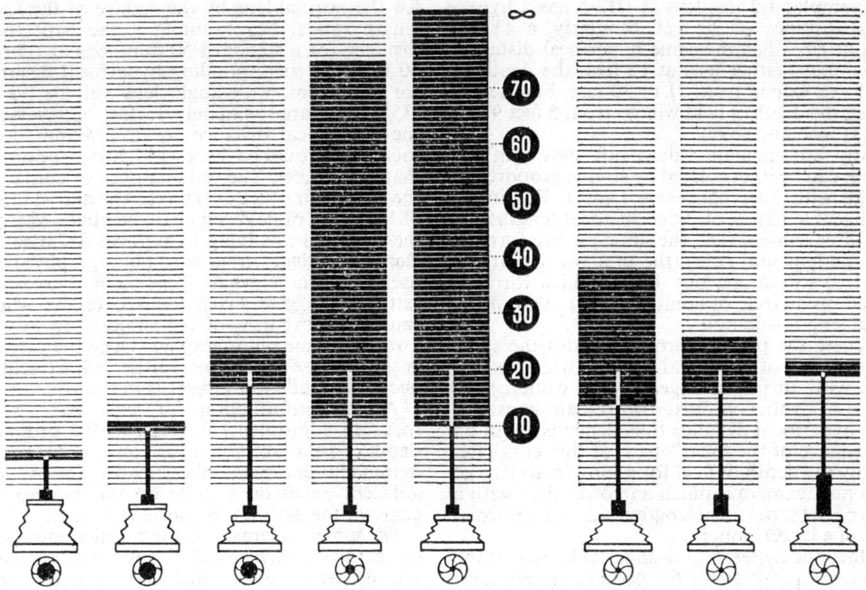

DEPTH OF FIELD IN PRACTICE. The extent of the depth of field depends on the focusing distance, the aperture, and the focal length of the lens. **First three columns:** *The depth obtained increases as the subject moves away from the camera.* **Fourth and fifth columns:** *With the lens focused on a given distance, the depth of field grows as the lens is stopped down.* **Last three columns:** *With a given distance and aperture, the use of lenses of increasing focal length again reduces the extent of the sharp zone available.*

the hyperfocal distance, the depth of field extends from the hyperfocal distance to one-third of the hyperfocal distance. At one-third of the hyperfocal distance, the depth of field extends from half to one-quarter, and at one-tenth of the hyperfocal distance the depth extends from one-ninth to one-eleventh.

The nearer the focused distance is to the hyperfocal distance, the greater the depth of field.

The hyperfocal distance depends on the same factors as the depth of field—standard of sharpness (circle of confusion), aperture and focal length of the lens:

$$H = 1000F/f$$

where: H = Hyperfocal distance
F = Focal length of lens
f = *f*-number of aperture

for a circle of confusion = 1/1000 of the focal length.

For a higher standard, e.g. 1/2000 of the focal length,

$$H = 2000F/f$$

For an absolute circle of confusion,

$$H = F^2/cf$$

where: H = Hyperfocal distance
F = Focal length of lens
c = Circle of confusion
f = *f*-number of aperture

All units of measurement must of course be the same throughout.

Simplified Depth of Field. For the same hyperfocal distance (fixed by the focal length and aperture of the lens) the depth of field values are always the same, and depend only on the distance at which the lens is focused. This simplifies depth of field calculations and tables. There is no need to work out the depth of field separately for every aperture and for every lens; it can be related directly to the hyperfocal distance.

In terms of the hyperfocal distance the near and far limits of the depth of field can be found as follows:

$$D_n = \frac{HD}{H + D}$$

$$D_f = \frac{HD}{H - D}$$

where: D_n = Near limit of depth of field
D_f = Far limit of depth of field
H = Hyperfocal distance
D = Focused distance

Used in conjunction with hyperfocal distance tables one depth of field table serves for all lenses and apertures. Once the hyperfocal distance corresponding to the focal length and aperture of the lens is known the depths of field at any focused distance can be read off from the tables.

Example: a 3 ins. lens at *f*/12·5 has a hyperfocal distance of 20 feet. Similarly, a 1½ ins. lens at *f*/6·3 has the same hyperfocal distance. If we focus either lens at 12 feet, the depth of field will extend from 7½ to 30 feet. Focused at 8 feet, the depth of field will go from 5 feet 9 ins. to 13½ feet, and so on.

Any intermediate values not shown in the tables can be interpolated by simple proportion.

Depth with Interchangeable Lenses. For interchangeable lenses of different focal lengths for the same camera (i.e., the same picture format) the fundamental rules (the first two equations quoted) still apply, but the simplified forms—based on a circle of confusion of 1/1000 of the focal length—do not.

Where the picture format remains the same irrespective of the focal length (which is the case with an interchangeable-lens camera) the circle of confusion must also remain constant. The hyperfocal distance thus depends again on the square of the focal length of the lens. That is why the depth of field for a long focus lens of a 35 mm. camera is much narrower than with a lens of the same focal length used as the normal lens of a larger camera.

Once the hyperfocal distance is known at the different apertures for the normal camera lens, the same hyperfocal distances for the other interchangeable lenses can be determined simply by multiplying the hyperfocal distance for the normal lens by the square of the focal length ratios. For example if the hyperfocal distance for a standard 50 mm. lens at *f* 5·6 is 30 feet, the corresponding hyperfocal distance for a 35 mm. wide-angle lens will be $30 \times (35/50)^2$ = approximately 15 feet. Conversely, the hyperfocal distance for a 135 mm. long focus lens would be $30 \times (135/50)^2$ = approximately 220 feet. The same applies of course to the values for other apertures. The actual limits of the depth of field can again be obtained from the equation involving hyperfocal distance.

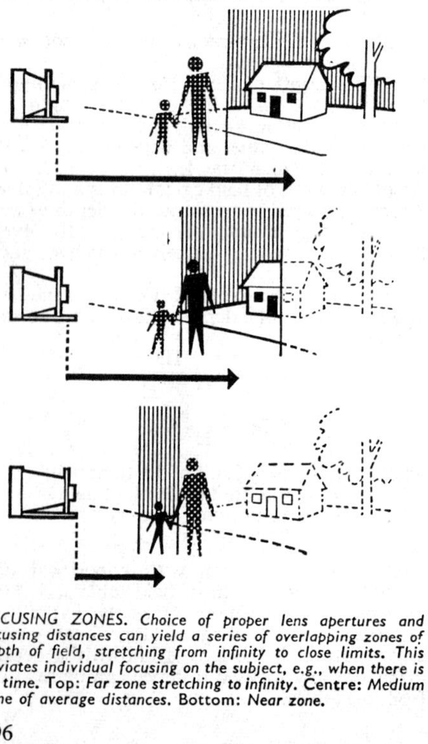

FOCUSING ZONES. Choice of proper lens apertures and focusing distances can yield a series of overlapping zones of depth of field, stretching from infinity to close limits. This obviates individual focusing on the subject, e.g., when there is no time. Top: *Far zone stretching to infinity.* Centre: *Medium zone of average distances.* Bottom: *Near zone.*

Zone Focusing. In action photography it is useful to know what focusing and aperture settings will give sharp focus over the whole range of movement of the subject towards or away from the photographer. Once the camera is set to these figures, no further adjustment is necessary while the subject is in the zone.

All the information for such fixed zone focusing adjustments is given in the ordinary depth of field tables, but for convenience many manufacturers print the figures for a few adjacent zones on a plate on the camera, or engrave the setting on the lens mount.

On some cameras, coloured indicating spots or arrows mark fixed focus and aperture settings which divide the whole range of the camera into no more than two or three zones. Focusing with such cameras is merely a matter of setting the lens to the appropriate Near, Average, or Distant Zone.

Close-up Depth of Field. Close-up pictures call for special consideration.

In copying, for instance, the subject (i.e., the original to be copied) is two-dimensional and there is no question of getting near and far objects sharp. The problem is to find the permissible error in setting up the camera and the subject, which is a matter of depth of focus.

In most close-ups the depth of field is so small that the zone of sharpness is in effect a flat plane, and the correct viewing distance depends only on the scale of reproduction on the final print. The smaller the negative the more it must be enlarged to give a standard scale of reproduction of, say, 1 : 1, so the circle of confusion must be correspondingly smaller.

For close-up work of this kind a suitable limiting diameter for the circle of confusion is about 0·025 mm. on full frame 35 mm. negatives, 0·0005 in. (0·012 mm.) on half frame 35 mm. negatives (as used in microcopying), 0·002 in. on 2¼ × 3¼ ins. (6 × 9 cm.) negatives, and so on. This is about equivalent to a standard of 1/2000 of the focal length under normal conditions.

While the normal depth of field equations still hold, a value based on the scale of reproduction is often called for.

Also, at such close distances the depth of field is practically the same behind as in front of the subject. The depth of field in front of and behind the plane of sharp focus is:—

$$d = fc(M + 1)/M^2$$

APPROXIMATE HYPERFOCAL DISTANCES (FEET OR FEET AND INCHES)

Focal Length of Lens in.	mm.	Hyperfocal Distance at Aperture f 1·5	2	2·2	2·8	3·5	4	4·5	5·6	6·3	8	9	11	12·5	16	18	22	25	32
1	25	55½	42	38	30	24	21	18½	15	13¼	10½	9–3	7–7	6–8	5–3	4–7	3–9	3–4	2–7
1½	37·5	83½	62½	57	44½	36	31	28	22	20	15½	14	11½	10	7–9	7–0	5–9	5–0	3–10
2	50	111	83½	76	59½	47½	42	37	30	26½	21	18½	15¼	13¼	10½	9–3	7–7	6–8	5–3
2½	62·5	139	104	95	74½	59½	52	46½	37	33	26	23	19	16¾	13	11¾	9–6	8–5	6–6
3	75	167	125	113	89½	71½	62½	55½	44½	40	31	28	23	20	15½	13¼	11½	10	7–9
3½	90	195	146	133	104	83½	73	65	52	46	36½	32½	26½	23	18¼	16¼	13¼	11¾	9–2
4	100	222	167	151	119	95	83½	74	59½	53	41½	37	30½	27	21	18½	15¼	13¼	10½
4½	112	250	187	170	134	107	94	83½	67	59½	47	42	34	30	23½	21	17	15	11¾
5	125	278	208	189	149	119	104	92½	74½	66	52½	46	38	33½	26	23	19	16¾	13
5½	137	—	230	208	164	131	114	102	82	73	57½	51	41½	37	29	25½	21	18½	14¼
6	160	—	250	227	179	143	125	111	89	79	62½	55½	45½	40	31	28	23	20	15¾
7	175	—	290	265	208	167	146	130	104	92½	73	65	53	46½	36½	32½	26½	23	18¼
8	200	—	333	305	238	190	167	148	119	106	83	74	61	53½	41½	37	30½	27	21
10	250	—	—	380	298	238	208	185	149	132	104	92½	76	67	52	46½	38	33½	26
12	300	—	—	455	357	285	250	222	179	159	125	111	91	80	62½	55½	45½	40	31

Standard of sharpness: circle of confusion = 1/1000 of the focal length. For a circle of confusion of 1/1500 of the focal length or for a 1½ × long focus or telephoto lens used with the same camera multiply the hyperfocal distance by 1½; for 1/2000 of the focal length or a 2× long focus or telephoto lens multiply by 2, and so on.

APPROXIMATE DEPTHS OF FIELD (FEET AND INCHES)

Hyperfocal Distance (Feet)	Depth of Field at Distance Setting of 2 ft. (0·6 m.)	3 ft. (0·9 m.)	4 ft. (1·2 m.)	5 ft. (1·5 m.)	6 ft. (1·8 m.)	8 ft. (2·4 m.)	10 ft. (3 m.)	12 ft. (3·6 m.)
3	1–2½ to 6–0	1–6 to ∞	1–8½ to ∞	1–10½ to ∞	2–0 to ∞	2–2¼ to ∞	2–3½ to ∞	2–5 to ∞
4	1–4 to 4–0	1–8½ to 12	2–0 to ∞	2–2¾ to ∞	2–5 to ∞	2–8 to ∞	2–10 to ∞	3–0 to ∞
6	1–6 to 3–0	2–0 to 9–0	2–5 to 12	2–9 to 30	3–0 to ∞	3–5 to ∞	3–9 to ∞	4–0 to ∞
8	1–7¼ to 3–8	2–2¼ to 4–10	2–8 to 8–0	3–1 to 13¼	3–5 to 24	4–0 to ∞	4–5 to ∞	4–9 to ∞
12	1–8½ to 2–4¾	2–5 to 4–0	3–0 to 6–0	3–6 to 8–6	4–0 to 12	4–10 to 24	5–5 to 60	6–0 to ∞
15	1–9 to 2–3½	2–6 to 3–9	3–2 to 5–5	3–9 to 7–6	4–3½ to 10	5–3 to 17¼	6–0 to 30	6–8 to 60
20	1–9¾ to 2–2½	2–7 to 3–6½	3–4 to 5–0	4–0 to 6–8	4–7½ to 9–4	5–9 to 13½	6–8 to 20	7–6 to 30
25	1–10¼ to 2–2	2–8 to 3–5	3–5½ to 4–9	4–2 to 6–3	4–10 to 7–11	6–1 to 11¾	7–2 to 16¾	8–1 to 23
40	1–11 to 2–1¼	2–9½ to 3–3	3–7½ to 4–5½	4–5½ to 5–9	5–2 to 7–1	6–8 to 10	8–0 to 13¼	9–3 to 17¼
50	1–11¼ to 2–1	2–10 to 3–2½	3–8½ to 4–4	4–6½ to 5–6	5–4 to 6–10	6–11 to 9–6	8–4 to 12½	9–8 to 15¾
75	1–11½ to 2–0¾	2–10½ to 3–1½	3–9½ to 4–3	4–8 to 5–4	5–7 to 6–6	7–3 to 8–11	8–10 to 11½	10¼ to 14¼
100	1–11½ to 2–0½	2–11 to 3–1	3–10 to 4–2	4–9 to 5–3	5–8 to 6–5	7–5 to 8–8	9–1 to 11¼	10¾ to 13½
150	1–11¾ to 2–0½	2–11 to 3–1	3–11 to 4–1½	4–10 to 5–2	5–9 to 6–3	7–7 to 8–6	9–4 to 10¾	11 to 13
200	1–11¾ to 2–0¼	2–11½ to 3–0½	3–11 to 4–1	4–10½ to 5–2	5–10 to 6–2	7–8 to 8–4	9–6 to 10½	11¼ to 12¾
300	2–0 to 2–0	2–11½ to 3–0½	3–11½ to 4–0½	4–11 to 5–1	5–10 to 6–2	7–9 to 8–3	9–8 to 10½	11½ to 12½
400	2–0 to 2–0	3–0 to 3–0½	3–11½ to 4–0½	4–11 to 5–1	5–11 to 6–1	7–10 to 8–2	9–9 to 10¼	11¾ to 12¼

APPROXIMATE DEPTHS OF FIELD (FEET AND INCHES)

Hyperfocal Distance (Feet)	Depth of Field at Distance Setting of 15 ft. (4·5 m.)	20 ft. (6 m.)	25 ft. (7·5 m.)	50 ft. (15 m.)	100 ft. (30 m.)	200 ft. (60 m.)	∞ ∞
3	2–6 to ∞	2–7 to ∞	2–8½ to ∞	2–9½ to ∞	2–11 to ∞	2–11½ to ∞	3–0 to ∞
4	3–2 to ∞	3–4 to ∞	3–5½ to ∞	3–8½ to ∞	3–10 to ∞	3–11 to ∞	4–0 to ∞
6	4–3½ to ∞	4–7½ to ∞	4–10 to ∞	5–4 to ∞	5–8 to ∞	5–10 to ∞	6–0 to ∞
8	5–3 to ∞	5–8½ to ∞	6–1 to ∞	6–11 to ∞	7–5 to ∞	7–8 to ∞	8–0 to ∞
12	6–8 to ∞	7–6 to ∞	8–1 to ∞	9–8 to ∞	10¾ to ∞	11¼ to ∞	12 to ∞
15	7–6 to ∞	8–7 to ∞	9–4 to ∞	11½ to ∞	13 to ∞	14 to ∞	15 to ∞
20	8–7 to 60	10 to ∞	11¼ to ∞	14¼ to ∞	16¾ to ∞	18¼ to ∞	20 to ∞
25	9–4 to 37½	11 to 100	12½ to ∞	16¾ to ∞	20 to ∞	21¼ to ∞	25 to ∞
40	11 to 24	13¼ to 40	15½ to 67	22½ to ∞	28½ to ∞	33½ to ∞	40 to ∞
50	11½ to 21½	14¼ to 33½	16¾ to 50	25 to ∞	33½ to ∞	40 to ∞	50 to ∞
75	12½ to 18¾	15¾ to 27	18¾ to 37½	30 to 150	43 to ∞	54½ to ∞	75 to ∞
100	13 to 17¾	16¾ to 25	20 to 33½	33½ to 100	50 to ∞	67 to ∞	100 to ∞
150	13¾ to 16¾	17¾ to 23	21½ to 30	37½ to 75	60 to 300	86 to ∞	150 to ∞
200	14 to 16¼	18¼ to 22	22½ to 28½	40 to 67	67 to 200	100 to ∞	200 to ∞
300	14¼ to 15¾	18¾ to 21½	23 to 27	43 to 60	75 to 150	120 to 600	300 to ∞
400	14½ to 15½	19 to 21	23½ to 26½	44½ to 57	80 to 133	133 to 400	400 to ∞

APPROXIMATE HYPERFOCAL DISTANCES (METRES)

Focal Length of Lens mm.	in.	Hyperfocal Distance at Aperture f 1·5	2	2·2	2·8	3·5	4	4·5	5·6	6·3	8	9	11	12·5	16	18	22	25	32
25	1	16·7	12·5	11·4	8·9	7·15	6·25	5·55	4·45	3·97	3·12	2·78	2·28	2·00	1·56	1·40	1·14	1·00	0·78
37·5	1½	25·0	18·8	17·1	13·4	10·7	9·40	8·35	6·70	5·95	4·67	4·17	3·41	3·00	2·34	2·08	1·71	1·50	1·17
50	2	33·3	25·0	22·8	17·8	14·3	12·5	11·1	8·90	7·95	6·25	5·55	4·55	4·00	3·12	2·78	2·28	2·00	1·56
62·5	2½	41·7	31·2	28·4	22·3	17·9	15·6	13·9	11·2	9·90	7·80	6·95	5·70	5·00	3·90	3·47	2·84	2·50	1·95
75	3	50	37·5	34·1	26·8	21·4	18·8	16·7	13·4	11·9	9·40	8·35	6·80	6·00	4·70	4·17	3·41	3·00	2·35
90	3½	60	45·0	41·0	32·1	25·7	22·5	20·0	16·1	14·3	11·2	10·0	8·20	7·20	5·65	5·00	4·10	3·60	2·81
100	4	67	50	45·5	35·7	28·6	25·0	22·2	17·9	15·9	12·5	11·1	9·10	8·00	6·25	5·55	4·55	4·00	3·12
112	4½	75	56	51	40·0	32·1	28·1	25·0	20·0	17·8	14·1	12·5	10·2	9·00	7·00	6·25	5·10	4·50	3·51
125	5	83	63	57	44·5	35·7	31·3	27·8	22·?	19·8	15·6	13·9	11·4	10·0	7·80	7·00	5·70	5·00	3·90
137	5½	—	69	63	49·0	39·3	34·4	30·5	24·6	21·8	17·2	15·3	12·5	11·0	8·60	7·65	6·25	5·50	4·30
150	6	—	75	68	53·6	43·0	37·5	33·3	26·8	23·8	18·8	16·7	13·6	12·0	9·35	8·35	6·80	6·00	4·70
175	7	—	88	80	63	50	43·7	39·0	31·1	27·8	21·8	19·4	15·9	14·0	10·9	9·70	7·95	7·00	5·45
200	8	—	100	91	72	58	40	44·5	35·8	31·8	25·0	22·2	18·2	16·0	12·5	11·1	9·10	8·00	6·25
250	10	—	—	114	89	72	63	56	44·5	39·7	31·2	27·8	22·8	20·0	15·6	14·0	11·4	10·0	7·80
300	12	—	—	136	107	86	75	67	54	47·6	37·6	33·3	27·2	24·0	18·7	16·7	13·6	12·0	9·40

Standard of sharpness: circle of confusion = 1/1000 of the focal length of the lens. For a circle of confusion of 1/1500 of the focal length or for a 1½× long focus or telephoto lens used with the same camera multiply the hyperfocal distance by 1½; for 1/2000 of the focal length or a 2× long focus or telephoto lens multiply by 2, and so on.

APPROXIMATE DEPTHS OF FIELD (METRES)

Hyperfocal Distance (Metres)	Depth of Field at Distance Setting of 0·6 m. (2 ft.)	0·75 m. (2½ ft.)	1 m. (3½ ft.)	1·2 m. (4 ft.)	1·5 m. (5 ft.)	2 m. (7 ft.)	2·5 m. (8½ ft.)	4 m. (13 ft.)
0·5	0·27 to ∞	0·30 to ∞	0·33 to ∞	0·35 to ∞	0·38 to ∞	0·40 to ∞	0·42 to ∞	0·44 to ∞
0·75	0·33 to 3·0	0·37 to ∞	0·43 to ∞	0·46 to ∞	0·50 to ∞	0·55 to ∞	0·58 to ∞	0·63 to ∞
1·0	0·37 to 1·50	0·43 to 3·0	0·50 to ∞	0·55 to ∞	0·60 to ∞	0·67 to ∞	0·71 to ∞	0·80 to ∞
1·5	0·43 to 1·00	0·50 to 1·50	0·60 to 3·0	0·67 to 6·0	0·75 to ∞	0·86 to ∞	0·94 to ∞	1·09 to ∞
2·0	0·46 to 0·86	0·55 to 1·30	0·67 to 2·00	0·75 to 3·0	0·86 to 6·0	1·00 to ∞	1.11 to ∞	1·33 to ∞
3·0	0·50 to 0·75	0·60 to 1·00	0·75 to 1·50	0·86 to 2·00	1·00 to 3·0	1·20 to 6·0	1·36 to 15·0	1·72 to ∞
4·0	0·52 to 0·71	0·63 to 0·92	0·80 to 1·33	0·92 to 1·72	1·09 to 2·40	1·33 to 4·0	1·54 to 6·7	2·00 to ∞
5·0	0·54 to 0·68	0·65 to 0·88	0·83 to 1·25	0·97 to 1·58	1·15 to 2·10	1·43 to 3·3	1·67 to 5·0	2·22 to 20
7·5	0·55 to 0·66	0·68 to 0·83	0·88 to 1·15	1·03 to 1·43	1·25 to 1·88	1·58 to 2·73	1·88 to 3·8	2·61 to 8·6
10	0·57 to 0·64	0·70 to 0·81	0·91 to 1·10	1·07 to 1·37	1·30 to 1·76	1·67 to 2·50	2·00 to 3·3	2·85 to 6·7
15	0·58 to 0·62	0·71 to 0·79	0·94 to 1·07	1·11 to 1·31	1·36 to 1·67	1·76 to 2·31	2·15 to 3·0	3·2 to 5·5
20	0·58 to 0·62	0·72 to 0·78	0·95 to 1·05	1·13 to 1·28	1·40 to 1·67	1·82 to 2·22	2·22 to 2·86	3·3 to 5·0
30	0·59 to 0·61	0·73 to 0·77	0·97 to 1·04	1·15 to 1·26	1·43 to 1·58	1·87 to 2·14	2·31 to 2·73	3·5 to 4·6
40	0·59 to 0·61	0·73 to 0·77	0·98 to 1·03	1·17 to 1·24	1·45 to 1·56	1·90 to 2·11	2·35 to 2·67	3·6 to 4·5
60	0·59 to 0·61	0·74 to 0·76	0·98 to 1·02	1·18 to 1·23	1·46 to 1·54	1·93 to 2·07	2·40 to 2·61	3·7 to 4·3
80	0·59 to 0·61	0·74 to 0·76	0·99 to 1·01	1·18 to 1·22	1·47 to 1·53	1·95 to 2·05	2·42 to 2·58	3·8 to 4·2
120	0·60 to 0·60	0·75 to 0·75	0·99 to 1·01	1·19 to 1·21	1·48 to 1·52	1·97 to 2·03	2·45 to 2·55	3·9 to 4·1
160	0·60 to 0·60	0·75 to 0·75	0·99 to 1·01	1·19 to 1·21	1·49 to 1·51	1·98 to 2·02	2·46 to 2·54	3·9 to 4·1

APPROXIMATE DEPTHS OF FIELD (METRES)

Hyperfocal Distance (Metres)	Depth of Field at Distance Setting of 5 m. (17 ft.)	7 m. (23 ft.)	10 m. (33 ft.)	15 m. (50 ft.)	25 m. (83 ft.)	40 m. (130 ft.)	50 m. (167 ft.)	∞ ∞
0·5	0·45 to ∞	0·47 to ∞	0·48 to ∞	0·48 to ∞	0·49 to ∞	0·49 to ∞	0·50 to ∞	0·50 to ∞
0·75	0·65 to ∞	0·68 to ∞	0·70 to ∞	0·72 to ∞	0·73 to ∞	0·74 to ∞	0·74 to ∞	0·75 to ∞
1·0	0·83 to ∞	0·88 to ∞	0·91 to ∞	0·94 to ∞	0·96 to ∞	0·98 to ∞	0·98 to ∞	1·00 to ∞
1·5	1·15 to ∞	1·24 to ∞	1·31 to ∞	1·36 to ∞	1·42 to ∞	1·45 to ∞	1·46 to ∞	1 50 to ∞
2·0	1·43 to ∞	1·55 to ∞	1·67 to ∞	1·76 to ∞	1·85 to ∞	1·90 to ∞	1·92 to ∞	2·00 to ∞
3·0	1·87 to ∞	2·10 to ∞	2·31 to ∞	2·50 to ∞	2·68 to ∞	2·79 to ∞	2·83 to ∞	3·0 to ∞
4·0	2·22 to ∞	2·56 to ∞	2·81 to ∞	3·2 to ∞	3·5 to ∞	3·6 to ∞	3·7 to ∞	4·0 to ∞
5·0	2·50 to ∞	2·91 to ∞	3·3 to ∞	3·8 to ∞	4·2 to ∞	4·5 to ∞	4·6 to ∞	5·0 to ∞
7·5	3·0 to 30	3·6 to 105	4·3 to ∞	5·0 to ∞	5·7 to ∞	6·3 to ∞	6·5 to ∞	7·5 to ∞
10	3·3 to 10	4·1 to 23·3	5·0 to ∞	6·0 to ∞	7·1 to ∞	8·0 to ∞	8·3 to ∞	10·0 to ∞
15	3·8 to 7·5	4·8 to 13·1	6·0 to 30	7·5 to ∞	9·4 to ∞	10·9 to ∞	11·5 to ∞	15·0 to ∞
20	4·0 to 6·7	5 2 to 10·8	6·7 to 20·0	8·6 to 60	11·1 to ∞	13·3 to ∞	14·3 to ∞	20·0 to ∞
30	4·3 to 6·0	5·4 to 9·1	7·5 to 15·0	10·0 to 30	13·6 to 75	17·2 to ∞	18·7 to ∞	30 to ∞
40	4·4 to 5·7	6·0 to 8·5	8·0 to 13·3	10·9 to 24·0	15·4 to 67	20·0 to ∞	22·2 to ∞	40 to ∞
60	4·6 to 5·5	6·3 to 7·9	8·6 to 12·0	12·0 to 20·0	17·6 to 43	24·0 to 120	27·3 to 300	60 to ∞
80	4·7 to 5·3	6·4 to 7·7	8·9 to 11·4	12·6 to 18·5	19·1 to 36	26·7 to 80	31 to 133	80 to ∞
120	4·8 to 5·2	6·6 to 7·4	9·2 to 10·9	13·3 to 17·2	20·7 to 32	30 to 60	35 to 86	120 to ∞
160	4·9 to 5·1	6·7 to 7·3	9·4 to 10·7	13·7 to 16·5	21·6 to 29·6	32 to 53	38 to 73	160 to ∞

CLOSE-UP DEPTH OF FIELD (INCHES)

Aperture f	Depth of Field for Scale of Reproduction 0·1 (1 : 10)	0·13 (1 : 8)	0·17 (1 : 6)	0·2 (1 : 5)	0·25 (1 : 4)	0·33 (1 : 3)	0·5 (1 : 2)	0·67 (2 : 3)	1 (1 : 1)	1·5 (3 : 2)	2 (2 : 1)	2·5 (5 : 2)	3 (3 : 1
2	0·22	0·14	0·084	0·060	0·040	0·024	0·012	0·0075	0·0040	0·0022	0·0015	0·0011	0·0009
2·2	0·24	0·16	0·092	0·066	0·044	0·026	0·013	0·0082	0·0044	0·0024	0·0016	0·0012	0·0010
2·8	0·31	0·20	0·12	0·084	0·056	0·034	0·017	0·010	0·0056	0·0031	0·0021	0·0016	0·0012
3·5	0·38	0 25	0·15	0·10	0·070	0·042	0·020	0·013	0·0070	0·0039	0·0026	0·0020	0·0016
4	0·44	0·29	0·17	0·12	0·080	0·048	0·024	0·015	0·0080	0·0044	0·0030	0·0022	0·0018
4·5	0·49	0·32	0·19	0·13	0·090	0·054	0·027	0·017	0·0090	0·0050	0·0034	0·0025	0·0020
5·6	0·62	0·40	0·23	0·17	0·11	0·067	0·034	0·021	0·011	0·0067	0·0042	0·0031	0·0025
6·3	0·69	0·46	0·26	0·19	0·13	0·077	0·038	0·024	0·013	0·0070	0·0047	0·0035	0·0028
8	0·88	0·58	0·34	0·24	0·16	0·096	0·048	0·030	0·016	0·0089	0·0060	0·0045	0·0036
9	0·99	0·65	0·38	0·27	0·18	0·11	0·054	0·034	0·018	0·010	0·0067	0·0050	0·0040
11	1·21	0·79	0·46	0·33	0·22	0·13	0·066	0·041	0·022	0·012	0·0082	0·0062	0·0049
12·5	1·37	0·90	0·52	0·37	0·25	0·15	0·075	0·047	0·025	0·014	0·0094	0·0070	0·0056
16	1·76	1·15	0·67	0·48	0·32	0·19	0·096	0·060	0·032	0·018	0·012	0·0090	0·0071
18	1·98	1·30	0·76	0·54	0·36	0·22	0·11	0·067	0·036	0·020	0·013	0·010	0·0080
22	2·42	1·58	0·92	0·66	0·44	0·26	0·13	0·082	0·044	0·024	0·016	0·012	0·0098
25	2·75	1·80	1·05	0·75	0·50	0·30	0·15	0·094	0·050	0·028	0·019	0·014	0·011
32	3·5	2·30	1·34	0·96	0·64	0·38	0·19	0·12	0·064	0·036	0·024	0·018	0·014
45	4·9	3·2	1·90	1·35	0·90	0·54	0·27	0·17	0·090	0·050	0·034	0·025	0·020

Circle of confusion 0·001 in. **The above is the depth of field at each side of the plane of sharp focus. The total depth twice these figures.**

CLOSE-UP DEPTH OF FIELD (MILLIMETRES)

Aperture f	Depth of Field for Scale of Reproduction 0·1 (1 : 10)	0·13 (1 : 8)	0·17 (1 : 6)	0·2 (1 : 5)	0·25 (1 : 4)	0·33 (1 : 3)	0·5 (1 : 2)	0·67 (2 : 3)	1 (1 : 1)	1·5 (3 : 2)	2 (2 : 1)	2·5 (5 : 2)	3 (3 : 1)
2	5·5	3·6	2·10	1·50	1·00	0·60	0·30	0·19	0·10	0·055	0·038	0·028	0·022
2·2	6·0	4·0	2·31	1·65	1·10	0·66	0·33	0·21	0·11	0·661	0·041	0·031	0·024
2·8	7·7	5·0	2·94	2·10	1·40	0·84	0·42	0·26	0·14	0·078	0·052	0·039	0·031
3·5	9·6	6·3	3·7	2·63	1·75	1·05	0·57	0·33	0·17	0·097	0·065	0·049	0·039
4	11·0	7·2	4·2	3·0	2·00	1·20	0·60	0·38	0·20	0·11	0·775	0·056	0·044
4·5	[illegible]	[illegible]	[illegible]	[illegible]	2·25	1·35	0·67	0·42	0·22	0·12	0·084	0·063	0·050
5·6	15·4	10·1	5·9	4·2	2·80	1·68	0·84	0·53	0·28	0·16	0·10	0·078	0·062
6·3	17·3	11·7	6·6	4·7	3·1	1·89	0·95	0·59	0·31	0·17	0·12	0·088	0·070
8	22·0	14·4	8·4	6·0	4·0	2·40	1·20	0·75	0·40	0·22	0·15	0·11	0·089
9	24·7	16·2	9·4	6·8	4·5	2·70	1·35	0·85	0·45	0·25	0·17	0·13	0·10
11	30	19·8	11·5	8·2	5·5	3·3	1·65	1·03	0·55	0·31	0·21	0·15	0·12
12·5	34	22·5	13·1	9·4	6·2	3·7	1·88	1·17	0·62	0·35	0·23	0·17	0·14
16	44	22·8	16·8	12·0	8·0	4·8	2·40	1·50	0·80	0·44	0·30	0·22	0·18
18	49	32	18·9	13·5	9·0	5·4	2·70	1·69	0·90	0·50	0·34	0·25	0·20
22	60	40	23·1	16·5	11·0	6·6	3·3	2·06	1·10	0·61	0·41	0·30	0·24
25	69	45	26·2	18·7	12·5	7·5	3·7	2·35	1·25	0·70	0·47	0·35	0·28
32	88	58	34	24·0	16·0	9·6	4·8	3·0	1·60	0·89	0·60	0·45	0·36
45	124	81	47	34	22·5	13·5	6·7	4·2	2·25	1·25	0·84	0·62	0·50

Circle of confusion 0·025 mm. **The above is the depth of field at each side of the plane of sharp focus. The total depth is twice these figures.**

where: d = depth of field behind or in front
M = Scale of reproduction (image size/object size)
f = *f*-number of aperture
c = diameter of circle of confusion
(all of these measurements in the same units).

This equation holds for close-up focusing with double-extension cameras or with extension tubes.

With positive supplementary lenses there is a change in relative aperture which must be allowed for by multiplying the *f*-number by a correction factor:

$$f_c = f \times S/(F + S)$$

where: f_c = *f*-number of combination
f = *f*-number of camera lens
S = Focal length of supplementary lens
F = Focal length of camera lens

To find the depth of field at close-up distances, ascertain the scale of reproduction for the focal length of the lens used and the focusing distance.

Then find the depth of field for that scale of reproduction and the aperture used in the close-up depth of field tables below.

Correct for supplementary lens if necessary.

Example: With a subject distance of 10 ins. and a 5 ins. lens the scale of reproduction is 1 : 1. If the lens is stopped down to *f* 8, the depth of field as shown in the table below will be 0·016 in. in each direction, or 0·032 in. in all (approximately 1/32 in.).

A 1 : 1 scale of reproduction will also result from using a 10 ins. supplementary lens with a 10 ins. camera lens. If the camera lens is stopped down to *f* 8, the correction factor for the *f*-number (for an object at infinity) is 10/(10 + 10), i.e., ½, and the effective *f*-number becomes *f* 4. The depth of field is then 0·008 in. in each direction, or 0·016 in. altogether (approximately 1/64 in.).

Limits of Sharpness. The depth of field calculations assume that the degree of sharpness corresponding to the chosen limiting circle of confusion is obtainable. There are, however, two reasons why this is not always possible.

(1) At large apertures the definition given by the lens may fall short of the assumed standard. That is because of aberrations which are usually most prominent at large apertures.

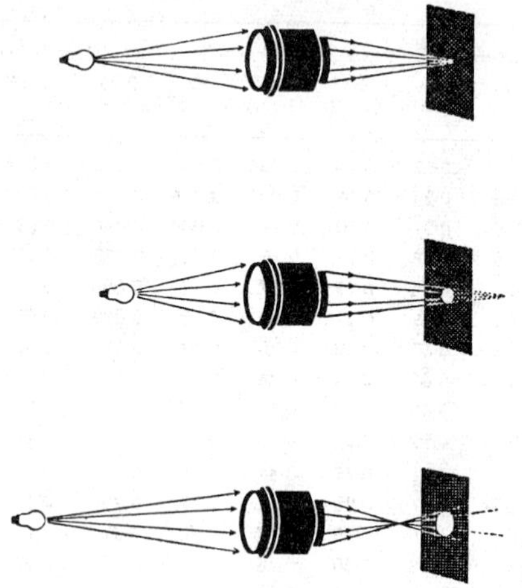

LIMIT OF SHARPNESS. Top: *When a lens is properly focused on an object, the rays of light meet in a point on the film or plate.* Centre: *If the object is nearer the lens, the rays meet farther away, and the plate records a blurred disc.* Bottom: *With the object farther away, the rays diverge again before they reach the plate, and also form a blur. This blur, or circle of confusion, is noticeable as a blur only if it exceeds a certain size. Below that size there is thus a range of distances nearer and farther away than the correct focusing distance, which still yield an acceptably sharp image. That range is the depth of field and can be controlled by judicious choice of apertures.*

(2) At very small apertures, the sharpness of the image is destroyed by the diffraction of light at the edges of the diaphragm. The smaller the aperture, the greater the diffraction. The actual depth of field therefore increases much more rapidly than the geometrical formula suggests, again with a limited and progressively inferior image sharpness.

So to keep within any adopted standard of sharpness, the lens must not be stopped down below a certain limiting aperture, depending on the circle of confusion chosen as the standard. L.A.M.

See also: *Aperture; Circle of confusion; Depth of focus; Optical calculations; Zone focusing.*

Books: *Photographic Optics*, by A. Cox (London); *Focal Focusing Chart*, by A. Cox (London).

DEPTH OF FOCUS

When the camera lens is focused on an object there is one position where the image is sharpest. The sharpness falls off as the film or plate is moved away from this position of exact focus. But there is a certain range of focusing movement within which it is not possible for the eye to detect the difference between a sharp and a very slightly blurred image. This zone of latitude is the depth of focus.

The depth of focus depends on the diameter of the circle of confusion, on the lens aperture, and on the distance between the lens and the image.

The depth of focus extends as far in front of the image plane as behind it.

It is given by the formula:

$$d = \frac{cvf}{F}$$

APPROXIMATE DEPTH OF FOCUS AT INFINITY SETTING (MILLIMETRES)

Hyperfocal Distance (Metres)	Depth of Focus for a Focal Length of 2·5 / 1	3·75 / 1½	5 / 2	6·25 / 2½	7·5 / 3	9 / 3½	10 / 4	11·2 / 4½	12·5 / 5	13·8 / 5½	15 / 6	17·5 / 7	20 / 8	25 / 10	30 cm. / 12 inches
0·5	1·3	2·8	5·2	—	—	—	—	—	—	—	—	—	—	—	—
0·75	0·84	1·9	3·3	—	—	—	—	—	—	—	—	—	—	—	—
1	0·63	1·4	2·5	3·9	—	—	—	—	—	—	—	—	—	—	—
1·5	0·42	0·93	1·7	2·6	3·8	5·4	—	—	—	—	—	—	—	—	—
2	0·32	0·70	1·3	1·9	2·8	4·0	5·0	6·3	—	—	—	—	—	—	—
3	0·21	0·47	0·83	1·3	1·9	2·7	3·3	4·2	5·2	6·3	7·5	—	—	—	—
4	0·16	0·35	0·63	1·0	1·4	2·0	2·5	3·1	3·9	4·7	5·6	7·7	10·0	—	—
5	0·13	0·28	0·50	0·78	1·1	1·6	2·0	2·5	3·1	3·8	4·5	6·1	8·0	12·5	—
7·5	0·083	0·19	0·33	0·52	0·75	1·1	1·3	1·9	2·4	2·9	3·5	4·7	5·3	8·3	13·8
10	0·063	0·14	0·25	0·39	0·56	0·81	1·0	1·3	1·6	1·9	2·2	3·1	4·2	6·0	9·0
15	0·042	0·093	0·17	0·26	0·38	0·54	0·67	0·84	1·0	1·3	1·5	2·0	2·7	4·2	6·0
20	0·031	0·070	0·13	0·19	0·28	0·40	0·50	0·63	0·78	0·95	1·1	1·5	2·0	3·1	4·5
30	0·021	0·047	0·083	0·13	0·19	0·27	0·33	0·42	0·52	0·63	0·75	1·0	1·3	2·1	3·0
40	0·016	0·035	0·063	0·10	0·14	0·20	0·25	0·31	0·39	0·47	0·56	0·77	1·0	1·6	2·2
60	0·010	0·023	0·042	0·065	0·094	0·13	0·17	0·21	0·26	0·31	0·37	0·51	0·67	1·0	1·5
80	0·008	0·018	0·032	0·049	0·070	0·10	0·12	0·16	0·19	0·24	0·28	0·38	0·50	0·78	1·1
120	0·005	0·012	0·021	0·033	0·047	0·067	0·083	0·11	0·13	0·16	0·19	0·26	0·33	0·52	0·75
160	0·004	0·009	0·016	0·024	0·035	0·051	0·062	0·079	0·097	0·12	0·14	0·19	0·25	0·39	0·56

This is the depth of focus at each side of the plane of sharpest focus. The total depth is twice as great.

APPROXIMATE DEPTH OF FOCUS AT INFINITY SETTING (INCHES)

Hyperfocal Distance (feet)	Depth of Focus for a Focal Length of 1 / 2·5	1½ / 3·75	2 / 5	2½ / 6·25	3 / 7·5	3½ / 9	4 / 10	4½ / 11·2	5 / 12·5	5½ / 13·8	6 / 15	7 / 17·5	8 / 20	10 / 25	12 inches / 30 cm.
3	0·028	0·063	0·11	—	—	—	—	—	—	—	—	—	—	—	—
4	0·021	0·047	0·083	0·13	—	—	—	—	—	—	—	—	—	—	—
6	0·014	0·031	0·056	0·087	0·13	0·17	—	—	—	—	—	—	—	—	—
8	0·011	0·023	0·042	0·065	0·094	0·13	0·17	—	—	—	—	—	—		
12	0·0070	0·016	0·028	0·043	0·063	0·085	0·11	0·14	0·17	0·21	0·25	0·34	—	—	—
15	0·0056	0·012	0·022	0·035	0·050	0·068	0·089	0·11	0·14	0·17	0·20	0·27	0·36	—	—
20	0·0042	0·0094	0·017	0·026	0·037	0·051	0·067	0·084	0·10	0·13	0·15	0·20	0·27	0·42	—
25	0·0033	0·0075	0·012	0·021	0·030	0·041	0·053	0·068	0·083	0·10	0·12	0·16	0·21	0·33	0·48
40	0·0021	0·0047	0·0083	0·013	0·019	0·025	0·033	0·042	0·052	0·063	0·075	0·10	0·13	0·21	0·30
50	0·0017	0·0037	0·0062	0·010	0·015	0·020	0·027	0·034	0·042	0·050	0·060	0·082	0·11	0·17	0·24
75	0·0011	0·0025	0·0044	0·0070	0·010	0·014	0·018	0·023	0·028	0·034	0·040	0·054	0·071	0·11	0·16
100	0·0008	0·0019	0·0033	0·0052	0·0075	0·010	0·013	0·017	0·021	0·025	0·030	0·041	0·053	0·083	0·12
150	0·0006	0·0012	0·0022	0·0035	0·0050	0·0068	0·0089	0·011	0·014	0·017	0·020	0·027	0·036	0·056	0·080
200	0·0004	0·0009	0·0017	0·0026	0·0037	0·0051	0·0067	0·0084	0·010	0·013	0·015	0·020	0·027	0·042	0·060
300	0·0003	0·0006	0·0011	0·0017	0·0025	0·0034	0·0044	0·0056	0·0070	0·0084	0·010	0·014	0·018	0·028	0·040
400	0·0002	0·0005	0·0008	0·0013	0·0019	0·0025	0·0033	0·0042	0·0052	0·0063	0·0075	0·010	0·013	0·021	0·030

This is the depth of focus at each side of the plane of sharpest focus. The total depth is twice as great.

CLOSE-UP DEPTH OF FOCUS (INCHES)

Aperture	Depth of Focus for Scale of Reproduction												
f	0·1 (1 : 10)	0·13 (1 : 8)	0·17 (1 : 6)	0·2 (1 : 5)	0·25 (1 : 4)	0·33 (1 : 3)	0·5 (1 : 2)	0·67 (2 : 3)	1 (1 : 1)	1·5 (3 : 2)	2 (2 : 1)	2·5 (5 : 2)	3 (3 : 1)
2	0·0022	0·0023	0·0023	0·0024	0·0025	0·0027	0·0030	0·0033	0·0040	0·0050	0·0060	0·0070	0·0080
2·2	0·0024	0·0025	0·0026	0·0026	0·0027	0·0028	0·0033	0·0037	0·0044	0·0055	0·0066	0·0077	0·0088
2·8	0·0031	0·0032	0·0033	0·0034	0·0035	0·0037	0·0042	0·0047	0·0056	0·0070	0·0084	0·0098	0·011
3·5	0·0038	0·0040	0·0041	0·0042	0·0044	0·0047	0·0052	0·0058	0·0070	0·0087	0·010	0·012	0·014
4	0·0044	0·0045	0·0047	0·0048	0·0050	0·0053	0·0060	0·0067	0·0080	0·010	0·012	0·014	0·016
4·5	0·0050	0·0051	0·0053	0·0054	0·0056	0·0060	0·0068	0·0075	0·0090	0·011	0·013	0·016	0·018
5·6	0·0062	0·0063	0·0065	0·0067	0·0070	0·0075	0·0084	0·0093	0·011	0·014	0·017	0·020	0·022
6·3	0·0069	0·0071	0·0074	0·0076	0·0079	0·0084	0·0095	0·010	0·013	0·016	0·019	0·022	0·025
8	0·0088	0·0090	0·0093	0·0096	0·010	0·011	0·012	0·013	0·016	0·020	0·024	0·028	0·032
9	0·0099	0·010	0·010	0·011	0·011	0·012	0·013	0·015	0·018	0·022	0·027	0·031	0·036
11	0·012	0·012	0·013	0·013	0·014	0·015	0·016	0·018	0·022	0·027	0·033	0·038	0·044
12·5	0·014	0·014	0·015	0·015	0·016	0·017	0·019	0·021	0·025	0·031	0·037	0·044	0·050
16	0·018	0·018	0·019	0·019	0·020	0·021	0·024	0·027	0·032	0·040	0·058	0·056	0·064
18	0·020	0·020	0·021	0·022	0·022	0·024	0·027	0·032	0·036	0·045	0·054	0·063	0·072
22	0·024	0·025	0·026	0·026	0·027	0·029	0·033	0·037	0·044	0·055	0·066	0·077	0·088
25	0·028	0·028	0·029	0·030	0·031	0·033	0·037	0·042	0·050	0·062	0·075	0·087	0·10
32	0·035	0·036	0·037	0·038	0·040	0·043	0·048	0·053	0·064	0·080	0·096	0·11	0·13
45	0·050	0·051	0·053	0·054	0·056	0·060	0·068	0·075	0·090	0·11	0·13	0·16	0·18

This is the depth of focus at each side of the plane of sharpest focus. The total depth is twice as great. Limiting circle of confusion: 0·001 inch.

CLOSE-UP DEPTH OF FOCUS (MILLIMETRES)

Aperture	Depth of Focus for Scale of Reproduction												
f	0·1 (1 : 10)	0·13 (1 : 8)	0·17 (1 : 6)	0·2 (1 : 5)	0·25 (1 : 4)	0·33 (1 : 3)	0·5 (1 : 2)	0·67 (2 : 3)	1 (1 : 1)	1·5 (3 : 2)	2 (2 : 1)	2·5 (5 : 2)	3 (3 : 1)
2	0·055	0·056	0·058	0·060	0·062	0·067	0·075	0·083	0·10	0·12	0·15	0·17	0·20
2·2	0·061	0·062	0·064	0·066	0·069	0·074	0·082	0·092	0·11	0·14	0·16	0·19	0·22
2·8	0·077	0·079	0·082	0·084	0·087	0·096	0·10	0·12	0·14	0·17	0·22	0·24	0·28
3·5	0·096	0·099	0·10	0·10	0·11	0·12	0·13	0·14	0·17	0·22	0·26	0·31	0·35
4	0·11	0·11	0·12	0·12	0·13	0·13	0.15	0·17	0·20	0·25	0·30	0·35	0·40
4·5	0·12	0·13	0·13	0·14	0·14	0·15	0·17	0·19	0·22	0·28	0·34	0·39	0·45
5·6	0·15	0·16	0·16	0·17	0·17	0·19	0·21	0·23	0·28	0·35	0·42	0·44	0·56
6·3	0·17	0·17	0·18	0·19	0·20	0·21	0·23	0·26	0·31	0·39	0·47	0·55	0·63
8	0·22	0·23	0·23	0·24	0·25	0·27	0·30	0·33	0·40	0·50	0·60	0·70	0·80
9	0·25	0·25	0·26	0·27	0·28	0·30	0·34	0·37	0·45	0·56	0·67	0·79	0·90
11	0·30	0·31	0·32	0·33	0·34	0·37	0·41	0·46	0·55	0·69	0·82	0·96	1·10
12·5	0·34	0·35	0·36	0·37	0·39	0·42	0·47	0·52	0·62	0·78	0·94	1·05	1·25
16	0·44	0·45	0·47	0·48	0·50	0·54	0·60	0·67	0·80	1·00	1·20	1·40	1·6
18	0·50	0·51	0·53	0·54	0·56	0·60	0·68	0·75	0·90	1·12	1·35	1·6	1·8
22	0·61	0·62	0·64	0·66	0.69	0·74	0·82	0·92	1·10	1·37	1·6	1·9	2·2
25	0·69	0·70	0·73	0·75	0·78	0·83	0·94	1·04	1·25	1·6	1·9	2·2	2·5
32	0·88	0·90	0·94	0·96	1·00	1·07	1·20	1·33	1·6	2·0	2·4	2·8	3·2
45	1·24	1·27	1·31	1·35	1·41	1·50	1·7	1·9	2·2	2·8	3·4	3·9	4·5

This is the depth of focus at each side of the plane of sharpest focus. The total depth is twice as great. Circle of confusion: 0·025 mm.

where: d = Depth of focus in either direction
c = Limiting circle of confusion
v = Distance between lens and image
f = *f*-number of lens aperture
F = Focal length of the lens

The total depth of focus, i.e. the total permissible film movement, is twice this value.

In terms of subject distance:

$$d = \frac{cDf}{D - F}$$

where: d = depth of focus in either direction
c = Limiting circle of confusion
D = Distance of subject
f = *f*-number of lens aperture
F = Focal length

The depth of focus is directly proportional to the *f*-number; the larger the aperture, the smaller the depth of focus.

Effect of Subject Distance and Focal Length. As the subject is brought nearer to the camera, the ratio D/(D—F) increases, and the depth of focus also increases. This is exactly opposite to the effect on the depth of field.

And as the focal length of the lens increases, the ratio of D/(D—F) also increases, and with it the depth of focus. This again is the opposite of the effect on the depth of field.

But the effect of the focal length is only appreciable at near focusing distances. If the subject distance is very much greater than the

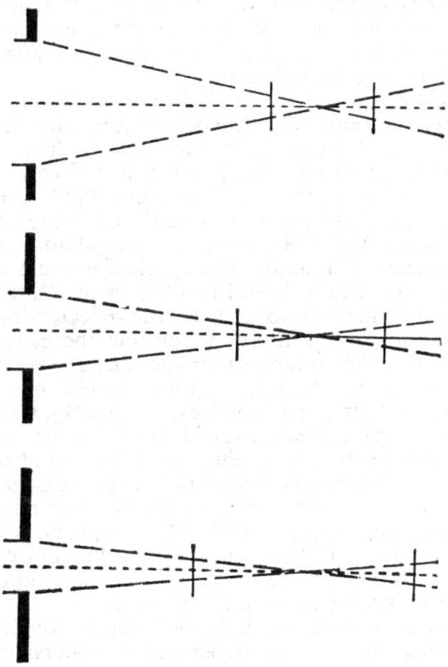

DEPTH OF FOCUS AND APERTURE. *Depth of focus increases the more the lens is stopped down because this decreases the angle of the cone of rays from the lens. At a small stop, therefore, the diameter of the cone reaches the limiting diameter of the circle of confusion more gradually and the film can be displaced more from the plane of maximum sharpness.*

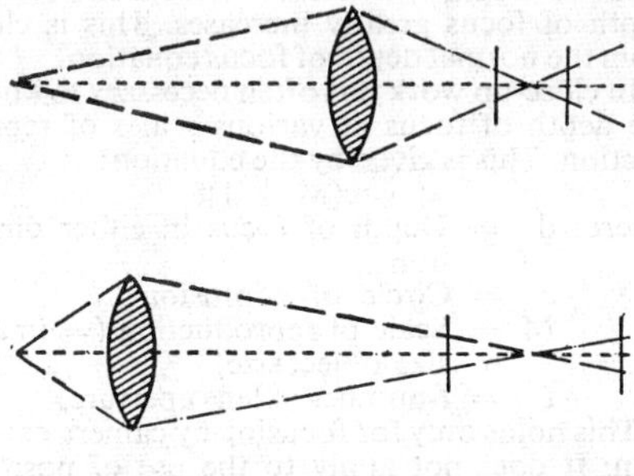

DEPTH OF FOCUS VARIATIONS. *Unlike depth of field, depth of focus increases for close subject distances because the cone of rays coming from the lens is longer and more pointed.*

focal length, the ratio D/(D—F) becomes nearly equal to 1. In that case the depth of focus becomes the product only of the circle of confusion and aperture, and is not affected by the focal length of the lens.

Standard of Sharpness. Finally, the stricter the standard of sharpness, i.e. the smaller the circle of confusion, the smaller the depth of focus.

If the diameter of the circle of confusion is 1/1000 of the focal length, the depth of focus is:

$$d = \frac{FDf}{1000\,(D - F)}$$

where: d = Depth of focus in either direction
F = Focal length of lens
D = Distance of focused subject
f = *f*-number of lens aperture

The circle of confusion is not a fixed quantity; it depends on the viewing distance of the final print, on the focal length of the lens, and on the scale of enlargement.

Hyperfocal Distance and Depth of Focus. The depth of focus is related to the hyperfocal distance by the formula:

$$d = \frac{F^2D}{H(D - F)}$$

where: d = Depth of focus in either direction
F = Focal length of the lens
D = Distance of focused subject
H = Hyperfocal distance of lens

The depth of focus given by any one of a range of lenses can be read off directly from the tabulated values of the formula in the tables.

All measurements must of course be in the same units.

To find the depth of focus of any lens set at infinity: look up the hyperfocal distance in the tables, then read off the corresponding depth of focus in the tables.

The tables are accurate within 10 per cent, provided the subject distance is more than 10 times the focal length. Thus for a 4 ins. lens we can use the figures down to a focusing distance of about 3½ feet; for a 5 cm. lens down to about 0·5 m.

Close-up Depth of Focus. At close distances the depth of focus greatly increases. This is clear from the normal depth of focus equation.

In close-up work it is often necessary to know the depth of focus at various scales of reproduction. This is given by the equation:

$$d = c(M + 1)f$$

where: d = Depth of focus in either direction
c = Circle of confusion
M = Scale of reproduction (= image size/object size)
f = *f*-number of lens aperture

This holds only for focusing by camera extension. It does not apply to the use of positive supplementary lenses when the effective *f*-number, and therefore the depth of focus, is decreased. But the same equation can be made to serve by multiplying the *f*-number by a correction factor:

$$f_s = f \times \frac{S}{F + S}$$

where: f_s = *f*-number of combination
f = *f*-number of camera lens
F = Focal length of camera lens
S = Focal length of supplementary lens

A range of values of the formula is tabulated above. The depth of focus of any normal lens used for close focusing can be read off from this table: first find the scale of reproduction for the particular lens and focusing distance, then read off the depth of focus for that scale of reproduction and the aperture used in the tables. (Correct if necessary for change in *f*-number by supplementary lenses.) L.A.M.

See also: *Aperture; Circle of confusion; Depth of field; Optical calculations.*
Book: *Photographic Optics*, by A. Cox (London).

DERMATITIS. Name given to a group of skin conditions. Sometimes produced by direct contact with certain processing chemicals.

See also: *Skin affections.*

DERMATOLOGICAL PHOTOGRAPHY. Photography used in the study of the normal and diseased skin. A considerable part of dermatological diagnosis depends upon the visual impression and therefore a useful photograph must portray a condition as observed by the clinician. The main points to be demonstrated are:

(*a*) Distribution—the location and spread of a disease.

(*b*) Shape of a lesion—discoid, circinate or plaque-like lesions.

(*c*) Configuration of edge—well defined, ill-defined or raised.

(*d*) Contour.

(*e*) Texture.

(*f*) Colour.

Penetration of the Skin by Visible and Invisible Radiations. The human skin is a translucent organ consisting of two main layers, the epidermis or surface layer and the dermis. Radiations in the ultra-violet, visible and infra-red bands penetrate the skin to varying degrees. The ultra-violet rays penetrate the surface only, visible light to the epidermo-dermal junction and infra-red into the dermis to the depth of the larger blood vessels. By using either an emulsion sensitive to a particular wavelength band or a light source emitting only one waveband, the photographer may select the "depth" at which the photograph is to be taken.

The definition of the skin surface will be affected by rays reflecting from the deeper layers, and the ultra-violet or blue wavebands are best used when the surface structure is to be portrayed. For the majority of lesions, orthochromatic emulsions produce adequate definition of surface and at the same time improve the contrast between pale red lesions, which predominate in skin disease, and the normal skin. When lesions are medium red in colour a normally balanced panchromatic emulsion is used. If dark red ulcerated lesions are present, a panchromatic emulsion with increased red sensitivity is preferable.

In selecting colour film for colour photography of the skin, preference should be given to an emulsion which has a good colour interpretation of red, as this is the predominant colour in diseased skin.

Lighting. Control of lighting conditions is essential both with tungsten lamps and flash. In the latter case it is preferable to fit a low voltage tungsten lamp to the reflector for modelling control. A directional light source such as a spotlight or small flash lamp in a reflector helps to show up the texture and contour of lesions. Heavy shadows ensuing from this beam should be filled in with a flood lamp or flash lamp with a large reflector. Back lighting can be useful to outline the edge of limbs which might otherwise merge into the background. Lighting ratios should not be greater than 3 : 1 when exposing colour film, but can be increased for black-and-white.

Backgrounds. A plain, unobstructed background is essential and dark backgrounds, such as black or grey, are preferable. For colour photography green or black backgrounds are recommended, the former can add colour contrast to pale erythematous lesions. It is important to keep backgrounds standard.

Special Techniques. Infra-red will demonstrate deep lying vascular structures or underside of scab-like lesions.

Electronic flash exposures with a U.V. transmitting filter over the lens (i.e. absorbing visible light) will emphasize pigmentary differences or changes in the keratin surface.

U.V. fluorescence photography can show fluorescence of organisms in the skin and hair which are diagnostic of some fungus diseases. In this case the source should be a U.V. lamp by itself and the lens fitted with a U.V. absorbing filter, so that the photograph only records the fluorescence.

Photomacrography of the living skin requires a special camera set-up with focal frames which fire an electronic flash on contact with the skin. Negative magnifications of up to × 5 can be obtained in this way. R.J.L.

See also: *Medical photography.*
Book: *Longmore's Medical Photography* (London).

DESENSITIZING. Certain dyes, called desensitizers, reduce the sensitivity of an exposed film or plate to light, without greatly affecting the latent image, So they can be used for developing negative materials by a much brighter safelight than usual. This is a great help in development by inspection and factorial development, though in some cases the actual factor is changed by the desensitizer action.

When they have been desensitized, slow orthochromatic materials can be developed by subdued white light, fast ortho films and plates by the sort of orange safelight normally used for bromide papers and fast panchromatic materials by a bright green safelight.

Red safelights are not suitable for desensitized panchromatic material because when certain desensitizers such as phenosafranine are present the red light attacks the latent image.

Colour of the dye has little direct influence on the degree of desensitization. The exact mechanism of the effect is not yet understood, but all the desensitizers may be considered as weak oxidants having the ability to trap photo-electrons that are released within the silver halide grains by the action of actinic light.

Desensitizers are best employed as a forebath. A few can even be added to the developer itself, but many have an adverse effect on the action of the developer. Fine grain developers of the ortho- or para-phenylene diamine type are not suitable for developing desensitized films or plates. Fortunately, these developers are mostly used for developing miniature films in tanks where there is no need for desensitization.

Method. In practice the film or plate is immersed in the desensitizer for 2–3 minutes in darkness, or by the normal safelight. It is then rinsed, transferred to the developer and developed by the appropriate bright safelight. Where the desensitizer is mixed with the developer, the bright safelight can be switched on after about 2–3 minutes development in total darkness.

Dyes. The useful desensitizing dyes are pinacryptol green, pinacryptol yellow, pinacryptol white, and phenosafranin. The first two should only be used as a forebath; the others may be used either as a forebath or in the developer. Phenosafranin has the disadvantage of staining the emulsion red.

A number of proprietary desensitizers are also on the market under various trade names.

An average working concentration is about 1 part in 5000 of water (or developer), usually obtained by diluting a 0·5 per cent stock solution.

Other Desensitizers. The oxidation products of certain developers have a desensitizing influence. For instance, a 1 per cent solution of metol (containing about 0·2 per cent each of sodium carbonate and potassium bromide), which has been oxidized by blowing air through the solution, or by long exposure to the air in an open dish, can be used as a desensitizing forebath. L.A.M.

DETECTIVE CAMERA. Term introduced in 1881 for a small camera disguised as some familiar object—e.g., a book or pocket watch—for taking pictures unseen by the subject.

See also: *Spy camera; Camera history.*

DEVELOPER IMPROVER. Several organic substances known as developer improvers are sometimes included in the developer either instead of, or in addition to, the restrainer. The chemicals used are benztriazole, 6-nitrobenzimidazole, and 3:5 - dichloro - 2 - amino pyridine. They have very high anti-fogging properties when present in the developer in small quantities—often less than 0·1 per cent.

Developer improvers do not greatly affect the activity or other properties of the developer; they merely keep down fog, including fog caused by lengthy or bad storage of the material.

Developer improvers are sold either in solid form or in solution under various trade names.

DEVELOPERS

A developer converts the invisible latent image formed during exposure into a visible form. It does this by changing into black metallic silver the silver salt compounds in the film or plate that have been affected by light.

Developers are sold either in ready-mixed powder forms or in made-up stock solutions which have only to be dissolved or diluted in water to be ready for use.

Many photographers choose to make up their own developer from one of the multitude of published formulae.

A developer for black-and-white films and papers usually consists of:—

(1) A developing agent: this does the actual work of converting the silver compounds in the sensitive material into a silver image. Some developers use more than one of these agents.

(2) A preservative: this stops the developer from deteriorating too rapidly.

DEVELOPING AGENTS COMPARED

Usual Name	*Other Names*	*Chemical Name and Formula*	*Solubility at Normal Temperature*			*Keeping Quality in Solutions*	
			Water	*Alkali Solutions*	*Sulphite Solutions*	*Weakly Alkaline*	*Strongly Alkaline*
Amidol	Dianol, Dolmi	2 : 4-diaminophenol hydrochloride $C_6H_3OH(NH_3Cl)_2$	Freely soluble	Freely soluble	Freely soluble	Does not keep	Does not keep
Chlorquinol	Adurol, Chlorhydroquinone	1-chloro-2 : 5-dihydroxybenzene $ClC_6H_3(OH)_2$	Fairly soluble	Fairly soluble	Fairly soluble	Good	Moderate
Glycin	Kodurol	p-hydroxyphenylamino acetic acid $OH,C_6H_4.NH(CH_2COOH)$	Almost insoluble	Freely soluble	Fairly soluble	Very good	Good
Hydroquinone	Quinol	P-dihydroxybenzene $C_6H_4(OH)_2$	Fairly soluble	Fairly soluble	Fairly soluble	Moderate	Poor
Meritol	(Proprietary developing agent)	Condensation product of para-phenylene diamine and o-dihydroxybenzene	Fairly soluble	Fairly soluble	Fairly soluble	Fairly good	Poor
Metol	Elon, Pertol, Pictol, Planetol, Rhodol, etc.	1-hydroxy-4-methylamino benzene sulphate (monomethyl-p-aminophenol sulphate) $OHC_6H_4NHCH_3HSO_4 \cdot \frac{1}{2}H_2SO_4$	Fairly soluble	Fairly soluble	Slightly soluble	Very good	Fairly good
Metol-hydroquinone	Metol-quinol, M.-Q.					Good	Moderate
Ortho-phenylene diamine	O.P.D.	1 : 2 diamino benzene $C_6H_4(NH_2)_2$	Fairly soluble	Fairly soluble	Fairly soluble	Good	Moderate
Para-amino phenol (base)		1-hydroxy-4-amino benzene $OH.C_6H_4NH_2$	Very soluble	Fairly soluble	Slightly soluble	Good	Good
Para-aminophenol hydrochloride		1-hydroxy-4-amino benzene hydrochloride $OH\ C_6H_4NH_3Cl$	Fairly soluble	Fairly soluble	Slightly soluble		
Para-phenylene diamine	P.P.D.	1 : 4-diamino benzene $C_6H_4(NH_2)_2$	Fairly soluble	Fairly soluble	Fairly soluble	Moderate	Moderate
Para-phenylene diamine hydrochloride		1 : 4-diamino benzene hydchloride $C_6H_4(NH_3Cl)_2$	Fairly soluble	Fairly soluble	Fairly soluble		
Phenidone	(Proprietary developing agent)	1-phenyl-3-pyrazolidone $C_6H_5N(CH_2)_2NHCO$	Slightly soluble	Fairly soluble	Fairly soluble	Good	Moderate
Pyro	Pyrogallic acid, pyrogallol	1 : 2 : 3-trihydroxybenzene $C_6H_3(OH)_3$	Freely soluble	Highly soluble	Freely soluble	Poor	Very poor
Pyrocatechin	Catechol, Catechin	1 : 2-dihydroxybenzene $C_6H_4(OH)_2$	Freely soluble	Highly soluble	Freely soluble	Poor	Very poor

DEVELOPING AGENTS COMPARED

Temperature Coefficient	*Inherent Grain*	*Development Speed (Activity)*	*Other Characteristics*	*Remarks*
Normal	Normal	Normal	Normal contrast. Will develop even in acid solution or in presence of hypo. Low fog formation	Amidol developers easily give pure blacks on bromide papers. Suitable for any normal developer, but not very popular owing to low keeping qualities
Fairly high	High	Normal	High contrast. Greatly resembles hydroquinone, but is more stable and less sensitive to temperature. Low fog formation	Warm tone developer for chlorobromide papers. Also suitable in any normal developer instead of hydroquinone
High	Normal	Low, except in strongly alkaline solution	Normal to soft gradation. Very low fog formation	Used in warm tone developers and in fine grain formulae. Very suitable for slow development methods as it produces practically no aerial fog
High	High	Faster the more strongly alkaline	High contrast, particularly with caustic soda or potash. High fog formation, hence needs lot of restrainer	Used as high contrast developer with caustic alkalies. Also in warm tone and normal developers
Normal	Low	Low	Fairly low contrast. Low fog formation	Used in fine grain developers. Has some characteristics of both paraphenylene diamine and pyrocatechin
Normal	Normal	High	Very low contrast. Moderate fog formation	Used in low contrast developers. Impurities in metol may cause skin poisoning with some people. Also used in many standard and fine grain formulae. See metol-hydroquinone
Normal	Normal	Normal	Average contrast; depends on relative proportions of metol and hydroquinone. Moderate fog	Mixture of developing agents; forms basis of many popular standard developers. Combines high activity of metol with contrast giving properties of hydroquinone
Normal	Low	Very low	Low contrast. Low fog formation	Used in fine grain developers. Similar to paraphenylene diamine, but does not stain
Normal	Normal	Normal	Low contrast. Low fog formation	Similar in some ways to metol, and used instead of the latter by people sensitive to metol poisoning. Also main constituent of many concentrated one-solution developers
Normal	Very low	Very low	Very low contrast. Moderately low fog	Used in many super-fine grain developers either alone or in combination with another developing agent (glycin, metol, etc.)
Normal	Normal	High	Soft gradation, resembles metol, but can be used in much lower concentrations	Used in combination with hydroquinone in developers of similar characteristics to metol-hydroquinone. Low toxicity and freedom from tendency to cause dermatitis
High	High	High	Fairly high contrast. Stains the image during development. A great deal of control is possible with pyro developers by altering bromide concentration	Traditional dish developer. Also used in pyro-metol developers
Very high	Normal	High	High contrast. Moderate fog formation. Stains and tans the image during development	Used in special tanning developers. Also ingredient of Meritol fine grain developers

(3) An alkali: most developing agents are only active in alkaline solution.

(4) A restrainer: this may be included to reduce fog during development.

Some fine-grain developers contain, in addition to the other ingredients, a solvent of silver salts. This is usually potassium thiocyanate or sodium thiosulphate (hypo) which dissolves some of the silver bromide in the emulsion so that the solution acts partially like a physical developer. To some extent the high concentration of sodium sulphite in most fine-grain developers does the same thing.

There are also one or two substances sometimes added which do not affect the characteristics of the developer as such, but help to give clean negatives. These include water softeners like Calgon (sodium hexametaphosphate), developer improvers and wetting agents.

Calgon prevents the formation of a precipitate in a developer solution made with hard water. This precipitate is chemically harmless, but it may settle on the film and stop the developer from reaching the image evenly. This may cause small light spots on the negative.

Developer improvers help to minimize fog. They are added to a developer in extremely small quantities and can be quite additional to the restrainer. They are particularly useful when stale materials are being developed.

Wetting agents are added to developers to ensure uniform wetting of the emulsion. This reduces the risk of white spots due to air bells.

The table above shows only the inherent activity of the developing agent, not how it behaves when made up into a developer.

NEGATIVE DEVELOPERS

Negative developers can be classified as:

(1) Normal contrast developers.

(2) High contrast developers.

(3) Low contrast developers.

(4) Formulae with special characteristics, such as high speed developers, tanning developers, etc.

(5) Various degrees of fine grain formulae.

The properties of any developer depend, however, as much on the relative proportions of the ingredients as on their chemical nature. The chemical process of development is much the same for all developers. The activity and other characteristics can be altered in many cases just as effectively by using more or less of, say, the alkali, as by changing from one alkali to another. Only a small range of chemicals is therefore necessary to produce developers of widely varying characteristics.

Metol-Hydroquinone Developers. The most popular normal negative developer is a combination of metol and hydroquinone, or M.Q. This is the standard formula recommended by most manufacturers of films and plates. M.Q. developers combine the characteristics of metol (high film speed and good gradation) with those of hydroquinone (high density and contrast).

Most M.Q. formula are of the following type:

Metol	8–35 grains	0·5– 2 grams
Sodium sulphite, anhyd.	½–2¼ ounces	12–60 grams
Hydroquinone	30–120 grains	2– 8 grams
Sodium carbonate, anhyd.	⅜–1¼ ounces	10–30 grams
Potassium bromide	2–20 grains	0·1– 1 grams
Water to make	40 ounces	1,000 c.cm.

This represents the relative proportion of the ingredients rather than the exact dilution. Usually the developers are made up in the form of more or less concentrated stock solutions which are then diluted about two to five times for dish development, and five to fifteen times for tank development. The development times naturally depend on the concentration as well as on the amount of agitation. The time for medium speed films and normal intermittent agitation is about eight to ten minutes at 65° F. (18·4° C.), with the above concentration.

These proportions are characteristic of many standard developers such as Focal Universal, Ilford ID-2, Kodak D.72, etc.

The normal metol : hydroquinone ratio is 1 : 4. This may be decreased to about 1 : 2 to produce a softer developer.

The usual sodium sulphite proportion does not vary much beyond the limits indicated, but occasionally a little sodium bisulphite or potassium metabisulphite is included.

The alkali concentration varies to a greater extent. It may fall to as low as one-quarter of the value given, though it is rarely higher than 1·5 per cent at normal tank strength. The alkali content greatly influences development time. This accounts to some extent for the wide range of different development times.

Potassium carbonate is sometimes used instead of the sodium salt. The equivalent concentration to produce the same pH at this level is very roughly the same. At high dilutions of the developer seven parts of potassium carbonate produce the same pH as about six parts of sodium carbonate. In their effect on the developer, one is as good as the other, but sodium carbonate is cheaper and more popular.

Sodium metaborate or Kodalk can be used in normal M.Q. developers with the following advantages: the energy of the developer is more easily controlled; there is less likelihood that aluminium carbonate will be precipitated from hardening fixers containing alum; no gas (carbon dioxide) is given off when the developer is neutralized in certain types of stop bath or acid fixer. This is particularly important in hot weather development where the emulsion easily forms blisters in any case.

(Sodium metaborate can also be made by mixing one part sodium hydroxide with about five parts borax, both being dissolved in water. This is equivalent to about seven parts of sodium metaborate.)

COMPARISON OF STANDARD METOL-HYDROQUINONE DEVELOPERS

	Ansco 61	*Defender* 53D	*Gevaert* G-201	*Focal Univer-sal*	*Ilford* ID-2	*Kodak* D-72	*Ansco* 47	*Defender* 3D	*Ilford* ID-34	*Kodak* D-61*a*	*Ansco* 48*M*	*Kodak* DK-50	*Kodak* DK-60*a*
Metol	1	1	0·4	0·5	0·7	0·5	0·75	0·8	0·6	0·8	2	2·5	2·5
Hydroquinone	2	4	1·5	1·8	2·7	2	1·5	1·5	2·5	1·5	1·5	2·5	2·5
Sodium sulphite, anhyd. ...	15	15	12	12·5	25	7·5	22·5	22·5	10	22·5	40	30	60
Sodium bisul-phite	—	—	—	—	—	—	0·5	0·5	2·5	0·5	—	—	—
Sodium carbonate, anhyd. ...	13	22	8	12·5	12·5	11	2·5	2·9	10	2·9	—	—	—
Sodium metaborate	—	—	—	—	—	—	—	—	—	—	10	—	—
Kodalk ...	—	—	—	—	—	—	—	—	—	—	—	10	20
Potassium bromide ...	1	0·6	0·5	0·16	0·6	0·3	0·4	0·43	0·3	0·4	0·5	0·5	0·5
Water to make	1,000	1,000	1,000	1,000	1,000	1,000	1,000	1,000	1,000	1,000	1,000	1,000	1,000
Tank strength ...	Above	Above	Above	Half above	Half above	Above	Above	Above	Above	Above	Above	Above	Above

The quantities given represent parts in 1,000 parts of water, i.e., grams per litre.

Caustic alkalis are rarely used in M.Q. developers.

The proportion of potassium bromide remains high to keep fog as low as possible, particularly with certain high speed films and plates. This slows down development, but is otherwise unimportant.

Phenidone-hydroquinone Formulae. As an alternative to metol, phenidone is used in a number of developers. Generally very much smaller quantities of phenidone are needed (around one-tenth of the metol concentration). The resulting developer has characteristics similar to a metol-hydroquinone formula, but has the advantage of not causing dermatitis in people whose skin is allergic to metol.

The phenidone-hydroquinone ratio is normally around 1 : 25 to 1 : 40; the other ingredients have the proportions and concentrations shown for the typical MQ formula above.

A typical negative developer formula (Ilford ID-67) is:—

Sodium sulphite, anhyd.	3	ounces	75	grams
Hydroquinone	140	grains	8	grams
Sodium carbonate, anhyd.	1½	ounces	37·5	grams
Phenidone	4½	grains	0·25	gram
Potassium bromide	35	grains	2	grams
Restrainer solution	290	minims	15	c.cm.
Water to make	40	ounces	1,000	c.cm.

Phenidone ... ...

The restrainer may be a solution containing 1 per cent each of benztriazole and sodium carbonate, or one of the various proprietary restrainer solutions on the market.

Dilution of the above developer is 1 : 2 up to 1 : 5. If the sodium sulphite concentration is reduced to about half and the potassium bromide concentration to about two-thirds, the developer can also be used for printing papers (contact papers at about full strength, enlarging papers diluted 1 : 1) and so becomes a universal developer.

Pyro-Soda Developers. Pyro is one of the traditional developing agents and is still popular with individual workers for developing plates by inspection. It tends to give thin, crisp negatives with excellent contact-printing quality. Unfortunately it also stains fingers and dishes (early photographers could generally be recognized by their blackened finger-nails).

Pyrogallic acid does not keep well in alkaline solutions so it is nearly always made up separately as a stock solution with potassium metabisulphite or sodium bisulphite.

The normal pyro concentration in the working solution is about 0·05–0·1 per cent. Some developers, however, contain up to 0·6 per cent.

The amount of sodium sulphite used varies between about 0·6 and 2·5 per cent. Pyro developers containing little sulphite stain the film yellow.

The alkali is generally sodium or potassium carbonate. Caustic alkalis have also been used, but they increase staining.

The proportion of potassium bromide in most normal formulae is about 0·1–1 gram per litre of working solution (2–20 grains per 40 ounces). Potassium bromide in a pyro developer reduces the effective speed of the negative material and it can be added during dish development by inspection to compensate for over-exposure.

The following is a typical pyro-soda developer formula (Ilford ID-1):

Stock solution A.				
Pyro	88	grains	5	grams
Potassium metabi-sulphite	22	grains	1·2	grams
Water to make	2	ounces	50	c.cm.
Stock solution B				
Sodium carbonate, anhyd.	¾	ounce	20	grams
Sodium sulphite, anhyd.	1	ounce	25	grams
Potassium bromide	10	grains	0·6	grams
Water to make	20	ounces	500	c.cm.

For dish development take 1 part A, 10 parts B, and 9 parts water. For tank development use only 5 parts B and 14 parts water to 1 part A. The average tank development time is 10–12 minutes at 65° F. (18·4° C.).

Glycin Developers. Glycin is a slow working developer which keeps very well in solution and is comparatively free from any tendency to aerial fog.

The following is a typical glycin developer formula:

Glycin	1½ ounces	37 grams
Sodium sulphite, anhyd.	1¼ ounces	30 grams
Potassium carbonate	7 ounces	175 grams
Water to make	40 ounces	1,000 c.cm.

For normal tank development dilute 6 times. Development takes about forty-five minutes at 65° F. (18·4° C.). For stand development dilute about fifty times, and develop for 4–5 hours.

Para-aminophenol Developers. Para-aminophenol is chemically related to metol and has similar characteristics as a developing agent. It is, however, free from any tendency to cause skin poisoning and can therefore be used as a substitute by photographers who are allergic to metol. A typical para-aminophenol hydroquinone developer is Kodak DK-93 :

Para-aminophenol hydrochloride	88 grains	5 grams
Sodium sulphite, anhyd.	1¼ ounces	30 grams
Hydroquinone	44 grains	2·5 grams
Kodalk	350 grains	20 grams
Potassium bromide	9 grains	0·5 grams
Water to make	40 ounces	1,000 c.cm.

Average development time is 10–20 minutes at 65° F. (18·4° C.).

Para-aminophenol is also the developing agent in many commercial highly concentrated developers.

High Contrast Developers. Hydroquinone is the principal developing agent for high contrast. So it is possible to increase the contrast given by a normal M.Q. developer by raising the proportion of hydroquinone to 6 : 1 or even 10 : 1, and using a solution 2-4 times stronger. In addition the bromide content of contrast developers is comparatively high.

Higher alkalinity of the developer also gives increased contrast, so sodium or potassium hydroxide may be used instead of sodium carbonate (about 1/10 the amount).

For even greater contrast hydroquinone is used alone with either sodium (or potassium) carbonate or sodium or potassium hydroxide as an alkali. The hydroquinone and alkali are usually made up in separate solutions and bisulphite or metabisulphite used as a preservative for the hydroquinone solution to give better keeping qualities. The caustic hydroquinone developers often contain a high concentration of potassium bromide (up to 10–12 grams per 1,000 c.cm. or ½ ounce per 40 ounces).

Hydroquinone developers without metol or other agents tend to have a long induction period, and develop the image simultaneously throughout the depth of the emulsion.

Extreme Contrast Developers. For certain graphic arts purposes images of maximum contrast and outline sharpness are required. These are obtained by developing ultra-contrasty films in extreme-contrast developers (usually based on hydroquinone) containing paraformaldehyde. These so-called lith type developers suffer rapid local exhaustion and thus yield special edge effects at boundaries between very bright and very dark image portions.

Exhausted developer from dark image areas diffuses across the boundary and holds back the density of neighbouring light areas. At the same time comparatively unused developer from thin image areas diffuses across to dense areas and develops up the edges of the latter to extra density. This yields an extra sharp

COMPARISON OF CONTRAST M.Q. DEVELOPERS

	Ansco 22	*Gevaert* G-203	*Kodak* D-163	*Defender* 9D*	*Ilford* ID-14	*Gevaert* G-201	*Ilford* ID-21	*Kodak* D-19	*Kodak* D-72	*Kodak* D-154
Metol	0·8	0·5	1·2	1	1·7	1·5	1·5	2·2	1·6	1·2
Hydroquinone	8	5	8·4	9	12·5	6	6·2	8·8	6	6
Sodium sulphite, anhyd. ...	40	50	38	75	75	50	25	100	22·5	22·5
Sodium carbonate, anhyd....	50	32	33	25	37	32	33	47	34	30
Potassium bromide ...	5	2	1·4	5	2	2	6·2	5	0·9	0·5
Water to make	1,000	1,000	1,000	1,000	1,000	1,000	1,000	1,000	1,000	1,000

*Also Kodak D-11.

COMPARISON OF HYDROQUINONE DEVELOPERS

	Ansco 70	*Gevaert* G-254	*Ilford* ID-13	*Kodak* D-8	*Kodak* D-153	*Kodak* D-178	*Ilford* ID-35	*Ilford* ID-38
Hydroquinone	12·5	55	12·5	15	12·5	30	19	9
Spdium sulphite, anhyd. ...	—	100	—	30	—	60	47	25
Sodium (or potassium) metabisulphite	12·5	—	12·5	—	12·5	—	—	—
Sodium hydroxide ...	18	—	—	12·5	25	12	—	—
Potassium hydroxide ...	—	25	25	—	—	—	—	—
Sodium carbonate anhyd. ...	—	—	—	—	—	—	—	27·5
Potassium carbonate ...	—	—	—	—	—	—	62	—
Potassium bromide ...	12·5	3	12·5	10	12·5	20	4·5	1·7
Water to make	1,000	1,000	1,000	1,000	1,000	1,000	1,000	1,000

All quantities are in grams per litre, or parts per 1,000.

boundary line, as is required for developing a line or half-tone dot image.

A typical formula is:—

Sodium sulphite, anhyd.	1½ ounces	30 grams
Paraformaldehyde	130 grains	7·5 grams
Potassium metabisulphite	45 grains	2·6 grams
Boric acid, cryst.	130 grains	7·5 grams
Hydroquinone	400 grains	22·5 grams
Potassium bromide	28 grains	1·6 grams
Water to make	40 ounces	1,000 c.cm.

After making up, this should be allowed to stand for about two hours before use.

Low Contrast Developers. Developers containing only metol or para-aminophenol as developing agent produce images of comparatively low contrast. The composition of the developer is similar to the standard M.Q. formula, but with more metol and without hydroquinone. Metol is more stable in solution and needs less sodium sulphite.

Developers of this kind tend to act as surface developers, i.e., building up the image first on the surface of the emulsion and only then going deeper. This property is useful for minimizing halation and similar effects which occur strongest in the deeper parts of the emulsion.

Kodak DK-15 is similar, but about three times as concentrated. It uses Kodalk instead of sodium carbonate and also contains some sodium sulphate to make it suitable for processing at high temperatures.

Many fine-grain developers give low contrast images.

COMPARISON OF METOL DEVELOPERS

	Ansco 120	*Gevaert G-253*	*Ilford ID-3*	*Ilford ID-15*	*Kodak D-165*
Metol	4·1	1·5	1·5	1·5	1·5
Sodium sulphite, anhyd.	12	10	6·3	10	6·3
Sodium carbonate, anhyd.	9·7	10	9	9	9
Potassium bromide	0·6	0·5	0·2	0·2	0·2
Water to make	1,000	1,000	1,000	1,000	1,000

Maximum Energy Developers. This type of developer is intended to bring out the maximum speed of a film or plate, particularly in cases of under-exposure.

Pyro-metol is a maximum energy developer which relies partly on the stain produced by pyrogallic acid to give printing density to the negative, and partly on the ability of the metol to develop an image even in greatly under-exposed regions of the film or plate.

The following is a formula for a pyro-metol developer:

Stock solution A		
Metol	35 grains	2 grams
Potassium metabisulphite (or sodium bisulphite)	70–100 grains	4–6 grams
Pyro	70–100 grains	4–6 grams
Water to make	20 ounces	500 c.cm.
Stock solution B		
Sodium carbonate, anhyd.	1½ ounces	37 grams
Water to make	20 ounces	500 c.cm.

For use take equal parts of A and B. Average development time is 6–8 minutes at 65° F. (18·4° C.).

Caustic metol-hydroquinone is a completely different type of high energy developer. This contains a high concentration of metol and hydroquinone (dissolved with the help of alcohol if necessary) and sodium hydroxide.

The following is the formula for a maximum energy developer:

Wood alcohol	2 ounces	50 c.cm.
Metol	240 grains	14 grams
Sodium sulphite, anhyd.	2 ounces	50 grams
Hydroquinone	240 grains	14 grams
Sodium hydroxide	160 grains	9 grams
Potassium bromide	160 grains	9 grams
Water to make	40 ounces	1,000 c.cm.

Development takes about 4–8 minutes at 65° F. (18·4° C.). The mixed developer hardly keeps at all.

If the sodium hydroxide is doubled, the development time can be reduced to about one minute. Some of the rapid developers are of the high energy type to utilize the maximum film speed.

The effective film speed is also increased by adding a small amount of hydrazine or its derivatives to the developer, with an anti-fogging agent. Such an energizing solution recommended by Kodak is :

Hydrazine dihydrochloride	100 grains	5·3 grams
6-nitrobenzimidazole nitrate (0·2%)	2⅔ ounces	67 c.cm.
Water to make	4 ounces	100 c.cm.

Three parts of this are added to about one hundred parts of a standard M.Q. developer. The developer and energizer mixture does not keep.

Concentrated Developers. Some developers can be made up very conveniently in strong stock solutions and diluted some twenty to sixty times for use.

All these formulae are based on para-aminophenol with potassium metabisulphite as preservative, and enough sodium hydroxide to bring them into solution.

A typical formula for a concentrated developer is:

Para-aminophenol (free base)	4 ounces	100 grams
Potassium metabisulphite	12 ounces	300 grams
Water	20 ounces	500 c.cm.

These ingredients are brought into solution with the least amount of concentrated sodium or potassium hydroxide solution and then made up to the final volume of 1,000 c.c. (40 ounces).

For normal tank development the solution is diluted about forty times. Development takes about fifteen to twenty minutes at 65° F. (18·4°C.).

The same principle works with metol-hydroquinone but not to such high concentration. A suitable formula is:

Wood alcohol	2 ounces	50 c.cm.
Potassium metabisulphite	35 grains	2 grams
Metol	¾ ounces	19 grams
Hydroquinone	1½ ounces	38 grams
Potassium metabisulphite	7 ounces	175 grams
Potassium bromide	160 grains	9 grams
Water to make	40 ounces	1,000 c.cm.

The potassium metabisulphite is dissolved in two lots. When the second lot has been dissolved as far as possible, just enough 50 per cent sodium or potassium hydroxide solution is added to dissolve all the precipitate in the solution. The bromide is then added, and the solution made up to the final volume.

For normal tank development this is diluted about twenty times.

Replenishers. Many developers can be used repeatedly but as each film takes some strength out of the solution subsequent development times have to be increased.

A better way of re-using developers is to replace some of the chemicals used up during development in the form of a replenisher.

The simplest replenisher would be concentrated developer stock solution, but all the ingredients of a developer are not used up at the same rate. The concentration of the developing agents and of the alkali decreases fairly rapidly whereas the potassium bromide content increases slightly. So a suitable replenisher is not identical with the developer; it contains the chemicals in the correct proportion to compensate for real losses during development. In use, about 15–25 c.cm. (½–1 ounce) of replenisher solution are added to 1,000 c.cm. (40 ounces) of developer working solution for each film developed. If this increases the solution volume too much, some of the old developer is poured off first.

Usually the composition of replenishers is:

Developing agents: about 1½–2 times the concentration in the working developer.

Sodium sulphite: same concéntration as in the developer.

Alkali: with alkalis like borax or metaborate, the replenisher will contain about 5–10 times as much as the developer. With carbonate alkalis the replenisher contains about 2–4 times as much as the working developer. Caustic soda or potash is also commonly used.

Restrainer: the majority of replenishers add no potassium bromide.

Other additions: neither developer improvers nor wetting agents are required in the replenisher. Solvent fine grain developer replenishers usually contain up to three times as much potassium thiocyanate or hypo as the developer itself.

Replenishers are only worth while with developers which keep well in solution. By using a replenisher, 1,000 c.cm. (40 ounces) of developer can be used for about twenty or more films. But pyro formulae and developers containing caustic soda as alkali deteriorate much too rapidly to justify replenishment.

There is no point in using a replenisher with dilute tank developers, nor with developing tanks which take less than about 400 c.cm. (15–16 ounces) of solution to develop one film. Under these conditions the developer is so exhausted after use that it is best thrown away.

Replenishers are best used in fairly concentrated solution (such as most fine-grain formulae) where it would be comparatively expensive to use fresh developer solution for each film.

High-speed Developers. Concentrated and highly alkaline developers will bring development times down to about ½–1 minute. They are usually derived from contrast developers of the metol-hydroquinone or caustic hydroquinone types which are already highly alkaline. With such strong solutions potassium bromide is usually not efficient enough to keep down fog, so a 0·1 per cent solution of phenosafranine is often added.

The presence of any salts not directly engaged in the development process slows down the development, so the amount of sodium sulphite is kept very low. In consequence these developers do not keep at all once they are mixed.

Such a developer may, for instance, be:

Sodium sulphite, anhydrous	52 grains	3 grams
Hydroquinone	52 grains	3 grams
Phenosafranin, 0·1 per cent	38 minims	2 c.cm.
Water to make	2 ounces	50 c.cm.

Just before use add an equal volume of 12 per cent sodium hydroxide solution.

A 10 per cent acetic acid stop bath is advisable immediately after development (which takes about forty-five seconds), followed by a rapid fixer.

Two bath methods of high speed development rely on very concentrated developing agents, followed by a short immersion in a caustic alkali solution. The following is a suitable formula:

Hydroquinone	2 ounces	50 grams
Sulphurous acid, 8 per cent solution	10 ounces	250 c.cm.
Phenosafranin, 0·1 per cent solution	¾ ounce	20 c.cm.
Water to make	40 ounces	1,000 c.cm.

The film or plate is immersed in the solution for 5–10 seconds and then given 3 seconds in a 30 per cent solution of sodium or potassium hydroxide, with 1 part formalin solution added to every 40 parts of alkali solution. Again a stop bath and rapid fixer should be used.

At high temperatures (up to 85° F. or 30° C.), development is complete after about 4–5 seconds in the developer, followed by 1–2 seconds in the alkali bath. There is little danger of damage to the emulsion from the high temperature and alkalinity of the solution (particularly if formalin is present in the second bath) because the emulsion has no time to swell appreciably.

After the high speed fixing bath the emulsion

is washed in running water for 30 seconds, and sandwiched in glycerine between two glass plates. It can then be enlarged immediately, and rewashed and dried normally later on.

Prints can be processed at high speed in much the same way. After the final 30-second wash they can be dried in front of a radiator or fire in about 30 seconds.

Tanning Developers. Certain developers tan the gelatin during development of the silver image. The degree of tanning (which hardens the gelatin and renders it insoluble) is proportional to the density of the silver deposit produced.

Most developing agents will, under suitable conditions, tan the gelatin. The effect is greatest with the polyhydroxy-benzenes, especially with pyrocatechin and pyrogallic acid.

One of the essentials of such a tanning developer is a low sulphite content, otherwise the tanning effect is largely lost. A fairly high alkalinity, and sometimes the addition of substances like alcohol, improve the tanning action. This is due to the polymerization of the oxidation products of the developing agent (quinone and quinone-like compounds) which are formed in the gelatin layer during development.

Tanning developers are employed for making gelatin matrices, i.e., images of hardened gelatin. These matrices are used in various processes—e.g., dye transfer and imbibition colour print processes, and photomechanical processes such as photolithography. Tanning developers are also sometimes used for controlling excessive contrast; where the emulsion has received most exposure, the increased tanning of the gelatin in those areas tends to retard development.

A typical tanning developer using pyrocatechin is:

Pyrocatechin	35 grains	2 grams
Sodium sulphite, 1 per cent solution	1 ounce	25 c.cm.
Sodium hydroxide, 1 per cent solution	2 ounces	50 c.cm.
Water to make	40 ounces	1,000 c.cm.

Average development time is 15-20 minutes at 65° F. (18·4° C.).

This developer must be made up immediately before use. Neither the sulphite nor the sodium hydroxide solutions keep well at this dilution, but are more convenient to measure when made up in this way.

A formula using pyrogallic acid is:

Stock solution A		
Pyrogallic acid	70 grains	4 grams
Sodium sulphite, anhydrous	175 grains	10 grams
Water to make	40 ounces	1,000 c.cm.
Stock solution B		
Sodium carbonate, anhydrous	485 grains	28 grams
Water to make	40 ounces	1,000 c.cm.

Equal parts of A and B are mixed immediately before use. Average development time is 6-10 minutes at 65° F. (18·4° C.).

A developer of this type using hydroquinone can be obtained by making up a standard caustic hydroquinone developer, but reducing the sulphite or metabisulphite content to about 10-20 per cent of the normal amount.

After development the unhardened gelatin can be washed out with warm water, leaving the tanned relief image. The silver image itself can also be bleached out either by a normal bleacher or by a tanning bleacher which further hardens the gelatin. The use of such a tanning bleacher before the removal of the unhardened gelatin will increase the contrast.

FINE GRAIN NEGATIVE DEVELOPERS

During normal development the image-forming silver in the sensitive emulsion tends to clump into comparatively large grains. These become unpleasantly obvious when the negative is sufficiently enlarged. Fine grain developers counteract this clumping tendency in three principal ways. There are thus three types of fine grain developer: normal developer, low energy developer and solvent developer.

Normal Developers. In the normal fine grain developer, the energy is reduced by lowering the alkalinity. The classic developer of this type is the metol-hydroquinone-borax formula (Kodak D-76, Defender 6D and Ilford ID-11):

Metol	35 grains	2 grams
Sodium sulphite, anhyd.	4 ounces	100 grams
Hydroquinone	88 grains	5 grams
Borax	34 grains	2 grams
Water to make	40 ounces	1,000 c.cm.

Average development times are about 10–20 minutes at 65° F. (18·4° C.).

There are three types of variation on this formula:

(1) The concentration of the developing agents may be varied. Some M.Q.-borax formulae contain less hydroquinone; others contain no hydroquinone at all, but have extra metol; others contain another developing agent such as glycin in place of the hydroquinone. The metol can also be replaced by 1/15 to 1/10 its weight of phenidone.

(2) The alkalinity may be increased by using an equal amount of metaborate or Kodalk instead of the borax, or by using more borax.

An increase in the borax or metaborate concentration will, other things being equal, slightly increase the grain and decrease the development times. This effect is particularly marked with Kodalk where a 2½ times increase in the alkali concentration cuts the development time by half.

(3) The alkalinity may be further stabilized, or even decreased by using a buffer mixture of borax and boric acid.

Thus the stabilized version of M.Q. borax contains 8 grams each of borax and boric

COMPARISON OF M.Q. BORAX FINE GRAIN DEVELOPERS

	Ansco 17	*Ansco* *17M*	*Defender* 4D	*Defender* 6D	*Edwal* 10	*Gevaert* *G-206*	*Ilford* *ID-11*	*Kodak* *D-76*	*Kodak* *D-76a*	*Kodak* *D-76d*	*Iford* *ID-68*
Metol	1·5	1·5	2	2	5	2	2	2	2	2	—
Phenidone	—	—	—	—	—	—	—	—	—	—	0·13
Hydroquinone	3	3	3	5	—	4	5	5	5	5	5
Glycin	—	—	—	—	5	—	—	—	—	—	—
Sodium sulphite anhyd. ...	80	80	75	98	100	100	100	100	100	100	85
Borax	3	—	5	2	10	2	2	2	2	8	7
Sodium metaborate	—	2	—	—	—	—	—	—	—	—	—
Boric acid	—	—	—	—	—	—	—	—	14	8	2
Potassium bromide	0·5	0·5	—	—	—	—	—	—	—	—	1
Water to make	1,000	1,000	1,000	1,000	1,000	1,000	1,000	1,000	1,000	1,000	1,000

The quantities are in grams per 1,000 c.cm. of solution, or parts per 1,000.

acid instead of 2 grams of borax alone. The grain and development time remain unchanged.

A somewhat different type of low alkalinity fine grain developer uses only metol and sodium sulphite. It is simply a 0·75–1·5 per cent solution of metol in 10 per cent sodium sulphite. The sulphite acts as alkali as well as preservative. The addition of 1–2 per cent of sodium bisulphite reduces the alkalinity still further and gives a buffered developer. Either of these formulae act as surface developers, building up the image first in the top of the emulsion layer before going deeper. This helps to increase the image resolution, quite apart from any improvement in grain. Over-development will, however, nullify this advantage.

Low Energy Developers. Developers of this type aim at reducing grain by using low energy developing agents in weakly alkaline solutions. (Para- and ortho-phenylene diamine are suitable developing agents, often used together with one or two normal agents.) The low activity of the developer decreases the effective film speed, and exposure may have to be increased as much as three times. The image contrast is usually low because the developers are incapable of producing high contrast and development is usually shortened to keep contrast low and grain subdued.

The basic superfine grain developer of this type is a 1 per cent solution of para-phenylene diamine in 9 per cent sodium sulphite. This is the Sease I formula. It gives very fine grain, but its activity is so low that the development time may be 45 minutes or more. In addition the effective film speed is reduced so seriously (up to 300 per cent extra exposure may be needed) that it is more sensible to use a slow fine grain film and develop in a normal fine grain developer. For this reason the formula is nearly always modified in one or more ways:

(1) A second developing agent may be added. In the Sease series this is glycin; Edwal 12 contains 0·6 per cent metol in addition. The Johnson fine grain developers use pyrocatechin as the second developing agent.

(2) An alkali may be added either to the plain para-phenylene diamine formula, or to its combination with a second developing agent. The alkali needs a fairly strong buffering activity. Trisodium phosphate is often used, and so is borax. The alkali concentration may be comparatively high.

Solvent Developers. In developers of this type, the fine grain is the result of adding a solvent of silver halide. This dissolves away part of the silver bromide grains during development, and in this way physically reduces clumping. At the same time some of this dissolved silver salt is developed and deposited as silver on the film, mainly on those parts of the emulsion which already carry a silver image. This is really a chemical process but it is known as physical development to distinguish it from the normal direct development of the silver salts to silver.

The common solvents are potassium thiocyanate, ammonium chloride and sodium thiosulphate (hypo).

Most of the action of a solvent fine grain developer is normal direct development, with some indirect physical development. But some developers rely almost entirely on depositing silver from the solution. In fact, they actually contain silver salts before development. These are the real physical developers.

The solvent type of fine grain developer removes silver from the emulsion so some of the latent image goes with it. This is another reason why the negative material needs extra exposure of anything from 50–100 per cent according to the type of emulsion.

Both solvent and low energy developers are classed as super fine grain developers.

Nearly all fine grain developers use a high concentration of sodium sulphite (8–10 per cent, as compared with an upper limit of 3 per cent for most normal negative developers). So fine grain developers keep well in solution, and can be used repeatedly. The high concentration of sulphite also helps by acting as a weak solvent.

One advantage of the solvent fine grain developers is that they give as fine grain as the low energy developing agents, but produce the same negative density and contrast quicker.

The formulae are mostly derived from the normal M.Q. borax and similar fine grain developers, with the addition of 0·1–0·3 per cent potassium thiocyanate or sodium thiosulphate. Most solvent fine grain developers call for slightly increased exposure.

Modern Fine-grain Developers. Both low-energy and solvent fine-grain developers are nowadays comparatively obsolete since the same combination of effective speed and fine-grain can be obtained by using a slower fine-grain film and developing this in a higher-energy developer.

A number of current proprietary fine-grain developers concentrate more on the use of mixtures of low and high energy developing agents in a moderately alkaline solution (achieved, for example, by the use of borax or metaborate). Such formulae fully utilize the emulsion speed of the film, yet yield reasonably fine-grain by avoiding the clumping of silver grains encountered with highly active M.Q. formulae.

A second advantage of a number of formulae of this kind is that they are concentrated solutions which can be diluted ten or more times for use. One lot of developer then serves for just one film and can be discarded afterwards.

High-definition Developers. With modern thin-emulsion films neither solvent nor low-energy developers yield the best results. The image there is easily attacked by the silver halide solvent action with adverse effects on negative quality. Moreover, such films of themselves yield appreciably finer grain than the average negative materials available at the time when special fine grain developers were highly popular.

With thin-emulsion films there is a greater risk of loss of image detail through lack of sharpness (due to image diffusion) rather than through graininess. Also such films tend to be contrasty and therefore require a soft working developer.

This has led to the emergence of high-definition developers which yield a soft gradation image that is concentrated in the surface of the emulsion rather than in depth. Developers of this kind are usually based on metol in an alkaline solution, or on para-aminophenol. Being surface developers, they yield maximum image sharpness, utilize the film speed to the greatest extent, and do not adversely affect the grain (which is fine anyway).

A typical formula, worked out by W. Beutler for slow films, is:—

Stock Solution A.		
Metol	175 grains	10 grams
Calgon	35 grains	2 grams
Sodium sulphite, anhyd.	2 ounces	50 grams
Water to make	40 ounces	1,000 c.cm.
Stock Solution B		
Calgon	35 grains	2 grams
Sodium carbonate, anhyd.	2 ounces	50 grams
Water to make	40 ounces	1,000 c.cm.

One part of each stock solution is mixed with ten parts of water.

A number of proprietary high-definition and fine-grain developers are based on similar principles. The developer may be made more active to yield higher contrast with medium speed (and hence less contrasty) films.

MONOBATH DEVELOPERS

Normally the developing and fixing processes are kept strictly separate. Under certain conditions the fixing chemicals can however be incorporated in the developer, so that both actions take place simultaneously. Monobath developers working on this principle have been known for a considerable time, but have only in recent years gained some practical importance. The earlier combinations of developer and fixer usually involved considerable loss of emulsion speed, fog and uneven action.

A number of photographic firms now market monobath developers that yield acceptable to good results.

A typical formula is:—

Hydroquinone	265 grains	15 grams
Sodium sulphite, anhyd.	2 ounces	50 grams
Phenidone	175 grains	10 grams
Potassium alum	320 grains	18 grams
Sodium hydroxide	320 grains	18 grams
Sodium thiosulphate, cryst.	2½ ounces	60 grams
Water to make	40 ounces	1,000 c.cm.

Since monobath development goes to completion, neither development time nor development temperature are particularly critical. As fixing proceeds together with development, the process is usually completed in 4 to 6 minutes; at higher temperatures both actions proceed faster and in colder solutions both take longer. There is thus no possibility of over-development.

On the other hand the balance between developing and fixing activity is fairly critical. A given monobath developer is thus suitable for only a limited range of negative materials, and different monobath formulae are usually necessary to deal with slow developing and fast developing emulsions. Some manufacturers have produced monobath developers in the form of two stock solutions, using them in different proportions for different films.

In practice, monobath development has appreciable advantages, especially in hot weather since the film spends only a short time in the processing baths. Also the use of a single solution—usually in concentrated form—is particularly practical for processing while travelling.

Against this high degree of convenience is the fact that monobath developers yield neither a very fine grain nor a very high degree of image sharpness in comparison with high-definition developers.

POSITIVE DEVELOPERS

The range of developing agents suitable for bromide and contact papers and lantern plates is limited. Developers with a tendency to stain

COMPARISON OF STANDARD POSITIVE M.Q. DEVELOPERS

	Defender 53D*	Focal Universal	Gevaert G-251*	Ilford ID-20*	Kodak D-163*	Ansco 103†	Defender 54D†	Gevaert G-252†	Ilford ID-36†	Kodak D-73†	Kodak D-158†
Metol	1·6	1	1·5	1·5	1·1	1·1	0·9	2·5	1·5	1·4	1·6
Hydroquinone	6	3·7	6	6	8·5	3·8	3·5	3·6	6·2	5·4	6·6
Sodium sulphite, anhyd. ...	22·5	25	25	25	38	15	13	25	25	20	25
Sodium carbonate, anhyd. ...	34	25	40	30	32	21	25	40	35	37	35
Potassium bromide	1	0·3	1	2	1·4	0·4	0·5	0·3	0·3	0·4	0·4
Water to make	1,000	1,000	1,000	1,000	1,000	1,000	1,000	1,000	1,000	1,000	1,000
Dilution for bromide paper ...	1 : 1	1 : 1	1 : 1	1 : 1	1 : 1	1 : 1	1 : 1	1 : 1	1 : 1	1 : 1	1 : 1
Dilution for contact paper ...	Undil.	Undil.	Undil.	Undil.	Undil.	Undil.	Undil.	Undil.	Undil.	Undil.	Undil.

*Mainly bromide paper and bromide lantern plate developer. †Mainly contact paper and contact plate developer. Quantities are in grams per 1,000 c.cm. of water, or parts per 1,000.

(e.g., pyro) cannot be used because the image colour of a print is important. The only agents of any importance for black tone development are metol, hydroquinone, amidol, and occasionally para-aminophenol. The agents used in warm tone developers (which can only be used with special chlorobromide papers and warm tone lantern slides) are chlorquinol, glycin, and sometimes pyro (with slides).

Positive prints do not call for low energy or fine grain developers; the grain size of a printing paper is of no importance—the image will not be enlarged, anyway. Even lantern slides have a grain that is very much finer than that of the negative.

The most popular formula is metol-hydroquinone with carbonate, and with glycin and (or) chlorquinol added for warm tone formulae. The only exceptions are certain special formulae for lantern plates or for high contrast, such as caustic hydroquinone, pyro, and certain warm tone developers containing ammonia or thiocarbamide for lantern slides. Very occasionally a physical developer is used for slides.

Characteristics. Positive developers can be classified as:

(1) Standard black tone bromide and contact paper developers. These also give black tones on bromide and contact lantern plates, and warm black tones on warm tone materials.

(2) High contrast and variable contrast developers for bromide papers and black tone (bromide) lantern plates. Contact (chloride) papers and plates do not respond so well to high contrast and variable contrast development. With warm tone (chlorobromide) papers and plates the contrast obtained is to some extent decided by the image colour aimed at.

(3) Warm tone developers. These are used almost exclusively with warm tone (chlorobromide) papers and lantern plates. Special warm tone developers for bromide papers have been published, but their action depends very greatly on the make and grade of the paper. Usually such warm tone bromide developers work with one or two makes only, and are not suitable for all black tone papers and plates. In any case the degree of control over the image colour is limited, and the tones are often distinctly unpleasant.

Positive M.Q. Developers. These developers are all based on the standard negative M.Q. formula.

Generally speaking, the main difference between negative and positive developers of this type is that while negative developers are mostly used fairly dilute (particularly for tank development), developers for bromide papers and black tone lantern plates are used in a much more concentrated solution. The developer for contact papers and chloride plates is used stronger still.

It is therefore possible to use one and the same formula for negatives, lantern slides, bromide and contact papers, by merely diluting the stock solution to suit. This is the basis of most of the so-called universal developers, like Focal Universal, Kodak D-163. The reason is simply that a standard M.Q. developer will develop almost any sensitive material.

There are several recognized departures from the standard formula.

First, papers have a tougher emulsion than negatives; it is less apt to swell or melt. So paper developers can be more alkaline and often contain up to 2 per cent sodium carbonate (or up to 4 per cent in the double strength contact paper developer).

Second, for really black tones, the potassium bromide concentration must be kept down to a minimum. This minimum amount of bromide needed to avoid fog varies from paper to paper, and also depends on the activity of the developer. That is partly the reason for the large number of formulae each specially recommended by a maker for his particular paper or papers.

Generally in bromide paper developers the potassium bromide content is near the lower limit of the range stated in the standard negatives M.Q. formula (about 0·07–0·1 per cent, or even less).

In some developers, particularly those designed for contact papers, the bromide concentration of the working solution is still less (0·04–0·06 per cent) even at the contact paper strength of the solution—i.e., when the concentration of all the other ingredients is twice as great as in the standard M.Q. formula. When the developers are diluted to normal strength for bromide papers or lantern plates, the amount of potassium bromide is only half as much.

With positive developers the use of a developer improver helps to give good black tones with the minimum of fog. The developer improvers are added in the proportion of about 1 part of the commercial solution to 100 parts of working developer. Some developers actually include a developer improver.

One typical formula is:

Metol	60 grains	3·5 grams
Sodium sulphite, anhyd.	1¼ ounces	35 grams
Hydroquinone	150 grains	8·6 grams
Sodium carbonate, anhyd.	2¼ ounces	63 grams
Potassium bromide	33 grains	1·9 grams
Developer improver	1¼ ounces	37 c.cm.
Wettol	190 minims	10 c.cm.
Water to make	40 ounces	1,000 c.cm.

The developer improver may be any one of a number of proprietary solutions on the market, or a solution containing 1 per cent benztriazole and 1 per cent sodium carbonate.

For use dilute 1 part with 2 parts water.

Positive Amidol Developers. Amidol was at one time very popular for bromide and contact papers, mainly because it very easily gives pure black image tones. But a standard M.Q. developer will give as good blacks if the potassium bromide concentration is kept low, or a developer improver is added. The disadvantages of amidol developers are that they do not keep well and must therefore be made up immediately before use, and that they stain the fingers.

AMIDOL DEVELOPERS

	Ansco 113	Defender 61D*	Ilford ID-22*	Kodak D-170*	Ilford ID-30†	Kodak D-162†
Amidol	6·6	3·8	6	4·5	6	4
Sodium sulphite, anhyd.	44	15	25	25	25	25
Potassium bromide	0·6	2·5	0·8	1	0·2	0·2
Water to make	1,000	1,000	1,000	1,000	1,000	1,000

*Mainly bromide paper developers.
†Mainly chloride paper developers.
Quantities are in grams per 1,000 c.cm., or parts per 1,000.

The developer itself is simply amidol, sodium sulphite, and potassium bromide. No alkali is needed; the low alkalinity of the sodium sulphite solution is quite sufficient.

High and Low Contrast Developers. The contrast obtainable with any one grade of bromide and contact papers or lantern plates can be slightly modified by using a high or low contrast developer.

The high contrast developers are much the same as the standard caustic hydroquinone or hydroquinone and carbonate developers used for negatives but the bromide content can be lower. Apart from their increased contrast, they differ from the normal M.Q. developers in three ways:

(1) The image colour is not a pure black, but slightly warm black because of the high bromide content of the hydroquinone developer. Even on standing, the working solution deteriorates rapidly and gives an unpleasant greenish black image. As the developer becomes exhausted, the effective speed of the paper decreases and the print needs a longer exposure.

(2) The solution very readily stains the paper base yellow-brown. This again is more marked as the developer deteriorates so forcing development beyond 2 minutes at 65° F. (18·4° C.) does more harm than good.

(3) The effective exposure latitude of the paper or lantern plate is increased. The positive can therefore be over-exposed by as much as 50 per cent (and development time reduced accordingly) without any falling off in print quality.

Low contrast is given by negative developers of the metol group. Again the potassium bromide content should be kept to a minimum. These developers keep well and do not stain but the printing exposure is critical; print quality deteriorates if the exposure is varied by more than about 10 per cent.

The contrast difference between caustic hydroquinone and standard M.Q. is equivalent to about half a paper grade.

Combinations of hydroquinone, M.Q., and metol formulae are used as variable contrast developers either separately in multi-dish methods of development, or mixed in varying proportions.

Warm Tone Developers. Most warm tone developers are suitable both for warm tone lantern plates and chlorobromide papers. They differ from normal black tone paper and slide developers in the following respects:

Developing agents: as metol tends to produce blue-black tones, it is used only in small amounts in warm tone developers; in many formulae it is left out altogether. But the amount of hydroquinone is fairly high, and is often supplemented by chlorquinol and glycin, both of which readily give warm image tones.

Preservative: the amount of sodium sulphite is more or less the same as in normal positive developers, although some hydroquinone formulae contain rather more sulphite to increase their keeping qualities. When they are made up in concentrated stock solutions (with the alkali separate) the hydroquinone stock solution many contain either potassium metabisulphite or citric acid to prevent premature oxidation.

Alkali: the proportion of alkali is the same as for normal M.Q. developers, but the actual concentration varies a great deal, since many

COMPARISON OF STANDARD WARM TONE DEVELOPERS

	Ansco 110	Ansco 115	Ansco 135	Defender 55D	Defender 58D	Gevaert G-261	Ilford ID-23	Ilford ID-24	Kodak D-16	Kodak D-32	Kodak D-155	Kodak D-166
Metol	—	—	0·8	0·8	—	—	0·25	—	—	—	0·4	0·6
Hydroquinone	3·8	2·4	3·3	3·3	—	3	3·1	3·4	9	3·5	4	4·2
Chlorquinol	—	—	—	—	4	—	3·1	3·4	—	—	—	—
Glycin	—	7·5	—	—	—	3	—	—	—	—	2·6	—
Sodium sulphite, anhyd.	9·5	22	12	12·5	16	20	25	31	25	3·1	22	12·5
Citric acid	—	—	—	—	—	—	—	—	—	0·3	—	—
Sodium carbonate, anhyd.	11	16	10	12·5	15·6	16	19	22	—	15	18	12·5
Sodium hydroxide ...	—	—	—	—	—	—	—	—	5	2·1	—	—
Potassium bromide ...	0·5	1	1·4–2·8	1·6–4	0·5	1	0·4	0·3–12	2	1·7	4	6·2
Water to make ...	1,000	1,000	1,000	1,000	1,000	1,000	1,000	1,000	1,000	1,000	1,000	1,000
Dilution	F.S.	F.S–2x	F.S.–2x	F.S.	2x	F.S.–4x	4–7x	2–30x	F.S.	F.S.	2–8x	2x
Colour	W	W–WB	W	W–S	W–B	BB–RB	W–S	W–R	W	W	BB–RB	RB

The quantities are in grams per 1,000 c.cm. or parts per 1,000.

F.S. = full strength. The dilution figure signifies the number of volumes to which the solution is diluted; e.g., 2x = diluted 1 : 1 to make 2 volumes.

Image colours: W = warm black, B = brown, WB = warm brown, BB = brown black, S = sepia, RB=red brown, R = red.

formulae may be diluted up to 10 or 20 times for the warmest tones.

Some warm tone developers even use ammonia in addition to sodium carbonate. This gives purplish brown tones. Many of the image colours obtained in this way are generally more suitable for lantern slides.

Restrainer: with most warm tone developers it is principally the potassium bromide that controls the image colour. The actual concentration of potassium bromide included varies from formula to formula. It may be as low as 0·03 per cent, or as high as 0·3 per cent (0·3–3 grams per litre).

The more potassium bromide there is in the developer, the warmer the tone. For this reason extra potassium bromide (as 10 per cent solution) is often added to the diluted developer. With certain formulae the bromide concentration of the working solution is kept more or less constant in this way, irrespective of the concentration of the other ingredients.

Potassium bromide lowers the effective speed of the chlorobromide paper or warm tone slide so for warmer tones much longer exposures are therefore needed with developers containing a lot of potassium bromide. The desired effect with any particular developer and make of paper can only be arrived at by experiment. The whole range of tones is, however, warmer with the more heavily restrained developers.

Special Warm Tone Developers. While practically all the standard warm tone developers will work equally well with both papers and lantern plates of the warm tone type, there are a few formulae mainly suitable for slides only.

Ammonia (purplish) tones are produced by adding ammonia or ammonium bromide and ammonium carbonate to the developer. The colour appears to its best advantage when projected or viewed by transmitted light.

The basic developer for this range of tones is a standard bromide paper developer with varying amounts of a 10 per cent solution of ammonium bromide and carbonate added as restrainer (1–10 parts restrainer to 20 parts developer).

Thiocarbamide tones, which range from grey to violet, can be obtained with the above developer by adding 1 part in 20 of a solution of 0·8 per cent thiocarbamide and 0·3 per cent ammonium bromide. A warm tone developer—preferably glycin or glycin-chlorquinol—used instead of the standard M.Q. and ammonium bromide/carbonate mixture is even better.

DILUTION OF DEVELOPERS

For reasons of convenience many modern developers are made up or sold in concentrated stock solutions requiring dilution before use. The degree of dilution depends on the initial concentration of the stock solution (which is limited by the solubility of the ingredients) and the purpose for which the developer is used. Thus the solution is generally diluted more for tank than dish development of negative materials or for development of papers.

The degree of dilution affects:

(1) The development time.
(2) The contrast.
(3) The exhaustion.
(4) The grain.
(5) The keeping qualities.

Development Time. The more a developer is diluted, the lower its activity (because of the lower concentration of the developing agents and alkali), and the longer the development time. Concentrated developers are therefore used where speed is important, and diluted solutions where the development rate is less vital, as in tank or stand development.

The increase in development time with dilution depends on the nature of the developer.

With standard M.Q. carbonate developers the time is proportional to dilution around the ideal concentration. However, the time increases more at higher dilutions and decreases less at much higher concentrations. With the buffered type of formulae the development time does not change so much with dilution. Thus with some fine grain developers of this kind a ten-fold dilution of the standard working solution may only result in treble the development time.

Contrast. In addition to the effect on development time, dilution tends to reduce the contrast obtainable. This is important with materials that are more or less developed to finality such as printing papers. The reason is that at higher dilutions the development rate falls off more rapidly as development progresses, than at lower dilutions where the development rate tends to be more constant.

Contrast developers are therefore used in more concentrated solutions than standard developers.

Exhaustion. At higher dilutions the exhaustion of the developer tends to be more complete than at lower dilutions. In practice that means that a smaller amount of stock solution can be used per film or other unit area of sensitive material in tank development with highly diluted developer. This method of development is therefore more economical than development with more concentrated solutions which are generally discarded long before they are exhausted.

The limit to the useful degree of dilution is set by the minimum amount of developing agents in the solution required per unit area of the material. When operating near or below this limit there is a risk that the working solution will exhaust itself before development has gone far enough, or at least that the activity will drop so much during development that the action cannot be accurately controlled.

Grain. Owing to the reduced activity of a diluted developer, the grain obtained is finer. In practice this is, however, not the ideal way of producing fine grain images, because it also gives lower contrast. The decrease in graininess is not very great if the image is developed to a higher contrast.

Keeping Qualities. Dilution seriously shortens the storage life of a developer. This is because of the lower concentration of preservative in the dilute solutions and because a given amount of oxidation makes proportionally greater inroads into the concentration of the developing agents. This is the main reason (apart from convenience) why developers are stored in concentrated stock solutions. After use, a diluted working solution is best discarded.

EXHAUSTION OF DEVELOPERS

The number of negatives or prints which a given amount of developer solution will develop depends on: the composition and concentration of the developer, the area of the emulsion surface (i.e., the negative or print size), and the density of the image developed (a dense negative or print takes more out of the solution than a thin one).

The combined effect of all these variables can be found out only by experience.

Negative Developers. The list below gives the developer requirements for typical batches of plates of various sizes when developed in a dish.

If the plates are developed one after the other the development time will have to be increased by about 5 per cent for each plate.

With normal tank developers made up by diluting concentrated stock solutions, allow about 16–20 ounces (400–500 c.cm.) per standard film of 70 sq. ins. (450 sq. cm.). Use the working solution once only. This film area corresponds approximately to an 8-exposure (2¼ × 3¼ ins. or 6 × 9 cm.) length of roll film or a 36-exposure length of 35 mm. film.

AMOUNT OF DEVELOPER REQUIRED

Batch	*Amount of Working Solution*
4 plates, 2¼ × 3¼ ins. (6 × 9 cm.)	3–4 ounces
2 quarter-plates, 3¼ × 4¼ ins. (9 × 12 cm.)	(80–100 c.cm.
2 half-plates, 4¾ × 6½ ins. (13 × 18 cm.)	6–8 ounces (160–200 c.cm.)
2 whole-plates, 6½ × 8½ ins. (18 × 24 cm.)	12–16 ounces (320–400 c.cm.)

For tank development in a small capacity tank, allow about 5–7 ounces (140–200 c.cm.) of working solution. Use once only.

With concentrated fine grain tank developers of the metol-hydroquinone-borax and other types of similar concentration, 16–20 ounces (400–500 c.cm.) will develop about two to four standard films. Again, the time has to be increased by about 5–10 per cent for each film developed.

With fine grain developers and commercial tank developers used in large bulk processing tanks replenishers can be added to make up for the lost developer activity.

Print Developers. When print developers are exhausted the image colour of the print may suffer and the longer development times necessary will cause staining of the paper.

The number of prints an average negative developer will develop depends on the size of the prints, as well as on the developer concentration.

Contact paper developers are generally used in comparatively strong solutions and they will usually do more work for a given volume than the weaker solutions used for enlarging papers. The exact exhaustion also depends on the proportion of dark tones in the prints; dark prints use up more developer than light ones. When a bromide or contact paper developer is

AVERAGE EXHAUSTION DATA FOR DEVELOPERS

Print Size		Approximate Number of Prints Developed by:	
ins.	cm.	*10 ounces (250–300 c.cm.) of Bromide Paper Developer	*20 ounces (500–600 c.cm.) of Bromide Paper Developer
2½ × 2½	6 × 6	50	100
2½ × 3½	6 × 9	35	70
3½ × 4½	9 × 12	18	35
3½ × 5½	9 × 14	15	30
4¾ × 6½	12 × 16	10	20
5 × 7	13 × 18	8	16
6½ × 8½	16 × 21	6	12
7 × 9½	18 × 24	3–4	7
8 × 10	20 × 25	3–4	7
10 × 12	25 × 30	2–3	5
12 × 15	30 × 38	—	3–4

*Only half this amount applies when using Contact Developer.

nearly exhausted the image-tones appear greenish-black, even after full development.

Once developer has been used, it does not keep well and should be thrown away. For economical working, prints are best planned in batches just large enough for each lot of developer made up. For this reason it is economical to deal with prints in uniform batches and to make up just sufficient developer to do one batch at a time. L.A.M.

See also: *Alkali; Chemicals; Cold weather; Colour film processing; Developer improver; Developers (obsolete); Developer testing; Developing agent; Development after fixing; Development history; Development theory; Fine grain technique; Hot weather processing; Latent image; Physical development; Rapid processing; Solutions; Tropical photography.*

Books: *Processing Chemistry*, by L. F. A. Mason (London); *Developing*, by C. I. Jacobson (London); *The Photo-Lab-index*, ed. by H. M. Lester (New York).

DEVELOPERS (OBSOLETE). A number of developing agents which were popular around fifty years ago have now fallen into disuse. Of these, the main agents are ferrous oxalate, edinol, eikonogen, metoquinone and ortol.

Ferrous Oxalate. This is the only inorganic developing agent which has ever been used to any great extent in practice. It is a clean-working developer which produces hardly any fog, but it is very sensitive to the presence of potassium bromide which materially decreases the effective emulsion speed. The developing agent itself, FeC_2O_4, is invariably prepared in solution by mixing ferrous sulphate with an excess amount of potassium oxlate solution:

A. Ferrous sulphate	5 ounces	125 grams
Sulphuric acid, 10 per cent solution	290 minims	15 c.cm.
(or citric acid	10 grains	1·2 grams)
Distilled water to make	20 ounces	500 c.cm.
B. Potassium oxalate (neutral)	10 ounces	250 grams
Distilled water to make	40 ounces	1,000 c.cm.

For use pour 1 part of A slowly into 3 parts of B, stirring well all the time. If the solutions are mixed the other way round, a yellow precipitate of ferrous oxalate is formed. This is insoluble in plain water and soluble only in potassium oxalate solution; once formed it is very difficult to dissolve.

A white deposit of calcium oxalate may be produced in the emulsion during fixing or washing if the water for the latter processes is hard. The deposit can be removed by a final rinse in a 0·5 per cent hydrochloric acid solution.

Edinol. Para-aminosaligenin, $NH_2C_6H_3OH.CH_2OH$. This is a derivative of para-aminophenol which it resembles in its action. It is fairly soft-working, and can be used as a developer with sulphite and carbonate. A suitable formula is:

Sodium sulphite, anhydrous	5 ounces	125 gram
Edinol	½ ounce	12 grams
Sodium carbonate, anhydrous	1½ ounces	37 grams
Water to make	40 ounces	1,000 c.cm.s

Eikonogen. Sodium 1-amino-2-naphthol-6-sulphonate, $NH_2C_{10}H_5OH.SO_3Na$. This is a derivative of ortho-aminonaphthol, and resembles metol to some extent. A related substance, 1-amino-2-naphthol-4-sulphonic acid, has also been used as a developing agent. Eikonogen can be used either by itself with sulphite and alkali in the same way as metol, or combined with other developing agents such as hydroquinone or pyro. A suitable eikonogen formula is:

Sodium sulphite, anhydrous	2 ounces	50 grams
Sodium carbonate, anhydrous	¾ ounce	19 grams
Eikonogen	1 ounce	25 grams
Water to make	40 ounces	1,000 c.cm.

Metoquinone. $(CH_3NH.C_6H_4OH)_2.C_6H_4(OH)_2$. This is a condensation product of 2 parts metol with 1 part hydroquinone. On dissolving in an alkaline solution of sulphite it acts like a metol-hydroquinone developer containing the two developing agents in a proportion of 2 : 1.

Ortol. $(NH_2C_6H_4OH)_2.C_6H_4(OH)_2$. This is a condensation product of 2 parts ortho-aminophenol with 1 part hydroquinone. A suitable formula for a concentrated developer is:

Potassium metabisulphite	¼ ounce	7 grams
Sodium sulphite, anhydrous	1½ ounces	37 grams
Sodium carbonate, anhydrous	1 ounce	25 grams
Ortol	½ ounce	12 grams
Water to make	40 ounces	1,000 c.cm.

For use dilute 1 part of the above with 10 parts of water. L.A.M.

See also: *Development history.*

DEVELOPER TESTING. A developer is tested (as distinct from analysing it) by observing the behaviour of a sensitive emulsion developed in it, and possibly by comparing this with the behaviour of another developer-emulsion combination. Such tests are carried out to plot time-gamma curves and establish sensitometric data and to obtain information about graininess, image structure and type, effect of variations of concentration, effect of different ingredients, temperature factors, etc. Much of this data is mainly of interest in the laboratory, and the testing procedures are laboratory methods.

Developer tests which a photographer is likely to want to carry out himself, may include:

(1) Comparison of contrast and rate of action of different developers and establishing an optimum development time for a new formula.

(2) Comparison of effective emulsion speed of different developers, or one developer at different development times.

(3) Comparison of graininess with different developers.

Basic Test Procedure. The practical part of most tests follows the same lines:

(1) Prepare two or more series of test exposures on a standard film. Each series should cover about seven exposures from one-eighth correct to eight times correct exposure. The subject should be one with a fairly long scale of tones from a reasonable amount of shadows to bright highlights; an indoor scene lit by lamps is best as it is easily reproducible and will not change in brightness with the weather. A suitable mark—e.g., a card with numbers which are changed for every shot—will help to identify the exposures given.

One good way of making the series is to use a 35 mm. miniature camera, and shoot off about three series on one cassette load, leaving a suitable number of empty frames between each set. The film can then be cut up at these empty frames before development.

As few cameras have really accurate shutters, the exposure for each shot of a series is best controlled by changing the aperture, keeping the shutter speed constant.

All the strips for a set of tests should be exposed at the same time.

(2) Develop one of the series or strips in a standard developer under standard conditions. This will serve as control.

(3) Develop the other series or strips in the developers, or the conditions, to be tested.

The evaluation of the results depends on the purpose for which the test is required.

Establishing Development Times. For this purpose a fair number of test series are required. If an approximate development time is known, a single test stage will give the necessary information; otherwise the test involves two stages.

To find an approximate time, first develop, say, five exposure sets for varying times. The range of times chosen will depend on the type of developer. For instance, a rapid developer with a medium speed film might be given times from 6 to 12 minutes, or a normal fine grain developer from 15 to 30 minutes. There should be a constant difference between the sets; thus the first set would be developed for 15 minutes, the second for 18, the third for 21 and so on.

Examination of the developed strips for contrast will then show which is nearest the correct time.

The negative selected should be the one which has received the minimum correct exposure—i.e., in which the shadow detail is just visible in the image.

To some extent the judgment will be arbitrary, and it may be desirable to make enlarger ments from the best negative to see whethe- the gradation is suitable for the paper and printing procedure to be used. Comparison with the control strip will also help.

All processing temperatures must, of course, be the same, and each strip should be developed in fresh developer, preferably diluted from the same batch of stock solution.

Comparison of Effective Emulsion Speed. This type of test may be useful for investigating the merits of finality development or claims for increased emulsion speed with maximum energy developers.

Here again a control strip is developed in a standard formula, while the other strips are developed in the formula to be tested. When investigating finality development, the test series is developed in the same solution as the control, but for an appropriately longer time.

To assess the gain in speed, the negatives corresponding to minimum correct exposure in the control and the test series should be compared. If these negatives have received the same exposure, the gain is obviously nil; where, say, the first correct negative in the control set corresponds to *f* 5·6 and in the test set to *f* 8, the gain is 1 stop, or 100 per cent. Intermediate stages can generally be judged by estimation.

Comparison of Graininess. Here the control strip is compared with a strip developed in the test developer for an optimum time as established separately. If that is not possible, the comparison strip is developed for a time recommended by the makers of the film or developer. The control developer should be a standard fine grain—e.g., M.Q. borax—formula, while the film should be a high speed coarse-grained type.

The negatives corresponding to the minimum correct exposure on each strip are compared for graininess. For this purpose, 20-diameter enlargements are made of a portion of the image which shows typical highlights, middle tones and shadows. The paper used should be a normal grade, glossy bromide paper in both cases; the control as well as the test film must therefore be developed to a suitable contrast for this paper grade.

It is important also to avoid enlarger vibration, as that will immediately destroy the definition of the graininess in the print.

Comparison of Definition. While the main factors determining the definition and resolution of an image are the optical system and the sensitive material, the developer employed does exert some influence. So does the degree of development. Thus if the image is developed mainly in the surface of the emulsion layer, the degree of spreading due to diffusion and irradiation is less than if the developer develops up the image throughout the depth of the layer (or if development is continued sufficiently long for the image to be developed in depth).

To test developers for their performance in this respect, the film selected is exposed to a standard resolution test target under specific conditions and different samples of the film so exposed developed under various conditions.

Resolution of the image can then be obtained by establishing the highest frequency of the test target still resolved (in lines per mm.) or by measuring the acutance of samples exposed for an acutance test.

If the test exposure contains images of resolution targets of different contrast or of a sinusoidal test target it becomes possible to plot a contrast transfer function of the developer-film combination. This transfer function of course also takes into account the performance of the optical system employed, but if all conditions are standardized apart from the developer, a comparison of the different transfer function curves shows up the differences in the definition performance of the developer.

A simpler and purely qualitative method of comparing the image definition obtained with different developers consists of exposing samples of a film to the image of a high contrast grid pattern. This consists of 20 to 40 lines per mm. with spaces between the lines equal to the line width itself. The comparison then merely involves a density measurement of sample areas of the image on the different films. These areas must correspond to similar positions in the image field.

The theory behind this method is that if the image on the film consists of very dense lines and virtually clear intervals, the integrated density reading is lower, the sharper the grid image is on the film. In theory completely opaque lines and completely transparent intervals should give an integrated density reading of 0·3 (in other words 50 per cent transmission). If the image is not perfectly sharp, the black lines tend to spread on the film, giving a higher integrated density reading.

While no silver image is ever completely opaque, this method still works in practice if the density difference between the black lines and the intervals is at least 2·0 and the fog density kept below a certain minimum level. The integrated density reading reflects both the resolution and the micro-contrast of the image, and so gives an indication of visual sharpness rather than pure resolution. This property, like acutance, is of practical significance. L.A.M.

See also: *Acutance; Grain; Resolving Power.*

DEVELOPING AGENT. Chemically, a developing agent is a reducing agent (not to be confused with a photographic reducer, used for decreasing the density of the silver image).

Developing agents can be classified according to their individual chemical properties, and to the way they behave in developers. The behaviour of any particular developing agent depends on the other ingredients present. Those affect its keeping quality, activity, and fine grain properties. So from a practical point of view it is better to consider the more common developer formulae and their agents at the same time.

Keeping Quality. Most developing agents are more rapidly oxidized in strongly alkaline solutions than in weakly alkaline or neutral ones.

Apart from this, however, some developing agents tend to keep better in solution than others, while some do not keep at all.

Development Speed. This depends largely on the alkalinity of the solution, the amount of restrainer, and the concentration of the developing agent. A table showing the inherent activity of various developing agents, is on pages 406–7.

The Watkins Factor does, however, offer a more accurate and less variable basis for comparing development speeds.

Effect on Emulsion. Some developing agents tend to produce images of low gamma while others are more suitable for use in high contrast developers giving high gamma.

The function of a developer should be only to act on the silver salts that have been exposed to light.

Unfortunately, most developing agents also affect unexposed silver salts, producing a silver deposit which has nothing to do with the photographic image. This is known as fog.

Much of the inherent tendency of a developer to form a fog depends on the developing agent. It is counteracted by adding a restrainer or a developer improver. The greater the fog formation, the more restrainer is needed.

The graininess in an emulsion is to some extent proportional to the activity of the agent. Low energy agents tend to make better fine grain developers. But with a suitable formula of controlled alkalinity and activity almost any developing agent can be used in a fine grain developer.

Effect of Temperature. The warmer the developer solution, the more rapidly it acts. The increase in development speed with temperature varies with the different developing agents. This sensitivity to temperature is stated in terms of a temperature coefficient. This is the ratio of the development times at two temperatures differing by 10° C. (18° F.). Thus, if a developer takes $7\frac{1}{2}$ minutes at 24° C. to develop a given film, and 15 minutes at 14° C. to give the same result, the ratio is $15/7\frac{1}{2}$, or 2·0. This is the temperature coefficient for that particular developer. The coefficient also depends to some extent on the composition of the whole developer. A normal value is about 1·8–2·2.

Developers also deteriorate more rapidly at higher temperatures, hence a developer solution should be kept cool. Some developing agents become virtually inactive at low temperatures.

Solubility. All developers contain the developing agent in solution, and most agents are soluble in water. Some are almost insoluble in water alone, and will dissolve only if the formula also contains alkali. Other agents (e.g., metol) will not dissolve readily if the solution already contains a lot of sodium sulphite. So metol is always dissolved in pure water and the sulphite added afterwards. Sometimes alcohol is added to the solution to increase the solubility of the developing agent. L.A.M.

See also: *Reducer*, *Developers*.
Book: *Developing*, by C. I. Jacobson (London).

DEVELOPING NEGATIVES

During exposure the light falling on the sensitive material affects the silver salts in the emulsion and forms a latent image. This latent image reacts with the developer during the development process which converts it into black visible silver.

This silver forms the final image which remains after the unused silver salts are dissolved and washed away in the fixing and washing baths that follow development.

There are two main methods of developing films and plates: by inspection or by time and temperature.

For the first, the material is usually developed in a dish in the darkroom by a suitable darkroom safelight.

For time and temperature development a dish may still be used; but it is usually more convenient to use a tank, particularly with roll and miniature films.

INSPECTION

This is the oldest way of developing plates or films. The sensitized material is immersed, emulsion up, in a dish of developer, and watched as the image appears. This is often called dish development.

It must be done by the light of a suitable safelight. High speed panchromatic materials are sensitive to even the dimmest darkroom safelight, and must only be examined intermittently.

Desensitizers can be used to lower the sensitivity of the emulsion and allow brighter safelights to be used.

Inspection development is comparatively little used in modern commercial processing. It does however demonstrate visibly the progress of development and permits a certain degree of control in special cases.

The method is primary suitable for plates and sheet films (which are handled largely in the same way).

Plates. As soon as the plate is in the developer the dish is gently rocked to keep the solution moving over the surface. For the first few moments (or minutes, with a slow working developer) there is a delay (the induction period) during which the solution penetrates into the gelatin layer.

Then, gradually, signs of the image begin to appear.

At first only the parts which have received the most exposure (the highlights in the negative) show up. Then the medium tones follow, and finally the shadows where the deposit of black metallic silver is least.

The image builds up density, and gets darker and darker, until no details are visible on the face of the plate as it lies in the dish. But when the plate is looked at from the back, the highlights and half tones appear black and dark grey, with the shadows light grey.

At this point development is complete. The plate is rinsed in clean water, and transferred to the fixing bath.

As long as the greater part of the image still looks white or light grey from the back, the negative is under-developed. If the plate is completely black all over it is over-developed.

The ability to judge the correct degree of development by the dim darkroom illumination comes from practice. As development tends to start at the surface, the first deposit of silver completely hides what is happening underneath.

But if the back of the plate is examined, the gradual building up of the image is easy to follow.

This process of depth development does not occur with some of the fine grain developers. With such developers examination of the unfixed negative from the back may be misleading. These developers are, however, rarely used for processing by inspection.

The complete steps in processing plates and sheet films by inspection are as follows:

(1) Immerse the plate or film, face up, in the dish of developer, making sure that the solution covers the whole surface at once.

(2) Dislodge any airbells from the surface of the emulsion.

(3) Rock the dish gently, but not too evenly, all the time.

(4) Inspect the negative from time to time,

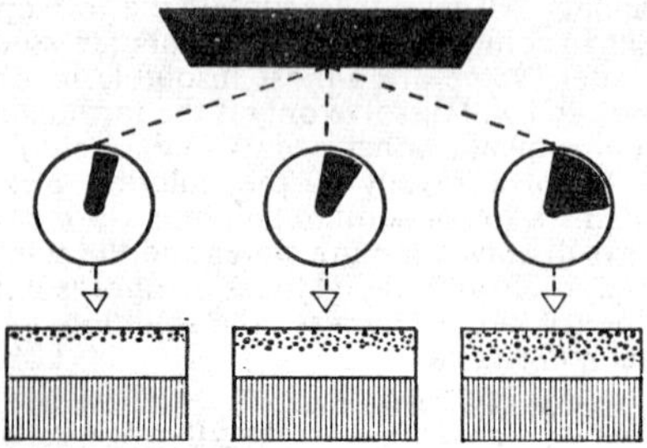

PROGRESS OF DEVELOPMENT. Development in most developers begins at the surface of the emulsion and progresses downwards with increasing development time.

lifting it out to see the image from the back, if necessary. Keep the dish covered with a card at all other times, and do not prolong the inspection, especially with fast panchromatic plates.

(5) When development has reached the required point, transfer the film or plate to a dish of clean water for 1 minute.

(6) Transfer the negative to the fixer.

(7) Rock the fixing dish occasionally, and leave the negative in it for about twice as long as the time required to discharge all traces of milkiness from the emulsion layer.

(8) Wash the negatives in running water for at least thirty minutes, or in eight changes of water of three to five minutes each. With plates take care they do not slide over each other in the washing dish.

(9) Lift the plates out of the washing water, drain, and stand up on edge in a dust-free place to dry. Alternatively, stand the plates in a drying rack. Sheet films should be hung up on a line by means of small film clips.

Rollfilms. Strips of roll film can in principle be developed in a similar way, by see-sawing the film through the developing dish. This method is, however, highly inconvenient and of no advantage over tank development, since individual control of negatives on the roll is impractical.

Control of Contrast. Development by inspection allows a certain degree of control over the final result. The length of development determines the contrast as well as the density of the negative.

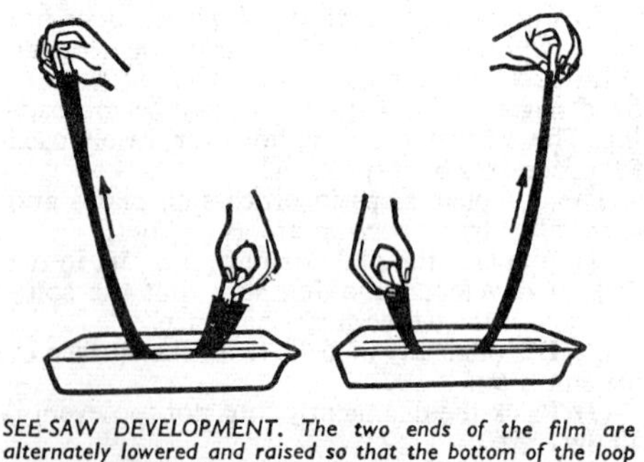

SEE-SAW DEVELOPMENT. The two ends of the film are alternately lowered and raised so that the bottom of the loop passes continually through the solution.

While it is difficult, if not impossible, to judge the contrast of a negative by safelight illumination, watching the negative density will give some idea of the progress of development.

Negatives can therefore be developed individually to different degrees of contrast to suit the subject, provided the exposure has been reasonably correct in the first place.

But to get the highest contrast it is better to use a contrast developer and to reduce the contrast of a contrasty subject it is better to use a soft developer or some kind of water-bath or two-bath development. These methods give the required contrast compensation without the excessive density produced by over-development, or the thinness which results from greatly reduced development.

Factorial Development. At a time when development tanks were still relatively unknown, photographers tried to put inspection development on a slightly more scientific basis. At the end of the last century A. Watkins showed that there was a relationship between the time of the first appearance of the image and the total time of development for approximately correct negatives. He found that the time of development could be calculated by noting the time it took for an image to appear and multiplying that by a factor. The so-called Watkins factor depends on the developer, but—within limits—neither on the temperature nor on the concentration of the developer, nor on the negative material. This method of development is known as factorial development. The Watkins factors vary from 4–5 for developers of the hydroquinone type to around 30 for metol.

Factorial development gives correct results only with correctly exposed negatives. Since modern films and plates contain sensitizing dyes which may affect the factor unpredictably, this method is not nowadays a very reliable one.

TIME AND TEMPERATURE

For any one film-developer combination the degree of development (under conditions of standard agitation) depends only on the temperature of the developer and on the time during which it is allowed to act. So if a film is developed in a given developer for a specified time and at a specified temperature, the resulting negatives will always be of the same quality. This system of time and temperature development is the basis of all modern tank procedures. With amateur tanks neither darkroom nor safelight is needed, since the tank is light tight and all operations—except for loading of the film—can be carried out in broad daylight.

Development Times. The various manufacturers recommend development times for their materials in certain standard developers. These times are intended mainly as a guide rather than as absolutely correct values. There is in fact no such thing as one correct development time; apart from variables such as agitation, etc., a great deal also depends on personal preference. For instance, some photographers may prefer comparatively thin negatives of low contrast which will enlarge well on a hard paper grade and need only short enlarging exposures; others again favour more vigorous negatives (especially when they are to be printed by contact). The range of correct development times is thus rather wide; comparison of different recommendations for essentially similar developers and negative materials may show variations of up to 100 per cent.

For that reason no development times are given here. The best procedure for establishing a suitable development time for a given developer-film combination is therefore to start from the maker's recommended figure, and modify it according to individual requirements. If the negatives turn out too contrasty the time can be reduced by 20–30 per cent; if they are too soft for the particular printing procedure employed, increase the time up to 50 per cent.

A number of variable factors have a bearing on the correct time of development: negative contrast; negative material; temperature; agitation.

Negative Contrast. The negative contrast aimed at is decided partly by personal taste, and partly by the printing process to be used.

Apart from individual preferences, miniature negatives are usually developed to a lower contrast (about gamma 0·7–0·8) than larger negatives (gamma 0·9–1·0). The reason for this is partly the finer grain obtained with shortened development, and partly the fact that fully developed negatives are denser and may require inconveniently long enlarging times. In addition a certain amount of contrast control may be desirable with a roll of film consisting of exposures of the same or similar subjects.

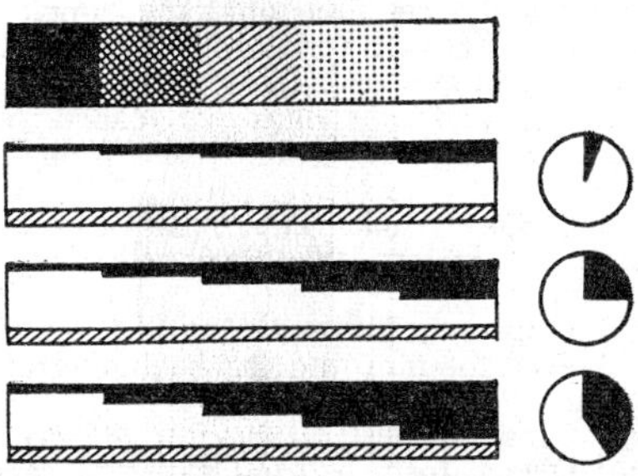

DEGREE OF DEVELOPMENT. *The longer the developing time, the blacker the densities become. They increase in proportion to the original exposure, yielding successively a soft, normal, and contrasty image of increasing tone separation.*

Negative Material. Different negative materials require different development times; some develop more slowly, others more rapidly. It is usual to divide negative materials into a number of development groups. Times are then quoted for materials of normal development speed, and modified accordingly for materials in any of the other groups.

Temperature. The temperature effect varies with the type of developer and is expressed as a temperature coefficient. This is the factor by which the development time has to be multiplied for a fall of 10° C. (18° F.) in the development temperature.

For most developers the temperature coefficient is in the neighbourhood of 1·9, though it may be as low as 1·5 or as high as 2·5. Strongly alkaline developers of the high contrast type such as caustic hydroquinone have high temperature coefficients. On the other hand, soft working developers like metol are much less affected by changes in temperature.

The useful range of development temperatures is about 60–75° F. (15–24° C.). At temperatures lower than 60° F. development times are inconveniently long; above 75° F. the gelatin emulsion becomes soft, and extra care is needed in handling the film to avoid injury to the gelatin. In fact, the use of hardeners or even special hot weather processing is advisable at high temperatures.

The table below gives the variations in development time for an average developer.

DEVELOPMENT TIMES AT VARIOUS TEMPERATURES

Time in Minutes at 65°F. (18·4°C.)	*Corresponding Time in Minutes at* 60°F. (15·5°C.)	62·5°F. (17°C.)	68°F. (20°C.)	70°F. (21°C.)	75°F. (24°C.)
5	6	5½	4½	4¼	3½
6	7¼	6½	5½	5	4¼
7	8½	7¾	6¼	6	5
8	9½	8¾	7¼	6¾	5½
9	11	10	8¼	7½	6¼
10	12	11	9	8½	7
11	13	12	10	9¼	7¾
12	14½	13	11	10	8½
13	16	14½	12	11	9
14	17	15½	13	12	9¾
15	18	16½	13½	12½	10½
16	19½	17½	14½	13½	11
17	20½	19	15½	14½	11½
18	22	20	16½	15	12½
19	23	21	17	16	13
20	24	22	18	17	14
21	25½	23½	19	17½	14½
22	26½	24½	20	18½	15½
23	28	25½	21	19½	16
24	29	26½	22	20	16½
25	30½	28	22½	21	17
26	31½	29	23½	22	18
27	33	30	24½	22½	18½
28	34	31	25½	23½	19½
29	35	32	26	24½	20

Agitation. By agitating the film in the developer, fresh solution is brought into contact with the emulsion, while the used-up solution is carried away and mixed up with the bulk of the developer in the dish or tank.

The film may be agitated all the time, or only intermittently. Constant agitation keeps the solution always on the move. It is essential with certain types of developing tank where the film is only partly immersed in the solution.

See-sawing a length of film through the solution is also a form of constant agitation for roll films.

In bulk processing tanks the solution is often agitated by a mechanical mixing device.

Intermittent agitation means agitation for a few seconds every two to three minutes. The solution is not mixed up as efficiently, therefore the film needs a longer development time. Usually the instructions with the developer quote the times for intermittent agitation; for constant agitation they can be shortened by 25–30 per cent.

Insufficient agitation (with intervals of more than about four to five minutes) may cause streaks of uneven density, light rims round dark image areas or dark areas with lighter spots in the centre.

Some degree of regular agitation during development is therefore essential. The shorter the over-all development time for the particular film and developer used, the greater the danger of uneven action through insufficient agitation.

With high speed developers and similar formulae the solution must never be allowed to stand still.

Too regular agitation may cause flow marks when partly exhausted developer flows over the emulsion surface in the same way and the same direction all the time, e.g., by rocking the dish or rotating the film reel always with the same motion. The developer dish, or film reel, should therefore be moved unevenly in all directions. If the film can be raised out of the solution and allowed to drain from time to time, so much the better.

Special Developing Techniques. The normal standard development technique serves on most occasions, but occasionally it does not bring out some particular aspect of the subject, or it does not make the most of a subject with some abnormal feature about it—e.g., extreme contrast or flatness. In such cases it is often possible to apply one of the many variations of the standard developing technique, although usually what is gained in one direction is lost in others. (The debit side of the balance is frequently overlooked, an oversight which is responsible for many exaggerated claims for unorthodox techniques or developers.)

There are, for example, non-standard techniques which aim at increasing film speed, improving grain, adjusting tone range, contrast, printing quality, etc. The commoner variants are development to finality, stand development, high speed development, two-bath and water-bath development, and the use of special forebaths.

TANK DEVELOPMENT

Most types of developing tank for amateur use consist of a light-tight container (usually made of black bakelite or other plastic) with some sort of holder inside for the films and plates.

Most roll film and 35 mm. film tanks carry the film wound up on a reel which may either be a spiral type or a reel with an apron.

Types of Tank. The spiral tank reel consists of two discs connected by a hollow spindle; a spiral groove runs on the inside faces of each disc until it reaches the spindle at the centre. The film is loaded into this spiral groove, which is the exact length to accommodate the entire film. Sometimes, the separation of the discs is adjustable so that films of different sizes (widths) can be processed.

Tanks are also made which allow the film to be loaded in daylight.

The basic construction of the apron tank is similar, except that the reel has no spiral groves. Instead, a celluloid or plastic band the same width as the film attaches to the centre spindle; the edges of this band have raised "dimples" embossed in alternate directions. The film is wound on to the spindle with the embossed band, the "dimples" of which touch the film at the edges only.

Some roll film and 35 mm. tanks are of the self-loading spiral type. In these tanks the cheeks on the reel can be moved through a small angle in relation to each other; the grooves are either moulded with projections which grip the film in one direction only, or the necessary one-way pressure is applied with the fingers. By turning the loose cheek to and fro, the film automatically feeds itself on to the reel once started in the spiral.

Most modern spiral reel tanks have cheeks of transparent plastic. This simplifies the reversal exposure when processing reversal colour films, making it unnecessary to remove the film from the tank reel. Otherwise reloading a wet film into a wet reel is tricky and liable to damage the film itself. Reloading a wet film into an apron tank is easier.

Plate and sheet film tanks are simple rectangular boxes with vertical grooves inside the body (sometimes as a removable rack) into which the plates or films are loaded. Sheet films are first loaded into stiff frames for rigidity.

Loading Spiral Tanks. The operations must be carried out in the dark. Their sequence is as follows:

(1) Prepare the film end.

(2) Feed the film into the beginning of the groove.

(3) Gradually push the film into the reel.

(4) Detach from backing paper or cassette spool.

(5) Put the reel into the tank.

After unrolling the beginning of the backing

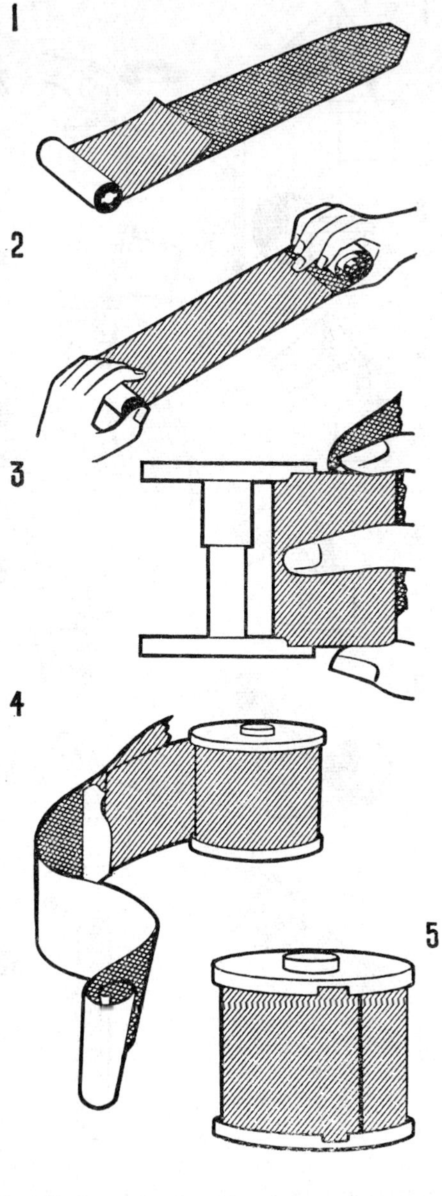

LOADING A ROLL FILM TANK. 1. Break the seal of the roll and unwind the backing paper (in total darkness) until the end of the film itself is reached. 2. Unroll a little more to have about 3–4 ins. free (but hold the spool to prevent it from unwinding altogether). 3. Guide the end of the film into the mouth of the spiral groove of the tank reel. 4. Feed the film into the reel until the other end (attached to the backing paper) is reached. A self-loading tank reel with rotating flanges greatly simplifies this operation. 5. Detach tne film from the backing paper and push the last bit into the reel.

paper, the film end may be prepared by bending in the extreme edge and sharply folding it back on itself. That gives the film a reinforced leading edge which is not so likely to stick in the groove. This trick is useful with roll films which are comparatively thin (3/1000 in.), but less necessary with the thicker 35 mm. films (5/1000 in.).

With 35 mm. film the specially shaped leader or trailer must be cut off.

The prepared end is pushed into the entrance of the spiral groove. (The reel must be absolutely dry, or the film may stick in the groove.)

The film is then fed into the groove gradually, bit by bit, unrolling the backing paper or pulling the film out of the cassette at the same time. If the going becomes difficult, the best way is to grip the outside of the film already on the reel with the whole hand, and gently rotate it to pull more film into the groove.

When all the film is on the reel, the other end must be detached from the backing paper. The tape with which the two are fastened should not stay on the film. With 35 mm. film the end is simply cut off as near to the cassette spool as possible.

Finally the last bit of film is pushed into the groove, the reel inserted into the tank body, and the lid closed.

There is also an alternative method. Here only the first few inches of the film are fed into the groove. The part of the film just outside the reel is then gently arched by pressing the two edges together.

While rotating the reel in the loading direction the leading end of the film is pushed right into the centre of the reel. The arched part of the film will then jump across the grooves. Where the reel has projecting lips at the entrance to the groove (as on some older designs of reel) the film must be eased past them at each revolution.

The reel is now rotated in the opposite direction, and as each part of the film reaches the inside of the spiral the pressure which holds it arched is released and it slips easily into tho groove.

With a self-loading tank the procedure is simplified, as it is only necessary to feed the beginning of the film into the groove. After that the self-loading mechanism takes over.

When the whole film is wound up in this way, the end is detached from the backing paper or cassette spool, pushed into the groove, and the reel inserted into the tank.

Loading Apron Tanks. To load an apron tank, the apron is first clipped to the centre spindle of the reel. The end of the film is then pushed under the same clip and the two are drawn around the spindle by rotating the reel.

When the film has wound on to the end, the backing paper or film cassette is cut away and the reel is rotated until the end of the apron is reached. This carries a clip by which it is secured to the rim of the reel. The reel is then dropped into the tank, the lid is fitted, and the rest of the operations can be carried out in daylight.

Loading Daylight Tanks. Each make of tank has its own method of loading, the differences lying mainly in the method of transferring the film from its spool or cassette on to the reel. The safest guide is to follow the usually very detailed maker's instructions.

Loading Plate and Sheet Film Tanks. According to the design of the tank, plates have to be fitted into the grooves of a plate rack, or into the tank itself. The important thing is to get each plate squarely into the right pair of grooves in the dark.

With some plate tanks a special loading jig is supplied. The jig is placed over the top of the plate rack or over the tank opening, when two guides on the jig align it with the correct grooves of the tank. The plates are then simply inserted through slits in the jig.

Sheet film is usually held in a frame made of stainless steel and then treated like a plate.

There are also special sheet film tanks, mainly to cater for the larger sizes of sheet film. Here again the film is inserted into stiff frame-like hangers. These are then suspended in a large rectangular tank of suitable size.

Processing in Tanks. Assuming the tank to be loaded with film or plates, the procedure is:

(1) Pour the developer into the tank.
(2) Agitate.
(3) Develop for the required time.
(4) Pour in rinse.
(5) Pour in first fixing bath.
(6) Pour in second fixing bath (if used).
(7) Wash.
(8) Remove from tank.
(9) Dry.

(1) The right amount of developer for the tank at the right temperature (between 60 and 75° F. or between 15 and 24° C.), prepared beforehand, is poured into the tank. A funnel is not necessary because most tanks already have funnel-shaped openings in the lid. The time when the developer is poured in is noted.

(2) Agitation should start as soon as the developer is in the tank. Normally that involves inverting the tank a few times—after closing the centre aperture. (Tanks have a special rod that is inserted through the centre opening and engages with the hollow spindle of the reel. This allows the reel to be rotated. Plate tanks have to be rocked or shaken.)

(3) Throughout the whole development period a close watch is kept on the temperature of the solution. It must not be allowed to vary more than a degree or so.

(4) The solution is agitated for about ten seconds every other minute and at the end of the development period it is poured out and the tank is immediately filled with water at the same temperature. Instead of washing water, a stop bath and a hardener may be used.

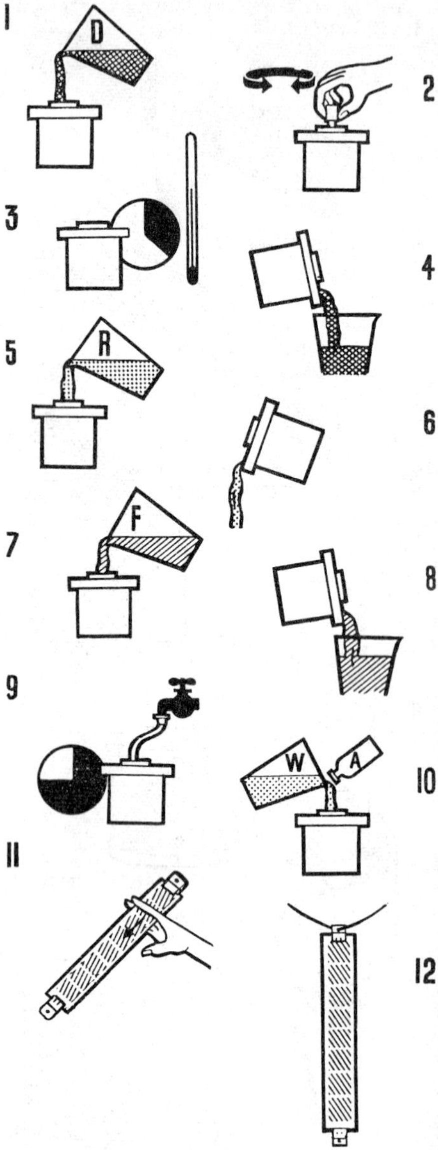

DEVELOPING FILMS. 1. Load the film into the tank, then pour in the developer. 2. Agitate the reel inside the tank, vigorously at first, and then at regular intervals. 3. Develop for the time recommended at the temperature of the solution. 4. Pour out the developer. 5. Fill the tank with clean water or with an acid hardening rinse. 6. Pour out the rinse. 7. Pour in the fixer. 8. Fix for the required time, then pour out (the fixer can be re-used until exhausted). 9. Wash, preferably by running water, for at least ¾ hour. 10. Give a last rinse with wetting agent and, if needed, a few drops of acid added. 11. Wipe the film free from surplus water. 12. Hang up to dry.

(5) The film is kept moving for about a minute and then the water is poured off and the fixing solution is poured in. The negatives are left in the fixer for at least ten minutes, again agitating periodically. It is then safe to open the tank and examine the film. When all the milkiness has gone from the film it should remain in the fixer for a further 10–15 minutes.

(6) A second bath of fresh fixer is often given to ensure really efficient fixation. The immersion time in this solution should be about ten minutes.

(7) The film or plates are then washed. The temperature of the washing water should be the same as that of the last fixing bath. Otherwise the temperature must be gradually reduced from the fixer to the washing water.

Roll films can be washed in the tank by passing a rubber tube from the water tap down the centre of the reel. Plates are best taken out of their tank and laid face up in a large dish, for most plate tanks do not allow efficient washing of the plates inside.

An alternative way of washing is to fill the tank with changes of fresh water, leaving it for 5 minutes each time. About six to eight such changes are sufficient.

After washing, the tank is filled with water again, and a few drops of wetting agent and, if necessary, dilute acetic or hydrochloric acid added.

(8) The film is then carefully unrolled from the reel, and a film clip fixed to each end. Most of the water can be removed by wiping it down but this entails some risk of scratching the wet emulsion. With plates the surplus water is best shaken off as far as possible.

(9) Finally, the films are hung up to dry in a dust-free place. Plates should be dried in special plate racks, while sheet films can be pinned up by one corner. L.A.M.

See also: *Agitation; Cine film processes; Cold weather; Colour film processing; Developers; Development after fixing; Densensitizing; Finality development; Fine grain technique; Fixing; Hot weather processing; Physical development; Rapid processing; Redevelopment; Reversal process; Stand development; Tanks; Tropical photography; Two-bath development.*

Books: *How to Do Your Own Film Processing*, by L. A. Mannheim; *Longmore's Medical Photography; Developing*, by C. I. Jacobson (all London).

DEVELOPING PRINTS

Prints are invariably developed by inspection by being immersed flat and face up in a dish of developer. During the development of an exposed print the developing solution converts the invisible latent image in the sensitive emulsion into a visible picture. The developed print is rinsed, and then fixed to dissolve away the unused silver salts. Finally, all soluble chemicals left in the paper are washed out and the print is dried.

Developers. The best developer to use depends to some extent on the type of printing paper.

Chloride contact papers are best developed in a contact paper developer containing a comparatively low amount of potassium bromide.

Bromide and chlorobromide enlarging papers are developed in a standard paper developer. Some of these serve equally well for bromide, chlorobromide and contact papers.

For warm tones on chlorobromide, contact, or enlarging papers, warm tone developers and special development methods are required.

Prints are generally developed by inspection in the light of a suitable safelight.

Processing. The sequence of operations in developing a print is:

(1) The print is immersed in the developer in such a way that the solution flows rapidly over the whole surface of the print. Small prints are simply pushed under the surface of the solution with the print forceps. With larger prints a satisfactory method is first to raise one end of the developing dish, so that all the solution collects at the other end, then hold the print, face up, and push one edge into the liquid. Finally release the end of the dish and allow the solution to flow in a wave over the whole of the print surface. Any air bubbles are immediately broken up with the forceps.

(2) The developing dish is gently rocked by raising and lowering one edge to keep the solution moving over the paper. Any corners of the paper which curl out of the solution during the first few seconds are pushed under with the forceps.

(3) When a print has had too little exposure, the image shows up very late and fails to reach a satisfactory depth of tone, no matter how long it is left in the developer. When a print has had too much exposure, the image flashes up in a few seconds and soon becomes too dark. The print always looks slightly darker under the orange safelight than it will look under normal lighting. If the safelight is olive-green, there will be less difference between the appearance of the print as seen in the darkroom and its appearance when it is finished and viewed by normal lighting.

(4) Development is continued until the image does not seem to go any darker; this point is usually reached after about two minutes, but it will depend on the developer in use (and its temperature) and also the type of paper. The print is then lifted out of the solution with the forceps, drained, and dropped, face down, into the rinse dish. If a chemical stop-bath is used the forceps should not be allowed to come into contact with the solution when transferring the print.

(5) The print is moved about in the rinse for about a quarter of a minute. If a chemical stop-bath is used instead of a plain water rinse, a

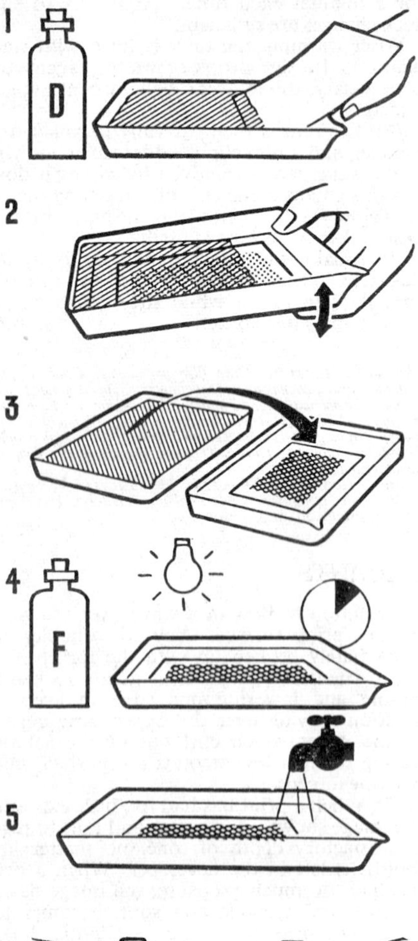

DEVELOPING PRINTS. I. *Slide the exposed sheet of paper into the developer.* 2. *Rock the dish during development until the image is fully developed out.* 3. *Transfer the print into an acid stop bath.* 4. *Transfer the print to the fixing bath. Leave for about 5 minutes (moving it about regularly). White light can be turned on after this stage.* 5. *Wash in running water or in frequent changes of still water.* 6. *Hang up (large prints) or lay down on clean muslin to dry (prints can also be dried by hot or cold glazing on glossy paper).*

separate pair of forceps should be used for handling the print in this bath and transferring it to the fixing bath.

(6) It is then lifted out, drained, and immersed, face down, in the fixing bath, where it is kept moving for about half a minute to make sure that the fixer reaches the whole print surface.

(7) When the print has been in the fixing bath for about a minute, the white light may be switched on, and the print examined. This will show whether the exposure was correct and whether the contrast is satisfactory. When further prints are immersed in the fixing bath, they should be pushed, face down, underneath the prints already there. The prints should be moved about from time to time to prevent them from sticking together and to bring them into contact with fresh solution.

(8) If two fixing baths are used, the print is lifted out of the first bath after about five minutes, and transferred to the second for the same length of time.

(9) Finally, the print is lifted out of the fixer and washed.

Rinse. The simplest rinse between development and fixing is plain water. This removes most of the developer from the prints before they pass to the fixing bath. It protects the fixer from excessive contamination and makes it last longer. The water should be changed frequently, as it, too, becomes contaminated with developer.

An acid rinse or stop-bath is better than plain water because it neutralizes the developer and arrests development in a second or two. And an acid rinse does not need to be changed as often as a plain water rinse.

The stop-bath may be a 5 per cent solution of potassium metabisulphite or sodium bisulphite, or a 2 per cent solution of citric or glacial acetic acid with about 1 per cent of sodium sulphite added.

Stronger stop-baths are not recommended, because they generate an appreciable amount of carbon dioxide gas in the emulsion when they react with the sodium carbonate of the developer. This gas may cause blisters on the print surface.

Development Time. The normal development time at 65–68° F. (18–20° C.) in one of the common paper developers is 45–60 seconds for chloride contact papers and 2–3 minutes for chlorobromide and bromide papers.

This is the minimum time the positive image needs to build up to its full depth of tone. Development usually takes more time in diluted solutions and less in strong solutions. With heavily restrained warm-tone developers, the development time with chlorobromide papers may be as much as half an hour.

Compensation in Development. When a print has been given too much or too little exposure, there is very little that can be done to put it right during development.

With a normal type of chloride or bromide paper, cutting short development to keep down the density of an over-exposed print usually flattens the contrasts and gives unpleasant greenish-black tones. The results of over-exposure can be improved by adding a restrainer or developer improver to the developer. But the print will never be as good as one that has been correctly exposed and fully developed.

Chlorobromide papers have more latitude, and development of over-exposed prints can be retarded by diluting the developer and adding a restrainer—e.g., potassium bromide. The development time is longer in a restrained and diluted developer, and can be stopped at any point. The shorter the development time under these conditions, the warmer the image tone. This method is therefore used in warm tone development and also to control density by three-dish development.

An under-exposed print is too light at the end of the normal development time, but the image is still gaining density (though very slowly); so slight under-exposure can be compensated by prolonging the development. This procedure often causes yellow developer stains on the paper, particularly if the developer contains a lot of hydroquinone.

Temperature. The best development temperature for papers is between 65 and 75° F. (18–24° C.). At temperatures below 60° F. (15° C.) the action of the developer is inconveniently slow; at temperatures above 75° F. (24° C.) the developer tends to stain the paper.

Variations of temperature between developer and rinse and between rinse and fixer do not matter as much as they do with films or plates; the emulsion is more firmly bonded to the paper base, and is not as likely to reticulate during processing.

Warm Tone Development. Chlorobromide papers developed in special warm tone developers can be made to yield a whole range of image tones from warm black to almost red. The tones depend on the make of paper and on the developer used, as well as on the relative exposure, dilution of developer, concentration of restrainer (potassium bromide or ammonium bromide) and development time.

The warm tone image has a much finer grain than the ordinary black tone image, although the graininess is not visible to the naked eye. During development, prints normally pass through the whole sequence of tones from reddish-brown to black, as they build up grain size and density.

In an energetic developer the transition of tones is too rapid, and by the time the image is dense enough to be acceptable, the grain size has already reached the point at which it gives black tones. By using a suitably diluted and restrained developer on the right kind of paper, the grain size can be kept down to the level that corresponds to the desired warmth of tone while density is built up. Chlorobromide papers are particularly suitable for this purpose, since they form an image from two different kinds of silver halide grains.

To compensate for the low density of the image at the warm tone stages, the exposure must be increased. The warmer the desired image tone, the more the exposure must be increased beyond normal.

So a warm tone image is one that has been over-exposed and then developed only until it reaches the right tone and density. This means that the relative exposure and development times have to be standardized if they are to produce exactly the desired image tone.

Exposure and Development. The exposure and development conditions must be determined separately for every developer-paper combination. This is best done by making comparative test strips.

First, an ordinary test strip is made on the chlorobromide paper, using a standard negative under set conditions of printing and processing. This test strip is fully developed—three minutes at 65–70° F. or 18–21° C.—in the warm tone developer. The developer is used at its normal concentration, corresponding to the coldest black it is capable of producing, without added potassium bromide. This strip is then used as a standard of reference to show the normal exposure required to give the nearest possible approach to a pure black-and-white print.

About half a dozen test strips are then prepared on the same paper. The exposure steps on each test strip are: normal (as found by the original strip), 1½, 2, 3, 4, 6, 9 and 12 times normal exposure. All six test strips are exposed under the same conditions, but developed separately in different concentrations of developer.

DEVELOPMENT OF TEST STRIPS

Strip	*Developer Dilution Times Normal*	*Extra Potassium Bromide (grams per 1,000 c.cm. working solution)*	*Development Time at 68°F. (20°C.)*
I	4	1·0	4½ minutes
II	8	1·5	8 minutes
III	12	1·5	12 minutes
IV	20	1·2	16 minutes
V	25	1·2	20 minutes
VI	30	1·2	25 minutes

This test strip series should show all image tones from warm black to the warmest the paper will give. If necessary, additional strips can be prepared to bridge any apparent gaps in the range.

The series also shows the exposure required to produce a print of the right density for each image tone.

To make a print of any required tone, the normal exposure is first found in the usual way.

The test strip showing the desired image tone is then selected from the series. This strip says at once which developer dilution and time of development will give the desired tone. Then, by referring to the exposure scale on the strip, the exposure to give the required density at that image tone can be arrived at.

The range of tones obtained and the exposures and developer dilutions required vary for each paper-developer combination. A new set of tests is necessary whenever the developer or the paper is changed.

The contrast of the prints decreases as the warmth of the tone increases, and a harder contrast grade is advisable when producing very warm tones.

Large Prints. When prints are too large for a normal developing dish, they may be developed in one of the following ways: by see-sawing each print like a roll film; by carrying out all the processing in a single large dish; by applying the processing solutions with a swab or spray.

See-saw development is useful if the print is just too large to lie flat in the largest dish available. If the dish is a little longer than the print is wide, the print can be held by the two short edges and developed by see-sawing it through the various solutions. The dishes need be no more than narrow troughs, provided they are long enough, and are sufficiently deep to allow a good covering of developer.

Single-dish development calls for only one dish big enough to hold the print. This dish is used for all the processing steps.

When the print has been exposed, it is laid flat in the dish and the developer is poured over it. As soon as the print is fully developed, the solution is poured off into a jug, and the rinse or stop-bath is poured in. After rinsing thoroughly, the dish is drained and the fixer is poured in and left for the usual length of time. Finally, the print is removed for washing and the dish is rinsed ready for the next print. (There is no difficulty in washing large prints in the bath or kitchen sink.)

Swab and spray development are methods for really large prints, like those used for making photo murals. With prints of this size the developer and other solutions are swabbed or sprayed over the print, and the picture is fixed and washed by the same means.

Because the surface of the print is exposed to the air most of the time, easily oxidized developers—e.g., those containing a high concentration of hydroquinone and alkali—are not suitable, as they are apt to produce stains and aerial fog. The best developers are those like glycin which work cleanly.

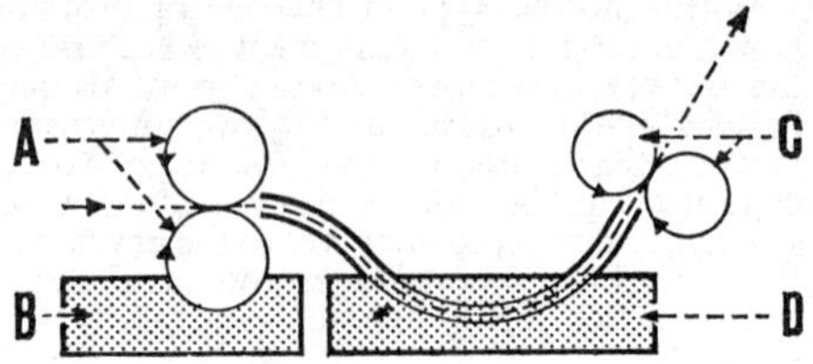

TWO-BATH DEVELOPING AND STABILIZING UNIT. The prints pass along the dotted line between developing rollers, which carry the activating solution to the print surface, and then through a stabilizer solution. They emerge merely damp from the final squeegee rollers. A, developing rollers. B, activator. C, squeegee rollers. D, stabilizer.

Print Processing Machines. As an alternative method of print development the use of motorized processing machines is gaining popularity. These pass the exposed sheet of paper in succession through a developer and a stabilizing bath, to deliver a ready processed print within about ten seconds. Motor- or hand-driven rollers carry the print through the solutions in a double trough and eliminate the need for processing dishes and even print washing (the paper leaves the processing machine in a semi-damp state).

The use of machines of this kind frequently calls for special papers incorporating some of the developing chemicals. The developer solution then consists only of an alkaline activating bath which enables development to be complete within the time (a few seconds) it takes the paper to pass through the developer solution.

Paper prints and enlargements produced in this way are reasonably stable for some months and even years.

For maximum stability for archive purposes refixing in a normal fixing bath is advisable.

L.A.M.

See also: *Cold weather; Colour printing technique; Developers; Dishes; Diffusion transfer; Giant enlargements; Glazing; Hot water bath development; Hot weather processing; Rapid processing; Redevelopment; Stabilization; Thorium nitrate forebath; Three-bath development; Toners; Tropical photography; Two-bath development.*

Books: *Complete Art of Printing and Enlarging*, by O. R. Croy (London); *Enlarging*, by C. I. Jacobson (London).

DEVELOPING TANK. Tank of plastic or stainless metal in which negative materials can usually be completely submerged in the developing solution for the period of development. Most developing tanks for amateur use are designed to permit all operations of processing (apart from the initial loading) to be carried out in normal light. Some even allow the loading to be done in daylight.

See also: *Tanks.*

DEVELOPMENT AFTER FIXING. During physical development an image is built up, by a process of silver intensification, on a very weak silver image first formed by normal chemical development.

Very minute amounts of silver are, however, already formed during the original exposure. These may by themselves form a foundation for a physically developed image. This makes it possible to fix a negative after exposure and then develop it in a physical developer by ordinary daylight.

The effective emulsion speed of the film or plate is greatly reduced and negatives need about 3–4 times the normal exposure. Some modern negative materials fail to react satisfactorily to this method.

A suitable developer is:

Stock solution A		
Potassium thiocyanate	1 ounce	25 grams
Silver nitrate	70 grains	4 grams
Sodium sulphite, anhydrous	½ ounce	12·5 grams
Sodium thiosulphate (hypo) crystals	88 grains	5 grams
Potassium bromide, 1 per cent solution	96 minims	5 c.cm.
Distilled water to make	8 ounces	200 c.cm.

Dissolve the silver nitrate and sodium sulphite separately in the smallest possible amount of distilled water. Slowly stir the sodium sulphite into the silver nitrate solution. When the precipitate which forms at first has redissolved, add the thiocyanate and the hypo, and finally the potassium bromide. Make up to the final volume with distilled water.

Stock solution B		
Metol	27 grains	1·5 grams
Sodium sulphite, anhydrous	½ ounce	12·5 grams
Distilled water to make	4 ounces	100 c.cm.

Just before use, mix 6 parts of A, 1 part of B, and 23 parts of water. Instead of B a commercial concentrated single solution developer of the para-aminophenol type may be used.

Development is very slow, and may take several hours. Since it is easily watched, this is not inconvenient.

After development is complete, the negative should be washed in running water for at least half an hour. L.A.M.

See also: *Physical development*.
Book: *Developing*, by C. I. Jacobson (London).

DEVELOPMENT CENTRE. Accumulation of silver atoms in a silver halide crystal formed by the action of light. Usually requires the presence of sensitivity specks in the crystal. The development centre or centres provide the point of attack for the developer to reduce the silver halide of the crystal to metallic silver and so form a visible image.

See also: *Latent image*.

DEVELOPMENT FOG. Fog formed by the development of unexposed grains of emulsion that occurs when the sensitized material is developed for too long a time or under unsuitable conditions.

See also: *Fogging; Fog level.*

DEVELOPMENT GROUP. Classification of a film according to its development time characteristics. Development groups are used in tables and calculators for working out the correct development time of any film with any developer. The group generally depends on the film speed, although there are exceptions when film-developer combinations behave in an unpredictable way. Thus such groupings are not infallible.

DEVELOPMENT HISTORY. Practicable photography began with the discovery of the phenomenon of development. Without the development process, all the energy needed to form a visible image must come from the light which falls on the sensitive surface during the exposure. In the absence of intense sources of illumination, far exceeding direct sunlight in power, prolonged exposures are necessary.

Although development is normally taken to cover only the conversion of latent images in silver halide into silver, N. Niépce used oil of lavender in 1822 to wash away the unexposed parts of his bitumen coatings.

The existence of a latent image and the possibility of its development at a later stage allows the image formation to be separated into two stages:

(1) The first stage, the exposure in the camera to form the latent image, requires very little energy and can be completed in a very short time.

(2) Almost all the work of forming the visible image is carried out in the development stage which can be postponed to a convenient time.

Development was discovered by Daguerre by a pure accident. Following some unsuccessful experiments during 1835, he placed an iodized silver plate in a cupboard. Some time later he found it bore a distinct image, though at the time when it was consigned to the cupboard it bore none. By a process of elimination he was able to attribute the image formation to the presence of mercury in the same cupboard. The mercury vapour had gradually condensed on the parts of the silver iodide which had been exposed in the camera.

In September 1840, Fox Talbot discovered how to develop a latent image on silver iodide paper using a solution of silver nitrate and gallic acid. He was enabled by this discovery to reduce his camera exposures from one hour to half a minute. This was, in fact, physical development.

In 1844, Robert Hunt introduced iron sulphate as a developer for silver halide paper materials and this developer played an important rôle in the early years of the collodion process of le Gray and Scott-Archer. Although pyrogallic acid, or pyrogallol, was shown by Regnault and Liebig to be an active developer and was suggested by Scott Archer himself for his process, the iron sulphate developer predominated for collodion wet plates.

Up to this time all development was of a physical character and was carried out in acid solution. Following the advent of the dry collodion plate, Mudd and Wardley found in 1861 that a plain ½ per cent solution of pyrogallol would serve as a good developer without the usual additions of silver nitrate and acid.

Alkaline Development. In 1862, Russell and Leahy discovered the bromided pyro-ammonia developer and thus initiated the alkaline "chemical" developer which was greatly to assist in the establishment of the silver bromide dry plates: first the collodion and albumen types and then the gelatin dry plate of Maddox.

The tanning development of gelatin emulsions by a pyro developer was demonstrated in 1881 by Warnerke.

A considerable advance in the technique of alkaline development came when Berkeley in 1882 improved the pyro developer by the addition of sodium sulphite which prevented the developer solution from quickly turning brown and minimized negative stain.

It is necessary to go back five years to note the discovery of the iron-oxalate developer by Carey-Lea in 1877, and two years to 1880 when Abney published the alkaline hydroquinone developer for gelatin-bromide negatives.

In 1880, too, Eder and Toth showed that pyrocatechin (catechol) would serve as an alkaline developer for dry plates. The next developers to be suggested were hydroxylamine, by Egli and Spiller in 1884, and phenylhydrazine, by Jacobsen in 1885.

The important developer discoveries which were to follow, up to the end of the century, sprang from the great expansion of the German dyestuffs industry. The outstanding figure in developer work at that time was Andresen who, noting the success of the polyhydroxybenzene compounds, turned his attention to analogous amino compounds. He discovered paraphenylenediamine, para-aminophenol and similar compounds as developers in 1888, Eikonogen in 1889, and in 1891 brought para-aminophenol on to the market as a very concentrated developer solution (Rodinal).

Although Andresen had claimed a very wide range of aminophenols within the scope of his patent, it was left to another, Bogisch, to search through the field again and to discover the specially valuable developing properties of the methylamino-phenols.

In 1891, the firm of Hauff, by whom Bogisch was employed, took out patents covering a series of aminophenols and marketed Metol. Later, by arrangement between Hauff and Agfa, to whom the master patent belonged, the same compound was marketed under the name Metol-Andresen.

More recent developing compounds of importance are listed in the table.

Developer Combinations. Of the developing agents available by the last few years of the nineteenth century, some were found to give desirable working qualities when used in pairs.

Chadwick, in 1891, used Eikonogen together with hydroquinone and this combination was sold in tabloid form by Burroughs Wellcome and Co. in 1895.

DATES IN DEVELOPMENT HISTORY

Date	*Developing Agent*	*Inventor*
1837	Mercury vapour	Daguerre
1840	Gallic acid	Fox Talbot
1844	Iron sulphate	Hunt
1851	Pyrogallol	Archer, Liebig, Regnault
1880	Hydroquinone	Abney
	Catechol	Eder and Toth
1884	Hydroxylamine	Egli and Spiller
1885	Phenylhydrazine	Jacobsen
1888	Para-phenylenediamine	Andresen
	Para-aminophenol	Andresen
1889	Eikonogen	Andresen
1891	Amidol	Andresen, Bogisch
	Metol	Andresen, Bogisch
	Glycin	Bogisch
	4-aminodiethylaniline	Hauff
1896	Ortho-aminophenol	Hauff
	Ortho-methylaminophenol	Hauff
1899	Chloro-hydroquinone (and bromo-)	Lüppo-Cramer
1904	Ortho-phenylenediamine	Lumière—Seyewetz
1914	4-amino-3 methyl diethylaniline	Fischer
1935	Ascorbic acid (vitamin C)	Mauer—Zapf
1940	Phenidone	Kendall

The metol-hydroquinone combination can be ascribed to no individual inventor but was in use in 1894 and was recommended by the Nepera Chemical Company in 1899 for use with its Velox gaslight paper. Gaslight paper helped to popularize the M.Q. developer.

In 1902, Lumière and Seyewetz patented the crystalline addition product of metol base and hydroquinone and sold it in 1903 as Metoquinone.

Chloranol consisted of metol and chlorohydroquinone and was patented by Lumière and Jugla in 1914.

Paraphenylenediamine together with hydroquinone formed the developing agent Hydramine of Hauff, 1896. Ortol of Hauff, 1896, was composed of hydroquinone with ortho-methylaminophenol, and another compound of the same date consisted of catechol and paraphenylenediamine. This last compound was marketed in 1938 as a fine grain developer, Meritol.

In 1935 Agfa introduced the Atomal developer which utilized hydroxyethyl-o-aminophenol and lately this compound together with glycin has formed the basis of Promicrol patented by John and Field.

The latest combined developer is Phenidone-hydroquinone which is based on l-phenyl-3-pyrazolidone discovered as a developer by Kendall in 1940 but not produced commercially until 1951.

Chromogenic Development. Several developing agents such as pyrogallol will produce a coloured dye image together with the silver image and some developers will react with their own oxidation products to give dye images, but none of these has proved of value in colour photography.

In 1913 Fischer found that suitable coupling agents could be added to paraphenylenediamine or para-aminophenol developer solutions to form dyes of the indophenol, indamine, indoaniline, azomethine, and indothiophenol classes. This type of coupling development forms the basis of the modern colour processes, though many systems incorporated the coupler in the emulsion. G.I.P.L.

See also: *Developers (obsolete); Discovery of photography.*

DEVELOPMENT PAPERS. Contact and enlarging papers coated with an emulsion which forms a latent image on exposure. The latent image is invisible until it is turned into metallic silver by being developed with suitable chemicals. Modern chloride, bromide, and chlorobromide papers are of this type.

See also: *Papers.*

DEVELOPMENT THEORY. Many of the chemical and physical changes that take place in the silver emulsion during processing are still not fully understood. This description gives in the simplest terms possible what appears to happen during development.

Chemical Development. In chemical development the essential function of a developing solution is to convert to metallic silver those silver halide emulsion grains that were partly decomposed by a previous exposure to light. This conversion is achieved by a process of chemical reduction that involves the transfer of electrons from solution into the silver halide grains, e.g.:

$$AgBr + \text{electron} = Ag + Br^-$$

or: silver bromide plus an electron gives metallic silver plus bromide ion.

The electrons required for this reaction are most conveniently obtained from aromatic organic reducing agents of the polyphenol (e.g., hydroquinone), aminophenol (e.g., metol), or polyamine (e.g., paraphenylenediamine) types.

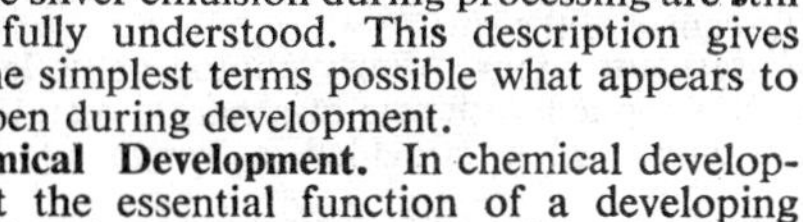

hydroquinone metol (base) paraphenylenediamine (base)

Particularly in the case of the phenolic reducing agents, the liberation of electrons is greatly facilitated by an increase in the alkalinity of the solution which induces the dissociation of the acidic hydroxyl groups. For this reason alkalis such as sodium carbonate, sodium borate, or sodium hydroxide are included in the developer solution. In some cases the emulsion is soaked in developing agent solution in one bath and is then transferred to an alkaline bath.

The addition of sulphite to the solution is necessary to inhibit the reaction of the reducing agent with atmospheric oxygen and to remove the developer oxidation products which would accumulate, retard development and stain.

In tanning developers, where it is the developer oxidation products that harden the gelatin, the sulphite concentration must be kept very low. Since sodium sulphite is itself a mild alkali it is not necessary to add a further alkali if a slow-acting metol developer is required; and amidol-sulphite solutions act quite rapidly at this low level of alkalinity.

The reducing action of the developing agent solution is exerted alike on both the exposed and unexposed grains and, in a sufficiently long time, all would become reduced to silver. The function of the latent image is catalytic; that is, those silver halide grains that carry a latent image are reduced much more rapidly than those that do not. The reduction of non-image grains causes a uniform veil, or fog, to be formed. Some fog can be tolerated in a negative but not in a print or diapositive.

The addition of small quantities of soluble bromide to the developer in many cases retards the reduction of the non-image grains more than that of the grains carrying latent image. The action of the soluble bromide is presumably connected with its ability to diminish the availability of silver ions at the surface of the silver halide grains. In some applications soluble bromide is not so efficacious as certain organic agents such as benzotriazole.

Physical Development. In physical development the enlargement of the latent image specks is achieved by the deposition upon them of silver that is not necessarily derived from the silver halide in the emulsion. A silver salt is added to the developer solution where it is slowly reduced to form a supersaturated silver solution. The presence of the latent image nuclei causes the silver to be deposited from solution, and the latent image specks become virtually silver plated. In time the physical developer will plate silver at random on to the walls of the dish and over the surface of the emulsion.

Since the rôle of the silver halide originally in the emulsion is to provide the latent image material during the exposure, it is not necessary afterwards. For this reason physical development can be effected after the exposed emulsion has been fixed out in a non-acid fixer. Development by the polyamine developers such as paraphenylenediamine is mainly physical because of their solvent action on silver halides.

The degree of apparent uniformity—the graininess of the developed image—depends primarily upon the characteristics of the emulsion used. However, for a given emulsion, the graininess of the image can be affected by the development conditions. In general, slow development to a low gamma yields a less degree of graininess, and the inclusion in the solution of solvents, such as thiocyanate or sulphite in high concentration, is beneficial.

The semi-physical development effected by paraphenylenediamine gives the finest-grained images, but with an effective loss of 70 to 80 per cent of the sensitivity of the emulsion. The metol fine-grain developers strike a compromise between loss of speed and fineness of grain. G.I.P.L.

See also: *Latent image.*
Books: *Processing Chemistry*, by L. F. A. Mason; *The Chemistry of Photographic Materials*, by G. F. Duffin; *Developing*, by C. I. Jacobson (all London); *Photography Theory and Practice* (English edition), by L. P. Clerc (London); *The Theory of the Photographic Process*, by C. E. K. Mees (New York).

DEVILLE, E. Dates unknown. Surveyor-general of Canada. Contributed to the theory of the half-tone screen (1895) and to photogrammetric restitution (1902).

DEXTRIN. Constituent of mountants.
Characteristics: Whitish powder.
Solubility: Slightly soluble in cold water, gives a milky solution if dissolved in boiling water. This solidifies on cooling.

DIACTINIC. Transparent to actinic light rays. An obsolete term applied to materials (e.g., filters) which allowed actinic light to pass.

DIAMINOPHENOL HYDROCHLORIDE. Developing agent; amidol. Sometimes also used as the free base, diaminophenol.

DIAPHRAGM

The diameter of a given bundle of light rays passing through a lens determines the brightness of the image formed. The maximum diameter of this bundle depends on the optical diameter of the lens. By placing plates with holes of various sizes, or a set of metal leaves to form an adjustable hole, in the light path of a lens it is possible to adjust the diameter of this bundle of rays, and hence the intensity of illumination of the image. The device with such a hole is known as a diaphragm. A small hole lets through less light than a large one, so the size of the hole can be used to alter the exposure.

Varying the size of this hole has two other effects. The first is that the hole in the diaphragm controls the angle of the cone of rays forming the image.

When the lens is focused sharply on the plate, the image of a point subject is produced by the tip of a cone of light rays coming from the back of the lens. If either the subject or the plate changes its position, the tip of the cone no longer coincides with the surface of the plate and the point image becomes a circular patch. The size of the patch depends on the angle of the cone and how far the tip is displaced behind or in front of the surface plate.

As the size of the hole in the diaphragm is reduced, the angle of the cone of rays becomes narrower and a change in the distance of the subject from the lens, or the lens from the plate produces a smaller change in the diameter of the image patch. So a small hole increases the depth of field covered sharply. And it also increases the range of backward and forward movement of the plate—i.e., the depth of focus—within which the lens will give a sharp image of a particular subject. In other words, focusing is made less critical when the hole in the diaphragm is small.

With cheap lenses the margins of the lens are not as completely corrected for aberrations as the central portion. In this second case, therefore, the diaphragm is so designed that even at its largest opening the diameter of the bundle of light rays passing through does not use the entire lens. Thus the image is formed only by the more highly corrected rays passing through the centre of the lens.

Waterhouse Stops. Early photographic lenses were supplied with sets of separate metal diaphragms which provided a range of hole

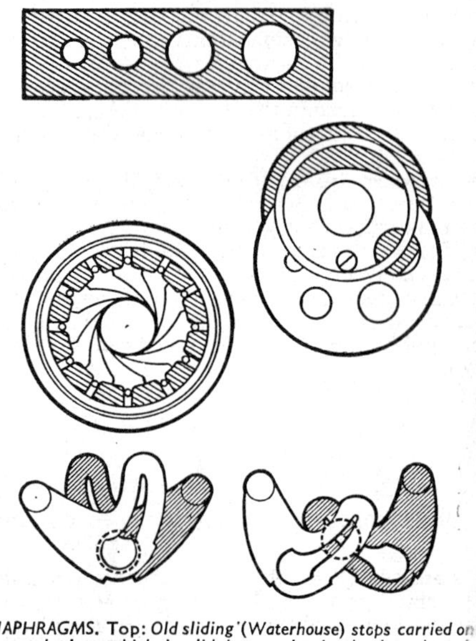

DIAPHRAGMS. Top: *Old sliding (Waterhouse) stops carried on a metal plate which is slid into a slot in the lens barrel.* Centre right: *Rotating stop plate.* Centre left: *Iris diaphragm consisting of a set of sliding blades.* Bottom: *Double-blade diaphragm positioned for a large opening* (left) *and a small opening* (right).

sizes. The diaphragms were called Waterhouse stops after John Waterhouse of Halifax who invented them in 1858. There was a slot in the lens barrel into which the diaphragms were inserted. Each diaphragm had a number referred to as a stop number.

Later Waterhouse stops were replaced by rotating stop plates. Since *c.* 1900 the iris diaphragm with variable apertures has been in general use, but the practice of referring to particular diaphragm openings as stops has persisted.

Process lenses still have, in addition to their iris diaphragm, a slot for Waterhouse stops, which often do not have circular openings because the shape of the aperture has an effect upon the dot formation produced by engraved half-tone screens.

Simple Diaphragms. The earliest type of diaphragm was simply a single hole in a plate that could be inserted between the lens components through a slit in the barrel of the lens mount—i.e., the Waterhouse stop. This idea was soon dropped because it meant that the photographer had to carry around a number of separate plates each drilled with a hole of a different size to give him a range of stops.

An improvement on this simple system is the stop plate carrying a number of holes arranged so that they can be brought into position by rotating or sliding the plate. This method is cheap and effective; it is still used on many inexpensive cameras.

Iris Diaphragm. Apart from the simpler cameras the type of adjustable stop now used almost universally for controlling the lens aperture is the iris diaphragm. The hole in this type is formed by a set of thin metal blades, mounted on a ring around the lens. Movement of a control lever on the outside of the lens mount closes the blades inwards or opens them out to make the hole smaller or larger. The *f*-number of the aperture is indicated on a scale alongside the control.

The greater the number of blades in an iris diaphragm, the more nearly the hole approaches a perfectly circular shape. At the same time, a truly circular hole is by no means essential for everyday photography.

In practice, the number of leaves in the diaphragm of the average camera lens covering the popular $2\frac{1}{4} \times 2\frac{1}{4}$ ins. format is ten.

This type of diaphragm is sometimes responsible for a peculiar form of highlight distortion. Bright points of light—e.g., stars in an astronomical telescope—appear as star-shaped patches. The number of points is related to the number of leaves in the diaphragm.

The iris diaphragm is generally an integral part of the lens/shutter assembly. Diaphragms are also used for controlling the amount of light passed by other optical systems—e.g., on the substage condenser of a microscope.

Double Blade Diaphragms. Another method of controlling the size of the lens opening is by the use of two movable diaphragm plates, each carrying a curved slot of decreasing width. The slots in the two plates overlap in the lens axis. Movement of the two plates brings either the broad end or the narrow end, or any intermediate position, of the slots to overlap in the centre of the lens, thus forming openings of different sizes. The shape of the opening usually also varies from approximately round at the largest aperture to more or less square at the smallest.

This type of diaphragm is used on inexpensive still and especially cine cameras, and also on cameras with automatic exposure control. There the movement of the diaphragm blades is controlled by coils energized by the current from an exposure meter cell.

Position of Diaphragm. In a lens which has only one element, the diaphragm may be placed in front or behind. The position affects the distortion of the image: if in front, it tends to produce barrel distortion and behind, pincushion distortion. It is generally placed in front of simple lenses.

In lens with several components, the diaphragm is almost always placed between two of the components—e.g., a symmetrical lens has the diaphragm inserted midway between each of the identical elements.

Effective Stop Diameter. The important thing about a stop is not its real diameter, but the diameter of the beam of light that it allows to pass through the lens to make the exposure. If the diaphragm is in front of the lens, then the two figures are the same—i.e., the diameter of the incident beam of light is equal to the diameter of the aperture in the diaphragm.

Where the diaphragm is placed between the lens components, the light incident on the front of the lens converges before it reaches the aperture. So in this case the diameter of the incident beam is greater than the diameter of the aperture through which it has to pass.

The diameter of the incident beam that just fills the aperture is known as the effective diameter of the diaphragm aperture, and the area of the beam is the effective aperture. In modern anastigmat lenses it may be 30 per cent greater than the actual diaphragm aperture.

The effective diameter of the diaphragm cannot be measured directly. It is measured in this way: the lens is focused on infinity and a sheet of opaque paper is substituted for the focusing screen; a pinhole is made in the paper opposite the centre of the lens; a sheet of sensitized paper is held in contact with the front of the lens, emulsion inwards; a bright light is held against the pinhole long enough to produce a black disc on the paper after development. The diameter of this exposed disc is the effective diameter of the stop.

Effects of Stopping Down. Apart from controlling the amount of light passing through the lens and governing the depth of field, changing the lens stop has several other effects.

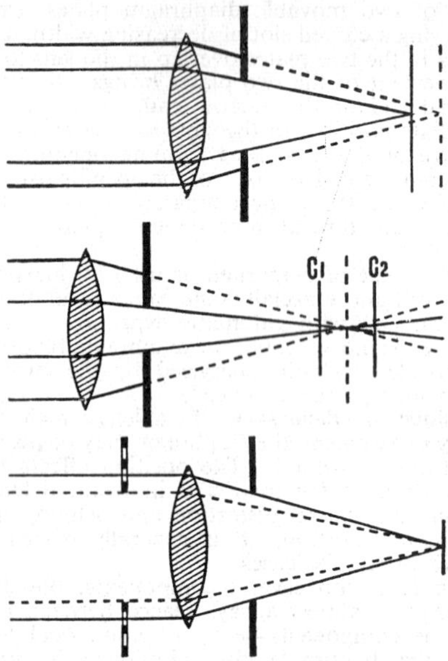

EFFECT OF LENS STOP. Top: *With simple lenses stopping down may change plane of sharpest focus by cutting out the uncorrected marginal rays.* Centre: *At smaller apertures the angle of the cone of rays is reduced, increasing the available depth of field.* Bottom: *The effective diameter of the stop depends on its position. In front of the lens it is identical with its actual diameter; behind, its effective diameter is larger.*

With both cheap lenses, and also many of better quality, stopping down may improve definition. This is because the correction of the lens for central rays is usually higher than for marginal rays. Stopping down there cuts out some of the marginal rays, leading to better image quality.

With modern high-speed lenses the correction is a compromise taking in both central and marginal rays. A certain amount of stopping down will still improve definition, but beyond the point of optimum definition no improvement is gained—in fact the image quality may deteriorate because the balance of correction between central and marginal rays may be upset.

Secondly, this change in balance between central and marginal corrections may also shift the plane of sharpest focus. The lens then in effect behaves as if it changed focal length during stopping down.

Thirdly, at very small openings the image definition deteriorates owing to diffraction. Light waves are slightly bent very near the edge of the diaphragm opening, leading to a certain amount of light scatter and unsharpness. At a small lens opening these deviated rays form a significant proportion of the light rays passing through the diaphragm and hence impair image sharpness.

Finally, a small diaphragm opening may improve the evenness of illumination in the image plane. This is likely to be the case where the lens mount itself tends to cut off some light (by vignetting). At a smaller diaphragm opening the reduced cone of light is no longer obstructed by parts of the lens mount and the effect of vignetting disappears. (Some falling off in illumination from the centre to the edges of the image always occurs. This is partly due to the longer path of oblique light rays from the lens to the image plane, and partly to the fact that the diaphragm opening as seen from any point near the edge of the image appears distorted—e.g., from round to oval.)

Relative Aperture. The effective diameter of the stop decides the amount of light that can pass through the lens to form an image. But the same amount of light can be spread over either a small brilliant image (with a lens of short focal length) or a large weak image (with a lens of long focal length).

So the important thing in deciding the brilliance of the image is not simply the size of the effective aperture. It is its size in relation to the focal length of the lens that matters. For this reason, the size of the aperture is always given as a fraction of the focal length—i.e., F/D where F is the focal length of the lens and D the effective diameter of the aperture. This fraction is called the relative aperture of the lens. All lenses with the same relative aperture in theory form images of equal brilliance of the same subject.

For example, a lens of focal length 12 ins. and an effective aperture of 1½ in. diameter would have a relative aperture of 8—i.e., 12/1½. Similarly a lens of focal length 6 ins. and an effective aperture of ¾ in. would also have a relative aperture of 8—i.e., 6/¾. So both lenses would give an image of equal brilliance of the same subject, and call for the same exposure.

The convenience of this method of stating the size of the stop is obvious. It means that the photographer has no need to know the actual diameter of the stop when he is adjusting his exposure; whether his camera is a 35 mm. miniature or a 8 × 10 ins. field camera, the same relative aperture number will give him the same exposure.

***f*-numbers.** The number indicating the relative aperture given by a particular stop is called the *f*-number, and its value is expressed by the letter *f* followed by the figure—e.g., *f* 4·5 which indicates that the effective diameter of the stop goes into the focal length of the lens 4·5 times. All lenses with the same *f*-number require equal exposures, no matter what the focal length of the lens or the diameter of the aperture may be. As the aperture increases, the *f*-number gets smaller; a big *f*-number—e.g., *f* 64—indicates a small aperture and a small *f*-number—e.g., *f* 2·0—indicates a large aperture. Where the lens has an adjustable diaphragm, the *f*-numbers are calculated for a number of stop sizes covering the whole range of the aperture adjustment. These are engraved on a scale alongside the diaphragm control.

Aperture Scales. The intensity of the light forming the image at any particular aperture (and hence the exposure) varies directly with the area of the aperture—i.e., with the square of its diameter D since

$$\text{Area} = \frac{\pi}{4} \times (\text{Diameter})^2$$

This is to say that it varies with

$$\left(\frac{1}{f\text{-number}}\right)^2$$

because the diameter of the stop

$$D = \frac{1}{f\text{-number} \times F}$$

(where F = focal length of lens)

So if the exposure is known for one aperture, it is possible to calculate the exposure given by any other by comparing the inverse squares of the *f*-numbers—i.e., the exposures at *f*-numbers, *f* x and *f* y will vary in the proportion

$$\frac{1}{(f\,x)^2} : \frac{1}{(f\,y)^2}$$

In other words, the exposure given at *f* 11 is half the exposure given at *f* 7·7 since $1/(11)^2$ is half $1/(7{\cdot}7)^2$.

It would be possible to choose any scale of *f*-numbers and work out the effect of changing from one to another by comparing the squares in this way. In practice the thing is made much easier by choosing the steps on the aperture scale so that a step up or down doubles or halves the exposure. So a scale starting with a maximum aperture of *f* x, would run: *f* x; $\sqrt{2}\,f$ x; 2 *f* x; $2\sqrt{2}\,f$ x, etc.—giving relative exposures of $1 : \frac{1}{2} : \frac{1}{4} : \frac{1}{8}$, etc.

The actual values on such a scale vary according to the starting point chosen, so it is possible to have any number of scales all based on progressive doubling of the exposure.

U.S. System. At first, practically every lens maker marked his diaphragms according to his own scale, but in 1881 the Royal Photographic Society proposed a Uniform System (hence U.S.) based on what was then the largest lens aperture; *f* 4. So the scale ran: *f* 4·0, *f* 5·6, *f* 8, *f* 11·3, etc. The Society recommended that each stop should be known by its relative exposure value, giving a scale of U.S. numbers: 1 (= *f* 4), 2 (= *f* 5·6), 3 (= *f* 8), 4 (= *f* 11·3), etc.

This system of marking was not much used in Britain or on the Continent, but it was used in America, particularly on the earlier Kodak cameras. It is being used again now for enlarger lenses where relative exposures are more important than actual aperture values.

Current Scales. Today all diaphragms are marked in *f*-numbers, but there are still three scales in existence since different lens manufacturers have based their scales on different maximum apertures.

In each case however the steps are chosen so that a change from one number to the next higher halves the image brightness and calls for twice the exposure.

The three current scales of *f*-numbers are shown in the table below. In some cases, where different manufacturers give slightly different figures, the less common is given in brackets below the usual figure.

Opposite each figure is shown the light-passing power of the aperture in relation to a standard of 10,000 units passed by an aperture of *f* 1.

British lens manufacturers mostly use the international series, 2, 2·8, 4, 5·6, 8, 11, 16, etc. which was adopted as the official British Standard in 1949, and which is now used throughout the world. European lens manufacturers at one time used to prefer the scale series, 2·3, 3·2, 4·5, 6·3, 9, 12·5, 18, etc.

Stops in this now obsolete Continental series pass about one-third less light than the nearest corresponding numbers on the British scale, but for all practical purposes the difference can be neglected. Often, in fact, a lens which has a full aperture that does not correspond to the

APERTURE SCALES COMPARED

f-number			*Relative light passing power*	*Decrease in light passing power (nearest 1/3 stops)*
1			10,000	0
	1·1		7,937	1
		1·2 (1·3)	6,300	2
1·4 (1·5)			5,000	3
	1·6		3,969	4
		1·8	3,150	5
2 (1·9)			2,500	6
	2·2		1,984	7
		2·5	1,575	8
2·8 (2·9)			1,250	9
	3·2		992	10
		3·5	788	11
4 (3·8)			625	12
	4·5		496	13
		5	417	14
5·6			312	15
	6·3		248	16
		7	209	17
8 (7·7)			156	18
	9		124	19
		10	104	20
11			79	21
	12·5		62	22
		14	52	23
16			40	24
	18		31	25
		20	26	26
22			20	27
	25		16	28
		28	13	29
32			9·9	30
	36		7·8	31
		40	6·5	32
45			5·0	33
	48 (50)		3·9	34

maker's normal scale may have one or more of the larger apertures numbered in the other series—e.g., 4·5, 5·6, 8, 11, 16.

T-Stops. Some of the light that would otherwise pass through the lens to form the image is always lost by reflection at the surfaces of the lens components. Coating the lens reduces this loss, but does not get rid of it altogether.

One lens manufacturer calibrates the stops in terms of the actual measured or calculated light transmission. The *f*-numbers thus obtained are called T (transmission)-stops. T-stops are always slightly higher than the corresponding *f*-number.

The system has been adopted principally for professional cine cameras. It is doubtful whether it has any real advantages for normal still photography. This is because most of the light supposedly lost by reflection does actually appear as general fog on the negative, so that it must in any case be taken into account as contributing to the exposure. On the other hand, the relative aperture may be incorrect from the light-gathering point of view for other reasons: the actual focal length may differ from the nominal value and the effective aperture may be different from the theoretical value. T-stop marking takes care of all these points. There is a further point that the theoretical *f*-number must be used in calculations of depth of field and hyperfocal distance which are based on the geometry of the lens and not on its actual light transmission. In practice, however, the differences in depth of field values derived from T-stops and from *f*-numbers are negligible.

Coupled Diaphragms. On some shutter and diaphragm assemblies, the diaphragm control is coupled to the shutter speed control in such a way that a change in the setting of one automatically changes the other to keep the resulting exposure the same. The coupling itself is adjustable so that the relationship between stop and shutter speed can be set to give any total value of exposure (known as the exposure value).

The idea is taken further in cameras coupled with a built-in exposure meter. The meter is pointed at the subject and the indicator needle is made to coincide with a reference mark. The control which does this automatically sets the diaphragm-shutter coupling ring to give the correct range of stop-speed values required for the prevailing light conditions. At that point it is still open to the photographer to select a particular shutter speed or stop (but not both) to serve his purpose.

In the so-called programmed shutter the diaphragm adjustment is linked with the shutter speed control, in such a way that adjustment yields a series of exposure value settings by progressively closing down the lens aperture and speeding up the shutter. This system does not provide alternative aperture-speed combinations for the same given exposure or exposure value. Some programmed shutters achieve this result by dispensing with a separate lens diaphragm altogether, and controlling the degree as well as the time of the opening of the shutter blades. Programmed shutters may or may not be coupled with an exposure meter system.

Automatic Exposure Control. On modern automatic or electric eye cameras the diaphragm adjustment is directly or indirectly controlled by an exposure meter—without even the need for making a meter needle coincide with any setting mark. In its simplest form —as used in a few cheap still and many cine cameras—the current from the exposure meter cell is used to move the blades of a double blade diaphragm to yield different openings.

More advanced electric eye systems, mainly on still cameras, employ a mechanical cam system which scans the position of the exposure meter needle. The cam in turn controls a spring drive that automatically closes down (or opens up) the lens diaphragm to an aperture corresponding to correct exposure. This takes place immediately before the shutter is released.

In more complex systems, especially in professional motion picture cameras, the photo cell may be used to control a motor driven diaphragm adjustment.

Automatic Diaphragm Control. Apart from meter-controlled diaphragm systems, preselector diaphragms are also used in single-lens reflex cameras. There the subject is first focused on a screen at the full aperture to get a bright image, and the exposure is made at the smaller aperture. The automatic mechanism here ensures that the diaphragm closes down to a pre-selected value (which may or may not be set by an exposure meter) just before the exposure. There are systems of varying complexity and automation for this purpose.

The simplest form is the straightforward manual pre-set diaphragm. The controls consist of the normal diaphragm setting ring and a second ring incorporating a mechanical stop and carrying the aperture scale (or setting index). During focusing the diaphragm ring is turned to its position for full aperture, while the selected working aperture is set on the second ring. Immediately before the exposure, turning the diaphragm ring as far as it will go sets the lens to the pre-selected stop. After the exposure the diaphragm ring is turned back again to the full lens opening.

Semi-automatic pre-selector systems are similar, again incorporating a second ring that controls a mechanical limit stop. The diaphragm ring is however in this case spring-driven and held tensioned at full aperture. On pressing the shutter release, a suitable linkage releases the spring, thus closing down the diaphragm to the pre-selected limit, an instant before the shutter opens. After every exposure the pre-selector mechanism needs retensioning by hand, which at the same time brings the diaphragm back to its full opening.

Fully automatic pre-selector diaphragms require no separate retensioning. The tensioning mechanism is often linked with the film transport (and shutter tensioning) so that working the transport movement brings all parts of the camera back to readiness for the next exposure. On reflex cameras where the mirror springs back after the exposure, the diaphragm may at the same time open again to full aperture as well.

Most of these forms of automatic diaphragm control are also available on reflex cameras with interchangeable lenses. This involves linking the automatic diaphragm release with the shutter release on the camera. Modern reflex cameras achieve this with an internal lever inside the lens mounting ring on the camera, arranged to engage a suitable control at the rear of the lens itself. Alternatively, the lens may carry its own release button to actuate the semi-automatic or automatic diaphragm. When the lens is mounted on the camera this button also bears on the camera release button to make the exposure an instant after the lens diaphragm has jumped to its pre-selected value. F.P. & L.A.M.

See also: *Aperture; Covering power; Exposure automation; Exposure value; Lens history; Shutter.*

DIAPOSITIVE. Positive picture on a transparent support intended for viewing by transmitted light—e.g., lantern slides and transparencies.

DIASCOPE. Optical projector for showing enlarged screen images of transparencies or diapositives.

DIATHERMIC MIRROR. Mirror with a coating that reflects visible light but transmits infra-red and heat rays. Used in light fittings and projector optical systems requiring a high concentration of light, where the heat also produced by the light source needs eliminating as far as possible.

See also *Heat filter; Mirror.*

DIAZO PROCESSES. Reproduction process in which a diazonium salt is destroyed by exposure to ultra-violet, violet or blue light. Materials using this system are also known as dyeline materials. An orginal document having a light-absorbing image (printing ink, pencil, Indian ink, aniline dye) is placed in contact with the dyeline material and the sandwich is exposed with the original facing a source of suitable light. The radiation is absorbed by the image but is transmitted by the non-image areas on to the diazonium salt in the surface of the underlying dyeline material. Exposure is continued until the diazonium salt is just completely bleached. When the sandwich is separated the dyeline material shows a positive copy of the original in unbleached diazonium salt. This is, of course, still light-sensitive and is pale yellow in colour. The diazo image is transformed into a coloured, visible, stable azo dye image by treatment with a suitable phenolic coupler. This step is usually termed development.

Because the dyeline process lacks the amplification stage of latent-image catalyzed reduction of silver halide that characterizes silver-based photography, the exposure times are relatively long and dyeline products are unsuitable for camera exposures. On the other hand, the low sensitivity allows handling in normal room lighting. Exposure to strong daylight, and especially sunlight, should be avoided. Special dyeline materials are made that are sufficiently sensitive to be exposed to a projected image, so long as an intense light source is used and the optics have a high ultra-violet transmission.

Applications. Diazo and dyeline products find their widest use in the direct copying of documents on a translucent base, such as engineering drawings on tracing paper, linen or matt-surfaced plastic film. In this application it is usual to adjust the exposure so that a trace of diazonium salt is left unbleached in the non-image areas. Development then gives a print having a line image on a weakly tinted background. This procedure ensures that no fine detail in the original drawing is lost by over-exposure.

Another field of increasing exploitation of dyeline copying is in commercial offices, both for convenience copying of correspondence and for systems of accounting, book-keeping, etc., whereby diazo reproduction is taking the place of expensive manual retyping. For this reason many business firms are ensuring that all stationery used has good translucency. Modern high-speed dyeline printing machines use 3–5 kilowatt high-pressure mercury vapour lamps. Alternatively a bank of actinic fluorescent tubes is arranged along the axis of a revolving glass cylinder against which the original and dyeline paper are held in close contact by moving fabric bands. The length of the exposure is adjusted by varying the rate of rotation of the cylinder. Many machines also incorporate a development section so that the paper can be exposed and processed to the final print at speeds of 50 feet per minute or more.

Materials. Two main classes of dyeline products are currently produced. The first is known as the two-component, dry or ammonia type, and the second as the one-component, semi-dry type.

The ammonia papers were the first to be produced commercially and were introduced by Kalle in 1923, being based on the ideas of Kögel. The essential components are a diazonium salt and coupler, both of low coupling power, that will couple only at high pH: an organic acid is added to the coating formulation to prevent premature coupling. Typical diazonium salts, often used as the zinc chloride double salts, are those derived from p-aminodiphenylamine, p-aminodimethylaniline, 2-amino-1-naphthol-5-sulphonic acid and p-aminophenylmorpholine, and couplers are 2,3-dihydroxynaphthalene-6-sulphonic acid, resorcinol and 2-hydroxynaphthalene-3,6-disulphonic acid. Development is effected by treating with warm, moist ammonia fumes.

The one-component or semi-dry products were put on the market by van der Grinten in 1927. The paper is coated with a diazonium salt and development consists of treating the surface after exposure with a thin layer of a solution of a phenol. Modern processing machines use special grooved rollers that will apply uniformly as little as 9 grams per sq. metre developer to the front and 7 grams to the back of the paper, which can therefore be quickly dried by brief passage through a heated zone. The diazonium salts frequently used cover a wide range of coupling activity. At one extreme is 4-N-benzyl-N-ethylphenyl diazonium zinc-chloride, that couples well only under alkaline conditions.

In practice a product coated with this diazo is developed by an alkaline solution of a phenol, borates being very commonly used. Other diazonium salts will couple under neutral or slightly acid conditions (pH below 4·5) and products sensitized with such diazos are developed by solutions containing large quantities of the sodium salts of weak acids such as citric, benzoic or formic. Diazos of this latter class are 2,5-diethoxy-4-phenylacetamidophenyl diazonium zinc-chloride, and 2,5-dimethoxy-4-p-tolylthiophenyldiazonium chloride. Phloroglucinol is almost universally used as the coupler, since it reacts readily with diazonium salts to give dark, almost black, dye mixtures. The acid developers are more resistant to discoloration due to aerial oxidation.

A special class of dyeline products is designed to produce prints that can themselves be used as masters for subsequent dyeline reproduction. These "intermediate" prints take the place of a valuable original drawing. The base for the dyeline coating is either translucent paper, or for special purposes, transparent linen or plastic film. Polyester film is gaining favour for its toughness and dimensional stability. The diazo and coupler are chosen to produce an image that will be acceptable visually but have a high absorption for U.V. and short-wavelength visible light. Resorcinol and its derivatives are frequently used for this purpose, as brown images are produced.

The diazo reproduction process is by nature one of fairly high contrast, so that reproduction of half-tone images is poor. Good reproduction on the usual type of paper, preferably baryta coated, is obtained if a screened positive is used as the original. Special products have, however,

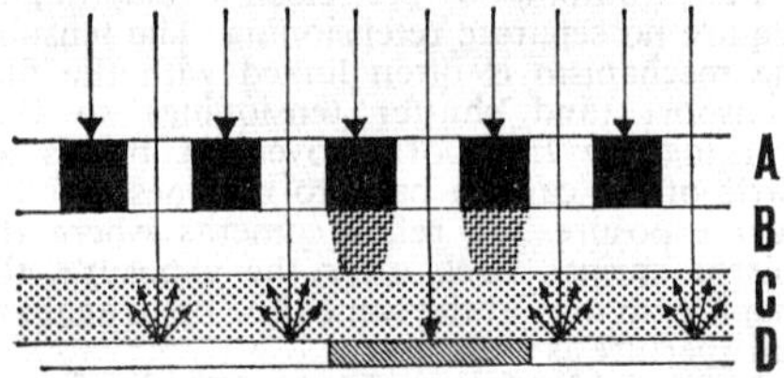

SCREENED DIAZO PRINTING. Diazo screen material carries a strippable screen layer in contact with the diazo layer. The plastic film support acts as a spacer between the original and the diazo layer. Light going through the screen is diffused and reflected back into the diazo layer from light areas of the original; dark areas reflect less or no light, so that in those parts the screen pattern is obtained after processing as a positive half-tone image. A, strippable screen layer. B, diazo layer. C, plastic film support. D, original (image shown shaded).

been developed for half-tone reproduction, but this is an unimportant use.

Preservation of Materials. Diazonium salts have a limited thermal stability, particularly in the presence of moisture, and the manufacturer ensures that the coated paper is produced as dry as possible, and that it is packed in moisture-proof wrapping. Nevertheless, some decomposition slowly occurs, and the practical life is 6–12 months. Storage in a refrigerator considerably extends the life. Opened packets deteriorate much faster. The most obvious change on keeping is a discoloration of the background and stale paper becomes unusable for this reason long before all the diazonium salt has become decomposed.

A recent development has been the introduction of dyeline papers of a truly dry nature. These are developed by heat after light exposure. Fundamentally, these thermal diazo papers are two-component ones in which the alkali for development is obtained by the action of heat, e.g. the decomposition of a chemical, such as the formation of ammonia from urea. The main difficulty is to produce a material that will give an intense image and have a reasonable shelf life.

Screens. Dyeline products normally give only copies of low contrast by reflex exposure, since most of the light is absorbed by the diazo (with consequent bleaching) before it can be selectively reflected by the non-image areas in the original. If, however, a fine carbon black screen is placed between the diazo and the light, the diazo beneath the dark elements of the screen is protected and is bleached only by light reflected and scattered in the non-image areas. No light is reflected by the image and the diazo remains unbleached in the image areas. Reflex copies of high contrast can therefore be obtained. The screen can be an integral part of the dyeline product, and is stripped away after exposure before development.

Scattering Images. Not typically a dyeline process, but one nevertheless utilizing a diazonium salt, is that of scattering photography. During bleaching by light a diazonium salt forms nitrogen gas as one of the decomposition products. The material for scattering images is a thermoplastic film containing a diazonium salt. On exposure, the nitrogen from the diazo forms micro-bubbles at high pressure. When, shortly after exposure, the film is heated the plastic softens and allows the nitrogen bubbles to expand. At the same time some crystallization of the polymer occurs and the result is a multicellular structure of 0·5–2·5 A. unit cell, which is highly light-scattering. Such an image will be equivalent to a density of at least 1·0 when inserted in the image plane of a projection system. The process gives a negative from a positive and vice-versa, and is used for making positives from silver microfilm negatives that can be projected on to highly-sensitive dyeline paper for positive prints. The image is fixed by exposure and heating to only a moderate temperature, so that the nitrogen diffuses out without distorting the polymer. D.J.F.

See also: *Document copying.*
Book: *Chemistry of Photographic Processes*, by G. F. Duffin (London).

3:5-DICHLORO-2-AMINO-PYRIDINE. Developer improver.

Formula and molecular weight; $C_5H_2N(Cl)_2NH_2$: 163.

Characteristic: White powder.

Solubility: Fairly soluble in water.

DICHROIC FOG. Kind of "bloom", sometimes seen on negatives, which looks purplish by transmitted light and greenish by reflected light. It is more obvious in the shadows than the highlights, and although there are many causes, the effect is always due to a deposit of colloidal silver.

The deposit can generally be wiped off the negative while wet, but once it dries on it may be stubborn. Brief treatment with Farmer's reducer will sometimes remove the fog, or it will generally yield to a bath in the following solution:

Thiourea	28 grains	1·4 grams
Citric acid	28 grains	1·4 grams
Water to make	5 ounces	125 c.cm.

Dichroic fog may be caused by excess of sulphite in a fine-grain developer; by hypo contamination of the developer; by traces of developer in the fixing bath or by insufficient fixing.

See also: *Faults; Fogging.*

DICKSON, WILLIAM KENNEDY LAURIE, 1860–1935. Scottish inventor, born in France. Was Edison's principal assistant in the development of the "optical phonograph" (1888), the Kinephonograph (1889), the Kinetoscope and Kinetograph (1891). Dickson made the first films for Edison. He left Edison in 1894 and joined Latham, with whom he developed the Panopticon in 1895. Left Latham and designed the Mutoscope.

DIETHYL-PARA-PHENYLENEDIAMINE HYDROCHLORIDE. Para-(diethylamino)-aniline hydrochloride. Developing agent for direct colour development with colour couplers.

Formula and molecular weight: $(C_2H_5)_2NC_6H_4NH_2HCl$; 200·5.

Characteristics: Whitish powder.

Solubility: Fairly soluble in water.

DIFFRACTION. Bending of light rays around opaque objects, specially apparent if the opaque object has a sharp edge.

It is customary to consider light as proceeding in straight lines, termed rectilinear propogation, and thus expect it to generate sharp shadows. However close examination of the shadow of an opaque edge illuminated by a plane wavefront shows that there is light present in the shadow which has seemingly been diverted around the edge. This spreading of light at the edge of obstacles is termed diffraction.

This phenomenon may be explained if the light propogation is considered as a wave motion. Each portion of the wavefront acts as a secondary source emitting a spherical wave and a new wavefront is formed as the common tangent to the spherical waves in the direction of propagation. When a wavefront meets an opaque edge the portion of the wavefront passing past the obstruction will spread behind due to the secondary spherical waves just inside the edge.

An example is an image-forming system with a finite aperture. The wavefront in the exit-pupil of a perfect lens imaging a point source will be a spherical wavefront with a sharply defined circular periphery. Due to this limitation of the wavefront the image of the point source is not an exact point but a diffraction disc of light, known as the Airy disc.

It is this disc of light (the actual image of a point source) that limits the resolving power of the optical system. If two object points are so closely spaced that the centres of their images are separated by the radius of the Airy disc then the two images will not be resolvable. For an

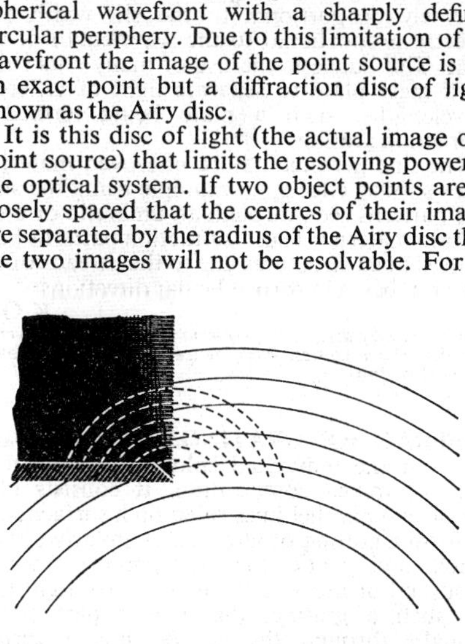

DIFFRACTION. When a wave front meets a knife edge, it generates a new wave front at the edge itself. This spreads round the obstacle, leading to some light being apparently bent round the edge.

optical system with a circular aperture limited only by diffraction the radius of the Airy disc is 0·61 times the mean wavelength of the image forming light divided by the numerical aperture. The resolving power, usually quoted in terms of lines per inch, is the reciprocal of the radius of the Airy disc.

For most photographic purposes it is agreed that the disc of confusion should be restricted to 1/2000 of the focal length of the taking lens if the picture is to be sharp in definition when viewed at the distance which gives correct perspective. For a lens of 2 ins. focal length this is a disc of confusion of 1/1000 in. diameter and the diameter of the Airy disc for a solely diffraction-limited lens will be of the same value when the stop number is 22. For higher stop numbers the Airy disc will increase in diameter proportionately and the image will no longer be satisfactory.

Depending on the state of correction of the taking lens it is possible that the above F/2000 condition is not satisfied unless the lens is stopped down by one or two stops from full aperture. For all stops between this satisfactory value and the stop when diffraction begins to predominate, the lens imagery will be satisfactory.

Where the standard of definition required is higher, the limiting aperture for acceptable freedom from diffraction becomes larger. For certain special purposes it may therefore be desirable not to stop down the lens beyond, say, *f* 11.

A slit aperture placed in an optical system will also give a diffraction pattern whose width is inversely proportional to the slit width and directly proportional to the wavelength of the light. This suggests the use of many slits ruled on a glass plate to separate light of different wavelengths, such a plate being termed a diffraction grating.

The fine periodic nature of woven materials gives similar effects, an example being the star-like patterns seen through an umbrella or a net curtain material. These diffraction patterns have cross shaped structures due to the threads of the material being in perpendicular directions.

K.G.B.

See also: *Airy disc; Lens performance; Resolving power.*
Books: *Wave Optics, Interference and Diffraction,* by B. Curry (London).

DIFFRACTION GRATING. Device for separating out the individual colours in a beam of light of mixed composition. It consists of a set of fine parallel lines ruled on a surface so as to form a grating of alternate, regularly spaced black and white (or transparent) lines, a thousand or more to the inch. When light falls on such a grating, the greater part passes straight through the spaces, but a certain proportion is bent by the phenomenon known as diffraction. The red rays are bent most and the violet least, which is the reverse of the action of a prism. But where the prism spreads out the violet and blue rays more than the red, the diffraction grating spreads the various frequencies evenly throughout the spectrum.

For most work—e.g., spectroscopy—a diffraction grating is better than a prism. At the same time, the spectrum given by a diffraction grating is less intense than one given by a prism because most of the light passes through the grating without being affected. Moreover, the prism forms one spectrum while the grating forms a series of spectra.

There are two sorts of diffraction grating; one consists of opaque lines ruled on glass (called a transmission grating) and the other consists of lines ruled on a metal surface (reflection grating). The spectrum of the first is seen by transmitted light, and the second by reflected light.

DIFFUSER. Any light-scattering medium placed in the path of a beam of light to soften its character. Undiffused light as it radiates from a light source is sometimes unsuitable for many types of photograph—e.g., portraiture, figure studies, still life—because it casts sharply defined shadows. While a hard light is excellent for emphasizing surface texture, roundness and subtleties of modelling call for a soft light which shifts the emphasis from the finer details to the broader conceptions of shape and form.

There are several ways of diffusing a light. The commonest is to furnish the lamp with a satin-finish or matt white reflector in place of a polished reflector, or to use no reflector at all. Reflections from matt surfaces are sufficiently diffused to avoid really sharp shadows.

Where even more diffusion is required or where the area of the reflector is small in relation to the source—e.g., in a spotlight—a diffusing screen is placed in front of the light. This screen may consist of a disc of fine white fabric or opal glass fastened over the opening of the lamp reflector, or a large screen of similar material supported on a stand between the lamp and the subject.

Flash-bulb reflectors are often fitted with a translucent plastic screen which serves the double purpose of diffuser and a protective shield if a flash bulb should explode on firing.

Diffusers may also be used over the camera or enlarging lens to soften the over-all definition of the picture. Diffusers of this type are usually clear glass inscribed with concentric circles, but very thin black chiffon or similar fabrics can be used quite effectively.

See also: *Lighting the subject; Soft focus; Texture screens.*

DIFFUSION. When light rays are reflected off or transmitted through a material in all directions, they are said to be diffused. White paint is a good diffuser because the light is reflected from a large number of microscopic flakes of

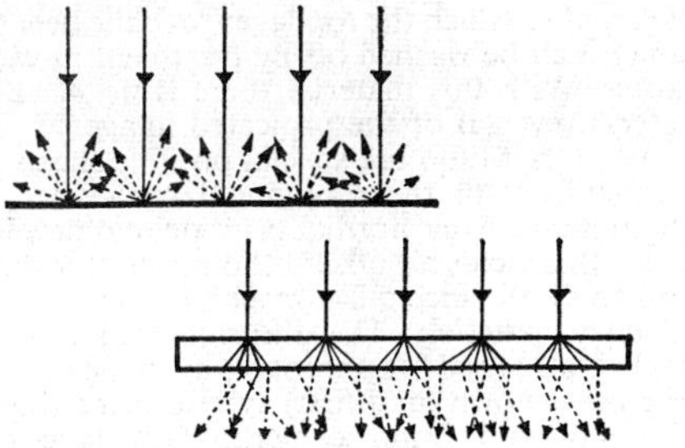

DIFFUSION. Top: *Light reflected from a rough surface is diffused by reflection in every possible direction.* Bottom: *Light passing through a translucent medium is diffused by scattering, yielding an evenly illuminated surface*

pigment, all lying in different directions, and therefore spreading rays at random. Milk is a good diffuser because it consists of large numbers of suspended globules of fat which behave in the same way as the flakes of pigment in the white paint.

Coloured materials will also diffuse light, but the fact that they appear coloured means that some of the light has been absorbed. So they are not as efficient in diffusing as white materials.

Soft focus lenses and diffusing discs may be used for either taking or enlarging photographs (or both) to diffuse and soften the image where extreme sharpness is undesirable. This effect is not really diffusion so much as refraction. What happens is that part of the light is used to produce a sharp normal image, and part is distorted by a refractive effect to produce one or more overlapping images. The combined effect is a softened but basically sharp image.

Opalescent diffusing screens are true diffusers and tend to spread light sources evenly and soften the sharpness of shadows.

DIFFUSION ATTACHMENT. Disc of lightly engraved or etched glass or transparent plastic. They are made to fit in front of the lens of a camera or enlarger to blur part of the image-forming rays and soften the definition of the picture. Chiffon and similar fabrics can also be used in the same way.

See also: *Soft focus.*

DIFFUSION TRANSFER

Photographic process in which a negative image is developed in contact with a suitable receiving layer, so that the unused silver salts diffuse into this layer, forming there a positive. The latter is then stripped off the negative which is usually discarded.

The process was invented almost simultaneously in 1939 by A. Rott of Gevaert and E. Weyde of Agfa. Intended initially as a method of rapid document copying, diffusion transfer systems have come to be applied also in instant-picture cameras.

Chemistry. The negative material for the diffusion transfer process consists of a normal sensitized silver halide layer containing also developer chemicals. The positive material is not light sensitive, contains development nuclei (which render unexposed silver halide or silver halide complexes developable) and a silver halide solvent such as sodium thiosulphate. The developer is usually an alkali solution, which may however also contain silver halide solvent. The primary role of this solution is to activate the developer in the negative layer.

During development the negative and the positive sheets are in close contact. The developer develops the negative image in the usual way. At the same time the unused silver halide in the negative layer is rendered soluble and diffuses into the positive layer. Under the influence of the development nuclei it is there reduced to metallic silver, forming a positive corresponding to the negative image. After development the two layers are peeled apart. Usually the negative is discarded, but can in certain cases be re-used for repeated transfers.

Photo-copying. During the transfer process the image is reversed left to right. To obtain a right-reading positive the negative must therefore be laterally reversed, which is the case with reflex printing and with contact printing through the back of a translucent original. The latter procedure usually loses some definition; hence reflex printing is the commonest method with diffusion transfer copying. Accordingly the contrast of the negative material must be very high, and this method is suitable only for copying line originals. Continuous tone originals need screening through a contact screen, which is at the best a make-shift procedure owing to the loss of definition and gradation.

The steps for making a diffusion-transfer copy are:—

(1) Place the negative paper face up on the top of a contact printing box and place the original on top of the negative.

(2) Expose for the appropriate time through the back of the negative paper.

(3) Feed the negative and a positive sheet into the developing solution. Mechanized developing machines enable the two sheets to be fed in, face to face, simultaneously, but separately. After passage through the solution—which soaks both layers—a pair of rollers squeegees the two papers into intimate contact.

(4) About half to one minute after the sandwich emerges from the processing machine, the negative and positive can be stripped apart.

Positive papers are also available coated on both sides, for producing double-sided copies.

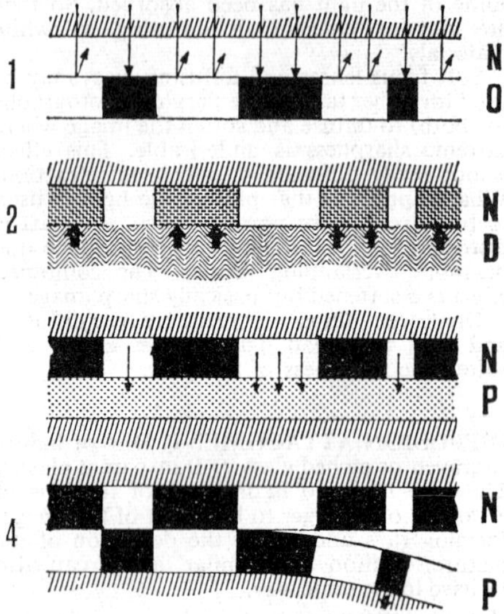

THE DIFFUSION TRANSFER PROCESS. 1. A negative paper N is placed in contact with the original O and exposed by the reflex method through the back of the negative paper. Light reflected from the white parts of the original produce a developable image in the negative layer. 2. The negative image developes in the developer D. 3. The developer-soaked negative is brought into intimate contact with a positive paper P. This is not sensitive to light, but contains special chemicals. These reduce unexposed silver salts diffusing into the positive from the negative. 4. The result is a positive image which can then be separated from the negative.

In this case a negative paper is exposed from each side of a double-sided original, and the two negative sheets fed through the processing machine together with the double-sided positive paper in between them.

Some types of diffusion transfer photocopying paper permit the production of several prints from one negative. In this case the negative paper is fed repeatedly through the developer, each time with a new positive sheet. Negative papers for this purpose have a higher silver halide content, leaving more halide available for diffusion, while the corresponding positive papers have fewer development centres. The positives obtained are usually grey rather than black, and about 8 to 10 positives can be produced from one negative.

High-speed negative papers for transfer diffusion copying can be printed by projection. In this case the original (for example a microfilm transparency or negative) must be placed in the enlarger reversed left-to-right to compensate for the lateral reversal taking place during transfer.

A special material for projection printing has the negative layer (sensitized) coated on top of the positive receiving layer. The negative layer in this case is a soft water-soluble gelatine layer. During processing the positive image is formed in the usual way in the lower non-sensitized layer, after which the top layer with the negative image can be washed off by treatment in warm water. With this material there is no need for lateral reversal of the projected image.

For offset litho use special positive sheets are available, with the receiving layer coated on aluminium. After normal printing and development this yields an offset litho master for direct use on small office offset machines.

Camera materials. The diffusion transfer principle is also used for photography in the camera, the best known (and first) system being that of the Polaroid Land camera. This takes the negative and the positive material in separate rolls or stacks of connected sheets. The negative is exposed in the image plane of the camera. The activating reagent for the developer (incorporated in the negative emulsion) is contained in small pods or capsules. The solution contains a jellifying agent to permit spreading between the negative and positive layers, without excessive contamination of the camera itself. After the exposure pulling a paper tab or leader draws the negative and positive between two steel rollers in the camera. This breaks the pod containing the activating chemicals and spreads them between the two sheets. After the appropriate processing time (from 10 seconds to 2 minutes, according to the type of material) the positive can be peeled off the negative.

Black-and-white print materials for the Polaroid Land camera use negative and positive papers; the effective emulsion speed may range from 200 up to 3,200 A.S.A. Special materials of this kind are available for oscillograph recording with speeds up to 10,000 A.S.A.

Alternative materials include a positive layer coated on film base for transparencies, and a negative coated on film base for subsequent recovery of the negative. After fixing and washing this film negative is then suitable for handling like any other normal negative, including for making enlargements.

Various other camera systems of diffusion transfer photography have been proposed.

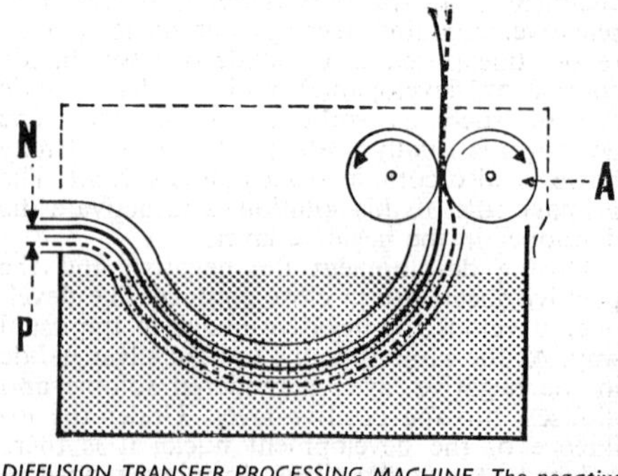

DIFFUSION TRANSFER PROCESSING MACHINE. The negative and positive papers pass through the developer (activator solution) and are pressed into contact on passing between the rollers. A, roller. N, exposed negative paper. P, non-sensitized positive paper.

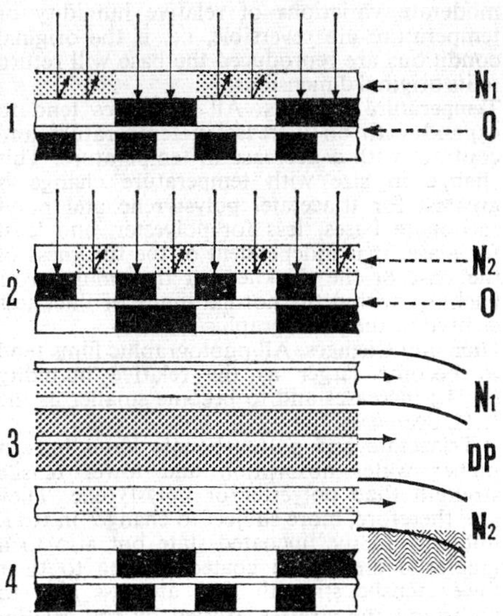

DOUBLE SIDED COPIES. 1. A negative paper N_1 is exposed by reflex printing to one side of the original O. 2. A second negative paper N_2 is exposed to the other side of the original. 3. The two negative papers are passed through the processing machine together with a double positive paper DP, carrying a positive layer on both sides. 4. The result is a double sided positive copy.

These include materials where the negative paper is spooled up on normal roll film spool and loaded into an ordinary roll film camera. When the whole negative strip is exposed, it is fed through an alkaline activating bath in contact with a non-sensitized positive strip. The resulting prints, after peeling off the negative, can then be cut up and handled like ordinary contact prints. The negative, if made on film, can be recovered for further use (enlarging, etc.) by a fixing treatment. This system is more versatile in that it is adaptable to conventional cameras and does not need a special camera design to allow for the processing; on the other hand the pictures are not available immediately.

Colour diffusion transfer. The Polacolor process, while producing positive colour prints by a kind of diffusion transfer, differs chemically from the black-and-white silver processes described so far.

The negative material here consists of three differentially sensitized layers (green-, red- and blue-sensitive), three layers containing special dye complexes and two spacing layers. Immediately below the top (blue-sensitive) emulsion layer is a layer containing yellow dye molecules linked to developer molecules. The green-sensitive layer has associated with it a magenta dye-developer layer, and the red-sensitive layer a cyan dye-developer layer.

The three sensitive layers analyse the subject colours in the usual manner of subtractive colour materials. During processing—initiated by spreading an activating jelly between the negative film and a positive paper—the dye developer molecules begin to diffuse into the neighbouring sensitized layers, and there develop the negative images. During development the dye molecule is fixed in the negative layer wherever a negative image is developed.

The unused dye developer molecules diffuse into the receiving layer of the positive material, forming a positive colour image there. The positive material also contains a layer of acid molecules which towards the end of processing neutralize the alkaline activator. When processing is complete, the positive print is peeled off the negative, and now carries a colour image consisting only of the dyes. The process depends greatly on the controlled diffusion of the dye developer molecules, achieved by the spacing layers and by careful balancing of the exposure and development times.

The process is also adaptable in principle to producing colour transparencies (by using a positive material coated on film base); the colour negative is not however recoverable.

L.A.M.

See also: *Document copying; Polaroid hand camera.*
Books: *Camera copying and reproduction*, by O. R. Croy; *Developing*, by C. I. Jacobson (London).

DIMENSIONAL STABILITY. Ability of a material to resist changes in size caused by changes in temperature and humidity, processing and ageing. It is normally expressed as a percentage change in the length or width of a material. The term dimensional stability is generally used in relation to emulsions coated on film base, as glass plates can for most practical purposes be regarded as being completely stable. Such very small changes as occur with these are mainly due to creep of the emulsion or substratum across the glass surface during processing rather than expansion or contraction of the glass itself.

With certain exceptions, photographic papers are subject to such large variations in size, as well as distortion, that they are not used where dimensional stability is important and, therefore, most photographic manufacturers do not publish figures for them. Some manufacturers supply special papers with waterproof base, or a laminated paper which has a thin metal foil sandwiched between two layers of paper. The latter type of paper has a high degree of stability.

Film has several advantages over glass; it is less liable to damage, easier to store and transport, and—in the graphic arts—permits the use of contact screens for halftone work. In addition, film is easier to handle, for example when making a composite negative or positive by patching up separate films or when making overlays. For these reasons, recent years have seen the introduction of more stable film bases. Although none of these is as stable as glass they are a considerable improvement over the commonly-used cellulose acetate. Any material, however, which is flexible is also capable of being stretched or compressed.

The factors affecting the dimensional stability of a photographic base are:—

(a) Its coefficient of thermal expansion, i.e., its ability to withstand changes in temperature.

(b) The degree to which it will absorb moisture and, therefore, its ability to withstand changes in relative humidity.

(c) Its tensile strength or ability to withstand the forces exerted by the gelatine coating which is itself subject to shrinkage and expansion.

(d) Its chemical inertness inasmuch as this affects the loss of plasticizer or other constituents during processing or ageing.

In practice, the dimensional change of a photographic material is the result of the sum of the temperature and humidity changes and the processing conditions. Changes in temperature invariably occur at a constant dew point and are therefore accompanied by changes in relative humidity. The changes may be additive or, more likely, may tend partially to cancel one another.

Changes in size due to ageing or processing are usually permanent, whereas changes due to

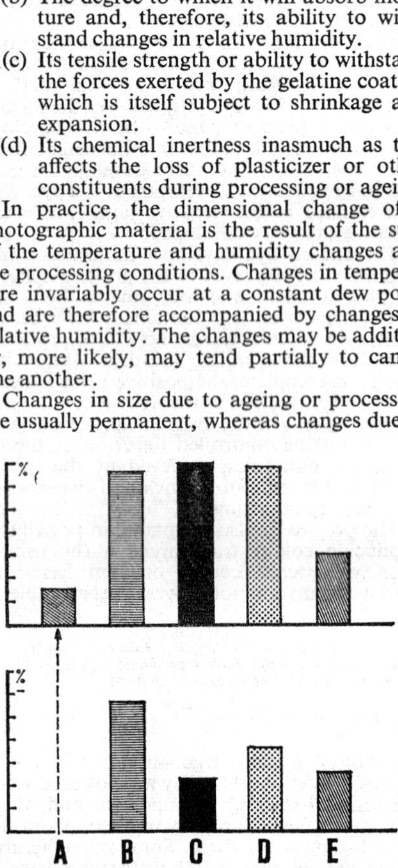

TEMPERATURE AND HUMIDITY CHANGES OF EMULSION SUPPORTS. Top: Average per cent change (relative) in size per degree C at constant relative humidity. Bottom: Average percentage change in size with changes in relative humidity for values between 20 per cent and 100 per cent, for the same emulsion and backing coated on different bases of similar thickness. A, glass (not affected by humidity change). B, triacetate. C, polystyrene. D, polycarbonate. E, polyester. Polystyrene has the best resistance to humidity changes, but is less satisfactory in temperature stability. Polyester to date provides the best compromise.

moderate variations of relative humidity or temperature are reversible, i.e. if the original conditions are reproduced the base will return to its original dimensions.

Temperature Changes. All the bases tend to expand with an increase in temperature and contract with a decrease in temperature. This change in size with temperature change is greatest for triacetate, polystyrene and polycarbonate bases, less for polyester, and least for glass. It is independent of the thickness of the base or the thickness of the emulsion or backing, within the normal limits of thickness as used in the photographic industry.

Humidity Changes. All photographic films tend to become larger as the relative humidity (R.H.) increases and to become smaller as the R.H. decreases.

Triacetate and polycarbonate bases have a higher water absorption and lower tensile strength than polyester or polystyrene. They are, therefore, more subject to changes in R.H. not only in the uncoated state but also to a greater degree when coated. Owing to their lower tensile strength they are less able to withstand the stresses exerted by the gelatin, when this expands with the absorption of water and contracts with the loss of water.

Both the thickness of the gelatin coating and the thickness of the base have an effect upon the dimensional stability. The thicker the gelatin coating the greater will be the force that it exerts upon the base. This is why continuous tone films which have a higher coating weight than high contrast films are coated on thicker bases, such as ·007 in. polyester or ·01 in. polystyrene.

Processing. Changes in size of a film due to processing are permanent and irreversible. Certain bases which are not so inert, such as triacetate, suffer a loss of plasticizer and solvent during processing and this results in a permanent size change. Even the inert film bases such as polyester which contains no plasticizer or solvent, however, show a permanent though smaller change due to the gelatin coatings. Processing can be regarded as an extreme change in R.H. which causes a permanent change in the gelatin coating, even when coated on glass. This change is usually a shrinkage which occurs with drying, but in the case of inert bases such as polyester is dependent upon the relative humidity to which the film was conditioned before processing. With a low R.H. before processing, and reconditioning the film to the same R.H. after processing, there will be an increase in size. A high R.H. equilibrium before and after processing will result in a shrinkage which is greater with increasing silver/gelatin ratio and with increasing area of developed silver. Thickness of gelatine coating and thickness of base also have to be taken into account.

Ageing. All films show a decrease in size with ageing. This change is caused by the shrinkage of the gelatin, gradually causing a plastic flow of the film base. In addition, bases such as acetate show a change due to loss of residual solvent.

Storage conditions play a large part in the degree and rate of change. High temperature will increase the plastic flow of the base. Low humidity storage tends to increase the compressive force exerted by the gelatin and, therefore, increases the shrinkage.

Certain conditions can be chosen which will minimize the change in size of a photographic image, or will help to keep the amount of change constant so that it can be allowed for.

Ideally the complete area in which films are to be handled should be air-conditioned. Temperatures from 14° C–20° C and R.H. from 50–60 per cent are about the optimum conditions.

Films should be conditioned to the temperature and R.H. of the working area by keeping in unsealed packages for at least 24 hours before use.

Excessive temperatures should be avoided during exposing as this can cause a permanent change in film size.

During processing the temperature and time in each bath, particularly during washing, should be controlled within reasonable limits.

The rate of drying can cause large variations in the change of size. On certain bases the processing shrinkage already referred to can become an expansion if the films are dried very quickly. Conditions of drying should, therefore, be standardized as far as possible.

Actual figures for the dimensional stability of the various bases available vary from one manufacturer to another. D.A.H.

See also: *Photogrammetry; Supports for emulsions.*

DIMETHYL GLYOXIME. Used in nickel toners, as it forms a red compound with nickel.

Formula and molecular weight: $(CH_3)_2(CNOH)_2$; 116.

Characteristics: White powder.

Solubility: Insoluble in water, fairly soluble in alcohol.

DIMETHYL-PARA-PHENYLENEDIAMINE HYDROCHLORIDE. Para-(dimethylamino) aniline hydrochloride. Developing agent for direct colour development with colour couplers.

Formula and molecular weight: $(CH_3)_2NC_6H_4NH_2HCl$; 172·7.

Characteristics: White powder. Similar to diethyl-para-phenylenediamine hydrochloride.

Solubility: Fairly soluble in water.

DIMMER. Electrical control which makes it possible to adjust the light given by a lamp or bank of lamps to any value up to and including the full normal brightness.

Dimmers usually consist of a number of open spirals of resistance wire connected in series with the light. The spirals are arranged so that movement of a control knob or arm short-circuits or disconnects them progressively until, in the extreme position, no effective dimming resistance is left in circuit and the full power of the supply is applied to the lamp. Alternatively, the circuit may start with a high resistance in series with it and movement of the control may add spirals in parallel, allowing more and more current to pass up to the maximum value.

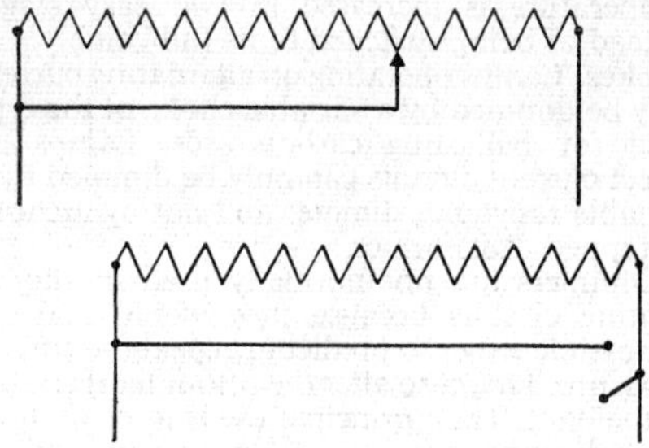

DIMMERS. Top: *Sliding resistor for adjusting the current taken up by the lamp, thus controlling its brightness.* Bottom: *Fixed dimming resistor with short-circuiting switch.*

Series-parallel Dimming. When the illumination is provided by more than one lamp of the same kind, a 2-pole double throw switch may be used to give one stage of dimming by connecting the lamps in either series or parallel. In the parallel position the lamps give their full brilliance; in the series position the voltage applied to each lamp is halved and the light is reduced to less than one-quarter.

The series-parallel arrangement is useful for increasing the working life of Photoflood lamps. Two lamps connected in series give enough light for focusing and arranging the subject and the switch has only to be thrown into the parallel position to make them give their maximum brilliance. This lengthens the life of the lamp in two ways; the full brilliance is used only for the brief taking interval, and not for the lengthy business of arranging the subject, and the filament lasts longer because its

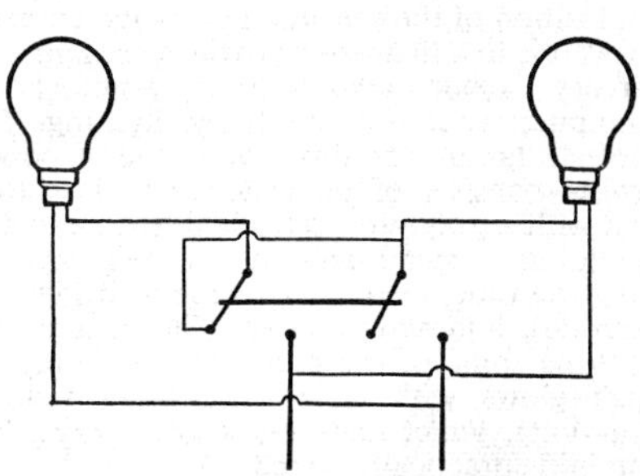

SERIES-PARALLEL SWITCH. With two switches wired up as shown and both open, the lamps are dimmed. They come to full brilliance on closing the switches (usually connected by a bar).

temperature is increased in two easy stages instead of being switched to its full value.
Chokes. Lamps operating on alternating current may be dimmed by a variable choke of the type used for ballasting carbon arcs. Lamps on direct current circuits can only be dimmed by a variable resistance dimmer and not by a choke or tapped transformer.

Dimmers are not normally used in subject lighting circuits because they would make it impossible either to predict or repeat the correct exposure. They also alter the colour temperature of the light. Their principal use is in projectors, for adjusting the screen brilliance, and in cinemas, for raising and lowering the house lights.

See also: *Electricity; Rheostat.*

D.I.N. SPEED. System of stating emulsion speeds, as laid down by the Deutsche Industrie Norm (German standards organization).

See also: *Speed of sensitized materials.*

DIOPTER. Unit expressing the power of a lens, mainly used by opticians. It is the reciprocal of the focal length in metres. When the power of a lens is stated in diopters (= d), its focal length is given by 100/d in centimetres, e.g., a lens with a power of 17 diopters has a focal length of 100/17 = 5·88 c.m. The power of supplementary close-up lenses used in photography is generally expressed in diopters.

See also: *Supplementary lens; Weights and measures.*

DIRECT CURRENT (D.C.). Term for electric current produced by a steady potential, e.g., generated by a dynamo or simple cell.

DIRECT VISION FINDER. Any form of viewfinder through which the subject is looked at directly, as with a telescope, instead of by reflection from a mirror.

DISCHARGE LAMP. Light source in which an electric current passes through a gas enclosed in a glass or quartz envelope.

The colour of the light is governed by the composition of the gas, and not by its temperature as it is in a filament or carbon arc lamp.
Mercury Vapour Lamp. Mercury vapour is the commonest source of discharge lighting. The simplest lamp of this type—the Cooper-Hewitt—consists of an evacuated glass tube fitted with a graphite electrode at each end and containing a small amount of mercury. By tilting the tube, mercury is made to bridge the electrodes, and start an arc. The heat of the arc turns some of the mercury into a vapour which glows with an intense light, rich in ultra-violet, violet and yellow-green rays, but containing practically no red.

The operating voltage depends on the length of the tube, and is usually below 100 volts, and a choke or resistance is necessary, as for most arc lamps. By connecting two lamps in series, less "ballast" is needed and the efficiency of the system is greatly improved.

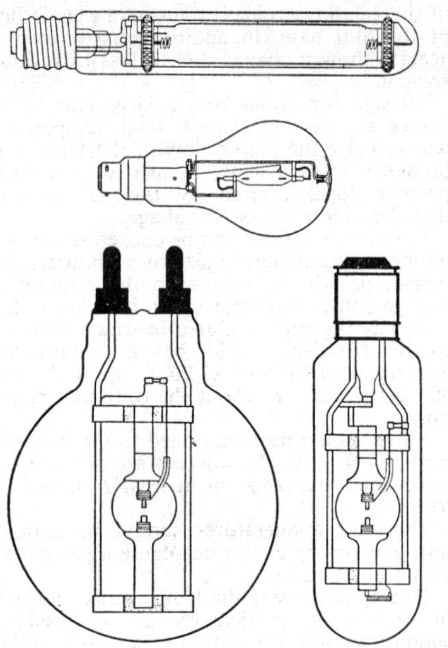

DISCHARGE LAMPS. Top: *250–400 watt high-pressure mercury vapour lamp. The discharge takes place inside a pressure tube sealed into the outer bulb.* Upper centre: *80–125 watt high-pressure mercury vapour lamp.* Bottom: *Two forms of compact source lamps for projectors and studio use.*

Low-pressure mercury vapour lamps are made to work off both D.C. and A.C. supplies. The visual efficiency is less than 20 lumens per watt, but the light is highly actinic. This type is used extensively in process work and plan copying.
Improved Types. If the pressure of the gas is increased, its resistance also increases. This means that the tube can be made shorter and still give the same—or more—light for the same electrical consumption. Lamps based on this principle have a short quartz tube an inch or more in length according to the power. The tube works at a pressure of 1 atmosphere and is surrounded by a partly evacuated bulb to cut down heat radiation and maintain the tube at an efficient working temperature. This type of lamp is made in ratings from 80 to 400 watts and has an efficiency up to 45 lumens per watt.

A modern development of these lamps is the tungsten-ballasted mercury vapour lamp. It also contains a tungsten filament which takes the initial surge of current when switching on and therefore does not need choke gear, so that the lamps can be plugged directly into the supply line. Another advantage of this type is that the tungsten filament, which fades down to a bright red colour as the mercury tube brightens, adds a useful amount of red to the light which is normally deficient at that end of the spectrum.

Lamps working at still higher pressures—up to 100 atmospheres—are nowadays widely used in photography. In the lower ratings—up to 1,000 watts—they are used for portraiture and spot lighting. More powerful lamps consuming several kilowatts are used in film studios. The operating voltage of such lamps may exceed 1,000, and the efficiency may be as high as 70 lumens per watt. As the concentrated discharge develops an appreciable amount of heat, which would even melt the quartz used, some form of water-cooling is necessary with the more powerful units.

A special type of high-pressure mercury vapour lamp known as the compact source lamp has been produced for projection in ratings from 2½ to 10 kilowatts at normal supply voltages.

These lamps consist of two electrodes, and sometimes a heated coil for striking, enclosed in a quartz bulb and surrounded by an evacuated bulb to maintain the discharge temperature. High working pressures are used, but normal voltages are sufficient since the impedance of the discharge is comparatively low.

Spectral Composition. The light of the mercury vapour lamp has no continuous spectral composition but consists instead of isolated spectrum lines. In the case of the Cooper-Hewitt lamps the visible output consists almost entirely of two lines in the violet, one in the yellow-green, and one in the yellow region of the spectrum. This results in very distorted colour reproduction when the lamp is used for general photography such as portrait work; skin defects are emphasized (being red in colour, any light source deficient in red will accentuate them).

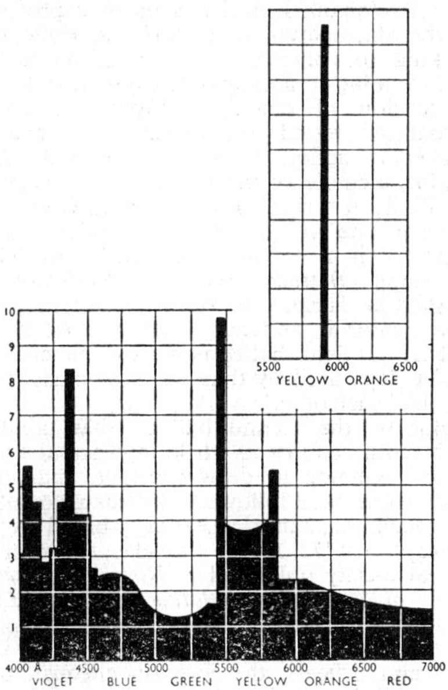

DISCHARGE LAMP SPECTRUM. A gas discharge lamp emits light of specific wavelengths only, and not a normal continuous spectrum. Top: Sodium lamp with single yellow waveband. Bottom: High-pressure mercury vapour lamp with pronounced individual lines but also a continuous background spectrum.

With high-pressure lamps the colour of the light is improved, since a low-intensity continuous background spectrum appears, including orange and red rays.

A further improvement in the mercury vapour lamp is produced by including some zinc or cadmium vapour in the tube. This increases the proportion of red rays and makes the light suitable for normal black-and-white, or even colour, filming.

Fluorescent lamps (below) are another type where the spectral composition has been balanced more evenly by additions to the interior of the lamp.

Sodium Vapour. Sodium vapour gives a light of similar electrical characteristics to mercury, but the colour produced is the familiar yellow of many of our road lighting systems. This lamp is suitable for photography with yellow-sensitive materials; the light is however monochromatic and contains only yellow. Hence all objects not reflecting yellow will appear black. The visual efficiency of a sodium lamp may be as high as 70 lumens per watt.

Xenon and Other Gases. Xenon high-pressure discharge lamps are fairly widely used for film and television studio illumination and also are gradually replacing arc lamps in motion picture projectors. Most electronic flash tubes are also in effect Xenon discharge lamps.

Like the high-pressure mercury lamp, the Xenon arc needs water cooling in the larger units.

They are used as illuminators in process camera work, for high-power lighting in high speed cinematography, etc. Small Xenon lamps have also appeared in narrow gauge cine projectors for home use.

Xenon lamps for projection purposes are usually compact source types; those for studio and graphic arts use may be strip sources in lengths up to 2 feet or more.

Xenon lamps also lend themselves to pulsed operation. In one unit for high-speed cinematography the lamp can operate for 2 seconds at a power of up to 7½ kW. Another pulsing Xenon lamp for motion picture projectors gives light pulses at a rate of 72 per second, in synchronism with the intermittent film transport, so that the projector needs no rotating shutter.

The visual efficiency of the Xenon arc is over 70 lumens per watt and the colour quality of the light is sufficiently similar to daylight to permit its use as an exposing light for photography with daylight type colour film.

Argon and Krypton have also been used

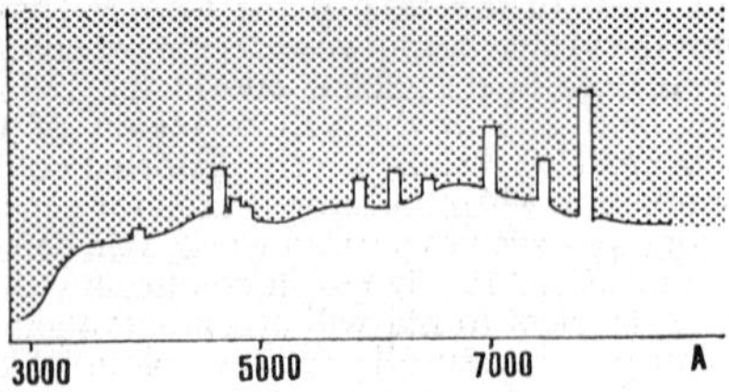

XENON LAMP SPECTRUM. *While the xenon lamp emits its radiation in discrete bands (the peaks in the curve) it also yields a somewhat lower level of radiation throughout the spectrum and is therefore suitable even for colour photography.*

either on their own or in combination with Xenon in high pressure discharge lamps.

Fluorescent Lamps. A large proportion of the radiations emitted by a mercury vapour lamp lies in the invisible, ultra-violet region—i.e., between 3,650 and 3,663 A. These radiations are useless for ordinary room lighting but they cause certain substances to fluoresce visibly. By coating the glass envelope with such substances the lamp is turned into a useful and highly efficient light source. This is the principle of the familiar tubular fluorescent lamp widely used for lighting public buildings and occasionally for domestic interiors.

As the frequency of the supply causes the light to fluctuate, certain precautions are necessary when making short instantaneous exposures by this form of illumination.

Fluorescent lamps, like other discharge lamps, must be connected to the supply through special control gear. F.P.

See also: *Fluorescent lamp; Light sources; Monochromatic illumination; Point source lamp; Spectrum.*

Books: *Fluorescent Lighting Manual*, by C. L. Amick (New York); *Fluorescent Lamps and Lighting*, by W. Elenbaas (Eindhoven); *Discharge Lamps for Photography*, by H. K. Bourne (London).

DISCOVERY OF PHOTOGRAPHY

Technical histories are histories of inventions—overflowing with argument about dates, names and precedence. These arguments are set off by historians who stress the importance of their own discoveries and are kept going by competing national vanities. But behind the smoke-screen of controversy there is a pattern of continuity, logic and progress. People, places and products matter less than we may think. In technological progress a man and his ideas are never quite alone. As many inventions are mutations of available knowledge, more than one person may hit on the same idea at the same time. The invention that matters most need be neither the first nor the best; it is the one that arrives at the right moment, strikes fertile conditions and leads to practical success.

The idea of photography is a synthesis of two lines of experience—both old. One is, that certain substances are sensitive to light; the other is, that light entering a dark room or box through a small hole will draw an inverted image of what is outside opposite to the hole. The chemical and optical lines of progress converge and where they intersect—photography is discovered.

LIGHT SENSITIVITY

It was noted thousands of years ago that foliage turned green in light and that light would produce purple dyes from the slime of some Mediterranean snails. It was common knowledge that certain colours were bleached by the sun. The respective functions of light, heat and air may not have been very clear to the ancients, and medieval alchemists added to the confusion by looking for mysterious chemical powers emanating from the heavenly bodies.

Early Experiments. Then in the seventeenth century the quest for exact knowledge based on experiments gradually made itself felt. In 1663 Robert Boyle, one of the founders of the Royal Society, described many chemical reactions and the resulting changes of colour. He mentioned that silver chloride turned dark under exposure; only he thought that this was caused by "air". In 1725 Johann Heinrich Schulze, a professor at the Universities of Altdorf and Halle, was looking for phenomena of phosphorescence and in doing so accidentally exposed to light a suspension of chalk in silver nitrate; he recognized, tested and correctly explained the darkening action so caused. In 1737 Jean Hellot, member of the Académie des Sciences de Paris, described invisible inks containing silver nitrate and gold chloride, which became legible when exposed to "air". In 1757, Giovanni Battista Beccaria, Professor of Physics at Turin, discovered the sensitivity of silver chloride and demonstrated that it was not air but light that turned his specimen blue-violet. It is unlikely that these scientists knew much of each other's work.

But by the second half of that fabulous eighteenth century its virile intellectual forces were beginning to draw together. Individual discoveries were followed by clear recognition of principles, planned research by broad literary surveys. In 1772, Joseph Priestley, fellow of the Royal Society, published his *History and Present State of Discoveries Relating to Vision, Light and Colour*, the first comprehensive description of the chemical action of light. In 1777, Carl Wilhelm Scheele, the Swedish chemist, published his dissertation *Aeris atque Ignis Examen Chemicum*, introducing the prismatic solar spectrum for the investigation of light sensitivity and ammonia as an agent for separating blackened from unchanged silver chloride.

In 1782, Jean Senebier, librarian of the city of Geneva, published his extensive experiments in colour sensitivity.

Thomas Wedgwood. The enthusiasm for experimenting and the number of experimenters grew rapidly. Among them was Thomas Wedgwood, fourth son of Josiah Wedgwood, the famous potter. In 1802 his friend Humphry Davy, later president of the Royal Society, published a report about Wedgwood's method of copying glass paintings or drawings on paper or leather which were treated by silver nitrate or silver chloride. He also described how silhouett es were made by the same method. (Jacques Alexandre César Charles, the versatile French Academician and popular lecturer, was said to have produced similar profiles by light before Wedgwood had done so, but there is no evidence that would confirm this belief.) Wedgwood had to keep his images drawn by light in darkness and could show them only by candle light; he found no way of making his work permanent—although Scheele's experience with ammonia should have been known among his friends. He was also unsuccessful in his attempts to make similar pictures in the camera obscura. There is an air of casualness about Wedgwood's efforts and Davy's summary of them. Wedgwood died three years after it was published, 34 years old. He touched the threshold beyond which photochemical experiments were made to serve graphic ends.

THE CAMERA IMAGE

Knowledge of the camera or, at least, of its principle has been ascribed to Mo Tzu in China of twenty-five centuries ago the Greek philosopher Aristotle (384–322 B.C.), the learned Arab Ibn al Haitam (965–1038), the English friar Roger Bacon (1214–94), the Hebrew scholar Levi ben Gershon (1288–1344) and others. But some of these assumptions are based on imaginative readings of barely more than sketchy remarks.

The Camera Obscura. That giant of the Renaissance Leonardo da Vinci (1452–1519), described the working of the camera obscura quite fully; only his account—confined to his manuscripts in mirror writing—did not become public property for centuries. An illustration of the camera obscura was actually published in 1545 yet the invention of the pinhole camera is still mostly linked with the name of Giovanni Baptista della Porta of Naples who gave a very clear description of it in his *Magia Naturalis sive de Miraculis Rerum Naturalium*, the first edition of which appeared in 1558.

In 1568, the Venetian nobleman Daniel Barbaro suggested the addition of a biconvex lens to the camera and of a diaphragm to improve the definition. Twenty years later, in a new edition of his work, Porta also mentioned the use of a lens as well as of a mirror to right the reversed image. In Porta's time a whole room was used as a camera obscura.

The seventeenth century, recognizing their usefulness to artists in drawing exact perspectives, brought portable cameras. The astronomer Johannes Kepler used one for topographical sketches in 1620, the Jesuit Athanasius Kircher illustrated a large portable model in 1646, the Praemonstratensian Johann Zahn described several small box reflex cameras in 1685.

Improvements. But it is the eighteenth century that made the camera obscura really popular; the improvements were both mechanical and optical. The shape was changed, among others, into that of tables, a sedan chair or a small tent that fastened to a drawing desk with the viewing piece turned into a reflecting periscope. Some cameras were equipped with a concave drawing board to compensate for marginal failings of the lens. Eventually, in 1812, William Hyde Wollaston replaced the old biconvex lens by a meniscus with a stop giving uniform distribution of reasonable sharpness over the whole field.

Now the image was coming out bright and sharp enough to evoke a desire in the draughtsman to catch this picture by means quicker than his hand. It certainly suggested this possibility to Fox Talbot in 1833 when he tried to sketch landscapes with the aid of a camera lucida on Lake Como (". . . the idea occurred to me . . . how charming it would be if it were possible to cause these natural images to imprint themselves durably, and remain fixed upon the paper!").

Retaining the Image. The vision of a mirrored view caught for good is probably as old as human imagination.

The Roman poet Publius Papinius Statius (40–96 A.D.) has it in some verses of *Silvae*.

In *Gyphantie*, a piece of science fiction from 1760, Charles-François Tiphaigne de la Roche romances about a subtle sticky substance which, coated on a piece of canvas, has only to be held before any object to depict it instantaneously.

Yet his were fancy dreams of magic at a time when down-to-earth practitioners of magic did not get beyond using silver solutions to make writing appear from nowhere or to turn a white man into a negro. Magicians' handbooks of the late eighteenth century have plenty of good advice to give on such subjects as well as entertaining uses of the camera obscura. However, when Talbot, returning from Italy, concentrated on the problem of marrying chemistry to the camera others were way ahead of him.

NIÉPCE

Joseph-Nicéphore Niépce of Chalon-sur-Saône, born 1765 into the well-to-do Burgundy bourgoisie, retired as a subaltern from a short army service and devoted himself to technical inventions. In 1813 he turned to lithography, which had recently been imported into France. Niépce could not draw and looked for ways of copying designs on to the lithographic stone. Thus in 1816 he became interested in the problem of producing images in a camera by light on sensitive materials.

Heliography. He made several cameras; one of them was the size of a jewel box. The dimensions of his cameras were determined by the focal length of the lenses he found. He fitted these in metal tubes and also used bellows. He improved definition by a stop and turned that into the iris diaphragm. He tried stone, glass, metals and paper for materials and sensitized these with a variety of substances following up references in the expanding literature of chemistry.

Niépce's progress was slow and groping; he worked for twenty years with admirable persistence yet apparently was handicapped by provincial isolation. His experiments were dominated and probably misdirected by the combined idea of a camera-made image that would also serve as a printing plate. Niépce called his process Heliography and applied that name both to images produced in the camera and to designs traced in contact from art work.

Limitations. His first camera images, in 1816, were weak negatives on paper treated with silver chloride and poorly fixed in nitric acid. He was aware of the problem of having to reverse the tones and preferred searching for alternative substances that would give him direct positives. After 1822, he settled down to an asphalt varnish on metal and glass and used a mixture of oils as a solvent for fixing the image. Niépce's varnish was not suited to recording tone values in camera views and, as sometimes a whole day was required for exposing these, the sun moving around the horizon also interfered with whatever modelling the material would yield. Niépce succeeded in taking a view in the camera in 1826, but his best heliographs were line subjects, engraved plates or prints made from engraved plates.

Thus Niépce invented a form of photography but he left it in a technical *cul-de-sac* waiting for Daguerre to find a way out.

DAGUERRE

Louis Jacques Mandé Daguerre, born 1787, was a gifted scenic painter and ingenious showman. His Diorama, opened in 1822, was acclaimed as one of the great attractions of Paris; it combined panoramic views with changing light effects and foreground props that blended into a colourful and realistic spectacle.

Daguerre, being interested in the interplay of light and perspective, made use of the camera obscura for sketching. He is supposed to have stimulated its improvement by the opticians, Vincent and Charles Chevalier, who equipped it with a double lens which they developed from Wollaston's meniscus. Daguerre discussed with the Chevaliers his ideas of catching the images in the camera obscura and even claimed to have succeeded in doing so.

Partnership with Niépce. In January 1826, on hearing about similar experiments by Niépce, the Chevaliers passed on his address to Daguerre. Daguerre wrote to Niépce with little loss of time; Niépce replied evasively. Twelve months later, in January 1827, Daguerre wrote again and Niépce now made careful inquiries about his correspondent. In August 1827, on his way through Paris, Niépce visited Daguerre and left impressed. He was now well over sixty, a tired provincial, financially much embarrassed, and in January 1828 he returned from London with a refusal of the Royal Society to take note of his invention as he was not ready to tell how he arrived at it. Daguerre was twenty-two years his junior, a very live Parisian, well off and successful in his undertakings.

By the end of 1829 the two arrived at a formal contract to run for ten years, agreeing to pool their secrets, mutually to improve upon their results, to publish their invention at the proper time under the joint names of Niépce and Daguerre and equally to share in the profits. To execute the agreement Daguerre came to Chalon and Niépce demonstrated his process. He also handed over a memorandum of detailed instructions how to use it. Daguerre does not seem to have disclosed anything; it is doubtful whether he had much to disclose at this stage.

The partners never saw each other again. Niépce died in 1833.

When Niépce and Daguerre signed their agreement in 1829 each seems to have expected too much from the other.

Niépce, who had been inclined to blame his optical equipment for slow progress, readily believed Daguerre's over-statements about his improvements to the camera obscura. Daguerre, whose chemical experience was sketchy, must have overrated the flexibility of Niépce's knowledge.

In any case when, in 1831, Daguerre repeatedly wrote to Niépce pressing for experiments with silver iodide, Niépce proved to be unreceptive. He thought he knew the substance from earlier trials, he was worried about getting negative images and he saw no way of fixing them. Eventually Daguerre must have realized that he was on his own and shut himself up in his laboratory.

Daguerre's Progress. Nobody was permitted to watch Daguerre's work and nobody really knows what he did except that for a few years he worked feverishly. His progress is related in a number of episodes which are colourful and plausible enough to have originated from a man with an obvious flair for effects.

One story tells how, in 1823, Daguerre noticed on a freshly painted scene the image of a tree which had been projected through a small hole in the window shutter; it was still there next day. This prompted him to try to retain images in a camera and during these trials he recalled that he had mixed iodine with the colours when painting that scene . . .

The second story tells about Mrs. Daguerre's visit, in 1827, to the celebrated organic chemist Jean Baptiste André Dumas to whom she complained that her husband was neglecting his business for the sake of dubious experiments. Dumas looked up Daguerre, but instead of discouraging him gave useful advice and promised expert help . . .

The third story is supposed to have happened in 1831; it relates to a silver spoon which, lying by chance on an iodized plate, left its perfect image on it . . .

The fourth story brings us up to 1835 when, having locked in a cupboard an insufficiently exposed iodized silver plate that had not shown traces of an image, on opening the cupboard Daguerre was surprised to see a fully developed image on it. The developing agent, found by eliminating one chemical after another that had been locked in with the plate, proved to be mercury . . .

The Process. Some of the materials used by Daguerre were used by Niépce before him; yet Daguerre's process was different. He employed silvered copper plates coated with vapours of iodine. He exposed the plate in the camera for fifteen to thirty minutes, producing an invisible or scarcely visible image. This latent image was developed by placing the plate over a vessel of heated mercury. Fine particles of the mercury clinging to the exposed regions of the silver iodide then formed a visible positive image. Finally Daguerre washed the plate in a hot solution of cooking salt, removing the unnecessary iodide and thus making the image permanent. The *Journal des Artistes* of 27th September 1835, claimed that he produced permanent images at that date, but in fact he found a way of fixing them only two years later.

It was also in 1835 that in view of his own progress Daguerre invited the son of his former partner, Jacques Marie Joseph Isidore Niépce, to accept a modified agreement changing the joint name from Niépce-Daguerre to Daguerre-Niépce. In 1837, demonstrating his latest results, he made the younger Niépce agree to drop his father's name altogether; the financial arrangements were to remain unaffected. In 1838 Daguerre's efforts to find money for a public company failed and so he turned to the influential scientist and politician Dominique François Arago for assistance in selling the invention to the French State. Rumours were helpfully afoot that "England offered to buy the secret of the process for 200,000 francs".

Official Recognition. On 7th January 1839 Arago reported to the Académie des Sciences about Daguerre's achievement. On the preceding evening the *Gazette de France* duly scooped the news. On 8th March the Diorama was destroyed by fire; Daguerre seemed ruined. On 15th June the Minister of the Interior, Tannegui Duchatel, presented to the Chamber of Deputies a bill granting Daguerre for life an annual pension of 6,000 francs and Niépce junior 4,000 francs. The bill was adopted by the Chamber on 3rd July based on the impressive exposition by Arago; the Peers adopted it on 30th July after a report by Joseph Louis Gay-Lussac, the great physicist.

On 19th August 1839 Arago, in a brilliant speech before the members of the Académie des Sciences and the Académie des Beaux Arts, described the technique in detail, sketched its history, forecast its future in the service of civilization and offered its free use to everybody. ("France has adopted this discovery and from the first has shown her pride in being able generously to donate it to the whole world".) Alas, on 14th August, Daguerre had secured a patent in England.

The building was overflowing with the cream of French intellectual life, outside Paris was buzzing with excitement and by then the whole world was listening. Ambassadors of the King of France presented daguerreotypes to courts abroad. Daguerre in return received decorations and presents from foreign sovereigns. Photography was officially born, ushered in by a beautiful gesture and in a setting of perfect stage management.

Dissent. There were dissenting voices long before the celebration. Francis Bauer, Fellow of the Royal Society, wrote a letter on 27th February 1839 to the Editor of the *Literary Gazette* of London to express the view that Daguerre's invention was the same as that of Nicéphore Niépce of whose work Bauer had proofs from 1827. Talbot wrote on 29th January to Arago and made a formal claim of priority to fixing images in the camera obscura. Hippolyte Bayard, a clerk in the Ministry of Finance, brought his own positive paper images to Arago three months before Daguerre's process was officially disclosed in August.

Daguerre's was really the wrong invention, yet launched so intelligently and with such force that other ideas could barely keep afloat in its wake. At a time when factories and railways were preparing great social changes which would soon produce mass demands for new arts, the daguerreotype presented an exquisite small silver image that was as unique and almost as laborious to duplicate as an artist's miniature painting. Niépce had been instinctively groping for the desirable solution when he tried to link his camera with lithography; but he became bogged down in the mutual incompatibility of the two techniques.

It was left to Talbot, a distinctly feudal figure, to grasp the essentials of such a democratic art and thus open the epoch of negative-positive photography—the functional method for producing with ease any number of prints from the original made in the camera.

TALBOT

William Henry Fox Talbot, born 1800, was the only son of William Davenport Talbot of Lacock Abbey in Wiltshire and Lady Elizabeth Fox Strangeways, eldest daughter of the second Earl of Ilchester. The Talbots had resided at the thirteenth century Abbey since the days of Henry VIII. Fox Talbot was educated at Harrow and Trinity College, Cambridge. He had a keen, exacting and versatile mind and developed interest in botany, mathematics, chemistry, physics, astronomy and archaeology. He wrote some tales in prose and verse which he hoped would be set to music and later learned works on classical research, on the book of Genesis and on English etymology. He had a wide knowledge of languages and travelled a great deal. For a couple of years he sat in Parliament and he was a Fellow of the Royal Society. Talbot's photographic activities cover merely a phase of his life and even whilst that phase lasted they were only a facet of it.

Photogenic Drawings. On some of his frequent and extensive Continental tours in the eighteen-twenties he had again and again attempted to make use of the camera obscura for sketching. Late in 1833 he became impatient about his own lack of success with the camera lucida (Wollaston's prismatic device for joint viewing and drawing the scene) and on his return to England in early 1834 he settled down to tackle the problem of capturing camera images by chemistry. Of Wedgwood's experiments he learned only later. It was suggested that he may have heard of Niépce's work as well. In any case by then the basic idea which Talbot pursued was to such an extent "in the air" that he could have hardly imagined it to be wholly original.

He started, though, from scratch. First he treated paper with silver chloride and on this sensitive material he stencilled in contact by light negative patterns of botanical specimens, lace and similar objects. He neutralized sufficiently the unexposed parts by washing the paper with a strong solution of common salt or potassium chloride to retain an image for some time. By 1835 he was able to produce tiny negatives by exposing them for half an hour in small locally-made cameras, which his wife referred to as "mousetraps".

When the news came from Paris in January 1839 about Daguerre's invention Talbot hastened to draw attention to samples of his "photogenic drawings" which happened to be in London; they were shown by Michael Faraday to the Members of the Royal Institution on 25th January. Talbot showed them to the Royal Society on 31st January and published his methods on 21st February under the title *Some Account of the Art of Photogenic Drawings; or, the Process by which Natural Objects may be made to Delineate themselves without the Aid of the Artist's Pencil.*

Talbot's friend, the astronomer Sir John Herschel, became particularly interested. In a letter to Talbot, dated 28th February, he made use of the word "photography". This word also appears on 25th February in the *Vossische Zeitung* of Berlin in a notice by the astronomer Johann von Maedler; Herschel and Maedler knew each other. The same word as a verb and its adjective "photographic" does crop up even earlier in Herschel's private writings. Herschel also coined the terms "negative" and "positive". More importantly it was Herschel who suggested sodium thiosulphate (hypo), the properties of which he discovered back in 1819 and which was to become the classic fixing agent.

With Herschel's consent Talbot wrote about this method of fixing on 15th March to the physicist Jean Baptiste Biot who published the letter in the *Comptes Rendus* of the Académie des Sciences; Daguerre made use of it.

Talbotype. At the same time Talbot sent news about having improved the sensitivity of his process by using silver bromide. In the autumn of 1840 he switched to silver iodide, also used by Daguerre, and under strikingly similar circumstances to those experienced by Daguerre he accidentally discovered that the exposure time could be greatly shortened by aiming only at a latent image and developing it after exposure. Talbot confined this technique to making negatives and continued to use silver chloride paper for printing out positives. He named his developer gallo-nitrate of silver and his new process Calotype, later Talbotype.

From 1841 onwards he protected his processes with a number of patents and demanded royalties for their use.

In 1844 he published *The Pencil of Nature,* which relates the story of his work, illustrating it impressively with a number of his own Calotypes: it is as telling a monument as any inventor could erect to himself.

Restrictions. Talbot's was the right invention—but not very happily put over. He claimed to have spent £7,000 on photography (about £28,000 in present value) and naturally hoped that his invention would make a fortune. Whilst all his scientific activities reveal him as an amateur in the best sense of that word, he rigidly watched over his photographic patents, claiming and patenting also improvements and discoveries made by others, and considering himself entitled to do so as the inventor of the calotype.

His real motives were probably less one-sided than the comments either of his admirers or of his critics. Pre-occupation with figures since early childhood and carefully balanced accounts of his personal expenditure suggest that he liked to know where he stood and was not indifferent towards money.

Talbot was bound to be indignant about Daguerre's cutting across his path, his lack of fairness in securing the protection of a British patent for the supposedly "free" French invention and, last but not least, about the greater commercial success of the competing French process.

Being concerned with respectability and appearances Talbot opened a photographic printing establishment for the making and sale of calotypes, and a commercial portrait studio —under the name of employees. Even when put under some pressure from the Presidents of the Royal Academy and the Royal Society who appealed to him to loosen his grip that threatened to handicap British progress, in a letter to *The Times* dated 30th July 1852 Talbot, whilst somewhat belatedly freeing his method of photography for scientists, artists and amateurs, still continued to insist on licensing portraiture for profit by the collodion process—the only worthwhile commercial application of camera work in those days. The relations between Talbot and some contemporary photographers were, in fact, such that when in 1853 plans were laid for the first Photographic Society in England, Talbot could not be persuaded to become its president.

Eventually, 1854, Talbot abandoned his patents once he lost his action against Silvester Laroche, a professional who practised the new collodion process. In the course of that case it must have been particularly disconcerting for Talbot to have to face the evidence of the Reverend Joseph Bancroft Reade, who proved that he had used a substance similar to Talbot's "gallo-nitrate of silver" before Talbot had it patented—allegedly knowing that the idea was not quite his own. (Reade used tannin as an accelerating influence in the blackening process before and during exposure whilst Talbot more decisively employed it for developing the latent image.)

BAYARD AND OTHERS

Another very disappointed man was Hippolyte Bayard of Paris. In 1840 he took a picture of himself in the nude captioning it thus "This corpse is M. Bayard . . . The Academy, the King and all who have seen his work admired it just as you do now. This brought him honour but not a penny. The Government who have done too much for M. Daguerre would do nothing for M. Bayard. So the wretch has drowned himself . . ."

The Late-comer. Born 1801, young Bayard watched his father making use of the sun for bleaching patterns on to ripening fruit through stencils. The boy repeated the game on paper which he dyed with safflower. He may have embarked on more serious experiments after 1837 but it is also possible that he only talked about them. He did begin work on 20th January 1839, i.e., after Arago's preliminary report on Daguerre and before Talbot's work became public.

Bayard fed into his camera moist paper which had been treated with silver chloride, then blackened by light and finally immersed in a weak solution of potassium iodide. He obtained a positive image bleached by the camera exposure, which took approximately an hour. He originally used potassium bromide for fixing. A couple of weeks after he started his experiments Bayard could show imperfect results to César Manuète Desprets of the Institut de France. Two months from his start he was handing around good prints. He brought them to Biot on 13th May and to Arago 20th May. He publicly exhibited thirty of them on 24th June.

By then, however, Arago was too much committed to Daguerre's cause, saw only complications arising from Bayard's intervention and cautioned him to be discreet. Bayard was a civil servant and his private interests were primarily artistic; he seems to have been timid and not much of a man of business. He laboured under the disadvantage of being a Frenchman in a France that had already made up its mind to celebrate a French invention of photography *à la* Daguerre. Because of that it was even overlooked that the pictorial accomplishments of Bayard's positive paper process temporarily surpassed those of Talbot, although it neither anticipated that of the Englishman nor in any way compromised the latter's basic merit of having developed the first workable method of producing negative-positive photographs.

Other Claimants. Even within its narrow field of achievement Bayard's priority was disputed. In France, Verignon and Lassaigne claimed to have developed similar processes before him. Fyfe is reported as having described a similar method before the Society of Arts in Edinburgh 17th April 1839.

On 13th April 1839, before the Bavarian Academy of Sciences in Munich, Franz von Kobell, a professor of mineralogy, and the optician Carl August von Steinheil presented negatives on paper sensitized with silver chloride, exposed damp in a camera made of cardboard and then fixed in hypo.

There were others as well, unnoticed by historians and unimportant to history. It is as well to remember that during the previous few years there had been notices in the press about Daguerre's experiments, some of them possibly calculated indiscretions. Besides his experiments, Daguerre was busy promoting his projects. First he was angling for private finance, then he started his subscription scheme for a public company and eventually he looked for help to the State. Since 1837 he talked increasingly often and, one might guess, quite volubly to scientists, artists, businessmen and politicians. He obtained permission to make pictures of famous monuments of the French capital. He took long exposures with his heavy apparatus perched on a vehicle surrounded by curious Parisian crowds. Stories spread to England, Germany, Russia and the United States. About his

"LA DAGUERREOTYPOMANIE." Théodore Maurisset's cartoon published in 1839, one of many which appeared at the time on the subject, reflecting the general excitement about this revolutionary process.

secrets he kept the people guessing and no doubt many of them spent much time in doing so. It was only to be expected that some of them would guess close enough and they were also bound to try out just how good their guesses were.

THE DAGUERREOTYPE ERA

The inventor who emerged too late will always attract sympathy, whilst recognition, success and, in particular, profits often blur the view of actual achievements and distract from real merits. Daguerre's showmanship and enterprise were destined to provoke doubts and even suspicion—yet they proved to be the motive power that sped photography into the world and around the globe.

Expansion. Within a few weeks from the release of the invention in Paris, daguerrean pictures were produced in England, Germany, Austria, Switzerland, Poland, Spain and the United States. A year later the process reached Japan and China. Daguerre's own handbook of instruction went into edition after edition in France, was published during the same year in English, German, Italian, Spanish and Swedish; the next year in Hungarian and Polish.

The original instructions and Daguerre's own equipment were in wide demand. Daguerre entrusted the sale of his apparatus to Susse Frere and Alphonse Giroux et Cie (Giroux, a relation, ran a stationer's business). Daguerre's camera weighed 50 kilograms (110 lbs.) and was sold for 400 frs. (£16—or £64 at present purchasing value). But before the end of 1839 the size and weight was reduced to one-third. Daguerre's plate measured 0·164 × 0·216 m., but cameras were also made one-third, one-quarter, one-sixth and one-eighth the original plate size. Plates cost between 1 and 4 frs. according to the size, auxilliary equipment 100 frs.

The image produced in the original camera was reversed left to right. It had a cold metallic sheen which caused reflections. It abounded in fine detail but was lacking in tone contrast. In strong sunshine 15 minutes exposure was needed; in less favourable light up to 30 minutes. Portraiture was torture; the portraits looked it.

Improvements. But the diffusion of information, stimulus and research that follows in the wake of major discoveries soon produced improvements. Mutations and duplications of closely related ideas are common at such stages in the development of any invention; claims of priority correspondingly frequent and on the whole irrelevant.

The sensitivity of the plates was increased by using a second halide (bromide or iodo-chloride of silver) as an accelerator. Such combined sensitizers were initiated by John Frederick Goddard and Antoine Frances Jean Claudet of London and Franz Kratochwila of Vienna.

In order to get stronger blacks, Baron Armand Pierre de Séguier devised better methods of polishing the metal plates; John Adams Whipple of Boston, Massachusetts, introduced machinery for the job.

Hypolite-Louis Fizeau of Paris turned the cold metallic glare of the daguerreotype into a deep purplish tone by adding gold to the fixing process; this also made the image more stable.

Johann Baptiste Isenring of St. Gallen in Switzerland found methods of retouching the eyes, gilding ornaments and colouring the face. So did others in England and France.

Prisms added to the lens reversed the image and thus normalized architectural views and other scenery (but again lengthened the exposure). Curiously enough, no great need for this reversal was felt by the sitters for portraits, who readily accepted the likeness they knew from their mirrors.

An ingenious camera with a concave mirror replacing the lens so shortened the exposure that it enabled Alexander Wolcott to open the world's first portrait studio in New York at the beginning of March 1840. One year later Richard Beard using the same camera opened the first photographic studio in Europe at the Polytechnic Institution, London, on 23rd March 1841.

The mathematician Josef Petzval of Vienna brought about a decisive turn in the fate of Daguerre's process by producing, in 1840, a new double lens with separate front and back components which, having an aperture of *f* 3·6, was thirty times faster than the Chevalier lenses originally used in Daguerre's cameras. These lenses, combined with the improved sensitizing methods and smaller plate sizes, reduced the time of exposure from a quarter of an hour to as little as 15 seconds. By 1841 Natterer in Vienna took military parades with an exposure of 1 second and Gaudin took Paris street scenes at one-tenth of a second—or so he said. Petzval's lenses were described abroad as "système Allemande" although neither the lens nor the designer came from Germany. The sales appeal of that phrase became so overwhelming that lens makers in other countries also engraved it on their mounts—sometimes irrespective of the performance of the product. Petzval got little financial benefit from his revolutionary design, quarrelled with those he thought deprived him of his dues and died a misanthrope. But he made portraiture a popular reality everywhere.

Portraiture. Cheap, quick and life-like portraiture was the key to the rapidly expanding boom of Daguerre's process. To have his image reproduced and perpetuated satisfied a deep urge in man. Sculptors, painters and lithographers could serve only the few who had the leisure to sit for them and the means of paying for a competent artist's skill. There were, of course, itinerant and back-street artists who would charge modest rates and produce quick results, but their work was mostly crude and the likenesses but sketchy. Technical tricks and tools to speed up the process and trace more exactly true likenesses had been sought for since the Renaissance and became numerous in the eighteenth century (silhouettes since 1760, the Physionotrace since 1789). The mid-nineteenth century did not at first work up a snobbish distinction between hand-made and not quite hand-made arts and so high and low alike rushed to have the capabilities of the camera tried out on their own faces. The artists themselves saw in it merely a new tool, an easier way to do the job and earn more money.

The protests of the intellectuals came later but soon enough. The poet Charles Baudelaire was among the first to mistake popular appeal for a sign of artistic inferiority. ("The whole filthy mob rushes in like a single Narcissus to gloat over their trivial images".) The "Daguerreomania" was now definitely on, to the delight both of those who profited by it and the others who would do so by lampooning it.

A New Industry. In the very first years of the daguerreotype era, among 20 camera artists operating in Hamburg, 13 used to be portrait painters and 2 lithographers. Of about 30 working in Berlin 16 used to be painters, 2 lithographers and 1 an engraver. Berlin, then a town of 300,000 inhabitants, cumulatively had 210 daguerreotypists established between the years 1839 and 1860. New York, then a city of 700,000, had in 1850 71 "daguerreotype galleries" employing 127 operators.

At the height of the daguerreotype business the best studio in Boston charged $5 (17s. 6d.; £3. 10s. today's value) for a large daguerreotype, but the standard price for a medium sized picture was $1. Towards the end of the boom prices slumped to 50, 25 and even 12½ cents. By that time the popular price in London was 1s. Richard Beard, a former coal merchant, is said to have amassed from his group of studios £40,000 (present value £160,000) during his first year of trading, the takings being sometimes £60 per day.

Some French daguerreotypists made 3,000 daguerreotypes per year. In 1849 the annual production of Paris amounted to 100,000 daguerrean pictures.

The commercial possibilities of what now would probably be termed "French postcards" were almost immediately discovered; pornography got itself attached to photography right from its tender youth.

But so has science. Astronomical photography, photomicrography, photogrammetry, stereoscopic photography and quite a few other fields of application were all approached in the daguerreotype era—just as Arago forecast that they should and would be.

But Louis Jacques Mandé Daguerre soon retired from the camera and to his painter's brush at the village of Bry-sur-Marne where he died in 1851, leaving behind a scenic fresco in the local church. He himself may not have succeeded in doing much for the art of the camera, but he certainly established its trade.

THE TALBOT HERITAGE

Daguerre's influence was widespread and short lived; Talbot's impact was isolated but penetrating.

Curiously enough both inventors managed also to retard progress. Daguerre's enthusiasm unwittingly misdirected the technique of photography for nearly two decades. Talbot's stubbornness no less effectively blocked its logical immediate development.

Drawbacks. The main restrictive effect was, of course, created by Talbot's own decision not to permit free and general use of his invention. His personal disclosures of it were confined to scientific papers and expensive pictorial publications (*The Pencil of Nature* 1844; *Sun Pictures of Scotland* 1845). Popular instruction books comparable with Daguerre's literature did not exist; the best English manual of the calotype was published as late as 1855. Talbot obtained patents and appointed agents abroad. His agents in the United States do not seem to have succeeded in selling a single licence. His agent in France was ineffectual in preventing infringements. Talbot's process did not flourish except in Scotland, where it was not patented.

Then there were technical difficulties although Talbot had an organization comparable with Daguerre's for supplying equipment. Paper that served both for negatives and prints posed problems; its texture, and sometimes even its watermarks, would imprint themselves on to the picture. Not only the depth but also the colour would vary between light violet and strong sepia. Outlines and detail were not very sharp and often disappeared in over-exposed highlights and under-exposed shadows. The negatives had to be taken in the strongest possible sunlight that in its turn would produce excessive contrast. Exposures were several minutes long and could only be gauged properly by a great deal of experience.

Yet some of its limitations were to become the very source of the sporadic but lasting triumphs of Talbot's process.

The small and precise daguerreotype looked just "a mirror with a memory" (Oliver Wendell Holmes). Registering every minute detail, fold and blemish with uniform, documentary emphasis, daguerreotypes have a quaint charm—heightened by the elaborate frames and boxes in which they are encased. The daguerreotypist, who so often was a trained painter of miniatures, aimed at a compressed and digested record of likeness without frills; he did not grope for expression and rarely caught "soul". Daguerreotypes looked more standardized and manufactured than most later products of photography.

D. O. Hill. Inevitably the calotypist worked more on his own, had to experiment and rely on individual ideas. The most outstanding of them was a Scotsman, David Octavius Hill. Born 1802, the son of a Perth bookseller, he became a painter, took a busy part in Edinburgh's art life and acted as Secretary of the Scottish Academy of Painting, Sculpture and Architecture. He painted romantic landscapes and drew scenic illustrations for engravings on Scottish subjects; his work was competent but insignificant.

In 1843 Hill set out to commemorate the "Act of Separation and Deed of Demission" by the Free Church of Scotland. It took him twenty-three years to finish this canvas of 4 feet 11 ins. by 11 feet 4 ins. depicting 470 personalities. (He asked for a fee of £3,000 and was paid £1,500—about £6,000 in present value.) The artistic merits of this work are negligible: it is just a hand-painted "photomontage". The task of assembling such a number of individual likenesses in a single composition would have defied a much greater painter than Hill and in any case he had no experience in painting portraits. Besides, the problem of securing sketches of so many people seemed formidable. But he thought of a new way of getting them and it is these "sketches" that for ever engraved his name in the records of history.

Hill had been encouraged by the physicist, Sir David Brewster, to collect his sketches with the camera and to enlist the assistance of Robert Adamson, who had recently opened a studio in Edinburgh. Between 1843 and 1848 Hill and Adamson produced hundreds of camera portraits, not only of members of the Free Church Assembly but also of other outstanding personalities of Edinburgh's society. They also photographed genre scenes in Scottish harbours and took architectural and other scenes.

The portraits remain outstanding as a sublimation of the fertile eighteenth century English and Scottish painting traditions and their vigorous translation into a new medium. Lack of definition and limited latitude are virtues, not weaknesses, in the pictures as visualized probably by Hill and taken mostly by Adamson. Their subjects are presented in terms of carefully distributed contrasts of tone. What there is of detail seems deliberately picked for emphasis. The poses may have had to be chosen to sustain an exposure of up to three minutes, but each one of them sums up a personality, stresses a mode of life and hints at much that was valid about the sitter beyond the confines of a single exposure. It is almost uncanny. This new vehicle photography had barely emerged from nowhere and, steered by an obvious traditionalist, it was not only picking up a trail at once but clearly thrusting towards new horizons. A mediocre Scottish painter became both its revolutionary pioneer and its greatest classic.

Robert Adamson left Edinburgh in 1847 and died a few months later, only 27 years old. Hill gave up photography and when he took it up again in the 1860s with another collaborator it was with much less success. The portraits which he took with Adamson received an "honorable mention" at the Great Exhibition of 1851. But there is no reference to Hill in the notices about the foundation of the Photographic Society of Scotland five years later. When he died, 1870, his obituaries did not as much as mention his photography and photographic journals took no notice of his death. His opus was unearthed after twenty years, to be universally recognized as the prodigious achievement of a great artist who succeeded in exploiting all the depth of expression his unwieldy tools would yield.

Hill's achievement in all its curious singularity has a threefold significance. It gives Talbot's process a sort of raison d'être. It demonstrates the prowess of this infant prodigy called camera. It established an early precedence for what is going to happen in the history of photography over and over again, where outsiders, amateurs and other non-professionals so often break into commercial trends and redirect them towards new aims.

Blanquart-Evrard. Next to Hill's, the most important use made of the Calotype process was by Louis Désiré Blanquart-Evrard. He failed to acknowledge his indebtedness to Talbot, but quickly recognized the value of the process in producing illustrative work. He exposed the paper sandwiched between glass. He smoothed it by a coating of egg white and milk whey (albumen), used the gold chloride formula of the daguerreotype to produce reliable brown tones, introduced a better developer and speeded up the printing process. In 1851 he built a printing works at Lille where he employed 30 to 40 assistants and is reported to have produced hundreds of thousands of prints. His first publication was the folio *Egypte, Nubie, Palestine et Syrie*, containing 125 prints from paper negatives taken on a Mediterranean trip by Maxime du Camp accompanied by Gustave Flaubert. Blanquart-Evrard also published impressive prints of monumental architecture studies by Henri le Secq and Charles Negre.

He was the first professional photographic publisher and the first to grasp that the camera could show more to more people than any pair of human eyes could ever hope to see without its aid.

Significance. Within a very few years of its birth photography was established as a trade, tried as a tool of art, used as a new medium of communication and thrust from many directions into the realm of science. Soon it was to father cinematography, that in turn became television's parent.

Photography and its offspring presented new visions of the world to the world and by doing so changed it.

Man who since the Middle Ages had grown increasingly introvert was made newly aware of his environment. His eyes were re-opened. Habits of verbal abstraction are now checked by a new wealth of visual impacts. Thinking is interrupted by seeing. Speculation is faced with documentation. The image has become partner and rival to the word.

The visual message, being direct and understood by everybody, enormously widened the scope of spreading knowledge and pleasure. Pictures may not always show facts, but thoughts do not always arrive at truth either. The mean of visual attraction may be on a lower level than that of intellectual appeal, but it does lay out an incomparably broader social foundation on which future generations should be able to build a great deal higher.

And that fact by itself makes photography the most significant invention since the discovery of printing. A.K.-K.

See also: *Camera history; Camera obscura; Chronology of photographic inventions; Development history; Historiography of photography; Lens history; Museums and Collections; Sensitized materials history.*

Books: *History of Photography*, by J. M. A. Eder (New York); *The History of Photography*, by H. and A. Gernsheim (Oxford); *The History of Photography*, by Beaumont Newhall (New York); *The History of the Discovery of Photography*, by Georges Pontonniée (New York); *The March of Photography*, by Erich Stenger (London); *A Source Book of Photographic History*, by W. Baier (London); *The Pictorial History of Photography*, by G. F. Pollack (New York).

DISDERI, ANDRÉ ADOLPHE EUGÈNE, 1819–90. French professional photographer and inventor. In 1864 produced a twin lens camera of which the upper lens was used as a finder only. Proposed use of photography for military purposes and drew up syllabus of instruction. In 1853 opened the largest studio in Paris and the following year introduced the photographic carte-de-visite which gave a tremendous stimulus to photography. Photographed many celebrities and published his photographs of celebrities from 1860 on. In 1861 was reputed to be the richest photographer in Europe. Opened branches in Toulon, Madrid, and London; died penniless, blind and deaf.

DISH DEVELOPMENT. Method of developing exposed sensitized materials by immersion in a shallow dish of the developing solution. This method enables the progress of development to be observed under a suitable safelight and controlled by adjustment of development time or sometimes the constitution of the developer.

See also: *Developing negatives; Developing prints.*

DISHES. Processing dishes are commonly used for processing single plates or sheet films. Prints are almost invariably processed in dishes.

The use of an open dish for negative development offers certain advantages over a tank. Development can be inspected under a suitable safelight. A relatively small quantity of solution is used and, after use, thrown away. Fresh solutions, uncontaminated by oxidation products, can thus be used for each negative. The solution can be kept moving to give even development. Rocking the dish is generally sufficient but, when quantitative measurements are to be made on the negative, as in photographic photometry or spectography, a brush can be used on the surface of the emulsion. For rapid processing at abnormally high temperatures, dishes are preferable since they can be readily heated, hold only relatively small volumes of solution and allow vigorous agitation during processing.

Dishes may be made of a variety of materials, the more common being xylonite, porcelain, plastic, enamelled steel and stainless steel.

Xylonite. This is one of the earlier types of moulded material used for making dishes. It has the great merit of being cheap and is reasonably resistant to processing chemicals. It does tend to stain, however, and the surface is not glass smooth and is somewhat difficult to clean. It is advisable to keep one particular dish for developer and another for fixer. The xylonite is usually of a fairly thin gauge and in the larger dishes it tends to buckle.

Porcelain. Porcelain was at one time the most popular material available for dishes principally because it is non-absorbent (at least when new) and has a glazed surface that is easy to clean and does not stain. On the other hand it is heavy and fragile and relatively expensive. After some time in use the glaze becomes covered with a fine network of cracks. When it reaches this stage it may absorb chemicals and cause staining.

Enamelled Steel. This is one of the most satisfactory of the cheaper materials available. It is as impervious to chemicals and as easy to clean as porcelain, but it is lighter and its surface does not tend to "crack" after a period of use. The one disadvantage of enamelled steel dishes is that the enamel chips off if the dish is knocked or dropped. Once the enamel is chipped off the inside of the dish, it exposes the metal to the action of the chemicals and the dish must be thrown away.

Plastics. Several types of plastics are now used for photographic dishes; they are generally cheap, light to handle, and easy to clean. According to the type of plastic used, they may be anything from tough and unbreakable to extremely fragile. As a rule, the harder the surface the more fragile the plastic, but some materials make up for their lack of surface hardness by being water-repellant and therefore stain-free and easy to clean.

Stainless Steel. This material is almost ideal from the functional angle, but it has the great drawback of being expensive. It is non-absorbent, easily cleaned, light in weight and virtually everlasting. While it is not affected by ordinary processing chemicals it may be attacked by some of the chemicals used for processing colour material.

Other Materials. In addition to the above, celluloid, ebonite, bakelite and glass have been used but have now been mostly superseded principally by enamelled steel and plastics. For very large prints it is customary to use shallow wooden trays, either lined with lead, or treated with acid-resisting paint or pitch.

Shapes. All processing dishes are rectangular and conform to the format of the ordinary negative and printing materials. Most—but not all—have a lip at one corner for pouring off the used solution. This lip is generally small, but on the more recent types of plastic dish there is a very generous lip.

The bottoms of dishes are shaped in some way to make it easy to insert a finger tip under the plate or print. Porcelain dishes usually have a depression at one corner and a groove along one side, but these are only effective with a plate or print that fills the dish. The best arrangement consists of two raised parallel ribs down the centre of the dish which hold the whole of the plate or print off the bottom even when it is very much smaller than the dish.

All dishes up to 8 × 10 ins. should be at least 1½ ins. (4 cm.) deep (and larger formats proportionately deeper) to prevent solutions from slopping over the edges. The sides should slope out slightly, and all the corners should be well rounded to make cleaning easy and to leave a gap to allow the edge of the material to be picked up.

Sizes. Most dishes of enamelled steel, plastics or stainless steel are available in a range of standard print sizes.

DISH SIZES

Ins.	*B.S. Size No.*	*cms.*
3½ × 4¾	1	9 × 12
5⅛ × 7⅛	2	13 × 18
6½ × 8½	3	16·5 × 21·6
*7 × 23		17·7 × 58·4
8 × 10	4	20·3 × 25·4
10 × 12	5	25·4 × 30·5
12 × 15	6	30 × 40
14 × 17		35·6 × 43·2
16 × 20	7	40 × 50
20 × 24	8	50 × 60
30 × 40		76 × 100

*Especially for strip postcards.

The size of a dish denotes the largest standard material that the dish will accommodate. This allows a slight gap or margin to spare, but in practice it is always advisable to buy one size larger than the largest material to be processed. Also, since there will generally be more plates or prints in the fixing dish than in the developing dish, the fixing dish should be the larger and deeper of the two. This distinction also makes it easy to tell one dish from the other in the darkroom.

If dishes of a sufficiently large size are not available and it is necessary to resort to see-saw processing, fairly deep dishes should at least be selected. G.B.M.

See also: *Developing negatives; Developing prints.*

DISODIUM PHOSPHATE. Alkali used in developers, principally in fine grain formulæ.

See also: *Sodium phosphate, dibasic.*

DISPERSION. Materials which have a refractive index which varies according to the wavelength of the light—i.e., bend the rays of some colours more than others—are said to disperse light rays. A prism placed in the path of a ray of white light bends the blue and violet rays more than the orange and red, so that it spreads out or disperses the colours as a continuous spectrum.

All transparent materials show dispersion. A lens is usually made of glass and is chosen to

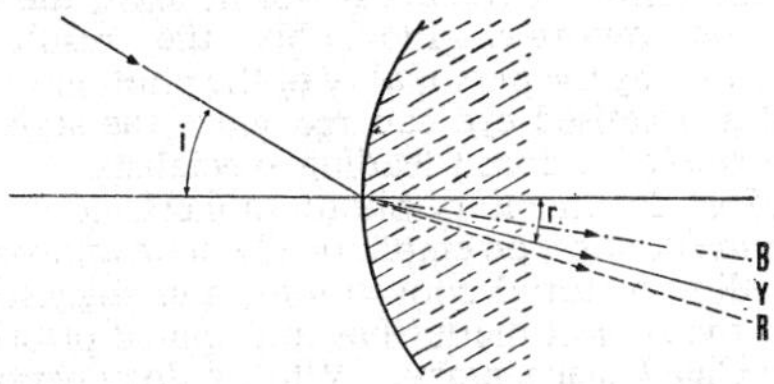

DISPERSION. A ray of white light entering a lens is split up into rays of its component colours, because the angle of refraction r for a given angle of incidence i varies with the wavelength of the light. This is the phenomenon of dispersion. The degree of dispersion depends on the type of glass the lens is made of, but not necessarily on its refractive index.

have a fairly high index of refraction to give it a high optical power for a shallow positive curvature. Unfortunately the unavoidable property of dispersion comes into the picture and the lens, instead of bringing the light rays to a point of sharp focus, spreads them out in a miniature spectrum. This effect is known as chromatic aberration; and designers of better quality lenses have to get rid of it by using more than one type of glass.

Glasses used by lens designers have to be made to very precise limits of refractive index and dispersion because the figure given by the manufacturer is the basis of the mathematical calculations used in designing the lens. R.B.M.

See also: *Aberrations of lenses; Lens.*

DISSOLVE. One of the most used transitions (next to the cut) from one scene to the next in a motion picture film. In a dissolve the second scene starts appearing faintly through the first scene at the beginning of the transition, then gets stronger and stronger while the first scene fades away. It can be produced in an amateur film by fading out one scene, winding the film back to the beginning of the fade in the camera, and then fading in the next scene. The superimposition of the fade-out and fade-in produce the dissolve. In professional motion pictures the dissolve is usually produced at the printing stage.

See also: *Cine technique.*

DISTANCE ESTIMATION. When a camera is focused by scale, the distance of the subject must be estimated or measured. One of the easiest ways of estimating the distance is to imagine how many times a 6 foot man could lie down flat between the camera and the subject. This method can become surprisingly accurate with practice.

The alternative is to measure or pace out the distance, taking the length of a comfortable stride as one yard.

It is useful to remember that an average person about 9 feet away just fills the longer side of the normal viewfinder. At 18 feet away he fills roughly half the finder, and so on. This is an approximate method, and the actual distances will vary a little with the camera used, and from finder to finder. So a few preliminary tests are advisable.

Where subject conditions make accurate estimation of the distance difficult (e.g., moving subjects) it is often better to use some zone focusing system, relying on depth of field.

See also: *Focusing rangefinders.*

DISTANCE GAUGE. Close-up accessory for certain cameras, especially miniatures, consisting of a rod or other fitting of fixed length which is attached to the camera and may carry a frame. When used in conjunction with specified supplementary lenses or extension tubes, the tip of the rod defines the exact camera-to-subject distance for a sharp image on the film. The frame outlines the field of view covered at that close up distance. Alternatively the gauge may consist of a set of extendable legs, the tips of which define both the subject plane and the field.

See also: *Close-ups; Macrophotography.*

DISTORTION. The camera lens normally forms an image which is a faithful representation in scale and perspective of the various elements of the scene in front of it. Any alteration in the shape or proportions of the normal image at any stage of the photographic process is known as distortion.

Some types of distortion are beyond the control of the camera user. One of the principal off-axis aberrations of lenses produces distortion of the image formed in the camera. When a lens suffers from distortion it reproduces straight lines away from the centre of the field as curves. Those lines whose middles curve inwards give what is known as pin-cushion distortion and those curving outwards barrel-distortion.

Other types of distortion can be deliberately produced by the photographer either for novel effects or to produce a more pleasing representation of the scene.

The use of a focal plane shutter when photographing fast moving subjects also gives rise to distortion, as different parts of the film are uncovered at different instants while the image changes in position.

Perspective distortion, which apparently exaggerates the relative proportions of objects close to the camera, is not in fact distortion; it is a true perspective image for the abnormally close viewpoint. This effect however is often used deliberately in such fields as, for example, architectural photography, for the sake of originality. Conversely the relative proportions of objects can be minimized by taking a more distant view-point and using a lens of longer focal length. This will still achieve the desired size of image on the negative.

See also: *Aberrations of lenses; Camera movements; Perspective; Shutter; Tricks.*

DISTORTION OF MATERIALS. Change in size of a photographic image due to expansion or shrinkage of the base support under the influence of temperature, relative humidity, processing or ageing.

See also: *Dimensional stability; Supports for emulsions.*

DIVERGING LENS. Lens which is thinner in the centre than at the edge. It has at least one concave surface. Rays of light passing through a diverging lens are refracted away from the optical axis. A diverging lens can only form a virtual image of the subject—i.e., one which cannot be focused on to a screen; it cannot form a real image as does a converging lens.

DOCUMENTARY PHOTOGRAPHY. At its lowest level, documentary photography is mere factual reporting, but in the hands of a master it can achieve the status of an art form.

Uses. Documentary photographs of the latter type are a popular feature in the illustrated magazines and even in the daily and weekly newspapers. Their principal value in this sphere is in bringing the reader a picture of the lives and customs of people normally outside his experience. Folk activities, peasant life, strange customs and crafts, are all a fruitful field for the documentary photographer.

At the same time, the famous documentary specialists find their material in less obvious and exotic quarters. These men are artists rather than globe-trotting reporters.

Documentary photography of this type has a very positive value in propaganda and education. Examples can be seen in the publications of political and religious groups, advertising literature and industrial house magazines.

Technique. Documentary photography is practised with all types of equipment from 8 × 10 ins. cameras down to sub-miniatures.

The aim is always to produce unselfconscious pictures, and most workers favour an unobtrusive camera and candid approach. Even so, some world-famous specialists use large stand cameras and achieve the unposed effect by skilful handling of the subject.

There is thus no established documentary technique; it is simply a matter of applying the principles of normal photography directed by the vision of the artist—or at least, the eye of the reporter. Inevitably the result is coloured by the personality of the photographer and a practised eye can recognize the style of any one of a dozen leading specialists.

Most documentary photographers, however, assume or affect an emphatically plain approach avoiding pictorial conventions, and suggesting that the subject matter has undisputed priority.

The Cine Documentary. While a documentary work may be no more than a single print, it is more often a series of prints, since it deals with life and movement. For the same reason that several prints are generally better than one, a cine film is better than a series of stills. The documentary film has in fact reached a high state of perfection in recent years. Public interest in this type of film is high; so much so that a number of documentary films have unexpectedly scored major successes in terms of the actual box office receipts.

See also: *Candid photography; Photo journalism; Picture series; Press photography.*

DOCUMENT COPYING

Photographic document copying is a method of making a factual copy, normally on paper, of an existing original or copy document. This can be printed, handwritten or drawn and may contain illustrations, both photographic (continuous tone) or reproductions (half-tone), or diagrams in monochrome and colour. There are various processes of document photography to copy such documents. Some have limitations and may not be able to deal adequately with all originals, especially those containing both illustrations and text, and some are unable to reproduce certain colours. As a general rule, these processes copy by contact printing and therefore produce same-size copies, but a few use optical methods and some can vary the size of the copy from that of the original.

The photo-copying methods belonging to the drawing office techniques (often called plan copying) are blue print (ferro-prussiate), diazo and true-to-scale (T.T.S.) and all these will print from a translucent master. Such masters are normally prepared by draughtsmen or tracers, but numerous methods of making them from opaque originals are also available. But most of the other document copying methods can make copies from opaque or double-sided originals and some do so directly, without making a negative intermediate.

PROCESSES

All these methods can be classified into five categories, namely:

(1) Plan copying, as previously outlined.
(2) Silver processes, which include negative-positive, reflex and direct positive processes.
(3) Transfer methods (diffusion and gelatin).
(4) Thermographic (heat) systems.
(5) Electro-photographic processes.

Blue Prints. The most familiar process is that commonly used for copying architects' and engineers' drawings and plans to give facsimile copies—i.e., copies of the same size as the original. An accurate tracing of the drawing is first made on a transparent or translucent paper or tissue. This is then placed on a sheet of sensitized paper with its back to the emulsion, and held in close contact with it by pressure. The sensitized paper is then exposed to light through the tracing and it receives an impression which may be visible and needs only to be fixed, or it may have to be developed. This depends upon the type of sensitized material used in the particular process.

In its simplest form, this process, known as the blue-print, cyanotype, or ferroprussiate process, was discovered by Sir John Herschel in 1842. It depends upon the fact that when light acts on certain sensitive ferric or iron salts, it can reduce them from the ferric to the ferrous state. In the well-known blue-print process the action of the light turns the treated paper blue, while the parts shaded from the action of the light by the lines of the drawing remain white. When the blue print is deep enough the whole print is well washed in water. This dissolves and washes away the unused ferric salt and fixes the image.

The blue print process uses ferro-prussiate paper, which is of low sensitivity and so is exposed in a photo-printing machine having a powerful light source. By using a translucent intermediate negative master, a positive copy having blue lines on a white ground is obtained and special chemicals can be used to strengthen the density of the blue image. The blue print is sometimes preferred by engineers because it will stand rough handling in the workshop, but generally the method is being superseded by diazo type. The ferro-gallate or black-line process giving blue-black lines on a white ground, was also used for some time.

Diazotype (Dyeline). Diazo paper uses a diazo

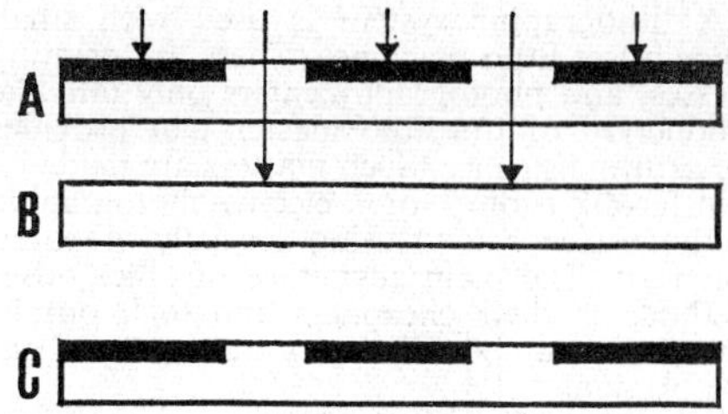

DIAZO PROCESS. The original for diazo duplication is printed, typed etc. on translucent paper. Light passing through this decomposes the diazo compound in the diazo paper. The unexposed compound is then coupled (developed) to form a dye image. This can be done in a solution, by ammonia vapour or by heated rollers. A, original. B, diazo paper. C, direct copy (positive-positive) after coupling.

type instead of iron salts and is exposed in the same type of apparatus. Being a positive method, it will produce directly from the normal translucent master a positive copy which could be one of many available tones. There are three types of material which differ in the method of development (coupling) and are generally described as semi-dry, ammonia or heat.

The semi-dry type is coupled with a single solution developer containing ammonia but emerges from the machine in a dryish state. The dry types are processed by exposure to ammonia vapour or heat. The necessity of extracting the ammonia fumes from the vapour is a small disadvantage, but the heat method is mobile although there is a tendency at present with this type to emit faint but unpleasant odours.

Diazo is extensively used in administrative offices for making carbon copies and for systems work. Small machines of elegant design have been introduced for this purpose, together with special translucent typing papers and diazo material of much better quality, which is able to produce a copy of good density. Because this material can be operated in daylight, is simple to use and is the cheapest form of sensitized material available, it has many advantages.

Lithographic Systems. The true-to-scale (T.T.S.) method requires more operational skill because it can be affected by temperature and humidity changes. In principle it is lithographic and is based on the premise that grease and water will not mix. The apparatus consists of three rollers which hold a continuous belt of linoleum and this is coated with an emulsion prepared by heating and known as the plate. When set, this emulsion receives an image from exposed but undeveloped ferro-prussiate paper, the image being inked and transferred manually to a receiving base which can be translucent, opaque, card or paper. The quality of the copy produced is of a high order and it is therefore a good method of making intermediates for diazo or other processes. The term "true-to-scale" is, however, a misnomer and should not be taken too literally.

A lithographic system is used with small office offset litho machines. These are printing presses, and photography enters only into the production of the litho master that provides the printing surface. Such masters are made by transfer-diffusion or electro-photographic methods (see below) using specially prepared materials. The main advantage of office offset methods is their cheapness and high output when large numbers of copies are required from one original.

Silver Processes. Most negative-positive processes, whether contact printing or photography in a copying camera, work on the same principle as normal negative-positive black-and-white photography. They thus involve the production of a negative and the printing of a positive. The negatives may be made on high-contrast films or plates (usually in a camera) or document papers (in a camera or a contact printer). This negative is then printed (usually by contact) on to a document paper to yield the positive. The negative and positive go through the usual processing stages of development, fixing and washing, followed by drying.

These processing stages tend to be lengthy and tedious, so modern negative-positive document processes frequently use rapid processing papers that incorporate the developing agent in the emulsion. Such papers are suitable for roller processing in developing machines; the paper there passes between a set of usually motor-driven rollers that apply a film of activator solution to the paper surface. This develops up the image within a few seconds. The paper then moves through a stabilizing bath and through a pair of squeegee rollers to emerge as a semi-dry print. Frequently the same paper is used for making the negative and the positive. Special papers are also available which give good copies from continuous-tone photographs and from normal photographic film negatives (also continuous-tone) in a fraction of the time required by traditional methods.

Self-positive papers use pre-fogged silver halide emulsions that give a positive copy by direct exposure. The exposure destroys the latent image fog by solarization or by the Herschel effect (if the exposure is carried out by orange, yellow or red light) so that the exposed portions remain white while the unexposed portions develop up black. Self-positive papers are usually processed by a development and stabilization technique.

Transfer Methods. These first make a negative from which a positive copy is transferred (without exposure) to a specially prepared paper.

The commonest method, diffusion transfer, uses a silver halide system. A sensitive negative paper is exposed to the original document, and processed in contact with a non-sensitive positive paper. As the negative image develops in the negative paper, the unexposed silver salts diffuse into the positive layer and are reduced by chemical fogging to a positive silver image. Normally transfer-diffusion yields one positive per working cycle (exposure of the negative and processing of the negative and positive papers). Several positive copies can be made from special negative materials by repeating the processing stage. The positive can also be on film (for making translucent masters for diazo copying), on metal plates for offset lithography, or on special materials to yield carbon masters for hectographic (spirit printing) duplicating processes.

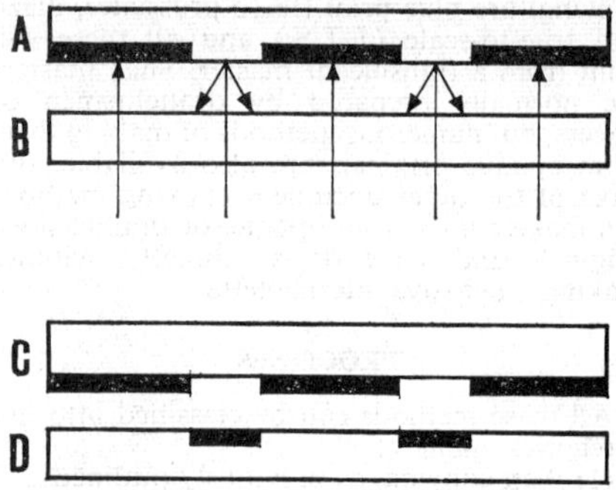

DIFFUSION TRANSFER. The negative is exposed by reflex copying and then passed in contact with a positive sheet through a developing solution. The unexposed sensitive compounds in the negative diffuse into the positive and there form a silver image. A, *original.* B, *undeveloped negative.* C, *positive.* D, *negative during development.*

Dye transfer uses a special emulsion which contains a dye, sensitive silver salts and developing chemicals. On activation in the processing solution these develop the silver salts, couple the dye and tan or harden the gelatin. In the

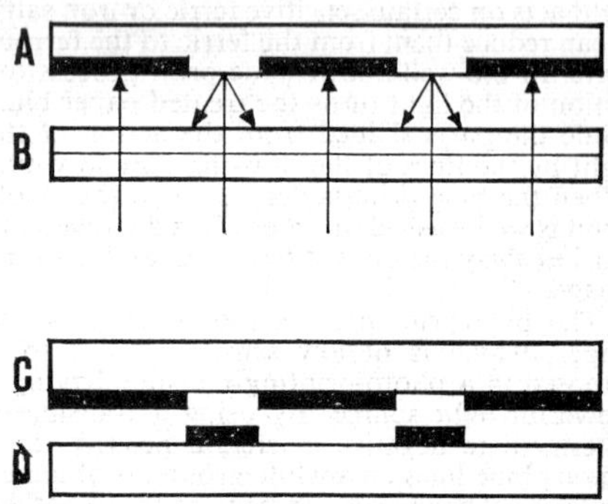

DYE TRANSFER. This uses a matrix with a gelatine coated surface which contains a dye and a mixture which can be activated to develop the exposed silver salts, couple the dye and tan or harden the gelatine. The matrix is exposed by reflection and after controlled development pressed against a positive paper. This picks up a thin layer of unhardened dyed gelatine from the unexposed areas, thus producing a positive image from a positive. A matrix may produce up to six copies by repeated transfer. A, original. B, matrix. C, copy. D, activated matrix.

resulting matrix exposed areas are hardened gelatin and the unexposed portions soft gelatin that can be transferred. After development the matrix is pressed against the positive paper which picks up a thin layer of unhardened dyed gelatin from the unexposed areas, thus producing a positive image. One matrix can give about six copies; the transfer is made in quick succession and so is cheap operationally.

As all transfer methods involve a lateral reversal of the image, they are usually employed with reflex printing methods where the negative image is reversed left to right at the exposure stage. Specially sensitive dye transfer papers also exist for exposure in a camera at an enlarged or reduced scale. These can also be used for making printing masters for offset lithography.

Thermographic Systems. These involve exposure by infra-red radiation, use no liquid chemicals, and produce instant dry copies. For a thermographic copy a heat-sensitive paper is exposed, by the reflex method, in contact with the original to be copied. The heat rays reflected

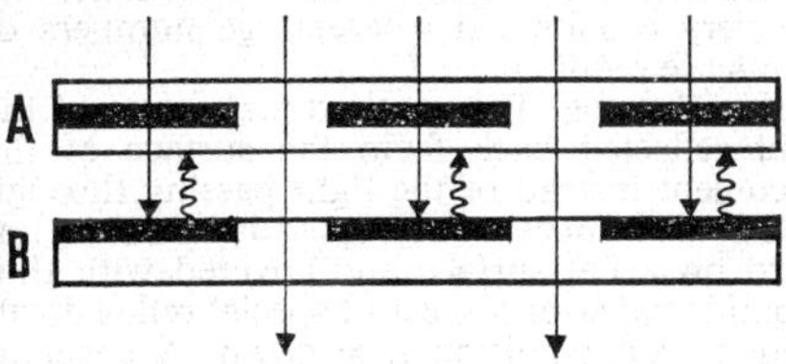

HEAT COPYING. The paper contains a heat sensitive compound and is backed up by the original. This sandwich is exposed to an infra-red source. The infra-red rays are absorbed by the carbon or metallic part of the image; the reflected heat forms an image on tho copy paper. A, heat sensitive paper. B, original reflecting heat and thus forming a positive copy from a positive original.

by the dark portions (text, etc.) of the original darken the thermographic paper, producing a positive image. To obtain a right-reading copy, the thermographic paper is exposed with its back in contact with the original; as a result the image is usually slightly unsharp. This process takes a few seconds, and is thus particularly suitable for automatic office copying machines which require no attention beyond feeding the original and the heat-sensitive paper through the unit.

An alternative system uses a carbon paper which is softened by heat and then transfers an image from the original on to normal paper. Another process uses a black paper with an opaque white coating. The heat melts this coating, allowing the black ground to show through as an image.

Since all heat processes depend on the reflection of infra-red radiation, they work only with originals printed or written in inorganic pigment inks. This covers typewritten and printed document material, but not originals written or printed in dye inks.

Electro-photographic Processes. These rely on the image-wise discharge of an electrically charged surface on exposure to light. The surface then carries an electrostatic image.

In xerography, the electrostatic image on a selenium coated plate is made visible by dusting with charged pigment particles. The pigment image is then transferred to ordinary paper and fused into the paper by heat treatment. This system is inexpensive to run, since the copies are made on ordinary paper; the equipment, however, is expensive.

A second system forms the electrostatic image on a zinc oxide coated paper which is then dusted with pigment powder and the latter fused into it.

Electrophotographic methods are not suitable for contact printing, since the electrostatic image would be destroyed on contact with another surface. The image must therefore be projected on to the plate or paper in a special camera, but copies can be made on any enlarged or reduced scale. Electro-photography is therefore used for re-enlarging microfilm negatives or positives. By appropriate choice of the electric charge of the plate and of the pigment powder, the process can yield a positive from a negative or a positive from a positive.

Electro-photographic methods are versatile because they can produce, in addition to document copies, masters for the diazo, hectographic and lithographic processes. They do not easily copy solid areas and for this reason are less suitable for continuous tone copying. New materials have improved these weaknesses, but further progress is still desirable.

Costing. The comparative costs of document copying methods depend on the actual cost of the material used for making the copy as well as on the investment in the equipment required and the efficiency and speed with which copies can be produced.

Material costs are lowest with the processes that produce the final copies on ordinary paper—offset litho and xerography. Offset litho is also capable of producing copies at a very fast rate, but requires a photographic litho master. Where many copies are required, the cost of the master per copy becomes negligible.

Xerography is rather slower since every copy requires a complete cycle of exposure, pigment processing, transfer and heat fusion. The high cost of equipment and operating time is balanced only by the extra versatility of the process; for mass output, xerography is primarily useful as a means of producing masters for the other processes. Automatic continuous flow xerography, utilizing a selenium drum, can appreciably speed up the operating cycle but needs even more expensive gear.

Plan copying methods, especially dyeline, use the cheapest sensitized material for the final copy. They are thus economical for a

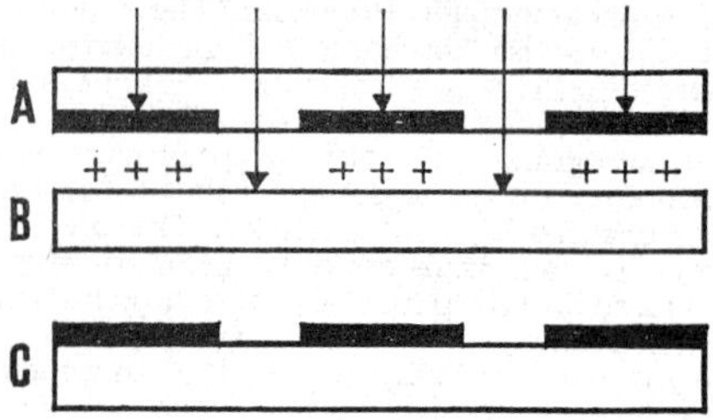

ELECTRO-PHOTOGRAPHIC COPYING. The original is optically projected on to a charged surface (selenium-coated metal or paper coated with resin and zinc oxide). Reflected light from the non-image areas of the original dissipate the charge, allowing an oppositely charged powder to adhere to the charged areas. The powder is fused on to the paper or transferred from the selenium to the paper and fused there. A, original. B, charged paper or metal surface. C, fused image copy. This yields a positive from a positive.

reasonable volume of multiple copies where the acquisition of offset litho machinery is not possible or desirable. Dyeline printing also requires a master—made photographically on translucent paper or film—and the speed of operation depends largely on the complexity of the exposing and processing machinery.

Thermographic processes are slightly more costly per copy (and the print cost is independent of the number of copies) but attractive for their simplicity.

Negative-positive (and direct positive) silver processes cost still more per copy, since the document paper is more expensive. For single copies they are comparatively slow, since—with negative-positive printing—both a negative and a positive must be made. One negative can however give any number of positives at a constant cost.

Transfer methods are slightly quicker for single copies since the final print is made in one stage. For a few copies, up to half a dozen, multiple transfer materials are suitable; for single copies and for high quality multiple copies, transfer processes are more expensive since every copy requires a negative and a positive sheet of paper.

TECHNIQUES

In office copying systems, prints are usually made by contact printing or reflex copying. This yields copies of the same size as the original. Where enlarged or reduced copies are required, the original must be re-photographed in a camera.

Contact Printing. Translucent originals, for example letters written on one side of the paper only, can be copied by transmitted light. This is suitable for silver and dyeline processes, and yields a negative or positive in one stage. If only a negative is required for reading, or with processes yielding a direct positive copy (self-positive papers, thermographic systems, etc.), the original must be placed with its back in contact with the sensitive surface of the copying material in order to yield a right-reading print. If the negative is used only in the intermediate stage for positive copies, the sensitive layer of the negative paper can be in contact with the image surface of the original.

Contact Printing from Tracings. In the modern drawing offices special printing machines are used for making prints from master tracings. These machines incorporate a stand supporting a hollow glass cylinder in which a fluorescent tube is mounted. The cylinder is turned slowly by an electric motor and at the same time the carbon arc travels continuously to and fro between the ends of the cylinder. The tracing and a sheet of blue or black line printing paper are wrapped around the cylinder and held in close contact by an outer covering of tightly stretched fine-wove canvas. The traversing speed of the tube can be adjusted to give exposures from 20 seconds to about 1½ minutes.

In large offices a modern printing unit, takes the print through automatically, developing it with ammonia vapour and delivering the finished copy ready for use. These machines are very economical where large numbers of prints are required.

Reflex Copying. This method makes use of the light reflected back from the surface of the document instead of the light passing through it. The document to be copied is laid face upward on a flat surface and covered with thin bromide paper or a sheet of special reflex document paper, emulsion face down. A sheet of plate glass is used to apply pressure to keep the two in close contact. Light is then allowed to pass through the back of the sensitive paper. This light is the same all over the copying paper, but more light is reflected back from the white paper of the document than from the dark print. So the copying paper in contact with the lighter parts of the document gets an extra dose of light—just enough to make it turn black on development while the printed parts remain white.

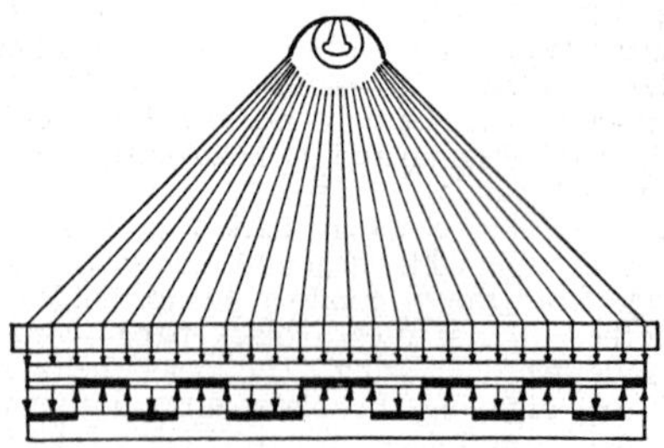

REFLEX COPYING. Light from above passes through a glass sheet and then through the sensitive paper laid emulsion side down on the document to be copied. Dark areas of the document (writing) absorb the light, white ones reflect it back on the emulsion, and thus produce a negative reversed image. Re-copying by the same process produces a positive.

Exposures are in the region of 5 to 25 seconds, the exact time being determined by making test exposures. The development, fixing, and washing of the print are carried out just as for a bromide print, and the final result is a negative of the document. Any number of positive prints can be made from this negative by contact printing in the usual way, through the negative on to a sheet of bromide paper.

Reflex copying is used both with negative-positive silver processes (usually employing special rapid processing papers with incorporated developing agent) and with transfer processes. Reflex copying is virtually the only contact printing system suitable for transfer materials since the image is reversed left-to-right during transfer and so needs the already reversed image produced by the reflex process. The same double reversal takes place in making the negative and from it the positive in the negative-positive process. Self-positive papers are not suitable for reflex copying since no second reversal of the image can take place.

Scale Copies. There are many cases in which it is important to have scale copies either larger than, equal in size to, or smaller than the original. Facsimile copies by reflex printing of very large plans would be unnecessarily bulky and expensive, while small documents are made much easier to read and handle if they are enlarged. Copies of this sort can be made in any type of camera from a miniature upwards, but half or whole plate sizes are generally used because they give usable prints direct from the negative without further enlargement.

The copying camera may be mounted vertically or horizontally and must have ample extension and a lens with a perfectly flat field and high resolution. Ordinary photographic films and plates may be used, but paper-based emulsions are more usual because they are cheaper and easier to store. These document papers are supplied in a wide range of thicknesses and emulsion speeds on either opaque or translucent paper bases.

A special stripping paper is also manufactured from which the hardened gelatin coating is stripped away after it has been exposed and processed. The resulting transparent negative can then be used for printing off any number of positives or it may be read as a positive through the back of the gelatin.

Camera copying is suitable for both negative-positive and transfer systems. When document papers are used for the negative, the positives are then produced by contact printing. Transfer materials need intense lighting of the original, since their speed is usually rather low. Camera copying is also the only way of making duplicates by electro-photographic processes.

For casual copying of originals of various sizes, reduced-scale film negatives can be made in an ordinary camera with appropriate close-up accessories; these negatives are then re-enlarged either to original or to a smaller size.

Microfilming. This is a special application of camera copying and involves reducing originals on to 35 mm. or 16 mm. film for particularly compact storage. The information from microfilm negatives is usually extracted by viewing in suitable viewers which project a magnified image on a screen. Microfilm negatives may also be enlarged on document papers when full size copies are required for reference or distribution.

Continuous-tone Originals. Most document copying materials yield an image of extreme contrast and are thus primarily suitable for reproducing text and line illustrations. Half-tone images—i.e. reproductions of photographs broken up into larger or smaller black dots as in a newspaper or book—can be copied in the same way but usually require shorter exposures to prevent the dot pattern of the image from blocking up. The copying material in this case must have a higher resolving power and—in contact copying—the original and copying paper must be in very intimate contact to reproduce the dots sharply.

Continuous-tone originals, such as photographic prints, are usually copied in a camera on to a soft working film (or plate). Duplicates can then be made by contact printing or enlarging the film on to a medium- or high-contrast bromide paper.

This is generally possible only with silver negative-positive systems. Continuous tone originals can be reproduced by transfer, dyeline and electro-photographic methods if they are first converted into a half-tone image. This involves making, for instance, a dyeline master on film by photographing the original with a half-tone contact screen in contact with the photographic film. A positive transparency made from this half-tone negative can then serve as the dyeline master. For electro-photographic reproduction the half-tone negative itself can be projected on to the xerographic plate or electro-photographic paper. Equally such a negative can be printed by contact or enlargement on to a high-contrast document paper for any number of positive half-tone prints.

With transfer processes, especially diffusion transfer papers, the half-tone negative is made by photographing the original in a copying camera, with the half-tone screen in contact with the negative paper. This is then processed in the usual way in contact with a positive paper to yield a positive half-tone print.

Half-tone contact screens are not suitable for contact printing processes, since the screen must separate the original from the sensitive material; hence this involves some loss of detail. Equally, half-tone screens are not suitable for reflex printing.

Special Techniques. The application of document photographing principles to stencil and jig making has been of the utmost value in speeding up production and shortening the

SENSITIZED MATERIALS USED IN DOCUMENT PHOTOGRAPHY

Type	Form	Colour Sensitivity	Speed	Contrast	Use
Films and Plates					
Contact	Sheet film	Blue only	Very slow	Extreme	Contact printing from negatives
Process	Sheet or plate	Blue only	Slow/med.	Med./high	General copying
Process Ortho	Sheet or plate	Ortho	Slow/med.	Med./high	Faded originals, etc.
Process Pan	Sheet or plate	Pan	Slow/med.	Med./high	Copying from colour
Microfilm	35 mm. film	Ortho or pan	Slow	Extreme	Microfilming and copying
Fast Microfilm	35 mm. film	Pan	Medium	High	Automatic microfilming
Positive F.G.	35 mm. film	Blue only	Slow	High	Printing microfilm negatives
Papers					
Contact Document and Reflex	Rolls or sheets	Ortho	Very slow	Extreme	Contact copying, reflex negatives printing document negatives
Medium Document	Rolls or sheets	Blue only	Slow	Medium	Camera copying, enlarging
Rapid Document	Rolls or sheets	Ortho	Med./fast	High	Camera copying, enlarging
Direct Positive	Sheets	Ortho	Slow	High	Direct positive copying by contact or reflex
Rapid processing	Sheets	Blue	Slow to fast	Med./high or extreme	Contact copying, reflex negatives, printing document and microfilm negatives; processed in roller processing units
Transfer-diffusion	Sheets	Ortho	Slow/med.	Extreme	Reflex copying
Dye transfer	Sheets	Blue	Slow	Extreme	Reflex copying
Dyeline	Rolls or sheets	Blue	Very slow	Extreme	Contact copying from positive translucent masters
Electro-photo	Rolls or sheets	Blue	Very slow	High	Enlarging microfilm negatives, camera copying
Thermographic	Sheets	Heat sens. only		High	Contact printing of documents in office machines

several procedures that intervene between the drawing office and planning and layout lofts in aircraft and marine engineering where large and costly templates, jigs, patterns, and stencils are used.

By an extension of the methods of document photography it has been possible to project on to sheet metal, plastic or other appropriate material the images, in exact scale, of the forms into which the material has to be cut. This method of marking off the work for machining does away with templates by printing or projecting an absolutely accurate photographic scale image on to the actual material to be used.

Originals that have become faded, stained, or even completely bleached or obliterated can often be rescued by photographic reproduction. The particular technique used depends upon the condition of the original.

Where the document or its contents are coloured, it is normal practice to increase the contrast of the final copy by using a panchromatic plate and a suitable filter. For coloured script, a panchromatic plate and a filter of the complementary colour are used to darken the tone of the script. For coloured paper, a filter of the same colour as the paper is used to make the paper reproduce in lighter tones.

It follows that for old documents where the script has faded and is yellow or brown but the background is reasonably white, an ordinary or panchromatic film with a blue filter will restore the contrast by darkening the tone of the script. Where the document itself has yellowed and gone darker, a high contrast panchromatic film with a deep yellow filter will give the necessary contrast by lightening the tones of the paper.

When blue prints have to be copied, panchromatic films or plates are used with a tricolour red filter to make the blue of the print come out black in the photograph. For Photostat paper which is a fairly deep yellow, a deep yellow filter is used to make the ground come out white.

Phototelegraphy. Phototelegraphy is another direct development of document photography. Here, the document is wrapped around a cylinder which rotates while a beam of light scans it from end to end. A photo-electric cell picks up the light reflected from each minute area of the document in turn, and changes it into electrical energy which can be sent along a wire or transmitted by radio. At the receiving end, the electrical impulses are used to control the strength of a similar beam of light traversing a sheet of sensitized paper mounted on a cylinder rotating in synchronism with the sending apparatus. The result is an image that can be developed out to a very fair copy of the original picture. Photographs transmitted in this way are often used in newspapers.

Illumination. In commercial copying equipment the lighting unit or units is an integral part of the apparatus. It is always designed to provide even illumination over the whole area to be photographed, and the light is designed to give a correctly exposed negative with the standard setting of lens, image distance and illumination. Some adjustment is possible, but it is generally better to stick to a standard setting and standard conditions.

In much of the commercial apparatus automatically controlled Photoflood lighting is used, the full power of the lights being on only during the actual period of exposure. As a general guide for work in a darkroom, four 100-watt lamps in reflectors at four corners of the table or support and 2 feet away from the surface upon which the document will rest should be adequate for ordinary copying. With fast panchromatic film or equivalent material and a lens aperture of f 11, the exposure will be about $\frac{1}{4}$ second. For micro-negative film the exposure will be about 2 seconds.

Copying equipment working with dyeline or other low-speed materials sometimes uses high-intensity mercury vapour lamps.

Special Lighting. Often there is no combination of film and filter that will make the written or printed matter stand out from its background when it is illuminated by the normal type of day or artificial lighting. When this happens, special forms of lighting like infra-red and ultra-violet will often give the right answer.

Infra-red light has been used successfully in photographing ancient manuscripts on leather and hide. It has also been successful in photographing charred and burned documents, in deciphering writing and printing obliterated by a censor, and in revealing the presence of chemically bleached writing on cheques, stock certificates and similar documents.

To find out if a document can be improved by infra-red light photography, it is first looked at through a deep red filter such as tricolor red. True infra-red filters are used for taking the actual photograph; they are useless for visual inspection. Infra-red negative stock must of course be used in the camera.

Ultra-violet light causes many inks and papers to fluoresce to a greater or less extent. This often makes it possible to photograph matter that has either completely faded or been intentionally bleached out or obliterated. The source of ultra-violet light is usually a quartz mercury vapour lamp used in conjunction with a filter known as a Wood's Glass filter which transmits only ultra-violet light and holds back all visible light.

The document is set up in front of the camera and photographed in complete darkness under the ultra-violet light. In this case there is no need for special negative material because the normal panchromatic plate is sensitive to the fluorescent light given off by the paper or script under this type of lighting.

Ultra-violet light has been used to reveal alterations on documents—forged dates, signatures, and the like—and for detecting the presence of otherwise invisible finger prints.

Sensitized Materials. There is quite a range of sensitized materials used in document photography. Plates, as a general rule, are not used.

For camera copying, the usual material is a contrasty process film, made in a variety of types and sizes by all main manufacturers of sensitive materials. They are further available in medium to high contrast, and blue-sensitive, orthochromatic or panchromatic.

Films for micro-copying (microfilming) are supplied in widths from 16 to 105 mm.; the gauges from 16 to 70 mm. may be perforated. Papers for copying cameras are supplied in rolls or sheets. Appropriate materials are supplied by the sponsors of the various special office copying systems such as diffusion, dyeline, electro-photography, etc.

The speed of particular types of material may vary with individual manufacturers; with equipment that is not precalibrated for automatic exposure, it is advisable to determine the correct exposure by actual tests on the sensitized material to be used.

EQUIPMENT

The apparatus used for document copying must on the one hand be able to handle the materials for a particular process, and on the other cope with a range of originals of different types. Since much of this equipment is used in offices by people not trained in photographic procedures, a certain degree of automation is also desirable to ensure correct exposure and processing of the document copying material.

For small-scale use there are two main types of copying equipment: flat-bed and rotary printers.

Flat Bed Systems. The basis of a flat bed copy-

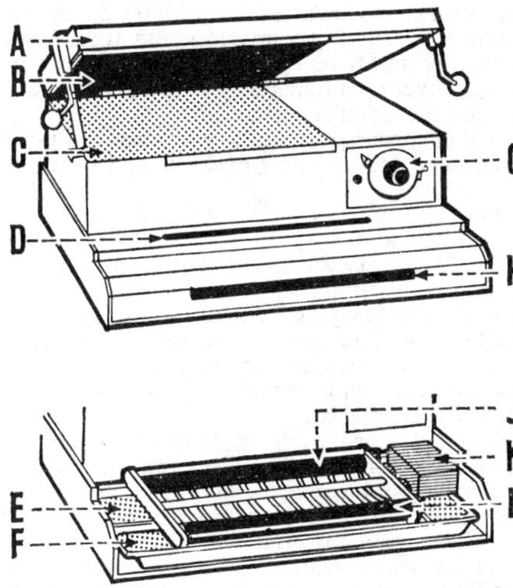

REFLEX OFFICE COPIER. Top: *The unit normally incorporates an exposing light box with built-in lamps and timer, plus a motor-driven roller processing unit.* Bottom: *The processing unit consists of the solution troughs for transfer diffusion or (as shown here) development and stabilization. A, lid. B, pressure plate. C, opal glass covering light box (the document paper and original are placed on this). D, delivery slot of processor. E, stabilizing bath. F, developing (activator) bath. G, timer. H, feed-in slot. J, squeegee rollers. K, drive motor. L, developing rollers.*

ing unit is a printing box with a number of lamps underneath a diffusing screen to illuminate evenly the full area of the document being copied. A lid lined with sponge rubber or foam plastic (or sometimes with a spring-loaded pressure plate) presses the original and document material into intimate contact.

The printing box may incorporate a mechanical or electronic timer connected to the lamps. Since the illumination is constant, the required exposure for any given process depends mainly on the sensitivity of the copying material; once this is determined for average originals, it only needs slight variation for different types of original. With two-stage copying processes, for example negative-positive, the two stages may require different exposures. On some copying machines these can be selected by push button controls once a basic exposure has been set.

Modern flat bed copiers for development and diffusion-transfer materials are frequently combined with a roller processing machine for the papers. This is motor driven and carries the paper through the processing solutions at a rate which ensures full development. The transport system for negative-positive processes differs from that for diffusion transfer materials, since the latter use only a single processing solution. In transfer processing machines the negative and positive papers are fed into the developing solution through separate slots which keep them apart until they are seized by the rollers and pressed into contact (once the emulsion layers are saturated with the developer solution). Usually the machine has three such slots to permit two negative sheets to be pressed into contact simultaneously with the two sides of a double-sided positive.

Roller processing units are also available separately for negative-positive and transfer systems.

Flat bed copiers are made in a number of sizes from about $8\frac{1}{2} \times 13$ ins. (foolscap) to 16×20 ins. or even larger. The flat bed copier is suitable for single sheet originals (e.g. letters) as well as for copying from periodicals and even books. Flat bed copiers for the dye transfer processes are similar in principle but have appropriately different arrangements for processing the material.

Some flat bed printers, especially those intended for making offset litho and other printing masters, have a vacuum unit to ensure absolutely perfect contact between the original and the copying material.

Rotary Copiers. These are more automatic than the flat bed machines and are suitable only for single sheet originals. The original and the copying paper is here fed into the machine past an exposure light (usually a strip lamp) or around an illuminated translucent drum. In either case the speed with which the material passes through the exposure station determines the actual exposure. This is matched to the sensitivity of the material used, but may with some machines be adjustable. After exposure the original and document paper are returned to the operator who then feeds the document paper through a motorized processing section. With some machines this is completely automatic and the machine itself delivers the original and the finished copy through separate outlets.

The rotary copier exists in various modifications for specific processes; for instance thermographic machines have an infra-red source for the exposure and no processing section.

Inexpensive contact printers may press the original and copying material into contact around the surface of a translucent plastic half cylinder containing the exposing lamps. This again is suitable only for single sheet originals, and the papers are pressed together with the aid of a flexible and spring tensioned apron. Such printers are slower to work, but suitable for more casual copying work.

Dyeline Machines. Small printers for diazo processes differ from negative-positive and transfer copiers in the light source (usually a mercury-vapour lamp since the dyeline paper is most sensitive in the ultra-violet spectral region) and in the processing arrangements. These may subject the paper to liquid treatment (to an ammonia solution—though some current materials are developed in liquids with a less objectionable smell), to ammonia gas or to heat. The exposure and processing sections may be separate, or combined into one unit.

Large dyeline printers used in drawing offices may handle papers up to 70 ins. wide. They consist of what is in effect a scaled up printing frame to take the original (usually a tracing or translucent master) and the dyeline paper. Exposure is to an ultra-violet source, after which the dyeline paper is processed.

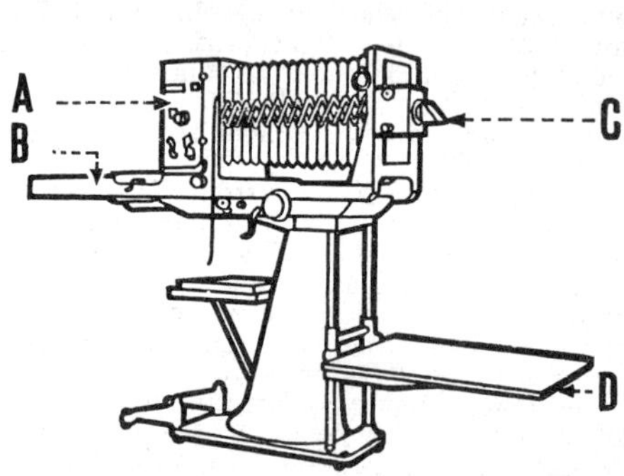

COPYING CAMERA. This is a large camera which holds paper in rolls up to 18 ins. wide and 350 ft long. The original is photographed via a mirror and lens system which produces a negative tonal copy on processing, which is quite correct reading. The developing and fixing trays are fitted to the rear of the camera, the paper being moved directly to the developer and transferred manually to the fixing bath. Washing and drying apparatus are quite separate. A, paper holder. B, processing trays. C, mirror and lens. D, copy holder.

Duplicating Machines. Hectographic, spirit printing and offset litho units are not photographic in operation, though they may use a photographic process for obtaining the printing master. The range is very wide and the machines vary in size and price according to the required output and format of duplicates.

Camera Copying. Large scale copying cameras can produce negatives or positives on any enlarged or reduced scale. A type of camera in use for many years—though now largely superseded by smaller and more convenient apparatus—takes the document paper in continuous rolls up to 350 feet long and 14 ins. wide. The paper is fed through the focal plane of the camera and cut off after every exposure. The sheet then passes through processing trays holding the developer and fixing bath. Such cameras are comparatively bulky and are often mounted on horizontal pedestals, photographing downwards on to a copy holder via a mirror or prism. This also reverses the image left to right, to yield a right-reading negative which can be used as it stands or contact printed to produce positive copies.

More modern document copying cameras frequently use a vertical layout, photographing downwards on to an easel with lights arranged to each side of it. The set-up is in effect an inverted vertical enlarger. The document copying material is placed in the top of a camera and may be held flat by a simple pressure plate or a vacuum board. Cameras of this kind are suitable both for making negatives by the normal negative-positive process and for exposure of the negative sheet of a transfer diffusion material. In the latter case no reversing mirror is needed, since a left-to-right reversal of the image takes place on transfer to the positive.

Special camera systems are also used for the electro-photographic processes, the camera being loaded with the selenium plate on which the electrostatic image is formed. The camera equipment there includes the apparatus for charging the plate and for dusting on the pigment powder and transfer to the final paper support.

Optical systems are also used for re-enlarging micro film negatives on to various types of document paper or on to sensitized material for forming masters for duplicating processes (offset litho, dyeline, etc.).

Continuous Flow Systems. These are fairly complex machines for the photographic production of large numbers of copies at comparatively low cost. They use electro-photographic and diazo systems.

Electro-photographic continuous flow machines work on the xerographic principle but with a selenium coated drum instead of a flat plate. The charging, exposure, powder processing, transfer and heat fusing stages are disposed around the circumference of the drum. The latter rotates continuously; the image is projected on the drum surface—at the exposure station—while the optical system scans the original document. At the end of the cycle the machine delivers the finished copy. The cycle can be repeated indefinitely for any number of copies at a rate of about four to six a minute. More advanced models may feed the paper in from continuous rolls and deliver copies at a paper web speed of up to 20 feet per minute.

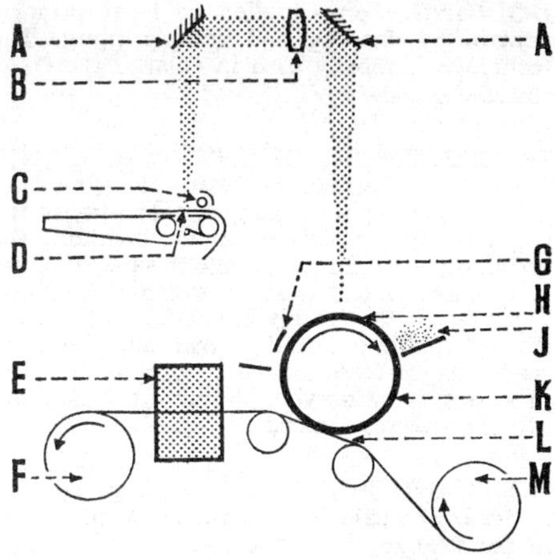

CONTINUOUS-FLOW XEROGRAPHY. The documents are fed continuously through the xerographic printer, the image being projected on to a rotating selenium drum. Various points of the latter carry the charging, exposing and powder processing stages. The powder image is then transferred on to a continuously moving paper web and passes through a heating stage to fuse the image on the paper. A, mirror. B, optical system. C, lighting. D, document on feed unit. E, heating stage. F, paper take-up. G, charging point of selenium drum K. H, exposing station. J, powder processing. L, transfer to paper. M, paper web feed.

Continuous flow diazo units also use rolls of dyeline paper. Here the original is fed into the machine which takes it, in contact with the dyeline paper, round an exposing drum (a translucent drum containing the exposure lights), after which the paper runs on into a development section (liquid or vapour or heat) while the original is returned to the operator. The machine may incorporate trimming equipment to deliver ready stacked sheets of copies. Where large numbers of copies of one document are required, the latter may be attached to the exposing drum. Where a number of documents are to be copied on a one-off basis, the originals are continuously fed into the machine. H.W.G., L.A.M. & H.R.V.

See also: *Copying; Diazo processes; Diffusion-transfer; Electrophotography; Fabric printing; Manuscripts and old documents; Micro-copying; Photo-telegraphy; Reversal materials; Thermography.*

Books: *Camera Copying and Reproduction*, by O. R. Croy (London); *Document Copying and Reproduction Processes*, H. R. Verry (London); *Electrophotography*, by R. M. Schaffert (London); *Micro-Copying Methods*, by H. R. Verry (London); *Xerography and Related Processes*, by J. H. Dessauer and H. E. Clark (London).

DODGING. Term applied to local printing control by shading and spot printing. The technique is mostly used in enlarging.

See also: *Enlarging*.

DOEBEREINER, JOHANN WOLFGANG, 1780–1849. German professor of chemistry and pioneer photo-chemist. He discovered many photo-chemical reactions, including the sensitivity to light of platinum salts, tincture of iodine, ferric and manganic oxalates. Discovered the catalysing effect of finely-divided platinum, and above all, the light sensitivity of ferric oxalate, upon which such processes as cyanotype, platinum printing and other processes depend.

DOGS AND PUPPIES. Some dogs take no notice of the camera; but others are interested in the strange object and must be allowed to sniff over the equipment and satisfy their curiosity before they are willing to co-operate. (It is advisable to leave the lens cap on to prevent the lens from being smudged by a wet nose or tongue.)

Handling. Dogs which have been over-petted are generally difficult subjects because they take little notice of commands to keep still, beg, etc.

It is always unwise to try to manage the animal and take the photograph too. The handling should be left to a helper who has been told in advance what sort of picture is required. The owner is not always the best person for this because the dog is apt to concentrate too much attention on him. The average dog can generally be relied upon for a certain amount of co-operation if treated sympathetically.

The Best Time. Just before his regular meal-time a dog is alert and willing to do everything he can to help if he knows that he will be rewarded with a tit-bit. After his meal, he will be less anxious to please, but he will also be less fidgety. Each time has its advantages and will result in pictures of quite a different character.

But the important thing is to let the dog please himself; no attempt should be made to force or threaten him or the resulting picture will be a sorry affair. If he cannot be tempted with a reward and refuses to respond to a normal command, he should be left for a time.

Setting. There must be complete freedom to shift the camera position, so it is better to work out-of-doors. If the dog is allowed to find a favourite spot in the garden, he is less likely to wander off than in strange surroundings. If he can be persuaded to stand or sit on a box or table, so much the better. This brings him up nearer to camera-level and prevents him from wandering too far afield. Animals always photograph best from a viewpoint more or less on their own level; high viewpoints make them appear stunted and out of proportion.

Attempts to introduce humour into animal pictures by dressing the subject up in hats or clothing merely inspire the contempt of anyone who cares either for animals or photography.

Background. The background should be kept as plain as possible, preferably contrasting in tone with the dog's colouring. It is a mistake to pose a black or brown dog against a background in shadow, or a light-coated dog against a white or sunlit wall. Bare brick walls and trellis-work fences should be avoided along with all backgrounds with an assertive pattern or texture. The mottled shadow pattern of foliage and trees, and the spotty effect of light-coloured flowers are examples of backgrounds that have ruined more than one good photograph.

A plain neutral-coloured piece of cloth or blanket without folds or creases makes an admirable background—dark for light subjects, and grey or cream for dark. The background should be supported about three to four feet away from the animal, so that it can be thrown out of focus and kept free from the dog's own shadow. If the camera has a fixed focus lens, even greater care must be taken to see that the background is free from folds or creases.

The blue sky makes an excellent background; to make use of it, the dog must be made to stand on a low wall or the roof of a shed. Under these circumstances a colour filter must be used to correct over-sensitivity of the material to blue, so as to give tone to the sky area. It is better to choose a cloudless blue part of the sky than to include distracting cloud shapes. When using the sky as a background, it is wise to watch out for neighbouring chimney-pots and wireless masts. The filter, of course, calls for an increase in exposure.

Lighting. Side lighting—i.e., with the light falling across the subject at about 45° from above and slightly in front—is best, as it gives good modelling and shows up the texture of the dog's coat. Flat or frontal lighting destroys texture and modelling because all the shadows are cast behind the subject, where they cannot be seen.

Back lighting—i.e., when the sun shines from behind the subject—creates a halo of brightly-lighted fur or hair around the subject. At the same time, it leaves the background in shadow so that the subject stands out sharply from it. Since it also leaves the front of the dog in shadow, the exposure must be increased some three to four times to record detail in the deep shadows. Back-lighting calls for a deep lens hood or some equally effective shield to prevent the sun from shining directly on the lens.

Reflectors. Direct sunlight is a very contrasty form of lighting unless it is accompanied by plenty of indirect light from other sources—white clouds, light-coloured walls and similar reflectors. If there is no such reflection, it must be provided or the range of contrast between the highlights and shadows of the subject will be too great for the sensitive material to record.

A white cloth or a sheet of white paper can be

made to reflect the sunlight into the shadows so long as it is not brought close enough to distract the subject. The reflection from a mirror or bright metal sheet is too harsh and directional. Properly handled, the reflector can be made to bring out detail wherever necessary. It can also be used to throw a catch-light into the eyes, or accentuate characteristic features.

Taking the Picture. The spot should be chosen and all the necessary preparations made before putting the dog in position, because animals become bored, and show it if they are kept waiting in one place for long. The lens aperture and shooting distance should be worked out to give the necessary depth of field with a shutter speed of 1/50 second or faster.

The dog should be photographed from a viewpoint at about the same level and from an angle at which he appears in three-quarter view —i.e., neither square on nor in sideways view, but somewhere between the two. The head should be in such a position that nose, eyes, and ears can all be kept in sharp focus.

It is most important to catch the dog when it is wearing a characteristic expression. Large breeds, like Alsatians, Great Danes, etc., should be taken when they are looking dignified and alert, whilst terriers are more suitably shown in a playful mood.

It is easier to deal with a sitting animal than one which is standing, and there is less risk of movement. But photographs taken to depict show points need standing positions which clearly show the typical features to advantage. This sort of photograph calls for the co-operation of an expert in the breed.

A suitable word from the owner just before the shutter clicks will make the subject prick his ears, open his eyes and generally look alert and keen. But such encouragement must be used sparingly or the word will either lose its magic, or excite the dog and make him uncontrollable.

Another person's help can often be very useful in attracting the dog's attention. They can hold a favourite toy or a fresh bone and often obtain a useful reaction, or if some completely strange plaything is presented at the critical moment, the result is often highly amusing and worth photographing. But the dog must not see the object until just before the shutter clicks or the element of surprise will be lost. And such tricks seldom work twice at one sitting.

Close-ups. Head-and-shoulder portraits of dogs require even greater care. There is always a risk of distortion in a close-up picture. Good quality lenses of normal focal length will allow considerable enlargement of a portion of a negative, and in practice it is better to work at 6 feet and enlarge the negative than risk distortion by coming close to the subject.

The most satisfactory way out is to use a long-focus lens as it gives a large-sized image at a reasonable distance. With the decreased depth of field of such lenses, accurate focusing is essential. Head studies of long-nosed dogs, like Borzois and collies, are difficult because the depth of field must be considerable to cover the entire head from nose to ears.

Here again three-quarter and profile studies are preferable to full-face portraits.

Action Photographs. Action photographs of dogs out-of-doors call for shutter speeds of anything from 1/200 to 1/500 second.

A running dog is best photographed by swinging the camera in the same direction as the movement, and making the exposure with the camera actually moving. If the dog is jumping over an obstacle, the shutter should be released at the top of the jump when he is moving neither up nor down.

An alternative way of photographing dogs in action is to use a synchronized speed flash or flashgun. With these aids, movement of almost any speed can be frozen.

Indoors. With a lens aperture of *f* 6·3, a fast panchromatic film, and a sunny room, good indoor photographs can be taken of any normally well-behaved dog. It is better to choose a room with plain distempered or papered walls and a floor-covering that is without a pattern. As much of the furniture as possible should be moved away to leave a reasonably clear area for working in.

Reflectors and/or artificial lighting should be brought in to relieve shadows and the subject should be posed as near the windows as possible, but away from shadows cast by the window-frame.

The camera should be mounted on a tripod or support and worked at the fastest shutter speed that conditions permit. If Photoflood lamps are used, they should be kept well away from the animal or it may be distressed by the intense light. Flash, on the other hand, does not often worry animals, and electronic flash, in particular, seems to go completely unnoticed.

Puppies. Most puppies are intensely active, and unless the subject is very young, or full and sleepy, the task of photographing it should be treated as action photography.

An old slipper or hat, held out by the assistant, can generally be relied upon to start something worth photographing, and a sudden strange noise will often make the puppy stop short in an interesting attitude. A sleepy pup cradled in a box or basket is easy to photograph and generally worth taking. It is particularly necessary when photographing tiny puppies to keep the camera down to their level.

Groups. Two or more animals together are very much more difficult to photograph than one. As soon as one looks right, the other is sure to move. If the dogs are free to roam about, it is wise to use as small a lens aperture as practicable to get the maximum depth of field. But the table-top technique is probably best, with willing hands to hold the dogs. The helpers should let the dogs go an instant before the shutter clicks, while at the same time another assistant makes a sudden noise or shouts to attract the heads around into the same direction. Such groups should never be attempted single-handed, nor even with only one helper. P.J.

See also: *Cats and kittens; Pets.*

DOLLOND, JOHN, 1706–61. English optician. Constructed achromatic eye-glasses and combinations of lenses for the telescope (1757). Biography by Kelly (3rd edition 1908).

DOLLY. Camera support mounted on wheels so that the camera—particularly a cine camera—can be more easily moved about in the studio or on level ground. Similar supports are also available for studio lamps.

In filming, the term also means the action of moving the camera—i.e., tracking with the subject.

DONISTHORPE, WORDSWORTH. Dates unknown. English barrister. With W. C. Croft patented (1889) a cinematograph apparatus. Worked on the dispersion process of colour photography (1875).

DONNÉ, ALFRED, 1801–78. French professor of medicine. Worked on the theory of the daguerreotype process and took one of the first (unsuccessful) portraits (1839). Produced daguerreotype photomicrographs in 1839. Devised the earliest method of printing from etched daguerreotypes 1839. In 1844 devised (with Foucault) a projection microscope.

DOPING PRINTS. Prints, especially on matt-surfaced or velvet papers, never look as brilliant or have such deep shadows when they are dry as when they are wet. This is the result of light scatter on the surface of the print.

The contrast range—and hence, the brilliance—of the print can be greatly improved by treating the surface of the paper with "dope", or oil reinforcement. Dope can be bought ready prepared from photographic dealers, or it can be made up as follows:

Turpentine	2 ounces	50 c.cm.
Mastic varnish	1 ounce	25 c.cm.
Linseed oil	1 ounce	25 c.cm.

The medium is rubbed well into the surface of the print with a piece of soft fluffless rag or the ball of the finger. The surface is then polished with a soft cloth until no streaks or marks are visible and the print is put aside to dry in a dust-free room. The disadvantage of this method is that drying is slow, and may take a couple of days or more. As a quicker alternative, a good brand of white shoe cream may be applied evenly and sparingly to the print with a soft cloth. The print is then polished with another cloth.

Once a print has been doped or polished, it is impossible to carry out any retouching; all blemishes should be removed before treating the surface. R.M.F.

See also: *Encaustic paste; Magilp; Waxing prints.*

DÖPPLER EFFECT. When a source of radiant energy moves towards or away from an observer, the observed frequency of the radiations is modified in proportion to the relative velocity of the source—e.g., as a locomotive with the whistle blowing passes through a station, the pitch of the whistle drops suddenly. When the engine is approaching the station, it adds its own speed to that of the sound waves travelling towards the observer. Immediately it passes the station the speed of the waves is reduced by the speed of the engine.

In astronomy this effect gives scientists information about the speed and direction of movement of stars. The frequency of the light waves from the star is modified according to its speed and whether it is coming towards or going away from the earth.

The effect is of interest in the recording of star spectra.

DOSIMETRY. Quantitative measurement of ionizing radiations, for example X- and gamma rays for the purpose of a continuous watch on personnel radiation hazards. These measurements can be carried out by instruments such as ionization dosemeters or by photographic means. The photographic method has the advantage over other detectors that it provides a permanent record, that it covers a much wider range of quantities of radiation and that it is particularly suitable for a large scale personnel monitoring survey. It has however the disadvantage that it does not provide a direct reading of the result, but requires processing and interpretation. The apparatus required consists of a film holder, the radiation monitoring film and suitable processing equipment.

Photographic Response to X- or Gamma Rays. The response of the photographic emulsion (usually X-ray film) depends on the quantity (or dose) of radiation to which it was exposed and also on the quality, i.e. the wavelength composition of the radiation applied. A film dosimeter should be capable of giving information on both quantity and quality of the radiation applied.

Quantity or dose. The units of quantity of radiation are the *röntgen* and the *rad.* The röntgen refers to a certain amount of ionization produced by the absorption of the radiation in air; the rad corresponds to the energy of 100 ergs absorbed per gram of any material. Both units are in practical use and differ numerically by about 6–10 per cent.

According to current recommendation the

maximum permissible dose to persons occupationally exposed to radiation can be expressed as a dose rate. It should not exceed 5 rads per year for whole body exposures. This corresponds to 0.1 rad per week or, within a 40-hour week, to 2.5 millirads per hour.

Before a monitor film can be used for interpretation, the particular batch of emulsion must be calibrated, i.e. the relationship between dose in röntgens (or rads) and photographic density for standard processing conditions must be known. This can be carried out by using a high precision dose- or doseratemeter.

The Film-badge System. This consists essentially of the film holder and the film packet. The film holder is a container into which the film packet is inserted. It is usually made of plastic with an open window and with areas covered by metal absorbers or filters. These may consist of duraluminium, tin, cadmium, copper, lead or other material. They serve (*a*) to smooth out to some extent the quality dependence of the film, and (*b*) to aid the interpreter in the assessment of the quality of radiation applied. The quality can be derived from the densities appearing on the film below the various filters after processing. The holder is usually provided with a means of attachment to clothing, so that it can be worn by the radiation worker.

The film packet is generally similar to a dental film packet, although the type of film selected may be different. As very low doses of radiation have to be recorded, the fastest type X-ray film is preferably used. Films having a fast emulsion coating on one side and a slow emulsion coating on the other side of the base have the advantage of covering an extremely wide range of doses of the order of 1: 10,000. Photographic dosemeters cover a wide range of modifications in regard to size, shape, type of filters and sensitized material. In some cases use is made of photographic papers instead of film.

Dosimetry of Electrons and Neutrons. The response of photographic film badges to electrons is greatly affected by the absorption of electrons by the wrapping material and corrections have to be made for this in any calibration. Neutrons, being uncharged particles, have no

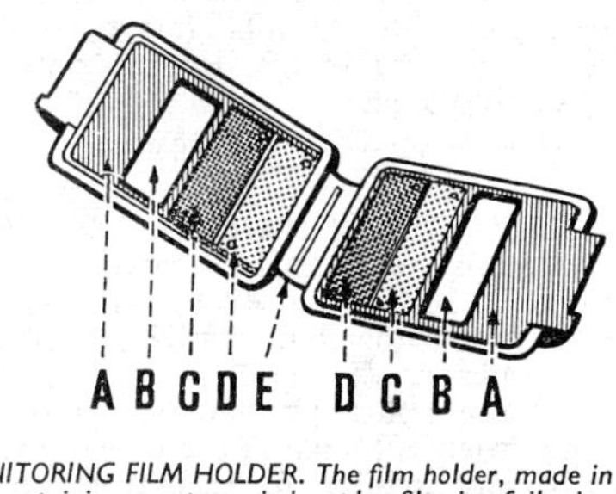

MONITORING FILM HOLDER. The film holder, made in plastic and containing an open window plus filtering foils, is worn by personnel subject to radiation hazard. The container holds a wrapped sheet of radiation monitoring film. A, plastic. B, clear window. C, cadmium foil filter. D, tin foil filter. E, hinge.

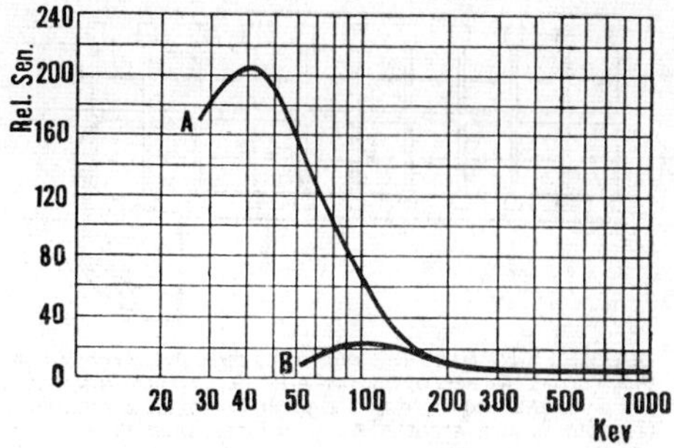

X-RAY FILM FOR DOSIMETRY. The speed of a normal X-ray film depends on the quality of the radiation (curve A—relative speed for a constant radiation dosage plotted against voltage of radiation). The maximum at 440 KeV corresponds to soft X-rays; for gamma radiation (200 KeV upwards) the sensitivity is less. By filtering the film with a 1 mm. thin metal foil a more even response to different radiation energies is obtained (curve B).

direct action on the photographic emulsion, but can be detected by indirect means, such as nuclear reactions caused either inside or outside the emulsion. In the case of so-called thermal neutrons (i.e. slow neutrons) the most efficient photographic response is achieved by placing certain foils of cadmium, indium, rhodium and others in contact with the emulsion or its wrapping.

In these foils electrons, and in some gamma rays, are released by neutron capture and thus photographic blackening can be achieved after processing. In the case of fast (high-energy) neutrons, hydrogen nuclei (protons) are set in motion by collision of fast neutrons with the gelatin of the emulsion. Using nuclear track plates or films, proton tracks will be formed and from their number per unit area conclusions can be drawn as to the neutron dose applied.

R.H.H.

Book: *Photographic Film Dosimetry*, by K. Becker (London).

DOUBLE-COATED FILM. Negative film—usually for the popular market—coated with two emulsions, one (the upper) fast, and the other slow.

The effect is to increase the tone range and exposure latitude of the film.

Double-coating is also used for some process materials to provide special scope for contrast control.

Not to be confused with bipack films, at one time used in colour photography.

See also: *Negative materials*.

DOUBLE-8 mm. Usual form in which 8 mm. cine film is made, exposed and processed. The film is actually 16 mm. wide, but with twice the number of perforations of regular 16 mm. film. It runs through the camera twice, being exposed only along half of its width the first time, and along the second half during the second run through. After processing the film is slit down

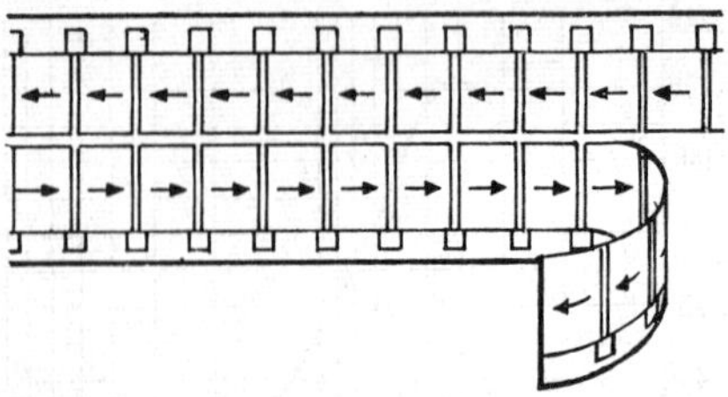

DOUBLE EIGHT FILM. This runs twice through the camera and thus carries two sets of images. After development the strip is slit down the middle to give two single lengths 8 mm. wide. (The film is here about 20 per cent larger than natural size.)

the middle to produce the 8 mm. width with its one perforation row. Super-8 film is usually packed in single 8 mm. widths.

See also: *Cine camera; Cine film; Cinematography.*

DOUBLE EXPOSURE. Intentional double or multiple exposures are often used for particular effects, foı trick photographs, and for making composite pictures for advertising purposes.

When the images are to be simply superimposed—e.g., for making a ghost photograph —one exposure is made after the other without changing the film or plate. In this case the highlights of each image are visible where they lie over the shadows of the other. Where two highlights coincide there is no second image since both appear as white paper on the finished print, and where two shadows coincide the result is simply a single tone of black.

Where each image is to appear separately on the film without any overlapping or ghost effect, the method is to mask off parts of the plate in turn while other parts of the plate are being exposed. The simplest way of doing this is to cover half of the lens at a time. This can be done with a lens cap, one half of which is cut away. The cap is adjusted so that one half of the picture seen in the focusing screen is masked off. The first exposure is made and then the cap is rotated through 180° so that it masks off the part of the plate already exposed. The second exposure is then made. In this way the whole of the plate is covered in two separate exposures. This is the standard method for making pictures in which the same person appears twice. It is essential for the half-mask to fit close up to the lens or the join will show.

In the older types of camera double exposure may occur accidentally and generally this is the photographer's fault through forgetting to wind the next exposure on. Otherwise it can be a fault in the film transport mechanism, or—very rarely—shutter bounce (when the shutter momentarily bounces open again after closing from the first exposure).

Most modern cameras have some double-exposure prevention device. On some models, however, this device can be disengaged so that double-exposures may be made when desired.

See also: *Film transport; Faults; Tricks and effects.*

DOUBLE EXPOSURE LOCK. Mechanism fitted on many modern roll film and 35 mm. cameras to prevent accidental double exposure. This is done by coupling the shutter wind or release to the film transport mechanism.

See also: *Film transport.*

DOUBLE EXTENSION. Amount by which the front of a field, technical or similar camera can be extended—e.g., to twice the focal length of the normal lens. It is used for taking close-ups and is also necessary when long focus lenses are used on a camera.

See also: *Extension of camera.*

DOUBLET. Name still sometimes given to a compound lens of the rapid rectilinear type but more commonly applied to the cemented combination of a positive crown lens with a negative flint lens.

DRAM OR DRACHM. Unit of weight equal to 60 grains ($\frac{1}{8}$ ounce) in apothecaries' measure, and $27\frac{11}{32}$ grains ($\frac{1}{16}$ ounce) in avoirdupois measure. Also denotes 60 minims ($\frac{1}{8}$ ounce) in fluid measure.

See also: *Weights and measures.*

DRAPER, JOHN WILLIAM, 1811–82. American professor of chemistry. Great investigator in the field of scientific photography. He was a pioneer daguerreotypist associated with Samuel F. B. Morse and also an independent discoverer of the law of photochemical absorption, since known as the Draper-Grotthuss Law. Probably also the first observer of the retrogression of the latent image in daguerreotypes. In 1840 he published the first paper on portrait photography, and took the first successful ones. Produced the first daguerreotype photographs of the moon and of the spectrum (1840). Collected works: *Scientific Memoirs* (1878).

DRAWINGS FROM PRINTS. It is possible to make a hand drawing with waterproof ink over the top of a photographic image and then to bleach away the image, leaving only the drawing on the paper.

See also: *Bleach-out process; Sketch photographs.*

DRIFFIELD, VERO CHARLES, 1848–1915. English chemist and amateur photographer. Undertook in 1890, with Hurter, the famous scientific study of the characteristics of photographic emulsions (H. and D. curve), and created photographic photometry. Both received the Progress Medal of the Royal Photographic Society in 1898 "for their work on the determination of the speed of plates". A Memorial volume of their collected papers was published by the R.P.S. in 1920.

DROP BASEBOARD. Term for the baseboard of a stand camera which is hinged to allow it to be swung down when a wide angle lens is in use so that it does not intrude on the wider field of such a lens. The drop baseboard is most commonly found on press cameras.

DROPPING BOTTLE. Glass bottle specially designed for pouring liquids drop by drop. The bottle has a ground-in glass stopper grooved half-way down on opposite sides. Corresponding grooves in the neck of the bottle enable the stopper to be turned to allow the liquid, when poured, to flow out in drops. Turning the stopper so that the grooves do not coincide closes the bottle.

Dropping bottles are useful for adding small quantities of solutions—e.g., wetting agent, developer improver, potassium bromide solution—to processing baths. They do not provide an accurate kind of measure.

DROPS. Drops are frequently used as a unit for measuring small quantities of liquid. But the size of a drop is by no means constant—it depends upon the nature of the liquid and the shape of the bottle lip or the dropper. For this reason the custom of regarding drops and minims as the same thing (or taking 20 drops as 1 c.cm.) is unreliable although it is near enough for most of the comparatively dilute aqueous solutions used in photographic processing.

DRYING

When negatives and prints have been processed they may be dried by various methods. The system used may largely be decided by the equipment available, although efficient drying is easily achieved without any special equipment, provided sensible care is taken.

Image density. The rate and other conditions of drying have some effect on the final density of the image. Hardening treatments during processing yield images of slightly lower final density, while rapid drying methods may produce higher densities. This is of practical importance only in scientific photography where the exact image density is quantitatively significant. In all cases however uneven drying tends to produce areas of different overall density, appearing especially pronounced when drops of water adhere to the emulsion surface and locally slow down drying.

PLATES AND FILMS

Once plate or film negative materials have been washed they should be dried as quickly as possible because all the time they stay wet they run the risk of being damaged.

To avoid drying marks, the negatives should be shaken and wiped free of water after washing. There are various types of chamois or viscose sponges available for wiping the surface of the emulsion, but unless they are kept absolutely clean and free of embedded dust, grit, etc., they may leave serious scratches down the whole length of a film or right across a plate.

It is advisable to put the film through a dish of water containing a few drops of wetting agent. When the film is hung up to dry, the water will drain off evenly.

It is perhaps safer to add a few drops of wetting agent to the last change of the washing water, and when the films are hung up to dry, the water will then drain off evenly without assistance.

Where the washing water is hard, a few drops of hydrochloric or acetic acid may also be added to the last lot of water. This will prevent scum deposits on drying.

Plates are best dried standing up in a wooden plate rack. There must be plenty of room for the air to get between neighbouring plates; if the plate rack grooves are set too close, only alternate grooves should be used.

Sheet films are best pinned up by one corner or hung from small film clips.

For roll and 35 mm. film, a normal film clip is fastened to one end, and the film hung up by it and weighted down by a second clip on the other end.

A number of modern drying units take the complete reel of a film tank and so make it unnecessary to touch the film until completely dry.

Drying Room. The room where the film is hung should be free from draughts or air currents that would stir up dust. For the same reason no one should be allowed to enter the room until the film is dry. While it is still wet, the gelatin surface has an uncanny affinity for stray dust particles, and once such particles become attached to the emulsion they are very difficult to remove. In summer it is wise to see that the drying room is also free from insects.

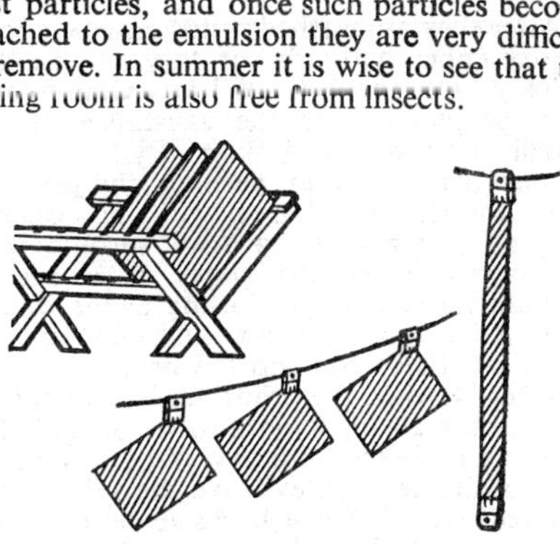

DRYING NEGATIVES. Left: *Plates are dried standing up on edge in a plate rack.* Centre: *Sheet films can be clipped on a line.* Right: *Roll and 35 mm. films should hang from a clip.*

Forced Drying. Negatives dry faster in a stream of warm air. Special film driers are available which blow heated air over the film surface and commercial drying cabinets generally incorporate a small heater unit and a fan for the same purpose. The air is usually filtered to remove any dust.

It is never wise to attempt to dry a negative in front of an electric radiator because this sort of heat dries the surface of the gelatin more rapidly than the deeper layers, causing the film to curl violently. This effect is overcome in commercial drying equipment by using specially filtered infra-red radiation of wavelength longer than 50,000 A. Heat rays of this wavelength are absorbed by the moisture in the film, but not by the gelatin layer.

High Speed Drying. Really rapid drying calls for more extreme measures.

The popular method among press photographers is to immerse the film for 3–4 minutes in 80 per cent (not stronger) pure alcohol or colourless industrial methylated spirit. This largely displaces the water in the emulsion, and the negative will then dry in 3–4 minutes in a current of air.

Another way is to harden the negative in a solution of 1 part formalin in 30–40 parts water. It can then be washed in one or two changes of nearly boiling water and dried in a few minutes in front of a fire in a current of air.

For temporary drying the wet negative may be soaked in a saturated solution of potassium carbonate for 1 minute, and wiped clean with a piece of chamois leather or viscose sponge. It can then be printed immediately. Such a negative must be rewashed and dried normally later on. When rewashing, it should first be immersed for 3–4 minutes in the minimum amount of water needed to cover it. Putting the negative in a large volume of water straight away may cause reticulation of the emulsion.

PRINTS

After they are washed prints may be allowed to dry either naturally, or they can be dried quickly with the help of heat. Glossy prints may be glazed with the assistance of heat, usually from a special glazer-dryer.

Natural Drying. After washing, the prints are well drained or blotted with photographic blotting paper to remove all surplus water. Photographic blotting paper is a specially manufactured type of blotting paper which is free from surface fluff and contains no chemicals likely to stain the print.

After blotting, the prints are laid singly, face down, on a clean, smooth towel or tablecloth, and left until they are dry. Woolly materials are unsuitable, as they leave the print surface covered with fluff which has to be rubbed off afterwards. Coloured materials are also unsuitable, because they are liable to stain the print surface if the colours are not fast.

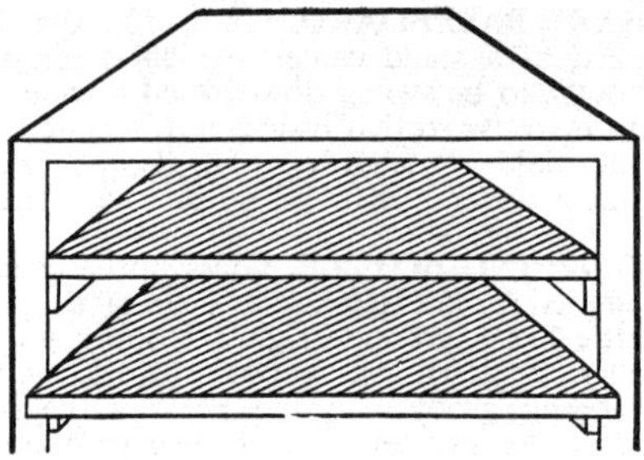

PRINT DRYING RACK. This consists of a series of frames with muslin stretched over them. The frames may be supported above each other in a covered rack to save space.

For large-scale drying, the prints may be laid on muslin stretched on wooden frames in a warm, dry atmosphere away from dust.

Alternatively, large pictures can be clipped together in pairs, back to back by one corner, and hung up to dry on a line. They may curl to some extent, but are easily flattened out afterwards (see below).

Heat Drying. Prints may be dried quickly by holding them in front of a fire or radiator. They must be kept moving, and turned over whenever they start to curl. They should not be allowed to become bone-dry, as the gelatin surface may then crack during flattening.

A glazing press can also be used to dry non-glossy prints quickly. Drying takes longer than for glazed prints, but it is still quicker than natural drying.

The glazing plate is taken out of the glazer, and the glazer is then heated up as for normal glazing. As soon as it is hot, it is switched off. The wet prints are laid on the glazing plate, face up, and all surplus water is blotted off. The plate is then replaced in the warm press, and the apron is fastened down over it. The prints thus dry fairly slowly in the press as it is cooling down. This method has the advantage that the drying speed is slowest at the time when rapid drying would be undesirable because it would cause the prints to curl violently.

The prints should be taken out of the press before they become bone-dry and brittle.

Flattening. Most prints, particularly large ones, tend to curl inwards during drying and must be flattened before mounting or filing. Very dry prints tend to curl more.

To flatten a print, it is laid, face down, on a soft but firm surface—e.g., a board or table covered with a layer of cloth. A ruler or straight edge is gently pressed diagonally across the back of the print and the print is drawn under it by the corner.

This will reverse the curl at one corner. The amount of curl taken out of the paper can be controlled by adjusting the weight on the ruler and the angle at which the print is drawn away from it. The greater the pressure on the ruler

STRAIGHTENING PRINTS. *Prints often curl during drying and can be straightened by drawing them backwards over the edge of a table or underneath a ruler.*

and the greater the angle between the paper and the board, the greater the anti-curling effect.

The same procedure is repeated for the other three corners. The whole print is then curled outwards, and will become flat if placed underneath a book for an hour or two.

The pressure of the ruler or straight edge must be applied with care, or the spot where it was pressed down may crease the print.

If the prints are too bone-dry, this way of flattening may crack the gelatin surface.

Most flattening methods, when applied to large prints, straighten the print out as a whole, but leave the edges wavy. This can be remedied in two ways:

(1) The print is laid face down on a clean sheet of paper. The back is slightly damped at the edges with a wet cloth or sponge until it just starts to go limp. A sheet of fairly absorbent paper—e.g., white blotting paper, or even a sheet of newspaper—is held in front of a fire or radiator until it is really dry and about to scorch. This sheet is then quickly laid over the back of the print, covered with a sheet of cardboard, and put under pressure for a few minutes. If necessary the process is repeated until the edges of the print are quite dry and flat.

(2) The back of the print is damped as before, and the print covered with a sheet of white blotting paper. It is then pressed by running a moderately hot household iron evenly over the whole print area. The print is finally covered with a sheet of cardboard and put under pressure until cold.

Glazed prints, if they curl at all, should be flattened by drawing them under a ruler (above), but taking care to avoid cracking the glazed surface; either of the wet methods would destroy the glaze. Usually, however, glazed prints flatten out of their own accord once they are cold. L.A.M.

See also: *Drying equipment; Drying mark; Faults.*

Books: *Developing*, and *Enlarging*, by C. I. Jacobson (London).

DRYING EQUIPMENT. Sensitized materials can be dried after processing by simply leaving them in a dry, dust-free atmosphere for several hours. Films are simply suspended from a film clip and weighted at the bottom end; plates are stood on edge in a plate rack, and papers are spread out, emulsion side up, on muslin stretched across an open frame.

But there are drawbacks to natural drying: it is slow, the soft emulsion is left exposed to dust and damage for a longer time; and in bulk processing plants it calls for a disproportionate amount of space.

Negative Materials. Drying can be speeded up by keeping a stream of dry air constantly moving over the surface of the emulsion. This is done by suspending the film in a drying cabinet.

Cabinets are usually designed to take a number of lengths of film at a time, but small types taking a single length are also made.

The principle is the same in every case. The cabinet is usually made in the form of a sheet metal cupboard with rods at the top from which the films can be suspended. An electric fan draws air into the cabinet through a dust filter and over an electric heating element. After passing around the films, the air is exhausted at the top of the cabinet.

The temperature inside the cabinet is indicated by a thermometer dial on the front and is regulated by an external switch either controlling a thermostat connected to the heating element, or switching one or more sections of the element in or out as required.

Cabinets of this type can be equipped with hanging rods for up to 100 lengths of roll film, or with racks for plates and cut film in sheaths.

In establishments where large quantities of processing are handled, it is common to use infra-red lamps to enforce rapid drying. These effectively dry the emulsion throughout its

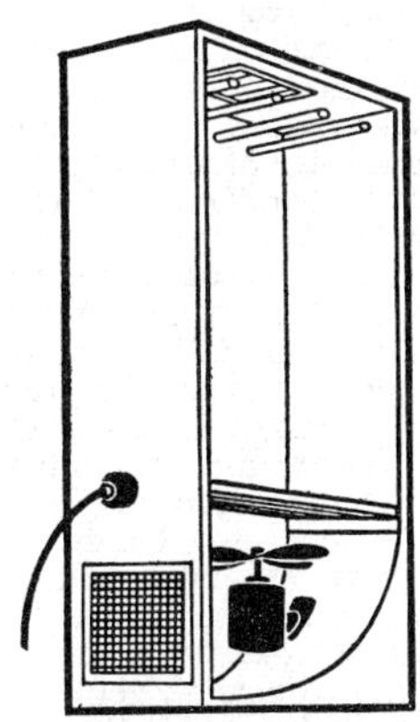

DRYING CABINET FOR FILMS. *The films are suspended above a low-power heating element, with a fan below to keep up a constant movement of warm air. A muslin screen filters the incoming air so that it is free from dust.*

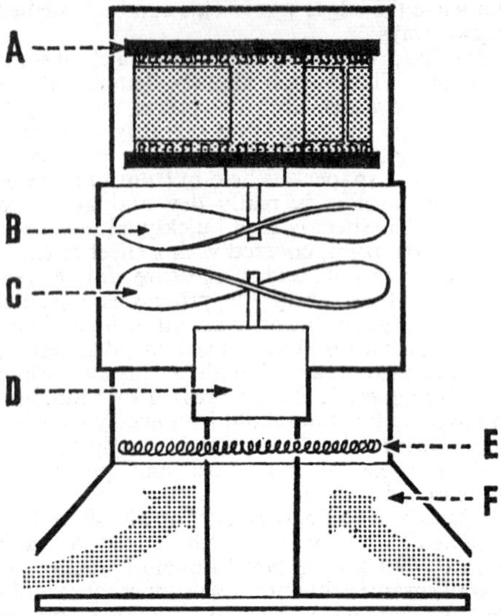

COMPACT 35 mm. FILM DRIER. This takes the film on its spiral reel direct from the developing tank. Similar driers are also available for other film widths from 16 mm. cine to 70 mm. film. The motor driven fan blows warm air through the film on the reel, while the air stream also rotates the reel. A, film on reel. B, fan blade connected to reel. C, motor driven drying fan. D, motor. E, heating coil. F, air stream.

depth in as little as two minutes; special precautions are, however, necessary to prevent the emulsion from melting.

Single-reel dryers take films on their spiral reels from developing tanks. Such dryers consist of a cylinderical housing into which the film reel is inserted from above, while a fan blows a current of air through from the bottom. The dryer may have a built-in heater for warm air drying. Single reel dryers of this type are available for the reels of normal roll film tanks, as well as for large capacity 35 mm. and 16 mm. cine film reels holding up to 100 feet of film. The main advantage of such a dryer is that the film need not be handled at all while wet, and is only unloaded from the reel when dry.

Prints. Special types of equipment are not generally needed for drying prints produced in quantities—i.e., in D. & P. work, in the picture postcard business, and for the press—as these are invariably finished in a hot glazer which automatically glazes and dries at the same time.

Special machines made for the purpose dry the prints between two bands of absorbent fabric which pass around a heated drum. The operation is continuous, the drum speed being controlled so that the prints stay long enough in the machine to emerge perfectly dry. Such machines usually have an output of the order of 200 8 × 10 prints per hour.

See also *Darkroom; Drying; Glazers and glazing machines; Processing machine.*

DRYING MARK. Marks left on the surface of the emulsion of a plate or film after the drying of isolated drops or streaks of water. The marks take the form of patches which are denser or less dense than the surrounding image and are left in the centre of a wet area that has taken longer to dry than the rest of the negative.

Marks of this kind show up on the final print and there is no cure for them once they are there. The only solution is to avoid them by either using a wetting agent in the final wash water, or wiping down the surface with a viscose sponge or chamois leather before drying the material. Either of these methods will prevent drops from forming on the surface.

It is also important to allow drying to proceed evenly; any attempt to speed matters after the negative has partly dried is always apt to leave uneven densities where the last patch dries off. This can be particularly important when matched negatives are required, as in colour printing with separation negatives. For this reason it is usual to allow separation negatives to dry naturally, all in the same position (relatively), so that any uneven densities due to drying will be in the same position on each negative.

Another type of mark which is often termed a drying mark is simply a limey deposit left after drying a negative that has been washed in hard water. Either of the above methods of preventing drops will work equally well here, or once the negative is perfectly dry, the marks can usually be rubbed away with a dry finger tip. Persistent lime deposits can be cleared by bathing the negative in a weak solution (1–2 per cent) of hydrochloric or acetic acid.

See also: *Faults.*

DRY MOUNTING. Popular way of sticking prints to a mount with shellac tissue (or similar material) with a hot iron or press.

See also: *Mountant; Mounting prints.*

DRY PLATE. In the earlier days of photography, when the wet collodion process was still popular, it was customary to refer to gelatin coated plates as dry plates. Nowadays all plates outside the photomechanical reproduction industry are gelatin coated and the term dry plate is no longer used.

DUBOSCQ, LOUIS JULES, 1817–86. French optical manufacturer. Used (1849, with Foucault) arc lamps for projection. Manufactured the Brewster stereoscope in 1851 and was probably the first Frenchman to make stereo photomicrographs (1857). Constructed apparatus for synthesizing motion from stereo series (1852), and for taking microphotographs. Built the Polyconograph camera for travellers (1861) which took 15 exposures on each plate.

DU CAMP, MAXIME, 1822–94. French writer. Used photography (calotypes) during his travels in the Near East and published in 1852 *Egypte, Nubie, Palestine et Syrie* illustrated with 125 photographic prints.

DUCOS DU HAURON, LOUIS, 1837–1920. French scientist, pioneer of colour photography. In *Les Couleurs en Photographie* (1869) he formulated the principles of the additive and most of the subtractive methods of colour reproduction, although owing to wrong primitive colours and the lack of panchromatic materials not all of these methods could be executed at that time. In 1878, with his brother Alcide Ducos du Hauron, he published a pamphlet, *Photographie des Couleurs,* in which their methods of producing colour photographs were described. Proposed in 1891 the anaglyph method of stereoscopy. Received in 1900 the Progress Medal of the Royal Photographic Society "for the initiation of three-colour heliochromy as a working process". Biography by G. Potonniée (Paris 1939).

DUFAY, LOUIS D. Dates unknown. French photographer. Originator of an additive colour process first introduced as the Omnicolour plate, later as Dufaycolor film. The process makes use of an integral ruled-screen tricolour filter. The original negative image is reversed in development to make a colour positive transparency. A negative-positive Dufaycolor film was also current at one time for motion picture use.

DUHRKOOP, RUDOLF, 1848–1918. German portrait photographer, active in Hamburg from 1883 onward and later also in Berlin. Treated portraiture as an art, and worked mostly in the gum print and bromoil processes of the fin de siècle period.

DULL WEATHER. Sunless, overcast days are bad for most sorts of photography—because of the character of the light, not so much because there is not enough of it. The trouble is that the illumination comes from all directions; it is completely diffused and so it casts no definite shadows. Without well-defined shadows there is nothing to give relief and modelling to the subject, so that it looks flat and lifeless.

Where there is no choice, and the photographer must make the best of the existing light, the contrast of the final print can be improved by giving the minimum exposure to the negative, and by developing for longer than the normal time. The use of flash equipment offers a third solution that is very popular with press photographers.

Dull weather has its uses, however, and it is not always desirable to adopt special techniques to increase the contrast of the final print. Where the photographer deliberately sets out to portray a mood of depression, sorrow, or wistfulness, dull weather gives him the most suitable lighting conditions.

Close-up photographs of white or very light flowers, like lilies and convolvuli, are best taken in low-contrast lighting of this nature, and a dull day is often better than a sunny one for outdoor portraiture.

Interiors are generally photographed in dull weather to avoid the very strong contrasts between sunlit windows and walls, and furniture in shadow. P.J.

See also: *Daylight.*

DUNCAN, CHARLES, 1870–1960. English photographer. Travelled all over the world taking pictures for Thomas Cook; photographed old masters and antiques for *The Connoisseur.* Autobiography: *A Photographic Pilgrim's Progress.*

DUPLEX SYSTEM. Proprietary system for running 9·5 mm. cine film in wide-screen as well as in normal form. Although interesting in its possibilities, it did not make much headway and disappeared again from the market. A similar system has also been proposed—but never realized—for an 8 mm. film.

The Duplex 9·5 mm. film has two perforation rows instead of one, the perforations still being positioned between the frames and not along the edges of the film. Both the camera and the projector must be designed for this type of film.

For normal filming the film runs through both camera and projector vertically, and the effect is the same as with the ordinary 9·5 mm. film. For wide-screen pictures the camera is held horizontally so that the film runs from side to side. The camera then only exposes the film along one half of the width, the images being beside each other instead of below each other. Once the film has run through the camera it is turned over and exposed a second time along the other edge.

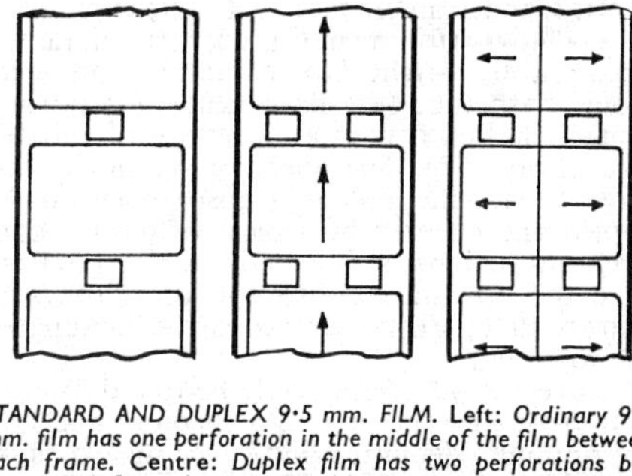

STANDARD AND DUPLEX 9·5 mm. FILM. Left: *Ordinary 9·5 mm. film has one perforation in the middle of the film between each frame.* Centre: *Duplex film has two perforations between each frame but is exposed in the same way as ordinary 9·5 mm. film.* Right: *For wide screen dual use the Duplex film runs through the camera twice horizontally and is slit down the middle after processing. Similar wide screen images have been proposed for 8 mm. film.*

After processing the film is slit along the middle and projected as the 4.75 mm.-wide strip running horizontally through the projector. The image has an aspect ratio of about 1·6 to 1, giving the wide-screen picture. On projectors designed for this system, the film transport mechanism and spools can swing over from vertical to horizontal running.

See also: *Cine film.*

DUPLICATE COLOUR SLIDES. Colour transparencies can be copied by the simple process of re-photographing the original colour transparency on a second piece of colour reversal film of either the same or a different make. Suitable means of re-photographing are either a camera adapted for same-size reproduction (by an extension bellows or extension tubes) or contact printing of the transparency (after removal from its mount) on to the fresh film.

A copy made in either of these ways is not truly a duplicate. Transparency materials are essentially of high contrast so re-photographing will generally give a copy of unusably high contrast with empty highlights and blocked shadows. In addition, since all colour materials use dyestuffs which are less than perfect, the colours in the copy will be somewhat degraded and shifted in hue. Colour errors which are acceptable or unnoticeable in the original transparency become apparent since they are now twice as bad and in addition direct comparison with the original transparency is possible.

Equipment and Technique. With proper methods, it is, however, possible to make acceptable copy transparencies using simple equipment. The contrast of the copy transparency can be lowered without too great a loss of colour saturation by the use of a suitable light source or a simple contrast reducing mask.

Electronic flash as a light source for copying of colour transparencies gives acceptable results. It can be used either for copying in a camera or for contact printing. Units specially designed for colour slide copying are commercially available.

If a 35 mm. camera fitted with extension tubes or a bellows attachment is used, this is mounted above a light-tight box containing an electronic flash tube (the flash head of a normal unit). The box carries an aperture of suitable size in one side. The aperture should be just larger than the colour transparency to be copied and covered by a piece of flashed opal glass on the inner side. In addition to the flash tube the box should also contain a tungsten lamp so that the transparency can be illuminated for focusing.

Correct exposure can only be found by trial and error. Set the lens to a medium aperture for optimum definition and (since the exposure time is controlled by the flash duration) set the shutter to 1/50 seconds (if a focal plane type) or any speed with a diaphragm type. Exposure can be varied by giving a number of flashes as well as by varying the lens aperture.

Contrast and Colour Control. Contrast of the copy transparency can be more precisely controlled by first exposing a thin, soft, black-and-white negative by contact with the original. After processing this is bound up with the original transparency in accurate register to form a contrast-reducing mask. Registration becomes easier if the negative is exposed through the back of the transparency and is thus slightly unsharp. Such an unsharp mask also gives copies of apparently enhanced sharpness. The contact mask will lower the contrast of the original to a degree which will depend on the contrast of the negative, whilst leaving the colour saturation unaltered.

In every case filters can provide overall colour correction.

The filters used should be the dilute complementaries of the colour cast to be corrected. A transparency which is too yellow in balance can be corrected by filtering the light source with a pale blue filter; a transparency which is too blue is improved with a pale yellow filter. The correction filter can be placed at any point between the light source and the transparency (or between the transparency and the lens when using a camera).

An alternative method of copying transparencies which is used by commercial colour laboratories involves the use of inter-negative film. In this method a colour and contrast corrected negative is first produced on a special colour negative film. Prints on positive copy film can then be made from this negative.

G.W.A.

See also: *Colour printing technique; Reversal materials.*

DUPLICATE NEGATIVES. Valuable negatives that might be difficult or impossible to replace are often duplicated for safety. There are several ways of making a duplicate, but the most reliable is by double contact printing on to ordinary lantern plates.

The negative to be duplicated is first printed in the normal way on to a lantern plate with the emulsion sides in contact. The lantern plate is processed to give a positive, and then when dry printed again on to a second plate, emulsion to emulsion. When developed, the second plate is the required duplicate negative. By suitably modifying the exposure, development, and grade of the plates, the character of the final negative may be controlled within wide limits. Ordinary (non-colour sensitive) process sheet film can be used in the same way.

Duplicate negatives can also be made by photographing an ordinary contact print or enlargement from the original negative, but the gradation and definition are never as good as in a duplicate made by double contact printing.

It is also possible to produce duplicates without an intermediate positive—i.e., by reversing the positive image produced under the negative in the first place. This is done either by bleaching out the positive image and then exposing and developing out the remaining silver bromide, or by substituting a dye image for the unexposed silver bromide and then removing all the silver.

Straightforward reversal processing of the film exposed through the negative also yields a duplicate negative. Reversal film specially designed for this purpose may be used, but many of the slow fine-grain negative films are equally suitable. Most manufacturers issue specific instructions for reversal processing of their negative materials. Such materials are not recommended for producing reversal positives (e.g. from positive transparencies), since the base tint of the negative film would also tint the positives.

Special film is available for making duplicate negatives without an intermediate positive. The film has a speed about equal to that of contact paper and when it is exposed under the negative it develops direct into another negative. Such materials may either be pre-fogged emulsions where the reversed image is produced by the destruction of the latent image fogging by exposure to yellow or orange light (Herschel effect); or they may be films with a diazotype (dyeline) layer which directly yields a negative image from a negative. Such dyeline films are widely used in producing duplicates from microfilm negatives.

Duplicate negatives have a special importance in cinematography where the original negative is never used for making 35 mm. release prints; the original is used for much 16 mm. work, but is sometimes duplicated in preparing special effects such as dissolves and fades. Amateur films are mostly made on reversal film, and if a copy should be required it is made by printing the original positive direct on to reversal stock.

See also: *Enlarged negatives; Optical aftertreatment; Reversal process.*

DUST. Dust is a nuisance at every stage in photography.

On the Lens. In time most camera lenses build up a coating of dust—in particular, box and folding hand cameras that have no protection in the form of a lens cap or hinged front. A dusty lens cannot form a sharp image and it tends to give veiled and foggy negatives. It can be cleaned by first removing the dust with a clean soft brush kept for the purpose, and then wiping the surface gently with a piece of clean chamois leather or lens cleaning tissue. Even better is a special anti-static cloth or brush; these do not induce static electricity, which causes dust to cling to the lens.

In the Camera. The opening and closing of bellows type cameras, in particular, sucks in dust that is always likely to settle on the back of the lens or on the surface of the emulsion. It is best removed by opening the back of the camera, loosening the dust with a soft brush, and blowing it out with blasts of air from a bicycle or car tyre pump—not by breathing moist air into it. Better still the dust may be sucked out with an ordinary vacuum cleaner. Care must of course, be taken not to distort and thus damage the bellows when doing this. Dust on the inner surface of the bellows can considerably lower the quality and the contrast of negatives. Bellows in such a condition should be re-blackened.

It is always easier to prevent dust from getting into the camera in the first place than to remove it once it is there. The camera should always be kept in its case when not in use; never opened longer or oftener than necessary and never carried loose in the pocket of a jacket or overcoat. It should also be given a thorough clean from time to time.

On the Sensitive Material. Specks of dust on the surface of a film or plate before exposure shield the emulsion and leave clear spots or pinholes on the negative. They can best be prevented by keeping dust out of the camera body (above) and film holders (below). Films or plates as they come from the factory can be assumed to be dust-free; they should not be allowed to lie face up in the darkroom when they are being loaded and the surface should never be touched with the fingers or brushed. If there is any reason to think that dust has settled on the surface of a film or plate, it should be shaken off by tapping the edge of the material lightly on the bench.

In Plate and Film Holders. If holders are carried around in a dirty case or cloth, dust creeps in at the edges and eventually finds its way on to the emulsion. While everything should be done to keep the holders away from dust, it is a good plan to draw the dark slides and blow out all the crevices at regular intervals. Particular attention should be paid to the velvet light trap. Dust harboured in it readily transfers itself to the dark slide and on to the emulsion.

Films are less likely than plates to acquire dust particles before exposure.

In the Darkroom. Two kinds of dust give trouble in the darkroom—ordinary dust and fine particles of photographic chemicals.

Ordinary dust should be reduced by the frequent use of a vacuum cleaner, not only on the floor, but on the bench and along the tops of cupboards. The darkroom floor should preferably be of non-dusting material without a carpet. The tops of processing solution bottles should always be wiped around before pouring out the solution to prevent any particles that have settled on the mouth of the bottle from being washed off into the solution. For the same reason all graduates and other vessels should either be kept upside down or covered when not in use; they should be allowed to dry naturally after washing and not be rubbed with a cloth, and they should always be rinsed under the tap before use.

Chemical dust is also created by allowing dishes that have contained solution to dry without first thoroughly washing them, or by neglecting to wash all places on the bench or floor where solutions have been spilled.

Chemical dust causes dark or clear spots or stains, according to its nature, on the emulsion surface of sensitized materials. It is generally the result of careless handling of darkroom chemicals, particularly fine substances like amidol. Here again prevention is easier than cure (since cure calls for tedious retouching). It is a sound rule to measure and mix all chemicals away from the darkroom.

It is wise not to dust or clean the enlarger or any other part of the darkroom immediately before use, as such action will leave dust particles floating about. The dust is potentially far more damaging in this state than when it is resting on top of the enlarger, shelves, etc. The darkroom should be cleaned at a time when no work is planned for some time afterwards.

In the Enlarger. Dust should not be allowed to accumulate on the enlarger lens or it will cause light scatter and consequent loss of sparkle and quality in every print made. This applies particularly to the back of the lens in vertical enlargers when the operation of the negative carrier and focusing movements causes dust to fall on to the lens.

Dust on the carrier glasses will show as white spots in prints and a similar effect will be produced if dust should be present on any glass surface near the carrier, such as a condenser or heat absorbing glass. Frequent cleaning of the entire optical system of enlargers is essential and carrier glasses should be cleaned and inspected for every negative enlarged. The simplest and most effective method of removing dust is, as mentioned above, to wipe the surfaces with anti-static cloth or one of the special brushes now available. Ordinary cloth is useless because of the static electricity induced by the rubbing. As a check that carrier glasses are clean, they should be focused on to a white paper on the easel, when any dust present can instantly be seen.

General Hints on Removing Dust. Most non-metallic surfaces—glass, plastic, celluloid, and the emulsion surface—tend to become electrically charged when they are rubbed with a cloth or brush. In a charged state they attract dust particles and hold on to them stubbornly. This is why dust can never be satisfactorily removed from such surfaces by rubbing; the more they are rubbed, the harder the dust clings. A single wipe over with a damp washleather is one very effective treatment. Any stubborn specks should then be removed separately with a corner of the washleather over the point of a pencil. Better still are the anti-static materials already mentioned above.

Dust is best cleaned out of crevices by blowing it out with a puff of air from a bellows. Miniature bellows can be bought which have the advantage of a very thin nozzle. Failing a small bellows, a cycle or motor tyre pump will do. Blowing moist breath into the part is worse than useless because it tends to mist over glass surfaces and start rusting in metal parts out of sight—e.g., in the shutter mechanism. F.P.

See also: *Faults; Retouching; Spots.*

DUSTING-ON PROCESS. Printing process using a support coated with a layer of gum arabic and a hygroscopic substance like honey, glucose or sugar—sensitized with a bichromate.

A coating of this nature becomes sticky with moisture that it absorbs from the atmosphere. But if light reaches any part of it, that part loses its stickiness. This is the effect that makes the process possible.

The sensitized surface is exposed to the light under a positive transparency. It is then left in the air but away from the light long enough to acquire the right degree of stickiness. The surface is then brushed over lightly with a powdered pigment of a suitable colour. Where the positive image has masked the surface, it remains sticky and holds the pigment. The operator is thus able to build up a positive picture which he can "fix" with the fixatif solution used by artists for crayon and pencil work.

This process can be used for printing a photographic image on materials like glass or ceramics by replacing the pigment with a powder that can subsequently be burned in or vitrified.

See also: *Printing processes* (*obsolete*); *Printing on special supports.*

DYE. Complex chemical used for colouring by molecular adsorption (as opposed to a pigment which colours by coating the substance with solid colouring matter). The actual colour of a dye is governed by the constitution of the dye molecule; the dyeing effect depends on the chemical nature of the substance being dyed.

Dye solutions can often be true solutions whereas liquid pigments are suspensions of solid colouring matter in a liquid vehicle.

Dyes are used in photography for colour sensitizing emulsions, for desensitizing emulsion, for toning transparencies and prints, to provide the coloured images in colour transparencies and prints and for retouching negatives and prints.

(1) As early as 1873, H. W. Vogel discovered that the use of a yellow anti-halation dye on collodion plates made them sensitive to green light. Subsequent experiments showed that a dye sensitizes a photographic emulsion to rays which are roughly of the same wavelength as those absorbed by the dye. The principle was progressively applied to gelatin emulsions as new dyes were developed and it is now standard

practice to sensitize plates and films to all colours of the spectrum—and it is even possible to make them sensitive to invisible radiations—e.g., infra-red.

Emulsions may be dye sensitized either by adding the dye to the liquid emulsion during manufacture, or by bathing the material in a dilute solution.

(2) Some dyes in solution—e.g., pinacryptol green—have the ability to make emulsions comparatively insensitive to actinic light. A plate or film immersed for a minute or so in such a desensitizing bath can subsequently be developed in yellow or weak white light.

(3) The silver image of a transparency or a print can be toned by first converting it into a colourless substance—e.g., silver iodide or cuprous iodide—(known as mordanting) and then immersing in a dye which is selectively absorbed on the substance and not by the gelatin. The result is that the image alone carries the colour. The dye toning process is also used in certain methods of making three-colour prints from separation negatives.

(4) The image of most colour transparencies, and prints made from colour negatives, consists of dyes. These are usually formed *in situ* during development by the interaction of oxidation products of the developer with colour couplers. As a result a dye image is formed in each emulsion layer together with the silver image, and after removal of the latter only the dye images remain.

(5) Water-soluble dyes of the same colour as the silver image—i.e., black, and the various cold and warm tones—can be used as an alternative to pigments for spotting out and retouching negatives and prints. The dye is absorbed into the gelatin and becomes part of the image instead of remaining on the surface. In the same way, dyes of various colours can be used to tint prints and transparencies by hand. L.A.M.

DYE DESTRUCTION PROCESS. Process for making colour prints with a tripack material which, however, already contains the final dyes (and not just dye couplers) in the emulsions before exposure. After exposure, a suitable bleaching agent, acting in conjunction with the silver image produced during the first development, bleaches the dye in the image areas in proportion to the silver densities present. If the silver image is a negative, this results in the formation of a positive dye image, and in this form the process can produce positives from colour transparencies without reversal. For negative-positive printing the image is in effect reversal-processed before reaching the dye bleaching stage.

See also: *Colour history; Colour materials; Colour printing processes.*

DYELINE. Printing process relying on the action of light on diazo compounds. Widely used in office copying procedures and for making duplicates of plans, maps, etc. The process yields a positive image from a positive and the prints are developed by treatment with ammonia vapour or certain organic solutions. The main advantage of the dyeline process is the cheapness of the material which is essentially a suitably impregnated paper.

See also: *Diazotype.*

DYE SENSITIZING. H. Vogel, in 1873 discovered that certain dyes were able to extend the colour sensitivity of a silver bromide emulsion. Dye sensitizers since developed have made it possible for manufacturers to sensitize emulsions to all visible rays and a wide range of invisible radiations, depending on the requirements of the emulsion.

See also: *Negative materials; Sensitized materials history; Sensitized materials manufacture; Sensitizer; Spectral sensitivity.*

DYE TRANSFER COLOUR PRINTS

The dye transfer method of making colour prints on paper is a subtractive imbibition process in which the dyes from three separately prepared images are transferred in register to one sheet of paper to form the colour print. A considerable amount of control is possible and it may be expected to give better results than integral image colour printing papers, especially with difficult subjects or when working from a transparency in which the colour rendering is faulty.

A good set of separation negatives (including the usual grey scales) is the first requirement for the normal process. They may be made from a transparency or taken direct from the subject. The procedure is as follows:

(1) Three relief matrices are made from the separation negatives either by contact printing or enlargement using specially prepared matrix film.

(2) The matrix film is exposed through the base and then developed in a tanning developer.

(3) The surplus emulsion which has not been tanned by the developer is now washed away in hot water leaving an image which looks very much like an ordinary positive made on sheet film except that it is in relief with the thickness of the gelatin varying from a trace in the light parts to a relatively heavy amount in the shadows.

(4) The three matrices are dyed up in cyan (blue green), magenta and yellow dyes respectively.

(5) They are then brought in turn into contact with a sheet of special transfer paper which absorbs the dye out of the matrices. The finished

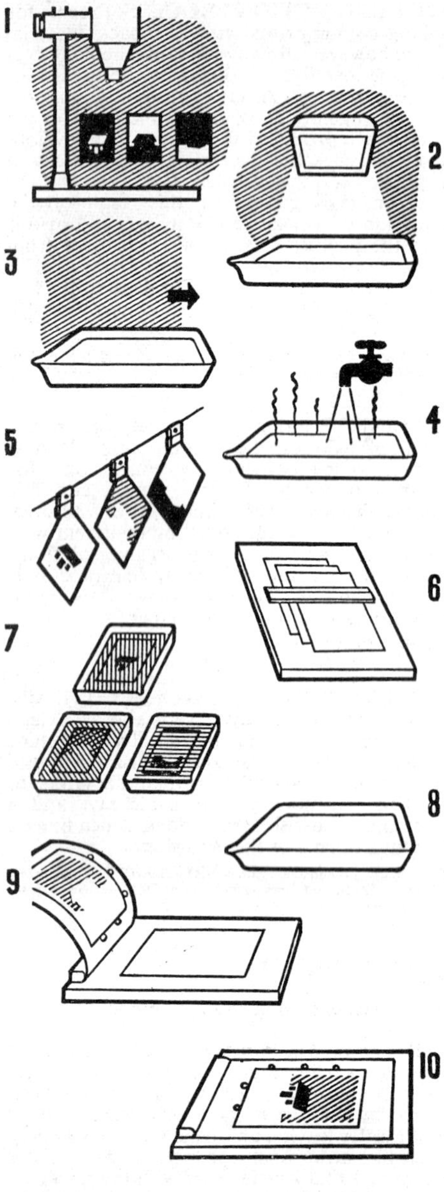

STAGES IN DYE-TRANSFER PRINTING. 1. *Make positives on matrix film by enlarging three-colour separation negatives in turn.* 2. *Develop the three matrices in a tanning developer.* 3. *Fix the matrices.* 4. *Remove untanned gelatin by treatment with hot water.* 5. *After chilling, hang up to dry.* 6. *Carefully register the three matrices, tape together along two edges, and trim the other two edges.* 7. *Expand the matrices and dye each in its appropriate dye bath.* 8. *Treat the dyed matrix in the acid rinse.* 9. *Position first matrix on transfer blanket and transfer dye image to final support.* 10. *Locate subsequent matrices and transfer in register with first image.*

print is therefore made up of three combined dye images.

The process divides itself conveniently into two operations: making the matrices and making the print.

Making the Matrices. Before starting to make the matrices three things have to be decided: the relative exposure for the three matrices which will give the correct colour rendering in the print; the actual exposures which will give a just perceptible density in the lightest parts of the matrices; the contrast to which they should be developed. The first of these is a problem common to all separate image colour-printing processes; it is dealt with by adjusting the printing exposures to give identical reproduction of the grey scales on each of the separation negatives. The second calls for the same technique as is used for finding the exposure for bromide prints—e.g., by test strips, exposure photometer, etc. The contrast to which the matrices are developed can be controlled by the composition of the developer and is largely a matter of experience, although the maximum density range of the separation negatives gives a fair guide. The vital factor in development is to make certain that each of the matrices receives exactly the same treatment. The easiest way to do this is to expose the matrices one after the other and to develop them either in separate trays on a rocker, or by the interleaving method in one dish. The danger is that all three matrices may be incorrectly exposed, whereas, if each matrix is treated separately, the fault can be found without wasting three sheets of film.

For amateur use there is much to be said in favour of separate development at room temperature (between 60 and 75° F.) with standardized agitation. As the speed of the matrix film varies with the development temperature it is wisest to find the exposure by the test strip method.

After development the matrices are put into an acetic acid stop bath and then washed off in hot water for about five minutes. The temperature of the hot water wash can be varied between 130 and 180° F. but must be nearly the same for the three matrices otherwise the expansion will not be the same and will cause trouble with registration.

Making the Print. The making of the actua colour print can be left until any convenient time. It is carried out by dyeing each matrix with the correct colour and then printing all three images by contact in register on to a single support.

One of the annoyances in the process is that the dye baths tend to go out of balance and have to be tested at intervals. The test is made by superimposing three dye images made from the same, or identical matrices. The resulting image should be grey. As the dyes are apt to give different degrees of contrast the test is best made with a step wedge matrix. If the dyes are found to be out of balance they have to be adjusted by adding acetic acid to whichever ones are low in contrast.

After dyeing up and before making the actual transfer the matrix is rinsed briefly in acetic acid to remove the surplus dye. It is here that the four normal controls can be applied. These are: the addition of sodium acetate to the rinse to reduce contrast in the middle and shadow tones; the addition of calgon to clear the highlights; reduction of contrast by reducing the strength of the acid rinse; increase in contrast, especially of the highlights, by adding extra acid to the rinse and carrying over more dye from the dye bath.

In each of these controls there are three variables: the composition of the rinse, e.g., the amount of chemical added; the rinsing time; the quantity of rinse used. It is only by carefully standardizing the procedure that consistent results can be obtained.

Another possible means of controlling contrast is to add acetic acid or ammonia to increase or reduce contrast of the dyes. Unfortunately the dyes are not equally sensitive to these additions and it is difficult to keep them in balance.

The three-dye images must, of course, be transferred in register. This may be done either by registering the matrices beforehand or at the time of transfer. With the first method the matrices are taped down in register over an illuminator where either two edges are trimmed or holes are punched to provide a mechanical guide for subsequent transfers. This method is ideal when several prints are to be made, but the extra time and apparatus needed is unnecessary when making only a few duplicate prints.

Individual registration is simple; the only apparatus needed is a strong clip and a thin sheet of celluloid. After transfer of the first colour image the print is covered with a sheet of celluloid and the next matrix registered on top of it. When the register is satisfactory the matrix and transfer paper are clipped together, the celluloid removed, and the matrix rolled or squeezed into contact. With practice the whole process should not take more than one or two minutes. The only disadvantage of the method is the possibility of scratching the matrix while registering it or while removing the celluloid.

Duplicate Prints. The process has the great advantage that after one good print has been made, duplicates can be produced quickly and cheaply. The average cost of one print only is rather more than for a similar carbro print because of the cost of the matrix film; but the actual cost of making duplicates is much less. One set of matrices can be used for up to 150 prints and the time taken for making each print varies from 10 to 40 minutes depending on the working conditions.

Dye Transfer v. Carbro. The chief disadvantage compared with carbro is the inability to make adjustments for bad register between the three-colour images. With careful work this should not cause any difficulty if the separation negatives are made from a transparency. If they are made direct from the subject the camera must be capable of exposing the three negatives in exactly the same plane and the lens must be well corrected for chromatic aberration.

At first sight the number of bottles and the space needed for working the process seem rather formidable, but the amateur who is not primarily concerned with making a print in the shortest time will only need space for five dishes of the size of print to be made, and access to a sink and hot water.

The control which can be applied during the process is considerable and this often tempts the novice to inaccurate work. In fact, a great degree of contrast control is possible, but this latitude often makes it more difficult to produce a good print unless the process is carefully controlled. The latitude in exposure of the matrices is on the other hand small and does not amount to much more than 5 per cent under or 15 per cent over the ideal exposure.

The latest development in the process is the panchromatic matrix film which makes it possible to produce colour prints direct from negative colour transparencies without the need for colour separation negatives. At first sight the process seems to have every advantage, but it is difficult to judge the quality of the transparency or to make a good colour print unless the transparencies are taken under standardized conditions or with a grey scale included in each picture. This fact may more than outweigh the saving in time for making separation negatives. H.

See also: *Colour print processes; Registering images.*

Books: *Colour Prints*, by J. H. Coote; *Dye Transfer Colour Prints*, by Viscount Hanworth; *Colour Photography in Practice*, by D. A. Spencer (all London).

E

EASEL. Flat board which, in photography, forms the surface to which printing papers are attached during enlarging. Nowadays, easels are mostly horizontal for vertical enlargers. Early horizontal enlargers used vertical easels.

See also: *Enlarger; Masking frame.*

EASTLAKE, SIR CHARLES LOCK, 1793–1865. English painter. Chief adviser to the Prince Consort and the Government on art. Director of the National Gallery and President of the Royal Academy, 1852; first President of the Royal Photographic Society, 1853–5.

EASTMAN, GEORGE, 1854–1932. American photographic manufacturer of Rochester, New York. His first products were plates for which he introduced machine coating in 1879. Introduced paper roll films in 1884 as negative material, in 1888 transparent film, and also in 1888 the "Kodak", the first of the now famous series of roll film cameras. In 1891 the daylight-loading film was introduced. The slogan, "You press the button, we do the rest," brought the taking of photographs to the millions, and popular photography spread over the civilized world. As a direct consequence, the Eastman Kodak Company provided a dramatic, almost fantastic example of growth and development in industry, and it has founded and supported a unique chain of research laboratories. Eastman's commercial manufacture of roll film provided the basic material for cinematography and made the huge film industry possible. Eastman introduced 16 mm. reversal film for amateur cinematography (and the necessary camera and projector) in 1923 and developed various colour photography processes to commercial application. Received the Progress Medal of the Royal Photographic Society in 1927.

Was one of America's greatest philanthopist. Eastman died by his own hand ("My work is done—why wait?"). Biography by C. W. Ackerman (Boston and New York 1930).

EBERHARD, GUSTAV, 1867–19??. German astrophysicist. Published numerous investigations on spectroscopy and on photographic photometry. Described the Eberhard effect (1926, 1931).

EBERHARD EFFECT. Dependence of the density of small areas of the developed image upon the size of the area. This is due to the diffusion of developer across the border between areas of differing exposure level, and to the increased density difference which occurs on a micro scale across the border compared

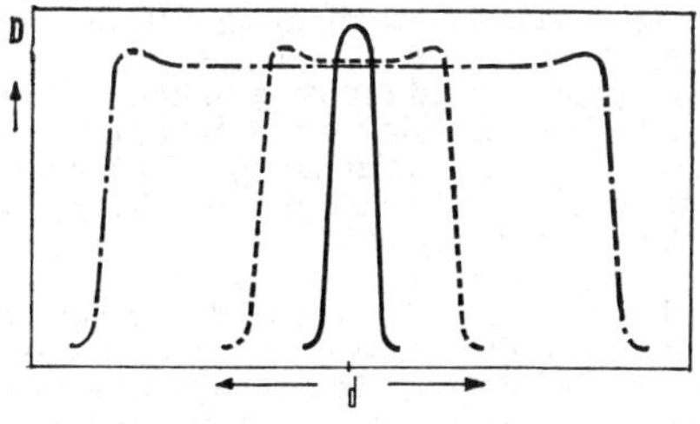

THE EBERHARD EFFECT. Microdensitometer traces show that the average density of small areas is higher than that of large areas. In all cases the density tends to rise near the edge of an area adjacent to a much thinner area. (d is the distance across the image.)

with that occurring on a macro scale between the centres of such areas. The smaller the areas in question, the greater will be the contribution towards their average densities of the border fringes which are subject to this enhanced density difference. The effect is observable for image sizes of less than approximately 1 mm.

See also: *Edge effect.*

EBONITE. Form of hard-vulcanized rubber, usually black, at one time much used in photography. It is relatively easy to mould or machine into camera parts and darkroom equipment, and it is not affected by photographic chemicals. Ebonite is, however, being more and more displaced by harder and tougher plastics. Nowadays its principal use in photography is for X-ray film processing tanks of up to 5 gallons capacity.

EBURNEUM PROCESS. Obsolete process brought out in 1865 by E. Burgess, a Norwich photographer. A positive image made by the carbon or collodion process was stripped and transferred to a support of imitation ivory. It was superseded by the Ivorytype which was much easier to produce.

ECOLOGICAL PHOTOGRAPHY. Ecology is the study of relationships under which living things exist together in their natural surroundings. Photographic records concerned with these studies include an appreciable quantity of straight-forward natural history pictures. But the quantitative assessment of ecological conditions usually involves a specialized application of aerial photography to investigate the distribution of plant life on a large scale.

The Nature Conservancy acquires Nature Reserves and is therefore concerned with ecology and conservation. These Reserves are managed under long-term management plans. Before management plans can be prepared, it is desirable to map the areas and also to carry out soil, vegetation, geological and physiographical surveys. Aerial photographic interpretation is playing an increasingly important role in carrying out these surveys.

The apparatus and exposing techniques used in ecological surveying are substantially the same as in normal aerial surveys. The amount of information obtainable depends largely on the skill of the interpreter.

Effective interpretation requires considerable experience with an extensive knowledge of the earth and life sciences, as well as a knowledge of photography. Interpreters with these dual qualifications are few and therefore it is necessary to find means of simplifying the interpreter's task by increasing the tonal contrasts on the photograph. Colour photography, filter and film combinations used wisely can greatly facilitate their work.

Films and Filter. The choice of film and filter combinations is of paramount importance in interpreting aerial photographs. This especially applies to the interpretation of vegetation where the tonal contrasts are usually low. The grey tones of the photograph are due almost entirely to the amount of light reflected by the object to the camera. Colour does affect the tone slightly but so little that it can be considered negligible. Light reflection depends on a number of contributory factors, the two most important being the position of the sun in relation to the camera and the texture of the surface being photographed. The tonal contrast between vegetation species under natural conditions is very limited. This especially applies to panchromatic film, which although sensitive to a wide range of colours, is not very sensitive to green light from foliage. A green filter is often used to remove both the red and blue light, thereby intensifying the contrasts in the green. Yellow filters are almost invariably used in aerial cameras. These filters cut out the blue light and this improves the haze penetration for maximum contrast on the photograph.

Infra-red film has two characteristics which assist the photographic interpreter. They are the unusual tonal rendering and its haze penetration, although the latter is not as efficient as is often supposed. Objects of high infra-red reflectance (chlorophyll content) appear very light in tone and those of low reflectance dark in tone. Extreme contrast is therefore obtained between some vegetation species. Infra-red is particularly helpful in the identification of trees, although deep shadows hide some detail. It is also of great value to physiographers and map makers because it enables the edges of water to be delineated with great accuracy, as clear water shows black on the photograph.

Colour photography has had limited use in Britain, partly because of greater cost and partly because of unfavourable weather conditions; reasonable colour rendering is possible only during a relatively limited period of the day, but high-speed colour emulsions can be valuable for specific purposes—e.g. to isolate areas of a particular species of grass from its associated vegetation. B.H.G.

See also: *Aerial survey.*

EDER, JOSEF MARIA, 1855–1944. Austrian scientist. Made outstanding contributions to photography, photochemistry and photomechanical work. In 1878 Eder won the great gold medal and one hundred ducats, the prize of the Wiener Photographische Gesellschaft (Photographic Society of Vienna), for his work on the action of chromic acid and chromates on gelatin and other organic substances. In 1879 his thesis on *The Chemical Action of Coloured Light* was published. In 1880 Eder became Professor of Chemistry at the Royal Technical School in Vienna. Soon after this he produced his comprehensive *Handbook of Photography*, which ran into many volumes and many editions. In 1889 he was appointed director of the Graphische Lehr- und Versuchsanstalt (Graphic Arts Institute), the greatest school of graphic art in the world. Finally, amongst other books, there is his great *History of Photography*, the last edition (German) of which appeared in 1932, and was translated by Edward Epstean in 1945. He received in 1884 the Progress Medal of the Royal Photographic Society for his work on silver chloride emulsions for gaslight development papers (1881). Biography by F. D. Worschak and O. Krumpel (Vienna 1955).

EDGE EFFECT. Generic term covering the various effects which can occur at the boundary between portions of the developed image of differing densities.

During development, the developer in the vicinity of a heavily exposed area becomes rapidly exhausted and the rate of development decreases. In the areas corresponding to less exposure adjacent to the heavily exposed area, however, the developer remains comparatively active. Due to lateral diffusion of this more active developer across the boundary between the two areas, a local increase in the rate of development takes place just inside the edge of the heavily exposed area, and this narrow strip develops to a higher density.

Similarly, in an area that has received only a low exposure, the developer remains relatively fresh and active. In adjacent areas corresponding to higher exposure levels, however, the developer becomes more and more exhausted as development proceeds. This relatively exhausted developer diffuses across the boundary into the lightly exposed area and lowers the activity of the solution at this point. Thus just inside the edge of the area of low density is produced a line of even lower density.

The net result of these two effects is the production, at the actual border between two areas of the type described, of a density difference greater than that which exists between the gross areas concerned. This enhanced edge contrast results in greater image sharpness, and is an effect which is in some instances sought after. The effect is emphasized by the use of a developing solution which exhausts rapidly during development, and by the avoidance of agitation.

See also: *Eberhard effect; Kostinsky effect; Mackie line.*

EDINOL. Para-aminosaligenin. Developing agent. Now obsolete for that purpose.

Formula and molecular weight: $NH_2C_6H_3OHCH_2OH$; 139.

Characteristics: White powder, similar to para-aminophenol in its developing action.

Solubility: Fairly soluble in water.

EDISON, THOMAS ALVA, 1847–1931. American inventor. Up to early in 1928, had taken out 1,033 patents among which were telephones, automatic telegraph systems, the phonograph, the first dictograph, incandescent electric lamps, and in 1891, the Kinetograph and Kinetoscope for taking and viewing serial pictures. Was probably the first to use on a practical scale film perforated at the sides, and was also the first to combine the projection of pictures with his phonograph to produce the first practicable, though far from perfect, sound film. Biography by F. L. Dyer and T. C. Martin (New York and London 1910).

EDITING

Arrangement of the individual shots of a cine film in the final order in which they are to be presented for the best filmic effect. This is the final stage in the production of a film. In a simple amateur film it involves sorting out the takes into their correct order, rejecting unsatisfactory ones, varying the length of shots to the mood of the film and joining all the resulting pieces together in the chosen order. In a more ambitious film judicious editing can control the tempo and even the story content of the film itself. In feature films editing is thus at least as important as scripting and direction.

Editing has a physical and a creative side. The first involves the sorting out of the shots and the splicing together. The second is concerned with cutting the shots to the right length, selecting the order in which they shall be joined and creating appropriate juxtaposition of scenes to convey ideas or impressions which may not be inherent in the individual scenes.

In the scripted film it is the editor's job to interpret the intentions of the script-writer and the director to produce the desired final effect.

Equipment. The main item of equipment is an animated viewer, preferably one that runs at a constant speed so that accurate timing of the shots is possible. Amateur viewers do not have this refinement. Many amateurs use their projectors instead, although inconvenient.

A good splicer is a necessity for joining the strips of film together with speed and accuracy. Film cement is the solvent used for what is actually welding and not a cementing operation.

Rewinders, spare spools and a winding-off plate are needed to enable film to be wound on and off the spools and generally handled with ease during the editing.

The sorting equipment often takes the form of a pin rack on which a row of film strips can be hung and sorted into order as required. Amateurs seem to favour an array of small tins numbered and stuck on to sheets of plywood of a convenient size. Storage tins serve to keep all the pieces of film identified and available.

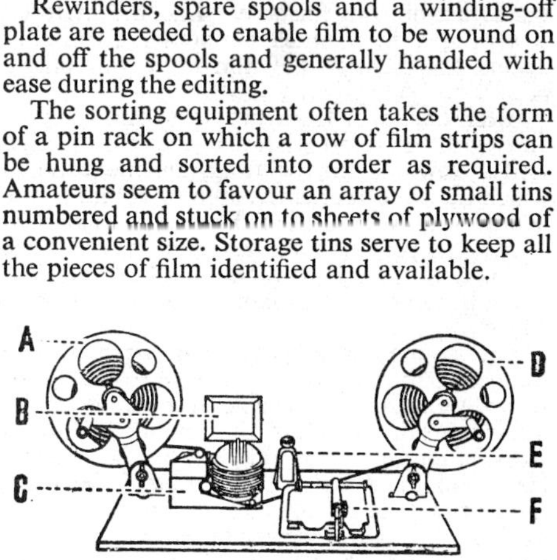

ANIMATED VIEWER. The different units are laid out on an editing bench for easy manipulation. A, normal feed spool. B, viewing screen. C, lamphouse. D, normal take-up spool with geared winding handle. E, film cement. F, splicer.

The smaller items of use in editing are cotton gloves, scissors, grease pencils, magnifying glass, duster, cleaning fluid, note pad and pencil.

Editing Procedure. Some 16 mm. and all 35 mm. users film with the negative-positive system. The first prints (or "rushes") made from the negative are projected as soon as possible, usually the day after shooting, and these are the positives that are handled in the cutting room in the building up of the cutting copy. Many narrow gauge users do their filming on reversal stock—in this case the editor should be given a reversal duplicate for cutting purposes in order to avoid the hazards of handling the original when doing the editing.

The editor has his own copy of the script. He is given a copy of the notes made by the director and copies of the continuity sheets prepared by the continuity assistant. During the showing of the rushes to the key members of the unit he also makes his own notes. He will thus know which are the best takes and will have noted any inferior sections which should be avoided if possible.

First the rolls of positive as received from the laboratory are broken down and the separate scenes are sorted out in numerical order to match the script. With the aid of his notes the editor immediately extracts the unwanted and faulty material. What remains is then joined into reels in script order and projected as a check that all the essential material for the film is there. If there is a sound track each reel will consist of two roils—separate picture and sound.

Then follows the rough-cut stage when all obviously unnecessary portions of the good takes are cut out—such as number boards, excess action and mistakes—and the material is arranged in a much condensed form that begins to resemble the film as it will finally be.

After approval from producer and director and with their suggestions, the editor then settles down to finer cutting, an operation in which he exercises his freedom of choice in the detailed consideration of the cuts from shot to shot. This process may last quite a long time and is a vital stage with continual experimenting to achieve just the right timings and general effects.

The sound, whether synchronized at the moment of filming or soon afterwards at a post-recording session, is whenever possible edited at the same time as the picture. Sound makes the work of editing more complex, but the specialized equipment helps to overcome the extra difficulties. The 35 mm. editor runs his lengths of film through a viewer or editing machine which not only shows a small animated picture but reproduces the sound from the sound track in synchronism. The narrow-gauge worker may use a specially adapted version of the 35 mm. editing machine, but in general most of his sound editing must be carried out on the sound projector.

In all but the most unpretentious films the next step will be the fitting of music and sound effects to the picture and the process of re-recording them with the existing sound track all on to one track.

After that the negatives or originals will be cut to match the cutting copy and a final print produced that has the sound and picture printed on to the same strip of film.

Silent films are naturally the easiest proposition. The amateur, especially, is not involved in such intense activity as the professional and his apparatus is simpler, but in principle his editing will conform to the same rules if he takes his film making seriously. When the fine-cut stage has been completed the main task that remains for him is to add the titles.

Editing Principles. The creative side of editing is one aspect of the expression of cinematographic art. While a number of principles have become established, largely by trial and error, as a sound basis for editing practice, they are guides rather than firm rules.

The art of editing in the main deals with continuity, timing and the creation of filmic relationships.

In its simplest terms continuity is concerned with keeping the film flowing smoothly without disturbing jumps or gaps.

Firstly, the editor has to choose the different types of shot available of a scene. Thus a long shot introduces the subject, while medium and close shots following after draw attention to specific details. The continuity of action depends also on the camera direction in a sequence. Thus when the camera moves in on a scene, the different shots should be from different viewpoints to avoid sudden jumps. Where such jumps are unavoidable, they can often be bridged by a cut-away—a shot of a different but related part of the subject.

Where the same action takes place in, say, a medium shot and a close shot, the two must be matched in joining, so that the action continues smoothly from the one to the other.

Cross-cutting or inter-cutting is a technique of linking two related and parallel actions in a film and incidentally also of heightening tension. The classic example is the rescue theme involving two basic sequences: the victim awaiting his or her end at the hands of the villain, and the rescuer hurrying to the scene. For cross-cutting the two shots are split up into short lengths and joined up alternately. The cross-cut part-shots are made progressively shorter and shorter to build up to the climax.

This is one of many aspects also of timing. Quite apart from the speed of movement of the subject in the film, the length of a shot conveys the rate of action. The shorter a shot appears on the screen, the faster the tempo or illusion of speed conveyed. Thus the impression of action can be derived from a rapid succession of very short shots without any significant inherent motion. Thus a flashing succession of newspaper headlines can suggest fast-moving events.

Inter-cutting can also stretch the length of an action. Thus a sequence of an accident lasting only a few seconds can be made to occupy two or three times its length in screen time by inter-cutting the event itself with the reactions of victims, onlookers, etc.

Conversely, editing can also compress time. For example if the story involves a transatlantic journey, a shot of the boat leaving the English port (established in previous shots) cut to the same boat with the Statue of Liberty in the background, tells the audience that the journey has taken place, without showing any of it in detail. Dissolves and fades are also classical means of bridging both space and time.

On a smaller scale the lapse of time—for example to eliminate a familiar and not particularly interesting action like dressing—can be bridged by a cut-away shot. In this case a close-up of a clock would do, and might also establish that the character is in a hurry.

The power of suggestion by the juxtaposition of seemingly unrelated shots to create a relationship is a further important tool of editing. Thus a shot of a character looking out of the window (filmed in the studio) followed by a scene of a cheering crowd taken from a newsreel conveys the impression that the character is watching the crowd.

Even the order in which the two shots are presented can give a different interpretation to the story: if the actor is seen first and the cheering crowd afterwards, the inference is that the crowd is cheering the hero. If the crowd shot precedes the actor, he may be taken to be a mere spectator. Similarly, shots of a burning model building, intercut with a racing fire engine and studio shots of rushing firemen, build up the story of a big fire.

While editing sound films is more complex, the scope of creative editing is greater still. Here sound effects can be added to the picture to create straightforward or contrasting relationships. They can also bridge visual points which would otherwise have to be linked by additional shots to preserve the continuity.

J.C. and L.A.M.

See also: *Cinematography; Cine technique; Splicing; Titling.*

Books: *How to Edit*, by H. Baddeley; *The Simple Art of Making Films*, by Tony Rose; *The Technique of Film Editing*, by K. Reisz (all London).

EDUCATION

EDUCATION. There are two and fundamentally different aspects of photographic education: on the one hand the use of photography as an educational aid, and the other the training of photographers.

Education with Photography. The application of photographic images to education covers cine films, and still films in the form of film strips and slides, as well as individual prints.

Cine films are available as specially planned films for class room showing with the aid of 8 mm. or 16 mm. projectors. Most of these are comparatively short but self-contained features, and are produced and planned as complete items covering a selected subject, being screened in the class room—usually after blacking-out—as part of a lecture or a complete lecture. Such films may run from ten minutes to half an hour in length and illustrate a subject, usually together with a commentary or other sound effects. In the 16 mm. gauge such films are generally mass produced and sold or hired to schools and other institutes.

Much the same applies to 8 mm. films, but there are two additional possibilities. The first is the production of the film by a teacher or a group of teachers. Here 8 mm. film has the advantage of cheapness both in material and in the camera equipment required. In most cases teacher-produced films are likely to cover mainly travel scenes, etc., though occasionally more ambitious projects may be handled. This is largely a matter of the cinematographic experience and skill of the teacher concerned.

The second possibility is the short and ultra-short concept film. Here single ideas are presented in filmic form, running on the average no more than two minutes. Such films may be shown on more or less frequent occasions in the course of a lecture. Their presentation involves special apparatus and systems, to permit immediate projection at any selected moment, much like drawing a diagram on a blackboard. Only the film can convey a moving idea—which may be either realistic or diagrammatic. Such concept films are generally mass-produced, and may be obtained by hire or purchase from suitable libraries.

Still pictures may be presented as film strips or slides. Film strips are available as self-contained series, sometimes with recorded tapes giving explanations and commentaries which may be played on tape recorders combined or linked with the film strip projector. Signals on the tape can be used to advance the projector automatically from one image to the next at the appropriate points in the narration. Slides are more flexible for presenting individual items, but here again a complete series may be linked with a recorded tape commentary.

Individual illustrations in the form of large transparencies can also be shown with overhead projectors. This has the advantage that the teacher or instructor can even draw diagrams on a transparent foil in the overhead projector, for presentation to the whole class in much the same way as on the blackboard.

Episcope projection of opaque illustrations (photographs, pictures from books, etc.) is less popular nowadays, mainly because the brightness of an episcope image is usually limited.

Education for Photography. Polytechnics and departments of technical colleges in Britain provide training courses for professional photographers. Usually they aim at preparing pupils for the IIP or City and Guilds examinations, which are regarded as standard qualifications for a photographer. In various countries in Europe there are special schools of technical college standard devoted entirely to photography.

All these schools and courses produce qualified practical photographers, and therefore provide instruction both in photographic theory and in actual picture taking. There are also courses at a number of institutes in Europe for theoretical photography. These are concerned with the academic training of photographic scientists and technicians, and lead to Ph.D. and similar degrees.

See also: *Examinations: Training and education in photography; Training of photographic technologists and scientists; Visial-aids.*

EFFLORESCENCE. Process by which a crystalline substance gives up its water of crystallization to the air and turns into the amorphous, powdery state. The white dust that forms on crystals of sodium carbonate (washing soda) when exposed to the air is an instance of efflorescence.

EIKONOGEN. Sodium 1-amino-2-naphthol-6-sulphonate. Developing agent. Now little used for that purpose.

Formula and molecular weight: $NH_2C_{10}H_5OHSO_3Na$; 261.

Characteristics: White powder. Somewhat resembles metol in its developing action.

Solubility: Slightly soluble in water at room temperature.

ELECTRIC CHARGES ON FILMS. The surface friction produced by handling or unrolling films may build up a charge of static electricity sufficient to create a spark or a visible glow discharge. This discharge is sometimes (but not often) intense enough to form a developable image on the film.

See also: *Static.*

ELECTRIC EYE. Common term, especially in advertising literature, for automatic exposure meter systems which directly control the lens aperture (and sometimes also the shutter speed setting) in response to the light falling on a photo cell.

See also: *Exposure automation.*

ELECTRICITY. Nowadays electricity is used in practically every department of photography—taking the photograph, processing the negative, and making the print or enlargement.

Most of the electricity used in these ways is for lighting; the remainder is for heating and for driving motors. The electricity supplied for all these purposes may be direct or alternating and its voltage may be around 110 volts or anywhere between 200 and 250 volts. The chances are, however, that it will be alternating current of 240 volts and a frequency of 50 cycles per second. This supply is gradually replacing all others throughout the British Isles. The standard American household supply is 115 volt, 60 cycle alternating current.

Changes or additions to the permanent wiring of buildings should preferably be made by a qualified electrical contractor. Any handyman who proposes to do the job himself should first study the relevant I.E.E. regulations on the subject. Copies of the rules applying to the particular job—e.g., permanent wiring, flexible cords, fuses—can usually be obtained from the local office of the British Electricity Authority or by application direct to the Institute of Electrical Engineers.

Equipment Circuits. There are two ways of connecting a number of pieces of equipment to an electricity supply—in series and in parallel.

When pieces of equipment are connected in series end-to-end, all the separate resistances of each item add together and their sum is the total resistance that the circuit offers to the flow of electricity from the supply. So ten pieces of the same kind of equipment offer ten times more resistance to the flow of electricity than a single piece of that equipment.

And the current that flows through the ten connected in series is exactly one tenth of the current that would flow if only one piece of equipment were there.

In the same way, two lamps connected in series pass only half as much current as one connected to the supply by itself. And since the voltage is the same in each case, the two lamps together consume only half as much electricity as one. But, of course, the amount of light is also greatly reduced.

When pieces of equipment are connected in parallel—i.e., side-by-side—each piece of equipment has the full voltage of the supply acting on it. So each separate item acts as though the others were not there. Each takes its full quota of current from the supply. So ten pieces of the same sort of equipment connected in parallel across the same supply take ten times more current than one. And the effective resistance of the whole circuit is one tenth of the resistance of a single item.

The general rule is expressed by the equation:

$$\frac{1}{R} = \frac{1}{r_1} + \frac{1}{r_2} + \frac{1}{r_3} + \text{etc.}$$

where R is the total effective resistance of the circuit and r_1, r_2, r_3 etc., are the separate resistances of each individual item.

Cells in Series and Parallel. When a number of cells are connected in series, end-to-end, the positive of one cell to the negative of the next, their combined voltage is equal to the sum of

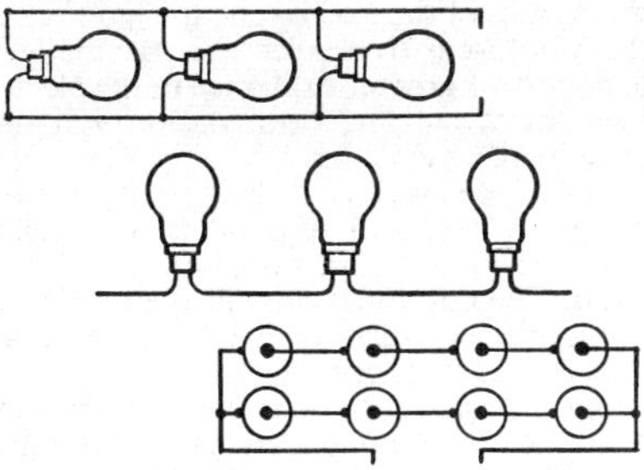

SERIES AND PARALLEL CONNEXION. Top: *Three lamps in parallel.* Centre: *Lamps in series.* Bottom: *Two batteries in parallel, each with four cells in series.*

the separate voltages. Ten 1·5-volt cells in series have a combined voltage of 10 × 1·5 = 15 volts.

The actual amount of the current that can be delivered by a cell is limited by its internal resistance, so that the current that can be delivered by a number of cells in series will be no greater than the current that can be delivered by a single cell.

When a number of cells are connected in parallel, with all the positive terminals connected together as one pole, and all the negatives as the other, the effect is exactly the same as if all the separate cells were merged into one. So ten 1·5-volt cells connected in parallel become one big cell, and the voltage remains unaltered at 1·5 volts.

Connected in this way each cell is still capable of delivering its full current, limited only by its internal resistance. So the ten cells in parallel are able to deliver ten times as much current as a single cell.

When cells are connected in series they give a higher voltage; in parallel, they deliver a higher current. There is a third type of circuit that gives an increase in both voltage and current. This method of connecting the cells is called series-parallel, and it is a combination of the other two.

Four 1·5-volt cells can be connected in series to make a 6-volt battery, and if they cannot deliver enough current for their purpose, an identical battery of four 1·5-volt cells can be connected in parallel with the first. The second battery will double the current available while the voltage will still be only 6 volts. This is an example of series-parallel connexion as it is often used for boosting batteries to supply extra flash bulbs on the same circuit.

The behaviour of cells in series and parallel is of great importance in photography with flash bulbs where a flash bulb that takes a relatively heavy current operates from small portable batteries. Delay or failure of the flash through weak batteries can be a very serious matter to the photographer.

Basic Calculations. The electrical units are interdependent, e.g., an electrical pressure of 1 volt applied to a circuit with a resistance of 1 ohm would send a current of 1 amp through the circuit.

The rate of power consumption in a circuit is given by the relationship:

$$\text{watts} = \text{volts} \times \text{amps.}$$

For instance, on a 250-volt mains supply, a lamp carrying 0·8 amps would consume 250 × 0·8 = 200 watts. The amount of electricity that would have to be paid for if this current were switched on for, say, 2,000 hours is given by the relationship:

$$\text{Kilowatt hours (units)} = \frac{\text{watts} \times \text{time (in hours)}}{1{,}000}$$

$$\text{or, in this case } \frac{200 \times 2{,}000}{1{,}000} = 400 \text{ units}$$

The relationship between volts, amps, and ohms, is usually expressed by the formula known as Ohm's Law:—

$$\text{Current in the circuit(C)} = \frac{\text{E(Voltage of the supply)}}{\text{R(Resistance of the circuit)}}$$

So if any two of these factors are known, the third can be found immediately.

For example if, say, a Photoflood lamp with a resistance of 190 ohms is connected to the 230 volt supply, the current that will flow in the lamp wiring will be:

$$\text{Current} = \frac{230}{190} = 1{\cdot}2 \text{ amps.}$$

The watts consumed by such a lamp will be: watts = volts × amps. = 230 × 1·2 = 275.

If such a lamp has a life of 2 hours, it will consume altogether 2 × 275, or 0·550 kilowatt hours, or units of electricity, i.e., just over half a unit while it lasts. F.P.

See also: *Batteries; Dimmer; Fuse; Rheistat; Series parallel; Switches; Transformer; Wiring.*

Books: *Electrical Facts for Photographers,* by R. H. Alder (London).

ELECTROFAX. Proprietary name for an electro photographic process depending on the image-wise discharge of an electrostatically charged dielectric layer coated on photographic paper. After treatment with charged pigment powder particles the electrostatic image becomes a visible pigment image which can be rendered permanent by the application of heat.

See also: *Electrophotography.*

ELECTROLUMINESCENCE. Direct conversion of electrical energy into visible light (instead of for example as a by-product of the heating effect of passing a current through a tungsten filament). A substance exhibiting electro-luminescence glows when subjected to an A.C. electric field.

The most promising electroluminescent substances to date are phosphorescent substances like zinc sulphide—closely related to the screen phosphors in cathode ray tubes.

The phosphor is distributed in an extremely fine powdered state over the surface of a base of metal or conducting glass. It may be suspended in a synthetic resin or a transparent vitreous enamel. A layer of conducting glass, resin or enamel, is applied over the phosphor. A final coat of transparent and electrically insulating enamel protects the surface and makes it safe to handle. When a voltage is applied to the outer conducting layers of the sandwich, the phosphor glows. The voltage required to produce luminescence varies according to the method and materials of construction, but may be anything from 50 to several thousand volts. The colour varies according to the phosphor used and the light output increases with frequency of the exciting current. A typical electroluminescent panel run at 350 volts and 400 c.p.s. gives a luminance of 22 feet-lamberts for a current consumption of 1.5 mA. per sq. in.

Photographic Applications. The soft green light emitted by the normal electroluminescent panel makes an excellent safelight for darkrooms where panchromatic films are handled. Electroluminescent lighting is used for darkroom clock faces, wall switch surrounds, instrument dials and pointers.

Photo-sensitive panels. A number of interesting and useful possibilities result from the combination of an electroluminescent layer with one that is photoconductive—i.e. that changes from an insulator to a conductor when light falls on it. When the two layers are sandwiched together between layers of electrically conducting glass carrying a sufficiently high voltage, a light directed on to the photosensitive layer causes it to conduct the electrical charge at that point. The electroluminescent layer is subjected to a voltage at that point and it glows. Even when the initial light is switched off, the glow from the electroluminescent layer continues to maintain the conductivity of the photosensitive layer, so the glow persists until the applied voltage is switched off. This basic phenomenon can be exploited in several ways:

1. *Image Intensification.* By adjusting the characteristics of the two layers, the light falling on the photosensitive layer can be made to excite a glow of greater brilliance in the electroluminescent layer. The combination will then act as an image intensifier.

2. *Image Converter.* The photosensitive layer can be made to respond to non-visible radiant energy—e.g. X-rays, ultra-violet or infra-red. The combination will then act as an image converter, turning an invisible image into one that can be seen.

3. *Image Storage Device.* Because the glow from the electroluminescent layer continues to excite the photoconductive layer in the areas of brightness even when the original image-forming rays are switched off, the panel will act as an image storage device that can hold the picture of transient phenomena as long as the applied voltage is maintained. Once the voltage is switched off the luminous glow disappears and the panel is ready to accept another image. In this way an electroluminescent panel can perform many of the functions of a photographic plate which needs neither developing nor fixing and can be re-used indefinitely. F.P.

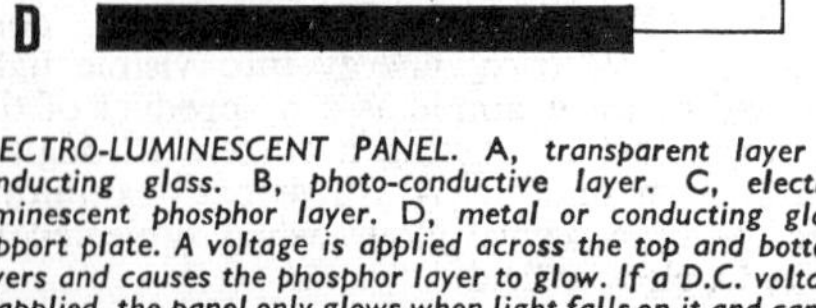

ELECTRO-LUMINESCENT PANEL. A, transparent layer of conducting glass. B, photo-conductive layer. C, electro-luminescent phosphor layer. D, metal or conducting glass support plate. A voltage is applied across the top and bottom layers and causes the phosphor layer to glow. If a D.C. voltage is applied, the panel only glows when light falls on it and can in this form be used as an image intensifier, and a short-term image retaining system.

See also: *Fluorescent lamp.*

ELECTROLYTIC DEVELOPMENT. System of processing a photographic image by the application of an electric voltage or field.

In photographic systems based on the use of silver salts, the essential step in image development is the deposition of silver. This is so even if the ultimate pigment is to be something other than silver, e.g., a dye.

In order to reduce silver salts to silver metal, it is necessary to donate an electron to each silver ion. In conventional developers, the source of electrons is the developing agent, usually an organic compound which, on giving up electrons, can pass into another, more stable, form. Since, therefore, the fundamental process involves the passage of electric current, it is possible to use electricity, more or less directly to carry out development.

So-called electrolytic development can be divided into two main categories; (a) generation of a developer by electrolysis, and (b) photoconductive systems.

Electrolysis. The first category applies to the development of conventional materials carrying an emulsion layer in which grains of silver halide are embedded in gelatin. Here the problem is one of application of the electrical energy available at a massive electrode to the latent image in each grain. The only practicable method is to use the electrical energy to generate a solution of low redox potential, i.e., a solution capable of reducing silver halides. It is this solution which diffuses into the emulsion layer and carries out the development along conventional lines. The electrolysis serves only to generate the reducing solution in the first instance and to regenerate it as it is used. It has been claimed that an image can be developed simply by placing an exposed layer in contact with a cathode in water, but this must have depended on the generation of some reducing agent, e.g., nascent hydrogen.

The nearest approach to a practicable system was produced by Roman who electrolysed a solution of a vanadium salt in hydrobromic acid. A solution of vanadous ion was formed in the cathode compartment and this solution could be pumped to the developing tank from which it returned for regeneration. The system could be used for the development of motion picture positive film.

Photoconductive Systems. These are of more interest at the present time inasmuch as they are adaptable to various processes for the recording and reproduction of information. In these systems the light image is converted in effect into an electrical image. In one form, a zinc oxide photoconductive emulsion is coated on a metallic support. After the xerographic exposure the layer carries a pattern of varying conductivity. This layer is wiped with a silver-bearing solution while a potential is applied between the wiping device and the metallic under-layer, the latter being the cathode. The local conductivity determines the intensity of the local current and, hence, the amount of silver deposited. In this way a silver image is built up electrolytically on the "conductivity image".

In another form, due to Berchtold, a sheet is made by cutting a slice from a bundle of innumerable, fine wires that are bonded together, but insulated from each other, by a resin. Such a sheet is highly conductive from front to rear but is a non-conductor sideways. A photoconductive (selenium) layer is coated on one side of the Berchtold sheet and a thin, transparent metal film is coated over the selenium to act as a contact. A wet, electrosensitive paper, which can carry a silver salt or a dye, is pressed against the rear side of the Berchtold layer by means of a conducting plate. If a voltage is applied between the front and rear contacts of the assembly while an image is thrown on the selenium layer, the electrosensitive paper will be affected by the current in contact with the micro-electrodes. The photosensitive part of the process is not expendable and can be used indefinitely.

Currently electrolytic systems of this kind are finding applications in document copying.

ELECTRONIC FLASH. Light source which utilizes the short duration flash of light produced by discharging a medium or high-voltage charge through a gas-filled tube.

See also: ***Flash (electronic).***

ELECTRONIC PRINTING

Method of making photographic contact prints and enlargements with a flying spot scanner instead of a continuous light source. An electron beam moves over the fluorescent face plate of a cathode ray tube in a regular pattern; the light spot produced exposes the negative point by point on to the printing paper. The main advantage over a continuous light source is that the light coming through the negative can be monitored continuously by a photo cell which can in turn influence the brightness of the light spot. That provides considerable scope for contrast control.

Photo-multiplier Monitoring. In electronic printing equipment the cathode ray tube is in virtually direct contact with the negative being printed. In a contact printer the positive material is in contact with the negative and a photo-multiplier tube records the light coming through the positive material.

In an electronic enlarger the cathode ray tube is in contact with the negative in the negative carrier. The enlarger lens there projects the image on to the printing paper on the enlarger baseboard in the usual way. The monitoring tube can again monitor the light coming through the printing paper; the more usual system is to have a beam splitter just below or just above the enlarger lens to divert a fraction of the light (about 5 per cent) to the photo tube.

In either case the photo-multiplier tube controls the intensity of the scanning spot and in this way the contrast of the resulting print. For this purpose the output of the photo-multiplier is fed into an inverse feed-back amplifier.

The feed-back amplifier is linked with the circuits controlling the intensity of the electron beam. When the photo tube registers a high light intensity (as the electron beam scans thin parts of the negative) the feed-back circuit reduces the intensity of the scanning spot. Hence thin parts receive less exposure than they would otherwise do. Conversely, when the photo tube output drops, the feed back amplifier increases the intensity of the electron beam so that dense parts of the negative receive more exposure. So what takes place is an automatic and simultaneous shading and over-printing of various parts of the negative according to their densities. This reduces the contrast of the print, so that even negatives of extreme tone range can yield acceptable prints on a normal or even hard grade of paper.

Put in another way, the negative is printed with a light source that corresponds to a luminous positive image. This takes over all the functions of an unsharp positive mask produced by conventional printing procedures, except that the light mask is formed automatically during the printing process and requires no preparatory work or processing. It is also auto-

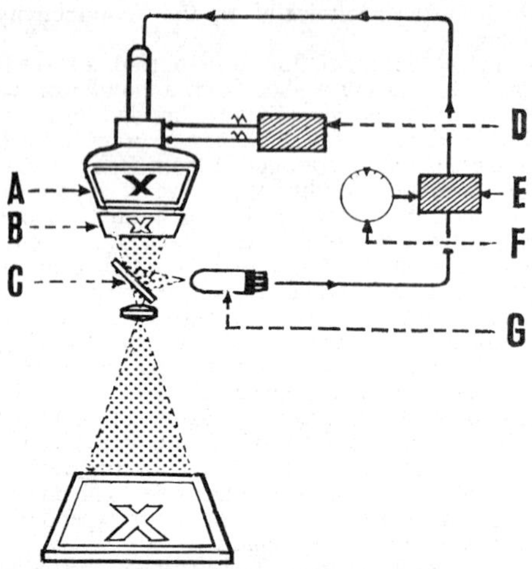

ELECTRONIC ENLARGER SYSTEM: A, cathode ray tube with flying spot scanner. B, negative. C, beam-splitting mirror. D, control of cathode ray tube. E, inverse feedback amplifier. F, contrast control adjustment (by feedback setting). G, monitoring photo tube. By additional control circuits this can also regulate the overall exposure by suitable brightness control of the flying spot.

matically related to the negative and in register with it.

Adjusting the feed-back of the amplifier controls both the mask density and contrast. If the feed-back is set to zero, the printing effect is the same as with a continuous light source: the scanning spot remains uniformly bright throughout the exposure.

Negative feed-back, as just explained, increases the scanning spot brightness when scanning dense negative areas and decreases it for thin negative areas. With 100 per cent negative feed-back the increases and decreases just compensate the negative densities; in theory this should give a completely uniform print density with no recognizable image. (In practice image outlines do remain visible owing to fringe effects, as explained below.)

The feed-back can also be made positive, so that the scanning spot gets brighter when the photo tube receives a lot of light through thin negative areas, and less intense when scanning denser areas. This would increase the contrast of the print; electronic printing thus provides a virtually unlimited degree of contrast control. In practice, electronic enlargers can handle negative density ranges up to about 3·0 and compress these to a range of 0·1 to 0·3 on a print. Even when a negative density range of 3·0 is reduced to 1·0 on the print, this corresponds to accommodating a subject brightness range of 1,000 : 1 (assuming a negative gamma of 1·0) within a print brightness range of 10 : 1.

Contrast Range and Detail Contrast. In theory a similar compression of brightness range would be possible by printing a contrasty negative on an extremely soft gradation paper. This would compress all tones uniformly and would equally reduce the detail contrast to an unnatural degree. The resulting print looks soft, even though it may have a full tone range from white to black.

Electronic printing gets over the difficulty by compressing the overall contrast without modifying the detail contrast to the same extent. The two can in fact be adjusted independently. This is important because a compressed tone range in a print where detail contrast is compressed along with overall contrast may make some detail invisible. Tone differences above the threshold of differentiation in the original may drop below this in the print.

Electronic tone contrast control thus corresponds more to tone separation.

This is achieved by controlling the size of the scanning spot as well as the inverse feed-back. While the amount of feed-back determines the overall contrast compression (corresponding to the mask percentage in masking), the spot size determines the unsharpness (and hence detail control) of the mask. The printer does not compensate for brightness variations over areas smaller than the scanning spot itself. It therefore preserves detail contrast both within highlights and within shadows, even when the overall exposure in negative shadows is increased and in highlights decreased.

The scanning spot size is chosen (by appropriate focusing of the electron beam) according to the requirements of the negative. A negative of great contrast but comparatively little fine detail would be printed with a large spot size and appreciable feed-back. A negative of lower overall contrast but with a great deal of fine detail (for instance an aerial photograph) would need a small spot size and moderate feed-back. In practice, spot sizes are held between 1/16 and ¼ in.

Fringing. The finite size of the scanning spot also leads to the fringing effect which is characteristic of electronic printing with a moving spot. As the spot scans a boundary of the image from light to dark, the light reaching the monitoring photo tube does not change suddenly but gradually. With the spot half way across the boundary, the light falling on the photo tube is half-way between that for the bright and for the dark areas. As however the feed-back circuit reacts instantly, the boundary of the shadow area begins to receive more exposure than the shadow region just traversed by the spot. Similarly the edge of the highlight area gets less exposure than the region covered by the spot immediately afterwards. This leads to a dark fringe at the border of the shadow area in the print and a light fringe at the border of the highlight area (where it is particularly noticeable). The width of the fringe depends on the time the spot takes to cross the boundary; with a fixed scanning speed it is greater the larger the spot.

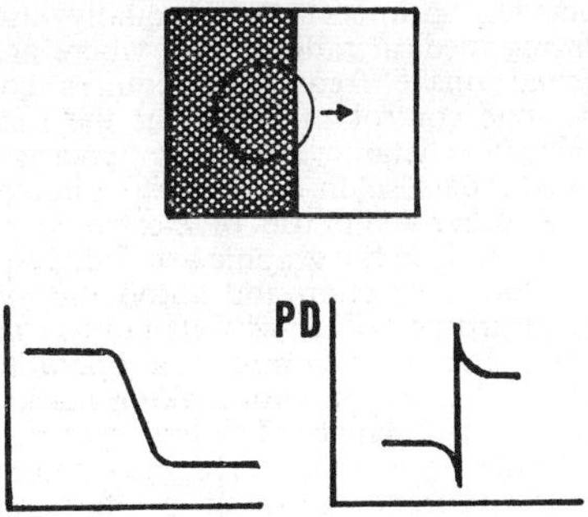

THE FRINGE EFFECT. When the scanning spot of the cathode ray tube moves from a dark shadow area to a light one (top), the feedback causes the spot brightness to drop gradually owing to the finite spot size (bottom left). The print density PD as a result decreases at the border of a light area, and increases at the border of a dark one. The smaller the scanning spot, the less the fringe effect, but the lower the detail contrast.

Scanning and Exposure. The actual scanning pattern that the flying spot traverses is not particularly important, provided all parts of the negative area are covered by the spot for the same time during a complete scanning cycle. The scanning pattern itself may either be a line scan (as in a television receiver) or an interlaced diagonal one. The total scanning cycle may last from 1/10 second to several seconds; one scanning cycle is the minimum exposure for a print.

The exposure control may be combined with the contrast control or it may be separate. In the first case the monitoring photo-multiplier tube automatically adjusts the brightness of the flying spot so as to yield a correct exposure for every part of the print during one scanning cycle (or a fixed number of complete cycles). Every exposure takes the same length of time and negatives of different density are compensated by appropriate changes in the spot intensity. The electronic printer then needs setting only to the paper sensitivity.

Instruments with separate control use a second photo-multiplier tube which automatically terminates the exposure after a predetermined amount of light has reached the tube. The exposure itself must then last for a whole number of scanning cycles. Alternatively the brightness range of the projected image on the baseboard can be measured by photo tubes before the exposure itself. These measurements establish the brightest highlight and darkest shadow tones and are then used to set both the required contrast control and the actual exposure on the control unit for the printer.

Electronic Colour Printers. The same principle can be used for producing colour prints from colour negatives. The negatives are printed by three successive exposures through a red, green and blue tri-colour filter. During each filter exposure the photo tube monitors the light through a complementary filter (i.e. cyan, magenta and yellow respectively). The photo tube output is in turn connected to three inverse feed-back amplifiers for the three printing colours. These are linked with three timers which control the individual printing exposures. The output of the monitoring photo tube provides automatic colour correction during each exposure. By appropriate adjustment of the control circuits any degree of over-riding colour correction can also be applied.

A similar arrangement is used for making colour separation positives from separation negatives, or for making separation negatives from colour transparencies. In the first case the negative and the printing material are changed between each exposure; in the second case the colour transparency remains in position but the three exposures are made on separate films. The inverse feed-back can then function both for automatic contrast masking and for colour masking. Colour masking is achieved by monitoring the flying spot through a non-complementary filter. For example, to correct the unwanted blue absorption of cyan dyes, the red filter exposure can be monitored through a yellow filter.

This use of an electronic printing system is mainly confined to graphic arts applications (where separation negatives are still employed almost exclusively). For this purpose there are also a number a more elaborate arrangements for complete colour correction in making separation positives from negatives. For

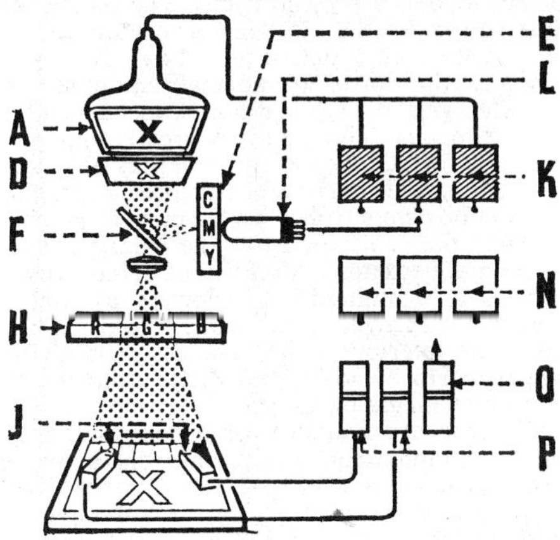

ELECTRONIC COLOUR PRINTING. An enlarger with a flying spot scanner tube can be adapted to colour printing by monitoring the light through complementary filters (C, M, Y—cyan, magenta and yellow), while the printing exposure takes place through the primary tricolour filters (R, G, B—red, green, blue). A, flying spot scanner tube. D, negative. E, complementary monitoring filters. F, beam-splitter mirror. H, printing filters. J, photometer probes for measuring the density range (prior to the exposure). K, colour feedback amplifiers. L, monitoring photo tube. N, exposure timers. O, matching density scale. P, photometers for setting the timers.

instance, in one electronic scanner all three separation negatives are scanned simultaneously by linked flying spots (or a single flying spot diverted to three negatives through beam splitters) while only one of the three negatives is being printed. The photo tubes that monitor the light coming through all three negatives (and hence react to colour errors recorded there) are then used to control the intensity of the flying spot which scans the negative actually being printed. In this way it is possible to introduce any degree of automatic colour masking and correction.

Applications. In the beginning electronic printers were developed for military purposes, especially the printing of aerial photographs. Such photographs have to be reproduced with maximum detail resolution, while the printing may require appreciable contrast control. Electronic printers prove equally useful in printing medical radiographs, where again the original image frequently requires considerable tone control to bring out the maximum significant detail. Electronic enlargers are widely used also in commercial photography.

The other main use of electronic printing equipment is in the graphic arts field, especially for colour correction and automatic masking. The electronic printer with its built-in masking control there saves appreciable operating time previously taken up with making masks in the conventional manner. Printing systems based on similar principles have also found their way into photo finishing printers and enlargers, and printing of motion picture positives in special fields. L.A.M.

See also: *Cathode ray tube traces; Contrast control; Enlarger Masking.*

ELECTRON MICROGRAPHY. Technique for recording photographically images formed by electrons in an electron microscope.

An electron microscope produces images by essentially the same process as the light microscope. However, the glass lenses in the latter are replaced by electromagnetic or electrostatic lenses (i.e. fields) and an electron beam provides the illumination. In an electron microscope electrons emitted from a heated tungsten filament are accelerated to a high velocity in an electric field of 50–100 kV. There are three "lenses" or fields: a condenser, an objective, and a projector lens. The condenser lens converges the electron beams on to the specimen which is placed just above the objective lens. Specimens are normally mounted on a metal grid containing numerous apertures approximately 70μ square. A greatly enlarged image is formed by the combination of objective and projector lenses, the precise magnification being controlled by the current passing through the projector lens. The enlarged electron image excites a fluorescent screen (used for visual examination); below is a camera system for recording the image photographically. The interior of the microscope is highly evacuated to avoid scatter of the electrons by collision with gas molecules.

Controls are available for precise alignment of the illuminating system and lenses, while both specimen and camera airlocks are fitted to save time when changing the specimen or photographic material. Provision is made for scanning the specimen for location of features of interest.

The major advantage of electronmicroscopy over light microscopy is the increased resolution. Whereas a resolution of 2,000 Å (about $\frac{1}{3}$ the wavelength of yellow light) is obtainable with a good light microscope, a figure of 5 to 10 Å is possible with most high-performance electron microscopes. This represents a useful magnification of one to two million times. The resolution obtained by no means represents the theoretical limit. This is not attained owing to aberrations in the lenses, primarily the objective, and the residual small instability in both high-voltage and lens current supplies.

Preparing Specimens. However good the microscope, for micrographs of high resolution the specimen must be thin enough to transmit a large fraction of the incident electrons without causing major changes in their velocity and direction. The specimen must also be stable and not contaminated at an excessive rate by the action of electrons on hydrocarbons present in the microscope system.

Methods of preparing suitable specimens vary according to the type of material and the nature of the examination to be made.

Microscopic particles are deposited on a very thin film of plastic (e.g. formvar or collodion), evaporated carbon or certain other inorganic materials and are viewed by direct transmission. This shows up shape, size and degree of aggregation of the particles (e.g., photographic emulsion grains, powders, virus particles).

Information on surface details of bulk material and microscopic particles (e.g. glass, fibres, metal surfaces) is obtained by replication. This involves coating a thin film of plastic, evaporated carbon or certain other inorganic materials directly on to the surface to be examined. The replica is removed either by stripping or by dissolving the specimen. It is then examined by direct transmission.

The microstructure of certain tissues and other bulk material can be conveniently studied by cutting ultra thin sections, the order of 200–600 Å thick, and viewing by direct transmission.

Many types of crystallographic material especially metals, may be examined by transmission, the specimen consisting of an ultra thin film produced by electropolishing or

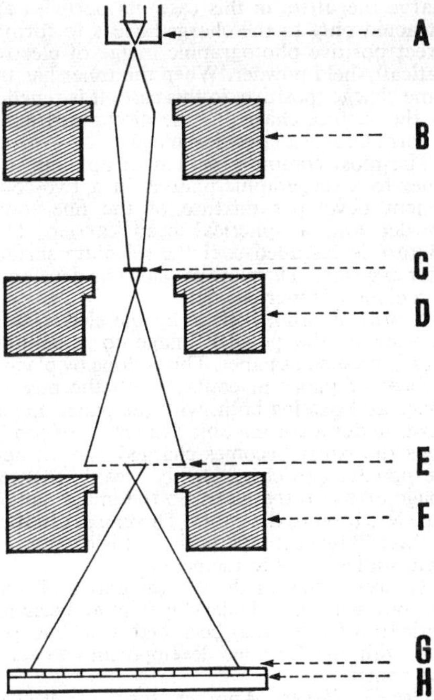

ELECTRON MICROSCOPE. An electron beam is focused through the specimen and the image projected in stages up to a fluorescent screen for observation, or a film for photography. Magnetic coils control the electron beam and so act as lenses in much the same way in which glass elements control a light beam. A, electron source. B, condenser lens. C, specimen. D, objective lens. E, intermediate image. F, projector lens. G, final image. H, fluorescent screen or photographic film.

chemical etching. Such specimens are more typical of the original bulk material than cut sections since incorporation of structural disorder, or the disturbance of existing disorder, is kept to a minimum.

Further information on the character of a specimen is frequently obtained by shadowing. This is the oblique deposition of a layer of material 10–30 Å thick having a high density to electrons (e.g. platinum/palladium and gold/palladium alloys) on to the specimen by evaporation in a vacuum. Raised areas of the specimen shield adjacent areas in a direction away from the source. These "shadows" are more transparent to electrons than the remainder of the specimen, and from their shape and size and angle of shadowing deductions can be made concerning the size and shape of surface features or the thickness and shape of particles.

Frequently thin sections of biological or animal tissue give insufficient contrast in the electron image. This can be overcome by treatment with solutions of inorganic materials such as uranyl acetate or phosphotungstic acid which are absorbed or react to different degrees with various constituents of the specimen, thus increasing their electron absorption.

Photographic Technique. For normal photographic records the plate or film is exposed directly to the electron beam. If a ciné record is required the fluorescent screen is photographed with a camera, but lower resolution is obtained.

All photographic emulsions give satisfactory results when exposed to electrons. There is a close correlation between speed to electrons and blue light, but not for white light since spectral sensitization is not operative in electron exposures. For most record purposes the fine grain process or lantern slide emulsions developed for recommended times in conventional MQ developers give the best compromise between speed and granularity.

Exposures must be correct; they are manually timed by raising and lowering the fluorescent screen, and range from 2 to 30 seconds.

Special Techniques. There are many other specialized techniques, e.g. the taking of stereoscopic pairs, the use of dark field microscopy and the recording of surface reflection micrographs. Some studies involve the use of special stages which enable the specimen to be heated, cooled, oriented or strained.

The desire of biologists to examine living tissue and of the metallurgist to examine specimens more directly typical of the bulk material is stimulating the development of instruments employing high accelerating voltages e.g. 300 kV to 1 mV, which gives greater electron penetration. R.B.F. & F.S.J.

See also: *Photomicrography*.

Books: *The Electron Microscope* by M. E. Haine and V. E. Cosslett (London, 1961); *Techniques for Electron Microscopy* by D. Kay (Oxford, 1961).

ELECTROPHOTOGRAPHY

Electrophotography comprises photographic processes in which the image is formed by a change in the electrical properties of the sensitive materials induced by the action of light. Photoconductivity is the basis for the electrophotographic processes in widespread use. In these processes, when the sensitive surface is exposed, the light-struck areas become conductive. If the surface has an electrical charge it is destroyed in the now photo-conductive areas, or if an electrical charge is applied after exposure, it is retained in the unexposed areas and not in the illuminated areas. Thus an invisible image is formed in which the shadows are represented by the charged and electrically resistant areas, and the highlights are represented by the uncharged and conductive areas. This invisible image is rendered visible by the attraction of developer to the charged area (direct-positive image) or by repulsion of developer to the conductive-uncharged area (reverse image).

Electrophotography provides a rapid means of reproducing direct-positive, reverse, line or continuous tone images. The results compare favourably with ordinary photographs taken by conventional methods. The main electrophotographic systems are xerography, Electrofax and persistent internal polarization.

Xerography. This is a dry process originated by C. F. Carlson. It is the oldest method of electrophotography and has been in use for a number of years, progressing from hand processing through the separate stages to make offset masters, to completely mechanized machines for office copying.

In xerography, a metal plate with a thin coating of vitreous selenium is sensitized by passing it under a corona charger (an array of wires at a potential of several thousand volts). This charger covers the plate with a layer of positive electrical charges.

The plate sensitized in this fashion can be exposed in the camera or by any of the normal printing methods. Its sensitivity is similar to orthochromatic film with a speed of 2.0 ASA to tungsten illumination.

The image on the plate is developed by dusting the plate with a pigmented powder, called a toner. When the toner has the opposite electrical charge (negative, in this case) its particles are attracted only to the charged areas to form a direct positive photographic image of electrostatically-held powder. When the toner has the same charge (positive, in this case) it is repelled by the surface charge to the discharged areas to form a reversal powder image.

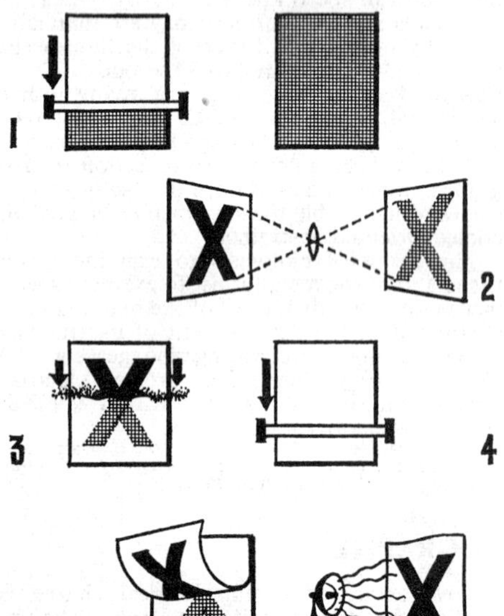

PRINCIPLE OF XEROGRAPHY. 1. A corona discharge from charging wires deposits a uniform electric charge on the surface of a selenium plate (or drum). 2. Exposure in the camera locally discharges the plate. 3. Treatment with fine powder reveals charged areas. 4. The plate is covered with paper and passed through the charging wires again. 5. This transfers the powder image from the plate to the paper. 6. Heat fuses the powder permanently into the paper support.

The most common method of applying the toner to a xerographic plate is in a two-component developer mixture of the fine toner powder and a spherical bead carrier. This mixture is cascaded over the selenium surface after exposure. The plate may also be developed by a cloud of toner powder.

A print is usually made by the electrostatic transfer of the powder image to a suitable substitute such as paper. This is done by placing a sheet of paper in contact with the powder image and passing both, with the paper uppermost, under a corona unit. Under those conditions the paper becomes charged and attracts the powder particles from the weakly-charged image areas on the plate. So the image simply transfers itself to the paper. The image can then be "fixed" by heating it until the fusible powder melts and adheres to the paper.

The resolution of the print is about 15 lines per mm. with cascade development and electrostatic transfer, but may be as high as 80 lines per mm. with powder cloud development and adhesive transfer.

Dielectric Paper. Another type of electrophotography utilizes a sensitive material of a photoconductive pigment dispersed in an insulating binder resin. A direct-printing electrophotographic paper is made by coating the mixture on paper. This paper can be charged, exposed, developed and fixed to produce a permanent print. The transfer step, necessary with a selenium plate, is obviated by use of this expendable medium.

The best known process of this type is called Electrofax.

Zinc oxide sensitized with dyes to extend its spectral response from the ultraviolet into the visible is commonly employed as the photoconductive pigment. A dye-sensitized Electrofax coating has a speed of about 2.0 ASA to tungsten light.

The coating is covered with a negative charge from a corona unit and exposed in the same fashion as the selenium plate in xerography. A technique devised for the development of zinc oxide paper employs a "magnetic brush". In this method of applying a toner-carrier mixture to the image surface, a magnetic iron powder is used as the carrier and the mixture manipulated with magnets to effect deposition of the toner. Another technique for use with pigment binder paper employs a toner dispersed in an insulating liquid such as a hydrocarbon or halogenated hydrocarbon. The sheet is immersed in the liquid and the fine toner particles are attracted to form the visible image. Fluid development is usually superior to carrier-toner in the develop-

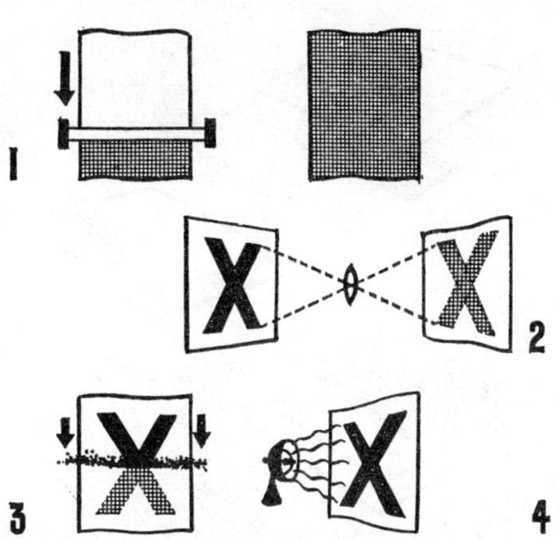

ELECTRO-PHOTOGRAPHY ON PAPER. 1. Paper with a zinc oxide or other dielectric coating receives a uniform charge from charging wires. 2. Exposure in the camera locally discharges the paper surface. 3. On treatment with pigment powder the latter adheres only to the charged areas. 4. Heat fuses this powder image to the support.

ment of continuous-tone images and for high resolution.

Another electrophotographic process utilizes a zinc oxide binder coated on an aluminium substrate, and develops by electrolytic reduction the conduction pattern formed. No charge layer is used, but development is carried out by applying a voltage to the exposed sheet while it contacts a solution of reducible ions such as silver. The ions are reduced to the metal in the light-exposed and conductive areas to yield a visible black reversed image.

Electro-plastic Systems. In the photo-plastic process a charge pattern image is formed on a thermo-plastic surface coated on a photoconductive layer. This charge pattern produces a stress in the plastic and—if the latter is softened by heating—permanently deforms the plastic in a micro-pattern. Under suitable illumination, this pattern can be viewed or projected as a photographic image.

Automation and Applications. Automatic machines based on both the xerographic and Electrofax methods are in use for copying of documents and drawings.

Both the magnetic brush and the liquid development techniques are being utilized in Electrofax office copies.

The use of toners with the same electrical charge as the surface charge to obtain reversed images has also been automated. Thus, xerography and Electrofax are applied to the production of positive enlargements from negative microfilm. High-output Xerographic machines utilize a selenium-coated drum which rotates continuously. Various stages around its circumference correspond to the charging, exposing, developing, and transfer phases. With it xerographic prints (e.g. from microfilms) can be made on continuous paper web. The electrolytic electrophotographic process machines are used extensively in this type of retrieval from microfilm.

Transfer of the xerographic image to an offset paper master as a convenient method to make such masters by photographic means was one of the earliest applications of xerography. The zinc oxide binder surface of Electrofax paper can also be converted to an offset master. Chemical treatment of the surface after fusing renders the background hydrophilic while fused toner images remain ink-receptive. Organic photoconductors in binders coated on aluminium are charged, exposed, developed and fixed in the same manner as the Electrofax coatings and are converted to a master by dissolving the background to expose the aluminium base.

The xerographic process is further used in radiography. Xeroradiographic prints are made by exposing selenium plates to X-rays, developing in semi-automatic equipment and transferring the image to sheets. The Electrofax process has been applied to computer write-out by exposing the paper by contact with a thin-window cathode ray tube. The binder system has also been used to form photo-resists by dissolving away the non-image areas to expose a suitable substrate.

As the zinc oxide binder can be repeat-processed it is capable of being developed in many colours. Using colour-separation masters, multicoloured line map prints are made by repeating the printing procedure using different-coloured liquid developers.

Persistent Internal Polarization. The formation of an image with internal charge separation within the photosensitive film is another electrophotographic process. It has been studied in the U.S.A. where it is called persistent internal polarization (P.I.P.) and in the Soviet Union where the sensitive material is called a photo-electret. By appropriate techniques this process can be made to produce either a direct or reverse image. As the charge pattern is internal and not on the surface as in xerography and Electrofax, it is relatively little affected by the atmosphere.

Although the scientific work on electro photography processes was carried out long before the introduction of xerography, the first successful process was made in 1948. Since then, new processes and especially new techniques of development have greatly expanded the applications into and beyond those of conventional photography. The fact that in the process the materials are not made light-sensitive until ready to be used offers a clear-cut advantage over the fogging reaction of silver halide system to stray radiation before use. The convenient electrophotographic printing procedures, coupled with their low cost, are thus extending beyond commercial copying, where they are essential. T.J.K.

See also: *Thermoplastic recording.*

Books: *Xerography and Related Processes*, by J. H. Dessauer and H. E. Clark; *Electrophotography*, by R. M. Schaffert (both London).

ELECTROPLANE CAMERA. Special type of cine camera in which one component of the lens is made to vibrate backwards and forwards in such a way that the focus of the lens is constantly shifting from the near foreground to the distance. At the same time, the lighting of the scene is switched in synchronism so that it shines only on that plane on which the lens is focused. The result is to give an optical illusion of extreme depth of field when the final positive film is projected and viewed in the normal way.

ELECTROTYPE. Copy of a relief printing surface. A mould is cast in a plastic from the original plate and is then electroplated with a metal, usually copper. When a sufficiently thick coating has been deposited, it is stripped off the mould, the back is filled solid with type metal, and the resulting electrotype is then used in printing in the same way as the original.

See also: *Photomechanical reproduction.*

ELEMENT. Basic substance which cannot be separated into component parts of simpler form by chemical methods of analysis.

Also term applied to single component of a lens system.

See also: *Chemicals; Chemical symbols.*

EMBOSSING PRINTS. Method of recessing the picture area of a print and leaving a raised white border in imitation of the "plate-sunk" effect incidental to printing with some types of line block.

The print is made on a sheet of paper big enough to leave a broad white margin all round. A rectangle of card about 1/16 in. thick is then cut so that it exactly coincides with the image. This piece of card is then placed on the print so that it covers the picture and is held in place with a dab of rubber solution. The two are then turned over on to a level surface.

At this point it is necessary to have a suitable tool, such as a blunt bone paper knife or the smooth end of a fountain pen. With this tool, the back of the print is worked down over the edge of the card by sliding the tool around the border and gradually increasing the pressure. Presently the paper will have a raised rectangle in the centre corresponding to the card underneath. When the print is turned face up and the card is removed from the front, the image will appear to be sunk into the surface. This is the effect produced by the pressure of the engraved or etched plates used for printing line reproductions. It is an effective way of finishing small unmounted prints—e.g., for sending out as Christmas or birthday greeting cards—and it has the advantage of giving the paper extra rigidity.

When a lot of prints are to be embossed in this way it helps to make a jig—i.e., a cut-out card just big enough to take the print, stuck down on to another card with the "plate" fixed in position in the middle. The print then fits automatically into the right position without measurement or any risk of slipping.

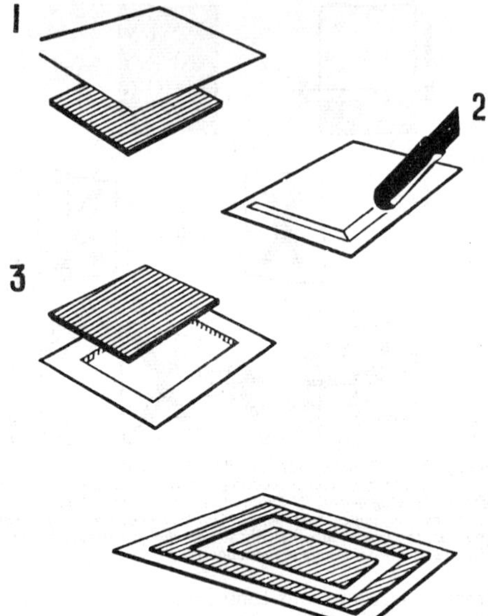

EMBOSSING A PLATE MARK. 1. Place the print over a sheet of stiff cardboard of the right size. 2. Press along the edges of the card from the back of the print. 3. The shape of the card is now imprinted on the front. Bottom: Two pieces of card stuck to a common base will produce a convenient jig.

When prints are to be mounted, the mount itself can be plate marked in the same way before or after the print is stuck down. In this case there is no need to leave a white border round the print because the necessary margin can be allowed for on the mount.

EMERSON, PETER HENRY, 1856–1936. British photographer, born in Cuba of an American father and an English mother. Settled in 1869 in England. Became photographer, advocate of "naturalistic" photography, and apostle of "art based upon scientific principles"; wrote a textbook, *Naturalistic Photography* (1889). He was opposed to retouching and "composite negatives". Advocated platinotype and photogravure as reproduction processes; favoured a moderate amount of soft focus. In 1891 he renounced the beliefs for which he had fought since before 1880 and declared that "photography was not Art".

EMULSION. An emulsion consists of a stable mass of finely divided particles of a substance uniformly dispersed in a liquid in which they remain suspended without dissolving. Each particle of the suspended substance is surrounded by a cushioning film of liquid which prevents it from joining particles around it.

The sensitive coating on photographic plates and films is a suspension of finely divided grains of silver bromide in gelatin.

See also: *Emulsion making; Gelatin; Sensitized materials manufacture.*

EMULSION MAKING. The manufacture of photographic emulsions involves several separate stages, each of which can give rise to a wide range of properties. It is by careful selection and combination of the treatments given that photographic emulsions of apparently distinct character are derived.

Precipitation. The photosensitive component of an emulsion is the microcrystalline phase of silver halide. Since silver chloride and silver bromide are crystallographically similar and can form mixed crystals in any proportion, grains varying continuously from pure silver chloride to pure silver bromide are obtainable. Silver iodide has a different crystal structure and is therefore only partially miscible with silver bromide. Consequently the practical upper limit of iodide concentration in a silver iodobromide emulsion is about 6 molar per cent (6 molecules iodide to 94 molecules bromide). Sensitivity generally increases from pure silver chloride to silver iodobromide, corresponding in practice to contact positive papers and negative camera films respectively.

The silver halide is formed by double decomposition from silver nitrate and alkali halide, e.g.:

$$AgNO_3 + KBr = AgBr + KNO_3$$

The precipitation is made in the presence of gelatin which acts as a protective colloid and allows a suspension of microcrystals to be obtained.

Ripening. Immediately after precipitation, the crystals are normally extremely fine, and since sensitivity is dependent upon grain size an emulsion thus prepared would be very slow. The ripening process is designed to increase the size of the grains to the desired level. Since crystal growth takes place by deposition from solution, and silver halides are very insoluble in water, solvents are added to increase their solubility by complex formation. These solvents are usually excess alkali halide and/or ammonia, and they may also be present during precipitation. Since silver chloride is more soluble than, say, silver iodobromide, the rate of grain growth of the former will be faster for given solvent conditions. Alternatively, silver chloride will generally require mild ripening conditions and silver iodobromide more vigorous conditions.

As with other chemical reactions, ripening proceeds at a greater rate with increase in temperature and this, together with control of ripening period, is used to control the grain size. Other factors, e.g. gelatin concentration, may also influence the course of ripening.

By the methods outlined above it is possible to arrive at emulsions of desired grain size characteristics, both with respect to mean grain size and spread of grain sizes. Thus a low speed, high contrast emulsion will require a narrow range of grain sizes with a small average, whereas a low contrast, high-speed emulsion will require the converse. Typical average grain diameters are, 0·1μ for positive emulsions, 0·3μ for high contrast "process" emulsions, 0·6μ for slow negative emulsions and up to 2μ for fast negative emulsions.

Washing. The salts left in solution after ripening (nitrate from precipitation and ripening solvents) can be harmful, particularly if present in sufficient quantity to crystallize out from the coated film. The emulsion is shredded after addition of gelatin to increase the concentration to 10–15 per cent and the soluble salts leached out by cold water. If the support is of paper, however, this may act as an absorbent for the salts and washing may not be required.

Digestion. Although to some extent the speed of the emulsion is determined by its grain size, it is obviously desirable to achieve maximum sensitivity of the grains. This is best done by heating the emulsions with a compound which is able to form silver sulphide specks at the grain surface. Such a compound may already be present as an impurity in the gelatin ("active" gelatin) or be deliberately added, e.g. sodium thiosulphate, thiourea (a few milligrams per kilogram of gelatin). Sensitivity first increases with time and then diminishes. Fog increases with time and an optimum point is reached when sufficient speed is obtained without excessive fog. This time is obviously dependent upon the temperature used and the concentration of sensitizer.

When maximum speed is desired, e.g. for high speed camera materials, use is made of gold sensitization. Small quantities of a complex gold salt (e.g. about 5 milligrams gold as ammonium aurothiocyanate per 100 grams silver nitrate) may be added before or after digestion. Such sensitization usually results in a reduction of contrast.

Stabilization. When the required sensitivity has been reached, the digestion process must be stopped. This is done by addition of a stabilizer which is usually a heterocyclic compound selected from certain classes, e.g. triazole, tetrazole, iminazole. Compounds of this type may also be used to control the tone of the image in silver chloride or silver chlorobromide emulsions.

Spectral Sensitivity. An emulsion prepared in the above manner has limited sensitivity, for example silver chloride is insensitive above about 400mμ, silver iodobromide about 500mμ. To extend this range optical sensitizers are used. These are dyes which stain the silver halide and are of such a type that the energy absorbed by them can be transferred to the crystal to form a latent image.

By this means the emulsion may be sensitized over the whole spectrum (panchromatic emulsions) or selected parts of the spectrum (orthochromatic, infra-red, colour emulsions).

Other Aspects. The above procedures are followed for all develop-out materials, together with various methods of controlling the physical properties of the emulsion. The actual constitution of the coated layer may, however, be modified to suit the emulsion to its use. For instance, positive paper emulsions, being of inherently high contrast are coated at relatively low coating weights (1–2 grams silver per square metre) whereas negative camera emulsions are coated at 4–8 grams silver per square metre. X-ray emulsions, being of high speed, are coarse grained and therefore of low covering power. High contrast is achieved by ensuring uniformity of grain size and by use of high coating weighs (10–15 grams silver per square metre). For direct X-ray exposure speed increases with coating weight and consequently values as high as 30 grams silver per square metre may be used.

For recording of ionizing particles, nuclear emulsions are used which are characterized by their high stopping power. Here the emulsion may contain over 80 per cent by weight of silver halide and be coated with dry thickness exceeding 1 mm. (conventional emulsions 5–15μ).

If high resolution is required, not only must the emulsion be of uniform fine grain but the effect of light scatter within the emulsion layer must be reduced. This is done by use of a thin emulsion layer or by use of a filter dye within the emulsion.

Print-out Papers. These are generally of low sensitivity and are capable of producing sufficiently high density without development. This is obtained by using silver chloride as the sensitive phase together with a halogen acceptor to remove the liberated chlorine which would otherwise retard the photolysis. Such a halogen acceptor can be silver nitrate, silver citrate or sodium nitrite.

This review of the principles involved in emulsion making is of necessity brief and generalized. Modern emulsion making is a specialized technology and while publications exist which provide recipes, it is certain that the results from these will be inferior to those from commercially available products. R.E.W.

See also: *Gelatin; Sensitized materials manufacture.*
Books: *Making and Coating Photographic Emulsions*, by V. L. Zelikman and S. M. Levi (London). *Film Coating Theory*, by B. V. Deryagin and S. M. Levi (London).

EMULSION REMOVAL. It may be desirable to remove the gelatin film from some old glass negatives or lantern slides. The glass can then be used for lantern slide cover glasses, for holding down film negatives in printing frames or even, in the very large sizes, for glazing greenhouses or garden frames.

When the negatives are very old the film is not easy to remove, and it is even more difficult to do this if the negative has been coated with protective varnish. The procedure is then as follows. Soak the negatives or lantern plates in cold water. (It is easier to remove the film after soaking for several hours.) Transfer the negatives to the following solution, heated up to nearly boiling point:

Caustic soda	4 ounces	100 grams
Water	40 ounces	1,000 c.cm.

This will remove the films very rapidly but is very corrosive if splashed on the skin or clothes. Transfer the glass to cold water, wash and dry.

Another method: soak the plates for some hours in water. Take a plate and place emulsion side upwards in a large photographic dish filled with hot water. After about a minute, take another plate of the same size with smooth edges, and use this as a scraper held at 45° to remove the film. It is quite easy to do this without risk of scratching the glass. Wash and dry the glass.

When making lantern slides, the easiest way to recover the glass from any waste plates is to take them from the fixing bath and dry without washing. After the film has dried it is usually possible to pull the film off the glass. If it does not come away readily, a short soaking will remove it.

The type of household bleaching fluid incorporating stabilized electrolytic sodium hypochlorite sold in bottles for domestic cleaning, etc., is also excellent for removing the emulsion from negatives. The fluid is simply applied at full strength to the emulsion surface and in a few moments the emulsion can be rubbed off. Concentrated hydrochloric acid can be used in the same way, but the fingers should not be immersed in the solution for any length of time.

All the above methods apply equally both to plate and film negatives. Special emulsions are available which permit easy removal. R.M.F.

See also: *Stripping.*

EMULSION SURFACE TEST. The emulsion side of negatives can generally be recognized by the slightly duller appearance. The difference is less noticeable in hardened or scratch-proofed negatives, but films so treated tend to curl towards the emulsion side when the negative is held in the palm of the hand. And when the emulsion is viewed obliquely against the light, the image can often be seen standing out in relief.

When the film is held up to the light and the negative image appears the right way around, the emulsion surface is then facing away from the observer. So if the subject includes any lettering, e.g., shop signs, notices, the emulsion side can be recognized at once.

There is no difficulty in recognizing the emulsion side of a plate negative because the difference between the gelatin and glass surfaces is generally obvious at once. If there is any doubt, the two sides should be touched in turn with the tip of the tongue; the emulsion side has an unmistakable tendency to stick.

The emulsion side of sheet-films also can be easily determined as all such material is notched. When a sheet is held so that the notch is at the top left-hand corner, then the emulsion side is facing away from the viewer.

Another way of finding out is to scratch the margin of the plate with a pin; only the emulsion side will scratch.

The emulsion side of most papers can be recognized by its sheen. Matt papers have very little sheen, but, like films, they tend to curl towards the emulsion side when laid on the palm of the hand. And if a corner of any paper is lightly bitten, the side that sticks to the teeth is the emulsion side.

ENAMEL AND CERAMIC PHOTOGRAPHS. There are various methods by which a photographic image may be made to appear permanently on a glazed porcelain surface. The principle is simple: the photographic silver image is replaced by an image consisting of a vitrifiable metallic oxide which will fire to the desired colour. This image is transferred to the unglazed pottery and thereafter the normal kilning and glazing proceeds as usual. Pictures applied in this way are absolutely permanent.

See also: *Ceramics; Dusting-on process; Printing on special supports.*

ENCAUSTIC PASTE. Mixture of roughly equal parts of white wax and oil of turpentine with half a part of dammar varnish, applied to the surface of a print (on rough or matt paper) to add depth to the shadows. The paste kills the "bloom" veiling the black areas of the image created by light scattered at the surface. Dopes for this purpose were popular before the advent of the many modern "lustre" surfaces, combining a grained or textured base with a fine surface gloss. It is doubtful if such surfaces are improved by treating with encaustic paste, megilp and similar preparations.

See also: *Doping prints.*

ENDOSCOPIC PHOTOGRAPHY. Technique of photography inside normally inaccessible cavities, using specially designed optical instruments and camera attachments.

Endoscopes. An endoscope (Endo-scope: from Greek, within + to examine) is an instrument, usually tubular, which can be passed into a narrow orifice for the purpose of examining internal structure. Until recently the connotation was exclusively medical, but endoscopic examination now forms part of industrial practice also.

In its simplest form an endoscope consists of a rigid tube open at both ends. This can be made to carry a miniature lighting system, either proximally to shine down the tube, or distally to illuminate the region beyond the tip. Although this is primarily an instrument for visual examination, provision can be made for the attachment of a motion picture, still, or television camera. Variations of this arrangement form the basis of the Brubaker (U.S.A.) photoendoscopic equipment. The term speculum is reserved for a simple endoscope when its length does not exceed its bore diameter by more than five times.

The present trend, however, is to use an endoscopic telescope in conjunction with a sheath, an open-ended endoscope, or a speculum which is usually made also to carry the lighting system. Light is generally piped towards the distal end of the telescope through a separate quartz rod. The Vulmirèe system (France), which was developed first, uses an incandescent light source for visual examination and cine photography, whereas the Storz equipment (Germany) favours electronic flash and is designed for still photography. The advantages of this basic optical arrangement are a wide angle of view, great

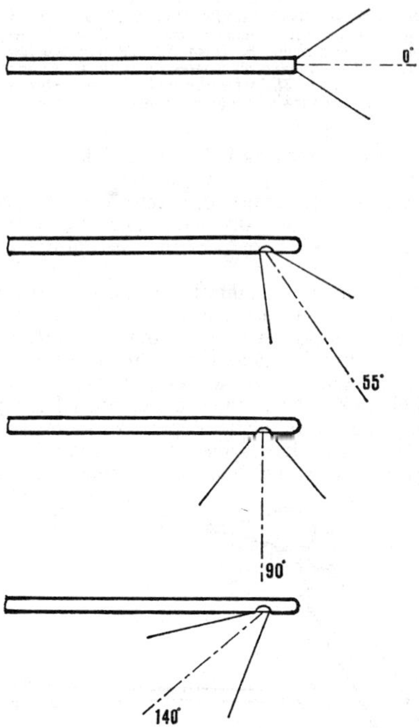

TYPES OF ENDOSCOPE. Top: *Direct endoscope.* Upper centre: *Forward system.* Lower centre: *Right-angle system.* Bottom: *Retrograde endoscope. The figures indicate the angle in degrees between the axis of the instrument and the viewing direction.*

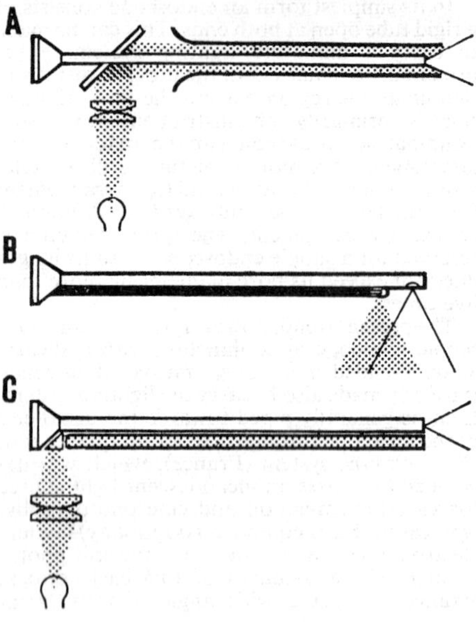

MAIN VIEWING AND ILLUMINATING SYSTEMS: A, proximal illumination used mainly in conjunction with wide-bore instrument or specula. B, Distal illumination provided by miniature filament lamp on a separate stalk. C, Quartz rod piped light system. The telescope and rod are embraced by another tube not shown. The light beams are shown shaded in each case.

tolerance in terms of depth of field, and the utilization of smaller and lighter cameras.

By definition, endoscopic telescopes are long in relation to their diameter. They range from 10 cm. in length by 2.8 mm. in diameter to 40 cm. in length by 6 mm. in diameter. They consist of a varying number of optical repeater systems to relay an image from the tip to an eyepiece—to which a camera may be attached—with the minimum loss of light. The terminal angle of view may be deflected by prisms.

Application. The laryngoscope, bronchoscope, oesophagoscope, gastroscope, cystoscope, proctoscope and sigmoidoscope are examples of instruments which can be passed into the natural orifices of the body, whereas the thoracoscope, peritoneoscope and culdoscope require a surgical incision in the chest, abdominal wall, or vaginal vault for their respective introduction. The auriscope is the shortest of the open-ended speculum type of instrument and the ophthalmoscope and retinal camera are endoscopes of a different order, there being no direct connection between the instrument and the body of the subject.

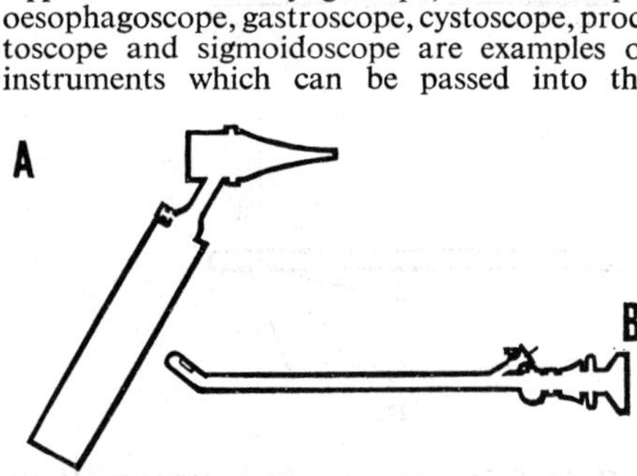

TYPICAL MEDICAL ENDOSCOPES. A, auriscopes or otoscopes. Speculum type with battery in handle. B, cystoscope with distal illumination and telescope optics. This is shown on the same scale as A.

Instruments such as the bronchoscope and sigmoidoscope consist essentially of a rigid tube of appropriate bore, carrying an obturator to facilitate introduction; small lamps are carried at the distal or proximal end to illuminate the orifice. Telescope optics are optional. More complex instruments such as the gastroscope and cystoscope invariably embody miniature telescopic optical systems so that a magnified or angled view can be obtained. The gastroscope is very often partly flexible at its tip.

Many of these instruments are immediately adaptable for industrial use. For example it is possible to examine car cylinders through the spark plug opening without removing the cylinder head; the inspection of pipes and drains for internal corrosion, scaling etc. can be carried out through inspection ports and this applies also to the testing of boilers. Sometimes, however, a more rugged construction is required and waterproof or other protective housings are a possibility.

In medical applications the endoscope is usually under the control of the surgeon because dexterity in the manipulation of the instrument is of the greatest importance. In such circumstances it may prove necessary to attach or remove a recording camera quickly. Such attachment should not be too permanent lest the safety of the patient or the convenience of the surgeon be in jeopardy. To this end a loose three-pin bayonet, or a cone or cup seating, whereby the camera can readily be applied to the telescope or endoscope with the fingers forms the general method; magnetic seatings have also been advocated. If any camera is attached to an endoscope it must be possible to view the field up to the moment of or during exposure; some form of reflex viewing is therefore necessary for both still and cine photography. It may also be necessary to provide a counterbalanced suspension for the entire equipment. Lighting must be cool and the operating voltage must be low or completely isolated to minimize the risk of electric shock. In industry these problems are simplified by the absence of such rigid requirements.

Fibre Optics. The quality of image resolution by fibre optics is probably not yet high enough to compete with conventional systems, but may be of some use in difficult situations; the definition obtainable is more on a level with television criteria. From a theoretical standpoint the application of fibre optics is confined to: (1) the design of flexible endoscopes, (2) guiding illumination along a tortuous path at the expense of considerable light loss, and (3) the connection of telescopes to heavy or bulky recording equipment.

It seems likely that developments along these lines will lend improvements to present techniques which were based initially on the concepts of Brubaker but are now being largely replaced by the more compact Vulmière system.

P.H.

See also: *Medical Photography.*
Books: *Longmore's Medical Photography* by Hansell and Ollerenshaw (London); *The Fundamentals of Still and Motion Picture Photography and Television through Endoscopic Telescopes* by Joseph D. Brubaker (Chicago).

ENERGIATYPE. A paper process invented by Robert Hunt in 1844; later called Ferrotype because protosulphate of iron was used as developer. It is not to be confused with the later kind of ferrotype or tintype.

ENGLAND, WILLIAM, ?–1896. English daguerreotype portraitist in London in the mid-1840s, later known for his views of Switzerland, the Tyrol and Italy. Took thousands of stereoscopic views for the London Stereoscopic Company 1854–63 in Ireland, America, France, Switzerland and at the International Exhibition in London, 1862. Invented the focal plane shutter with variable slit 1861.

ENLARGED NEGATIVE. A number of special printing processes are too slow for anything but contact printing. If large prints are to be made from small negatives by any of these processes, they cannot be made by enlarging direct on to the sensitive material. An enlarged negative must be made, and from it contact prints made on to the particular type of printing material.

Contact printing may also be the faster method for making large numbers of prints from one negative. For very large production runs it may even become necessary to print simultaneously from several duplicate negatives.

Enlarged negatives can be made either via a positive transparency or by direct reversal.

Positive Transparency Method. Here a diapositive is made by contact printing the negative in the normal way. The positive is then projected in the negative carrier of the enlarger on to a slow non-colour-sensitive film or plate which is then processed to give the enlarged negative. This method allows considerable adjustment to be made in the contrast and density of the duplicate, while part only of the image may be enlarged if the rest is not required.

If a same-size duplicate negative is required, it is simply printed by contact from the transparency. Enlarged or duplicate paper negatives may also be made in this way.

Direct Reversal Method. Enlarged negatives can alternatively be made by projecting the negative on to an ordinary plate or film and processing the enlargement to give a reversed image. At the same time it is generally more satisfactory to use materials specially designed for reversal processing.

The original negative must be inserted the wrong way round in the negative carrier of the enlarger so that the emulsion side faces away from the lens. If this is not done, the enlarged negative will have the image reversed left to right, and be unsuitable for contact printing.

The image on duplicate negatives made by direct reversal and by contact will be the wrong way round in any case, unless the negative is turned over for printing, and in the case of contact printing the image will lose some sharpness because the emulsion surfaces of negative and duplicating material are not in direct contact.

See also: *Duplicate negatives.*

ENLARGER

In the early days of photography, photographic prints were mostly made by exposing a piece of sensitized paper in contact with the negative. This produced a positive image the same size as the negative image—satisfactory at a time when negatives were big. Nowadays, with the popularity of small negative sizes, big prints are made by optical projection printing.

The apparatus in which the printing process is carried out is called an enlarger. This projects an image of the negative on to the sensitized paper so that, by adjusting the distance between lens and paper, it is possible to obtain almost any degree of enlargement.

There are two principal kinds of enlarger: vertical and horizontal. The vertical type is used almost exclusively in everyday photography; the horizontal type was popular in the early days but is too cumbersome for normal printing; it is now used only in certain specialized branches of photography. There is a third, fixed focus, type of enlarging device which has an extremely limited scope and is practically obsolete.

All enlargers employ the same working principle. The negative is brightly lighted and its image is focused by a lens on to a sheet of sensitized paper. After a sufficient time of exposure, the paper is processed to give an enlarged positive image. The different degrees of enlargement are obtained by adjusting the distance between the lens and the negative and the lens and the paper. As with a camera, the two are inter-related (the lens-to-paper distance corresponds to the image distance in the camera, and the lens-to-negative distance to the object distance) for a given focal length of lens.

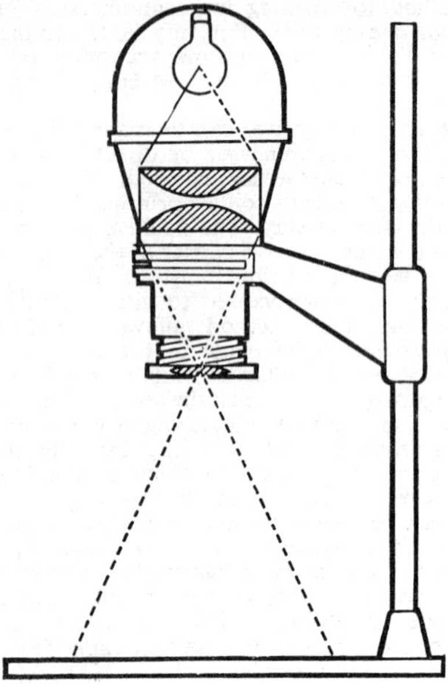

VERTICAL ENLARGER. The optical system of the conventional vertical enlarger consists of a lamp (usually opal or sprayed) and a condenser which spreads the illumination evenly over the negative in the negative carrier. The negative is projected on the baseboard by the lens. Raising or lowering this entire unit alters degree of enlargement; movement of lens controls sharpness.

To obtain a high degree of enlargement involves raising the enlarger so that the lens is a long distance above the baseboard; for a smaller degree of enlargement, the lens is brought nearer to the paper surface (and at the same time farther away from the negative). For enlarged printing the lense-to-paper distance is always bigger than the lens-to-negative distance. The scale of enlargement follows the normal rules for optical magnification: $M = v/u$. (v = lens-paper distance; u = lens-negative distance.)

All enlargers have certain basic parts in common—i.e., a means of illuminating the negative, a holder for the negative, a lens (usually with provision for focusing) and a paper holder. The first three items are usually assembled in a single unit which is moved bodily towards or away from the paper holder or easel to change the scale of enlargement. The lens is focused either independently or automatically to produce a sharp image on the easel.

ILLUMINATING THE NEGATIVE

The problem common to all types of enlarger is that of illuminating the negative brightly but evenly. If the illumination is weak, it is difficult to focus the enlarged image visually, and the exposure becomes unduly long; if it is uneven, the paper will be exposed unevenly and give a patchy print with some parts over- and some under-exposed. There are various ways of achieving correct results.

The negative image is made up of varying densities of silver deposit distributed in a layer of virtually transparent gelatin. When rays of light fall on it, some are blocked by the silver deposit, and some pass straight through the clear parts of the negative. If the light rays reach the negative in the form of a directed beam, the rays passing through remain mostly inside a fairly narrow cone; if the original rays are diffused, they will strike the negative from all angles and will fan out in all directions equally when they emerge from the other side. In this case there will be no bright central cone of rays.

When an illuminated negative is being projected in the enlarger, the only useful rays emitted by the light source are those that fall on the enlarger lens. Any rays of light that do not strike the lens are useless. Rays that are already directed towards the lens before they reach the negative will generally fall somewhere on the lens and pass through to form an image. Even though the rays will mostly emerge in a cone and not in a parallel beam, the greater part of the cone will fall inside the area of the lens.

If the rays are not pointing in the direction of the lens when they fall on the negative, the cone of light leaving the negative will either miss the lens altogether, or only the outer fringe will fall on it. So, viewed from the point of view of the lens, the image patch illuminated by the ray will either appear dark or at least very much less bright than if the ray had originally been pointing towards the lens. But if all the rays used for illuminating the negative are directed towards the lens, then, when viewed from the lens, the negative will appear uniformly brightly illuminated.

Direct Illumination. So one effective way of illuminating the negative is to ensure that all rays of light from the source converge in a cone with its apex at the lens. All the rays from the source that pass through the negative will then fall on the lens and help to form an image on the easel. This condition is achieved by placing a suitably shaped reflector behind the light source, and a condenser, composed of one or more lenses, between the source and the negative. The reflector redirects all the rays travelling away from the negative back to the source. The condenser collects the direct light rays from the source and the indirect rays from the reflector and concentrates them into a cone of light. With a perfect reflector and condenser system and a point source of light, the tip of the light cone lies inside the disc of the lens and no light is wasted.

In practice the cone of transmitted light is never focused exactly on the back of the lens because such an arrangement would call for re-focusing for every change in the scale of enlargement with its consequent alteration of the negative/lens separation. The light is made to converge on a point beyond the lens so that the section of the cone of light at the plane of the lens is somewhat wider than the disc on the lens. This wastes some of the light but it allows the lens to be focused over a wide range of scales of enlargement without any need to re-focus the illumination.

Diffused Illumination. On the other hand, if the rays from the light source are completely diffused in all directions, then the rays that pass through the negative will emerge in all directions and not in a concentrated cone. Every illuminated image patch will radiate light uniformly in all directions like a tiny point source. Some of this light will fall on the lens and pass through to form an image. And whether the image patch is at the centre of the negative or at the edge, the total amount of light it radiates will be the same, and the proportion that falls on the lens will be the same. So, once again, when looked at from the lens, the negative will appear uniformly illuminated.

This type of illumination can be achieved by placing a diffuser between the light source and the negative and employing a reflector with a matt diffusing surface behind the source. But whereas in a perfect condenser system all the light from the source is usefully employed, in this second system only a proportion of the total light leaving the source falls on the lens and helps to form an image. In practice this may be only a few per cent of the light passing through the negative; hence it is very much less efficient to illuminate the negative by diffused light than by direct light.

Callier Effect. There is one more important difference between direct and diffused negative illumination. When direct light falls on the negative image it passes straight through the clear parts, but where the rays strike even a thin density they are broken up and scattered by the silver grains so that they emerge in all directions as though they had been diffused by a sheet of opal or ground glass. This means that the highlight areas of the print—i.e., under the denser parts of the negative—receive less light than they should, so they appear whiter. In other words direct illumination increases the contrast of the negative, leading to more contrasty prints.

This selective scattering of the light by the denser parts of the negative is known as the Callier effect after its discoverer. No such effect takes place with diffused illumination of the negative since the rays falling on the clear areas of the negative are already diffused.

A scratch or other physical blemish on the

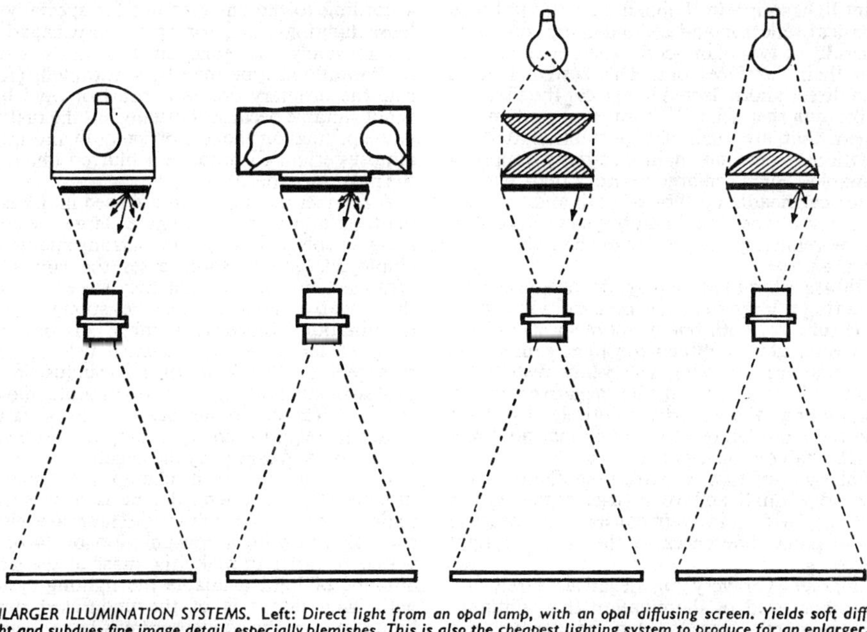

ENLARGER ILLUMINATION SYSTEMS. **Left:** *Direct light from an opal lamp, with an opal diffusing screen. Yields soft diffused light and subdues fine image detail, especially blemishes. This is also the cheapest lighting system to produce for an enlarger, and is therefore found in a number of inexpensive models.* **Centre left:** *Indirect diffused light from two opal lamps with an opal diffuser. The usual system for enlargers for large negatives. Some modern enlargers replace the bulbs by a grid-shaped fluorescent tube.* **Centre right:** *Direct light from clear lamp and condenser. Yields intense light, increased image contrast, and shows up finest detail. However, needs constant refocusing of lamp.* **Right:** *Direct light with opal lamp and condenser. The most common system in use, with most of the advantages but not the drawbacks of the clear lamp and condenser.*

negative scatters the light just as the silver grains do, and direct illumination produces a similar exaggeration of the contrast. So scratches and blemishes and grain—all tend to be rendered more obvious in the print by direct than by diffused illumination. For the same reason, since the resolution of detail depends on the contrast between adjacent light and dark areas, direct illumination, which exaggerates tone contrast, automatically gives better resolution of detail than diffused illumination.

The Callier effect is greatly reduced with negatives not containing silver grains, such as the dye images of colour films. This is because the dyes are transparent, and not opaque particles like silver.

Lighting Systems Compared. To sum up the differences: direct illumination makes more efficient use of the light available; it increases contrast and gives better resolution. But it makes grain, scratches and blemishes more obvious. Diffused illumination gives softer contrast (which is sometimes an advantage) and lessens the effect of grain, scratches and blemishes. But it is relatively less efficient and calls for either a much more powerful light source or longer exposures than direct lighting.

In practice an enlarger may have either direct, diffused, or semi-diffused illumination.

Direct illumination is produced by using a point-light source in conjunction with a polished spherical reflector and a condenser, consisting generally of two plano-convex lenses assembled with their flat faces out. This system focuses both direct and reflected images of the filament in the enlarger lens. It is mostly used in enlargers that are run off a portable battery in situations where no main electricity supply is available. Such enlargers are fitted with a motor-car headlamp type of bulb and the high efficiency of the illumination system makes up for the relatively low power of the light (24 or 36 watts).

Diffuse illumination may be produced: by mounting a large diameter opal bulb in a matt white reflector with one or more sheets of opal or frosted glass between the light source and the negative; by using an evenly distributed source of light behind the negative; or by illuminating the negative entirely by light reflected from lamps screened so that no direct rays can fall on the negative.

Enlargers of this type are manufactured for both very small and very large negatives but they vary widely in their construction because of the great differences in the types of light source.

Enlargers for very big negatives rarely use any other kind of illumination than diffused. This is partly because big condensers are expensive (and more difficult to make) and partly because the heat from the enlarger lamp may crack even thick condenser lenses.

Semi-Diffused Illumination. In this type of illumination, the direct light of a condenser arrangement is partially diffused so that its optical adjustment is no longer critical. The light is therefore made appreciably larger than a point source (generally an opal bulb) with a white reflector, rather than a polished one. The greater part of the light is still directed in a broad, slightly convergent beam on to and around the back of the lens, but the diffusion is sufficient to smooth out the effect of incorrect focusing of the light source and condenser. It also prevents an out-of-focus image of the light source itself—e.g., the filament of a tungsten lamp—from being superimposed on the image of the negative on the easel.

Semi-diffused illumination is the most popular type for normal enlarging because it combines most of the advantages of both direct and diffused systems while the characteristic disadvantages of each tend to cancel out.

Light Sources. The light source in an enlarger may be one of the following types: tungsten filament (lamps), high pressure mercury arc, vapour discharge tubes, non-electric sources of various types.

Tungsten filament lamps with fully diffusing opal or frosted glass bulbs are the commonest form of light source for semi-diffused illuminating systems. They provide a bright sphere of practically uniform illumination. Consumption ranges from 60 to 150 watts according to the enlarger but for specially intense lighting—e.g., for very dense negatives, or unusually big enlargements—an over-run or Photoflood type may be substituted. (As a rule the ordinary domestic pearl or opal bulb is not suitable because it usually has the maker's name printed on the end of the bulb and this is almost certain to produce a blurred image on the printing paper.)

A number of such bulbs are used in diffused-light enlargers to illuminate large negative areas evenly. Various lamp arrangements are employed, and in some cases the lamps are arranged movably so that they sweep all over the negative area to provide specially even illumination. Since a number of powerful tungsten lamps generate a great deal of heat, enlargers of this kind (usually industrial or professional models) may incorporate blower cooling. There, an airstream from a fan is blown through the enlarger lamp house to cool the lamps, and especially the negative.

Lamps of the car headlamp type may be fitted in enlargers intended to be used away from main supplies of electricity. These are either 6 or 12 volt with a consumption of 24 or 36 watts. In order to make the most of the small amount of light available the lighting system with such lamps is of the undiffused direct light type and the condenser must be re-focused for each change in the scale of enlargement.

The normal type of projection lamp bulb is also used as the light source in a number of enlargers—mostly for miniature negatives. In this type of lamp the filaments are accurately positioned and closely bunched while their rigid mounting prevents them from moving out of the focal plane of the condenser and reflector system. Projection lamps have a tubular instead of a pear-shaped envelope so that they can be mounted close to the condenser and reflector. This arrangement makes the maximum use of the available light.

Because of the filament arrangement these lamps must be mounted vertically, so if they are used in a vertical enlarger the lamphouse must be specially designed and include a mirror to direct the light downwards.

High pressure mercury arc lamps utilize an arc struck between tungsten electrodes in a small glass or quartz tube containing mercury vapour with a trace of argon. The tube is enclosed in a partly evacuated outer glass envelope to limit the heat loss and maintain the pressure at the arc. Such lamps have a very high luminous efficiency and are excellent for making very big enlargements—e.g., photomurals—where an extremely brilliant light source is necessary to keep exposures within reasonable limits. A disadvantage is that once the lamp is switched off the arc cannot be struck again until it has been allowed to cool off for several minutes. In common with all arc lamps high pressure mercury vapour lamps must be operated through a current-limiting choke. They have a life of about 1,500 hours.

Mercury vapour, fluorescent and cold cathode tubes are used principally for large negative sizes that cannot be evenly illuminated by any of the more compact light sources. A number of tubes are arranged side by side, or a single long tube is bent to form a grid so that the whole of the negative area is covered. The result is an even light which requires little further diffusion. Cold cathode tubes in particular are excellent light sources used in this way. Their light is practically cold, so the lamp house can be reduced to a mere shallow box; they can be used intermittently with no delay after switching on, and the light is highly actinic. An 80 or 100 watt lamp is ample for most purposes and the useful life may be 6,000 hours or more.

Where no electricity supply or battery power is available, practically any of the traditional non-electric sources may be used—e.g., coal gas, acetylene and other gas burners, and paraffin wick and mantle type lamps. Such sources are generally more suitable for burning vertically to allow for a ventilating chimney immediately over them. For this reason they are almost always used only in horizontal enlargers.

CONSTRUCTION

An enlarger is made up of two principal units, the head assembly and the easel. The negative is held in a carrier in the head assembly and projected on to a sheet of sensitized paper on the easel. The two units are mounted on a track consisting of a rigid tube or a pair of guide rails so that the distance between them can be adjusted to vary the enlargement scale.

The whole arrangement may be designed to work with the guide rails horizontal or vertical. In the vertical arrangement, the track forms the support for the head assembly and is rigidly mounted on the easel to form a base; in the horizontal arrangement the two units merely rest on the track.

The head assembly consists of the lamphouse, the negative carrier, the lens with its focusing mechanism.

Lamphouse. The lamphouse is a ventilated metal box which houses the light source, a shaped reflector and a condenser unit or diffusing screen or both. The reflector is usually spherical and the light source is placed at its centre of curvature so that all rays falling on the surface are reflected back through the source. In a direct illumination system the reflector is polished, but for diffused and semi-diffused systems it is painted matt white or satin finished.

Most enlargers of the direct or semi-diffused type have condensers with a focal length approximately half that of the enlarger lens. As the condenser must be big enough to cover the whole negative this requirement produces an abnormally thick lens in the larger formats if a single lens is used. In practice, single condenser lenses are used only for 35 mm. and other small negatives. It is customary to use a plano-convex lens with the flat face towards the negative. (It may even be pressed into contact with the negative to keep it flat.)

For negative formats with a bigger diagonal than about 2 ins. two plano convex lenses are employed with their flat faces outwards. These are equivalent to a single lens of twice the thickness, but it is easier to make two separate plano-convex lenses than one double convex. The double assembly suffers less from aberrations, and it is less likely to crack under the heat of the light source.

In direct and semi-diffused lighting systems, the position of the lamp is usually adjustable with respect to the condenser so that the light can be roughly focused to concentrate the greater part of the rays on the back of the enlarger lens.

The typical vertical enlarger lamphouse is constructed in the form of a cylinder with a hemispherical cap which forms or houses the reflector. The lamp is mounted on the end of a metal tube which enters the top of the cap through a friction-tight bush. This enables the lamp to be moved up or down to focus the beam. In practice the lamp is moved about with no negative in the carrier until the baseboard is evenly illuminated.

A vertical enlarger lamphouse designed to work with a projection type lamp has a hori-

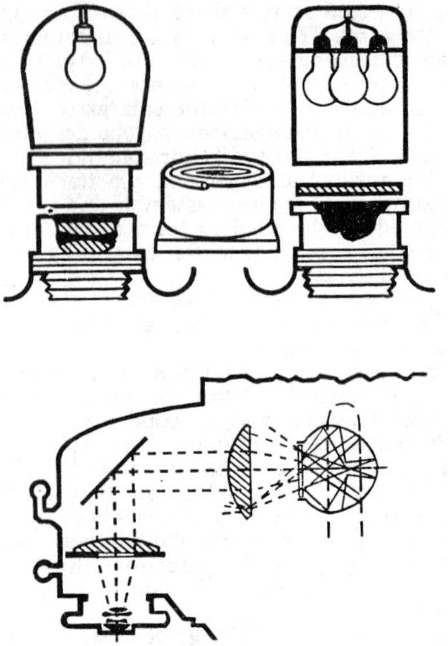

LAMPHOUSE SYSTEMS. Top left: *Orthodox and popular system carrying the lamp on an adjustable rod, held above the condenser.* Top centre: *Cathode tube grid for diffused illumination, as used in large-size professional models.* Top right: *Multi-lamp unit with opal glass diffuser.* Bottom: *Modern system utilizing a projector lamp surrounded by a special reflector, with 45° mirror to deflect light to negative.*

zontal lamp chamber in which the lamp is mounted vertically in front of a reflector which projects the light on to a 45° mirror. From the mirror the light is reflected down on to the condenser. This arrangement is necessary because projector type bulbs must be worked in a vertical position, and it also makes for more efficient cooling.

On a horizontal enlarger the lamphouse is a rectangular box with a ventilating chimney on top. One end of the box is taken up by the condenser, and the lamp and reflector are mounted on a movable slide in front of it so that the light can be focused to give uniform illumination. This type of lamphouse is convenient for non-electric light sources such as gas or oil burners.

The simple types of diffused light lamphouse are similar in construction to the normal condenser type. The condenser unit is replaced by one or more sheets of opal glass and the reflector is given a matt finish instead of being polished. As the light is almost always an opal lamp and not a point source, the shape of the reflector generally departs from the conventional spherical section so as to give even illumination right to the edges. The lamp is mounted in a fixed holder; there is no need for it to be adjustable. This system is only suitable for small or medium sized negatives—e.g., 35 mm. up to 2¼ × 3¼ ins.

Another basic type of diffusing lamphouse has a number of lamps—or a single, shaped tubular lamp—arranged in a trough around the outside so that the light is directed on to a central dome-shaped reflector. Light from the reflector alone is allowed to fall on the negative, the direct rays from the lamps being cut off by the inside wall of the trough. Lamphouses of this type can be made to evenly illuminate negatives of anything up to quarter-plate size—i.e., 3¼ × 4¼ ins. (9 × 12 cm.).

Finally there is the flat, box-shaped lamphouse that houses a grid of tubular mercury vapour—or more recently, the popular cold cathode-discharge lamps. In this type the grid extends over the whole of the negative area and the tubes are practically touching. The result is an evenly distributed light right to the edges of the negative even without the addition of a diffuser. Since the light given out by this type of lamp is practically cold, ventilation is no problem and the lamphouse is little more than a flat box just big enough to hold the grid. The necessary control gear may be built in to the lamphouse or it may be supplied as a separate unit.

Lamphouses of this type are suitable for negatives of all sizes, but particularly for the larger formats.

Negative Carrier. The function of the negative carrier is to hold the negative flat and parallel to the lens plane at the right distance from the lens.

The carrier usually fits into a slot in the enlarger body and supports the negative in front of the condenser. With some carriers it is possible to rotate the negative by turning a knob on the end. This enables the operator to square up the image with the sides of the paper easel. The movement is particularly valuable on horizontal enlargers where it is not always easy to adjust the paper position.

Plates to be enlarged are held in a frame type carrier in which the plate lies in a shallow rebate around the inside of the frame and is retained by spring clips. Or it may be held in a double frame, hinged to open like a book.

Films may be sandwiched between sheets of glass in a frame, or they may be held by the edges only in a glassless carrier. The glass sandwich type of holder is used for film negatives of all sizes, but the glassless type is suitable only for miniature or small negatives. Glassless carriers have the advantage of offering only two film surfaces for dust to cling to, whereas in the sandwich type there are four glass surfaces in addition. But any film larger than 2¼ × 2¼ ins. is apt to buckle under the heat of the lamp and go out of focus—and even miniature negatives may give trouble if the ventilation is inadequate or the lamp is left switched on too long.

Film negative holders normally incorporate a mask, or one half of the frame is shaped to serve the same purpose. The opening in the mask is the exact size of a single frame without allowance for the clear margin. This ensures that no unnecessary light passes through the negative to form a possible source of stray reflections that might fog the enlarged image.

In some enlargers the negative stage has built-in adjustable masks which permit masking down of the image area as required.

In certain miniature enlargers the negative is simply held against the lower face of the condenser or diffuser by a metal plate cut away to mask the negative area. This arrangement helps to keep the negative flat but may defeat its own object unless the glass surface is well cooled by efficient ventilation.

Roll film is often handled in the complete roll, and 35 mm. film either in the roll or in strips of six or nine frames. So the negative holder, while keeping the film under pressure for enlarging, must permit it to be moved on to a new frame without being removed from the enlarger. In the cheapest instruments the two halves of the carrier are simply held together by a spring with just enough pressure to allow the film to be drawn between them. With this arrangement there is always the risk of scratching from particles of grit or rough spots on the face of the carrier in contact with the film. Better enlargers have a lever which releases the pressure while the film is being moved.

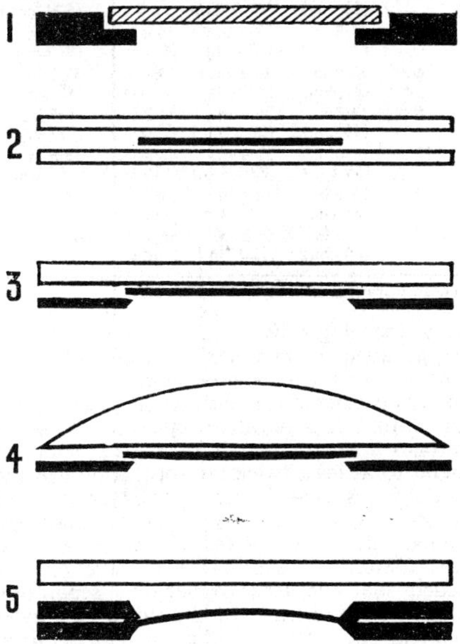

NEGATIVE CARRIERS. 1. Open frame carrier for plates. 2. Double glass sandwich for films; keeps the negative flat, but collects dust on four surfaces. 3. Single glass plate and metal frame. 4. Condenser and metal frame; often used in 35 mm. enlargers. 5. Glassless with heat-absorbing glass; limited to small negative sizes as larger films distort.

In most enlargers the negative carrier is inserted through a light-trapped opening and is held rigidly in the beam of the light in the focal plane of the lens. Some of the more expensive instruments, however, have a pivoted housing for the carrier which enables the negative to be tilted at an angle in conjunction with an opposite tilt of the easel to correct for converging verticals. This device gives a more accurate correction than a simple tilt of the easel (which corrects the convergence but elongates the subject unnaturally and puts part of the image out of focus).

Lens. The lens is connected to the negative carrier housing by a bellows or a telescopic metal sleeve so that it can be moved to and fro for focusing the image.

The focal length of the lens depends upon the format of the negative to be covered. Generally the focal length is equal to or somewhat longer than the diameter of the condenser and also about equal to the diagonal measurement of the negative in question—i.e. equal to the focal length of the normal camera lens for that size.

The maximum aperture of the lens must give sufficient illumination on the paper to focus by when an average negative is being enlarged to the fullest extent possible.

While interchangeable lenses of certain cameras (mainly miniatures) can be used for enlarging, the use of a high-quality enlarging lens is preferable for two reasons. One is that camera lenses are usually corrected to give best definition at high reduction—with the lens focused on distant subjects and with a comparatively short lens-to-image distance. In enlarging it is the object distance (negative to lens) which is small and the image distance comparatively greater. The second reason is that high-speed camera lenses may contain several cemented elements which can get damaged by the heat from the enlarger light. This is liable to occur with enlargers using a powerful directed light source.

The lens may be equipped with click-stop adjustment—i.e., the diaphragm is controlled by a lever which engages with a click that can be felt and heard at each stop value.

A red or orange glass filter may be fitted in front of the lens in such a way that it can be moved into the light beam to hold back the actinic rays. This allows the position of the sensitized paper to be adjusted with the enlarger switched on. The filter is normally hinged or pivoted in front of the lens. On some of the large professional models it slides behind the lens.

Nowadays practically all enlarger lenses are bloomed to cut down internal reflections and increase the image contrast.

Focusing Mechanism. The image is focused by moving the lens closer to, or farther away from, the negative. This calls for a light-tight sleeve between the lens and the negative and a mechanism for adjusting the separation.

TYPICAL LENS AND CONDENSER SIZES

Negative Size ins.	Negative Size cm. (approx.)	Diagonal of negative or minimum diameter of condenser ins.	Diagonal of negative or minimum diameter of condenser cm.	Focal length of lens ins.	Focal length of lens cm.	Maximum aperture
¼ plate (3¼ × 4¼)	9 × 12	6	15	5–6½	13–15	4–5·6
2¼ × 3¼	6 × 9	4½	11	4–5	10–13	3·5–4·5
2¼ × 2¼	6 × 6	3½	8·5	3–4	7·5–10	3·5–4·5
1 × 1½	2·4 × 3·6	1¾	4·3	2–3	5–7·5	3·5

The minimum condenser diameter for an enlarger with a tilting negative carrier is always about 25 per cent larger than for a non-tilting type to ensure that the full negative area is always well within the light beam.

In the simplest arrangement the lens is coupled to the head assembly by a pair of metal tubes which telescope one into the other. A pin in the inner tube engages with a helical slot in the outer, so that turning the lens moves it in or out. In a more elaborate version of this system the pin and slot is replaced by a multi-start helical thread, and this may be further modified so that the lens does not rotate. These systems provide a limited extension for one particular focal length only.

In the more elaborate units, the lens panel is mounted on a rack and pinion slide or a set of rails and is connected to the enlarger head by a flexible opaque bellows. This arrangement permits any desired length of extension and can be designed to provide the necessary range of movement for lenses of several focal lengths. It is popular for enlargers intended for more than one size of negative—e.g., 35 mm. (2 ins. lens) and 2¼ × 2¼ or 2¼ × 3¼ ins. (3½ ins. lens). This method also allows for some degree of reduction.

The movement of the lens panel may be controlled by the rack-and-pinion mechanism mentioned above, or by a friction drive. Some models incorporate a micrometer screw for fine focusing. A locking screw may also be provided to set the focusing position.

The maximum scale of enlargement is obtained when the lens is racked back so that the negative lies very near its back focus; this results in an almost infinite scale of enlargement. As the lens is moved farther away from the negative the scale gets progressively smaller until the separation is approximately equal to twice the back focus. At that point the negative is reproduced 1 : 1—i.e., the negative and projected images are the same size. Further extension gives a reduced image.

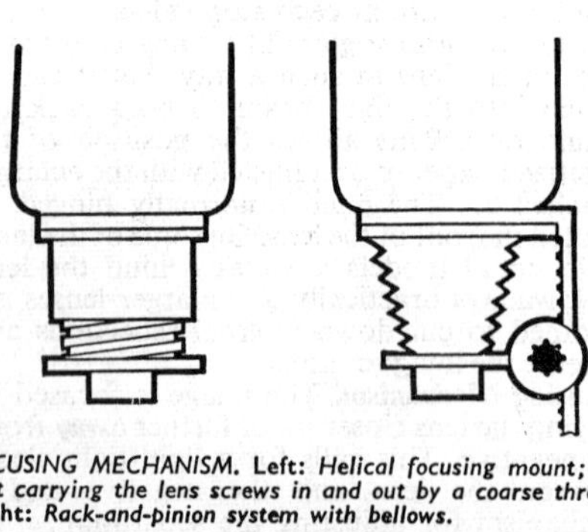

FOCUSING MECHANISM. **Left:** *Helical focusing mount; the unit carrying the lens screws in and out by a coarse thread.* **Right:** *Rack-and-pinion system with bellows.*

LENS-TO-BASEBOARD DISTANCES AT VARIOUS MAGNIFICATIONS

Magnification (Diameters)	Focal Length of Lens 2 ins. (5 cm.)	3 ins. (7·5 cm.)	4⅛ ins. (10·5 cm.)	5⅝ ins. (13·5 cm.)
1	3·9 in. (10 cm.)	5·9 in. (15 cm.)	8·3 in. (21 cm.)	10·6 in. (27 cm.)
2	5·9 in. (15 cm.)	8·9 in. (22·5 cm.)	12·4 in. (31·5 cm.)	15·9 in. (40 cm.)
3	7·9 in. (20 cm.)	11·8 in. (30 cm.)	16·5 in. (42 cm.)	21·2 in. (54 cm.)
4	9·9 in. (25 cm.)	14·8 in. (37·5 cm.)	20·7 in. (52·5 cm.)	26·6 in. (67·5 cm.)
5	11·8 in. (30 cm.)	17·7 in. (45 cm.)	24·8 in. (63 cm.)	31·9 in. (81 cm.)
6	13·8 in. (35 cm.)	20·7 in. (52.5 cm.)	29 in. (74 cm.)	37·2 in. (95 cm.)
8	17·7 in. (45 cm.)	26·6 in. (67·5 cm.)	37·2 in. (95 cm.)	49·9 in. (127 cm.)
10	21·6 in. (55 cm.)	32·4 in. (82·5 cm.)	45·5 in. (116 cm.)	58·5 in. (148 cm.)
12	25·6 in. (65 cm.)	38·4 in. (97·5 cm.)	53·7 in. (136 cm.)	69 in. (175 cm.)
15	31·5 in. (80 cm.)	47·2 in. (120 cm.)	66·1 in. (168 cm.)	85 in. (216 cm.)

These distances are slightly rounded off.

The focusing range is normally designed to give a range of enlargement from infinity to same size with a particular lens. If more than one lens is catered for, the focusing range must start with a lens negative separation equal to the back focus of the shortest focus lens and extend to at least twice the focal length of the longest focus lens—i.e., in an enlarger designed for both 35 mm. and 2¼ × 3¼ ins. negatives it must extend from 2 ins. (5 cm.) to at least 7 ins. (18 cm.). If such extreme reductions are needed that the lens to negative separation with the normal lens would need to be greater than can be achieved, then a shorter focal length lens should be used instead.

Automatic Focusing. The lens panel may be racked along the slide by a hand-operated knob or lever, or it may be coupled to the head raising and lowering mechanism so that it auto-

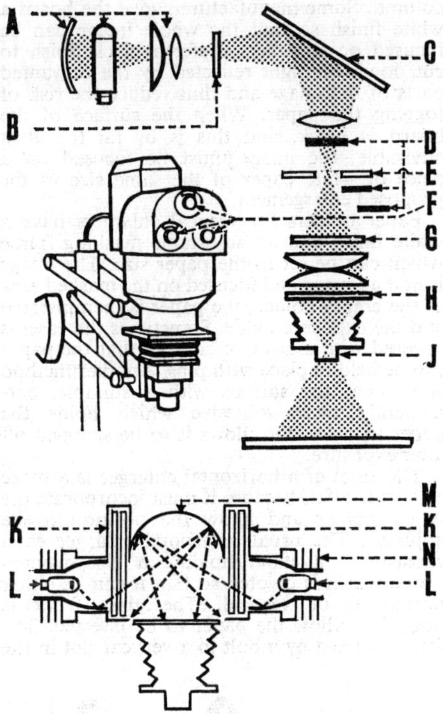

COLOUR MIXING HEADS. Top: *Colour control in the beam. The light from a projection-type lamp is focused on to a slit and subtractive filters inserted separately into the beam to any desired extent. The filters are controlled by setting knobs on the lamphouse. This modified light beam then reaches the negative through a diffuser.* A, *reflector and condenser system of projection lamp.* B, *heat filter.* C, *mirror.* D, *subtractive filters.* E, *slit.* F, *filter controls on lamphouse.* G, *diffuser.* H, *negative.* J, *enlarging lens.* Bottom: *Light control with integrating sphere.* K, *graduated adjustable subtractive filters.* L, *lamp.* M, *integrating spheres.* N, *cooling fin on lamphouses.*

matically moves the panel to give the correct extension for any degree of enlargement—i.e., according to the distance of the enlarger head from the easel. This is done by moving the lens panel by a rocker arm riding on a suitably shaped cam or profile plate attached to the fixed part of the enlarger.

Rangefinder Focusing. The enlarging head may also incorporate a ranging device. This is usually some form of split-image system in which two separate images—either of the negative, or of an illuminated slit—are projected on to the baseboard and are brought into coincidence by moving the lens focusing control. When the images coincide, the negative will be sharply focused. It is easier to focus this way than by trying to decide when the image is at its sharpest.

The method most generally used is to insert a mask behind the lens so that each half of the lens forms its own image of an illuminated slit mounted in the focal plane. When the two halves of the slit are brought into register the lens is sharply focused. The slit image is then moved out of the way and the negative carrier slid into position, bringing the negative into the focal plane previously occupied by the slit.

Magnifier Focusing. Any of the normal types of focusing magnifier may be used for determining when the image is sharp. The basic principle with all such systems is to reflect part of the enlarged image on to a viewing screen with a magnifying lens above it. When the magnified screen image appears sharp, the image on the easel will also be sharp. Magnifiers of this type are not usually supplied with the enlarger; they are bought as an accessory.

Magnifiers of this type must be very accurately made really to be an advantage; a slight difference in the distance between the mirror and screen and mirror and baseboard will result in inaccurate focusing on the final print.

Instrument Focusing. The sharpness of the image on the baseboard can also be checked with a photo-electric sharpness meter. This compares the contrast of the sharp image with the contrast when it is unsharp. A special high-contrast target (such as a black-and-white screen pattern) is used in the negative carrier during focusing. The image is projected on a special photocell system, and the sharpest position found by observing the point where the reading of a measuring instrument is at its highest.

Raising and Lowering Gear. The scale of enlargement is adjusted by altering the distance between the enlarger head and the easel. In the commonest vertical arrangement upright metal tubular columns or girders, anchored to the easel, support the head and form a track on which it can be slid up or down. It can be locked in any position by a clamp, and its weight may be counterbalanced by a cable passing over a pulley and attached to a suitable lead counterweight suspended inside the column.

The use of a weight and pulley is the most

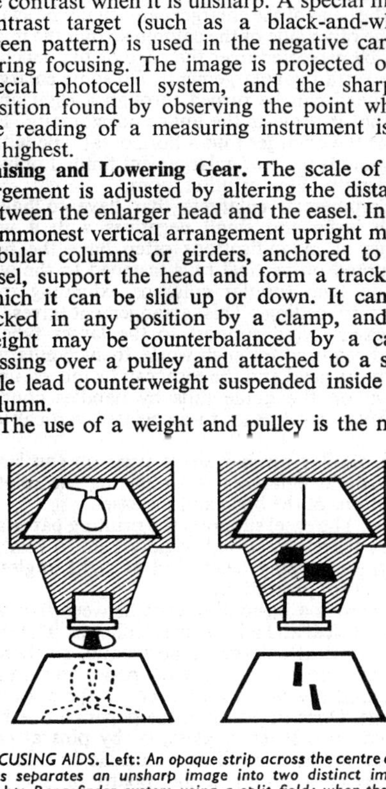

FOCUSING AIDS. Left: *An opaque strip across the centre of the lens separates an unsharp image into two distinct images.* Right: *Rangefinder system using a split field; when the split line fuses on the baseboard, the image is sharp.*

simple system of providing a counterbalance, but enlargers also employ systems using springs instead.

The head may be raised by a rack and pinion mechanism or a friction wheel drive. Another method is to move it up and down by a head screw and nut in the same way as the tool rest of a lathe is moved along the bed.

On big enlargers the raising and lowering adjustment of the head is controlled by hand wheels at the front of the baseboard. On some advanced models the movement may even be motor driven.

In some enlargers the head is attached to the column by two sets of parallel links pivoted at each end so as to allow the head to move up and down while its axis remains vertical (this is commonly called a parallelogram movement). The weight of the head is counter-balanced so that very little effort is needed to raise it. A clamping screw holds the head firmly once it has been set at the desired height. This arrangement usually incorporates an automatic focusing mechanism coupled to the link motion in such a way that the image on the easel remains sharp over the whole range of movement of the head.

The head of the enlarger can usually be swung to the other side of the column so that the image can be projected over the edge of the bench to give extra big enlargements. On some enlargers the head can be turned through a right angle so that it projects the image horizontally. This turns the enlarger into a horizontal type with a maximum scale of enlargement limited only by the length of the printing room.

Commercial enlargers that have to handle both large and small negatives with a wide range of magnifications often have the head mounted on vertical rails fastened to the wall, or on twin tubular columns extending from floor to ceiling. This arrangement gives the necessary rigid support to the larger type of head required for such work. In such enlargers both the head and the easel can be moved up or down on the guide rails by handles coupled through bevel gears to associated vertical lead screws.

A similar mechanism is used on the larger horizontal enlargers for adjusting the relative positions of the body and the easel.

Easel. The easel supports the printing paper and must therefore be flat, rigidly held in relation to the enlarging head, and at right angles to the axis of projection.

There is a basic difference between the easel of a vertical and a horizontal enlarger. The easel of a vertical enlarger is horizontal, so there is no difficulty in arranging the paper on it in any required position with respect to the projected image. Once in position, the paper can be held down by a sheet of glass, or by pins at each corner or by a masking frame. So in this case the easel is simply a flat board or sheet of metal which also forms the support for the column. Some manufacturers give the board a white finish so that the whole image can be focused on it, others prefer a black finish to cut down the light reflected by the unwanted parts of the image and thus reduce the risk of fogging the paper. When the surface of the board is black, and this is by far the most advisable, the image must be focused on a piece of white paper of the same size as the proposed enlargement.

Paper holders for vertical enlargers have a white finish and an adjustable masking frame which can be set to the paper size. The image is first adjusted and focused on the masked area of the empty holder; the paper is then inserted and the exposure made. Sometimes the easel is covered with a layer of cork so that the paper can be held in place with pins. Another method is to coat the surface with a suitable, permanently tacky adhesive which holds the paper in place but allows it to be stripped off after exposure.

The easel of a horizontal enlarger is a more elaborate affair because it must incorporate the paper holder and allow the position to be adjusted. The usual horizontal enlarger easel consists of a stand holding a glass-fronted frame which is pivoted so that it can be either horizontal or vertical. The glass front is hinged to allow the paper to be inserted. The frame is fixed by a bolt in a vertical slot in the

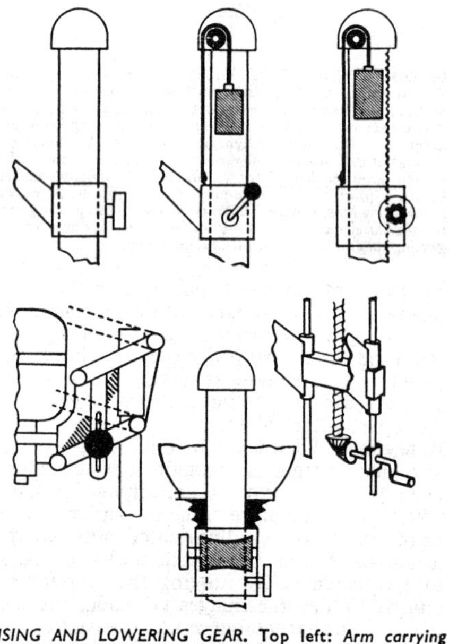

RAISING AND LOWERING GEAR. Top left: *Arm carrying lamphouse moves up and down column, being clamped in place.* Top centre: *Sliding supporting arm with counterweight inside column.* Top right: *Counter-weighted lamphouse arm moved by rack and pinion drive.* Bottom left: *Spring-loaded parallelogram system.* Bottom centre: *Friction drive on column.* Bottom right: *Lamphouse arm raised by screw gear.*

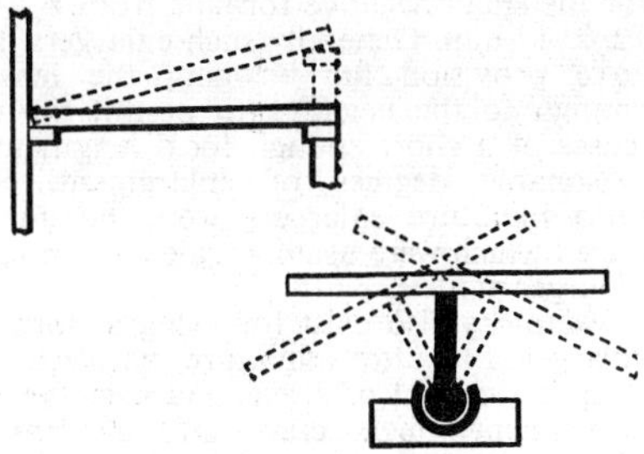

TILTING EASELS. Left: *Loose paper board tilted by raising support at one side.* Right: *Paper holder mounted on a universal ball-and-socket joint for free tilt in all directions.*

stand; when the bolt is slackened off, the frame can be raised, lowered or turned to bring the image into the required position. The front of the easel is white with black lines marking out the boundaries of the various sizes of printing paper.

To use an easel of this kind, it is first placed at the right distance to give a sharp image of the required size. The frame locking-bolt is slackened off and the frame moved about until the image occupies the rectangle corresponding to the paper size. The frame is locked, swung horizontal and the glass front opened. A piece of paper is laid in position and the frame is closed and returned to the vertical for the exposure.

It may be locked at any intermediate angle for correcting converging verticals.

This type of easel may also be faced with cork or adhesive instead of glass to allow the paper to be pinned or stuck in position.

An adhesive support for the paper during enlarging can be made by pouring a layer of jelly into a shallow tray of suitable size. The ready-made composition used in "jellygraph" duplicators is excellent for the purpose, or a home-made substitute can be prepared as follows:

Soften 1 ounce (25 grams) of leaf gelatin by soaking in water for 24 hours. Drain off the water, pour over 6 ounces (150 c.cm.) of glycerine and heat in a double saucepan until the gelatin dissolves. A few drops of oil of cloves is added to prevent the mixture from deteriorating. The warm gelatin is poured into the tray and left to set. When cold the surface will be tacky enough to hold single weight paper firmly in position but it is not so successful with double weight papers.

After a time the surface may lose its tackiness by being covered by a film of grease or dust, but it can be restored by simply melting the mixture and allowing it to set again. This type of paper holder is particularly useful with horizontal enlargers where the paper must be held on a vertical surface.

Horizontal Enlarger. In principle there is no difference between a vertical and a horizontal enlarger. The horizontal arrangement was the first to be used, being simply an adaptation of the existing "magic lantern" type of slide projector.

Horizontal enlargers are not nowadays manufactured for the smaller and more popular negative size because the layout takes up much more space than the vertical arrangement, the paper is more difficult to control, and the image is not so easy to examine. The only remaining horizontal enlargers are those specially made for the larger plate formats—e.g., in photomechanical work, document copying and commercial and professional studios—where the weight of the assembly makes the vertical construction unnecessarily expensive.

Even so, because of the greater convenience in use and saving of space, the vertical arrangement is often preferred, notwithstanding the extra cost, even for the largest instruments.

The lamphouse of the typical horizontal enlarger is simply an oblong box housing a reflector, light source and condenser. A light-trapped chimney on top of the box carries off the hot air.

This type of house incidentally allows a variety of non-electric light sources to be used. The negative carrier on a horizontal enlarger is often equipped with movements that allow the negative to be rotated to level up the horizon and fitted to correct converging verticals. The lens panel is connected to the body by a bellows and is moved forward or backward by a rack and pinion mechanism.

The whole assembly is attached to the end of a pair of wooden or metal rails which carry the easel and keep it at right angles to the projection axis. Adjustment of the scale of the image is carried out by sliding the easel along the guide rails.

Horizontal enlarger design has remained practically unchanged for the past fifty years. The demand for this type is not great enough to justify the cost of press tools or moulds so the few that are made are hand built on the old pattern

Even when the horizontal type of arrangement is more convenient—e.g., for making photomurals or other giant size enlargements—most workers prefer to use the head of a vertical enlarger to project the image and, in fact, many vertical enlargers are designed to allow the head to be turned through 90° for this purpose.

Fixed Focus Enlarger. The fixed focus enlarger (now practically obsolete) is simply a vertical enlarger designed to gives a standard size of enlargement—e.g., post-card—from one particular size of negative. It is usually made of the cheapest materials and components and built down to a price. The object of this type of instrument was originally to overcome the reluctance of the ordinary non-technical snapshooter to buy a miniature camera. The negatives produced disappointingly tiny contact prints and this type of user was not prepared to go to either the trouble or expense of making enlargements. The fixed focus enlarger gave him pictures big enough for his purpose with no more trouble than contact prints.

The arrangement was simply a long conical light-tight box with an opal lamp and diffuser at the narrow end and a paper holder at the opposite end. There was a slot for the negative holder and the lens was mounted in the centre of a suitably placed light-tight partition inside the box.

Enlargers used for commercial mass printing from specific negative formats are frequently also fixed-focus types. They are however appreciably more complex in the provisions for holding the paper (often in continuous rolls) and for automatic exposure measurement and timing.

Daylight Enlarger. A special type of fixed-focus enlarger has no lamphouse, its place being taken by a sheet of opal diffusing glass. When pointed towards the daylight an enlarged image of the negative is projected on to a sheet of bromide —or even print-out-paper—in the opposite end of the box. The paper is inserted and removed in the darkroom.

Some larger models were even built into a darkened room with the negative receiving daylight through a hole in the wall. Enlargers of this type are not made today.

Enlarging Adaptors. Various forms of adaptor have been manufactured in the past for converting a plate camera into an enlarger. The usual design consists of a lamphouse, condenser and negative holder which is attached to the camera in place of the plate back. The focusing extension of the camera must be great enough to allow a range of movement up to a distance of twice the focal length of the lens from the negative.

The advent of the cold cathode grid light source for enlarging has made it possible to design a compact adaptor in which the lamphouse is reduced to a shallow square box housing the discharge tube.

Another type of adaptor, now obsolete, enabled a plate camera body to be fitted to an aperture in the window blind of the darkroom. A sloping white-painted board outside the window reflected the sky light through the aperture and illuminated the negative in a holder fitted in the back of the camera. This arrangement went out of use as soon as electric lamps became available, but it is still employed by workers remote from electricity supply services.

Like all daylight printing methods it suffered from the variable nature of the light and the fact that no work could be done during the hours of darkness.

Ultra-miniature Enlargers. When ultra-miniature cameras first appeared, existing enlargers for 35 mm. and larger negatives were unsuitable for the small negative formats from 8 × 11 to 12 × 17 mm. Generally such enlargers did not have provision for bringing the lens near enough to the negative to permit the use of lenses of a short enough focal length to yield reasonable degrees of enlargment. Special ultra-miniature enlargers were therefore produced which were usually scaled-down versions of larger models.

Adaptors also exist for using a normal-size enlarger for ultra-miniature negatives. Such adaptors consist of a lens and negative carrier attachment which screws into the lens panel of the enlarger in place of the normal lens. The light beam from the enlarger illuminates the negative in its own ultra-miniature carrier, with the enlarging lens being part of the adaptor to focus the negative image in turn on the baseboard.

The lens in this case has its own focusing adjustment. The negative carrier takes ultra-miniature negatives in strips, and can accommodate image sizes frequently down to the frame size of 8 mm. cine film.

Enlargers for Colour. For making enlargements from colour negatives on to positive colour papers (or from colour transparencies on to reversal papers) the enlarger must provide some control of the colour of the printing light to correct the colour rendering of the final print. In their simplest form, colour

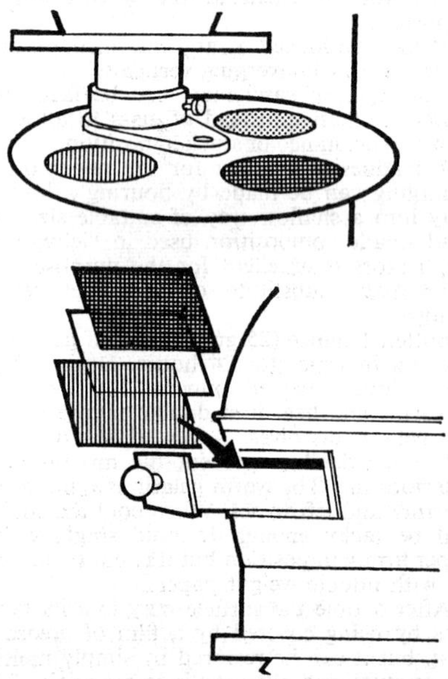

SIMPLE COLOUR ENLARGING SYSTEMS: Top: *Filter turret fitted in front of enlarger lens for three-exposure printing. The turret carries three tri-colour (red, green and blue) filters which can be brought into position as required.* Bottom: *Filter drawer for subtractive printing. Printing filters of various strengths fit into a drawer between the negative and the light source to modify the colour of the printing light.*

enlargers incorporate a pull-out drawer in the lamp house to permit the placing of appropriately-coloured cyan, magenta or yellow filter foils of varying densities between the lamp and the condenser. The filter foils are available in standard sizes (for example 12 × 12 cm.). The correction is achieved by combining two or more filter foils of one or two colours and similar or different densities. The actual filter combination required depends on the colour of the enlarging light, on the characteristics of the printing paper and the colour balance of the negative. This is determined in each case by preliminary tests, and the colour print is made by a single exposure by the appropriately-filtered light.

An alternative method of making colour prints is to expose the paper three times through a red, green and blue filter respectively. The relative exposures through the three filters determine the colour rendering—again settled by preliminary tests. The filters in this case are mounted in front of the lens, either in a filter slide or a filter turret. Most normal enlargers can be adapted for this method of printing.

Enlargers for colour printing must have a light source with a continuous spectrum and a consistent colour quality. Tungsten lamps are the most usual enlarger lighting for colour, and are generally run through a voltage control circuit to ensure a constant operating voltage (and hence constant colour temperature) for consistent results. Certain types of fluorescent lighting have, however, also been used for colour enlarging.

More advanced colour enlargers may provide semi-automatic or automatic systems of light control in a special lamp house which can be used—interchangeably with the normal lamphouse—on standard enlargers. Such enlargers may use either the three exposure or the single exposure printing system.

For three-exposure printing, the lamphouse incorporates a red, green and blue filter which are switched in front of the enlarger lamp in turn (there may even be three enlarger lamps, each with its own filter). The filters are controlled by electric circuits, which move them into position for pre-determined times according to controls set on the enlarger head or on a separate timer unit. The latter may in turn be controlled by various photo cell systems that measure the intensity and colour quality of the light coming through the enlarger lens.

Colour enlargers for single exposure printing may incorporate one or more light sources with graduated cyan, magenta and yellow filters in front of them. External controls vary the position of each filter in the light beam, thereby adjusting the amount of each colour component of the light. This filtered light is mixed in an integrating sphere or reflector arrangement to provide thoroughly diffused light of uniform colour. In semi-automatic and automatic units, the filtering selection is controlled from a photo-electric measuring system which measures the colour of the light coming through the enlarger lens.

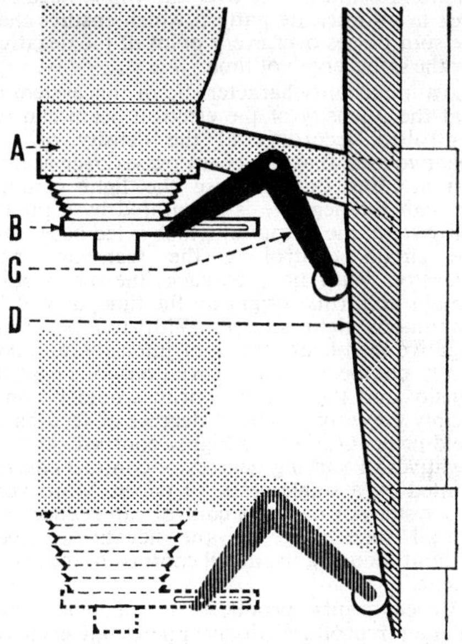

PRINCIPLE OF AUTOMATIC FOCUSING. The position of the lens panel is governed by a lever following a cam. At high degrees of enlargement (top) this raises the lens panel; at low magnifications (bottom) it lowers the lens panel to increase the lens-to-negative distance for sharp focus. A, enlarger head. B, lens panel with lens. C, control lever. D, focusing cam.

Colour enlargers using an integrating and colour mixing system of this nature inevitably have to be diffused light enlargers. For colour work this is no drawback; even a directed light enlarger would yield no greater image contrast owing to the very small light scattering ability of the dye images. Comparatively intense lighting is, however, necessary to yield practical exposures, since part of the light is filtered out by the colour control filters. Colour enlargers therefore generally work with one or more high intensity lamps. They also need adequate ventilation to prevent overheating which would damage both the filters and the colour negative.

Enlargers for commercial colour printing may incorporate considerably more complex lighting and colour control arrangements.

Electronic Enlargers. The enlargers described so far use a continuous and uniform light source that simultaneously and evenly illuminates the whole negative area. By contrast, so-called electronic enlargers use a flying spot scanner—an electron beam moving over the surface of a cathode ray tube—as the light source. This exposes the image spot by spot, like the scanning system of a television screen, and not all at once. During the exposure the scanning beam moves over the whole negative area in an intricate pattern which ensures that the spot passes over every point of the negative for the same length of time.

An important characteristic of this system is that the intensity of the electron beam can be controlled according to the density of the negative area that the beam passes over at any instant. For this purpose the light coming through the negative is monitored by a photo cell system, the output of which is fed back into the circuit controlling the scanning spot intensity. With no feed-back, the spot would remain uniformly bright all the time, as with a continuous light source. With 100 per cent negative feed-back, the spot intensity automatically compensates for variations of negative density and the resulting print would be uniformly grey. Intermediate degrees of feed-back yield prints of lower or higher contrast, so that negatives of varying degrees of contrast can be printed on a single grade of paper. Moreover, this system of contrast control can reduce the overall tone range of the negative image, without reducing the detail contrast to the same extent.

Hence prints produced in an electronic enlarger reproduce all the fine detail even of very contrasty negatives without exceeding the contrast range of the paper. The effect of the flying spot scanner exposure is that of an unsharp contrast reducing mask—virtually a luminous positive low contrast image of the negative being printed, and produced during the printing exposure.

The photo cell system can also register the total light passing through the negative and cut off the exposure when sufficient light has passed to produce a correctly exposed print. Both the contrast (feed-back) and exposure level are usually adjusted by calibrated settings of the control circuits.

Electronic colour enlargers working on similar principles use in addition cyan, magenta and yellow filters in front of the monitoring photo cell, and red, green and blue filters in front of the enlarger lens for three successive exposures. The cyan, magenta and yellow filters in turn analyse the red, green and blue content of the light coming through the colour negative and control the exposure accordingly through the three printing filters.

F.P. and L.A.M.

See also: *Callier effect; Cold cathode lamp; Colour Printing Technique; Contrast Control; Electronic printing; Enlarging; Enlarging filter; Enlarging lens; Paper holder; Voltage stabilizer.*

Books: *Enlarging*, by C. I. Jacobson (London); *The Complete Art of Printing and Enlarging*, by O. R. Croy (London).

ENLARGING

Projection printing or enlarging on development papers requires suitable equipment.

Exposures must be found by trial or measurement, and the prints go through a standard routine of development and fixing.

Large quantities of enlargements of the same size can be made from the same, or similar, negatives at about the same rate as they can be printed by contact. But the rate is far less for prints of different sizes from different negatives where separate exposure tests are necessary.

The prints can be made in any size or format, and from the whole or part of the negative. In projection printing the appearance of the finished print can be controlled within wide limits by the choice of paper surfaces, individual technique, and the use of accessory equipment. Special devices, like image distortion and combination printing, can be employed.

The image colour can be black, or warm-black to red-brown with warm-tone papers. Toning processes can produce any other colour.

Enlarging papers are made in a number of grades, so that acceptable prints can be made from negatives of almost any contrast.

Enlarged prints are made by projecting the negative image on to the printing paper through an enlarger. Many enlargers can be used for same-size printing, and some even for reduction, e.g., for making lantern slides or film strips from larger negatives.

Enlargements are usually made on bromide or fast chlorobromide paper. Such papers are sensitive to ordinary artificial lighting or daylight and must be handled and processed only under an orange or olive-green safelight.

SETTING UP THE ENLARGER

The steps in making enlargements fall into two basic operations: setting up the enlarger, and making the exposure—after which the print is processed.

When setting up the enlarger, it is quicker and better to follow a routine procedure.

Sorting the Negatives. Negatives for enlarging must be selected even more critically than for contact printing. Slight unsharpness that passes unnoticed in a contact print shows up distinctly in an enlargement, and pinholes, grain, scratches, and other flaws become much more noticeable. A good negative, having had the correct exposure and correct development, will be of moderate density and contrast, and generally of fairly fine grain. Excessive negative exposure and development tend to increase contrast and grain. Grossly under-exposed negatives cannot yield perfect enlargements as essential shadow detail is missing.

The contrast in a negative can be controlled in enlarging by choosing the appropriate grade of paper, and the print density is, of course, determined by length of exposure.

By carefully examining the negatives under a

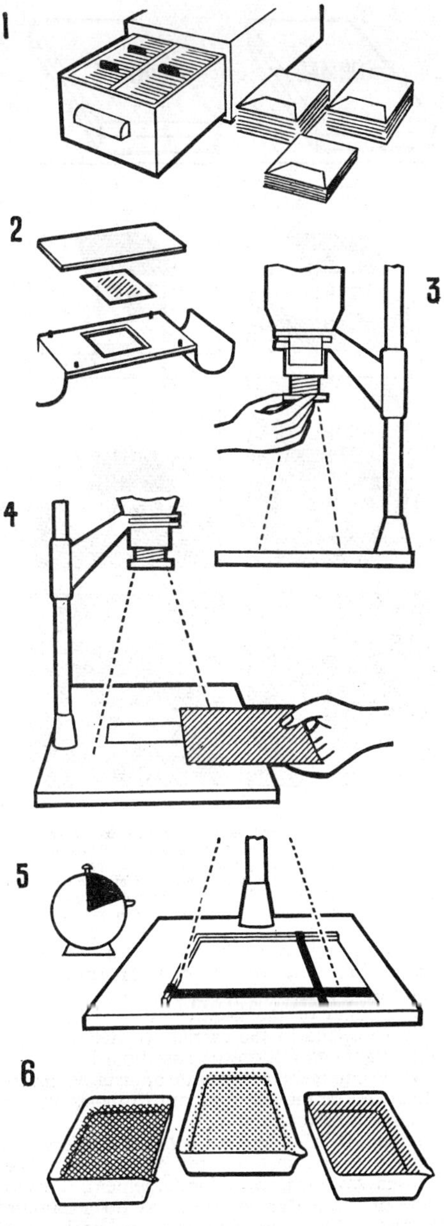

SEQUENCE OF ENLARGING STAGES. 1. Sort out the negatives, discarding any that are unsuitable for enlarging due to obvious faults. 2. Place the selected negative in the carrier. 3. Adjust magnification and focus of the enlarger. 4. Make a test strip to establish the correct exposure necessary. 5. Locate printing paper on baseboard (e.g., by means of a suitable masking frame), and expose for the time suggested by the test strip after development. 6. Process the exposed enlargement (develop, rinse, fix, wash, and dry).

magnifying glass, or even by projecting them in the enlarger on to a white sheet on the baseboard, they can then be graded in contrast and density groups.

NEGATIVE CONTRAST AND PAPER GRADE

Negative Contrast	*Appearance of Highlights*	*Negative Shadows*	*Paper Grade*
Very low	Very similar in density usually fairly thin		Ultra hard
Low	Medium	Thin	Hard
Medium	Normally dense or Very dense	Thin but full of detail Medium	Normal
High	Dense to very dense	Fairly thin but full of detail	Soft
Extreme	Very dense, almost blocked up	Thin, may lack detail altogether	Extra soft (negative frequently unprintable)

Under-exposed negatives appear thin and soft (unless greatly forced during development) since the maximum image density is low by comparison with the fog level. Printing on a hard paper grade will distort the tone scale, but is the only way of producing a reasonably presentable print.

For judging both contrast and density a useful aid is a collection of standard negatives of the appropriate contrast groups and of four or five degrees of density. After ascertaining the correct paper grade and a relative exposure for each, these can be mounted between glass. Such a set will serve for direct comparison with any new negative to be printed. In estimating the negative quality, the film or plate is best observed against a well diffused light source (e.g., a light box or the light reflected from a pale coloured wall), holding the negative at the same distance from the light every time. Observation against a direct source (such as a bare lamp) is misleading.

An alternative method of judging the image is to lay the negative down on a page of a book printed on white paper. A normal negative of average contrast should show good tone and detail, with some density even in the thinnest parts, and the text underneath should be just readable through the densest areas. Similar examination of other negatives will show whether the highlight density (which can be used to estimate exposure) is higher or lower. It will also show whether the contrast (as indicated by relative density of highlights and shadows) is greater or less.

Usually it will be possible after some experience to dispense with most of these aids to judging negatives.

Loading the Negative Carrier. Before loading, both the negative and the glasses (if any) of the negative carrier should be cleaned to remove as much dust as possible.

NEGATIVE AND PAPER CONTRAST. To obtain a well-graded print, the paper contrast used must be matched to the negative gradation. Top left: *Contrasty negative with very dense highlights and thin or very thin shadows. This needs a soft paper grade to accommodate the full density range. With a normal paper the result would be distinctly hard, and with a hard paper the image would be almost devoid of tonal gradation.* Centre left: *Normal negative showing average tone scale. This will yield the best print on medium or normal paper grade.* Bottom left: *Soft negative with comparatively small tone differences between even the extremes of light and shade in the subject. A hard (and sometimes extra hard) paper grade is required to separate the tone values.*

Glass surfaces are best cleaned with a soft cloth or chamois leather. If the glass is breathed on and the film of moisture appears quite even without any streaks or patterns, the surface can be accepted as being clean. Rubbing and polishing glass has the undesirable effect of charging it with static electricity and making it attract dust particles. So immediately before use, the negative and each glass surface is gently dusted with a clean, soft brush specially kept for the purpose, and then wiped over with the dry ball of the thumb. This discharges any static electricity on the surface, and removes the last traces of dust.

The details of the loading procedure depend on the design of the negative carrier.

The negatives are inserted with the emulsion side facing the lens. If they are loaded wrong side down, the projected image will be reversed left to right, but it will not suffer in any other way. The image may be deliberately reversed in this way for better pictorial effect so long as the picture does not contain anything that obviously reveals the trick, e.g., lettering.

The edges of the negative should be completely masked in the carrier so that no stray white light can fall on the base-board.

Some enlargers incorporate adjustable masks which can be moved in or out to mask off the negative area required.

Focusing. With focusing enlargers where the image size is adjustable, the picture has to be focused for each size of enlargement. This is done in much the same way as on a camera by moving the lens forwards and backwards until the image appears sharp.

The image is observed on a sheet of white paper held exactly where the bromide paper will be held when the enlargement is being made. It is absolutely necessary for the surface of the white paper to occupy the same position as the emulsion of the bromide paper, so that the image on the final print will be sharp. If any type of paper holder or masking frame is used for the bromide paper, it must also be used to hold the paper on which the image is focused.

The enlarger head is moved up or down the column until the projected image of the negative (or the selected part of the negative) is about the right size. The lens is then focused until the image is reasonably sharp, and the size of the picture is checked. If it is too large, the enlarger head is lowered; if too small, it is moved up the column and the image is again focused sharply.

For fine focusing, the lens is moved until the image just passes the point of maximum sharpness. This position is noted and the lens is moved in the opposite direction until the image again just begins to be unsharp. By successive up and down movement, these limits of sharpness are brought as close together as possible. Finally, the lens is set half-way between the two limits.

There are several devices available for getting over the difficulty of focusing dense or flat negatives where the image is either too faint or not contrasty enough to be seen clearly. Such devices are called focusing aids.

There are two kinds of focusing aid: separate accessories and built-in devices.

The accessories include focusing negatives and ground glass viewing devices.

A focusing negative is a piece of black film carrying fine transparent lines, crosses and circles, extending over the whole negative area. The clear, distinct lines are easy to focus, and once the position of maximum sharpness has been found, the negative is removed and the film or plate to be enlarged is inserted in the carrier in its place.

Provided that nothing is allowed to shift when the substitution is made, the resulting enlargement should be the sharpest that can be made from the negative.

Screen focusing devices usually consist of a stand which holds a mirror and a ground glass screen. When the stand is placed on the surface of the paper on the enlarger easel or in the paper carrier, the mirror reflects part of the image on to the ground glass screen. The screen is set at a convenient angle for viewing, and the dimensions of the device are arranged so that the distance travelled by a ray of light falling on the ground glass surface is exactly equal to the distance it would have travelled if it had been allowed to fall on the surface of the paper. So if the image is sharply focused on the screen, it will also be sharp on the print. Most focusing devices of this sort incorporate a focusing magnifier.

Another focusing accessory consists of a strip of black paper mounted in a holder to fit over the centre of the lens. The two halves of the lens each form distinct images. These two images merge into one when the negative is accurately focused.

Some enlargers have a calibrated scale on the column, and a second scale on the lens mount or on the focusing rack. When both scales are set to the same value, the lens is sharply focused. The enlarger head is set for the required magnification against the scale on the column, and the lens is then set to the same figure.

Other enlargers are equipped with a built-in rangefinder. This projects two spots of light on to the baseboard and the image is focused by adjusting the lens until the spots coincide.

Sharpness meters can also be used to determine the position of optimum sharpness of the image on the baseboard. The image is here focused on a photo cell or set of photo cells, and the output observed on the measuring instrument while adjusting the lens focus. The maximum reading of the measuring needle indicates the position of maximum sharpness.

Automatic enlargers cut out the whole procedure of focusing. They are set once only for the lens to be used. After that, raising and lowering the lamphouse increases or decreases the picture size, and focuses the image at the same time. L.A.M.

Same-size or Reduced Prints. The enlarger can be used not only for its basic purpose of producing larger size pictures but also to make prints of the same size as the original negative, or even to a reduced scale.

This method is valuable when it is necessary to be able to hold back or bring out certain portions of the negative by dodging or any other method of control to which contact printing does not lend itself. In enlargers which have bellows or focusing mounts where the total extension between lens and negative is twice that of the focal length of the lens—e.g., giving a 6 ins. extension with a 3 ins. lens—it is possible, without any further aids, to project an image the same size as the negative on to the baseboard. Many enlargers, however, do not cater for this long extension. In most instances this can be overcome by inserting an extension tube between lens and lens panel.

If a reduced size print is required—e.g., when making photographic miniatures—an even longer extension is necessary. While it is possible to provide this by inserting an extension tube between the lens and lens panel, the method may be limited by either the diameter of the lens mount or of the extension tube. One of these may cut off the corners so that the whole of the original negative is not reproduced. The amount of extension from negative to lens required when reducing the negative to give an image one-half smaller is equal to three times the focal length of the lens. To reduce a negative to one-third of its original size an extension of four times the focal length is needed.

The easiest way of getting reductions without extension tubes is to use a lens of shorter focal length. This needs less extension and has sufficient covering power for the whole negative when used at more than double-extension.

The above methods of printing same-size or on a reduced scale are useful when making transparencies from larger negatives—e.g., to suit a miniature projector. It is only necessary to substitute lantern plates or positive film for the paper on the baseboard. W.D.E.

EXPOSURE

In enlarging, the correct exposure is usually found by the aid of a test strip, a step wedge, a photometer, or by density measurements carried out on the negative.

Test Strips. Exposure is generally determined by making a test strip. For this, a piece of the actual paper on which the enlargement is to be made is exposed in sections on the enlarger base. Each section is given a longer exposure than the one before, and the correct exposure is based on the appearance of the developed strip of paper.

Procedure. (1) The enlarger is adjusted for the required print size, and the negative is focused in the normal way.

(2) The orange filter is placed over the enlarger lens and a strip of the enlarging paper to be used is held flat on the base-board where it includes the most important part of the image. (A part should be selected with as full a range of tones as possible, from the brightest highlights to the deepest shadows, over the whole length of the strip.)

(3) The orange filter is swung out of the way, and the whole strip exposed for, say, 2 seconds.

(4) The enlarger lamp is switched off, and half an inch or so of the strip is covered with an opaque card.

(5) The enlarger lamp is then switched on again for 2 more seconds.

(6) The above procedure is then repeated with increasing exposures of 4, 8, 16, 32, etc., seconds until the whole strip has been covered.

It is not necessary to switch off the enlarger between the test exposures. The opaque card can be moved along the strip at the required intervals, so long as the enlarging paper is not touched.

The whole strip has now received a series of exposures of 2, 4, 8, 16, 32, etc., seconds. Thus each section of the strip receives twice the exposure of the section before it.

After processing the strip should show one correctly exposed step. Its exposure time is then the correct exposure for the final print.

Where the correct exposure lies between two consecutive steps, a second test strip may be made to cover this range only. Thus between 8 and 16 seconds, the intermediate steps would be 10, 12, and 14 seconds. This second strip pinpoints the exact exposure.

But if all the sections are too light, a further test strip must be made, starting at the last exposure on the old strip. If this gives an inconveniently long time for the correct exposure, the light must be increased by either

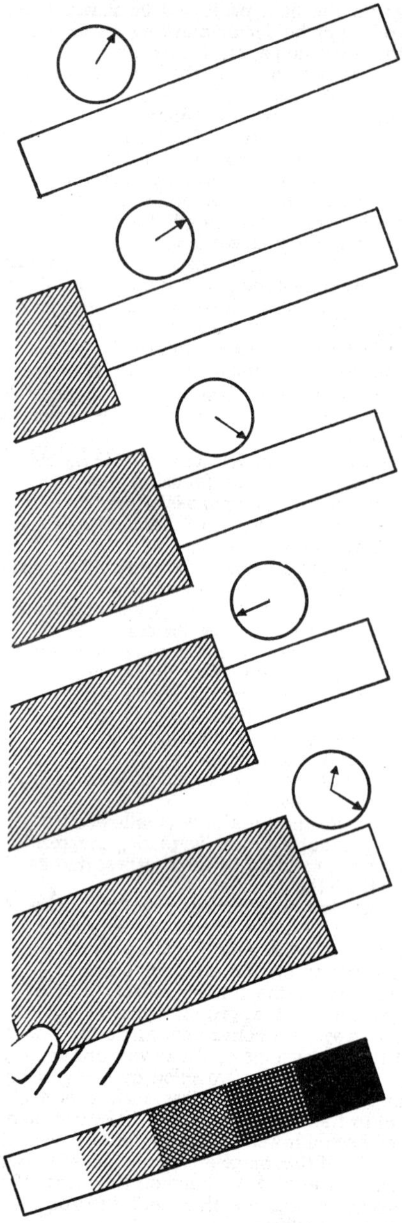

MAKING A TEST STRIP. Arrange a strip of the enlarging paper on the baseboard so that it covers a picture area with a wide range of image tones. Expose for 5 seconds, and cover up a small section of the strip with an opaque card. After a total of 10 seconds cover a further section of the strip, following up after 20, 40 and 80 seconds from the beginning of the exposure. After processing the strip should show one section of reasonably correct exposure. A further strip covering a narrow exposure range will pinpoint the exact time required.

increasing the lens aperture or using a more powerful lamp in the enlarger.

If all the sections are too dark, there is no point in making another test strip with exposures shorter than 2 seconds. Exposures as short as this are normally much too short to be timed conveniently or accurately. The solution is to use a smaller lens aperture or a weaker lamp, and adjust matters to give a correct exposure of suitable duration.

Where the approximate exposure is known from experience, or can be arrived at by other means, one strip covering a narrower range is sufficient. For instance, where the exposure is known to lie between 8 and 32 seconds, use test steps of 8, 10, 13, 16, 20, 26 and 32 seconds.

Step Wedge Tests. The test strip can be produced in one operation by using a step wedge in the shape of a strip, circle, or square of film printed with sections of progressively increasing density. The density of each section is arranged so that the paper under it receives a known proportion of the total exposure.

The wedge is placed over a piece of the enlarging paper on the base-board, and the negative image is projected on to it for a fixed time—e.g., 1 minute. After processing the test piece should show one correctly exposed section. Once the section is known, the equivalent exposure that it has received can be easily calculated.

On some step wedges used for enlarging, each step is printed with the appropriate fraction of the total exposure that it gives. Sometimes each step carries a code letter which shows on a calculator disc the exposure equivalent to any over-all exposure.

If all the steps on the exposed test piece are too light or too dark, the over-all exposure is increased or decreased accordingly.

Exposure by Photometers. A grease-spot photometer can be made to give the correct enlarging exposure and paper grade. For this purpose it must be used in conjunction with a set of specially prepared grey scales, one of which must be made for every grade and type of paper used.

To prepare a grey scale, the enlarger is first adjusted (with no negative in the carrier) to give a standard photometer brightness reading on the easel.

A strip of each enlarging paper is then given a series of exposure steps as for a test strip. In this case each step is given about 1½ times the exposure of the one before: 2, 3, 4, 6, 8, 12, 16, etc., to 360 seconds. After processing, the strip should show a full range of tones from white paper to completely black. If the steps are all too light or too dark, the brightness of the illumination used for the grey scales will have to be increased or decreased accordingly.

The grey scales can also be made by exposing the paper under a step wedge, although most enlarging step wedges do not cover a wide enough range.

The grey scale is mounted on a card, and each step is marked with a tone factor. The tone factor is calculated by dividing the standard light strength as measured by the photometer by the exposure in seconds that the section has been given.

The contrast range covered by the strip and full particulars of the make and grade of the paper are written on the card. The contrast range is determined by taking the tone factor for the lightest tone—the first grey step—and dividing into the tone factor of the first full black: i.e., the contrast range is the ratio:

$$\frac{\text{Exposure given to first full black}}{\text{Exposure given to first grey scale step.}}$$

Determining Paper Grade. The negative is focused in the enlarger in the usual way and the brightness of the lightest and darkest parts of the projected image is measured with the photometer.

The ratio of these two brightness readings is a measure of the contrast range of the negative.

The grade of paper needed to include all the tones of the negative must have the same contrast range as the negative. So once the ratio has been measured, it is only necessary to pick out the grey scale with the same contrast range; the grade of paper on which the grey scale is printed will be the correct grade for the negative in the enlarger.

When using variable contrast papers, determination of the paper grade in effect involves finding the best printing filter to use, or the most suitable ratio of exposures through the two maximum and minimum contrast filters.

Alternatively it is possible to use a contrasty paper grade for all negatives, and to soften the gradation by a controlled preliminary fogging exposure. This technique is known as flashing. It is fairly widely used in commercial photo finishing, where automatic measuring systems determine the amount of flashing required for a particular gradation. In the amateur darkroom, flashing is less useful, since the pre-exposure must be exactly determined, and also be uniform over the whole print area. The degree of flashing for any required print contrast there must be determined by careful tests.

Determining Exposure. To find the exposure needed, the photometer reading for the darkest part of the negative is divided by the tone factor for the first grey step on the grey scale. The very lightest parts of the negative will then print black, with all the intermediate tones naturally falling into place.

For example, say the readings of a directly calibrated photometer for a negative are 120 in the darkest and 10 in the lightest areas. The contrast range is 12 : 1, and a paper with a grey scale which indicates such a contrast range—probably a grade 3 or hard paper—would be chosen for the print. If its highest tone factor is, say, 8, the correct exposure time needed is 120 ÷ 8, or 15 seconds.

A new set of grey scales must be printed for each new batch of paper, because paper speeds may vary from batch to batch. And it is advisable to print new scales after every year or so that the paper is stored to allow for loss of speed with age.

Once the grey scales have been made, there is no need to make test strips or trial exposures. The method mainly relies on the accuracy of the photometer readings, and on rigidly standardized processing—i.e., full development in an accurately made-up standard developer. The same developer must be used for the grey scales as for the actual prints.

Use of Density Measurements. The correct exposure can be arrived at by measuring the range of densities in the negative by a densitometer, or specially adapted photo-electric exposure meter.

Density measurements made in this way can be used to determine both the exposure and the grade of paper needed to give an acceptable print from any negative. A set of standard test prints is made with a negative of known density and contrast. The standard prints relate the densities of the negative to actual print exposures at a given magnification and lens stop. From this information exposures at different magnifications and lens stops are calculated.

The relationship between the measured densities of the standard negative and any specimen negative is the same as the relationship between the exposure needed for any particular print in the prepared series and a similar print made from the specimen negative.

But the conditions in which such density measurements are made are quite different from those in an enlarger. The effective contrast range of the negative, for instance, is not necessarily the same as that measured by the densitometer. The difference depends largely on the enlarger illumination used; it may even call for a different grade of paper from that indicated by the densitometer readings.

Exposures at Different Magnifications. A change from one degree of enlargement to another demands a corresponding change in exposure time. The exposure must be increased when the scale of enlargement is increased, and vice versa.

The table below gives approximate values at different magnifications. It applies to normal condenser enlargers with semi-diffused lighting. With enlargers which have fully diffused illumination, the variation of the exposure time with the degree of enlargement is slightly greater.

Effect of Lens Stops. When the time for the correct enlarging exposure is inconveniently short for accurate timing—e.g., if it is 5 seconds or less—it can be extended by using a smaller lens stop. And exposure times at high magnifications can sometimes be shortened by opening up the lens aperture. But the change in exposure cannot be calculated by using the *f*-numbers marked on the aperture ring.

The optical system of some enlargers tends to direct more light through the centre than through the outer regions of the enlarging lens For this reason the light passed by the lens does not always vary in strict proportion with the size of the stop.

Some enlarging lenses have the stops marked in direct values: 1, 2, 4, 8, 16, in accordance with the actual measured amount of light that they pass at the various diaphragm settings. The numbers show the relative exposure at each stop without further calculation.

Where the lens stops are marked in *f*-numbers (which are not directly proportional to the light passed) the exposure times at the various stops can be found from the table above.

This table applies only to enlargers with semi-diffused or fully diffused illumination. With condenser enlargers which have a point

APPROXIMATE EXPOSURES AT DIFFERENT MAGNIFICATIONS

Relative Exposures in Seconds at Degrees of Enlargement													
1½×	*2×*	*2½×*	*3×*	*4×*	*5×*	*6×*	*7×*	*8×*	*10×*	*12×*	*14×*	*16×*	*20×*
—	—	—	5	8	12	15	18	25	36	50	60	80	115
—	—	—	6	10	14	18	23	30	43	60	75	95	140
—	—	5½	7	11	16	21	26	34	49	70	85	110	160
—	4½	6½	8	12½	19	24	30	38	58	80	95	130	185
—	5	7½	9	15	22	27	33	45	65	90	110	145	210
—	5½	8	10	16	24	30	38	50	72	100	125	160	230
—	6	9	11	18	26	33	41	55	80	110	140	175	255
—	6½	9½	12	19	28	36	45	60	85	120	155	190	4½ m.
—	7	10½	13	21	30	39	48	63	90	130	165	205	5 m.
5½	7½	11	14	22	32	42	52	68	100	140	175	220	5½ m.
5¾	8¼	12	15	24	34	45	54	75	110	150	185	240	5¾ m.
7½	11	16	20	32	48	60	75	100	145	200	250	5½ m.	7¾ m.
9½	13	20	25	40	60	75	90	125	180	250	5 m.	6½ m.	9½ m.
11	17	25	30	50	70	90	115	150	210	5 m.	6 m.	8 m.	11 m.
13	20	28	35	55	85	105	130	175	240	5¾ m.	7 m.	9½ m.	13 m.
15	23	33	40	64	96	120	150	200	4¾ m.	6¾ m.	8 m.	11 m.	15 m.
18	28	40	50	80	120	150	180	250	6 m.	8½ m.	10 m.	13 m.	18 m.
23	33	48	60	100	145	180	230	5 m.	7¼ m.	10 m.	12 m.	16 m.	23 m.
33	50	70	90	150	215	4½ m.	5¾ m.	7½ m.	11 m.	15 m.	18 m.	24 m.	33 m.
45	65	95	120	200	4¾ m.	6 m.	7¾ m.	10 m.	14½ m.	20 m.	24 m.	32 m.	45 m.
65	100	140	180	5m.	7¼ m.	9 m.	11½ m.	15 m.	22 m.	30 m.	36 m.	48 m.	65 m.

EXPOSURES AT DIFFERENT APERTURES

				Relative Exposure in Seconds at					
f 3·5	f 4	f 4·5	f 5·6	f 6·3	f 8	f 9	f 11	f 12·5	f 16
¾	1	1¼	2	2½	4	5	8	10	16
1¼	1½	1¾	3	3½	6	7½	12	15	24
1¾	2	2½	4	5	8	10	16	20	32
2	2½	3¼	5	6¼	10	12½	20	25	40
2¼	3	3¾	6	7½	12	15	24	30	48
3¾	3½	4½	7	9	14	17½	28	36	56
3	4	5	8	10	16	20	32	40	64
4	5	6¼	10	12½	20	25	40	50	80
4¾	6	7½	12	15	24	30	48	60	96
5½	7	9	14	17½	28	35	56	70	112
6½	8	10	16	20	32	40	64	80	130
7	9	11½	18	23	36	45	72	92	145
9½	12	15	24	30	48	60	96	120	190
12	15	19	30	38	60	75	120	150	240
16	20	25	40	50	80	100	160	200	5¼ m.
19	25	31	50	62	100	125	200	250	6¾ m.
24	30	38	60	75	120	150	240	5 m.	8 m.
30	40	50	80	100	160	200	5¼ m.	6¾ m.	10¾ m.
38	50	65	100	130	200	4¼ m.	6¾ m.	8¾ m.	13½ m.
48	60	75	120	150	240	5 m.	8 m.	10 m.	16 m.

source of illumination the differences are smaller, and also depend on the degree of enlargement.

Checking the Paper Grade. The approximate paper grade is decided beforehand when the negatives are classified, but it must be checked by actual trial.

Once the correct exposure has been found by the method chosen, a small test piece or even the test strip itself will show whether the grade chosen is suitable. Different grades of the same paper usually require different exposures, so a second test strip will usually have to be made to find the correct exposure for the new grade.

Automatic Exposure Systems. Various semi-automatic and fully automatic exposure measuring devices can simplify the job of determining enlarging exposures. Most of them are based on the measurement of the light reaching the enlarging paper, and frequently incorporate some means of switching the enlarger lighting off when the paper has received enough exposure. The measurement takes into account the full negative area, and so requires certain over-riding adjustments for predominantly light or predominantly dark negatives. Also, separate adjustment is necessary for different contrast grades and paper types. The measuring arrangement generally uses a system of photo-multiplier cells.

A simple method is to measure the light reflected from the enlarging paper during the exposure on to a photo-multiplier cell. This may be fixed to or near the edge of the masking frame and oriented so as to cover the whole paper surface. The sensitivity of the photo-multiplier cell circuit is adjustable to allow for various paper speeds, and after preliminary calibration this system yields reasonably reliable exposures. On switching on the enlarger lamp for the exposure, the photo cell starts to charge up a capacitor system. When the latter is fully charged, a leakage current activates a further circuit which switches the lamp off again. Exposure control units of this kind must also be adjusted for the reflectivity of the paper surface, but automatically take into account the lens aperture and degree of magnification. Generally the photo cell is set to cover a specific paper size.

A second method uses a special masking frame with a transparent or translucent top, with one or more photo-multiplier cells arranged underneath. These cells then measure the light coming through the enlarging paper, and again control the exposure accordingly. In addition to adjustments for paper sensitivity, these masking frames have to be calibrated for the transmission characteristics of the paper base (single or double weight, etc.) and also for the print size—where this is adjustable. Masking frames of this kind may include sets of photo cells behind red, green and blue colour filters and so control the exposure or the filtration when making colour enlargements.

A third system measures the light coming through the enlarger lens—either via a beam splitter built into the enlarger or by a swing-out photo cell placed in position for measurement only. In the first case a proportion of the light is constantly diverted to a photo-multiplier cell and again controls the enlarging exposure as before. In the second case the photo cell is used for measurement of the light intensity only, and the reading transferred directly or indirectly to a suitable timer. This system requires a calibration adjustment for the paper speed and also for the degree of magnification, since the light flux through the lens does not vary a great deal with a change of enlargement.

Making the Exposure. The exposure is conveniently made by switching the enlarger light on and then switching it off at the end of the required time.

The paper is placed in the paper holder or

masking frame and positioned with the lamp switched on and the orange filter in front of the lens. The lamp is then switched off, the orange filter swung out of the way, and the lamp switched on again for the exposure.

The most convenient way of controlling the exposure is by a timer wired into the enlarger lamp circuit. This is set to the appropriate exposure time, and the exposure started by pressing a switch or button. The enlarger light then remains on while the timer runs down, and automatically switches off at the end of the exposure. The automatic exposure control systems mentioned above include a timer (usually electronic) for this purpose.

With almost all enlargers there is a risk of vibration, certainly at high degrees of magnification. And if the enlarger head is moving in relation to the baseboard during the exposure, the picture will not be sharp. So it is advisable to allow time for any vibration to die down before switching on. For the same reason neither the enlarger nor the table it stands on should be touched or shaken during the exposure. Professionals, who use sturdy equipment and handle a great deal of work, expose by removing and replacing the orange filter, rather than switching the enlarger on and off. Continued switching is time-consuming, and also considerably shortens the life of the lamp.

Vibration can also be caused by people walking about in the room or on the same floor during the exposure (particularly if the darkroom is on an upper floor). It may even be necessary to cease enlarging temporarily while heavy traffic is passing in the street outside.

Shading and Spot Printing. If areas of the negative are too dark or light to print satisfactorily, the exposure that the paper receives may be selectively modified.

This is commonly done by cutting out cardboard of the desired shape and holding it over the paper during part of the exposure in the case of shading, or using it as a mask for additional local exposure in the case of spot printing. With both methods, the card must be moved slightly during the exposure to avoid printing as a sharp edge. The hands can also be used for shading.

With electronic enlargers shading can take place automatically to pre-determined degrees.

LARGE PRINTS

Most enlargers are designed to give enlargements of at least 12 × 15 ins. from the whole negative with the lamphouse at the top of the column. This size is as big as most amateurs normally wish to print, but occasionally they may want to make an enlargement bigger than this, or to enlarge only a part of the negative to this size. In either event, if 12 × 15 ins. is the normal limit of the enlarger, some special arrangement must be made to increase the scale of reproduction. There are two ways of doing this: by using a lens of shorter-than-normal focal length, and by increasing the lens-to-paper distance.

Short Focus Lens. The scale of enlargement can be increased by substituting, say, a 2 ins. (5 cm.) lens in an enlarger normally designed for a 3 or 4 ins. (7·5 or 10 cm.) lens. But it must be possible to bring the lens to within 2 ins. of the negative, and if the focusing scale does not permit this, the new lens must be fitted in a recessed adaptor on the lens panel. Often this arrangement means that the iris diaphragm control is out of reach of the fingers and a special adaptor has to be added to make adjustment possible.

But even when the negative can be focused, if the new lens is a normal type its shorter focal length will also mean that it will cover a smaller area of the negative. So the arrangement will not work if the whole of the negative is to be enlarged. To enlarge the whole negative, a wide angle lens would be required—i.e., one with a covering power greater than that of a normal lens of the same focal length.

The arrangement has other drawbacks—e.g., the focal length of the condenser will no longer accord with the new negative-to-lens separation, and the lamp may be too close to the condenser. In theory these factors upset the even illumination of the negative, but in practice the lamp can generally be adjusted, or provided with extra diffusion, to give reasonably uniform lighting over the whole area.

A 2 ins. (5 cm.) lens used in this way in an enlarger normally designed for a 3 ins. (7·5 cm.) lens would give a picture 1½ times bigger than normal each way—i.e., instead of a 12 × 15 ins. picture, the new size would be 18 × 22½ ins. (46 × 57 cm.).

Increasing Lens-Paper Separation. The height of the enlarger column sets the limit on the size of the enlargement that can be made with a given lens. But there is no limit to the size so long as the paper can be placed far enough away from the lens. Most enlargers are designed so that the enlarger head can be swung around through 180° on its column to allow the image to be projected off the baseboard for this purpose. The baseboard is weighted—e.g., with a pile of books—and the enlarger is placed near the edge of the table or bench with the head pointing down at the floor. The paper is then either exposed on the floor or, if the separation is too great, on a box or similar flat object.

Some enlargers are made so that the head can be turned to project the image horizontally. In this case the paper is either pinned to at board or sheet of card against the wall, or held in the type of easel manufactured for use with horizontal enlargers.

When big enlargements are made in this way, the risk of blur from vibration is greatly increased because the effect of relative movement between lens and paper increases in proportion to the separation and because lens and paper are independently supported instead of being on a single rigid structure.

The moral is simply to take every possible precaution against movement during exposure. If the lamp switch is on the enlarger instead of in series with the flex, there is always a risk that switching on will set the head vibrating. So it is wise to hold a piece of black card in front of (but not touching) the lens for some seconds after switching on the lamp, and then to make the exposure by uncovering and covering the lens with the card when everything has settled down. And unless the darkroom has a concrete or similar type of solid floor, no one should walk about during the exposure.

Exposure. The exposure for big enlargements is apt to be lengthy because the intensity of the light falls off in proportion to the square of the distance between the lens and paper—i.e., doubling the distance calls for a four times increase in exposure. During a long exposure there is a greater risk of accidental movement, and any light leaks or unsafe darkroom lighting are more likely to fog the paper. So it is as well to keep the exposure as short as possible by other means.

One way is to make the exposure at the widest lens aperture that will give sharp definition over the whole picture area. If only a part of the negative is being enlarged in the centre of the field it may not be necessary to stop the lens down as far as when enlarging the whole of the negative. This is because the definition of the lens over the central part of the field is fairly even; falling off only becomes serious towards the outer region of the field.

Another useful expedient is to increase the power of the light by fitting a more powerful high-efficiency lamp.

When using a light as intense as this in the enlarger lamphouse, greater care than usual must be observed in sealing off light leaks and covering up any light or polished surfaces that might reflect the glare from the printing paper back on to it again. It is also advisable to have the paper resting on a sheet of matt black paper to absorb any light falling outside the actual image area. Finally it should be remembered that it will be unwise to trust the red safe screen normally used in front of the enlarger lens for checking the image.

PROCESSING

When the print has been exposed under the enlarger, it only remains for it to be processed in dishes.

The sequence involved in processing the print is: develop, rinse, fix, wash, dry. The actual solutions employed may vary according to certain requirements, but in general they will consist of a standard print developer, plain water or an acid stop-bath for the rinse, and an acid-fixer for the fixing.

When dealing with large prints, a special procedure is used if the correct size dishes are not available for processing. L.A.M.

See also: *Colour printing technique; Combination printing; Converging verticals; Developing prints; Electronic printing; Enlarger; Flashing; Focusing (automatic) Giant enlargements; Photomurals; Shading and spot printing.*

Books: *How to Make Your Own Enlargements*, by L. Gaunt (London); *Enlarging*, by C. I. Jacobson (London); *The Complete Art of Printing and Enlarging*, by O. R. Croy (London).

ENLARGING BACK. Attachment which is fitted in place of the plate holder on a plate camera, enabling the camera to be used as an enlarger. The device consists of a negative carrier and an illuminating system including a diffuser and lamphouse. A modern version of this fitting employs a cold cathode grid illuminant which cuts down the bulk and eliminates the diffuser.

ENLARGING FILTER. Enlargers of all types are usually fitted with a coloured filter. This can be slipped on to the lens or brought in front of it so that the position of the image on the bromide paper can be checked before exposure without fogging the paper.

The filter is made of glass, gelatin, or plastic, coloured either a deep brown or medium red. The material may be fitted into a slip-on mount, like a colour filter, or it may be hinged or pivoted so that it can be swung into position without touching the lens (this is the better method as there is no risk of moving the lens).

Such filters are not safe over a long period; they are intended only for a brief check, nor should the image be focused on the paper with the filter in position because the lens may not focus the visible, non-actinic red rays and the actinic blue rays in the same plane.

The holder for the filter may allow the filter to be changed. This is useful when special filters are needed for other reasons—e.g., correcting colour balance when making colour prints.

Colour filters which slip over the lens of the enlarger are used when enlarging on variable contrast paper. This type of paper yields soft or hard print gradation according to the colour of the printing light. Filters for contrast control are normally of gelatine and are supplied in a cardboard holder, but for regular use they can be bought mounted between glass in metal mounts.

See also: *Colour printing technique; Enlarger; Variable contrast papers.*

ENLARGING LENS. The lens used in an enlarger must be capable of giving a critically sharp image with a perfectly flat field. A lens intended for colour enlargements must, of course, be fully colour-corrected. Good colour correction is still desirable for black-and-white enlarging as well, since chromatic aberrations lead to unsharpness. This is particularly important with enlargers using tungsten and cold cathode lighting alternately.

With enlargers working exclusively with mercury vapour or other lamps rich in ultra-violet radiation, the lens should be corrected for these short wavelengths. Otherwise the lens may have to be refocused after visual focusing (since the eye is most sensitive to the medium wavelengths of the visible spectrum). On some specialist enlargers this is done automatically, for example when a yellow viewing filter is swung out of the optical path.

Using the Camera Lens. While some camera lenses will perform reasonably well as enlarging lenses, they are never ideal. A camera lens is computed for short lens–film and long lens–subject distances. This is exactly opposite to the enlarger lens which works with long lens–film (paper) and short lens–subject (negative) distances. Sometimes an improvement in image quality can therefore be obtained by mounting the camera lens back to front in the enlarger, with the front (from the camera aspect) facing the negative.

A three or four element anastigmat is generally satisfactory for normal enlarging. If a camera lens is used to save the cost of a separate lens, it should be of symmetrical or almost symmetrical construction. At the same time, lenses of high-class miniature cameras are frequently used and give good enough results for all normal requirements.

Focal Length. The enlarging lens must be able to cover the full negative area with which it is intended to be used. Generally an enlarging lens of the same focal length as the normal camera lens (in other words roughly equal to the diagonal of the negative format) is satisfactory.

A slightly longer focal length—for example 60 mm. instead of 50 for 35 mm. negatives, or 90 mm. instead of 75 mm. for 2¼ × 2¼ ins. square formats and so on—is frequently popular. By not extracting the maximum covering power of the lens it is frequently possible to obtain a better average definition, especially at full aperture.

Where an enlarger is used for several negative sizes, for example 24 × 36 mm. and 2¼ × 2¼ ins., it is more convenient to have two enlarging lenses, one for each format. While the longer focus lens could be used for the smaller negative size, it necessarily limits the maximum degree of magnification obtainable from a given enlarger.

A short focus enlarging lens can on the other hand be used with larger negative sizes where very low degrees of enlargement or even reductions are required. Thus a 50 mm. enlarging lens can under these conditions be used safely with a 2¼ × 2¼ ins. or even a 2¼ × 3¼ ins. negative. At the great lens-to-negative distances required for same size copying or reduction the image angle covered by the lens is reduced.

Heating Up. When a condenser is used to focus the light from the lamp on the lens, it also concentrates the heat on it. This is one reason why it is better to use a lens specially designed for enlarging instead of one primarily intended for taking the photograph. Enlarging lenses are made without cemented components, whereas one or more glasses of a camera lens are frequently stuck together with cement which is softened by even a moderate rise in temperature. The risk to the lens can be eliminated by interposing a heat-arresting filter between the lamp and the condenser. A heat filter, composed of slips of Chance ON20 glass, safeguards the lens and is also a great help in preventing the heat from buckling or damaging the negative. This risk is greater with older lenses; some of the more modern types, even if cemented, use cements of greater heat resistance.

Maximum Aperture. The maximum aperture of the enlarging lens depends on the scale of enlargement it will be used for. Miniature negatives need very much more enlargement than, say, half-plate negatives, and so the lens used needs to be able to pass more light to keep exposure reasonably short when making big pictures. For this reason 35 mm. enlargers are commonly equipped with *f* 3·5 lenses, while enlargers for 2¼ square, or 2¼ × 3¼ negatives have *f* 4·5 lenses and half-plate and over sizes have lenses often no more than *f* 6·3.

Stops. At times it is necessary to stop the enlarging lens down from the maximum aperture, and since in the darkroom it is difficult to see the stop values, so-called click-stops are now common. The iris setting ring is fitted with a leaf spring which clicks into notches corresponding with the various aperture settings. Any value can thus be located by the feel or the sound of the clicks as the ring is turned from the fully open position.

The stop values on an enlarging lens are not usually calibrated as for a camera lens—i.e., *f*5·6, *f*8, *f*11, and so on—since this would mean that the first step on the majority of lenses would be less than a whole stop. This would be confusing, so the lens is sometimes marked 1, 2x, 4x, 8x, and so on, each step calling for twice the exposure of the one before.

In a condenser enlarger with a small light source, stopping down the enlarging lens does not necessarily bring about a proportional decrease in illumination on the easel. This is because the condenser frequently concentrates the light on to an area in the lens that is smaller than the maximum aperture. In other words, only a small central portion of the lens is being used. Until the area of the iris is reduced below the illuminated area of the lens, there is no significant reduction in the illumination on the easel.

Coated Lenses. All modern enlarging lenses are

now bloomed (coated on their glass-air surfaces with metallic fluorides) to reduce the amount of light lost by reflection and to prevent degradation of the image by scattered rays. When the lens is not coated, this stray light spreads an even veil of fog over the whole print which makes little difference to the shadows but greys the highlights and flattens the contrasts of the print. The increased contrast given by a coated lens is not enough to affect the paper grade which will be used, but it improves the definition of the print. This is because the contrast of the finer details is not blotted out by light scattered inside the lens.

See also: *Aberrations of lenses; Coated lens; Lens.*

ENPRINT. Fixed-ratio enlargement produced from the whole of a negative in an automatic large-output printer. Since the production of enprints requires little or no more time than that of contact prints, photo-finishers often offer such fixed-ratio enlargements at the same (or only slightly higher) prices as contact prints—and well below the price of normal enlargements. Originally enprints were offered as $3\frac{1}{2} \times 4\frac{1}{2}$ ins. enlargements from 24×36 mm. miniature negatives; nowadays similar fixed-ratio enlargements are available from a number of other negative sizes up to about $2\frac{1}{4}$ in. square and under various trade names.

See also: *Sizes and packings; Wholesale photo-finishing.*

EPIDIASCOPE. Projector which combines the functions of an episcope and a diascope in the one apparatus.

See also: *Projector (still).*

EPISCOPE. Optical projector for showing an enlarged image of small opaque objects, photographs and printed matter. By employing an episcope, a lecturer or teacher can show small objects conveniently to audiences.

See also: *Projector (still).*

EPSTEAN, EDWARD, 1868–1945. American photo-engraver. Writer and translator of articles and books on the history of photography, collector of books on photography and photomechanical work. Donated this collection to the Library of Columbia University, New York.

EQUIVALENT FOCUS. Distance between the nodal point and its corresponding principal focal point on the axis of a lens. Same as focal length.

EQUIVALENT WEIGHT. Term used in chemistry to denote the weight of any given substance that will either unite with one gram of hydrogen or displace it from a compound. The equivalent weight of any substance can be either measured or deduced, and it is possible, by equating the equivalent weights of the chemicals taking part in a reaction, to describe both the reaction and its products in quantitative terms.

See also: *Chemical calculations.*

EREMIN, YURIY PETROVICH, 1881–1948. Russian landscape photographer, from 1931 lecturer at the Moscow Architectural Institute and the Institute of Artistic Design. Documented Moscow's architectural reconstruction.

ERFURTH, HUGO, 1874–1948. German portrait photographer in Dresden from 1896 onwards, making only gum prints and later oil-pigment prints. In the 1920s he was the most celebrated photographer of the German intelligentsia. Moved to Cologne in 1934, where he was active until bombed out in 1943; his work survived in a bank safe.

ERNEMANN, HEINRICH, 1850–1928. German camera manufacturer. He induced the German government to endow a professorship of photographic chemistry at the Technical College in Dresden (filled by Dr. Robert Luther). His firm (and others) were merged, in 1926, into the Zeiss-Ikon A.G. In 1925 introduced the Ermanox camera with the Ernostar f1·8 lens, then the fastest camera.

ETHER. Diethyl ether; sulphuric ether. Solvent for many purposes, including making collodion emulsions.

Formula and molecular weight: $(C_2H_5)_2O$; 75.

Characteristics: Volatile pleasant smelling liquid. Very inflammable.

Solubility: Partially miscible with water.

ETHYLENE DIAMINE TETRA-ACETIC ACID. (E.D.T.A.) Used in developers, usually as the tri- or tetra-sodium salt, to prevent hard water precipitates. It is more stable than Calgon in strongly alkaline solutions. Also used in the compounding of combined bleaching and fixing baths (blix baths) from hypo and the ferric-EDTA complex. Sold under various trade names, e.g. Sequestrol (Geigy).

Formula and molecular weight: $(NC_2H_4N)(CH_2COOH)_4$; 292.

Characteristics: White crystalline powder. One of a group of chelating compounds which are able to form stable, soluble compounds with a number of metals such as calcium, magnesium, iron, etc. Certain metals, such as iron and copper, if present as impurities in developer chemicals, will tend to act as oxidation catalysts in the presence of E.D.T.A. and can cause accelerated exhaustion and aerial fogging.

Solubility: Sparingly soluble in acid solutions below pH 6 and more freely soluble in alkaline solutions.

EVANS, FREDERICK HENRY, 1852–1943. English professional photographer from 1898, previously a bookseller in the City of London and an amateur photographer for sixteen years. Member of the Linked Ring, 1901. Best known for his photographs of English cathedrals and French chateaux taken for *Country Life*. Also took many fine portraits of his artistic and literary friends. Gave up photography when platinum paper became unobtainable during World War I.

EVER-READY CASE. Carrying case, usually leather or plastic, which can be opened to give access to all the camera controls without being removed from the camera. The base of the case is generally fastened to the camera by a knurled screw which fits the tripod bush on the camera body and is threaded to take a tripod screw.

See also: *Case*.

EVERSET SHUTTER. Generic term for all shutters of the type—common on cheaper cameras—in which one movement of the release control tensions the shutter operating spring and then fires the shutter so that no separate cocking control is needed.

EVIDENCE BY PHOTOGRAPHS. The purposes for which photographs are produced in evidence in legal proceedings are broadly:

(1) To provide a record of something that it would otherwise be difficult or impossible for the Court to see—e.g., scenes of serious crime as they appear before anything has been disturbed; vehicles which have been involved in collisions, photographed before any repair or salvage has been done; roads, buildings and other distant places which may be many miles from the Court; injuries which will have healed and be no longer visible by the time of litigation; copies of documents, the originals of which cannot be made available.

(2) To present the results of an experiment or test (in many cases made in a laboratory) that it is not practicable to demonstrate in Court—e.g., photomicrographs; photographs taken by ultra-violet fluorescence; photographs illustrating some test or measurement; comparison photographs demonstrating the similarities between finger-prints, footprints, tool marks, fired bullets or cartridge cases, etc.

(3) To show something that only photography can reveal (infra-red photographs and photographs taken with unusual filters).

This article is not concerned with how such photographs are taken; it deals only with their use as evidence in the Courts.

Presentation. There must be no handwork on either negatives or prints to be used in evidence. Simple shading control during printing is permissible, and often indeed essential. But there must be no spotting, even where the filling up of a pinhole or two could not possibly lead to any misrepresentation. It is a legal axiom that not only must justice be done, but that it shall be manifest to all present that justice is being done. In the same way photographs should not only be free from faking; it must be clear that there could be no possibility of faking.

First and foremost, this calls for clean working in the darkroom. But if in spite of it the negative or print shows a flaw, the flaw must remain, and must be disregarded or explained in Court as need arises.

There should be no reconstruction of the scene for photography. No attempt should be made to replace anything inadvertently displaced. The photographer should simply record what is in front of his camera and explain, if necessary, any discrepancy between what the photograph shows and what witnesses describe.

There is no room for ingenious technical tricks in this field; the mere fact that photographs are being used to illustrate points of evidence implies a belief that the camera cannot lie, so all work should be as straightforward as possible. (Photographs produced under laboratory conditions of subjects normally invisible to the human eye are, of course, excepted). And the photographer should always assume that, while the members of the Court may know everything about the law, they know little or nothing about the technicalities of photography.

The following example illustrates this point: in a legal dispute over a road accident, one side produced photographs showing the view which the defendant would have had round a bend if he had been driving on his correct side of the road. The pictures were taken with a lens of normal focal length (i.e., 45–50° angle of view). The other side objected that there were only thirteen white-line dashes visible in the pictures, although it was agreed that there were in fact seventeen dashes between the camera position and the turn of the road. It required considerable explanation to establish that the photographs were a fair and accurate representation of what the defendant should have seen, and that no photograph could ever show all the foreground in front of the camera position unless the lens had an angle of view of 180 degrees.

Similar difficulties can also arise from the use of photographs of scenes artificially lit—e.g. streets at night. A lay court will expect the photographs to show neither more nor less than an eye-witness could have seen. They will not usually be able to appreciate the effect of the cumulative sensitivity of the photographic emulsion, by which the amount of detail recorded in a photograph of a dimly lit scene depends entirely on the exposure time.

Court Procedure. The usual procedure for taking the evidence of a photographer witness runs something like this. First he takes the oath (unless he has conscientious objections to doing so, when a form of declaration may be substituted) and gives his name and profession or official status. He then states where, when and of what he took the photographs, that he made prints, which he produces, and that the negatives are in his possession. (He should take the negatives with him for inspection if requested, but they are rarely asked for.) He will then describe what the photograph, or each photograph of a series, shows. Finally, he may be cross-examined to bring out any point that the other side feels will be to their advantage. (The "other side" simply means the side that did not call him as a witness).

There is no recognized standard type of print for use in evidence. Single prints may be mounted on thin card with a brief descriptive caption. A series of photographs may be printed with a wide margin on the left-hand side, and then numbered and bound together bookwise in a light cover with a page of descriptive captions at the beginning.

The size of prints which should be produced depends on two things: perspective considerations, and the fact that they will be examined in the hand. The perspective should be as far as possible correct when the picture is held at the normal viewing distance. Prints should also be of a size convenient to hold. The size meeting both requirements best is generally something between 5 × 4 ins. and 12 × 10 ins.

The number of copies produced depends on the type of Court hearing the case. Where there is no jury—e.g., in a Magistrates' Court, County Court or before a Coroner sitting without a jury—four copies should suffice: one copy for each side in the case, one for the Bench and one for the witness himself to refer to. In trials by jury twelve copies are needed.

Captions. Captions must be rigidly factual and descriptive; there must be nothing to imply the conclusion which one side or the other is seeking to establish. In a road accident case it would be proper to caption a picture:

"View along the Weston–Eastwich road, taken from point A looking in the direction of B." A caption which added, ". . . showing that B is clearly visible from the correct side of the road at A," would be objected to by the Court, and the photograph might be declared inadmissible as evidence.

With photographs of a more technical type, such as are used to illustrate comparisons of finger-prints, tool marks, etc., it is permissible and indeed often necessary to indicate the important points by means of, for example, ruled lines or arrows. But if any of the indicating lines unavoidably obscure any detail, even unessential detail, there should always be a plain unmarked copy of the photograph for reference. H.J.Wa.

See also: *Crime photography; Forgeries; Police photography; Traffic photography.*
Book: *Photographic Evidence*, by C. C. Scott (Kansas City).

EXAMINATIONS. A few British professional and educational organizations hold annual examinations in photography. These examinations are usually intended for professional photographers but amateurs are also eligible. Such examinations confer certain diplomas or similar qualifications; they are not, however, needed by any person for carrying on photographic work.

The most important photographic examinations in Great Britain are those of the Institute of Incorporated Photographers and the City and Guilds of London Institute. Employers normally accept a pass in either of these examinations as evidence of ability and some firms make it an essential qualification.

Institute of Incorporated Photographers. The examinations of the I.I.P. are held in three grades: preliminary, intermediate and final, and they must all be taken in that order. The standard required for each grade is high and carries a practical test in addition to the normal written paper. The preliminary candidate must have an elementary knowledge of photography and be able to demonstrate it in practical tests. The intermediate examination requires a more thorough knowledge and practical ability in the general techniques of photography. The final examination may be taken in any of four different sections as follows: Commercial, Industrial and Advertising; Portraiture and Pictorial; Scientific Applications of Photography; Medical Photography. A high theoretical and practical standard is required in the section taken. In special circumstances candidates may be exempt from the preliminary and intermediate examinations.

Candidates must be capable of expressing themselves clearly and have a good general education. Past results show that many candidates fail because they lack normal ability in these respects.

City and Guilds of London Institute. These examinations fall into two grades—intermediate and final. The standard is approximately the same as that for the I.B.P. examinations. The intermediate consists of a written paper and a practical test; the final comprises a common written paper, a second written paper in either of two basic sections, and a practical test. A successful candidate who obtains a first-class certificate in at least one section of the final may later qualify for a full technological certificate by passing suitable external examinations in chemistry, physics and (sometimes) mathematics.

Other Examinations. There are examinations of a similar nature in other countries. More common, however, are the internal examinations of the various schools and colleges of photography throughout the world. These examinations are generally recognized in the absence of tests conducted by independent professional bodies. They follow a general pattern, but for internal examinations, attendance at the school is normally essential. In Britain, the examination at one of these schools carries exemption for the I.I.P. preliminary.

Various specialized bodies also hold examinations in their own fields—e.g., radiography, photo-mechanical printing and the photographic retail trade, etc.

As a suitable qualification, employers will sometimes accept Fellowship or Associateship of one of the leading photographic societies or institutes. This may again entail an examination, but normally the award is based on inspection of specimens of the photographer's work submitted, and on his past achievements.

Before offering himself for examination, the candidate should make sure that his experience is adequate. He may have taken a course in photography, not recognized by the examining body, which may not have covered an essential aspect. All examining bodies publish a syllabus of their examinations and sometimes copies of previous test papers. A close study of these will show the candidate if any gaps exist in his studies.

American Practice. There is no standard examination system comparable to that existing in England. The closest thing to it is the Civil Service Job classification breakdown which calls for tests in proficiency in the various categories. As a rule, these tests are based upon local arrangements for examinations; there is no completely standardized national procedure for this. J.D.C.

See also: *Careers in photography; Training and education in photography; Training of photographic technologists and scientists.*

EXHAUSTION OF SOLUTIONS. The useful life of developing and fixing solutions is limited by a number of factors: the active ingredients become used up or they automatically decompose in the working solution, the solution becomes contaminated with impurities produced by the decomposition of the active ingredients, and certain solutions (developers) are oxidized by contact with the atmospheric air or air dissolved in the water.

Nothing can prevent the active ingredients from being used up during the actual processing, but the other causes of exhaustion can often be lessened by taking suitable precautions.

Water used in making up developers should be boiled for about five minutes to get rid of dissolved air. Chemicals should be bought in the smallest pack that will serve normal needs—it is not economical to buy in bulk if the chemicals deteriorate through being stored for a long time in an opened container. Most developers, silver salt solutions and ferric salts keep better in dark—e.g., amber or brown—bottles which prevent the contents from being decomposed by actinic light rays.

Developer stock solutions stay fresh longer if they can be kept out of contact with the air. There are several ways of doing this:

One method is to float a film of paraffin on the top of the solution and to draw off the developer through a glass tube that dips under the surface. In this case fresh solution must be added before the oil falls to the level of the open end of the tube.

Another method is to drop glass marbles into the bottle as the solution becomes used up so as to bring the level up to the neck of the bottle. This reduces the surface of the liquid in contact with the air in the bottle and so retards oxidation.

With modern plastic bottles it is easy to squeeze the bottle itself to reduce the air space to a minimum. The bottle cap must of course fit tightly. Plastic bottles are less recommended for long-term storage owing to their porosity.

See also: *Developers; Fixing.*

EXHIBITIONS AND SALONS. Photographic salons and exhibitions are either confined to members of a particular society, or are exhibitions which are open to all photographers anywhere. Practically all active photographic societies run an annual exhibition of their own members' work either on the society's own premises or in some local hall. The bigger societies which promote open and international exhibitions always arrange for an adequate display in local halls or art galleries. Such exhibitions to be successful and gain general acceptance must comply with rules laid down by The Royal Photographic Society in this country or by The Photographic Society of America in the United States. The conditions are becoming fairly well standardized throughout the world so that entrants may be assured of a fair deal.

Briefly, it is stipulated that there must be no restricted classes, that the accepted entries shall be properly shown in suitable premises and that conditions should allow for the proper display of at least 150 prints. Exhibitors are thus assured that their prints will be suitably displayed if accepted.

The generally accepted conditions are:

(1) Entries shall be limited to not more than four prints from any individual (an exception to this is the London Salon which allows six prints).

(2) All pictures shall be submitted to the judges or selectors.

(3) The exhibition is not limited to special types of photographer—e.g., to men or women professionals, miniature photographers, or any other arbitrary classification less than the entire photographic body.

(4) The promoters must publish and furnish a complete catalogue of prints accepted and hung.

Entry forms must be provided and as names and addresses of successful entrants are nearly always published in the exhibition catalogue, regular exhibitors are in the habit of receiving entrance forms from exhibitions all over the world. Would-be exhibitors who have not yet been so successful can find from the photographic press regular notes about all the important exhibitions at home and abroad.

Mounts. Most exhibitions limit the size of the mount to 16 × 20 ins., although a few such as the London Salon will accept up to 20 × 25 ins. 16 × 20 ins. is a good standard size and it is the size that should be adopted by entrants who wish to send their pictures to a number of different exhibitions. This size will always be accepted and much smaller mounts are not generally desired. If, for example, the pictures are displayed under glass it is necessary that the mount should fit the size of the glass used and this is generally made to cover the 16 × 20 ins. mounts.

Entrance Fees. It is customary to ask for an entrance fee—e.g., five shillings or a dollar—to cover the expenses of running the exhibition. One of the few exceptions to this practice is the Annual Exhibition of The Royal Photographic Society for which no entrance fee is required. People sometimes object that exhibitions should not charge an entry fee as they are asking for work from exhibitors, but on the other hand very few exhibitions charge an admission fee and there is the cost of the hire of the hall or gallery to pay, the cost of printing the catalogue, and many other incidental expenses. In any event, few societies are able to make an exhibition pay, and practically all are prepared to lose money and look upon the cost as good publicity for the society.

Organization. Organizing a photographic exhibition is no easy task. Once it has been decided when it is to be held and the necessary hall has been booked, entrance forms have to be sent out all over the world, and the necessary publicity notices have to be sent to the photographic press. Arrangements must also be made for the reception of the parcels, their unpacking, checking, and counter-checking with the entrance forms. The importance of adequate space for this work is often overlooked, for it is by no means unusual for a society to receive several hundred parcels of 16 × 20 ins. prints and anything from 1,000 to 1,500 individual pictures. As all of these pictures will have to be returned in the original wrapping, a proper system has to be worked out for unwrapping, marking the packing paper to correspond with the numberings on the print, checking the entrance forms, taking out the postal orders, cheques and other forms of entry fee, and arranging the pictures for the judges or selectors. It is most important for all the wrappings to be safely stored for the month or two which will elapse between receipt and re-dispatch.

Arrangements have to be made to notify suitable selectors in advance of the judging date and provide accommodation for them. In most cases a panel of selectors is chosen, consisting of three or perhaps five members so as to make a majority vote possible. On the day of judging, facilities must be provided for placing the prints in proper lighting in front of the judges and for recording their decisions. Once the selection has been made, the forms have to be marked and the pictures transported to the place where they are to be hung. Arranging the pictures on the walls is a highly skilled task which should be entrusted to someone with experience because the appearance of an exhibition can be made or marred by the way in which the pictures are displayed. Pictures which are likely to conflict with one another must be kept well apart and a general harmonious arrangement of all the prints aimed at. Once the order of pictures has been decided upon they must be numbered and a catalogue prepared. The catalogue must then be printed and dispatched to successful entrants at once. It is customary to have the exhibition opened by some well-known person and, if the society is to benefit from the publicity of this exhibition, necessary arrangements must be made to have it reported in the local newspapers and the national photographic press. P.W.H.

See also: *Judges for exhibitions; Judging exhibition entries.*
Books: *Exhibition Prints*, by A. D. Bensusan (London); *Photographic Exhibitor's Reference Book*, by F. Harris (London).

EXPEDITION PHOTOGRAPHY. Expeditions, no matter what their nature or purpose, must always be recorded comprehensively by a competent photographer. All the requirements of the organizers of the venture must be met—in still and cine, and in black and white and colour. The photographer must further bear in mind possible uses for picture and film material by the popular press, for television, and for lectures. Photographs frequently prove excellent fund-raisers, to help finance either the venture they depict, or future expeditions.

It is essential to gain a clear idea, well before the expedition leaves, of the precise nature of the photographic records which are required, and the particular purposes to which they will eventually be put. Adequate documentation and illustration is an essential part of the responsibility of the explorer to posterity.

A small venture can afford to leave its photography in the hands of one who is primarily there as a geologist, naturalist, or doctor, and who regards the taking of pictures as just another job, but the high standards demanded today make the inclusion of an expert cameraman essential in any expedition of importance. It may even be found necessary to have one man for stills work, and another for movie work.

It is prudent to obtain a clear understanding (preferably in writing) before departing, of the copyright position; normally copyright of all photographs taken on the expedition should belong to the expedition as a whole, and would be vested in the leader, or the financial sponsor at home. Agreement should be reached at an early stage on the procedure to be adopted if, for instance, the doctor wanted to use his own unofficial pictures to illustrate a lecture, or a book on his experiences; preliminary discussion and formal decision on such matters before departure can avoid much hard feeling later.

EQUIPMENT

The time and trouble spent on selecting, designing and testing the equipment to be taken will be amply repaid on location; repairs or modifications then are likely to prove difficult or unreliable, or even impossible. Early discussion with the manufacturers on the best type of equipment for the conditions likely to be encountered is strongly recommended.

Complete re-lubrication of all camera moving parts might be necessary if sub-zero temperatures are to be met and this is a task for the expert. Lack of this precaution may cause shutters to become sluggish, or not to work at all, when very cold; lens mounts and aperture adjusting rings may become impossible to turn, because of the great increase in viscosity of ordinary lubricants below zero.

Modifications in camera design may also be found advisable. Some controls, for instance, cannot be manipulated with any ease by thickly-gloved hands, and this difficulty must be overcome.

If work is to be done at extreme temperatures, whether high or low, attention should be paid to dimensional clearances in the film path, or the film movement mechanism. The coefficients of expansion of dissimilar materials are unlikely to be identical, and adequate clearances at + 20° C. may prove entirely inadequate at − 40° C. This is the type of problem which the well-informed manufacturer would be able to handle effectively. Special machining of the interior may well prove essential, if the equipment is to work reliably.

Even if the equipment is of the highest standard, it is still essential to test it under conditions as near as possible to those likely to be encountered.

Practical experience with cameras which have been cooled to sub-zero temperatures should be gained if an expedition to the Arctic is proposed—and several hours must be allowed for the cooling procedure, as it takes a surprisingly long time for the interior of a camera to get really cold.

Even more important is the testing of any apparatus to be used under water. Pressures well in excess of the maximum likely to be met should be imposed upon all equipment, as a substantial safety margin is advisable. This precaution should be observed with all equipment to be taken under water, and not merely with the photographic apparatus.

Cameras and other instruments to be used in caves must be protected from moisture. This applies particularly to any electrical gear.

Expedition conditions are frequently such that the photographer works under considerable physical and mental strain. At extreme temperature human reactions tend to be slow, and at high altitudes the relative lack of oxygen impedes clarity of thought. When submerged, the photographer has so many other things to think about that there is a grave risk of his attention becoming distracted from his main purpose. For these reasons it is important that the camera controls should be simple to use. As much as possible of the operation should become mechanical, leaving little for the photographer to do, except to select the subject and press the button. To this end, much thought in simplifying the controls of the instrument before departure will be amply rewarded.

SENSITIZED MATERIAL

Careful consideration should be given as early as possible to the types of sensitive material to be employed, and again the experienced guidance of the manufacturers should be enlisted.

At low temperatures, the sensitivity of the emulsion decreases, and its colour balance may alter; provision for dealing with this first alteration is simple, if the extent is known. The shift of colour balance is more difficult to predict, and a series of practical trials may be necessary; the appropriate correcting filter for any given circumstances may then be determined.

At high altitudes, the ultra-violet content of the light tends to be much higher than near sea level, and adequate U.V. filtration is important.

Another difficulty to be expected at low temperatures is the tendency of the film base to stiffen, and become brittle, and even to snap if sharply stressed. This factor should be remembered when selecting equipment; cameras which employ sharp bends in the film track will almost certainly lead to trouble, and modifications in design may become necessary.

In hot climates changes in sensitivity, and in colour balance, are again to be expected, but a much greater problem arises through the inherent instability of the emulsion under hot and humid conditions. At above 40°C., emulsion stocks can become useless in a few weeks; they produce the faults which are normally associated with out-of-date films, i.e., high fog-level, low emulsion speed, low contrast, and low maximum density and, with colour films, a pronounced colour shift. The obvious precautions are to keep the films in their original sealed tropical wrappings as long as possible, to store them after use in a dry atmosphere (by the use of silica gel, in a closed container) and to maintain them, both before and after exposure, at as low a temperature as can be achieved.

This advice to keep emulsions cool may seem vain in an arid desert at 50° C., but much may be accomplished by the use of wide-mouthed vacuum jars, or even by burying the stocks in the sand, as it is very much cooler two feet below the surface. The location must, of course, be carefully marked.

If adequate cooling and prompt processing cannot be assured, then it would be unwise to attempt to make any use of infra-red material, as this has notoriously poor keeping qualities under these conditions.

The relative instability of the latent image in any emulsion kept at high temperatures must never be over-looked. Since apparent under-exposure is the most obvious result of this instability, the experimentally-minded photographer, if faced with the utter impracticability of cool storage and prompt processing, might care to investigate the effects of deliberate over-exposure. It would, however, require luck to gain complete satisfaction in this way.

PROCESSING

The precautions to be taken after exposure are also of great importance, and plans should be made, if possible, to process films with the minimum of delay, and to subject them to minimal temperatures during the inevitable storage period. The provision of a temporary darkroom at base should be considered, before the expedition departs. It is highly desirable to make arrangements to fly exposed colour films home, for the more carefully controlled treatment they demand. However, a technique has been worked out, for certain stocks, which enables the first development to be performed in the field, so that the effects of subsequent storage become relatively unimportant.

Provision should be made if possible, to process an occasional test-film on the site, to enable both the condition of the material and also the technical quality of the photography to be assessed at regular intervals. If films cannot be processed on location, and all material is sent home for development, then the expedition photographer should be sent regular reports on the condition and technical quality of his work. D.H.O.J.

See also: *Archaeology, Caves, Cold weather, Geology, Hot weather Processing, Marine photography, Mountains, Polar photography, Tropical photography, Under-water photography.*

Book: *Photography on Expeditions*, by D. H. O. John (London).

EXPIRY DATE. Date stamped on cartons of roll films and packings of 35 mm. film to indicate the date by which the material should be developed. Generally this is 2–2½ years from the date of coating for black and white materials, and about 1½ years for colour materials; the manufacturer's guarantee of the film does not normally extend beyond the expiry date, although the material can generally be used for considerably longer.

See also: *Keeping qualities of materials.*

EXPLODED VIEWS. Photographs of machines or similar constructions dismantled into their parts and laid out so as to correspond as closely as possible to the way in which they are assembled.

Exploded views of articles consisting of an assembly of smaller parts have a special function in engineering. They show clearly how separate components are assembled, and indicate details of internal construction that may not be obvious in a complete unit. In advertising, exploded views help to emphasize important selling points about the way the article is made up and are of special value for instructional and educational use in the form of illustrations in literature or as lantern slides or film strips.

Drawings (line or wash) do not always carry the conviction of a genuine photograph; but on the other hand, special care must be taken when using a camera in order to avoid such faults as distortion, heavy and featureless shadows and wrong relative size of components. A number of small errors of this sort may collectively render an otherwise good photograph useless as a functional illustration.

It is as well for the photographer to possess some technical knowledge of the type of subject to be exploded, and special skill is called for in arranging and lighting the groups to make the picture tell the right story.

General Arrangements. The final print must show the exploded parts and nothing else. A plain background must be used and the components arranged on it in the order of assembly to avoid an horizontal line produced by the join of a vertical background with the table-top. The main item should be placed in such a position in relation to the camera that the other parts can be added with as little overlapping as possible.

Side elevations (sometimes used in drawings) are generally avoided because of the way in which the perspective of successive parts increases with their distance from the centre. As an example, a washer directly in front of the camera and edge-on appears as a straight line; when additional (and similar) washers are added on each side, the images progressively become more circular. In practice the centre washer can be turned slightly (technically known as "cheating") to show the third dimension, and the more distant washers can be turned in the opposite direction to lessen the effect.

It is generally necessary to suspend small parts by fine threads (that can be touched out afterwards) or use small pieces of clay or plasticine to fix them in position on the background. Sufficient separation between parts must be left to show their individual identity; but by varying the distance between certain groups it is possible to indicate the order of assembly. Sufficient perspective should be given to show the third dimension without causing one part to hide another, and the assembly as a whole should not recede in too steep general perspective or it may falsely indicate a reduction in size for the more distant parts. It is common practice to include all small screws or bolts, not as if they were suspended in mid-air, but arranged in groups near their functional position, so that their use is easily inferred.

Lighting. One of the commonest failings in photographs of mechanical components is the introduction of conflicting shadows by the use of complicated lighting arrangements. Forward shadows (back lighting) are sometimes effective, but in general, hard lighting must be avoided and special care taken to ensure that no shadow links one part with another. As a rule, soft frontal lighting is safest as it is often impossible to cancel out shadows from one side by balancing the light from the other. In a few cases it may be necessary to light the group by reflected light from a large sheet of white paper, or by turning the lamps on to a white ceiling. When dealing with polished metal surfaces it may be helpful to photograph the subject within a tent constructed from tracing paper or other translucent material, otherwise any polished surface that does not reflect light will appear black. Shadowless lighting can be achieved by placing the parts on a sheet of translucent material which is illuminated indirectly from below—i.e., by reflected light from a white sheet of paper below this surface.

Camera. For this work a camera with a complete range of camera movements is essential. Skill in the use of camera movements can be effective in overcoming difficulties in focus and perspective. Longer-than-normal focus lenses are most helpful up to a point, and tend to flatten perspective, which, in most cases, is an advantage. Normal and shorter-than-normal (wide angle) lenses increase perspective and are notoriously unsatisfactory when used close to small subjects arranged on the table-top.

One other cause of failure is sometimes produced by vibration. In general it will be necessary to employ a small lens stop, which in turn will mean longer exposures. Even the smallest amount of subject movement can spoil such an assembly, and it pays to take a great deal of trouble to ensure that both camera and subject are rigidly supported so that relative movement is impossible. **B.A.**

See also: *Industrial photography.*

Books: *Photographing Machinery*, by B. Alfieri (London); *Basic Photography*, by M. Langford (London).

EXPLOSIONS. In the study of explosions, photography can show what happens at the centre: the initiation point and speed of detonation; the velocity of the gaseous and solid products. Photography can also measure the blast: the speed, shape and behaviour of shock waves; their propagation from variously shaped charges; and the effects of their collision or concentration.

Of the precautions necessary to protect operators and equipment, the most obvious is to get as far away as possible and use telephoto lenses.

This may be many miles for nuclear explosions, a remotely-controlled camera, being mounted in a tower, protected from blast, and photographing through a telescope. Distances can be shorter for conventional explosives, while for close work the explosions are contained within blast walls or pits and observed through periscope systems. Extreme close-ups are possible with miniature charges, housed in metal or reinforced wooden boxes with replaceable armoured-glass windows.

The photographs are generally taken by multiple exposures in still or lenticular plate cameras or with high speed framing cameras taking up to several million frames per second. Rates of expansion may be measured with streak cameras.

Most explosions produce light, which can be used for illumination but is often a disadvantage since the flash obscures the subject. It is usual therefore to underexpose for the explosive flash by using small apertures and high shutter speeds, then to flood the subject with light from other sources, so that the non-luminous gases, solids and shock waves are properly recorded.

These sources must obviously be powerful and may comprise electronic flash, argon bombs, spark gaps and rare gas discharge tubes.

A single flash, used with a high-speed framing camera, must have a flat uniform output so that all the frames are evenly exposed. In open flash methods, where flash duration determines exposure time, this may be under 0·01 microseconds, several flashes, controlled by delay timers, producing multiple exposures.

Smoke obscuring an explosion can be penetrated by a high-intensity X-ray flash beamed through the subject from a point source to make a shadowgraph on a film opposite. Here the flash heads are protected behind heavy wooden "windows" transparent to X-rays, while instead of a camera, only an X-ray film holder is needed, designed to blow away without damage.

Several other methods produce shadowgraphs, and often involve simplified schlieren techniques to record blast waves. They may be made by incident light from behind the explosion or by reflected light against a screen. Blast can also be observed by the apparent distortion of straight lines or posts, due to the differential refractive index of air at the wave front.

Where camera shutters are called for, electro-optical devices such as the Kerr cell provide the microsecond exposures required. They are not completely opaque, so are usually supplemented by rapid-action mechanical shutters; or opening immediately before the explosion, another closing afterwards.

Such systems require precise programming and synchronization, both with the explosion and the flash. A typical sequence is for the shutter to initiate the explosion which then triggers the flash.

Alternatively, the explosion may trigger the shutter, and the shutter the flash. Photocells, explosive switches, shutter synchronizing contacts and electronic timers provide triggering impulses. P.B.M.

See also: *High-speed cinematography, High-speed photography; Schlieren photography.*

EXPOSURE

Exposure in the scientific sense is the product of the intensity of light and the time during which the light acts. In terms of practical photography it means the particular combination of lens aperture and shutter speed that must be chosen to give a good negative.

The aim of the photographer is almost always to achieve a satisfactory print or transparency by giving the least possible camera exposure. This may be for three reasons: to get a print of the best possible quality by keeping the actual value of the exposure low; to keep the duration of exposure short, to arrest movement; and, by keeping the lens aperture small, to give the greatest possible depth of field. The last reason does not always apply: there are times when a large aperture is used deliberately to reduce the depth of field.

In some instances the brightness range of a subject is too wide to be recorded by the photographic process. This applies in particular to colour photography: in a scene with a wide range of tones only a limited group will be reproduced in correct colours—hence the recommendation of choosing "flat" subjects for colour photographs. Where it is impossible to do anything to reduce the brightness range of such a subject by suitable lighting, the photographer must make up his mind which group of tones he wants to come out in true colours and then expose for that group.

An under-exposed photograph lacks detail in the areas corresponding to the shadows; over-exposure causes the highlights to become flat and "burnt out".

In black-and-white photography by the negative-positive process, there is a range of exposures that will give satisfactory prints, reproducing quite faithfully the tones of the original scene. This range is a measure of the latitude of the emulsion. But the art of correct exposure is to give the least possible exposure that will show the required amount of detail in the shadow portions of the final print. Resolution will then be at a maximum, and grain at a minimum. Hence the old saying "expose for the shadows and the rest will take care of itself".

Reversal—and in particular colour reversal—materials have much less latitude than those used for negative-positive work. For direct colour transparencies there is only one correct exposure, and deviations one stop higher or lower give only just acceptable results.

Blackening Effect. Apart from the above term, the scientific definition of exposure denotes the total amount of light falling on the unit area of the photographic material, i.e., the product of the intensity of the light and the time during which it acts. The important thing is to know how much exposure produces how much blackening after development.

In general the blackening of the image increases with increasing exposure, but it is not always constant for the same total exposure. When sensitive materials are given the same total exposure throughout a range of varying values of time and light intensity, it is found that for each material there is one combination of time and intensity that produces most blackening. A reduction in intensity and corresponding increase in time, or an increase in intensity and a corresponding reduction in

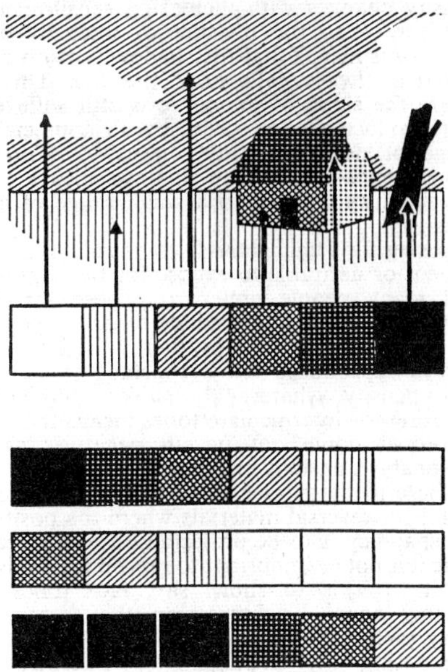

EXPOSURE AND TONE SCALE. *An average subject will, with correct exposure, record every tone value by a corresponding reversed density on the negative. Under-exposure records only the highlights, and all shadow tones remain clear without detail. Over-exposure records the shadows, but all brighter tones become uniform black.*

time, will decrease the blackening. In other words changing the intensity and time from their optimum value, lowers the photographic sensitivity. This effect is called the reciprocity failure.

The blackening also changes if the exposure is given intermittently. This happens because an intermittent exposure works in the same way as a long exposure at a lower intensity. If the same exposure is given continuously to one piece of material and intermittently to another, the total time taken by the exposure is greater for the intermittent case and the average intensity is decreased. The result is similar to an exposure given at lower intensity and as a rule causes the sensitivity to decrease. The effect is important in the design of instruments for sensitometry.

Exposing for Shadows. The light image that falls on the sensitive material in the camera consists of a series of patches of different degrees of brilliance. All these patches receive the same time of exposure, but because the intensity of the light on them varies, they are reproduced as areas of different densities on processing. These densities are faint for the shadow portions and heavy for the highlights.

When camera exposures are correct, the shadows in all scenes are reproduced as areas of similar low densities. On the characteristic curve which shows how the negative density changes with increasing sensitometric exposure—i.e., the product of intensity and time—this means that the shadow portion of a negative always falls near the toe. This is where the slope of the curve is still sufficient to reproduce shadow detail by an appreciable range of densities in the negative. Correct camera exposure for the shadows means that the shadows are tied to a small range of low densities.

Exposing for Highlights. It is not always convenient or desirable to expose for the shadows and many people prefer to expose for the highlights or the intermediate tones. Exposing for the highlights means that a highlight is always reproduced at a constant (relatively high) density, whatever the subject. Similarly, exposing for intermediate tones means that the mid-tones come out on all negatives as a reasonably constant medium density.

These methods of exposure are particularly useful for reversal materials where the positive transparency is to be projected on to a screen. Here it is not as important to reproduce shadow detail as it is to show, say, face tones at roughly equal brightness on the screen in every picture.

Exposing for highlights or intermediate tones rather than for the shadows is also useful where the tone range is too wide to be taken care of by the print. In such cases it pays to sacrifice detail in the shadows rather than in the medium tones and highlights.

Before the photographer can choose between exposing for the shadows or the highlights, he must be able to estimate or measure the tone range of his subject. What follows is concerned with the method of using various devices to determine the correct exposure for any particular set of conditions.

Estimating Exposure. With modern photographic black-and-white materials, a standard exposure of 1/50 second at *f* 11 will give a reasonable negative for a surprisingly large number of amateur snapshots. This is because most of such photographs are taken in bright sunny weather and the subject is generally in the nature of a group of people in the open or something involving streets and buildings.

With average subjects of this sort, any variation in lighting is usually well within the latitude of the film, so one standard exposure will serve on most occasions. A few simple and easy-to-remember rules cover the occasions when the standard exposure is not suitable. Thus: very early or very late in the day, multiply the standard exposure by ten; in shade, multiply it by two to four; in bright open places like the beach, or on mountains, divide it by two to four.

Rough and ready rules of this sort cannot guarantee first class results every time, but when they are combined with a little experience they will nearly always give a negative that can be made to yield a satisfactory print.

Indirect Methods. Under this heading come calculators, tables, and all methods of fixing the exposure without direct reference to the actual subject in front of the camera. It is possible to fix the exposure in this way because the strength of the light at any particular place and time is roughly the same every year.

So it is possible to give a reasonably accurate figure for exposure once the following facts are known: the geographical latitude, the time of the day and year, the weather, the type of subject, and the speed of the sensitive material.

The geographical latitude determines the height of the sun, and thus the amount of light that reaches the subject. This figure can be stated accurately for any particular place.

The time of the day and year also determines the height of the sun. The time of year is usually given simply as the month. The time of day should be given in solar time and not in British Summer Time or any similar artificial time. Local times may vary from solar time by as much as three quarters of an hour since it is generally inconvenient for local time to vary from one part of a country to another. The errors introduced into the estimation of exposure by using the wrong time of day are small around midday but may become quite large when the sun is low.

The weather is the remaining factor that decides the strength of the light falling on the subject. This cannot be measured accurately and depends upon how the individual photographer interprets a number of conditions.

The brightest possible weather condition is a clear sun with white clouds. The clouds help to reflect light to the earth's surface, thus lightening the shadows. Rather less light comes from sun in a clear blue sky which gives very strong shadows.

Shadows are one of the most important helps in estimating weather conditions. They are deepest for the "Clear sun", slightly weaker for the "Sun and white clouds" which do not obscure the sunlight, definitely watery for "Hazy sun" and absent for both "Dull" and "Very dull" weather conditions. The last two classifications probably cause more trouble than the others because they cover a much greater range of possible light values than any of the others. Only experience can teach which one to apply.

These factors—latitude, time and weather—govern the light that falls on the subject.

The subject, according to its reflecting powers, governs the amount of light reflected into the camera to form the image. So the subject is classified according to its light-reflecting properties.

There are two current ways of classifying the subject. In one the subject is classified by its character: "open sea", "distant landscape", "landscape with light foreground", "landscape with medium foreground", "close up subjects", etc. In the other, the subject is classified by its tone: light, medium, or dark, sometimes its distance from the camera, and often the direction of the light falling on it. The existence of these two subject scales is responsible for much of the difficulty that people experience in using exposure tables, charts and calculators.

The user of an exposure chart or guide must remember that the same subject at different distances from the camera and under different conditions may require different exposures. Thus, a figure in silhouette against the sky some distance away is interesting for its shape, and not for its shadow detail, for the simple reason that the eye looking at the original scene is not capable of resolving the detail. But the same figure closer to the camera will call for a longer exposure because in this case there will be visible shadow detail to record.

There is another reason why a distant figure requires less exposure. A certain amount of light is always scattered by haze in the air between the distant subject and the camera. This scattered light tends to lighten the shadows and to destroy the detail in them, both to the human eye and the photographic material.

The direction of the light may also have a considerable effect on the final exposure. Exposure against the light must always be heavier than exposure with the light, unless the light source itself forms part of the against-the-light scene, e.g., in a sunset. When photographing directly with the light, the shadows are invisible both to the eye and the camera. This is a high-key photograph for which exposure is always relatively small.

By contrast, when photographing directly against light, all the shadows are directed to the eye and the camera, so shadow detail is extremely important and exposure must be relatively heavy. There is one exception to this rule: when photographing a silhouette effect, shadow detail is not only unimportant, it is undesirable, so the subject needs only a short exposure.

Direct Methods. The above methods arrive at the exposure without actually measuring the strength of the light falling on or reflected from the subject. But provided that certain precautions are observed actual measurement of the light gives more accurate results. There are several ways of measuring the light that exposes the sensitized material in the camera:

(1) Actinometers measure the strength by the time the light takes to darken a test patch of sensitized material to a standard depth.

(2) Extinction meters relate the strength of the light to the faintest image that the eye can see through a neutral density wedge.

(3) Integrating exposure meters measure the cumulative effect of the light on a photo-electric cell over a period.

(4) Photo-electric meters indicate the light strength in terms of the electrical signal generated by a photo-electric cell directed either at the subject from the camera, or at the light source from the subject.

(5) Photometers measure the relative brilliance of a typical area of the subject.

The use of these instruments to determine exposure varies according to their different principles and is discussed under the individual headings.

Calibration. There are many reasons why it pays to go to some trouble to calibrate an exposure meter. The first is that no two meters, even meters of the same make and model, can be expected to respond in exactly the same way. The same can be said with even more truth of the camera shutters which eventually decide the exposure. Finally, individual photographers have their own ideas of what is a well-exposed negative.

The meter should be used to find the exposure for a series of photographs of widely different subjects. Each subject should be photographed at one quarter, one half, twice and four times the indicated exposure.

Inspection of the developed film will show at once which value of the indicated exposure gives the desired type of negative. If the exposure required is consistently greater or less than the indicated reading, the easiest way of making the correction is to adopt a correspondingly higher or lower speed figure for the negative material. This applies in particular when using automatic cameras.

The meter must be calibrated separately for

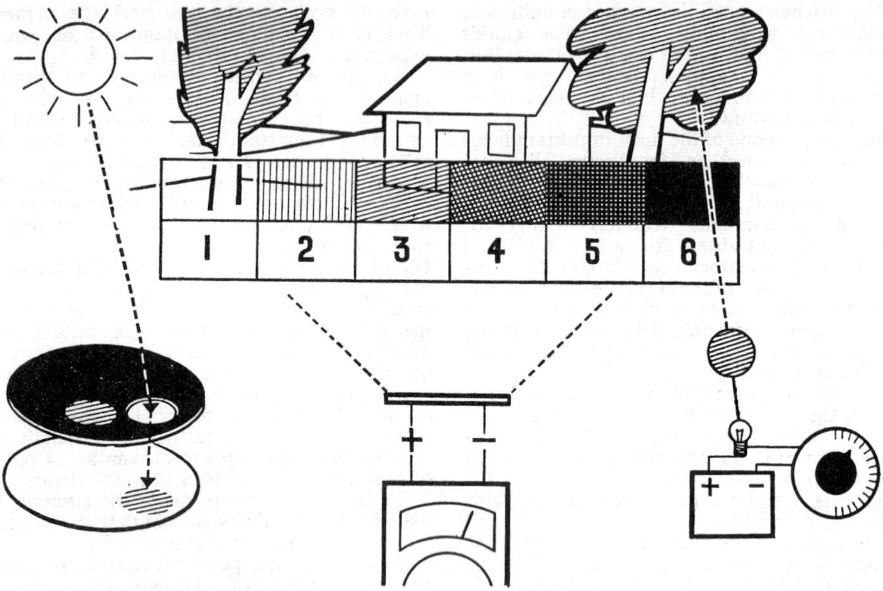

EXPOSURE METER SYSTEMS. Left: *Measurement of strength of light by an actinometer. This establishes the time taken by a piece of sensitive paper to darken under the influence of light so that it matches a predetermined depth of tone. Correct exposure is derived from this time.* Centre: *Measurement of average brightness of light reflected from subject by means of an extinction meter (observation of detail through a scale of numbers of increasing density), or by a photo-electric meter.* Right: *Measurement of brightness of individual subject areas by matching against them an illuminated area in a photometer. Adjustment of lamp illuminating the comparison area gives value for brightness. (The lamp is calibrated periodically to provide a zero point.)*

any of the other methods of use—incident light, substitution (i.e., taking readings on near-by objects of the same tone as a more distant subject), etc. It must also be re-calibrated if the photographer makes any change in the developer or development technique.

The speed figure finally adopted should allow for the fact that an under-exposed negative is no good, whereas an over-exposed negative will at least give an acceptable print. So, for general work, it is wise to adopt a speed figure of one half of that which would, according to the calibration, give the exact exposure. This recommendation is good for black-and-white negative and positive materials and even for colour where separate negatives and positives are used. It does not apply to reversal materials and normal colour materials, where only the exact exposure is admissible.

Minimum Correct Exposure. For best possible image quality the minimum exact exposure is desirable even with black-and-white negative materials. This applies especially to miniature and ultra-miniature camera films on the one hand, and to materials of very high sensitivity on the other. While under-exposure obviously leads to loss of image detail in the shadows, over-exposure beyond the minimum level necessary (in other words making use of the exposure latitude) tends to coarsen the image grain, and decrease definition. It does so by bringing out more of the image in the depth of the emulsion and so increasing the visible effect of light scatter and irradiation.

For these reasons the above-mentioned safety factor (adopting one-half of the speed figure that would give the exact exposure when setting the film speed on the exposure meter) is not in the interest of image quality, but only of convenience.

Such a safety factor is equally undesirable with ultra-speed negative materials. In this case the same image deteriorating factors already mentioned play their part to an enhanced degree. Further, the purpose of an ultra-speed film is to permit the shortest possible exposures in unfavourable light conditions. Hence the use of a speed reducing safety factor is a waste of versatility.

At one time standards of film speed rating and of exposure meter calibration between them more or less automatically led to exposures with such a safety factor. Methods of film speed measurement were internationally revised from 1960 onwards, and current standard film speed ratings, when used with a normally calibrated exposure meter, yield virtually the minimum correct exposure for the material.

Development Technique. The type of developer and the technique of development influence the effective speed of the sensitized material—

and hence the exposure it needs. Some fine grain developers, for instance, greatly reduce the effective speed and may require four times the normal exposure to be given. The exposure must be made with this fact in mind.

Exposure (Light) Values. The result of using any exposure calculator or meter is to obtain a number which indicates the camera exposure necessary to get a satisfactory picture with a material of given speed. This number has been called the exposure value.

The exposure recommendation embodied in an exposure value reading can be translated into a range of lens apertures and shutter settings—e.g., 1/25 second at *f*8, 1/50 second at *f*5·6, or 1/100 second at *f*4. These combinations all represent the same exposure value or photographic exposure.

It is decidedly desirable to be able to vary either shutter speeds or lens apertures in such a manner that the exposure value is automatically kept constant. This is realized in a number of modern shutters. Here a special exposure-value control is set to the number which has been obtained either from an exposure table or a suitably calibrated meter. This control is coupled to the diaphragm. Once it has been set, any change in the shutter speed setting will automatically alter the diaphragm so that the total amount of light passed on pressing the shutter release remains constant. Thus the shutter cannot be re-set without altering the diaphragm opening, but the diaphragm can be changed independently of the shutter, thus providing the same degree of flexibility as with the normal types of shutter.

This cross-coupling feature of the exposure value system survives in cameras with built-in coupled exposure meters. As, however, the process of reading off and transferring an exposure value is eliminated, the exposure value scale becomes unnecessary. The meter setting control simultaneously sets also a correct aperture-speed combination; alternative combinations are then selected as before by the cross-coupled controls.

Choice of Speed-Stop Combination. All systems of exposure measurement give exposure recommendations in the form of a series of shutter speeds and lens apertures. The photographer must make the final decision about which particular combination of aperture and shutter speed he will use: a small aperture and a slow speed, or a large aperture and a fast speed? There is no meter that will give the answer. Provided that the effective exposure remains constant, any one of a number of speed-stop combinations can be chosen.

There are two conflicting requirements to be reconciled. One is the need to stop down the lens to gain the utmost depth of field; the other is the need to keep the exposure time short so that a moving object or camera shake will not blur the picture.

When a camera is focused for a certain distance, objects in a zone extending both in front of and behind the plane on which the camera is focused also appear reasonably sharp. This zone is the depth of field.

The depth of field is governed, among other things, by the size of the lens aperture; the smaller the aperture, the greater the depth of field. So in the final speed-stop combination, the stop must be small enough to give a depth of field that will include all the objects that must appear sharp in the finished photograph.

But if the object is moving during exposure, the image on the film will also be moving all the time the shutter is open. This means that no matter how quickly the shutter opens and closes the image is bound to be blurred. But the faster the shutter speed, the smaller the blur will be. So in deciding on the speed-stop combination, the speed must be fast enough to reduce the blur to such a small amount that it will not show in the final print. The permissible amount of blur is decided by the diameter of the circle of confusion.

Finally, if the camera is to be held in the hand, there is the need to guard against camera shake. For the majority of photographers and cameras, this means that to get a really sharp picture, the shutter speed should be 1/50 second or even shorter. Some people can hold a camera steady for as much as 1/5 second, but this ability is rare.

The stop and speed finally chosen must reconcile the above—not always compatible—

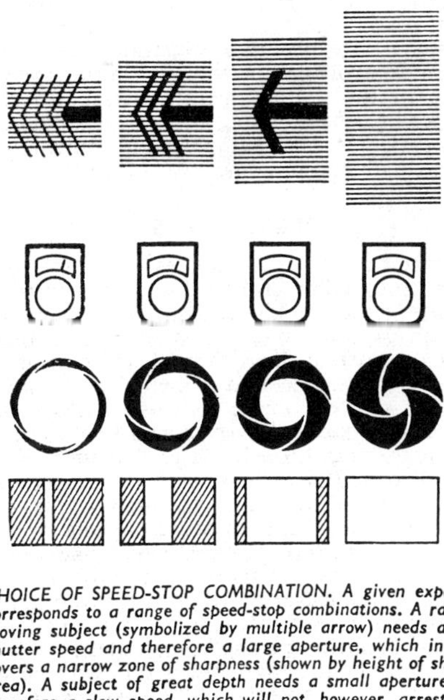

CHOICE OF SPEED-STOP COMBINATION. A given exposure corresponds to a range of speed-stop combinations. A rapidly moving subject (symbolized by multiple arrow) needs a fast shutter speed and therefore a large aperture, which in turn covers a narrow zone of sharpness (shown by height of shaded area). A subject of great depth needs a small aperture and therefore a slow speed, which will not, however, arrest fast movement. Choice of best settings may need a compromise.

requirements either by a compromise or by deliberately ignoring one of the factors.

Certain simple cameras with coupled meter relieve the photographer of this worry by providing only one aperture-speed combination for each exposure setting—usually a compromise which would be used in the majority of cases. Cameras with such a so-called programmed shutter may not even have aperture or speed scales at all, and make picture taking technique simpler—at the expense of some versatility in special conditions, where exceptional depth of field or high speed may be needed. W.F.B.

See also: *Actinometers; Aperture; Characteristic curve; Contrast; Exposure meter; Exposure tables; Exposure value; Extinction meter; Photometer; Reciprocity law; Sensitometry; Speed of sensitized materials.*

Books: *Successful Exposure in Photography*, and *The Right Exposure Every Time*, both by L. A. Mannheim (London); ***Exposure***, **by W. F. Berg (London);** ***Exposure Meters and Practical Exposure Control***, **by J. F. Dunne (London).**

EXPOSURE AUTOMATION

The photographic process depends on the action of a specific amount of light on a sensitive material. This quantity of light can be measured by various means. So it is not surprising that designers of photo equipment have tried at various times to make light measuring devices directly regulate the mechanisms and adjustments that control the exposure in a camera or other photographic system. Exposure automation is the term applied to set-ups which do this purely by the action of light and without direct interference by the operator. (The latter must however make appropriate calibrating adjustments to set the system for specific operating conditions such as the speed of the sensitive material employed.)

This characteristic distinguishes automatic exposure control systems from semi-automatic ones, where the exposure is set by the physical adjustment of a measuring device to take an exposure reading.

Systems of exposure automation may in a camera control the aperture or shutter speed setting or the shutter operation itself. In an enlarger or other printing equipment an automatic system is usually employed to control the printing light or in some cases a mechanical shutter.

STILL CAMERAS

Aperture Control. The simplest form of exposure automation in a still camera controls the size of the lens diaphragm by the current from an exposure meter. Where in a normal meter the current flows through the coil of a measuring instrument, it is here led through one or a pair of minute coils associated with feather-light metal leaves. According to the current flowing through the coil, these leaves are pulled across each other to a greater or lesser extent. The leaves themselves are so designed that their movement results in a larger or smaller aperture through which the light reaches the lens. As more light falls on the meter cell, the leaves move in such a way that the opening formed by them becomes smaller; with less light the opening becomes larger.

Exposure control systems of this kind have been fitted in inexpensive still cameras as well as cine cameras. They allow only a limited amount of calibration adjustment for film speeds (see below) and usually are associated with a single speed shutter.

Servo—Control Setting Systems. A more complicated but also more versatile approach is to use a mechanical scanning system to determine the position of the indicating needle of an exposure meter built into the camera. This scanning mechanism is then linked to a shutter or aperture setting drive.

In practice this is how such a system works. When the photographer begins to press the release button of the camera, a stepped cam moves down over the exposure meter needle. The end position of this cam depends on the position of the meter needle, and hence on the light falling on the photo cell. Further pressure on the release button then frees a spring which drives the aperture setting ring until it is stopped by the scanning cam. The point at which the aperture setting ring—controlling the iris diaphragm—stops depends in turn on the position of the scanning cam and hence directly on the amount of light which reached the meter cell. Alternatively, the camera release button may directly drive the aperture setting ring until the scanning mechanism stops it.

A number of variations of design exist. Thus

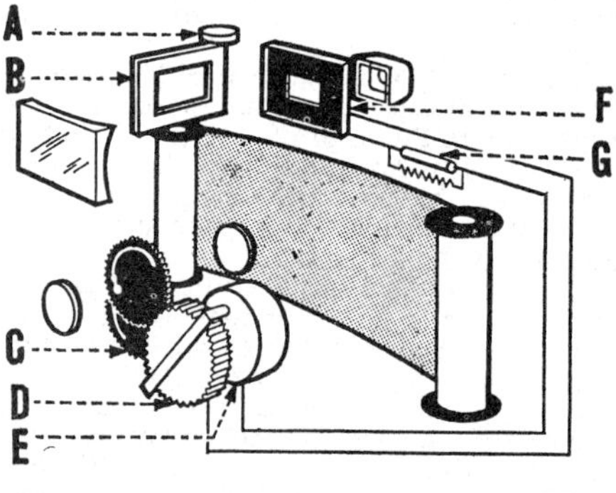

SIMPLE ELECTRIC EYE STILL CAMERA SYSTEM: A, *film speed setting control; places a mask* B *in front of meter cell to cut down light for slower films.* C, *aperture plates worked by meter gear* D, *to increase or decrease light going through lens.* E, *meter coil.* F, *meter cell inside finder.* G, *thermistor and resistor circuit to compensate for temperature variations.*

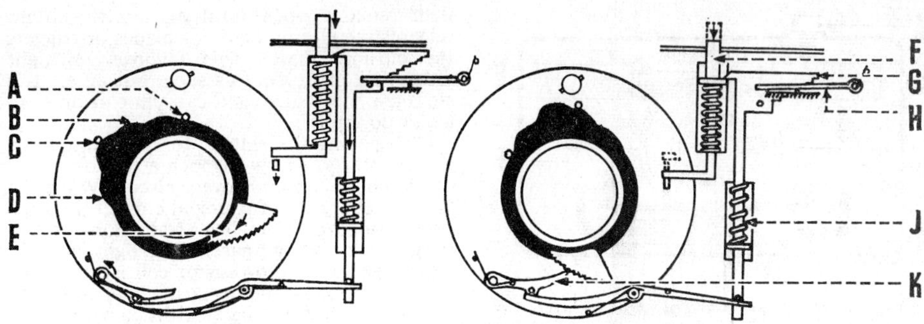

AUTOMATIC SELF-POWERED SHUTTER SETTING. 1. Before release. 2. During releasing, before the exposure. Pressing the release actuates a spring which makes the scanning segment of the exposure meter lock the meter needle in its position. The movement of the control cams for shutter speeds and apertures is arrested in a position corresponding to the needle position. A, cam follower for shutter speed settings. B, shutter speed control cam. C, cam follower for aperture settings. D, aperture cam. E, arresting segment of programming ring. F, release button. G, scanning segment of exposure meter. H, meter needle. J, spring of scanning segment. K. lever linkage engaging programming ring.

the servo drive may be linked to both the aperture and the shutter speed adjustment (as in programmed shutters), or it may adjust the shutter speed when the aperture setting range has reached its limit (or vice versa). More elaborate, but less reliable, servo systems have also been proposed. In one the exposure meter was linked to a thin plate that partially covered a perforated plate and so obstructed the flow of air from a kind of dashpot to a greater or lesser extent. This air flow was then used to regulate the shutter setting.

Yet another system uses a small electric motor to drive a setting pointer to a position where it lines up with the meter needle. The same motor then also drives the aperture and shutter speed adjusting rings to set a correct exposure time. Motorized servo systems are however more common in cine cameras.

Calibration Adjustments. Like a straightforward exposure meter, automatic control systems need adjustment for films of different speed. In the simplest designs this can be done by putting neutral density filters in front of the meter cell or by obscuring part of the cell surface and so adjusting the effective sensitivity.

In servo systems involving a coupling between the exposure meter and the exposure setting mechanism, the calibration adjustment is built into the coupling.

In such systems it is frequently possible to pre-select a shutter speed and let the scanning mechanism adjust only the lens aperture. In the simplest case the shutter speed selection itself also acts as a film speed setting: the camera works with slow shutter speeds for the slowest films and with the highest shutter speeds for fast films. Where there is a shutter speed adjustment on top of the film speed setting, this merely involves having a wider adjustment range to accommodate the two.

Exposure Signals. Since a camera with auto-

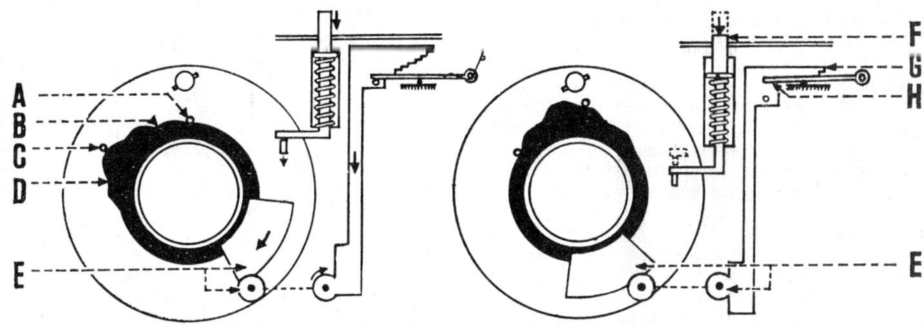

AUTOMATIC SHUTTER DRIVEN EXPOSURE SETTING. Pressing the release drives the shutter cam, controlling the aperture and speed settings, until the scanning cam of the meter is arrested by the meter needle. 1. Before release. 2. During releasing, immediately before the exposure. A, cam follower for shutter speed settings. B, shutter speed setting cam. C, cam follower for aperture settings. D, aperture cam. E, gear sector and gearing. F, release button. G, needle scanning segment. H, exposure meter needle.

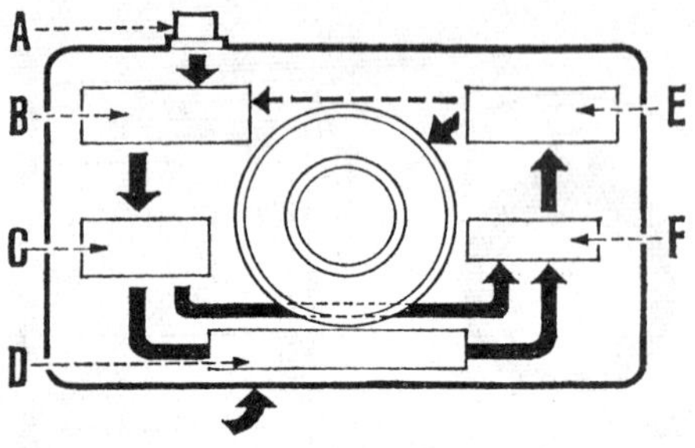

SCHEME OF DIRECT EXPOSURE CONTROL OF SHUTTER. On pressing the release a suitable device (e.g. a release magnet) opens the shutter. It also initiates light measurement via a photo cell. When an appropriate amount of light has reached the cell, a further control element closes the shutter. A, release. B, release magnet. C, impulse generator or other initiating system for light measurement. D, photo cell. E, closing mechanism of shutter. F, measuring unit of photo cell—for instance a switch comparing the impulse direct from the generator C with a second impulse weakened by the photo resistor cell.

matic exposure control sets itself according to the prevailing light, some indication is necessary for the user to show whether the light conditions are adequate for a correct exposure. This is usually done by coloured signals visible in the camera viewfinder. When a red signal appears, or persists on taking the first pressure on the release button, the light is insufficient for the exposure range provided and no correct exposure is possible. If the signal is green or turns green on pressing the button, the prevailing light conditions are suitable for a correct exposure.

On cameras permitting a pre-selection of shutter speeds, a red signal can be turned green by choosing a suitable slower shutter speed. In some cases the signal is given by a pointer visible in the camera finder, which moves from a green to a red sector (or vice versa), and may also indicate the actual lens aperture selected by the automatic mechanism. Alternatively, a green and red aperture scale may move past a fixed index mark.

Yet another method of providing an exposure signal is the blocking of the release button when the light is unsuitable for a correct exposure.

In conditions where the automatic system is incapable of setting a right exposure, either flash or time exposures are usually possible by disengaging the automatic control. With some cameras such disengagement permits the setting of any non-automatic exposure combination.

Direct Shutter Control. While the systems so far described rely on light measurement with an instrument which is coupled with the aperture and/or shutter controls, a more direct approach to shutter operation is also possible. Here the current from the photo cell is used to regulate the closing of the shutter, after the latter has opened for the exposure. The exposure commences by the opening of the shutter, which then remains open until an electric circuit releases the closing blade or blades or triggers the closing phase. This involves no light measuring instrument as such, but an electric storage system such as a capacitor or an electronic detector.

On opening the shutter for the exposure a current starts to flow through a photo-resistor cell to charge up the storage circuit. When the latter is charged, an overspill current (in some cases running through a transistor system) triggers the closing phase. If the light intensity falling on the photo-resistor cell is great, the charging period—and hence the interval before the closing phase is triggered—is comparatively short. In poor light the resistance of the photo-resistor increases and the current takes longer to charge up the electronic storage system, thus leading to a long exposure time. In practice, control systems of this nature can cover exposure times from 1/1000 second (or even shorter) to several minutes in a continuous range.

If the measuring cell is situated so that it receives the light through the camera lens, the exposure time automatically allows for the lens aperture in use.

The charging time can also be varied by appropriate adjustment of the electric circuit; this provides for different film speed settings. Where the photo-resistor cell does not receive the light coming through the lens, a switching adjustment has to be linked with the aperture control so as to select different response sensitivities at different aperture settings.

In view of the cumulative action of such an exposure system, it can respond to very low lighting levels. Hence the usual stop/go signals are rarely necessary. Instead a special signal may be provided to warn the user that the exposure time will be long enough to necessitate the use of a tripod.

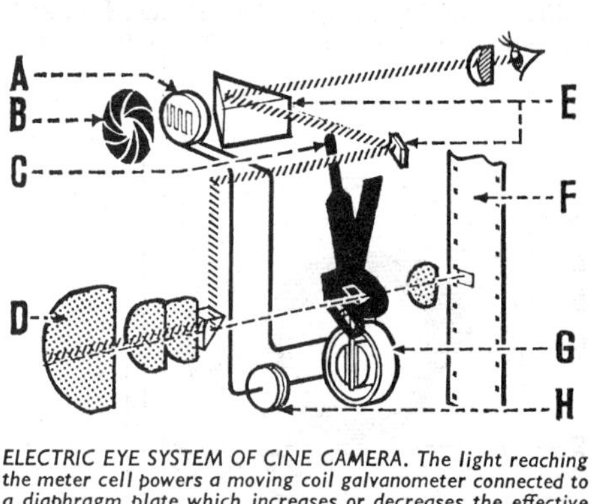

ELECTRIC EYE SYSTEM OF CINE CAMERA. The light reaching the meter cell powers a moving coil galvanometer connected to a diaphragm plate which increases or decreases the effective lens aperture. A pointer forming part of the plate may be visible in the viewfinder, and indicate the actual aperture selected by the system. A, cadmium sulphide cell. B, iris diaphragm (controlling light reaching cell, used for film speed setting). C, pointer. D, camera lens. E, finder system. F, film. G, galvanometer. H, mercury battery.

CINE CAMERAS

Since the light reflected from the subject may change during, for instance, a panning shot the electric eye system should monitor and adjust the camera exposure setting continuously. This is perfectly straightforward with direct aperture control systems and to a more limited extent with motor-controlled servo systems.

Direct Aperture Control. Most of the simpler automatic or "electric eye" cameras utilize a straightforward aperture control system relying on the movement of light-weight blades with special cut-out slots. These blades are moved in front of the lens (or between the lens elements) by a coil or coils energized by the current passing through the meter cell. This system is therefore similar to that described for simple still cameras. There is usually a limited adjustment for different film speeds, achieved by partially obstructing the meter cell to a greater or lesser extent. Sometimes a range of framing speeds of the cine camera may also be accommodated in this way. Where a cadmium sulphide photo resistor cell is used, the adjustments may also involve the electric circuit.

Where the photo cell receives the light coming through the lens (usually by a beam splitter somewhere in the optical path or a reflecting shutter), linking the cell to the aperture control provides an automation system which also corrects its own errors. As the light falling on the photo cell depends partly on the light passed through the lens aperture, any tolerances or inaccuracies in the aperture calibration and in the mechanical linkages directly affects the meter cell as well. This will then readjust the aperture accordingly.

Motorized Servo Systems. In a number of cine cameras the current from the exposure meter circuit is balanced against a reference current in a bridge circuit. Any unbalance actuates a servo motor which drives the aperture adjustment to increase or decrease the lens opening until the point of balance is regained. Adjustment of the lens opening in this way may modify the resistance of the balancing circuit (and so allow for the greater or lesser amount of light falling on the photo cell). Or the light may reach the photo cell through the lens aperture, so that the motorized adjustment of the latter again automatically provides the point of balance—and hence correct exposure. The servo motor is switched on just before the camera starts running, and remains switched on during filming so as to follow all changes in light intensity.

PRINTING AND ENLARGING

Amateur Enlarging. Normal enlarging procedures use comparatively long exposure times (several seconds upwards) so that photo multiplier cell circuits become the most convenient means of controlling the exposure itself. In most systems a photo multiplier cell starts registering the light the moment the enlarger is switched on for the exposure. When sufficient light—reflected from the enlarging paper, going through the paper or diverted from the light beam going through the lens—has reached the cell, an associated relay automatically switches off the enlarger illumination. The associated electronic circuits have to be adjusted for the speed of the enlarging paper, and its surface reflection or transmission characteristics—according to whether the light is registered by reflection from the paper surface or through the paper base. The meter cells, however, automatically allow for the magnification and the lens opening of the enlarger. Where the light is measured in the enlarger beam (for example via a beam splitter) the magnification must be set in addition.

Commercial Printers. Printers and enlargers

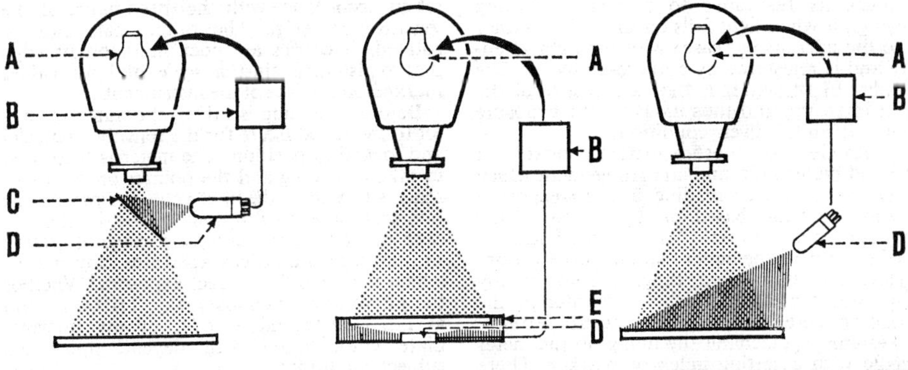

EXPOSURE CONTROL SYSTEMS IN ENLARGERS. Left: *Photo cell with beam splitter monitors light coming through enlarger lens and switches enlarger lamp off after the required exposure time.* Centre: *The photo cell is built into a special masking frame and measures the light coming through the enlarging paper.* Right: *The photo cell measures the light reflected from the enlarging paper on the baseboard. A, enlarger lamp. B, timer control unit. C, beam splitter. D, photo cell. E, special masking frame.*

used in photo-finishing laboratories largely utilize automatic exposure control systems of similar type. Since in commercial printing it is the output speed that matters, the photo cell system may be linked to the circuit that adjusts the printing light intensity rather than the duration for which it is switched on. In electronic printers using a flying spot scanner the measuring photo cell may control the intensity of the electron beam and so yield different printing exposures, according to the density of the negative, during a given exposure time.

Intensity control of the printing light is also used in continuous motion picture printers for producing positive prints from the negative film. Here the films run through the printer at a uniform rate, so variations of exposure time are impossible. L.A.M.

See also: *Cine camera; Diaphragm; Electronic printing; Enlarging: Exposure; Shutter.*

EXPOSURE CALCULATOR. Device which simply puts all the information contained in an exposure table into a more convenient form. It generally takes the shape of a slide rule with logarithmic numbers plotted in the form of scales, either along straight slides or circular discs. The most common calculators are really not pure calculators at all, but a combination of tables and a slide rule in straight or circular form.

Book: *Focal Exposure Chart*, by W. D. Emanuel (London).

EXPOSURE COUNTER. Mechanical feature on miniature and some roll-film cameras to indicate how many frames of a load of film have been exposed.

See also: *Film transport; Miniature camera.*

EXPOSURE DATA. Information printed as a caption beneath a published photograph, in photographic magazines and journals, giving details of the shutter speed, aperture, etc., used when the photograph was taken. Often based on recollection or reconstruction.

EXPOSURE METER

Instrument which measures light intensity and indicates this in terms of exposure settings for a camera. The light may be registered and measured in a variety of ways, such as the visible darkening of a piece of sensitized paper; comparison with the brightness of a standard light source; and by the effect on a photo-electric cell. Exposure meters using the latter system are most widely used and are also known as photo-electric meters.

The photo-electric exposure meter is designed around a so-called barrier layer or a photo-resistor cell. The former is a small metallic looking disc which generates an electric current when light falls on it. The photo resistor cell changes its resistance to a current flowing through it when light falls on the cell. In either case the current increases with the light intensity and is registered by a micro-ammeter. The reading of the latter is then a measure of the light intensity, and thus indicates the exposure required under given conditions.

Photo-electric exposure meters consist of a cell and meter unit, and an arrangement which converts the meter reading into an exposure recommendation based on the speed of the photographic material. Various methods are used to convert the meter reading into exposure settings for the camera, ranging from directly calibrated scales on which the pointer of the micro-ammeter shows actual exposure settings, to systems of mechanically lining up the meter needle with a setting index or marker. There the mechanical movement required also adjusts a scale showing exposure settings or adjusts the exposure settings of the camera itself.

DESIGN

The design of photo-electric meters varies from one make to another. Although all employing the same principle, meters differ in methods of use, particularly in translating the needle reading into terms of shutter speed, aperture and film speed.

Calculators. The original and simplest device for converting the meter reading into an exposure recommendation consists of two circular concentric scales, the outer marked in the whole range of shutter speeds and the inner in the whole range of lens apertures. The inner scale is normally fixed, but it can be unlocked and set in accordance with the speed figure of the sensitive material. The outer scale can be rotated. It carries an index mark on its edge that registers against a scale of light values marked on the face of the instrument.

Before a reading is taken, the outer scale is set to the speed figure for the sensitive material and locked in position. The meter is then used to take a reading and the pointer on the inner scale is turned to the light value indicated. The two scales are now in their correct relation to each other to give a complete range of related shutter speeds and lens apertures, any one of which will give the correct exposure. Whether the photographer chooses, say, 1/1000 second at *f* 2, 1/30 second at *f* 11, or any intermediate combination, will depend upon the subject conditions.

Setting Markers. Most current meter types have an auxiliary needle which is set to the indication of the meter. This act of setting automatically

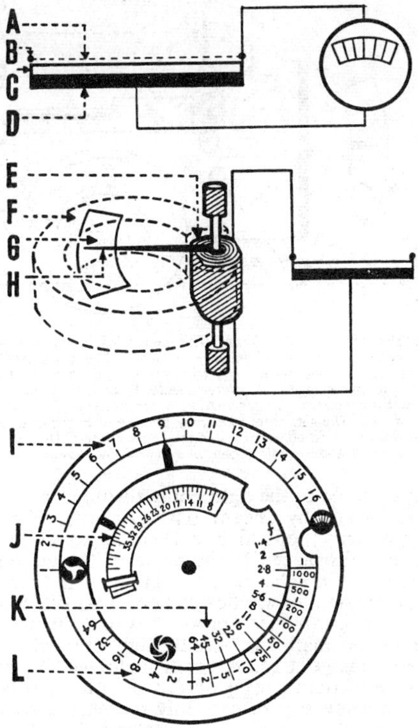

PHOTO-ELECTRIC EXPOSURE METER. The meter consists essentially of three units: the photo-electric cell, a micro-ammeter, and some kind of calculator. Top: The photo-electric cell generates a current when light falls on it. A, very thin transparent layer of gold. B, contact ring. C, selenium layer. D, iron disc. Centre: The micro-ammeter measures the current produced by the cell. E, coil. F, magnet. G, scale. H, pointer. Bottom: The calculator converts the scale readings into exposure values. I, scale readings. J, film speeds. K, apertures. L, shutter speeds. Some meters have much more simplified calculators or are direct reading once a suitable scale has been selected for the film speed used.

adjusts the calculator, allowing the correct exposure to be read off the calculator scale which has been adjusted for the speed of the material in use.

In another method, the meter needle is brought to the same point on the scale each time by a diaphragm, variable flap, or similar device which alters the amount of light reaching the photo-electric cell. Both methods are favoured with meters which are incorporated into cameras, since the act of setting adjusts the lens aperture and/or shutter speed of the camera at the same time. Some meters introduce variable electric resistances to bring the needle of the meter to the fixed reference point. More sensitive meters than those incorporating barrier-layer types of cell are those based on photo-resistors; these are substances, the electric resistance of which drops with increasing light flux falling upon them. The meter unit then consists of the photoresistor, a small battery and a micro-ammeter, and its sensitivity may be of the order of 1,000 times higher than that of the barrier-layer type.

Reflected Light Meters. The commonest type of photo-electric meter measures the total amount of light which comes from the scene to be photographed. While it is the brightness of the shadows which determines correct photographic exposure in most scenes, the proportion of light to dark is fairly constant and the total amount of light reflected from the scene gives a reasonably accurate indication of exposure. At the same time, the reading needs to be interpreted according to the nature of the subject.

Since the reflected light meter is meant to measure the amount of light coming from the scene to be photographed, it must not receive light from areas outside the scene. It is usual to define an "angle of acceptance" which is the angle of the cone of light received by the meter.

Incident Light Meters. An alternative type of meter measures the light incident on a scene instead of that reflected from it. The principle of operation of such incident light meters is effectively that of measuring the brightest spot on a scene, this bright spot consisting of a piece of white translucent opal material—the receptor—which is arranged in front of the photo-electric cell. Such a meter is therefore used by holding it so that it receives the same light which reaches the subject.

The receptor has two functions; first, to take in all the light falling on those parts of the subject that face the camera, and second to present this light as a uniform bright surface to the meter cell. Incident light meters therefore do not rely on a limited acceptance angle; in fact the greater the acceptance angle the better.

The ideal receptor for an incident light meter is a hemispherical or conical translucent body placed in front of the cell. This covers the light reaching the subject over an angle of about 180°. Flat diffusing reflectors placed in front of the photo cell are almost, but not quite, as efficient. The light absorption of the diffusing receptor is usually so adjusted that the incident light meter yields the same exposure reading with an average subject as a normal reflected light meters, these being designed without its diffusing attachment). Apart from the receptor, the construction of the incident light meter is identical with that of the reflected light meter.

Built-in Meters. Many modern cameras have an exposure meter built into the body, and often coupled with the exposure settings. Built-in meters are usually highly compact versions of normal separate meters; cadmium sulphide cell systems permit particularly compact design.

Meters built into the camera are usually reflected light meters the meter being designed to take in the same angle of view as the lens and to be used from the camera position. This makes for convenience in handling: the meter is always at hand and ready to use, and cannot be left behind. On the other hand the versatility of use is reduced; it is less convenient to move the camera about to take close-up or other special readings, and a built-in meter in a camera becomes particularly inconvenient for incident light measurement. Often no incident light diffuser is in fact supplied with such a meter.

The orthodox design of a built-in meter places the meter cell either around the lens (in fixed lens cameras), near the viewfinder window, or in some other convenient position on the camera front. Meters in this way measure the light reflected from the subject. Various attempts have also been made to measure on the one hand the light coming through the lens, and on the other the brightness of the image formed on the film or—in reflex cameras—on the viewing screen. Through-the-lens exposure measurement has the advantage, especially with interchangeable lens cameras, that the meter cell receives light from exactly the same subject field as is taken in by the lens. The acceptance angle of the meter is thus modified accordingly when a lens of different focal length is used.

Designs along these lines have used (a) a swing-away meter cell behind the lens (which moves out of the way immediately before the exposure), (b) meter elements built into the mirror of a reflex camera, (c) meter cells to

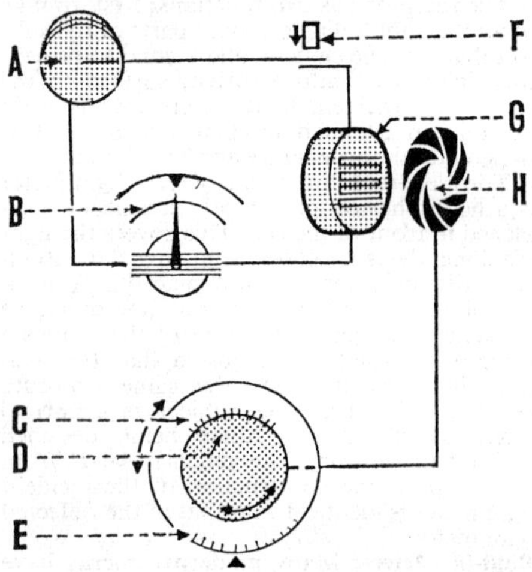

ZERO SETTING CADMIUM SULPHIDE METER. The cadmium sulphide cell changes its electric resistance according to the amount of light falling on it and so controls a galvanometer in the circuit. In a zero setting system the light is kept constant by an iris diaphragm or similar device. A, battery of the meter circuit. B, galvanometer with meter needle lined up opposite a setting index. C, shutter speed scale. D, aperture scale (usually linked with the iris diaphragm H). E, film speed settings. The functions of the controls C and D may be directly linked to aperture shutter speed settings in the camera. F, switch (to prevent drain on the battery when not needed). G, cadmium sulphide cell.

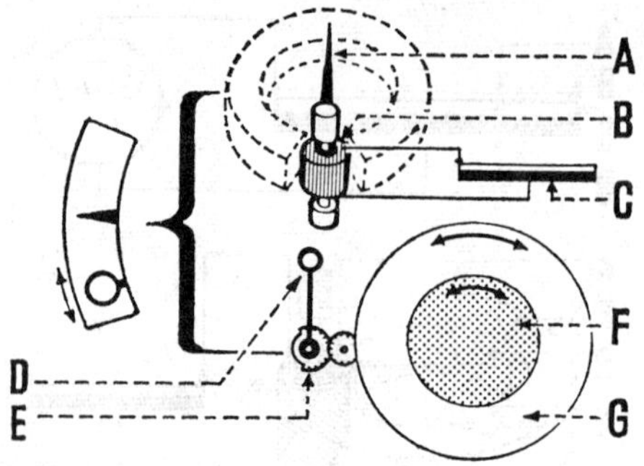

NEEDLE MATCHING SYSTEM. The measuring instrument and the calculator are mechanically separate. The calculator is linked to a second needle or setting marker which is lined up with the meter needle. A, meter needle. B, coil. C, photo cell. D, setting marker. E, gearing. F, aperture ring. G, shutter speed ring. The setting marker may be geared to the aperture and shutter speed adjustments on the camera itself. The setting marker and meter needle appear in a common window.

which some of the light going through the lens was diverted by a beam splitter or other reflecting systems (popular in cine cameras), and (d) meter cells placed in the plane of the viewing screen or in some other position in the viewfinder system of a reflex camera. One difficulty with the last-named approach is that the meter should normally measure the light reaching the whole image area. A small meter cell in the centre of the focusing screen or other part of the viewing system can only measure a small portion of the subject. The meter cell then acts like a photometer and the exposure settings must be arrived at by appropriate spot readings.

If the exposure meter is behind the lens, the light reaching the meter automatically also allows for the lens aperture. Similarly, filters placed in front of the lens hold back light also from the meter cell (though the meter is not necessarily affected in the same way as the film by the filter).

Reflex cameras that keep the lens aperture fully open during viewing and focusing require a separate allowance for, or coupling with, the aperture adjustment.

Automatic Exposure Control. In the majority of built-in camera meters the setting control of the meter is mechanically coupled with the exposure settings (aperture and speed) of the camera. The operation of taking an exposure reading therefore directly sets the exposure as well. Such coupled meters are often advertised as automatic, though they still need a manual operation of lining up the meter needle with a fixed or movable pointer. They are more correctly described as semi-automatic.

In fully automatic exposure systems the meter mechanism is directly linked to the lens aperture (or shutter speed setting), adjusting the latter, directly or indirectly, according to the light falling on the photo cell.

The simplest mechanism uses two slotted feather-weight aperture discs geared to the meter coil. Movement of the slots relative to each other enlarges or reduces the effective lens aperture. In other systems the meter mechanism controls the movement of a mechanical stop or the meter needle is scanned by a mechanical device. The lens is then stopped down to the appropriate point by a spring-loaded pre-selector iris on pressing the shutter release. The pre-selector system may also work the shutter speed adjustment instead. Some cameras also employ miniature servo motors for the same purpose.

CHARACTERISTICS

Some of the performance characteristics of exposure meters have been standardized by appropriate standards of the British Standards Institution (B.S. No. 1383:1947) and of the American Standards Association (PH2.12–1957). These cover acceptance angles, sensitivity and colour sensitivity.

Acceptance Angle. The angle of acceptance decides the amount of light reaching a meter. It therefore also influences the apparent sensitivity of the meter, and has at times been made a misleading selling point. The above-mentioned standards specify the measurement of the acceptance angle. In practice the meter should not have an angle very much greater than the angle of view of the standard camera lens.

While it matters comparatively little with a subject of uniform brightness distribution (such as an open view) whether the meter takes in a very wide or a narrow view, a wide acceptance angle can cause the meter to give misleading readings under a variety of subject conditions. These include scenes where strong light sources or strong reflections appear just outside the field of view and may thus influence the meter reading. With a wide acceptance angle it is also more difficult to take reliable close-up readings.

Limiting Devices. Various devices are used to limit the acceptance angle. The simplest method is to arrange for the cell to be situated at the bottom of a "well". A single well must be deep to be effective, but the depth can be decreased by using many small wells, subdividing the whole area of the cell.

Another method is to arrange a lens in front of the cell to focus an image of the scene on the cell surface. In this way a rather larger amount of light reaches the cell than by the method of using a simple well; and the meter can be made much slimmer. Many meters use a multicellular lens which looks rather like an enlarged version of an insect eye.

Sensitivity and Measuring Range. Response sensitivity of exposure meters has been standardized by the British Standard Specification No. 1383: 1947 which requires a minimum deflection of 2 mm. of the meter needle for a change of 2 : 1 in brightness.

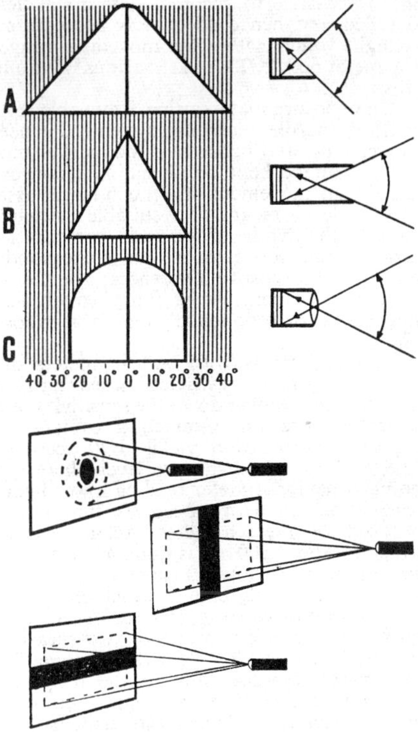

ACCEPTANCE ANGLE. Top: *The angle of acceptance of the meter is usually limited by a hood in front of the cell. The curves shown are plots of the amount of light reaching the cell from the various directions when the meter is pointed at a uniformly illuminated surface. A, shallow hood covering a wide angle. B, deep hood covering a narrow angle. C, shallow hood with lens. This can still cover a narrow angle but receives more light than the corresponding hood alone, and is more compact.* Bottom: *To measure the acceptance angle, bring the meter up to a suitable test chart until the needle deflection is constant when the cell covers a well defined area.*

The actual sensitivity of the meter to light depends on the meter design and also on the measuring range provided. High-quality selenium cell meters may react to lighting levels as low as 0.1 foot candle. An inexpensive meter of this kind built into a camera may have its lowest response level up to ten times as high; on the other hand a cadmium sulphide photo resistor cell meter may be 50 to 100 times as sensitive as the best selenium cell instrument. The highest lighting level likely to be encountered in normal photography is between 1,000 and 2,000 foot candles, so that a high-class selenium cell meter would have to cover a lighting range of around 20,000 : 1, and a cadmium sulphide cell of about 2,000,000 : 1.

Covering such range provides certain difficulties in obtaining the readings. It is often physically impossible to accommodate a sufficiently detailed and hence adequately accurate scale of light values within the movement range of the meter needle. There are various ways out of this.

On exposure meters using a movable setting marker (needle matching system) the needle movement—and hence also the marker movement—can be kept non-linear and converted to a linear movement of a much larger associated scale by means of a suitable mechanical cam. In the middle of the scale, comparatively large needle movements thus correspond to fairly small exposure differences; at either end of the movement range fairly small needle movement corresponds to appreciable exposure changes.

Simple cameras with built-in meters often have a very much restricted measuring range, achieved by keeping down the sensitivity of the meter. Where for instance a camera with exposure automation yields 1/30 second at *f*2·8 as its lowest exposure level, there is no point in having a meter reading lower lighting levels than this with the slowest film likely to be used. So there a meter with a measuring range of about 500 : 1 is quite adequate. This corresponds to about 10 exposure value intervals, which are easy enough to handle.

The commonest way of covering a big exposure range is by means of two measuring ranges. One is the low light range, going from the lowest response limit of the meter to a level about 200 times as high. This covers mainly indoor and artificial light subjects. At higher lighting levels the meter cell is covered by a baffle which lets through about 1 per cent of the light, thus cutting down the effective sensitivity of the meter by a factor of 100. This at the same time increases the exposure range 100-fold, leading to a total range of 20,000 : 1. With the baffle in front of the cell the meter is then used for measuring brilliant daylight scenes.

With cadmium sulphide cell systems either a baffle or an electrical resistance can be used to cut down the meter sensitivity.

The exposure scale of the meter then carries two series of aperture (or shutter speed) figures, one for use with the baffle and the other without.

Some meters also take a booster cell to increase the sensitivity in very poor light.

Spectral Sensitivity. Photo-electric cells have about the same spectral sensitivity as the human eye. They have the same peak in the yellow-green, but are more sensitive to blue and red than the human eye. Meters vary considerably in their spectral sensitivity, particularly in their sensitivity to blue, and the American Standard Specification No. PH52.12–1957 dealing with meters allows quite wide limits for this.

The colour sensitivity of the cell is quite unlike that of any photographic material, even when colour filters are used to correct the colour sensitivity of the material. Thus, when using the meter on highly coloured scenes, it is possible for exposure recommendations to be misleading. This applies particularly to photographs of sea and sky scenes of the intense blue that sometimes occurs in southern countries. Under normal conditions, however, the discrepancies between the colour sensitivities of the cell and the photographic material are not important since the colour of the average scene is fairly constant.

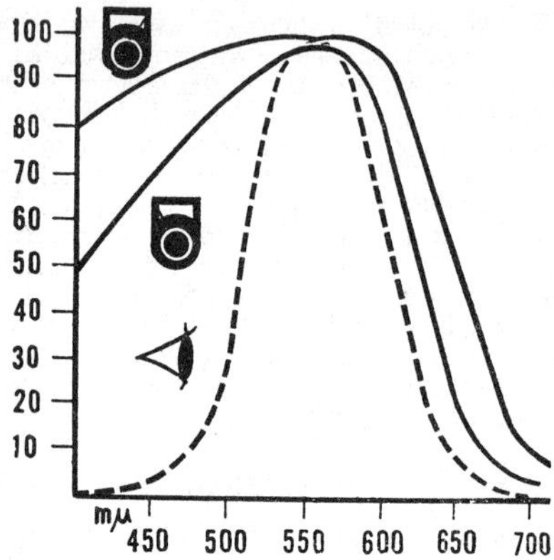

SPECTRAL SENSITIVITY. The curves for photo-electric meters (solid) are not always the same as for the eye (dotted). The upper and lower limits of sensitivity shown are those acceptable to the American Standards Association.

METHOD OF USE

The figure indicated by the meter is an indication of the strength of the light, and no more. Before it can be used for calculating an exposure, two other factors have to be taken into account: what sort of negative the photographer wants to produce—dense, medium, or thin—and how the meter is used—i.e., what fraction of the light coming from the subject is actually measured by the meter.

Adjustment for Type of Negative. No matter which of the various methods of use the photographer adopts, he aims at getting the type of negative that suits his particular style of work. This type of negative may be either denser or thinner than the negative that would result from following the exposure indicated by the meter. But the exposure must be increased to get a dense negative and decreased to get a thin one. So the photographer must correct his meter reading so that, if necessary, he will consistently under- or over-expose his negatives by the right amount to give him the negatives he likes. The easiest way to do this is to calibrate the meter and adopt a suitably revised speed for the sensitive material.

Adjustment for Method of Use. There are several different ways of using a photo-electric exposure meter; some depend on personal

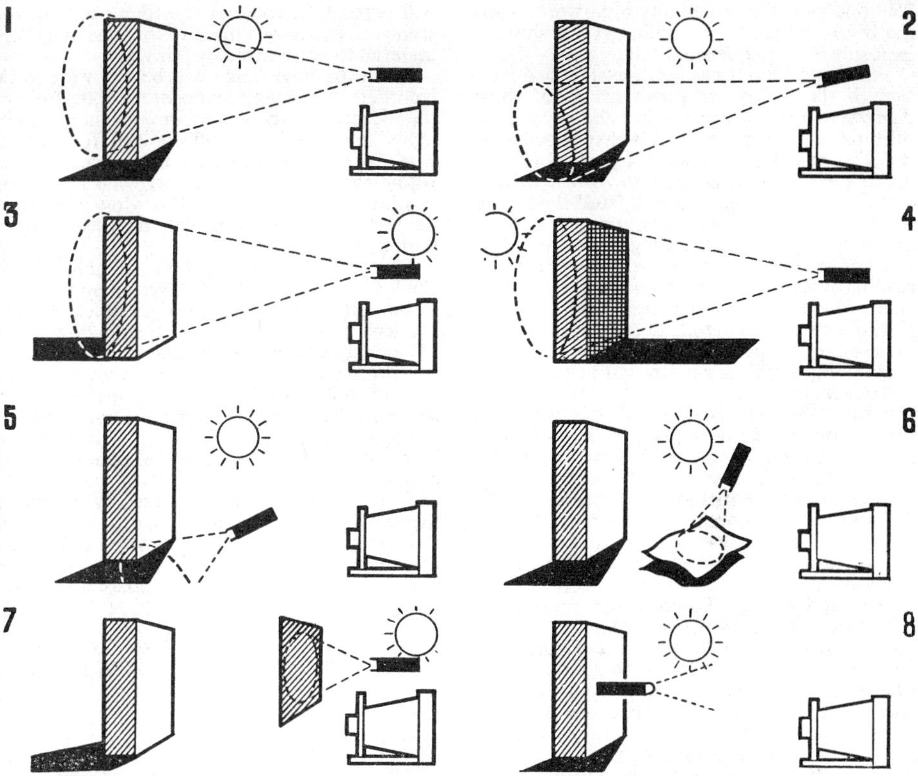

USING AN EXPOSURE METER. 1, Over-all reflected light readings taken from the camera position with the meter pointing at the subject. 2. The meter should point slightly downwards to exclude large sky areas. 3. If the light is fully behind the camera, the exposure indicated may be halved. 4. Against the light, the exposure indicated should be doubled. 5. If the shadows are important, the meter reading should cover them at close range. 6. For a highlight reading the meter receives the light from a white surface lit by the main light. The correct exposure is then about eight times the figure indicated. 7. In substitution readings the meter is pointed at a grey card or other object chosen to match the important part of the scene. The substitute object must also receive the same light as the actual subject. 8. For incident light readings the meter points towards the camera from the subject, and thus measures the light reaching the subject. The meter cell must be fitted with a suitable diffusing cone or screen.

choice, others are dictated by special circumstances. Each method measures a different fraction of the light reflected from or falling on the subject, and for any given set of lighting conditions each will give a different reading. But in the end, they must all lead to the same exposure, so each method has its own factor by which the reading must be multiplied before it will give the right answer.

The conversion factors are only approximate. They will help the photographer who wishes to change temporarily from one method of use to another. But if he wishes to adopt the new method permanently, he should calibrate his meter afresh to arrive at the exact speed figure for his purpose.

Average Scene Brightness. The most common method is to measure the average brightness of the scene to be photographed. To do this the meter is pointed at the scene from the camera position, looking at the scene in the same direction as the camera. This method is sometimes described as measuring the light reflected from the scene. Used in this way, the meter is accurate enough for a surprisingly large number of subjects. So long as the direct rays of the sun are prevented from falling on the light-sensitive surface of the meter, the exposures given should be satisfactory for most general purposes, using the speed figure given by the maker of the film or plate. But this method does not give reliable readings when the picture includes the sky.

Tilting the Meter. In ordinary black-and-white photography, at least, it is common practice to expose for the shadows. But, using the above method of measuring the average scene brightness, two scenes with the same shadow brightness which included different amounts of sky would give two different exposure readings.

The one with the largest sky area would show the bigger reading and would get too short an exposure in consequence.

For this reason it has become standard practice to tilt the meter down at about thirty degrees to the vertical so that the sky has no part in the meter reading. It is easy to see when the sky has been excluded by watching the reading as the meter is tilted down. The reading will first decrease quite rapidly and then reach a steady value when further tilting makes no difference. This steady reading is the one to use.

Since excluding the sky reduces the meter reading, it is necessary to compensate for the change by adopting a new figure for the speed of the sensitive material. On an average type of subject, a speed figure double that used when pointing the meter straight at the subject will be satisfactory.

Close-up Method. If, instead of using the meter from the camera position and tilting it down, it is brought close to the important part of the subject, the sky is automatically cut out. This is a favourite method with many workers and so long as a preliminary set of calibrating exposures is run off, it gives consistently good results. It has the disadvantage that it takes no account of the effect of haze between the camera and subject. To allow for the brighter tone value of the haze, the exposure must be shorter by about half, or the normal speed figure for the material should be doubled.

Allowing for Tone Range. The speed figures supplied with the meter for normal use from the camera position are worked out for an average type of subject. If the tone range of the subject is especially wide, the shadows will be deeper and should therefore have a longer exposure than in normal subjects. But the highlights will be brighter, thus giving a higher meter reading and indicating an even shorter exposure. So that to put matters right, the subject should be given double the indicated exposure, or the normal speed figure for the sensitive material should be halved.

Similarly, if the subject has an unusually narrow range of tones, the indicated exposure should be halved, or the normal speed figure for the sensitive material should be doubled.

The above recommendations apply to negative materials. With a reversal film the highlights must remain correctly exposed even if the subject has a wide tone range and no doubling of the exposure is therefore desirable. If the tone range is unusually narrow, however, a correctly exposed reversal transparency will appear rather dull and dark because there are no bright highlights. In this case the indicated exposure should be doubled, or the normal speed figure for the sensitive material halved. In practice this applies especially to outdoor exposures on colour film in dull weather, mist, etc.

When the exposure is based on the shadow brightness, what happens to the highlights will depend on whether the tone range of the scene is narrow enough for the negative material to be able to cope with it.

Often the tone range will be too wide to be recorded by a single exposure, so that either the deepest shadows or the brightest highlights have to be sacrificed. So, for critical work, the meter must be used to measure the whole tone range. If it is too wide it may be possible to correct it by throwing additional light into the shadows with a reflector or a flash bulb.

The tone range is the ratio of highlight to shadow brightness. For an average subject it is of the order of 100 : 1. But light is scattered in the lens and dispersed into the shadow areas of the image on the sensitive surface. This compresses the tone range by an amount that depends on such things as the properties of the lens and the distribution of light and shade in the scene. A tone range of 100 : 1 in the scene is compressed under average conditions to about 30 : 1.

The tone range is measured by taking a reading first on a highlight in the scene, and then on some shadow detail. When doing this the angle of acceptance of the meter must be completely filled by the portion of the scene to be measured.

FILM SPEED FACTORS FOR METER METHODS

Measurement taken	*Multiply normal film speed rating by*
Average scene brightness	1
Meter tilted at 30 degrees	2
Close-up	2
Close-up for subjects of wide tone range	$\frac{1}{2}$
Close-up for subjects of narrow tone range	2
Shadow brightness	16
Artificial highlight	$\frac{1}{8}$

When shadow detail is an essential part of the picture, the shadow brightness can be measured direct by bringing the meter close up to a representative shadow part of the subject. The angle of view of the meter must be completely filled by the shadow portion. When the meter is used in this way it will give a very much smaller deflection than if it had been pointed from the camera position. So the speed figure for the sensitive material should be multiplied by 16 (or the exposure divided by 16 if the speed scale of the meter will not show such a large figure).

Artificial Highlights. All the methods of using a photo-electric meter described so far have been concerned with correct rendering of shadow detail without reference to what happens to the rest of the picture.

For subjects of wide tone range this may mean that the highlights are beyond the range of the material so that they will appear "burnt out", i.e., lacking in detail. An alternative method which guards against this failing is to use the meter to measure the brightness of an artificial highlight. This may either be a highlight present in the scene itself, or one which is placed there in the form of, say, a white handkerchief.

The exposure given is based upon this highlight reading in such a way that in the photograph the density of the strongest highlight is always the same, no matter what the subject or its lighting. In other words, the strongest highlight is "tied" to the top of the characteristic curve just as in the shadow method the weakest shadow is "tied" to the toe of the curve.

If an ordinary photo-electric meter is used to measure the brightest highlight in this way, it will give a very high reading and recommend a very short exposure. The exposure recommended will, in fact, be about six times too short so the indicated time must be multiplied by six, or the normal speed figure divided by six, to get a correctly exposed negative. As in all other methods discussed, the exact multiplying factor can only be found by actual experiment.

Using Incident Light Meters. It is but a small step from measuring the brightness of an artificial highlight to incorporating it in the meter itself, as is done in incident light (or highlight) meters.

Incident light meters were originally designed to be pointed at the light, but many photographers now prefer to point the meter at the camera. A hemi-spherical receptor in this position is illuminated in much the same way, for example, as a human face. The exact shape of the receptor apparently does not matter much as long as it adds up evenly all the light falling on it. Some meters use cones or inverted cones, or even flat screens.

It is sometimes necessary—with artificial light in particular—to set up an incident light meter where the illumination on it is representative of that falling on the subject. As a rule, however, it is sufficiently accurate to take a meter reading from the camera position with the meter pointed towards the lens.

Exposure by incident light meter gives highlights of constant density. This is an advantage in any work, particularly for colour materials where faulty exposure not only produces faulty tones, but also upsets the colour balance. It is again an advantage in any work where the tones should appear much the same from one negative to the next, because the brilliance of the highlights (and especially of the skin tones in colour photography) establishes the general tone of the picture.

Duplex Method. Possibly the most critical way of using an incident light meter is the Duplex method. It involves two meter readings: one with the meter facing the camera, the other facing the strongest light falling on the subject. The camera exposure derived from these readings is the *f*-number which is half-way between the *f*-numbers corresponding to the two readings. In most instances an incident light meter in which the light collector is conical or hemispherical in shape will give an exposure recommendation which is the same as that given by the Duplex method.

Key Tone Method. The most popular types of photo-electric meter aim at correct exposures for the shadows, and highlight meters are only concerned with the highlights. But there is an intermediate method designed to give an exposure that will reproduce the medium tones at a constant density.

In this case, instead of pointing the meter at the subject, it is pointed at an artificial medium tone such as a piece of grey card. The reading obtained will of course depend on the reflection density of the card and must be calibrated separately for any particular tint.

This key tone method of measuring the amount of light from a grey card can be applied to any selected tone, darker or lighter than the medium grey. Some photographers carry about with them a series of grey cards from which they select one matching the brightness they are interested in.

They find it easy to stand at the camera position and pick out the card that comes nearest to the brilliance of the important part of the scene. Once they have decided which card to use they simply point the meter at it and take a reading. This saves them the trouble of walking over to the subject with the meter—a course that is not, in any case, always open where the subject is inaccessible.

Incident Light Method with Normal Meters. Sometimes the light is so weak that the normal meter does not give a reading when pointed at the subject in the usual way. When this happens the meter can be used as an incident light meter and pointed at the light instead of the subject.

The ordinary meter is not calibrated for use in this way, and no rigid rule can be laid down to suit every meter since not all meters respond in a similar fashion. In any case, the nature of the subject itself influences the final exposure.

But as a first rough guide the meter reading should be reduced 100 to 1,000 times. Even so, it will still be wise to give exposures of ¼ and 4 times, as well as the estimated exposure, to be certain of a good result.

Limitations of Built-in-Meters. Exposure meters built into a camera are more awkward to use for special reading methods (close-up, highlight, duplex, incident light) which involve taking a reading from a position away from the camera view point. With an electric-eye camera where the meter sets the exposure controls immediately before the exposure, readings other than straightforward reflected light readings of the actual subject become virtually impossible. The meter cell has to take in the actual view recorded on the film during the reading. Direct reflected light readings give reasonably reliable exposures in about 90 per cent of all picture taking occasions. There are two ways of dealing with some of the remaining 10 per cent.

(1) Where the exposure has to be halved, doubled, etc. for subjects of unusual tone range, the film speed setting can be adjusted instead. Thus to halve the effective exposure the film speed setting on the meter is doubled (A.S.A. scale). But remember to reset the film speed adjustment to its normal rating again afterwards.

(2) Close-up readings and readings with the meter tilted slightly downwards can be made by depressing the release button slightly during the taking of the reading. That takes up the first pressure and locks the exposure selection mechanism. The release button must be kept partly depressed in this way while taking up the position for the actual picture, and then pressed through fully. This is fairly easy when the camera only has to be tilted slightly between the reading and the exposure; it becomes more awkward when movement of the camera from one position to another—with the release button partly depressed—is involved.

W.F.B. (L.A.M.)

See also: *Actinometer; Cine Camera; Exposure; Exposure automation; Exposure tables; Extinction meters; Photo cell; Photometer.*

Books: *Successful Exposure in Photography, The Right Exposure Every Time*, and *How to Use Your Exposure Meter*, all by L. A. Mannheim (London); *Exposure*, by W. F. Berg (London); *Exposure Meters and Practical Exposure Control*, by J. F. Dunn (London).

EXPOSURE TABLES

Tables listing all the factors which affect the exposure. These are: the weather; the subject; the time (of the day and the year) and the latitude, which determine the altitude of the sun; the altitude; the speed or sensitivity of the photographic material; the filter factor.

A figure is given to each. The user has to estimate the various conditions, and add up all the figures. The answer is a figure opposite to which the exposure can be read off in a table.

Alternatively, all this information may be conveniently contained on an exposure calculator, e.g., in the form of a slide rule.

The basic idea underlying the design of all exposure tables and similar devices is that the various factors listed are independent of one another and that any one can vary without affecting the rest. But this assumption is only approximate and for this reason alone the results are not as accurate as those obtained with meters that measure the actual light on the spot. Still, for most purposes the latitude of the photographic materials is enough to take care of any errors.

The main errors in the estimation of individual factors occur with weather conditions and subject brightness. While full sunshine is reasonably predictable in strength (except in industrial areas where smoke haze cuts out quite a lot of light), dull weather may vary widely according to the thickness of the cloud layer. Similarly, subjects in normal light are straightforward enough; scenes in the shade, especially under trees, etc., are rather undefinable in brightness (e.g., the trees may be nearly bare or full of leaves).

The following list shows by comparison what sort of accuracy can be expected from such methods of measuring exposure. It proves that quite satisfactory results can be obtained by the use of exposure tables or calculators.

Exposure tables, charts and calculators	45%
Extinction meters	87·8%
Photo-electric meters	97·3%

Standard Exposure Tables. The tables which follow were at one time the basis of a British Standard (No. 935:1948), and are typical of the conditions that have to be taken into account.

To use the tables, add together the numbers corresponding to the appropriate conditions in tables 1 and 2. Then subtract from the total the number from table 3 corresponding to the speed of the film or plate used. The resulting number will enable appropriate values of exposure time and lens aperture to be obtained from Table 4, as shown in the example given below.

Example.—A photograph of a near-by, unshaded subject is to be taken at 4 p.m. B.S.T. in September, at a latitude of 50° N., the sun being obscured by slight haze. The material being used has a speed of 40 A.S.A.:

from Table 1e—lat. 50° N.: Sept., 3 p.m. G.M.T. (4 p.m. B.S.T.)	1
from Table 2—hazy sun; near-by, unshaded subject	58
Total	59
For 40 A.S.A. (see Table 3) subtract ..	27
Final number	32

Suppose that the photograph is to be taken at *f* 8. In Table 4, in the line corresponding to *f* 8, the number 32 does not occur; therefore take the nearest figure, which is 31, and the exposure time indicated is 1/30 second.

Care should be taken in using the figures for the light indices (tables 1a–1g) early and late in the day. The figures have been worked out for the middle of the respective months, and may not be reliable earlier or later in the month at the first or last hours of the day. The times used are local solar times, not official local time.

W.F.B. (L.A.M.)

See also: *Actinometer; Exposure; Exposure Meter; Extinction meter; Photometer.*

Books: *The Right Exposure Every Time*, by L. A. Mannheim (London); *Exposure*, by W. F. Berg (London).

TABLE 1a. LATITUDE 0–15°

EQUATOR.—Central Africa, Northern Congo, Cameroons, Victoria Nyanza, Northern Brazil, Amazon River, Southern Guiana, Southern Venezuela, Ecuador, Southern Colombia, Southern East Indies.

0–15° N. Central Africa, Somaliland, Abyssinia, Southern Sudan, Guinea, Northern Guiana, Northern Venezuela, Panama, Northern East Indies, Southern Philippines, Ceylon, Southern India, Central America.

0–15° S. Portuguese West Africa, Southern Congo, Nyasa Lake, Central Brazil, Northern Bolivia, Northern Peru, Southern East Indies, Extreme North of Australia.

0–15° *North*	*Noon*	11 1	10 2	9 3	8 4	7 *a.m.* 5 *p.m.*	0–15° *South*
June	0	0	0	0	1	2	December
July May	0	0	0	0	1	2	January November
August April	0	0	0	0	1	2	February October
September March	0	0	0	0	1	3	March September
October February	0	0	0	0	1	3	April August
November January	0	0	0	1	1	4	May July
December	0	0	0	1	1	4	June

TABLE 1b. LATITUDE 15–25°

15–25° N. Southern Egypt, Northern Sudan, Sahara, Northern Philippines, South Mexico, Southern India, Indo-China, South Arabia.

15–25° S. Northern Australia, Madagascar, Northern half of Union of South Africa, Southern Peru, Bolivia, Southern Brazil, Paraguay.

15–25° *North*	*Noon*	11 1	10 2	9 3	8 4	7 5	6 *a.m.* 6 *p.m.*	15–25° *South*
June	0	0	0	0	1	2	5	December
July May	0	0	0	0	1	2	6	January November
August April	0	0	0	0	1	2	10	February October
September March	0	0	0	0	1	3	—	March September
October February	0	0	0	1	1	4	—	April August
November January	0	0	0	1	2	7	—	May July
December	0	0	1	1	2	8	—	June

TABLE 1c. LATITUDE 25–35°

25–35° N. Tripoli, Tunis, Morocco, Southern States of U.S.A., Southern Japan, Southern China, Northern India, Persia, Palestine, Southern Mediterranean, Northern Egypt.

25–35° S. Southern half of Union of South Africa, Uruguay, Northern Argentine, Central Australia.

25–35° *North*	*Noon*	11 1	10 2	9 3	8 4	7 5	6 *a.m.* 6 *p.m.*	25–35° *South*
June	0	0	0	0	1	1	4	December
July May	0	0	0	0	1	1	4	January November
August April	0	0	0	0	1	2	7	February October
September March	0	0	0	1	1	3	—	March September
October February	0	0	0	1	2	6	—	April August
November January	0	1	1	1	3	—	—	May July
December	1	1	1	2	4	—	—	June

TABLE 1d. LATITUDE 35–45°

35–45° N. Southern France, Spain, Portugal, Central States of U.S.A., Central Japan, Central China, Caspian Sea, Black Sea, Northern Mediterranean, Turkey, Greece, Italy.
35–45° S. Central Argentine, Southern Australia, New Zealand and Northern Patagonia.

35–45° *North*	*Noon*	11 1	10 2	9 3	8 4	7 5	6 6	5 *a.m.* 7 *p.m.*	35–45° *South*
June	0	0	0	0	1	1	3	9	December
July May	0	0	0	0	1	1	3	—	January November
August April	0	0	0	0	1	2	6	—	February October
September March	0	0	0	1	1	4	—	—	March September
October February	1	1	1	1	3	10	—	—	April August
November January	1	1	1	2	5	—	—	—	May July
December	1	1	2	3	7	—	—	—	June

TABLE 1e. LATITUDE 45–55°

45–55° N. **England and Wales, Eire**, North of France, Northern States of U.S.A., Southern Canada, Northern Japan, Northern China, Southern Russia, Germany.
45–55° S. Southern Argentine, Falkland Islands.

45–55° *North*	*Noon*	11 1	10 2	9 3	8 4	7 5	6 6	5 *a.m.* 7 *p.m.*	45–55° *South*
June	0	0	0	0	1	1	2	5	December
July May	0	0	0	0	1	1	2	7	January November
August April	0	0	0	1	1	2	5	—	February October
September March	0	1	1	1	2	4	—	—	March September
October February	1	1	1	2	4	—	—	—	April August
November January	2	2	3	4	—	—	—	—	May July
December	2	2	4	6	—	—	—	—	June

TABLE 1f. LATITUDE 55–65°

55–65° N. **Scotland**, Central Canada, Southern Alaska, Central Russia, Southern Scandinavia, Denmark, Iceland.
55–65° S.

55–65° *North*	*Noon*	11 1	10 2	9 3	8 4	7 5	6 6	5 7	4 *a.m.* 8 *p.m.*	55–65° *South*
June	0	0	0	0	1	1	2	3	6	December
July May	0	0	0	1	1	1	2	4	10	January November
August April	0	0	1	1	1	2	4	—	—	February October
September March	1	1	1	2	3	6	—	—	—	March September
October February	2	2	3	4	9	—	—	—	—	April August
November January	4	5	7	—	—	—	—	—	—	May July
December	6	7	—	—	—	—	—	—	—	June

TABLE 1g. LATITUDE 65–75°

65–75° N. Central Greenland, Northern Alaska, Northern Russia, Northern Scandinavia.
65–75° S.

65–75° *North*	*Noon*	11 1	10 2	9 3	8 4	7 5	6 6	5 7	4 8	3 9	2 10	1 *a.m.* 11 *p.m.*	*Mid-night*	65–75° *South*
June	0	0	0	1	1	1	2	2	3	5	7	9	10	December
July May	0	1	1	1	1	1	2	3	5	7	—	—	—	January November
August April	1	1	1	1	2	2	4	7	—	—	—	—	—	February October
September March	2	2	2	3	4	8	—	—	—	—	—	—	—	March September
October February	5	5	7	—	—	—	—	—	—	—	—	—	—	April August

TABLE 2a. SCENE INDICES FOR BLACK-AND-WHITE NEGATIVE MATERIAL

Land and marine scenes

Scene	Distant	Semi-distant	Near-by			Close-up		
Shade condition			Un-Shaded	Light Shade	Heavy Shade	Un-Shaded	Light Shade	Heavy Shade
Clear sun	*49	*52	*55	58	61	*58	61	64
Hazy sun	*52	*55	*58	61	64	*61	64	67
Cloudy, bright...	55	58	61	64	67	64	67	70
Cloudy, dull	58	61	64	67	70	67	70	73

*Side lighting, add 3; back lighting, add 6.

Sky conditions.—Clear sun—scene illuminated by direct obscured sunlight, but light clouds may be present.

Hazy sun—the sun's light weakened by a blanket of haze, but the sun's disc still visible. Haze may weaken sunlight early or late in the day, even if the sky is clear overhead.

Cloudy, bright—the sun obscured by light cloud, but its approximate position discernible by a bright area in the sky.

Cloudy, dull—the sky nearly or completely filled with heavy cloud.

Shaded scenes.—Light shade—scene shielded from direct rays of sun but illuminated by about half of the open sky.

Heavy shade—scene shielded not only from sun, but from most of the sky (e.g., scenes in narrow streets).

Scene structure.—Scenes are classified in Table 2a according to distance, because subject distance provides a fair index of lowest scene brightness; as the subject distance increases, dark parts of the scene appear to merge with lighter areas, raising the lowest scene brightness.

Semi-distant objects, photographed with a long-focus lens to yield the effect of a close-up, should be treated as a close-up.

Reversal films do not relate exposure so directly to lowest scene brightness and the classification of distance is omitted from Table 2b. it may be noted, however, that the scene indices given in Table 2b for side and back lighting may be varied by 3; decreasing the index increases the dramatic light effect, increasing the index yields more detail in shadow areas.

TABLE 2b. SCENE INDICES FOR REVERSAL MATERIAL*

Shade condition	Front Light	Side Light	Back Light for		Light Shade	Heavy Shade
			Lighting Effect	Shadow Detail		
Clear sun	52	55	55	58	61	64
Hazy sun	55	57	57	59	—	—
Cloudy, bright	58	—	—	—	—	—
Cloudy, dull	61	—	—	—	—	—

*These indices are for average coloured subjects. For light coloured subjects, subtract 2 from the index; for dark coloured subjects, add 2 to the index.

TABLE 3. INDICES FOR FILM SPEED (TO BE SUBTRACTED)

Speed	ASA	10	12	16	20	25	32	40	50	64	80	100	125	160	200
	DIN	11	12	13	14	15	16	17	18	19	20	21	22	23	24
	ASA log		2°			3°			4°			5°			6°
Index		21	22	23	24	25	26	27	28	29	30	31	32	33	34

Speed	ASA	250	320	400	500	640	800	1,000	1,300	1,600	2,000	2,500	3,200
	DIN	25	26	27	28	29	30	31	32	33	34	35	36
	ASA log			7°			8°			9°			10°
Index		35	36	37	38	39	40	41	42	43	44	45	46

TABLE 4a. EXPOSURE VALUE REQUIRED

Exposure Value	–4	–3	–2	–1	0	1	2	3	4	5	6	7	8	9	10
Total Index	76	73	70	67	64	61	58	55	52	49	46	43	40	37	34

Exposure Value	11	12	13	14	15	16	17	18	19	20	21	22	23
Total Index	31	28	25	22	19	16	13	10	7	4	1	–2	–5

Table 4b is for cameras with shutters scaled in exposure values. With both tables 4a, and 4b, intermediate index totals correspond to intermediate exposures, e.g. a total of 35 or 36 would indicate an exposure value required of 9½.

TABLE 4b. EXPOSURE REQUIRED

Aperture	*Shutter Speed in Seconds*																	
	1/2000	1/1000	1/500	1/250	1/125	1/60	1/30	1/15	1/8	1/4	1/2	1	2	4	8	16	32	60
f36	0	3	6	9	12	15	18	21	24	27	30	33	36	39	42	45	48	51
f32	1	4	7	10	13	16	19	22	25	28	31	34	37	40	43	46	49	52
f25	3	6	9	12	15	18	21	24	27	30	33	36	39	42	45	48	51	54
f22	4	7	10	13	16	19	22	25	28	31	34	37	40	43	46	49	52	55
f18	6	9	12	15	18	21	24	27	30	33	36	39	42	45	48	51	54	57
f16	7	10	13	16	19	22	25	28	31	34	37	40	43	46	49	52	55	58
f12·5	9	12	15	18	21	24	27	30	33	36	39	42	45	48	51	54	57	60
f11	10	13	16	19	22	25	28	31	34	37	40	43	46	49	52	55	58	61
f9	12	15	18	21	24	27	30	33	36	39	42	45	48	51	54	57	60	63
f8 or **f7·7**	13	16	19	22	25	28	31	34	37	40	43	46	49	52	55	58	61	64
f6·3	15	18	21	24	27	31	33	36	39	42	45	48	51	54	57	60	63	66
f5·6	16	19	22	25	28	31	34	37	40	43	46	49	52	55	58	61	64	67
f4·5	18	21	24	27	30	33	36	39	42	45	48	51	54	57	60	63	66	69
f4 or **f3·8**	19	22	25	28	31	34	37	40	43	46	49	52	55	58	61	64	67	70
f3·5	20	23	26	29	32	35	38	41	44	47	50	53	56	59	62	65	68	71
f2·9 or **f2·8**	22	25	28	31	34	37	40	43	46	49	52	55	58	61	64	67	70	73
f2·5	23	26	29	32	35	38	41	44	47	50	53	56	59	62	65	68	71	74
f2 or **f1·9**	25	28	31	34	37	40	43	46	49	52	55	58	61	64	67	70	73	76
f1·5 or **f1·4**	28	31	34	37	40	43	46	49	52	55	58	61	64	67	70	73	76	79

EXPOSURE VALUE. Setting on certain shutters to give exposures as ranges of aperture-shutter speed combinations by a single setting. The setting in effect controls the actual light reaching the film, irrespective of the combination of aperture and shutter speed, as long as the exposure value setting remains the same. Exposure values are sometimes also referred to as light values.

Some exposure meters show their readings as exposure values; these may again be either arbitrary, or expressed in photometric units such as foot candles, which have to be transferred to a calculator disc.

Shutter Values. The exposure value scale on various modern shutters combines apertures and shutter speeds in a single exposure setting. The exposure values usually run from 2 to 18, and each exposure value represents a given amount of light passed through the lens. The higher the exposure value, the less the light passed through; or, from the point of view of setting exposures, the brighter the subject, the higher the exposure value that must be set.

This system reduces exposure to its simplest terms. All the photographer has to do is to find the exposure value for the prevailing conditions from a suitably calibrated meter or table, and set the shutter scale to the same value.

The scale of exposure values is so arranged

EXPOSURES AND EXPOSURE VALUES

Exposure Value	*Shutter Speed (Seconds) at Aperture*									
	*f*1·4	*f*2	*f*2·8	*f*4	*f*5·6	*f*8	*f*11	*f*16	*f*22	*f*32
−1	4	8	¼m.	½m.	1m.	2m.	4m.	8m.	¼h.	½h.
0	2	4	8	¼m.	½m.	1m.	2m.	4m.	8m.	¼h.
1	1	2	4	8	¼m.	½m.	1m.	2m.	4m.	8m
2	1/2	1	2	4	8	¼m.	½m.	1m.	2m.	4m.
3	1/4	1/2	1	2	4	8	¼m.	½m.	1m.	2m.
4	1/8	1/4	1/2	1	2	4	8	¼m.	½m.	1m.
5	1/15	1/8	1/4	1/2	1	2	4	8	¼m.	½m.
6	1/30	1/15	1/8	1/4	1/2	1	2	4	8	¼m.
7	1/60	1/30	1/15	1/8	1/4	1/2	1	2	4	8
8	1/125	1/60	1/30	1/15	1/8	1/4	1/2	1	2	4
9	1/250	1/125	1/60	1/30	1/15	1/8	1/4	1/2	1	2
10	1/500	1/250	1/125	1/60	1/30	1/15	1/8	1/4	1/2	1
11		1/500	1/250	1/125	1/60	1/30	1/15	1/8	1/4	1/2
12			1/500	1/250	1/125	1/60	1/30	1/15	1/8	1/4
13				1/500	1/250	1/125	1/60	1/30	1/15	1/8
14					1/500	1/250	1/125	1/60	1/30	1/15
15						1/500	1/250	1/125	1/60	1/30
16							1/500	1/250	1/125	1/60
17								1/500	1/250	1/125
18									1/500	1/250
19										1/500

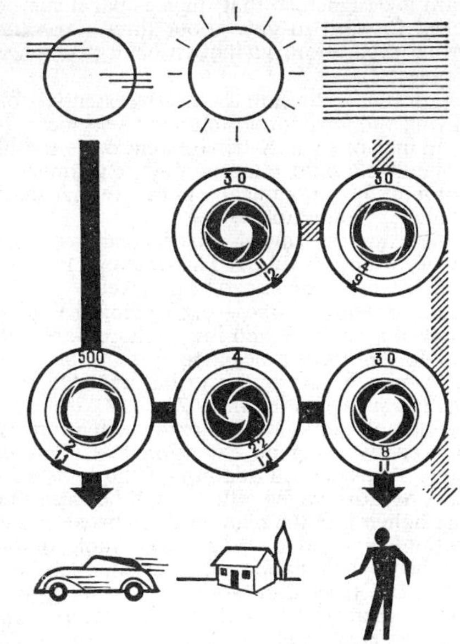

EXPOSURE VALUES. A shutter scaled in light values has coupled aperture and speed settings. Bottom: for a given light value (e.g., 11) setting a fast speed automatically opens the aperture, or stopping down sets a slow speed, so as to keep the exposure constant. To change the exposure for different light conditions, simply alter the light value setting.

that each number is equivalent to half the exposure of the next lower number, and twice the exposure of the next higher number.

Any given exposure value covers a range of aperture shutter-speed combinations that yield the same exposure. For instance, the exposure value 13 gives a range of aperture/shutter-speed combinations—e.g., 1/500 second at *f* 4, 1/250 second at *f* 5·6, or 1/125 second at *f* 8, or 1/60 second at *f* 11—each of which gives the same amount of exposure. Similarly, an exposure value of 9 gives 1 second at *f* 22, ½ second at *f* 16, ¼ second at *f* 11, ⅛ second at *f* 8, etc.—again all the same thing in terms of the amount of light received by the film during the exposure.

The aperture and shutter speed settings on the shutter are coupled to the exposure value control; once an exposure value has been set, stopping down the aperture—e.g., to increase the depth of field—will automatically adjust the shutter to a correspondingly slower speed. Alternatively, setting the shutter to a fast speed (e.g., to cope with rapid subject movement) automatically opens up the diaphragm aperture.

The exposure therefore remains automatically constant for that exposure value if either shutter or diaphragm settings are altered.

The exposure value system can also be used for estimating exposures by rule-of-thumb. Thus an average subject in bright weather may need an exposure value setting of 12. For a brilliant subject it would be correct to advance the setting to 13; or if the sun goes in, to reduce it to 11 or 10, and so on.

In the same way filter factors can be considered as reducing the exposure value: for a factor of 2 the exposure value must be reduced by 1, for a factor of 3 by 1½, for a factor of 4 by 2, and so on.

Mathematical Equation. The exposure values E.V. of the exposure value scaled shutters are given by the equation:

$$2^{EV} = f^2 \times 1/t$$

where f is the aperture number, and t the exposure time in seconds. This can also be written as:

$$\text{E.V.} = \log_2 (f^2 \times 1/t)$$

or, more conveniently,

$$\text{E.V.} = 3{\cdot}322 \times \log(f^2 \times 1/t)$$

For instance with a shutter speed of ⅛ second ($1/t=8$) and aperture *f* 16 we get:

$$\text{E.V.} = 3{\cdot}322 \times \log(16^2 \times 8) = 3.322 \times 3{\cdot}311 = 11.$$

Therefore this combination corresponds to an exposure value of 11. A combination of ¼ second at *f* 22 would give $3{\cdot}322 \times \log(22^2 \times 4)$ which is $3{\cdot}322 \times 3{\cdot}288$, or 10·93, again approximately 11. Similarly, an equivalent combination of 1/250 second at *f* 2·8 gives an exposure value of $3{\cdot}322 \times \log(2{\cdot}8^2 \times 250) = 3{\cdot}322 \times 3{\cdot}292 = 10{\cdot}95$. All the speeds on the shutter are chosen so that they give exact or very near whole numbers when combined with the standard aperture numbers in the exposure value equation.

The exposure value scale is logarithmic, and can thus cover enormous exposure ranges with a comparatively limited series of numbers. A scale from 2 to 18 is sufficient for all normal photographic purposes, since it represents exposures from 1 second at *f* 2 to 1/500 second at *f* 22; in other words a range of over 1: 60,000.

A scale from about −16 to +22—less than 40 numbers—corresponding to a range of about 1: 250,000,000,000 would cover every conceivable exposure ever made, from an astronomical photograph taken on a long winter night of 18 hours to extreme speed shots of around 1/10,000,000 second.

Exposure values also fit into the Apex system, where the exposure value is the sum of similar single-digit unit values for the lens aperture and the shutter speed.

Cameras with coupled exposure meters work on the same system, including the cross-coupled aperture and speed settings, but have no exposure value scale. The latter becomes unnecessary where no transferring of readings is required from the meter to the shutter. L.A.M.

See also: *Apex system; Diaphragm; Exposure; Exposure Tables; Shutter.*

EXTENSION BELLOWS. Attachment for miniature and other rigid cameras with interchangeable lenses, providing a variable degree of lens extension for close-up and macrophotography. Usually—though not invariably—extension bellows are produced for specific camera models. Since the variable extension requires visual focusing, they are used with reflex cameras or in conjunction with reflex housings or attachments. The focusing range with extension bellows (using a lens of appropriate focal length) frequently covers distances from infinity down to same-size reproduction,

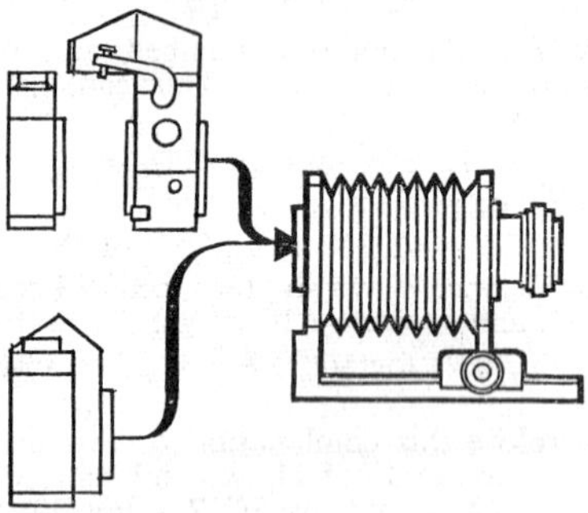

BELLOWS ATTACHMENT. This provides a variable extension for close-up work over a continuous range of distances. Bellows units are usually intended for miniature reflex or rangefinder miniatures in conjunction with a reflex housing. Mostly they are designed for specific camera models.

and often even larger. Various adaptor rings or fittings take different lens units.

The bellows attachment generally consists of a fixed rear member which attaches to the camera and a movable front unit which runs along two or more rails. In many cases the whole bellows attachment can also move forward and backward when mounted on a tripod. The movements are controlled by rack and pinion systems. For reference purposes one or more of the rails may carry numerical scales to enable a given camera setting to be repeated without lengthy experimenting.

See also: *Close-ups; Extension of camera; Extension tube.*

EXTENSION OF CAMERA. Strictly, the maximum distance to which the lens can be moved in front of the sensitive material. The term is also generally applied to the extent of the focusing movement where this involves travel of the lens as a whole. Normally the amount of extension available allows the lens to be focused on objects down to about 3 feet.

With a double extension camera, a second sliding member on the baseboard allows the lens panel to be racked forward to give twice the normal lens-film separation. This permits objects to be photographed the same size.

With a triple extension camera, the baseboard is designed so that the lens panel can be racked forward to give about three times the normal separation, giving an enlarged image of objects.

A double extension is mostly essential for copying and for photographing objects measuring an inch or two. A triple extension is useful for really close-up photography—e.g., macrophotography, botanical photography and philatelic photography.

The extension can also be increased by the use of extension tubes. This method is only possible when the lens can be removed.

Long extension tubes—e.g., giving the equivalent of a double extension or more—are not used for the larger negative formats because they would be too bulky to be practical. For negative sizes greater than $2\frac{1}{4} \times 2\frac{1}{4}$ ins. any considerable amount of extra extension is achieved by using an extension box. This is simply a box-shaped adaptor of the same size as the negative which either fits in between the plate holder and the plate back or between the lens and lens panel. Where the whole outfit must be reasonably portable, the extension box may be constructed in the form of a collapsible bellows. Extensions of this type are used on plate cameras when working with lenses of abnormally long focal length.

Extension and *f*-number. The marked value of the stop can be accepted as the actual effective aperture so long as the object is farther away than ten times the focal length of the lens. In practice this condition usually holds good within the focusing range provided by the normal single extension. But when a double or triple extension is used, the effective value of the aperture is appreciably reduced.

The practical rule for finding the new value of the *f*-number is to multiply the marked *f*-number by $N/(N-1)$ where N is the number of times the object distance is greater than the focal length of the lens.

For example, if the object is one foot away when it is focused sharply by a 6 ins. lens (so that $N = 2$), then the effective aperture at any *f*-number is given by *f*-number $\times\ 2/(2-1)$—i.e., *f*-number $\times$ 2. So that a stop marked *f* 4 would have to be regarded as *f* 8 when working out the exposure.

Extension and Magnification. The final purpose of devices for increasing the camera extension is to obtain a bigger image than the normal close limit of focusing will permit. If the scale on which the subject is to be reproduced is S (i.e., size of the object $\times$ S = size of image) then the amount of extension required is given by $f \times S$.

For example, if the object is to be reproduced to a scale of half life size with a lens of 2 ins. focal length, the length of extension required would be $E = 2 \times \frac{1}{2} = 1$ in.

Extension and Image Distance. When a lens of focal length f is required to work at a distance u from the subject, the amount d it

APPROXIMATE OBJECT DISTANCES (INCHES) AND SCALES OF REPRODUCTION

Focal Length of Lens in.	cm.	*Object Distances for Scale of Reproduction* 0·1 (1 : 10)	0·13 (1 : 8)	0·17 (1 : 6)	0·2 (1 : 5)	0·25 (1 : 4)	0·33 (1 : 3)	0·5 (1 : 2)	0·67 (2 : 3)	1 (1 : 1)	1·5 (3 : 2)	2 (2 : 1)	2·5 (5 : 2)	3 (3 : 1)
2	5	22	18	14	12	10	8	6	5	4	3¼	3	2¾	2¾
2½	6·25	27½	22½	17½	15	12½	10	7½	6¼	5	4¼	3¾	3½	3¼
3	7·5	33	27	21	18	15	12	9	7½	6	5	4½	4¼	4
3½	9	38½	31½	24½	21	17½	14	10½	8¾	7	5¾	5⅜	4⅞	4¾
4	10	44	36	28	24	20	16	12	10	8	6⅝	6	5⅝	5¼
4½	11·2	49½	40½	31½	27	22½	18	13½	11¼	9	7½	6¾	6⅜	6
5	12·5	55	45	35	30	25	20	15	12½	10	8⅜	7½	7	6¾
5½	13·7	60½	49½	38½	33	27½	22	16½	13¾	11	9¼	8⅜	7¾	7⅜
6	15	66	54	42	36	30	24	18	15	12	10	9	8½	8
7	17·5	77	63	49	42	35	28	21	17½	14	11¾	10½	9¾	9¼
8	20	88	72	56	48	40	32	24	20	16	13¼	12	11¼	10¾
10	25	110	90	70	60	50	40	30	25	20	16⅝	15	14	13¼
12	30	132	108	84	72	60	48	36	30	24	20	18	16¾	16

APPROXIMATE OBJECT DISTANCES (CENTIMETRES) AND SCALES OF REPRODUCTION

Focal Length of Lens cm.	in.	*Object Distances for Scale of Reproduction* 0·1 (1 : 10)	0·13 (1 : 8)	0·17 (1 : 6)	0·2 (1 : 5)	0·25 (1 : 4)	0·33 (1 : 3)	0·5 (1 : 2)	0·67 (2 : 3)	1 (1 : 1)	1·5 (3 : 2)	2 (2 : 1)	2·5 (5 : 2)	3 (3 : 1)
5	2	55	45	35	30	25	20	15	12·5	10	8·3	7·5	7	6·7
6·25	2½	69	56	44	37·5	31·2	25	18·8	15·6	12·5	10·4	9·4	8·8	8·3
7·5	3	83	68	53	45	37·5	30	22·5	18·8	15	12·5	11·3	10·6	10
9	3½	99	81	63	54	45	36	27	22·5	18	15	13·5	12·6	12
10	4	110	90	70	60	50	40	30	25	20	16·7	15	14	13·3
11·2	4½	123	101	78	67	56	45	33·6	28	22·4	18·7	16·8	15·6	15
12·5	5	137	113	88	75	63	50	37·5	31·2	25	20·8	18·8	17·5	16·7
13·7	5½	150	124	96	82	69	55	41	34·8	27·4	22·8	20·5	19·2	18·2
15	6	165	136	105	90	75	60	45	37·5	30	25	22·5	21	20
17·5	7	192	158	122	105	88	70	53	44	35	29·2	26·7	24·4	23·3
20	8	220	180	140	120	100	88	60	50	40	33·3	30	28	26·7
25	10	275	225	175	150	125	100	75	63	50	42	37·5	35	33·3
30	12	330	270	210	180	150	120	90	75	60	50	45	42	40

must be racked forward from the infinity mark is given by the equation:

$$d = \frac{f^2}{u - f}$$

For example, if a 10 cm. lens is to photograph an object 0·5 m. (50 cm.) away, the camera extension must be:

$$\frac{100}{50–10} = \frac{100}{40} = 2{\cdot}5 \text{ cm.}$$

Extension and Image Size. The size v of the image of an object of size u formed at a given extension d by a lens of focal length f is:

$$v = \frac{d\,u}{f}$$

For example, a 5 cm. lens focused on an object 1·5 cm. high with an extension of 2 cm. will produce an image

$$= \frac{2 \times 1{\cdot}5}{5} = 0{\cdot}6 \text{ cm. high.}$$

That is, two-fifths the original size. F.P.

See also: *Bellows; Close-ups; Extension tube; Macrophotography; Optical calculations; Reflex attachment; Technical camera.*

EXTENSION TUBE. Tube of specific length threaded to take the camera lens at one end and to screw into the mount at the other. Extension tubes are used to increase the working separation between the lens and the sensitive material to give a magnified image. They are used mainly for miniature and small cameras where they do not need to be of any great diameter. On larger cameras the extension is usually provided in the form of double or triple extension bellows.

Extension tubes are usually available in standard lengths: ½ in., 1 in., 2 ins., etc. They extend the focusing range in a number of fairly short zones. The focusing movement of the lens itself helps to bridge the gap between one fixed extension and the next.

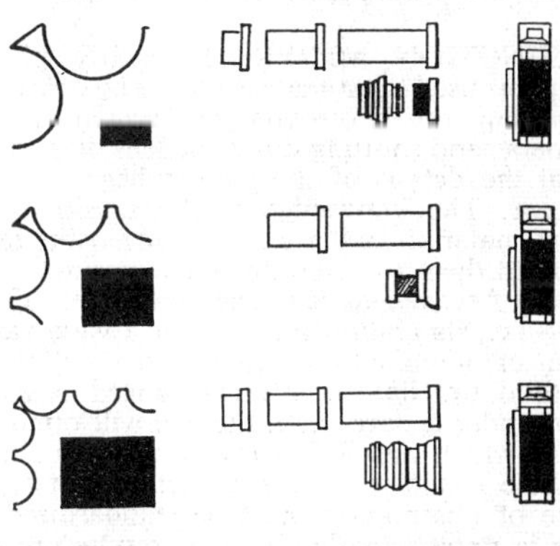

EXTENSION TUBES. A set of tubes used with an interchangeable lens camera will yield a range of magnifications up to several times natural size. For greatest magnification use most tubes and shortest focus lens.

APPROXIMATE SCALE OF REPRODUCTION WITH EXTENSION TUBES

Focal Length of Camera Lens		*Scale of Reproduction with Extension Tube of*									
in.	*cm.*	*½ in. 1·3 cm.*	*¾ in. 1·9 cm.*	*1 in. 2·5 cm.*	*1½ in. 3·8 cm.*	*2 in. 5 cm.*	*2½ in. 6·3 cm.*	*3 in. 7·5 cm.*	*3½ in. 9 cm.*	*4 in. 10 cm.*	*6 in. 15 cm.*
2	5	0·25–0·31	0·38–0·44	0·5–0·56	0·75–0·81	1·00–1·06	1·25	1·5	1·75	2	3
2½	6·25	0·2–0·27	0·3–0·37	0·4–0·47	0·6–0·67	0·80–0·87	1·00–1·07	1·2	1·4	1·6	2·4
3	7·5	0·17–0·25	0·25–0·33	0·33–0·41	0·5–0·58	0·67–0·75	0·83–0·91	1·00–1·08	1·16	1·33	2
3½	9	0·14–0·24	0·21–0·31	0·29–0·39	0·43–0·53	0·57–0·67	0·72–0·82	0·86–0·96	1·00–1·10	1·14	1·7
4	10	0·12–0·24	0·19–0·31	0·25–0·37	0·37–0·49	0·50–0·62	0·63–0·75	0·75–0·87	0·87–0·99	1·00–1·12	1·5
4½	11·2	0·11–0·26	0·17–0·32	0·22–0·37	0·33–0·48	0·44–0·59	0·55–0·70	0·67–0·82	0·78–0·93	0·89–1·04	1·33
5	12·5	0·10–0·17	0·15–0·22	0·20–0·27	0·27–0·34	0·40–0·47	0·50–0·57	0·60–0·67	0·70–0·77	0·80–0·87	1·2
5½	13·7	0·09–0·16	0·14–0·21	0·18–0·25	0·27–0·34	0·36–0·43	0·45–0·52	0·55–0·62	0·64–0·71	0·73–0·81	1·1
6	15	—	0·12–0·21	0·16–0·25	0·20–0·29	0·33–0·42	0·42–0·51	0·50–0·59	0·57–0·66	0·67–0·76	1
7	17·5	—	0·11–0·22	0·14–0·25	0·21–0·32	0·29–0·40	0·36–0·47	0·43–0·54	0·50–0·61	0·57–0·68	0·85–0·96
8	20	—	—	0·12–0·18	0·19–0·25	0·25–0·31	0·31–0·37	0·38–0·44	0·44–0·50	0·50–0·56	0·75–0·81
10	25	—	—	—	0·15–0·22	0·20–0·27	0·25–0·32	0·30–0·37	0·35–0·42	0·40–0·47	0·60–0·67
12	30	—	—	—	—	0·17–0·26	0·21–0·30	0·25–0·34	0·29–0·38	0·33–0·42	0·50–0·59

There is not much point in using short extension tubes with long focus lenses, as the focusing extension is often longer than the extension tube.

All the calculations for normal camera extension apply equally to extension tubes. The *f*-number correction is also the same.

The third table above provides a range of scales of reproduction for each lens-extension tube combination. The lower limit of the range applies to the lens when focused on infinity, the higher value to the lens focused on a close-up subject. The nearest focusing point of the lens alone is taken as about 3 feet (1 metre) for lenses up to 4½ ins. focal length, about 6 feet (2 metres) for 5–7 ins. lenses, and about 12 feet (4 metres) for 8–12 ins. lenses. The longer focus lenses are considered more as narrow angle lenses for miniature cameras, since the relatively narrow extension tubes would prevent them from covering the plate of a larger camera.

With the very long extension tubes the actual range of scales of reproduction obtainable by varying the focusing setting is comparatively limited. So only one value is given in the table above. L.A.M.

See also: *Close-ups; Extension bellows; Extension of camera; Macrophotography; Optical calculations.*

EXTINCTION METER. The early photographers used to estimate exposure by observing the image on the ground-glass screen of their cameras and shutting down the lens diaphragm until the details of the picture began to disappear. The knowledge of the sensitivity of their material and a simple conversion table allowed them to determine the exposure.

Visual Accommodation. The adaptation of the eye—i.e., its ability to see under a wide variation of illuminations—makes the extinction method unreliable. After prolonged observation under a dark cloth the eye will often see detail which was at first invisible. So to obtain reliable results with any extinction meter, the time of observation must be standardized. If this is done remarkably good results are obtained, even by different observers using the same meter.

Construction. Since the majority of cameras have no ground-glass screen, separate extinction meters have been developed. Many modern extinction meters use a wedge of graduated density with a series of figures behind it. This is mounted in a tube or a flat tablet through which the scene to be photographed is observed. The last figure which can still be seen is referred to a table, adjusted for the film speed, from which the exposure is read off directly.

Another type of extinction meter is simply a small camera in which a lens throws an image of the scene on to a small screen and in which the diaphragm is changed until the detail disappears. In another rather similar instrument the scene is used to illuminate a piece of opal glass against which a distinctive figure is placed. The diaphragm is altered until the figure becomes indistinct and the diaphragm reading gives the exposure directly from a table.

At one time, a diaphragm attachment was available for twin lens reflex cameras enabling the viewing screen of the camera to be used as an extinction meter.

Using Extinction Meters. The commonest source of error in using an extinction meter is to take too long over a reading. The remarkable powers of accommodation of the human eye will make it adjust itself to see faint figures in the meter that were invisible at the start. The extinction meter is probably the only instrument that gives more accurate results when it is used quickly and carelessly than when it is treated deliberately.

So with all extinction meters which are held close against the eye, the time of measurement must be standardized. This is not necessary with the type of meter which is held well away from the eye since in this case the general state of adaption of the eye is not altered. This type of meter is probably less reliable than the others and is normally available only in rather cheaply made forms.

Since the extinction meter makes use of a rather ill-defined property of the human eye, it is not surprising to find that different users obtain different results with the same type of meter. So every user should calibrate his own meter to suit his personal characteristics. An extinction meter calibrated in this way can be a very reliable instrument.

Most extinction meters are designed to accept light from the whole of the scene. This fact may give rise to misleading readings. The meters should never be pointed directly at the sun or they will give meaningless readings. W.F.B.

See also: *Actinometer; Exposure; Exposure meter; Exposure tables; Photometer.*
Books: *How to Use Your Exposure Meter,* by L. A. Mannheim (London); *Exposure,* by W. F. Berg (London).

EYE. Light-sensitive organ of vision. The human eye is a very highly developed organ; it is, in fact, part of the brain. In the embryo it begins as an outgrowth from the rudimentary brain, and later develops into the eye. While the eyes of animals and birds are more highly developed for special purposes, the human eye is probably the best for all-round performance.

Anatomy of the Eye. Like the camera, the eye has a lens, a shutter (the eyelid), a variable aperture stop or diaphragm (the pupil), and a light-sensitive surface (the retina).

The lens consists of a transparent elastic substance whose curvature—and consequently, focal length—are controlled by a ring of muscles round its edge. The eye is focused on objects at different distances by altering the focal length of the lens. The shape of the cornea however, contributes to the total focusing power. Due to the fact that the eye is filled with fluid, the internal and external focal lengths are different. For distant vision, the former is 22 mm. the latter, 17 mm. The diameter of the lens is about 10 mm. and the angle of view is between 150° and 170° depending on the orientation.

The lens itself produces considerable chromatic and other aberrations, but the interpre-

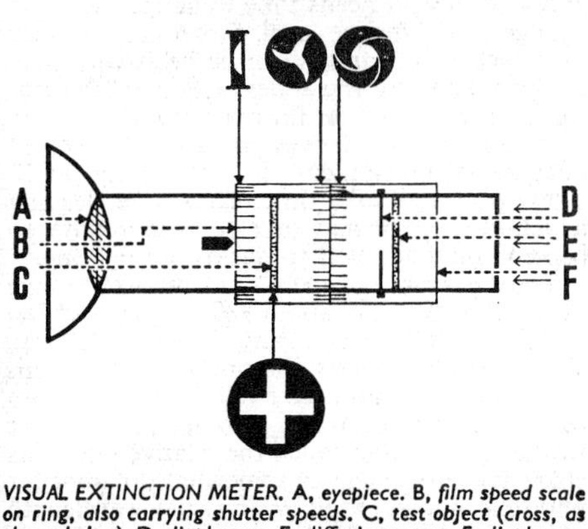

VISUAL EXTINCTION METER. A, eyepiece. B, film speed scale on ring, also carrying shutter speeds. C, test object (cross, as shown below). D, diaphragm. E, diffusing screen. F, diaphragm adjusting ring carrying also aperture scale. In use the diaphragm is closed down until test object just disappears with meter pointed at subject; scales then indicate exposure settings.

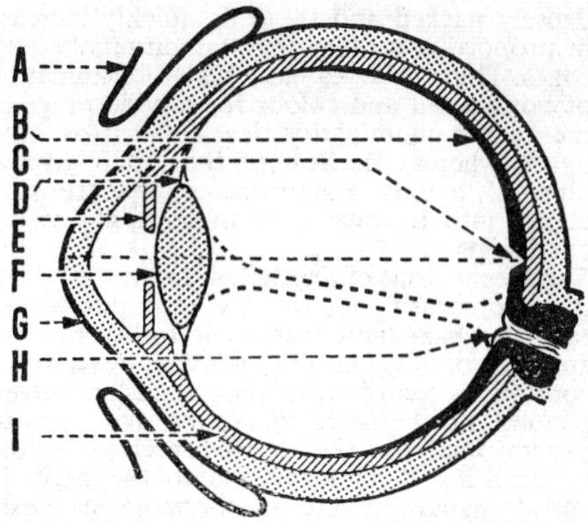

THE HUMAN EYE. A, conjunctiva. B, retina. C, fovea centralis (yellow spot). D, ciliary muscles controlling the lens. E, iris. F, lens. G, cornea. H, blind spot, being the ending of the optic nerve; is virtually insensitive. I, choroid.

tive mechanism of the brain normally allows for these defects. Spherical aberration, coma and astigmatism, although present, are reduced by the design of the eye, and the considerable field curvature is offset by the curvature of the retina.

The pupil of the eye is a variable aperture situated in front of the lens and composed of a ring of muscle which lies in the coloured part of the eye. This muscle controls the size of the lens aperture and gives it a normal working range of about *f* 4 to *f* 8 (·15 in. to ·075 in. diameter, or 3·8 mm. to 1·9 mm.). The eye can adjust quickly between these limits, but, given time, it can cover a range from *f* 1·8 to *f* 12 (·33 in. to ·05 in. diameter, or 8·5 mm. to 1·25 mm.).

The size of the pupil is controlled by a reflex action from the brain; it automatically contracts in bright light and dilates in dull.

The Retina. The area of the inner wall of the eye ball is divided into a multitude of tiny light-sensitive cells which comprise the retina. These cells feed signals about the light falling on them to the brain via the optic nerve. There appear to be two different kinds of cell: firstly, the cones, which are used mainly in day vision and have high discrimination of colour and intensity level; and secondly the rods, which are mainly used in night vision, are much more sensitive but have no colour discrimination. There are about 7 million cones and about 120 million rods in the eye but they appear to be interconnected in large groups since there are only 800,000 nerve fibres in the optic nerve.

The rods and cones are not uniformly distributed over the area of the retina. A small area on the optic axis of the eye called the fovea is very rich in cones and has few if any rods present. Here the cones are spaced at intervals corresponding to about 700/mm. As we move out from this area, the cells become much less densely packed and the rods quickly increase in proportion so that they far outnumber the cones. Thus the fovea has a high discrimination of both detail and colour for a wide range of the higher illumination levels (such as daylight), whereas the rest of the retina gives a blurred, mainly monochrome, picture and comes into its own only under night vision conditions.

The Mechanisms of the Retina. Over most of its range of sensitivity, the eye responds to the rate at which light reaches it rather than to amounts of light; in other words it deals with a continuous record rather than an instantaneous picture. Furthermore the retina and nervous system analyses the image received before transmitting the information to the brain in coded signals. There is therefore a closer analogy with a television system than with a photographic system.

If, however, a very short flash (less than $\frac{1}{5}$ second) of light enters the eye, the retina no longer responds to the rate at which light is arriving, but as if the amount of light in the flash had been spread uniformly over $\frac{1}{5}$ second. This is termed the persistence of vision. It implies that if a series of flashes arrives at a rate higher than about 5/second they will seem to merge together into a continuous illumination.

The signals sent to the brain by the retinal cells record only changes in light intensity. If the light is uniform very little information is transmitted. This method of analysis determines the way the eye is used. The head is moved to direct the optic axis towards the interesting part of the scene whose image then falls on the foveal region. The muscles holding the eye introduce a small continuous involuntary wobble so that the image is continually moved over the fovea. The resulting varying light intensities on the cones are a highly important part of normal vision. If the "wobbles"

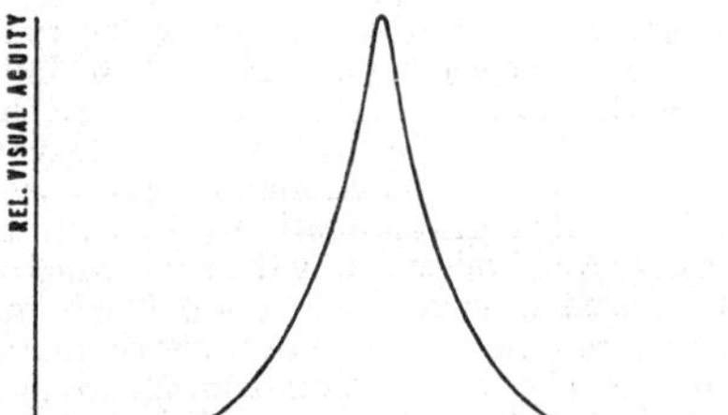

VARIATION OF DETAIL DISCRIMINATION ACROSS RETINA. The visual acuity (sensitivity to detail) is highest in day. vision when the cones are active (solid curve). The peak corresponds to the fovea and the gap to the so-called blind spot. For night vision when the rods are in use, acuity is very much reduced (dotted curve). It is in fact lowest in the fovea since this part of the retina contains hardly any rods.

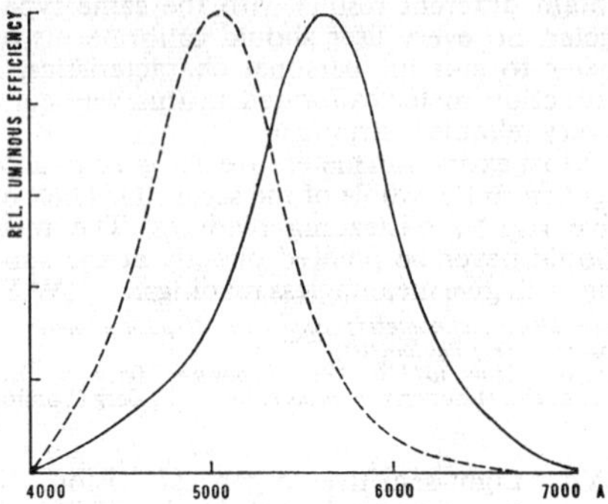

RELATIVE LUMINOUS EFFICIENCY CURVES. The solid curve shows the luminous efficiency of the cones in the retina which operate for day vision; the dotted curve corresponds to the rods which operate for night vision.

are stopped by an artificial optical device in front of the eye, normal vision soon disappears.

Sensitivity. The mechanisms of vision operate over an enormous range of brightness (properly called luminance) from about 1,000,000 ft.-lamberts to 1/1,000,000 ft.-lambert—a range of 10^{12} : 1. The upper limit (10^6 ft.-lamberts) is set by the pain induced by intensely high lumniances. A pure white object in strong sunlight might have a luminance of 10,000 ft-lamberts, in artificial room lighting it might be 10 ft.-lamberts and in moonlight it might be 1/100 ft.-lambert.

The cones in the retina, with their high discrimination of colour and detail, act over the range 10^6 to 10^{-3} ft.-lamberts, and the rods overlap this with their range 10 to 10^{-6} ft.-lambert. The lower region where only rods are effective is called scotopic vision, the upper where only cones respond is photopic vision, and the region of overlap, mesopic vision.

The eye cannot see all this range of luminances at once; it needs time to adapt itself to a change in luminance level. Small changes (say by a factor of 5 or 10) can be dealt with relatively quickly by the reflex action of the iris, just as one might alter the aperture of a camera lens. But larger changes involve complicated adaptation phenomena in the retina itself. These processes take time. Thus, after leaving a well-lit room, it may take 10 to 15 minutes to become fully adapted to darkroom lighting.

The eye is sensitive to rays of light over a range of wavelengths from about 380 mμ to about 750 mμ although the useful range is 400 to 700 mμ. The sensitivity varies continuously over this range with a pronounced maximum in the green region of the spectrum, and its plot against wavelength is often called the relative luminous efficiency curve. This is somewhat different for photopic and scotopic vision, the wavelength of maximum sensitivity varying from 555 mμ for the former to 510 mμ for the latter F.J.B.W.

See also: *Colour vision; Vision.*

EYE CAMERA. Special motion-picture camera designed to record the movements of a sitter's eyes when looking at a picture for the first time. The film is analysed to show the order in which the various items and points of interest in the picture attract the attention of the viewer. Records of this type are made in the course of research into the functional value of picture composition and layout.

The term is also applied to the special cameras used in medicine and dentistry for taking close-up pictures—e.g. of the human eye or mouth. Cameras of this type are designed for operation by nurses and assistants without a particular photographic training. The camera is a self-contained unit which may include a built-in flashgun or electronic flash head, some device for centring the subject at a fixed distance from the camera lens and a handle or handles incorporating the shutter release.

On some cameras of this type the subject is located by four pointers attached to the camera body which fix the distance and mark the corners of the field covered by the lens. Or the camera may project a pair of light beams which converge on the subject when it is at the correct distance.

The lens works at a fixed distance setting and once the aperture is adjusted for the speed of the film it never needs to be changed. The duration of the electronic flash is short enough to freeze any movement of the subject or camera and the shutter speed is short enough to prevent a secondary image from being recorded by the ambient lighting.

This type of camera is also used for making record photographs and copying in a number of applications where the size of the subject does not vary much. Since the flash source is mounted close to the camera lens, the lighting tends to be flat and shadowless, but does show full detail of the subject.

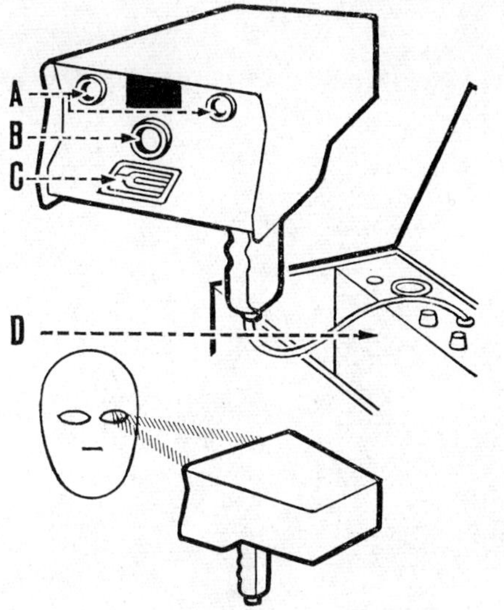

EYE CAMERA: Top: *The camera has a lens set to a close subject distance and projects two rangefinder beams converging on the same distance. An electronic flash tube below the lens provides the illumination.* A, *rangefinder windows.* B, *lens.* C, *flash tube.* D, *power pack.* Bottom: *When the light beams from the rangefinder windows exactly converge on the subject, the latter is at the correct distance for a sharp picture.*

See also: *Close-ups; Macrophotography; Medical Photography; Ophthalmic photography.*
Book: *Longmore's Medical Photography* (London).

F

FABRE, CHARLES, 1867–1945. French physicist. Director of the Institut d'Optique (Institute of Optics), Paris. Writer of books on photography and optics. Biography: *Charles Fabre, Jubilé scientifique, 3, December* 1937 (Paris 1938).

FABRIC PRINTING. There are several ways of making a photographic print on fabric. In some methods—e.g., diazotype—the image is produced by dyeing the actual fabric; in others, the fabric is simply used as a support for a silver or pigment image.

Commercial materials. Several manufacturers produce sensitized fabrics for various purposes, ranging from tracing cloths to continuous-tone photo linens.

Sensitized tracing cloth is used in drawing offices to make master positives for dyeline or diazo printing. It consists of a silver chloride or bromide emulsion coated on a varnished or otherwise water-proofed translucent cloth base. The material is exposed and processed like any other document copying material. The finished print usually takes pencil or ink to permit modifications or corrections to the drawing before mass reproduction.

Continuous-tone sensitized fabrics are intended for contact printing or enlarging like any normal photographic printing paper. The emulsion here is impregnated into the fabric (e.g. linen). The latter is semi-stiff when dry, but goes limp on immersion in the processing solutions. It can thus be folded up, making it possible to develop large prints in medium sized dishes. Exposure and processing follow the normal procedure; for drying the material may have to be stretched on frames to ensure a really flat print. Alternatively the dried linen print can be re-wetted and ironed with a household iron set to low heat. (The intermediate drying is usually desirable to ensure that the image is firmly embedded in the fabric.)

Home-made Formulae. There are various formulae for preparing fabrics for daylight or artificial light printing. Most of them yield a contact speed material too slow for projection printing. If the fabric area to be covered is larger than the negative chosen for a print, an enlarged negative must first be made. The negative and fabric are in this case placed in a printing frame for the exposure.

Printing-out Method. The fabric to be printed is first well washed (preferably boiled), immersed in a sizing solution for 3 minutes, and hung up to dry. A formula for a sizing solution is:

Arrowroot	88 grains	5 grams
Sodium chloride	88 grains	5 grams
Glacial acetic acid	¾ ounce	20 c.cm.
Tannic acid	88 grains	5 grams
Distilled water to make	10 ounces	250 c.cm.

The tannic acid is dissolved by boiling in about half the water. The arrowroot is made into a paste with a few drops of water, and the rest of the water is brought to the boil and poured on to it. The other ingredients are then dissolved, and the tannic acid solution is added.

After sizing, the material is immersed for a few minutes in a 10 per cent solution of silver nitrate in distilled water containing 0·1 per cent nitric acid, and dried in total darkness.

It can then be printed out, toned, fixed, etc., like ordinary printing-out paper. The fabric should be handled carefully in the printing frame to prevent it from moving and thus printing double. As a further safeguard, it may be stuck down temporarily with rubber solution mountant on to a sheet of stiff paper.

Diazotype Printing. Diazotype images are based on the decomposition of dye-forming compounds by light. In diazo printing, the final dye image produced after coupling is faintest in the areas of the print that have received the greatest exposure. So a positive transparency must be used to give a positive image in the print.

The fabric to be printed on is soaked in a hot (60-70° C.) 1 per cent solution of primuline (not a primuline dye) for five to ten minutes. During that time it is worked about with a glass rod to ensure thorough penetration and even distribution of the primuline. It is then drained, rinsed in cold water, and hung up, preferably in a current of warm air, until it is quite dry.

It is next immersed (in the dark) in a sensitizer solution containing 0·4 per cent sodium nitrite and 0·6 per cent oxalic acid. This solution must be made up cold and used immediately. The fabric is then blotted between two sheets of blotting paper, and exposed straight away in a printing frame behind the positive transparency.

The material is exposed until a small test piece, exposed directly to the printing light alongside the frame, gives only a slight colour when it is developed.

Still in the darkroom, the exposed fabric is then soaked in one of the developer solutions below, and then thoroughly washed for about thirty minutes.

After exposure the transparency should be washed to remove traces of sensitizer left on it by the moist fabric.

DIAZOTYPE DEVELOPERS

Chemicals	*Concentration of Ingredients and Colours*					
	Yellow	*Orange*	*Brown*	*Red*	*Purple**	*Blue*
Phenol	1%	—	—	—	—	—
Resorcinol	—	0·6%	—	—	—	—
Sodium hydroxide	—	0·9%	—	0·8%	—	—
Pyrogallol	—	—	1·2%	—	—	—
Beta-naphthol	—	—	—	0·6%	—	—
Alpha-naphthylamine	—	—	—	—	1·2%	—
Oxalic acid	—	—	—	—	0·1%	—
Eikonogen	—	—	—	—	—	1·2%

*After this developer the material should be rinsed in a 5 per cent tartaric acid solution, and dried without further washing. Any remaining naphthylamine can be removed from the hands and dishes by rinsing with the sodium nitrite and oxalic acid solution followed by clean water.

Iron-Silver Printing. This method makes use of the action of light in reducing ferric salts to the ferrous state in which they can be made to interact with silver nitrate to produce a silver image. It gives positives from negatives.

The material is immersed, in the dark or by a red safelight, in the following solution and poked about thoroughly with a glass rod for five minutes:

Ferric oxalate	4 ounces	100 grams
Oxalic acid	½ ounce	12·5 grams
Silver nitrate	1½ ounces	37·5 grams
Distilled water to make	20 ounces	500 c.cm.

The oxalic acid and oxalate are dissolved in about half of the water, and the silver nitrate in the rest. The solutions are then mixed when required. The sensitizer is kept in the dark.

After immersion the fabric is drained, dried, and ironed out smooth, all by the light of a yellow safelight of the kind used for contact papers. It is printed out until the image appears about half as dark as required, washed for ten minutes in the darkroom in running water, and fixed in a 5 to 10 per cent plain sodium thiosulphate solution. The process is completed by washing and drying the fabric.

Photo-Mordant Dye-Printing. Fabrics may also be printed by using the photographic image as a basis for a dye. One way of doing this is to wash the fabric well and then sensitize it with a solution of potassium or ammonium bichromate. After drying, the fabric is placed under the negative and exposed to daylight until a faint image appears. This image consists of chromium hydroxide which acts as a mordant for such dyes as alizarine and anthracine. The fabric is simply boiled in a solution of the appropriate dye, cleared in a solution of sodium hypochlorite, washed in soap and water, and dried.

Dyes may also be formed in the fabric by using the chromic acid available in the image to oxidize an organic compound.

Images formed in either of these ways are as permanent as the dyes used. The principle can also be applied to prints on a paper base.

Artificial Light. Fabrics like canvas can be sensitized with an emulsion fast enough for contact or projection printing by artificial light.

All operations, including the preparation of the fabric, must be carried out in total darkness or by a deep red safelight. (Absorption of light by the fabric, even before coating, is said to fog the material.)

The fabric is soaked in soap-suds for about twelve hours, washed well in running water, and dried in the dark. It is then sprinkled with water and ironed with a moderately hot iron. It is next soaked in the presensitizing solution for 20 minutes. A suitable presensitizer is:

Potassium bromide	70 grains	4 grams
Cadmium bromide	½ ounce	12·5 grams
Potassium iodide	70 grains	4 grams
Distilled water to make	40 ounces	1,000 c.cm.

The fabric is then dried and sensitized in the following solution:

Silver nitrate	125 grains	7 grams
Citric acid	30 grains	1·8 grams
Distilled water to make	10 ounces	250 c.cm.

For sensitizing, the fabric is best laid in the bottom of a clean dish. The sensitizer is then poured over it, and any air bubbles are broken up with a glass rod. After about five minutes, the solution is poured off and the fabric is dried in the dark.

The sensitized material can then be treated in the same way as an ordinary bromide development paper. The speed of the emulsion varies with the temperature and other working conditions, so exposures have to be found by trial. L.A.M.

See also: *Printing on special supports; Stripping.*
Book: *The Complete Art of Printing and Enlarging,* by O. R. Croy (London).

FACTORIAL DEVELOPMENT. Method of developing by inspection in which the time to the first appearance of the image is multiplied by a factor to give the total development time.

The method was first suggested by A. Watkins, in 1893, for which reason the factor is generally referred to as the Watkins Factor. For those who develop by inspection, the method is reliable, although with fast panchromatic materials there is always a risk of fogging as the plate is scrutinized. Each developer has its own factor.

See also: *Developing negatives.*

FADE. Way of starting or finishing a scene in a motion picture film by making it emerge gradually out of darkness (fade in), or darkening it on the screen to complete blackness (fade out).

A fade-out should usually be followed by a fade-in of the next scene. This type of transition implies a passage of time or a change of subject, much like a chapter in a book.

A fade-out superimposed on a fade-in produces a dissolve which often implies a change of location without a change of time, or a transition between related ideas.

In the camera a fade-out can be produced by gradually closing down the lens aperture or reducing the shutter opening of the camera. A fade-in is obtained in the reverse manner. In professional motion picture practice these effects are produced by the film laboratory during printing.

See also: *Cine technique.*

FADING. Disappearance or discoloration of silver image in negatives and prints, especially the latter. It is due to insufficient washing or occasionally to insufficient fixing. With prints, fading may also be due to the use of mountants containing acid or traces of certain oxidizing agents such as free chlorine.

A slightly faded image can sometimes be restored by bleaching it in a permanganate bleacher containing potassium bromide or chloride, followed by a rinse in sodium bisulphite solution, a good wash, and redevelopment in a normal developer (e.g., M.Q. or Amidol). Alternatively, it may be copied on a blue-sensitive film or plate which will tend to reproduce the yellowish faded areas almost as dark as the unfaded ones.

For maximum protection against fading, prints may be toned with sulphide or gold; they are then less easily attacked by oxidizing agents.

See also: *Faults.*

FAHRENHEIT SCALE. System of thermometry (called after its originator, the German physicist G. D. Fahrenheit) based on the mistaken assumption about the coldest temperature attainable. Fahrenheit based his scale on the temperature of a particular freezing mixture, and with that as his zero, he graduated the mercury thermometer to register the boiling point of water as 212°. On this scale, the melting point of ice occurs at 32°. In spite of its unsound basis, the Fahrenheit scale has conveniently spaced degrees for most practical purposes. Until recently the standard temperature scale in Great Britain, it is now gradually being replaced by the Centigrade scale.

See also: *Thermometer; Weights and measures.*

FARADAY, MICHAEL, 1791–1867. English scientist. Showed for the first time a few of Fox Talbot's photogenic drawings at the Royal Institution on 25th January, 1839. He also gave the first public description of the process on this occasion.

Faraday was also interested in "the apparent animate movement of objects from the inanimate", by which he meant the optical illusion caused by the rotation of the phenakistiscope or zoetrope. Biography by S. P. Thompson (1898).

FARMER, ERNEST HOWARD. 1860(?)–1944. English photographer. First Head of the Photographic Department of the Regent Street Polytechnic, London. Invented reducing bath (1883) (Farmer's Reducer) consisting of hypo to which is added ferricyanide of potash. Described in 1894 the reaction between the silver image of a bromide print, bichromate, and gelatin, which is the basis of the bromoil process.

FARMER'S REDUCER. Name sometimes given to the potassium ferricyanide reducer after the originator, E. Howard Farmer.

FASHION PHOTOGRAPHY

Fashions in fashion photography change almost as fast as they do in clothes. In the past few years this branch of photography has been revolutionized in Britain, America and France.

There is a changed market for the fashion industry, and thus for the fashion photographer and the publishers of his pictures. When the glossy fashion magazines started, their market and that of the couture houses which produced fashion, was the rich, sophisticated, middle-aged woman. Her tastes and habits dictated the style of fashion photographs and fashion magazines up to during and just after World War II. Then, the points of interest about a garment were the texture of the cloth, its hang, its decoration, and the setting it was seen in. Techniques—lighting, cameras—were adapted to show these details.

But since the War, and particularly within the last five years, there has been a great change. A vast new market has appeared: the young wage-earning woman. Her demands are rather different. She wants smart cheap clothes for everyday wear. She is unimpressed by embroidery, decoration, fussiness. So the manufacturer discards the elaborate fabrics and the costly handwork and turns to large volume output. To keep the turnover up, styles must change frequently—hence the importance of the two Collections a year in Paris, London and Italy. Clothes must become obsolete before the next batch is ready for the shops, so very distinctive "looks" follow each other with extreme rapidity.

Today the single, 200 guinea "creation" for the rich woman is almost irrelevant as a commercial article. The change is reflected in today's fashion pictures.

Speaking very generally, there is no time today to appreciate the finer details of a garment. It is made, sold and discarded quickly. So the fashion photographer's function is not so much to create an accurate and attractive visual record of the garment, which will be appraised by critical and well-informed buyers, as to project romance and excitement, to suggest that the clothes have magic sewn into them. Wit and artistic sensitivity have become more important in fashion photography than the traditional expertise of the photographer.

Fashion photography divides broadly into two aspects: editorial and advertising.

EDITORIAL

This branch of fashion photography began in the glossy magazines. Basically the editorial purposes of a fashion photograph are: (i) to inform the reader that a garment exists; (ii) to express a preference—this dress, of the many available has been chosen because the editors feel it embodies an important fashion point; (iii) to contribute to the excitement of the magazine; (iv) to stimulate its manufacturer or seller into buying space in the publication.

Organization. All serious fashion magazines create their own illustrations, that is, "hand out" pictures are never used. On a large magazine a great deal happens before the photograph is taken. The fashion editors on the staff keep a continuous watch on the market, and report trends to the editor. Two or three months before publication day the editor, the art director and the fashion editor decide what kinds of clothes they are going to show, and roughly plan the make up of the magazine: say eight pages of Paris Collections, a spread of cheap winter coats, four pages of ski-ing clothes, and so on. A fashion editor is assigned to each feature, who chooses the actual clothes to be photographed, and gets them together in the office, with all the necessary accessories—hats, shoes gloves and jewellery. The art director assigns a photographer, who, with the fashion editor, engages models. The photographer may work full time for the magazine, or be under contract to produce so many pages a year or he may be a freelance. The art director will also specify whether the photographs are to be taken in the studio or on location.

Photographers. With the change in the emphasis of fashion journalism and photography, photographers have also changed. Five years ago, the successful photographer was typically a mature, artistic, sophisticated man. There are still one or two who survive, but the majority now are in their mid-twenties, from a variety of backgrounds—the east end of London, or a university—but they are in touch with their time and the all-important world of the young, and they know how the young women of our time want to look.

As well as this being in touch with the undercurrents of fashion, he must combine a number of talents. He must be a stage director—for part of his job will be the creation of dramatic tableaux. He must be able to make contact with his sitters—whether professional models, or celebrities, and be able to put them immediately at their ease. He must be able to dissect the fashion qualities of a garment, to analyze its line, texture and mood, and be able to devise poses to bring these out. He has to have a feeling for pictorial composition, be able to flatter the girl and the garment, and create a two-dimensional image that is a harmonious whole in itself. He must be able to put some new quality into every picture he takes. And finally he has to be a competent photographic technician, if no more—though this is perhaps the least important part of the job.

Models. Models are now younger, less supercilious, sexier. The professional demands on them are no less, perhaps even more, and the rewards are immense but, as ever, girls with the right qualities are very scarce. A good model is half way to a good fashion picture, and many photographers work with a very small number of girls, perhaps only one, so that the essential sympathy and understanding between them is maintained. A model and photographer working well together show an almost balletic harmony.

Poses. With the use of miniature cameras, fast films and high aperture lenses, it is no longer necessary for the model to hold a rigid pose. Often she is made to move about while the pictures are taken, and in general her positions will be much more natural and easy than in the traditional fashion picture. As the photographer shoots, the well-trained model will make small alterations in the positions of her head and limbs at the rate of perhaps one a second for exposure after exposure.

Composition. If it is possible to generalize about so diffuse a subject, the modern trend in fashion pictures is towards two-dimensional-ness. This in turn is part of a larger trend towards simplifying the composition of the pictures, eliminating details and strengthening the basic shapes. The best contemporary fashion pictures seldom try to suggest a real three-dimensional space in which the model exists. Perhaps this can be traced to the influence of the Impressionists and their followers, perhaps it is due to modern layout which demands much greater adaptability to form the elements of a page make-up.

The illusion of depth—a "hole in the page"—which can look up, or down, or be shallower or deeper, only makes pictures more difficult to place beside each other.

Lighting. Nowadays light is seldom used to achieve three-dimensional modelling, but to create pleasing patterns on the printed page. The apparatus of front, three-quarter and back lighting of the classical set-up has been discarded. In its place there is a very simple, standard set-up which can be used in the studio or found on location with ease. In the studio it imitates window light: a bank, placed to one side of the model, between 45° and 90° to the photographer. The model is placed within 4 or 5 feet of a plain mid-tone background, so that her lit side stands out against a darker wall, and her unlit side is seen in silhouette against a brighter wall. If the picture is full length, a paper backdrop is used which rounds off the corner between the floor and wall, and in effect suspends the model in a dimensionless, shadowless, limbo.

This set-up can be seen in almost any issue of any fashion magazine and, of course, it has numerous adaptations.

Another common arrangement uses an even density of light in all directions, from above, behind, below, in front and both sides. This is particularly effective in pictures of diaphonous lingerie, producing a composition of delicate mid-tones against a white ground.

Another trend is to use a ring light round the camera lens, producing a perfectly even flat light, with shadows round the outer edges of the face, limbs, and so on.

Whatever the lighting set-up used, multiple shadows, such as mar much cinema lighting, are never seen, and single shadows very seldom, and then only when they form a definite part of the composition.

On location this approach to lighting means that pictures can be taken under overcast skies, in shadow, inside rooms, or in flat sunlight with equal success. But English weather being what it is, this last seldom occurs.

When light is available on location, artificial aids are seldom used; that is, reflectors, or fill in-flash. The gain in picture quality is offset by the complication of the equipment and the possible ugliness of the results. With most locations, it is always possible to move to find better light.

Equipment. Most fashion photographs are taken with 35 mm. or 2¼ ins. square reflex cameras, and a variety of lenses. Tripods are not often used, and high speed films allow handheld shooting even in the studio. Coarse grain is often needed to give a romantic quality to the finished picture and, in any case, with the fine blocks of a glossy magazine it is no drawback. Lenses from 18 mm. to 1,000 mm. are used with 35 mm. cameras to give either gross distortion or extreme differential focusing and flattening of the image.

Controls. The basic aim of the high fashion picture is the creation of a mood and an atmosphere.

The photograph is only a means to this end, and a great deal of the effect is obtained in the darkroom. To simplify the composition of the picture, contrast is often manipulated, either by sharply increasing it—sometimes so far as to eliminate the mid-tones altogether—or by reducing it. Definition is often decreased, either by using extreme high speed films and force development to produce "golf-ball" grain, or by using a long exposure and moving the camera, or by moving the printing paper under the enlarger. Every photographer has his own favourite techniques, and again generalization is impossible, but it is safe to say that no fashion photographer can afford to neglect dark-room techniques as, say, the photojournalist can.

Locations. Generally it is the decision of the art director whether the pictures are to be taken on location or in the studio, but as pictures tend to concentrate more and more on the clothes and less on their surroundings, the two kinds of picture, which were once very distinct, are beginning to merge. When clothes are photographed outside the studio it is usual now to show them in suitable or neutral settings. Not long ago it was fashionable to use surrealistic backgrounds—e.g. evening dresses photographed in gas-works—but this trend is now over.

Locations in fact present magazines with embarrasing problems. The new fashion picture often requires only the simplest of props—a tree trunk or a spray of leaves may provide all the setting that is needed. But editors are beseiged with offers of free travel from airlines in return for publicity given in the editorial matter.

Thus it may well be cheaper to send a photographer, two models and a fashion editor for three weeks to Japan, than to send the same party for three days in the Cotswolds. But when the Japanese photographs are published, they either look fussy and old-fashioned by showing too much local colour or they show so little of Japan that the airline is dissatisfied.

The foregoing remarks apply to the glossy fashion magazines. They can be extended to fashion photography in other publications by observing that the less sophisticated the surrounding matter, and the less the space available for the fashion pictures, the simpler they will be; also, the poorer the quality of printing, the better the photographs have to be, in newspapers, there is no room for extreme high or low key, and definition must be high.

ADVERTISING

As in editorial work there are graduations of technique and interest. Generally though fashion advertising pictures will show the garment more clearly than in editorial pictures; they will attempt to create less of a mood; technique, lighting and models will be more ordinary.

Larger cameras are quite often used and, since an advertising picture has to be suitable for a number of different publications, quality has to be higher. Less freedom is allowed the photographer in choosing his model, pose and lighting—he often has to work from a sketch prepared by the advertising agency. P.La.

See also: *Lighting the subject.*
Book: *Lighting for Photography*, by W. Nurnberg (London).

FAULTS

The photographic process involves so many variable factors that faults are bound to occur sometimes in its working. These faults may show up in a variety of ways.

Identifying Faults. Before any attempt is made to trace a fault, whether by tables or personal investigation, the fault must first be identified; trouble-shooting charts are useless unless the user knows what type of fault he is trying to track down.

Often the nature of the fault seems clear, but even so, it is better not to jump to a hasty conclusion. The defect should be examined closely—through a magnifier if necessary—and all peculiarities, however slight, should be noted. Any unusual circumstances which can be related to the fault, even remotely, should be borne in mind—e.g., very cold weather, a stock clearance bottle of developer, a sticking dark slide.

Tracing Faults. The more common photographic faults will be found in the tables that follow. The major faults which appear in these tables are also dealt with in detail in separate articles and information on these in the tables is therefore abbreviated. However, faults can be caused by so many things that it is almost impossible to anticipate every one.

If a fault cannot be traced in the tables, systematic investigation should reveal the cause.

First it is necessary to establish the stage of the process at which the fault occurred. This narrows the field and eliminates any causes outside that particular stage—e.g., to establish if horizontal black lines across a print occurred when the print was made, the negative is first examined. If this reveals similar, but clear lines, the print-making stage is at once cleared from blame. Black marks on the negative that only appear in the picture area and not in the border left by the rebate in the camera indicate that the trouble could have only happened when the film was in the camera with that part of the negative in position in the film aperture. By further investigations, other possibilities are similarly eliminated until the source of the trouble is revealed.

Often, however, such a methodical system points to more than one answer. A fault may arise at any one of several stages—e.g., excessive grain might be caused by faulty film material, over-exposure, unsuitable developer, high processing temperatures, etc. When this happens it may be possible to eliminate some causes—e.g., over-exposure would show other signs as well—but failing this, the only course available is to check each stage.

Similarly, faults in colour balance in a colour print may be due to errors in the colour correction during printing, or they may already be inherent in the colour negative (for example lack of balance due to mixed lighting). Such faults are more difficult to establish directly from the negative, and only become apparent if correction of a fault in a colour print inevitably gives rise to other errors in colour rendering in subsequent prints.

If no satisfactory solution can be found in this way, photographic publishers and manufacturers are usually willing to give advice if full details are provided.

FAULTS IN BLACK-AND-WHITE NEGATIVES

Description	*Cause*	*Treatment*
Image lacks detail in shadows	Under-exposure	Density of any faint detail present can be improved by intensification
Image weak and flat, but reasonable detail in shadows	Under-development	Intensification or make a contrasty duplicate negative (optical intensification)
Image dense, shadows veiled, highlights very flat but dense. Border of negative clear	Over-exposure	Subtractive reduction
Image hard, with particularly dense highlights but good contrast and detail in shadows	Over-development	Proportional reduction

FAULTS IN BLACK-AND-WHITE NEGATIVES

Description	*Cause*	*Treatment*
Image extremely thin, very little detail, especially in shadows, very weak contrast	Under-exposure and under-development	Difficult to improve. Intensification or make a contrasty duplicate negative (optical intensification)
Image hard with very little detail in shadows	Under-exposure and over-development	Super-proportional reduction to reduce highlights, then intensify to improve shadows
Very little contrast and absence of strong densities but plenty of detail in negative	Over-exposure and under-development	Proportional reduction to lower over-all density, then intensify to increase contrast
Image extremely dense all over with no clear celluloid or glass except around rebate	Over-exposure and probably over-development	Proportional reduction
Image fades after some time and/or turns brown	Insufficient washing and/or fixing	If deterioration not too advanced, bleach in halogenizing bleacher and redevelop in ordinary developer
Image and emulsion vanish during after-treatment in certain solutions	Excessive immersion times in acidified baths, or use of baths too strongly acid	None
Partial or complete reversal of negative to a positive image	Accidental fogging, possibly due to unsafe darkroom light, before or during development	None
Negative shows over-all veil, particularly in shadows; no clear celluloid or glass visible		
(1) Borders of negative perfectly clear	Over-all fogging occurring when negative was in camera. Caused by faulty shutter or light leak	Subtractive reduction to clear shadows
(2) Borders of negative also veiled	Over-all fogging due to excessive development time; developer temperature too high; developer exhausted; developer incorrectly mixed; unsafe darkroom light; light leaks in camera back, film packing, darkroom or processing tank; faulty sensitized material due to staleness, packing, etc.; or aerial fog due to excessive exposure to air during development	Subtractive reduction to clear shadows
Over-all fog on negative (including borders) which appears red by transmitted light and blue by reflected light	Dichroic fog, due to developer exhaustion or contamination with hypo; developer too warm or development time too long; too high sulphite content in fine-grain developers; fixing bath exhausted or too warm; fixing bath contains excess of carried-over developer due to exhausted stop-bath or insufficient rinse; or negative exposed to strong white light before completely fixed	Bathe negative in solution of thiourea and citric acid (1–2 per cent each) in water
Image very dull and clear areas obscured; back of negative has brownish appearance	Insufficient fixing	Refix and wash
Over-all fog which is yellow by transmitted light	Similar to dichroic fog	As for dichroic fog
Edge of negative shows black fog, tailing off towards centre of negative	Light leak due to faulty plate-holder or camera back; roll film loosely wound or exposed excessively to daylight during loading, unloading or storage; or red window in camera unsafe	Use abrasive reducer or retouch on print
Numbers and other markings from roll film backing paper appear as image on film	Badly stored or stale film	In slight cases, retouch negative or print
Irregular, forked or tree-like wavy black lines, sometimes resembling lightning	Fogging caused by static electricity discharged when roll film is unrolled quickly in very dry weather.	Retouch on negative or print
Black lines or smudges over image, but not on borders of negative	Pinhole in bellows or camera body	In very slight cases retouch print

FAULTS IN BLACK-AND-WHITE NEGATIVES

Description	Cause	Treatment
On films only, very short black lines, right-angled and thinner at ends; sometimes coinciding with identical depression on film surface	Cinch mark, due to film buckling	Retouch on negative or print
Image has light streaks or lines around the edges of dark areas or running in straight lines from dark areas; or, similarly, dark lines on light areas	Mackie lines, due to insufficient agitation of developer, causing local concentration of exhausted developer	In slight cases, retouch negative or print
Density irregular and streaky	Insufficiently mixed or unstirred developer; or insufficient agitation during development	None
Thin, dark or light straight scratches on negative	Abrasion marks or scratches, due to dust, grit or faulty film loading. Light scratches usually occur before exposure, dark scratches after exposure	In slight cases, negative or print may be retouched
Large, irregular, light areas with sharp edges	Unequal flow of developer during processing, causing uneven development	None
Irregular areas of higher density	Variations in developer concentration or temperature, due to addition of developer during development	None
Irregularly defined light areas, often contaminated with scum	Developer standing unprotected from dust and air, causing a scum of oxidation products which locally retard development	None
Regular wave-like markings on dish developed negatives	Insufficient agitation during development	None
Speckled and mealy unevenness of image density	Deterioration resulting from damp storage or staleness of material, or use of old developer	None
Gradual increase in density towards one end of negative; with plates, sometimes very noticeable at edges	(1) Uneven drying (2) Faulty focal plane shutter	Depending on extent, negative may be "shaded" during printing
Edges of high densities (bright parts of subject) spreading into lower densities	Irradiation, due to light scatter in emulsion	In slight cases, retouch negative or print
Halos or rings around very high densities (usually light sources in the picture)	Halation, due to reflection of light from negative support, particularly with glass plates	In slight cases, retouch negative or print
Curved lines of increased density; may be accompanied by star-shaped images	Lens flare	Use abrasive reducer or retouch on negative or print
Finger marks		
(1) Light marks	Negative contaminated with grease or fixer from fingers before development	In slight cases, retouch negative or print
(2) Dark marks	Negative touched before development, with moist finger-tips, or fingers contaminated with developer	In slight cases, retouch negative or print
Small, usually well defined, light or dark spots	Ordinary or chemical dust on negative before development	Retouch negative or print
Clear, light, circular spots with soft edges	Air-bells on negative during early stages of development	Retouch negative or print
Small dark or light flecks of irregular shape	Particles of undissolved chemicals in developer	Retouch negative or print
Small, light, undefined brownish specks	Over-used and insufficiently agitated developer	Retouch negative or print
Light or dark drop-like densities, sometimes with darker edges	Drying marks, due to persistent drops of water when remainder of negative is dry; or splashes of water when negative is dry	Sometimes removed by thorough re-washing; wipe away all surplus water before drying again
Black specks of foreign matter in emulsion	Grit in processing solutions (particularly washing water); or dust in drying	Do not attempt to wipe away persistent specks; retouch print
Small, clear holes in emulsion	Black specks in emulsion which have been dislodged and washed or wiped away	Spot out hole with blocking out medium; retouch print

FAULTS IN BLACK-AND-WHITE NEGATIVES

Description	*Cause*	*Treatment*
Holes or hollows in surface of negative	Due to insects, bacteria or moulds, particularly during long drying times in warm, moist weather	In slight cases, retouch negative or print
Small blisters in emulsion	Usually due to transferring negative from highly alkaline developer to strongly acid stop bath or fixer	Prick blisters with needle before dry, or retouch print
Emulsion leaving negative support (liable to occur with plates), particularly at edges	Frilling; often a manufacturing fault but may also be caused by same reasons as blisters; or by excessive handling of edges with warm fingers	Carefully replace emulsion before dry
Image distorted and emulsion melting	Excessively high temperatures during processing or drying	None
Surface of negative dries with a fine grain-like pattern in emulsion surface. Often feels rough when dry	Reticulation, due to sudden changes in temperature between different processing solutions	None. When gelatin anti-curl backing on roll film reticulates, it will not usually show on prints
Negative appears opalescent and yellowish-white	Sulphur from decomposing fixer, due to excess of acid; high temperature; or insufficient sulphite in acetic acid fixers	Harden in 1 per cent formalin and then bathe in 10 per cent solution of sodium sulphite at about 100° F.; wash well
Fine-grained white deposit	Chalky residue from very hard water	Bathe negative in 2 per cent solution of acetic acid; wash well
On wet negatives, shiny deposit which dries with a white powdery appearance	Precipitate of aluminium sulphite, due to fixing bath losing acidity	Harden in 1 per cent formalin and then bathe in 5 per cent solution of sodium carbonate; wash well
Surface of negative has frosty deposit	Insufficient washing after fixing	Re-wash
Streaky markings similar to appearance of oil on water	Dirty or contaminated dishes or tanks used in processing	Bathe negative in 1–2 per cent solution of acetic acid
Red, blue or green tinge on negative support	Anti-halation backing, which should normally disappear during processing	If colour is deep or uneven, soak negative in 5 per cent sodium sulphite solution

FAULTS IN BLACK-AND-WHITE PRINTS

Description	*Cause*	*Treatment*
Flat tones showing insufficient contrast but adequate detail in highlights	(1) Contrast grade of paper too soft for negative used (2) Print was over-exposed and under-developed (3) Developer too dilute or too cold	No satisfactory remedy, make new print
Tones too light but show evidence of correct contrast	(1) Paper was under-exposed (2) Developer exhausted (3) Print was slightly under-developed	No satisfactory remedy, make new print
Tones very contrasty with little detail in highlights; shadows dense	(1) Contrast grade of paper too hard for negative used (2) In slight cases, print under-exposed and over-developed	No satisfactory remedy, make new print
Over-all heavy tones with veiled highlights	(1) Print was over-exposed (2) Developer too warm (3) Severe over-development	No satisfactory remedy, make new print
Tones show good contrast and quality but highlights not clean	(1) Print slightly over-exposed and/or over-developed (2) Very slight fogging	Very short immersion in Farmer's reducer
Image and print borders fogged all over	(1) Unsafe darkroom safelight (2) Unsafe enlarger safelight filter (3) Stray white light from enlarger or, during exposure, reflection from light walls, etc. (4) Darkroom not thoroughly blacked-out (5) Paper stale (6) Developer contained insufficient potassium bromide, was too warm or too concentrated	No satisfactory remedy, make new print
Tones are a greenish-black colour	(1) Chloride and bromide prints under-developed (2) Chlorobromide prints over-developed (3) Developer exhausted (4) Excess of potassium bromide in developer	Print may be toned if colour is objectionable

FAULTS IN BLACK-AND-WHITE PRINTS

Description	*Cause*	*Treatment*
Prints turn yellow on keeping	(1) Incompletely fixed or insufficient washing after fixing (2) With mounted prints, mountant used contains free acid, or impurities present in mount	Bleach in concentrated halogenizing bleacher and redevelop in ordinary developer
Prints turn yellow or brown during processing	(1) Excessive development (2) Developer contaminated with hypo or containing insufficient potassium bromide (3) Developer too warm (4) Insufficient rinse between developing and fixing (5) Fixer insufficiently acid, possibly due to developer contamination	Try developer stain remover
Yellow or brown stains	(1) Inefficient agitation or prints stuck together during fixing (2) Exposure to white light before fixing was complete (3) Insufficient rinse between developing and fixing (4) Exhausted or insufficiently acid fixer	Bathe in appropriate stain remover after refixing and rewashing in cases (1) and (4)
Brown marks on hot glazed prints	Insufficient fixing or washing	In some cases, bleach in a potassium permanganate bleacher and then redevelop, wash and dry
Sharply defined patches of uneven density	Print not completely covered by developer at beginning of development	None
Blotchy, uneven density with very flat contrast	Development time much too short	None
Surface has white streaky scum	(1) Dirty dishes (2) Hard water used in washing	Immerse in bath of 2 per cent acetic acid
Small, black spots, clearly defined; no corresponding spots on negative	(1) Undissolved chemicals in developer (2) Air bubbles in rinse, stop bath of fixer	Retouch
Small white spots, clearly defined; no corresponding spots on negative	(1) Air bubbles in developer (2) Small impurities in paper	Retouch
Irregular and concentric ring shaped markings	Newton's rings, due to minute gap between negative and enlarger carrier glasses	Retouch
Blisters in emulsion	(1) Differences in temperatures of processing solutions too great (2) Differences in alkalinity of developer and acidity of stop bath or fixer too great	In slight cases, remove loose emulsion and retouch with medium thickened with gum
Fine, short black lines	Stress marks, due to pressure or creasing on emulsion	Retouch
Mottled tones, although print was fully developed	Stale paper	None, but stale paper can be used if a developer improver is first added to developer
Greenish image tones	Under-development, exhausted developer, or stale paper	Tone in direct gold tones; can be avoided by use of developer improver
Black or fogged edges	Paper stored in damp or very warm room, or near gas stove or fire from which gas has reached paper	None, but in slight cases paper can sometimes be used if developer improver is first added to developer
Image reversed left to right	Negative placed wrong way up in enlarger or printing frame	None

FAULTS IN THE PICTURE

Description	*Cause*	*Treatment*
Image blurred all over	(1) Incorrect focusing of camera or enlarger (2) Camera or enlarger shake	None
Image consistently blurred in one area	Film plane not flat, due to opening bellows too sharply with film cameras; distorted camera back or plate holders; faulty loading of plates or films; or weak or broken pressure plate in roll film cameras	None

FAULTS IN THE PICTURE

Description	*Cause*	*Treatment*
Moving objects blurred	Shutter speed too slow to arrest movement	None
Near objects sharp, far objects blurred, or vice versa	Insufficient depth of field, due to using too large lens aperture	None, but with some subjects it can be used intentionally to subdue unwanted parts of picture
Image blurred and flat in contrast all over	Lens misted over, due to condensation when cold lens is brought into contact with warm air. Dirty lens	None
Picture confined to circle shape, image dark and blurred	Lens not extended on cameras fitted with collapsible mount	None
Out-of-focus blob intruding into picture area on some or all pictures	Holding camera so that hand or fingers obstruct lens; flap of ever-ready case obstructing lens; or wide-angle lens used in plate camera without wide-angle rack	None. In slight cases, exclude in masking of enlargement
Dark corners	Unsuitable lens which does not cover complete negative area. If edges are pronounced, lens hood of wrong design used, or rectangular lens hood fitted on incorrectly.	In slight cases, exclude in masking of enlargement or shade during printing
Framing of subject consistently inaccurate	(1) Viewfinder mounting out of adjustment (2) Wrong viewfinder for lens fitted on camera (3) Faulty sighting with frame type viewfinders (4) Failure to compensate for parallax errors	None
Two different images in same picture	Double-exposure	None
Moving parts of subject appear in picture twice as double image	Shutter bounce, due to shutter bouncing momentarily open after initial exposure	None, but have shutter examined by reliable camera repairer
Dimensions of subject distorted	(1) Foreshortening, due to using viewpoint too close to subject (2) Distortion due to lens	None, but in the case of (1), a more distant viewpoint should have been chosen
Clouds in skies fail to register clearly	Relative over-exposure of sky which could be avoided by using filter	Give sky extra exposure during printing

FAULTS IN COLOUR TRANSPARENCIES (POSITIVES)

Description	*Cause*	*Treatment*
Image too light and colours weak	(1) Over-exposure (2) Incorrect processing, especially if excess of one colour is present in tri-pack materials	Only (1) can be treated, in the case of additive transparencies, by intensification
Image too dark	(1) Under-exposure (2) Incorrect processing, especially if colour rendering is inaccurate	Additive transparencies may be improved by reduction. With tri-pack materials, manufacturer may suggest satisfactory formulae for treatment
Highlights burnt-out or shadows blocked-up	(1) Brightness range of subject too great (2) Incorrect processing	None
Image excessively orange in colour	(1) Daylight film used in artificial light. This is most likely cause (2) Incorrect processing	Try binding pale blue filter tint over transparency
Image excessively blue in colour	(1) Artificial light film used in daylight. This is most likely cause (2) Incorrect processing	Try binding pale pink filter tint over transparency
Image consists almost entirely of one colour only	(1) Monochrome filter left over camera lens (2) Incorrect processing (3) Coloured light used (4) Sheet film exposed through base	None

FAULTS IN COLOUR TRANSPARENCIES (POSITIVES)

Description	Cause	Treatment
Some colours correct, others unaccountably false	(1) Uneven or mixed lighting used for subject (2) Vapour discharge lighting used for subject (3) Colour casts reflected from objects near subject	None

FAULTS IN COLOUR NEGATIVES

Description	Cause	Treatment
Image too light	(1) Under-exposure (2) Under-development of film (3) Exhausted or wrongly compounded developer (4) Stale film	None
Image too dark	(1) Over-exposure (2) Over-development or excessive temperature	None
Image very contrasty	Over-development	None
Areas of uneven density	Insufficient agitation during development	None
Stains and blotches	(1) Exhausted developer (2) Insufficient agitation (3) Contamination of bleacher or bleach-fixer by developer (4) Fogging or reflections in camera	None
Strong overall stain	(1) Orange stain: dye mask in film (2) Other stains: incorrectly compounded developer	(1) This is normal for certain films and is not a fault, hence requires no treatment. The dye mask colour however makes the already difficult job of estimating colours in a negative virtually impossible. (2) None
Negatives clear very slowly in bleacher, or not at all.	Exhausted bleacher	Repeat processing stages starting with bleaching in fresh solutions
Overall opalescent yellowish appearance of film	Insufficient time in bleacher or exhausted solution	Repeat processing steps starting with bleacher in fresh solution
Grainy appearance on film surface	Reticulation due to processing solutions being at different temperatures	None
Clear spots on film	(1) Dust in the camera (2) Undissolved particles or sediments in the developer adhering to the film, or air bells due to insufficient agitation at start of development	With care may be retouched on the print, or even on the negative in the case of large formats
Scratches	Careless handling of film when hanging up to dry or during processing	Retouching in negative or print

FAULTS IN COLOUR PRINTS FROM COLOUR NEGATIVES

Description	Cause	Remedy for subsequent prints
Image too dark	(1) Over-exposure (2) Over-development or too warm developer	(1) Reduce total exposure (or three exposures in proportion for three-exposure printing if colour balance is correct) (2) Check solution temperatures and carefully observe recommended colour development time
Image too light	(1) Under-exposure (2) Under-development (3) Exhausted or too cold developer	(1) Increase total exposure and check that exposure increase of filter factors has been allowed for (2) Keep to recommended development time (3) Check development temperature and use fresh solutions
Image shows predominant overall tint	(1) Incorrect filtering in single-exposure white-light printing (2) Incorrect balance of exposures in three-exposure additive printing	(1) Add or increase filter density of colour corresponding to cast (or decrease filter densities of other two colours); in case of predominant casts of primary colour (red, green or blue), decrease filter density of complementary colour or increase densities of other two colours (2) Increase exposure through tricolour filter of same colour as cast or decrease exposure through other two filters; in case of complementary casts (yellow, magenta or cyan) decrease exposure through complementary primary filter (blue, green or red respectively) or increase exposure through other two filters

FAULTS IN COLOUR PRINTS FROM COLOUR NEGATIVES

Description	*Cause*	*Remedy for subsequent prints*
Cyan cast or dirty yellows which cannot be removed by filtering correction or requires disproportionately heavy cyan filtering or short red filter exposure	Infra-red radiation affects red sensitive layer of paper	Use heat filter or infra-red absorbing filter between lamp and negative in enlarger
Colours wrong only in highlights and/or shadows, or specific areas of picture, impossible to correct fully by adjusting printing exposures	(1) Negative under- or over-exposed (2) Negative under- or over-developed (3) Subject exposed by mixed lighting (for example daylight and artificial light)	(1) Measure exposures carefully (2) Keep to recommended development of colour negative, especially development time (3) Always illuminate the subject only with light sources of the same colour temperature
Final print does not correspond in depth or colour balance to test on which it was based	(1) Test and final print made under different conditions (2) Unnoticed variations in enlarger lighting, processing solutions, etc. (3) Different batches of colour paper used for test and for final print	(1) Keep careful notes of filtration and exposure for all tests and check that optimum values established for test are also repeated for final print (2) Use a voltage stabilizer in the enlarger lighting circuit; develop final print in same developer as used in test, and check that developer is not getting progressively exhausted
Image dull and lifeless	(1) Negative too soft (2) Print under-developed or processed in exhausted developer	(1) Use a contrasty grade of colour paper (if available), or make both test and final print with longer development, adjusting filtration accordingly to correct change in colour balance (2) Increase development time and/or replace developer if necessary
Image harsh and contrasty, detailless highlights and heavy shadows	(1) Negative too contrasty or under-exposed (2) Print over-developed	(1) No remedy, though sometimes reduction of development time for test and for final print with corresponding adjustment of filtration may produce passable result (2) Decrease development time
White borders and highlights stained or veiled	(1) Stale or exhausted developer (2) Excessively high processing temperature (3) Insufficient rinsing before bleach fix bath (4) Stray light in darkroom or paper exposed too near to safe-light (5) Exposure to white light before bleaching is complete (6) Incorrectly compounded developer	(1) Use fresh developer (2) Observe recommended processing temperature (6) Use maker's recommended processing chemicals. If this does not help, add 0·1% hydroxylamine hydrochloride to colour developer
Highlights stained, but borders clean	(1) Over-exposure of print (2) Stray light during exposure in enlarger	(1) Reduce exposure (and adjust filtering if necessary)
Irregular patches or streaks	(1) Prints sticking together in processing solutions (2) Prints handled with fingers during processing (3) Fogging by light	(1) Keep prints moving about continuously, especially in colour developer (2) Avoid contaminating solutions with each other (3) Use maker's recommended safe-light and avoid examining prints too close to safe-light before completion of bleach
Discolouration of parts or whole of print when dry	(1) Insufficient fixing or exhausted bleach fix bath (2) Prints hot glazed or dry mounted by heat	No remedy
Colour fringes and blurred outlines near edges of print	Enlarger lens or (if fault present on negative) camera lens not sufficiently corrected for colour	No remedy
Colour fringes and blurred outline all over print	(1) Enlarger vibration during exposure (2) Enlarger or paper slightly moved when changing filters	(1) After last touching enlarger wait for a few seconds before starting exposure (2) Check enlarger set-up for rigidity

FAULTS IN REVERSAL COLOUR PRINTS FROM TRANSPARENCIES

Description	*Cause*	*Remedy for subsequent prints*
Image too light	(1) If associated with blank highlights: over-exposure (2) Associated with high contrast: excessive first development (3) Associated with low contrast: insufficient colour development	(1) Reduce printing exposure (or exposures when using additive printing method) (2) Check temperature and time of first development (3) Check temperature and time of colour development (also use fresh solutions)

FAULTS IN REVERSAL COLOUR PRINTS FROM TRANSPARENCIES

Description	*Cause*	*Remedy for subsequent prints*
Image too dense	(1) With blocked-up shadows: under-exposure (2) With low contrast: insufficient first development (3) With high contrast: excessive colour development	(1) Increase exposure (through all three filters in proportion when using additive printing method) (2) Observe recommended time and temperature in first developer, and use fresh solutions
Image too contrasty with blank highlights and blocked shadows, usually also with falsified colours in parts of picture	Transparency too contrasty for printing	No simple remedy; reduce contrast of transparency by masking or make prints by negative positive process via an inter-negative
Overall colour cast	(1) Incorrect filter combination with white-light printing method (2) Incorrect balance of exposures through tricolour filters with three-exposure printing (3) Fogging in darkroom	(1) Reduce filter density of same colour as cast or increase filter densities of other two colours; in case of primary (red, green or blue) casts increase filter density of complementary colour (cyan, magenta or yellow respectively) or reduce filter density of other two colours (2) Decrease exposure through tricolour filter of same colour as cast or increase exposure through other two filters; in case of magenta, cyan or yellow casts increase exposure through complementary primary filter (green, red or blue)
Generally degraded colours	Transparency unsuitable for printing owing to colour faults or inherent colour cast which cause further deterioration in reversal print	No remedy

FAULTS IN GLAZING AND MOUNTING

Description	*Cause*	*Treatment*
Prints stick and cannot be stripped when dry	(1) Glazing sheet was dirty (2) Glazing solution exhausted	Soak in water to remove and then re-glaze
Unglazed spots on surface of glazed print	(1) Air bells trapped between print and glazing sheet (2) Squeegee applied too heavily or too lightly (3) Dirty glazing sheet	Soak in water and then re-glaze
Glazed surface shows oyster-shell markings	Uneven drying	Soak in water and then re-glaze
Glazed surface covered in fine scratches	Glazing sheet covered in fine scratches	Soak in water and re-glaze print on new glazing sheet
Prints refuse to glaze	Prints left for too long in an acid hardening fixer	None
Wet mounted prints will not stick	(1) Too little, or more likely too much mountant used (2) Print applied to mount before mountant was tacky	Remove immediately, clean and re-mount
Dry mounted prints will not stick		
(1) Tissue sticking to mount, but not to print	(1) Too much heat	Adjust temperature or time of pressing
(2) Tissue sticking to print but not to mount	(2) Insufficient heat or too short pressure	
Dry mounted prints only sticking in places	Uneven application of heat or pressure	When due to too much heat, it is better to allow print to cool for a few seconds before trying again
Dry mounting tissue will not stick to either print or mount	(1) Stale tissue (2) Iron or press not hot enough	(1) None (2) Check equipment
Dry mounting tissue protruding from edges of mounted print	(1) Print not absolutely dry before mounting (2) If tissue was trimmed after tacking on to print, trimmer blade may have pulled tissue and print unevenly, trimming more of latter	May be removed by carefully trimming with razor blade, then pulling cut strip free of mount
Mounted prints have slightly uneven surface or light creases	Too much wet mountant or creases in dry mounting tissue	In some cases, print may be removed, cleaned and re-mounted
Cockling—i.e., small areas of print leaving mount and swelling into irregular hump shapes	Moisture in print or mount before dry mounting	If print can be removed: clean, wash and dry thoroughly; ensure mount is perfectly dry; then re-mount print. Sometimes, with dry mounting, can be removed by applying heat with heavy pressure

FAULTS IN GLAZING AND MOUNTING

Description	Cause	Treatment
Surface or image of print damaged by dry mounting	(1) Type of print unsuitable for dry mounting—e.g., carbro prints, tri-pack colour prints	(1) None
	(2) Glazed prints dry mounted with textured surface between print and heat source	(2) Remove print and re-glaze and re-mount
Dye retouching shows after mounting print	Colour of dye changed by heat; dye retouching should be done after mounting	Some improvement possible by retouching again
Colour prints partially or wholly discoloured after glazing or mounting	Colours attacked by heat	None

J.D.C. & L.A.M.

See also: *Aberrations of lenses; Abrasion marks; Air bells; Blisters; Camera shake; Contact printing; Converging verticals; Covering power; Developing negatives; Developing prints; Dichroic fog; Distortion; Double exposure; Drying marks; Enlarging; Fading; Fingerprints; Flare; Fogging; Frilling; Ghost; Grain; Halation; Irradiation; Movement; Newton's rings; Over-development; Over-exposure; Reticulation; Solarization; Spots; Stains; Under-development; Under-exposure.*

Books: *Colour Films*, by C. L. Thomson; *Developing* and *Enlarging*, both by C. I. Jacobson (all London).

FEDERATION INTERNATIONALE DE L'ART PHOTOGRAPHIQUE. International movement officially launched in 1947 with the broad aims of raising the status of amateur photography as a common cultural link between nations, improving photographic techniques by the interchange of ideas, and contributing, by its international associations, to world peace.

There are affiliated associations in a large number of countries spread over both hemispheres, and representatives meet and exchange views at periodical gatherings. The Fédération organizes a biennial international photographic exhibition and issues monthly bulletins to members. These bulletins draw attention to matters of technical interest in the photographic press and give members regular reports on the work of the branches and committees.

Prints from the biennial salon are circulated for exhibition to affiliated associations throughout the world with the object of letting not only photographers, but the ordinary citizen of each nation learn something of the activities of other nations in a language that all can understand. The Fédération publishes a yearbook of representative photographs following the biennial exhibition.

FEERTYPE. Process invented by Dr. Adolf Feer in 1889. Paper is sensitized with a diazo compound which forms a coloured salt when exposed to light. The paper is exposed under a negative and washed to remove the unused chemicals, leaving a positive image in the selected colour.

See also: *Printing processes (obsolete).*

FEES. Prints accepted for publication in the press or any form of illustrated literature—e.g., travel guides, catalogues, brochures—are paid for at rates which vary according to the publication, size of reproduction, etc.

Prices charged for taking photographs—e.g., for commercial or private customers—depend on the type of work undertaken, the number and size of the prints supplied, and numerous other factors such as local supply and demand.

See also: *Costing prices of commercial photographs; Reproduction fees; Selling photographs.*

FENTON, ROGER, 1819–69. English barrister; from 1847 till 1862 architectural, landscape and portrait photographer. Founder and first Honorary Secretary of the (Royal) photographic Society (1853). The first to photograph war scenes under fire and to take the actual battlefields. Sailed from England in February 1855 as accredited war photographer for the Crimean War and took during the spring of 1855 over 350 photographs. Biography by H. & A. Gernsheim (London 1955).

FERGUSON, WILLIAM BATE, 1853–1937. English scientist; member of the Research Laboratory of Ilford Limited. With B. E. Howard he worked out the time and temperature method of development (1906). Received the Progress Medal of the Royal Photographic Society in 1914 for his researches, discoveries and publications on the physics and chemistry of photography. He edited the *Hurter and Driffield Memorial Volume* of the R.P.S. (London, 1920).

FERRI. (Slang.) Abbreviated term for potassium ferricyanide; usually applied to reducers containing ferricyanide.

FERRIC ALUM. Ferric ammonium sulphate. Constituent of some blue and blue-green toners and reducers.

See also: *Alum, Iron.*

FERRIC AMMONIUM CITRATE. Used in certain sensitizers and toners.

Characteristics: Brown or green crystals of slightly variable composition. The brown variety is the more basic one. Both are slightly sensitive to the action of light.

Solubility: 20–30 parts in 100 parts of water at room temperature.

FERRIC AMMONIUM OXALATE. Used in certain sensitizers (e.g. in the blue-print process) and toners.

Formula and molecular weight: $Fe_2(C_2O_4)_3 \cdot 3(NH_4)_2(C_2O_4) \cdot 6H_2O$; 856.

Characteristics: Green crystals. The salt, particularly in solution, is sensitive to light.

Solubility: Highly soluble in water at room temperature.

FERRIC AMMONIUM SULPHATE. Constituent of some reducers and toners for blue and blue-green images.

See also: *Alum, Iron.*

FERRIC CHLORIDE. Iron chloride. Used in reducers, various iron (blue print) and iron-silver processes and as etching mordant for copper, for processes such as mezzotints and photomechanical work.

Formula and molecular weight: $FeCl_3 \cdot 6H_2O$; 270. Also available as the anhydrous salt $FeCl_3$; 162.

Characteristics: Yellow very deliquescent crystals.

Solubility: Highly soluble in water. The anhydrous salt is also freely soluble in alcohol, ether, and other organic solvents.

FERRIC OXALATE. Used in certain sensitizers (e.g. in the blue-print process) and toners.

Formula and molecular weight: $Fe_2(C_2O_4)_3 \cdot 6H_2O$; 484.

Characteristics: Greenish crystals. The solution is sensitive to light.

Solubility: Fairly soluble in water.

FERRIC POTASSIUM OXALATE. Used in sensitizers for iron processes; also in blue toners.

Formula and molecular weight: $Fe_2(C_2O_4)_3 \cdot 3K_2(C_2O_4)_2 \cdot 6H_2O$; 982.

Characteristics: Green crystals. Solution sensitive to light.

Solubility: Fairly soluble in water.

FERRO-PRUSSIATE PROCESS. Little used for ordinary photographic printing; it is mainly employed for reproducing plans and drawings, which are commonly called blue prints. The paper is sized and then sensitized in equal parts of the following solution, mixed just before use:

Stock solution A		
Ferric ammonium citrate	10 ounces	250 grams
Water	40 ounces	1,000 c.cm.
Stock solution B		
Potassium ferricyanide	3½ ounces	90 grams
Water	40 ounces	1,000 c.cm.

The made-up solution should be kept in the dark.

The paper is pinned to a board and a liberal quantity of the sensitizing solution is applied with a brush or clean sponge. After sensitizing, the paper is dried in the dark.

The dried paper is held in contact with the plan, drawing or negative to be printed and exposed to daylight or arc light. Printing is slow and should be carried on until the shadows attain a bronze tint. The blue print is developed by simply soaking it in two or three changes of clean water.

See also: *Document copying.*

FERROTYPE PROCESS. Method for making direct positive prints in the camera on black or chocolate enamelled iron plates (called ferrotype plates). A variation of the Ambrotype, introduced by A. A. Martin in 1853 and very popular in the United states in the 1860s. In Europe chiefly used by street, beach and itinerant photographers. The plates were originally coated by the wet collodion process but dry plates were later manufactured.

The ferrotype process was popular for while-you-wait photography since there was no time wasted in drying the base. Since the original image is converted into a positive and there is no intermediate negative, it is reversed from left to right—a fact that the while-you-wait photographer rarely pointed out to his customers.

FERROTYPING. American name for the process in which photographic prints on bromide and contact paper are glazed by squeegeeing them into contact with a highly polished surface. At one time the most convenient surface was provided by sheets of japanned iron (ferrotype plates) specially produced for the purpose. Nowadays sheets of highly-polished chromium-plated steel, stainless steel, or plastic, are favoured because they are less easily damaged than ferrotype plates.

FERROUS OXALATE. Used as a developing agent, but now obsolete.

Formula and molecular weight: FeC_2O_4; 144.

Characteristics: Yellow salt. Generally prepared in solution by adding ferrous sulphate solution to excess potassium oxalate solution. The solutions do not keep well.

Solubility: Insoluble in water, but soluble in solutions of potassium oxalate.

FERROUS SULPHATE. Green vitriol. Used in certain clearing and stain removing baths.

Formula and molecular weight: $FeSO_4.7H_2O$; 278.

Characteristics: Green crystals.

Solubility: Freely soluble in water at room temperature. Does not keep well in solution.

FIBRE-OPTICAL SYSTEM. Optical system using glass fibres and fibre arrays as light-guides and for relaying images.

Light-guides. A light-guide is a transparent rod having polished surfaces coated, whenever practical, with material of lower refractive index to protect them. Light entering the end face is refracted towards the axis. Light incident within a certain cone becomes sufficiently oblique to the cylindrical surface to be totally internally reflected there by the refractive index difference between core and coating. It is thus trapped and zigzags down the core of the light-guide until it is refracted out of the far end.

Such guides still transmit light when they are bent over radii more than a few times the diameter of the guide itself. They also transmit well, down to a diameter of ten wavelengths or less, given a coating of two wavelengths' thickness.

Light may therefore be guided down flexible fibres of optical glass a few ten-thousandths of an inch across and several feet long. In these guides there are several hundred reflections per inch along the length. Light would be almost completely absorbed during this number of reflections from a metal mirror.

Image-transmitting Fibre Bundles. Consider a close-packed array of fibrous light-guides in which the position of each fibre in the array is the same at its two ends. If a pattern of light be displayed on one end, each fibre will transmit the light from a small area of the pattern to the second end and the image will be reassembled

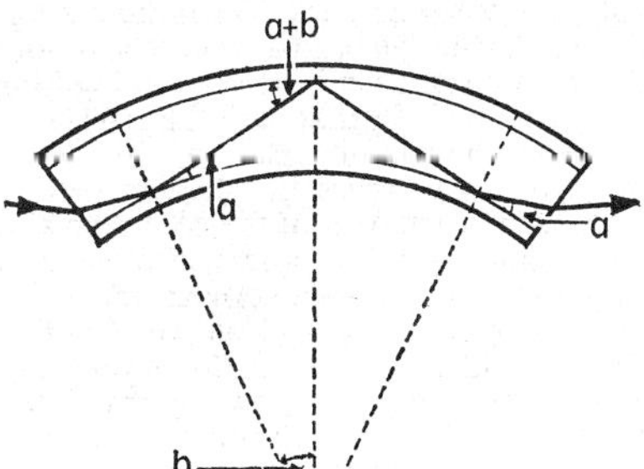

LIGHT GUIDE PRINCIPLE. A ray passing through a curved light guide of circular cross section is internally reflected, as long as the angle (a) on the inside face of the light guide is within the angle of total internal reflection. With a curved light guide, the combined angle (a + b)—where (b) is the angle of curvature—must be within the angle of total reflection. This is usually the case as long as the guide is not curved into a radius of less than about three times its diameter.

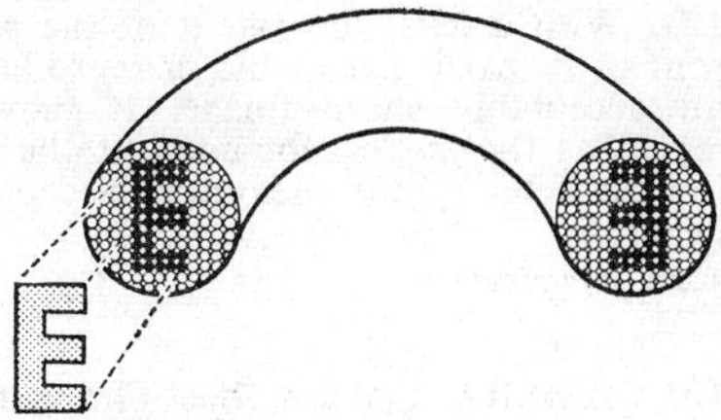

IMAGE FORMATION. In a fibre optical system each fibre end picks up and transmits the average light falling on its cross section. A fibre bundle can transmit a picture broken up into image elements corresponding to the fibre elements. (The object is here seen from the front; the fibre bundle however is seen from behind—hence the apparent reversal.)

on that surface. The principle is much the same as that used in television, with the important difference that, because the light from every point of the incident image is transmitted all the time, no storage of energy or scanning device is needed.

Such a bundle of fibres can be very flexible, because the fibres need to be constrained to a definite order only near their ends. With an objective lens on one end and an eyepiece or camera on the other, such a bundle forms an endoscope for inspecting cavities inaccessible in a direct line from outside. Light may be transmitted into the cavity through fibres around the image-transmitting area. Such endoscopes have application to industrial and medical inspection.

Simple bundles are made of fibres between 0.0005 and 0.002 in. in diameter and are capable of resolving up to 20 lines per mm. on their end faces. More elaborate bundles are being developed in which each of these fibres contains a matrix of high refractive index cores surrounded by low refractive index material. With these the resolving power may rise to 100 lines per mm.

Fibre Faceplates. When an image is formed on a phosphor layer inside a cathode-ray tube or image-intensifier, a photographic emulsion cannot be placed in immediate contact with it. Definition is lost because the light spreads sideways in its passage across the thickness of the glass envelope. Lenses can transfer to a film situated elsewhere only a few per cent of the light emitted at all angles by the phosphor. They also add to the length of the whole system.

A mosaic screen of parallel glass fibres (fused together so as to be vacuum-tight, but remaining optically separated by their coatings) therefore replaces the simple glass window. This redisplays the image on the accessible outer surface of the tube, transferring as much as three-quarters of the light from the phosphor.

Such fibre screens can be made with one surface curved to fit the image surface of the electron optics, so acting as field flatteners.

R.W.J.U.

See also: *Cathode ray tube traces.*

FIELD. With a lens, the extent of the scene in front of it that it is capable of reproducing as an acceptably sharp image. It may be expressed as the angle subtended at the lens by the diameter of the maximum acceptable image circle.

See also: *Covering power.*

FIELD CAMERA. Plate or sheet film camera for commercial and professional photographers, sufficiently portable to be taken on location for field work. The field camera is the forerunner of the modern technical camera, and has many of the adjustments and controls of the latter.

Field cameras were usually made in quarter-plate to whole-plate sizes ($3\frac{1}{4} \times 4\frac{1}{4}$ ins. to $6\frac{1}{2} \times 8\frac{1}{2}$ ins.) Larger formats are no longer portable enough to qualify for consideration as field cameras; they are studio models.

Construction. Most field cameras are made of wood and are of the folding baseboard design. A shallow box-like camera back is hinged to a baseboard. The box is the camera back and takes a focusing screen or plate or film holders. Usually the focusing screen runs in guides and has to be removed before the film or plate holder can be inserted. Frequently the film holder can be inserted in two ways, horizontally or upright.

On some models the carrier for the screen and film holders rotates to permit rapid switching over from horizontal to upright shots, or to arrange the picture format in any intermediate position.

The baseboard carries extending runners rather like a flat bed technical camera to provide a double or sometimes even triple focusing extension. The lens panel moves along these runners and is connected to the camera back by a set of bellows.

In addition to holding the back at right angles to the baseboard, struts usually allow it to be fixed at an angle to the vertical as well. This gives a limited tilting movement about the horizontal axis. Sometimes there is also a lateral swing to bring one side closer to the lens than the other.

The lens panel is hinged to give both lateral and vertical swings and in addition incorporates rising and cross front movements.

The Tripod. The field camera is invariably used on a tripod, so this frequently forms an essential part of it. In some cases the camera baseboard is itself the tripod head, with the legs attached to the baseboard.

The tripod normally consists of collapsible wood or metal legs and tends to be bulky to ensure the necessary rigidity for supporting the appreciable weight of the camera.

Lenses and Shutters. The lens of the field camera is usually interchangeable and mounted on lens panels which clip into the front standard. Focal lengths vary according to the camera format. Field cameras used to be equipped with convertible lenses as well, where two components of a lens unit could either be used together or separately—giving longer focal lengths in the latter case.

On the more modern versions the shutter is built into the lens and changed together with the latter. That means that every lens has its own diaphragm shutter associated with it, which, of course, makes the acquisition of an optical outfit more expensive.

As the older types of field camera were designed for use with slow sensitive materials and time exposures at small apertures, they are not always fitted with shutters. Sometimes the lens cap is considered adequate for controlling the exposure. Alternatively, a variety of shutters attach to the front of the lens (blade or roller blind types) and are then changed over from one lens to the other when the lenses themselves are changed.

Field cameras also exist with behind lens shutters, but generally internal shutters of any kind are rare.

Focusing. The field camera is always focused visually on a ground glass screen fitted in place of the plate holder. The screen holder is so designed that the ground surface lies in the exact plane occupied by the emulsion surface of the plate or film when the latter is in the camera.

Sometimes the ground glass screen carries a clear spot with a reference mark (hairline cross) in the image plane for aerial focusing. This can even be home-made: it involves drawing a cross in pencil in the centre of a ground glass screen, and cementing a small round microscope cover glass over it with Canada balsalm. The latter fills in the ground surface, rendering it completely transparent, while the pencil cross provides the visual indication of the plane in which the image must be focused.

Usually the focusing movement is a rack and pinion mechanism which pushes the lens standard forward or back. On more elaborate field cameras the camera back may be mounted on a separate standard instead of being hinged to the baseboard. In that case the back is also movable on the baseboard.

Scope. For many years the field camera was the standard instrument of the commercial and technical photographer, and has most of the scope—even if not all the convenience—of the modern technical camera. Compared with the latter, the field camera was often more limited in its camera movements and adjustability.

Its main fields of application were landscape photography, architectural and industrial work as well as close-up, medical and scientific photography.

While wooden field cameras are often regarded nowadays as museum pieces, they were made with surprising precision and are still found in the second-hand camera markets.

See also: *Camera movements; Extension of camera; Technical camera.*

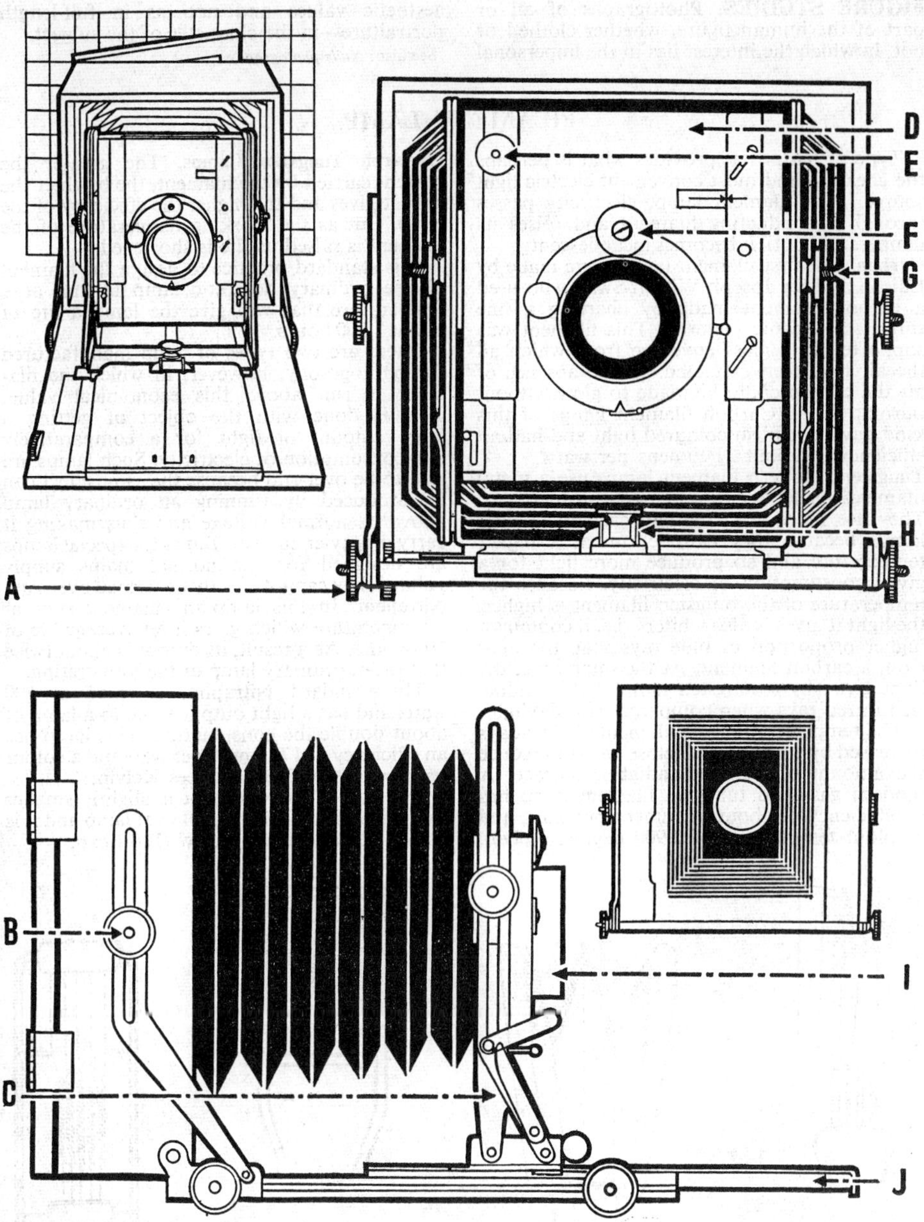

FIELD CAMERA. Top left: *General view against* 1 *in. square grid.* Top right: *Front view.* Bottom left: *Side view.* Bottom right: *Rear view with back removed.*

The main features of the camera are: A, knob for locking movement of back along baseboard. B, swing back control, also drops baseboard when required for use with wide-angle lenses. C, bracket supporting lens standard. D, removable lens panel, carrying lens and shutter, held in position by a sliding ledge at the side. E, rising front control; two opposing arrows on the front of the panel indicate when the lens is centred. F, shutter speed control (many of the older field cameras have air brake shutters, in which case the speed dial is linked to the air brake cylinder). G, knob locking lens panel swing. Two catches on the front of the panel engage when the lens panel is at right angles to the baseboard. H, catch to lock movement of lens standard along baseboard. I, lens. J, baseboard extension.

FIGURE STUDIES. Photographs of all or part of the human figure, whether clothed or not, in which the interest lies in the impersonal aesthetic values and not—as in full-length portraiture—in the character of the subject.

See also: *Nude photography.*

FILAMENT LAMP

Type of lamp that provides what is perhaps the cheapest and most convenient electric light source. In a filament lamp, electricity passes through a conducting filament and raises its temperature until it becomes incandescent.

Carbon. The first filament lamps were made by Edison and Sir Joseph Wilson Swan who used a carbon filament made by charring a fine thread of organic material. This filament was supported in a glass envelope from which all the air had been extracted. In the absence of air the carbon could be made to glow without burning away. Carbon filament lamps of this kind gave a reddish coloured light and had an efficiency of about 2½ lumens per watt.

Tungsten. Modern filament lamps use a metal filament—generally tungsten—in an inert gas. They are more efficient than carbon filament lamps because they can be run at a higher temperature and so produce more light for a given consumption of electricity. Because the temperature of the tungsten filament is higher, the light it gives is also whiter—i.e. it contains a higher proportion of blue rays than the light from a carbon filament. At the same time, the light given by a tungsten lamp is still unduly rich in red rays when compared with daylight.

The temperature of the filament is nowadays increased by coiling it in a close spiral to reduce the amount of the free radiating surface. A modern gas-filled tungsten filament lamp has an efficiency of about 13 lumens per watt, and a colour temperature of 2,900 degrees Kelvin.

Over-run Tungsten Lamps. The greater the current carried by the filament, the brighter the light it gives and the higher the efficiency of the lamp. But as the working temperature of the filament is raised its life is shortened.

It is standard practice to design the filament of the ordinary domestic lamp to run at a temperature that will give the lamp a life of about 1,000 hours.

There are two types of lamp manufactured for photography, however, in which the filament is run above this economical value. This is done with the object of getting a large amount of light for a comparatively low consumption of electricity. Such lamps are said to be over-run because the same effect can be produced by running an ordinary lamp above its normal voltage and thus making it carry a heavier current. But as the special lamps are designed for the normal mains supply voltage they cannot strictly be termed over-run.

Nitraphot. In this lamp the filament runs at a temperature which gives it an average life of 100 hours. As a result, its output is about twice that of an ordinary lamp of the same rating.

The standard Nitraphot is rated at 500 watts and has a light output equal to a lamp of about double the consumption. This lamp has an efficiency of 17 lumens per watt and a colour temperature of 3,200 degrees Kelvin.

The bulb of a Nitraphot is slightly smaller than that of an ordinary 500 watt lamp and it is fitted with an Edison Screw (E.S.) cap.

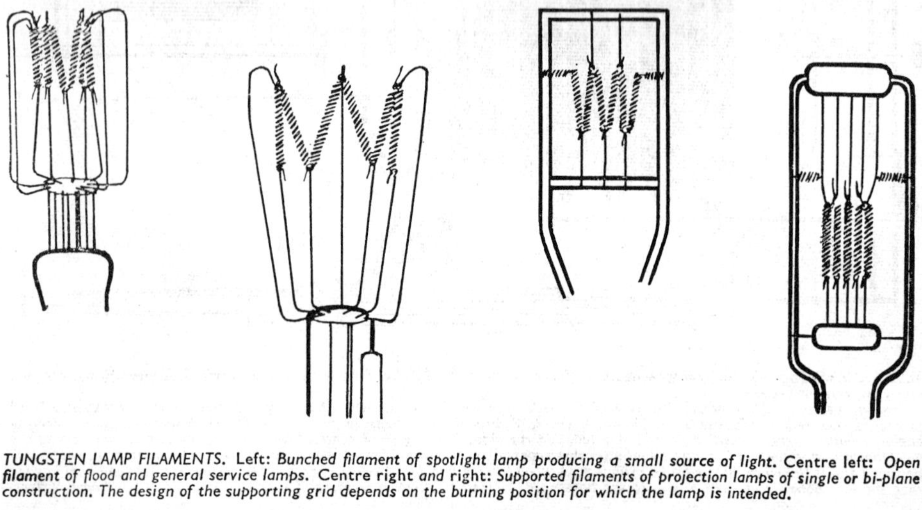

TUNGSTEN LAMP FILAMENTS. Left: *Bunched filament of spotlight lamp producing a small source of light.* Centre left: *Open filament of flood and general service lamps.* Centre right and right: *Supported filaments of projection lamps of single or bi-plane construction. The design of the supporting grid depends on the burning position for which the lamp is intended.*

Photoflood. This type of lamp is so severely over-run that it has a life of only a few hours, but it gives a very brilliant light in relation to its consumption. There are two sizes in common use.

The No. 1 consumes 275 watts and is equal to a normal lamp of 800 watts. This lamp lasts for about two hours when used intermittently—i.e., when it is never switched on for more than a minute or two at a time.

The No. 2 Photoflood consumes 500 watts and is equal to a normal lamp of 1,600 watts. This lamp lasts for about 7 hours when used intermittently.

Photoflood lamps have an efficiency of about 36 lumens per watt and a colour temperature of 3,400 degrees Kelvin.

Photoflood lamps are popular for amateur still and cine photography indoors because they are cheap to replace and enable comparatively ambitious effects to be achieved with the current from a domestic supply point.

Projection Lamps. Special filament lamps are manufactured for use in conjunction with optical projection systems—e.g., in spot lamps and still or motion picture projectors. These lamps are made as small as possible, and the filament is designed to provide a very compact area which will remain rigidly in focus.

One way of reducing the size of the source is to design the lamp for 110 volts instead of the normal 200–250 volts. Halving the working volts enables the filament to be made proportionately shorter and thicker, which gives a brighter, whiter, and smaller source. Such lamps must be used in conjunction with a voltage-reducing transformer or resistor.

Small, high intensity, lamps of this type tend to run very hot and must have ample—often forced—ventilation. In addition, unless specially designed, they must always be mounted upright or the hot filament may sag out of shape (or one coil may even come into contact with another and cause a short circuit). The working position of projection-type lamps is always marked on the bulb.

Projection lamps generally have pre-focus caps, which are flanged and fitted in such a way that the distance from the filament to the locating flange is the same for all lamps of that type—i.e., so that the light source always lies on the optical axis of the projector.

Integral Reflector Lamps. Lamp reflectors tend to be awkwardly large, add to the weight and cost of the lamp support and quickly tarnish and collect dust. For these reasons external metal reflectors are being replaced by reflectors inside the lamp itself where they stay bright, take up no extra space and add very little to the cost of the lamp.

Where the optical performance does not need to be very precise, as in the photoflood lamps, it is enough to silver part of the inside of the glass bulb. Such photoflood lamps have a modified bulb shape and part of the glass is

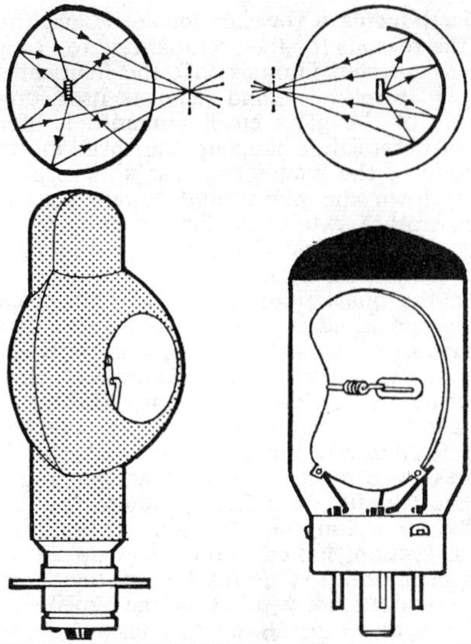

INTERNAL REFLECTOR LAMPS. Left: *The fully silvered bulb has its back blown out to an ellipsoidal reflector and the front to a spherical reflector with only a window for the focused light beam. The spherical front reflector thus utilises virtually all the light from the filament.* Right: *An internal proximity reflector has a similar function and makes an external mirror in the lighting system unnecessary. Both these lamps concentrate their light into the plane of the transparency or film without the need for separate condensers.*

mirror-coated inside to act as a reflector. This concentrates approximately 75 per cent of the light in a cone with an angle of 60°. Integral reflector lamps of this type are manufactured in the two regular sizes: 275 watts (3 hour life) and 500 watts (10 hour life). The colour temperature is adjusted to suit the A type colour materials.

Internal reflector lamps are also coming into increasing use in projectors, especially for cine. Here one of the problems is to get the maximum amount of light concentrated on the area of the gate. Even with a precisely-aligned reflector and condenser system, a good deal of the light spills around the edges of the light collectors or falls uselessly around the target area. Some types of projection lamp therefore use a small, precision reflector mounted inside the lamp and close to the filament. The same result can also be achieved by shaping the projection lamp envelope to act as a reflector, the glass being silvered inside. In this position the reflector intercepts a much higher proportion of the light emitted by the filament and will focus it accurately on the gate. In an 8 mm. or 16 mm. projector this arrangement does not even need a condenser system. Such lamps often also use a low-voltage filament, with a further gain in luminous efficiency. The combined increase in light output is so great that an integral reflector lamp will give a screen brightness equal to an orthodox system of five times the power.

Quartz-Iodine (Tungsten-Iodine) Lamp. One of the reasons for the eventual failure of a gas-filled tungsten lamp is that the filament gradually evaporates (and deposits itself on the inside of the glass envelope) until it breaks. Even before that happens, the shrinking cross section of the filament increases its resistance, cuts down the power and lowers its colour temperature, while the film of tungsten that settles on the inside of the glass reduces the useful light still further.

In the quartz-iodine lamp a small trace of iodine enclosed in the envelope prevents this from happening. When the lamp is switched on, the iodine vaporises and combines with the tungsten deposit on the envelope to form tungsten iodide. This decomposes in contact with the incandescent filament into tungsten, which unites once more with the filament, and free iodine vapour. As a result, the envelope stays clean, the tungsten filament maintains its cross section, lumen output goes up 20 to 30 per cent and the useful life of the lamp is roughly doubled. In addition, the lumen output and colour temperature stay constant throughout the life of the lamp—an important advantage in photographic studio lighting.

The reaction between the iodine vapour and the tungsten deposit on the envelope will only take place if the temperature is above 250°C.

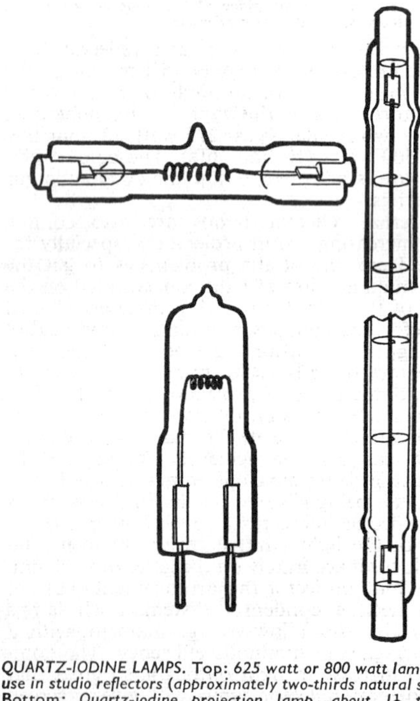

QUARTZ-IODINE LAMPS. Top: *625 watt or 800 watt lamp for use in studio reflectors (approximately two-thirds natural size).* Bottom: *Quartz-iodine projection lamp, about 1½ times natural size.* Right: *1,500 watt strip form lamp for floodlighting. This is about 10 ins. long.*

This is ensured by enclosing the filament in a narrow quartz or high-melting-point glass tube instead of the conventional glass bulb. So a quartz-iodine lamp is very much smaller than a normal tungsten lamp of the same power and can be used with smaller and nearer reflectors. Quartz-iodine lamps for general lighting have a straight tube with a contact at each end.

Quartz-iodine lamps are used for general studio floodlighting in place of ordinary tungsten filament lamps, the units being very much smaller and easier to arrange and store than orthodox floods of the equivalent power.

Quartz-iodine floodlights are available in all the normal voltage ranges from 12 up to 240 and in a range of ratings from 100 to 10,000 watts. The portable lamps of up to 1,000 watts use a short tube and can be operated at any angle without ill effects. The more powerful studio lamps have a longer tube which is apt to blacken at one end if it is worked at more than 4 degrees to the horizontal.

Quartz-iodine photographic floodlights may be of the over-run type. This means that the filament is designed to give an abnormal amount of light at the cost of a much shorter life.

Normal quartz-iodine floodlights—from 1,000 to 10,000 watts—have a guaranteed life of up to 2,000 hours with a luminous efficiency of 22 lumens per watt and a colour temperature of 3,000°K. Quartz-iodine photographic floods—usually below 1,000 watts—have a luminous efficiency of 34 or more lumens per watt and a colour temperature of 3,400°K, but the life drops to 12 hours. Even so, the performance and life of a quartz-iodine photographic flood are greater than for an ordinary photoflood, although the cost is much higher.

This type of lamp is also fitted in the compact, camera-mounted, portable lighting units used for amateur still and cine photography. The small quartz tube requires only a tiny reflector, and a typical 800 watt lamp will give a light equal to three 500 watt reflector photofloods and last twice as long.

As the light source is a horizontal tube, the spread of light is greatest in the horizontal plane and least in the vertical plane. This suits the photographer on most occasions, but for the special jobs that require a greater vertical spread, the lamp unit can be swivelled at right angles. Some types of quartz-iodine lamps are produced as sealed beam units inside specially designed reflectors resembling a motor car headlamp. When the lamp burns out the whole reflector is replaced.

Single-ended quartz iodine lamps are further used in slide and film projectors where the extremely compact type of light source makes it possible to employ a much more efficient optical system. For this purpose the filament is over-run, giving a colour temperature of some 3,300°K. and an efficiency of about 28 lumens per watt with a life of at least 50 hours.

Quartz-iodine lamps run at a much higher temperature than normal tungsten lamps and must be left for as long as 10 minutes to cool down before they can be handled. On switching off, the iodine vapour produces a pink glow and it can be taken that as long as this glow is visible, the lamp will be much too hot to touch.

If the tube of the lamp is touched with greasy fingers when it is cold, the marks will produce opaque black spots when it heats up. These can be cleaned off with methylated spirit but it is wiser to polish the tube clean with a dry rag before use if it has been handled at all. F.P.

See also: *Lamp caps and fittings; Light sources; Projection principles.*

Books: *Modern Electric Lamps*, by D. A. Clarke; *Atlas Lighting Manual*, by H. Hewitt and H. C. Weston; *Electric Lamps*, by the British Lighting Council (all London).

FILING NEGATIVES AND PRINTS

Any filing system for negatives should serve two main purposes. Firstly, it has to hold the negatives themselves together with the relevant technical and other data and, secondly, it should allow any negative to be found at a moment's notice.

Every serious photographer should, therefore, have a storage system for negatives and data (probably including a contact print for identification) and an indexing system. Both must allow for indefinite expansion. There must always be room for more without having to reclassify the system.

Storing Negatives. The basis of the storage system for negatives will be negative albums. Often it is most practical to make such an album from transparent or translucent envelopes, punched along one side so that they can be inserted into a loose leaf binder. Or it may be found more convenient to stick a narrow strip of paper to one edge of each envelope and punch this.

The envelopes are numbered, and the negatives are inserted in the order in which they were exposed. For additional safety it may be as well to write the number of the negative on the margin of the film with a very fine pen. Some photographers even add their names.

This will do for single film negatives of most conventional sizes. 35 mm. negatives are best stored in lengths of four or six exposures in envelopes of suitable size and shape. The envelopes containing a strip can still be numbered consecutively, while each negative receives the envelope number and suffix. Thus envelope No. 485 would contain negatives 485a, 485b, 485c, and so on.

Plates are best kept in special plate boxes which are grooved to keep each plate separate, and the grooves are numbered.

Where there are several negative sizes which will not go into one type of album, they may be accommodated in several albums and have the negatives numbered separately in each. The numbers will be preceded by a letter, so that, say, all the numbers from A1 onwards will be 35 mm. negatives, from B1 onwards V.P., $2\frac{1}{4} \times 2\frac{1}{4}$ ins. and $2\frac{1}{4} \times 3\frac{1}{4}$ ins. films, and the series beginning with C1 may be half-plate glass negatives.

Photographers who specialize in one particular subject can start a series D1—e.g., for portraits—or go further and subdivide and use double prefixes.

But although some people are quite at home in a complicated system of this sort, others find it very confusing. Anyone who finds such a system confusing should make negative numbers depend on their size and chronological order only.

As an alternative to making the envelopes up into a negative album, they can simply be kept loose, in their numerical order, in boxes.

Negative Data. Most negative albums have no adequate provision for entering exposure and other details. Therefore these should be entered separately on data sheets.

The data are the usual details about subject, location, aperture, shutter speed, time of day or other illumination data, sensitive material, filters, date, and development.

Each data sheet carries the same number as the corresponding negative, and has a contact print attached to it. With several negative sizes the tidiest plan is to make an enlargement of the whole of every negative so that all the prints are the same size (which will depend on the largest negative size). The prints need not be technically perfect, as they serve mainly for identification. On the prints should be written details of where shading or overprinting is needed during enlargement, and what parts of the negative are to be used.

Further details to go on the data sheets are: the most suitable grade of enlarging paper, whether and where the picture has been submitted, published, sold, etc.

A typical layout for a data sheet may look something like this:—

DATA SHEET

Neg. Number...........	Roll........... Exp. No............
Contact print or standard size enlargement	Subject ....
	Exp............... sec., *f*...........
	Plate/ film ....
	Lighting ....
	Time of day......... Date.........
	Filter......... Camera............
	Lens......... cm./in., *f*............
	Developer ....
	min. at ....°F./°C.
	After-treatment ....
	Suitable printing paper...........
	Notes ....
	....

Indexing. These data sheets can serve as their own index. The most straightforward method is to file them under subject headings arranged alphabetically in a loose leaf book. Thus there will be headings like Children, Portraits, Industrial, etc., with the data sheets inserted under them in numerical order.

If the number of sheets under any heading becomes too great for convenient handling, it will be necessary to subdivide the headings. Portraits can be subdivided into the names of sitters (arranged alphabetically). Travel into geographical regions and place names, and so on.

An index also needs extensive cross-references. These can be dealt with by preparing several data sheets of the same negative where required, and inserting them under the several appropriate headings. To save time and materials, blank sheets should be inserted under the alternative headings. These sheets will only bear the negative number or numbers, and a reference where the data sheet is to be found. For instance, under Table Top there might be a sheet marked: "C255, C256, A327, A335—see Advertising Photos."

In this way the indexing can be made independent of the storage system. The negative numbers only identify the negatives, and the two systems are self-contained. To use the index look up the subject and subdivision, run through the sheets there to find the required picture, note the negative number and then take the corresponding negative out of the file or box. This is the only occasion when the negatives themselves are handled.

An alternative method of filing negatives makes use of a card index. Each negative is stored separately in an envelope as before (or strips of four or six negatives to one envelope in the case of 35 mm. film). Each envelope is then stuck on the back of a card.

This system needs a separate index, which can also be a card index, or else a loose leaf binder. However, instead of data sheets there is one sheet or card for each subject. On this all the relevant negative numbers are entered. No subdivisions are used; the sheets are simply labelled like the index in a book. Thus when the sheet Villages becomes too full for easy reference, it is replaced by several sheets: Villages—Kent, Villages—Sussex, Villages—Wales, etc.

The Decimal System. There is yet another system, which can be adapted for photographic filing and applied to data sheets and cards.

The subjects are divided into not more than ten groups, each covering as broad a field as possible. The groups are assigned single key numbers from 0 to 9—e.g., 0 for Portraits, 1 for Sports, 2 for Travel, etc.

Each group is then subdivided, and each subdivision given a second key number. Thus, under Travel, 21 might mean Europe, 22: Asia, 23: North America, etc.

Further subdivision produces 211: British Isles, 212: Western Europe, 213: Central Europe, 214: Scandinavia, etc., while the next stage would be 2121: Holland, 2122: Belgium, 2123: France, etc.

A key list of these numbers is prepared to show which key and subsidiary key numbers refer to which classifications. It is also necessary to prepare an index card or sheet for each negative, bearing the negative number and classification number. If several negatives fall into the same category, they all come on to one index card. If any card becomes too full, it is replaced by a number of new ones, carrying the division a stage farther.

Some negatives may occur on several index cards under different classification numbers. Thus St. Paul's Cathedral would fit into both 2114 (Travel — Europe — British Isles — London), and 5413 (Architecture—Church Buildings—Cathedrals—Seventeenth Century).

Instead of index cards the data sheets of the negatives can be used. The classification number is written on them in addition to the negative number. Entries under alternative headings are dealt with by blank sheets bearing only the two numbers.

The index cards or data sheets are kept in numerical order according to their classification numbers, and not by the negative numbers. The order of the classification numbers depends on the actual digits, from left to right, and not on the length or numerical value of the complete number. Thus 4216 would come before 638, not after.

To find a negative of any subject: look up the subject classification in the key list, find the corresponding index card or sheet, and get the negative number from it, take out the corresponding negative from the album or file.

Print Records. A processing record of every print made enables identical further prints to be made at a later date.

The method is simple. Before each piece of bromide paper is exposed in the enlarger, it is numbered on the back with a soft pencil using very slight pressure to avoid pressure marks on the face of the print. The pencil marks will not wash off during normal processing, and when the print is finished the reference numbers can be gone over with Indian ink or a rubber stamp for greater permanence.

For the actual records it is convenient to use a large note-book ruled in columns. As each print is made, the following details are entered in the book: print number, negative number, paper type and grade (usually the manufacturer's code number will do for this), the maker's batch number, and the exact degree of enlargement.

If the enlarger column is marked in inches or centimetres, it is easy to note the height of a convenient part of the lamphouse above the baseboard. All enlargements with the lamphouse in this position will have the same degree of enlargement. Otherwise to get the exact degree of enlargement it is necessary to measure the exact width of the mask in the negative carrier—or draw or scratch two lines a fixed distance apart on an unimportant part of the negative—and to measure the corresponding size of the image of these two lines on the baseboard.

Further details required are: focal length of enlarging lens (only if there are several lenses for different negative sizes), aperture, exposure time, developer, shading or other control, print size, after-treatment, and a column for any other notes (what the print is used for etc.). If a photometer is used for determining exposures, there should also be a column for the photometer readings.

To make a duplicate of any print, it is only necessary to look up the print number in the record book, select the paper, set the enlarger and expose, etc., as indicated.

Print Files. The print files are much simpler, as prints will be taken out as well as put in. The files will consist of the finished prints on hand before submitting them to journals or papers.

The prints are sorted out according to subjects, and put into large envelopes (about 8 × 10 ins. to accommodate the largest print likely to be used) or boxes, with the name of the subject written on the outside. Containers are filed alphabetically. If a print comes under several headings, it is filed under the most appropriate. A sheet of paper with a note—e.g., See also print No. XX, Churches—as applicable, is filed under the alternative subjects. It is best to stick to this method even if there are several prints from the same negative, rather than to distribute them in various containers. The prints are then all in one place, and a check can be kept on how many prints there are from one negative in stock.

It is also possible to use the decimal system already described for prints using the same key list as for the negatives. The prints have the classification numbers put on the back, and are stored in the order of these numbers. Alternative classifications are dealt with as before by blank sheets with appropriate cross-references. If possible, it is best to avoid overlapping classifications.

There is no need to go to a very elaborate system, as the quantities in stock are much more likely to be easily manageable than the negative files. If, however, there are too many, it is satisfactory to store them in the order of their negative numbers and use the negative indexing system to find a print of any particular subject.

All prints should be numbered with the negative number as well as with the print number, so that reference can be made to the negative files and processing records for all technical details.

Prints Sent Out. It is essential to keep a record of all prints submitted for publication.

For this a loose leaf book will serve, devoting a sheet to each negative number from which prints have been sent out. The sheets are inserted in numerical order.

On each sheet is entered the date when a print was submitted, its number, where it was sent, when it was accepted (or returned), published, the size of the published picture, and when it was paid for (including amount received).

If several prints are sent out, these details are noted about every one of them. It is also necessary to state whether the negative or copyright is sold, or if the prints have been submitted through an agent. If a contact print from the negative is stuck on each sheet, there is no need to refer back to the negative file to identify the print.

This record also needs two indexes. In the subject index under each heading is entered the negative from which prints have been submitted, accepted, or rejected, and also where they were submitted. This subject index can also be based on the decimal system as for negatives.

The second index is one of magazines and papers to which pictures have been submitted. Under each paper is entered the negative number, subject, date of submission, and publication.

With these two indexes it is possible to see at a glance what markets have been covered with which subjects.

Storing Transparencies. Transparencies comprise lantern slides, film slides (film transparencies bound between glass), unbound transparencies, and film strips. These may be either black and white or colour.

Slides and film slides may be stored loose in boxes, with the slides numbered for index reference. Each box may hold about 50–100 slides. Special slides boxes are made which hold the slides in separate grooves for easy access and sometimes have a carrying handle for portability. Large collections can be stored in drawer cabinets made up of such boxes.

Unbound transparencies, especially in colour, are best kept in transparent envelopes; this permits viewing without risk of damage. Film strips are stored rolled up in cardboard or metal tins.

Indexing Transparencies. Any of the indexing systems for negatives and prints can be used, according to the size of the collection and needs of the owner.

Advanced systems like detailed subject subdivisions or the decimal classification will prove most useful when the transparencies are required for a stock file for publication. In that case it is best to have separate files for unique transparencies (e.g., original colour transparencies produced in the camera on reversal film), and for duplicates or positives printed from an original or negative. The former would then be indexed like a negative file and the latter like a print file with references to the originals or negatives.

Slides illustrating a lecture are best kept together as sets, though individual numbering will enable them to be referred to in the common index with other transparencies whenever an individual slide is required.

With film strips a set is joined together anyway so the sets can be filed by subject.

It is absolutely vital for a filing system, once started, to be kept up to date. There are no half measures: it is either complete and accurate, or it is useless. **L.A.M.**

See also: *Film storage.*
Book: *Camera Copying and Reproduction*, by O. R. Croy (London).

FILL-IN LIGHT. Any light directed on to the subject from a lamp or reflector to illuminate shadows cast by the principal light on the subject. Without some such form of shadow illumination, those parts of the subject not lighted by the principal light would appear in the finished photograph as dense black areas with little detail.

For normal subject lighting, the power of the fill-in illumination of the shadow areas is usually one-third of that given by the principal lighting.

See also: *Against the light; Flash technique; Lighting the subject; Portraiture.*

FILM CLIP. Metal (usually stainless steel) or plastic clip attached to the ends of roll films so that they can be hung up to dry. The clip for the top end hangs by a hole or a hook and the lower clip may be weighted to keep the film taut. As the ends of a wet film are slippery, the jaws of film clips are spiked or serrated and held closed by a strong spring. A normal type of clip—e.g., a bulldog paper clip—is apt to slip and the film must be handled specially carefully when dipped in.

Metal film clips should always be made of rustproof material.

See also: *Developing negatives.*

FILM COUNTER. Counting mechanism in many modern cameras, especially 35 mm. models, which shows the number of exposures already made on a length of film, or the number of frames still available.

See also: *Film transport; Miniature camera.*

FILMING. Term broadly covering all the activities of making motion picture films with narrow gauge (including amateur) and standard cine camera.

See also: *Cine camera; Cinematography; Cine techniques; Editing; Projection practice; Projector* (*cine*).

FILM PACK. Standard form of pack in which negative material may be bought. It consists of a number (usually twelve) of separate sections of film assembled in such a way that they can be exposed singly in an adaptor fitted in place of the plate holder of a plate camera and removed singly for processing. Film pack material is generally rather more flimsy than cut film and should therefore be processed in a dish, as the cut film holders for tank development are not suitable.

See also: *Film transport.*

FILMS. Sensitized materials in the form of an emulsion coated on a flexible base—e.g., celluloid or plastic. Films for still photography are available in four standard forms: roll films, perforated film, sheet film and film packs.

History. Long before the introduction of celluloid film, attempts were made to replace glass plates by a less heavy and fragile support.

Stripping film, which could be peeled off the glass plate, was mentioned by Frederick Scott Archer in his communication of the collodion process in March 1851. Other methods of stripping the negative from its glass or paper support were put forward by Alexander Parkes (1856), C. M. Ferrier (1857), Leon Warnerke (1875), Thiébault (1883), Georges Balagny (1883). George Eastman introduced stripping film for use in the Eastman/Walker roller-slide in 1886.

Flexible sheet film needing no support was patented by Alexander Parkes in 1856. Though his film of several layers of plain collodion never found application, with this idea Parkes laid the foundation for many flexible film supports which were devised mostly in the 1880s, the most important of which was celluloid, a material invented by Parkes in 1861. Various films of gelatine and/or collodion were devised by H. J. Palmer and F. Wilde (1878), Ferrier (1880), E. Stebbing (1881), Alfred Pumphrey (1881) and others. Celluloid film was first manufactured in thin (1/100 in.) clear sheets for John Carbutt, an English photographer in Philadelphia, who in 1888 introduced emulsion-coated celluloid sheet film.

The Rev. Hannibal Goodwin of Newark applied for a patent on 2 May 1887 for a very thin transparent rollfilm made of celluloid, which was not granted until 1898. Meanwhile the Eastman Company had captured the market with a similar rollfilm patented by their employee H. M. Reichenbach, 1889. This eventually led to a lawsuit lasting twelve years; in 1914 the Eastman Kodak Co., who by that time had a virtual world monopoly in rollfilms, settled with the Ansco Co.—owners of Goodwin's patent—for five million dollars.

With this transparent rollfilm invented by Goodwin and popularized by George Eastman, the modern era of photography began, the only change being the replacement of inflammable celluloid by non-inflammable cellulose acetate about 1930. H. & A.G.

Roll Films. These are mounted on a numbered opaque paper backing and wrapped on a wood or metal spool. The commonest size is 2½ ins. wide film which yields eight pictures measuring 2¼ × 3¼ ins., twelve pictures 2¼ × 2¼ ins. or sixteen pictures 1⅝ × 2¼ ins., according to the camera.

Perforated Film. The original film used in 35 mm. miniature cameras was simply standard 35 mm. cine negative film. Nowadays perforated 35 mm. film is manufactured specially for still cameras. It differs from the cine film in a number of respects—in particular in the type of coating. The perforations enable the film to be moved exactly one frame at a time by the film transport mechanism, thus eliminating the numbered paper backing. The standard picture size for cameras taking perforated 35 mm. film is 24 × 36 mm.; a number of cameras use smaller formats such as 18 × 24 mm. (half-frame) or 24 × 24 mm. Perforated film in larger sizes is used in special scientific and technical photography—e.g., in aerial survey and instrument recording cameras, as well as for document recording. Established standard widths there are 70 mm. and 105 mm.

Perforated 70 mm. film is also becoming popular for professional and technical cameras and even special roll film cameras. The film yields pictures 2¼ ins. wide ty 2¼ to 2⅞ ins. long (according to camera) and is usually packed in cassettes holding up to 15 feet of film.

Certain hybrids between roll film and perforated film—a kind of perforated roll film—also exist for special cameras. In one case the film is rolled up in special cassettes together with a backing paper, and a perforation (one per picture frame) serves to arrest the transport movement of the camera rather than to advance the film directly.

Perforated and unperforated 16 mm. film is used in sub-miniature or ultra miniature cameras yielding pictures from 10 × 14 mm. to 12 × 17 mm. The film is usually packed in single (sometimes double) cassettes without a backing paper. Either the film perforation or a gearing with the take-up spool of the cassette serves to advance the film. One camera even uses unperforated 9.5. mm. film.

Sheet Film. This, known also as flat film and cut film, is manufactured in all normal plate negative sizes. The base is thicker than that used for roll films and there is no paper backing. Sheet film is loaded into special film holders which hold the film flat and in the correct focal plane in the camera. Sheet films can also be loaded into plate holders in the same way as a plate, but a piece of black card of the same size as the film must be placed behind it to make up the thickness. Film sheaths are also available to hold sheet film in plate holders; these are flat pieces of metal with grooves to hold the film edge in place, and go into the plate holder in the same way as a glass plate.

As there is very little difference between the feel of the back and the front of sheet film, it is notched in manufacture so that the emulsion side can be identified. When the film is held with the long side vertical and the notches are at the right hand end of the top edge, the emulsion surface is facing towards the operator.

Film Packs. A film pack is a convenient package of twelve pieces of cut film attached to paper separators with numbered tabs in such a way that each film in turn can be brought into position for exposure in the camera. This method of packing has the advantage of allowing individual films to be extracted and processed without waiting for the whole pack to be exposed. Film packs are now virtually obsolescent; they were always more expensive exposure than the equivalent size in sheet film.

The film pack is exposed in a special film pack adapter which fits into the plate holder grooves in the back of a plate camera.

Cine Films. The sensitized materials used in cinematography are normally available in four gauges: 8 mm., 9·5 mm., 16 mm., and 35 mm.; for special techniques, such as wide screen projection, larger gauges such as 65 mm. are used.

Cine films are perforated down one or both sides. The size and number of perforations depends on the type of film and gauge.

Spools, magazines or cassettes are employed for loading cine film into the camera.

See also: *Cine film; Colour materials; Film transport; Negative materials; Perforation; Reversal materials; Sizes and packings; Supports for emulsions; Transparency materials.*

FILM SETTING. Method of assembling printed texts etc. on a film by printing individual letters from appropriate negatives. The film carrying the completed text is then used for making a printing plate by photo-mechanical means. In film setting machines the selection of the letter negatives and their alignment, exposure of the film etc., proceed automatically at high speed, being controlled by a keyboard somewhat like a typewriter keyboard.

See also: *Photomechanical reproduction; Photo-typesetting.*

FILM STORAGE. Sensitized materials gradually deteriorate from the moment of manufacure. This deterioration usually involves loss of speed, change of contrast, increase of background fog and—in the case of colour materials—a change of colour balance. Manufacturers therefore mark film cartons and packages with an expiry date (usually about two years from the date of coating) during which time they guarantee the specified characteristics of the film under proper conditions of storage. In practice films may remain serviceable for a considerably longer time.

The storage conditions naturally influence the keeping qualities of the film. Particularly harmful are—in order of importance—radio-activity and harmful radiations (of importance in atomic energy establishments and radiographic departments of hospitals), fumes, high humidity and high temperatures.

Storage Conditions. Ideally films should be stored in a cool, dry and adequately ventilated place free from radio-activity, and industrial solvents and vapours (formaldehyde, hydrogen sulphide, sulphur dioxide, motor exhaust gases, ammonia, mercury vapour, turpentine and other solvent vapours). The ideal temperature is 40 to 50° F (4 to 10° C) and the ideal relative humidity 40 to 60 per cent.

Tropical Storage. Manufacturers are increasingly going over to packing films in sealed metal foil or plastic wrappings—originally used for materials intended for tropical climates. Such wrappings effectively protect the film against all outside influences (apart from temperature) as long as the wrapping is unopened.

While the film is in such hermetically sealed wrappings, it can also be kept in a refrigerator for longer storage life. This is desirable in hot weather and climates, especially if the film stock must retain consistent characteristics for some time. (For example in documentary coverage with colour film, where large amounts of film of the same batch number are obtained.)

Films stored in a refrigerator must be allowed to reach the prevailing temperature again before the package is opened; otherwise moisture may condense on the cold film and ruin it. For the same reason open film (once taken out of its sealed package) must not be replaced in the refrigerator. Once the package is unsealed, the film is best stored in a metal box containing a dehydrating agent (e.g. activated silica gel) to provide a dry atmosphere. Preferably films should be exposed and processed within a few hours of loading into the camera.

In tropical climates moulds are also liable to attack the film when in the camera.

Processed Films. Negatives and coloured transparencies should be stored in a reasonably cool and dry place, as far as possible away from fumes such as hydrogen sulphide, etc. Excessively warm and excessively dry conditions may render the film brittle; with cine films stored in cans it is usually desirable to have a humidifier pad inside the can to keep the film supple.

In the tropics the greatest danger is attack from insects, bacteria and moulds. Films are best kept in envelopes (preferably treated with fungicide) with a closing flap. Periodic inspection is desirable.

Storage Regulations. Modern cellulose acetate and polyester films present no special fire risk. The old-fashioned cellulose nitrate films however were highly inflammable and subject to special regulations for storage (in accordance with the Factory and Workshops Acts in Britain, and corresponding regulations in other countries). These regulations limit the quantity of nitrate film that can be stored in a single cupboard or drawer, and also specify the type of building required.

Cellulose nitrate films are only likely to exist in old files and motion picture stores; in view of the risks and special regulations, organizations holding such stocks are gradually preparing duplicate negatives or transparencies on safety-base films. L.A.M.

See also: *Keeping qualities of materials.*

FILM STRIPS

A film strip is a series of positive transparencies printed on a single strip of 35 mm. perforated film. Film strips may be viewed in a transparency viewer, but they are primarily intended for projection.

A film strip offers a convenient method for showing a number of pictures which have some common link—e.g., the illustrations for a lantern lecture, a series of pictures to advertise a manufacturer's product, or a planned, step-by-step educational sequence accompanied by either a recorded talk, or a set commentary read by the teacher or instructor. Many manufacturers equip their representatives with portable projectors.

Film Strip Printers. The positive prints are made either by contact or projection on to special perforated or unperforated 35 mm. film coated with a slow bromide emulsion of medium contrast. They may also be made directly on to reversal material in the camera although this method is only worth while in special circumstances.

It is usual to make contact prints from 35 mm. negatives for reference only; serious film strips are almost always made by projection printing because it permits more control of the picture and allows strips to be made by reduction from negatives larger than 35 mm.

Contact printers are usually in the form of a frame similar to the body of a 35 mm. camera with the lens panel removed. The spool chambers are big enough to hold at least five feet of positive film, and the picture aperture can be masked for single or double frames. A hinged gate over the picture aperture holds any selected negative in position for printing. After exposure the winding knob automatically advances the positive film to the next frame. The printing light may be an ordinary electric light bulb suspended above the printing table and connected through a dimming resistance for controlling the brilliance.

There are also semi-automatic printers for turning out numbers of film strips from the same copy negative.

Projection printers are necessarily more complicated than contact printers because they must allow for the separate adjustment and refocusing of each new negative. In this case the film holder is generally mounted on a slide on which there is also a focusing stage. The white surface of the focusing stage is at exactly the same height from the carrier as the surface of the emulsion in the gate of the film holder.

The image is focused on the stage with the slide at one end of the track. It is then moved to the other end, bringing the film gate into the exact position previously occupied by the focusing stage. This system allows the negative frames to be printed in any order, or orientation and with any individual degree of magnification.

A 35 mm. camera can also be used as a projection printer by loading it with positive film and fitting an extension tube to give 1:1 reproduction. The film to be printed is held flat in a negative carrier over a cut-out aperture in the top of a box. The light from a lamp inside the box diffused by a sheet of opal glass illuminates the negative and provides the printing light. The camera is held on a copying stand at the correct height over the negative and the exposure is made with the camera shutter.

Special close-up gear of this kind exists for a number of miniature cameras. It usually consists of suitable extension tubes and/or supplementary lenses, together with a copying stand or distance gauge that automatically supports the camera at the right distance from the negative (or transparency). The latter then only has to be placed on top of a light box.

Printing Technique. Film strips must be planned as a whole because they are printed on one continuous length of film. Once printed, the order can only be changed by cutting and splicing the film, which is rarely desirable nor always practicable. Otherwise the whole strip must be reprinted.

The same positive material is used for the whole strip, so the negatives must all be of similar contrast. This is easy enough where film strips are printed from a set of negatives specially prepared for the purpose. For convenience, the negatives should be in the correct order, and all of the same size. When taken on miniature film, the negative film itself can be processed to produce a direct reversal positive film strip.

There are four principal steps in printing a film strip:

(1) Selecting the negatives and arranging them in the order in which they are to appear on the strip. This includes negatives of titles and captions.

(2) Making test exposures to find the exposure time for each negative. A careful note is taken of the exposure time, and amount of magnification or reduction to be used for each negative.

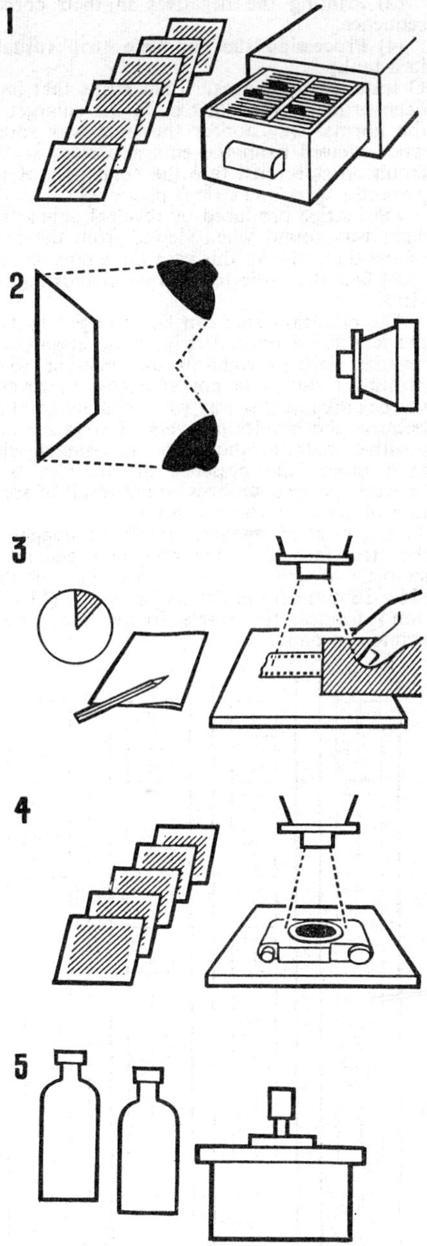

FILM STRIP PRINTING. ***1.*** ***Select negatives and place in correct order for final sequence of images on strip. 2. Prepare titles and copy them by photographing (preferably on same size of negatives as used for rest of strip). 3. Make test exposures for each negative and note down exposure times in sequence. 4. Print the negatives on to the strip in their final order. 5. Process, wash, and dry the exposed positive strip.***

(3) Printing the negatives in their correct sequence.

(4) Processing the complete strip (usually in a tank).

Orientation of Images. Film strips that have been printed by contact or in the enlarger in the normal way appear the right way round when viewed from the emulsion side: so the emulsion side must face the condenser of the projector when the strip is projected.

Film strips produced by reversal appear the right way round when viewed from the back of the film. So in this case the emulsion side must face the projector lens when showing the strip.

The emulsion side can be arranged to face the lens for a normal strip, if the negative is reversed left to right in the enlarger when printing. But it is not advisable to reverse the negative in this way when contact printing because the emulsion layers of negative and positive material should be in contact with each other; the presence of the film base between the two surfaces would result in some loss of sharpness in the print.

Arrangement of Images. If all the images of the strip face the same way, arrangement is simple. Where a strip contains double (24 x 36 mm.) frames as well as single (18 x 24 mm.) frames, the single frames are always printed in pairs.

Where the strip contains horizontal and vertical single and double frames, the work of the projectionist is made easier if the frames are printed so that the film strip holder needs only to be rotated through 90° when passing from one type of frame to any other. Some projectors may permit a 90° movement.

Thus all vertical double frames and horizontal single frames (vertical strip) should be printed facing one way, and follow underneath each other when the images are viewed the right way round. All horizontal double, and vertical single, frames (horizontal strip) should also be printed facing in one direction, following each other from left to right when the strip is looked at the right way up and the right way round.

The standard type of strip has a blank 9 ins. leader, with a frame marked "start" before the first frame of the strip, and it finishes with a frame marked "end", and a blank 9 ins. trailer.

Processing. It is not practicable to process individual frames separately, and the whole strip is best processed in a developing tank in the same way as 35 mm. film. Such tanks take time to fill and empty in the usual way, and as development times are short, it is wise to fill the tank with the developer first and then immerse the reel in it; this is better than pouring the developer into the tank after the

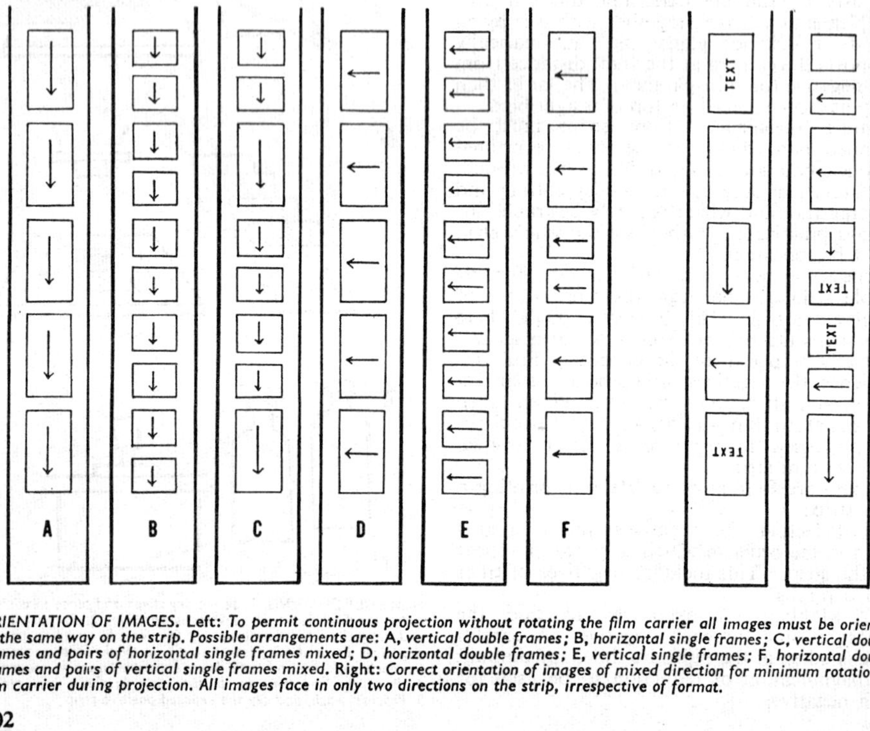

ORIENTATION OF IMAGES. Left: To permit continuous projection without rotating the film carrier all images must be oriented in the same way on the strip. Possible arrangements are: A, vertical double frames; B, horizontal single frames; C, vertical double frames and pairs of horizontal single frames mixed; D, horizontal double frames; E, vertical single frames; F, horizontal double frames and pairs of vertical single frames mixed. Right: Correct orientation of images of mixed direction for minimum rotation of film carrier during projection. All images face in only two directions on the strip, irrespective of format.

reel is there. After development the reel is lifted out in the darkroom and immersed in a dish containing the rinse or stop bath. The tank is meanwhile emptied and the fixer poured in. The reel is then replaced in the tank for fixing and subsequent washing.

The temperature should be constant between developer and rinse, and between rinse and fixer. An acid hardening fixer is advisable.

Single test exposures can be treated like small prints and developed in a dish.

All processing of the sensitive film strip should be carried out by the light of an orange safelight of the type used for bromide paper.

Projection. Film strips are projected in suitably adapted miniature projectors. These incorporate a film strip holder—which may be interchangeable with a slide holder—with feed and take-up spools to take film strips up to 10 feet in length. The individual frames are brought into position by turning the spool knobs and the film is held flat in the gate by a pressure plate. The whole holder rotates through 90° or 360° (according to the model) to project upright and horizontal frames. Better quality projectors include some means of releasing the pressure plate during film transport so as to reduce the risk of scratches, and may also feature interchangeable gate masks for screening 18 × 24 mm. as well as 24 × 36 mm. frames.

Commercial Film Strip Production. The general principles of printing film strips commercially are similar to those of amateur production as described. The equipment and procedure are, however, different, since the amateur methods would be too slow and uneconomical.

Commercial practice generally utilizes semi-automatic or automatic printing machines and continuous processing units of the type employed in motion picture laboratories. The negative for a commercial film strip may be passed through the printer over and over again in a closed loop, or else a number of negative strips joined end to end to print a whole film strip series in one run.

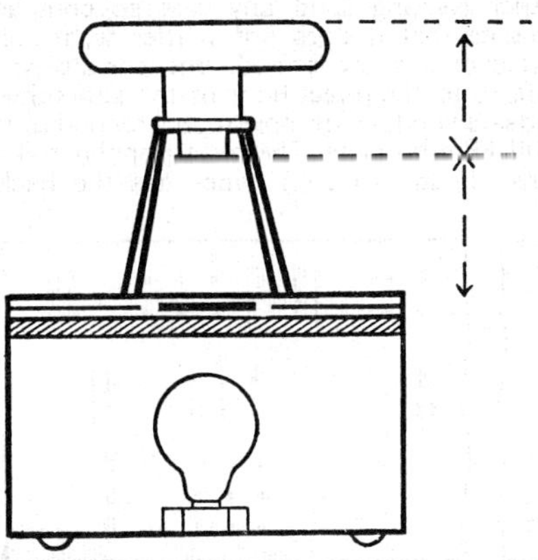

FILM STRIPS BY COPYING. A miniature camera with a 1 : 1 extension tube and distance gauge can copy 35 mm. negatives (or transparencies) directly over a light box.

Availability and Uses. The increasing interest in the use of visual aids in education has made the film strip very popular in recent years. There are now numerous firms who specialize in the making of film strips, and most of these publish catalogues.

Many photographic organizations also provide some kind of film strip service, usually in the form of a free lending library. Film strips covering both technical and pictorial aspects of photography are available, nearly always supplemented with a scripted commentary. More recently, particularly in the U.S.A., film strips have also been issued with tape recorded commentaries to accompany them.

A number of film strip projectors exist combined with a built-in gramophone record player. The strips are then issued complete with a pre-recorded disc. This is more practical with film strips than, for instance, with slide sets, since the strip already carries the images permanently connected.

Other sources of film strips are the publicity departments of major industrial concerns, government bodies and public services.

Film Strips in Colour. Sets of colour illustrations such as paintings and the like, can be turned into colour film strip series by direct photography in a miniature camera on reversal colour film. Since paintings (but also miniatures, objets d'art, etc.) are best photographed by artificial light for maximum control of exposure conditions, an artificial light type colour film is required. The exposures should—as far as possible—be the same (determined after preliminary tests). Where mixed horizontal and vertical views are used, be sure to turn the camera round to only one vertical position, corresponding to the proper orientation of a mixed film strip (see above).

The subject distances (and hence magnifications) can of course vary according to the size of the original. The colour reproduction is also liable to vary somewhat from a true rendering; colour films are rarely able to reproduce artificial pigments with absolute fidelity. In principle a colour negative film can also be used, for printing the positive on to a transparency film. This enables a number of strips to be made of the same negative strip. On the other hand printing control is more difficult and colour negatives contact printed on to positive colour film are liable to lose sharpness owing to the number of emulsion layers involved. Projection printing gives better results. A colour duplicating film can also be used where many copies of this strip are subsequently required.

Film strips can also be made from a set of colour slides. This calls for an inter-negative film or a soft gradation transparency duplicating film. The first yields colour negatives which again have to be printed on to a positive film. The second yields direct colour transparencies with only a comparatively slight loss of colour saturation and tone gradation. (Ordinary colour reversal films invariably yield duplicates of excessive contrast and degraded colours.)

Commercial filmstrips in colour are also available; some outstanding examples deal with well known paintings. L.A.M.

See also: *Contact printer; Contact printing; Slide diagrams; Slides; Printing materials; Projection practice; Projector (still); Visual aids.*

Books: *An Introduction to Filmstrips*, by H. R. and I. W. Dance (London); *Projecting Slides*, by C. W. Long and M. K. Kidd (London); *Making Lantern Slides and Filmstrips*, by C. Douglas Milner (London).

FILM TRANSPORT

Any camera which takes sensitive material in pre-packed loads for a number of exposures must have an accurate, simple and foolproof method of replacing an exposed piece of film in the image plane by an unexposed one. In roll film and 35 mm. cameras this involves winding the film on through an appropriate length corresponding to the space taken up on the film by the image. The transport system has to provide for the smooth advance of the film, give an indication of the amount of film advanced, and show the number of exposures of the load already used up.

The design of the transport system depends on the kind of material employed—roll film, perforated 35 mm. or other film, and so on.

ROLL FILM CAMERAS

On inexpensive roll film cameras (also on most older models) the film transport consists simply of a knob on a shaft which engages the take-up spool inside the camera. The beginning of the film is attached to this spool during loading. The degree to which the film has to be advanced from picture to picture is established by the numbering on the backing paper of the film, and observed through a window in the back of the camera.

Backing Paper. Roll films are backed with an opaque paper strip which carries numbers corresponding to the number of negative spaces on the film. The backing paper is a little wider than the film and extends beyond it at each end. The leading end of the film is attached to the backing paper by a piece of adhesive tape, and there is enough spare paper ahead of it to attach to the take-up spool in daylight without exposing the end of the film.

As the backing paper is wound on to the take-up spool the first thing that is seen through the safe-window in the back of the camera is the outline of a hand. This indicates that the end of the film is just about to enter the negative area. From this point onwards it is not safe to open the camera or release the shutter.

After the hand there may be a series of warning dots followed at last by the figure (1). This indicates that the film is now in position for the first exposure. As each exposure is made, the film is wound on to the next number. When the last frame has been exposed, the film is wound completely on to the take-up spool, leaving its own spool empty and ready to transfer to the take-up chamber.

The backing paper is commonly either black paper with a red or yellow back and black figures, or black paper with a green back and white figures. The safe window in the back of the camera may be either red or green. At one time it was usual to buy orthochromatic film on red backing paper, and panchromatic on green. Red windows were used for orthochromatic films and green windows were either substituted for the red or stuck over the top when a panchromatic film was in the camera. Experience seems to show, however, that the paper backing is in any case so completely opaque that it does not matter what colour paper or window is used. But it is always unsafe to let the direct light of the sun shine on the safe window for more than a second or two.

Roll Film Formats. The most popular roll film size—2¼ ins. (6 cm.) wide—has the backing

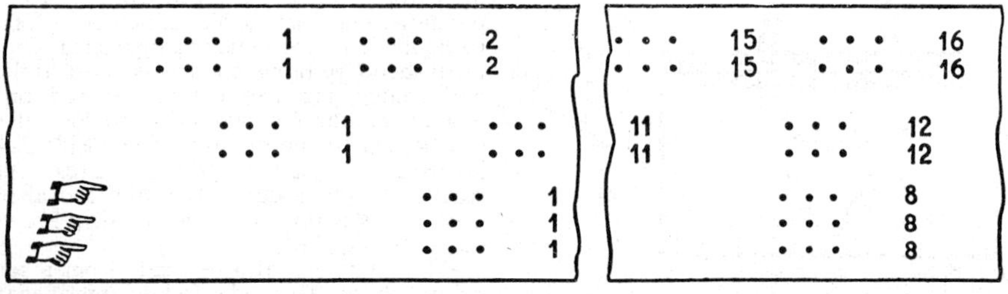

BACKING PAPER NUMBERS. The backing paper of practically all roll films carries a series of numbers to show how far the film must be wound on to the next exposure. The numbers are observed through the red window in the camera back. Films yielding various numbers of exposures of alternative sizes (e.g., 120 or 620 film) carry several sets of numbers.

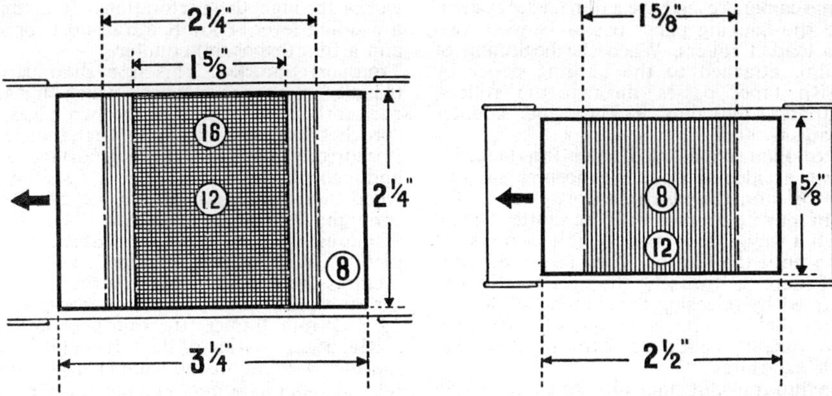

FILM WINDOWS. The positions of the windows in the back of a roll film camera are laid down by a British Standard, and correspond to the location of the numbers on the backing paper of the film. Two-size cameras have the appropriate two film windows. The arrows indicate the direction of film travel, the shaded areas the alternative smaller picture sizes. On most modern roll film cameras a shutter keeps the film window closed to protect it against strong light, except when the film is actually being wound on. Left: Size 120 and 620 film for 8 exposures 2¼ × 3¼ ins. or 12 exposures 2¼ × 2¼ ins., or 16 exposures 1⅝ × 2¼ ins. Right: Size 127 film for 8 exposures 2½ × 2½ ins. or 12 exposures 1⅝ × 1⅝ ins. Some cameras take 16 exposures 1¼ × 1⅝ ins.

paper numbered down the left edge to give 8 exposures to the roll, down the centre to give 12 exposures and down the right edge to give 16 exposures. So the position of the safety window in the back of the camera indicates the format of the picture. When the window is at the left side the camera takes the oblong format—2¼ × 3¼ ins. (6 × 9 cm.). When the window is in the centre, the camera takes the square format—2¼ × 2¼ ins. (6 × 6 cm.); and with the window at the right the camera takes the 16-on size—i.e., 1⅝ × 2¼ ins. (4·5 × 6 cm.). As an alternative to the right-hand window, older cameras or adapters that take these half sizes have two safety windows. Each number on the backing paper appears twice—once behind each window—and an exposure is made with it in each position. In this way 16 exposures are made on the normal 8-exposure film.

The backing paper for 1¾ ins. (4·5 cm.) wide film is numbered down the centre for the oblong 1⅝ × 2½ ins. (4 × 6·5 cm.) format of 8 exposures and down the left hand edge for the square 1⅝ × 1⅝ ins. (4 × 4 cm.) size. The film windows in the back of the cameras taking this film size are positioned accordingly. For the 16-on size of 1¼ × 1⅝ ins. (3 × 4 cm.) the camera carries two windows along the centre line of the back, each number being wound to the first and second window in turn.

Some cameras are equipped to take more than one size of picture on the same film. In particular, a number of cameras will take either 8 or 16 exposures on the standard roll of film. The back of such cameras carries an extra window that is only used when taking the half-size frame, and a half-size mask that is slipped into the negative aperture when the film is loaded. There are also cameras which will take all these formats; they have half-size and square masks and three safety windows. All cameras that give optional formats have some arrangement for masking the viewfinder or indicating the boundaries of the optional sizes.

Automatic Film Transport. More expensive modern roll film cameras (in particular the better-class single-lens and twin lens reflexes) have provision for advancing the film without reference to the numbering on the backing paper. The film transport itself may still be a knob or else a winding lever or crank. The transport movement automatically locks when it has advanced the film by exactly one frame interval.

There are several systems. In the simplest case the film is loaded in the normal way and wound on until the figure 1 on the backing paper appears in the safety window. From that point the window is covered and the amount of film used at any time indicated on a visible counter. The film counter (and the locking system) may be driven by a spring-loaded wheel bearing on the edge of the film inside the camera, or it may simply be coupled to the film transport shaft. In the latter case some compensating device usually allows for the growing diameter of the take-up spool as the film winds up on it, to ensure reasonably even spacing of the frames on the film.

Many current roll film cameras have no safety window at all. There the procedure during loading is to wind the backing paper until a reference mark on it comes opposite a locating mark in the camera back. The latter is then closed and the film wound until the transport locks. The distance from the reference mark to the first film frame is standardized, so that a given number of turns of the film transport shaft automatically brings the first exposure on the film into position by the time the transport locks.

Some cameras even have a film feeder system, where the backing paper passes between two spring-loaded rollers. When the beginning of the film, attached to the backing paper by adhesive tape, passes through the rollers, it initiates the film locking and counter mechanism.

Coupled Film Wind and Double Interlock. To prevent accidental double exposures on one frame of film, the film transport system of a camera may be coupled with the shutter release in such a way, that it is impossible to press the release until the film has been advanced after an exposure. Conversely, the transport lock is only freed by releasing the shutter. With such cameras there is often some way of putting the device out of action to permit intentional double exposures.

The film transport may also be coupled with the shutter to tension the latter automatically during the advance of the film.

Normally cameras with any form of double exposure lock only have a single B setting for time exposures on the shutter. Long time exposures must be made with a cable release and a locking arrangement. The T setting, available on shutters for cameras without double exposure lock, cannot be used in this case, since it requires a second pressure on the shutter release to close the shutter. The double exposure lock makes such a second pressure impossible until after the film has been advanced.

In addition to coupling the film and shutter setting operations, some cameras also have a device which operates when the setting knob has not been turned on after the last exposure. An internal metal shutter automatically masks the viewfinder after the shutter release is pressed, warning the user that he must wind on the film before he can take another photograph.

MINIATURE CAMERAS

As 35 mm. film has no backing paper, its frames cannot be visibly numbered as they are with roll film. On the other hand the film is perforated, and the transport system usually

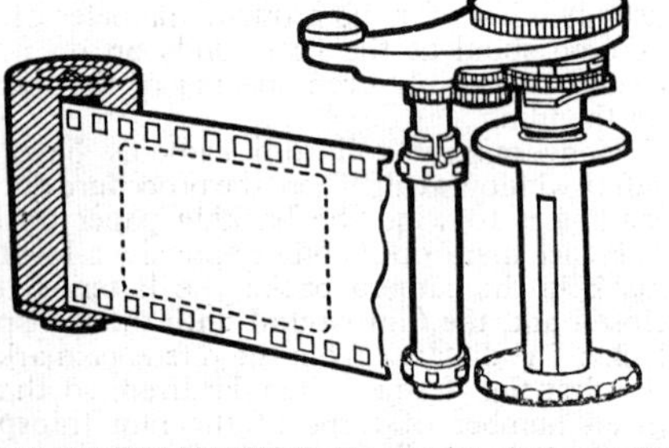

MINIATURE FILM TRANSPORT. The sprocket wheel, geared to the take-up spool, pulls the film out of the cassette. On modern cameras a rapid winding lever works the mechanism.

works through the perforations. It comprises a winding lever or knob, a transport sprocket, and a film (exposure) counter.

Transport Sprocket. This is a shaft carrying (usually) two sprocket wheels which engage the perforations of the film at the two edges. The sprocket thus pulls the film through the camera. It is directly linked with the transport movement and incorporates a lock which arrests the shaft when the right length of film has been pulled through. With the standard 24 × 26 mm. frame, eight film perforations correspond to one picture. With cameras taking 24 × 24 mm. pictures the transport lock is arranged to stop the movement after six perforations, and for 18 × 24 mm. frames after four perforations.

The usual location of the sprocket is between the film window in the camera back and the take-up spool. Some cameras have a single transport sprocket set in the film track. In either case the sprocket only rotates in one direction, but can be disengaged from the transport movement to turn freely in either direction for rewinding the film.

The Lever Wind. On most current 35 mm. cameras the film transport control is a lever operated by a flick of the thumb or finger. Frequently this folds into the camera when not in use.

Usually a single swing of the winding lever advances the film by the right amount for one exposure. On some cameras the film can also be advanced by a number of short swings, the lever locking after a transport movement corresponding to one film frame. On cameras which have this provision, the lever must be pulled up to its fullest extent to complete the transport movement. The lever then flies back under spring action to be ready for the next transport cycle.

Miniature cameras of older design have a transport knob instead of a lever. This requires turning, but is geared again to the sprocket and take-up spool.

A number of unorthodox film transport movements have also appeared from time to time: levers on the camera front, plungers in the top or at the side etc.

The winding movement is also linked to the take-up spool which winds up the film. This linkage is usually a friction drive so that the spool can slip on its shaft as it gradually gets full, to keep its movement in step with the transport sprocket. A few cameras have used a direct gearing between the transport lever or knob and the take-up spool, dispensing with the sprocket. In this case the spacing of the frames on a film is uneven, since the spool diameter increases as more film is wound on it.

The Film Counter. In a 35 mm. camera a mechanical counter keeps track of the number of frames exposed of a cassette load of film. This film counter or exposure counter is geared to either the transport sprocket or the shutter tensioning mechanism. It consists of a graduated disc which rotates by one division as the film is

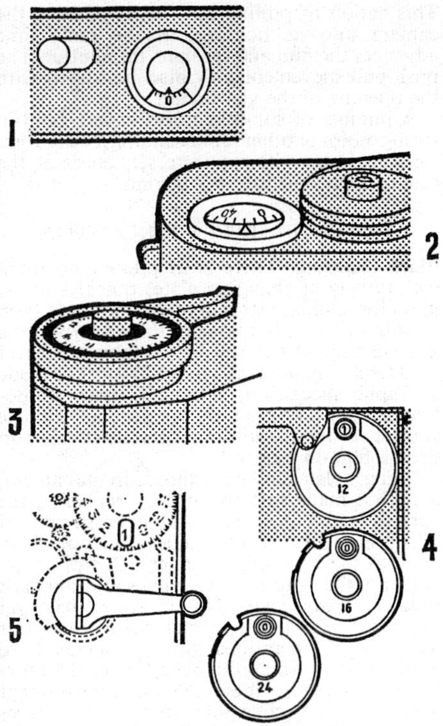

FILM COUNTERS. 1. *Film counter of miniature camera, reset to zero by a setting button.* 2. *Automatic film counter, resets itself to zero on opening the camera after unloading.* 3. *Counter in transport lever, manually set at any time.* 4. *Interchangeable film counters for certain multi-format twin-lens reflexes.* 5. *Film counter of twin-lens reflex coupled with transport crank.*

advanced, or turns through about one revolution ending up with one division further—as read off opposite an index mark.

On most cameras the film counter is reset to its starting point after loading a film. The film itself is usually loaded in light-tight cassettes, with the beginning of the film protruding. After threading the film into the camera the exposed film leader has to be wound up on the take-up spool with the camera back closed, to ensure that a fresh and unfogged piece of film comes into position behind the film window for the first exposure. This involves making two or three "blind" exposures beforehand. Many film counters, when set to a starting index, automatically show when sufficient film has been advanced to take up the fogged beginning.

More advanced and expensive 35 mm. cameras may have an automatically resetting film counter. Here the counter returns to 0 on its own on opening the camera back or withdrawing the take-up spool (where this is removable).

There are two common counting systems: from 0 upwards (to about 40) to show the number of frames exposed; or from 36 backwards to No. 1, to show the number of frames still left on the film. In the latter case the counter must be started at a point corresponding to the actual length of film in the cassette. Beyond that the difference between the two systems is merely a matter of calibration and personal preference.

Double Exposure and Film Lock. On most current 35 mm. cameras the film transport is linked with the shutter tensioning mechanism, so that accidental double exposures become impossible. The film transport lock is in turn linked with the shutter release, to prevent blank frames (the advance of a film without taking a picture).

Deliberate double exposures are usually possible by disengaging the transport sprocket (by means of a reversing button or clutch) while working the transport lever. This tensions the shutter without advancing the film. Some camera designs incorporate a film release button to release the transport lock for deliberate film advance without exposures; this is useful when a partly exposed film has to be reloaded. The double exposure and film release functions may also be provided by various other means.

On cameras with interchangeable magazines the film transport mechanism may be interlinked with certain other safety locks to prevent inadvertent removal of the camera magazine without first closing the film apparature.

Rewinding. Since the film is wound on to the take-up spool from a light-tight cassette but does not itself have any backing paper or other protection, it must be rewound into the cassette when the whole film length is exposed. Otherwise opening the camera would fog and therefore ruin the whole film. The rewind mechanism on a 35 mm. camera consists of a knob which may have a built-in folding crank. This acts on a shaft which engages the centre spool of the cassette. To permit winding, the transport sprocket must first be disconnected to enable it to turn backwards. This is achieved with a reversing clutch in the form of a button, lever or similar device. The film is then rewound against the resistance of the friction drive of the take-up spool.

On a few cameras the film counter also moves backwards to its starting point during rewinding; usually however the film counter mechanism is disengaged as well. The point at which rewinding should be stopped can generally be felt by the resistance as the film leader is pulled off the take-up spool. Further rewinding pulls the film leader into the cassette. This is of no great importance if the complete film was exposed, but rewinding should stop before the film leader is withdrawn into the cassette if a partly exposed film has to be reloaded again.

A few cameras can work with a light-tight take-up cassette instead of the bare spool; in this case no rewinding is necessary and the full take-up cassette can be unloaded from the camera at the end of the film. Certain rapid loading systems, employing a completely enclosed film magazine with light-tight feed and take-up compartments also need no rewinding.

Rapid Winding Systems. Quick-operating devices are available for some 35 mm. cameras to transport the film even faster than is possible with the normal lever wind. These accessories are useful for action photography and wherever a large number of exposures must be made rapidly.

For even higher operating speeds the film can be automatically advanced after each exposure by a spring motor attachment. With this device it is possible to make exposures at the rate of 2 or 3 per second. One or two makes of camera have such a spring—or electric—motor built in, and after the first pressure on the shutter release button, exposures are made automatically much faster than they could be made by hand. This type of camera will make a series of still photographs of an action lasting only a second or two.

In a fully motorized camera the motor automatically transports the film and retensions the shutter, leaving everything ready for the next exposure. When fully wound, a spring motor may be able to operate the complete exposure sequence a dozen or two dozen times. Electric motor film transport systems or attachments are usually also suitable for remote control. In this case the camera can be set up—for example for nature photography—and operated some distance away by the photographer who only has to press a button at the end of a remote control cable. This releases the shutter and actuates the transport and tensioning cycle. Remote control systems of this kind can even operate by wireless signals.

ULTRA-MINIATURES

Still cameras taking 16 mm. or even narrower film are generally loaded with single or double cassettes. With the single cassette the transport system is in the main a scaled-down version of a 35 mm. film transport. A sprocket here acts on the perforations of the film in the usual way. If the camera uses unperforated film, the transport mechanism acts through the take-up spool core. Ultra miniatures using a single cassette need rewinding of the film as with a 35 mm. camera.

In a number of cases, especially with two-cassette systems, the take-up cassette contains no spool. The film is then simply pushed into the take-up chamber where it curls up, guided by an internal spring. This design makes the cassette more compact, and is also found on a few 35 mm. cameras.

Ultra-miniatures may have a transport lever or knob, but a number of other systems are also popular. One of these is a push-pull action: the camera is pulled out of a housing which normally protects the lens and finder opening. This action of pulling out (or of pushing the camera into its housing) at the same time advances the film and tensions the shutter. The push-pull movement may also be linked with the opening of the viewfinder.

A number of sub-miniatures have a built-in spring motor or other rapid winding mechanism. This again permits exposure sequences at the rate of 2 or 3 pictures per second.

PLATE AND SHEET FILM CAMERAS

Plate Changing. There is at present no satisfactory way of changing plates quickly. Many ideas for loading plates in magazines have been tried in the past, but none of them survived the introduction of the roll film. Plates are never used when a number of exposures must be made in rapid succession, although press photographers, by long practice, can generally change a plate in about the same time as it takes to wind a film on by hand.

When a plate has been exposed in the camera, it is first made safe to light by replacing the dark slide and then the plate holder is removed by sliding it out of the guides that hold it in place at the back of the camera. A fresh plate holder is then slid into the guides and the dark slide is withdrawn. No inventor has yet succeeded in improving materially upon this sequence of operations. Some cameras have plate backs in which the plate holder is simply clipped in place; these are an improvement over those that slide into guides. Some have sections of the guides cut away so that the plate holder has only to slide about one quarter of the length of the guide.

All types of plate changing magazine are handicapped by the weight of the plates. Even if such magazines could be made to work satisfactorily, no one would willingly carry around up to a dozen plates in his camera when a single roll of film would serve the same purpose.

Sheet Film. Single sheets of film, which are in many cases replacing plates, are used in much the same way as the latter. They are loaded in sheet film holders with a dark slide, and the loading and changing procedure is as slow as it is with plates. A loaded sheet film holder is however lighter. Magazine operations with sheet film involves the use of a special packing, the film pack.

Film Packs. The film pack usually holds twelve exposures in the form of pieces of film cut to the plate size of the camera. Each piece of film is attached to a strip of black paper which forms both a backing paper and a leader by which the film is pulled into place before exposure. The films are packed card-fashion in a shallow metal box, and the paper leaders are folded back around the bottom edge of a separator so that when one of the leaders is pulled, the film is drawn down one side of the separator and up in front of the other.

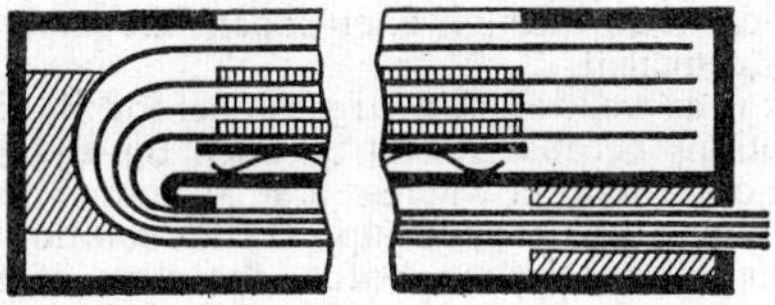

CROSS-SECTION OF FILM PACK. A pile of 12 sheet films (3 shown here), each attached by one edge to a long strip of black paper, is placed inside a flat metal box behind a film aperture. The paper leaders go round behind the pressure plate, with the ends projecting as tabs outside the edge of the box. After exposure, the top film is moved out of the way round the back by pulling the paper tab, leaving the next film in taking position.

Film Pack Adapter (F.P.A.). A metal container, called a film pack adapter, holds the film pack and is attached to the back of the camera like a plate holder. To make an exposure the dark slide is withdrawn from the film pack adapter and the first paper leader is pulled up and torn off.

This draws away the outer sheet of black paper and uncovers the first film. When this has been exposed the second paper leader is pulled out.

This pulls the exposed film around to the back of the pack and uncovers the second film. The backing paper behind each film prevents light from passing through it during exposure and affecting the next film.

The film pack can be changed quicker than a roll film, and it has the added advantage that once the dark slide is replaced, the half-used pack can be removed from the camera, leaving it free to be used with either a focusing screen or another type of sensitive material. On the debit side; the film pack is the dearest way of buying sensitive material; it adds to the bulk of the camera, and the dark slide is another item to think—or forget—about.

SPECIAL TRANSPORT SYSTEMS

Large-Capacity Cameras. Certain types of industrial and commercial cameras have to carry large film loads of several hundred exposures to cope with a high work rate, without the loss of time involved in frequent cassette or roll changing. In principle these transport systems work on the same lines as the automatic transport of a 35 mm. camera: the film is wound on by a fixed distance after an exposure while the shutter is retensioned. Usually the transport system is driven by an electric motor which may be remote-controlled. Examples are cameras used in aerial photography, instrument recording, microfilm copying, etc.

Continuous Transport. For some types of photography the film in the camera is advanced not intermittently, frame by frame, but continuously. This is the case in all systems where the subject also moves past the camera at the same time. This permits exposures at a high rate and is used in such fields as aerial photography and mapping, microfilming, etc.

The main requirements of the transport system are a smooth movement of the film at a constant speed, and accurate matching of the film movement with the speed at which the object or objects flow past the camera. No shutter tensioning of film counting is involved, and the exposure takes place through a narrow slit immediately in front of the film.

Instant Pictures. In a few special camera systems processing takes place inside the camera and yields finished prints. In this case the film is advanced by pulling a protruding leader out of the camera after every exposure. The material is arrested at appropriate intervals by a mechanical device inside the camera.

F.P. & L.A.M.

See also: *Films; Miniature camera; Remote control.*

FILTER BAG. Suitably dyed transparent plastic bag used to modify the colour of light sources, especially flash bulbs. The normal range of dye colours includes pale yellow and pale blue to lower or raise the colour temperature of flash bulbs for work with artificial light or daylight type colour films respectively. Special deep red or almost opaque filter bags have also been used for infra-red photography; with such a bag the flash bulb or lamp emits practically no visible light at all and can be used for infra-red photography at night without attracting attention. In use, the filter bag is simply fitted over the bulb before the exposure.

FILTERING. Filtering removes undissolved particles and other solid impurities from a solution. These may be present in freshly made solutions, and also accumulate as a sediment in developers, etc., during use. If not filtered out, they may settle on the emulsion during processing and either cause clear spots by preventing the developer from reaching the surface, or black spots if the particle consists of undissolved active chemical.

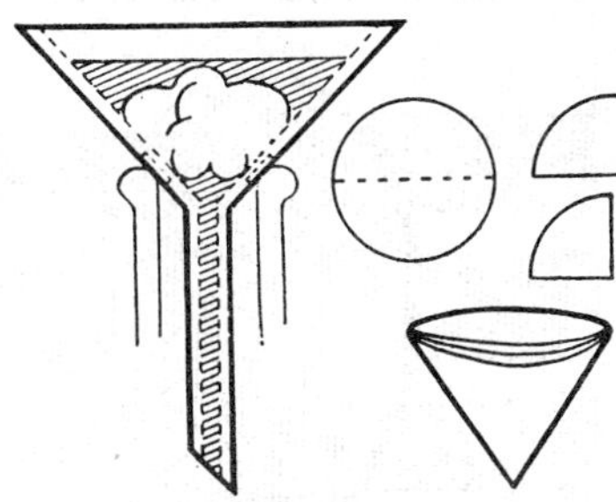

FILTER FUNNEL. Either cotton wool or filter paper will hold back solid particles. The paper is folded to form a cone which fits into the funnel and is supported by the sides.

A solution is filtered by being poured through a funnel lined with a porous material—e.g., a wad of cotton wool, or a cone made from a circular disc of the special filter paper that can be bought for the purpose. The finer the pores, the more efficient the filtering action—i.e., the smaller the particles of solid matter that will be removed.

But for the same surface area of the filtering material, the solution takes longer to pass through a fine material than one with more open pores.

Whenever possible, solutions should be filtered while hot, as the process is then quicker. Solutions run through cotton wool more rapidly than through filter paper, but paper is more efficient.

It is as well to use a large funnel for filtering solutions because it can be filled up and left without attention whereas a small funnel must be constantly topped up. If the solution is being filtered into a bottle, the stem of the funnel can be inserted in the neck of the bottle and no support is needed. But if the solution is being filtered into a dish, the funnel should be mounted on a filter stand. This consists of a metal ring or a clamp to hold the funnel, clamped to the vertical pillar of an ordinary chemistry laboratory retort stand.

See also: *Chemicals; Solutions.*

FILTERS

Filters are used in photography to modify the light falling on the subject or passing through the camera lens. Thus we have camera filters which are coloured discs of glass or gelatin placed on the lens, and lamp filters which are placed in front of the light source.

As their name implies, filters actually stop some of the light falling on them, so that the colour or nature of the light which passes through is changed. With coloured filters, the light which they pass is governed by the actual colour of the filter: thus a yellow filter will pass red and green light, but tend to stop blue light. In this way, by controlling the proportions of colour coming from the subject (or light source), the relative tones in the negative and final photograph can be modified.

There are five main reasons for wanting to use a filter:

(1) To correct the imperfect colour sensitivity of the film, and make it translate the subject into tones of grey of more or less the same brightness as the colours appear to the eye. Such filters are called correction filters.

(2) To distort (brighten or darken) the reproduction of certain colours for special effects. This intentional distortion is used in pictorial photography and in certain commercial and scientific work. Filters used in this way are called effect filters or contrast filters.

(3) To change the colour temperature of the light. This is important for colour photography. Such filters are called colour compensating filters.

(4) To take pictures by the light of a single colour only; mainly for scientific and other special purposes such as telephotography by infra-red light. In principle, most of these filters are extreme cases of the contrast filters mentioned above.

(5) To take photographs by polarized light, or to cut out the effect of polarized light in photographs. Such filters are called polarizing filters.

Gelatin Filters. The commonest camera filters are thin pieces of dyed gelatin. They are easiest to make accurately, and offer the widest range of colours.

Gelatin filters are cut to the required size and used in front of the lens in special holders. For better protection the gelatin discs may even be mounted between the lens elements themselves.

As gelatin filters are very sensitive to mechanical injury, they should be carefully protected against scratches, dirt, dust, grease and moisture. The gelatin surface must never be touched with the fingers as it is almost impossible to clean.

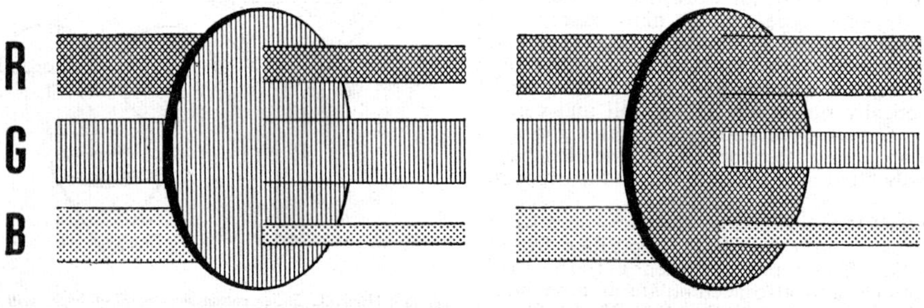

ACTION OF A FILTER. Filters act by holding back some of the light reaching them. They pass almost all the light of their own colour. but appreciably reduce the intensity of light rays of different colours. As a result they also call for extra exposure.

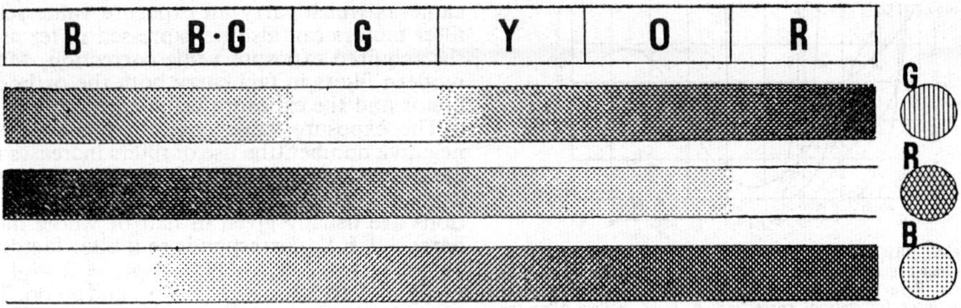

FILTER ABSORPTION. A filter usually has its maximum transmission in one particular region of the spectrum. It will lighten objects of its own colour, and darken others to an increasing degree the farther away they are on the spectrum scale. Thus a red filter will darken blues most, as blue is at the other end of the spectrum. A green filter will darken both blue and red.

Gelatin Glass Sandwich. Where gelatin filters have to be constantly handled, they are usually cemented between glass. The workmanship of the cementing and the quality of the glass are important. If the glass surfaces of the filter are not absolutely flat and parallel the thickness of the glass may upset the optical properties of the lens. The danger is less with unmounted gelatin filters which are so thin that they have practically no effect.

The best filters for high quality scientific work are prepared as optical flats of guaranteed accuracy. For most normal photography where such high precision is not needed, the gelatin is usually cemented between flat and polished plate glass.

Cemented filters must be treated with the same care as camera lenses. They should never be allowed to get wet or dirty. Although the edges of the filter sandwich are usually sealed, moisture and cleaning fluids will in time penetrate and affect the cement or the gelatin. Filter holders which exert uneven pressure on cemented filters upset their optical parallelism, particularly of those of the highest quality.

Some modern cemented filters have dyed cellulose sheets instead of gelatin.

Glass Filters. Much more robust filters are made from mass coloured glass. Such filters consist of glass only. While they are not so easily damaged, the range of colours in which they can be manufactured is more limited. The optical accuracy of the best solid glass filters can be as high as that of an optical flat.

Lamp Filters. Filters are also made in the form of lamp filters which are large thick sheets of dyed gelatin. They are placed over the lamp or other light source, and not over the camera lens. As they cannot affect the performance of the lens, their optical quality does not matter and they do not call for any special care in handling.

Where the subject is illuminated by a number of lamps, it is generally more economical and easier to make any colour correction by using a filter in front of the camera lens. But lamp filters are useful for dealing with only part of the lighting—e.g., for special effects in colour photography. Sometimes a lamp filter is used to limit the colour of the lighting to one particular wavelength as in photography by infra-red and ultra-violet light.

Liquid Filters. For scientific work special filter effects may be required which are not obtainable or practical with normal ready-made gelatin or glass filters. There it is often most convenient to use solutions of certain coloured chemicals in suitable glass cells. The thickness of the cell (i.e., the thickness of solution placed in the path of the light) determines the degree of absorption, while the choice of substance gives the necessary absorption characteristics. Liquid filters may be also used for absorbing heat.

Chemicals used for liquid colour filters are largely inorganic salts, for instance potassium bichromate (orange), ammoniacal copper sulphate (blue-violet), copper sulphate or nitrate (blue-green), iodine in carbon tetrachloride (violet) or in potassium iodide solution (brown), various nickel and cobalt salts, etc. Visually colourless solutions (e.g., sodium nitrite) are also used as ultra-violet filters. Suitable combinations of several solutions—usually in separate cells—can isolate very narrow bands of the spectrum.

The main advantage of liquid filters—apart from their versatility and cheapness—is the exact reproducibility of characteristics. These depend on variables which are easily controlled with great precision; namely cell thickness, and composition and strength of the solutions used.

Absorption Curves. Filter characteristics are specified by absorption curves which plot the filter density against the wavelength of the light. Such a curve therefore shows the wavelength of maximum transmission (i.e., the predominant colour of the filter) as well as the amount of light of other colours passed through. The individual points of the curve indicate relative—i.e., percentage—absorption and transmission of each wavelength. The shape of the curve corresponds to the absorption with a light source of equal energy throughout the

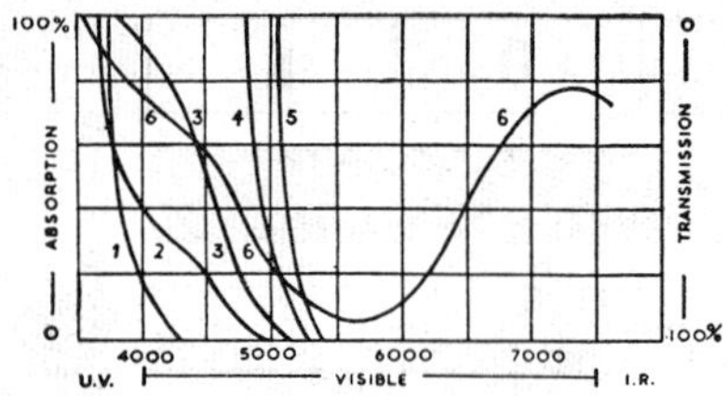

TYPICAL ABSORPTION CURVES OF CORRECTION FILTERS. 1. Colourless ultra-violet filter for high altitudes and colour film. 2. Pale yellow filter for slight correction. 3. Medium yellow filter for normal correction. 4. Deep yellow filter for over-correction. 5. Very deep yellow (minus blue) filter, often used in photomicrography. 6. Yellow-green filter.

spectral range covered, though the curve is rarely obtained by direct measurement with such a source.

Absorption curves are significant with separation filters for three-colour photography, where they help in correctly matching a set of filters for the most suitable overlap in absorption ranges. In other fields, too, the curve is a help in selecting the best filter for a particular purpose.

Filter Factors. Filters affect the colour of light they pass by holding back some of it. So they call for a longer camera exposure to compensate for the light lost. In practice the exposure time is multiplied by a filter factor.

Usually filter factors—other than the lowest values—are rounded off to the nearest whole number. This is accurate enough for everything except certain sensitometric work. In ordinary photography the difference between factors of 3·8 and 4 or 4·2 is too small to matter. In fact, errors up to 20 per cent can be ignored.

Instead of increasing the exposure time, we can widen the lens aperture. Here again, setting the aperture to the nearest half stop is quite accurate enough for ordinary photography.

The filter factor is not a constant number. It depends on: the colour and depth of the filter; the colour sensitivity of the film or plate; the colour of the light; the colour of the subject.

So to know the factor of a particular filter, we must also know the colour sensitivity of the film and the colour of the light. For really accurate tone reproduction we may have to know the colour of the subject as well.

The filter colours may vary from that of pale yellow or green correction filters to that of deep effect and monochromatic ones. Generally, the deeper the colour of the filter, the more light it holds back, and the higher its factor.

Filter manufacturers always provide a set of tables giving filter factors for the various conditions normally anticipated.

Filter Factors and Exposure Values. For cameras which carry an exposure value scale, filter factors can also be expressed in terms of the required exposure value correction. Many modern filters in fact carry both the orthodox factor and the exposure value correction.

The exposure value correction is always a negative number (the use of filters increases the exposure required and hence reduces the exposure value setting). Exposure value corrections are usually given in half or whole numbers: $-\frac{1}{2}$ E.V. corresponds to a filter factor of $1\frac{1}{2}\times$, $-1 = 2\times$, $-1\frac{1}{2} = 3\times$, $-2 = 4\times$, $-2\frac{1}{2} = 5$ to $6\times$, $-3 = 8\times$, and so on. The exposure value correction also corresponds to the number of lens stops by which the lens must be opened up when using the filter.

Negative Material. The colour sensitivity of the film plays its part because some films are excessively sensitive to certain colours and thus need lower filter factors for filters of those colours. We can classify films and plates according to their colour sensitivity as follows:

Blue sensitive materials are sensitive to ultra-violet, violet, and blue light only. They are rarely used with filters, and then only with pale yellow ones, which have a high factor when used with this type of material.

Slow orthochromatic materials are sensitive to ultra-violet, violet, blue, blue-green, and slightly to green.

Fast orthochromatic materials are sensitive to yellow-green and yellow as well, but to a much greater extent to violet and blue. With both slow and fast orthochromatic materials yellow and yellow-green filters (to reduce the excessive blue sensitivity) are the only ones that can be used.

Red panchromatic materials are sensitive to all colours of the spectrum, and excessively sensitive to orange and red. They thus need comparatively low filter factors with red filters.

Correct panchromatic materials are also sensitive to all colours, but their sensitivity is more balanced. Thus they are not excessively red sensitive, and need higher filter factors with red filters.

Infra-red materials are highly sensitive to violet and blue, almost insensitive to blue-green, green, yellow, and orange, and sensitive to red and infra-red. They are used with special deep red or infra-red filters only.

Colour of the Light. The colour of light depends on the light source. Daylight is more blue (has a higher colour temperature) than tungsten electric light. Thus in daylight blue filters need lower filter factors, and red filters higher.

With artificial light, which is comparatively redder, red filters need lower factors

APERTURE CHANGES WITH FILTER FACTORS

For a factor	$1\frac{1}{2}$	2	3	4	5–6	7–9	10–13	14–18	20–27	30–35	
Increase aperture by	$\frac{1}{2}$	1	$1\frac{1}{2}$	2	$2\frac{1}{2}$	3	$3\frac{1}{2}$	4	$4\frac{1}{2}$	5	stops

and blue filters higher factors than by daylight.

For really accurate scientific or reproduction work and also for colour photography, the smaller variations in colour between different types of artificial light must also be allowed for—e.g. low power household lamps, high power studio lamps, Photofloods, etc. Even

SUMMARY OF CORRECTION FILTERS

Filter	*Daylight Factor* LO	O	P	R	*Art. Light Factor* O	P	R	*Effect*
Ultra-violet	1½	1¼	1¼	1¼	*	*	*	Used to darken sky at high altitudes
Pale-yellow	2	1½	1½	1½	1½	1½	1½	Slightly darkens sky tones
Medium-yellow	3	2	1½	1½	2	1½	1½	Darkens sky; bring out clouds
Deep yellow	5	3	2½	2	2½	2½	2	Very full correction of sky tone
Yellow-green	4	2½	2	2	2½	1½	1½	Slightly darkens sky, also darkens red with pan film
Pale green	*	3	3	3	3	2½	2½	Correct tones on pan film for outdoor portraits, etc.
Medium green	*	*	3½	4	*	3	3½	Fully correct tone reproduction on red pan
Pale blue		*	*	*	*	1½	1½	Correct tones on pan films in artificial light

LO = slow orthochromatic, O = fast orthochromatic, P = correct panchromatic, R = red panchromatic.
* There is no point in using this film-filter combination under these circumstances.

the colour temperature of daylight depends on the time of day and on the weather. But for all normal work these variations can be ignored.

Special light sources such as arc lamps, gas or mercury vapour discharge tubes, etc., have their own filter factors. They are usually published by the manufacturer of the film or plate likely to be used under such conditions.

Correction Filters. Nearly all orthochromatic and panchromatic materials are excessively sensitive to blue. This means that in subjects like landscapes, the blue sky is much too light in the final print. The tone difference between the sky and white clouds is also lost.

Both these troubles can be overcome by using a yellow filter. This colour works equally well with orthochromatic and panchromatic materials. The depth of the filter used depends on the amount of correction required.

Some correction of orange and red tones, in addition to blue, may also be desired with panchromatic materials, especially those of increased red sensitivity. Thus, outdoor portraits of sun-tanned faces will look too pale without a filter to reduce the excessive red sensitivity. Moreover, for recording and similar work, completely correct tone reproduction is often wanted. Here a yellow-green or green correction filter should be used with panchromatic materials.

As long as these filters are not too deep, they can be used with orthochromatic as well as panchromatic materials. With orthochromatic materials they are no better than yellow filters, because there is no red sensitivity to correct. But they are useful when we wish to carry as few filters as possible and have to work with both ortho and pan materials. A yellow-green or pale green correction filter is the best all-round filter for such occasions.

The excessive red sensitivity of some panchromatic films and plates also makes itself felt in photography by tungsten filament electric lamps. In portraits especially, flesh tones are reproduced too pale, red lips hardly show at all unless heavily made up with lipstick, and blue eyes become too dark. A pale blue or blue-green correction filter is therefore used.

Orthochromatic films are always highly sensitive to blue light and therefore never need a blue filter.

At very high altitudes the blue sky tends to reproduce as a fairly deep tone on panchromatic film and in mountain and Alpine photography a pale yellow filter gives all the correction necessary. At altitudes above 5,000 feet a special colourless U.V. (ultra-violet) filter is often used.

Sky Filters. Yellow and yellow-green correction filters are often loosely called sky filters because they are commonly used to darken the sky tones. But this name is more especially applied to the graduated correction filters. These are filter discs of which the top half is coloured, and gradually fades away towards a colourless lower half. They are used in landscape photography when only the sky is to be toned down, while the foreground requires no correction.

The light from the sky passes through the top (coloured) half of the filter, and the light from the subject through the bottom (colourless) half. As the camera exposure is given to suit the foreground, these graduated filters do not need increased exposures, and have a factor of 1.

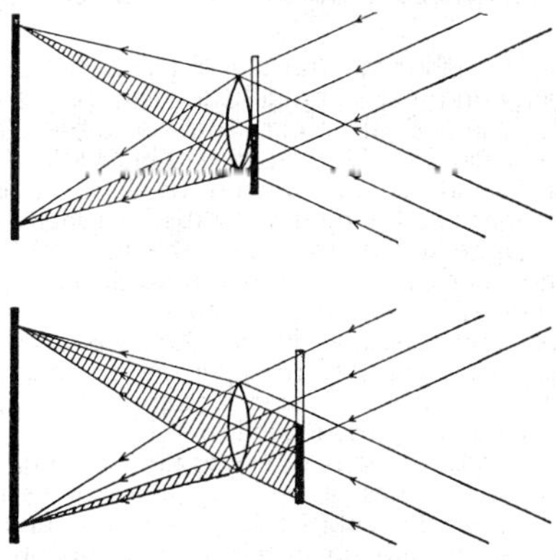

GRADUATED SKY FILTER. Top: *This should be mounted a little way in front of the lens to be effective so that rays from the sky are filtered, but rays from the foreground are not.* Bottom: *Mounted close to the lens it is unable to separate the rays from sky and foreground, and filters both equally.*

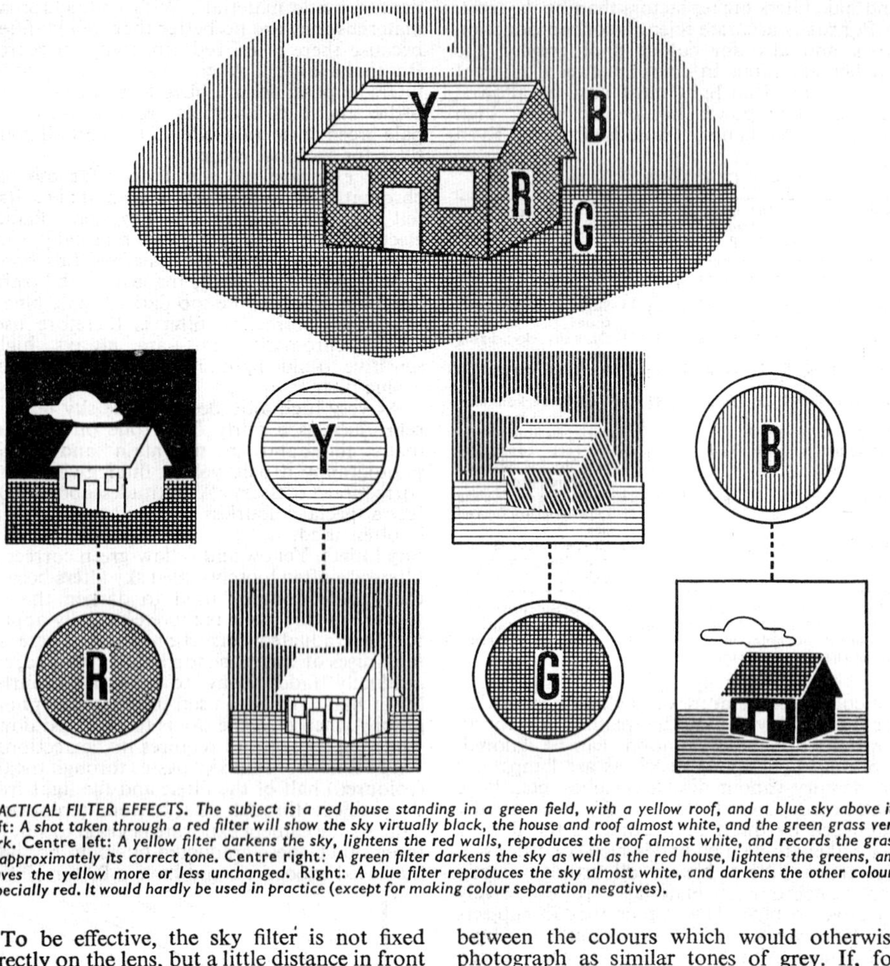

PRACTICAL FILTER EFFECTS. The subject is a red house standing in a green field, with a yellow roof, and a blue sky above it. **Left:** *A shot taken through a red filter will show the sky virtually black, the house and roof almost white, and the green grass very dark.* **Centre left:** *A yellow filter darkens the sky, lightens the red walls, reproduces the roof almost white, and records the grass in approximately its correct tone.* **Centre right:** *A green filter darkens the sky as well as the red house, lightens the greens, and leaves the yellow more or less unchanged.* **Right:** *A blue filter reproduces the sky almost white, and darkens the other colours especially red. It would hardly be used in practice (except for making colour separation negatives).*

To be effective, the sky filter is not fixed directly on the lens, but a little distance in front of it, in a special holder. Some of these filters are square and fit in holders which allow the glass to slide up and down in front of the lens. The centre of the filter can thus be adjusted to the height of the sky line in the picture. This adjustment can only be made accurately on a camera with a focusing screen.

Effect and Contrast Filters. Orange and red effect filters are used to carry the modification of sky tones beyond correct reproduction, to give specially dark dramatic skies.

In addition, filters of the various colours are made to emphasize the brightness of certain colours in the subject (e.g. flowers), and to bring out or play down certain tones in copying and other fields such as photomicrography for which special sets of contrast filters are made.

Here the task of the effect and contrast filters is mainly to increase the tone contrast between the colours which would otherwise photograph as similar tones of grey. If, for instance, red and green are reproduced in their correct brightness on normal panchromatic film, they both look about the same. To differentiate between them, we can use either a red filter which will lighten the red and darken the green, or a green filter which will darken the red and lighten the green. Deep yellow will, of course, also act as a contrast filter.

Contrast filters are almost exclusively used with panchromatic materials.

We can best visualize the effect of these filters by laying out the main spectrum colours in their natural order: blue—blue-green—green—yellow—orange—red.

Any filter will darken the tones of colours farthest away from its own colour on the spectrum scale. At the same time, the filter will lighten the tone of objects of its own colour, and it will make objects of neighbouring colours come out slightly lighter. Thus blue will be reproduced much darker with a red filter than with a yellow; green will come out as a much lighter tone with a green filter.

SUMMARY OF CONTRAST AND EFFECT FILTERS

Filter	*Daylight Factor* P	R	*Art. Light Factor* P	R	*Effect*
Blue ...	5	7	10	15	Darkens red, yellow lightens blue
Green ...	6	7	6	7	Darkens red, orange, slightly darkens blue, lightens green
Deep yellow	2½	2	2½	2	Darkens blue, lightens yellow, orange
Orange ...	5	4	2½	2½	Darkens blue, green, lightens orange, red
Light red ...	7	4	4	3	Darkens blue, green, lightens orange red. Generally increases image contrast
Deep red ...	16	8	8	6	As light red, but with greater effect

P = correct panchromatic, R = red panchromatic.

Colour Separation Filters. Separation negatives for making colour prints are taken in sets of three through the tricolour filters, red, green, and blue.

Normal tricolour filter sets are used to make colour separation negatives directly from the subject. The filters are more or less the same as the contrast filters of the same colour.

Special narrow-cut tricolour filter sets are used to make indirect colour separation negatives from colour transparencies. These filters pass light of practically only their own colour.

With either type of tricolour filter the filter factors must be highly accurate, as even small errors upset the colour balance of the final colour print. The exact colour temperature of the light source is also much more important. These factors are generally supplied by manufacturers for specific films or plates, since general factors as used with correction and contrast filters are not accurate enough.

Filters for Colour Film. In colour photography the main function of the filter is to correct the colour rendering of the picture by matching the colour balance of the film to the colour quality of the lighting, or by compensating for inherent deficiencies of certain light sources. In colour print making filters are also used to control the colour rendering and balance of the finished print.

Filters for colour work are mostly very much paler than for black-and-white films, and they are confined to rather fewer colours. The normal correction and effect filters for black-and-white film cannot be used with colour film; they would yield a colour picture with a very strong predominant colour tint of the filter itself. (But strong filters are used sometimes for special pictorial colour effects.)

The four main groups of filters used in colour photography: light balancing filters, conversion filters, colour compensating (C.C), filters and general correction filters.

Light balancing filters modify the colour temperature of the light passing through the filter. They are available usually in two series: bluish (to raise the colour temperature) and reddish or amber (to lower the colour temperature). Such filters are sometimes calibrated in the actual decamired shift reproducing the colour temperature: for instance an R3 filter makes the light passing through redder by 3 decamireds.

Conversion filters are a special kind of light balancing filter to adapt specific types of colour film to a different colour temperature. Usually they are recommended by the makers of a given colour film for matching that film with a given light source. The effect of such conversion filters may vary slightly with colour films of different makes though of the same nominal colour balance. The commonest filter of this kind is the type A filter used for adapting artificial light (photo-flood) type colour film to shooting in daylight.

Colour compensating filters have specific absorptions in one or two colours to compensate for colour deficiencies or irregular balance of special light sources (such as fluorescent tubes, electronic flash, etc.). These filters are also used for controlling the light when making colour prints or transparencies by the negative/positive or reversal systems.

The commonest C.C. filters are cyan, magenta and yellow series in various densities from 0·05 to about 0·9. These densities are equivalent densities to light of the complementary colour. The filter designation in most cases indicates both the colour and the equivalent density by a code: thus a CC20M filter is a magenta filter with a density (to green light) of 0·20. C.C. filters also exist in red, green and blue. These are equivalent to combinations of the corresponding complementary colours.

General correction filters for colour include haze and skylight filters, which mainly absorb ultraviolet and a certain amount of blue light. They correct for the excessive blueness for colour pictures taken of distant subjects in the presence of atmospheric haze, and shots in cloudy weather.

Filter Factors for Colour. Filters for colour photography rarely have high factors, since the correction required is usually fairly small. Light balancing and conversion filters of the reddish or amber series usually have factors between 1¼ and 2 ×; bluish filters have generally higher factors (up to 3 or 4 ×). The factors for haze and skylight filters usually vary between 1 and 1½ ×.

Haze Filters. Detail in distant landscape views is generally obscured by atmospheric haze. Haze is simply light scattered by the water vapour and dust particles in the atmosphere, and it is mostly composed of blue and ultraviolet light—hence its blue colour.

Most photographic materials are unduly sensitive to light of this colour and they tend to over-emphasize the effect of the haze in the finished photograph. So they lose even detail that the eye could distinguish when the photograph was taken. But the haze can be greatly reduced by using a yellow filter which holds back the scattered blue and ultra-violet light.

This penetration of the haze increases as the filter cuts out more and more of the blue end of the spectrum, therefore it is more marked still with orange and red filters.

This bluish haze is also objectionable in colour photographs of distant scenery. Here, a colourless, ultra-violet-absorbing, or a very pale yellow filter may be used. But anything more pronounced than the palest yellow completely falsifies the colour balance.

Infra-red Filters. An extreme degree of haze penetration is possible with infra-red film used with an infra-red filter. This combination cuts out all visible rays and passes only infra-red light which is not scattered at all by the atmosphere. We can thus get extremely clear pictures of long distance shots. This technique is therefore greatly used in aerial photography and survey.

The infra-red filter makes skies come out jet black. Green leaves, plants, etc., reproduce almost white because the chlorophyll (green colouring matter) in the plants strongly reflects infra-red light. The result resembles snow photographs by moonlight.

Infra-red filters (with infra-red film) find much use in medical photography and research; e.g. for photographs to show the structure of the blood vessels below the skin surface.

Various other materials, including semi-charred carbon, reflect infra-red light. Thus charred documents can be deciphered if they are photographed by infra-red light, because the charred paper still reproduces as a light tone.

Invisible Light Filters. Infra-red lamp filters can be used over a light source such as a flash bulb for night photography by invisible light. The infra-red filter cuts off all visible illumination, passing only the invisible infra-red which lights up the subject and records the image on the film.

Invisible light photography is also possible with ultra-violet light. Many substances glow in the dark when exposed to invisible ultra-violet rays. This fluorescence is used to distinguish between different inks and pigments which look identical in ordinary light. Ultra-violet light photography is thus a valuable tool in examining paintings and documents for suspected forgeries.

Here again the ultra-violet filter is placed over a special ultra-violet lamp, cutting out all visible light. (This filter must not be confused with the colourless ultra-violet filter used over the camera lens for mountain and high altitude photography. The colourless filter absorbs ultra-violet light, while the lamp filter absorbs everything except ultra-violet light.) When photographing by ultra-violet light, a colourless filter is usually placed over the camera lens, since in most cases we want to record not the ultra-violet light itself, but only the fluorescence.

Spectrum Filters. These are contrast filters used in scientific work. They transmit only a narrow band of wavelengths—in effect only one colour of the spectrum.

For even greater accuracy, the scientist uses interference filters. These are colour filters which derive their colouring not from any dye or pigment, but purely from optical interference. The colours are therefore very pure (i.e., the transmission bands much narrower than with conventional dyed filters), and are not liable to fade or deteriorate.

Polarizing Filters. These can cut out light which has undergone an orientational change known as polarization.

Light is polarized when it is reflected from the surface of any transparent medium. This includes glass, water, the glaze on china and pottery, varnishes, polishes, etc. The amount of polarization depends on the exact angle at which the light is reflected from the surface. Metallic surfaces do not polarize light on reflection.

The light from the blue sky, particularly at right angles to the position of the sun, is also polarized.

The two main uses of polarizing filters therefore are to reduce or eliminate surface reflections when photographing subjects behind glass, under water, etc., and to darken blue skies without affecting the rest of the scene. This property is particularly useful in colour photography, where any other type of filter would upset the colour balance of the film.

The polarizing filters sold for photography are usually sheets of a polarizing substance prepared in plastic form, and mounted between

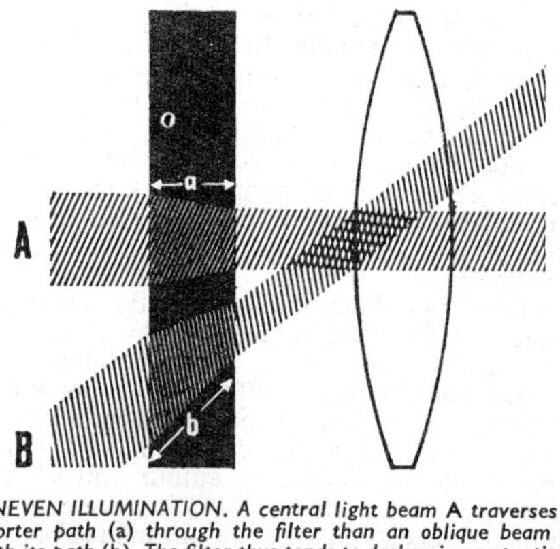

UNEVEN ILLUMINATION. A central light beam A *traverses a shorter path* (a) *through the filter than an oblique beam* B *with its path* (b). *The filter thus tends to darken image portions near the edge of the field more than in the centre.*

glass in much the same way as gelatin filters. The position of a polarizing filter on the lens is important; it is effective only if its polar axis is at right angles to that of the light it is intended to hold back. In practice, the subject is looked at through the polarizing filter and the filter is rotated until the objectionable reflections are reduced to a minimum. There is usually a handle, or some kind of reference marks on the mount so that the filter can be fitted to the lens in the position that gives the best visual result.

Optical Limitations of Filters. Lenses are usually corrected for use on their own; the addition of any other optical element, even a filter, is liable to impair the maximum performance of the lens. Imperfections of the filter (lack of parallelism of the filter surfaces, flaws, also dust and dirt) can cause loss of sharpness and loss of contrast. For high quality work filters must therefore be as precisely made as the lens itself.

The thickness of the filter, especially with a lens having a wide image angle (where an appreciable portion of the light rays go through the filter obliquely) enhances the falling off of illumination towards the edges of the field. Oblique rays have a longer path to travel through the filter, and thus more of the light is absorbed. At close range the difference in the light path through the filter glass at the centre and at the edges of the field may also upset the optical correction of the lens, and cause a slight degree of distortion. This is greater with glass filters than with the thinner gelatin filters; for highly accurate work the latter are therefore preferable.

Generally it is best to avoid using a filter where it is not really necessary. In particular the practice of keeping an ultra-violet filter permanently on the lens for protection is not in the interest of maximum lens performance. L.A.M.

See also: *Clouds; Colour materials; Colour printing technique; Colour temperature; Heat filter; Negative materials; Polarized light; Spectral sensitivity; Spectrography; Tone rendering.*

Books: *How to Choose and Use Film and Filter*, by L. A. Mannheim (London); *Colour Films*, by C. L. Thomson (London); *Exposure*, by W. F. Berg (London); *Camera Techniques*, by H. J. Walls (London); *Filter Practice*, by H. Clauss and H. Mensel (London).

FINALITY DEVELOPMENT. Procedure of developing an image until all the silver salts affected by light have been reduced to silver.

In normal negative development the action of the developer is stopped at a specified stage, depending somewhat arbitrarily on the contrast and density required. The time taken to reach this stage depends on the developer, but is a matter of minutes and has to be carefully controlled in each case to yield consistent results.

In finality development the action of the developer is carried on as far as it will go, until the image reaches gamma infinity. In practice that involves developing the negative material for up to hours in the normal developer.

Finality development has the following results:

(1) The effective emulsion speed increases because all the exposed silver salt is reduced instead of only part of it. The actual gain in speed in terms of shadow detail reproduced is in the region of 60–120 per cent, although exaggerated claims have put it as high as 800 per cent.

(2) The image contrast greatly increases. As a result, the density in the medium tones and highlights becomes very great, and the slightest over-exposure yields an almost unprintable negative.

(3) The fog level increases.

(4) The grain increases; this procedure is not, therefore very suitable for miniature work.

(5) The exact development time is not critical.

In practice, the disadvantages under 2, 3 and 4 greatly outweigh the advantage of extra speed. Finality development was, however, specified as standard procedure in the old (1931) DIN standard of film speed ratings. This was one of the reasons why that particular DIN standard was never accepted for international use.

Finality development is nevertheless popular for papers, where the increase in contrast and density is not so great, and the image grain does not matter. In this case finality development has the added advantage of yielding superior image tones (except with some chlorobromide papers). Development must, however, still be stopped before the appearance of fog although fogging is generally delayed by suitable restrainers or developer improvers.

At normal temperatures, chloride papers are developed to finality after about 1–1½ minutes, chlorobromide and bromide papers after about 3–4 minutes. L.A.M.

See also: *Developing negatives; Stand development.*

FINDER. Abbreviated name for viewfinder. A sighting device built in or attached to a camera through which can be seen an area of the subject corresponding more or less exactly to the image focused on the sensitized material during exposure.

See also: *Viewfinders.*

FINE GRAIN TECHNIQUE. Whenever negatives are printed by enlargement, the fineness of the grain in the image is important. This applies especially to miniature work where the negative sizes of around 1 × 1½ ins. often require considerable magnification. Coarse grain shows itself in an uneven, broken-up appearance of the image tones—particularly in the medium densities—which is generally unpleasant. One of the aims of miniature technique is therefore the production of fine grained images.

The factors affecting the fineness of the grain in the finished picture are:

(1) Choice of negative material.
(2) Exposure of the negative.
(3) Choice of negative developer.
(4) Degree of negative development.
(5) Choice of enlarging illumination.
(6) Choice of paper.
(7) Degree of enlargement.

In brief, the best results are obtained by using a fine grain film, correct exposure, normal fine grain development to a moderate gamma, and enlargement in an enlarger with semi-diffused or diffused lighting on a normal paper grade.

Negative Materials. The most important factor determining the graininess of the final image is the size of the silver halide grains in the negative emulsion. Special fine grain materials are available for miniature photography. They tend to be comparatively slow, but should be chosen whenever the extra speed of hypersensitive materials is not essential.

Provided the intrinsic grain of the emulsion is sufficiently fine, the other points of fine grain technique are considerably less important. They can, however, help to keep down the graininess with what would otherwise be a comparatively coarse grained film.

Negative Exposure. Over-exposure tends to increase the grain of the negative image, even though the latitude of the emulsion may be able to cope with excessive exposures. To make the most of the limited fineness of grain with faster emulsions, the exposure should be the minimum required to yield adequate shadow detail, but no more. The use of some exposure guide—preferably a meter—for really accurate exposures is therefore desirable. Current ASA speed ratings are designed to yield minimum correct exposures; for most consistent results calibration of the exposure meter to the film is advisable.

Fine-Grain Developers. Energetic developers tend to clump the silver grains in the image into larger aggregates. Fine grain developers generally are formulae of curtailed activity, using either a low concentration of alkali (or else weak alkalis) to keep the pH down, and/or low-energy developing agents, and/or silver solvents.

Many of these developers also lower the effective emulsion speed, especially with fast films. Generally it is better to use a moderate fine grain developer with a fine grain film than an ultra-fine grain developer with a fast coarse grained film. The effective speed of the latter combination may be no higher than of the former, and the contrast is likely to be lower for the same degree of graininess. High-speed negative materials are intended for occasions where subject conditions demand the maximum speed without regard to graininess.

Modern thin emulsion fine grain films can safely be developed in a normal (non-fine grain) developer without the risk of undue grain, but for maximum acutance and definition special developers are desirable with such films. These developers also bring out the full film speed.

Degree of Development. The average grain size in most negative materials increases as development progresses. The lower the image contrast produced—i.e., the shorter the development time—the finer the grain. Part of the function of a fine-grain developer is in fact to keep down the contrast. Fine grain development is to some extent a compromise between the lowest negative gamma that will give a good print and the largest grain that can be tolerated. Miniature negatives are therefore best developed to a maximum gamma of 0·7–0·8. Over-development, especially when coupled with over-exposure, greatly increases the grain.

But low-contrast development is not the complete answer to fine grain. While a low-contrast negative will have an image of comparatively fine grain, it needs a hard paper to produce a satisfactory print. On the other hand, a hard paper tends to emphasize the grain of the negative. The most useful fine grain developers are therefore those made up of ingredients which tend to produce reasonably fine grain with reasonably good contrast.

Enlarging Illumination. The ideal enlargers for miniature negatives are those using a light source that is either semi-diffused (opal lamp and condenser) or fully diffused (lamps and diffuser or cold cathode grid). Condenser enlargers with a point source of light tend to emphasize the grain of the negative image, making it much more prominent in the print.

Choice of Enlarging Paper. Contrasty papers accentuate the grain in the negative, soft papers tend to subdue it. However, the final graininess—the combined effect of negative grain and contrast of printing paper—does not normally vary a great deal between an under-developed negative printed on a hard paper and a more liberally developed (but not over-developed) negative printed on a normal paper. A negative developed to a moderately low gamma and printed on a hard paper does, however, utilize the film's effective speed better.

The paper surface also affects the graininess of the final image. Smooth surfaced (especially glossy) papers tend to show up the grain, rough surfaced papers tend to lose the image grain in the paper texture. Large prints of considerable magnification are therefore best made on rough paper.

Degree of Enlargement. Obviously the more the negative image is enlarged, the more prominent the grain will be. The photographer who likes to leave plenty of "air" around the subject gets a smaller image than the one who takes the trouble to arrange the subject in the finder so that it fills the whole of the negative area right up to the margins. The first worker may have to enlarge his negatives up to twice as much as the other, and in consequence get consistently grainier prints in spite of all his fine grain technique. So, where it can be done without prejudice to correct perspective and depth of field, the image of the subject should always occupy the whole of the available negative area. L.A.M.

See also: *Developers; Grain; Negative materials; Physical development; Resolving power; Surface development.*
Books: *Do Your Own Film Processing*, by L. A. Mannheim; *Developing*, by C. I. Jacobson; *35 mm. Photo Technique*, by H. S. Newcombe (all London).

FINGERPRINT PHOTOGRAPHY. Technique of photographing finger print impressions on a variety of objects and surfaces; widely used in police work to establish the identity of suspects and to obtain and present evidence.

The basis of the use of fingerprints in crime detection is the fact that the ridge pattern of the skin on the fingers of a person is unique and permanent throughout his life.

Like many other branches of photography, fingerprint recording is a specialist job and requires a knowledge of the science of fingerprints by the operator to enable him to produce the result the expert needs to reach his findings.

There are four types of finger impression which it is sometimes necessary to photograph. They are:—

(1) Latent impressions—sometimes invisible to the naked eye.

(2) Moulded impressions—left behind after the fingers have been coated with some substance such as blood.

(3) "Pick-up" impressions—where the ridges of the fingers have picked up dust or powder and left behind the pattern in the dust.

(4) Inked impressions—taken on a form or card for record purposes.

The satisfactory recording, photographically, of a latent fingerprint found at the scene of a crime depends to a large extent on the correct choice of powder used to develop the impression when it is found. For the other types it is usually only a question of choosing the correct film and illumination.

Equipment. The equipment used varies according to the location and article on which the impressions are found. Many are in difficult positions requiring some ingenuity on the part of the operator to get his camera into position.

Where conditions are suitable, a technical double extension camera (stand camera) enables the impression to be recorded in actual size. Special fingerprint cameras have the lens set at a fixed distance from the film and the impression, again to give an actual size image. The format of technical cameras is usually 4 × 5 or 4¾ × 6½ ins.; fingerprint cameras take 2½ × 3½ in. pictures except for one or two specially made instruments which are 4 × 5 ins. or thereabouts. Occasionally 35 mm. cameras are used but are not generally favoured.

Illumination can be by daylight, photoflood, other tungsten light or by flash. The fingerprint camera is normally battery-operated using four built-in bulbs. Some of the specially made instruments use electronic flash.

When dealing with light impressions on a dark background, a camera fitted with a prism or optical mirror can be used. This eliminates some of the intermediate stages in the preparation of the final print. Filters can be of great assistance in rendering the background darker or lighter.

Inked impressions on cards or forms can be copied satisfactorily on a process camera or by means of xerography in the case of good clear impressions.

Materials. As the aim is, in most cases, to increase contrast, process film is widely used but panchromatic and orthochromatic films are as popular. For a very weak low contrast mark, a "lith" type material as used in graphic arts may occasionally be used coupled with extreme contrast development. The choice depends to a large extent on the strength of the impression and the colour of the background.

Technique. With a developed (e.g. powder-dusted) latent impression or a moulded or pick-up one, the choice of camera depends very much on where the impression is and what it is on. If a non-portable surface is in the range of the technical camera or fingerprint camera, either can be used. If there is a sequence of impressions or part of a palm, the larger format of the technical camera may be required, to record the whole area on one negative.

The technical camera is set up with the bellows extended to give an image size of 1 : 1. When focusing the camera, the surface on which the impression appears, the lens panel and the focusing screen should, wherever possible, be parallel to avoid distortion. Accurate focusing is essential in view of the shallow depth of field at this extension. On curved surfaces or surfaces which slant away from the camera, a small aperture is necessary to render the whole of the impression sharp.

With daylight, a white card used as a reflector may assist in evening up the illumination when the light is coming from one side only. Portable artificial light sources should be moved into such a position that the image is shown at its best on the focusing screen. On strongly-patterned, but shiny surfaces, the impression may be made visible by deliberately lighting the surface so that the latter reflects glare into the camera lens. The impression then stands out against the glare. Oblique lighting may be necessary to show up certain moulded impressions (such as those found in soft putty) or pick-ups. Flash bulbs should not be too powerful and can be used to advantage at an oblique angle. They are to be preferred when dealing with anything like soft putty, as the heat from the photoflood or other lamp may melt the subject.

If it is not possible to record the impression in actual size, some sort of a marker photographed with it provides a guide for future enlarging.

The fingerprint camera is normally hand held, with or without a template in position over the impression to be photographed. At the front of the camera, legs or an enclosed mouth rest against the surface being photographed. Some cameras require a time exposure as their source of illumination is low. In this case the camera must be held perfectly still during the exposure. Modern fingerprint cameras with more powerful light sources permit exposures of 1/30th second or faster at small enough apertures to bring impressions on a curved surface within the available depth of field. Usually both aperture and shutter speed are fixed to yield a correctly exposed picture with a specified light source and film speed.

Impressions revealed by the use of a light powder can be photographed with a camera fitted with a prism or an optical mirror direct on to bromide paper. This gives a print which is right way round but with the tones reversed, i.e. the impression appears dark in colour on a light-coloured background. If several prints are required, then several exposures have to be made. This is simplified by using a special multiple sliding back holding a long narrow piece of bromide paper. This method saves time and material in the production of the prints.

Portable articles on which fingerprints are found are taken from the scene to the studio where better facilities are generally available to deal with them.

The copying of inked impressions from forms or cards is still best done by orthodox photography using any stand, fingerprint or process camera.

Special treatment is required when dealing with mirrors. An impression can be seen on the surface of the glass and another on the silvered back of the mirror and one appears superimposed on the other. A dark powder is used to reveal the impression and a piece of ground glass is placed ground side down on the impression. The mark on the surface of the mirror is clearly visible but the one on the silvering of the mirror cannot be seen. The photograph is then taken in the same way as any other print.

Weak marks on glass can sometimes be dealt with better by illuminating them from behind and photographing them in silhouette against the light.

Impression on a multi-coloured surface are dealt with by using a powder which flouresces under an ultra-violet lamp. Only the flourescing ridges of the pattern are recorded and the colours of the background completely eliminated.

Printing. No matter whether the fingerprint has been revealed with a light-coloured powder or a dark powder the resulting print must show the impression as a dark pattern on a light background. This is to permit comparison of both impressions in the same form (i.e. the fingerprint form with its black inked impression and a photograph showing black impressions on a light base).

Where the fingerprint has been developed by the use of a black powder, all that is required is a straightforward print from the negative on a suitable grade of paper. If a light-coloured powder has been used, a transparency made from the negative serves to produce subsequent prints. The transparency is made by contact printing the negative on to a piece of slow process film with the negative placed upside down in the carrier. A thin-based film (e.g. 3/1000 in.) is used for the negative to minimize any loss of definition due to the two emulsions not being in close contact. The use of a prism or mirror does away with the need to make a transparency or even a negative.

Photographing the impression in actual size in the first instance simplifies subsequent processing, as prints or transparencies can all be made by contact printing.

Contact prints can be made on bromide or contact papers. Many operators prefer the former in the belief that it records more of the range of tones sometimes found in the impression or that it provides a better balance between the weak and the strong areas. H.Sc.

See also: *Crime photography; Evidence by photography; Police photography.*

Book: *Photography in Crime Detection*, by J. A. Radley (London).

FINGERPRINTS. To the photographer, fingerprints may be either a nuisance or they may actually form his subject—e.g., in crime detection and police records. In the first place they are to be avoided, in the second, the aim is to get as clean a photographic image of them as possible.

Prevention. When the fingers press on the surface of a sensitized material they may leave behind prints that become visible later. There are three kinds of contamination which may be responsible: grease, perspiration and processing chemicals.

(1) The natural oil on the skin may leave a greasy, water-repellent imprint. This has no effect on a finished negative or print, but if it is left on the surface of the material before it is processed, it prevents the processing solution from acting on the emulsion. If made on an undeveloped negative the result is a light imprint after processing. A greasy fingerprint on the developed negative will hold back the

fixer and allow the emulsion to get blacker, so the imprint will show as a black mark on the half-tones.

A greasy finger-mark on a print before development will hold back the developer and later show as a white imprint on the half-tones and shadows.

A greasy imprint made on the surface of a print after development and before fixing will blacken over the half-tones where it has been exposed to light but it will remain clear over the highlights. Later, however, if the print is exposed to daylight for any length of time, such imprints will gradually darken and become visible. The remedy here is to wash hands in soap and hot water immediately before a processing session, and then refrain as far as possible from touching the sensitive surfaces. Forceps should always be used for handling prints.

(2) Imprints of another kind are caused by perspiration on the fingers. The moisture is usually slightly acid and, although it is immediately dissolved and rendered harmless by the aqueous processing solutions, it can set up a chemical action with visible consequences if transferred to the surface of a dry negative or print. This type of imprint may appear either white on half-tones and shadows or dark or discoloured on half-tones and highlights. People whose hands tend to perspire unduly should rinse them in water frequently during processing to dilute the dissolved matter in the perspiration. They should avoid handling sensitized materials before processing and use a self-loading spiral tank for developing films.

(3) The fingers can become contaminated by processing chemicals when manipulating plates and prints in the various solutions. The ideal prevention is to use forceps or, if that is impossible, rinse the fingers in clear water before as well as after putting them in the solution.

Contamination can also occur by indirect means—i.e., from touching objects that have been handled on a previous occasion with contaminated fingers—e.g., taps, bottles and stoppers, door-knobs, and particularly towels. This is a further reason for rinsing the fingers immediately after they have been in contact with a solution. For the same reason it is always wise to wash off any dribbles of solution that run down the side of a liquid chemical bottle.

Colour materials also will finger-mark from any of the above causes and the results show equally over all colours, shadows and highlights. The imprint may be black, clear or coloured depending upon the type of material and the stage at which the fingers touch the surface.

Photography. The grease or perspiration that is normally present on the fingers leaves an invisible print on everything it touches. It is possible to render these imprints visible by dusting with a suitable powder. Photographic records of them can then be made. Fingerprint photography forms an important branch of criminal investigation.

Fingerprints taken for police records may later have to be photographed so that they can be projected in enlarged form for comparison with prints found elsewhere. The photographic technique here is simply a matter of making a normal black-and-white record of a clear black on white impression. F.P.

See also: *Faults; Fingerprint photography.*

FINLAY, C. L., 18??–1936. English inventor of mosaic screen additive colour photography process: Thames plate (1908) with integral screen, Paget (1913) with separate screen.

FINLAY COLOUR. Additive colour process in which the ruled mosaic screen was separate from the monochrome negative. It had the advantage that copies could be made by normal black-and-white technique and then turned into colour transparencies by binding each copy up with its own colour screen.

See also: *Colour photography: History.*

FIREWORK DISPLAYS. There are two ways of photographing fireworks: with a time exposure, and with an instantaneous exposure.

Fireworks by Themselves. If the camera is set up on a tripod and the shutter is left open, the light of each firework will register on the film from start to finish. The shutter can be closed after a single firework has gone off, or left open to register the pattern left by several. This technique is also effective for photographing the lightning flashes in the course of a thunderstorm after dark.

The most effective way of collecting a number of firework patterns over a period of time is to close the shutter after each principal event and open it just before the next. This prevents the picture from being spoiled by registering minor fireworks and casual light sources. If this method is adopted, the shutter should be operated by a very long cable release so that the operator's arm can rest; waiting for several minutes with the finger at the ready on the shutter release can be very tiring.

Instantaneous Exposure. Short shutter exposures are only possible when the display produces a great volume of light that lasts for some seconds. At such times, very effective pictures can be made of the faces of the spectators or of silhouettes against the light of fireworks or bonfires. This type of subject calls for the maximum speed that can be obtained by combining large lens apertures with high speed pan films, and processing that puts film speed before every other consideration.

Flash. Both flash bulbs and flash powder can be used to improve firework pictures. The use of flash bulbs or electronic flash to light up people in the immediate foreground calls for double exposure technique. The most effective method is to set the camera on a tripod, focus the lens on the foreground, and stop down so far that the picture will be very much under-exposed. This exposure must be made in between fireworks or the trails will be blurred. Open flash is used to save having to reset the shutter for the fireworks.

After making the first exposure at the best moment for recording an interesting foreground, the shutter is closed, the lens refocused on infinity, and the firework trails are taken in the normal way by opening the shutter on B. The resulting picture will show the fireworks exploding in the sky and reproduce enough of the foreground—e.g., the backs of spectators—to give more depth and interest than would normally appear.

Sensitized Material. For all work of this type only the fastest panchromatic films are of any use—particularly as the subjects are largely red, which would not come out on an orthochromatic film.

Most colour materials are fast enough to record the coloured patterns of the fireworks, and the results are frequently enchanting. Here, too, it is worth-while experimenting with flash to record the foreground objects.

When tracing the fireworks upon the film, using a time exposure, an aperture of *f* 8 with a film of speed 32 ASA would normally be suitable.

When examining the results, it is as well to remember that this type of subject always gives a disappointing looking negative in black-and-white. The only reliable way of judging whether the negative is worth printing is to print it.

Most firework negatives should be printed on a hard, glossy or lustre surfaced paper so that every trail or spark appears as white as possible.

For those who enjoy using photo tints, firework prints—and lantern slides—offer plenty of scope and the effects have at least a pleasing novelty that even the most hardened purist would hesitate to condemn. L.V.

See also: *Light sources; Light sources in the picture.*

FISCHER, NICOLAS WOLFGANG, 1782–1850. Austrian Professor of medicine, University of Breslau. Discovered the light sensitivity of silver albuminate and published a pamphlet in 1814 on the action of light on silver chloride. In 1818 he described how to precipitate metallic silver from silver chloride by the action of zinc, a reaction afterwards made use of in the recovery of silver from photographic residues and waste.

FISH. There is no great difficulty involved in photographing river and sea fish under controlled conditions in a suitable glass-sided tank. This may be either a fair-sized tank with a movable glass plate with which the subject can be penned into a narrow space, or the tank itself may be made narrow enough in the first place. A convenient tank for this is 2 feet long by 18 ins. high by as little as 4 ins. in depth.

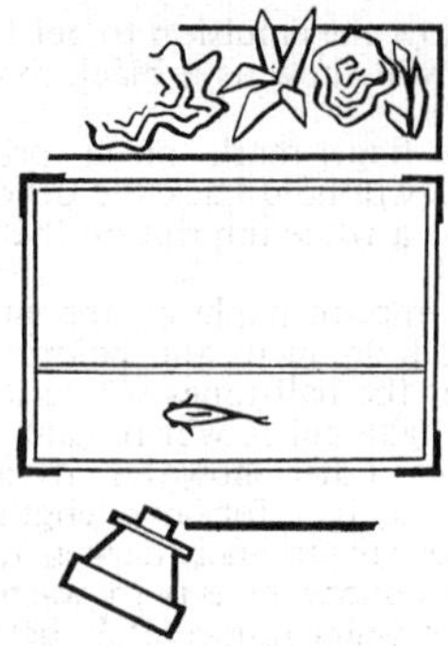

AQUARIUM SET-UP. The fish is confined to a narrow zone by a sheet of glass in the water. Black paper surrounds the lens to eliminate reflections from aquarium wall. Backgrounds can be set up behind to appear as if they were inside.

When photographing small fish, the lens is generally brought up so close to the subject that there is very little depth of field. Under these conditions, the narrower the tank the better, because fish have a tantalizing habit of keeping beyond the zone of sharp focus if they are allowed the necessary freedom.

The tank must be fitted with the best polished plate glass, front and back, to prevent optical distortion. All stones, sand and water-weed that form the scenery inside the tank must be well washed. Unless this is done, the soil and dirt will be stirred up by the movement of the fish and degrade the quality of the picture. For the same reason, the fish and the water are always allowed to settle for an hour or so before taking any photographs.

Avoiding Reflections. The camera is never set up exactly square with the front of the tank or it would photograph its own reflection as well as the fish. It is moved about 2 ins. to the side so that the lens-subject line is at an angle to the front of the tank. The small shift out of the square-on position makes practically no difference to the focusing of the camera, but it keeps its reflection out of the picture. All other reflections from the front of the camera and its surroundings are cut off by supporting a sheet of dull black paper or fabric opposite the front of the tank and just clear of the camera lens.

These precautions are, however, not necessary if the camera is brought up to the glass so that the lens hood actually touches it.

Fish may be photographed successfully indoors with artificial light. Again care must be taken to arrange the tank and lighting so that no reflections are shown in the glass, for a tank filled with water can act as a very effective mirror. Most of the front reflections may be obviated with a black cloth as already mentioned. Two floodlights set up above the camera and about 3 feet each side of it give good lighting, but the whole set-up should be carefully examined on the focusing screen before making an exposure. Another method is to place the tank close to a north window, with one floodlight on the opposite side.

Electronic flash or flash bulbs shining down from above the water surface can also be used.

Suitable Subjects. The majority of fresh water fish may be photographed in a tank of the type described above. It is better to choose small specimens; a fish 6 ins. long will make a better picture than one twice the size.

Any scenery included in the tank should be of the type usually associated with the fish to be photographed. When making records of a fish that spends most of its life at the bottom of the stream, for instance, the bottom, with characteristic weeds, should be shown in the picture. If the fish is a surface feeder, there is no need to show the floor of the tank. Since both the front and back of the tank are glass, extra scenery, and particularly the background, can be most conveniently built up outside and if they are thrown slightly out of focus it helps to give a feeling of depth to the picture.

Fish in Aquariums. The photography of the larger sea fish in public aquariums calls for special equipment and facilities not available to the ordinary photographer. A considerable amount of floodlighting is used and special precautions are taken to avoid reflections. Generally, there is an anti-reflection hood extending from the lens right up to the face of the glass, and part of the illumination may be provided by flash bulbs sealed into glass containers submerged with the fish. O.G.P.

See also: *Pets; Underwater photography.*

FISH EYE LENS. Extreme wide-angle lens covering an angle of view of around or nearly 180°. Lenses of this kind can thus take in a horizon-to-horizon view and yield a circular image. This shows appreciable wide-angle distortion (barrel distortion) since the circumference of the image circle with a 180° angle corresponds to a straight line. Only lines going through the centre of the subject field are reproduced straight (but still foreshortened towards the edge of the field).

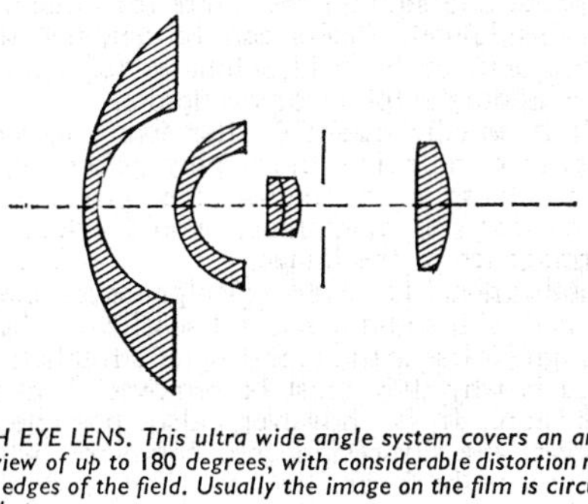

FISH EYE LENS. This ultra wide angle system covers an angle of view of up to 180 degrees, with considerable distortion near the edges of the field. Usually the image on the film is circular in shape.

To take in the extreme angle, fish-eye lenses have a highly curved protruding front element resembling a fish eye—hence the name. They are used for various kinds of trick photography as well as for recording the complete sky when pointing directly upwards. Fish eye lenses usually have a maximum aperture around *f* 8. The extremely short focal length makes focusing unnecessary, since the depth of field covers all distances from infinity down to a few inches. They are available for 35 mm. and occasionally larger still cameras and for cine cameras.

See also: *Lens history; Wide-angle lens.*

FISH GLUE PROCESS. Process in which a photographic image is produced by printing from a negative on to a metal plate coated with a light-sensitive layer of ammonium bichromate dissolved in fish glue. Where the light has acted, the glue is rendered insoluble and the plate is developed by washing away the soluble glue. The image is heat-hardened to form an enamel-like acid resist and the plate is then etched to produce the relief-printing image. The process is used in photo-mechanical reproduction. The plate is subsequently etched to form a relief printing image, the insoluble glue forming a resist. The process is used in photomechanical reproduction.

FIXED FOCUS. On many of the cheaper cameras, there is no focusing movement and the lens is at a fixed distance from the film. With such fixed focus cameras, all objects beyond about 7 feet from the camera are in reasonably sharp focus.

The success of fixed focus relies on the depth of field of the lens. For cheapness, such cameras are fitted with lenses of relatively small aperture and proportionately great depth of field. A typical example would be a 2¼ × 2¼ ins. box camera fitted with a 7·5 cm. lens with a working aperture of *f* 11. If such a lens is focused on 23 feet, the depth of field at *f* 11 extends from 11½ feet to infinity. In practice the definition would be considered sufficiently good from about 8 feet. It is thus possible to use such a camera for making reasonably good pictures by simply avoiding subjects closer than 8–10 feet. Many cameras of this type can be used closer than this because they are fitted with lenses with a working aperture even smaller than *f* 11. Others have a lens of shorter focus—e.g., 5 cm.—with a correspondingly greater depth of field.

Fixed Focus Setting. Any focusing camera can be converted at once into a fixed focus camera for snapshooting by focusing the lens at the hyperfocal distance for the aperture in use. If the lens has a depth of field scale, setting it for fixed focus working simply means adjusting the focus so that the distant limit of the depth of field for the working aperture coincides with the infinity mark. Obviously, the smaller the working aperture, the closer the near limit of the zone of sharp focus to the camera.

By adding supplementary lenses fixed focus cameras can be used for subjects closer to the camera than the near limit of the depth of field.

Technique. Fixed focus makes the camera easy to use and eliminates the risk of getting the subject out of focus. But while it gives a sharp picture of the subject (so long as it is not too close), it also gives a sharp picture of everything else more than 8 feet or so in front of the camera. This general sharpness is sometimes undesirable. It is all very well for a street scene or a landscape. But trouble arises when the real subject of the picture is viewed against an unimportant or ugly background.

How to Subdue the Background. The owner of a focusing camera can deal with the unwanted background by differential focusing. This cannot be done with a fixed focus camera, so the background must be played down in one of the following ways.

By choosing a plain background such as the sky, or a stretch of sand, grass, or snow. In this case the eye has nothing to focus on but the subject, so it no longer tends to wander away from the point.

By shooting from a low viewpoint. This cuts out the competing objects near the subject and shows it against more distant things which may be lighter in tone and may also be out of focus. Best of all, it may show the subject against the plain blue tone of the sky.

By choosing a contrasting background. A light-toned subject will stand out if it is photographed against a dark-toned background or an area in shadow. In the same way, a dark subject will stand out from a light-toned background.

By swinging the camera. If the camera follows a subject that is moving across the line of view, the subject will be sharp and the background blurred.

By using flash. So long as the background is considerably farther away from the camera than the subject, it will be too dark to attract attention. This method can only be applied in poor light, and it works only if the background itself is comparatively dark in tone. F.P.

See also: *Depth of field; Zone focusing.*

FIXING

When negatives or prints have been developed, they should be given a thorough rinse in water or a stop bath. This is to remove or neutralize the surplus developer on the emulsion. They must then be fixed.

THEORY

Fixing is the part of the photographic process in which the unexposed and unreduced silver halide is removed to render the image stable in white light and also to remove the turbidity. This is done by dissolving the silver halide in a solution of a substance that can form soluble silver salts. A wide range of substances is available ranging from cyanides, and thiocyanates, which act quickly, to sulphite and a strong solution of potassium bromide, which are slowest in action. For general purposes sodium thiosulphate (hypo) is the most effective agent.

When silver halides dissolve in thiosulphate solutions they form a series of silver-thiosulphates $Na_x\,Ag_y\,(S_2O_3)_{(x+y)/2}$ according to the conditions. Only two complexes are important in ordinary fixation; in both these $y = 1$ and x is either 3 or 5. When the hypo concentration is below 25 per cent, $x = 3$. Above 50 per cent, conditions favour $x = 5$. Between these ranges, both forms will be present in substantial proportions.

It is better to fix in baths of low silver content because this favours the more rapid removal of the silver-thiosulphate complexes in the subsequent washing. These conditions are usually achieved in practice by fixing in two consecutive baths.

It is undesirable for most purposes to use a plain hypo solution for fixing, especially if the films or plate are transferred direct from the developer to the fixer, because the developer will continue to act in the presence of the fixer. This often results in the formation of a dichroic silver deposit as a stain. This trouble can be avoided by using an acid fixing bath.

The addition of an acid, even one as weak as acetic acid, to a solution of sodium thiosulphate, causes decomposition and the deposition of sulphur. This decomposition can be avoided by having bisulphite present. A simple acid fixer can be compounded from sodium thiosulphate and sodium bisulphite (or potassium metabisulphite). Fixers can be acidified with acetic acid or by acid sulphates (e.g., alum) after adding sulphite or bisulphite.

It is equally essential after fixing to wash negatives and prints thoroughly before drying them. This is to remove the products of fixing that, if remaining, would affect the permanence of the image. G.I.P.L.

Stabilization. The silver-thiosulphate complexes formed with sodium are still sensitive to light, though far less so than the original silver halides. That is why they must be removed from the emulsion. It is, however, also possible to convert the silver salts into complexes which are virtually insensitive to light and thus sufficiently stable to remain in the emulsion.

The advantage of such stabilizing agents is that the fixing process can be very greatly shortened—to a few seconds under favourable conditions—and subsequent washing becomes unnecessary. Suitable complexes for this purpose are formed with agents like thiourea and its derivatives, certain thiocyanates and a number of other substances.

An image stabilized in this way is not as permanent as a properly fixed image. Prolonged action of light and of heat (especially in the presence of moisture) will cause eventual decomposition and deposition of silver to form stains. For that reason stabilization is of particular use in document copying where copies are normally stored in conditions of darkness and are mostly needed for short-term use. A stabilized image can be made fully permanent by refixing—possibly at a more convenient time than the original occasion of copying and processing—in a normal fixing bath.

FIXING NEGATIVES

It is specially important for negatives to be completely fixed because a spoiled negative cannot be replaced and many types of after-treatment will ruin a negative in which there are still traces of silver salts.

Sodium thiosulphate or hypo is the most common fixing agent. Certain other chemicals may be added to the fixing bath to prevent staining, to harden the gelatin, or fulfil some purpose other than fixing.

Acid Fixers. The presence of an acid in the fixing bath ends the process of development at once by neutralizing any alkali carried over from the developer. This prevents the developer from continuing to act locally and form stains on the negative during fixation.

As most acids by themselves decompose sodium thiosulphate, sodium sulphite must also be present to act as a preservative. The same purpose is served by using a bisulphite as the acid salt.

The simplest type of formula is that for bisulphite acid fixing bath:

Sodium thiosulphate	8–12 ounces	200–300 grams
Sodium bisulphite (or potassium metabisulphite)	1 ounce	25 grams
Water to make ...	40 ounces	1,000 c.cm.

This can be made up as (A), a 50 per cent stock solution of sodium thiosulphate, and (B), a 10 per cent stock solution of potassium metabisulphite. For use, two parts of A are mixed with one of B.

A solution containing 5 per cent sodium sulphite and 1·5 per cent concentrated sulphuric acid can be used instead of the potassium metabisulphite stock solution.

Another alternative for the acid stock solution is 10 per cent sodium sulphite with 4 per cent of acetic acid added. This solution is mixed with an equal volume of water and added to the same amount of 50 per cent hypo solution.

In all cases (including acid hardening fixers) the solutions must be mixed cold.

Acid Hardening Fixers. Hardening fixers use either chrome alum or potash alum as hardening agent. The former has the greater hardening power, but the latter keeps better.

The formula for chrome alum acid hardening fixer is:

Stock solution A.		
Sodium thiosulphate (hypo)	24 ounces	600 grams
Potassium metabisulphite ...	2 ounces	50 grams
Water to make	40 ounces	1,000 c.cm.
Stock solution B.		
Chrome alum	1 ounce	25 grams
Water to make	40 ounces	1,000 c.cm.

Equal parts of A and B are mixed just before use. The solution will not keep longer than a day or two.

The formula for potash alum acid hardening fixer is:

Stock solution A.		
Sodium thiosulphate (hypo)	12 ounces	300 grams
Water to make	40 ounces	1,000 c.cm.
Stock solution B.		
Sodium sulphite, anhydrous	3 ounces	75 grams
Glacial acetic acid	2¼ ounces	63 c.cm.
Boric acid	1½ ounces	38 grams
Potash alum	3 ounces	75 grams
Water to make	40 ounces	1,000 c.cm.

For use, 3 parts of A are mixed with 1 part of B.

With acid and acid hardening fixing baths it is safe to expose the material to white light as soon as the fixer has started to act, as development is arrested almost at once. But it is never wise to subject unfixed material to strong light for long because it slows down fixation and the action of the light may deposit silver on the emulsion as a thin fog.

Rate of Fixing. The speed with which a fixing bath acts depends on: the sodium thiosulphate concentration, the exhaustion of the solution, the negative material, and the temperature of the solution.

The rate of fixing increases with increasing sodium thiosulphate concentration up to a maximum strength of about 35–40 per cent.

Above that the gain in fixing speed with increasing concentration is less significant, and 30 per cent is a convenient optimum concentration.

The presence of certain other chemicals decreases the fixing rate. Acid fixers and acid hardening fixers work slower than plain solutions containing the same amount of sodium thiosulphate.

As the solution becomes exhausted, the thiosulphate concentration, and thus the fixing speed, decreases. Some of the silver compounds, particularly iodide, in a partly exhausted fixing bath also slow down the action of the bath.

The fixing time also depends to some extent on the negative material. Thick emulsions—particularly double coated ones—contain more silver bromide than thin ones, and take longer to fix. Many high speed emulsions contain a proportion of silver iodide, which is much less soluble than silver bromide, and consequently takes longer to fix. For the same reason, when the silver bromide is converted into silver iodide as in physical development, fixing takes a longer time than usual.

As a rule, papers fix very rapidly. This is because the emulsion layer is thinner and also because, in the case of chloride and chloro-bromide papers, the emulsion contains silver chloride which is much more soluble than silver bromide.

The effect of temperature on the fixer is the same as on all other solutions: the warmer the bath, the quicker it acts. At high temperatures a hardening fixer is essential.

For really rapid fixing there are special high speed fixing baths.

Exhaustion of the Fixer. The area of negative material that can be fixed in a given volume of fixing solution is mostly limited by accumulation of iodide and dilution, caused by water carried over by the negative.

Both these effects slow down the fixing rate. In addition the increasing concentration of silver compounds makes fixation progressively difficult, until the point is reached when the material cannot be completely fixed even if the bath is replenished by adding sodium thiosulphate. The high silver content of exhausted fixing baths also leads to staining of materials.

EXHAUSTION OF FIXING BATH

Negative Size in.	cm.	Number of Negatives per 40 ounces (1,000 c.cm.)
1 × 1½	2·4 × 3·6	1,000
1⅛ × 1⅝	3 × 4	1,000
1⅝ × 2½ (V.P.)	4 × 6·5	480
2¼ × 2¼	6 × 6	360
2¼ × 3¼	6 × 9	240
2½ × 4¼	6·5 × 11	180
3¼ × 4¼ (¼-plate)	9 × 12	140
3½ × 5½ (postcard)	9·5 × 14	100
4¾ × 6½ (½-plate)	12 × 16	75
6½ × 8½ (whole plate)	16 × 22	40
8 × 10	20 × 25	30

Film Size or Number	Number of rolls per 40 ounces (1,000 c.cm.)
35 mm. film, 18–20 exposures	60
35 mm. film, 36 exposures	30
No. 828 or 888 (Bantam) rolls	100
Standard No. 27 or 127 (V.P.) rolls	60
Standard No. 20, 120, 62, 620, or Z20 rolls	30
Standard No. 16, 116, 616, or Z16 rolls	20

The acid in an acid fixing bath is gradually neutralized by the developer, which is normally alkaline, carried over in the emulsion of the negative. This will happen sooner when the intermediate rinse is not as thorough as it should be. If the acid in the fixing bath is exhausted before the fixer its action can be restored by adding further acid stock solution.

The number of films or plates that can be fixed in a given quantity of fixer depends finally on the type of the negative material, and on the amount of silver bromide left behind in the emulsion after exposure and development. Thin emulsions films and dense negatives will exhaust the bath less than under-exposed and under-developed negatives where a lot of unused silver bromide remains to be dissolved away.

As a maximum, 40 ounces of fixing solution containing 25 per cent of sodium thiosulphate will fix about 2400 square inches of material (about 1·5 square metres per 1000 c.cm.).

This is given in the tables above in terms of the number of actual negatives that the bath will fix.

Testing for Exhaustion. The speed with which the milkiness of the negative clears gives some indication of the approach of exhaustion. When the clearing time is more than three times as long as with the fresh bath, the solution is best thrown away.

A more exact test is to add about 20 minims (1 c.cm.) of 5 per cent potassium iodide solution to 1 ounce (25 c.cm.) of the fixing bath. If the solution goes cloudy and does not clear on shaking, the fixer is exhausted.

The acidity of the solution can be checked by dipping a piece of blue litmus paper into it. The litmus paper will turn red immediately if the bath is sufficiently acid.

Keeping Properties. Freshly mixed fixing solution will keep almost indefinitely; fixing baths are not exhausted by storage as are developers. In a partly exhausted fixing bath, especially if it is left exposed to light, the silver compounds gradually decompose, forming a blackish sludge of silver. If the acidity is low, a white deposit of aluminium hydroxide may also form from the potash alum present. All these precipitates should be filtered off before use.

Hardening fixers with chrome alum lose their hardening properties within 1–2 days.

Two-Bath Fixing. To guarantee thorough fixation, the normal fixing bath may be followed by a freshly made up bath of the same solution. This second solution removes any last traces of silver salts that may be left in the emulsion when the first bath nears exhaustion, The second fixer stays practically fresh all the time.

When the first bath is thrown away, the second one takes its place, and a fresh second bath is made up. This system uses no more fixer than the single bath method, but it cuts out the risk of incomplete fixation.

High Speed Fixing. As a quick fixer for negatives, ammonium thiosulphate may be used instead of the sodium compound. In practice this is often done by adding 1 part of ammonium chloride (sal-ammoniac) for every 5–7 parts of sodium thiosulphate contained in the solution. The chloride attacks stainless steel, however, in which case ammonium sulphate may be used instead.

Such fixing baths act in about one-quarter to one-half the time needed with normal fixing solutions but they are said to affect the permanence of the silver image.

Ammonium thiosulphate is nowadays widely used in commercial rapid fixing baths. A suitable formula is:

Ammonium thiosulphate	8	ounces	200	grams
Sodium sulphite, anhydrous	265	grains	15	grams
Acetic acid, 28%	1¼	ounces	55	c.cm.
Boric acid	130	grains	7·5	grams
Potassium alum	265	grains	25	grams
Water to make	40	ounces	1,000	c.cm.

The chemicals are dissolved in the order listed in about three-quarters of the final volume of water.

Commercial rapid fixing baths frequently are sold in two stock solutions, one containing the ammonium thiosulphate and the other the sulphite and hardening agent.

A 40 per cent ammonium thiocyanate solution containing 5 per cent formalin may be used. This will fix in about half a minute and even less if it is used hot (up to 120°F. or 50°C.). During the very short immersion times the gelatin is not likely to soften appreciably.

The high fixing rate of ammonium thiocyanate makes it suitable for low temperature processing. The solution has a very low freezing point and it takes only a few minutes to fix even below 32°F. (0°C.).

The use of a small proportion of thiosulphate with thiocyanate fixers helps to stabilize the images if not washed after clearing.

Potassium cyanide is another chemical with rapid fixing properties but it is rarely used because it is very poisonous. It has the added disadvantage of attacking the silver image and acting like a weak negative reducer.

FIXING PRINTS

Fixing is as necessary for prints as for negatives—i.e., it removes unused silver compounds still present in the print after development. These silver compounds would in time tend to discolour and mar the print. Thorough fixing is also important if the print is to be toned later.

Fixers. The commonest fixing bath for prints is an acid fixer containing sodium thiosulphate (hypo), and an acid, together with sodium sulphite or an acid salt like potassium metabisulphite. The acid immediately neutralizes any alkali carried over from the developer, particularly if a plain water rinse is used instead of an acid stop bath. It arrests the action of the developer and so prevents stains. It also lengthens the useful life of the fixer.

The formula given for negatives may be used, but only half the amount of sodium thiosulphate is necessary for prints.

Prints should never be fixed in a bath which has been used for negatives. Such a fixer may contain enough silver salts to stain prints, even when it has no obvious effect on films or plates. There is also the risk that the prints will be stained by backing dye dissolved from the back of films or plates.

Acid hardening fixers are not usually necessary for prints; the gelatin layer of printing papers is not as sensitive to temperature changes or physical injury as the emulsion of a film or plate. A hardening fixer may, however, be advisable if the prints are to be hot glazed. In that case the acid hardening fixers for negatives are suitable, but at half strength.

Fixing Time. The usual time necessary for completely fixing chloride contact papers is about one minute, and for chlorobromide and bromide papers about two minutes, all at 65–70°F. (18–21°C.). When the bath is nearly exhausted, fixing may take two or three times as long.

When prints are left in a fresh acid fixing bath, particularly if it is strongly acid, the solution may in time dissolve away some of the fainter highlight detail. And after prolonged fixing the paper fibre absorbs acid, which subsequent washing may not remove completely. This acid can in time attack the silver image.

Prints should therefore not stay in the fixing bath for longer than about ten minutes.

Two-Bath Fixing. As a safeguard against incomplete fixing, two similar fixing baths may be used. The first bath does the main work of fixing, and the second removes any traces of silver salts left in the emulsion. The first bath gets exhausted in the normal time, while the second stays fresh much longer.

When the first bath is exhausted, it is thrown away; its place is taken by the second bath, and a new second bath is made up.

The amount of fixing salts used up is no more than for a single fixing bath, but the two baths ensure complete fixing, even when the first bath is nearly exhausted.

Rapid Fixing. The fixing time of prints is already very short, so the special high-speed fixers used for negatives are rarely needed for normal photography.

The only rapid fixer of practical importance for prints is ammonium thiosulphate. It may be used as a 20 per cent solution with 3 per cent potassium metabisulphite added. Alternatively, it can be made by adding ammonium chloride to a normal acid fixing bath (1 part to every 6 parts of sodium thiosulphate in the fixer).

At normal processing temperatures, this solution will fix prints in about 15–20 seconds.

Exhaustion. During fixing, the silver salts removed from the prints accumulate in the fixing bath. With print fixers, the concentration of silver salts usually becomes too great to permit complete fixation before all the thiosulphate in the bath is actually used up. This is why nothing much is gained by using print fixers as concentrated as negative fixers; they would still stop working after about the same number of prints.

Print fixers should never be overworked. They may still appear to act satisfactorily, when they are in fact spent. A fresh fixing bath is particularly essential for print permanence.

The number of prints which 1,000 c.cm. or about 40 ounces of fixer containing 20–25 per cent sodium thiosulphate will fix is approximately the same as the number of negatives. The actual work done by the fixing bath is less, because printing papers contain less silver than negative materials. But the amount of dissolved silver which the bath can contain and still work satisfactorily is also lower.

The keeping qualities of print fixers are much the same as those of negative fixers.

Tests for Fixing and Exhaustion. Chloride and cream-tinted bromide papers do not change colour in the fixing bath. Ordinary bromide prints turn from pale yellow to white. This, however, is not a complete guide for fixing.

A better test is to drain and wash a print after fixing, and touch one corner with a crystal of sodium sulphide (or place a small drop of 10 per cent sodium sulphide solution on it). If the spot tested goes brown, the print still contains silver salts and is not completely fixed. If the test gives a brown stain, even after prolonged fixing (5 minutes or more), the fixing bath is exhausted.

The fixing bath itself may be tested directly by the potassium iodide method, while blue litmus paper will check acidity. L.A.M.

See also: *Cold weather; Exhaustion of solutions; Hot weather processing; Hypo eliminator; Rapid processing; Stabilisation; Tropical photography.*

Books: *Developing*, by C. I. Jacobson (London); *A Textbook of Photographic Chemistry*, by D. H. O. John and G. T. J. Field (London); *Processing Chemistry*, by L. F. A. Mason (London).

FIZEAU, HIPPOLYTE LOUIS, 1819–96. French physicist. Invented gold toning of daguerreotypes in 1840. Inventor of a very successful process of etching daguerreotypes for mechanical printing (1841). Studied (1844, with Foucault) the action of various light sources and of the spectrum on daguerreotype plates and found reciprocity law failures. Took first pictures of the sun on daguerreotype (1845, with Foucault).

FLARE. Bright patch (sometimes also a ring or other irregular pattern) on the print caused by strong light reflections inside the lens. The light may be any bright source inside or on the edge of the picture area, or even a bright rim on the edge of the lens mount or hood.

All lenses tend to form flare spots, although by good design the worst of them can be made to fall outside the limits of the negative normally covered by the lens.

Almost every camera lens will form one or more flare spots if it is pointed towards a light source. These spots show up on the focusing screen and it is a good plan to study the performance of a lens in this way for those occasions when the picture area includes a light source. The quickest way of making such a check is to set the camera up in the darkroom in front of a lighted candle. The position of the flare is noted as the lens is aimed directly at the flame and also to each side of it. It is often possible to arrange the picture so that the flare falls on a light area of the subject.

The position of flare on the viewing screen of a twin lens reflex is no guide to the position of the flare formed by the taking lens. With such cameras, and all cameras without focusing screens, it is necessary to open the back and examine the actual image formed by the taking lens on an improvised screen.

Coating the lens greatly reduces the tendency to form flare.

See also: *Coated lens; Faults; Ghost; Halation; Irradiation.*

FLASH BULB

Flash bulbs are artificial light sources which generate light by the rapid combustion of certain metals in oxygen. The whole process takes place inside a glass bulb resembling an electric light bulb and yields a short but intense flash of light. The bulb is fired electrically, and can be used once only. Flash bulbs are therefore sometimes described as "expendable flashes".

History. The first forerunner of the flash bulb was a magnesium lamp by Chauffour in 1893, which electrically fired magnesium ribbon in a glass bulb filled with oxygen. It was designed exclusively for underwater photography. The first flash bulb proper was invented by Vierkötter in 1925 and consisted of a charge of flash powder inside a light bulb fired by a filament covered with magnesium powder. The combustion took place in a vacuum and the bulb gave a light output of over 100,000 lumen seconds. In 1925, Vierkötter produced an improved version filled with low pressure oxygen. In 1929, Ostermeier made the first commercial flash bulb (Vacublitz) filled with aluminium foil. These were marketed in Germany. In 1934, Phillips produced the first wire filled flash bulbs in Holland. Foil- and wire-filled bulbs began to be made in America under licence in 1930 and 1936 respectively where they were soon produced in far greater numbers than they ever had been in Europe.

Construction. The flash bulb consists of a glass envelope shaped like an electric light bulb the end of which carries two electrical contacts.

Inside the bulb, the contacts are joined to a thin wire filament covered with a primer of explosive paste. The rest of the envelope is filled with a loose mass of fine aluminium wire in an atmosphere of oxygen at a reduced pressure.

When a battery is connected across the terminals, a current flows through the filament which becomes incandescent (burning out almost immediately), and ignites the explosive primer. This in turn ignites the aluminium foil or wire, which burns, generating a flash of intense light.

Modern ultra-small flash bulbs use zirconium wire filling, providing the same light output from a smaller quantity of wire. At one time flash bulbs also used to be filled with aluminium foil or with an extra large quantity of primer paste and no wire at all.

The electrical current to fire the bulb is generally derived directly or indirectly from a dry battery. The minimum voltage required is 3 volts, and the maximum with most bulbs is 30 volts (as supplied by a battery-capacitor unit).

Certain large bulbs are designed to be fired from 110 or 250 volt mains, but unless this is specially stated on the package, mains firing can be very dangerous. Where a bulb is designed for mains firing, it must be connected directly to the supply, and not wired through a synchronizer or synchronized shutter.

The usual way of firing bulbs in present-day flash guns is by a battery-capacitor circuit, where a 15 to 30 volt battery charges up a capacitor which in turn discharges through the flash bulb on closing the firing circuit. Such a capacitor can also fire several bulbs (usually up to three or four) which are generally connected in series—wired so that the current passes through one bulb after the other. To fire larger numbers of bulbs booster boxes are sometimes used, employing a high tension battery of 90 to 300 volts; this can then fire several dozen flash bulbs at a time.

Low voltage batteries can also be used for direct firing, but are very rarely used nowadays, since the synchronizing reliability depends very much on the freshness of the batteries.

Flash bulbs can also be fired without electrical connexions at all, simply by being placed very close to another bulb fired in the normal way. In this case the light energy of the first flash sets off the combustion in the second bulb.

Characteristics. Flash bulbs differ mainly in: the amount of light they give, the time they take to light up, the duration of the flash and the colour of the light. Standard information on each of these points is therefore given with each flash bulb. It may be given in the form of a graph showing the variation of light intensity with time, from the instant when the firing circuit is closed to the end of the flash.

Peak brightness defines the maximum light flux in lumens when the flash is at its brightest, and corresponds to the highest point or peak of the graph. A small bulb may have peak illumination of 500,000 lumens, while large bulbs go up to 6 million lumens.

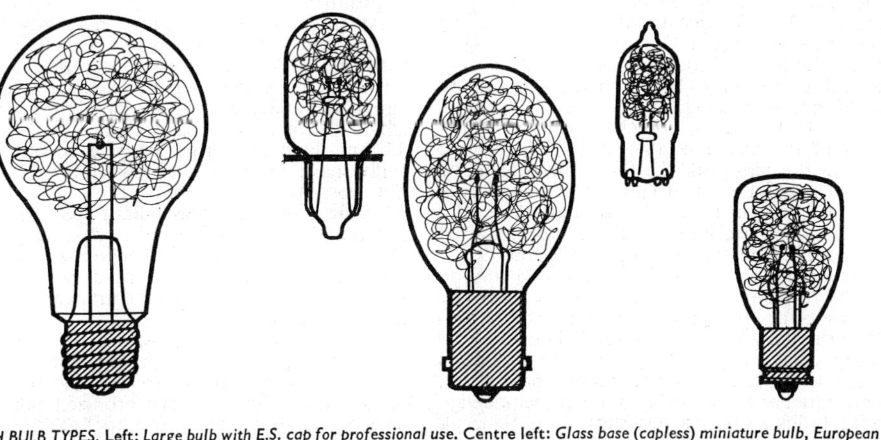

FLASH BULB TYPES. Left: *Large bulb with E.S. cap for professional use.* Centre left: *Glass base (capless) miniature bulb, European type.* Centre: *Small bulb with S.C.C. cap, usually of focal plane type.* Centre right: *Ultra-miniature all glass capless bulb (American type).* Right: *Miniature American bulb with metal cap. All the bulbs are shown approximately two-thirds natural size except fo the large E.S. capped bulb which is half natural size. They are also all available with a blue coating for colour photography.*

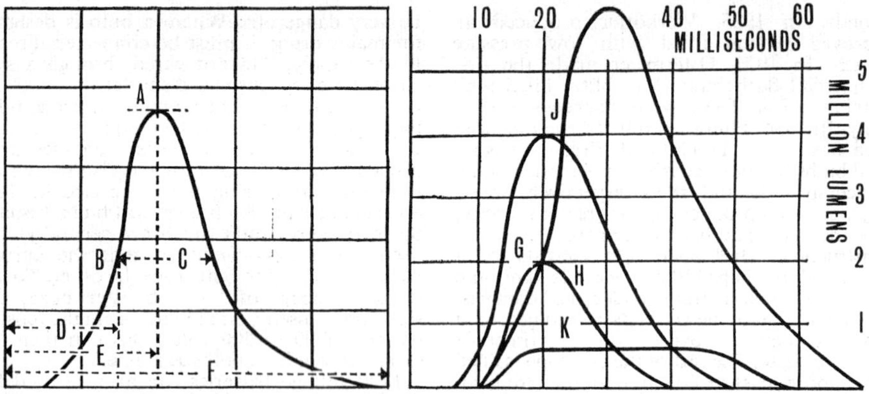

FLASH CURVES. Left: *The data obtainable from a flash curve are:* A, *peak intensity;* B, *half-peak intensity;* C, *half-peak duration;* D, *half-peak delay;* E, *peak delay;* F, *total duration.* Right: *The typical curves represent:* G, *medium class M bulb;* H, *small class M wire-filled bulb;* J, *large class S bulb;* K, *focal plane bulb.*

The amount of light (as distinct from its intensity) that a bulb yields is measured in lumen-seconds. This depends both on the intensity and the duration of the flash and determines the exposure that the flash produces on the photographic material. Large bulbs have a high light output (over 100,000 lumen seconds), while small bulbs may provide as little as 5,000 lumen-seconds. The same total light output may be generated by either a very intense flash of comparatively short duration, or a less intense flash of longer duration.

In terms of the flash graph, the total light output is the area below the curve.

With most flash bulbs the light intensity rises up to a definite peak and then falls away again. Photographically, the useful part of the flash occurs while it is above half its maximum intensity. The half-peak duration extends from the instant when the curve first rises to half its peak level to the instant when it falls back to the same level again.

The half-peak duration may be quite short—e.g., 5 milliseconds (1/200 second)—or as long as 50 milliseconds (1/20 second).

The half-peak duration does not give any idea of the shape of the peak of the curve. Above the half-peak the flash may rise to its full brightness in a sharp point and immediately fall off again, or it may remain fairly level at a lower value. The latter type of flash is required for focal plane shutter synchronization.

The fact that the flash takes a little time to reach its full brilliance after firing must be allowed for in synchronization. The actual time the flash takes to develop is known as the firing delay or time lag; it can be measured either to the peak or to the first half-peak point, and stated as a peak delay or a half-peak delay. With bulbs yielding a very short flash, the peak is more easily timed, while the half-peak delay only is significant with long-flash bulbs.

Most wire-filled flash bulbs have a colour temperature of about 3800°K and therefore need suitable compensating filters when used for colour photography.

Some bulbs are available with a tinted protective lacquer to raise the colour temperature for matching with daylight type colour film. Daylight flash bulbs carry a blue coating and have an effective colour temperature of 6000°K. Blue bulbs are useful for fill-in flash in daylight.

Flash bulbs for amateur photography are made with two kinds of fittings for appropriate flash guns. Both are known as capless bulbs, as distinct from the larger types which have a metal cap. The European capless bulbs have the terminal wires emerging from the glass base on opposite sides; the American A.G. type has a flattened base with the terminals emerging directly from the bottom and bent up in opposite directions.

American miniature bulbs frequently have a metal cap with a centre contact (U.S. miniature or U.S.M.). The cap is here held in the flash gun by a spring that engages a recessed rim at the end. Small focal plane bulbs have a small centre contact bayonet cap (S.C.C.) of 15 mm. diameter. Large size professional flash bulbs have an Edison Screw (E.S.) 27 mm. diameter cap.

Classification. As most flash photography involves the use of synchronized flash, the firing characteristics affecting synchronization are important in selecting the right bulbs for a given job. Flash bulbs are accordingly grouped in several classes.

Class M (Medium) bulbs form the majority on the market. The peak delay is about 18–22 milliseconds. The bulbs are suitable for synchronization with X- synchronized shutters at slow shutter speeds (up to 1/30 second) or with M-synchronized shutters and delayed release synchronizers at all shutter

TYPES OF FLASH BULB

Type	Output Thousand Lumen-Seconds	Peak Million Lumens	Half-Peak Delay Milli-Seconds	Peak Delay Milli-Seconds	Half-Peak Duration Milli-Seconds	Cap
Clear bulbs (Colour temperature 3800°K)						
Miniature and Small Class M	5·5–10	0·5–0·8	12–14	18–20	8–10	Capless or U.S.M.
Medium Class M	16–20	1·2–1·4	15–16	20	12–15	Capless
Medium Class M	22–30	1·2–1·8	13–15	20	12–16	E.S.
Large Class M	50–75	2·8–5·0	12–15	20–24	14–20	E.S.
Class S	95–110	3·6–5·2	19–25	30	17–22	E.S.
Small Class FP	15	0·5–0·6	15–17	—	25–30	S.C.C.
Large Class FP	45–80	1·0–1·6	17–18	—	45–55	E.S.
Blue tinted bulbs (colour temperature 6000°K)						
Small and Medium Class M	7–9	0·5–0·7	14–16	18–20	12–15	Capless
Large Class M	27–30	1·4–1·8	12–15	20–24	14–20	E.S.
Class S	45–50	1·8–2·5	19–25	30	17–22	E.S.
Small Class FP	7	0·25	17	—	25	S.C.C.
Large Class FP	22	0·5	18	—	45	E.S.

speeds up to 1/500 second. The light output ranges from 5,000 to 75,000 lumen-seconds, and peak intensities from 500,000 to 5,000,000 lumens. The half-peak duration varies from 8–10 milliseconds for the small bulbs to 16 (and occasionally 20) milliseconds for the large. Some bulbs in this class may be extra-small in size. E.S. caps are fitted on most types; most small bulbs are capless.

Class S (Slow) bulbs have as their main feature high light output (around 100,000 lumen-seconds; peak intensity around 5,000,000 lumens) and long firing delay (30 milliseconds to peak). These bulbs are not really intended for synchronization. With most shutters the speed is limited to 1/10 or 1/25 second, but with fully synchronized shutters and adjustable delayed release synchronizers (which can allow for the greater firing delay) faster speeds are possible. At fast speeds however, a lot of the light output of the bulb is wasted, and it is more practical to use a class M bulb of the same peak intensity but shorter duration. The half-peak duration of class S bulbs is about 20 milliseconds. All these bulbs have E.S. caps, and they are the only type of flash bulb that can be fired from a 110 to 250 volt mains supply.

Class FP (Focal Plane) bulbs are specially designed for synchronization with focal plane shutters at all speeds. They are distinguished by a long and fairly flat peak; the light remains reasonably even during the time of travel of the focal plane shutter. The peak intensity (about 500,000 to 2,000,000 lumens) is rather lower relative to the light output (15,000 to 80,000 lumen-seconds) than with class M bulbs of similar size. The half-peak duration is 25–30 milliseconds with the smaller bulbs for miniature cameras, and 45–50 milliseconds with the larger bulbs for focal plane shutters on press cameras. S.C.C. or E.S. caps are fitted.

Special Flash Bulbs. Flash bulbs are also used in high-speed cinematography to provide the intense light necessary for very short periods. Special lamps of this kind exist with a flash duration of over 1½ seconds. This is adequate for most high-speed sequences. For example, one second covers the exposure—at 2,000 frames per second—of 25 feet of film. (In practice, this is a 50 feet load, for the other 25 feet have to run through the camera first while the latter accelerates to this framing speed.)

For longer filming sequences—for example, 6–8 seconds—it is possible to link several such extra long peak bulbs to fire in succession. Provided the flashes are suitably synchronized, the lighting should be steady.

Unusual flash bulbs proposed include discharge-firing types, where the flash is initiated by a spark passing through the primer paste on the terminals to the wire filling between the terminals. This type of bulb has a high internal resistance, no connecting filament, and needs a high voltage but low intensity firing current. Such bulbs are intended for firing by a piezo-electric impulse, in other words a minute current of several hundred volts produced by mechanical impact on crystals of certain substances. Piezo-electric firing systems have been suggested for building into camera shutters; their main merit is that they are compact and need no external current.

Safety Measures. Despite all precautions, a bulb may burst on firing with consequent risk of injury to the people around. So all bulbs are now covered with a layer of cellulose lacquer (clear on clear bulbs, tinted on colour bulbs) to reduce the risk of flying glass splinters.

The most likely cause of bursting is the entry of air into the low-pressure atmosphere of oxygen. Some makes therefore carry an indicator spot inside the bulb. This spot changes colour if any leakage has taken place.

An additional precaution in use is to have a safety screen of wire mesh or transparent plastic in front of the flash bulb when taking portraits and especially pictures of children. Such shields are particularly desirable when the flash is fired at close range. L.A.M.

See also: *Flash (electronic); Flash equipment; Flash synchronization; Flash technique; Light sources.*

Books: *Successful Flash Photography*, by L. A. Mannheim (London).

FLASH CAMERA. Camera type, often of simple box camera construction, incorporating a small flash gun taking miniature flash bulbs. On box cameras of this kind part of the front of the camera body is usually shaped in the form of a metallized reflector with a central socket for the bulb. The camera further takes batteries in a special compartment to fire the bulb, and the firing circuit is wired internally to the shutter.

Miniature (35 mm.) flash cameras follow two main designs. One incorporates a tiny reflector in the space normally taken up (sometimes on alternative versions of the same camera) by an exposure meter. Here an ultra-small flash bulb is pushed in from the top, and the reflector spreads the light just sufficiently to cover the subject field taken in by the camera.

The alternative design has the socket for the flash bulb in the top of the body. This may be covered by a folding flap which on opening also acts as the flash reflector. Alternatively, the reflector may rise up out of the camera top. In either case the light spread is calculated to throw some light directly on the subject and some up at the ceiling for bounce flash.

The power for the firing of the bulb is provided in most cases by one or two pen-light miniature cells which fit into a special camera compartment. One or two cameras use miniature mercury button cells together with a transistor circuit to step up the cell voltage sufficiently to charge a capacitor.

Where the camera has an adjustable lens aperture it can be used over a range of subject distances, according to the guide number of the flash (usually only one type of flash bulb can be used). With cameras where the lens aperture is restricted to one or two settings, the flash distance is similarly restricted.

Special flash cameras for scientific purposes are frequently designed to operate at specific close-up distances. There the proximity of the flash to the lens provides virtually shadowless illumination. Some cameras of this type use a built-in electronic flash unit—occasionally in the form of a ring light surrounding the lens. This type of equipment is finding increasing use in ophthalmic, dental and pathological photography. Here the flash intensity, close-up setting of the camera (with or without supplementary lenses) and the film speed to be used are matched to achieve correctly exposed and fully sharp results at the prescribed setting. L.A.M.

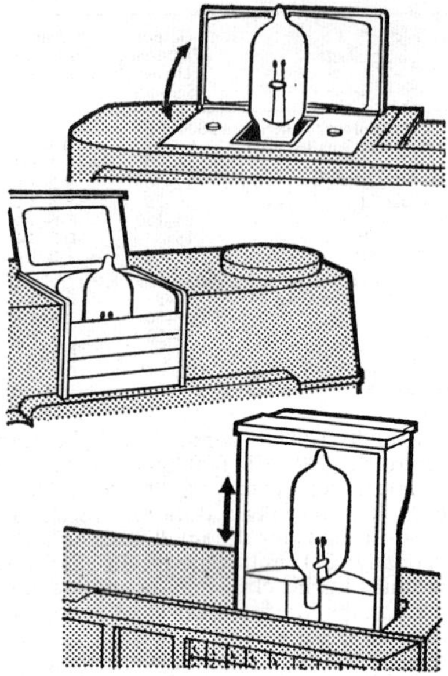

BUILT IN MINIATURE FLASHGUNS. The ultra-small AGI type flash bulb permits incorporation of flashguns in 35 mm. miniature cameras. The commonest designs are the hinged reflector with the bulb socket in the camera top or in a compartment on the front (top *and* centre), *and the pop-up reflector plus lamp socket* (bottom).

See also: *Box camera; Close-up camera; Eye camera.*

FLASH CAPSULE. Now obsolete. Measured quantity of flash powder which used to be supplied in the form of a capsule to plug into a flash gun.

See also: *Flash powder.*

FLASHED OPAL. Glass in the form of a clear sheet with a thin opal coating on one side. It is used as a diffuser in front of the lamps in contact printers and enlargers.

FLASH (ELECTRONIC)

Electronic flash, as used in photography, is produced by an instantaneous electric discharge between two electrodes in a gas-filled glass bulb (called a tube).

In practice, the electrical energy for the discharge is stored in a condenser (otherwise known as a capacitor).

The main components of an electronic flash unit are therefore: the power supply, the capacitor, the triggering circuit, and the flash tube itself usually with a reflector.

The power supply in portable equipment is often packed with the capacitor in one unit—the power pack. The triggering circuit and the flash tube make a second unit which may resemble a flash gun in appearance.

Miniature portable electronic flash units contain everything in a single lampholder, including the power supply, capacitor and flash tube. The smallest types are compact enough to fit into the accessory shoe of a miniature camera like a normal flash gun for bulbs.

The Power Supply. The electric current used to charge the capacitor must be a direct current of the appropriate voltage for the tube (usually around 2000 volts, depending on the type, but often as low as 250 volts or less with low-voltage tubes).

Direct current at a suitable voltage can rarely be produced directly; it is mostly derived from a comparatively low voltage source, and stepped up. The following systems are in use:

(1) Accumulator with vibrator, transformer and rectifier circuit; or with transistor system.

(2) A.C. mains with transformer and rectifier.

(3) High-tension drybatteries.

Accumulator Units. An accumulator (usually of the lead-acid or nickel-cadmium type) provides an efficient power supply, and is reasonably portable. Its main disadvantage is its weight, to which has to be added the weight of the transformer and other units required with it.

An advantage is that accumulators can be recharged fairly easily. Some electronic flash units utilize very small light-weight accumulators and incorporate a charging circuit. The number of flashes obtainable from an accumulator charge is small in that case, but the unit is intended for frequent recharging by simply plugging it into the mains supply.

The accumulator provides current at about 2–6 volts, according to the number of cells used. To convert this into the necessary high voltage involves three operations:

(1) A vibrator or make-and-break circuit (similar in principle to the circuit of an electric bell or buzzer) converts the continuous direct current into an intermittent one.

(2) The low-voltage intermittent current is converted into a high-voltage alternating current in the following way. The intermittent current is fed into the primary winding of a transformer where the interruptions in its flow produce a fluctuating magnetic field. This induces an alternating current in the secondary winding of the transformer. The value of the induced voltage depends on the ratio of the number of turns in each winding. So the transformer can be made to convert the low-voltage intermittent current into alternating current at any desired voltage.

(3) A rectifier converts the high-voltage alternating current (which is useless for charging a capacitor) into a high-voltage direct current. The rectifier may be of the radio valve type or a metal rectifier. In either case it acts by letting the current pass in only one direction; a simple rectifier thus yields an intermittent direct current. With a suitable circuit of two rectifiers a smooth current is obtainable.

Modern electronic flash units replace the vibrator circuit and rectifier with a transistor system, saving weight and bulk.

Mains Supply. High-powered electronic flash units have a heavy current consumption and accumulator operation would be inconvenient. As these units do not have to be portable, they can be supplied directly from the mains.

Generally the mains voltage is still too low and has to be stepped up. With A.C. mains this requires a transformer and a rectifier, while with D.C. mains a vibrator or transistor—as for an accumulator unit—is also necessary.

Units employing low-voltage flash tubes can be charged directly from a D.C. mains supply, or through a rectifier from an A.C. supply. With alternating current, two rectifiers can be connected in a doubling circuit which doubles the voltage output, and can be made to yield 500 volts from a 250 volt mains supply.

Frequently portable units also permit mains operation, usually as an alternative to battery operation. In most cases the unit incorporates a charging circuit which automatically recharges the accumulator when connected to the mains.

High-tension Batteries. These charge the capacitor directly through a resistor, and need neither vibrator, transformer nor rectifier. The weight of the power pack is therefore much lower, especially when it consists of light-weight hearing aid batteries, but operation is more expensive as the batteries have to be replaced completely. High-tension operation is more suitable for low-voltage tubes (250-500 volts although it has been used up to 800 volts) since the bulk of the batteries required to produce higher voltages would be too great.

The Capacitor. This stores electric power supplied by the transformer and rectifier or by the high-tension battery.

In principle, a capacitor consists of two layers of metal foil each connected to a terminal of the power supply and separated by a thin layer of insulating material.

The amount of electrical energy that a capacitor will store depends only on its capacity

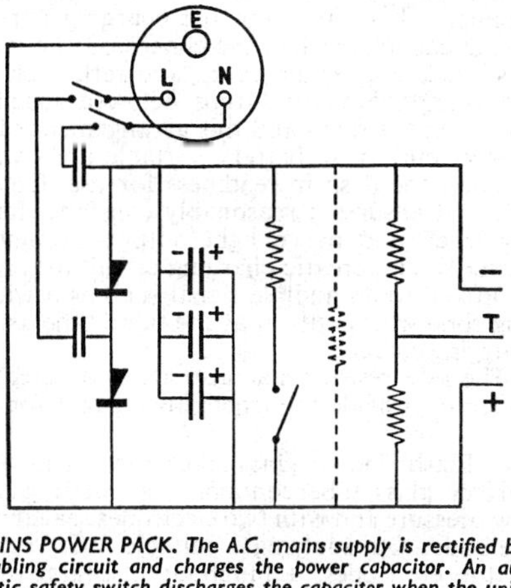

MAINS POWER PACK. The A.C. mains supply is rectified by a doubling circuit and charges the power capacitor. An automatic safety switch discharges the capacitor when the unit is opened. E, earth. L, live. N, neutral. T, trigger lead.

and not on the strength of the charging current. So a small current can be made to charge up a large capacitor with a considerable amount of electrical energy. The intense flash is the result of discharging in a fraction of a second all the electrical energy that has built up inside the capacitor over the whole of the charging period. This, of course is the basic principle.

Capacitors are rated in microfarads according to their capacity, and in volts D.C. according to their maximum working voltage. Electronic flash unit capacitors range from 30 to 200 microfarads, at working voltages up to 2500 volts. The plates of these capacitors are usually aluminium foil, separated by thin oiled-paper insulation. Such capacitors are known as paper-type capacitors.

An alternative type of capacitor uses a liquid electrolyte (generally a solution of an alkali). Electrolytic capacitors are smaller and cheaper for the same capacity than the paper-oil type, but are not as durable. Moreover, an electrolytic capacitor must always be connected into the circuit with its + pole wired to the + pole of the charging circuit; reversing the polarity of the capacitor ruins it. Paper insulated capacitors can be connected without regard to polarity.

Electrolytic capacitors are suitable only for low-voltage flash tubes and are rated up to 500 volts with capacities of up to 600 microfarads for electronic flash units.

Electrolytic capacitors tend to lose their storage ability when left lying unused for long periods. They then become "depolarized" and have to be reformed by being kept under charge for several hours. The main result of a partially depolarized capacitor is reduced capacity and greater leakage rate, and hence lower available flash energy. If an electronic flash unit using electrolytic capacitors is in only intermittent use, it is advisable to keep the capacitor under charge for about one quarter hour once a week.

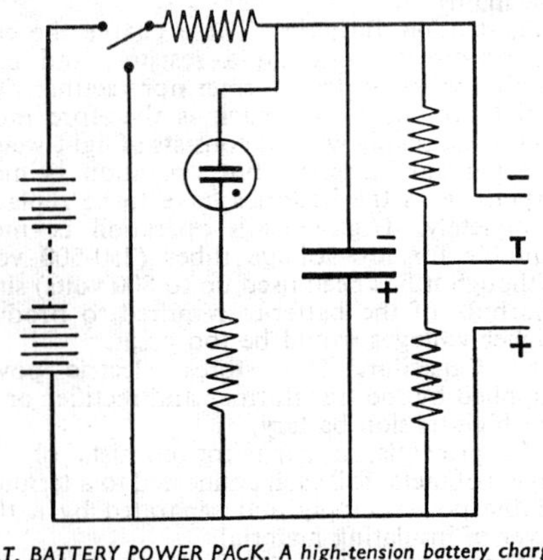

H.T. BATTERY POWER PACK. A high-tension battery charges the power capacitor directly through a resistor, controlled by a two-way charge-discharge switch. T, trigger lead.

Wiring of capacitors follows the same principle as wiring of batteries. Connecting a number in series proportionately increases the voltage they can carry but only with the capacity of a single unit; connecting them in parallel increases their total effective capacity, but only for the voltage of a single unit.

Charging Time. The capacitor takes some time to charge up—on commercial electronic flash units from 5 to 15 seconds. The charging time depends both on the size of the capacitor, and on the charging current. The latter is determined by the resistance of the charging circuit.

Small sets powered by high-tension batteries with small capacitors charge up quickly; large units with big capacitors and transformer circuits take longer.

Leak Resistors. In most electronic flash units the two sides of the capacitor are connected through a high resistance of 1 to 10 megohms. This arrangment is largely a safety measure to prevent accidental shocks as it gradually empties the capacitor it left in a charged condition. The rate of leakage is not sufficient to affect the normal operation of the flash unit.

The leakage current can also be used to indicate the state of the charge on the main capacitor, since it is proportional to the charge. For this purpose a micro-ammeter may be connected in series, or a small neon bulb joined in parallel. The latter lights up when the leakage current exceeds a certain value, and thus indicates when the capacitor is charged. Generally this neon bulb is housed in the flash head itself with the tube.

In practice the point at which the neon bulb lights up is somewhat below the maximum capacitor charge. Hence if the capacitor is allowed to charge up for a few seconds longer, the flash obtained is more powerful.

As this may well lead to inconsistent output, better type modern electronic flash units incorporate transistorized automatic charging circuits. These interrupt the charging current to the capacitor when the charge on the latter has reached a certain value, and restart charging when the charge drops below a specified level. On the one hand this arrangement saves battery current in battery portable units when keeping the flash in readiness for use. On the other, it ensures a reasonably constant charging level, and hence light output—for every flash. The automatic charging circuit may also be linked to an audible signal such as a whistling tone which cuts in at the same time as the charging current.

The leak resistor may also act as a potentiometer to provide the triggering current for the flash.

The Flash Tube. This takes the form of a narrow glass tube containing an inert gas at low pressure and with two electrodes sealed into the ends. On applying a suitable high voltage to the electrodes, a luminous discharge takes place through the gas.

The pressure of the gas determines the resistance of the tube and the voltage required for the discharge, while the nature of the gas determines the colour of the light. Most electronic flash tubes contain krypton, xenon, and small quantities of other rare gases, balanced to yield a light approaching daylight in colour.

The tube itself may have various shapes: straight, coiled, or U-shaped. The straight and U-shapes are common for smaller flash units, while the more powerful tubes are of the coiled type. The energy dissipated at the peak of the flash is very great, and the tubes are generally made of quartz or special glass to withstand the strain.

Electronic flash tubes are also made in the form of a ring to encircle the lens to give shadowless illumination, particularly for dental and medical photographs. In such specialized fields it is often an advantage to eliminate shadows so that the tone and colour values of the subject can be studied independently.

Flash Tube Fittings. There are no standardized flash tube fittings as there are standard lamp caps for flash bulbs, since each maker of flash units uses his own way of mounting the tube in the unit. Some tubes are available simply with soldering tags, while others use one or more of the various radio valve bases such as 3-pin, 4-pin or international octal. A few tubes are even made with standard or small Edison screw fittings.

Flash Reflectors. Electronic flash tubes are invariably mounted in a reflector to concentrate the maximum amount of light on the subject. Reflector shapes vary enormously both according to the size of the flash unit and the purpose for which it is intended.

Small portable units have often a trough-shaped reflector with a straight flash tube. This provides the highest light efficiency (important to make up for the comparatively low power of such units) but spreads the light over a comparatively limited angle. Frequently this only just covers the angle of view of a normal camera lens; when used with wide angle-lenses such flash units may only illuminate the centre of the subject area. In any case the flash needs accurate aiming.

Larger portable and semi-professional units often have a reflector of lower efficiency but more even light spread and less harsh illumination. The reflector may with these units be adjustable to provide a narrow or a wide beam of light to choice.

Commercial studio electronic flash units use lamp sets of a great variety of shapes and sizes for broad to narrow floods of light and even highly directional reflectors for spot lighting effects.

Some makers of electronic flash outfits tint the reflector (or the protective plastic screen in front of it) faintly yellow to make the colour quality of the flash match more closely the colour of average daylight when used with daylight type colour films.

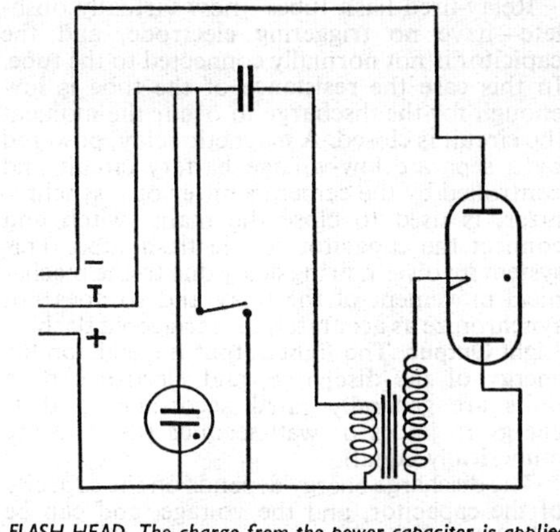

FLASH HEAD. The charge from the power capacitor is applied across the tube. Closing the firing switch generates a pulse in the coil which triggers the discharge.

Firing the Flash. The two main methods are: a triggering electrode and a magnetic relay.

Most modern electronic flash units use the triggering electrode, since it is much safer and more convenient in practice.

In this type of circuit the main capacitor is connected permanently across the flash tube terminals. The resistance of the gas in the tube is normally too high to permit a direct discharge. For firing the flash there is a third electrode—usually a coil of wire wound round the outside of the tube. An instantaneous triggering voltage of 5,000 to 10,000 volts applied to this electrode ionizes the gas in the tube, thus lowering its resistance and allowing the capacitor to discharge its energy through the tube in the form of a flash of light.

The triggering voltage is usually supplied by a small induction coil energized by a suitable voltage picked off a potentiometer across the main capacitors. The triggering switch may be the synchronizing contacts inside the camera shutter or a synchronizer or a press button on the flash head, or any type of remote switching device. The whole process of triggering takes a few microseconds and is, from the point of view of synchronization, virtually instantaneous.

An alternative method uses a cold cathode valve to control the triggering current. The triggering voltage is permanently connected to the anode and cathode of the valve and the actual firing impulse is released by a low voltage applied to the grid of the valve. The advantage of this arrangement is that no high-tension current passes through the triggering switch at all, and that a very weak current can trigger the flash. This permits operation by impulses from a microphone or photocell, and is the basis of many automatic triggering methods for complex studio set-ups and also ultra high-speed work.

Relay-fired flash tubes—now virtually obsolete—have no triggering electrode, and the capicitor is not normally connected to the tube. In this case the resistance of the tube is low enough for the discharge to occur the moment the circuit is closed. A magnetic relay, powered by a separate low-voltage battery circuit and controlled by the camera shutter or a synchronizer, is used to close the main switch and connect the capacitor to the flash tube. This system involves a firing delay due to the mechanical movement of the relay and so does not synchronize as accurately as a triggered flash.

Light Output. The light output depends on the energy of the discharge, and electronic flash units are generally rated according to their energy in joules or watt-seconds (the two are numerically equal).

The discharge energy depends on the capacity of the capacitor, and the voltage, and can be calculated from the equation:

$$J = \frac{1}{2} \times C \times (kV)^2$$

where J is the energy in joules, C the capacity in microfarads, and kV the voltage in kilovolts.

Thus a discharge from a 66 microfarad capacitor charged at 2500 volts (2·5 kilovolts) would give a flash of $\frac{1}{2} \times 66 \times (2{\cdot}5)^2 = 206$ joules (or watt-seconds). Not all this energy is necessarily converted into light, as there are losses in line resistance, and leakage.

Flash tubes are designed to withstand the energy of a particular intensity of discharge and are rated according to the discharge voltage and power loading they can carry.

The light-output in lumen-seconds depends on the efficiency of the tube and of the discharge circuit in lumens per watt. Generally a 200 joule tube has an output of 7 to 10 lumen-seconds and gives a roughly equivalent exposure to a normal flash bulb of similar output. The effectiveness of the flash unit for exposure purposes depends on the light output, the power, as well as the reflector efficiency. Since electronic flash tubes are usually associated only with the reflector in which they are fitted, it is customary to quote guide numbers for the whole flash unit. These are—unless inflated by advertising claims—a more reliable indication of the performance of a flash unit than the discharge energy in joules or watt seconds. In other words, when two units of say 25 and 35 watt-seconds are compared, the 25 watt-second flash reflector may conceivably be so efficient that the smaller unit actually provides greater illumination on the subject.

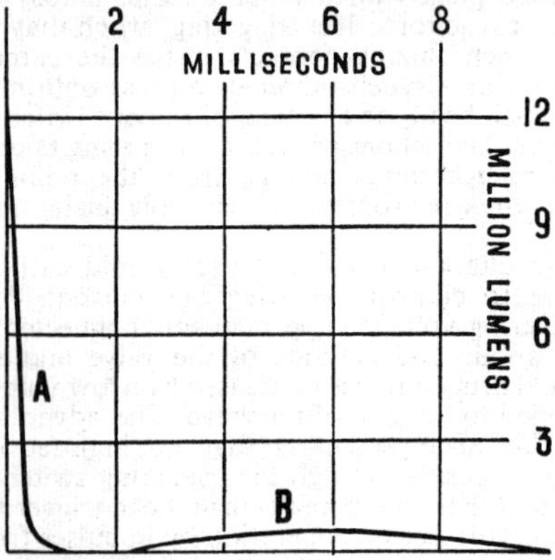

ELECTRONIC FLASH CURVE. A: The electronic flash is very brief and intense, with virtually no firing delay. B: By comparison even the fastest flash bulb is very slow in firing with a small peak intensity but considerably longer duration.

There are industrial standards (for example the German standard DIN 19011) which precisely specify the measurement of a guide number for electronic flash units. The method is similar to film speed measurement and involves comparison exposures of a film sample behind a grey scale.

Exposures are made on the one hand to a standard light source and on the other to the flash to be measured. The guide number is calculated from the grey scale steps behind which the film shows, after development, a minimum density of 0.1 above fog.

The main feature of these guide numbers is that they are accurately reproducible and not subject to optimistic advertising claims. The standard lays down a room factor (reflected light from walls indoors) and other details to be taken into account in measurement. The recycling time and the number of flashes obtainable from a set of batteries or an accumulator charge are also the subject of specific measurement methods.

Flash Duration. With normal electronic flash units the duration of the flash varies between 1/300 second and 1/5,000 second. With a 100 joule unit operating at 2,000 volts, it is around 1/3,000 second; portable units of 50 to 60 joules operating at 250 volts have a flash duration of between 1/500 and 1/1,000 second. This is as short as, or shorter than, the shortest exposure time that most normal camera shutters can give. Electronic flash is therefore specially suitable for high-speed action photography, and can arrest most fast movement.

Mathematically, the flash duration can be calculated from the equation:

$$t = C \times R$$

where t = duration in microseconds
C = capacity in microfarads
R = tube resistance in ohms.

The tube resistance is the apparent resistance during the discharge, and is about 5 to 6 ohms for a tube operating at 2000 volts.

Thus a 200 joule tube operating at 2·5 kilovolts (as described before) with a 66 microfarad capacitor would have a flash duration of 400 microseconds, or 1/2,500 second. But the resistance of the wires connecting the capacitor to the tube must also be taken into account.

Other Characteristics. In addition to light output and flash duration the following data

TYPES OF FLASH TUBE

Type	*Joules*	*Recommended Voltage*	*Recommended Capacitor (mfd.)*	*Output Lumen-seconds*	*Effective Duration seconds*
Small low voltage	35–50	200–250	1,000	2,000	1/300
Medium low voltage	75–100	450–500	600	4,000	1/500
Medium voltage	100–200	1,000–1,300	200–400	4,000–8,000	1/1000–1/3000
Small high voltage	100–300	2,500	30–100	4,000–13,000	1/3000–1/5000
Studio	400–1,000	2,500	150–300	18,000–50,000	1/800–1/2000
Large studio	10,000–16,000	3,000–4,000	2,000–3,000	500,000 and over	1/300
Microflash	35–50	7,000–7,500	1–2	1,000	1/1000000

are also of interest in photography.

The firing delay of instantaneous firing discharges is of the order of 20 to 30 microseconds, or about one fiftieth that of the fastest firing (class F) flash bulb. This type of flash is defined as class X flash for purposes of synchronization. With relay-fired units the delay is similar to class F flash—5 to 10 milliseconds.

The peak intensity is proportional to the square of the voltage discharging through the tube. Its practical importance is not very great, since the total light output matters more. The peak intensities of normal flash tubes vary from 10,000,000 to 50,000,000 lumens.

The colour temperature depends on the mixture of the gases in the tube. Strictly speaking, it is not really possible to specify a colour temperature with flash tubes since the spectrum is not a continuous one. For photographic purposes, however, most modern flash tubes behave like a light source with a colour temperature of around 6000°K and are thus suitable for colour photography with daylight type colour films.

Types of Flash Unit. Electronic flash units on the market are classified according to their portability as well as their power supply.

Battery-portable units carry the power source (accumulator and anciliary gear, or high-tension battery) and the capacitor in a power-pack—usually a plastic case with a shoulder-strap. The flash tube and triggering circuit are built into a flash head similar to a normal flash gun which carries the reflector and can be fixed to the camera by the accessory shoe or tripod bush. These units weigh from 2½ to 15 lbs., and have a power loading of 35–200 joules.

Mains-portable units are similar, but are designed for connexion to a mains supply, and are thus not self-contained. Some units provide for alternative battery and mains operation.

Miniature battery units, working from small nickel-cadmium accumulators or hearing aid batteries, combine all components in a single compact unit that fits directly on the camera. Outfits of this kind weigh about 1–2 lbs, and have a power loading of 20 to 35 joules.

Studio electronic flash units nearly always derive their power entirely from the mains supply. They resemble a studio floodlamp unit in appearance, the power pack being built into the base of the lamp stand or into a separate case on wheels. One power pack may supply several lamp units in various combinations. Many of these units incorporate a filament lamp of 100 to 500 watts, to serve as guide light when setting up the lighting scheme. The power may be from 500 to 20,000 joules, 1,000 joule units being the most common type.

Safety Measures. The high voltage of the capacitor charge of most electronic flash units is capable of giving a fatal shock, and special precautions are necessary in handling this type of equipment.

It is unwise at any time to open the case of an electronic flash unit while it is still connected to the mains, or until it has been left disconnected for some minutes. Even then, before doing any repairs or testing, the main terminals of the capacitor should be short-circuited with the blade of a well-insulated screwdriver.

The unit should never be left where it is likely to get wet, or even damp, because the danger of shock from moist surfaces is greatly multiplied. And connecting plugs should never be pulled out, or the tube removed from its holder while the set is switched on.

One built-in safety measure is the leak resistor connected across the main capacitor. This ensures that the charge drains away when the set is not in use. Generally an additional discharge circuit consisting of a resistance of about 5000 ohms is provided and is automatically connected across the capacitor on switching the unit off. This immediately dissipates the charge.

Similar discharge resistors are often fitted to the connecting plugs of the cable from the power pack to the flash holder. Pulling out the plug then automatically discharges the set, ruling out any possibility of shock from the exposed connexions.

All exposed metal parts of a flash unit should be earthed.

Apart from normal photographic flash units, various types of electronic flash have been designed for specialized applications. The most important of these are micro-flash, stroboscopic, and X-ray flash units.

Micro-flash. The duration of an electronic flash is already very brief but by special design it is possible to achieve even shorter flashes for ultra-high-speed photography in such fields as ballistics.

As the flash duration depends largely on the capacity of the main capacitor, micro-flash units use very small capacitors of the order of 1 to 2 microfarads. The voltage has to be correspondingly high (generally at least 7,500 volts) to yield sufficient light. Even so, microflash units rarely exceed 35 to 50 joules, while the flash duration is of the order of one microsecond (1/1,000,000 second).

Stroboscopic Flash. If the triggering circuit of a flash unit is permanently closed, the tube will discharge automatically every time the charge on the capacitor reaches a sufficiently high level. The unit will thus flash continuously at a rate depending on the characteristics of the charging circuit.

Units of this type are used for stroboscopic examination of moving objects—e.g., machinery.

The faster the capacitor charges up, the greater the rate of flashing. Stroboscopic units therefore have comparatively small capacitors and a high-output power supply. As the power is generally derived from the mains, this type of flash requires a large transformer and a correspondingly heavy rectifier.

The flash tube itself must also be specially designed for stroboscopic work, as a normal tube would not stand up to a series of discharges in rapid succession. The flashing rate may be up to several hundred flashes a second.

X-ray Flash. The principle of electronic flash can also be applied to X-ray tubes by passing an instantaneous discharge through the X-ray tube instead of a continuous current. Like a normal electronic flash unit, an X-ray flash unit requires a power supply, a capacitor, and a firing circuit for operation, though the detailed circuits are different. The voltages are up to 200 times as high, while the flash duration is of the order of 1 microsecond. L.A.M.

See also: *Flash bulbs; Flash equipment; Flash Synchronization; Flash technique; High speed photography; Stroboscopic flash.*

Books: *"Flash!", Seeing the Unseen by Ultra-high-speed Photography*, by H. E. Edgerton and J. R. Killian (Boston).

FLASH EQUIPMENT

In normal flash photography the bulbs are usually fired in a flash holder or flash gun which may be attached to the camera and connected with the shutter by a firing lead.

Originally flash guns were flash holders incorporating a synchronizer. As nowadays all camera shutters are internally synchronized, such synchronizers have disappeared from use, and the terms flash holder and flash gun are used interchangeably.

The basic components of a flash holder are:

(1) A battery case to hold the batteries which fire the flash.

(2) A lamp holder to take the bulb.

(3) A reflector to concentrate the light.

(4) A firing switch or outlet to an external switch to complete the flash circuit.

Many flash holders also incorporate:

(5) A test circuit to test the bulbs and the firing circuit.

(6) A bracket or suitable arrangement for mounting the holder on the camera.

(7) A firing lead and plug to connect to the camera shutter.

Nowadays many flash guns incorporate these features (or at least the first four) in very compact units, sometimes no bigger than a match box.

Battery Cases. Early flash guns used to take two or three 1·5 volt dry cells (to yield a firing voltage of 3 or 4·5 volts) in a comparatively bulky battery case resembling a large pocket torch. Nowadays direct firing by low voltage batteries (which had to yield a high current to be reliable) has given place to battery-capacitor firing, using a small anode battery of 15 to 30 volts and a capacitor. This arrangement is considerably more compact, and is sometimes even built into the camera itself.

The battery may measure as little as ½ × ½ × 1 in., and is usually of the type used in various kinds of hearing aid. It charges a capacitor (mostly a 50–100 microfarad electrolytic type) which on closing the firing circuit fires the flash bulb.

The advantage of this system is that the charge of the capacitor is constant even when the battery is nearly exhausted, because a failing battery merely takes longer to charge the capacitor. So a capacitor always fires the bulbs reliably. There is no appreciable direct drain on the battery; only enough electricity is used each time for the flash itself, and the current is sufficient to fire several bulbs at once. The battery thus lasts much longer and as a result, battery-capacitor flash guns are rapidly displacing ordinary battery flash guns.

In practice, the charging circuit also incorporates a resistor of several thousand ohms. This increases the charging time to several seconds instead of a fraction of a second, but prevents any sizeable discharge of the battery directly through the bulb. In most battery-capacitor circuits the bulb itself completes the charging circuit so that the capacitor does not charge up until the bulb is inserted.

Since an electrolytic capacitor is used (a paper type of the same capacity would be much too bulky) the battery must always be inserted the same way round. Usually the battery case is marked with appropriate symbols to indicate the position of the + and − poles of the battery as a guide to fitting a new battery.

The battery and capacitor assembly generally take up no more room than two or three cells of a standard battery. It is therefore possible to convert a plain battery model to capacitor operation by making a suitable

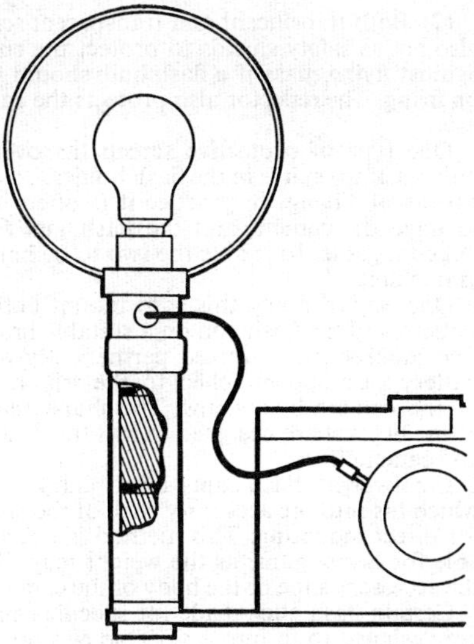

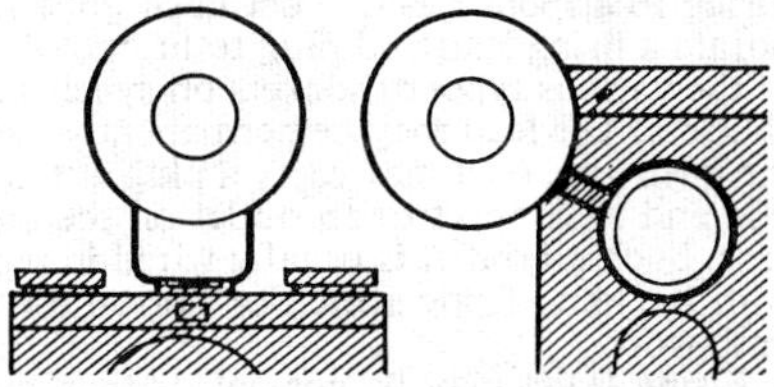

FLASH GUNS. **Top:** *The orthodox flash gun contains all the electrical circuit inside a tube which carries the flash bulb socket and the reflector. A flash cable connects the gun to the shutter. The gun is usually joined to the camera by a bracket.* **Bottom left:** *Many modern guns are small enough to clip into the accessory shoe. The firing connexion is either a flash cable or a contact in the accessory shoe.* **Bottom right:** *Some flash guns have an arm with bayonet ring for mounting on the finder lens of certain twin-lens reflex cameras.*

insert to hold the battery and capacitor circuit, and loading this into the battery case in place of the normal batteries.

The details of construction will of course depend on the dimensions of the battery case.

Lamp Holders. Amateur flash guns usually take capless glass-base flash bulbs. There are two types: the European one where the glass base is more or less round, and the American (A.G. type) with a rather smaller oblong base. In both cases the terminal wires from the filament inside the flash bulb protrude on opposite sides of the glass base. The two types of bulb cannot normally be used in the same flash gun, especially as the American type capless bulb is designed for ultra-miniature flash guns.

In America miniature metal-capped bulbs are also used (but not the European glass based bulb) and flash guns carry a corresponding lamp holder.

Professional flash guns have one of two types of lamp holder:

(1) To take small centre contact bayonet cap bulbs with a cap diameter of 15 mm. (S.C.C., A.S.C.C., or B15s).

(2) To take Edison screw cap bulbs with a cap diameter of 27 mm. (E.S., or E27).

The S.C.C. lamp holders are intended for the smaller types of focal plane (class FP) bulbs and the E.S. lamp holders for all large professional flash bulbs. The flash guns for the latter are also comparatively larger.

At one time bulbs with S.C.C. cap were the standard amateur flash bulb, and a number of simple flash guns taking this type of bulb still exist. These can be converted to take the European type of glassless bulb with the aid of a simple adaptor.

Adaptors are also available for fitting S.C.C. bulbs into E.S. lamp holders and vice versa.

Most lampholders on flash guns incorporate an ejector to remove the hot fired bulb from the gun without touching it by hand. This is simply a button which bears against the centre of the glass base from underneath, and pushes the lamp out of its fitting when required. On S.C.C. lamp holders the mechanism consists of a spring-loaded contact and a catch which normally retains the bulb in the holder against the thrust of the spring. On pulling aside the catch, the spring-loaded centre contact ejects the bulb.

Reflectors. The flash bulb emits its light in all directions, and if used by itself only about 8 per cent of the light would reach the subject directly. The reflector collects some of the remaining 92 per cent of wasted light and redirects it towards the subject. The reflector is thus of great importance in assessing the effective light of the flash reaching the subject. The simplest types of reflector may utilize about 15 to 20 per cent, and the most efficient type about 60 to 70 per cent. With the same flash bulb in two different reflectors the exposure may thus vary by the equivalent of over two stops. That is one reason why guide numbers for flash exposures as given by bulb manufacturers are necessarily vague and subject to considerable modification in practice.

The efficiency of a reflector depends on its size relative to the flash bulb, as well as its shape and surface.

The larger the reflector, the more light it collects. For the same effect, a large bulb needs a larger reflector than a small bulb.

Amateur flash guns for miniature bulbs usually have a reflector of about 3 inches in diameter, though for higher efficiency a reflector size at least twice as large is better. The smallest guns for the American A.G. capless bulbs have minute reflectors only 1 to 2 inches in diameter (especially where such guns are built into the camera itself).

Professional flash guns for large bulbs have reflectors from 7 or 8 to 16 inches in diameter.

For compactness some flash guns use a folding fan type reflector which opens out to about 3 to 4 inches in diameter, but folds together into a space no larger than 1 × 1½ inches on the front of the gun.

The simplest reflector is a plain white or silvered disc held behind the bulb. Most flash holders, however, have curved reflectors which provide a more concentrated and even flood of light.

The most efficient reflector shape is the parabola which concentrates the light into a beam. This beam may be parallel, converging or diverging, according to whether the bulb is at the focus of the parabola, in front of the focal point, or behind it. Some reflectors of this type are adjustable to permit movement of the bulb to vary the angle covered. Many reflectors also provide for a vertical adjustment to centre bulbs of different sizes.

The reflector surface may be polished or satin finished. A polished reflector is the more efficient, but this type can give rise to extra brilliant "hot spots" or to areas of uneven illumination if the reflector shape is not formed with sufficient precision. A satin-finished reflector throws a little less light on to the subject, but diffuses it more evenly. This effect is still greater with matt white reflectors. Grained reflector surfaces are similar to the satin finish for diffusion, but are claimed to be more efficient.

Screens. Some flash guns are fitted with transparent or translucent plastic screens. These clip over the reflector or are screwed on and made to swing out of the way to allow the bulb to be inserted. The screens serve two purposes.

(1) The translucent type acts as a diffuser, softening the light from the flash and spreading it more evenly over the subject area. This involves a loss of light and therefore calls for more exposure.

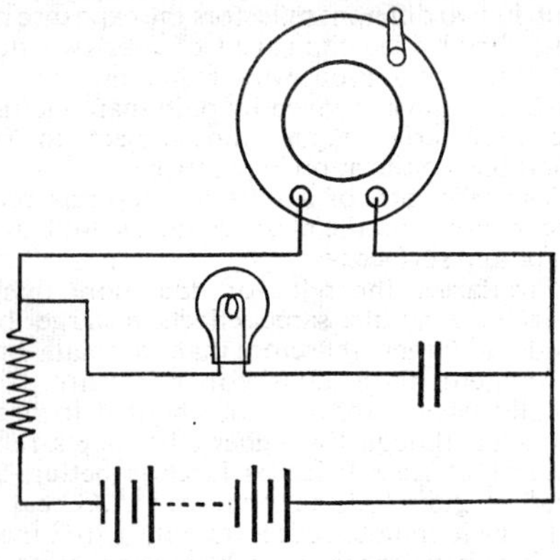

CAPACITOR FLASH CIRCUIT. An anode battery charges the capacitor through a resistor and the flash bulb filament. On closing the shutter contact, the capacitor is short-circuited across the bulb and discharges, thus instantly firing it.

(2) Both translucent and transparent screens also act as safety shields to protect the subject against flying glass if a flash bulb should burst on firing. The reflector also protects the camera user.

One type of protective screen fits over the bulb itself when it is in the flash holder.

Means of Fixing. In practice it is often useful to have the camera and the flash gun firmly linked together to enable the two to be handled as one unit.

One way of doing this is to mount both the camera and the flash gun on a suitable bracket. The bracket may be fixed permanently to the battery case and attached to the tripod bush of the camera by a screw. The shape and size of the bracket are designed to suit the flash gun and camera in use.

Lightweight flash guns often carry a foot which fits into the accessory shoe of the camera for direct mounting. This method is not advisable for heavy guns, as the weight may distort the accessory shoe or the body of the camera.

Certain flash guns made for specific cameras are designed to fit into a suitable clip or screw on to the camera concerned.

Firing Plugs and Lead. Most flash guns incorporate a firing lead and plug to fit one or other of the various types of sockets on synchronized shutters. The lead may be permanently fixed to the flash gun, or it may carry a plug at the flash gun end to connect to an outlet on the battery case. Such an outlet is usually wired in parallel with the hand firing switch or button for open flash work.

Firing leads may be coaxial or twin cables. With a coaxial cable the outer lead is the earth connexion. The plugs, both for the shutter and for the battery case, are generally designed to fit only one way to prevent the insulated terminal of the battery case from being connected to the earth terminal on the shutter and so firing the bulb as soon as it is inserted in the lamp holder.

Originally manufacturers used various designs of flash socket on their shutters. With the exception of the first type listed below, which is nowadays the standard flash fitting, most of the others have become obsolete.

(1) Coaxial plug and socket of 3 mm. diameter. This is used on most Continental cameras, as well as on many British makes. The plug may be in line with or at right angles to the cable. Some cameras have a special catch which prevents the plug from being accidentally pulled out of the shutter.

(2) Coaxial plug and socket of 3·8 mm. diameter.

(3) A.S.A. bayonet plug and socket. These are used for some American cameras. The socket on the shutter resembles a tiny bayonet lamp cap, while the plug corresponds to the bayonet socket.

(4) A.S.A. twin pins and sockets (bipost). The shutter carries the pins, while a female plug

makes the connexion. These also are used for some American cameras.

(5) Non-standard types include various forms of two-pin and three-pin plugs, coaxial bayonet, etc.

With certain cameras the electrical connexion is fitted inside the clip which holds the flash gun to the camera, or inside the accessory shoe to which the gun is attached. In both cases the flash gun must be specially designed for the camera concerned, and no external firing lead is required.

Permissible Current. The various electrical connexions and contacts must withstand the maximum firing current likely to pass through them.

The better makes of flash cable and contacts are tested to an instantaneous voltage of the order of 600 volts and can thus carry the primary trigger surge of electronic flash units. The maximum steady current loading for periods not exceeding 1/10 second, is usually around 10 amps at 24 volts.

Test Lamps. Some of the more elaborate flash guns incorporate testing circuits with a built-in test lamp. This lamp enables the user to check the working of the synchro-contacts of the shutter, the connexions and firing lead, the battery, and the flash bulb itself. A separate test lamp (e.g., 6 volt motor-car bulb) inserted in the lamp holder in place of the flash bulb will also test all the above points except the flash bulb.

Extension Units. For multiple flash work it may be necessary to synchronize several flash bulbs to one synchronizer or shutter. The usual way of doing this is to use extension units which are linked to the main unit and shutter by means of extension cables. Several systems are possible for doing this:

(1) One battery or battery-capacitor unit fires several bulbs. The bulbs are connected in parallel or series (according to the circuit) and the extension units are plugged into a suitable extension socket on the master unit. The current must be strong enough to fire all bulbs simultaneously.

(2) One battery charges several capacitors each of which fires one bulb. The main unit is a normal capacitor flash gun, while the extension units are similar but with an extra connecting socket and no battery. The usual scheme is for the extension units to be connected directly to the shutter, and the main unit plugged into the extension unit. Alternatively the leads from each unit are plugged into a multi-way adapter which in turn plugs into the shutter. This is normally the most reliable system.

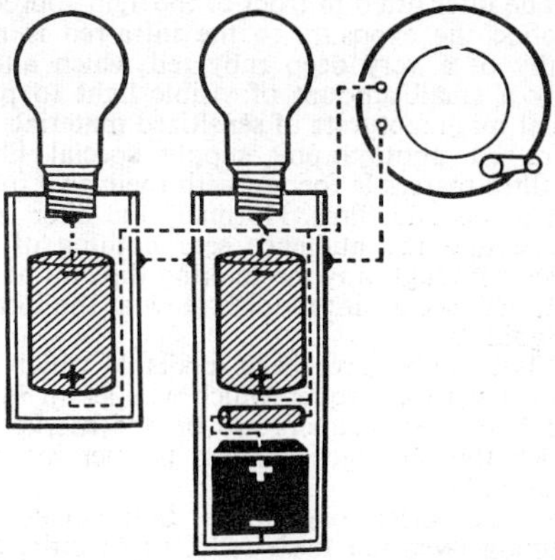

EXTENSION FLASH. One battery of a capacitor flash gun can power one or more extension units as well, each with its own capacitor. The units are here connected in parallel.

(3) Several separate capacitor flash guns, each with its own battery unit, are connected in parallel to the shutter or synchronizer. This again connects via a multi-way adapter to the various guns. The units are thus independent of each other as far as power is concerned. There is, however, the risk that the capacitor of one unit may fire the bulb of another when the latter is connected. It is advisable therefore to have a rectifier in the circuit of each gun to prevent its current from going into another circuit (e.g., if the resistance of the circuit of one of the other guns in smaller). L.A.M.

See also: *Batteries; Flash bulb; Flash camera; Flash (electronic); Flash for infra-red; Flash powder; Flash synchronization; Stroboscopic flash.*

Books. *Successful Flash Photography*, by L. A. Mannheim (London).

FLASH FOR INFRA-RED. Infra-red flash has been used for a variety of special purposes, including press photography in the blackout and in situations where a bright flash would be disturbing, candid photography, detection of criminals in the dark, cinema audience-reaction studies, unobtrusive instrument recording in aircraft, dark adaptation and other studies of the eye, and photography of industrial operations in the dark—e.g., sensitized material manufacture and processing.

Equipment. Infra-red flash photographs can be taken with ordinary cameras and flash attachments, the special requirements being a filter in front of the flash bulb to confine the rays to the infra-red, and a film sensitive to infra-red rays. In some cases it may be necessary to make a slight adjustment of the focus in the form of a slight forward shift as though the subject were somewhat nearer than its actual distance. Many modern lenses have an infra-red datum line marked on the focusing mount.

The light from ordinary flash bulbs and electronic flash tubes is rich in infra-red rays and both of these light sources are suitable for infra-red flash pictures.

The filter fitted in front of the light source to confine the exposure to the infra-red is normally of a very deep ruby red which allows only a small amount of visible light to pass. Most manufacturers of sensitized materials for infra-red photography supply special filters or filter materials for use with their own infra-red plates and films. Normally the filter is so dense that the audience sees nothing of the flash, although anyone looking directly at the bulb will see a dim reddish glow at the instant of ignition.

Flash bulbs have been available which are coated with a lacquer which acts as an infra-red filter and lacquers are on the market by which the photographer can lacquer his own clear bulbs.

Plastic sheet bags have been made for slipping over the flash reflector or bulb, and having the characteristics of infra-red filters. It is a simple matter to fix a sheet of infra-red filter gelatin over the reflector, but it should be protected by sandwiching it between stout sheets of transparent plastic.

In arrangements where the flash is entirely independent of the camera, frames carrying the filters can be placed over the lights, or darkroom lamphouses can be used if an infra-red filter is inserted in place of the normal safelight glass and a sheet of tin or aluminium foil curved around the back of the bulb inside to increase the reflectance. Booths or other devices with infra-red filters covering them may also be constructed to house the lamps. In such cases the filters should be between sheets of glass since the heat of the flash might tend to destroy the brittle gelatin filters.

Technique. In infra-red flash photography, the subject should be unaware of being photographed, but it is practically impossible to make flash pictures in such a way that the subject could not in any circumstances see a flash. If there is enough illumination to give a reasonable exposure, anyone looking directly at the reflector when the flash is fired, will see a deep red flash. So the attention of the subject must always be drawn away from the light source, as in a theatre or cinema, where everyone is looking at the stage or screen, or the lighting must be indirect—e.g., "bounced" from the ceiling or a wall—or the exposure must be made through an existing and permanent red light, such as part of the illumination, or an exit sign. In direct infra-red flash photography, it is best to use a reflector which gives well-diffused rather than concentrated illumination.

The actual exposure depends on the nature and shape of the reflector and it is not possible to give precise data. But with the infra-red material and flash bulbs available through the normal sources of supply there is no difficulty in achieving a flash factor of 50 or even 100 for an exposure time of 1/50 second. The flash factor is the product of the *f*-number and the lamp-to-subject distance in feet. Detailed tables are given by the manufacturers.

Certain special films are available in 35 mm. form with greatly enhanced infra-red sensitivity which are particularly suited for infra-red flash photography. The guide number is as high as 200 with suitable flash bulbs, indicating a four times increase in infra-red speed.

Satisfactory audience-reaction photographs have been made in an auditorium 35 × 65 feet by placing five 1,000 joule electronic flash units in a booth at the side, pointing the reflectors to the ceiling, and covering them with sheets of infra-red filter between glass. Using indirect illumination of the audience and the earlier type of infra-red film, good exposures of about one hundred people in the middle of the room were obtained at *f* 3·5, the effective exposure time being 1/10,000 second.

For photographing a long row of operators spooling very fast aerial film in total darkness, twelve large flash bulbs were used, one suspended over each operator and machine, and two extra for supplementary illumination. All the flashes were fired together. Using the earlier infra-red film, a good picture was made at *f* 16 with open flash. W.C.

See also: *Infra-red photography.*

FLASH GUN. Camera accessory consisting of a battery case, lamp socket, and reflector, for firing flash bulbs. It may incorporate a synchronizing mechanism, or may be designed to be fired by the flash contacts built into most camera shutters.

See also: *Flash equipment.*

FLASHING. Method of varying the contrast of a photographic paper by using a brief auxiliary fogging exposure in addition to the printing exposure. First suggested by G. B. Harrison and J. Mitchell in 1950.

A paper suitable for projection printing is basically a high contrast bromide paper with a very sharp-toed characteristic curve and which, without a flash exposure, is suitable for printing very thin negatives. Introduction of an auxiliary exposure reduces contrast by lengthening the toe; the higher the proportion of flash to basic printing exposure the softer will be the result. The process is limited by the production of visible fog. The same principle may be adopted for control of contrast of contact papers.

It is immaterial whether the flashing exposure is applied prior to, simultaneously with, or subsequent to the printing exposure.

Altering gradation by the use of flashing techniques is simpler than using the so-called variable contrast papers which vary in contrast according to the wavelength of the light source to which they are exposed, and hence require a set of suitable coloured filters to exploit their possibilities to the fullest extent.

Although flashing papers will provide acceptable prints from a fairly wide range of negatives, the use of this method to deal with the very hard negative tends to give a print of rather veiled appearance; this is inherent in the method.

A typical flashing projection printer is provided with push-button controls for the auxiliary exposure, the appropriate button being selected according to the negative density range; the printing exposure is automatically controlled in the usual way by a photo-cell.

This technique of automatic printing is widely used in commercial photo-finishing establishments. H.P.

Flashing is also a standard technique in the making of photo-mechanical half-tone negatives where it is used to form a minute dense core in the centre of each dot, which helps to form a firm "shadow" (small dot) formation and is lost in the lighter tones which are represented by larger dots.

Local Flashing. The term "flashing" is also applied to a technique used to blacken parts of the print by exposing them to the direct action of white light. This is straightforward local fogging, and has no relevance to contrast control.

For example, certain types of photograph look more satisfactory when surrounded by a black border.

This is produced by flashing a narrow edging all around the exposed print. The centre of the print is shielded with a rectangle of black card, the negative is removed from the enlarger, and the light is switched on for a second or so in order to fog the border. The black card must be in good contact with the paper.

Selected areas of the exposed print may also be darkened by going over them with a small spot of light projected from a suitable pocket torch. The light should be just strong enough to produce blackening in about 5–10 seconds. Throughout the flashing process, the enlarger light is left switched on, with the orange filter in front of the lens, so that the image can be seen on the paper to guide the operation.

See also: ***Latensification; Shading and spot printing; Vignetting.***

FLASH POWDER. Until the introduction of the flash bulb the only self-contained light source for indoor night photography was afforded by burning magnesium, either in the form of ribbon, foil, or more generally, powder. The first two are obsolete, as is the older form of magnesium powder. This was used in a special flash lamp burning methylated spirit, the powder being blown into the flame by air pressure from a bulb.

In its later form the magnesium powder is mixed with chemicals which make it easier to burn.

Powder of this type is used in a flash tray in which it is ignited by a spark generated by a wheel and flint. It can also be ignited by touch paper and an ordinary match, by plunging a lighted taper fixed to a stick into the heap of powder, or by an electrically-heated element.

Flash powder has largely been superseded by the modern flash bulb, and the electronic flash unit, but there are occasions when its use offers advantages—e.g., when the illumination is needed for a much larger area than that covered by the largest flash bulb. Greater illumination can be provided by a big quantity of powder spread in a trail instead of a heap. Used for this purpose, it is rather cheaper than the large-size flash bulbs.

Commercial flash powders are supplied in double containers. One holds the magnesium powder, the other the igniting compound. In the interests of safety these two powders must be kept apart in their containers until required for use.

After they are mixed, great care must be taken not to allow the powder to become damp or it may form lumps. These must be removed before using or there will be a risk of their jumping from the tray and causing damage or injury.

Flash powder must be carefully handled or it may produce severe burns and cause damage, and even start a fire. In some cases its use is prohibited.

Finally, it is not suitable to be used in connexion with the flash synchronized shutters.

Safety Precautions. Flash powder must be treated like an explosive. Never pour the powder directly from the container to the tray of the flash lamp. First place upon a sheet of paper, thence on the tray, and replace the lid or cork in the container immediately.

Should the powder fail to fire on the tray, pour it back on to the paper before investigating the cause. If the touch paper seems to have failed, do not approach or disturb it until it really has gone out. Keep the tray well away from curtains and furniture, and remove or cover anything valuable.

Ignition. The flash lamp is necessary to fire flash at a precise instant; otherwise a large tray of metal will do on which the required quantity of powder can be spread. The exposure is made by placing the touch paper supplied in the powder and touching it with a lighted taper on a stick.

When a large quantity of the powder is to be fired, it may be divided into a number of smaller flashes at different points. This is rather slow, but with interiors speed is not important.

Quantity. The quantity of powder necessary depends upon the distance from the subject, the tone of the walls, whether light or dark, the make of powder, and the speed of the film. As a rough guide to form the basis for further trials, using a fast panchromatic film, with the subject at 12 feet, and medium-light walls, 20 to 30 grains of the powder will be sufficient.

Flash Capsule. Flash powder also used to be

supplied in a capsule form which enabled it to be used as easily as a flash bulb. The capsule contained a measured quantity of smokeless flash powder formed around a wire filament connected to a pair of contacts which could be plugged into a cap for inserting in a flash holder in the same way as a bulb. When the firing battery was switched on, the wire heated up and fired the powder. R.M.F.

See also: *Flash technique; Light sources.*

FLASH SYNCHRONIZATION

There are two basic methods of taking a photograph by flashlight: by firing the flash while the shutter is open, and by opening the shutter while the flash is alight.

Open Flash. The first method is the one practised with flash powder illumination; it is also used with flash bulbs, when it is called open flash exposure. With this method all the light of the flash is used in making the exposure, but, as the shutter has to remain open for a period that will include the whole time of ignition, it is not normally suitable for fast exposures. A charge of flash powder, for example, might take several seconds to burn, and the shutter would have to remain open for the whole of that time. A flash of this type can be used only with still subjects.

Flash bulbs have a much briefer period of ignition, so the shutter only needs to be open for a shorter time.

The open flash technique can still be adopted—i.e., the shutter is opened as for a brief time exposure, the flash is fired by hand, and the shutter is closed. It is possible in this way to leave the shutter open for as little as half a second. Even so, the method is only suitable for still subjects.

Synchronized Flash. The second and the modern way of taking flash photographs is to synchronize the firing of the flash with the shutter movement in such a way that the maximum brilliance of the flash coincides with the instant when the shutter of the camera is fully open at an instantaneous shutter speed. This synchronization of the flash with the shutter must allow both for the time the shutter takes to open and for the time the flash takes to reach its full brightness or peak. It can be achieved in several ways:

(1) By means of a flash firing contact built into the shutter.

(2) By means of a mechanical synchronizer coupled to the shutter.

(3) By means of an electro-magnetic shutte release connected to the flash circuit (magnetic synchronizer).

Synchronized Diaphragm Shutters. In a synchronized diaphragm shutter—i.e., all between-lens shutters and most behind-the-lens shutters other than focal plane types—a flash contact connected to a moving part of the mechanism closes the firing circuit of the flash at the right moment. This may be either when the shutter blades are fully or nearly fully open, or some time (usually about 1/60 second) before they begin to open. In the latter case a delayed action release mechanism may be incorporated to delay the shutter opening for a time corresponding to the firing delay of the bulb. There are thus several kinds of synchronization:

(1) X-synchronization fires the flash at the instant the shutter blades are fully open. The flash contact is generally a simple make-and-break contact connected to the ring or lever actuating the blades. X-synchronization is suitable for electronic flash which has no firing delay, i.e., where the flash reaches its peak almost the instant the firing circuit closes. Any shutter speed up to the fastest (1/300 or 1/500 second) can be used.

Class M, and S flash bulbs can also be synchronized with an X-synchronized shutter, provided the shutter speed is slow enough to cover the firing delay of these flash bulbs, as well as an appreciable part of the half-peak duration. That is usually a shutter speed of 1/25 to 1/30 second.

(2) F-synchronization closes the firing circuit

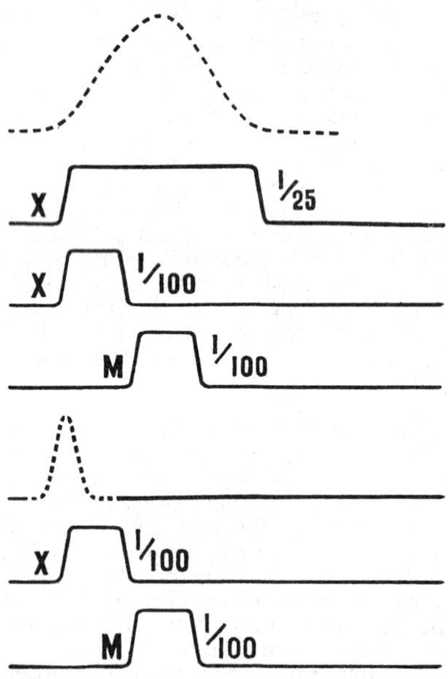

SYNCHRONIZATION. Top: A class M bulb (dotted curve) can be synchronized with an X type shutter at 1/25 second to catch all the light. At 1/100 second the shutter closes too early, but with M synchronization any speed captures the peak of the flash. Bottom: With electronic flash (dotted curve) only X-synchronization is possible; when set to M the shutter only begins to open after the flash is finished.

when the shutter blades are half open, i.e., about seven to eight milliseconds (1/125 to 1/150 second) before they are fully open. The flash contact is of the same type as that of X-synchronized shutters. F-synchronization used to be intended for the now obsolete Class F flash bulbs as the peak brilliance of the bulb coincided with the full opening of the shutter. Class M and S bulbs can still only be synchronized at slow shutter speeds. F-synchronization can also be used with the obsolete relay-fired electronic flash units which have a firing delay of about eight milliseconds due to the operation of the relay switch, but normal (Class X) electronic flash cannot usually be synchronized.

This type of synchronization was at one time provided on diaphragm shutters and is still built into some of the cheaper shutters, as found on box cameras, where the shutter takes fairly long to open (irrespective of the time it stays open).

Both X and F-synchronization are based on the fact that the exposure time is longer than the flash duration. The shutter speed therefore plays no part in exposure determination unless the flash is mixed with other light such as daylight.

(3) M-synchronization closes the firing circuit about sixteen to eighteen milliseconds (roughly 1/60 second) before the shutter is fully open—well before the blades start to move. The flash contact is closed by the release mechanism, but an escapement (generally shared with the mechanism of the self-timer) delays the opening of the blades for the required time.

This synchronization is suitable for class M flash bulbs (which have a half-peak delay of 16 milliseconds) used at all shutter speeds up to the fastest (1/300 to 1/500 second). Class S bulbs still need slow shutter speeds, while M-synchronization is not suitable at all for electronic flash.

With M-synchronization the exposure time is shorter than the flash duration, which is why the shutter opening must be timed so that the blades open and close while the flash is at its peak. The faster the shutter speed, the less of the light is utilized, and exposure determination (e.g., by guide numbers) must therefore take the speed into account.

M-synchronized shutters usually also incorporate X synchronization. They are often referred to as speed-synchronized or (incorrectly) as fully synchronized.

(4) XM-synchronized shutters have two flash contacts built in, one for X and one for M-synchronization. A synchronizing lever selects the type of synchronization required, or the shutter may have two separate flash outlets. Many precision shutters are of this kind and are thus equipped to tackle a wide variety of flash subjects.

(5) MF and XMF-synchronized shutters incorporate the different types of synchronization indicated by the respective letters. They are similar to XM-synchronized shutters, but rare nowadays as class F bulbs are no longer made.

(6) Full synchronization denotes that the delay of the synchronizing mechanism is continuously adjustable instead of being limited to two (or three) fixed settings. The term is also used incorrectly for XM-synchronized shutters.

The adjustment of fully synchronized shutters may range from 0 to 25 or even 30 milliseconds, thus covering even Class S flash bulbs. In practice full synchronization is of no special advantage any longer, since virtually all flash bulbs have standardized firing delays according to their class.

SHUTTER SPEEDS AND SYNCHRONIZATION

Flash Class	*Setting on Diaphragm Shutters*		*Focal Plane Shutters*	
	X	*M*	*Simult. Release*	*Delayed Release*
X	1/500*	—	1/60†	—
F	1/100	—	1/30‡	—
M (small)	1/30	1/500*	1/30‡	—
M (large)	1/30	1/500*	1/30‡	1/60†
S	1/15	1/30	1/10	1/125
FP	1/30	1/500*	1/30‡	1/1000*

The speeds are the top speeds that can be synchronized with the given settings and flash bulbs. Older cameras may have settings of 1/25, 1/50, etc. second instead of 1/30, 1/60, etc.

*or fastest speed of shutter.

†1/30 second on larger cameras.

‡1/15 second on larger cameras.

Synchronized Focal Plane Shutters. In a focal plane shutter the synchronizing contact may be closed by a cam on the blind roller.

The problem of focal plane shutter synchronization is somewhat different from that of diaphragm shutter synchronization because the shutter exposes the negative in successive strips instead of all at once. For synchronization at high shutter speeds the peak of the flash must coincide with the moment when the blind begins to uncover the film, and last long enough to cover the full time it takes the blind to travel across the focal plane. This time is usually 20–50 milliseconds irrespective of the shutter speed set. The focal plane shutters of miniature cameras take the shorter time quoted; shutters of larger cameras—e.g., press cameras —take longer.

For that reason special long-flash bulbs (Class FP) are used for focal plane shutter synchronization. Bulbs of this type have a half-peak duration of 20 milliseconds or more, and the peak itself is comparatively flat, yielding an even level of light throughout the flash.

Class M and S bulbs, even with their long half-peak duration, are less suitable for this purpose as the flash changes appreciably in intensity above the half peak, and the negative

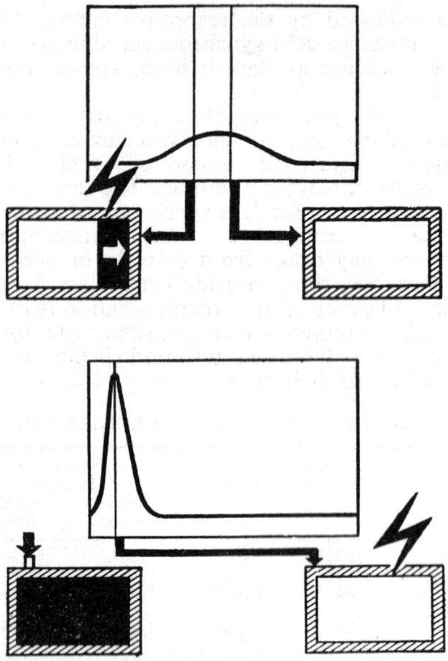

FOCAL PLANE SYNCHRONIZATION. Top: *An ordinary flash bulb must be fired just before the shutter is open so that the full opening coincides with the peak.* Bottom: *An electronic flash is fired only when the shutter is fully open. In both cases a slow shutter speed must be used which simultaneously uncovers the whole frame; long peak bulbs permit fast speeds as the flash lasts while the shutter runs down.*

is not exposed evenly. Generally one or both ends, corresponding to the beginning and end of the travel of the blind, are under-exposed as compared with the centre.

All types of flash other than FP bulbs are therefore best synchronized with a shutter speed at which the blinds are momentarily open to the full size of the negative. This is generally 1/50 second with miniature cameras, and 1/20 or 1/25 second with larger types. At higher speeds than this, the blinds close up to form a slit which travels across the negative and exposes it piecemeal.

Some focal plane shutters incorporate two contacts wired in series, both of which must be closed before the flash can be fired. One contact is the normal synchronizing contact, while the other is connected to the release button. This prevents the firing circuit from firing any bulbs in the flash gun while the shutter is being tensioned. This release contact is additional to any further synchronizing contacts that may be built in, and works in series with any of them.

As with diaphragm shutters, there are several kinds of synchronization for focal plane shutters.

(1) Simultaneous release fires the flash when the first blind has reached the end of its travel. It corresponds to X-synchronization on diaphragm shutters. Provided the set speed is 1/60 second (often marked by a special symbol), or 1/15 to 1/30 second, or slower, with larger cameras, the negative is fully uncovered at this instant. This kind of synchronization is thus suitable for electronic flash at these speeds, or for most types of flash bulbs (all classes) at the next slower speed (1/30 or 1/15 second). The slower speed is required with bulbs to allow for the firing delay.

(2) Delayed release closes the firing circuit about sixteen to twenty milliseconds before the first blind begins to uncover the film. This corresponds to M-synchronization with between-lens shutters, but is suitable only for class FP bulbs with which any shutter speed can be used. Class M bulbs, even at slow speeds, cannot be synchronized satisfactorily since the flash is over before the first blind has finished travelling. Some class S bulbs can be synchronized under these conditions, but again the exposure may be uneven over the negative, especially with larger cameras.

On certain cameras the delayed release closes the firing circuit 18 milliseconds before the first blind has reached the end of its travel. This limits the use of FP bulbs to slow shutter speeds, but permits synchronization of class M bulbs at intermediate speeds (up to about 1/100 second on miniature cameras).

Electronic flash (class X) cannot be synchronized with a delayed release contact.

(3) Multiple synchronized focal plane shutters incorporate both simultaneous and delayed release contacts, analogous to an XM-synchronized diaphragm shutter. Generally, however, no synchronizing lever is provided, and the two contacts are wired to separate outlet sockets. Sometimes the shutter speed dial is coupled with the synchronization so that it brings in the simultaneous contact at speeds of 1/50 second and slower, and the delayed release contact at faster speeds.

(4) Fully synchronized focal plane shutters have a dial or similar device for setting the firing delay. The figures on the dial do not necessarily denote delay times.

Wiring of the Shutter. The flash contacts inside the shutter are wired through to a suitable flash socket on the shutter housing or on the camera body. In most cases one side of the circuit is earthed to the metal parts of the shutter, while the other is insulated from it. It is therefore important to preserve the correct polarity when connecting the flash plug to the shutter. This is generally achieved by the use of the now standard co-axial flash socket; the outer contact is earthed, and the inner contact is insulated.

With a symmetrical two-pin socket both contacts must be insulated. Alternatively, the two pins may be different in size or arranged so that the plug can only be inserted one way round, preventing a short circuit which would prematurely fire the flash. Unsymmetrical three-pin sockets serve the same purpose, the third pin acting as a locating device.

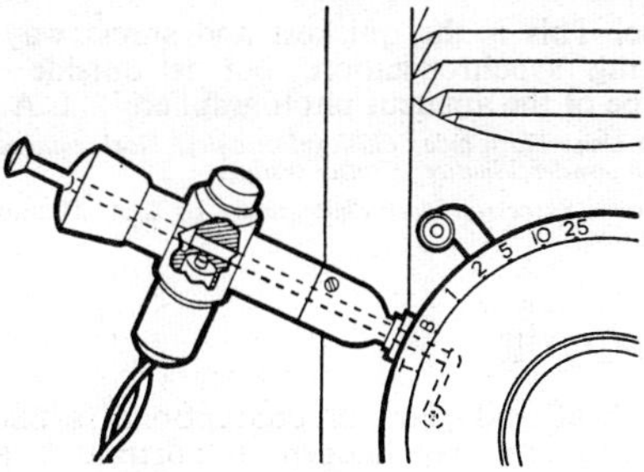

MECHANICAL SYNCHRONIZER. *A contact inside the plunger unit closes the firing circuit as the shutter opens.*

Mechanical Synchronizers. Older shutters without flash contacts can be synchronized by means of an external synchronizer. The simplest form is the mechanical synchronizer which makes the contact inside an attachment which screws into the cable release socket or is linked with the release button of the camera.

The firing circuit is closed during the movement of the plunger or button of the attachment as it releases the shutter. A simple adjustment moves the point of contact to adapt the synchronizer to different shutters.

Most mechanical synchronizers are of the simultaneous release type, corresponding to X or F-synchronization. Delayed release is difficult to time reliably.

Magnetic Synchronizers. These release the shutter by the action of a solenoid which operates the release lever. The energizing current for the solenoid also fires the flash.

The solenoid is a coil of insulated wire surrounding a soft iron core. When a current passes through the wire, the coil becomes a powerful magnet which can be made to attract a suitable iron armature connected to the shutter release. Passing a current through the solenoid therefore operates the release and trips the shutter.

The solenoid itself simply operates an electrically controlled remote shutter release. To turn it into a magnetic synchronizer, a flash circuit is wired in parallel with it, and powered

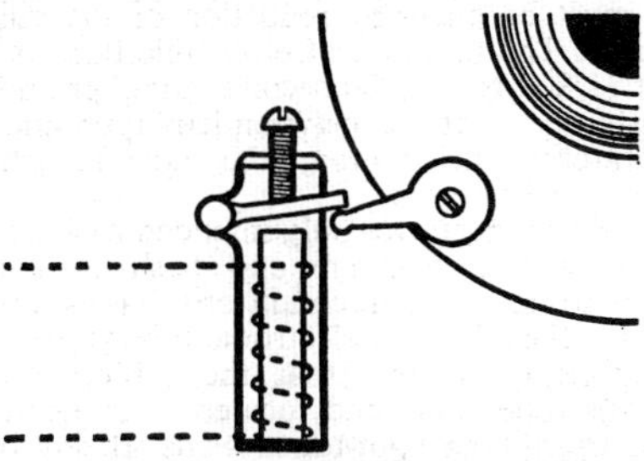

MAGNETIC SYNCHRONIZER. *The firing current for the flash actuates the armature of the magnet and thus trips the shutter. A set screw adjusts the timing and controls synchronization.*

from the same battery. Closing the circuit therefore fires the flash as well as releasing the shutter.

Magnetic synchronizers are always delayed release synchronizers, since the firing circuit must close before the shutter can be released. The delay depends on the distance the armature has to travel before it trips the shutter, and is usually adjustable.

Magnetic synchronizers need a fairly strong current for their operation, and the batteries used must be capable of giving at least three amperes.

Since the solenoid depends on a steady flow of current, it is not normally suitable for operation from a battery-capacitor unit, which only supplies a current pulse.

Testing Synchronization. Testing methods fall into two groups: visual and photographic.

Visual methods depend on observing the flash or a suitable test lamp through the shutter —or next to it—during release.

The simplest method works well with X-synchronized diaphragm shutters using an electronic flash unit. Direct the flash at a light wall and point the camera at the same wall. Open the camera back and release the shutter (with electronic flash connected to the synchronizing outlet).

If the synchronization is correct, the shutter appears fully open when the flash fires, a full size bright circle being visible through the camera back.

Incorrect synchronization will show the silhouette of the partly open shutter blades against the bright circle—or no bright circle at all. The lens aperture should, of course, be wide open.

The same system works with focal plane shutters, but here incorrect synchronization results in only part of the negative area appearing illuminated.

Photographic methods involve directly or indirectly photographing the flash. With diaphragm shutters, a direct photograph of the flash bulb, taken at a small aperture and synchronized with the shutter, will show the filament or charge in the process of combustion. The exact stage of this combustion in the picture will indicate whether the flash is early or late in firing.

With focal plane shutters, a picture of the flash itself is not so useful, as it is at a different stage in different parts of the negative. A better arrangement is to photograph a white area illuminated by the flash, so that it fills the negative.

If the exposed area appears uniformly lit, the synchronization is in order. If one or the other end of the negative appears lighter than the other, synchronization is incorrect. The extent and position of the lighter area, relative to the movement of the shutter, then indicates whether the flash is early or late, and by approximately how much.

A falling off in density at both ends of the negative indicates that the flash was too short for the shutter speed employed.

Professional testing methods involve the use of oscillograph equipment, and record the closing of the flash contact and the shutter opening as traces on the screen of a cathode ray tube. This is the quickest and surest way of testing synchronization, but is outside the scope of the amateur photographer. L.A.M.

See also: *Flash bulb; Flash (electronic); Flash equipment; Flash powder; Shutter; Shutter testing.*

Book: *Successful Flash Photography*, by L. A. Mannheim (London).

FLASH TECHNIQUE

Modern flash equipment has greatly increased the scope of photography under almost all conditions. The flash bulb and the electronic flash unit will not only get pictures in the dark; they can be used in conjunction with numerous other light sources, particularly with daylight. Thus in weak light flash enables instantaneous exposures to be used at small lens apertures, thus giving greater depth of field than would otherwise be possible. Flash can also be used to simulate sunlight on a dull day. In strong sunlight flash can be used to lighten the shadow areas of the subject and so even up excessive contrasts such as are beyond the range of colour film.

Flash can be used to stop fast movement, and moving subjects can be brought within the scope of slower films, particularly colour emulsions. The fastest movement can be completely arrested by electronic flash.

Subjects which are disturbed or harmed by the heat and glare of conventional artificial lighting, as may happen in portraiture and when photographing flowers and insects, can safely be photographed with flash.

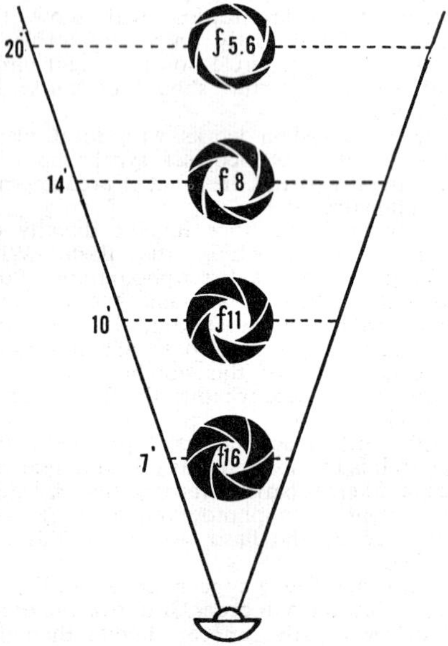

GUIDE NUMBERS. The flash distance and f-number for correct exposure are inversely proportional, and for any one film and flash bulb their product is constant. This product is the guide number and serves for quick exposure calculation. Correct aperture is guide number divided by distance. (Guide number is here 110, so at 10 feet aperture should be f 11.)

Guide Numbers. Exposure with flash is usually determined by means of the guide numbers quoted by flash bulb and electronic flash unit manufacturers. These numbers are so computed as to take account of the power of the flash and the speed of the film, and are the product of the lens aperture (*f* value) and the lamp-to-subject distance in feet. To find the correct lens aperture it is therefore only necessary to divide the guide number of the bulb in use by the known lamp-to-subject distance. Conversely, the correct lamp-to-subject distance for a given lens aperture can be ascertained by dividing the guide number by the known aperture.

The guide number system takes into account the fact that the amount of light reaching the subject varies inversely as the square of the distance from the light source. When altering the flash-to-subject distance it is therefore important to remember that, if this distance is doubled, only one quarter of the light reaches the camera; and if it is halved, the amount of light is increased four times. In the former case, the lens aperture will have to be opened up two whole stops, and in the latter, it must be stopped down by the same amount.

The normal guide numbers apply only in cases where the camera shutter remains open throughout the duration of the flash, which with flash bulbs varies between 1/25 and 1/200 second. If the shutter only remains open during part of the flash, less light is able to affect the film during the exposure, and allowance for this must be made by reduction of the guide number. The duration of electronic flash is so brief (1/800 to 1/2,000 second) that, provided the shutter is of the between-lens type and is accurately synchronized, the normal guide number can be used.

Guide numbers are only intended as a guide to the correct aperture and flash-to-subject distance under average conditions. It is assumed that the flash bulb or electronic tube is used in an efficient reflector, that the subject is of average tone value and located in a medium sized room with light-toned walls. If any one of these assumptions does not apply, it may be necessary to open or close the lens aperture by one whole stop or more. With colour film (and particularly with reversal emulsions), which has much less exposure latitude, it is advisable to aim at correct exposure to within half a stop either way.

Camera Guide Number Scales. Some cameras carry an additional scale, used in conjunction with the distance or aperture scale, to indicate correct flash exposures.

There are two systems. The first is a simple scale of guide numbers, mounted on the aperture ring or on a ring next to the aperture scale. On setting the guide number to be used against the distance (either on the focusing ring or on a secondary distance scale), an index shows the corresponding required lens aperture. Alternatively, the index may have to be set against the distance and the aperture read off for each guide number.

The second type found on certain cameras goes one step further to make the procedure semi-automatic. Here the act of setting the guide number on the scale couples the aperture control with the focusing movement. Focusing on a closer distance automatically closes down the lens aperture, focusing on more distant subjects opens up the aperture. In this case the focusing movement is also restricted to a certain range which corresponds to the distances over which the aperture range of the camera can yield correct exposures with the selected guide number. Thus with a guide number of 220 and a smallest stop of *f*/22, 10 feet is the nearest distance at which a correct exposure is possible. So when set for this method of operation with a guide number of 220, the camera will not focus down below 10 feet.

Adjusting Exposure. If use of the normal guide numbers results in consistently wrongly exposed results, the numbers can be converted according to the following table, which provides adjustment of exposure between two times and one half of that resulting from direct calculation.

When a second flash as a fill-in light is placed appreciably farther away from the subject than the key light, the guide number of the latter can remain unchanged for purposes of calculating the exposure. But if two flash bulbs are used in the same position, the guide number should be multiplied by 1·4. When three lamps

ADJUSTMENT OF EXPOSURE

Change of lens aperture:	*Multiply Guide Number by:*
1 stop larger	·70
½ stop larger	·84
1 stop smaller	1·4
½ stop smaller	1·2

of the same type are used, multiply the guide number by 1·7, and with four, multiply by 2.

As a general rule, where two flashes of different strength are used at different distances, the exposure can be worked out as follows. Find the aperture required separately for each flash from the guide number and the distance. If the resulting apertures are the same, use one lens stop smaller. If the apertures differ by ½ to 1½ lens stops, use half a stop smaller than the smaller aperture (higher *f*-number). If the difference is more than two lens stops, use the smaller stop for the exposure and ignore the larger one.

With three flashes combine the exposure for the two more powerful ones in the way just described, and compare the resulting stop with the lens stop for the third flash. Then combine again in the same way. With four or more flashes proceed similarly; usually the effect of these additional flashes—if weaker than the first three flashes—can be ignored.

As it is impossible to judge lighting effects with flash in advance, Photoflood lamps can sometimes be used to advantage for this purpose. If an exposure meter reading is taken of the experimental set-up, it is then possible by trial and error to work out the relative power of the two forms of lighting and so calculate the flash exposure by this means. It is advisable to keep a record of each lighting set-up including the exact position of the lamps and their distance from the subject. This makes it easy to reproduce the same conditions at will and so ensure correct exposure, which is particularly critical with colour film.

There are special exposure meters that can read flash exposures. With electronic flash, taking a reading involves setting up all the flash units and firing them in synchronism while taking an exposure reading. With flash bulbs this is however expensive, since it means wasting one set of bulbs for the exposure reading, and then using a further set for the actual exposure.

Flash at Camera. The simplest and handiest position for a single flash gun is at the camera, to which it can be secured by a bracket. This is an excellent position when flash is used as auxiliary lighting to soften shadows in daylight, but has certain disadvantages when flash is the sole light source. Results are somewhat

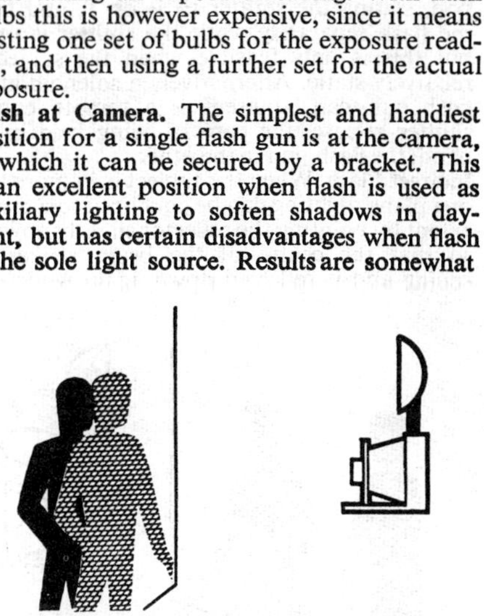

FLASH ON THE CAMERA. The easiest way of taking flash shots, as both camera and flash holder are firmly connected and freely mobile. However, it yields flat lighting and a heavy shadow close behind the subject.

USING A FILL-IN REFLECTOR. A white reflecting screen on the shadow side of the subject will lighten and soften the hard shadow cast by a single flash bulb.

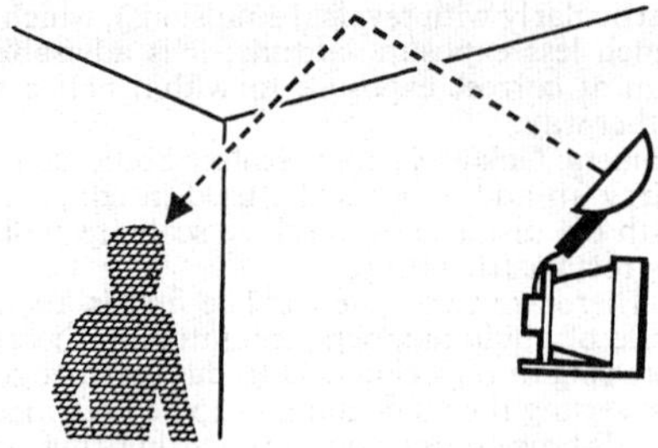

BOUNCE FLASH. The flash is directed at the ceiling and walls so that the subject is lit by soft diffused light, without harsh shadows. The exposure must allow for lower light.

flat and lacking in modelling, a sharp black shadow is cast behind the subject, whose eyes appear with unnaturally light pupils, due to an internal reflection in the retina. With colour film the same reflection causes the eyes to appear red.

Modelling can be improved by placing a reflecting screen to one side of the subject, using for example a white card or sheet or a board covered with silver paper, placed in such a position that the flash is reflected on to the shadow side of the subject.

Flash Remote from Camera. The drawbacks mentioned above are eliminated if the flash can be placed slightly above and to one side of the camera.

If the flash lead is not long enough for the purpose, it may be possible to use the open flash technique, securing the camera to a tripod, setting the shutter to B or T, and firing the flash separately while the shutter is open; but this is only feasible when the subject is relatively static. Alternatively a solenoid at the flash gun can be used as a remote control shutter release, the camera being held in one hand, the flash gun in the other.

Indirect Flash. When the subject is in more than one plane, lighting can be equalized to a certain extent by pointing the reflector slightly upwards so that the centre of the beam strikes the ceiling and is reflected down again while only the edge of the beam strikes the foreground of the subject directly. This method calls for a lens aperture two stops larger than normal. If there is a considerable distance between foreground and background, the reflector should be turned right away from the former and directed straight towards the latter. The lens aperture in this case should be calculated for the background.

Flash can also be directed away from the subject altogether and aimed at light walls and ceiling, so as to bounce back to the subject—hence the term bounce flash. This method gives an agreeably diffused result, but with colour film it can obviously only be used when the walls and ceiling are dead white. The lens aperture must be two to three stops larger than when direct flash is used.

Flash Without Reflector. Very natural lighting effects can be obtained in interiors by dispensing altogether with the reflector. In small light-coloured rooms the lens aperture required will be one stop larger than normal; bigger rooms call for one and a half stops more, and dark-toned rooms two and a half stops. Never use flash bulbs at the camera without reflector; serious injury may result in the event of a bulb bursting.

In general, when working with only one flash bulb, make as much use as possible of the reflecting power of walls to lighten shadows.

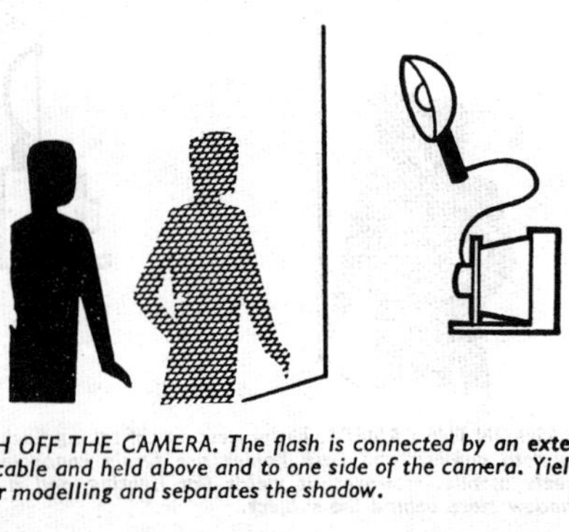

FLASH OFF THE CAMERA. The flash is connected by an extension cable and held above and to one side of the camera. Yields better modelling and separates the shadow.

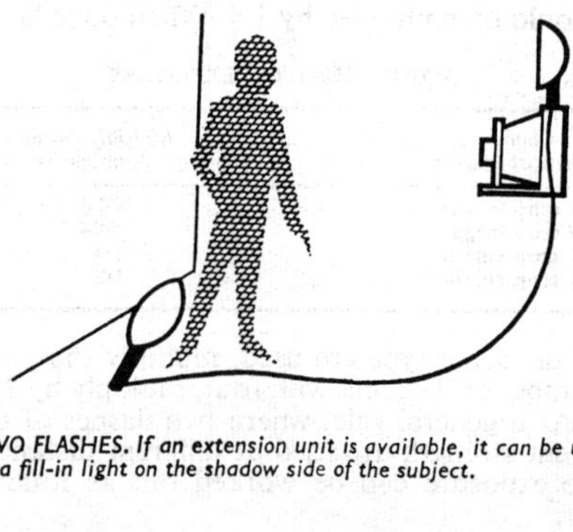

TWO FLASHES. If an extension unit is available, it can be used as a fill-in light on the shadow side of the subject.

Take care that the image of the bulb itself is not reflected back to the lens by mirrors, windows or other reflecting surfaces.

Two Flashes. The quality of results is greatly improved if more than one flash unit is available. Though the outfit becomes more cumbersome a second flash-gun is a real advantage, even in action and press photography. It is often possible to persuade a bystander to hold the second gun; or it can be fixed in position with a universal clamp.

With a single flash gun the range of contrast between highlight and shadow areas of the subject may well be beyond the range of the paper on which the negative is printed. With a second flash, we immediately have full control of contrast. The normal procedure when using two flash bulbs of equal power is to have one flash gun remote from the camera but nearer the subject as main light, and the second flash-gun at the camera as a fill-in. In this way the fill-in flash softens the shadows cast by the main light without casting any of its own. Alternatively the relative power of the flashes can be adjusted by using different types of bulb for main light and fill-in.

Placing Fill-in Flash. Contrast adjustment will normally be effected by moving the main flash relative to the subject, while the camera with its fill-in flash remains stationary. As the light reaching the subject varies as the square of the distance between it and the flash bulb, a convenient way of memorizing the appropriate flash-to-subject distances for given contrast ratios is to refer to the *f*-numbers engraved on the camera lens diaphragm. According to the calibration, these will either read 4, 5·6, 8, 11, 16 and 22, or 3·5, 4·5, 6·3, 9, 12·5 and 18. As is well known, each number represents an aperture admitting half the amount of light of its left-hand and double that of its right-hand neighbour. If an illumination ratio of 2 to 1 is required and the camera and fill-in flash are placed at 11 feet from the subject, the main flash gun should be placed at 8 feet. If a ratio of 3 to 1 is desired, respective distances should be 11 and 6·3 feet; and for a ratio of 4 to 1, 11 and 5·6 feet.

Three or More Flashes. When using more than two flashes, the best method of adjusting the lighting is to replace the flash bulbs temporarily by Photofloods in appropriate light sockets and reflectors and judge the effect visually, calculating the exposure by the method already described.

Though it might appear less complicated to dispense with flash altogether and make the exposure by Photoflood light, actually flash offers very definite advantages which outweigh the extra trouble involved. First and foremost, the much higher light output of flash bulbs makes it possible to use smaller apertures and thus obtain greater depth of field. Then the ability to synchronize the flash to the instantaneous shutter speeds makes it unnecessary for models to remain motionless, which is an obvious advantage when children or animals are involved, and results in much more natural and relaxed poses. For colour portraiture, with the comparatively slow films at present available, flash is almost indispensable to those who do not have access to an elaborately equipped professional studio.

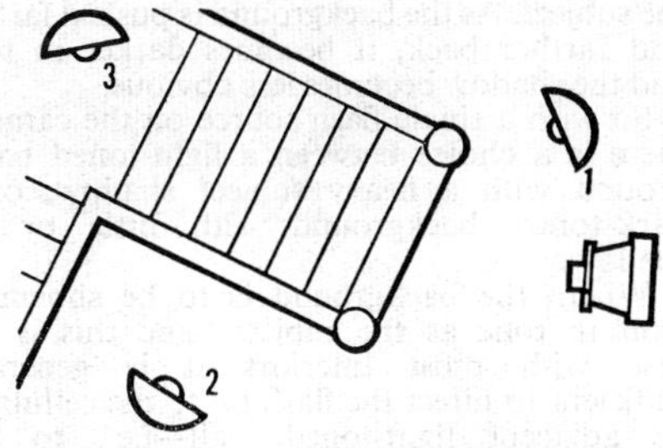

MULTIPLE FLASH. One flash holder can illuminate a large room if the camera is set up on a tripod, the shutter opened for a time exposure, and the flashes fired in succession from various points. Avoid directing any flash into the lens.

Multiple Successive Flashes. The ability to fire successive flashes with electronic units can be turned to good account with static subjects such as interiors, architecture, and copying.

The flash tube can be recharged and fired from different positions to provide modelling and soften or entirely eliminate shadows. Unless flash is the sole light source it is necessary to close the camera shutter during the recharging period. Flash bulbs can also be fired in succession, but if a large number are required, the cost becomes prohibitive.

Flash and the Background. In the simplest case of a single flash on the camera, the subject casts a shadow on the background. The shadow is just visible from the camera position as a black border around the half of the subject on the opposite side of the camera to the flash—or below it where the flash is mounted immediately above the lens. The closer the subject is placed to the background, the more obvious this fringe of shadow becomes. This is because the background receives almost as much light as

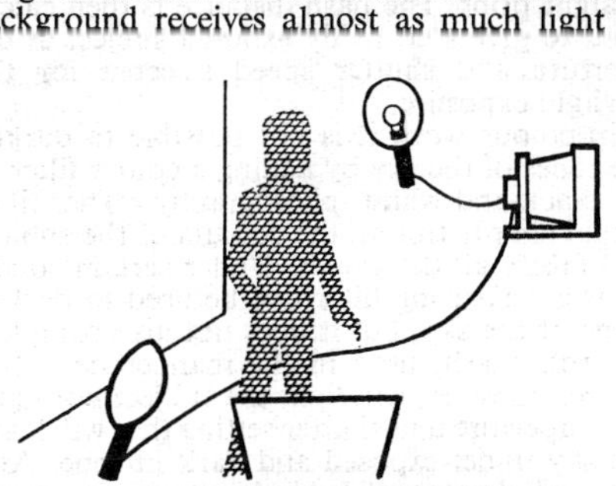

EXTENSION FLASH WITH REFLECTING SCREEN. With a reflecting screen used to fill in harsh shadows, the extension flash can be used as an effect light on the background.

the subject. As the background is pushed farther and farther back, it becomes darker in tone and the shadow becomes less obvious.

So with a single flash source on the camera, there is a choice between a light-toned background with a heavy subject shadow, or a dark-toned background with little or no shadow.

Where the background is to be shown as light in tone as the subject—and this is the case with most interiors—it is generally sufficient to direct the flash on to the ceiling or an adjacent light-toned wall—i.e., to use "bounce" flash technique. The flash is then fairly evenly distributed over the subject and the background (so long as they are only a few feet apart) and no hard shadows will appear anywhere. The drawback here is that the lighting is often too soft; it lacks depth and relief and gives no control of the relative tones of subject and background. It is satisfactory, however, for subjects of general interest where the subject is no more important than its surroundings.

For greater control it is necessary to use a separate flash—or Photoflood lighting—on the background. Illuminated in this way, the background can be made to reproduce as light in tone as wished by bringing the background light closer. The background illumination may be directed from immediately behind the subject and concentrated on the area where the shadow will fall from the principal light mounted on the camera. Or the principal light may be moved up and to the side so that it no longer casts a shadow on the background. (In this case a fill-in flash or white surfaced reflector will have to be used to light up the shadow side of the subject.)

Out of doors, supplementary flash lighting may be used for the reverse effect—i.e., to brighten the tone of the subject in relation to its background. This technique is often useful when the subject is dark in tone and is seen against a sunlit background. The normal camera exposure is adjusted so that the background will be under-exposed and come out fairly dark in a straight print. The flash distance is then calculated to give a correctly exposed subject at the aperture and shutter speed selected for the daylight exposure.

In colour work it is not possible to darken the tones of the sky by adding a colour filter as in black-and-white photography. The filter would simply tint all the colours of the subject and falsify all the values. Under certain conditions a polarizing filter can be used to darken some of the sky, but it does not give complete control. Flash, used in the manner described above, however, will light the subject normally at an aperture and shutter setting that will leave the sky under-exposed and dark in tone. And although the sky will be darker than in a normal exposure made without flash, its colour and the colours of the rest of the subject will remain reasonably the same. With flash bulbs it will be necessary to use a suitable blue filter over the lamp or a special blue-dipped daylight bulb to correct the colour temperature. No filter is required with electronic flash since this has approximately the same colour temperature as daylight.

For colour shots with flash, colour films designed for short exposures must be used rather than the special types produced for long exposure times in the studio. The use of the long type of colour film—even with correction filters—may lead to loss of speed and changes in colour rendering, owing to reciprocity-failure effects.

Flash and Daylight. Flash can be used as an auxiliary light source to soften hard shadows resulting from exposures in bright sunlight. This is particularly useful with colour film, which can only record a limited contrast range. The technique is often known as synchro-sunlight.

When calculating the correct balance between daylight and fill-in flash it should be remembered that with flash bulbs, the slower instantaneous shutter speeds (e.g., 1/25 second) when the flash duration is shorter than the shutter opening, have no effect on the power of the flash. The full flash intensity is only reduced when the shutter speed is faster than the flash. When this takes place depends on the flash duration, which varies according to the type of bulb from 1/25 to 1/200 second. The duration of electronic flash discharge is always shorter than the fastest shutter speed with which it can be synchronized.

Generally therefore the shutter speed only influences the amount of daylight reaching the film, and not the flash intensity. Thus an exposure of 1/50 second will admit only half the daylight admitted by a 1/25 second exposure, without reducing the flash intensity.

Balancing Daylight and Flash. The desired balance between daylight and fill-in flash can therefore be obtained by varying either the shutter speed (affecting the daylight), or the flash-to-subject distance (affecting the flash intensity.

SYNCHRO-SUNLIGHT. A synchronized flash can serve as fill-in light in outdoor shots against the sun. The exposure must be carefully balanced not to overlight the shadows.

The flash intensity can also be altered by using a bulb of different power or by placing a piece of cloth over the bulb. (With colour film, such cloth must be dead white so that it does not affect colours.)

The normal procedure in practice is to set shutter speed and lens aperture as for an exposure to daylight only. Next divide the guide number of the flash bulb to be used by the lens aperture already selected. The resulting figure gives the flash-to-subject distance at which daylight and flash will illuminate the subject with equal intensity. If a relative intensity of 4 to 1 between daylight and flash is desired, the flash-to-subject distance must be doubled.

Simulating Sunlight. To simulate sunlight in dull weather, or to use flash as an effect light, reduce the flash-to-subject distance by half again, so that you are exposing according to the guide number. The flash then becomes the main light.

When using colour film, balanced for daylight, the flash bulb must be blue-tinted so that it will match the colour temperature of daylight.

Electronic flash, being of approximately the same colour temperature as daylight, is quite suitable for the purpose without modification.

H.A.S.

See also: *Colour technique; Fill-in light; Portraiture outdoors; Stroboscopic flash; Underwater photography.*

Books: *Successful Flash Photography*, by L. A. Mannheim (London); *Synchro-flash Photography*, by G. L. Wakefield and N. W. Smith (London).

FLAT FILM. Type of sensitized film material, also known as cut or sheet film. It is available in normal plate sizes and, to minimize buckling, usually has a thicker base than roll film.

See also: *Films.*

FLATNESS. Term applied to photographic prints exhibiting no clear black or white tones; lack of contrast.

FLATTENING PRINTS. Prints on drying tend to curl towards the emulsion surface. They can be flattened by various methods—e.g., pressure or stretching over the edge of a board in the reverse direction of the curl.

See also: *Drying.*

FLEXICHROME PROCESS. Obsolete commercial process for producing coloured prints from black-and-white negatives. The negative is printed on a special stripping material. This is developed in a tanning developer which hardens the gelatin in the image areas. The unhardened gelatin is then washed off in hot water leaving a swollen relief image.

A bleacher is applied to remove the silver image and the material is soaked in a grey dye which is absorbed by the gelatin matrix in proportion to the thickness of the gelatin and thus in proportion to the original silver deposit. The gelatin layer is then stripped and transferred to a paper support.

Special coloured dyes are next applied to the grey image. These dyes displace the grey dye and leave a colour image which varies in depth of tint according to the amount of gelatin available. In the same way, a second colour may be applied to displace the first.

The process allows the photographer to exercise almost unlimited control over the final result. At the same time, the colours will "take" only in proportion to the thickness of the gelatin image, so that the process remains photographic.

Flexichrome used to be popular for producing colour originals for magazine illustration from ordinary negatives, and was also practised by some portrait photographers and pictorial workers for exhibition.

See also: *Colouring prints.*

FLOODLIGHT. Source of main general lighting. A floodlight may consist of one or more quartz iodine or 500-watt tungsten filament lamps mounted in a white enamelled reflector shaped like a shallow box or trough. The light from such a unit spreads evenly over the subject and does not cast hard shadows.

Floodlights may hang from the ceiling, stand on the floor, or be mounted on an adjustable stand on wheels. In portrait studios the floodlights are generally part of the permanent fittings. For amateur use, a Photoflood in a shallow matt-surfaced reflector makes a useful floodlight.

See also: *Lighting equipment.*

FLOWERS. Photographs of flowers are almost always better when they are taken against a plain background, and for tall flowers like hollyhocks, sunflowers, etc., the best background of all is the blue sky. Sunshine is not essential but it helps to add modelling and texture.

Most flowers require a 2× or 3× yellow or green filter and panchromatic film. The shutter speed must be kept short enough to deal with movement of the subject in the wind.

If the natural background is fussy, the subject should be photographed against a large piece of card, light or dark grey to suit, supported behind the plant or fastened on to it by paper clips. With short-stemmed plants the background card may be pinned between wooden stakes pushed into the ground. The card may be curved behind the chosen flower to protect it from the wind.

The colour of the background should be chosen to contrast with the subject, a light flower being given a darker background and vice versa. It is generally better to take a few flowers well grouped than a large mass of blooms of the same sort. The whole group, including the foremost petals, must be sharply focused.

Backlighting emphasizes the delicacy and transparency of flower petals and foliage, and sidelighting shows up the texture. Over-dark shadow areas should be lightened by holding or supporting a piece of white card or cloth so that it reflects the principal light into the dark parts.

Flowers Indoors. When cut flowers are photographed indoors, much of the appeal depends on careful arrangement and choice of setting. The container should be in keeping with the type of flower; dainty flowers need delicate, fragile containers; large, sturdy blooms look best in a substantial jug or vase. And as the flowers are the important thing, the container should be plain and unobtrusive. For the same reason furniture and other ornaments should be kept out of the picture.

Grouping flowers attractively calls for some skill and flair; it is generally better to avoid symmetrical arrangements and the blooms should not be bunched in the dead centre of the picture or spread out at regular intervals.

Small Flowers. Very small flowers call for true close-up technique. They can be held in position with Plasticine and photographed by artificial lighting but the work must proceed quickly because the flowers wilt when they are out of water and under the warm lighting. Alternatively a sheet of heat-resisting glass can be placed in front of each lamp to minimize the heating effect. The grouping of such small subjects must be kept as nearly in one plane as possible because of the extremely shallow depth of field of the lens at such close distances.

Lighting indoors should be made to look as natural as possible—i.e., it should fall from above and to one side. Often the best lighting arrangement comes from standing the flowers on a window ledge—but not in direct sunshine —and brightening up the shadow side with a reflector.

Equipment. The best camera for flower photography is either a plate camera with back focusing, or a single lens reflex. Both of these cameras have focusing screens on which the picture can be comfortably studied, and can be fitted with long focus lenses which give big pictures without the distortion of a too-close viewpoint. Miniature cameras—particularly those with special close-up magnifying reflex attachments—can also be used. Finally, almost any camera with a supplementary lens of about 1–2 diopters can be made to turn out good flower photographs within its own limitations.

Sensitized Material. There is no room for artistic falsification of tone values in flower photography. All the colours of the subject must be reproduced in their true brilliance or the photograph will look false to anyone who knows anything about flowers. So the subject demands a medium speed panchromatic film with perhaps a 2× green filter.

This means that deep red, blue, or mauve flowers and their foliage will all reproduce in more or less identical shades of grey. The result is that such flowers are lost amongst the foliage. It is of course possible to lighten the colours of the flowers with a suitable filter, but only by falsifying the tone values. If the photographer simply wants to make flower pictures, he should avoid flowers of such deep shades and use lighter coloured varieties.

Colour film is the only material that will do justice to flowers, and there are few better subjects for turning into transparencies. Here, as elsewhere, the photographer must resist the temptation to cram as many highly coloured blooms as possible into the picture. The best effects are undoubtedly achieved in subtle tones of the same colour or combinations of pastel shades. The colours will be true to life only if the subject is given the exact exposure. This means that all exposures should be measured with an accurately calibrated exposure meter. Where there is a mixture of colours in the scene, the exposure may have to be measured within inches of the most important colour to ensure that it is reproduced correctly. P.J.

See also: *Botany; Gardens.*
Book: *How to Shoot Outdoor Colour*, by G. Wells and F. Smith (London).

FLOW PHOTOGRAPHY. Method of studying phenomena of a transient nature which are difficult to observe normally. This covers subjects whose activity is too fast for the eye and brain to perceive as discrete parts of a sequence, or subjects whose rate of change are too slow for the retinal memory to carry an image in the brain long enough to compare with the next impression.

Still Photography of Liquids. Water flow can be recorded and studied with the still camera, to show surface and depth flow, and the effect of water flow on, say, the bed of a river. For this work one might employ candles, electric floats; positive, neutral, or negative buoyancy particles; dyes; or air bubbles. The last, due to their positive buoyancy should be restricted to use in water tunnels where the water flows upward through the working section. Using candle floats (birthday cake), or electric floats (comprising a miniature battery and miniature bulb in a buoyant waterproof case) and an interrupter in front of the camera lens, it is possible to record the direction and velocity of surface flow. For depth flow a deep draught float is needed.

The interrupter could be a five-bladed fan revolving at 1 rev per 10 seconds to give exposures of 1 second every other 1 second. The camera shutter is left open for the appropriate time. The result shows a series of varying length tracks indicating the velocity and direction of the floats over the water. Without the interrupter the result shows various flow patterns.

Under water or through a glass-sided tank, neutral and negative buoyancy particles will show flow stream phenomena between the surface and the disturbed flow associated with the bed of the model or river, etc. Dyes and air bubbles coupled with short or long exposures show dispersion of flow, and limits of disturbances.

Still Photography of Gases. Since gases are compressible it is possible to make use of optical techniques which are sensitive to variations of refractive index in the gas. Schlieren and interferometric systems make it possible to record shock waves and other disturbances associated with gases passing an object at high speed. The sensitivity of a Schlieren system can be varied; e.g. by altering the position of the source or knife-edge or the use of graded glass filters. Colour materials can be used with a multi-colour filter, say red, green and blue, where green is in the middle and gives the picture its background colour. Light rays passing through the system may be bent due to refraction and pass through the red or blue portions indicating high or low densities according to the orientation of the filter.

The interferometric method shows refractive index variations as fringes, and each fringe indicates a specific value.

A multi-spark light source can be used in conjunction with a Schlieren system. The spark sources are arranged in a circle and fired successively. This produces a plate with a series of small images in a circle. The duration of the spark and the interval between the sparks are measured in micro-seconds. Thus it is possible to study a hypersonic shock front for a brief instant.

Long exposures, or many short exposures on one plate, in conjunction with back-illuminated smoke or vapour, show density variations and total limits of pollution from chimneys and ship funnels. They can equally show up the flow of air over car bodies, ships, aircraft, buildings and industrial sites.

Time-lapse Cinematography in Liquids. Time-lapse is one of the few methods of recording phenomena that enable us to analyze and visualize changes occurring too slowly to see. By making the interval between exposures many seconds, minutes, or even hours, the action is speeded up many times on projection at normal speed (16 or 24 frames per second)—e.g. a plant growing from seed to full bloom in three minutes.

To study flow in liquids, the same floats or particles already mentioned may be used. By projecting the film at normal speed it is possible to visualize form, etc. and also to analyze the film frame by frame. This is most easily done with a special analyzer which measures the position of a float against a fixed datum and transfers the information to punched tape. Later the punched tape can set out the information in tabular form, or draw graphs of the float behaviour and flow velocities. It is also at this stage that variables can be fed into the computer to save further mathematics later.

Time-lapse will show the effects of flow, e.g. the drift of a sand-bar due to water flowing over it.

A method of studying flow on the full scale, e.g. currents, etc., inside a large harbour, is to place suitable radar reflectors in the water. These are picked up by the aerial and imaged on the radar screen as dots. A cine camera arranged to take a picture (e.g. one a minute) of the radar screen, might—on projection at normal speed—show in the harbour unsuspected currents, eddies, etc.

Time-lapse Cinematography in Gases. By taking a picture of the sky every few seconds meteorologists have been able to study cloud formations, etc. An automatic iris control fitted to the camera lens coupled to the exposure meter ensures even exposure throughout the film.

Normal and High-speed Cinematography in liquids. Normal and high speed cine are considered together, since it is not unreasonable to film a model at model speed for analysis and to film the same model at a higher speed to simulate the full scale for visualization, e.g. a model boat in waves. The relationship between camera projection speed is given by the equation:

$\sqrt{\text{model scale}} \times$ projection speed f.p.s. = camera speed f.p.s.

For example, for a boat model on a scale of 1 : 64, we have:

$\sqrt{64} = 8 \times 24$ f.p.s. = 192 f.p.s.

Thus a film shot at 192 f.p.s. will give the impression of a full size ship at sea if it is projected at 24 f.p.s.

Suitable markers (e.g. lighted torch bulbs) fixed to the model and to rigid points in the background will provide datum points in the picture area from which measurements may be taken, or allow assessment on an automatic analyzer.

Neutrally buoyant particles placed in the flow stream ahead of a model will show the stream lines and disturbed regions round a ship's hull, piers, etc. With two cameras mounted at 90° to each other it is possible to plot X, Y and Z co-ordinates of the positions of these markers frame by frame for a three-dimensional representation.

Underwater records are obtained by filming through a glass window. Alternatively, one end of a periscope may be placed in the water, with the camera fixed to the other end above water. Other possibilities are an underwater camera—

either fixed, or remotely controlled with a closed-circuit T.V. viewfinder.

Stroboscopic light sources are valuable for studying oscillatory or cyclic phenomena. Putting the stroboscopic light source slightly out of phase with the subject, e.g. a propellor, produces a form of mock high-speed cinematography if filmed at normal speed.

Normal and High-speed Cinematography in Gases. Cine cameras in conjunction with Schlieren systems enable flow patterns and shock waves to be studied. The camera taking speed is ascertained linearly in relation to smoke plumes or streams in a wind tunnel. For example, a factory chimney model may be set up with a ratio of full scale to model scale wind velocities of 5 : 1. For projection at 24 f.p.s., a camera speed 120 f.p.s. will give full scale simulation.

The combustion of fuel in engines can be studied with a high-speed camera, filming a special engine with a toughened glass cylinder wall or cylinder head.

High-speed cameras usually have a built-in time base film marker, which leaves a series of dashes on the margin of the film. The frequency of the marker might be 1,000 c.p.s. in which case a film showing one dash per three frames would have been taken at 3,000 f.p.s. A.W.G.

See also: *High-speed cinematography; Schlieren photography; Time lapse photography.*

FLUORESCENCE. As a rule, when materials absorb light they convert it into kinetic energy, or heat. But some substances re-emit part of it as light of a longer wavelength. Such substances are said to fluoresce. Dyes, like fluorescein, eosin, and rhodamine, will fluoresce in solution. Zinc and cadmium sulphides, suitably activated with other metals, will also fluoresce. These, or similar substances, are used for coating the glass tubes of fluorescent lights. Certain substances will absorb invisible rays of ultra-violet light and convert it into light of a visible wavelength. Sometimes the radiation goes on even after the exciting light has been removed. This kind of radiation is known as phosphorescence.

The familiar fluorescent lighting is produced by coating a mercury vapour tube with substances that fluoresce under the influence of ultra-violet radiation. The invisible ultra-violet rays are converted into visible light, and by a suitable choice of coating material the colour of the resultant light can be varied throughout a range of "daylight" tints.

Fluorescence enables some materials to be photographed by invisible radiations when other non-fluorescing substances remain invisible. Among substances that fluoresce, some do so more strongly than others. This provides a valuable means of creating contrast differences which do not exist in ordinary light.

Apart from its use in radiography (fluorography) fluorescence photography can reveal things that would otherwise be invisible—e.g., the presence of two kinds of ink, which fluoresce differently, in a falsified document. The technique is used in a number of different fields of photography and scientific observation.

See also: *Fluorescent lamp; Forgeries; Luminescence; Ultra-violet and fluorescence photography.*

FLUORESCENT INK. Printing ink containing a fluorescent pigment or dye that absorbs short wavelength, ultra-violet and blue-violet rays (to which the eye is comparatively insensitive) and re-radiates their energy as light rays of longer wavelength (yellow to red). On dull, grey days or in poor light, such pigments (largely used for poster work) appear brighter to the eye than a non-fluorescing pigment and also to the photographic emulsion of those high-speed panchromatic films which are over-sensitive to red. This effect can be compensated for, if necessary, by the use of a filter complementary in colour to the fluorescent ink.

See also: *Paintings and drawings.*

FLUORESCENT LAMP. Type of vapour discharge lamp in which the glass envelope is coated with a substance capable of effective fluorescence. The rays emitted by the lamp cause this coating to fluoresce.

Fluorescent lamps are normally tubular with a contact cap at each end; this cap may be bi-pin, bayonet or raised contact. For convenience the tubes are usually mounted or clipped into holders which also serve as trough reflectors or diffusers.

Quality of Light. By choice of the coating substances used, fluorescent tubes can be made to emit light of different but very constant colour temperatures; the more commonly manufactured range is natural daylight and warm-white. For accurate colour-matching, a special tube is also made.

Fluorescent lights are very efficient in their use of electricity because they turn so little of it into wasteful heat. A tube consuming 40 watts gives about the same amount of light as an ordinary domestic tungsten filament lamp of 150 watts. The light is soft, low intensity, and diffused, since the tube may be anything from two to eight feet in length. For this reason the photographic applications are limited to general illumination, in particular, of the copy board in process and document photography, and of such subjects as oil paintings.

Operation. Fluorescent lamps may be run off either an A.C. or D.C. mains supply.

On A.C. the tubes will flicker at double the frequency of the supply and are not suitable for instantaneous exposures of a shorter duration than two cycles—i.e., of anything less than 1/25 second with a 50 cycles per second supply. If a steady light is essential—as in cinematography—three tubes may be connected across the phases of a three-phase supply.

Special control gear is required with all arc and vapour discharge lighting because the internal resistance is not constant. For D.C. operation a ballast resistance must be connected in the circuit and, since this accounts for about half the total consumption, it lowers the efficiency of the lamp. On an A.C. supply the lamp must be connected in series with a choke to limit the current. Where the supply voltage is too high or too low for the tube, a reactive transformer is used in place of the choke. In addition to a choke or transformer, the circuit may also need one or more condensers to compensate for the inductive load and maintain the power factor of the supply.

There are two ways of starting up a fluorescent lamp. The commonest method is by passing a starting current through a heating filament inside the tube, and the other by applying an initial voltage high enough to ionize the gas in the tube. The first type of tube is called a hot cathode; it is used for practically all domestic and public building illumination and for subject lighting in photography. The second, known as cold cathode illumination, is used in photography principally as an enlarging light in the form of a compact grid extending over the whole of the negative area.

Luminous Panel Light. Certain substances emit visible light rays when placed in a high voltage alternating current field. Materials of this type form the basis of a special type of luminous panel lighting. A layer of the material is sandwiched between a layer of metal foil and a layer of transparent conducting plastic. The metal side of the sandwich is supported on an insulating base and the transparent side is covered with a sheet of glass or transparent insulating plastic. When an alternating current at 230 volts is applied to the conducting layers, the intermediate coating glows with a brightness of about ½ foot lambert. The light emission can be increased by applying a higher voltage, and the colour of the light can be varied according to the composition of the coating material or by incorporating a suitable filter in the transparent layer. The total amount of light available depends on the area of the panel.

Luminous panel lighting has been applied on a limited scale as a safelight for both panchromatic materials and enlarging papers, and for luminous notices and clock faces in photographic material coating rooms. This form of lighting has the advantage of being practically cold and therefore exceptionally efficient (because practically all the electrical energy is converted into light).

Such luminescent panels are also coming into use as a light source for contact printing and copying. In a luminescent contact printer the negative and the positive paper are pressed into contact with the luminous panel. Although the actual brightness of the panel is not very great, the light output is fully utilized by the direct contact with the printing material and hence reasonable printing exposures become possible.

Luminous panels on their own have also been used for document copying. Here again the panel is pressed into contact with the sensitive material and the document or printed matter to be copied. If the latter is a page in a book, the panel—which is thin enough for the purpose—may be placed inside the book and the latter closed to provide the necessary contact pressure. Printing here is by the usual reflex process, with the front of the document paper in contact with the original and the back pressed against the luminous panel. Both the conventional transfer and negative-positive copying processes are suitable for this purpose. F.P.

See also: *Discharge lamp; Electroluminescence; Light sources.*

Books: *Fluorescent Lamps and Lighting*, by W. Elenbaas (Eindhoven); *Fluorescent Lighting Manual*, by C. L. Amick (New York); *Fluorescent and Other Gaseous Discharge Lamps*, by W. E. Forsythe and E. Q. Adams (New York).

FLUORESCENT WHITES. Some photographic papers are treated in manufacture with a fluorescent dye to increase the effective reflectivity of the base. This adds brilliance to the highlights of the photograph, particularly when it is viewed in daylight. This is because daylight is rich in the ultra-violet light which causes the dye to fluoresce.

There is one great disadvantage of this type of base. If photographs made on it are intended for reproduction, the highlights must not be worked up with ordinary process white. A special type of fluorescent process white must be used, otherwise the retouching will show clearly on the finished half-tone reproduction although invisible in the original print.

FLUOROGRAPHY. Process of photographing the visible image produced by the impact of invisible radiations on a fluorescent screen. There are two methods of fluorography—contact and optical. In the contact method, the sensitive photographic material is pressed into contact with the screen. In the optical method, the image is photographed by setting up a camera in front of the screen.

The commonest application of optical fluorography is in mass miniature radiography.

FLUOROTYPE. A paper process invented by Robert Hunt in 1844. The paper was prepared with sodium fluoride and developed with protosulphate of iron. In spite of the reasonably short exposure of half a minute, the fluorotype did not come into general use.

FLYING SPOT SCANNER. System in which an electron beam scans an image by moving over it in the form of a spot of light. This spot can expose an image on a sensitive surface if the spot is made to vary in intensity on going through a normal negative or positive. Such a set-up is used in electronic printing systems. If the spot of the electron beam is made to vary in intensity while passing over different points of a phosphorescent layer, an image can be made visible to an observer in front of the screen. The scanning speed must then be high enough to cover the whole image within the period of persistence of vision. This is the use of a flying spot scanner in assembling an image on a television picture tube.

See also: *Electronic printing; Television.*

***f*-NUMBER.** Numerical expression of the relative aperture of a lens at its different stops. The *f*-number is equal to the focal length divided by the effective diameter of the lens opening, and is written in various forms—e.g., *f* 8, *f*/8, 1 : 8, etc. When focused on infinity, all lenses stopped down to the same *f*-number produce images of equal illumination (apart from differences due to varying reflection losses). For a given shutter speed, a given *f*-number always corresponds to the same exposure.

See also: *Aperture; Diaphragm.*

FOCAL LENGTH. Term in optics referring to the distance from the plane in which the lens forms an image of objects at infinity to the node of emission. It is important in photography because it decides the scale of the image (the longer the focal length, the larger the image) and the extension of the camera (a longer focal length calls for a greater amount of camera extension). More precisely, it is the equivalent focal length defined by the relation $f = y/\tan \beta$ where β is the angular subtense of a small object of image height y.

The focal length of a lens is usually engraved on the mount thus: $f = 90$ mm. This should not be confused with the *f*-number.

See also: *Lens.*

FOCAL PLANE. Imaginary plane on which a lens forms a sharp image when correctly focused. For sharp pictures the emulsion surface of the plate or film in a camera must be accurately positioned in the focal plane of the taking lens.

See also: *Lens.*

FOCAL PLANE SHUTTER. Shutter consisting of one or more roller blinds of fabric or metal, having a generally variable slit which moves across the back of the camera when the release is pressed and exposes the sensitized material progressively. This type of shutter is mounted as close as possible to the surface of the sensitized material—i.e., to the focal plane. In that position it is most efficient, and also gives consistent exposures with lenses of varying focal lengths. The focal plane shutter was invented by William England, 1861; re-introduced by B. J. Edwards, 1879, and included in the camera patented by Ottomar Anschütz, 1888, to whom its invention is usually erroneously attributed.

See also: *Shutter.*

FOCAL POINT. Point of intersection of all rays of light transmitted by a lens from a given object point. When the object is at infinity, and the incident rays are parallel to the lens axis, the image is the principal focal point and lies on the axis of the lens.

See also: *Lens.*

FOCUS. Point of convergence. In optics it is the point on the optical axis of a lens or mirror through which all rays of light parallel to the axis pass after refraction (lens) or reflection (mirror).

See also: *Lens; Mirror.*

FOCUSING

The camera lens forms an image of the objects in front of it on the film or plate in the back of the camera, but the image is sharp only when the distance between the lens and the film is correct. What the correct distance is depends on the focal length of the lens and how far away the object is. Adjusting the camera to this distance is known as focusing.

All but the cheapest cameras have some means of focusing. Very cheap box cameras have fixed focus lenses. They work at small apertures and so have sufficient depth of field —from about 10 feet to infinity—to give a sharp picture without focusing. Such cameras are only intended to photograph subjects more than 10 feet away, except when using supplementary lenses.

There are two ways of adjusting the image sharpness: by changing the actual distance between the lens and the film, and by changing the focal length of the lens.

There are several ways of changing the lens-film distance.

On most miniature cameras, for example, it is done by a helical focusing mount in which the lens can be rotated to screw it towards or away from the film.

In most plate cameras and many twin lens reflexes the whole lens panel is moved forward or backward by a rack and pinion.

In many field and technical cameras the actual film plane is moved forward or backward for focusing.

Again, the focal length of the lens may be altered by increasing or decreasing the separation between the lens elements. Usually the front cell of the lens moves in or out in a screwed mount, and the method is known as front cell focusing. Many of the simpler modern cameras have front cell focusing.

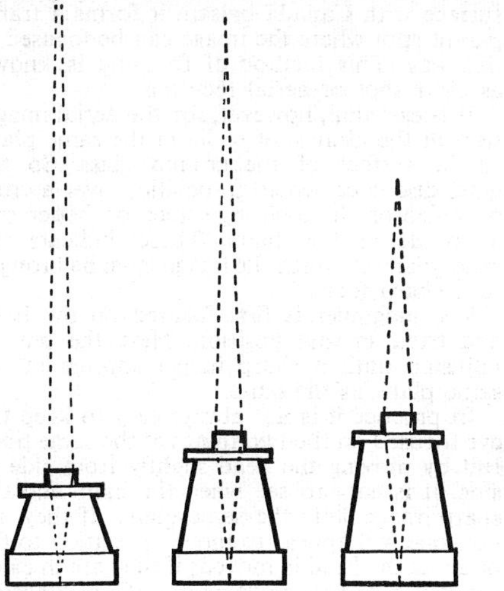

PRINCIPLE OF FOCUSING. When a lens is focused at infinity, rays from a very distant object point converge to a corresponding image point a given distance behind the lens in the focal plane. As the object point is brought nearer, the image point moves farther away from the lens: the lens-film distance must be increased to obtain a sharp image of close objects.

In effect the use of positive supplementary lenses is simply another way of shortening the focal length of the camera lens thus enabling it to focus on close-up objects within the limits of the camera extension.

Direct Screen Focusing. In one of the oldest and surest methods of focusing, the image is observed on a ground glass focusing screen in the back of the camera, and the lens is adjusted until the subject is sharp. The ground glass screen is then replaced by the sensitive material in the film or plate holder. It is essential for the surface of the sensitive material to lie in exactly the same plane as the ground surface of the focusing screen. This calls for accurate construction of the plate holder and focusing screen slide. Given this, direct screen focusing is extremely reliable.

The focusing screen not only shows the sharpness of the image, but also the exact picture that will be photographed on the negative. It is, therefore, a means of composing as well as focusing.

The drawbacks of direct screen focusing are that it is slow and the image appears upside down on the screen. As the image must not move while the focusing screen slide is changed for the plate or film holder, it is limited to more or less still subjects, and the camera must be mounted on a tripod.

Reflex Focusing. In the reflex camera the image is focused on a horizontal screen set in the top of the camera. The image is projected on to it by a mirror behind the lens. When the shutter release is pressed the mirror swings out of the way and immediately afterwards the shutter opens and exposes the film or plate in the back of the camera. The time lag between focusing and exposure is reduced to the time taken by the mirror to move out of the way—a fraction of a second.

Reflex focusing has all the advantages of the ordinary focusing screen but once the image is focused the exposure can be made right away. This means that the camera can be held in the hand and used for taking moving subjects.

Reflex cameras have one great drawback: it is not always convenient to use the full aperture of the lens, and when it has to be stopped down beyond about *f* 6·3 the picture on the screen might become too dim to be of any use.

This disadvantage has been overcome in most modern cameras by use of a pre-set iris control. The aperture required for the exposure is selected but not actually adjusted. This allows the lens to be used at full aperture for focusing. When the shutter release is then pressed, the iris automatically closes to the selected aperture before the shutter opens.

All the focusing aids used with the ordinary focusing camera can be applied to the reflex screen.

Reflex cameras usually have a built-in focusing magnifier, while a collapsible hood takes the place of the focusing cloth.

A further feature of a reflex screen image is that it is reversed left-to-right. This mainly affects the use of the screen for arranging the subject and following movement. The majority of current miniature reflexes incorporate an optical reversing system such as a pentaprism to yield an upright and right-way-round image for eye-level viewing.

Some miniature single-lens reflex designs have also succeeded in eliminating yet another disadvantage of the moving mirror system: the disappearing image. In these models the mirror

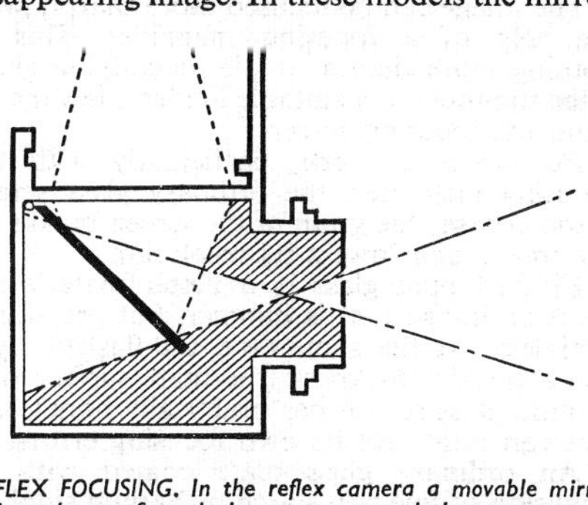

REFLEX FOCUSING. In the reflex camera a movable mirror deflects the rays from the lens on to a ground glass screen at the top. This shows the exact sharpness of the image as it would appear on the film. The mirror swings away for exposure.

returns to its viewing position immediately after the exposure so that the subject continues to be visible once the picture is taken.

Twin-Lens Reflex Focusing. Twin-lens reflex cameras really consist of two camera systems built one above the other. The bottom camera is used solely for taking the picture, while the top one is used quite independently for viewing and focusing. The two camera lenses are, however, geared together, so that what is sharp on the upper reflex focusing screen will also be sharp on the film behind the lower lens.

The twin lens principle has several advantages: there is no time lag between focusing and exposure; the focusing (and viewing) lens always remains at full aperture (in fact, the aperture of the focusing lens is often larger than that of the camera lens, giving more critical focusing and a brighter image).

But there are also drawbacks: as there are two lenses, focusing accessories such as supplementary lenses must be bought in matched

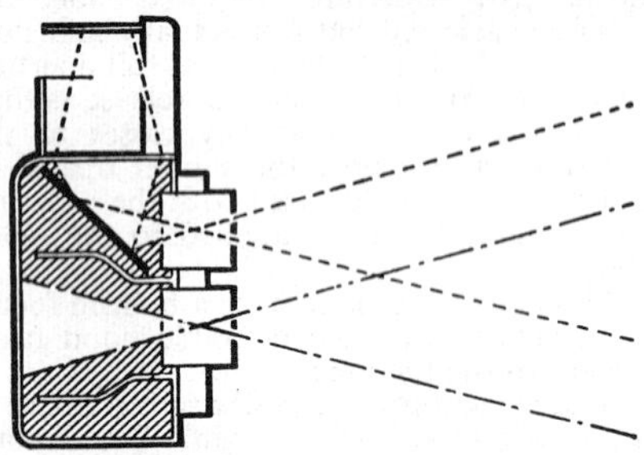

TWIN LENS REFLEX FOCUSING. Instead of the movable mirror of the single lens reflex, the twin lens reflex has a second focusing camera with a matched lens coupled to that of the taking camera so that both are focused in unison.

pairs instead of singly; the twin-lens reflex camera is subject to parallax error.

Aids to Screen Focusing. There are a number of aids to exact focusing on a screen.

The image can be focused more sharply with the help of a focusing magnifier. This is nothing more than a simple magnifying glass, often mounted in a suitable holder a few inches from the focusing screen.

For accurate work, particularly with the focusing magnifier, the ordinary glass screen is too coarse; the grain of the screen breaks up the image and lowers its resolution.

Flashed opal glass is a better material; it gives an image that is dimmer, but practically grainless. At the same time, the flashed layer has a certain thickness (the matt surface of a ground glass screen has practically none), and this can introduce its own focusing errors.

An ordinary glass plate coated with an emulsion of fine rice starch or barium sulphate in gelatin also gives a very fine screen.

Clear Spot (Aerial) Focusing. The image formed in the plane of the focusing screen can be observed directly through a magnifier without any assistance from the screen itself. If a disc of clear glass is cemented on the ground surface with Canada balsam it forms a transparent spot where the image can be focused in this way. This method of focusing is known as clear spot or aerial focusing.

It is essential, however, for the aerial image, seen in the clear spot to lie in the same plane as the surface of the ground glass. So the glass disc is cemented in position over a cross pencilled on the focusing screen or better still a couple of fine hairs. These indicate the exact plane in which the image must be brought to a sharp focus.

The magnifier is first focused on the hairs and fixed in that position. Next the lens is adjusted until a sharp image appears in the same plane as the hairs.

In practice it is not always easy to keep the eye focused on the two things at the same time. But by moving the head slightly from side to side, it is easy to see when the hairs and the sharp image lie in the same plane. If they are not, one will appear to move in relation to the other as the head is moved; if they are in exact register, they will move as one. This extension of aerial focusing is known as parallax focusing.

A pair of crossed glass wedges, recessed into the focusing screen so that the crossing point is exactly level with the focusing surface, can show up such a parallax effect automatically. When the image is not exactly focused, it appears split across the dividing line between the wedges, with the two halves displaced relative to each other. Focusing in this way is visually the same as with a split-image rangefinder. Many current miniature reflex cameras incorporate this feature.

Double Image Focusing. When the lens is not focused properly, different parts form images in slightly different positions on the screen. The separate images may be fairly sharp but they all combine to form a blur.

If a strip of opaque material is placed across the centre of the lens, about one-third of the lens diameter in width, it will leave two images, formed by the left hand and right hand segments of the lens. These images will be sharper than the image formed by the whole lens, but they will be out of register unless the lens is correctly focused.

The strip must be as close to the front surface of the lens as possible. The strip must, of course be removed before the exposure.

It is nearly always better to focus with the camera lens at its largest aperture, and stop down for the exposure later if necessary. At large apertures the depth of focus is small, so the transition from sharp to unsharp is much more rapid. The image is brighter and easier to see. An exception to this method is when using certain lenses, that shift their focus when the aperture is decreased.

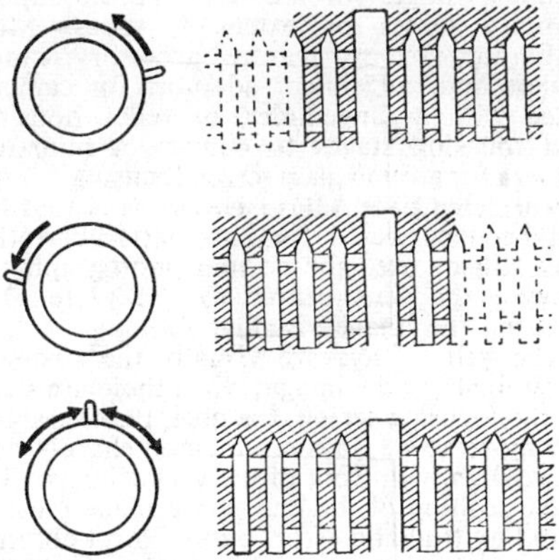

BRACKETING. Top: Observe the image on the focusing screen and start from a setting with the near limit of the subject unsharp. Centre: Continue until the far part begins to become unsharp. Bottom: Set the focusing control midway between the positions corresponding to the beginning of unsharpness at each end.

Bracketing. The image often appears reasonably sharp anywhere inside a certain range of focusing movement. This is because the resolution of the screen and possibly the eyesight of the operator limit the degree of sharpness that can be appreciated. For most practical purposes it is enough to narrow down this range to a minimum by trial and then to set the focusing adjustment midway between the front and rear limits within which the image looks equally sharp.

This method of focusing is called bracketing.

Accurate focusing must take into account the whole of the image, not merely the central area. No lens is good enough to give the same standard of sharpness all over the plate; it is often impossible to focus the image sharply at the centre and the edges of the screen at the same time. When this happens, the bracketing method is used to set the lens midway between the positions that give maximum sharpness to the centre and the edges of the image. In this position over-all sharpness can then be achieved by stopping down the lens.

Focusing by Scale. Nearly every camera has a focusing scale, even if it also has a focusing screen or a rangefinder. The scale is useful for action photographs.

Where the whole lens is moved for focusing, the distance through which it must be moved forward from its infinity position to focus it on a nearer object is given by the equation.

$$d = \frac{F^2}{u - F}$$

where d = forward shift of lens from infinity position
F = focal length of lens
u = distance of object from lens.

In practice the exact focal length of the camera lens is not very simple to measure. The value marked on the lens itself is usually good enough, but is not always accurate. Individual lenses may deviate slightly from their nominal focal length.

By focusing first on infinity—i.e., a very distant object—and on an object say 3 feet away, the exact focal length can be calculated from the equation:

$$F = \frac{dU}{V}$$

where F = focal length of camera lens
d = lens extension for close-up focus (i.e., difference between infinity and close-up settings)
U = size of object
V = size of image.

All measurements must be in the same units; e.g., all in inches, or all in centimetres.

With front-cell focusing lenses, calibration of the scale is much more complicated, and depends very much on the construction of the lens. The amateur can only calibrate or check the focusing scale of such a lens by trial and error.

In most cases the lens-to-film distance is so small in comparison with the subject distance, that it does not matter whether the subject distance is measured from the front or the back of the camera. But for accurate work, particularly for close-ups, the subject distance should be measured from the front of the lens, or still more exactly from the forward nodal point (approximately one-third of the thickness of the lens from its front surface).

A few cameras, however, carry a warning that all distances must be measured from the plane of the film.

Where the camera is focused by scale, the actual subject distance must either be measured separately, or estimated (e.g., by assessing the size of an average person in the viewfinder).

With action subjects where the distance may change suddenly, the focusing scale is best used in conjunction with a depth of field indicator to select suitable zone focus settings rather than exact subject distances.

Rangefinder Focusing. In focusing by range-

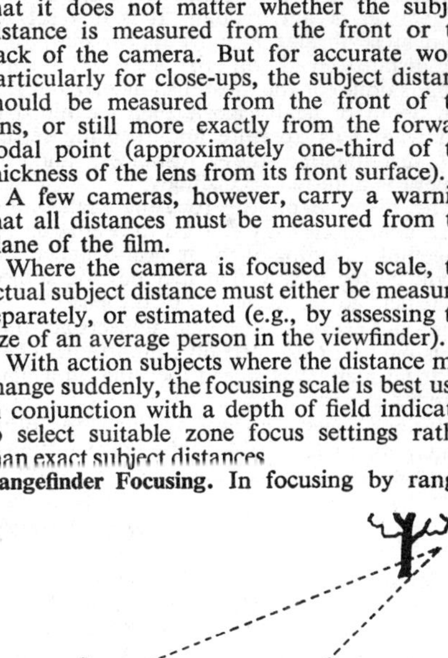

RANGEFINDER FOCUSING. The rangefinder measures the subject distance by the convergence of two rays from the ends of a fixed base, as observed by the coincidence of two images.

finder, the photographer either sets the lens focusing scale to the distance given by the rangefinder (where a supplementary, or non-coupled rangefinder is used) or the act of adjusting the rangefinder automatically focuses the lens on the subject (where the camera has coupled rangefinder focusing).

In any case, the rangefinder must first be adjusted to read the subject distance. Generally this means adjusting the lens (with a coupled rangefinder) or moving a lever or turning a wheel to produce coincidence between either two overlapping images or two halves of a split image. The point of coincidence is easy to recognize in good light but it may be impossible when the light is poor.

It always helps if the part of the subject to be focused on includes a well-defined edge—e.g., the profile of a face, a gatepost, or the branch of a tree. If the important area contains no such clear line, as for instance when the subject is a fluffy kitten on a plain carpet, then the finder should be ranged on the edge of some other convenient object at the same distance.

Where, in addition to the subject itself, objects in front and behind must also be rendered sharply, a single reading of the rangefinder is not enough. It is first set for the nearest object that must appear sharp, and the distance read off the scale. The same thing is done for the most distant sharp object. The depth of field scale is then examined to find what lens aperture and focusing setting will just include these distances, and the camera is adjusted accordingly.

The distance on the focusing scale under these conditions may not be the same as the rangefinder reading of the subject distance, but it will be the setting that includes the subject in the field governed sharply by the lens.

Close-up Focusing. With near subjects accurate focusing becomes specially important. There the ideal method is screen focusing on a plate or reflex camera.

Various accessories are available for other camera types to establish the correct subject distance. The simplest may take the form of distance gauges for use with specific supplementary lenses or extension tubes. More elaborate systems feature accessory rangefinders and rangefinder adaptors for cameras fitted with a rangefinder, or reflex housings and focusing stages to convert a miniature camera for ground glass screen focusing.

Uncorrected Lens Adjustment. A lens that has not been corrected for chromatic aberration may still be used for normal photography by moving the lens nearer to the plate after focusing the image visually.

The yellow rays are visually the strongest when focusing the image; when these are sharp on the focusing screen, the blue, photographically more active, rays are focused sharply on a plane slightly in front of the visual image. The exact position of the blue-sharp plane must be found by trial and error, but for most chromatically uncorrected lenses it is displaced approximately 1/40th of the distance of the plate from the lens.

To arrive at a reasonable point for starting experiments, the image should be focused visually—preferably with a magnifier.

The camera extension is then measured, and the lens moved back from 1/30 to 1/40 of the distance. A series of exposures should then be made at, in front of, and behind this distance to find the point of maximum sharpness. For the final exposure the lens should be worked at a very small aperture.

As only the image formed by the blue rays is in focus, all the other unsharp rays must be prevented from registering on the plate. This means either using a plate that is insensitive to all but blue rays—i.e., a blue-sensitive process type plate, or fitting a deep blue filter in front of the lens to allow only blue rays to reach the plate. Either of these measures makes for long exposures so that the method is only suitable for still subjects. L.A.M.

See also: *Circle of confusion; Close-ups; Depth of field; Depth of focus; Distance estimation; Extension of camera; Focusing (automatic); Focusing mechanism; Focusing stage; Reflex attachment; Supplementary lens; Zone focusing.*

Books: *Optics*, by A. Cox (London); *Focusing*, by E. S. Bomback (London).

FOCUSING (AUTOMATIC)

Method of adjusting the sharpness of an image without visually judging it. Strictly speaking, fully automatic focusing also involves a system of getting an unsharp image sharp on its own by suitable servo circuits. This term is sometimes also applied to the various systems of rangefinder coupled focusing in which setting the rangefinder for a given distance automatically focuses the lens.

This again however is not really automatic, since it involves both visual judgement (though not directly of image sharpness) and manual control.

Automatic focusing systems require two stages: measurement of the sharpness of the image, and a provision for evaluating the measurement and using it to adjust a focusing mechanism for maximum sharpness.

Of the various sharpness measurement methods proposed the only one in practical use nowadays relies on measurement of image contrast. The contrast of a sharp image of a given subject is higher than the contrast of the same image when it is out of focus.

An instrument which is able to measure the light distribution in the image, and compares

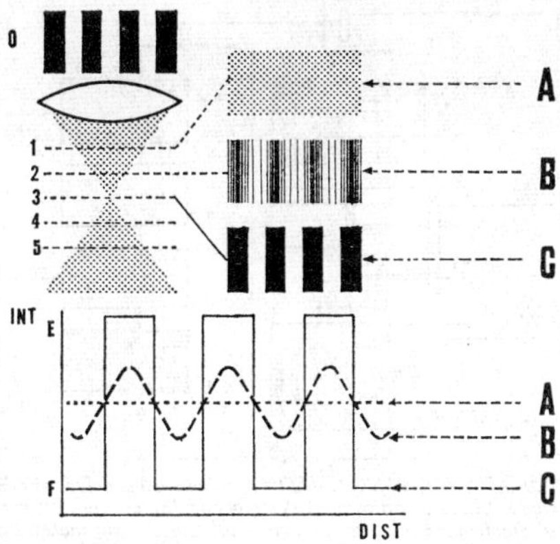

IMAGE SHARPNESS AND LIGHT DISTRIBUTION. Top: *When the image of a pattern O is projected on a screen, it appears sharpest (C) in the focal plane (3) of the lens. In the plane 2 (and 4) the image is partly blurred (B); in the completely defocused planes 1 and 5 it tends to be a uniform area of brightness (A).* Bottom: *The brightness distribution of these images, if measured in a plane across the bars, corresponds to the three curves A, B, and C, oscillating to varying degrees between E and F about the mean level A.*

for a sharp and an unsharp picture, can therefore also measure sharpness. This may be done with a dectector which registers independently adjacent light intensities of an image, and which does so in a non-linear fashion. That is to say, it must react disproportionately to different light intensities, like a photographic emulsion or a cadmium sulphide photo conductor. Such a photo conductor is in fact used in virtually all current automatic focusing systems.

On plotting the output of the cell while the image passes from unsharp through the plane of maximum sharpness to unsharp again, the current follows a fairly straight curve with a pronounced dip as the image becomes sharp, and a rise as the image becomes unsharp again.

Measuring Instruments. This change in output of the photo cell can be used to monitor the sharpness of an image without actually observing it. One instrument of this kind uses two photo cells. One of them has a diffusing screen in front of it to compare the output corresponding to a completely defocused image with the output of the second cell responding to the change in sharpness. The two cells are connected in a bridge circuit, the output of which feeds a galvanometer or ammeter. The deflection on the meter dial is greatest when the image is at its point of maximum sharpness.

Instruments of this nature exist both for camera focusing and for enlarging. In the first case a probe containing the photo cells is placed in the image plane of the camera (or on the focusing screen); in the second case it is placed on the enlarger baseboard. To ensure maximum contrast of the sharp subject (and hence greatest sensitivity of the instrument) an artificial subject or target is sometimes used, consisting of a uniform pattern of black-and-white lines or dots, such as a ruled halftone screen. This is placed in the negative carrier of the enlarger or in the subject plane of the camera during the focusing operation.

Automatic Ranging Systems. This principle of sharpness measurement can be extended in

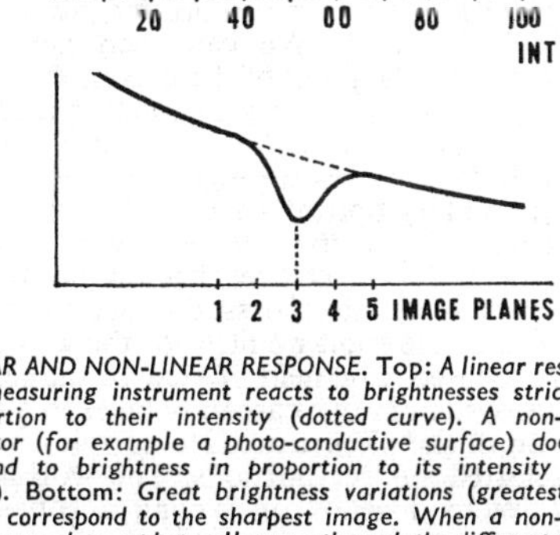

NEAR AND NON-LINEAR RESPONSE. Top: *A linear respond-g measuring instrument reacts to brightnesses strictly in oportion to their intensity (dotted curve). A non-linear ceptor (for example a photo-conductive surface) does not spond to brightness in proportion to its intensity (solid rve).* Bottom: *Great brightness variations (greatest con-st) correspond to the sharpest image. When a non-linear eptor such as a photocell moves through the different planes an image from unsharp through sharp to unsharp again, the put drops distinctly in the plane of maximum sharpness.*

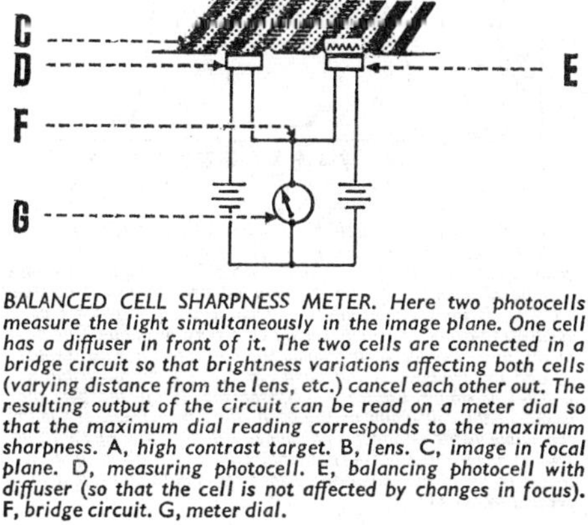

BALANCED CELL SHARPNESS METER. Here two photocells measure the light simultaneously in the image plane. One cell has a diffuser in front of it. The two cells are connected in a bridge circuit so that brightness variations affecting both cells (varying distance from the lens, etc.) cancel each other out. The resulting output of the circuit can be read on a meter dial so that the maximum dial reading corresponds to the maximum sharpness. A, high contrast target. B, lens. C, image in focal plane. D, measuring photocell. E, balancing photocell with diffuser (so that the cell is not affected by changes in focus). F, bridge circuit. G, meter dial.

various ways to provide a fully automatic focusing system. There the output of the photo conductor cell is continuously monitored and used to drive a servo motor. The latter adjusts the focus to maximum sharpness.

One method is to have two such photo cell systems positioned in front of and behind the plane of maximum sharpness. The differential outputs of the two sensing systems are in turn connected in a bridge circuit. When the plane of maximum sharpness is exactly in the middle between the two sensing systems, their output is exactly balanced. As the plane comes nearer the one or the other cell pair, this system produces a signal. The sign of the signal depends on the direction in which the plane of maximum sharpness has moved. This signal can then be used to drive a servo motor to restore the image plane (for example by refocusing the lens) into the exact halfway position.

A second ranging system uses a single photo

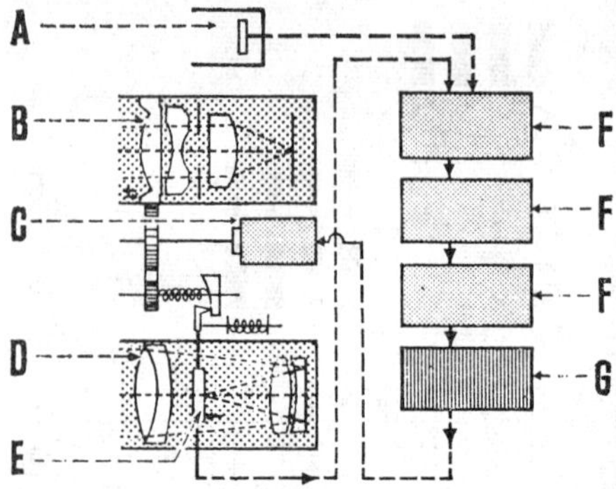

AUTOMATIC FOCUSING CAMERA. A photocell is driven through the focal plane of an optical system and the cell output monitored electronically so as to switch off the driving motor when the cell has reached the plane of maximum sharpness. The motor also shifts the camera lens to produce a sharp image on the film. A, *automatic gain control (to compensate for subjects of different brightness).* B, *optical system of camera.* C, *driving motor for focusing movement (camera lens and photocell).* D, *optical system of photocell.* E, *cadmium sulphide cell.* F, *electronic stages of system.* G, *switching stage to control motor.*

cell (or cluster of cells) which oscillates to and fro at a fixed frequency through a series of planes which includes the plane of maximum sharpness. During each passage through the sharp plane, the output drop of the cell shows itself as a signal pulse. If the sharp plane is exactly in the centre of the oscillation range, the output takes the form of a signal of a regular pulse frequency. When the sharpest plane is not in the middle of the oscillation range, the pulse signal becomes unsymmetrical. The rate of change of the signal frequency is then measured and used to activate a servo moto which readjusts the position of the lens or the image plane. This permits focusing not only o a static but also of a moving image.

A third way is to monitor the output of single photo-conductive cell as it passes from position of maximum unsharpness to maxi mum sharpness. While the output drop in thi case is fairly gradual, electronic differentiatic can yield a steep curve indicating the rate change of output. We can then determir fairly reliably the point of the maximum rate change—i.e. the point of maximum sharpness

Such a system has been used in a camer The cell there moves through the image plan of a secondary optical system, with the mov ment geared to the motor-driven focusi movement. As the cell reaches the position maximum image sharpness, the focusing mot cuts out at the right point, with the lens acc rately focused. The field of view of the sensi optical system in this case only takes in a sm portion of the picture area, to permit focusi on selected subject portions. The working this system depends on having a subject sufficient detail contrast within the image ar being measured. L.A.

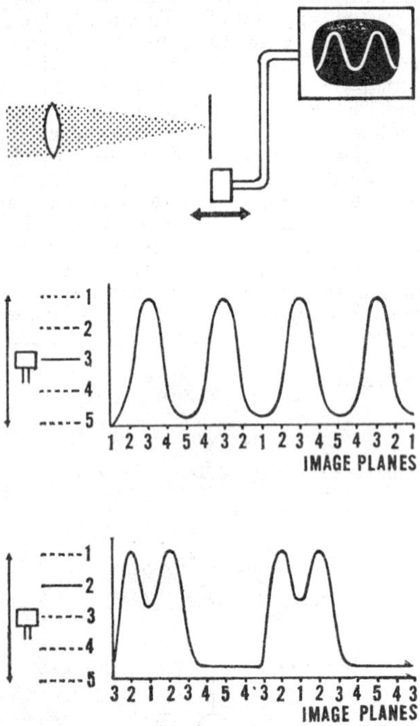

THE OSCILLATING OPTICAL RANGER. Top: *If a photocell oscillates to and fro across the focal plane of an image, its drop in output in the focal plane gives rise to a signal pulse that can be observed as a wave form on an oscilloscope screen.* Centre: *If the focal plane is exactly in the mid-point 3 of the oscillating cycle, the wave form has a regular shape. The peaks of the pulses occur at the positions corresponding to the plane 3, and the troughs at the positions corresponding to planes 1 and 5 (maximum unsharpness).* Bottom: *If the sharpest plane is not in the mid-point of the oscillating movement of the cell, the wave form changes and shows up the lack of sharpness. This change in the pulse frequency can also drive for instance a motor to move the lens and restore maximum sharpness in the central image plane.*

FOCUSING CLOTH. Square of opaque cloth, often lined with black velvet, which is draped around the focusing screen and over the head and shoulders of the photographer to shut out all light but that of the image on the ground glass. When not in use it provides a convenient light-tight wrapper for the loaded plate holder.

FOCUSING HOOD. Folding hood, usually of metal, surrounding the focusing screen of waist-level reflex cameras to shield off light for a clearer view of the screen image. Often incorporates a fold-away magnifying lens (focusing magnifier) to permit precise focusing of an enlarged centre section.

FOCUSING MECHANISM

All cameras (apart from the fixed focus types) must have some means of focusing—i.e. of setting the lens at the correct distance from the plate to give a sharp image of the subject to be photographed. There are several ways of doing this; any one of them to be efficient must fulfil the following requirements: it must be free from backlash and play, and it must maintain the lens rigidly in a plane parallel with that of the sensitive material.

Most, but not all, focusing arrangements are associated with a scale that shows, for every position of the lens, the distance in feet or metres from the camera at which objects will be sharp on the negative.

Rack and Pinion Focusing. Most of the older types of camera, and almost all cameras taking half-plate or larger negatives, focus by rack and pinion. The lens, in such cases, is attached to a panel strutted at right angles to a movable slide running in guides on the baseboard. A brass strip let in to the underside of the slide carries teeth that engage with a pinion turned by a thumb-wheel at the side of the baseboard. Turning the thumb-wheel moves the slide in or out. On large cameras the slide is moved by a rack and pinion at each side of the baseboard to prevent it from binding in the guides.

Generally, the lens panel can be set up and ocked anywhere along the slide, and once it is oughly in position the rack and pinion ocusing is used for the final adjustment. This rrangement has the added merit of doubling he available camera extension and also roviding for lenses of long and short focal ngth.

Rack and pinion focusing is satisfactory here the amount of lens movement is relatively rge—i.e. on say, a whole-plate field camera here the lens moves forward nearly four ches from the infinity position to focus on feet. But it is not accurate enough to be tisfactory on smaller cameras where, for stance, a 2 ins. lens moves forward only a ttle more than one tenth of an inch from finity to focus on 3 feet. To give comparably ne adjustment on the smaller camera the ameter and tooth sizes of the pinion would ve to be scaled down so far that the parts ould not be strong enough for their work.

ther Methods of Shifting the Lens Panel. On ocess and copying cameras, the rack and nion arrangement is sometimes discarded for nut and leadscrew combination, where the nut is attached to the slide and the leadscrew is turned by a handle or handwheel at the back of the camera. With this set-up, the operator can stay in one position for focusing irrespective of the position of the lens panel on the slide.

On some hand cameras where the focusing scale extends only from infinity down to 3 or 4 feet, the lens panel is moved by a lever pivoted on the baseboard. The operating arm of the lever is relatively long and moves over a calibrated focusing scale. In this way it is possible to have an open and easily read scale even though the movement of the lens is no more than a small fraction of an inch.

Focusing Lens Mount. Where the amount of focusing movement required is no more than an inch or so and on all cameras where there is no separate lens panel (such as on most miniatures and self-erecting cameras) the focusing lens mount is the most convenient arrangement.

The simplest focusing mount consists of two closely fitting sections of tube sliding one inside the other, one holding the lens, and the other being flanged to attach to the front panel. This type of simple mount is good enough for visual focusing, but the backwards and forwards movement is too small to allow for accurate calibration.

But if a helical groove is cut in one tube and

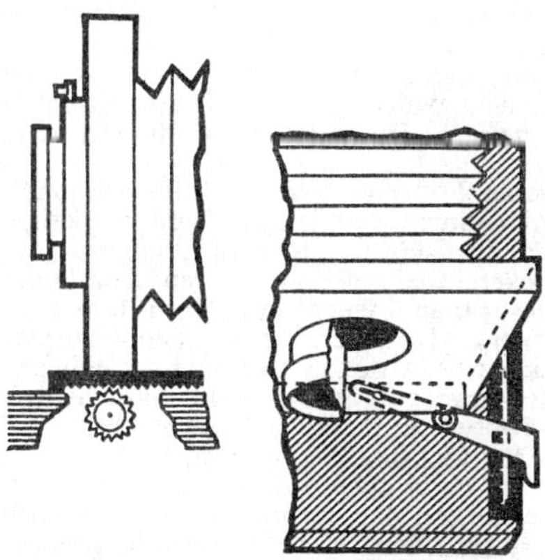

FOCUSING MOVEMENT. Left: A rack-and-pinion movement racks the lens panel forward and backward. Right: An alternative system moves this panel by a pivoted focusing lever.

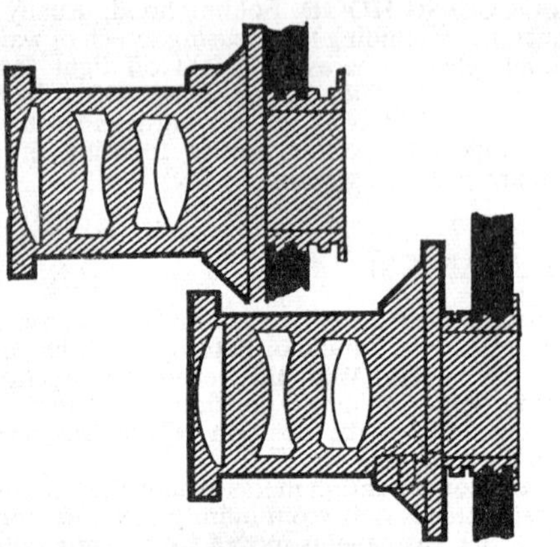

HELICAL FOCUSING MOUNT. The lens unit moves forward as a whole in a screw mount running in a similar thread in the camera body. Easy to couple to rangefinders. Left: *Infinity position.* Right: *Focused on near distance.*

made to engage with a pin in the other, a large rotary movement of the lens can be made to produce a relatively small forward shift. This is the principle of the helical focusing mount which is widely used in one form or another.

The helical focusing mounts on modern cameras may have either three helical grooves, or the groove and pin arrangement may be replaced by the more accurate, but more expensive, multiple-start screw-thread.

As the rotary movement of the helical screw is so much greater than the forward shift, it is possible to mark off a very accurate scale around the edge of the mount—the shift from infinity down to 3 feet generally covering about three parts of a full turn.

A pure rotary movement of a helical screw mount also rotates the lens with it—an inconvenient feature from the point of view of reading aperture scales, fitting rectangular lens hoods, etc. Most modern helical focusing mounts therefore incorporate a rectilinear movement which rotates only the mount but not the lens. The latter moves directly in and out.

Close-up Focusing Mounts. With cameras of normal dimensions it is difficult to design a helical focusing mount with a single screwed lens barrel that will focus the lens from infinity to closer than 3 feet (1 m.). This is a matter of geometry. However, some manufacturers—particularly in the 35 mm. and sub-miniature fields—have got over this difficulty by using two concentric threaded tubes interconnected with the focusing ring or arm in such a way that the outer tube unscrews fully before the inner one can start turning. When the focusing ring is turned back, the inner tube screws in first and the outer one next. The extra extension provided in this way allows a 35 mm. camera with a 5 cm. lens to be focused down to as close as 4 in. (10 cm.). This type of mount is more suited to miniature reflex cameras than to cameras with a separate viewfinder because it becomes impossible to compensate for parallax and difficult to focus by measurement at such close quarters.

Front Element Focusing. Many hand cameras—particularly the less expensive types—employ front element focusing in which the focal length of the lens is altered instead of its distance from the sensitive surface. This is done by moving the front element of the lens forward, and for this purpose the lens mount is split and the front element is set in a helical mount. As the front element of the lens is rotated, the separation between the lens elements increases, and the focal length shortens. The back element of the lens does not move, but as the lens now has a shorter focal length, it is now in front of its infinity position —i.e., it is focused on a closer distance.

The focusing scale is engraved around the rim of the lens, where a relatively large amount of rotation corresponds to a small shift of focus.

Although front element focusing is cheap and convenient, it is not by any means ideal. Its greatest drawback is its effect on the aberrations of the lens. These cannot be highly corrected over the whole range of lens element separations and an average value of correction, which is always something below the best, must be accepted.

Front element focusing complicates the use of such things as rectangular lens hoods graduated filters and polarizing filters.

Focusing Back. It is not always necessary or desirable to focus a lens by moving it backwards or forwards. Alternatively, the sensitized material itself may be moved closer or farther from the lens to obtain the correct focus. This arrangement is made possible on some cameras by means of a focusing back—

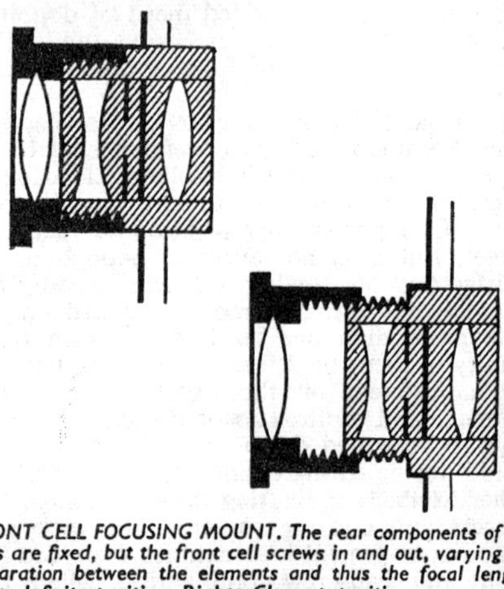

FRONT CELL FOCUSING MOUNT. The rear components of the lens are fixed, but the front cell screws in and out, varying the separation between the elements and thus the focal length. Left: *Infinity position.* Right: *Close-up position.*

that is, the back of the camera can be moved to alter the lens-to-emulsion distance. This is effected by a normal rack and pinion mechanism; the rack used for focusing the lens normally is commonly made long enough to serve the back mechanism as well.

The principal advantage of the focusing back is that the image size on the focusing screen does not undergo great changes of size when brought in and out of focus—as occurs when focusing the lens itself. This is of importance when setting-up the camera to reproduce a particular image size: the camera distance is adjusted until the image on the screen is the required size and can be then focused with the back without upsetting the image size very much.

Most field and technical cameras are fitted with a focusing back.

The Focusing Scale. There is no standard system for calibrating focusing scales. One very good reason for this is that the number of divisions of the scale that are necessary between infinity and the closest focusing distance varies according to the focal length of the lens.

At the same aperture, a lens of 50 mm. focal length has twice the depth of field of a lens of 100 mm. focal length, so that it needs only half the number of subdivisions of its focusing scale for the same accuracy in focusing.

Very short focus lenses can, in fact, dispense with a focusing scale altogether. Instead, they can be focused quite satisfactorily in zones. The whole focusing scale is split into two or three zones, classified as near and distant, or near, midway and distant, and generally marked on the scale with coloured spots.

Long focus lenses, on the other hand, cover a relatively narrow depth of field and call for a focusing scale calibrated in short steps up to distances of several hundreds of feet in order to be sufficiently accurate.

Calibration. Where the focusing scale of a camera has been damaged or lost, it is a fairly simple matter to calibrate a substitute, especially where the focusing adjustment has an infinity stop—as it usually has. (If there is no stop, the infinity position can be found by putting a piece of ground glass in place of the plate or film, and focusing visually on a distant row of trees or buildings.)

The distance (d) that a lens of focal length (f) must be moved forward from the infinity position to focus sharply on any object at distance (u), is given by the formula:

$$d = \frac{f^2}{u - f}$$

If f is in inches, d and u must also be in inches.

Depth of Field Scale. Many cameras, in addition to a focusing scale, carry an indicator that shows the near and far limits of the depth of field of the lens at any particular aperture. The depth is often shown on the half of the focusing scale carrying the reference arrow. On each side of the arrow are scales corresponding to the lens aperture numbers. The positions of the pairs of aperture numbers indicate the limits of the depth of field on the adjacent focusing scale. For example, with a 90 mm. lens, focused on 10 feet, the upper scale would show *f* 8 on the depth scale opposite 8 feet on the focusing scale, and on the lower scale (on the other side of the focusing reference arrow) *f* 8 on the depth scale would appear opposite 12 feet on the focusing scale. This system of marking shows at a glance that when the lens is focused on 10 feet and stopped down to *f* 8 the depth of field extends from 8 feet to 12 feet. The depths at other apertures are shown in the same way opposite the relevant aperture values.

Where the depth of field is not shown directly on the lens mount in this way, it is often given in the form of a table engraved on a metal plate and attached to the side or back of the camera body. F.P.

See also: *Camera movements; Focusing.*

FOCUSING SCALE. Scale of distances marked on the focusing mechanism of a camera. When a particular distance is opposite the scale reference arrow, the camera lens is focused sharply on all objects at that distance.

FOCUSING SCREEN. Sheet of ground glass mounted in a frame and fixed in the camera in such a position that the camera lens forms a visible image on its ground surface.

In the normal types of screen-focusing cameras, the screen can be slid or swung away to allow the sensitive material to take its place. In the reflex camera the screen is fixed in the top of the camera and the rays of light passing through the lens are reflected on to it by a hinged mirror.

In each case the ground surface of the screen lies at exactly the same distance from the lens (measured along the optical axis) as the surface of the sensitized material. So that when the image is focused on the screen, an identical image will be formed on the surface of the sensitized material.

The clarity of the image on the screen is greatly affected by the texture of the ground surface. A coarsely ground screen gives a bright and well-distributed image, but it does not permit really fine focusing because the definition is limited by the granularity of the screen. A finely ground screen will give critical definition, but the illumination tends more and more to be confined to a small area of the screen lying on a line from the observer's eye to the back of the lens.

To counteract the drop in image brightness

from the centre to the edges of a focusing screen, a field lens is often used. This is a very thin Fresnel lens which evenly distributes the screen brightness.

In place of ground glass, some modern screens have a specially grained texture—often impressed in a plastic sheet. A combination of this with the Fresnel field lens principle has been used in some cameras to produce a particularly fine and evenly brilliant screen image.

Focusing screens often have pencilled grid lines to aid composition.

See also: *Focusing.*

FOCUSING STAGE. Accessory for miniature cameras and other models with interchangeable lenses (except reflex cameras) to permit accurate focusing on a ground glass screen when taking close-ups or photomicrographs.

Construction. The main parts of a focusing stage are:

(1) A ground glass screen housing, usually with a focusing magnifier.

(2) A camera holder.

(3) Provision for holding the lens, usually with a focusing mount incorporated.

(4) Provision for moving the ground glass screen housing and the camera body alternately into position behind the camera lens.

Focusing stages are always designed for specific cameras, but the principle is the same in each case. The ground glass screen housing is mounted on a movable platform together with the camera (without its lens). The housing is accurately matched to the camera, so that the distance from the subject to the ground glass screen is exactly the same as the distance from the subject to the focal plane of the camera.

The focusing magnifier may be a vertical or an inclined or right-angle one. Vertical magnifiers yield an upright but laterally reversed image; inclined and right-angle magnifiers incorporate a mirror or prism reflecting system above the ground glass screen and form an upright image which may be laterally reversed or right-way-round. Since focusing stages are mostly used for copying, the orientation of the image is not vitally important and the viewing direction is largely a matter of convenience with the particular set-up. In most cases the magnifiers are interchangeable.

For normal work, the power of the magnifier is about 4–5× but for fine focusing, high power magnifiers (up to 30×) are available.

The camera holder keeps the camera firmly fixed and accurately aligned alongside the focusing screen on the movable platform.

The movable platform permits either the camera or the focusing screen to be brought into position over the lens. The movement may be sliding or rotary. Instruments of the latter type are often called focusing turrets. Some types of focusing stage dispense with the movable platform altogether; the ground glass screen housing or the camera is simply screwed or clamped in position on the lens housing.

Using the Focusing Stage. The focusing stage is generally set up so that the camera points vertically downwards, though it may equally well be used horizontally. Camera and lens are fitted in position; the ground glass focusing screen is normally a part of the stage. With the ground glass screen housing in position behind the camera lens, the image is focused on the screen. The camera body is now moved over to replace the screen housing behind the lens. This usually requires a simple movement of the sliding or rotating platform. The exposure is then made by releasing the camera shutter in the normal way.

Owing to the unavoidable separation be- tween the lens flange and the camera, the lens cannot normally be focused on infinity. Special ways of mounting the lens—e.g., by fitting a short barrel to a lens of long focal length—are adopted to overcome this limitation.

Focusing stages are more often used fo close-up photography and copying, where th extra extension (about ½ in.) is necessary anyway Additional extension tubes or bellows extensio units are employed to obtain larger scales c reproduction. Focusing is carried out with th normal focusing barrel of the lens, or a speci focusing mount between the lens and the stag

The focusing stage is often the basis of more extensive copying outfit incorporating a easel, support pillar, and lighting system.

By coupling the focusing stage to the tul of a microscope with an adapter, it can l used for photomicrography. When the sta forms part of a camera system, a suitab adapter is generally listed as an option fitting. Usually a clear glass "aerial" focusi screen is also available. L.A.M

See also: *Close-ups; Reflex housing.*

FOCUSING TURRET. Form of focusi stage, with the ground glass screen holder a the camera mounted on a rotating, instead a sliding platform.

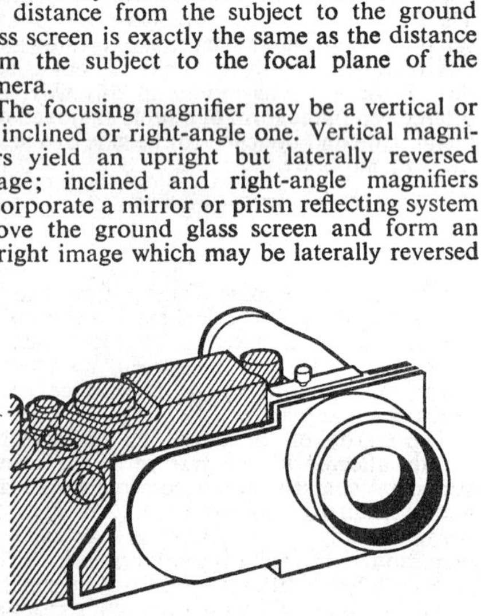

FOCUSING STAGE. *A sliding platform carries the camera together with a ground-glass screen housing. After focusing on the screen the camera moves into place behind the lens.*

FOG (ATMOSPHERIC). Mist is composed entirely of water-vapour particles and occurs in both town and country; fog is mist plus smoke (or more rarely, fine dust), so it occurs almost always in densely populated and industrial areas. The smoke particles are dark brown or black, therefore they tend to absorb all the light that falls on them. Under these circumstances, even deep filters make little improvement to the clarity of the picture since even the red and infra-red rays are absorbed.

Photography in fog, as in mist, is a matter of broad effects. In fog, however, the lights of street lamps, torches, vehicles and shops, offer possibilities which are often exploited for pictorial effects.

Exposure. There are no very definite rules for exposure in fog. So much depends upon the kind of result aimed at, the nature of the subject and its distance from the camera.

The exposure is not critical; a very wide range of values will give acceptable results. In general, exposures tend to be lengthy and it is advisable to use a fast panchromatic film.

Technique. To derive the greatest pictorial benefit from fog and mist, the following technique is advised.

Keep one object, preferably dark-toned, in the immediate foreground of the picture, as this helps to accentuate recession of planes and adds a "stepping-stone" for the eye into the picture beyond. An archway, statue, column, tree, etc., is suitable for this. If the fog is insufficiently dense to divide near-at-hand planes, use a diffusion disc on the lens. No exposure increase is needed. Should the subsequent print still exhibit too much definition and clarity, reprint, using the diffusion disc on the enlarging lens. Contrasts are flattened by this treatment, and paper one grade harder is required to counter the lack of dark tones.

The sun, in trying to break through, gives heightened effect on moist roof-tops and masonry. A viewpoint which gives oblique back-lighting is recommended. This produces unusual colour pictures too.

Photographing lighted street lamps, shop-windows, etc., in colour in fog during the day is mixing light of different colour-temperatures; however, it comes off very well in many instances.

Since visibility is limited, subject-interest should not be farther than 30–50 feet away from the camera, according to density and lighting. Use of flash, unless much nearer the object than the camera, has little to commend it. A strong beam of light, as from car headlamps, can give dramatic spotlight if falling at a suitable angle on to a person, obelisk, building, etc.

Limitations. Contact prints cannot do justice to pictures taken in fog and mist; consequently such negatives must always be enlarged. One or two in an album or collection are attractive and "different", more are apt to depress and defeat their own purpose.

It is best to have a definite location and subject in mind before venturing with a camera in the fog—two or three hours wandering is conducive to chills and colds. Remember that mist and fog are damp, and prolonged exposure to them will mean that the camera has to be dried afterwards. P.J.

See also: *Mist*.

FOGGING. Fog or veil is any visible deposit or density in the negative or print not forming part of the photographic image. Several kinds of fog can be distinguished.

(1) Optical fog, caused by the action of unwanted light.

(2) Development fog caused by excessive silver deposit during processing.

(3) Chemical fog caused by unwanted reactions in the processing solutions (e.g., aerial fog).

Optical Fog. Unwanted light may reach the sensitive material:

(1) During loading or unloading the camera.
(2) In the camera itself.
(3) During handling in the darkroom.
(4) Due to unsafe darkroom illumination.

(1) Fog can be caused during loading and unloading of roll films if the spool is allowed to unroll itself partly after being unsealed or before being sealed up. The light then penetrates round the edges of the film and gives rise to patches of density extending from the edge and getting lighter towards the middle of the film.

Loading or unloading a film in brilliant sunshine carries a similar risk; this may also show itself on the first few frames of a 35 mm. film if the cassette opening is not absolutely light-tight—e.g., because the velvet is worn.

Plate holders may similarly develop light leaks, especially through a faulty velvet light-trap. These produce thin streaks of fog across the negative, heaviest near the end where the dark slide is withdrawn.

To avoid fog from these causes, roll films should be loaded and unloaded carefully in subdued light or in the shade; 35 mm. cassettes should not be reloaded indefinitely (except special types designed for the purpose); and plate holders should be examined periodically for leaks and have the velvet brushed or renewed if necessary.

(2) Light leaks in the camera itself produce irregular fog patches on the negative. The shape and position of these patches will usually give some clue to the location of the light leak.

The most common leaks occur in the bellows of folding cameras, but may also be traced to badly mounted lenses, or to an unsafe film window or to damage to the camera.

Scattered light and flare inside the lens also produce fog. This may be a thin general fog, or it may form patterns and patches due to reflections inside the lens when photographing against the light. Both types are greatly reduced with coated lenses.

A dirty lens also scatters light.

All these types of fog (except those arising from leaks in the camera back) are characterized by the fact that they extend only to the limits of the image frame, but not into the margins of the film or plate which are normally clear.

(3) Stray light due to leaks, flare, and scattering can reach the paper on the enlarger baseboard during enlarging. This is analogous to stray light in the camera, and can be prevented by seeing that the enlarger lamphouse is light-tight, by avoiding bright reflecting surfaces on or near the enlarger, by masking the negative in the negative carrier to the actual area required, and by using a coated enlarging lens.

(4) Unsafe darkroom illumination due either to use of a safelight which is too bright or of the wrong colour: too long exposure of the material to it will produce over-all fog or uneven patches. Generally materials are more likely to be fogged before processing than after they have been in the developer for some time.

Fog of this nature is avoided by using the correct type and strength of safelight and keeping the material at the recommended distance from it. Panchromatic films and plates which are to be developed by inspection are best desensitized first.

Unsafe darkroom lighting also gives rise to several peculiar fog phenomena, such as the Sabatier effect and partial reversal of the image, while an unsafe red safelight may even destroy part of the latent image.

Development Fog. Development fog is the result of the development of unexposed silver halide grains. A special kind of development is also caused by the deposition of silver from the developer.

All emulsions contain a certain proportion of silver halide grains which are reduced to silver without exposure to light. This is the basic fog density and depends on the nature of the emulsion (e.g., high-speed panchromatic films usually have a higher basic fog density than slow process plates). It can, however, be influenced by storage conditions, by the activity of the developer, and by the use of restrainers or developer improvers.

Unsuitable storage increases the number of developable but unexposed silver halide crystals—often unevenly—and causes local or general fog.

Overactive developer, protracted development (including development to finality), and too warm a developer reduce a proportion of the unexposed silver halide crystals and usually give rise to general fog.

Potassium bromide reduces the activity of the developer and inhibits the development of unexposed salts. Developer improvers have a similar effect and will often also counteract the tendency of stale materials to become fogged.

Certain fine grain developers, particularly those incorporating a silver solvent, deposit silver on the image from the solution. A certain amount of this silver also tends to be deposited all over the film or plate as a general fog.

Chemical Fog. This covers all types of fog caused by side reactions in the developer and fixer. Two general groups can be distinguished (both may occur together):

(1) Silver deposits.

(2) Chemical stains.

(1) Silver deposits may be grey or coloured, depending on the fineness of the deposit. One well-known form is dichroic fog, which appears reddish by transmitted light and bluish green by reflected light. The main causes are:

Silver deposition from the developer: this is similar to the silver deposited by solvent developers, but it is more uneven and may be caused by contamination of the developer by fixing salts.

Silver deposition from the fixer: this may be the result of an excessive concentration of silver salts in an exhausted fixer. Alternatively, if the fixer is not sufficiently acid (because large amounts of developer have been carried over into it) parts of the image will go on developing in the fixer and produce solvent fog.

To prevent this type of fog, the fixer must not be overworked and special care should be taken to avoid contamination of the developer by fixing chemicals. A good rinse between development and fixing will prevent contamination of the fixing solution by developer.

(2) Chemical stains can also form in the developer or fixer. The commonest cause is aerial oxidation of the developer, from too much exposure to air during development. The oxidation products of certain developing agents (notably hydroquinone, pyro, and pyrocatechin) are strongly coloured and have marked staining properties. Such stains are particularly disturbing with prints, as the paper base easily takes and shows the stain. A general stain in a negative is less troublesome; in fact some developers (e.g., pyro-metol) are designed purposely to produce such a stain to improve printing quality.

Usually, however, such stains are not popular on negatives since they make estimation of printing exposures difficult. Also, stains of this nature are frequently uneven, and will thus show up in the form of patches on the print.

To avoid aerial oxidation, the material must not be exposed too much to the air during development, and prints must not be overdeveloped (certain contact papers are specially sensitive in this respect). Developers with too low a sulphite content have the greatest tendency to produce stains.

Aerial oxidation can also occur in the fixing bath if it is not sufficiently acid—e.g., because of carried-over developer. A stop-bath between development and fixing (with prints an acid stop-bath) helps to prevent this fog from forming.

Too warm or too exhausted a fixing bath, especially one of the acid-hardening type, may produce uneven fog by toning parts of the image brown. This reaction is similar to that of the hypo-alum toner.

With colour films and colour prints, incorrect processing (wrong developer, wrong temperature, etc.), can produce both silver deposits and colour fogging—often noticed as a colour cast in highlights.

Removal of Fog. Uniform fog, provided it is not too strong, is not usually troublesome in negatives. If it is due to silver deposits, it can generally be removed by a weak reducer. For prints, a thiocarbamide reducer can be used; for negatives, Farmer's reducer is usually most convenient.

Chemical fog caused by oxidation will often disappear if the image is bleached in a permanganate bleacher and redeveloped in a non-staining developer (but not a fine-grain formula).

Restoring Unused Fogged Materials. Any plates, films and bromide paper that have been accidentally exposed to light may be made usable by soaking for five minutes in either of the following solutions.

Chromic acid	110 grains	6·2 grams
Potassium bromide	220 grains	12·5 grams
Water to	40 ounces	1,000 c.cm.

Potassium bichromate	160 grains	8·3 grams
Hydrochloric acid	1 ounce	25 c.cm.
Water to	40 ounces	1,000 c.cm.

The treatment should be followed by thorough washing and drying and the whole process must be carried out in darkness. The speed of the material should be rated at 1/5 to 1/10 of the original figure. L.A.M.

See also: *Faults; Keeping qualities of materials.*

FOG LEVEL. Basic density of fog produced by development of a sensitized material without exposure. The minimum density of the photographic image produced by exposure must be sufficiently above the fog level to show detail. Paper prints should have no fog level at all, or the highlights will appear degraded.

See also: *Fogging; Speed of sensitized materials.*

FOLDING CAMERA

All the early cameras had to fold because they used such large plates that a rigid body would have presented serious problems in transport and storage. But today the term is restricted by usage to cameras of the quarter-plate (3¼ × 4¼ ins.) or smaller sizes.

Sizes. The majority of folding cameras in the popular class take size 120 or 620 roll film and yield 8 exposures 2¼ × 3¼ ins. (6 × 9 cm.) or 2¼ × 2¼ ins. (6 × 6 cm.). A number of models are also made for 16 exposures 1⅝ × 2¼ ins. (4·5 × 6 cm.) on the same size of film. Certain types will take two picture sizes, e.g., 2¼ × 3¼ ins. and 2¼ × 2¼ ins. or 2¼ × 3¼ ins. and 1⅝ × 2¼ ins., by means of a thin metal mask inserted into the film aperture. Sometimes this mask is built in, and folds away into the film chambers.

The next most popular film is size 127 for 8 exposures 1⅝ × 2½ ins. (4 × 6·5 cm.). Reduced picture sizes (e.g., 12 exposures 1⅝ × 1⅝ ins. or 4 × 4 cm. and 16 exposures 1⅛ × 1⅝ ins. or 3 × 4 cm.) are rarer but exist. Cameras taking other roll film sizes (e.g., 116, 616, 29, etc.) are virtually obsolete.

The folding camera is today mainly a thing of the past, and only very few models are made, although the folding press-type of cameras in the 4 × 5 ins., ¼ plate, and 6 × 9 cm. formats are still popular. The latter can be regarded as mainly professional equipment.

An appreciable number of precision folding cameras use 35 mm. miniature film. These are, however, miniature cameras in all respects and will not be considered here.

The simplest type of folding camera is simply a box camera designed to go into a smaller space. Each manufacturer added his own improvements to this basic type, but most of the major developments are common to all makes.

Construction. The classical folding camera is a shallow box, one side of which hinges outwards and can be locked at right angles to the body to form a baseboard. This baseboard supports the lens and shutter when open, and protects them when closed. A light tight bellows joins the lens panel to the body. The back of the body has either an attachment for plate holders and a focusing screen, or if it is a roll film camera, spool chambers, a pressure plate, film guide, and number windows. Some folding cameras are made to take both films and plates.

The lens panel can be moved to and fro to focus the lens. Focusing is generally from 3 feet to infinity, but on double and triple extension types the baseboard telescopes out to give increased extension for close-up photographs.

Most folding cameras for popular use have a self-erecting lens mounting that springs into position automatically when the baseboard is released. Once in position the lens is focused either by rotating it in a helical focusing mount, or moving it bodily by a lever acting on the self-erecting mechanism. Sometimes there is no hinged baseboard and the lens panel simply springs forward into position on a lazy-tongs device.

Folding cameras for technical or close-up work, however, still retain the baseboard and

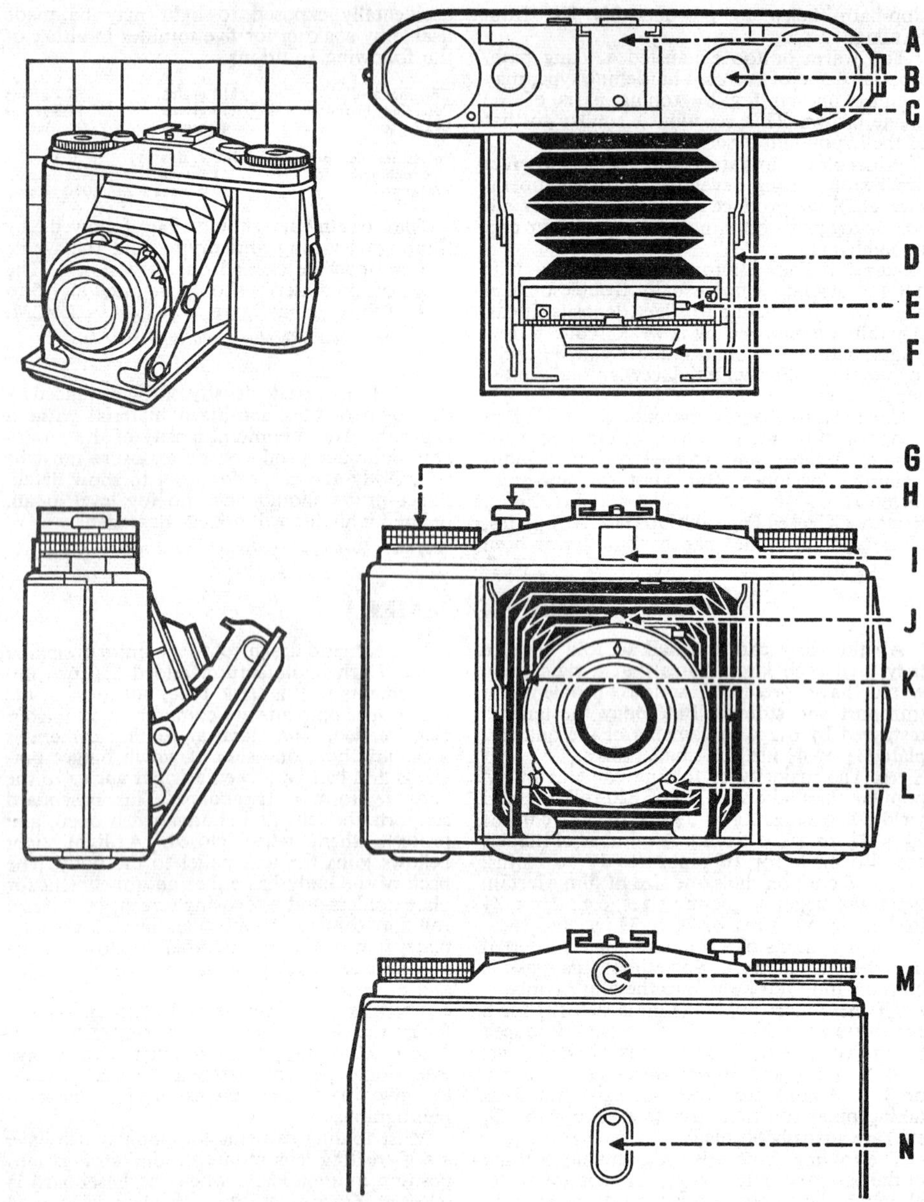

TYPICAL FOLDING ROLL FILM CAMERA. **Top left:** *General view to scale against grid of 1 in. squares.* **Centre left:** *Side view with camera half closed.* **Top right:** *Top view.* **Centre right:** *Front view.* **Bottom right:** *Rear view.*

The main features and controls are: A, accessory shoe. B, front opening button. C, depth of field calculator (is often engraved on the front mount next to the distance scale). D, strut holding the lens panel rigid. E, flash socket. F, front cell focusing mount. with distance scale. G, film transport knob. H, release button. I, viewfinder. J, shutter tensioning lever. K, cable release socket. L, synchronizing lever and self-timer (exists in various forms and positions according to shutter model). M, viewfinder eyepiece. N, film window for observation of backing paper during winding. Usually covered with a safety shutter.

guide rails because it is not possible to incorporate the necessary double extension and rising and cross front in a self-erecting lens panel.

Optical Equipment. Normal folding roll film cameras are fitted with 3- or 4-element lenses of triplet construction and maximum apertures from *f* 6·3 (*f* 8 or 11 in simplest models) to *f* 3·5 or *f* 2·8. Certain precision cameras of this type may also feature more advanced 5-element lens systems.

Normally the lens is not interchangeable (folding, press and field cameras are exceptions), but the standard models of most makes are available with a choice of lenses and shutters to suit different price levels.

Many folding cameras in the expensive class have a built-in rangefinder coupled to the lens through the self-erecting mechanism. Adjustment of the image of a particular object in the rangefinder automatically focuses the lens on that object.

The rangefinder is adjusted by a lever or knob on the camera body.

On a number of models, a rangefinder is built in but not coupled to the lens. Measurement of distance and focusing involves two operations in such a case.

The better folding cameras almost always have direct-vision optical viewfinders.

Some cameras with coupled rangefinder focusing use the same window for rangefinder and viewfinder. In this way no time is wasted in shifting and refocusing the eye between adjusting the rangefinder and taking the photograph.

Shutter Equipment. Amateur folding cameras almost invariably carry a between-lens diaphragm shutter. The simplest versions have 2- or 3-speed types with X-synchronization, advanced models feature 4-speed to 10-speed shutters with maximum speeds up to 1/500 second and XM-synchronization.

On the earliest models the shutter release was always a part of the actual shutter mechanism. It was often a small projecting lever, difficult to locate, and by no means standardized in its method of operation. In more recent years the shutters of the better folding cameras were released by pressure on a substantial button or plunger on the body of the camera.

Film Transport. The film is wound on by a knob or key, usually directly connected to the shaft which engages the film take-up spool. On more modern cameras a film lock automatically arrests the transport mechanism when the right amount of film has been wound on, and also works a mechanical film counter. On most models, however, a red window in the camera back serves for observation of the numbers on the backing paper to determine the correct position of the film from one exposure to the next.

The risk of making a double exposure is avoided by mechanically coupling the shutter release and film winding mechanism so that the shutter cannot be released for a new exposure until the film has been wound on to the next frame. Such coupling is not a simple matter and it is found only on the more expensive folding cameras.

Scope. The convenience of the folding camera has been exploited in practically every field of photography. It has a close rival in the miniature camera which is equally adaptable. There is very little to choose between the actual bulk of the two cameras. The most powerful argument in favour of the miniature camera is the increased versatility of the expensive precision models with their wide range of special accessories. The larger folding camera can advance the argument that it takes a bigger picture and is less troubled by dust, scratches, grain in the film, and inevitable deterioration in the quality of the image when greatly enlarged. F.P.

See also: *Films.*
Books: *How to Choose the Camera You Need*, by W. D. Emanuel and L. Gaunt; *Camera Techniques*, by H. J. Walls (both London).

FOOD. There is a steady demand from advertising agents and the cookery page editors of women's magazines for photographs of prepared dishes. Such photography may look easy but it is beset by as many difficulties as any other specialized branch of illustration.

Sensitized Material. While there are still examples of the use of black-and-white material for photographing food, the majority of such pictures are now taken in colour because the colours of an expertly cooked dish and of the garnishes, sauces, table linen and decorations provide much of their attraction.

The need for first class definition in the final reproduction rules out 35 mm. colour which tends to lose some sharpness in enlargement and reproduction.

Most workers in this field use an integral tripack cut film which they process in the studio darkroom.

Camera. Specialists almost invariably use a professional plate or sheet-film camera, and a negative size which may vary from ¼ plate to 10 × 8 ins. The negative material used today is mainly of the sheet-film type. Anything smaller than ¼ plate, is likely to involve a loss of quality intolerable for this type of work.

This work calls for a selection of lenses of varying focal lengths since the subject may be a single dish or a laid table. Large apertures are not important since there is no objection to moderately long exposures. But all lenses must be fully colour corrected and capable of giving critical sharpness right to the picture edges.

Lighting. Artificial lighting is always used because it is always under control and it allows conditions to be reproduced exactly if a retake is necessary. Normal lighting equipment is suitable provided that all the light sources are of the same colour temperature and adjusted to the colour material in use, by filters if necessary. Fairly powerful sources are used to keep the exposure down to reasonable limits, but none of the lamps must be close or powerful enough to dry up the food during focusing and exposure.

Preparing the Photograph. Food pictures which accompany "How-to-do-it" magazine articles or advertisement copy must be true to colour and sharp all over since the photograph must give an exact impression of the subject.

A good food picture is the result of careful planning and team work between photographic and culinary experts. The food must be both perfectly cooked and perfectly displayed if it is to appeal to the magazine reader.

As a rule, the photographer has little freedom to choose the picture; he works to a scheme decided upon after long and often tedious editorial conferences. And as most periodicals are planned several months ahead, he may have to provide special delicacies which are out of season together with suitable floral decorations. Fortunately such things are easier to obtain nowadays thanks to deep frozen food and flowers by air.

At the same time, the photographer learns by experience that certain foodstuffs do not photograph well and he must be able to suggest reasonable substitutes.

If the photographer is not himself an expert in the kitchen, he must work in conjunction with an experienced chef who can look after both the preparation and display of the food to be photographed.

The specialist food photographer works in a combined studio-kitchen or kitchen-studio, the main thing being to have the photographic apparatus on the spot where the food is being cooked. He must decide in advance on the correct style and arrangement of table linen, silver, china, flowers, etc. Close attention to such details is vital because the final picture must survive the critical inspection of expert housewives and cooks.

Taking the Picture. Once all the necessary preparations have been made, a rehearsal should first be staged with a dummy dish and a trial exposure actually run off to make sure that everything is as it should be. Retakes in this type of work can waste a great deal of time and money.

If the trial exposure is satisfactory, the picture is arranged and then, just before making the exposure, the real dish, freshly prepared, is slipped into position. In this way the food appears fresh and attractive; nothing looks worse than a meal that has been standing for some time. And under the powerful lighting normally used, the colour and general appearance of food can change very quickly.

The arrangements must be organized to run smoothly and without crises. Careful timing is called for; most dishes take some time to bring to the point when they are ready for serving, and a dried up vegetable or a cold, greasy-looking meat dish can ruin the picture, discourage the cook and put everyone concerned to the trouble of a retake. H.G.C.

FOOT (FT.). Lineal measure equal to 12 ins. or 1/3 yard. The foot measure is common to all nations and periods being based on the length of the human foot from which it takes its name.

See also: *Weights and measures.*

FOOT CANDLE. Unit of illumination representing the light intensity over a surface one foot away from a standard candle. It is equal to one lumen per square foot.

See also: *Light units.*

FOOT LAMBERT. Unit of luminance representing the brightness of a perfectly reflecting and diffusing surface when the light falling on it has an intensity of one foot candle. Also known as an effective foot candle.

By international agreement luminance is now expressed as luminous intensity per unit area, e.g., in candles per square metre or square foot.

See also: *Light units.*

FOOTAGE COUNTER. Measuring device in a cine camera to show the length of film of a load already used up. Usually the mechanism is very simple and relies on measuring the thickness of the film wound up on the take-up spool, giving an approximate indication of the exposed length.

See also: *Cine camera.*

FORCEPS. Various types of forceps are used in photography for manipulating prints in the developer, stop bath, fixer, and other solutions to keep the fingers dry and prevent them from becoming contaminated with solution and spreading stains. Forceps also reduce the risk of contamination of the various solutions with each other, and save a great deal of the time taken up by rinsing the hands or fingers before handling unexposed papers again.

Print forceps may be of various sizes (the largest are generally most practical) and are available in plastic or stainless steel.

One pair of forceps is required for the developer and another for stop bath and fixer, to avoid contamination. In practice the print is lifted from the developer with the developer forceps, dropped into the stop bath without immersing the forceps, and handled from then on with the fixing forceps.

To prevent accidental interchange, the two pairs of forceps should be different in size, type, or colour.

Plastic forceps are preferable for the fixer, since stainless steel collects a silver deposit from the silver in the fixing bath and becomes discoloured in time. The silver can usually be removed by immersion in an ordinary negative reducer (e.g., permanganate), or by polishing with a silver polish.

See also: *Print paddle.*

FOREGROUND. The part of the scene that lies nearest the camera. The great depth of field of a miniature camera makes it possible to include very near objects in the foreground and still render them sharp. With larger formats this may only be possible by stopping the lens right down and giving a very long exposure. If the camera is fitted with movements, the front can be swung to bring the foreground into sharp focus without stopping down the lens.

A blurred or out of focus foreground is generally unpleasing, but there are occasions when it helps to add depth to the picture by pushing the subject back into a more appropriate perspective. A similar effect can be achieved by the use of heavy, dark tones in the foreground to frame the more distant subject—e.g., by shooting from under an archway or through overhanging foliage.

See also: *Camera movements.*

FORENSIC PHOTOGRAPHY. Application of photography to the law, criminal investigation, police court proceedings and the like.

See also: *Crime photography; Evidence by photographs; Fingerprint photography; Forgeries; Police photography.*

FORESHORTENING. Appearance of an object when viewed obliquely. A relatively long and slender object—e.g., an arm or leg—when seen at an oblique angle by the camera gives an image which is by comparison short and thick. The size of near parts of the subject may appear greatly exaggerated if, in addition, the viewpoint is close.

See also: *Perspective.*

FORGERIES. Photography has made possible a number of techniques for detecting forgeries.

Forgery may be defined here as the fraudulent making or alteration of any document with intent to deceive or, as the old common-law definition runs, "to the prejudice of another man's right". "Documents" in this context may be widely construed to mean not only letters, receipts, cheques, wills, registers, business papers, certificates of ownership, passports, etc., but also paper money, books and even pictures.

Forgery can of course be a serious criminal offence; its detection and proof therefore call for careful expert examination of the questioned documents. Photographic methods constitute a generally useful and sometimes essential part of this examination.

We may distinguish between documents which are:

(1) Wholly false, or

(2) Partly genuine and partly false, where the forgery has consisted in making an erasure and/or alteration and/or addition to some material part.

In general, photography is of less importance in the examination of documents which are wholly false, in which the chemical and physical analysis of papers and inks is in most cases of relatively greater importance.

Technically speaking, the expert examination necessary may also be divided into:

(1) Handwriting comparisons, and

(2) All other types of examination.

The reason for this distinction is that the study of handwriting (including typewriting) to test genuineness or determine authorship is a very specialized expertise, acquired only by considerable experience and with only a limited number of practitioners, whereas the other types of examination depend much more on the application of techniques familiar to every competent forensic scientist.

The uses of photography in this field fall broadly into three categories.

(1) The permanent recording for the examiner, or illustration for a court, of visual findings made during the examination. This commonly requires photography at magnifications of, say, 1 to 20 times, sometimes with the aid of colour filters. Sometimes also colour photography is necessary for a proper display of the significant findings.

(2) The duplication of parts of documents as a stage in their comparison. This applies particularly to handwriting examinations. The expert's normal procedure there is to copy the questioned and control documents at rather over natural size (sometimes a direct photocopying process suffices), cut out and assemble all the examples occurring of each individual letter and stroke formation, and examine analytically whatever regularly recurring characteristics may be revealed by this process.

(3) The use of photography as a scientific tool in its own right for gaining information obtainable in no other way. This normally involves photography using an isolated region of the spectrum, possibly one outside the visible range. The following photographic techniques may all be used for these purposes.

Straight Macrophotography. Usually this is done in black-and-white or (less commonly) colour. Apart from its use in handwriting comparison already mentioned, this is used for the demonstration of "awkward" joins produced by additions and alterations. Magnified recording of stroke intersections can show the order of writing, as a wet ink stroke which crosses one already dry diffuses into the latter. (This does not apply to ball point.) In the comparison of genuine and forged paper money and certain official documents the forgery reveals itself partly by the fact that its printing lacks the clarity and crisp quality of that on the genuine document.

Specimens of handwriting and typewriting may also be compared by using superimposed transparencies. The complete identity of two signatures is a proof that one is forged and was copied from the other by tracing, since no one ever makes two completely identical signatures. The same process will reveal those small irregularities of alignment and spacing which characterize a particular typewriter.

Filters. The use of complementary and contrast filters in the visible region can: isolate material from a distracting background (cancellation marks on stamps or writing on a stained or closely printed surface; or render pale coloured writing as dark as possible; differentiate between two closely similar but not identical hues. As already mentioned, the most satisfactory result is sometimes obtainable only with colour photography.

Photography by Infra-red. The uses of this can be broadly divided into:

(*a*) utilization of the fact that infra-red penetrates paper better than visible light;

(*b*) utilization of the differing infra-red reflectivities of visually similar dyes and pigments.

The first may for example render incompletely erased pencil strokes legible, the incomplete erasure having left some graphite particles just under the surface of the paper. The second is perhaps the most useful single technique in distinguishing fraudulent alterations; even if these are so skilfully done as to be visually unnoticeable, the chances are that the inks of the original and added entries will have different infra-red reflectivities. In some cases, also, certain otherwise difficult colours (e.g. browns) can be filtered out using infra-red.

Ultra-violet Photography. "Straight" ultra-violet photography is not often utilized. It requires only a source of ultra-violet radiation (normally a mercury-in-quartz arc) and a filter transmitting ultra-violet only. All emulsions are highly sensitive to ultra-violet. Such photography is of course limited to the 360–400 mμ region when ordinary photographic lenses are used, since glass absorbs wavelengths less than about 360 mμ. However, even with this limitation, it can be useful for differentiating certain visually indistinguishable pigments.

The examination and photography of fluorescence (i.e., the re-emission of absorbed ultra-violet as visible radiation) is on the other hand much used in this field. It is particularly valuable for revealing ink writing which has faded or has been obliterated by chemical means. The fluorescence of the paper base is affected by the chemicals which have soaked into the paper from the ink.

Since fluorescence is always very faint compared with the exciting radiation, it is essential to illuminate the document with ultra-violet only (ultra-violet lamp with Wood's glass screen) and to let only visible light reach the emulsion by using an ultra-violet-absorbing filter on the camera. The exposures will be long, and panchromatic emulsions should be used as the fluorescence may be of any colour.

Shadowgraphy with X-rays. This could of course be considered not strictly a photographic process, since it does not require a camera or lens, and since whatever it reveals could in many cases have been seen on a fluorescent X-ray screen. However, photography on X-ray film is normally more convenient. This is a standard procedure in testing the authenticity of paintings, since it may reveal earlier brush-work which has been painted over.

An allied technique uses a slightly radioactive sheet material instead of X-rays—e.g., acrylic polymer sheet containing carbon-14. The radioactive sheet, the document and a piece of X-ray film are clamped together as a sandwich and a rather long exposure given (hours or days). This may be used to reveal the structure of paper and similar materials which are more or less opaque to light but completely transparent to normal X-rays. It has been used to detect philatelic frauds by showing up otherwise invisible gummed joints, etc. H.J.Wa.

See also: *Crime photography; Evidence by photographs; Police photography.*

Books: *Evidential Documents*, by J. V. P. Conway (Springfield, Illinois); *Suspect Documents, their Scientific Examination*, and *Forgery Detection*, both by Wilson R. Harrison (London); *Questioned Document Problems*, by Albert S. Osborn (Albany, New York).

FORMALIN. Formaldehyde solution. Used as a hardening and tanning agent, especially for scratch-proofing.

Characteristics: Colourless liquid with unpleasant pungent smell. Formalin is a 40 per cent solution of formaldehyde gas (HCHO, molecular weight 30) in water.

FORMAT. Size of a book with particular reference to the relative proportions of the lengths of the sides. The depth of the page is always stated first—e.g., the format of this book is $9\frac{1}{4} \times 6\frac{3}{4}$ ins. Also applied to photographs.

FORMULA. In photography the term applies to the recipe of ingredients required to make up a particular processing solution—e.g., a developer.

It also applies to the symbols indicating the molecular composition of a chemical—e.g., $AgNO_3$ (silver nitrate).

See also: *Chemical symbols; Solutions.*

FOUCAULT, JEAN BERNARD LÉON, 1819–68. French physicist. One of the foremost photographic scientists in the first quarter century of photography. Constructed (1844, with Donné) a projection microscope, and (1849, with Duboscq) an electric arc lamp. Studied (1844, with Fizeau) the effect of light on daguerreotype plates, and produced the first daguerreotype photograph of the sun in 1845. Collected works edited by Gabriel and Bertrand (Paris 1878).

FOURIER ANALYSIS. Mathematical technique which can greatly simplify the analysis of complex waveforms, used in the evaluation of optical images and of recording systems. A waveform may be considered as the sum of its individual frequency components (sine and cosine waves), and the effects of any system by which it is transmitted or recorded may then be expressed at any frequency as a simple scale factor.

Book: *Fourier Methods in Optical Image Evaluation*, by H. E. Linfoot (London).

FOX TALBOT, WILLIAM HENRY, 1800–77. English archaeologist, chemist, linguist, mathematician and inventor of photography.

See also: *Talbot, William Henry Fox.*

FRAME. Single image in a series on a strip of film. Applied mainly to cinematography (for example in defining frame size and camera speed in frames per second), but also used in still photography where for instance a single exposure on a 35 mm. film is often referred to as a frame. As this frame size is normally 24 × 36 mm., an 18 × 24 mm. image size is often referred to as a half-frame.

The frame is also the boundary in the viewfinder which outlines the limits of the field-of-view of the camera. Framing a shot means aligning it within this viewfinder frame, so that the latter takes in just the subject field required.

FRAME COUNTER. Precision measuring mechanism built into certain more advanced cine cameras to show the actual number of individual frames exposed during the run of a film. It is used for accurate location of a particular spot on a film, when rewinding for double exposures, dissolves and other effects.

See also: *Cine camera.*

FRAMING PHOTOGRAPHS. There are two principal reasons for putting a photograph in a frame: so that it can be hung up easily and seen to better advantage, and to protect it from dust, damp and finger-marks while it is displayed.

The amateur, unless he is an experienced carpenter, is best advised not to attempt to make a frame. The result will almost certainly be a waste of time and timber.

A photograph that deserves a frame will usually justify the expense of having it professionally framed.

It is of course possible to buy frames ready-made. All the photographer has to do then is fix the photograph in the frame—not quite such a difficult task.

The actual designs of frames vary considerably. In its simplest form a frame can be made of quite plain wooden moulding; a more elaborate type may consist of a number of ornate borders of machine decoration. The modern trend however is to use a simple, narrow frame with plain sides.

Choosing the Frame. The frame should normally be plain and unobtrusive. Its purpose is to isolate the photograph from its surroundings so that the picture will attract and hold the attention. An ornate or showy frame only draws attention away from the picture or at least clashes with it.

If the photograph is to fill the whole space in the frame, the frame should be a wide one; otherwise the picture will not be sufficiently isolated. If the photograph is to appear on a mount, then the frame may be narrow, because in this case, the mount itself can provide a border. Generally, the bigger the mount in relation to the photograph, the narrower can be the frame.

The glass should be of the kind intended specially for framing pictures; this is much thinner than the glass used for glazing. The quality of this glass varies widely, so it should be carefully inspected for flaws and waviness. These faults show up much more over a black and white photograph than over a coloured painting.

It is a good plan to run a narrow strip of gummed paper around the edge of the glass and the inside of the frame before fitting the picture in position. This prevents dust and damp from entering from the glass side of the frame.

Framing the Photograph. Before starting the actual framing, the print should first be mounted. The mount must be fairly thick and fit the frame exactly. If a cut-out mount is used it must be backed with card of the same size. The glass in the frame must be cleaned thoroughly—especially on the inside surface. The mounted photograph is placed in the frame and secured along all edges with heavy gummed paper; this is stuck with half its width over the back of the mount and the remainder over the back of the frame.

Cord or wire for hanging the picture should be attached to the frame and not the mount. This is best done by fixing a screw eye into each side of the frame. The cord is attached to the eyes and kept short so that it does not show above the top edge of the frame (unless the picture is to hang from a picture rail). Normally it is better to hang it from a hook held by one of the special hard steel pins which can be driven into plaster or wood without leaving an ugly hole.

Passe-partout. A cheaper way of framing photographs is to use passe-partout binding.

For this type of framing it is essential for the print to be on a mount of least 1/16 in. thick.

A piece of picture glass must be cut to exactly the same size as the mount or the mount may be cut to match the size of the glass, using the glass as a template.

The mount and glass are then bound together

with the passe-partout tape. The latter can be used in one complete length, or cut into four strips—one for each side. This second method is the easier, especially when dealing with large pictures. The passe-partout in either case must be mitred on the corners of the picture.

In cases where it is ever likely to be necessary to remove the print from its frame or passe-partout binding, a backing card or board should be used. The backing is cut to the same size as the mount and fitted over it before applying the binding.

Stick-on hangers can be obtained for attaching the cord or wire to the back of the mount. These are surprisingly efficient although they may not look reliable. Another type has a ring on a split pin; the pin is pushed through the mount and secured by bending flat on the other side. If this system is used, obviously the mount must be backed with a second piece of cardboard; the split pin is then only pushed through the rear card. It is of course essential with this method to attach the ring before binding together the glass, mount and backing card.

Whether using a proper frame or passe-partout binding, some thought should be given to the colour. This should contrast with the wall on which it is to hang; if it blends into the wall its purpose (which is to isolate the picture) is defeated.

J.D.C.

See also: *Mounts*.

FRANCE

Like all inventions, photography was the culminating result of patient work in many countries extending over a period of centuries. Nevertheless, it is certain that an image of the exterior world was permanently recorded on a light-sensitive surface for the first time in France.

Early Experiments. Nicéphore Niépce (1765–1833) was the first man to succeed in recording an image in a camera obscura. In May 1816 he photographed buildings on his property near Châlon-sur-Saône on sheets of paper impregnated with silver chloride, but was unable to fix them. He experimented with various materials such as paper, glass and metal (zinc and copper). In 1822 he obtained recognizable images on glass coated with bitumen of Judea, though not in a camera obscura.

Between 1823 and 1826 he improved the process, which he named heliography and in which the image was obtained on a bitumen coated plate of zinc or pewter. After exposure the unaffected bitumen layer was dissolved away. Copies of line drawings obtained by superimposition could be satisfactorily etched and used as printing plates. Camera views, however, were too faint to be transformed into printing plates. Niépce's first view from nature was taken from his window in 1826. The following year he submitted a communication on heliography together with specimens to the Royal Society, London, who could not give it recognition because he did not disclose manipulatory details. Niépce was also the first man to suggest the use of an iris diaphragm to correct lens aberrations and eliminate marginal rays.

While Niépce's final object was to make printing plates of camera pictures, L. J. M. Daguerre (1787–1851) aimed simply at camera pictures, and it was his meeting with Niépce that led to his success. In 1827 Daguerre made Niépce's acquaintance through the optician C. L. Chevalier and on the 14th December 1829 they went into partnership.

The Daguerreotype. Both during Niépce's lifetime and after his death the process was gradually improved, finally resulting in the daguerreotype: a copper plate was coated with silver iodide and the photographic image was developed by the action of mercury vapour and fixed by what is today known as sodium thiosulphate. On 7th January 1839 Arago announced the invention to the Académie des Sciences. The French government then voted legislation for the purchase of the process "in order to offer it as a free gift to the world", and on 19th August 1839 Arago repeated his announcement and described the invention at a formal session of the Institut de France.

The daguerreotype process had a number of disadvantages, the principal one being that only a single copy of each exposure could be obtained. In February 1839, Hippolyte Bayard (1801–87) invented a negative-positive paper process. However, his direct positive process on paper, invented the following month, was much better. Unfortunately, it did not win commercial success and Bayard's name was forgotten for many years.

Later Developments. W. H. Fox Talbot's process was an improvement on the daguerreotype; it was perfected in France by Louis Blanquart-Evrard in 1846 and Gustave Le Gray in 1849. The latter obtained excellent transparent negatives by the use of dry waxed paper.

In 1851 Blanquart-Evrard opened an "imprimerie photographique" (photographic printing works). He made large numbers of photographic prints which were published in albums thus being one of the very first workers to undertake systematic mechanization of the photographic process.

In 1847, Claude Niepce de Saint Victor (1805–70), a cousin of Nicéphore Niépce, invented the albumen process. The sensitized surface consisted of silver nitrate, the developer was gallic acid and the fixer sodium thiosulphate.

In 1849, Le Gray suggested the possibility of using collodion in photography—a solution of gun cotton in ether, but the process was first invented by Frederick Scott Archer who published it in 1851. During the siege of Paris in 1870 the collodion process was used to make microfilms for sending messages across the enemy lines by carrier pigeon.

At this period numerous French experimenters developed various mechanical processes for the reproduction of photographs, e.g., Hippolyte Fizeau (1841)—Niépce de Saint Victor (1853) and Alphonse Poitevin (1855). In 1878, Petit invented a system of producing dots for the half-tone process. About 1858 Nadar succeeded in making instantaneous exposures by the light of a Bunsen battery.

Colour Photography. The next major invention after Niépce's to which Frenchmen can lay claim is that of colour photography. In his communication of February 1891 to the Académie des Sciences (Academy of Sciences) Gabriel Lippmann (1845–1921) explained the principles of the interference method.

But colour photography had been invented already in 1869 by two men working independently: Charles Cros (1842–88) and Louis Ducos du Hauron (1837–1920). In February of that year these two experimenters announced the principles of three-colour photography. They produced the first colour images by the superimposition of three plates.

In 1907 the brothers Auguste and Louis Lumière produced their autochrome plate, using a pigment process which made additive colour photography a commercial proposition.

Applications of Photography. France has made an important contribution to the study of movement by photography, which led ultimately to the motion picture. E. J. Marey (1830–1904) made a "fusil photographique" (photographic gun) in 1882. In 1883 he founded in Paris a centre for studying movement, and this still exists today under the name Institut Marey. In 1888 he perfected his process by making a "chronophotographe à pellicule" (film chronophotograph).

Finally on the 13th February 1895 the brothers Lumière announced their invention of the process which they christened "cinematograph".

Experiments in stereoscopy began very early, as is shown by the fact that daguerreotype pairs are in existence for stereo viewing. In 1891 Ducos du Hauron had perfected the anaglyph method of stereo viewing.

In connexion with the scientific applications of photography the following names deserve mention:

In 1861, Colonel Aimé Laussedat invented the process of photogrammetry, enabling linear measurements to be taken over great distances by photographic means.

In 1919, Poivilliers invented the stereotopograph, an apparatus working on stereoscopic principles for the measurement of heights from aerial photographs.

The history of photomicrography goes back to 1840. In that year Dr. Alfred Donné produced daguerreotypes with a microscope, and in 1845 he published the first book illustrated by photomicrographs.

In 1882 the brothers Paul and Prosper Henry made an important contribution to astronomical photography by devising a photographic method of recording the transit of Venus from the Pic du Midi observatory. As a result, an international union was formed for mapping the skies.

In the same year, Alphonse Bertillon took out a patent for a system of photographic identification, which is still in use in most countries of the world for criminal record purposes.

In 1908 Edouard Belin invented a method of transmitting pictures by wire. In 1914 the first news photograph to be transmitted by "belinogramme" was published. It is now in regular use in newspaper offices.

The Art of Photography. In its early days, photography was mainly used for portraiture. Most daguerreotypes were portraits. Some remarkable views of Paris were made by this process between 1840 and 1850, many of them by Bayard.

Gustave Le Gray and Blanquart-Evrard later made a number of landscapes and views of monuments. But the long exposure time required made the taking of action pictures difficult. Maxime Du Camp travelled in Egypt, Nubia, Palestine and Syria between 1849 and 1851 and brought back some remarkable "photographic drawings", as he called them.

Many amateurs took up photography as a hobby; amongst them Victor Hugo, his sons and his son-in-law Auguste Vacquerie made a fine collection of pictures in Jersey and Guernsey from 1853 to 1870.

Photography made its official public début in a section of the World Exhibition of 1855 in Paris.

During the whole of this period photographers concentrated on a realistic portrayal of their subjects, and the results, particularly in the case of Bayard's work, were generally not lacking in charm. Portraits were very conventional and, in general style resembled the work of contemporary painters. The outstanding pioneers were the Nadar family, foremost among them Felix Tournachon, called Nadar (1820–1910). Together with his brother he opened a studio in 1852 which was later taken over by his son Paul. Nadar was a curious type, a man-about-town, inventor and friend of Jules Verne; he went in for ballooning and all sorts of other adventurous experiments besides his photographic activities. His portraits, which depict the celebrities of a whole period of history, have a very distinctive and vigorous style and they gave rise to much controversy on the question of whether photography could be called an art, in the course of which Charles Baudelaire took his stand against photography while Eugène Delacroix was in favour of it.

Under the Second Empire (1852–70) another photographer, André Disderi, achieved great success with multiple portrait photographs in small format, which became known under the name "Carte de Visite". So popular did these become that as Napoleon III was riding through Paris at the head of his troops on the way to the war in Italy in 1859, he did not hesitate to stop the procession in order to have his portrait taken by the "Master" as a memento of the great occasion. At the end of the Crimean war (1855) photographic history was made when a photograph was taken instantaneously at the victory parade in Paris.

In the same period, two Parisian workers are worthy of special mention: the portrait photographer Adam Salomon and the landscape photographer Adolphe Braun; the latter also took excellent studies of contemporary Parisian life.

Later, serious photography, like original art, underwent an eclipse for a number of years, though it did become a popular pursuit among amateurs. Paul Nadar made some excellent portraits of important personalities; particularly worthy of mention are the serial studies in expression of General Boulanger and the chemist Michel Chevreul, which were published in the newspaper *Le Figaro* in 1886/7. But photographs as a whole became a bad imitation of the worst painting.

Puyo and Robert Demachy made some good pictorial studies with soft-focus effects. At the turn of the century a swarm of photographic clubs encouraged the spread of this rather over-rated technique.

One photographer was in a class by himself and a forerunner of the modern school: his name was Eugène Atget (1856–1927). Though he enjoyed no commercial success, he produced straight photographs without any special effects. Apart from their considerable documentary value, Atget's photographs still seem very fine today and have a certain naïve charm.

After the end of the first World War the most prominent French photographers were Madame Laure Albin-Guillot and Emmanuel Sougez. The latter is particularly noteworthy for having introduced the conception of pure photography into France. His work, the best of which probably consists of still life studies, shows how beautiful reality can be when presented completely without affectation and without recourse to retouching.

Amongst the older generation today, some work displays a stark realism; at the other extreme we find a whole school of surrealists.

There is also a school of journalistic photography, many of whose members produce work tending more to the illustrative than the documentary. Some aim to produce work that alternates between the poetic and the ironical.

A number of photographers of foreign origin have settled in France, and their influence has often been considerable.

Societies. Paris houses the oldest photographic society in the world, the Société Française de Photographie (French Photographic Society), founded in 1851 under the name of Société Héliographique and renamed in 1854. The Society brings together the most eminent personalities in French photography in the scientific, technical, industrial and artistic fields. It has a fine collection of apparatus and prints going back to the earliest times.

The French photographic industry is organized in an association, the Syndicat Général des Industries Photographiques (General Syndicate of Photographic Industries). The users of photography, professionals and dealers, belong to the Confédération Française de la Photographie (French Confederation of Photography). There is also the Syndicat National du Cinéma Sub-standard (National Substandard Cine Association), while some professional photographers belong to the Fédération Française de la Photographie (French Federation of Photography).

Amateur societies and clubs are organized in the Fédération Nationale des Sociétés Photographiques de France (National Federation of French Photographic Societies), which is a member of the Fédération Internationale d'Art Photographique (International Federation of Photographic Art). Several of these societies and clubs hold important international exhibitions.

Exhibitions. Every second year an exhibition of photographic equipment and products is organized in Paris, the principal French manufacturers and the French representatives of foreign manufacturers taking part. Since 1955 this exhibition, covering photography, cinematography, and the optical industry, has been held alternately in Paris and abroad. Its Paris home is the Grand Palais des Champs Elysées. Named the Biennale Photo Cinéma Optique the exhibition has an average of about 300,000 visitors, and there are some 300 exhibitors from all over the world. The 1957 Biennale was held in Washington, U.S.A. A number of important meetings of an artistic and cultural nature are held during the exhibition.

A Salon National is held annually in the Mansard gallery at the Bibliothéque Nationale (National Library). It is confined to photographers of French nationality or resident in France and gives the opportunity of appraising the standard of French photography. After being shown in Paris, the prints from the Salon National are displayed in most of the larger provincial towns. The Coupe de France is an annual inter-club competition, the best photographs being exhibited in Paris and the provinces. The Centre International de la Photographie organizes international shows roughly once every three years and these shows als tour the provinces. The Cabinet des Estampes of the Bibliotheque Nationale arranges exhibitions on specific subjects.

A number of scientific associations exist, corresponding to the numerous applications of photography. The majority of these associations concern themselves with both still photography and cinematography.

Study Centres. Two organizations take the lead in photographic research and teaching: the Institut d'Optique (Optical Institute) and the Ecole Nationale Technique de Photographie et de Cinéma (National Technical College of Photography and Cinematography).

The Conservatoire National des Arts et Métiers (Museum and School of Arts and Crafts) has a very fine collection of cameras and numerous daguerreotypes, and prints dating from the early period of photography. The Museum at Chalon-sur-Saône has some interesting items relating to Nièpce, and a fine collection of old cameras.

The Cabinet des Estampes (print-room) at the Bibliothèque Nationale contains the world's largest library of photographic prints, including numerous early works—amongst them the Nadar collection. As a copy of all photographs published must legally be deposited at the Bibliothèque, and many exhibition prints are likewise deposited, in particular those from the Salon National and the Coupe de France, this collection is constantly being added to. It provides the world's largest pictorial review of the whole history of photography, comprising more than 2 million prints with 500,000 now being catalogued. Prints are available for inspection by interested persons and copies may be taken from them. They are often loaned for display in exhibitions illustrating particular subjects.

The Société Française de Photographie also has a collection of historic photographs which may be viewed on request. It holds regular exhibitions in its gallery. There are a number of photographic print libraries open to the public (sometimes on application) in Government offices and other institutions, e.g., the Archives de France (Record Office), the Présidence du Conseil (Cabinet), the Documentation Française and the Centre National de la Recherche Scientifique.

Numerous French organizations and bodies make considerable use of photography thus contributing towards its progress, among them the Centre National de la Recherche Scientifique (National Centre of Scientific Research) the Institut Pasteur (Pasteur Institute), the Museum d' Histoire Naturelle, the Commissariat Général à l'Energie Atomique, the École Polytechnique, the various astronomical observatories and the Sorbonne University.

The Ministry of Education has set up a department of audio-visual studies as part of the Institut Pedagogique National. The department has its own laboratories and studios, and provides photographs and undertakes all kinds of research for educational purposes, besides giving instruction in photography for teachers of all grades.

During the Biennale exhibition of May 1955 it was resolved to set up an International Centre of Photography and Cinematography, with headquarters in Paris. Forty-three countries are represented in this organization, together with large international associations concerned with photography. It takes advantage of the presence in Paris of UNESCO and works in close collaboration with the United Nations.

The Centre organizes international exhibitions on general or specific subjects. It calls international conferences every two years, and these are held alternately in France and abroad. It also brings together international experts according to their work, especially in the following fields; studies in applied techniques, documentation studies, studies in picture psychology and sociology.

Publications. The following are the principal photographic publications in France: *Le Photographe* (official organ of professional photographers); *Sciences et Industries Photographiques* (technical journal); *L'Officiel de la Photographie* (organ of the Fédération Nationale des Sociétés); and the amateur magazines *Photo Cinéma*, *Photo Ciné Revue* and *Photo Monde*. P.S.

Industry. The French photographic industry has shown considerable expansion during the past five years in the value of production. There are at present nearly a hundred producers established mainly in the Paris and Lyons areas. These together provide employment for an estimated 12,000 persons, the majority of whom are concerned with the manufacture of sensitized materials.

Among the equipment sector, which consists of about ninety producers with some 2,000 employees, the manufacture of still cameras is the most important single activity, whilst the production of cine apparatus and lenses has been undertaken for many years. Leading companies in this sphere are Ercsam, Heurtier and Pathe, whilst the firms of Angenieux and Berthiot are well known for their high quality lenses. Foca, Sem, Malik, Indo-Fex, and S.F.O.M. are the chief manufacturers of still cameras and projectors.

There are currently about seven producers of sensitized materials and chemicals, employing approximately 10,000 workpeople, the principal item of production being photographic paper. The leading company in the field is Kodak-Pathe, with 5,750 employees in 1963, whilst the other major producer is Lumiere, employing some 1,000 workpeople in 1963, both of these manufacturing sensitized paper, cine and roll film on a large scale. Other major producers are Gevaert (Pont-a-Marcq), Bauchet and Guilleminot which manufacture sensitized paper and film.

The industry is served by a trade association, the Syndicat General des Industries Photographiques et Cinematographiques Substandard (General Syndicate of the Photographic and Narrow Gauge Cinematographic Industries), which performs the function of representing manufacturers' interests in all commercial and Governmental matters.

The output of still cameras reached 628,000 units with a value of NF 40·4 million in 1963, the majority of which were accounted for by box-type models, whilst still projector production totalled 80,000 units worth NF 13·2 million. Within the total of other equipment and parts both cine cameras and projectors have increased in output since 1959, a large proportion of these being exported in contrast to still equipment, the bulk of which is taken up by the domestic market.

Sensitized material production has risen overall by more than 50 per cent in both volume and value during the last five years to reach 48·0 million square metres with a value of NF 422·8 million in 1963.

Approximately three-quarters of the total quantity in this year was accounted for by sensitized paper, with 35·7 million square metres, the output of which has shown substantial growth in the period, whilst still film has also been increasing in quantity, due mainly to graphic arts material, to a total of 4·0 million square metres in 1963. The output of X-ray film has remained fairly steady in the region of 3 million square metres each year since 1959, and cine film production has fluctuated about the 5 million square metres mark each year during the period.

Imports. The value of imported photographic goods has risen by over five times between 1959 and 1963 to a total of NF 226·4 million in the latter year.

Two-thirds of this figure is composed of apparatus, among which still cameras, particularly the miniature type, have gained an important place.

In 1963, imports of all types of still cameras reached a total of 467,100 units with a value of NF 41·0 million, the majority of which were supplied by West Germany with 206,100 cameras worth NF 27·6 million.

Imports of cine cameras have also shown a sharp increase in both quantity and value during recent years.

The bulk of these are currently amateur 8 mm. models, totalling 55,700 units valued at NF 14·3 million in 1963, of which the United States supplied 19,700 cameras worth NF 4·0 million and Switzerland 9,500 cameras with a value of NF 2·9 million. Cine projector imports have also expanded considerably, the 8 mm. models reaching 43,400 units worth NF 10.6 million during 1963, the chief suppliers being Austria (15,000 projectors worth NF 3·2 million) and Switzerland (7,900 projectors valued at NF 1·8 million).

Among other apparatus, photocopy equipment has become a major item, the chief sources in 1963 being (in NF million), the United Kingdom (20·3), West Germany (5·2) and the Netherlands (3·8).

Sensitized material and chemical imports have shown considerable growth over the last five years, although not holding such an important place as those of apparatus.

The principal item imported is photographic paper, totalling 9·3 million square metres with a value of NF 22·8 million in 1963, the main sources of supply being West Germany with 3·4 million square metres worth NF 7·7 million, and the Netherlands with 2·4 million square metres valued at NF 5·6 million. The Dutch paper is primarily photo-copying material for

Among other important products is roll film, office-copying machines and systems, the larger part of which is composed of black-and-white stock (totalling 0·6 million square metres worth NF 10·2 million in 1963), of which Belgium supplied 0·5 million square metres valued at NF 5·0 million.

Exports. Exports of equipment and materials have expanded by almost a half in value since 1959, reaching NF 158·2 million in 1963, the larger part of which is accounted for by sensitized products.

Among apparatus exported, the leading place is held by lenses destined for use with still and cine equipment. These amounted to NF 15·8 million in 1963, the largest purchaser being Switzerland, which took NF 9·4 million's worth during the year.

Cine cameras have gained an important position in recent years, reaching 5,400 units with a value of NF 6·1 million in 1963. Of this total, 8 and 9·5 mm. gauge models accounted for 4,300 units worth NF 2·2 million, the largest single market for which was the United States of America with 1,700 cameras valued at NF 1· million.

Exports of cine projectors have also recorded an increase since 1959 to 1,900 units worth NF 7·6 million, the larger part being accounted for by the 16 mm. type (1,100 projectors with a value of NF 6·8 million in 1963).

Still camera exports have fallen in value during the past five years to a 1963 total of 71,000 units worth NF 3·1 million. The majority of these are of the amateur type (70,000 cameras worth NF 2·0 million), and important purchasers are West Germany and Algeria.

The value of sensitized materials exported has recorded a considerable increase since 195 to reach a total of NF 101·5 million in 196 The leading product, cine film, totalled NF 44· million during the year, of which colour stock accounted for NF 27·5 million. The major purchasers of the colour reversal type are West Germany and Italy.

Another important item is sensitized paper exports of which have more than doubled in both quantity and value over the period to reach

FRENCH PRODUCTION OF PHOTOGRAPHIC EQUIPMENT

NF Million

(Manufacturers' Net Selling Price ex-works)

	1959	1960	1961	1962	1963
Still cameras	30·7	37·1	43·8	42·7	18·2
Still projectors	} 41·7	41·9	17·3	20·5	18·3
Processing and finishing equipment			6·3	5·6	8·5
Other equipment and parts			22·2	25·0	28·4
Total still apparatus	72·4	79·0	89·6	93·8	73·4
Lenses and cine equipment	n/a	n/a	34·9	36·9	35·3
Total production	n/a	n/a	124·5	130·7	108·7

FRENCH PRODUCTION OF SENSITIZED MATERIAL

Thousand Sq. Metres

	1959	1960	1961	1962	1963
Sensitized paper	19,068	21,324	25,332	33,924	35,724
X-ray films	2,664	3,084	3,252	2,832	2,880
Still film- Miniature:					
Black and white	276	281 }	457	514	589
Colour	136	176 }			
Roll	1,186	1,357	1,285	1,322	1,261
Sheet	1,279	1,610	1,631	1,765	2,143
	2,877	3,424	3,373	3,601	3,993
Plates	35	30	26	23	23
Cine film: 35 mm. Positive:					
Black and white	2,695	2,695 }	4,102	3,329	3,542
Colour	1,063	1,213 }			
35 mm. Negative:					
Black and white	437	553 }	810	901	796
Colour	118	115 }			
16 mm. and smaller gauge Positive:					
Black and white	91	98 }	649	733	1,066
Colour	522	634 }			
	4,926	5,308	5,561	4,963	5,404
Total sensitized material	29,570	33,170	37,544	45,343	48,024

FRENCH IMPORTS OF PHOTOGRAPHIC GOODS

NF Million c.i.f.

	1959	1960	1961	1962	1963
Still cameras	4·1	6·3	26·7	33·8	41·0
Cine cameras	4·8	4·3	9·7	16·8	18·0
Still projectors, enlargers and reducers	1·6	4·7	6·6	9·7	10·9
Cine projectors	3·3	3·0	6·8	9·9	14·7
Lenses	0·6	0·7	4·0	6·3	6·1
Photocopy equipment	1·8	4·7	9·5	19·5	36·7
Other equipment and parts	4·3	10·2	14·2	18·5	19·5
Total apparatus	20·5	33·9	77·5	114·5	146·9
Sensitized paper	5·8	9·3	15·1	19·7	22·8
X-ray film and plates	1·9	3·3	4·3	8·5	14·8
Roll film	3·2	5·2	6·7	8·2	10·2
Plates and sheet film	3·9	5·3	8·3	11·0	12·2
Cine film	7·4	7·9	13·6	15·1	15·7
Chemicals	0·9	1·5	2·3	3·5	3·8
Total sensitized material and chemicals	23·1	32·5	50·3	66·0	79·5
Total imports	43·6	66·4	127·8	180·5	226·4
(£ million)	3·2	4·9	9·3	13·1	16·4

FRENCH EXPORTS OF PHOTOGRAPHIC GOODS
NF Million f.o.b.

	1959	1960	1961	1962	1963
Still cameras	4·7	2·4	3·6	3·2	3·1
Cine cameras	4·6	5·4	5·9	9·9	6·1
Still projectors, enlargers and reducers	2·6	2·5	2·8	2·6	2·0
Cine projectors	2·9	3·2	3·6	5·4	7·6
Lenses	8·8	10·1	13·9	20·2	15·8
Photocopy equipment	1·9	3·5	4·8	4·7	4·1
Other equipment and parts	12·4	17·2	19·1	17·8	18·0
Total apparatus	37·9	44·3	53·7	63·8	56·7
Sensitized paper	11·9	19·6	18·3	19·6	24·4
X-ray film and plates	10·8	12·7	12·5	9·7	12·1
Roll film	9.9	11·0	10·1	9·4	9·6
Cine film	36·6	37·6	41·7	38·9	44·2
Plates and sheet film	4·8	6·6	7·5	8·2	9·0
Chemicals	1·3	1·9	2·0	2·0	2·2
Total sensitized material and chemicals	75·3	89·4	92·1	87·8	101·5
Total exports	113·2	133·7	145·8	151·6	158·2
(£ million)	8·3	9·8	10·6	11·0	11·5

FRENCH EXPORT MARKETS FOR SENSITIZED MATERIALS 1963

Country	*Sensitized Paper NF Million f.o.b.*	*X-ray Film NF Million f.o.b.*	*Roll Film NF Million f.o.b.*	*Cine Film NF Million f.o.b.*
European Common Market:				
Belgium	7·6	0·1	0·1	0·3
W. Germany	2·0	4·9	0·0	10·4
Italy	2·7	2·8	1·9	16·9
Netherlands	0·2	—	0·3	0·2
E.C.M. Total	12·5	7·8	3·2	27·8
European Free Trade Area:				
Austria	—	—	0·2	0·7
Denmark	—	—	0·4	0·7
Finland	—	—	—	1·1
Norway	—	—	0·1	0·1
Portugal	0·1	—	—	—
Sweden	0·1	—	0·1	0·6
Switzerland	0·4	—	0·3	0·6
U.K.	1·0	—	0·2	1·0
E.F.T.A. total	1·6	—	1·3	4·8
Algeria	2·5	—	0·9	1·0
Morocco	1·0	1·7	0·3	1·2
Tunisia	0·7	0·4	0·1	0·3
U.S.A.	1·4	—	0·9	0·6
All other countries	4·7	2·1	2·9	8·5
Total exports	24·4	12·1	9·6	44·2

***Total of Colour Reversal Film, all gauges.**

20·2 million square metres with a value of NF 24·4 million in 1963.

Leading markets for sensitized paper are Belgium, taking 8·7 million square metres worth NF 5·0 million, and West Germany, which received 2·5 millions square metres valued at NF 1·4 million during the year.

Among exports of the remaining product[s] roll film totalled 619,000 square metres with [a] value of NF 9·4 million in 1963, the bulk [of] this being composed of black-and-white stoc[k] (601,000 square metres worth NF 8·9 million[)] with Italy and the United States as chief market[s].

G.R.H. & A.G.[B.]

FRAUNHOFER, JOSEPH VON, 1787–1826. German optician. After an early struggle became assistant and later partner in an important optical works. In 1814 he discovered a method of calculating a spherically and chromatically corrected objective. He also invented a machine for polishing large and mathematically accurate spherical surfaces. Between 1814 and 1817 he determined accurately the dark lines in the solar spectrum, since known by his name. He was the first to study diffraction with a grating and was responsible for producing high quality optical glass. Biography by M. von Rohr (1929).

FREELANCE PHOTOGRAPHY

Practically every publication that prints photographs accepts a certain proportion of them from freelance workers. This means in theory that the scope of the freelance photographer is unlimited. In practice this is not altogether true, because the freelance photographer who tries to compete with the full-time agency and press photographers is unlikely to get very far. The regular photographers employed by the agencies and newspaper offices have facilities for transport, communication and publication that no freelance could afford. Furthermore they have a regular salary to tide them over their occasional failures whereas every time the freelance draws a blank, his expenses are a dead loss.

Scope for the Freelance. So in actual fact the freelance must narrow down his field of activities to the subjects which for one reason or another do not attract the agencies and the press. Very broadly this means avoiding anything of a topical or live news character and going after subjects of general and lasting interest. It means relying on inspiration and imagination more than the sporting calendar, the passenger lists of luxury liners and airlines and the police news.

This, at any rate, is where he makes a start; later on, if his work becomes known, he may be commissioned by editors who want his particular style of work on a topical or feature story. When that happens he may find himself working alongside the staff photographers but not in competition with them.

Even when he takes his camera along to photograph events of topical interest, the freelance approaches them from a different angle. While the regular cameramen are getting their pictures of the show itself, making sure of at least one good shot of every important person, and one of the principal event of the occasion, the freelance is looking around for the incidental stories—the faces of the onlookers, the lost child, or the puppy that slips through the police cordon at the crucial instant.

This does not mean that the freelance should ignore the regular subjects, only that he should keep his eyes open for everything else, too.

But most of the time the freelance concentrates on feature material where he has a chance of selling a number of pictures rather than one lucky shot. He keeps his ears and his eyes open for the ideas that it would not be worth the while of the pressmen to follow up, and when he travels he does so with the object of making a series of salable pictures. These he may send to appropriate weekly or monthly magazines, to travel agencies, or transport companies—e.g., airline, shipping, road, and railway.

Many freelance workers specialize in one particular kind of picture—e.g., animals, high speed flash, or landscapes—but the narrower their field of specialization, the less chance there is of their being able to make a living at it. By specializing and becoming known, there are many opportunities for the spare time freelance to make a profitable sideline of his hobby. But this is a different matter from turning it into a career.

Other Freelance Outlets. The scope of the freelance photographer is not limited to working for publication. There are other directions in which he can make money with his camera. He may make a profitable sideline of at-home portraiture, particularly of young children, a branch which few professionals trouble about. The professional naturally prefers people to come to his studio where he has everything standardized and under control. He is also spared the waste of time in packing and transporting his equipment. The freelance, on the other hand, has no studio anyway and does not have to add the extra time and trouble of working away from home to his costs. Some freelance workers with a flair for this type of photography have eventually opened up their own businesses.

There is also an outlet for technically sound freelance photographers in the commercial field. Small businesses and manufacturers of all kinds need photographs of their products for advertising, publicity, and sales literature. Very often the number of pictures they use in a year is too small to be worth the attention of a regular commercial photographer. By getting in touch with such small factories, shop-keepers and the like, the freelance can build up a useful connexion, but his main occupation must allow him to visit his various clients during their working hours.

There are also seasonal openings for freelances with their own cameras to work as beach and street photographers. But such jobs are strictly part-time work and not genuine freelancing. The photographer in such cases has to rely on the organization of the company that employs him; he could not afford to run the whole business himself.

Freelancing as a Career. Freelance photography is not a difficult field to break into and anyone with a camera, a few rolls of film and a good story idea, may succeed at the first attempt. But to sell enough pictures month after month to earn a living, is a much more difficult matter.

The competition in the freelance field is keen because photography nowadays is a mass art. In the old days, people used to buy box cameras and content themselves with taking family snaps. Now they buy expensive cameras—often later models than most professionals possess—and start trying to earn spare-time money by submitting pictures to magazines and newspapers. When they get a few pictures published they often throw over their regular business to become full-time freelances. Most of them drop by the wayside because they mistake the ability to take a picture with the imagination to create or recognize one. They seldom develop a "picture sense", without which no freelance can go very far.

Some people are born with a picture sense, but most beginners have to acquire a picture sense the hard way—by working at it.

Anyone who wishes to become a successful freelance must study thousands of pictures other people have taken and analyse their weak and strong points. They have to take thousands of pictures themselves and learn to adopt the same critical, objective attitude to their own efforts. Most of all, they have to take their own cherished pictures around to editors and listen to their unflattering comments. Then finally they have to go back and try again until they begin to master the formula for turning out shots with above-average human interest.

Few people have the necessary tenacity. Most photographers who attempt freelance work discover that it gives them at best an erratic and unpredictable income and a lot of frustration. Economic pressure forces most of them to find a niche in a steadier, if less romantic branch of the profession. They open portrait studios, take baby pictures, go in for advertising or commercial work, or join the staff of some magazine or newspaper. A few strong individualists who prefer working on their own stick at it long enough to make a success of it.

For the average photographer freelancing is not a practical way of making a living. Even so, some of the best known and most successful photographers in the world do all their work on a strictly freelance basis and have done so from the beginning.

Submitting Prints. The prints should be dry mounted on card if possible so that they do not become dog-eared by being handled. Presentation is very important in any graphic art field and if a photographer does not respect his own work other people are not likely to take care of it either.

No freelance should ever show contact prints. Contacts should never leave the darkroom.

Every picture benefits from enlarging and trimming, while good retouching and finishing can improve the appearance of specimens enormously. Anyone who finds difficulty with this end of the job would be wise to get a specialist to finish his prints properly.

Every freelance should make up a rubber stamp with his name, address and phone number, and use it on the back of each print. It establishes identity, ensures the return of work and tells the editor whom to contact in the event of an assignment.

The freelance who wants to find his way into the high fee paying field must first carefully study the markets. Far too many send unsuitable material to an editor who then feels that his time is being wasted. A paper's or magazine's editorial policy should be examined until the freelance is thoroughly conversant with that journal's style. Photographs submitted should be sufficiently varied in format to allow the make-up department of the journal to arrange an interesting lay-out.

It is most important to send in prints at the time when the editor is looking for that particular subject. If the photographer is simply working through his print library and submitting likely specimens to suitable publications, the subjects must be appropriate to the issues then in preparation. With monthly magazines in particular, it should be remembered that the copy is made up several months in advance of publication. So there is no point in trying to sell Christmas pictures in December.

It must be made clear, when selling a picture, what rights are being offered—e.g., "First exclusive world rights", "First exclusive British rights", or "First Australian rights" or just "Publication rights". In the majority of cases the editor is only interested in normal publication rights. However, when exclusive rights are offered, the fee should be high enough to compensate for the sales that are being missed by ruling out other markets Editors seldom insist on buying world rights or transfer of copyright, but when they do they must be prepared to pay for it. Anyone unfamiliar with the marketing procedure should get in touch with a reputable photo agency who will handle the work on a commission basis if it has merit.

Basic Rules. Many budding freelances are discouraged because of their ignorance of the elementary requirements of the work. They may be able to take pictures, but that is all. Unfortunately there are other conditions to be complied with before the pictures can come to the notice of the right editor, be accepted, published and paid for.

It always pays to see editors in person whenever possible, rather than to submi work through the post or through an agent.

There is nothing to be learned from a rejection slip but an editor's personal comments can be very illuminating—and helpful.

No one should pester editors for assignments at the outset. The way to attract their attention is to show them some good original picture stories; once their confidence is established, assignments will come automatically. But no editor wants to entrust an important job to an unknown, unproved cameraman. This explains why specimens are so important.

When a photographer has a good story idea, but cannot see the editor to discuss it personally, his best course is to put it down on paper and submit it in writing, rather than take it up with subordinates. A good editor can recognize a story at once and he may offer an assignment. If the idea is worth using the editor will be glad to pay for it and, even if the author's picture-taking ability is not up to standard, the editor may buy the idea and have one of his staff carry it out. The fact that the idea has been submitted in writing is protection against anyone using it.

It is a waste of time for a beginner to try to break into the top markets first unless he has some really outstanding pictures. He will be much more likely to get encouragement from editors of smaller publications which cannot afford to employ their own staff of photographers. Such editors are always looking for new talent. And the big markets will still be there when he is ready for them.

The freelance should never be shy about asking for credit-lines. A photographer has the same reason for wanting to be identified with his work as a writer. But unless the editor is asked at the time he may forget it and apologize later. The time to raise the question (and that of fees!) is before publication. H.B.

See also: *Agencies; Photo-journalism; Reproduction fees; Selling photographs.*

Books: *How to Take Photographs that Editors Will Buy*, by R. Spillman (London); *The Business of Photography*, by C. Abel (London).

FREQUENCY. Number of complete events of a regular cyclic sequence, usually counted over one second. The term is especially applied to the wave propagation of the various forms of energy—e.g., electromagnetic and acoustic radiations, as well as regular mechanical movement.

The frequency is inversely proportional to the wavelength and proportional to the velocity of the radiation, e.g., if the velocity of a wave is 1,000 feet per second and its wavelength (measured from the crest of one wave to the corresponding point on the next) is 10 feet, then in every 1,000 feet there will be 1,000/10 = 100 complete waves. This number of waves is radiated every second, so the frequency is 100 cycles per second.

In the general case, the frequency n of a radiation travelling with a velocity of V and a wavelength λ is given by the equation

$$n = \frac{V}{\lambda}$$

V and λ must both be measured in the same units, i.e., metres per second and metres, or feet per second and feet.

It is more convenient to describe very short waves—e.g., of light and X-radiations—in terms of their wavelength in Angstrom Units. Radiations of longer wavelength—e.g., in the radio band—may be expressed in either wavelength or frequency. Much slower cyclic events are more conveniently expressed in terms of frequency.

Electrical energy is generally distributed as an alternating current—i.e., a current which alternates in a regular sequence between a maximum positive and a maximum negative value. The frequency of the alternations in Great Britain is almost always 50 cycles per second, but there are a few local exceptions. Elsewhere in the world the frequency of alternating current supplies varies but is mostly between 40 and 60 cycles.

There is no risk of damage involved in connecting heating equipment designed for one frequency to a supply of a different frequency, but the performance of motors and electronic equipment will be affected and in some cases the equipment may suffer.

Sound Frequency. The pitch of a tone depends on the frequency of the vibrations of the sound wave carrying it. In sound recording and reproducing equipment (e.g., tape recorders, amplifiers for sound cine projectors) the frequency range gives an idea of the performance—with certain qualifications. An average 8 mm. sound projector can reproduce a frequency range of about 80 to 6,000 cycles per second, a high quality tape recorder might cover 40 to 20,000 cycles, assuming loud speaker equipment, etc., that can handle such a range.

Spatial Frequency. In measurements of lens performance and resolution the line spacing of a test target (i.e., line pattern) is often expressed as a spatial frequency. Thus a target of 50 lines per mm. has a spatial frequency of 50. L.A.M.

See also: *Cycle; Wavelength.*

FREQUENCY RESPONSE. Term derived from audio engineering, indicating how an optical system or a sensitive emulsion (or a combination of both) reproduces detail of various fineness. The latter is specified in terms of spacial frequency, referring to the number of lines per mm. of a resolution test target. The number of lines resolved is plotted against the contrast of the image; the complete graph is

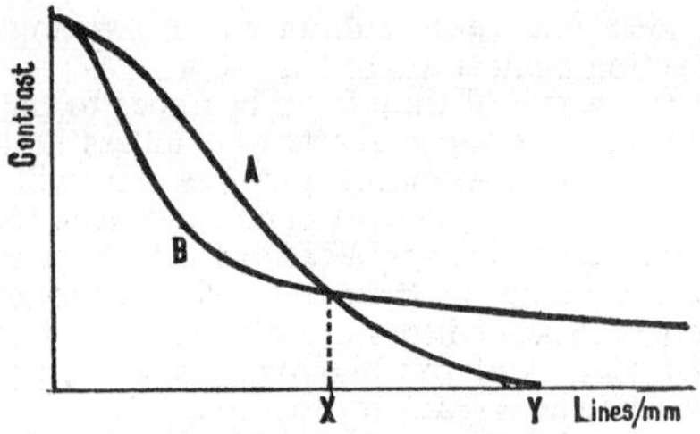

FREQUENCY RESPONSE CURVE. This plots the contrast of an image obtained by a lens or lens and film system against the lines per millimetre resolved. Curve A indicates a high contrast at low spacial frequencies but a limited ultimate resolution. This corresponds to an image which looks brilliant and apparently sharp, but cannot resolve fine detail. Curve B represents lower contrast but higher limiting resolution. The image may look fuzzier but in fact shows more detail. For detail corresponding to a lower spacial frequency than the point X curve A represents better image quality despite its limiting resolution at Y.

known as the transfer function or contrast transfer function, and provides a fairly good indication of the performance of the system.

See also: *Information in the image; Lens performance.*

FRESNEL LENS. Short focus condenser lens of a special design used principally in spotlights. While a normal short focus condenser lens is thick and heavy, the Fresnel lens is relatively thin and light so it is less apt to crack from the heat of the lamp, which can be quite considerable.

The lens consists of a series of concentric stepped rings, each one being a section of a convex surface. The combined effect of all the rings is the same as that of a single lens of normal shape with the same diameter and curvature. Sometimes the lens is moulded in one piece, but for more accurate optical equipment it may be assembled from a number of optically ground curved prisms.

Fresnel lens patterns, embossed in plastic, are also used to distribute the image brightness over the area of a focusing screen. The screen may itself be the plane surface of the lens.

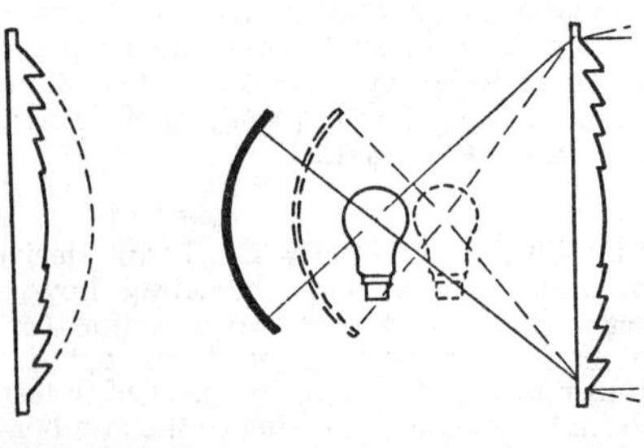

FRESNEL LENS. Used as a condenser lens in spotlights, it consists of a thin stepped disc with each step having the curvature of a much thicker lens.

FRESSON PROCESS. Early carbon printing process in which no transfer to a final support was necessary. It was similar to the Artigue process, but used a different type of paper.

See also: *Printing processes (Obsolete).*

FRIEDMAN, JOSEPH F., 1899–1956. American research chemist born in Russia. Emigrated to the United States in 1904, took Ph.D. in chemistry at Chicago University and joined the Ansco Corporation in 1943. Co-inventor with Edwin Land of the polaroid process. Dr. Friedman also played a leading part in the development of Technicolor. His *History of Colour Photography* (1944) is a standard work.

FRIESE-GREENE, WILLIAM, 1855–1921. English professional photographer, and one of the pioneers in motion picture photography. Patented (with Evans) a "machine camera" in 1889 which took instantaneous photographs in rapid succession. He was a prolific inventor; his patents include a two-colour additive method of colour cinematography. Biography by Ray Allister (London 1948).

FRILLING. Fault caused when an emulsion separates from the edges of the support, generally because of some abnormality in the temperature or chemical composition of the processing solution. The loose gelatin swells and expands, causing it to crinkle and form a characteristic frill.

Frilling seldom occurs with modern sensitized materials. It is usually the result of very high temperatures in processing—and violent changes in solution temperature. Another possible cause of frilling is the use of strong acid rinses or stop-baths immediately after the film has been in a strongly alkaline developer. Excessively long immersion times in solutions can be another cause.

Frilling is also troublesome in carbro printing during the development or transfer of the gelatin relief image. The cause in this case may be high temperatures, excessive immersion times, or an insufficiently wide border on the original bromide print.

See also: *Blisters; Faults; Hot weather processing; Tropical photography.*

FRITH, FRANCIS, 1822–98. English landscape photographer from 1853 onward; partner in a Liverpool printing firm which moved to Reigate in 1859, when a photographic printing department was added. Frith produced a large number of photographically-illustrated books with views taken on three extensive tours of Egypt and the Near East 1856–60. Also issued portfolios of photographs of many West European countries. Frith & Co. became the largest photographic publishers of English and foreign views in the world, their catalogues listing several hundred thousand pictures.

FRONT ELEMENT FOCUSING. Method of focusing in which the front component of the lens is moved in or out to focus the image.

FROST. The photographic possibilities of frost are mostly confined to pictures of trees and bushes white with rime and to records of the frost patterns on window-panes. With the first type of subject a filter must be used when blue sky is included in the picture or the frost-whitened branches and twigs will tend to merge with the sky above the line of the horizon. As frosty subjects of this sort are generally photographed in dull, overcast weather (the rime quickly disappears in sunshine), the contrast of the negative may have to be exaggerated by prolonging development.

Frost patterns on window-panes are best photographed from the inside against a background formed by supporting a sheet of black cloth or dull black paper at 45° outside the window. The room should, at the same time, be darkened except for the one window to guard against unwanted reflections.

The most beautiful effects with frost patterns are obtained by photographing them as large as possible. This can be done by normal close-up or macrophotography techniques.

For photography in frosty weather the lens of the camera should be at the same temperature as the air, or it will mist over and give a blurred image. For this reason the camera should not be carried in an inside pocket if it is to be used for taking photographs in the open air. A warm camera taken out in cold air tends to form mist on the inside of the lens. The lens of a cold camera, brought into a warm room, tends to mist on the outside. P.J.

See also: *Cold weather; Snow.*

FROST DEFORMATION PROCESS. Process employing a photoconducting thermoplastic film on which an electrostatic image is produced and which upon heating is deformed by the electrostatic change in the corresponding image areas to form a light-scattering image.

See also: *Thermoplastic recording.*

FUNNEL. Aid to pouring liquids into narrow-necked bottles. It consists of a hollow stem, widening out into a flared conical shape.

Funnels may be made of glass or plastic; the latter less breakable. Pliable plastic funnels (polythene) are particularly useful.

The neck of the funnel should be ribbed, so as to leave an air space when it is inserted into the neck of a bottle.

Funnels can also be used for filtering; in that case a filter paper or a plug of cotton wool is inserted in the top.

Most conical funnels flare out at an angle of 60°. The size is specified by the diameter at the top. For most photographic purposes a 4 ins. (10 cm.) funnel is a suitable size.

FUSE. Safety device consisting of a deliberate weak link in an electric circuit designed to break the circuit when the current exceeds the safe working level. Usually the fuse is a short length of easily fusible wire (hence its name) inserted in the circuit in the form of a cartridge, or wired across a pair of contacts fitted to a porcelain holder.

If the circuit becomes overloaded (either through a short-circuit of bare wires touching, or through connecting across it more lamps or other gear than it can carry) the wiring heats up. The fuse then melts and breaks the circuit before any other parts of the wiring can heat up sufficiently to do any damage.

Open wire fuses can be repaired easily by just replacing the molten piece of wire. The correct gauge of fuse wire should always be used. With cartridge fuses the whole cartridge must be replaced. If the fuse blows again immediately, the wiring of the circuit or the appliances in it must be checked for faults.

FUSE RATINGS

Type of Circuit or Appliance	*Type of Fuse*	*Rating (Amps.)*	*Location of Fuse*
Amplifiers, and other individual low power appliances	Cartridge	2–3	In appliance
Projectors and other more powerful units	Cartridge	5	In projector
House lighting circuits	Porcelain	5	In fuse box
Power circuits	Porcelain	10–15	In fuse box, sometimes in power socket
Main collective house circuits	Porcelain	20–60	In special fuse box; sometimes these are service fuses and must be repaired only by service staff of electricity suppliers
Individual power units (heaters, etc.) with fused plugs	Cartridge	3–13 (according to appliance)	In plug

When replacing or repairing fuses, the rating must never exceed that of the blown fuse. If a heavier fuse is inserted in the circuit, it may be able to carry a greater load than the rest of the wiring and is then no longer a weak link; i.e., it is useless as a safety device.

Common fuse ratings are shown in the table above. The heavier fuses and those covering the lighting and power circuits of a flat or house are generally installed in fuse boxes with a main switch. The box can only be opened when the circuit is switched off. In the case of cartridge fuses or fuses in individual appliances the equipment must be disconnected from the electricity supply (not just switched off) before attempting to replace or repair the fuse. L.A.M.

See also: *Wiring.*

G

GAIN (PROJECTION). Measure of the useful brilliance emitted by a projection screen in relation to the intensity of the illumination falling on it. When a perfect diffuse reflector or (back projection) transmitter, illuminated by a light intensity of one foot candle, has an apparent brightness of one foot lambert, its gain is unity in all directions.

As no screen will diffuse or reflect all the light falling on it, the gain of a screen is always less than unity, although directional screens may have a gain of more than unity over a narrow zone immediately in front.

See also: *Projection principles; Screens for projection.*

GALVANOGRAPHY. Term applied to photographic processes in which a photographically-produced relief image is given a conducting coating of black lead and then heavily electroplated to form a shell. This then can be backed up with type metal to form a printing plate for photomechanical reproduction.

GAMMA

Gamma is a sensitometric quantity derived from the characteristic curve of photographic emulsions. It is loosely interpreted as a measure of the contrast reproduced in a negative image, i.e., the ratio of negative contrast to original subject contrast for a given range of tone values. Such a comparison is based on the assumption that the brightness values of the subject tones can be accurately measured, and that the corresponding negative densities fall within the straight-line portion of the characteristic curve—a condition rarely fully satisfied in practice. Gamma is often also quoted as a yardstick for the degree of development, especially when film manufacturers recommend development times for their materials. In that sense it can be said that negatives developed to the same gamma show comparable tone reproduction.

In technical terms, gamma (written γ) stands for the slope or inclination of the straight line part of the characteristic curve. Mathematically it is defined exactly like the slope of a railway line—i.e., as the ratio of the height gained to the distance travelled in a horizontal direction. In our case the height is density (D) and the base is the log of the exposure (E). So $\gamma = D/\log E$.

Gamma is important to the photographer because it tells him how his photographic material will respond to a change in exposure. In a material with a low gamma, a small change of exposure will produce a small change in density; but the same small change of exposure with a high-gamma material will produce a much larger change in density.

Materials capable of widely different gamma values are available. The particular type of material that the photographer chooses depends on how he intends to translate differences in brightness of the object—and hence differences in exposure—into differences in density on the film, plate, or paper. If the brightness differences—i.e., the tone range of the subject—are small, he chooses a material of inherently high gamma. If they are large and he wishes to record the total range of exposures, he chooses a low gamma material.

Negative Materials. In practice each of the main groups of negative materials has its own very definite gamma characteristics.

Normal photographic subjects call for films and plates with a gamma value of around 1·0, varying from 0·6 to 1·5. This enables the emulsion to record the wide range of tones which occurs in an outdoor scene. When printed on a material of similar gamma value, the print will be as true a reproduction of the original as is possible. The exact condition for this occurs when gamma (negative) × gamma (positive) = 1. This is known as the Goldberg condition.

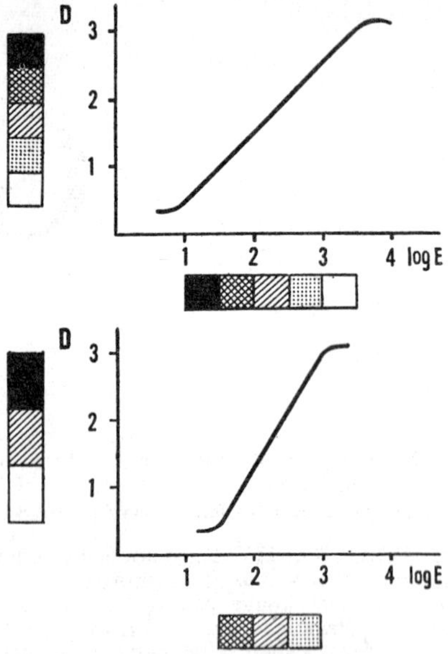

GAMMA AND CONTRAST. Top: *A negative of gamma unity reproduces the subject tones as corresponding image opacities, with the image being of the same contrast as the subject.* Bottom: *A negative of high gamma reproduces only a limited range of subject tones, and exaggerates the contrast.*

X-ray records have to be made on very sensitive films of high gamma. Here the image has inherently very low contrast, and a high gamma material enables the radiographer to reproduce the small differences in log brightness as much larger changes in density, thus increasing the sensitivity of the method. X-ray films may reach gamma values of well above 3.

Document copying and process work require materials of even higher gamma values. Here it is only necessary for the photographic material to record two tones: black and white. Often a typed letter represents quite small differences in log exposure which are turned into large differences in density in the copy.

For process work, the differentiation between black and white must be as great as possible. Materials with gamma values between 5 and 10 are available for this purpose.

Printing Papers. To match the films and plates mentioned, there are photographic papers with a correspondingly wide range of gamma values. Gamma values are not, however, commonly quoted for papers, since it is not usual to confine the image to the straight-line portion of the characteristic curve. Average slope figures are used instead. These may be taken, for example, between two points on the curve where the slope is one-tenth of the average slope.

Most papers for the printing of snapshots, etc., are divided into 4 to 6 grades, the slopes of which match the range of the negative materials used for this purpose. Papers described as soft or normal will make a good print from a negative developed to a gamma of around 1 and their slope figure is of the order of 2. Very soft papers have slope figures of half that value, and very hard papers may have slopes half again as great. Papers for document and process work are harder still and slope figures are not as a rule quoted.

These references to the slope values of papers must not be misunderstood. As a rule a paper is not chosen because it will give satisfactory tone reproduction from a given negative, but simply because its useful exposure scale is identical with the density range of the negative. More exactly, the paper is chosen because its useful exposure scale covers that part of the density range which the photographer wishes to reproduce in his print. This choice ensures that he will make use of the full range of tones from black to white that the paper is capable of reproducing. It means also that the relative tone values, far from being faithfully reproduced, are in most instances falsified.

Gamma Control by Processing. The gamma value of a photographic material is not fixed and can often be varied within wide limits by the method of development and the choice of developer. In general, gamma is low for short times of development or if the temperature of the developer is low. As the time of development is increased, gamma values rise until they reach an optimum value which is known as γ_∞ (gamma infinity). The plot of gamma against time of development is known as a gamma-time curve; such curves are often plotted for a range of times of development.

The type of developer used also affects gamma. The highest gamma values are obtained with very alkaline developers containing hydroquinone and sometimes no other developing agent. Fine-grain developers, such as buffered borax metol-hydroquinone developers or developers based on p-phenylene-diamine, will give low gamma figures. By a suitable choice of developers and developing conditions gamma values varying between 1·5 and 3 may be obtained, for example, on the film used for document micro-copying.

Information on the relationship between gamma and development is generally stated in the manufacturers' descriptive literature, but where high accuracy is required such information must not be taken too literally. There may be variations between batches of the material and the conditions of development, agitation, etc., may not be the same—e.g., in the development of separation negatives.

Separation negatives must be developed to the same gamma figure but gamma often changes with the colour of the exposing light. So any natural variation in gamma with the colour of the filter must be compensated for by changing the time of development. This means

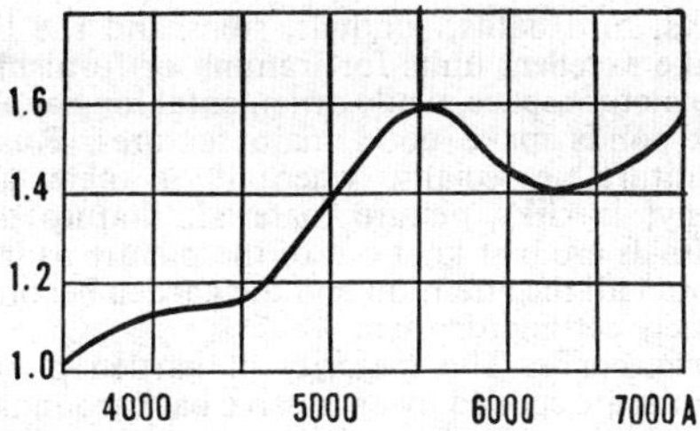

GAMMA AND LIGHT COLOUR. The gamma obtained on a panchromatic material varies with the colour (wavelength) of light. This variation may be as much as 0·4 from blue to red.

that the photographer must determine gamma-time curves under his own conditions which he must keep rigidly constant throughout.

Gamma and Wavelength. With many materials gamma changes with the colour of the exposing light. Materials behave differently in this respect and no general rule can be given, but usually the gamma value for a material exposed through a blue filter is lower than those for green and red filters. For ultra-violet light, gamma figures drop even lower, a fact which causes considerable difficulties in spectrographic work. The change is caused by the difference in absorption of the light by the photographic emulsion. With ultra-violet the absorption is so strong that only the top layer of the emulsion receives any light at all.

The only way of preparing a sensitive material which has uniform gamma throughout the spectrum is to add to the emulsion a screening dye with an absorption spectrum complementary to that of the emulsion, so that the total absorption is constant at all wavelengths. This yields a constant-gamma material.

Variable-gamma Materials. Many workers like a printing paper with a gamma or slope which they can alter at will so that they can print negatives of widely different contrasts without having to use different grades of paper. Papers of this type are in fact marketed by several manufacturers. They are prepared from a mixture of two emulsions, one hard and the other soft. One is dye-sensitized—e.g., to blue and green light—whereas the other emulsion may be sensitive to blue light only. By changing the colour of the enlarging light with different colour filters the proportion of the total exposure given to each emulsion can be controlled at will. With some materials a range of specific filters is supplied, one for each contrast grade. Alternatively, a single filter may be used to give one extreme of contrast, and the material exposed partly with and partly without the filter for intermediate gradations. In this way a selection can be made from the two emulsions, giving a range of contrasts equal to that normally covered by four or five separate paper grades. W.F.B.

See also: *Characteristic curve; Developing negatives; Sensitometry; Spectral sensitivity.*

Books: *Exposure*, by W. F. Berg (London); *General Sensitometry*, by Yu. N. Gorokhouskii and T. M. Levenberg (London).

GAMMA RADIOGRAPHY. Radiography using gamma rays, instead of the normal X-rays. Gamma rays are emitted by radioactive bodies and have a shorter wavelength than X-rays—i.e., around 10^{-10} cm.—they are therefore more penetrating. This form of radiography is now extensively used in industrial non-destructive examination of castings, welds, and moulded parts.

The most widely employed sources of gamma radiation are radium and the gamma-active isotopes produced in atomic reactors. These are supplied in the form of cylinders of 1 to 6 mm. of isotopes of metals such as antimony, cobalt, iridium and tantalum. The gamma energy of the source and its other properties—e.g., half-life and activity per grain—are specified according to the type of specimen to be inspected.

One of the principal advantages of isotopes as a source is that they can be used in positions where it would be impossible to accommodate an X-ray tube—e.g., inside pipes and on welded girder-work.

The technique is simple: an X-ray type of film in a holder, usually sandwiched between lead foil intensifying screens, is placed against the subject, and the isotope is uncovered on the opposite side so that the radiations pass through the subject before falling on the film. By suitably choosing the characteristics of the source and the time of exposure, the resulting image reveals variations in the internal structure of the specimen—e.g., flaws, scale inclusions, and cavities in welds and castings.

Where a number of small items have to be examined in this way, the isotope is simply surrounded by any number of specimens, each one associated with its own film and screens. All are then exposed at the same time.

Isotopes for gamma radiography are dangerous to handle; they are normally supplied in lead bombs which are opened from a distance with tongs or other remote handling equipment only when everything has been placed in position and no one is left within range. The exposure is made by uncovering the isotope and leaving it—often for a day or more—in a locked room surrounded by the specimens to be radiographed. At the end of the calculated exposure time the bomb is again closed up before anyone is allowed near the specimens.

Gamma radiations are used in the treatment of cancer but not normally for the radiographic examination of the body.

See also: *Industrial radiography.*

GAMMA RAYS. Radio-active radiations similar to X-rays but of a shorter wavelength and hence greater penetrating power. Gamma radiography with isotopes is now an important department of industrial radiography.

See also: *Spectrum.*

GARDENS. Any attempt at conveying the size of a large garden generally results in extensive areas of uninteresting foreground. The massed flowers and main features of the small garden—lily pond, pergola, rockery, etc.—are generally laid out on a more modest scale that is easier for the normal angle lens to handle.

Viewpoint. Generally it is better to choose characteristic portions of the garden rather than try to include the whole scene, and all stiff and formal effects should be avoided. Nothing is less attractive than a square-on photograph of a flower bed or border where a narrow uniform strip of flowers stretches from one side of the picture to the other. Straight-down-the-garden-path pictures are equally dead looking.

A viewpoint on a corner usually gives a picture in which the space is filled in a much more interesting fashion; paths then lead into the scene at an angle, and the varied masses of light and shade formed by groups of flowers can be balanced to make a pleasing and original composition.

A camera at eye-level will lessen the importance of the closest blooms, but indicate size of bed and general layout. For distances of 10–20 feet, waist-level viewpoints are often best; much depends upon the varying heights of the flowers themselves, whilst for small border clumps camera heights of 1 to 2 feet at distances of 4 to 8 feet are advised. Dead-overhead viewpoints of selected, even-growing masses, provide interesting "pattern" pictures.

It is often possible to find an effective viewpoint in which a view of the garden is seen from indoors through an open french window or door. And interesting photographs can often be taken from a first floor window—a viewpoint which, in many cases, has the advantage of excluding the sky.

Garden Plants, Trees, Shrubs. Rockeries and massed blooms are easier to arrange in the picture space than isolated clumps of assorted specimens.

In black-and-white pictures, the flowers of the lightest tones should be placed to provide the main point of interest, but in making colour photographs the place of interest should go to the most brilliant colour.

Plants that carry all their interest at the top—like standard roses—are not easy subjects unless they are photographed close up. Flowering shrubs and fruit trees are good space-fillers for upright or square picture shapes. Dark shrubs and laurels make poor subjects lacking interest but are excellent for backgrounds.

Garden Furniture. Garden features like archways, bird baths, sundials, steps and the like make excellent units for framing or furnishing the picture space, while ornamental loggias and lily ponds make good main features. Rustic furniture, especially when it is new and shiny, is risky picture material. Statues and animals are best kept out of the picture as they attract all the attention and the garden becomes a mere setting for them.

Backgrounds. The majority of garden photographs are spoiled by aggressive backgrounds—white painted trelliswork, wire-netting, railings and ugly fencing. It is usually difficult to avoid this type of background but it is possible to play it down—it can be blurred by selective focusing and the picture can be taken when the background is in shadow.

On the other hand, close-boarded fencing, wattle and old roughcast walls can actually add decorative interest to the picture if the lighting is oblique and shows up the texture.

The sky, usually the best of backgrounds, is rarely effective in a garden photograph and is generally better cut out altogether by a suitable choice of viewpoint.

Many garden pictures are marred by details that could easily have been avoided before the picture was taken—dead or withered blooms near the camera, stakes and canes, name-tags and odd tools. Daisies that pass unnoticed on the lawn come out as disagreeable white spots on the print.

Lighting. The best lighting is provided by slightly diffused sunshine with plenty of reflection from white clouds. This type of lighting gives modelling and relief without dense black shadows.

Flat, high angle or fully frontal lighting all tend to destroy modelling and depth, while back lighting is rarely effective, particularly when it shines through a network of small branches.

Filters are not required unless the flowers are tall, and are photographed against a blue sky background, when a 2× or 3× yellow, green, or yellow-green filter may be used with advantage. An orange filter gives greater tonal contrast, but at the expense of over-dark foliage. Broad lawns and open scenes which include the sky also need a filter.

Early morning and evening sunshine is useful for spotlighting individual features, but it should always be supplemented by a white reflector or the shadows will be solid black.

Camera. All the normal types of camera are suitable for garden photography. Even a fixed focus camera will turn out attractive garden pictures, but it calls for even more attention to the background than usual. A focusing camera has the advantage that it can be made to throw the background out of focus. **P.J.**

See also: *Botany; Flowers; Trees.*

Book: *How to Shoot Outdoor Colour*, by G. Wells and F. Smith (London).

GARDNER, ALEXANDER, 1821–82. American photographer born in Scotland. Emigrated in 1856 and assisted Matthew Brady, first in his New York studio and from 1858 as manager of his new Washington studio. During American Civil War, worked first on Brady's staff of war photographers then in document and map copying department of the Federal Army of the Potomac. At the same time, Gardner and other former employees of Brady took some of the most famous Civil War photographs. One hundred of these were published in *Gardner's Photographic Sketch Book of the War*, (1865). In 1863 Gardner opened his own portrait studio in Washington. In 1868 he became official photographer to the Union Pacific Railroad Company. In 1873 he made a portrait collection of criminals for the Washington police.

GAS BURST AGITATION. System of stirring up processing solutions in tanks by means of nitrogen gas led into, and released near the bottom of, the tank in periodic bursts.

See also: *Agitation.*

GASLIGHT PAPER. Old name for contact paper, derived from the fact that it was designed for printing by artificial light, and introduced at a time when gas light was the normal artificial illumination.

See also: *Papers.*

GATE. Part of the cine camera and cine projector in which the film is held during the exposure or projection respectively of each frame. The term is derived from the construction of the gate in early and many current instruments, where a hinged plate or frame with an aperture swings into position over the film.

The gate incorporates guides to keep the film flat and in alignment, to avoid gate float or unsteadiness in the moving picture during projection. The guides hold the film by the edge only to prevent scratching. The gate is one of the most precisely machined parts of the camera or projector, and must be polished to a highly smooth finish. The aperture in the gate, the so-called gate mask, defines the area of the film projected on the screen. The dimensions of the gate mask are laid down by international standards.

By derivation, the picture stage in a slide projector is frequently also known as the gate, though its mechanical construction differs appreciably from that of a cine projector.

See also: *Cine camera; Projector (cine).*

GAUDIN, MARC ANTOINE AUGUSTINE, 1804–80. French photographer, optician, and scientist. Improved the daguerreotype process. Achieved the first instantaneous exposures in 1841. Investigated (1841) the Becquerel effect. Studied gun cotton (1847) and silver halide collodion emulsions (1853, 1861). Proposed (1853) potassium cyanide as fixing agent, gelatin and other substances as image carriers. Described (1861) an improved collodion emulsion for negatives.

GAUMONT, LÉON, 1863–1946. French pioneer of the cinema. In 1895 became director of the Comptoir Général de Photographie and marketed the Demeny-Decaux Chronophotographe. This was followed by other cine apparatus for professionals and amateurs and by still cameras, development machines, synchronized sound-and-film apparatus (Chronophone 1910), three-colour additive cine projection apparatus (Chronochrome 1912), and the Ideal Sonore sound film method (1929). Important film producer.

GAUSS POINT. Point on the optical axis of an objective situated at a distance from the focus equal to the focal length of the equivalent simple lens. There are two such points, one for the lens used normally, and one when it is reversed. Also commonly known as the nodal points.

See also: *Lens.*

GAY-LUSSAC, JOSEPH-LOUIS, 1778–1850. French chemist and physicist. Experimented on photochemical reactions, among them that between chlorine and hydrogen (1809). In 1839 he made the report on the discoveries of Daguerre for the Chamber of Peers at the same time as Arago reported for the Chamber of Deputies. Suggested, with Arago, the possibility of producing maps photographically.

GELATIN. A naturally occurring protein used as the binding medium for the silver halide crystals in the common type of photographic emulsion. It is isolated from animal skins and bones by various processes, which have considerable influence on the properties of the product.

It is insoluble in cold water but will absorb about 10 per cent of its weight of water, with consequent swelling. Hot solutions will set to a rigid jelly at certain temperatures, and this is one of the properties which is useful in the manufacture of photographic emulsions. It is not, however, a definite chemical compound, and a given sample of gelatin will contain molecules of various molecular weights, ranging from about 20,000 to over 100,000, and of various amino-acid compositions.

As a protein it has both acid and basic properties. It can also combine under certain conditions with some heavy metal ions to form insoluble products. Some photomechanical processes are based on the light-induced hardening of gelatin. For example, the action of light on a gel containing potassium dichromate leads to the formation of a compound which is insoluble in water. Washing with warm water removes the unexposed areas, leaving a relief image. Gelatin is also hardened by other agents, including formaldehyde and the oxidation products of some developers.

An important photographic property of gelatin is its protective colloid action towards the silver halide crystals during the preparation of an emulsion. If aqueous solutions of silver nitrate and a potassium halide are mixed, the solution becomes opalescent, but the gelatin prevents the crystals from growing sufficiently to be precipitated. In the absence of gelatin the silver halide is precipitated as crystals which are too large for photographic purposes.

A property of gelatin directly connected with the actual photographic process is its role as a halogen acceptor. The action of light on a silver halide leads to the formation of atoms of silver and of the halogen, which can recombine to form the original silver halide and so nullify the effect of the light. This recombination can be prevented by the addition of compounds which combine with the halogen, and gelatin itself is such a compound.

One of the earliest materials used as a binder for silver halides was collodion, or cellulose nitrate, on which the "wet collodion process" was based. In 1871, however, gelatin was found to be a good binding medium, and had the advantage over collodion that emulsions made with gelatin were much more sensitive, and were also sensitive when dry. Not all gelatins however gave equally sensitive emulsions, and the "good" photographic gelatins were found to contain a few parts per million of certain sulphur compounds. The addition of such compounds, for example allyl thiourea, to "poor" photographic gelatins enabled sensitive emulsions to be obtained. Nowadays photographic gelatins are of high purity and the sensitizing compounds are added in controlled amounts.

Synthetic polymers, such as polyvinyl alcohol and polyamides have been proposed as substitutes for gelatin, but none has proved completely satisfactory as yet. Where wet processing is not required, as for some print out-recording papers, materials such as polystyrene have been used as the binding medium. E.M.

See also: *Emulsions; Sensitized materials manufacture.*
Book: *Photographic Gelatin*, by F. G. Clegg and R. J. Croome (London).

GELATINO-. Prefix once applied to certain photographic processes and materials—e.g., gelatino-bromide process, gelatino-chloride paper—indicating the employment of an emulsion of light-sensitive chemicals in gelatin. The use of the prefix was more general at the time when gelatin emulsions were replacing collodion for normal photography.

GELATIN-SUGAR PROCESS. Daylight printing process in which the action of light produces selective hardening of a sensitized layer of pigmented gelatin and sugar coated on a paper base. The image is developed by washing away the unaffected material with hot water. The process is almost identical with the gum bichromate process.

GENRE. Term for the type of photograph that depicts people in their normal environment of work or leisure.

The essence of genre is its unaffected portrayal of the everyday life of average men and women. Photography of this type is only successful when the whole feeling is natural and unforced.

GENTHE, ARNOLD, 1868–1942. American portrait photographer born in Germany. Settled in San Francisco 1895, took up photography as a hobby and in 1897 opened a professional portrait studio. Made excellent pictures of the Chinese quarter and of the earthquake and fire in 1906; also of his many travels in foreign countries. Moved in 1911 to New York where he was prominent as a photographer of stage and film actors and other celebrities. Autobiography: *As I Remember* (1937).

GEOLOGY. Geology is the study of the rocks which make up the earth's crust and includes as sub-divisions mineralogy, palaeontology, crystallography, and the specialized study of metal ores. Technically a rock is any substance forming part of the earth's crust; even such substances as coal and peat are rocks to the geologist.

Field Work. Any part of the globe where natural rock is exposed is a suitable site for a geological investigation. Nowadays a great deal of the geologist's work consists of laboratory examination of specimens collected in the field and his records, as a result, consist largely of photomicrographs and macrographs of material specially prepared.

These records are not complete, however without a certain amount of descriptive illustration of the locality and site from which the material was obtained. To complete the record photographs are usually desirable to show the topography of the terrain, the general configuration, bedding, etc., and direction and angle of dip (if any) of the exposed rocks. Photographs of any usual or unusual features are taken which, used in conjunction with a map, give the reader a fair idea of what he would actually see on arrival at the site.

The photographic apparatus carried should be as small and light as possible in view of the other equipment the geologist must carry. Any good film camera is suitable. A reflex is useful but a miniature or folding roll film camera is more convenient. It must be borne in mind that a certain amount of scrambling over rough ground may have to be done and a secure, well fitting camera case is essential.

Preparation and Technique. When photo

graphing large expanses of exposed rock, it is obviously not practical to put in a great deal of preparatory work on the surface, but a fair amount of help can often be obtained from local conditions, if used to best advantage. On a weathered surface, for instance, the harder rocks or layers, being less worn, protrude beyond the surface of softer rocks. Cast shadows will show up such formations in strong relief at certain times of the day.

Some idea of the size of the rock exposure should be recorded by including in the photograph a familiar object of known size. The geologist's hammer will serve if it will be sufficiently visible in the photograph; for more extensive exposures, a walking stick painted alternately black and white in 6 ins. sections along its length is commonly used.

When a site has been selected—e.g., a cliff face showing structures such as bedding—it is not advisable to photograph it in direct frontal lighting, as a good deal of the bedding which is faint will not record well. When possible, the site should be photographed when the light is falling at an oblique angle, thus forming a certain amount of shadow and exaggerating the structure. From small structures which have been badly weathered, loose grains and vegetable matter should be brushed away to reduce confusion in the photograph.

Where colour variation is the main feature in a rock exposure, it is generally necessary to use a panchromatic emulsion with a suitable filter to record colour differences faithfully. 35 mm. colour film has also been employed successfully for this purpose. The primary object of photography in the geological field is the production of a true and accurate record; the claims of an aesthetically pleasing picture and observance of the laws of composition must always be of secondary importance.

It is advisable where possible to take several shots of a site from different angles. These give a far better idea of the general lay-out than one comprehensive view.

Panoramas. Reasonably pleasing panoramic views can be produced without the aid of a special camera if certain basic points are observed. These are:—

(1) The number of exposures must be adequate to cover the area being photographed and an overlap of at least 25 per cent is normally allowed from one exposure to the next.

(2) The camera should be rotated through the necessary arc on an imaginary axis passing through the lens assembly. Mounting on a tripod is usually accurate enough for most purposes.

(3) The base of the focal plane should be in the same horizontal plane throughout.

(4) Exposures should be made in rapid succession, care being taken that moving clouds or other factors do not cause a variation in lighting conditions.

(5) The prints should all be made at the same time, each print being carefully developed to the same density as its immediate neighbours. They are assembled into a panorama by mounting on cardboard and roughly tearing the overlapping print at the matching point so that a chamfered edge is obtained. The tear is so made that the chamfer occurs on the underside of the overlapping prints. By using this procedure the uneven edge of the tear will normally be almost invisible if the prints are well matched for density.

Hand Specimens. Where the macroscopic structure of the specimen is of interest, the face to be photographed should be as flat as possible and, if necessary, should be ground flat or polished. This is done by grinding with successively finer grades of abrasives, preferably in powder form moistened to a paste, on a flat surface until the required degree of detail in the surface of the specimen is visible. With a fossil, however, the shape and any signs of distortion are as important as the surface features and there should be no trimming or grinding of the surface. It should, however, be cleaned up to show the surface features to best advantage. Particles of the sedimentary rock surrounding the specimen should be picked out with a scriber or similar instrument, and the surface to be photographed brushed clean before photographing.

Sometimes only the one surface is exposed and the specimen is left attached to a portion of its groundmass; at other times the specimen is picked clean and photographed against a black or white background. The treatment is a matter of individual preference, but it also depends on the nature of the specimen—e.g., it may be necessary to keep the specimen on its groundmass in order to keep it whole, just sufficient of the surface being picked and scrubbed clean to show the detail.

Photomicrography. If the rock structure is too fine to record macroscopically, a thin section can be prepared and magnified by transmitted light through a microscope.

The preparation of the thin section requires considerable skill and patience. One side of a small piece of rock is ground with fine carborundum powder to give it a very smooth surface. It is then mounted on a glass slide with Canada Balsam. The other side of the specimen is then ground with successively finer abrasives until it reaches the required degree of thinness. By polarizing the light through the microscope the thickness of the section is indicated by the colour of the crystals. From this colour the thickness can easily be determined with the aid of the polarization colour scale.

Finally the thin section must be covered with Canada Balsam and then protected by a cover glass.

Photographing the Thin Section. A petrological microscope is necessary for photographing thin sections since a great deal more detail and structure can sometimes be seen by using polarized light.

Photomicrography is a skilled art and it is normally desirable to use an instrument specially designed for this work. Some 35 mm. cameras can be used and the manufacturers supply a special adaptor for this purpose. Provided certain rules are observed, it is possible to arrange a suitable plate camera and a microscope for satisfactory results.

Normally petrological work does not call for a high degree of magnification, but the mining geologist frequently uses very powerful objectives when examining polished ore-minerals, and photographing these specimens with the aid of a vertical illuminator can be very difficult, even with a specially constructed instrument.

All types of emulsion are used in this work, ranging from process to panchromatic, depending on the degree of contrast required and the presence or absence of colour in the specimen. M.C.K. & D.G.W.

See also: *Archaeology; Metallography.*

GERMANY

An important part in the invention and progress of photography has been played by Germany, and today photography in the Deutsche Bundesrepublik (German Federal Republic, West Germany), as well as being a recreation for millions of amateurs and an indispensable aid to most branches of science, engineering, and commerce, provides the basis for a flourishing industry with a world-wide reputation.

Discoveries in Chemistry. 1193–1280: Albertus Magnus (Albert von Bollstädt, a Dominican monk), one of the most learned men of the middle ages, was the first to draw attention to the blackening of silver salts when in contact with organic substances. ("A silver nitrate solution dyes human skin black, and the dye is difficult to remove.")

1565: Georg Fabricius was the first to discover the existence of silver chloride as a natural mineral and to describe it.

1683–1737: Kaspar Neumann was the first to give details of the sensitivity of mercury salts to light.

1725: Johann Heinrich Schulze discovered that silver salts are sensitive to light. Published his discovery, 1727.

1782: A. Hagemann gave the first definite details of the sensitivity of gum to light.

1801: Johann Wilhelm Ritter discovered the ultra-violet rays of the spectrum and their darkening effect on silver chloride and was the first to observe the conflicting effects of red and violet rays.

1814: Nicolaus Wolfgang Fischer discovered that silver albuminate is sensitive to light and ascertained further facts about the light sensitivity of silver compounds.

1819: Theodor Freiherr von Grotthuss enunciated the law of photo-chemical absorption and elaborated the theory of chemical reactions to light.

1831: Johann Wolfgang Doebereiner discovered the high light sensitivity of iron oxide and manganese oxide when combined with oxalic acid.

1832: Dr. Gustav Suckow discovered that potassium bichromate becomes light-sensitive when mixed with an organic-substance.

1839. 13th April: Franz von Kobell, together with Carl August von Steinheil, presented to the Bayerische Akademie der Wissenschaften (Bavarian Academy of Science) paper negatives made in a cylindrical camera.

1839: Albrecht Breyer invented the reflex method of photographic copying of opaque originals without a camera.

1873: Hermann Wilhelm Vogel discovered optical sensitization. It was of fundamental importance both for correct tonal rendering in monochrome and in colour photography.

1891: Momme Andresen introduced the Rodinal developer on which many later developers are based.

1895: Wilhelm Conrad Röntgen discovered X-rays and invented radiography.

1901: Heinricus Lüppo-Cramer invented desensitization.

1903: Adolf Miethe and Arthur Traube discovered isocyanine dyes and introduced them into photographic practice. They are of special importance for colour sensitization and paved the way for the modern panchromatic plate.

1904: Otto Perutz succeeded in preparing the first panchromatic emulsion to incorporate a permanent colour sensitizer. Additional sensitizing dyes were introduced by Ernst König, Benno Homolka and other chemists at the Höchst dye factory.

1905: Adolf Tellkampf and Arthur Traube jointly invented the photographic tracing process known as "Fotoldruck".

1912 (1913, 1914): Dr. Rudolf Fischer took out basic patents for the subtractive colour process using intregal tripack film and paper, and involving colour development.

1917: Gustav Kögel invented the most widely used modern plan copying process—the Ozalid process.

1920: Henricus Lüppo-Cramer introduced desensitizers which made photographic development possible in relatively bright light.

1920: Robert Schuloff discovered the green sensitizer.

1936: Introduction of integral tripack colour process (Agfacolor) based on the work of Dr. Gustav Wilmanns and his associates, and using non-diffusing couplers.

1949: Introduction of document copying process, based on the work of Dr. Edith Weyde, giving a direct positive from the original by image diffusion.

Camera Developments. 1671: Athanasius Kircher, a Jesuit, improved the laterna magica.

1676: First portable reflex box camera obscura described by Johann Christoph Sturm, Nürnberg.

1685: Small box reflex cameras made by Johann Zahn.

1841: Peter Wilhelm Friedrich Voigtländer produced in Vienna the first daguerreotype camera made entirely of metal. He also founded the firm of Voigtländer in Braunschweig (1849).

1914: Oskar Barnack completed the first two models of the Leica camera.

1925: Introduction of the Leica camera, marking the beginning of the miniature era.

1928: Reinhold Heidecke introduced the Rolleiflex camera: the first precision twin-lens reflex.

1936: The first single lens miniature reflex camera, the Kine-Exakta, was constructed at the Ihagee Camera Works in Dresden.

Development of Optics. 1840: Peter Wilhelm Friedrich Voigtländer pioneered the production of the Petzval lens.

1865: Carl August von Steinheil, working with his son Adolf, manufactured the distortion-free, symmetrical Periskop lens; and in the following year Adolf Steinheil computed the Aplanat lens, which was free from distortion and dispersion.

1886: A glass technology laboratory was founded by Friedrich Otto Schott at the instigation of Ernst Abbe for the production of special glass for lens manufacture. This work laid the foundation of modern lens design. Abbe associated himself with Carl Zeiss' workshop and later produced the anastigmat lens invented by Paul Rudolph.

1889: Paul Rudolph computed an unsymmetrical anastigmat giving fine definition to the edge of the image. The Zeiss Protar and Tessar lenses were developed from this lens.

1893: Emil von Hoegh, in the firm of C. P. Goerz, constructed the first symmetrical (Goerz) double anastigmat lens, the Dagor.

1902: Christian Bruns and Friedrich Deckel constructed the Compound shutter.

1912: Friedrich Deckel devised the Compur shutter, the most universally used between-the-lens shutter in the world today.

1924: The Elmar, the first special anastigmat lens for the Leica, was computed by Professor Dr. Max Berek.

Other Developments. 1839: 25th February: the word Photographie was first used in the journal *Vossische Zeitung* by Johann von Maedler.

1882: Georg Meisenbach introduced the direct reproduction of half-tone images by means of a cross-line screen. Meisenbach's Autotypie was the first photographic half-tone process used commercially in book printing.

1925: Dr. Vierkötter introduced the safe and smokeless vacuum flash bulb with tin foil for electric ignition.

Industry. The German photographic industry is almost as old as photography itself. As soon as the necessity for the manufacture of photographic products arose, existing plants and laboratories took up the work and increased their sphere of activity.

The following dates denote the origin of a number of firms whose activities reflect practically the whole German photographic development.

Voigtländer & Sohn (founded in Vienna in 1756), in Braunschweig since 1849—lenses, cameras.

Dr. Carl Schleussner, Frankfurt-on-Main (since 1830), supplied raw materials for the wet collodion process in 1860.

C. A. Steinheil, Munich (since 1837)—lenses.

Carl Zeiss, Jena (since 1846)—lenses.

E. Schering, Berlin (since 1851) began the production of photographic chemicals in 1854.

Eduard Liesegang, Düsseldorf (since 1854)—projectors, enlargers and photographic papers.

Trapp & Münch, Friedberg, Hessen (since 1861) produced photographic salts and albumen papers.

Zeiss-Ikon (a merger of the firms of Zeiss, Ika, Ernemann, Contessa, Nettel, Goerz), formerly in Dresden, now in Stuttgart and Kiel, traces its origin back to Richard Hüttig's camera workshop in Berlin in 1862—lenses, cameras.

Ernst Leitz, Wetzlar, was founded by Carl Kellner and taken over by Ernst Leitz, sen. in 1869, the name being changed at the same time by the new owner—lenses, cameras, microscopes.

C. P. Goerz, Berlin (since 1886)—lenses. After the first World War the firm passed into the hands of Carl Zeiss, Jena.

Agfa Aktiengesellschaft für Anilinfabrikation was established in 1867 and began production of photographic developers in 1889.

Mimosa, A. G., formerly in Dresden, now in Kiel, was founded in 1889 in Cologne-Ehrenfeld—photographic materials.

Industry and the World Wars. Except for a monopoly of the production and supply of sensitizing dye stuffs, the German photographic industry before 1914 developed on similar lines to that of other European countries. After the First World War, in spite of the confiscation of many patents and the losses resulting from the economic collapse, German industry made a remarkable recovery for the following reasons:

(1) Manufacturers had to look for new products and markets. Big firms incorporated old crafts and hand-made goods to be manufactured on industrial mass-production lines.

(2) New inventions were exploited for this drive (miniature camera, flash bulbs).

(3) The industry publicized photography on a scale not previously attempted in Europe.

The Second World War ended all this and the industry, in addition to suffering considerable damage, was afterwards divided into two parts, one in the Deutsche Bundesrepublik (German Federal Republic, West Germany) and the other in Deutsche Demokratische Republik (German Democratic Republic, East Germany).

WEST GERMANY

Reconstruction began at once in the Federal Republic. In the first ten post-war years its photographic industry again became stable and successful. Three important factors were:

(1) The revival of the Verband der Deutschen Photographischen Industrie e.V. (German Photographic Manufacturers' Association) in Frankfurt in 1946 and its intelligent and energetic leadership.

(2) The initiation of Photokina, the International Photo and Cine Exhibition, Cologne, which in 1950 for the first time displayed to visitors from all over the world the new German photographic products and their applications. From 1951 Photokina became international both in its trade fair and its cultural section. It is now recognized as the most important exhibition of photographic products and activities in the field of photography. The 1963 Photokina attracted 619 industrial exhibitors from 22 countries, and many more nations contributed to the picture exhibitions. The fair occupied 760,000 square feet in nine halls and there were more than 202,000 visitors during the ten days of opening.

(3) The foundation of the Gesellschaft zur Förderung der Photographie e.V. (Society for the Advancement of Photography), Frankfurt, an organization financed voluntarily and jointly by industry and dealers. It has created a centre supplying the press, free of charge, with articles on photography and good pictures, it runs a general advertising campaign for the benefit of photography, and engages in many other promotional activities.

There are over 10,000 retailers in the German Federal Republic (West Germany) and their interests are represented by Deutscher Photo- und Kinohändler-Bund (German Photographic Dealers Association) at Hamburg. Besides the usual photographic supplies and processing services, many dealers run their own lectures and instruction courses. A 1962 survey revealed that 16 million adults in West Germany are camera owners.

Societies and Clubs. The leading German photographic society is the Deutsche Gesellschaft für Photographie e.V. (The Photographic Society of Germany) in Cologne. Membership is by invitation only, and it is open to persons who have distinguished themselves in the service, and furthered the progress, of photography. The Society organizes international congresses and awards an annual prize for cultural achievement in photography (which includes a sum of D.M. 10,000).

The top practising photographers are organized in the Gesellschaft Deutscher Lichtbildner (Society of German Photographers). Members must submit work for each annual exhibition. The bulk of the professional photographers is embraced by Centralverband des Deutschen Photographenhandwerks (Federal Association of the Photographic Craft) counting about 6,000 members.

Press and magazine photographers are organized in the Deutscher Journalistenverband (German Journalists' Association) at Bonn.

The amateur clubs—about 550—are federated to the Verband Deutscher Amateurphotographen-Vereine e.V. (Association of German Amateur Photographic Clubs) which in its turn is affiliated to the Fédération Internationale de l'Art Photographique.

Schools. The leading school of photography, the Höhere Fachschule für Photographie (High School for Photography) was built and opened in Cologne in 1954. It is sponsored and financed jointly by the State North-Rhine-Westphalia, the City of Cologne, the photographic industry and trade. It enjoys the reputation of being the best equipped place for studying and practising photography. It has three sections, one for instructing dealers and dealers' employees, one for creative and colour work and one for research. It covers 45,000 square feet and occupies six floors.

Four more schools, Bayerische Staatslehranstalt für Lichtbildwesen (Bavarian State School of Photographic Arts) in Munich, Folkwangschule (Folkwang-School), in Essen, Staatliche Hochschule für Bildende Künste (State School for Art) in Hamburg and Letteschule (Lette School) in Berlin endeavour to educate pupils mainly in the creative side of photography. There are also a number of private schools, many of which specialize either in particular techniques—e.g., colour—or follow definite artistic trends. The Faschschule des Deutschen Photo- und Kinohandels (School of the German Photo- and Cine Trade), founded 1957 in Kiel, is organized by the Dealers Association and controlled by the Ministry of Culture of Schleswig-Holstein. In three yearly courses it teaches technical chemical and sales knowledge to dealers' employees.

Publications. Since the early days of photography a firm in Halle has published a wide range of standard works of scientific and artistic value intended both for professional and amateur photographers. This firm now publishes in Düsseldorf the monthly magazine *Foto Prisma* for professionals and *Foto Rundschau* for amateurs. A Munich Company which specializes in the publication of photographic books is also responsible for *Foto Magazin*, the largest German amateur photographic magazine, and the dealers' magazine *Der Fotohändler*.

The official organ of the photographic dealers, Inpho, is published in Düsseldorf. A more technical magazine, MFM (Monatliche Fototechnische Mitteilungen), is published in Ludwigsburg, and the weekly Photo-Presse, published in Hannoversch-Münden, serves the interests of the professional photographer.

A Berlin publishing house specializing in periodicals for the radio, photographic and motion picture industries, publishes the monthly magazine *Photo-Technik und Wirtschaft*, dealing mainly with industrial and trade matters. In Stuttgart, Dr. Wolf Strache revived in 1955 the famous pre-war yearbook *Das Deutsche Lichtbild*.

Dr. Erich Stenger of Kreuzwertheim-am-Main was responsible for an important scientific and historical collection, which bears his name and which, after his death in 1957, was bought by the Agfa Works in Leverkusen. In 1963 the Stadt (City) Museum of Munich took over the large photographic (mainly optical), collection of Dr. Rudolf Loher, opening the first photographic museum in West Germany.

Artistic Trends. The first important exhibition aimed at raising the level of pictorial photography was organized in Hamburg in 1893 by Ernst Juhl and Alfred Lichtwark. The principal participating photographers were Rudolf Dührkoop of Hamburg, Nicola Perscheid of Berlin, Wilhelm Weimer of Darmstadt, and Hugo Erfurth of Dresden, later Cologne. Another outstanding international exhibition was staged at Dresden in 1909.

After the First World War two main trends developed, both of which had a profound influence on photography everywhere.

One was the abstract movement, derived from the famous Bauhaus in Dessau, later Weimar, which became the pioneering centre of modern art where photography was recognized and encouraged as a means of artistic expression. Well known personalities from many lands—e.g., the Hungarian, Laszlo Moholy-Nagy, and the Austrian, Herbert Bayer—were attracted to the Bauhaus, worked there, and carried the Bauhaus spirit abroad. The use of photography for publicity, as in posters, murals and book-covers, was one of the particular achievements of the Bauhaus Group. The outstanding event of the period was the exhibition *Film und Photo*, Stuttgart, 1929.

A typical lone worker of the same period was Albert Renger-Patzsch who specialized in the portrayal of simple objects, hitherto overlooked by photographers, often in a new large close-up treatment. His rendering of these objects came to be known as the Neue Sachlichkeit (New Objectivity) and it revealed both the abstract and the factual beauty of our everyday surroundings.

Another factor was the rise of photojournalism. In this the publishing house of Ullstein, Berlin, with the magazines *Berliner Illustrirte*, *Uhu* and *Die Dame* paved the way for photo journalism as we know it today. Dr. Erich Salomon was a singularly outstanding photographer in the field of candid photography and his remarkable studies of political personalities, taken with inferior materials and the available light without the victim's knowledge, introduced a new era in photographic reporting. Many other photographers who have since achieved world fame, such as Alfred Eisenstaedt, first obtained recognition through their work for Ullstein. Leading illustrated magazines the world over owe much to the ideas and initiative of these German photographers and publishers.

Dr. Paul Wolff did much to rescue photography from stiff formality in his pioneer use of the miniature camera for his dynamic and unorthodox pictures. His work inspired amateurs and professionals alike to look at people and the world around from a new angle.

The main aim of photographers in West Germany since the war has been to catch up with the progress made in the western world during the pre-war years, to learn to use new equipment and techniques—e.g., electronic flash—and to resume the experimental work of the period between the wars—e.g., abstract and impressionist trends like "fotoform" and "subjektive fotografie" but with a modern approach.

A most hopeful proof of the new vision of modern times has been afforded by the Arbeitsgemeinschaft Jugend Photographiert exhibitions, organized by Photokina since 1954. Photographs selected from many thousands of entries have shown that the German Federal Republic is again contributing greatly in the field of creative photography. From the prizewinners have emerged a new crop of professionals. Many of these are photo-journalists, who have founded an international group known as European Magazine Photographers, to which nationals of other European countries are also invited.

The West German Government established in 1961 an annual Youth Photo Prize, the sum of DM. 10,000 being presented to the winning pictures. L.F.G.

Industry Today. The West German photographic industry is one of the oldest established and most highly developed in Europe. Its primary interest has traditionally been in the manufacture of equipment, in particular high quality cameras and lenses, although in recent years the production of sensitized materials has come to gain an important place.

There are at present 157 establishments undertaking the manufacture of photographic goods, together providing employment for approximately 50,000 persons, and these are located principally in the southern part of the country, leading centres of production being Stuttgart and Munich. Of the total number of establishments, 141 with about 38,000 employees are concerned with equipment production whilst the remaining 16 establishments and 12,000 work people undertake the manufacture of sensitized goods and chemicals.

The most important single manufacturer is Agfa-Gevaert, A.G., a subsidiary of the industrial combine Farbenfabriken Bayer, which has about 11,000 employees and a turnover of DM. 425 million in 1962. This company produces still and cine equipment and a wide range of sensitized materials, and its output of still cameras was announced to have reached one million units per annum in 1963.

In addition to Agfa, the largest companies in the equipment field are Franke & Heidecke ("Rollei" trade mark), Kodak, Leitz, Voigtlander, and Zeiss Ikon, being together responsible for almost the entire industry's output of still cameras. Among these, Kodak specialize in the production of 35 mm. still cameras, having 2,050 employees and a turnover in 1963 of DM. 135 million, whilst Zeiss Ikon are well known in the sphere of optical goods in addition to the manufacture of miniature cameras and lenses, with some 5,300 employees and a turnover of DM. 100 million in 1963.

The sensitized materials sector is headed by Agfa, with Perutz (also controlled by Farbenfabriken Bayer) as the second largest photochemical producer, making all types of film, plates and chemicals, having some 1,500 employees. The formerly independent Agfa Company merged with the major Belgian producer, Gevaert, during 1964 to form the largest photographic manufacturing group in Europe. The combined annual turnover in 1963 was about £108 million sterling.

Another important manufacturer of sensitized products is Schleussner ("Adox" trade mark), which also produces still cameras and projectors, with about 2,000 employees. This company has now become a fully owned subsidiary of DuPont, the leading United States chemical producer.

There are four trade associations, membership of which is voluntary, covering between them practically the whole equipment and photo-chemical sectors of the industry. The principal organizations are the Verband der Deutschen Feinmechanischen und Optischen Industrie and the Verband der Deutschen Photographischen Industrie, which represent the interests of photographic apparatus manufacturers with regard to all trade matters.

An important influence on the development of the industry has been the Photokina, an international trade fair started in 1950 and held thereafter at varying intervals in Cologne, which forms a focal point for new ideas and products.

Production. The production of photographic goods has shown a steady growth in output during the last five years, although since 1960 the effects of Japanese competition in the important field of still cameras has been reflected in reduced production and exports. The industry also undertakes the manufacture of a wide variety of accessories and lenses, whilst the output of cine equipment, which has hitherto been relatively small in keeping with the domestic market for such products, is now being expanded to provide increased supplies in the future.

All types of sensitized materials are produced, the output of which has risen considerably in volume since 1959, and this group now accounts for about two-fifths of the total value of photographic goods production. Important items are sensitized paper and roll film, both of these having expanded noticeably during the period.

Imports. The value of imported photographic products has recorded a substantial gain during the last five years, reaching a total of DM. 193·6 million in 1963, of which approximately a half was composed of sensitized materials and chemicals.

Among apparatus, imports of still cameras have more than trebled in value during the period, the bulk of these being accounted for by the amateur type. In 1963 supplies came principally from Japan, with DM. 4·0 million, and the U.S.A. with DM. 5·7 million's worth.

The largest item, cine equipment, has shown an impressive rise in value to DM. 17·9 million in 1963, of which DM. 15·1 million was composed of 8 mm. cameras and projectors. The main sources of the latter were Japan (DM. 7·3 million), and Switzerland (DM. 4·0 million).

Sensitized material imports have increased considerably in value between 1959 and 1963, reaching DM. 102·2 million in the latter year. The greatest expansion has occurred among sensitized paper, imports of which have nearly trebled in value to DM. 18·7 million, due mainly to increased imports from Belgium, which supplied DM. 12·0 million's worth in 1963. The leading product throughout the period has been roll and cine film, together totalling DM. 46·1 million in 1963, the principal sources of supply being Belgium (DM. 10·1 million), Italy (DM. 10·2 million), and the U.S.A. (DM. 11·5 million). Belgium is also the main supplier of plates, X-ray, and sheet film, with DM. 16·7 million's worth out of the 1963 total imports of DM. 32·9 million.

Exports. A more moderate rate of increase than among imports has been shown by exported photographic equipment and materials, although their value is more than treble that of the former. Of the DM. 650·1 million total value of exports in 1963, the largest share was accounted for by apparatus, in particular still cameras, the exports of which rose to a peak in 1960 but have since declined somewhat with the growing competition of Japanese producers in world markets. The largest single customer is the United States, whilst the other E.E.C. members and the E.F.T.A. countries are important markets.

Still projectors, enlargers and reducers together comprise the next most important item exported, with a total value of DM. 65·3 million in 1963, of which Switzerland received DM. 7·1 million, Sweden DM. 8·3 million, and the U.K. DM. 6·0 million's worth. Exports of cine cameras and projectors have shown little growth in value during the last five years, although in 1963 8 mm. equipment production started to expand and should be reflected by increased exports in the future. This type accounted in 1963 for DM. 23·3 million of the DM. 42·5 million total, a major customer for which the U.S.A. taking DM. 4·5 millions' worth during the year.

Sensitized material exports have risen by 75 per cent in value since 1959 to reach DM. 221·1 million in 1963. Among the leading products is photographic paper, totalling DM. 93·2 million during the year, the chief destinations of which were the U.S.A. (DM. 11·1 million), Italy (DM. 8·3 million), and France (DM. 6·2 million).

Roll and cine film are also leading products, together accounting for DM. 85·1 million in 1963, of which Switzerland received DM. 8·5 million and the U.S.A. DM. 5·5 million, whilst exports of plates, X-ray, and sheet film have trebled in value since 1959 to reach DM. 28·8 million, France receiving DM. 4·6 million's worth.

G.R.H. & A.G.B.

WEST GERMAN PRODUCTION OF PHOTO EQUIPMENT

Quantity (000s)	1959	1960	1961	1962	1963
Technical cameras	5	8	5	3	4
Other cameras	2,766	2,731	3,005	2,577	2,473
Camera bodies	82	73	42	33	41
Total cameras	2,853	2,812	3,052	2,613	2,518
Lenses	2,141	2,037	1,977	1,749	2,155
Photocopiers	29	37	37	36	50
Exposure meters...	1,299	1,503	1,386	1,334	1,257
Still projectors and viewers	317	382	502	752	719
8 and 16 mm. cine projectors	70	56	56	69	62
Value (*DM million*)	1959	1960	1961	1962	1963
Technical cameras	6·1	8·5	7·6	5·2	6·5
Other cameras	243·0	279·0	277·4	227·6	225·4
Camera bodies	15·8	15·9	11·6	10·1	12·0
Total cameras	264·9	303·4	296·6	242·9	243·9
Lenses	76·6	77·2	79·9	66·1	72·3
Photocopiers	12·9	16·2	16·6	19·9	23·6
Exposure meters...	29·9	34·5	31·4	30·9	34·1
Still projectors and viewers ...	40·3	54·3	76·0	73·9	71·7
8 and 16 mm. cine projectors	18·9	17·3	21·5	27·5	26·2

WEST GERMAN PRODUCTION OF PHOTOGRAPHIC GOODS

DM, Million (Manufacturers Net Selling Price ex-Works)

	1959	1960	1961	1962	1963
Still cameras	264·9	303·4	296·6	242·9	243·9
Still projectors and viewers	40·3	54·3	76·3	74·0	71·7
Cine projectors	23·9	23·2	27·0	30·8	29·2
Lenses	76·6	77·2	79·9	66·1	72·3
Other equipment and parts	210·2	240·8	249·1	246·6	287·8
Total apparatus	615·9	698·9	728·9	660·4	704·9
Sensitized paper	169·5	184·4	205·1	190·1	218·1
Roll film	118·6	106·5	124·0	145·7	144·2
Cine film, sheet film and plates ...	92·8	98·9	106·7	120·6	128·0
Chemicals	16·7	19·4	25·5	24·3	23·8
Total sensitized materials and chemicals	397·6	409·2	461·3	480·7	514·1
Total production	1,013·5	1,108·1	1,190·2	1,141·1	1,219·0
(£ million)	90·5	98·9	106·3	101·9	109·2

WEST GERMAN PRODUCTION OF SENSITIZED MATERIAL

Thousand Sq. Metres

	1959	1960	1961	1962	1963
Sensitized paper	65,888	71,540	78,516	70,874	83,003
Roll film	3,472	3,200	3,716	3,977	3,804
Sheet film and filmpacks	553	611	692	778	258*
Total	69,913	75,351	82,924	75,639	87,065

*(Figure not comparable with earlier years due to altered coverage)

WEST GERMAN IMPORTS OF PHOTOGRAPHIC GOODS

DM. Million c.i.f.

	1959	1960	1961	1962	1963
Still cameras	5·5	8·1	8·5	11·3	17·5
Cine cameras and projectors	9·3	13·2	17·8	18·2	17·9
Still projectors, enlargers and reducers	2·7	2·3	3·4	3·2	5·5
Lenses	5·1	6·6	6·7	6·6	7·5
Other equipment and parts	5·4	12·4	14·7	22·8	43·0
Total apparatus	28·0	42·6	51·1	62·1	91·4
Sensitized paper	6·9	8·7	12·1	17·1	18·7
Roll and cine film	23·8	29·0	39·0	47·4	46·1
Plates, X-ray film and sheet film ...	20·2	23·1	24·7	28·9	32·9
Chemicals	1·4	2·2	2·8	3·8	4·5
Total sensitized material and chemicals	52·3	63·0	78·6	97·2	102·2
Total imports	80·3	105·6	129·7	159·3	193·6
(£ million)	6·8	9·0	11·6	14·2	17·2

WEST GERMAN EXPORT MARKETS FOR AMATEUR STILL CAMERAS 1963

Country	*No. (000s)*	*Value (DM. Million)*
European Common Market:		
Belgium/Luxembourg	63·6	3·2
France	209·9	22·0
Italy	141·4	12·3
Netherlands	191·2	7·4
E.C.M. Total	606·1	44·9
European Free Trade Area:		
Austria	59·7	5·6
Denmark	49·4	3·6
Finland	29·5	1·0
Norway	28·0	1·7
Portugal	6·3	0·4
Sweden	84·6	6·5
Switzerland	128·6	12·3
U.K.	129·6	10·1
E.F.T.A. Total	515·7	41·2
Australia	32·1	2·5
Brazil	11·0	0·9
Canada	19·7	2·4
Hong Kong	15·8	4·2
Mexico	29·4	2·5
New Zealand	17·1	0·8
Union of S. Africa	21·4	1·6
U.S.A.	146·6	25·3
All other countries	245·8	22·3
Total exports	1,660·7	148·6

WEST GERMAN EXPORTS OF PHOTOGRAPHIC GOODS
DM. Million f.o.b.

	1959	1960	1961	1962	1963
Still cameras	169·8	179·0	164·2	160·3	156·8
Cine cameras and projectors	33·9	31·5	31·9	31·2	42·5
Still projectors, enlargers and reducers	32·7	44·8	58·0	59·5	65·3
Lenses	45·0	43·9	46·6	46·1	50·3
Other equipment and parts	95·0	88·9	90·9	104·6	114·1
Total apparatus	376·4	388·1	391·6	401·7	429·0
Sensitized paper	64·3	69·7	75·5	75·3	93·2
Roll and cine film	47·1	54·1	67·2	70·4	85·1
Plates, X-ray and sheet film	9·3	11·5	16·3	21·3	28·8
Chemicals	7·1	8·7	10·6	11·5	14·0
Total sensitized material and chemicals	127·8	144·0	169·6	178·5	221·1
Total exports	504·2	532·1	561·2	580·2	650·1
(£ million)	42·0	44·3	50·1	51·8	58·0

EAST GERMANY

Organizations. Professional photographers are organized in the appropriate sections of the Handwerkskammer (Chamber of Crafts) and the Verband der Deutschen Journalisten (Union of German Journalists). Young amateurs join the photographic association of their school, youth club or the youth group at their place of work. Older amateurs can join photo-groups sponsored by their Trade Unions or clubs of the Deutsche Kulturbund (German Cultural Federation). There is an elected leadership for all groups, circles and clubs in the Zentrale Kommission Fotografie—ZKF (Central Photographic Commission), an organ of the Deutsche Kulturbund, with sections for colour photography, youth photography, exhibitions, tuition, etc. The ZKF represents East German amateurs in working with societies from other countries and with international bodies. Branch organizations of the ZKF, known as Bezirks-und Kreiskommissionen für Fotografie (Local Commissions for Photography), are found in various districts; their leadership is elected. In these districts, there are regular courses of instruction. A postal study course keeps the group leaders up-to-date in their photographic and instructional knowledge.

Characteristic of East German organizations is the close co-operation between amateurs, professional photographers and photo-journalists in the running of the various commissions, courses and groups. This is why there is no need for special photographic societies.

The ZKF has established an archive of the best photographs by East German photographers, and also runs the Museum für Fotografie (Museum for Photography) in Dresden; it publishes the periodical *Fotografie*.

Exhibitions. There is a widespread and lively interest in picture exhibitions. In schools, clubs, firms and neighbourhood areas, there are often small exhibitions of defined subject content to which anyone can contribute. Besides these, there are annual or biennial progress shows of the best pictures in black-and-white and colour. Societies from other parts of the country are invited to submit work for consideration in these. Every other year, the Deutsche Fotoschau (German Photo Show), and every third year the Berliner Internationale Fotoausstellung—Bifota (Berlin International Photo Exhibition), take place. In 1962, nearly 10,000 photographs were contributed to the latter from forty-one countries.

Newspapers, periodicals and social organizations often hold photographic competitions, the results of which either go on exhibition or are published.

Schools. Besides an engineering school (Ingenieurschule für Feinwerktechnik) in Jena, in which photographic and other subjects are covered, and the Deutschen Hochschule für Filmkunst (German High School for Cinema Art) in Berlin-Babelsberg, there is a specific training centre for photography in Leipzig. This has grown from the Fachschule für fotomechanische Vervielfaltigungsverfahren (Trade College for Photomechanical Processes) founded in 1893 at the then Leipziger Kunstakademie (Leipzig Academy of Art). Since 1945, the Abteilung Fotografik at the College der Hochschule für Grafik und Buchkunst (Department of Photographic Design at the College of Graphic and Printing Arts) had been operating a five year course leading to a final diploma (Hochabschluss). This course provides training in photography as an art form and its application in conjunction with typography and design in, for example, lay-out, poster work, publicity, advertising and in illustrated periodicals. Since 1963, this department has also run a three year specialist class for applied photography.

Publications. The most important photographic publications are issued by the Fachbuchverlag, Leipzig, which absorbed the former VEB Fotokinoverlag, Halle. The latter, with 125 years of publishing history behind it, published about thirty different photographic books annually, as well as the monthly periodicals *Fotografie* and *Fotokino-Magazin* (Phot-Cine Magazine), the former on the advanced, the latter at beginners' level. The press is also responsible for the export magazine *Jenaer* Rundschau (Jena Review). A Berlin publisher caters for still and cine amateurs with technical leanings through its monthly *Bild und Ton* (*Picture and Sound*). Scientific, medical and other fields of applications are being covered by several publishers specializing in the subjects in question. Furthermore, nearly every firm in the photographic industry produces its own detailed information sheets.

Trends. The overriding significance of photography lies in its pre-eminence as a medium of pictorial communication, aesthetic reflection of reality and expression of human experience.

East German photographers accept an obligation towards those traditions and pioneers of photography who were particularly concerned with human dignity and the beauty of environment. They made their own contributions to photographic history by research into the long neglected area of working class photography. The social criticism and evidence of these pictures about the life of workers set a pattern for a human, committed and positive photographic outlook. This outlook is opposed to formal experiments, abstraction and decadence, which are looked upon as emotional negations and anxiety in the face of realities, reflecting distortion of human values. Reaching beyond the preoccupation with the formal elements of picture making, there is a search for picture content that will reflect human ideals as defined by the country's social order with an emphasis on peace and international understanding.

Interest is centred in three areas: colour photography, which is predominantly used as a reversal process; the camera work applied to photographic journalism, which is also accessible to amateurs, who are often concerned with the pictorial documentation of their local and industrial environment; and youth photography, which is sponsored by the youth movement and the schools alike (50 per cent of all schools have their own photographic work groups).

Industry. The optical industry and camera works, centred in Jena and Dresden, were largely destroyed during the war, but have since been reconstructed under public ownership. The old works were considerably enlarged and expanded in line with national economic planning.

East Germany specializes today above all in making quality single lens reflex cameras, interchangeable lenses, and apparatus for scientific photography and laboratory technology. Besides a number of old established but continually developed makes, various automatic cameras have also been designed, and all kinds of photographic accessories (exposure meters, projectors, flash apparatus and so forth) and cameras in the middle and lower price brackets (especially for beginners' photography) have come into production. In keeping with growing needs, there has been considerable increase in the production of films and chemicals since 1945.

Products are sold in specialist state-run and private shops, which offer a great deal of service to customers. Photographic goods may also be obtained through mail order houses.

The East German photographic industry exhibits annually at the Leipzig Fairs and in a special Fair, devoted to film, radio, photography and television in Dresden. Export is handled by a state organization through agencies in most countries. The whole photographic industry is of particular export importance to East Germany. A.N. & B.B.

Production. During 1963 the output of photographic equipment and materials has shown a small overall increase, although production of several times within the total has remained static or fallen off. The East German industry has clearly not succeeded in maintaining its growth of the previous year, due most probably to labour shortages and other economic difficulties in the country.

Still cameras continued to be the most important products among apparatus, and provided the only increases during the year. The output of reflex cameras gained 14·1 per cent in quantity, compared with only 2·3 per cent in 1962, to reach the highest figure yet attained with 139,600 units. Production of miniature types, although rising by 10 per cent to 192,900 units, showed a slackening in the growth of nearly 30 per cent achieved during the previous year and output has not yet regained the level reached in 1960. Other equipment almost exactly reversed its gain of 1962 with a fall of 13·3 per cent to 187,200 units.

The value of photocopy apparatus produced in 1963 dropped further to DM 56·8 million, whilst other equipment and parts showed a small fall to DM 25·1 million in contrast with its previously rising value.

Sensitized material output, excluding film base, rose by 2·5 per cent to a total of 27·6 million square metres in 1963, compared with an increase of 5 per cent a year earlier.

The leading product continued to be sensitized paper, which showed a slightly increased rate of growth with a rise of 4·6 per cent to reach 11·5 million square metres. The output of negative still and cine film of both black-and-white and colour types, however, remained unchanged at the 1962 level.

Production of positive cine film achieved a rise overall from 13·4 to 13·6 million square metres, whilst altering in composition. Black-and-white stock again increased slightly during

EAST GERMAN PRODUCTION OF PHOTOGRAPHIC EQUIPMENT

	1959	1960	1961	1962	1963
Thousand Units					
Reflex cameras	153·0	98·5	118·1	122·3	139·6
Miniature cameras	249·2	243·9	133·8	175·3	192·9
Other equipment	155·8	187·6	189·5	215·9	187·2
Total	558·0	530·0	441·5	513·5	519·7
DM Million					
Photocopy apparatus	56·7	61·9	62·3	58·5	56·8
Other equipment and parts ..	35·8	25·0	25·7	28·1	25·1

EAST GERMAN PRODUCTION OF SENSITIZED MATERIAL
Thousand Square Metres

	1959	1960	1961	1962	1963
Film base	24,403	25,318	23,959	26,273	27,844
Sensitized paper	11,520	11,250	10,668	11,006	11,508
Still and cine film:					
Negative black and white ..	2,520	2,430	2,253	2,178	2,176
Negative colour	274	323	306	342	342
Cine film:					
Positive black and white ..	9,007	8,917	8,674	10,168	10,697
Positive colour	4,107	3,778	3,715	3,198	2,851
Total paper and film	27,428	26,698	25,616	25,892	27,574

EAST GERMAN EXPORTS OF PHOTOGRAPHIC EQUIPMENT
Thousand Units

	1959	1960	1961	1962	1963
Reflex cameras	84·5	99·0	83·0	80·9	82·3
Other equipment	161·3	175·0	159·7	148·3	133·5
Total	245·8	274·0	242·7	229·2	215·8

EAST GERMAN EXPORTS OF SENSITIZED MATERIAL

	1959	1960	1961	1962	1963
Film base	706	468	640	632	340
Sensitized paper	2,931	2,517	3,109	2,508	2,553
Still and cine film:					
Black and white	8,812	7,858	9,948	9,435	10,006
Colour	3,676	3,456	3,398	3,045	2,749
Total paper and film	15,419	13,831	16,455	14,988	15,308

the year, and now accounts for nearly 80 per cent of the group's total, but the production of colour material declined noticeably.

The output of raw film base rose to a smaller extent than in 1962 to reach 27·8 million square metres, practically all of which was taken by the domestic market.

Exports. East German exports of photographic equipment continued their overall decline during 1963, although reflex cameras showed a marginal increase to 82,300 units. Since the destination of exports is not shown in the trade returns it is difficult to know where changes have occurred, but some information can be gathered from other countries' import statistics. Thus, in 1963 the U.K. received 23,900 still cameras from East German producers, compared with 15,200 units during the previous year, the bulk of these in both cases being composed of miniature models. Still cameras imported by Yugoslavia, however, fell heavily from 13,200 to 4,300 units over the same period, whilst Hong Kong purchases rose slightly from 1,800 to 2,600 units.

Exports of sensitized materials (excluding film base) totalled 15·3 million square metres in 1963, a rise of 2·1 per cent from 1962 contrasting with a fall of almost one-tenth which took place in the latter year. Both sensitized paper and black-and-white still and cine film contributed to this increase, but colour still and cine stock dropped by 9·7 per cent in volume to continue its decline of recent years. Raw film base exports also fell off in 1963.

The leading market for all types of film is the U.S.S.R., whilst other important purchasers include Eastern Bloc members, India, and Hong Kong.

G.R.H. & A.G.B.

GEVAERT, LIÉVIN, 1868–1935. Belgian photographer. Founder (1894) of the N.V. Gevaert Photo-Producten works in Antwerp. Gevaert manufactured originally photographic papers but soon started manufacture of all types of film.

GHOST. Gratuitous image of a light source formed by double reflection between the internal surfaces of the front and back components of a lens. Ghost images are troublesome if they are sharp, and lens designers try to arrange matters so that they will be diffused over the whole focal plane and therefore less obvious. Most lenses produce ghost images if a brilliant light source is included in the field of view.

Such images are a particular kind of flare spot in the form of an actual image, generally distorted, of the light source. A window will often appear as a ghost image in an interior, or a bright patch of sky in an outdoor scene. The ghost is usually inverted and occurs on the opposite side of the picture to the true image.

See also: *Coated lens; Flare; Faults; Halation; Irradiation.*

GHOST PHOTOGRAPH. Trick effect produced by making two exposures of the same scene on the same plate. The first exposure is made of the empty scene, and the second with someone present—e.g., sitting in a chair. The print shows the sitter as a transparent figure through which the background can be seen.

It is equally effective, and sometimes more convenient to photograph the person against a dead black background and then make a normal exposure on the same plate of the scene in which the ghost is to appear. In this case the position of the ghost must be marked on the focusing screen so that it can be correctly placed in its setting in the second exposure.

The same technique can be used with a cine camera to produce a moving picture of a ghost.

See also: *Spirit photography; Tricks and effects.*

GIANT ENLARGEMENT. The following procedure is reasonably adequate for the production of the occasional large (but not too large) print. As a means of producing giant enlargements in variety and quantity, however, both the equipment and the procedure would be unsuitable. There is very little standard equipment available that measures up to the requirements of the professional maker of large prints. Enlargers, processing tanks, etc., must of necessity be purpose-designed and purpose-made.

A correctly exposed and developed negative should enlarge up to twenty diameters without too serious a loss of resolution. To produce prints of this proportion involves no basic departure from normal enlarging technique. The main consideration is that of equipment—what new equipment must be obtained and how best to modify and use existing equipment, particularly in relation to the space available.

The mounting of the enlarger lens, as well as the control of parallelism of negative holder, lens panel, and projection board, must be done carefully and accurately.

Enlarger. When considering the enlarger it is important to remember that the light source will be much farther away than usual from the printing base. A light source of the double condenser type makes focusing easier, minimizes loss of contrast and cuts down the length of exposure.

Where there is enough working space the ideal instrument is undoubtedly the horizontal enlarger. Few models are now made, but second-hand ones can be obtained and are relatively inexpensive. A second-hand quarter-plate or half-plate field camera will serve very well as the basic component of a home-made horizontal enlarger.

It is possible to convert some vertical enlargers to horizontal operation. In a number of instances the lamphouse assembly can be swung and locked in the horizontal position, and there is also available a prismatic attachment that fits on to the enlarger lens and throws an image at right angles to it. In other instances the lamphouse assembly can be completely detached and secured in a home-made horizontal cradle.

It is possible, however, on many vertical enlargers to throw a larger-than-normal image

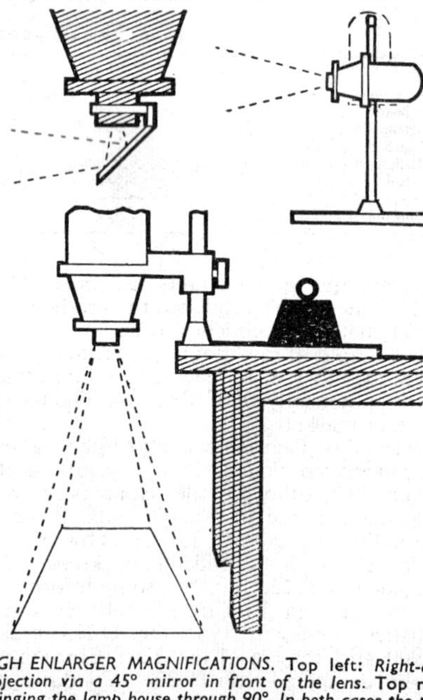

HIGH ENLARGER MAGNIFICATIONS. Top left: *Right-angle projection via a 45° mirror in front of the lens.* Top right: *Swinging the lamp house through 90°. In both cases the paper is pinned to a suitable wall.* Bottom: *Projecting on to the floor*

vertically by mounting the lamphouse assembly on the outside of the column and projecting the image over the edge of the bench on to the floor. The baseboard must be securely weighted down—e.g., with a pile of books.

Equipment. Outsize dishes are not essential. Three dishes, 16 × 20 ins. can be used to develop, rinse and fix prints of a considerably larger size. An ordinary domestic bath, provided the water is changed frequently, will serve very well as a washing tank.

Bromide and chloro-bromide paper can be bought in rolls in two standard widths, 20 ins. and 40 ins., the normal length being 25 feet.

The following items will be needed for cutting the paper:

(1) A sharp-pointed, short-bladed knife.

(2) A metal or hardwood straightedge.

(3) A strip of hard, smooth material (sheet metal, sheet plastic or glass) as a cutting base.

Because it is packed in a tight roll the paper will retain a spring-like curl until it has been well immersed in developer. The normal masking frame, therefore, besides being too small, would be inadequate to hold the paper firmly in position. It can best be secured to the base by darkroom pins.

Printing. The printing of giant enlargements calls for no departure from the normal procedure in the preparation of negatives. However, greater care than usual must be taken to keep negatives free from dust, scratches and other blemishes. Proper attention to cleanliness in processing can save hours of afterwork on the print.

In the following outline of print-making technique it will be assumed that the object is to produce an enlargement 20 × 30 ins.

Cut off a strip of paper from a 20 ins. roll and make the normal exposure test.

Cut a 32 ins. length of paper off the roll. Pin this length into position along the two 20 ins. edges. Do the same along the 32 ins. edges without piercing the paper—i.e., so that only the head of the pin presses down on it.

The scale of enlargement required will make a long exposure time necessary. For this reason, the strongest illumination that can be safely used should be fitted in the enlarger.

The considerable loss of contrast due to distance of projection when making a large print from a comparatively small negative makes necessary the use of high-contrast emulsions, which generally means results which are not of the highest possible quality. The professional avoids this difficulty by making from the original negative a near-perfect "reproduction print", using normal dodging and printing-in techniques. This print should be rich in detail in all tones (i.e. not too contrasty) as it will be copied to make a negative of suitable size, density, and contrast from which the ultimate enlargement will then be made.

Instead of thus making a copy negative, it is possible to use a pre-solarized film of a type supplied by a number of leading manufacturers. This material enables the production of a negative directly from a negative. By using an enlarger, printing control is still possible and the negative can be made to the required size, as before.

Processing. When the paper has been unpinned after exposure, it will spring back into a tightly-wound scroll 20 ins. in length.

Grasp the centre of the scroll with one hand and with the other slacken off about 12 ins. of paper, holding it tautly away from the scroll.

Feed the leading edge of the paper into the 20 ins. side of the developing dish. Press it firmly to the bottom and make sure it is quickly immersed by vigorous movements of the hand from side to side. Slacken off a further length of paper and repeat the process, while at the same time rolling up the paper from the leading edge until the whole area of paper has been drawn through the developer and thoroughly immersed.

Continue coiling and uncoiling the paper, first from one end and then from the other, until development is complete.

The print will now have become limp and easy to handle. Fold it in half, emulsion outwards (making sure it does not crease), rinse it in a water stop bath, then transfer it to the fixing bath.

To avoid uneven development it is essential to carry out the initial immersion in developer as quickly as possible. The beginner may find it easier to soak the paper in water after exposure until it is limp and easy to handle. The paper should then be wound once again into a scroll, drained, and fed into the developer by the method already described. If the final print consists of two or more separate pieces, exposed either in a single projection or in several projections, the professional excludes visual judgement in developing the sections, relying entirely on time and temperature control.

The print, having been fixed and washed, should be secured to a line by clothes pegs and left to dry

Prints which are made in sections should be dried in such a manner that each section undergoes identical conditions with the others. If hung up, they should all hang in the same way, so that any stretch in one occurs also to the others, to the same extent, and in the same direction.

When dry the print is trimmed to size and knifed and spotted. Particular care is given to spotting out the white marks left by the pin heads along the edge of the print. D.M.

See also: *Enlarging; Photo mural.*

GLAISHER, JAMES, 1809–1903. British physicist and meteorologist. Made valuable contributions to photographic chemistry, emulsion making, actinometry, spectrophotography and colour analysis. Third president of the Royal Photographic Society, holding the longest term of office (1869–74, 1875–92).

GLAMOUR PHOTOGRAPHY. Whatever it means in other applications, the word "glamour" means to the photographer the result of certain tricks and devices deliberately employed to enhance the specifically physical qualities in the portrait of a beautiful woman. Most of these techniques originated in the motion picture studios. In most cases, the aim is frankly to create an appearance of refined sex appeal.

Sex appeal is basic to glamour, but so is refinement. Any suggestion of vulgarity in a picture, and all glamour is ruined.

Much trickery has been invented by photographers in recent years to accomplish the basic purpose of the glamour photo—to make the subject look more desirable. Most of the trickery involves sublimating the earthy elements of sex appeal, and translating them into classic or stylized terms and interpretations. The best glamour picture is always an idealization of feminine beauty as well as a subtle compliment to the feminine appeal of the subject.

Unless the subject is a flawless beauty, however, with features such as no artist could improve, the glamour photographer must do more than merely record his subject's beauty in specific detail. Instead, he may show very little which is specific, and let his picture consist of a few vivid impressions.

Glamour photography incorporates a number of specialized techniques to arrive at the desired effect. Here are the principal considerations.

(1) The model. The model must have at least the basis of glamour, in the form of either beauty or sex appeal, and the more obvious these qualities, the easier the task of creating a glamorous picture.

(2) The setting. Take the pictures against a glamorous background, perhaps suggesting a life of ease and wealth. Remember, the object is to make your sitter look desirable. The setting should have an intimate and inviting air. In the studio, such a setting can be suggested with a simple prop.

(3) The costume. Intimate garments such as nightdresses and negligees have the highest glamour potential for studio pictures, whereas bathing suits and other sports costumes are favoured for outdoor shooting.

(4) The pose. Try to show the best features of a girl's figure in silhouette, put special emphasis on a good bust line and on trim hips and legs; then pose her face to give a flattering view of her hair, lips and eyes.

(5) Expression. For the intimate studio glamour shot, the ideal expression requires the girl to look pleasantly approachable and is arrived at usually by experiment—e.g., she may be asked to try sly, seductive smiles, to part her lips invitingly, to think about something good to eat, to try sidelong glances or direct gazes. This is the most difficult aspect of glamour photography in the studio. Working outdoors with the model in play clothes, it is much easier. It is necessary simply to have her look happy and friendly, as if she were naturally full of fun —and the quality will be there.

(6) Lighting. All the standard techniques of lighting for portraiture and figure work can be well applied in glamour photography. Basically, the theory is to emphasize the subject's most interesting features with highlights, and to subdue or conceal her other features with shadows.

(7) Make-up. This is especially important for indoor glamour work. Make-up artfully applied can do much more than cover complexion defects: it can accomplish some of the same purpose as lighting, stressing the good features and subduing, or better still correcting, less pleasing features.

(8) Retouching. Here again, the main application is for indoor close-up studies. The retoucher can work magic on a portrait type of glamour shot, removing all defects in the complexion, adding highlights where they will do the most good, and generally rounding off even the bone structure of the model. Backgrounds also can be subdued if necessary. If the photographer lacks the skill to do the retouching personally, there are professional retouchers who will do the job for him.

(9) Diffusion. Here is a trick which anyone can apply quickly and easily to glamourize his subject.

Simply place a diffusion attachment in front of the lens when shooting the picture, or, if no such attachment is available, use a piece of black gauze. The effect is to show all the main lines of the picture with apparent clarity, while all the undesirable details (complexion defects, hair on arms, and wrinkles) are obscured. It is important to accomplish the diffusion in the camera, rather than in the enlarging process. Camera diffusion spreads the highlights and lightens the shadows; enlarger diffusion, on the other hand, spreads the black accents and shadow areas and degrades or greys the highlights.

(10) Use good taste. It is treacherous work to seek consciously to emphasize sex appeal and there is a constant danger that the picture may become offensive. Remember that, while the sitter may wish to appear glamorous and will co-operate to that end, she will certainly not thank the photographer for making her look cheap or vulgar. E.M.H.

See also: *Make-up for photography; Models; Nude photography; Portraiture.*

Books: *Glamour in Your Lens*, by J. Macgregor; *Posing for the Camera*, by H. Shepard and L. Meyer; *Posing Patterns*, by L. E. Broome (London): *How to Photograph Women*, by Peter Gowland (New York).

GLARE. In a camera, non-image forming light which may originate from scattering at dust particles, scratches and damaged lens coatings, and from reflections at lens surfaces, the iris diaphragm and lens shoulders, etc. This light usually results in an overall fog, or veil, at the image plane, though in some cases ghosting of the bright parts of the image may occur. Fortunately the tolerances of veiling glare in a camera are much larger than in a visual optical system, since the image contrast can often be maintained by suitable adjustments during exposure and development. However steps must be taken to reduce glare so far as is possible. Since reflections at the iris diaphragm frequently account for a large proportion of glare, stopping down the aperture is not necessarily beneficial; thus it should be ensured that the diaphragm as well as the bellows, lens shoulders, etc. in a system are matt black in order to minimise reflectivity. Lens surfaces should be kept free from dust and protected from scratching. The most effective step towards the elimination of glare is the use of coated lenses, particularly in systems having several air-glass surfaces.

Outside the camera glare refers to the reflection of intense light into the camera lens from highly reflective surfaces (e.g. water, glass, etc.). In outdoor photography this glare is avoidable only by changing the camera viewpoint. With artificial light, repositioning the light source or sources helps.

In copying set-ups glare reflected from the surface of the original being copied (even if not shiny) may reach the lens and produce a "hot spot", i.e. specially dense image patch on the negative. The remedy is correct setting up of the lamps at a suitable angle.

See also: *Coated lens; Document copying.*

GLASS. Transparent brittle substance made by the fusion of certain silicates. Physically it is a supercooled liquid, with no regular solid structure such as a crystal lattice. Different mixtures of various metallic silicates produce glasses of different optical properties. Apart from its use for the manufacture of lenses, glass is used in photography as a support for negative and positive emulsions (plates and slides), for glazing printing and picture frames, and for graduates, processing dishes, tanks, funnels and bottles for the various chemical solutions. Of the solutions that attack glass, hydrofluoric acid and certain of its salts are the only ones likely to be used by the photographer.

The greatest disadvantage of glass for articles in regular use is its brittle nature. A number of items of photographic glassware are nowadays made of a special hard glass similar to the familiar glass ovenware, but this material tends to make a clumsy, heavy article. For such purposes, plastics—in particular the flexible polythene varieties—are becoming increasingly used.

The hard surface of glass can be made to take an extremely high polish, however, and in this respect it probably has no equal at a competitive price. Because it gives a higher quality glaze than any other material, many photographers and a number of commercial printers use plate glass sheets for glazing prints in spite of its inconvenience.

See also: *Optical glass.*

GLASS (PHOTO-SENSITIVE). Glass containing certain heavy metal compounds can under exposure to ultra-violet radiation undergo an electronic modification of the metal ions. On subsequent heat treatment this leads to a change in the colour or the transparency of the exposed areas. In such glass it is therefore possible to produce photographic images. The first photo-sensitive glass of this kind was announced in 1947 in America.

There are three main kinds of photo-chemical action in glass: metallic image formation, non-metallic crystal formation and differential solubility.

Photo-sensitive glass for metallic image formation contains compounds of gold, silver, copper or palladium. It may also contain optical sensitizers such as cerium to extend the spectral absorption range; copper and silver act as their own sensitizers in this sense.

The glass is exposed through a normal black-and-white negative to ultra-violet rays between 3,000 and 3,500 A in wavelength, usually to a high intensity mercury vapour lamp or carbon arc. The sensitivity is very low, requiring exposures of the order of ten minutes. The image is formed throughout the depth of the glass with ultra violet rays of longer wavelength, but only in the surface if very shortwave radiations are used.

After exposure the image is developed by heating the glass to around 600°C. The image colour depends on the exposure and may vary from blue to brown, according to the metallic compound incorporated. Development takes about half an hour to one hour. The image itself is due to the conversion of the metal ions into sub-microscopic metal particles. Such images are virtually grainless and of very high resolving power. Cooling of the glass fixes the image which is then virtually permanent unless the glass is heated to its softening point again.

Photo-sensitive glass of this type is used for making graticules, photographic reproductions and photographically decorated glass ware.

In certain types of glass the metal particles produced by the exposure can act as nuclei for the growth of non-metallic crystals. The image then is an area of greater scattering power or opacity than the original glass itself.

Photo-sensitive glass with differential solubility yields a milky opalescent image after exposure and heat treatment. This image is preferentially soluble in glass etching solvents like hydrofluoric acid. Exposed and developed areas of the glass etch up to fifty times faster than unexposed areas, leading to direct glass etching in depth. Such images are much sharper and more detailed than indirectly photo etched glass with a resist produced by conventional photographic methods. L.A.M.

GLASS PICTURES. Photographic images can be etched on glass by using a gum bichromate print as an intermediate. The process gives a "print" from a positive transparency, the image appearing as a positive when it is viewed against a dark background.

The Negative. Sized paper is first coated with a solution of one part each of gum arabic, sugar, and ammonium bichromate in eight parts of water and allowed to dry in the dark. The prepared paper is then exposed to daylight behind a positive transparency.

The sensitized layer of gum is hardened and made less sticky by the action of light, so, at the end of exposure, the areas behind the clear parts of the transparency are not as sticky as the areas behind the denser parts. While the surface is in this condition, it is dusted with fine bitumen powder; the powder sticks to the lightly-exposed areas more readily than to the heavily-exposed areas. The result is a direct positive of the transparency in bitumen dust on the sensitized paper.

This image is pressed into contact with the glass plate which has been previously warmed. The warmth softens the bitumen dust and makes it stick to the glass surface. When the paper support is soaked with water and peeled off, the bitumen powder is left and forms a positive image resist of the original transparency.

Etching. The glass is then exposed to the etching action of hydrofluoric vapour. Where the surface is protected by the bitumen dust, no etching takes place; elsewhere the glass is attacked by the acid and the result is a photographic image. The bitumen dust is finally washed off with petrol.

This is a reversal-negative process; the gum-bichromate-bitumen image being a reversal stage. The etched image is negative but acts as a positive since the dense areas scatter light and appear bright instead of dark. Hence the need for a positive original.

This method produces images which are an integral part of the glass. Image layers can be produced on top of the glass by any of the methods described for printing on metals.

Photographs may also be printed on a glass surface by any of the regular methods of printing on solid supports—i.e., the glass may be coated with a sensitized emulsion and printed under a negative by contact or enlarging; or it may have an image transferred to it by the transfer coating process; or it may take the place of the final support in the carbon or carbro processes.

Pictures in Glass. By an entirely different process photographic images can be formed inside the glass.

This process makes use of a special type of photosensitive glass containing colloidal silver or other metal in suspension. The glass is exposed through a negative to a strong source of ultra-violet light. This converts the colloidal metals in the glass into a form in which they can be precipitated and rendered visible. The actual precipitation is brought about by heating the glass to just below its annealing temperature. The result is a visible image embedded in the glass. L.A.M.

See also: *Printing on special supports; Stripping.*

GLASSWARE. Glass is a transparent material, therefore the shape of anything made of glass can only be clearly defined by lighting it so that it stands out as a dark shape against a light background or as a brightly lit shape against a dark background. With some subjects it is effective to combine both methods.

Silhouette Treatment. With the first type of lighting some of the sparkle of the glass may be lost but every detail of the shape will be perfectly clear. This is a great advantage if the photograph is to be reproduced. Glass most suitable for this treatment should be tall rather than squat—i.e., vases, decanters, jugs, etc.—and it should not have too much decoration.

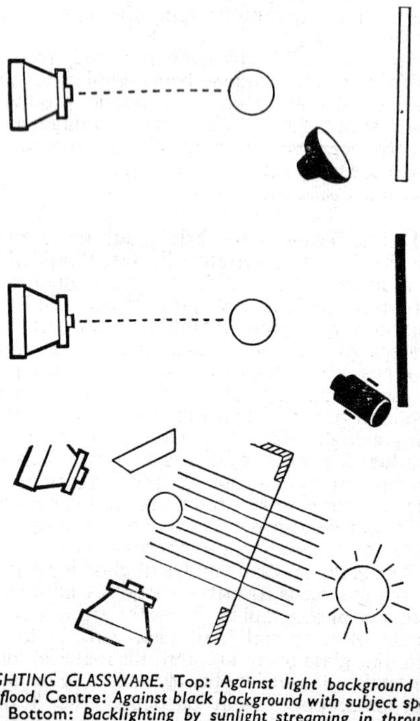

LIGHTING GLASSWARE. Top: *Against light background lit by flood.* Centre: *Against black background with subject spot-lit.* Bottom: *Backlighting by sunlight streaming in through window; reflecting screen used on shadow side. The two camera positions yield a white and a black background respectively.*

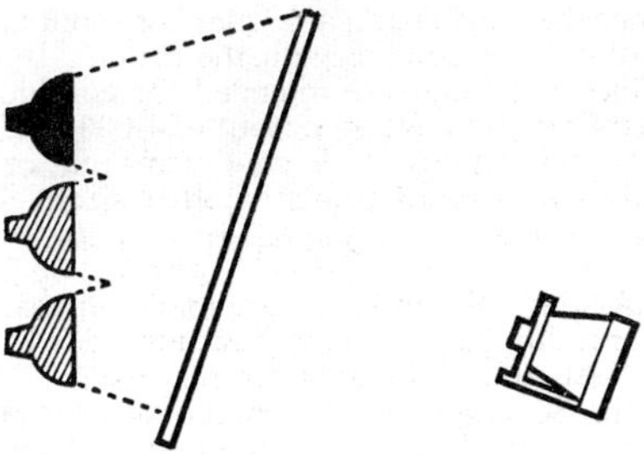

ETCHED GLASS IMAGES. *To take an etched motif in a window, the back is evenly lit by several lamps, or one lamp moved into successive positions during the course of the exposure. The camera should point squarely at the window.*

For this type of photograph the subject is placed, preferably on a plain surface or on a sheet of glass, near the edge farthest from the camera. This arrangement prevents the background line from cutting through the subject. A vertical white background, several feet behind the subject, is lit by spot or floodlight. No light is allowed to fall on the subject, therefore it stands out in silhouette. Variations in the background are sometimes made by casting shadows across it. This calls for care to avoid confusing the outline of the subject. For the same reason, textured backgrounds are avoided as they tend to give a false texture to the subject itself.

The best viewpoint for this type of photograph is nearly level with the base of the subject.

Highlight Treatment. It is more difficult to photograph glass so that it appears light against a dark background and the technique varies to some extent with the type of glass. Cut or engraved glass usually looks best against a dark background, more especially a plain surface that does not fight with the pattern on the subject. And it should curve upwards behind the subject so that there is no horizontal edge to cut the photograph in two. Tracing paper laid over a sheet of plate glass provides one suitable type of background as it enables the subject to be lit from below—a form of lighting which is often necessary.

The main modelling light is kept low and almost behind the subject. It is usually provided by a spotlight to ensure that light falls only on the subject, leaving the background in shadow. Where this light does not illuminate the subject sufficiently a second spotlight underneath the background is directed vertically upwards on to the subject; this light helps to outline such parts as the rim of a glass or vase or it may be used to pick up highlights in the base and stem. The viewpoint for this type of photograph can be anything up to 60° above the horizontal, but experienced workers tend to keep it lower.

Combined Silhouette and Highlight Treatment. A combination of light and dark background tones is usual with flatter subjects like ash-trays, shallow bowls and dishes. The subject is placed on a sheet of clear plate glass which is supported 3 or 4 feet above a plain white background. This white background is then lit by a spotlight adjusted until the areas of highlight and shadow look right when seen from the camera viewpoint. The final effect is that part of the subject is silhouetted against a light area, while parts like the rims and upper edges stand out as highlights against the shadow area of the background. A second light source above the plate glass is occasionally directed on to the subject itself to add extra highlights and sparkle.

Equipment. Really high class work for catalogue and "glossy" magazine reproduction is almost always done with a large format plate camera using a long focus lens stopped right down to give the necessary depth of field.

As with most commercial photography, a lens with a long focal length is generally chosen to avoid close-up distortion.

Normally, when photographing glass, best results are obtained from panchromatic material which helps to prevent excessive contrast in the negative. D.L.H.

See also: *Ceramics; Silverware.*
Book: *Lighting for Photography*, by W. Nurnberg (London).

GLAUBER'S SALT. Another name for sodium sulphate decahydrate. Used to prevent swelling of emulsions in hot weather processing.

GLAZERS AND GLAZING MACHINES. Prints may be glazed by using a glazing sheet on its own or in conjunction with a glazing machine. In the latter case, heat is applied to the glazing sheet to speed up the process.

For really large scale glazing where high speed is important, rotary glazers incorporating their own glazing surface may be used.

Glazing Sheets. Any smoothly polished, hard surface is suitable for glazing. Special sheets for glazing can be bought; these available are usually enamelled metal or chromium, the latter being more expensive but less likely to scratch easily.

Glass is a suitable substitute for a chromium glazing sheet, and it is almost universal opinion that plate glass will give a much better glaze than any other material; it does, however, need meticulous and regular polishing. Prints may even be glazed by squeegeeing them direct on to windows provided that the glass is perfectly smooth and clean.

Glazed tiles are suitable for glazing prints, although the surface of the tile must be free from cracks if a smooth glaze is to be obtained.

Glass and tiles are not suitable for use in a flat bed glazer where heat is applied. It is invariable practice to use chromium glazing sheets in machines.

Flat Bed Glazer. A flat bed glazer consists of a shallow metal box with a slightly curved top. This forms the bed of the glazer; it is heated by

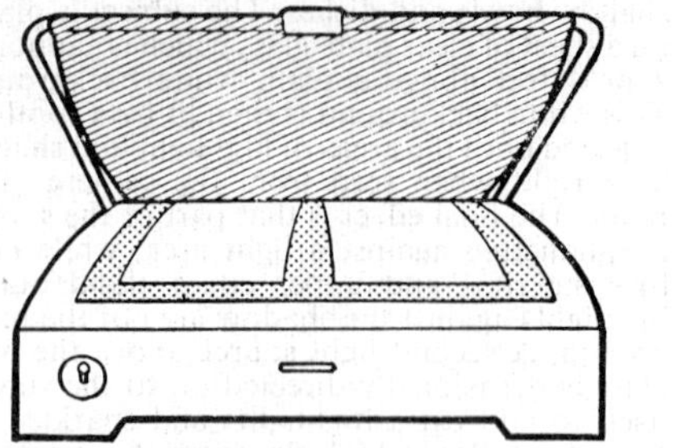

FLAT BED GLAZER. The prints are squeegeed on the glazing sheet which in turn is heated from below. A fabric apron keeps the prints pressed against the sheet.

electric elements inside the box, the temperature being controlled by a switch and thermostat.

The bed is made to accommodate a standard-sized glazing sheet. An open frame carrying a pressure cloth of fabric is fastened over the bed so that the pressure cloth holds the glazing sheet in close contact with the top of the bed.

In use the heat is switched on while the wet prints are being squeegeed on to the glazing sheet. The glazing sheet is then laid on top of the bed and the pressure sheet is stretched over it and fastened down. After a period of 5 to 10 minutes, the prints dry and can be removed from the press.

Glazers of this type are made to take from one to four 8 × 10 ins. prints and consume from 300 to 1,000 watts depending on the size.

Rotary Glazer. Wholesale photo-finishers, photographic agencies and others turning out large quantities of prints use power-driven automatic rotary glazing machines.

In these machines the glazing sheet takes the form of a large cylinder mounted on a horizontal spindle and turned by a small electric motor. The wet prints are fed under a squeegee roller at the base of the rotating drum and held in contact with the polished surface by an endless band of fabric.

Electric elements heat the drum from the inside, and the rate of rotation is adjusted so that the prints emerge dry and glazed from under the fabric band after making about half a revolution in contact with the drum.

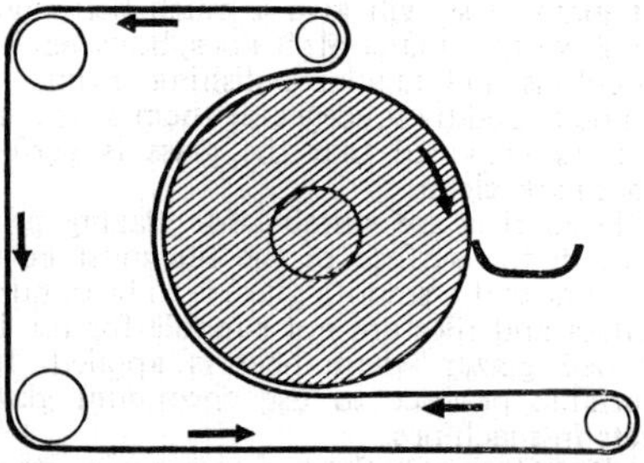

ROTARY GLAZER. The prints are applied to the glazing surface of a rotating heated drum, and pressed in contact by the continuous apron. One revolution glazes and dries the prints.

Such machines are intended for continuous operation and have an output of 1,000 to 2,000 2½ × 3½ ins. (6 × 9 cm.) per hour or an equivalent area of prints of other sizes. F.P.

See also: *Darkroom; Drying equipment; Glazing.*

GLAZING. When prints are made on gaslight or bromide paper, a glossy surface adds to the contrast and gives a better rendering of fine detail. So a glossy surface should always be used for small contact prints and all photographs intended for reproduction. Glossy paper as supplied by the manufacturer dries with a shiny surface, which can be further improved by glazing. The wet print can be glazed by squeegeeing it on to a clean polished surface.

For small scale glazing, almost any highly polished surface will serve, but a sheet of polished plate glass, free from scratches, gives the highest gloss. Many workers use chromium plated steel sheets, and thick celluloid and glazed pulp boards are also used. These replace the once-popular ferrotype plates made of enamelled iron which were cheap but had the disadvantage that they were easily damaged.

Preparing the Surface. The glazing surface must be clean. The best way of cleaning it is to rub it vigorously with a pad of cotton wool or clean rag soaked in methylated spirit. The surface is then dusted with French chalk and polished with a clean dry cloth. Or the surface may be polished with the following mixture:

Beeswax	2 ounces	50 grams
Turpentine	40 ounces	1,000 c.cm.

A few drops are applied over the whole surface with a piece of clean rag, after which it is polished with a clean dry cloth.

Glazing Solution. When prints are simply squeegeed on to a polished surface they often have an annoying tendency to stick. This tendency can be overcome and a much higher gloss given to the surface by using a glazing solution.

For those who like to make up their own, the following formula is offered:

Prepared ox gall	½ ounce	12 c.cm.
Water about	40 ounces	1,000 c.cm.

The prints are soaked in the solution for about five minutes before glazing.

There are a number of proprietary glazing solutions on the market. These solutions may be applied direct to the glazing surface in their concentrated form, or they may be diluted to make a bath in which the prints are soaked.

After the prints have been allowed to soak in the dilute solution they are removed one at a time and the surface is lightly swabbed with cotton wool to remove scum or small particles of grit. The print is lowered to the glazing surface one end first so that no air is trapped between the print and the glazing surface. If the concentrated solution is used it is rubbed lightly over the polished surface and the prints are laid on it straight from the final washing water.

To ensure even glazing, the prints are best squeegeed on to the plate once they are in position. A suitable method is to cover the prints on the plate with a rubber sheet and gently squeegee from one end to the other with a flat or (preferably) roller squeegee. After that the plate is drained, the rubber sheet replaced and squeegeed again more firmly. Finally it is advisable to blot off all surplus water from the back of the prints before letting them dry or inserting the plate into a hot glazer. When completely dry the prints should fall off.

In wholesale photo-finishing the prints are squeegeed on to a polished chromium-plated drum or sheet and dried off in a few minutes by artificial heat. These hard-worked surfaces are kept clean by washing them frequently with hot water and soap.

It is not necessary with modern papers to harden the prints before glazing as the emulsions are sufficiently hardened during manufacture. As a precaution, a fixing-hardening bath should be used in the tropics or in very hot weather. Prints that have not been hardened take the highest gloss.

Faults in Glazing. If good quality paper is used and the glazing surface is kept scrupulously clean most glazing troubles will be avoided. There are, however, some faults that have other causes:

Prints refuse to strip after drying: generally caused by a dirty surface, or exhausted glazing solution. If the trouble is very frequent, it can be avoided by drying the prints first, and then soaking them for ten minutes before glazing.

Small unglazed specks: squeegeeing is to blame and is failing to make perfect contact between the print and the glazing surface. Or small particles of grit in the washing water may be clinging to the print and keeping it out of contact with the glazing surface.

Oyster shell markings: these are caused by patchy drying through the uneven application of heat, or changes of temperature during the drying period.

Glazing Matt Surface Prints. A matt surface shows less detail than a glossy surface and it also has a shorter range of contrast. For this reason, matt surfaced prints are preferred for pictorial work where a slight suppression of hard detail is often an advantage.

But a matt surface print can be glazed when necessary; it may be that the negative is not available and the print is needed for photo-mechanical reproduction. Glazing is also helpful when a print upon a matt surface paper is to be copied.

It is never possible to produce a very high glaze on a matt surface print, but it can be made about equal to the surface of a glossy print before glazing.

A matt print can be given a gloss with a

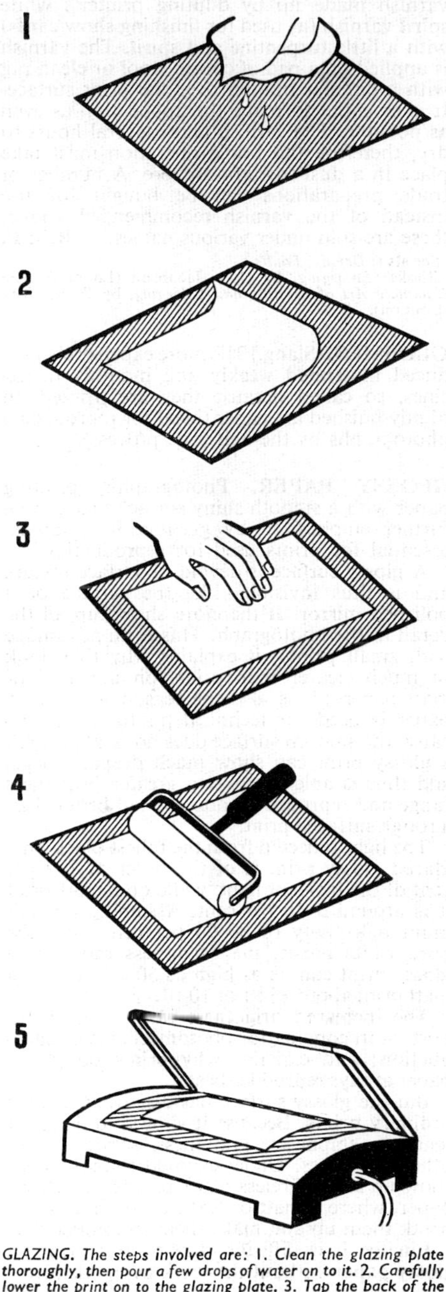

GLAZING. The steps involved are: 1. Clean the glazing plate thoroughly, then pour a few drops of water on to it. 2. Carefully lower the print on to the glazing plate. 3. Tap the back of the print with the fingers to expel any air. 4. Squeegee down, first in one direction, then in the other 5. Place the glazing plate into the hot glazer, and fasten down the cloth apron, then leave until dry. For cold glazing blot off surplus water, and leave the plate to stand on edge until dry.

varnish made up by diluting printer's white spirit varnish (as used for finishing show cards) with a little turpentine and spirit. The varnish is applied on a pad of cotton wool or clean rag with a circular motion over the print surface. It is essential to make the application as even as possible. The varnish takes several hours to dry, therefore the whole operation must take place in a dust-free atmosphere. A number of trade preparations can be bought for use instead of the varnish recommended above; these are sold under various names. R.M.F.

See also: *Drying; Faults.*
Books: *Enlarging*, by C. I. Jacobson (London); *The Complete Art of Printing and Enlarging*, by O. R. Croy (London).

GLOSSIES. (Slang.) The more expensively produced illustrated weekly and monthly magazines, so called because they are printed on highly finished art paper (ideal for reproducing photographs by the half-tone process).

GLOSSY PAPER. Photographic printing paper with a smooth shiny surface that can be further emphasized by glazing. It is almost essential for prints used for reproduction.

A glossy surface is devoid of surface texture and is thus invisible—like the surface of a polished mirror. It therefore shows up all the detail in the photograph. This is an advantage with small prints; it explains why they look so much clearer than prints on unglazed or matt papers; it is also the reason why glossy paper is used for technical photographs. Because the smooth surface does not scatter light, a glossy print can show much deeper blacks, and thus is able to cover a greater brightness range and reproduce shadow detail better than a rough surfaced print.

The light reflected from the fullest black on a glazed glossy print is of the order of 1–2 per cent of the incident light, while on a matt print it is around 5–10 per cent. Assuming a maximum reflectivity of 80–90 per cent from the pure white paper, the brightness range of a glossy print can be as high as 50 : 1, and of a matt print about 15 : 1 or 10 : 1.

The increased brightness range is also important in copying for photomechanical reproduction; and explains why prints on glossy paper always reproduce best.

But the glossy surface has disadvantages for ordinary prints. Because it shows up detail, it tends to emphasize graininess, scratches and other blemishes. Under-exposed shadow areas show up as featureless black patches on glossy paper, where a matt or textured surface would break them up and make them less noticeable.

Finally, it is difficult to retouch a print on glossy paper without the work being visible.

GLUE PROCESS. An obsolete variation of the gum pigment process for making prints. The paper is coated with pigmented glue, sensitized with potassium bichromate and the soluble glue washed off in warm water after the print has been exposed through the paper stock.

GLYCERINE. Used to thicken reducers, etc., for local application; also when enlarging scratched negatives.

Formula and molecular weight: $(CH_2OH)_2 CHOH$; 92.

Characteristics: Colourless oily liquid with sweet taste.

Solubility: Mixes with water in all proportions.

GLYCIN. Para-hydroxyphenyl glycine; Kodurol. Developing agent.

Formula and molecular weight: $C_6H_4(OH)(NHCH_2CO_2H)$; 167.

Characteristics: White powder.

Solubility: Practically insoluble in water, but freely soluble in solutions containing sodium carbonate or other alkali.

GODDARD, JOHN FREDERICK, 1795–1866. Science lecturer in London; was employed by Beard and Wolcott to speed up the daguerreotype process by chemical means. Published his successful application of bromine as an accelerating sensitiser on 12 December 1840. This shortened the exposure from about 15 minutes to about 2 minutes, thus constituting a major improvement in the daguerreotype process.

GOERZ, CARL PAUL, 1854–1923. German optician. Founder (1886) of the C.P. Goerz A.G. works in Berlin-Friedenau. Manufactured cameras from 1888 and the famous Dagor double anastigmat lens (calculated by von Hoegh) from 1893; also the wide-angle Hypergon (1900), cine apparatus, optical instruments and telescopes. Introduced race finish photography in 1922. The factory later merged into the Zeiss-Ikon A. G. combine.

GOLDBERG WEDGE. Sensitometric wedge giving a continuously increasing density from one end to the other. It consists of a layer of gelatin which is thicker at one end than the other—i.e., wedge shaped. The gelatin is evenly tinted and, because of its varying thickness, gives a changing density. The gelatin is sandwiched between glass.

See also: *Sensitometry; Wedge.*

GOLD CHLORIDE. Chlorauric acid. Used in gold toners.

Formula and molecular weight: $AuCl_3.HCl.4H_2O$; 412.

Characteristics: Very hygroscopic yellow crystals.

Solubility: Highly soluble in water.

GOLDEN SECTION. (Sectio aurea.) Art term derived from interpretation of classic architecture and referring to the division of a line in such a way that the smaller division is to the longer as the longer to the whole. The resulting proportions have been claimed to yield ideal rectangular picture shapes. Loose usage of the term led to the belief that this point of division could also be used as a fulcrum of pictorial composition by only approximating the calculations involved.

See also: *Composition.*

GOLD POTASSIUM CHLORIDE. Used in gold toning baths.

Formula and molecular weight: $KAuCl_4.2H_2O$; 416.

Characteristics: yellow deliquescent crystals. Should be kept in the dark in sealed tubes, or as a 1 per cent solution in distilled water.

Solubility: Highly soluble in water.

GOLD SODIUM CHLORIDE. Gold sodiochloride. Used in gold toners instead of gold chloride, where a neutral solution instead of the acid one produced by gold chloride is required.

Formula and molecular weight: $AuCl_3.NaCl.2H_2O$; 428.

Characteristics: Hygroscopic yellow crystals.

Solubility: Very soluble in water.

GOODWIN, HANNIBAL, 1822–1900. American clergyman and amateur photographer. Filed in 1887 a patent application for celluloid roll-film. The patent was granted in 1898 and confirmed in 1914 after a long lawsuit against the Eastman Kodak Company.

GOST SPEED. Film speed rating in use in the Soviet Union and some eastern European countries. GOST speeds are quoted on an arithmetical scale and are fairly similar to the A.S.A. arithmetical speed numbers. The GOST speed criterion is based on the exposure point corresponding to a specified minimum density above base plus fog of the film.

See also: *Speed of Sensitized Materials.*

GRADATION. Term referring to the tone scale or contrast range of a developed image. An image which shows a large number of intermediate tones of grey between the extreme dark and light tones is said to have a soft gradation; if there are only a few recognizable shades of grey between the extreme dark and light tones, it is said to have a hard gradation.

The gradation of an emulsion is at once obvious from the steepness of its characteristic curve. An emulsion with a steep curve will possess a harder gradation than one with a shallow curve. If both emulsions receive the same range of regularly increasing exposures, the hard emulsion reaches maximum density in a few steep steps while the softer must be taken up a greater number of shallow steps to reach the same density.

Generally speaking, where the subject contains a large number of tones which must be reproduced to convey a true impression of its appearance—e.g., a rounded surface—a soft gradation sensitive material must be used. Where there are only a few intermediate tones —e.g., in a silhouette or in deliberate posterization—a hard gradation is more suitable.

There is a fairly constant relationship between speed and gradation: slow materials tend to have a hard gradation and fast materials a soft gradation.

Speed is generally the important factor in deciding the choice of a negative material, but development papers are classed according to their gradation, and speed in this case is unimportant.

Development and developers exercise a definite effect on the gradation of an emulsion. Broadly speaking, extended and concentrated development tends to give hard gradation and curtailed and dilute development gives soft gradation. F.P.

See also: *Contrast; Negative materials.*

GRADE. Classification of the inherent contrast in printing papers (which are usually obtainable in a range of different contrast grades).

See also: *Papers.*

GRADIENT. Slope of the characteristic curve of the emulsion at any point, and a measure of the contrast of the image at that point.

GRADUATE. Graduated vessel—usually of glass—for measuring the volume of liquids. Normally scaled with etched or painted figures.

See also: *Measure.*

GRAIN. The normal negative image consists of black metallic silver distributed more or less in proportion to the light that acted on the material during the exposure. Areas that received a relatively great exposure show a greater image density than those receiving less exposure. But the silver forming the image is not as evenly distributed as it appears to the naked eye. Examination under the microscope reveals that it is made up of irregularly shaped clumps of black silver. These clumps are formed by groups of the original exposed silver bromide grains after development has changed them into black metallic silver. While the grains appear to have grown together to form the larger masses of silver visible under the microscope they do not as a rule touch, and the apparent clumps are the result of the random distribution of the original halide crystals in the photographic emulsion.

In addition to being visible under the microscope, the grains also become apparent in the print when the negative is enlarged. The grains then tend to break up the image and give it an unpleasant, mealy appearance. This is the reason why so much attention is given to the elimination or reduction of grain by manufacturers and photographers.

Speaking generally, the grain size of an undeveloped photographic emulsion increases with its sensitivity so that fast emulsions have a coarser grain than slow ones. The grain in a developed emulsion also tends to increase with the size of the crystals in the undeveloped emulsion and as a rule it increases with the development time.

Graininess and Granularity. The graininess of a developed photographic material refers to the appearance of the finished photograph and may be defined by the distance from which a photograph has to be viewed to make the grainy impression disappear completely. On the other hand granularity is physical and may be defined in terms of the local changes in density when a narrow pencil of light scans the test patch. Both granularity and graininess are subjects of a great deal of present-day research, one of the aims being to define a method of determining the granularity which will link it with the grainy appearance of the picture. The problem has largely been solved in recent years.

There have been many attempts to define granularity, of which that proposed by Selwyn in 1935 has found general acceptance. If the density of a piece of uniformly exposed and processed photographic film or paper is measured in a densitometer, in which the diameter d of the measuring aperture can be varied, density values taken at different points will differ, more so the smaller the size of d. This is because the population and size of the developed grains is not completely uniform. The density values measured with any one aperture will cluster around the average value, which would be obtained with a very large aperture, and will in general follow a bell-shape Gaussian distribution, the shape of which is determined if the spread σ is known. 2σ is the range between which 68 per cent of all measured values will lie; 999 out of 1,000 measurements will lie between 6σ.

Selwyn's relationship between the value of σ for a given average density and the diameter d of the measuring aperture is simply:

$$\sigma \times d = \text{constant}$$

The constant is termed granularity.

In practice, granularity is determined on a scanning microdensitometer, the fluctuating output of which is fed into a so-called rms (root mean square) voltmeter, on which the values of σ may be read directly.

The granularity increases with density, since with higher density the number of grains per unit area, and thus their fluctuation, is greater. In order to describe completely the granularity of a material, an array of curves, plotting granularity against density, would be necessary. The arrays vary for different emulsions and conditions of processing, but they are sufficiently similar to allow the granularity of a material to be represented by a single figure with reasonable accuracy.

For this purpose the granularity value for a certain density is conveniently chosen. Two conventions have been proposed and termed Class I and Class II granularity.

Class I granularity is used for the routine evaluation of negative materials of one type and is simply the granularity at a density of 0·8 (including fog, but excluding base).

For a granularity figure which is more representative of what might be seen in a print, Class II data are necessary. Here the negative is developed so that the average slope of the characteristic curve falls between the limits laid down for the determination of speed; this would be 0·65 ± 0·05 for amateur black-and-white films, and around 0·8 for commercial films. Granularity is then determined at a density of 1·2. In this way a good correlation between granularity and the graininess of the print is established.

The Print. Graininess in the print is always worst in the middle tones. The grains of paper emulsions are always much finer than those of negative emulsions and the positive image is—where graininess is a consideration—usually an enlargement of the negative; consequently, the graininess of the print is essentially due to the negative image. The print graininess is low in the highlights where a small change in density in the negative is reproduced by a much smaller change in density in the print because here the emulsion is being used in the region of the toe of the characteristic curve for the paper.

Graininess is also reduced in the deep shadows, since here the eye is not as sensitive to small changes in density.

Granularity in the print is affected by the granularities of both negative and positive materials. When the scanning aperture is small, the granularity in the print will be essentially that of the positive material. For larger scanning apertures, the granularity will be more nearly governed by that of the negative, and will represent more nearly what is perceived by the eye as graininess. The granularity in the print measured with larger apertures also depends on the contrast of the positive material, since densities will be simply proportional to the density of the negative multiplied with the slope of the positive curve. Since for an average good photograph the Goldberg condition applies:

$$\gamma\,\text{neg} \times \gamma\,\text{pos} = 1{\cdot}5$$

one may define as print granularity the negative granularity divided by γ neg.

The Negative. The graininess of the negative increases continuously with density provided the light flux behind the negative, i.e. on the viewing side, is kept constant. A heavily exposed and consequently dense negative therefore yields a much more grainy print than one normally exposed.

This makes it important to choose the camera exposure correctly.

Similarly a print from a foggy material tends to be a good deal more grainy than one from a clean negative.

The fact that the photographic image is built up from a large number of small grains limits the degree of enlargement which can be used in practice, for two reasons. One is that the grainy appearance is often considered unpleasant, the other that the presence of grain limits the resolving power of a photographic material so that the picture appears blurred if enlarged unduly.

The concepts behind these well-known facts have been refined in recent years. Granularity has the same properties as the "noise" in a wireless receiver; i.e. it obscures the intelligibility of the signal. In order for a signal to be clear—and this can only be stated within given probability limits—it must exceed the noise the more the greater the certainty required that the signal is genuine.

For a small area of density to be clearly distinguishable from its surrounding areas, it must exceed these so much that only a few density steps may be clearly distinguishable within the tone range provided by a photographic material. For larger areas, more steps will be available. The detailed discussion of this situation is the task of information theory. W.F.B.

See also: *Characteristic curve; Developing negatives; Fine grain technique; Information in the image; Latent image; Negative materials; Speed of sensitized materials.*

Books: *Exposure*, by W. F. Berg (London); *Photographic emulsions*, by E. J. Wall (London).

GRAIN (WEIGHT). Smallest unit of weight used in all British and American systems of measurement.

See also: *Weights and measures.*

GRAININESS. Mealy appearance of the printed image caused by the clumping together of the silver grains forming the negative.

See also: *Grain.*

GRAM. (French: Gramme). Unit of weight in the metric system related to the other units by powers of ten.

See also: *Weights and Measures.*

GRANULARITY. Term referring to the structure of the sensitive emulsion as represented by the measured variation in the distribution of an apparently uniform silver deposit. It is a scientific concept whereas graininess is a subjective impression; a highly granular, but very dense deposit would appear less grainy than a lighter deposit which was observed by the eye at a correspondingly more discriminating level of visual acuity.

See also: *Grain.*

GRAPHIC ARTS. A comparatively modern term covering all the processes of making reproductions of illustrations for printing purposes. The hand- or machine-setting of the type characters to form the text is not usually included in the term, nor is the machine-printing of the end product. An exception is the processing of the product (i.e. the exposed photographic film) of photo-composing (filmsetter) machines.

Camera Work. Graphic arts cameras vary in size and complexity from small simple ones providing a maximum negative size of 12 × 15 ins. to large, high-precision motor-driven equipment which may weigh several tons and give negatives of 40 × 60 ins. or more.

Originals (generally called copy) for reproduction may be photographs, drawings or paintings, while colour copy is often in the form of colour transparencies. The copy is placed on the copyboard of the camera and lit evenly with arc lamps or, in the case of transparencies, illuminated from behind, then photographed to obtain one or more negatives. High-precision enlargers may also be used for this work, especially when involving colour transparencies. The initial continuous-tone separation work is done on special contact-printing boxes, which use a vacuum to hold the original in close contact with the photographic film. In common with the cameras and enlargers, they also have mechanical or electronic timers for controlling exposure, mechanical drive to change the colour filters, and pin register for locating copy, film, colour, masks, etc.

Black-on-white pen-and-ink line drawings are usually photographed on very high contrast (mainly silver chloride) photographic emulsions of the lith type, where the lines are recorded in the negative as completely clear transparent areas in a dense, opaque ground. For letterpress plate-making, the negative is contact-printed on zinc (sometimes on copper) in a light-sensitive colloid substance which can then be baked to form an acid-resisting enamel. The metal surrounding the lines is then etched, leaving the raised letterpress image.

For lithography, the line negative can be contact-printed on the litho metal (commonly anodized aluminium) to form a grease-based ink image. When the plate is stretched tightly around the plate cylinder of the press and damped with water, the image area accepts the ink, which is repelled by the damp areas. This image is then transferred to the offset blanket (a tightly-stretched layer of mid-hard rubber on the offset cylinder) and from there to the paper by pressure; hence the term "offset-lithography".

For photogravure, a line positive is used to

form a light-hardened gelatin negative image on the copper cylinder. The gelatin resist, thickest in the white areas and thinnest in the black, is etched in ferric chloride solution, which penetrates the thin gelatin layer in the areas of the lines of the image, but is too viscous to penetrate the thicker areas. Thus the lines are etched into the copper. The screen pattern of thin lines at surface level, which support the scraping action of the doctor blade when printing, is separately exposed when making the carbon print.

Tone Illustrations. For pictures containing a tone range, the half-tone dot formation negatives for letterpress work are made on graphic arts cameras with a half-tone screen supported in front of the film. After processing the latter is printed on metal in an acid-resisting enamel, much as for line work, and possibly etched by the powderless technique. But it is usually necessary to include, at some stage of the work, an adjustment of the tones of the picture, to correct the inherent shortcomings of the photographic stage and to allow for the requirements of the printing method.

For lithography, the plates may be contact-printed on the metal from line and half-tone negatives. More often the plates are made from line and half-tone positives, which, in the case of the half-tone images, may well have been prepared by photographing continuous-tone negatives through a half-tone contact screen. In that case, the retouching for tone correction is usually carried out partly at the negative stage (chiefly for the light tones) and completed on the half-tone positives.

For photogravure, positives are required for making the etch resist on the cylinders, and for conventional photogravure the positives are in continuous tone. The retouching is therefore in continuous tone for both negative and positive and it is convenient to treat the dense tones of each; that is, the light tones of the negative and the dark tones of the positive, which are appropriately strengthened with neutral grey dye.

Dot Etching Retouching. The retouching of half-tone negatives or positives is rather different from continuous tone retouching, because the size of the dots needs to be altered. Dye retouching is unsuitable because it would affect the amount of exposure of the dot formations and thus the printability of the plate. Dot images have to be "all-or-nothing"—to print or not to print. Hence the dot size must be reduced—usually with ferricyanide-hypo which may be applied locally with a brush. Alternatively, parts of the picture may be covered with a cellulose varnish and the whole bathed in the reducer solution. Knife scraping and abrasive reduction are other, less-used, methods.

Colour separation images require not only tone correction but colour correction. This work requires great skill but it tends to spoil colour photographs, making them resemble paintings. To avoid this, photographic colour-masking techniques effect at least some of the necessary correction, making less retouching necessary. Electronic colour correction is an even more automatic system of masking; it is particularly effective for providing the necessary quality in the black printer plate and in reducing the weight of the colours in the neutral colours, which are best made with the black printer.

F.H.S.

See also: *Half-tone screen; Lith film; Masking; Photomechanical reproduction; Photo typesetting.*

GRATICULE. Pattern of fine lines superimposed on the visible image in a piece of optical equipment to facilitate sighting, aligning or measuring. Known in America as reticule or reticle. There are many kinds for varied applications, including such optical aids as microscope stage micrometers, divided circles employed in theodolites and other surveying instruments, master templates for profile projection, scales for linear measurement, and cross-lines grids as used in the focal plane of instruments such as telescopes, microscopes and binoculars.

A graticule is usually considered to be made up of two parts. One is the support or blank, usually of glass but sometimes of metal or plastic material, and the other is the pattern. According to the nature of the pattern and its relationship to the associated optical or mechanical device, the graticule may be used for a variety of purposes including measurement of length, angle, area, volume, or for reference when positioning an instrument.

The uses of graticules in everyday photography are not extensive but one important application is in aerial survey cameras which employ a so-called register plate on which is a graticule pattern reproduced on each photograph as an aid to subsequent measurement. Again, in assessing projection-lenses a resolution-test graticule is often used thereby enabling the viewer to make some estimate of the lens performance.

Modern graticule-making processes are very numerous; they range from those employing photography almost exclusively to those which mainly involve mechanical techniques. Etching applied to glass itself or to metal films on glass is frequently employed. Modern developments have taken advantage of the many benefits to be derived from the use of vacuum-evaporated or sputtered metal films.

Since in most applications clarity of vision is of prime importance photographic processes for producing the final pattern have not, in general, been of the conventional gelatine-emulsion type. The use of alternatives—bichromated fish-glue, polyvinyl alcohol, shellac, and more recently synthetic photo-sensitive materials—has been widespread because, among other attributes, they possess the valuable property that after exposure and development the unexposed areas are completely removed. Nevertheless, sometimes high-resolution gelatine emulsions of the Lippmann type are employed for the final pattern but in such cases exceptional care must be taken to ensure cleanliness.

In many processes which are mainly based on photography the original almost invisible colloid image is intensified by introducing darkening agents such as lead sulphide, colloidal silver, or certain dyes. For more permanent patterns it is often necessary to etch the lines and figures in the glass surface after first preparing on it a protective metal or other stencil. Other processes involve heat treatment of a metal or the introduction of metal into the glass itself under the influence of an electric field. More recently, permanent graticule patterns have been prepared by using photo-sensitive glass which yields the appropriate image after exposure to ultra-violet radiation followed by heating. E.Bo.

GREETING CARDS. The photographer has the opportunity of producing original greeting cards which will appeal more to the recipients than the mass-produced cards sold in the shops. He can make Christmas cards, birthday cards, baby arrival cards, bookmarks, gift cards, wedding congratulation cards, wedding anniversary cards, or "get well" cards. They can be produced in many ways and range from straightforward prints to those made by photomontage, or from black-and-white to colour.

Simple Cards. The simplest greeting card can be made by pen lettering a greeting with opaque ink on the back of a film negative, and then printing by contact or enlarging. This can be done by anyone, and is easier than trying to write a greeting in reverse on the film side of the negative, or making a copy negative of the combined greeting and illustration.

Another simple method is to put the photographs in commercial printed folders. This, however, can hardly be claimed as making greeting cards.

A development of this idea—and one which is attractive—is to use master masks for the greeting part of the card. These can be obtained commercially and with a wide range of designs, but it is not difficult to make them at home. An original is first drawn in Indian ink on white paper to conform with the size of the finished cards. A space is left where the illustration is to be printed later. The surround can be decorated with a fine line or fancy bordering, or the whole area can be left completely blank, so that the picture has a white border when printed later.

A copy is now made on a process or line film and developed in a maximum contrast developer to obtain a good contrasty negative. This copy can be made on one of the special thin base process films, and then later an opening can be cut out to allow the film negative to be inserted below. The combination when printed by contact will be perfectly sharp. The size of the copy negative will depend on the size that the finished card is to be. If the illustration is 2 × 3 ins., then half plate size would be suitable for the master mask negative. This idea is also useful in that a few copies can be printed from each of a selection of different negatives, instead of having to keep to the same illustration on every card.

After the film is dry, the opening for the subject negative is cut out, and the negative taped in place. A piece of cellulose tape is ideal for this purpose and, as the background negative has great density, it will not show. The opening for the illustration negative can be cut with a sharp penknife or a razor blade, and a steel rule and set square. It is important that the edges are clean, square and straight.

If glass plates are used for the background negative instead of film, the window for the illustration negative can be cleared by scraping off all the gelatin emulsion from the glass base. When the film is placed on this and printed by contact, there will be some very slight softening of the background design unless a great deal of pressure is used to secure optical contact. The cards can be printed from such a multiple negative, however, by projection in the enlarger. It may be necessary to shade either the illustration or background negative during exposure to give the same printing density to each part.

Folded Cards. A variation on the above method of making photographic greeting cards is to use a much larger sheet of photographic paper for the greeting card background and then double or four-fold it, to improve on the ordinary flat greeting card. This looks effective, but, if expense is a consideration, the large size photographic paper needed for this method is costly. For example, to produce a four-fold card size 4 × 5 ins. needs a piece of 8 × 10 ins. photographic paper for each card.

If a two-fold or a four-fold card is made it can have a touch of colour added with a bow or knot of ribbon tied at the side. Selecting the colour of the ribbon to blend with the picture can add extra appeal—e.g., by combining a tartan ribbon with a highland scene, or using regimental colours on a soldier's camp photograph.

The manufacture of the two-fold or the four-fold card does not require much special description, but there are one or two points to watch. First, the illustration and greeting must be so placed on the paper that they come into correct alignment when trimmed and folded. Again, the prints should not be dried by heat, otherwise the gelatin emulsion may crack when folded. While single or double weight photographic paper can be used, the larger sizes of single weight paper may crease when wet unless handled very carefully.

Where greeting cards are to be made by the more complicated methods—e.g., photomontage or photograms—it is better to prepare a key original and make a copy negative from this, for the making of a number of photomontage or photogram cards by direct methods can be costly and tedious.

Anyone not an expert at lettering, can use stencils, nursery, cut-out letters, cine titling letters, or draw the letters with a pantagraph, to word the greeting. Another way is to use the actual handwritten signatures of the people in the photograph, or to put the greeting in script lettering. Illustrations for the background can be traced, stencilled, or caricatured, methods which call for very little drawing skill.

Mounting and Colouring. Where photographs are to be mounted on to board to make a greeting card, they should be dry mounted and not wet mounted, to avoid cockle. Cover boards can be purchased from artists' sundriesmen. They should tone with the colour of the photograph to show the picture off to best advantage. Boards which are ornate in finish, or glaring in colour, should not be used. Pastel shades are generally preferable. A white board is best for black-and-white prints and ivory or cream toned board for the sepia and warm tones. Tinted base photographic papers can also be used with advantage for some subjects—e.g., photographs of sunsets, cornfields, autumn scenes, or golden sands.

Colour can be introduced either by using actual paper colour prints, or by making use of the less costly hand-colouring methods. Where a number of hand-coloured prints are to be made of one subject it pays to cut stencils out of waxed paper for the principal objects, or the "rubbing-on" oil colouring method can be used. Here, all one part of all the cards is coloured at one time—e.g., skies or sea. This method also saves colouring time. Colour can be introduced into the background of the cards by drawing something in the illustration suitable for tinting—e.g., holly and berries, a lantern, the traditional robin, bluebirds, thumbnail sketches of flowers, windmills, etc.

If a large number of greeting cards are required in colour, genuine colour photographs made by the dye transfer colour process are quicker to produce and relatively cheap.

Suitable Subjects. As subject matter for the illustrations, either general pictures can be chosen or those with a personal interest. For Christmas cards, such subjects as snow scenes and old buildings and churches are suitable. For birthday cards beautiful scenes or subjects which have some personal associations for the recipient are quite apt. For baby arrival cards and wedding anniversary cards, have a family motif. A good camera study of the church where the pair were wed would provide something which no commercial card could give. Gift greeting cards should be of pleasant, decorative subjects. Greeting bookmarks offer unlimited scope for both ideas and subjects, and "get well" cards can be treated seriously or in lighter vein. Humorous table top photographs or happy snapshots can do much to cheer up a depressed patient.

All types of photographic greeting can be improved by a little originality in presentation.

Deckled edged mounts can be made by clamping the mounting card (in bulk) between two pieces of plywood or hardboard, the same size as the cardboard, in a vice and drawing the point of a large nail cross-cross over the edges of the card.

Hand torn edges can be made by placing the card on a flat surface, putting a steel straight-edge on the top of it, and pulling the surplus of the paper edge upwards, while holding the steel rule very firmly on top of the photographic paper or mounting card. At least 1 in. extra must be allowed all round the board to grip while tearing. A.Pg.

See also: *Lettering; Mounts; Title; Tricks and effects.*
Books: *The Complete Art of Printing and Enlarging*, and *Design by Photography*, by O. R. Croy (London).

GRENZ RAYS. Soft X-rays used for the industrial radiography of textiles and materials with too small a range of densities to give an image with normal X-rays.

GREY SCALE. Print or transparency consisting of a series of grey tones extending in regular steps of increasing depth of tone from white (or clear) to black (or opaque). Grey scales are used as test objects in sensitometry and for adjusting exposure and development to give true colour balance in making three colour separation negatives.

See also: *Wedge.*

GROTTHUSS, THEODOR FREIHERR VON, 1785–1822. German pioneer in photochemistry. Discovered (1817) the basic law, that, "only the absorbed light rays are active in the production of chemical changes. This law, the Grotthuss law of photochemical absorption, was named after him. It was later forgotten and re-discovered by Draper (1841). Among his numerous photochemical experiments are some on the bleaching of dyes by light of different colours.

GROUND GLASS. Plate or sheet glass which has been abraded on one face to give a translucent, light-arresting surface on which a visible image can be formed. It is the standard material for making focusing screens for field and reflex cameras.

See also: *Focusing; Focusing screen.*

GROUPS. The fashion in group portraiture has changed more than in any other branch of photography. The florid "draping" of the sitters in groups of forty or fifty years ago has now given place to neat, formal posing.

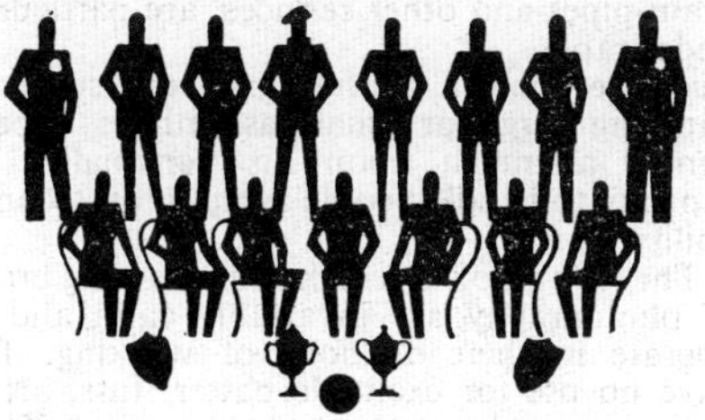

POSED FORMAL GROUP. The regular arrangement ensures that everybody is well visible in the picture. Such photographs of sport teams should include reserve players and club officials as well as trophies in the foreground.

Large Groups. Groups involving 100 sitters or more are usually taken with a panoramic camera. This gives a negative 6 ins. or more in depth and as long as the group demands. Prints up to 3 feet in length are not uncommon. This type of work is largely in the hands of specialist firms who act for other photographers on a trade basis.

Medium Groups. Groups up to 100 sitters are commonly taken on a ½-plate camera and the negative subsequently enlarged to the required size. Such plates (or cut-films) are normally processed in fine-grain developer to ensure maximum, grain-free resolution of each sitter's face. In order to make maximum use of the plate area and to produce a print of pleasing proportions, it is customary to "bank-up" the group as much as possible. In a typical arrangement for a school group of 100 persons, 20 of the smallest pupils would be sitting cross-legged on the ground; 20 (staff and senior pupils) sitting in chairs; 20 standing immediately behind them; 20 standing on forms; and a final 20 standing on tables forming the rear rank.

The photographer is careful when marshalling the group to arrange for even balance of height in all ranks—normally he puts the tallest pupils in the centre and allows the ranks to tail away to the sides of the group. A group always looks better if all the members adopt the same pose: either all arms should be folded or all hands should rest in laps; all ankles should be close together or all knees crossed. The exact arrangement rests with the photographer, but uniformity is essential.

Smaller Groups. Into this category come such assemblies as cricket XI's, football teams and other such sporting groups. The usual arrangement for an "eleven" makes use of five chairs only. The captain sits in the centre chair with two players on each side of him. The six remaining members of the team stand behind the seated five covering the gaps between the sitters. Trophies, if any, are placed on the ground in front of the captain.

Occasionally teams wish to be photographed holding their sports gear—e.g., tennis racquets, hockey sticks or rifles in the case of a shooting eight. Particular care is then taken to get the members to hold their gear in a uniform manner or the final effect will be ragged and untidy.

Family Groups. Here are included groups taken on the diverse occasions on which families foregather to celebrate some notable event. Strict formality is not appropriate to such groups; the photographer aims at achieving a sense of relaxed, comfortable unity. It is customary for the central characters to be seated with the women seated beside them. The men stand and the children are seated on rugs on the floor.

Informal Groups. Such seemingly simple pictures call for the highest degree of technical and creative ability. The formal group is a comparatively simple matter—sitters are posed in conventional fashion and all eyes look towards the camera. In the informal group, the reverse should be the case. There should be some central point of interest to create a sense of artificial unity. A high viewpoint is desirable in many cases so that the camera may appear to be, as it were, an eavesdropper to some notable occasion.

To cite examples: a trophy has been presented to a golfer and the picture shows, from a 7-foot viewpoint, a re-staging of the presentation. All involved are closely gathered around and vigorously clapping hands are seen. Such a treatment is more effective than a formal line-up. A similar theme from a different viewpoint would be treated as follows: the wedding cake is about to be cut and all concerned are grouped closely around the table. The photographer exposes from a very low viewpoint to stress the style and elegance of the gowns worn by the ladies.

Whatever the subject chosen for an informal group, care should be taken that none of the sitters appears to be aware of the camera's existence. This means adopting a technique similar to that for candid photography, in which various methods are used for concealing the camera—e.g., aiming the camera "blind" from inside an open coat.

With many informal groups it is important to capture the atmosphere of the occasion too.

STUDIO FAMILY GROUP. A suitable settee holds the group together. The adults sit "backwards" to keep their legs on the far side of the settee. Toys add atmosphere. Relaxed poses and expressions are important.

POSED INFORMAL GROUP. The arrangement of the figures is less regular, but still carefully set out. Bodies are posed so that they tend to face inwards to the centre of the picture.

This may mean including in the picture certain objects that are a feature of the surroundings, choosing the position of backgrounds carefully, and selecting the right moment for making the exposure. The choice of viewpoint, influencing the composition, can also play an important part in getting the atmosphere—for instance, a group of chorus girls may look more agile and full of movement if the picture has strong diagonal composition rather than static, horizontal emphasis.

It is, of course, possible to "stage" an informal group beforehand. This involves careful handling and direction of the people if the effect is to look natural. Usually it is better to work out a rough arrangement and make the exposure when the people have relaxed and forgotten the photographer's intentions.

Commercial studios often arrange informal groups for advertising pictures. These are frequently taken in the studio—even if the setting is an outdoor one. In this case, professional models are mostly used, experienced in posing naturally. Photographs of this type are more complex in setting-up than the final picture suggests, and involve much patience and attention to minor details. But the same advice as for other informal groups still applies.

Lighting. Large groups indoors are best lit evenly from both sides, so that shadows more or less cancel each other out. Such lighting can consist of floods or flash.

When working outdoors it is wise to avoid direct sunlight, as this will cause unpleasant strained expressions and also cast harsh shadows. Soft daylight from a clouded sky is ideal. Diffused sunlight helps to give the picture some sparkle.

Backgrounds. In all groups the position in relation to the background is important. The chosen site must be level in a "left-to-right" plane and selected so that it avoids unsightly distraction in the background. A smooth, distant line of trees is almost ideal. Buildings in the background are avoided unless there is some very good reason for including them. The backs of buildings, with their unsightly drain-pipes and other services, are particularly undesirable.

Audience Control. A photographer's success in handling large or small assemblies depends almost as much upon his personality and capacity to handle people as upon his technical ability.

The men who succeed in this exacting branch of photography are invariably calm and deliberate in their methods of working. They have no use for excitable patter, fuss, artistic antics or ponderous humour. G.Cg.

See also: *Banquets; Weddings.*
Book: *Group Photography*, by G. Catling (London).

GUIDE NUMBER. Alternative name for the flash factor used in arriving at the correct exposure to give with a particular make of flash bulb. The number (quoted by the makers for a particular flash bulb, shutter speed, and sensitized material) divided by the subject distance, gives the recommended aperture.

Example: the correct *f*-number to use when taking a subject 10 feet away with a flash bulb of guide number 120 is

$$\frac{120}{10} = 12.$$

This assumes that a normal reflector is used.

Specific methods of measuring guide numbers as laid down in certain industrial standards. These involve comparing the densities obtained on a film exposed to a standard light source and to the flash under test. With flash bulbs the light is measured of the bare bulb and the resulting guide number multiplied by 2, assuming a reflector factor of 4. With electronic flash the measurement automatically takes into account the power of the flash reflector. The amount of light likely to be reflected from the walls of a room under normal conditions is also allowed for in the standard measurement method.

See also: *Flash technique.*

GUM ARABIC. Gummy substance obtained from the stem of the acacia and also occurring in the fruits of certain other plants. It is soluble in water, but becomes insoluble when treated with certain bichromates and exposed to light. Used in mountants, for sizing paper before sensitizing for certain processes, and in the gum bichromate process.

GUM BICHROMATE PROCESS. Contact printing process in which the image is formed on a coating of sensitized gum containing a suitably coloured pigment. Exposure to light renders the gum insoluble, so when the exposed paper is "developed" in running water the pigmented shadows remain while the highlights are washed clean.

The gum bichromate process can be made to yield prints of great beauty but it has been practically superseded in recent years by bromoil and bromoil transfer. At the same time the gum print has a distinctive character of its own. There have been attempts at revival, although the technique is not easy to master.

Materials. The process calls for only a few inexpensive materials. These are: potassium bichromate 10 per cent solution; white gum arabic; moist water colours in tubes, or dry powder colours; paper for sensitizing (Whatman Cartridge, or good sized paper); one or two brushes, including mop, camel hair brush, and a 2½ ins. flat brush for smoothing (good quality paint brushes will serve).

Sensitizer. One ounce of gum arabic is dissolved in 3 ounces of water, preferably by placing the gum in a muslin bag suspended in water. The potassium bichromate is a 10 per cent solution made up with hot water. This is simply a saturated solution and it is near enough if there are always some undissolved crystals at the bottom.

The sensitizing solution varies according to the technique of the worker, but a mixture of equal parts of the gum and bichromate solutions will be found satisfactory. One part of ivory black or a rather smaller quantity of tube colour are added to every four parts of this solution. The two must be mixed with a glass rod until all trace of grittiness disappears.

Coating. The paper is pinned to a large drawing board over two or three thicknesses of blotting paper. The sensitizing solution is then applied with the coating brush in a series of brisk overlapping strokes from side to side. Then the smoothing brush is used to smooth out the coating with similar strokes from top to bottom. If necessary, any ridges still showing are broken up by going over the surface again with lighter strokes from left to right. When the coating is correctly carried out the coating will be smooth. The brushing is very lightly kept up until the pigment mixture has begun to set. The paper is then left in a warm dark place until it is dry.

When the coating works too stiffly and it is not possible to smooth out the brush lines, it is an indication that there is too much gum. If on the other hand the coating takes too long to set and small transparent patches appear, there is too little gum in proportion to the amount of pigment.

The paper is not sensitive until it is almost dry, therefore the coating can be done in bright artificial light.

The negative that prints best is one that is rather thin without strong contrasts, of the type that is regarded as ideal for enlarging.

Exposure. The print is made by contact with the negative in an ordinary printing frame by daylight or colour arc. There is only a slight visible image during printing and it is no guide to printing depth. As a guide, a piece of P.O.P. or self toning paper is put out at the same time under a negative of similar density and contrast. When the P.O.P. print is of the right depth for

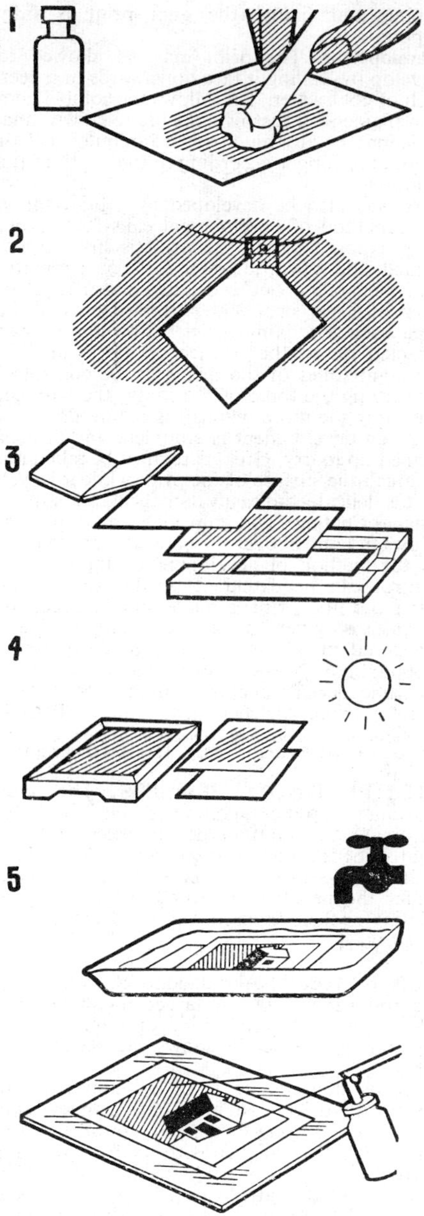

GUM BICHROMATE PROCESS. 1. Apply the mixture of gum bichromate solution and pigment to the paper by a suitable brush. Work in subdued light. 2. Hang the paper up to dry in the dark. 3. Place the paper in a printing frame with the negative. 4. Print out by daylight, using a P.O.P. paper at the same time as a check on the progress of the exposure. 5. Develop by floating face down in a dish of cold water, or attach to a glass plate and wash off soluble gum by an artist's spray.

toning and fixing, the gum print is fully exposed.

Development. The print may be allowed to develop by leaving it face downwards in a deep dish of cold water. This allows the soluble gum and pigment to dissolve out in their own time. This may be anything from 15 minutes to 5 or 6 hours or longer according to the depth of the printing.

It may also be developed by a light spray. This method of development calls for a more fully exposed print. The print is first placed, coated side down, in water for 3 or 4 minutes and then taken out and laid upon a sheet of glass of larger size. Water from a spray of the type used for fixing artists' drawings is then directed on to the surface of the print. The highlight tones of the print can be controlled by varying the force of the spray; the stronger the spray the more pigment is removed.

When development is complete, the print is pinned up to dry. Great care must be taken not to touch the surface of the print while it is wet, as it is delicate and easily damaged at this stage.

Double Coating. The gum bichromate process allows of control over the contrasts of the print by the method of development adopted. This control can be extended by double coating. After the first print is made and finished, the surface is given a second coating, dried, registered under the negative and exposed a second time. The second development allows the same measure of control and can be made to yield a stronger, richer print. R.M.F.

See also: *Control processes, Impressionistic photography.*
Book: *Pigment Printing*, by G. L. Hawkins (London).

GUM PLATINUM PROCESS. The gum platinum process combines the gum bichromate and platinotype processes. A fairly light print is made upon platinum paper. This is then gum-sensitized and the print is registered under the negative a second time. The gum coating is then exposed and developed in the usual way.

GUN CAMERA. Aerial gunnery camera for recording strikes on the target in combat, and for use without ammunition in gunnery training practices, when it shows the accuracy of aim.

In its commonest form, it comprises an electrically-driven 16 mm. cine camera with, in the past a long-focus, but nowadays ultra wide-angle lens, aligned with the free guns in a turret or fixed forward guns in a fighter or bomber. In the latter case, the gun camera is also used to record rocket and dive bombing attacks, although these may also be covered by fast recycling still cameras on larger film.

The camera is actuated by the firing or release button, and usually continues running for some seconds after ceasing fire, to give the projectiles time to reach the target. During these seconds, the firing aircraft often starts violent evasive action so that, unless it has an ultra-wide-angle lens, the camera loses the target. Various counters to this have been tried, from devices to deflect the optical axis, to backward-pointing cameras for use during the pull-out from ground-strafing attacks. There remains the difficulty of seeing damage through flying debris.

Another gun camera, used in training, photographs through the actual gunsight, recording the sighting graticule as well as the target. Its pictures are particularly easy to assess.

In most cases the films are assessed by single-frame projection on a gridded screen. Where the target is another aircraft, the shadow from an identical model of the type is matched to the target silhouette. This shows the target's attitude and direction, and also the range. The increasing use of colour film has made assessment much easier.

Gun cameras are not confined only to aircraft. They are often used in close-range anti-aircraft gunnery, and similar instruments may be fitted to missile-tracking radar aerials.

P.B.M.

See also: *Air force photography.*

GURNEY-MOTT THEORY. A theory of the way in which a latent image is formed in a sensitized silver emulsion by the action of light, proposed by R. W. Gurney and N. F. Mott in 1938.

It is generally accepted that exposure to light causes photolytic decomposition of the silver halide grains in an emulsion and deposits minute specks of metallic silver which form nuclei for the developer to start work on. As a result of the presence of one such speck on the surface of a grain of the halide, the developer is able to convert the whole grain into metallic silver.

The Gurney-Mott theory explains the mechanism of this photolytic action. It suggests that a quantum of light releases an electron from a halide ion, and that this electron attaches itself to an existing silver speck or nucleus. The electron carries a negative charge and this enables it to attract an unattached silver ion from the interior of the crystal and thus add a silver atom to the speck. In this way the speck builds up to form a starting point for the subsequent action of the developer.

See also: *Latent image.*

H

HALATION. A photograph of a bright source of light is often surrounded by a sharply defined ring with a diffused outer edge. This effect is called halation and is produced by the fan of scattered light reflected from the back of the support of the photographic material.

Halation is minimized by reducing the support thickness, which decreases the diameter of the ring of reflected rays. This ring is smaller with films than with plates because the film

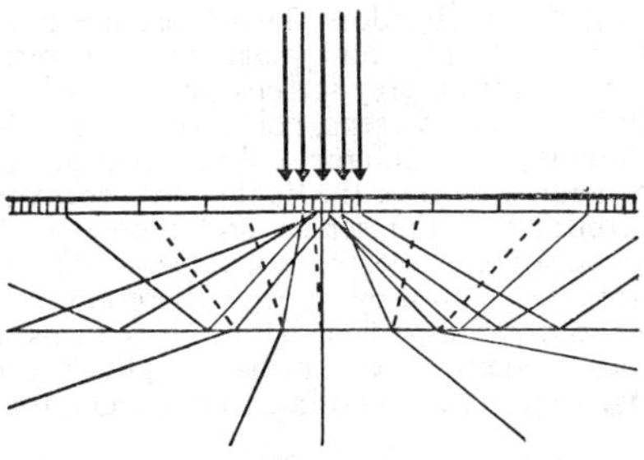

HALATION. Light scattered in the emulsion is reflected from the bottom of the support and at a certain angle forms a halo round bright image points (e.g., light sources in the picture).

base is thinner. It is also reduced by using a dyed backing layer to absorb the rays which penetrate to the rear of the support.

Certain films are coated on a grey base which absorbs a lot of the light that normally reaches the emulsion by reflection from the back of the support.

Halation commonly occurs in night photographs where street lamps are included in the picture. Daylight interiors which show unshielded windows also suffer from halation; this occurs in the dark areas around the window.

Over-exposure also promotes the appearance of halation since the comparatively weak halo image then becomes developable. That is also why it is so apparent around brilliant over-exposed image spots.

See also: *Faults; Flare; Ghost; Irradiation.*

HALF-FRAME CAMERA. Miniature camera taking 18 × 24 mm. pictures, yielding up to 72 exposures from a standard 35 mm. cassette. The term "half-frame" refers to the fact that the picture area is half of the full 24 × 36 mm. frame size. Cameras of this kind are popular as compact pocket models.

See also: *Miniature camera.*

HALF-PLATE. Standard format for negatives and prints, measuring $4\frac{3}{4} \times 6\frac{1}{2}$ ins.

A half-plate is not half of a whole plate; that size, $4\frac{1}{4} \times 6\frac{1}{2}$ ins., is not used in photography.

See also: *Sizes and packings.*

HALF-TONE BLOCK. Photomechanical printing block in which the printing areas are formed by a pattern of raised dots. The dots are small or large according to the tone of the original. Highlights are normally represented by very small dots which may, however, be removed by deep-etching or engraving if necessary. The shadow dots join to produce almost solid areas. An optical illusion of intermediate grey tones is produced by the varying proportions of dot and white paper corresponding to the half tones of the original subject. Half-tone reproduction is used for reproducing photographs and other continuous tone originals.

See also: *Photomechanical reproduction; Scaling for reproduction.*

HALF-TONES. Grey tones intermediate between shadows and highlights in a picture. Sometimes also called middle tones.

HALF-TONE SCREEN. Line or dot pattern used to convert the continuous tones of varying darkness in a photograph, etc., into a discontinuous pattern of constant density but varying area. In a half-tone image lighter or darker tones are produced by smaller or larger dots on lines—which, through being uniformly spaced thus occupy a greater or lesser proportion of a given unit area. This conversion into a half-tone image is necessary in lithographic or letterpress printing from continuous tone originals, because litho plate or letterpress block can imprint only a uniform ink density on paper.

Half-tone images can be produced in many ways; the most usual is to convert the continuous tone into a regular dot pattern. For this purpose, a glass cross-line screen can be used in a process camera to make screen negatives or positives, and a film contact screen can be used in a camera, in a contact box, or on an enlarger baseboard.

Cross-Line Screens. These are made from glass and carry equally-spaced parallel lines which have been etched and filled with an opaque substance. Two glasses are cemented together so that the lines cross at 90°. Screen rulings range from 50 up to about 200 lines per inch.

In letterpress printing the choice of screen ruling is determined by the paper upon which the illustration is to be printed, while fine screens (120 upwards) are generally used for photolithographic reproduction. The finer the screen, the finer the detail that can be reproduced but the smoother must also be the paper to ensure even tone.

The screen is held in the camera a few millimetres in front of the image plane, and light passing through the apertures formed by the lines exposes a dot formation.

Theory of Dot Formation. The penumbral theory (Clerc), suggests that the screen ruling casts a shadow, and that when the screen is at correct distance, the light is most intense at a point on the photographic material midway between two lines. Between these is an area of penumbra which is related to the intensity of light reflected from the original at any given point. If the original reflects little light, a small dot is built up; if it reflects more, the dot spreads. It is also suggested (Fruwith and Mertle) that the diffraction of light behind the sharp-edged opaque rulings contributes to the build-up of a graded dot.

Half-tone exposures in photo-engraving are adjusted so that a small isolated dot in the shadow areas, and a chessboard pattern dot in the highlight areas, is produced. The screen ruling, exact distance between the screen and the film, and the size (and shape) of the lens aperture must be carefully matched to obtain the correct dot formation.

For photolithography and for powderless etching processes in photo-engraving, it is necessary to expose for a formation of higher contrast. Normally these formations are obtained by giving a general exposure to obtain detail, and a "flash" exposure, using a small lens opening, to build up the density of shadow dots.

Contact Screens. These are on flexible material and are used in direct contact with photographic film. The screen elements are made up of variable density dots. Two types are available: neutral grey, and magenta. For direct colour separation grey screens must be used.

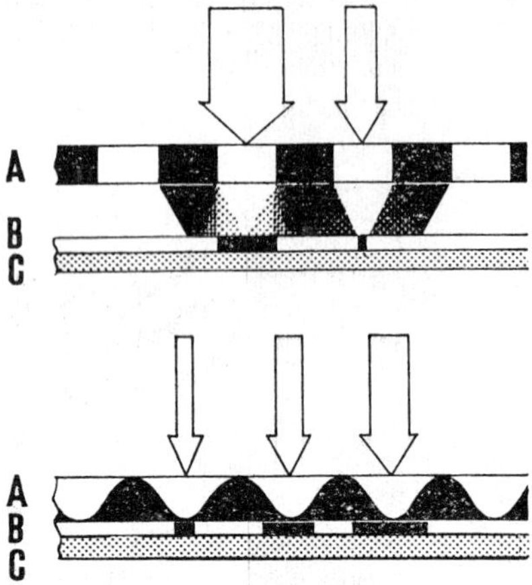

DOT FORMATION WITH HALF-TONE SCREENS. Top: *Cross-line screen. The screen, fitted slightly in front of the film, casts unsharp shadows on the film. Where light from highlights reaches the film the penumbra spreading into the intervals between the lines becomes bright enough to record on the emulsion as a large dot. In the shadows only the centre of the space is light enough to record as a small dot. The dots are sharp because the ultra high contrast film records no intermediate tone gradations.* Bottom: *The contact screen already has an unsharp dot pattern and is used in contact with the emulsion.* A, *screen.* B, *emulsion.* C, *support.*

With magenta screens, contrast is controlled by using lights of different colour for exposing. A magenta filter over the light source decreases the contrast of the screen and increases the contrast of the image, while a yellow filter does the reverse. With grey screens, the contrast of result can be modified by using screens of various gradation, i.e., normal, bright, flat, or by varying flash exposures or development times.

Special Screens. Gauze and linen screens have been used to reproduce continuous-tone originals. These are wire, or bolting cloth, stretched taut, and used in the camera in the same way as a cross-line screen. Single line screens which break the image into parallel lines of varying thicknesses and screens with single wavy lines have also been used for special effects. Screens in which the elements take the form of grains have been used in the past, and there have been experiments with screens in which lines cross at 60°.

Relative Advantages. The cross-line screen has certain faults which affect tone reproduction. However, it is a traditional instrument, is easily used and is very suitable for conventional photo-engraving. The contact screen is less expensive, is easily stored, but may be damaged by grit, scratching or solutions. It is claimed that, because the contact screen is used in direct contact with photographic material, resolution is higher than with cross-line screens. Further, contact screens can be designed to produce images with a specified tone range.

In practice, contact screens are used more frequently in photo-lithography than in photo-engraving, although it is considered that they are very suitable for making the negatives required for powderless etching and for photo-polymer plates. F.G.W.

See also: *Half-tone screen reproduction; Photomechanical reproduction.*
Book: *Photographs and the Printer*, by F. H. Smith (London.)

HALF-TONE SCREEN REPRODUCTION. When half-tone prints are to be reproduced by a half-tone process the prints must be re-photographed through a half-tone screen held in the camera in the usual way. Because of this, two uniformly distributed sets of dots, those of the print and those of the screen, are superimposed and a pattern, known as moiré, is created. It has been shown that if the dots coincided no pattern would result, but this is rarely possible, and precautions are always taken when re-screening to reduce moiré so far as is possible.

Several methods have been suggested, all of which involve diffusion of the image. A rough way of doing this is to focus the image so that it is slightly unsharp. Another is to focus the image sharp, but to vibrate the camera slightly and regularly during the exposure.

A more reliable method is to hold a piece of thick, clear film parallel to the front component of the lens and to vibrate this regularly during the exposure. Other suggestions have been to cover the original in the camera copyholder with a piece of matt film, or to substitute a grain screen for the normal cross-line screen in the camera when re-photographing.

Probably the most effective method is to place the original in the copyholder so that the lines of dots form an angle of 30° with those of the screen ruling. (A 30° separation is maintained in four-colour half-tone printing in order to reduce moiré). If a circular screen can be used in the camera this can be rotated to an angle which appears to reduce moiré to the greatest extent.

Optical apparatus has been designed to help solve this problem. Prominent is an anti-flare housing which fits on the lens panel and within which a rotating and rocking glass disc produces a controlled displacement of the image. The selection of cams to drive the glass is determined by screen ruling.

Optical tone filters are also designed to reduce moiré. On the surface of each filter, which is used in front of the lens, thin film discs of a coating (similar to lens coating) form a random pattern which deflects rays and produces a more continuous pattern out of the original half-tone dot pattern. Different filters have different proportions of coated and uncoated area. The choice of the correct filter depends upon magnification and the screen ruling of the original.

It is also possible to reduce moiré when reproducing half-tone proofs on a special electronic engraving machine. F.G.W.

HALOGEN. Collective term for the elements fluorine, chlorine, bromine, and iodine. Compounds of the latter three, with silver, form the main light-sensitive substances used in photography. Binary compounds of a halogen with another element—e.g., a metal—are known as halides.

See also: *Silver halide.*

HAND CAMERA. Name given to the types of camera that are normally used without a tripod —e.g., cameras for negative sizes of ¼-plate or smaller, with instantaneous shutters and focusing scale, rangefinder, or reflex focusing. The term was heard more often in the early days of photography when the stand camera was in more general use.

H. & D. SPEED. System of speed rating named from the initials of the surnames of two scientists, Hurter and Driffield. They collaborated in establishing the first generally-used system of measuring the sensitivity or speed of sensitized materials. The H. & D. system is based on the measured densities resulting from a range of exposures. The relation between H. & D. speeds is arithmetical—i.e., a film or plate rated as 1,000 H. & D. is twice as fast as one with a speed of 500 H. & D.

See also: *Speed of sensitized materials.*

HAND-OR-STAND CAMERA. Term, no longer in general use, describing a type of camera intermediate between the hand and stand types. It was usually a ¼-plate folding camera with rising and cross front and sometimes a small amount of vertical swing back camera movements.

In construction and purpose it was an attempt to bridge the gap between the two basic types. It was intended to be small enough to be carried unobtrusively and used for instantaneous exposures (therefore it was equipped with a shutter) and yet have enough of the features of the field camera to find favour with the serious photographer of landscapes and architecture, for which purposes it was equipped with a focusing screen as well as a focusing scale.

By and large the hand-or-stand camera never succeeded in satisfying either class of user and it is no longer made, although it is still occasionally listed in the second-hand market.

See also: *Camera history.*

HANDS. A hand is a very human and live thing and a photograph must reveal these qualities or it will fail in its purpose. A great deal can be learned about the character of people by watching how they use their hands in everyday activities such as picking up a cup of tea, doing up a parcel and striking a match. The correct technique in photographing hands is to avoid a deliberate pose; the hands must be allowed to fall naturally into a relaxed position in character with the sitter. Hands are an essential part of a portrait, and must be made to fit into the composition in such a way that the eye sees the picture as a whole. A picture of a pianist, for example, should never give the impression that the hands were "placed".

While men are rarely conscious of their hands, women are more likely to be.

Distortion. The first and main aim in photographing hands must be to avoid distortion. This calls for a long focus lens and if possible a camera with swing back. The swing back is used to bring the plane of the hand as near as possible to the focal plane of the camera.

Lighting. Point source lighting—e.g., one or two 500 watt spots—is needed to bring out the contours and the texture of both men's and women's hands. The key light must be oblique and the shadow side filled in—e.g., with a diffused 500 watt flood. One spot and one flood are the minimum necessary. The fill-in light must be just strong enough to soften the hard shadows thrown by the key light, and no more. Where there is strong back lighting, care must be taken in balancing the lights or flare will occur on the edges of the fingers causing loss of texture.

When photographing men's hands the fill-in light should be only just strong enough to reveal detail in the shadows, while allowing sufficient contrast to produce a texture effect and perhaps even allow the hairs to stand out. With women's hands the fill-in light should be somewhat stronger so as to give a softer effect.

Special attention must be given to the background. Any surface that reflects highlights must be avoided unless the hands are actually engaged in working on a polished object, where such reflections are unavoidable.

Shadows. Great care must be taken to avoid awkward shadows cast by the fingers, which in extreme cases may give the impression that the sitter has four hands. In order to avoid shadows on the background the hands should be an appreciable distance away from it. With a back-lighted subject, the frontal fill-in light must be strong enough to kill the shadows cast towards the camera by the key lighting. There could, for example, be heavy shadows cast where a hand is resting on a book, which has been strongly back-lighted to bring out texture and provide modelling. Here the fill-in light could conveniently be improvised by using a small mirror to reflect some of the back light.

Hand Position and Camera Angle. Where hands are included in portraits they must never look posed, but at the same time it is necessary to pay attention to their angle in relation to the camera. The hands must be relaxed, with the fingers extended naturally. A view of the back of the hand with the fingers folding into the palm is ugly and meaningless. Hands should also neither be photographed with fingers extended towards the lens, nor so placed that the wrist is nearest the lens while the fingers point away from it.

It follows from the foregoing that unless a deliberately distorted view is required, unusual camera angles should be avoided.

Special Cases. When photographing a broad hand with the back presented to the camera, it is a good plan to place the index finger and the second finger so that they meet, making a division between these and the third finger—the third and fourth being slightly parted. This arrangement gives an impression of length. The lighting should fall at an oblique angle—at least 45° to the flat surface of the hand.

Where hands are to be clasped, intertwined fingers must be avoided at all costs. Occasionally the clasped hand is essential for certain effects—e.g., in dramatic poses. In such instances the technique is to take the left hand upraised—e.g., to the level of the chin—with the tips of the fingers of the other hand resting lightly in the palm, or vice versa. In lighting this arrangement, the key light for the face will provide adequate modelling without any additional lighting from the front, while the spread of the back-light—e.g., a 1,000 watt spot—will cover both head and hands, producing a strong feeling of depth.

Choice of Film and Paper. Medium speed panchromatic film is preferable for this work; processed in a fine grain developer it gives the desired amount of contrast. For printing, a glossy surface is preferable to the usual art paper, as a smooth surface does not conflict with the texture of the hands. Men's hands should be printed somewhat darker than women's. E.C.

See also: *Portraiture*.

Book: *Lighting for Portraiture*, by W. Nurnberg (London).

HANFSTÄNGL, FRANZ, 1804–77. Bavarian lithographer and portrait photographer. Established a lithographic business in Munich, 1834. In 1853 he opened a portrait studio and commenced photographic art publishing. His son, Edgar Hanfstangl, was the first to introduce pigment printing on a large scale in Germany.

HARDENER. Chemical—usually potash alum, chrome alum or formalin—used to toughen the gelatin emulsion (generally of films and plates, but sometimes of papers) to render it less liable to physical damage by scratching or rubbing.

Some hardeners may be incorporated in the fixing bath; others may be used as a separate bath after normal processing. Each has its own characteristics. The latter method is more usually employed in special circumstances—for instance, processing cine films, etc.

See also: *Hardening bath; Scratch-proofing.*

HARDENING BATH. High solution temperatures—above 75°F.—tend to make photographic emulsions turn dangerously soft or even melt and run off the support. So in hot weather processing, and also when using highly alkaline developers (e.g. the Meritol-caustic type), a hardening bath is essential. This may be used after the rinse or stop bath, or it can be combined with the stop bath. Some hardening solutions can even be used before development.

The most common hardening agents are chrome alum and potash alum. The former is more effective, but the latter keeps its hardening properties better in solution.

Formalin is also useful as a hardener, though its action on the gelatin is not of the same kind as that of the alum hardeners. Finally, tannic acid hardeners increase the actual physical hardness of the gelatin and can be used to make the film resistant to scratches and abrasion; they are only used after washing.

Chrome Alum. The simplest way to harden films or plates after development is to immerse them for about five minutes in a 3–5 per cent solution of chrome alum.

In hot weather, the solution should also contain about 10–15 per cent of sodium sulphate crystals (Glauber's salt) to prevent undue swelling of the gelatin.

All chrome alum hardeners (including acid hardeners and acid hardening fixers containing chrome alum) will harden as long as the solution looks purplish by transmitted light. When it starts to turn green, it should be thrown away. In acid solution the hardening properties disappear within a few days, particularly if acid sulphites are present. That is why acid hardeners (combined stop bath and hardener) and acid hardening fixers with chrome alum do not keep.

The presence of citric, acetic, and other organic acids also reduces the hardening action even of a fresh chrome alum bath. So negatives must be well rinsed between any stop bath and a chrome alum hardening bath.

An effective acid hardener (which is also a stop bath) consists of equal parts of 2 per cent solutions of chrome alum and potassium metabisulphite. This is best prepared just before it is needed, and discarded after use.

Potash Alum. Although chrome alum is the more effective hardener, potash alum keeps much better even though the solution must be acid to prevent precipitation of aluminium hydroxide by any alkali carried over from the developer.

A suitable potash alum hardening stop bath consists of a solution containing 2 per cent of potash alum, and 1 per cent each of sodium sulphite and glacial acetic acid.

A special potash alum acid hardener stock solution may be added to a plain hypo solution to convert it into an acid hardening fixer.

Formalin. A formalin hardening bath for use after development consists of 1 part formalin (40 per cent) solution diluted with 80 parts of water. The negative should be soaked in this bath for about 5 minutes.

Alkaline formalin hardeners are also useful before aftertreatment such as intensification, etc. The hardening is said to be more uniform than with acid alum hardeners, and there is less risk of uneven action of the solution used in the various aftertreatment processes.

A suitable solution contains 1 per cent each of formalin and sodium carbonate. Addition of about 10 per cent of sodium sulphate (Glauber's salt) prevents the alkali from swelling the emulsion. As the formalin takes some little time to harden the gelatin, a certain amount of swelling would otherwise be bound to occur.

Hardening Before Development. When processing at temperatures above 85°F. (30°C.) the films or plates are best immersed before development in a hardener of the following type:

Formalin solution (40 per cent)	1 ounce	25 c.cm.
Sodium sulphate, crystals	8 ounces	200 grams
Water to make	40 ounces	1,000 c.cm.

The material is immersed in the hardener for 3–4 minutes, rinsed in a 15 per cent sodium sulphate solution, and transferred to the developer. The hardened gelatin slows down the penetration of the processing solutions, so somewhat longer development is needed.

Chrome alum hardeners are not as suitable for use before development because they are definitely acid and thus reduce the activity of the developer. L.A.M.

See also: *Fixing; Hot weather processing; Scratch-proofing; Tropical photography.*

Book: *Developing*, by C. I. Jacobson (London).

HARDENING FIXER. Fixing bath, particularly one for fixing negatives, that contains a hardener so that the emulsion is hardened and rendered less liable to injury in subsequent washing, drying and handling.

Proprietary packs of fixer are usually available with and without the hardener.

HARD WATER. When developer is made up with hard water, the calcium and magnesium salts responsible for the hardness react with the sodium carbonate of the developer and precipitate a white deposit. (Although the sodium carbonate content is slightly reduced by the reaction, the amount is negligible.)

The precipitate has no undesirable chemical effects but it may settle on the surface of the emulsion and form clear spots or pinholes where it has prevented the developer from acting.

Calgon added to the water in the proportion of 1 gram to every litre of water prevents the formation of an insoluble precipitate. It has no ill effect on the working of the developer apart from increasing its alkalinity very slightly.

The deposit may also be removed by allowing the developer to stand until the milkiness settles and then decanting the clear solution.

In large-scale processing laboratories it may be worth while installing ion-exchange water softeners. Even small units can be fitted into the supply line at a suitable point.

See also: *Faults; Water.*

HARRISCOLOR PROCESS. An obsolete two-colour subtractive process for cinematography. Prints from colour-separation negatives are made on single-coated film printed first through the back, processed, and blue-toned with iron. The residual emulsion on the print is subsequently printed, processed and red-toned.

HAZE. Suspension of minute particles of dust and other matter in the atmosphere which scatters light and thus tends to obscure detail and degrade tones in distant views. The scattering is greatest for the blue component of the light and the red component penetrates the haze most readily; the haze itself therefore appears bluish, and its effect can be reduced by photographing the scene with a blue-absorbing —e.g., yellow, orange, or red—filter.

Haze is often of value in pictorial work by assisting in the differentiation of planes.

See also: *Fog (atmospheric); Mist.*

HEAT FILTER. Optical device for removing heat radiation from a light beam without appreciably affecting the light output.

The common sources of light used in projectors and similar apparatus emit a high proportion of their total radiation output as infra-red or heat radiation which consists of wavelengths greater than 0·7 micron. For example the proportion of the total radiation emitted as heat by tungsten projector lamps operating at 2800°K and 3400°K is approximately 94 per cent and 86 per cent respectively. For carbon arcs at 3800°K and 6000°K the proportion is 80 per cent and 48 per cent respectively. High-pressure mercury vapour lamps have a high luminous effeciency but still only emit 20 per cent as visible radiation, although the 1 kw. water cooled version will give 2½ times the light output with less than half the heating effect of a 1 kw. tungsten lamp. The xenon arc is similar in heat output to the carbon arc and has a group of strong monochromatic lines in the region 0·8 to 1·0 micron. Fluorescent mercury lamps with either hot or cold cathodes emit negligible infra-red, but are of limited use.

The heat radiation must be prevented from reaching the optical components and, in particular, the photographic image. Here the heat filter permits visible radiation to reach the optical system but absorbs, or otherwise diverts, the heat radiation.

Transmission Filters. The simplest heat absorbing filter is a thickness of water, about 1 cm., in a glass cell. This, however, only absorbs beyond 1·4 micron and so is restricted to low-temperature sources. The addition of 2 per cent by weight of copper sulphate ($CuSO_4 \cdot 5H_2O$) to the water reduces the transmission limit to 0·8 micron but introduces a slight red deficiency in the transmitted light. More conveniently, heat absorbing heat resisting glass filters can be obtained. For example, a 4 mm. thickness of Chance type HR1 glass transmits a total of 82 per cent of the visible and only 7 per cent of the heat radiation from a tungsten lamp operating at 2854°K.

It is not possible with absorbing filters to obtain a sharp transmission cut-off on the long-wavelength side and, therefore, to obtain adequate visible transmission, some infra-red transmission must be allowed. Further, the absorbed radiation heats the filter causing a liquid one to evaporate and may cause a glass one to crack.

Mirrors. An entirely different type of filter, in principle based upon the interference of light waves, has ideal transmission properties. This filter consists of a multiple sandwich of alternate layers of transparent materials of high and low refractive index, deposited by evapora-

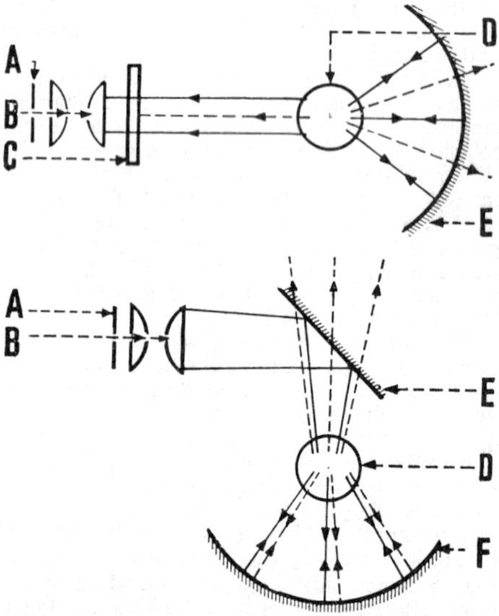

COLD MIRRORS IN PROJECTION SET-UPS. Top: *The lamp is backed by a light-reflecting but infra-red transmitting mirror. An additional heat filter in the light path protects the transparency.* Bottom: *An alternative set-up uses a normal metallized reflector behind the lamp, but an infra-red transmitting mirror in the light path to reflect the light through 90°. A, film or transparency. B, condenser system. C, heat filter. D, light source. E, cold mirror (infra-red transmitting, light-reflecting). F, fully reflecting mirror. The solid lines (shown as far as the first condenser) are light rays, the dotted lines infra-red radiation.*

tion upon a glass base-plate. The thickness of each layer is a fraction of a wavelength of light. Such a multiple layer can be designed to either transmit the visible and reflect the infra-red or vice-versa. The former is an infra-red mirror or heat filter and the latter is popularly known as a "cold" mirror or diathermic mirror. The transition from transmitting to reflecting is very rapid.

In use, an infra-red mirror is placed between the source and the optical system and a concave cold mirror used behind the source. Alternatively, light from the source can be reflected into the optical system, by a plane cold mirror placed at 45° to the optic axis. This has the advantage that no infra-red mirror is required and that the usual metallic concave mirror can be used behind the source. Multiple-layer mirrors of both types designed for normal incidence reflect 98 per cent and transmit 85 per cent of the incident radiation at the relevant wavelengths. The performance of those designed for use at 45° is not quite so good however.

Projection lanterns used for illumination are not usually equipped with heat filters with the exception of those used in photomicrography. Here it is often necessary to protect the speciment from undue heat.

The use of a heat filter of one type or another does not lessen the heating effects brought about by conduction and convection. These must be minimized by correct design and adequate cooling. G.G.T.

See also: *Diathermic mirror.*

HEAT RECORDING

While in most photographic processes the image is formed by light on a light-sensitive layer, thermo-sensitive systems can record images formed by heat radiation. This is useful for detecting objects in the absence of visible light, and also for establishing and measuring temperature distribution and differences in a variety of research applications.

Dectectors. An important element in any heat recording system is the detector. This is the device which converts the energy of the radiation falling on it into another form, which may be electrical, photographic or visual. Detectors are classed as either photon or thermal detectors.

The first type responds to the wavelength of the radiation and the second to the intensity of the heat source, ignoring the colour (spectral content). Both types of detector can be either image-forming, i.e. a real picture of the object is formed, or non-image-forming in which only a signal is produced. This latter method is widely used in military applications where the signal so obtained can be applied in a variety of ways, e.g. to control the movements of a guided missile. Only image-forming systems where a resultant picture is viewed or photographed are considered here.

Photon Detectors. In these the incidence of a small bundle of light energy (photon) on the detector releases an electron so that an electric signal, whose magnitude depends on the object being recorded, is available. Since these photons may be in the infra-red range they can be called heat rays. However in this class of detectors it is not the heating effect which forms a picture, but the photon effect; in other words, it is not the intensity of the heat which is important but its colour.

Examples of cameras and viewers of this type are the image convertor, the image dissector, the image orthicon, the vidicon and the metascope.

In the image convertor the infra-red (invisible) image formed on the photo-cell causes electrons to be accelerated down the tube. These electrons strike a phosphor film at the end of the tube and cause it to glow, giving a visible image. The image so formed is reduced in size and is upside down. A suitable viewer magnifies and erects the image or a camera can photograph it.

The image dissector is a television camera capable of providing a television picture using infra-red radiation. Because of fundamental limitations (noise—i.e. a high level of spurious signals not originating from the object) it has not been used very much in infra-red applications. This device, which is a scanning system, can be used in very low levels of illumination. Pictures can be obtained for scenes in bright moonlight. An image intensifier version of this device has been used to detect celestial bodies, such as the satellites of Jupiter, during daylight.

The metascope is a near infra-red viewing device. An infra-red lamp illuminates the target which is then focused by a mirror system on to a phosphor screen. This device is used at a wavelength of 1·3 microns. There are certain military applications for driving at night and it can also be used in signalling systems.

Thermal Imaging Detectors. These are of two main types—the evaporograph and the absorp-

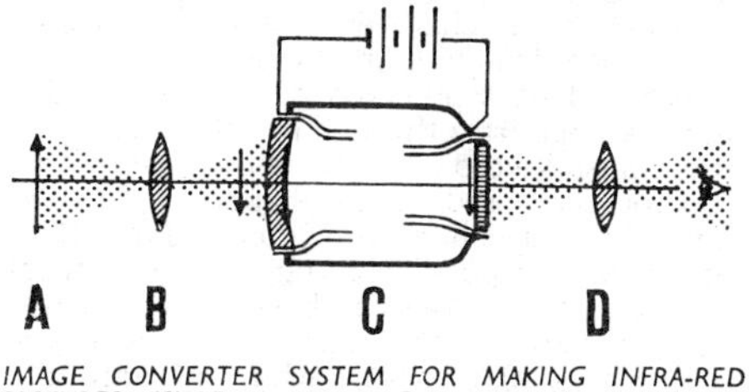

IMAGE CONVERTER SYSTEM FOR MAKING INFRA-RED PICTURES VISIBLE. A, *object.* B, *camera lens.* C, *image converter reproducing the infra-red image projected on the photo cathode as a visible image on the fluorescent screen.* D, *viewing lens.*

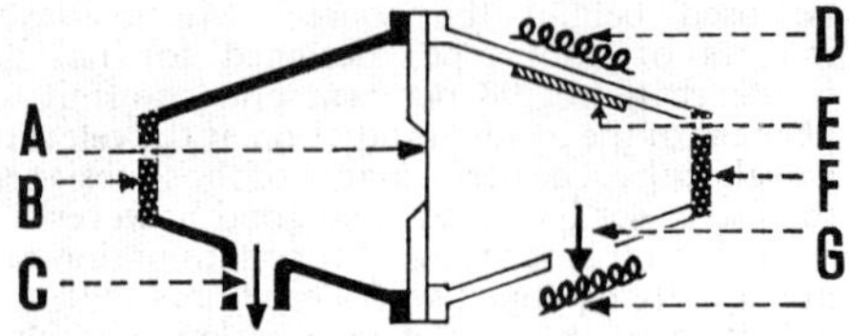

EVAPOGRAPH CAMERA. This consists of two evacuated chambers, one containing oil vapour and separated by a membrane having a layer of gold black. When an infra-red image is focused on the front of the membrane, the oil film deposited on its rear evaporates image wise, forming a picture which can be photographed through a glass window at the end of the oil chamber. A, evaporation membrane. B, infra-red transmitting sodium chloride window. C, vacuum line to pump. D, heating coils. E, absorbent pad of oil to keep the chamber saturated with oil vapour. F, conducting glass window.

tion edge image convertor. The evaporograph consists of a vacuum pump, cell, optics and camera. It relies on the evaporation of an oil film from a membrane on which an infra-red heat picture is projected.

The camera itself consists of a chamber that can be evacuated to a pressure of about 0·01 mm. The heat image is focused through an infra-red transparent sodium chloride window on to the black side of the membrane. The oil film on the membrane is evaporated from the warmer area of the image and condenses on the colder area, resulting in an oil film of varying thickness corresponding to the heating effect of the image. By optical interference this thin film of oil appears in various colours which are then photographed in the usual way by a camera and colour-sensitive photographic film.

The operation of the absorption edge image convertor depends on the fact that certain semi-conductors, such as amorphous selenium, change their absorption and hence their transmission of light when locally heated. Any object which is even only slightly warm will, when focused on this selenium film, cause the film to transmit less light.

In a practical set-up infra-red rays from the object are focused by means of a parabolic mirror through the infra-red transmitting rock salt window on to the selenium film. This is 0·001 mm. thick and is backed with a chromium film to improve the absorption characteristics. Convection losses of heat are reduced by evacuating the chamber. The projected image heats the selenium film and decreases its transmission. This is observed by shining a sodium lamp via a mirror through the film and viewing the transmitted picture. Parts of the film which do not receive radiation will appear light yellow and the image itself will appear dark. It has been claimed possible to photograph a teapot in the dark at a considerable distance when the temperature is only about 60° C. It is even possible to image objects which are only 10° C above their surroundings.

Photographic Thermometry. When an object is photographed in the usual way using, say, panchromatic film, the negative shows a variation in brightness corresponding to the variation in brightness of the object. Using a densitometer, which will measure the transmission, one can measure the actual brightness of the object at various points, from the photographic information.

In the same way, it is possible to record the infra-red radiation from the object on infra-red film. In other words, the heat emissivity at various points on the object are recorded as corresponding density variations on the film. This, in effect, is photographic thermometry.

Fast infra-red film is used for the photography of hot objects in the dark to determine the temperature distribution over the surface, and studies of temperatures ranging from 250° C to 500° C are possible. The film should have a peak sensitivity of about 8,500 A. The hot object should not be polished but dark, and high-aperture lenses are necessary.

Colour Photography. Heat records on colour film are possible with temperature-sensitive paints when a minimum of apparatus is required. They give a record of surface temperature changes in the form of a pattern of coloured isotherms. A permanent record may be made by still colour photography or changes over a period may be recorded on colour film in a cine camera.

Certain paint compounds not only change colour at one temperature but change to various colours throughout a range. For example, a blue paint compound may change to bright green, then yellow and eventually to light brown as the object is heated.

Colour changes are also a function of time. Although the immediate effect occurs in one or two seconds, changes may continue for several hours. Exposure times of the order of ten minutes are required and it is possible with careful work to measure changes of colour equivalent to temperature changes of 10° C.

Pyrometry. This is the measurement of the temperature of a body by studying the nature of its radiation. Both optical and photographic techniques are available.

In optical pyrometry the human eye is the detector of radiation and relies on its ability to estimate degrees of brightness. The surface whose temperature is to be measured is viewed side by side with a source whose temperature is adjustable. Both the brightness and the colour of the comparison source are used to calibrate the temperature. The pyrometer consists of a telescope or viewer that shows the two sources alongside each other in a split-field eyepiece. The intensity and colour of the comparison source may be adjusted by rotating various colour filters and choppers in front of that part of the viewer which projects the comparison source, until a perfect match of colour and intensity is obtained. The rotation necessary is marked off on a temperature scale.

The job of the eye in the pyrometer may b

taken over by a camera. For low temperatures, infra-red film is necessary since most of the radiation occurs in this range. For high temperatures (and pyrometry is concerned mainly with this), the furnace or hot object emits light which can be recorded by panchromatic film. The brightness of the light emitted by the body increases with rising temperature (although the two are not directly proportional).

The procedure consists of photographing tungsten-strip lamps burning at various calibrated temperatures. This photographic strip (which looks rather like a graded density wedge) is the master sample. The hot body is photographed throughout a range of exposure times and aperture stops.

Such a range of exposures is necessary because the exposure required for optimum density differentiation of temperature variations depends on the temperature itself. Hence a range of master samples, also exposed at different times and with different apertures, is used.

The densities of the negative are measured by standard photographic densitometric techniques. Comparison of the density of the test negative with the densities of a master sample strip exposed for the same time and at the same aperture then indicates temperature of the hot body. This technique can under favourable conditions estimate temperatures with an accuracy of plus or minus 3° C at 1500° C. It has been applied to the measurement of the temperature of furnace walls, hot metal ingots, welding methods, and—with certain modifications—of radio valves and superheated-steam lines.

Indirect Scanning Methods. Certain semi-conductors are particularly sensitive to temperature changes and greatly alter in their conducting properties as they get slightly warmer or colder. They are the basis of a num-

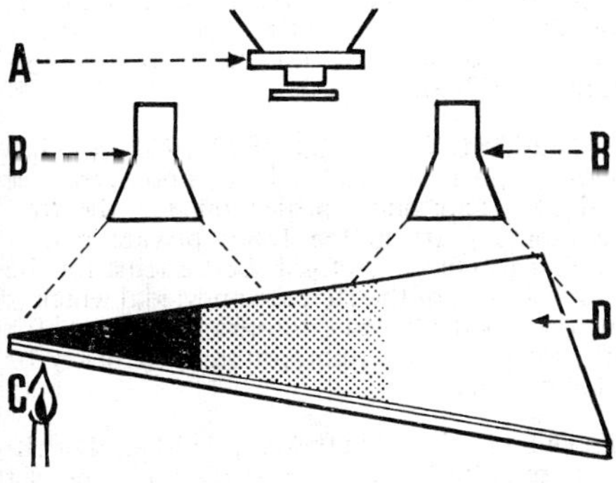

DIRECT PHOTOGRAPHY WITH HEAT SENSITIVE PHOSPHOR. A, camera with ultra-violet absorbing filter over lens. B, ultra-violet lamps. C, hot end of object. D, cold end of object. The ultra-violet radiation makes the phosphor glow, heat tends to subdue the glow. If photographed under suitably standard conditions, temperatures and temperature gradients can be determined by measuring the photographic densities of the object on the negative.

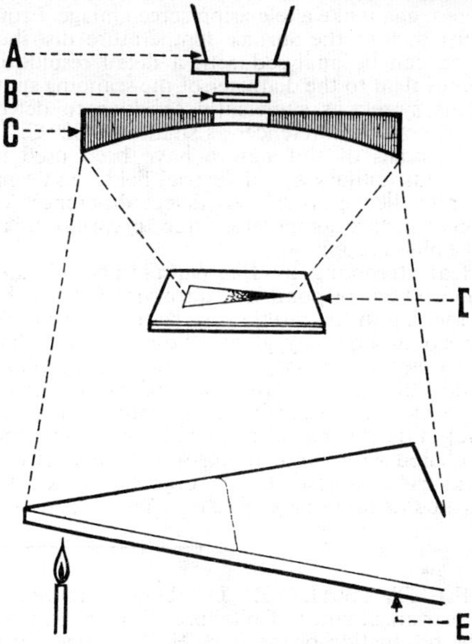

INDIRECT PHOSPHORESCENCE PHOTOGRAPHY. An image of a heated object with its temperature gradients is projected by a mirror lens on to a phosphor-coated screen irradiated by ultra-violet rays. The camera then photographs this image on the screen. A, camera. B, ultra-violet absorbing filter. C, paraboloid mirror. D, layer of phosphor on screen. E, heated object.

ber of industrial temperature measuring instruments. The most sensitive of them will also react to the radiant heat in the image of an object projected on the cell and can thus be used for infra-red or heat photography.

Such a recording system works in a manner analogous to a television link: it involves optical scanning of the object being examined or measured and reconverting the current fluctuations produced in the semi-conductor into signals that can reconstitute a picture. In practice a mirror optical system (which does not absorb infra-red rays like a glass lens) focuses an image of the object on to the photo-conductive cell. The optical system moves so that the cell scans the image from top to bottom and side to side like the image on a television screen.

The intensity of heat radiation from the object modulates the current in the cell circuit; this is amplified and in turn modulates the intensity of a projected light beam. The latter scans over a photographic film in synchronism with the scanning pattern of the mirror system in the "camera". The film thus records the temperature variations in different parts of the object as different densities. For instance, on the film warm subject areas reproduce darker than cold areas. The final black-and-white picture is made up of a series of closely spaced lines, again like a television screen image. From this picture the surface temperature distribution can be analysed with a detail resolution equivalent to the diameter of the scanning spot. This system is sufficiently sensitive to detect temperature differences as small as 0·02° C.

Systems of this nature have been used in various military and industrial fields, and even for medical purposes to detect differences of body surface temperature due to variations of the blood supply.

Heat Recording by Phosphorescence. Certain phosphors respond to ultra-violet light by radiating in the visible spectrum. For example the phosphor may glow yellow or red. It has also been found that the temperature of the phosphor affects the brightness which decreases rapidly with increasing temperature. So if an object is coated with such a phosphor and radiated with the light from a mercury arc it will glow yellow. If the object is gradually heated its brightness will drop. With a constant value of ultra-violet radiation, the brightness may change by as much as 10 per cent per ° F and the light emitted lies in the photographic region.

The method is applied in two ways:—the phosphor may be coated directly on to the heated object or an image of the hot object may be focused on to a phosphor-coated screen.

The primary advantage of such a system is that it simultaneously shows a "thermal map" of the surface. For quantitative measurements of the heat distribution the photographic recording is analysed by densitometry.

For the projection method a large-aperture mirror optical system is required because of the relatively low output of the phosphors. The arrangement consists of an ultra-violet source, the mirror lens, a camera, and a filter to exclude ultra-violet radiation from the photographic film. It is possible to record an entire temperature pattern with one exposure. J.N.W

See also: *Image amplifier; Infra-red; Thermography.*

HECHT, WALTER D. Dates unknown. Austrian scientist. Collaborated with Eder in the production of the Eder-Hecht wedge sensitometer. This, when used with a standard quantity of magnesium ribbon as light source, was widely employed in sensitometry on the Continent until the introduction of the D.I.N. system.

HECTO-. Prefix used in the metric system denoting one hundred—e.g., hectogram, hectolitre. Little used in everyday practice.

HEGE, WALTER, 1893–1955. German photographer whose fine surveys of Naumburg, Bamberg and other German cathedrals in the late 1920s, and of the Athenian Acropolis and Olympia in the 1930s, are an important contribution to art history and the photography of architecture and sculpture. In 1930 Hege was appointed professor of photography at the art school in Weimar, but gave up the position in 1935 to devote himself entirely to photography and a number of art-historical and nature films. After World War II, Hege worked on art books and lectured on colour photography.

HELIOGRAPHY. Name given by Nicéphore Niépce to his photographic process for producing camera pictures and for copying engravings by superposition. The world's first photograph from nature was a heliograph of the view from a window at Niépce's house Gras, near Châlon-sur-Saône, taken in summer 1826. It was made on a polished pewter plate coated with bitumen of Judea, which hardens under the influence of light. After an exposure of about eight hours in the camera obscura, the parts of the bitumen which remained soluble were washed away with a mixture of oil of lavender and white petroleum. The result was a permanent direct positive picture in which the lights were represented by bitumen and the shades by bare metal.

In the case of copies of line-engravings, the plate could be etched and printed from.

See also: *Photomechanical reproduction.*

HELIOTYPE. A modified collotype introduced by the London portrait photographer Ernest Edwards, December 1869, and worked in partnership with R. L. Kidd.

HELMHOLTZ, HERMANN VON, 1821-94. German physicist and physiologist. Photographed the solar spectrum, improved Brewster's stereoscope, contributed notably to the theory of colour and was partly responsible for the Young-Helmholtz theory of colour sensitivity of the eye. Biography by Königsberger (1902–3).

HENDERSON, A. L. 18??–1900. English photographer. Specialized in photoceramics and in background projection for theatres. For some years in the 1880s private photographer to Queen Victoria. Left a trust for the establishment of the Henderson Medal which is awarded by the Royal Photographic Society for the best papers on photographic chemistry or related subjects.

HENNEMAN, NIKOLAAS, 1813–?. Dutchman, one-time valet to Fox Talbot, who set him up as manager of the first photographic printing establishment, in Reading, 1843, and in a Calotype portrait studio in Regent Street, London, in 1847. In 1851, Henneman made the prints of photographs of the Great Exhibition for the illustrated Reports of the Juries.

About 1857, his studio was closed down and in the 1860s he was employed in Oliver Sarony's studio in Scarborough.

HERSCHEL EFFECT. Ability of red or infra-red radiations to bleach out a latent image previously formed in an emulsion otherwise insensitive to this region of the spectrum. The effect is most marked with the pure silver bromide emulsion of certain development papers and in particular with those that contain an excess of soluble bromide.

HERSCHEL, FRIEDRICH WILHELM (SIR WILLIAM), 1738–1822. German astronomer living in England. Investigated (1800) the properties of the solar spectrum and discovered the infra-red rays. Devised means of measuring and recording infra-red. Biography by Sime (1900).

HERSCHEL, SIR JOHN FREDERICK WILLIAM, 1792–1871. English astronomer and pioneer photographic chemist, and independent inventor of a photographic process on glass (1839). Son of Friedrich Wilhelm Herschel. Coined the word "photography" (1839) and introduced the use of the words "negative" and "positive" as well as "snapshot". Discovered (1819) the thiosulphates and proposed in 1839 the sodium salt hypo as photographic fixing agent. Investigated the photochemical action of the solar spectrum. Discovered the different spectral sensitivities of the silver halides, chloride and bromide, and of silver nitrate. Discovered the Herschel effect, the name given to the fading caused by infra-red radiation falling on an exposed image, which has been the starting-point for much modern research on the nature of the photographic latent image. Discovered the bleaching effect of mercuric chloride upon the developed silver image, and the restoration by immersing such a bleached print in a solution of hypo; this is the basis of the mercury intensification process and also of the so-called magic pictures which produce an image when breathed upon. Herschel described the light sensitivity of the citrates and tartrates of iron and of the double iron-ammonium citrates. Invented cyanotype, the first blue print (1842). Also discovered and described the argentotype process, in which iron salts were used to precipitate silver under the action of light, and suggested microfilm documentation.

HIDE. Camouflaged screen or hut from which animals and birds can be photographed without the photographer revealing his presence.

See also: *Big game; Birds.*

HIGH CONTRAST DEVELOPER. Class of developers which are intended to produce images of higher contrast than normal.

HIGH KEY. All work in photography is based on tones—dark, middle and light. In normal work some of the tones from each group are used; the print thus ranging in tone from black, through the middle tones and the lighter ones to white. High key work is concerned with light and occasionally middle tones, leaving out the lower ones altogether. Only certain subjects are really suitable for this type of treatment.

The subjects that are best so treated are those which are to convey cheerfulness, lightness and delicacy, such as in portraiture of children and nudes, and in landscapes, atmospheric effects, early morning scenes and glancing light on water. It is wise to choose a subject which has no strong shadows, for full black is not going to be used. A high key print starts in the camera; it is not made by trying to turn an ordinary print into one of high key. It is dependent on lighting and subject matter. The ideal model for high key portraiture, for example, would have fair hair and a light skin, light blue eyes, and lightly made-up lips, and should be wearing light clothes.

Lighting. The lighting must be full and soft. For portraiture it is very necessary to have the subject in a light room with plenty of light reflected from all surfaces and a white sheet on the floor for the same purpose. There should be full frontal soft lighting from the direction

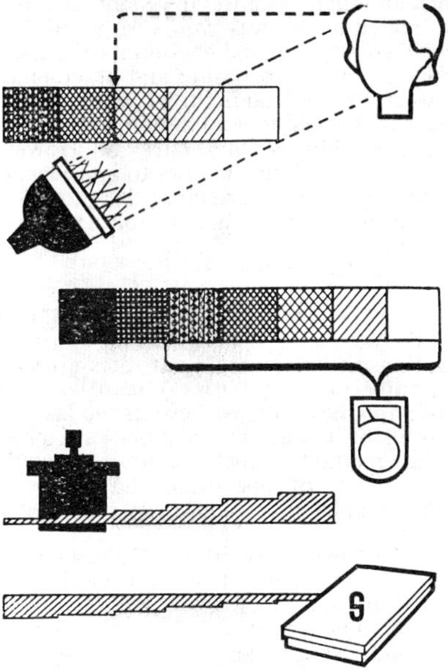

HIGH KEY TECHNIQUE. The elements in the production of a high key picture are: Top: Choose a light subject and light it softly to avoid heavy shadows. Upper centre: Expose for the highlights. Lower centre: Develop to a low contrast in a soft-working developer. Bottom: Print on a soft paper grade to preserve full detail in the highlights.

of the camera and no shadows cast by the features, or they will photograph too dark. It is sometimes wise to place the lights so that one of the sources, quite near the camera viewpoint acts as a model light, casting some shadows. The remaining lights should then be placed so as very nearly to "kill" these shadows. They should not, of course, be permitted to cast counter shadows. Soft light is best obtained by allowing plenty of space between the lights and subject, for distance is a good diffuser. Diffusing screens of various types are also useful. Having obtained this result, it is then necessary to illuminate the background independently to raise its tone to either a shade lighter than the highest lights on the subject or a shade darker; the latter is usually best. If the background were left unlighted, even though white, it would photograph a mid-grey owing to its greater distance from the light source.

In daylight it is necessary to choose a day when conditions create a soft, diffused lighting, and the surroundings should help to give the feeling of lightness.

It is quite possible to use this type of lighting to give a high key effect to any suitable subject that demands the quality of delicacy and lightness, but in landscape work it is wise to wait for the right conditions—i.e., where the subject has no strong shadows. Misty, atmospheric subjects are the popular ones for such treatment, although against the light effects in mist treated in this way can be very satisfying.

Technique. It is essential to continue the special treatment through exposure and development. For work by artificial light, where all shadows have been reduced to a minimum, the exposure is not critical but should be fairly full. However, as there are no deep shadows to consider, full exposure for a high key subject is not going to be as long as that for a similar subject with a full tonal range.

Providing that enough exposure is given to produce a deposit of silver through the whole image, the type of negative wanted is largely a matter of personal preference. With outdoor scenes, however, it is desirable to give enough exposure to lighten any of the darker tones to avoid contrast.

In developing, it is essential to secure a softly-graded negative with all the highlight tones separated—a negative which will yield the soft, delicate print essential to high key reproduction. This can best be obtained by the use of a dilute developer.

More people fail in making the print than at any of the other stages. The correct procedure is to choose as soft a paper as possible and print it so that all the importance is on the highlight details while the darkest tone is no deeper than a middle grey. It is surprising how dark even this can look by contrast. The paper chosen should not, of course, be so soft as to make the print appear flat and dull overall. If the high-key print lacks sparkle, much of its purpose is lost.

The choice of paper base colour and surface can also make a big difference. A white base paper will enhance the general effect of delicacy; a cream base paper looks yellow and faded.

The aim in high key work is to achieve softness, lightness and delicacy, and all efforts must be directed to that end. D.A.

See also: *Lighting the subject; Low key.*

HIGHLIGHT EXPOSURE METER. Type of exposure meter that measures the intensity of the light falling on the subject.

See also: *Exposure meter.*

HIGHLIGHTS. Bright parts of the subject which are reproduced as the densest areas in the negative and as the lightest areas in the print or transparency.

HIGH SPEED CINEMATOGRAPHY

Motion picture recording at rates exceeding about fifty pictures (frames) per second, to photograph movement and events too fast to be perceived by visual or straightforward photographic methods. Such recording requires special means of operation, and high speed cinematography is accordingly classified as follows:

(1) High speed, at 50 to 500 frames per second, using intermittent film motion and mechanical shuttering;

(2) Very high speed, at 500 to 100,000 frames per second, using moving film and image compensation;

(3) Ultra high speed, at 100,000 to 10 million frames per second, using a stationary film with moving image systems;

(4) Super high speed, in excess of 10 million frames per second.

The boundaries between systems are not precise as some cameras employ combinations of techniques. This arrangement fits in with common practice in the main and also the terms used follow the same pattern as those used in the classification of radio frequencies although, of course, frames and cycles do not agree. It is unfortunate that superlatives were used too readily, that is, before the cameras in the last two groups became possible.

High speed cameras of group (1) are specially developed versions of normal cine-cameras, being limited in speed by the physical strength of the film, for the stresses imposed by starting and stopping between exposures can exceed the breaking point of the film base causing tears first at the sprocket holes. The fastest cameras operate at up to 300 f.p.s. on 35 mm. film and 450 f.p.s. on 16 mm. film.

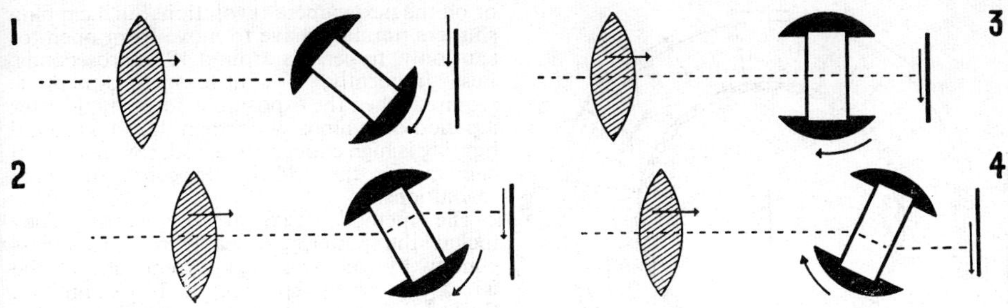

MOVING PRISM HIGH SPEED CAMERA. A rotating prism acts as shutter and keeps the image sharply in position on the moving film. 1. End piece of prism cuts off light. 2. Prism opens light path; image is displaced upwards by refraction. 3. The image follows the moving film. 4. After maximum downward displacement, the prism cuts off the light again.

In the cameras of group (2) the film is moved smoothly and continuously and some form of compensation is required to avoid image blur during the exposure.

In cameras of groups (3) and (4) the film is stationary and the image only moves.

Rotating Prism Compensation. Between the lens and the film is placed a rotating prism of regular form (parallel sided block, cube or octagon). By refraction, the light forming the image is moved during the exposure in the same direction and at the same speed as the film, the prism and film drives being suitably geared and driven from a common motor. Thus, during exposure, there is no relative motion between the image and the film. Suitable masks are fitted to the prism to limit the exposure time to the period during which compensation is possible. Masks with smaller apertures are also used when still shorter exposure times are required.

The physical strength of the film base again sets the limit and the maximum rates achieved are 20,000 f.p.s. with 8 mm. frames and 10,000 f.p.s. with 16 mm. frames. To maintain an economical camera life, it is common to restrict framing rates to 80 per cent of the maximum value.

Rotating Drum Systems. Some improvement in speed can be made at the expense of total recording time by attaching a length of film to a rotating drum so that the film now has support and the limit is set by the strength of the drum.

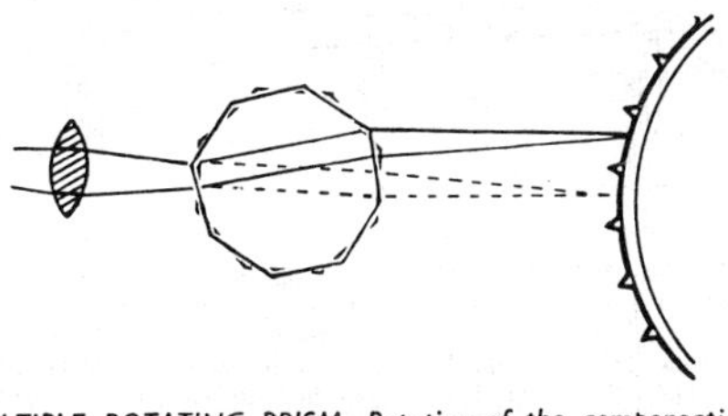

MULTIPLE ROTATING PRISM. Rotation of the compensating prism keeps the image stationary on the moving film as it passes over the sprocket. Successive frames are exposed as each prism surface faces the lens in turn. Prism and film movement are synchronized. Maximum camera speed is 10,000 frames per second on 8 mm. film.

Using a rotating mirror for image compensation, a commercially available camera gives 224 frames of 7·5 × 10·2 mm. at 26,000 f.p.s.

Using two concentric drums, one carrying the film and the other a number of lenses, frames of 5 × 8 mm. have been recorded at 40,000 f.p.s.

With a single objective lens and a rotating drum, it is also possible to synchronize a series of very short duration light flashes to the drum motion, so that the brief exposure of each flash is so short as to give no appreciable image movement. This permits framing rates up to about 100,000 f.p.s.

A complex installation of five drums and fifty spark gaps, all synchronized, produced 2,500 frames 25 × 20 mm. at 250,000 f.p.s. and 5,000 frames 25 × 10 mm. at 500,000 f.p.s.

Multiple-lens Cameras. To exceed the framing rates possible with moving film, various methods exist in which the image is moved along a stationary film.

The subject is imaged very near the surface of a rotating mirror (commonly about 25 mm. width). This image can then be re-imaged on to the film by an arc of lenses. A relay lens between the objective lens and the mirror images the stop of the objective on to the stops of the lenses on the arc. As the mirror rotates, the image-forming light beam moves from lens to lens on the arc producing a series of images on the film, with negligible image-to-film motion during each exposure. Exposure time of individual frames can be varied by changing the aperture stops.

Suitable mirrors can be rotated up to 80 per cent of their bursting velocity due to centrifugal forces, and when made of suitable materials, e.g. high tensile steel, rotation speeds of 500,000 r.p.m. have been achieved.

With individual frame sizes of 8 × 8 mms., all speeds up to 10,000,000 f.p.s. are possible. By changing the lenses used, larger frames at lower rates can be recorded using the same camera and mirror. The number of frames recorded is commonly between 30 and 200.

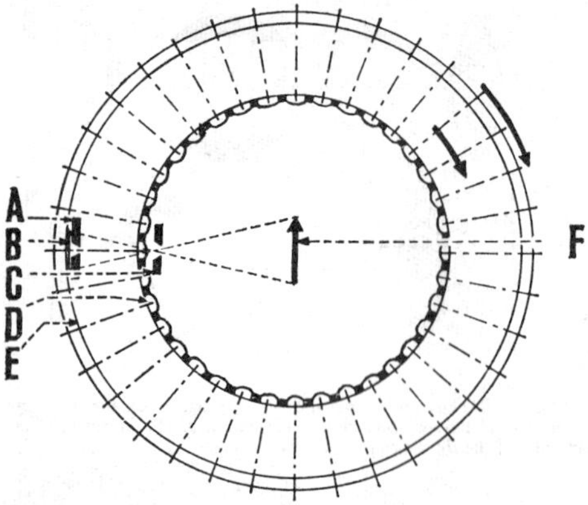

ROTATING DRUM CAMERA. The film is fastened to the inside of a drum which further carries a ring of lenses, so that each lens covers one frame on the film. The image of the subject is projected into the centre of tho drum and from there directed to the circumference. As the drum rotates at a high speed, each moving lens picks up the image and keeps it stationary on the individual film frame behind the lens until the next lens moves into position. A stationary between-lens shutter cuts off the light after one revolution of the drum. A, stationary focal plane shutter (slit through which image is projected). B, final image, moves with film. C, stationary diaphragm shutter. D, moving lenses in drum. E, moving film in drum. F, intermediate image projected into centre of drum.

Use of two rotating mirrors and multiple reflections can produce still higher rates.

With a complex system of lenses, mirrors and prisms, combined with a rotating mirror, a form of focal plane shuttering can be achieved and is claimed to give shorter exposure times without loss of aperture.

To increase the number of frames in a sequence and to reduce synchronization problems, it is sometimes advantageous to keep the camera continuously recording. Thus irrespective of mirror position, an exposure is being made at all times somewhere on the film track. This can be accomplished by using multiple optical paths in one camera or by synchronizing two or more cameras together.

Adding a rotating drum mechanism to a rotating mirror system can increase the number of frames recorded without reducing frame rate excessively but the camera becomes more complex and expensive.

Many of these cameras do not produce records which can readily be projected as cine-films, although skilful printing can make this possible. However when the number of frames is not large analysis is normally done frame by frame.

Capping Shutters. When photographing self-luminous subjects with high speed cameras which depend on polarized light (Faraday or Kerr effects) or on rotating drums or mirrors, a capping shutter is necessary if the luminosity continues after the camera exposure. Otherwise the image can record by polarizer break-through or on the next camera revolution. Such capping shutters normally have to move from open-to-shut only, in periods around 10 microseconds. Less frequently, a shut-to-open system is required when the exposure is to be made after a period of luminosity or when the background lighting is high enough to record, say on a drum camera, by the additive exposure of many revolutions.

The methods used are numerous. They include the sputtering of lead wire on to a glass window by passing a high current through the wire; explosively spraying a black mixture (lamp black, say, plus a greasy binder) on to a window; explosively shattering a glass window sandwiched between two sheets of perspex such that it becomes relatively opaque by multiple scattering. Various tricks with very thin metal foils have been proposed which give opening or closing shutters depending on arrangement—all involve some electrical discharge circuit to produce the necessary energy, usually demanding many thousands of volts and large capacitors.

Passing a discharge through a thin metal layer forming a mirror, so removing the metal layer can also be used as either type of shutter.

Multiple Cameras. A series of pictures taken in sequence by a number of single-shot cameras constitutes a form of cine-photography. The advantages lie in the increased information per frame, flexibility in frame interval and exposure time, shorter exposure times and reduction of image movement.

An assembly of twelve Kerr cell cameras has been so used, each at 0·1 microsecond exposure.

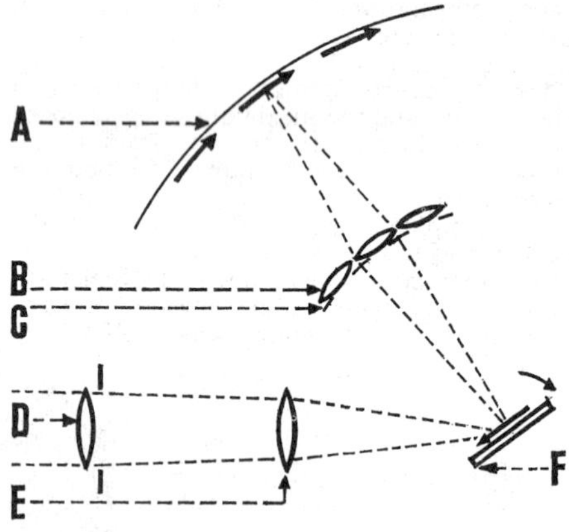

ROTATING MIRROR CAMERA. The objective lens images the event via a relay lens on to the surface of the rotating mirror. The relay lens images the aperture of the objective lens successively on to miniature lenses which form further images of the subject on the film. Owing to the rotation of the mirror the reflected beam travels in the direction shown by the arrows, exposing successive frames. With two rotating mirrors the travel speed of the beam can be very greatly increased. A, film. B, miniature lenses forming final images. C, lens stops. D, objective lens with lens stop. E, relay lens. F, rotating mirror.

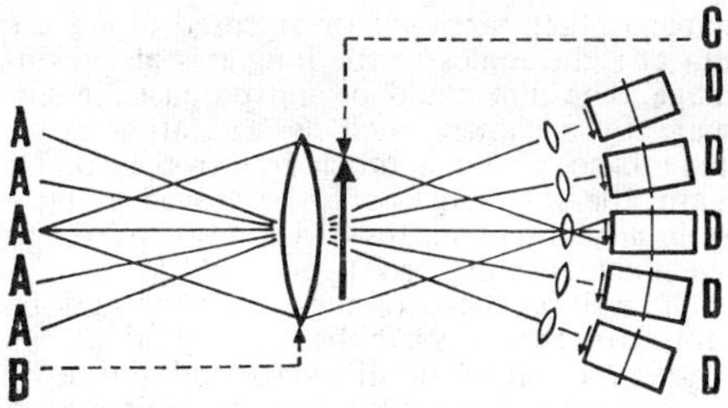

MULTIPLE DRUM MULTIPLE SPARK CAMERAS. Each drum has its own lens and spark series. As each spark is imaged on the aperture of only one camera lens, it cannot affect the other four cameras. A, sparks. B, field lens. C, subject. D, drum cameras.

When triggered in immediate succession, this set-up can give a rate of 10,000,000 f.p.s. Kerr cell shutters have been operated at still shorter exposures down to 5 millimicroseconds ($^5/1{,}000{,}000{,}000$ of a second). Similarly Faraday shutters and image tubes can be used.

For non-luminous events, a battery of cameras each with its own spark light source has been used at up to 100,000 f.p.s.

Some difficulty in analysis occurs in that these cameras all have slightly different viewpoints. To some extent this is overcome by using mirrors to bring the viewpoints closer. An alternative method is to make one camera with a sectional lens, that is, the lens is divided into segments which are then displaced a little from each other. Arranging a rotating disc with a single aperture causes each lenslet to be exposed in turn, giving a series equal in number to the number of lenslets. Such a camera, which is simple and relatively cheap, takes eight frames at 6,000 f.p.s.

Image dissection is also a kind of multiple exposure system, but with more complex recording and analysis requirements.

Optical Delay Camera. The extremely short exposures possible with a Kerr cell permit a further technique. The light from the subject is relayed to the camera via a system of mirrors such that six possible paths exist, all of different lengths. Light only travels 1 foot in 1 millimicrosecond. Thus, if the path length difference between adjacent cameras is 5 feet, then at exposures of 5 milli-microseconds, the light for one exposure is different from the light used in the next exposure. All six paths can be made to pass through the same Kerr cell, so that one shutter action produces six separate images corresponding to six separate time periods in the subject. Such pictures of exploding wires have been made at 5 milli-microsecond exposures and with frame intervals of 15 milli-microseconds, equivalent to 66,000,000 f.p.s.

Image Tube Framing Cameras. The shuttering and deflection properties of image tubes can be used to produce a short sequence of pictures on one tube. Image tubes can already be used to produce shorter exposures and higher deflection rates than other techniques and also can even intensify the light available making it possible to record events previously not sufficiently bright. A commercially-available camera gives three frames, each 17 × 25 mms., with exposures as short as 5 milli-microseconds at intervals of 50 milli-microseconds or 20,000,000 f.p.s. Other cameras give six or eight pictures at up to 1,000,000 f.p.s.

Timing. Cameras with moving film require timing marks on the film to assist in interpretation. In many applications the film never achieves a steady velocity and is often still accelerating when the magazine is exhausted. So small marks recorded at regular time intervals along the film edge indicate the film velocity (more precisely, the time resolution) at any point of the film. These timing marks can be produced by a variety of techniques which depend on some independent stable frequency source.

A tuning fork is such a source and its vibrations can be maintained by an electro-magnet powered by, say, a 12-volt battery. A ray of light shines via a slot mounted on the fork, the fork's vibtations causing the slot to interrupt the beam regularly, thus a 500 cycles per second fork produces 1,000 marks per second on the film; the camera rate is readily computed, e.g., 4 marks per frame—260 f.p.s., 4 frames per mark—4,000 f.p.s.

With the advent of transistors and miniaturization in electronics, quite compact units

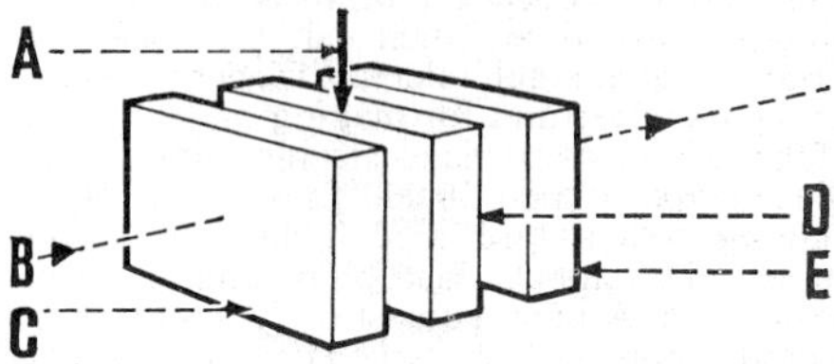

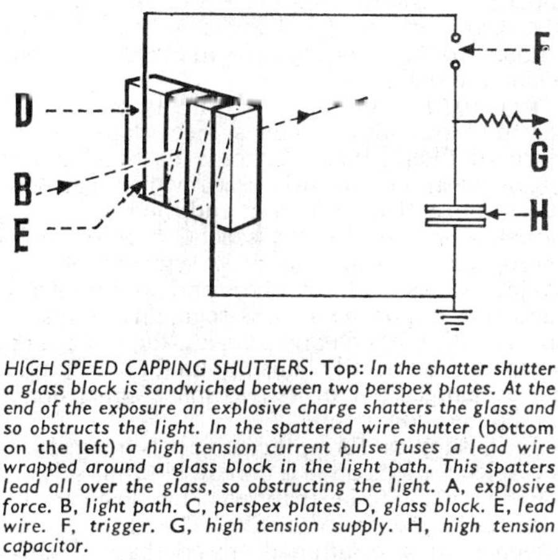

HIGH SPEED CAPPING SHUTTERS. Top: *In the shatter shutter a glass block is sandwiched between two perspex plates. At the end of the exposure an explosive charge shatters the glass and so obstructs the light. In the spattered wire shutter (bottom on the left) a high tension current pulse fuses a lead wire wrapped around a glass block in the light path. This spatters lead all over the glass, so obstructing the light. A, explosive force. B, light path. C, perspex plates. D, glass block. E, lead wire. F, trigger. G, high tension supply. H, high tension capacitor.*

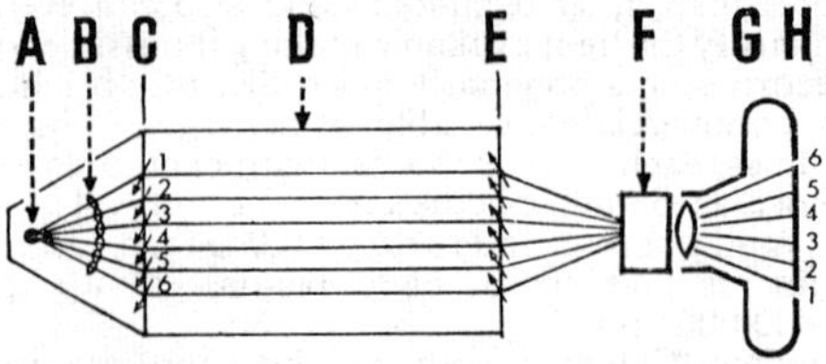

OPTICAL DELAY CAMERA. The light from the event (from the subject) passes through a series of field lenses into an optical time delay tank. There it is reflected between a series of mirrors in the first and second mirror plane. It thus reaches the film plane of the camera at different instants separated by the time taken by the light to traverse different lengths of the optical path. This yields a series of successive images 1 to 6 on the film. A, *subject.* B, *field lenses.* C, *first mirror plane.* D, *optical time delay tank.* E, *second mirror plane.* F, *Kerr cell shutter.* G, *35 mm. miniature camera.* H, *film plane.*

using crystal oscillators and dividing circuits to drive argon, neon or xenon lamps recording on the film through a narrow slit can be made. They have the advantages that the duration of the mark and its rate can be adjusted to suit requirements.

Particularly with cameras at rates below 1,000 f.p.s. a neat method is to include a clock (dial or digital) in the field of view so that, during projection, it can be observed at the same time as the record.

In the ultra high speed class with rotating mirrors, it is customary to direct a beam of light via the mirror to a path alongside the film where it can, by suitable masks, make marks on the film. Simultaneously the beam is received by a photo-electric tube which can then send out electrical pulses suitable for triggering events or light sources and for making measurements of the camera speed via electronic equipment.

Illumination. Up to 50,000 f.p.s., a variety of lighting systems are used including tungsten lamps, photofloods, flash bulbs and electronic flash tubes. A more sophisticated source is the pulsed xenon arc lamp which can be driven for short periods into peak powers of 10 kilowatts and produces enough light over an area of 1 square foot for photography at 20,000 f.p.s. on black-and-white film.

For higher framing speeds, many events are self-luminous but when subject lighting is required then either electronic flash tubes or argon flash bombs are used, which can be synchronized to the camera and the event.

Streak Cameras. In this special type of high speed camera a very narrow segment of the subject is selected for analysis and continuously successive exposures of this segment are made so that *all* visual changes during the recording period are observed. Thus the final photograph has exchanged one of its two space dimensions for a dimension in time.

In its simplest form, the subject is imaged by the objective lens on to a narrow slit thus selecting the portion to be analyzed. This slit-image is then re-imaged on to the film. The film may be attached to a rotating drum so that the slit-image is then streaked or smeared along the film at right angles to the long axis of the slit-image. The film could be moved past the slit-image by a camera such as a rotating prism camera, so giving a much longer record. Or again, the slit image could be moved along a stationary film by means of a rotating mirror to give higher writing speeds.

The writing speed of these cameras may be quoted in two ways, either the speed of the relative motion of the slit-image and film or by the time it takes for the slit-image to be moved by an amount equal to its own width. One is usually given in mm. per microsecond and the other in seconds or fractions of seconds. The latter is a better indication of performance and is the time-resolution of the camera.

The fastest rotating mirror cameras give writing speeds up to 40 mm. per microsecond (equivalent to 100,000 feet per second or nearly 100 times the speed of sound—Mach 100) with resolution times as short as $2{\cdot}5 \times 10^{-9}$ seconds (or 2·5 divided by 1,000,000,000). Rotating drums are unlikely to exceed supersonic peripheral speeds and so are at least 100 times slower.

Many streak cameras are not continuously viewing, hence it is necessary to synchronize the event to the period during which a record can be taken. By use of multiple optical paths within the camera, or multiple cameras, this can be overcome. A rotating element designed to achieve this with fewer components uses both refraction and reflection within a rotating element of twin prisms. As these must be transparent, only lower speeds can be used as the prism material will not have as high a tensile strength as metal mirrors.

Image tubes can also be used as streak cameras with relatively shorter recording periods. A slit-image of the subject is imaged on the photo-cathode. By electronic deflection within the tube, a streak image then appears on the screen. Such tubes have achieved speeds of 1,000 mm. per microsecond and resolution times of $0{\cdot}2 \times 10^{-9}$ seconds—1/5 of a milli-microsecond.

Claims have been made of resolution times approaching 10^{-11} seconds and the estimated limit is probably around 10^{-13} seconds. The time for light to travel one wave-length is only about 10^{-15} seconds, so it is apparent that absolute physical limits are being approached.

"Photo-Finish" Streak Cameras. One technique has been used which does not fit into either streak or framing categories but is somewhere between the two. If a streak camera is used to photograph a missile whose image velocity is equal to the streak speed, each portion of the missile will be photographed in turn and the film will show a still picture of the whole of the missile. This is a streak image of the changes in a given slit in space but is also equivalent to a very short exposure with a focal plane shutter.

It is also a high-speed version of the race-

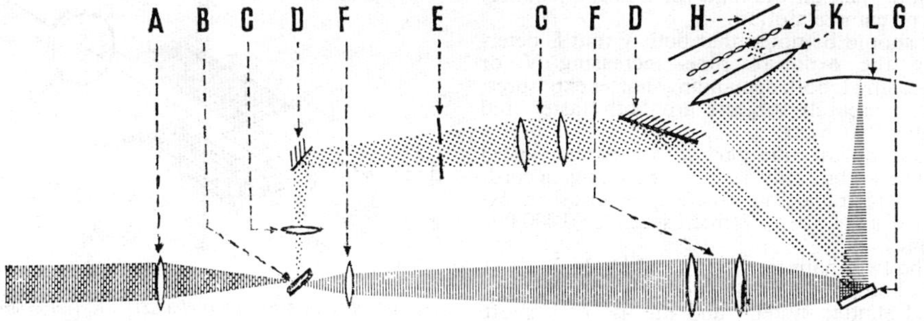

SIMULTANEOUS STREAK AND FRAMING CAMERA. A beam splitter divides the image rays from the objective lens along two paths to a rotating mirror. This in turn projects the images on the one hand through the imaging lenses of a framing camera, and on the other on to the film plane of a streak camera. A, objective lens. B, slit and beam splitting relay mirror. C, framing field and relay lenses. D, fixed relay mirrors. E, primary framing stop. F, streak camera field and relay lenses. G, rotating mirror. H, framing camera film plane. J, imaging lenses and secondary stops. K, framing camera projection lens. L, streak camera film plane.

track photo-finish camera. It is very useful in missile studies, as it can give a high-resolution picture of a missile in flight whose length is so much greater than its girth, where a more normal optical system could not.

Multiple-Slit Streak Cameras. If instead of a single slit, an array of parallel slits is used, a record sampling more portions of the event is obtained. The recording period is, of course, very much reduced, as the streak image of one slit should not overlap that of the next slit. Such a system, when the number of slits is large, becomes an image dissection system and operates as a form of framing or cine-camera.

Streak versus Framing Cameras. These cameras are not so much comparable as complementary. A framing camera records two spatial dimensions but only samples in time, thus it is possible for a change to occur between two frames which is not evident in either frame. A streak camera records one spatial dimension (i.e. a sample of space) continuously in time, such that no change in time within the space sample is missed.

The streak camera is ideal for studying events of uniform growth or decay, an expanding sphere or cylindrical discharge, when only the rate of growth is to be measured. But its results can easily be misinterpreted.

Combinations of framing cameras and streak cameras or streak cameras with a number of single shot cameras produce the maximum amount of information with the minimum possible errors and the minimum of equipment. One combined camera is commercially available giving up to twenty-three frames, 6 × 12 mm., at 8,000,000 frames per second and a streak record for 4·5 microseconds at 27·6 mm. per microsecond with a possible resolution time of 7·7 milli-microseconds.

Applications. Up to 20,000 frames/second, the subjects for high-speed cinematography cover a wide variety of fields in both industry and research including motors, relays, switches, textile machinery, internal combustion studies, rockets, biology and medicine.

At higher speeds, the uses are more likely to be in scientific research on explosives (both conventional and nuclear) plasma studies particularly for atomic energy fusion reactors, supersonic research for aircraft and missiles and studies of the fracture of materials, e.g., shattering of glass in car windscreens. G.H.L.

See also: *Flow photography; Image dissection; High speed photography.*

BOOKS. Proceedings of the International Congresses on High Speed Photography held in Washington (1952); Paris (1954); London (1956); Cologne (1958); Washington (1960); The Hague (1962); and Zurich (1965); *High Speed Photography*, by F. R. Saxe (London).

HIGH SPEED PHOTOGRAPHY

Exposure times of mechanical shutters in still cameras generally have a shortest limit of around 1/1000 second. Photography with shorter exposures is classed as high speed work, and nearly always requires special techniques.

To arrest motion in a picture, either the shortest exposure possible is used or the camera is moved in synchronism with the subject (permitting the background to blur).

A formula for the optimum exposure time is:

$$T = \frac{L}{500 \times V} \text{ seconds}$$

where L is the largest subject dimension to be recorded and V is the subject speed in the same units as L per second. For example a car moving at 68 m.p.h. (100 feet per second) would probably have a longest frame dimension, L, of 50 feet. Thus $T = \frac{50}{500 \times 100}$ or $T = 1/1000$

second—just on the limit for a normal shutter and no camera motion.

It should be noted that both V and L determine the exposure time—increasing V or decreasing L both demanding shorter exposures. Many research subjects are both faster and smaller than a car.

Even an air rifle dart with a similar speed (100 f.p.s.) but only an inch or so long, according to formula and practice; requires an exposure as short as 2 microseconds (0·000,002 second).

The two main methods of taking high speed still photographs are the use of a suitable high speed shutter system, and the use of a short duration flash while the shutter is open.

High speed shutter systems may be magneto-optical, electro-optical, or electronic units using image converter tubes.

High speed flash systems may utilize electronic flash discharges or sparks, or special discharges such as X-ray flashes for high speed radiography.

Magneto-optical Shutter. This utilizes the Faraday effect, i.e., the rotation of the plane of polarization of light passing through a transparent medium in a magnetic field.

To use the Faraday effect, a suitable medium in a magnetic coil is placed between crossed polarizers. Dense flint glass is generally used, since it shows considerable rotation of the plane of polarization for a given magnetic field and is convenient to handle. With no current in the coil, there is no magnetic field, no rotation of the plane of polarization occurs and so no light is transmitted. When a suitable current is applied (often 1,000 amperes needing 10,000 volts) the plane of polarization is rotated until it agrees with the second polarizer, and the maximum light is transmitted. This current can be supplied by discharging a capacitor through the coil using a spark gap as a switch. The time of the discharge depends on the capacitor size, the voltage, and the number of turns in the coil. Exposure times down to 1 microsecond have been achieved.

In practice, a cylinder of the glass is placed in front of the lens and co-axially with it, together with the crossed polarizers, while the coil surrounds the cylinder.

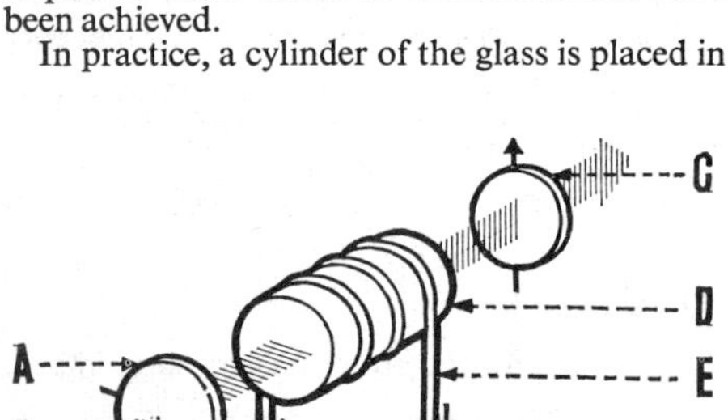

ELECTRO-MAGNETIC (FARADAY) SHUTTER. A, *first polarizer.* B, *light ray from subject.* C, *second polarizer.* D, *glass cylinder.* E, *coil. The second polarizer passes light only when the glass cylinder rotates the plane of polarization under the influence of the magnetic field. This is created by a high-voltage pulse passed through the coil.*

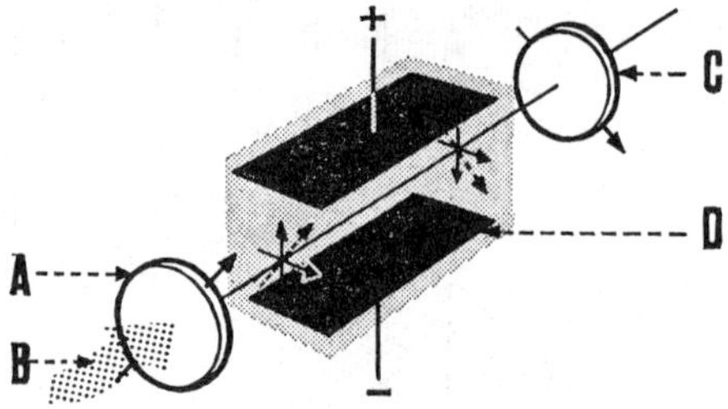

ELECTRO-OPTICAL SHUTTER (KERR CELL). A, *first polarizer.* B, *light beam from subject.* C, *second polarizer.* D, *electrodes in liquid cell. The liquid becomes birefringent on application of a high-voltage pulse to the electrodes.*

The capacitor discharge may be controlled by a spark gap, the spark being in turn actuated by the subject itself (i.e., the shock wave or the flash from an explosion under examination).

Electro-Optical Shutter (Kerr Cell Shutter). Kerr discovered that certain mediums in the presence of an electric field become bi-refringent, that is, light polarized in one plane has a different velocity in the medium to light polarized in a plane at right angles.

The usual Kerr cell shutter consists of a glass cell fitted with electrodes and filled with nitrobenzene, and placed between two polarizers. The whole assembly may be mounted in front of the camera lens or within the optical system. The first polarizer is set such that its polarizing plane is at 45° to the cell plates and the electric field. Plane-polarized light entering the cell becomes circularly polarized, i.e. has two equal components each at 45° to the original plane of polarization, 90° to each other.

If a suitable voltage is applied to the plates (commonly near 20,000 volts), the phase difference produced by the differing velocities is such that on recombining at the second polarizer, the resultant plane-polarized beam is in agreement with this polarizer. With no voltage applied, no phase change occurs and the resultant beam is polarized such that no light is transmitted. Thus by switching the voltage on and off, a shutter is produced and exposure times down to a few thousandths of a microsecond (1/200,000,000 second) are possible.

Image Tubes. A variety of image tubes exist including image converters, orthicons, image-orthicons and image-intensifiers. Some of these can act as high speed shutters. Metals, particularly caesium, have the property that, in a suitable electric field, they emit electrons when irradiated with light. Also certain materials exist which emit light when bombarded by electrons (e.g. a television screen).

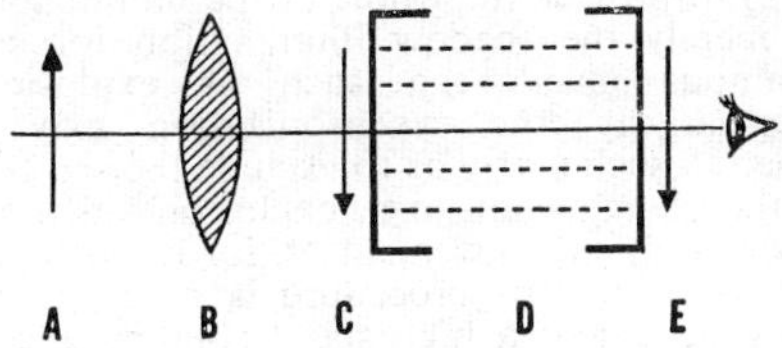

HOLST IMAGE CONVERTER. The main elements are: A, subject; B, lens; C, semi-transparent photo-cathode; D, electron stream; E, viewing screen observed by eye. As the image of the subject is formed on the photo-cathode the latter emits electrons. These travel through the evacuated tube and excite the fluorescent screen to produce an image there.

A combination of these into one evacuated tube can produce an image converter. The light receiver is called a photo-cathode, and the light emitter is the screen. An image is focused on the photo-cathode by a lens. By applying suitable electric and magnetic fields the image can be faithfully reproduced on the screen, which in turn can be photographed by a recording camera. With no voltage applied to the tube, no picture is produced. Thus by turning the electric field on and off, the tube acts as a shutter. The voltages employed are usually between 6,000 and 25,000 volts. In some tubes, other electrodes (grid electrodes) are inserted which can control the tube with lower voltages (e.g. about 300 volts). Additionally, by suitable electrodes or magnetic coils, the image on the screen can be deflected, permitting other techniques.

The orthicon or iconoscope uses as the light receiving element a mosaic of photo-emissive elements. These store an image which is later scanned off by an electron beam. The image-orthicon has a photo-cathode, similar to the image converter. The electrons from the cathode strike a storage plate which is again scanned later. An image-intensifier has a photo-cathode and a number of secondary-emissive surfaces, the electron beam passing from one to another in turn and being amplified before reaching

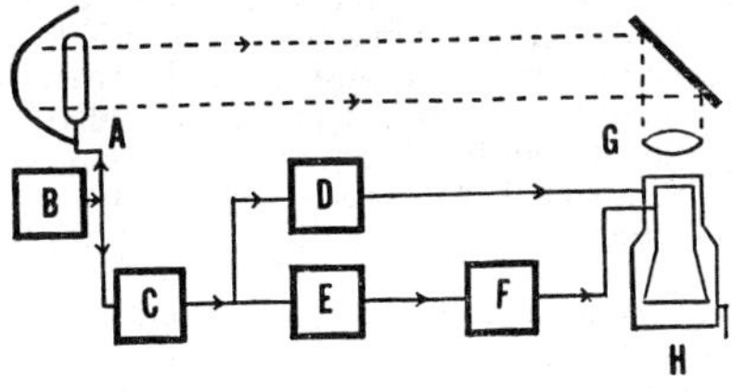

IMAGE CONVERTER SET-UP. This system is used for single photographs of non-luminous objects which are illuminated by a flash tube. The image converter shutter is synchronized with the peak of the flash. A, electronic flash tube powered in the usual way. B, trigger circuit. C and E, variable delay circuits D and F, thyratron switches. G, lens. H, image converter

either a screen or a storage plate. The gains possible in intensifiers, to date, are several thousandfold. An image converter can give some light intensification, e.g. at 25,000 volts, the light gain in a simple tube can be ten times.

Using image converters, exposures as short as 0·005 microseconds have been achieved, and it is in these devices, particularly, that further advances will be made. The physical limits are probably 1/10,000 of present results.

Flash Photography. The Electronic flash is now standard photographic equipment and is used not only when the available light is too weak but, with an open shutter, to arrest motion due to the shortness of the flash. Normal photographic electronic flashes have durations in the region of 100 to 1,000 microseconds, which is too long for most high-speed subjects. But with special tubes, capacitors and high voltages (7,000 volts and more), durations of a few microseconds or so are quite possible.

As the exposure time gets shorter, the brightness of the light source must increase. The threshold of fast films has been measured at very short exposures to be 10^{-3} metre-candle-second. This means, allowing for all the likely losses, that the subject illumination needs be about 1 metre-candle-second or 1,000,000 lumens per square metre for 1 microsecond, equivalent to a source of 1,000,000 candles at a distance of 1 metre.

The light sources capable of producing such high levels normally used in high speed photography include special electronic flash tubes, sparks and flash bombs. Sparks are produced by capacitor discharges across tungsten, steel, or copper electrodes. They have been made with durations as short as 7 millimicroseconds (7/1000 microseconds). In the microsecond region, an improved spark gap has argon gas blown into it via a hole drilled along the axis of one electrode giving a longer gap for the same voltage, more light output and a repeatable line source suitable particularly for schlieren photography.

A special type of spark is obtained by the explosion of thin metal wires. The wire is connected in series with a spark gap, wired in parallel with a bank of capacitors. These are charged to 15,000 to 60,000 volts. As the voltage rises, the spark gap breaks down, and the wire explosively vaporizes with a brilliant flash. Generally, however, explosions are more likely to be events studied by high speed photography than light sources for other subjects.

High speed radiography units have been designed for exposures as short as 0·03 microseconds. A high vacuum X-ray tube is used for the purpose and is fed by 500,000–2,000,000 volt pulses at several thousand amps.

Flash bombs are tubes containing argon with a transparent end-window and a small explosive charge at the opposite end. The explosive, when detonated, produces a luminous shock-wave in the gas which produces a high light level for periods from fractions of a microsecond up to many microseconds, depending on the geometrical structure.

The latest light source is the giant pulsed laser, giving extremely energetic flashes as short as 30 milli-microseconds, concentrated in a very narrow spectral band and in a very parallel beam. It allows photography of the surfaces of other highly luminous surfaces as the subject light can be ignored by a narrow-band filter set to accept the laser light only.

With these various devices, high speed still pictures can be taken, using normal cameras with open shutters, always assuming that the background light level is not excessive.

Stroboscopic Operation. Using a still camera and a flashing light source (stroboscope) a subject can be photographed in a series of positions in one picture. Alternatively, by synchronizing the flashes with a subject with a repetitive motion, a succession of exposures can be superimposed of the subject in the same position.

The first method, with the flashes at regular intervals, is useful in studying movement sequences (e.g. athletes, dancers, etc.). The second method serves to study mechanical motion when it is not possible to produce enough light to take a single short-duration exposure or when it is desirable to check that the subject is truly repetitive. Stroboscopic flash units are available at speeds up to many thousand flashes per second in some special cases, but more commonly operate at around 100 per second.

Lighting Technique. High speed photographs taken with electronic or optical shutters need a high level of illumination. Self-luminious subjects, such as flash discharges, explosions, etc., usually themselves provide sufficient illumination for high speed photographs. Non-luminous subjects, as well as those where the self-illumination is not to be recorded, must be lit up separately. There the light source may be either a constant-intensity lamp, or a flash unit synchronized with the shutter.

Suitable constant-intensity sources are high pressure mercury vapour lamps. They are available in units up to 1,000 watts, with an arc length of about ½ in. With an appropriate reflector, lighting levels up to 50,000 foot-candles can be achieved over an area 5 ins. in diameter, 8 ins. from the lamp.

Pulsed xenon arc lamps can reach light intensities ten times as high. The lamp usually burns at 50 watts, and is pulsed up to 10,000 watts for periods of 2–3 seconds.

In many cases the high instantaneous intensity of electronic flash or spark units makes them more suitable as high-speed light sources, even when an electronic or optical shutter is used. The light usually still lasts longer than the effective shutter opening, so that the latter controls the exposure time.

The light may be synchronized to the shutter so as to provide the whole of the illumination, or merely the shadow illumination for self-luminous phenomena such as explosions. Alternatively, the synchronization may be timed in such a way as to exclude the self-light of the phenomenon (e.g., explosion flash) and record only the movement of, for instance, the shock wave. This procedure is possible only with the aid of a high speed shutter. Such a shutter is also useful for eliminating other extraneous light. That may include the "tail" of the high speed flash discharge itself; even a 2-microsecond discharge takes about 15 or more microseconds to die away completely. Such a comparatively low intensity light may, without a Kerr cell or similar shutter, record on the film owing to its duration, and blur or obscure the main subject.

Synchronization. It is equally important not only to select the correct exposure time but also the correct exposure instant. Various techniques are possible and each new experiment may suggest or demand a new method.

When the event is self-luminous, a photoelectric trigger is most commonly used. It has the particular advantage that the trigger sensitivity level can be so adjusted that the exposure is not made until the subject is sufficiently bright to be recordable. For exposures at later instants, the same trigger may be used and electronic delay circuits inserted before a start signal is passed to the shutter.

Should the event and the camera both be capable of precise triggering, suitable signals can be generated from a common source thus allowing any possible time arrangement to be selected. An important factor is to know the response time of all components. This becomes increasingly important as time intervals get shorter, reaching eventually conditions in which it is essential to make due allowance for the transit time of light. In one nanosecond (0·000,000,001 second) light only travels approximately one foot.

When the event is non-luminous, some light source is necessary. Especially at the shorter exposures this is some form of electrical discharge started by an electrical signal. The event must then be made to activate some electrical circuit and in a sense be made to take its own picture. For example it can be arranged that a light beam to a photocell is interrupted, a projectile can break a wire or pierce a screen to give a signal. At lower speeds, relays can be used. Each subject usually suggests its own solution.

On supersonic projectiles or events initiated with a bang, an acoustic pick-up can be a very elegant device. For example, when photographing the action of guns and their projectiles, a pick-up to detect the initiating explosion or the projectile's shock wave can be used to start the light source and, when necessary, the shutter. Merely moving the pick-up along and below the line of flight can trigger exposures at different instants, by knowing the distance the pick-up was moved and comparing it with the change in the projectile's position in the pictures, the projectile's velocity can be readily given as a multiple of the speed of sound.

More elaborate arrangements are sometimes required, including, for example, coincidence circuits when two parts of an experiment are not precisely controllable. But normally relatively straightforward optical, acoustic or mechanical devices will suffice.

Applications. High speed still cameras have been used in almost every field of scientific and industrial research, permitting the study of subjects whose changes are far too rapid for the unaided human eye. It is true that it is in weapon research that most advances have been made but it is one of the bonuses of warlike activities that high speed photography has been developed to its, already, very high standard and that it is now available for more peaceful ones.

It has been a valuable tool in the studies of motors, complex mechanical processes, medicine and the peaceful uses of atomic energy.

Exposure Effects. Exposures of short duration at high light intensities are subject to reciprocity failure. The effective sensitivity and contrast of most materials reaches a minimum around exposures of 10 microseconds. Below that the reciprocity law is usually valid, and the sensitivity of blue-sensitive emulsions remains independent of the exposure time. With optically (dye-) sensitized emulsions a further loss of speed occurs around 1 microsecond exposures, and is believed to be due to the sensitizing dyes. Above 10 microseconds the speed and contrast increase, showing a maximum at 10,000 microseconds (1/100 second). This is generally the optimum exposure time from the point of view of speed for most materials. L.A.M. & G.H.L.

See also: *Ballistic photography; Chronophotography; Flash (electronic); High speed cinematography; Image amplifier; Image dissection; Spark photography; Stroboscopic flash.*

Books: *High Speed Photography*, by F. R. Saxe (London); *Progress in Photography* (Vol. II), ed. by D. A. Spencer (London); *The Photographic Study of Rapid Events*, by W. D. Chesterman (Oxford); *"Flash!" Seeing the Unseen by Ultra-high Speed Photography*, by H. E. Edgerton and J. R. Killian (Boston); Proceedings of the various International Congresses on High-Speed Photography held in Washington (1952); Paris (1954), London (1956), Cologne (1958), Washington (1960), The Hague (1962), Zurich (1965), etc., information also available from the British National Committee for High Speed Photography, through the Royal Photographic Society.

HIGH SPEED PROCESSING. Special formulae and techniques enable negatives and prints to be processed in an extremely short time. High speed processing is used principally: in photofinish race recording, when a finished print can be produced less than two minutes after the photograph was taken; in newspaper offices and picture agencies, for dealing with urgent topical news photographs; and in hospitals, where medical X-ray photographs are often processed at high speed to show the results of an operation. Instrument recording cameras may incorporate high speed processing features for immediate access to results.

See also: *Rapid access system; Rapid processing.*
Book: *Developing*, by C. I. Jacobson (London).

HILL CLOUD LENS. Camera lens designed by R. Hill for photographing cloud formations over the whole sky; the camera therefore has to cover an angle of 180° with the lens pointing vertically upwards. Any surrounding vertical objects such as trees appear on the plate laid out in a radial direction. Distortion is corrected by using the same camera as a projector, with the image on the developed plate projected on to a hemispherical screen. In the lens a moderately wide angle lens (covering about 60°) is placed behind the stop, and in front is a negative meniscus component, arranged such that an object field of 180° is compressed into the smaller (60°) field of the rear element.

See also: *Fish eye lens; Lens history.*

HILL, DAVID OCTAVIUS, 1802–70. Scottish painter and photographer. He was the first to demonstrate the artistic potentialities of photography. In 1843 he decided to paint the first General Assembly of the Free Church of Scotland.

This involved a total of some 470 portraits of delegates and it was to assist him in this task that he called in the aid of calotype. This was in collaboration with Robert Adamson, who was recommended to him by Sir David Brewster. The collaboration lasted until the illness of Adamson in 1847, and to this period belongs the best work Hill did.

It is generally conceded that some of his portraits comprise work which has never been excelled.

Biography by H. Schwarz (English translation: London and New York 1932).

HINE, LEWIS WICKES, 1874–1940. American sociologist who took up photography in 1905 to illustrate his work, documenting the miserable lives of poor European immigrants, and of Pittsburg iron and steel workers, 1908. As staff photographer to the National Child Labour Committee, Hine exposed in 1911, shocking conditions which contributed to the passing of the child labour law. During and after World War I he worked for the American Red Cross.

Hine also made a photographic record of the erection of the Empire State Building, in New York, 1931.

HIRE PURCHASE. Hire purchase trading in Britain is regulated by the Hire Purchase Act, 1938, the short Act of 1954 and regulations made from time to time by the Board of Trade under emergency powers. The Act now applies to all hire purchase transactions in photographic or other goods in which the hire purchase price does not exceed £300 (not £100, as prior to August, 1954.) Further, the Advertisements (Hire Purchase) Act, 1957 lays down certain minimum information which must be included in all advertisements to which this Act applies wherein goods are advertised as being available on hire purchase or credit sale.

Under a hire purchase agreement goods are hired to the customer who is granted an option to purchase, which is taken up when he completes the last payment. Until that last payment is made the goods belong to the dealer.

Terms. The extent to which hire purchase terms are available on photographic goods is increasing and a large number of dealers now offer them. The terms vary. The majority of photographic dealers use those recommended by the Photographic Dealers Association to its members, namely a deposit of 50 per cent of the hire purchase price with interest on the balance of 5 per cent for nine months, and 7½ per cent for longer than nine months. Legal minimum requirements laid down in June, 1962 are a 10 per cent deposit, and balance payable in not more than three years.

Very few photographic dealers will offer hire purchase terms on goods priced below £5 or will extend the payment period beyond two years. Instalment payments are usually made either monthly or weekly.

The financial status and stability of an applicant for hire purchase terms are of utmost importance to the dealer. It is usual to present the would-be purchaser with a proposal form for signature, and for details of his residence, occupation, etc., and banker's reference.

The Agreement. In the case of goods to which the Act applies, a hire purchase agreement must be entered into—over the signatures of the dealer (called the "Owner") and the customer (called the "Hirer").

Before the agreement is signed the Owner must declare in writing, to the prospective Hirer, the cash price of the goods. The agreement itself must contain the cash price, the hire purchase price, the amount of each instalment and when payable, and a list of the goods to which the agreement relates, sufficient to identify them. If the goods are second-hand, this should be stated so as to avoid the statutory implication that they are of "merchantable quality". Further, the agreement must contain a prominent notice in the terms prescribed by the Schedule of the Act, which sets out:

(1) The right of the Hirer to terminate the agreement at any time before the final instalment becomes due (subject to certain conditions of payment to the Owner).

(2) Restriction of the Owner's right to recover goods from the Hirer, without his consent, once the Hirer has paid one-third of the hire purchase price. If that has been paid the Owner may only recover the goods by obtaining an order in the County Court.

Finally, once the agreement has been signed, a copy must be sent to the Hirer within seven days.

(Agreement forms, printed and detailed, which, when filled in and signed, conform with all provisions of the Hire Purchase Acts, may be obtained ready printed.)

Method of Financing. The method of financing the hire purchase system of trading is important and must be considered carefully by the dealer. He may finance it himself or enlist the services of a finance house. If he finances the deal himself, he will require an amount of extra capital which may be considerable. The custom with some dealers is to finance the smaller transactions themselves and to hand the bigger deals to a finance house. The finance houses operate various methods, but dealers usually prefer an arrangement in which they sell the equipment to the finance house who then hires or sells it to the customer. Where this arrangement is used, the finance house, under its agreement with the dealer, usually has a right of recourse against him if the customer defaults. A.F.A. (A.J.W.)

American Practice. Hire purchase is much more widespread in the U.S.A. (there known as time payments). Terms are not so closely con-controlled by law, and vary throughout the country. General terms, subject to a thorough credit investigation of the purchaser, are: 10 to 20 per cent deposit, with the balance payable over 3 months to 1 year. The interest charge is about 6 per cent on the unpaid balance.

Many dealers do not finance the sale themselves; once their bank or financing agency has approved the sale, they receive the full purchase price from that bank or financing agency. The latter then collects the monthly payments from the purchaser.

HISTORIOGRAPHY OF PHOTOGRAPHY. The first task of any history of photography is to relate technical progress. Beyond that it has to take note of the men who use the techniques and the work they accomplish. The history of photography is also a history of photographers and photographs. Side by side with the technical record unfold the stories of photography as an industry and as an art form.

Technical progress in photography stems from three sources: chemistry, optics and engineering. Whenever developments from one angle overtake the others a new challenge ensues for advance all round. This pattern of technical interactions is complex enough to discourage professional historians. Histories of photography are, as a rule, compiled by students of photography who, as historians, are mostly amateurs.

As photography is not much more than a century old a fair amount of source material is accessible, some of it still unpublished or just waiting to be discovered. Old photographs and historic equipment offer self-documenting evidence and confront the student with *embarras de richesse*. There are but few safe short cuts of organized basic information; what we have of contemporary or near-contemporary writing befogs rather than illuminates the origins and decisive turns of the art. The historian, being forced to accomplish first a great deal of original field research, is in danger of getting prematurely exhausted and myopic. Unskilled or limited research is easily satisfied with its own results and proceeds to vest its guiding interests in them; dates and names that are but pegs on which to hang a tale will then be mistaken for the whole story.

Thus histories of photography are sometimes bogged down at elementary levels; the assembling of data and cataloguing them according to controversial priorities. Preoccupation with details is more frequent than constructive integration of available knowledge, or imaginative interpretation of trends.

In particular, insufficient attention has so far been paid to cause-and-effect relationships between technical changes and modes of picture-making. The rejuvenating impact of advancing photographic tools—which steadily broaden the photographer's view of his subject matter and periodically vary his methods of pictorial articulation—awaits much study. Short of it, any grand synthesis of the history of photography is difficult to conceive.

Component histories of lenses, sensitized materials, cameras, etc., are more easily assembled although still not always simple to follow. It is a weakness of specialist historians to delight in expounding subtle distinctions or to overstress freakish phenomena—neither of which is, as a rule, relevant to normal developments. Yet this is not over-important.

The historiography of photography is badly in need of more scholarly monographs, dispassionate biographies and thoroughly researched local and national histories. The immediate task of new historians lies in this sphere of working to limited objectives and presenting balanced views of specific areas rather than in staking comprehensive claims of coverage based on incomplete or uneven material. A.K.-K.

See also: *Artificial light history; Camera history; Chronology of photographic inventions; Cine history; Colour photography history; Development history; Discovery of photography; Lens history; Literature on photography; Museums and collections; Sensitized materials history.*

Books: *A Source Book of Photographic History*, by W. Baier (London); *History of Photography*, by H. and A. Gernsheim (Oxford): *The History of Photography*, by B. Newall (New York); *The History of the Discovery of Photography*, by G. Potonniée (New York): *The March of Photography*, by E. Stenger (London).

HOEGH, EMIL VON, 1865–1915. German lens designer. Calculated in 1892 one of the most important modern lens systems, the symmetrical double anastigmat, manufactured by C. P. Goerz, Berlin, and later named Dagor.

HOLD-ALL. Leather, fabric or plastic carrying bag—usually fitted with a shoulder strap—designed to take the whole of a photographer's equipment of camera, exposure meter, interchangeable lenses, filters, lens hoods, spare films, etc.

See also: *Cases.*

HOLDING THE CAMERA. Blur caused by camera shake is one of the commonest faults in negatives exposed with the camera held in the hand. This is understandable because most amateur snapshots are taken with shutter speeds of 1/25 and 1/50 second. Most people think that they can hold a camera steady for such short intervals. But actual tests show that shake is common at all shutter speeds up to 1/150 second. Even so, it is fairly easy to learn to hold a camera steady at 1/25 second.

Gripping the Instrument. The exact way of getting hold of the camera body will depend on the type of the camera used, on its size relative to your hands, and on personal preference. The grip to aim at in every case is one which permits pressing of the release—and operation of other essential controls, e.g., rangefinder focusing knob where appropriate—without detracting from the support which the hands should give the camera. This is often a matter of trial and error; try different holds and see how comfortably you can release the shutter while still holding the camera still.

Cameras with eye-level finders are best pressed against cheek and forehead for steadiness, with the hands gripping each end of the body. With a folding model, the fingers of one hand may, if there is room, curl round one side of the baseboard, or else be tucked behind it. In either case, keep all fingers clear of the lens and rangefinder windows (if any). If the camera is in an ever-ready case, use the third and little fingers of one hand to keep the lid or flap of the case out of the way of the lens when necessary.

A camera strap (or the strap of the ever-ready case) can increase steadiness, too. For that purpose, shorten the strap until the camera lies well up on the chest with the strap hanging round the neck. Then bring the camera up to the eye, and hold it so that one or both hands are inside the loop of the strap. A certain amount of outward pressure will then tension the strap and keep the camera firmly pulled against the face.

The normal hold for a camera with waist-level finder—box cameras as well as reflex models—is to press it against the chest keeping both arms and elbows close to the body. If the camera or case has a neck strap, it is again

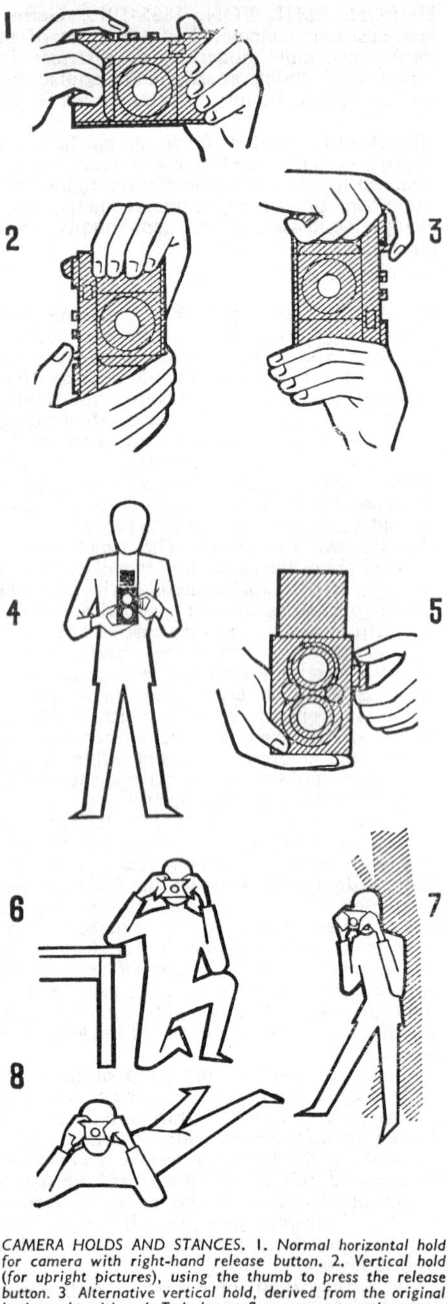

CAMERA HOLDS AND STANCES. *1. Normal horizontal hold for camera with right-hand release button. 2. Vertical hold (for upright pictures), using the thumb to press the release button. 3 Alternative vertical hold, derived from the original horizontal position. 4. Twin-lens reflex camera on a short strap round the neck. 5. The camera should still be steadied by the hands. 6. A table will serve as a support for steadying the hands while holding the camera at slow shutter speeds. 7. Leaning against a tree for support. 8. Floor-level camera position.*

useful to shorten this so as to bring the camera up as high as possible for comfortable viewing. Pulling against the strap will aid in holding the camera steady.

Finally, release the shutter smoothly. To do this you must practise until your finger can take up the first idle motion of the release quite quickly, and then squeeze steadily until the shutter trips. Do not try to release the shutter with one sharp movement or you will amost certainly jerk the camera. Aim at maintaining a uniform pressure throughout, without worrying about when the actual release will occur.

Supports. When making a fairly lengthy exposure, look around for something to brace your body against or rest your elbows upon. By using a tree, the side of a building, a lamp post, the top of a car or a gate, you can greatly reduce the risk of camera shake from the start. It is even better if you can hold the camera itself against something solid like a tree or lamp post.

If you are forced to work without some such support, go down on one knee and brace your elbow against the other, like a marksman. If this is impossible, or if it results in an unsatisfactory viewpoint, then at least stand with your feet well apart and your heels flat on the ground.

Breathing. Some people find it easier to breathe out and then release the shutter, others breathe in and release the shutter while they hold their breath. Both of these methods lead to trouble if you have to wait before you make the exposure. The best way is simply to breathe naturally throughout the release.

Finally, waste no time over the operation. Do not prepare to release the shutter until you are quite ready to shoot, then proceed without hesitating. If you spend needless seconds at the ready you will stiffen up, start worrying about camera shake, and inevitably produce it.

See also: *Camera shake; Camera support*

HOLLAND. Photography was originally introduced into Holland by itinerant exponents of the daguerreotype, and there are several good specimens of their work in private and official collections. Since that time the growth of photography has kept pace with that in other European countries in the scientific, professional and amateur fields. The popularity of photography in Holland owes a great deal to the use of the camera to illustrate the country's vast amount of public and private contact with overseas.

Science and Industry. Dutch scientists have made a number of important contributions to photographic progress—principally in the optical field. The invention of the microscope by Zacharias Janse of Middelburg in 1590 and of the Dutch telescope by J. Lippershey of the same town in 1608 can both be regarded as essential preliminaries to the optics of photography. Snellius, a professor of Leyden, discovered the laws of refraction in 1625 or earlier, and in 1678 the famous Dutch scientist, Huygens, originated the wave theory of light.

In recent years Prof. Dr. A. Bouwers has made further contributions in mirror optics and wide screen cinema projection, while the popular Albada viewfinders are the invention of the Dutch optician General L. E. W. van Albada, who also made remarkable discoveries in wide-angle stereophotography.

The Universitaire Film (Academical Film Centre) at the University of Utrecht has an international reputation in the field of medical cinematography.

In industry Holland has one manufacturer of sensitized materials—roll films, printing papers and X-ray films—and photographic chemicals. In addition there is a world famous organization in Eindhoven which exports flash bulbs, studio lamps of all kinds, and cinematograph projectors for both 16 and 35 mm. films. This great industrial concern has also been closely concerned in the development of the electron microscope and its application to scientific and industrial problems. The best-known Dutch optical firm at Delft is famous for its optical mirror systems and has produced a new wide-screen method for cinematography without chromatic aberrations—of importance with colour films. They also make equipment for mass miniature radiography for export.

Holland has several small factories making screens, and various small pieces of equipment. At Utrecht there is a famous factory making microscopes, and at Venlo glass is produced. Several manufacturers of colour films have their own processing laboratories in the country.

Holland is a very photo- and film-minded country, and imports of equipment and materials are considerable. By far the highest total percentage of all imports at present are from Belgium and West Germany. The U.S.A. leads in imports of 8 and 16 mm. colour cine-films.

Organizations. There is the Nederlandse Vereniging voor Wetenschappelijke Fotografie (Netherlands Society for Scientific Photography) which is associated with the International Scientific Film Association, and the Nederlandse Vereniging van Fabrikanten en Importeurs van Fotografische Producten (Netherlands Society of Producers and Importers of Photographic Products). The retail trade has its own Bond Nederlandse Fotohandelaren (Society of Photographic Dealers) and the professional photographers are organized under the Nederlandse Fotografen Patroonsvereniging (Society of Master Photographers). The Nederlandse Fotovakschool (Netherlands School for Professional Photography) is the only educational institution authorized by the Ministry of Education to hold examinations and award qualifications in technical and commercial photographic subjects. Press photographers also have their own professional organization.

The various trade associations organize joint propaganda drives, using instructive cinema films—e.g., to promote the sales of cameras and flash equipment.

The Dutch Society of Press Photographers holds an annual World Press Contest at the Hague. Entries of press photographs from all over the world are exhibited on this occasion, and high cash prizes awarded to the winners. The first prize is £1,000 and a free trip to Holland for the occasion.

In the University of Leyden, part of the Rijks Prenten Cabinet (Government Print Collection) consists of the famous "Gregoire" collection of daguerreotypes and historical photographs.

The amateur movement in both still and cine photography is very strong. There are 110 camera clubs in the Bond van Nederlandse Amateur Fotografen Verenigingen (Federation of Dutch Amateur clubs), while sixty-six amateur cine clubs are united in the Nederlandse Organisatie van Amateur Film-clubs, or N.O.V.A. (Netherlands Organization for Amateur Cine Clubs).

The Nederlandse Amateur Fotografen Vereniging (Netherlands Amateur Photographic Society) of Amsterdam, founded in 1887, is the oldest photographic society on the continent of Europe.

Dutch photographers are divided in their approach between the romantic and the modern as sponsored by such organizations as the Netherlands Fotografen Kunstkring or N.F.K. (Netherlands Society of Photographic Artists) and the Gebonden Kunsten Federatie or G.K.F. (Federation of Modern Craftsmen).

Publications. Holland produces a fair amount of photographic literature. Periodicals include the magazine *Focus*, established in 1914, and the younger *Foto*. Amateur cinematography is catered for by *Smalfilm*, while *De Fotohandel* and *Foto Mercuur* are directed at the photo-dealers. A large number of periodicals and books in English, German and French about photography are imported, since a very high proportion of Dutch photographers are able to read these languages. Books and articles by Dutch writers are also regularly translated into other languages. D.B.

HOLLYER, FREDERICK, 1837–1933. English portrait photographer and specialist in reproductions of paintings. Best known for his platinotype portraits of artists and authors taken in their own surroundings 1890–1914.

HOLMES, OLIVER WENDELL, 1809–94. American writer and amateur photographer. Collector of stereographs and designer of an open type stereoscope (1861).

HOME-MADE EQUIPMENT. Every photographer makes use of home-made accessories at some time or other, even if it is no more than a card clipped to the side of a lamp reflector to deflect part of the light.

TOOLS AND MATERIALS FOR MAKING EQUIPMENT

Article	*Used for*	*Type or Size*	*Remarks*
Abrasives	Wood	Sand or glass paper	Available in grades of coarseness; use a cork sanding block for flat work
	Metal	Emery paper or cloth	Keep old and used pieces for final polishing. Available in various grades
Adhesives	Paper, cloth	Rubber gum	—
	Wood	Cold glue (various makes)	Follow simple directions on tin or tube, glue under pressure whenever possible
Bending bars	Bending sheet and strip metal	2 bars, $9 \times \frac{1}{2} \times \frac{1}{4}$ in.	Make up from mild steel bar; fit a $\frac{1}{4}$ in. bolt and nut at each extreme end
Chisels	Wood	$\frac{1}{8}$, $\frac{3}{8}$, $\frac{3}{4}$ in. suitable	Preserve original edge angle when sharpening. Use mallet and not hammer to drive
	Metal	$\frac{1}{4}$ and $\frac{1}{2}$ in. suitable	Cut along vice jaws or bending bars with a shearing action
Dies	Cutting screw thread on rods, bolts, etc.	$\frac{13}{16}$ in. diam. dies cut threads up to $\frac{1}{4}$ in.	Taper off end of work to be threaded. Cut with $\frac{1}{2}$ turn forward and $\frac{1}{4}$ turn back
Files	Metal	6–8 in. medium double cut	Use with slow, steady strokes; relieve pressure on return stroke
Glass cutters	Cutting sheet glass	Diamond	Practice gives the "feel". Steady, confident strokes give cleanest cut. Support glass to be cut adequately along the whole length
		Wheel	Adequate for amateur requirements
Punches	Centring for drills	Centre punch	Place point at acute angle to locate spot, then tip to vertical to strike
	Setting pins and nails	Flat punch, nail punch	Hit squarely
Saws	Wood	Hand saws	Not recommended for small work
		Tenon	The best for small work
		Coping	For cutting round curves. Do not "force" cut
	Metal	Hack saws	Fine tooth blade for sheet and tube work (24 points per in.) Coarser (18) for heavier bar material
Soldering	Brass, copper, tin	Tinman's solder in sticks	Spirit or paste flux suitable. Tin pieces before "sweating" together over flame
	Electrical work	Resin-cored solder only	Hot iron, minimum of solder on joint
Taps	Threading holes	Taper, second or plug available in all sizes	Procedure as for dies. In blind holes, start with taper and "bottom" with plug tap
Vices	Wood	Various sizes	Steel frame type with renewable wood jaws preferred
	Metal		Make a pair of soft "clams" or false jaws, to hold finished work. Do not rivet on the jaw edges; the jaws are hardened and will chip

One of the main reasons for making a piece of equipment is that it will then exactly fill the particular need. Another equally important reason may be the cost of the manufactured article. This is especially so with elaborate and expensive items like enlargers. Here, the time spent on assembling, testing and adjusting is not something which the home builder has to charge against the cost. The manufacturer cannot help but pass this expense on to the purchaser.

Dependent on the equipment and ability of the builder, the home-produced article may or may not be as presentable or efficient as the factory-made tool. The finish is merely a matter of taste on the part of the builder. Relative efficiency is a thing which the user will have to consider when deciding whether to build or purchase. Many items of equipment are available in different standards of accuracy, and with the normal facilities of the home builder he cannot hope to produce an article comparable to the best commercial product.

This is the case with colour filters, to take one example. Many manufacturers offer a range of filters made from optical glass dyed in the mass, and worked to the same precision limits as apply to lenses. They are therefore expensive, and are used where the utmost scientific accuracy is necessary. However, where the requirements are not so exacting, a less accurate and much cheaper type may be perfectly satisfactory. Such an alternative range is often offered by the same manufacturer.

It is here that the home-made article begins to be a worthwhile proposition. Correctly coloured filter gelatin is available very cheaply, and it makes the production of a set of cemented glass filters in a range of colours a practical project. Gelatin filters can be made exactly to the user's requirements for size, shape and colour with next to no equipment. Even coloured plastic sheet has been used successfully.

Bellows. New or replacement bellows for enlargers, etc., can be made up quite well by most handymen. Some patience is required here in marking out, although this only entails simple geometry.

Materials required for first-class job are:

(1) Thin black leather "skiver", which can be obtained from any leather shop. This is not very expensive and is more supple than most other materials that have been used.

(2) Some very thin material—such as that from which a good quality pocket handkerchief is made.

(3) For stiffening purposes ordinary drawing paper between the leather and material will help to keep the folds in the finished bellows.

(4) Rubber solution as an adhesive.

Quite good bellows can be made up from ordinary stout brown wrapping paper for those who cannot undertake the more elaborate job.

Unorthodox Materials. Home-made colour filters are only one example of how the home builder can profitably consider substitute materials. While camera bodies today are largely die-cast in light alloys, not long ago they were made of wood and wood is still an excellent stand-by for the photographic handyman. There is no reason to be shy of trying unorthodox or unconventional materials. In these days of plastics, resin-bonded plywood and super-strong cold-setting glues, he has available a host of easily-worked materials denied to the earlier worker in ordinary wood.

To take another example, commercially produced lens hoods are generally metal spinnings but an equally strong and in some ways superior hood is easily made from thin plywood. Most of the principal metal components of an enlarger can be made of wood, and have been so constructed in hundreds of successful cases.

Many commercial components—e.g., enlarger parts, tripod heads, rangefinder bodies and the like—are castings, often in light alloys, but it is nearly always possible for the metal worker to fabricate similar items from sheet and rod. Brass is generally a suitable material, as it is easily soldered, but, if there are no soldiered joints, aluminium is an easier metal to work.

Given only an elementary equipment of hacksaws and files, quite complicated items can be built up without any professional skill.

Apart from soldering (which is not everyones' forte) quite a large proportion of home-made metal equipment can be assembled with small brass screws and nuts. These are obtainable from any electrical or ironmonger's shop—sized from No. 1 B.A. to No. 6 B.A. would be found very useful. Many uses can also be found for aluminium or copper rivets.

Sheet metal work is something else which the home constructor needs no elaborate plant to tackle successfully. While the manufacturer uses bending and folding machines to make his shapes, the home worker can produce a satisfactory job by bending his sheet between two bars held in a vice and cutting out his shapes with hand snips. Enlarger bodies, dishes, reflectors, lenshoods and darkroom lamps are just a few of the sheet metal items which are readily produced by "one-off" hand methods.

Hand-shaping methods can be applied to many other simple things—e.g., plastic print tongs and paddles, wooden masking frames, drying racks and metal dishes.

Adapting Scrap. Many home-made articles can be produced from scrap or discarded material. Many examples will spring to mind of scrap articles and materials which need little work to adapt them for photographic purposes: roller squeegees from rubber hose and a length of wooden broom handle; enlarger lamphouses from paint or other cans; reflectors from aluminium pudding basins; safelights from coloured glass jars—the list is endless. And there is no limit to the ingenuity which can be brought to bear in recognizing a photographic use for improbable material.

The great attraction in this pursuit lies in the fact that the final product often costs literally nothing. In some cases the appearance of the finished article may betray the original purpose of the material used, but if the user is satisfied with its performance that is all that matters.

Ideal Facilities. At the other end of the scale comes the amateur who can call on the resources of an enthusiast's workshop, containing machine tools and precision equipment. Many an amateur engineer and craftsman has built up such an outfit. Access to or the possession of a lathe opens the door to precision machining comparable with the best from any factory. The amateur's methods will be different, and his production will take longer, but his results will be in no way inferior to the factory-made article. Indeed, they may well be better and more adapted by individual needs.

The importance to photography of the home constructor has been well recognized in the number of books and magazines giving designs, instructions and hints on building. These provide valuable information and short cuts for making serviceable equipment with little or no skill or training. And there is nothing to be ashamed of in home-made equipment; after all, the prototype of every successful new device was once a "home-made" project.

Tools. While many articles can be made with the simplest tools such as are found in every household, the scope of the amateur can be greatly extended by adding a few others as included in the list above. Generally, the cost will be saved on the first few items made with their assistance. L.C.M.

See also: *Adhesive; Bellows; Blacking.*

Books: *Hints, Tips and Gadgets for Amateur Photographers* (London); *Making an Enlarger*, by H. van Wadenoyen and J. Holtam (London).

HOME MOVIES. Motion-picture equivalent of the snapshot album, made possible by the introduction of a ciné camera film which is developed to a direct positive by the reversal system. The film used in the camera can be projected after processing, thus effecting a considerable economy in operating costs.

See also: *Cinematography*.

HOMOLKA, BENNO, 1860–1925. Austrian chemist. Worked in Höchster Farbwerke, near Frankfurt, specializing in dyestuffs and photographic chemistry. In 1904 discovered pinacyanol, the red sensitizer, and pinacryptol green, the desensitizer. Also invented a dye-print process.

HOOKE, ROBERT, 1635–1703. English physcist and experimenter; assistant of Robert Boyle. Reported on a projection system in a light room 1668. Designed a conical portable camera obscura, 1680. Hooke made many contributions to microscopy.

HORGAN, STEPHEN H., 1854–1941. American photographer and writer on the photomechanical process. One of the pioneers of the use of half-tone blocks for newspaper printing, the first being published in the New York *Daily Graphic* in 1880.

HORIZON. Line where the earth and sky appear to meet. The position of the horizon in a landscape determines where the interest of the picture lies—if it is nearer the top of the picture than the bottom, the interest is in the landscape, and vice versa. That is why it should never run across the middle of the photograph, because that automatically splits the picture into two equal parts which compete for the observer's attention.

In a strong landscape such as occurs in hilly country, the horizon will be in the upper half, or, indeed, it may with advantage be eliminated by trimming away the sky. But in fen country, for example, or by the sea, a low horizon will carry a sky which dominates the scene with its cloud formation.

There are three different ways of adjusting the position of the horizon: by tilting the camera up or down; by keeping the camera level and raising or lowering the front; and by keeping the camera level and later leaving out part of the top or bottom of the negative during printing or enlarging.

Tilting the camera up or down introduces some distortion of the perspective of the picture. This passes unnoticed in a picture made up of rounded and irregular-shaped masses. But if there are any strong vertical lines—e.g., straight, tall trees, or the sides of high buildings—the tilt may be unpleasantly obvious. It is also impossible to join up the separate prints of a panorama series made with a tilted camera.

The ideal way of controlling the interest and adjusting the level of the horizon in the picture is to use a camera with a rising and falling front. This method does not distort any vertical lines in the picture and shows a natural perspective of trees and buildings.

If a landscape is photographed with a camera without a rising front, verticals will not be distorted so long as the camera is kept absolutely level. In this case the horizon will run across the centre of the negative, but there is no reason for it to run across the centre of the print. The top or bottom of a contact print can easily be trimmed away to give the picture the required balance, and in an enlargement the photographer is free to please himself about the position of the horizon. This is the only method open for making panoramas with a camera that has no rising front.

The camera viewpoint will determine how much is included in the foreground and how the foreground objects are laid out in the picture. Thus a low viewpoint will include only very near objects on a large scale, and these will hide anything farther away. If these subjects are taller than the height of the camera, they will also obscure the sky line and often exclude the sky altogether.

Conversely, a high viewpoint will not include any particularly near details, and provides an over-all view over medium distance objects, with the sky line formed by the limit of the field of vision. The higher the viewpoint, the farther this limit, and the less will appear of closer subjects.

The straight line of the horizon should generally be interrupted before it meets the edge of the picture. This is merely a matter of choosing a viewpoint that makes use of a suitable tree, or similar natural object. F.P.

See also: *Landscapes.*

HORIZONTAL ENLARGER. Earliest type of enlarger; an obvious extension of the old magic lantern, substituting a negative for the slide, and a piece of sensitized paper for the projection screen. This type of enlarger occupies more space than the modern vertical enlarger, but it is still widely used in making giant or mural enlargements.

HORSES. Successful photography of horses requires patience and a knowledge of horses. It is essential to understand how to show the points of different breeds correctly, both when stationary or in motion. It is useless to photograph a Thoroughbred in the same way as, say, a Shire horse or a Hackney.

Types. A light horse such as a Thoroughbred, Hunter or Hack must be photographed with the camera at a right angle to the subject to show its lines and lightness to full advantage.

The photograph may be taken from either side, although the near (left) side is the more conventional, but the horse must be posed with the foreleg nearest the camera slightly in front of the far foreleg, and the nearer hind leg a few inches behind the far hind leg. The tail must be shown in the normal position for the breed; in the case of an Arab it is carried almost like a plume. The head should be turned very slightly towards the camera with both ears pricked forwards and clearly visible. The neck should be flexed to show the crest (top of the neck)

HORSES FOR SHOW. Left: *Camera position (represented by shaded square) for photographing a race-horse to the best advantage.* Right: *Corresponding ideal position for a heavy horse, to show up strength and solid build.*

to advantage and there should be no suggestion that the horse pokes its nose forwards. The mane should be on the off (right) side of the neck and if shown should be plaited and not blowing about in the wind.

A heavy horse, such as a Shire or Clydesdale, may be photographed from a slight angle to show the powerful hind quarters to advantage. In this case the legs should be level, to convey the impression of strength and solidity.

The neck must be well bent to show its powerful line and high crest. There again the head should be turned slightly towards the camera so that intelligence in the eye and breadth of forehead may be shown.

A Hackney (trotting horse) should be photographed with the hind legs well back, so that the horse "covers a lot of ground" and conveys the impression of latent power.

When photographing groups of moorland or New Forest ponies, which cannot be posed, it pays to wait until two or three of the animals nearest the camera are in typical positions.

When a mare and foal are being photographed together the foal should stand very close to its dam on the side nearer the camera, with its face no farther forward than the mare's shoulder.

Equipment. The apparatus most favoured by animal photographers is of two kinds, either a reflex camera or a press type with a lens of slightly longer focal length than standard. A press camera should have a large direct vision viewfinder in which the subject can be watched easily for any sudden movement, such as the flick of an ear or a swish of the tail, either of which could spoil the result. A quiet shutter is important, because any sudden noise, however slight, is liable to cause a horse to turn its head or move an ear. The shutter must be capable of fast speeds in order to catch movement in jumping and other action shots.

Fast panchromatic material is used to give good colour rendering and to permit the use of fast shutter speeds. Filters are useful when photographing a bay horse with black points or when taking groups of ponies ranging in colour from chestnut to dark brown. A medium yellow filter is used when taking a grey horse, either jumping or whenever it is seen against the sky.

Lighting. The most suitable lighting for all types is given by rather hazy sunshine at about 45° to the subject, as such lighting will show the coat to advantage and the shape of the animal, without dark patches of shadow.

Horses Standing Still. Great care is necessary in choosing a suitable location to photograph a horse because it is most important that the animal should be standing on level ground and the background be as plain as possible. Precautions are taken to see that fences, stable doors or the branches of trees do not link up with the outline of the horse in any way. If the horse is to be shown unmounted, it should have either a headcollar or a bridle, with the reins resting on the withers. It should be held with a leading rein by a groom standing in front of the horse and well away from it. Where a horse and rider are shown, no one else is necessary to hold the horse, as the rider will control it and position it correctly. The camera is held so that it is at about the height of the horse's shoulder and must on no account be tilted up or the subject will be in very bad perspective.

It is usual for the animal to be in its summer coat and well groomed to bring out the sheen, which is a sign of good condition; the hooves should be oiled.

HORSES IN ACTION. Top: *The best time to photograph a jumping horse is at the take-off, or half-way across the hurdle. For the latter the jump must be clean; the take-off is nearly always reasonably safe.* Bottom: *A horse landing does not look elegant unless his back feet are well up, nor when he has touched down and is regaining his stride.*

Horses in Motion. Jumping pictures, whether in the show ring or on the racecourse, are either taken with the camera placed very low down at an angle of 45° to the landing side of the jump, or from the side of the jump with the camera a little higher for a solid jump, such as a wall. The exposure must be made in the split second when the horse is over the middle of the jump.

Hackneys, when trotting, are photographed at right angles to the camera to show all four legs clearly and with the leading foreleg at the top of its action, as this "high stepping" appearance is the animal's most important characteristic.

Groups of riders—e.g., at a hunt—are taken from a three-quarter front view in order to show the faces of both the riders and their mounts.

Negatives should be very fully exposed to show all possible detail, and development time curtailed to produce a well graduated negative, full of detail. Over-development leads to a loss of quality in the rendering of the coat.

As this is a technical subject there must be no retouching or faking, the object being to portray the subject accurately but to the best advantage. R.R.

See also: *Movement; Sport.*

HORSLEY-HINTON, ALFRED, 1863–1908. English amateur landscape photographer, follower of P. H. Emerson's Naturalistic movement, and one of the leading spirits in the Linked Ring Brotherhood, founded by him in 1892. Horsley-Hinton was an influential authority on artistic photography and editor of *The Amateur Photographer* from 1893 until his death. He wrote a number of text-books, of which *Artistic Landscape Photography* is the most important.

HOT WATER BATH DEVELOPMENT. By using a bath of hot water it is possible to reduce the contrast of a print by as much as one paper grade.

In principle, the process is similar to water bath development of negatives.

After exposure in the usual way, the print is immersed in a standard developer until the image begins to appear. It is then transferred immediately to a dish of hot water at about 130–140°F. (55–60°C.), and allowed to develop out. If the image does not build up sufficient density, the print is returned to the developer for half a minute or so, and then returned to the hot water bath. This process may be repeated until the print is fully developed.

The decrease in contrast obtained depends on how much development the print received before transfer to the hot water. The sooner it is transferred, and the warmer the water, the softer the result will be. The temperature of the developer itself should be kept below about 60°F. (16°C.), so that the emulsion has a chance to become evenly charged with developer solution before any appreciable development takes place. If the emulsion is not uniformly saturated, the image may develop up unevenly in the water bath.

The water bath soon goes cold and becomes contaminated with developer, so it should be renewed every few minutes.

This method is not suitable for unhardened paper emulsions—e.g., as used for carbro and bromoil.

See also: *Contrast control; Water bath development.*

HOT WEATHER PROCESSING. Photography in the tropics demands a special technique of its own. But normal hot weather conditions (as experienced in fairly temperate climates) present no great difficulties except those concerning processing.

Normal processing temperatures range from about 60 to 75° F. (15–24° C.). If the processing solutions are warmer than 75° F. (24° C.) special precautions are then necessary. At these temperatures the gelatin emulsion takes up a lot of water and swells considerably. It becomes soft, highly sensitive to mechanical injury, and it may even melt and run off the support.

So in hot weather it is important to prevent the gelatin from swelling unduly, and to harden the emulsion layer to reduce the risk of damage from careless handling.

A change of one degree above, say, 75° F. has much more effect than a change of the same amount below 65° F., so accurate temperature control is even more important than usual—not only to keep the temperature of each bath constant, but to see that there is no variation between baths.

Swelling of Emulsion. To prevent the gelatin from swelling, the time that the film or plate stays in the various solutions—particularly before hardening—should be kept as short as possible. Low energy developers need very long development times and should be avoided, and a processing technique should be adopted that does not include any form of pre-soaking. The fixer solutions should be fresh to reduce fixing times, and washing should not exceed 30 minutes. At temperatures between 75° and 95° F. (24–35° C.) the washing process takes much less time, anyway.

Apart from these general precautions certain changes must be made in the developer.

All developers containing a high concentration of alkalis, such as sodium carbonate or hydroxide, soften the gelatin layer to some extent and are unsuitable for hot weather development. The most useful type of developer follows the metol-hydroquinone-borax formula.

Further, a neutral salt, such as sodium sulphate, reduces the absorption of water by the gelatin. So, although the salt generally increases the development time, it is customary to add about 4 ounces (100 grams) of sodium sulphate crystals (Glauber's salt) to every 40 ounces (1,000 c.cm.) of working developer.

With a metol-hydroquinone-borax formula the development times are modified thus:

HOT WEATHER TIMES

Temperature of Developer °F.	°C.	*Time of Development compared with time without Sulphate*
75	24·0	As for 65° F. (18·5° C.)
80	26·5	As for 70° F. (21·0° C.)
85	29·5	Half time for 65° F. (18·5° C.)
90	32·5	Half time for 70° F. (21·0° C.)

Hardening the Emulsion. One—or both—of the following expedients will serve to harden the gelatin: an acid hardening fixer may be followed (after a brief rinse in an acid stop bath) by a special hardener. A suitable formula for a chrome alum hardener is:

Chrome alum	1¼ ounces	32 grams
Sodium sulphate, crystals	5 ounces	125 grams
Water to make	40 ounces	1,000 c.cm.

This hardener is effective as long as the solution looks bluish purple.

If the weather is very hot, with solution temperatures above 95° F. (35° C.) a formalin hardener may be used before development. This will tan the gelatin and prevent it from softening. After a good rinse the material is then developed in any tropical developer.

Technique. Apart from controlling the nature and temperature of the processing solutions, hot weather processing calls for some adjustment of technique.

It is most important to avoid handling the sensitized materials with hot sweaty fingers before processing. Unnecessary contact with the emulsion contaminates the surface and leaves fingermarks behind. Plates and films should be handled by the edges only. It is often difficult to observe this rule when threading a film into a spiral developing spool, and anyone who intends to do a lot of hot weather processing should buy one of the self-loading tanks or one of the daylight loading types.

Once the emulsion is wet, it is even more important to avoid touching it with the fingers. The ends of films developed by see-sawing should be held with stainless steel clips of the type used for hanging up films to dry. If the ends are held in the fingers the gelatin will melt and run.

In very hot weather it is as well to carry out all processing at the time of day when the temperature is lowest—i.e., in the early morning. At that time both the air temperature and the temperature of the water supply are generally low enough to permit processing without any special precautions. If the processing is being undertaken in a daylight developing film tank away from darkroom facilities it is a simple matter to load the tank during darkness the night before and carry out the processing next morning.

During hot weather there are more flies and dust about than usual. This means that the negatives must be dried off as quickly as possible in a current of filtered air—e.g., in a dry cupboard with ample ventilation holes at top and bottom, covered with muslin. L.A.M.

See also: *Hardening bath; Tropical photography.*

Books: *Camera in the Tropics*, by G. C. Dodwell; *Photography on expeditions*, by D. H. O. John (both London).

HÜBL, ARTHUR BARON VON, 1852–1932. Austrian officer. Studied chemistry; became director of the Militärgeographisches Institut (Military Geographical Institute), Vienna. Made valuable contributions to photogrammetry, especially as applied to map making, and to photomechanical processes, colour photography and colour printing. Published in 1882 a reliable process for the preparation of platinotype paper (invented by Willis). Worked on emulsions and sensitometry, invented multiple gum printing (1898), and improved the ozotype process (1903).

HUGHES, ALICE. Dates unknown. English photographer of royalty and society women from *c.* 1887 on, originally as studies for her father, a Court painter. Her fine platinotype prints are the epitome of an elegant age. Miss Hughes sold her London studio in 1910; opened up again *c.* 1916 and remained active until *c.* 1925. Autobiography: *My Father and I* (1923).

HUNGARY

Photography has deep roots in Hungary; in 1840, a year after Daguerre's invention was made public in Paris, the first daguerreotypes were displayed at the Magyar Tudomanyos Akádémia (Hungarian Academy of Sciences). Soon several daguerrotype studios were working in the capital of the country. The first of these studios was opened by Jakab Marastoni (1804–60) in June 1841. The calotype (talbotype) process did not come into general use in Hungary, but Archer's wet collodion process rapidly achieved popularity and soon more studios were opened in smaller towns. Many of the contemporary photographers were portrait painters who—though in most cases only temporarily—exchanged the brush for the camera. József Borsos (1821–83) was the most outstanding of these. Károly Szathmáry Pap (1813–87), painter and photographer, pioneered photographic journalism: as early as 1854 he took pictures in the Crimean war.

In 1880 there were 257 independent photographers and 215 journeymen operating in Hungary, and Hungarian photographers were entering international shows, including the first world photographic exhibition held in New York.

In the last decades of the nineteenth century the development of photography as an art was greatly helped by the growing number of amateurs, some of whom proved to be photographers of outstanding artistic ability.

Hungarians also contributed to camera technique. Josef Petzval, the Hungarian professor, designed the first modern rapid portrait photographic lens. He was born in Szepesbela in 1807, became an eminent mathematician, and at the age of nineteen taught at the University of Pest. In 1840 he was invited to the University of Vienna, where he took charge of the mathematics department. Here he became interested in the optics of photography, and set out to design a system of lenses which would have a much greater speed than Daguerre's f14 achromatic objective. He computed a successful four-piece asymmetric f3·6 objective which was a revolutionary step forward, particularly in portrait photography where it reduced exposures to the order of a second or two instead of minutes.

Petzval's objectives were prepared and marketed by the Viennese optical firm of Voigtländer. His work gave a tremendous impetus to the science and art of photography by making the designing of lenses a matter for exact optical and mathematical computation. The Wiener Akademie der Wissenschaften (Vienna Academy of Science) and several other scientific associations made him an honorary member of their organizations and when he died in 1891 a street was named after him in Vienna. Today his name is known all over the world and in Hungary photographers revere his memory.

Ferenc Veress was the first person in Hungary to work on the problem of colour photography. His colour-sensitive emulsion was praised at the World's Fair in Paris in 1900 by A. Londe, head of the section on photography, and by Professors H. W. Vogel in Berlin and J. Eder in Vienna. Ferenc Veress tried out more than 500 colour emulsions, and colour contact prints made by him, preserved in the Magyar Nemzeti Múzeum (Hungarian National Múseum), have retained their bright colours to this day.

Photographic Industry. In Hungary a considerable photographic industry has developed since the end of the Second World War. The firm Forte established in Vac by the American Kodak company in 1923 made only photographic paper. It was damaged during the war, but since then it has been rebuilt, modernized, and made suitable for producing other photographic materials. This factory now makes photographic chemicals, photographic paper, plates and films and sensitized materials for industrial and medical radiography. The production of negative and reversal colour film started in 1963. The factory exports quantities of photographic materials, especially to the Eastern countries and to the tropics.

After 1945, the manufacture of cameras was started by a Hungarian optical works, beginning with the development of a roll film camera. In 1954, in addition to simple cameras, the mass production of roll film cameras of a more advanced type was launched. Optical instruments are also made.

Other Hungarian factories and precision engineering shops manufacture enlargers, laboratory materials and equipment, mainly for home consumption, and a number of chemical factories are engaged in turning out the basic materials for photographic chemicals.

The development of the Hungarian photographic industry is the result of continuing large-scale research work conducted since 1945, primarily in the laboratories of the various photographic manufacturers as well as in independent research laboratories. 1963 research led to the production of thin emulsions of high sensitivity.

Associations. The foremost organization of both professionals and amateurs is the Magyar Fotomüveszek Szövetsége (Association of Hungarian Photo Artists) in Budapest; a member of the Fédération Internationale de l'Art Photographie (FIAP). An organization of scientific experts called the Optical and Kinotechnical Association has made an immense contribution to the development of the Hungarian photographic industry. It maintains a large number of departments and research committees which embrace practically every branch of photochemical and optical science.

Hungarian photographic artists have a separate organization within the Magyar Képző- és Iparmüvészek Szövetsége (Association of Fine and Commercial Artists).

Photo-reporters working for the press are members of the photographic division of the organizations. Many press photographers are employed by the Magyar Távirati Iroda (Hungarian News Agency) which supplies photographs to all kinds of Hungarian and foreign publications. The most outstanding photographic artists and photo-reporters are engaged in training the younger generation of photographers. These experts hold lectures and arrange debates at which they analyse the artistic and technical problems of photography and criticize the pictures exhibited. The Hungarian daily papers and illustrated periodicals constantly publish the finest works of photographic artists and reporters.

The amateur movement in Hungarian photography is particularly strong. In a country of 10 million people the number of amateur photographers is more than a quarter of a million and still increasing. Most amateur photographers today are members of photographic groups in their own place of employment and there is hardly a factory or institution without such groups. Amateur photographers in the provinces are usually members of groups organized in the cultural centres which are to be found in most villages and small towns, and the primary, secondary and vocational schools usually have photographic groups for pupils. There are about 2,000 such groups in Hungary.

There are twenty-eight amateur organizations throughout the country. The oldest of these is the Magyar Dolgozók Országos Müvészfényképezö Egyesülete (Hungarian Workers' Association of Artistic Photographers), recently renamed the Budapest Photo Club, with about 1,000 active members. This organization, formed in 1899, consists mostly of people who are not members of a factory, trade union or similar photo club.

Within the national organizations of amateur photographers, the factory photo clubs and the photographic circles of the provincial cultural centres, planned, artistic training is carried on. Annual national exhibitions of photography are held and several times a year local exhibitions are arranged, and there are lectures by experts and artists. Various organizations support the development of photographic art by sponsoring photographic contests. Amateurs also take part in international events.

Exhibitions. The first large-scale international photographic exhibition was arranged at Budapest in 1910, when about 2,000 pictures were submitted from all parts of the world. In the following decades exhibitions were held occasionally: international, national exhibitions, one-man shows etc. Since 1957 exhibitions have been regular and uninterrupted: five international exhibitions in Budapest and several salons displaying the art of a particular artist or groups of artists. On an exchange basis, works from many countries was exhibited in Budapest, collections of Hungarian works having been shown in Poland, France, The Soviet Union, German Democratic Republic, Czechoslovakia, etc. Most of the exhibitions were arranged by the Magyar Fotomüvészek Szövetsége (Association of Hungarian Photo Artists. The Magyar Ujságirók Szövetsége (Union of Hungarian Journalists) arranged two international Interpress exhibitions. The Magyar Távirati Iroda (Hungarian News Agency) used to exhibit the works of its photographers. Professional portraitists united in Fényszöv arranged in 1959 and in 1963 international exhibitions of portrait photography. Some of the exhibitions were organized around a particular theme for example, the Fifth International Exhibition "Women in the 20th Century".

Besides the exhibitions shown in the capital, several others are regularly arranged in country towns by the local clubs.

Art. Hungarians have contributed substantially to photographic art. From the period of daguerreotypes onwards, outstanding photographs have been preserved, showing the artistic value of masters like József Borses (1821–83), Károly Koller (1823–90), Antal Simonyi (1821–92) Sándor Strelisky (1851–1922) and Ede Ellinger. However, real striving after artistic expression began around the end of the century with the advent of amateurism. Thereafter many landscapes, city sights and genre pictures were taken. As in the West, from 1895 to 1925, pictorialism (imitation of paintings) was the main trend but already at that time there were some signs of photographic realism in Hungary.

The thirties brought the renaissance of Hungarian photographic art. Most characteristic of this period were the new themes: everyday happenings in the lives of simple people, typical figures of the masses, scenes with animals and landscapes. Some Hungarian photographers were fighting hard (for social progress.

Significant members of this group are: Kata Sugár (1910–43), Kata Kàlmán and Klára Langer. The latter has been awarded by the Hungarian Government the title of "Merited Artist". Several photographers have achieved world-wide success winning awards at many international exhibitions and competitions. Best known in this respect are: P. F. Angelo and Károly Gink.

Several gifted Hungarian photographers spent most of their life abroad, enriching with their works the culture of some western country: Làszlò Moholy-Nagy, Brassai (Gyula Halász), André Kertész, Robert Capa (Endre Friedmann) etc.

For the last thirty or forty years Hungarian photographers have been winning awards at international photographic exhibitions. Over a hundred gold medals and first prize certificates, won at the big international shows, testify to the high standards of their art. In 1939, 1,496 pictures were accepted from 100 Hungarian photographers, and were displayed at 70 international exhibitions. The 1940 *American Annual of Photography* carried a list of the world's best photographic artists and among the first ten were two Hungarian photographers. In our day, too, Hungarian workers are frequently awarded first prizes in international exhibitions. Aladar Szekely, Rudolf Balogh and Frigyes Haller and Ernö Vadas were outstanding photographic masters of the first half of this century.

Aladar Szekely (1870–1940), was a pioneer who strove to express the personality and character of his models. He discarded the artificial poses and the conventional studio scenery which created a false atmosphere and produced realistic portraits imbued with profound humanity. He published an album *Authors and Artists* followed by a volume of photos about the Hungarian poet Endre Ady.

Rudolf Balogh (1879–1944), a great figure in Hungarian artistic photography, established an entirely new style. He initiated themes based on the life of the Hungarian people, characteristic Hungarian landscapes, folk activities and ethnographic characteristics. His work won many international prizes. He died in 1944 at the age of 65.

Frigyes Haller (1902–54) scored many successes at international exhibitions with his bright village and peasant genre pictures and lively urban scenes. His greatest merit, however, lay in his work as a critic and teacher. He taught an entire generation of artistic photographers, and his pupils following in his footsteps are winning new honours at exhibitions at home and abroad. Haller was one of the finest and most highly cultured photographic aesthetes of his day and was for several years the President of the Budapest Photo Club.

Ernö Vadas (1899–1962) had a significant influence on Hungarian photography and was a master of themes showing everyday events in the lives of simple people; he also took landscapes and animals. After World War II, he began to apply new perspectives, mostly bird's eye views, to his work in cities, villages, factories, etc. He was the President of the Association of Hungarian Photo Artists.

In Press photography, Károly Escher and János Reismann achieved great success. Dr. Zoltán Tildy and Károly Koffán are noted for animal and bird photography while Dr. Ernö Vajda has become world-famous for his landscape and plant pictures.

Publications. In the nineteenth century, books on photography were published infrequently and they were mostly translations of foreign works.

Recently, however, a great number of books have appeared, some on the history of photography and others showing the latest developments in the form of Year Books.

Several magazines appeared at the beginning of the century, the first being *Az Amatör* (*The Amateur*). From 1921 onwards *Fotomüvészeti Hirek* (*Photo Art News*) gave technical and artistic advice to amateurs, while *Magyar Fotográfia* (*Hungarian Photography*) catered for professionals. *Fotoélet* (*Photo Life*) was launched in 1931. The current monthly for the amateur movement is *Foto* which, besides having articles on many aspects of photography, shows the best works of Hungarian and foreign artists. The Association of Hungarian Photo Artists publishes for its members a quarterly paper *Fényképmüvészti Tájékoztato* (*Review of Photo Art*) containing historical and aesthetical essays on photography.

Present Trends. In recent years, in addition to black-and-white photography, marked progress has been made in colour. The advent of colour materials of Hungarian manufacture and the introduction of special colour exhibitions has brought about a great increase in the popular interest in this field.

Present-day photographs by Hungarian artists are characterized by a feeling for reality, emotional depth, a healthy, optimistic view of life and simple technique. Hungarian photographic artists and amateurs draw the themes of their works from the everyday life of the ordinary people. They visit the workers in the factories, the peasants on co-operative farms, men of science in their laboratories and record their lives in the midst of their pursuits or out in the open on their days of rest.

They are interested first of all in the portrayal of reality, in the artistic presentation of life and not in the little tricks of photo-technique, the interplay of lights and shadows for their own sake, the tinsel of formal elements and artificial themes. To them, the main themes of photographic art are man and human creation. E.V. (I.H. & P.B.)

HUNT, ROBERT, 1807–87. English mineralogist. Outstanding research worker in the early days of photographic science. Writer of the earliest general treatise on photography and several important manuals. A founder-member of the (Royal) Photographic Society. Discovered (1844) that ferrous sulphate is a useful developing agent. Investigated the influence of the spectrum on light-sensitive substances (1843). Proposed several non-silver printing processes. Devised one of the earliest actinometers (1845).

HURLEY, FRANK, *c.* 1885–1962. Australian photographer and polar explorer. Official photographer to five Antarctic expeditions and in both World Wars. Best known for his pictorial record of Sir Ernest Shackleton's Antarctic expedition 1914–16.

HURTER, FERDINAND, 1844–98. Swiss chemist. Worked (from 1867) in an English chemical factory and was a colleague of V. C. Driffield. These two scientists and amateur photographers, not satisfied with empirical methods for measuring the speed of plates, established the first reliable approach to accurate sensitometry. His first work was to devise actinometers that would measure light intensity. Thereafter, in collaboration with Vero C. Driffield, worked out the method of measuring the sensitivity of photographic plates, involving the measurement of light intensities and the resultant densities. They also worked on the theory of the latent image, its development and the control of the development factor (gamma). Published a full description of the apparatus and methods in 1890.

The Royal Photographic Society granted Hurter and Driffield the Progress Medal in 1898 and instituted in 1918 the H. and D. Memorial Lectures. Collected works: *H. and D. Memorial Volume*, edited by W. B. Ferguson (London, 1920).

HUTTON, KURT (Kurt Hübschmann), 1893–1960. English photographer born in Strasbourg, (then) Germany. Opened a portrait studio in Berlin, 1923. 1929–34 worked as photographer to German illustrated weeklies, then settled in England, and became staff photographer to *Picture Post* from its foundation in 1938 until 1955.

HYALOTYPE. (Greek hyalo = glass). Name given by Frederick and William Langenheim of Philadelphia to their albumen-on-glass lantern slides in 1849. Hyalotypie was also the name given by F. A. W. Netto of Leipzig to the cliché-verre technique, 1840.

HYDRATE. Term applied at one time only to a salt formed by replacing one of the atoms in water by a metal.

Such a salt is nowadays called an hydroxide and the term hydrate is commonly reserved for compounds containing water—e.g., water of crystallization.

See also: *Chemicals; Crystal.*

HYDROCHLORIC ACID. Muriatic acid; spirits of salt; hydrogen chloride solution. Used in various bleaching baths.

Formula and molecular weight: HCl; 36·5.

Characteristics: Concentrated hydrochloric acid is a solution of about 32 per cent (by weight) hydrogen chloride gas in water. It is a colourless liquid with a strong pungent smell.

HYDROFLUORIC ACID. Hydrogen fluoride solution. Used in stripping solutions to remove gelatin emulsion layers from glass plates.

Formula and molecular weight: HF; 20.

Characteristics: Commercial hydrofluoric acid is a 50 per cent solution of hydrogen fluoride in water.

It is a colourless fuming liquid, and attacks glass. It is therefore kept in guttapercha, paraffin wax or polythene bottles. The solution is also very corrosive to the skin. The fumes must not be inhaled.

HYDROGEN PEROXIDE. Used in hypo eliminators.

Formula and molecular weight: H_2O_2; 34.

Characteristics: The commercial hydrogen peroxide is a solution in water, labelled in terms of volumes indicating how much oxygen gas it can give off. A 10-volume solution contains about 3 per cent pure hydrogen peroxide, a 20-volume solution about 6 per cent.

HYDROQUINONE. Hydrochinon; hydrochinone; quinol; 1 : 4-dihydroxybenzene. Used as a developing agent.

Formula and molecular weight: $C_6H_4(OH)_2$; 110.

Characteristics: White or pale buff coloured needle crystals.

Solubility: Slightly soluble in water at room temperature. Freely soluble in hot water.

HYDROXIDE. Chemical compound which contains one or more of the hydroxyl radicals; chemical formula, OH—e.g., sodium hydroxide ($NaOH$).

HYGROSCOPIC. Term indicating that a substance absorbs moisture from the atmosphere—e.g., common salt which becomes damp when exposed to the air.

HYPERFOCAL DISTANCE. When a lens is focused on infinity, the depth of field extends from infinity to a point nearer the camera. The distance from the camera to this near limit of the sharp field is called the hyperfocal distance.

If the lens is focused on the hyperfocal distance, the depth of field extends from half the hyperfocal distance to infinity.

See also: *Depth of field; Zone focusing.*

HYPERSENSITIZING. Term applied to any method of increasing the effective speed of an emulsion after manufacture and before exposure in the camera.

There are three ways of doing this: the film or plate may be immersed in a hypersensitizing solution; it may be exposed to the hypersensitizing action of certain vapours; or it may be hypersensitized by exposure to light. Some of these methods can also be used after exposure in the camera.

The Hypersensitizing Bath. Normally a silver bromide emulsion contains a certain amount of free potassium bromide. This reduces the sensitivity, but it is included to make the emulsion keep in store. A bath in distilled water removes most of the free bromide, and thus increases the effective emulsion speed, but of course films or plates treated in this way must be used as soon as possible after the treatment. This applies to all hypersensitization.

Bathing in an alkaline solution such as diluted ammonia (e.g., 1 part 0·880 ammonia, 8 parts pure alcohol, 24 parts water) is still better, and a 1 per cent solution of triethanolamine is equally effective.

There is an even greater increase in speed if the bath contains ammoniacal silver nitrate, say 1 per cent silver nitrate in distilled water with ammonia solution added until the precipitate formed at first is just redissolved. As this method increases the emulsion fog, the developer must contain a developer improver or anti-fogging agent.

In every case the film or plate is immersed (in total darkness) in the solution, drained, and quickly dried in the dark. The speed increase is about 50 per cent (with plain water) to 200 per cent (with silver nitrate).

Exposure to Vapours. When the film or plate is hypersensitized by vapour, it is placed in a closed container with a little mercury or sulphurous acid solution. There is no need to unwrap the material, as long as the packing does not contain any metal. A closed box of plates or spool of paper-backed film can be hypersensitized in a sealed jar containing a few drops of mercury or potassium metabisulphite solution in a small dish. The mercury or sulphur dioxide vapour soon fills the jar, and penetrates the wrapping.

This treatment takes about 24 hours and although the vapour pressure (particularly of mercury) is very low at ordinary temperature, it is quite effective.

The results of vapour treatment are not very consistent; different materials require different periods of exposure to the vapour for the best effect. Some makes and types respond more readily than others. The speed increase varies from one make of sensitive material to another, and even sometimes with the same make.

Pre-exposure. Exposing the sensitized material very briefly to white light before exposure in the camera—i.e., flashing—increases the effective speed by giving the emulsion a threshold exposure, i.e., overcoming the inertia. Usually the fog is also increased.

The easiest way is to give the film a short initial exposure in the camera with the lens pointed at an evenly illuminated out-of-focus surface. The strength of the light and the aperture should be adjusted so that when the shutter is set to its fastest speed the exposure is about 1/200 to 1/100 that needed just to produce an image.

The extent of the increase in speed depends on the negative material used, and the most suitable level of illumination must be found by experiment. L.A.M.

See also: *Latensification*.
Book: *Developing*, by C. I. Jacobson (London.)

HYPERSTEREOSCOPY. Stereoscopic photography carried out with abnormal separation of the two viewpoints—i.e., from points more than the normal interpupillary distance (2½ ins.) apart.

If the viewpoints are separated by twice the interpupillary distance, the impression created by viewing the finished stereoscopic pair in a suitable viewer will be equivalent to looking at a half size scale model of the subject at half the actual distance from the camera to the original subject.

In other words, the subject will appear to be the same size as if it had been taken on a normal stereoscopic pair, but the stereoscopic effect will be doubled.

Terrestrial. Beyond a certain distance—about 200 feet—objects no longer produce a stereoscopic impression because the angle subtended by the observer's eyes is so small that both eyes see substantially the same picture. But if two photographs of the scene are taken from viewpoints sufficiently far apart (determined by the viewpoint distance), the resulting pictures will look stereoscopic.

In this way, distant mountain ranges can be made to stand out in relief by taking hyperstereoscopic photographs from two widely spaced camera positions.

Aerial. Consecutive photographs from an aerial survey series form hyperstereoscopic pairs because the respective viewpoints may be separated by a mile or more along the track of the aircraft. When such photographs are looked at in a stereoscopic viewer, the observer sees in effect a perfect scale model from a distance of a few feet.

This technique is employed in interpreting aerial reconnaissance photographs in wartime when it often shows up otherwise perfectly camouflaged buildings.

Astronomical. The same principle applies to astronomical photographs taken at consecutive instants of time. Because the earth is moving through space, the photographs are taken from viewpoints separated by a distance proportional to the time between one exposure and the next. Here again, seen in the viewer, the effect is that of a three-dimensional scale model. F.P.

HYPO. Every-day name for the fixing agent, sodium thiosulphate. Derived from the old and incorrect name, hyposulphite of soda. Nearly always used with an acid.

HYPO ELIMINATOR. Solution which removes small traces of hypo (sodium thiosulphate) from negatives and, more often, prints. By using a hypo eliminator, the washing time can be greatly curtailed—an advantage in high speed processing. A satisfactory formula is as follows:

Hydrogen peroxide (10 volumes)	5 ounces	125 c.cm.
Ammonia solution (0·880)	190 minims	10 c.cm.
Water to make	40 ounces	1,000 c.cm.

This formula converts any remaining traces of sodium thiosulphate into sodium sulphate, after which the hydrogen peroxide decomposes into water and the ammonia evaporates. There are no other solid compounds left in the paper as there are when potassium permanganate is used as a hypo eliminator.

The above solution is used after the prints have been washed for at least thirty minutes. The prints are immersed one by one, until the whole batch is in the solution. They must not be allowed to stick together, and should be turned over all the time. After ten minutes the prints are returned to the washing water for a further ten minutes, and then dried.

If the prints are left in the hypo eliminator for more than ten minutes, it may slightly reduce the density of the lightest highlights.

1,000 c.cm. or 40 ounces of hypo eliminator are enough for about 120–150 6 × 9 cm. (2¼ × 3¼ ins.) prints or the equivalent in other sizes. The mixed solution does not keep, and is discarded after use. L.A.M.

HYPOSULPHITE OF SODA (HYPO). Incorrect but widely used name for sodium thiosulphate, the most common fixing agent for photographic sensitized materials. Discovered by Sir John Herschel, in 1819, as a solvent for silver salts, and recommended by him twenty years later as a photographic fixing agent.

I

ICE RINK. Many colourful and spectacular ice rink productions are presented each year in addition to regular ice hockey matches, races, and normal skating events at the many ice rinks throughout the world. There are also touring ice shows that visit the theatres with suitable stage accommodation and facilities. All these spectacles are well worth photographing. Permission should, however, always be obtained.

Ice rink photography is considered the most difficult of all artificially lit stage subjects because of the size of the rinks and the area which can be covered by the skaters. The speed of the performers and the quick changes in the colour and power of the lighting all add to the difficulties of the photographer. Solo skaters move at high speed and rarely do their stunts in the same place twice, so it is almost impossible to keep them in focus.

A miniature camera is undoubtedly the most suitable for ice rink photography, especially because of the faster lenses available and the greater depth of field obtainable. Working at full aperture with fast film, sufficiently high shutter speeds can usually be chosen to freeze the movement of skaters, if not too close.

Electronic flash is widely used for publicity photographs at skating rinks. Such work is mostly done by official photographers and no attempt at using flash should be made without first obtaining permission. Most normal flash equipment is only powerful enough for closer shots of individual skaters, and for general views it is still necessary to resort to available light techniques. For stopping really fast action, however, electronic flash is essential.

The popular method of getting sharp pictures is to set the lens to cover a favourable zone, and concentrate on capturing interesting shots in that area. A preliminary visit helps in giving the photographer a chance to note where the pictures occur and select a suitable seat. The lighting is contrasty and a soft two-bath metol developer can be recommended for this and similar subjects. L.V.

See also: *Circus, Dancing, Theatre.*

I.C.I. STANDARDS. Alternative name for the C.I.E. standards of colour description. A system of reference for colours.

See also: *Colour measurement.*

IDENTITY PHOTOGRAPHS. The idea of using identity photographs on passports, shooting licences, etc., was first put forward in 1851 by Dodero, a Marseilles photographer. Identity photographs were first used in America by the Chicago & Milwaukee Railway Co. for season ticket holders in 1861.

Today photographic portraits are used for official records of identity on such documents as military identity cards, passes, and passports. Standard conditions are set out in detail in the British Standard Specifications (B.S. 967: 54) of which the main requirements are as follows:

All photographs must be taken full face with the head uncovered and with the camera not less than 4 feet from the subject. The head must occupy no less than half the print height.

Prints must be made on single weight paper with a surface that will take writing in ink. Prints for passports must in addition be on ordinary (non-waterproof) printing paper that can be mounted in the passport with adhesive.

Passport photographs must have a trimmed size of between 1½ × 2 ins. and 2 × 2½ ins. and two copies are required.

For all other identity photographs the permissible trimmed sizes are 24 × 36 mm. and 38 × 38 mm. with a tolerance of plus or minus one millimetre on each dimension.

In the U.S.A. passport photographs have to be 2½ ins. square, glossy, and two copies are required. Other identification pictures for licences issued by police departments (chauffeur's licence, etc.) are to be round, 1½ ins. in diameter, two prints being required. Photographs for the armed services must be taken by the service issuing the identification card, and are 1½ × 1½ ins. in size.

Colour prints are now also admissible as identity pictures in the U.S.A. if they satisfy the other conditions mentioned above.

ILLUMINATION. In photography, illumination is the distribution of light from one or more sources over the subject being photographed and over the surface of the sensitive material. The illumination of the subject depends upon the type, power, position, and colour of the photographic lighting, and the illumination of the negative material depends upon all these, being modified by the size of the lens aperture and the scale of reproduction. But in each case the illumination obeys the same laws and is measured in the same units.

Intensity and Brightness. The intensity of a light can be measured in two ways—by measuring the actual energy present, or by comparing its brilliance with that of a standard light source.

To measure the energy of a light source it is made to shine on the blackened surface of a thermopile—a device which converts heat into electricity. The light is absorbed by the black surface and transformed into heat. The thermopile then transforms the heat into electricity and the electricity is measured with a suitable meter. This method is used for determining the efficiency of light sources—i.e., how much of the fuel consumed by the source goes into useless heat and how much into light.

The original standard of light was the flame of a specified type of candle burning at a specified rate and the illumination it produced at a distance of a foot is still known as a foot-candle of illumination.

The old standard candle gave a surprisingly uniform light, but in the course of time science has continued to change it for more accurate references. No matter what new reference is adopted, however, it is always quoted in terms of the brightness of the original candle.

The present standard source of light is taken as that given out by platinum at its temperature of solidification; and in order to preserve continuity with former standard candles it has been agreed that it should be regarded as having a surface brightness of so many "old" candles per unit area.

Luminous Flux. The light intensity is often also stated in terms of the luminous flux. The unit is the lumen, and is comparatively familiar to photographers who concern themselves with the efficiency of lamps. The lumen is the flux on a unit surface at a unit distance from a light source of one candle-power. If we imagine a sphere of unit radius surrounding the source, all the light flux falls uniformly on it, and each unit of area receives 1 lumen. As the surface area of the sphere is 4π units, a source of 1 candle power emits 4π, or about 12·6 lumens.

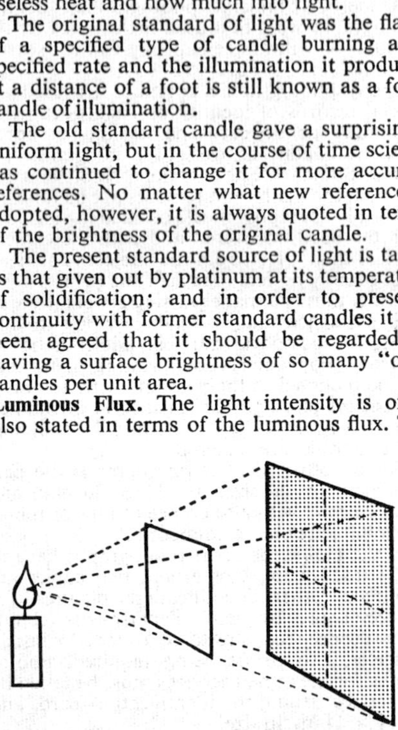

INVERSE SQUARE LAW. At 2 feet from the source a unit area receives only one-quarter the light it gets at 1 foot, as the same amount of light spreads over 4 times the area.

The Inverse Square Law. The inverse square law states that the illumination given by a light source varies inversely as the square of the distance from the source—i.e., the intensity depends on the area of the surface over which the flux is spread. This can be understood by seeing what happens when a cone of light falls on the same surface at different distances from the source. The total amount of light in the cone is always the same, no matter what the distance, but the cross sectional area of the cone at any point, and hence the area over which the light is spread, increases with the square of the distance from the source.

So the illumination in foot-candles at a distance D from a source of candle power C is equal to $C/(D \times D)$, i.e., C/D^2. If the light does not fall perpendicularly on the receiving plane then the intensity is reduced by a factor equal to the cosine of the angle between the incident beam and the normal or perpendicular to the plane.

The law is of fundamental importance in photography because it governs the change in exposure required when a light source is moved towards or away from the subject (or when the enlarger lamphouse is raised or lowered).

The law applies only to illumination by uniformly diverging rays from a point source—it does not apply to parallel rays which theoretically give the same illumination at any distance.

Brightness of a Light Source. Foot-candles are units of intensity of illumination and not of the brightness of a surface. The illumination of a surface is the luminous flux reaching it per unit area: the lumen per square foot was termed the foot-candle. The brightness or luminance of a surface depends on the amount of light falling on it and the proportion of this reflected, and is expressed by its luminous intensity per unit area (candles per square foot). The brightness has also been given by multiplying the illumination by the reflector factor, and expressing it in lumens per square foot (foot-lamberts).

Illumination of Image Plane. The aim of every means of illumination used in photography is to produce an image on the sensitive material in the camera. The degree of illumination of the image is governed by the brightness of the light falling on the object being photographed, the reflectivity of the object, and the relative aperture of the lens. If the brightness, B, of the object is measured in candles per square foot, then the illumination in the image plane, I, is given by the formula:

$$I = \frac{\pi B}{4f^2}$$

where f is the relative aperture of the lens.

If the emulsion had not been put in the camera but it had been put in contact with the object plane, the intensity of the illumination would be πB and then the ratio of the two intensities is 1 : $4f^2$. This ratio is of importance for finding the relation between the exposure necessary for printing by contact, and optical projection (enlargement). R.B.M.

See also: *Colour temperature; Lighting the subject; Light sources; Light units; Reflectors.*

Book: *Principles of Illumination*, by H. Cotton (London).

IMAGE. Rays of light reflected from an object and focused on a flat surface by a lens or mirror form patches of light on the surface corresponding to the bright parts of the subject. These patches of light together form a visible two-dimensional representation of the object. This recognizable picture of the object is called a real image.

The image may be formed on a plane in space, and not on a surface (e.g., in the plane of a clear screen in clear spot focusing). It is then called an aerial image.

When the eye looks at an object through a telescope, an optical viewfinder, or in a mirror, it sees an image of the object. This image cannot be focused and viewed on a screen, like a real image. This type of image is known as a virtual image.

In photography there is a third type of image—the invisible change in the emulsion produced when a real image is focused on it for the duration of the exposure. This image must be developed before it can be seen. It is called a latent image.

IMAGE AMPLIFIER. Electronic system for intensifying an optical image by conversion into electron streams followed by reconversion into light. In its simplest form it consists of an evacuated glass envelope with a semi-transparent photo-cathode at one end and a luminescent screen, similar to the viewing screen of a television receiver, at the other. The photo-cathode receives the primary optical image and reproduces it as a pattern of emitted electrons. These are accelerated away from the photo-cathode and focused on the luminescent screen by an electron lens (a suitably shaped electric field, or combination of electric and magnetic fields). The screen emits light under electron bombardment, and re-converts the electron image to a visible light image.

The Light Gain. The light amplification may be expressed as the photon gain, which is the ratio of the number of photons emitted by the screen to the number of photons incident on the photo-cathode. It depends on the quantum yield of

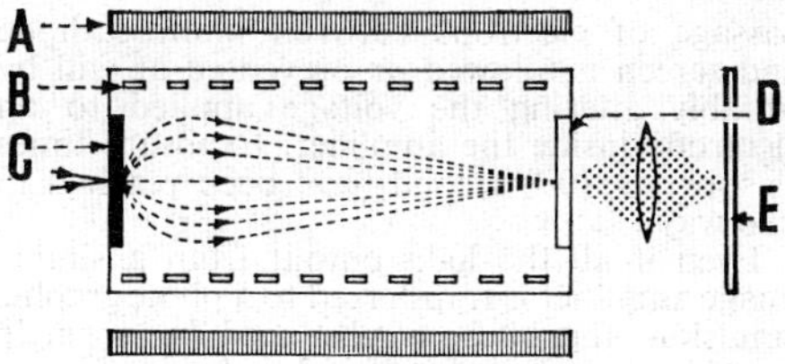

SIMPLE IMAGE AMPLIFIER. A, solenoid to produce focusing magnetic field. B, accelerating electrodes. C, photo-cathode. D, luminescent screen. E, photographic emulsion. The electron stream is shown dotted.

the photo-cathode for light of the spectral distribution of the primary image, on the accelerating voltage applied between the photo-cathode and the screen, and on the efficiency with which the screen converts the energy of the electrons to light.

The photon gain of simple image amplifiers may be as high as 100 for blue and ultra-violet light. Gains in red and infra-red are lower because of the lower quantum yield of available photo-cathodes. The resolution may be 50 line pairs per mm., comparable to that of fast photographic emulsions.

When the output of an amplifier is recorded photographically by using a lens to transfer the image from the screen to a photographic emulsion, the effective light gain is less than the photon gain, because the screen emits light in all forward directions, and the fastest lenses cannot accept more than about 25 per cent. This waste of light is avoided in some amplifiers which have the screen on the inside of a very thin mica window. The emulsion can be pressed into contact with the outside of the window, and substantially all the light output is used with little loss of definition. An alternative method is to use a fibre-optical window. This directly transmits the light from the screen to a photographic emulsion in contact with the window, with virtually no loss.

Amplifiers of this kind have been particularly successful in infra-red recording. Photographic emulsions are very insensitive at these wavelengths, and the conversion of the infra-red image to a visible image allows exposure times to be reduced by a factor of 1000 or more.

High-Sensitivity Systems. By using an electron lens which demagnifies the electron image, the light *per unit area* at the screen can be increased by a factor of 50 or more, over and above the photon gain. An amplifier of this kind, with a fast objective forming the input image, forms the basis of a camera of extremely high speed, and has enabled photography of starlit scenes with less than 1 second exposure. The same principle has been used in connection with the photography of faint X-ray fluoroscopic images.

The conversion of the optical image to an electron image also makes it possible to use image amplifiers as high speed shutters. The passage of electrons between photo-cathode and screen is allowed or prevented at will by suitably altering the voltage applied to an electrode inside the amplifier. Exposure times as short as 10^{-9} second have been reached in this way.

Even if all the light output from a simple image amplifier is transferred to a photographic emulsion, the photons produced by a single photoelectron are far too few to produce a detectable image above fog. The higher photon gain needed to achieve this is obtained by incorporating stages of image-preserving electron multiplication between the photo-cathode and the screen. By this means amplifiers have been made with photon gains exceeding 10^5, and resolution of 30 line pairs per millimetre. They have been used in a variety of applications to photograph images which are so faint and transient that only a small number of photons is available. W.L.W.

See also: *Electronic Photography, Photo-cathode*

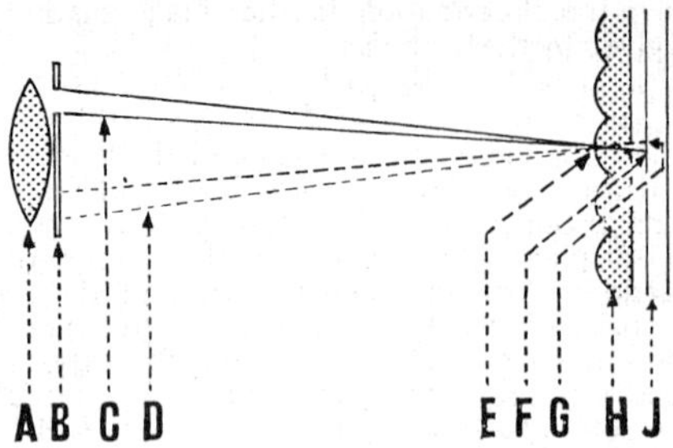

LENTICULAR IMAGE DISSECTION: The lens images the event on to the lenticular plate. The aperture of a rotating perforated disc (Nipkow disc) is imaged on to the photographic plate by one of the lens elements of the lenticular plate. When the aperture of the rotating disc has moved to a new position (dotted) the event is imaged also in a new position on the plate. All other lens elements of the lenticular plate produce similar images. A, main lens. B, Nipkow disc. C, aperture in first position. D, aperture in second position. E, single element of lenticular plate H. F, image in first position. G, image in second position. J, photographic plate.

IMAGE DISSECTION. To take a series of exposures in one camera (as in cinematography) it is normally necessary to move either the film or the image by one picture width between exposures.

To do this rapidly requires both complicated and expensive equipment. However, if the image is divided into very many image points and these points separated by a number of point widths, then it is only necessary to move the image relative to the film by one point width to produce the next picture in the sequence. With one point width of, for example, one thousandth of the frame width, the required motion between frames has been reduced by this factor. If the separation between points is ten times the point width, then at least ten frames in sequence are possible and if points have this separation in both directions, then there can be up to 100 frames in a sequence.

With such an array of points, the simplest straight line motion is at an angle to the array and can, without overlap, readily give fifty frames.

The final record is a scrambled picture of all the different frames and requires un-scrambling

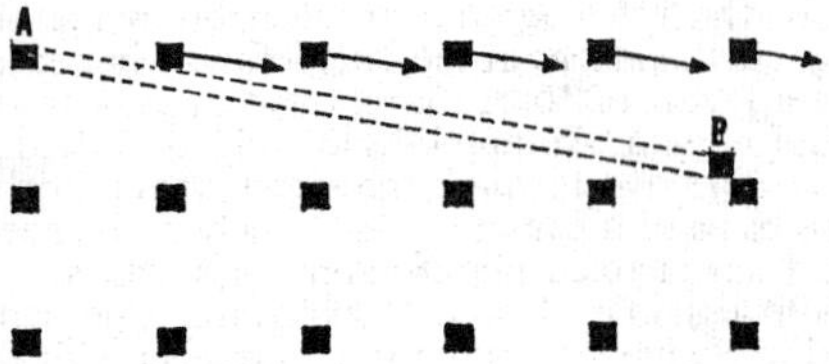

PORTION OF A DISSECTED IMAGE. The image points are shown in the starting position. During a fifty-frame track the point A moves to the point B in its fiftieth position. All other points move simultaneously in the same manner.

before the frames can be seen separately. Fortunately it is possible to use the camera as its own un-scrambler.

These cameras can either be made of equal speed to other cameras at much lower cost or can give higher speeds than were possible by other cameras because of physical limitations. Variations in technique occur in the methods of dissection and image displacement.

Dissection. The pattern into which the image is to be dissected can be made as a grid having a small hole for each image point. The object can be imaged on to the grid and the transmitted dissected image then re-imaged on to a photographic plate or film. Assuming the holes to be arranged in a regular manner such that holes are ten hole widths apart, the total area of the holes to the total area of the grid will be 1:100. Thus only 1 per cent of the light will be available after the grid.

A more economical method in most cases is to make the grid of an assembly of very small lenses, each of which images the aperture of the objective lens on to the photographic emulsion as a small circle. Almost all the light reaching the small lenses is then transmitted. This system is more efficient while the *f*-number of the small lenses is better than that of the objective lens.

Such arrays of small lenses are available. In fact, film has been made with such arrays attached. It is more common to produce these as lenticular plates which consist of many semi-cylindrical lenses side by side. Two such plates, arranged with the cylindrical axes at right angles, are equivalent to an array of small lenses.

A single lenticular plate produces an image made up of narrow parallel lines. This can be used, but only permits deflection at right angles to these lines and the number of frames in sequence will be much reduced. Such a system is then very similar to a streak camera with multiple slits.

Deflection. The simplest technique uses a single lenticular plate (or array of parallel slits) and a simple mechanical movement of the photographic plate such as releasing it and allowing it to fall under gravity or driving it along with a spring or electro-magnet. The exposure must be limited to the time for the plate to move the distance between two image lines. Such a camera has produced ten frames at 100,000 f.p.s.

Using a similar dissection system and moving the image by a rotating mirror as in a streak camera, ten frames have been taken at 100,000,000 f.p.s.

With a crossed pair of lenticular plates, and a Nipkow disc fitted to the objective lens to produce a moving lens aperture, as many as 300 frames have been recorded in a single sequence at up to 100,000 f.p.s.

Image Tubes. Combining image dissection with all the advantages offered by the image tube gives the maximum possibilities. An image tube was specially designed having a metal layer with a suitable pattern of holes deposited on its photo-cathode so that additional optics were eliminated. This tube has recorded fifty frames in one microsecond that is, at 50,000,000 f.p.s. Given a suitable subject, it is capable of up to 1,000,000,000 f.p.s., and when the light intensification possibilities of such tubes are fully available, even higher speeds should be possible.

Fibre Optics. Long flexible bundles and face plates made of very fine glass fibres (diameters of less than 0·001 in. are possible) are opening up new optical techniques.

They could be readily applied as image dissectors with the added advantage that they can be used to re-arrange an image into a different pattern. Thus, in theory, one end of a fibre optics bundle could have the fibres tightly packed to accept almost all of an image, while at the other end the fibres could be arranged in one long single line of fibres. The length of such a line makes this impractical but with a number of parallel lines adequately spaced, the requirement for an image dissection camera is achieved. In the other systems, the image was, of necessity, only being sampled, but with fibre optics the chance of detail being missed is reduced almost to zero.

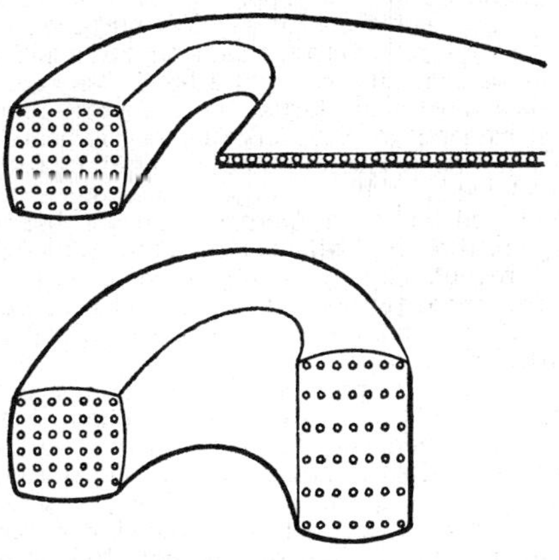

FIBRE OPTICAL DISSECTION SYSTEMS: Two possible arrangements are shown, one giving a single line image (top) and the other a spaced multiple line image (bottom), both from a compact original image. The same bundle used in reverse reconstitutes the image.

A camera with a fibre-optic dissector and a moving plate has produced twelve frames at 500,000 f.p.s. Image dissection techniques are also proposed to make high speed cine-radiography possible, a new tool of value to medicine.

Applications. Applications, to date, have been in studying rapid mechanical motions, projectiles and explosives for example. Image dissection cameras are ideally suited for applications in cine-microscopy and have been used in the study of liquids and particles such as coal dust.

Apart from high speed recording techniques image dissection is also suitable for storing a number of images on one film or plate—e.g. in document recording. The camera set-up is then quite simple—using a sampling grid with holes or a lenticular plate—since the only movement involved is the shift of the grid by one point or element interval from one exposure to the next. For reading, the multiple image is placed behind the same grid, and the latter gradually moved to assemble one image after another in turn.

G.H.L.

See also: *Fibre-optical system; High-speed cinematography; Lenticular system.*

Books: *Image Dissection in High-speed Photography* by J. S. Courtney-Pratt; also the Proceedings of the International Congresses on High Speed Photography.

IMAGE MOTION COMPENSATION (IMC). Method of obtaining a sharp image of a moving object on a film, by advancing the latter in step with the image movement. Used in a variety of fields where subject or camera movement is unavoidable, such as aerial photography, high speed recording, etc.

See also: *Aerial survey, Moving image photography.*

IMBIBITION PROCESS. Any process for making prints by the selective absorption of dyes on a gelatin or similar surface.

See also: *Colour print processes; Dye transfer colour prints.*

IMPRESSIONISTIC PHOTOGRAPHY. In emulation of Impressionist paintings, some pictorial photographers in the early 1890s strove to achieve similar effects by blurring the image by optical means and printing or enlarging it on coarse paper. Soon, however, the Impressionists were no longer satisfied with these more or less legitimate means. The gum bichromate printing process introduced in

1894 by A. Rouillé-Ladévêze enabled the photographer to modify the image to such an extent that it could no longer be considered a reproduction of nature made by the camera. By adding different pigments and using rough drawing paper, the print could be given the appearance of a red chalk or charcoal drawing. By exposing the negative initially to coarse canvas the photograph could be made to imitate a reproduction of a painting. Thus Impressionism in photography dwindled into empty aestheticism.

See also: *Gum bichromate process.*

INCANDESCENT LAMP. Term generally applied to light sources in which the light is emitted by a glowing metal filament heated to incandescence by an electric current, in particular, tungsten filament lamps. Also sometimes applied to arc lamps and flash bulbs.

See also: *Filament lamp; Light sources.*

INCH (IN.). Lineal measure equal to 1/12 of a foot or 1/36 of a yard. In 1305 Edward I standardized the inch as equal to "three grains of barley, dry and round". Nowadays it is taken as the 36th part of the standard yard measure maintained at the Exchequer offices in London.

See also: *Weights and measures.*

INCIDENT LIGHT METER. Exposure meter which measures the intensity of the light falling on the subject (not that reflected from it).

See also: *Exposure meter.*

INDIA. Photography is not a part of the traditional Indian cultural heritage, having been introduced to India by a few European professionals. Amateurs were mainly influenced at first by British magazines, and later, by the flow of photographic literature from all parts of the globe.

History. Arab Pundits knew the basic optical principle underlying photography as early as A.D. 1038 and, according to references found in Amsu Bodhini Shastra, by Maharshi Bharadwaja, much of the art of photography, and its scientific principles, was known at the times of the Vedas thousands of years ago.

Although no direct part was taken by India in the invention of photography, Messrs. Bourne & Shepherd opened a studio in Calcutta in 1840—surely one of the earliest established in the world.

The early wet-plate process of photography was unknown in India, but after the invention of the dry-plate process European professionals introduced photography into the country.

The first pictorial photographer of outstanding merit in India was Shapur Bhedwar. He worked during the late nineteenth and early twentieth centuries, when modern electrical aids were not available and studios and enlargers had to depend on daylight. Bhedwar was also the first Indian member of the London Salon, and several of his pictorial studies were exhibited in England, which was then the Mecca of Pictorial Photographers. J. S. Tarapore soon entered into a healthy rivalry with him, and this really opened the eyes of Indian camera-users to the possibilities of pictorial photography.

Industry. Until recently there was only one photographic factory, at Belgium, and its sole products were field-cameras in teak wood. This factory survived for some forty years, until Japanese competition slowed down its progress. The import of small hand cameras for plates and film packs from Germany, and the growing demand for more handy equipment eventually stopped the manufacture of field cameras in India.

Up to the end of World War II, the demand for photographic goods in India was very small, and many of the materials required could be imported easily and cheaply. In 1942 certain import restrictions were introduced, and after the war India began to become more photography-conscious. Both these factors boosted the demand for photographic goods and, after Indian Independence in 1947, some industrialists considered setting up factories in the country. A beginning was made with manufacturing photographic accessories such as albums, storage boxes, printing frames, leather cases and a few other small items.

In 1956, the import situation became serious when licences were cut down, but mid-1957 all imports of photographic equipment were banned. Only sensitive materials were allowed in to a limited extent. Even that quota was cut down from time to time and was, in 1963, practically nil. The supply of x-ray films is, of course, sufficient for medical purposes.

Today simple German cameras are made in Baroda under licence, and a few private companies have also started the manufacture of box cameras. A new large factory is nearing completion at Thana, Maharashtra State, some twenty miles from Bombay; this too is the joint venture of a German and an Indian organization, and will give India its first factory for manufacturing photographic papers. A Government factory at Ootacamund will, with French collaboration, manufacture roll films. Processing chemicals are also manufactured.

Although the prospects for the photographic industry in India are bright, certain high quality components are not produced in the country and this hampers the manufacture of lenses, shutters, and the like. The use of the full capacity of the factories is therefore dependent on availability of imports. Yet some photographic apparatus such as enlargers, slide-projectors, printing machines, trimmers, storing boxes for slides, flash bulbs, slide holders, masking frames, drymounting tissue papers etc. is manufactured in a few larger cities.

Colour photography is practised on a limited scale in India. The reversal (transparency) process is favourite with the amateurs, and 35 mm. and $2\frac{1}{4} \times 2\frac{1}{4}$ are the most popular sizes used. Very few do their own processing. There are well equipped processing laboratories in Bombay, Calcutta, Delhi, Madras, and a few other big cities. Colour printing is very scarce and there is much room for improvement.

Societies. About 1920 the Photographic Society of India was founded in Calcutta, but due to lack of enthusiasm its life was short. It was not until 1932 that a small band of very devoted and competent pictorial photographers formed themselves into an exclusive group called the Camera Pictorialists of Bombay. Their aims were the advancement of pictorial photography and its recognition in India as a branch of the fine arts.

The enthusiastic activities of this group resulted in a rapid growth of photographic clubs and societies, of which there are today about forty. It also pioneered the first all-India exhibition in 1933 followed by three international ventures in 1935, 1937 and 1939, where the cream of the world's pictorial workers exhibited.

In 1937, the Photographic Society of India was established in Bombay and is now the largest and most progressive photographic group in the country.

In 1953, some of the photoclubs joined together to form The Federation of Indian Photography, with its headquarters in Bangalore it is now affiliated to the F.I.A.P.

Another amalgamated group, formed in 1956, is the Indian Academy of Photography, with its headquarters in New Delhi. Dr. S. Radhakrishnan, President of the Republic of India, is the Patron of this Academy. It is the parent society of some of the best known and oldest societies in India. It is not affiliated to any foreign society.

In addition to the forty well-established photographic societies, transient clubs number up to one hundred. Some of these are affiliated to the R.P.S., P.S.A., F.I.P.P., or the I.A.P.

There are professional photographers' associations, each working on behalf of its own branch of the profession.

The Amateur Cine Society of India is the largest cine society in the country. It has held several successful film contests.

Exhibitions. The Simla Fine Art Society and the Bombay Art Society, which were founded in 1872 and 1888 respectively, laid the foundation for pictorial exhibitions in India by admitting pictorial camera studies to their annual art exhibitions, and contributed towards its promotion by offering attractive awards. In 1947, the Bombay Art Society, for the first time in its history, made H.E. The Governor of Bombay's Prize also available, in rotation, for photographic art at its annual exhibition.

Lord and Lady Canning opened the Exhibition of the Photographic Society on 2nd February 1856 in Bombay, perhaps the first photographic exhibition in the world, which opened on 22nd December 1852 at the rooms of the Society of Arts.

Publications. One of the main outlets for pictorial photography in India is the Press. For a long period *The Times of India's Illustrated Weekly* was the only paper in India that encouraged appreciation of pictorial sense. Later the *Kaiser-I-Hind*, a local Anglo-Gujarati weekly, and *Sakal*, a Marathi weekly, made a substantial contribution towards the advancement of photographic art. Purely photographic magazines have also played their part, and prominent among these were (but are no longer published) *Camera in the Tropics*, and *Indian Photography and Cinematography*.

There are now no regular publications on photography as such. Domestic ones solely devoted to photography, have never been able to compete with foreign periodicals entering the country. Enthusiasts, have from time to time brought out Indian annuals of photography. A few individual photographic societies also have their own monthly journals.

Training. As the need for the development of photography as a profession has become evident, courses in photographic techniques are now being held in many cities. Poona has a Government sponsored training centre, The Film Institute of India; students are eventually absorbed by India's large film industry.

Trends. The availability of modern miniature cameras, fast lenses and emulsions, have revolutionized photography in India to some extent. The new generation of amateurs tend to favour the candid type of work, rather than traditional pictorialism. There are, however, many amateurs who continue to follow the old school.

With the rapid industrial progress in India during the past ten years, photography has helped the engineer, the doctor and the technologist. Most of the large scale industries in India, such as, iron and steel, textiles, chemicals, and many others, as well as medicine, rely on photographic research in one form or another. Almost all of them have their laboratories, where photography plays a prominent part.

The film industry, which is one of the largest in the world, is fully equipped to meet the most modern demands. In recent years, photography has become the backbone of advertising and the Press, and plays an important part in business.

Photography plays a prominent part in visual education. Some of the advanced schools, colleges, and technical institutes have their individual photographic societies, and laboratories. Lectures and demonstrations are accompanied by projection of slides or cine-films. The U.S.S.R. Information Service, the British Information Service, and the United States

Information Service are helpful in spreading education and knowledge not only to students but also to the public in general. These services, and the Government's own cine-units, are giving valuable service in spreading knowledge to the thousands of villages of India and helping to remove illiteracy. J.N.U.

INDICATOR CHEMICAL. Substance which by a change of colour shows variations in the acidity or alkalinity of solutions, or detect the presence of silver or other compounds.

Three main types of chemical indicators are used in photography:

(1) pH (acidity) indicators,
(2) silver indicators, and
(3) hypo indicators.

Hypo indicators are generally used only in solution, pH indicators may be used in solution or in the form of test papers impregnated with the indicator, while silver indicators are generally used as test papers.

pH Indicators. These are organic dyes, usually weak acids or alkalis, which change colour at different degrees of acidity (i.e., at different pH values) of a solution.

The best-known pH indicator is litmus which is red in acid solutions of a pH below 5 and blue in alkaline solutions of a pH greater than 8. The point of change is therefore near the point of neutrality (pH 7).

Litmus is generally used in the form of litmus test papers for testing the acidity of stop baths and acid fixing baths. If the paper turns blue or even just mauve on insertion in the bath, the acidity is spent and the acid stop bath needs renewal. In the case of the fixing bath more acid has to be added, provided the fixing capacity is not exhausted as well.

Some indicators may be added in solution to provide a continuous check on the state of the bath. An example is bromcresol purple which is yellow below a pH of 5, and turns purple at a pH above 7. The change is slightly on the acid side which is an advantage since it provides a safety margin. Furthermore, the change in colour of a stop bath containing this indicator is easy to see by the yellow or orange light of a darkroom lamp. Some commercially available stop baths do in fact contain such an indicator. Some indicators which change at the right pH are not visible by a darkroom safelight (e.g., yellow to red) and therefore not of practical use with stop baths.

Bromthymol blue, changing colour from yellow to blue between pH 6 and 8 is also suitable for this purpose. Most other indicator dyes change colour at a much lower or higher pH and are thus not much use with stop baths. They may, however, still be useful for checking the state of special processing solutions.

In practice most indicators are used in a 0·01 to 0·05 per cent solution in water or aqueous alcohol, and added to the solution to be tested at about ten drops per ounce.

Silver Indicators. These show the silver content of fixing baths, thus giving an idea of the state of exhaustion. Silver indicators are available in the form of indicator papers which are impregnated with a substance that precipitates the silver as a brown stain. The silver content of the bath is assessed by the time the test paper takes to discolour. Fixing baths containing more than 4 grams of silver per litre (35 grains per pint) for use with negatives or 1·5 grams per litre (14 grains per pint) for prints are no longer able to dissolve the unused silver compounds in the emulsion completely and must be discarded.

Hypo Indicators. These show whether washing is complete, or whether any hypo is left in the emulsion of the negative or print. The most common indicator for this purpose is a 0·1 per cent solution of potassium permanganate containing 0·1 per cent sodium carbonate. To test for hypo, drain the negatives or prints, and collect about ½ ounce (10–20 c.cm.) of the drainings in a test tube or glass measure. Add one drop of the permanganate solution; if the colour persists, the water is free from hypo. If the colour is discharged, make a blank test with washing water direct from the tap. Hypo is present if the colour is discharged more rapidly in the drainings than in the control water (more than one drop may have to be added). The blank test or control is necessary because the washing water may contain substances other than hypo which also decolourize potassium permanganate solution. L.A.M.

INDICATOR DYES

Indicator	*Acid State* Colour	pH	*Alkaline State* Colour	pH
Malachite green	Yellow	0	Blue-green	2
	Blue	11	Colourless	13
Crystal violet	Green	0	Violet	2
Thymol blue	Red	1·2	Yellow	2·8
	Yellow	8	Blue	9·6
Bromphenol blue	Yellow	3	Blue	4·6
Methyl orange	Red	3·1	Yellow	4·4
Bromcresol green	Yellow	3·8	Blue	5·5
Methyl red	Red	4·2	Yellow	6·3
Para-nitro-phenol	Colourless	5	Yellow	7
Litmus	Red	5	Blue	8
Bromcresol purple	Yellow	5·2	Purple	6·8
Bromthymol blue	Yellow	6	Blue	7·6
Phenol red	Yellow	6·5	Red	9
Cresol red	Yellow	7·2	Red	8·8
Naphthol-phthalein	Red	7·3	Blue	8·7
Thymol blue	Yellow	8	Blue	9·6
Phenol-phthalein	Colourless	8·3	Red	10·5
Thymol-phthalein	Colourless	9·3	Blue	10·5

INDUCTION PERIOD. Interval elapsing between the immersion of a negative in the developer and the appearance of the first sign of an image. Important in factorial development.

See also: *Developing negatives.*

INDUSTRIAL PHOTOGRAPHY. Any application of photography or photographic processes used in an industrial organization. Such uses may be in research and development, for improving production efficiency, or in promoting sales and service, as well as the photography of industrial installations themselves.

Every facet of photography is used in industry: as an illustrative medium for advertising and publicity; as a medium of communication; to arrest motion or integrate the effect of long periods of time; to provide changes of scale; to observe, record and measure in places where human beings would be limited by danger, fatigue, climate or toxic substances; and to duplicate photographic records for wide distribution. Cinematography is used as a time microscope with a magnification of 800 times or more or to reduce time to an infinite extent.

Applications include:

(a) The entirely functional: e.g., non-destructive testing of components by radiography; scientific analysis by high-speed cine film; time and motion study by time lapse photography; reproduction of engineering drawings and commercial documents; communication of information by photo-drawings; measurement of dimensions by photogrammetry; the duplication of written and illustrated matter by photolithography.

(b) The illustrative and communicative: e.g., in sales literature, safety propaganda, technical reports and for customer information.

(c) As a recording medium; e.g., in electron-microscopy, micro-recording, and oscilloscope trace photography.

The industrial photographer should therefore have a wide range of basic skills plus special training in his particular industry. His knowledge should cover all processes from cinematography, sound recording and photolithography to colour processing, photomicrography, and photographic instrumentation.

Many industrial organizations employ staff photographers. Some have a centralized photographic and reproduction service while others have a number of units each responsible to a particular department. These units may be small or large, specialized or general.

The advantages to a company using an internal photographic service are: That the photographer can be called upon at short notice, and is familiar with the company's products. Such a unit is not subject to divided loyalty, introduces no security risk, can be tailor-made to the company's requirements; may be more economical than an external service; provides quick delivery; and retains possession of negatives without extra charge. Industrial organizations with or without staff photographers may use the services of free-lance photographers and outside studios. Such services are available for illustrative photography of high impact value and pictorial appeal, for technical and scientific applications and for document recording and reproduction. R.F.

See also: *Cathode ray tube traces; Commercial photography, Document copying; Electrophotography; Exploded views, High-speed photography; Industrial radiography; Infra-red photography; Instrument recording; Machinery; Motion Study; Oscillograph recording; Photo-elasticity; Photo-micrography; Schlieren photography; Templates.*

INDUSTRIAL RADIOGRAPHY. Röntgen discovered X-rays in 1895, and in the same year he used his discovery to make an X-ray picture of the scale weights in a wooden box. This was the first industrial radiograph. Since that date X-rays have become an indispensable tool in a great many branches of industry—quite apart from their important uses in medicine and scientific research.

Non-destructive Testing. The principal use of radiography in industry is for non-destructive testing—i.e., examining castings, welds, and finished components for internal flaws without breaking them open or destroying them.

A weld or a casting may conceal a defect which will pass the usual physical tests and inspection, but cause the part to fail after some use. Previously such flaws could not be detected without destroying the part. This meant that no matter how many samples were taken, or what proportion of the finished articles were tested to destruction, the soundness of the part put into actual use had to be taken on trust. The risk was considerable in the case of highly stressed objects like flywheels, turbine shafts and aeroplane parts.

Radiography now makes it possible to check the internal soundness of anything from a weld or a light alloy casting to heavy iron

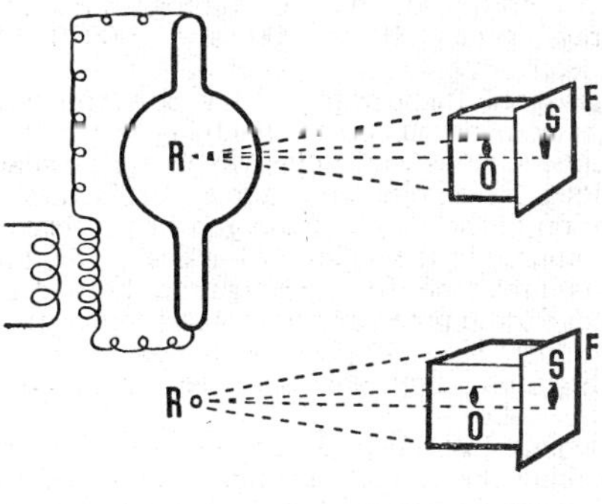

INDUSTRIAL RADIOGRAPHIC METHODS. Top: *X-ray tube as the source.* Bottom: *Radioactive substance used as source of gamma rays. In both cases the film records a shadow of the flaw (voids, cracks) in castings, welds, etc. R, source, O, flaw. S, shadow of flaw. F, X-ray film in contact with casting being radiographed for examination.*

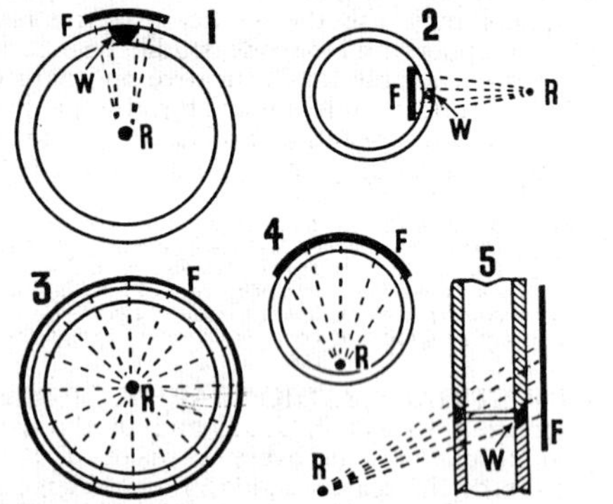

PIPE WELD RADIOGRAPHY. 1. and 2. Longitudinal welds, exposed from the inside and outside respectively. The film must be long enough to cover the section of the weld to be investigated. 3. and 4. Circumferential welds. 5. Circumferential weld exposed through both pipe walls, producing oval shadow on film. R, radioactive source. F, film, W, weld.

castings several feet through. X-ray inspection is an accepted routine test of the welding that now often takes the place of rivets in the construction of bridges, ships, boilers and pipes. It is also used to test the fitness of trainees for employment as welders.

Portable X-ray units are used in the building trade to show the position of the reinforcing bars in concrete structures when machinery or plumbing has to be installed after erection of the building. In this way the holes for the pipes or holding-down bolts can be drilled to miss the metal and so avoid weakening the structure.

Quality of X-rays. The energy of radiation applied in industrial radiography covers a very wide range from about 20 kilovolts to several million volts. The choice for the correct voltage depends on the atomic composition and thickness of the specimen to be radiographed. For instance, steel thicknesses of 1 in. and 6 ins. may be radiographed with X-rays generated at 200 and 1,0000 kV. respectively.

High-speed Radiography. Very powerful units have been developed for studying fast-moving events, such as explosions which are normally hidden in smoke and flame, projectiles in motion in the barrel of the gun or just leaving the muzzle in the region of the flash. The same units are used for investigating how dense fluids flow in pipes, and how metal cracks under a sudden blow. Exposures in this branch of radiography may be as short as one millionth of a second.

Assembly Inspection. X-rays are also used for checking the correct assembly and spacing of the parts in many manufactured products—particularly where a number of components are moulded into a plastic body and cannot be easily inspected.

Manufacturers also use permanent and continuous X-ray inspection to verify the contents of packaged products and to detect the presence of foreign bodies in foodstuffs.

Soft X-rays (Grenz Rays). The generation of soft X-rays (i.e. of long wavelength) has made it possible to widen the field of application.

Hard X-rays will only show tone differences in the developed image if there are correspondingly well-marked differences in thickness or opacity in the subject. Soft X-rays, however, will show up comparatively small differences and so can be used to examine and record the internal structure of such substances as leather, wood and even glass. The soft rays are also useful for examining living plants and other biological specimens.

Gamma Rays. Gamma rays have a shorter wavelength than X-rays and so are more penetrating. They are given off by radio-active substances. At one time radium and its gaseous decay product, radon, were the chief sources available, but nowadays atomic piles are being used to produce a number of suitable radioactive isotopes, such as Cobalt 60, Iridium 192, Caesium 137, and others.

Gamma radiography is carried out by placing a source at a distance on one side of the specimen and a sensitive film on the other. When two lengths of pipe are welded together, for example, a radiograph of the weld is obtained by sliding a source along the inside of the tube until it lies at the centre of the weld. A sensitive film is wrapped around the tube at the weld and left for the required exposure time. The duration of the exposure may range from minutes to several hours, depending on the thickness of the metal, the diameter of the pipe, and the activity of the source. When the film is unwrapped and developed, it shows a radiograph of the weld in a straight line.

The radio-active source may consist of a cylindrical isotope only 2 or 3 mm. in height and diameter and is usually kept in a lead container. Sources of this size can be introduced into places that would be completely out of reach of the normal X-ray tube—e.g., inside pipes and drilled holes, to examine welds, cracks, and shrinkage. Because the source is relatively small, it produces well-defined radiographs.

Gamma radiography with isotopes is flexible, easy to carry out, and very cheap when compared with the initial and operating costs of an X-ray unit doing the same class of work.

Stereo-Radiography. The localization of defects can be achieved by stereographic methods. This may be done by calculating the depth of the defect within a specimen from measurements taken on a film on which two radiographic exposures were made in succession, each with the X-ray tube in a different position with respect to the specimen.

Assessment of Radiographic Quality. The technical quality of a radiograph depends on contrast and definition. The ability with which an industrial radiograph provides information concerning the internal conditions of a metal specimen can be assessed semi-quantitatively by measuring the sensitivity of fault detection. This is defined as the percentage sensitivity = d/t × 100 where d is the smallest detectable thickness and t is the total thickness of the specimen. Thus the lower the percentage value the higher is the sensitivity. The device used for this semi-quantitative assessment is a penetrameter. It may consist of a thin step wedge of the same material as the specimen being examined, or it may consist of a sheet of metal of a given thickness having notches of various depths. Also wires of different diameters embedded in plastic or rubber may be used. The percentage sensitivity is determined by establishing the smallest thickness of material which can easily be revealed by the radiograph. With X-and gamma rays a sensitivity of ½–2 per cent is attainable, depending on the material and thickness of the specimen and particularly on the quality of radiation applied.

Protection. Any person engaged in the field of industrial radiology has to take adequate protection against the hazards. In industry radiation protection is strictly controlled by statutory regulations made under the Factories Act in the United Kingdom. Similar regulations are in force in many other countries. The guiding principle for the operator is not to exceed a certain maximum permissible dose or quantity of radiation. He must be able to protect himself from the radiation behind a sufficient thickness of lead, bricks or other suitable material and by a safe distance from the source, and also to measure the dose he receives in a specified time. This is carried out by means of ionization dose rate meters and by wearing photographic dosimeters (film badges).

G.T.S. and R.H.H.

See also: *Dosimetry; Gamma radiography; High speed photography; Intensifying screen; Medical radiography; X-ray crystal analysis; X-ray film.*

Books: *Handbook of Industrial Radiography*, by J. A. Crowther (London); *A Further Handbook of Industrial Radiography*, by W. A. Wiltshire (London).

INERTIA. Particular value of exposure arrived at by continuing the straight line portion of the

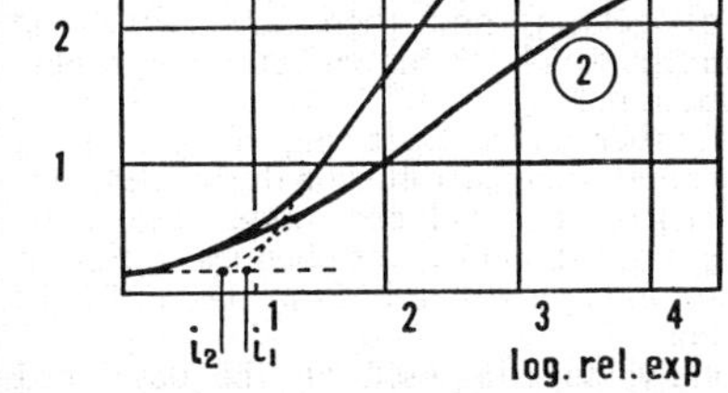

INERTIA. This is the point where the continuation of the straight line part of the curve cuts the fog level line, and is the basis for H. & D. speeds. It is, however, misleading, as a fast film 1 may have a higher inertia i_1 than a slow film 2.

characteristic curve to meet the exposure axis. It is constant for any given combination of emulsion and developer and is the figure used in determining H. and D. speeds.

When the speed of an emulsion is measured by the H. and D. system, the figure increases as the developing time is extended. At the same time the inertia falls to a lower value. The effect is due to the presence of free bromide in either the developer or the emulsion.

See also: *Speed of sensitized materials.*

INFECTIOUS DEVELOPMENT. Effect that can be observed in some negative images, particularly in process-type emulsions developed in a hydroquinone developer with a low concentration of sulphite. One of the oxidation products of the developer accelerates the action of the partly exhausted solution. The result is that lightly exposed parts of the image —e.g., the fringes of lines exposed by irradiation or halation—may develop to a visible image. This effect may cause blocking up in line negatives having fine close lines.

INFINITY (∞). Mathematical term for a dimension or quantity of sufficient size to be unaffected by finite variations. Also a position on the focusing scale of a camera lens. With the lens set to this position, the sensitized material is in its rear focal plane, and receives a sharp image of very distant objects.

Although only infinitely distant objects are perfectly sharp, all objects extending forward to the hyperfocal distance for the particular lens are rendered sufficiently sharp for most purposes. For checking the infinity setting of a lens visually on a focusing screen, the object at infinity may in practice be the horizon, or an object at 1,000 yards or so from the camera.

INFORMATION AND INQUIRIES. The answers to most technical questions about photography are usually to be found in books and in technical journals dealing with the special subject. Where the information is insufficient, authors, who can usually be approached through their publishers, will generally give the reader fuller details.

Non-technical information is sometimes more difficult to obtain, the chief difficulty being that the inquirer does not know where to turn and wastes time in trying the wrong source of information. For instance, manufacturers of cameras will not usually enter into correspondence with individual photographers, as they prefer them to deal with the retailer or agent. Chemical manufacturers cannot be expected to divulge the composition of proprietary solutions where the formulae have not been published, nor can this type of information be found elsewhere.

On the other hand, most information can be obtained from the editorial departments of photographic magazines. Focal Press have a special Information Department, which will answer inquiries direct. A self-addressed, stamped envelope for a reply should always accompany the query.

The following list is intended to serve as a guide to those seeking information in Britain about some aspect of photography. Where more than one source of information is given, the most likely is listed first.

Books. Rare or out-of-print: Librarian, Royal Photographic Society of Great Britain. Current publications: photographic publishers, booksellers, photographic dealers. In Great Britain, any public library can obtain books available in other public libraries.

Advice on specific literature is also given by Focal Press.

Cameras and Accessories. Photographic dealers. Special books are published that deal with one make only—e.g., Focal Press *Camera Guide* and *Camera Way* series on individual camera techniques.

Clubs and Societies. Royal Photographic Society; Photographic Alliance of Great Britain; local public libraries; local newspapers; local photographic dealers.

Legal. General information: H.M. Stationery Office. Photographic: *The Market for Photographs*. Literary: *The Writers' and Artists' Year Book*.

Customs. Full information can be obtained from H.M. Customs and Excise. Leading photographic dealers are usually prepared to assist by answering queries on this subject.

Employment. Local branch of Ministry of Labour. Institute of British Photographers. Vacancies are often advertised in *The British Journal of Photography* and other photographic magazines.

Examinations. Examinations to test various grades of proficiency are held by the Institute of British Photographers and by the City and Guilds of London Institute, Department of Technology.

Fellowship and Associateship. Particulars as to qualifications are obtainable from the Secretary of the relevant society. Honorary titles are sometimes awarded in recognition of special services.

Photographic Technique. Books and magazines supply information of varying degrees of technicality. The Focal Press Information Department answers specific queries by post.

Models. Model agencies. *International Model* Year Book (London). Sometimes art schools can put inquirers in touch with models. Some photographic classes occasionally employ models. Classified advertisements in photographic magazines.

Manufacturers' Instructions. For new apparatus, dealers and agents. For second-hand apparatus, manufacturers' instructions are often not available. Many popular cameras are dealt with in Focal *Camera Guides* and *Camera Way* books.

Photographic Permits. Appropriate authority or public relations office. It should not be assumed that permission to photograph will necessarily be granted—e.g., at some forms of public entertainment copyright difficulties preclude the issuing of permits.

Training. Professional: schools and colleges. For time tables and fees, apply to Principals of Institutes in question. If in doubt whether photographic instruction is supplied in any particular district, inquire from the local Education Officer. Examinations: Institute of British Photographers; City and Guilds of London Institute, Department of Technology.

Amateurs: clubs and societies. Evening classes of educational bodies—e.g., London County Council (information from local Education Officer or public library).

Selling Pictures. Photographic and press agencies; magazines, publishers. Publications: *The Market for Photographs*; *The Writers' and Artists' Year Book*; *Willing's Press Guide*. B.M.

See also: *Books on photography; Literature on photography; Museums and collections; Print libraries.*

INFORMATION IN THE IMAGE. Photographic efficiency and image quality are of particular concern in those applications where it is necessary to record as much information as possible in the photographic image. Astronomy and aerial photography are fields where the best possible photographic performance is often essential. Stemming from these and other closely-related scientific fields, new techniques have evolved with which photographic performance may be measured. These techniques have the same basis as those used to evaluate electrical communication systems, namely the Fourier analysis of complex waveforms.

By Fourier analysis a waveform may be resolved into a series of elementary (sine and cosine) waves of varying amplitude and frequency. The effects on the waveform of a transmitting or receiving medium may then be expressed as a simple scale factor for each frequency component. The frequency response of an amplifier provides a familiar example of this, and temporal frequency in the electrical case is replaced by spatial frequency when discussing optical and photographic image quality.

The photographic layer consists of a more or less random distribution of grains in a less dense layer of gelatin, and there are two basic mechanisms which inhibit perfect photographic reproduction.

Firstly there is the scattering of light within the layer and chemical diffusion during development. These are included under the same heading since the effects of chemical diffusion are often indistinguishable from those of light scattering.

Secondly, superimposed on the developed image are those fluctuations in density which are due to the discrete granular structure.

Spread Function. A useful representation of

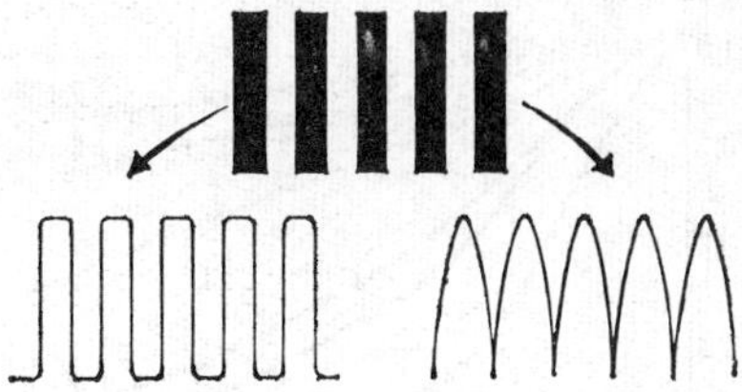

TYPES OF SPREAD FUNCTION. *An original pattern of black and white lines can—according to the characteristics of the image recording system—yield either a squarish spread function (left) or a more pointed type (right). The squarish function yields greater visual sharpness (acutance); the second type of image spread is capable of greater resolution but decreased apparent sharpness.*

light scattering is provided by the spread function. This is the term given to the image density distribution which results from exposure to a pin-point of light.

The spread function is the basic unit of all image formation since any exposure can be considered as a two-dimensional array of point sources, each of which will give rise to an image density distribution which is determined by the spread function. The final image will be given by the summation of the spread functions, each having the same shape, but with size depending on the exposure intensity at that point and the density-exposure characteristics of the photographic layer. The imaging properties of a lens system or an aperture can also be represented by a spread function.

Although the spread function is such a fundamental concept, it has limitations due to difficulties of manipulation. For example, when a point of light is imaged by a lens onto a film, each element of the lens spread function will be further spread by the photographic layer. Prediction of the final image is then difficult, while the problem of deducing the exact form of the exposure from observation of the image is virtually impossible. It is to overcome these problems of image evaluation that Fourier analysis is necessary, and this leads to the concept of the transfer function.

Transfer Function. The transfer function may be obtained by analyzing the spread function into spatial frequency components. By definition it may also be obtained by measuring the amplitudes of the images of sinusoidal patterns. For these images the sinusoidal shape is maintained, and the amplitude at any frequency is merely changed by a scale factor which is the value of the transfer function at that frequency. It is this property of the transfer function which makes it such a useful concept. For example, the overall transfer function of a camera system is merely the product of the transfer functions of the lens and the film.

The transfer function, also known as the contrast transfer function, is a two-dimensional graph. It is a plot of spatial frequency against contrast. Its shape shows how efficiently a lens (or film) system reproduces an image.

Granularity. Fourier analysis also proves useful for the description of photographic granularity. If a uniformly exposed and developed film is scanned with a densitometer which has a long, narrow aperture, then a one-dimensional density waveform will result which represents the distribution of the developed grains.

Granularity is very much a statistical property, and one useful method of statistical analysis is provided by the *autocorrelation function.* This measures the degree of correlation between the granularity fluctuation at any point of an image with those at every other point of the image. Although it provides a complete and fundamental description of granularity, like the spread function it is not always convenient when evaluating photographic image quality, and the *noise power spectrum* has distinct advantages.

The noise power spectrum results when the granularity waveform is resolved into spatial frequency components. It bears the same relationship to the autocorrelation function as the transfer function does to the spread function, and it may therefore be obtained by Fourier analysis of the autocorrelation function. It may also be obtained directly from the granularity waveform by a "power spectrum" analysis. The term "power" originates from the field of electrical communications and signifies that the mean-square fluctuation is measured rather than the mean fluctuation, since the latter is neces-

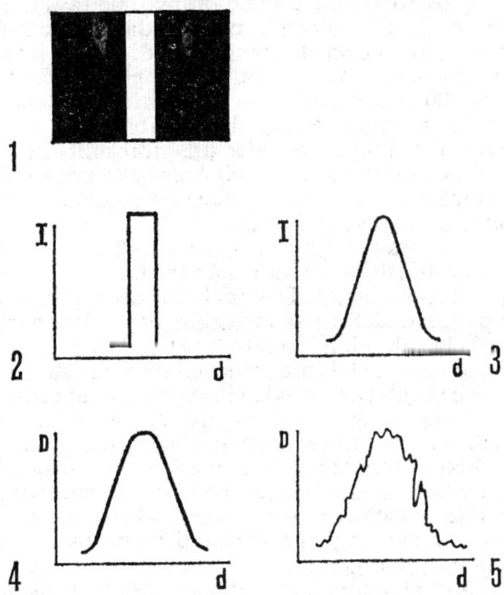

SPREAD FUNCTIONS. *1. Original of a white line on a black background. 2. On plotting or intensity (I) against the distance (d) across the image formed by a lens this should yield a "square" curve. 3. In practice the image formed by the lens on a film tends to spread, giving this typical spread function curve. 4. The photographic process causes further spreading; here density D is plotted against distance (d) across the image. 5. The final density output, with superimposed noise due to granularity is even more distorted.*

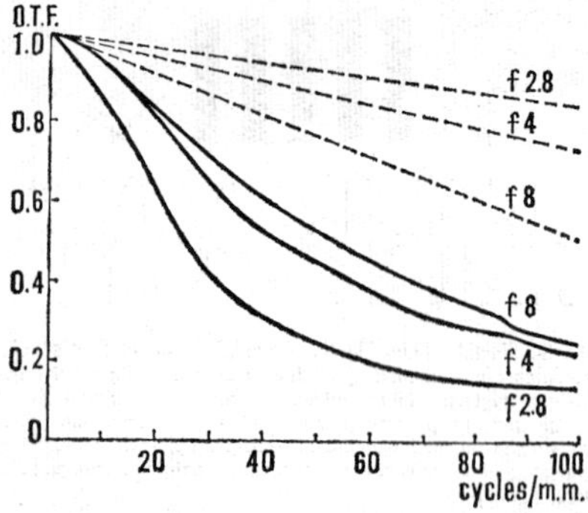

OPTICAL TRANSFER FUNCTION. This plot of image contrast against resolved spacial frequency shows how well a lens (or lens and photographic) system performs. The dotted lines indicate the response of a perfect lens at different apertures, limited only by diffraction. The solid curves show typical practical responses. The higher the OTF value at any point the greater the information recording capacity.

sarily zero. In electrical applications the mean-square fluctuation of a current or voltage wave-form corresponds to electrical power.

Quantum Efficiency. There is a further interesting way of measuring the size of the granularity fluctuations. These fluctuations can never be less than the limit imposed by the fluctuations in the exposure which are due to the quantum nature of light. The ratio of these quantum fluctuations to those due to granularity is used to measure how closely granularity approaches its lowest value. This ratio is sometimes termed the *equivalent quantum efficiency*.

Different films may have widely differing combinations of transfer function and noise power spectrum, and so will have a wide range of imaging characteristics. This may cause confusion when selecting a film for normal visual purposes such as portrait photography. But for scientific purposes the choice is relatively straightforward, since the scientists' criterion is usually that the photographic image should have the highest possible signal-to-noise ratio. The signal size is determined by the transfer function and the noise by the noise power spectrum.

Information Theory. It is this consideration of signal-to-noise ratio which leads to the evaluation of photographic image by information theory. Information theory is the general name given to a series of theorems which define how much information can be obtained from a signal in the presence of a given signal-to-noise ratio. It happens that a measure can be given to "amount" of information which is intuitively satisfying and has scientific rigour. The unit of information is the binary digit or "bit", and corresponds to the amount of information which is conveyed by a "yes" or "no" answer when both were equally probable.

The *information capacity* defines the amount of information which may be stored in the photographic image. Calculations show that one square centimetre of film may be capable of storing more than a million bits of information. All these yes or no answers are about the intensity distribution of the exposure. The more questions that must be answered, the better the film that is needed. R.Sh.

See also: *Grain; Lens performance; Resolving power.*

Books: *Mathematical Foundations of Information Theory,* by A. I. Kinchin (New York); *The Theory of the Photographic Process,* by C. E. K. Mees (New York); *Fourier Methods in Optical Image Evaluation,* by E. H. Linfoot London).

INFORMATION THEORY. General name given to theorems relating to the transmission and recording of information and which permit the evaluation of the efficiency of recording and reproducing systems, and the choice of systems for particular purposes.

The basis of these theorems is that "amount" of information can be given a precise scientific definition which is in keeping with intuitive ideas. The natural units of information fall on a binary scale, and the ability of a system to transmit or store information is sometimes termed the information capacity.

See also: *Information in the image.*

INFRACHROMATIC. Term sometimes used to describe emulsions sensitive to infra-red radiations, as used for infra-red photography.

INFRA-RED (I.R.). Band of invisible rays that occurs in the spectrum beyond the visible red region is known as the infra-red. These rays, which have a wavelength of roughly 7,000 to 150,000 Angstrom units penetrate fog, mist and haze more readily than visible light rays and the longer wavelengths transmit heat. Infra-red rays up to 12,000 Angstrom units will produce a developable image on specially sensitized photographic plates.

Long-wave infra-red rays also affect certain photo-resistor cells and can thus be used to see and record objects in visual darkness. Glass is opaque to infra-red radiation beyond certain wavelengths but rock-salt and some semi-conductor metals are infra-red transparent.

The sun's rays are rich in infra-red radiation, and any body heated to just below red heat will act as a source of artificial infra-red rays.

Flash bulbs are another convenient source of infra-red radiation. They can be screened with a filter which removes all the visible light rays and allows only the infra-red to be used for taking photographs.

Distant scenes normally obscured by mist or haze can often be photographed successfully by using a plate specially sensitized to infra-red rays and adding a filter over the lens to cut out all but the infra-red region of the spectrum.

See also: *Flash for infra-red; Infra-red photography; Spectral sensitivity; Spectrography; Spectrum.*

INFRA-RED PHOTOGRAPHY

Infra-red photography is of interest to the amateur and commercial photographer and to scientists and technologists because it permits results to be obtained which are not possible with panchromatic films. In its practice there is scarcely any difference between infra-red and normal photography. The same cameras and light sources can be used, together with the common developers and fixing solutions. The main difference lies in the choice of film and the need for filters.

TECHNIQUE

The peculiarities of infra-red photography lie in the ability of the film to "see" what the eye cannot see (permitting, for instance, photography in the dark); in the fact that many materials reflect and transmit infra-red in a manner different from the way in which they behave towards visible light; in the ability of infra-red radiation to penetrate certain kinds of haze in the air so that photographs can be taken of distant objects which cannot be seen or photographed on normal films; and in the ability to photograph hot objects by the heat rays which they emit. These properties permit infra-red photography to be used as an important adjunct to photography by normal light.

The infra-red lies "below the red"—that is, it includes all radiations between wavelengths just beyond those of the deepest red end of the visible spectrum, and the invisible heat waves which can be felt but not seen. But though we are able to double our range of vision in the spectrum by using infra-red photography, there is a limit to the part of the infra-red which can be photographed.

Photographs by infra-red were made in the last century, but it was not until the early nineteen-thirties that it became possible to carry out infra-red photography with the ease and certainty of ordinary photography. Just as orthochromatic and panchromatic films owe their response to green and green-and-red light to the inclusion of certain dyes, infra-red-sensitive materials owe their response to special infra-red-dye sensitizers. The earlier infra-red-sensitive materials were quite low in speed, but with recent improvements in the methods of making photographic emulsions and the discovery of new dyes, it has been possible to make available for general use infra-red films which are at least as sensitive as the fast panchromatic films.

Infra-red photographs are indispensable to astronomers, physicists and other scientists and have permitted many important discoveries to be made. It was, in fact, with these discoveries in view that most of the development of infra-red films was originated, but out of this highly specialized development have come films of far broader interest. The visible spectrum ends to all practical purposes at a wavelength of 7,600 A., but films have been made, and are used by scientists, which respond to radiations beyond 12,000 A. Those which are used for general infra-red photography, however, must have good speed and in practice they respond to about 9,000 A. with a maximum response at a wavelength somewhat longer than 8,000 A.

Infra-red sensitive materials need special storage, and refrigeration is often advised. This is partly because the sensitizing dyes employed tend to be less stable, and partly because by their very nature infra-red emulsions are sensitive to heat radiations of the shorter wavelengths so that mere storage in darkness is not always sufficient.

Infra-red Sources. The light sources commonly used in photography are suitable for infra-red materials. Most of them depend for their visible light on heating certain materials to visible incandescence, but a great part of their radiation is nevertheless in the invisible infra-red region. The tungsten filament lamps of the normal and over-run (Photoflood) variety and the foil- or wire-filled flash lamps are especially suitable, since they have the peak of their emission in the part of the infra-red where the films have their maximum sensitivity. The natural incandescent light source is the sun. It is clear from its warming influence that it is a great source of infra-red, and, in fact, it is this rich source of radiations that makes outdoor infra-red photography possible.

In addition to the sources which depend on the heating of materials, there are some which depend on an electric discharge through gases in a tube. There are many such sources, but some, like the fluorescent tubes used for lighting, are not suitable for infra-red work. Electronic flash tubes are most satisfactory.

It is always necessary to use a filter for infra-red photography because all the normal films are highly sensitive to blue light and, if no filter were used, the infra-red effect would be largely masked by the blue exposure. The common yellow filters may be used, but it is best to use the standard red filters of the tri-colour type. More pronounced effects may be obtained by using deep red filters. Filters require an increase of exposure which is greater the deeper the filter.

Equipment. Normal cameras can be used for infra-red photography, but there are two points to be checked: the bellows, film holders and draw slides, although black, must be tested to ensure that they do not transmit infra-red; the focus of the lens in the infra-red may be different from that in the visible spectrum.

The camera and its attachments can be tested by loading film in the camera or in the film holder, searching it closely with a bright electric lamp, and developing the film. Absence of fog indicates safety. The presence of fog means that the bellows (i.e., the camera) must be changed, or the film holder must be replaced by one that is opaque to infra-red. Some manufacturers test their holders for infra-red and guarantee them safe.

Most lenses do not have their best focus in the infra-red when they are focused for visible light. Usually it is necessary to rack the lens forward slightly, as if focusing on a nearer object. The exact distance to give a sharp image must be determined by a few tests.

For development, standard developers may be used. Fixing, washing and drying are as for normal photographic materials.

Lighting. Outdoor photography is done with the sun and skylight as the natural sources of infra-red, and a filter is used over the camera lens to confine the exposure to the longer wavelengths.

The normal rules for subject lighting are followed, although compensation must be made for the fact that the photographic contrast of scenes in the infra-red is higher than for ordinary photography.

In the studio with artificial lighting, normal tungsten filament lamps, photoflash-type lamps, Photoflood-type lamps, and electronic discharge tubes are all satisfactory. The lighting is generally flat from the front.

Filters must be used on the lens, or, to make pictures "in the dark", on the light source. Infra-red polarizing filters are manufactured for use over light sources and over the camera lens for special polarized-light photography in the infra-red. As in the case of all filters, those used over the lens must be chosen so that they will not impair the definition of the lens.

APPLICATIONS

For many years, the popular conception of infra-red photography was that it was a way of seeing through fog. This was largely the result of the sensational handling of the subject by the press. In fact, infra-red photography does not provide a way of seeing through fog, although it can improve visibility through certain kinds of haze.

Long-distance Photography. This property is particularly important for long-distance photography on the ground where the detail of distant objects is often obscured by haze, and for high altitude and especially oblique photography from the air. Actually, infra-red photography does not always result in a very marked increase in the range of vision, but it may increase the contrast of the distant subjects and thus the amount of detail which can be seen. This produces the effect of greater penetration.

For long-distance infra-red photography the modern fast infra-red film is used with a red filter equivalent to a tricolour red. By this means the "atmosphere" so dear to many landscape photographers is entirely lost, and the general effect is one of enhanced and distorted contrast. Grass and leaves of deciduous trees appear white as if covered with snow, although the coniferous trees generally appear dark. The clear sky is very dark, particularly away from the horizon and away from the sun. Clouds appear white and, in the case of heavy cumulus clouds, stand out strikingly against the dark sky. High, wispy clouds which may be invisible in an ordinary photograph may show up clearly. Water generally appears black, unless it is turbid with suspended matter.

Portraiture. Portraiture, outdoors or inside, is not satisfactory by infra-red unless for an occasional bizarre effect because by infra-red flesh has a chalky appearance, red lips are very pale, and eyes appear as dark spots. Sometimes, however, doctors have infra-red photographs made for diagnostic purposes. The infra-red tends to penetrate the turbid skin and show up the veins lying close to the surface, and from their appearance certain conclusions can be drawn.

Survey and Reconnaissance. In infra-red photography outdoors from the ground or from the air, grasses and the foliage of deciduous trees appear white because of the transparency of the green colouring matter, chlorophyll, in the infra-red, and the high reflectance of the cellulosic structure of the grass and leaves. The result is that the infra-red contrast of a landscape or an aerial terrain may be quite different from the visual contrast, and this, over and above enhanced penetration resulting from longer wavelength, may help to increase visibility of distant objects. Further, it helps in aerial survey and reconnaissance to distinguish deciduous trees and grass from coniferous, diseased and dead trees and burnt grass which tend to appear dark in an infra-red photograph.

In war, the infra-red is of potential use for camouflage detection, especially in areas where the day-by-day erection of camouflage cannot be observed. Most green paints, which match green foliage visually in the summer, are strong absorbers in the infra-red and appear dark in an infra-red picture, while natural green foliage photographs as white. Some paints have been developed in which the spectral reflectance is high in the infra-red and they are more difficult to detect by infra-red photography. Special multilayer colour films have even been made and used in which one layer responds to the infra-red and produces a distorted-colour picture in which infra-red-absorbing camouflage shows up as an abnormal colour.

In practice, infra-red photography has been used in forest survey to distinguish between stands of coniferous and deciduous trees, and it has even been stated that if a yellow instead of a red filter is used it is possible to differentiate between different types of conifers.

An important application of infra-red aerial photography in war is its use for determining depth of water and detecting underwater obstacles off potential landing places on enemy coastlines. It has been claimed that by comparing infra-red and visible-light photographs in average coastal waters, depths to 20 feet can be determined to an accuracy of 10 per cent. It has been suggested also that the method might be applied to construction of charts, study of sandbars, silting of navigable channels, control of erosion and pollution, charting of currents and the study of marine life. Seaweed surveys have been made by infra-red photography.

In the Dark. Since the infra-red is invisible, photography in total darkness can be readily carried out if infra-red film is used and the light source is covered with an infra-red-transmitting filter.

Infra-red flash photography has been used in this way for many special purposes, such as press photography outdoors in the blackout during war and in other situations where a bright flash would be disturbing (or betray the photographer), detection of intruders and criminals in the dark, audience-reaction studies in lectures, training-film projection and motion-picture theatres, unobtrusive instrument recording in aircraft, dark adaptation studies of the eye, and photography of industrial operations which are carried out in the dark, as in the photographic industry.

Usually, filters which permit the passage of the minimum of visible light are used—and it must be recognized that deep red light can be very visible in the dark. The filters may be placed in holders over the reflectors, or the flash bulb or the jackets of electronic discharge tubes may be dipped in a dark, infra-red-transmitting lacquer. Normal safelight lamps provide a good lamphouse if an infra-red filter is used in the place of the safelight glass. Suitably dyed plastic bags have also been used as covers for flash bulbs.

In audience-reaction studies by flash, it is usually sufficient to cover flash lamps and reflectors with an infra-red filter, and point them to the ceiling as for bounce-flash to produce illumination which is invisible to the audience. (If a person looks directly at a flash lamp covered with a deep infra-red filter, he will be conscious of a red flash.)

A few deep ruby bulbs among the cove lights or in the exit signs of rooms can often provide enough unobtrusive illumination for short-exposure infra-red photography.

With the modern high-speed infra-red film, short exposures will give good photographs at the normal low level of illumination pertaining in motion picture theatres, night clubs, and so on. Under low levels of tungsten-filament lamp illumination, the fast infra-red film will permit fuller exposures than the fastest panchromatic films for the same exposure values.

Document Inspection. Infra-red photography has found several applications in criminological investigations and it is a standard tool in many laboratories for the study of faded, burnt, worn, dirty or altered documents; the differentiation between pigments, dyes and inks which may appear indistinguishable to the eye; examination and identification of cloth, fibres and hair; detection of secret writing, and a variety of other special applications.

In the study of documents, infra-red photography was dramatically used in the early thirties for showing up printing underlying the obliterating ink of censors which, although black to the eye, happened to be transparent to infra-red. In this respect, infra-red photography is an important adjunct to ultra-violet photography in document examination. Printed matter, engravings and photographs which have become indecipherable through dirt or age can frequently be revealed, mechanical or chemical erasure can often be determined even if overwritten, provided, of course, the overwriting is in an ink transparent to infra-red, and even writing originally on documents charred in fires has been made readable provided the charring has not gone too far. Infra-red photography has taken a place with chemical, ultra-violet and X-ray study in determining the authenticity of works of art.

Scientific and Industrial. Infra-red photography is one of the accepted tools of the applied photographer, and the imaginative worker should have little difficulty in recognizing its possibilities and enlarging its field. Among the special uses, mention can be made of: plant pathology, in the study of plant diseases where there is change in pigment or cellular material; palaeobotany, particularly in coal petrology; in the textile field for detection of irregularities in dyeing or weaving of cloth or damage to the fibres, particularly where the material is dyed a dark tone and visual examination is impossible; in photomicrography of deeply pigmented tissues and thick sections to show enhanced details of internal structure; in the study of furnaces while they are operating.

Since heated objects emit infra-red radiation, in some circumstances they may be photographed on infra-red films by their own rays. The surface temperature of hot objects such as hot plates, engine parts, stoves, cooling ingots and castings, has been studied in this way. The useful range is limited, however, because at temperatures below about 400° C. exposure times become impracticably long.

Infra-red photography is a tool of rapidly increasing importance in astronomy and spectrography. Through its means, hundreds of new lines have been recorded in the spectra of the elements, much has been learned of the composition of the stars and of the atmospheres of the planets, new stars have been discovered and the haze of nebulae has been penetrated to bring us a fuller knowledge of the universe. W.C.

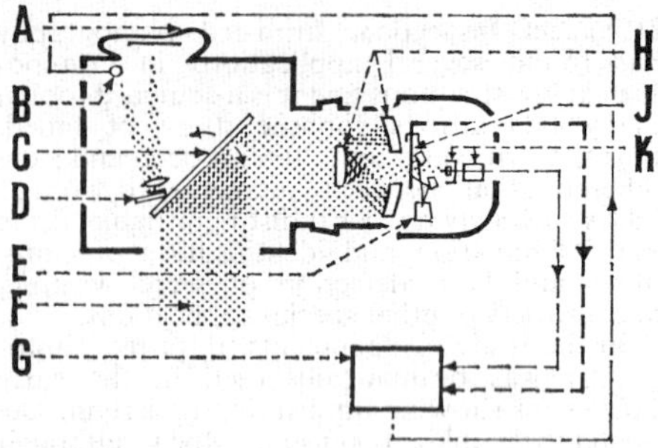

INFRA-RED SCANNING CAMERA. This uses a movable mirror to scan the subject, directing the infra-red rays on to a photo-electric detector. The output of this modulates the intensity of a glow tube, the light of which is focused on to the film by a mirror moving with the main scanning mirror. This reconstitutes a visible photographic image. A, camera back. B, glow tube. C, target scanning mirror. D, image scanning mirror. E, black body temperature reference. F, radiation beam from object. G, amplifier and electronic control system. H, mirror optical system. J, chopper to produce signal pulses. K, infra-red detector and pre-amplifier.

While direct infra-red photography is limited to a maximum wavelength of about 13,500 A, the recording of longer wave radiations is possible by certain other methods. For instance at very low temperatures the latent photographic image can be partially destroyed by long-wave radiations (the Herschel effect) of wavelengths extending beyond the 13,500 A limit. This is a somewhat indirect—and not yet practical—way of utilizing long-wave infra-red rays.

Thermography. While normal infra-red photographic techniques use radiation reflected from the subject to form images, thermography utilizes infra-red radiation naturally emitted by objects at all temperatures above absolute zero. Further, it works with infra-red receptors other than photographic emulsions, such as certain types of photo-electric and transistor cells. This extends the sensitivity range to wavelengths beyond 300,000 A and is suitable even for recording heat radiations from objects emitted at 0° C.

A thermographic recording camera consists of a moving mirror and mirror lens system which between them scan the subject being recorded, focusing the heat rays on to an infra-red detector. The scan moves over the subject area in a television-like pattern.

Radiation from successive small scanned areas is converted into an electrical signal which is amplified and used to modulate the light emission from a glow tube. An image scanning mirror, synchronized with the subject scanning mirror, directs the beam from the glow tube over a sheet of photographic material, reproducing the television-type scanning pattern. The intensities of infra-red radiation received by the detector from each subject point are thus reproduced by different exposures of various points of the photographic material during the scan.

The resulting thermo-photograph indicates a true spacial relationship of the radiation temperatures of the various details in the scanned areas. Tones ranging from black to white indicate increasing radiation temperatures. The actual temperature at every part of the image can be assessed from a grey scale recorded in the photograph, corresponding to calibrated radiation temperatures. This provides a possibility of analyzing the temperature of image portions in the picture with accuracy of a fraction of a degree C.

Applications of such thermographic scanning techniques include the analysis of military targets and backgrounds (the presence of vehicles on the ground can for instance be detected by the higher temperature of their engines), monitoring industrial process installations, testing heating and cooling systems, investigating the degree of heating-up of engines, electrical components and other machinery in use, as well as medical and biological investigations.

In the last case slight variations in the skin temperature of a subject may reveal hidden tumours and other clinical conditions.

See also: *Aerial survey; Air Force photography; Flash for infra-red; Heat recording; Negative materials; Spectrography; Ultra-violet and fluorescence photography.*

Books: ***Infra-red and Ultra-violet Photography*, by Eastman Kodak (Rochester, U.S.A.); *Photography by Infra-red*, by W. Clark (London).**

INKS FOR PHOTOGRAPHS. It is necessary to choose a suitable ink for writing numbers or data on the emulsion of negatives or prints. Ordinary writing ink should not be used for reference numbers or data on the margins of negatives; there is a tendency for the ink to run; it is not easy to make it take on the smooth surface, and some inks tend to come off after drying.

The best ink for this purpose is black drawing ink which should be applied with a fine pen of the kind used for map drawing.

Ordinary commercial writing inks can be used for writing data on the back of prints. Most of these inks are permanent but if the print has not been thoroughly washed they may bleach, especially if the print is stored under damp conditions. Some of the inks used in ball point pens are not permanent, especially if exposed to strong light for long periods. A signature on the print should always be written in black drawing ink.

The choice of ink for titling prints mounted in albums depends upon the tone of the pages. On a dark toned page the white ink sold by photographic dealers for this purpose should be used. If the white looks a little too glaring it may be diluted with a trace of black ink. This will give a grey tint which can be attractive on dark grey or black paper. For light toned or white pages black ink looks best.

If required, a waterproof ink can be made up as follows:

Add ½ ounce of borax to 4 ounces of hot water and heat together with ½ ounce of shellac until most of it has dissolved. Pour off the clear liquid and mix with enough Indian ink or lamp-black to leave a good black impression.

Glossy surfaced papers and hardened negative emulsions are reluctant to accept inks and water-colours. The difficulty can be overcome by adding a drop of prepared ox gall (obtainable from any artists' supplier) to the medium, or by applying a smear of the preparation to the surface to be written or painted on. R.M.F.

See also: *Lettering; Titles.*

INSECTS. It is assumed that anyone who is so interested in insects that he wants to make photographs of them will also know something of their habits; if not, it will pay him to find out. In this branch of photography few things help more than a knowledge of what the subject will do next.

Stalking insects in the field with, say, a reflex camera is good fun and it sometimes results in a good photograph. But in the open air it is difficult to keep the subject in one place, and there is usually a wind to make matters worse. It is also unlikely that the subject will be kind enough to pose against a suitable background. Far better results are obtained by bringing the insects indoors or, better still, by rearing them from the egg or caterpillar stage and taking their photographs by artificial light under controlled conditions.

Camera. The single lens reflex or a miniature with a close-up reflex attachment is the best camera for insect photography. The twin lens reflex and the conventional hand camera are unsatisfactory for such close-up work because they both suffer from parallax. As the aim is generally to get as large an image as possible on the negative, the camera must have a very long extension or be fitted with a supplementary lens to permit close-up work.

Lighting. For insects like butterflies, dragonflies and large caterpillars, two Photofloods will give all the light needed. They should be controlled by a series-parallel switch so that they can be run in series during the business of arranging and focusing the subject and then instantly switched to full brilliance just before making the exposure. Smaller insects are best illuminated by a focused spotlight, and restless subjects may have to be taken with a flash.

Some insects, in particular dragonflies, are quite docile in the dim light, but will fly as soon as the bright lights are switched on. Photographs of such subjects are best taken by focusing in as dim a light as possible and then making the exposure with a synchronized flash bulb.

Flash bulbs are also used to arrest slow movement like the crawling of a caterpillar. The flash bulb is held at about 4 feet from the subject for a lens aperture of *f* 26 and is kept well to one side to throw up the modelling of the subject in sharp relief.

High speed electronic flash lighting may be used to arrest the movement of insects that normally move about quickly and demand very short exposures and proportionately brilliant lighting. The duration of the flash is so short—anything down to 1/10,000 second—that movement ceases to be a problem.

Exposure. As in other branches of natural history photography, every detail must be critically sharp. This means small apertures and long exposures. It is impossible to give hard and fast rules about exposure, but a time of about one second at *f* 22 on a fast pan film or plate with two 275 watt Photofloods at 3 and 5 feet from the subject should give a reasonable negative.

Negative Material. The choice of negative material generally calls for a compromise between two opposing techniques: using a fast, but relatively grainy emulsion, and using a fine-grain, but relatively slow emulsion. If the image can be made large enough on the negative in the first place, then it is better to use a very fast panchromatic film or plate, since there will not be enough enlargement to show the grain.

But if it is impossible to get a big image on the plate to start with—either because the subject tends to move about, or because the camera cannot be focused close enough—then it is better to use a slower but finer grained emulsion that will stand the necessary extra amount of enlargement. A smaller image on the plate means that a shorter exposure can be given, and this helps to compensate for the slower speed of the negative material.

Colour. There is no special difficulty about photographing insects in colour. The technique is the same as that used for monochrome photography except that lighting, filters and exposures need even greater care than usual. It is even more difficult to give any guidance on exposures for colour than for black-and-white. The problem can be solved only by experiment, particularly as even individual makes of colour film vary from one batch to the next.

Large Insects. The larger insects are persuaded to take up a position on suitably arranged foliage set up in front of the camera. Butterflies are fairly docile in artificial light, and a patient operator can generally succeed in getting the subject to rest where he wants it by going to work slowly and gently. The members of some species will expand their wings in the light of the Photofloods; others—those that do not naturally rest with their wings open—may refuse, and have to be photographed with them folded.

With this type of subject there is a strong

temptation to use a wide lens aperture and a fast shutter speed to arrest movement. But the best photographs are made by the photographer who has the patience to wait until the insect is quite still and then to give a relatively long exposure. In this way the lens aperture can be kept small enough to give the necessary depth of field. If the insect is free to move, the image must be constantly examined through the focusing magnifier to keep it focused within the limits of critical sharpness.

Small Insects. The photography of small objects like eggs, young caterpillars and grubs of insects has to be handled by the special technique of low-power photomicrography. A magnification of about five times life size on the negative is generally sufficient. With larger magnifications than this it is difficult to obtain the necessary depth of field, even at very small apertures.

The enlarged image is best produced by using a short focus lens and a long camera extension. A camera with a normal extension may be mounted with its back to the lens panel of another with a long bellows. The subject of such large magnifications calls for a small spot of concentrated light.

Underwater Insects. Water-dwelling insects and pond life may be photographed either from above or from the side. When they are photographed from above, the insects are placed in a shallow glass dish with just enough water for them to swim in. This restricts their movement and keeps them within the limits of the depth of field of the lens.

When water insects are photographed from the side, they are confined in a very narrow Perspex or glass tank which leaves them freedom to move up and down, but not from front to back. This arrangement greatly simplifies focusing.

The field covered by the lens can be marked out on the side of the tank, so that once the lens is set to include the full width of the tank in its depth of field there is no further need to watch the image in the focusing screen. The photographer simply waits for the insect to move into the picture area, switches on the lights and makes the exposure.

This type of set-up leaves the photographer free to use a background of the most suitable tone to show up the subject and even to employ back lighting against a dark ground, perhaps the best type of illumination for dealing with the almost transparent hairs and antennae possessed by many water-dwelling insects.

When the subject is enclosed between glass walls in this way, the image on the focusing screen must always be examined very carefully for unwanted reflections with the full lighting switched on.

Life Cycles. Perhaps the most interesting insect photographs are those that show a series of events, like the successive stages in the pupation of a caterpillar, or a butterfly emerging from its chrysalis. This type of record calls for careful preparation because most metamorphoses of this sort are over very quickly. The camera, lighting and plate holders must all be ready for instant use, and the photographer must be prepared to keep up a continuous watch so as not to miss the start of the process.

Controlling Live Subjects. It is fairly common practice to control the movement of insects by stupefying them with carbon tetrachloride or ether, or chilling them in a refrigerator to slow down their rate of movement. This should be done only as a last resort with insects like silver fish and cockroaches that move at great speed and seek dark corners. In such cases the method of control should always be stated—e.g., etherized, temperature controlled, etc.

Photography in the Wild. In spite of the many difficulties inherent in attempts to photograph insects in the field, there are occasions when this is desirable.

When the insects are at rest, a tripod should be used; all preparations such as the setting up of the camera and the removal of unwanted foliage require great care in order not to disturb the insect. For the photography of insects feeding on flowers, and of other moving subjects, it is usually necessary to hold the camera in the hand, but sometimes a single support resting on the ground can be utilized to give extra steadiness. A compromise must be reached between a short enough exposure to arrest slight movement and a small enough aperture to give sufficient depth of field.

For this work, a single-lens reflex is desirable, and it is best to bring the subject into focus by moving the camera as a whole nearer to or farther away from the subject, rather than to use the normal focusing system of moving the lens in relation to the plate or film. If a reflex is not available, a rod may be attached to the camera so that this rod projects from the front, but is just out of the field of view. The camera, with its focus correctly set, is then moved bodily to or from the subject until the end of the rod is in line with the insect, whereupon the exposure is made.

When photographing moving insects with the camera held in the hand, it is as well to make several exposures of the same subject, as it is very probable that some will be spoilt by the movement of the insect or by camera shake.

Dead Insects. What has been said above refers to the photography of living insects. Photographs of dead insects should be made against a plain background. The subject should never, for example, be attached to a flower to make it appear to be alive. There are, unfortunately, in books on natural history, many such faked photographs, and in some of them the corpses are arranged in positions that the living insects would never assume in nature. Photographs like these bring discredit on natural history photographers and are greatly to be deprecated; if the insect is dead, it is better to be honest about it and mount it like a museum specimen. No skill in disguising the withered shell of a dead insect can ever imitate the living reality. S.B.

See also: *Close-ups; Macrophotography.*
Books: *All About Photographing Insects*, by G. E. Hyde (London); *Entomological Photography in Practice*, by E. F. Linssen (London).

INSTANTANEOUS EXPOSURE. Old-fashioned term for an exposure made with an automatic shutter speed of 1/25 second or less. Nowadays it has no significance except that the initial "I" is still much used to indicate the automatic exposures on shutter controls.

INSTANT PHOTOGRAPHY. Photographic system that yields a finished picture immediately after the exposure—usually by processing within the camera.

See also: *Polaroid Land camera; While-you-Wait Photography.*

INSTANT RETURN MIRROR. Mirror system in single-lens reflex cameras where the mirror returns automatically into the viewing and focusing position immediately after the exposure. This may be coupled with an automatic diaphragm mechanism which at the same time re-opens the lens iris to its maximum aperture. Systems of this kind are usually built into cameras with focal plane shutters; they are rare with a diaphragm shutter.

See also: *Reflex camera.*

INSTITUTE OF AMATEUR CINEMATOGRAPHERS. Organization founded in 1932 to help amateur cinematographers by disseminating technical information and providing facilities for the exchange of views and experiences. All duties in the organization are performed in an honorary capacity.

The Institute has its headquarters in England, but its membership includes amateur cinematographers and a large number of societies and clubs in many parts of the world.

Membership is open to all *bona fide* amateurs who either practise or are interested in cinematography, and the council awards Fellowships to members in consideration of outstanding film productions or for services rendered to the movement.

Special publications, exclusive to members and affiliated societies, are issued at frequent intervals and the Institute provides technical advice, film and book libraries, and film and script criticism. Membership carries with it privileges for cheaper technical services and a low-fee lending library of outstanding amateur films. The annual film competition is the oldest of its kind in the world, and the prizewinning films are presented at the annual London Amateur Film Festival, sponsored by the Institute.

INSTITUTE OF INCORPORATED PHOTOGRAPHERS. Formerly the Institute of British Photographers. Was founded in 1901 to represent all who practise photography as a profession in any field and to establish for the profession a reputation commensurate with the importance of photography in the public service. It has played a major part in the development of professional photography in the United Kingdom during the last half-century. The Institute is frequently consulted by government departments and its recommendations contributed to the framing of the sections bearing on photographic matters in the Copyright Acts of 1911 and 1956.

The affairs of the Institute are regulated by its Memorandum and Articles of Association and the Companies Act, 1948. In 1960, the Institute was reconstituted as a professional institution with a new main object "To improve the technical knowledge and professional status of persons engaged in photography and ancillary occupations and to achieve uniformly high standards of professional conduct". The organization had previously been part trade association—a situation which was incompatible with the holding of professional examinations and the general development of photography as a profession.

The Institute is governed by a Council which is elected by the members and comprises twenty persons who are either Fellows, Honorary Fellows, or Associates. A number of standing or ad-hoc Committees are in operation, advising the Council on matters including the following spheres: advertising photography, cine photography, education, finance, industrial photography, medical photography, membership, professional fees, Photographic Convention, Register of Incorporated Photographers, qualifications, press photography and general practice.

Over one hundred meetings are held annually in many parts of the United Kingdom under the auspices of the Institute, ranging from those organized by the eighteen local centres to national conferences such as the Photographic Convention and the annual Congress which are of more general appeal. The meetings of the Institute are an important means of diffusing knowledge on photographic and professional matters and provide valuable opportunities for the interchange of ideas on topics of current interest.

The Institute publishes a monthly journal, *Record*, and has issued a series of documents on professional matters including *Standards of Practice for Commissioning Advertising Photography, Conditions of Engagement and Scale of Professional Fees for Commercial and Industrial Photography, Terms for the Employment of Professional Models, Fees and Conditions for granting Reproduction Rights in Copyright Photographs, Insurances for Photographers and Job Evaluation for Industrial Staff Photographers.*

The Institute also issues a *Register of Incorporated Photographers*, and a manual entitled *The Professional Photographer in Practice* has recently been published under its auspices.

Membership, which is worldwide, is open only to those persons professionally engaged in photography or an ancillary occupation and admission may be gained either through the examination system or under what are termed the Direct Entry regulations which require the submission of work.

The examinations of the Institute are held at three levels and all have a strong practical bias. Associateships and Fellowships have been awarded by the Institute since 1938 and those holding these titles are entitled to the designatory letters A.I.I.P. and F.I.I.P. respectively.

INSTRUMENT RECORDING. Photography offers a convenient means of recording the readings of scientific instruments. It releases the human observer, and at the same time provides a permanent record of the readings of every instrument throughout the duration of the experiment.

There are two principal ways of recording instrument readings with photographic material: dial instrument records—e.g., photographs taken of the normal instrument dials; and trace records—e.g., of instruments in which a moving spot of light takes the place of the calibrated dial and pointer.

Dial Instrument Record. The instrument dials are grouped closely together and a cine camera fitted with a wide angle lens is focused on them so that they fill the negative area. As a rule the aim is to read the instruments at regular intervals over a long period of time. This is done by setting the shutter release of the camera for single shot working and making the exposures by means of a solenoid-operated shutter release energized at regular intervals by a contactor clock.

The instrument dials are illuminated by diffused lighting so that the pointers do not case confusing shadows, and the light sources are placed so that they are not reflected in the dials. Where possible the glass is removed from the front of the instrument to reduce the risk of reflection of either the light source or the camera itself in the polished surface.

The dials are preferably finished in dead black with white figures and pointer as this combination gives a clear image and less risk of trouble with reflections.

If the exposures are made at fairly short intervals, the lighting is left switched on all the time. This is wasteful if the exposures are made at intervals of an hour or so. In this case the solenoid is equipped with switch contacts to switch on the light just before the shutter itself is tripped by the mechanism; the entire operation is then automatic.

The whole set-up is enclosed to keep out extraneous light and enable a standard exposure to be given.

If the instruments can be strongly lighted, microfilm is used for its high resolution; this is specially necessary when there are a large number of instruments in the picture.

Self-Contained Units. Very often the camera has to record instruments inside a self-contained piece of apparatus out of reach of a power supply—e.g., in a meteorological balloon. In this case the power for lighting the instrument dials and driving the camera is supplied from a small storage battery. If the test is to last for a long period, the camera may be operated by clockwork to save the drain on the battery. The battery can be made to last even longer by cutting down the power of the illumination until it just gives a readable image on the fastest panchromatic film. This is only possible when the grain of the emulsion does not make it difficult to read the instrument to the required degree of accuracy.

In self-contained units of this type, the size of the unit is kept down by using an arrangement of mirrors to give the required optical separation between the lens and the subject within the space available. The mirrors are of the surface silvered type to avoid getting double images with each mirror used.

Trace Records. Photographing normal instrument dials is at best a make-shift arrangement that is only justified where the instruments are already performing a distinct function—e.g., for records of the instrument panels in an aircraft. Where the sole purpose of the instruments is to provide a permanent record, it is far more satisfactory to get rid of the dial and substitute a moving light spot for the pointer. The light spot can then make its own record directly on a moving strip of sensitized film or paper without any need for a camera. If necessary, the moving elements of a number of instruments can be grouped to record simultaneously on the same piece of sensitized material. Or each instrument may make a separate record.

How the light spot is produced depends upon circumstances. In the simplest case the moving element of the instrument carried a mirror. A narrow beam of light falls on the mirror and is reflected on to the sensitized material. When the mirror rotates, the light moves across the sensitized material and leaves a developable trace.

The light trace may also be produced in a cathode ray tube. In this case its movement may be recorded directly on sensitized material, or it may fall on a fluorescent screen and leave a luminous trace for camera recording.

Instrument records of this type, where the instrument is specially designed for the purpose, are made nowadays inside projectiles and in other places unsuitable either a human observer or conventional instruments. F.P.

See also: *Cathode ray tube traces; Oscillograph recording; Time lapse photography.*

INSURANCE

In so far as a camera, with its appurtenances, is part of the equpment of a professional photographer's studio it will be insurable with the stock in trade, fixtures, fittings and other contents of the studio.

Similarly, the camera in private ownership is insurable with furniture, clothing and personal effects in the owner's private dwelling house or flat.

In both cases it will be insured whilst in the premises described in the policy against such perils as are fully set out in the policy itself.

Householder's Policy. The ordinary householder's policy does not insure a camera for more than 5 per cent of the declared full value of the contents. So the owner of a valuable camera, or, say, a cine-projector, should make sure that the value of each separate article in his photographic equipment does not exceed 5 per cent of the full value of the contents of his house. If it does, he must so declare it, have his equipment specially described in the policy of insurance, and insured specifically for such amounts as he considers reasonable.

This means that if the total contents of his house are insured for £2,000, a camera or other item of photographic equipment valued at more than £100 should be specially insured.

A comprehensive policy on the contents of a private house or flat provides for the temporary removal of personal effects, including cameras. Up to 15 per cent of the total value of the contents will be held insured against a limited number of perils while temporarily removed to other dwelling houses or other buildings in which the owner or his family may be residing. In other words, when the owner goes off on holiday the things he takes with him, including his camera, will be insured against certain risks whilst on the premises of the hotel in which he is to spend his holiday.

But fifteen per cent of the total value of the contents of a house does not go very far if the owner's wife has packed her smartest frocks, a few of her best furs and a selection of her choicest jewellery. There will be very little left of the permitted amount of insurance to take care of expensive cameras.

Note also that ordinary insurances are restricted to stated locations. The camera is insured only while it is in the house or flat, in the studio, or in the hotel.

All Risks Insurance. The operator often takes his apparatus to the scene which he wishes to photograph, and, unless he is a studio-portrait photographer, cannot wait for his subjects to visit him. The real need of the camera owner is a form of insurance which will protect his camera against loss or damage, anywhere.

This is available. It is known as "All Risks" insurance, but although its name implies that every form of loss or damage is covered, this is not strictly true. It does not cover wear and tear, mechanical breakdown, or electrical defects. It may exclude customs confiscation or detention. It sometimes excludes theft from unlocked motor cars. Loss or damage by rioters, strikers, and other forms of malicious damage are usually excluded, as is all damage or loss occasioned by war, invasion, foreign enemy and the like. Earthquake (outside Britain), or looting or pillage following earthquake, are other excluded perils.

These exceptions apart, the purpose of an All Risks insurance on cameras is to provide insurance protection against loss or damage, however and wherever caused. It may contain a clause which provides that the owner shall bear a small share—usually ten shillings or so—of claims in respect of damage to flash bulbs, which are particularly susceptible to damage, and special clauses dealing with damage or breakage of lenses are sometimes encountered.

Premium rates vary within fairly wide limits. Amateur photographers can obtain All Risks insurance on a camera worth £50 for as little as 10s. 6d. per annum if the cover is restricted to the United Kingdom, but for world-wide policies required by professional and press photographers, annual premium rates will rise to as much as 45s. per cent of the value of the apparatus to be insured.

In the United States of America the average premium rate is $1.50 per annum per $100 value insured. Usually there is a premium discount over a 3-year period, with an effective rate of 2½ times the annual rate.

Valuation. It is desirable that the camera or other equipment should be independently valued before it is insured. Apart from determining the real value of the camera, it enables the owner to describe it accurately. This description is of great importance if the camera should later be lost or stolen.

It is important that the full value of the equipment should be insured, because the policy usually contains an average clause. This provides that in any case where the value of any camera or apparatus exceeds the sum insured on it, the insured will be considered as being his own insurer for the difference between the real value and the sum insured, and shall share the loss accordingly. It is sometimes possible, by arrangement with the insurance company, to have this clause deleted, but only if the company is satisfied that the full value of the photographic equipment is insured.

Claims. The policy of insurance is a contract of indemnity. This means that the insurance company or underwriter undertakes to make good, by cash payment, reinstatement or repair, the intrinsic loss suffered by the owner as a result of the loss or damage from the insured peril. It is not an undertaking to pay a stated sum: the amount payable is measured by the extent of the loss suffered, with a maximum of the sum insured which is set out in the policy itself.

If there should be any dispute between the parties about the settlement of a claim under the policy, it is referred to the decision of an arbitrator, and the policy contains an arbitration clause which fully protects the interests of both insured and insurer.

It is rare for the arbitration clause to be invoked. Claims usually are negotiated by independent assessors, who do not hesitate to enlist the services of experts when dealing with losses of such specialized property as photographic equipment and cameras.

All Risks insurances on cameras and photographic equipment are transacted by a number of insurance companies and by underwriters at Lloyd's. The office or offices handling the ordinary insurances of the photographer, either amateur or professional, will generally quote for his camera insurance, or will at least put him in touch with an insurer transacting the business.

Plate Glass Insurance. Apart from the insurance of his cameras and photographic equipment, there are other risks against which the owner of a professional studio should seek insurance protection.

If the studio forms part of the ground floor of the premises, it is more than likely it will have display windows, and the accidental breakage of the plate glass in windows, doors, fanlights and other parts of his shop front as well as neon signs, glass counters, display cabinets, and the like should be covered by insurance.

Motor Car Insurance. The motor car used by a professional photographer to make visits to the scene of his operations must be insured to conform with the provisions of the Road Traffic Acts, but the compulsory cover prescribed by law is restricted to insurance against liability for personal injuries to third parties. It is desirable that motor vehicles should be insured in such a way that the third party indemnity is extended to cover damage to the property of third parties as well as injuries to the person. Accidental damage to the motor vehicle itself should also be insured, including damage by fire, loss or damage by theft, and malicious damage as well as risk of injuries to passengers.

Motor insurers provide what is known as a "comprehensive" policy which is wholly admirable. It gives wide cover and deals effectively with the hazards of motoring. It should be noted that mechanical breakdowns of motor cars are not normally insurable.

The professional photographer who uses his private car for professional purposes is advised to examine his existing motor insurance policy, and in particular, the certificate of insurance which is issued to him by his insurance company. A motor car used solely for social, domestic and pleasure purposes is insurable at a lower rate of premium than one used for professional purposes. And while it is true that the user of a private car who insures it as such is permitted in the terms of his policy (and certificate of insurance) to make what are called "business calls in person", it is unlikely that a private motor car used for the purposes of a professional photographer is insurable at "private car" premium rates.

Where the car owner merely uses his car to take him to and from the studio, this will be treated as "private purposes", and the lower premium rate will be charged, but if, in addition, he uses his vehicle to visit churches to take the inevitable wedding group photographs, if, in fact, his motor car is used to convey apparatus and operator to places where work is to be done, the motor car should be insured as a private car used for business purposes and the appropriate premium for such use paid.

It is a serious offence at law to use a motor car on the roads without proper insurance, and the importance of ensuring that motor vehicle insurance is in order cannot be over-emphasized. Any photographer who allows an employee to use his car for business should see that the employee is correctly insured and, of course, that he holds a current driving licence.

Liability Insurance. Other forms of insurance to which the professional photographer should give attention include the important "liability" insurances.

As an employer he is liable at common law for the consequences of injuries to people in his employment caused by his negligence. He is also liable for injuries to people in his employment caused by the negligence of their fellow employees. To deal with this liability he must take out an employer's liability policy. This will cover all his employees, both regular and casual, and a modest premium based on the total wages expended in each year is charged for cover which fully meets all contingencies under this heading.

Apart from liabilities to employees, the professional photographer owes duties to members of the public, to the people who visit his premises as clients or otherwise and to the people who pass by. He is required to exercise reasonable care in the practice of his profession so that no harm shall come to them through his negligence. His studio must be maintained in a safe condition, well lighted, in proper repair, and when he, or members of his staff, leave the studio on "outside" work, care must be taken to see that no damage or injury is done to third parties even when the latter do everything possible to impede photographic work by crowding in on the subject.

These public, or Third Party liabilities, are taken care of by public liability policies which usually contain a limit of indemnity in respect of any one accident. As awards in damages by the courts to injured third parties are tending to rise, it is unwise to fix the limit of indemnity at less than £25,000. A.S.W.

INTAGLIO PROCESS. Photomechanical printing process in which the printing area of the block is incised into the metal and the non-printing areas stand up. When the block is inked up and then wiped off, the sunk areas are left filled with ink which prints as an image when pressed on to the paper. The most common example of this is photogravure.

INTEGRAL TRIPACK. Set of three emulsions and a filter layer coated on to a single film and used for making subtractive negatives or transparencies in an ordinary camera, or for colour prints when coated on a white base.

See also: *Colour film principles; Colour materials.*

INTEGRATING EXPOSURE METER. Meter which adds up the amount of light coming from the scene during exposure; all other meters measure the amount of light received at any one instant and, if the amount of light varies, the meter reading varies with it. Integrating meters can only be used for making long exposures, therefore they are not much used by the ordinary photographer. They are employed principally in process work where exposure times often run into many minutes and where arc lamp illumination may vary considerably during the period of exposure.

The term is also sometimes applied (incorrectly) to ordinary reflected light meters and sometimes to incident light meters, by virtue of the fact that both these meters "integrate" the light reaching them from a wide angle and give an average reading. The use of the term "integrating" is confusing in this connection and therefore best avoided.

See also: *Actinometer.*

INTEGRATION TO GREY. Principle of assessing the colour balance of a colour negative for printing. It assumes that the overall colour of average subjects in nature is more or less white or grey. This implies that if the colours in the image formed by an outdoor subject are completely mixed up (e.g. by complete diffusion) the result would be a fairly neutral colour tone.

The principle is applied in automatic colour printing, where the intensity of the red, green and blue components of completely diffused light derived from a colour negative are measured. The measurements are related to the red, green and blue content of the printing light to establish the required exposures (or the necessary printing filter pack) to yield a correctly balanced print.

In practice not all subjects integrate to grey. For example outdoor scenes showing large expanses of grass or woodland yield a predominant greenish tone on integration; a colour negative from such a scene, when printed by integration to grey, will therefore yield too magenta a print. Indoor shots including large areas of a single colour will exhibit similar anomalies. When printing such colour negatives, an over-riding allowance must therefore be made to obtain a correctly balanced print.

In integrating to grey, the measurements must also be taken with reference to an assumed outdoor colour balance. Under these conditions indoor scenes taken by artificial light or flash need a different allowance in integration.

See also: *Colour printing technique.*

INTENSIFICATION. Process of strengthening a photographic image which is too weak to give a satisfactory picture. It is possible to intensify both positive and negative images.

Negatives. Most methods of chemical intensifying either add another metal to the silver image, or make the image less transparent to actinic light. In either case the effective density increase is proportional to the existing density of the image. So the highlights are intensified more than the shadows and the over-all contrast of the negative is increased.

No chemical intensification can add much density to very faint images, and it cannot therefore greatly improve under-exposed negatives.

Most intensifiers (except perhaps silver) increase the image grain.

Chromium Intensification. Chromium intensification consists of first bleaching the image, and then redeveloping it. Bleaching is carried out in a halogenizing bath containing potassium bichromate.

Stock solution A		
Potassium bichromate	¼ ounce	12·5 grams
Water to make	10 ounces	250 c.cm.
Stock solution B		
Hydrochloric acid concentrated	¼ ounce	12·5 c.cm.
Water to make	10 ounces	250 c.cm.

The working solutions depend on the degree of intensification required as follows:

CHROMIUM INTENSIFIERS PROPORTIONS

Intensification	Proportions of A	B	Water
Slight	10 parts	10 parts	5 parts
Medium	10 parts	4 parts	11 parts
Strong	10 parts	2 parts	13 parts

The negative is left in the working solution until it has bleached right through and turned yellow-buff. It is then rinsed in a 2·5 per cent potassium metabisulphite solution, until the yellow stain has completely disappeared. It is then rinsed in water and transferred to a 3 per cent borax solution. Finally, the negative is redeveloped in an ordinary (not fine grain) metol-hydroquinone developer.

Mercury Intensification. In mercury intensification the image is first bleached and then darkened.

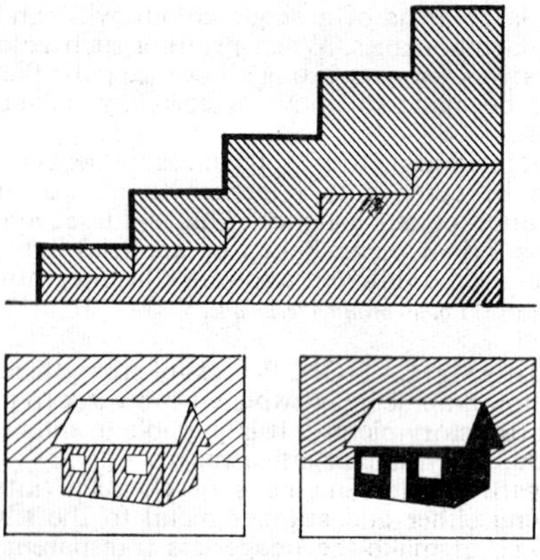

INTENSIFICATION. This consists of adding density to the image by chemical action, and usually increases also the over-all contrast of the negative. The density can only be added to existing image areas, not to clear parts.

The bleach consists of a 2·5 per cent mercuric chloride solution containing about 1·5 per cent hydrochloric acid. After bleaching the negative is rinsed in three or four successive baths of 3 per cent hydrochloric acid (a few minutes in each), washed in water, and then treated with a suitable darkener.

A number of chemicals are capable of darkening the image. Those below are arranged in order of increasing intensifying power:

(1) A 10 per cent sodium sulphite solution.

(2) A normal metol-hydroquinone or amidol developer.

(3) 0·880 ammonia solution diluted 1 : 20.

(4) A 2·5 per cent solution of sodium thioantimoniate (Schlippe's salt) containing 1 per cent ammonia.

(5) A 1·5 per cent sodium sulphide solution.

(Both the mercuric chloride and the sodium thiosantimonate are intensely poisonous.)

Mercury intensification is uncertain and it does not form a permanent image.

Silver Intensifier. This is really a physical developer used on the fixed film. It does not appreciably increase the negative grain.

One of the earliest formulae was that of Wellington which used a mixture of silver nitrate, ammoniun thiocyanate, and hypo, with pyro to trigger the reaction. A more modern and cleaner formula is the following:

Stock solution A		
Sodium sulphite, anhydrous	2 ounces	50 grams
Distilled water	40 ounces	1,000 c.cm.
Stock solution B		
Silver nitrate	2 ounces	50 grams
Distilled water	40 ounces	1,000 c.cm.
Stock solution C		
Sodium thiosulphate, crystals	4 ounces	100 grams
Distilled water	40 ounces	1,000 c.cm.
Stock solution D		
Metol	140 grains	8 grams
Sodium sulphite, anhydrous	88 grains	5 grams
Distilled water	40 ounces	1,000 c.cm.

For use, stir 1 part of A slowly into 1 part of B and add 1 part of C. The white precipitate first formed should dissolve at this stage. Then stir in 4 parts of D. Immerse the negatives until sufficiently intensified, and then leave in a 20 per cent plain fixing bath for about ten minutes before washing. The working solution of the intensifier does not keep.

Toning Intensifiers. It is possible to intensify a negative image by sulphide-toning it by the method used for toning prints. Here the intensifying effect is partly the result of turning the image yellow and thereby making it much more opaque to the active rays in the printing light. The toned image is permanent.

Negatives can also be intensified in this way if the image is coloured by dye-toning it with a red dye, or by developing the bleached negative in a colour-coupling developer for red.

Optical Intensification. Apart from changing the silver image of the negative chemically, very thin negatives can be copied in such a way that the visual effect of the image is enhanced. This involves re-photographing the film or plate by low-level side light or dark-ground illumination. The individual silver grains, though not producing sufficient density to record tone detail on printing by transmitted light, then strongly scatter the illuminating light and so show up a reasonably strong image. This is particularly distinct if the silver is first converted to silver halide by a suitable bleaching process. The negative image then appears as a positive, and on re-copying yields a further negative of improved density and contrast.

An alternative way of optical intensification uses a complex reflecting system that directs the light to and fro through the negative up to five or more times before the light beam reaches the printing paper. During each passage through the negative the optical image is thus reinforced to show up improved density and contrast. Owing to the requirements of such a set-up, the multiple reflection leads to some loss of definition.

Prints. Negative intensifiers like chromium and mercury, which actually add neutral image-density, can be made to work on prints. The results are, however, inferior, partly because the reducer usually has an effect on the image-colour, and partly because the print-image does not in any case offer much scope for intensification. Again, it is usually quicker to make a fresh print.

Staining and similar chromatic intensifiers which intensify the negative by changing its colour and thus making it more opaque to actinic light, are used as print-toners, but not for intensifying prints.

Certain toning processes themselves also intensify the original image. L.A.M.

See also: *Optical aftertreatment.*

Books: *Developing*, by C. I. Jacobson; *The Complete Art of Printing and Enlarging*, by O. R. Croy (both London).

INTENSIFIER, IMAGE. Term generally applied to photo-electronic systems which convert very faint optical images into more intense ones. The faint image is projected on to a photo-cathode; the electron streams emitted by the latter are amplified electrically to form a brighter image on a receiving screen or on the film.

See also: *Image amplifier.*

INTENSIFYING SCREEN. Foil or plate transparent to X-rays, carrying a substance that fluoresces when struck by X-rays or certain other invisible radiations. Used mainly in radiography in contact with the film which is much more sensitive to the visible light emission of the screen than to the X-ray beam. A given radiographic exposure therefore yields a greater image density—though the screen grain may reduce definition.

See also: *X-ray films.*

INTENSITY SCALE. Scale of exposures in which the steps differ in the intensity of the light received but not in the time of exposure—e.g., the scale resulting from exposing through a step wedge and not by subjecting each step to an increasing time of exposure at a constant intensity of illumination. This is the type of scale favoured for sensitometry.

INTERCHANGEABLE LENS. Lens system which can be removed from a camera to permit the use of alternative lenses of different focal lengths and/or apertures. Popular on better types of miniature camera, but also on many professional and nearly all technical camera systems.

See also: *Lens interchangeable.*

INTERFERENCE. As light is a wave motion the waves may be out of phase, with the "crests" of one set coinciding with the "troughs" of another from the same source. When this occurs, the light is extinguished. This most commonly occurs by reflection in a thin film; if its thickness is comparable with the wavelength of the light, light waves reflected from the back may be out of phase with light reaching the front, and interference will occur.

The anti-reflection coating of a lens produces an example of interference: the light rays reflected from the surface of the glass are out of phase with the rays reflected from the surface of the coating (the thickness of the coating is controlled to make this happen). The result is that the individual reflected rays cancel out and there is no reflection from the lens surface.

Interference commonly leads to the production of colours. If the light is initially white, only those wavelengths are extinguished for which the thickness of the film produces interference, and the residual light is coloured. Newton's rings are an example of this effect.

Lippmann's process for making colour photographs relies on stationary waves produced by interference.

See also: *Diffraction.*

INTERIORS. Photographing the interior of a building has much in common with the photography of architecture in general. The aim is to produce an acceptable representation of the scene as it appears to the eye, with no violent or apparently false perspectives, no featureless shadows and no "burned out" highlights.

Equipment. The most satisfactory apparatus, containing all the requisite movements, is the field or stand camera. Most of the modern 4 × 5 ins. cameras are based on the original design of the early stand camera, including swing back (both vertical and horizontal) and the independent front with swing, rise and fall. Cameras without an adequate rising front impose definite limitations and it is nearly always better to use a camera possessing all the normal movements.

A tripod which can be used up to a height of 5 feet 10 ins., or any intermediary height down to 2 feet, is of inestimable value. The lower positions facilitate the photography of details on the sides of table tombs, font bases and similar subjects situated at low levels.

When using roll film it is advisable to have three roll film holders. One should be devoted exclusively to interiors and very contrastive subjects, one for normal exteriors and similar well-lit subjects and the third for flat exteriors, tablets, ledger stones, and other shots lacking in contrast. Processing can then be adjusted to suit the particular requirements of each roll (e.g., for soft contrast subjects, increased development should be given).

Two small pieces of lighted candle or a pocket torch are useful for judging the extreme edge of the field where otherwise it would be too dark to focus visually.

Viewpoint. For general interiors it is as well to use as distant a viewpoint as possible, and the longest focal length that will include the required area of the subject. Where the space is confined and it is necessary to use an extreme wide angle lens close to the subject, the viewpoint must be selected with great care or else the perspective will be so violent that the resulting photograph will look more like a tunnel than a room.

But there are, of course, instances when the creative photographer can put to good use the wide angle technique, with its exaggerated perspective. Many examples of such work, where distortion of perspective is both obvious and at the same time pleasing, can be found in architectural books and journals.

The height of the camera is important: a low viewpoint gives a good ceiling or roof perspective but foreshortens the floor and vice versa. When photographing a low cloister, for instance, where the interest is in the vaulted roof, a viewpoint about 2 feet from the floor with the lens well raised would show all the details in good perspective. Photographing this same view with the tripod extended and the lens below centre would produce a fine photograph of the floor, but the roof would be in such sharp perspective that its features would be lost.

To obtain correct drawing without appreciable distortion, only the centre of the field of the lens should be used, therefore the longer the focus of the lens in relation to the plate the better.

Lighting. Most interiors, with the exception of cinemas, theatres and other darkened buildings, can be photographed by natural lighting, but it is generally better to supplement it with some form of artificial light.

Large interiors—e.g., of a theatre or cinema —call for either a battery of 500 watt floodlights, a number of flash bulbs or a really large dose of flash powder. Flash powder is cheaper than bulbs, but it is advisable to use it only where there is a good extractor fan to suck the resulting clouds of smoke from the building.

Flash bulbs are the most convenient form of lighting, but for big buildings they are rather costly. Experienced workers often photograph large interiors by the light of consecutive flashes covering the whole field, but the flash position must be carefully chosen to prevent the result from looking patchy.

Flood lamps have the advantage that they allow the final effect to be judged before exposure. Also, flood lamps can be tied to a pole and hoisted aloft in dark church roofs. A suitable power-supply is, however, necessary.

Shops, showrooms and well lighted buildings can be satisfactorily photographed by their own lighting. Electronic flash is not as a rule powerful enough for general interiors, but can be very useful for close-ups and details.

When artificial light sources are used to supplement existing lighting or daylight, they must not be allowed to cast defined shadows, particularly when they are used to lighten up shadows in a photograph taken against the light.

Professional photographers are often asked to take factory or school interiors without interrupting the normal activity of the people present. A reflex camera with a four-way swing front, a $f4{\cdot}5$ lens and a fast panchromatic plate form an excellent combination for this type of work. With brightly lighted subjects of this type it is often possible to give an exposure of 1/15 sec. at $f11$ even when photographing against the source of light. However, to get really good results, the photographer should have the building, and those present in it, under his control during the assignment.

Windows are difficult to reproduce by normal methods. If the exposure is long enough to show detail in the interior, it over-exposes the windows and reproduces them as blank areas of paper surrounded by halation. The trouble can be partly cured by using fill-in lighting for the interior.

Sometimes it is possible to set up the camera and make an exposure in daylight for the windows only, making a second exposure for the interior by artificial light after sunset. Where the camera can be left undisturbed for the required time, this is an excellent way of solving the problem.

Setting Up. The camera is set up at the chosen viewpoint and adjusted to give a sharp image over the required depth of field, with all vertical lines parallel to the sides of the picture and all horizontal planes in satisfactory perspective. Where the camera has no camera movements, it may be necessary to accept some distortion which it may be possible to correct in part during enlarging.

Focusing. Focusing follows the pattern for normal photographs of architecture—that is, everything should be critically sharp (except when special effects are required). Particular attention should therefore be paid to depth of field, ensuring that the lens is stopped down sufficiently to render every plane in the subject sharp.

When photographing interiors the vertical swing front is used more than when taking the outsides of buildings. Flights of stairs, for instance, call for vertical swing to bring the whole length inside the depth of field of the lens. while details of floors and roofs are much easier to record with the aid of this camera movement.

The relatively poor lighting of interiors is apt to make focusing difficult. But it is quite easy to focus on a small electric torch, or even a candle flame held by an assistant or simply propped on a table or chair.

When it is impossible for the photographer to get behind the camera to focus, the method of working is simple: he measures the distance from the subject to the position of the camera. Then he focuses on a piece of printed matter at a similar distance from the camera, in a place where he can conveniently view the screen. The next step is to put the camera into the required position. The subject is now sharply focused on the film.

Sensitized Materials. As it is often desirable to give individual development, sheet films and plates are preferable to roll-films. With roll films and tank development control of gradation is confined to choice of developer and development technique.

With interiors in particular, it is unlikely that a roll film would give a series of negatives that would compare with the same subjects photographed on a dozen sheet films and developed individually.

Medium speed panchromatic plates are ideal because they have a fine grain and the contrast can be regulated by varying the length of development. They should, of course, have anti-halation backing.

The best plate for architectural interiors is the double coated type consisting of a fast emulsion coated on a slow contrasty emulsion. These plates lend themselves to the abnormal contrasts so often encountered when photographing the inside of buildings where light sources—windows or artificial lights—appear in the picture.

Exposure. When photographing interiors, a photo-electric exposure meter is invaluable, although the photometer type of meter, with its narrow acceptance angle which makes it possible to take an accurate reading on an isolated shadow, can be even more useful.

With interiors the reading of any type of meter must be regarded as a guide and not an infallible indication. It is generally safe to take the meter reading on the darkest part of the subject (the roof of the building or its darkest corner) and then multiply the exposure indicated by 3. When photographing details against a contrasting background—e.g., bronze statuary against white marble—the meter should be read close up to the subject to exclude the effect of the background.

Stained glass windows can generally be well rendered by measuring the exposure necessary to photograph them from outside, and giving from 6 to 10 times this exposure inside. This generous exposure, followed by development curtailed to one-third the normal time for general interiors, will give excellent negatives.

For dark woodwork detail it is practically impossible to over-expose, but great care must be taken in developing to avoid high contrast, otherwise the highlights will print white and the shadows black instead of in proportionate tones of grey.

As a very rough guide for general interiors where the lighting is normal, using a medium speed material and a stop of *f* 32, an exposure of well under 30 seconds should be ample.

With architectural interiors, the old rule of exposing for the shadows and developing for the highlights will nearly always prove to be reliable.

Development. Development calls for just as much care as exposure: the aim in the one case is to give enough exposure for the shadows to show detail, in the other case to give enough development to show detail in the highlights but not so much that they become clogged and look "burned out" in the print.

The sort of negative aimed at is one which will give a full scale of tones when printed on normal bromide paper. L.H.-F.

See also: *Architecture; Camera movements.*

Book: *Basic Photography*, by M. Langford (London.)

INTERMITTENCY EFFECT. Phenomenon met with in sensitometry having reference to the fact that a large number of brief separate exposures do not produce the same effect on the sensitive material as the theoretically equivalent single exposure.

INTERNATIONAL CANDLE. Basic unit of illumination; the light output of a spermaceti wax candle burning at the rate of 120 grains of wax per hour.

See also: ***Light units.***

INTERNEGATIVE FILM. Specially designed type of colour negative film adapted for the production of colour prints from colour transparencies. Such colour prints may be either on paper base or on clear film base. In the latter case the resulting product is a further colour transparency which may be either the same size as the original or an enlargement or reduction. To avoid the inevitable loss of colour saturation associated with copying colour transparencies, modern internegative film generally incorporated built-in coloured coupler masking.

Although it is generally desirable in copying or printing colour transparencies to maintain or even lower the overall contrast of the process, the contrast of the highlight parts of the subject will need to be maintained or raised. This requirement results naturally from the shape of the normal characteristic curve of photographic sensitive materials. One commercially available internegative film maintains the overall contrast of the copy negative by producing a characteristic curve with the required slope in the mid tones but enhances the contrast of the highlights in the copy. To accomplish this, the characteristic curve of the internegative material has a shoulder of increasing gradient in place of the more usual decreasing gradient.

With such a material incorporating both coloured coupler masking and tone scale correction excellent colour prints and duplicate transparencies can be made from colour slides.

INTERSECTION OF THIRDS. Popular art term decribing the division of the picture area into thirds by two horizontal and two vertical lines. The points where these lines cross have been claimed to be particularly suited for placing the picture's "centre of interest". This recommendation means, of course, turning into an over-simplified positive assertion the truth that neither a strictly symmetrical position nor one near the edges is likely to be the best for most subjects.

See also: ***Composition; Golden section (secto aurea).***

INVERSE SQUARE LAW. When a surface is illuminated by a point source of light, the intensity of illumination produced is inversely proportional to the square of the distance separating them.

This is the law of inverse squares; it is easier to understand from a practical example. If the distance between the light and the surface is doubled, the intensity of illumination is reduced to one quarter; at three times the distance, it is reduced to one-ninth and so on.

See also: ***Illumination.***

INVERTED TELEPHOTO LENS. Lens designed to have a short focal length but a relatively long back focus—i.e., a short focus lens which works at a relatively long distance from the plate. The construction consists of a group of diverging lenses followed by a converging group.

Lenses of this type are used in one method of colour cinematography where room must be left between the lens and the film to accommodate a prism system.

See also: *Wide angle lens.*

INVERT (INTAGLIO) HALF-TONE. Photomechanical reproduction process similar in some respects to photogravure. The image is etched into the cylinder to comparatively equal depth and the tones are produced by a half-tone dot formation. The Dultgen process is a well-known form.

INVISIBLE RAYS. Visible light forms only a small proportion of the range of electromagnetic vibration. The whole range extends from radiations with a wavelength of several miles to those with a wavelength of less than a hundredth of an Angström unit (one ten-millionth of a millimetre).

Over this immense range of radiations only a small fraction (from about 4,000 A to about 7,000 A—i.e., violet to red) is visible as light.

Above and below the visible spectrum the radiations take the form of invisible rays. Below—i.e., longer than—the visible red rays there are infra-red, heat, and "wireless" waves, all of which are invisible. Above—i.e., shorter than—the visible violet rays, there are invisible ultra-violet rays, X-rays and gamma rays.

Some of these forms of radiation—e g., the infra-red, ultra-violet, X- and gamma rays—although invisible, are able to produce a developable image in a photographic emulsion. Furthermore, these radiations all behave in characteristic ways which differ from visible light—e.g., X- and gamma rays will penetrate opaque objects, ultra-violet rays cause certain substances to fluoresce visibly, and infra-red rays will penetrate fog and haze more readily than visible light.

Thus invisible rays can be utilized for certain types of photography that would not be possible with normal visible light.

Illustrations: Plate Section XV.

Book: *The Measurement of Optical Radiations*, by G. Bauer (London).

IODINE. Used in reducers and bleachers.

Formula and molecular weight: I_2; 254.

Characteristics: Black crystals. Attacks rubber and cork, therefore keep in glass stoppered bottles.

Solubility: Very slightly soluble in water, freely soluble in potassium iodide solutions, also in alcohol and ether.

IRIS DIAPHRAGM. Standard type of adjustable lens stop. It consists of a number of thin metal or vulcanite leaves arranged to form a roughly circular aperture. The aperture size can be varied by movement of a lever.

See also: *Diaphragm.*

IRIS MOUNT. Type of lens mount that stands out from the lens panel with the iris diaphragm adjustment well clear.

See also: *Lens mounts.*

IRON ALUM. Chemical used in some toners and reducers. Available as purple crystals which give a yellow solution.

See also: *Alum, Iron.*

IRON PRINTING PROCESSES. Various largely obsolete methods of making prints from negatives by the conversion of certain ferric to ferrous salts on exposure to light. Same as iron salt processes. Blue prints are made by an iron printing process.

IRON SALT PROCESSES. Photographic processes utilizing the action of light on ferric salts which are reduced to the corresponding ferrous salt. The latter acts as an intermediary in forming a visible image. Thus it can be converted into Prussian blue (blue print process) or used to reduce silver salts in the emulsion to silver (Kallitype and other iron-silver processes) or to reduce palladium or platinium salts (palladiotype and platinotype).

In a variation of the blue print process (Pellet process) the unexposed iron salts are converted into Prussian blue to form the image.

Yet another type of iron salt process is based on insoluble coloured compounds formed by ferric salts with gallic or tannic acid (ferro-gallic and ferro-tannic process).

See also: *Fabric printing; Ferro-prussiate process; Printing processes (obsolete).*

IRRADIATION. Rays of light falling on the sensitive emulsion tend to be scattered by reflection from silver halide crystals as they travel through it. This scattering effect is called

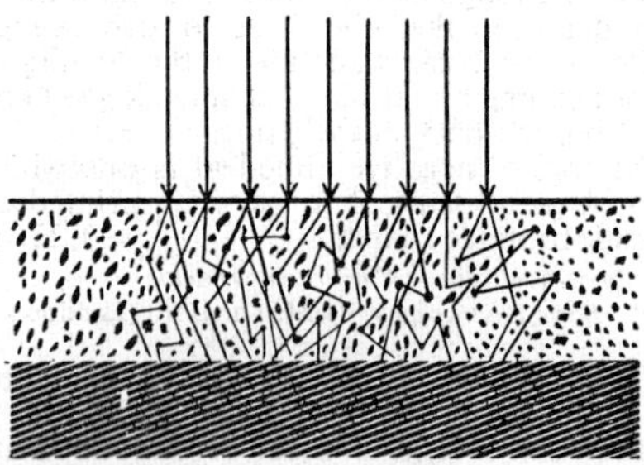

IRRADIATION. The scattering of light by the grains in the emulsion causes the image to spread sideways from the directly exposed area, resulting in a diffused surround.

irradiation; it is caused by the physical structure of the emulsion.

The effect of irradiation is to spread the image and destroy definition—i.e., it lowers the resolving power of the emulsion and is one of the reasons why thick emulsions have a lower resolving power than thin.

Irradiation increases the effect of halation—i.e., image spread caused by light rays scattered within the emulsion layer after reflection from the surface of the support.

The effects of irradiation can be minimized by processing emulsions in a surface developer. In this way, the image is confined to the surface of the emulsion and spreads less severely.

See also: *Faults; Flare; Ghost; Halation; Negative materials.*

ISENRING, JOHANN BAPTIST, 1796–1860. Swiss painter, engraver, daguerreotypist and photographer. In December, 1839, Isenring produced daguerreotypes of his native town, St. Gallen, Switzerland, and from May 1840 took experimental portraits of friends. In July 1841 opened a studio in Munich. He was one of the first to retouch and colour daguerreotypes. Biography by E. Stenger (Berlin 1931).

ISOCHROMATIC. Term much used at one time to describe emulsions which were more or less sensitive to colours in proportion to their visual intensity. It covered many types of sensitized materials, some of which were sensitive to all colours, and some to only a part of the visible spectrum. It has now been dropped in favour of orthochromatic and panchromatic.

ISOTOPES. Atoms of an element having different atomic masses (i.e., different individual atomic weights). As long as they still have the same atomic number, they are chemically the same element and behave like it in practically all respects. They cannot be separated by normal methods of chemical analysis but only by physical means which depend on atomic weights. For example, heavy hydrogen (atomic mass 2) is an isotope of ordinary hydrogen (atomic mass 1).

Most elements have several natural isotopes. Artificial isotopes may be of photographic use as they are often radioactive. These are produced in an atomic reactor and are available as a convenient source of gamma rays. They are used in industrial radiography for the non-destructive testing of welds and castings.

See also: *Gamma radiography.*

ISRAEL. Photography came to the land of the Bible many years before the creation of the modern state of Israel. It played an important part in documenting the events leading to the establishment of the nation's statehood, and has been used to record the history of the last fifty years. Photographs were a publicity tool serving to inform the outside world of the events of the time, as well as of the aspirations which caused those events. The pictures had to be realistic enough to convince, and also sufficiently beautiful to catch the eye and fire the imagination of the viewer.

Industry and Commerce. No photographic equipment is manufactured in Israel although items such as studio tripods and light stands are sometimes built to special order. Bromide and gaslight papers are produced by two local firms, but they are not of the highest standard. Most foreign brands are now readily available, though prices are somewhat higher than abroad. Cameras and projectors displayed in the shops are mostly Japanese, U.S., German and Russian (available in that order).

Commercial photography is quite new and there are only three commercial studios worthy of the name in the whole country as yet. The first large scale D & P laboratory was opened in Jaffa only a short time ago. It accepts black-and-white work from all over the country, mostly through a chain of supermarkets. All photographic shops and most studios accept amateur work which they develop and print themselves. Most of the darkrooms are small, one-man affairs. Drugstores and chemists generally do not sell films nor accept D & P. Colour processing is done by some shops and at least one manufacturer's agent in Tel Aviv accepts Kodak film for processing in the U.K.

Exhibitions and Archives. Photographic exhibitions are held from time to time, and are usually one-man shows. Most take place at the Artists' Club, Jerusalem, or at the Journalists' House, Tel Aviv. A few competitions have been sponsored by the Prime Minister's Office, and the entries exhibited to the public.

Today the Jewish Agency's Jerusalem office contains the greatest photographic archive in the country, some 75,000 negatives including work by Kluger, Lerski and other masters. Another large photo archive is that of the Government Press Division's Tel Aviv office. It was started in 1948 and now contains about 50,000 negatives. Both these archives serve the general public for a nominal fee.

Associations. Photo-journalists are organized in the Irgun Tsalamei Haitonut (Press Photographers' Union), which is very exclusive. Illustrated magazines buy pictures from all sources, although Government Press Cards are distributed only on recommendation by the Union.

Studio owners belong to the Igud Baalej Melaha (Artisans' Organization), while the Igud Hasoharim (Dealers' Organization) represents photographic retailers.

There is little contact between amateurs, photo-journalists and portrait photographers. Amateurs are numerous but quite unorganized. So far all attempts to start photo clubs have been short-lived.

Training. Little proof of qualifications is required of newcomers to the trade and formal training is rare. When a lack of industrial photographers was felt, the Government sponsored, in 1960, a two-year course for photographers' assistants in Jerusalem. One school for girls, at Haifa, also runs photographic courses.

Publications. The two million Israelis read about 150 illustrated magazines including one for women, three movie journals and three armed service magazines (published by the Army, Navy and Air Force). There are also publications on electronics, architecture, engineering, motoring and the theatre. There are some weeklies of general interest in Hebrew as well as in numerous foreign languages. Many of these employ staff photographers, but all accept freelance work when suitable matter is offered.

Photographic magazines on the news stands include many publications from the U.S.A. and Great Britain. Photographic literature is available mostly in English, German and French.

Photography Today. Visitors to Israel will find a diversity of subject matter. The beaches of Tel Aviv and the long Mediterranean shore are a photographer's paradise. Antique ruins and statuary are to be found at most historical places mentioned in the Scriptures. Many villages are peopled by new immigrants from some seventy countries, who still live according to their native customs and frequently wear the costume of their country of origin. Roughly half of the country is desert land, roamed by Bedouin and their herds of sheep and cattle. Eilath on the Red Sea offers the world's best underwater photography. The Government Tourist Office publishes a weekly information booklet on parades, shows and festivals of interest to the tourist. It is distributed free at hotels and Tourist Information Centres.

It should be borne in mind that the light in Israel is harsh, with exceedingly black shadows and blinding highlights, and thus many newcomers under-expose their black-and-white pictures and over-expose their colour slides.

During the three months of winter, December to February, torrential rains may be expected and these constitute a real danger to equipment. The hot winds of spring and autumn make films and paper curl up, and load them with static electricity. These winds also carry, from the Arabian desert, sand which can penetrate even tightly closed camera bags. It is therefore a good idea always to have some polythene bags ready, in which to stow equipment and films against rain and wind.

In the heat of summer, cameras and films should be kept out of the sun (and such places as a car's glove compartments) as much as possible. Colour films should be kept in refrigeration unless used within a few weeks, and processed as soon as possible after exposure. Films should be removed from the refrigerator a few hours before shooting to avoid condensation of moisture, and cans should not be sealed airtight after exposure.

Visitors should bear in mind that orthodox Jews object, on religious grounds, to being photographed, and Bedouin frequently expect a tip for posing. Certain military installations must not be photographed, and these are marked clearly by signboards in both Hebrew and English. W.H.

ITALY

Photography in Italy has roots which can be related to the sixteenth century. At that time, the camera obscura, the origin of which is very remote, was described by Leonardo da Vinci in the *Codice Atlantico*. Later, in 1568, the Venetian nobleman, Daniello Barbaro, improved it by putting a lens where there had formerly been a small hole through which passed the rays forming the image. The scientist, G. B. della Porta of Naples described it in his book *Magia Naturalis* (1558) and as a result its invention was often wrongly attributed to him. Nevertheless his name has always been linked with the history of the camera obscura as he made it widely known.

In the early days of photography proper, daguerreotypes were made in Italy, but few details are available as at the time Italy was split up into a number of small states which were not united until 1870.

Already, about 1853, there were excellent photographers, some using the collodion process, others albumen, among them Luigi Bardi in Florence, Ponti in Venice and Anderson in Rome—(the firm established by Anderson still exists today). At about this period, Robert MacPherson worked with great success for over twenty years in Rome. The Alinari brothers established a considerable business in Florence in the 1850s with views and architecture, later reproducing on collodion plates the principal works of painting and sculpture in Florence, Rome, the Vatican and elsewhere in Italy.

At the beginning of the present century they opened a collotype establishment for printing three-colour reproductions of the principal paintings in Italy, which have been exported throughout the world.

Italy has had some outstanding photographers. Typical of these was Sella, who produced excellent mountain studies on collodion plates with a telephoto lens. He was attached as official photographer to a number of high mountain expeditions including the Himalayan expedition to Karakorum (K2) led by the Duke of Abruzzi, and to Alaska, the Caucasus, etc.

Pesce of Naples had an excellent collection of historic cameras and lenses. Other equipment or parts, both photographic and cinematographic, are on display in the Milan Science and Technology Museum.

The Italian photographic plate industry started its activity in Milan during 1886, thanks to Michele Cappelli. During the first Italian Photographic Exhibition held in Florence in 1887, the dry silver bromide plate was presented to the public and placed on the market. Approximately 20–30 sq. metres of plate were manufactured daily during the first years, while as early as 1910 the daily average had reached 300 sq. metres. After passing through the austerity of the war period, and its shortage of glass supplies, the daily production of plates increased to 500 sq. metres in 1926. The manufacture of printing paper in Italy dates back to 1905, and by about 1914 100,000 sq. metres were produced monthly. Two plants for the manufacture of photographic and cinema films were established in 1919 at Ferrania, a locality among the mountains near Savona. One manufactured the celluloid base and the other the emulsion. The true and proper manufacture of photographic and cine negative and positive film started after 1920. Industrial production of radiographic films, filmpacks and 16 mm. sub-standard films started between 1926 and 1930. Colour films and paper for colour prints have been produced since 1949.

According to a survey carried out among the middle classes, 60 per cent of Italian families own a stills camera, and 6 per cent have a movie camera and projector. Of the films used for stills photography, 42 per cent are made in Italy.

The number of photographic dealers has greatly increased since the late war and now totals over 5,000. They are supplied by about thirty-five wholesalers. Very few of these dealers are now equipped to do their own finishing; they call on the services of numerous independent photo-finishers, some of whom are fully equipped for colour.

Professional Training. At present, twelve schools of photography offer professional training to those who intend to take up photography as a career. Half of these are state-run while the others are private concerns. The State schools include those of Milan, Turin, Florence and Rome, where the courses last seven years, while in the others the period of study ranges from 2 to 4 years. Once they have their diplomas, the students from these schools easily find good jobs, particularly in large concerns with photographic laboratories. However, no one wishing to set up in business as a photographer is required to produce evidence of his qualifications or general education, but only needs to fulfil certain formalities of an administrative character. There is no organization linking professional photographers together for technical or instructional purposes.

The number of professional portrait photographers has greatly decreased in recent years, while industrial, publicity, press and action photography has become correspondingly more popular.

Amateur Clubs and Societies. The leading association of Italian photographers in the nineteenth century was the now defunct Società Fotografica Italiana (Italian Photographic Society) of Florence. Its president was the late G. Pizzighelli, who was well known for his researches in photographic optics and platinum paper. The society with the largest membership today is the Società Fotografica Subalpina (Subalpine Photographic Society) of Turin, which also dates from the past century. In the past five or six years there has been a considerable increase in the number of societies throughout the country, which now total 200, with an aggregate membership of over 9,000. They are particularly active in the organization of exhibitions. Many are autonomous, while others form part of the Circoli Ricreativi per i Lavoratori (Workers' Recreation Clubs) which have been formed for the employees of all the larger firms. The majority of photographic societies are affiliated to the Federazione Italiana Associazioni Fotografiche (Federation of Italian Photographic Associations). Two photographic societies broke away from the Federation with the declared aim of raising the standard of pictorial photography. One is now extremely active, especially in organizing exhibitions of members' work.

The older-established societies generally have large premises for meetings, and equipment for still and sub-standard cine projection; some also have a completely equipped darkroom for the use of members. Owing to the high rents now prevailing, most of the more recently formed societies have no proper headquarters, meetings being held at cafés and in members' own homes.

Publications. Five photographic journals, all monthly, are published in Italy. *Ferrania, Fotografia* and *Progresso Fotografico* in Milan; *La Gazzeta della Fotografia* at Palermo, and *Il Corriere Fotografico* at Turin. Italian editions of the West German *Foto-Magazin* and the American *Popular Photography* are also published. There are three news-sheets on photography: *Il Commercio Fotografico* (monthly) and *Punto Focale* (monthly) of Rome, and *Il Giornale del Fotografo* (quarterly) of Trieste. Three publishing companies specialize in books on professional and amateur photography. A.O. (C.M.)

Industry. The photographic industy in Italy is comparatively small and concerned mainly with the production of sensitized materials. The number of manufacturing establishments in operation in 1963 was approximately 25, with a total employment figure in the region of 7,500 persons. Of this figure, the equipment sector accounts for the majority of producers, whilst sensitized materials manufacturers number about half a dozen, although the two sectors have about equal numbers of employees.

The industry is principally located in Northern Italy, centring on Milan, whilst there are also several producers around Florence and Rome.

Among leading manufacturers of equipment is Bencini of Milan, producers of inexpensive still cameras particularly for export markets, Durst at Bolzano, well known for their enlarging equipment and also still cameras, and Ferrania Apparechi Foto, makers of still cameras and accessories. In the field of projection apparatus, the firms of Microtecnica and Silma, both located at Turin, produce cine projectors, whilst Malinverno, established at Milan, manufactures stilll projectors.

The sensitized materials sector is dominated by the Ferrania Company of Milan, makers of all types of photographic film and paper, whilst there are in addition a small number of manufacturers of plates and paper. In 1963, Ferrania had a sales figure of £17 million, and the number of persons employed was in the region of 4,700.

During 1964, Ferrania was acquired by the Minnesota Mining and Manufacturing Company of America, who also control Bauchet, the French film and paper producer.

Since the sectors of the photographic industry form only a minor part of the optical goods and chemical industries, separate statistics of production are difficult to obtain. However, the following figures have been estimated for the value of output over the last five years, and demonstrate the predominance of sensitized materials.

No breakdown of this data is available, but in the light of export statistics the production of cine projection equipment appears to be a leading activity among apparatus manufacturing. Sensitized materials have shown a moderate growth in output since 1959, among which cine film probably now accounts for about one-third of the total value, whilst X-ray film is also an important product.

Imports. Evidence of a growing domestic market for photographic goods is provided by the threefold increase in the value of imports which has taken place since 1959, reaching a total of Lire 21,214 million in 1963.

Almost half of the total is now accounted for by apparatus, among which still cameras are a leading product. The majority of these are of the amateur type (Lire 2881 million in 1963), the principal supplier of which is West Germany (Lire 1,772 million during 1963).

Imports of cine equipment have followed a sharply rising trend over the last five years, the larger part of which is attributable to cameras. Among these the 8 mm. type predominates, totalling Lire 2,382 million in 1963, the main sources of which were (in million of Lire), Japan (1,229), Switzerland (514), and West Germany (341). Cine projection apparatus is also composed mainly of 8 mm. gauge models (Lire 1,480 million in 1963), of which Austria supplied Lire 301 million's worth and Switzerland Lire 205 million's worth.

The leading place among imports of sensitized material is held by cine film, which accounted for 35 per cent of the Lire 10,932 million total in 1963.

The greater part of this item is supplied by France (Lire 2,032 million), whilst the U.S.A. (Lire 655 million), Belgium (Lire 577 million), and West Germany (Lire 509 million) are also important sources.

Plates and still film together form another main product among imports, totalling Lire 3,407 million in 1963, of which the roll film part was valued at Lire 2,203 million. France is the leading supplier of this, with Lire 816 million, whilst West Germany was responsible for Lire 603 million's worth and the U.S.A. for Lire 399 million's worth.

Another important material is sensitized paper, which has more than doubled in value over the past five years to reach Lire 2,966 million in 1963. West Germany is the principal source of this, with Lire 1,275 million during the year, whilst France supplied Lire 410 million, and Belgium Lire 485 million's worth.

Exports. The value of exported photographic equipment and materials has expanded two and a half times since 1959, to a total of Lire 13,418 million in 1963.

In keeping with the Italian industry's pattern of production, about a third of total exports is represented by apparatus, among which cine projection equipment holds the leading position. Exports of the latter more than quadrupled since 1959, reaching Lire 1,985 million during 1963, consisting of 16 mm. and larger gauge projectors worth Lire 1,257 million (of which the U.S.A. was the chief purchaser, taking Lire 329 million's worth), together with 8 mm. equipment totalling Lire 728 million (the principal market for which was France with Lire 135 million's worth).

Exports of cine cameras are small in comparison, and consist mainly of 16 mm. and larger gauge models. Still cameras, however, form an important item, the bulk of which are of the less expensive amateur type (Lire 272 million in 1963), the principal markets for the latter being the U.K. (Lire 84 million), and West Germany (Lire 62 million).

Among other still and cine apparatus, the largest group is composed of still projectors, enlargers and reducers, together totalling Lire 1,137 million in 1963, of which Lire 225 million reached the U.S.A.

Sensitized materials exported have shown a substantial rise in value in recent years to reach Lire 8,851 million in 1963. The most important product is cine film, totalling Lire 3,541 million during 1963, for which the largest single purchaser is West Germany, receiving stock worth Lire 1,485 million, whilst Yugoslavia took Lire 104 million's worth.

The other leading product group is composed of still film and plates, worth Lire 4,772 million in 1963, among which the largest part is due to sheet film (Lire 3,683 million). The main purchaser of this latter material is France (Lire 819 million), whilst the U.S.A. (Lire 711 million) and West Germany (Lire 426 million) are also important markets. Among other items, roll film accounted for Lire 1,018 million during 1963, of which Yugoslavia received Lire 348 million's worth. G.R.H. & A.G.B.

ESTIMATED ITALIAN PRODUCTION OF PHOTOGRAPHIC GOODS

Milliards * of Lire

(Manufacturers' Net Selling Price ex-works)

	1959	1960	1961	1962	1963
Still and cine equipment	2·4	2·4	3·0	3·7	4·5
Sensitized materials	13 4	15·6	16·0	19·6	22·0
Total production	15·8	18·0	19·0	23·3	26·5
(£ million)	9·1	10·3	10·9	13·4	15·2

*Thousand million.

ITALIAN IMPORTS OF PHOTOGRAPHIC GOODS

Million Lire c.i.f.

	1959	1960	1961	1962	1963
Still cameras	1,330·0	1,725·8	2,036·4	2,519·8	2,912·8
Cine cameras and parts	919·0	1,390·3	1,550·5	2,130·9	2,674·6
Cine projectors and parts	410·5	596·6	938·5	1,042·3	1,679·3
Still projectors, enlargers and reducers	860·3 }	1,242·6 }	1,954·6 }	827·6	815·1
Other still and cine apparatus and parts				1,997·7	2,200·0
Total apparatus	3,519·8	4,955·3	6,480·0	8,518·3	10,281·8
Sensitized paper	1,291·4	1,528·4	1,951·5	2,315·4	2,966·3
Roll film	1,441·9 }	1,162·3	1,525·2	1,717·8	2,203·0
Plates and sheet film		690·9	699·7	1,011·2	1,204·3
Cine film	1,746·4	1,648·6	3,324·7	3,810·8	3,845·8
Chemicals	208·2	244·4	468·2	593·9	712·6
Total sensitized materials and chemicals	4,687·9	5,274·6	7,969·3	9,449·1	10,932·0
Total imports	8,207·7	10,229·9	14,449·3	17,967·4	21,213·8
(£ million)	4·7	5·9	8·3	10·4	12·2

ITALIAN EXPORTS OF PHOTOGRAPHIC GOODS

Million Lire f.o.b.

	1959	1960	1961	1962	1963
Still cameras	274·2	293·4	424·6	387·0	299·2
Cine cameras and parts	13·4	77·9	164·7	125·0	245·9
Cine projectors and parts	369·7	760·8	859·4	997·9	1,984·5
Still projectors, enlargers and reducers	852·6 }	964·1 }	1,329·6 }	1,080·9	1,136·9
Other still and cine apparatus and parts				646·3	900·3
Total apparatus	1,509·9	2,096·2	2,778·3	3,237·1	4,566·8
Sensitized paper	190·2	258·9	322·3	464·1	412·5
Roll film	1,706·8 }	2,042·2 }	358·2	973·7	1,018·3
Plates and sheet film			2,921·0	2,756·7	3,753·4
Cine film	1,802·7	2,317·7	3,416·7	4,027·0	3,541·3
Chemicals	35·0	74·6	125·9	105·9	125·5
Total sensitized materials and chemicals	3,734·7	4,693·4	7,144·1	8,327·4	8,851·0
Total exports	5,244·6	6,789·6	9,922·4	11,564·5	13,417·8
(£ million)	3·0	3·9	5·7	6·6	7·7

ITALIAN EXPORT MARKETS FOR SENSITIZED MATERIALS, 1963
Million Lire f.o.b.

Country	35 mm. Cine Film	Roll Film	Sensitized Paper	X-ray Film
European Common Market:				
Belgium	11	22	13	63
France	155	213	72	779
W. Germany	1,382	72	15	411
Netherlands	2	2	4	101
E.C.M. Total	1,550	309	104	1,354
European Free Trade Area:				
Austria	44	10	3	62
Denmark	16	10	3	67
Finland	2	5	2	12
Norway	—	10	—	31
Portugal	16	17	12	87
Sweden	16	22	4	129
Switzerland	10	25	11	42
U.K.	16	22	1	1
E.F.T.A. Total	120	121	36	431
Argentine	66	4	2	40
Brazil	48	2	10	4
Czechoslovakia	40	9	7	—
Hong Kong	111	10	1	2
U.S.A.	230	18	2	692
Yugoslavia	80	99	10	273
All other countries	509	446	241	931
Total exports	2,754	1,018	413	3,727

IVES, CHARLES, 1902–62. American photographic scientist and engineer. Pioneered research in motion picture engineering and developed new methods and techniques for rapid processing in high temperature solutions. From 1940, supervised the Photographic Chemistry Department of Eastman Kodak; 1956–7 member of the Department of Military Photography, and later Technical Staff Assistant to the Head of the Applied Photography Division.

IVES, FREDERIC EUGENE, 1856–1937. American photographer and inventor. Among his 70 U.S. patents are many concerning improvements in the half-tone process, the most important being an improved half-tone screen, 1885.

For colour photography with three separation negatives, Ives invented the Photochromoscope camera, 1891, and viewing instrument, 1892—later called Kromskop. He also brought out stereo versions of these, and the three-lens lantern Kromskop for projection in colour, 1895. Whilst his first colour camera worked with a repeating back, several later versions were one-shot colour cameras. Ives also invented a 3-D cine camera working on the Anaglyphic system, and the parallax stereogram process requiring no viewing instrument, 1903.

The latter was re-introduced after World War II as Vita Vision. Progress Medal of the R.P.S. 1903. Autobiography, Philadelphia 1928; biography by L. W. Sipley (Philadelphia 1956).
H. & A.G.

IVORYTYPE. Photographic positive intended to imitate a painting on ivory. The method of preparation (no longer practised) was as follows. A print on plain salted paper was stretched face up on a sheet of glass and coloured by hand. The print was then impregnated with wax, stripped from the glass and squeegeed face down on to a hot sheet of polished glass. The paper base, then translucent (owing to the wax), was finally backed up with white or ivory-tinted paper.

The same name was sometimes applied to transparencies prepared by the Eberneum process, which gave a similar effect. This method was more troublesome and never achieved the popularity of the Ivorytype.

See also: *Eburneum process; Printing processes (obsolete).*

J

JANSSEN, PIERRE JULES CÉSAR, 1824–1907. French astrophysicist. Invented in 1874 a "photographic revolver" with focal plane shutter for the chronophotographic study of the Venus transit. Photographed the sun in 1877, the big comet of 1881 and sun spots in 1885. Worked on solarization (1880) and found in 1881 the reciprocity law failure. Received the Progress Medal of the Royal Photographic Society in 1906 "for his eminent services to photography, especially in its applications to solar and spectroscopic investigations".

JAPAN

Photography is used extensively in Japan in science and industry, and besides its normal commercial applications, it also has a very strong amateur following.

History. The first camera to be introduced into Japan was a daguerreotype apparatus brought through Dutch trade to Nagasaki in 1840 at a time when Japan's foreign commerce was limited almost wholly to that with Holland. A Nagasaki chemist, Toshinojo Ueno, acquired this daguerreotype outfit and began a study of photography, and today he is considered the father of photography in Japan.

Ueno's photograph of the powerful feudal lord Prince Nariakira Shimazu taken on 1st June 1841, in Kagoshima in southern Kyushu, is recorded as the first in Japanese history. The use of the camera for portraits and scenic photographs spread rapidly among the Japanese, who displayed considerable skill in the new medium of art. A number of students of photography emerged from among the great feudal lords, including Prince Shimazu and other nobles and leading scholars of the day as well as from among the common people, many of whom in the early period had regarded the new instrument with superstitious fear.

Photography spread in the various treaty ports and through the influence of the foreign residents until finally foreign trading firms and Japanese middlemen in Nagasaki and Yokohama began to deal in photographic supplies. Two major Japanese importers of photographic material were established by Tokichi Asanuma and Rokuemon Konishi, both of whom began as middlemen during this early period.

The pioneers in professional photography in Japan were Renjo Shimooka, who began his work in Yokohama in 1862, and Hikoma Ueno, who opened his studio in Nagasaki in the same year. Subsequently eminent pupils of these two pioneers began to attract attention, and with the advance of technical skill throughout the country photography developed into a highly respected profession and art.

During the Meiji Period (1868–1911) photographic research organizations were established by the Englishman W. K. Burton, Kazumasa Ogawa, Seibei Kajima and others.

This period saw the beginning of Japan's photographic industry and the first appearance of photographic publications. Amateur photography spread rapidly, and the foundation for today's popular and highly developed photographic culture, industry and profession was established. S.U.

Science. Photographic science and technical training have been incorporated in the curricula of such Japanese educational institutions as Kyoto and Chiba universities, where a photographic science department has been established. Tokyo Photographic Junior College has departments in photographic industry and photo-engraving, while the Chiba and Nihon universities offer motion picture courses.

Three of the four major photographic equipment and supply manufacturing companies in Japan maintain independent research laboratories, while the fourth has a research department. There are more than 300 independent photographic research workers, among whom are a number of university professors and graduate students.

Only a few of the many camera manufacturing companies have their own research laboratories and their staffs are relatively small. However, the ability of the research workers in the lens-making companies and in universities is recognized throughout the world.

The Society of Scientific Photography of Japan, established in 1926, and the Optical Science Section of the Society of Applied Physics of Japan, founded in 1952, have a total membership representing 90 per cent of the aforementioned scientists and research workers and regularly hold lectures, discussions and other meetings and publish periodicals. During 1954, over 150 research papers and reports on various aspects of photography were published.

In assessing the comparative levels of Japan's photographic science and technology, one can note that the quality of Japanese sensitive materials is the most highly advanced, followed by optical science. Cameras and accessories come third on the list.

The war and the confused situation of the immediate post-war years were major handicaps which hampered Japanese photographic research workers and cut them off from activities in foreign countries. However, various projects were undertaken. One group conducted a research into the reaction mechanism of chemical sensitizers and their adsorption to the silver halide surface by means of electrochemical methods.

Other groups sought to account theoretically for various anomalous effects of sensitizing. The research laboratory of one leading photographic manufacturer conducted studies to explain the mechanism of latent image formation and chemical sensitizing by experiments on the failure of the reciprocity law. Since 1941, research has been carried out on the ripening inhibitor contained in photographic gelatin. Natural and artificial ripening inhibitors for bromide emulsion, including known and unknown constituents, were tested and more than 130 inhibitors were synthesized. A part of these results was presented to the International Congress of Photography in 1953. Some excellent inhibitors were discovered by these researches—e.g., vitamin B_1 and penicillin G and their derivatives—and a convenient process was worked out for testing inhibiting power by the silver potential titration of the fog inhibitors.

An electron microscope cinematograph film dealing with the change of silver salt under exposure was successfully made by the same research laboratory referred to above. An ingenious process for manufacturing silver halide emulsion continuously by mixing and overflowing, followed by washing, was developed. Another firm introduced an improved process for manufacturing a new kind of printing paper in which polyvinyl alcohol is used as a vehicle instead of gelatin. This photographic paper has been placed on the market. S.F.

Amateur Photography. About 3 million people own cameras in Japan, and of these about 60,000 are members of nearly 1,500 clubs, societies and other amateur photographic organizations established throughout the country. Of these organizations, there are eighty-eight amateur photographic clubs which have over one hundred members each, and twelve which have over 1,000 members each. The largest society is the All-Japan Association of Photographic Societies which has a membership of about 10,000 and is sponsored by the newspaper *Asahi Shimbun*. This organization was established in 1926, and among its annual activities are the International Photographic Salon and the All-Japan Photographic Salon.

Photographic contests are held regularly and are extremely popular. Among the outstanding ones is the Fuji Photo Contest sponsored by the Fuji Film Company, and in 1955 this had nearly 114,000 entries. In Tokyo there is an average of seven photographic exhibitions every month.

There are nine photographic magazines, of which the *Asahi Camera*, with a circulation of 180,000, is the largest. The *Asahi Camera*, *Camera Mainichi*, and *Sankei Camera* are published by newspaper companies. In addition, there are the *Nippon Camera*, *ARS Camera*, and other widely circulated photographic magazines. *ARS Camera*, which was first published in 1921, is the oldest amateur photographic magazine in Japan. There are several photographic annuals, such as the *Asahi Camera Annual*, *ARS Photographic Annual* and others. Lectures and books are also widely published.

Pictorial Photography. The carbon process was introduced into Japan about 1877 following research and experiments by Matsusaburo Yokoyama, but it was not widely adopted. However, after the introduction of gum prints and through exhibitions of foreign photography, from about 1894 Japanese photographers began to take an increased interest in pictorial photography.

With the establishment of the Tokyo Shashin Kenkukai and the Nihon Shashinkai (Photographic Society) in 1907, the trend toward pictorial photography became more marked. In 1915 a photographic course was included in the curriculum of the Tokyo Fine Arts School. In 1921 the magazine *Photographic Arts* was published.

Until about 1925 the various photographic trends apparent in English photographic periodicals and annuals exerted considerable influence in Japan. This phase produced such pictorial photographers as Shinzo Fukuhara.

The German trend of pictorial representation influenced Japanese photography from about

1930, but only for a short time; otherwise the tradition remained unchanged. Following the Pacific War, the influence of American journalistic photography with its emphasis on realism and humanism of the social scene became a major trend in the creative art and there was an increase of interest in workers like Ihee Kimura and Ken Domon, who are representative of the pictorial photographers of this period.

The International Photographic Salon is the major photographic exhibition in Japan and is displayed in Tokyo and other cities; the 15th was held in 1955. Other important annual photographic exhibitions are the Japan Photographic Salon, and the Japan Photographic Art Exhibitions. In addition, a large number of photographic collections are printed and published each year. S.K.

Industry. The photographic industry in Japan—established mainly in the Tokyo area—has shown impressive expansion in output over the last five years, although the number of producers has remained relatively steady. In 1963 there were about fifty companies engaged in the production of photographic equipment and materials, many of which are also manufacturers of non-photographic goods, together providing employment for approximately 25,000 people.

By far the larger part of the industry has traditionally been occupied with the production of equipment and parts. Some thirty-five companies are mainly concerned with cameras, a further four are fully integrated manufacturers of lenses, whilst the remainder, apart from those which compose the sensitized goods sector (the two most important of which also manufacture cameras), are firms undertaking the production of accessories and parts.

There are six leading equipment manufacturers, which together account for about 50 per cent of the industry's total camera production, and an indication of their size may be gained from the number of their employees in 1963, which were as follows: Asahi Optical (1,300), Canon Camera (3,600), Minolta (1,800), Nippon Kogaku (2,200), Olympus Optical (600), and Yashica (2,600). Among these, during 1963 the Canon Camera Company produced over 450,000 still cameras and about 50,000 interchangeable lenses, whilst Yashica's output of still cameras was in the region of 700,000 units in addition to 150,000 cine cameras. Olympus Optical, who account for about 80 per cent of the industry's total production of "half-frame" 35 mm. still cameras, produced about 750,000 units.

The sensitized goods section of the industry has not been so highly developed and, although manufacturing all types of material, lags considerably behind that of other countries in the West, particularly in the field of colour film. Estimated total employment in this sector of the industry during 1963 was between 5,000 and 10,000 persons.

The leading manufacturer is the Fuji Photo Film Company, employing 5,900 persons in 1963, which produces still and cine film, X-ray film, sensitized paper, photographic chemicals and optical glass, and has in recent years begun the production of cameras and projection equipment. This company supplies about three-quarters of the domestic market for all types of film and approximately one-third of that for sensitized paper.

The other major producer of sensitized goods is the Konishiroku Photo Company, which has 3,600 employees and manufactures all types of film and plates and sensitized paper, in addition to cameras. There are two other main companies engaged in the manufacture of photographic paper, one of which also produces film, in addition to a small number of firms active in the photocopying field.

The industry has always been dependent to a large extent upon exports. Its development has been facilitated by the unrestricted entry of photographic goods allowed by the United States, which takes about one-third in value of total Japanese photographic exports, and more recently by the relaxation of tariff and quota barriers against Japanese equipment by several European countries. Government assistance in the form of company taxation relief and loans at low interest rates granted to export industries has also aided expansion.

There are three trade associations serving the photographic industry, the most important of which is the Japan Camera Industry Association. This is financed by its forty-five members, who account for about 90 per cent of the producers of cameras, lenses and accessories, and in addition to the usual functions of such bodies provides services in connection with design, research, and sales promotion work. A second organization, the Japan Camera Inspection Institute, has the function of inspecting and maintaining standards of quality in equipment destined for export markets.

Production. The value of Japanese production of photographic goods has more than doubled over the five years from 1959 to reach a total of $291·5 million in 1963.

The larger part of output is composed of equipment, totalling $199·1 million in the latter year, among which the most important items are still and cine cameras. An impressive rise in both quantity and value of these has taken place.

The type of still camera manufactured has changed since 1959 to emphasize the leading place held by 35 mm. and latterly the smaller format models, which together now compose the bulk of output, whilst the twin-lens reflex type has declined and the production of folding models has almost ceased.

The majority of cine cameras are 8 mm. models, and the large proportion of output taken by export markets is shown by the figures for 1963 when about three-quarters of total production in both quantity and value was sent abroad.

Among other items of importance are lenses, and still and cine projectors, the output of the latter having been relatively low up to the present but is now undergoing expansion.

Sensitized goods, although accounting for less than one third of the total photographic production value in 1963, have shown considerable growth over the past five years.

Output of the largest single item, sensitized paper, has doubled in both quantity and value since 1959 to reach 37·6 million square metres worth $28·6 million in 1963, nearly all of which is due to black-and-white material. X-ray film has also shown expansion during the period to 3·6 million square metres with a value of $16·8 million in the latter year.

Production of still film and plates has expanded by about two-thirds in total over the past five years to reach 5·8 million square metres worth $33·6 million in 1963, almost all of which is composed of black-and-white material. In contrast, output of cine film has recorded only a small amount of growth, reaching 4·4 million square metres with a value of $13·4 million during 1963. The larger part of this is also accounted for by black-and-white stock, although the production of colour material is gradually increasing.

Imports. Imports of photographic equipment and materials by Japan are comparatively small in relation to the volume of goods exported. This is to a large extent the result of tariff and quota restrictions, coupled in the case of equipment with the expansion of the domestic industry in recent years, although during 1964 such restrictions are being released somewhat. Nevertheless, the total value of imported products has rather more than doubled in the five years since 1959 to reach $16·8 million in 1963.

Less than one-third of the total is attributable to apparatus, among which the largest single item is cine cameras. Imports of these totalled 2,150 units worth $470,000 in 1963, of which 16 mm. gauge models comprised 1,100 cameras valued at $445,000. The principal sources of the latter are the U.S.A., with 600 units worth $315,000 and Switzerland with 400 cameras valued at $47,000.

Still cameras represent the remaining product of importance, totalling 2,200 units with a value of $337,000 in 1963, the leading supplier of which is West Germany (600 cameras with a value of $88,000).

Sensitized materials comprise the larger part of imports and provided the bulk of their growth since 1959. Within the total value of $11·7 million in 1963, the largest product is cine film, worth $5·1 million. Among this product the major items are 35 mm. colour negative and positive stock (12·1 million metres worth $2·0 million), and 16 mm. black-and-white negative and positive material (27·3 million metres with a value of $1·4 million). The principal supplier of both types of film is the U.S.A., which accounted for 9·9 million metres valued at $1·7 million of the 35 mm. gauge stock, and 25·6 million metres worth $1·3 million of 16 mm. stock.

The other item of importance which has recorded considerable growth is roll film and plates, together totalling $4·4 million in 1963. The largest single products within the total are 35 mm. negative and reversal type film, reaching a total of 7,000 square metres with a value of $170,000, of which the U.S.A. supplied 3,500 square metres worth $95,000 and West Germany 3,000 square metres worth $66,000.

Exports. The value of photographic exports has recorded a substantial increase during the past five years, almost trebling since 1959 to reach a total of $91·5 million in 1963.

In keeping with the pattern of production, by far the greater part of exports is attributable to equipment which now accounts for 94 per cent of the total value, at $86·2 million. Still cameras are the leading product, exports of which reached 1·3 million units with a value of $36·0 million in 1963, and the accompanying table shows that the majority are 35 mm. models. The largest single market is the United States, taking 370,300 cameras worth $10·8 million during the year, with West Germany in second place (126,000 units valued at $2·9 million) and the United Kingdom third with 58,900 cameras worth $1·4 million.

The other major item which has shown an impressive rate of expansion in the period is cine cameras, exports of which attained 620,500 units with a value of $26·2 million in 1963. Nearly all of these are 8 mm. models (618,000 units worth $21·6 million), the principal purchasers being the U.S.A. with 300,700 cameras valued at $8·9 million, and West Germany with 50,100 cameras with a value of $2·1 million.

Among other products exported, photographic lenses hold an important place with 465,700 units worth $6·4 million in 1963, of which the United States received 231,000 lenses valued at 2·2 million.

Exports of cine projectors have also risen and almost quadrupled over the period, reaching 86,800 units with a value of $5·0 million during 1963. Of these, 8 mm. models accounted for 84,800 projectors worth $2·6 million, the principal markets being the United Kingdom (19,300 projectors worth $518,000) and West Germany (16,300 units valued at $476,000).

Still projector exports, although comparatively small, have risen since 1959 to reach a total of 131,000 units worth $1·3 million in 1963. Large purchasers of these are the United States, receiving 11,200 projectors worth $96,000, and the United Kingdom with 9,000 units valued at $81,000 during the year.

Exports of sensitized materials have expanded considerably over the past five years, although representing only a small part of total photographic goods, to reach $5·3 million in 1963.

The major product is sensitized paper,

totalling 7·1 million square metres with a value of £1·9 million, the larger part of which is composed of photocopy material. Principal markets for all types of paper in 1963 were the U.S.A. (3·2 million square metres worth $907,000), Formosa (0·5 million square metres valued at $280,000), and Sweden (0·5 million square metres worth $90,000).

Roll film and plates together account for the major part of the remaining materials exported, with 650,000 square metres worth $1·4 million in 1963. The largest item within the total is black-and-white roll film (493,000 square metres valued at $1·0 million), of which the leading purchasers are Brazil, taking 64,000 square metres valued at $196,000, and Hong Kong with 41,000 square metres worth $137,000.

Among other sensitized materials exported, cine films reached 17·4 million metres with a value of $862,000, of which 35 mm. black-and-white stock accounted for 10·2 million metres worth $426,000, the principal market being China with 4·4 million metres worth $151,000 in 1963.

G.R.H. & A.G.B.

JAPANESE PRODUCTION OF STILL AND CINE CAMERAS

	1959	1960	1961	1962	1963
Still Cameras					
Quantity (000s)	1,844	1,859	2,355	3,123	4,085
Value ($ million)	51·2	51·7	65·8	92·7	120·2
Cine Cameras					
Quantity (000s)	278	595	685	818	809
Value ($ million)	8·5	19·4	23·0	29·4	32·1

JAPANESE PRODUCTION OF PHOTOGRAPHIC GOODS

$ U.S. Million (Manufacturers' Net Selling Price ex-works)

	1959	1960	1961	1962	1963
Still cameras	51·2	51·7	65·8	92·7	120·2
Cine cameras	8·5	19·4	23·0	29·4	32·1
Still projectors	1·1	1·3	1·8	1·8	1·8
Cine projectors	5·1	6·4	6·4	8·8	8·5
Lenses	5·6	7·7	10·6	14·2	17·1
Other equipment and parts	9·7	9·7	11·6	16·4	19·4
Total apparatus	81·2	96·2	119·2	163·3	199·1
Sensitized paper	13·3	15·0	16·6	21·4	28·6
X-ray film...	10·8	10·3	12·6	14·6	16·8
Still films and plates	21·4	29·5	28·5	32·4	33·6
Cine film	9·2	7·9	8·0	7·5	13·4
Total sensitized material	54·7	62·7	65·7	75·9	92·4
Total production	135·9	158·9	184·9	239·2	291·5
(£ million)	48·5	56·8	66·0	85·4	104·1

JAPANESE PRODUCTION OF SENSITIZED MATERIAL

Thousand Sq. Metres

	1959	1960	1961	1962	1963
Sensitized paper	17,624	20,538	23,861	31,218	37,585
X-ray film...	2,153	2,383	2,597	3,041	3,602
Still film:					
Black and white	3,152	3,923	4,057	4,844	5,500
Colour	59	58	77	166	238
	3,211	3,981	4,134	5,010	5,738
Plates	161	156	133	149	127
Cine film:					
Black and white	4,227	3,473	3,438	3,225	3,113
Colour	1,061	1,502	1,462	1,229	1,297
	5,288	4,975	4,900	4,454	4,410
Total sensitized material	28,437	32,033	35,625	43,872	51,462

JAPANESE IMPORTS OF PHOTOGRAPHIC GOODS

$ U.S. Thousand c.i.f.

	1959	1960	1961	1962	1963
Still cameras	148	344	468	276	337
Cine cameras	326	326	452	525	470
Still projectors	71	245	396	215	377
Cine projectors	75	119	70	100	297
Lenses	78	137	240	235	400
Other equipment and parts	882	1,071	3,156	2,077	3,200
Total apparatus	1,580	2,242	4,872	3,428	5,081
Sensitized paper	166	163	303	536	1,102
X-ray film...	78	93	233	511	565
Still film and plates	911	1,071	1,564	2,415	4,392
Cine film	3,864	5,161	4,414	4,475	5,114
Chemicals	44	136	150	242	559
Total sensitized material and chemicals	5,063	6,624	6,664	8,179	11,732
Total imports	6,643	8,866	11,536	11,607	16,813
(£ million)	2·4	3·2	4·1	4·1	6·0

JAPANESE EXPORT MARKETS FOR STILL AND CINE CAMERAS. 1963

Country	*Still Cameras* No. (000's)	*Still Cameras* Value ($000's f.o.b.)	*8mm. Cine Cameras* No. (000's)	*8mm. Cine Cameras* Value ($000's f.o.b.)
European Common Market:				
Belgium	15·2	558·8	3·2	138·3
France	3·0	102·2	3·0	142·8
W. Germany	149·5	3,774·5	54·8	2,360·0
Italy	7·6	185·8	25·0	879·0
Netherlands	2·8	43·4	0·2	6·8
E.C.M. Total	178·1	4,664·7	86·2	3,526·9
European Free Trade Area:				
Austria	0·9	18·6	0·2	10·8
Denmark	2·8	80·9	1·9	81·8
Finland	1·3	26·1	0·7	31·6
Norway	2·1	49·2	0·8	36·5
Portugal	1·6	34·1	0·3	11·7
Sweden	23·3	608·5	3·9	141·4
Switzerland	85·2	2,474·1	59·5	2,911·0
U.K.	57·2	1,803·3	12·4	430·1
E.F.T.A. Total	174·4	5,094·8	79·7	3,654·9
Aden	13·3	372·0	7·8	300·0
Australia	32·2	902·1	8·3	324·4
Brazil	19·0	393·0	1·0	39·5
Canada	34·8	1,024·7	13·9	539·2
Hong Kong	55·9	1,639·8	9·3	580·1
Singapore	29·0	846·0	5·0	263·1
Union of S. Africa	22·7	633·8	7·5	316·2
U.S.A.	339·3	10,584·6	143·1	6,071·4
All other countries	279·0	9,794·5	52·7	2,509·9
Total exports	1,177·7	35,950·0	414·5	17,809·4

JAPANESE EXPORTS OF STILL CAMERAS BY TYPE

Type	1959 No. (000s)	1959 Value ($000s f.o.b.)	1963 No. (000s)	1963 Value ($000s f.o.b.)
35 mm. Focal Plane Shutter	44·4	2,905	186·3	11,454
35 mm. non-F.P. shutter	413·0	8,696	819·3	19,504
Twin-lens reflex	130·6	2,386	109·6	2,843
Folding	2·3	37	1·2	53
Others	99·0	1,031	135·4	2,096
Total	689·5	15,055	1,251·8	35,960

JAPANESE EXPORTS OF PHOTOGRAPHIC GOODS

$ U.S. Thousand f.o.b.

	1959	1960	1961	1962	1963
Still cameras	15,055	16,212	19,969	29,394	35,950
Cine cameras	4,801	11,960	15,873	21,783	26,200
Still projectors	639	839	1,450	1,220	1,335
Cine projectors	1,332	2,343	2,348	3,095	4,989
Lenses	3,163	3,634	4,126	4,698	6,365
Other equipment and parts	6,108	6,582	9,656	12,937	11,392
Total apparatus	31,098	41,570	53,422	73,127	86,231
Sensitized paper	739	655	959	2,085	1,885
X-ray film...	127	190	339	678	977
Still film and plates	632	885	1,053	1,282	1,436
Cine film	102	254	378	693	862
Chemicals	36	26	44	84	107
Total sensitized material and chemicals	1,636	2,010	2,773	4,822	5,267
Total exports	32,734	43,580	56,195	77,949	91,498
(£ million)	11·7	15·6	20·1	27·8	32·7

JENA GLASS. Name for the optical glasses made in the Thuringian town of Jena.

As photography developed during the last century, it started the search for glasses with a much lower dispersion in relation to refractive index than had previously been required for optical work.

Glasses of this specification were first produced by Otto Schott and Ernst Abbe, working together. They joined forces in 1886 with Carl Zeiss and with the backing of the Prussian government started the large scale manufacture of the new optical glasses at Jena.

In a short time they gained almost a monopoly of the optical glass trade. The Jena factories continued to enjoy this monopoly until World War I. Since that time the optical glass industry developed in Great Britain has ended the exclusive supremacy of the Jena factories in that field.

See also: *Optical glass.*

JENKINS, C. FRANCIS, 1867–1934. American pioneer of cinematography. Founder and first President (1916) of the Society of Motion Picture Engineers. Designed in 1893–94 the Phantascope projector and camera with continuous movement and optical compensation. Prolific inventor holding more than 300 patents; in later years active in the field of radio and television. Autobiography: *The Boyhood of an Inventor* (1931).

JEWELLERY. Photographs of jewellery fall into two distinct categories; real jewellery and imitation, or costume, jewellery. Imitation jewellery is photographed for pictorial or catalogue purposes only, so it is easier to deal with, presents no special technical difficulties and can be handled by the ordinary studio.

Photographing real jewellery, because of its extreme value, is highly specialized work. There are many uses for such photographs, but it is a fairly limited market for photographers. First, because jewellery of this expensive class is also frail, jewellers are not very keen to let it be taken away from their premises. Most of the really important jewellers, in fact, have a photographic department on the premises run by their own photographic staff. Some set aside a room for the use of an outside photographer who may keep his own apparatus set up in position for use at any time.

A knowledge of gemmology and the crystallography and structure of precious stones is a great advantage for the photographer.

Jewellery in this category is photographed for many purposes apart from the pictorial and advertising interest. In one large firm all jewels that are regarded as important new creations are photographed so that the pictures may be interchanged between the various branches all over the world, so that designs and trends may be maintained at a high standard. Albums of these photographs are kept in the showrooms to assist clients in choosing designs around which their old and out-of-date jewellery can be remodelled. The photograph of the original jewel before remodelling can be of great assistance to the designer when making up the new design. Photographs are taken of these original jewels to keep a record of the stones before they are unmounted in case of any dispute afterwards.

Important jewellers hold photographic records of all pieces of jewellery made by them to hand to the police in the event of loss or burglary.

Photographs are also taken of large and specimen stones on offer in the London or other markets. These are then sent to overseas branches for the consideration of likely purchasers as it is not advisable to keep sending stones backwards and forwards unless something definite is demanded. An X-ray photograph is sometimes taken of a pearl where there is a doubt about its being cultured. The pattern on the film caused by the deflection of the rays shows at once whether it is genuine or not. Photomicrographs taken of faults and inclusions enable a stone to be identified. Stones have actually been identified in a court of law by means of such a photograph taken years previously. A gem is often identified solely by photographs, and photography does provide a useful supplement to the normal records kept by the jeweller of each important stone.

Photographs are taken from a number of angles, including plan, from behind, and side views. These photographs are printed the same size and attached to the record. Enlargements may be made to indicate precisely the position of flaws and surface imperfections, and in some cases photomicrographs are made of superficial irregularities with the stone immersed in oil. The oil, which is highly refractive, creates an intense black line around the edge of the stone, clearly outlining the shape of the indentation.

Another photographic means of identifying stones is to record the pattern produced on a sensitive emulsion when a narrow pencil of light falls on the stone from a given angle. An extension of this method is to record the pattern of fluorescence produced by the diamonds in a piece of jewellery exposed to filtered ultra-violet light.

Diamonds are sometimes coloured artificially by exposing them to the radiations of radium. Where this has been done, it can be detected by putting the diamond in contact with a photographic plate in the dark for a period of several hours—if the stone has been radium treated, an image shows on the plate. X-ray records are also used to identify the details of the mount in which the gems are set.

Jewellery designers make considerable use of photography for enlarging or reducing a design, thereby saving themselves hours of work.

Equipment. Jewellery is usually photographed same size—as it loses a certain amount of definition and brilliance if enlarged. The best camera for this work is a 8 × 10 ins. process camera on rails mounted on trestles. A vertical easel covered with thick lino is used for pinning up jewels for photographing because they are displayed to better advantage hanging than lying flat.

An 18 ins. process lens is used as it is essential to have a long focus lens to keep the camera well away from the jewels and so allow sufficient space between lens and easel for manœuvring the lights. A vertical set-up on the principle of a copying camera is also useful for jewels that have to be laid down flat.

For general and perspective work, a whole plate stand camera with colour-corrected anastigmat lens is used. If it should be necessary to reduce the size of a jewel in taking the photograph, the amount of reduction is kept to the absolute minimum.

Lighting. The most suitable type of illumination is given by a floodlight with a general diffusing type of reflector, fitted with a 1500 watt tungsten light. A pointed or concentrated light certainly does not give as good results.

When lighting jewellery, the main source must always be supplemented with reflectors to fill in dark hollows in the stones which would otherwise debase their apparent quality. Small movable white reflectors on stands, adjustable white roller blinds on two sides opposite the light and a large white card reflector around the lens are used to highlight the reflecting surfaces of the jewels.

The main light source shines at an angle of 45° on to the subject.

Materials. Most coloured precious stones, such as emeralds, rubies and sapphires, reproduce darker on orthochromatic emulsions; and stones are usually more valuable when the colour is deeper. For this reason ortho films are often preferred. But a panchromatic emulsion is used when a fault or inclusion is to be photographed.

Colour transparencies of jewellery are made on reversal colour film, using tungsten light and a correction filter. It is usual for salesmen to take out 2 × 2 ins. colour slides and a viewer to avoid carrying too much jewellery.

Background. The detail and beauty in stones is always more apparent when the background is kept simple. A fine-grained linen paper—of a suitable shade—is generally sufficient.

To avoid shadows which may obscure the detail of the delicate mountings, the piece may be fixed to a sheet of glass with cellulose cement. The glass is then supported a few inches away from the background.

The finished print is made on a brilliant black-and-white bromide paper and is always glazed to add sparkle. A.C.K.

See also: *Glassware; Silverware.*

JOHNSTON, JOHN DUDLEY, 1868–1955. English amateur photographer. Honorary Secretary, Curator and Editor of the Royal Photographic Society. A founder of the Photographic Alliance of Great Britain and a well-known lecturer, exhibitor and exhibition judge. President of the R.P.S. 1923–25 and 1929–31; received their Progress Medal in 1951.

JOIN-UP. Panoramic and wide angle views can be made by joining several pictures, each of which contains part of the scene. The prints used must overlap each other to allow for trimming to exact line of joining. For minimum distortion, strict rules must be observed when shooting.

See also: *Panorama.*

JOLY, JOHN, 1857–1933. Dublin professor. Was a pioneer in screen-plate colour photography. His linear screen-plate process (1893) was the forerunner of many other additive colour processes, such as the Thames, the Paget, and the Dufay plate and film, to mention but the more important.

JONES, HENRY CHAPMAN, 1855(?)–1932. English chemist and photographer. President of the Royal Photographic Society, 1912–14. Instructor in, and writer of books on, photography. In 1901 introduced a sensitometer, under the name of the Chapman Jones Plate Tester, that was widely used by photographic manufacturers in England. It replaced the Warnerke sensitometer.

JONES, LOYD ANCILE, 1884–1954. American physicist and electrical engineer. Joined the Research Laboratories of the Eastman Kodak Company on their establishment in 1912 and worked there (later as Head of the Physics Division) until his retirement a fortnight before his death. Specialized in sensitometry and the measurement of the properties of photographic emulsions. Received in 1948 the Progress Medal of the Royal Photographic Society "for his great and outstanding contributions to photographic science and practice, notably in the field of sensitometry".

JOULE. Unit of energy, equal to 10,000,000 ergs, or 1 watt-second. It is named after J. P. Joule (1818–89) who first formulated a mechanical equivalent of heat energy. In photography joules are used to measure the energy discharged in an electronic flash tube. In terms of light output of an electronic flash, 1 joule is approximately equivalent to 40 lumen-seconds.

See also: *Light units.*

JUDGES FOR EXHIBITIONS. Most photographic exhibitions are judged by qualified persons having no direct interest in the organization holding the exhibition.

In Britain, the various area Federations of amateur photographic societies maintain panels of judges; these are listed collectively in a publication of the central body, the Photographic Alliance. All of these judges are available to affiliated societies. In some cases, specialists are listed for exhibitions devoted to particular subjects (e.g., nature photography).

In other cases, it is common for well-known photographers, picture magazine editors, and artists to be invited to judge photographic exhibitions. Provided that the exhibition is of a reasonable standard and size, a courteous letter will frequently produce agreement to judge an exhibition; but adequate notice must always be given.

Invited judges sometimes prefer to have the exhibition entries sent to their home or office for selection; others may visit the organizers' headquarters for the purpose. Always, however, selection is carried out at least a few days before the exhibition is due to open so that the catalogue can be prepared and printed. Frequently the judge will also attend either the opening of the exhibition or a meeting at which a commentary on the prints is given and any lantern slide entries are projected.

At all times, the decisions of the judge must be sincerely respected.

JUDGING EXHIBITION ENTRIES. Every year thousands of photographs are submitted to pictorial photographic exhibitions by entrants who hope that their pictures will be considered good enough to be accepted for hanging. The experts chosen by the organizers to perform the task of selection are called judges (and sometimes also selectors, jurors, or critics). Their function is to consider the pictures submitted and select those which in their opinion possess the highest merits.

The selection of a picture for hanging is not decided by the whim of the judge; there are a number of systems for arriving at a fair assessment of the merits of a print. There is usually a simple explanation for any apparent anomalies. For example, an entrant may have his picture accepted by one well-known exhibition and yet rejected by another. In such circumstances he may quite well feel that the judges cannot be right in both cases and that he has a legitimate grievance. But he may not be competing against prints of the same standard on both occasions, and the best at one exhibition may be the second best at the other.

The number of prints accepted varies from one exhibition to another. Any pictorial exhibition approved by the Royal Photographic Society of Great Britain in this country, and by the Photographic Society of America in America displays all accepted prints so that they can be properly seen by the general public. But the space available varies from large wall spaces in a first-class art gallery to comparatively few screens in a small local hall. The organizers of the exhibition at their preliminary meeting have to weigh the space available for their final display against the total entry they are likely to receive and the proportion of this entry it is advisable or possible to hang.

The average open international exhibition ("open" in this case meaning that the exhibition is not confined to members of a particular society or club) receives on the average from 500 to 1,000 entries from all over the world. Bigger exhibitions, like the Royal Photographic Society's Annual Pictorial Exhibition and that of the London Salon of Photography, may receive anything from four to six thousand prints. As in neither of these cases will more than two or three hundred prints be hung in the gallery space available, the percentage of entries displayed is naturally small. Acceptance at such exhibitions is therefore more highly prized. The average provincial exhibition usually hangs one in three or one in four of the prints submitted, unless, as occasionally happens, the standard is unusually low. So it follows that a print rejected from the major salons may quite well be accepted in a less well-patronized exhibition.

Sometimes there is only one judge, but generally a panel of judges makes the selection. A panel consists of an uneven number; say three, five, or occasionally seven, so that in a case of dispute they can return a majority verdict.

A good judge must be an experienced critic, able to make up his mind quickly. He should be able to see at a glance whether a photograph shows evidence of original thought or is merely a slavish copy of somebody else's previous exhibition success. And he should know what constitutes a generally accepted standard of good pictorial photography. The judge must also be free from prejudices against certain types of subject; he should be open to new ideas and not hide-bound by tradition. Even so, in the end much depends upon his personal taste in pictures.

It is not always appreciated that judges give their time freely in return for nothing more than their out-of-pocket expenses, and there are not many experienced judges who have enough leisure to travel from one part of the country to another on this arduous work. The soundness of a judge can be proved only by the resulting exhibition and while it is not possible for the judge to please everybody, he soon loses his reputation if the majority of experienced photographers consistently disagrees with his selection.

Amateur photographic organizations in some countries maintain lists of judges prepared to give their services free (excepting expenses).

The A, B and C Method. There are various methods of judging, but the most popular is what is known as the A, B and C method. Under this system there is open discussion between the members of the panel. The pictures are put up one by one in front of the panel and each member in turn classifies the picture as A, B or C. A signifies that they consider it well up to the standard required, B that it is not quite equal to the A standard, but is worthy of further consideration and discussion, and C that it is below standard. If with a panel of three there is one A and two B's, the print will go into the B category and come up for further consideration, whereas if there are two A's and one B, the print will be classified as A. C prints are not generally viewed a second time.

When the first round has been completed, the total number of A's and B's is counted and if for example the organizers have a thousand entries and decide to hang two hundred and fifty pictures and there are only one hundred A's, then the B's are gone through again and the best of these selected. If the total number of A's and B's does not reach the desired number on this round, the organizers will have to put up with a smaller number, and this often happens. For example, the gallery accommodation may be sufficient for two hundred and fifty pictures, but the total A's and B's may not add up to more than two hundred. In such a case only two hundred pictures will be hung.

Marking Method. Some judges prefer to allot marks, say from one to ten, to each print, according to their opinion of its merits. The total number of prints selected for the exhibition is then made up by going down the list of each category and taking all the tens, then all the nines, and so on down until the total number desired is reached. The objection to such a system of judging is that it is impossible to attach numbers to abstract qualities with any degree of accuracy.

Push Button Method. In the third system, known generally as Push Button judging, each judge has an electrical device consisting of three push buttons, or a three-position switch for the A, B and C categories. The voting is done secretly and one judge does not know which way the others are voting. Those who advocate this method claim that it prevents any one judge from dominating the others, but those who oppose it say that it is fairer to the entrant to have a free discussion amongst the judges on the merits of the picture and that in any case a judge who allows himself to be dominated by others will not maintain his reputation very long.

Points Method. Another idea that has been tried without much success has been to provide each picture with a sheet listing a large number of qualities to which the judges allot points. The various qualities may reach as many as twenty and include among other things, technique, presentation, originality of ideas, composition, and emotional impact. With twenty such categories and a maximum of ten marks in each category, the ideal or perfect print would have a possible two hundred marks. The difficulty is that no one can dissociate the various qualities that go to make a good picture, or say that any one quality is more important than another. Finally, even if the judges were skilful enough to consider each print in this way, to do it conscientiously would take up too much time.

Popular Vote. Occasionally attempts have been made to submit the judgement to popular vote, leaving the visitors to decide which prints are the best. The usual procedure is to give each visitor a form in which, for example, he is asked to place say the ten best prints in order of merit. Unfortunately in this as in many other branches of life the mere counting of heads does not produce wisdom. Certain types of pictures are " sure-fire winners " in popular voting. Young and helpless looking kittens and puppies; children with their hands in jam-pots and with jam-covered faces; cats in baskets; kittens in boots; very small children with a highly sentimental or somewhat sugary appeal.

Then there are certain conventions that the general public expects the photographer to observe. For example, the sharper the picture, the better; a broad treatment of tones, or any picture which faithfully reproduces the soft mistiness of evening light or early morning fog will be peered at closely and dismissed as " not sharp".

No method of judging prints is perfect, but it is generally agreed that the A, B and C system is the best available and it has now been adopted for most of the exhibitions in this country and abroad. P.W.H.

See also: *Criticism of photographs; Exhibitions and salons; Judges for exhibitions; Visual appeal.*

K

KALLITYPE. Obsolete printing process introduced by Dr. W. J. Nichol in 1899, based on Herschel's argentotype. Similar in appearance to platinotype but forming an image in metallic silver instead of platinum.

See also: *Obsolete printing processes.*

KALVAR PROCESS. Proprietary name of a low-speed heat-developable light-scattering photographic system. The image is produced by the decomposition of diazo compounds in a plastic layer, liberating nitrogen. On heating, the nitrogen forms minute bubbles in the plastic layer, producing a vesicular light-scattering image.

See also: *Diazo processes.*

KÄSEBIER, GERTRUDE, 1852–1934. American pictorial photographer. Founder member of the Photo-Secession, 1902.

KATAPOSITIVE. Positive printed on an opaque base to be viewed by reflected light as opposed to a diapositive which is printed on a transparent base to be viewed by transmitted light. The term is rarely used to-day.

KATUSHEV, JAKOV MATVEEVICH, 1885–1958. Russian specialist in scientific photography. Consultant to the Central Laboratory of Cinematographic Chemistry 1928–33 and Professor of Photographic Chemistry in the Moscow Polygraphic Institute 1937–46. Received Order of Lenin for scientific research and educational activities.

KEEPING QUALITIES OF MATERIALS. The length of time sensitive materials stay fresh depends on the storage conditions, the chemical properties of the support and the dyes used for colour sensitizing.

Storage Conditions. Certain conditions make negative materials deteriorate rapidly: moisture or high atmospheric humidity, high temperatures and the presence in the atmosphere of fumes and gases (e.g., sodium sulphide, or fumes from burning fires) which react with the silver halide in the emulsion.

Sensitized materials keep best in a cool, dry place, well away from smoke, fumes and volatile substances. Special precautions are required in tropical and other extreme climatic conditions. The effects of unsuitable storage conditions on sensitized emulsions are loss of speed, sometimes with increase of contrast, increased fog level and mottling, streaks, and patches caused by the uneven action of damp, fumes or high temperature.

Chemical Properties of the Support. Most emulsions lose speed when kept for any length of time. Plate materials suffer least; under normal conditions they may remain usable for many years. With films the plasticisers added to the base may affect the emulsion characteristics in time. This risk is virtually absent with modern polyester and similar supports. In the days of cellulose nitrate films the products of the gradual decomposition of cellulose nitrate also interfered with the emulsion and the latent image during storage. The paper base of printing papers has no adverse effect on keeping qualities, though these materials, too, lose speed and change in contrast on storage.

Sensitizing Dyes. In orthochromatic and panchromatic negative materials the sensitizing dyes tend to decompose very slowly. This lowers the speed and alters the colour sensitivity. In colour films of the multilayer type the decomposition of the sensitizing dyes may upset the colour balance.

Ordinary roll films will keep their speed and general characteristics up to about two years from the date of manufacture, so long as they are properly stored. An expiry date is usually stamped on the film carton.

Expiry Date. Film materials are generally still usable even after the expiry date, although they will not be as fast as when new. Special precautions—e.g., using a developer improver—should, however, be taken when developing stale materials in order to keep down fog. Colour films are less likely to be usable if good colour rendering is required as the emulsion characteristics in the individual layers are liable to change at different rates. This tends to affect the colour balance during storage.

The dyes used for sensitizing negative emulsions to the more extreme infra-red radiations are very unstable and in consequence the materials are usable only up to a few weeks from the date of sensitizing. Manufacturers often make such emulsions to special order only. Even then the films or plates may have to be stored with solid carbon dioxide at temperatures of around −80° C. to slow down the rate of deterioration. Low temperature storage is also important in order to eliminate fogging from heat radiations which may affect infra-red materials of long-range sensitivity.

Because plates and printing papers keep better than films, expiry dates are not usually given for them. Papers especially keep well for several years if properly stored, and the loss of speed rarely matters. Here, too, special precautions—e.g., use of developer improvers—are advisable when processing old material.

Some scientific materials have to be hypersensitized before use to give them the highest possible speed. Such hypersensitized materials will only retain their extra speed for a few days. After that time they lose their speed again, but increase in fog density.

Exposed Materials. Long storage of materials after exposure and before development may partially destroy the latent image and thus reduce the effective speed of the emulsion. This applies particularly to high-speed emulsions. Exposed materials should therefore be developed as soon as possible. This is specially important with colour materials where a delay of development of more than two months may noticeably alter the results. L.A.M.

See also: *Faults; Film storage.*

KEIGHLEY, ALEXANDER, 1861–1947. English amateur pictorial photographer. Exhibited at the Photographic Society (later Royal) in 1889 and afterwards in many European and American cities. Many of his most famous pictures were taken with a box camera, disguised in brown paper. Started print-making *c.* 1883 with albumen paper and later used Platinotype and carbon. Often made exhibition pictures from enlarged 16 × 20 in. negatives printed by contact in daylight. Biography: *Alexander Keighley 1883–1947* (Royal Photographic Society, London, 1947).

KEITH, THOMAS, 1827–95. Scottish gynaecologist and surgeon in Edinburgh and amateur photographer 1853–56. Dr. Keith made splendid architectural and landscape photographs of Edinburgh and the surrounding country by the waxed paper process.

KELLER-DORIAN, ALBERT, 18??–1924. French photo-engraver and manufacturer of textile embossing rollers. Pioneer of an additive three-colour cinematography process using lenticular films (with minute embossed lenses on one side of the film support, and with a panchromatic emulsion on the other). Through a three-colour filter separation of colours, images are formed in the emulsion, which are reversed in development; when projected with a similar filter they produce a picture in natural colours (1908). The Keller-Dorian process was improved by Berthon and became known as the K.D.B. process, the rights of which were acquired by the Eastman Kodak Company. They improved and marketed it in 1928 under the name Kodacolor. Another version was the Siemens-Berthon process and the Agfa amateur colour film.

KELVIN DEGREES. Units on the absolute scale of temperature, numerically equal to degrees Centigrade plus 273. Named after W. T. Kelvin (1824–1907) and used for measuring colour temperatures of light sources.

See also: *Colour temperature.*

KENNETT, RICHARD, 1817–96. English amateur photographer. Produced and patented stable silver bromide gelatin pellicles in 1873. These could be kept and, when swelled and melted in water, used to coat glass plates, thus replacing collodion. In 1874 he produced dry silver bromide gelatin plates which were manufactured, 1876.

KEPLER, JOHANNES, 1571–1630. German astronomer. Discoverer of the inverse square law relating the intensity of light with the distance it travels from the source. Founder of modern optics by his book, *Dioptrice*, published in Augsburg, 1611. Enunciated the laws of planetary motion. For topographical sketches used, before 1620, a tent camera obscura with a revolvable lens tube projecting through the top, which projected images on to the paper. Biography by D. Brewster (8th ed. 1874).

KERR CELL. "Shutter" used in high speed photography consisting of a glass tank filled with a fluid which, in conjunction with a pair of polarizing screens, will allow light to pass through it when a high voltage is applied. This permits exposure times of less than 1/1,000,000 second.

KEY. A photograph can be an acceptable picture without including the whole of the tone scale from black to white that is usually present in photographs. The image may consist of only a few shades of grey selected from any part of the tone scale.

If tones are taken from the lower end of the scale the picture will be made up of black and a few shades of dark grey with only small areas of highlights. The general impression will be one of sombre gloom. An image of this type is said to be in a low key.

If the image is formed by a few pale grey tints and white paper, it will convey an impression of lightness and delicacy. Such a picture is said to be in a high key.

The photographer may set out deliberately to restrict the number of tones in the final picture. He may wish to produce either a high or a low key picture to suit the special character of the subject.

A true high or low key picture is produced by restricting the tones of the original image—i.e., by manipulation or selection of the lighting. A portrait of a fair-haired subject wearing light-toned clothing and photographed against a white background by soft lighting from the front will produce a high key picture without any departure from normal exposing and processing technique. The delicate atmosphere of such a picture cannot be imitated by taking an ordinary photograph and falsifying its tones in development or exposure or by any kind of aftertreatment of the print.

In the same way, a true low key picture can be produced only as a result of low key lighting; it is not a print that has been deliberately over-exposed or over-developed.

See also: *High key; Low key.*

K-FILTERS. Colour filters made by dyeing gelatin with the dye Filter Yellow K. Once necessary with early pan and ortho materials.

KILO. Prefix signifying a thousand in the metric system—e.g., kilogram, kilometre. Seldom needed in photography.

See also: *Weights and measures.*

KILOWATT. One thousand watts. Sometimes a more convenient practical unit of electrical power than the watt.

KILOWATT HOUR. Commercial unit of electricity consumption; the amount of power used in one hour by an apparatus rated at 1,000 watts or one kilowatt.

KINEMATOGRAPH. (Greek, kinema = movement.) More correct, but less common, term for cinematograph.

KIRCHER, ATHANASIUS, 1601–80. German mathematician and scholar. Gave descriptions of the camera obscura 1646, magic lanterns and other optical instruments 1671. Is often named (erroneously) as the inventor of the optical lantern.

KITES. Cameras suspended from kites have been used for taking aerial photographs of the ground. A series of exposures may be made automatically by a clockwork mechanism which transports the film and sets the shutter, or a single exposure may be made by setting a clockwork or wind-driven delay mechanism before the flight.

See also: *Balloons and kites.*

KLIČ, KAREL, 1841–1926. Czech painter and newspaper artist. Creator of modern photogravure; he combined the use of the half-tone screen, carbon tissue, aquatint grain and iron chloride etchants (1879). Also introduced rotogravure—the process of printing intaglio cross-line screen images from cylinders on high-speed rotary presses (1895). Biography *Karl Klietsch* by K. Albert (Vienna 1927).

KNIFING. Removal of opaque blemishes from a negative or print by scraping with a knife.

The action of knifing actually removes fine layers of emulsion until the density has been reduced by the required amount.

See also: *Retouching.*

KOBELL, FRANZ VON, 1803–75. German mineralogist, Professor at University of Munich. He and Steinheil were the first in Germany (1839), to make paper negatives by their own process, similar to Talbot's. Also made photomicrographs at an early date and invented the galvanographic (non-photographic) reproduction of paintings in 1840. Biography by A. Dreyer (Munich 1904).

KODACHROME. Pioneer subtractive process of colour photography worked out by L. D. Mannes and L. Godowsky and introduced by Kodak in 1935. It makes use of an integral tripack in which the emulsions contain no colour formers but are subjected to individual dye development.

See also: *Colour history.*

KODACOLOR. Originally an additive method of producing colour photographs with the aid of a lenticular screen embossed on the front of the film. It was first made available commercially in 1928 on 16 mm. cine film. The name is now applied to a negative-positive process for making subtractive colour prints.

See also: *Cine history; Colour photography history.*

KODAK. Trade name of Kodak Limited, coined by George Eastman, founder of the organization, in 1888 and first registered in Great Britain on May 3rd, 1888, then in the United States on September 4th, 1888, and subsequently in many other countries. It was first used in connexion with Eastman's original box-type camera. Eastman was determined to adopt a trade mark that could not be misspelled or mispronounced, in any language, or infringed or copied by anyone. He wanted a strong word that could be registered and that everyone could remember and associate with the products of his manufacture. He based it on the letter K, which attracted him because it was "firm and unyielding" and easily pronounced, and by a process of association and elimination he arrived at the word Kodak.

See also: *Camera history.*

KODALK. Proprietary alkali of the sodium metaborate type used in developers.

Characteristics: White crystalline powder.

Solubility: Freely soluble in water at room temperature.

KÖNIG, ERNST, 1869–1924. German dye-chemist. Producer of many sensitizers of which orthochrome, pinaverdol, pinachrome and others have played an immensely valuable part in ortho- and panchromatic photography (from 1904), and the di- and tri-carbocyanines (1923) have extended the sensitivity into the extreme red and near infra-red. Pure filter dyes were also produced by König and he also investigated the desensitizers discovered by Lüppo-Cramer. He invented pinachromy by leuco bases, and was author of a number of books on colour photography.

KOSTINSKY EFFECT. Phenomenon of development resulting in the shifting of the relative position of two image points situated very close to each other. The shift is the result of the change in the concentration of active—or exhausted—developer ingredients immediately between the points.

Where two heavily exposed point images are developing in a solution without agitation, they rapidly exhaust the developer in contact with them. When that happens, they depend for further increase in density on the fresh solution that diffuses from the emulsion around them. If the points are close together, the solution in between them becomes exhausted at a very early stage and development ceases where the points are closest.

On the outside of each point, however, the images can go on developing because a supply of fresh developer is constantly diffusing through the emulsion from unexposed areas to take the place of the exhausted solution.

The result of this selective action is that the nearer borders of each point stop developing while the outer borders continue, so the regions of maximum density move away from each other and the point images appear farther apart than they would have done if development had been even all over.

The opposite effect occurs when the negative image consists of clear spots on a dense ground. In this case, the solution lying immediately between the points becomes exhausted and fresh solution diffuses inwards from the clear spots. The supply is limited, however, and the demands of the heavily exposed space between the points is great. So the adjacent borders of the image points cannot get enough fresh developer to satisfy their needs, and the margins remain clear instead of developing to a visible density. The demands for fresh solution at the outer edges of the spots are only about half as great, so the image develops at full strength right up to the edge. The net effect of all this is that the clear spots grow on their nearer borders and so appear closer.

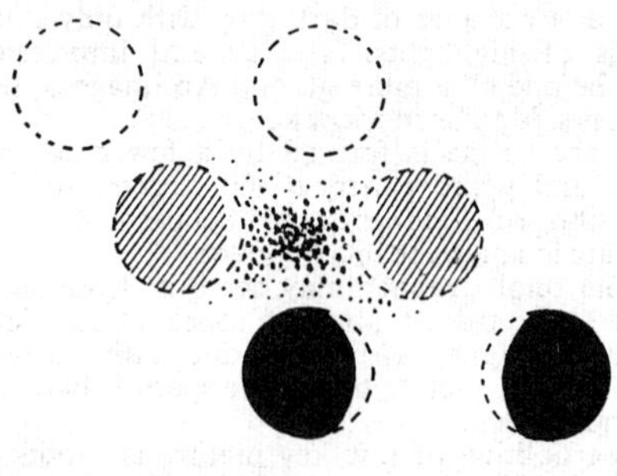

KOSTINSKY EFFECT. Developer between closely spaced heavy image points gets exhausted and under-develops areas closest to each other, so that image points appear to move apart. The opposite occurs with clear spots on a dark ground.

Image movements of this type cause errors in certain scientific measurements based on photography and produce distortion of the sound record in cinematography. F.P.

See also: *Eberhard effect.*

KROMSKOP. Instrument (invented by F. E. Ives, 1892) for viewing monochrome colour separation transparencies through a system of mirrors and colour filters to produce a natural colour picture by additive synthesis. The mirrors were arranged so that the images of the separation transparencies were all combined in register. The transparencies were illuminated by a single source and the single or stereosopic colour image viewed through eyepiece lenses.

KRONE, HERMANN, 1827–1916. German landscape and portrait photographer, and professor of scientific photography at the Dresden Polytechnic (later Technical College). Investigated multiple solarization in 1851. Worked for a time with Taupenot's collodio-albumen process, and experimented with the Lippmann colour process.

KÜHN, HEINRICH, 1868–1944. Austrian pictorialist of German descent. Studied science and medicine in Dresden. From 1888 Kühn devoted himself entirely to photography. Introduced, 1897, the multiple gum bichromate process; also made glue prints. Prominent exhibitor of portraits and landscapes, and a member of the Linked Ring.

L

LABELS. Made-up chemicals are usually well enough labelled by the manufacturer for the label to remain readable throughout the life of the contents. But trouble arises with bottles containing stock solutions. These are handled more than most bottles—often with wet hands—and the inevitable dribbles of solution soon discolour and loosen ordinary gummed labels.

The two essentials for such labels are that they should be clear and be proofed against wet solutions so that they will stay clear. The label should be boldly printed on white paper in ordinary draughtsman's waterproof indian ink. It can then be made solution-proof, in one of two ways.

For a temporary job, the label should be held in place with transparent cellulose adhesive tape. The tape should cover the whole label and overlap generously on to the bottle all round so that dribbles of solution cannot get at the label. This method is useful for labels that carry the date of making up; they are easy to strip off and renew.

Permanent labels are best attached with a fairly thick solution of celluloid in amyl acetate—made by shredding old film, cleaned of its emulsion, into the solvent and leaving it for a day or two. The bottle should be polished clean and dry, and then given a coat of the celluloid varnish. When this coat is dry, the back of the label is brushed with the varnish and pressed on to the bottle. After half an hour, the front of the label and the side of the bottle for half an inch around are coated with the varnish. This makes a permanent label which will withstand handling, solutions, and washing in cold water. It will stand up to hot water long enough to allow the bottle to be washed out.

Perspex dissolved in chloroform or ethyl acetate makes an even harder and more water repellant coating.

Another temporary label can be made by writing on the glass with one of the special pencils sold for the purpose, and covering the letters with cellulose tape. Glass writing inks are also available from laboratory suppliers in various colours (including white for dark bottles). They will not wash off and can be removed only by hard scrubbing. Another alternative is the special self-adhesive tape with a white, matt "write-on" surface. The title is printed on the matt side and the section of tape simply pressed on to the bottle.

But no matter which system is used, the solution should be labelled right away to avoid mistakes. Relying on memory is bound to lead to errors.

It is a good plan to mark bottles with labels cut to a distinctive shape—e.g., round for negative developer, triangular for fixer, long and narrow for paper developer. As soon as the shapes become familiar, it is no longer necessary to read the label. This can be useful when working only by safelight.

Another valuable aid to quick and certain recognition is to store the various processing solutions in bottles of different shapes and sizes.

This may produce an untidy array of bottles on the shelf, but it makes the solutions so much easier to recognize by sight, and it is the only way of recognizing them by touch in the dark.

Where small objects such as nails, screws, etc., are kept in boxes, place a sample of each kind on a sheet of bromide paper under the enlarger and make a silhouette print. This makes a label that is easier to see at a glance than a printed description. F.P.

See also: *Bottles.*

LAINER, ALEXANDER, 1858–1923. Austrian chemist, professor at Salzburg and Vienna. Recommended in 1889 (independently of M. Andresen) the acid fixing bath for the removal of fog. Worked on platinum prints and discovered in 1896 the Lainer effect (acceleration of developer action by potassium iodide). Established (1900) in Vienna a dry plate and paper factory (later Herlango).

LAMBERT. Unit of brightness applied to the measurement of the luminosity of illuminated surfaces and of incandescent bodies. One lambert is equal to a rate of light emission or reflection of one lumen per square centimetre—i.e., 10,000 lumens per square metre.

In scientific work the brightness of a surface is expressed by a measure of the light emitted or reflected by it, and termed its luminance in terms of its luminous intensity per unit area (candles per square inch).

See also: *Light units.*

LAMP. Artificial light source, usually electrically powered. The lamp is the source itself (e.g. the light bulb), and is used in various lighting units such as reflectors or spot lamps for lighting up the photographic subject, or incorporated in an optical system for projection etc.

Photographic Lighting. Most lamps for lighting the subject are of the incandescent tungsten filament type. They range in power from 100 watts up to 10,000 watts, and may be used in lamp reflectors or lamp units for flood lighting, spot lighting, etc. Special miniature spot lamps are employed in photomicrography and other scientific fields.

Other lamp types for lighting the subject include fluorescent and vapour discharge lamps. With some of these the output of invisible radiation (e.g. ultra-violet rays) is used for special photographic techniques.

Discontinuous lamps—flash bulbs, electronic flash and spark sources—are employed both in normal photography and in high speed work. Carbon arcs used to be popular for high intensity lighting, but are nowadays nearly completely displaced by high-power incandescent filament lamps.

Projection. Most projector lamps, especially for still and narrow-gauge cine projectors, are again incandescent filament lamps. High-power motion picture projectors may use carbon arcs or—in modern units—concentrated gas arcs.

For highest light efficiency, lamps of this kind frequently use the nearest approach to a point source with a suitably designed condenser or other optical system.

Small size transparency viewers employ either normal filament lamps or—in the pocket type viewer—battery-powered miniature lamps. Large-size display cabinets for transparencies frequently use fluorescent tubes of selected colour quality as the light source.

Printing and Enlarging. Lamps in contact printers and printing boxes are nearly always groups of filament lamps. Small amateur printers may use a single 15 watt lamp; large size printing boxes may take up to a dozen lamps of 60 or more watts each.

Enlarger lamps are mostly filament lamps of 75 to 250 watts; some enlargers use over-run projection lamps with a suitable optical system. Special forms of discharge lamps (cold cathode lighting) sometimes provide a uniform area of diffused light.

Printers and enlargers for photo-mechanical reproduction may work with mercury vapour or other discharge lamps. These yield a high intensity of ultra-violet radiation to which certain graphic arts and copying materials (e.g. pre-sensitized litho plates or dyeline papers) are specially sensitive.

Darkroom Illumination. While conventional darkroom lamps use low power filament bulbs (mostly 25 to 40 watts), the spectral characteristics of sodium vapour lamps and of electroluminescent panels have been employed to provide a light safe for certain types of sensitive material.

See also: *Carbon arc lamp; Discharge lamp; Electricity; Enlarger; Filament lamp; Flash bulb; Flash* (*electronic*); *Fluorescent lamp; Lamp caps and fittings; Light sources; Light units; Neon lamp; Point source lamp; Projection principles; Projector* (*cine*); *Projector* (*still*); *Safelight; Viewers.*

LAMP CAPS AND FITTINGS

When ordering or replacing lamps for use in photographic lighting or apparatus, the photographer needs to know certain details of construction to avoid mistakes.

There are British Standard Specifications dealing with the basic types of electric lamp. Most lamp manufacturers list these standard types under various trade code descriptions, and most also manufacture non-standardized types for special purposes.

When ordering ordinary tungsten filament lamps for any purpose, the minimum essential data to be given are: voltage, watts consumption, type of glass envelope—i.e., clear, frosted or opal—and type of cap.

When ordering other types—e.g., projector bulbs, studio spotlights, fluorescent and vapour discharge lamps, and flash bulbs and tubes—it may be necessary to state further particulars—e.g., A.C. or D.C. supply, and the maker's code description.

Both electric incandescent and flash bulbs are fitted with caps that allow them to be secured in a lampholder and connected to the electricity supply. These caps differ in both design and size. The following descriptions cover the lamp caps most used in photography, but there are many others which are used less often for specialized applications.

In addition to the different types of cap connexions listed below, various adaptors are made to enable one type of cap to be used in a socket of different type.

Bayonet Caps. The bayonet type of cap is pushed straight into the socket and turned clockwise to secure it, two studs on the body

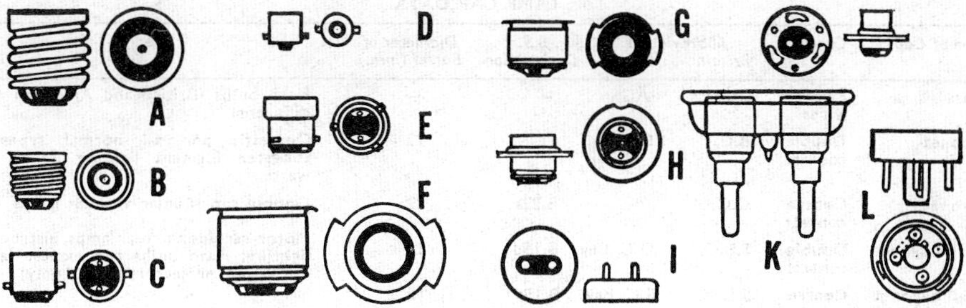

LAMP CAPS. A, *giant Edison screw (G.E.S.).* B, *Edison screw (E.S.).* C, *double bayonet cap (B.C.).* D, *small centre contact bayonet cap (S.C.C.).* E, *three-pin bayonet cap.* F, *mogul prefocus.* G, *medium prefocus (P.F.).* H, *double contact prefocus (motor car headlight type).* I, *two pin.* J, *double contact prefocus with slotted flange.* K, *bipost.* L, *4-pin base.*

of the cap engaging with shaped slots in the socket.

There are two normal double contact sizes—the large double contact bayonet cap (B.C.) and a smaller double contact version (S.B.C.). The filament in B.C. and S.B.C. lamps is connected to two insulated contacts on the base.

Single contact lamps have a single central contact and the circuit is completed through the wall of the cap itself. There are single centre contact versions of the large cap (C.C.) and the small (S.C.C.). There is also a miniature centre contact cap (M.C.C.).

The B.C. is used in Great Britain (but not in America) for domestic and some photographic lamps of relatively low consumption (up to 150 watts). The smaller type of cap is used in flash bulbs, motor car lamps and in instrument lighting circuits. The miniature cap is used on pilot and indicator lamps.

In the U.S.A. the S.B.C. and S.C.C. caps are also known as double contact bayonet candelabra (D.C. Bay.) and single contact bayonet candelabra (S.C. Bay.) respectively.

Screw Caps. There are four common types of screw cap, the miniature Edison screw (M.E.S.), the small Edison screw (S.E.S.), the Edison screw (E.S.),and Goliath Edison screw (G.E.S.). The M.E.S. cap is used on flashlamp bulbs, pilot and indicator lights; the S.E.S. on miners' handlamp bulbs; the E.S. cap for all domestic lamps in America, for large flash bulbs and for various lamps of the order of 200 watts in Great Britain, and the G.E.S. cap for lamps of up to 1,500 watts. In addition, a very small cap is used for flash unit test lights.

American equivalents of S.E.S., E.S. and G.E.S. caps are candelabra (Cand.), medium screw (Med.) and Mogul (Mog.) respectively.

Prefocus Caps. Where the filament of the lamp must always be at a fixed distance and position in relation to the base of the support (as in projection lamps) the bulb is fitted with a special type of bayonet cap known as a prefocus cap. There are several types of prefocus cap varying in the method of engaging the contacts and positioning the filament. Three of the caps are used in photographic projectors.

In the first and most usual type there is a single centre contact and the lamp is held in the socket by a pair of flanged projections which engage with corresponding slots in the socket and fix the height of the filament in relation to the base. There are two sizes of prefocus cap of this type, the large prefocus and the medium prefocus.

American designations are Mogul prefocus (Mg. Pf.) and medium prefocus (Md. Pf.).

A second type with double or single contacts and smaller than the above, has a circular locating flange or disc with a single cut away portion which engages with a projecting key in the lamp socket. This type is principally used in motor car headlamps.

The third and smallest prefocus cap is surrounded by a circular flange in which there are three elongated tapering holes; these lock with a bayonet action over three corresponding studs on the face of the socket. This type also has a double or single contact cap (D.C. Pref. and S.C. Pref. respectively in America) and is mostly used in some sound projector exciter lamps.

A modern projector lamp fitting which also locates the filament accurately is the four-pin valve base. This is pushed straight into the corresponding socket. The pins are 1 mm. thick and spaced 15/32 inch or 12.5 mm. apart in a square. The lamp carries a keyed central locating peg which ensures that the filament is corrrectly orientated and that the right pins go into the right sockets. (Only two of the pins carry the current.)

Bipost Fitting. A special type of two-pin cap is used for heavy current lamps in film studio work. These caps, of which there are two common sizes, have contacts formed by two metal posts which are continued inside the lamp as the filament supports. Lamps of this type will carry currents up to 10 kW.

Special Caps. In addition to the above types of cap, there are a number of special fittings used on lamps which have a more limited application in photography.

There are several types of fitting for double

LAMP CAP DATA

Type of Cap	*Contacts*	*Abbreviation Britain*	*Abbreviation U.S.A.*	*B.S. Designation*	*Diameter of Barrel (mm.)*	*Uses*
Glass	Two wires	Capless	AG	—	—	Flash bulbs (British and American type different)
Bayonet	Double contact	B.C.	B22		22	Domestic and all normal types of tungsten filament lighting up to 150 watts
Bayonet	Centre contact	C.C.		B.22s	22	Double cap tubular vacuum lamps
Small bayonet	Double contact	S.B.C.	D.C. Bay.	B.15d	15	Motor-car side and rear lamps, instrument lighting. Flash bulbs and exciter lamps for 35 mm. projectors (S.C.C. only)
Small bayonet	Centre contact	S.C.C.	S.C. Bay.	B.15s	15	,,
Miniature bayonet	Centre contact	M.C.C.	Min. Bay.	B.9s	9	Pilot indicator lights
3-pin bayonet	Double contact	3-pin B.C.		B.22m	22	Mercury, sodium, etc., vapour discharge bulbs with choke connected in circuit
3-pin bayonet	Centre contact	3-pin C.C.		B.21s	21	,,
4-pin valve	4-pins (2 carrying current)			G17q	29	Projectors (pin-spacing 12·5 mm.) for lamps up to 1200 watts
2-pin glass	2-pins	—	—	—	—	Tungsten iod ne projector lamps (½ in. pin spacing)
Goliath Edison Screw	Centre contact	G.E.S.	Mog.	E.40	40	Tungsten flood lights and vapour discharge lamps up to 1,000 watts
Medium Edison Screw	Centre contact	E.S.	Med.	E.27	25	Tungsten and projector lamps up to 250 watts. Large flash bulbs
Small Edison Screw	Centre contact	S.E.S.	Cand.	E.14	14	Miner's lamps and similar portable battery lamps
Miniature Edison Screw	Centre contact	M.E.S.		E.10	10	Pilot and indicator lights
Bipost	2-pin		Md. Bip.		—	Studio spotlights up to 5 kW. (distance between centres of pin = 1½ in. and ⅞ in.)
Prefocus large	Single contact		Mg. Pf.	P.40	40	Projectors
Prefocus medium	Single contact		Md. Pf.	P.28	28	Motor-car headlamps
Prefocus disc	Single or double contact			P.22s. or d.	22	,,
Prefocus disc	Single or double contact		S.C. Pref. or D.C. Pref.	P.15s. or d.	15	Projectors, exciter lamps and motor-car headlamps
Bi-pin	Double contact			G.24	35} 23}	Fluorescent tubes. Dimensions refer to barrel diameter. Pin spacing = 13 mm.

ended tubular filament lamps—some have an insulated centre contact cap at each end, others have plain metal caps which also form the contacts. A third type of "straight light" tubular lamp fitting consists of metal posts projecting from the back of the tube 38 mm. from each end. This arrangement is adopted so that when the tube is in position in the appropriate type of socket, the contacts and sockets cannot be seen from the front.

Tubular fluorescent lamps are made with two different types of end fitting, Bi-pin (up to 4 feet) and B.C. (8 feet).

Pear-shaped fluorescent and vapour discharge lamps have a three-pin bayonet cap which prevents them from being accidentally plugged into an ordinary tungsten lamp holder. This is necessary because fluorescent lamps must be connected to the supply through a choke; if a vapour discharge lamp could be plugged into an ordinary supply point it would short circuit the supply and almost certainly destroy the tube. Sodium vapour discharge lamps are fitted with ceramic bayonet caps.

Flash bulb caps may be either of the S.C.C. bayonet type (in the smaller sizes) or E.S. type (larger sizes). Modern miniature flash bulbs often carry no cap at all, contact being made directly with the two wires entering the glass.

Electronic flash tubes have caps corresponding to radio valves—e.g., 4-pin, Octal or UX—or in some cases the tube may be supplied with soldering tags only.

Lamp Socket Adaptors. The lamp socket is normally part of the apparatus in which the lamp is used, and there are naturally sockets corresponding to all the types of lamp cap described above. As a general rule any lamp

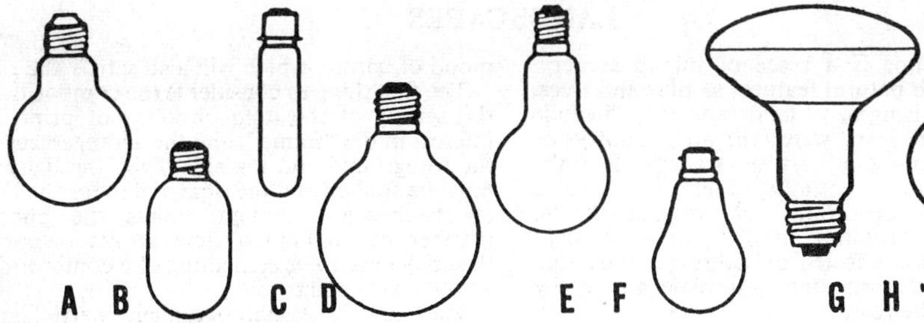

BULB SHAPES. Approximately 1/6 natural size. A, class E projector lamp (for use with epidiascopes and spotlights). B, wide tube class A1 projector lamp (for use in lantern slide and similar projectors). C, narrow tube class A1 projector lamp (for cine and miniature projectors). D, floodlamp. E, high power studio lamp. F, household lamp. G, integral reflector photoflood. H, integral reflector projection lamp. The blown out bulb shape acts as a reflector and is internally silvered.

plugged into a particular socket should have a cap that corresponds with it, and the differences in shape and size do in fact ensure this. However, there are adaptors which enable one type of lamp cap to be plugged into another type of socket. Such adaptors are always of a type that will permit a lamp to be plugged into a socket designed for one of higher consumption, never the other way round. (If a low power socket were adapted to take a high power lamp, the fuse would blow when the lamp was switched on.)

The only exception to this rule applies to adaptors for flash bulbs. In this case there is no objection to the use of a large bulb in a small socket, and suitable adaptors are available for making all sizes of flash bulb and socket completely interchangeable.

The most useful lamp socket adaptors are the G.E.S. to E.S., the G.E.S. to B.C., the E.S. to B.C. and the B.C. to E.S. In addition, for flash bulbs, there are E.S. to S.C.C. and S.C.C. to E.S., as well as capless to S.C.C.

Lamp Shapes. The shape of the glass envelope surrounding the light source varies according to the function of the lamp and the disposition and loading of the filament. As far as possible, the glass is kept well away from the filament to keep its temperature down. This is the reason for the pear shape of the normal type of electric light bulb as used in domestic lighting, photographic flood lighting, enlargers, and similar situations where there are no other factors to dictate another shape.

Fluorescent and cold cathode lamps give off very little heat and can therefore take the shape of a narrow tube.

Quartz-iodine lamps are also tubular and very narrow. The smallest are about ½ inch in diameter and 3 inches long; large types may be up to 3 feet long and about 1 inch in diameter.

The mercury and sodium vapour type of discharge lamp may also be tubular, but in this case the narrow tube containing the vapour is generally surrounded by a larger glass envelope. The space between the two is evacuated to prevent loss of heat from the inner tube and thus maintain the lamp at a more efficient working temperature.

Tubular lamps may be either single or double ended—i.e., they may have both contacts at one end, or one at each. Generally such lamps are double ended because a socket at each end gives a more secure mounting. Light sources of this type may be manufactured in any shape to suit the working conditions—e.g., circular tubes are used in the integrating sphere type of enlarger lamphouse.

Specially shaped glass envelopes are used in certain integral reflector lamps. Reflector flood lamps have a squashed pear shape, with the area of glass immediately behind the filament silvered. This acts as its own reflector.

Internally-silvered projection lamps have a blown-out bulb. The portion immediately behind the filament is ellipsoid in shape, acting —with the internal silvering— as a more or less parabolic reflector. The front of the lamp is also blown out in a spherical shape and internally silvered except for a central window. This silvered area reflects the light from the filament back to the filament and from there to the rear reflector. This ensures that very little light is lost, and all comes out as a more or less parallel beam through the front window. F.P.

See also: *Filament lamp; Flash bulb; Light sources.*
Book: *Modern Electric Lamps*, by D. A. Clarke (London).

LAMPHOUSE. Light-tight (and usually ventilated) chamber holding the light source in projection apparatus—e.g. an enlarger, lantern slide or film strip projector or cinematograph film projector.

The size and shape of a lamphouse is governed by the particular function of the equipment it is part of and the lamp it has to accommodate.

See also: ***Cinematography; Enlarger; Projectors (still).***

LANDSCAPES

A landscape is a piece of inland scenery. Besides such natural features as hills and trees, fields and hedges, a landscape may include water (river, lake, waterfall) and countryside objects forming a part of scenery, like the village church, cottages, inn, farms; also agricultural operations of various kinds, ploughing, harvesting, etc. The appearance of a landscape is affected by natural effects, e.g., sunshine and storm, light and shade, and by the seasons of the year.

The landscape photographer is presented with the raw materials, and he has to make his picture from them. He does so through a selection of viewpoints by movement within the chosen scene, and by a choice of lighting and seasonal conditions suited to his purpose. His object, generally, is to produce a picture portraying some characteristic of the scene or mood of nature, which will also satisfy the eye.

The first thing to consider is the composition, the setting of the item or items of principal interest in the "frame", and the arrangement of the foreground and the sky. Then the lighting must be studied and the time of day for the best effect chosen, although, unless the photographer has unlimited time at his disposal, this may have to be something of a compromise with the viewpoint.

The landscape photographer, particularly if he is on tour, has to make the best of things as he finds them, or note the required conditions and return another day.

The Principal Interest. A landscape, like any other kind of picture, needs a principal item of interest to which the eye naturally travels. This, as a rule, lies in the distance or middle distance.

The main item of interest may be emphasized

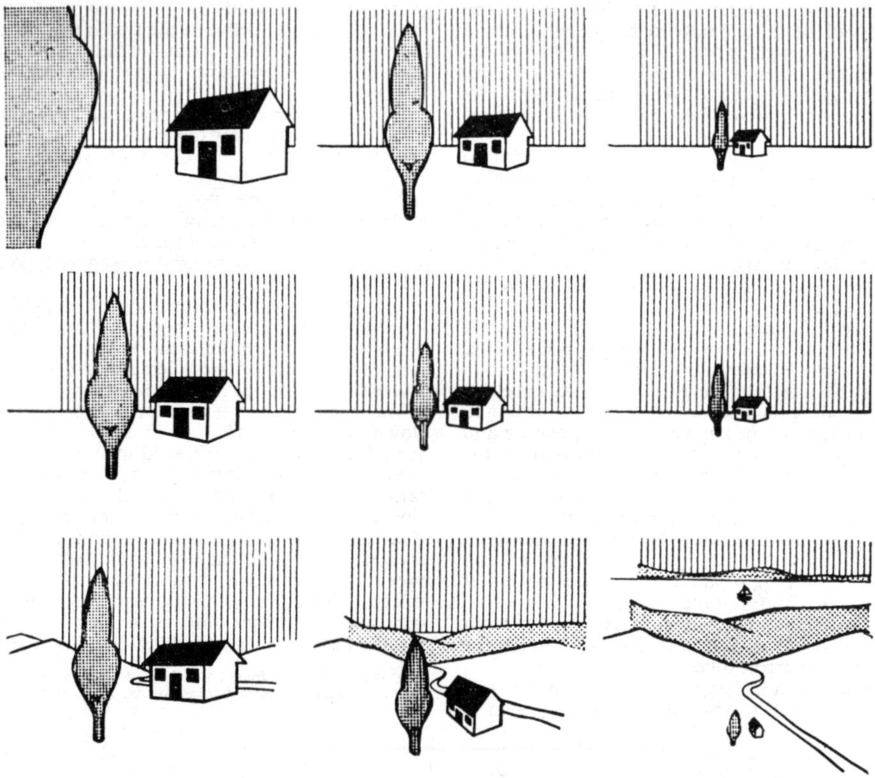

HEIGHT AND DISTANCE OF VIEWPOINT. Top row: *The closer the viewpoint, the more near objects are emphasized at the expense of distant ones. At close range the foreground will only be partly included.* Centre row: *Use of a long focus lens from a more distant viewpoint will still yield large images but without the excessive distortion. As the distance increases further, the proportions will not change appreciably.* Bottom row: *The height of the viewpoint determines the relative placing of foreground and background. From a low viewpoint the foreground and middle distance will fill most of the view. As the camera position goes up higher, the foreground begins to drop away and more distant parts of the scene come into view. From a still higher point the picture becomes virtually an aerial photograph, with no foreground and the increasingly distant planes spread out before the camera.*

by adjusting its tone scale and placing in the final picture. It is not often possible to emphasize it by differential focusing, because most lenses render the middle distance and distance equally sharp at all apertures.

But there is often some atmospheric haze in the air and this splits up the picture into more or less clearly defined planes. This haze lightens the tone and softens the edges of shapes in the far distance. There is less haze between the closer planes and the camera, so the nearer they are, the clearer and sharper they look. And the shadows of nearer objects have a thinner veil of mist over them, so they appear blacker than more distant objects.

This natural separation of the picture planes can be reduced by the use of colour filters.

But the interest must also be prevented from being led out of the picture. Straight, horizontal lines—fences, roads, waterways, and the horizon itself—tend either to divide the picture or to direct the eye out of it. But so long as the line is broken or hidden by some other object before it reaches the margin of the picture, it loses most of its distracting power.

One useful way of keeping the interest within the boundaries of a landscape picture is to frame it with dark foreground objects—e.g., a gateway or overhanging trees.

The Sky. The sky is generally part of any landscape picture, and, in fact, it may provide the main interest. But no matter how much of the picture space it occupies, or whether it plays a principal or secondary part, it must consist of something more than an area of blank paper. Photographic films and plates are almost always over-sensitive to blue light, so unless special precautions are taken, the final print will show no difference in tone between the blue sky and the white clouds. So the sky will be blank and the photograph "baldheaded".

Sky tones can be correctly reproduced—or if

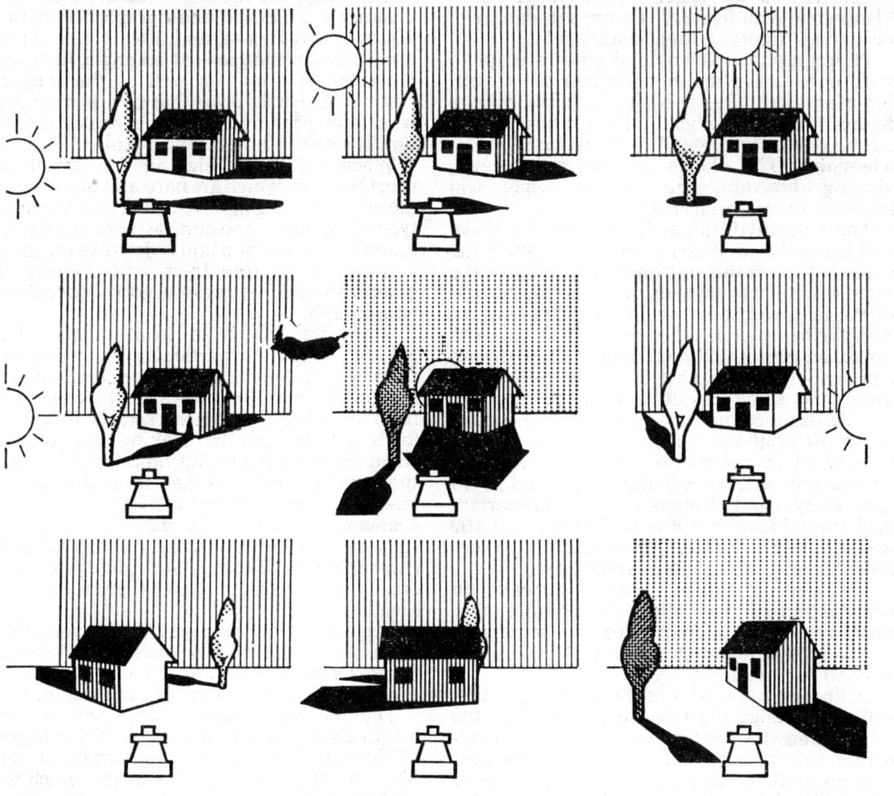

DIRECTION AND ANGLE OF LIGHT. Top row: The length and depth of the shadows depend on the height of the sun. Low morning or evening sun yields long and luminous shadows; as the sun rises in the sky the shadows shorten, and at midday they are short and intense. Centre row: The direction of the sun determines the direction of the shadows and extent of the shadow area. Side lighting casts shadows across the subject, and shows up modelling and texture. Back lighting throws shadows towards the camera which sees the subject almost as a silhouette. Front lighting casts all shadows behind the subject, and produces evenly lit, flat results. Bottom row: The different lighting effects are obtainable also by taking the camera round the subject.

necessary exaggerated—in several ways.

(1) By using a colour filter that holds back the excess of blue rays;

(2) By a special graduated sky filter that filters the light from the sky area only;

(3) By printing in clouds from a separate negative during enlargement; or

(4) By using a polarizing filter to darken a blue sky. This is a particularly useful method when using colour film, where a coloured filter for such a purpose is obviously out of the question.

Figures in Landscapes. Human figures may be included in a landscape photograph, but they should fit naturally into the scene and not be on a big enough scale to become the main interest. Generally speaking, if it is possible to read the expression on the face of any person in the picture, that person will automatically become the subject and the picture will no longer be a landscape study. When people are included, they attract attention out of all proportion to the space they occupy, and a "landscape with figures" is always in danger of becoming "figures with a landscape".

Photographs in towns and cities are apt to look unnatural without people in the streets, but even there, the photographer who includes figures must remember that there is always a strong risk of their stealing the picture.

Viewpoint. The choice of camera distance, viewing direction, and camera height will influence the pictorial effect.

The camera distance determines the scale, and hence the emphasis of the main part of the subject. It does this not only in relation to the picture area (this depends on the angle of view of the lens, too) but also in relation to nearer and farther picture elements. Thus at close range, an object will dominate the foreground and tend to dwarf anything behind. Farther away, the same object will merge into the middle distance or background and lose most of its pictorial significance. (Distance here is considered in comparison to the object size. Thus a gate or fence is in the foreground 10–15 feet away, and becomes an unimportant background feature if it is farther than 50–100 feet. A house at that distance would still be very much the foreground of a larger view.)

The viewing direction largely depends on what aspect of the subject the photographer wants to show. Different sides of a static subject will show different lighting effects, especially early or late in the day.

Deliberate movement of the camera position will also change the relationship between the subject and the background. With near landscapes this can be used to shift prominent features of the skyline (e.g., hills, etc.) to one or other side of the main pictorial interest. In this way foreground objects can be made to cover up disturbing features in the middle distance.

The camera height will determine how much the camera sees of near and distant parts of the scene. A high viewpoint—a bird's eye view—shows the landscape against the ground, with middle distance and background well separated. The nearest features drop out of the picture altogether, and there is little difference in the sharpness of the various picture planes.

From a low viewpoint the foreground crowds into the picture and dominates or even obscures much of what is behind. It also places the nearer features against the sky, and so isolates them from their surroundings.

Lighting. The strength, direction and quality of the light on the scene can make very vital differences in the character of the final picture. There is no need for the photographer to accept these conditions as he finds them; he can control them by choosing the season, the time of day and the weather that will give him the type of lighting he wants. Beyond this, the effect of the lighting at any particular time will depend on whether the camera looks at the scene from the north, south, east or west.

Generally, the most satisfactory lighting for a summer landscape photograph occurs in the hours before 11 a.m. and after 3 p.m. At such times the sunshine falls at an angle, lighting up the sides of things as well as the tops, and casting longish shadows that give them strength, relief and proportion. In the winter the ideal time is between 11 a.m. and 3 p.m.

When the sun is high in the sky, it casts short shadows which are hard and black because there is very little light reflected into them. This type of lighting is too contrasty for good photographic reproduction and it destroys the natural appearance of roundness and solidity that objects should possess in a good monochrome photograph.

When the sun is in front of the photographer, it gives a particular effect known as "contre jour" lighting. In this kind of lighting, shadows are cast towards the camera and most of what the camera sees is in shadow. Such lighting can be striking and dramatic when there are important objects fairly near to the camera, but it is apt to make an open landscape look dull and lifeless.

Camera. Any good-class camera can successfully by used for landscape photography. For controlling perspective, and giving scope for creative composition, interchangeable lenses are invaluable.

Fixed-focus cameras are not very suitable for open landscapes, because their zone of sharpest focus is generally around 20 to 30 feet and they give very poor definition of very distant parts.

The normal camera lens has a focal length about equal to the length of the diagonal of the negative—i.e., it has an angle of about 53°. This type of lens takes in too much foreground and sky for landscape photography where the interest does not lie nearer than the middle distance. It is better to use a lens of longer focal length—i.e., with a narrower angle of view—so that the near foreground and the

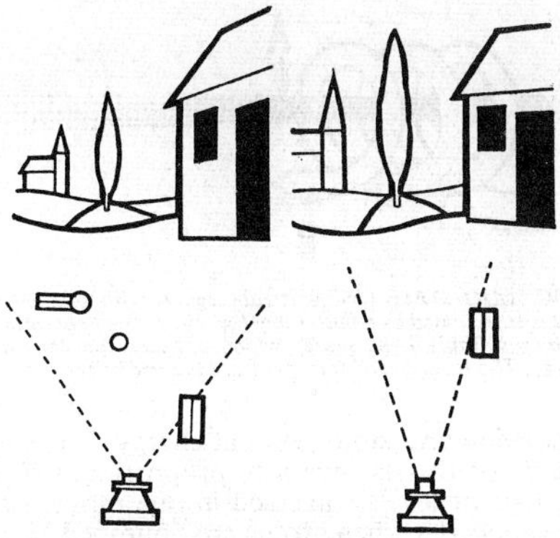

PERSPECTIVE AND FOCAL LENGTH. **Left:** *A short focus lens from a close viewpoint emphasizes perspective and depth.* **Right:** *A long focus lens from farther away compresses depth and subdues perspective making objects look flat.*

upper sky area are cut out and the important part of the scene fills the negative.

Where the camera lens is not interchangeable for one of longer focus, the unwanted foreground or sky can be left out when enlarging the negative, or trimmed off the finished print.

A lens with a shorter-than-normal focal length—i.e., a wide-angle lens—is sometimes used in landscape photography for special effects, such as exaggerating the size of near objects or conveying an impression of spaciousness.

Exposure. In open landscape photography the lens is usually stopped down as far as possible to give the maximum sharpness and depth of field.

This means that the shutter speed often is the limiting factor. On calm days it is possible to work at shutter speeds of 1/25th second, or even longer if the camera is on a tripod. But if there is a slight breeze, a shutter speed of at least 1/100th second must be used to prevent the movement of branches and eaves from showing, and on a really windy day, speeds up to 1/500th second may be necessary.

If there are people moving about in the scene, nothing slower than 1/100th second should be used.

The direction of the light falling on the scene must also be considered in conjunction with any exposure aid. As a rough guide, the exposure recommendation should be doubled if the light is shining directly from the side, and multiplied by four when shining towards the camera.

If there is running or falling water in the scene, nothing faster than 1/250th second should be used, or the water will look frozen.

Sensitized Material. Medium speed panchromatic film, processed with fine-grain technique, will give big enlargements of good quality.

The really fast emulsions are only necessary in dull weather and very early and late in the day.

Filters and Lens Hood. For general use a medium yellow filter is all that is required to brighten the greens and retain the sky. For dramatic effects, when greater contrast is desired or a bold cloudscape emphasized, an orange filter can be used.

Care should be exercised in the use of a green filter. In spring, when the foliage is light, a green filter will bleach the tone values out of recognition and produce an effect of snow. But later in the year when the greens have darkened, a 2× green filter is invaluable in bringing life to an otherwise dull scene.

For the reason that a person shades his eyes when looking at a distant scene, a lens hood should be regarded as an essential. And in "contre jour" photography it is helpful to hold, or get someone else to hold if the camera is not on a tripod, a map, book or folded paper at a suitable distance from the lens to cast a shadow on it, and thus extend the efficacy of the hood.

Care must be taken to avoid cutting off a part of the picture through the mask being too near.

Colour. Open landscapes taken in colour tend to be disappointing because of the over-all monotonous green, particularly in high summer, with its distant haze.

Spring and autumn are the best times for colour work; spring with its blossom and varying shades of green, and autumn with its galaxy of tints. But it is important to have an appropriate object of principal interest—e.g., a grey stone church, a black and white half-timbered cottage or one of tinted plaster and thatch, a carpet of bluebells, a field of poppies, even a simple white field-gate—to act as a foil to the green landscape. P.J.

See also: *Mountains; Panorama.*

Book: *The Photography of Scenery*, by C. D. Milner (London).

LANGENHEIM, FRIEDRICH, 1809–79. German photographer, brother of Wilhelm Langenheim and a relative of F. Voigtländer. Established with Wilhelm a studio in Philadelphia; introduced the Petzval-Voigtländer lens to the U.S.A. The two brothers are always linked historically.

LANGENHEIM, WILHELM, 1807–74. German photographer, relative of F. Voigtländer. Emigrated to Philadelphia, U.S.A. with his brother Friedrich and founded there in 1840 a Talbotype and daguerreotype studio (still in existence a century later under the name of C. W. Briggs Co.). Introduced the Petzval-

Voigtländer lens to the U.S.A. in 1842–3. In 1847 purchased the American rights of the Talbotype patent from Fox Talbot and photographed the Niagara Falls in 1845 on 5 daguerreotypes. From 1850 onwards sold lantern slides (Hyalotypes) which were made by the albumen process of Niepce de St. Victor.

LANTERN PLATE. Photographic plate coated with slow bromide, chlorobromide or chloride emulsions specially made for the production of positive transparencies (lantern slides) from negatives.

See also: *Printing materials; Sizes and packings.*

LANTERN SLIDE. Positive transparency mounted between glass or in a suitable slide frame for projection in a slide projector. The term derives from the time when still projectors were known as magic lanterns.

See also: *Slide (transparency).*

LAROCHE, SILVESTER. 1809–86. French-Canadian portrait photographer with studio in Oxford Street, London. His resistance to Talbot's injunction became a test case in December 1854 when it was proved that Archer's collodion process did not come under Talbot's Calotype patent. As a result, photography with collodion was free for all to practice.

LARSEN, LISA. Dates unknown. American photographer born and educated in Germany. Spoke French, English, German, and some Danish and Russian. Began photographic career in New York and in 1949 joined *Life* as photo-journalist. 1953 Woman Photogapher of the Year; won Matthew Brady Award same year. Career cut short in early life through fatal illness.

LASER. Light source emitting continuous or intermittent light pulses in the form of coherent waves of great concentration and intensity. Also known as optical maser. Laser stands for light amplification by stimulated emission of radiation.

The light produced is extremely pure in colour: its line width is less than 0.03 A—narrower than any known spectral standard. The light waves are coherent: instead of the usual random radiations, they radiate in phase and so can be controlled like radio waves. Output is emitted in a 5-mm. wide beam diverging by less than 0.5°. At 10 kilowatts, this is some 5,000,000,000 times brighter than light of the same wavelength in a similar beam from the sun.

Due to its coherence, it can be focused to a spot of only 10^{-8} sq. cm.

Operating Principle. Lasers utilize the ability of atoms to switch energy levels by respectively absorbing and emitting light energy. The wavelength of this is inversely proportional to the number of levels traversed in one move. Laser action occurs when atoms are "pumped" into a metastable energy level, with intense short-wave light. If they are stimulated into returning to the lowest energy level, the light waves emitted are coherent and have the laser emission characteristics described above. Stimulated from an outside source, the laser behaves as an amplifier; stimulated by its own spontaneous emission, it generates light.

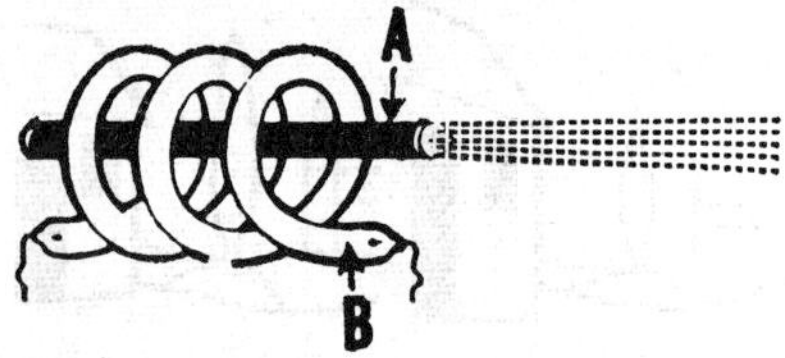

BASIC SOLID STATE LASER. A ruby rod A with semi-silvered ends is surrounded by a flash tube (for intermittent operation) or another suitable light source which provides the "pumping" energy. The laser beam emerges from the end of the rod.

For this, the laser is positioned between two facing mirrors, one partially silvered. Spontaneously emitted light waves are trapped between the mirrors and oscillate through the laser, stimulating further emission which finally penetrates the partial mirror.

Laser Construction. Lasers may be solid, liquid or gaseous and may operate at low absolute temperatures. The first and commonest laser comprises a synthetic ruby rod (aluminium oxide containing chromium), with silvered ends, pumped by electronic flash. This excites the chromium atoms, which emit red light of 0.05 A, bandwidth. Input is some 10 kilowatts and maximum output 1 kilowatt, lasting only as long as the flash. Various other fluorescent metallic crystals are also suitable. Improved pumping methods, using mercury arc lamps, have decreased input and produced continuous emission. Some continuous lasers use helium and neon gas in an ordinary discharge tube with a 50-watt R.F. pumping current. They emit 10 milliwatts, in five infra-red wavelengths, some only 10^{-6} A wide.

Liquid lasers contain benzene or other organic ring liquids. Pumped by ruby lasers, they operate on stimulated Raman scattering—emission wavelength depending on molecular vibrations. They emit red or infra-red radiations and show tunable characteristics. The latest developments are injection lasers, using semi-conductors pumped by ordinary current.

Applications. Lasers are still in their infancy and their usefulness depends on future developments.

Their frequency range may eventually include the whole visible spectrum, leading to variable tuning and bandwidth, besides light amplification. This offers the photographer a powerful cool light source, tunable to single pure colours or wide mixtures. Available in flash or continuous form, the narrow beam can be condensed to a minute spot.

The latter property has already been exploited in photomicrography. The high-intensity microsecond flash will assist high speed photography and may even make astronomic flash exposures possible, while the spectral purity of the light may make safelights and filters unnecessary. P.B.M.

LATENSIFICATION. Latent image intensification, often called latensification, is very similar to hypersensitizing, but is carried out after exposure and before the negative is developed.

There are three methods of treatment: with a chemical solution; with a chemical vapour; and with actinic light.

Chemical Baths. The baths used for hypersensitizing work equally well as latensifiers. Triethanolamine is particularly effective.

A 0·5 per cent solution of potassium metabisulphite or sodium bisulphite will give an increase in effective speed that can be increased still further by adding an equal quantity of sodium sulphite to the solution.

Treating films or plates after exposure has the advantage that there is no need to dry them again before development. Latensification can, in fact, be the first step in processing, although the speed gain is said to be better if the material is dried after latensification.

Vapour Treatment. Vapour treatment with mercury or sulphur dioxide is exactly the same as for hypersensitizing.

Light Treatment. Latensification by exposure to light does not take place in the same way as hypersensitizing by the same means. At first sight it would seem that the only effect of an additional very small exposure after the true exposure would be to add an imperceptible amount of fog. But, in fact, the effective emulsion speed is appreciably increased.

This is because the light at such very low intensities acts selectively. It is not strong enough to form many new developable image points, but not too weak to go on building up already existing parts of the latent image. By suitable adjustment of the latensification exposure, the latent image exposure can in fact be raised to the upper limit of the threshold exposure for the material in question. A certain amount of fog is also formed, but the gain in the image is much greater.

A darkroom lamp makes a suitable light source, and a dark green panchromatic safelight with a 15 watt bulb has been recommended. The material should be exposed about 15 feet away from the light for anything between 30 minutes and 2 hours. L.A.M.

See also: *Hypersensitizing.*
Book: *Developing*, by C. I. Jacobson (London).

LATENT IMAGE. Exposure to light in a camera does not as a rule produce a visible change in the silver halide grains which, suspended in gelatin, make up the photographic emulsion with which films, plates, and papers are coated. But it does bring about an invisible change which can be revealed by the photographic developer. The invisible image which was present before development is called the latent image, the term latent standing for hidden or invisible. Apart from the process of development, science knows of no other way of revealing the presence of the latent image.

Sensitivity Specks. If exposure is increased very much beyond the level necessary to produce a latent image, the emulsion crystals or grains print out as a visible image because the continued action of light causes the silver bromide to disintegrate into bromine and metallic silver, the latter appearing on the grain surface in the form of small specks. This suggests that the latent image itself is present as small specks of metallic silver on the grain surface.

These specks of silver are extremely small. Their size can be worked out approximately from the ratio of the exposures required for print-out and latent image formation. Such calculations indicate that the mass of a latent image speck is only one part in a hundred million of the whole of the silver bromide grain.

Small as they are, these specks are of fundamental importance in modern theories of latent image formation.

On development, these specks of silver which were formed during exposure act as points of attack for the developer. Starting at the specks, the grains are changed by the developer to black metallic silver, yielding a visible image.

Since the latent image specks are so small that they escape even the most sensitive direct methods known to science, it is no wonder that any theory on the mode of formation of the

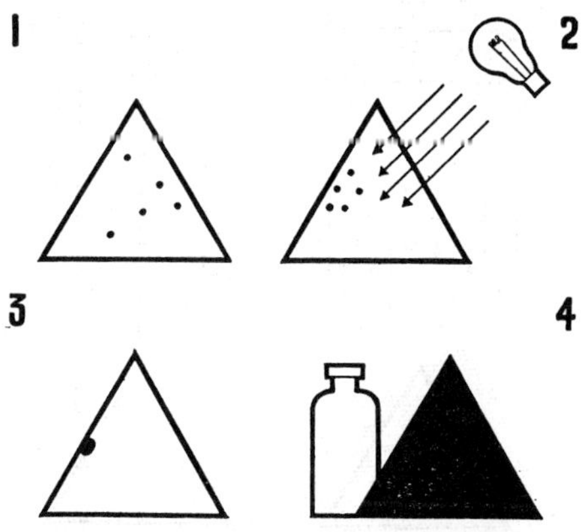

IMAGE FORMATION. 1. Silver specks in the silver halide grain. 2. Condensation after continued exposure to light. 3. Developable sub-image. 4. Developed silver image.

latent image is necessarily speculative, the more so the greater the detail the theory attempts to cover. The general features of any theory are, however, established in broad outlines.

Stages of Formation. The latent image in any one grain is formed in several stages. That means, in terms of the silver speck hypothesis, that several atoms of silver are required to form the latent image. Light (in the form of light quanta; i.e., small parcels of light energy) is absorbed and forms silver anywhere in the silver halide crystal. To form a metallic speck, these silver atoms must flock together. Thus a coagulation or condensation process is an essential feature of latent image formation.

Condensation phenomena are well known in nature and follow well-defined laws. Examples are the formation of water drops to make fog or rain, the growth of snow-flakes and other crystals from solution. When the process of growth is slow, the crystals or particles are few, but large in size; if the process is fast, many but small particles are formed.

Rate of Growth. In the photographic emulsion grain, the rate of growth of the latent image specks is governed by the rate at which light reaches the grain, in other words, by the intensity of exposure. In comparing equal exposures (i.e., in which the product intensity × time is constant) for varying intensities and times, many of the resulting changes in photographic behaviour are accounted for by a straightforward assumption. This postulates that latent image specks formed at low intensities and long times of exposure are few and large and those formed at high intensities and short times numerous and small. One of the most important differences is in the rate of development: flash exposures require longer development than normal snapshots to get full sensitivity and contrast.

During the build-up of the latent image there

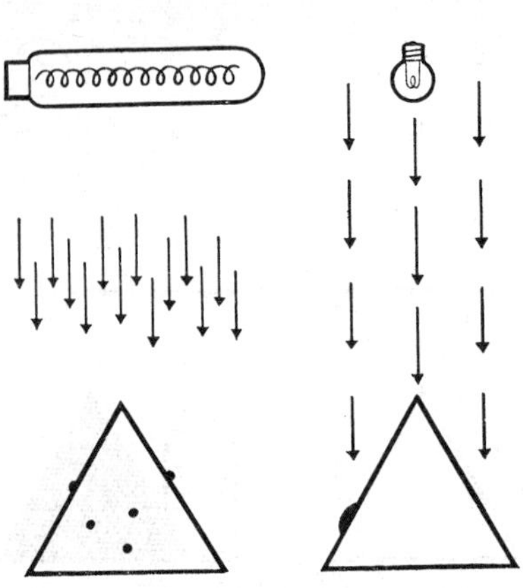

LATENT IMAGE AND LIGHT INTENSITY. **Left:** *Short exposure to high intensity forms many small sensitivity specks.* **Right:** *Long exposure to weak light forms few but large specks.*

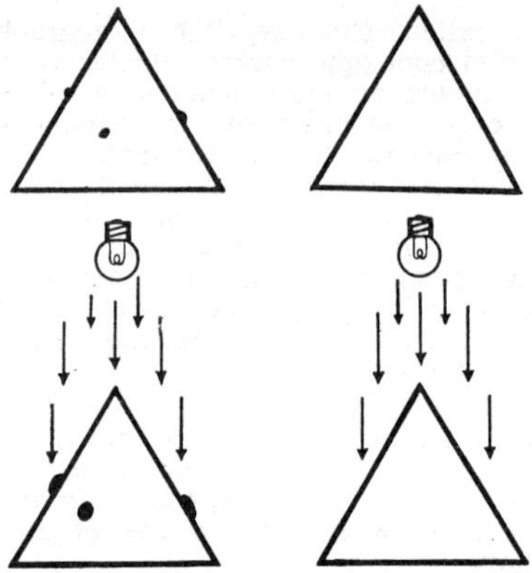

LATENSIFICATION. **Left:** *Exposure to very weak light will build up existing sub-image specks* **Right:** *The same weak exposure will not form new sub-image specks.*

occurs a stage at which a stable result of exposure is produced which is not yet developable and therefore—by definition—not a latent image. This has been termed the sub-image. The result of a short-time exposure is the production of numerous grains carrying sub-image specks; this causes a considerable sensitization of the emulsion towards a subsequent exposure of longer duration to light of low intensity. Conversely, a low-intensity exposure can build up an image-wise sub-image to a fully developable latent image (latensification).

Reciprocity Failure. The first stage in latent image formation is the most difficult: at low intensities of light, it is the sub-image stage which is inefficient and responsible for the loss of sensitivity for low intensity—long time—exposures (low intensity reciprocity failure).

Crystal growth is much helped by the presence of nuclei on which a start is made. In latent image formation these are provided on the grain surfaces by the sensitivity specks. The nuclei act by holding the first atoms in place; once a reasonable number of atoms have aggregated, the speck is stable. At high intensities the problem of stability does not arise since the specks are built up to stable size quickly. Then even quite ineffective nuclei become operative, and in a photographic grain these are often situated inside where the developer cannot reach them.

The internal image centres come into action if a developer containing a solvent for the silver halide is used. If any surface image present is first bleached away by an oxidizer, the internal image can be studied on its own. With commercial emulsions, the sensitivity on internal development is, as a rule, less than that for surface development, especially for long-time exposures at low intensities.

The formation of the internal image as well as the small size of the latent image specks accounts for the sensitivity loss at high intensities and short times of exposure (high intensity reciprocity failure).

Other Latent Image Aspects. The more detailed theories of latent image formation have to take into account that absorption of light in silver halides produces both silver and halogen atoms. If given a chance these two will recombine to reform silver halide, so that the effect of exposure is lost. The process of chemical sensitization to add speed, is variously regarded as providing places where the silver atoms can collect—the silver-sulphide sensitivity centres of Sheppard and of Gurney and Mott—or chemical acceptors for the halogen atoms (Hickman, Mitchell).

None of the details of the various processes postulated are known with any certainty. There is little doubt, however, that the very first result of the absorption of light in a silver halide crystal is to split an electron from one of the halide ions. This spare electron is freely mobile through the crystal and, in effect, corresponds to the movement of a silver atom through the crystal, since, if the electron stayed with any of the silver ions in the crystal, a silver atom would be formed. If an electron is caught so as to be held tightly somewhere on the crystal, a silver ion could move up to it and form a silver atom there. This appears to be the function of the silver sulphide sensitivity speck, according to Gurney and Mott.

Their theory therefore comprises both an electronic and an ionic movement of which only the second strongly depends on temperature. This accounts for the complicated way in which the sensitivity of photographic emulsions depends on temperature. W.F.B.

See also: *Development theory; Gurney-Mott theory; Photochemistry.*

Books: *Photographic Sensitivity* (London); *Progress in Photography* (2 vols.), ed. by D. A. Spencer (London); *Theory of the Photographic Process*, by C. E. K. Mees (New York).

LATERAL REVERSAL. Term applied to an image indicating that it has been reversed from right to left. The image of an object seen in a mirror is inverted in this way, so is the image on the ground glass screen of a reflex camera.

Some optical direct viewfinders give a laterally reversed image. A normal negative is both vertically and laterally inverted in the camera, but right and left, and top and bottom, are restored in printing by contact or projection.

Photomechanical printing from blocks does not correct for lateral inversion, therefore the original negative must be made through a reversing prism with the camera at right angles to the copy.

Prints made by the carbon process are laterally inverted unless the negative is first reversed or the print is made by double transfer.

LATITUDE. Term indicating the range of exposure over which a photographic emulsion will yield an acceptable negative. Latitude is directly related to the length of the straight line portion of the characteristic curve and the contrast range of the subject. Generally speaking, the faster the film and the less the subject contrast the greater the latitude.

Very slow films and colour films have practically no latitude; the exposure must lie within very narrow limits (less than half a stop) to achieve perfect results.

Modern high-speed films, on the other hand, have an overall exposure latitude of about 8 stops for an average subject; within that range the negative will be acceptable.

See also: ***Exposure; Negative materials.***

LAUSSEDAT, AIMÉ, 1819–1907. French officer of the Engineers' Corps. Pioneer of photogrammetry which he introduced into practice *c.* 1851. He designed or invented many early photogrammetric instruments. In 1861 he produced plans of parts of Paris from photographs taken from the roof of the Polytechnic College and from the tower of St. Sulpice. Same year was appointed Captain of the phototopographic army brigade. Retired in 1879 and became director of Conservatoire des Arts et Metiers in Paris. Wrote a number of books on photogrammetry.

LAUSTE, EUGÈNE AUGUSTIN, 1857–1935. French inventor. Pioneer of cinematography and sound film. Worked from 1887 to 1892 in Edison's laboratory as assistant to W. K. L. Dickson on the Kinetoscope and from 1894 with Major W. Latham on a projection apparatus for moving films. In 1895 constructed his first (wide film) projector, the Eidoloscope, also cameras and a printer. In 1896 joined the American Biograph Co. From 1904 experimented on the recording and reproduction of synchronized sound and movement (pioneer patent of 1907) and showed his first sound film in Brixton in 1910. Made many sound films before the outbreak of the first World War.

LEAD ACETATE. Sugar of lead. Formerly used in combined toning and fixing baths for daylight printing papers, also in certain toners and intensifiers.

Formula and molecular weight: $Pb(CH_3CO_2)_2.3H_2O$; 379.

Characteristics: White crystalline powder, resembling sugar. It is strongly poisonous.

Solubility: Highly soluble in water at room temperature.

LEADER. Beginning of a miniature film used for loading the film into the camera. This is always fogged, and the film has to be advanced by two or three frames after closing the camera in order to bring the first section of unexposed film into position behind the lens. With most miniature cameras the film leader has to be specially trimmed to shape for loading. Film

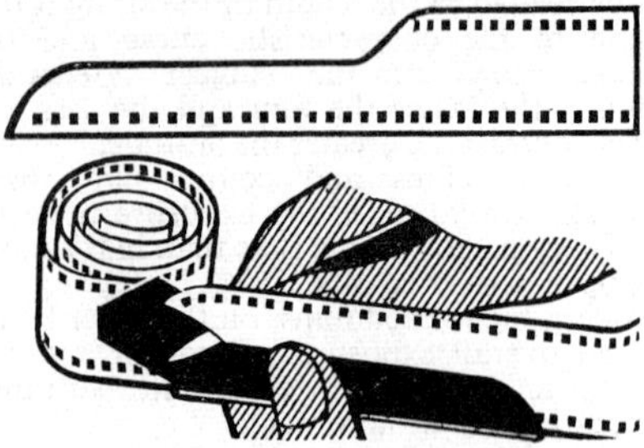

TEMPLATE FOR FILM LEADER. Some miniature cameras need specially shaped film leaders for loading; these may be cut accurately with the aid of a suitable template.

supplied in cassettes or refills has the beginning already suitably trimmed. Special trimming templates are available for trimming film cut from bulk lengths.

With narrow-gauge cine film the film leader is the initial length used for threading the film through the transport mechanism and on to the take-up spool of the cine camera or projector.

Such a leader, especially on 16 mm. release prints, carries certain reference information about the film as well as cue frames for the precise location of the film in the projection gate. This permits a smooth start to be made to a performance.

The required location and nature of such cues, and the length of the leader, are laid down by various national and international standards. This applies also to 35 mm. motion picture film. Similar leaders are provided on still film strips produced for educational and instruction purposes.

LEAD NITRATE. Formerly used in combined toning and fixing baths for daylight printing papers, also in certain toners and intensifiers.

Formula and molecular weight: $Pb(NO_3)_2$; 331.

Characteristics: White crystals. Strongly poisonous.

Solubility: Highly soluble in water at room temperature.

LECTURES. Still and cine projection can be used to illustrate lectures for entertainment, education and instruction. The simplest arrangement is to have a projectionist projecting slides when required (and signalled by the lecturer) to show specific points made. For more elaborate arrangements the projector can be remote operated by the lecturer himself, or the lecture recorded on tape which also controls the slide projector by means of cueing signals.

See also: *Projection practice; Visual aids.*

LE GRAY, GUSTAVE, 1820–62. French painter and photographer, inventor of the waxed paper process 1851. A founder member of the Societé Héliographique. Took a large number of photographs for the Commission for Historic Monuments, 1851. Famous for his instantaneous seascapes, 1856 and wrote several *manuels* of photography. Retired from photography early in 1861 because of the carte-de-visite vogue, and became professor of drawing in Cairo.

LEICA. First modern 35 mm. miniature camera. An unnamed prototype was built by Oscar Barnack in 1914 for his private use, to take photographs on the standard film used in commercial cinematography. In 1924 Ernst Leitz of Wetzlar started manufacturing the camera. The first model had a fixed lens, a focal plane shutter, coupled film transport and shutter tensioning. It used 35 mm. film in special cassettes which were inserted into the camera in daylight. Subsequent models retained the same principles and basic construction, but incorporated various refinements such as interchangeable lenses, coupled rangefinder focusing, an extended range of shutter speeds, and flash synchronization.

The Leica also originated miniature photography which, owing to the unusual technical requirements and novel scope of that type of camera, existed for a long time largely as a separate photographic movement.

The Leica was also the first camera to be used as the basis of a camera system planned to cover practically every field of photography by the use of an extensive range of accessories and specialized photographic equipment. Although many miniature cameras have followed this lead, the Leica system is to-day still one of the most comprehensive and versatile.

See also: *Camera history; Miniature camera; Miniature camera technique.*

LEITZ, ERNST, 1843–1920. German optician. Established in 1869 the Ernst Leitz optical works in Wetzlar. Specialized in the manufacture of microscopes, prism binoculars and other fine-mechanical instruments. Biography: *Jubilee Book*, 1849–1949, edited by A. Berg (Wetzlar 1949).

LEITZ, Dr. ERNST, 1871–1956. German, son of Ernst Leitz, who succeeded his father as head of the Leitz works in Wetzlar, Germany, in 1920. Expanded production lines to include industrial optical measuring instruments and photographic equipment. Manufactured the Leica camera (invented by Oskar Barnack 1913) from 1924. Biography: *Jubilee Book*, 1849–1949, edited by A. Berg (Wetzlar 1949).

LENDVAI-DIRCKSEN, ERNA, 1884–1962. German portrait photographer. Her most important work was the large number of volumes of *Das Deutsche Volksgesicht* publised in the 1930s, consisting of portraits of typical German peasants in their regional costumes.

LENS

Optical device for forming an image of an object. In photography, the lens always consists of a portion of a transparent medium—e.g., glass or plastic—bounded by either two curved surfaces, or one curved and one plane.

When rays of light pass from one transparent medium into another, their direction of travel usually changes—i.e., they undergo refraction. When they pass at an angle into a denser medium, then the rays are bent towards the normal (a line perpendicular to the surface at the point of entry of the ray). If the rays are themselves normal to the surface they suffer no refraction. When the rays emerge from the denser medium, they are bent away from the normal unless themselves normal to the surface.

If the surfaces of the denser medium are parallel, then the refraction at entry is reversed at exit so that when the ray comes out, it is still parallel to its original direction but displaced to one side or the other by refraction inside the medium. But if the surfaces of the medium are not parallel, then the refraction at entry will no longer be exactly reversed at exit, and the incident and emergent rays will no longer be parallel. In other words, the light ray after passing through the medium will be bent at an angle to its original direction; and the greater the angle between the surfaces of the medium, the more it will be bent. This is the basic principle of the lens.

Simple Function. A lens is simply a glass disc shaped so that rays from a point object on one side pass through and are bent either inwards towards a common point on the other side of the lens, or outwards as though radiating from a point closer to the lens than the object. This is achieved by shaping one or both faces of the disc like a portion of the surface of a sphere.

If the curves are so arranged that the lens is thicker in the middle than at the edges, parallel rays of light falling on it will converge to a point on the opposite side. It is then called a converging lens.

If the curves are so arranged that the lens is thinner in the middle than at the edges, parallel rays of light falling on it, when seen from the other side, will appear to radiate from a point. It is then called a diverging lens.

Since any object can be considered as a collection of points, then a converging lens, by producing point images of all the object points, forms an image similar in all respects to the object. If a light reflecting surface is placed in the plane of the image points, a visible image of the object will be formed on it. This, in fact, is the type of image formed on the focusing screen of the camera and which registers as a latent image on the sensitized material during exposure. Such an image is said to be a real image.

A diverging lens does not form a real image; the rays of light from the object are not bent inwards after refraction: they open out so that they appear to come from a point nearer than the object. This point is merely apparent; it is not real. So the image formed by all the points is called a virtual image; it cannot be made visible on a reflecting surface like a real image.

Lens Shapes. The behaviour of the lens varies according to the number of curved surfaces and the direction of the curvature. When both surfaces are convex or there is one convex and one flat face, the lens is always a converging type. When both are concave, or there is one concave and one flat side, the lens is always a diverging type. When there is one concave and one convex side the lens may be either converging or diverging depending on the relative curvature. If both sides have the same radius of curvature the lens is neither diverging nor converging. Each type has its special uses and each has a distinguishing name:

TYPES OF LENSES

Plane	Convex	Concave	Name	Thickest at	Effect
1	1		Plano-convex	Middle	Converging
1		1	Plano-concave	Edges	Diverging
	2		Bi-convex	Middle	Converging
		2	Bi-concave	Edges	Diverging
	1	1	Concavo-convex	Middle	Converging
	1	1	Concavo-convex	Edges	Diverging

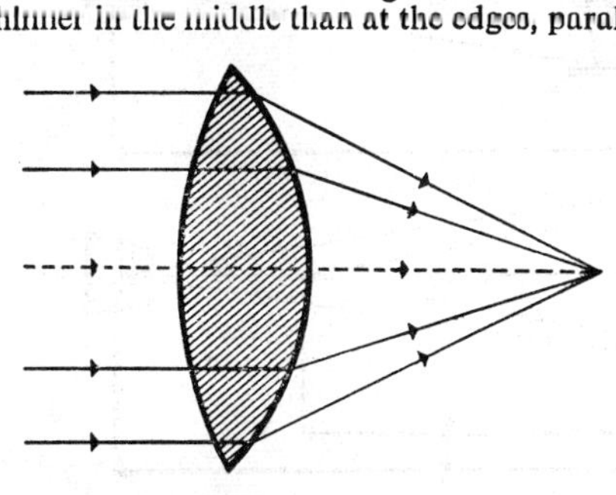

FOCUS OF CONVERGING LENS. A converging lens bends parallel rays of light so that they converge to a point of focus situated at a fixed distance behind the lens.

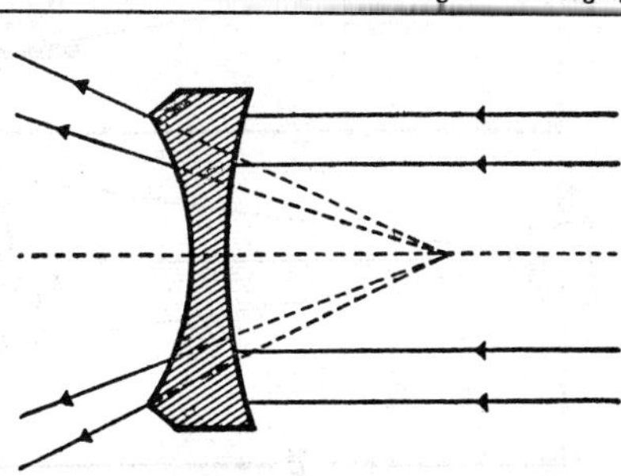

FOCUS OF DIVERGING LENS. A diverging lens bends parallel rays of light so that they diverge and appear to come from a point of focus in front of the lens.

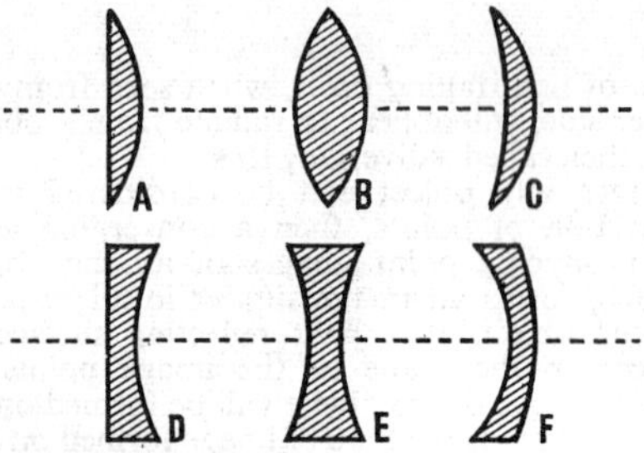

LENS SHAPES. Top: Converging lenses. A, plano-convex. B, bi-convex. C, concavo-convex (meniscus). Bottom: Diverging. D, plano-concave. E, bi-concave. F, concavo-convex

In practice lenses may be either simple—i.e., a single shaped glass—or compound—i.e., an assembly of glasses with characteristics intended to balance out certain inherent shortcomings of the simple components.

A simple lens is known as a lens element. A combination of two or more elements which still makes up only a part of a complete optical system is a lens component. This term is also applied to separate uncemented lenses used on their own in an optical system. Thus a camera lens may for instance consist of four elements; if two of those elements are cemented together and the other two stand free, the lens is said to have three components.

There are a number of basic optical relationships which, with certain modifications, apply to both simple and compound lenses.

The Principal Focus. The straight line about which the lens or objective is symmetrical is called the axis. The centres of curvature of all the curved faces of the lens lie on the axis.

When rays of light pass through a converging lens parallel to its axis, as though coming from a point very far away—i.e., at infinity—they are bent inwards and meet at a point on the axis. This point is called the principal focus of the lens. Light passing through a diverging lens in the same way is bent outwards so that it appears to come from an imaginary point on the other side of the lens. This point also is called the principal focus of the lens.

The principal focus of a converging lens is a real point and the image formed by the lens is a real image that can be focused on a screen. The principal focus of a diverging lens is not a real point in the same sense, and the image cannot be made visible by projecting it on to a screen. This type is known as a virtual image.

If the rays of light pass through the lens in the opposite direction, they reveal a second principal focus on the other side of the lens. The planes at right-angles to the lens axis which pass through the principal foci are called the principal focal planes of the lens. Both the principal foci lie on the axis of the lens.

They are also known as the front principal focus and back principal focus, according to the orientation of the lens. By standard definition the front of a camera lens is that end which is intended to face the object being photographed, while the back of a camera lens is the end which faces the sensitized material and the focal plane of the camera. When the front of the lens faces an infinitely distant object, the light rays from the latter converge at the back principal focus. Hence the other principal focus is the front principal focus.

The front of an enlarger lens is that end of it which is intended to face the baseboard of the enlarger, while the back is the end facing the negative being enlarged. The principal foci are therefore disposed accordingly.

Usually lenses have plane or spherical surfaces since these are easy shapes to grind and polish, but sometimes non-spherical (aspheric) surfaces are used.

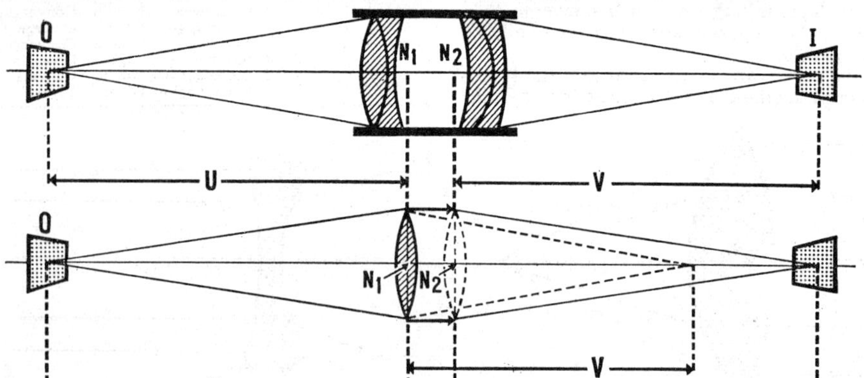

NODAL POINTS. A composite lens behaves as if it were a simple lens which receives the light at one point and discharges it at another. These two points are the nodal points, and the dotted lines passing through them are the principal planes. O, object. I, image. N_1, front nodal point. N_2, rear nodal point. U, object distance. V, image distance.

Focal Length. The distance from the centre of the lens to the principal focus is known as the focal length of the lens. This is the distance at which all parallel rays of light—i.e., those coming from a very distant object—converge to form a sharp image on a film or plate. If the object is closer, then the image will be formed farther away from the lens.

If the object is at infinity, the lens-plate distance that gives the sharpest image is equal to the focal length of the lens. If the object is closer than infinity, the required distance is always greater than the focal length. Finding the distance that gives the sharpest image of a particular object is known as focusing the lens.

If the focal length of the lens is known, the lens-plate distance can be calculated from the basic equation:

$$\frac{1}{u} + \frac{1}{v} = \frac{1}{F}$$

where u = distance from the object to the lens
v = distance from the image to the lens
F = focal length of the lens.

This equation is not very convenient for calculations and adaptations are more often used.

Equivalent Focus. Like simple lenses, compound lenses—i.e., with more than one component—have their principal foci found in the same manner by determining the position of the image-point of a distant object on the axis of the lens.

Simple lenses are treated as though they have no thickness, and it is assumed that the object-lens and lens-image distances are measured from the centre of the lens. Compound lenses cannot be treated as though they have no thickness, and in calculations on compound lenses, the object distance is measured from one fixed point in the lens and the image distance measured from another. These points are called the nodal points of a lens and the planes through them at right-angles to the axis are called the principal planes. The distance from the second nodal point to the rear principal focus is the equivalent focal length of the lens, the distance from the first nodal point to the front principal focus is also the equivalent focal length of the lens.

The equivalent focal length governs the size and position of the image formed in the same way as the focal length governs the size and position of the image formed by a simple lens.

For all optical calculations such as those based on the equation

$$\frac{1}{u} + \frac{1}{v} = \frac{1}{F}$$

the object and image distances are measured from the nodal points, and the value given to F is the equivalent focal length of the lens.

When the lens is part of a camera, it is more convenient to define the image position for a distant object in terms of the distance from the surface of the rear element to the rear principal focus. This distance is called the back focal distance of a lens. There is also the corresponding front focal distance.

Nominal Focal Length. While lenses for cameras, etc., are normally designed for specific focal lengths, the actual focal length of a series of similar lenses as manufactured may vary slightly owing to manufacturing tolerances. In most cases (except where photographs taken with the lens require precise dimensional valuation) this is unimportant provided the other lens calibrations (especially the distance scale) are adjusted accordingly. Thus a standard lens for a 35 mm. miniature camera may be listed as having a focal length of 50 mm.; the lenses actually produced may range in focal length from 48 to 52 mm., and the lens maker does not generally mark such deviations. By international standards the acceptable measured focal length may be within plus or minus 4 per cent of its marked nominal value.

Focal Length and Field. If a comparatively large image is required, a lens of long focal length is used, since this produces a larger image than one of short focal length. At the same time, for a given negative size, the larger the image formed by the lens, the smaller the proportion of the field which will be recorded on the negative; so in practice the final choice of focal length is controlled by the amount of the field that the lens is to cover, or its angle of view.

The exact relation of the field covered to the focal length of the lens is given by the equation

$$2F \tan \frac{\theta}{2} = d$$

where F = equivalent focal length of the lens
θ = total angle of field subtended by the object at the lens
d = diagonal of the negative.

The field of photographic lenses generally lies between 15° and 140°. Within these limits lens designers have produced a wide variety of lenses, each one with a particular purpose to fulfil. Thus a lens which is designed for a 30° field gives poor results over a 50° field, and although a 50° angle lens has quite a good performance over a 30° field, it is never as good as a lens designed to cover 30° and no more.

The angle of a lens gives very little indication of the general type of construction—e.g., Speed Panchro lenses of various designs have fields of from 14° to 53°, while triplets vary from 25° to 70°.

However, for fields greater than 70°, special wide-angle lenses are made. These are generally variations of the symmetrical type of construction. A good modern wide-angle lens may cover an angle of as much as 100° at *f*32. Such a lens has a maximum aperture of *f* 6·3, which is, however, used only for focusing. Special wide aperture lenses are available to cover 70° to 80° at apertures as great as *f* 4·5.

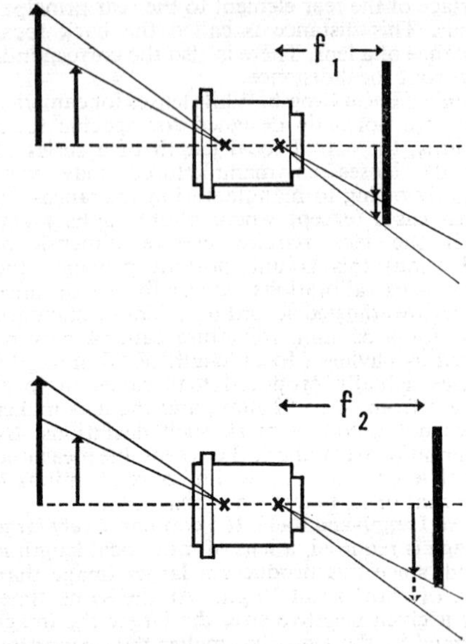

FOCAL LENGTH AND SCALE. Top: *A lens of short focal length, f_1 produces a smaller image of an object a given distance away, but covers a greater part of that object.* Bottom: *A lens of longer focal length f_2 yields a larger image of the same object at the same distance in front of the lens, but on a similar negative size will include a smaller object area.*

Normally a lens of very long focal length requires a long camera extension. This is undesirable, and has led to the development of the telephoto lens which has a long focal length, but which works with a normal amount of camera extension.

The power of a telephoto lens is defined as the ratio of its equivalent focal length to its back focal length. This ratio is increased by making the lens of two widely separated components. The front component causes the light to converge, while the back component is divergent. In the most modern designs, both the front and rear components are quite complex.

Telephoto lenses of early forms suffered from severe distortion, but this was considerably reduced in later designs. An average modern telephoto lens will work at an aperture of *f* 5·6 and cover a field of 15° to 30°.

Aberrations. All lenses suffer more or less from basic shortcomings inherent in their design and construction which prevent them from forming a perfect image of the subject. These imperfections, known as aberrations, are of a number of distinct kinds, and are eliminated or reduced by the lens designer in various ways—e.g., by introducing extra lens components with equal and opposing aberrations.

The process of reducing the aberrations is known as correcting the lens. and the further it is carried the more expensive it becomes. In practice a lens is corrected only just enough to meet the specification laid down; to go beyond that point would simply multiply the cost of the lens to no good purpose.

It is because of the lens aberrations that the distinction between the front and the back of a lens becomes significant. Lenses are corrected for the conditions under which they are intended to be used. Thus a camera lens is normally used in such a way that the front focal distance is very much greater than the back focal distance, in other words with reasonably distant objects. When these conditions are reversed, for example in macrophotography where the image distance becomes appreciably greater than the object distance, the lens may no longer yield its best performance. In such a case it is sometimes useful to reverse the lens, and use it with its front facing the sensitized material and the back facing the object.

Due to the residual aberrations in the lens, the focal length as measured with light passing from the front to the back may be different from the value obtained with the light passing from the back to the front. The former value is known as the back focal length, and the latter value the front focal length. When the term focal length is used without qualification, it is normally understood to mean the back focal length.

An enlarging lens, by the same token, is normally intended to be used with the front focal distance (lens-to-enlarging paper) greater than the back focal distance (lens-to-negative). Special lenses, for example for process work, are often corrected to perform best under conditions where the front and back focal distances are more or less the same or do not exceed a ratio of about 3 : 1 or 4 : 1.

Calibrated and Computed Focal Length. The focal length of a lens also determines the image scale, in other words the magnification under specific conditions. If the lens suffers from distortion—one of the various aberrations—the image scale is no longer constant over the whole of the image. It may then be necessary, in certain applications of a lens, to assume an adjusted value of the focal length. This is chosen to distribute the distortion in a manner best suited to the conditions under which the image is valuated. This adjusted value is known as the calibrated focal length. It is used mainly in the calculations involved in measuring dimensions from the image formed by the lens under specified conditions.

When complete constructional data of a lens design (such as the curvatures, refractive indices, thicknesses and separation of lens elements) are available, the focal length of a lens can be completed mathematically as the distance from the rear nodal point to the principal focus. This is the focal length for a light beam of negligible diameter parallel with the axis of the lens. Such a limitation to a narrow axial light ray eliminates the influence of aberrations on the focal length; the resulting value for the latter may be different from the actual focal length measured.

Aperture and Diaphragm. It is generally necessary to be able to control the amount of light that passes through a lens—e.g., as a means of adjusting the exposure. This is done by placing a diaphragm with a hole in it in the path of the light rays. The size of the aperture can be varied according to the amount of light that is to be allowed to fall on the sensitized material to make the exposure. The control which alters the size is calibrated in *f*-numbers. The *f*-number at any setting is the ratio of focal length to effective optical diameter (which is not necessarily the same as the physical diameter of the lens opening); it is thus a relative measure of the aperture. The brilliance of the image varies in proportion to the *f*-number; it does not depend upon the actual size of the lens or its diameter. So all lenses set to the same *f*-number form images of equal brilliance and hence call for the same exposure, no matter what their focal length.

The diaphragm has another function: it controls the depth of the field of sharp definition within which any focused object lies. The smaller the aperture, the greater the depth of field sharply covered.

The aperture also gives control of certain types of aberration, and all lenses require a diaphragm to restrict their useful area in some degree for this reason.

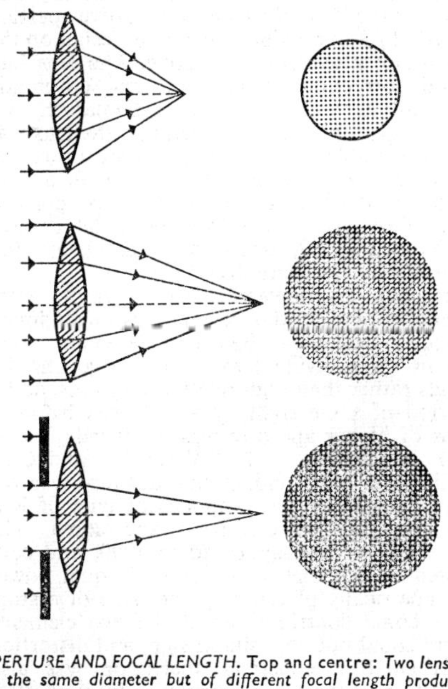

APERTURE AND FOCAL LENGTH. Top and centre: *Two lenses of the same diameter but of different focal length produce images of different sizes and therefore of different brightness.* Bottom: *As the lens is stopped down, the brightness of the image decreases still further.*

Coating. When a lens is composed of a number of elements, the amount of light lost through reflection at each of the surfaces may add up to a serious amount of the image-forming rays. Most of this loss is prevented by the process of coating or blooming the lens surfaces with a thin film of a suitable metallic fluoride. Practically all lenses are now treated in this way.

Special Lenses. In some cases lenses are required with special characteristics—according to the particular work the lens is intended for. Lenses designed for such purposes are usually named according to their function—e.g., enlarging lens, process lens, projection lens.

There are also lenses which are designed to produce special effects. These involve quite unusual constructions rather than just differences in normal characteristics. Instances are: the Hill cloud lens, designed to cover a field of 180 degrees so that the sky can be photographed on one negative; the variable focus lens, used to get continously variable focal lengths, and thus image sizes; the Fresnel lens—a simple lens of unusual shape—intended for spotlamps.

Engraved Data. All lenses of any value have certain information on the the rim of the mount.

A recommended standard form of marking lists the focal length, maximum aperture, name of lens maker, name of lens, and maker's serial number. For example: 50 mm. *f* 2·8 Smith Anastigmat No. 12876. Certain alternative marking systems are also in use, including the town of the lens maker, and different ways of expressing the maximum aperture (e.g. as a ratio of aperture diameter to focal length): G. Smith, Leeds, Anastigmat, *f* = 50 mm., 1 : 2·8, No. 12876.

The iris diaphragm control is also marked with the different aperture settings, normally chosen from the international standard aperaperture range. Sometimes the maximum aperture does not fall into the international standard range; the older continental aperture range is, however, now obsolete.

Where the lens has a distance scale, this may be marked in either feet or metres or both. A standard recommendation is that the distance figures should follow the series H, H/2, H/3, H/4, etc. (H is the hyperfocal distance at full aperture, i.e. the nearest distance within the depth of field range when the lens is focused on infinity.) Lenses do not necessarily follow this recommendation. D.P.C. (L.A.M.)

See also: *Aberrations of lenses; Aperture; Circle of confusion; Coated lens; Covering power; Diaphragm; Lens care; Lens history; Lens, interchangeable; Lens manufacture; Lens mount; Lens performance; Lens testing; Optical calculations: Optical glass.*

Books: *Introduction to Applied Optics* (2 vols.), by L. C. Martin (London); *Photographic Optics*, and *A System of Optical Design*, by A. Cox (London); *Fourier Methods in Optical Image Evaluation*, by E. H. Linfort (London); *Measurement of Optical Radiations*, by G. Bauer (London).

LENS CAP. Opaque metal, rubber or plastic cover slipped over the camera lens to protect it. Interchangeable lenses are usually supplied with a cap for both ends, for complete protection, when the lenses are detached from the camera. With early cameras which had no shutters the photographer made the exposure by uncapping the lens for the required time. This method is still used for making long exposures with stand cameras.

Some hand cameras have a baseboard which covers the lens when the camera is folded, in which case no lens cap is required.

LENS CARE. The optical glass used for making lenses is usually much softer than other kinds. It is not easy to clean once it has been allowed to get dirty; everything should be done to keep it clean in the first place. If the lens has a lens cap, it should be left on unless the camera is actually being used; if the lens is detachable, it should have a cap over each end when it is out of the camera. Folding cameras should never be left open when they are not in use.

The lens surfaces should never be touched with the fingers. Fingers leave greasy prints that dust will stick to, and may leave acid perspiration that will actually destroy the high polish.

Dry dust is best removed from the lens with a soft camel hair brush or by blowing it off with an air jet from a watchmaker's bellows —not by breathing on it.

Grease and finger marks may be removed by wiping the surface gently with a piece of soft, well-washed cotton rag or wash-leather. The special tissues made for cleaning lenses are excellent; they can be used once and thrown away, so that there is no danger to the lens surface from dust and grit picked up and embedded in the tissue.

A stubborn film over the lens may call for a drop of methylated spirit on the cleaning tissue. No spirit should actually flow on to the lens or it may penetrate between the cemented components and loosen them.

When, through wrong cleaning methods, a lens has lost its high polish, it should be returned to the maker for repolishing; no one but a specialist should tackle this job which requires precision equipment.

LENS HISTORY

The lens in its simplest form has been known for hundreds of years; references to the use of a magnifying glass occur in writings of the eleventh century. The first use of a lens in a camera, however, was probably in the camera obscura of Cardano in 1550; the lens was of biconvex form and made of crown glass. Chemical recording of the image was unknown; it was simply viewed on a screen, often a marble table-top. No attempts were made to correct the lens for the various optical aberrations, and the photographic lens really originated in 1812, when the English physician, Wollaston, first recognized one of the most important principles in lens design—that certain aberrations could be controlled by suitably positioning a stop with respect to the lens.

The Wollaston meniscus gave reasonable definition at *f* 16 over a field not greater than 50°. More simple meniscus photographic lenses have been made than of any other form, for use in fixed focus box cameras. Wollaston's lens was uncorrected for colour, and the achromatized meniscus was introduced by Chevalier in 1829, the principle of combining a positive lens of crown glass with a weaker negative element of flint glass having been already applied to the telescope object glass.

The period 1840–60 saw the birth and growth of photography, as in those years great strides were made in the chemical recording of the image formed by the camera lens. The advantage of a fast lens—that is, a lens of large relative aperture—was now obvious, and ensuing developments had this aim, together with improved definition over a larger angular field.

Petzval Lens. Recognizing the need for lenses of increased speed, the Hungarian mathematician J. Petzval, in 1840 made a complete break from the traditional lens type of his day and evolved the Petzval lens with a relative aperture rated at *f* 3·6, possibly basing his design on the Fraunhofer telescope objective. The lens had good correction of spherical and chromatic aberrations and coma, but astigmatism and field curvature were considerable; the lens was most suitable, therefore, for portraiture requiring high aperture but small field and established itself at once as the best portrait lens all over the world. Many years later it was developed for high-speed taking and projection lenses in cinematography.

Orthodox Development. Few attempts were made to develop the Petzval lens on its introduction, possibly because designers were intent on evolving lenses with large angular fields rather than high relative apertures, and a period of some eighty years elapsed before a lens of higher aperture was produced.

Following the lead of Wollaston and Chevalier, the next fundamental advance was the introduction of the symmetrical type of lens to give reduced distortion and coma. The Goerz Hypergon appeared in 1900, and consisted of two meniscus lenses of equal power symmetrically placed on either side of a stop. The coma contributions of the two elements were equal but opposite in sign, and distortion was small if spherical aberration was also small—that is, at low aperture ratios. Spherical aberration was considerable, chromatic aber-

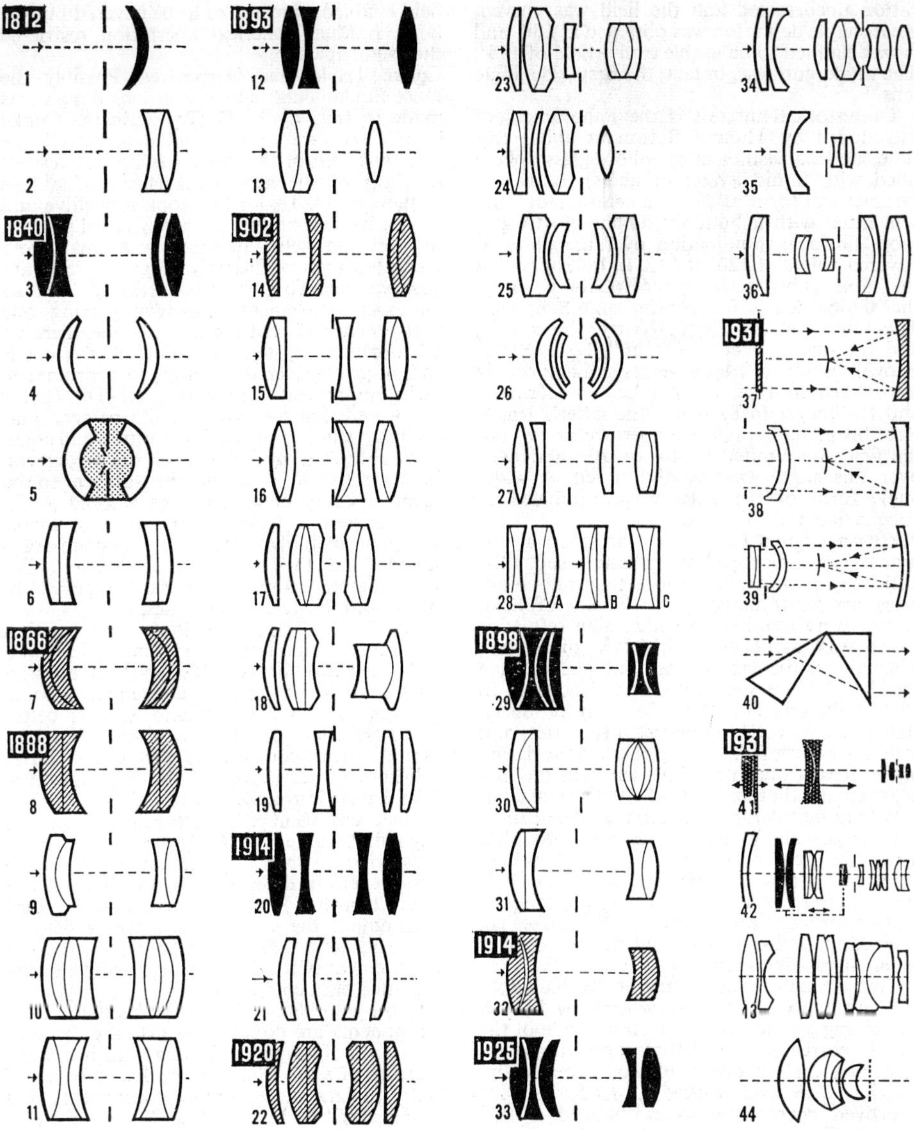

DEVELOPMENT OF PHOTOGRAPHIC LENS DESIGN. Simple lenses: 1. *Wollaston Meniscus.* 2. *Achromatized Meniscus.* (Early symmetrical designs.) 3. *Petzval lens.* 4. *Goerz Hypergon.* 5. *Liquid filled Panorama lens.* 6. *Globe lens.* 7. *Rapid Rectilinear, the best-known symmetrical lens.* 8. *Ross Concentric.* 9. *Zeiss Protar.* 10. *Double Protar.* 11. *Goerz Dagor.* Anastigmat lenses: 12. *Original Cooke Triplet.* 13. *Aldis Anastigmat.* 14. *Zeiss Tessar, the best known triplet derivative.* 15. *Dallmeyer Pentac.* 16. *Leitz Hektor.* 17. *Zeiss Sonnar f2.* 18. *Zeiss Sonnar f1·5.* 19. *Cooke Series X.* Modern high-speed symmetrical types. 20. *Cooke Aviar.* 21. *Ross Homocentric.* 22. *Cooke Series O.* 23. *Warmisham. f1·3.* 24. *Lee f1·0.* 25. *Zeiss Orthometar.* 26. *Zeiss Topogon wide angle.* 27. *Zeiss f0·85.* Telephoto and inverted telephoto lenses: 28. *Telephoto attachments for normal camera lenses.* 29. *Zeiss Telephoto.* 30. *Dallmeyer Adon.* 31. *Busch Bistelar.* 32. *Booth Telephoto.* 33. *Distortionless telephoto lens.* 34. *Inverted telephoto design.* 35. *Hill Cloud lens.* 36. *Watson variable focus zoom lens.* Mirror lens systems: 37. *Schmidt.* 38. *Bouwers and Maksutov.* 39. *Linfoot and Hawkins.* Anamorphic system: 40. *Brewster's prisms.* Zoom lenses: 41. *Busch Vario-Glankar.* 42. *Zoomar for 35 mm. cameras.* Aspheric lenses: 43. *Elgeet Golden Navitar.* 44. *Yoshido f 0·519.*

ration uncorrected and the field was curved. Reasonable definition was obtained at *f* 30, and it was claimed to be usable over a field of 135°. The Hypergon was, in fact, the first wide-angle lens.

Of historical interest is the Panorama lens introduced by Thomas Sutton at about this time, and consisting of a hollow glass sphere filled with liquid (water or acids). It was of symmetrical form as it had a central stop, and was fitted with a butterfly diaphragm to give more uniform illumination over the plate. It covered a field of 120° at *f* 12. In 1862 appeared the first achromatic symmetrical-type lens, the Globe lens of Harrison and Schnitzer. The outer glass surfaces were concentric, and the lens covered 92° at *f* 17·5. Better known, however, are the lenses simultaneously introduced in 1866 by Steinheil in Germany and Dallmeyer in England; the latter's Rapid Rectilinear had a relative aperture of *f* 8, but the field was limited to 40° or 50°, and even over this angle was heavily curved. A wide-angle form of the R.R. was introduced to cover a field of 100° at *f* 15.

Early Anastigmat Period. Lens designers at this time were handicapped by limitations set by the optical glasses available: to satisfy the conditions for achromatism and flatness of field, glasses were required in which high refractive index was associated with low dispersion, whereas, in the crown and flint glasses then used, high index went with high dispersion. In 1888 Abbe and Schott in Germany produced stable glasses with the desired properties, and soon afterwards Schroeder published the design of the Ross Concentric. This lens was the first anastigmat—that is, the field (70°) was reasonably flat and tolerably free from astigmatism—and consisted of two achromatic doublets using the new barium crown glass of high index and a new flint glass of low index. This combination, however, was the opposite of that required for correction of spherical aberration, and the lens could be used only at the small relative aperture of *f* 16.

This difficulty was overcome by Rudolph, of Zeiss, who in 1890 introduced the Protar, combining an old achromat on one side of the stop to correct spherical aberration with a new achromat on the other side to correct astigmatism. The lens worked at *f* 7·5. In 1894 improved corrections were obtained in the Double Protar, in which the old and new achromats were cemented and duplicated on either side of the stop. Von Hoegh of Goerz achieved a similar result with six elements in the Dagor, working at *f* 6·8 over a field of 70°.

The symmetrical form was retained at this time to reduce coma and distortion, but each half could be used separately, with poorer correction, giving lenses of different focal lengths. The complete lens was then called a convertible. These symmetrical type anastigmats had similar properties; the relatively flat field enabled wide angles to be covered but the large residual spherical aberration restricted the useful aperture.

Cooke Triplet and Derivatives. Possibly the most fundamental advance in lens design was made in 1893 by H. D. Taylor in the Cooke lens. Curvature of the field depends on the resultant sum of the powers (with due respect to sign) of the separate surfaces of a lens system. H. D. Taylor first took a positive and a negative lens of approximately equal powers, so that the field curvature was small, and separated them sufficiently to give a resultant positive lens. To correct distortion he divided the positive element into two, placing one part on each side of the negative member. To achromatize a system for points off the axis it had been considered essential to achromatize each component separately; H. D. Taylor discovered that the system could be achromatized as a whole. Spherical aberration and coma were corrected by suitable choice of curves, and astigmatism by placing the stop near to the negative component. The lens worked at *f* 3 and covered a field of 40°; it was, therefore, considerably faster than contemporary designs—e.g., the Dagor—but covered a smaller field. Later, more symmetrical types were introduced to cover wider angles.

The Cooke Triplet provided a construction capable of adaptation in symmetrical and unsymmetrical forms. Most modern anastigmats are derived from it, the tendency being towards the symmetrical form with its better field correction. In the Aldis Anastigmat at *f* 4·5 the front components were cemented. A well-known derivative of the triplet is the Zeiss Tessar introduced by Rudolph in 1902; a doublet rear member is employed, and the lens originally worked at *f* 4·5 over a 50° field. The aperture has been increased to *f* 3·5 without serious deterioration in definition. The Leitz Elmar is a Tessar type, as also is the Kern Anastigmat; the latter is used over a field of 65°. The Ross Xpres can be regarded as a development of the Tessar type, having a triple-cemented back component.

In the Voigtländer Heliar the front and back components are doublets, as also are those of the Booth (Dallmeyer) Pentac with a relative aperture of *f* 2·9. The Hektor, due to Berek of Leitz, utilizes three doublet components and works at *f* 1·9. In the Zeiss Sonnar *f* 2 and *f* 1·5 lenses, triple-cemented components are used; the angle covered is 48°. In a relatively simple derivative of the triplet, Lee (1926) increased the aperture to *f* 2·5 by using a double-back component in the Taylor-Hobson Cooke Series X.

Quadruplet Types. Whilst H. D. Taylor in the Cooke Triplet had actually divided the positive lens into two components, he also suggested that the negative lens could be similarly divided and the two parts placed on either side of the stop, the resultant system being of a symmetrical type, more suited for larger angular fields. This form was developed by Goerz in their Celor and Dogmar lenses (1898) and further developed by Warmisham about 1914 into the Taylor-Hobson Aviar working at *f* 4·5. In the Ross Homocentric all four components are of meniscus form, and the lens works at *f* 5·6 over a 50° field.

The demands of the cinema industry and of the miniature camera user were for faster lenses with a high standard of definition. By employing doublet negative lenses, Rudolph increased the aperture to *f* 3·5 in the Zeiss Planar, and in 1920 Lee increased it still further to *f* 2·0 in the Taylor-Hobson Series O. The Leitz Summar, the Schneider Xenar and the Zeiss Biotar are all of similar construction. Departure from symmetry and increase in complexity are shown in the *f* 1·3 lens, due to Warmisham, and the *f* 1·0 lens, due to Lee.

The quadruplet lens with meniscus components is particularly adaptable to a wide-angle form at a moderate aperture. The Zeiss Orthometar introduced in 1931 and the Ross Wide-angle Xpres have a relative aperture of *f* 4 and cover a field of 70° with very good distortion correction. The Zeiss Topogon, introduced in 1933, covers 90° at *f* 6·3, and a similar form is made by Bausch and Lomb as the Metrogon. The Ross *f* 5·6 wide-angle survey lens has the rear positive component divided into two elements.

Petzval Derivatives. The Petzval type of lens, characterized by large aperture but small field, remained largely undeveloped for many years. Designers were at first attracted more by the potentially better field correction of the symmetrical type, and later by the even greater potentialities of the triplet construction. For portraiture, however, the lens remained popular for a long time. The demand of the cinema industry for projection lenses with large apertures but relatively small angular fields ultimately aroused interest in its capabilities. Most cinema projection lenses are of the Petzval type with relative apertures up to *f* 1·9. When larger angular fields are involved, as in back projection, the lenses are usually triplet derivatives.

As a high speed taking lens for cinematography the type was developed by Warmisham to *f* 1·3; and for radiography to *f* 0·85 by Zeiss and to *f* 0·83 by Lee.

Telephoto Lens. The telephoto principle was applied by Barlow to telescopes in 1834, and to photographic lenses in 1891 simultaneously by Dallmeyer in England, Duboscq in France and Miethe in Germany. Dallmeyer first introduced a complete telephoto lens, consisting of an achromatic positive component and a triple negative. This was too slow, and was superseded in 1892 by a symmetrical triple negative lens to be used with a (positive) portrait lens of the Petzval type. A year later he introduced a negative attachment consisting of a pair of negative doublets and considered particularly suitable for medium magnifications. Several firms marketed their own versions of these negative attachments. Variation of focal length of the combination was obtained by altering the separation of the negative attachment and the positive lens with which it was used. These combination lenses were unsatisfactory, since field curvature and astigmatism were over-corrected (that is, the field was convex towards the lens), pincushion distortion was prominent and oblique colour was objectionable.

In 1898 Rudolph of Zeiss reverted to the original aim of Dallmeyer and introduced a telephoto lens complete. Dallmeyer in 1901, followed by Plaubel in 1905, introduced another type of attachment in the form of a low-power Gallilean telescope. The Dallmeyer attachment was called the Adon and had a magnification of three; it was not completely telescopic, and could be used alone as a telephoto lens of small aperture. This type of attachment was considered particularly suitable for use with the small lenses fitted to hand cameras with limited bellows extension. In more recent times a Gallilean attachment has been introduced by Schneider, and is mainly intended for use with lenses working with 8 mm. and 9·5 mm. film.

Fixed Focus Telephotos. The first successful fixed focus telephoto lens, capable of replacing the usual lens on the hand camera and giving a larger picture with the same bellows extension, was probably the Bistelar introduced by Busch in 1906. It had a relative aperture of *f* 7 or *f* 9, depending on the focal length, and a telephoto effect of two, and was followed by the Zeiss Magnar of *f* 10 and a telephoto effect of three.

In 1910 Dallmeyer produced the Grandac, and in 1912 Booth offered the New Large Adons with relative apertures of *f* 4·5 and *f* 5·6 and of similar form to the Bistelar. The Telecentric of Ross, with apertures of *f* 5·4 and *f* 6·8, had a triple positive component. These telephoto lenses were free from many of the defects of the variable magnification types, but still suffered from astigmation and distortion.

Anastigmat Telephotos. The first anastigmat telephoto lens was designed by Booth in 1914 with an aperture of *f* 6; it was corrected for flatness of field and astigmatism, as well as for spherical and chromatic aberrations, and in definition was a considerable improvement on its predecessors. In 1920 the same designer lightly simplified the construction by using a cemented positive component, and the lens was made by Dallmeyer as the Dallon with an aperture of *f* 5·6 and a telephoto effect of two. The Zeiss Teletessar at *f* 6·3 also consisted of two cemented components, whilst the Ross Teleros at *f* 5·5 has a triple cemented negative component. By separating the positive component into two elements (one cemented) Lee increased the aperture to f3·3.

Distortion had been considered an inherent drawback of the telephoto construction, and was present in all of the types mentioned. The first distortion-free telephoto lens was designed by Lee, with an aperture of f5 and a telephoto effect of 2·3. The field covered by anastigmatelephoto lenses is usually 30°. For cinematography, the aperture of the telephoto lens has been advanced to f1·65.

Inverted Telephoto Lens. In the telephoto construction with the positive component in front of the negative the equivalent focal length is greater than the back focal distance: with the negative component in front of the positive the focal length is less than the back focal distance. For a given focal length, therefore, the inverted telephoto construction gives a greater back clearance, a convenient arrangement if it is necessary to interpose other optical elements between the lens and the focal plane. In some systems of colour cinematography, for example, a beam splitting prism has to be accommodated in this space. One such inverted telephoto design, due to Lee, is used for photography in Technicolor; a lens of 1½ ins. focal length covers the 35 mm. film frame, with an aperture initially f2 and later increased to f1·7.

Another effect of the inverted telephoto construction is a widening of the angular field, and this property has been utilized by R. Hill in his Cloud lens for photography of the whole sky, the field covered being 180° at a very small aperture; distortion is deliberately large. The principle of the Hill lens was extended by Zeiss in the Pleon and Sphaerogon lenses patented about 1935. The Elgeet Company has recently introduced a 13 mm. f1·5 reversed telephoto lens for 16 mm. cameras.

Recent Developments. The post-war years have seen great activity in photographic lens design, and possibly the most important reason for this is the introduction of new optical glasses. High-index glasses with low dispersion have been introduced, the use of which enables flatter curves to be employed on the lens surfaces, with reduced zonal spherical aberration and astigmatism.

No fundamentally new design has been introduced, but improved developments have been made of most of the already existing types. Successors of the triplet construction with a meniscus element in the front air space (originally due to Ernemann) include the Kodak Cine Ektar f1·9 lenses and the Kern Switar and Pizar lenses of f1·4. From the quadruplet design the following lenses have evolved: the Wray Lustrar f4·5; the Summarit f1·5 and Summarex f1·5 for the Leica camera; the Taylor-Hobson Amotal f2 for the Foton camera and the Ivotal f1·4 for 16 mm. cinematography; the Dallmeyer Septac f1·5; the Kodak Aero Ektars f2·5; the Ross Xtralux f2; the Wray Unilite f2 and Cine Unilite f1·9 and f1·0 copying lens. An f0·71 lens of unusual form with a very short back focus has also been produced.

Performance of the Petzval type has been improved by the use of crown glass of high index, and there has been a revival of the use with this lens of a negative component close to the focal plane to correct field curvature in a 2 ins. f1·5 projection lens, by Schade, to cover 14°. This arrangement was originally used by Piazzi Smyth in 1866.

Aspheric Lenses. The advantage of using a non-spherical refracting surface, to correct aberration in a lens system, has been known since the seventeenth century. However, in the succeeding two centuries this advantage was hardly exploited, since little was understood as regards the aberration of optical systems and furthermore the ability to manufacture such surfaces was not developed.

In the early part of the present century, however, interest was renewed as the photographic lens, using conventional spherical surfaces, underwent significant development.

Between 1908 and 1922 Carl Zeiss took out several patents with respect to both the design and manufacture of lenses using aspheric surfaces. Earlier attempts by Zeiss to manufacture aspherics were mainly concerned with the moulding of components rather than the then almost traditional hand figuring methods.

The full significance of the use of aspheric surfaces, other than conic-sections, was demonstrated when Berthardt Schmidt introduced the system that now bears his name.

Like many previous steps forward his invention was of a simple nature but it opened up a whole new line of thinking to the photographic art. Schmidt's discovery comprised of only a thin aspheric corrector plate and a spherical mirror but the essence of his discovery lay in the positioning of the corrector plate relative to the mirror such that it contributed to both the axial and oblique aberration corrections.

Stemming from this discovery have come many modifications including the Dioptric Schmidt type of system which replaces the spherical mirror by a more conventional refracting member. Such derivatives have yielded lenses of very high relative apertures of f0·7, covering substantial fields of view.

Since the end of the 1950s manufacturing techniques developed to provide surfaces of larger deviations of figure than the earlier corrector plates, whilst achieving an accuracy of figure equal to that of more conventional spherical surfaces. As a result the aspheric surface can either be used to simplify an existing type of construction yet yield an equivalent performance, or such a surface may enable the optical designer to achieve a result hitherto difficult to obtain by using spherical surfaces only.

Examples include the 12 mm. *f* 1·2 Elgeet Golden Navitar and the 5·8 mm. Kinoptic lens for the 16 mm. format. For the 35 mm. cine format the 18 mm. and 25 mm. *f* 2·0 Cooke Speed Panchro Series III lenses both have an aspheric components, as have the Varotal III and Varotal V 10 : 1 zoom lenses for television cameras.

Zoom Lenses. The scale of the picture taken with a camera depends upon the focal length of the lens used; from a given viewpoint to photograph a particular object at different magnifications necessitates the use of different lenses. In cinematography a need arose for a lens of continuously variable focal length, within certain limits, so that close-ups could be taken, the image remaining sharply focused on the film as its size was altered. Projection lenses of variable power were described as early as 1902, but taking lenses of this class seem to have been developed from 1930 onwards. The earliest of these included the Bell and Howell Cooke Varo, the Busch Vario-Glaukar and the Astro Transfokator. The latter was an afocal zoom attachment intended to be used in front of the normal motion-picture lens. These lenses represent the first known deliberately designed zoom lenses and in every case some kind of cam was used (i.e. mechanical compensation) to hold the image in constant focus while the focal length was being changed.

Due possibly to the economic conditions of the film industry in the 1930s, activity in the design of zoom lenses subsided until 1946, when F. G. Back introduced the Zoomar lens for 16 mm. cameras. This lens involved a succession of positive components in which the alternate second and fourth components were coupled together and moved linearly along the lens axis to vary focal length. Such a system yielded an approximation to a constant image plane position during change of focal length (optical compensation).

Zoom lens design developed rapidly since then and currently zoom lenses are available for microscopy, cine and still photography, television and telescopic applications. The economic advantages of optical compensated zooms against the performance advantages of mechanically compensated zooms has in the past tended to cause manufacturers to produce only one type or the other dependent on their type of market. However, current trends involving a wider specification on range of focal lengths and aperture suggest that the mechanically compensated zoom holds the better promise for future development.

The majority of current optically compensated zoom lenses achieve a maximum focal length to minimum focal length ratio of about 4 : 1, but the Zoomar Company also introduced a lens having a 7 : 1 ratio for television cameras. In 1950 H. H. Hopkins described a mechanically compensated zoom lens having a 5 : 1 ratio whilst in 1960 Rank Taylor Hobson introduced the Varotal III zoom lens involving three moving members having a dual range of 10 : 1 in two 5 : 1 steps. The changeover from one focal range to the other being achieved by a fourth moving member occupying one of two positions.

Since 1960 focal ratios extended even further with the introduction of mechanically compensated zoom lenses for cine and television by Angenieux and Rank Taylor Hobson having continuous ratios of 10 : 1. Such lenses are optically very complex and required intense development of their associated mechanics. However, today such zoom lenses are attaining a level of performance equivalent to the range of fixed focal length lenses they are intended to replace. Generally at least two components of the lens system are moved in relation to the focal plane and to each other, and precision cams giving non-linear axial movements are necessary.

Mirror Lenses. A disadvantage of the orthodox lens with refracting elements—a dioptric system—is that colour correction is never perfect, there being always a residual secondary spectrum. The attraction of an all-reflecting (catoptric) system with its complete freedom from chromatic aberration is therefore obvious, particularly in the utilization of actinic ultraviolet radiation or in extending the sensitivity of the emulsion to the infra-red. Whilst mirror-lens combinations had been used previously, the arrangement used by Schmidt in 1931 was probably the first successful photographic lens of this type. Coma and astigmatism of the spherical mirror were eliminated over a curved field by placing a stop at the centre of curvature, and spherical aberration was corrected by a plate of suitable aspheric form in the plane of the stop. The system is strictly catadioptric, since refraction occurs at the correcting plate, but even so, chromatic aberration is negligible. A relative aperture greater than *f* 1·0 was obtained, but the field was steeply curved.

The aspheric correcting plate was replaced by Gabor with a meniscus element using spherical surfaces, and independently by Bouwers and Maksutov with a concentric or near-concentric meniscus of appreciable thickness. An achromatized corrector consisting of a meniscus plate and a doublet aspheric plate has been proposed by Linfoot and Hawkins, on the grounds of greatly improved definition, whilst Wynne has suggested a double meniscus corrector.

The main application of the Schmidt lens and its developments has been in astronomy, where extremely high apertures are required. For X-ray fluorography, a Danish firm has made a 4 ins. *f* 0·75 for 35 mm. film and a 6 ins. *f* 0·85 for 70 mm. film. The lens has also been used for television projection, the curved end of the cathode-ray tube matching the curved field of the lens. Aspheric correcting plates of moulded gelatin have been described.

Ultra-Wide Aperture Photographic Systems. In general such systems fall into one of two types. They are either the derivatives of the Schmidt system (reflecting or dioptric); or extensions of the double Gaussian, Triplet or Petzval types. The advantages in the choice of any system evolves round the field coverage requirement and whether the image plane need be flat.

Mirror Schmidt systems of apertures around *f* 0·7 have been used extensively in television projection, whilst the Dioptric Schmidt has found a major application in cine radiography work. The advantage here lies in the fact that the focal plane lies external to the lens and does not vignet a central area of the lens aperture as in the mirror Schmidt system. The focal plane is still curved but in the opposite direction to the classical Schmidt and the film is curved to suit. Examples from the patent literature by Warmisham and Wreathall show apertures of *f* 1·0 covering 32° of field and *f* 0·7 covering 22½° of field.

Derivatives from the Double Gaussian, Triple and Petzval types of construction have found application in cine photography, cine radiography and cine projection. The two former types yield essentially a flat field but of reduced angle compared with the dioptric Schmidt.

Examples for 16 mm. cine include *f* 0·95 lenses by Angenieux and S. O. M. Berthiot, these being basically extended Double Gaussian systems. Triplet derivatives include an *f* 0·75 covering a 14° field by Schade and the *f* 0·519 lens by Yoshido, the latter using an aspheric surface to achieve the extreme aperture. Kaprelian has shown an example of an extended Petzval type of construction having a relative aperture of *f* 0·7 covering a field of 34°.

Non-Reflecting Films. At every air-glass surface of a refracting lens system approximately 5 per cent of the light incident there is reflected; the precise amount depends on the refractive index of the glass (and on the degree of polish and cleanness of the surface). In a complicated system, therefore, as much as 50 per cent of the incident light may be reflected backwards and forwards between the surfaces, a good proportion finally emerging and spreading over the image plane. The light transmission of the lens is thus reduced, and from the point of view of speed its rated *f*-number is incorrect: for an unbloomed lens of eight air-glass surfaces the effect is equal to approximately half a stop. For this reason it has been proposed that aperture markings should be photometrically calibrated, and some manufacturers are, in fact, engraving "T-numbers" on their lenses. The effect of the non-image forming light falling on the focal plane may be more serious than the fall in transmission, however, because the contrast of the image may be seriously reduced.

H. D. Taylor in 1892 first recorded that the surface reflection at an air-glass surface was reduced if the surface had become naturally bloomed by atmospheric action. For light of a given wavelength, if a layer of a suitable refracting material is deposited on the surface with a thickness of one-quarter wavelength, destructive interference occurs between light reflected at the air-film, film-glass surfaces and the amount of light reflected is reduced. Ideally, the refractive index of the film, for the wavelength considered, is equal to the square root of that of the glass on which it is deposited. Magnesium fluoride has a suitable refractive index, the value for the helium *d* line being 1·390. This substance is deposited by vacuum evaporation to the requisite thickness. The process is most commonly employed for the production of non-reflecting films, and was first described by Cartwright in 1939.

For white light, the film thickness cannot be ideally correct for all wavelengths and a compromise is adopted; as an average value, reflected light is reduced to 1 per cent per surface. With a complex photographic lens the increase in light transmission is appreciable, and a bloomed lens of eight air-glass surfaces is only about one-quarter stop slower than an ideal lens of the rated *f*-number (compared with one-half stop for an unbloomed lens of similar construction). The increase in contrast, due to the reduction in reflected light, is quite considerable.

Anamorphotic Systems. In general, within the limits set by any inherent distortion, the photographic lens forms an image geometrically similar to the object. Trapezoidal distortion can, of course, be introduced by a suitable tilt of the plane of the photographic plate or film and of the screen of an enlarger. Descriptions of optical systems which will deliberately distort the object by giving different magnifications in two mutually perpendicular meridians date back for a hundred years. Brewster's teinoscope consisted of a refracting prism system, whilst G. Chrétien proposed the use of cylindrical lenses in three-colour photography, and combinations of cylindrical reflecting and refracting surfaces have been suggested.

Apart from its application to comic pictures, anamorphosis has been little used in photography up to recent times; with the advent of wide-screen cinematography, however, there has been a revived interest. With an anamorphic attachment fitted to the taking lens, a scene of approximately twice the usual width is compressed into the standard film width, whilst there is little change in the vertical direction; on projection, a similar attachment expands the picture in width to fill the wide screen. A refracting prism system of the Brewster type has been developed by Taylor, Taylor & Hobson; the Delrama anamorph of De Oude Delft is a reflecting system employing a spherical and a cylindrical mirror. K.J.H. (P.A.M.)

See also: *Anamorphotic lens; Aspheric lens; Camera history; Coated lens; Lens; Zoom lens.*

Books: *Handbook of Photography*, by Henney and Dudley (New York); *Photographic Optics*, by A. Cox (London).

LENS HOOD. Shield fitted to prevent rays of light from outside the picture area from falling on the lens. Although rays outside the angle of view of the lens do not affect the sensitive surface directly, they set up internal reflections in the lens and may cause ghost images and a general degradation of the contrast of the negative.

There is no need to use a lens hood with the sort of box cameras which have the lens set well back into the body—often behind the diaphragm—but a hood is almost a necessity on the more expensive cameras with wide aperture lenses which are usually without any protection from stray light.

The simplest type of lens hood is no more than a short piece of tube, blackened on the inside to reduce surface reflections. It is made to slip directly on to the rim of the lens or filter mount. Some cameras have a screw fitting.

Most lens hoods are made funnel-shaped with the object of providing a relatively deep hood without cutting off the corners. The angle of the funnel conforms more or less to the angle of view of a lens of normal focal length. With this shape of hood there is always a risk of unwanted reflections from the inside surface if it is at all smooth or bright. The cheaper hoods of this sort are painted with dead black paint; in the more expensive types there are fine ridges machined on the inside surface to reduce reflections still further.

The most efficient hoods are those which conform to the shape of a square or round box with an aperture cut on the side opposite the lens. With this sort of shape there is no risk of internal reflection as none of the surfaces is so inclined that it can reflect incoming light rays on to the front of the lens. The aperture of the hood should, ideally, be of the same proportions as the negative; it should also be as small as possible without actually obstructing the angle of view of the lens (the distance between the lens and the hood aperture will govern this). Unfortunately, such hoods cannot be used with lenses which have front cell focusing unless the hood is put on after focusing.

Most lens hoods for hand and miniature cameras are of rigid metal construction; a few are collapsible. Larger hoods for studio and field cameras are often in the form of a rectangular bellows which can be extended to the limit where it starts to cut off the corners of the image on the focusing screen.

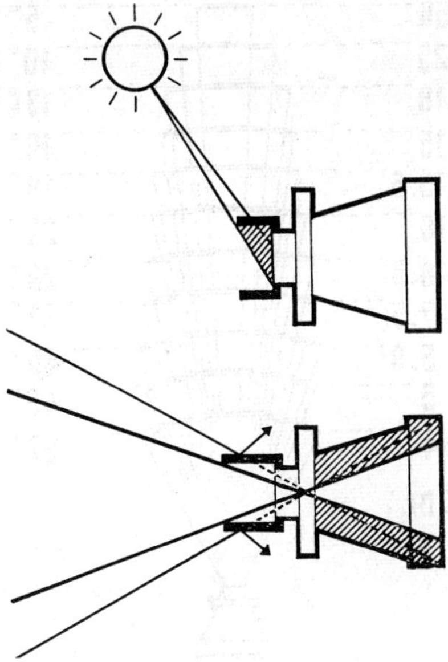

LENS HOODS. Top: The main purpose of the hood is to shield the lens from light sources outside the picture area, which might otherwise cause reflections inside the lens. Bottom: For maximum efficiency the hood should cut down the cone of light accepted by the lens so that it just covers the negative size.

When choosing a lens hood for a camera with a rising front, it is important to check it for cut-off with the front raised to the limit. A hood which just fails to cut the corners of the image with the lens central will cut off the bottom corners when the front is raised.

The shorter the focal length of the lens, the shallower the hood that can be used with it. Telephoto lenses, because of their narrow angle of view, can be fitted with very deep and efficient hoods which are a great help in producing a brilliant, fog-free image.

LENS, INTERCHANGEABLE

Lens system fitted into a camera (or projector or enlarger) in such a way as to be easily removable and changed against alternative lenses of different focal length or aperture. The use of interchangeable lenses extends the scope of a camera; such lens systems are widely used in better-class cine, miniature and reflex cameras as well as in many professional camera types. On many cine cameras variable focus (zoom) lenses serve the same function without the need for lens changing.

Optical Considerations. Alternative focal lengths provide a range of angles of view and image scales. They are frequently specified in relation to the normal focal length of the lens for a given picture format. This is accepted as being approximately the same as the diagonal of the picture format. Thus the diagonal of a theoretical 24 × 36 mm. format (in practice the image produced on the film is often slightly smaller) would be about 43 mm.; usually the normal focal length for a miniature camera is

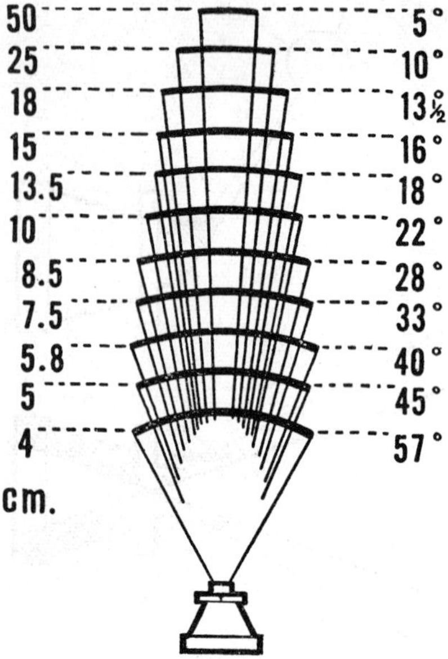

INTERCHANGEABLE LENSES. Certain cameras, notably precision models taking 24 × 36 mm. negatives, often take a whole series of interchangeable lenses of various focal lengths to cover a range of image angles. A typical series may include lenses from 3·5 or 4 cm. wide-angle to 40 or 50 cm. telephoto units with angles from around 60° down to 5°.

45 or 50 mm. The diagonal angle of view when the focal length is the same as the image diagonal is about 53°. The normal (or "standard") focal lengths for different image sizes are shown in a table below:

STANDARD FOCAL LENGTHS

Nominal Image Format *mm.*	*ins.*	*Diagonal* *mm.*	*Normal Focal Length* *mm.*	*ins.*
8 × 11		13·6	15	⅗
10 × 14		17·2	17–20	⅝–¾
12 × 17		20·8	20–25	¾–1
18 × 24	¾ × 1	30	30	1¼
24 × 24	1 × 1	34	35–40	1⅜–1⅝
24 × 36	1 × 1½	43	45–50	1¾–2
28 × 40	1⅛ × 1⅝	49	50	2
cm.				
4 × 4	1⅝ × 1⅝	56·5	55–60	2¼–2⅜
4.5 × 6	1¾ × 2¼	73 }	75	3
4 × 6·5	1⅝ × 2½	76 }		
6 × 6	2¼ × 2¼	81	75–80	3–3¼
6 × 9	2¼ × 3¼	100	100–110	4–4¼
9 × 12	4 × 5	150	150	6
12 × 16	4¾ × 6½	200	200	8
13 × 18	5 × 7	225	200–250	8–10
10 × 21	6½ × 8½	265	250–300	10–12
18 × 24	7 × 9½	300	300	12
20 × 25	8 × 10	320	300–380	12–15

Lenses of shorter focal length, yielding a wider angle of view, are generally referred to as wide-angle lenses; longer focal lengths are long focus or telephoto lenses. (The two terms are used interchangeably. Strictly speaking there is a difference in the optical construction—though not in the use.)

The shorter the focal length of a lens, the smaller the scale of the image taken in, but the wider the field covered. A wide angle lens is thus useful for taking in a large view in confined quarters where it is not possible to get far enough away from the subject to include it all with a normal lens. The image usually shows an enhanced perspective effect, because the wide view tends to take in a considerable portion of foreground, and such foreground objects loom large in the picture compared to the more distant ones.

The longer the focal length, the larger the image scale but the smaller the view covered. Long-focus and tele lenses therefore are useful for showing up detail of distant objects on a large scale. The perspective in long focus views appears subdued, largely because such a view frequently takes in only a small image field of objects situated at a distance where there is less graduation in depth. In fact, the perspective depends only on the camera view point and from a given viewpoint the geometrical perspective of any portion of the image taken in by a wide-angle normal and long-focus lens is identical but the apparent perspective in viewing the print is dependent on the viewing distance.

Short focus lenses yield a considerably greater depth of field at any given subject distance and lens aperture than long focus lenses. The depth of focus, at a given relative aperture, is however the same. When related to the image scale (magnification or reduction) a depth of field, however, is also constant irrespective of the focal length. That is to say, the depth of field for a subject taken with a long-focus lens is the same as the depth for the same subject taken with a shorter focus lens at a sufficiently closer distance to reproduce the subject in the same scale on the film.

While a normal or even wide-angle lens for a given picture format can be used as a long-focus lens for a smaller format, the reverse is not practical. This is because a wide-angle lens must be designed with a suitable covering power to cover the field of the image format with which it is used. A normal lens for a given format is not usually able to cover the wider field required of a wide-angle lens for a larger format, even though both may be the same focal length.

Wide angle lenses for reflex cameras are also subject to design limitations posed by the need for an adequate back focus to clear the reflex mirror. This calls for inverted telephoto type constructions.

Convertible and Interchangeable Lenses. There are three main ways (apart from zoom lenses) of providing alternative focal lengths for a camera:

(1) By changing the whole lens;

(2) By changing part of the lens;

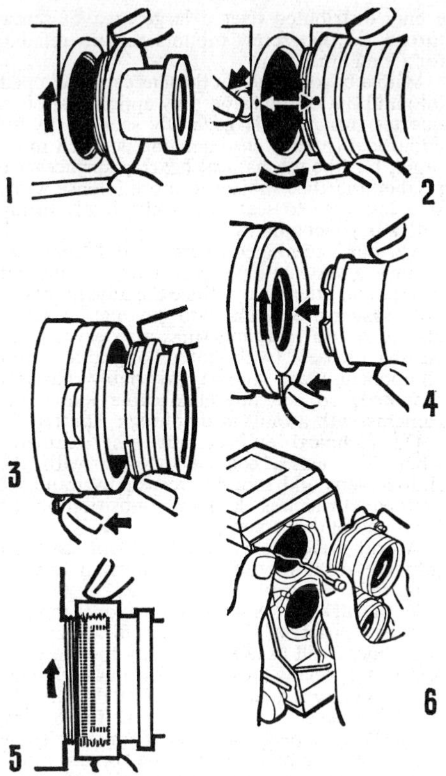

LENS CHANGING SYSTEMS. 1. Standard screw-in mount. 2. Bayonet mount in focal plane shutter camera. 3. Bayonet mount holding lens in front of diaphragm shutter. 4. Bayonet mount of convertible lens system (with rear lens component built in behind the shutter). 5. Interchangeable lens mount with retaining ring. 6. Interchangeable twin-lens panel.

(3) By fitting lens attachments in front of the fixed lens of the camera.

Fully interchangeable lenses are invariably mounted in front of the camera shutter. With focal plane shutter cameras this presents no problem, since the shutter itself is well behind the lens. The range of focal lengths available for such a camera is usually limited only by the physical dimensions of the camera mount and by considerations of usefulness. Angles of view may range from nearly 180° to 1 or 2°, focal lengths (with for instance a 35 mm. miniature camera) from 8 or 10 mm. (fish-eye lenses) to several metres.

Practical ranges found on professional cameras often go from 20 to 1,000 mm.; for amateur use about 28 to 200 mm. is adequate.

With diaphragm shutter cameras the construction becomes a little more complicated. Here the interchangeable lens range needs a reasonable constant back focus to avoid vignetting if the diaphragm shutter is too far behind the rear of the lens. Focal lengths are thus more limited, but again nearly always adequate for amateur use. The limitation is given by the minimum back focus to enable the lens to fit in front of the shutter (for wide-angle lenses) and by the dimensions of the shutter (for long-focus lenses).

An alternative, with diaphragm shutters, is to change the lens together with the shutter. This requires a separate shutter for each focal length, but is frequently used with professional and technical cameras.

In a convertible lens system for alternative focal lengths part of the lens—the rear element or elements—remain fixed in the camera, while only the front units are interchanged. The different front units together with the fixed rear component then yield the alternative focal lengths. The rear unit is mostly mounted behind the shutter, a diaphragm-type built into the camera.

This particular type of interchangeable lens system is nowadays only used on a few 35 mm. miniature cameras; it is somewhat

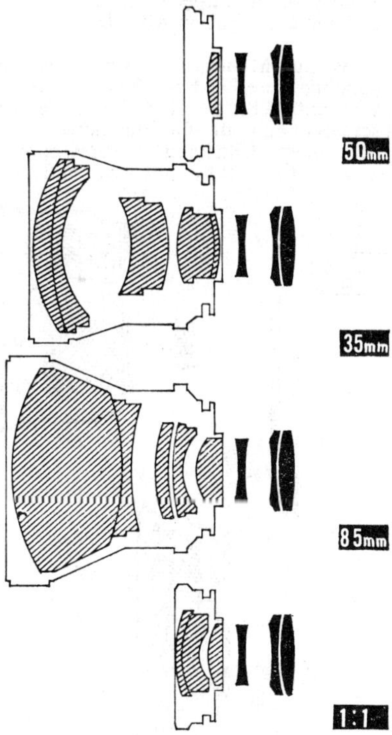

CONVERTIBLE LENS SYSTEM. The rear elements of the standard lens (top) are permanently built into the camera—usually with the shutter—but the front element is interchangeable. Alternative front elements can be combined with the rear components to produce a wide angle or telephoto lens system or even a lens unit permanently focused on a given close-up distance, for instance for same-size copying.

limited in the range of focal lengths available (usually in a ratio of 1 : 3), and also poses special design problems.

The design problems with convertible lenses of this kind are considerable, and both the wide angle and the tele constructions are usually considerably more complex than they would be with a fully interchangeable lens of similar focal length.

More old-fashioned types of convertible lens for field cameras consist of two components or groups of elements which can be used either singly or together to give different focal lengths.

A related design idea is the use of afocal lens attachments which fit in front of the complete (and firmly built-in) camera lens to convert the latter into a wide-angle or tele lens unit. Such conversion attachments have been produced for a number of cheaper miniature cameras, as well as for twin-lens reflexes where changing both the finder and the taking lens is not always practical. In the simpler versions they sometimes involve certain quality compromises, since it is not easy to design a camera lens so that it gives optimum performance on its own as well as in combination with the lens attachment.

Lens Changing Mounts. The commonest ways of changing lenses is by a bayonet or by a screw mount. With a screw mount the whole lens screws in and out of a fitted flange in front of the camera. Changing of screw-mounted lenses tends to be slow, but wear and tear is evenly distributed over a large area of screw thread, so that screw mounts remain reliable for a long time.

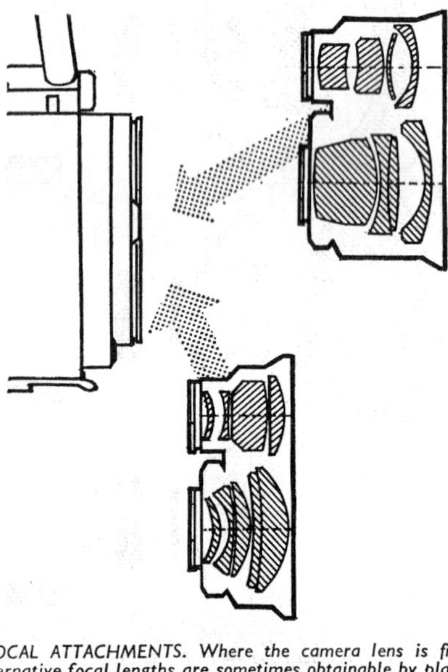

AFOCAL ATTACHMENTS. Where the camera lens is fixed, alternative focal lengths are sometimes obtainable by placing magnifying or reducing attachments in front of the camera lens. This system is used for some twin-lens reflex models where for instance a wide angle attachment (top) can convert a 75 mm. camera lens into an effective 52 mm. one, while a tele attachment (bottom) can increase the focal length to around 115 mm.

With a bayonet mount the lens can be clipped into and out of the camera by approximately a quarter turn, and is generally secured by an additional catch. Lens changing is much more rapid, but the mechanical bearing surfaces are smaller and thus subject to more intense wear. It is less easy to seat a lens absolutely firmly with a bayonet mount.

Various combined screw and bayonet mounting systems have been tried at different times in an attempt to utilize the advantages of both systems. Systems with bayonet or screw clamping and retaining rings also exist. Convertible lenses (with interchangeable front components) are almost invariably bayonet mounted; so are interchangeable lenses for cameras with a built-in diaphragm shutter.

On technical and certain other cameras where the lenses change together with the shutter, lens and shutter are usually mounted on a panel which is clamped in position on the front unit of the camera.

Afocal attachments may screw or bayonet into the front mount of the normal camera lens.

With still cameras screw and bayonet mounts exist in a variety of sizes and specifications, and the camera will usually take only the lenses of a series specially designed for it. With cine cameras there is greater uniformity through the use of standard mounts: type D for 8 mm. cameras and type C for 16 mm. cameras. These are screw mounts, in which the thread, pitch, and the flange-to-film plane distance are accurately specified.

For rapid lens changing cine cameras often carry two or three focal lengths on a lens turret. Rotation of the turret brings the alternative focal lengths into place in front of the film. Turret cine cameras are however giving way to an increasing extent to cine cameras with zoom lenses. Turret systems have also been tried with still cameras, but have not been successful owing to the great bulk of a turret with two or three lenses.

A special kind of interchangeable lens mount is the sub-assembly mount where the camera takes one or more interchangeable units consisting of a focusing mount and aperture controls. The lenses themselves then fit interchangeably into this sub-assembly mount. The advantage is less bulky interchangeable lenses, since the latter consist only of the lens unit proper without a separate focusing mount for each focal length.

Rangefinder and Exposure Meter Coupling. Cameras in which the image is focused on a ground glass screen (technical and reflex models) need no special focusing provisions for the different focal lengths. Where the same focusing mount or focusing scale is used for different focal lengths, the distance figures are however not correct for all of them. Full interchangeable lenses on rangefinder cameras usually carry a differential coupling which moves the rangefinder mechanism through the same distance corresponding to a given focus change of the lens. Thus long focus lenses, which require an appreciable focusing travel for setting to nearer distances, still work correctly with a built-in camera rangefinder.

The same arrangement is not always possible with convertible lenses, since the focusing movement is also built into the camera and remains constant. A given focusing travel (extension of the front lens unit) thus corresponds to different distance settings according to the focal length of the lens. For that reason convertible lens systems have fallen largely out of use with rangefinder cameras; rangefinder focusing there involves focusing with the rangefinder in the normal way, reading off a distance figure on the standard distance scale and re-setting the focusing mount to a corresponding distance figure on another scale.

Exposure coupling—with cameras having built-in semi-automatic or automatic exposure meters—is however easier with a convertible lens system. There the shutter is firmly built into the camera and is easily linked with the iris diaphragm of the lens (sometimes even the iris is built into the camera). The same applies to interchangeable lenses that fit into the front of a shutter built into the camera. The coupling there usually takes place via some system of preselector diaphragm controlled by the exposure meter or by the meter setting control.

Complete exposure coupling with cameras taking fully interchangeable lenses in front of a focal plane shutter is rare, since such cameras are not normally intended for the user who requires exposure automation. Exposure coupling with focal plane shutter cameras is mostly confined to coupling the shutter speed setting control with the exposure meter, the aperture being pre-set and pre-selected separately on the lens. On one or two cameras however the aperture control on the lens can also be linked with the exposure meter.

Exposure coupling problems with technical cameras rarely arise, as these cameras are not normally fitted with built-in exposure meters.

Viewfinders. Cameras other than single lens reflex and technical types need supplementary viewfinders or viewfinder adaptations to show the field of view taken in with the different focal lengths. Camera makers produce various types of multiple finder for this purpose; on many modern rangefinder cameras the viewfinder carries additional image frame lines to show the field of view with selected alternative focal lengths. For the very long focal lengths rangefinder cameras often use a reflex housing for accurate focusing and viewing. Twin-lens reflex cameras need no special viewing provision, as long as the alternative lenses (or lens attachments) are available in matched pairs to take the place of, or fit over, the finder lens as well as the taking lens.

Projectors and Enlargers. The lenses of slide projectors and often also cine projectors are interchangeable to provide alternative screen image sizes for a given projection distance, or alternative projection distances for a given screen size. The longer the focal length, the smaller the screen image size (other things being equal), or the longer the throw required for a given screen image size. Short focus projection lenses thus are useful when projecting in small rooms; long focus lenses are essential for large halls. The lenses are usually fitted in screw mounts or in sliding tubes.

Most enlargers—especially those taking a range of negative sizes—have interchangeable lenses. The changing mounts are usually screw threads, since speed of lens changing is of no importance. The focal lengths of enlarging lens used are usually slightly longer than the corresponding normal camera lenses, to ensure adequate coverage. The problem of alternative focal lengths for a single negative size rarely arises—except possibly for high degrees of reduction where a shorter focus lens (shorter than that normally used for a given negative size) permits a higher scale of reduction without excessive lens to negative distances. At such great extensions even a short focus lens will cover the negative area adequately. Multi-format enlargers sometimes carry two or even three lenses mounted on a turret or sliding mount for more convenient lens changing.

L.A.M.

See also: *Afocal lens; Convertible lens; Lens; Projection principles.*

LENS MANUFACTURE

The manufacture of a photographic lens is the product of the combined skills of the mathematician who designs it and the craftsmen who fashion the lens elements and mounting.

There is probably no product which demands higher precision in all stages of design and production than a photographic lens and the industry is one in which the craftsman still holds a premier place.

For a lens system to be satisfactory for use as a photographic lens it should be free to a very high degree from chromatic and spherical aberrations, from coma, astigmatism, and distortion, and it should have a flat field so that it will produce a sharply defined image of a given object plane over the whole of the image plane it is intended to cover.

It is in practice impossible to eliminate all aberrations completely; the designer's task is to reduce them all simultaneously to the

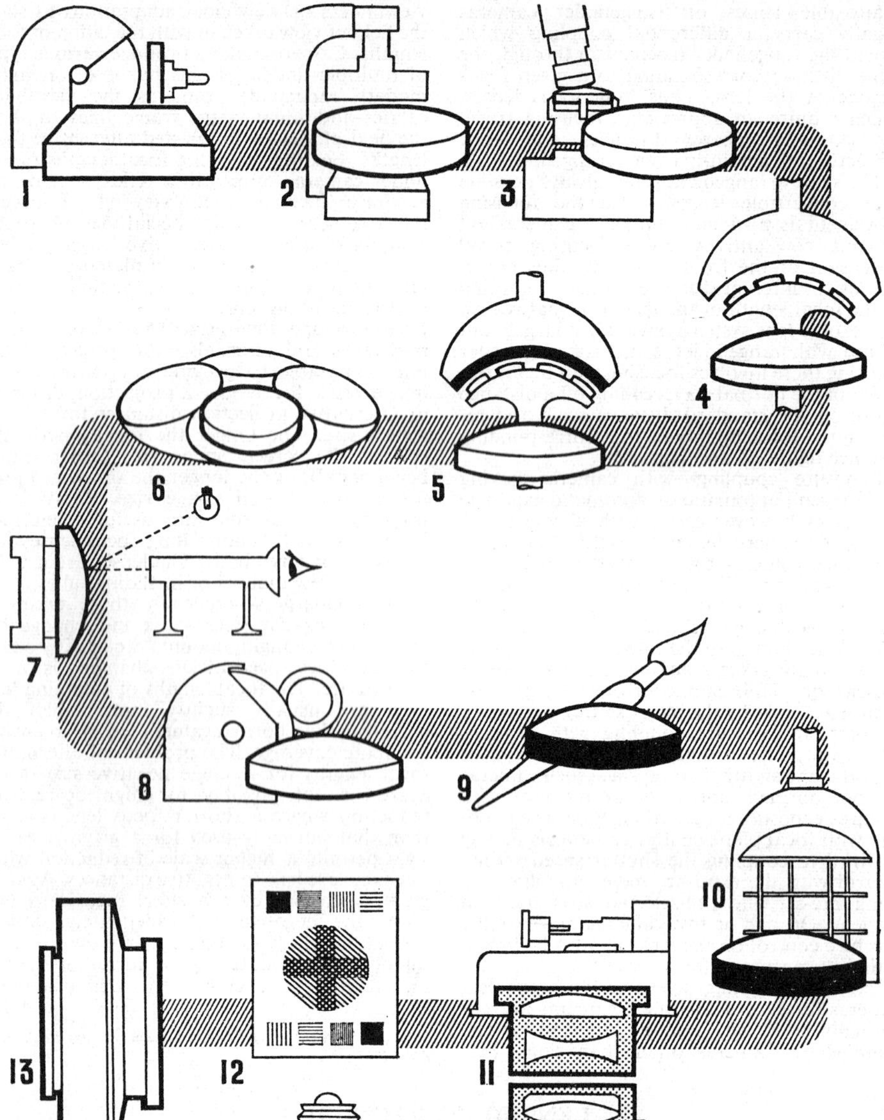

***LENS MANUFACTURE.** The precise operations and the way in which they are carried out vary from one lens manufacturer to another, and indeed from one lens design to the next. The sequence of the major steps does however follow this pattern: 1. Cutting of blocks of raw glass into plates for moulding. 2. Moulding the blanks. 3. Smoothing (rough grinding) the blanks. 4. Fine grinding the blanks (mounted together on a grinding tool). 5. Polishing. The polishing head is fitted with a pitch casting which carries engraved in it a series of grooves. These are filled with jeweller's rouge (a special form of iron oxide). 6. Testing curvature with test plate. The appearance of Newton's rings when the test plate is in contact with the lens surface indicates how accurately the surface is finished. 7. Centring lens for edging. 8. Edge grinding. 9. Blackening of edges. 10. Coating of surfaces. 11. Mounting lens units in mount. 12. Final testing. 13. Finished lens mounted in diaphragm shutter.*

greatest extent possible. This task becomes progressively more difficult as the aperture of the lens is increased.

To meet these requirements the designer has at his disposal about a hundred types of optical glass with varying refractive indices and dispersive characteristics from which he can make the lens elements which he combines to form photographic objectives. Each element can be made with an infinite range of radii and with thicknesses and separations which are only limited by practical considerations of weight and convenience. He has also to keep in mind when combining a number of elements that if his combination is disproportionately long with respect to its diameter the loss of light due to vignetting at the margin of his image may become serious.

Designing. The design of the lens thus presents a complex mathematical problem with many variables. The classical method of going about it used to be to work out the general form of a lens from known principles, using logarithms and slide rules, followed by laborious ray tracing and subsequent modifications. Even when done with calculating machines, this ray tracing frequently involved many months of work before the design of the lens could be finalized.

The modern way is to do the job with electronic computers. These permit both far more elaborate ray tracing systems and provide the required answer within hours instead of months. There it is possible to try out a considerable number of alternative specifications to arrive at the one which meets the requirements best.

Lens designs evolved in this way are nowadays frequently tested by instruments that show the performance of a lens in the form of contrast transfer functions, which can be read off directly from the face of a cathode ray tube.

Optical Glass. Optical glass differs from ordinary sheet glass or plate glass in two essential ways. Firstly it must be homogeneous throughout its thickness and free from internal strain, and secondly its ingredients must be very carefully controlled so that the properties of the glass—e.g., its refractive index and dispersion—meet the requirements of the lens maker.

Optical glass is composed essentially of a very pure variety of sand specially chosen for its freedom from iron. This is mixed with one or more of a wide variety of metallic oxides chosen to give the glass the particular properties required. Lead oxide produces some flint glasses which generally have a relatively high refractive index and a high dispersion. Barium oxide gives a range of glasses having a low dispersion compared with their refractive index which are invaluable for producing lenses with anastigmatic flattening of the field. The mixture is melted in a fire clay crucible at a temperature of about 900° C. and constantly stirred until no further "gassing" occurs and all bubbles are removed. When the crucible cools down the glass tends to break up into pieces which are subsequently re-heated, pressed into plates and then annealed by being cooled down very slowly, over a matter of several days, to remove internal stresses. Some of the very modern and rarer types of optical glass are now melted in platinum crucibles and this results in purer and cleaner melts.

The resulting plates of optical glass are the materials which the lens maker works on for making his lens systems, which usually consist of not less than three and frequently as many as six or even seven separate glass elements. In most cases each of the elements is made of a different kind of optical glass.

If only a small number of lenses is to be made it is usual to cut up plates of glass, but where large numbers of a type are to be made it is more usual to use moulded blanks.

Plates of glass are cut up on a circular saw, the glass plate being held between vice jaws faced with hard wood pads, and fed up to the saw. The saw blade may be made of soft iron, about 10 to 12 ins. diameter with tiny diamond splinters embedded in the edge but diamond-bonded wheels are now more common.

After the glass has been cut into squares of a suitable size and thickness each piece is formed into a rough circle by shanking or nibbling off the corners with a crude tool known as a pair of shanks. The rough discs of glass so formed are then cemented together into a stick or roll and formed into a smooth cylinder either by rolling on a rotating flat steel plate with a mixture of emery powder and water, or by the more modern method of putting the roll through a centreless grinding machine. The circular blanks are then taken apart and are ready to be ground on their faces to the required curvature on each side. The circular blanks are now often cut from the plate by means of diamond-bonded trepanning tools.

Where large numbers of one type of lens are to be worked, this stage of the work is bypassed by the use of moulded blanks of the required diameter and having on them roughly the correct radii. These are supplied by the glass maker who makes them by cutting the original plates up into pieces of the necessary weight, reheating until they are sufficiently plastic to be pressed between iron tools of the required radius, and then reannealing. Mouldings naturally save the lens maker a good deal of work and waste less of the expensive optical glass.

Modern methods of continuous smelting of optical glass yield moulded blanks directly from the melt. Here the ingredients for the glass are fed continuously into one end of a platinum trough, and the glass passes through various temperature controlled stirring stages, to be tapped at the other end of the trough. The tapping takes place automatically, releasing a drop of molten optical glass of pre-determined weight (controlled by the temperature and the viscosity of the glass). The glass drops drop into suitable moulds where they are pre-moulded into blanks, followed by annealing. This manufacturing procedure is virtually automatic, but is economical only when large numbers of similar blanks (batches of the order of 10,000 or over) are required. Frequently these blanks can pass directly to the grinding and polishing stage.

Roughing the Blanks. Until recently all these grinding operations were performed by presenting the glass disc by hand to iron tools ground to the required radius of curvature; if a convex surface was required it was ground on a concave tool and vice versa.

To-day the preliminary work of forming the curves is more often done on a spherical milling machine with a cup-shaped diamond-impregnated tool. The spindle carrying the tool is capable of being rotated over a considerable angle. When the axis of the tool is in line with that of the lens blank, a flat surface will be generated. If the axis of the tool is inclined at an angle to that of the lens blank, a spherical surface is generated on the glass, and as the angle between the two axes increases so the radius of curvature on the lens is shortened. The tool revolves at great speed and is supplied with a copious flow of liquid coolant, while the lens blank is rotated very slowly below it. A convex curve is produced by cutting with the inside edge of the cup wheel, and a concave curve with the outside edge.

After the required curves have been milled or "roughed" on the two faces, the lens blank is ready to be ground and polished.

Grinding. First, the lenses have a pad of pitch stuck on to one side. They are then held by hand and applied individually to a spherical tool similar to that used for hand roughing, and ground against the rotating tool with a medium grade of emery or carborundum powder in water.

Next the lenses are "blocked"—i.e., they are stuck by means of the pitch backing on to a spherical tool of suitable radius. The tool is heated, to melt the pitch and make it stick, and a number of lenses are fixed to it in this way and put through the subsequent stages of grinding and polishing at the same time. It is obvious that the number of lenses which can be worked at one time depends on the diameter and the deepness of curvature. A single lens with a deep curve approaching a hemisphere might have to be worked by itself, while the number of lenses with very shallow curves and plane surfaces that can be polished is limited only by the relative diameter of the lens to that of the tool.

The iron tools on which the grinding is done are first turned as accurately as possible to the required curve and then ground together in pairs, one convex and one concave and accurately measured from time to time on a spherometer—an instrument that measures the radius of curvature. The grinding process is adjusted until finally the tools have the required radius and a very fine finish.

The block of lenses is then screwed to a vertical spindle on which it can be slowly rotated. The iron tool oscillates over it and the block is fed in stages with successively finer grades of abrasive powder in water. After each grinding stage the block is thoroughly cleaned and examined to see that all the pits left by previous grindings with coarser abrasives have been removed. After the block has been finally cleaned from the finest grinding it is ready for polishing.

Polishing. The polishing is done on the same type of machine and in the same way as the grinding, but the polisher consists of a metal spherical tool coated with an even layer of pitch in which are cut grooves which retain the polishing medium—generally rouge (iron oxide) or cerium oxide. The pitch backing allows the lenses to "give" and so keep in intimate contact with the polisher.

As polishing proceeds, the lenses are examined from time to time to see that the action is even over the surface of the block. If it is found that the edges are polishing faster than the centre, the stroke of the arm carrying the polisher is shortened so that more work is put on to the central area, and vice versa. Sometimes the grooves in the polisher are widened at some particular zone for a similar reason. From this it can be seen that the polishing of lenses to a high degree of precision is very skilled work needing true craftsmanship.

The time for polishing varies considerably according to a number of considerations—e.g., the size of the block and the working temperature. Polishing is preferably carried out in an air-conditioned room, free from dust and at a set even temperature. Two to three hours is a reasonable time for a block of about six to eight inches diameter.

Checking. The block of lenses is passed as satisfactory when examination by a powerful magnifying glass shows that all traces of greyness have disappeared, and when it appears truly spherical and of the correct radius when a test plate is applied.

The test plate consists of a glass or quartz block polished flat on one side and with the required radius of curvature accurately polished on the other. Test plates are always made in pairs, concave and convex, ground and polished together, and tested from time to time during manufacture on a very accurate spherometer.

To use the test plate, both it and the lens to which it is to be applied are carefully cleaned to remove all surface grease and dust, and then placed in contact. When the surfaces in contact are viewed by a monochromatic light source—e.g., from a sodium lamp—any lack of perfect contact between them is seen as a pattern of light and dark bands, each dark band representing a separation between the glass surfaces of one half a wavelength of the light employed—about ·0003 mm. For a surface to be considered good enough for a good photographic lens not more than three such bands are tolerated, from which it will be appreciated that the spherical surfaces are accurate to a very high degree indeed.

Edging. When a lens has been finally polished on both sides it is edged. This consists of grinding the edge of the lens to the specified diameter, at the same time ensuring that the mechanical axis of the lens coincides with the optical axis. The optical axis is the imaginary line joining the centres of the two spheres of which the curved surfaces of the lens are parts. A photographic lens consists of a series of elements, each of which must be truly mounted about a common optical axis. The performance of the lens must be symmetrical about the axis, and this only occurs if all the elements are themselves truly centred on it. So the operation of edging is highly important and must be performed with the utmost care and precision.

The edging is done by first cementing the lens to an accurately made brass ring chuck which is then revolved slowly. As it turns, the operator observes the image of a small light source such as a lamp filament reflected on the surface of the lens. If the image moves, the operator alters the position of the lens by small transverse movements on the chuck. (The cement is previously softened by heating in a gas flame.) He continues to adjust the position by trial and error, until the images of the source, as seen by reflection at the back and front lens surfaces, remain stationary as the lens revolves.

The spindle, chuck and lens are then fitted into a grinding machine where the lens is slowly fed up to a rapidly revolving carborundum wheel or a copper lap impregnated with diamonds. Accurate adjustable stops on the machine determine the amount to be ground off the edge, and when the operation is complete the lens is removed from the chuck, and after cleaning is ready to be mounted.

Coating. Today nearly all photographic lenses have their surfaces coated or bloomed before being mounted. This process has the effect of increasing the light transmission of the lens and also of reducing the amount of light reflected back as stray light within the camera.

The material used for coating is magnesium fluoride. The lenses and the magnesium fluoride are placed in a vacuum chamber suitably separated. The chamber is then exhausted to a very high vacuum by means of a backing pump and an oil diffusion pump. The fluoride is then melted by an electric filament heater which causes it to evaporate and it condenses as a fine film on the lens surfaces. The thickness of the coating is controlled by watching the reflection of a lamp on the lens surfaces. When the colour of the film indicates that the thickness is correct the current is cut off and the vacuum broken. A purple film gives maximum transmission in the yellow green of the visible spectrum.

Mounting. The lens elements are now ready for mounting. The mount itself has to be turned to high precision limits, and particular attention must be paid to the truth of all the screw threads. It is also important for each glass to be held securely but without any pressure which might produce distortion, especially with changes of temperature.

Great accuracy is needed in mounting the elements so that the separations are exactly as calculated by the designer and the position of the iris diaphragm relative to the elements is correct.

Testing. Testing is the last, but far from least important, stage in the production of lens.

There are several methods of testing. In one, an image of an illuminated test object is produced by the lens and carefully examined with a high power microscope. Another method consists of using the lens to project on a screen a greatly enlarged image of a test object and then studying the image in detail. Two types of error are looked for—symmetrical, which may be due to the design or the construction, and unsymmetrical, which arises from slight inaccuracies in making or mounting the lens elements. Photographic tests are also commonly used, as well as interferometric tests at the design stage. A.W.S.

See also: *Information in the image; Lens mounts; Lens performance; Lens testing; Optical design; Optical glass.*

Books: *A System of Optical Design*, by A. Cox (London); *Photographic Optics*, by G. Franke; *Fundamentals of Optical Engineering*, by D. H. Jacobs (London); *Lens Work for Amateurs*, by H. Orford (revised by A. W. Lockett) (London); *Prism and Lens Making*, by F. Twyman (London).

LENS MOUNT. Metal housing which holds the assembly of glasses making up lenses. It may also house the iris diaphragms and may be incorporated in the construction of between-lens shutters. Lens mounts on some cameras are fixed and non-interchangeable.

There are five principal types of lens mount: standard, sunk, collapsible, between-lens shutter types and focusing mount. There are also several varieties of the last-named type.

On a standard mount, the flange screw is near the back of the lens barrel so that most of the lens projects from the front of the panel when it is screwed home. This type of mount is sometimes called an iris mount because it is convenient for the fitting and adjustment of the iris diaphragm. The standard type of mount adds its length to the existing camera extension and is useful in keeping down the over-all dimensions of the camera. Lenses for stand and studio cameras usually have standard mounts.

On a sunk mount, the flange screw is placed as close to the front end of the lens barrel as possible, leaving only a narrow rim to take the lens hood or filters. This type of mount is fitted on reflex cameras; it looks neater and gives the lens better protection than the standard mount, but it calls for more camera extension. The iris diaphragm in this case is controlled by a rotating ring on the front of the mount.

A collapsible mount is simply a standard mount that allows the lens to be pushed back into the body of the camera when not in use. It forms a convenient way of providing the necessary camera extension for the short focus lenses of miniature cameras. The lens barrel can be locked in the forward working position by pulling it out and turning it to engage a bayonet catch. This type of mount may also incorporate the focusing mechanism.

The between-lens shutter mount is found on the majority of hand cameras. The lens components are actually fitted by the manufacturer so that in effect the lens is sunk into the shutter mechanism and the blades of the shutter and the leaves of the iris diaphragm lie between the front and back components.

All British lens mounts, unless otherwise stated, are screwed with R.P.S. standard threads:

STANDARD FLANGE THREADS

Screw Diameter	1·000	1·250	1·500	2·000	2·250	2·500	2·750
Threads per inch	24	24	24	24	24	24	24
Screw Diameter	3·000	3·500	4·000	4·500	5·000	6·000	
Threads per inch	24	12	12	12	12	12	

Mounts made on the Continent usually have metric threads.

Cine Camera Mounts. Most cine cameras have a standard screw mount and the lenses are usually interchangeable. Two widely used standard mounts are the Type C for 16 mm. cameras and the Type D for 8 mm. The Type C mount has a diameter of 25·4 mm. (1 in.) with thirty-two threads per in. and a flange-to-film distance of 17·5 mm. (0·69 in.). The Type D mount for 8 mm. cameras has a diameter of 15·8 mm. (0·632 in.) with again a pitch of thirty-two threads per in. and a distance of 12·3 mm. (0·484 in.) from the lens flange to the film. There are, however, also a number of non-standard mounts for specific cine cameras which can then be used only with the lenses designed for them.

Bayonet Mount. On a number of cameras, especially 35 mm. miniatures and roll film reflexes, the lens is fitted to the camera by a bayonet mount. This permits quick lens changing, since only a quarter turn or half a turn is required to position the lens. Usually a catch secures the lens in position, and has to be pressed or disengaged before the lens can be removed or changed.

Bayonet mounts are rarely standardized, partly because the flange-to-film distance tends to vary from camera to camera. So even when one lens fits the mount of another camera, there is no guarantee that the lens will be positioned at the correct lens-to-film distance. Sometimes different cameras using bayonet-mounted lenses by the same manufacturer deliberately have different "keys" to the bayonet mount to prevent a lens from being used on a camera other than that for which it has been designed and set.

On some 35 mm. rangefinder cameras the bayonet mount engages a mechanism which automatically changes the image frame of the viewfinder to match the focal length of the lens.

Focusing Mounts. In a focusing mount the lens barrel is mounted in a housing which is screwed or slotted to allow the lens to be focused by sliding or turning it in the mount. The most usual system is the helical mount which screws the lens in and out by means of a coarse screw thread.

Occasionally lenses may have a dual focusing mount with two (or even more) helical threads and separate lens tubes for each to give a greater range of lens-to-film distances. This is useful for extra close focusing and such macro mounts—available for certain miniature still and narrow-gauge cine cameras—permit focusing down to a few inches.

Filter Mounts. The front mount of a camera lens if frequently designed to hold filters securely. The simplest way is to finish the front mount with a plain rim which then takes push-on filter fittings. These are easy to attach or remove, but not very secure.

The majority of camera lenses carry a screw thread on the front for screw-in filters in various standard diameters. A few cameras have a bayonet front mount to take bayonet mounted filters which are quick to attach or remove, yet secure in position.

The front lens mount is equally intended for taking supplementary lenses in their appropriate mounts or fittings. F.P. (L.A.M.)

See also: *Diaphragm; Focusing mechanism; Lens, interchangeable; Lens manufacture.*

LENS PERFORMANCE. Any assessment of lens performance is an attempt to determine the faithfulness of the image forming process carried out by a lens, that is, the degree to which an image corresponds to a perfect two-dimensional projection of the object. This correspondence is never perfect because the image-forming process is limited by the diffraction of light at the finite apertures used. In considering the reproduction of images—and the information contained in them—we have to deal with the shortcomings of elements other than just the lens. However, this section will be concerned with the lens image forming properties only and will briefly indicate some of the methods used to assess performance.

Telescopes and Star Images. Although primarily concerned with photographic optics, the earliest assessment of lens performance was concerned with astronomical telescopes and the images of stars. Aberrations in the lens led to diffuse and complex images with the pure aberrations being characterized by distinctive light distributions in the image. An assessment of the state of correction was based on the percentage of the energy falling in successively larger circular areas of the image. This did not always correlate with the observed performance, an obvious exception being a comatic flare which is quite disturbing in any subjective test.

Interferometry. Interferometric tests of photographic lenses, as carried out on a Twyman-Green type interferometer, give the departure from a true sphere of the wavefront leaving the lens. If the wavefront is truly spherical, or deviates from a sphere by not more than a quarter of a wavelength, then the image as formed by the lens for that particular field angle and focal plane will be indistinguishable from the theortically best available.

However, not all lenses are well corrected and even well corrected lenses have aberrations when used off axis. It is then necessary to assess the possible performance of a lens from the more complex interferogram. A criterion that suggests itself is to determine the portion of the wavefront that lies within the aberration tolerance (Raleigh Tolerance of $\frac{1}{4}\lambda$), as clearly the light from this region forms the bright core of the image whereas the remaining wavefront region forms a diffuse patch. Attempts to correlate such a criterion with an assessment of performance derived from test chart images have not however proved to be very satisfactory. It is not possible to predict the resolution limit but only to confirm that the percentage area of the wavefront well corrected is to some extent proportional to the logarithm of the resolution limit.

Edge Gradient and Acutance. One approach involving the actual image formed by a lens considers the image of an edge between two regions of unequal illumination. For a perfect lens (this term implying diffraction limitations only) the image is an intensity curve which decreases monotonically and is of finite slope. The slope at the mean intensity is termed the edge gradient. In practice this gradient decreases as the lens aberration increases. The effect of this decrease of edge gradient is evident in the images of extended objects as blurred edges and a general unsharpness.

The practical measurement of the edge gradient requires a knowledge of the magnification of the image, as increased magnification will decrease the edge gradient in the same manner as increased aberration. With this precaution it is possible to compare the performance of lenses under equal conditions. This test does not have the same sensitivity to image degradation as other tests to be described or subjective assessments. One difficulty is that the eye and the receptor used in the physical measurement of the edge gradient do not respond in the same way to variations in light intensity.

A criterion taking this point into account is the acutance. A high acutance value indicates that the lens forms sharp images of edges but not necessarily good images of point sources. It implies that the image of a point source is compact without a halo of light about it. Thus a high acutance is not the same as a high resolution limit since a point image with a sharp centre spot and a diffuse halo (e.g. with spherical aberration) would give a high resolution limit but a low acutance value.

The acutance is more concerned with low spatial frequency detail, as implied in its use of an edge as object, and so is a quality criterion covering only a part of the workings range of a lens. If later components in the optical train restrict the spatial frequency range in use then the value in assessment of performance increases.

Resolution Limit. This is the limit of resolvable detail, as determined from test charts. The form of test charts is very variable, each generation correcting the faults of the previous one. Thus there has been a progression from multiline gratings to two-line Cobb charts, in an effort to eliminate spurious resolution, and a return to three or four-bar gratings. Gratings with continuously varying periods in the form of straight lines or sectors of a star have also been used.

However, most of the test charts used have been such that their images do not have the same form of intensity distribution as the object. This leads to difficulties in the measurement of the resolution limit due to the lack of resemblance. It further makes the decision of continued recognizability difficult and partially accounts for the variation of results between observers using the same test chart and for the greater divergence between the values obtained using different forms of chart.

The resolution limit criterion is also unsatisfactory in that the single-number performance figure derived is based solely on the high spatial frequency performance of the lens with no consideration at all given to the imaging properties at low spatial frequencies. This fault is made more serious by indications that it is the low spatial frequencies that are operative in any subjective assessment.

Transfer Functions. To overcome some of the above objections, lens designers now work with the optical transfer function. This function gives the ratio of the image contrast of a grating

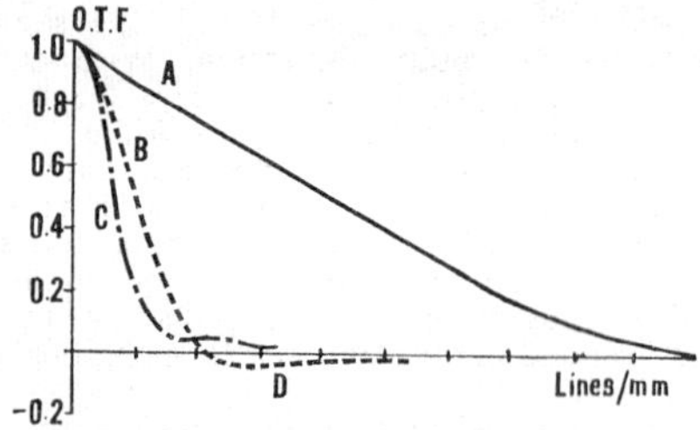

EFFECT OF ABERRATION ON OPTICAL TRANSFER FUNCTION. A, transfer function curve for perfect lens. B, curve of lens defocused by 0·004 inch at f5. This already reduces the resolution limit to one-fifth of the value for a perfect lens. Reversal of contrast takes place in the region D. C, curve for lens with primary spherical aberration, partially corrected by a change of focus to give the same order of marginal aberration as in curve B.

of sinusoidal transparency to the object contrast as a function of spatial frequency. Sinusoidal here means that the transparency distribution of the grating from line to line follows the form of a sine wave. Spatial frequency is the reciprocal of the grating period expressed in lines per millimetre.

The spatial frequency Re, corresponding to the resolution limit of the lens as derived from classical theory, is given by

Re = 2 (Numerical Aperture)/Wavelength.

For an *f* 5 system Re = 400 lines/mm.

Two facts should be noted if this function is to be adopted for use in the assessment of lens performance. Firstly, any object, represented by a two dimensional incoherent intensity distribution, may be analysed into a set of Fourier terms. Each term of this set represents a sinusoidal transmission grating of a particular periodicity and may be transferred to the image plane using the optical transfer function where the set is reconstituted to form the final image. This procedure illustrates the use of a sinusoidal grating.

Secondly, whatever the form of the spread function formed by the lens, the sinusoidal grating will always be imaged as such a grating but with reduced contrast, possibly zero, and laterally shifted from the Gaussian image position.

It will be seen from the optical transfer function of a perfect lens that at all spatial frequencies above zero lines per millimetre the lens performance is impaired, a necessary condition imposed by diffraction at the edge of the lens opening. The aberrations in the lens reduce the optical transfer curve below the optimium. It is never possible for it to rise above the curve for the perfect lens.

Attempts have been made to derive a single figure of merit from such curves. An early attempt integrated the area under the curve and derived the equivalent band-pass, between zero spatial frequency and same value Re with unit contrast transfer, that would give the same integrated area. The upper spatial frequency limit Re is then taken as the figure of merit.

This method does not allow for the shape of the junction between the curves, so that several curves with equal integrated areas beneath them and, therefore, the same equivalent bandpass, appear to have equal performance when this is not so. Account must be taken of the relative importance of various spatial frequency regions and this is dependent on the receptor used in any final assessment and the spatial frequency spectrum of the object. If, however, an assessment is to be made of the lens alone, then the integrated square of the optical transfer function is a satisfactory assessment criterion.

All the above criteria can be measured for individual points in the image field only. If the assessment is to be applied to the whole image field then it is necessary to determine the mean of all the values representing individual points in the field. If some particular region in the field is considered of greater importance, for example the central region, it may be necessary to weight this region accordingly before determining the mean value of the performance figure for the whole image field.

K.G.B.

See also: *Aberrations of lenses; Acutance; Information in the image; Lens testing; Optical design; Resolving power.*

Books: *A System of Optical Design*, by A. Cox; *Photographic Optics*, by G. Franke (both London).

LENS TESTING

The lens manufacturer uses highly specialized and very expensive equipment to make his own tests and in many instances he is more concerned with tests which assist him to maintain a consistently high standard of production rather than the tests which are of interest to the amateur user.

Before embarking on an extensive series of lens tests the average user must have a clear idea of the type of information he is likely to obtain.

He must understand that perfection cannot be expected from his lens and that performance will vary from one type of lens to another according to the type of work for which it was designed, its relative aperture and field of view and its cost.

An early understanding that some residual lens aberrations are theoretically inevitable will encourage him to concentrate on a search for the more obvious errors which may be caused by faulty manufacture, accidental damage or inaccuracies in mounting the lens on the camera.

The presence of different types of residual lens aberrations in different lenses means that their relative performance can be expected to vary as the conditions of use are changed. Similarly, the performance of a motor car varies under different conditions of use. An attempt to assess either performance by a single figure of merit can be quite valueless and often misleading. Rather than rely on a lens assessment such as resolving power which requires expert knowledge, experience and elaborate equipment, it is preferable to allow a good lens to speak for itself in doing the type of work for which it was designed.

The visual and photographic tests described will assist the photographer to understand the limitations of his lens and suggest ways of minimizing their effects in practice.

The defects which are most likely to impair the sharpness and clarity of a photograph can be divided into two main categories. The first concerns the quality of the lens by itself and the second concerns the mechanical precision with which the lens and sensitive emulsion are positioned in the camera.

Whilst a qualitative test of the lens by itself requires special equipment together with some technical knowledge of photographic optics and experience in interpreting results, the user can conduct simple tests which will enable him to recognise lens errors and to see their effects.

Before testing for the presence of aberrations, second-hand lenses should be examined for gross errors arising from accidental damage or unskilled repair work.

Examine the metal components of the lens to see whether parts have been bent, screw thread and seating shoulders bruised or whether lens elements are loose in their cells. The lens elements of complex photographic lenses have to be centred about a common optical axis to an accuracy of a few ten-thousandths of an inch and quite small displacements are sufficient to spoil performance.

Centring Errors. The effects of centring errors on definition can be seen by using the lens at full aperture to form an image of a brightly illuminated pin-hole in an opaque metal sheet or card. The pin-hole must be placed as closely as possible to the axis of the lens, and, in the case of camera lenses, at a distance from the lens exceeding twelve times the focal length.

View the image on fine ground glass using a high-powered magnifier carefully focused on the greyed surface which should face the lens. As the lens is racked slightly inside and outside focus the image should remain circular in shape. If it shows an uneven shape with a flare to one side the lens is badly out of centre.

To make the test more critical make a lens holder which permits it to rotate smoothly about its own optical axis, maintaining focusing distance and a stationary image. Irregularities in the image due to lack of centring will be seen to rotate with the lens.

Mounting Errors. The performance of the best lens can be ruined by inaccurate mounting on the camera. The emulsion must be at the correct focusing distance from the lens, it must be flat and it must be at right angles to the optical axis of the lens. The precision of these mechanical settings must be increased as the quality of the lens is increased and as wider relative aperture reduces depth of focus.

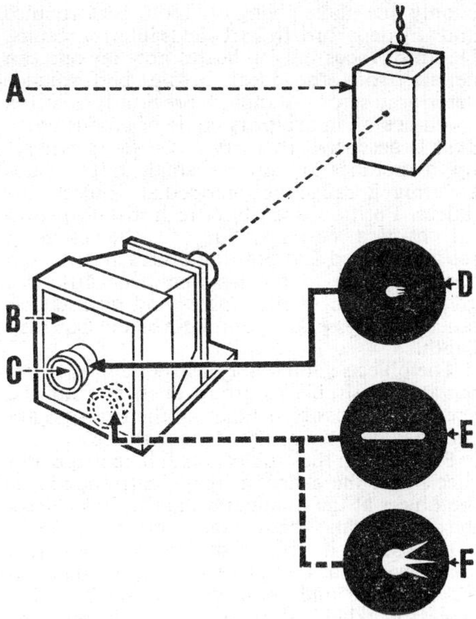

TESTING FOR CENTRING. Examination of the image of a pinhole in the centre of the ground glass screen will show whether the lens is correctly centred. An image near the edges may also show up other faults. A, illuminated pinhole. B, ground glass screen. C, magnifier. D, image point with lens out of centre. E, astigmatism. F, coma.

A first class lens provides its best performance in one flat plane at right angles to its axis passing through the best focusing position on axis. The emulsion is placed in a cone of light proceeding from the lens aperture to an axial image point. In the absence of aberration the ideal point image of a point object will only be recorded when the emulsion is precisely at the tip of the cone. An emulsion in other positions will intersect the cone and finite circular areas known as discs of confusion will be recorded instead of points.

The angle of this cone is inversely proportional to *f*-number and the diameter of the disc of confusion will be the error of focusing divided by the *f*-number. Thus an error in focusing of ·004 in. with an *f* 2 lens produces a circle of confusion of ·002 in. A similar argument shows that squareness of the emulsion plane is equally important for image points out in the field of view.

It is dangerous to relate mounting errors too rigidly to depth of focus or depth of field. There is only one plane giving optimum performance and any departure from it yields inferior results. Depth of focus only indicates how far one can depart from the ideal before performance deteriorates to a standard previously selected by choice of an arbitrary circle of confusion.

Test Procedures. To carry out a test series set up a number of similar small test objects covering a range of intermediate object distances. Position one object at a distance from the camera corresponding to the focusing setting selected and position others at distances corresponding to the far and near distances given in depth of field tables and also at distances at, say, ½, 1, 1½ and 2 times the expected depth.

The objects should not obstruct one another, but it is desirable to group them as near the lens axis as possible so that all their images are clustered near the centre of the negative.

Photograph these objects at full aperture and check that the sharpest image corresponds to the object at the nominal distance. This checks the axial distance between lens and emulsion.

Now set up a series of objects in a flat plane, taking care that this plane is at right angles to the lens axis and positioned at the best distance for axial performance as determined by the previous test. The objects should be distributed over the complete field of view in some symmetrical pattern centred about the lens axis.

Photograph these objects at full aperture and check that any variation of image quality is distributed symmetrically around the axis of the lens. A variation of definition in different parts of the negative may be an indication of lens quality, but, if these variations are symmetrically disposed about the centre, the emulsion is likely to be flat and at right angles to the lens axis to a satisfactory degree.

Resolving Power. Although experienced interpretation of resolution does assist the manufacturer to attain a consistently high standard of production, it does not necessarily provide any simple and straightforward comparison between the quality of different lenses. Naturally, the manufacturer does make such comparisons, but the assessment of resolving power is only a part of this procedure and it takes a very extensive form covering assessment under a variety of conditions.

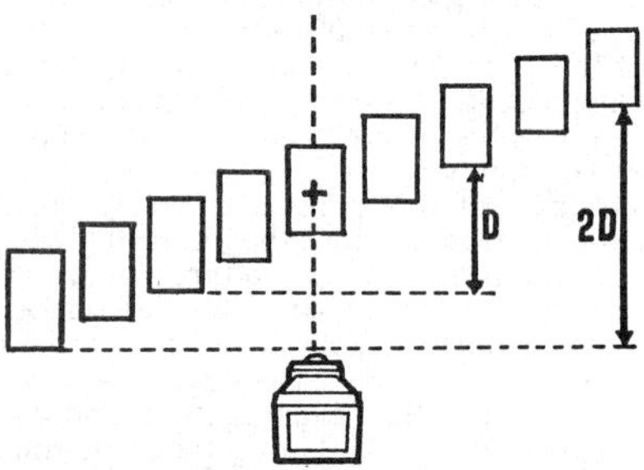

TESTING FOR OPTIMUM FOCUS. Layout of test objects (e.g., test charts) at various distances in front of camera. D, total depth of field at nominally focused distance.

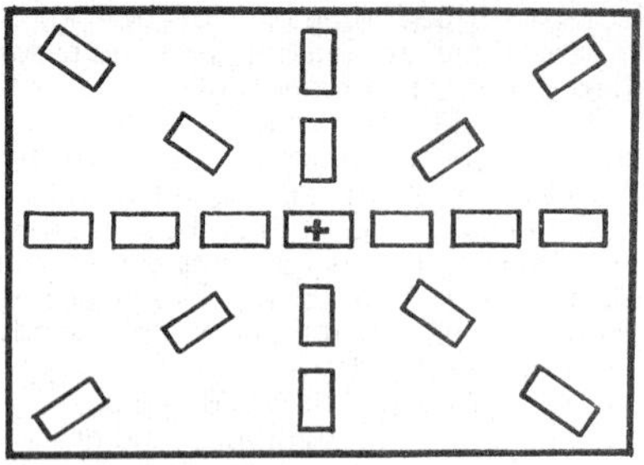

TESTING FOR COVERAGE OVER FIELD. Layout of test charts over the field of the lens to check definition and compare the edges and corners of the field of view with the centre.

Resolution depends on a considerable number of factors. Apart from the magnitude of the residual lens aberrations, resolution depends on the conditions of test. It varies particularly with the layout of the test objects, the contrast of the object, the grain and sensitivity of the emulsion, exposure and development conditions and with the criteria adopted for focusing. In comparing two lenses, for example, it often happens that one appears to resolve better than the other when the object being photographed has detail of high contrast, whereas the second lens has a higher resolving power than the first when the contrast of the object is low. Similarly, it may happen that the lens which yields the higher resolving power is not the better for definition.

It is important to realize this difference between resolution and definition. Resolution is concerned solely with ability to record very fine detail whereas definition, or acutance as it is called in some quarters, is concerned with the sharpness or lack of "fuzz" at a boundary between larger areas of different density. In some lenses having appreciable aberration the best focusing position for maximum resolution is not necessarily the best for maximum definition.

The best way to test the performance of a lens and its mounting on the camera is to take a series of photographs of various types of subjects. A careful choice of subjects and conditions will give the amateur a much clearer and more satisfying indication of performance than any attempt to assess resolution by a test procedure which is necessarily complex, difficult to perform without experience, and liable to yield misleading results.

If photographs of different types of subjects show a consistent error and the lens appears to be in good condition, the camera should be tested to see whether the lens is badly mounted.

In those cameras which allow fine ground glass to be positioned in the focal plane,

LENS TESTING CHART. Chart for copying in natural size. A number of such charts laid out along the diagonals of the field of view covered by the lens will give some information about the covering power. Photographed at a distance of 10 feet, the medium group of lines marked M give images which with a 1 in. lens are 1/1200 in. wide, with a 2 ins. lens 1/600 in. and so on up to 1/150 in. with an 8 ins. lens. The clear spaces between the lines are the same width as the lines. The finer sets give lines with 1/2 and 1/3 of this separation, and the coarser sets have separations $1\frac{1}{3}$, 2, and $2\frac{1}{2}$ times that of the M set. The circular dots are the same diameter as the thickness of the medium lines. Charts should be made up which are both negative and positive copies of this chart and photographed under the same conditions. In the negative charts, especially those set up near the corners of the field, the images of the dots will, on examination with a powerful magnifier show up any coma flares, and spread due to lateral chromatic aberration.

mounting errors may sometimes be discovered visually with the aid of a high-powered magnifier. In all cases, however, it is preferable to make tests photographically, especially with roll film cameras where the precise position taken up by the film is not well defined by the mechanical construction.

Spherical Aberration. This aberration concerns object and image points lying on the axis of the lens and takes its name from the fact that it must always be present in a lens made from a single piece of glass if its polished surfaces are spherical in shape. It refers to the way in which rays of light passing through the outermost annular zones of the aperture of a single lens converge to cross its axis nearer to the lens than do the rays passing through other zones nearer the centre.

To test for spherical aberration, set up the centring test described for centring errors taking care that the pin hole is bright enough to give sufficient light in the aberration fringes of the image and that the fringe is not lost by the effects of stray light falling on the ground glass. To give the maximum information about aberrations the pin hole should be as small as is consistent with enough light getting through to make all aspects of its image easily visible. Coloured fringes due to chromatic aberration should be removed by filters between lamp and pin hole. It is preferable to choose a filter transmitting light in the centre of the photographic spectrum, but change of filters can be used to assess variation of spherical aberration with colour.

The first impression of the presence of spherical aberration will be a vagueness about the position of best focus at full aperture. With a very well corrected lens the image comes into focus crisply, suddenly and definitely.

Next, stop the lens down by means of its iris and note whether the presence of spherical aberration is sufficient to change the position of best focus. Shift of focus on stopping down is present in a large variety of lenses and can amount to as much as 0·2 per cent of the focal length.

With the lens fixed in its best focusing position at full aperture, note the size and intensity of the aberration fringe round the image and see what changes occur on stopping down. If spherical aberration is corrected, the presence of residual zonal spherical aberration is likely to be most easily recognized when a relatively large but faint fringe is rapidly removed (well within one stop from full aperture) leaving a much smaller but brighter fringe which is removed at a slower rate on further stopping down. A lens which provides one large fringe which reduces steadily in size all the way down to small stops without any change of character may not be corrected at all for spherical aberration.

An interplay between two types of fringes is sometimes apparent in a corrected lens with zonal errors when the lens at full aperture is very slightly racked inside and outside the best focusing position.

Axial Chromatic Aberration. This is a failure of the lens to bring light rays of different wavelengths to a common focus. This can cause poor definition.

If the lens is intended for use throughout the visible spectrum, remove any filters before making a test. The effects of this aberration may be confused by any variation of spherical aberration with colour, but the appearance of the image should be in accordance with the best compromise correction for chromatic aberration over the whole lens aperture. This may show up in a corrected lens as a purplish or greenish fringe round the image. Blue fringes of importance photographically are difficult to see when using tungsten illumination deficient in blue light, and the eye and a photographic emulsion respond to different colours to varying degrees.

Find the best focusing position for a number of apertures, first in red light and then in blue light. If the focusing position, aperture for aperture, is always farther from or nearer to the lens in one colour compared with another, the correction of chromatic aberration may be suspected.

Astigmatism. This is an aberration affecting off-axis image points in such a way that the definition of a line image at one position in the field of view will vary according to the inclination of the line with respect to the centre of the field.

To test for astigmatism visually, arrange an illuminated pin hole as an object, and make provision whereby the pin hole, and thus its image, may be moved away from the axial position towards the edge of the field of view.

Take care that the pin hole moves in only one flat plane and that this plane and the ground glass are at right angles to the lens axis. Leave the focus set at the best position for the centre of the field and see how off-axis aberrations combine to give complex and irregular image patches as the pin hole is moved away from the axial position. The extent of the field of view that can be examined in this way will be limited by the effects of oblique illumination on to ground glass and by the construction of the magnifier, but it is possible to check that at all points equidistant from the lens axis the image has the same appearance. This symmetry of performance around the lens axis is another indication of a centred lens.

Astigmatism is rarely the only aberration present and the appearance of the image is complicated by the presence of field curvature, coma and transverse chromatic aberration. To test for astigmatism under these conditions, refocus the lens viewing a pin hole image well out in the field of view. Any tendency towards two distinct focusing positions, one giving minimum aberration fringes in one direction with respect to the lens axis and the other giving minimum fringes in a direction at right angles to the first, may be recognized as the effect of astigmatism. The distance between these focusing positions is an approximate measure of the astigmatism that is present.

The confusing effects of coma may be removed in some cases by stopping the lens down, but it should be realized that, whilst the magnitude of astigmatism is the same at all stops, its apparent effect is smaller at small stops.

Repeat this test at different positions in the field of view and note whether there is a field position where astigmatism has been corrected.

Coma. This is another aberration concerning off-axis image points, and shows itself by the fact that a lens without any correction for coma reproduces object points as an unsymmetrical light patch flaring away to one side like the tail of a comet.

Comatic aberrations are usually of a complex nature and all that the average user can do is to examine the image at different points in the field and to notice the variations of the effects of coma at different positions and stops.

Curvature of Field. In the absence of all other aberrations, curvature of field will be an aberration which does not impair definition provided the sensitive emulsion can be placed on a surface which is curved (usually concave to the lens) to the extent demanded by this aberration. If a lens suffering from curvature of field is used with a flat emulsion definition will deteriorate steadily away from the centre of the field as the dimensional displacement of the emulsion from the ideal curved focal surface exceeds the depth of focus of the lens.

The test described above for astigmatism will show quite clearly the effects of curvature of field. If the ground glass has to be moved from the best focusing position on axis to achieve a better focusing position for a point well out in the field, then curvature is present. In some cases the way in which astigmatism and curvature are off-set against one another can be seen. In the case of anastigmat lenses try to find a position in the field of view where both these errors are absent.

Transverse Chromatic Aberration. This is a fault which causes the scale of the image to depend on the colour of the light and appears as a coloured fringe around the image.

In the above tests dealing with off-axis image points, any colour fringing extending along a line passing through the centre of the field will be due to this aberration. Since its extent and effects are the same at all apertures the confusing effects of other aberrations may be avoided by stopping the lens down to a sufficient degree. The presence of a complete spectrum of coloured images shows that this aberration is not corrected. Residual errors in a corrected lens, however, may show as purplish and greenish fringes on either side of the image.

Special Conditions. The tests described above should be adequate in the case of most lenses used on still cameras, but cine lenses and miniature camera lenses present difficulties in connexion with short focal lengths yielding very high standards of definition on small negative areas. Although the ground glass and magnifier may be replaced by a microscope-viewing system of suitable numerical aperture, the mechanical difficulties will be too severe for most users and photographic testing is to be preferred.

The photographic tests for mounting errors described above may be used for this purpose if the test objects are chosen to show three types of definition. These are the definition of a point, the definition of a radial line or edge stretching out like a spoke from the centre of the field and the definition of a line running transversely across the radial lines like the rim of a wheel.

Test objects consisting of printed matter interspersed between radial and transverse lines are quite suitable for this purpose. A series of photographs taken at slightly different focusing positions and at different relative apertures, with and without colour filters, will suggest themselves to the user who has studied the visual tests described above.

Avoid the use of contrasty emulsions and contrasty developers which will hide important aberration fringes, and make certain that the test objects are evenly illuminated and square on to the camera.

Unless the user is experienced in the assessment of lens performance, he should not assume that the presence of aberration is an avoidable fault in the lens. The appearance of a visual image must be considered in conjunction with the power of the magnifier and any doubts should be checked by photographic tests on a correctly adjusted camera.

It is impossible to lay down any hard and fast

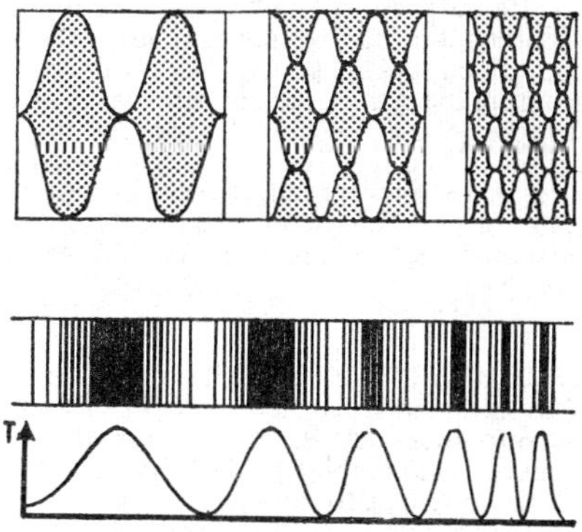

SINUSOIDAL TEST TARGETS. Top: Template type transparent target. This oscillates in transfer function measuring instruments, so providing a sinusoidal density change of the image being scanned. Bottom: Sinusoidal target with continuously varying spatial frequency.

rules concerning permissible amounts of aberrations. If a lens fails to give satisfaction when used for the purpose for which it was designed, it is best to refer to a reputable authority. Correction of faulty lenses should be left to the manufacturer who has the original design information and the correct tools and fixtures for such work. G.H.C.

Optical Transfer Function. This function is primarily concerned with describing the imaging properties of a lens in such a manner that these properties may be included with those of associated components in the imaging system to give the overall imaging performance. However, the function may be used to describe the lens performance alone.

The optical transfer function relates the ratio of image contrast to object contrast of a sinusoidally transparent grating as a function of the reciprocal of the period, termed spatial frequency, of such gratings. The function covers the complete spatial frequency range from zero lines per millimetre, for example, for a blank surface, to the classical resolution limit of the lens under test. The contrast ratio is normalized to unity at zero lines per millimetre, thus taking into account the photometric properties of the lens. Contrast is defined to be Michaelson contrast or visibility.

The sinusoidally transparent objects necessary for such an examination may take several forms. One of the earliest used is a series of apertures of sinusoidal form profile cut in an opaque sheet. If this is scanned by a slit the object is of the correct sinusoidal form with a d.c. bias. A series of such objects would be required to cover the complete spatial frequency range and frequently a set of such apertures of varying periodicity is placed on the periphery of an object drum. More than one period of each spatial frequency must be present to reduce errors in results obtained with such objects at higher spatial frequencies.

An alternative form of object is a sinusoidally transparent grating of continuously varying period. Such gratings may be produced by photographing a suitable cut out mask, of the form previously described, moved in the direction of the length of the lines during exposure, or by photographing the fringe pattern of a modified Michaelson interferometer.

Such an object grating, used for the measurement of the optical transfer function, is suitably illuminated from a white light source, and is moved continously in the object plane of the test lens. The image of the grating moves across the fixed scanning slit S. The light passing through the scanning slit is collected into a photo-detecter and the output from this is measured and possibly also continuously recorded.

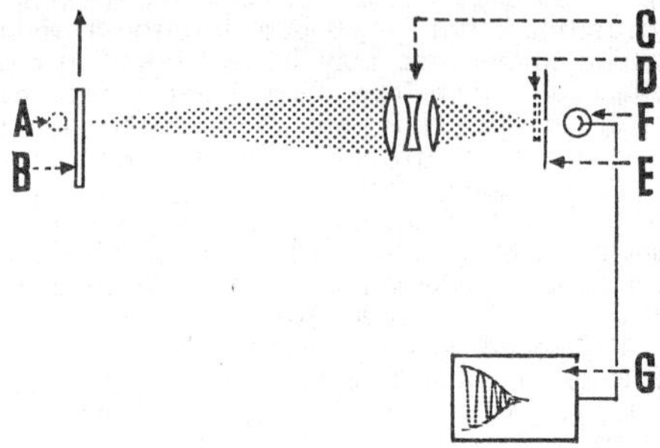

OPTICAL TRANSFER FUNCTION MEASUREMENT. A, *light source.* B, *grating.* C, *test lens.* D, *image of grating.* E, *scanning slit.* F, *photocell.* G, *recorded signal. The illuminated grating moves continuously in the object plane of the test lens so that the grating image moves across the fixed scanning slit. The light passing through the slit is collected by a photo detector and the output of the latter measured and if required continuously recorded.*

The recorded signal will oscillate with the movement of the grating and the contrast in the image may be evaluated as a function of spatial frequency in the image plane of the lens.

If the object grating is of unit contrast for all spatial frequencies the contrast variation recorded as a function of spatial frequency is the required function. This requires a narrow scanning slit, its width being about one-tenth of the finest grating period it measures. The change of spatial frequency in the grating must also be slow to counteract the influence of one period of each spatial frequency only being present. With these conditions fulfilled it is possible to obtain the optical transfer function of a lens to sufficient accuracy. K.G.B.

See also: *Aberrations of lenses; Faults; Information in the image; Lens performance; Optical design; Second-hand equipment.*

Books: *A System of Optical Design* and *Optics*, by A. Cox (London); *Measurement of Optical Radiations*, by G. Bauer (London).

LENTICULAR SYSTEM. Photographic process using a lenticular screen to break up an image into linear or area components which are subsequently recombined. The purpose of such splitting up is usually to accommodate two or more images interspersed in each other on the same area, Under some circumstances this greatly facilitates recording as compared with methods that record such images on separate sheets of sensitive material. Applications include a now obsolete system of colour photography, stereo photography, and image dissection for high-speed recording, etc.

Lenticular Screen. The main component of all lenticular systems is the lenticular screen itself, consisting of a transparent support embossed with a regular pattern of lens surfaces. Usually these are cylindrical running across the screen in one direction as strips. Nowadays such screens are usually made of plastic.

By embossing a second set of linear elements at right angles to the first, a truly lenticular screen—i.e. consisting of spherical lens elements—is obtained. The same effect is possible by combining two suitably matched screens at right angles.

When this screen is placed in front of, and in contact with, the surface on which the camera lens projects an image, the individual lenticular elements break up the image into lines or points. They concentrate the elements into a smaller area, leaving spaces between them. Additional images can be recorded in these spaces by slightly displacing the lens laterally or by moving the object in front of the lens.

The record then contains a series of interlaced images which can be reconstituted by observation through a similar lenticular screen. In any one position of the screen relative to the image all the components belonging to one image are visible again as a continuous picture. On moving the viewpoint (e.g. the eye) or on shifting the screen relative to the multiple image, a different set of image components become visible, assembling a new picture.

Lenticular Colour Photography. One of the first systems of additive colour photography for motion pictures used a normal panchromatic emulsion embossed with a regular texture of cylindrical lenses. This lenticular structure was impressed on the film base, and the taking lens of the camera was fitted with a filter consisting of three bands of red, green and blue. On exposing the film through the base each cylindrical lens element focused a red, green and blue band (from the filters) on to the emulsion.

In each image of the filter, the red, green or blue area will only receive an exposure if that colour is reflected from that particular area of the subject. Thus a blue image point will produce a dense patch on development under the image of the blue filter band, but not under the images of the other filter bands which remain clear. On reversal processing the patch corresponding to the exposed area becomes transparent while the rest of the area covered by the tiny lens becomes opaque.

The film is projected through the same type of coloured filter. The white light passing through the clear patch is focused again by the lenticle through the corresponding colour band of the filter on to the screen.

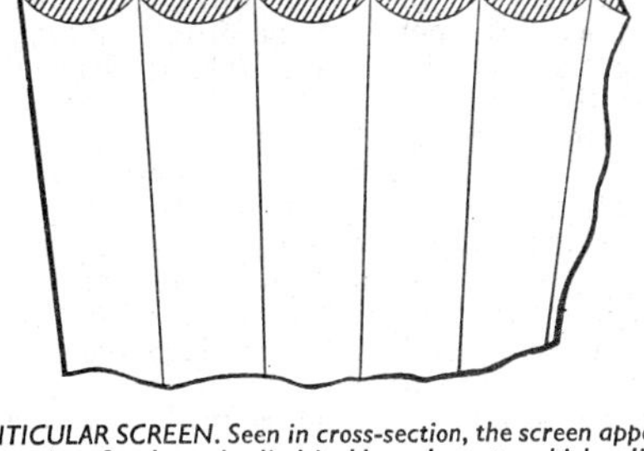

LENTICULAR SCREEN. Seen in cross-section, the screen appears as a series of embossed cylindrical lens elements, which split up the image into strip elements.

All image points of the various colours of the scene are synthesized by the taking filter in this way and reproduced as more or less clear patches on the developed emulsion. On projection (with the emulsion facing away from the lens) these patches transmit white light. Its intensity is proportional to the brightness of the original subject at each point. The light is again directed by the lenticular screen through the corresponding colour bands of the projection filter and recombined to form an image in the colours of the original subject.

Stereo Photography. In one system of stereoscopic photography two lenses focus two images (corresponding to the left eye and right eye views) on to the same film through a lenticular screen. At the correct spacing of the lenses relative to the curvature of the lenticular elements the screen records the two images as a series of interlaced strips. After printing as a positive, the individual images are separated by viewing through a similar screen. The stereoscopic effect is limited to a fairly small viewing angle in front of the picture, but it is produced without the use of a special viewer or spectacles.

This type of stereo image is known as a parallax stereogram. In a more elaborate system a single taking lens moves during the exposure, while the film with the lenticular screen is moved in synchronism. If the movements are appropriately matched, the strip elements record a continuous change of viewpoint from the left-to the right-eye view. This yields a so-called parallax panoramagram. Reconstitution of the image takes place in the same way as with the stereogram, but the stereoscopic effect is continuous over a much wider viewing angle. A system of this kind has even been used for printing three-dimensional images by a mass printing process. The printed reproduction is covered with a layer of plastic into which the lenticular elements are impressed by a suitably machined heated roller.

Image Dissection. A lenticular plate can be used to split up and multiply an image for high-speed photography. With picture sequences at fast rates the necessary movement of the film to record one frame after another runs up against appreciable mechanical problems. If a lenticular system is used, the images can be broken up and recorded side by side in an interspersed fashion on the same sensitive surface; the required movement is then very much reduced, making greatly increased exposure speeds possible. A similar lenticular plate is used to "unscramble" the composite image frame by frame. This merely requires successive very slight movements of the lenticular plate relative to the image, to bring the lenticular elements into register with all the image points belonging to one picture.

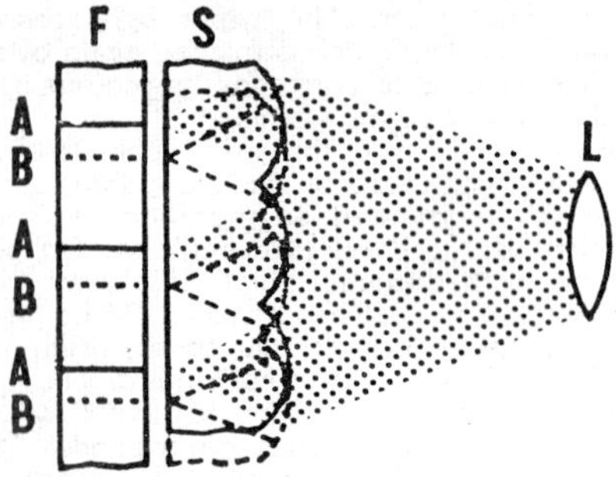

MULTIPLE IMAGE FORMATION. The lens L projects its image via the lenticular screen S on the film F. The screen breaks up the image into line elements A with spaces in between. If the screen is moved (dotted) a new set of line elements B is formed in the spaces between the first set. With two crossed screens, dot elements are formed.

Multiple Image Storage. The same principle can also be used for micro-recording. Here a number of separate originals—for example pages of a book—are photographed in succession on to the same sheet of film. A lenticular screen is placed in front of the film. The image is thus split up and concentrated into dots which may occupy only a fraction of the space of the film surface.

For successive exposures the lenticular screen is slightly displaced, forming subsequent dot images in the spaces between the original one. This therefore yields the interlaced pattern of a multiple picture.

To unscramble the composite record it is then only necessary to project the developed negative through the lenticular screen. Progressive movement of the screen relative to the developed film projects successive images (the different originals copied in turn) on to the viewing screen or on successive sheets of copying paper.

Whenever lenticular dissection is used, this naturally limits the resolution of the image by the spacing of the lenticular elements. But the latter also expand optically the image points and thus produce a continuous and undistorted picture on projection. F.P., (L.A.M.)

See also: *Colour history; Image dissector; Parallax stereogram.*

LE PRINCE, LOUIS AIMÉ AUGUSTIN, 1842–90. French artist, chemist, inventor and pioneer of cinematography. Introduced to photography by Daguerre, friend of his father, saw Muybridge photographs in 1879 and started work on motion picture projection while living at Leeds. His U.S. patent application of 1886 (granted 1888) described a camera or projector with one or several lenses, use of flexible film, intermittent movement and obscuration. In 1888 took 12 to 20 pictures per second on film 2⅛ ins. wide (first sensitized paper film, later gelatin stripping film and finally celluloid-base film). His first cine film projections were witnessed in Leeds 1889 and Paris 1890. He disappeared without trace, on a journey, at Dijon, 16th September, 1890.

LEREBOURS, NOEL-MARIE-PAYMAL, French optician, publisher and photographer. Made some of the lenses for the original (official) daguerreotype apparatus put on the market 1839. Published in conjunction with Susse frères an edition of Daguerre's manuel, 1839. Sent artists to many countries to take daguerreotype views to illustrate *Excursions Daguerriennes* 1841–3. Introduced quarter-plate daguerreotype cameras with achromatic lenses for portraiture, 1840. Wrote several manuals on the daguerreotype process. Founder member of the Société Héliographique. Published, with others, under the title *Lithophotographie* (1853) large views with good half-tone using albumen on lithographic stone.

LE ROY, JEAN ACME, 1854–1932. American pioneer of cinematography. Worked on a device for projecting Edison Kinetoscope pictures on to a screen for a large audience and exhibited moving pictures in New York on 5th February, 1894. Exploited his invention as a showman, but never patented or commercialized it.

LERSKI. HELMAR, 1871–1956. Swiss actor who took up portrait photography in America in 1911. 1915–27, cameraman for German motion picture companies, then returned to portrait photography, first in Berlin, after 1931 in Israel. Specialized in close-ups of simple people, changing their expression and apparent personality by subtle lighting.

LE SECQ. HENRI (Jean-Louis-Henri Le Secq Destournelles), 1818–82. French painter and photographer. Chiefly known for his architectural photographs taken for the French Government, 1851. Also photographed for Blanquart-Evrard. Gave up photography in the late 1850s, and formed a museum of wrought-iron work at Rouen.

LETTERING. Some prints need a title—e.g., to indicate the name of the sitter in a portrait or of the place in a landscape. It is usual to letter the words by hand on the mount below the print, and many an otherwise good piece of work is ruined in the process. Photographers too often do not realize that hand lettering is an exacting form of skilled craftsmanship that takes years to master. Fortunately there are several ways in which the amateur can make a neat, workmanlike job of lettering a title without being an artist.

The only equipment necessary is a steel rule, H and HB pencils, and a sandpaper block. Pencils softer than HB should not be used or the lettering will rub off when the print is handled or filed with other photographs. The sandpaper block is needed for keeping the point of the pencil sharp.

Simple Lettering. The simplest method is to fit the letters into rectangles roughly half as wide as they are high. All the letters of the alphabet can be fitted into such rectangles with the exception of M and W which need a little extra width. The style of the letters should be kept simple and the uprights should be at right angles to the base. Each line must be made with a single stroke of the pencil; it spoils the appearance of the letter to go over it more than once. If plain block letters look too stark, they can be improved by adding light serifs to produce a Roman type of lettering, but any further embellishment should be avoided. After all, the photograph is the important thing and the title should not compete with it for attention.

The title should be lettered on the mount, starting below the left-hand corner of the print and with about half an inch separating the bottom edge of the print and the top of the letters. First it is as well to letter the title between ruled lines on a spare sheet of paper and experiment with the letter spacing until the words look right and are easy to read with no ugly gaps or crowded letters. The back of the paper can then be rubbed over with a soft pencil and the lettering traced lightly through into place on the mount with a hard pencil. This takes care of the spacing and permits the title to be lettered on to the mount with confidence. Where the photographer wishes to include his own name, he should letter it in the same style as the title. An ordinary signature on the mount looks pretentious.

Other Methods. The ordinary draughtsman's celluloid stencils can be used for pencil lettering. But the mount should be pinned down on to a drawing board and the stencil positioned with a tee-square.

A title typed on a sheet of paper, cut out and pasted on to the mount can look neat and effective. It is generally better to use suitably tinted paper; white is apt to make the mount or the print look grubby.

Prints in albums may be titled by any of the above methods, although cut out typewritten words are usually more suitable as the pages of the album are usually black. White ink may also be used for tilting on black pages but mistakes are difficult to erase.

Photographic Christmas cards may be lettered by hand, but if a number are being turned out it is easier to draw out the lettering, photograph it, and print it in position along with the negative. There is no reason why old Christmas card titles should not be cut up to form the required words. The ready-made lettering on thin adhesive paper used by commercial artists may also be used for making up the title or greeting message to be photographed. P.C.P.

See also: *Caption; Greeting cards; Inks for photographs; Slide diagrams; Titles.*

LETTERING LAYOUT. Letters should fit into equal rectangles, except for specially wide ones (e.g., MW). Titles should be lettered simply and kept short.

LETTERPRESS. Method of mechanical printing from a raised surface which carries the ink. Derived from the fact that in the oldest and still most widespread printing process for text matter the letters stand up above the general surface. Half-tone blocks with a dot structure of raised dots are equally referred to as letterpress blocks.

See also: *Photomechanical reproduction.*

LEVELS AND PLUMBS. Devices to aid accurate horizontal and vertical alignment of cameras and other equipment. Levels are necessary in all kinds of survey photography, architectural work, stereoscopic photography, panoramas, etc. Plumbs are used in copying set-ups, architectural and surveying photography, etc.

A level (spirit level) consists essentially of a sealed glass tube filled with spirit, leaving a small air bubble. As long as the tube is truly horizontal, the bubble will lie exactly between reference marks in the middle. If the tube is inclined along its length, the bubble will move towards the upper end. This arrangement is very sensitive to the slightest inclination.

The level may be a separate accessory mounted in a suitable metal or wooden block with its under-surface absolutely parallel to the tube, or it may be built into the camera or other equipment. Built-in spirit levels are more commonly found on older-type stand cameras, and sometimes on modern tripods.

Some types of spirit level consist of a horizontal round cell instead of a tube. On truly level positioning the bubble should be in the middle of the flat top of the cell. This level shows departures from horizontal in any direction, not only along one axis as in the case of the tube level.

A plumb is a small weight attached to the end of a thin cord. When the weight hangs freely from a fixed point, the cord provides a truly vertical reference—e.g., for setting up the camera accurately over a particular point on the ground.

A vertical spirit level, i.e., one set in one limb of an accurate T-square or similar instrument, will also serve for vertical alignment.

In some cameras a levelling indicator is visible in the viewfinder in the form of two pointers. When these are in line, the camera is level. L.A.M.

LEVY-DORN, MAX, 18??–1929. German roentgenologist. Made the first stereo-radiographs (1897) and the first cine-radiographs, also the first complete X-ray installation in a hospital in Berlin (1906). Stressed, as early as 1903, the need for protection from X-rays.

LIESEGANG, FRANZ PAUL, 1873–1946. German manufacturer of projection apparatus, son of Paul Johann Liesegang. Wrote much on the history of projection and on the forerunners of cinematography.

LIESEGANG, FRIEDRICH WILHELM EDUARD. 1803–69. German photographer. Founded the firm of Liesegang (Elberfeld, 1854, later Düsseldorf) which was carried on by his son and grandsons, eventually to merge, as the oldest component, into the Agfa concern. Manufactured at first projection apparatus; from 1857 on, also calotype materials and photographic papers.

LIESEGANG, PAUL EDUARD JOHANN, 1838–96. German photographic manufacturer, son of Friedrich Liesegang. Suggested green safelights for the darkroom in 1861. Introduced the Lilliput camera (3 × 3 cm.) in 1882. Founded several periodicals: *Photographisches Archiv* in 1860, *Photograph scher Almanach* in 1861, and *Laterna Magica* in 1877.

LIESEGANG, RAPHAEL EDUARD JULIUS, 1869–1947. German colloid chemist, son of Paul Johann Liesegang. Published experimental work on electrolytic development on photochromy, and on the bleach-out process. Described the principle of lenticular screen colour photography processes (1896). Author of many books and articles.

LIGHT. Light is merely one of the many forms of electro-magnetic radiation encountered in nature. Such radiations occur as radio, radar, radiant heat and infra-red therapeutic rays, visible light, ultra-violet rays, X-rays, and the gamma rays caused by the breakdown of certain atoms.

Electro-magnetic radiation will travel across empty space and needs no material conductor. All forms of it have the same velocity in vacuum—3×10^8 metres, or 186,000 miles, per second. The velocity in air is practically the same.

Corpuscular and Wave Theories. The first attempt to explain the nature of light was Newton's corpuscular theory. He postulated that light consisted of small corpuscles shot out from the light source rather like bullets from a machine gun. This explained reasonably well many of the optical phenomena known at the time. At about the same time Huygens suggested that light was a wave motion. The discovery, during the eighteenth century, of diffraction and inteference effects, and later the assumption that light was one of a number of electro-magnetic radiations, eventually gave preference to the wave theory as the more likely explanation of the nature of light.

The present view is that both aspects of light appear to have a degree of simultaneous existence. In other words, light can behave either as waves or as particles. The former are needed to explain diffraction, interference and the colour of different kinds of light; the latter are associated with energy aspects, including the interaction of light with matter to initiate chemical reactions. The formation of a photographic image in an emulsion is the most obvious example.

Wave Characteristics. When regarded as consisting of waves, electro magnetic radiation is associated with a definite frequency governed by the following relationship:

$$\text{Velocity} = \text{frequency} \times \text{wavelength.}$$

The frequency of a certain hue of blue-green light is, for example, 6×10^{14} cycles per second, and its wavelength is therefore $0{\cdot}5 \times 10^{-6}$ metres.

Physical interaction between light and material bodies leads to the phenomena of reflection, refraction and dispersion. Straightforward reflection, with the light ray thrown back from a surface virtually unchanged, is a characteristic of the action of mirrors. When light passes from one transparent medium to another—e.g., air to glass—its velocity and therefore its wavelength are changed. This is refraction, the phenomena to which the action of a lens is due.

The wavelength of light is so short that it must be measured in special units. The commonest is the Angstrom unit, which has as its symbol the capital A. There are ten million Angstrom units to the millimetre. It is sometimes more convenient to use larger units like the micron which is the thousandth part of a millimetre, or the millimicron (mμ), which is the millionth part of a millimetre—i.e., one micron = 1000 millimicrons = 10,000 A.

The nanometer (nm.) is a more modern term for the millimicron.

So $0{\cdot}5 \times 10^{-6}$ metres, the wavelength of blue-green light, would be written as 500mμ or nm, or 5000A.

X-rays have such a short wavelength that

they are measured in x-units of which there are ten thousand million in a millimetre.

Photography is concerned with wavelengths ranging from a few mμ (X-ray range) to 1500 mμ.

Light as Energy. Like matter, light can be subdivided quantitatively until we arrive at the smallest basic unit. This is the photon and corresponds—at least in conception—to Newton's corpuscles. A photon is an elementary energy packet with one so-called quantum of energy. The size of this quantum depends on the wavelength of the light, and is defined as:

$$E = h\nu$$

where his Plank's constant and ν the frequency.

The effect of light in terms of exposure of a photographic emulsion is associated with light quanta. It is the energy of the light which initiates the reactions leading to the conversion of silver halide into metallic silver forming the image. R.B.M. & L.A.M.

See also: *Angstrom; Colour; Light units; Spectrum.*

LIGHT GUIDE. Optical unit which can carry light rays along paths other than straight lines. Usually the term refers to glass rods or bundles of glass fibres which can be bent into any shape, but which ensure that a light ray entering the rod at one end will always emerge at the other. The light guide carries the light along its tortuous path by a series of successive internal reflections. As long as the guide is not bent to too small a radius virtually no light will be able to escape.

See also: *Fibre optical systems.*

LIGHTING EQUIPMENT

In the early days of photography the sun was the only source of photographic lighting; with the development of electric lighting, photographers were quick to appreciate the advantages of a source that they could control. To-day as much, if not more, photography is carried out by electric light as by daylight, and the equipment available ranges from small portable units to permanent installations capable of creating any type of lighting effect and powerful enough to permit exposures of short fractions of a second.

The most popular sources of electric lighting are tungsten filament lamps of the normal and overrun (Photoflood) types. These are cheap, convenient and very suitable for most amateur and professional needs. Arc lamps of various types, vapour discharge and fluorescent lamps and flash, both bulb and electronic, have their special fields of operation.

Controls. Before a light source can be usefully employed for photography, there must be some means of controlling it.

The power of the light is controlled in the simplest fashion by the switch which turns it full on or off. It may be further controlled by inserting a dimming resistance in the circuit, or by series-parallel switching arrangements; by this means the power of the light is continuously variable between full power and off (dimming resistance), or adjustable to give full power, reduced power of a fixed value, and off (series-parallel switching).

The concentration of the light is controlled by shaped reflectors behind the light source, or lenses (spotlights) in front. In the case of reflectors, the light is spread and, depending on the surface of the reflector, somewhat diffused; this is useful for providing general illumination of fairly large areas. With a spotlight, the light can be concentrated on a smaller area and used to get crisper results with harder shadows. When maximum diffusion is required, a diffuser is fitted in front of either reflector or spotlight.

The colour of the light may be controlled by either the choice of the source, or the use of colour filters over the light. Such measures are only required in colour photography where colour temperature is of great importance.

The position of the light source is, in practice, the most fundamental of all controls. To enable convenient alteration of the light position, various supports, brackets, and stands are made so that the direction and angle of the light is adjustable with ease.

Amateur Outfit. Few amateurs have the space to spare for ambitious lighting equipment. In most cases the gear must be packed away after the photographic session, therefore it needs to be of the type that dismantles easily or telescopes into a small space.

While it is true that good pictures can be made with only one lamp, its range of effects is limited and exposures are apt to be unduly long. A good working compromise between such austerity and the elaborate set-up of the professional is provided by the following:

(1) Modelling light: one Photoflood or 500 watt tungsten lamp in reflector on an adjustable telescopic tripod stand. The adjustment should allow the lamp to be placed anywhere from four to eight feet from the floor. This lamp provides the main source of light; where it is not enough to cover the subject it can be supplemented by the fill-in light.

(2) Fill-in light: one Photoflood or 500 watt tungsten lamp in a shallow reflector on a small hook-or-stand support or on a clamp fitting. This light is normally required for illuminating the shadows, but it may be replaced by a flat white reflector so that it can be added to boost the modelling light—e.g., to cover a broader area.

(3) Background light: one 150 watt tungsten lamp in a deep reflector or fitted into a reading

lamp holder with a suitable cone of white paper to concentrate the light on the background. This light is used for controlling the general or local tone of the background, particularly in portraiture. It may also be used for back lighting to create a bright rim of light around the subject.

(4) Spotlight: although a spotlight is much more expensive than any of the other forms of lighting, it opens up a wide choice of special effects both in portraiture and still life, and it can be packed into a small space. For these reasons the serious amateur would complete his equipment with a 250 watt focusing spotlight with fittings for barndoor, snoot, lampfilter and diffuser.

TYPICAL LIGHTING INSTALLATIONS

Equipment	*Watts*
Commercial Studio	
(Equipped for six cameras on still-life subjects)	
18 250 watt lamps in deep reflectors	4,500
12 500 watt spotlights	6,000
1 1,000 watt flood	1,000
1 7-amp enclosed arc (on 200 v.)	1,400
Total	12,900
Large Portrait Studio	
4 Mercury vapour tubes, 200 watt	800
5 1,000 watt lamps in white-painted trough	5,000
2 1,000 watt floods	2,000
2 1,000 watt spotlights	2,000
2 500 watt spotlights	1,000
2 250 watt floor lights	500
1 1,000 watt spotlight with mirror reflector only	1,000
Total	12,300
Small Portrait Studio	
1 400 watt Mercury vapour tube, 2 500 watt Nitraphot — in movable trough from ceiling, diffused	1,400
1 120 watt Mercury vapour tube, 1 500 watt Nitraphot — in small white trough, diffused. Floor stand with counter-weighted arm	620
1 1,000 watt Daylight lamp in convertible studio light. Floor stand with large matt aluminium bowl. Direct or indirect light	1,000
1 500 watt Nitraphot with shallow blue-tinted bowl. Direct or indirect. Floor stand	500
1 1,000 watt Spotlight	1,000
1 500 watt Spotlight	500
Total	5,020
Studio of Daily Pictorial Newspaper	
1 2,000 watt spotlight	2,000
2 1,000 watt spotlight	2,000
1 20 amp. arc spotlight (on 230 v.)	4,600
3 1,000 watt floods, in 18 in. painted reflectors	3 000
Total	11,600
Back Projection Equipment	
1 50 amp. lantern with water cell (on 230 v.)	11,500
6 250 watt overhead spotlights	1,500
2 250 watt floor spotlights	500
Total	13,500
Commercial Sub-standard Cine Studio	
(This lighting equipment permits shooting on a 30 feet wide set at *f* 2 or *f* 2·5)	
50 275 watt Photofloods, overhead	13,750
8 500 watt ceiling floods	4,000
2 500 watt spotlights	1,000
2 1,000 watt spotlights	2,000
4 275 watt Photofloods in metal reflectors	1,100
Total	21,850

The total power consumption of all this equipment would not overload the normal room power supply.

A desirable extra is a switch—preferably the foot-operated type—connected to the two Photoflood lamps so that they can be switched in series for focusing and arranging the subject, or in parallel to give their maximum brilliance for the exposure.

There are several types of collapsible reflector on the market for amateurs who have no room to store rigid reflectors. These reflectors are never as good as the normal type, but they are satisfactory so long as they have a matt and not polished reflecting surface.

Commercial Studio. The lighting equipment of the commercial studio is dictated by the type of work undertaken. A studio specializing in one particular branch—e.g., portraiture or catalogue illustration—may be relatively lavish with equipment suited to that type of work but poorly equipped to handle general subjects. It may suit one studio better to have a number of mobile units and little in the way of fixed lighting, whereas another may need to have a number of fixed lights controlled by separate switches on a central panel.

The basic set-up in most studios consists broadly of the following:

(1) General lighting: banks of tungsten, mercury-vapour or fluorescent lamps in trough-type reflectors. These are suspended or built in above the camera position to direct a flood of general light forward on to the subject. They may be adjustable for height and position, and they are generally arranged in a number of separately switched units. The total power consumed by the general lighting of an average studio is often more than 10,000 watts.

(2) Effect lights: any number of tungsten lamps from 250 to 1,000 watts, mounted on adjustable mobile stands. These lights are brought up as required to give general or local modelling, fill-in, background and back lighting. The shape and size of the reflectors vary according to the purpose of the lamp—e.g., shallow and broad for supplementary general lighting, or deep and narrow for illuminating smaller areas.

(3) Spots: any number of fixed and focusing spotlights mounted on adjustable mobile stands with provision for inserting masks, filters, barndoors, hoods, snoots, etc.

The spotlights, because of their power and flexibility (in position, and in breadth and shape of beam) tend to be used for all the principal work of modelling. Light sources may be either projector-type tungsten lamps, or carbon, mercury or high intensity arcs, ranging from 500 to 2,000 watts or more.

In studios which specialize in popular and "while-you-wait" portraiture the principal lights are often fixed arrangements disposed around the walls and over and in front of the sitter's chair. Any of a number of types of subject lighting can be produced by pressing the appropriate switches. This is quicker than moving lights about, and the exposures can be standardized. With very little training an operator can take charge of such a studio and turn out a portrait to a set pattern in a few minutes. The photographs produced by such lighting arrangements are apt to be stereotyped.

Control Boxes. Where a number of independent lamp units in a photographic studio are used at the same time, they may conveniently be controlled from a lighting control box. In its simplest form this is a sheet metal box on rubber feet or castors, with two or more switched socket outlets connected in parallel to a tough rubber-sheathed cable and plug which connects the box to the mains socket outlet. This allows two or more lighting units (depending on the number of outlets provided) to be plugged in and switched on and off individually. It also allows the lighting to be controlled from anywhere in the studio irrespective of the position of the permanent socket outlet.

In addition to the switches on each of the separate sockets, the control box may provide other facilities. Certain outlets (e.g. for photoflood lamps) may be associated with a series/parallel switch allowing pairs of lamps to be run in series while the subject is being arranged and then switched to parallel working for maximum brilliance during exposure.

One or more socket outlets on the box may be connected to a variable resistance controlled by a hand wheel on the outside of the box to allow a bank of floodlights or a spotlight to be dimmed at will.

The control box usually incorporates a pilot lamp which lights up when the box is connected to the mains to show that outlets are live.

Another type of box, popular for amateur studio lighting, allows an even number of photofloods to be connected as two groups in series or parallel by pressing a foot switch mounted on the box.

Voltage Control. Variations in the voltage of the electricity supply used for studio lighting affect both the total light output and the colour temperature of the lamps. This is particularly important in colour photography, where a supply of constant voltage is essential.

One means of providing this is an auto-transformer connected to the alternating current supply. Tappings on the secondary winding are connected to a continuously variable control so that the transformer ratio can be altered to give a constant output voltage for every value of supply volts between the upper and lower permissible limits of supply voltage variation. A meter across the secondary of the transformer allows the voltage to be kept at a constant figure by turning the control up or down as required. This type of control can be used for either keeping the supply at a constant voltage or for providing a range of output voltages from a given supply.

Where the only purpose is to maintain a constant voltage the job can be done automatically by a constant voltage transformer (CVT). This makes use of a control circuit consisting of a separate coil wound on a saturated iron core and connected in opposition to the main winding. This is done in such a way that an increase in the supply voltage is cancelled out by an opposing voltage of the same value in the control winding while a drop in the supply volts is compensated for by a greater drop in the opposing volts. Once again the net result is that the useful voltage of the transformer remains unchanged.

As an alternative, the voltage can be controlled by a variable resistor (rheostat) wired in series with the lamp circuit. The resistance required at any one time depends on the resistance of the rest of the circuit, hence rheostats must be selected individually for the requirements of the equipment being controlled.

A rheostat has the advantage that it can be used with direct (D.C.) as well as alternating current (A.C.), and rheostats are sometimes employed as dimmers where a continuous variability of light output is needed. However, when the voltage is reduced with a rheostat, an appreciable amount of current is wasted and converted into heat. For that reason transformers are preferred to resistors in virtually all cases where an A.C. supply is available.

Colour Temperature Control. To provide direct control of the colour temperature of tungsten filament studio lighting, an auto transformer may be used with a series of tappings covering a range of voltages from some 50 per cent below to 50 per cent above the normal supply voltage. A rotary or sliding control allows any voltage to be applied to the lamps according to the colour temperature required, a higher voltage raising both the colour temperature and the level of the lighting and vice versa.

One form of control box designed specially for still and cine photographic studios combines all the separate pieces of equipment required for colour temperature control: auto-transformer, voltage selector and indicators for showing colour temperature and lighting level. This type of unit helps to extend the useful life of photographic lamps since it allows them to be under-run while the subject is being arranged and then brought up to the required colour temperature for the actual exposure. The saving can be considerable when it is remembered that the life of a 1000 watt studio flood varies from 2,000 hours under normal conditions to 12 when over-run to give a colour temperature of 3,400°K. F.P.

See also: *Batteries; Diffuser; Dimmer; Electricity; Light sources; Light supports; Reflector; Spotlight; Studio; Switches; Wiring.*

Books: *Lighting for Photography* and *Lighting for Portraiture*, by W. Nurnberg (London); *The Business of Photography*, by C. Abel (London).

LIGHTING THE SUBJECT

The photographer who works out of doors in daylight must take the subject lighting as he finds it. Up to a point he can select the kind of lighting he wants from variations in direction, intensity and character of the sun's light in the course of the day, or the season, or the weather. He can modify the lighting of near subjects by the use of a reflector or supplementary flash. But he has no control over the principal lighting and in fact he has little wish to exercise control since the natural lighting is part of the character of the subject.

Things are different when he comes to take a picture indoors. Here he has complete control of the lighting of the subject. This time, he not only has to select, arrange and compose his picture material—he has to light it. And whereas the lighting of a picture that was obviously taken out of doors is accepted by most people without challenge, the lighting of an indoor photograph is as fair a ground for the critic as the composition. To a very great extent, the lighting is the picture; and the success of the result depends upon understanding how much to use and where to put it.

The tools are simple and few; there is the spotlight, which gives a hard and relatively narrow beam of light; the floodlight, which gives a broad, soft, general illumination; and various types of light reflector—e.g., white or aluminium-surfaced flat boards—which reflect more or less diffused light on to the subject.

The effect of any single light on the appearance of the subject can be controlled in a number of ways.

(1) The angle of the rays falling on the subject may be varied—i.e., the lamp may shine either horizontally, or from vertically above or below, or from any intermediate angle.

(2) The position of the light may be varied, making the lamp shine from in front, from behind, or from any intermediate point.

(3) The distance from the subject may be changed from near to far.

(4) The power of the lamp may be varied according to the consumption or type of the light source chosen.

(5) The softness and spread of the rays may be changed by letting them shine through a translucent diffusing screen—e.g., of stretched white muslin.

(6) The colour temperature may be controlled by the use of tinted filters in front of the lamp.

All these various forms of control can be exercised with a single light source. But any number of single sources may be used at the same time, and each one may be varied in any of the six ways independently of the others i.e., with two lamps, there are six ways in which the effect of either lamp may be altered, associated with six states of the other. So that with only two lamps in use, there are thirty-six ranges of lighting effect, and since each range of effects offers anything up to another dozen recognizably different effects, the photographer is faced with an embarrassing number of alternatives in arranging the lighting. And as the number of lights increases, things get more complicated.

So a haphazard experimental approach is out of the question. Correct subject lighting follows as a logical result only when the photographer can visualize the characteristic effects of the basic lighting conditions; he has already in his mind a picture of how he wants the subject to look, and he places his lights to give the effect corresponding to his mental picture.

It does not follow, however, that by employing all of these lighting effects a better picture will be achieved.

BASIC LIGHTING CONDITIONS

It is first of all necessary to recognize the basic lighting conditions. Five lighting conditions give characteristic effects which can be easily recognized. These are when the light, in relation to the subject (not the camera), comes from front, side, behind, top, below. (In practice these basic conditions are rarely used; it is more usual to combine the separate characters of two or more by adopting a compromise.)

Frontal Lighting, When the light shines on to the front of the subject it casts no definite shadows that are visible from the camera position. So the subject appears completely without relief; it is recognizable only by its shape against the background, and by its own differences of colour and tone. Where the surface of the subject curves away from the camera, however —e.g., along the sides of the cheeks and neck in a full-face portrait—less of the light is reflected back to the camera, so the tone is proportionately darker.

Frontal lighting is called "flat" or "shadowless" lighting; it can give a high key picture of

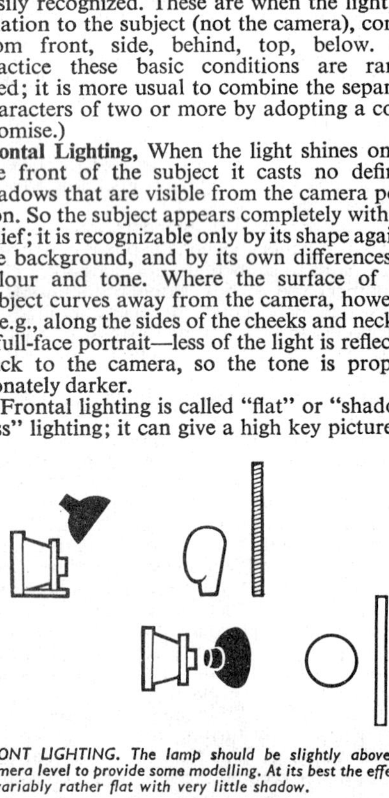

FRONT LIGHTING. The lamp should be slightly above the camera level to provide some modelling. At its best the effect is invariably rather flat with very little shadow.

subjects which have no strong colour or tone contrasts of their own; it is the lighting generally recommended for straightforward colour photography because the illumination of all the colours is substantially equal and therefore the colours are more likely to come within the exposure range of the colour film.

Frontal lighting is often encountered in everyday life when the sun shines from behind the observer, so it is accepted as normal and calls up no special associations.

Side Lighting. When the light shines from one side on to the subject, it illuminates that side brightly and leaves the other in complete darkness. The result is to make the subject appear to stand out in strong relief because any feature that projects—e.g., the nose in a full-face portrait—reproduces as a highlight with an associated strong black area. The black area consists of the unlighted part of the projection and the shadow cast by the projection on the adjacent surface. A depression in the surface of the subject shows up as a highlight on the edge farthest from the light source and a shadow area on the same side as the light source.

Side lighting is common in normal, everyday experience and is easily accepted as natural.

Top Lighting. When the light shines from directly over the subject it casts all shadows vertically downwards over the subject. Generally, it highlights the features in the same way as side lighting—i.e., it emphasizes projecting features and depressions by reproducing them as areas of associated strong light and shade. Top lighting is not a normal effect away from the tropics, so it looks unnatural even when produced by artificial light.

Lighting from Below. When the light shines from below, the shadows are cast upwards. Projecting features are brightly lighted on the lower side and the upper surfaces are in shadow. The shadow on the side of projecting features joins up to the cast shadow to make a single black area.

This effect never occurs in nature so it looks completely false. The most familiar example of such lighting is the light of theatre

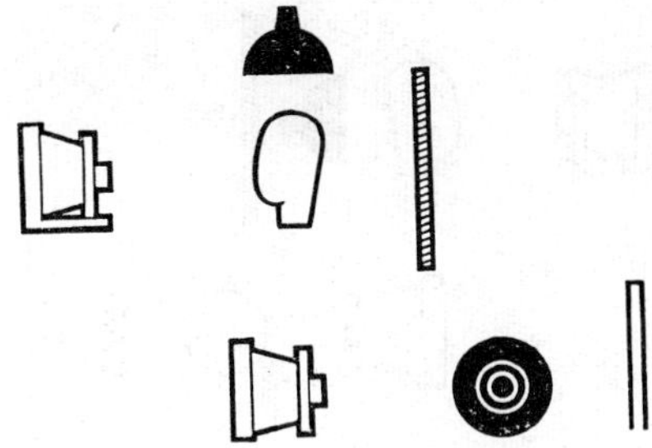

TOP LIGHTING. Throws features facing the camera into relief. All shadows are cast downwards, and the effect is not a natural one. It can, however, be used as effect lighting.

stage footlights so, possibly by association, lighting from below has come to suggest drama and artificiality.

Lighting from Behind. When the light shines from behind the subject, all shadows are cast towards the observer. If the light is directly behind, the only indication of its presence may be a brilliant rim of light produced when the rays are reflected from surface roughnesses, fibres or hairs, the general effect looking something like an eclipse of the sun. This rim of light becomes wider as the lamp moves around towards the side, top, or bottom of the subject.

Rim lighting and backlighting are common in subjects seen against the rising or setting sun, so they are not unnatural in their effect. At the same time they leave the front of the subject completely in shadow, so they are rarely used as a sole source of photographic lighting. When they are used, it is as effect lights and not as principal illumination.

The Basic Lights. There is no limit to the number of lights that can be employed or the number of ways in which they can be arranged to provide a scheme of subject lighting. But there are only four functions that any particular light source can fulfil (although one light source may answer more than one of these purposes at a time)—i.e., it may be the principal (or modelling) light, a fill-in light, an effect light or it may light up the background.

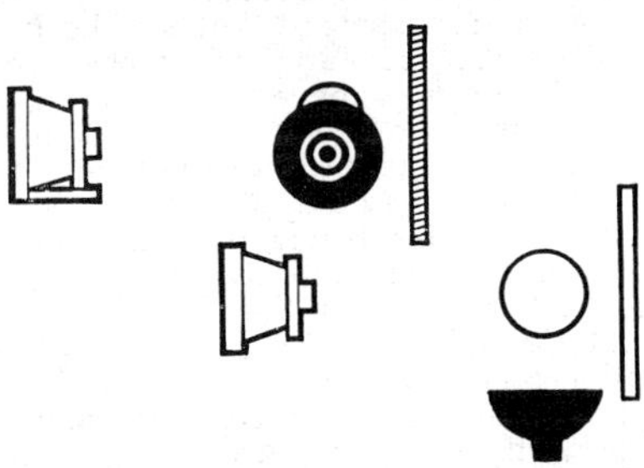

SIDE LIGHTING. Throws features facing the camera into bold relief. Brilliantly lights one side of the subject, leaving the other in deep shadow, unless filled in by reflectors.

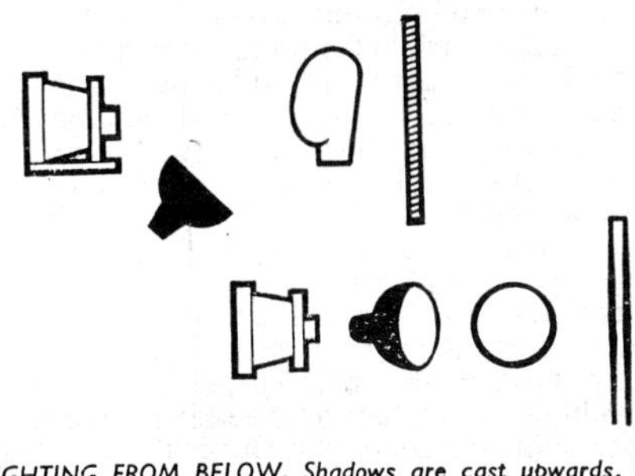

LIGHTING FROM BELOW. Shadows are cast upwards, producing theatrical and dramatic effects. The light should not be too low, otherwise the result looks weird.

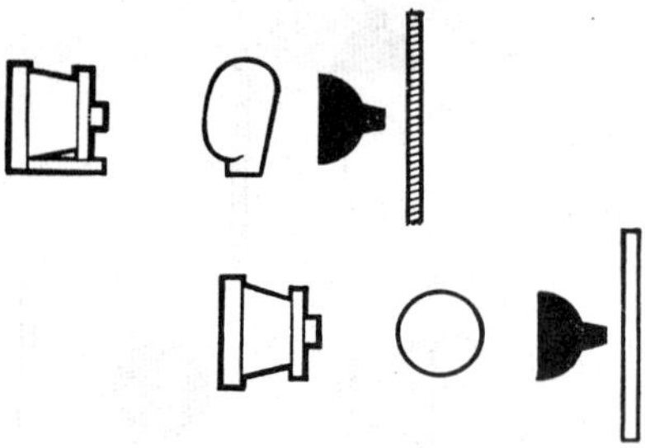

REAR LIGHTING. Leaves the subject as a detailless silhouette, usually with a brilliant rim of light outlining the subject shape. Mostly used as effect lighting only.

Principal Light. In every scheme there is one light that dominates all others; it creates the most noticeable highlights and it casts the most important shadows. This is the light that "holds the picture together" and its power and position must be adjusted so that its purpose is obvious. If any other light in the system is strong enough to compete with the principal light it will divide the interest, possibly produce conflicting shadows, and the picture will generally be confusing and incoherent.

Fill-in Light. The job of the fill-in light is to illuminate the shadows cast by the principal light.

The photographic emulsion is not able to register satisfactorily such a great contrast range as the human eye, so while the eye may still be able to discern detail in the shadows cast by the principal light, the camera might reproduce them as featureless black patches. So more light must be directed into the shadows to bring the over-all contrast of the subject within the range of the sensitized material.

A reasonable balance between highlight and shadow illumination is achieved when the shadow brightness is approximately one-third of the highlight. This condition may be created by adjusting the distance between the fill-in light and the subject, or by altering the power of the light sources.

The position of the fill-in light must be chosen so that it illuminates the shadows cast by the principal light but casts no new shadows that would be visible from the camera. The fill-in light is almost always provided by a floodlight giving a soft and evenly distributed illumination.

Reflectors of various types can be used in place of fill-in lights, both out of doors in sunshine and in the studio. While not being quite as versatile as a fill-in light, these are often more portable and convenient.

Effect Light. An effect light is one that is directed on to the subject with the aim of producing a distinctive highlight—e.g., a bright rim, a highlight on hair or cheek in a portrait—or of adding extra sparkle and brilliance. One of the most common effect lights is a light placed behind the subject and shining towards the camera to give a halo of light around the subject. Any number of such effect lights may be introduced, and they may be provided by spotlights, floods, or even reflectors. Mirrors are particularly useful for this kind of illumination.

Background Light. The background of subjects photographed in daylight is a matter for selection; it may be empty or furnished with objects, it may be brightly lighted or in shadow according to how the photographer takes the picture, but in every case, it is already there. In artificial light photographs, the photographer creates the background; nothing will appear in the finished picture unless he has put it there and then either intentionally or accidentally directed light on to it.

The photographer may choose a flat, plain, neutral tinted background for his subject, or it may be curved, textured or printed with a pattern; but the amount that appears in the finished picture, and the brilliance of its tone in relation to the subject, is a matter of lighting. It is possible to concentrate on the subject lighting and trust to luck that whatever incidental light falls on the background will produce a passable effect. (This is the characteristic approach of the beginner who forgets about the background until it turns up in the finished print to remind him unpleasantly of his oversight in arranging the subject.)

In all serious work, however, the background is separately lighted by a specially arranged background light. This may be supplied from a spot or a flood, depending on whether the light is to be spread evenly or concentrated.

Where a spotlight is used, the shape of the patch of light projected on to the background may be altered by manipulating the focusing adjustment, by altering the angle at which the beam strikes the surface and by employing masks (e.g., "barndoors") in front of the spotlight so as to cast a shadow.

Where a floodlight is used, the area illuminated can be controlled by turning the lamp and reflector and by supporting a shaped piece of card in front of the light to cast a shadow over part of the surface.

The broad object of background lighting is to create a tone contrast between the background and the part of the subject that appears against it. In the simplest case, where the subject is lighted from one side, the background is lighted so that the shadow side of the subject is seen against a light tone and the lighted side against a dark tone. It may be brightly lighted all over so that the entire edge of the subject stands out in a darker tone than the background. Or it may be given a low level of illumination so that all the edges of the subject appear brighter by contrast. The range of effects is infinite and completely under control, but it can never be left to chance; the background lighting is as much a part of the picture as the subject lighting itself.

Reflectors. Where there is already one source of illumination, the effect of a second source can be created within limits by using suitably disposed reflecting surfaces. These generally consist of flat boards painted with white enamel or aluminium paint, stretched sheets of white fabric or even sheets of white drawing paper. A reflector of this type can be arranged so that it intercepts rays from a light source and reflects them in another direction. A reflector with a perfectly polished surface reflects the light in straight lines, in the same way as a mirror, whereas the more usual matt white surface scatters the rays and behaves like a weak and very diffused floodlight.

Most of the basic lighting conditions can be reproduced approximately by suitably placed reflectors so long as the position and angle of the true light source lends itself to the arrangement. For instance, when the light source is in front of the subject, a reflector can be arranged to direct part of its rays on to the sides, top, bottom, or back of the subject—but not on to the front.

A reflector may also be used to reflect all the light from the source so that no direct rays from the source fall on the subject. This may be done to create a softly diffused source of light where the direct rays would give too hard an effect—as when a flash is "bounced" off a light-coloured wall or ceiling.

A reflector may thus be regarded as a weak and diffused supplementary light source. Its feeble character can generally be offset by placing it close to the subject, and even its diffusion can be overcome by using a mirror instead of a white surface.

SHADOWS

Shadows are inseparable from almost every lighting arrangement. The only position from which a lamp casts no shadows that will come out in the photograph is from the position of the camera itself. From all other positions, every light source casts a shadow that may or may not reproduce in the print. Generally speaking any light source in the lighting arrangement is bound to cast shadows that reproduce in the print—otherwise it would not be doing a useful job. And it is a safe rule that any shadow that can be detected in the actual set-up will appear darker in the final print.

The quality, size and shape of the shadow vary according to the effective area of the light source, the size of the object casting the shadow, and the relative distances of the source and the object from the surface on which the shadow is cast.

Quality of the Shadow. The light source affects the quality—i.e., tone and sharpness—of the shadow according to the effective area of the source.

A point source of light without a reflector will cast on to a surface perfectly sharp

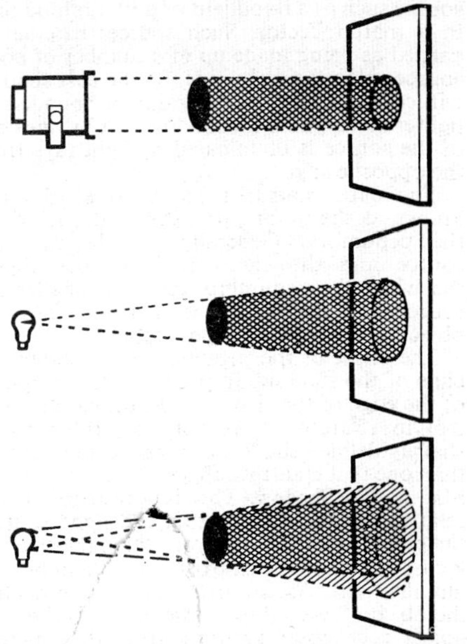

SHADOW FORMATION. Top: *A parallel beam from a spotlamp casts a sharp shadow of the subject in approximately the same size.* Centre: *A point source casts a sharp shadow larger than the subject.* Bottom: *A larger source casts a diffused shadow of a dark core and a blurred lighter rim (a penumbra).*

shadows of any opaque object; this applies no matter what distance there is between light source, object and surface. But the closer the light is brought to the object, the bigger the scale of the shadow, and the closer the surface is brought to the object, the smaller the shadow. In every case, the shadow will be equally dense all over.

A source of parallel light rays—e.g., a spotlight—casts a sharp shadow that is always exactly the same size as the object, no matter how far the object and the surface may be from the source. Here again, the shadow density is the same from edge to edge.

What happens with a source of a definite size—e.g., a pearl Photoflood bulb—can be understood by visualizing the shadows cast by two light sources side by side. Each source will cast a separate shadow of the object on the background and if the lamps are far enough apart, rays from the lamp on the one side will fall on the shadow cast by the lamp on the other. So each shadow will be lighter in tone than if the second lamp had not been there.

If the lamps are brought closer together, the cast shadows will approach each other and finally meet and overlap. Where the two overlap there will be an area of deep shadow receiving no illumination at all, flanked by areas of lighter tone where the shadow cast by one lamp is illuminated by the other.

The same thing happens with a broad light

source such as a floodlight or a Photoflood bulb in a matt reflector. Such sources can be regarded as being made up of a number of point sources placed side by side. So the cast shadow will consist of a dark central area bounded by lighter areas where the shadow cast by one edge of the source is illuminated by light rays from the opposite edge.

The dark centre of such cast shadows is known as the umbra, and the lighter fringe as the penumbra. Generally, the broader the source and the closer it is to the object, the wider the penumbra. And the smaller the source and the farther it is away from the object, the narrower the penumbra

The effect of the penumbra is to soften the edge of the shadow. In practice, the sharpness of the edge of the shadow calls for careful control in relation to its purpose; generally, a sharply defined shadow attracts more attention than one that is softly diffused.

Size of the Shadow. This is governed by the relative distances separating the light source, the object and the surface on which the shadow is cast (except in the case of the spotlight type of illumination). As the light source approaches the object, the shadow grows bigger and as the object gets closer to the surface the shadow grows smaller. By adjusting the relative position of subject, light source and background in this way, the photographer can introduce background shadows of the subject—or any suitably shaped object—into the picture. The shadows can be either large and dominating, or small and subordinate according to the arrangement employed.

Shape of the Shadow. If the background is flat and at right angles to the beam of light, the cast shadow is a scale reproduction of the projection of the object. Since this type of shadow is simply a straightforward silhouette of the subject, it is uninteresting and rarely worth including in the picture. But when the shadow falls at an oblique angle, it becomes distorted—generally in some interesting way—and can make a valuable contribution to the composition. The distortion can be varied still further by curving the background in any direction. Here again the range of effects obtainable extends the power of the photographer to express his ideas in terms of light and shade.

Shadow and Subject. There are generally two surfaces on which the lighting casts a shadow of the subject: the background and the supporting surface—e.g., the table or bench, or the floor on which the subject stands. In some circumstances both these shadows will appear in the picture, in others, only one of them, and both can if necessary be removed by various means.

Shadows on the Background. When the subject is lighted from the front, its shadow will fall directly behind it on the background and the camera will not see it. When the light shines at an oblique angle from the side or above the subject the shadow will fall too far away to be included in the picture area. In all such cases the photographer can ignore the shadow of the subject when he is arranging the background. If there is an unwanted area of shadow cast by oblique lighting, he can make it fall outside the picture area by simply moving the subject away from the background.

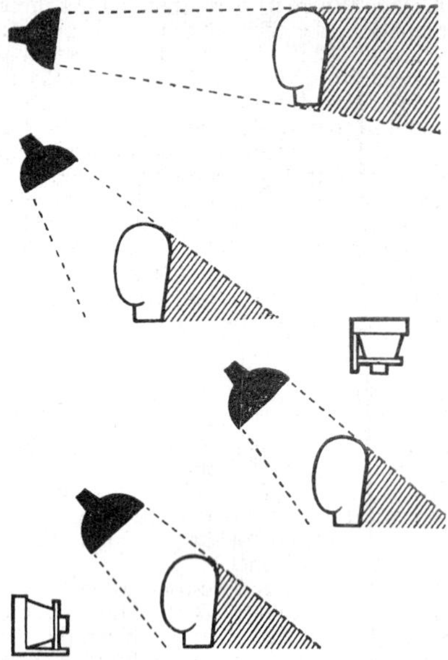

SHADOW POSITION. Top: *With front lighting the shadow is cast directly behind the subject, while top lighting or side lighting moves the shadow down or to one side respectively. This helps to separate subject and shadow.* Bottom: *With a given light position, a level camera position includes less of the shadow in the picture than a high camera angle.*

But in all intermediate lighting positions the shadow of the subject falls inside the picture area. It is part of the picture, and as such, it is subject to all the methods of control described above—i.e., it can be varied in density, sharpness, size and shape. In addition, the shape of the shadow can be altered by turning the subject in relation to the light and the background, and its tone can be made lighter by directing a light on to it from the front or side. If necessary, the whole shadow or only a part of it can be made to disappear by lighting it in this way with a flood or spotlight.

Shadows on the Support. Normally the shadow on the support—e.g., the floor—is not so much in evidence as the part that falls on the background. There are two reasons for this: because the camera sees less of the horizontal surface of the support than of the vertical surface of the background, and for the psychological reason that the shadow on the support is accepted as a natural part of the subject. Everything resting on a surface (as most things do) casts a shadow on that surface and in fact this kind of shadow is more likely to be missed when it is not there than noticed when it is.

When the camera looks down on the subject, the shadow on the support becomes more noticeable and it can be made even more so by appropriate lighting—e.g., low-angle lighting will produce unusually long shadows.

But there are occasions when the shadow is just a distraction—e.g., in catalogue illustrations and the like. To attempt to "kill" the shadow with another light might simply add another shadow on the opposite side of the subject. The correct method, whenever practicable, is to rest the subject horizontally on a sheet of glass supported above a sheet of white paper. This white paper is in effect a background and may be illuminated either by a special light or by the subject lighting. If the subject lighting provides the "background" light then it must fall at such an angle that the cast shadow lies outside the picture area.

This method of supporting the subject also makes it possible to light it from below to get rid of shadow areas in which the detail would otherwise be lost—an important point in illustrations of products for catalogues, advertisements and instruction books.

STANDARD LIGHTING ARRANGEMENTS

Lighting to the photographer is what brush and paints are to the artist and one should be as free to express himself in his own way with his materials as the other. The last thing a photographer should seek to do is to copy slavishly any so-called standardized method of lighting. At the same time, while he ought not to be concerned with any such arrangement as a goal, he can save himself a lot of trouble and experiment by using it as a point of departure from which to develop.

Broadly speaking, the subject lighting may be arranged to reproduce either modelling or texture. In addition there are a number of special effects: silhouettes, shadowless lighting, high and low-key lighting, lighting for the photography of polished surfaces, and sometimes lighting for a given contrast range.

Lighting for Modelling. In normal straightforward photography of people, solid objects, and three-dimensional subjects in general, the aim of the lighting is to convey the depth and roundness of the subject. It does this by making one side of the subject light and the other dark, with a gradual transition from one to the other where the surface is curved. It creates planes of differing tone values in which the nearer parts of the subject contrast with those farther away. As the natural light out of doors shines downwards, the most natural effect is achieved by making the principal light shine downwards on the subject too.

So the basic position for the principal light

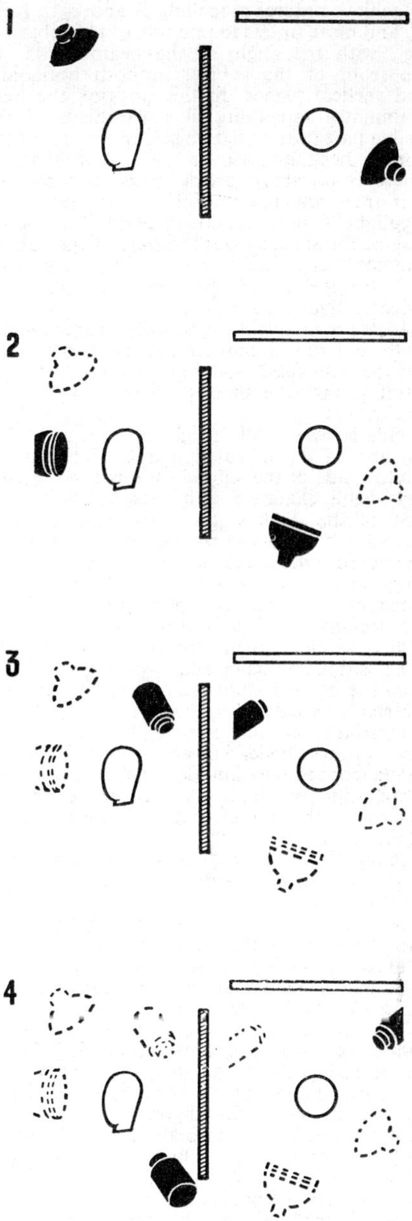

BUILDING UP A LIGHTING SET-UP. 1. The principal light is a floodlamp to one side of the camera, shining down on the subject. This provides the main modelling. 2. A diffused flood is added as a fill-in light from the front and on the same level as the subject. 3. A spotlamp is introduced to light up the background. 4. A second spotlamp acting as effect light (in this case a side-rim light) completes the basic set-up.

to achieve natural modelling is above, in front of, and more or less to one side of the subject—e.g., with the angle of the beam at 45° to the front of the subject in both horizontal and vertical planes. In this position the light illuminates something like two-thirds of the visible part of rounded objects like a ball or the human head and leaves the rest in shadow. There is no sharp dividing line between the two areas and the smooth change over from highlight to deep shadow suggests the roundness of the surface. But the part of the subject that receives no light will be completely black and featureless. In practice, enough stray reflected light usually reaches even the deepest shadows to enable the eye to detect some detail there, but the illumination is much too weak for the sensitized material to record it. The result is that the shadows print out as solid blacks.

This is why a fill-in light is necessary. The purpose of the fill-in light is to light up the shadow side of the subject (but without casting any visible shadows of its own). Raising the level of the shadow illumination in this way reveals enough detail to make the dark tones interesting and narrows the over-all contrast range so that all the tones can register in proportionate densities in the photograph.

Normally the best position for the fill-in light is as close to the camera as possible. In this position it casts no shadows that are visible from the camera station and it illuminates all the shadows cast by the modelling light that are visible from that point. A reflector can be made to serve instead of an actual light source but its scope is more limited and for it to provide enough illumination it may have to be brought so close to the subject that it intrudes into the picture area.

Lighting the Background. At this point the subject will appear naturally lighted with highlights on the top and one side and more or less luminous shadow areas on the opposite side and below. It will look nicely rounded, but the shadow side will tend to be lost against the dark background and in a portrait the hair will tend to look lifeless.

The tone of the background must now be adjusted by adding background lighting. There are no rigid rules about this; the subject may call for a plain neutral tone provided by a broadly dispersed floodlight, or any of the light and shade effects obtainable with a spotlight or masked flood. This stage in the light is dictated by the photographer's mental picture of how he wants his subject to look. Generally the level of the background illumination should be below that of the subject, but there are occasions when a brightly illuminated background and a comparatively dark subject may be more appropriate.

Lighting the background will make the edge of the shadow side of the subject visible and for many subjects the lighting will now be satisfactory. But while the modelling of the subject will be satisfactory up to a point, the general appearance will suggest that it consists of only the front half stuck on to the background—i.e., an orange against a white background would look like half an orange with its flat surface against a white wall.

A final suggestion of roundness and depth can be given by the use of back lighting. The back light is placed so that its rays strike the side of the subject and create a rim of light which is visible from the camera position. (The lens must be screened from the rays of the back light by a deep hood and the light itself must be carefully controlled to prevent any rays from striking the lens.)

The bright rim of light, fading towards the front of the subject, creates an illusion of depth that was previously absent. It pulls the subject away from the background and makes it clear that it occupies a position in space between the observer and the plane of the background.

The lighting set-up now provides a sound system of basic lighting for modelling that will serve for most subjects. By varying the power and character of the lighting the general effect can be varied within wide limits without departing from the basic principles.

Profile Lighting. Where the interest of the subject lies in the outline, there are two methods of lighting that will emphasize it—profile lighting, which shows a bright outline of the subject against a dark background, and silhouette lighting which shows a dark or even black outline against a comparatively much lighter or white background.

Profile lighting is often employed in portraiture where this aspect of the sitter's face is the most characteristic. It is created by a back light so placed that its rays strike the edge of the subject at an angle of about 45° from behind and above. This illuminates a narrow band around one side of the subject and fades into darkness towards the front. The background may be lighted to throw up the shadow side of the subject while leaving the lighted profile to appear against a dark background. Or the whole shadow side of the subject may be left to merge into an unlighted background to emphasize the outline of light around the opposite side.

This lighting technique is effective for nude figures studies.

Although the bright outline is the interesting feature in this type of lighting, it is generally advisable to show something of the rest of the subject. This can conveniently be done by throwing back some of the profile lighting with a reflector facing away from the camera. The same result can be arrived at by using a very subdued fill-in light from the camera position.

Profile lighting is frequently used as an effect in conjunction with straightforward modelling lighting, particularly in glamour portraiture. In this case it may be used from one side or both.

Silhouette Lighting. In silhouette lighting the subject is shown as a more or less solid black shape against a light background. The object here is to illuminate the background brightly, and the subject not at all.

One effective way of doing this is to pose the subject against a thin and uncreased white sheet or tablecloth and make the exposure by firing a flash directed at the back of the sheet, opposite the subject.

The same effect can be produced by lighting a white background with a floodlight placed behind the subject. With this method there is more risk that stray light will be reflected on to the front of the subject.

Here again, the silhouette technique is often combined with other lighting so that while the main lighting emphasizes the significant profile, the secondary lights pick out other features necessary to the final effect.

LIGHTING DIFFERENT MATERIALS

Surface texture is one of the most important and revealing characteristics of many materials, and it expresses itself in the form of uniform unevenness of one sort or another. This unevenness is a manifestation on the surface of the material of its essential grain or structure. A polished surface has no texture in the accepted sense of the word, yet the fact that it can take and retain a polish suggests a dense, hard structure—e.g., glass, chromium plate, diamond. But the rough splintered face of sawn timber speaks directly of its fibrous texture, just as the hairy nap on a piece of Harris tweed or the smooth sheen of a piece of silk conveys a clear impression of the physical nature of the material.

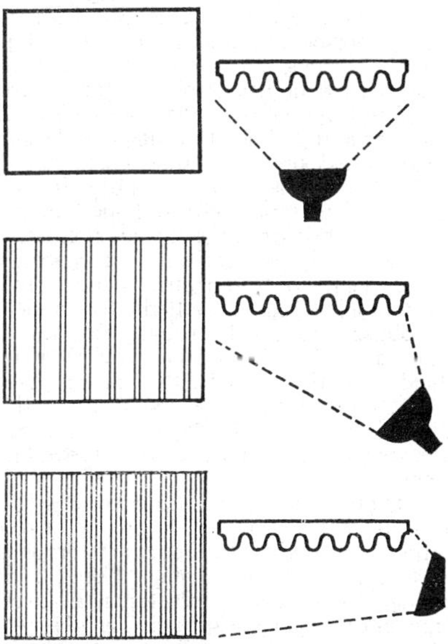

LIGHTING FOR TEXTURE. Top: Very low oblique lighting emphasizes texture by casting pronounced shadows of surface detail and unevennesses. Centre: Oblique lighting shows up texture naturally by providing normal modelling. Bottom: Front lighting illuminates all the shadows evenly, and thus kills the texture effect, making the subject look completely flat.

In many subjects texture is as important as shape and modelling. It is an essential feature in photographs of materials—e.g., fabrics, leather, wood—and it is also vital in portraiture, where the appearance of the skin—smooth and rounded or rough and wrinkled—says so much about the age and character of the subject. The texture of food, fruit and flowers seen close up can evoke many of the actual physical sensations of the real thing.

Broadly speaking, texture is revealed by side lighting and obliterated by frontal lighting. When the light strikes the surface at such an acute angle that it just skims the surface, the tiniest projection will be brilliantly lighted on one side and cast exaggerated shadows on the other. If the light comes from a point source, or from a parallel beam—e.g., from a focusing spotlight—the cast shadow will be sharp and incisive. The surface texture will be emphasized and sharpened by this treatment, whereas, if the light is diffused, the cast shadows will be blurred and the texture softened.

The type and angle of lighting used to bring out the surface texture will depend on the material being photographed. The hard, low-angle lighting that would be quite suitable for a piece of canvas would make velvet look bristly and stiff instead of silky and pliable.

A fill-in light is just as necessary here as in normal lighting. Without some illumination in the shadows, they would appear inky black in the print. This might be an advantage with subjects that call for the greatest possible emphasis of texture, but generally the photograph will be better for some shadow illumination provided by a fill-in light or reflector shining from the camera position. Again, the more powerful the fill-in light the less prominent the texture will be.

One of the problems of reproducing texture arises when it is to appear simply as an extra in a normal photograph. In such cases the modelling and the texture lighting would combine to give an unnatural or positively displeasing effect. Draped materials, for example, would call for careful modelling to show the soft roundness and depth of the folds. To introduce strong cross lighting for texture would defeat the object of the principal lighting.

In such cases the answer is to arrange the main lighting to give the best modelling over most of the subject and then to indicate the texture by a suitably placed rim light. Where the near margin of the rim lighting just begins to fade out, the rays glance off the surface of the subject at an extremely acute angle and emphasize the texture strongly. When this technique is applied to a sufficiently strong point in the picture, the eye accepts the suggestion of texture as applying to all the subject.

Polish. Strictly speaking, polished surfaces show nothing of the texture of the material, but if the polish is an essential feature of the subject, it must be reproduced in the photograph. Polish is indicated by light reflections, but there is nothing gained by allowing extensive areas of the surface of the subject to be taken up with uninteresting patches of light toned reflections or distorted images of the light sources. A few odd catchlights are all that is necessary to show that the surface is polished. With many subjects considerable skill and a measure of compromise are required to reduce the surface reflections to a catchlight or two. The effect can be judged only by examining the image on the focusing screen or by viewing the subject from the same position as the camera lens. It may be necessary to move the principal lighting to get rid of major reflections and introduce a small, carefully controlled spot to provide the catchlight on the polished surface.

When a polished surface has a pattern—e.g., the grain of wood—reflections mask the pattern and are therefore undesirable. Moving the lights to get rid of the reflection may produce bad subject lighting, and any of the tricks of dulling the surface would obliterate the pattern or marking.

In these circumstances the answer may be to use a polarizing screen in front of the lens. This can be rotated until the unwanted reflections (which consist mainly of polarized rays) are absorbed; the surface markings will then be visible. A polarizing screen has no effect on light reflected from metal surfaces, however, since these rays are not polarized when reflected unless the light source itself is polarized.

Sheen. Many materials have a glossy surface combined with a rough or irregular texture—e.g., skin, silk, ornamental leather, feathers, and fur. In such materials the polished surfaces are neither large nor flat enough to produce directional reflection, but they combine to give a sheen or gloss to the material. This sheen is a characteristic that must be preserved in the photograph but it cannot be captured by texture lighting. Oblique lighting emphasizes the physical texture but falls at the wrong angle to reflect the glossiness. On the other hand, a broad soft flood of light from a high, frontal angle will give plenty of reflection but tend to flatten the physical texture of the surface. If it is necessary to reproduce both, the answer is to arrange the principal lighting to emphasize the sheen and a rim light to suggest the texture.

Translucence. Some materials, while they are by no means transparent, will allow light to disperse through them—e.g., flesh, hair, many of the plastics, flower petals and leaves. This translucence can only be revealed by rays travelling towards the camera. Such rays strike the subject and penetrate the surface, lighting it up with a luminous glow that radiates in the direction of the camera. When the lights fall on the front of the subject this glow is invisible and the material looks dead.

Materials. The standard techniques employed for rendering the texture of a number of common materials are described briefly below.

Damask and patterned materials: normal lighting to give modelling to the folds with a contrast filter for the pattern.

Flowers, foliage, fruit: soft, normal lighting with effect lighting from behind and above to show translucence and add rim light to pick out shape and texture. Opaque subjects are best lighted from the side with plenty of soft fill-in lighting.

Glassware: (1) for catalogue-style illustrations, normal lighting using a point source—e.g., a carbon arc spot—as the principal light to give sharply defined shadow. The shadow must not be exaggerated and a careful watch must be kept for excessive reflections and catchlights.

(2) Experimental treatment making use of exaggerated patterns of shadows and highlights created by refraction in the glass. The background or support on which the shadows are cast needs to be close to or touching the subject. Here again much can be done with point source illumination.

(3) Silhouette treatment created by photographing the subject against a lighted background. The effect is more delicate if the subject is supported horizontally on a sheet of glass above the "background" and it may be further elaborated by the use of effect lights to create highlights brighter than the background tone while the main outline of the object is darker.

Leather: (1) smooth finishes—e.g., calf—are best lighted with a combination of soft normal lighting and as much hard oblique lighting as is needed to bring out the surface texture.

(2) Grained, glossy finishes—e.g., crocodile—call for plenty of soft frontal light to reproduce the shine plus a spotlight for modelling the object. Grained, rough finishes like natural crocodile, lizard and snakeskin may be lighted normally, to show the natural pattern, or cross-lighted to show the texture. If the pattern is to show clearly, there must be no shine visible on the surface when viewed from the camera.

(3) Hide: the rough-textured "pigskin" finish is best reproduced by point-source lighting from slightly to the rear (to emphasize the texture) and as much diffused frontal lighting as is necessary to register the over-all shape and form of the subject. The glossy type of finish can be given similar treatment except that the texture lighting should be softer to give more prominence to surface reflections from the rounded pattern than to the shadows. An effect light—usually a small spot—will generally be necessary to add life and modelling.

Its position will depend upon the shape of the subject which might be a pair of gloves, a handbag, or a suitcase.

(4) Suède: this must be carefully brushed in the direction of the pile and then given hard lighting for texture with a point source—e.g., a carbon arc—in combination with softer general lighting to show the shape of the subject and its basic feeling of limpness.

Tweed: this calls for contrasty oblique lighting—e.g., from a carbon arc spot—to pick up the texture and some soft frontal light to show the pattern. If the material is draped, it may be necessary to suggest the texture with rim lighting and devote the principal lighting to modelling the folds and recording the pattern.

Silk: a soft flood of frontal light will bring out the shimmer and gloss of the material, but an effect spotlight should also be added to introduce life and modelling and prevent the photograph from being flat and humdrum.

Paper: the lighting will vary according to the type of paper and the reason for photographing it: backlighting for translucent papers; cross lighting for rough textured papers; frontal lighting for patterned papers.

Pottery: the unglazed variety is best given normal but hard lighting to bring out the essential roughness of its texture. An effect spotlight can be usefully included to add a highlight. Glazed pottery needs soft lighting with no reflections, except possibly a catchlight, otherwise the beauty of the pattern will be obscured.

Delicate china should be given backlighting to reproduce its translucence with enough frontal light to show the shape and pattern.

Silver and polished metal ware: this generally calls for a very special technique. The basic requirement is a soft and widespread source of illumination which will form no definite reflections on the surface of the subject. The light of a flood may be "bounced" off a large sheet of white paper; the subject may be set up in a muslin tent illuminated from outside by one or more floodlights while the camera peeps in through a hole. There are various tricks for dulling the surface—e.g., with putty—to kill reflections, but these can easily destroy the real character of the metal.

Wood: wooden objects are usually finished smooth or even polished, so they have no real surface texture.

At the same time, however, the principal object of polishing is to reveal the beauty of the grain, so the grain must be reproduced in the photograph. Normal lighting for modelling is satisfactory, but the rendering of the grain can be improved by using a colour filter in front of the lens. The colour filter should never be deeper than the palest tint of the wood—i.e., a pale yellow for such woods as light oak and sycamore, a deep yellow or orange for walnut, and a light red for mahogany.

CONTRAST AND EXPOSURE

When the lighting on the subject has been arranged to give the required effect, other photographic considerations must be made; these are principally concerned with subject contrast and exposure.

Contrast. In taking pictures by artificial light, the subject contrast is completely under the control of the photographer. His principal light sets the level of the highlight illumination and his fill-in light fixes the level at the other end of the scale. It is not enough to build up a lighting scheme that gives the desired visual impression; he must know what is going to happen to the highlights and shadows when they have been translated into the corresponding dark and light grey tones of the print. How they will reproduce will be governed by his handling of the successive variables of negative exposure and development and the contrast grade, exposure and development of the print.

All this points to the need for careful records of the lighting arrangements and rigid control and recording of all the subsequent stages. Without a great deal of experience in handling subject lighting it is impossible to predict the final effect of any lighting set-up. The professional can acquire the experience fairly quickly; the amateur can avoid a lot of wasted exposures by making detailed notes of each set-up and its results.

Once the lights are in position, the contrast can be controlled by moving them closer to the subject or taking them farther away. As a starting point, the principal lighting should be three to four times as bright as the fill-in lighting. If both sources are of equal power and are broadly diffused, the fill-in light should be twice as far from the subject as the principal light. For colour photography, the fill-in light needs to be closer to give low contrast.

This basic arrangement assumes that the photographer wants to cover all the tones of the subject from the deepest shadow to the brightest highlight in an evenly graded black to white scale of tones in the print. If, however, he wishes to bunch all the deep greys together into a mass of solid black to create a special effect, while allowing the middle and light tones to reproduce normally, he can move the fill-in lamp farther away and put the subject contrast outside the range of the sensitized material. Or a similar effect may be achieved by underexposing slightly. He will then finish up with a print in which all the feebler subject brightnesses were too weak to register on the negative and thus print out dead black.

With the same lighting contrast, a longer exposure would allow the shadow detail to record normally but clog up the highlights so that they would reproduce as blank white paper. This effect is never deliberately sought as clear white paper has none of the aesthetic possibilities of equally blank dark areas.

Correct adjustment of the contrast range established by any lighting set-up is just as important as the placing of the lights; unless it is deliberately planned, the appearance of the final print will be a matter of guesswork.

Exposure. There is no one "correct" exposure for any given subject and lighting arrangement. It is true that if the contrast of the lighting is adjusted so that it exactly corresponds to the gradation of the sensitized material, only one exposure value will give a faithful representation of the subject brightnesses in the print. But very often the photographer does not want a faithful representation. He may wish to compress the dark or the light tones; he may wish to produce a high or low key picture. The photographer has at his disposal a number of variables which enable him to control the appearance of the final print. Exposure is one of these and it is no more possible to lay down hard and fast rules about it than to tell a painter how he should apply his colours.

The only really satisfactory answer is to use a photo-electric exposure meter and take readings on or at the parts of the subject that are to appear fully exposed in the negative. If the subject is more or less even in tone all over, a reading from the camera position will be satisfactory, but if the important tones occupy only a small part of the picture—e.g., the face in a portrait—then the meter must be taken right up to the subject so that it sees only that part.

If the separate readings of the required highlight and shadow luminosities show that the range is too great for the material, then the lighting must be adjusted to decrease the contrast.

Frequently, however, the contrast will lie well within the range of the material, and there will in that case be a range of exposures that will give proportional reproduction of the tone values of the subject.

The normal type of photo-electric meter may not be sensitive enough to give a reading on the shadow areas even if the meter is fitted with a low-light scale. When this happens, the answer is to use a highlight reading as a basis for calculating the exposure. If, for example, the contrast range is known to be short, then the highlight reading can safely be used without risk of under-exposing the shadows.

While actual measurement with an exposure meter is the only reliable method of arriving at the exposure, there are various published tables that enable a fair estimate to be based on the set-up data.

The factors that have to be considered in using the tables are:

(1) Subject: light or dark; colour.

(2) Room: light or dark walls and ceiling; small or large—i.e., reflecting surfaces close to or far away from subject.

(3) Lights: number, power, types of reflector, distances from subject and angles at which light strikes subject; colour temperature (for colour photography).

(4) Sensitized material: speed; colour sensitivity; latitude.

The tables give an arbitrary figure to each of these factors and the total figure for each lighting set-up, subject and sensitized material corresponds to an exposure value stated in the final column.

A condensed table of this type is given below. It gives exposures for a single Photoflood bulb (or equivalent 500 watt light source) in a wide-angle, matt surfaced reflector, for fast panchromatic film. The table refers to subjects of average brightness; exposures should be doubled for dark subjects and halved for light ones. It assumes that the lamp is used in the basic modelling position—i.e., 45° above, and on the side-front of the subject with a fill-in light or reflector directed from the camera position.

SHUTTER SPEEDS

Distance of Lamp from Subject (feet)	*Aperture*							
	2	2·8	4	5·6	8	11	16	22
3	1/400	1/200	1/100	1/50	1/25	1/10	1/5	½
4	1/200	1/100	1/50	1/25	1/10	1/5	½	1
6	1/100	1/50	1/25	1/10	1/5	½	1	2
8	1/50	1/25	1/10	1/5	½	1	2	4
11	1/25	1/10	1/5	½	1	2	4	8
16	1/10	1/5	½	1	2	4	8	16
22	1/5	½	1	2	4	8	16	32

The speed for any other lamp can be calculated from its effective light output—e.g., with two Photoflood bulbs the exposure would be halved—i.e., from 1/25 to 1/50 second.

Effect of Inverse Square Law. Under certain ideal conditions it would be possible to say what change should be made in the exposure if the lights were moved closer to or farther away from the subject. According to the inverse square law, moving a lamp to double the distance calls for a four times increase in exposure; to three times the distance, a nine times increase, and so on. But in practice it is not possible to apply this rule rigidly.

First of all, the law applies only to light radiating from a point source—like the spokes of a wheel. It does not apply to the haphazard bundle of light rays thrown out by a pearl Photoflood bulb in an arbitrarily curved reflector that may be either matt, polished, or something in between. With such a light source it is possible for a proportion of the rays at either the centre or around the edges of the beam to remain more or less at the same intensity as the distance is increased while the rest of the rays fall off in intensity according to the law of inverse squares. Everything depends on the shape of the reflector, the type of bulb, and the position of the bulb in the reflector.

Then the law takes no account of reflections from other surfaces, and in practice these can contribute a great deal to the illumination of the subject. When a lamp in a reflector is close to the subject all its light may be concentrated in a hard patch of light, leaving much of the subject only weakly illuminated. If the same lamp is moved twice as far away, a proportion of the rays will then fall on the surroundings. If these are light in tone and happen to be suitably placed, they may reflect more light than before on to the parts that were only weakly illuminated. Under these circumstances it might not be necessary to give the exposure anything like as much as the four times increase demanded by the law of inverse squares.

Finally, with a spotlight focused to give a parallel beam, the law of inverse squares does not apply at all. If the beam is truly parallel, it will have the same intensity, and call for the same exposure, at 50 feet as at 5. In fact, a large diameter spot focused to give a converging beam would, up to a point, call for a decrease in exposure as it was moved farther away from the subject. This ignores though, any light loss due to dust in the air.

All this points to the impossibility of laying down reliable rules for calculating the exposure required with any given set-up. The professional worker does have a chance to standardize certain lighting arrangements and find out by trial and error the correct exposure for each. By constantly using the same lighting equipment, he can usually judge correct exposures merely by experience. But for the amateur the most reliable guide is a photo-electric exposure meter and a book of systematic notes. F.P.

See also: *Against the light; Colour technique; Flash technique; Lighting equipment; Portraiture; Tricks and effects.*

Books: *Lighting for Photography*, by W. Nurnberg (London); *Lighting for Portraiture*, by W. Nurnberg (London); *Pictures by Indoor Lighting*, by H. van Wadenoyen (London); *Camera Portraiture*, by A. Williams (London).

LIGHTNING. Flashes of lightning are not difficult to photograph. The light of the flash is bright enough to give a good image on a medium-speed film with an aperture of *f* 11. This means that a box camera can be used as well as any other type.

The camera is simply supported on a tripod and aimed at the part of the horizon where the flashes are most frequent. The shutter is set at B, held open until a flash occurs, and then released. It may also be left open long enough to record a number of flashes, but in this case each successive flash adds to the illumination of the scene generally. The landscape and clouds show up more and more clearly, and cause the lightning itself to lose intensity by comparison.

Photographing from an open bedroom window will provide a skyline of rooftops in semi-silhouette, a fitting setting for the subject. If there are distant spires or trees, etc., which can be included, so much the better.

The wider the aperture of the lens, the more the details of the scene are reproduced. It is generally an advantage for the picture to show something of the landscape and clouds, for which reason it is advisable to use an aperture of about *f* 6·3 on medium or fast pan films, although a very much smaller stop will reproduce the lightning. In principle, however, most of the scene will appear as a silhouette, so the camera should be pointed at something which has a characteristic outline.

LIGHT SOURCES

The sun and the moon are the only natural light sources normally employed by the photographer. Moonlight itself is of little practical value as a light source, and is only of use for special effects. The sun is, for general photography, the most important and widely used of all light sources.

The sun is not considered in this survey of light sources since the technical considerations of natural light are sufficiently vast to be dealt with elsewhere, separately, in a number of different contexts. Since the sun is also the only useful natural light source, no alternative choice of lighting can be considered.

Apart from the sun (and sometimes the moon) the other light sources the photographer has available are those created artificially by electricity or combustion. Electric light sources are usually the more convenient to use, although at times non-electric sources of lighting may be necessary or offer some special advantage.

History. From the earliest days of photography attempts were made to take pictures by artificial light in order to shorten the exposure, and to enable portraits to be taken on dull days. The first artificial source employed was oxy-hydrogen light (limelight), used by L. L. B. Ibbetson of the Royal Polytechnic Institution, London, in 1839–40 for daguerreotypes taken with a microscope. Limelight proved too harsh for portraiture. The same applies to John Moule's Photogen patented in 1857, which burned a pyrotechnic compound, with a screen of blue glass to protect the sitter from the glare.

A more satisfactory source was magnesium wire or ribbon, the actinic qualities of which were observed by Prof. Robert Wilhelm Bunsen of Heidelberg and his assistant (Sir) Henry Roscoe, 1859. The first portrait by this light was taken on 22 February 1864 by Alfred Brothers of Manchester, the sitter being Roscoe. Brothers also demonstrated its use in a Derbyshire coalmine soon afterwards. Flash-light, or magnesium in powder form, first proposed by W. White in 1865, only came into use in 1886, but smoke formation was a great drawback. In 1892–3 Chaufour designed an electric photoflash lamp of coiled magnesium wire in a glass globe, ignited by a Bunsen battery. The disadvantage was that the globe frequently burst through the expansion of the oxygen in it. This was avoided in Dr. Paul Vierkoetter's vacuum flash-bulb, 1925. The modern safety flash-bulb was born, and has been in general use since 1929 when Johannes Ostermeier substituted aluminium wire or foil for magnesium, in oxygen in a sealed bulb.

Some of the micro-daguerreotypes of medical specimens taken by Léon Foucault and published as engravings by Alfred Donné in 1845, were illuminated by a carbon arc lamp. Fox Talbot demonstrated photography by the flash of a Leyden battery in 1851. For portraiture, the carbon arc lamp powered by a Bunsen battery was experimented with by Gaudin and Delamarre in France 1854. Levitzky in St. Petersburg and Nadar in Paris installed this lighting system in their studios in 1857 and 1858 respectively. Nadar photographed the sewers and catacombs beneath Paris by the same light system in 1861–2—the first photographs ever taken underground. The first portrait studio in the world using electric light exclusively was opened in London in 1877 by Henry van der Weyde, an American. He used a 6,000 candle-power carbon arc powered from a dynamo, and a lighthouse lens in the 4 ft. reflector. Public electricity supply did not become generally available until the early years of the twentieth century.

A few provincial studios used gas lighting augmented by reflectors, one of the first being P. Maitland Laws of Newcastle-on-Tyne, *c*. 1880. The heat generated limited its use.

H. & A.G.

ELECTRIC

Five main types of electric light source are used to provide photographic lighting: tungsten filament, arc, vapour discharge, fluorescent and flash discharge. Although flash bulbs are fired by an electric current, the light they give out is produced by combustion of a magnesium mixture in the form of foil, wire, or paste, so they do not rank as normal electric light sources.

The principal variables in electric light sources are: consumption (measured in watts), total light output (measured in lumens), efficiency (measured in lumens per watt), colour temperature (measured in Kelvin degrees) and life (measured in hours).

Tungsten Filament. The light emitted by a tungsten filament lamp is produced by passing an electric current through a tungsten filament enclosed in a glass envelope (usually filled with an inert gas). The wire is heated to incandescence by the passage of the current. Light output and colour temperature depend upon the design of the filament and the applied voltage. Several types are used in photography.

First there is the ordinary domestic lamp normally supplied with a bayonet cap and a frosted or pearl bulb in powers up to 150 watts. Filament lamps of higher power are supplied with screw caps and the glass envelope is normally too large for domestic fittings.

Then there are abnormally bright lamps of the domestic type in which the extra light is obtained by over-running the filament—i.e., by applying the usual domestic supply voltage to a filament which would normally be run on a lower voltage. The result is a lamp that gives an abnormally bright light but has an abnormally short life. These lamps vary in performance and life according to the degree of overloading.

The third type of filament lamp is designed for use in projectors and spotlights. In such lamps the filaments are designed to conform to the optical requirements of the projection system and may be bunched to provide an approximate point source, or spread out to give a flat, uniformly bright grid. Size, consumption, output and method of connecting vary widely according to the type of equipment in which they are used.

A further and more modern kind is the quartz-iodine lamp which contains the filament in a narrow quartz (or high-melting glass) tube together with a trace of iodine vapour. This type of lamp may be over-run and yields a light of constant intensity and characteristics throughout its burning life (which is longer than that of normal filament lamps).

Filament lamps are suitable for both A.C. and D.C. supplies.

Arc. There are several types of arc lamp, the light in all cases being produced by the passage of an electric current across a gap between two electrodes.

The commonest type is the carbon arc lamp, with electrodes made of carbon which may be either open to the air (open type) or almost completely enclosed in a glass globe (enclosed type). The carbon electrodes may be of plain carbon run at a low current density (low intensity arc) or cored with certain metallic salts and run at a high current density (high intensity arc). Carbon arcs are used in spotlights, floodlights, and particularly for film studio lighting where the near-daylight colour temperature of the high intensity arc is welcome.

There are both D.C. and A.C. carbon arcs.

Another type of light source based on the arc principle is the concentrated source in which the current passes across the gap between two

ELECTRIC LIGHT SOURCES COMPARED

Light Source	Watts Consumption, up to	Total Light Output in Lumens, up to	Efficiency, Lumens per Watt	Colour Temperature °K	Average Life (Hours)
Filament					
Domestic type	150	2,000	13	2,600–2,800	1,000
Over-run	500	18,000	36	3,400	2–6
Over-run for colour photography	10,000	360,000	36	3,100–3,200	10–50
Quartz-iodine	10,000	220,000	22	3,200	2,000
Quartz-iodine (over-run) ...	1,000	34,000	34	3,400–3,450	15
Projection	1,000	27,000	27	3,100	50
Arc					
Low intensity carbon ...	2,000	30,000	15	2,700–3,000	—
High intensity carbon ...	14,000	420,000	30	5,400–6,400	—
Concentrated gas arc ...	250	10,000	45	3,400	—
Mercury and other Vapour Discharge					
Cooper-Hewitt	266	6,500	24	5 bands, 3,650A. to 5,791A.	1,000 or longer
Medium pressure (4 atmos.)	240	10,500	43	Similar to Cooper-Hewitt but with inclusion of some red rays	
High pressure (75 atmos.)	1,000	65,000	65		
Sodium	140	10,000	70	—	2,500
Fluorescent	125	5,750	46	3,700–4,800	1,000

No comparative figures are possible with flash.

metallic beads enclosed in a glass bulb containing mercury vapour. In a third type the arc takes place in an argon-filled bulb and the light is emitted by a coating of molten zirconium on one of the electrodes. These point sources are highly efficient and are used in certain types of projector and in scientific optical equipment. They are manufactured for both A.C. and D.C. supplies.

Vapour Discharge. Vapour discharge lamps are based on the fact that an electric current passed through certain gases at reduced pressure causes them to emit light. The wavelength of the light emitted depends on the nature of the gas, and the light is produced by a phenomenon known as resonance vibration which is not necessarily accompanied by the generation of any considerable amount of heat (as occurs in filament and arc lamps).

The basic lamp of this type is the Cooper-Hewitt mercury vapour lamp, usually consisting of an evacuated glass tube (up to 6 feet long) with an electrode at each end and a small amount of mercury to supply the vapour. These lamps give an intense blue-violet light which is very actinic to blue-sensitive emulsions and is particularly suitable for blockmaking and for printing on document materials.

There are also medium and high pressure versions of this type of lamp. These tend to have a higher efficiency and are much smaller than the low pressure Cooper-Hewitt type.

The design of high pressure mercury vapour lamps varies widely; they are used in many fields of photography from small projector lamps and spotlights to high power film studio spots and floods. As the light frequencies generated are restricted to one or two sharply defined bands, they are useless for colour photography. The physical characteristics, contacts and supply requirements of such lights vary as widely as the applications.

In addition to mercury, other substances, notably sodium, are used in vapour discharge.

Sodium lamps are familiar as the bright yellow street lamps but they may also be used as photographic lighting. Other gases, such as neon, give light which has applications in fields like advertising display, but is not used in photography.

Fluorescent. Fluorescent light is simply a development of the vapour discharge principle. The fluorescent lamp is a vapour (generally mercury) discharge lamp coated with a substance that fluoresces under the action of the rays emitted by the vapour. The lamp is mostly manufactured in the form of a tube with a contact cap—usually bi-pin, bayonet or raised contact—at each end. The colour temperature of the light depends on the nature of the coating materials; most manufacturers list three colours—natural, daylight, and warm-white. A special colour-matching shade is also available.

Flash Discharge. Electronic flash tubes are another special application of the vapour discharge principle. The tube in this case is normally only a few inches in length and it is filled with a mixture of rare gases—e.g., argon, krypton or xenon. The discharge is of extremely short duration and the energy dissipated is correspondingly high; the result is a brilliant flash of light. The energy is provided by building up a charge of electricity in a condenser connected across the terminals of a main supply or a suitable battery. Some tubes are designed to flash as soon as the voltage is applied; others are permanently connected across the supply terminals and flashed by applying a "triggering" voltage.

The brilliance of the flash depends on the type of tube and the quantity of electricity discharged through it from the condenser. The output is measured in watt-seconds or joules and may range from less than fifty to many thousands.

As the light is cold, intensely bright, and almost identical in colour temperature with daylight, it is widely used for flash photography, not only of moving subjects, but even for serious portraiture, copying, photomicrography, underwater photography and medical and dental photography. F.P.

Colour Characteristics. Tungsten filament lamps range in colour temperature from 2,600 to 3,400°K. The over-run lamps for colour photography—3,400°K—are suitable for type A artificial light reversal colour film. Standard studio lamps for colour photography burn at 3,200°K and are used with type B or tungsten artificial light colour film. Quartz-iodine lamps exist in both kinds, the 3,400°K ones being over-run to a greater degree than those burning at 3,200°K.

Ordinary household lamps, with their low colour temperature, are not really suitable for colour photography except with very heavy compensating filtration.

Most other light sources (arcs, discharge lamps, fluorescent lamps, electronic flash) have no continuous spectrum of uniform or uniformly varying energy distribution; certain spectral bands tend to predominate. Where the source has an underlying emission in all the main spectral bands, the colour quality can be approximately specified by an equivalent colour temperature rating. With most colour films (but not all) such sources will give results similar to those obtained by exposure by a light of the appropriate colour temperature.

NON-ELECTRIC

Non-electric artificial light sources are used in photography either because no electricity supply is available, or because of their special characteristics.

In the first case they can serve to illuminate the negative in an enlarger, the slide in a projector, the subject to be photographed, or they can be used as the light source in contact printing. In the second case they may be used because of certain technical advantages they possess.

Any artificial light source can be used for photography providing sufficient exposure is given.

Most non-electric sources are inconvenient, however, and in practice the photographer who has no electricity available only reverts to gas or pressure oil-lamps, or one of the other sources described below, because of their occasional usefulness.

PRINCIPAL NON-ELECTRIC LIGHT SOURCES

Source	*Main Application*	*Equivalent Electric Lamp Wattage (Panchromatic Material)*
Gas:		
No. 1 Mantle	General	25[1]
No. 2 Mantle	(studio lighting;	40[1]
No. 3 Mantle	enlarging;	50[1]
No. 4 Mantle	projection)	75[1]
Acetylene:		
3-jet burner	Enlarging and projection	About 200 Depends upon jet size and gas pressure[2]
Oil:		
Wick	Photomicrography	½-in. wick = 4, 1-in. wick = 10, large circular wick = up to 40[3]
Incandescent	General	60
Pressure	General	150–200
Candle	Special effects; calibration of photometer exposure meter	1[3]
Magnesium ribbon	Still-life; contact printing; night photography	250
Flash bulbs	Night photography; illumination of shadow areas in sunlight, etc.	Depends on bulb size. Average over duration of flash—30–200 kilowatts

[1]Varies with make and age of mantle.
[2]About two-thirds of the wattage of electric lamp giving same visual illumination.
[3]Very deficient in blue.

Gas. By this is generally meant coal-gas, consisting almost entirely of a mixture of hydrogen, methane and carbon monoxide, all of which burn with a bluish non-luminous flame. To obtain a good light the gas flame is made to heat a fragile fabric of metallic oxides (mainly thoria). This "mantle" is heated to incandescence and gives out a light approaching that of an electric lamp. The amount of light given out depends upon the size of the mantle, its composition and age, and on the gas supply.

A mixture of butane and propane is sold in cylinders for isolated houses, caravans, etc., and it behaves very much like coal-gas.

Acetylene. This is a gas unlike coal-gas in that it gives a brilliant light when burnt as a simple jet without the need of a mantle. Acetylene is normally generated by dripping water on to calcium carbide in a specially designed container and hence is a useful illuminant for portable lamps. It is little used nowadays, but some country houses have been piped for its use.

The arrangement for enlargers or projectors generally comprises three jets formed as a unit. Mixed with oxygen, acetylene gives an intensely hot jet which was used to heat a block of lime to produce genuine lime-light. This brilliant but temperamental light source was at one time popular in "magic lanterns" and for theatre stage spotlighting. It went out of favour when the more convenient electric arcs and filament lamps were introduced.

Oil-lamps. These take three forms: simple wick, incandescent, pressure.

The first utilizes a wick up which oil is drawn by capillary attraction and is vaporized by the heat of the flame. Carbon particles ("soot") rising from the wick in the updraught separate out by breakdown of the hydrocarbon (oil) and are then rendered incandescent—thereby providing the light source. This source is extremely deficient in blue.

The second type is designed to mix enough air with the oil vapour to give complete combustion. The flame is hotter, but no longer luminous. It has a mantle added and this becomes incandescent in the heat of burning oil vapour in a similar manner to the gas mantle.

The third type also utilizes a mantle but it is heated by gas formed from oil vapour passing through tubes preheated by the heat from the mantle flame and under pressure obtained by manual operation of an air-pump attached to the oil reservoir. In principle it is the familiar blow-lamp with the addition of a mantle.

Candle. The candle, once burning has started, functions like a small wick-type oil-lamp. The paraffin wax of which it is composed melts and the liquid is drawn up the wick to burn with separation of carbon which glows and gives out light.

Magnesium. This is an inflammable metal which is rather difficult to light due to a thin tough coating of oxide which forms on it when stored, but once lit it burns with a great heat and vivid blue-white flame more akin to the light given by a Photoflood than by a normal electric lamp. A cloud of dense white smoke (magnesium oxide) is produced; this is harmless but messy. There are also various formulae for flash powders which incorporate powdered magnesium and other chemicals.

All the above sources require oxygen to function and adequate ventilation must be provided both to supply the needs of the source and to remove fumes.

Flash bulbs. Although these are fired electrically, they can be classified as non-electric light sources because the flash is caused by combustion. The mechanics and technique of flash are, however, a normal aspect of photography.

APPLICATIONS OF NON-ELECTRIC SOURCES

Many of the non-electric light sources can be used in photography, and in specialized cases may be preferable.

Contact Printing and Projection. Contact printing may very conveniently be carried out by household incandescent oil-lamps, gaslight or any of the other sources listed above. The use of such sources presents no particular problems in contact printing; gaslight papers, now better known as contact papers, were so named because gaslight was the commonest printing light at one time.

With one or two exceptions, vertical enlargers are made for use with electric light only and cannot be adapted very easily for non-electric sources. Horizontal enlargers are easily adapted. Most models of this latter type have a roomy lamphouse sufficiently well ventilated to take illuminants such as incandescent gas, a pressure oil lamp, or an acetylene burner. A diffuser of ground glass or flashed opal must be inserted between the light source and the condenser. This prevents an image of the source from appearing in the print and avoids the need for adjusting the position of the light each time the degree of magnification is altered.

A horizontal enlarger of this design may also be used for projecting slides. The diffuser can with advantage be omitted in this case as the human eye is not very critical of uneven illumination and the degree of magnification is not so frequently changed. The light source is adjusted by first roughly focusing the projected image, next removing the slide and centralizing the light, and finally moving the source backwards or forwards to give even illumination.

A special projector for miniature slides and film strips is made by at least one manufacturer. Two models are available to take, respectively, one of two well-known pressure oil-lamps. The lamp is instantly removable when required for ordinary lighting purposes.

Subject Lighting. When taking photographs by the light from non-electric sources, exposure is the major problem. Photo-electric exposure meters can be relied upon when using panchromatic material—if the lighting is strong enough to produce a reading. Quite often, though, even the incident light method of taking a reading will not give a satisfactory indication, or the lighting may be so unevenly distributed that the reading of an integrating meter is meaningless. The correct exposure must then be found experimentally.

Orthochromatic materials are not very suitable for work with these light sources, owing to the low colour temperature which leads to an appreciable speed loss in emulsions not sensitive to red.

With coal-gas lighting, the usual exposure calculating tables may be used. A single burner fitted with a No. 1 (bijou) mantle is equivalent to a 25 watt electric lamp; a No. 2 (medium) mantle should be rated as 40 watts and a No. 3 mantle as 50 watts.

Exposures using incandescent oil-lamps may be similarly calculated on the basis of 60 watts per lamp, or for the more powerful pressure type, 150–200 watts.

Magnesium ribbon burns with a very actinic light and two pieces about four inches long, burning at 6–8 feet from the subject in an average size room, enable $f\,8$ to be used for the approximate burning time of $\frac{1}{2}$ second.

Owing to the relative inflexibility of non-electric lighting systems, advantage should be taken of reflectors. A large mirror or white sheet can supply, in effect, a second light source. Reflector fittings such as are used for electric lighting cannot readily be contrived, but mirrors hung or stood behind the source can increase the useful illumination on the subject.

Street Lighting. Some street lighting is still provided by incandescent gaslight, but even within the circle of light cast by the lamp, exposures of 10 seconds at $f/2{\cdot}8$ (five minutes with a box camera) will be required. Lamp-posts in narrow streets and alleys can make good night pictures, especially if a figure is included and there is a wet pavement to add interesting reflections. To obtain an adequate outline of the surrounding walls an exposure of one minute at $f/2{\cdot}8$ will be necessary. Anti-halo films or backed plates are essential, and a lens hood must be used to exclude all unwanted direct lights. A coated lens is a definite advantage.

If the photographer can look down on such an alley, the top shade of the lamp will screen the light source and reduce flare and halation.

Equally interesting pictures can be taken with relatively short exposures by using the twilight to illuminate the shadows. For example, an effective picture of a lamp-lighter adjusting a lamp could be obtained at dusk with ½ second exposure at $f/2{\cdot}8$. Alternatively flashlight may be used to help out the street lighting so long as it is relatively subdued and does not betray itself by producing additional shadows of its own.

Firelight. A group sitting near a camp fire may be photographed by the light of the fire using a short time exposure. The size of fire, its rate of burning and the distance of the group from it may all vary so widely that it is impossible to suggest exposure times; but as there is no exact criterion for the amount of detail required, the exposure is not critical. It is best not to include the flames in the picture, although a much smaller fire arranged nearer the group may be safely included. With a ring of people around a bonfire, a very effective picture can be obtained by throwing a little flash powder—wrapped up in paper—into the fire.

The camera should be shielded from the direct flash by the people around the fire.

Weak Light Sources. Matches, candles, etc., can be used successfully in making a picture, but the occasions when they constitute the only source of photographic lighting are rare.

With one candle at 15 ins. from a person's face, and a second at 3 feet distance, an exposure of 20 seconds at $f/2{\cdot}8$ is ample. And since the intensity of light varies inversely with the square of its distance from the source, the exposure may be shortened by bringing the light closer to the subjects. With two candles at about 12 ins. from the subject, a well-exposed negative can be obtained in 1 second at $f/2{\cdot}8$ (30 seconds with a box camera). Screens must be used if the candles are not to appear in the picture; if included, they must be close to the subject or they will appear unnaturally bright by comparison with the illumination of the subject.

A lighted match held at the end of a cigarette will light the face characteristically and produce an interesting portrait. The match should be struck upwards and the camera shutter opened immediately the match is at the cigarette end and while the actual head is still burning. Both the subject and the camera must be kept steady while the match head flares up; the shutter should then be closed. An aperture of $f/4{\cdot}5$ should be used with a fast panchromatic film.

An excellent detailed picture of an ordinary open type fireplace may be obtained by the light of the burning coals in about half an hour at $f/3{\cdot}5$.

Ultra-speed Films. All the above exposure suggestions are based on a medium speed film of 80 to 160 ASA. Materials of extreme speed (800 ASA or more) permit even hand-held instantaneous exposures at full aperture (1/15 to 1/30 second), especially when processed for maximum emulsion speed.

Infra-red. Infra-red radiation for photography is normally obtained electrically, but, using special infra-red film, pictures can be taken in the dark by the "light" from a hot stove or even a couple of hot flat-irons. As lenses are not normally corrected for infra-red, the change in focus must be borne in mind. Some cameras have an auxiliary focusing index for infra-red work. An approximation may be found by focusing visually on a ground-glass screen with a tri-colour red filter over the lens, or the lens may be moved forward about 1/250 of its focal length.

Specialized Uses. Certain non-electric light sources are valuable in their own right for technical reasons. The modern wax candle and nightlight are very consistent and steady in their light output, and one make of photometer exposure meter is intended to be calibrated against such a source. Another valuable source is the wick type oil-lamp which finds wide application in photomicrography of fine detail.

Its steady flame is virtually structureless and can be focused in the plane of the microscope slide without contributing extraneous detail to the final image. R.E.

See also: *Carbon arc lamp; Colour temperature; Daylight; Discharge lamps; Filament lamps; Flash bulbs; Flash (electronic); Flash powder; Fluorescent lamp; Lamp caps and fittings; Light units; Moonlight; Neon lamp; Point source lamp; Sunlight; Zirconium lamp.*

Books: *Principles of Lighting*, by W. R. Stevens (London); *Modern Lighting Technique*, by H. Hewitt (London).

LIGHT SOURCES IN THE PICTURE

Although camera owners are often advised to have the light coming from behind or from one side of them, many effective pictures can be secured by deliberately including the light source in the photograph.

With the camera in this position, however, the full strength of the light will be falling on the lens. Special care is therefore called for to avoid diffusion, fogging and flare in camera and lens; and special exposure and processing techniques are required.

Avoiding Reflections. The internal condition of the camera is important. All metal parts and also the surfaces of fabric bellows should be totally non-reflecting. This is true for general photography, of course, but when shooting directly into the light there must be absolutely no possibility of internal reflections degrading the negative. Similar attention should be given to the lens mount and to the interior of the lens hood, both the often unsuspected cause of veiled negatives.

Cleanliness of the lens is also of prime importance. The slightest deposit of dust on any of the surfaces will result in veiled negatives. There will also be the risk that a brilliant source of light will introduce reflections between the components of a lens and cause circular flare patches to appear in the negative. In addition, fogging can be caused by light scattered within a lens. To a certain extent, and depending upon the design of the lens, use of a coated optical system will help to prevent such troubles.

A lens hood should always be used for taking photographs into the light. It may at first seem incongruous to try and shield the lens from a light source that is intentionally pointed at the lens; there is, nevertheless, the possibility of reflections arising from surfaces outside the picture area. Indoors, these may well come from mirrors or the glass of framed pictures. Even the highly polished surface of a door or table top can reflect sufficient light into the lens to reduce the brilliance of a negative. Out of doors, reflections may come from shop windows, car bodies and windscreens and from puddles.

Reducing Contrast. With the light source occupying a prominent position in the picture, other parts of the subject will be in shadow and this excessive contrast is perhaps the greatest danger with this form of photography. The ideal method of overcoming this hazard is by using a *shadowless* fill-in floodlight to bring the level of the general illumination up to the required brilliance. Note the accent on shadowless. This is important because in practically all pictures in which a light source is included the effect desired is for this to be accepted as the only light. For example, a still-life subject of a book, spectacles and oil lamp or a candle would be entirely ruined if the appearance of shadows made it obvious that the scene was lit, not by the candle or reading lamp, but by studio type floodlighting; the effect would at once be artificial.

For all indoor photographs in which table lamps, candles, matches are included, it is a simple matter to provide this over-all fill-in illumination with either diffused flash or diffused floodlight. An effective method of doing this is by turning the reflector towards the ceiling. When working out of doors the prevailing conditions have to be accepted and steps should be taken to reduce the excessive contrast by employing a modified form of processing. It is also possible to secure softer negatives by exposing at dusk.

Exposure. As with all forms of photography, calculation of the exposure is of the greatest importance. It is obvious, however, that in this case normal methods of estimating exposure cannot be employed. If, for example, the exposure were to be calculated from the brilliance of the light source the surroundings would be considerably under-exposed; conversely, if the rule "expose for the shadows" were obeyed, then the light source would be hopelessly over-exposed. Therefore, it is necessary to strike a balance between the two extremes that will produce the desired result, bearing in mind the main point of interest.

Although for street scenes at night the fastest film should be used, for still-life sets indoors it is advisable to employ a medium speed plate or film. This is because the slower emulsion is likely to give a more evenly balanced result with detail at both ends of the scale than superspeed emulsions.

Processing. With high contrast being the chief feature of against-the-light photography, processing must be controlled to avoid increasing this contrast further with the resultant loss of detail. A good negative of a candle picture—or for that matter any photograph in which a light source is deliberately made the centre of attraction—should show detail in all areas, even in the flame or lamp filament. Generally, it is sufficient if the normal developer is used but with the time of immersion reduced to about two-thirds of that usually given. For some subjects, however, it may be necessary to resort to a soft working developer.

The Sun. The sun itself can be included in the field of view when it is suitably filtered through thick mist or fog. If using an integrating exposure meter, take one reading facing the sun and another away from it; then choose a compromise exposure so as to obtain any detail desired in the darker parts of the picture.

In bright weather fine pictorial effects are obtained by photographing the reflection of the sun as seen on a sheet of water. Contrasts here will be high, and shadow detail will probably have to be sacrificed.

Sunsets are not difficult to photograph provided care is taken to expose while the sun is partially veiled by cloud. Best results are obtained when there is plenty of thin cloud and during the last thirty minutes before the sun actually sets. Using fast panchromatic film and no filter, an exposure of 1/100 second at *f* 16 or *f* 22 should produce a satisfactory result, but it is impossible to give a definite figure because the intensity of the light is constantly changing. If a meter reading is taken this can usually be halved.

A suggested exposure for colour film of speed 50 ASA is 1/50 second at *f* 11. The ideal method of working is to make three test exposures at three different apertures. A useful range would be 1/50 second at *f* 8, *f* 11, and *f* 16.

The Moon and Moonlight. With a cloudless sky, a full moon can be included in the picture with an exposure of about 1 second at *f* 2 using the fastest panchromatic film. Best results are secured when overlooking water so that the path of reflected light can be included.

Landscapes can be taken by moonlight, but in this case the moon itself should not appear in the picture because the long exposure necessary would show it as a blur owing to the relative movement of moon and earth. On a clear night with a full moon an exposure of 20 minutes at *f* 4·5 on fast panchromatic film should secure detail in all but the darkest objects. The movement of the moon and the long exposure make it impossible to render sharply defined shadows.

To photograph the moon itself a telephoto lens is necessary because with a lens of normal focal length the image will be extremely small. Exposures for a full moon in a cloudless sky, on the fastest panchromatic film, range from 1/5 second at *f* 3·5. Long exposures are impracticable because the moon then appears in the negative as an elongated blur of light.

Interiors. The pictorial effect of interiors is often enhanced when part or all of the normal room lighting is included in the field of view. This lighting often takes the form of a shaded standard or table lamp. If the domestic bulb is used in the lamp, the general level of room lighting will have to be low, otherwise the contrasts will be insufficient and the effect will be lost. The alternative solution is to replace the ordinary bulb by a Photoflood, keeping the over-all room lighting at a moderate level.

Assuming the light source to be included in the picture is shaded, contrast should not be excessive; but it is advisable to take exposure meter readings on the light source (as it will appear in the picture) and the main subject area, choosing an exposure that will ensure that the former is not "burnt out" while sufficient detail is retained in the latter.

Fires and Firelight. Effective photographs can be secured of groups around the fire. These are of two types: those showing the actual flames and those showing the effect of firelight.

To include the flames in the picture, a large aperture lens is necessary and with fast panchromatic film exposures need to be about ½ second at *f* 3·5 with one Photoflood directed at the ceiling to provide over-all diffused illumination.

The effect of firelight can be simulated by placing a floodlamp in the fireplace and combining this with general illumination as above. In this case exposures can be much shorter—about 1/20 second at *f* 4·5—or alternatively, flash bulbs can be used: one in the fireplace and one aimed at the ceiling. If the fireplace appears in the picture, a person or piece of furniture should be placed between the lamp and the camera to prevent the direct rays of the light from reaching the lens. Processing should be controlled to avoid excessive contrast.

Candles and Matches. Excellent photographs can be made of lighted candles as, for instance, on a Christmas or birthday cake, at *f* 4·5 on fast panchromatic film with an exposure of 1/2 second. To avoid movement of the flame the subject should be well screened from draughts and no movements made in the room immediately before exposing.

Because modern plates and films have an anti-halation coating on the back to prevent the spreading out of brilliant lights in the image, it is almost impossible to secure the attractive glowing halo as pictured by artists. To record this halo unbacked plates should be used. For a picture of a man lighting a pipe with a match an exposure of about 1/5 second at *f* 3·5 with fast panchromatic film should secure detail in the flame and in the face.

The suggested exposures given above are for use when candles or matches form the sole light source. A useful reduction of contrast, without loss of realism, can be obtained by using diffused fill-in illumination, but this must be shadowless.

An alternative method, of particular benefit to those who do not possess wide-aperture lenses, is to use concentrated artificial light, e.g., a Photoflood in reflector, or small spot, so placed as to shine from the same direction as the candles or matches. Exposures can then of course be much shorter, or smaller apertures can be used.

Fireworks. Any camera from humble box to expensive miniature can be used for photographing fireworks. All that is necessary is to have the camera on a firm support and to point it at the part of the sky where rockets are bursting and to keep the shutter open while the show lasts. The more bursts included on one film the more effective will be the result. A fast panchromatic film should be used. Catherine wheels and roman candle type fireworks make attractive pictures if the shutter is left open for about five seconds. When long exposures are given care should be taken to avoid including people in the foreground.

Street Scenes at Night. A large aperture lens is

essential for photographing street scenes at night to allow a short exposure to be given which will stop movement. A brilliant subject like Piccadilly Circus can be recorded with an exposure of 1/20 second at *f* 3·5 on fast panchromatic film. For less brilliant scenes a short time exposure can be given with the camera on a rigid support, but the lens should be covered while cars are passing or their lights will be recorded as white lines. Street scenes at night after rain make attractive studies. The camera should be kept dry and the lens protected from raindrops and splashes. A pocket torch is a help when setting controls. As long as they keep moving pedestrians will not be recorded during long exposures.

Street lamps included in the field of view will inevitably cause halation. With larger negative sizes this can be reduced by physical abrasion, but a little can safely be left to give the impression of a shining source.

Modern highway illumination often employs sodium or mercury lamps. With a rigid tripod and the fastest panchromatic film, any type of camera may be used. Lamps near the camera should be screened from the lens by a tree or sign. Exposures range from about 45 seconds at *f* 11 for sodium lamps and about 5 seconds at *f* 11 with mercury lamps.

Illuminated Signs. Almost any camera can be used for photographing illuminated signs provided a rigid support is used. Flashing signs which tell a pictorial story progressively can be fully recorded only by keeping the shutter open for one complete sequence. Moving traffic and pedestrians cannot therefore be included. On the fastest panchromatic film exposures range from about 1/50 second at *f* 3·5.

Double Exposure. When the actual unshielded source of light is to appear in the picture it will be much brighter than the scene it illuminates. So the problem is to give an exposure that will be sufficient for the detail in the scene, but not too much for the light source. The lighting contrast is usually much greater than the latitude of the emulsion, so it is not possible to give a single exposure that will include both the source and the rest of the subject. But it is often possible to solve the problem by giving two exposures, a short one for the source and a longer one for the surroundings. The source is switched off during the second exposure, during which the scene is lit by some other suitable source, outside the picture area.

In a picture including a lighted candle the camera is set up on a tripod and the first exposure is made with the candle lighted. For this, a time of 1/25 second at *f* 5·6 on fast panchromatic film may be about right. The candle is then blown out and the subject is lit by a diffused even lighting that casts no definite shadows. The second exposure is then given to register the details of the rest of the scene. This exposure will vary according to the strength of the lighting.

The same principle can be applied to outdoor scenes including street lamps or other forms of public lighting. In this case the camera is set up on the tripod in daylight. Some time before the lamps are switched on, and while there is still some daylight remaining, the first exposure is given. This must be long enough to register a fairly thin image of the scene. The shutter is then closed and the camera left in the same position until after dark. A short exposure is then given with the lights switched on and the resulting negative shows the lights and just enough detail in the scene to complete the picture.

The moon calls for special treatment. If a moonlit scene is given a long enough exposure to show the landscape, the moon, because of its movement in relation to the earth, appears as an elongated patch of white paper. The answer is to give an initial exposure long enough to record the details of the landscape, arranging the picture so that the moon is just off the edge throughout. At the end of the first exposure, a mark is made on the focusing screen where the moon is to appear. The camera is then moved so that the image of the moon coincides with the mark, and the second exposure is given. This is just long enough to record the image of the moon without blur. The position chosen for the image in the picture should be reasonably close to the actual position of the moon during the first exposure or the shadows will appear unnatural. And as the moon always looks unnaturally small in a photograph, it is an advantage if the image can be recorded with a 2× or 3× telephoto lens while the landscape is photographed with a normal lens. J.Bl.

See also: *Firework displays; Flare; Ghost; Halation; Irradiation; Light sources; Moonlight; Shop windows; Sunrise and sunset.*

LIGHT SUPPORTS. Fixed lamps are used only in some commercial portrait studios where every subject is taken with the same basic lighting. Most other branches of photography call for some form of mobile, adjustable lamp support.

Designs vary considerably, especially on the matter of adjustable joints for reflectors. However, most supports can be classified as follows: clamp holders, collapsible stands, heavy duty stands, booms, and ceiling fittings.

Clamp Holders. The clamp-on type of lamp support is particularly convenient for the amateur worker. It is simply a bulb-holder and reflector fitted to a powerful spring clamp with jaws that can be opened with a scissors action

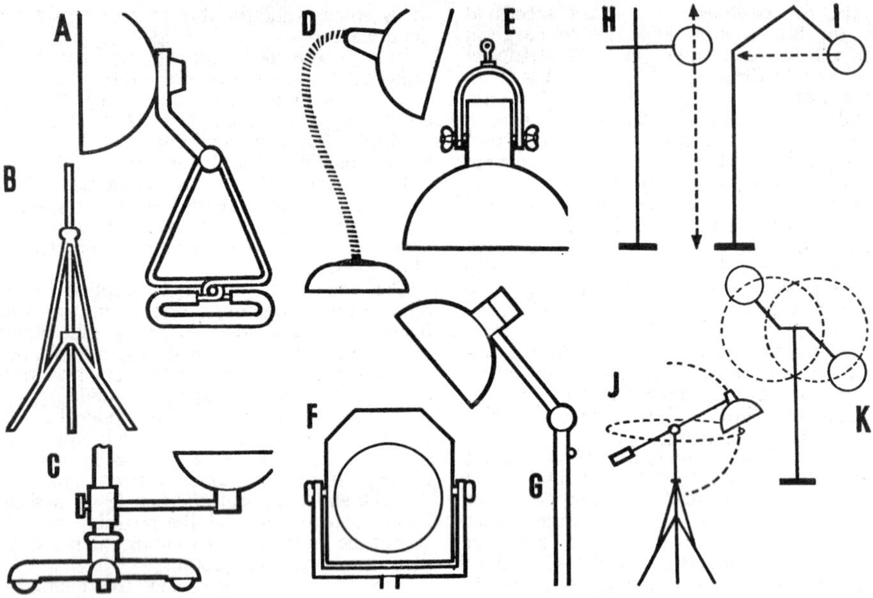

LAMP HOLDERS AND SUPPORTS. Left: *Examples of lamp bases. A, clamp-on unit. B, tripod (used with light telescopic stands). C, studio base with ball-bearing casters and movable low-angle bracket.* Centre: *Methods of lamp adjustment. D, flexible arm; universally adjustable, but not very rigid. E, hanging fitting for flood lamp. F, pivoted support for spotlamp. G, hinged arm; may be clamped at any angle.* Right: *Range of movement of individual light sources on stands. H, arm on column; provides vertical and angular adjustment. I, hinged arm providing lateral adjustment. J, boom light for top and under lighting; keep the light stand itself well away from the subject. K, twin unit with sources adjustable in any position to each other.*

and clamped on to any convenient object—e.g., the seat or back of a chair, the edge of a table, a book shelf or curtain rail. A clamp of this sort provides a cheap substitute for an adjustable stand; it takes up less room and is often more rigid.

Collapsible Stands. The simplest support is a small metal base with a stem to which the lamp reflector is attached by a universal joint that can be clamped by a butterfly nut to hold the reflector at any desired angle; alternatively, a swan-neck tubing may be used. This type of support has little or no adjustment for height, but it can be placed on the floor, a chair, a table, or even hooked on to the picture rail.

For regular use the reflector and adjustable clamp are obtainable mounted on a telescopic tube with a collapsible tripod stand. This type of unit gives height adjustment from about three to eight feet from the ground. Such supports are steady enough for Photoflood bulbs in light aluminium reflectors.

Heavy Duty Stands. Stands to hold high power studio lights and spotlights need to be of heavier calibre than the collapsible stand. The base is a substantial cast ring or spider mounted on heavy duty castors and broad enough to prevent the unit from being accidentally pushed over even when fully extended.

With the exception of lightweight collapsible stands for single Photoflood bulbs, most studio lighting stands are mounted on wheels which allow the lamps to be easily pushed into position or cleared out of the way when they are not wanted.

Booms. It is not always convenient to have the light vertically over the base of the support. This is avoided by using a cross-arm or boom, clamped to the centre pillar of the support and holding a lamp and reflector unit at the end; the other end is usually counterbalanced. The boom enables the lamp to be brought close up to the subject or even over it, while the stand itself can be kept out of the way.

Ceiling Fittings. The ceiling provides a convenient support for fixed-position lighting units. In commercial and portrait studios it is customary to suspend banks of floodlights from the ceiling for providing general illumination of the subject area. The whole bank may be controlled by a dimmer or by switching some of the lamps off, and the assembly may be counterweighted to allow it to be raised and lowered. Ceiling units have the merit of eliminating stands and trailing cables and the disadvantage of being fixed.

One type of lamp support makes use of the ceiling to hold the stand firm. The column of the stand has a circular pad at the top and bottom and it is simply extended and braced between the floor and ceiling with a quick-release toggle device. Once the column is fixed in position it will support a reasonably heavy lamp and reflector without any risk of toppling over. This arrangement permits the lamp to be held anywhere between floor and ceiling level; it can be moved about the studio, and it eliminates the usual heavy, space-monopolizing stand. F.P.

See also: *Lighting equipment.*

LIGHT TRAP. Opening in a wall or partition shaped so that no light rays can pass through from one side or the other.

Rooms. Many darkrooms—particularly in busy press offices, commercial photographic establishments and hospitals—have light trapped entrances so that the staff can pass freely in and out while sensitized materials are being handled. Windows and ventilation holes in the darkroom are also fitted with light traps.

Light traps for these purposes consist of a passage incorporating an angle or bend shaped so that all rays of light coming directly from the source are blocked by the walls. This does not prevent light from leaking around the source by repeated reflection from the walls of the trap, so it is essential for all surfaces to be painted with a dead black paint that will prevent light scatter.

The simplest form of light-trapped darkroom entrance is a thick curtain hung right across the door. Anyone entering closes the door before drawing the curtain to one side and going into the room. This arrangement relies on the co-operation of the person entering for which reason most professional darkrooms have permanently baffled or shaped entrances.

Lamphouses. The ventilation passages in the lamphouses of enlargers and projectors are light trapped to prevent the light source from shining through into the room.

Lamphouse light traps usually consist of a ring of holes in the body of the lamphouse masked by a baffle carrying the equivalent number of holes placed so that there are no direct paths from the source into the room. Dead black paint on the inner surfaces absorbs any reflected light.

Ventilators. Another kind of light trap, very essential in darkrooms, is the ventilator type. This must be capable of supplying the room with fresh air (and allowing the exit of stale air) without admitting any light.

Generally the ventilator is included in the black-out arrangement of the windows—e.g., a hole in the black-out is light trapped by staggered wooden slats that overlap each other. At least one commercial make of window black-out incorporates this system.

Special light-trapped electric fans are also made for maintaining fresh air supplies in darkrooms. The fans are designed to fit snugly into a circular hole in the window black-out.

Sensitized Material Containers. A different type of trap is incorporated in holders for sensitized materials—e.g., plate and cut film holders, film packs and cassettes. In the holders for plates and flat film, there is always a slit to accept the dark slide, while in cassettes there is a slit through which the film leader projects. These slits are sealed by strips of black velvet or fabric.

In time, the light seals on re-usable equipment tend to lose their resilience and allow light to leak past. When—or preferably before—this happens, it is easy to strip off the material and glue a fresh piece in place of it.

The velvet light seals on the consumable type of film cassette cannot be renewed without pulling the case to pieces. These cassettes are not, however, intended to be used more than once; they are frequently reloaded without giving trouble, but the velvet of the light seal is apt to pick up dust and grit which may leave scratches along the whole length of the emulsion. Reloadable cassettes are designed so that when the camera is closed over them the light seal springs open and does not rub the film.

See also: *Darkroom.*

LIGHT UNITS

The units used for measuring light may be divided into those for measuring its colour or quality and those for measuring its strength or intensity.

COLOUR AND QUALITY

Pure spectrum colours are usually denoted by their wavelength. The wavelength of electromagnetic radiation may be expressed in any convenient units of length. Various other units are also used to measure colour quality.

Wavelength. Radio waves are usually measured in metres, but for visible radiation (light) the metre is much too big a unit so the millionth part of a metre is ordinarily employed. This unit is called the micron, and is written μ. Even this unit is inconveniently large for some purposes and the thousandth part of it (or millionth part of a millimetre) is commonly used. This is the millimicron, written mμ. Another unit ten times smaller is also often used. This is named after the Swedish physicist Angstrom, and is written A. Thus, the wavelength of a certain red, for example, may be denoted as 0·6438μ, 643·8mμ, or 6438A.

Colour Temperature. Pure spectrum colours are exceedingly rare in practice, however, and the near-white colour of lighting usually encountered in photography is more conveniently specified in terms of the temperature of a perfectly radiating body giving out light which matches that under consideration. Colour temperature is measured in degrees Kelvin

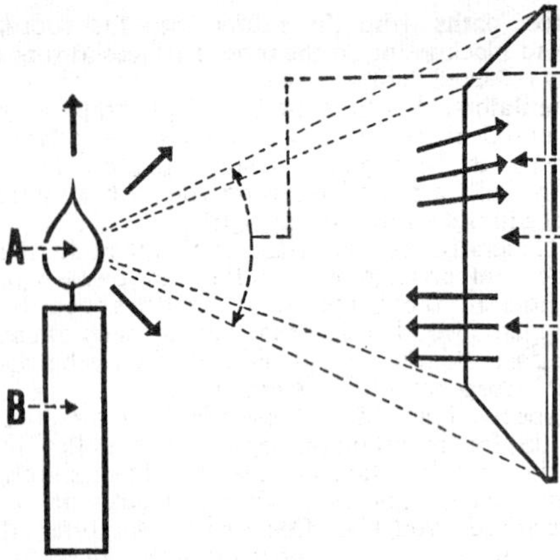

LIGHT UNITS. A, luminous intensity or candle power, which is the amount of light emitted in all directions by a candle made to standard specifications. B, light source of one candela. C, luminous flux; the unit is the lumen which is the light received by a unit area at unit distance from a source of 1 candela. The total emission of this source is thus 12·57 lumens. D, illumination, which is the luminous flux per unit area and is measured in lux (lumens per square metre) or lumens per square foot (previously known as foot candles). E, reflectivity or reflection factor, which is the proportion of the incident light reflected by a surface. F, luminance or intensity of light reflected by a surface. Measured in nit (candelas per square metre) or in foot lamberts. A perfect diffusely reflecting surface (reflectivity 100 per cent) illuminated by 1 lumen per square foot will reflect 1 foot lambert; illuminated by 1 lux it reflects 1 nit.

which have the same magnitude as degrees Centigrade but are based on absolute zero (−273°C.) instead of the melting point of ice.

QUANTITY AND INTENSITY

Units of quantity and intensity of light cannot be put on an absolute physical basis because of the psycho-physiological factors involved. The units used depend upon the conditions under which measurement is being considered.

At different times and in different countries various terms and units have arisen which have never been satisfactorily nor thoroughly standardized. Hence the reader will find considerable variation of nomenclature in literature dealing with light. Preference is here given to those terms adopted by the International Commission on Illumination, but reference is made to other popular or well-established terms. The table provided below compares these and also some more obscure terms that may be met with. (The International Commission on Illumination is generally known by the abbreviation C.I.E., for the official French name Commission Internationale de l'Eclairage. The initials of the translated name, I.C.I., commonly used in America, are deprecated because they conflict with a well-known British trade-mark and because the initials of the name translated into other languages are different.)

There are four basic concepts of particular interest in photography. These are: the power of light sources; the intensity of light falling on a subject; the intensity of light reflected from a subject; and total light energy emitted or received.

Direct use is made of these concepts, respectively, when determining exposure by: the use of calculating tables involving lamp powers and distances; the incident-light method with a photo-electric meter; the common reflected-light technique using an integrating photo-electric meter, photometer, or other light meter; the use of flash bulb data, and also for test strips and plotting characteristic curves.

Knowledge of light units is not essential when making exposure measurements using empirically calibrated meters, and the photographer may never need to work out in actual figures the relation between lumens, foot lamberts, candelas, etc. But a knowledge of the basic principles underlying photometric units is always of value.

Power of Light Sources. The international standard of light emission was originally based on a specified wax candle burning under specified conditions. Hence the term candle power arose and the unit was called the international candle.

Candela: today, the international unit is the candela. It is of such magnitude that the luminance of a perfectly radiating body at the melting point of platinum is 60 candelas per sq. cm. This standard was proposed by the United States in 1936 and has now been adopted because it can be more accurately reproduced than the standards heretofore in use (i.e., the wax candle and special lamps).

Candle power: this is now defined as the light-radiating capacity of a source in a given direction in terms of the luminous intensity expressed in candelas. For all practical purposes the candela is the same as the international candle, from which it differs by less than 2 per cent.

Candle: in the Unites States the use of the term candle is being continued and it now applies to the new unit as defined above; if distinction is necessary the terms new candle and old candle are used.

Luminous flux and lumen: the luminous flux or light flow that falls on 1 square foot of surface 1 foot away from a point source of 1 candle power is called the lumen. If a uniform point source of 1 candela is placed at the centre of a hollow sphere of 1 foot radius every square foot of the inside of the sphere—and there are 4π (12·57) of them—will receive 1 lumen. Thus a uniform source of 1 candela emits 1 lumen per unit solid angle and provides a total luminous flux of 12·57 lumens.

Mean spherical candle power: the light-giving power of a lamp has been commonly expressed in the past in terms of mean spherical candle power (M.S.C.P.). This is the average value of the luminous intensity in all directions and when multiplied by 4π (12·57) gives the total lumens supplied by the lamp. The average luminous intensity in all horizontal directions from a source (mean horizontal candle power or M.H.C.P.) is sometimes used when no light is radiated over a substantial angle (such as in the usual incandescent gaslight fitting) and where the M.S.C.P. would not, therefore, give a fair representation of the useful luminous intensity.

Efficiency (lumens per watt): the efficiency of an electric lamp has been quoted in the past in watts per candle power. This is responsible for the name "half-watt" being applied to tungsten lighting—from the rather optimistic estimate of efficiency of ½-watt per candle power. Actually this efficiency is only attained by over-run lamps (such as class A1 projection lamps). Lamp efficiency is now expressed in lumens per watt. If all the power supplied to an electric lamp were converted into white light the maximum theoretical efficiency would be about 220 lumens per watt. Fluorescent lamps have been made to supply up to 70 lumens per watt. A normal 100 watt lamp has an efficiency of 15 lumens per watt, which means that only 7 per cent of the energy supplied is obtained as light—the remaining 93 per cent is wasted as heat! Even a Photoflood has only double this efficiency. The practical photographer thinking in terms of 500 watt spots or 2 kilowatt broadsides is completely ignoring the many factors that intervene between the input of current to his lamps and the reception of light by his lens. To bridge this gap we must now consider the further concepts listed above.

Intensity of Light Falling on a Subject. A point source would cause exactly the same amount of light to fall on 1 square foot of surface situated at a distance of 1 foot as would fall on 1 square metre at a distance of 1 metre. Each subtends unit solid angle at the source and if the source is of 1 candela each receives a luminous flux of 1 lumen. But in one case the luminous flux is spread over a square foot and in the other it is spread over a square metre. Since 1 square metre equals 10·764 square feet, the illumination in the first case will, therefore, be 10·764 times that of the latter.

Lux, lumen/square foot and foot candle: the illumination on a surface at 1 metre distant from a point source of 1 candela is, of course, 1 lumen per square metre. This is the international unit of illumination and is called the lux. The illumination at 1 foot from a source of

TERMS OF LIGHT MEASUREMENT

Popular Concept	*Technical Term*	*Symbol*	*Unit*	*Abbreviation*
Colour	Wavelength	λ	Micron (= 10^{-6} m.) Millimicron (= 10^{-6} mm.) Angstrom (= 10^{-7} mm.)	μ mμ A
Strength of light sources	*Luminous intensity Candle-power	I	*Candela; candle (U.S.A.) International candle (obsolete)	cd; c.
	*Luminous flux Pharos	F or Φ	*Lumen	lm.
	Efficiency (of a light source)	η	Lumens per watt †Watts per candle power $= \frac{12\cdot57}{E/w.}$	lm/w. w/c.p.
Intensity of light falling on a subject	*Illumination Luminous flux density Pharosage	E	*Lux; †metre-candle (= lm/sq.m.) Phot (= lm/sq. cm.) Lumen per square foot; †foot candle Nox (= 10^{-3} lx.) Luxon (ophthalmic use only)	lx; m.c. ph. lm/sq.ft.; f.c. — —
Intensity of light reflected from a subject, or emitted from a surface	*Luminance †Brightness Helios	L B H	*Nit (= cd/sq.m.) †Stilb (= cd/sq. cm.) Candela per sq. in. Candela per sq. ft. Foot lambert; †apparent foot candle; †effective foot candle (= $1/\pi$ cd/sq.ft.) Blondel; *apostilb (= $1/\pi$ cd/sq. m.) †Skot (= 10^{-3} asb.) †Lambert (= $1/\pi$ cd/sq. cm.) †Millilambert (= 10^{-3} L)	nt. sb. cd/sq. in. cd/sq. ft. ft. L. (foot lambert) asb. (apostilb) L. mL.
Total energy	*Luminous energy †Quantity of light	Q	Lumen-second; lumerg Lumen-hour Lux-second Photon (used in theory only)	lm-sec. lm-hr. lx-sec. —

*International terms †Deprecated terms

1 candela is 1 lumen per square foot (10·764 lux) and this is the popular British unit. It was previously known as a foot candle. General room lighting for reading and writing is usually of the order of 5 lumens per square foot.

Luxon: the luxon is a special unit used in considering light within the eye. It is the illumination on the retina produced by viewing a surface having a luminance of 1 candela per square metre when the pupil area is 1 square millimetre. This unit was originally termed the photon, which term now has a quite different application (see below).

Intensity of Light Reflected from a Subject. Luminance, or surface brightness, depends upon the power of reflection of a surface and must not be confused with the preceding concept of illumination, which is simply a measure of the incident light.

Nit and foot lambert: the international unit of luminance is the nit which is 1 candela per square metre. A common British unit, however, is the foot lambert which is defined as the luminance of a uniform diffuser emitting 1 lumen per square foot.

Reflection factor: the reflection factor of a surface is the ratio of the luminous flux reflected to the luminous flux incident; it is always less than unity.

Luminance factor: the luminance factor of a surface is the ratio of luminance to illumination. It always refers to specified conditions of illumination and angle of view. The luminance factor may greatly exceed unity in certain directions. For example, with 10 lumens per square foot illumination, a glass-beaded screen may produce a luminance of 50 foot lamberts when viewed from near the projector axis (i.e., luminance factor = 5) but less than 5 foot lamberts from more than 30° (i.e., luminance factor = ·5 or less).

Candelas per unit area: many light sources being extensive in area cannot be considered as point sources and the concept of luminance is also applied to them. Such high intensity surfaces are usually measured in candelas per square inch.

Luminosity: the term brightness is preferably avoided because of possible ambiguity due to the very broad meaning of the term in its popular non-technical use. It may be used in non-quantitative statements, but refers to the visual sensation and not the photometric quantity. The international term is luminosity. Thus the luminance of a surface may be doubled yet it would be permissible to say that the luminosity (or brightness) had not doubled since the sensation called brightness is generally judged not to be doubled.

Total Light Energy. Lumen second: luminous energy or quantity of light is measured in lumen seconds, also known as lumergs. One lumen second is the quantity of light given out in 1 second by a source generating 1 lumen of light power. This unit is used when considering the light given by a flash bulb. 1,000 lumen seconds is mathematically the same as 1 lumen shining for 1,000 seconds or 10,000 lumens for one-tenth of a second. In photographic practice this relationship is modified by the reciprocity failure of sensitive materials. Flash bulbs may be accurately compared by their total light output when considering open-flash technique, but not when using them synchronized with high shutter speeds since two bulbs having the same output in lumen seconds might have quite different durations of flash, and hence different peak flux in lumens.

Lux second: the lumen second may also be used when dealing with the effect of light falling on a surface for a measured length of time (photometric exposure) but the surface-area must also be specified. For this purpose a more convenient unit is the lux second which is one lumerg per square metre.

Photon: the photon is the quantum or smallest quantity of light energy that can be emitted or received. It is inversely proportional to wavelength and the photon of ultra-violet light is twice the magnitude of that at the red end of the spectrum. It is of importance in theoretical physics but of no direct practical utility. R.E.

See also: *Illumination; Light sources.*
Books: *Photometry*, by J. W. T. Walsh; *Principles of Illumination*, by H. Cotton (both London).

LIGHT VALUE. Generally, this is a term applied to arbitrary numbers assigned to subject and lighting conditions in exposure tables and some exposure meters.

Many exposure tables tabulate separately the various exposure conditions (time of day and year, weather, subject, film speed, etc.). Each set of conditions carries an arbitrary number; these numbers are added up to give a light value which is referred to a final table from which the correct exposure is read off.

Exposure values, as used on shutters, are also frequently referred to as light values.

See also: *Exposure value.*

LINE BLOCK. Photomechanical printing block consisting of a thin metal (usually zinc) plate in which the non-printing areas around the image are etched away, leaving a raised printing surface corresponding to the original image. There are only two possible tones in a line block: the full intensity of the ink, and the unprinted areas of the paper. There is no gradation in the tones of a line block impression. Line blocks are used for reproducing diagrams and black and white originals; they are not suitable for reproducing photographs and originals in continuous tone.

See also: *Photomechanical reproduction.*

LINE COPY An original document or drawing consisting essentially of two tones (e.g. black and white) without intermediate tones.

LINE OF BEAUTY. Term originally used by the English artist, William Hogarth (1697–1764), to relate gracefulness to a curved S-shaped line. He discussed it in his essay, *Analysis of Beauty*, and employed its snake-like form in some of his paintings. It is one of the elements of picture construction sometimes used by photographers who try to compose their work in the manner of academic painters.

LINKED RING BROTHERHOOD. A secession from the Photographic Society of Great Britain in 1892, under the leadership of A. Horsley-Hinton. The Linked Ring included all the younger pictorial photographers of note: George Davison, J. Craig Annan, Frederick H. Evans, Alexander Keighley and others.

LIPPMANN, GABRIEL, 1845–1921. French physicist, professor at the Sorbonne. Originated the interference method of direct colour photography (1891), which produces fixable, accurate and excellent colour photographs but is difficult to carry out. Received the Royal Photographic Society Progress Medal in 1897 and the Nobel Prize in 1908. Suggested the lenticular film method of additive colour photography.

LIPPMANN EMULSION. Sensitive emulsion containing very little silver halide in extremely fine grains so as to be practically transparent. It is characterized by almost complete absence of any scattering phenomena, and was used by G. Lippmann (1845–1921) for his colour process.

LIPPMANN PROCESS. Obsolete method of making pictures in natural colours invented by Professor G. Lippmann. The interesting feature of this process was that the colours were produced without using either dyes or pigments; they were the result of the interference of the light rays forming the image. (Interference is the effect responsible for the familiar colours in an oily film on the surface of water).

Lippmann took his pictures on a panchromatic plate with its emulsion surface facing away from the lens and in contact with liquid mercury. Light rays passing through the emulsion were reflected back by the mercury. The reflected rays interfered with the incident rays and formed a latent image which varied in depth according to the colour (wavelength) of the rays.

When this image was developed and viewed at a particular angle in contact with the surface of a mirror a faint picture was visible in natural colours.

The technical difficulties of the process are immense, and no one has succeeded in carrying it beyond the stage of a scientific novelty.

LITERATURE ON PHOTOGRAPHY. Bibliographies of specialized photographic literature can be found in numerous publications.

The subject catalogue and supplement of the Royal Photographic Society's library list books and pamphlets in the library up to the year 1952. Many of them are in foreign languages. The catalogue has about 180 subject classifications, with an alphabetical index.

Proceedings of the Centenary Conference 1953 of the Royal Photographic Society include a section which deals with 82 contemporary books on various applications of photography.

The *Periodicals Catalogue* of the Royal Photographic Society covers periodicals in the Society's library up to the end of 1955. These include all important British and American journals and many in foreign languages.

The *British Journal Photographic Almanac* publishes an abridged list of text books covering most general and certain special fields of photography. It mainly confines itself to current books, but also carries a limited selection of older publications.

Progress in Photography Volumes I (1940–50), II (1951–54) and III (1955–58) give an international survey of discoveries and new techniques in the field of photography for the years covered. Bibliographies are appended to most of the articles.

This publication was continued in 1959 as a quarterly review entitled *Perspective*, containing news of progress in photography, cinematography, sound and image recording.

The Library Association of Britain has published two guides to photographic literature. *Some Tools for Reference and Information Work in Photography* (1953), by R. S. Schultze, covers some fifty reference works, arranged according to type of publication. *A Reader's Guide to Books on Photography* (1954) applies only to British books and American books published in Britain. It includes many of the more elementary books, but also has an advanced section, *Special Applications and Subjects*.

The Society of Chemical Industry's Annual Reports on the Progress of Applied Chemistry include a section, *Photographic Materials and Processes*, which has a comprehensive bibliography on photographic work from the chemical point of view.

In the United States, the catalogue of the Epstean Collection, *A Catalogue of the Epstean Collection on the History and Science of Photography and its Applications especially to the Graphic Arts*, now in the library of Columbia University, covers books up to 1937. Over 1,400 titles are listed, grouped in 84 categories.

An article, "Photography's 100 Best Books", was published in the November, 1954, issue of *Photography* (U.S.A.). It gives a brief indication of the scope and standard of each title mentioned. The books quoted are mostly English or American, with a few major works in other languages.

The *Photo-Lab-Index* (U.S.A.) contains *A Guide to the Literature of Photography and Related Subjects* (1943) by A. Boni. It lists 3,000 books, pamphlets and articles in journals, drawn from various countries. They are listed alphabetically by author and subject.

The periodical literature of photography is covered by three abstracting journals: *Photographic Abstracts* published by the Royal Photographic Society, *Science et Industries Photographiques* (Photographic Science and Industries) (France), and *Abstracts of Photographic Science and Engineering Literature* published by the Columbia University in the City of New York, Department of Graphics, School of Engineering and Applied Science, with the editorial co-operation of the Society of Photographic Scientists and Engineers. This Journal, which first appeared in March 1962, superseded the *Eastman Kodak Monthly Abstract Bulletin*, U.S.A., and *Ansco Abstracts*, U.S.A., both of which ceased publication in December 1961. The *Proceedings of the Centenary Conference* 1953 of the Royal Photographic Society include a report on other abstracting journals which cover photographic literature.

The first three of these abstracting journals deal with the more important patents in the field of photography. A full list of British patents will be found in *Abridgments of British Patent Specifications Group XX, Photography.*

The *World List of Scientific Periodicals* lists photographic journals published between 1900 and 1950, with the British libraries in which copies may be found. The British Museum has in preparation the *British Union Catalogue of Periodicals*, which will include many photographic journals.

Special scientific and technical books can in Britain be obtained on loan through one of the interlending library services. The National Central Library acts as an interlending centre for most public libraries and many specialist libraries, including that of the Royal Photographic Society and the research libraries of Ilford Ltd. and Kodak Ltd. The library of the London Science Museum, which itself has a collection of photographic books, acts as an interlending centre for scientific libraries. The Finsbury Central Library has the largest collection of books, pamphlets and periodicals on photography of any Public Library in the United Kingdom. This Metropolitan Special Collection is housed in St. Luke's Branch Library, Lever St., E.C.1.

A new international bibliographic guide to general and specialized literature on photographic processes, techniques, theory, chemistry, physics, apparatus, materials and applications, industry, history, biography, aesthetics, etc., entitled *Photographic Literature*, and edited by Albert Boni, was published in December 1962 by Morgan and Morgan, Inc., Publishers, New York. N.W. (G.L.)

See also: *Books on photography; Information and inquiries; Museums and collections; Periodicals; Standardization in photography.*

LITH FILM. Thinly coated emulsions which produce extremely sharp images of high density and contrast; used in the graphic arts for making line and screen images, and in some phototypesetting systems. The characteristic curve of a lith film has a steeply rising straight-line portion, and a very short toe. When developed in a formaldehyde-hydroquinone developer, fully-exposed areas develop slowly at first, then, by a process of infectious development, rapidly increase in density.

Orthocromatic lith films are used for making black-and-white line and screen negatives and positives, in the camera or by contact. Half-tone images, made with either a cross-line or a contact screen, have well-formed dots which are suitable for reduction by dot etching. For normal images, the film is developed at 68° F. for a period of 2–3 minutes, but it may be developed by inspection when the growth of the image is checked by placing a safelight under a transparent dish, or by suspending one over and close to a white dish.

A special technique with lith films is fine-line development, used for copying screen proofs, type matter and thin lines. The developing dish is agitated for 20 seconds, then allowed to remain stationary for the remainder of the development period. Reaction products prevent the veiling over of fine detail. Another technique is the etch-bleach reversal process for making positives, where the gelatin and silver of a developed image area are dissolved, the unexposed silver halide is exposed to white light, and the film re-developed to form a positive image.

Lith films with a panchromatic sensitivity are used for making colour line and direct colour half-tone separation negatives. F.G.W.

See also: *Developers; Graphic Arts; Photomechanical reproduction.*

LITHOGRAPHY. Method of mechanical printing from a flat surface where the printing areas accept greasy ink while the non-printing areas are dampened with water and repel the ink. The printing surface nowadays usually is a suitably grained zinc or aluminium plate, and the image is produced on it mostly by photographic means.

See also: *Photo-litho masters; Photomechanical reproduction.*

LITMUS. Vegetable colouring matter used as an indicator in chemistry; it is turned red by most acids and blue by alkalis. It is available as a solid or a solution in water, or in rolls or books of litmus impregnated paper.

See also: *Indicator chemical.*

LITRE (LITER). The standard of capacity in the metric system. One litre is for all practical purposes equal to 1,000 cubic centimetres.

See also: *Weights and measures.*

LIVER OF SULPHUR. Chemical used in some toners; potassium sulphide. It is of variable composition.

LOCAL CONTROL. Any method of controlling the appearance of a limited area of the photographic image—e.g., to darken or lighten its tone. The term is applied to local control of development of the negative or positive, and dodging, shading and spot printing during exposure in making enlargements or contact prints.

LÖCHERER, ALOIS, 1815–62. German pharmaceutist, from 1847 professional portrait photographer in Munich. Noted for his photo-story of the transport and setting up of the gigantic statue of Bavaria in 1850, and for Calotype portraits of celebrities published in 1854. Löcherer also made genre and nude studies for artists and reproductions of engravings in the Bavarian royal collection, 1855.

LOCKING CABLE RELEASE. A release which, by means of a locking screw or similar device, enables a shutter to be locked in the open position when it is set at "B".

LOGARITHM. Mathematical function used for making certain calculations easier. It is the "power" to which the "base" number (usually 10) has to be raised to equal a given number. For example, $100 = 10^2$, therefore the logarithm of 100 is 2·0. Its principal use in photography is to reduce the characteristic curve of the emulsion to a manageable form. The curve is required to show the effect of increasing exposure in a series of steps, each one representing a constant multiple of the value of the one before. The first step, for instance, may be from 1 to 10, while the last is from 10,000 to 100,000 exposure units.

If the curve were plotted normally, it would not be possible to give equal importance to the first and last steps (and in sensitometry they each have the same weight). The effect of using logarithms of the numbers instead of the numbers themselves is to give the same spacing along the exposure axis to every comparable step from the lowest to the highest values.

Some film speed systems use a logarithmic scale, where an increase of 10 degrees represents a ten-fold rise in speed. This means that a very great range of relative sensitivities can be accommodated on a comparatively short scale of numbers.

Logarithms to the base 2 are used in the formula for establishing exposure value numbers and also ASA log film speed degrees. Here every increase of 1 in the index number (exposure value or log speed) corresponds to a doubling of the basic value (exposure, speed).

LOGETRONIC. Proprietary name of first electronic printing and enlarging units utilizing a flying spot scanner.

LONG FOCUS. Description of a lens indicating that its focal length is longer than the diagonal of the negative it is designed to cover—i.e., longer than the focal length of a normal lens.

See also: *Lens; Telephoto lens.*

LONGMORE, T. A., 1905–57. English radiographer and medical photographer. Pioneered methods of training of workers in these fields and the standardized approach to clinical photography. Author of the first comprehensive text on these subjects (1944) which is now a standard work in worldwide use.

LONG SHOT. Large view in a cine film which takes in the entire scene or setting in which the action of a particular sequence takes place. It is generally a distant view—a landscape, street scene or the like out-of-doors, or the whole of a room in an indoor setting. Any human figures in a long shot would occupy less than one-third of the height of the frame.

LONG TOM. (Slang.) Abnormally high-powered telephoto lens of the type used by press photographers for taking sports pictures from a distance—e.g., at cricket matches and boat races. So called because of its resemblance to an old type of long-barrelled naval gun of that name.

LOW CONTRAST DEVELOPER. Certain developers tend to produce an image of lower contrast than others. They may be used deliberately to play down the over-harsh contrasts of certain subjects, or they may be employed because of other qualities associated with such developers—e.g., fine grain.

LOW KEY. Effect produced by a predominance of dark tones in a picture. Low key pictures have a heavy and dark character with a particular richness of shadow detail.

A low key photograph is a product of the right subject and suitable lighting; the effect cannot be successfully imitated with an unsuitable negative by any method of printing or aftertreatment.

The subject should be one which lends itself to dramatic, mysterious or rugged treatment and the actual tones should be inherently dark. A fair complexioned woman with blonde hair would be a bad choice for a low key study.

The lighting should limit the important parts of the subject to a narrow range of dark tones while at the same time providing good modelling. Unimportant areas and the background may be considerably darker than the subject, to give it emphasis. To relieve the over-all heaviness of tone there should be at least one small highlight area.

The exposure must be calculated specially for the tone range over the important part of the subject—not necessarily over the whole picture.

Under-exposure will result in loss of detail in the important tones; over-exposure will give better shadow rendering, but may burn out the highlights.

Development of the negative should be normal, but in a developer that favours good shadow detail—e.g., a metol sulphite developer. Under-development will cause loss of contrast, with murky and flat tones; over-development will increase contrast and destroy gradation in the smaller but quite important highlight areas.

The print should be made on a paper that is known to give good separation of the darker tones and which is capable of producing deep, rich blacks. Glossy and velours papers are excellent in this latter respect.

The only satisfactory way of arriving at the correct exposure is to make one or more test strips over the important areas, develop and fix them fully, and examine them by white light.

Without a lot of practice, it is almost impossible to select the correct strip under the darkroom safelight; the usual tendency is to err on the side of under-exposure.

When the print has been processed, care must be taken to preserve a clean, bright surface when it is dried. Hard water or insufficient washing can both lead to dulling of the surface sheen, which in turn causes a loss of contrast in the very dark tone areas. A final rinse in a weak solution of acetic acid usually prevents this trouble. J.D.C.

See also: *High key.*

LUMEN. Unit of light flux; it is the power of light falling on 1 square metre of a hollow sphere of 1 metre radius at the centre of which is a light source of 1 candle power. The total light flux emitted by the unit source is 4π lumens, 4π (= 12·57) being the area of the sphere of unit radius.

See also: *Light units.*

LUMEN SECOND. Unit used in measuring the total light output of a source—e.g., for stating the amount of light given out by a flash bulb, or the amount of light falling on a test piece of sensitized material. The quantity of light in lumen seconds is equal to the power of the source, expressed in lumens, multiplied by the duration in seconds.

See also: *Light units.*

LUMIÈRE, AUGUSTE, 1862–1954. French scientist, son of Antoine Lumière. With his brother, Louis, manufactured dry plates and later roll film. The two of them together did valuable work in photographic chemistry, colour photography and cinematography. Auguste was also famous in the sphere of biological science: founded the Clinique Auguste Lumière and devoted much of latter life to work in the physiological laboratory. Historically, his name is nearly always linked with his brother, Louis.

LUMIÈRE, LOUIS, 1864–1948. French scientist, son of Antoine Lumière who founded a dry-plate factory in Lyon in 1882. With his brother Auguste assisting, the factory expanded and started roll film manufacture in 1887.

Together with his brother, he carried out and published much scientific and technical work on photographic subjects. Louis and Auguste (alone and with Seyewetz) made valuable contributions to photographic chemistry, especially development (1891), and received the Progress Medal of the Royal Photographic Society in 1909. They produced Lippmann photographs in 1892, the Autochrome screen plate for colour photography in 1903 (commercially in 1906) and demonstrated publicly their Cinematographe camera and projector in 1895, using an intermittent (claw) movement with perforated 35 mm. film; the design followed in nearly all modern cine apparatus. In 1900 the two brothers demonstrated the Photorama, a 360° panoramic projection device. Biography by Bessy and Duca (Paris 1948).

LUMINANCE. Property of emitting light either as a source, or by reflection. The intrinsic luminance of a body (the amount of light coming from a unit area) is measured in foot lamberts—i.e., the intensity of the light falling on the surface measured in foot candles multiplied by the reflection factor of the surface.

See also: *Light units.*

LUMINESCENCE. Visible glow of certain substances when subjected to stimulation by electro-magnetic radiation, electric fields or heat. The general term luminescence also embraces phenomena like electro-luminescence, fluorescence and phosphorescence.

See also: *Electroluminescence; Fluorescent lamp.*

LUMINOUS FLUX. Intensity of light (i.e., visible radiation emitted by a source) or in common usage a measure of the power of the source. It is measured in lumens.

See also: *Light units.*

LUNAR CAUSTIC. Commercial form of silver nitrate, the chemical which enters into the composition of most of the sensitized materials used in photography.

LÜPPO-CRAMER, HINRICUS, 1871–1943. German scientist and writer on photography. A life-long worker in the photographic industry. first with Schering, then as manager of the Schleussner dry-plate factory, and later as technical director of the Kranseder dry-plate factory at Munich and as director of research in the Deutsche Gelatinefabrik A.G., Schweinfurt. He worked on hydroquinone-derivative developers (1902) and on desensitizers and in 1920 introduced his system of developing in bright light. Contributed notably to knowledge of colloid chemistry in relation to photography and to the latent image theory. The Royal Photographic Society granted him their Progress Medal in 1932.

LUX. Measure of the illumination over a surface.

A level of illumination of one lux means that one lumen of light is distributed over every square metre of surface. In other words, it is the illumination per square metre of a surface at a distance of 1 metre from a point source of 1 candela.

See also: ***Light units.***

LUX SECOND. Unit based on the lux, used in sensitometry for measuring exposure.

It is the amount of light that falls in one second on a surface illuminated with an intensity of one lux.

See also: ***Light units.***

M

MACH, ERNST, 1838–1916. Austrian physicist. The invention of X-ray stereoscopy is attributed to him. He suggested to Eder and Valenta that, by shifting the X-ray tube, it should be possible to produce stereoscopic radiographs. This was found to be correct and the first pictures so produced appeared in 1896. Mach was the first to use the electric spark as a light source in photographing projectiles 1885–6. He was also the first to use a rotating sector wheel in measuring sensitivity and was also a pioneer in stroboscopy and in time-lapse photography. Biography by A. Lampa (1918).

MACHINERY. Machines are photographed (a) to record their movements for analysis (b) to record their design or operating features for technical or catalogue illustration (c) for advertising purposes. As a preliminary to photography, a knowledge of the purpose of the machine and its method of operation, together with a clear understanding of its most important features, is necessary.

Motion Analysis. Movement is usually analyzed by "freezing" it by high speed photography, but sometimes the solution may involve integrating time, as in time-lapse photography.

The factors determining the method to be used and the equipment required are:

(1) the type of motion to be studied—intermittent, continuous or self-activated;

(2) speed of movement, often calculated or estimated from the frequency of movement and the distance travelled or based on design information;

(3) the amount of blur that can be tolerated in the photographic image (a relatively large degree of blur can often provide sufficient information);

(4) type of illumination—self-luminous or requiring front, side or rear lighting, diffuse or directional lighting, of continuous, intermittent or brief duration. On these factors depends the exposure necessary and hence the type of shutter, camera and form of lighting.

Some problems can be solved by using ordinary equipment. For others it is often necessary to use special apparatus, including:—magneto-optical shutters for exposure times of 2 to 20 microseconds; electro-optical Kerr cell shutters for times of 0·007 to 7 microseconds; electronic shutters (image converter tubes) for about 0·02 microseconds, or oconoscopes for about 2 microseconds. Illumination by electronic flash is suitable for exposures down to 1 microsecond, while sparks produced by capaciter discharges cover 1 to 0·1 microsecond. High-intensity continuous sources such as high-pressure mercury vapour lamps and xenon tubes may be needed for high-speed cinematography. Photo-cell, mechanical, electronic and acoustic triggering systems provide synchronization of the exposure with the event. Repetitive flash sources or shutter operation is used for stroboscopic recording. Other special equipment for high-speed work includes streak cameras, optical compensation (rotating drum and prism) cameras; image switching cameras and image sampling cameras.

Record Photography. This calls for the selection of emulsions and filters to show fine details and reveal differences of the surface textures and of the colours of the various materials and finishes that are used. Slow panchromatic emulsions are generally satisfactory but orthochromatic emulsions are often a better choice to show texture when the resultant reproduction of colour values is acceptable. An even spread of light is necessary for illuminating basic detail, but should be supplemented by directional lighting to create shadows that reveal shape and form and to produce reflected highlights to show up the nature and finish of the material. The most suitable form of lighting depends upon particular circumstances. At times the prevailing light for basic illumination, can be supplemented by a simple arrangement of lamps to fill-in dark areas and to create contrast. At other times existing light from unsuitable angles must be swamped by special illumination.

If the absence of movement and of extrane-

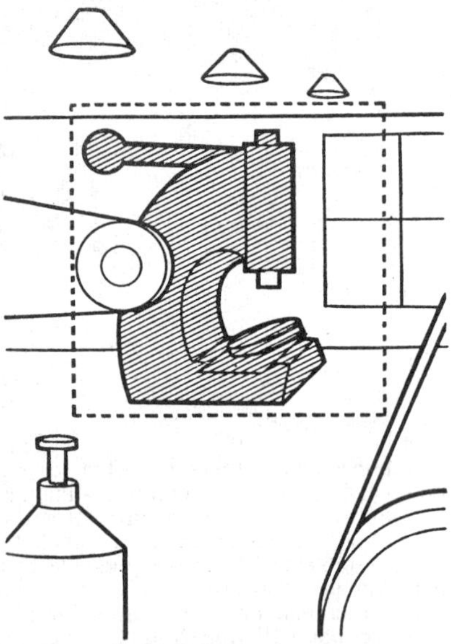

MACHINERY. When photographing inside a factory there are a number of points to watch, outside the picture area as well as in it. These include lamps that may cause flare, windows in the background (usually need covering up), vibration due to other machines (stop them if necessary), reflections from bright steel parts, and polished surfaces just outside the picture area.

ous light permits a sufficiently long exposure time, the area can be "painted" with light from a moving lamp to provide all required illumination. The direction of movement of the lamp needs preliminary testing to avoid unwanted glare from flat reflecting surfaces.

When long exposures are not possible due to vibrations, strong partial lighting, movement, etc., the light intensity must be increased. The problem then becomes mainly the physical one of installing lamps in sufficient quantity and in appropriate positions. The type of light source is less important.

Machinery with rotating parts in motion can often be photographed with long exposure times, the image blur giving a good impression of movement. Flash illumination can be used during such exposure to produce a superimposed sharper image of the moving part.

When figures are included in the picture, either because they cannot be removed or because they are required to give scale or to demonstrate a method of operation, pose them naturally and doing the right thing technically. Do not let them appear more important than the machine, and do not let them obscure any important part of the machine. Consult the safety officer to check that machine guards are correctly adjusted.

Look for objects that are out of place and for untidy details that may appear in the picture; have them removed from the scene.

Confine retouching to the obliteration of blemishes in the machinery and to the elimination of backgrounds by blocking-out. The latter task is much facilitated by paying attention to the background before exposing.

Backgrounds can be suppressed by controlling the illumination so that details are either underexposed or overexposed. Or get somebody to support a sheet behind the machine and to move it horizontally and vertically during the exposure to form a blurred image that isolates the machine from its surroundings. Select the camera angle so that the required details appear in the photograph clearly and without distortion and so that the required information is conveyed accurately.

Advertising. Any creative photography of machinery requires a different approach to produce a visual impact within the frame of the picture. This may mean selecting contrasting patterns and shapes and tonal areas, or bringing them out by suitable lighting. Unusual camera angles can emphasize perspective effects. When machine operators are included, choose the lighting and camera position to concentrate interest on either the operator or the machine. Generally the techniques required are pictorial rather than industrial—the impression rather than the technical information counts. R.F.

See also: *High-speed cinematography; High-speed photography; Industrial photography.*

MACKENZIE - WISHART ENVELOPE. Elaboration of the normal type of plate holder, enabling plates to be changed in daylight. Plates are loaded into light-tight envelopes which are then inserted in daylight into a special plate holder. The plate holder is so constructed that when the sheath is withdrawn, it opens the envelope and allows the plate to be exposed. Replacing the sheath closes the envelope again and allows it to be exchanged without a visit to the darkroom. With one of the special holders, the number of plates that can be changed and exposed is limited only by the number of loaded envelopes.

The envelopes are no longer made and are only to be had on the second-hand market.

MACKIE LINE. Clear line surrounding an intense highlight on a developed emulsion. It is the result of contamination of the solution by the waste products of development released in the highlight region. Only occurs when the developer is given insufficient agitation.

See also: *Eberhard effect; Edge effect; Kostinsky effect.*

MACPHERSON, ROBERT, 1811–72. Scottish surgeon in Edinburgh. Settled in Rome in early 1840s and became a painter. Changed to photography in 1851 and became the leading architectural photographer in Italy, particularly celebrated for his views of Rome and of sculpture in the Vatican. Devised a process of photolithography, 1855.

MACROPHOTOGRAPHY

Technique of taking larger-than-life photographs with ordinary camera lenses. Macrophotography ends and photomicrography (with microscope lenses) begins, at about 10× diameters.

Magnified pictures taken in this way are extremely useful for study and records in medicine, entomology, science and philately. They are also of striking general interest when of familiar small objects on an enlarged scale.

There are two ways of taking macrophotographs: by using a normal camera and lens with an abnormally long extension tube or bellows, and by using a large—e.g., whole plate, or 8 × 10 ins.—camera with an abnormally short focus lens. In either of these ways, an enlarged image of the subject is focused on the plate and photographed.

BASIC SYSTEM

The main consideration with a basic system is the type of equipment and lighting to use. Certain cameras and lenses are more suitable for macrophotography than others.

Focusing. The size of the camera for macrophotography is unimportant; anything from a miniature to a large plate camera will do. But it is essential for it to be equipped with a screen for visual focusing. It is possible to do ordinary close-up photography without focusing visually by measuring the lens-subject distance, or fixing it by a rod or other gauge. But in macro work, the lens is much closer to the subject than in ordinary close-ups. The depth of field is so shallow that the distance cannot be measured with sufficient accuracy; often a difference of 1/100 in. is enough to throw the subject right out of focus.

For subjects that are likely to move between focusing and exposure, a reflex camera or one with a reflex attachment is essential. If the subject is stationary, any plate camera or plate back attachment that will also take a focusing screen may be used.

CAMERA EXTENSION REQUIRED FOR MACROPHOTOGRAPHY

Focal Length of Lens		*Plate-Lens Separation*					
		Scale 1 : 1		*Scale 1 : 5*		*Scale 1 : 10*	
in.	*cm.*	*in.*	*cm.*	*in.*	*cm.*	*in.*	*cm.*
1	2·54	2	5·08	6	15·24	11	27·94
2	5·08	4	10·16	12	30·48	22	55·88
3	7·62	6	15·24	18	45·72	33	83·82
4	10·16	8	20·32	24	60·96	44	111·76
5	12·70	10	25·40	30	76·20	55	139·70
6	15·24	12	30·48	36	91·44	66	167·64

In the general case, for a lens of focal length = F and a magnification = M (times the size of the subject), the plate-lens separation S, measured from the back nodal point of the lens, is given by the equation:

$$S = F \times (M + 1)$$

The distance from the lens to the subject, measured from the front nodal point of the lens, is given by the equation:

$$D = F \times (1 + \frac{1}{M})$$

Advantages of Back Focusing. A camera with back focusing is a great advantage in macro work. This arrangement allows the lens to remain fixed in relation to the subject while the back is removed to focus the image. In this way the scale of magnification remains constant and focusing becomes more sensitive. When the image is focused by moving the lens only, comparatively large lens shifts may—especially near same-size reproduction—not lead to any plainly visible change of sharpness over a fairly wide range. The transition from sharp to unsharp becomes even more rapid at close range when the whole camera—with the lens and film fixed relative to each other—is moved nearer to or farther away from the subject. Such an abrupt change of sharpness is generally regarded as facilitating exact focusing.

Shutter. Instantaneous shutter speeds are not required in normal macro work; the brilliance of the image is so weakened by the amount of magnification that only time exposures are possible. All that is needed is a shutter that can be operated through a long flexible release without transmitting any vibration to the camera. Even this is not always necessary; the exposure may be made by switching the light on and off in an otherwise dark room. The exposure may also be made with a black card held in front of the lens after the dark slide has been drawn.

The situation is different when photographing moving specimens with electronic flash illumination. In this case the short duration of the flash is the thing that arrests the movement. As it lasts for less than 1/1000 second it can be synchronized to work with a fast shutter speed. By doing this, the shutter is not open long enough for the normal room illumination to form a secondary image. So for this particular technique it is an advantage to have a shutter synchronized for electronic flash at 1/100 second or less.

Lens. Most better quality lenses are perfectly suitable for macro-photography. Symmetrical lens designs are generally preferable. The main requirement for a macro-camera is adequate lens extension. Even for same-size reproduction the lens must be moved out to double its infinity distance from the image plane. This usually requires special close-up gear, such as extension tubes or extension bellows. They follow the general principles of such equipment for close-up photography, except that in macro work the over-all extension must be even greater to yield appreciable magnifications.

Normally camera lenses are computed for best definition at short image distances and long subject distances. In macro-photography these conditions are reversed, especially with unsymmetrical lens designs. At high magnifications it may be an advantage to use a normal camera lens reversed so that the rear component faces the object. This sometimes has the disadvantage that the diaphragm can only be adjusted from within the camera.

A standard lens also has the drawback that the large camera extension required for magnifications much above same-size may make it difficult to keep the whole camera set-up rigid. Any vibration is in addition magnified by the increased lens to film distance.

For this reason special short-focus macro lenses are available. For miniature cameras these may have focal lengths of 1 in. or less and are intended for specific magnification ranges. (A 1-in. lens, even with a normal covering angle of some 60°, will cover a 24 × 36 mm. negative area comfortably at same-size reproduction since the lens is then 2 ins. from the film. The same lens would not be able to cover that negative area when focused on infinity.)

Macro lenses need not have a large maximum aperture since the lens is usually stopped down appreciably to give a reasonable depth of field.

Depth of field in macro-photography depends, for a given standard of sharpness, only on the lens aperture and on the magnification. For a given scale of magnification it is therefore immaterial whether a short focus or a long focus lens is used. Where it is impossible to approach the object really closely, a lens of longer focal length has however the advantage that it gives the required scale of reproduction at a longer lens-to-object distance.

Stop Values. When a normal camera lens is used for macrophotography at abnormal plate-lens separations, the marked stop values no longer apply. The amount of light passing through the lens is the same irrespective of the camera extension, but at abnormal extensions it has to spread over a much bigger picture area. This means that for a given stop diameter, the bigger the picture, the weaker the image. And a weaker image needs a longer exposure.

The effective aperture, Ae, for a lens of focal length F with a rated aperture of Ar when working at an extension E, can be found from the equation:

$$Ae = \frac{E}{F} \times Ar$$

For example, if a 2 ins. lens stopped down to *f* 8 is being used to form an image magnified 5 × life size, the camera extension will be approximately 12 ins. So the effective aperture will be given by:

$$Ae = \frac{12}{2} \times \frac{8}{1} = 48$$

—i.e., *f* 48

The small *f*-value of a lens working at the great extensions common in macrophotography explains why the photographer must either use a very intense light source or be content to take pictures of still subjects which will allow him to give long exposures.

Of the two alternatives intense lighting is preferable, especially for colour photography. With very long exposures reciprocity failure may make it difficult to estimate the correct exposure time. For this reason flash lighting (see below) is ideal.

Camera Types. The two most suitable systems for macro-photography are on the one hand technical cameras with their double or triple bellows extensions, and on the other miniature and roll film reflex cameras with interchangeable lenses. A camera with a fixed lens, while usable for ordinary close-up photography with the aid of supplementary lenses, is not suitable for macro-photography. Here the power of supplementary lens that would be required to achieve magnifications above same-size reproduction usually leads to unacceptable deterioration of image quality.

For high magnifications with a technical camera a short focus lens is usually ideal. As already mentioned, this arrangement permits shorter camera extensions for a given magnification.

With miniature cameras the necessary lens-to-film separation can be achieved by extension tubes or extension bellows inserted between the lens unit and the camera body. If the camera is not itself a reflex model, a reflex housing, focusing stage or other focusing aid permitting visual focusing on a ground glass screen is required.

The same considerations apply to roll film reflexes (which also must take interchangeable lenses). In this case the scale of the set-up is necessarily also more bulky. Extension tubes and bellows for high magnifications are more readily available for 35 mm. miniature cameras.

Miniature cameras also have certain limitations. On the one hand the film capacity is an advantage where whole picture series are to be taken of the same subject or of similar subjects under identical conditions. But if conditions vary from frame to frame, it becomes more difficult to give uniformly correct exposures. Where only one or two photographs are to be taken, cutting off and processing short lengths of 35 mm. film tends to be cumbersome.

Normal Lighting. For still subjects almost any form of artificial light will provide satisfactory illumination; the power is unimportant since the exposure can be prolonged indefinitely. For preference, however, the light should be in the form of a beam of parallel rays of light rather than a flood of soft diffused rays. The appeal of macro pictures is in the normally invisible detail that they show, so that a fairly hard cross lighting is usually best.

An enlarger lamphouse with the negative carrier and lens removed makes an excellent source of illumination, and equally good pictures can be made with a good focusing hand torch. The light from the torch is generally better for being slightly diffused through a sheet of ground or opal glass.

When a hard light is shone across the specimen to emphasize detail, a second fill-in light or a white card reflector must be used to throw some illumination on to the shadow side of the subject. If this is not done, the parts in shadow will be completely black and featureless.

If an enlarger lamphouse is used it is often useful to fit a Photoflood in place of the normal lamp. The extra brilliance of the Photoflood compensates for the weak illumination of the enlarged image; it makes focusing easier and shortens the exposure. At the same time it generates a great deal of heat and must be used with caution, particularly on live insects or delicate plants.

Flash Bulb. Flash bulbs are a useful form of illumination for special subjects—e.g., for natural history photographs of living specimens in the field. The bulb can be brought close enough to the subject to make instantaneous exposures possible and as there are no weighty batteries involved, the set-up can be made easily portable.

It is often an advantage to use the flash bulb without a reflector when working close to the subject in this way. The reflector gives too broad a source of light when it is only a few inches away from the subject. This means that the shadows have no definite edges, and crisp detail is absent. Using the bulb alone reduces the width of the source and gives better rendering of detail and surface texture.

Flash bulb illumination is apt to be expensive for this type of work where a large number of exposures may have to be made to get a few acceptable pictures.

If the camera can be supported, open flash technique may be used at very small apertures, but with the popular type of midget flash bulb it is even possible to make synchronized exposures of 1/25 second and shorter on medium speed pan film at lens apertures of *f* 11 and smaller.

Electronic Flash. Electronic flash is perhaps the most suitable of all forms of illumination for macrophotography. It is bright enough to enable small apertures to be used, and its short duration arrests the movement of the liveliest insect. It is equally good even for still subjects because it cuts out any risk of camera shake during exposure. The cost per exposure is negligible, and the size of the source is small enough to give good definition at normal working distances.

A 100 joule flash will provide enough illumination to give fully exposed negatives magnified 10 × life size on medium speed pan film at apertures as small as *f* 36.

Ring Illuminator. A special type of electronic flash tube for very close working is manufactured in the form of a ring. The ring is big enough to encircle the lens of most cameras. A ring illuminator of this kind produces virtually no shadows since from the point of view of the lens looking through it, the light falling on the subject is uniform from all directions. For this reason the detail of the subject must consist of well-defined tone or colour differences or it will not show up at all. Any subject where the interest is mainly texture or detail in relief is better photographed with side-lighting. Ring illumination is useful in photographing cavities and it is the standard lighting for deep intra-oral dental photography.

Transmitted Light. Macro subjects are usually opaque, and so are normally illuminated from the front. But, particularly in laboratory work, it is often necessary to photograph specially prepared sections—e.g., of botanical or pathological specimens. Such sections—and other semi-transparent subjects like insects' wings—are best photographed by transmitted light. In this case there is no need for the exact methods of subject illumination that are necessary in photomicrography. Any of the light sources used for reflected light illumination may be used for transmitted light simply by directing them on to the back of the subject (which is held in position on a sheet of glass, or mounted like a microscope specimen).

A wide variety of effects can be obtained by changing the character and angle of the illumination. If the light shines directly from behind the subject, all the opaque parts will appear in silhouette and the result will be a strong black image without half tones. If the light shines obliquely on the subject the outlines of details will appear to glow against a dark background. (The appearance is similar to that produced by dark ground illumination in photomicrography.)

It is never a good arrangement to shine an undiffused light directly on to the back of the specimen as any uninterrupted rays fall on the lens and create flare and light fog. Where the details are to appear in silhouette, it is sufficient to place a sheet of white paper behind the subject and light it with a spot of light so that the subject is seen against a brilliant white ground.

With any system of transmitted light, there is a risk of halation and irradiation. These effects destroy the definition of fine detail, so they should be minimized by using backed plates and the minimum exposure.

SUBJECTS

The technique of macrophotography may vary according to the subject. There are two important divisions—still subjects and moving subjects.

Still Subjects. For still subjects there is no need for either reflex focusing or brilliant illumination.

The camera should be fixed firmly to the same table top or optical bench as the support for the subject to minimize the effect of vibration.

The set-up may be arranged vertically or horizontally. A horizontal arrangement is more suitable when a comparatively large and heavy plate camera body is used to provide the necessary bellows extension. A vertical arrangement is more suitable for a set-up consisting of say, a ¼-plate camera body and a long extension tube.

The light source, screened to prevent direct rays from falling on the lens, is mounted on an adjustable support so that it can be directed on to the subject from any angle.

In use, the subject is mounted in front of the lens and focused sharply on the screen with the lens at full aperture. The subject-lens distance is adjusted to give an image of the required size, and the lamp is moved about until the illumination is satisfactory. For most subjects the light should shine at an angle of about 45° from the front and to one side of the subject. A piece of white card is then held or fixed to reflect light on to the shadow side of the subject and adjusted until the shadow detail looks bright enough on the focusing screen.

The lens is then stopped down—generally to its smallest aperture—and the exposure made by covering and uncovering it with a piece of black card (after withdrawing the dark slide from the film holder if a technical camera is used).

Moving Subjects. For moving subjects, reflex focusing is essential. Some workers use an ordinary reflex camera, others prefer a miniature reflex or a miniature camera with a reflex attachment. The miniatures are handier for work in the field, but for extreme magnifications a miniature lens in a normal reflex camera is probably the best combination.

If the set-up is to be reasonably portable, the extension—and magnification—must be kept to the minimum. In this case the flash bulb or tube must be mounted on the camera; side lighting is out of the question. This means that some crispness of definition must be sacrificed.

A stationary arrangement can be constructed with a longer extension and this will handle greater magnifications. It also allows the flash to be placed to one side of the subject with a reflector to illuminate the shadows. This arrangement, whether vertical or horizontal, is suitable for photographing living insects, etc.

When an electronic flash tube is mounted on the camera in a portable set-up, the lens stop is dictated by the distance of the flash from the subject. With a stationary set-up, any stop may be used because the exposure can be regulated by moving the flash relative to the camera.

When working with flash it is generally necessary to employ an independent focusing light. This is because the magnified image is too feeble to be focused easily by the normal room or day light. A good electric torch, mounted on the side of the flash head and focused on the subject gives enough light.

In use, the focusing light is switched on and the lens opened out to full aperture while the subject is arranged and focused. When the image is satisfactory in position and sharpness, the lens is closed down to the working value and the release is pressed.

It is always a great help if the movement of the subject can be limited in some way so that once its image has been focused on the screen it cannot move either out of focus or off the plate. If this can be done, as soon as the image has been focused sharply, the lens can be stopped down to its working aperture and the subject observed directly. This enables the photographer to press the shutter release the moment the subject is in a favourable position instead of having to risk losing the picture at the critical instant. F.P.

See also: *Close-ups; Depth of field; Extension of camera; Focusing stage; Optical calculations; Photomicrography; Reflex attachment; Supplementary lenses.*

Books: *Camera Close-up*, by O. R. Croy, and *Camera Copying and Reproduction*, by O. R. Croy (both London).

MADDOX, RICHARD LEACH, 1816–1902. English doctor and amateur photographer. Inventor of the first workable silver bromide gelatin emulsion (1871). He made his ideas public "to point the way", without patenting this revolutionary process and passed his last years in straitened circumstances. Received the Progress Medal of the Royal Photographic Society in 1901 for his invention, which, with improvements by others, had laid the foundation of the dry plate and film industry.

MAGAZINE. Light-tight container for holding lengths of film or (in older versions) stacks of sheet film or plates. It is designed to be attached to or inserted in a still or cine camera, exposed, and then removed in normal light. A magazine therefore allows a camera to be operated continuously without the need for a darkroom for reloading, and it affords complete protection of the sensitized material before and after exposure. In many cases the film can be removed from the camera in the middle of its run with the loss of no or only a single exposure. Such a magazine incorporates a dark slide or similar device for rendering the film opening light-tight while the magazine is removed or interchanged against another.

35 mm. Magazines. Several 35 mm. cameras are designed for use with interchangeable magazines or magazine backs. These are almost invariably designed for a specific camera type and feature

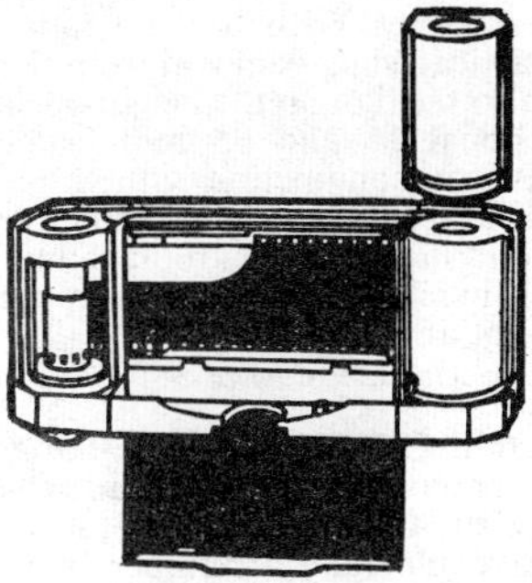

INTERCHANGEABLE 35 mm. CAMERA BACK. This type of magazine locates the film accurately in the focal plane of the camera and permits changing in mid-film and reloading without rewinding. Safety interlocks to prevent fogging of film however make manipulation slightly involved.

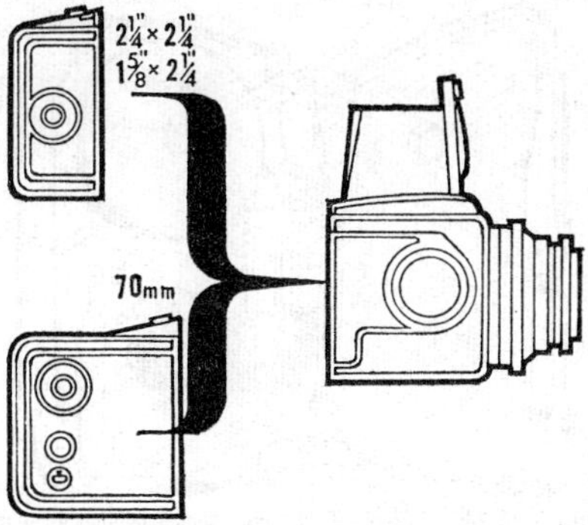

INTERCHANGEABLE ROLL FILM MAGAZINES. The use of several pre-loaded magazines with a roll film camera designed for them permits a rapid interchange of film types as well as picture formats such as 2¼ × 2¼ inches, 1⅝ × 2¼ inches, etc., or a choice of roll film and 70 mm. film.

a film feed chamber, take-up chamber (with built-in spool), film track, and frame counting mechanism. A film indicator is usually also fitted. The magazine is loaded with 35 mm. film in standard cassettes, and on insertion in the camera engages the film transport mechanism of the latter. Interchanging magazines in the middle of the film is a simple operation, and the operative steps involve closing the film aperture with a dark slide or similar device.

In a 35 mm. camera the film plane must be correctly located within a small fraction of a mm., so this aspect calls for special precision in the design of the magazines as well as the camera. Two common systems are automatic film track location and mechanical camera location. In the first a pressure plate pushes the film forward into the film track of the camera itself when the magazine is loaded into the camera. This always positions the film correctly, but calls for fairly complicated design. Mechanical location systems rely on the magazine making contact with a specific and accurately machined part of the camera, locating studs or the frames of camera body and magazine respectively, usually serving this purpose. In this case the magazine has its own film track, and the precision of location is subject to wear and tear of the location surfaces, incidence of dust, grit, etc.

The magazines frequently incorporate safety interlocks to prevent accidental removal from the camera while the film aperture is open, accidental operation of the camera while the film aperture is closed, accidental opening of the magazine in daylight, etc.

Working with magazine backs permits rapid changing over from one film-type to another, and also makes possible rapid continuation of work at the end of a film, provided a second pre-loaded magazine is available. The use of magazine backs is sensible only if at least two such magazines are available.

Bulk Magazines. Magazine backs holding from 35 to 100 or more feet of film are available for certain 35 mm. cameras. These are intended for long shooting sequences where frequent reloading would be inconvenient (e.g. instrument recording, micro filming, etc.). Here the magazine is loaded directly from bulk film. It incorporates safety features similar to those of the normal interchangeable magazine-back, to prevent accidental fogging of the film.

In automatic recording cameras bulk magazines frequently are linked with automatic film transport devices.

Roll Film Magazines. These are available for specific roll film cameras (usually taking size 120 or perforated 70 mm. film) and also for press and technical cameras. In the latter case they are also known as roll film adaptors, and fit on to the camera back in place of the usual sheet film or plate holder. Roll film magazines for specific cameras again have to be designed individually for the camera. In principle they work in the same way as 35 mm. magazine backs, but usually have their own film track and are located by the fitting together of accurately machined surfaces on the camera and the magazine respectively. They are loaded with normal rolls of film (mostly size 120) or with cassettes of 70 mm. film.

Sheet Film Magazines. Magazines to hold six units of 4 × 5 ins. sheet film are made by one of the leading manufacturers of press-type cameras, to fit their own and similar press and technical cameras. All or any of the sheets may be removed for processing at any time without the remainder being wasted. Magazines holding different types of film can be interchanged easily and rapidly. Such magazines are particularly useful when it is necessary to keep exposed material in a particular order.

Cine Magazines. These are produced for cine cameras in gauges from 8 to 35 mm., and are available in various types. Standard magazines for certain 8, 9.5 and 16 mm. cameras are sold pre-loaded with film and are returned to the

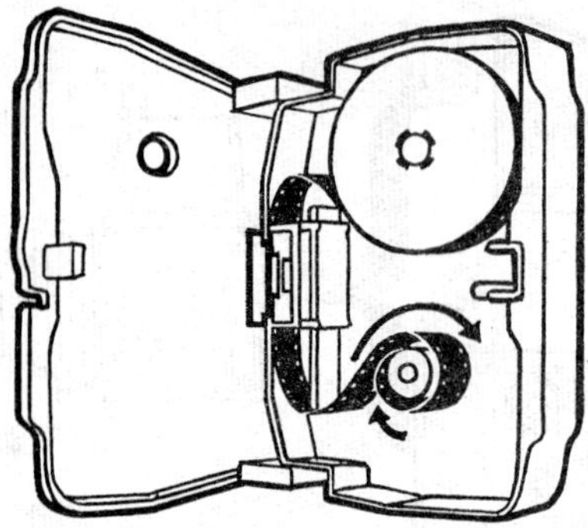

8 mm. CINE MAGAZINE. This takes spools of double-8 mm. film and is then easily inserted or changed round inside the cine camera, eliminating the need for elaborate loading manipulations.

maker for processing. The magazines may incorporate their own film transport movement to link up with the camera drive, or they may let the film run out of the magazine in a loop through a light-tight channel, for threading into the camera film track, where it engages with the film claw or other transport mechanism. With double 8 mm. cameras the use of a magazine speeds up the change over between the first and the second run of the film. With a few cameras the whole magazine turns over with part of the camera body, thus eliminating reloading.

Certain 8 mm. and 16 mm. magazines can be loaded by the user from standard 25 foot or larger spools. With 16 mm. and 9.5 mm. gauge the magazine itself can usually be removed from the camera in mid-film, with the loss of only a few frames. With double 8 mm. film this is not practical, since removal from the camera will fog also the film on the neighbouring track.

Professional 16 and 35 mm. cameras are mostly used in conjunction with robust magazines which are loaded from bulk supplies by the user and incorporate a footage indicator to show how much film remains unexposed in the magazine at any time. L.A.M.

See also: *Cassette, Cine Camera.*

MAGAZINE CAMERA. Before the advent of roll film, many attempts were made to design a camera with a plate magazine incorporated so that it could take a number of exposures without the inconvenience of changing plate holders.

Some of these cameras held a dozen plates in a holder which kept the foremost plate pushed forward into the focal plane. After exposure, operation of a lever released the front plate and allowed it to fall into a storage chamber in the base of the camera. The next plate was then pushed into position automatically by a spring. These devices were temperamental and unpredictable in action and the attempts to improve them were abandoned on the arrival of the roll film.

A recent series of roll film cameras are designed to take magazines which are merely dropped into the camera, making it immediately ready for action. No spooling, rewinding, etc., are necessary, and the magazines are discarded on processing the film.

See also: *Camera history.*

MAGAZINE PHOTOGRAPHY

The United States has exerted a strong influence on the styles and methods applied to magazine publishing in other countries. There is a world-wide tendency to follow the established patterns of American practice, on which this article is based.

Since the big picture magazines emphasize the picture story, it probably could be called the basis for magazine photography. It is the picture story which has been popularized most in latter years, and it is the picture story approach which holds the interest of most of the younger photographers.

Most general magazines which feature text also use pictures and many have one or more picture stories in every issue. The specialized, trade publication is also a major market.

Despite the growing, widespread interest in pictures for publication, the increasing number of photographers in the magazine field makes it a highly competitive one.

The Photographer. The average magazine photographer is a freelance. In the United States the majority are concentrated in New York City, with the second largest group deployed in and around Hollywood, but most major cities have one or more photographers who use them as bases for specializing in magazine photography in regions often consisting of several states.

Many magazine photographers have studios, with thousands of dollars worth of equipment. They usually specialize editorially in fashions or portraits or home suggestions, and may do most of their work for advertisers. Many others work with only a few small cameras but, since the advent of electronic flash, a travelling photographer may be so weighted down with paraphernalia that transportation becomes a serious problem.

To light up the whole arena when a circus appeared at Madison Square Garden in New York City, one photographer using colour employed 2,800 pounds of electronic flash equipment. A radio remote control set-up was also used, with a transmitter on his camera and a radio receiver activating the electronic flash. Flash bulbs are seldom used by the magazine photographer, whilst much of his work is still done in natural light.

Many magazine photographers spend most of their time travelling. In one ten-day period, one freelance reported having shot hockey in New York one night, baseball near San Francisco the next morning, football at Los Angeles the next afternoon, another baseball game on the Mexican border the next day, a horse race at Santa Anita the next morning, another football match at Los Angeles the following day, and three ski stories at Sun Valley the next three days. Then, following a quick flight to New York and back to Los Angeles, he covered a travel story on the Pacific, en route to Hawaii.

The American freelance may be independent, or be represented by a picture agency. He usually works for a day rate, which may or may not be supplemented by a space rate, or for a story guarantee. Almost always, expenses are in addition to payment for his time. Although most photographers naturally prefer not to, some will work on speculation. They may or may not be members of the American Society of Magazine Photographers, which began as a professional society but changed its status to that of a union; it is seeking acceptance of a code which has as one of its central features a minimum daily rate of payment for magazine photographers.

The Pictures Wanted. Generalizations are difficult because magazines use all types of photographs, but a great deal of work—both black-and-white and colour—is done in 35 mm. and 120 size. In colour, the editors favour as large a transparency as the subject matter allows for best results, but not at the expense of editorial impact.

With its beginning in 1936, the pioneering American picture magazine, *Life*, presented this credo: "To see life, to see the world, to eye-witness great events, to watch the faces of the poor and the gestures of the proud, to see strange things, machines, armies, multitudes, shadows in the jungle and on the moon; to see man's work, his paintings, towers and discoveries; to see things dangerous to come to; the women that men love and many children; to see and to take pleasure in seeing; to see and be amazed; to see and be instructed; thus to see, and be shown, is now the will and new expectancy of half mankind."

Since then, television has whetted even more the desire of a visual-minded public to see as much as it can. And although the public is absorbed vitally in important events, it is just as interested when photography gives significance and meaning to the familiar, the commonplace, the ordinary. A magazine photographer can find dramatic, exciting, moving pictures in everyday life and day to day events.

Magazines of this type see pictures from all the major news services and are a focal point for submissions from the agencies and established photographers—but they are usually interested in contributions from anyone who thinks he has a printable picture, amateurs not excepted. In general they should be addressed to the Contributions Editor, but in the event of a major news break the nearest staff representative of the magazine could be contacted. If pictures which are submitted are trimmed, it is always a good idea to send a set of contact prints too.

Despite the general emphasis on the picture story, many magazines are also a market for outstanding single pictures. In Great Britain and other countries many of the leading illustrated magazines, have, in fact, ceased publication in recent years. There is today a greater demand for material suitable for specialized publications, rather than for the general features magazines, and also a greater market for single pictures and small features, instead of photo-stories.

Working Organization. The bulk of work for the leading American magazine is done by thirty-four staff photographers (thirty-two actually staff, two more on a contract basis for their exclusive services) and many freelances, some of whose work appears so regularly that they come to be identified with the magazine. Of the thirty-four, the biggest single group is located at the home office in New York. The others are distributed among the various domestic and foreign bureaux. Los Angeles and Chicago, for example, each have four staff photographers; Washington and Paris, three each.

Each staff photographer owns his basic equipment, but receives an annual amortization allowance. Many photographers own much in excess of their basic equipment, and the magazine also has a pool of equipment available to photographers working on assignment.

Some magazine photographers take their place among the great journalists of this day. In covering news, when the magazine is necessarily faced with following the newspapers, besides trying for better pictures, emphasis is placed on comprehensive, analytical, penetrative reportage. A picture taken from a fresh angle may be the difference. Frequently, effective layout is the answer.

Most photographers customarily work from a script presented by one of the magazine's special departments (e.g., religion, science, modern living, fashion, etc.) into which the editorial staff is divided. This script presents, as clearly and specifically as possible, the reason for the story, major angles to be developed and, when possible or necessary, definite picture ideas.

The script generally is merely advisory or suggestive. It gives the photographer a working basis for his story, but does not bind him to specific details (except in certain cases, such as a science or fashion story, where a particular point must be made in a particular way; and even in these instances the photographer may devise a better way).

All departmental scripts are presented to the picture department and the picture editor holds daily meetings with his assignment staff to consider them. When a script is approved, it becomes the function of the picture editor, with the help of his aides, to select the photographer.

Depending upon the time and scene of the story, the picture editor makes every effort to choose, from among his available staff and freelance photographers, the right man or woman for the story being considered. Some photographers complain that, as a result, there is too much "typing" of photographers. This complaint certainly has some basis but does not take into account one prominent photographer's comparison of photography to spaghetti: "It all depends on the sauce." In this case, the "sauce" is the photographer's style.

Many magazine photographers consider themselves artists—and a few are. Better reproduction facilities permit a magazine to regard a photographer's mental and emotional approach as much more important than the mechanical. A certain technical competency is assumed, but a technically poor picture which tells a story is preferable to a better quality picture which does not.

Editorial Routine. A magazine photographer is generally accompanied on an assignment by a reporter, i.e., a researcher, a junior editorial worker who helps with arrangements and writes captions. Frequently the photographer never sees the pictures he takes until they are published. The film is sent by express delivery to the darkroom, where it is processed and delivered to the negative readers who edit the contact prints into story form; this is done with the help of the photographer and/or researcher, if either can be present, or from captions. Although most of the bureaux have their own processing facilities, the average story—in the interest of speed in handling—is sent to the central office, which returns a set of contact prints to the photographer.

The photographer may have taken hundreds even thousands of pictures, of which only relatively a few have any chance of being used, even when the story is successful. This overshooting has a purpose: greater slectivity.

In the form it is presented to the editors for consideration and possible layout, the picture story is printed full scale on glossy, single-weight paper in sizes of 8 × 10 ins. or 11 × 14 ins. for the more outstanding pictures.

A picture story ranges from a page and a half to the "photographic essay", a major presentation, of usually seven to nine pages, either in black-and-white or colour, often with the photographer's credit line. In making photographs the photographer should supply some pictures of vertical, and some of horizontal, format. Each feature should have a key picture —one picture which pinpoints the essence of the whole story.

Normally, the magazine's last black-and-white pages close four days before publication. A black-and-white cover would have been selected a week before and the first pages of an issue usually close nine days before publication.

The bulk of the colour pages are scheduled about nine weeks in advance, but up to six pages an issue may be allotted to "fast colour", with a normal sixteen-day closing. This development offers the opportunity of presenting stories, as well as single pictures in colour on a quick schedule. R.Md.

See also: *Freelance photography; Photo journalism; Picture series; Press photography; Reproduction fees; Selling photographs.*

Books: *How to Take Photographs that Editors will Buy*, by R. Spillman (London); *Pictorial Journalism*, by L. Vitnay, J. Mills and R. Ellard (New York); *The Technique of the Picture Story*, by D. D. Mich and E. Eberman (New York).

MAGENTA. Secondary colour, of a purplish-red shade, complementary to green. Sometimes also called minus-green—i.e., white with the green component removed leaving only blue and red. Magenta dyes are used in subtractive colour synthesis to colour the image formed in the green-sensitive layer of an integral tripack colour film.

MAGENTA CONTACT SCREEN. A contact screen composed of magenta dyed dots of variable density, used for making half-tone negatives in the camera.

See also: *Half-tone screen.*

MAGIC LANTERN. Early name for a slide projector. The term is still sometimes used by non-photographers.

See also: *Projector (still).*

MAGIC PICTURES. One-time popular photographic novelty. It consisted of a sheet of apparently plain paper on which an image was made to appear by warming the paper or breathing on it.

One method of making such pictures was to bleach out a fixed, but not toned, albumen paper print. By pressing this into contact with a piece of cloth or blotting paper damped with hypo solution, the image could be made to reappear.

See also: *Printing processes (obsolete).*

MAGILP. Mixture of linseed oil and mastic varnish, originally an oil-painter's medium, but also used as a thinner in the oil and bromoil processes and for imparting a gloss to matt or textured surface bromide prints. Also known as megilp.

See also: *Doping prints.*

MAGNESIUM. Used as ribbon or wire to provide a cheap and convenient source of light for still photography and contact printing. The ribbon, 100 cm. of which weighs about 1 gram, can be lighted with a match. It burns with an intensely actinic light—and, unfortunately, a considerable amount of white smoke.

As it oxidizes rapidly, the ribbon is stored in an air-tight box, just as much as required being pulled through a slit.

The light from a given length is reasonably constant, so that an exposure can be repeated exactly by burning the same length of ribbon at the same distance.

The exposure required for magnesium ribbon depends upon the weight of magnesium burned. 1 gram of ribbon (about 5 ins. of ribbon 1/12 in. wide) gives a light with a guide number of approximately 30 when used with a medium speed pan film. The approximate length of magnesium to give a fully exposed picture of a subject in a room with light walls and ceiling is shown in the table below.

Where the length is too great to be handled easily, the ribbon can be roughly woven into a square mat and lighted at one corner.

EXPOSURES FOR MAGNESIUM RIBBON

Distance	*Inches of Magnesium Ribbon Required at Aperture f 8*		
Feet	*Film Speed Slow*	*Film Speed Medium*	*Film Speed Fast*
5	10	5	2½
10	30	15	7½
13	45	22½	11½

At one time, magnesium was also very commonly used in powder form as a constituent of flash powder. A typical formula is as follows: magnesium powder, 6 parts; potassium chlorate, 4 parts; manganese dioxide, 1 part.

Flash photograph is now nearly always done with flash bulbs or electronic flash.

See also: *Flash powder; Light sources.*

MAGNET FILM. Narrow gauge or full width (35 mm.) cine film with standard perforations, fully coated with a magnetic oxide layer instead of a photographic emulsion. Also known as magnetic film. Used for live sound recording in professional motion picture practice. The perforation of the film ensures exact and consistent synchronization of the sound with the picture.

The sound recording from the magnetic film may be transcribed (after appropriate editing of picture and sound) on to an optical sound track to be combined with the image on the final release prints.

Alternatively a magnet-film with the final sound track may be run through the reproducing equipment in synchronism with the film being projected.

See also: *Sound recording; Sound synchronization.*

MAGNETIC STRIPE. Narrow strip of a magnetic iron oxide layer applied near the edge along narrow gauge cine films. The stripe serves for recording and playing back a sound track accompanying the film. It is applied in various ways, either during the manufacture of the film (immediately after coating the emulsion) or after processing.

See also: *Sound recording; Sound synchronization.*

MAGNETIC TAPE. Sound recording medium in magnetic tape recorders. It consists of a flexible base of cellulose acetate, PVC or polyester, ¼ in. wide and 0·002 in. or less thick, coated with magnetic iron oxide—normally on one side only. The tape is supplied wound on to plastic reels in standard sizes and lengths.

The playing (or recording) time of a spool of tape on a particular machine depends on the recording speed of the machine, the number of tracks recorded over the width of the tape and the thickness of the tape (since the thinner the tape the more can be wound on a spool of a given size).

Tape is manufactured in four different thicknesses, known from their relative playing times per spool as Standard, Long Play, Double Play and Treble Play. The thicker tapes are more robust, less troubled by "print through" (transference of the record between layers on the spool), and cheaper for a given length than the thinner tapes. However, the thinner tapes allow more material to be recorded on a given spool size. For example a 7 in. spool can hold up to 3,6000 feet of treble play tape—giving a total playing time of 6 hours at 3¾ ins. per second on a twin-track recorder. Being more flexible, thin tapes can be held in closer contact with the face of the recording head. On quarter track recorders where the actual width of the recording is less than 1/16 in., poor contact between tape and head seriously affects the reproduction quality. For this reason, Standard tape is never used on quarter track recorders where quality is important, and in any case this tape is virtually obsolete.

Perforated Tape. Some tape attachments for coupling a tape recorder to a cine projector employ perforated tape to ensure accurate synchronization between sound track and picture. This tape is perforated to run over the standard 8 mm. cine film sprockets. The perforations occupy half the effective width of the tape so the sound track may consist of a single half track recording or two quarter track recordings.

Magnetic Film. For sound recording in motion picture studios a 16 mm. or 35 mm. perforated film is sometimes used, carrying a magnetic oxide coating in place of the photographic emulsion. This has the advantages of perforated tape, and also permits recording with a wider track or with several tracks on the same film. The film is edited together with the picture film and then serves for the production of a final release print of a motion picture with an optical sound track.

Striped Cine Film. The functions of recording tape and cine film can be combined by coating the magnetic iron oxide on to the edge of the photographic film. The coating is applied in various ways—e.g. as a full stripe (0·1 in. wide) on the unperforated edge of single perforated film or as a half stripe (0·03 in. wide) on both sides of single perforated film. A balancing stripe is in most cases applied to the opposite edge of the film to the one carrying the record to enable the film to be spooled up evenly. Film of this type is used principally for the production of the final married print of a sound and picture squence separately recorded.

Video Tape. This is a special type of magnetic tape used for recording the television picture signal. Because of the greater amount of information it has to carry, video tape must be wider and run at a higher speed than normal sound recording tape. A typical 2 ins. wide tape runs through the video tape recorder at 6–12 ins. per second. The signal is, however, recorded across the tape by a fast rotating set of recording heads. The effective head-to-tape speed may be as high as 1,500 ins. per second for broadcast standards, or as low as 200 ins. per second for closed-circuit TV requirements. Where the information to be stored is simpler, or a lower standard of picture quality is acceptable, a narrower tape may be employed—e.g. 1 in. wide. Even ¼ in. tape with a linear speed of 120 ins. per second has been used for video recording. F.P.

See also: *Sound recording; Tape recorder.*
Books: *Tape Recording and Reproduction*, by A. A. McWilliams; *The All-in-One Tape Recorder Book*, by J. M. Lloyd (both London).

MAGNIFICATION

Relationship, usually represented by the letter M, between the size of the object photographed and the image of it formed by the lens. The sizes of the object and image are proportional to their distances from the nearer nodal points of the lens. If the image distance is v, and the object distance u, then

$$M = \frac{v}{u}.$$

Basic Equations. In normal photography where the object distance u is large compared with the image distance v, the image is very much smaller than the object and M is a fraction. For this reason the relationship between the two is often expressed as a reduction, r, which is the reciprocal of magnification—i.e.,

$$r = \frac{1}{M}.$$

On the other hand, it is equally common practice to regard M as expressing the ratio of the image and object sizes without any implication of actual increase in size—i.e., when M is less than unity, the image is reduced; when it is larger than unity, the image is enlarged.

The magnification can also be expressed in terms of the focal length, *f*, of the lens:

Taking the fundamental equation for an object distance u and image distance v,

$$\frac{1}{F} = \frac{1}{u} + \frac{1}{v}$$

and multiplying both sides by v, then

$$\frac{v}{F} = v\left(\frac{1}{u} + \frac{1}{v}\right) = \frac{v}{u} + 1$$

or

$$v = F\left(\frac{v}{u} + 1\right)$$

and since

$$M = \frac{v}{u} \text{ (above)}$$

then

$$(1)\ v = F(M + 1)$$

and

$$(2)\ u = \frac{F}{M}(M + 1) = F\left(1 + \frac{1}{M}\right)$$

These formulae for v and u are the fundamental formulae of magnification and they also

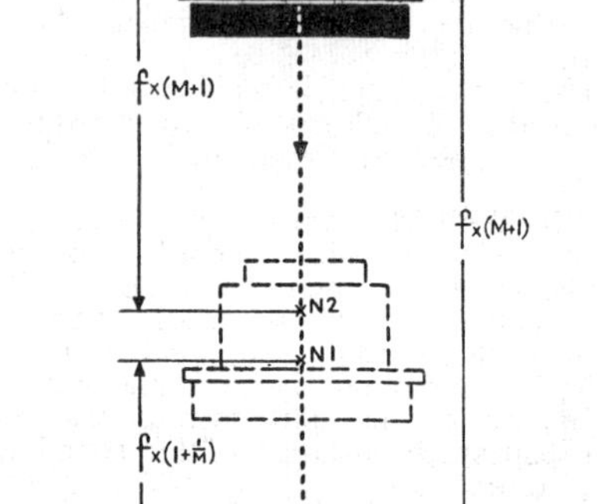

ENLARGER MAGNIFICATION. The relative positions of negative and paper depend on the focal length of the lens and the magnification employed. With a given separation of negative and enlargement there are two lens positions at which a sharp image is formed; in one case an enlarged image, and in the other a reduced one (dotted lens). The factors are: f, focal length; M, magnification; N_1 and N_2, nodal points.

represent the relative separations that will produce a magnification = M. From them it is possible to calculate every image and object size and the separation relationships encountered in photography. Some typical examples are worked out below.

(1) What is the maximum scale of reproduction that can be obtained with a 6 ins. lens and a camera extension of 9 ins.?

$$(v = 9, F = 6)$$

from equation (1)

$$9 = 6\,(M + 1)$$

therefore

$$M = \frac{9}{6} - 1 = \frac{1}{2}$$

In other words, using the full extension, the negative image will be half the length or width of the object.

(2) An enlarger has a maximum negative-lens separation of 5 ins. What is the smallest image it will give with a 3 ins. lens?

(u = 5—since the negative is the "object" — F = 3)

From the equation we get:

$$u = F\left(1 + \frac{1}{M}\right)$$

$$5 = 3\left(1 + \frac{1}{M}\right)$$

$$\frac{1}{M} = \frac{5}{3} - 1 = \frac{2}{3}$$

therefore M = 1·5.

In other words, the image on the enlarger easel will be one and a half times as long or as wide as the negative.

(3) A 5 cm. lens is to be used for macro-photography up to a magnification of 10 ×. What total lens-film separation (v) must be allowed for?

$$(F = 5, M = 10)$$

from equation (1)

$$v = 5\,(10 + 1) = 55 \text{ cm.}$$

This (less the maximum extended length of the camera) would be the length of the extension tube required to give an image enlarged 10× with the lens in question.

In practice, the image and object distances in all the above examples must be measured from the nearest nodal point of the lens, and the distance separating the nodal points must be added to the over-all length of the resulting set-up.

If the expressions for object and image distances are added, the result will be the total separation of the object and image planes (neglecting the relatively small separation of the nodal points)—i.e.,

$$v + u = F(M + 1) + F\left(1 + \frac{1}{M}\right)$$

$$= F\left(M + 1 + 1 + \frac{1}{M}\right)$$

$$= F\left(M + \frac{1}{M} + 2\right)$$

For any given total distance of v + u, there are two values of M that will satisfy the equation. This simply means that there are two possible positions of the lens which will give a sharp image; one which gives a magnified image, and the other reduced image. The only exception is where the lens is midway between the planes and M = 1; in other words, when the object and image sizes and the corresponding separations are equal.

Exposure Factor. The light passing power of a lens depends both on the actual lens diameter and on the distance from the lens opening to the image plane (i.e. the film). In normal photography of distant to reasonably close subjects the lens-to-film distance does not vary a great deal, and for exposure purposes we can assume this to be approximately equal to the focal length of the lens. It thus becomes possible to specify the light-passing power of the lens opening in terms of the *f*-number—i.e., the ratio of the focal length to the optical lens diameter. This is convenient because it provides a way of calibrating the lens openings of any lens in terms of their exposure effect.

With very near subjects, photographed at magnifications of 0·1 and larger, this approximation is no longer accurate enough. The exposure must then be increased by a factor which is the square of the actual lens-to-film distance divided by the square of the focal length of the lens.

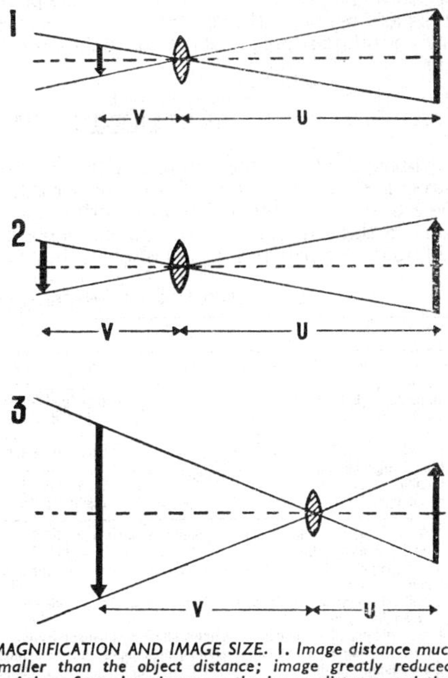

MAGNIFICATION AND IMAGE SIZE. 1. Image distance much smaller than the object distance; image greatly reduced. 2. A long focus lens increases the image distance and thus yields a larger image. 3. Reducing object distance and increasing image distance also increases scale of reproduction. In all cases: u, object distance; v, image distance.

In practice this exposure factor can also be expressed in terms of the magnification, which yields a very simple equation:

$$\text{Exposure factor} = (M + 1)^2$$

Thus for a close-up in half natural size ($M = 0{\cdot}5$) we get an exposure factor of $(0{\cdot}5 + 1)^2 = 2\frac{1}{4}$. For full size reproduction the exposure factor becomes $(1 + 1)^2$, or 4 X. At high magnifications it increases very rapidly.

This exposure factor applies only where the lens is extended beyond the infinity position for close focusing. With supplementary lenses the effective lens-to-film distance remains almost unchanged and no exposure increase is necessary.

Magnifying Devices. The magnification of the image is decided by the ratio of the object and image distances, so any device that either shortens the object distance or lengthens the image distance, will increase the magnification.

(1) The distance of the object from the lens can be shortened easily by bringing the lens closer to the subject. When this is done, to maintain sharp focus, the distance between the lens and the image plane must be correspondingly increased—e.g., by camera extension or by adding extension tubes. If the lens has a relatively long focal length, this increase will be considerable. To reduce the extension to practical proportions at large magnifications the focal length of the lens must be kept as short as possible, either by fitting another lens, or by adding a supplementary lens to the camera lens.

For macrophotography up to a magnification of about 10×, 5 cm. is the longest practicable length. In dealing with microscopic objects the lens may have a focal length as short as 1/12 in. and work almost in contact with the subject.

(2) Where it is not practicable to bring the lens close to the subject, the magnification can be increased by increasing the lens-image separation—i.e., by using a lens of longer focal length or (up to moderate amounts) by adding a negative supplementary lens. In this case magnification is a relative term; the image is generally much smaller than the object, but larger than with the normal camera lens.

The telephoto lens is a special case because it gives a magnified image without requiring any increase in camera extension over a normal lens. The explanation is simply that the construction of the telephoto lens brings the rear nodal point in front of the lens. As the image-lens separation is measured to the rear nodal point, the effective distance is much greater than the physical separation or the separation given by a normal lens.

The magnifying effect of a telephoto lens is not quoted in terms of the ratio of image to object size; it is expressed as the power of the lens, a figure (written 2×, 3× and so on) that indicates how many times in linear measurement the telephoto image is bigger than the image of a normal lens with the same back focus. F.P.

See also: *Close-ups; Macrophotography; Optical calculations.*

MAKE-UP FOR PHOTOGRAPHY

Make-up serves the photographer in two ways: it eliminates the need for most or all of the retouching on the negative and print, and it adds to the attraction of the model. Make-up will obliterate patchy skin tones, blemishes and lines, and it can also change the shape of the whole face or of individual features.

Corrective Make-Up. Corrective make-up is make-up applied to change the proportions of the face—e.g., to narrow a too-broad nose or

MAKE-UP FOR BLACK-AND-WHITE PHOTOGRAPHS (ON PANCHROMATIC MATERIAL)

	WOMEN	*MEN*
Pan-stik or Pan-cake make-up	6N	9N
Powder	Translucent	Tech. Special
Eyeshadow (Lining colour)	6 (Blue-grey) or 22 (brown)	22 (Brown)
Pencil	Light Brown, brown brownish-black, black	Brown, brownish-black
Lipstick	Sparkling scarlet	———
Moist rouge (lips)	———	T-3
Mascara	Brownish-black, black	Optional
Dry rouge	Light Tech.	Dark Tech.

Women with dark or tanned complexion, use 7N Pan-stik or Pan-cake. Men with dark or tanned complexion use 10N Pan-stik or Pan-cake.

To highlight areas of face, use Pan-stik or Pan-cake several shades lighter than foundation colour: 3N for women; 6N for men.

To shadow areas of face, use Pan-stik or Pan-cake several shades darker than foundation colour: 9N for women; Light Egyptian for men.

Erace may also be used to lighten shadows under eyes.

Women may also use blue, green, or blue-green eyeshadow sticks as well as 6 or 22 lining colour. Available with or without iridescence.

Lighter shade than Sparkling Scarlet lipstick is Clear Clear Red; darker shade is Brilliant Red.

Mascara is available in Cake, Cream or Wand.

Dry rouge is used to give more natural appearance to completed make-up.

Pan-cake may also be used as body make-up.

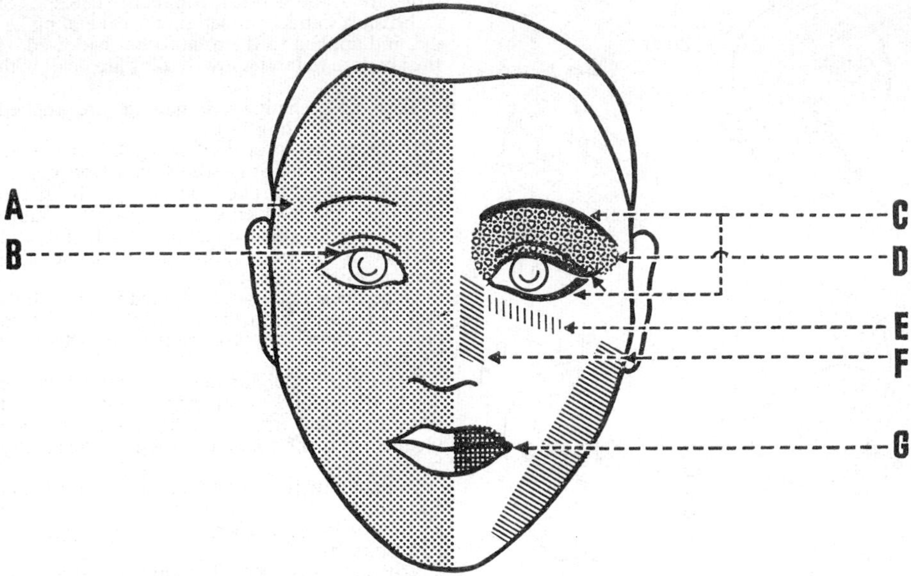

MAKE-UP CHART. A, foundation; determines the general tint of the complexion. This is also applied to all visible body portions (e.g., neck, shoulders, etc.). B, eyelash make-up; usually black for dark complexions, left untouched for light faces. C, pencil for eyebrows and eyes. D, eye shadow. E, highlights (under eyes, also around mouth and nose; generally applied with a wet brush, and blended into the foundation with a dry brush). F, shading; often applied initially underneath foundation, the latter then being rubbed away to bring up the shadows. G, lip-stick or lip paint; may be applied with a brush.

jaw, or to remove a double chin. This is done by applying suitable highlights and shadows with make-up of the right colour. It is important to use make-up colours that look right to the sensitive material in the camera.

Colours that merely look right to the eye may reproduce in a completely different tone in the photograph.

It has become common practice for normal portraiture to use panchromatic films or plates in conjunction with the range of make-up colours specially produced for the purpose, such as Max Factor Panchromatic Make-up. The brightness of each shade of make-up colour is indicated by a number.

A highlight is normally made three or four shades lighter than the foundation colour, and a shadow, three or four shades darker. For example, if the foundation colour or basic skin tone has been applied in 6N Pan-Cake, then 2N or 3N would be used for the highlight and 9N or 10N for the shadow.

Application of the Make-Up. The face to be made up must be clean, dry, and free from any grease or oil. The shadow areas are first filled in with Pan-Cake shadows; using a sponge for

MAKE-UP CHART FOR COLOUR PHOTOGRAPHY (EASTMAN FILM 5251 (35 mm.) and EKTACHROME 7255 (16 mm.))

	WOMEN	*MEN*
Pan-stik	7-25B; 7-25BC; 7-25CN	7-25EN; 7-25FN; 7-25GN
Powder	Translucent	Tech. special
Creme rouge	Flame	None
Dry rouge	Flame or Light Tech.	Dark Tech.
Moist rouge (Lips)	———	T-3, T-2 (Dark)
Lipstick	7-22; 7-22B	———
Eyeshadow (Lining colour)	7-34 (Grey-brown) No. 3 (brown)	No. 3 Brown
Eyeshadow Stik	1R. Brown, 1R. Grey	———
Fluid Eyeliner	Brown, brownish-black; black	———
Eyebrow pencil	Light Brown, brown, brownish-black	Brown, brownish-black
Mascara	Brown, brownish-black or black	Optional

Apply Pan-stik foundation and blend very thin.
Eyeshadow colours other than grey or brown may be used for more dramatic eye effect.
Pan-cake shades, if preferred to Pan-stik, are 7-25A, B, C, D, E, F, G, H. Pan-cake may also be used as body make-up if necessary.
For Kodacolor (stills) and for Ektacolor (stills) use Society shades of make-up. Select Pan-stik or Pan-cake foundation one shade darker than natural skin tone. Lipstick shades in medium pink, clear red or orange tones photograph best. Dark blue-reds are to be avoided.

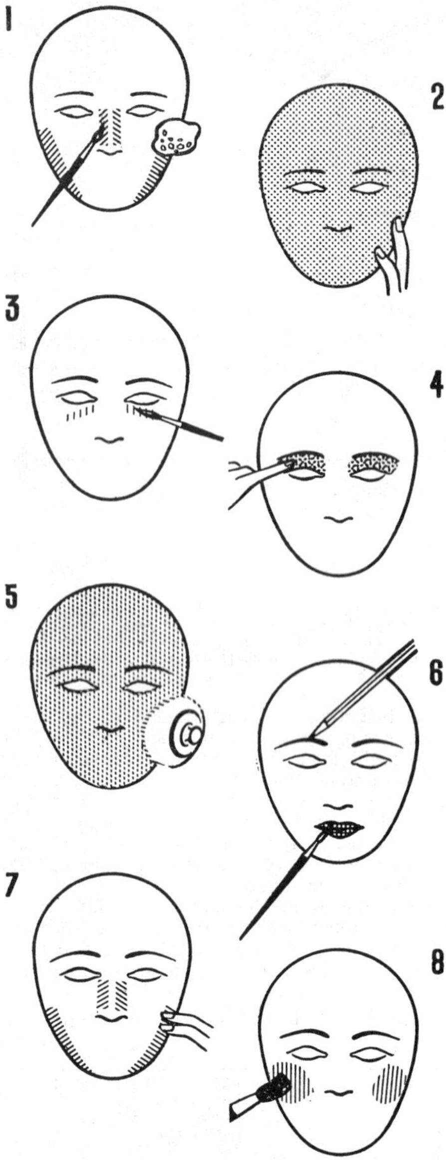

APPLYING MAKE-UP. 1. Apply basic shadows with make-up cake (except eye shadow). Use brush for small areas, sponge for large ones. 2. Cover whole face with foundation cream. 3. Apply highlights (under eyes, lines around mouth or nose) with wet brush, and blend into foundation with clean dry brush. 4. Moisten eyelids with cream and apply eye shadow with finger tips. 5. Powder over make-up with powder puff. 6. Line eyebrows with eyebrow pencil and apply eye-lash make-up. Also make up lips using a suitable thin brush. 7. Rub shadow areas with finger tips (as applied in the beginning) to make them show up through the foundation. 8. Apply cheek rouge with a brush and blend carefully into face tone.

large areas and a brush for small. The sponge or brush is wetted, rubbed on the cake of make-up, and applied to the area to be shadowed. If the jaws are to be shadowed, they are dealt with first.

Then nose shadows, if needed, are applied with a flat blending brush.

If the make-up is used in stick form—e.g., Max Factor Pan-Stik—the foundation colour is put on first, stroked on straight from the container and blended with fingertips or damp sponge to a smooth even texture. Highlights and shadows are applied afterwards with blending brushes or the finger tips.

Once the shadows are all filled in, the whole face is covered with the Pan-Cake foundation colour, completely concealing the shadows previously applied.

Pan-Cake highlights are next applied with a wet brush under the eyes and in the lines around the mouth and nose. The highlights are then blended into the foundation with a clean, dry brush.

Eye shadow is next applied, the eyelids being first moistened with a little cream so that the eye shadow will blend to an even tone with the Pan-Cake base.

When the make-up is completely dry, it is lightly powdered and buffed all over with a powder puff to produce a smooth skin texture and remove excess Pan-Cake which dries as surplus residue.

The whole face is now brushed with a powder brush.

To complete the make-up, the eyes are lined with pencil, eyebrows are made-up, eye-lash make-up is applied, and finally, lipstick.

At this point the shadow areas are lightly rubbed with the finger tips. This makes the shadows appear according to the amount of rubbing, giving complete control of the final effect. Normally the shadows are rubbed until they are only just visible, in which state they will appear completely natural in the photograph. The effect is best judged by looking into a mirror at arm's length. If the shadow is just barely visible, it will photograph well.

The general effect is enhanced by brushing dry rouge sparingly on to the cheeks: a light shade (Max Factor Light Tech.) for women, and a darker shade (Max Factor Dark Tech.) for men.

This improves the modelling of the cheekbones and adds sparkle to the eyes.

Make-up that has been on for some time and lost its freshness may be restored by powdering over with a powder two or three shades lighter than the one originally used.

A model with a dry skin can prevent cake make-up from looking dry and flaky by using Invisible foundation cream before making up the face.

Photographic Make-up Kit. The following list gives the composition of a suitable kit for photographic make-up.

3 Pan-Cake, Pan-Stik or S.S. Grease (one for base, one for highlighting and one for shadowing).

1 White Lining (No. 12)
1 No. 6 Lining (dark grey)
1 No. 25 Face Powder
1 Moist Rouge (No. 2 Special)
1 Eyebrow Pencil
1 Eyelash Make-up
1 Cleansing Lotion or Cream
1 Skin Freshener
1 Powder Puff or Powder Brush
1 Lip Brush
1 Sponge

If colour film is to be televised in black-and-white, use a minimum of cheek rouge. The foundation colour chosen may be a shade darker than the normal skin tone. M.F.

Books: *Photographic Make-up*, by J. Emerald; *The Technique of Film and Television Make-up*, by V. J. R. Kehoe (both London).

MANLY, THOMAS, 18??–1932. English photographer. Improved in 1898 the Marion transfer process for gum-bichromate pigment prints and patented it under the name Ozotype. In 1905 he described his Ozobrome process in which a gelatin silver bromide image is transferred by contact to a pigment paper. From this H. F. Farmer worked out the carbro process (1919) which was put on the market by the Autotype Company.

MANUSCRIPTS AND OLD DOCUMENTS

Normal document copying is carried out with special permanent equipment. But the photographer who is interested in copying old and rare documents in museums, libraries, churches, and private collections, almost always has to go to his subject and photograph it under unfavourable conditions with the equipment that he takes along with him. This is a special field of photography, and even of document photography.

Typical subjects include old photographs, maps, plans, drawings, etchings, printed books, and manuscripts, each presenting a different problem, and needing to be treated in a different way. Usually the photographs are required for illustrated articles or books on the subject they depict, as examples of a period in art or history, or for study and records. These subjects are usually found in public or private collections—and anyone privileged to photograph them should treat these old collections with the greatest of care, first because they are usually unique and for that reason above price, and secondly in fairness to other photographers.

Equipment. Sharpness and detail are most essential for this type of photograph. And while there is much to be said for the miniature camera, a 8 × 10 ins. or whole-plate camera is the most suitable for this sort of work—preferably of the square-bellows, back-rack-and-pinion focusing type. If miniature equipment is used it should be in the form of the single-lens reflex, which allows the use of interchangeable lenses, and of extension tubes or bellows.

For good definition, an anastigmatic lens of *f*7·7 at full aperture is all that is required as the subject is still and can be given a long exposure.

The most convenient focal length depends on the scale to be copied. Often a same-size picture is required; sometimes an even larger scale is wanted.

The following table indicates the camera extension necessary for various scales of reproduction:

SCALE AND CAMERA EXTENSION

Lens-to-Film Distance		*Scale of Image*
¼ more than focal length	=	¼ object size
½ more than focal length	=	½ object size
double focal length	=	same size
3 times focal length	=	twice scale
4 times focal length	=	3 times scale

Filters are essential for copying coloured, tinted, or faded originals. A 2× yellow and if possible a tricolour set of four should be included in the equipment, preferably of gelatin sealed between optical flats. The filters are best held behind the lens in a three-sided frame screwed or clipped to the lens panel. Final focusing should always be done with the filter in position, to avoid errors due to possible lens aberrations.

Because of the need for accurate focusing, a magnifier is most necessary—either one of the proprietary instruments or one of the linen tester type which does a good job of work and has the advantage of being pocket size.

An ordinary tripod should be avoided if possible because the camera must be dead square with the original, and even when the tripod has rubber feet it is an awkward thing to set accurately on a polished floor. The most useful thing is the centre pillar type of stand. This means extra weight to carry around, but the legs are completely adjustable for uneven surfaces and the head can be set to the exact height required with the turn of the handle. The tilting top also gives any angle required—even vertical—which means that the subject can be laid flat on the floor and photographed from above.

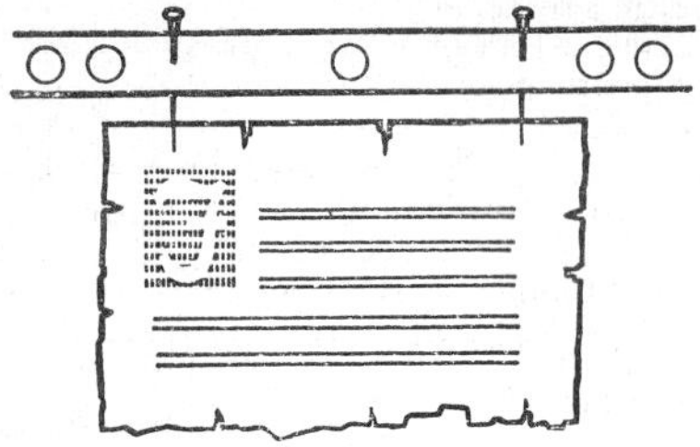

HOLDING OLD DOCUMENTS. *Two lengths of tape (at top and bottom) are pinned to an easel and long pins stuck through the tape to press flat against the document. This holds the document securely in position without any risk of damage—an important point with old manuscripts.*

Odd items of equipment that are useful for this work include: drawing pins, household pins (including very large ones if possible), strong tape about $\frac{3}{4}$ in. wide (bookbinder's tape is ideal if obtainable), a pair of clamps and at least four large bulldog clips with strips of chamois leather stuck over the jaws to prevent damage and grip more firmly; a couple of wooden wedges, about four inches long, one inch wide, and about one inch at the thick end, are useful for levelling; a length of good stout string and a pocket tape measure for checking distances. At some time all these items will be needed in setting up the various subjects dealt with. Improvisation is a valuable art.

The photographs may be taken on plates or film. The normal subject will usually be covered by a slow panchromatic or orthochromatic material, but sometimes a process or pan-process is better. This does not mean that all these types have to be carried around. It is nearly always possible to find out in advance which will be needed.

Setting Up. This is the operation that calls for improvisation. A given subject may be unmounted and it may not be permissible to stick pins through it. Even so it has to be placed so that it may be photographed. In the studio this can be done on an easel; away from home, anything flat that drawing pins can be stuck into—e.g., the side of a large carton, a wooden box, or even a pastry board—will do. Two pieces of tape are placed tightly across this easel in a horizontal position with two drawing pins at each end (one always comes out at the wrong moment). The distance between the tapes should be about $\frac{1}{2}$ in. greater than the height of the subject. Household pins are then pushed through the tape bringing the points toward edges of the subject to be photographed until the tip of the pins just press flat against it. One pin in each corner (and another in the centre of the tape if the subject is large) will hold the subject without damage, while the ends of the pins—if they show at all—can easily be touched out on the negative.

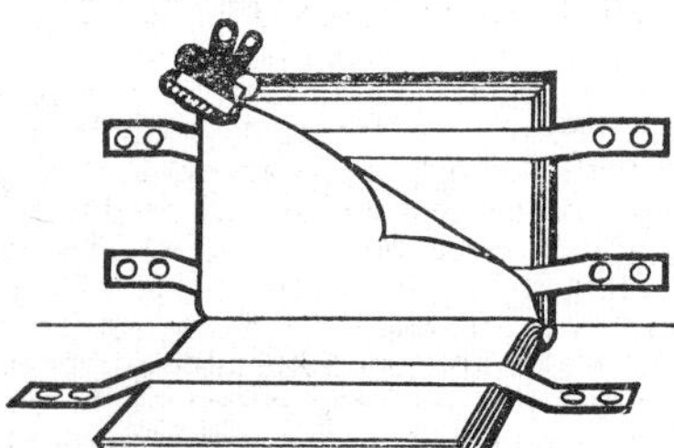

HOLDING OLD BOOKS. *The book page to be copied is set up vertically against the easel, taped down a few pages behind the one required, and the latter clipped with bulldog clips. The horizontal page not being taken is also taped down.*

For a book or manuscript page, a board as above is required, the book being opened at right angles with the page to be photographed in a vertical position against the board. Two or three tapes are pinned to the board across the vertical portion of the book, one or two pages behind the one to be photographed, the upper tape being about one inch from the edge. The page or folio to be copied is then laid open and clipped to those held by the tapes with a bulldog clip at each corner. To get the page to lie flat, as the clips are being fixed at each corner, place a piece of clean paper over the page to prevent rubbing or contact with the hand and gently press from the centre of the book diagonally towards each corner to remove any trace of a bulge. If the horizontal page not being taken tends to get in the way, tape will hold it down. A piece of black paper should always be laid on this page to cut out reflected light and prevent uneven illumination At home, a flat subject is best held in an old printing frame, or, if it is a book, in one of the special bookholders made for the purpose. Finally the wedges are used to set the easel absolutely vertical in all aspects.

Lighting. With flat subjects lighting presents no special problems. A pair of lamps are the best source since it is risky to rely on daylight. The lightweight music-stand type of lighting stand is excellent. For the light source, three 200 watt lamps (one spare) of the interior sprayed type are suitable. A length of cable and an assortment of plugs and adapters should be carried.

When working away from home it is as well to find out in advance the voltage where the work is to be done.

One lamp should be placed on either side of the camera and spaced according to the size of the subject to give even illumination over the whole surface. Great care must be taken to avoid reflections from subjects with shiny surfaces; if the lamps are placed too close to the camera there is a danger of reflection from the lamps on the surface of the subject. If reflections do appear, the lamps must be moved out sideways until the reflections vanish.

On the other hand, with an uneven or cockled chart or map, the lamps should be placed as close to the camera as possible to avoid shadows from the irregularities in the surface.

The lamps must never be placed so close to the original that the heat from them will cause the subject to curl, as, apart from the risk of damage, there is the possibility of the subject moving during the actual exposure.

Exposure. Hard and fast rules cannot be laid down and much depends on experience. There are so many factors to be considered—type and colour of original, reduction scale, type of plate, and lighting—that each subject must be decided on its merits. Even with an exposure meter, allowance must be made when the subject is close. When copying close up, the exposure indicated by the meter should be multiplied by

$$\frac{(\text{camera extension})^2}{(\text{focal length of lens})^2}$$

to arrive at the correct exposure. This is usually more convenient than recalculating the *f*-numbers, which alter when the lens is used at greater than normal extension.

For example: if the meter indicates an exposure of 10 seconds at a particular stop when photographing same-size with a lens of 8 ins. focal length (when the camera extension would be 16 ins.) the correct exposure would be

$$10 \times 16^2/8^2 = 10 \times 4 = 40 \text{ seconds}$$

Old Photographs. These may be of any colour from a good glossy black-and-white to a faded yellow ghost image, or even a combination of both. They may be flat, curled, or even badly cracked and crumpled. In this case, it is as well to try to persuade the owner to allow the picture to be pressed or even reglazed. If reglazing is undertaken the print should be given a light wipe over with cotton wool moistened with carbon tetrachloride to remove any greasy fingermarks.

For faded yellow prints or daguerreotypes a slow process or fine-grain ordinary plate will often work wonders. Orthochromatic plates are usually good for copying the black-and-white and brown tones. Really dark brown or red-brown prints, like dark oil paintings, should always be given a little more exposure than seems necessary.

Maps and Plans. These may be printed in plain line, coloured or tinted. The technique will depend on the colour to be brought out, and its background. In most cases a pan plate with a filter the same colour as the background will produce the desired effect. If the subject is badly stained, but the aim is to reproduce mainly the drawing then a pan-process plate and a red filter are called for.

If a map has to be taken in sections and the resulting prints joined together, the correct technique to avoid difficulty in matching up at the joins is as follows: take the first half, then turn the map upside down and take the second half, without moving camera, focus, or easel. Take four quarters in this order: top left quarter, bottom right quarter; reverse map; take top right quarter, and finally bottom left quarter, keeping the map on the same plane and moving nothing else. A slight overlap should always be allowed at all joining edges.

Manuscripts. Manuscripts to be copied may be paper or vellum and of these, those on vellum usually present the most difficulty. Often they are highly illuminated with a large amount of beaten gold leaf ornamentation. A fine gradation pan plate with a 2× yellow filter will give adequate colour correction. For the gold leaf, the secret is to capture sheen without shine, and a little manœuvring of the lamps is usually all that is necessary. Most owners of this type of manuscript forbid the use of glass over illuminations because of the risk of damage.

Vellum often contains holes which need backing up with white paper. It is important for the backing paper to cover the whole page or it will show through as a light patch. Vellum is also sometimes slightly transparent, and allows any writing to show through from the back. A sheet of black paper behind it will often help, but it will darken the whole sheet and the exposure must be increased accordingly. Lastly, one side of a sheet of vellum is often very different in colour from the other and if several folios are being taken, it may be necessary to allow for this by varying the exposure.

Processing. Normal processing techniques will cover most of the subjects mentioned above. Either contrast or normal developers are appropriate. Tank development may be used for a long series of similar subjects but dish development should always be employed for the more special ones. After two-thirds of the normal development time has gone by, the appearance of the image should be checked visually. The finished print is intended to reproduce the original faithfully and in some cases to bring back to vivid life an image that is fading away. A glance at the right moment will show whether to give a little extra development to increase the contrast or to stop before losing a delicate tint present in the original. D.A.W.

See also: *Antiques; Copying; Document copying; Macrophotography; Microcopying; Paintings and drawings; Philately; Supplementary lens.*

Book: *Camera Copying and Reproduction*, by O. R. Croy (London).

MAPPING. Production of maps by scientifically planned survey flights. This is one of many photographic methods of recording measurements.

See also: *Aerial survey; Photogrammetry.*

MAREY, ETIENNE JULES, 1830–1904. French physiologist. In 1882, began to use photography to analyse animal and human movements. Muybridge and Marey kept one another acquainted with their experiments and Marey invented a whole series of methods for registering complicated and fleeting physiological actions, notably the ophygmograph for heart action, a photographic gun (1882) for taking serial pictures of birds in flight; and in 1887 his "chronophotograph"; the second model, designed by Marey and Demeny in 1890, used flexible film and is a forerunner of the modern cinematograph camera. In 1892 Demeny and Marey devised the Phonoscope, a projection apparatus for synthesizing motion. Marey also invented slow motion cameras (700 images per second in the 1894 model).

MARINE PHOTOGRAPHY. The sea and its harbours and traffic of ships, yachts and small boats offers a wealth of subjects to the photographer. All these subjects are apart from ordinary photography as the sailor is apart from those who live and work ashore.

The photographer's approach to marine subjects is essentially pictorial; if his results do not convey something of the poetry of the sea and the drama of the life afloat, then he will have missed the most significant feature.

Cameras. The most convenient type of camera for marine photography is the miniature, but it does not necessarily produce the best pictures. It is small, which is an advantage when moving around in small boats, and it will make a series of pictures in rapid succession—a useful point when following groups of small racing yachts.

Then, although there is no difficulty in filling the small viewfinder of a miniature when standing on shore, it is not easy to do it accurately afloat. A slight error in lining up the subject may call for serious cuts in the picture area to get the horizon straight again in the print.

Finally, the negative given by the miniature is uncomfortably small, and when it is enlarged to a reasonable size, such valuable details as the delicate tracery of rigging and the texture of a bellying sail are apt to disappear.

For the serious professional a quarter-plate or even a 4 × 5 ins. camera is a more satisfactory instrument. The smaller amount of enlargement required for the bigger negative means that the essential detail can be preserved even when the subject and camera may both be moving.

Whatever type of camera is used, it must have a direct vision finder. It is extremely difficult to stand up in a boat and look down into a reflex or waist-level finder. With a direct finder, the eye observes the subject naturally and there is no difficulty in maintaining balance. It is always wise to allow a margin around the subject in the viewfinder so that the picture can be trimmed to make the horizon level. This tell-tale horizontal line in marine pictures shows slight errors in lining up that would never be noticed in a normal photograph.

The shutter must be speeded to at least 1/500 second for taking photographs of ships and yachts from another boat. It must be remembered that the camera is on a moving platform. (The fastest sailing boat never moves at more than 14 to 16 knots. At this speed, an exposure of 1/100 second would suffice to arrest the movement if the camera were operated from a steady platform, but a much higher speed is called for when the picture is being taken from another boat.)

One final essential is a really deep lens hood. This serves the double purpose of protecting the lens from flying spray and cutting off stray reflections from the surface of the water.

Filters. A good range of filters is an advantage because of the great variation of tone in sea and sky. It is very easy to under- or over-correct, and any errors will show up immediately.

When photographing sails against the sky, they must appear in their true tone relationship. Seascapes generally need more correction than landscapes, but too deep a filter kills atmosphere and loses the beauty of soft gradation.

A pale green filter and a series of yellow (pale, medium and deep) cover all the normal requirements of the marine photographer. The most convenient form of filter holder is the type that screws into the back of the lens hood; in small craft the photographer must be prepared to "step lively" and a slip-on filter is apt to fall off in the process.

Sensitized Materials. The miniaturist should use nothing faster than a medium speed panchromatic film so that he can retain the advantage of fine grain. In summer the light may occasionally be strong enough to use a slow panchromatic film where there is not a lot of subject or camera movement, but generally it is wiser to use the faster film and cut down the exposure time. The user of a quarter-plate or 5 × 4 ins. camera can afford to stick to the fastest panchromatic plates because at the normal scale of enlargement the coarser grain will not be troublesome.

Seascapes. The light at the seaside is very rich in actinic rays and the scene is mainly made up of blue sea and sky and light-coloured stretches of sand. So everything makes for very short exposures. It is usually advisable to give one-half to one-quarter the exposure that would be given to a normal subject inland.

Photographs taken on the open sea call for even shorter exposures—one-quarter to one-eighth of the exposure for a normal inland view.

Breaking waves and moving water generally should be photographed at a shutter speed slow enough to show a slight blur—i.e., at not more than 1/200 second. A faster shutter speed than this arrests the movement completely and makes the water look stationary and "frozen".

Flat lighting is useless for catching the sparkle of the spray and breaking waves; the right time to take such pictures is when the sun is shining across, or even slightly towards the subject.

Seascapes call for careful development: over-developing burns out the delicate highlights—e.g., of sunshine on white sails and hulls. A soft-working developer and a leaning towards under- rather than over-development gives the right sort of result.

Harbours. Ports and harbours offer a choice of subject material of such variety that it is impossible to give any general guidance. Since the water is still, however, it is always worth while to study the reflections of ships and spars and cordage for possible pictures. And the subject, whatever it is, should be simplified as far as possible. General views of harbours and craft are apt to be very confused and busy with distracting detail. Exposure should generally be based on the shadows.

When boats are packed tightly together in a harbour, a higher viewpoint will help to sort out the general effect of confusion. Side or back lighting also helps to separate the craft more and prevent their details merging.

There is nearly always plenty of activity worth photographing in harbours. Jobs such as the cleaning of boats or repairing of fishing nets are part of the everyday life in some harbours, and as such are an essential part of the atmosphere.

Equally important as subjects are the local characters to be found in many harbours. It is usually better to photograph these people by candid techniques rather than to ask them to pose.

Yachts. Sailing boats at sea form the most attractive of all marine pictures. These are subjects that can be found at every seaside resort, but the best opportunities occur where there are regular regattas. For the few shillings that it costs to hire a small motor launch, the photographer can make his own choice of groups or single boats, on a long tack with the lee-scuppers awash, or sailing before the wind with spinnakers hoisted and the crews taking it easy. But do not forget that powered vessels must, if necessary, give way to yachts.

Small boat sailing is a highly popular pastime and all around the coast weekly regattas are held by the various yacht clubs. It is not unusual to see starts of twenty-five to thirty ships on the Solent, the Clyde, and many seaside towns, so the photographer who wishes to specialize in this type of subject never need travel far to find his material.

If the photographer is working from a small boat, he will be unable to keep up with the race and should take up a permanent position either at the start or somewhere along the course where the yachts can be expected to sail fairly close.

A point somewhere on the course is generally to be preferred as small boats are apt to be regarded as a nuisance around the start.

Sunshine is essential in yachting pictures; without it, the sails instead of looking white and sparkling will look drab and dead. There is no particular angle from which yachts look best, but when a boat is lying well over it is better to be on the lee side so that the picture will show the deck and crew and not just the bottom of the hull.

So long as the photographer can get as close to the subject as he wishes, there is no point in using a long focus lens. A normal angle lens fairly close to the subject increases its size in proportion to the background whereas a long focus lens makes it look less imposing in proportion.

Shooting from a small boat is a knack that has to be learned. The photographer must stand with feet apart in line with the direction of the motion of the boat. He must keep his balance without supporting himself against a mast or the gunwale because that would increase the movement of the camera. And although the lens hood will keep a certain amount of spray off the lens, the camera should be held in the lee of the body except when shooting.

The best advice that can be given to anyone who contemplates taking photographs from a yacht during a race is—don't. At such times anyone but a working member of the crew is unwelcome, and in any case passengers have their work cut out in keeping out of the way.

Subject Contrast. Photographs near or on the sea tend to be excessively contrasty because of the brilliance of the light and the absence of reflected light in the shadow areas. The contrast can be reduced by exposing fully and cutting the development time, or by using a soft-working developer.

It is not possible to reduce the contrasts of scene by providing a reflector to throw light into the shadow areas. But the same effect can be arrived at by taking the photograph on a day when there are large masses of sunlit white cloud in the sky. At such times the contrasts of the scene are very much lower than when the brilliant sun shines out of a cloudless blue sky.

Exposure. The only safe guide to exposure with subjects at sea in sunshine is an exposure meter and even that must be checked by experiment before the reading can be accepted. On an average sunny day, exposure on a fast pan film with a medium filter will be of the order of 1/100 second at *f* 11, but it will vary according to conditions to considerably more or less than this figure.

Finally, as both sand and salt water are bad for fine mechanisms and exposed metal, the camera should be kept covered up or in its case and taken out only for the shortest possible time when it is absolutely necessary. F.W.B.

See also: *Seaside; Underwater photography.*
Book: *Camera Afloat*, by H. S. Newcombe (London).

MARKETING PHOTOGRAPHIC GOODS. The marketing of photographic goods has made rapid strides in the past decade, and the methods of marketing have also varied considerably. The turnover of all kinds of cameras, equipment and sensitive materials has also advanced and the availability and selection of cameras and accessories has extended considerably.

Imports come from Japan, Germany and a dozen other countries. It is estimated that there are over 12 million camera users in Britain, spending over £60 million a year on amateur photography alone. Marketing, therefore, is now a complex operation.

The Photographic Manufacturer. The larger manufacturers supply the photographic dealers direct, either via road or rail, or from warehouses distributed throughout the country.

Photographic manufacturers employ divisional sales representatives to call on dealers throughout the country, and link this up with trade advertising, exhibitions and sales training courses in the selling and use of their more specialized equipment. (British manufacturers sometimes get a price advantage over their foreign competitors, because there is no duty on home produced goods, but both carry the same purchase tax. On the other hand, British manufacturers may have to import raw materials at a greater cost than the price prevailing in some foreign countries, and labour costs may be higher in Britain.)

The manufacturer has many things to consider besides competitive quality and price. For example, with sensitized photographic materials he must also consider right storage, suitable packing and transport and the expiry date of his wares. Any one of these being faulty could have a disastrous effect on his marketing. Providing instruction booklets or leaflets with practically every item is essential. Continued sales depend on proper use of the equipment and material supplied, and therefore photographic manufacturers spend large sums on booklets, lecturers, advertising and sales and technical training to ensure that the dealer who sells them and the customer who uses them know the essentials. The manufacturer who invents and introduces new products may have to build up a possible market for these goods, and this raises total marketing costs.

Some manufacturers may supply the larger dealers direct, and the wholesalers will supply the smaller firms. The marketing of a product by a manufacturer has to be considered well in advance, for, not only must he consider the best time of the year to market the goods, but also the design, price, window displays and counter cards, advertising, packing, quantity to make, and sales policy and distribution.

The Photographic Importer. The job of marketing photographic goods is both comprehensive and skilled, and calls for business acumen of the highest order. Imports from abroad have to go through Customs to be cleared and the appropriate duty paid. In Britain and the United States there are a large number of photographic importers who handle vast quantities of goods during the year, and also pay the duty and see that the right amount of purchase tax is added to the price and collected by them for transfer to the proper authorities.

The photographic importer, having obtained the goods through Customs clearance, then advertises through the trade press and sends sales representatives round to the photographic dealers to get orders for the goods he has to sell. To make the distribution fair and ensure a reasonable livelihood for the importer, the manufacturer abroad usually only has one importer for his wares in each country. When the importer only has a small sales staff, he may also market his wares through photographic wholesalers. The task of the importer is important, for by stocking the various new lines as they become available he is helping to ensure that both professional and amateur photographers have the opportunity to purchase practically any of the cameras and equipment made throughout the world. A good knowledge of photography and photographic equipment is necessary if the importer is going to be able to recognize what will sell and what is really new on the market. He must also decide how much any piece of equipment should sell for to give both him and the photographic retailer a reasonable profit. Importers pay visits to camera works abroad to see the latest models being made, and to discuss with the manufacturers future likely demands and markets. (The photographic importers in Britain have their own Association—The Photographic Importers Association—which looks after their interests and welfare.)

While the average-sized photographic importer may only handle the goods, the larger ones may employ skilled staff to check the goods and skilled repairers for any equipment which proves faulty and which is covered by their guarantee. The photographic importer also has heavy insurance and security charges to meet.

The Wholesaler. Wholesalers throughout the country offer a good service to the photographic dealer, and particularly to the smaller dealer, who is in this way enabled to keep a small but comprehensive stock, and obtain quick local deliveries. The wholesaler fulfils another useful function in taking a considerable quantity of the maker's stock, thus helping the manufacturer's storage problems. Photographic wholesalers usually have a sales staff to cover specific areas but the larger ones cover the whole country.

The Retailer. It is through the photographic retailer that the final link in the marketing chain comes to fruition, for he meets the members of the public and is in direct touch with public demand. The photographic dealer and his assistants have to be knowledgeable and qualified people. Whereas there used to be but a few pieces of equipment to sell, today the dealer needs to know something about the many different types of cameras and accessories available and their manifold uses.

Photography is now used for many purposes, including medicine, science, aerial work, research, publicity, forensic medicine, astronomy, photomicrography, micro-photography, document copying and many more. The photographic dealer is expected to be able to advise the prospective customer on the equipment he needs, show him how to use it correctly, criticise his results, and help him to obtain improved results.

The photographic dealers in Britain are divided into three groups: the purely photographic dealer (who sells photographic equipment only); the chemist; and the photographer/dealer. The sizes of the firms may vary and whereas one dealer may sell films and the cheaper range of cameras and equipment, another may sell every type of equipment, and have projection theatres and special departments and assistants for specialized goods. The processing of films is now rarely done by the photographic dealer, but by a "finisher" who collects and delivers films for processing, and deals with both black and white and colour materials.

Staff training for dealers' assistants has become a necessity today. In Britain, assistants can take courses sponsored by The Photographic Dealers Association and a number of the larger manufacturers, or take courses at technical colleges in London and the provinces.

The cameras and equipment the photographic dealer has to sell today are supplied in the greater part from abroad, and nearly all this equipment carries a fairly high import duty and purchase tax. Thus the selling prices of equipment are higher in Britain than in most countries abroad. In Britain, photographic films and paper are made by several large firms in bulk, and some specialized photographic equipment is made, but the popular-sized still cameras and cine cameras and projectors are made abroad.

There are fewer photographic retailers per head of the population in Britain than elsewhere.

For example in Britain there are approximately only twenty photo dealers to each million of the population, while in Switzerland there are about 150. Other countries of Europe, e.g. West Germany, France, Denmark, Sweden, Belgium and Holland, have about sixty photo retailers to every one million of their population.

The business of selling cameras and equipment is still booming, and providing there is no recession of trade—or unemployment—it should continue for a large number of years. The production of photographic journals and text books has also undoubtedly increased the trade of the photographic retailer. A.Pg.

See also: *Production of photographic goods.*

MARKING UP. Technique of indicating final size and other details (cutting, retouching, etc.) on an illustration to be reproduced for blockmaking or other purposes.

See also: *Scaling for reproduction.*

MARTIN, PAUL, 1864–1942. English wood-engraver and photographer, known chiefly for his candid shots taken in the 1890s with a concealed hand camera in London streets and at the seaside. 1900–1908, worked as a free-lance press photographer; thereafter as a process engraver in London. His photographs of London by night, 1895–6, were the first of their kind. *Victorian Snapshots* by Paul Martin, (1939).

MARVILLE, CHARLES. Dates unknown. French photographer to Museums. In the1850s he recorded the quarters of Paris to be pulled down to make way for the boulevards of Haussmann.

MASER, OPTICAL. Light source of special characteristics depending on stimulated emission of radiation. Maser stands for microwave amplification by stimulated emission of radiation, derived from micro wave communication, but also applicable to light waves—hence the description "optical" maser. Nowadays generally known as a laser.

MASKING

Photographic masking of any kind is essentially the technique of interposing a mask—something which will reduce the quantity of light—in the path of the light forming an image and thus modifying the photographic result. As a simple example, holding the hands, or a piece of card, over part of the image in an enlarger for part of the exposure, masks the print. However, it is more usual in these days to limit the term to the use of a photographic image for that purpose and to use the word mask whether the image is positive, negative, a colour separation record or, indeed, a coloured record.

MONOCHROME MASKING

The method is used in this case almost invariably to reduce contrast, although not necessarily over the whole of the image. Consider the simplest case; a negative may have a good tone range, but too high contrast for convenient printing on the required material. The way to solve the problem is to make a very

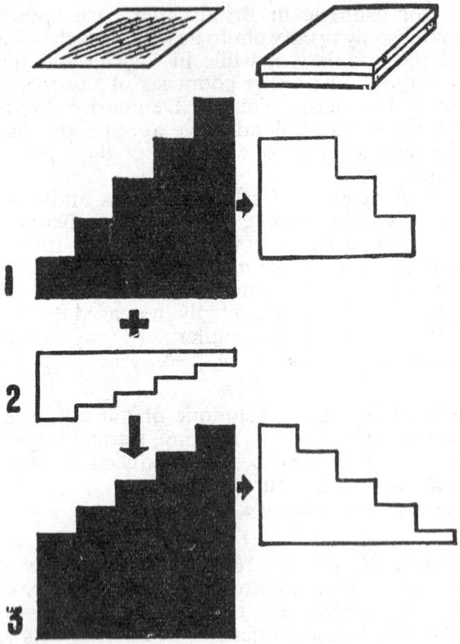

MASKING FOR CONTRAST. 1. A very contrasty negative may not record all tones on printing paper (limited contrast range shown by height of "staircase"). 2. To overcome this a low-contrast positive is made. 3. This is bound up with the negative, and reduces the tone range, enabling the paper to reproduce all the tones within its own contrast range.

weak and soft, but full-toned, positive contact print from the hard negative on another plate or film. When dry, this is bound into contact in accurate register with it. The mask will then tend to cancel out the negative and it will have the effect of considerably reducing the contrast of the negative, which may be very useful.

Selective Masking. In industrial photographs where often a machine, or other object, has to be photographed in a workshop against an obtrusive background, masking is a useful alternative to blocking. It permits the presence of the other machines to be shown in subdued tones.

The method consists of painting out the machine with opaque on the original negative. This is then contact printed on a plate and under-developed so that only the shadow tones of the background machines are weakly recorded. When the mask is dried and bound back in register with the negative (from which the opaque has been washed off), it prints with the shadow tones of the background reduced in weight. The background is still visible but is less obtrusive and the important machine stands out satisfactorily. The mask can be made to a suitable density to cancel out the background to any required extent.

Masking for Tone Control. Often the important tones of a photograph lie at the highlight end of the range (a bridal dress, for example) or at the shadow end (a railway engine, for example).

The negative of a predominantly light toned subject will usually record the highlight detail best on a more contrasty grade of bromide paper, but the shadow tones are then much too black. This can be overcome by masking the shadow tones by means of an under-exposed mask which records only the lower end of the range. A print is then possible on the more contrasty bromide paper which stretches out the highlight range considerably. Of course there is a commensurate loss of shadow detail, but at least the shadows are no longer too dark and the gain in highlight quality can be quite remarkable. (Uneven flattening in tone range can even be an advantage in some cases.)

At the other end of the scale, boosting the shadow quality is best tackled in two stages. The initial negative of the subject should be fairly soft and thin; that is to say, fully exposed and not fully developed.

The first stage of the masking operation is the production of a fairly soft contact positive on film of much the same quality as used for highlight improvement above. When that positive is dried a plate is exposed against the back of it and developed to produce a soft and very thin laterally reversed negative having detail in the shadow tones and an over-all even density of about 0·3 in all lighter tones. When

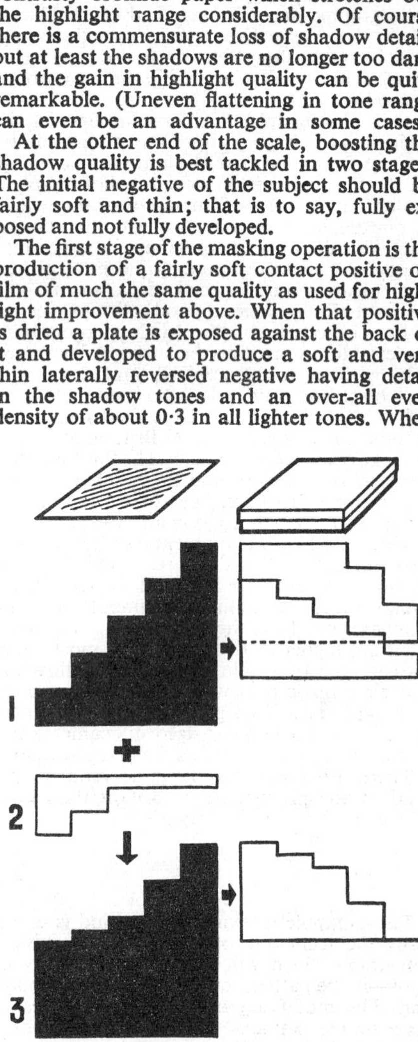

MASKING FOR TONE CONTROL. 1. Contrasty negative may record full tone range on soft paper, but result looks dull. On normal paper rendering is better, but some tones (e.g., shadows) are lost. 2. A low contrast positive mask of shadows only is made. 3. Bound up with the negative, this compresses shadow gradation without flattening highlights. Uneven compression of tone range yields better print.

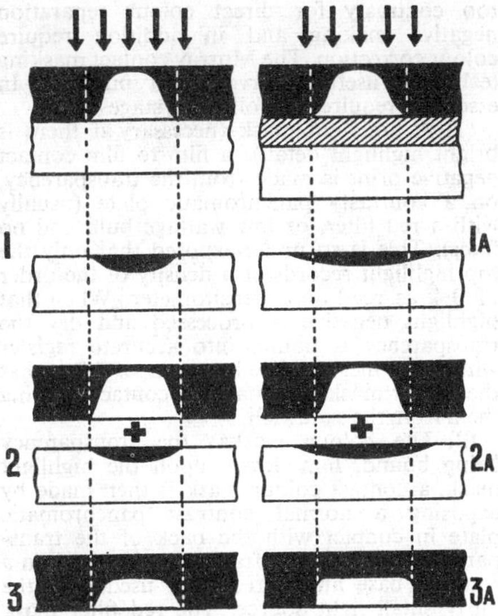

SHARP AND UNSHARP MASKING. Left: Sharp mask. 1. Mask made by contact exposure. 2. Negative and mask bound up. 3. Resultant density reduced. Right: Unsharp mask. 1A. Mask exposed with spacer between negative and positive. Blurred mask outlines larger than on negative. 2A. Negative and mask bound up. 3A. Density reduced; blurred mask outlines also cancel slight edge blurring of negative image, resulting in an appearance of increased sharpness.

the negative mask is bound up in register with the original negative the shadow tones are emphasized by the mask and all higher tones are evenly raised in density. There is thus a gain in total density range and the gain in quality is all in the lower end of the tone scale. This is an exception to the generality of masking effects because of the increase rather than decrease in density range; it works very well in practice.

Unsharp Masking. Most masking methods may be used with sharp or unsharp masks. It is claimed that unsharp masks improve the rendition of detail, which may seem a contradiction. What happens is that the blurred outline of the unsharp mask ("thickened" would be a better term) tends to cancel the slight blurring of the outlines that is present even in a nominally sharp negative. In effect it outlines each small piece of detail so that it does, in fact, become more plainly visible. In practice the visual effect is a general one of apparently greater sharpness; the outlining is too delicate to be seen in detail with the naked eye.

A suitable method of making the mask is to use a fairly small light source and to suspend it, at an angle, some distance above the contact frame in which the mask is printed. A thin sheet of transparent plastic (a spacer) is used between the negative or positive from which the mask is being printed. The printing frame revolves on a rotating table throughout the exposure.

COLOUR MASKING

Any colour print or transparency can be masked, when it is copied, in any of the ways described for monochrome masking above, but there is also the possibility of colour masking which offers considerable advantages in improving the colour reproduction and compensating for deficiencies in the colours used.

In theory, the yellow, magenta-red, and cyan-blue tricolour printing colours forming the image of a colour transparency or colour print should each perfectly transmit (or reflect) two-thirds of the spectrum of white light. The yellow should transmit the whole of the green and red from about 4,900 Angstrom units to the limit of vision in the red (around 6,800 A.), completely absorbing the remainder. The magenta-red should transmit the whole of the blue and the whole of the red from before the beginning of the blue (around 4,000 A.) to about 5,100 A., and from about 5,900 A. to about 6,800 A. at the visual end of the red, with a gap of absorption between them. Finally the cyan-blue should transmit the whole of the blue and green from about 4,000 A., to about 6,100 A. at the beginning of the orange-red, absorbing the remainder.

Unhappily only the yellow printing colour is really efficient; the magenta-red is fairly efficient in its red transmission, but very poor in the blue, also it does not absorb green as

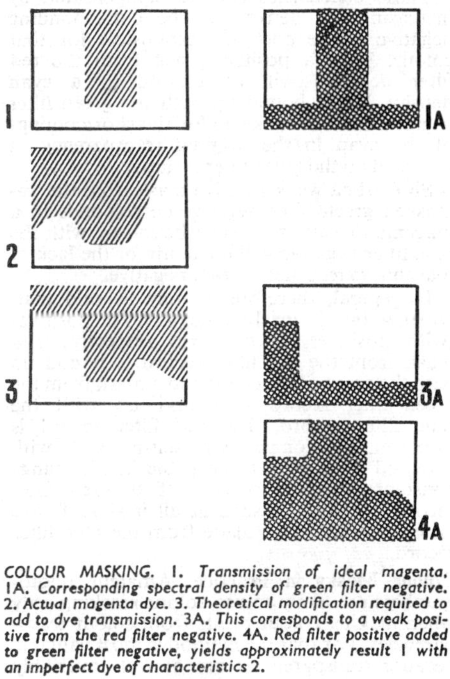

COLOUR MASKING. 1. Transmission of ideal magenta. 1A. Corresponding spectral density of green filter negative. 2. Actual magenta dye. 3. Theoretical modification required to add to dye transmission. 3A. This corresponds to a weak positive from the red filter negative. 4A. Red filter positive added to green filter negative, yields approximately result 1 with an imperfect dye of characteristics 2.

well as it should. The cyan-blue transmits blue only fairly well and green poorly; indeed that is the reason why magenta is commonly called "red" for it is not blue enough and cyan "blue" because it is not green enough. One might suppose that a magenta-red might be improved by adding blue to it, but that would reduce the red reflection and cannot be done. For much the same reason cyan-blue cannot be much improved. There are better magentas and cyans, but unfortunately they are fugitive and cannot be used.

In terms of percentage, on a basis of calling the total amount of blue, green and red reflected by good white paper 100 per cent of each, an excellent yellow may transmit 90 per cent each of green and red. However, magenta-red transmits 90 per cent of red and less than 40 per cent of blue, while cyan-blue transmits less than 75 per cent of blue and not more than 50 per cent of green. It is basically for this reason that the colour printings of any colour reproduction need to be colour corrected either by hand or by colour masking.

The colour filters used to take the colour separations or the responses of the three layers of transparency materials are qualitatively quite efficient and their quantitative faults can be dealt with by increasing the exposure time or the sensitivity of tripack layers; it is the printing colours which are the difficulty.

These deficiencies can be compensated by increasing the density of the corresponding negative image portions responsible for that colour. Thus a positive mask from the red filter negative will correspond to a cyan density. When bound up with the green filter negative this will make up for the shortcomings of the cyan in the original transparency as recorded on the green filter negative.

Similarly a weak positive mask from the unmasked green filter negative corresponds to a magenta density, and when bound up with the blue filter negative will make up for the lack of magenta as recorded on that negative.

In general, therefore, it may be said that, whether one is masking negatives or positives (with positives or negatives respectively) a mask from the red filter record is bound up with the green filter record and a mask from the green filter record is bound up with the blue filter record. The red filter record is rarely masked (or perhaps "self-masked" with a record from itself to keep the density range constant with the others). If the red filter record is colour masked at all it should have an extra weak mask made from the blue filter record.

Colour Masks in Practice. Amateur photographers are likely to use colour masking only when making colour prints from colour transparencies. For that the method is most valuable because transparencies are almost invariably too contrasty for direct colour separation negative making and in addition require colour correction. The Murray contact masking technique usefully serves both purposes. In essence it requires the following stages.

(1) A highlight mask (necessary if there is bright highlight detail): a film to film contact negative print is made from the transparency, on a contrasty panchromatic plate (usually with a red filter, or low wattage bulb and no filter). This is so under-exposed that only the top highlight records at a density of the order of 0·4, as read on a densitometer. When that highlight negative is processed and dry the transparency is bound into accurate register with it by means of cellulose tape. After that the colour masks are made by contact printing them from the sandwich.

(2) The colour masks: the transparency being bound, film down, upon the highlight mask, a contact colour mask is then made by exposing a normal contrast panchromatic plate in contact with the back of the transparency. The printing frame should stand on a rotating base and a red filter used over the fairly small light source. The red filter mask should have a minimum density of about 0·2 and the range from minimum to maximum density should be about 40 per cent of the density range of the transparency.

This is repeated with a green filter record which should have a minimum density range of about 0·2 and a range of about 30 per cent of that of the transparency.

(3) The separations: when the two colour masks are made the highlight mask is no longer required and is dispensed with.

The transparency is then bound into register first with the red filter colour mask (so that it is thus film side out). With that combination of transparency plus mask, both the red filter (cyan-blue printer) and the green filter (magenta-red printer) separation negatives are made by contact or projection on normal contrast panchromatic plates.

If a four-colour print is required (that is, with a black printing in addition to the colours) a fairly successful result can usually be obtained by using a yellow-green filter through the transparency plus red filter mask combination. The black printer is, however, always rather a problem and is best treated according to the main colours of the subject, often by exposures through each tricolour filter in turn, with modification of the timings.

In the case of the red filter exposure, for the cyan-blue printer separation, the mask serves only to reduce the high contrast of the transparency to a convenient level. The mask is itself a red filter record, and the result is said to be "self-masked".

But in the case of the green filter exposure, for the magenta-red printer separation, the effect of the mask is different. It usefully lowers the contrast range, but it also modifies the result selectively. In fact, it will be found that the blue and green colours of the transparency are recorded in the negative as much denser than usual, while red colours are recorded more transparent than usual. Therefore the magenta-red print from the negative is strong in the red colours and much weaker in the blues and greens—which is precisely what is required and would have to be contrived by hand retouching if not by this automatic photographic means.

Once the cyan-blue and the magenta-red printer negatives (and perhaps the black printer) are made, the transparency is removed from the red filter colour mask and registered on the green filter mask, that combination being used when making the blue filter (yellow printer) negative.

The density range of the separation negatives is, of course, governed by the printing process for which they are made (carbro, dye transfer, etc.). In general the minimum density should not be much less than about 0·5 and the range from minimum to maximum probably of the order of 1·0 for dye transfer and rather less for carbro according to the type of sensitizer that is being used.

Step Wedges and Colour Patches. Among very experienced colour workers there is some controversy about whether a step wedge laid beside the transparency should be recorded identically in each record of a colour set. In any case a black-to-white step wedge should always be included and colour patches are also useful, both for identifying the separations and for checking the success of the printing. A very expert worker might vary the contrast of the record of the wedge in one or more of the separations, especially if he wished to make some modification of the effect in the print. But until a great deal of experience is gained, it is best to aim to make all the separations so that the density range and quality of the record of the step wedge is as precisely matched as possible throughout the set.

Masking Positives or Negatives. The above description relates to the masking of positive transparencies with negative masks. Fortunately the same technique works equally well for the masking of negative subjects with positive masks. For example, a colour negative for one of the colour negative-positive processes can be masked exactly as described for a positive colour transparency. That is, a red filter mask is used for making both the red filter positive (in this case) and the green filter positive (perhaps also the black printer), while a green filter mask is used when making the blue filter positive separation.

Furthermore the same basic principle can be used for a set of separation negatives which are to be colour masked when making positives (for example for printing the reliefs for dye transfer).

A negative range of around 1·0 is suitable for printing dye-transfer positive reliefs, depending somewhat on whether printing is by contact or projection and the type of enlarger in the latter case. (For contact printing the masked negatives the initial negatives must be on film, which is quite possible in these days of dimensionally stable film bases, and the masks must be made in contact with the backs of the negatives.)

The masked negatives must have a range of 1·0 which means that the initial separation negatives must be a good deal more contrasty to allow for the reduction of contrast by the masks. Suitable figures may be 40 per cent of the contrast range for the red filter mask and 30 per cent for the green filter mask. To suit those requirements the initial red filter and green filter negatives should have a range of 1·6 (say from 0·3 to 1·9) while the blue filter negative should have a range of 1·4 (from 0·5 to 1·9).

The two colour masks (preferably unsharp) are then made by contact from the red filter and green filter negatives, the mask from the red filter negative having a range from 0·8 to 0·2 and the mask from the green filter negative having a range from 0·6 to 0·2. (The densities are quoted as from shadow to highlight so that the effect of negative plus positive is obtained by simple addition.) The red filter mask is then bound up into register, first with its own red filter initial negative (thus self-masking it) and printed, then with the green filter negative and printed. The green filter mask is used on the blue filter negative. The total density range of negative plus mask totals in each case from 1·1 to 2·1 which is 0·6 higher than the normal colour separation negative range of from 0·5 to 1·5. So the masked negatives will require four times the normal exposure (antilog 0·6).

PROFESSIONAL MASKING

An extremely wide range of masking methods, both for monochrome and colour are used in practically all branches of professional photography. One absolutely invariable feature of all such work is the extremely precise sensitometric and densitometric control which is always employed: empirical methods are entirely inadequate.

The cine industry uses masking to a great extent. Processes range from simple back projection methods, where the actors performing before the projected background mask out portions of it (because they are solid), to all the intricate matte processes. These are used in combining two or more negatives of parts of the complete picture when making the prints for projection in the cinemas. By these methods it is possible to avoid sending actors on expensive location shooting.

Complete colour masking controls are also employed in the preparation of colour motion pictures.

Masking in the Graphic Arts. Masking is used to a continually increasing extent in the processes employed in the preparation of printing surfaces of all kinds.

The basic essential for many of the processes is mechanical means of inserting and replacing masks and other negatives or positives into precise register so that a number of different masking exposures may be made on one photographic plate. The best known of these is a special make of masking camera on which a great many different masking techniques can be undertaken with certainty of precise register. On this camera a negative or positive of large size can be contact printed with light which proceeds from a much smaller mask so that the exposing light is itself masked. The contact print is therefore modified as required for colour masking or for such purposes as the introduction of lettering into the picture or the formation of a montage. A different approach to somewhat similar problems is one make of supplementary camera, which can be fitted to a process camera and projects masking images into the image plane via a transparent pellicle mirror.

Integral Masks. One firm's integral mask invention as used in their own colour negative-positive process is useful both for graphic arts and for pure photographic colour print making.

According to this idea the middle (magenta-red printer) layer of an integral tripack transparency material contains a weak orange-yellow dye and the bottom (cyan-blue printer) layer contains a weak red dye before the material is exposed. The colour-development of the material dissipates the dyes in the situation of the respective images. The result is a colour negative which already has its green filter record and red filter record positive masks automatically within it. Therefore when the colour negative is printed by contact or projection on integral tripack material on a white base it yields an automatically colour-masked print. If colour separation positives are made they also are colour masked. Those separation positives may be used for graphic arts reproduction or, if they are in the form of dye-transfer reliefs, for paper print making.

Coloured Masks. A development of colour-masking principles is the idea of making the colour masks themselves coloured, so that they have different effects upon the colour separations according to the colour of the filter used to make the exposure through the mask.

One of the first of these methods used three bi-pack films, each effectively two-thirds of a tri-pack colour transparency film. Thus one film is sensitive to blue and green light in its two layers, and colour-develops to yellow and magenta respectively; the second film is sensitive to blue and red light and colour-develops to yellow and cyan; the third is sensitive to green and red light and colour-develops to magenta and cyan respectively.

Consider the first of those bi-pack films, when contact printed beneath a positive colour transparency. The top, yellow layer of the transparency is "minus blue" and so provides a negative image in the top, blue-sensitive, layer of the bi-pack material. Similarly the middle, magenta-red, layer of the transparency is "minus green" and so provides a negative image in the lower, green sensitive, layer of the bi-pack material. But the top layer of the bi-pack develops to a yellow negative image which cancels out the yellow positive image of the transparency; similarly the lower layer of the bi-pack colour-develops to a magenta-red negative image which cancels out the magenta-red positive image of the transparency. Thus, when the processed and dried bi-pack colour negative mask image is bound up in register with the initial positive colour transparency, it cancels out the yellow and magenta-red image layers and the combined result appears as a black positive image of the cyan-blue layer of the transparency in a red ground. This can be photographed with red light to produce a colour separation cyan-blue printer image of the cyan-blue layer of the colour transparency, cleanly separated from the yellow and magenta-red layers. Similarly, by using the other two bi-pack film masks, one can obtain a colour separation of the yellow and of the magenta-red layers of the transparency.

A further step are the current tri-pack colour masking films, the three (sometimes more) layers of which develop to the appropriate colours to serve as masks according to the colours of the filters used to make the colour separation images. A simplified example shows the principle.

Suppose a colour tri-pack film material is sensitive, as usual, to blue light in the top layer, to green light in the middle layer and to red light in the bottom layer. But suppose that the three layers colour develop, not to the normal yellow, magenta and cyan colours in the respective layers, as with a normal colour transparency material. Instead, the colours produced are cyan, yellow and magenta, at around half normal strength. If such a material is contact printed beneath a positive colour transparency and developed directly to a colour negative image, the result looks peculiar, but provides a colour-mask which fairly efficiently colour corrects all three of a tri-colour separation set and, indeed, also the black printer of a four-colour set.

Thus the bottom, magenta coloured negative image (formed by red light from the transparency) acts as a colour-mask for the green filter separation negative. This is because magenta is a "minus green" colour, and the yellow and cyan-blue image layers above it would not affect that separation because they both freely transmit green light.

Similarly it can be shown that the yellow middle image layer of the coloured mask (a "green filter" record) masks the blue filter (yellow printer) separation, again according to the general principles of colour-masking, and the top (blue filter) cyan coloured image layer acts as a mask for the red filter (cyan printer) separation.

This description shows the general principle of the coloured colour-mask systems. In practice the various makes of the film are somewhat more involved, and they can be used for colour-masking both contact and projection separations of colour transparencies, and also for the camera colour separation of opaque coloured copy.

Electronic Colour Masking. A great deal of research is going on in the use of electronic principles for colour masking.

The methods rely on the fact that photo-electric cells can be made which will respond to some or all of the visible spectrum. It is therefore possible to scan a colour object, print or transparency, with three photo-cells each sensitive to a different portion of the spectrum. They can then each control a fluctuating light spot and produce three or four colour separation negatives or positives.

It is possible to cause part of the impulse from one or more of the cells to be shunted into the circuit of the others and so to introduce colour masking to a practically unlimited extent. Such methods can be used for pure colour photography, of course, but also the impulses can be caused to engrave printing plates direct.

The principle is used in a variety of ways. For instance, there is a device used on a process camera by which an opaque coloured copy, on the camera copyboard, is scanned by a moving spot of light. The intensity of this spot is controlled by the instantaneous response of three photo-electric cells examining the quantities of red, green and blue light reflected by the different colours of the copy. During the scan, which takes several minutes to cover the copy, the camera lens is left open. A colour filter over the lens ensures that one colour-masked colour separation negative is recorded on a panchromatic film or plate in the image plane. The computer portion of the apparatus is then set for the next colour separation, and the copy is light-scanned again, and so on for all four of the separation set.

Another method resembles that of photo-telegraphy but in this case a colour transparency is placed around the transparent cylinder of the portion which may be considered the "sending" part of the apparatus. The cylinder revolves and a very small spot of light shines through the transparency from within the cylinder. The light head slowly moves along the axis, so that the whole of the transparency is "scanned" by the light spot. Three photo-cells measure the quantity and colour of the light spot transmitted through the transparency. The electronic computer portion of the apparatus introduces colour-masking, under-colour removal, tone correction, etc. and feeds four lamp heads. These project four separately modulated points of light upon four films placed around the cylinder at the other end upon which the yellow, magenta-red, cyan-blue and black printer separation negatives are formed in unison with the track of the scanning spot.

Again, there are electronic machines which scan opaque coloured copy or a colour transparency and simultaneously engrave, into a plastic or a metal, a colour-masked, colour separation, half-tone printing plate, the operation being repeated for each of the four separations. If the image is engraved into a transparent plastic having an opaque or a red surface, it may then be printed-down upon a litho plate for that process.

A difficulty of the above three methods is that the various machines usually cannot cope with opaque and transparent subjects of all sizes from miniature transparencies perhaps to a large carpet, for example. Some inventors have approached the problem by designing apparatus into which one places an uncorrected set of continuous tone separation negatives or positives and, by the electronic scanning system, a continuous tone or a half-tone screened set of colour separation positives or negatives is produced, either one at a time or all four simultaneously. F.H.S.

See also: *Electronic printing; Photomechanical reproduction; Registering images; Separation negatives.*

Books: *Amateur Carbro Colour Prints*, by Viscount Hanworth (London); *Colour Prints*, by J. H. Coote (London); *Amateur Dye Transfer Colour Prints*, by Viscount Hanworth (London); *Colour Photography in Practice*, by D. A. Spencer (London).

MASKING FRAME. Paper holder for enlarging to keep the paper in position on the easel. Usually also produces a white border round the print. The border and the size of the image area framed may be adjustable.

See also: *Masks; Paper holder.*

MASKS. The appearance of most prints is enhanced by a neat white margin produced by masking off the border of the paper during printing. If the print is made on double weight paper large enough to leave a wide border, it can take the place of the usual mount. The normal mask is rectangular or square in shape, although at one time other shapes were popular (notably circles and ovals).

White Borders. A permanent mask can be built up on the negative with strips of adhesive black paper. Passe partout binding or lantern slide binding strips will serve for this purpose. Care must be taken to keep the corners truly square by checking with a set-square and ruler.

Ready-made masks cut from opaque black paper can be obtained in various shapes, and of different sizes for both prints and lantern slides. These are mostly used for contact printing. If numbers of prints are being made it is a good plan to attach the mask to the front of the negative with transparent self-adhesive tape. This saves constant readjustment each time a fresh piece of paper is inserted.

For contact printing, transparent ruby masks cut from celluloid are available and are preferable to paper masks, being more durable and easily handled. They also facilitate positioning of negative, paper and mask in the contact printer because the actual details of the negative can then still be seen through the mask.

These masks can also be used in contact with the negative when enlarging.

Lantern slide masks are supplied with openings of various sizes and shapes, but domes, cushions and similar fancy shapes are not in favour to-day. These masks are generally made of black paper, although some makes—particularly in 35 mm. size—are also available in metal foil. Masks are also supplied ruled with white guide lines so that the user can cut the shape he desires to suit the subject.

The simplest method of masking a lantern slide is to build up the rectangle with strips of lantern slide binding before adding the cover glass. Or the mask may be attached to the cover glass in the same way. This method affords additional protection against breakage.

Masking Frames. When making enlargements the usual method of masking is to use a masking frame on the baseboard of the enlarger. A masking frame consists of a metal frame made from strips of metal from ⅛ to ¼ in. wide. The frame is hinged to a baseboard so that it can be lifted up to allow the paper to be placed in position. Metal pegs or guide strips are fastened to the baseboard along two adjacent sides of the paper area to allow the sheet of printing paper to be accurately located under the mask. The baseboard is usually

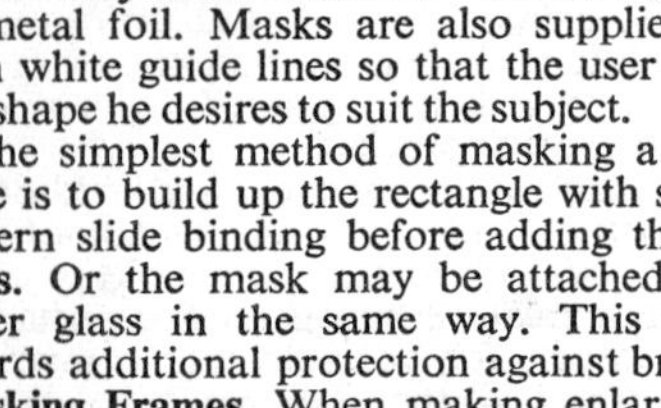

MASKING FRAME. The sliding strips adjust the picture area, and the hinged frame holds the paper in position on the baseboard. Adjustable margin stops may be provided.

painted a dull white and the frame a dead black so that the image can easily be focused and composed before inserting the paper.

Adjustable masking frames of this type have two sliding members of black metal strip which form two adjacent sides of the mask. The other sides are formed by the metal frame that carries the sliding strips. Any size of mask up to the full size of the frame can be produced by adjusting the position of the strips on the side members of the frame. They are so mounted that, while they will slide easily, they always lie square to each other and parallel to the corresponding rigid sides of the frame.

On some models, the paper locating strips are themselves also adjustable so that the width of the border can be varied.

Frames of this type are made heavy enough to stay in position on the enlarger baseboard once they have been set to mask the projected image. They serve the double purpose of masking the paper and holding it flat. Some workers use fixed masks consisting of glass plates of a range of different sizes with a border formed with strips of passe partout binding.

Black Borders. It is sometimes effective to give the print a narrow black border. All that is necessary for this is a sheet of metal or a glass plate covered with opaque black paper a quarter of an inch or so smaller in length and width than the finished print. After the print has been exposed, the opaque mask is laid in position under pressure leaving a uniform margin of the print exposed all round, and the white light switched on for a few seconds. The same result can be obtained by trimming the negative to the bare image size and printing it with a clear strip showing along all four sides of the paper, but the result is not as sharp as that given by a mask.

Trick Masks. In addition to the normal type of mask which gives a white border, various trick shapes are used when appropriate.

Oval and circular masks are useful when the corners of the negative are unsharp. A circular mask is easily marked out with a compass, but an oval mask calls for a special technique. The easiest way to mark out an oval is to use two pins stuck into a board at about three-quarters to two-thirds the length of the negative apart. A loop of string is placed over the pins and the length adjusted until a pencil held upright inside the loop cannot quite reach to the top or bottom edge of the print. If the pencil is now moved around so as to keep the loop of string taut it will trace out an oval. The dimensions of the oval can be adjusted by varying the spacing of the pins and the length of the loop. Once the oval has been drawn on the masking paper in this way, the blank can be cut out with a sharp razor blade.

Other shapes popular for trick effects are keyholes, "binocular" openings, and jagged edges suggesting that the subject has broken through the paper. Binocular openings in the form of two overlapping circles might be used to give atmosphere to a long narrow landscape view of a horse or yacht race so long as the mask does not cover an important part of the subject.

Many of these fancy shapes are more effective if they are reversed—i.e., to give a black instead of a white masked portion. This is printed in the same way as a black border, using the blank cut out from the centre of the mask to cover the centre of the print during the second exposure to white light.

There are various other ways of producing special masking effects. One effective method is to cover the glass of the frame with sand, sugar, rice, cotton wool or any loose substance, and clear away the area to be printed. This leaves a rough edge with a texture depending on the substance used. The same idea can be used to produce a black border by clearing the masking substance away from the edges and giving a second exposure to white light.

If a soft edge to any type of mask used in contact printing is required, it can be produced by supporting the mask on a sheet of glass above the printing surface. The softness can then be imparted by moving the light during printing. In enlarging the light cannot be moved to soften the edge and the simplest method is to use a mask under the negative, interposing a sheet of glass between the two to throw the mask out of focus.

Cutting Masks. It is not difficult to cut masks; the requirements are thin black paper—e.g., the wrappings from bromide paper—a sheet of glass, a steel rule and set-square, and a print trimming knife. The knife must have a keen cutting edge or the paper will be torn rather than cut. The method is as follows:

Place the paper on the glass and with the steel rule and square mark out the rectangle to be cut with a pencil line. It is a good plan to lay the negative or slide over a sheet of graph paper; the limits of the edges required can then easily be noted with a pencil mark and transferred to the black paper. To cut the mask, rest the knife against the edge of the steel rule, press it firmly down at the beginning of the cut, hold it straight, and cut along the whole of one side of the mask in a single stroke. If the cut is made in more than one stroke there will almost certainly be an irregularity in the edge of the mask. R.M.F.

See also: *Paper holder*.

MASS MINIATURE RADIOGRAPHY. Mass miniature radiography is a special branch of medical radiography used for making a preliminary X-ray examination of the chests of large numbers of people. It aims at detecting the signs of pulmonary tuberculosis or other chest diseases in its early stages so that the disease can be tackled in good time.

Method. To take a miniature radiograph, the subject is placed with his back to an X-ray tube and his chest against a fluorescent screen. When the X-ray tube is switched on it produces a visible radiographic image of the subject's chest on the screen by fluorescence. This image is photographed on 35 mm. film by a miniature camera focused on the screen. If the miniature radiograph shows any dubious areas the subject is then given a normal full-scale radiographic inspection using large-size X-ray films and equipment.

Equipment. The first equipment used consisted essentially of an X-ray unit, fluorescent screens, and a camera using 35 mm. film, all three being connected together by a light-tight hood; the camera usually had a lens of aperture $f1{\cdot}5$. This type of equipment has largely been replaced by units with a mirror-lens system, and an effective aperture about $f0{\cdot}75$, in order to reduce the X-ray exposure to the patient. The size of film employed for recording the fluorescent image is now mainly 70 mm. or 100 mm. in width, in single square sheets or in rolls. The fluorescent screen emits either yellow-green, or blue, light so that the sensitivity of the film must be selected accordingly.

Equipment for mass miniature radiography is designed to be relatively easy to transport and operate. Many M.M.R. units are completely self-contained, and have their own transportable generators and darkrooms.

Subject Identification. Each person to be examined is given an identification card on which he fills in his name and address. This card is inserted in a slot in the unit, and the serial number thereon is projected optically on to the film during the exposure. It is therefore impossible for anyone to be associated with the wrong radiograph.

The staff operating the equipment work from behind screens to shield them from unnecessary exposure to the X-rays. A trained team can deal with as many as 600 people per day.

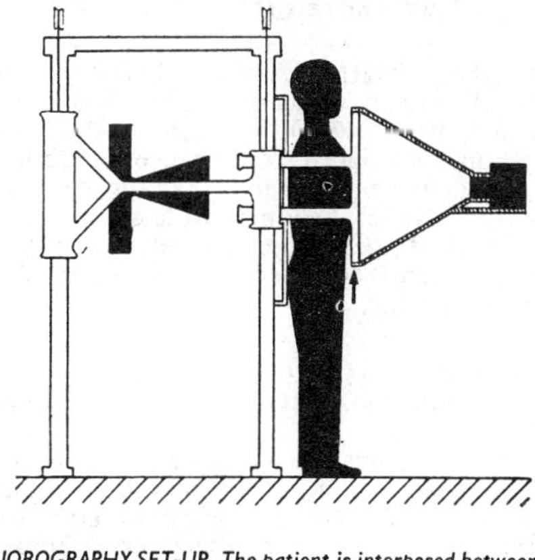

FLUOROGRAPHY SET-UP. The patient is interposed between the X-ray tube and a fluorescent screen which is in turn photographed with a miniature camera.

Organization. In addition to the actual operating staff, an M.M.R. service needs its own administration for arranging mass examinations, identifying negatives, interpreting the results and following up positive cases.

M.M.R. teams regularly tour the country and it is possible for anyone who wishes to have his chest radiographed at the recommended 12-monthly intervals. G.T.S. (C.S.)

See also: *Medical radiography.*

MASTIC VARNISH. Varnish (principally for negatives) prepared from the resin exuded from the stems of the Pistacia Lentiscus. The resin is insoluble in water but soluble in alcohol and a number of other spirits and oils.

MATRIX. Relief image used for making dye or pigment prints by transferring colour from the raised parts of the relief to a suitable paper or other base. In photography, gelatin matrices are employed for special processes such as dye transfer colour printing. These matrices are generally made by developing an exposed silver halide emulsion in gelatin with a tanning developer. Alternatively, a photographic image in a gelatin emulsion may be bleached in a tanning bleacher. Either method selectively hardens the gelatin, allowing the unhardened parts to be washed out in warm water, leaving the gelatin relief.

It is usual for the exposure to be made through the base when making a matrix.

See also: *Dye transfer colour prints.*

MAXWELL, JAMES CLERK, 1831–79. Scottish mathematician, physicist and natural philosopher. Author of valuable investigations on the perception of colour. Demonstrated the fundamental basis of three-colour photography (1861). His researches established the electro-magnetic theory of light. Biography by Campbell and Garnett (2nd edition 1884).

MAYALL, JOHN JABEZ EDWIN, 1810–1901. American daguerreotypist. Worked in Philadelphia (1842–6), London (1847–66) and Brighton. Introduced from France the albumen glass negative process in 1851. Also patented Ivorytype, 1855. Leading professional in the carte-de-visite period, which he made popular in England in 1860. Introduced the carte album in the same year.

MEASURE. Graduated containers for measuring volumes of liquids, and sometimes solids.

Most measures are glass or plastic beakers or tumblers with a scale marked down one side. They may be cylindrical or conical in shape; in the latter case the interval between the scale divisions is necessarily greater near the bottom than at the top.

Measures for large volumes may be in the form of plastic or enamelled jugs.

Scales. Measures may be graduated in metric units (cubic centimetres or millilitres and litres) or in Imperial fluid measure (minims, drachms, fluid ounces, or pints). Sometimes a measure carries graduations in both systems.

Sizes. Small measures hold 1 or 2 ounces, subdivided in drachms and tens of minims; or 25–50 c.cm., subdivided in c.cm. (scientific laboratory glass-ware is almost always graduated in millilitres). They are used to measure out stock solutions for dilution, etc. Intermediate sizes hold 4 to 20 ounces (or 100 to 500 c.cm.) to measure out larger quantities such as working solutions required to fill a developing tank. The largest measures may hold 40 or more ounces (1 litre upwards) for filling larger tanks.

Accuracy. Cylindrical measures in the small sizes are expected to be accurate to the nearest 0·1 c.cm. or to the nearest 2 minims. With a 100 c.cm. or 4 ounce measure an accuracy of 1 c.cm. or ¼ drachm is quite sufficient, and many measures are considerably less accurate than that. An accurate 500 c.cm. or pint measure should measure to the nearest 5 c.cm. or drachm.

The accuracy depends both on the shape of the measure and on the method of calibration. The most accurate types are the cylindrical and conical measures calibrated individually by an engraved scale. Tumbler and other measures with etched or painted scales are still reasonably accurate for most photographic purposes.

Plastic measures with embossed graduations are as accurate as any glass measures of the same size, as the moulds can be made with a high degree of precision.

Enamelled jugs with painted or enamelled-on scales cannot be expected to measure more exactly than to the nearest ½ ounce in the 20-ounce size.

Burettes and pipettes may be used for measuring small volumes with particular accuracy.

Reading the Scale. The correct level from which to read a volume of liquid is the bottom of the

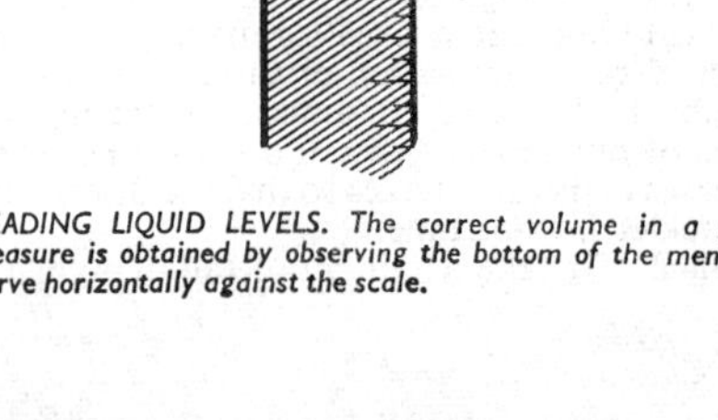

READING LIQUID LEVELS. The correct volume in a glass measure is obtained by observing the bottom of the meniscus curve horizontally against the scale.

meniscus of liquid—i.e., the lowest level of liquid in the middle of the measure. The measure must be held level, and the scale read with the eye at the same level as the liquid surface. With opaque measures the bottom of the meniscus may have to be estimated.

Solid Measures. Powdered chemicals can also be measured out in glass or plastic measures if the volume of a given weight of powder is first found by trial. Once established, the relationship between volume and weight will hold only for that particular chemical in that form. For instance acid fixing powder will need a different measure from hypo crystals.

Solid measures of this type are available for various chemicals (e.g., fixing salt) which need not be weighed too accurately and can be made from old wide-mouthed bottles by affixing a suitable calibration mark (e.g., a strip of gummed paper) at a level corresponding to, say, 1 or 2 ounces. Sometimes the lids of tins in which chemicals are packed are made to hold a specified amount of the contents.

See also: *Weights and measures.*

MEDICAL PHOTOGRAPHY

Medical photography is essentially factual. Its primary function is to put on permanent record the salient visible features of the case without embellishment, exaggeration, distortion, deliberate obliteration of detail or recourse to any artistic or spectacular effects. It embraces the whole field of photography and includes a few methods or techniques which are solely medical. The competent medical photographer must be skilled above the average in all aspects of photographic practice and he must possess a general knowledge of human anatomy, elementary physiology, elementary histology, a general knowledge of the terms used in pathology and bacteriology, a general knowledge of medical and surgical terms and their significance, a knowledge of the common processes of treatment, and experience in hospital etiquette.

The apparatus must be the best of its kind and comprehensive because no one type can be expected to cope with all classes of medical photography.

Medical photographs are useless unless they are properly classified and recorded in detail. All data concerning the type of case, the exact methods used to take the negative, and how it was subsequently printed, must be noted and filed. Proper keeping of records is an essential part of medical photographic practice. T.A.L.

History. The first application of photography to medicine was made within six months of the publication of the daguerreotype process, by Alfred Donné of Paris, who made photomicrographs of sections of bones and teeth, February 1840. The earliest publication with engravings copied from photographs of medical specimens is the Atlas accompanying Donné's *Cours de Microscopie* (1845). After the introduction of the wet collodion process, books on photomicrography, including medical subjects, were published by Lionel S. Beale (1857), J. W. Draper (1858) and Josef Gerlach (1863). By the time Albert Londe published *La Photographie Moderne* (1888), with a chapter on medical photography, fast gelatine dry plates and achromatic microscope lenses and condensers had eased the task of the photomicrographist.

The first known use of photography for comparative medical records occurred in 1852. H. W. Berend had photographs taken of his orthopaedic cases in Berlin before and after treatment: the same year Hugh Welch Diamond began photographing mental patients at the Surrey County Asylum. In France, Dr. Baillarger photographed cretins in 1851.

In 1861 J. Ganz started taking stereoscopic photographs for T. Billroth at the Chirurgical Clinic in Zürich, and some were published in 1867. Five years previously the earliest medical publications illustrated with photographs from life had appeared: *Album de Photographies Pathologiques* and *Mécanisme de la Physionogie Humaine*, both by G. B. Duchenne, the founder of electrotherapy, who photographed his patients from 1856 onward.

A. B. Squire, a London surgeon, published hand-coloured photographs of the main types of skin diseases, 1864–66, and A. Hardy in Paris published a dermatological iconography with photographs by A. de Montméja in 1868. For this kind of work the over-sensitivity to blue and violet of the early colour-blind plates had the advantage of exaggerating the texture and discoloration of the skin.

Thousands of photographs of wounds inflicted during the Civil War 1861–65 are preserved at the National Library of Medicine, Washington. *The Photographic Review of Medicine and Surgery*, started by F. F. Maury and L. A. Duhring in 1870 in Philadelphia, was the first medical journal illustrated with actual photographs.

From 1882 onwards great progress was made in medical photography by Albert Londe at the Hospice de la Salpêtrière for nervous diseases. Unco-operative patients were photographed with a small reflex camera worn on a strap round the neck, with which the photographer approached them unobtrusively. This camera is described in Londe's book *La Photographie Medicale* (1893), the earliest specialized treatise on this subject. In post-mortem photography the affected organs were taken with the camera fixed vertically over a table, and the lens stopped down to obtain greatest depth of field. Londe used chronophotography to analyse the movements of epileptics and cripples with a 12-lens apparatus and electrically operated shutters. The twelve photographs were taken on one plate.

Specialized techniques were devised for endoscopic photography. The first good photographs of the larynx were taken by J. N. Germak in 1862 and still better ones were published by Lennox Browne in 1883. He used a 10,000 candle-power carbon arc lamp, with reflector, as illuminant, the light being concentrated through two series of plano-convex lenses, with a plane mirror to reflect the light into the larynx.

The exposure was ¼ second. In 1885 the living human retina was successfully photographed by William Thomas Jackman and J. D. Webster.

Walter Woodbury described and illustrated the Photogastroscope in 1890, but the first practical gastroscopic camera was constructed by the German doctors Lange and Meltzing in 1898. The 2 in. long camera, containing a film 1/5th ins. wide and 20 ins. long, was supplied with light by a small electric lamp. The first photographs of the interior of the bladder were published by Max Nitze, inventor of the first satisfactory photographic cystoscope, in 1894. The following year the discovery of X-rays by Prof. W. C. Röntgen and the introduction of the Lumiére brothers' cinematograph apparatus opened new fields in medical illustration. A.G.

General Photography. Although the very sick or seriously injured patient may be under the direct care of the medical or nursing staff, while the case is in the medical photographic department the photographer is in some measure responsible. He must possess sufficient knowledge of the medical aspects of the case to appreciate the limitations imposed in handling it. He is not normally expected to remove or replace splints or dressings but must be capable of doing so should the need arise, and also be competent to deal with an emergency until medical aid arrives.

General record photographs are standardized in scale of reproduction, angle of view, lighting direction and contrast, and colour rendering. Where the contours are to be shown, photographs are taken from at least two different directions, usually at right angles. Surface lesions are generally photographed first from far enough away to show the lesion in relation to adjacent anatomical landmarks, and then close up to show detail.

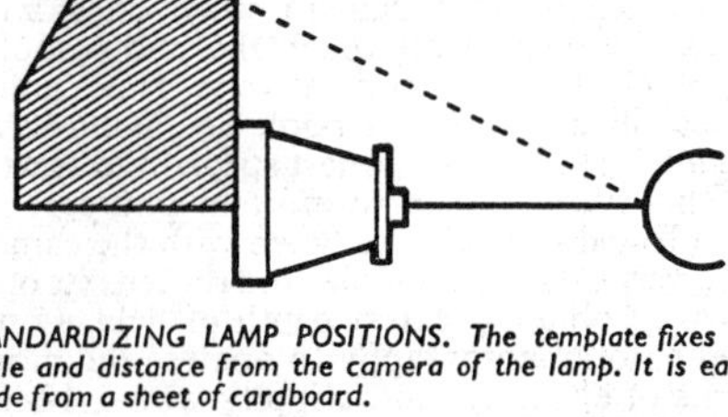

STANDARDIZING LAMP POSITIONS. The template fixes the angle and distance from the camera of the lamp. It is easily made from a sheet of cardboard.

General record photography demands not only familiarity with all types of camera and lighting unit, but also the ability to produce comparable standard views of the same and similar cases with different apparatus and not always in the same place or under exactly the same conditions.

Materials. Black-and-white photography still holds pride of place in medical work, but an increasing proportion of photographs are now made in colour. Medium-speed orthochromatic films are most generally useful, with recourse to panchromatic materials where heavy areas of red preponderate (e.g. fresh specimens, operation fields, etc.). In colour photography, 35 mm. reversal transparencies have found the greatest favour. Since each exposure yields but a single original it is wise to make paired exposures so that one can be retained in much the same way that a negative is normally kept. It seems likely that the use of colour negative material for all classes of work will eventually be adopted.

Infra-red. Infra-red radiation is capable of deeper penetration into human tissue than white light; moreover, it is reflected back rather diffusely. If there is some structure immediately below the skin surface which has a higher absorbing property than the surrounding tissue it is recorded on the infra-red photograph as a shadow. For example, the superficial blood vessels appear as a dark pattern against a pale background. For this reason infra-red photography of the superficial venous system can be helpful in examining changes due to varicose conditions, obstruction, heart disease, and pregnancy.

In dermatology infra-red can reveal changes taking place under a superficial scab. In ophthalmic work it has a particular value in enabling photographs of the pupil of the eye to be taken through a cornea opaque to visible light. It may also be used to measure the way the diameter of the pupil changes in response to various levels of illumination since it has no apparent effect on pupillary diameter.

For gross specimens infra-red has been found useful in showing up deposits or tumour tissues not normally differentiated by ordinary photography. To this end specimens may also be injected with special substances which reflect or absorb infra-red radiations, thus introducing differentiation from a photographic standpoint.

The camera and light sources commonly in use in the medical photographic department are suitable for infra-red work in conjunction with the appropriate filters and sensitized material. The almost opaque filters used to filter off unwanted visible rays may be used over the camera lens or the sources of radiation, depending on the type of work being undertaken. As there will be a difference between the visible and infra-red focus of any camera lens, it is advisable to achieve a preliminary focus through a deep red filter, before the infra-red filter is placed in position.

Stereoscopic Work. Three-dimensional photographs are always useful for indicating depth, especially when views at right angles are impracticable. Cameras with twin lenses are not commonly found in the equipment of the medical photographic department so the photographer must often improvise. This means that he must fully understand the principles of stereoscopy and be capable of adapting his standard apparatus. He must also be familiar with the methods of presenting and viewing stereoscopic pairs.

The medical photographer is usually expected to understand the fundamentals of stereoscopic radiography and the operation of the Wheatstone and binocular stereoscopes for viewing radiographs.

Cinematography. Nowadays considerable use is made in hospitals of cinematograph films—usually 16 mm.—of operating techniques and other suitable subjects. The medical photographer needs to understand the principles of cinematography and be familiar with the handling of up-to-date 16 mm. cinematograph cameras and accessories. He should have experience with black-and-white and colour materials in a studio, in daylight, and under the special conditions of lighting in the operating theatre. He also needs a practical knowledge of script preparation, continuity, camera angles, movements, cutting, animation, titling and editing. It is an advantage for him to be acquainted with special applications like cinemicrography, time lapse and high speed work, as well as sound recording and reproduction.

Dermatology. Skin photographs are used to demonstrate such conditions as inflammatory changes, ulcerations, eruptions and burns. It is usual to take general views to show the location of the lesion and close-ups to show its details.

When a series is taken, standardization of technique is most important. If the purpose of the photographs is to evaluate a change in the colour of the skin, they are made on panchromatic material with correction filters, or colour material with a control of normal skin taken under the same conditions of lighting and exposure. On the other hand, greater contrast may be deliberately produced by using orthochromatic materials or panchromatic plates and films with contrast filters, or the skin texture may be accentuated by employing harsh oblique illumination. Subcutaneous conditions can sometimes be demonstrated by infra-red technique.

Ophthalmology. The external eye can be photographed to demonstrate injury, disease and loss of muscular function. It is desirable to have an apparatus for fixing the head. A small view camera mounted on an ophthalmoscope stand is very useful; otherwise a camera with visual focusing is generally to be preferred—e.g., a miniature reflex.

Special apparatus is used for macrographic study of the iris and vascular network. Part of the internal surface of the eye (the retina) may be photographed through the dilated pupil. The instrument used for this purpose is a combined ophthalmoscope and camera.

Black-and-white, infra-red and colour materials are all used in eye photography.

Dental Photography. Dental photography may be used for general teaching, or for compiling records of unusual and rare conditions, or for making a graphic record of progress during treatment, or for psychological value—i.e., to prove that a particular treatment is having the desired effect. Photographs taken in comparative youth provide a useful guide when contours are to be built up with dentures in later life.

Although most cameras are suitable for this work, it is found best to have a specially designed small and compact camera unit and a simple illuminating system permanently fixed on a bracket table at the chairside. The medical photographer may only be asked to provide technical assistance, exposures being made by the dental surgeon as the need arises. Since colour changes are so important a factor in dealing with teeth and gums, colour materials are definitely to be preferred to black-and-white.

Orthopaedics. The principal use of photography in orthopaedics is to supplement written descriptions. It also provides a permanent record of progress and, where the improvement is gradual and prolonged, it is a means of encouragement to the patient.

Deformities, either congenital or acquired, are photographed with supplementary views to show limitation of function. The details of joint movement generally and the demonstration of gaits are subjects for cinematography.

Plastic Surgery. The value of photography in plastic surgery cannot be over-estimated. Not only are routine records required to show the

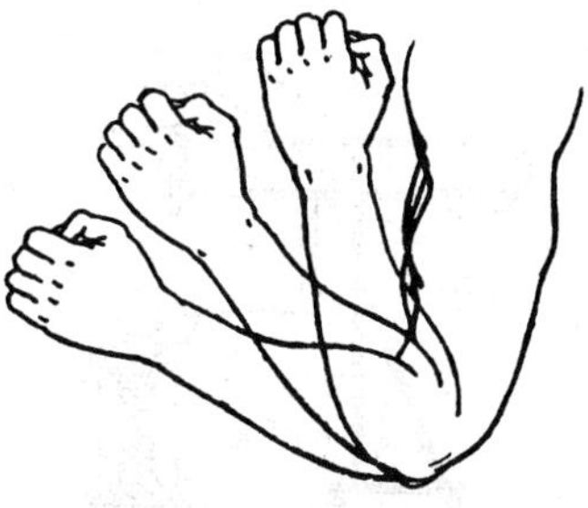

MULTIPLE EXPOSURE TECHNIQUE. Two or more exposures on the same negative may show the limits of a body movement for comparison with normal range of movement. This is useful for illustrating progress after injuries or operations.

protracted stages of almost any plastic repair, but such records may be called upon in the event of litigation. These photographs also play an important part in maintaining the morale of patients who often become discouraged during a long series of operations with little result to see in the intervening phases.

Because so many views are generally required, the miniature camera is often chosen for this work. This approach demands exacting darkroom work to obtain the necessary quality. On the other hand, it makes for portability and interchangeability of equipment to be used in the studio, the operating theatre and the consulting room.

General Surgery. Work in the operating theatre requires not only high technical photographic skill but patience, ingenuity, and ability to work in a team. Knowledge of surgical routine is an advantage as much time is saved if the photographer knows at which stages he is required to take pictures and can be prepared accordingly. The photographer who works in the operating theatre must understand and carry out the rules of asepsis. He must also take precautions against explosions of anaesthetic vapours arising out of electric sparks from his lamps or switches.

Broadly speaking, photographic work in the operating theatre is divided into still, endoscopic, and motion picture. Still pictures mainly show results whilst motion pictures show how the job is done; both may be made at the same operation.

Still photography includes pre-incision views, the operative field at various stages, demonstration of special features, and the final result. There is a difference of opinion regarding the most suitable type of camera, but the choice usually rests between the miniature which can be carried easily into position, as required, and gives a surgeon's eye view, or the larger camera on a heavy duty stand which is either left permanently in one position or wheeled up as required. Sometimes such a stand has a platform, carrying its own flood-lighting units, which raises the photographer well above the operating table

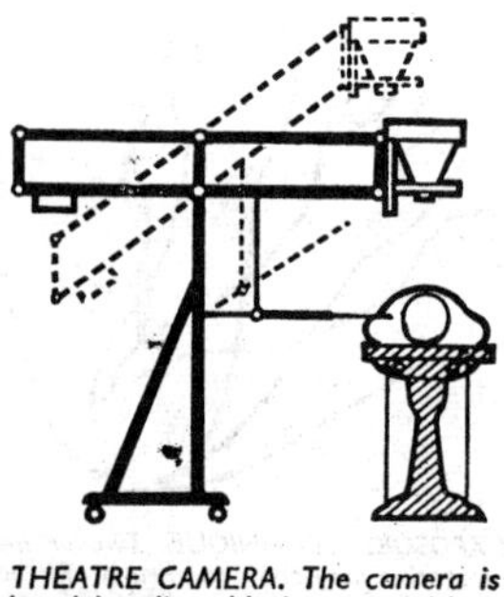

OPERATING THEATRE CAMERA. The camera is fixed on an upright stand and is adjustable in any position with positive locking once adjusted. The focus is fixed in line with the pointer.

level. Otherwise the choice of illumination is flash bulbs or electronic flash (suitably protected) as these permit the use of small lens apertures and provide desirable focusing tolerance. They also give the high intensity light required by colour processes without the undue heat of tungsten or photoflood lamps. For cinematography, the heat from continuous light sources may constitute a problem. In order to overcome this, there are some designs of lamps which make use of cold mirror reflectors, i.e., surface coatings which reflect light, but transmit heat. An additional heat filter is sometimes placed in the light path.

Generally speaking, anything which can be seen by means of an endoscope (the instrument used for internal examination through the body orifices) can be successfully photographed. Broadly there are two varieties of endoscopic examination: where the orifice can be widened to permit direct photography; the illumination is usually directional light from spotlights or focused sources, or the speculum used for dilatation may carry a small illumination source at its distal end. Where the endoscope itself carries an optical system and its own illumination, such as the bronchoscope or cystoscope, with a miniature camera attached to the viewing end. By using a special reflex focusing mechanism the field can be selected without removing the camera. Either black-and-white or colour materials may be used for endoscopic photography.

Ultra-miniature cameras have been devised for introduction into the body. One of these is the stomach camera. Such instruments carry their own light source, also exceedingly small, and may be positioned by X-ray screening.

Motion pictures of surgical operations are made for a number of different purposes:

(1) They may be made for general medical exhibition and will be rather lengthy, showing an operation in detail. Such films may include pre-operative investigation, pre-medication data, anaesthetic technique, and post-operative handling and treatment. This type of film may be made up from shots of several patients, so long as the fact is not apparent.

(2) The film may emphasize some particular aspect, such as the nursing and after care of the case for showing to an audience of students.

(3) The film may be a "short" to demonstrate only one special feature, such as a method of injection or suturing. A collection of such shorts is considered very useful for teaching and individual items are easily replaced or revised.

(4) The film may be made to show a new development in operative technique to an audience of specialists; all normal or generally accepted actions are omitted from this type of film.

Motion pictures may be made in black-and-white or in colour but the tendency today

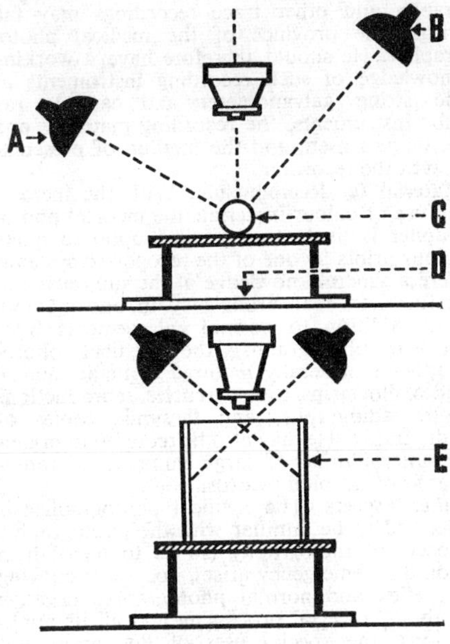

PHOTOGRAPHING SPECIMENS. Top: *General purpose layout. A, modelling lamp. B, fill-in light. C, specimen supported on glass. D, white background.* Bottom: *Set up for even diffused light. E, white cylinder. The light reaches the subject by reflection from the inside of the cylinder.*

is towards colour, except perhaps for the instructional short. Most films are "silent" with suitably printed captions, but more ambitious productions have a sound track added professionally, or by magnetic striping.

Gross Specimens. The subjects in this class of work may be preserved in jars or be fresh from the operating theatre or post-mortem room (where the photography in fact, is often done). A knowledge of asepsis, antisepsis, and general handling of specimens is required of the photographer dealing with such specimens. The preparation of the specimen and the method of presentation may be directed by the pathologist, but a skilled medical photographer is capable of handling routine specimens unaided.

He must know the effects that various immersion fluids have on colour retention, as it has a bearing on his choice of negative materials, illumination and filters.

For black-and-white work, he may use either orthochromatic or panchromatic stock with contrast filters to accentuate red contrast, or panchromatic material with selective filtration to increase contrast, or to produce maximum detail in one selected colour. He may also use colour material. The background in this work is usually of the shadowless type for black-and-white, and a plain colour for colour work.

When the specimen is in a jar, the photographer often has to exercise a considerable amount of ingenuity in order to show the important features, to control specular reflections, and to eliminate unwanted reflections.

Instruments. For the successful photography of instruments and apparatus a good general knowledge of photography is necessary, but the photographer must also know the function of each article so that it may be depicted lucidly. For example, if the parts are disconnected they must be shown in their correct position for assembly.

Bacteriological Cultures. Cultures may be infective or delicate and in either case will call for knowledge and care in handling lest also they become contaminated with other organisms through exposure to the air. The cultures may be on transparent, translucent or opaque media, in test tubes or in Petri dishes. The photographs may be taken by frontal, front oblique, transmitted, or oblique back lighting, on varieties of black-and-white materials with or without filters, and on colour materials.

Macrophotography. Small specimens of all kinds are photographed by direct camera enlargement up to ×10 or even ×20. The medical photographer must therefore be familiar with apparatus for macrophotography, or be able to adapt standard apparatus for such work. He must understand the use of all the standard types of lighting: incident or transmitted light, diffused or direct, and be familiar with the methods of mounting and presenting specimens.

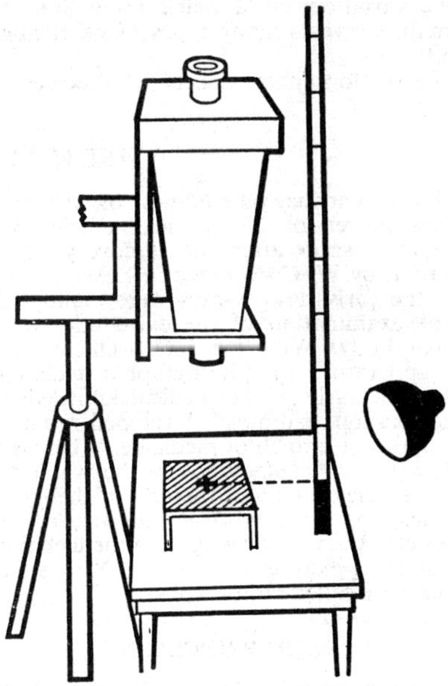

CLINICAL MACROPHOTOGRAPHY. The subject is arranged on a low table to keep the focusing screen conveniently low. A high-power magnifier serves for examination of the image.

Photomicrography. This is an important field of medical photography although it may not be performed in every photographic department.

The work may be done in the pathological, bacteriological, histological, or any department which normally makes visual examinations with the microscope. On the other hand all photomicrographic work may be centralized in one specialist department doing nothing else.

Generally speaking, the medical photographer is expected to possess a working knowledge of the microscope, including types and uses of different objectives and oculars, and sub-stage condensers. He must be familiar with all the common light sources for black-and-white and colour materials, and must also understand the use of special forms of lighting—e.g., dark ground, polarized, ultra-violet and phase contrast.

The experienced medical photographer has at least an elementary knowledge of staining and mounting and is capable of recognizing good and bad slides, as well as elementary tissues.

Trace Recordings. An increasing number of photographic trace recordings are used in modern diagnostic practice. The principal recording instrument is the electrocardiograph. In this the variations in electric potential brought about by heart action are traced by a spot of light on a moving strip of sensitized paper. The developed record shows the cyclic variations as a black wavy line. This technique reveals many types of heart abnormality.

The photographic aspects of electrocardiography and other trace recordings may fall within the province of the medical photographer. He should therefore have a working knowledge of such recording instruments as the string galvanometer and cathode ray tube instruments, the recording materials and techniques used, and the method of presentation of the recording.

Material for Reproduction. With the increasing use of colour materials the medical photographer is likely to be called upon to make colour prints by one of the recognized systems. Here a general knowledge of the subject is less useful than real proficiency in one method.

In addition to normal enlargements from medical photographs, the medical photographer is generally required to make lantern slides, film strips, transparencies, reproductions from radiographs, and facsimile copies of radiographs. He may also be required to organize microfilming of large numbers of radiographs or hospital records.

Other Aspects. The medical photographer is expected to be familiar with the photographic aspects of radiography (so as to be of help should an emergency arise), document copying by reflex and normal photographic negative methods, reversal processing and all its applications, the special uses of fine grain and physical developers, and tropical processing generally. He must have a knowledge of the principles of design and lighting of a studio and darkroom, and the care and maintenance of the electrical equipment. T.A.L. (P.H.)

See also: *Dental photography; Dermatological photography; Endoscopic photography; Mass miniature radiography; Medical radiography; Ophthalmic photography.*

Book: *Longmore's Medical Photography*, by P. Hansell and R. Ollerenshaw (London).

MEDICAL RADIOGRAPHY

The diagnostic possibilities of the penetrating power of X-rays in medicine were recognized soon after the discovery of this radiation by W. C. Röntgen in 1895.

In the early days X-rays were mainly used in the examination of the extremities of the human body. With the improvement of X-ray apparatus, of photographic techniques and particularly of the medical knowledge of interpretation, radiology developed rapidly to an important branch of medicine. Whereas the term *medical radiology* covers the whole field of X-ray diagnosis and therapy in theory and practice, the term *medical radiography* comprises chiefly the knowledge and the techniques of obtaining pictures by means of X-rays. Such pictures are called radiographs.

BASIC PRINCIPLES

In the production of a radiograph X-rays are emitted from the focal spot of an X-ray tube. The radiation passes through the object to be radiographed and is absorbed differentially within the various regions of the object. The degree of absorption depends on the atomic composition of the object, on its density and its thickness. The radiation transmitted by the object impinges on the photographic material (X-ray film) behind the object. Processing of the X-ray film yields a radiograph in which the shadow images of the various layers comprising the thickness of the object are superimposed. The radiograph is then inspected on a viewing box for interpretation by the radiologist.

At a very early time of the development of X-ray diagnostic techniques it was realized that certain organs, such as the stomach and colon can be made visible radiographically, despite the small difference of opacity in these organs. This is done by giving the patient a drink of radio-opaque substances, such as liquids containing barium. Later it was shown that by means of injections and oral applications of radio-opaque substances other

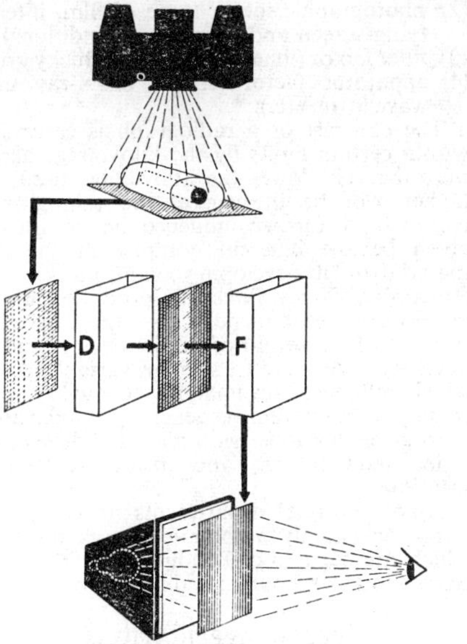

RADIOGRAPHIC CYCLE. Top: *The image is formed by exposure of the film to X-rays passing through (and being partially absorbed by) the subject.* Centre: *Development and fixation transforms the latent image into a permanent visible one.* Bottom: *The radiograph is viewed by transmitted light against a light box, usually as a negative.*

organs, such as the gall bladder, kidney and urinary tract can be made visible. The principle of this procedure is to introduce elements of high atomic number into the body which increase its subject contrast.

Geometry of Image Formation. The ideal condition for radiographic image formation would be a point projection where each point of the object is reproduced as such in the plane of the image. Owing to the finite size of the focal spot of an X-ray tube this ideal cannot be attained in practice and penumbra effects impair the definition in radiography. Geometrical sharpness depends on the size of the focal spot and on the focus-object film distances according to the equation:

$$P = \frac{F.A}{B}$$

where P = width of penumbra, F = diameter of focal spot, A = object-film distance and B = focus-object distance. Hence unsharpness increases with increasing focal spot diameter, with decreasing focus-object distance and with increasing object-film distances. All radiographic procedures are aimed at improving the sharpness of the image.

The X-ray Tube. This essentially consists of a highly evacuated glass vessel provided with 2 electrodes, the cathode and the anode. The cathode is the filament heated to incandescence during the operation of the tube, and produces a stream of electrons when a high voltage is applied across the two electrodes. The electrons impinge on a tungsten disc embedded in the anode which is usually made of copper. The impact of electrons on the anode produces X-rays.

The anode disc is inclined against the axis of the X-ray tube in such a way that the geometrically effective focal area is reduced with respect to the effective area bombarded by the electrons. This also reduces the optically effective focal spot area.

The X-ray tube voltages used in medical radiography range from about 40–125 kilovolts. Tubes vary in focal spot size and loading capacities. Rotating anode tubes are frequently used in hospital X-ray departments. By subjecting the anode to 3,000 revolutions per minute the focal spot of the tungsten disc is never bombarded to electrons for more than a fraction of a second, making heat dissipation easier than in stationary anode tubes. Hence a considerably smaller focal spot can be obtained with the same loading as for a stationary tube, leading to improved sharpness.

Materials. In the radiography of extremities (hand, arm, foot and legs) where the utmost sharpness of bone structure is often required, the exposures are usually taken on non-screen X-ray films, (i.e. double-coated films of specially high X-ray sensitivity). These may be supplied in light-tight paper envelopes, so as to avoid the insertion of film into cassettes in the darkroom.

The majority of exposures, such as pelvis, abdomen, chest and often skull radiographs are taken on X-ray films sandwiched between a pair of salt intensifying screens. These are layers of calcium-tungstate or barium lead sulphate coated on a flexible support of cardboard or plastic material. When struck by X-rays these screens emit a bluish fluorescent

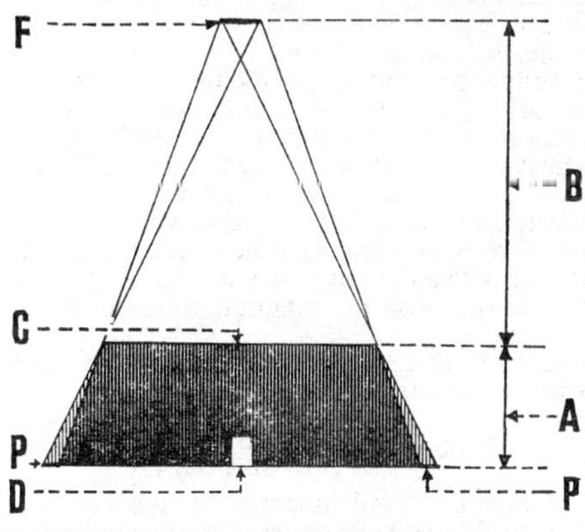

RADIOGRAPHIC IMAGE FORMATION. *Shadow projection from a focal spot of finite size, always produces a penumbra P. This can be reduced by making the ratio B/A as large as possible. (The distances from the focal spot F to the object C and the film D respectively). The penumbra P is also smaller with a smaller focal spot.*

light to which the silver bromide grains of the film are particularly sensitive.

With the application of intensifying screens exposure times can be very much reduced if compared under the same radiation conditions with those required for non-screen films and this is the main reason for their use. On the other hand screens impair definition because of the diffusion of their fluorescent light in all directions.

Potter Bucky Grids. X-rays suffer a considerable amount of a scatter when they pass through the thicker parts of the human body. This scattered radiation can be very troublesome in radiography as it impairs the image quality. The most effective method of reducing scattered radiation is the use of a Potter-Bucky grid or diaphragm which is placed between the object and the film during exposure. The grid consists of a series of thin lead strips placed lengthwise side by side and held together, but separated from one another, e.g., by plastic material or wood. The primary radiation emanating from the focal spot of the X-ray tube passes unhindered between the lead strips, whilst the scattered radiation, being random in direction, is absorbed by the lead strips and thus prevented from reaching the X-ray film. As part of the primary radiation is absorbed by the lead lamella and would thus produce shadows on the film, the Potter-Bucky grid is moved during the exposure. There are also stationary grids available in which the lead strips (or lamella) are made fairly thin and the distances between them are considerably smaller than in the movable grids. Although the stationary type of grid leaves shadows in the radiograph, these are less troublesome, when the radiograph is inspected at a distance of about 2 feet.

Fluoroscopy. In certain radiological investigations the patient's chest or abdomen has to be screened by means of a fluorescent screen and a permanent record is not necessarily made. Screening or fluoroscopy is an aid either to follow peristaltic movements of the stomach or colon, to observe the patient's heart in motion or the behaviour of certain organs whilst the patient or the organ moves. Fluoroscopy is carried out with special screens containing mostly a layer of zinc sulphide activated by cadmium or other substances. The screen is usually covered with a lead glass to reduce the dose of radiation received by the radiologist.

Exposure. The many factors influencing exposure can be summarized as follows:

$$\text{Exposure time} \propto \frac{\text{distance}^2}{\text{mA} \,.\, \text{KV}^n} \times \text{F}$$

where mA = milliampere, i.e. tube current, KV = kilovoltage across the X-ray tube, n depends on the range of kilovoltages applied and F is a function of the following factors:

(1) Object factor (thickness, atomic weight and density of the object);
(2) photographic factor (type of film, intensifying screen and processing conditions);
(3) filter factor (inherent filtration, Bucky grid);
(4) apparatus factor (circuit of X-ray unit, wave form, etc.).

The contrast of a radiograph is controlled within certain limits by the kilovoltage across the tube. The lower the kilovoltage used, the higher will be in general the radiographic contrast. A further influence on contrast is given be the inherent contrast of the film material and its developing conditions.

Processing. X-ray films are processed in the developers recommended by the manufacturers, such as metol-hydroquinone or phenidon developers. As there are so many variable factors which influence the final image quality, processing of X-ray films is generally standardized for a given developer with regard to developing time, temperature, and using replenisher solutions'

Larger hospital departments nowadays use semi- or entirely automatic processing units which are capable of handling 100–200 larger sized films per hour including drying.

SPECIAL APPLICATIONS

Medical radiography now covers such a wide field of specialized examination methods that only a few of these can be briefly discussed here.

Stereoradiography. If two consecutive radiographs are made and the X-ray tube is displaced by about 6.5 cm. (the inter-pupillary distance) between the exposures, the resulting radiographs form a stereoscopic pair. When these are viewed on a special viewer (e.g. Wheatstone stereoscope) the various parts of the body appear in three dimensions. The method can be used for the localization of foreign bodies or for the determination of certain displacements within the human body etc.

Planigraphy or Tomography. Where a cavity, for example in the lung or a foreign object is obscured in the ordinary radiograph due to the superimposition of the many layers of the object, the image can be made to stand out clearly by the special technique of planigraphy or tomography. In this method the tube and the film are moved during exposure in opposite directions about a common axis. If this lies in the plane of the cavity or foreign body to be examined, all the layers above and below the selected plane will appear blurred in the radiograph but everything in the selected plane will be relatively sharp. Several exposures are usually made before the plane in question is found.

The distances from the front or rear of the patient with respect to the selected plane can be arrived at by this method.

Miniature mass radiography. For chest examinations of a whole population, photographs of the fluorescent screen image are taken on special 35 mm. or 70 mm. square sized films or on

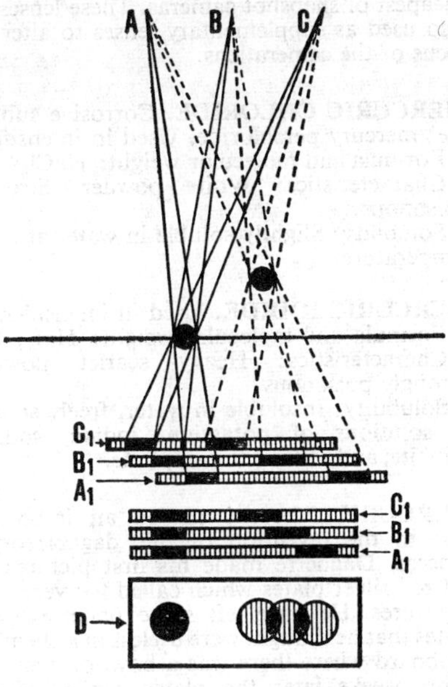

TOMOGRAPHY. The X-ray tube and film are moved in opposite directions about an imaginary common pivot during exposure. Everything in the plane of this pivot will then be sharp, everything else blurred. A, B, C, successive tube positions. A_1, B_1, C_1, successive film positions (shown above each other for greater clarity). D, resultant image.

other formats. Specialized equipment is available for this purpose and some of the apparatus is moved in special mass radiography vans to areas where such examinations are intended.

Dental Radiography. In the radiography of teeth small X-ray tubes with fine focal spots are built into easily controllable X-ray units. The small dental film (an X-ray type film specially wrapped in standardized size) is generally held in place by the patient. Because of the small area of the focal spot and the fact that teeth and jaw lie close to the film, good definition can be obtained.

Image Intensifier. One of the difficulties of fluoroscopy is the relatively low brightness level of the fluorescent image. A solution of this problem was found by the introduction of image intensifiers. In one version of such a device X-rays strike a photo-emitting surface and the electrons from this surface are accelerated upon a luminescent phosphor. In the early time of development of image intensifiers a small fluorescent image was inspected with binocular viewers. The popular size of present intensifier images is 7 and 9 ins. The brightness intensification is at least 1,000 times and this permits a considerable reduction in X-ray dose compared with that used in conventional fluoroscopy. One of the main advantages of the intensifiers is that the radiologist can view the image in a non-darkened room, and that they can be used in conjunction with cine cameras and closed-circuit television systems.

Cine-radiography. X-ray cine films were produced very early in the development of radiology and were used either directly or indirectly. The direct method was relatively little in use, as large film cassettes had to be moved in rapid succession. In the indirect method the fluoroscopic image was photographed by a cine camera. Both methods have been abandoned now, because of the high X-ray doses involved. With the introduction of image intensifiers it is now the intensified image that is filmed. This has revolutionized cineradiography and led to a new diagnostic tool of great importance.

Closed circuit television. Using X-ray television combined with image intensifiers the radiologist can conduct his examination by remote control outside the X-ray room. Hence he is removed from the exposure to radiation and he can discuss the picture with his colleagues and can dictate his findings without alarming the patient. A further advantage is the possibility that a large audience can watch the fluoroscopic examination whilst remote from the examination room. In the search for still lower irradiation doses and greater convenience image storage tubes are being introduced. In these the televised image can be held for up to 10 minutes after the X-ray source has been switched off.

Because of the potential hazards involved when working with X-rays these new methods are an important development as they contribute considerably to the use of lower radiation dose levels in the interest of patient, radiographer and radiologist. R.H.H.

See also: *Autoradiography; Industrial radiography; Mass miniature radiography; X-ray film.*

Book: *Longmore's Medical Photography*, by P. Hansell and R. Ollerenshaw (London).

MEDIUM SHOT. Shot in cinema production which covers a limited area of a scene in which mainly localized action takes place. This is usually a section of the whole setting of a sequence. When a medium shot involves people, it usually corresponds to a view taking in a small group or single figures shown in full length.

As a medium shot is the intermediate between a long shot and a close-up, its scope is rather broad and usually considered in relation to long shots and close-ups in the same sequence. Medium shots can be sub-classified into medium long shots—taking in a larger view but not large enough for a long shot proper—and medium close-ups, for example of a half-length figure.

See also *Cine technique*.

MEES, CHARLES EDWARD KENNETH, 1882–1960. English born photographic chemist and physicist. In 1903 Dr. Kenneth Mees, in collaboration with Samuel E. Sheppard, repeated and amplified Hurter and Driffield's work on sensitometry, their results being published in 1907. In 1908 Dr. Mees joined the English photographic manufacturing firm of Wratten and Wainright, and in 1913 established, at George Eastman's request, the Research Laboratories of Eastman Kodak Ltd. at Rochester, U.S.A. which he directed for over forty years. The establishment of a photographic museum at George Eastman House in 1947 was also due to Mees. Author of *The Theory of the Photographic Process* (1942) and other books.

MEGA-. Prefix signifying one million, e.g. megacycles, megohm. Applied mainly to electrical units.

MEGILP. Dope for increasing the richness of the blacks of a print. Mixture of linseed oil and mastic varnish. Also known as magilp.

MEISENBACH, GEORG, 1841–1912. German engraver and art publisher. Founded (Munich 1873) the first zincographic etching establishment in Germany. Started experiments with half-tone screens in 1879 and introduced commercially half-tone screen Autotypie for book printing in 1882. Opened a branch in London, 1884.

MELAINOTYPE. (From Greek melas = black.) Name generally given to Ambrotypes on the Continent.

See also: *Ambryotype.*

MÉLIÈS, GEORGES, 1861–1938. French conjurer (director of the Théâtre Robert-Houdin, Paris). Bought in 1895 a Bioscope from R. W. Paul, decided to make and sell his own films in 1896 and became one of the most famous early film directors, specializing in trick and fantastic films. He ended as a stall-holder in the Gare Montparnasse. Biography by Bessy and Duca (Paris 1945), including an autobiographical section.

MENISCUS LENS. Simplest lens used in photography. It is a single disc of glass, thicker in the centre than at the edges, and it has one face convex and the other concave.

The meniscus shape reduces spherical aberration somewhat, and a small stop in front or behind still more, while the combination of meniscus shape and small stop keeps astigmatism within tolerable limits. The lens is otherwise uncorrected for any of the aberrations.

Single meniscus lenses are found only in the cheapest of snapshot cameras. These lenses are also used as supplementary lenses to alter the focus of the camera lens.

MERCURIC CHLORIDE. Corrosive sublimate; mercury perchloride. Used in intensifiers.

Formula and molecular weight: $HgCl_2$; 271.

Characteristics: White powder. Strongly poisonous.

Solubility: Slightly soluble in water at room temperature.

MERCURIC IODIDE. Used in intensifiers.

Formula and molecular weight: HgI_2; 454.

Characteristics: Heavy scarlet powder. Strongly poisonous.

Solubility: Insoluble in water, freely soluble in solutions of potassium iodide, sodium sulphite, and sodium thiosulphate.

MERCURY. Mercury played an important part in the invention of the daguerreotype process. Daguerre made his first pictures on iodized silver plates which called for very long exposures. He once left some under-exposed plates that he thought were useless in a chemical cupboard where there was a bowl of mercury. Some weeks later the plates carried visible images. Daguerre concluded that the images had been intensified by the mercury vapour.

Today the photographic uses of mercury are limited to the method of hypersensitizing negative materials by exposure to mercury vapour either before or after exposure. The method is uncertain in its effects, but under suitable circumstances it may double the effective speed of the emulsion.

The vapour given off by a drop or two of mercury accidentally spilt in the darkroom can create a puzzling and persistent nuisance. It will penetrate most forms of packing and can completely change the characteristics of any sensitized materials stored in the room. Once it has contaminated a porous surface it is almost impossible to get rid of it.

Mercury vapour in a workroom can also act as a long-term poison so the liquid should be handled with extreme caution and kept in a well stoppered bottle.

One of the principal uses for mercury is as the indicating liquid in thermometers, but mercury thermometers are not easy to read under the darkroom safelight. In thermometers intended for use in photography, the mercury is nowadays replaced by a blue indicating liquid (generally dyed alcohol) which looks black by red or amber safelight illumination.

Many types of mirror are produced by coating a mercury amalgam on to glass or metal, but here also mercury is losing favour. In photography its place is being taken by amalgams of harder metals—particularly for surface-silvered mirrors.

A pool of mercury was used as the backing of the emulsion in the Lippmann colour process for colour images by diffraction.

Another unusual photographic application of mercury is in the so-called photography of perfumes. A speck of the odiferous substance is floated on the surface of a vessel of mercury. After a time, a photograph taken in brilliant cross lighting reveals patterns on the polished surface. These patterns vary according to the nature of the substance and they are only present when it has a recognizable perfume.

Mercury vapour is also used as the luminous gas in certain discharge lamps.

Formula and atomic weight: Hg. 201.

Characteristics: Heavy light grey metal, liquid at normal temperatures. Solidifies at −38° F. (−39° C.). Boiling point 670° F. (357° C.).

MERCURY PERCHLORIDE. Chemical used in intensifiers; mercuric chloride. It is very poisonous.

MERCURY VAPOUR LAMP. Special type of arc lamp in which the glowing arc is maintained in an evacuated glass tube containing vaporized mercury.

See also: *Discharge lamp.*

MERITOL. Proprietary developing agent.

Formula and molecular weight: $C_6H_4(OH)_2$. $C_6H_4(NH_2)_2$; 218.

Characteristics: White crystalline powder.

Solubility: Fairly soluble in water at room temperature.

MESSTER, OSKAR, 1866–1943. German pioneer of cinematography, 1896, the motion picture film industry, and of aerial photography. Designed apparatus and accessories for cinematograph cameras and projectors, and for aerial photography. Donated his collection on the history of motion pictures to the Deutsches Museum in Munich. Autobiography (Berlin, 1936).

METALLOGRAPHY

Metallurgists study the structure of metals with the aid of an optical or electro nmicroscope and frequently use a camera in conjunction with it to record their observations.

SPECIMEN PREPARATION

The metal specimen must, first be prepared in such a way as to reveal clearly the details to be recorded.

Preparation of the Surface. A specimen of suitable size for examination is cut from the object in question taking care not to strain it in the process, otherwise structures may be developed which are related to the strain and not to the original structure of the metal. The specimen is then polished and etched or treated by electrolytic polishing and etching.

The specimen is usually filed, or ground flat on a wheel, before being rubbed down on emery papers of increasing fineness. It is usual to rub down the surface in such a way that the previous lines of scratches are removed by the succeeding scratches of the next finer polishing process—i.e., by working in a direction at right angles to the existing scratches. In this way the effects of the previous scratches can be completely removed. It is usual to use emery papers numbers G, M, F, 0, 00, 000 in succession; the use of papers finer than 000 is generally of little value, and in some cases, particularly for electrolytic polishing, 00 emery paper is sufficiently fine. To remove these last fine scratches, the specimen is then polished on a revolving pad covered with either selvyt cloth, special cloth or velvet, to which the polishing powder is added, the pad being kept moist with running water. This last stage of the polishing process induces the formation of a thin layer of "flowed" metal on the surface, the irregularities of the scratched surface being partly rubbed away and partly filled up.

The powders employed in polishing depend upon the nature of the metal and are usually oxides of aluminium, magnesium and chromium, and in particular cases, diamond dust. For soft metals, such as aluminium, magnesium, zinc, etc., alumina, which is levigated to get a particularly fine particle size, is used. For hard metals, such as steels, chromic oxide is used; and for very hard substances, like tungsten carbide, fine diamond dust.

Etching. After the specimen is polished, the actual structure of the metal is revealed by the process of etching. This consists of dipping the specimen in an alkaline or acid solution for a few seconds or more, thereby allowing an attack on the surface to take place, which first removes any film of flowed metal; the attack is of a chemical nature. If the metal is homogeneous, then the etching reagent will dissolve

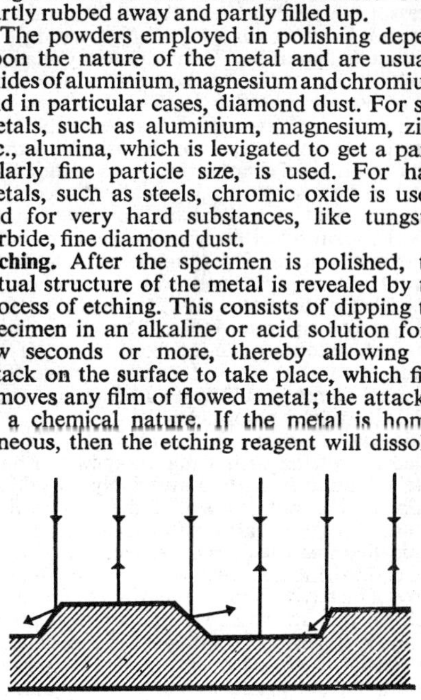

EFFECT OF ETCHING. This dissolves away different solid phases, crystal face or constituents at varying rates, and thus produces hollows or crystal patterns that are characteristic of the parts treated. When vertically illuminated and examined under a microscope, the various surface areas reflect the light in different ways and thus show up structure.

away the surface of each crystal of which the metal is composed, in amounts depending on their respective orientations. Under vertical illumination the junctions of these crystal faces will appear dark, since the surfaces of adjacent crystals have been differently attacked. But if the specimen is not homogeneous and contains one or more constituents or phases, then an etching reagent is chosen which will attack one or other of the constituents so that it can be identified in the presence of others by its difference in colour. The exact choice of etching reagent depends upon its chemical relationship with the metal or alloy.

Electrolytic Polishing. Instead of the above standard methods of polishing and etching, electrolytic polishing offers a particular advantage when a metal is soft and therefore not easy to polish. The specimen is prepared as before down to 00 emery paper and then is made the anode in a suitable electrolyte so that the surface of the specimen is removed by the passage of the current. Electrolytic etching may be carried out in the same bath, but using a lower voltage.

Thin films suitable for transmission electron microscopy are prepared in the same way from material which has already been thinned down to 200 microns or less. The final thickness necessary for adequate electron transmission must not be more than 1000A (0.1 micron).

Vacuum Evaporation. Thin films may also be formed by deposition of the metal on to a suitable substrate, but this method does not give specimens which are representative of the bulk material. The most common deposition method is that of thermal evaporation in a vacuum. The metal to be evaporated is usually placed in a basket or boat of molybdenum or tungsten, or if in the form of a thin wire, is wound round a tungsten filament. Under high vacuum, a heavy current (20 amperes or more) passed through the basket or filament causes the metal to melt and finally evaporate. It is collected on a substrate of suitable material from which, if required, it can be removed at a later stage. The two most common substrates used are mica and rock salt (sodium chloride).

Sectioning. Thin specimens may also be obtained by cutting with a microtome from bulk material, but this method invariably introduces strain, and is not suitable if the specimen is required for deformation or dislocation studies.

Replication and Shadowing. A method of wide application in transmission electron microscopy which overcomes the disadvantage of thick opaque specimens is that of replication. Briefly, it consists of coating the specimen with a material such as formvar or evaporated carbon, so that the latter takes up the surface topography of the specimen. The replica is then removed, either by stripping or by destruction of the specimen.

The visibility of surface details can be improved by shadowing. This is the deposition of a thin film of metal obliquely on to the object by evaporation in vacuo. Regions of the specimen screened by projections show up as areas of higher electron transmission so that heights and depths of projections and depressions respectively can be evaluated from a knowledge of the angle of deposition.

Metals widely used for shadowing are gold, palladium or an alloy of the two. A carbon/platinum mixture gives a deposit almost free of structure.

RECORDING METHODS

Photomicrography. A photographic record of what can be seen under the microscope is of immense value in the study of metals. Different ways of illuminating the specimen have resulted in the development of new techniques for further elucidating metal structures. Permanent records of the appearance of the specimen can then be carried out by a suitably modified technique of photomicrography. There are a number of such techniques in normal use.

(1) Vertical illumination: in the examination of opaque bodies, such as metals, a device known as the vertical illuminator is generally used. This consists of a plane glass reflector mounted immediately behind (above) the objective and reflecting light on to the specimen. The light passes by reflection from the specimen, back through the vertical illuminator to form the final image. The source of light is often a small electric bulb with a very finely-ground glass diffusing screen between it and the glass reflector. If the source of light is so placed that its distance and that of the focal plane of the eyepiece from the reflector are equal, then "critical illumination" is obtained. This is the best mode of illumination possible.

Both achromatic and apochromatic objectives can be used. When work of a really critical nature is undertaken—e.g., as in high-power photomicrography—apochromatic objectives are best; but for routine observations, the less expensive achromatic objectives, if of good quality, are quite satisfactory. When apochromatic lenses are used, the red, green and blue rays are united at the same focus, thus eliminating the secondary spectrum and producing an image free from colour interference. Achromatic objectives are not free from secondary spectrum.

For metallographic work, the most useful objectives are 25, 16, 8, 4 and 2 mm. (uncorrected for cover glass). The 2 mm. objective is an oil immersion lens with a numerical aperture (N.A.) of about 1·4. In order to get the best results from any objective, the magnification of the image should not exceed 1,000 $\times$ N.A. For convenience a chart can be constructed relating camera extension to magnification for each objective.

Since the resolving power of an objective is directly proportional to its N.A. and inversely proportional to the wavelength of light used

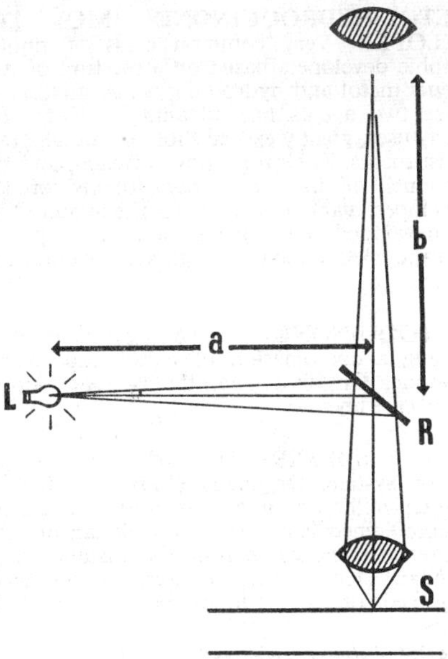

VERTICAL ILLUMINATOR. L. small light source. R, plane glass reflector. S, subject. a, distance from lamp to reflector. b. Distance from reflector to focal plane of eyepiece. The light from the lamp is reflected on to the subject. and reflected back through the illuminator into the eyepiece. For critical illumination the lamp distance from the reflector should equal the distance of the reflector from the focal plane of the eyepiece; the subject then appears at its brightest.

(i.e., it varies as N.A./λ) it follows that the larger the N.A., the greater the resolution or the finer the structure seen; the shorter the wavelength the greater the resolution. Hence it is usual in photomicrography to use filters passing wavelengths from 400 to 600 mμ in conjunction with backed process plates (blue-sensitive) or orthochromatic plates (blue- and green-sensitive).

(2) Oblique (dark ground) illumination: in certain circumstances structures may be more clearly revealed if the light falls obliquely on to the metallic surface before entering the objective. Special objectives are made with annular mirrors built into them, so that the external source of light falling on the mirror is reflected on to the metal surface and then back through the objective to the eyepiece.

(3) Polarized light: structural features can be observed on polished metallic specimens under polarized light which are not discernible under vertical or oblique illumination or by the aid of phase-contrast microscopy—e.g., a ray of polarized light may be unaffected by the metal itself, but de-polarized by impurities which would therefore appear dark. An ordinary microscope can be fitted with the necessary polarizing devices.

(4) Phase-contrast microscope: this type of microscope depends on the difference in phase of the light reflected from constituents which lie at different levels in the specimen surface and gives clear distinctions between constituents which are otherwise difficult to distinguish photographically. The alterations required to convert an ordinary microscope into a phase-contrast microscope are simple in principle.

Electron Micrography. The primary advantage of the electron microscope over the optical microscope is its much higher resolving power (about 200 times greater for a modern high resolution instrument), so that much finer details of specimen structure can be recorded. In particular the presence of dislocations can be detected in transmission microscopy of thin films.

Three modes of examination of metals by the electron microscope can be distinguished.

(1) Electron transmission: This is the method most generally employed and is applied to the examination of replicas or thin films. The latter must not exceed 1,000 A in thickness for examination in conventional instruments.

(2) Electron reflection: An image can be formed by collecting electrons scattered from a metal surface on which an electron beam is incident at a small angle. This reflection method has the advantage that it can be used with bulk metal specimens. The resolution obtainable is low for an electron microscope, but may be as much as 10 times that of the optical microscope using vertical illumination.

(3) Electron diffraction: Electrons are diffracted by a crystal lattice in a manner analogous to X-rays. In modern electron microscopes it is possible to make a simple adjustment of the optical system to enable a diffraction pattern of a small area of a specimen to be imaged on the screen of the microscope or recorded photographically. The crystal structure of the material on which the electrons are incident can be deduced from the pattern, an operation which may serve to identify chemically the area under examination.

Investigations with X-rays. Micrography by X-rays (radiography) permits the investigation of large metallic objects. By this means it is possible to determine the positions of holes, impurities, flaws, etc., in the interior of the metal.

For a more detailed structural examination, microradiography may be employed. This allows the discovery of cracks, microscopic inclusions and crystal shape.

X-ray crystallography provides a means of investigation of the atomic structure of metals.

G.C.F., M.L.V.G. & F.S.J.

See also: *Electron micrography; Gamma radiography; Geology; Industrial photography; Industrial radiography; Photomicrography; Polarized light; X-ray crystal analysis; X-rays.*

Books: *Handbook of Industrial Radiography*, by J. A. Crowther (London); *Infra-Red and Ultra-Violet Photography*, Eastman Kodak Company (Rochester, N.Y., U.S.A.); *The Electron Microscope (The Present State of the Art)*, by M. E. Haine and V. E. Cosslett (London); *Techniques for Electron Microscopy*, by D. Kay (Oxford).

METAL PRINT. Metal and other solid surfaces can be treated so that they can be printed with a photographic image. The printing may be carried out by contact printing by daylight or artificial light, or by enlarging.

After varnishing to prevent chemical action between the processing solutions and the bare metal, a photograph can be applied to it by any of the following methods:

(1) The surface may be brushed over with a sensitized coating or emulsion, exposed under the negative by contact or in the enlarger, and then developed.

(2) It may have the image transferred to it by the method of transfer coating.

(3) It may take the place of the final support in the carbon or carbro processes.

See also: *Printing on special supports; Stripping.*

METER. Instrument with a moving pointer, counter or scale used for measuring. In photography, special meters may be used for determining the strength of the light when making exposures in the camera (exposure meter) and printing from the negative (generally a photometer); for measuring the colour temperature of artificial light sources (colour temperature meter); for measuring distances (telemeter—i.e., rangefinder); for measuring the temperature of solutions (thermometer); and for measuring the current and voltage in electrical equipment (ammeter, voltmeter).

Also U.S. spelling for metre.

METHYL ALCOHOL. Wood spirit. Simplest alcohol. In the commercially pure state it may be used as a substitute for ordinary alcohol—e.g., for rapid drying of negatives.

Formula and molecular weight: CH_3OH; 34.

Characteristics: Volatile clear liquid. Poisonous.

Solubility: Miscible with water, ethyl alcohol acetone, and most other solvents in all proportions.

METHYLATED SPIRIT. Commercial form of alcohol, denatured by various additions to render it undrinkable (pyridine, various oils), and usually dyed purple to draw attention to its denatured state.

METOL. Elon; Genol; Photol; Pictol; Planetol; Rhodol; monomethyl para-aminophenol sulphate; para-hydroxy-monomethylaniline sulphate; 1-hydroxy-4-methylamino benzene sulphate. Developing agent.

Formula and molecular weight: $(CH_3NHC_6H_4OH)_2.H_2SO_4$; 344.

Characteristics: White crystalline powder. It may cause skin poisoning with some people allergic to it.

Solubility: Fairly soluble in water at room temperature. Less soluble in sulphite and alkali solutions.

METOL-HYDROQUINONE (MQ) DEVELOPER. Very common class of photographic developers based on a mixture of the agents metol and hydroquinone. A mixture of these two agents has advantages which, for many uses, greatly exceed those of the separate constituents. The proportions of each, and the amounts of the other constituents of the developer, vary according to the material to be developed—e.g., plates, films, or papers—and the effect aimed at—e.g., soft or contrasty image.

METOQUINONE. Developing agent containing a combination of metol and hydroquinone. On dissolving, the two developing agents are liberated.

METRE (METER). Unit of length in the metric system. Originally (1801) specified as one-ten-millionth of a quarter of the earth's circumference (equator-to-pole distance). The reference standard is now the distance at a standard temperature between two marks engraved on a platinum bar preserved in Paris.

See also: *Weights and measures.*

METRE CANDLE. Unit of illumination more usually known as the candle metre and also as the lux.

See also: *Light units.*

METRIC SYSTEM. System of weights and measures laid down in 1801 by the French Republic and now almost universally employed for scientific measurements on account of its arithmetical convenience, as all units are interrelated in terms of 10 or its powers.

The basic unit is the metre, originally specified as one-ten-millionth of the pole-to-equator distance. When this distance was measured more accurately and found not to agree with the original metre, the latter was referred to a standard metal bar preserved in Paris (copies of which were preserved in other national capitals). More recently the metre has been redefined in terms of the wavelength of a specified spectral line.

Length in the metric system is measured in kilometres (1,000 metres), metres, centimetres (1/100 of a metre) and millimetres (1/1000 of a metre).

Area is measured in squares of the above units and also in units of 1 hectare (10,000 square metres) and 1 are (100 square metres).

Volume is measured in cubic centimetres (c.cm.) and multiples and sub-multiples of the litre (the volume of 1 kilogram of pure water at 4° C.).

Millilitres and cubic centimetres are virtually (though not absolutely, due to a slight difference in definition) identical in size.

Weight is measured in terms of the gram

(originally specified as the weight of one cubic centimetre of water at 4° C. but now referring to a standard weight preserved in Paris). Other units are the milligram (1/1000 gram) and kilogram (1,000 grams).

An Act of Parliament, passed in 1864, made the system legal in Great Britain. The inch equals 2.54 cms.

Imperial weights and measures are now defined by reference to the metric standards.

See also: *Weights and measures.*

MICRO. Prefix indicating one millionth part of any unit. Also used as a prefix to indicate anything small or minute or used for examination or measurement of small objects.

MICRO-COPYING

Method of photographing original documents, drawings, etc. at a greatly reduced scale on film or paper. Being considerably less bulky, such micro-copies save valuable storage space. When required they can be read by optical apparatus or printed on sensitized material to produce a copy of the size required. Micro-copying methods, usually on film, serve many purposes, but were originally intended to preserve in reduced form valuable documents against their loss by fire or other causes. They can also be used for other security and preservation purposes, because micro-film, correctly processed and stored, will outlast all the normal types of paper.

In 1853 Sir John Herschel suggested copying reference works, public records etc., on microscopically small negatives and using an enlarger to reproduce them on a readable scale. His proposal had to wait 85 years for realization, for microfilm documentation was only begun on a large scale under the threat of war about 1938, when the most valuable books in the British Museum were copied on 35 mm. film, and some newspapers had microfilm copies made of their old volumes. Banks also began adopting the system for their records.

Any number of duplicates of the original microfilm can of course be made and, thanks to this process, copies of rare books and records have been made available to students and libraries all over the world. It may well be that the libraries of the future will consist of collections of microfilm copies of books, manuscripts and newspapers, occupying something like one hundredth or less of the space and capable of yielding further copies at quite a small cost compared with other means of reprinting or reproduction.

The airgraph service used during the last war is an interesting application of micro-copying. A form upon which a message of some 700 words could be written by the sender was copied on microfilm, sent overseas, and then reproduced photographically at its destination for a cost of less than one penny per letter.

The main kinds of micro-copy in current use are continuous roll microfilm, unitized film cards, and micro cards.

Microfilming. The straightforward system of micro-copying of documents, etc. is to photograph them one by one on a roll of 16 or 35 mm. film. The 16 mm. gauge is more compact from the point of view of convenience in storing, but 35 mm. film may be necessary where very large originals are to be copied (for example plans and drawings). To preserve the maximum amount of detail in big engineering drawings, 70 mm. or even 105 mm. wide film is sometimes used.

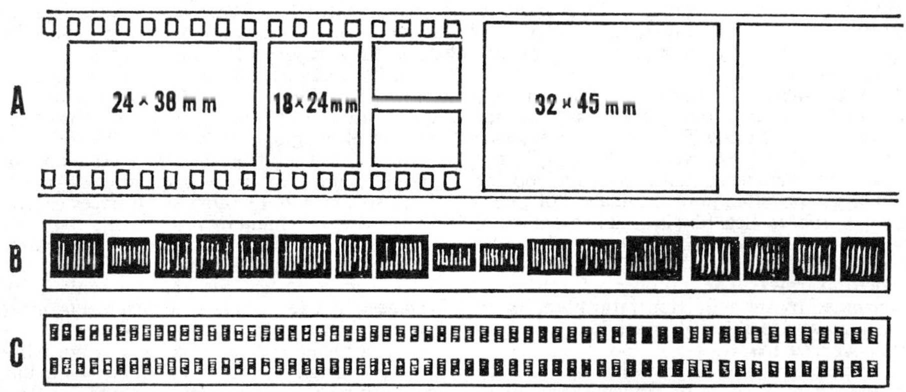

MICROFILM SIZES. A, 35 mm. film. This can be used either perforated or unperforated. Perforated film provides a choice of full frame 24 × 36 mm., half-frame 18 × 24 mm. or smaller split sizes. The standard image area on unperforated film is 32 × 45 mm. used for drawing offices. B, 16 mm. film at a 1 : 19 reduction, for documents, correspondence, etc. C, 16 mm. microfilm used at 1 : 40 reduction, providing four times as great a capacity per film length. The 16 mm. sizes are both unperforated.

Continuous microfilm is the most suitable method for micro-copying great volumes of coherent material, such as books, complete files of correspondence, etc. The reduction ratios employed vary from about 1 : 10 (on 35 mm. film) to 1 : 40 (16 mm.).

For locating individual items on a continuous roll of microfilm (usually up to 100 feet of film length) a separate indexing system is usually necessary. This specifies the roll on which any item is recorded, and the position of the required information on that roll. Special film readers reel the film through at high speed, while suitable locating marks between the frames enable the position of any one to be determined with reasonable accuracy.

Unitized Film Cards. Where a roll of microfilm covers a large number of disconnected items, keeping track of them requires an indexing system appreciably larger in bulk than the microfilm record. It then becomes more practical to file individual microfilm frames or groups of frames (short strips) as separate units.

The current means of such unitized treatment is the film card: an index card 3 × 5 ins., 3¼ × 7½ ins. (tabulator card format) or in one of several other reasonably standard sizes. The card carries a cut out aperture in which the microfilm frame or strip is located. The film may be held in the aperture either by a transparent and very thin plastic envelope, or by transparent adhesive tape. The mounting procedure is usually mechanized and there are various machines which cut the individual frames from the continuous microfilm roll and automatically place them in the aperture card at the rate of some hundreds or even thousands an hour. This is important in organizations using film cards in bulk, to avoid the mounting operation from becoming a handling bottleneck.

The microfilm cards can then be filed in any group or subject filing system. An alternative method is to use tabulating punched cards as aperture cards, retrieving any individual microfilm image by high speed mechanical sorting systems.

Microfilm Sheet and Micro-cards Sheet film is also used for microcopying, partly to avoid the inconvenience of using film in roll form for such purposes as the publication of out of print books and periodicals, and partly for copying very large and often poor originals. For publication purposes sheet film is used in a camera of the step-and-repeat type, which holds a flat sheet of film on which small images are recorded in rows. The number of images is determined by the reduction ratio used, about 80 being average although much higher recordings have been made.

The negative obtained in this way can be printed on to a sheet of transparent film to produce a positive, or on to opaque card 3 × 5 ins. These are micro cards. Similar opaque cards exist in many forms. One type for instance contains 100 images on a 6 × 9-inch card and is printed by lithography from 35 mm. film. Another kind, made especially for the legal profession, contains 200 images on each side of a card 6½ × 8½ ins.

Micro-positives can also be made on an opaque strip of sensitized material, printed from a negative roll. The opaque strip or tape has an adhesive backing which permits it to be cut in any required length and attached to any card, additions being made as required.

Micro-cards of this kind are useful as they allow direct filing of appreciable amounts of information on a single card. Also, since any number of positive micro-copies can be made, cross filing and multiple filing is greatly simplified. This method is cumulative, so that any number of new items can be added as required.

Such a system is thus a halfway house between roll microfilm and unitized single frame filing. (But some types of unitized aperture cards will also take up to a dozen microfilm negative images in strips.)

An important variant of microfilming is the recording on microfilm both of text and of a binary code. This code is used to search electronically at extremely high speeds, and—with some apparatus—to produce a readable copy automatically from the film. For certain applications it is even more convenient to encode the whole information on the microfilm in a binary system.

Equipment. Microcopying is carried out by various forms of special miniature camera equipment with film magazines to carry several hundred feet of film. The camera is generally mounted above a table with a flexible system of lighting so that a range of sizes of document can be evenly illuminated. In advanced apparatus, the setting and release of the shutter, the control of the lighting, and even the turning of the pages of the book are synchronized and remote controlled.

Many normal miniature cameras also can be adapted for microcopying with specially designed accessories for this purpose.

The cameras used for roll microfilming on a large scale are generally of two types, planetary and continuous. Planetary cameras record a set number of images, usually 1,600, 800 or 600 on a roll of 100 feet. The camera moves along an upright column to give varying sizes of reductions, so that two or more small pages can be photographed simultaneously. This type of apparatus is normally suitable for single sheets, large or small, books and magazines. After every exposure, the film is automatically advanced to a new frame. The documents must, however, be moved manually.

The continuous type of camera is normally limited to single sheets, about 12 ins. wide, but these can be of any reasonable length. They can be fed into the camera by hand, or mechanically at very high speeds. The exposing unit automatically illuminates and photographically

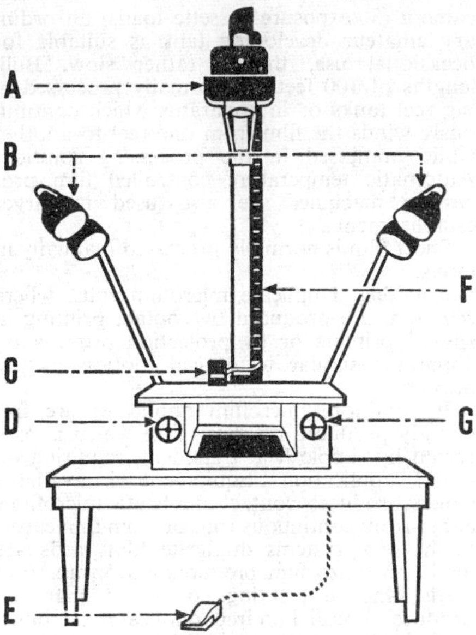

MICRO-FILMING CAMERA. A, camera unit. B, fixed lamps. C, exposure meter on movable arm to swing over document to be copied, and providing standard exposure times for all originals after calibration. D, exposure control wheel, acts by varying the light intensity. E, foot switch operating the shutter. F, reduction scale on column. G, camera control wheel for raising and lowering camera unit, also adjusts lens focus.

records the original, repeating this operation with each document. By this means a variable number of documents can be recorded on the film, the actual number depending on the length of the original and other conditions.

In the most advanced continuous microfilm cameras the film constantly moves past a narrow slit, while the originals are carried on a moving band past the camera below the lens. The movement of the film and of the originals is synchronized, so that the camera continuously records a sharp image on the film. Cameras of this kind can microfilm documents at the rate of several thousand an hour.

Automatic microfilming cameras also exist for producing film cards with the micro-copy ready mounted in an aperture card. The copying unit is here loaded with stacks of aperture cards carrying unexposed microfilm. The cards are moved into the image plane of the camera one by one for the exposure, and then pass through a developing, fixing, washing and drying sequence. The solutions are sprayed on the film surface. After processing and drying, the machine delivers a film card ready for filing. The whole operation is automatic, and the operator only has to place the originals on the copying table and press a button. The output speed is not very high, but this system eliminates the manual work of processing, cutting and mounting the microfilm.

Some cameras can photograph both sides of a double-sided document simultaneously on the same film. The original here is suspended between two mirrors which reflect the image from each side into the lens of the microfilm camera. With more conventional microfilm cameras, recording of two sides of a document involves turning the original over between one exposure and the next, or making a double exposure with the original placed in alternate halves of the copying easel. The latter has to be covered with black non-reflecting material to avoid fogging of the half of the film not being exposed.

Cameras for large size (sheet film) microfilming are often of the horizontal type. They hold the original on a vertical easel, while the camera runs on rails to permit variable reductions. The easel may be illuminated from the front as well as from the rear (through a translucent panel), for copying opaque as well as transparent or translucent originals.

Films and Processing Microfilm used for micro copying must have very fine grain, usually high contrast and high resolving power because the image is naturally very small. Several of the major film manufacturers provide special emulsions for microfilming in 16 and 35 mm. perforated or unperforated form. The unperforated type permits a more efficient utilization of the film size.

70 mm. film for microfilming has similar characteristics, while sheet films, where used, are usually fine grain process emulsions.

Microfilms are generally available both as blue sensitive (non-colour sensitized) and as panchromatic emulsions. The latter are preferable when multi-colour originals have to be copied; non-colour sensitized emulsions are more frequently used for making positive copies.

Microfilms are normally processed to the

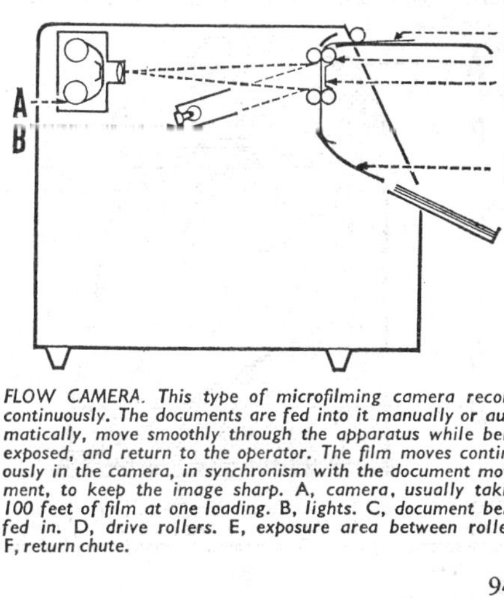

FLOW CAMERA. This type of microfilming camera records continuously. The documents are fed into it manually or automatically, move smoothly through the apparatus while being exposed, and return to the operator. The film moves continuously in the camera, in synchronism with the document movement, to keep the image sharp. A, camera, usually taking 100 feet of film at one loading. B, lights. C, document being fed in. D, drive rollers. E, exposure area between rollers. F, return chute.

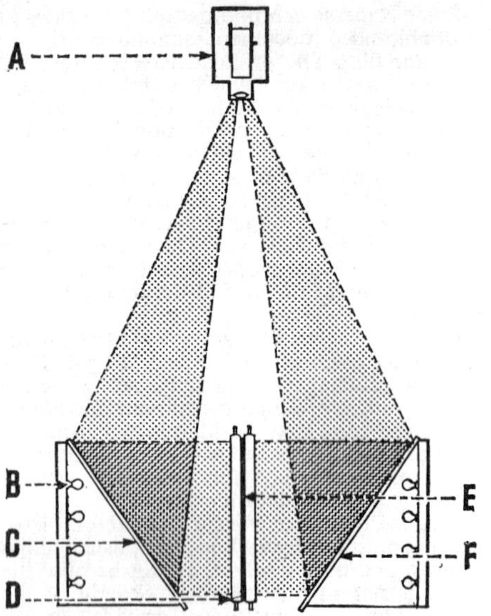

DUAL MICROFILMING. This serves for simultaneously copying both sides of a document. The front and the back views are reflected by mirrors to the camera, recording both sides side by side on the film. A, camera. B, panel of lamps to illuminate document. C, front mirror. D, document (edge on view). E, drive rollers. F, rear mirror.

highest possible contrast (gamma values of around 2·5 to 4) for easy readability. Softer development may be required if continuous tone or half-tone originals are copied, but it is rarely possible to allow for this with large output cameras, unless such originals are presorted and copied together.

All sizes of roll microfilm can be developed manually, using reels to hold the film. For microfilms made in an ordinary miniature camera (36-exposure cassette loads) an ordinary amateur developing tank is suitable for occasional use, though rather slow. Bulk lengths of 100 feet are normally processed in big reel tanks or in apparatus which continuously winds the film from one reel to another while immersed in the processing solution. Automatic temperature controlled film processing machines are also used in larger establishments.

Sheet film is normally processed manually in tanks.

Duplicating Duplicate microfilm rolls, where required, are produced by contact printing in special printers or by projection printing on apparatus similar to optical motion picture printers.

In addition, microfilm duplicates are frequently produced on diazo film which is processed by simple heat treatment, requiring no liquid application. Equipment is available which produces contact duplicate microfilms either from continuous rolls or from film cards. With some systems duplicate film cards are made on diazo film pre-mounted on aperture cards, and in printing equipment that can produce several hundred duplicates an hour. Such duplicates are required for cross filing and also for distributing, for instance microfilmed engineering drawings, in a number of departments of a manufacturing firm.

Readers and Printers For direct reference to microfilm negatives special viewing apparatus is available. These microfilm readers usually take the form of a table viewer with a back projection screen of 18 ins. square or more. The film, on a spool, is fitted into the viewer and threaded on to a take-up spool in much the same way as in a cine projector. Operation of a control knob or lever moves the film to bring a new frame on to the screen. In addition such viewers may incorporate a rapid film transport to bring a frame from, say, the middle of a reel

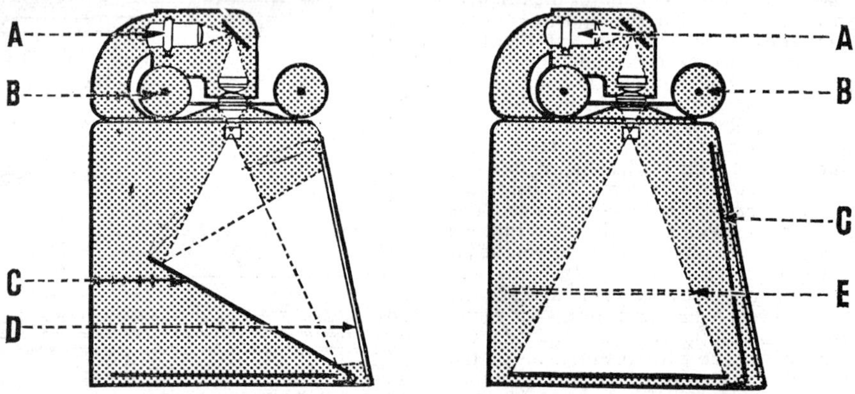

MICROFILM READER. This projects the image optically for viewing and sometimes also permits photographic enlargements by direct projection. Left: Mirror in reading position. Right: mirror in enlarging position. A, projection lamp. B, film. C, mirror. D, reading screen. E, paper holder in raised position for small prints.

quickly into position. In advanced units built-in sensing devices can automatically bring any required frame into the picture gate on adjustment of a selector mechanism.

Mostly microfilm viewers enlarge the microfilm image from 12 to 20 times, to bring it back to approximately original size. Viewers for unitized film cards, and also for micro-cards work on similar principles; a micro-card viewer illuminates the card by reflected light.

Combined camera and viewing equipment is also made, so that the apparatus can be used for both copying and viewing. On the other hand microfilm viewers frequently incorporate their own printing system for re-enlarging microfilm frames. The required frame is located on the viewer screen, after which a swinging mirror re-directs the projected beam on to a sensitive paper surface for making the enlargement. Reader printers mostly use electrophotographic, electro-lytic or diazo materials for the re-enlargement, and the positive copy emerges from the printer within a minute or so, ready processed. The largest units can re-enlarge microfilms to the size of big engineering drawings.

Microfilm negatives can also be enlarged on to document copying paper—usually of the fast projection type—for processing in the conventional way (development, fixing, etc.) or in motorized rapid roller processing units.

Continuous flow electro photographic printers, using the drum principle, exist for re-enlarging microfilms in bulk. The film moves continuously through a projector which projects the enlarged image on to a rotating selenium drum. This is charged, exposed, treated with charged pigment powder, and the image transferred to ordinary paper at various points around the circumference of the drum. The paper enlargements then emerge in a continuous strip or web and such units can produce full size enlargements at a web speed of 20 or more feet per minute.

This particular system has been used in reproducing photographically out-of-print books, rare editions, etc.

Zinc oxide coated paper is also used in electrophotographic printers. Here the paper is charged and image exposed, developed and fused on the paper itself, so that no selenium plate or drum is required. H.W.G., L.A.M. & H.R.V.

See also: *Copying; Diazo processes; Document copying; Electrophotography; Thermography.*

Books: *Camera Copying and Reproduction*, by O. R. Croy (London); *Micro-Copying Methods*, by H. R. Verry (London).

MICROFILMING Technique of copying documents etc., on a greatly reduced scale on 16 or 35 mm. (or sometimes wider) film for compact filing and storage. Microfilms may be stored in roll form or as unitized film cards, carrying microfilm frames or strips of frames in pockets or in a cut-out aperture.

The microfilm material used is usually a panchromatic emulsion of very fine grain and high resolving power. In the 16 and 35 mm. gauges it may be perforated or unperforated.

See also: *Micro-copying.*

MICROFLASH. Electronic flash of extremely short duration (of the order of 1/1,000,000 second) used for the photography of specially fast moving subjects—e.g., in ballistics, design of machinery, aircraft research, and others.

See also: *High speed photography.*

MICRON. One millionth of a metre, also sometimes known as micrometre. The term is most commonly used in expressing the wavelength of light.

See also: *Light units.*

MICROPHOTOGRAPH. Photograph made on a greatly reduced scale; the opposite of a photomicrograph. J. B. Dancer produced microphotographs 1/16 in. in diameter, taken on microscope slides, to be viewed in a powerful microscope, May 1853. They became fashionable six years later. Microphotographs were also sometimes set as jewellery under thick plano-convex Stanhope lenses, and in fancy goods such as penholders. In October 1857 Sir David Brewster forecast that microphotographic copies of secret dispatches might be placed in a space no larger than a fullstop, and transmitted by post. During the Russo-Japanese War, 1904–5, it was stated that Japanese spies carried microphotographic reports under their nails, in their nostrils, and in their stomachs in ivory containers that resisted the gastric juices. Present-day microfilming is the modern application of microphotographs which can under special circumstances reduce documents several hundredfold.

MICROPHOTOGRAPHY. Term used on the Continent for the technique of making large photographs of very small objects through the microscope. In Britain and the U.S.A. this technique is known as photomicrography, while the term microphotography is applied to the reverse method—i.e., of making microscopically small photographs of objects by optical reduction.

MICRORADIOGRAPHY. Special radiographic technique for examining the internal structure of thin specimens at high magnifications. The technique is also known as X-ray micrography.

To make a microradiograph, a plane section

of the specimen up to 1 mm. thick is placed in contact with the emulsion of an ultra-fine grain film or plate and exposed to a source of X-rays.

The image in the emulsion is then enlarged up to 300 times for examination, and shows the internal structure of the specimen.

The technique depends on the use of the finest grained emulsion possible. Before modern ultra-fine grain materials became available, Lippmann emulsions were employed.

MIETHE, ADOLF, 1862–1927. German astrophysicist. From 1899 Professor at the Technische Hochschule (Technical University) in Berlin as successor of H. W. Vogel. Earlier a director of Voigtländer & Söhne. Calculated an aplanat lens in 1888 and introduced a tele objective in 1891. In 1902 discovered new isocyanine sensitizing dyestuffs (with A. Traube); the Aethyl-Rot was used in the panchromatic Perutz Perchromo plates in 1904. Also worked on three-colour additive projection and printing.

MILE. (Lat. mille, a thousand.) Lineal measure equal to 1,760 yards, or 5,280 feet. Originally 1,000 paces or roughly 1,680 yards.

The nautical mile, standardized by the British Admiralty, is approximately one minute of arc of the earth's surface and equals 6,080 feet. This is also the British Geographical mile.

The International Geographical mile is one-fifteenth of a degree of the earth's circumference measured at the equator and equals approximately 4·61 statute miles—i.e., miles of 5,280 feet.

See also: *Weights and measures.*

MILLI-. Prefix signifying a thousandth part of. It is used principally in the metric system—e.g., millimetre, millilitre, milligram.

See also: *Weights and measures.*

MILLIMICRON. Thousandth part of a micron, used principally to measure the wavelength of light rays.

See also: *Light units.*

MINIATURE CAMERA

Term applied generally to camera using perforated (or sometimes unperforated) 35 mm. cine film as negative material for still photography.

A wider view, based on an official definition by the Royal Photographic Society, includes cameras taking negatives up to 2¼ ins. square. However, there are many constructional principles and points of technique which specifically apply to 35 mm. cameras and are not shared by roll film cameras (even for negative sizes below 2¼ ins. square). For that reason the term miniature camera is here used only for cameras taking 35 mm. film.

Precision. Miniature negatives have to be enlarged to yield prints of a reasonable size and this magnifies any shortcomings of definition. So the camera has to be constructed with greater mechanical precision and fitted with lenses of a higher standard of sharpness than a larger camera of comparable quality.

The degree of enlargement necessary also emphasizes grain and physical faults in the negative. The films therefore require much more careful treatment, and often special fine grain processing. From the start, miniature cameras were thus associated with a particular miniature technique of photography. In due course this not only influenced technique with larger film sizes but was also responsible for great advances in film manufacture.

Film. The majority of 35 mm. miniature cameras take 24 × 36 mm. (approx. 1 × 1½ in.) negatives, but some take 24 × 24 mm. negatives, and a few, odd sizes like 24 × 32 mm. The 24 × 32 mm. size is an attempt to get away from the stablished 24 × 36 mm. format.

A second fairly popular size is the 18 × 24 mm. half-frame (i.e. half the full 24 × 36 mm. frame). This reduced picture area permits an appreciable scaling down of the camera size, leading to more compact cameras.

As 35 mm. film has no backing paper, film transport has to be entirely mechanical. The perforations of the film are utilized for this purpose. The transport system therefore usually consists of a sprocket wheel or shaft which engages the perforations and pulls the film past the picture aperture. This is coupled to a take-up spool (often by a slipping clutch system to compensate for the varying thickness of the spool as more film is wound on it). A transport lock comes into operation when sufficient film has been advanced for the next shot. At the same time the transport sprocket also works a film counter.

The film is carried in light-tight cassettes which are loaded into the camera in daylight. When the whole film is exposed it usually has to be rewound into its cassette before unloading, but in some cameras a second cassette takes the exposed film. In that case no rewinding is necessary.

Some miniature cameras take interchangeable backs or interchangeable magazines to permit rapid loading or film changing in the middle of a length. The interchangeable magazines themselves are loaded with standard cassettes of film, and carry suitable means of locating the film in exactly the focal plane of the camera. Usually dark slides and other safety devices prevent fogging of the film while switching over magazines.

Lenses. The majority of miniature cameras

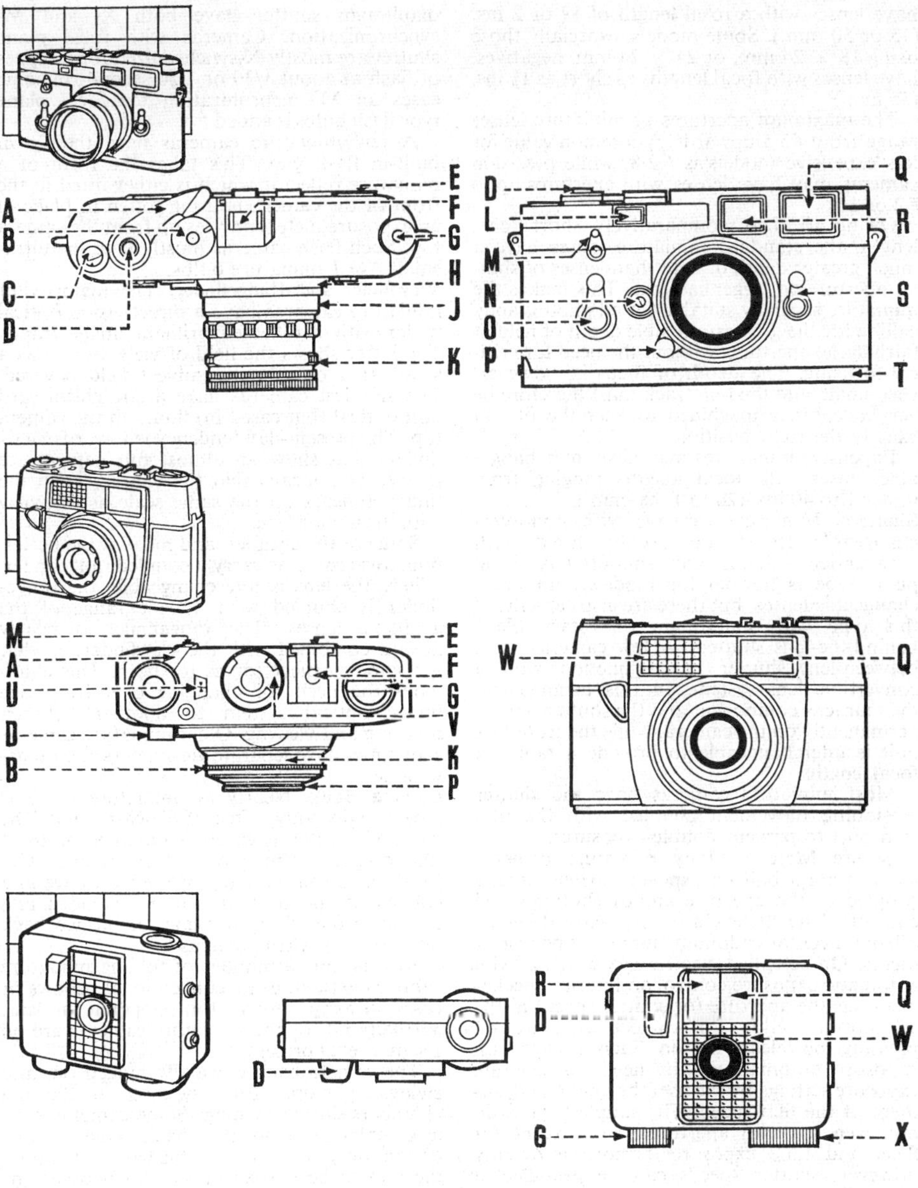

MAIN TYPES OF 35 mm. MINIATURES. Top: *high precision camera with interchangeable lenses, coupled rangefinder and focal plane shutter. It may sometimes have a built-in exposure meter and usually has a wide range of accessories.* Centre: *Popular miniature camera. This has a fixed lens, diaphragm shutter and usually a built-in exposure meter (frequently with automatic exposure control). More advanced models may have a coupled rangefinder and sometimes even interchangeable lenses.* Bottom: *Half-frame camera taking 18 × 24 mm. pictures. Some of these are scaled-down 24 × 36 mm. models, while others are specially compact designs incorporating an exposure meter, motorized film drive and simplest possible operation.*

A, *film transport.* B, *shutter speed setting.* C, *film counter.* D, *release.* E, *finder eyepiece.* F, *accessory shoe (may incorporate a flash contact).* G, *rewind knob or crank.* H, *lens changing mount.* J, *distance scale.* K, *aperture setting.* L, *rangefinder window.* M, *reversing clutch.* N, *lens changing catch.* O, *self timer.* P, *focusing lever or mount.* Q, *viewfinder window.* R, *illuminating window for finder frames.* S, *finder frame selector for interchangeable lenses.* T, *removable base plate.* U, *exposure meter needle (often visible in finder.)* V, *film speed setting of exposure meter.* W, *meter cell.* X, *motor winding knob.*

have lenses with a focal length of 1¾ or 2 ins. (45 or 50 mm.). Some models, especially those using 18 × 24 mm. or 24 × 24 mm. negatives, have lenses with focal lengths as short as 1⅜ ins. (35 mm.).

The maximum apertures of miniature lenses range from $f\,3{\cdot}5$ upwards. A common value for less expensive models is $f\,2{\cdot}8$, while precision cameras may have lenses with apertures up to $f\,2$ or 1·5.

Owing to their comparatively short focal length, the standard miniature lenses have a much greater depth of field than lenses of similar aperture on larger cameras. This makes the miniature specially suitable for all photography calling for the greatest possible depth of field at fairly large apertures without the need for critical focusing. The depth of focus is, however, very small, and the film track must therefore be very accurately machined to keep the film in exactly the right position.

Expensive miniatures may have interchangeable lenses with focal lengths ranging from under 1 to 40 ins. (20 to 1,000 mm.).

Shutters. Miniature cameras with non-interchangeable lenses are usually fitted with between-lens diaphragm shutters. A focal plane type is normal for models with interchangeable lenses, but there are also cameras of this type with diaphragm or rotary blade behind-the-lens shutters. A few cameras use a between-lens shutter in conjunction with a convertible lens system. In this arrangement the rear lens component and the shutter remain permanently on the camera, while the front lens unit is interchangeable to provide a range of focal lengths.

Most miniature cameras have the shutter tensioning movement coupled with the film transport to prevent double exposures.

Exposure Meters. Many miniature cameras incorporate a built-in exposure meter, usually coupled to the aperture and/or shutter speed controls. The meters are compactly designed selenium cell or cadmium sulphide type instruments. On so-called electric eye cameras with automatic exposure control an internal mechanism sets the aperture (or shutter speed) to the appropriate value for correct exposure on pressing the release button. Such models with exposure automation thus need no separate exposure settings by the user, beyond an adjustment of the film speed. The simplest types do not even have an aperture scale (except for flash and time exposures); more frequently however, shutter speeds can be pre-selected manually.

On interchangeable-lens cameras with focal plane shutter the exposure meter is frequently an accessory that clips on to the camera, but still couples with the shutter speed (and occasionally also the aperture) adjustment.

Flash. Simple miniature cameras and models with automatic exposure control are usually X-synchronized for flash; some models with diaphragm shutter have both X- and M-synchronization. Cameras with focal plane shutter are mostly X-synchronized for all types of flash at about 1/30 or 1/50 second; in some cases an M-synchronization for focal plane type flash bulbs is added.

A few miniature cameras have their own built-in flash gun. This takes the form of a miniature reflector which is either fitted in the front of the camera housing (often in place of an exposure meter) or rises up from the camera top. Such flash cameras mostly take the ultra-small AG 1 miniature bulbs.

Viewfinder and Rangefinder. The majority of miniature cameras have a direct-vision optical finder with a reflected brilliant image frame. The latter shows the field of view as well as a small area outside the subject field covered. The simplest cameras have a straightforward non optical (but cased-in) finder in the camera top. The present-day tendency is towards viewfinders that show an almost full natural-size image. This means that the view through the finder appears on the same scale as the view with the unaided eye.

Some of the cheaper, and most of the better, miniature cameras carry a coupled rangefinder. Where the lens is interchangeable, the rangefinder is coupled with a wide range of the available lenses. The rangefinder is nearly always combined with the viewfinder, so that the same eyepiece serves for both. This eliminates the need for changing over from one finder optical system to another between focusing and viewing. Older miniature cameras sometimes have separate viewfinders and rangefinders.

Camera Body. Nearly all miniature cameras have a rigid body, often of die-cast metal. The lens and shutter assembly is mounted on front (diaphragm shutter); on a few cameras with focal plane shutter some of the lenses are in a collapsible mount that can be extended and locked in the taking position, or pushed back into the body when not in use.

At one time a number of folding miniature cameras existed, using conventional bellows or cased in arrangements for extending the lens. Virtually no folding miniature cameras are on the market at present.

The camera back is usually hinged on, and swings right open for easy access to the film chambers during loading. Some cameras have a completely removable back. One or two designs only have a removable base and require the film to be pushed into a slot between the film aperture and the rear body panel. This can slightly slow down loading.

Types of Miniature Camera. 35 mm. cameras can be classified according to the basic characteristics.

Simple miniatures usually have a three- or four-speed between-lens shutter, simple viewfinder and little or no additional refinements.

Medium-class miniature cameras often incorporate a coupled—and frequently automatic—exposure meter, a wider range of shutter speeds, and better quality lenses. Occasionally medium class miniatures have a rangefinder, but this is rarer than an exposure meter. More advanced types feature both exposure meters and rangefinders, and vary mainly in the versatility of their shutter speed range and exposure automation.

Precision miniatures are made to the highest standards of fine-mechanical engineering. They usually have coupled rangefinder focusing, frequently with interchangeable lenses, and with a focal-plane or behind-the-lens diaphragm shutter. An exposure meter may also be built in.

The half-frame miniature, taking 18 × 24 mm. pictures, is often a scaled down variant of the full size miniature. Half-frame cameras tend to be comparatively simple models, sometimes with fixed focus lens, and automatic exposure control. They are mainly intended as a pocket companion camera, though some models exist with specifications approaching those of a full-size precision miniature.

The motorized miniature incorporates a spring motor which permits rapid sequence shots at a rate of up to 8 shots a second, with as many as 24 shots at one winding. The motor automatically retensions the shutter, transports the film and advances the film counter after every exposure.

A built-in battery-powered electric motor may do the same job.

The miniature reflex was originally built on the same lines as the larger single-lens reflex cameras, but has now been developed into a distinct type of its own. Present-day models often have an eye-level viewing arrangement for the focusing screen, showing an upright and, in some cases, right-way-round image. 35 mm. twin-lens reflex cameras are very rare.

Miniature Systems. Some of the more expensive precision miniatures have developed into an extensive and highly versatile system of equipment. Apart from the normal camera accessories (filters, lens hoods, cable releases, cases, etc.), this system includes specialized apparatus designed to adapt the camera for particular types of work.

Some of the accessories extend the scope of the camera. This group covers interchangeable lenses and viewfinders to go with them, simple close-up attachments, extension tubes and rapid winding gear.

Other more complex equipment converts cameras into specialized photographic tools for particular purposes, e.g., copying outfits, microscope attachments for photomicrography, reflex housings for tele- and close-up photography and stereo adapters for stereo photography. The basic unit of the system is in each case the miniature camera itself.

Miniature Conversion. A few roll film cameras of larger sizes can be adapted for using 35 mm. film by suitable conversion outfits. This may combine some advantages of the particular camera employed (e.g. reflex focusing in the case of a twin-lens reflex) with the economy of 35 mm. photography. However, the performance of the lens in relation to the size of the negative is not always as good as that of a lens specially computed for a genuine miniature; conversion can also only be successful if the camera to be adapted is made with sufficient precision to meet the demands of miniature technique. L.A.M.

See also: *Camera history; Exposure automation; Flash camera; Leica; Magazine; Miniature camera technique; Quick-fire camera; Ultra-miniature camera.*

Books: *Cameras*, by W. D. Emanuel and A. Matheson (London); *The Contax Way*, by H. Freytag (London); *The Leica Way*, by A. Matheson (London); *The Retina Way*, by O. R. Croy (London); *How to Choose and Use Your 35 mm. Camera*, by L. Gaunt (London).

MINIATURE CAMERA TECHNIQUE

The 35 mm. miniature camera, in spite of its limitations, is an extremely versatile instrument. The system of photography to which it has given rise is capable of good work within its own sphere.

The modern miniature camera has unique advantages, and a technique of its own. It is light and compact, so it is easy to carry and handle. It usually has a lens of wider aperture than that on bigger cameras and in some instances faster shutter speeds, while it holds negative material for a large number of exposures. In most miniatures it is possible to cut off and process part of a film, leaving the rest in the camera. There are miniatures with coupled rangefinders for precise and easy focusing and others with reflex focusing on a ground-glass screen. Most important of all, the best miniatures will take a range of lenses and accessories which make them almost universal instruments.

The eye-level miniature reflex with its pentaprism viewing system provides a direct view of the screen image at eye level. It is particularly easy to use with lenses of various (and variable) focal lengths and over focusing ranges from infinity down to a few inches.

But the miniature has no camera movements—i.e., front and back swing and tilt, rise and cross front, triple and double extension—and this explains some of its limitations in specialized commercial photography. However, by the use of extension tubes the absence of great bellows extension can be overcome.

The small negative calls for first-class material, exact exposure and development, and scrupulously clean processing.

Faultless photographs can only be pro-

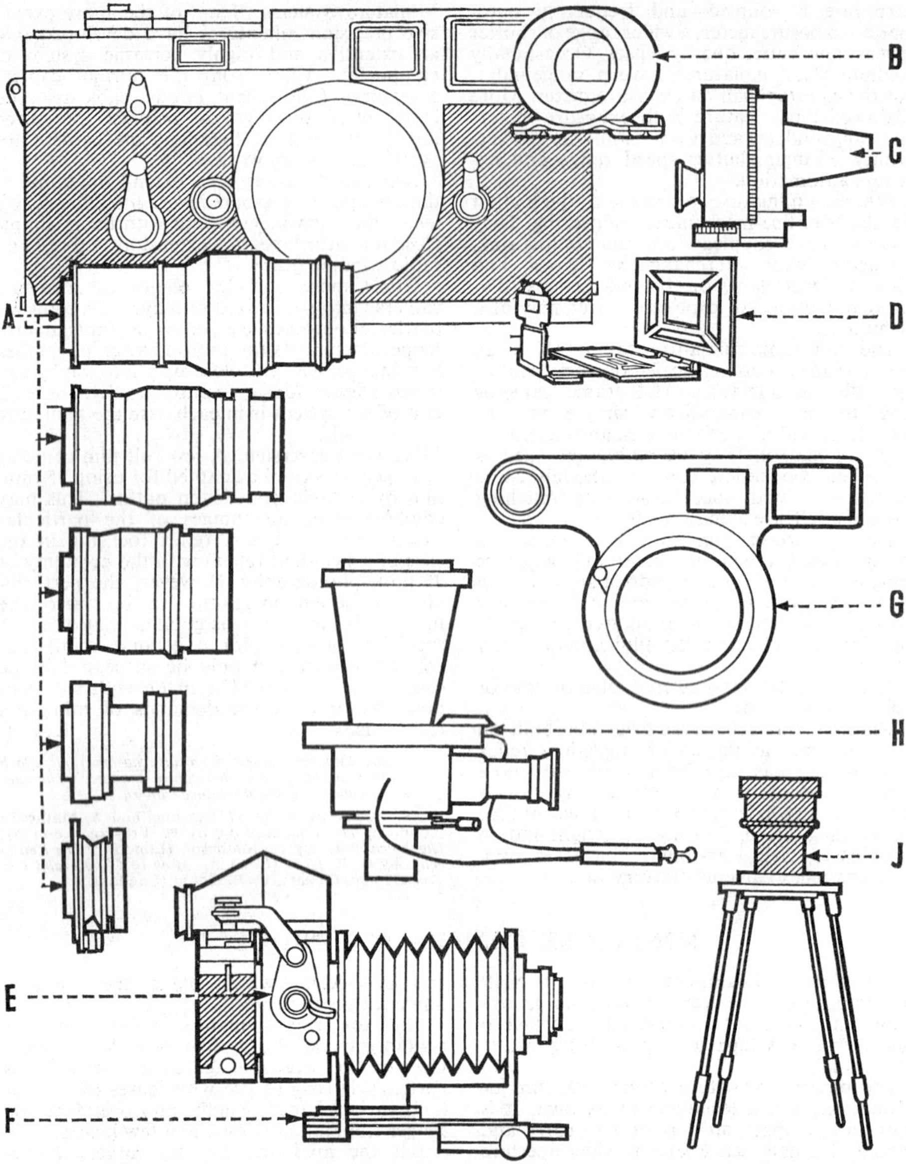

THE MINIATURE SYSTEM. The complete range of accessories of a precision miniature extends its scope to many applications. The most advanced 35 mm. miniatures thus comprise an almost self-contained system of equipment, to control the field of view, the scale of the subject, as well as operation.

A, interchangeable lenses may include focal lengths from around 1 in. for wide angle shots to 40 ins. or more for telephotography. B, C, and D, special finders for the alternative focal lengths clip into the accessory shoe on the camera. But many 35 mm. cameras indicate the fields of view of a selected range of lenses already in the viewfinder. E, a reflex housing in effect converts the rangefinder miniature camera into an eye-level reflex and is particularly useful for accurate focusing in telephotography with the longer focal lengths as well as for close-up work. F, extension bellows cover a continuous range of very near distances. G, subjects closer than about 3 ft. can often be covered with focusing adaptors which fit between the camera and lens, and also automatically adjust the rangefinder. H, micro adaptors permit the camera to be mounted on a microscope for direct photomicrography. A focusing eyepiece permits observation of the subject during exposure. J, distance gauges used with the extension tubes or supplementary lenses ensure rapid and exact placement of camera and subject when copying at selected fixed scales.

duced by the miniature camera system with great care and skill, not merely in processing, but right from the start—even before the picture is taken.

Cleanliness is an essential and calls for care in handling, carrying, storing and maintaining the camera and all its accessories and lenses.

Shutter Operation. Miniature negatives require a greater degree of enlargement than bigger sizes, so it is more essential to avoid camera shake. There are many satisfactory methods of holding the camera; the important thing is for everyone to find out which method gives him a firm, steady—and comfortable—grip on the camera and a simple, smooth pressure on the shutter-release. Each photographer must develop a shake-free shutter-release technique. A gentle squeeze with the most sensitive touch at the actual moment of release should be the aim. This calls for practice. Precise definition and dead-sharp negatives are impossible if there is the slightest trace of camera shake.

While it is possible, with some practice, to hold a camera reasonably steady during exposure times of 1/30 second or even longer, the high shutter speeds in nearly every case do give noticeably sharper results. Tests have shown that even with the standard lens many photographers get a small amount of camera shake—noticeable as unsharpness in big enlargements—at shutter speeds slower than 1/125 to 1/250 second. When no other considerations (such as a small aperture for adequate depth of field) make a slow shutter speed imperative, it is best to use the fastest available speed that gives a correct exposure.

For exposures with slow shutter speeds it is best to support the body and the arms against something solid for greater steadiness while releasing the shutter. A rigid support like a tripod is also useful, provided it is really rigid. The average light weight tripod tends to vibrate, and the "kick" of a focal plane shutter, and especially of the mirror of a single lens reflex camera may be enough to cause a slight image blurr even when the camera is mounted on the tripod. Often in such cases, and with shutter speeds between 1/15 and 1/60 second, the camera can be held more steadily in the hand (if well supported) since this tends to subdue camera vibration.

Using the Viewfinder. The eye should be as close as possible to the rear window of the viewfinder and must look straight through it. With simple optical viewfinders of the older type (without a brilliant luminous frame) oblique viewing gives misleading image boundaries. Even with brilliant-frame or bright-line finders the eye should be centred behind the eyepiece to take in the full image frame. Beyond this, the main problem is that of parallax—the slightly different fields of view taken in by lens and viewfinder owing to their different positions. With the standard lens only close shots of under 10 feet are affected. On some miniatures the necessary compensation is made automatically; in others the photographer must remember that his lens is looking from a viewpoint so much above, below or to the side of the finder, and make his own allowances.

When lenses of different focal lengths are used on miniature cameras, a variable type of viewfinder is often employed. This accepts various masks or allows other adjustments so that the viewfinder shows exactly the same field of view as the selected lens.

If the finder built into the camera has a multiple image frame for different focal lengths, the view taken in by lenses of various focal lengths can be checked directly, even before the lenses are changed.

In a single-lens reflex camera the screen visible in the finder shows more or less exactly the view taken in by the lens and no alternative finders are needed. Nor is there any parallax problem, since the lens itself forms the finder image.

Focusing. Correct focusing is an important part of successful miniature camera technique and can be achieved in more ways than one. Rangefinder focusing is a straightforward routine and, when the finder is coupled, the lens is automatically adjusted for focus as the images in the finder come together. It is usually more convenient to focus on some clearly defined edge in the picture if there is one. If the double image cannot be seen in the rangefinder window the chances are that the camera is not pointing straight at the subject unless the subject happens to be already sharply focused.

Another method of focusing is to estimate the distance and set the lens to the mark on the scale. This is useful when pictures must be taken unobtrusively and when to hold the camera up some time for focusing would attract attention. The greater depth of field given by miniature lenses makes such focusing techniques possible.

Zone focusing is a variation of this routine. The lens is pre-set at a distance where the depth of field at the aperture being used covers the near and far limits of subject movement. Pictures can then be taken without stopping to focus whenever the subject comes inside the zone of sharpness. This method is more successful at small lens apertures as the zone is then wider.

The image with the miniature reflex is focused on the screen. In some reflexes the sharpness of the picture can be checked over the whole screen area. More usually however only a central circle or ring actually shows the picture going sharp or unsharp as the lens is focused.

In addition many eye-level reflexes incorporate an optical wedge in the centre of the screen which acts as a split-image rangefinder. Here the image in the centre of field appears broken up across a horizontal dividing line, with the upper and lower halves displaced sideways relative to each other. As the lens is set

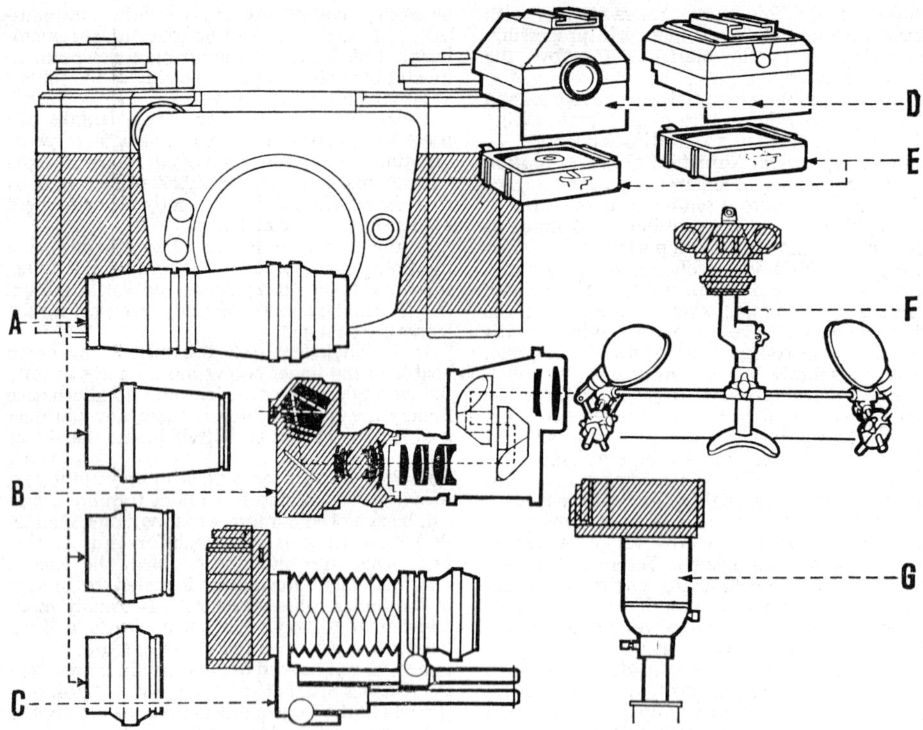

A MINIATURE REFLEX SYSTEM. As the reflex camera incorporates a universal viewing and focusing system, it needs fewer accessories to achieve the same scope as a rangefinder miniature. A, interchangeable lenses cover focal lengths from about 1 in. upwards. Ultra wide-angle lenses may be fitted in the camera if the mirror is swung out of the way. The reflex system is also ideal for zoom lenses. B, telescopic and other afocal attachments provide the equivalent of long focus lenses when fitted in front of the standard camera lens. C, extension bellows provide increased lens-to-film distances for close up and macrophotography. D, interchangeable finder units for eye-level or waist-level viewing. E, interchangeable focusing screen for alternative focusing possibilities with wedge rangefinders, fully matted screens, aerial images etc. F, copying or close-up stands (with or without lighting units). G, microscope adapter.

to its position of maximum sharpness for the subject focused, the two halves of the image join up and become continuous.

With some reflex focusing systems the image is broken up more irregularly in the central field, but again becomes smooth and continuous when the lens is sharply focused.

Rangefinder miniature cameras with interchangeable lenses and focal plane shutter can frequently be used with special reflex housings which in effect convert the camera into a reflex type. This is useful with long-focus lenses (since focusing on the screen is then more accurate than focusing by rangefinder), and for close-ups where the rangefinder cannot be used.

Grain. Compared to other larger negative sizes, miniature negatives require much more enlargement when prints are made. For this reason it is essential that adequate precautions should be taken to avoid objectionable grain in the image.

Where speed is unimportant, the slower fine grain negative materials are the better choice. However, to retain the fine grain, it is equally important to keep exposure to a minimum; over-exposure leads to coarser grain.

By using special fine-grain developers, further improvement results (although over-development will cause coarser grain). Most genuine fine-grain developers lower the film speed, although quite a few developers have been introduced which, besides giving finer grain, are actually claimed to increase the film speed.

The miniature photographer therefore can use a combination of film, exposure and development which will give perfectly adequate fineness in the grain. With the materials now available, grain presents no serious problems.

Exposure. On many modern miniatures exposure measurement is made convenient by a built-in exposure meter. Usually this is best suited for average reflected light readings, with the camera pointing at the subject. The exposure meter is in most cases designed to take in approximately the same view as the camera lens. If the meter indication is visible in the viewfinder and the meter coupled with the aperture and shutter controls, it is only necessary to point the camera at the subject and adjust the controls until the meter needle matches an appropriate index mark or pointer.

Such readings are suitable for approximately 90 per cent of all subjects encountered; special corrections may have to be made for against-the-light shots, subjects of exceptional contrast or (in the case of reversal colour films) of exceptionally low contrast, and subjects of predominantly light or predominantly dark tones. Often it is most convenient to allow for this by an adjustment of the film speed setting.

With fully automatic exposure meters the meter mechanism itself sets the exposure control when the release button is depressed for taking the picture. The advantage of this system is that no separate exposure readings are required, but such exposure automation is often also less flexible when specific combinations of aperture and shutter speed have to be selected for depth of field or subject movement.

Choice of Equipment. The range of lenses and accessories of the miniature camera makes it possible to carry out a wide variety of work and to overcome some of the drawbacks of the absence of camera movements and the small negative. Excellent work can be done with one lens, but it is the extra equipment which gives miniature photography its enormous scope and advantages.

A suggested basic outfit ought to include at least the camera and its normal lens, with an ever-ready case, lens hood, cable release, tripod, and a photo-electric exposure meter (where this is not already built in). This equipment could be the starting point; the aim being to add other items when necessary. To some photographers the most important extra would be a longer-than-normal focal length lens; others would want a wide-angle lens or a lens of the widest aperture. Most good miniature manufacturers list a number of lenses of short, normal, long and extra-long focal length. The quality is not necessarily constant and the cost varies accordingly; but for miniature cameras, the higher quality lenses should be an automatic choice.

Close-ups. In the field of close-up photography the miniature camera is particularly efficient and there are accessories—e.g., supplementary lenses, extension tubes, and special equipment —which permit copying and close-ups to be made to a very high standard. A difficulty is that the viewfinder is no longer accurate at these short distances, but it is overcome by the use of special distance gauges, copying stands, and frame finders. With a miniature camera, hand-held close-ups are practicable and accurate range finding is often possible at close quarters by the use of a separate rangefinder or by an addition to the existing rangefinder.

With miniature reflex cameras close-up photography is much less of a problem. If the lens is not interchangeable, supplementary lenses can be used to extend the focusing range downwards. With interchangeable lenses closer distances are obtainable by extension tubes. In either case the screen of the camera finder system shows the actual subject area taken in as well as the exact sharpness of the image.

In close-up photography with increased lens-to-film distances (e.g. by the use of extension tubes or extension bellows) an exposure correction is required which depends on the scale of reproduction of the image on the film. The correction becomes significant at magnifications greater than about 0·12 (1/8 reduction) and rises astronomically in macro-photography at scales larger than same size. No such correction is normally necessary when supplementary lenses are used.

In all close-up photography the depth of field is very limited and it may often be necessary to work with small apertures and correspondingly increased exposure times. Under those conditions special exposure increases may be necessary to compensate for reciprocity failure effects, particularly with colour film.

The Miniature Approach. Good prints from small negatives are a result of several things:

(1) Correct exposure, because the small negative is not really suitable for after-work, so exposure faults are not easily remedied later.

(2) Perfect definition, because unsharpness can be particularly noticeable when magnified by enlargement.

(3) Standardized processing to cope with a long length of film containing up to 36 negatives, often covering a variety of subjects and lighting conditions.

(4) Maximum use of negative area, because with such a small frame available none of it should be wasted.

Tackling the subject with these points in mind wins half the battle. The other half concerns the choice of subjects.

The wide aperture lens makes hand-held shots practicable in poor light and makes possible easier night-time shots indoors and out as well as stage photography by ordinary stage lighting; circus pictures; seaside illuminations, etc. The convenience of the small-size camera makes it ideal for hand-held photography of subjects such as children and animals that are always on the move and do not stay still; for unposed natural pictures of every-day life; and for tours, holidays, trips abroad, and family events. The large number of exposures available at one loading is an invaluable asset in producing picture sequences.

The interchangeable lens system permits the subject matter to fill the negative every time, and provides the means for specialized work in industry, medicine, research. The system also enables good portrait, architectural and landscape photography to be undertaken, in spite of there being certain limitations in these fields.

The very great depth of field of the standard and wide-angle lenses is a useful feature; though it can also be a drawback where differential focusing is needed to isolate a subject from its surroundings.

The secret of good miniature camera work lies in thoroughly understanding the equipment and exploiting its advantages. A.H. (L.A.M.)

See also: *Fine grain technique; Focusing stage; Miniature camera; Reflex attachment; Reflex camera.*

Books: *How to Choose and Use Your 35 mm. Camera*, by L. Gaunt (London); *My Way With the Miniature*, by L. Vining (London); *35 mm. Photo Technique*, by H. S. Newcombe (London).

MINIM. Smallest of all units for measuring liquids. One minim is equal to $\frac{1}{60}$ drachm, or approximately an average drop. 20 minims correspond to about 1 c.cm.

See also: *Weights and measures.*

MIRED. Contraction of "micro-reciprocal degrees". The mired value of a light source is the colour temperature in degrees K divided into 1,000,000.

See also: *Colour temperature.*

MIRROR. Polished surface which reflects rays of light falling on it in accordance with certain optical laws.

Mirrors may be either flat or curved. They may be made of metal or glass.

Glass Mirrors. A glass mirror consists of a sheet of plate glass coated with silver on one side. For ordinary purposes, the glass side of the silver is used as the reflecting surface and the outside of the silver coating is covered with a protective coat of paint. Mirrors of this type keep their brilliance indefinitely because the silver coating is completely sealed. The only disadvantage is that the surface of the glass itself acts as a mirror and gives a weak secondary reflected image. Normally this is not troublesome, but it cannot be tolerated in optical and photographic apparatus for which front surface (often surface silvered) mirrors must be used.

Surface Silvered Mirrors. In this type of mirror, the polished glass is coated with silver, usually by cathodic sputtering. The glass in this case is merely a support for the silver film and does not protect it. Usually a very thin coating of transparent varnish is sprayed on to the silver

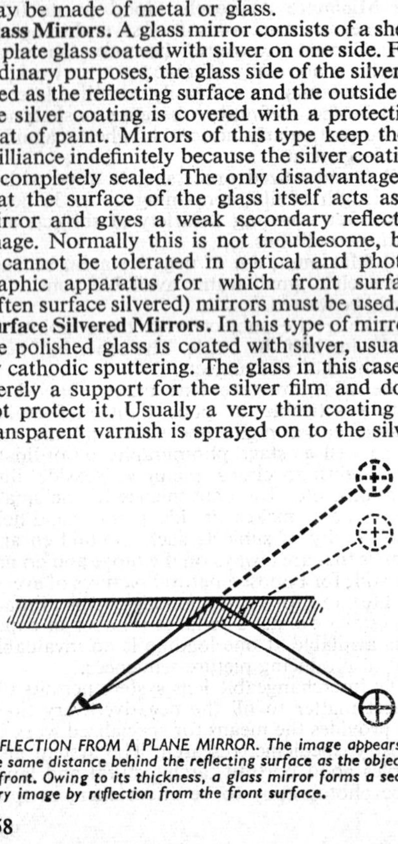

REFLECTION FROM A PLANE MIRROR. *The image appears at the same distance behind the reflecting surface as the object is in front. Owing to its thickness, a glass mirror forms a secondary image by reflection from the front surface.*

to prevent it from being oxidized quickly, but even so, surface silvered mirrors tend to lose their brilliance and have to be resilvered from time to time. Silver is nowadays replaced by other metals and alloys, particularly aluminium deposited on the glass support by vacuum evaporation. In photography, the principal use for front surface mirrors is for the viewing mirrors in single and twin lens reflex cameras.

Half-Silvered Mirrors. The metallic coating on a glass mirror can be made so thin that it reflects only part of the light falling on it and allows the remainder to pass through without interruption. Half-silvered mirrors of this type are used in some optical rangefinders, visual exposure meters and microscope illuminators.

Diathermic Mirrors. Normally the silvering of a mirror reflects visible light as well as invisible ultra-violet and infra-red radiation equally. Reflecting layers can however also be made selective in their action, reflecting certain wavelengths of radiation and letting others pass through. This is analogous to the action of a dichroic filter. With selective mirrors, the aim is usually to reflect visible light but to let through infra-red radiation, and hence heat. Such mirrors are known as diathermic; typically they may reflect up to 90 per cent of the light (compared with over 95 per cent for fully reflecting mirrors) and about 10 per cent of the infra-red component.

Diathermic mirrors are used mainly to direct light into places where the attendant heat radiation is undesirable—for example as lamp reflectors in projectors, spotlights, etc. Some present-day projection lamps with integral or proximity reflectors use diathermic mirrors. In a projector this means that the radiated heat reaching the transparency or film is greatly reduced and consists almost only of heat coming directly from the lamp filament, while the light includes that concentrated by the mirror.

The mirror layer can equally be designed to reflect heat and transmit visible light. It would then act somewhat like a heat filter, with the advantage that it does not itself heat up to anywhere near the same extent.

Metal Mirrors. Mirrors may be made by giving a high polish to a metal surface by burnishing. The metal principally used is stainless steel but mirrors are also made by plating chromium on to a support of another metal. It is difficult and therefore very costly to produce a perfectly flat metal surface, and the reflecting power of

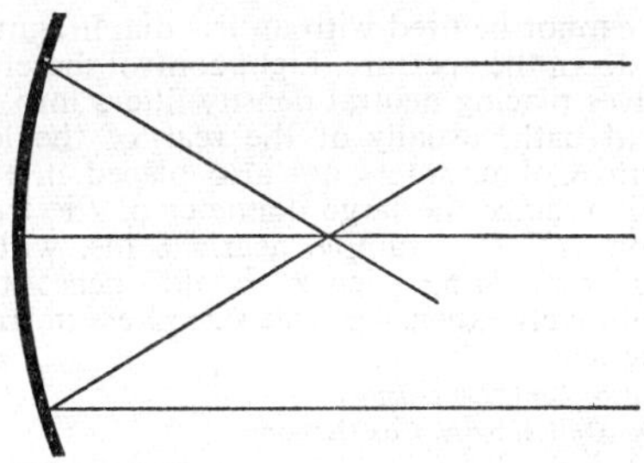

REFLECTION FROM A CURVED MIRROR. Parallel rays of light are reflected back to a focal point. This is situated at a distance equal to half the radius of curvature.

the polished metals is not as high as that of the evaporated metal on glass mirrors. So the optical quality of commercially produced metal mirrors is rarely as high as that of glass and they are only used in places where a glass mirror might get broken by shock or heat—e.g., in some camera viewfinders and in the lamphouses of enlargers and projectors.

Curved Mirrors. Curved mirrors are used to serve the same purpose as a lens in many kinds of optical apparatus. A curved mirror surface behaves in much the same way as a lens with the advantage that it is completely free from chromatic aberration. A concave mirror behaves like a convex lens—i.e., it brings parallel rays of light falling on it to a real focus (but on the same side as the light source). A convex mirror behaves like a concave lens—i.e., it brings parallel rays of light falling on it to a virtual focus (behind the mirror).

Curved mirrors intended for forming an image are usually spherical—i.e., the surface is part of the surface of a sphere. A parabolic shape is used for mirrors intended to reflect a parallel beam of light—e.g., in lamphouses, motor car headlamps and searchlights. When the light source (which is as nearly a point as possible) is placed at the focus of the parabola, the resultant reflected beam is parallel and as wide as the diameter of the opening. Parabolic mirrors have the advantage that they can be extended to reflect most of the source's rays.

Cleaning and Resilvering. Ordinary mirrors may be cleaned by washing the glass with warm soapy water and drying with a chamois leather.

Surface silvered mirrors cannot be cleaned by ordinary means. The most that the user can safely do is to remove dust from the surface with a soft camel hair brush. Surface aluminized mirrors can be washed with a detergent.

When the silvering of a mirror becomes too dull or scratched it can be renewed. It is not worth the amateur's while to resilver a mirror.

See also: *Beam splitter; Heat filter; Mirror lens; Reflector.*

MIRROR LENS. Image-forming system using curved mirrors instead of transparent glass elements. Usually mirror optical systems use glass elements—which are easier to make in the high precision required—to provide the surface for subsequent silvering. Newton investigated the properties of both lenses and mirrors and came to the conclusion that lenses could never be freed from chromatic aberration. So he decided that it was better to use mirrors than lenses in constructing telescopes. His conclusions were later found to be incorrect, but mirrors are, nevertheless, more suitable for astronomical telescopes than lenses, because they produce much better axial images.

History. Spherical mirrors were first used in place of lenses in microscopes by Robert Barker in 1736. They were soon replaced by achromatic lenses, but in recent years a number of reflecting microscopes have been constructed under a grant from the Nuffield Foundation. These microscopes have aspherical mirrors of aluminized speculum metal for both condenser and objective and they give a numerical aperture of 0·65. They are used principally for infra-red and ultra-violet microspectroscopy for which they are particularly suited by their freedom from chromatic aberration. Reflecting microscopes employing spherical surfaces are now obtainable commercially.

The 200 ins. mirror in the astronomical telescope at the Mt. Palomar observatory in California is the largest in the world.

Advantages. An *f* 4 concave mirror will produce a practically perfect axial image, while if the mirror is made parabolic, even larger apertures are possible. By contrast, an achromatic lens telescope objective is limited to a maximum aperture of *f* 5 by its residual chromatic aberration. In astronomy, where the highest standard of definition is required, even the parabolic mirror is usually limited to *f* 5.

Disadvantages. Mirrors have the constructional disadvantage that they form an image which is in the path of some of the incident light, so blocking the centre of the initial beam and slightly reducing the aperture of the mirror.

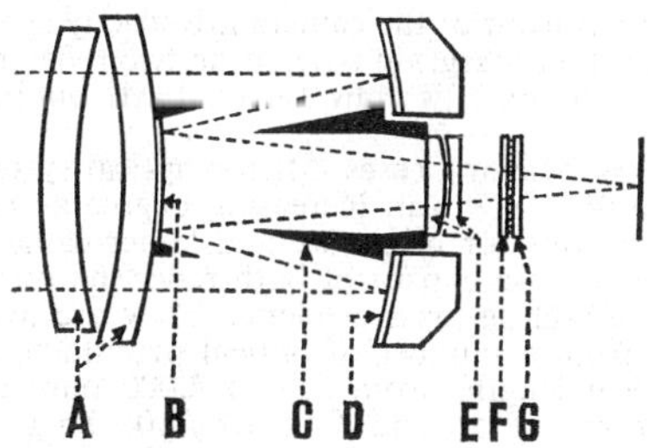

TELEPHOTO MIRROR LENS. Typical of long focus mirror optical systems for miniature still cameras, this lens uses both refracting and reflecting elements. A, front corrector elements. B, convex secondary reflector. C, cone to keep out stray light. D. primary concave reflector. E, refracting lens elements. F, neutral density filters (the only means of controlling light intensity since the lens aperture is fixed). G, colour filters if needed. Lenses of this kind are produced in focal lengths from 500 mm. upwards.

Sometimes a second mirror is used to reflect the light and form the final image behind a hole in the centre of the first mirror.

Astronomical telescopes are normally used for photographing heavenly bodies, as well as for observing them; so in this specialized field, it can be said that mirrors have been used as photographic lenses for some time. Unfortunately, because of the large off-axis aberrations, telescope systems, whether made with lenses or mirrors, cover only a very small field, so they are not suitable for normal photography where a very much wider angular field is required.

Schmidt Mirror Lens. Bernhard Schmidt in 1931 developed a mirror lens which overcame many of the disadvantages of the telescopic system. The Schmidt camera employs a spherical mirror to form an image on the photographic film.

The aberrations of the mirror are compensated by a glass plate with an aspheric surface mounted at the centre of curvature of the mirror. The curvature of field is, however, very large and has to be overcome by using a special curved photographic plate. When this is done, it is possible to cover a field of 24° and work satisfactorily at an aperture of *f* 1·0, which is an extremely good performance for such a simple optical system.

Schmidt-type lenses have been improved by using spherical and plane instead of aspherical surfaces, and it is now possible to produce a mirror lens system with a flat field at *f* 1·0.

Extreme speed mirror lenses, with apertures to about f/0.6—are used for fluorography and other applications involving the photography of fluorescent screens. At this high aperture the field is no longer quite flat, and this can be allowed for either by incorporating a curved fluorescent screen in the camera employing such lens or by holding the film in a slightly curved plane.

By astronomical terminology the Schmidt camera has the photographic plate right inside the mirror system. Mirror lens systems for fitting in front of the camera are strictly speaking of the Cassegrain type, using two reflectors. The light path is thus folded back on itself twice.

Camera Mirror Lenses. Mirror optical systems are also used to an increasing degree as telephoto units for miniature and larger cameras. Usually these consist of both reflecting mirror and refracting glass elements. They are available from a number of optical manufacturers in focal lengths from 500 to 2,000 mm. and apertures of around *f* 4.5 to *f* 10. They are intended mainly for use with reflex cameras, since the long focal length necessitates very accurate focusing. Also the camera and lens assembly must be mounted on a very firm support; at the telephoto magnification of 20 × with a 1,000 mm. lens the very slightest vibration leads to unsharp pictures.

One characteristic of mirror lenses is that they cannot be used with an iris diaphragm to stop down the aperture. Light control therefore involves placing neutral density filters into the optical path, usually at the rear of the lens system. Colour filters are also placed in that position, since the large diameter of the front component (for example nearly 8 ins. with a 1,000 mm. f 5.6 lens) would necessitate prohibitively expensive large-size filters in front of the lens.

See also: *Lens; Lens history.*

Book: *Optics*, by A. Cox (London).

MIRROR PHOTOGRAPHS. There is nothing difficult about taking a photograph of a scene reflected in a mirror. What the eye sees when it looks at the mirror from the camera position will come out in the photograph exactly as though the objects reflected were being photographed directly, but there are one or two points that must be remembered.

Focusing. The reflection in the mirror must be regarded as being as far behind the mirror as the object is in front of it, so the lens must be focused on that distance plus the distance from the camera to the mirror. It must not be focused on the surface of the mirror. For example, suppose the photographer is taking his own portrait and he is 3 feet from the mirror with his camera 1 foot in front of him. His reflection will be 3 feet behind the mirror and, as the camera is 2 feet in front, then the total distance from the camera to the reflection will be 5 feet. Accordingly the camera lens must be focused on that distance.

This simple relation only holds good so long as the camera is looking straight into the mirror. If it is pointed at an angle, to photograph objects lying to the right or left or above or below, then the focused distance must be equal to the distance from the camera to the object via the mirror measured along an imaginary line joining the two—i.e., along the path of a ray of light which proceeds from the object to the mirror and is reflected from the surface into the camera lens. This distance can be found by measuring from the object up to the surface of the mirror in the direction of the

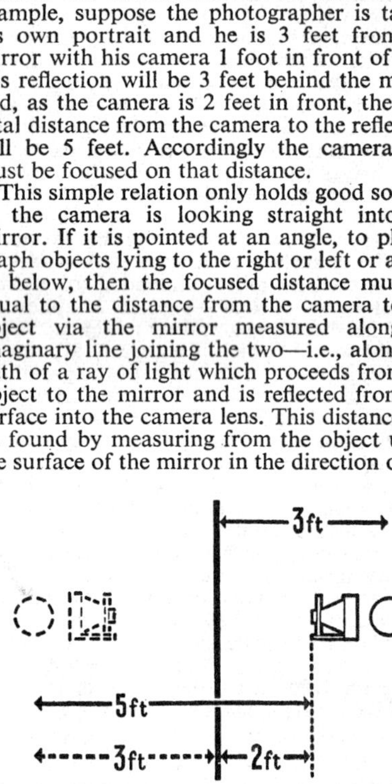

FOCUSING IN A MIRROR. The total focusing distance is the sum of the camera-mirror and of the mirror-subject distances, i.e. the subject is focused as if it were behind the mirror.

reflection of the camera, and from that point on the surface of the mirror to the camera itself.

With rangefinder or screen focusing, so long as the reflection and not the surface of the mirror is focused on, the image will be sharp, no matter what the angle of the line of sight or the distances of subject and camera from the mirror.

Effect of Glass. The ordinary domestic "looking-glass" is silvered on the back of the glass, and there is always a dim secondary reflection from the surface of the glass in addition to the bright reflection from the silvered surface. At a distance of several feet, this double reflection is not usually noticeable, but it may be objectionable in close-up pictures. The answer then is to use a polarizing filter in front of the lens, turning it until the surface image is no longer visible in the focusing screen—or when viewed directly through the filter.

The image reflected by the mirror is reversed from left to right. This does not matter with some subjects but it is undesirable in portraits and pictures where there is any printed matter. If the photograph is intended to look what it is, a mirror reflection, then there is no need to do anything further about it, otherwise the negative must be printed with the emulsion side away from the paper to cancel out one reversal with another.

Photographs may be taken of subjects reflected in the surface of distorting mirrors of the type set up in amusement arcades and fairgrounds. Here the same rule holds good as for plane mirrors, but it is not possible to fix the position of the reflection accurately. With rangefinder or screen focusing the lens should be focused on the important part of the reflection—e.g., the face—and a small stop used to bring as much of the rest as possible into focus. With a scale focusing camera, all that can be done is to set the lens as for photographing in a plane mirror and then use the smallest practicable stop to extend the depth of field sufficiently to cover any errors.

Where the mirror has a regular curve—e.g., spherical or parabolic—the image can be focused directly by rangefinder or screen in the normal way. But scale focusing by any of the methods described above is useless because the image may lie farther away from or closer to the surface of the mirror than the object reflected. The position of the reflection can be found from the fundamental optical equation if the focal length of the mirror is known. The focusing scale can then be set for the total mirror-reflection and mirror-camera distances. F.P.

See also: *Self-portraits; Tricks and effects.*

MISONNE, LEONARD, ?–1943. Belgian pictorialist specializing in gum prints and bromoils of stormy landscapes with contre-jour effects.

MISSILE PHOTOGRAPHY. Rockets, satellites and other missiles are photographed in flight to obtain information about their behaviour and data for future improvements, etc. Such recordings can be made from the ground or from the missile itself. Photographs from missiles are also used for reconnaissance and survey work.

From the Ground. Two objectives of missile photography from the ground are the recording of events, and the acquisition of metric information. Event photography is performed by tracking telescopes, except for short duration events which may be done by high-speed cameras. Metric information of a missile may be obtained by cinetheodolites, or by ballistic cameras.

Event Photography. This is primarily concerned with the recognition of missile events such as changes in the plume and attitude of the missile and the changes of position from image to image, as well as its appearance. In the U.S., long focal-length telescopes have been used with 70 mm. cine cameras. Frame rates and focal lengths vary, depending on the problem being studied.

In one camera, for example, five effective focal lengths are provided from 100 to 500 ins. with corresponding fields from 1½° to ¼°. Pointing accuracy of 2 milliradians is obtained by two operators, one for elevation and one for azimuth, using stiff-stick controls capable of 10° per second tracking velocity. For remote operation, radar range signals are fed into the electronic computer and thence to the instrument to keep it on track by converting radar data to azimuth and elevation co-ordinates, as well as positioning the glass wedges for automatic focus control at all focal lengths.

The use of a 70 mm. cine camera provides time-coded sequential photographs of missiles in flight with high resolution. Tracking telescopes are typically mounted in a 16 feet diameter astrodome 30 feet above ground level to avoid atmospheric turbulence, and in an atmosphere air conditioned at 50 per cent relative humidity and within 5°F. of ambient temperature for protection against sudden changes of temperature.

Metric Photography. This is concerned with the accurate location and trajectory of missiles. Both cine cameras and plate cameras are used. Cinetheodolites are surveyors' theodolites to which a 35 mm. cine camera has been added to eliminate the surveyor's visual sight, while the missile is being tracked.

A true line of sight can be determined from the azimuth and elevation readings by noting the position of the target relative to the cross-hairs. Focal lengths in current use vary from the 12 to 180 ins. With two or more instruments arranged on accurately surveyed base lines a spatial position can be determined. Range accuracy is 15 to 30 seconds of arc.

For greater distances, a satellite tracking

camera is used as a ballistic camera. For such a camera, a large field of view and high optical speed are of prime importance. The lens is usually a mirror-optical system, a typical set-up may cover 5° × 30° with a 20 ins. focal length at *f* 1 and a film scale of 406 seconds of arc per mm. In operation, the camera alternately tracks the satellite and the stars at the rate of sidereal motion parallel to the satellite orbit, thereby exposing the satellite on a background of faint stars on a single film. The satellite's position is noted in relation to the exact centre of the middle break in the star trails.

For accurate trajectory measurements of missiles where resolution is secondary to freedom from coma and distortion, a more compact ballistic camera with a 12 ins. lens yields an accuracy of 2 seconds of arc over a 33° field of view. Where range is not limited by the brightness of the light source, the simplicity and portability of such a camera has led to its widespread use. An external capping shutter avoids the vibration and hot air turbulence caused by an electric motor operating an internal shutter.

An ideal triangulation system would provide a 45° cut angle from the base line between stations to the missile. If the angle from one station is given, the optimum angle from the other one is such that the sum shall be 90°. Three stations at the apices of an equilateral triangle close to 50 nautical miles with the launch, or nadir of the missile at the centre of the triangle and a fourth camera close (1 n. mile) to the centre provides an optimum set-up, although only occasionally achieved in a trajectory.

From Missiles. Photography from missiles may be for surveillance of the sky for other missiles or satellites, or for reconnaissance of the ground. The prime requirement is area coverage. TIROS (Television and Infra-red Observation Satellite) is a good example of a spin stabilized satellite camera. Originally a 104°-field lens and later at 76° wide-angle lens covering 200,000 square miles was used to expose a 5/8 in. diameter Vidicon tube for 1/500 second, the image scanned in two seconds with 500 lines, and stored on reversible metallized magnetic tape for transmission on command.

The received image is photographed with a 35 mm. camera on a kinescope, and cloud depiction charts are prepared from the positive-film for distribution on facsimile circuits. In spite of many steps in the process, it has been possible to resolve ice floes in Canadian east coast waters.

The use of modulation transfer functions for each factor affecting the quality of an aerial photograph greatly aids the design of a photographic system. The lens function may be multiplied by the film emulsion function, the vibration function, the function for uncompensated linear image motion, the atmospheric wavefront disturbance due to boundary layer effects and "seeing" transfer function, to yield an overall performance curve, in which each individual factor may be varied to determine its particular effect on the system.

Photo-transmission systems from a drone or missile normally include a standard reconnaissance camera, either frame, strip or panoramic. This is followed by an in-flight rapid processor-slack box. The developed film is drawn past a high resolution film scanner and a high frequency airborne data link transmitter. The ground station receives the TV message and records it on 5-ins. film which is processed within a few seconds and viewed on a light table.

Due to the fact that the aerial camera can register more than twenty times as much information as the best TV camera, the transmission of this information cannot be continuous. Several minutes are required to read out each photograph, even at 20 megacycle bandwidths.

A special scanning tube has to be used to dissipate heat from the phosphor, The phosphor is deposited on a metal drum which acts as a heat sink and rotates the phosphor scanning area.

Rapid Processing. Silver halide films may be processed in 30 seconds or less by monobaths of low or high viscosity using spray cell, porous applicator, viscous applicator, or web methods of applying the monobath to the emulsion. Xerographic processes with a special film can achieve 200 lines/mm. resolution at an ASA speed of 20 with no liquid processing required. With the possible exception of photopolymers, this is the most promising type of film, where expense is no consideration, for photography from missiles.

Scope of Missile Photography. In its early history, missile photography was almost exclusively high-speed photography. As film speeds increased and instruments and their

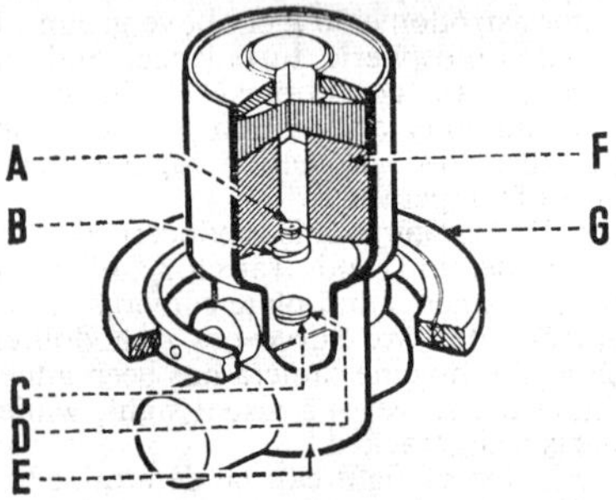

CAMERA ORIENTATION CONTROL. A star tracker of this type may be used to keep a camera system in a missile vertical for accurate aiming. The sensing system permits lining up and following a fifth magnitude star within two seconds of arc accuracy. It works by photo-electrically monitoring the light coming through the optical tracking system. A, transfer lens. B, nutation wedge. C, condenser lens. D, gratitule surface. E, photo-multiplier cathode. F, solid catodioptric telescope. G, gimbal mounting.

associated environmental control improved, the short exposures needed to achieve high resolution became less important. Photographs above the atmosphere looking up have achieved 1/10 second of arc angular resolution, and looking down they have exceeded one second of arc.

Looking up from the ground, the wave-front deformation of the turbulent atmosphere limits the resolution to two seconds of arc, except under unusual conditions. The seeing strength factor is approximately proportional to square root of path length and inversely proportional to cube root of path elevation. Boundary layer turbulence from a high speed missile is proportional to atmospheric density and Mach number (speed relative to the speed of sound) and must be used as an additional factor in photography from missiles in the atmosphere. L.H.N.

See also: *Ballistic photography; Rapid access processing.*

MIST. Particles of water vapour suspended in the atmosphere. These particles tend to reflect the rays at the blue end of the spectrum, but they are more transparent to the rays at the red—and particularly the infra-red—end. The result is that light shining on to mist looks bluish-white, and light shining through mist looks yellow or orange.

When an object is seen through mist, the mist reflects the blue and violet rays and only allows the rays towards the red and infra-red end of the spectrum to pass through. At the same time, light falls on the mist from the observer's side and the opposite thing happens: the redder rays pass through the mist and away from the observer, while the violet and blue rays are reflected back towards the observer. So the object appears veiled with a bluish-white haze.

To the camera, matters are even worse, because the normal photographic emulsion is more sensitive to the blue end of the spectrum than to the red. So that in the ordinary photograph of a misty scene, the effect of the mist is much more noticeable than it is to the eye.

The effect of mist can be minimized by using a filter which holds back the violet and blue rays. Yellow, orange, and red filters are suitable; the deeper the tint of the filter, the less the mist will show in the photograph. If the mist is to be removed as completely as possible, as in the tele-photography of distant objects, a special infra-red sensitive emulsion is used in conjunction with an infra-red filter.

Pictorial Effect. Mist suppresses the details of the scene and leaves only the main masses visible.

It divides the whole scene into clearly defined planes, with near foreground objects appearing as very dark shapes, while those farther away become rapidly lighter in tone until the more distant simply merge into the general whiteness of the mist itself.

So mist photography is usually concerned with broad, simple ideas.

Exposure. Because the mist scatters light into the shadow areas of the subject, photographs taken in mist should be given a shorter exposure than indicated by any of the normal types of exposure meter or calculator (these guides to exposure always assume normally-lighted shadows).

The beauty of a mist photograph is in its softness; any tendency to over—or even full—exposure produces a negative with too much contrast for this class of picture. For the same reasons, development should never be continued beyond the recommended time, and a soft-working developer should be chosen. P.J.

See also: *Fog (atmospheric).*

MODELLING. Gradual variation of tone that helps to display the shape of an object. In two-dimensional photography the solid form of the subject can only be suggested—in particular by controlling the way in which the light strikes the subject. With the right arrangement of lighting, the form of the subject can be represented in terms of light and shade to give a more or less convincing impression of three-dimensional reality.

A ball illuminated exactly from the front will show little variation in tone and will look flat like a disc. If the light is moved more to one side, a shadow will appear on the other side of the ball; this shadow will follow the curve of the ball and so show its form more clearly. A light used in this way is often called a modelling light.

See also: *Lighting the subject.*

MODELS. The best way to obtain the services of a model, if the work justifies the expenditure of money for model fees or if it can be afforded, is to go to a model agency and hire a professional.

Agency Models. At the agency files show pictures of hundreds of models, all kinds of characters, men, women and children and young girls. The abilities and photogenic qualities of a model are easier to judge accurately by studying pictures than by interviewing the model in person. So it is better to consult the files of a large model agency for the talent required.

Even the amateur, if he has his heart set on producing really good pictures of living models, would do well to consider hiring a model from an agency. He will have to pay her for about two hours of work as much as the average working man earns in eight hours, but the result will almost certainly be far better pictures than he can hope to obtain with amateur models. If that were not the case, photographers would not spend sums running into thousands of pounds annually to hire professional models. If amateurs were as good as the trained professional, there would be no room for the professional model.

Other Sources. Those who have no access to a model agency or who feel the expense is not justified need not despair. Good models of all kinds can be found in every city, town and village. It is only necessary for the photographer to seek them out, get them interested in posing, and find out by making test pictures which of them have the photogenic qualities he needs.

Character models of all kinds, including the glamorous girls who are always in greatest demand among photographers, can be found in the amateur theatrical groups. These amateur actors are willing and even eager, almost without exception, to be amateur models as well. And they are likely to have the unselfconscious qualities which make them at ease in front of the camera—a most important consideration.

For graceful girls with beautiful figures the most likely place to try is the local dancing school. Here the photographer is almost certain to find suitable models for bathing suit and figure studies among the dancing pupils. Poise and confidence are drilled into the dancing students from childhood, and the exercise of dancing practice gives them not only gracefulness, but the highly essential control of the pose.

They are also taught to control their facial expressions, and especially to be able to smile on command. False modesty is no part of their make-up; the dance student accepts naturally the sort of scanty costume that other girls would find embarrassing.

Dancers, too, will often pose in return for no other fee than a set of prints. It helps a dancer to check her technique to see herself in pictures, and if she ever becomes a professional she can made good use of them. But do not think of dance students only in terms of dance pictures; they are likely to be good models for any kind of photography.

Selection. Not all pretty girls make good models, so the photographer would do well to scrutinize the prospective model closely, with the following items in mind, before he commits himself to photographing her. Has she good teeth and pretty eyes? These are the features most difficult to retouch. Does she smile easily and naturally? A pleasant smile is a model's greatest asset. Has she a good bust line? This is an element of importance for bathing suit shots. Are her ankles trim? Trim ankles usually indicate a girl who will pose gracefully. Does she use her hands in a delicate way?

This is an important element in registering feminine charm on film. Is her posture good? The girl who habitually holds her head up and carries herself confidently will require less coaching before the camera than the one who slumps.

Is she self-confident? Humility is no virtue in a model. The best model is aware of her appeal, proud of her beauty, and happy about displaying it for the camera. Can she face bright lights without squinting? Note her reaction to sunlight; beware if she wears sunglasses.

Finally, does the idea of modelling appeal to her? Unless she is eager to co-operate, it is better to forget her. She should not be coaxed; no matter how beautiful she may be, she will be hard to photograph unless she co-operates whole-heartedly.

To determine the real extent of her interest, the photographer should give the prospective model his telephone number and wait for her to call for an appointment, rather than call her himself. This will weed out any whose interest is too mild.

Model Release. When a satisfactory model is found, she should sign a model release if the pictures are intended for publication.

The form approved by the Institute of Incorporated Practitioners in Advertising and the Institute of Incorporated Photographers contains the following details: name of photographer; negative series number; name of advertising agency or client; order number; product or service; and date of sitting.

This is followed by the body of the form, which reads:

In consideration of the sum of £.......... which I agree shall be paid to me (or my agent) for posing for the photograph(s) referred to above, I permit............................ and its licensees or assignees to use the photograph(s) and/or drawings therefrom and any other reproductions or adaptations thereof either complete or in part, alone or in conjunction with any wording and/or drawings for all advertising and editorial purposes in connection with the above product or service.

Unless otherwise agreed the photograph(s) and any drawings or adaptations thereof shall be deemed to represent an imaginary person.

I understand that I do not own the copyright of the photograph(s).

*I am over 21 years of age.

*I am not over 21 years of age but my parent/guardian has agreed to my undertaking this work and to the use of my photograph(s) as mentioned above.

*Delete as applicable.

This is followed by the name (in capitals) of the model, his or her signature and address, the date, and the signature of a witness.

A copy of the form must be supplied to the advertising agency/client with the photograph(s).

If the model is under legal age, the parent or guardian must also sign the release to signify his consent. E.M.H.

See also: *Copyright and the photographer.*

Book: *International Model Year Book* (London).

MODELS (SCALE)

There are two ways of photographing a scale model: so that it looks like a model, or like the real thing.

Natural Representation. The photographer who aims at a natural representation makes no attempt to disguise the fact that the subject is a miniature replica of the real thing. So he can employ the normal techniques that apply when photographing other small objects—e.g., coins, ceramics, glass and silverware.

The common denominator in photographing most small things is a long focus lens, and so long as the camera can be equipped with one, the negative size matters little—a miniature camera is no better than a quarter- or half-plate instrument. The principal advantage of the miniature is its greater depth of field, but, as the subject is still, the lens of a bigger camera can be stopped down until the whole subject is sharp. Where the camera has no provision for fitting a long-focus lens, the only alternative is to take a small image on the negative and enlarge that part only.

The important point is that the camera should not be brought so close to the subject that the natural proportions become distorted. For this reason the subject should, preferably, be taken with its longest axis parallel to the focal plane—e.g., ships and locomotives should be taken in side view and not three-quarters or end-on. This is not, however, vital in the case of scale models photographed at a distance in proportion to their size since the perspective then remains the same. A side view will on the other hand bring out more of the detail which is often the main inducement to photograph a model.

As a rule the model-maker is proud of his work and, far from wishing to disguise its smallness, he may want to draw attention to the minute detail and the skill required to execute it. For this reason it may be useful to include another object in the picture for comparison—e.g., a tiny model might be supported on a match box. And although small objects are generally photographed with the camera down on their level, in this case a somewhat higher viewpoint adds to the miniature aspect of the subject.

Artificial light is generally more satisfactory than daylight because it can be controlled and the conditions can be exactly reproduced if there has to be a retake. It is always safe to use natural angle lighting—i.e., at 45° from the front and side—with plenty of fill-in light from a reflector or a second lamp near the camera. Reflections from bright metal or painted surfaces should be watched for and dealt with either by dulling them with a touch of putty, or by moving the lights.

The setting of the subject is best left absolutely plain apart from anything introduced to give the scale, and the background should be well below the general tone of the subject and free from distracting shadows.

Exposure should be determined by a photoelectric meter in the first place, but the best results will be obtained by actual trial, making exposures of longer and shorter duration than the meter reading.

In all other respects the subject can be treated normally. What is wanted is the presentation of a straightforward "catalogue illustration" because the interest is all in the detail and workmanship of the subject, and not in the photography.

Illusion of Reality. The difficult part of model photography comes when the miniature replica has to appear as the full-size object. The deception is rarely successful, particularly when the model is on a very much reduced scale. It is almost impossible to present a picture of a scale model to the eye so that its appearance is identical with that of the real thing. There are several reasons for this, the principal ones being bound up with the perspective, the definition and the tones.

(1) Theoretically the perspective of a particular model can only look right if the viewing distance is scaled down in proportion to the scale of the model. Suppose the problem is to take a photograph of a 1 : 30 scale model of a motor-car 15 feet long so that the observer will imagine he is looking at a photograph of the car itself. It would be reasonable to allow the subject in each case to fill two-thirds of the width of the picture. The real motor-car, photographed with a camera with a normal lens—i.e., of about 50° angular field—would fill two-thirds of the whole picture width from a distance of approximately 30 feet. To give the same perspective, the scale model

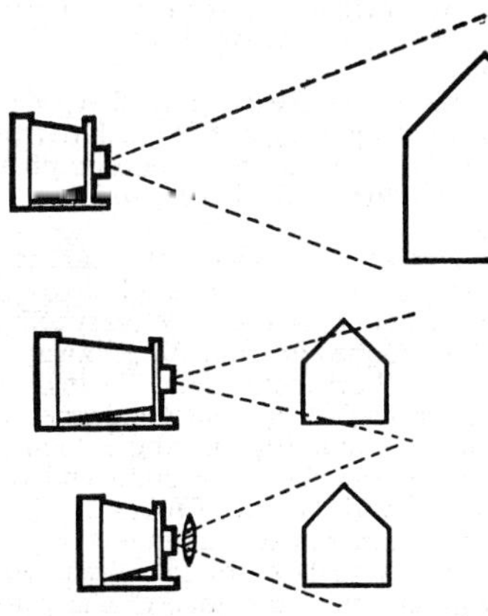

PERSPECTIVE IN SCALE MODELS. For correct perspective the model size as well as taking distance must be in proportion. To cover adequate angle, close-up lenses may be necessary.

would have to be photographed from a distance of 1 foot. At this distance the lens would have to cover a picture width of 9 ins. so that the 6 ins. long model would occupy two-thirds of the picture.

But, with a normal lens, the angular field of about 50° only holds good so long as the subject is more than ten times the focal length distant. For close-up work, the lens must be racked out beyond the normal focusing range, and the further it is racked out the smaller its field becomes.

This diminution of the field of view is more pronounced, the longer the focal length of the lens. In theory it would at a distance of 1 foot be negligible only if the lens has a focal length of about 1 in. Lenses of such short focal length are only available for subminiature cameras.

In practice it will be necessary either to put up with a certain loss of angle of view (thus the model of the car would fill more than two-thirds of the negative), or to restore the normal field covered by using a supplementary close-up lens and leaving the camera lens set to infinity. The focal length of the camera lens then becomes immaterial, except in that it determines the image scale on the negative. The larger the negative image, the less enlarging it will need subsequently to yield natural perspective at a normal viewing distance of 10 ins.

If the scale of the model is larger, e.g. 1 : 15 instead of 1 : 30—1 foot long instead of 6 ins.—natural perspective would be achieved by photographing it from a distance of 2 feet. Or, by accepting a viewpoint twice as far away—60 feet instead of 30 feet (and accepting the perspective rendering that goes with this increased distance)—the 1 : 30 model could again be taken at 2 feet.

In the general case of a 1/x scale model, if this is photographed to reproduce the perspective of the original as seen at distance D feet, the camera must be placed at D/x feet from the model.

(2) In most cases the zone of sharp definition in the model photograph is much narrower than it would be in the real photograph it is seeking to imitate, and this inevitably reveals the deception.

Going back to the model motor-car: it was shown that a 1 : 15 scale model photographed 2 feet away would have the same perspective as the real thing looked at from 30 feet away. But even with a short-focus (e.g. 2 ins.) lens stopped down to *f* 22, the sharply defined field would start at 1 foot 10½ ins. and extend only to 3 feet 9 ins. So the foreground and background would both be out of focus.

If the same camera had really taken a photograph of the actual motor-car, even with a much larger aperture, the zone of sharp focus would have extended appreciably in front of and behind the subject—e.g., at *f* 8 the scene would be sharp from about 14 feet to infinity—an effect which it would be impossible to reproduce in the model set-up.

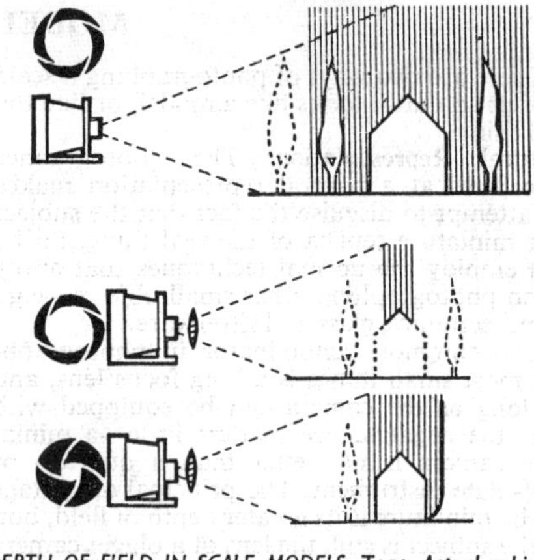

DEPTH OF FIELD IN SCALE MODELS. **Top:** *Actual subjects allow substantial depth.* **Centre:** *Close-up depth is limited.* **Bottom:** *So use small apertures and near backgrounds.*

Sometimes it is possible to get around the difficulty by using a background set close behind the subject, but in that case the background scenery itself will be so close to the observer that its perspective too will have to be dealt with.

The fact is that it is generally impossible to reproduce the natural falling off in sharpness in the model set-up and this shortcoming will reveal the deception even if the perspective has been reproduced perfectly.

(3) Out of doors there is generally some haziness in the atmosphere that tends to soften and obscure objects progressively as they get farther away from the observer. The eye comes to accept paler tones and softer edges as indicating distance and sharp definition, and strong contrasts of light and shade as indicating nearness. This separation of the planes of the picture is a characteristic of outdoor subjects that is not normally present in model set-ups.

The Practical Approach. Although it is theoretically impossible to photograph a model set-up so that it looks exactly like the real thing, up to a point in practice some reasonably satisfying approximations can be achieved. Anyone who is aware of the limitations has a better chance of achieving something worth while.

The subject should always be photographed from the eye level of an imaginary 6 foot person scaled down in proportion to the model—e.g., a 1 : 50 scale model should be taken with the camera lens 6 × 12/50 ins.—i.e., about 1½ ins. above the ground level of the model.

The viewpoint should be as close as possible to the subject. At the same time it should not be so close that the resulting depth of field with the lens stopped right down is unnaturally restricted.

In this respect, photographing models differs from the photography of other small objects. It is the usual practice in such cases to use a long-focus lens and a distant viewpoint with the object of avoiding strong perspective effects. But the perspective which would look overdrawn on a full-scale subject photographed from a close-up viewpoint looks normal on a scaled-down model. This is because the subject distance is seen in relation to the scale of the model. In a photograph of a full-scale dog-kennel, a subject distance of 3 feet from the camera would appear too close, but with a 1 : 20 scale model doll's house in place of the dog kennel, the subject distance would be equivalent to 3 × 20 = 60 feet and would no longer appear uncomfortably short.

If a scale model is photographed from a distant viewpoint with a long focus lens, the perspective is unpleasantly flattened exactly as it is when a full-scale subject is photographed with a high-power telephoto lens.

If possible the background should be flat, and close enough to the subject to be included in the depth of field. A suitably scaled down picture of hoardings, a brick wall, a screen of tall trees, either painted or photographed, can lend an amazing feeling of realism. Any such artificial background must, of course, have its lighting at the same angle as the lighting on the subject.

Reduced-Scale Optical Systems. Special viewing instruments based on endoscopes can be used for viewing a scale model from the proportionate distance and eye-level required to present the subject in its natural full-scale perspective. Here the optical system of a short focus telescope is mounted in a narrow tube terminating at a right angled prism. The viewer is set up so that the prism occupies the position of the eye of an imaginary model-sized observer. A real observer looking into the viewer gets a true impression of what the model would look like if it were scaled up to full size.

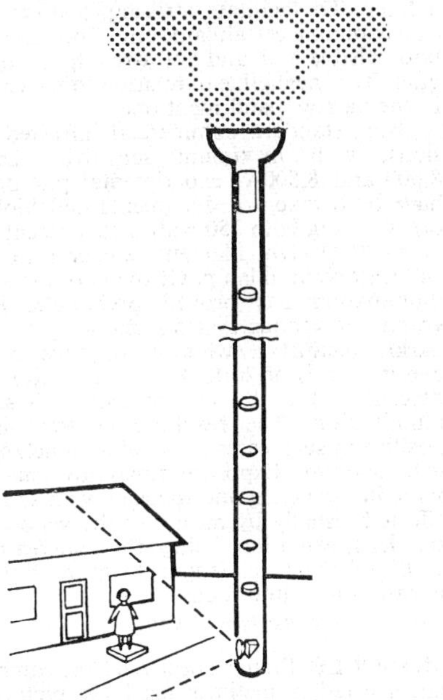

MODELSCOPE. The optical system of an endoscope, with a roof prism at the base, records close-up views of models as if a scaled-down observer were standing in the midst of the model scene. The camera (preferably reflex) is mounted on the unit.

If the eyepiece is replaced by a camera—preferably a reflex—the image can be focused and recorded just as it can with a microscope or telescope. This type of viewer is useful for photographing architectural models in life-size perspective. Another use is for taking still or moving pictures of model sets to project as a background to live action shots. The realism of the effect is derived from the fact that the primary image forming lens has a very short focal length.

This has the same relationship to the dimensions of the scale model as the lens of a normal camera has to a life-size set up. Accordingly both perspective and depth of field assume realistic values.

Camera Movements. A number of the difficulties of photographing models can be lessened by the use of camera movements.

The swing back helps to correct the distortion produced by looking down on the model. Even if the camera has no swing back, the same result can be achieved by keeping the camera level and photographing the subject on the top half of the negative. Without one of the means of correction it is impossible to show the top and front of the model at the same time without apparent distortion.

The swing front is used to extend the depth of field so that more of the foreground can be reasonably sharp. By swinging the front forward, the distance between the lens and the top of the plate is increased. This is equivalent to racking the lens forward to focus the nearer foreground.

The lateral swings are also useful for correcting any unpleasant convergence of horizontal lines when the camera looks at the subject from one side—i.e., so that the photograph will show something of the side as well as the front of the model.

Illumination. Lighting should present the subject in more contrasty tones than the background. The background should be softly lighted by a reflector or a well-diffused flood so that none of the shadows are as black as those of the subject itself.

If it is possible to introduce a third plane intermediate in both tone and position, it will enhance the feeling of depth and greatly improve the result.

When lighting for a natural effect, the modelling light should be kept at a distance so that its rays have something of the parallel character of sunlight. Otherwise the distribution of shadows will look unreal.

Everything that has been said above assumes that the model itself is a perfect miniature of the real thing right down to the last visible detail. If the finish of any part is in any way below standard—e.g., showing an obviously abnormally thick layer of paint or a mark left by an apparently giant cutting tool—all the care and calculation spent on getting the perspective, sharpness, tones and lighting right will be wasted. F.P.

See also: *Close-ups; Perspective; Table top.*

MOHOLY-NAGY, LASZLO, 1895–1946. Hungarian painter and photographer; professor at the Bauhaus. Experimenter in abstract photography and photomontage, 1922–28. He produced "photograms" by placing three dimensional objects on light sensitive paper and so produced contours, shadows and, in the case of translucent objects, texture also.

MONCKHOVEN, DÉSIRÉ CHARLES EMANUEL VAN, 1834–82. Belgian chemist and photographer. One of the foremost photographic scientists of the nineteenth century. Working in Vienna (1867–70) he built a famous studio for the photographer Rabending where life-size portrait enlargements were made. In 1878 perfected the preparation of silver bromide gelatin emulsions in the presence of ammonia. In 1879 improved the manufacture of dry plates and sold emulsion to dry plate factories. Later established in Belgium a factory for pigment papers. Also worked on spectral analysis and astronomy and improved the solar enlarger to which he added artificial illumination. Wrote on photographic chemistry and optics.

MONOBATH. Single solution combining developing and fixing chemicals, for simple and rapid processing. Monobaths for various materials are available commercially.

See also: *Developers.*

MONOCHROMATIC ILLUMINATION. Lighting with rays of virtually a single wavelength, used to eliminate chromatic aberration when maximum sharpness is required—generally in scientific photography and photomicrography. The sodium vapour and mercury vapour lamps are the commonest source of such light. The sodium lamp is now improved in its efficiency: one hundred lumens per watt, or more with lamps above one hundred watts. This increased efficiency is partly due to care in manufacture, but mainly to a good distribution of sodium along the lamp, due to the fact that pins placed in the inner tube allow sodium drops to collect at strategic points as the lamp cools.

Besides, the inner sodium discharge tube is enclosed in a larger tubular bulb, the two being separated by a vacuum, and this reduces the effect of ambient temperature.

Uses. The photographic lens can only accurately focus light of one or two particular wavelengths. It is true that at medium apertures the focusing setting provides sufficiently sharp focus for light of all wavelengths to which panchromatic emulsions are sensitive. Yet at these same apertures, photography by infra-red or ultra-violet radiation calls for an adjustment of the setting obtained by visual focusing, and at wide apertures such as *f* 2 the setting may not even be the same for red and blue. It follows therefore that, other things being equal, a record photograph exposed by monochromatic light will be sharper than if a wider band of the spectrum is used. In this way, all chromatic aberration is eliminated.

Sodium light is particularly useful for the photography of transparent or translucent objects, as modelling and depth are particularly well rendered without objectionable parasitical reflections.

Sodium and Infra-Red. As the sodium vapour lamp also emits rays in the infra-red band between 8,183 and 8,194 A., it can be used for infra-red photography provided that the visible rays are suitably eliminated by means of a filter. To facilitate striking, sodium lamps also contain a certain amount of neon, emitting both visible red and infra-red light, but the quantity is negligible in relation to the emission in the narrow band mentioned.

Using standard commercial infra-red emulsions, with maximum sensitivity between 8,000 and 8,500A. experimental photographs have been taken of documents and biological organs using both 250 watt incandescent lamps at 3,000° Kelvin filament temperature and a 140 watt sodium lamp. Of the two, the sodium illumination has proved preferable. Photographs of venous systems, showing the veins darkly against a white arm, gave negative density ratios of 20 to 45 per cent with incandescent and 20 to 75 per cent with sodium illumination. The resulting contrast in the positive in such cases is therefore much greater with sodium. Exposure times are much the same in each case, and focusing with sodium is effected visually by means of the yellow light, the lens extension being then increased by 1/200 of the focal length so as to bring the infra-red rays into focus. M.D.

See also: *Discharge lamp; Laser.*

MONOCLE. Simple spectacle lens, completely uncorrected, sometimes used for pictorial or portrait photography where the soft definition of the image is not undesirable. Otherwise used as a supplementary for changing the focal length of the camera lens.

MONOCULAR TELESCOPE. Originally developed for visual observation, monoculars are now made adaptable for photographic purposes. They enable a normal camera lens to be converted into a lens of longer focal length, the increase being dependent on the magnification of the monocular. In construction, monoculars resemble one half of a conventional pair of prism binoculars.

MONOPACK. Special form of sensitized material used in colour photography. It is a tripack of emulsions with intervening filter layers, coated on a common base. Each emulsion records one third of the spectrum. Also referred to as an integral tripack.
See also: *Colour materials.*

MONTAGE. Composite photograph made by cutting out parts of a number of prints and mounting them on a common support. The technique is used for producing trick photographs, making panoramas and aerial survey mosaics, and assembling photomurals.

Type of Prints. Some types of montage are purposely intended to appear as a collection of combined prints—producing a pattern or over-all impression. For this type of montage, the prints used do not have to conform to any critical requirements. However, for work in which the joins must be invisible (e.g., aerial survey), certain precautions must be taken in making the separate prints.

The scale of enlargement must be exactly the same for each print in the montage. This is best done by marking on the enlarger baseboard points where recognizable details appear in adjacent negatives (i.e., the overlapping edges of each picture); when the reference points of one negative have been thus recorded, and the print made, the enlargement of the next negative is adjusted until the matching details exactly fit the reference marks on the baseboard.

The illumination in the enlarger must be also perfectly even; falling off at the corners, producing light corners on the prints, makes it difficult to match the tones of adjacent prints.

All prints for the one montage should be made on the same paper, and to make the mounting as inconspicuous as possible, it is better to use single weight paper.

Cutting the Prints. It is important for the cut edges to be clean and smooth. This calls for sharp scissors or a thin, new safety razor blade. When a safety razor blade is used, the edge can be cut obliquely, so that the emulsion surface is undercut and comes to a thin edge, which can be stuck down more easily and so does not show up on the finished picture.

If both prints have similar types of blurred background—e.g., foliage—they can often be merged better if the foreground print (i.e., the one to be pasted on top of the other) is

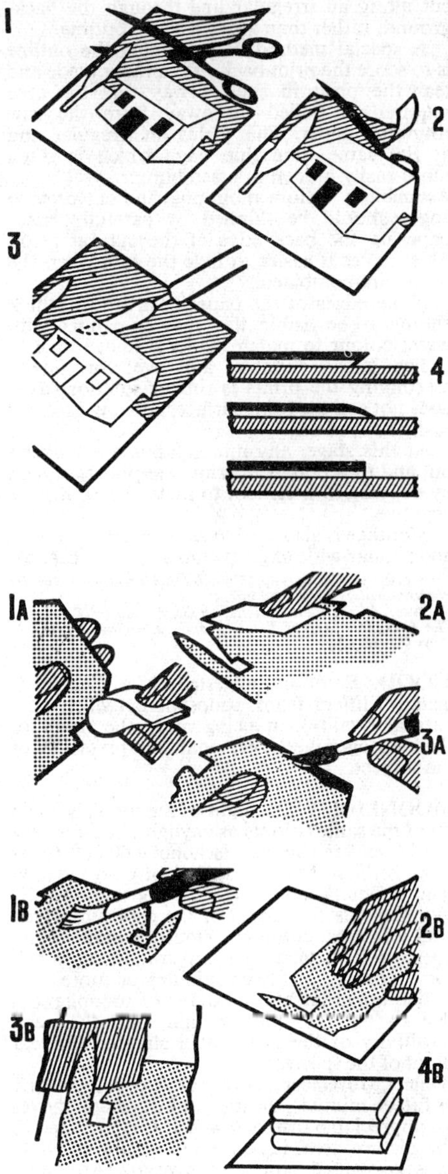

CUT-AND-PASTE MONTAGE. Top: *Methods of cutting.* 1. *When cutting with scissors, first trim away bulk of unwanted area.* 2. *Then cut accurately along outlines.* 3. *With a knife, move the print past the bent blade.* 4. *This gives an oblique cut and is easily stuck down flat. A scissor cut always leaves a thick edge.* Centre: *Preparing the outlines.* 1A. *Smooth out jagged edges.* 2A. *Thin edge with sandpaper.* 3A. *Paint with water colour.* Bottom: *Pasting.* 1B. *Cover back of cut-out with adhesive.* 2B. *Stick down on background.* 3B. *Smear thin details with adhesive on paper from underneath and stick down.* 4B. *Put under pressure until dry.*

cut along an irregular line through the background, rather than along a sharp outline.

A special method of producing the outline is to score the print back with a razor blade and tear the print in such a way that the area required is pulled up away from the surrounding paper. This yields an irregular and at the same time thin edge which is often more easily hidden in assembling.

Assembling. Before mounting, the cut or torn edges should be thinned by carefully sandpapering the back edge of the cut-out print. This makes it easier to hide the join when the prints are combined.

If the edges of the paper base are still thick enough to be visible, they may be painted with water colour to match the local image tone.

The most suitable adhesive for finally assembling the prints is rubber solution, as it does not soil the print surface, and any excess is easily rubbed off.

At this stage, any minor faults are touched out and the composite print is re-photographed by copying in a camera to make a new master negative.

Montage is also used to assemble panoramas, and similar wide angle pictures. L.A.M.

See also: *Aerial survey; Combination printing; Murals; Photo murals; Tricks and effects.*

Books: *All the Photo-Tricks*, by Edwin Smith (London); *The Complete Art of Printing and Enlarging*, by O. R. Croy (London).

MOOD. Emotion suggested by a picture. It usually differs from atmosphere (which is a natural quality), in being more the result of the photographer's personal interpretation of the subject.

MOONLIGHT. The light of the moon is made up of the same colours as daylight, but it is very much weaker (the sun is about 600,000 times more brilliant than the full moon). So the only real difficulty about taking photographs by moonlight is the length of the exposure. Even with a fast panchromatic emulsion and a wide-aperture lens, pictures of open landscapes call for exposures of fifteen minutes or more.

The apparent bluish colour of moonlight is due to the fact that in dim light the peak sensitivity of the eye is at a slightly different point of the spectrum.

Subject Movement. During an exposure as long as fifteen minutes, the image of the moon moves an appreciable distance, and if it is included in the picture area, it comes out as a long white sausage-shaped blur on the print. (Any stars in the picture area will appear as lines of light and not as points.) This is no drawback if the moon itself is kept out of the picture, but in that case there is a danger that the resulting photograph will be merely a flat and toneless view without anything to indicate that it was taken by moonlight.

Method. There are several photographic methods of including an image of the moon in a moonlight picture. The simplest method is to photograph the scene just after sunset a few days before full moon when the moon rises early in the evening. The exposure for the twilit landscape is about right for the moon itself so the moon may be included in the picture, and photographed at the same time as the scene.

Other methods rely on making two separate exposures; one for the landscape from which the image of the moon is excluded, and the other for the moon in which the exposure is so short that even if a part of the landscape is included, it does not register. The negative of the moon may be made by a second exposure on the same film or plate as the landscape, or it may be made separately and printed in afterwards like a cloud negative. The advantage with a combination print is that the same negative of the moon can be used for a number of different landscapes; the position of the shadows in the scene should, however, match the position in which the moon is printed.

Lens. If the moon is photographed with a normal angle lens, the resulting image is apt to look disappointingly small. For the picture to look natural, the image of the moon must be enlarged up to three times the size that it would normally appear in the print. This can be done by making the exposure for the moon through a long focus or telephoto lens with a power of 2× or 3×, while the exposure for the landscape is made through a normal angle lens. When the two negatives are combined and printed, the eye accepts the result as natural.

Alternatively, the moon may be taken through the normal lens, an enlarged duplicate negative made, and the latter combined with the landscape negative.

Exposure. Open landscapes under a full moon need an exposure of about 2–5 minutes at $f\,3{\cdot}5$ on the fastest panchromatic emulsion.

The light given by the half-moon is less than half of the light of the full moon; exposures under the half-moon should be at least three times the full moon figure—i.e., 6–15 minutes at $f\,3{\cdot}5$ on fast panchromatic film.

The above exposures will generally produce very dark shadows which help to convey the atmosphere of a moonlit scene. If the picture is to show the same amount of shadow detail as a daylight photograph, the exposures must be multiplied by four.

Equipment. Any camera that will take a time exposure is suitable for making pictures by moonlight. It should be mounted on a very steady tripod because of the long exposure times that have to be given. Preferably the camera should have both normal and long focus lenses so that the moon can be photographed separately on an enlarged scale for combination printing.

The lens aperture should be at least $f\,3{\cdot}5$ or the exposure times will be inconveniently long. The effect of stray lights and reflections is greatly increased by the darkness and the very long exposure, so the lens should be equipped with a really efficient lens hood. A piece of black card is useful for holding in front of the lens while lights of road vehicles are crossing the picture area.

Sensitive Material. While orthochromatic and slow materials can be used in moonlight photography, the fastest panchromatic plates and films are better, because they make for shorter exposures, less risk of movement during exposure, and consequently sharper pictures. F.P.

Image Amplifiers While normally instantaneous exposures by moonlight are impractical, they become possible with electronic aids such as image amplifiers. Since the amplification factors of such systems are well over 1000-fold, exposure times of a fraction of a second can record more detail than is visible under the lighting conditions by the naked eye.

Even cinematography is possible under such circumstances. A typical set-up might consist of a large-aperture lens (usually faster than *f*1) that projects an image on to the photo-cathode of the image amplifier, and another lens system that re-photographs the image from the screen of the amplifier tube on to 16 mm. film. An appreciable brightness gain also results from the fact that a large image is reduced to a small one and so concentrates the same energy into a smaller area, leading to a higher relative intensity.

See also: *Light sources in the picture.*

MORDANT. Substance which adsorbs and holds dyes specially strongly. In photography, mordants are used for dye-toning and certain colour printing processes where the silver image is converted into a mordant image in a mordanting bath, and then treated with a dye solution. The name mordant is also generally used for etching solutions used by block makers.

MORSE, SAMUEL FINLEY BREESE, 1791–1872. American painter and one of the inventors of the electro-telegraph. Met Daguerre in Paris and had a daguerreotype camera made late September 1839. One of the first in the United States to attempt portraits (September/October 1839, with exposures of 10 to 20 minutes). For a short time had a daguerreotype studio in New York (1840) in collaboration with Draper but soon abandoned photography.

MOTION PICTURE PHOTOGRAPHY. Alternative term for cinematography, especially on the professional level.

See also: *Cinematography.*

MOTION STUDY. In modern industry great importance is placed on economy in movement patterns of men and machines engaged in production. The design of the product and its method of use; machine and tool design, workplace layout and manual operations, all demand the achievement of the most efficient utilization of both effort and time.

The aspect of work study which deals with movement economy is known as motion study and is one of the most important activities of the work study engineer. It involves the analysis of movements into their basic functional elements. The work of the motion study engineer is to observe, or in the case of a new development to determine, the essential movements. He then analyses the records to work out the best movement patterns.

Motion study employs many specialized techniques of which photography is but one. The camera is especially useful in three main directions:

(1) For the analysis of movements which are either too rapid or minute to be appreciated readily by visual observation. This is known as **micromotion** study.

(2) For recording the paths of movement involved in a particular operation. This type of photographic record is known as the **chronocyclegraph.**

(3) For making general pictorial records of machines, tools, layouts and the like.

Micromotion Study. Micromotion employs a cine camera, usually taking 16 mm. films. The method is to take a cine film of an operation and include in the picture a chronometer graduated to read in 1/2000th of a minute. The resultant positive is projected frame by frame and examined to determine the points in time at which one basic element of movement merges into the next. This analysis is tabulated and studied and supplies the data necessary for the engineer to develop an improved method. The essential points to be observed in taking the film are to avoid heavy shadow areas and to

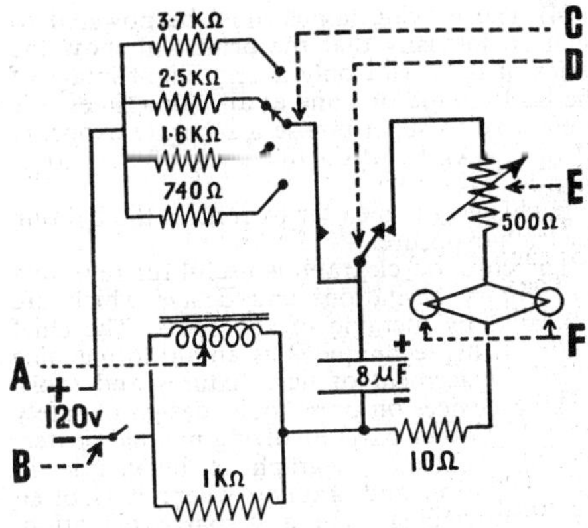

CIRCUIT FOR CHRONOCYCLEGRAPH INTERRUPTOR. A, high-speed relay coil. B, main switch. C, speed selector switch controlling flashing rate. D, make-and-break. E, variable resistance controlling lamp brightness. F, lamps.

fill the frame from a viewpoint which will permit all movements to be seen clearly.

Chronocyclegraphs. Micromotion is primarily concerned with the separate basic elements and particular movements and their relevance to the whole operation, but it very frequently happens that the path and amplitude of a complete movement which is a composite of many basic elements must be considered as a whole in a single picture.

This is where the still camera has the considerable advantage in that it makes it possible to record a complete trace of the movement—i.e., a chronocyclegraph. The technique consists in attaching to the moving member a small lamp which is made to flash at predetermined intervals by a suitable interrupter in the lighting circuit.

The lamps are switched on while the part is in motion and a normal brief exposure is made of such duration that it covers the whole cycle time. The resultant print shows the paths of movement as a series of white dots.

The interrupter apparatus is so designed that the intervals between the flashes may progress in steps from 8 to 30 per second. The current supply to the lamps is also variable so that intensity of illumination may be controlled for the various flash rates. A further characteristic is that the lamps light up at full brilliancy and die off slowly so that the flash produces a pear shaped image. The pointed end of the image indicates the direction of travel.

The following points must be observed in order to obtain the best results:

(1) The picture should fill the available area.

(2) A small aperture must be used in order to obtain a considerable depth of field.

(3) The photographer must be alert to recognize the beginning and end of the movement to be recorded. He must keep the shutter open only long enough to include the actual movement.

(4) The moving lamps must be powered to such an intensity that the print will show the series of dots with only a very slight image of the background or none at all. Sometimes it is preferable to superimpose a separate exposure of the background on the track of the movement.

In this way it is easier to control the lighting for each exposure.

The chronocyclegraph is useful for revealing momentary hesitations and delays which are indicated by merging of the spots. The chief value of this technique is as an aid in machine design, placement of jigs, fixtures and tools, locking devices on press tools, design of safety gates and other parts involving manual dexterity. The chronocyclegraph is invaluable in demonstration and training where it is often more convincing than a verbal explanation. **This technique gains considerably in value if a** stereo camera is used and the prints viewed stereoscopically. Relative distances are more easily defined, and movements in the plane of the taking lens axis are easy to resolve. J.H.

See also: *Chronophotography; Stroboscopic flash.*
Books: *Motion and Time Study,* by R. M. Barnes (London); *The Purpose and Practice of Motion Study,* by A. G. Shaw (London).

MOTOR RACING. Photographers who are also keen on watching motor racing naturally want to take pictures. Such pictures can be interesting and attractive records and are not difficult to get.

Equipment. Equipment must be specialized to some extent. A simple camera will only take pictures of racing cars when they are standing still; when they are moving they can only be taken with a camera with a good lens and shutter. The lens should have an aperture of at least $f\,3{\cdot}5$ and the shutter must be speeded to 1/500 second or faster. Normally the fastest panchromatic film should be used, but when a really bright day is expected it may be possible to use a slower fine grain film.

The most suitable camera for all-round photography of racing cars has a coupled rangefinder and direct (preferably frame-type) viewfinder. An interchangeable long focus lens is useful for the occasions when it is impossible to get close to the track. But if the photographer is content to limit his choice of subject and plan his course of action beforehand, he can still get good pictures with an ordinary camera without these refinements. In either

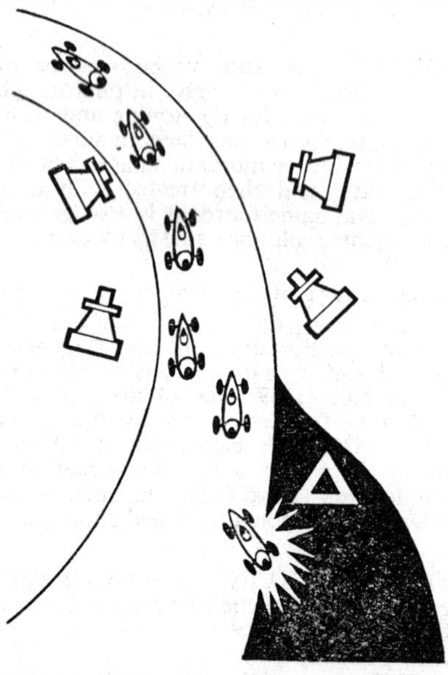

RACE TRACK SHOTS. Cameras with normal focus lenses are suitable for broadside or side-front shots on the inside and outside of bends (upper pair of cameras). Long focus lenses (lower pair of cameras) permit three-quarter or front views from farther back. Beware of the crash area just beyond the bend.

case, larger cameras are less suitable.

Position. The best position for taking the pictures of the race in progress is on the inside of the track at about 25 feet from the subject. As a rule, however, special permission (usually reserved for the press or the official race photographers) must be obtained for shooting from this position. At club meetings where attendance is restricted to invitation, the photographer can usually wander about as he pleases. Otherwise under ordinary racing conditions, the best position is as close to the track as possible.

The ideal position is on the track on the inside of a bend with the sun shining from behind the camera or across the line of fire but certainly not into the lens. The more light there is shining directly on the subject, the smaller the aperture that can be employed, and the greater the depth of field in which the subject will appear sharp. A shutter speed of 1/500 second should be used wherever possible and an exposure meter reading taken to give the aperture for this speed. (The meter must be aimed at the ground, not the sky.)

The photographer then stands as close to the bend as possible and sights and focuses on the line taken by the cars as they come towards him around the corner. The cars will naturally corner as close to the inside of the bend as possible and the photographer must take care that in his eagerness to get a picture he does not impede the view of the oncoming drivers.

As the car approaches it is kept in the centre of the viewfinder by swinging the camera, and the shutter is released as it passes over the spot previously focused on. The swinging movement of the camera must be smooth and steady and it should "follow through" even after pressing the shutter. Once this technique is mastered, sharp pictures can be obtained even with a shutter speed as low as 1/125 second unless the shot is taken on a bend.

(It is not generally realized that a car passing straight across the line of fire can be arrested with a slower shutter speed than one approaching direct or in three-quarters view. This is because the subject can be held absolutely stationary in the viewfinder when it is moving past the camera, but when it moves towards the camera, even though it is stationary in the finder, it is growing bigger as it approaches.)

By swinging the camera, the subject is made to appear sharp, but the background will of course be blurred. This is actually desirable in most shots as it increases the impression of speed.

For the photographer who has to go along with the crowd and do all his shooting from the public enclosures, the best position is on the outside of a wide bend. This is where a great deal of exciting duelling occurs and here drivers can often be caught fighting to regain control of a skidding car. At the same time it is advisable to keep away from the point on the outside of the bend directly in line with the oncoming cars because that is the point where crashes are most likely to end up.

There are also good pictures to be had from positions along the straight, but these call for very skilful panning to keep the cars steady in the finder when they may be travelling at well over 100 m.p.h. As conditions are constantly changing during the course of the race, it is as well to keep away from the grandstands which restrict freedom of movement.

At big crowded meetings, the photographer who is not adept at fighting his way to the ropes will find it a great advantage to get hold of a box to stand on.

Motor Cycles. Most of what has been said about motor cars applies to motor cycles, but there are one or two points of difference that need to be watched. Firstly, the motor cycle is a smaller target and calls for a closer viewpoint or a lens of longer focus than would be satisfactory for photographing cars.

Then the technique of cornering is different because the motor cycle has to lean over to counteract the centrifugal force. Generally speaking the racing car tends to corner wider and at a higher speed, relying on drifting and controlled skidding as much as steering for changing direction. The motor cyclist hugs the inside of the bend and leans over as far as possible just short of skidding. This makes the motor cycle a more impressive subject because the angle of the machine gives an added impression of speed that is absent with four wheels.

The best point for photography is on the outside of a sharp bend where the machines will be going relatively slowly and leaning well over. It is better to shoot a machine leaning away from the camera than towards it—i.e., shoot from the outside of the bend not the inside. This is because more light falls on the upper side of the machine and rider and more detail comes out in the photograph.

A broadside view from a low viewpoint is generally better than a three-quarter shot, but it calls for expert camera swinging and timing.

The corner is also the best place for shooting sidecar races. For such subjects it is important to take up a position on the sidecar side of the machines irrespective of whether this is on the outside or the inside of the bend otherwise the antics of the passenger will be masked by the machine and rider.

Club Meetings. The more homely club meetings can be just as interesting as international events and provide excellent practice for the photographer who wants to gain experience in motor racing photography. The cars at such events may not be the fastest, but they call for exactly the same technique as the star entries at international races.

There is a lot of interest to be had in photographing the various types of car present in the paddock. A stroll around usually reveals somebody taking his car to pieces or doing some last-minute tuning. For this type of shot there is no need for a camera with an expensive specification. The people in the paddock as well as the cars make an interesting study; the enthusiasts who attend motor race meetings are a colourful and distinct type and a keen student of human nature will certainly want to include some representative specimens in his day's bag. G.S.D.

See also: *Movement; Sport.*

MOTTLE. Print defect characterized by non-uniform density differences.

See also: *Faults.*

MOUNTAINS

Mountain photography includes two different types of work, with a certain amount of common ground. The first type is the photography of mountains and is an extension of landscape photography; the second is photography on and from mountains—i.e., mountaineering photography and work above the snow line. Photography on high mountains, as distinct from Alpine views, was undertaken under great difficulties by pioneers in this field—for example, Auguste Bisson, Samuel Bourne, and Aimé Civiale.

Equipment. The mountain climber has usually many things to carry and it can be quite a problem, especially in high alpine work, to cut weight down. The camera is only part of the total equipment, and not the most essential part. It competes with food, clothing, maps and instruments, ice-axe, rope, and perhaps pitons and crampons, so on most climbs only a light camera can be justified. The walker, especially in Britain, has more freedom, but generally prefers light equipment. Weight includes the instrument, accessories and supplies of material, so roll film cameras or miniatures are preferable to larger models.

Type of Camera. Either 35 mm. cameras or small folding cameras with picture sizes up to 2¼ ins. square are most suitable. They should be sturdy, well cased to ensure resistance to rough work, and simple in operation. A self-erecting front and a shutter release that can be worked with gloved hands are helpful. It should be possible to fit a filter on the lens and leave it in position. Automatic film counting is invaluable in alpine work where the dazzle from snow makes it almost impossible to read the numbers through the window. But if windows must be used, they should have metal covers to prevent fogging.

The light is usually bright to brilliant so extreme aperture lenses are not required; a good f4·5 or even an f6·3 anastigmat is all that is needed. High shutter speeds too are unnecessary, for almost all work can be done at 1/100 second.

For landscapes a rangefinder is not necessary; it is useful for climbing shots. Experienced climbers may dispense even with this since they acquire good judgement of distances by rope handling.

A light yellow filter of factor approx. 1½× is enough for most shots, including snow. A contrast filter can be added if desired, for occasional dramatic effects.

Sensitized Materials. For all black-and-white mountain photography, panchromatic material is best. For miniature cameras, slow, fine grain panchromatic material; for large sizes—i.e., 2¼ × 3¼ ins.—fast film can be used if desired since grain is less important, although the faster film will give lower contrast, and mountain scenes are often themselves of very low contrast.

All normal colour processes are suitable for mountain photography. But a haze filter at least must be added in the mountains where there is more ultra-violet and blue light, especially in snow or at high altitudes.

At times a somewhat stronger filter than the colourless u.v. is useful, whilst some workers recommend and use artificial light material corrected for daylight by adding the yellow or

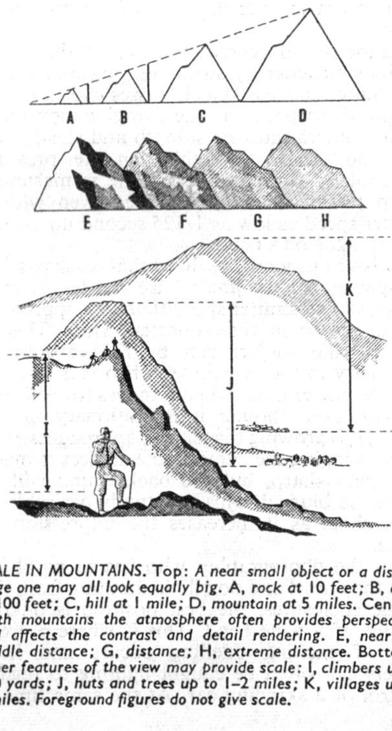

SCALE IN MOUNTAINS. Top: *A near small object or a distant large one may all look equally big. A, rock at 10 feet; B, crag at 100 feet; C, hill at 1 mile; D, mountain at 5 miles.* Centre: *With mountains the atmosphere often provides perspective and affects the contrast and detail rendering. E, near; F, middle distance; G, distance; H, extreme distance.* Bottom: *Other features of the view may provide scale: I, climbers up to 300 yards; J, huts and trees up to 1–2 miles; K, villages up to 5 miles. Foreground figures do not give scale.*

salmon pink filter recommended by the makers. This eliminates excess violet and u.v.

Telephoto Work. Telephoto work is a fascinating extension of normal mountain photography. For 35 mm. instruments, lenses of 13·5 or even 18 cm. can be used, but the equivalents in $2\frac{1}{2} \times 3\frac{1}{2}$ ins. cameras are very heavy and bulky. No rangefinder coupling is necessary because all subjects are at infinity. A medium yellow filter will give good results in the high Alps, whilst a red filter may be advisable in the greater haze of this country.

Lighting and Weather. Clear sunny weather with some light clouds in the sky is good for general record work, but the finest pictures occur in unsettled conditions, with the sun breaking through storm cloud or mist sweeping across hillsides.

Lighting and weather cannot be controlled, but with a map and a note of light directions at different times of day according to season, some idea can be obtained of the range of effects that may be expected in clear weather. But it is often necessary to study an important subject under varying conditions before its possibilities can be properly understood.

Most mountain landscapes are at their best in the early morning or late evening, when long shadows reveal and accentuate contours. This is particularly so in the high Alps, for the modulations of the snow and glaciers are then seen to the greatest advantage. But the top lighting of noonday, useless for ordinary landscape, is valuable when taking steep north faces, with the light glancing down.

Lighting for Colour. The clear rarified air of the mountains produces more pronounced contrast between sunlit and shadowed parts especially on high ridges. So unless there are clouds or haze present to diffuse the harsh sunlight and reflect light into the shadows, care must be taken to avoid large shadow areas.

As with monochrome, early morning and late evening lightings are often best. The weaker, yet warmer—i.e., more orange—light contrasts effectively with the blue or bluish shadow areas, yet each is bright enough to come within the restricted range of the colour material.

Seasons for Colour. In low hills such as in Britain, much of the charm of colour is in the contrast between foreground and distance. The heavy monotonous greens of high summer do not make attractive colour pictures. There is much more beauty in the subtler tints of late autumn, winter and early spring.

The same conditions apply, to some extent, in alpine regions although with evergreens there is less variety in the green of the foliage. In the foreground shots, however, from the end of the snows until early summer, the varied hues of the

LIGHTING IN MOUNTAINS. Top left: *Side light is the most useful lighting as the shadows indicate the shape of the hills and there is usually sufficient haze to give an impression of depth.* Top right: *Front light with the sun behind the camera produces little modelling, and atmospheric haze is least effective with this kind of lighting. It is, however, very suitable for colour photography.* Bottom left: *Back lighting with the camera more or less facing the sun gives contrasty and dramatic effects. It emphasizes receding planes, and tends to suppress detail.* Bottom right: *Top light from the midday sun in the summer is rarely suitable for open views, but may be successful for illuminating steep rock faces which will then show the same kind of modelling as by side light.*

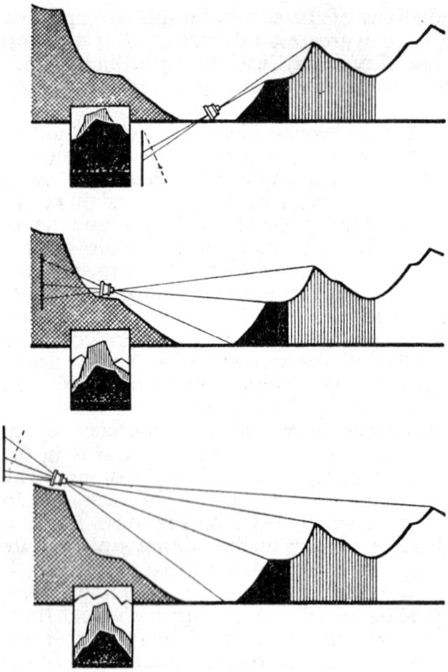

VIEWPOINTS. Top: *Shooting upwards from a valley makes the foothill almost obscure the summit.* Centre: *Halfway up the other side of the valley the foothill, summit, and background are in proper proportion.* Bottom: *Too high a viewpoint dwarfs a mountain as more distant peaks tower above it.*

alpine plants, the gentian, the alpenrose, the saxifrages lend a special attraction to colour photographs of mountain scenery.

Viewpoint. Viewpoints on valley or plain level have the disadvantage that when the mountain subject appears in reasonable proportion it is too small and distant to be effective, whilst a near view, being practically below the subject, seriously distorts the perspective. Photographs taken from hill tops often miss the valleys and lower slopes. Viewpoints at intermediate levels offer the best chance of a good picture. When the foreground slopes away sharply it also opens up the valley vistas and it is surprising how the view expands even with a relatively small gain of height above the valley floor.

Climbing Shots. The recording of technical detail on rock climbs and on major mountaineering expeditions is an important part of mountain photography. For normal work with single lens cameras, extreme angles of tilt either up or down should be avoided. The best climbing shots are taken either from opposite the route by someone not in the party or among the party itself on traverses (i.e., horizontal movements across rock) or on ridges.

Stereoscopic Work. Stereoscopic photographs do not suffer from this limitation of tilt, and in fact pictures looking steeply up or down have astonishing realism and drama. But apart from climbing shots, stereo pictures of general views are unimpressive so long as the normal lens separation is used. There is, in fact, virtually no impression of depth.

Effective stereo shots of distant peaks are best taken with abnormal lens separation, sometimes with the left and right pictures made from as much as 100 feet or more apart. For such "long base" shots settled conditions without moving cloud shadows are essential. But such pairs reveal third dimensional detail unobtainable in any other way, and the mountain looks like a model a few feet away with all depth accentuated. C.D.M.

See also: *Caves; Geology; Landscapes.*

Books: *The Photography of Scenery*, by C. D. Milner (London); *The Mountain Scene*, by F. S. Smythe (London).

MOUNTANT. The important thing about an adhesive used for mounting prints is that it must not contain any chemicals which might attack the photographic image. For this reason, glue and gum should never be used. Many suitable types of adhesive are available: starch paste, dextrine, gelatin, rubber solution and dry mounting tissue.

Starch Paste. This is made from ordinary starch powder. A quantity, say two tablespoons, of the powder is placed in a basin. About a dessert-spoonful of water is added and the starch is worked into a stiff paste free from lumps.

Boiling water is added (with constant stirring) to bring it up to the consistency of thick cream and at this stage is put into a saucepan and boiled for about two minutes.

A few drops of carbolic acid solution are then added and the mixture allowed to cool. When it sets, the skin which will have formed on the top is removed.

Dextrin. Dextrin mountant is best bought ready made and is used in exactly the same way as starch paste.

Here is a formula for those who prefer to make it themselves:

Dextrin (white)	14 ounces	350 grams
Water	20 ounces	500 c.cm.
Oil of wintergreen	5 minims	0·3 c.cm.
Oil of cloves	5 minims	0·3 c.cm.

The water is placed in a double saucepan over heat and when it has reached a temperature of 160° F. (this is critical) the dextrin is stirred in slowly. When it has dissolved, the preserving oils are added with constant stirring. This paste should be kept a week or two before using to permit it to mature.

Gelatin. Gelatin mountant has the advantage that the solvent evaporates rapidly so there is no danger of cockling the mount or print. One part of ordinary cooking gelatin is covered with water and allowed to swell. This takes some time and the water must be changed several times during the period. The water is poured off and the soaked gelatin is heated in a water bath until it melts. Twelve parts of methylated spirit are added slowly while the mixture is stirred. It is then poured into a large wide mouth jar and covered. The solution sets when cool. It is melted again before use by standing the jar in hot water. It is applied quickly and evenly with a soft brush to the back of the print which is then placed as quickly as possible on the mount.

Rubber Solution. This adhesive has rapidly taken the place of the older mountants. It has the advantages of being very clean and easy to use. It adheres strongly, but if necessary, the print can be easily stripped from the mount. It is non-cockling. Several brands of adhesive rubber solution are marketed specially for photographic mounting.

Only these rubber mountants should be used for photographic prints; other rubber solutions may contain substances which are harmful to the photographic image in various ways.

One drawback of rubber solution is that the rubber layer between the print and the mount gradually tends to lose its binding properties. The print may then start peeling off.

Dry Mounting Tissue. Dry mountants that cause adhesion when heat is applied are the least messy to use. Tissue impregnated with shellac is the most common and is usually sold in sheets or rolls interleaved with waterproof paper. Provided the tissue is used properly, really perfect contact between the print and mount can be achieved. The layer of shellac will also serve to protect the print from any impurities in the mount. Dry mounting tissue must always be stored in a cool and dry place to prevent it from deteriorating.

Solutions of thermoplastic synthetic resins are also used for dry mounting. These are brushed or sprayed on to the back of the print, and the latter can then be treated—once the resin is dry—like a print with dry mounting tissue already attached. R.M.F.

See also: *Adhesive.*

MOUNTING PRINTS

The print and its mount should always be truly rectangular. This can be ensured by careful trimming.

Trimming. The best trimmer is the desk type, operated by pressing down the top platform. In this type the necessary guide stop and margin indicator are incorporated. For prints up to 5 × 7 ins. the type of trimming board that has a push-down knife or guillotine is satisfactory.

Prints can, of course, be trimmed with a straight edge, using a set square as a guide. The print is placed upon a sheet of glass and cut through at one stroke with a very sharp knife or a safety razor blade in one of the holders sold for the purpose.

If the print has a white margin it is best to use as a guide a glass or Perspex straight edge through which the edges of the print can be seen: it is then easy to trim the print exactly parallel to the edge of the mask. When trimming a print without a white margin, the knife should be inclined slightly outwards. This will avoid leaving a white edge showing around the print.

Paste Mounting. The print is placed face down on a sheet of clean paper or an old newspaper. The mountant—e.g., starch paste—is then rubbed on the back of the dry print with the fingers or a brush (a discarded shaving brush is ideal). The brush is wiped across the print on all four sides on to the paper. This ensures that the edges, and especially the corners, are coated with paste. At this stage the print must not be allowed to slip over the paper because it is important to keep the paste off the front surface.

The print is next lifted carefully and placed in the correct position on the mount. This position should be indicated beforehand by light pencil marks. The best way is to attach the top edge, and then carefully lower the rest of the print into place to prevent air pockets between the print and the mount. A sheet of clean paper is then put over the print and pressure applied with a squeegee. The soiled paper upon which the print was pasted should be replaced by a fresh sheet for the next print.

Wet Mounting. Prints may be paste mounted when still wet if the excess water from the final wash is first removed by squeegeeing the prints face down on a sheet of glass.

The mountant is applied to the top print. It is peeled off, placed in position on the mount, covered with a sheet of white paper, and rubbed down into place. The surface of the print is then dried with a clean cloth, covered with a fresh sheet of paper and squeegeed firmly into contact with the mount.

Wet mounting should be avoided because when prints are mounted wet there is always a risk that the mount will cockle on drying.

Rubber Gum Mounting. The print is trimmed and its place on the mount is marked by two light pencil dots at the two top corners or by a rectangle ruled in pencil.

An ample supply of rubber solution is taken up on a palette knife and spread evenly over the rectangle of the mount, carrying it well beyond the dimensions of the print. The back of the print is then covered in the same way and both surfaces are allowed to dry for about a quarter of an hour.

The print is then laid in position, covered with a sheet of waxed paper, and squeegeed

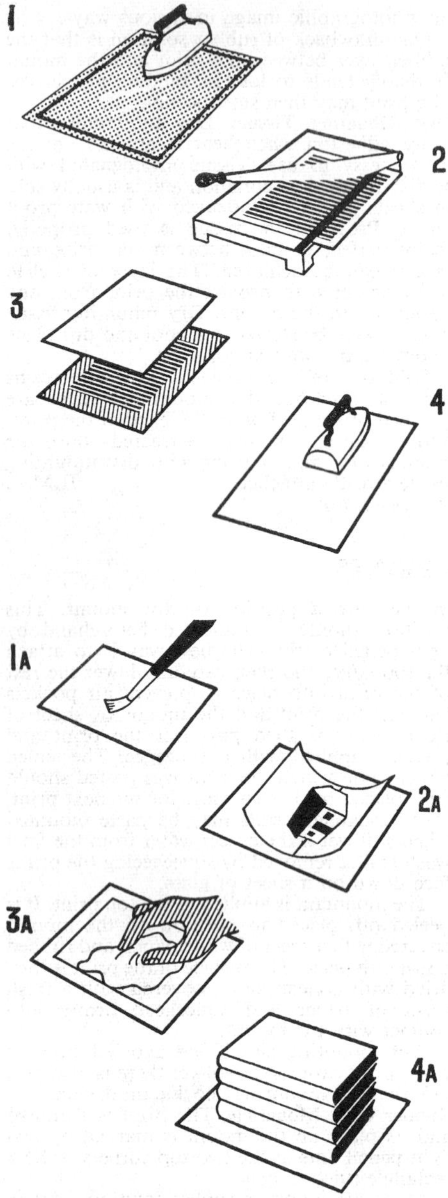

METHODS OF MOUNTING PRINTS. *Top sequence: Basic stages in dry mounting with tissue. 1. Tack down a sheet of tissue. (slightly larger than print) to the back of the print with a hot iron. 2. Trim edges of tissue and print. 3. Place on mount and cover with clean paper. 4. Apply hot iron. Bottom sequence: Main steps in wet mounting with paste. 1A. Rub paste well into back of wet print. 2A. Carefully lower print on to mount. 3A. Wipe over to press down. 4A. Put print and mount under pressure for several hours.*

into contact with the mount. Once contact between the print and mount has been made, the print should not be moved. After about ten minutes the surplus solution around the edges can be rubbed away with the fingertip without harming the mount or the surface of the print.

Dry Mounting. Professional and commercial workers attach prints to the mount with dry mounting tissue. This is paper coated on each side with shellac which melts when heated in the dry mounting press and holds print and mount together. Prints mounted in this way do not cockle and the surface is left smooth and free from irregularities. The method is permanent because the tissue effectively stops any chemical impurities in the mount from getting through to the print.

A special heated press is necessary for dry mounting prints in any quantity and especially for the larger sizes. These presses are heated by gas or electricity and maintain the correct temperature for melting the tissue. The two faces of the press are very strongly constructed so that they will give the necessary high and uniform pressure all over the print.

A sheet of tissue, larger all round than the finished size of the print, is placed over the back of the print. The tissue is then touched down with a small heated fixing iron applied around the centre of the print. This melts the tissue and fixes it temporarily in position on the back of the print. Prints must always be quite dry before being mounted. There must be no dirt or grit on the print or the cover plate or it will indent the surface and ruin the print.

The print with the tissue attached is then trimmed and the tissue is touched down on the mount with the iron at all four corners. The tissue is then stuck to the print in the centre and to the mount at the corners, so that it holds the print in position ready for the mounting operation.

The bed plate of the machine is heated to 140°F. as indicated on the thermometer. The print, face up, is placed in the press with a metal cover plate on top of it and the pressure is applied for three or four seconds (prints on double weight paper need longer time). The print is removed and while still hot is bent outwards to remove the slight curl left by the press. If the print comes out of the press attached to the tissue and not to the mount it generally means that the press is not hot enough. If it is attached to the mount and not to the print the press is too hot.

It is possible to dry-mount a print with an ordinary domestic flat iron. The hot iron is first used to touch the tissue down on to the centre of the print and then, after trimming, on to the mount, exactly as the small fixing iron is used above. The mount and print are then laid on a smooth surface and a thin metal plate is laid on the print to act as a separator.

Pressure is applied with the hot iron over the whole area of the print until it is firmly stuck to

DRY MOUNTING EQUIPMENT. Left: *Electrically heated dry mounting press for professional use.* Right: *Touching iron (may be electric) for attaching tissue to prints.*

the mount. This method is fairly satisfactory for small prints but less effective for larger sizes. It is not as certain in action as a real dry mounting press.

If a dry mounting solution is applied to the back of the print it is only necessary to let the solution dry thoroughly. The print is then placed in position on the mount, inserted in the dry mounting press and placed under pressure and heat as with dry mounting tissue.

Tape Mounting. Double-sided adhesive cellulose tape is also a neat means of mounting prints. When placing large prints on exhibition mounts it is advisable to attach only the top edge of the print. Smaller prints intended for albums can be mounted by two strips of adhesive tape attached at the top and bottom edges.

This is the procedure. Place the print face up on the mount and mark the exact position of the two upper corners with fine pencil marks. Then turn the print over on its face. Pull a length of adhesive double sided tape off its roll and stick it, with the free adhesive side downwards, on the back of the print near the top edge. If the print is to be fixed at the bottom as well, repeat with the second strip near the bottom edge. Next peel off the protective paper from the top of the adhesive strip, exposing the second adhesive surface. Place the print in position on the mount, holding it so that the top edge only touches the mount at an angle, with the adhesive surface of the strip well clear of the mount.

Align the print carefully with the pencil marks made before, and gently lower the paper on the mount so that the adhesive strip makes contact with the mounting board. Then press down firmly, running a soft cloth, (or a clean finger) along the print surface above the adhesive strip.

Print location must be correct the first time; once the adhesive strip has made contact with the mount it is rarely possible to move the print again without damaging the photograph or the mount or both.

Mounting for Exhibition. Prints may be mounted for exhibition by any of the methods described, but there are generally special conditions to be fulfilled. Most exhibition authorities stipulate mounts of certain standard sizes—20 × 25 ins., 16 × 20 ins. and 12 × 15 ins. are common. It is also usual to stipulate that the mount must be white or of light tone.

The size of the print is usually left open, but in most cases it will be found that the smaller the print the larger the apparent size of the mount. For example, a print 3 × 4 ins. would look well upon a mount 8 × 10 ins., but a print 6 × 8 ins. would look as if it had insufficient margin.

With the exception of the commercial mount where the position of the print is already marked, the photographer is free to decide the position of the print for himself. It is generally agreed that with vertical prints there should be more space at the sides of the print than along the top. And with horizontal prints there should be more space at the bottom than at the top; in fact a print mounted in the centre often has the appearance of dropping off the mount.

The best position should be found by experiment and marked before the print is finally mounted.

Fancy Mounts. Many workers attempt to enhance the appearance of their work by various more or less dubious embellishments such as mounting the print on a tinted mount or giving it a deckle edge. Such extras are almost always obtrusive and compete with the picture.

Pencil lines lightly ruled on the mount can be very effective, particularly if there are white areas running off the edge of the print. But the lines must be neat, rectangular and lightly drawn.

Titles and signatures on the mount are better left out unless the worker is skilled in the art of lettering. Unskilled lettering can easily ruin the appearance of an otherwise excellent print. R.M.F.

See also: *Embossing prints; Framing photographs; Mountants; Mounts.*

Books: *Enlarging*, by C. I. Jacobson (London); *The Complete Art of Printing and Enlarging*, by O. R. Croy (London).

MOUNTS. The main purpose of a mount is to isolate the photograph from its surroundings. Interest is then concentrated on the photograph and not attracted by detail outside the picture area. It also serves to provide the photograph with a stiff support.

The simplest mount is just a plain piece of card, coloured cream, ivory or white. A thickness of 1/16 in. is normally adequate.

Simple mounts are usually best, and exhibition prints are nearly always displayed on quite plain mounts varying only in surface and colour.

Colour. The image colour of a photograph can range from a cold blue-black to a warm reddish brown—or it can be toned to a variety of other colours. The actual paper base of the print can also vary from white to deep cream. So the colour of the mount must be chosen to blend with the print. A cream based print will tend to look degraded on a white mount; and although a warm toned print may be put on a white mount, it generally looks better on ivory or cream. Dark coloured mounts are best avoided, the only exception being for colour photographs, which gain in brilliance when mounted on a black or dark grey card.

Surface. Mounts are also available with different surface textures—e.g., smooth, egg shell, rough or with a linen texture.

Choice of surface is less important than colour but it does have some influence on the final appearance of a photograph. Pictures of a strong, bold character look better on rough mounts, and light delicate subjects need a smoother surface.

Size. If the mount is required only as a support, it is better to buy a size slightly larger than the print. The surplus border is then trimmed off after the print has been mounted. This leaves a cleaner edge.

When the mount is used to provide a border as well, there is no practical limit to the size that may be used, but the normal practice is to choose a size that will provide about 2 ins. border along the top and sides with rather more along the bottom. However, as mounts are made in standard sizes, the width of the border can only be varied by trimming the mount or altering the dimensions of the print. So unless a very wide border is required, a 12 × 15 ins. print is used on a 16 × 20 ins. mount or trimmed down slightly and used on a 12 × 15 ins. mount. A variation sometimes effective is to mount a horizontal print on a larger size vertical mount with the print near the top edge.

The standard sizes of mounts are the same as those for prints, but mounts are not normally made smaller than 6½ × 8½ ins. Most exhibitions have a rule that prints should be on mounts of a certain size. This is usually 16 × 20 ins., or 12 × 15 ins. but with minor exhibitions smaller sizes are often accepted.

Quality. It is important to choose a mount made specially for photographic prints. Ordinary mounting card may contain chemical impurities that will affect the permanence of the photographic image. Poor quality mounts also have a tendency to split along the edges, particularly when heat is applied—as in dry mounting.

Variations. Other variations on the simple mount are also used. The picture area can be cut out, and the edges of the opening bevelled. The picture is then mounted behind the opening so that it is framed by the mount. (Cutting out this aperture requires considerable skill.

Mounts can also be obtained with plate markings—i.e. with a sunken area slightly bigger than the print to be used.

Artists' drawing paper may also be used as a mount when firmness is not important. The most suitable is hand made paper, which has a very pleasing surface texture.

Coloured paper of a contrasting shade to the mount can also be introduced between the print and the mount. The paper is cut slightly larger than the print so that it will show a narrow border. With the right subject this narrow border between print and mount can look very effective. These papers, called border tints, are usually coated with shellac on one side for dry mounting.

Folder Mounts. In its simplest form, the folder mount is a piece of thin card folded in half. The picture is mounted on one half of the card, while the remaining half serves as a cover.

Folder mounts embody much the same features as the plain card mount, but are mostly made in smaller sizes and of much thinner card. They are also made in various colours with mottled and grained effects, deckled and torn edges and even with silver or gilt border lines.

To avoid the use of a mountant, folder mounts usually incorporate some easy system of attaching the print—e.g. four diagonal slits for the corners of the print (a method commonly used in albums). Another type has a cut-out frame stuck over the mount along three edges, the fourth being left unstuck so that the print can be slipped in.

Folder mounts are mostly used for small prints that are to be carried in the pocket or handbag. They protect photographs from surface marks and help to keep them flat. J.D.C.

See also: *Albums; Greeting cards; Mounting prints.*

MOVEMENT

If the subject of a photograph is moving as the shutter is released, its image will move over the surface of the film or plate in the camera. No matter how slowly the subject is travelling or how quickly the shutter opens and closes, the image is bound to move to some extent. Because of this movement the negative and any prints made from it will show blur. And if the negative is enlarged, the blur will be increased in proportion.

Even so, the blur is not bound to be visible in the finished print. In fact, if it is less than 1/100 in. wide the average human eye will not be able to detect it. With this as a starting point it is possible to lay down conditions for photographing moving subjects in such a way that the final picture will appear to be perfectly sharp—i.e., so that the actual blur will not be greater than 1/100 in. on the print.

Slowest Shutter Speed. If the negative is to be

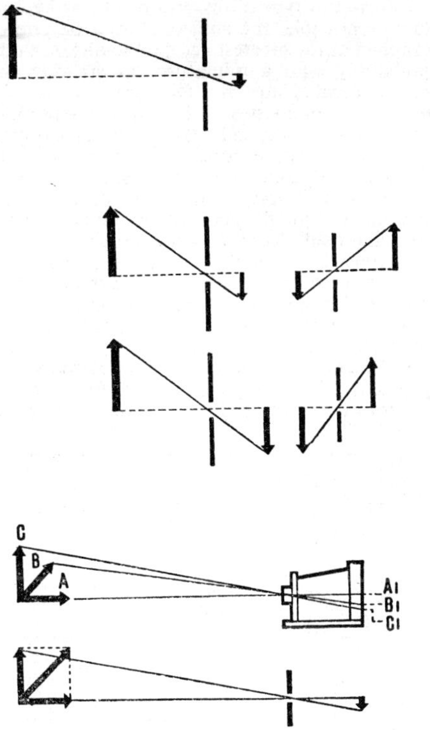

OBJECT AND IMAGE MOVEMENT. Top: *The image movement is smallest for a distant moving object.* Centre left: *At a given distance a short focus lens gives a smaller image (with less movement) than a longer focus lens.* Centre right: *If, however, both negatives are enlarged to the same image size, the blur through image movement will be the same.* Bottom: *Object movement along the camera axis causes the smallest image movement, across the axis the greatest. Movement at an angle is intermediate in effect, as it can be resolved into a movement along the camera axis and one at right angles.*

used for making contact prints which will be viewed at the normal distance of 10 inches with the unaided eye, the condition to be fulfilled is simple: the image must not move more than 1/100 in. while the shutter is open.

In this case it is only necessary to calculate the time the image takes to move 1/100 in. on the plate, and the answer will be the slowest shutter speed that will arrest the movement of the subject.

For example, a racing motor car is to be photographed passing across the field of view 300 feet away at 120 miles per hour (176 feet per second), the negative is to be taken through a 10 ins. lens, printed by contact, and viewed from 10 ins. (these conditions give natural viewing). As the car (300 feet away from the lens) moves 176 feet in one second, the image (10 ins. away from the lens) moves (by simple proportion) (176/300) × 10 ins. = 5·86 ins. It moves this distance in 1 second, so that it will move 1/100 in. in 1/(5·86 × 100) or, roughly, 1/600 second.

This is the lowest permissible shutter speed (l.p.s.) that will arrest the movement of the subject at that particular speed, distance and direction of movement. For other shutter speeds, distances and directions, the corresponding lowest permissible shutter speed follows by simple reasoning.

If the racing car is travelling at 240 instead of 120 miles per hour, the slowest permissible shutter speed would be 1/600 ÷ 2 = 1/1200 second.

If the racing car is being photographed at 100 feet instead of 300 feet, the slowest permissible shutter speed would be 1/600 ÷ 3 = 1/1800 second.

Movement across the field is most marked; movement towards the camera becomes merely a relatively slow change in size. So if the car were moving towards or away from the camera instead of across the field of view, the l.p.s. could be made three times as long—i.e., 1/600 × 3 = 1/200 second. And if the car had been moving obliquely, at 45° to the camera-subject line, the l.p.s. can be twice as long—i.e., 1/600 × 2 = 1/300 second.

Effect of Focal Length. The above theory holds good for contact prints from negatives made with a 10 ins. lens and viewed at a distance of 10 ins. But if the negative is enlarged, then the blur is enlarged with it, and blur that was not noticeable in a contact print might become visible in an enlargement viewed at the same distance. So for every diameter the negative is enlarged, the l.p.s. must be made shorter in proportion—i.e., if the negative is to be enlarged four times, the above l.p.s. must be divided by four.

Generally, the amount of enlargement that must be given to a negative to produce a natural-looking print is found by dividing the focal length of the taking lens (in inches) into the focal length of the lens that gives a natural print by contact—i.e., a 10 ins. lens.

A negative taken through a 2 ins. lens must be enlarged 10/2 = 5 times to give a natural print. But the amount of blur in the negative is smaller in the same proportion because the film is 2 ins. away from the lens instead of 10. So that the l.p.s. for any particular moving subject is the same irrespective of the focal length of the taking lens.

But this is true only as long as the negative is enlarged to make a print for natural viewing at 10 ins. Very often, however, the subject is deliberately magnified beyond its natural print size either by taking the negative through a telephoto lens, or by giving extra enlargement to a negative taken through a lens of normal focal length (as when a 35 mm. negative, taken with a 2 ins. lens is enlarged 10 times).

If prints made in this way are viewed from a distance of 10 ins. they look so many times "larger than life" and any blur caused by subject movement is magnified in the same proportion. Theoretically, the l.p.s. should be made correspondingly shorter—i.e., a negative taken through a 2 ins. lens and over-enlarged 10 times should be taken at half the l.p.s. for the same negative enlarged naturally five times.

Complex Motion. What has been said about simple motion can rarely be applied in practice because so many subjects involve more than one kind of motion. If a man is being photographed walking past the camera, a fairly slow shutter speed will arrest the movement of his body, but something very much faster would have to be used to give a sharp image of his hands and feet in the middle of their swing. The shutter speed that would arrest the movement of a railway engine would still leave the spokes of the wheels blurred.

In fact, almost every form of movement met with in practice is complex, and no photographer could hope to work out the exact minimum shutter speed necessary to arrest the movement of a given subject in time for the answer to be of any use. But a knowledge of the theory is a useful guide when adopting the more practical approach described below.

Focal Plane Shutter Distortion. In a camera with a focal plane shutter movement of the subject (or the camera) may distort the image on the negative even at shutter speeds fast enough to prevent blur. This is because the moving slit of the shutter exposes the negative piecemeal and not simultaneously all over. The speed of the shutter is the time taken to uncover and cover a single point on the surface. This may be so short that the image does not move enough to show blur in the print. But the total time taken by the shutter to expose the whole surface in this way is much longer and the image may also be moving as the slit moves across the focal plane. In that case each successive image strip exposed through the slit will be displaced further in the direction of the image movement, and the subject will appear to be sloping in that direction in the finished print. The wheels of a motor-car travelling past the camera will appear to be sloping either forwards or backwards if the camera is held so that the shutter travels from bottom to top or top to bottom respectively. If it travels from side to side, the image will be either compressed or elongated in the direction of movemet (i.e. round wheels appear oval).

The extent of the distortion depends on the time it takes for the slit to traverse the image. So the closer the subject to the camera—i.e., the bigger the image it forms—the worse the distortion. And the longer it takes the slit to complete its travel—i.e., the slower the speed of the blind or the farther it has to go—the greater the distortion. This explains why, under similar conditions, there is less distortion with a miniature focal plane shutter than with a larger type where the slit has a greater distance to travel.

Where this type of distortion is to be kept as low as possible, the subject should be photographed at the greatest acceptable distance and preferably with a miniature camera. On the larger focal plane shutter cameras the slit width and blind speed—i.e., spring tension—are adjusted independently. With such instruments it is possible to give the same exposure with several combinations of slit width and blind speed. Where there is a choice, that using the highest blind speed in conjunction with the widest slit gives the least distortion.

It must be remembered that what has been said above relates only to the image distortion produced by movement. The blur is controlled independently and in exactly the same way as with a between-lens shutter.

The Practical Approach. In practice the whole problem of photographing moving subjects is very much simpler than the theory suggests.

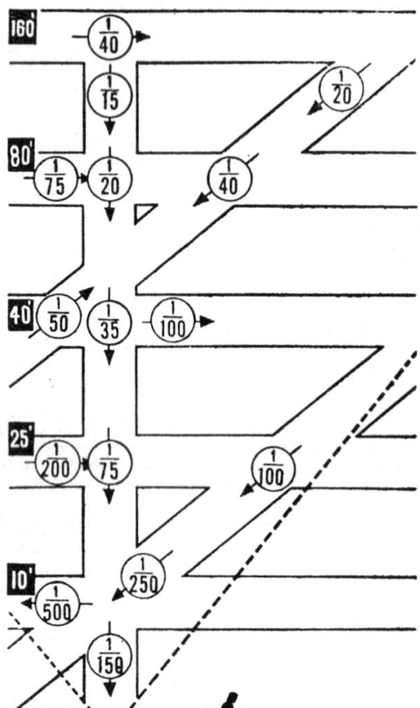

MAXIMUM EXPOSURE TO ARREST MOVEMENT. The figures indicate the slowest speeds that can be used when photographing a cyclist moving at about 15 m.p.h. at various distances from the camera. A subject moving at twice this speed (i.e., at 30 m.p.h.) will need twice as fast a shutter speed (half the exposure time—e.g., 1/100 second instead of 1/50), while the exposure can be doubled for a subject moving at half the speed (7–8 m.p.h.). The shutter speeds are of course approximate, and not necessarily marked on every shutter. Use either a suitable in-between setting, or where this is not possible, the next faster speed marked on the shutter scale.

The experienced photographer avoids movement blur in one or more of four ways:

(1) By shooting from a distance and angle that will minimize blur.

(2) By using the fastest shutter speed possible, irrespective of the theoretical l.p.s.

(3) By making the exposure at a "dead" point in the motion.

(4) By swinging the camera to keep pace with the movement of the subject.

Distance and Angle. The closer the camera is to the subject, the greater the blur on the negative, and at any distance, the blur is worst when the subject is moving across the field of view and least noticeable when the subject is moving towards or away from the camera. So if conditions preclude the use of fast shutter speeds the photographer should not work close to a moving subject, and he should choose a position where the direction of movement is not directly across the field of view, but at about 45° to the line of sight. (A head-on photograph would show even less blur, but the three-quarter view gives a better impression.)

Shutter Speed. From actual experience it is possible to make a list of the l.p.s. values that apply to a number of representative subjects, moving at various distances and direction. These values take into account any secondary movement of the subject (e.g., the arms and legs of a walking person, which move faster than the body) as well as its speed so they are more reliable than a theoretical list based on speed alone.

At the same time, the photographer cannot carry such a list in his head, and he will rarely have time to refer to it. In practice he simply uses the fastest shutter speed that the circumstances—lighting, lens aperture and film speed—will permit. In fact, within limits, the actual shutter speed used is less important than the other practical considerations governing the amount of blur.

The table below gives, however, a set of suitable exposure times.

The Dead Point. A swinging pendulum comes to a stop at each end of its swing; an athlete or a horse and rider at the highest point of a jump are moving neither up nor down; there is an instant in walking and running when the hands and feet are stationary. In almost every form of complex movement, there are "dead" points where the movement halts. These are the instants that the practised photographer looks for if he is out to avoid blur in his pictures.

It is not enough merely to be able to recognize the dead point; the photographer must learn to allow for the time lag in both the human and the camera mechanism, or he will find that he is making his exposures after the critical instant has passed.

Swinging the Camera. One of the most effective ways of minimizing the blur caused by straightforward movement is called "swinging the camera". This simply means that the camera is kept pointing at the subject as it moves past, the image of the subject being held in the centre of the viewfinder up to and during the instant of exposure.

When the camera swings with the subject in this way the subject will come out sharp on the negative, and the background will be badly blurred. But as a rule the blur, far from being objectionable, helps to separate the subject from the background and actually increases the impression of speed. Swinging the camera will render completely sharp only those objects which are solid units, such as cars, trains, etc., but there will still be evident a blurring in arms and legs of subjects such as human beings, as well as the limbs of animals.

General Technique. An experienced photo-

PRACTICAL SHUTTER SPEEDS FOR MOVING OBJECTS

Distance		*Exposure Time (seconds) for Object Speed of*								
ft.	*m.*	2 3·2	4 6·4	8 13	16 25	30 50	60 100	120 200	250 400	500 *m.p.h.* 800 *km.p.h.*
1	0·3	1/3000	—	—	—	—	—	—	—	—
2	0·6	1/1500	1/3000	—	—	—	—	—	—	—
4	1·2	1/750	1/1500	1/3000	—	—	—	—	—	—
8	2·4	1/400	1/750	1/1500	1/3000	—	—	—	—	—
16	4·8	1/200	1/400	1/750	1/1500	1/3000	—	—	—	—
32	10	1/100	1/200	1/400	1/750	1/1500	1/3000	—	—	—
64	20	1/50	1/100	1/200	1/400	1/750	1/1500	1/3000	—	—
125	40	1/25	1/50	1/100	1/200	1/400	1/750	1/1500	1/3000	—
250	80	1/10	1/25	1/50	1/100	1/200	1/400	1/750	1/1500	1/3000
500	150	1/5	1/10	1/25	1/50	1/100	1/200	1/400	1/750	1/1500
miles	*km.*									
¼	0·4	1/2	1/5	1/10	1/25	1/50	1/100	1/200	1/400	1/750
½	0·8	1	1/2	1/5	1/10	1/25	1/50	1/100	1/200	1/400
1	1·6	2	1	1/2	1/5	1/10	1/25	1/50	1/100	1/200

These shutter speeds apply for objects moving across the line of sight of the camera. For movement diagonally towards or away from the camera, the exposures may be 1½ times as long (e.g., 1/500 second instead of 1/750). For movement directly towards or away from the camera 3 times as long exposures may be given (e.g., 1/250 second instead of 1/750).

The standard of definition is 1/1000 of the focal length of the normal camera lens. With long focus or telephoto lenses the speeds must be multiplied (or the exposure *time* must be *divided*) by the power of the telephoto lens, e.g., 1/100 second becomes 1/200 with a 2× telephoto lens.

grapher who wanted to photograph, say, a horse and rider clearing a hurdle, would go to work like this:

(1) He would choose a station which his subject would approach in three-quarter view and at which, at the instant of clearing the hurdle, it would fill about half of the picture space.

(2) Having decided what exposure to give he would achieve it with the lens aperture that would give a depth of field that just included the jump and no more. This would permit the fastest possible shutter speed in the prevailing lighting conditions.

(3) He would get rid of horizontal blur by swinging the camera as the horse and rider approached.

(4) He would get rid of vertical blur by clicking the shutter as horse and rider reach the highest point of the jump.

Deliberate Blur. Blur in a photograph of a moving subject is not always objectionable; there are times when a certain amount actually improves the picture. In fact the only photographs in which blur is completely out of place are scientific records of movements too fast for the eye to follow. In pictorial work it is rarely advisable to show both subject and its background perfectly sharp even when it is possible. If both are sharp the photograph simply looks as though the subject were stationary. F.P.

See also: *Camera shake; Chronophotography; High speed cinematography; High speed photography; Holding the camera; Motion study; Moving image photography; Sport; Stroboscopic flash; Time lapse photography.*

MOVIE. Originally slang term for motion picture, now applied to amateur aspects of cinematography. Thus an amateur cine camera is frequently referred to as a movie camera, and amateur filming as movie making.

(See also: *Cine camera; Cinematography; Cine technique.*

MOVING IMAGE PHOTOGRAPHY. Technique of stopping movement of an image relative to the film, caused by the object or camera moving during exposure. Also known as image motion compensation (IMC).

Fast shutter speeds, the conventional means of "freezing" motion, inevitably produce a blur, however minute, since the image moves while the shutter is open. The shortest exposure time can at best keep the blur within given limits of sharpness, depending on image speed. Often the fast (and grainy) film and wide lens apertures entailed themselves limit resolution. However rapidly the object or camera moves, optium resolution demands the complete elimination of all image motion.

This can be achieved by moving the film in step with the image or by moving the optical system. The speed and direction of image motion are assessed and the film made to travel at this speed, while the camera is aligned to match the direction. In this way, the image remains virtually motionless on the film. Alternatively, the lens may be swung or the whole camera panned to follow the object. The latter is of course the standard method of following movement and is, quite properly, a method of image motion compensation since, if accurately carried out, it causes the image to remain in one position on the film. But panning is not always convenient or even possible with heavy or complex apparatus.

Various means of moving the film are used. They include pulling it by the edges, holding it by suction to a moving platen or moving the back of the camera. The simplest method, however, utilizes the film transport mechanism, augmented by additional controls.

Image speed at the focal plane is calculated from the object or camera speed of movement multiplied by image scale—a function of lens extension (or focal length, if at infinity) and lens-to-object distance. Thus, if an object moving at 100 ft, per second is reproduced on the film at a scale of 1:50, the image travels at $100 \times \frac{1}{50} = 2$ ft. per second and the film must maintain this speed during exposure to record a motionless image.

Variations of the IMC Principle. The above applies to single-shot and intermittent (or repeating) cameras. The principle is taken further in continuous strip photography, exemplified by the Sonne aerial camera. Such instruments are shutterless, the film being

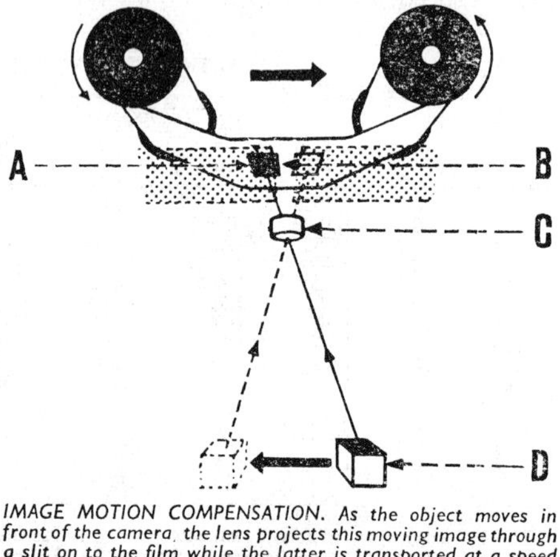

IMAGE MOTION COMPENSATION. As the object moves in front of the camera, the lens projects this moving image through a slit on to the film while the latter is transported at a speed corresponding to the image movement on the film itself. A, image on film. B, slit. C, lens. D, object. The initial position of the object and image are shown solid. A subsequent position during the movement is shown dotted.

exposed through a stationary slit across the focal plane. It is drawn continuously past this at the calculated image speed, the width of the slit being adjusted to give the correct exposure. So the effect is that of a focal plane shutter, but the result is a continuous record of objects passing the camera lens. The optical slit need not be at the focal plane: it may be placed in front of the lens or in the object plane. It can even be dispensed with entirely and a bar of light from a projector focused on the object plane instead.

Factors Affecting Resolution. Since the image is virtually stationary, resolution should be the best of which the lens and film are capable. But this depends on several conditions. Object movement can be in one direction only: the film cannot for instance follow the vertical and random movements of a runner's legs as well as his horizontal travel. Again, any inclination towards or away from the camera will alter image scale and with it, image speed. For the same reason, the object must be in one plane, parallel to the film: it can have little or no depth because any background will be blurred, as in panned action shots of racing cars, etc. Even if this were not so, changes in perspective would still cause slight unsharpness.

Any form of moving film compensation for motion therefore demands precise operation. In aerial photography for example, accurate flying is essential, since any change in aircraft height, speed or attitude will affect these critical factors. Some limitations have to be accepted: the inherent difference in scale between the centre and edges of a photograph is one. Another is when the object is not parallel with the film: in this case the film movement is usually matched to the most important zone in the image.

Applications. IMC is used extensively in modern aerial photography, where high aircraft speeds and the extremes of resolution required make it mandatory. Standard air cameras can be adapted fairly easily by the addition of special IMC magazines; continuous strip photography requires a specialized instrument. With either, the focal length of the lens and the aircraft height and speed determine film movement, and these are often monitored and fed in electronically.

A variation of the continuous strip is a proposed aerial television system, in which the camera makes single line scans only, across the flight path below the aircraft. At the receiving end, the lines are printed off the tube face on to a moving strip of sensitized material. This advances at a speed equivalent to that of the aircraft, so building up a continuous picture of the ground underneath it. The principle is very similar to that of radio-facsimile transmission of pictures and documents.

Conventional document and microfilm copying methods often utilize moving image techniques to increase output; an image of the moving original or microfilm is projected through a slit on to sensitized material moving at a corresponding speed.

The Sonne principle has been applied to highway survey in America, using a 35-mm. continuous strip camera mounted on a boom on a truck. Image scale is predetermined by the camera height above the road and film movement is servo-controlled from the truck's speed. Floodlights on the truck give strong directional lighting to show up surface defects.

Race finish cameras work in the opposite way, the camera being stationary and pointing along the finishing line at the moving contestants. Blurred leg movements and any distortion due to inaccurately matched image speed is acceptable since the finishing line is a fixed point.

Moving image cameras are also used in industrial photography to examine regular surfaces, either by visible light or by X-rays. If the object is flat it is traversed past the lens; if it is cylindrical, it is rotated. In the same way, certain dental X-ray cameras take a panoramic view of complete sets of teeth on moving film.

Normal panoramic cameras may either themselves rotate about the rear nodal point of the lens or may sometimes have a rotating prism in front of the lens, scanning an arc. In either case the film is moved past a focal plane slit at a speed corresponding to the rotation. Alternatively, the lens and slit may turn together, wiping an image on to a stationary strip of film curved about the common centre of rotation. In panoramic work there is no change in perspective, so the object can have unlimited depth.

The moving image technique is useful in high speed photography, particularly in cinematography where conventional intermittent mechanisms would impose too much stress on the film. Instead, many high-speed cine cameras run the film continuously past the gate while keeping the image stationary on the film by means of a parallel-sided prism in place of a shutter disc. This turns in synchronism with film movement, deflecting the image rays throughout exposure in the direction of film travel. Even projectiles and rockets can be stopped in flight by the sweep image camera. It makes a single exposure on a strip of film curved about a rotating mirror which sweeps an image of a slit, and of the object passing it, along the film. Although not primarily concerned with making an accurate reproduction, when carefully set up, the camera can produce undistorted pictures of ultra-fast-moving objects. P.B.M.

See also: *Aerial survey; High-speed cinematography; Panoramic camera; Photo-finish camera; Periphery photography.*

MUCK SPREADING. (Slang.) Term of disparagement for the bromoil and oil pigment finishing processes, or any oil finishing.

MUNSELL SYSTEM Classification system of colours, based on a three-dimensional solid representation of a selected set of colour hues, their desaturation to white and degradation to black.

See also: *Colour measurement.*

MURALS. In the main, murals can be copied in the same way as ordinary paintings. But murals are usually much larger and may have to be taken in sections.

Since the process is apt to be tedious, before adopting it a careful check should be made to see if a suitable viewpoint for a single exposure has been overlooked (e.g., from outside the building, looking through an open window). When an immovable object (e.g. a lighting fixture or pillar), prevents a dead-on view, the camera may be placed off centre, if it is provided with movements for correcting parallels. If such a camera is not available, an ordinary off centre shot may be tried; sometimes it is possible to correct the resulting distortion at the printing stage.

If, however, there is no easy way, the mural must be photographed in sections. The prints are then assembled as a photomontage which is subsequently re-photographed. (The method cannot be used for murals in relief—e.g., sculptures—because the change of viewpoint causes a change in the perspective from section to section.)

Taking the Photographs. By using the focusing screen or viewfinder on the camera, a rough guess is first made of the number of separate exposures likely to be needed to cover the whole mural. All the exposures must be made at the same distance from the wall and the sections should be mapped out so that each fills as much of the available negative space as possible.

There may be only a single row or several rows—e.g., two rows of two, or one row of four. If the mural is an odd shape—e.g., on the wall of a stairway—it may have to be covered in rows of unequal numbers. The section plan must in any case allow for plenty of overlap.

Method. The camera is set up dead square with the wall at the end of a row so that the outside edge of the painting lies just within the field of view. An assistant then sticks a small white spot of paper on the mural just inside the field of view on the other side of the viewfinder or focusing screen. All four corners of the section are indicated on the mural in this way with white spots of paper lightly stuck to the surface.

A length of wire is then knotted to indicate the distance between the mural and a suitable reference point on the camera. (String is apt to stretch.)

After the first photograph has been taken, the camera is moved along the row until it covers the second section as indicated by the white spots marking the inner edge of the first section. Two new spots are then added to mark the inner edge of the new section. Before exposure the camera is checked to see that it is square with the wall and at the correct distance from it. The procedure is repeated until all sections in the row have been photographed.

For subsequent rows it is only necessary to attach one extra spot to mark each section. This also ensures the overlap between the two rows as well as between the two sections.

The lighting must be even over the whole of each section and must be identical for each exposure. Slightly darker or lighter corners can be passed off in a single print, but in a montage they show up the joins at once.

If it is necessary to use a filter to improve the colour contrast, it must be used for every exposure and not simply for the sections where the colour rendering requires it.

Making the Prints. The print from each negative should be as large as possible—although if more than a dozen are involved, the complete montage may be difficult to handle with prints much above quarter-plate size.

The enlargements should be made on single weight glossy paper with the enlarger perfectly square with the easel so that the scale of enlargement is the same for each edge of the negative. The illumination must be even to avoid dark or light corners on the prints. The enlargements may be numbered on the back in sequence with the negatives. Until a satisfactory set of prints has been produced, the enlarger should be left locked in its position on the column.

Apart from these precautions, no special technique is necessary until the prints are processed.

When the number of prints involved is not too great, it is best to process and dry all the prints together. Otherwise the agitation and development time and temperature for each print must be rigidly controlled to give identical tones.

Mounting the Prints. The montage is best assembled by dry mounting.

The base consists of a sheet of card or stiff paper large enough to accept all the prints and still leave a small border. Dry mounting tissue is tacked on to the back of each print and they are then trimmed just enough to leave clean edges. A print from a corner of the mural is placed on the corresponding corner of the card, leaving a slight border, and mounted with a dry mounting iron.

The remaining prints are mounted in the order in which the negatives were made, each overlapping its neighbour to give a continuous picture. If the detail on one print does not correspond with the next a fresh print of slightly smaller or larger scale must be made. This can be done by marking on a sheet of paper the actual position in which certain details should come, laying it on the enlarger easel, and adjusting the scale of enlargement to fit the marks. In practice, slight errors in matching the edges pass unnoticed.

Next the montage is retouched. The white reference spots are touched out with dye and any discrepancies in the tones of adjoining prints corrected by building up the lighter tone by the edges with powdered graphite.

The montage is then copied, using normal copying technique and adjusting the lighting to minimize the effect of the joins. Soft, frontal lighting is best, so long as it is even and does not cause reflections on the print surface or shadows from the joins. J.D.C.

See also: *Montage; Photo murals.*

MUSEUMS AND COLLECTIONS

Photographic progress—technical as well as artistic—has been traced in a number of museums and collections throughout the world. Several national photographic societies have collection of apparatus and pictures from the beginnings of photography, and there also exist a number of famous private collections.

Great Britain. The photographic department of the Science Museum, London, has a fine collection of apparatus, both still and cinematographic—though many of its rarities are on loan from the Royal Photographic Society and the Will Day collection of cinematography. There are also on display photographs; these are not shown as pictures in their own right but only as specimens of various processes.

The photographs at the Victoria and Albert Museum, London, are not on display, but the collection can be used by the public. It lacks any historical sequence, being rather a conglomeration of photographs presented from time to time: chiefly portraits, topography, architecture, and art reproductions.

The Permanent Collection of the Royal Photographic Society of Great Britain, London, falls into three divisions: apparatus, books, and photographs. The collection of apparatus, including the items on view in the Science Museum, is, like the library, probably unrivalled. Both can be studied by members, and by students on special request.

A library catalogue was published in 1939 and a supplement in 1953. The collection of photographs contains large numbers of the Old Masters of photography in Britain, and the work of late-Victorian and Edwardian pictorialists is particularly well represented. Unfortunately the photographs have not yet (1964) been catalogued, and like the library and apparatus, are not on view to the public.

The photographic department of the Imperial War Museum, London, comprises about two million photographs from World War I onward, well indexed and available to the public.

The Kodak Museum at Harrow has an excellent technical collection largely devoted to apparatus, both still and cinematographic, and including the forerunners of cinematography. The range of exhibits includes the more important developments in the science of photography and examples of some of its industrial and other applications. The exhibits are not restricted to Kodak products, though naturally these form the largest proportion. The museum contains sections on colour photography and photomechanical printing. An informative illustrated catalogue was published in 1947.

One of the largest picture libraries in the world is the *Radio Times–Hulton Picture Library*, incorporating, in addition to modern photographs, the extensive historical collections formed by several private collectors, and the negative stocks of a few leading nineteenth-century photographic firms. The photographs are mainly collected from the "news" point of view, and include portraits of celebrities. They are classified in four sections: portraits, historical, topographical, and modern, in order to make them easily available to the public.

The National Buildings Record, London, formed at the beginning of the last war, contains about 800,000 photographs of English buildings and sculpture of historic interest.

Eadweard Muybridge bequeathed to the Public Library at Kingston-on-Thames much of his pioneer work (both negatives and published photographs) on the motion of human beings and animals, including his projection apparatus the zoopraxiscope, and his cuttings books.

The Reference Library, Birmingham, contains 22,000 social-documentary photographs and 14,000 negatives of the late-Victorian and Edwardian period taken by Sir Benjamin Stone and other members of the National Photographic Record Association.

The Scottish National Portrait Gallery, Edinburgh, possesses about 1,000 calotypes by Hill and Adamson, and several hundred of their negatives. The Edinburgh Public Library has about 450 calotypes by Hill and Adamson.

The Royal Scottish Museum, Edinburgh, has a few pieces of Talbot's apparatus, and calotypes by Talbot and John and Robert Adamson.

Other European Collections. There is no public museum in Europe devoted to the permanent display of photographs, much to the detriment of the appreciation of the past achievements of photography and its present-day applications in almost every sphere of civilized life. Nor does there yet exist an organized catalogue of photographic collections though it has been proposed by H. and A. Gernsheim that one should be published under the auspices of U.N.E.S.C.O.

Austria: The late Dr. J. M. Eder formed an historical collection, from the scientific point of view, at the Graphische Lehr- und Versuchs-Anstalt, Vienna. The catalogue of the library, published in 1936, lists 7,500 books.

Belgium: The Musée de l'Histoire des Sciences (Science History Museum), Ghent, has a small photographic department chiefly consisting of apparatus, books and other relics of the Belgian chemist D. van Monckhoven.

Czechoslovakia: The National Technical Museum, Prague, has a large photographic and cinematographic department, dealing chiefly with the technical aspect.

Denmark: The Royal Library and the Royal Academy of Fine Arts in Copenhagen have large collections of historical architectural photographs—40,000 and 30,000 respectively. The Royal Library has, in addition, collections of photographic portraits and views.

France: In consequence of an arrangement whereby photographers can deposit copies of their pictures for preservation, the Bibliothèque Nationale (National Library) Paris, has numerically probably the largest collection of photographs in existence, which runs into several million. They are indexed under subjects and photographers' names. None are on display: in other words, they form an extensive pictorial archive. Foreign photographers are scarcely represented, and even leading nineteenth-century French photographers like Nadar are represented mainly by small published pictures (carte-de-visite and cabinet size). In 1955, the Georges Sirot Collection formed an important acquisition of exhibition prints and since then, other French private collections have been purchased.

A large and important historical collection, chiefly of apparatus relating to photography and cinematography and examples of processes, is on permanent display at the Conservatoire National des Arts et Métiers, Paris, which published a catalogue of its collection in 1949.

The Musée Carnavalet—the museum of the City of Paris—assembles photographic documents of Paris, and possesses many fine pictures by Marville and Atget.

Historically, the most important French collection is that of the Société Française de Photographie (French Photographic Society). It illustrates the development of photography in France both as an art and as a science from Niépce and Daguerre on, including apparatus. No catalogue has been published. Unfortunately the collection has been in storage for many years.

The Archives des Documentations Photographiques (Record Office of Photographic Documents) at the Palais-Royal, Paris, have a substantial collection of photographs of all kinds of subjects, including a large number of Atget's negatives.

The Archives Photographiques d'Art et d'Histoire (Record Office of Artistic and Historical Photographs), Paris, collect photographs of architecture, public monuments, and art reproductions.

An important private collection of about 1,500 daguerreotypes and some apparatus and books, assembled by Albert Gilles of Paris, was dispersed after his death and acquired partly by the Bibliothèque Nationale and partly by André Jammes, who assembled the most important private collection in France.

The Musée Denon, an art museum at Châlon-sur-Saône, possesses Nicéphore Niépce's apparatus and a number of his photo-etched plates (heliographs) and other relics relating to the pioneer of photography.

Germany: The Deutsches Museum (German Museum), Munich, used to have a good department before World War II devoted to the technical development of photography and some of its scientific applications, as well as an excellent range of cameras. The Museum was damaged during the war and the collection is still (1962) in storage.

A photographic museum was founded at the Stadt Museum (City Museum), Munich, in 1962, initially to house the collection of about 1,000 lenses and optical instruments of Dr. Loher. The Stadt Museum having been very active during the last few years in arranging photographic exhibitions, it is expected that the photographic museum will eventually also comprise photographs.

The Staatliche Landesbildstelle, Hamburg, owns, besides old photographs of Hamburg, an important collection of daguerreotypes by Hermann Biow, Carl Ferdinand Stelzner, and others, which were formerly in the city's art museum. In addition, there are collections of pictorial photographs *c.* 1900 formed by Dr. Ernst Juhl, and of contemporary German photography formed by Fritz Kempe.

There is a small museum of historical photography at Dresden, containing about 600 photographs by Hermann Krone, Hugo Erfurth, Franz Fiedler and other Dresden photographers.

The Staatliche Lichtbildstelle, Marburg, is a centre for photographs of German architecture and sculpture, analogous to the National Buildings Record, London.

The historical collection of Dr. Erich Stenger was acquired in 1955 by Agfa and will form the basis of a photographic museum at Leverkusen. Apart from photographs and apparatus, the collection contains an excellent library of books and journals.

A photo-historical collection is in process of formation at the Folkwangschule, Essen-Werden.

Holland: The Print Department of the University of Leiden has begun to build up a collection of photographs for which the Gregory collection of daguerreotypes and stereopictures forms the nucleus.

Following the showing of the Gernsheim

exhibition in 1956, the Print Department of the Stedelijk Museum, Amsterdam, decided to collect contemporary photographs of aesthetic value, as did several German museums mentioned above.

Italy: The Department of Antiquities and Fine Arts of the Town Council of Rome, and the Historical Museum of the Risorgimento in Rome, have photographic archives.

Sweden: The Technical Museum, Stockholm, has a collection of equipment and examples of processes. H. & A.G.

U.S. Collections. There are but two museums in the United States solely devoted to photography; the American Museum of Photography in Philadelphia, and the George Eastman House in Rochester, New York. The former is rich in the work of Philadelphia pioneers, particularly the brothers Frederick and Wilhelm Langenheim. In addition to a large collection of photographs, the museum owns many examples of apparatus.

The George Eastman House in Rochester, N.Y., occupies the palatial mansion built by George Eastman in 1905. To the existing building, a theatre seating 550 was added in 1950 for the presentation of moving pictures. The Eastman House contains an extensive collection documenting the entire history of photography from Daguerre to the present day. More than a thousand cameras, including examples of the earliest type, make up the technical division of the collection; the photographic collection (incorporating the collections of Gabriel Cromer and Alden Scott Boyer) contains outstanding examples of the work of daguerreotypists, calotypists, and later photographers. In co-operation with the Museum of Modern Art, the George Eastman House possesses the largest collection of moving pictures in the United States.

The photographic division of the Smithsonian Institution, Washington, D.C., contains perhaps the largest collection of early American equipment, including cameras used by the pioneers, Alexander Wolcott, Samuel F. B. Morse and John W. Draper, early motion picture equiment, and apparatus illustrating the development of sound recording.

Two important collections of photographs are to be found in Washington in the Library of Congress and the National Archives. Both of these institutions possess negatives of the famous Civil War photographs collected by Mathew B. Brady. To the Library of Congress there have come, year after year, photographs and early cine films entered for copyright, and as a result the collection is perhaps the richest in the United States. The National Archives own negatives and prints taken for government projects, including the remarkable early photographs of the U.S. West, taken shortly after the Civil War.

The Gernsheim Collection has been largely built up to illustrate the history and aesthetic development of photography. The 35,000 photographs range from the world's first, by Nicéphore Niépce, to the work of leading contemporary photographers and include exceptionally fine collections of the classics of photography, chiefly British, though leading nineteenth-century French, Italian and American photographers are represented. The collection also comprises apparatus (including forerunners of cinematography), autograph MSS, and a library of about 3,500 nineteenth-century books and journals and all important publications on the pre-history of photography from the sixteenth century on. An illustrated catalogue entitled *Masterpieces of Victorian Photography from the Gernsheim Collection* was published by the Arts Council of Great Britain on the occasion of an exhibition at the Victoria and Albert Museum in 1951. Since then, an enlarged international selection has been shown in art museums of many West European countries, and in the U.S.A. under the title *120 Years of Photography*. In each case an illustrated catalogue was published the most expensive being that of the Detroit, U.S.A. exhibition (1963) with 1,020 items. The Gernsheim Collection is now owned by the University of Texas.

A few art museums have collected examples of artistic photography, notably the Museum of Modern Art and the Metropolitan Museum of Art, both in New York. Other museums are following the trend established by these New York institutions, and small collections are housed at the Chicago Art Institute, the San Francisco Art Museum, the Museum of Fine Arts in Boston, and other leading museums.

The important Epstean Collection of books and periodicals relating to the history and science of photography and photomechanical printing is at Columbia University, New York.

Historical societies often possess photographs by the thousand. Of the many predominant in the field, three can be mentioned in particular: the New York Historical Society, New York; the American Antiquarian Society, Worcester, Mass.; the Minnesota Historical Society, St. Paul, Minn.

One of the fields which seems to have attracted most individual collectors in America is that of stereoscopic photography. A review of American collections (by L. Dexter) in *Eye to Eye*, the journal of the Graphic History Society of America, revealed no fewer than twenty-one collections of sufficient importance to be described.

Two collectors have specialized in daguerreotypes: Mrs. Zelda P. Mackay of San Francisco, and Ray Phillips of Los Angeles.

An interesting collection of early cinematographic apparatus has been formed by Mr. D. Malkames in Tuckahoe, N.Y.

There is a growing interest in the history of photography, and other collections will no doubt be formed in course of time. B.N.

MUSGER, AUGUST, 1869–1929. Austrian priest and professor of drawing and mathematics. Described in 1905 optical compensation (using a mirror polygon) for cinematography which led in 1914 to the manufacture of the Ernemann Zeitlupe (a high speed cine camera).

MUYBRIDGE, EADWEARD JAMES, 1830–1904. English photographer. Worked mainly on the West Coast of the U.S.A., taking landscapes. Produced the first serial photographs of humans and animals in motion by successive exposures (with 12 to 36 cameras) at regular intervals. Synthesized motion from these serial photographs in his Zoogyroscope (1879) and later in his Zoopraxiscope. Published 780 plates (1887) illustrating the human figure and animals in motion, taken under the auspices of the University of Pennsylvania. His work (1877–85) greatly stimulated other investigators in chronophotography, analysis and synthesis of motion, and in early cinematography. Retired in 1900 to his native town of Kingston-on-Thames; it is there that a collection devoted to his work is housed in the Public Library.

N

NADAR (Gaspard Félix Tournachon) 1820–1910. French portrait photographer, previously caricaturist, writer and aeronaut. Became professional photographer in Paris, 1853, in collaboration with his brother Adrien and opened his own studio, 1858. Nadar photographed nearly all the great men active in Paris during the Second Empire and the Third Republic. Among the earliest to try electric light for portraiture, Nadar pioneered many fields; he was the first to photograph from a balloon, 1858, and underground (catacombs and sewers of Paris) 1861–2, and made the first photo-interview, with the centenarian chemist, M. E. Chevreul, 1886. Nadar patented balloon photography for ordnance survey, 1858. His exploits with his balloon *le Géant*, 1863, the largest in Europe, made Nadar an international figure.

During the Siege of Paris, 1870–1, he commanded an observation balloon corps and provided a balloon postal service. In 1874 Nadar lent his studio for the first exhibition of the Impressionist painters. Author of a number of books, including *Quand j'étais photographe* (1899).

See also: *Aerial survey; Balloons and kites.*

NAMIAS, RODOLFO, 1867–1938. Italian chemist. Made many valuable contributions to photography, especially in the field of colour work (mordant dye pictures). Evolved the permanganate reducer. Also established the photographic periodical, *Il Progresso Fotografico.*

NAPPELBAUM, MOISEI, SOLOMONOVITCH, 1869–1958. Russian portrait photographer. Opened his own studio in St. Petersburg, 1910 after working in Warsaw, Moscow and U.S.A. Made the best-known portrait of Lenin, and of many Communist Party Leaders, scientists and artists.

The only Soviet photographer to merit the title Outstanding Artist of the R.S.F.S.R. (Soviet Socialist Republics.)

NARROW GAUGE CINE FILM. Film of reduced width—as compared with the standard gauge 35 mm. motion picture film—for amateur cinematography. The most used gauges are 8 and 16 mm., and—to a lesser extent—9.5 mm. Also known as sub-standard cine film, though this term is going out of use.

See also: *Cine film.*

NATIONAL BUILDINGS RECORD. Independent organization which receives a grant from the Treasury through the Ministry of Public Building and Works. It was set up in 1941 when historic buildings throughout the country were being destroyed or damaged by enemy action.

It maintains a very large library of photographs of the historic architecture of England and Wales, which is arranged topographically in order of counties in such a way that any building, if it is represented in the collections, can be looked up at once.

There is also a considerable collection of measured drawings and specialized indexes of various kinds. The library is open to members of the public for reference purposes at all normal hours.

The Record deals with all types of buildings, domestic, ecclesiastical, and civic, dating from the earliest times to about the middle of the nineteenth century, after which its files are more selective. The collections are concerned not only with general views but also with details of interior decoration, fittings, and sculpture, the ecclesiastical sections being particularly rich in this respect.

The photographic library now contains over half a million prints. There is a small staff of permanent photographers who are at present engaged on specific programmes with a view to covering eventually the whole of England and Wales. Speaking generally, the important traditional architecture of the great cities and of the smaller towns has been covered, but there are many buildings of less importance in rural areas which are not yet represented. Considerable attention is now being given to nineteenth century buildings.

One of the chief current duties of the Record is to ensure that adequate photographs and measured drawings are made—if they do not already exist—of all buildings of any merit that are likely to be demolished. Photographs of many important buildings demolished during the earlier years of this century are still required.

Photographs of a purely picturesque nature are not wanted. The Record is primarily interested in obtaining photographs which give a good and complete account of a structure both inside and out, with special reference to any significant architectural detail. It is also interested in photographs showing groups of old buildings in their natural setting, particularly in villages and towns where development may ultimately affect their appearance.

The collection of negatives numbers approximately 325,000, a great many of which have been given or bequeathed by interested persons. Specialized collections are purchased from time to time as funds permit. Prints are supplied to the public at normal charges. In those cases where negatives are deposited on indefinite loan the copyright of the lender is strictly observed. The Record is a non-profit making institution whose collections and facilities are freely available to all those who wish to make use of them.

See also: *Print libraries.*

NATIONAL PHYSICAL LABORATORY. British establishment founded in 1900 for the purpose of "standardizing and verifying instruments, for testing materials and for the determination of physical constants". The National Physical Laboratory, which is now part of the Department of Scientific and Industrial Research, covers 70 acres and employs more than 1,300 people. It is divided into nine divisions: Aerodynamics, Applied Physics, Autonomics, Basic Physics, Light, Mathematics, Metallurgy, Ship and Standards. In addition to its original objectives N.P.L. now carries out a considerable programme of fundamental research work and undertakes investigations of special problems.

The Central Photographic Section is responsible for a large part of the general and cinematic photography of the laboratory, including cinematography. It also gives advice on photography within the laboratory and to other government departments, industry, etc., and it is also part of the responsibilities of the section to provide photocopying service using photostat and microfilm processes and including plan-copying of engineering drawings. The section contains a lofty studio, 20 × 35 feet, with a gallery. Lighting, all of which is controlled from a mobile switching trolley, totals 60 kW and includes high-speed flash equipment. The studio is fitted with 10 feet wide roller backgrounds in different colours, as well as hanging curtains.

A variety of still cameras is available, from an industrial model extendible to 40 ins., to normal 35 mm. equipment. Both 35 mm. and 16 mm. cine cameras are used, as well as special high-speed cameras operating up to 8,000 frames per second.

The Section is also responsible for the installation and maintenance of all still and cine projectors in Glazebrook Hall, the Laboratory's new conference centre. The wide use of 2kW Xenon enclosed arc lamps is a notable feature of the equipment.

Photography is also widely used for specialized scientific purposes by a number of Divisions throughout the Laboratory. For example, in the Aerodynamics Division, various techniques of high speed photography over a wide range of framing rates are used for flow-visualization in wind-tunnels, shock-tubes and a water-tunnel, and for studies of boundary-layer transition. Cine films have proved of great value in making known the results of aerodynamic research, and there is a Divisional library of such films.

NATURALISTIC PHOTOGRAPHY. A book by Dr. P. H. Emerson appeared under this title in 1889. It was a manifesto of his belief in a return to nature for subject matter, pictorial photography having become bogged down in the artificiality of studio compositions. Emerson had some followers, but most of these became addicts of impressionistic photography a few years later.

NATURE PHOTOGRAPHY. Since the earliest days of photography the camera has been used extensively for making both pictorial and scientific records of nature subjects. To-day the camera is used to supplement the naturalist's notebook, to photograph plants, animals and insects not easily observed because they are too small, or appear only after dark, or because they are normally inaccessible to the student; while the camera has brought back pictures of the marine life and vegetation under the sea.

See also: *Animals; Big game; Biology; Birds; Botany; Fish; Flowers; Insects; Pets; Reptiles; Trees; Underwater photography; Zoo.*

NAVAL PHOTOGRAPHY. Photography in the Fleet Air Arm is limited to use of standard cameras as employed in the R.A.F. In training, for the calibration of weapons, and particularly in research, many special cameras are used both by the Royal Navy and by the Royal Naval Scientific Service.

Operational and Training Cameras. The cameras for training purposes include many standard commercial and R.A.F. cameras; they are used principally for the recording of data or for accurate measurement of results of various operations. Cameras used at sea have to withstand exposure to bad weather, salt spray and continuous salt air and must necessarily be very robust and reliable. Simplicity coupled with very extensive weather-proofing are therefore the main criteria governing the design. Also the cameras have often to be used in restricted space and under very noisy conditions by relatively unskilled personnel, and the photographs must be taken at precise times when many other events are occurring. Because of this they are very strongly constructed and all operating levers, etc., are of generous proportions.

Dial Camera. The dial camera is used for purely recording purposes. This camera takes daylight loading roll film 2½ ins. wide, sufficient for some forty exposures on a 2¼ ins. square format. The lens works at f4·5 and has an accurate focusing scale from infinity to 18 ins. A very large brilliant viewfinder with parallax adjustment, which can be swung for viewing from above or from one side, is fitted. The camera can be fired by hand or by a built-in solenoid. The shutter is of the pre-set type, the pre-setting being done by the film wind mechanism which is actuated by a cord wound on a drum. This camera is often arranged to take photographs of the various dials, clocks and computing mechanisms in ships at the instant of gunfire or the launching of other projectiles.

Ballistic Recording Cameras. Much of the other purely naval photography consists in recording the falling of projectiles in the sea or the position of bursts of anti-aircraft fire in relation to targets. For these purposes the cameras are purely instantaneous recording measuring instruments, the lenses being calibrated for focal length and the resulting photographs measured and used for plotting purposes. A watch is also photographed at the instant of exposure to give the necessary time base for subsequent plotting.

One camera of this type is known as the Low Angle Marking Camera. The film used is 70 mm. wide and the picture size is 2¼ × 6 ins. The shutter is of the louvre type, situated immediately behind the lens. A prism is fitted over half the lens, having a constant deviation of 80° so that each exposure actually takes two fields of view, one of the firing ship and the other of the fall of the projectiles. The camera is fitted with a sighting arrangement so that it can be kept aligned on the firing ship; a side mirror in the sight enables the operator to see the fall of shot so that he can make his exposure at the correct time. The film is transported and the shutter set by actuating a lever. Since this camera is generally used on the deck of a small ship it is fitted to the tripod so that it can swing freely and is partly stabilized by a heavy pendulum so that it remains upright while the ship rolls.

For recording high-angle gun fire against aircraft, slightly modified cine cameras are normally used by the Navy. One camera which is particularly favoured is eminently suitable for general use at sea because of the ease with which it can be adapted for many purposes and fitted with lenses, watches, counters, single shot devices, etc. The magazines are easily loaded and are fitted directly to the camera; there is no need to thread the film in the gate.

Photo-Theodolite. Photo-theodolites of a special type are used for calibrating various weapons. These are normally film cameras using film 5½ ins. wide with a fixed lens of 5 ins. focal length at f4. During exposure the film is pressed against a register glass which is ruled in accurate squares and the lens is carefully calibrated for distortion. The pattern of squares is thus transferred to the film at the instant of exposure and the position of any missile or splash can be measured in relation to the grid with considerable accuracy.

Two such photo-theodolites are used synchronously to mark the exact position of the missile in space. The cameras are mounted on levelling screws and are aligned accurately in fixed directions so that by simple trigonometry or graphical plotting the exact position of the missile can be calculated. The cameras are electrically fired in synchronism from a central observation point and the film is wound on and the shutter re-set by an electric motor. The camera is capable of recording at the rate of two pictures per second but, where more rapid repetition rates are required, as many as six such cameras may be rigidly mounted together in one framework, all actuated by a single main driving shaft with cams to fire each shutter in sequence.

Torpedo Camera. Another important camera is used to photograph the flight in air of torpedoes dropped from aircraft to determine the trajectory. In this camera, which uses plates 24 ins. long by 8 ins. wide, or in another version film 12 ins. wide and 36 ins. long, the exposures are made by means of a louvre shutter which is actuated by the movement of the sighting telescope. A moving mask in the focal plane covers the whole plate except for a narrow strip. This mask is connected to the sighting telescope in such a way that when the torpedo is centred in the telescope, the image in the camera coincides with the opening of the slit. The shutter opens each time the slit moves its own width. In this way a composite photograph is made showing the torpedo in successive instants during its flight and from this it is possible to analyse the trajectory.

Underwater Cameras. Underwater photography naturally plays a large part in Naval research and for this purpose several still and cine cameras in specially designed watertight cases are used with various types of lighting and flash gear. These cameras are used by free divers or frogmen. Cameras are also fitted in the hulls of ships to photograph underwater phenomena and some of these are of very special design although the actual film winding arrangements may be of normal types. It frequently happens that such cameras are driven synchronously with some machinery and exposed by means of stroboscopic electronic flash equipment. In all such underwater cameras the lenses are of special design for underwater use and are frequently of periscopic or telescopic types having relatively short focal lengths with a very long extension.

Cine Cameras. The need to record the images on television and radar cathode ray tubes has resulted in the development of special cine cameras having a long exposure with a very fast film change mechanism. In such cameras, the film may be exposed for 1 second and then moved into position for the next exposure in 1/100 second.

General Research. Research work for the Navy is carried out by the R.N.S.S. who maintain establishments dealing with all the multitudinous aspects of Naval work. Photography plays a very large part in much of the research work for which many special cameras have been devised. The cameras, however, although having novel features are normally of fairly well-known types. In particular image stabilized high-speed cameras of the drum type are largely used, some of them having repetition rates up to 50,000 pictures per second, each picture being 1 in. square. J.H.-D.

See also: *Air Force photography.*

NEGATIVE. In photography, an image on a film, plate or paper in which light or transparent areas represent the dark tones of the original subject and dark or opaque areas represent the light tones of the original.

The term negative is also used in electricity to define a contact which has a lower potential of electromotive force than an associated contact (the positive) or point connected in the same circuit.

NEGATIVE CARD. Name for a form of negative material, now little used, consisting of a negative emulsion coated on to a stiff card support. Positive prints are made by projecting the negative in a type of episcope, or by treating the card support with oil to make it translucent.

See also: *Paper negative; Supports for emulsions; While-you-wait photography.*

NEGATIVE CARRIER. Frame in which the negative is supported for projection in an enlarger.

An open frame (glassless carrier) is used for plates and miniature negatives, but the larger roll and cut films have to be sandwiched between glass plates.

NEGATIVE LENS. Lens which is thinner at the centre than at the edge, and which can form only a virtual image of the subject.

See also: *Diverging lens.*

NEGATIVE MATERIALS

Photographic negative materials essentially consist of a sensitive emulsion—which is nearly always silver bromide in gelatin—with or without other additions, e.g., silver iodide to increase speed and colour sensitivity. This emulsion is coated on a suitable film, glass, or (more rarely) paper, support.

There is a wide range of negative materials available to the photographer. One negative material may differ from another in any of the following respects:

(1) Speed.
(2) Colour sensitivity.
(3) Gradation.
(4) Graininess.
(5) Resolving power.
(6) Development characteristics.
(7) Physical characteristics of the emulsion.
(8) Nature of the support.

These properties determine the usefulness of a material for any particular purpose.

Speed. The extent to which an emulsion is light sensitive is usually referred to as its speed. This can be actually measured and expressed numerically in a number of different ways.

Originally the speed figures were derived from the minimum exposure needed to produce a certain degree of blackening on the film or plate after development. At one stage, comparative speeds were also arrived at by making a series of exposures of typical subjects, and examining the prints from the negatives to find the minimum exposure which will give acceptable prints.

Most current speed criteria are based on the exposure at which the characteristic curve reaches a certain minimum gradient.

Speed Systems. There are two practical types of speed system: arithmetical and logarithmic.

Arithmetical systems express the speed in cardinal numbers which are directly proportional to the sensitivity of the material and inversely proportional to the exposure needed for any subject. A film, for instance, with an arithmetical speed of 200 is twice as fast as one with a speed of 100 (measured in the same system). The slower film, therefore, needs twice as much exposure as the faster to give the same result. The most widely used current arithmetical system is the ASA speed (American Standards Association, also used in a corresponding British Standard). The Russian GOST standard uses a similar system, and so

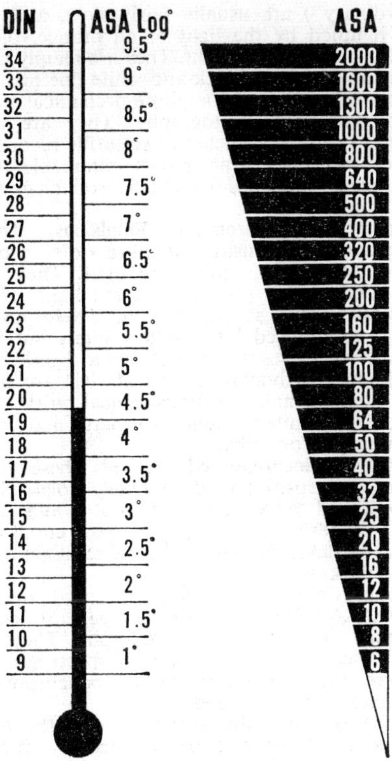

FILM SPEED SYSTEMS. In the arithmetic ASA system film speed is directly proportional to the speed number. In logarithmic systems the speed doubles for every constant increase in the speed number. Thus in the DIN range every increase of 2/3 corresponds to a doubling of the speed. On the ASA log scale every increase of 1° doubles the film speed. These three speed systems are fully interconvertible, being based on the same speed criterion. There are also other arithmetic speed systems with values similar to ASA.

do a number of now obsolete speed criteria.

Logarithmic systems express the speed in numbers which are logarithmic functions of the actual speed. In most of these systems every increase of approximately 3 doubles the sensitivity; every decrease of 3 halves it. For instance, a material with a speed of 26 needs twice the exposure for the same subject as a material with a speed of 29, measured on the same scale.

Arithmetical systems result in large cumbersome speed figures, whereas logarithmic systems give the same values in a much more convenient form. A range of 1 to 10,000 in arithmetical speed numbers corresponds to only 0 to 40 in logarithmic degrees.

The most widely used logarithmic system is the DIN scale current on the European Continent.

An alternative logarithmic scale (ASA log) increases the speed number by 1° for every doubling of the speed. Frequently ASA arithmetical and log numbers are quoted together· for example ASA100/5°.

Materials of 25–400 ASA are mostly used for general photography. The fastest types (which may have speeds up to 800 or 1,200 ASA) are specially suitable for photography by artificial light or under poor lighting conditions.

Films and plates of 25–100 ASA are used in commercial and technical photography, where, for much of the work, a fast material is not essential, but where the characteristic advantages of the slower types—e.g., fine grain and high contrast—are more important for many of the subjects tackled.

Even slower materials are used for copying and reproduction and for photomechanical work where the high contrast required necessarily goes with low sensitivity. This is more of an advantage than a drawback, because it allows the material to be handled by a reasonably bright safelight during loading into the camera and processing.

The slowest plates of all are manufactured for making lantern slides and as such are not negative materials.

Speed Variations. The effective speed of a material also depends to some extent on how the exposure is given—i.e., by a bright light acting for a short time, or a weak light acting for a long time.

According to the classical reciprocity law (formulated in 1876), the exposure required to produce a given density or range of densities on development should depend only on the product of the intensity of light multiplied by the exposure time. The law does, in fact, hold for normal light intensities and exposures of the order of 1 second. But with very intense lighting and short exposure times (less than 1/1000 second) —e.g., in electronic flash photography and also in high speed spark photography—the effective speed of the material may be lower because of reciprocity failure.

This loss of speed (often accompanied by reduced contrast) is not related to the normal emulsion speed; it is more marked with some makes of material than others. Some materials are therefore more suitable for photography under such conditions than others of the same nominal speed.

A similar loss of speed arises at low levels of light intensity, requiring exposure times longer than 1 second, as in artificial light photography. The reciprocity failure can become very marked at extremely low lighting levels, as in astronomical photography. Frequently special emulsions are made for such applications, designed to yield their maximum film speed at low lighting levels.

The effective emulsion speed of films and plates also depends on the conditions of development.

Colour Sensitivity. A straightforward silver halide emulsion is sensitive over the approximate wavelength range 2,000A.–5,000A.—i.e.,

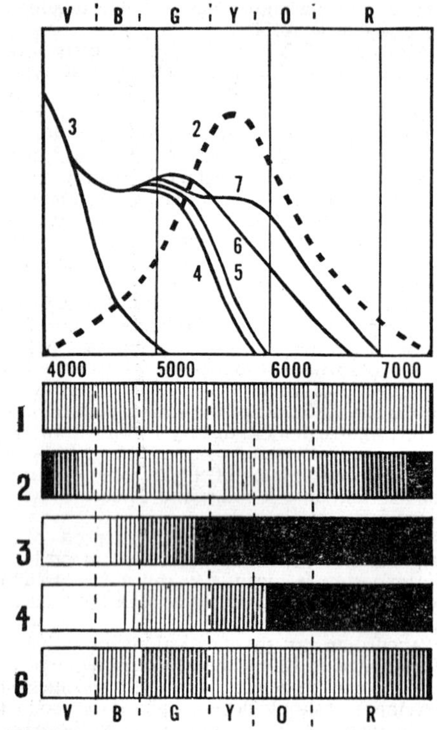

COLOUR SENSITIVITY. Top: Sensitivity of different types of emulsion over the visible spectrum, compared with the sensitivity of the eye (dotted curve). Bottom: Relative brightness of spectrum colours as the eye or the different emulsions would "see" them. In both cases, the curves and tones indicate: 1. Spectrum of uniform energy throughout (on which the other tone scales are based). 2. Sensitivity of the eye. 3. Ordinary blue-sensitive emulsion. 4. Slow orthochromatic emulsion. 5. Fast ortho material. 6. Normal panchromatic emulsion. 7. High speed pan emulsion with increased red sensitivity.

to the ultra-violet and, in the visible region, to violet, blue and (slightly) to blue-green. It is insensitive to longer wavelengths—i.e., to yellow, orange, red and infra-red.

The limit of sensitivity at the short-wavelength end in the ultra-violet is fixed by the ultra-violet absorption of gelatin, but can be extended in certain ways (see below). The limit of sensitivity at the long-wavelength end depends on the properties of the silver halides themselves, but can be extended by means of sensitizing dyes. Different dyes extend the limit by different amounts, and the various types of colour-sensitive emulsions thus produced are described in the following sections.

This extension of sensitivity by means of dyes does not, in general, affect the ultra-violet sensitivity of the emulsion, or alter its limit. Also, all colour-sensitive emulsions remain more sensitive to blue-violet than to any other colour.

Non-colour-sensitive materials coated with unsensitized emulsions (sometimes known as "ordinary") are usually fairly slow, and can be handled by the light of an orange, olive-green or brown safelight. They are mainly used for copying from black-and-white line or half-tone originals, and for photomechanical work and document photography. They are also used in certain branches of scientific recording where certain characteristics obtainable only with such emulsions (e.g., high resolving power) are important.

Slightly Orthochromatic Emulsions. These emulsions are sensitive to ultra-violet, violet, blue, blue-green, and green light. They can be handled and processed in red light. Their excessive blue sensitivity can, where necessary, be reduced by a yellow filter, but this greatly decreases the speed, often to $\frac{1}{4}$ or less.

Slightly orthochromatic materials are used for copying and photomechanical work, and also to a limited extent in general and commercial photography.

Highly Orthochromatic Emulsions. These emulsions are sensitive to ultra-violet, violet, blue-green, and yellow light, but not orange and red. Materials coated with these emulsions must be developed by a deep red safelight or in total darkness.

These emulsions range in speed up to 500 ASA. They are used in general, commercial and technical photography. They are also used in those branches of spectrography where the red sensitivity of panchromatic materials is not required.

The excessive blue sensitivity of orthochromatic materials can be lessened by yellow filters which, however, reduce the effective speed (filter factor $1\frac{1}{2}$–$3\times$). A yellow filter shifts the zone of highest sensitivity from the blue region of the spectrum into the region where the human eye is most sensitive. This shift is taken advantage of in photomicrography where the image must be perfectly sharp. Using a yellow filter with a highly orthochromatic plate ensures that the rays by which the visible image is focused will also be the rays which form the photographic image.

If it is necessary in photomicrography to use a panchromatic emulsion to reproduce colours at the red end of the spectrum, the microscope must be fitted with an expensive objective with a very high degree of colour correction (apochromatic objectives).

Certain fast orthochromatic films are especially made for fluorography—i.e., the photography of X-ray images and cathode ray traces on fluorescent screens.

Panchromatic Materials. These emulsions are sensitive to ultra-violet and all wavelengths of the visible spectrum. Their sensitivity is usually lowest to blue-green, and they may therefore be developed by the light of a deep green safelight. But even this safelight must be used for very brief periods of inspection only; it is safer to handle and process panchromatic films and plates in total darkness.

Practically all films and plates for general photography, including amateur work, are nowadays panchromatic.

Usually the sensitization results in a reproduction of colour values in tones of grey which corresponds more or less to the visual brightnesses of these colours—except for the blue and violet region of the spectrum. Such materials are known as correct panchromatic (also orthopanchromatic). Certain special materials, and generally the ultra high-speed films and plates, have an increased red sensitivity and are known as red-panchromatic. This is useful in artificial light photography, since such materials lose little or no speed when exposed by tungsten filament lighting which is richer in red radiation and daylight (see below).

Panchromatic plates and films range in speed from very slow—for process work, copying coloured originals, spectrography, and photomicrography of stained specimens, etc.—to the very fast materials of about 500–1,200 ASA which are used in general photography.

While panchromatic materials are sensitive to light over the whole visible spectrum, their relative sensitivity to individual colours is not the same as that of the human eye. The human eye is more sensitive to yellow light than to blue and red, whereas panchromatic films and plates are more sensitive to violet and blue light, and (in the case of red-panchromatic materials) to orange and red, than to yellow.

Filters may be used to correct the over-sensitivity of an emulsion to any particular colour, and also to alter the colour response to give contrast between areas that differ in colour, but which appear equally bright to the eye. The loss of speed of panchromatic materials when used with filters varies with the type of filter.

Infra-red-sensitive Emulsions. These emulsions are made by treating the ordinary emulsion with special dyes. This type of emulsion is sensitive to violet, blue, and blue-green light, usually insensitive to green and yellow, and sensitive to red and invisible infra-red rays. Infra-red-sensitive emulsions can generally be developed by the light of a special green safe light.

Although the emulsion obtained in this way is sensitive to infra-red rays, it is still much more sensitive to blue. When these special emulsions are used for taking photographs by infra-red light, therefore, the unwanted blue rays must be held back by a filter. Suitable filters are made of deep red or black glass which transmits only red and infra-red radiation. Infra-red-sensitive plates and films are, therefore, comparatively slow (80 ASA and lower) when they are used for photographing by infra-red radition alone.

Infra-red-sensitive materials are used extensively in aerial photography and photogrammetry. Infra-red radiation is not appreciably scattered by mist and haze. So it is possible to get clear pictures of distant landscapes on infra-red materials when ordinary photographic emulsions would be useless. Infra-red films and plates are also used in technical, industrial, and forensic photography, astronomical photography, and spectrography.

The range of wavelengths to which an infra-red emulsion is sensitive depends upon the sensitizing dyes employed. Infra-red materials for normal use are sensitized to wavelengths up to 8,800 A. (The sensitivity of panchromatic materials ends at about 6,600 A.) Scientific materials may be sensitized to wavelengths up to 10,000 A., while special sensitization has produced emulsions sensitive to wavelengths up to 12,000 A. for astronomical photography and infra-red spectrography. Such materials do not keep well.

Short Ultra-violet-sensitive Emulsions. These emulsions are used in scientific work, mainly ultra-violet spectrography.

All ordinary emulsions are sensitive to ultra-violet wavelengths down to about 2,000 A. But at short wavelengths the gelatin begins to absorb the radiations, and below about 2,300 A. this absorption seriously decreases the effective sensitivity. It also means that at short wavelengths the image is confined to the surface of the emulsion layer, and is therefore of low contrast. The resulting variation of contrast with wavelength is a disadvantage in spectrography. Hence for this work so-called "uniform-gamma" emulsions are sometimes used, in which the relatively excessive penetration of the emulsion by the longer ultra-violet wavelengths is prevented by the addition of suitable substances to the emulsion.

Sensitive materials to record the very short wavelengths in the region in which gelatin absorbs can be prepared in three ways. (1) By making emulsions which contain little or no gelatin (Schumann plates). (2) By making the silver halide grains project above the surface of the gelatin layer, thus reducing absorption by the gelatin to a minimum. Such emulsions are very easily damaged. (3) By sensitizing or coating the film or plate with fluorescent materials which glow with visible light under the action of ultra-violet rays. The fluorescent material may have to be removed before processing.

Daylight and Artificial Light Speeds. The spectral composition of daylight is appreciably different from that of artificial light such as tungsten illumination. The former contains comparatively equal proportions of radiations of the various wavelengths of the visible spectrum, while tungsten lighting is comparatively poor in blue and violet rays and correspondingly richer in orange to red (and infra-red) radiation.

This difference naturally also affects the speed of the various kinds of sensitive material. Owing to the low blue content of tungsten illumination the effective speed of a non-colour sensitized emulsion may be one-half or even less of the effective daylight speed. A normal ortho emulsion may also lose one-third to one-half of its daylight speed when exposed by tungsten lighting.

Many manufacturers quote a separate speed figure for tungsten illumination with their materials. The standards for emulsion speed measurement lay down only one way of determining the speed figure, using a light source approximating daylight in quality. Tungsten speed figures are therefore only an approximate indication of how the material is likely to behave.

With panchromatic emulsions the value of a tungsten speed figure is doubtful, since it depends also on the method of exposure determination employed. Thus some film makers list panchromatic materials with a tungsten speed the same as their daylight speed, while others quote tungsten figures of about 30 per cent below the daylight speeds. Where an exposure meter is used for measuring the lighting level it is best, for accurate exposure determination, to find the effective film speed by a series of test exposures with that meter under the required lighting conditions.

Gradation and Contrast. Negative materials may differ from each other in respect of contrast and gradation. These two terms are commonly used synonymously and, though always associated in one way in practice, are quite distinct in theory. Contrast is the slope of the characteristic curve—that is, the rate at which density changes with exposure (or, properly, its logarithm). With a material of high contrast a small increase in exposure will produce a large increase in density and vice versa. In addition to this, however, the total exposure range covered by the useful portion of the characteristic curve is also a practically important property of any material. It determines the total brightness scale—that is, tone range or subject contrast—which the material can faithfully reproduce. Strictly, the term gradation should refer to this property.

With actual materials, high contrast is always associated with a small useful exposure range or a short scale of gradation; low contrast with a long scale of gradation. Moreover, both properties are associated with emulsion speed: slow emulsions tend to yield high contrast (hard emulsions) and to have a short exposure range: fast emulsions tend to yield low contrast (being soft-working) and a long exposure range.

These associations depend upon the fundamental properties of the silver halide crystals (large crystals being more light sensitive than small ones), upon the processes used in emulsion manufacture, and upon the properties necessary if emulsions are to be useful in practice.

Slow, very contrasty materials are used in photomechanical reproduction and for copying line originals where the only tones present are black and white. Similar, though less extremely slow and contrasty, materials are used for rendering the low-contrast subjects frequently found in astronomy, photomicrography, etc. Fairly high contrast is also needed in copying continuous tone originals, in aerial photography, and in many technical and scientific applications.

The sensitive materials employed in everyday general and commercial photography are mostly of medium contrast.

Low contrast is an undesirable characteristic in any sensitive material. It is never welcome for its own sake, but where extreme speed is important, some lowering of contrast must be accepted.

Gradation and Exposure Latitude. The emulsions of most medium-speed to fast materials can reproduce a tone range, or subject brightness scale, much greater than that of most normal subjects. The exposure times given in the camera may therefore in many cases be increased by several hundred per cent above the minimum without the exposure given to any part of the image exceeding that which the emulsion can reproduce. This is the exposure latitude.

The permissible latitude therefore depends upon the tone range of the subject and upon the gradation (useful exposure range) of the emulsion. The more contrasty the subject, and/or the less the useful exposure range of the emulsion, the less the latitude. For the same reason, and quite apart from the emulsion speed demanded, it is essential to use fast materials—that is, those which have associated with their speed a long useful exposure range—when the subject contrast is high.

The exposure range, and therefore the latitude, of comparatively contrasty materials can

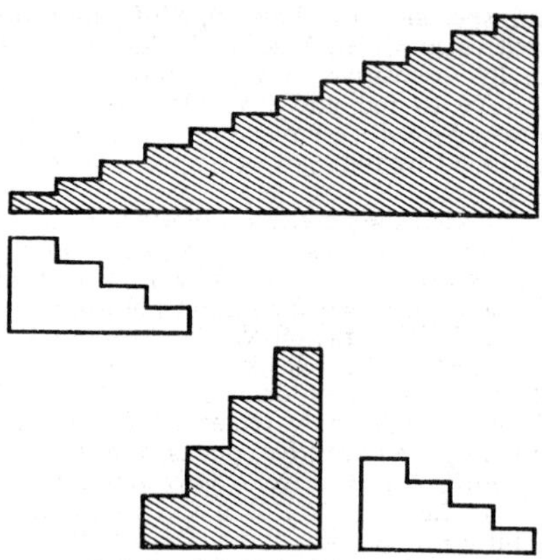

GRADATION AND EXPOSURE LATITUDE. Top: Soft-gradation emulsion (long tone scale) can accommodate subject with great latitude. Bottom: Contrasty (short tone scale) material has little or no latitude and needs exact exposure.

be increased by double coating with two emulsions—a fast and a slow layer.

Contrast Variation. Although the contrast of the developed image depends primarily on the emulsion, it can be influenced by other factors.

The most important of these is the developer and development time used. Some developer formulae inherently give higher contrast than others; and with any developer, increasing the development time increases the contrast up to a certain maximum.

Contrast also varies with the colour of the light. With panchromatic materials, sensitive to all colours, it is generally lower for blue than for red or reddish light. Hence a negative made through a blue filter has, other things being equal, a lower contrast than one made through a red filter.

This has to be taken into account when making colour separation negatives in colour photography. In practice the difference may be adjusted by giving the blue-filtered negatives longer development than the others. But there are also special materials which allow all three separation negatives to be developed for the same length of time to give images of equal contrast.

The contrast range of an emulsion may also change when the intensity of the exposing light is very high or very low (e.g., when pictures are taken by high speed electronic flash).

Graininess. The developed photographic image is not continuous; it consists of small particles of opaque, black silver, embedded in the gelatin layer. These particles become visible when the negative is greatly magnified. The tones no longer appear smooth and homogeneous, they look broken up and grainy. This graininess may show up unpleasantly in enlarged prints.

How far a negative may be enlarged before the grain becomes objectionable depends first of all on the size of the original silver halide crystals in the emulsion. The original size of the grains varies from one type and make of sensitive material to another, so the graininess of the image depends on the choice of sensitive material, among other things. A number of secondary factors also affect the grain size—e.g., the developer used, negative exposure and contrast, printing illumination and print surface.

Graininess and Speed. There is no regular connexion between the size of the silver halide crystal and the amount of light that it takes to make it developable. A large crystal generally needs no more light than a small one. So the larger the crystals, the more image silver is produced by the same amount of exposure; i.e., in effect the larger crystals tend to be more sensitive than the small ones. It follows that the coarse grained emulsions tend to be fast, and the fine grained materials slow.

The emulsions used in normal photography vary from medium to fairly fine grain. The choice of material is often a compromise be-

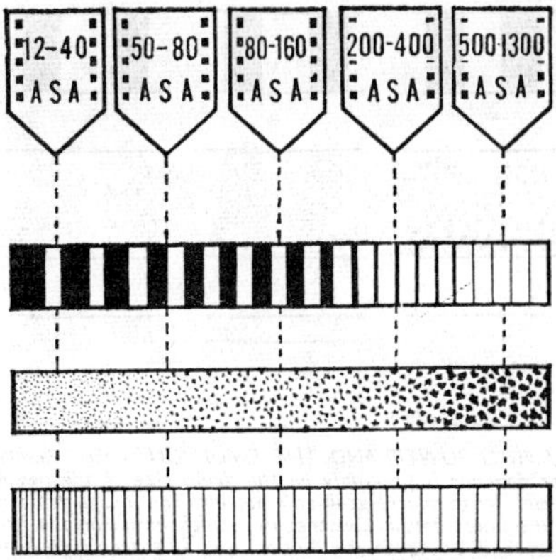

FILM CHARACTERISTICS. Top row: *Speed.* Second row: *Contrast; tends to decrease with faster films.* Third row: *Grain; tends to increase with speed.* Bottom row: *Resolving power; is generally higher with slower emulsions.*

tween a number of factors, among which speed and grain figure largely. Miniature camera photography in particular calls for the finer grain types of film, because the negatives have to be enlarged on a proportionately bigger scale.

Special fine grain negative materials are manufactured for photomicrography, micro-copying, and spectrography, all of which call for an unusually high standard of resolution.

Completely grainless emulsions which are slower even than contact papers are used for radiography, for making graticules, and wherever the resolving power must be as fine as possible.

Resolving Power. The resolving power of a photographic emulsion is its ability to separate small image detail. It is governed by the thickness and turbidity of the emulsion and by its graininess.

The effective resolution of the image also depends on a number of other factors apart from the sensitive material, such as subject contrast, exposure, the developer and technique of development, and the resolving power of the optical system.

The relative resolving power of sensitive materials is measured by photographing a test chart consisting of sets of lines spaced at varying distances apart under controlled conditions of illumination, exposure, and development. The resolving power of the emulsion is expressed as the greatest number of lines per millimetre that can be distinctly seen (under suitable magnification) on the negative of the test chart. This figure, of course, represents the maximum number of lines that the emulsion is able to resolve under ideal conditions of testing.

A photographic emulsion of silver halide and gelatin is a turbid medium, so it tends to scatter light that passes through it. As a result,

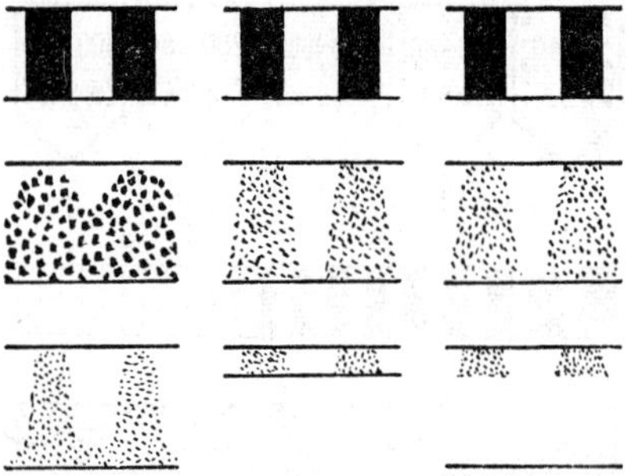

RESOLVING POWER AND THE EMULSION. Left: *Resolving power depends appreciably on the grain size of the negative material. With coarse-grained emulsions the image grain may hide the finest detail.* Centre: *In a thick emulsion the image may spread owing to irradiation and in extreme cases may obliterate the smallest image detail altogether. A thin emulsion layer thus increases resolving power.* Right: *In the same way surface development which confines the image to the surface of the emulsion layer, improves effective resolution as if the emulsion itself were thinly coated.*

the silver image spreads outside the limits of the optical image. The more turbid and the thicker the emulsion, the more the light spreads and destroys the definition of fine detail. This results in lowered resolving power.

Over-exposure increases the amount of scattered light that registers on the sensitive emulsion, and also increases the depth of the image in the emulsion layer, with a corresponding loss of resolving power in each case. Over-development of the negative material destroys resolution by emphasizing the light scatter and coarsening the image.

The greater the emulsion grain, the lower the resolving power. This explains why the fast double-coated materials used in general photography have a comparatively low resolving power. The slower, fine grain emulsions have much higher resolving power. Special thin emulsions are used to give the necessary high resolving power for such work as aerial photography, photomicrography and spectrography.

For very high resolution—e.g., as in making graticules—Lippmann type plates are used. These have a very thin and almost grainless emulsion layer which is practically transparent. The silver halide content of the emulsion is comparatively low, so that such materials are often slower than the slowest lantern plates or contact papers.

Contrast Transfer Function and Acutance. In practice it is not only the resolving power that is important, but also the contrast of fine image detail. This affects the visual sharpness of an image quite apart from its resolution. Thus it is possible to have a film of high resolving power but low detail contrast, yielding an image which appears less sharp than one obtained on a film of lower resolving power but high contrast.

A more complete specification of the film performance in respect of image detail is therefore given by the so-called contrast transfer function. This is a curve which plots the resolving power against the image contrast. A practical significance of the transfer function is that it permits the selection of a sensitive material for any specific purpose, taking into account the degree of practical resolution required. Thus where it is important to resolve very fine detail irrespective of the contrast performance, a film of high resolving power at low contrast is necessary. On the other hand where extreme fine detail is less important, a film of higher contrast in the low resolution range of the transfer function is the better choice.

Another measure of the sharpness of the image obtainable on the film is the acutance. This is a physical measure of the transition across an image boundary between a light and a dark area. The steeper this transition, the sharper such an outline appears. This property is also known as the edge contrast or outline contrast of the image. It is not directly related to the resolving power; it is possible to have a low resolution film of high acutance and vice versa. Generally thin emulsion films yield images of the highest acutance.

Development Characteristics. The characteristics of a negative material must always be considered in conjunction with the developer that is to be used with it. So for any sensitive material it is necessary to know something about its actual rate of development; its potential maximum contrast; and the effect of developers on its effective speed. All these factors depend on the developer as much as on the negative material.

The rate of development in a particular developer is often highest for the slower sensitive materials; the fastest panchromatic materials generally require the longest development times. Special fast-developing high-speed films and plates are, however, manufactured for press photography and other work where rapid processing is required.

While the relative image contrast (gamma)

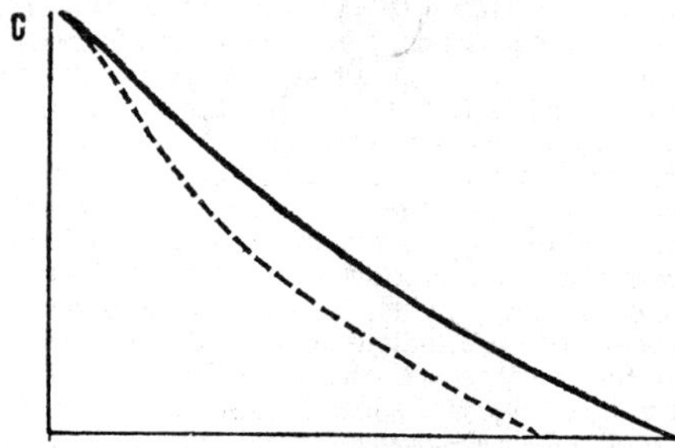

CONTRAST TRANSFER UNCTION. *Instead of specifying the resolving power as such, the transfer function plots resolution against image contrast and so gives a more comprehensive picture of the performance of the material. The solid curve represents an ideal material; in practice the best emulsions still fall rather below that (dotted curve).*

TYPES OF NEGATIVE MATERIAL

Type	Speed °B.S.I. ASA	DIN	Available as	Grada-tion	Grain	Develop-ment Speed	Uses and Notes
Panchromatic Emulsions							
Ultra-speed	800–1200	30–32	P, S, R, M	m.	n.	II–III	Press, poor light, artificial light
High speed	200–500	24–28	P, S, R, M	m. or s.	n.	IV	Press, poor light, artificial light, general photography
High speed aerial	100–200	21–24	P, Perforated roll	h.	n. or mf.	III	Highly red-sensitive. Specially designed for aerial photography. Sometimes made on special non-shrink supports
Fluoro-graphic	100–200	21–24	M. Perforated roll	m.	mf.	II	Specially designed for mass miniature radiography and fluorography with yellow-green screens
Medium speed	64–160	19–23	P, S, R, M	m.	n. or mf.	III	General photography scientific, colour separation
Fine grain	25–64	15–19	P, S, R, M	m. or h.	f.	II	General photography, scientific
Extra fine grain	16–32	13–16	M	h. or eh.	ef.	I	Miniature work in good light, micro-filming
Fast re-cording	—	—	S, M, perforated roll	m. or h.	mf.	S	Recording traces on fluorescent screens, fast moving lightspots, etc.
Process	—	—	P, S	eh. or uh.	f. or ef.	S	Screen and line separation negatives from colour originals, photomicrography, scientific
Micro-copying	10–20	11–14	M	eh.	ef.	I or S	Microfilming of coloured originals, scientific
Orthochromatic Emulsions							
High speed	200–400	24–27	P, S	m.	n.	III	Press, general photography, portraiture, medical
Medium speed	80–160	20–23	P, S, R	m.	n.	III	General and commercial photography
Fluoro-graphic	—	—	M	h.	n.	II	Specially designed for mass miniature radiography and fluorography with green screens
Rapid	40–80	17–20	P, S	m.	mf. or n.	II	General, technical, scientific, and commercial photography, copying of continuous tone originals, especially faded or yellowish images
Slow	12–25	12–15	P, S	m. or h.	f.	I	Scientific, photomicrography, copying continuous tone originals
Recording	—	—	S, M, Pp, perforated roll	eh.	mf.	S	Recording traces on fluorescent screens, moving light spots
Process	—	—	P, S	h. or eh.	f. or ef.	S	Copying of line or screen originals, photomicrography, etc.
Litho process	—	—	P, S, Pp	uh.	ef.	S	Specially designed for photo mechanical screen reproduction, producing extra sharp dot quality—e.g., for lithography
Maximum resolution	—	—	P	uh.	uf.	S	Extremely thin emulsion, used for making graticules and scientific work requiring utmost resolution
Blue-Sensitive Emulsions							
Medium speed	20–40	14–17	P, S	m.	n.	III	Commercial and scientific
Fluoro-graphic	—	—	S	h.	n.	II	Fluorography with blue screens
Commercial	8–12	10–12	P, S	m. or h.	mf. or f.	I	Copying, scientific
Process	—	—	P, S	h. or eh.	f. or ef.	S	Copying line or screen originals, scientific, photomicrography
Recording	—	—	S, Pp	m. or h.	f.	S	Recording traces on blue cathode ray screens, recording moving light spots—e.g., mirror galvanometer readings
Miniature positive	—	—	M	h. or eh.	ef.		As a negative material for the microfilming of monochrome originals; scientific
Infra-red Emulsions (speeds are with infra-red filter)							
Rapid	50–100	10–21	P, S, M	m.	n.	I–II	Long distance work, aerial survey, haze penetration
Slow	4–12	8–12	P	h.	mf.	I	Scientific
Process	—	—	P	h.	mf. or f.	S	Special copying techniques

Speed: Where no speeds are given, exposures are determined by tests for the particular set-up in use—e.g., with process materials. The materials are arranged approximately in descending order of speed. Gradation: S—soft; m—medium; h—hard; eh—extra hard; uh—ultra-hard; Grain: n—normal; mf—medium fine; f—fine; ef—extra fine; uf—ultra-fine or grainless. Availability: P—plates; S—sheet film; R—roll film; M—35 mm. miniature film; Pp—paper. Development Speed: I—very rapid; II—rapid; III—average; IV—slow; V—very slow; S—material requiring special developers and/or development procedure.

This list does not cover a large number of scientific materials for specific purposes which are made to special order only.

depends largely on the developer and development time, the maximum contrast (gamma infinity) obtainable in a given developer depends on the inherent gradation of the negative material.

Development in many fine grain developers lowers the effective emulsion speed of the material. Most fine grain developers lower the effective speed of the faster materials more than they do the slow. This difference is very marked with some ultra-fine grain developers. Many ultra-fine grain developers in fact offer little or no advantage with most types of high-speed materials over, say, ordinary fine grain developers with medium speed materials.

Physical Characteristics. The thickness of the emulsion coating varies according to the purpose of the material. Thick emulsions contain more silver salts than thin, so they are capable of yielding a greater maximum density, but their resolving power is correspondingly lowered. Very thin emulsions are used for work calling for the maximum amount of resolution—e.g., for fine line reproduction and making graticules.

Certain negative materials are prepared in such a way that the emulsion can be stripped off the support and, if necessary, transferred to another. These usually have a hardened emulsion coated on top of a layer of clear, unhardened gelatin. The soft gelatin makes it easy to strip the emulsion off its support. In addition, the emulsion may have a supercoat of adhesive to allow it to be transferred and stuck to another support. Such materials are known as transfer materials.

The material may also be given a final protective coating of clear unsensitized gelatin, known as a supercoat. This layer protects the sensitized emulsion from physical injury—e.g., scratching and abrasion—which tends to produce dark stress marks in the developed emulsion, even though the material itself is not visibly marked. In transfer materials, the supercoat may consist of a layer of adhesive to attach the emulsion to its new support.

Double-coated Emulsions. Many sensitive materials—particularly films for the popular photographic market—have two emulsion layers: a slow one coated directly on the support, and a fast one coated on top of the slow emulsion.

During exposure the fast emulsion layer is affected first. Further exposure affects the lower layer and forms a latent image there, so increasing the effective tone range of the material.

Double-coated materials are particularly suitable for general amateur photography because of their wide exposure latitude. Their main disadvantage is that double coating increases the thickness of the emulsion layer and gives rise to increased irradiation, so that double-coated materials have a comparatively low resolving power.

Supports. The base on which the emulsion is coated is mostly either a glass plate or a flexible film of cellulose acetate or other plastic. Glass plates are used for accurate work such as photogrammetry where dimensional stability is important, but special transparent plastic bases have been developed which are almost as stable as glass in this respect.

In addition to these transparent bases, several opaque materials are used as supports for negative emulsions. These include card (for the cheap sensitized material known as negative card), paper (for cheap paper roll films), and sheet metal (for templates).

Most negative materials are treated with some form of backing or sub-coating of dye or pigment to absorb the light scattered within the emulsion layer and reflected from the base.

Roll films and cut films are also coated with a gelatin backing to counteract the tendency of this type of support to curl in towards the emulsion layer. The anti-halo dye or pigment is usually put in this anti-curl coating. L.A.M.

See also: *Acutance; Cine film; Colour materials; Dimensional stability; Films; Gradation; Grain; Information in the image; Keeping qualities of materials; Resolving power; Sensitized materials history; Sensitized materials manufacture; Sizes and packings; Spectral sensitivity; Speed of sensitized materials; Supports for emulsions.*

Books: *Developing*, by C. I. Jacobson (London); *Exposure*, by W. F. Berg (London); *Sensitometry*, by L. Lobel and M. Dubois (London).

NEGATIVE-POSITIVE VIEWER. Viewing system that permits a negative to be observed as a positive image (and vice versa). Viewers of this kind rely either on fluorescence quenching by infra-red radiation, or on electronic reversal of an image on a television screen.

See also: *Viewers.*

NEGATIVES FROM COLOUR SLIDES. It is often useful to be able to make black-and-white prints from colour transparencies. Since the transparency is a positive it is first necessary to make an ordinary monochrome negative from it; this negative is then used to make the black-and-white print (or a monochrome slide).

The kind of negative to be made depends upon the original transparency and the sort of print required. There are negatives—e.g., 35 mm. and other miniature sizes—made by contact printing, from which prints are made in the enlarger; small negatives made by projection—e.g., to 2¼ ins. square and 2¼ × 3¼ ins.—from miniature transparencies, to print by contact or projection; and enlarged negatives from which prints are made by contact.

Contact Negatives. These are best made on to a medium speed panchromatic film; 35 mm. films are printed in a 35 mm. printer of the type used for making film strips or miniature slides.

The disadvantage of the method is that it is difficult to prevent dust and scratches from being reproduced on the negative. These blemishes are subsequently magnified if the final prints are made by enlargement.

Contact prints from roll film transparencies —i.e., $2\frac{1}{4} \times 2\frac{1}{4}$ ins. or $2\frac{1}{4} \times 3\frac{1}{4}$ ins.—are made in the same way as lantern slides but on medium speed panchromatic film or plates instead of lantern plates.

Negatives by Projection. It is sometimes more convenient to make enlarged negatives from 35 mm. transparencies; the conventional lantern slide sizes are easier to handle and retouch than miniature slides made by contact. In this case the colour transparency is held in the negative carrier of the enlarger. The negative is then made by normal enlarging technique on to a suitable plate or section of film held in a frame on the enlarger baseboard.

Because the sensitive material is so much faster than bromide paper, every trace of stray light must be cut out and all operations conducted by the correct type of safelight.

If enlargements are made from negatives projected in this way, there is a further loss of definition that does not occur in the first method described, but greater freedom from mechanical defects.

Where large black-and-white prints, all of the same size, are required, an economical method may be to make an enlarged negative from which full size contact prints can be run off. Compared with the other methods, this way is too expensive in negative material to be used for making only two or three prints of one size.

Whenever the copy negative is to be printed by contact, the fact that it will be reversed from left to right must be allowed for by reversing the colour transparency in the enlarger. This enables the negative to be printed with the emulsion in contact with the printing paper. The precaution is unnecessary with negatives for enlarging because they can be reversed without affecting the definition of the prints.

Sensitized Material. The copy negative should generally be made on panchromatic material, although orthochromatic and non-colour sensitive material may be quite suitable for certain subjects just as it is in making monochrome negatives direct from the original subject. In the same way, filters may be used over the printing light to vary the relative brilliance of the colours, or to correct the colour temperature of the light source.

Medium speed material is essential to give fine grain in negatives that have to be considerably enlarged. For negatives for contact printing, faster pan films of lower contrast may be used with advantage.

The contrast of most colour transparencies is appreciable, so the film must be developed in a soft-working solution and processed to give a soft negative. The correct method is best found empirically. Any normal negative developer at half the usual strength can be used, and a few test exposures at various development times will soon give an acceptable and workable level of contrast.

Printing Light. The colour temperature of the printing light influences the tones of the negative. Tungsten lighting, for instance, being rich in red rays, will tend to lighten the tone of red and yellow objects and darken the tone of blue. This effect can be corrected by the use of a green-blue, half-watt-to-daylight correction filter in front of the lamp. Mercury vapour lighting, being very rich in blue and ultra-violet, lightens blue tones and darkens red. The spectrum of cold cathode lighting is very similar to that of daylight.

The foregoing comments apply to the making of positive prints from reversal colour transparencies; photographers who expect to require many monochrome prints may prefer to use one of the colour negative systems.

It should also be borne in mind that reversal papers are on the market which enable positives to be made without an intermediate negative. C.D.M.

See also: *Duplicate negatives; Separation negatives.*

NÈGRE, CHARLES, 1820–79. French painter and photographer. Began to calotype *c.* 1850, opened a portrait studio in Paris, but is chiefly known for architectural and genre photographs. Founder member of the Société Française de Photographie. Improved various photomechanical processes by which he obtained excellent half-tone pictures: heliogravure, 1854, collotype, 1856, and photogalvanography, 1861.

NEO-COCCINE. Red dye used in negative retouching. It stains the gelatin and prevents some of the actinic light from passing through and affecting the printing paper. A faint, even tint can thus be given to thin parts of the negative to prevent them from printing too dark during enlarging.

NEON LAMP. Glass bulb filled with neon gas at very low pressure and having two electrodes connected to a conventional type of cap. When the lamp is on, the gas between the electrodes glows with a pinkish light.

Lamps of this type are not bright enough for normal lighting, but they have a number of specialized uses.

Because the current consumption is very low, the light is practically cold, and there is no filament to break or burn out. Neon lamps are used as pilot lights, indicators, warning lamps and in situations where absolute reliability is essential and low current consumption is an advantage.

Because the light "strikes" as soon as the voltage is applied, and goes out immediately without any afterglow on switching off, neon lamps are ideal for use in stroboscopes.

The neon arc strikes as soon as the applied voltage exceeds roughly 80 volts. This property is made use of in electronic timers. In these instruments the neon lamp operates a switch and breaks the enlarger lamp circuit. It does this when the voltage of an associated bank of condensers rises to a particular value; this charging period corresponds to the desired exposure time.

NEWMAN, ARTHUR SAMUEL, 1862–1943. English engineer, designer of photographic and cinematographic apparatus. Devised in 1889 his first diaphragm shutter; in 1890 the Celeritas shutter (with Simpson's of Clerkenwell, London). Manufactured with the firms of Newman and Guardia (till 1908) and Newman and Sinclair several cameras, e.g., the Standard, Nydia, Sybil, Centum (100 exposures on unperforated 35 mm. film), N & S Reflex camera. Made early cine apparatus for Friese-Greene, the Viventoscope projector for Blair (1897), supplied films and gave cine shows. In 1909 advocated standardization of motion picture films. In 1911 made the first N & S cine camera and later the clockwork Auto Kine camera. He worked as consultant for photographic and cine companies. A founder member of the British Kinematograph Society. Received the Progress Medal of the Royal Photographic Society in 1936. The Newman Lecture and Plaque were established in his memory.

NEWS PHOTOGRAPHY. Section of press photography concerned with taking pictures of topical events.

See also: *Freelance photography; Magazine photography; Photo journalism; Picture series; Press photography.*

NEWTON'S RINGS. Concentric bands of coloured light sometimes seen around the areas where two transparent surfaces are not quite in contact. The rings are the result of interference and occur when the separation between the surfaces is of the same order as the wavelength of light.

In enlarging, Newton's rings can form between the surface of the negative and the glass negative carrier plate, when they produce alternate light and dark bands on the print. The cure, if the negative size is small enough to permit it, is to use a glassless carrier. With larger negatives a mask around the edge, sufficient to take the negative (usually the back) out of contact with the glass, will often put things right. In extreme cases, the negative may be pressed on to a pool of glycerine on the glass so that all air bubbles are excluded. A print made in this way will show no sign of rings. Newton's rings are more likely to occur with negatives which are not completely dry, or under conditions of high humidity.

Newton's rings can also be a nuisance in film slide projection when they form between the shiny side of the film and the cover glass in contact with it. It is difficult—and often impossible—to get rid of Newton's rings between glass-to-glass or glass-to-film surfaces. But if one of the glass surfaces is lightly etched with acid to give a fine matt texture the trouble cannot occur. Glasses treated in this way are available for mounting transparencies and also for glass negative carriers. Although it is only necessary to use treated glass for one half of the sandwich, in practice both halves are treated so that it does not matter if the glasses are interchanged. The grain of the etching is so fine that it does not show itself either on the projected image or the print.

The appearance of Newton's rings between two flat or curved glass surfaces is used as a check for flatness or regularity of curvature in fitting and inspecting optical components.

See also: *Faults.*

NEW ZEALAND. The photographic history of New Zealand has been much affected by its geography. Although only a small country, with a population of under three million, it spreads over a distance of about 1,100 miles from north to south. Difficulty of access resulted in there being at least two independent pioneering photographic groups in the nineteenth century. The two oldest towns, photographically speaking, are Auckland and Dunedin. Dunedin, well towards the south, was founded by Scots, whereas Englishmen were the main settlers in Auckland.

Old records show that in 1898 the Auckland Photo Club, as it was then called, was already holding annual exhibitions and the President of the club was the then Governor of New Zealand, Lord Onslow. This Vice-Regal patronage has continued—in fact, within the last year the Auckland Photographic Society, which started as the Auckland Photo Club, has won the Cup, donated by another Governor General.

The Dunedin Photographic Society's recorded history pre-dates that of the Auckland Society's earliest records and it is therefore safe to say that there was in New Zealand by 1960—only twenty years after the country became a Crown Colony—a well-established photographic group of both professional and amateur photographers. It is interesting to note that the professionals and amateurs have worked together in harmony in the photographic societies, clubs and groups in New Zealand throughout the photographic history of the country.

By 1860 there was at least one photographic retailer, Mr. H. Holland, in Victoria Street, Auckland, and Waterworths, another early dealer, was in business prior to 1880. A third name mentioned in early records is Harringtons.

Photography had an early influence on the paintings of a well-known New Zealand painter, Goldie. Many of his paintings were done on a sensitized base, and this undoubtedly accounts for many of his astoundingly accurate portraits of famous Maori chieftains.

During the past thirty years the Auckland Public Library has collected and filed old photographs of early Auckland. This service, has helped greatly in establishing the accuracy of historical comments.

By 1913 Sucklings had purchased a large shop in Queen Street, Auckland's main thoroughfare. Sucklings offered a four-hour developing and printing service for roll films and plates. This, perhaps more than anything else, helped to popularize the idea of amateur photography in New Zealand, and when Kodak (N.Z.) Limited, was established in the early 1920's, taking over Suckling's shop in Queen Street, New Zealand photography could be said to have come of age.

In a recent survey it was estimated that the average New Zealand home has three cameras, one of which is 35 mm. There is a movie camera in every tenth home and a projector of one type or another in every eighth home. (In spite of the of the fact that the importation of cameras into the country is subject to Government control.)

Industry. In recent years almost all of the major photographic companies in the world have established representatives in New Zealand, and several manufacturers have opened colour film processing laboratories.

At the present time New Zealand manufactures 16 mm. sound projectors, 35 mm. slide projectors in two types, some twenty or thirty types of screen, simple hand illuminating viewers, slide boxes in a multitude of types, but no cameras.

New Zealand can service almost any camera, and there is available a wide range of brands of film, black-and-white, colour negative and colour reversal and most of these can be processed in New Zealand.

Societies and Clubs. Because of the difficult terrain that separates the principal cities in New Zealand, it was not until around the late 1930's that there was much direct communication between all the main centres. Now, with better roads and a good air service, New Zealanders are extremely travel-conscious, and this has led in 1954 to the establishment of the Photographic Society of New Zealand. Initially a gathering of camera clubs who came together to advance photography as an art, it has now been extended to include over 100 clubs. The Society has succeeded in welding together the clubs throughout the length and breadth of New Zealand. It holds an annual conference and it publishes its own magazine *New Zealand Camera*.

At about the same time, the professional photographers in New Zealand established the Professional Photographers' Association, which has groups for dealers, commercial and portrait photographers, etc. Members do their utmost to maintain standards of ethics and work quality; they run schools and lecture programmes and there is hardly a professional photographer in New Zealand who is not a member.

Publications. Apart from *New Zealand Camera*, there is a bi-monthly trade journal *New Zealand Studio*.

A number of the individual camera clubs publish their own journal.

There have been numerous publications of photographs, mainly in colour, of the impressive New Zealand scene. One of these, prepared by Ken Bigwood, is a two-volume publication called *Pictorial New Zealand*. Another, entitled *New Zealand Birds*, by Geoff. Moon, A.R.P.S. shows the native bird life.

Schools. Although there are night school courses and apprenticeship systems, New Zealand has no examinations or accepted professional standards. An amateur can learn from the clubs and societies, or can enrol for one of the two- or three-year courses conducted one night a week at various colleges by many of the country's leading photographers, but these vary greatly in coverage and standards. On the commercial side there is also no training programme. The Professional Photographers' Association is trying to standardize procedure, and it may not be long before recognized photographic courses will be available through the technical colleges.

Application. New Zealand photographers are employed in hospitals, universities and government organizations, particularly in the Department of Scientific and Industrial Research. Many of these have done original scientific work in the photographic field. A camera recently produced by Mr. S. Rumsey of the Dominion Physical Laboratory in Wellington is designed to take a picture of the visible sky, all at one time. The lens of this camera, which is of Mr. Rumsey's own design, is considered by many to be a major break-through in this specialized field.

In the field of medical photography, New Zealand has established a reputation for its visual training systems, usually with colour slides, which are used in training doctors and nurses, and also in the recording of certain heart and lung operations—in some cases the first of their kind in the world.

Photography is being increasingly used by industry. Many firms microfilm their documents, and a photographic record of company and product progress is common. Larger organizations publish illustrated house magazines.

New Zealand, as a small country, has always lacked capital investment, and often it has been left to the government to pioneer certain aspects of photographic work. The New Zealand National Film Unit, for example, has for many years been producing top quality films, and has won a number of awards for its productions.

This is a field where private enterprise has difficulty in competing.

The government also maintains a large publicity department where advertising for the tourist trade is created. R.A.H.

NICKEL NITRATE. Used in certain toners.

Formula and molecular weight: $Ni(NO_3)_2 \cdot 6H_2O$; 291.

Characteristics: Green hygroscopic salt.

Solubility: About 50 parts in 100 parts of water at room temperature.

NIÉPCE DE ST. VICTOR, CLAUDE FÉLIX ABEL, 1805–70. French cavalry officer and inventor of the albumen-on-glass process, 1847. Modified the heliograph (bitumen process) of his cousin Nicéphore Niépce into a heliogravure photomechanical printing process on bitumen-coated steel plates which gave good half-tone prints, 1855. Following Edmond Becquerel's ideas Niépce de St. Victor succeeded by about 1860 in taking natural colour photographs (called heliochromy) on silver plates by the interference method. They could not be fixed and could only be viewed in dim light. Gabriel Lippmann solved the problem in 1891. Biography by R. Colson (Paris 1898).

NIÉPCE, ISIDORE, 1805–68. French, son of Joseph Nicéphore Niépce whom Isidore succeeded (1833) in the partnership with Daguerre. Does not appear to have made any important scientific or technical contribution to the further work of Daguerre, but took part in the commercial exploitation of the Niépce-Daguerre inventions. In 1839 obtained a pension from the French Government for the disclosure of the daguerreotype and heliographic processes. Published a brochure against Daguerre, 1841.

NIÉPCE, JOSEPH NICÉPHORE, 1765–1833. French amateur scientist. Invented photography in the camera, and heliogravure, both of which he called heliography. Experimented with lithography and tried to produce paper images and litho printing blocks (on stone, metals, glass) by copying drawings through the action of light and subsequent etching. Used a camera (with lens) as early as 1816, and also contact copying without a camera. In 1816 obtained negatives by the action of light on paper (imbibed with silver chloride), but could only partially fix them. After experimenting with a number of substances, achieved satisfactory results by coating a pewter plate with bitumen, exposing it (by contact, or for hours in a camera), washing away with lavender oil and petroleum the shadow portions which had not been hardened by light, and thus obtaining a permanent, direct-positive picture. He was the first to use an iris diaphragm and bellows in his cameras. 1822 is probably the year in which he obtained the first successful, permanent image by contact copying; the first surviving photograph from nature, taken in a camera, dates probably from 1826. Niépce used in 1829 silvered metal plates for his bitumen photographs and darkened the shadows (bare silver) by iodine fumes; this method was described in detail in his supplement to the partnership agreement which he concluded with Daguerre in 1829, and doubtless contributed to Daguerre's discovery of the light sensitivity of iodized silver plates (1831). Niépce's contact with Daguerre had been established in 1826 through the Parisian optician, Charles Chevalier, from whom both bought lenses and optical apparatus.

The subsequent history of the Niépce-Daguerre invention will be found in the biographical notice on Daguerre. Niépce died in 1833, and his son Isidore became Daguerre's partner. Biographies by Fouque (Paris 1867) and H. & A. Gernsheim (London 1956).

See also: *Discovery of photography.*

NIEPCEOTYPE. Name occasionally and incorrectly given to J. N. Niépce's heliography, and Abel Niépce de St. Victor's albumen-on-glass process.

See also: *Discovery of photography.*

NIGHT PHOTOGRAPHY

Photography in one way or another can still be carried on out of doors after the sun has set, by the afterglow in the sky in the period of dusk, by existing street, shop-window, or floodlighting, and by moonlight. The photographer can supplement these effects to a greater or less extent by adding lighting of his own—generally flash.

Equipment. The camera, its accessories, and a firm tripod form the principal items of equipment. There is one important addition that does not usually form part of a photographer's kit—a pocket torch.

There are some after-sunset pictures that can be taken with any camera and others that call for something special. If the lighting is so feeble that the photograph needs a time exposure, with the camera mounted on a tripod or other support, then one camera is as good as another. Night scenes were photographed in this way as early as the beginning of the century with lenses and sensitized materials that were impossibly slow by modern standards. So for time exposures, even a box camera will serve; the quality of the results being as good as its simple lens and limited adjustments will permit. The subject in all such cases must of course be still.

Moving subjects call for a lens with an aperture of f3·5 or preferably f2·0 with accurate focusing and a shutter giving automatic exposures down to half a second. With such subjects it is not essential for the camera to be on a tripod, but some form of support is usually desirable. This is the field of the miniature camera and, in fact, the advent of this type of instrument, together with modern high-speed panchromatic emulsion, was directly responsible for the popularity of night photography.

Whatever the camera, it must have a large, clear viewfinder, plainly marked scales, and a deep lens hood. Automatic film transport is a desirable feature as it is difficult to see film numbering at night. A bloomed lens is an advantage when there are artificial lights facing the camera.

Lighting. Apart from photographs taken by dusk, moonlight, or lightning, all photography at night relies on the presence of some form of artificial lighting. This may consist of isolated street lamps or of masses of lights in busy city centres, in arcades and fair grounds, or of the special concentrations of lights at the foot of floodlit buildings. The subject may be the pattern made by the lights themselves, or an illuminated scene in which the light sources may or may not appear.

Contrast. In every case, the special character of the illumination has to be considered and, in the main, this means its peculiarity of falling off abruptly at a short distance away from the source. This is the result of the operation of the inverse square law. Artificial lighting indoors is either concentrated by reflectors or so much of it is reflected back on to the subject by walls and ceiling that the law applies only in a very modified fashion. But outdoor lighting usually radiates in all directions and is not often reflected back in useful amounts by light surfaces nearby. So the intensity of outdoor lighting at, say 20 feet from the source is only quarter the intensity at 10 feet, at 30 feet it is one-ninth, at 40 feet it is one-sixteenth and so on.

The consequence is that the picture tends to concentrate itself into localized pools of light immediately surrounding the source, with featureless areas of black in between. This is not what the eye, with its tremendous range of sensitivities, sees when it looks at the same scene. So the picture is apt to be disappointing and not at all what the photographer hoped to get when he made the exposure.

There are several things that can be done to improve the picture. The first is to choose scenes in which there are light-reflecting surfaces—buildings, foliage and flowers, even vehicles—to make the most of what light there is in the faintly illuminated areas. Secondly, the highlights in the picture can be doubled by choosing a rainy night—preferably just after a shower—when they are reflected in the wet pavement. It is often possible to make use of the surface of a river or lake in the same way, while snow-covered ground and light dusty road surfaces are equally helpful.

The violent contrasts of the scene can be reduced still further by keeping the light source out of the picture. This can often be done by choosing a viewpoint where the sources lie outside the field of the lens. Another solution is to choose a position where a foreground object—e.g., the trunk of a tree or a lamp-post—comes between the light source and the camera. This trick is particularly effective when the obstruction has a recognizable shape—e.g., a piece of statuary or even a human figure.

It also helps to use a low-contrast, sensitized material—i.e., one of soft gradation. This point is generally taken care of automatically, since the night photographer almost always uses fast panchromatic materials which tend in any case to have a soft gradation.

For the same reason, exposure should be on the generous side to get the maximum shadow detail, and development can safely be cut by 25 per cent to prevent the highlights from blocking up.

Using Flash. When flash is used out of doors at night there are no light surroundings, like ceilings and walls, to act as reflectors and throw indirect light back on to the subject and its setting. So flashlight in the open falls off rapidly and illuminates nothing beyond the actual subject. (It also illuminates any foreground objects very brightly so that they can quite easily steal the picture.)

This characteristic offers a valuable means of completely isolating a particular subject. Anything photographed in this way prints out against a solid black background. If the background and setting are necessary, the easiest way to handle the problem is to shoot around dusk, when there is just enough sky light remaining to record the setting while the shutter is open for the flash. Usually, the best way to do this is to light the subject by open flash technique and leave the shutter open long enough to give the required exposure to the landscape. Many novel effects can be created by this method of combining flash and evening light.

Halation. When the light sources appear in the picture, they are bound to show some sign of halation. This causes the image of the source itself to spread, and surrounds it with a halo of scattered light. As a rule, the longer the exposure and development, the worse the halation. It is also worse with unbacked materials, but nowadays any plate or film likely to be used for night photography is backed or treated against halation by the manufacturer.

Halation can be avoided by cutting out the light source, or it can be kept to a minimum by giving no more exposure or development than is absolutely necessary. Development in a surface developer is also an excellent way of avoiding halation. Finally, it can be lessened by rubbing down the affected areas of the negative with a fine abrasive paste like Baskett's reducer. But it must be remembered that a trace of halation adds a natural radiance to the light sources that can be most effective.

Flare. Most lenses produce some form of flare, but there is generally some position of the source where flare is avoidable. These positions can be found by setting up a candle or lighted torch bulb in a darkroom, focusing the lens on it, and noting the effect of shifting the image about on the focusing screen. Any serious flare noted can be avoided by keeping the light source away from the corresponding position of the test light in the picture area.

The position of a flare patch on the focusing screen of a twin lens reflex is no indication that the taking lens will form a similar patch on the negative. Flare in most lenses is generally worst at small apertures. Coating a lens reduces the likelihood of flare; any photographer who intends to take up night photography can always have the lens of his camera coated reasonably cheaply if it is not coated already.

Focusing. At night it is difficult to focus by any of the regular methods, because the subject is too dim to be seen in a focusing screen or a rangefinder field, and the light sources are too intense. The camera can be focused on a torch held near the subject; another method is to measure or pace out the distance, or, if necessary, estimate it. Slight errors in focusing are in any case less noticeable in a night photograph. As a rough rule the lens should be focused on a point about one-third of the way from the front edge of the depth of the field to be covered sharply, but it is easier to turn it right to the infinity stop for scenes where the interest lies beyond about 50 feet.

Exposure. There is no single correct exposure for any night subject; everything depends on what the final print has to show. The minimum exposure that will register the light sources would be adequate where the picture consisted of a pattern of lights only, but as soon as the negative must include an image of objects or scenery illuminated by the light source, then it becomes necessary to give a longer exposure. And the longer the exposure, the more of the scene will appear in the final print.

At the same time, there is no point in giving such a lengthy exposure that the details of the subject will be as plain as in a daylight photograph, because without some proportion of completely black shadow the photograph would lose the essential after-dark atmosphere. So the best exposure will generally lie somewhere between the above extremes. In practice this allows a great deal of latitude which is extended still further at the printing stage. By choosing a hard or soft paper, the amount of shadow detail in the final print can be widely controlled.

All that can be said about exposure is that it should be based on a selective exposure meter reading, interpreted in the light of experience and experiment and study of data published with night photographs taken under similar conditions. A rough starting point is indicated for each of the various classes of subject that are dealt with below.

Most night pictures are time exposures made with the camera mounted on a steady support or a tripod. In addition to the usual precautions taken to ensure that the camera does not move during the exposure, a special technique is necessary if there are passers-by or lighted vehicles moving in front of the camera. A piece of black card is held in front of the lens while the interruption is going on and removed as soon as it is over. This procedure can be repeated ad lib., so long as a check is kept to see that the total time for which the lens is uncovered adds up to the required exposure. However, people walking fairly quickly across the picture can be neglected; they will not affect the image.

Sensitized Materials. Since even the brightest of night scenes is dull compared with daylight, only the fastest panchromatic materials can be considered. Even for time exposures it is better to use fast panchromatic materials for the sake of their softer gradation.

Often the fastest materials are still not fast enough—particularly for use in a camera with an $f4{\cdot}5$ or smaller lens. At such times it is worth-while hypersensitizing the film or plate. This can be done either before or after exposure. There are various methods of hypersensitizing which can be expected to double or treble the speed of the material for a few days at least.

Processing and Printing. As the subject contrasts are generally too extreme to be recorded by the film, development should aim at low contrast—e.g., by curtailing the development time up to 25 per cent, by employing two bath or water bath development, or (preferably) by developing in a compensating fine-grain formula. Fireworks and similar light trace patterns, however, are rendered best by development in an ordinary M.Q. developer to get extra contrast.

The typical night photograph is best printed on a fairly hard grade of paper to bring out the shadow detail and to also reproduce good quality blacks.

Landscapes. The best time for taking open scenery and panoramas is immediately after the sun has set, and as soon as the buildings and streets are lighted up. At such times there is still enough sky light left to show up silhouettes of buildings and the line of the horizon, while lighted windows and street lamps stand out against an inky-black background.

Open landscapes with plenty of lighted roads and buildings—or even public illuminations—make excellent pictures taken in this way. Sometimes a better result is achieved by making two exposures, one just after the sun has set, and another very much later, when all the lights have been switched on. The camera must of course be left untouched on the tripod between exposures.

The exposure for the landscape immediately after sunset should be 4–8 seconds at *f* 8 on fast panchromatic film. If a second exposure is given later to get more lights in the picture, the same exposure should be given.

Streets. In the older parts of most towns there are lamplit street corners, archways, flights of steps and alleys that make excellent picture material. Generally all that is needed is a suitable figure either under the lamp or silhouetted against its light and the result is an atmospheric picture that can be taken even with a box camera. If the light is reflected in a wet pavement, so much the better. Exposure will be anything from 5–10 minutes at *f* 11.

Street scenes in the busier part of the town call for instantaneous exposures to arrest movement. Fortunately the lighting is more powerful in such places and, by shooting when the subject is well within the full brilliance of the lamp, exposures as short as 1/25 at *f* 3·5 will give a reasonable image. At this speed the movement should be towards or away from the camera. By taking the moving figures in silhouette against a brightly lighted surface, or against lights reflected in a wet pavement, even higher speeds may be used.

Many of the latest types of mercury or sodium vapour street lamps enable much shorter exposures than this to be given, particularly if the subject or group is standing directly under the lamp and silhouetted against pavement reflections from other lights. A single figure can be very effectively photographed looking at a newspaper under the street light, because the newspaper acts as a light reflector to improve the illumination of the face.

Shop Windows. Lighted shop windows give an opportunity for two types of picture: the display figures and window-dressing arrangements and people looking at the window.

(1) As the lighting is all directed away from the observer, there is no danger of flare or halation and the results are usually excellent. One or two window shoppers silhouetted against the light help to make things more interesting. Exposures will depend on the number and power of the lamps (which vary widely) and a photo-electric meter reading is the only answer.

(2) A camera station well to the side is necessary to show the faces of the spectators as well as something of what they are looking at. This gives the photographer a chance to steady the camera against the side of the building and take a time exposure. As the faces are lit only by reflected light, the exposure must be as long as can be given without showing movement. Since the subjects are interested in what they are looking at, it is often possible to give an exposure as long as 5 seconds, but even at 1 second the result may be satisfactory.

Busy Centres. Every city has its Piccadilly Circus where there is a general blaze of light from street lamps, lighted shop windows, advertising signs and the like. Here there are interesting groups of people, newspaper and flower sellers, traffic and lively action. The possible subjects are endless and the light is generally strong enough to make snapshooting possible. Even so, the longest practicable exposure should always be given at the greatest stop that will give the required depth of field. A tripod will be out of the question, but there is never any shortage of lamp-posts, pillar boxes etc., to serve as a camera rest.

For general views of the centre, it is a good idea to shoot when traffic is stationary, just as the traffic lights are changing over. This is the right time also to catch pedestrians waiting to cross and who are generally standing under the brightest illumination.

Normal shots around centres of this type should be given an exposure of 1/10 second at *f* 2, but by carefully timing the instant of exposure it is possible to get pictures which show little movement at ½ second at *f* 4·5.

A novel type of picture results from stopping down the lens and giving a long time exposure during which the lights of moving vehicles register as parallel streaks of light. For this type of shot the camera should be on a tripod and if possible looking down on the centre.

Fairs and Markets. Pictures around fairs at night call for a variety of techniques because the subjects and lighting vary so widely. Generally speaking, there is not much point in trying to get pictures in the poorly lighted spots. The fun of the fair is in its life and action, and there are usually a number of vantage points around the more elaborate mechanical entertainments where the lighting is bright enough for instantaneous exposures. Here, the technique recommended for busy city centres will apply and exposures will be about the same.

Brightly lighted stalls like shooting galleries, houp-la, and coconut shies can be tackled like shop windows—i.e., showing the spectators silhouetted against the brightly lighted background of prizes, or shooting from the side to catch their expressions. Markets with their stalls, cheap-jacks and their audiences can be dealt with in the same way.

One great advantage of the individual lighting in such places is that, by taking a little trouble over the viewpoint, it is generally possible to isolate the subject completely against a dark background.

Finally, the long exposure technique applied to roundabouts, switch-backs or the Big Wheel will often yield delightful light-trace patterns if the shutter is left open for one or more revolutions.

Floodlit Buildings. These are ideal subjects for simple cameras, since they can be given a long time exposure without any fear of movement. The building should first be studied from every angle to find the most effective viewpoint—it is rarely the flatly-lighted view from straight in front that shows the subject to the best advantage. With a camera that has no rising front movement it will usually be necessary to choose a distant viewpoint to keep the tilt of the camera as small as possible. At the same time, a violent tilt from a close-up viewpoint may be effective.

The subject should be framed if possible—e.g., by shooting from between trees or foliage, or the bold shapes of other buildings—to give a feeling of depth and distance. A foreground figure silhouetted against the light is an alternative device for imparting space.

Exposures with this type of subject vary with the intensity of the floodlighting and the colour of the building. Average conditions call for about one second at *f* 8 for white lighting; when the lighting is coloured, it is generally weaker, and will require up to four times the white light exposure.

Colour. It is possible to make colour pictures of after-dark scenes which include artificial lighting, but it is difficult or impossible to get the colour values exact. If the lighting is a mixture—as it usually is—of tungsten, fluorescent and gas discharge light sources, no single type of colour film can reproduce them all faithfully at one and the same time. The result, whether you use colour film corrected for artificial or daylight, is bound to be a compromise. This does not always mean that the result will be unpleasant, but it will not be an exact reproduction of the colours of each light source.

The problem is worse when there is still some daylight left in the sky and you want to reproduce it as well as the street lighting, advertising signs, lighted house and shop windows and so on. A professional photographer will generally use artificial light material and concentrate on getting good, clear colours in the man-made lighting. Amateurs usually want to shoot only one or two frames on this kind of subject and are not prepared to load up with an A-type colour film when the greater part of it would be wasted. Those who shoot off a few night exposures on a daylight colour film must put up with a certain amount of colour distortion, however carefully they gauge the exposure.

The extreme lighting contrast of night photographs also means that the dark parts of the picture are invariably underexposed—they are, after all, night. Similarly, light sources in the picture are over-exposed, While this does not necessarily detract from the pictorial effect, it does mean that on colour film a certain amount of distortion of the shadow and brightest tones occurs. With long exposure times, reciprocity failure effects lead to further colour changes. In practice, the results are usually still acceptable, especially as few people can accurately remember—or even reliably see—the colour values in a dimly lit scene.

Paradoxically, accurate exposure in night photographs is not so critical, because the exposure level mainly affects the degree of gloom or brightness in the final colour transparency. Accordingly, the effects obtained vary with the exposure, but the result is an authentic night shot over a surprising range of exposure levels. Without this pseudo-latitude successful night photography would be impossible.

Rain. Taking photographs in the rain at night is an unpleasant procedure, but it is always worth while going out with a camera immediately after a shower when the wet pavements and roads reflect all the lights of street lamps, shop windows, advertising signs and so on. The pavement reflections also form a bright background to figures and objects that would otherwise be lost. The extra reflected light enables exposures for most subjects to be cut by about a half.

Snow. A covering of snow is an even better reflector which also adds a touch of novelty to the most ordinary scene. Powerful lights near the camera should be avoided as they will create a glare in the foreground that nothing else in the picture will be able to compete with. Shooting during a snow-fall gives the picture a misty, soft-focus appearance.

Exposures can be cut to half the normal value or even less.

The use of a lens hood is desirable to protect the lens against falling snow.

Mist and Fog. Both these conditions affect the subject in the same way: they obliterate distant lights and put a luminous halo around the ones nearer the camera. So they tend to isolate and simplify the subject and lend it an atmosphere of mystery or unreality. Here again, a well-placed figure can often have a dramatic impact.

There is no need to increase the normal exposure—in fact, particularly in mist, it may be possible to reduce it. This is because the mist acts as a reflector and actually increases the illumination of the shadows. F.P.

See also: *Faults; Flash technique; Light sources in the picture; Moonlight; Shop windows.*

NITRAPHOT. Slightly over-run type of tungsten filament lamp. Similar to a large size (No.2) over-run Photoflood but having a longer working life.

See also: *Filament lamps.*

NITRATE FILM. Standard 35 mm. commercial cine film which used to be coated on a base of cellulose nitrate. It is no longer used because it was highly inflammable.

See also: *Supports for emulsions.*

NITRIC ACID. Aqua fortis. Used in dish cleaners, and is the etching mordant for zinc and magnesium in photomechanical work.

Formula and molecular weight: HNO_3; 63.

Characteristics: Highly corrosive yellowish liquid (colourless if pure).

Solubility: Mixes with water in all proportions.

NODAL POINT. A ray of light enters a lens at a definite angle with the optical axis and leaves the lens also at an angle with the axis which is, in general, different from the first. However, sometimes the angle for the entering ray is equal to that of the emergent ray. For this condition the position where the entering ray would cut the optical axis is called the first nodal point, and the position on the axis from which the emergent ray appears to come is called the second nodal point.

In general when there is the same medium on both sides of the lens, usually air, the nodal points are where the principal planes cross the axis; but in cases when there are different media surrounding the lens, as in an oil-immersion microscope objective or the human eye, the nodal points are not in the principal planes of the lens.

See also: *Lens.*

NOISE POWER SPECTRUM. The fluctuations in a photographic image which are due to granularity may be represented by a density distribution with the characteristics of a random waveform. Fourier analysis of this waveform into spatial frequency components leads to the noise power spectrum. This term originates from the field of electrical communications, and the name *Wiener* spectrum is sometimes alternatively used.

See also: *Information in the image.*

NORWAY. Norway has not contributed to the invention and development of photography, nor has the country any photographic industry. The population is only 3 millions and the basis for such an industry would be too small to be firmly established.

The Past. The first daguerreotypes made in Norway date from 1842—three years after Daguerre made his invention public. It is known that in 1842 a daguerreotypist named Knudsen settled in Oslo, the capital of Norway (then Kristiania). From 1843 there are daguerreotypes from Bergen signed Iversen, and these two may be considered as Norway's first portrait daguerreotypists. These was also a calotype portrait photographer, H. T. Winther, who opened a studio in Oslo in 1842.

The most prominent portrait photographers using the wet plate process were L. Szacinski and O. Nyblin of Oslo (Kristiania). The former, who was an immigrant Pole, was also by appointment photographer to the Royal Court.

Amateur photography has developed greatly during the last decades. The first known amateur photographer was the clergyman, Ole Tobias Olsen, who became interested in photography early in the 1860's. He used a wet plate camera of considerable weight and size, and with this equipment he made several thousand photographs of various parts of Norway. Most of these are of historical interest and are housed in the Norwegian Technical Museum at Oslo, where his photographic equipment may also be seen including his darkroom tent and complete wet plate outfit. Bearing in mind that Norway—although beautiful—is very mountainous and difficult to penetrate, especially when there were no railways, no automobiles and very poor roads, Ole Tobias Olsen's performance is unique not only from a photographic point of view, but also as an athletic achievement.

The first amateur club, Amatørfotografen (The Amateur Photographers), was founded in Oslo in 1888 and lasted until 1907. The members used mostly dry plates. It numbered about 100–150 members, the best known being Frithjof Nansen, the Norwegian polar explorer and scientist, in later years famous for his international work for the refugees after the first World War on behalf of the League of Nations.

Present-day Activities. In 1921 a new amateur club was founded, named Kristiania Kamera Klub (now Oslo Kamera Klub). Membership has grown from 100–150 to over 700. In 1927 this club together with two other clubs, Rjukan Kamera Klub and Fredrikstad Kamera Klub, founded the Norsk Selskap for Fotografi (Norwegian Photographic Society) which is the central organization of Norwegian camera clubs.

The number of clubs is now seventy-seven and the total membership about 4,000. The Norsk Selskap for Fotografi publishes the journal *Kamera*, and arranges yearly photographic competitions. The N.S.F. is a member of F.I.A.P. and of P.S.A.

The professional photographers in Norway are organized in Norges Fotografforbund (The Norwegian Photographers' Association) founded in 1894, now numbering about 300 members. Their official publication is the *Norsk Fotografisk Tidsskrift.*

Photography is not introduced as a subject in any of the high schools or universities of Norway, but the university of Oslo has a laboratory for photographic research. Photographic education is included as a special subject in the Craftsman Schools (for apprentices) at Oslo and Trondheim and the Arts and Crafts School at Oslo. Moreover, members of the Norges Fotografforbund lecture on photography, especially to candidates to the Master Craftsman Certificate.

Pictorial photography, which has developed considerably in recent years in Norway, is following the general world trend. Norwegian photographers are active in international exhibitions and competitions. The country's largest camera club, the Oslo Kamera Klubb, celebrated its fortieth anniversary in 1961 by holding an international exhibition in the Museum of Applied Art in Oslo. R.Mo.

NOTCH. Sheet film is notched to ensure loading the material into the film holder with the emulsion facing the right way. When the notch is at the top left-hand side of the sheet, the emulsion side is facing away. Various shapes of notching indicate different types of emulsion.

See also: *Sizes and packings.*

NUCLEAR PHYSICS. The photographic detection of atomic particles, such as electrons, protons, alpha particles, neutrons, mesons and others has greatly contributed to our fundamental knowledge in atomic and nuclear physics. With nuclear track plates it is possible to trace the paths of fast charged particles, when these travel through the photographic emulsion. They leave rows of developed silver grains, i.e. tracks in the layer which can be identified under microscopic inspection.

History. O. Mügge (1909) was the first who observed charged particle tracks in photographic emulsions and Kinoshita (1910) working in Rutherford's laboratory made systematic studies of the photographic recording of alpha particle tracks. M. Blau and H. Wambacher (1920–1930) investigated the use of photographic emulsions for protons and for nuclear disintegrations. The climax of this development was reached when in 1945–9 C. F. Powell and his collaborators in Bristol refined the technique by using a new type of nuclear track plate supplied by the photographic manufacturers and discovered the existence of pi-and mu-mesons. Powell received the Nobel prize for this work.

Types of Elementary Particles. A proton is the nucleus of the hydrogen atom with a positive charge of unity.

An electron has a negative charge with a mass 1840 times smaller than that of the proton.

A neutron has no charge and is one of the building stones of the nuclei of atoms. It has practically the same mass as the proton.

The alpha particle with its double positive charge is the nucleus of the helium atom; it has 4 times the mass of a proton.

Mesons are also elementary particles and there exist 11 varieties of them; these differ in mass, charge and some are uncharged. They have masses 200–1000 times that of the electron. Mesons are believed to be responsible for the exchange interactions between protons and neutrons in the atomic nucleus.

When a fast charged particle passes through the silver bromide grains of a photographic emulsion it loses energy by the ionization it causes within the grains. The nature of the tracks depends on the number of ionizations the particle causes per unit length of path. The rate of production of ions by a particle is proportional to the square of its charge and inversely proportional to the square of its velocity. Hence a heavy particle of a given energy, for example an alpha particle with its double charge and its relatively low speed because of its great mass, produces so many ionizations that it renders all grains developable along its path. A proton of the same energy has a higher velocity because of its smaller mass and leaves a few gaps between the developable grains in its initial path. Towards the end of the trail, when the proton is slowing down the grain density increases. Hence different particles may be distinguished by the change in grain density per unit length of path. Furthermore the discrimination is often facilitated by the amount of scattering a particle suffers due to interactions with atoms through which it passes.

A neutron, having no charge, causes no ionizations and leaves no tracks in the emulsion. If it collides, however, with a hydrogen nucleus in the gelatine of the emulsion, this may take up the energy of the neutron and from the lengths of the proton tracks thus obtained, conclusions can be drawn as to the energies of the colliding neutrons.

Nuclear Track Plates. One of the outstanding properties of nuclear track plates is their proportion of silver bromide to gelatin which is much higher than in normal emulsions. Hence the grains are very closely packed, thus permitting the formation of continuous rows of grains. In order to record the full range of particles, emulsion coatings may be up to 1 mm. thick. As very high energy particles may have ranges of the order of several centimeters in the emulsion stacks of plates may be superimposed. Such stacks of up to 50 cm. thickness have been carried in balloons in tests reaching an altitude of 100,000 feet. In this experiment 300 kg. of silver bromide emulsion was used and 20 tons of hypo was required for the fixing of the emulsion.

Processing. The processing of nuclear track plates having emulsion thicknesses up to 50–100 microns is almost conventional apart from slightly increased bathing times. Usually metolhydroquinone developers are used.

For plates of 200, 600 and 1000 microns, the so-called temperature cycle method is used. The plates are soaked in the developer solution at 5°C. so that the developer can diffuse to the depth of the layer without acting on the latent image. The temperature is then raised to about 25°C. and development takes place simultaneously throughout. The soaking and development time may, depending on emulsion thickness, take up to 10 hours; the fixing time 100 hours and washing including drying up to 10 days. A 1000-micron thick emulsion layer thus requires about a fortnight for its complete processing.

Inspection. For the detailed inspection of tracks formed in the emulsion high power microscopes with oil immersion objectives are required. In order to obtain a pictorial record of three-dimensional nuclear events a great number of photo-micrographs have to be made at different focusing levels and several hundreds may be required for one event. Finally the various photomicrographs are arranged in a mosaic.

Cloud-, Bubble- and Spark Chambers. Other detectors of charged particles which use photographic recording methods are the well established Wilson cloud chambers. In these the particles leave a visible track of condensed droplets in passing through super saturated water vapour. The trail of drops is photographed by synchronized electronic flash with a stereoscopic camera.

More recently huge and very powerful detection instruments have been developed such as the bubble and the spark chambers. In the former liquid hydrogen or propane is used at a temperature above the normal boiling point of the liquid, but boiling is prevented by applying a high pressure. This is reduced for a short time so that the liquid remains super-heated. If a charged particle passes through the super-heated liquid it causes a local heating along its path thus forming visible bubbles for a few milliseconds. These can be photographed by illuminating the bubbles in a dark field. In one investigation 50,000–100,000 pictures may be needed with two cameras to select and reconstruct nuclear events in space.

In the spark chamber the track of a pre-selected particle reveals itself as a train of sparks caused by an electrical breakdown between plates. This is triggered by the ionizing particle.

Spark track photographs can be taken from mirrors set at 45- which allow a three-dimentional reconstruction of the event.

Autoradiography. The photographic emulsion, e.g. fast X-ray film, is often an alternative choice to the use of Geiger counters particularly when very weak radioactive samples have to be measured and when many samples have to be measured simultaneously. R.H.H.

See also: *Autoradiography; Bubble chamber photography; Dosimetry.*

Books: *The Study of Elementary Particles by the Photographic Method*, by C. F. Powell, P. H. Fowler, and D. H. Perkins (London); *Nuclear Physics in Photographs*, by C. F. Powell and G. P. M. Occhialini (Oxford).

NUDE PHOTOGRAPHY

In photographing the nude there are two particular difficulties to overcome: one, the naked presence of human body and the built in realism of the photographic process. The mechanical aspect of photography contrasts strongly with all the other graphic arts and the basic manual skills they involve. For this reason, at earlier periods, attempts were made to suppress blatant reality via various special printing processes such as bromoil, bromoil transfer, pigment ana gum bichromate, and the like. Although perfectly valid in their own way, these processes led to increasingly laboured and unsatisfactory imitations of Romantic or impressionist painting, purely to escape the lens's "reality", not only in the hazardous field of nude photography, but also in many others.

The lens of course "sees" objectively. The essence of the camera and at the same time its most obvious application is the straightforward record. The lens registers everything before it indiscriminately. The human eye, however, is different. It corrects and selects according to requirement, even governing its own exposure. It freely appraises and arranges, in an effort to bring the elements of choice, emphasis and suppression to the various parts of the scene.

The close relationship between eye and lens reveals the futility of borrowing techniques from the other graphic arts. Through this partnership of precise optical reproduction and human selectivity, still photography and, in particular, the cinema have become a peculiarly typical medium of expression of our time. In the hands of the talented, the camera becomes an integral part of life, portraying facts in detail but colouring them with vision. Yet the understanding of its operational intricacies is far from simple, requiring much ability, taste and discrimination, and in no sphere of photography is this more true than in nude photography.

Nude photographs have a right to existence and publication only if they satisfy basic aesthetic and pictorial requirements. For this reason they are considered difficult even from a professional point of view. The photographer himself therefore has to be his own best critic, and uncompromising in his attitude to everything which detracts from the artistic image of the picture or which could be held to be sheer eroticism.

But the nude does not need justification as it presents anyone of reasonable artistic sensibility with one of the most poetic motifs conceivable. Approached in the right spirit and with developed critical judgment, the nude should create no more problems than any other photographic field.

Equipment. Formerly there was no question that the best negative size for nude photography was also the biggest. Even today the professional who still prefers the large-format camera for studio work has every reason for his choice. The big ground-glass allows easy control of detail. The resulting contact prints or colour transparencies require the minimum extra work for reproduction purposes. Furthermore, modern cameras of the large-format type have become handier and more streamlined, and weight and size are no longer serious obstacles for their hand-held use in fields demanding more mobility.

Many specialists in nude photography prefer the $2\frac{1}{4} \times 2\frac{1}{4}$ ins. twin- or single-lens reflex camera. The construction of these cameras requires the square negative for optimum use of the design's potential, and this apparent necessity in fact proves in many ways a virtue. The square format is adaptable and permits the choice of vertical or horizontal arrangements at will. The negative is big enough to allow sectional enlargements to be made from it without noticeable loss of quality, yet at the same time the camera is compact and simple in operation. Composition and sharp focus are facilitated by the bright ground-glass screen. The fact that the screen image is reversed from left to right is no real disadvantage, and one gets used to it very quickly in practice. If necessary, pentaprism devices are available, which will give a laterally correct, upright image. The 6×6 cm. reflex is in fact the ideal instrument for most general work, and the negative a good compromise between large and small.

35 mm. technique has been transformed by the arrival of ultra-fine grain, thin-emulsion films, with their excellent resolution and contour sharpness. Their appearance enabled the miniature camera to compete on far more favourable terms with larger cameras, while retaining its own special advantages. The changes brought about were even more widespread than those heralded by the earlier introduction of fast emulsions, and affected the development of nearly all black-and-white films in all sizes. It is, however, a paradox that the thin-emulsion film provided some of the best-known photographers of the "nouvelle vogue" with an excuse for expressing their "disgust at so much sharpness" and producing pictures, among them nudes, in which graininess became a virtue.

The thin-emulsion film really comes into its own in combination with the miniature single-lens reflex, with automatic diaphragm and exposure coupling, pentaprism and instant-return mirror, an instrument which has gained tremendous popularity since the beginning of the sixties. The single-lens reflex finder has many advantages, and probably is the nearest approach to the ideal. It gives a large, bright, upright and laterally correct image, totally devoid of parallax, and showing the exact field and sharp focus of any lens without the need for an appropriate additional finder.

These developments have made the present-day 35 mm. camera almost universal in its scope. In the field of nude photography, especially by available light, its outstanding advantage lies in its ease of handling and complete mobility. But it is a precision instrument demanding a specialized and disciplined approach if above-average results are to be achieved, and slipshod work will merely produce pictures which might as well have been taken with the simplest snapshot camera. Many current 35 mm. cameras embody automatic exposure control. This is quite all right for casual record photography, but in order for it to be acceptable for serious work, there must be provision for manual operation as well.

Choice of a camera depands, of course, not only on technical specification, but also on one's own personal preference and custom. A reasonable basic outfit could consist of the following: a 35 mm. camera for available-light nudes, dynamic action, and all cases where quick shooting is necessary, using films (according to requirement) of any speed between 16 and 800 ASA; a single-lens reflex with normal, wide-angle, and medium-tele lenses would probably be ideal, and a $2\frac{1}{4} \times 2\frac{1}{4}$ ins. twin- or single-lens reflex expressly for colour.

Materials. Present-day miniature films of medium sensitivity have a smooth gradation, and can be used for most general subjects. However, in spite of their rather steeper gradation, ultra-fine grain emulsions are to be recommended whenever the finest detail is required. These slow, contrasty films must not be overexposed, and are best if uprated to the next higher sensitivity (for example, 20 ASA becomes 40). They should be developed in energetic but soft-working developers—the very opposite of super fine-grain formulae—which give maximum emulsion speed consistent with acceptable gradation. Such developers properly used with a correctly-exposed film of the right kind will soon demonstrate their efficacy in subduing excessive contrast, grain, and blocked-up shadows. At the same time they enhance contour-sharpness and graphic effect.

In colour, reversal films are best reserved for projection. Colour prints from reversal films only give satisfactory results as a rule from transparencies of soft, even hues.

The versatile negative colour process has much to recommend it. If the user has neither the time nor the equipment to carry out the processing himself, with its rather complicated filtration steps, he would be well advised to work with a recognized processing laboratory. This is certainly not cheap, but generally results in consistent and reliable prints which in the long run save time and trouble.

Models. Finding a model can be something of a problem, but not so fraught with difficulties as one might think. Certainly no one who is really serious about nude photography will consider it impossible to solve. It should not, for example, be too difficult to persuade your wife or girl-friend to pose, and members of your own family are always potential models. If the results are good, they will probably be exhibited or at least shown to friends, and this may well bring in more recruits. Before you take any photographs, however, it is important to explain to your model exactly what you intend to do with the pictures, especially if there is a possibility of their being published. In this case the model's permission is essential (and, if she is under legal age, her parent or guardian's too), preferably in the form of a written model release.

Working with professional models is always a pleasure. Although they may not necessarily know the ultimate purpose of the photographs —something which is primarily the business of the photographer, they are quicker to grasp an idea and interpret and react to instructions, and this saves time and effort. Make-up, hairstyles and accessories are the tools of their trade, together with the ability to achieve absolute accuracy in pose and expression before the camera, the prerequisite of carefully arranged artificial-light nude work in the studio. The open air nude, on the other hand, is generally far more relaxed and unposed.

The attractiveness of a young girl often lies purely in her personal charm. This makes it difficult to set out a list of desirable attributes. Furthermore it would be folly to expect to find perfection in shoulders, arms (especially), hands, legs, breasts and back in most women. Such combination is rare, and each model usually has an imperfection of some sort.

The photographer of the nude has a more difficult task than the painter. He must be more critical of his model and more able to recognize her particularly attractive features. He cannot alter in the same way as the artist with brush or pen; he must omit, arrange, distribute, emphasize and conceal. The model's appearance as a whole is of far greater importance than minor bodily deficiencies, and she must move well. In addition, the photographer needs at least an element of psychological skill in his directing ability.

The Nude Out of Doors. Nudity in a photograph must have a *raison d'être*. It must not contradict its surroundings. The most convincing, and perhaps the simplest, setting is out of doors. Today, as a reflection of human longing for fresh air, space, and sunshine, naturist camps and reservations have grown in popularity and number, and have made the notion of open air nudity readily acceptable.

To photograph your model in motion, you can simply allow her to wander about at will, while you follow her movements with the camera and expose at the right moment. But there is a likelihood that this will result in perhaps one breathtaking success along with thirty-five mediocre shots, which would strain even the most dynamic concept of miniature technique. So it is much better to keep everything under your control. The best kind of directing is an almost impreceptible prompting, so that the model herself does not know the precise instant of each shot. You can suggest she runs into the water or over the sand dunes. Directing need not be complicated, but it must be there, and the photographer should dictate the right moment rather than leave it to chance. The results should still be totally spontaneous.

In summer the most favourable time for photography is up to 11 o'clock in the morning and between 4 and 6 o'clock in the afternoon. Slanting rays of sunlight have a charm all of their own, which is equally valid for black-and-white and for colour film, with which the lower colour temperature gives a pleasantly warm cast. Strong sidelight is particularly effective with nudes, giving simple modelling to the lines of arms or hips, or delineating the thrust of shapely breasts. Lightening the shadows is mostly superfluous with black and white, but necessary for colour.

Working against the light usually guarantees striking effects in black-and-white, and also in colour photography, provided that measures are taken to brighten the shadows via white reflectors or fill-in flash. The against-the-light technique adds brilliance to almost any picture, with gleaming highlights side by side with dark shadows, and, in the case of nude photography, the naked form acquires an embracing luminosity. Strong contrasts and deep shadows appear, and the elements of the picture gain unity. When backlighting comes from water, it creates reflections all round the model.

Backlighting is also the most effective form of lighting when the nude figure is to be an integral part of a landscape. Precise exposure is essential. The old rule of thumb to give three times the exposure is a guide, but may also mislead. It is essential to decide what are the most vital parts of the picture; e.g., a girl's body which must remain dark apart from back-lit contours demands a different exposure from a model moving about in the sea. If you are in doubt, you should expose for the shadows, taking a close-up reading with the exposure meter, and shielding it from the direct rays of the sun. When necessary, the incident light method, as opposed to reflected, can be adopted.

Flat frontal lighting is still the safest for colour, but only used for black-and-white nude photography if particularly graphic effects are required with orange or red filters, or alternatively high-key effects in diffuse light with a blue filter.

Diffuse veiled sunlight is particularly valuable. It gives abundant illumination, especially by the sea, and when light mist or haze is present, backlit photographs taken in it acquire a beautiful luminosity. If the mist is fairly dense and a telephoto lens is used, to give a reasonably large image, the sun itself may be included in the picture. A lightly overcast sky can be ideal for certain kinds of colour work, and noticeably beneficial to skin tones. Cloudy conditions, however, are unsuitable for black-and-white, unless you are aiming at special effects, such as high key, since, in the normal way, the dull light will have an adverse effect on flesh tones, producing a rather ugly result.

One further vital point must be made in connection with the nude outdoors: it is essential to ensure that the model has no unsightly patches on her body caused by sunbathing in a swimsuit.

Indoors. Choice of motif is more limited in this case, but there are still many interesting possibilities, such as window lighting, bedroom and bathroom scenes, and the girl can be given something appropriate to do, such as arranging her hair in the mirror. Ordinary domestic surroundings, with tables, sideboards, armchairs etc. are but rarely happy choices; the nude in these surroundings looks quite out of place or even an object of ridicule. But the inspired photographer may be able to use even such unlikely contrasts for surrealist effects.

Large rolls of white or grey paper may be hung on the wall of a room to make a neutral background. A tripod is absolutely necessary, and any difficulties are likely to involve lighting.

Contrasts between the area of the window and the dark parts of the room are always considerable, unless a daylight studio with lighting from three sides and above is available. These contrasts need to be reduced, and the best method is the old professional trick of using reflecting screens and folding sheets of aluminium foil which can be moved about until the desired amount of fill-in illumination is achieved. Aluminium sheets throw a rather metallic gleam on the skin. Carefully controlled, this gleam is generally desirable, especially with slightly oily skin. If this is better avoided, however, particularly for colour work, large sheets of matt white paper or linen are preferable.

Artificial Light. The composed nude photograph almost invariably requires artificial light. The most usual and convenient form of illumination is obtained from Photoflood lamps. They are inexpensive and easy to control and position. Basic equipment for the average amateur studio might consist of the following: one general-purpose floodlamp, one diffused flood, and one spotlight. However, it is quite possible to make do with the diffused flood and the spotlight alone. Each must be mounted on its own movable stand. A considerable amount can be achieved with such a simple set-up.

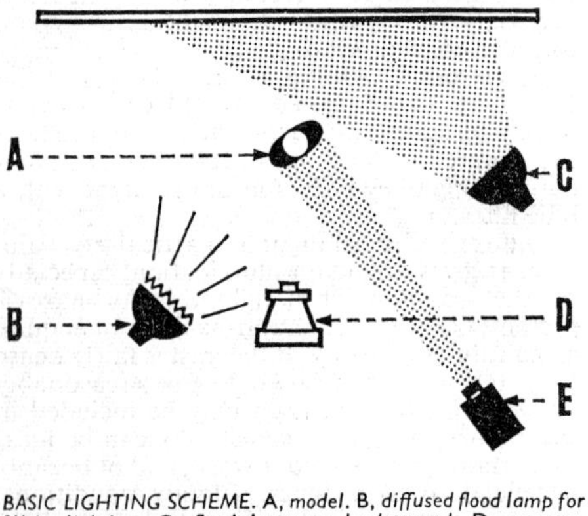

BASIC LIGHTING SCHEME. A, model. B, diffused flood lamp for fill-in lighting. C, flood lamp on background. D, camera. E, spot lamp for key lighting.

In positioning the lights, the photographer should bear in mind that there is nothing to be gained by using more than necessary. In fact, where two lights are enough, a third may well spoil the effect. But the third light can prove particularly useful for illuminating the background.

Both main lights should be kept fairly well back from the model, and they should never compete with each other. The principal illumination must come from one direction to give modelling, while the other lamp lightens the shadows.

As with natural light, aluminium foil sheets can often be used to good effect.

Background. In all indoor work the line formed at the angle of floor and wall can prove a nuisance. The problem can be solved with a roll of white or grey background paper roughly 6 yards long and 3 or 4 yards wide. This is hung from the wall in a gentle curve to the floor, and the model stands, sits or lies on one end. It can be illuminated according to taste, and forms what is known as a seamless background.

Colour film calls for fine pastel skin tones. Often, however, skin tones can be far from perfect in actual fact. A half-stop extra exposure will make them paler and more idealized. A good compromise is to use the half-stop extra with reflected illumination from light neutral grey paper, which will counteract any tendency for the colours to get too washed out. It is essential, of course, to see that the colour temperature of the lighting is correct for the film in use.

Flash. Electronic flash and bulbs can be used to give direct frontal light from the camera position, and this can prove reasonably satisfactory with colour. With black-and-white, however, it is better to have camera and flash apparatus separated by one or more extension cables or even slave units. Extension cables should preferably not exceed 9 feet in length. The flash is then virtually independent of the camera and much the same rules apply as for lighting with photofloods.

The simplest standard flash lighting for nudes is from above and to one side. The flashgun can either be mounted on a stand or an assistant can hold it.

Bounce flash is excellent for soft, even lighting. The flashgun is pointed up at an angle to the ceiling (which must be white or light-coloured), and the exposure increased by one to two and a half stops. A certain amount of trial and error may be necessary. This kind of illumination is virtually shadowless, and ideal for high-key work. It may be found worthwhile

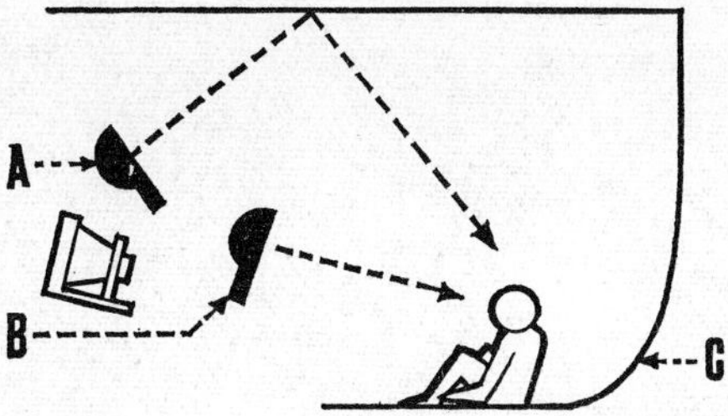

BOUNCE FLASH ARRANGEMENT. A, *bounce flash reflected from ceiling on to model.* B, *flash off the camera for key lighting.* C, *paper roll background.*

to supplement it, however, with direct light from a second flash head.

Here again the exposure must be found by experience.

A disadvantage with flash work is that it is difficult to judge the final effect of the lighting, although some idea may be gained by using tungsten lighting to simulate the flash. Therefore there are certain types of nude shot which suggest themselves more readily for flash, such as the model under a shower or in a foam or sauna bath. The short duration of electronic flash will freeze any movement and the light itself gives a good rendering of wet skin, while the inherent nature of the subject promises spontaneity.

It is possible to use mixed flash and tungsten illumination for black-and-white work, but it is more difficult with colour, because of the different colour temperatures of the two forms of lighting. Colour conversion filters (on the lights) and considerable experimenting may be necessary.

Flash is useful as a fill-in light out of doors for against-the-light shots. The brilliance of the flash must be reduced, however, either by varying the flash-to-subject distance or by using a handkerchief or similar cloth in front of the reflector, otherwise the picture will look completely unnatural because of the predominance of frontal lighting. H.R

See also: *Glamour photography; Hands; Lighting the subject; Models; Portraiture.*
Book: *The Photographer and the Nude,* by H. Rittlinge (London).

O

OBERNETTER, JOHANN BAPTIST, 1840–87. German photographer and photographic manufacturer. Introduced commercially, 1868, under the name of Aristotypie the collodion silver chloride paper of G. W. Simpson, and invented *c.* 1886 a photogravure process on copper plates called Lichtkupferdruck. Obernetter also introduced *c.* 1870 a collotype process similar to Josef Albert's. In the 1880s his firm made good orthochromatic dry plates.

OBJECTIVE. Lens, in an optical instrument, that normally faces the object. More often applied to the lenses of telescopes and microscopes than cameras.

OCUP, PETER ADOLFOVICH, 1883–1963. Russian press photographer and Master of Photographic Art. Established a portrait gallery of famous writers and artists, including Tolstoi, Chekhov and Gorkij. Made photographic documentation of Russo-Japanese War 1904–5, and of the Revolution. Took a series of portraits of Lenin, collected under the title Leniniana. Received the Order of the Red Banner of Zabour and the Order of Lenin.

OFFICE COPYING Photographic method of making same-size copies of original documents, letters, drawings, etc. There are various processes in use, utilizing either silver halide systems or electro-photographic, heat and other systems. Office copying machines are usually designed for simplest possible handling, with automatic exposure control and processing, to allow use by comparatively unskilled staff.

See also: *Document copying: Document copying equipment.*

OHM. Unit of resistance in electricity, often denoted by the symbol Ω. For convenience, other units also used are the kilohm (1,000 ohms) and the megohm (1,000,000 ohms).

See also: *Electricity.*

OIL PRINTING. Printing process introduced in 1904 in which oil pigment was applied, by brush, to a gelatin relief image.

See also: *Printing processes (obsolete).*

OIL REINFORCEMENT. Oil pigment can be used to reinforce the tones of prints on matt or rough surfaced papers. The method makes it possible to alter the tonal range of the print. Shadows can be made deeper in tone or highlights made lighter, or too insistent details suppressed. Many photographers question the use of such after-treatment on the grounds that it is not a photographic process, but for those who wish to experiment, here is the procedure:

The print must be quite dry for this work. It is laid face up on a smooth surface and the emulson side is rubbed over with print dope. This can be bought ready prepared or made up from a mixture of:

Turpentine	2 ounces	50 c.cm.
Mastic varnish	1 ounce	25 c.cm.
Linseed oil	1 ounce	25 c.cm.

The materials must be of good quality and purchased from an artists' colourman.

After the print has been treated with this medium, care being taken to go over the whole surface, the pigment can be applied in the form of good quality artists' oil colours or those supplied for the bromoil process.

The colours must harmonize with the colour of the print. Brown-black suitable for bromide prints on warm tone papers can be made by mixing black with a little burnt sienna.

A little of the pigment is squeezed from the tube on a sheet of clean glass and mixed with a small quantity of the medium. The mixing is done with the flat blade of a knife and care taken not to have the colour too strong.

A plug of cotton wool is rubbed into the mixture and lightly rubbed on to the print to apply the colour as evenly as possible. A second piece of clean cotton wool is then used to work the colour well into the print, and to remove any surplus.

The colour should be even; if there are patches of uneven density or colour the whole will have to be wiped off with cotton wool saturated with the medium or pure turpentine, and a fresh start made.

Any modification of the print may then be proceeded with. Using a tuft of cotton wool lightly moistened with the medium, the highlights that are to be reduced are wiped over. Any points of light, such as shafts of sunshine or the catch lights in the eyes of a portrait, are brightened up with cotton wool wrapped round the point of a small stick. Cloud form can be emphasized by using larger pieces of cotton wool. Any sharp outlines left can be smoothed out by gentle rubbing with clean wool. Local areas can be made darker with more pigment applied on cotton wool on a pointed stick.

R.M.F.

See also: *Doping prints; Retouching.*

ONE SHOT COLOUR CAMERA. Special type of camera used for making three-colour separation negatives with a single exposure.

See also: *Colour camera.*

OPACITY. Anything which transmits (e.g., a photographic silver deposit) 1/xth part of the light falling on it, is said to have an opacity of x. The reciprocal 1/x is called the transparency, and the logarithm log x is known as the density.

See also: *Density.*

OPAL GLASS. Glass which has a translucent opal tint throughout its thickness, as opposed to flashed opal which is opal on one surface only.

It is used as a diffuser in enlargers, photometers and other optical systems.

OPAL LAMP. Ordinary filament lamp with an opal glass bulb. When maximum diffusion is required, this bulb is better than those which are merely sprayed or frosted on one surface.

OPALOTYPE. Obsolete, once popular, type of print made by transferring a carbon process image on to translucent opal glass or by printing direct on to a bromide emulsion ready coated on opal glass. Such prints were often finished by mounting them on plush in gilt frames or with elaborate coloured borders.

See also: *Printing processes (obsolete).*

OPEN FLASH. Flash technique in which the camera shutter is held open while the flash is fired. In open flash, the duration of the exposure is governed by the burning time of the flash bulb or flash powder.

See also: *Flash synchronization.*

OPHTHALMIC PHOTOGRAPHY

Elementary ophthalmic photography is within the province of nearly every medical photographer. When the needs of the ophthalmologist alone are to be met, certain specialized cameras and apparatus rather than routine studio equipment are essential.

Choice of general equipment is largely determined by the areas to be covered which, in the main, range from head-and-shoulders to part of a single eye. By virtue of this restricted range, much useful work can be done with 35 mm. cameras and materials; many small units favour this approach since the use of monochrome and colour reversal stock almost eliminates darkroom work. The following notes apply equally to large or small negative sizes.

Success in ophthalmic photography depends upon attention to the following four points:

(1) Comfortable, but firm fixation of the head of the patient and in turn the direction of gaze of the eye. The former is achieved by a headrest; the latter by providing a fixation point—usually a small, bright, movable light.

(2) The use of intense short duration illumination (i.e., flash) to minimize discomfort, arrest movement, and permit the use of the smallest possible lens aperture. The last two features are particularly important when working at a scale of 1 : 1 or greater, as will often be the case.

(3) The control of reflections. The surface of the eye presents a wet, shiny surface which inconveniently reflects an image of any light source directed upon it or any bright object placed in its vicinity. The aim is therefore to use a single light source in dim surroundings; the size and position of the single reflex which results is carefully adjusted to avoid obscuring a point of interest.

OPHTHALMIC TABLE. Universal table for ophthalmic photography with head rest for patient and central geometric stand for positioning camera equipment.

(4) Whenever possible, photographs should be made to a definite scale, which will depend upon the area involved. Thus scale may vary from a reduction of 1 : 3 for a head and shoulders to a magnification of 3 : 1 in the case of a single eye. These figures relate to a finished print size of $4\frac{3}{4} \times 6\frac{1}{2}$ ins. which is usually adequate.

Sensitized materials employed in ophthalmic photography are strictly conventional; uses are to be found for 35 mm. colour materials. In the larger sizes medium speed orthochromatic film is recommended for general purposes, with high speed softer contrast panchromatic film on occasions.

General Technique. External photography of the patient falls broadly into the following categories, on which choice of equipment is based:

(1) Head-and-shoulders—to show general appearance, head tilt or ocular torticollis.

(2) Face only—to show distribution of lesions in relation to the eyes. Here the avoidance of distortion is of importance.

(3) Both eyes—mainly to compare one eye with another or to demonstrate a bilateral lesion in moderate detail only. This is also the standard pose for eye movement studies in which the nine positions of gaze are frequently recorded.

(4) Single eye: a close-up view of one eye only is invariably required to record adequately all but gross lesions of the external eye and anterior chamber. A magnification of at least $\times$ 2 to $\times$ 3 should be presented in the final print, and it is therefore desirable that at least half this magnification is obtained at the negative stage.

(5) Macrophotography of the eye: it will sometimes be necessary to make photographs of a portion of the eye to simulate the appearance through a low-power microscope. This is better achieved by using a short focus lens at great extension rather than through a compound microscope.

All the above work may be carried out with any camera capable of triple extension—by means of bellows or extension tubes—together with a range of lenses. For example, a 4 $\times$ 5 ins. camera should be provided with lenses of 2, 5 and 8 ins. focal length, whereas 35, 50 and 90 mm. lenses will be found most suitable for use with a 35 mm. camera.

A single lens reflex is probably ideal for these purposes, but conventional film and plate cameras will serve almost equally well. As camera requirements may be expressed in terms of a series of fixed extensions, much can be done with simple fixed-focus equipment which facilitates standardization of results.

Other necessities comprise a specially constructed table on which are mounted a headrest for the patient at one end, and a movable geometric stand which acts as the camera support. This latter may be slid on guide rails the whole length of the table and thus, with a range of lenses, all the areas referred to above can be embraced by this excursion; the stand itself also provides fine adjustment of camera position in three planes.

Lighting. As already indicated, some form of flash illumination should be employed. As such sources are likely to be used in fairly close proximity to the eye, they must be well shielded.

A single 500 watt focusing spotlight is sufficient to provide general illumination for setting up and focusing as well as for cinematography when the use of a single light source is desirable. If electronic flash is chosen as the exposing source one flash head capable of dissipating 200 joules is essential and two may be an advantage on occasion.

Reflectors should be detachable to restrict the effective size of the light source at close quarters.

At all times the source of illumination should be kept at the greatest possible distance from the subject consistent with correct exposure at a nominal lens aperture of *f* 32. With a medium-speed orthochromatic material this distance will vary from about 3 feet for a whole head to 15 ins. for a single eye and even less for macrophotography.

Modern battery/mains operated electronic flash units having compact flash heads have sufficient light output for photography of the eye in colour, provided they are used close to the subject; moreover, their spectral emission is well matched to daylight colour film. Marked lens apertures will now range from *f* 16 to *f* 11 as the position of the light source—held, say, at the lens panel—and the degree of camera extension vary over distances from 1 to 3 feet.

Conditions more nearly approaching the desiderata outlined above may be attained by the use of expendable flash bulbs; this method is, however, relatively expensive.

Cinematography. The motion picture has a special place in ophthalmic photography. In the studio it is used for the study of eye movements, pupillometry and in the recording of such phenomena as nystagmus. In many instances use may be made of the time-compression and time-extension properties (fast and slow motion) of this medium.

In the operating theatre cinematography can record and reproduce a view of surgical procedures which is otherwise seen only by the surgeon himself.

The use of two 500 watt spotlights, each 6 feet distant, permits exposures on artificial light 16 mm. reversal colour film at 16 f.p.s. using a lens aperture of *f* 5·6. This level of illumination does not distress patient or surgeon and does not promote undue drying of exposed surfaces. For such work, the camera should be mounted about 30 ins. vertically above the face of the supine patient and roughly on a level with the chin: at this distance a 15 mm.

lens will accept a general view and a 102 mm. lens enables the frame to be filled by a single eye. Apart from these special points standard cinematographic practice is observed.

Retinography. Photography of the fundus oculi can, as yet, only be undertaken with specialized equipment such as a fundus camera. Briefly, the periphery of a large aspheric objective directs near-axial illumination through the pupil of the eye, thus illuminating a portion of the retina. The emergent rays are brought to a focus by the central portion of the objective to form an aerial image of the retina which is picked up and magnified by a small reflex camera integral with the instrument.

Light sources for this type of work have varied from the carbon arc to an incandescent lamp with an annular filament. The latest development in this respect is the use of electronic flash.

Keratography. This is a technique for producing photographs which provide information regarding the surface and curvature of the cornea. By suitable calibration of the photo-keratoscope which makes these records, the radius of curvature of different parts of the cornea may be directly measured. The method is also of value in the diagnosis of early keratoconus.

Essentially, a large illuminated target is placed a short distance in front of the eye under examination; a fixed-focus camera whose lens mount protrudes through the centre of the target, records the image of this target, which is mirrored by the cornea. The system is so arranged that when the target, which carries a geometric pattern, is at its optimum working distance the reflected image will also be in sharp focus at the film plane. Commonly the illuminated target consists of a series of concentric rings which, when reflected by a perfectly spherical surface of appropriate radius, appear to be equally spaced and uniform in thickness. In keratography, the size of the rings bears a definite relationship to the radius of curvature of the cornea and local surface irregularities are indicated by breaks in the rings or distortion of the image.

Various other experimental techniques have been tried from time to time. Stereoscopic photography, for example, has its applications and ingenious cameras have been designed for this purpose. This technique has been used to photograph the angle of the anterior chamber as seen through a gonioscope.

In gonioscopy, however, as in the recording of biomicroscopic appearances with slit-lamp illumination, the ophthalmic artist has pride of place. P.H.

See also: *Medical photography*.

Books: *A System of Ophthalmic Illustration*, by P. Hansell (Oxford and Springfield, U.S.A.); *Longmore's Medical Photography* (London).

OPTICAL AFTERTREATMENT. Optical methods of negative aftertreatment are used for adjusting contrast and for intensification.

Copying to Adjust Contrast. If a negative lacks contrast, one way out of the difficulty is to make a duplicate and add extra contrast in the process of duplication. The same process can be used to produce a duplicate with a lower contrast than the original if the original negative has too much contrast. This avoids ruining the negative during aftertreatment.

The duplicate negative may be made by contact printing or copying in the camera. In each case it is first used to produce a positive transparency from which the duplicate is obtained by repeating the same process.

The contrast of the original negative is decreased in the duplicate by using a soft gradation lantern plate (for contact printing) or a soft negative material in the camera (during copying). A soft working developer for the intermediate positive and/or the duplicate negative will have the same effect.

The contrast of excessively soft negatives is increased by using a hard lantern plate or process material in the camera, and/or developing in a contrast developer of the caustic hydroquinone type. This is the method often adopted for making tone separation negatives.

Copying for Intensification. Grossly underexposed negatives often have only a very faint ghost image in the shadow regions. This is much too thin either to print in the normal way or to respond to intensification.

But even a thin image shows up clearly if it is viewed against a dark or black background by oblique lighting. The visual effect of the image is considerably intensified in this way.

This effect can be utilized by copying the negative by this type of lighting. The principle of the lighting set-up is similar to that of dark ground illumination in photomicrography.

The negative to be copied is mounted in a frame with an aperture just large enough to

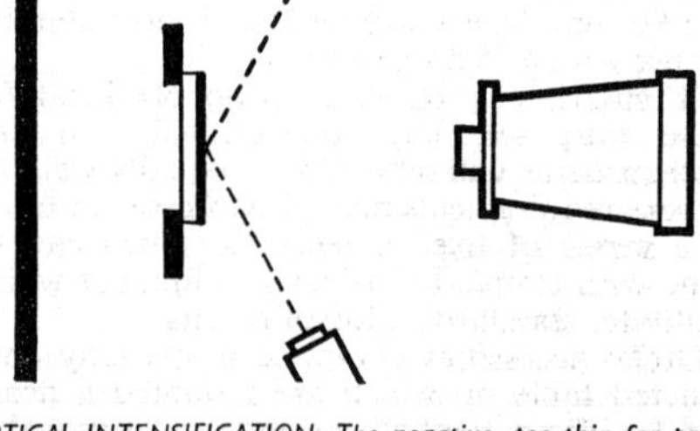

OPTICAL INTENSIFICATION. The negative, too thin for printing, is halogenized, and copied in front of a black background, by oblique illumination. The copy negative shows greatly improved shadow detail brought out by scattered light.

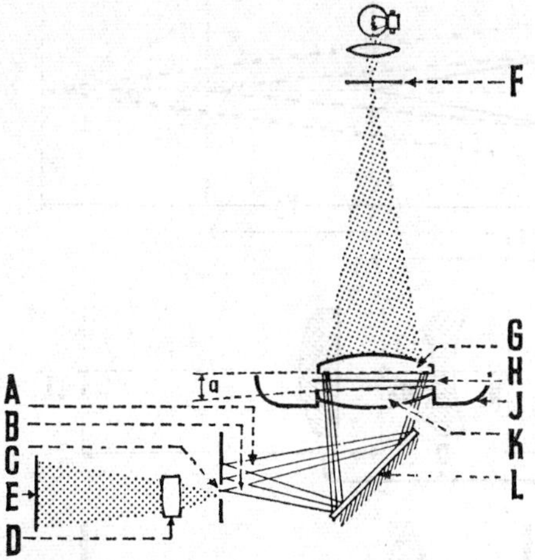

OPTICAL INTENSIFIER LAYOUT. The negative is placed between the semi-reflecting plane surfaces of two plano-convex lenses at a small angle (a) to each other. A light beam projected through this system forms, in addition to a primary image, a series of successive images due to inter-reflection between the plane lens surfaces. These successive images largely overlap and reinforce the primary image and can, by positioning a slit, be projected on to the printing paper as an optically intensified image. The number of intensifying reflections possible is limited by the loss of definition occasioned through incomplete superimposition of the reinforcing images. A, first converging point of image beam. B, final converging point of image beam. C, screen with aperture. D, projection lens. E, printing paper. F, diaphragm in front of light source (point source). G, upper condenser lens with semi-reflecting plane surface. H, negative. J, oil bath (to eliminate additional interface reflections). K, lower condenser lens. L, movable mirror to locate any of the beam converging points in the screen aperture C.

take the negative. The frame is set up a short distance in front of a black background—e.g., a piece of black velvet, or a door opened on to a darkened room—and illuminated by a pair of spotlights, one pointing from each side of the negative and striking it at an oblique angle from the front. The camera is set up opposite the negative so that it looks through it at the black background.

Under these conditions the faint image will appear as a positive image, and straightforward copying in the camera will directly produce a duplicate negative. This negative usually shows shadow tones in sufficient strength for normal printing or enlarging.

The visual effect of the image can, if necessary, be further increased by bleaching the original negative in a halogenizing bleacher—e.g., copper sulphate. Alternatively, this optical intensification effect can be further amplified optically. This involves a fairly complex set-up of lenses and mirrors which reflect the light to and fro through the photographic image, building up an optical image of increasing contrast and density. L.A.M.

See also: *Contrast control; Duplicate negatives; Intensification; Masking.*

OPTICAL AXIS. Axis of symmetry of an optical system—e.g., of an assembly of glasses making up an objective or a projection system.

See also: *Lens.*

OPTICAL BENCH. Basic laboratory device for measuring the performance of lenses. It consists of a long wooden or metal track carrying accurately centred supports for a light source, the lens under test, and a screen on which the image of the light source formed by the lens can be examined.

The optical bench may also be used for photometry by mounting a Bunsen photometer in the centre of the bench with light sources for comparison mounted on slides at either side.

The performance and characteristics of concave mirrors may be investigated on the optical bench in the same way as lenses except that the screen on which the image is formed is on the same side of the mirror as the object (usually a fine wire grid in front of a diffused light source). In this case the screen must be shielded from the direct rays of the source so that it receives only the image forming rays from the mirror. So the screen must lie slightly to one side of, or above, the optical axis and the angle of the reflected ray must be taken into account when measuring the various separations.

These are the basic essentials, but there is provision for adding other measuring or optical equipment. The distances between all the items can be varied, without moving them off the common optical axis.

OPTICAL CALCULATIONS

The symbols used in the following equations are collected and defined as follows:

F = Focal length of a complete lens system (total focal length).

f_1 = Focal length of the first component of a two-component system.

f_2 = Focal length of the second component of a two-component system.

P = Power of a lens in diopters.

S = Axial separation between two components.

u = Conjugate distance between the front nodal point of a lens and the object (in an enlarger the object is the negative).

v = Conjugate distance between the rear nodal point of a lens and the image (in an enlarger the image is the copy board).

D = Distance between object and image ignoring nodal point separation = sum of conjugate distances = $(u + v)$.

O = Size of object (linear).

I = Size of image (linear).

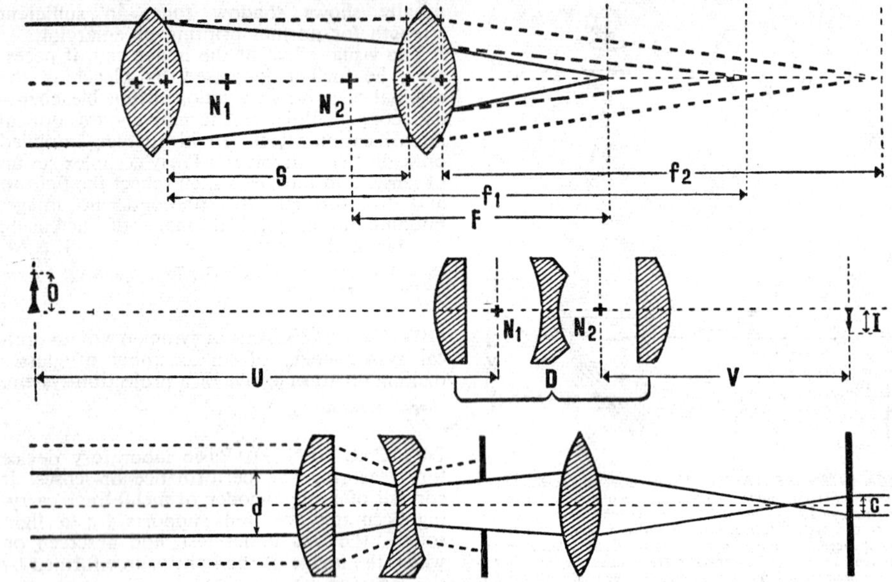

SYMBOLS AND DIMENSIONS. Top: *Lens combinations.* N_1, *front nodal point of combination.* N_2, *rear nodal point of combination. S, separation between components (from rear nodal point of first to front nodal point of second unit).* f_1, *focal length of front component.* f_2, *focal length of rear component. F, effective focal length of complete lens system.* Centre: *Object and image. O, object size. I, image size. U, object distance (to front nodal point). V, image distance (from rear nodal point). D, object-image distance, ignoring nodal point separation.* Bottom: *Aperture and sharpness. d, effective clear diameter of lens. C, circle of confusion.*

M = Optical magnification = $\dfrac{\text{size of image}}{\text{size of object}} = \dfrac{I}{O} = \dfrac{v}{u} = \dfrac{1}{R}$

R = Optical reduction $= \dfrac{\text{size of object}}{\text{size of image}} = \dfrac{O}{I} = \dfrac{u}{v} = \dfrac{1}{M}$

d = Effective clear diameter of a lens.

n = Relative aperture or *f*-number $= \dfrac{F}{d}$

c = Diameter of circle of confusion.

H = Hyperfocal distance.

L = Movement of lens system along its axis.

POSITIONS OF LENSES AND FOCAL PLANES

The fundamental equation is

$$\frac{1}{F} = \frac{1}{u} + \frac{1}{v}$$

Several rearrangements of the parent equation are possible but the following are selected as being the most convenient in use.

Focal Length. The focal length required by conditions of use can be defined in terms of any two of the following factors: object distance, image distance, magnification or reduction, sum of conjugate distances.

$$F = \frac{uv}{(u + v)} = \frac{u}{(R + 1)} = \frac{Rv}{(R + 1)}$$

$$= \frac{Mu}{(M + 1)} = \frac{v}{(M + 1)} = \frac{DR}{(R + 1)^2}$$

$$= \frac{DM}{(M + 1)^2}$$

Object Distance (e.g., closest distance in copying, lens negative distance in enlarging). This can be determined if two of the following factors are known: focal length, image distance, magnification or reduction, sum of the conjugate distances.

$$u = \frac{vF}{(v - F)} = F(R + 1) = \frac{F(M + 1)}{M}$$

$$= \frac{DR}{(R + 1)} = \frac{D}{(M + 1)}$$

Image Distance (e.g., total camera extension in copying, lens-paper distance in enlarging). This can be determined if two of the following factors are known: focal length, object distance, magnification or reduction, sum of the conjugate distances.

$$v = \frac{uF}{(u - F)} = \frac{F(R + 1)}{R} = F(M + 1)$$

$$= \frac{D}{(R + 1)} = \frac{DM}{(M + 1)}$$

Magnification or Reduction. This can be determined from two of the following factors: object distance or size, image distance or size, focal length or sum of conjugate distances:

$$M = \frac{I}{O} = \frac{v}{u} = \frac{(v - F)}{F} = \frac{v}{(D - v)}$$
$$= \frac{F}{(u - F)} = \frac{(D - u)}{u}$$
$$R = \frac{O}{I} = \frac{u}{v} = \frac{F}{(v - F)} = \frac{(D - v)}{v}$$
$$= \frac{(u - F)}{F} = \frac{u}{(D - u)}$$

Conjugate Sum (e.g., total height of a negative above the baseboard in an enlarger). This requires two of the following factors: focal length, object distance, image distance, magnification, or reduction.

$$D = \frac{F(M + 1)^2}{M} = \frac{F(R + 1)^2}{R} = \frac{v(M + 1)}{M}$$
$$= v(R + 1) = u(M + 1) = \frac{u(R + 1)}{R}$$
$$= u + v$$

Object Size (e.g., field size covered in close-ups or copying). This can be determined from the magnification or reduction and the image size, or from the latter with two of the following factors: image distance, object distance, sum of conjugate distances, focal length.

$$O = IR = \frac{I}{M} = \frac{IF}{(v - F)} = \frac{I(D - v)}{v}$$
$$= \frac{I(u - F)}{F} = \frac{Iu}{(D - u)} = \frac{Iu}{v}$$

Image Size (e.g., screen dimensions required in projection). This can be determined from the magnification or reduction and the object size, or the latter together with two of the following factors: image distance, object distance, sum of conjugate distances, focal length.

$$I = \frac{O}{R} = OM = \frac{Ov}{u} = \frac{O(V - F)}{F}$$
$$= \frac{Ov}{(D - v)} = \frac{OF}{(u - F)} = \frac{O(D - u)}{u}$$

Space does not permit examples for every variation of the fundamental equation but the following will indicate the manner of use.

Example 1. What focal length of lens is required to form a half-size image of an object if object-image distance is fixed at 9 ins.?

$$D = 9; R = 2$$
$$F = \frac{DR}{(R + 1)^2} = \frac{9 \times 2}{(2 + 1)^2} = \frac{18}{9} = 2 \text{ ins.}$$

Example 2. What are the conjugate distances of a 4 ins. focal length enlarging lens working at 5 times enlargement?

$$F = 4; M = 5$$
$$u = \frac{F(M + 1)}{M} = \frac{4(5 + 1)}{5} = \frac{24}{5} = 4{\cdot}8 \text{ ins.}$$
$$v = F(M + 1) = 4(5 + 1) = 24{\cdot}0 \text{ ins.}$$

Example 3. What is the focal length of a lens with conjugate distances 30 and 6 ins.?

$$u = 30; v = 6$$
$$F = \frac{uv}{(u + v)} = \frac{30 \times 6}{30 + 6} = \frac{180}{36} = 5 \text{ ins.}$$

Example 4. What is the distance between a lens and its image plane if the focal length is 20 ins. and the object 10 feet from the lens?

$$F = 20; u = 120$$
$$v = \frac{uF}{u - F} = \frac{120 \times 20}{120 - 20} = \frac{2400}{100} = 24 \text{ ins.}$$

Example 5. What is the maximum magnification possible in an enlarger with a baseboard lens distance of 36 ins. and a focal length of 4 ins.?

$$v = 36; F = 4$$
$$M = \frac{(v - F)}{F} = \frac{(36 - 4)}{4} = 8 \text{ diameters}$$

Example 6. What is the field width covered when copying with a 5 ins. lens at 30 ins. from the object, on a negative $2\frac{1}{4}$ ins. wide?

$$F = 5; u = 30; I = 2\tfrac{1}{4}$$
$$O = \frac{I(u - F)}{F} = \frac{2\frac{1}{4}(30 - 5)}{5} = 2\tfrac{1}{4} \times 5$$
$$= 11\tfrac{1}{4} \text{ ins.}$$

COMBINATIONS OF LENSES

Total Focal Length (e.g., effective focal length with a supplementary lens). The focal length of a combination of two lens components is dependent on the focal length of each component and the separation between them.

$$F = \frac{f_1 f_2}{(f_1 + f_2 - S)}$$

Focal Length of One Component. If the focal length of one component (e.g., camera lens) of a two-component system is fixed, the focal length of the other (e.g., supplementary lens) needed for a given total focal length for the combination is defined by the equations:

$$f_1 = \frac{F(f_2 - S)}{(f_2 - F)}$$
$$f_2 = \frac{F(f_1 - S)}{(f_1 - F)}$$

These focal lengths are positive when the lens is collective, and negative when dispersive.

If the focal length of one component (e.g., the supplementary lens) is known in diopters, the equation becomes:

$$f_1 = \frac{F(100/P - S)}{(100/P - F)} \text{ cm.}$$

or:

$$f_1 = \frac{F(39{\cdot}37/P - S)}{(39{\cdot}37/P - F)} \text{ ins.}$$

Axial Separation. The separation between two lens components of given focal length which will provide a given total focal length for the combination is defined by the equation:

$$S = f_1 + f_2 - \frac{f_1 f_2}{F}$$

Here again the focal lengths are positive for collective, and negative for dispersive, lenses.

If only approximate focal lengths are required in the previous equations and the two components are placed very close together (e.g., a supplementary lens mounted on a camera lens) S can be ignored, and the previous equations become:

$$F = \frac{f_1 f_2}{(f_1 + f_2)}$$

$$f_1 = \frac{F f_2}{f_2 - F}$$

and so on.

Object Distance with Supplementary Lenses. When the camera lens is focused on infinity, and a supplementary lens f_2 is added, the object distance for a sharp image becomes:

$$u = f_2$$

The distance is thus independent of f_1 or F, or S. This does not apply when the camera lens is focused on nearer objects; in that case the combination has to be calculated.

Example 7. What is the focal length of the combination of two collective lenses with focal lengths 4 ins. and 9 ins. if the separation between their nodal points is 1 in.?

$$f_1 = +4;\ f_2 = +9;\ S = 1$$

$$F = \frac{f_1 f_2}{(f_1 + f_2 - S)} = \frac{4 \times 9}{(4 + 9 - 1)} = \frac{36}{12} = +3 \text{ ins.}$$

Example 8. What is F in example 7 if the 9 ins. lens is dispersive?

$$f_1 = +4;\ f_2 = -9;\ S = 1$$

$$F = \frac{f_1 f_2}{(f_1 + f_2 - S)} = \frac{4 \times (-9)}{(4 + (-9) - 1)}$$

$$= \frac{-36}{-6} = +6 \text{ ins.}$$

Example 9. In a telephoto system comprising a front collective lens of 5 ins. focal length and a rear dispersive lens of 6 ins. what is the separation between their nodal points to produce a 10 ins. combination?

$$f_1 = +5;\ f_2 = -6;\ F = +10$$

$$S = f_1 + f_2 - \frac{f_1 f_2}{F} = (+5) + (-6)$$

$$- \frac{5 \times (-6)}{+10} = -1 - \frac{-30}{+10} = -1 + 3$$

$$= 2 \text{ ins.}$$

Example 10. What is the maximum focal length of supplementary lens required to shorten the focal length of a 4 ins. lens so that it can focus on an object 10 ins. from the lens maintaining a camera extension or v dimension less than 5 ins.?

$f_2 = 4;\ u = 10;\ v = 5$; assume S to be negligible

$$F = \frac{uv}{(u + v)} = \frac{10 \times 5}{10 + 5} = \frac{50}{15} = 3\tfrac{1}{3} \text{ ins.}$$

$$f_1 = \frac{F(f_2 - S)}{(f_2 - F)} = \frac{3\frac{1}{3}(4 - 0)}{4 - 3\frac{1}{3}} = \frac{13\frac{1}{3}}{\frac{2}{3}} = \frac{40}{3} \times \frac{3}{2} = 20 \text{ ins.}$$

FOCUSING MOVEMENTS

Lens Movement. The forward movement of a lens to focus sharply a near object can be defined in terms of the distance of the object and the focal length of the lens. This movement is measured from the position in which the lens focuses sharply an infinitely distant object.

Object Distance. If a lens is moved forward by a known amount from the position in which it focuses an infinitely distant object, the position of an object which will be sharply focused can be defined in terms of this movement and the focal length of the lens. In the case of cameras where object distances are intended to be measured from the focal plane instead of the node of the lens, the object distance can be denoted by the symbol Up.

When u, the object distance, is measured from the front nodal point of the lens system:

$$L = \frac{F^2}{(u - F)} = \frac{FI}{O} = \frac{Fv}{u}$$

$$\text{and } u = \frac{F(F + L)}{L}$$

When Up is measured from the focal plane:

$$L = \frac{Up - 2F - \sqrt{Up^2 - 4FUp}}{2}$$

$$\text{and } Up = \frac{(F + L)^2}{L}$$

An approximate equation which can be used for most practical purposes is:

$$L = \frac{F^2}{(Up - 2F)}$$

Magnification and Camera Extension. The maximum enlargement obtainable from a given camera extension (i.e., focusing movement in front of the infinity position) is:

$$M = \frac{L}{F}$$

Conversely, the extension required for a given magnification is:

$$L = FM$$

Example 11. In a camera where object distances are measured from the lens node, what is the focusing movement from the infinity position for a 6 ins. lens to focus at 8 feet?

$$F = 6;\ u = 96$$

$$L = \frac{F^2}{(u - F)} = \frac{6^2}{(96 - 6)} = \frac{36}{90} = 0{\cdot}4 \text{ ins.}$$

Example 12. In a camera where object distances are measured from the focal plane, what is the forward focusing movement of a 5 ins. lens to focus an object 3 feet?

$$F = 5;\ Up = 36$$

$$L = \frac{Up - 2F - \sqrt{Up^2 - 4FUp}}{2}$$

$$= \frac{36 - 10 - \sqrt{1296 - 720}}{2}$$

$$= \frac{26 - \sqrt{576}}{2}$$

$$= \frac{26 - 24}{2}$$

$$= 1 \text{ in.}$$

The approximate equation gives:

$$L = \frac{F^2}{(Up - 2F)} = \frac{25}{36 - 10} = \frac{25}{26}$$

$$= 0{\cdot}962 \text{ ins.}$$

Example 13. What is the forward focusing movement from the infinity position required for copying at a reduction of 3× (i.e., a magnification of $\frac{1}{3}$) with a 12 cm. lens?

$$F = 12;\ M = \tfrac{1}{3}$$

$$L = FM = 12 \times \tfrac{1}{3} = 4 \text{ cm.}$$

Focusing Movements with Supplementary Lenses. If the total focal length of the combination of a camera lens and a supplementary lens is computed as described in the section on Combinations of Lenses, all the equations in the other sections can be used to determine the behaviour of supplementary lenses.

Alternatively, the following simplified procedure will yield information on the focusing movements using supplementary lenses.

When a camera is focused on infinity, the distance focused on when a supplementary lens is put in front of the camera lens is equal to the focal length of the supplementary lens. This distance should be measured from the supplementary lens—strictly, from its nodal point, but the front surface is usually good enough.

The focal length in inches of a supplementary lens whose power is expressed in diopters is equivalent to 39·37 divided by the number of diopters.

Thus the distance focused on when a two-diopter positive supplementary lens is positioned in front of a camera lens focused on infinity is 39·37 ÷ 2 = 19·7 ins.

When a camera is not focused on infinity, the new object distance, Us, when using a supplementary lens can be determined if the focal length of the supplementary lens, f_2, and the focused object distance, u, of the camera lens alone is known.

Assuming negligible separation between the two lenses, the equation is:

$$Us = \frac{uf_2}{u + f_2}$$

Example 14. A camera is sharply focused on an object 5 feet from its lens. If a 1 diopter positive supplementary lens is placed in front of the camera lens, what is the new object distance which is sharply in focus?

$$u = 60;\ f_2 = 39{\cdot}37$$

$$Us = \frac{uf_2}{u + f_2} = \frac{60 \times 39{\cdot}37}{60 + 39{\cdot}37} = \frac{2362{\cdot}2}{99{\cdot}37}$$

$$= 23\tfrac{3}{4} \text{ ins.}$$

DEPTH OF FOCUS AND DEPTH OF FIELD

The numerous equations under the above heading are all based on the assumption that some degradation of definition is permissible in the recorded images of objects not sharply in focus. The criterion of permissible unsharpness is the circle of confusion.

Circle of Confusion. The diameter of the circle of confusion defines in an arbitrary manner the degradation of definition that can be accepted for a particular purpose. A perfect lens can record pin-point images of pin-point objects at the distance sharply focused upon. Point objects at other distances are recorded as circular patches, or circles of confusion, instead of true point images. In photographs to be viewed at an eye distance of about 10 ins. it is considered that circles of confusion less than ·005 in. in diameter are indistinguishable from true points. The circle of confusion which can be allowed in the negative should be chosen with respect to the degree of subsequent enlargement, if any, and to the probable viewing distance of the print. Many depth of field tables are based on the acceptance of a circle of confusion whose diameter is equivalent to the focal length of the lens divided by 1,000.

Depth of Focus. The permissible tolerances in the distance between a lens and the sensitive emulsion, i.e., depth of focus, are:

$$\text{Permissible tolerance} = cn$$

For near objects this becomes:

$$\text{Permissible tolerance} = \frac{cvn}{F} = \frac{cun}{(u - F)}$$

These tolerances for depth of focus ignore any considerations of depth of field and should therefore be reduced as much as possible.

Near and Far Limits. The size of the circle of confusion and thus the degradation of definition increases gradually as the object distance is increased or decreased from the ideal sharply focused distance. If the degree of degradation of definition that can be permitted is defined in terms of the size of the circle of confusion, the near and far limits and thus the depth of field within which the object distance may vary can be determined from equations involving knowledge of the focal length of the camera lens, the relative aperture at which it will be used, and the object distance at which it is sharply focused.

The fundamental equations for depth of field are:

U_1 = Nearest distance in focus

$$= \frac{Fu(F + cn)}{(F^2 + ucn)}$$

U_2 = Farthest distance in focus

$$= \frac{Fu(F - cn)}{(F^2 - ucn)}$$

Taking c as 1/1000 of the focal length, the equations become

$$U_1 = \frac{uF(1 + n/1000)}{(F + nu/1000)}$$

$$U_2 = \frac{uF(1 - n/1000)}{(F - nu/1000)}$$

Ignoring the fraction n/1000 as negligible compared with 1 further simplification gives the approximate equations:

$$U_1 = \frac{uF}{F + nu/1000}$$

$$U_2 = \frac{uF}{F - nu/1000}$$

For most practical purposes especially where u is large relative to F, depth calculations may be simplified by first computing the hyperfocal distance and proceeding from there.

Hyperfocal Distance. This is the nearest distance in focus when the lens is precisely focused for an infinitely distant object.

$$H = F + \frac{F^2}{nc}$$

U_1 = Nearest distance in focus

$$= \frac{Hu}{(H + u - F)} \text{ or approx.} = \frac{Hu}{(H + u)}$$

U_2 = Farthest distance in focus

$$= \frac{Hu}{(H - u - F)} \text{ or approx.} = \frac{Hu}{(H - u)}$$

If the nearest and farthest distances U_1 and U_2 are fixed then the lens must be focused on a distance u such that:

$$u = \frac{2U_1 U_2}{(U_1 + U_2)}$$

and the aperture must not be wider than

$$n = \frac{(U_2 - U_1)F^2}{2U_1 U_2 C}$$

when u is large compared with F.

Interchangeable Lenses. The simplification of the depth-of-field equation based on a circle of confusion of F/1000 is only valid for a lens of normal focal length, i.e., approximately equivalent to the image diagonal.

This condition is not fulfilled with interchangeable lenses of different focal lengths used on one and the same camera where negatives are used to about the same degree. Here calculations can conveniently be based on the equation for the hyperfocal distance:

$$H = F + F^2/nc$$

Usually F can be neglected compared with F^2/nc and H becomes proportional to F^2.

If this hyperfocal distance H_s is known for the standard focal length F_s, the value H_x for a different focal length F_x is given by:

$$H_x = H_s (F_x/F_s)^2$$

Depth of field limits for the alternative lens can then be recalculated from its hyperfocal distance.

Total Depth. When the lens is set at short copying conjugate distances the depth of field is small and it is sufficient to compute the total depth of field between the near and far limits instead of each limit separately.

Total depth (far to near)

$$= (u_2 - u_1) = \frac{2cn(M+1)}{M^2} = 2cnR(R + 1)$$

Example 15. A miniature camera mounts interchangeable *f* 2 lenses. What is the permissible tolerance on the distance between lens and emulsion if the circle of confusion must not exceed ·002 in.?

$$c = \cdot002;\ n = 2$$

$$\text{Tolerance} = cn = \cdot002 \times 2 = \cdot004 \text{ ins.}$$

To maintain this tolerance with any of the lenses the seating face on the camera and also the shoulders on each individual lens must be held to not more than half this, i.e., ·002 ins.

These tolerances must be reduced appreciably if the performance expected from the use of depth of field tables is to be obtained.

Example 16. What is the depth of field of a 2 ins. lens working at *f* 4 and focused on an object at 3 feet (36 ins.) from the lens if the permissible circle of confusion is ·002 ins.?

$$u = 36;\ F = 2;\ n = 4;\ c = \cdot002$$

$$u_1 = \frac{Fu(F + cn)}{(F^2 + ucn)} = \frac{2 \times 36(2 + \cdot008)}{4 + \cdot288}$$

$$= \frac{72 \times 2\cdot008}{4\cdot288} = \frac{144\cdot576}{4\cdot288} = 33\cdot7 \text{ ins.}$$

$$u_2 = \frac{Fu(F - cn)}{(F^2 - ucn)} = \frac{2 \times 36(2 - \cdot008)}{4 - \cdot288}$$

$$= \frac{72 \times 1\cdot992}{3\cdot712} = \frac{143\cdot424}{3\cdot712} = 38.6 \text{ ins.}$$

The approximate method gives:

$$H = F + \frac{F^2}{nc} = 2 + \frac{4}{\cdot008} = 502 \text{ ins.}$$

and

$$u_1 = \frac{Hu}{(H + u)} = \frac{502 \times 36}{502 + 36} = \frac{18072}{538}$$

$$= 33\cdot6 \text{ ins.}$$

$$u_2 = \frac{Hu}{(H - u)} = \frac{502 \times 36}{502 - 36} = \frac{18072}{466}$$

$$= 38\cdot8 \text{ ins.}$$

Example 17. If two objects are positioned 4 feet and 5 feet from a 5 ins. lens, what object distance should be focused upon and what is the widest relative aperture to yield a standard of definition corresponding to a circle of confusion of ·004 ins.?

$$u_1 = 48;\ u_2 = 60;\ F = 5;\ c = \cdot004$$

$$u = \frac{2u_1\, u_2}{(u_1 + u_2)} = \frac{2 \times 48 \times 60}{48 + 60} = \frac{5760}{108}$$

$$= 53\tfrac{1}{3} \text{ ins.}$$

and

$$n = \frac{(u_2 - u_1)F^2}{2u_1\, u_2\, c} = \frac{(60 - 48) \times 5^2}{2 \times 48 \times 60 \times \cdot004}$$

$$= \frac{12 \times 25}{96 \times \cdot24} = \frac{300}{23{\cdot}04} = f\,13$$

Example 18. How much depth of field can be expected when a lens working at $f\,4$ copies an object at half size reduction and the criterion for sharpness corresponds to a circle of confusion of ·002 ins.?

$$c = \cdot002;\ n = 4;\ R = 2$$

$$\text{Total depth} = (u_2 - u_1) = 2cnR(R + 1)$$

$$= 2 \times \cdot002 \times 4 \times 2(2 + 1)$$

$$= \cdot032 \times 3$$

$$= \cdot096 \text{ ins.}$$

G.H.C.

See also: *Aperture; Depth of field; Depth of focus; Extension of camera; Extension tubes; Magnification; Supplementary lenses.*

Books: ***Introduction to Applied Optics* (2 vols), by L. C. Martin (London); *Photographic Optics*, by A. Cox (London).**

OPTICAL CEMENT. Adhesive used to bind elements. The oldest is Canada Balsam, but now synthetic resins are used. An optical cement must have a suitable refractive index to match that of the glasses being cemented, as well as resistance to variations in humidity, temperature, etc.

See also: *Adhesive.*

OPTICAL CONTACT. Contact between two surfaces signifying that there is no space left between them; when two transparent surfaces —e.g., the components of a lens—are in optical contact the join thereafter behaves as though it were a single surface separating the two components. Absolutely matching surfaces in contact may adhere without adhesive.

OPTICAL DESIGN

As generally understood, optical design comprises two related activities, which may be described as optical engineering, and as optical design proper.

Optical Engineering. This field covers those studies which indicate the general form of the system to be developed. In earlier times this was a comparatively straightforward matter, but with the application of optical systems to a wider range of use it has taken on an entirely new complexion.

The general framework of an optical system must be developed so that it will fulfill conditions such as the following:

(*a*) An image of adequate size should be formed on a desired receptor. This may be the retina of the human eye, a photographic plate, a sensitive surface of a xerographic type, a photo-electric surface in a television tube, an infra-red or an ultra-violet detector used in a scanning mode;

(*b*) An image of adequate brightness must be formed, so that any of the devices listed above will provide a signal of suitable strength;

(*c*) The system must fit into a stipulated space, which may often require that it be folded with prisms and mirrors;

(*d*) The system must be capable of being fabricated from materials which will transmit desired spectral regions and which will stand up to severe environmental conditions. Such materials must also be available in sizes which will satisfy the needs of the system.

The application of photographic and allied techniques to the wide range of problems in the modern world means that a photographic lens is only one special kind of optical system; many other types may be needed to obtain desired results.

Designing the System. Once the general form of an optical system is determined, whether it be a simple photographic lens or a really complex system, the next stage is to design it in such a way that it provides an image of acceptable quality. This is optical design proper.

In order to provide perfect image quality a lens or optical system must bend the light rays from an object point so that they converge in an image point, and this point must lie in the proper place. The failure of a lens to provide this perfect performance is due to the presence of aberrations. The following basic types of aberration are recognized.

(*a*) Spherical aberration, which gives a circular patch of light instead of a point image.

(*b*) Coma, which produces a light patch flaring away like the tail of a comet.

(*c*) Astigmatism, which produces a light patch that can be changed to a short vertical or horizontal line by shifting the focal plane.

(*d*) Chromatic aberrations, which produce colour fringing of either a symmetrical or of an unsymmetrical form.

(*e*) Distortion, which causes a square object to be reproduced either as an approximate square with concave (pin cushion) or convex (barrel) shaped sides.

It is the responsibility of the optical designer to correct these aberrations so as to obtain the best possible quality. A considerable body of mathematical theory has been built up which analyses the character of these aberrations.

The simplest group are known as the Seidel or third order aberrations, and fairly simple mathematical expressions are available to deal with these. Beyond this point it is necessary to trace rays of light through the lens. At each surface it is necessary to calculate first where a ray encounters it, and after that to calculate in detail how it is refracted at that surface.

Rays and Ray Tracing. There are two distinct classes of rays, meridional rays and skew rays. A meridional ray is one which lies in a plane containing the axis of symmetry of the lens. A ray which does not lie entirely in this plane is a skew ray.

The tracing of meridional rays is a much simpler and less time consuming activity than tracing skew rays. As a rough estimate it can be stated that, in order to trace a meridional ray with a simple desk calculator, the time needed is about two minutes for each surface that the ray encounters. To trace a skew ray under the same conditions requires about 15 minutes, and the chances of arithmetical errors are much greater.

In the days when all ray tracing was done with desk calculators the labour of designing was very heavy. It was the rare exception to trace skew rays, and the total number of rays which were traced in the intermediate design stages would not normally exceed about fifteen. At the present time all serious optical design is carried out on electronic computers. On even a slow machine the time to trace a skew ray across a surface is about $1\frac{1}{2}$ seconds (compare this with 15 minutes!) while on a reasonably fast machine the time is less than $\frac{1}{4}$ second. About ninety-six rays will be traced at the intermediate design stages.

At the final evaluation stages it is not at all unusual to trace about a thousand rays or more. The intersection points of these rays in the focal plane of the system are plotted (using automatic machines) to produce a spot diagram, comprising points or spots whenever a ray meets the focal plane. From these spot diagrams it is possible to obtain a much clearer idea of what the lens performance will be.

The advent of electronic computers in the past few years has made a revolution in lens design. The laborious process of ray-tracing once set the pace for design development. This is no longer the case and the emphasis has been shifted towards the development of higher and higher levels of performance. In particular the use of computers has made it possible to design zoom lenses of outstanding performance, an achievement that would have been much more difficult if ray-tracing still had to be carried on by hand.

Automatic Design. This is logically the next step. An electronic computer is capable of making decisions when it is given a suitable criterion by which to operate. If then a computer carries out a ray-tracing programme it can examine the results it obtains, it can make suitable changes in the radii, separations and thicknesses of lenses which should improve the overall performance, and then go into a new ray-tracing programme. This cycle can be repeated until it seems that no further improvement is possible.

Such a process of automatic lens design is no longer a visionary goal. Several programmes directed towards this end are now in use throughout the world. They still need improvement in order to produce results which will satisfy the human designers employing them, but it appears that this is a matter of detailed refinement rather than of establishing new principles. A.C.

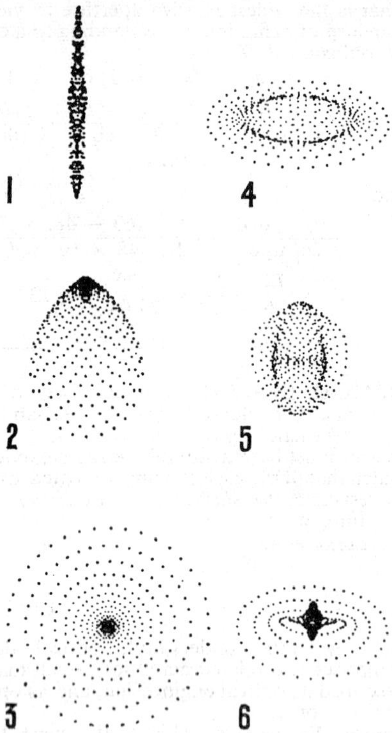

SPOT DIAGRAMS. Rays can be traced through a lens by a computer which directly prints out the distribution of the different rays corresponding to an object point in the image plane. This then indicates the shape and image spot takes in the presence of various optical aberrations. 1. Shape of the light patch in the presence of astigmatism. 2. Coma. 3. Spherical aberration. Each point represents the intersection of a light ray with the image plane. 4, 5, and 6. Light patches obtained with a mixture of aberrations. These light patches were obtained with a fairly simple 8 in. lens and are magnified 300 times (2, 3, and 6) or 1,000 times (1, 4, and 5).

See also: *Aberrations of lenses; Lens performance; Lens testing.*

Books: *Optics*, by Arthur Cox (London); *A System of Optical Design*, by Arthur Cox (London).

OPTICAL FLAT. Filter and other optical element made to a specially high degree of accuracy with both surfaces absolutely flat and parallel. Filters used for particularly exact photography must be optical flats but they are unnecessarily accurate—and expensive—for the requirements of normal photography.

In practice, minute errors still occur in the accuracy of optical flats. Departures from flatness are expressed in terms of light wavelengths, and from parallelism in terms of minutes or seconds of arc.

OPTICAL GLASS

The modern photographic lens uses a variety of glasses for its construction and in providing glasses of the appropriate properties, many stringent requirements have to be met.

Ideally a lens should reproduce a point as a point and a straight line as a straight line in the plane of the photographic plate even though the various parts of the subject are at differing distances from the lens. In practice a simple lens departs from the ideal, first because defects arise from the fact that the lens surfaces are spherical in form and secondly because glass itself disperses white light into its components and thus produces a spectrum instead of a point image at its focus.

In practice the various defects or aberrations of the simple lens are reduced to a minimum by a proper choice of glasses with the appropriate curvatures used in combination with one another, but at its best the composite lens is a compromise between one aberration and another. The larger the aperture and the greater the curvature of the lens surfaces the more difficult it becomes to reduce the aberrations, and the evolution of the modern photographic lens has depended largely on the development of optical glass.

Use of Different Glasses. The production of a spectrum at the lens focus (chromatic aberration) was considered by Newton to be the chief obstacle to his design of refracting telescope. Between the years 1726 and 1760, Chester Moor Hall and John Dollond showed that chromatic aberration could be reduced by combining a positive lens of a crown glass (low refractive index and low dispersion) with a negative lens of a flint glass (high refractive index and high dispersion). With the glasses existing at that time it was possible to achromatize only for two wavelengths and there remained a residual chromatic aberration called secondary spectrum.

The history of the development of optical glass is largely concerned with attempts to produce glasses to give a reduced secondary spectrum. The important effects of boron and barium in the composition of the glass were discovered by Sir George Stokes and the Rev. Vernon Harcourt in England, and the effects of many other elements were explored systematically by Abbe and Schott in Germany. Many of the glasses so developed had to be abandoned but a few glasses were sufficiently stable to be produced commercially and by their use the apochromatic lens which achromatized at three wavelengths became possible. Such glasses had relatively low refractive indices so that their use was restricted.

Modern glass type for optical purposes include a wide variety of different crown and flint glasses. The term crown glass still remains as a generic description for low dispersion glasses (V-value above 50 to 55) while flint generally applies to glasses of a V value below 50. In addition glasses with a refractive index below about 1.6 are frequently referred to as "light" and glasses with a higher refractive index as "dense". Glasses above 1.7 are extra dense types. Frequently the main addition to the glass melt is also indicated, so that there are barium crown and barium flint glasses, zinc and borosilicate crowns, lanthanum crowns and flints, etc.

While early optical glasses mainly comprised a range from low to medium-refracting low-dispersion crowns up to highly refracting and high-dispersion flints, the development of modern glass types brought highly refracting glasses of medium and even low dispersion into the range. These include rare-earth glasses like lanthanum flints and crowns and thoria glasses.

The value of high-refractive crown glasses is that they can be combined with low refractive glasses of the same dispersion value (the light crown types) to provide colour correction over a wide range of wavelengths.

Homogeneity. The glasses available to Moor Hall and Dollond were found to be inhomogeneous and the sharp variations in composition, which appeared as striations, rendered the glasses unsuitable for optical purposes. The process of eliminating the striations by stirring the glass in its molten state was investigated by a Swiss, P. L. Guinard, in the years 1768 onwards, and by Faraday at the Royal Institution, and this process forms the basis for modern optical glass making technique. Faraday at the Royal Institution conducted experiments on homogenizing, the glass being melted in platinum trays. The methods of Guinand were first tried out on a production scale in England in 1848.

The methods followed in the manufacture of optical glass have developed rapidly over the last 25 years but in all processes stirring is a necessary phase to eliminate inhomogeneities. Prior to World War II, stirring techniques were very similar in principle to those developed by Guinard. At this time, each glass was melted individually in a crucible or pot and usually the

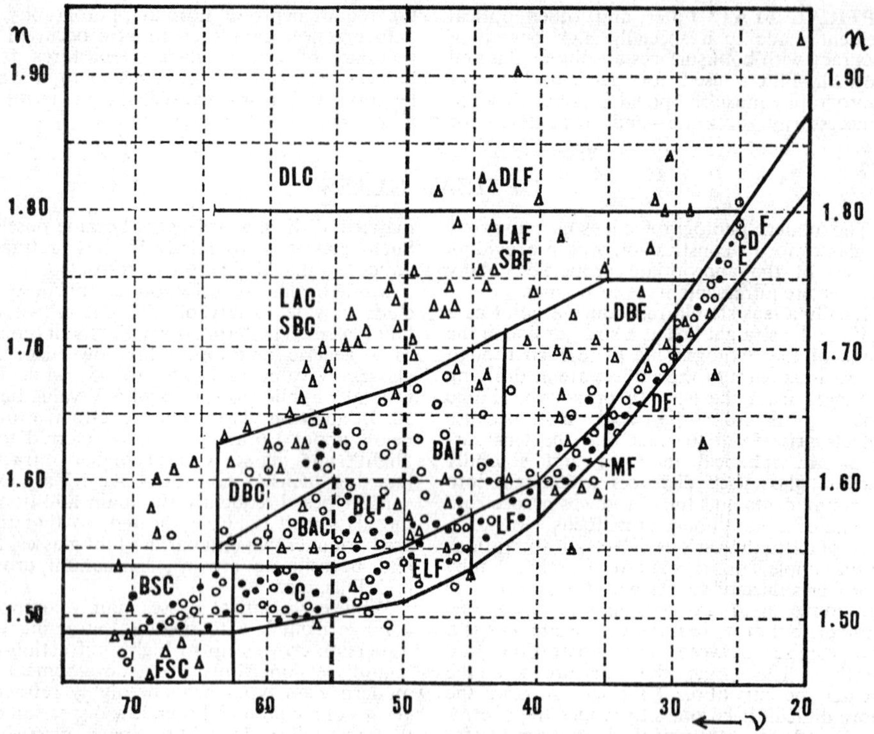

OPTICAL GLASS TYPES. Glasses are specified by their refractive index (n) and dispersion (ν). The diagram shows available glass types of different dispersions and refractive indices. By tradition, glasses with an Abbe number above about 50 to 55 are termed crown glasses (to the left of the heavy broken line); glasses of higher dispersion (lower ν value) are flint glasses. The divisions indicate the most commonly used terms for the glass types, partly specifying also their chemical composition. Crown glasses: FSC, fluorosilicate crown; BSC, borosilicate crown; C, crown; BAC, barium crown; DBC, dense barium crown; SBC special barium crown; LAC, lanthanum crown; DLC, dense lanthanum crown. Flint glasses: ELF, extra light flint; LF, light flint; MF, medium flint; DF, dense flint; EDF, extra dense flint; BLF, barium light flint; BAF, barium flint; DBF, dense barium flint; SBF, special barium flint; LAF lanthanum flint; DLF, dense lanthanum flint. The solid circles represent glasses available at around the turn of the century; open circles indicate glasses developed up to about 1950 and open triangles are glass types produced since then. These include virtually all the lanthanum glasses as well as rare earth types.

whole pot was allowed to cool and crack up. The cooling was carried out under controlled conditions so that large lumps of glass were obtained and these rough lumps were re-softened and moulded into slabs in order to allow a critical quality examination to be carried out. Glass defective either because of bubble inclusions or striations was cut away and the resultant glass sold either as slabs or as re-moulded lens blanks.

At a later date, the pots were preserved and the contents cast into a large block. Such a block was kept whole by careful cooling and when cold it was polished to allow optically good parts to be located. These were cut away by diamond charged saws and finally processed to the shape required. By either of these processes, precise control of optical properties, homogeneity and bubble quality was difficult, with problems arising from failure to re-produce melting and stirring conditions precisely. Whilst these problems have not been entirely resolved by the modern continuous process, the new technique is such as to allow a greater degree of control to be imposed on all phases of the process with the consequent advantages of reproducibility of optical properties and improved glass quality. In this process, raw materials are fed continuously into a melting unit of special design in which the glass is successively melted and fined off to remove solid and baseous inclusions. Finally it is stirred and formed continuously into bars, blocks or pressings suitable for processing to the optical designers final requirement.

Reference has already been made to the importance of chemical homogeneity and the necessity of stirring to remove sharp variations in composition. The homogeneity of glass is also dependent on a past thermal history and it is important that this should be uniform throughout the glass in order to achieve uniformity of glass structure and hence uniformity of optical properties. To achieve these necessary requirements all optical glass is annealed to special cooling schedules.

In the most modern method of manufacture, raw materials are fed continuously into a melting unit of special design. In this unit, the glass is successively melted, fined off to remove solid and gaseous inclusions, stirred, and finally continuously formed into bars, blocks and pressings suitable for processing to the optical designers' final requirements.

Requirements of Optical Glass. The precise requirements of optical glass are:

(1) The glass must have accurately defined values of refractive index and dispersion, repeatable from one melt to another for the same type of glass.

(2) The glass must be uniform in respect of both chemical composition and physical state. Uniformity of chemical composition is achieved in the stirring process and physical uniformity, that is, uniformity of refractive index in a glass of a uniform chemical composition, is achieved by precise annealing.

(3) It must transmit as much light as possible and must not absorb selectively light of different wavelengths.

(4) It must be as free as possible from bubbles.

(5) It must be as resistant as possible to the action of water vapour and other gases present in the atmosphere.

(6) It must be as resistant as possible to the action of acidic materials used in grinding and polishing.

PROPERTIES OF TYPICAL OPTICAL GLASSES

Type	*Refractive Index (nd)*	*V**
Fluor crowns, borosilicate crowns and hard crowns	1·465–1·525	70–59
Barium crowns	1·552–1·657	63–51
Barium flints	1·551–1·700	55–41
Flints	1·542–1·927	47–21
Special rare earth glasses	1·642–1·744	58–44

*The *V*-value is a measure of the dispersion; it is a function of refractive index *nd* (helium *d* line) and the difference in refractive index for the hydrogen *C–F* lines. The lower the *V*-value the higher the dispersion.

On account of the rather unusual chemical composition of many optical glasses it is often not possible to achieve all these requirements in a particular glass, but clearly the optical properties and homogeneity must take first place and the other requirements must follow. It is not always possible, for example, to achieve a glass of the highest durability because its chemical composition has to be such as to give abnormal optical characteristics. In such a case, it may be necessary to accommodate the component at an internal position in the lens so that its surfaces are protected.

Manufacturing Methods. The pot process is being increasingly superseded by continuous melting processes. These enable the optical properties of glass to be maintained to very close limits over long periods of time and also make it possible to cast large slabs of optical quality glass which are needed for such purposes as protective viewing windows in atomic energy facilities and similar situations.

As with the classical pot process, stirring is an essential feature of this new technique, as compositional variations, arising for example from selective volatilization of the constituents and from corrosion of refractory material, must be removed from optical glass. This continuous process enables glass of a high degree of homogeneity and freedom from bubbles to be produced. The main commercial problem remaining is that the quantities produced are large and it therefore becomes necessary to limit the number of types of optical glass that can be manufactured in this way to those that are needed in substantial quantities.

The continuous process is subject to patent protection and also considerable secrecy as to the exact details of manufacture. In general terms the glass is melted in a rectangular bath, heated largely by electrical conduction, and the melted but bubbly glass is then submitted to a "fining" process for which equipment made in platinum is used. After the glass has travelled through the fining system it is stirred mechanically, again in platinum apparatus, in such a way that every portion of the glass is subjected to the same treatment and the uniformity of the product is thereby assured. With electric heating, the temperature in the various stages of the process can be controlled over long periods of manufacture to within 1 or 2° C. and the effects of volatilization, which were so serious under older pot methods, have virtually been eliminated.

The continuous stream of glass which emerges from the stirring equipment may be converted into individual pressings, appropriate to the lenses that are to be made. This involves the use of automatic pressing machines fed from the glass stream with drops of controlled weight. Alternatively, the glass may be converted into continuous strips from which a variety of lens or prism shapes can be produced, or can be flowed into large moulds in order to produce large slabs. Slabs produced by this method in the normal course of manufacture can be up to 6 feet square and 1 foot thick and can weigh a ton or more—a far cry from the haphazard selection of small pieces of bottle glass which were all that were available to the earliest optical designers.

The use of platinum equipment has eliminated the contamination that used to occur by the solution of the refractory materials used as a container, and also has simplified the removal of bubbles, many of which were produced by the interaction of the constituents used for the glass and the ceramic pot.

At the same time, there are some technical limitations to the continuous method. Glasses with extreme optical properties can give rise to trouble due to volatilization and refractory solution. The colour of a glass is primarily determined by the choice of raw material and by impurities taken into the glass as a result of refractory corrosion. The chemical elements used for the manufacture of the more extreme glasses are more liable to give rise to difficulties over both volatilization and refractory corrosion. Consequently, with these glasses, the technique involving the use of pure platinum crucibles heated in electric furnaces has been maintained. Materials such as lanthana, thoria, cadmium, tungsten, are used in the manufacture of these special glasses in order to produce the wide range of optical characteristics required for specialized design.

The use of platinum crucibles in electric furnaces also continues to some extent for types of optical glass that are needed in relatively small quantities, for instance, in the manufacture of microscope lenses. The cost of manufacture in this way is notably higher than that of the production of optical glass by the larger scale continuous process.

Both methods of manufacture—the continuous process and the platinum pot process—are under continual investigation in the research laboratories and in the works of optical glass manufacturers, with particular effort being directed towards the problem of improving control of both glass quality and optical properties. It is anticipated that this requirement for tighter control will become progressively more important as a result of increasingly tight specifications being imposed on the optical properties by new and more sophisticated lens design. At the same time, new and improved glasses are being developed and it is clear that this work, in due course of time, can lead to some redundancy in existing glass types.

W.M.H. & H.C.

See also: *Glass; Lens history; Lens manufacture.*

OPTICAL SENSITIZING. Method of extending the spectral sensitivity of an emulsion by the use of dyes. The basis of all ortho and pan emulsions. Either the emulsion or the coated plate may be treated, and the extra sensitivity depends on the dye, e.g., eosine will sensitize an "ordinary" blue-sensitive plate to yellow light, cyanine will sensitize it to red. Other dyes can sensitize emulsions to invisible radiations (infra-red rays). The patent literature lists many sensitizing dyes.

See also: *Negative materials; Sensitizer; Spectral sensitivity.*

OPTICAL WEDGE. Device used in sensitometry to measure the effect of a range of exposures on a laboratory sample of sensitized material. It generally consists of a strip of material—glass, celluloid or plastic—covered with a pigment or developed silver emulsion layer which is clear at one end and gradually becomes opaque towards the other. The transition from transparent to opaque may take place smoothly (continuous wedge), or in regular steps, in which case it is called a step wedge.

See also: *Sensitometry; Wedge.*

OPTICS. Branch of science concerned with the behaviour of light. There are three main divisions of the subject: physical, physiological and geometrical. The first is concerned with the composition and character of light. The second is concerned with light in its relation to human vision. The third division deals with the geometry of the paths followed by light rays under particular conditions—e.g., when passing through a lens or after reflection from a polished surface.

See also: *Colour; Illumination; Image; Lens; Light; Mirrors; Optical calculations; Projection principles; Refraction; Reflection; Vision.*

ORDINARY EMULSION. Term used to describe a "colour-blind" or blue-sensitive emulsion to distinguish it from an orthochromatic or panchromatic emulsion.

See also: *Negative materials; Sensitized materials history.*

ORTHOCHROMATIC. Materials which are sensitized to all the visible spectrum with the exception of the deep orange and red rays. Orthochromatic films used to be popularly known as chrome films, but this suffix is now also applied to panchromatic and to colour films.

See also: *Infrachromatic; Negative materials; Ordinary emulsion; Orthopanchromatic; Panchromatic; Spectral sensitivity.*

ORTHOPANCHROMATIC. (Correct pan). Term used to describe materials sensitized to all colours of the visible spectrum but with an evenly balanced red sensitivity. In this they differ from the excessive red sensitivity of some high-speed panchromatic emulsions which are produced to give speed as a first consideration.

ORTHO-PHENYLENE DIAMINE. 1 : 2-diaminobenzene; ortho-aminoaniline. Fine grain developing agent.

Formula and molecular weight: $C_6H_4(NH_2)_2$; 108.

Characteristics: White crystalline powder.

Solubility: Fairly soluble in water.

ORTOL. Developing agent. Now largely obsolete.

Formula and molecular weight: $2(CH_3NHC_6H_4OH)_2H_2SO_4.C_6H_4(OH)_2$; 798.

Characteristics: White powder. In developers acts somewhat like metol-hydroquinone.

Solubility: Fairly soluble in water.

OSCILLOGRAPH RECORDING

Technique of recording in graphic form, physical changes against time. Photo-oscillography is the technique of oscillograph recording using light beams or electron beams to expose traces on photographic materials.

History. The first known recording instrument appeared about 300 years ago in Oxford, England, and was used to record temperature as it varied with time. Early recording techniques included the use of a needle to scratch through a smoked surface, previously applied to glass or paper. During its evolution numerous marking methods and materials were tried as increased mechanization of the nineteenth and twentieth centuries created the need for accurate recording media. The development of recorders continued for many years to be centred around mechanical, ink, pressure, heat, and electric marking methods.

In 1867, a photothermograph and a photobarograph were developed and used by the Royal Meteorological Society. These instruments, the first to record photographically, may be seen at the Science Museum, London. In 1897, Sir J. J. Thomson at the Cavendish Laboratory in Cambridge, using a cathode-ray tube for his experiments, measured the ratio of the charge of the electron to its mass. In the same year, Professor F. Braun applied a cathode-ray tube to industrial testing, intiating what is now a universal instrumentation technique.

In the following years, two photographic-marking methods evolved: light-beam recording and electron-beam recording. The acceptance of the photographic-marking methods occurred because of inherent limitations of mechanical, ink, pressure, heat, and electric marking equipment. The frequency response of typical mechanical-stylus recorders was in the 0–100 cycles per second range. In addition, restrictions of size and stylus swing limited the number of information channels and the amplitude that could be recorded on paper of a given width. The advent of photo-oscillography overcame the low-frequency-response limitation; present-day light-beam recorders are capable of frequencies to above 10^4, and electron-beam recorders to above 10^8. Photo-oscillography also made it possible to overlap traces so that as many as sixty channels might be recorded on 12-ins. wide sensitized paper.

Sensing Devices. Physical qualities such as temperature, force, stress, light level, fluid flow, acceleration, displacement, and vibration can be measured by first converting the energy into an electrical signal. In that intermediate form, the energy can be transmitted, processed and converted into measurable data. The sensing device is called a transducer and examples are thermocouples, strain gauges, differential transformers and accelerometers.

Transducers may be active or passive. The active transducer is a calibrated device for converting a small sample of some form of mechanical energy into analogous electrical energy. Examples are piezo-electric and magnetic transducers.

Passive transducers act as valves to control the flow of electrical energy from an outside source. Strain gauges and resistance thermometers are examples of passive transducers.

Transducer output-signal variations, when linked to the oscillograph recorder, cause corresponding variations in sweep of the recording beam. The beam exposes the photo-sensitive material moving continuously past the recording point at a constant rate, leaving a path that is the final trace on the record.

Light-beam Oscillography. Light-beam oscillographs are usually of the mirror-galvanometer type. Light from a source within the oscillograph is reflected from a tiny mirror mounted on the rotating coil of the galvanometer. The coil with its mirror twists in proportion to the electrical current, and the light reflected from the mirror falls on the moving photo-sensitive material. After calibration, the mirror position (and hence the trace) is a measure of the exact amount of current.

Although the mirror-galvanometer is perhaps the most critical component in the light-beam oscillograph, recording cannot occur unless other components are provided. Besides the light-sensitive material and a transport to drive it through the recorder, the light source, the trace optical system, and time reference and other optical systems must be installed in a light-tight enclosure provided with suitable controls. The resulting unique "camera" is the photographic light-beam oscillograph.

These instruments are of two kinds: develop-out and print-out recorders. Develop-out oscillographs use a tungsten lamp to record on

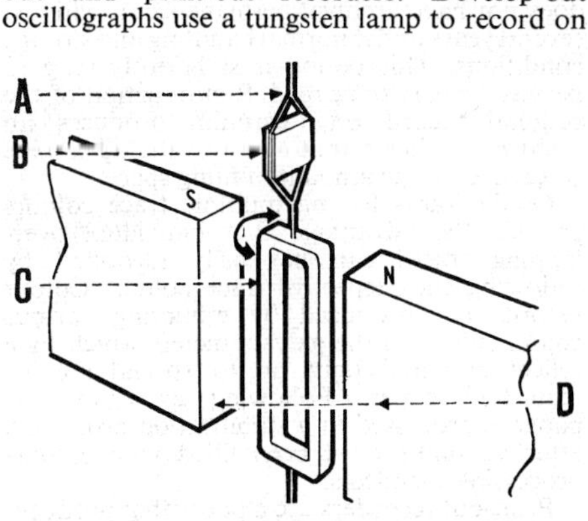

GALVANOMETER OSCILLOSCOPE. A, suspension. B, mirror. C, coil. D, magnet poles. Current pulses going through the coil cause the latter to turn in the magnetic field; a light beam directed on the mirror moves with the latter and can thus trace a record of the movement on a sensitised paper or film.

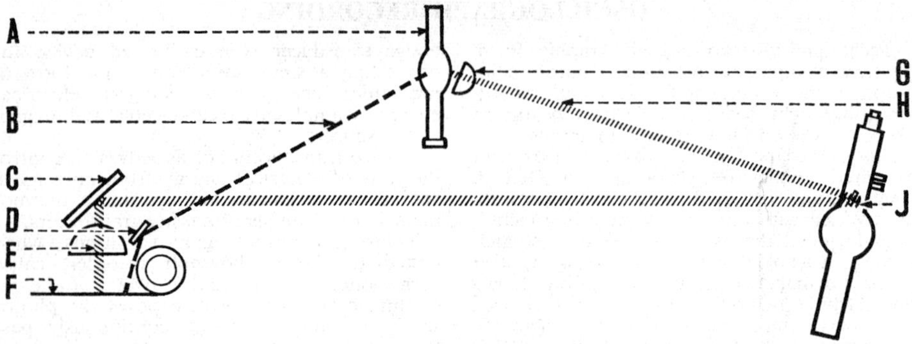

GALVANOMETER OSCILLOGRAPH RECORDING SYSTEM. A, *light source.* B, *light beam for recording reference grid on the paper or film.* C, *main recording mirror.* D, *grid recording mirror.* E, *main recording lens.* F, *surface of recording paper.* G, *collector lens.* H, *main recording beam.* J, *galvanometer mirror.*

film or paper which must then be chemically processed to produce a visible image. Develop-out sensitized materials must be handled in a darkroom under recommended safelight conditions; this applies to loading and unloading the oscillograph magazine, as well as to processing the material.

Develop-out records are generally in long roll form which precludes processing by hand methods. Rewind batch-type processors, normally used to develop aerial films, were once widely used for processing oscillograms. Since the mid-1950s, compact, continuous stabilization processors have been used. Table-top stabilization processors develop, stabilize and dry records at a nominal 6 feet per minute. Floor models operate at speeds up to 200 feet per minute. The wash cycle normal to standard photographic processing is omitted, thus speeding up access time. Although the record does not have archival permanence, it will last several years under normal handling and storage conditions. This period is sufficiently long to permit the data to be read. Reproduction of the original record on microfilm produces an archival duplicate that also has the advantage of saving a large amount of filing space.

Oscillograms having multiple trace colours provide the advantage that individual overlapping traces can be easily identified by following the respective trace colour. Colour records are produced by mounting various colour filters on the galvanometers which then reflect coloured light on to special colour-recording paper. Following exposure, the paper is processed in a stabilization processor, after its tanks have been filled with colour-processing chemicals.

Print-out recorders use a paper that produces a visible trace without chemical processing. Print-out paper is relatively slow, as compared with develop-out paper; therefore, a higher intensity light source must be used in the recorder. Usually, the lamp is a mercury or xenon arc lamp, but improvements in the sensitivity of print-out materials have enabled high-intensity incandescent lamps to be used in some lower writing-speed applications.

After the traces are exposed, the paper must be given a second exposure, called latensification, to make the latent image visible. The second exposure can be made to room light, but most modern oscillographs are equipped with built-in latensification lamps. Subsequent exposure of the record to ultra-violet or excessive, prolonged or repeated exposure to room light will degrade and eventually obliterate the trace image. The record will also be rendered useless if it is duplicated with reproduction equipment having high-intensity light sources. However, the image of some print-out papers can be given a chemical-stabilization treatment, which renders it permanent and capable of reproduction in any equipment.

Emulsions, whether develop-out or print-out, are coated on both light-weight and heavy-weight paper base. The lightweight paper allows duplicates to be made on standard diazo and similar office reproduction machines.

Electron-beam Oscillography. The advantage of electron-beam recording over other recording methods is that the electron displacement is virtually inertialess. Therefore, very high-frequency recording is possible, and performance is believed to be the highest of any existing recording method. Recording can take place either directly by exposure from electrons, or indirectly by photographing the luminiscent trace on the face of a cathode-ray tube.

Direct Exposure. In practicing the direct-exposure method, special electron-sensitive recording film is used; its emulsion is particularly sensitive to electron energy and relatively less sensitive to photon energy. The sensitized material is placed inside the special recording tube, the tube is evacuated, the electron-beam

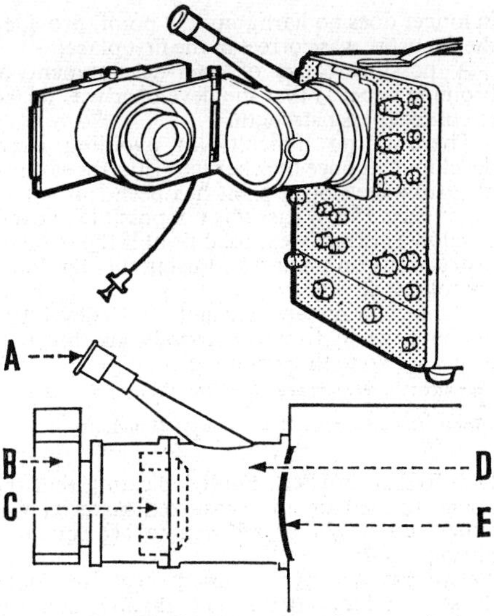

OSCILLOSCOPE CAMERA. This consists of a camera back and lens, usually with shutter, set for a close-up distance and mounted on the front of the oscilloscope screen to record the trace on the latter. Top: General view with camera unit swung away on the light excluding tube. Bottom: Side view. A, eyepiece for visual observation of the oscilloscope screen. B, camera back. C, lens and shutter unit. D, light-excluding hood, also holds the camera lens at the correct close-up distance. E, oscilloscope screen.

exposure is made, the tube is returned to normal pressure and opened, and the film is removed and processed.

Indirect Exposure Exposure by light emitted by the phosphorescent screen of a cathode-ray tube, when its phosphor coating is activated by electrons, is the more common technique. The widespread use of oscilloscopes and improvements resulting in high-speed, high-resolution sensitized materials, popularized phosphorescent-screen cathode-ray oscilloscope photography. Today, the oscilloscope is used in nearly every laboratory, and single-shot recording cameras are available to fit nearly every make of instrument. It is easy to make and to assess photo-oscillograms made on such equipment, particularly when instant-processing materials are used as the recording medium. Because of its universal use, the oscilloscope and the recording of single shot cathode-ray tube images has been well documented in electronic and photographic literature.

The standard technique for photographing the cathode-ray trace uses an oscilloscope camera, but a still camera can also be employed. In either case the camera is focused on the cathode-ray tube screen and is usually mounted at the end of a light-excluding hood. The camera can record either continuous events in which the electron-beam follows the same repetitive path for a period exceeding the exposure time; or separate events in which the beam makes one trace only per single transient.

In the first case, the camera exposure can be as long as necessary to obtain a satisfactory record. In the second case the shutter is opened just before the trace occurs and is kept open for its duration. This may necessitate a fairly wide lens aperture and fast film, particularly at high writing speeds of the spot. Special instant processing materials are available for oscilloscope photography that have speeds as high as 10,000 A.S.A. equivalent, which greatly simplifies the exposure problems.

Special cameras are available for recording transient events that occur at random, and/or which may be prolonged and vary continuously. In this case the film moves continuously at a constant speed thus providing the time axis (just as paper is transported in the light-beam oscillograph). Therefore, the time base of the oscilloscope is switched off so that the spot moves only in the amplitude scan.

Transient events can also be photographed in a similar manner on a drum camera. Cameras of this type are capable of very high film-travel speeds. H.W.H.

See also: *Cathode ray tube traces; Instrument recording.*

Books: *The Photography of Cathode Ray Tube Traces*, by N. Hendry and W. Nethercot (London); *The Photographic Recording of Cathode Ray Tube Traces*, by R. J. Hercock (London). *The Cathode Ray Oscillograph in Industry*, by W. Wilson (London); *Techniques of Photo-Recording from Cathode-Ray Tubes* (Du Mont Laboratories, Division of Fairchild Camera and Instrument Corporation); *Manual of Oscillography*, by Howard W. Hoadley (London).

OSTWALD, WILHELM, 1853–1932. German professor of physical chemistry at Leipzig University. Introduced a new theory of colour and published his Colour Atlas which embodied a system of colour tabulation commencing with pure hues to which were added varying proportions of black and white. Invented the Katatype process for colour prints (1902). Proposed theories on the chemical development of the latent image and on the ripening and growth of the grains in an emulsion. He received the Nobel Prize for Chemistry in 1909. Autobiography: 1926/27.

OVER-DEVELOPMENT. Fault indicating that the developing solution was allowed to act too long on the sensitized material to produce an acceptable result.

Negatives. An over-developed negative looks dense and contrasty. The shadow detail is well graded, but the highlights are clogged and extremely dense. Such a negative can easily be distinguished from one that has been over-exposed but correctly developed. The over-exposed negative is dense but flat, and even in the darkest areas the shadows are veiled.

A normal print from an over-developed

negative is harsh and contrasty; the shadows print out in solid black and the highlights are chalky and featureless.

Assuming that the developer is correctly mixed and fresh, over-development may be caused by errors in working out the development time for the particular combination of sensitized material and developer. Or it may be due to faulty timing or to over-agitation. Other possible causes are that the developer concentration was too high, or the development time did not allow for a temperature change.

When developing by inspection, over-development may be due to the photographer's inability to judge the appearance of a fully developed negative.

An over-developed negative may be bleached and redeveloped, as in intensifying, or it may be reduced by a suitable reducing bath. The exact treatment adopted will depend on the contrast of the negative, a different method being adopted according to whether the contrast is to be unchanged or reduced.

If the amount of over-development is slight, it may be sufficient to print on a soft grade of paper. In this case the increased density of the negative is partially offset by the higher speed of the soft paper.

Over-development can be avoided by mixing and using the developer according to the maker's instructions and by using a reliable developing chart which takes into account the differences in all the current kinds of sensitized material and developer, and makes allowances for temperature, agitation, age, and exhaustion of the solution.

When developing by time and temperature it is always advisable to work to a darkroom timer that rings a bell at the end of the required development period.

When developing by inspection it is generally safe to stop development when the shadow detail can be seen showing through the back of the emulsion.

Over-development should always be avoided with miniature negatives as it produces an increase in the grain size.

Prints. When a print has been correctly exposed, development for considerably longer than the recommended period has no serious effects; it may, in fact, give a richer print. If continued for too long the print will become fogged and stained and the image colour will alter.

Over-development with normal exposure can be distinguished from over-exposure with normal development without much trouble. The over-exposed print very rapidly builds up density in the developer and is more or less black and obscured before there is time to do anything about it.

Since prints are developed by inspection, the only reason for continuing beyond the correct time is inability to recognize a fully developed print. The tendency is in any case to take the print out of the developer too soon; leaving it in longer does no harm, up to a point, provided the exposure was correct in the first place.

A possible cause of over-development of bromide paper is that the developer was mixed to contact paper strength.

There is no difficulty in avoiding over-development once the characteristic appearance of a fully developed print has been learned by experience. In any case it is not possible to over-develop a correctly exposed print if the maker's instructions on mixing and using the developer are followed.

Contact prints are normally fully developed in about twenty to thirty seconds, and bromide prints in two to three minutes. F.P.

See also: *Developing negatives; Developing prints; Faults; Over-exposure.*
Book: *Developing*, by C. I. Jacobson (London).

OVER-EXPOSURE. Fault indicating that the image formed on the sensitized material was either too bright or allowed to act too long during exposure.

Negatives. An over-exposed negative looks dense and blocked up. The shadow detail is all there, but it is usually veiled over with a deposit of silver while the highlight areas look completely opaque.

In extreme cases the whole negative looks black with only a hint of an image in the deepest shadows.

An over-exposed negative can be distinguished easily from one that has been correctly exposed but over-developed. The over-exposed negative tends to be flat and lacking in contrast, because all the tones, given long enough exposure, would record as dense deposits while the brightest highlights could add no further density. The over-developed negative tends to be excessively contrasty. This is because the highlights go on developing at a disproportionately greater rate than the shadows.

A print from an over-exposed negative, on a normal grade of paper, tends to be flat and muddy. It may, however, yield an acceptable print on a harder grade of paper, but in either case the excessive density means that it will take a long time to print.

Over-exposure may be caused by inaccurate estimation or measurement of the prevailing light, by incorrect setting of the shutter speed and lens aperture, or by a faulty exposure meter or camera shutter. In flash photography it may be the result of an error in calculating the stop and subject distance from the flash factor. It may even be caused by using a developer that increases the effective film speed.

The commonest cause of over-exposure is undoubtedly human error. The photographer either does not use the meter correctly, or interprets its reading literally instead of making allowances for such factors as subject contrast, filter, film speed, light direction and colour temperature, and type of developer to be used. Owners of single lens reflex cameras frequently focus at full aperture and then forget to stop down before exposure.

Over-exposure in extremely cold weather is often the result of shutter lag caused by coagulated oil or, in focal plane shutters, by hardening of the roller blind fabric.

Generally speaking, over-exposure is much less serious than under-exposure because all the detail is there, although over-exposure tends to increase the grain—an important consideration in miniature work.

An over-exposed negative can usually be made to yield a satisfactory print on a suitably hard grade of paper. The principal drawback is the long printing time, but there are two ways of avoiding this. One method is to replace the normal printing or enlarger lamp with a Photoflood, in which case the enlarger should always be switched off immediately after the exposure, because of the great heat generated; the other is simply to reduce the negative with Farmer's reducer. As this reducer tends to increase contrast, it partially corrects the flatness of the negative tones at the same time.

Over-exposure can be prevented by using a reliable exposure meter and making due allowances for all the relevant subject, lighting and processing conditions. If, in spite of this, negatives are regularly over-exposed, either the meter or the camera shutter may need checking.

In cold weather the camera should be kept in a warm place and not left in an unheated room where the low temperature may slow down the shutter and give over-exposed negatives.

Prints. An over-exposed print has a heavy, sombre appearance; the highlights are veiled over and the shadows are completely blocked-up areas of deep black tone.

It is not possible to mistake a blackened, over-exposed print for one that has been correctly exposed but over-developed. Even a print that has been developed for two or three times the normal length of time will look little darker than one that has been given the correct development if the exposure was correct; after a very long time the print will become covered with general fog, and there may be stains in places, but up to that point over-development has no other effect.

Some workers over-expose prints habitually, consciously cultivating a gloomy, low-key style of expression.

The cause of accidental over-exposure is usually failure to switch off the light in time, either through faulty timing or carelessness. It may occur in enlarging when the lens is opened up for focusing the image accurately and not subsequently stopped down to the correct aperture for exposure.

Over-exposure may also be the result of changing from one batch of paper to another of the same specification but of higher speed due to manufacturing differences.

Over-exposure can be avoided by making careful tests before exposure on strips of the actual printing paper. Further tests should be made after any major change in magnification or if the lens is stopped down after making the test exposure. (In many enlargers the optical conditions are such that when the lens is stopped down, the light passing through is not reduced in proportion to the aperture numbers.) F.P.

See also: *Exposure; Faults; Over-development.*
Book: *Exposure*, by W. F. Berg (London).

OVERHEAD PROJECTOR. Still projector for projecting large-size transparencies, including diagrams and plans drawn on a transparent or translucent support. The transparency is placed on a horizontal illuminated light box and a lens projects the image forward via a 45° mirror mounted some distance above the light box. Used for certain educational and industrial purposes.

See also: *Projector (still).*

OVER-RUN LAMP. Lamp which is deliberately run on a higher voltage—and at a higher temperature—than normal—e.g., Photofloods. This greatly increases the light output and efficiency, raises the colour temperature, but appreciably shortens their life (often to a matter of a few hours).

See also: *Filament lamp.*

OXALIC ACID. Used in certain toners, also used as a preservative of pyro solution.

Formula and molecular weight: $(CO_2H)_2 . 2H_2O$; 126.

Characteristics: White crystals. It is poisonous.

Solubility: Fairly soluble in water at room temperature.

OX GALL. Commercially prepared liquid used by photographic retouchers and artists for treating a glossy surface—e.g., of a print or slide—to make it accept pigments applied with a brush. Also used in glazing prints.

OXIDATION. Chemical reaction increasing the proportion of oxygen or similar electronegative components of a chemical compound (or decreasing the proportion of hydrogen or similar electropositive components). Thus conversion of a silver image into silver halide by a ferricyanide and bromide bleacher consists of oxidation of the silver to silver bromide (here bromine is the electronegative component):

$$K_3Fe(CN)_6 + KBr + Ag = K_4Fe(CN)_6 + AgBr$$

Every oxidation is necessarily accompanied by a reduction, i.e., the increase of the proportion of the electropositive component of a compound. In the above example potassium ferricyanide $K_3Fe(CN)_6$ is at the same time reduced to ferrocyanide $K_4Fe(CN)_6$, potassium (K) being the electropositive component.

Similarly, during development silver halide is reduced to silver while the developing agent is oxidized.

OZALID PROCESS. One of the diazotype printing processes used principally for making positive prints from tracings of draughtsman's drawings.

Special paper, sensitized with a diazo compound and a colour coupler is exposed behind an ordinary line tracing and developed either with ammonia vapour or by damping it slightly with a dilute alkali.

The image normally develops as reddish-brown lines on a faintly purple tinted ground.

As the paper is not processed wet, it does not change dimensions and the print is reasonably true to size.

See also: *Document photography; Printing processes (Obsolete).*

OZOBROME PROCESS. Fore-runner of the carbro process. In common with carbro the process gives pigment prints direct from bromide contact prints or enlargements without the action of light. It was invented by Thomas Manly in 1905.

See also: *Printing processes (Obsolete).*

OZOTYPE PROCESS. Method of pigment printing invented by Thomas Manly, 1899, later replaced by the ozobrome process. One of the many processes which finally led to present-day carbro printing.

See also: *Printing processes (Obsolete).*

P

PACKING AND SENDING PHOTOGRAPHS. Of the thousands of prints and transparencies that are sent through the post, a high proportion arrive in bad condition—bent, creased, dog-eared, and in some cases even torn or broken. Often valuable photographs are worthless when they are received at their destination. Too often the Post Office gets the blame when in fact the trouble lies entirely with the sender.

A finished picture represents such an expenditure of time and trouble that it is worth while taking the little extra trouble in packing it to ensure that it reaches its destination in first-class condition.

Unmounted Prints. The method of packing depends largely on the number being sent and their size. Small prints will usually go into an envelope with stiffeners to keep them flat. The envelope must be of the strongest type; some of the cheaper varieties tear very easily. Stiffeners can conveniently be cut from strong cardboard and should be slightly larger than the prints themselves. One is placed on each side of the prints and the sandwich is held together with rubber bands.

Larger prints usually have to be parcelled and the packing materials should be selected carefully. The prints themselves should be sandwiched between strong cardboard stiffeners, while the corners and edges should be protected with corrugated cardboard held in place with string or adhesive tape. The parcel is then wrapped up in strong brown paper and fastened securely with string. Cellulose tape is also useful for holding down the edges of the parcel and prevents the paper from being torn off in transit.

An alternative method is to roll the print up and pack it in a stiff cardboard tube. This protects the prints against almost all risks of injury short of deliberate damage.

Mounted Prints. The mount itself is stiff, but that does not mean that packing can be skimped. Where there are several prints they should be packed in pairs face to face, with the surfaces protected from abrasion by tissue or other soft paper. The stiffeners should be of heavy board just larger than the mounts and corrugated cardboard should be used around the edges. (It is often advisable to use a sheet of thin plywood in place of one of the stiffeners if the prints are very large, very valuable, or have to be sent overseas.) The parcel is finally wrapped in stout brown paper and well tied and sealed.

Exhibition prints are often sent in specially constructed flat cases made of hardboard or plywood. This inevitably increases the weight of the parcel, but affords much greater protection for the contents and provides a convenient storage place for prints at other times.

The address should always be written clearly in block capitals and in addition, the name and address of the sender should be plainly stated. It is also important that the name and address of the sender is written either on the enclosed prints or on a covering letter included with them. This enables the prints to be returned to the sender if the outer wrapping should be torn off or obliterated.

How the prints are sent through the post will depend on the number being sent. A parcel of up to four mounted prints, size 16 × 20 ins., can be sent by letter post. Larger numbers may be sent by parcel post up to a weight limit of 15 lbs. Letter post is usually the quicker of the two. Either way the parcels can be registered.

When prints are sent overseas the sender may have to complete a customs declaration form stating the contents of the parcel and the estimated value. This is then stuck on the outside of the package. Prints conveyed by parcel post abroad cannot be registered but they can be insured, which amounts to the same thing. Details of postage, registration or insurance can be obtained from any Post Office.

In the U.S.A. prints may be mailed either 1st class or 4th class. In the latter case the package must be open for inspection (equivalent to British printed matter postage).

Transparencies. In most cases the value of a transparency—especially in colour—is greater than that of a print and consequently careful packing is even more important to avoid possible damage or loss. It is usual for transparencies to be either mounted between glass or else slipped into a transparent paper or plastic bag for protection. Even cardboard-mounted 35 mm. transparencies should be protected in such an envelope.

It is almost impossible to pack glass slides in such a way that breakage is impossible, but it is quite easy to take precautions that reduce the risk to a minimum. The secret is in packing the slides tightly. A stout cardboard box should be used, into which the slides fit exactly. Pack any vacant space tightly with folded newspaper or corrugated cardboard. Leaves of paper between the slides can be included, but are not essential. The lid of the box must not be neglected. This too should be padded out so that when the lid is on the slides cannot move in any direction. Plenty of corrugated cardboard should then be bound round the box in both directions before the usual outside packing is completed.

Although a strong cardboard box will give adequate protection in most cases, a wooden, or metal box may be more suitable if slides are being dispatched in quantity. The interior of such a box should first be lined completely with some sort of padding, and the slides packed tightly as before. But batches of more than six in one box will be safer if wads of corrugated cardboard are interspersed at intervals.

Transparencies without glass mounts will also travel best in a strong cardboard box. Care should be taken to see that they remain flat in the box, and plenty of padding with folded newspaper will give added protection.

Quick Delivery. From time to time photographs have to be got away in a rush, and the normal postal service would be too slow. One quick way is to take the parcel or packet to the railway station for despatch by train. British Railways will accept parcels for dispatch by the first available train on either weekdays or Sundays. The parcel is clearly marked "To be called for" at the distant station and the addressee is advised by telephone or telegram to meet the train and collect the parcel.

Photographs may also be sent by express post all the way to their destination, but this is quite a costly business if they have far to go. A cheaper way is to send them by express Railex. In this case the Post Office dispatches the parcel by messenger to the station where it is put on the first available train. It is collected at the station nearest to its destination by another messenger and taken to the address. The weight limit for parcels sent in this way is 1 lb.

A parcel for an overseas address can also be delivered by express post when it reaches the immediate neighbourhood of its destination. This service is available in certain countries only; it is arranged through the Post Office where full particulars can be obtained of the places served and approximate delivery time.

No special express service is available for photographic prints in America, but the Western Union messenger service is often used for this purpose within localities and the various air express services throughout the country will dispatch such materials from coast to coast within hours. G.W.P.

PAINTING AND PHOTOGRAPHY. The influence of painting on photography became apparent from the moment the daguerreotype and calotype processes were published. During past centuries many artists had striven to copy objects as realistically as possible and in 1839 many considered photography as an easy, quick and cheap means towards this end. Paul Delaroche, France's most celebrated artist at the time, summed up this attitude with his classic remark on first seeing a daguerreotype: "From today, painting is dead!"

A high proportion of early photographers were former artists, e.g. D. O. Hill, Henry Collen, Charles Nègre, William Lake Price, O. G. Rejlander, etc. Though admittedly not of front rank, they brought to photography an artistic training in composition and an eye for picturesque effects, and until the mid 1850's they treated subject matter in a straightforward and realistic way. It was only when art critics reviewing photographic exhibitions taunted photography with being a mechanical reproductive process, denying it creative possibilities, that many ambitious photographers tried to prove the opposite by making photographic pictures in subject matter and styles borrowed from painting, and quite outside the scope of photography. Leading spirits in this conscious attempt to raise photography to Fine Art were William Lake Price, O. G. Rejlander and H. P. Robinson. The majority of their fanciful compositions, made by combination printing or photomontage, looked at best like reproductions of paintings, but never like the achievements of an independent art. Yet this desire for recognition as artists led astray many otherwise excellent photographers like Hill and Julia Margaret Cameron, who achieved art when they were least conscious of it.

The photographic Fine Art movement had its centre in the Photographic Society of London, and from here pictorialism gradually spread to wherever photographers met in organizations, and competed in exhibitions. Their exhibition pictures, made as relaxation from the everyday routine work of portrait photography, usually demonstrated extraordinary technical skill, but outside these circles few considered them as Art.

The first open rebellion against artificial picture making was P. H. Emerson's *Naturalistic Photography* (1889), which came at a time when similar movements appeared in literature, drama, and other cultural fields. Soon after this the majority of the new generation of landscape photographers came under the influence of Impressionist painting. The understandable wish to dissociate themselves from the masses of new amateurs who were mere snapshooters led to the use of elaborate techniques such as the gum process, which made possible considerable manual interference with the photographic image, thus proving that photography was not mechanical, and frequently resulting in pictures imitating graphic techniques. The aesthetic movement in photography which began in 1892 with the *Linked Ring*, and received great impetus from the American *Photo-Secession* ten years later, was a world phenomenon closely related to the *art nouveau* movement in outlook. This kinship led to the acceptance of photography by many art groups of the period in joint exhibitions, especially in Germany and Austria.

Other *fin-de-siècle* photographers considered it the highest aim of photographic art to make fake paintings in the style of Old Masters.

During and after World War I, under the influence of new movements in art, there sprang up at around the same time numerous similar trends in photography: *abstract photography* (Vortographs by Alvin Langdon Coburn, Schadographs by Christian Schad, Rayographs by Man Ray, and photograms by Moholy-Nagy); *Neue Sachlichkeit* (New Realism) photographs by Albert Renger-Patzsch, Edward Weston and others; *surrealist* photographs by Winifred Casson, Peter Rose Pulham, Angus McBean and others. Of these three movements, only *Neue Sachlichkeit*, which was really a return to the sharp objective photography of the earliest period, with special emphasis on close-ups, has left a lasting mark on the further development of photography.

After World War II the acceptance of abstract art, now no longer experimental but world-wide, also gave photography a new look, largely under the leadership of Otto Steinert's *Subjective Photography*. Over-stressing of the graphic element often led to an attendant loss of photographic halftone and texture—photography's greatest assets. While certain attempts to bring photography into line with contemporary tendencies in art are obvious, the effects are today normally achieved by purely photographic means, and are certainly preferable to the outdated pictorialism still in favour with the older members of many photographic clubs and societies. Photographic documentation and reportage has by its very nature always remained outside the influence of aesthetic trends and painting, and has probably for this very reason developed into one of the most important fields of original expression in photography. H.G.

Book: *Creative Photography: Aesthetic Trends* 1839–1960 by Helmut Gernsheim (1962), London.

PAINTINGS AND DRAWINGS

Photographic reproductions of works of art are used for study by people who do not have access to the original, for magazine and book illustration and for records for art historians and picture restorers. For all these purposes the photograph must be sharp and clear, and as close to the original as possible. In addition to reproductions of whole paintings, some exciting results can be produced with exact size details or even by photographs taken 3–4 times magnified to show the artist's technique, brush strokes and craquelure.

Equipment. The studio equipment necessary is simple, but it must be accurate and strong: a firmly-built square bellows camera to take plates or sheet film, if possible not smaller than whole plate; a first-class modern copying lens with a slot to take gelatin filters; a solid stand with easy raising and lowering movement and an upright easel fixed to a steel frame. The easel must move up and down, and the frame must have a sideways movement. Accessories needed are a lens hood, focusing magnifier, light meter, filters (these can be bought comparatively cheaply and allow the photographer to invest in a large variety for experiment), and the normal darkroom equipment. It is most important to have a dark matt background, preferably black.

Artificial lighting is by far the most satisfactory because its constant nature gives a good basis for judging exposures and makes reflections easier to control. The most effective and reliable permanent lighting set-up is provided by a pair of upright parabolic trough reflectors fitted with 500 watt photo pearl bulbs every 18 inches. The troughs, up to eight feet high, painted white inside and mounted on a stand with rubber wheels or castors, can be easily moved about and give a good even light to cover the biggest pictures.

If, however, the work has to be done in a house or sale room, shallow bowl reflectors on a portable stand taking 500 watt bulbs are quite satisfactory. This type of lighting is suitable tot the amateur.

Sensitized Materials. Sheet films or plates can be used; the latter are dimensionally more stable. But modern polyester film bases are sufficiently stable for most practical purposes.

A soft gradation panchromatic film is best for paintings, and for drawings a slightly more contrasty emulsion. Even with black-and-white drawings it is advisable to use a panchromatic plate. Any discoloration or stains in the paper can then be made visible or subdued by the use of appropriate filters.

When photographing an oil painting, the aim is to produce a soft, clear, fully-exposed negative with plenty of detail in the shadows; with a drawing or water colour a slightly more contrasty, clean-looking negative is required.

Either dish or tank development may be

used, but the negatives must be kept free from grit and blemishes, particularly when they are to be used to show condition, as no retouching must be done at all.

Prints should be made on a smooth surface paper, either unglazed glossy or semi-matt. The expert never uses paper with an obtrusive surface or sepia tone; he knows that he is reproducing a picture, not making one.

Method. The picture to be photographed is set up as upright as possible on the easel, or fixed to some solid foundation if in a sale room or house. The image is then checked to see that it is square on the ground glass and includes the whole of the painting. It is next focused carefully with a focusing magnifier, and the lens is stopped down sufficiently to ensure good definition over the complete area of the plate. Focusing is carried out finally with the filter in position if a filter is to be used. The photographer now stands with his head in front of, and level with, the lens, and examines the picture for reflections, moving the lights about until there are none, and always checking up from the level of the lens.

Next, using a photo-electric meter, a reading is taken on a darkish part of the picture (if it is an oil painting), and the exposure is worked out, allowing for the filter if any. It is generally safe to double the resulting calculation with rich oil paintings as they absorb so much light, and it is important to reproduce every detail. For drawings, a general meter reading should be taken from a distance of about 12 ins. It is easy to cast a shadow over the picture without noticing it, so every care should be taken to prevent this happening.

With a drawing, accuracy of exposure and development is more important than with an oil painting, as the difference between charcoal, pencil and crayon is so slight. Failure to appreciate this fact explains why so many reproductions of fine pencil drawings look as though they have been done with pen and ink. The whole character of the original can easily be lost unless special attention is given to this point at every stage, including the printing.

Whenever possible the picture should be photographed out of its frame so that all the edges show. The edges must never be trimmed off or the result will look more like a picture postcard than a reproduction of a masterpiece.

Filters. Modern films have such a balanced colour sensitivity that filters are not often needed. However, there are times when a filter helps. In photographing a portrait it is quite useful to use the type of filter that is slightly greenish-blue and gives good modelling in the features. This filter can also be used to advantage in photographing pictures during cleaning as it shows up dirty patches of varnish very clearly. Another filter which is useful is the orange-red. This is a help when the picture is so dark and dirty that detail cannot be got in any other way. A yellow-green and a set of yellow filters are sometimes helpful in photographing landscapes to get detail in the blue skies.

Printing. Printing the negative, if everything has gone according to plan, is straightforward. If, however, it has been impossible to avoid reflections, the resulting dense areas on the negative may be rubbed down with metal polish or reduced with ferricyanide. And if parts of the negative print up too dark, tissue paper cut exactly to the shape of the thin patches may be put on the ground glass of the printing box under the negative, or the thin parts may be held back in the usual way during enlarging. These controls should not be allowed to falsify tones too much.

Care must be observed in finishing the print not to do any spotting that removes part of the picture. With good technique, it should not be necessary to do any finishing at all.

Slides are another form of reproduction in great demand by lecturers on art history. The best method to get slides of first-class quality is to make them through a camera on to a lantern plate. The negative is set up in a frame with an even light behind it and photographed in the normal way.

Infra-red, ultra-violet, X-ray and micro-reproduction of pictures all have their uses for the art historian and restorer, but this work is a branch of scientific photography and lies outside the scope of the photographer who simply wishes to make a good reproduction of a painting or drawing. V.C.W.

Colour. If colour transparences are required, colour reversal film should be used. When colour prints are wanted, negative colour film should be used, and although this type of film is more accommodating to different kinds of lighting, the best results are obtained when exposures are made under the conditions for which the film is balanced. "Universal" negative colour films are usually balanced for clear flash-bulbs (about 4,000° K.), and a Wratten 82C filter is appropriate when the original is lit by tungsten lamps. Weak colour-negatives will not give good prints, so it is safer to err on the side of over-exposure. For many subjects negative colour film as intended for normal photography is perfectly suitable, but certain subjects containing areas of pure white or lettering may appear to lack sparkle when the colour prints are made. The low contrast required for good tonal rendering of the picture as a whole then allows white to become a little degraded. In such cases it may be advisable to use an inter-negative type of colour film, especially made for copying coloured originals, transparencies or prints. The characteristic curves of inter-negative emulsions steepen considerably at the highlight end of the scale and excellently clean colour prints can be made from a properly exposed internegative. The production of an internegative requires the use of a preliminary test-chart of a range of neutral greys, from a dense black to pure white paper, upon which one or more trial exposures are made. After processing, the tricolor densities are measured with a densitometer and their curves drawn against the reflection-densities of the test-chart. The correct relative exposures which place the highlights upon the steeper portion of each curve can then be deduced with any correction-filtering which may be seen to be required. This may seem laborious, but important subjects will merit the extra trouble.

Pigment Characteristics. In Nature, colours may sometimes have a high degree of purity (i.e. a low proportion of rays of wavelengths outside the band of the predominant hue). This is because phenomena such as refraction, diffraction and polarization tend to be highly selective in the way in which different wavelengths of light are emitted.

Ordinary pigments, however, act upon the light which falls upon them simply by absorbing much of the light of wavelengths other than those of the principal colour. The efficiency of this is low in that the purity, saturation and brightness of the best of pigments are imperfect. In other words, some of the wavelengths which should be entirely absorbed are, in fact, reflected whilst wavelengths which should be reflected are partially absorbed. Hence, even the brightest blue pigment will reflect some yellow, orange and red rays and at the same time fail to reflect more than about 40 per cent of the blue rays. Blue pigments are particularly poor in this respect, but to some degree this is true of all artists' colours, particularly when ground in oil or varnish medium. The result is an inability of mixtures of pigments to match exactly certain colours in Nature. Expert painters overcome this by clever juxtapositioning of colours or manipulation of the technique itself, to produce an illusion of greater brilliance and saturation of colour than is there in fact. This may be readily proved by holding a colourman's sample chart of tube pigments close to any oil-painting.

So paintings are nearly always a compromise in faithful colour rendering, even though the painter may choose from a hundred prepared pigments—and whether he was conscious of this at the time or not. When the photographer comes to reproduce the painting, the limitations go further. Firstly, all practicable processes of colour photography employ only three channels of colour analysis, plus, in certain work, a grey channel to reinforce the light and shade values.

These are achieved either by filtering or by selective sensitization of a three-layer colour film. In either case there is some unwanted reaction to red values by the blue and green channels. Blue is slightly recorded in the red and green channels and some green rays are recorded in the red and blue channels.

These faults of recording can be corrected by independent masks, for work which merits the very time-consuming use of such techniques. In certain negative colour processes integral coloured masks, which are automatically formed in one or more image layers during processing, substantially reduce faults due to defects of recording and deficiencies of the dye-stuffs which form the three negative images. In reversal transparency processes, however, such automatic masking is not possible because the colour film upon which the exposure is made forms also the final positive. In such cases and in any other process where masking is not incorporated the degradation of colour rendering will be the sum of any faults of selectivity between the three channels and the dyes or pigments which form the ultimate print or transparency.

Dyes produce visual colours by selectively absorbing wavelengths other than those of their own colour. Most positive colour processes nowadays use yellow, magenta and cyan dyes which should in theory completely absorb blue, green and red respectively.

In practice dyes do not behave in this ideal way and have to be balanced one with another in purity and exposure-density response. As a result complete fidelity in reproduction is impossible; some degradation is inevitable. In making their compromises, colour-film makers aim to record neutral greys fairly accurately, together with plausible flesh-tones and foliage-colours. But colours which cannot be matched by varying combinations of density of the three primaries in a given system (i.e. the cyan, yellow and magenta dyes of the film) cannot be accurately reproduced.

In graphic arts techniques, gross errors are often reduced by retouching intermediate images or by correction-masking; but in the end the limitations are imposed by the defects of the inks or dyes used for the ultimate printing.

Fortunately, many works of art can be reproduced without such faults being very obvious, particularly when the reproduction is smaller in size than the original. Subjectively, the eye tends to consider colours to be brighter and more pure when the image becomes smaller and this offsets the reproduction losses to a large degree.

Fluorescing Agents. In recent years, fluorescing agents have been introduced to the textile and paper trades to produce so-called "whiter than white" effects. These substances absorb ultra-violet radiation present in the illumination, and re-emit it as visible light. This can lead to a treated sample of white material reflecting 15 to 20 per cent more visible light than an untreated one, giving the illusion of greater brightness.

Unfortunately, the extra light may not be equally distributed over the visible spectrum. Though a black-and-white photograph of an artist's drawing or painting may be quite satisfactory, so long as he has not made corrections with ordinary white pigment, a colour film may record the particular tint of the fluorescing substance quite differently from any pigment whites. Even in monochrome, any retouching, with visually matching white paint, shows up.

The only cure is to photograph such paintings with strict exclusion of ultra-violet rays. Ordinary household lamps of low wattage and general purpose floodlamps give off too little ultra-violet, in relation to the longer wavelengths, to cause trouble. But sunlight, electronic flash, arc-light, high-efficiency tungsten lamps and most kinds of fluorescent and discharge-lamps used in photography all require UV-absorbing filters in front of them. Even photoflood lamps emit enough ultra-violet to cause trouble when highly fluorescing substances are present in the original. (A camera-filter will not cure such trouble as the fluorescence has already occurred in the original before the light reaches the camera lens).

Old drawings and paintings and modern ones for which only the finest hand-made papers, panels or canvasses have been used, rarely contain fluorescing substances. The UV-absorbing nature of the binders in oilcolours would destroy the effect even if present. A number of modern pigments formed from *dyestuffs*, however, fluoresce when exposed to a source of ultra-violet radiation. For display and show-card work this effect is deliberately exploited in "Dayglow" type paints. In cheaper decorators' colours and whenever the illusion of great brilliance is more important than other qualities, such fluorescence is welcome, even though it cannot be reproduced by photographic means. In some cases, the fluorescence inherent in a particular dye may be of a different colour from that which it normally exhibits. The cure, as in the case of fluorescent papers and textiles, is to exclude ultra-violet rays from the original when photographing it, whether in monochrome or colour.

The presence of fluorescing substances may be detected by exposing the subject to the light from a "black-light" UV-lamp in a darkened room. An unmistakeable glow shows up the offending areas. Artists who are preparing work especially for reproduction owe it to their clients to use only art-boards free from "brightening agents" and either to use only the finest artists' quality water colours of the traditional kind, or to submit the poster and designers' colour charts to this test before admitting them to their studio. The cost of retouching and correction of defects in process engravers' establishments is considerable. Process engravers themselves, because of the especially UV rich illumination which they prefer, owe it to their clients to test originals before copying and to employ light-filtering where this is needed. J.V.

See also: *Copying; Manuscripts and old documents; Murals.*

PALLADIOTYPE. Printing process introduced *c.* 1916 capable of very delicate tone rendering. Special paper was used and the final image consisted of palladium. The process was very similar to platinotype.

See also: *Printing processes (obsolete).*

PAN AND TILT HEAD. Tripod attachment which permits the camera to be turned around its vertical axis and to be tilted in a vertical plane. Elaborate pan and tilt heads are used mainly in cinematography where they permit smooth movement of the camera during shooting, but are equally useful for accurate setting up in certain fields of still photography. The head may carry scales showing the horizontal and vertical rotation in degrees.

See also: *Camera supports.*

PANCHROMATIC. Sensitized materials with a response to all colours of the visible spectrum, including the red rays.

See also: *Infrachromatic; Negative materials; Ordinary emulsion; Orthochromatic; Orthopanchromatic; Spectral sensitivity.*

PANCHROMATIC VISION (P.V.) FILTER. Viewing filter through which the scene appears more or less in monochrome as a panchromatic emulsion would reproduce it. The effect of using filters in front of the lens can be approximated by adding them to the P.V. filter.

See also: *Tone rendering.*

PANNING. Term often used to indicate the swinging of the camera following a moving object, in order to record a sharp image.

See also: *Swinging the camera.*

PANORAMA. Picture that presents a continuous view of the landscape. In practice such comprehensive pictures are generally built up from a series of separate photographs covering the horizon, in slightly overlapping sections. There are also special panoramic cameras designed to scan any arc of the horizon and record the picture on a single continuous length of film.

Precautions. When a normal camera is used for making panoramas, certain special precautions must be taken or the adjacent photographs will not dovetail into one another without leaving a visible join.

(1) The camera must be mounted on a tripod. This is to make sure that all the photographs are taken from the same viewpoint. If the viewpoint moves between one exposure and the next, the perspective of the two photographs is no longer the same, and it is not possible to find a common boundary for joining them.

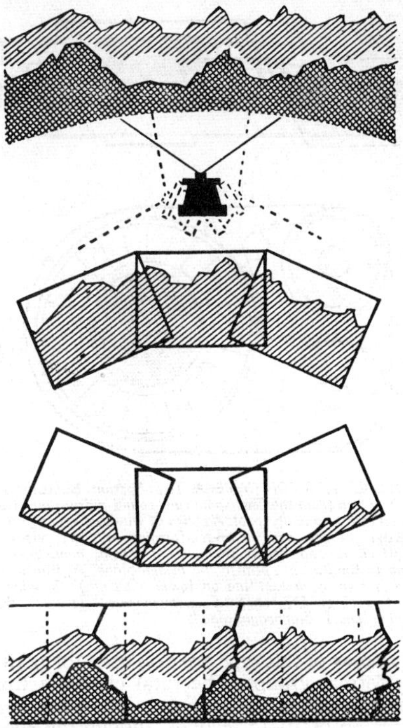

PANORAMAS. Top: *In taking a panorama the camera should be firmly mounted, and rotated through a fixed angle (preferably about the front nodal point of the lens), after each exposure. Successive views must overlap to permit matching up later on.* Centre: *The camera must be mounted absolutely level, otherwise the horizon line will appear to curve down (camera pointed up) or up (camera pointed down).* Bottom: *One way of hiding the joins in the assembled print is to follow the outlines of prominent features of the view in cutting.*

(2) The camera should preferably be mounted on a special head marked off to show the successive camera positions (for the particular lens in use) that will cover the horizon in sections with a suitable overlap between each section.

(3) The camera must be absolutely level—i.e., the focal plane of the lens must be vertical—or the adjoining edges of the separate pictures will not match. For this reason many of the panoramic heads incorporate a spirit level.

(4) If the horizon is to appear above or below the centre of the picture, the adjustment must be made by using the rising and falling front, if the camera has one, or by trimming the prints; it cannot be done by tilting the camera.

Panorama Tripod Heads. It is possible to take panoramas with the camera mounted on an ordinary tripod head by loosening the clamping screw and turning the camera through the required angle between exposures. After taking the first exposure looking at, say, the extreme right hand side of the view, the camera is swung around until a point on the horizon that previously appeared just inside the left hand edge of the finder now lies just inside the right hand edge. The second exposure will then overlap the first by a sufficient margin. Further exposures are made in the same way until the whole of the required field has been covered.

Special tripod heads are available for taking panoramas. These make it easy to turn the camera through the exact angle between exposures. The head generally incorporates a spirit level and is in two parts, a fixed base which screws on to the top of the tripod and a swivelling head which can be locked in any position. A scale ring graduated from 0° to 360° around the base shows how far the camera has been turned. With a lens taking in an angle of, say 50°, the head would be set at zero with the camera at the right hand end of the picture and for subsequent exposures moved to 45°, 90°, 135°, and so on, covering the picture area with an overlap of 5° between exposures.

There are special panorama heads which indicate the correct position for each exposure by an audible click. These are designed to work with a lens of a particular focal length, or they may have a set of interchangeable rings which suit a number of different lenses. Automatic heads of this type cut down the time between exposures and so make it possible to deal with subjects where there is some movement—e.g., of clouds or traffic—in the scene.

Suitable Subjects. The subject chosen for a panorama should have plenty of variety; mountains, docks and riverside buildings seen from the opposite side of the river and famous views are the right sort of material. Scenes with a straight, unbroken horizon line are generally not worth taking. Moving objects call for extra care to see that they do not appear on two sections of the same panorama. There is an unavoidable loss of time between one exposure and the next, and in that time a moving object can pass from one section into the next. On windy days, when the clouds are moving and changing shape quickly, it may be impossible to make a satisfactory panorama showing clouds.

Vertical Panoramas. In making a vertical panorama, the camera is mounted on a tripod fitted with a tilting head and swung in a vertical arc. Subjects for this type of panorama are rare because the whole field can generally be included in a single exposure with a lens of normal focal length. The view from the top of a cliff that falls vertically away from the camera makes a suitable subject if, for instance, there are streets and houses in the valley immediately below the camera and mountain scenery facing it.

Mounting Panoramas. Provided the prints are made from correctly made negatives there is no difficulty in joining up and mounting adjacent pictures. The prints should be first arranged side by side and one lapped over the other until the edge of the upper print registers with the scene in the lower. The overlapping edge of the top print is then lightly stuck down on the lower, preferably with india rubber solution.

When dry, both prints are cut through together down a line midway between the overlapping edges. This means that when the prints are mounted edge to edge, the two sides of the join will register perfectly and be almost invisible.

An even less conspicuous way of matching up is to follow prominent contours of the scene when cutting one edge (e.g., the left-hand one) of each print (except the left-hand print of the series). Provided these contours all lie within the overlap area of the adjoining print, they will tend to hide the join.

When mounting, the edges should be sandpapered down from the back as thin as possible before assembly. The base of the one print and the surface of the edge area of the adjoining print are then smeared with rubber gun, and the top print placed in position. Any excess gum can be rubbed off the surface when dry. P.J.

See also: *Hill cloud lens; Montage; Panoramic camera.*

PANORAMIC CAMERA. Special type of camera for making a continuous record of a wider section of the horizon than could normally be photographed in a single exposure. The lens scans the view from side to side and exposes a long narrow strip of film through a moving slit.

In practice the lens is made to rotate about its rear nodal point together with the whole camera on a tripod head, while a governor-controlled motor winds the film past the slit from one spool to the other. The film movement is carefully synchronized with the camera rotation, so that the image remains in effect stationary with respect to the film. The film itself is supplied in special lengths to suit the camera. With a suitable tripod mounting, such a panoramic camera can cover a complete circle of 360°.

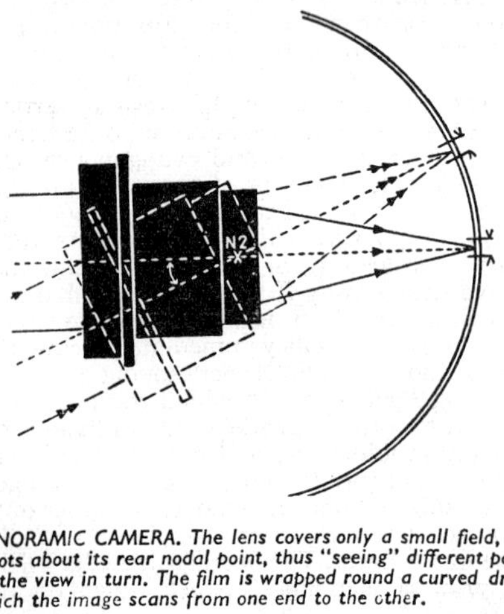

PANORAMIC CAMERA. The lens covers only a small field, but pivots about its rear nodal point, thus "seeing" different parts of the view in turn. The film is wrapped round a curved drum which the image scans from one end to the other.

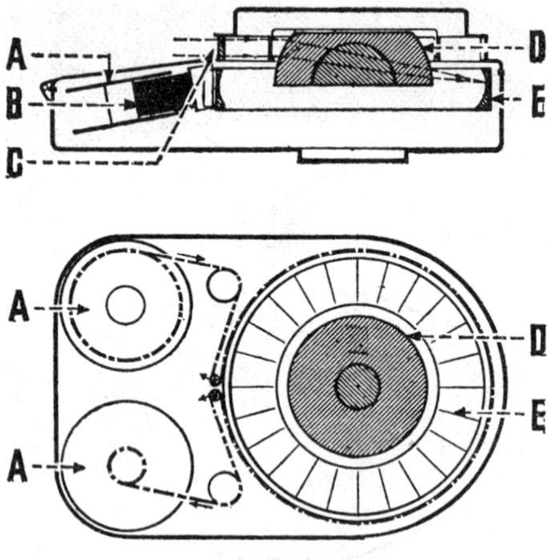

355 DEGREE HORIZON CAMERA. Top: Section. Bottom: Top view. The film from the feed spool runs round a drum and from there on to the take-up spool. A series of windows separated by diaphragm blades takes in image-forming rays over virtually the full circle and directs them via a special hemispherical lens on to the 35 mm. film in the image plane. A, film spool. B film (shown by broken line on lower diagram). C, window (one of a ring). D, concentric hemispherical lens. E, field flattening lens. F, diaphragm blade.

Cameras of this type are also used for taking long formal groups. The total exposure time from one end of the picture to the other is in the region of 10–15 seconds, though the effective exposure of each point in the picture may be around 1/10 second. This could be varied by making the slit narrower or wider, but in practice a simpler method of exposure control is adjustment of the aperture.

In the case of a group shot the whole group should keep still while the camera is running. Double exposure stunts are, however possible by one or two members at the first extreme edge running round the back to the other end of the group as soon as the camera view has passed them.

Special panoramic cameras have also been developed which cover a view up to 355° in a single simultaneous exposure on a curved strip of film with a special optical system.

See also: *Fish-eye lens; Hill cloud lens; Photo finish camera.*

PANTONE. Special type of dry litho photomechanical reproduction process developed from about 1924 onwards by Ronald Trist. This process was lithographic in principle but printed without water on ink-repelling areas.

A sheet of copper or, more commonly, copper-plated steel, was plated thinly with chromium. An acid resist image in line or half-tone was formed on the chromium which was then etched away down to the copper base in the non-image areas. The bare copper was silver plated and then treated with mercury which amalgamated with the silver, after which the plate was ready for the printing machine.

Printing ink will not take on a mercury-silver amalgam, but it takes very well on chromium; thus the plate could be rolled up with printing ink and used for direct or offset printing.

At its inception, Pantone appeared to be most promising. It would print on litho or letterpress machines, and the inventor devised many ingenious and useful developments of the basic principle; but it failed for such reasons as the tremendous printing pressure necessary on letterpress machines, the difficulty of replenishing the mercury by addition to the ink, objection to the poisonous nature of the mercury, litigation over patent rights, and scarcity of mercury. The commercial development of chromium plating was prompted by its use in the Pantone process.

PAPER GRADE. Most types of development paper are manufactured in a range of numbered grades. Each number corresponds to a particular type of image contrast—e.g., hard, normal, soft. Although there is no general agreement between manufacturers, most tend to favour five grades—No. 1 (Soft), No. 2 (Normal), No. 3 (Hard), No. 4 (Extra Hard) and occasionally, No. 5 (Ultra Hard). It does not follow that the same grades of different manufacturers all produce the same contrast.

By selecting paper of the correct grade it is possible to produce acceptable prints from negatives with contrast scales ranging from very soft to very hard.

A normal negative will yield the best print on a normal paper grade. However, a soft and flat negative will require a hard paper grade to put adequate brilliance into the print, similarly a contrasty negative needs a soft paper grade to record the extremes of th etone scale and produce an acceptable print.

See also: *Contact printing; Contrast; Enlarging; Papers; Variable contrast paper*.

PAPER HOLDER. On a horizontal enlarger the easel and paper holder are combined, but on a vertical enlarger the paper holder is usually a separate accessory.

Adhesive Holders. One of the earliest devices for holding down the paper was a shallow tray filled with a glycerin-gelatin jelly. The paper was simply pressed on to the surface of the jelly and held by adhesion. Nowadays the same effect can be produced by a double-sided adhesive sheet. One modern enlarger with a pressed steel baseboard is supplied with two magnetized strips which hold the paper firmly by its edges. Nowadays, however, most photographers favour the type of paper holder which is also a masking frame. These are made in both fixed and adjustable types and by virtue of the fact that they hold the paper by its edges, invariably give a print with white margins.

Masking Frames. The fixed size masking frame, which is commonly made in postcard and two-on postcard sizes, is simply an inverted shallow box the same size as the paper with a further box which fits neatly over it. The bottom of the outer box is cut away to form the picture aperture with a narrow rebate $\frac{1}{8}$ or $\frac{3}{16}$ in. wide all round. The printing paper is placed on the lower half, and the upper section is fitted over it. This holds the paper flat and leaves it exposed except for the margins.

Masks of this sort are made in both metal and moulded plastic.

Adjustable Types. Adjustable masking frames usually consist of either a metal or wooden baseboard to which is hinged an open frame of metal of square section which carries two metal masking strips. These strips are made to slide along adjacent sides of the frame so that they can be adjusted to enclose a rectangular picture space of any dimensions up to the full size of the holder. When a sheet of paper is placed between the board and the frame, the frame itself masks the paper on two adjacent edges, and the adjustable strips mask the remaining two.

The width of the white border is fixed by the width of the metal frame, and the adjustable strips are set so that they mask off the same width along their sides of the paper. Some types are made that permit actual adjustment of the border width. The paper is located under the mask by pressing it against two stops as the mask is lowered.

Masking frames for enlargements without white border frequently use a hinged glass plate in place of the frame, to hold the paper flat. Locating stops at the side of the frame help in positioning the paper. The glass plate must naturally be kept absolutely clean and free from dust, as well as from scratches. The glass itself is a flawless sheet of flat plate glass.

Borderless enlargements can also be produced on a simple board with two adjacent raised edges acting as positioning guides.

Exposure Control Systems. Advanced paper holders for commercial printing and enlarging incorporate a photo multiplier system that measures the light coming through the paper in the holder and is used to control enlarging exposures. The surface of the holder is in this case a translucent sheet of glass, with one or more photo-multiplier cells underneath it. These integrate the light coming through (in other words measure the average light intensity) and thus automatically allow for different negative densities and different degrees of enlargement. The photo-multiplier cell usually controls a relay that switches off the enlarger light (the cells having been activated by switching on) when a predetermined amount of light has reached the photo-multiplier cell.

The exposure control must be calibrated according to the translucency of the paper base and the paper speed; a further adjustment is also necessary for different print sizes. This arises because the photo-multiplier cells measure the light coming through a given area of the paper holder surface; if this area is reduced by masking down for smaller print sizes, the light reaching the cell naturally also decreases (although the light intensity on the paper may remain the same).

Similar automatic paper holders for colour enlarging may incorporate three sets of photo-multiplier cells with red, green and blue filters respectively in front of them. Each cell then monitors the appropriate colour component, and controls the exposure by light of that colour, or the movement of filters in the enlarging lamp house. G.W.A. & L.A.M.

See also: *Masks.*
Book: *Enlarging*, by C. I. Jacobson (London).

PAPER NEGATIVE. One of the cheapest forms of negative material consists of a negative emulsion coated on a paper base, known as negative card.

Many cameras produced for use by "while-you-wait" photographers are designed for this material. The card is loaded into sheaths and exposed in the camera in the normal way. After processing it can be printed in a number of ways: by being re-photographed in the original camera on to a similar card; by treating it with oil to make it translucent and then printing by transmitted light; by printing it by reflected light in a special enlarger working on the principle of the episcope; or by reflex printing. The last method is commonly used in document photography, especially in office copying machines.

Instant-picture cameras based on transfer diffusion also use a paper-based negative. This, however, is discarded after processing, where its function is to produce the positive image by transfer.

Paper negatives have been favoured by pictorialists and portrait photographers because they are so much easier to retouch than films or plates. It might sometimes be worth while to make an enlarged paper negative from a miniature negative that calls for retouching. In this way any number of prints can be made from the finished negative.

See also: *Supports for emulsions; While-you-wait photography.*

PAPERS

The final form of a large proportion of photographic records and pictures is the paper print. It is produced in most cases by exposing a sheet of sensitized paper in contact with a negative, or to the image of a negative projected by an enlarger.

The main types of paper for general photography are development papers which form a latent image that must be developed before it becomes visible. These are used in pictorial, technical, commercial, and professional photography. Special emulsions in this group are used for document photography and recording.

Another type of paper which used to be widely popular half a century ago is daylight paper. This forms an image on direct exposure, and needs no development. These papers are known as either print or printing out papers and were formerly employed for pictorial photography, but are now almost entirely confined to making proofs; within this type also come a number of specialized materials for plan-copying and similar purposes.

DEVELOPMENT PAPERS

Development papers vary in three important respects:

(1) Type of emulsion—i.e., chloride, bromide, chlorobromide, etc. (This also governs the development characteristics and image tones obtainable.)

(2) Contrast range—i.e., soft, normal, hard.

(3) Physical characteristics—i.e., thickness, surface finish, colour, etc.

The sensitive layer of most development papers consists of silver chloride, silver bromide, or a mixture of the two, in a gelatin emulsion. The relative proportions of silver chloride and silver bromide determine both the sensitivity and the image tone of the paper. Certain papers also contain a small proportion of other salts such as silver iodide.

With one or two exceptions—e.g., variable contrast papers—printing papers are not colour-sensitized. They are sensitive only to blue and violet rays.

There are four types of emulsion commonly used for coating development papers: silver chloride for contact papers, slow chlorobromide for warm tone contact papers, silver bromide for enlarging papers, and fast chlorobromide for warm tone enlarging papers.

Silver Chloride Papers. These are papers in which the emulsion contains silver chloride only.

The sensitivity of these papers is comparatively low, and they are generally suitable only for contact printing in a printing frame or printing box. Because of this limitation they are therefore referred to as contact papers, or by their older name of gaslight papers.

They are so slow to artificial light that they may be handled by subdued artificial light, or by a bright yellow safelight.

Chloride papers give blue-black tones in a metol-hydroquinone developer and take about 45 seconds at 65–68° F. (18–20° C.) to develop.

Slow Chlorobromide Papers. These papers have an emulsion that consists mainly of silver chloride, with a smaller proportion of silver bromide.

They are up to ten times as sensitive as silver chloride contact papers, and are generally suitable for contact printing, although they are sometimes used for enlarging.

As silver bromide is more sensitive than silver chloride, these papers must be handled under an orange or yellow-green safelight.

Slow chlorobromide papers give warm black tones on development in a metol-hydroquinone developer. They can, however, produce brown-black to warm brown and even reddish images by over-exposure followed by development in a dilute and heavily restrained developer.

Silver Bromide Papers. These papers are known as plain bromide papers. The emulsion contains only silver bromide with perhaps a trace of silver iodide.

Bromide papers are the most sensitive of all printing materials (100 to 1000 times as sensitive as chloride contact papers). They are mainly used for making enlargements where the intensity of the printing light is generally very much less than for contact printing.

As all bromide papers are sensitive enough to be fogged by even weak white light, they must be handled only by olive-green or orange darkroom illumination. Certain bromide papers are double coated and slightly colour sensitive. These can only be handled safely by the light of an orange safelight.

Bromide papers give pure black prints on development in metol-hydroquinone. Development usually takes about 1½–2½ minutes at 65–68° F. (18–20° C.).

Fast Chlorobromide Papers. Fast chlorobromide papers have an emulsion that contains mainly silver bromide, with a smaller amount of silver chloride and sometimes a little silver iodide. These papers are nearly, though not quite, as sensitive as ordinary bromide papers. They are mainly used for making enlargements, and need about 1½ to 4 times as much exposure as bromide papers.

Some fast chlorobromide papers are sensitive to green light as well as to blue and violet. An olive-green darkroom light is therefore not always safe enough; orange is better.

Fast chlorobromide papers give warm-black image tones on direct development in metol-hydroquinone paper developers. The usual development time is about 2–3 minutes at 65–68° F. (18–20° C.). Prolonged development of over-exposed prints in a dilute and heavily restrained developer gives warm brown and sepia tones as it does with the slow chlorobromide emulsion. The range of tones available is, however, not as wide with the slow chlorobromide emulsions.

Automatic Development Papers. Papers intended for automatic roller processing machines frequently incorporate also hydroquinone or another developing agent in the emulsion. During development the paper then passes

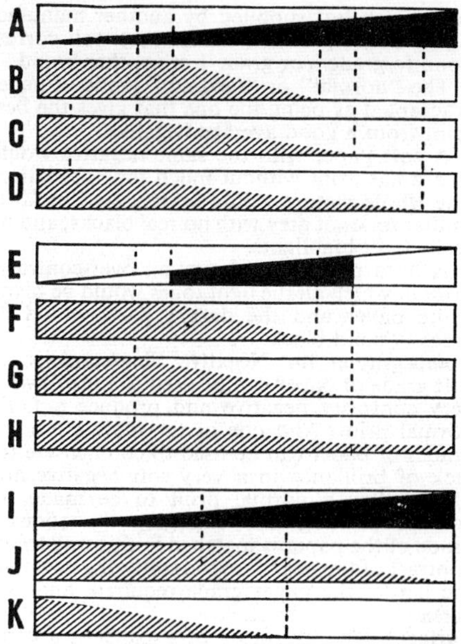

PAPER GRADES. Top: A, negative with long scale of tones from black to transparent. B, printed on hard paper; C, on medium paper; D, on soft paper. These papers reproduce respectively a short range, most, or all of the negative tones, spreading them from black to white. Centre: E, negative of short scale of tones. F, printed on hard; G, on medium; H, on soft paper. The negative tones are increased in contrast, reproduced correctly, or flattened respectively. Bottom: I, negative with long scale of tones. J, printed on hard paper to secure tone separation and highlight detail at the expense of the shadows. K, exposed to secure shadow detail at the expense of highlights.

through an alkaline activator bath which in conjunction with the developing agent develops the image on the spot within a few seconds. Papers of this kind may have either chloride or bromide emulsions.

Contrast Grades. Most development papers are made in several contrast grades. A range of such grades is necessary so that the photographer can select the most suitable to make the best possible print from every type of negative—from soft to hard.

The paper grades are usually classified by each maker according to his own ideas. There is as yet no standard system, but most of those in use at present conform broadly to the following scale:

No. 0: extra soft
No. 1: soft
No. 2: normal, or medium
No. 3: vigorous, hard, or contrasty
No. 4: extra vigorous, extra hard, or extra contrasty
No. 5: ultra vigorous, ultra hard, or ultra contrasty.

A particular grade of paper made by one manufacturer does not necessarily correspond to a paper carrying the same number or description but supplied by another manufacturer. Grade 2 of one maker may well correspond to grade 3 or grade 1 of another brand.

The "normal" or "medium" grade of paper is accepted as being the one that gives the best print from a good average negative.

A soft paper with the same negative would give a flat print without much tone gradation. The whole picture would be rendered in fairly similar tones of grey with no real blacks, and no really light highlights.

A hard paper would give an over-contrasty print in which all the light tones would be blank white paper and the darker parts would be areas of solid black.

Compensating for Negative Shortcomings. A soft grade of paper can be used to tone down a very contrasty negative and produce a fairly normal print. And one of the more contrasty grades of paper can be used to compensate for lack of brilliance in a very soft negative and again allow a normal print to be made. In contact printing and enlarging the contrast grade of the paper is matched in this way to the contrast of the negative; the softer the negative, the harder the paper grade required, and vice versa.

Surfaces. The "surface" of a paper is a combination of its texture and its finish.

Its texture may be smooth, fine grained, or rough. In addition, there are several artificial surfaces which are made to resemble fabrics like silk, rayon, canvas, etc.

Its finish may be glossy, semi-glossy (described under various names such as semi-matt, half-matt, velvet, lustre, etc.) and matt.

A glossy finish can only be produced in conjunction with a very smooth surface texture. A glossy print can, in addition, be glazed after processing to give a mirror-like shiny surface. A matt finish must also have a very smooth surface texture; but a semi-glossy or lustre finish may be used for every type of texture.

Enlarging papers and, to a lesser extent, contact papers, are made with a variety of surfaces. This variety enables the photographer to enhance the artistic effect of a picture by choosing a suitable paper to match the mood or character of the subject.

The Paper Base. The paper support that carries the sensitive layer may be either white or tinted, generally ivory or cream. Some modern white-base papers incorporate a fluorescent dye to increase the luminosity of the whites.

Most papers are made in a range of base thicknesses; the thinnest is light-weight (air mail), the thin standard base is single-weight, and the thick standard is double-weight. The thickest base of all is treble-weight, which is a comparatively heavy card.

Some types of paper base may be translucent, especially for document copying purposes. These are generally thin bases.

Papers intended for high speed processing often have a waterproof base. This is to minimize the absorption of solutions and so speed up the drying.

Printing paper manufacturers have their own code letters and figures which they print on the pack to indicate the contrast grade, surface, base tint, and paper thickness of the contents.

Choice of Papers. The selection of a print surface and paper tint for any particular print is largely a matter of taste. Prints intended for reproduction are an exception; they are always supplied on glossy or glazed white-based paper.

Apart from this special requirement, the photographer can generally contribute something to the pictorial appeal of his work by selecting a paper with the right characteristics.

A cream paper base may help to impart a suggestion of warmth and sunniness to the picture, and a warm-tone image tends to enhance the impression. Traditional or dignified types of subject, as well as character portraits, often look better on cream or ivory tinted papers.

Essentially black-and-white subjects—e.g., snow, mountains, machines—are generally handled better by a white based paper.

The smoother the paper surface, the more it shows up fine image detail. The smoothest glossy surface, particularly when glazed, is therefore specially suitable for scientific and technical subjects, small contact prints and prints for copying and reproduction.

A rough surface breaks up empty image areas, so it livens up broad subjects and big enlargements where there is comparatively little fine image detail. And rough paper also helps to hide graininess in big enlargements.

Artificially embossed surface textures such as rayon, silk, etc., have a similar effect.

Maximum Tone Range. The paper tint, image colour, and surface, affect the visual tone range of the print.

The visual range is narrower when the paper base is tinted than when it is white, because the tint darkens the lightest tones. A tinted paper surface reflects only 30–50 per cent of the light falling on it, compared with up to 90 per cent for a really white surface.

A warm tone image has a shorter range than a pure black one because it can never reproduce such deep shadows.

Semi-matt and rough surfaces scatter the light falling on them. This also increases the visual brightness of the shadows and so narrows the available range of tones. (This explains why matt prints often appear to lack brilliance and why they look brighter and more contrasty when they are wet than they do after they have dried.)

The maximum visual contrast is given by a pure black-and-white glazed glossy print. The white is the brightest possible, while the smooth surface and black tones ensure low reflectivity in the darkest image areas.

But the photographer who goes in for unusual

paper surfaces runs the risk of distracting attention from the subject matter of his photograph. Most photographs, in fact, look best when printed on a non-committal, semi-matt, print surface, on white or ivory-white paper.

Document Copying Papers. Papers for office and document copying purposes have special characteristics for the job they are required to do. They are mostly slow in speed and ultra-hard in contrast. Papers for office copying machines are frequently intended for reflex printing and can be handled by ordinary subdued daylight or artificial light. Mostly they are silver chloride papers and may incorporate a developing agent in the emulsion for rapid processing systems. Faster emulsions are also available for enlarging microfilm negatives; papers of this kind have to be handled in a darkroom by appropriate safelight illumination.

Some document copying processes follow special systems and may use pre-fogged papers (which yield a positive image from a positive on exposure to yellow light which destroys the existing latent image) or a diffusion-transfer process.

Document papers are mostly coated on a lightweight paper base and are matt surfaced. Thicker paper bases are also available, as well as translucent bases. The latter have a structure reminiscent of grease-proof paper, and are intended for making document copies that are to serve as master negatives or positives for subsequent mass printing by dye-line and similar processes.

COLOUR PAPERS

Papers for making colour prints are mainly of two types: positive and reversal.

Positive Colour Papers. These are intended for making colour prints from colour negatives. They invariably work on the subtractive colour principle and consist of three sensitive layers sensitized to blue, green and red light respectively. The layers also incorporate complementary colour couplers. As a result the blue-sensitive layer yields a yellow image on development in a colour developer; the green-sensitive layer yields a magenta image and the red-sensitive layer a cyan image. The three layers record the colour components of the negative image and reverse both the tone and the colour values of the latter.

Sometimes the sensitization of the three layers of a positive colour paper is specially matched to the characteristics of the dyes in the colour negative material of the same manufacturer. In extreme cases the nominally red sensitive layer may for instance be sensitized mainly to infra-red and the deep red end of the spectrum. This tends to improve colour separation and reproduction.

Processing involves colour development, a stop bath and bleaching of the silver image plus fixing out of all the silver salts. Manufacturers usually issue specific instructions and processing kits for colour papers of their own make; universal colour processing formulae for positive colour papers also exist. Usually the printing procedure requires some modification when non-standard formulae are used. In any case development time and temperature in the colour developer are very critical.

The exact colour balance of the print is also controlled deliberately by adjusting the colour of the printing light with the aid of appropriate filters.

Reversal Colour Papers. These yield a positive colour print from a positive transparency and require reversal processing. This resembles the procedure for reversal colour films. The processing solutions and procedure must be accurately matched to the particular make of colour paper employed; the manufacturer issues full instructions and kits where the paper is released for processing and printing by photographers. (Some reversal colour papers are released only to colour finishing laboratories appointed by the manufacturer.)

The structure of the reversal colour paper is very similar to that of a positive colour paper, and formation of the colour image proceeds in the same way as in the case of the reversal colour film. Colour rendering can be controlled within limits by modifying the printing light with filters.

Special Colour Papers. Some types of colour paper, both of the positive and of the reversal kind, work by a dye-destruction process. Here the paper already contains finished dyes in the three emulsion layers; these dyes are destroyed image-wise during processing and so leave a dye image after removal of the developed silver image and silver salts.

Special colour papers, sometimes with only two colour-forming layers, are also used in certain recording applications.

DAYLIGHT PAPERS

At one time daylight papers were very widely used for making positive prints. They form a visible density directly on exposure to actinic light, whereas development papers do not show the effects of exposure until treated with a developer.

The sensitivity of daylight papers is very much less than that of development papers, and they are nowadays very little used except in cases when their slow speed and special characteristics are an advantage.

Recording Papers. The main current use of "daylight" papers is to record traces of moving spots of light, for example as reflected from the mirror of a galvanometer or other recording instrument. The paper there moves continuously past the location of the moving light spot; the vibration of the latter thus exposes a trace on the moving strip of paper. This image is visible immediately after the exposure, but darkens on further exposure to a controlled light source.

Printing-Out Papers. Print-out papers of the more old-fashioned kind are usually coated with a sensitive emulsion of silver chloride in gelatin or collodion. Their sensitivity is so low that they may safely be handled in ordinary artificial light, and even in subdued daylight.

Print-out papers usually give a reddish image which is toned with gold or platinum before fixing to produce sepia to brown prints. Ordinary print-out papers used to be employed by professional photographers for proof prints.

Some so-called self-toning daylight papers incorporate the gold toning salts in the emulsion. Such papers tone automatically in the fixing bath and do not therefore need a separate toner.

Special Daylight Materials. Other daylight printing materials rely on the sensitivity of either ferric salts or of colloids like bichromated gelatin.

In iron printing papers the action of light reduces ferric salts to the corresponding ferrous compounds. These in turn can be made to form usable images in a number of ways.

In the obsolete platinotype and palladiotype papers, the ferrous compounds reduced platinum or palladium salts to the corresponding metals, thus producing permanent platinum or palladium images in attractive tones of brown.

In the same way the ferrous compounds can be made to react with silver salts to form a black silver image (Kallitype process).

A blue image can be produced by making the ferrous salts react with potassium ferricyanide to precipitate Turnbull's Blue as in blue-print papers.

In certain types of blue-print paper the images are formed from the unexposed ferric salt. These papers therefore give negative prints from negatives, and positives from positive transparencies.

Iron printing out materials are mostly used for photomechanical reproduction of drawings.

Bichromated colloid (gelatin, gum, glues) materials make use of the tanning action of bichromates on colloids that have been exposed to light. This renders the gelatin or gum insoluble in the exposed parts of the print. The rest remains soluble and can be washed off with hot water to leave a relief image.

If the gelatin coat is coloured with a pigment a visible pigment image is left after washing. This is the principle of the carbon and carbro processes.

The tanned gelatin image also possesses the property of taking up greasy inks in proportion to its degree of exposure and consequent hardening. This method of pigmenting the image is the basis of the oil (and derived bromoil) processes.

Bichromated colloids are extensively used in photo-engraving to produce etching resists and other images by the action of light.

Other daylight printing materials utilize the action of light on certain diazo compounds. After exposure, these compounds lose their ability to produce dyes by coupling. So by treating the material with suitable couplers after exposing it under the negative, the unexposed parts of the image form a dye image but the exposed parts remain unaffected. This produces a negative image from a negative and a positive from a positive. Like blue-print papers, diazotype materials are mainly used for photomechanical reproduction. L.A.M.

See also: *Colour materials; Contrast; Document copying; Glossy paper; Keeping qualities of materials; Paper grade; Printing materials; Sizes and packings; Supports for emulsions; Variable contrast paper.*

Books: *Enlarging*, by C. I. Jacobson (London); *The Complete Art of Printing and Enlarging*, by O. R. Croy (London).

PARA-AMINOPHENOL. (Free base) 1-amino-4-hydroxy-benzene. Developing agent used in certain concentrated developers.

Formula and molecular weight: $NH_2C_6H_4OH$; 109.

Characteristics: White crystalline powder.

Solubility: Slightly soluble in water at room temperature. Freely soluble in alkaline solutions, especially caustic alkalis. Very slightly soluble in sulphite solutions.

PARA-AMINOPHENOL HYDROCHLORIDE. 1-amino-4-hydroxybenzene hydrochloride. Developing agent.

Formula and molecular weight: $NH_2HClC_6H_4OH$; 145·5.

Characteristics: White powder.

Solubility: Fairly soluble in water at room temperature. Freely soluble in alkali solutions. Very slightly soluble in sulphite solutions.

PARABOLIC MIRROR. Silvered glass or metal reflector of which the cross-section along the axis is a parabola. This shape has the property that if a point source of light is placed at its geometrical focus, the light reflected by the mirror will be a parallel beam. In practice, this condition can never be completely achieved because there is no such thing as a point source.

Parabolic mirrors can be used in spotlights instead of the conventional condenser lens. Parabolic reflectors also have special uses.

See also: *Mirror; Reflector.*

PARALLAX. Apparent alteration in the relative position of two objects that occurs when there is a change in the viewpoint of the observer. Also, term generally applied to the difference between two viewpoints. Because of parallax, the picture seen in the viewfinder of a camera

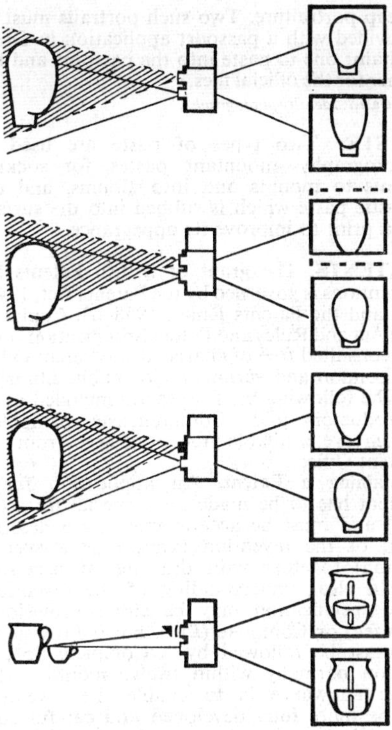

TWIN LENS CAMERA PARALLAX. Where the viewfinder is in a different position from the camera lens, it necessarily covers a slightly different view. At near distances this may lead to parts of the subject (e.g., the top of a portrait head) being cut off in the negative. Top: Uncorrected parallax error in twin lens reflex camera. Upper centre: Correction by masking the finder so that it always covers less than the film. Lower centre: Tilting the finder lens downwards at near distances. Bottom: Optical wedge, deflecting finder beam for close-ups; the field covered becomes the same, though the viewpoint does not.

is not the same as the picture seen by the lens, some inches below or to the side.

The effect is not serious at distances beyond about 6 feet, but closer than this it is necessary to allow for the difference in viewpoint. This may be done by reducing the finder field to take in only the subject area covered all the time, by correction marks in the finder, or by a mechanical shift of the finder field (with a movable mask or by tilting the finder). Automatic parallax correction systems have such a mechanical adjustment coupled with the focusing movement of the lens.

Any such compensating system can only correct for the change in image field occasioned by the parallax error; it cannot correct for the different perspective caused by the displaced viewpoint. In practice this is significant only at very close subject distances, as in macrophotography. If, for example, the camera lens is looking directly at the face of a small cube, the top of the cube will not appear in the picture but even when the viewfinder is tilted the picture it sees will still include the top of the cube.

The only type of viewfinder free from parallax error is the focusing screen of the single lens reflex camera.

See also: *Viewfinder*.

PARALLAX STEREOGRAM. Composite positive print or transparency prepared in such a way that when viewed from a particular distance with both eyes a stereoscopic picture results. The stereogram actually consists of a combined stereoscopic pair with each picture broken up into very fine strips. The pictures are printed so that the strips of left and right hand pictures alternate. This is then covered by a suitable grid. The spacing between picture strips and grid lines, and the separation of the two, are so chosen that when looked at from the correct viewing distance each eye sees its own picture strips while its view of the other set is blocked by the grid lines.

Parallax stereograms can be printed from a normal stereoscopic pair or they can be made directly from the subject. In one typical system the original negative is made in a camera on a single plate behind a suitable grid. The camera has two lenses, separated by the interpupillary distance, and associated with a pair of prisms. The prisms bend the incoming rays of light towards the centre so that the images formed by the lenses are superimposed instead of lying side by side.

If the right and left images formed on the plate by the grid were printed in the normal way, the resulting picture when viewed would be pseudoscopic in exactly the same way as a normal stereoscopic pair printed without transposition. The correct viewing conditions are achieved, however, by simply shifting the viewing screen the width of one line to the side. This produces the same effect as trans-

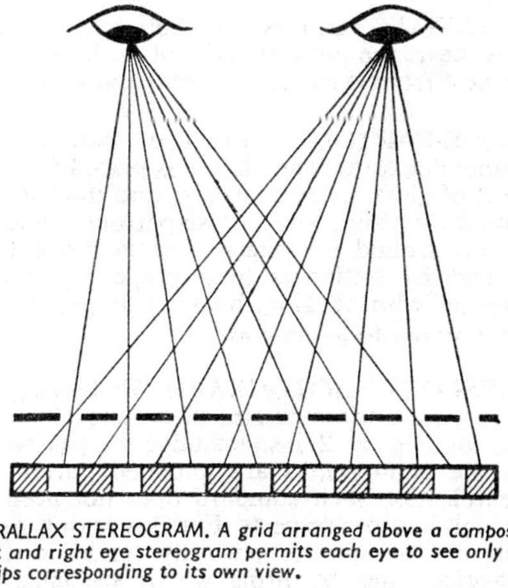

PARALLAX STEREOGRAM. A grid arranged above a composite left and right eye stereogram permits each eye to see only the strips corresponding to its own view.

position in printing a normal stereoscopic pair.

The prisms used with the taking lenses in this method produce lateral inversion which is inevitable in the interest of simplicity.

The principle of the parallax stereogram has been applied to numerous stereoscopic systems the first being that of F. E. Ives, 1903. Instead of grid lines, a screen of lenticular lenses may be used, and by making numerous exposures from slightly differing viewpoints it is possible to produce a picture which presents a changing aspect when the viewpoint is moved.

See also: *Lenticular system; Stereoscopic photography.*

PARA-PHENYLENE DIAMINE. (Free base.) 1 : 4-diaminobenzene. Developing agent in fine grain developers.

Formula and molecular weight. $C_6H_4(NH_2)_2$; 108.

Characteristics: Brownish white powder.

Solubility: Fairly soluble in water at room temperature.

PARA-PHENYLENE DIAMINE HYDROCHLORIDE. 1 : 4-diaminobenzene hydrochloride. Developing agent in fine grain developers.

Formula and molecular weight: $C_6H_4(NH_2HCl)_2$; 181.

Characteristics: White powder. More stable than para-phenylene diamine.

Solubility: Fairly soluble in water at room temperature.

PARAPHOTOGRAPHY. Collective term embracing image-forming systems based on other than the silver halide process. It covers thermography, electrophotography, photochromism, diazotypy, photopolymerization, and magnetic image recording, which are sometimes also referred to as unconventional, unorthodox, or just non-silver systems.

PARAXIAL. Term used in optics to describe rays near the optical axis of a lens. Rays farthest from the axis are termed abaxial.

PASSE-PARTOUT. Gummed material for framing pictures. The picture is placed facing a sheet of glass of the same size and the two are bound together by passe-partout binding passed around the edges. The gummed tape is sold by stationers in a range of colours together with backing, hanging loops, etc.

See also: *Framing photographs.*

PASSPORT PHOTOGRAPH. Full-face portrait with a trimmed size of between 2 × 1½ ins. and 2½ × 2 ins. included in passports. As the photographs are intended purely for identification their standard does not need to be high in other respects. For this reason passport photography is generally carried out at cut-price rates by firms which specialize in cheap portraiture. Two such portraits must be provided with a passport application in Great Britain; one to paste into the passport and the other for the official files.

See also: *Identity photographs.*

PASTE. Two types of paste are used in photography—mountant pastes, for sticking prints to mounts and into albums, and encaustic paste which is rubbed into the surface of a print to improve its appearance.

PATENTS. The grant of British Patents for inventions is governed by the Patents Act, 1949–57, and the Patents Rules, 1958–63. Copies of the Act and Rules and Patent Specifications may be consulted free of charge at the Patent Office in London and various major public libraries.

The following brief notes are intended as an introductory guide. Comprehensive pamphlets which are available free of charge from the Patent Office.

Obtaining a Patent. An application for a Patent has to be made on a special form (£1 fee) and must be accompanied by a description of the invention (known as a specification) together with drawings if necessary for a clear understanding of the invention. The specification may be either Provisional (no fee) or Complete (£10), but if Provisional it must be followed by a Complete Specification normally within twelve months. This time allowance is to enable the invention to be more fully developed and careful consideration to be given to the drafting of a Complete Specification. This must be a full and detailed description from which a competent workman could carry the invention into practical effect using only the document and drawings (if any) and without further directions. The Provisional Specification, on the other hand, need only disclose the basic features of the invention, although as much as is possible should be described. The filing of an application (£1) with a Provisional Specification establishes a priority date and gives the inventor an option, so to speak, on a consequent Patent, and thereby forms an interim measure of protection. A Provisional Specification does not constitute a Patent but merely forms a legal basis for the earlier date (the priority date) to be accorded to a subsequent Complete Specification. A Provisional Specification is not published, and its contents remain confidential to the Patent Office until a Complete Specification is filed, accepted and published.

No models are required nowadays. Patents are granted on payment of a sealing fee of £3 following satisfactory examination of the documents. Thus the basic cost of obtaining a Patent is £14 (£1 on application, £10 on filing a Complete Specification, and £3 sealing fee) and this covers the first four years from the date of filing the Complete Specification.

Thereafter annual renewal fees are payable (if it is desired to keep the Patent in force) until its final expiry after sixteen years.

The granting of a Patent does not imply that there is any commercial value in the invention, and it should be realized that few of the 30,000 inventions patented annually ever become a financial success.

Value. A Complete Specification must end with a summary of the main novel feature of the invention and any important subsidiary features. This should be in the form of clear, brief statements (known as claims).

The value of a Patent depends mainly on the wording and validity of the claims of the final Complete Specification. Specialists, known as Patent Agents, are usually employed to draft at least the Complete Specification. Obtaining a really useful Patent can be an expensive enterprise and the would-be patentee is strongly advised, before filing a Complete Specification, to consult a reputable manufacturer as to the probable value of his invention. Some manufacturers advertise their willingness to consider "gadgets" and give experienced advice as to the best course to adopt. The reputation of such advertisers should be checked before disclosing what might be a valuable idea, and the prior filing of an application for a Patent with a Provisional Specification is always a wise precaution. As a Provisional Specification must be followed by a Complete Specification within twelve months, a manufacturer should be approached well before the end of that period to leave enough time to proceed with the application.

Patent Rights Abroad. If it is desired to obtain Patent rights outside the United Kingdom application has to be made to the government of the country in which protection is desired, be it one of Her Majesty's Dominions or a foreign state. The procedure varies considerably and prospective applicants are advised to consult a registered Patent Agent. Under the terms of an international convention an application for a Patent in most countries abroad, if made within twelve months of the date of application in the United Kingdom, has the same effect as if made on the day the application was made in the United Kingdom. A list of such countries is given in the free literature available from the London Patent Office.

Applying for foreign Patents can be an expensive procedure, and an international search is to be recommended. This might reveal lack of novelty in the invention and thereby save the costs of making applications. A Patent Agent can arrange for a search to be made.

Patents in the U.S.A. The United States of America offers a particularly wide market for photographic equipment and its Patent procedure is therefore worthy of special attention.

An application for a Patent by a citizen of the United States has to be made with a formal Petition (this usually includes a Power of Attorney in favour of a registered patent attorney or agent) and Oath of inventorship, accompanied by a full description of the invention is the form of a specification (including a drawing in every instance where the nature of the case admits of illustration) and Claims (stating precisely what is to be protected). A filing fee of $30 (exclusive of attorney's fees) must be paid. On approval a further fee of $30 is payable before a Patent is granted. There are no further fees or taxes and the protection lasts for 17 years from the date of issue.

The most vital part of the application is the Claims and as these have to be filed in the first instance it is desirable to employ a patent attorney. Names of attorneys may be obtained from the nearest Patent Law Association. This will, of course, increase the cost of obtaining a Patent. It is wise to have a search made before applying for a Patent as the filing fee is not returnable if no Patent is granted.

There are several good books published giving practical advice on all aspects of inventorship. These may be consulted in public libraries or obtained from the Patent Office in Washington.

Should the cost of obtaining a Patent be a deterrent there is an alternative course available. This is to make a complete written description, in duplicate, of the objects and purpose of the invention, describing its construction and operation in detail, including whatever drawings or sketches assist in understanding it. One copy of the papers should be signed and dated by the inventor and two reliable witnesses who understand the invention. The signed copy should then be carefully preserved and the unsigned copy sent to a likely manufacturer. This procedure is even more valuable if the invention has first been put into practice and successfully operated before two witnesses who understand it. Only manufacturers of repute should be consulted as there is no legal redress until a patent (for which the above document is valid proof of inventorship) has been obtained.

R.E.

PAUL, ROBERT WILLIAM, 1869–1943. English engineer and scientific instrument maker. Pioneer of cinematography. Copied Edison's Kinetoscope in 1894. Built in 1895 a cine camera, and in 1896 the first English cine projector, the "Theatrograph". Made the first English films; later he made (with Professor Silvanus Thompson) animated films for teaching. In 1912 he abandoned cinematography and devoted himself entirely to precision instrument building. His works were taken over in 1920 by the Cambridge Instrument Company.

PEARL LAMP. Ordinary filament lamp with a bulb sprayed or frosted to diffuse the light. Not to be confused with the opal type of lamp.

PELLET PROCESS. Specialized application of the iron salt process for giving positive blue-line copies of line tracings—e.g., engineering drawings and architects' plans.

See also: *Printing processes (obsolete)*.

PELLICLE (PELLICULE). Thin film used as semi-reflecting surface in one-shot colour cameras. Also word sometimes used for the thin layer of emulsion covering a plate or film. The term occurred in the early Pellicle processes, in which commercially prepared emulsion was used for coating dry plates at home.

See also: *Beam splitter.*

PENTAPRISM. A five-sided glass prism inserted between the focusing screen and the viewing eyepiece of an eye-level reflex camera. Also sometimes called roof prism. Whereas a simple 45° reflecting mirror gives an image of the reflex screen that is either upside down or

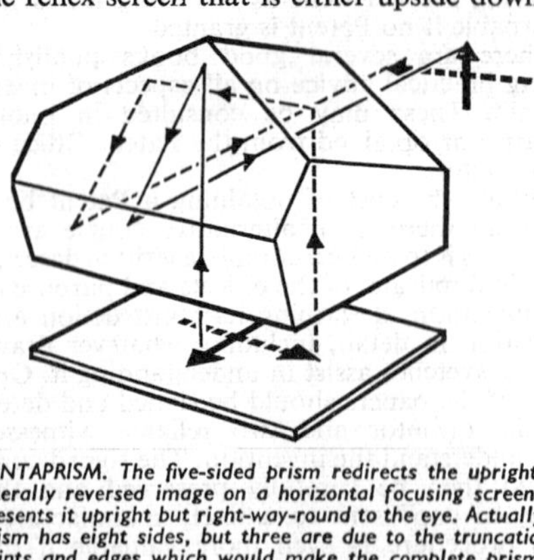

PENTAPRISM. The five-sided prism redirects the upright but laterally reversed image on a horizontal focusing screen and presents it upright but right-way-round to the eye. Actually the prism has eight sides, but three are due to the truncation of points and edges which would make the complete prism too bulky.

laterally reversed (left-to-right), the pentaprism, by introducing extra reflecting surfaces, shows the subject both upright and with the right and left sides in their proper place.

See also: *Viewfinder; Reflex camera.*

PER CENT (%). Parts per hundred. Way of stating concentrations of solutions, proportions, ratios, etc. Thus a 10 per cent solution of a chemical contains 10 parts of the chemical in every 100 of solution.

See also: *Chemical calculations; Solutions; Weights and measures.*

PERCENTAGE SOLUTION. Solution containing a given quantity of a dissolved substance in a stated volume of the solvent—i.e., 10 parts of a 10 per cent solution of, say, potassium bromide contain one part of the bromide.

Usually a percentage figure refers to weight of solid per volume of solution (w/t), but in certain cases (e.g., with liquids like alcohol) a volume per volume (V/V) solution may be meant. For accurate work weight of solid per weight of solution (W/W) may be specified.

See also: *Chemical calculations; Solutions.*

PERFORATION. Small symmetrical shaped holes spaced at regular intervals down the length of special sensitized materials. The holes may be in a single row down one side or the middle of a film or they may run lengthwise down both sides of the film.

Perforations are mostly employed for cine films, although miniature and sub-miniature still cameras use perforated film in the standard cine gauges.

The main purpose of perforations is to provide a means of film transit (at the same time frequently operating a counter which measures the amount of film used). Another important function of perforations, in cinematography, is to ensure that the film shall move and come to rest with great precision in relation to the significant parts of the mechanism—e.g., the picture gate. Owing to the considerable reduction and re-enlargement of cinematograph pictures in taking and projection, perforations must be of extreme accuracy: errors in the dimensions or positions of the holes would cause unsteady pictures and rapid wear of the various mechanisms. All these dimensions (over-all width of film, separation of perforations, size of perforation, radius of corner, etc.) together with the permissible tolerances have been laid down in British Standard No. 677, parts 1 to 3, and in the comparable standards of the U.S.A. and other countries.

History. The standard negative perforation devised by Edison was the first to be used, and from about 1910 onwards was used for all motion picture film, both negative and positive.

In 1925 the present standard positive perforation was adopted when tests had shown that the slightly greater height and rounded corners gave a longer life to release prints during repeated projection.

This shape of perforation was completely satisfactory for projectors, which are sprocket driven, but it was unsuitable for camera equipment where the perforation is also used to locate the film in the gate. All existing cameras had register pins which fitted the negative perforation with the necessary accuracy, and the studios were unwilling to change all the cameras in use to the slightly larger and differently shaped pin required by the positive perforations.

Up to the present time, all 35 mm. negative, master positive, and duplicate negative stock is perforated with the barrel-shaped Bell-Howell hole, whilst all black-and-white release positive stock has the rectangular positive perforation.

Types of Perforation. 8 mm., 9·5 mm. and 16 mm. cine films all have the same type of perfora-

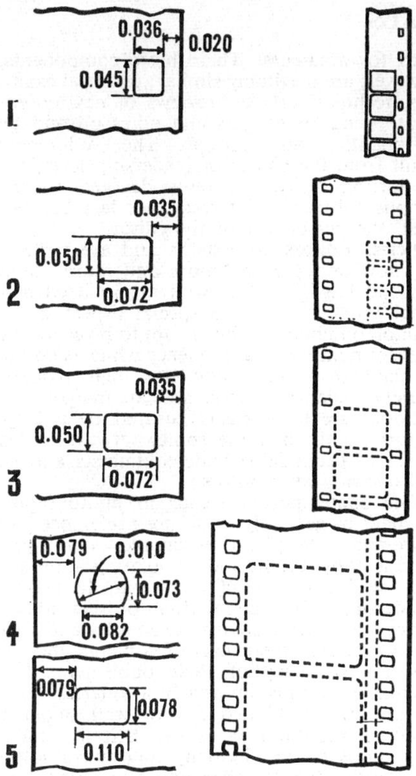

PERFORATION TYPES AND SIZES. 1. Super-8 mm. film. 2. Standard double-8 mm. film. 3. Standard 16 mm. film. 4. Negative perforation of standard 35 mm. film. 5. Positive perforation of standard 35 mm. film. All dimensions are in inches.

tion—a rectangular hole with rounded corners. This is used for reversal and for negative and positive stock.

On 8 mm. film, the perforations are spaced at single frame intervals along one side of the film. Double-run 8 mm. has the same perforations but down each side of a 16 mm. wide film. In practice, the film is run through the camera twice, using only half the film width each time. The film is subsequently split down the middle, when it assumes the dimensions, etc., of normal 8 mm. film. Super-8 mm. film has a smaller and almost square perforation hole.

On 9·5 mm. film, the perforations run down the centre of the film between the framing of adjacent pictures. A special type of 9·5 mm. film has two central perforations.

16 mm. film has perforations along both sides with silent film and down only one side with sound film; they are spaced at single frame intervals.

The shape of perforations for 35 mm. films depends upon the type of film:

(1) The standard perforation for cine negative films is circular, but with flats at the top and bottom. This is employed for ordinary 35 mm. negative that runs vertically in both camera and projector, in which case there are four pairs of perforations per frame. In one special wide screen projection system, where the film runs horizontally with the longer dimension of the negative image along the length of the film, there are eight pairs of perforations per frame. Another system uses six pairs of perforations per frame, with the film running vertically.

The same perforation shape is also used in 35 mm. film prepared for use in miniature still cameras.

(2) The standard perforation for 35 mm. positive cine films is rectangular with a straight top, bottom and sides, and rounded corners; it is spaced the same as for normal 35 mm. negative film. This shape of the perforation is similar to that used on all narrow gauge films and gives a satisfactory standard of accuracy combined with long working life in projectors. 35 mm. "still" positive films also have this kind of perforation (except positives processed by reversal from the original exposure).

(3) Positive films having the widescreen pictures taken through an anamorphic lens have perforations similar to (2), but rectangular and slightly smaller in dimensions; this is to make room for the greater horizontal dimensions of the image and to accommodate the magnetic sound tracks that are coated alongside the visuals.

The 70 mm. and 80 mm. perforated films are also used in special cameras for research work, in particular for the study of aircraft and guided missiles.

Colour Film. The perforation of colour film is more complicated. For processes where three coloured part images have to be registered in successive operations on printers which, like the camera, have register pins, negative perforations have always been used. Integral colour processes, which do not call for such precise registration, use negative-perforated camera stock and positive-perforated print stock.

Other Sensitized Materials. In various types of recording devices employing sensitized film or paper, perforations serve both to feed the strip and often to time the recordings. A recent British Standard specification, No. 1193: 1954, standardizes the perforations on film and paper, in various widths from 16 mm. to 80 mm. With the exception of 16 mm., all perforations are identical with the standard 35 mm. positive perforation.

Certain magnetic sound recording tapes have circular perforations to facilitate accurate synchronism of the sound recording with the picture, although such tape is prone to expansion and contraction throughout its life. G.H.S.

See also: *Cine film; Film transport.*

PERIODICALS

The function of the periodical press is to report on new developments, to keep readers stimulated and to serve dealers and manufacturers as advertising media in their particular market.

Photographic periodicals are produced by independent publishers, photographic societies and trade associations as well as by major manufacturers and even retailers. Their revenue is derived only in part from what the readers pay for the publication whilst the other, sometimes the larger, part is contributed by advertisers or sponsors.

Types. Most periodicals are published monthly. Some, however, appear quarterly, bi-monthly and a few are weeklies. Prior to the second World War, at least two German journals were published twice weekly. The frequency of publication is, as a rule, due to commercial considerations and is of greater concern to advertisers than to readers. The actual volume of genuinely new and important information of photographic interest is limited and thus the inevitably repetitive element of the editorial contents is as a rule inversely proportionate to the length of the period that separates two consecutive issues of any periodical.

Photographic periodicals may be classified as technical, pictorial or trade journals although the three groups often overlap. Most amateur magazines gravitate towards the middle group.

Standards, methods and policies vary a great deal. Readers' interest always ranks as the first consideration in selecting and angling editorial material—but first of all must come the decision as to what sphere of readers the publication in question is to attract. It is at this critical stage that the interests of advertisers, organizations or sponsors are taken into account and where the basic compromise is reached between serving the public and paying the publication's bread and butter expenses.

Periodicals professing lofty aims and maintaining an aggressive independence have a notoriously low chance of survival. This is particularly true of those crusading, from time to time, for some artistic principle.

High level technical journals often have to be specially underpinned by the photographic industry, by learned associations or simply by the selfless enthusiasm of their editors and contributors. The higher their level the lower are the fees they can pay for contributions.

Trade and market journals are sometimes linked to associations whose interests they represent, in which case a fair proportion of their contents is bound to be devoted to business politics. Independent publications have to aim at a brighter editorial approach to compensate for some first hand news which may be monopolized by the "official organs".

The greatest variety of approach to and presentation of the subject is found among magazines for amateurs. Their basic components, however, are inevitably similar; pictorial examples, technical articles, reviews of new equipment, ideas for gadgets and notes on and by personalities and groups. The differences result from the choice of readership level, the proportion of the contents devoted to the various spheres of interest, and, last but not least, the personality of the editor.

Policies. Editors are experts and are reputed to know and preach "more and more about less and less". Editors with pronounced personal interests in some special aspect of or opinion on photography are apt to revert to the relevant points with a frequency which is bound to slant their work and tire their readers. Variety, balanced views, and an invigorating outlook, are more easily ensured by a large editorial staff than the somewhat out-moded device of perennial outside columnists and so called steady contributors.

The contemporary trends in photographic journalism generally indicate preference for more and better pictorial examples, simple and concise copy, bright and stimulating approach and free interchange of views with the reader.

Readership. The circulation of periodicals varies greatly from country to country and also, of course, with their level of contents and the business efficiency of their publishers. Top circulation claims approach a quarter of a million in the United States, well under a hundred thousand in Great Britain, under fifty thousand in Germany and much lower in France, Italy, Sweden and Switzerland.

The advertiser is less interested in absolute figures than in the relative size of specific market coverage any periodical may offer and whether this coverage is expansive, stationary, receding or seasonally fluctuating.

For the reader, what matters is not how many other people read the same magazine but whether he likes what he reads.

Obviously the reader's choice of periodical reading will be conditioned first and foremost by his own interests. If these are pictorial, he will know only too well what type of pictures and so which pictorial magazine he prefers. If his interests are technical, his own level of understanding will be as sure a guide as any, and he will soon learn to distinguish journalistic bluff hiding behind terminological fireworks from genuine information. If his hobby is centred on clubs, exhibitions and societies, he has only to look for the journals which provide most of that type of news. Specimen copies are supplied free as a rule and the choice is so wide that there is no reason for any photographer not to have the periodical he needs, likes and deserves. A.K.-K.

See also: *Books on photography; Literature on photography.*

Book: *Royal Photographic Society Periodicals Catalogue* (London).

PRINCIPAL PHOTOGRAPHIC PERIODICALS OF THE WORLD

Title	Frequency
*Amateur and General—Still and Cine (*organ of amateur organization)*	
Amateur Cine World, London	W
Amateur Photographer, London	W
Amateur Photography * Johannesburg, S.A.	M
Ashahi Camera, Tokyo, Japan	M
Australian Photography, Sidney, Australia	M
Bild und Ton, Berlin, E. Germany	M
Camera 35, New York, N.Y., U.S.A.	B
Camera, Lucerne, Switzerland	M
Camerart, Tokyo, Japan	B
Camera Magazine, London	M
Camera Owner, London	M
Camera World, Sidney, Australia	M
Cine Amateur, * Paris, France	M
Cine Camera, London	M
Cinema Pratique, Paris, France	B
Colour Photography, London	M
Eight Millimetre Magazine, London	M
Eight Millimetre Movie Maker and Cine Camera, London	M
Ferrania (Ferrania), Milan, Italy	B
Film-Kreis, * Der, Munich, Germany	M
Focus, * Haarlem, Holland	F
Fotografia, Warsaw, Poland	M
Fotografie, * Leipzig, E. Germany	M
Fotografie, Prague, Czechoslovakia	M
Fotokino-Magazin, Leipzig, E. Germany	M
Foto-Magasinet, Copenhagen, Denmark	M
Foto-Magazin, Munich, Germany	M
Foto-Prisma, Düsseldorf, Germany	M
Galerie, Die, Vienna, Austria	B
Good Photography, London	M
Leica Fotografie (Leitz), Frankfurt, Germany	B
Leica Photography (Leitz), New York, N.Y., U.S.A.	Q
L'Officiel de la Photographie et du Cinema, * Paris, France	M
Modern Photography, New York, N.Y., U.S.A.	M
Nordisk Tidskrift für Fotografi, Gotenburg, Sweden	M
Photo-Amateur, * Liestal/BL, Switzerland	M
Photo-Cinema, Paris, France	M
Photo-Cine-Revue, Paris, France	M
Photographic Journal, The, * London	M
Photographics New Zealand, Auckland, N.Z.	M
Photography, London	M
Photography and Travel, Johannesburg, S.A.	M
Photo-Japan, Tokyo, Japan	B
Photo-News, Guildford, Surrey	W
Photo-Tribune, Antwerp, Belgium	M
Popular Photography, New York, N.Y., U.S.A.	M
Practical Photography, Guildford, Surrey	M
P.S.A. Journal, The, * Chicago, Illinois, U.S.A.	M
Rolleigrafie, Munich, Germany	B
Schmalfilm, Berlin, Germany	M
Schweizer Schmalfilm, Winterthur, Switzerland	M
Sovietskoje Foto, Moscow, U.S.S.R.	M
Spiegelreflex-Praxis, Düsseldorf, Germany	Q
U.S. Camera and Travel, New York, U.S.A.	M
*Professional—partly also for trade and industry (*including official organ of professional or industrial association—see also Scientific)*	
American Cinematographer, Hollywood, California, U.S.A.	M
British Journal of Photography, London	W
Camera-Europhot, * Lucerne, Switzerland	Q
Canadian Industrial Photography, Toronto, Canada	B
Film-Technikum, Weisbaden, Germany	M
Film User, London	M
Foto-Blickpunkt, Viersen, Germany	M
IBP Record, * London	M
Industrial and Commercial Photographer, London	M
Industrial Photography and Commercial Camera, Sidney, Australia	B
Industrial Photography, New York, N.Y., U.S.A.	M
Industrial Reproduction, New York, N.Y., U.S.A.	Q
Industrial Screen, London	M
International Photographer, Los Angeles, California, U.S.A.	M
International Photo-Technique (Linhof), Munich, Germany	Q
Kinomechanik, * Moscow, U.S.S.R.	M
Le Photographe, Paris, France	M
MFM—Moderne Fototechnik Ludwigsburg, Germany	M
National Press, New York, N.Y., U.S.A.	M
Norsk Fag Foto, * Oslo, Norway	M
Österreichische Photo-Zeitung, Vienna, Austria	M
Photo Presse, * Hanover, Germany	W
Photo – Schweizerische Photorundschau, * Berne, Switzerland	F
Photo-Technik und Wirtschaft, * Berlin, Germany	M
PMI—Photo Methods of Industry, New York, N.Y., U.S.A.	M
Professional Photographer, The, * Milwaukee, Wisconsin, U.S.A.	M
Reproduction Methods for Business and Industry, New York, N.Y., U.S.A.	M
Research Film, Göttingen, Germany	SA
Svensk Fotografisk Tidskrift, * Stockholm, Sweden	M
*Scientific (*Partly also organ of professional association, mainly presenting original papers and abstracts)*	
Abstracts of Photographic Science and Engineering Literature, New York, N.Y., U.S.A.	M
British Kinematography, * London	M
Journal of Photographic Science, * London	B
Journal of the Biological Photographic Association, * Boston, Mass., U.S.A.	Q
Journal of the Optical Society of America, * New York, N.Y., U.S.A.	M
Journal of the SMPTE, * Pa, U.S.A.	M
Journal of the Society of Photographic Instrumentation Engineers, * Hollywood, California, U.S.A.	Q
Journal of the Society of Scientific Photography of Japan, * Tokyo, Japan	Q
Jurnal Nauknoi i Prikladnoi Fotografii i Kinematografii, Moscow, U.S.S.R.	M
Kino-Technik, * Berlin, Germany	M
Medical and Bilogical Illustration, London	Q
Perspective, London	Q
Photogrammetric Engineering, * Wisconsin, U.S.A.	B
Photographic Abstracts, London	Q
Photographische Korrespondenz, Darmstadt, Germany	M
Photographic Science and Engineering, * Easton, Pa., U.S.A.	B
Photographie and Wissenschaft (Agfa), Leverkusen, Germany	I
Repro-Graphie, * Bad Homburg, Germany	B
Revue D'Optique, * Paris, France	M
Science et Industries Photographiques, Paris, France	M
Tecknika Kino i Televidenya, * Moscow, U.S.S.R.	M
Zeitschrift für Wissenschaftliche Photographie, Leipzig, E. Germany	M
*Trade (Mainly for dealers or distributors, including official organ *)*	
Australian Photo Dealer, * Sidney, Australia	M
Der Foto-Händler, Munich, Germany	F
Der Photo Markt, * Düsseldorf, Germany	M
Der Schweizer Photo-Händler, * Zurich, Switzerland	M
Fotonyheterna, Stockholm, Sweden	M
Inpho, * Düsseldorf, Germany	F
Japan Camera Times, Tokyo, Japan	M
Japan Camera Trade News, Tokyo, Japan	M
Norsk Fag Foto, * Oslo, Norway	M
Photo Dealer, New York, N.Y., U.S.A.	M
Photographic Retailer, The, * London	M
Photo Trade, Toronto, Canada	M
Photo Trade News, Sidney, Australia	M
Photo Trade of Japan, Tokyo, Japan	B
Photo Trade World, London	M
PTN Photographic Trade News, New York, N.Y., U.S.A.	F

Frequency: B = bi-monthly; F = fortnightly; I = irregular; M = monthly; Q = quarterly; SA = semi-annually; W = weekly

PERIPHERY PHOTOGRAPHY Technique of photographing 360° all-round views of objects to obtain a flat "rolled out" picture. Such pictures may be obtained either by rotating the object in front of a camera fitted with a moving film system, or by surrounding the object by an all round optical system that records the whole circumference simultaneously.

Periphery Camera System For this method the object is set up on a rotating turntable opposite the camera. The latter has a travelling back to move the film during the exposure. The lens projects the image of the object on to the film through a narrow slit, which ensures that only that part of the object directly facing the lens axis is recorded at any instant.

The film holder and turntable are driven by motors synchronized to keep the image sharp during the exposure. The speed of the turntable and of the film holder are linked in such a way that the film travels through its full length while the turntable rotates through one revolution.

A camera of this kind can handle objects of sizes from several feet in diameter down to a fraction of an inch (for example a pencil surface). The scale of reproduction must be accurately related to the size of the object and hence to its distance in front of the camera lens. For that reason a camera set-up of this nature can only deal adequately with objects of approximately cylindrical shape where all surface features are more or less in one plane.

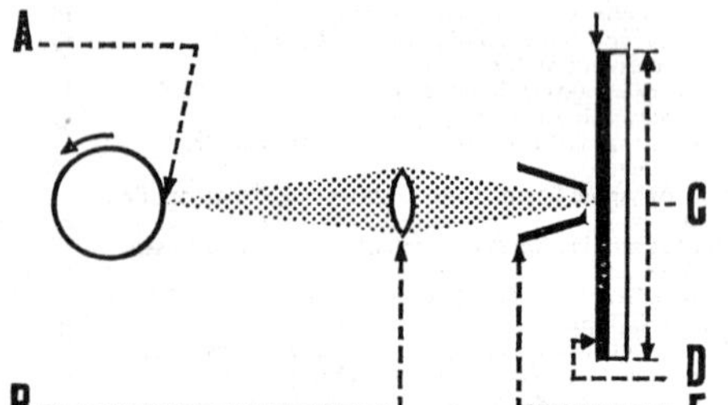

SET-UPS FOR PERIPHERY PHOTOGRAPHY. A, portion of cylinder or other round object imaged on the film. B, lens. C, extent of plate movement. D, travelling plate. E, slit. The distance between the object's surface—either the outside (top) or the inside (bottom)—and the lens is matched with the lens-to-plate distance and the plate movement so that the latter covers exactly the external or internal circumference of the object.

Parts of the object that project appreciably from, or recede behind, that plane would move at a different speed during the rotation of the turntable and therefore lead to a distortion (and often unsharpness) of the image on the film. The degree of unsharpness under these conditions depends on the slit width. At its narrowest setting this is usually around 0·003 in., which gives sharp but still distorted images of object features whose movement speed did not exactly match the film movement.

The resulting image on the film covers a full 360° view of the circumference of the object, but in a flat plane. This is frequently more suitable for the examination of surface details than any orthodox view could be.

Panoramic Mirror Systems Instruments that provide a simultaneous all-round view of the circumference of an object use a panoramic mirror surrounding the object itself. The latter goes through the centre of the mirror, which then projects an image of this circumference view through a further mirror system into an eyepiece or on to a photographic film. Compared with a periphery camera, this arrangement needs no moving film, but is limited to smaller objects. (In practice mirror systems like this exist taking objects up to a diameter of about 4 ins.) The image produced is distorted, consisting of a flat ring, the circular inner and outer boundaries of which define the limits of the field of view along the axis. Since the image itself is stationary, an additional dimension of movement can be brought in to record different sections along the length of the object with a cine camera. This technique has been used to check the production of continuous lengths of items like steel cables, machined rods, etc.

The circular image can be converted into a flat one by a reversing optical system that projects the periphery image in a curved plane surrounding the optical system. If a photographic film or paper is wound around this image plane, it can be unrolled after exposure and processing to yield a flat and rolled-out but undistorted view of the object. The light source for such photography is usually a ring light surrounding the panoramic mirror of the taking system. By linking the optical rectifying system directly with the panoramic mirror unit, a flat rolled-out image can be produced by a single exposure. L.A.M.

See also: *Moving image photography.*

PERMITS TO PHOTOGRAPH. With certain exceptions people are allowed to photograph anything to which they have obtained legal access. The exceptions are subjects coming under the Official Secrets Act, certain subjects covered by copyright, anything obscene, views in the Law Courts, and the Houses of Parliament. For the latter, permission may be granted upon request but only when the House is in recess.

When a photographer wishes to take photographs on private property or in a private building, requesting permission is a matter of good manners. It is always as well to ask before starting to take pictures if there is any doubt. Permission should be asked to take pictures inside cathedrals, churches, historic buildings or private houses, and if possible the permit should be obtained in advance of the visit. A letter to the dean of a cathedral, the vicar of a church, or the custodian or owner of private property will usually have the desired effect.

A stamped addressed envelope should always be enclosed as a gesture of courtesy.

No one is normally given permission to take photographs in a church whilst a service is being conducted. The one exception is at a wedding ceremony, but in that case the vicar must always be consulted first.

In some parks and public places it is necessary to have permission to erect a tripod. This is because a tripod constitutes an obstruction and there are often bye-laws controlling or prohibiting their use. Anyone who particularly wants to use a tripod in such places should apply for permission to the Park Superintendent.

No one needs permission to take photographs of other people provided the photographs truthfully represent the subject and are not libellous. Even so it is only courteous to ask permission to photograph a stranger, and usually an offer to supply a print is all that is required to procure a willing sitter.

Photographs may not be taken in the cinema, at the theatre or the circus without permission. The cinema and theatre have the question of copyright to consider and although photography is occasionally permitted at the theatre the photographer must make sure of it first. The circus is a different matter. Permission is usually given readily, and one circus even has a competition for the best photographs taken during a performance. The only provisions are that camera users must not stand up, or interfere with anyone else's enjoyment, nor must they use flash except at a special photo-call.

Photography is also prohibited in most military and atomic energy installations.

Regulations abroad generally follow the same pattern as in Britain, though in a number of countries restrictions are more severe and may include strategic factories, bridges, port installations, etc. G.W.P.

See also: *Copyright and the photographer.*
Book: *Photographers and the Law*, D. Charles (London).

PERSPECTIVE

Perspective is a graphic device which artists use to reproduce the appearance of scenes and solid objects on a flat surface—i.e., for representing a three-dimensional subject in two dimensions. There is more than one method of perspective representation, but the method generally employed by draughtsmen and photographers is the simplest. This supposes that the observer is looking, with one eye, through a window pane at the scene, from a fixed viewpoint opposite the centre of the pane. Seen in this way, the outlines of the various objects in the scene can be imagined as being drawn on the glass, and if the whole scene were actually filled in with a pencil or brush, the result would be a picture in two dimensional perspective. This imaginary window will be useful in explanations that follow; it must be remembered, however, that this is only an artificial device for convenient description.

PRINCIPLES

There are three things that the photographer needs to know about perspective: the basic principles, perspective as it is interpreted by the camera and the subjective aspect—i.e., the mental impressions produced by the perspective picture.

The picture that would be produced by drawing over the outlines of the scene framed in an imaginary window pane has only two dimensions; it shows the relative positions of points in the field of view embraced by the frame, but not their distance from the observer. Distance in the perspective picture is conveyed by size. A football would appear as a circular disc, which would grow proportionately smaller as the football went farther and farther away. One football twice as far away from the observer as another would be represented by a disc of half the diameter. Similarly the apparent separation between parallel lines—e.g., the edges of a road or the rails of a railway—running away from the observer gets less and less as they recede; at one point twice as far from the observer as another, the separation is half as much; three times farther away, it is one-third as much, and so on. On the other hand, the separation between parallel lines running across the scene parallel to the window pane is to all intents and purposes constant because all points are at the same distance from the plane of the window.

Viewpoint and Picture Plane. An object in the scene subtends the same angle at the viewpoint as its perspective image on the picture plane—i.e., on the window pane. Both the object and its image may therefore be regarded as the bases of similar triangles whose common apex is the eye of the observer at the fixed viewpoint. So in a perspective view, the size of the object is to the size of the image as the distance of the object from the viewpoint is to the distance of the image from the viewpoint.

In other words, when the observer traces on the window pane the outline of an object seen through it, the closer his eye is to the window,

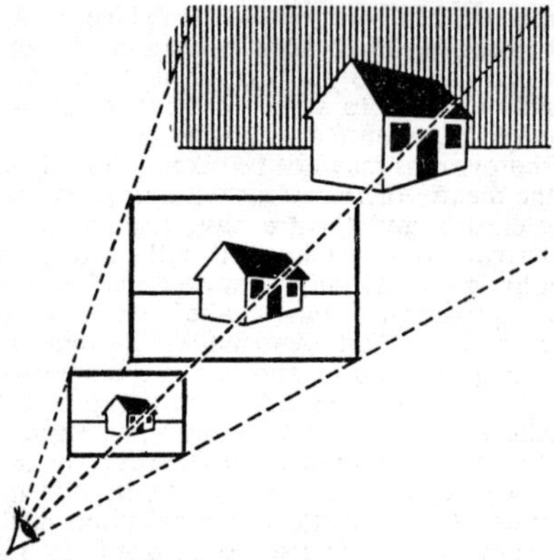

SCALE AND PERSPECTIVE. Moving the picture plane farther away from the eye of the observer increases the scale of the image but does not alter its perspective.

the smaller the outline image will be in relation to the pane, and the farther away, the larger. (If the window is so far away from the viewpoint that it actually touches the object, then the image on the screen will be the same size as the object.)

But as the eye of the observer comes closer to the window, and the relative sizes of things get smaller, more things are included in the area enclosed by the frame. And as the observer backs away from the window, the relative size of things seen through it increases and fewer things are included in the area enclosed by the frame.

With a short distance between the eye and the pane, the eye can survey a wide angle of the scene and see a large number of objects which form relatively small images in the picture plane. With a long distance between the eye and the pane, the eye is forced to observe only a few objects within a narrow angle of view. But since the frame is the same size in each case, the few objects that fill it in the second case will appear correspondingly larger in scale than in the wide-angle view.

In practice the closest distance at which the average human eye likes to look at things is 8 to 10 inches. So normal scenes are drawn (or photographed) as though traced on a picture plane about ten inches away from the eye.

Size and Viewpoint. In the perspective picture it is the relative size of objects in the scene that fixes their position. Of two objects of the same actual size the one that traces the smaller image will be the farthest away. But their relative sizes—i.e., how much bigger one is than the other—will depend on how far they are away from the observer.

If the observer is looking along a line of evenly spaced telegraph poles with his back against the first, then the third pole, which is twice as far away as the second, will be half the height. The fourth pole will be only $1\frac{1}{2}$ times as far away as the third so it will be only 2/3 the height. The 25th pole will be 23/24 of the height of the 24th, and so on.

The same relationship will apply to equidistant points on an object—i.e., the farther they are from the viewpoint the closer they approach each other. And their separation is greater near the viewpoint—e.g., from 1/2 to 2/3, etc.—than it is farther away—e.g., from 24/25 to 25/26, etc. So the first telegraph pole in the row appears very much taller than the second and the 24th appears very little larger than the 25th, yet all the poles are the same size and are equally spaced.

The change in relative size depends solely on the distance from the viewpoint; it does not matter how large or how small an area of the scene is included. If the frame of the picture

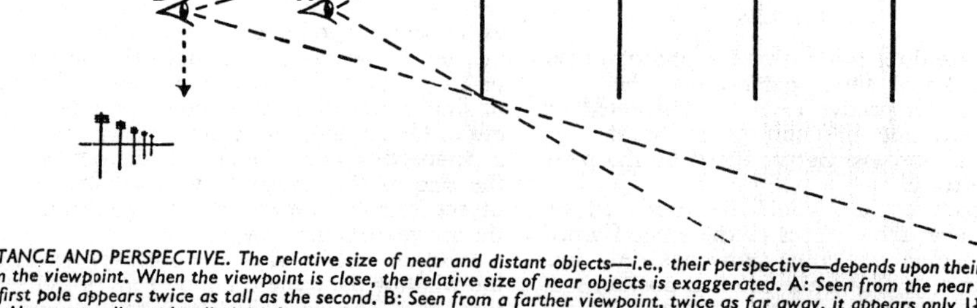

DISTANCE AND PERSPECTIVE. The relative size of near and distant objects—i.e., their perspective—depends upon their distance from the viewpoint. When the viewpoint is close, the relative size of near objects is exaggerated. A: Seen from the near viewpoint the first pole appears twice as tall as the second. B: Seen from a farther viewpoint, twice as far away, it appears only $1\frac{1}{2}$ times as tall. Also, succeeding poles diminish less in size, and the distance between them in the image decreases only slowly.

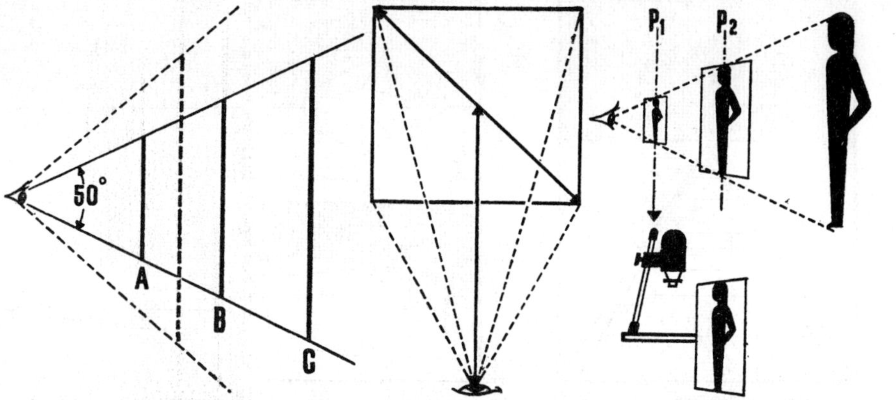

SIZE OF FRAME AND SCALE. Left: *Pictures must be viewed at such a distance that they lie within the angle of view of the eye, i.e., 50°. I they lie outside (broken line) they call for an abnormal amount of eye movement.* Centre: *To lie inside the comfortable viewing angle the picture must be looked at from a distance not less than its diagonal.* Right: *A small picture from a near viewpoint looks the same when enlarged to twice the size at twice the distance.*

cuts off the first twenty-three poles, the second pole in the picture—i.e., No. 25—will still be only 23/24ths the height of the pole in front, and not half the height.

This change in the relative size of objects receding from the viewpoint is the thing that fixes the perspective of the picture. So the perspective is a function of viewpoint only, and is not influenced by any other factor.

Size of Frame. With the eye at any given distance from the picture plane, the size of the frame does not affect the real size of the image. A small frame will include only a correspondingly small part of the scene while a larger frame would include more, but the size of the images traced on the picture plane would be the same in each case.

There is no limit to the amount of the scene that can be included in the picture; it is only necessary to go on increasing the size of the frame. But in practice the human eye can only take in objects included in an angle of about 50°. So the normal practice is to limit the size of the finished picture to dimensions that fall within the angle of 50° covered by the eye from its most convenient viewing distance. This means that the diagonal of the picture should be about equal to the viewing distance—e.g., a picture looked at from 10 ins. away should not normally exceed 6 × 8 ins.

Scale. The size of the image traced by an object on the picture plane in relation to the size of the frame increases as the picture plane is moved farther away from the viewpoint and towards the scene. If the separation of the viewpoint from the picture plane is doubled, the size of objects included in the picture will be doubled also. But the same result can be produced by enlarging the original picture to twice the size. In both cases the perspective of the picture remains unaltered and there is nothing in the appearance of the image of a particular object to show whether its scale has been enlarged (or reduced) by altering the viewpoint-picture plane separation or by keeping the separation constant, tracing out the scene, and then enlarging the result. In each case the scale of the objects and their perspective will be the same.

Looking from its fixed viewpoint, there is nothing to indicate to the eye (apart from the purely physical difficulty of focusing very close objects) whether it is looking at a miniature perspective picture 1 in. away, or one ten times as high and wide but 10 ins. away, or one 100 times bigger but 100 ins. away. Neither the scale nor the perspective nor the area enclosed by the frame are changed by making the same proportionate change in both viewing distance and frame dimensions. So the artist can please himself whether he paints a miniature portrait to be looked at one foot away or a larger picture to be looked at from a proportionately greater distance.

Horizon. If the picture plane is vertical and the viewpoint is opposite the centre of the frame (the basic assumption in normal plane perspective) then the horizontal line joining the viewpoint to the centre of the frame if produced will meet the horizon, and the line of the horizon will bisect the resulting picture. If the horizon is above or below the centre of a picture this is either the result of deliberately omitting part of the picture, above or below the horizon, or of a departure from the conditions of normal plane perspective.

The position of objects in the picture in relation to the horizon is an important factor that indicates their proportions. Assuming a normal viewpoint of about 5 feet from the ground, then all points at that height lying directly between the viewpoint and the horizon will be on the level of the horizon. Wherever the

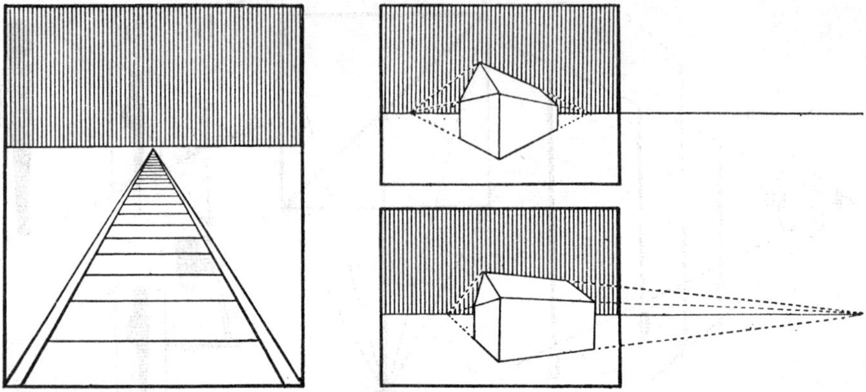

VANISHING POINTS. Parallel lines receding directly from the observer appear to meet at a point (called the vanishing point) on the horizon in the centre of the picture plane. **Right:** *When the parallel lines are at an angle to the horizon, the vanishing point lies off the centre or even outside the picture plane. The horizon itself corresponds to the height of the observer above the level of the lowest part of the subject. It can be raised or lowered at will.*

line of the horizon intersects objects in the picture, the point of intersection will be 5 feet above the ground. So the height at which the horizon intersects any objects in the picture is also an indication of the height of the viewpoint above normal ground level.

Vanishing Point. In a perspective reproduction of a scene, all parallel lines which appear to converge will continue to approach each other in direct proportion to their distance from the observer, until finally they meet at a point on the horizon. This point is known as a vanishing point; there is one such point for all parallel lines lying in the same plane.

For all parallel lines normal to the picture plane the vanishing point lies on the horizon at the centre of the picture. As the parallels incline to one side or the other of the normal, the vanishing point moves right or left until finally it passes outside the frame. When this happens, the vanishing point will still lie on the horizon, but the horizon will have to be extended outside the frame. The limit is reached when the lines lie parallel to the picture plane and the vanishing point lies at an infinite distance along the extended line of the horizon.

Oblique Perspective. Everything that has been said above assumes that the picture plane is vertical, but the same principles can be applied to show what happens when the picture plane is tilted.

From the point of view of an observer drawing over the outlines of the scene looked at through a window pane, it makes no difference whether the picture plane is vertical or tilted. The picture that he traces on a sloping pane will still fit over the scene as he sees it so long as he looks at it from the same viewpoint and holds the frame at the same angle.

But if the picture traced under these condi-

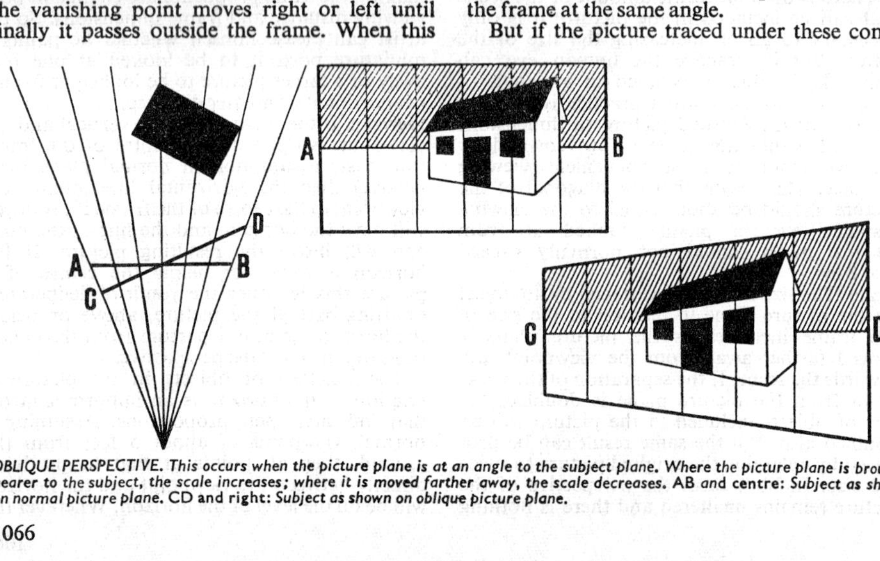

OBLIQUE PERSPECTIVE. This occurs when the picture plane is at an angle to the subject plane. Where the picture plane is brought nearer to the subject, the scale increases; where it is moved farther away, the scale decreases. AB *and centre: Subject as shown on normal picture plane.* CD *and right: Subject as shown on oblique picture plane.*

tions is looked at in the normal way—i.e., held up straight—the relative proportions of the various objects will be different from those given in a normal plane perspective.

Suppose the picture had been traced with the top of the frame tilted forward towards the scene. Then the picture plane would be closer to objects in the upper half of the scene than in the lower half, and the top would be farther away from the viewpoint. This means that the objects in the upper half of the scene would be reproduced on a larger scale than those in the lower half, and vertical parallel lines would therefore appear wider apart at the top than at the bottom of the picture. They would only appear parallel if the picture were sloped away from the observer at the original angle at which it was tilted.

In addition to increasing the scale of all objects in the upper half of the picture, the forward tilt of the picture plane would also increase the perspective of the objects in proportion to their increased nearness. (The difference in size between perspective images of objects increases progressively as they approach the picture plane as illustrated by the telegraph pole example above.)

Similarly inclining the picture plane to make the top slope away from the scene would result in a picture in which vertical parallel lines would be wider apart at the bottom than the top, and the perspective of objects in the lower part of the picture would be increased.

Parallel lines running across the picture from side to side would remain parallel on the tilted picture plane but their linear separation on the plane would increase with increasing tilt—in other words the scene would be stretched out from top to bottom, a smaller amount of the scene being drawn out to fill the full height of the frame. A similar thing occurs when a circle of light shines on to an oblique surface: the circle becomes elongated and requires a larger area to accommodate its full circumference.

So tilting the picture plane has three effects on the appearance of the perspective picture:

(1) It enlarges the scale of objects progressively as they are brought closer by the tilt.

(2) It increases the perspective of objects brought closer by the tilt.

(3) It elongates the subject in the direction of the tilt.

Panoramic Perspective. It has already been said that there is no limit to the amount of the scene that can be included in the picture by making the frame larger—or by bringing the viewpoint closer to the picture plane. This is all very well in theory, but in practice there are objections because the human eye is unable to embrace an angle greater than about 50° and pictures covering such a wide field are apt to appear distorted when looked at in the normal way.

The apparent distortion results from the fact that the picture plane is only really normal—i.e., square on—to the line of sight at the very centre. The picture plane becomes increasingly oblique to the line of sight away from the centre, and the bigger the frame, the more oblique will be the angle of the picture plane at the edge of the picture.

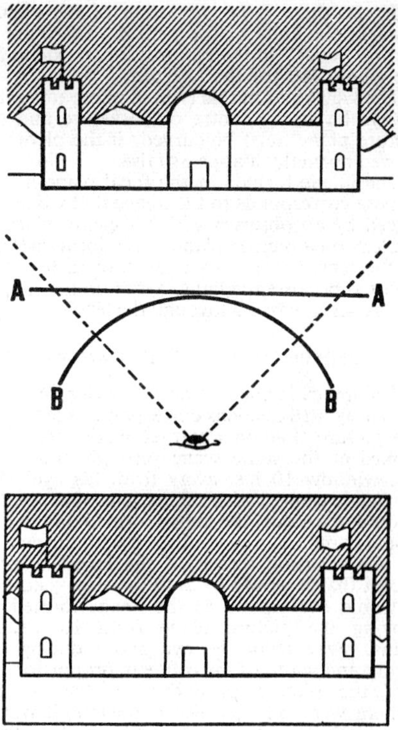

PANORAMIC PERSPECTIVE. AA: At the centre, the picture plane lies at right angles to the line of sight and objects are represented in normal perspective. Towards the edges of the picture, the plane becomes more and more oblique and produces more and more of the characteristic distortion. When the picture plane is wide in proportion to its distance from the eye of the observer—i.e., in a panoramic view—the distortion is proportionately great. BB: To present a live impression of the scene it is necessary to use a curved picture plane with the eye situated at its centre curvature.

It was shown above that when the picture plane is at an oblique angle the result is to enlarge, elongate and increase the perspective of the objects traced out on it. All these effects occur towards the edges of the field covered by the picture plane and are more pronounced as the edges get farther away from the centre.

Curved Panoramic Perspective. There is another way of arriving at a perspective view of an abnormally large scene which has none of the above objections. This is to trace out the scene on a curved picture plane with the viewpoint at its centre of curvature. In this case the picture plane is always normal to the line of sight, so there is no distortion of the picture towards the ends of the picture. There are in fact no "ends" to the picture; it can extend for the whole 360° of the horizon.

The viewpoint in this type of panorama must always be at the centre of curvature and the picture plane must be curved; if the picture is viewed normally it appears false.

The image formed at the focal plane in the camera corresponds to the image that would be traced by an observer with his eye in place of the lens on a picture plane at the focal distance of the lens. (Except that the camera image is inverted because the light rays from the subject cross over in passing through the lens.)

PERSPECTIVE IN THE CAMERA

The image formed on an 8 × 10 ins. focusing screen by a 10 ins. lens corresponds exactly to the picture that an observer would see if he looked at the same scene through an 8 × 10 ins. window 10 ins. away from his eye. If a miniature camera with a 2 ins. lens were pointed at the same scene, the image formed on the film would compare in every respect with the picture on the focusing screen of the larger camera except that it would be 2/10 the size—i.e., 1/5. It would correspond to the image formed by moving the picture plane from 10 ins. to 2 ins. away from the eye and reducing the height and width of the frame in proportion.

If the picture produced by the smaller camera were enlarged five diameters, it would be identical with the 8 × 10 ins. picture. Similarly a camera with a 5 ins. lens would give a picture which would be identical with the 8 × 10 in. pictures after 2× enlargement.

Normal Lenses. The general aim in taking a photograph is to recreate the impression of the original scene. While the original scene would be related to a picture plane of any size at any distance from the observer, people mostly hold things about 10 ins. away for comfortable viewing and their eyes have an angle of view of about 50°. So a photograph should reproduce the effect of looking at the original scene through a frame with a diagonal of 10 ins. and held 10 ins. away from the eye. This gives an angle of view of approximately 50°.

These conditions can be met directly by a camera with a 10 ins. lens working on a plate with a diagonal of about 10 ins.—e.g., a whole plate ($6\frac{1}{2} \times 8\frac{1}{2}$ ins.) camera. A contact print from such a camera held 10 ins. away from the eye would reproduce the effect of looking at the original scene through a $6\frac{1}{2} \times 8\frac{1}{2}$ ins. window. The same conditions can be met by subsequently enlarging the negative from a 2 ins. lens five times, a 3 ins. lens 3·3 times and so on.

In each case, to fulfil the required viewing conditions, the focal length of the lens must be approximately equal to the diagonal of the plate or film covered. Such a lens is regarded as

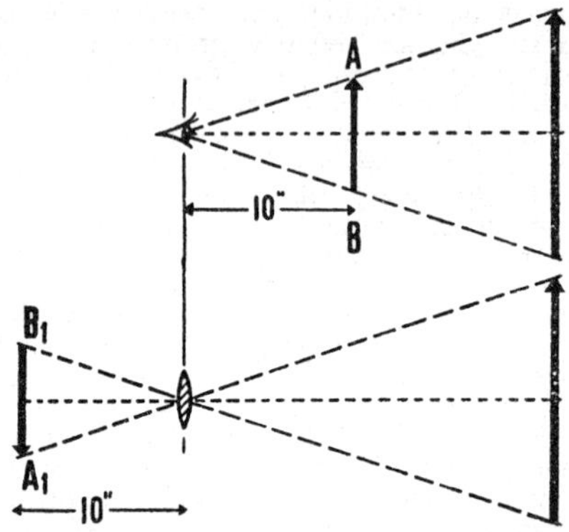

CAMERA PERSPECTIVE. The camera image formed by a 10 ins. lens corresponds to the picture traced out by an observer on a window pane in front of him, 10 ins. from his eye. AB picture plane. A_1B_1, camera image.

the normal lens for that particular negative format. The normal lens for a 35 mm. miniature camera—negative $1 \times 1\frac{1}{2}$ ins. (24 × 36 mm.), diagonal approx. 1·75 ins. (43 mm.)—is $1\frac{3}{4}$–2 ins. (4·5–5 cm.); for $2\frac{1}{4}$ ins. (6 cm.) square cameras—diagonal approx. $3\frac{1}{2}$ ins. (80 mm.)—is 3–$3\frac{3}{8}$ ins. (7·5–8·5 cm.) and so on.

Any camera with a normal lens of focal length $= f$ will give a picture in natural perspective for viewing at 10 ins. if the negative is enlarged $10/f$ times.

Generally, if the normal lens of the camera is replaced by one of n times its focal length, then the image size will be multiplied n times on the new negative while an identical picture can also be produced by enlarging the original negative n times. If the whole of the "normal" negative is enlarged, the dimensions of the picture will, in addition, be n times the dimensions of the long-focus picture.

Long-focus Lenses. If the picture plane is moved farther away from the observer's eye, it includes less and less of the scene but the perspective remains unchanged; 20 ins. away from the eye, the frame will include only half the picture width that it included at 10 ins. But all the objects in the field included will be twice as big in relation to the frame as they were before.

The same effect is produced by doubling the focal length of the camera lens. A long-focus 20 ins. lens will cover only half the field of the normal 10 ins. lens, so everything will be reproduced at twice the previous size. But the perspective of the picture will not be altered. So that the same picture would result from enlarging the corresponding part of the negative taken by the 10 ins. lens (and covering the larger view) to twice the size.

This holds good equally for all other sizes of camera and standard lenses when the focal length of the lens is doubled without altering the viewpoint.

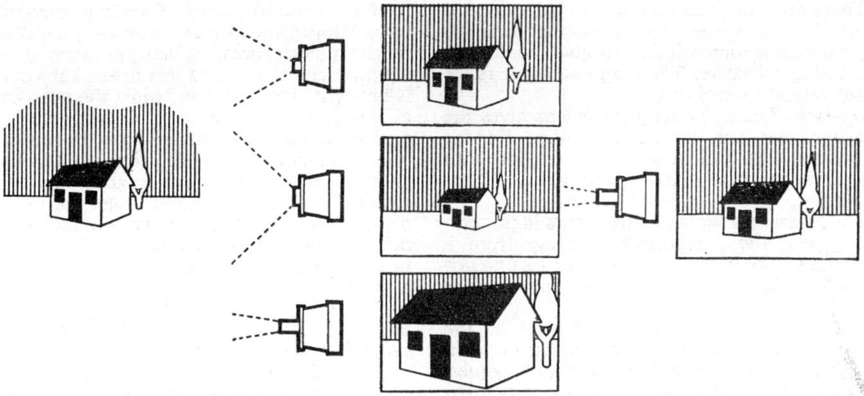

FOCAL LENGTH AND PERSPECTIVE. The focal length of the lens alters the size, but not the perspective of the image, provided the subject-camera distance remains the same all the time. Top: Subject taken with normal lens. Centre: Taken with wide-angle lens from same viewpoint. Bottom: Taken with long-focus lens from same viewpoint. Right: Subject taken with long-focus lens from more distant viewpoint gives normal sized image with flatter perspective.

Wide-Angle Lenses. If the normal picture plane is brought closer to the eye—e.g., to 5 instead of 10 ins.—the resulting picture will include twice as much of the scene. All the objects in the scene will be half the original size in relation to the window, but their perspective will remain the same, except that panoramic perspective distortion, as mentioned earlier, will increase as the angle of view widens.

This corresponds to what happens when a lens with a focal length shorter than normal—i.e., a wide-angle lens—is used in the camera. As for the long focus lens, if the normal lens is replaced by one of $1/n$th the focal length, then all objects in the field will be reduced to $1/n$th the normal image size. An identical image of the objects common to the normal and wide-angle negatives can be produced by reducing the scale of the normal image by $1/n$th, but the dimensions of the resulting picture would only be $1/n$th the dimensions of the wide angle picture.

CONTROLLING PERSPECTIVE

It will be clear from what has been said above that there is no direct connexion between the perspective of the picture and the focal length of the lens. From the same viewpoint, all lenses give a picture with the same perspective. If one lens gives a smaller image than another, the situation can be restored by enlarging. And no matter what the size of the camera or the focal length of the lens, the resulting negative, or part of it, can always be enlarged to produce a print which, when viewed from the normal distance of 10 ins., will present the original scene in its natural perspective. When using long focal length lenses and small negative material, such correct prints would prove impracticably small, however.

The main object in using a long focus (or telephoto) lens is to get a bigger picture of the subject on the negative when it is not possible to get it by going closer to the subject.

The main object in using a short focus (or wide angle) lens is to take in more of the subject when it is not possible to include everything by moving farther away.

Neither of these functions is primarily concerned with the effect of the change on the perspective of the picture.

But there are occasions when the sole object of using these lenses is to achieve a particular effect of perspective and not because it is impossible to get close enough to include less, or far enough away to include more of the scene with a normal angle lens.

Normal Lens. Within limits, there is no need to use a special lens to control the perspective of the picture. Perspective is controlled by viewpoint and nothing else.

By coming close to the subject, the picture can be given the violent type of perspective in which objects get smaller very rapidly as they go farther away. The disadvantage of this method is that as the camera comes closer, the field covered by the lens shrinks in proportion, so less of the scene appears in the picture.

By moving farther away from the subject, the picture can be given the type of perspective in which objects get smaller very slowly as they recede. The disadvantage of this method is that, as the camera goes farther away, the size of the image gets smaller and the picture includes more and more of the scene for which the photographer has no use. It is possible to get over the difficulty by masking off the unwanted part of the scene and enlarging the remaining important part to the required size. Up to a point this is an excellent, plan, but where only a small portion of the negative has to be greatly enlarged, the loss of quality is a serious drawback.

There are thus definite limits to the extent to which it is possible, with a normal lens, to choose a viewpoint which will give the desired effect of perspective. The long and short focus lenses extend these limits.

Long-focus Lens. The long-focus lens gives the photographer the large-scale image that is associated with a close-up viewpoint, but with the gentle perspective associated with a distant viewpoint.

The classic example of this occurs in portraiture where the perspective resulting from a close viewpoint is apt to be neither pleasing nor characteristic. There are two separate reasons for this.

(1) The closer the sitter is placed to the camera, the greater the disproportion between the distances of near and far features becomes. When the head of an average sitter is pointed towards the camera, the tip of his nose is about 6 ins. nearer the lens than his ears. So that if the camera is 4½ feet away from the sitter's nose, that feature is 1/10 nearer the camera than the ears and will be reproduced on a scale 1/10 larger than the ears. A more distant viewpoint—up to 15 feet—restores the proportions to acceptable standards while the use of a lens, with a focal length of two or three times normal enlarges the image to the same size as would be produced by a close-up viewpoint with a normal lens.

(2) As the lens is brought closer to the head of the sitter it looks at it from a wider and wider angle, and sees less and less of the sides of the face. So the features in the centre of the face—e.g., nose, mouth, eyes—appear larger in proportion to the whole area seen by the camera. A normal mouth is about one half as wide as the face, but from a close viewpoint it may appear as much as two-thirds the width, and the ears may not even show at all. Such a portrait would give a false impression of the actual appearance of the face.

Here again, a more distance viewpoint restores the natural proportion, and the long-focus lens enlarges the image as needed.

Wherever it is necessary to prevent objects nearer the camera from over-shadowing those farther away, the standard technique is, first, to choose a viewpoint far enough away to show the near and far objects in the desired perspective proportions, and then to magnify the important area of the scene until it fills the whole picture area. This technique applies equally to portraiture, still life, landscape and architectural photography, and it should not be confused with the use of a long-focus lens simply to produce a bigger image than would be given by a normal lens—e.g., in sports photography and astronomy.

Wide-angle Lens. A near viewpoint gives a perspective in which the size of near objects is increased in relation to those farther away. Where this effect is undesirable, the answer is to adopt a more distant viewpoint, but on occasion it offers a useful means of lending importance to the subject and of playing down other objects in the picture area. A lens of normal focal length includes less and less of the subject as it is brought closer, and so, while the perspective of the close viewpoint may be satisfactory, the amount of the scene included may be too small.

A wide-angle lens—i.e., one which has a shorter-than-normal focal length—enables the camera to be brought close to get the required perspective, but it also covers a larger-than-normal field, so that the whole of the subject included with its angle of up to 100° is the same as that covered by a normal lens with its angle of about 50° working from a correspondingly greater distance. The field is the same, but the perspective is different.

This technique is used for making a particular object or part of the subject—e.g., a building—appear very much larger than the background and objects immediately behind it. This use of a wide-angle lens to control perspective is quite distinct from its use from a near viewpoint to cover a larger area of the field than could be photographed with a normal lens—e.g., in making a record of the interior of a building where the camera distance is limited.

Camera Movements. Finally, perspective can be controlled in the camera by the use of the various camera movements. The principle is the same for all the movements of the back of the camera—i.e., vertical and lateral swings—and for movements of the back in relation to the front—i.e., rise and cross front movements. (It does not apply to front swings used alone because these simply swing the focal plane of the lens to increase its effective depth of field to cover an oblique subject.)

Every camera movement of this type used to control perspective does it by altering the angle between the subject plane and the plane of the sensitized material—i.e., of the back of the camera. And the principle employed in all cases is that: all parallel lines in the scene that are also parallel with the plane of the back of the camera reproduce as parallel lines; all parallel lines in the scene that run at an angle to the back of the camera converge to a vanishing point on the horizon (either in the picture or outside it on the extended line of the horizon); and the more the parallel lines in the scene slope away from the camera, the closer the vanishing point lies to the optical centre of the picture. The optical centre is the point on the horizon immediately opposite the camera—i.e., on a straight line from the camera, normal to the plane of the sensitized material.

Correcting Horizontals. For example, if the camera is looking on a corner of a building, the sides of the building will each be at an angle to the back of the camera. So the side walls will each appear to converge to a vanishing point on the horizon. Swinging the back of the camera to bring it more nearly parallel to one of the sides will reduce the perspective of that side and in-

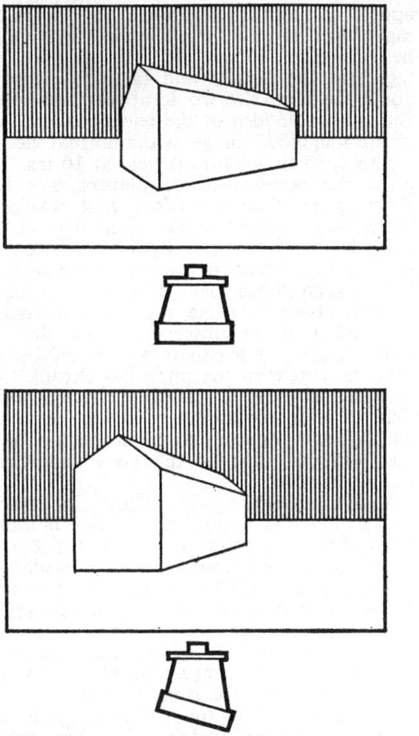

EFFECT OF LATERAL SWING. Top: *The camera back is in its normal position, i.e., at right angles to the lens axis, and the resulting image corresponds to what the eye would see from the same viewpoint.* Bottom: *The back of the camera has been swung to bring it more nearly parallel to the left side of the building. This has the effect of reducing the convergence of parallel horizontal lines on that side. At the same time the convergence of the horizontal parallels on the side at right angles increases. The perspective is now abnormal but for some purposes the effect may in fact be more pleasing or even necessary.*

crease the perspective of the other. On the one side, the parallels will open out and the vanishing point will move farther away from the optical centre of the picture while on the other side they will converge at a steeper angle and the vanishing point will move closer to the optical centre. So the perspective of one side is changed at the expense of the other.

This method of perspective control is applied chiefly in architectural and industrial photography where two aspects of the same subject are to be shown in a single picture by taking a three-quarter view. It may be produced by simply swinging the back of the camera, but it can also result from the use of the cross front—i.e., the lens is moved to the side and the whole camera turned until the back is at the required angle to the subject.

The principle is also applied—though less often—to figure studies to give more pleasing proportions.

Correcting Verticals. The same principles apply to the correction of the perspective of vertical lines. In this case, when the parallel lines make an angle with the back of the camera—e.g., when the camera is pointed up at a tall building—the lines converge to a vanishing point on the vertical line through the optical centre. As the back of the camera is brought more nearly vertical by the use of forward swing of the back, the vanishing point moves farther and farther away and the verticals become more nearly parallel. When the back itself is vertical then the vanishing point will be infinitely far away and all vertical lines in the camera image will be parallel to the upright sides.

The same procedure applies equally to the less usual case in which the camera is pointed down as when taking a photograph of one tall building from the top of another. Here the verticals in the subject will converge to a vanishing point below the picture and will become parallel when the camera back is swung backwards to bring it vertical. In both the above cases overall sharpness is regained by swinging the lens-panel into a position parallel with the film plane.

When either vertical or horizontal perspective is altered in this way, the movement that restores the parallels also elongates the subject in the same direction.

PERSPECTIVE ANOMALIES

A perspective picture is essentially a picture seen or photographed from one particular viewpoint. The picture that the observer traces on the window pane will exactly correspond to the scene in front of the frame only so long as he keeps his eye in the same position. If he looks at his tracing from nearer or farther away or from one side or the other, the tracing will no longer correspond to what he sees through the glass.

In the same way, the picture drawn by the camera is only correct when seen from the viewpoint of the lens, or when an enlarged reproduction is seen from a proportionately distant viewpoint.

If a negative is taken with a normal lens of f ins., then a contact print to appear in natural perspective must be viewed with the eye f ins. away. If it is enlarged n times, then the enlargement must be looked at from $f \times n$ ins. away.

Thus, a contact print from a normal negative taken with a 10 ins. lens must be looked at from a viewpoint 10 ins. away, opposite the centre of the print. A contact print from a negative made with a 2 ins. lens would have to be looked at from 2 ins. away. As the unaided eye could not focus an image as close as this, the negative must be enlarged—e.g., 5 diameters—and the enlargement looked at from $2 \times 5 = 10$ ins. away.

An exhibition print, made from the same negative, enlarged twenty times would have to be looked at from a distance of 2 × 20 = 40 ins. A quarter of the negative enlarged on the same scale would still have to be looked at from the same distance to appear in natural perspective.

If these conditions are met, then it does not matter from what angle or viewpoint the original negative was made, nor whether it was made with a telephoto, normal, or wide-angle lens. The picture will look right. There will be no distortion of the perspective.

But these ideal conditions rarely hold in practice; they are more often than not violated either in ignorance or deliberately, and in either case the perspective of the picture will look unnatural. The eye will tolerate a large amount of some types of perspective distortion but even small amounts of other types will offend it. Two principal classes of perspective distortion occur in photography; due to incorrect viewing distance and due to incorrect viewing angle.

Incorrect Viewing Distance. In some cases the observer does not know the correct viewing distance for the picture he is looking at, and may even be misled into choosing the wrong one. If a print is put into his hand, he will automatically hold it 10 or 12 ins. away. He might then be looking at a contact print from a negative made with a 10 ins. lens (in which case he will be right) or a 15× enlargement from a negative made with a 2 ins. lens (which he ought to look at from 15 × 2 = 30 ins.) or a contact print from a negative made with a 6 ins. wide-angle lens (which he should hold 6 ins. away). All the prints could be the same size, so that there would be nothing to indicate why only one of them looked natural.

But often the choice of the wrong viewpoint is deliberate. The commonest example of this is in pictures taken with a telephoto lens with the object of showing a large image of a distant subject—e.g., of a batsman at the wicket. To appear in natural perspective, such a picture made with, say, a 40 ins. telephoto lens should be viewed at 40 ins. At that distance, the fact that the bowler and batsman appeared about the same size would appear reasonable. But the whole idea of the telephoto lens is to give a magnified image at the normal viewing distance, so the picture is held at 10 ins. This gives the appearance of players, not four times larger than life (since that would be incredible) but four times closer. And at that close distance the observer still expects to see a noticeable difference in the sizes of the near and far batsmen. Since they nevertheless persist in looking about the same size, the unconscious conclusion of the observer is that they are much closer together than they should be, and that the length of the pitch has shrunk from 22 yards to about 5. The same effect is noticeable in telephotographs of rowing eights in which stroke appears to be sitting on bow's feet with the rest of the crew sandwiched tightly between.

Again, a picture taken with a wide-angle lens would give natural perspective if looked at from the viewpoint of the lens. If a 8 × 10 ins. contact print from a negative made with a 6 ins. wide-angle lens were looked at from 6 ins. away it would look perfectly natural, but at 10 ins. the ground appears to be sloping steeply and the scene has an opened-out look that is highly unnatural. Here, however, the choice of the lens is dictated solely by the problem of getting the widest possible field of view from an inescapably close camera position, and false perspective is the price that has to be paid.

The false appearance cannot be removed simply by enlarging until the correct viewing distance increases to 10 ins., because then the eye 10 ins. away is trying to take in a picture that extends beyond the normal 50°

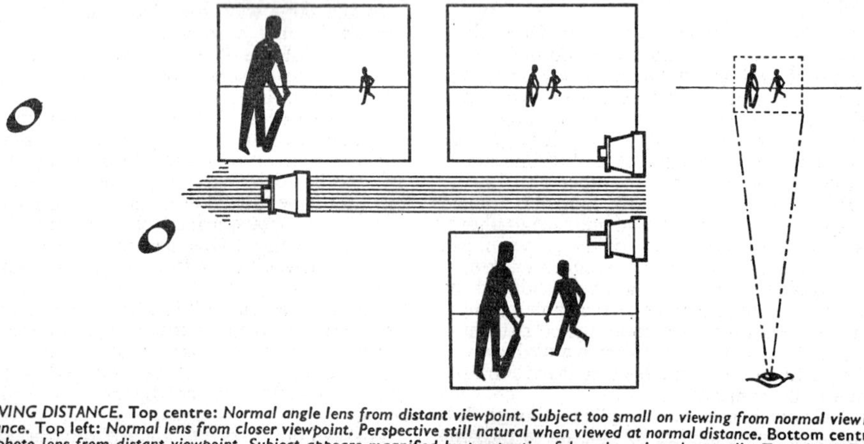

VIEWING DISTANCE. Top centre: *Normal angle lens from distant viewpoint. Subject too small on viewing from normal viewing distance.* Top left: *Normal lens from closer viewpoint. Perspective still natural when viewed at normal distance.* Bottom centre: *Telephoto lens from distant viewpoint. Subject appears magnified by perspective false when viewed normally.* Top right: *Telephoto picture observed at greater distance acquires natural perspective but observer loses benefit of magnification.*

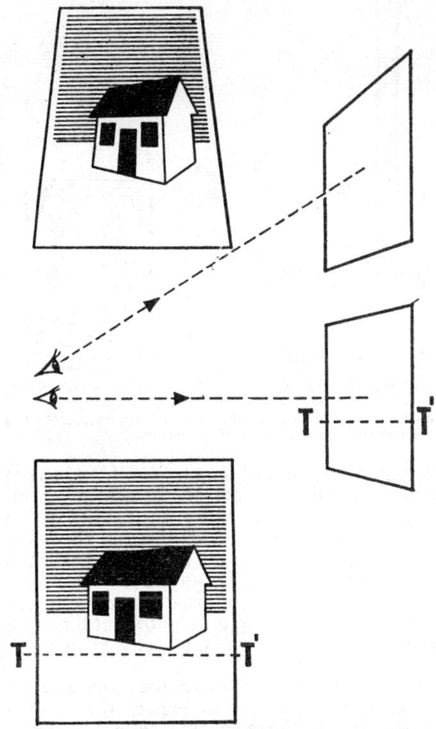

VIEWING ANGLE. For natural appearance the picture should be looked at on a level corresponding to the relative height of the camera lens at the time of exposure. If it is seen from a lower level—e.g., if it is hanging high up on the wall—it will appear in false perspective. Tilting the picture forward counteracts this. TT^1: When part of the picture is trimmed away for effect, the viewpoint does not change. It does not move to a point opposite the geometric centre of the print. The picture should therefore be viewed from its original viewpoint, even if this is opposite a portion outside the print area.

field of vision. No matter how much the wide-angle picture is enlarged, the observer will tend to back away from it until it subtends an angle of about 50° at his eye, and under those conditions the perspective will look wrong.

Incorrect Viewing Angle. Even when the eye looks at a print from the correct distance, the perspective will still look wrong unless it also looks from the correct angle—i.e., from a point corresponding to the position of the lens at the time of exposure. Generally, this is a point directly opposite the centre of the horizon on the negative. This condition again is often violated with the result that the observer feels that there is something wrong, without always realising that he is looking from the wrong angle.

Anyone looking at an enlargement from the right or left half of a negative will automatically hold it straight in front of him with his eyes opposite the centre of the print. In fact, to get the correct perspective, he ought to look at the print through one eye from a point opposite the edge along which the cut was made—i.e., opposite the centre of the original complete picture.

Similarly, when the lower or upper part of a print is trimmed away, the eyes of the observer should still look at it from a point opposite the horizon, whether the trim has left it at the top or bottom of the resulting picture.

The fact that people usually look at a photograph with both eyes means that the resulting effect of perspective is at best only a compromise. However, the difference in the angles from which each eye sees the picture gets less as the viewing distance increases; this is one of the reasons why a big enlargement is so much more pleasing to look at than a small print.

Marginal Distortion. A further cause of apparently incorrect perspective in photographs arises because the photographic recording surface is flat whereas the retina of the eye is curved.

The perspective images of objects at the edge of a picture, particularly of a wide-angle photograph, are drawn out because the picture plane is flat and so becomes increasingly oblique away from the centre of the field. But the image projected on the retina of the eye by objects in the field of vision falls on a spherical surface, with all points the same distance from the eye, so that the perspective is no different at the edges from what it is at the centre.

To a man standing in front of a row of identical posts stretching away to either side, the apparent diameter of the posts gets gradually smaller as they get farther away. But to a camera from the same position, the posts will actually get wider and wider away from the centre because each successive post records

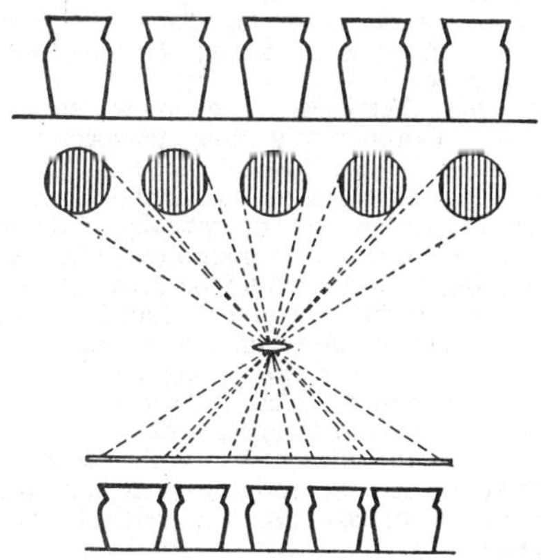

MARGINAL DISTORTION. The images of solid objects are progressively distorted towards the margins of the plate because the off-centre rays meet the plate at an increasingly oblique angle. Distances along radial lines increase here.

its image at a point farther from the lens and on an increasingly oblique surface. The result of this disparity between the visual and photographic perspective is that the eye rejects as false the drawn-out perspective of objects towards the edges of wide-angle photographs.

This is really panoramic perspective distortion in another guise. The significant point is that it affects only solid objects (which are seen from a different direction at the edge of the field than at the centre). If the subject is effectively in a flat plane, the oblique view of marginal areas from the lens position foreshortens apparent dimensions in that plane; this foreshortening exactly compensates for the elongation produced in the image plane.

This marginal distortion is a function of the lens angle, and has nothing whatever to do with lens construction or correction. Nor must it be confused with image distortion (such as pin-cushion or barrel distortion) introduced by the lens construction itself, though pin-cushion distortion may appear similar.

The marginal distortion due to purely perspective effects can become disturbing in wide-angle shots taken with a lens covering an image angle over about 75° on the picture format. It appears particularly pronounced where objects of obvious depth are arranged near the edges of the picture—for example heads in portrait groups, tables or other furniture of large surface expanse in interiors, etc. The greater the depth of the object relative to the width of the subject field, the more obvious the distortion effect. Such objects are therefore best kept away from the edges of the picture.

The same marginal distortion is liable to appear when vertical or horizontal sliding movements (rising or cross-front) are employed on a technical camera. In this case the image on the film in effect represents in part the marginal area of a much larger picture format corresponding to a wide-angle view. Such distortion can however be minimized, under suitable conditions, by tilting or swinging the camera back.

Converging Verticals. The most familiar example of apparently false perspective in photographs is the inward slope of vertical lines—e.g., the sides of buildings—when the camera is pointed up. Theoretically, if the eye takes up a position corresponding to the lens, the sloping verticals would be acceptable because the perspective would in fact be true to life. In practice, however, for a number of reasons, the eye will not accept the fact that receding vertical parallels converge although it is quite prepared to accept the effect in the horizontal plane. One reason is that when the eye looks up at a tall building, there is nothing to indicate directly that the verticals of it converge.

But when the image appears printed on a rectangular piece of paper, the eye can no longer ignore the inward slope. Again, the eye takes in a vertical building by scanning it from bottom to top, and does not consciously compare the width of the bottom with that of the top because it does not see them at the same time (except when they are so far away that the convergence is negligible anyway). But on the reduced scale of even a large photograph, the eye can take in the top and bottom at the same time and cannot escape the fact that they are unequal in width.

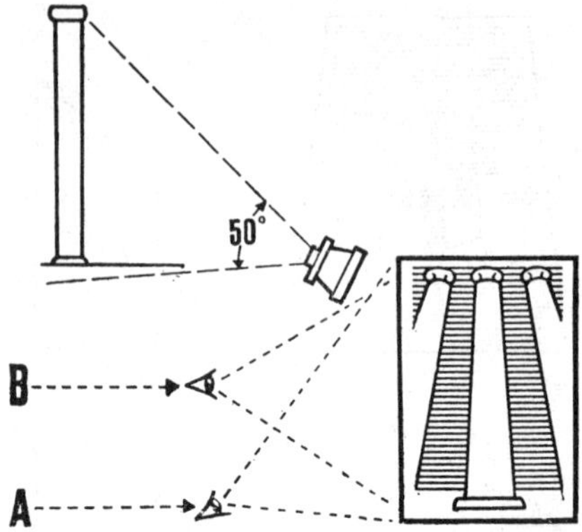

CONVERGING VERTICALS. A: A picture with converging verticals looks natural if viewed from a point corresponding to the position of the camera. B: It only appears unnatural when the eye is opposite the centre of the frame.

The result is that to satisfy the psychological dislike of vertical convergence, the photographer has to "correct" the natural perspective by swinging the back of the camera. In so doing he actually falsifies the real facts but the important thing is that the parallel verticals look right and the photograph becomes acceptable.

With a given camera tilt the convergence of verticals appears much more pronounced with a wide-angle lens than a long focus one. This difference is also largely psychological, since the perspective of the view taken in by the long-focus lens (assuming the same camera viewpoint) is identical with the perspective of the corresponding centre section of the wide-angle view. In the latter however the very much greater convergence of lines near the edge of the picture intensifies the general impression of "falling over backwards" of the building. In the long focus view, taking in a smaller subject field, the maximum angle between the converging lines at each edge of the picture is smaller and the total effect therefore less extreme.

Compensation. From what has been said above it will appear that most photographs are looked at from wrong viewpoints, distances or angles, and with two eyes instead of one; so that in theory, few people should be able to look at a photograph without finding the perspective unnatural. In practice, however, there is always a certain amount of unconscious compensation.

From long experience the eye becomes educated to accept a wide measure of disparity between the real and the reproduction and, fortunately for most photographers, there is rarely any need to worry about getting exact perspective.

At the same time, when dealing with straightforward subjects, the more the photographic conditions comply with the requirements of natural perspective, the better the result will be in the finished picture. F.P.

See also: *Camera movements; Circle of confusion; Converging verticals; Psychology of vision.*

PERUTZ, OTTO, 1847–1922. German photographic manufacturer. Started as a chemist, bought in 1880 a chemical and pharmaceutical firm in Munich, and began manufacturing silver bromide gelatin dry plates based on Obernetter's formulae. After Vogel's discovery of sensitizing dyestuffs Perutz manufactured orthochromatic and later panchromatic plates. The first was the Vogel-Obernetter silver eosin plate (1884, still manufactured in 1955). Perutz also manufactured silver chloride plates for lantern slides, sheet films, and plates for aerial photography and for photo-mechanical work, antihalation plates. Introduced at an early date controlled processes, testing and research laboratories and built a film factory in 1922 from which issued 35 mm. films, fine grain films, X-ray films, etc. (*Denkschrift zum 75-jährigen Perutz-Jubiläum,* Munich 1955.)

PESTLE AND MORTAR. Heavy porcelain or earthenware bowl (the mortar) and a pear-shaped ball of similar material fitted with a handle (the pestle) used for grinding up chemicals to make a powder.

PETS. The photograph of a cat or a dog is not merely a representation of an animal; it is a picture of an individual. Every domestic pet has its distinct personality and, unless this is made clear, the photograph is a failure. The photographer must be prepared to study his subject and then set out to capture the attitude and expression that will convey its character in unmistakable terms. This sort of photography involves an appreciable amount of patience, understanding—and luck.

The Camera. Stand cameras are unsuitable; an animal cannot be expected to wait patiently in one place whilst all the necessary adjustments are made.

Many photographers who specialize in this field prefer a reflex camera, as the subject can be watched and focused on the screen right up till the moment of exposure. The reflex camera also has the advantage that it lends itself to use at a low viewpoint.

Others favour the miniature camera, because of its great depth of field and interchangeable lenses of large aperture.

There is something to be said for both, but no matter what camera is chosen, it should be equipped with a large aperture lens or lenses, so that it can be operated at reasonably fast shutter speeds. All controls should be easy to work and the camera generally should be handled without any apparent hurry or fuss that might disturb or excite the subject.

The Lens. A long-focus lens has the advantage of giving a large image of the subject from a relatively distant viewpoint and it helps to cut out distracting background objects. At the same time, it lacks the depth of field of the normal angle lens and calls for a lot of fidgeting with the focus if the subject is moving about. The best compromise is a lens with a focal length of rather more than the diagonal of the picture—e.g., a 5 cm. lens on a 35 mm. camera or a 4 ins. (10 cm.) lens on the popular $2\frac{1}{4} \times 2\frac{1}{4}$ ins. (6×6 cm.) size.

The Sensitized Material. Only panchromatic materials should be used because it is absolutely necessary to show the colour and markings of the subject in true tone relation to the surroundings. Generally, only the fastest films and plates are worth considering, but if the lighting conditions are good enough and the subject is not likely to be restless, a medium speed panchromatic emulsion will give the best pictures. The fast film, having a coarser grain, will stand less enlargement. This, in its turn, calls for a closer viewpoint to give a bigger image, and very close viewpoints have a number of disadvantages.

Rabbits. No attempt should ever be made to photograph a rabbit in its hutch. The closed-in top and sides prevent effective lighting for modelling and texture. Its coat must look like fur; this means side-lighting and accurate focusing. A garden table and a suitably plain cloth held to form a background is better than working on the ground, unless the rabbit is white and will therefore stand out against it.

Closely-cut grass also makes quite a suitable setting, but a sky background is unnatural. The shutter speed should not be slower than 1/50 second, and the camera should be at the same level as the rabbit if possible. No filter is needed.

Birds in Cages. If the cage has a glass-edged base or internal hanging-mirror, these should be removed in case they catch the light and reflect back into the lens. It is best to stand or hang the cage in the garden. Backgrounds must be plain; dark-toned for light birds and vice versa. Watch out for the sun rays striking the plating on bars and fitments. Side or oblique backlighting is effective. A supplementary lens is needed if a long-focus lens is not available—this means shallow depth of field. A sky background is very suitable if it is well-filtered, i.e., by a 3× yellow, green or orange filter.

Photographing in a cage indoors in the window is possible but more difficult. If the camera is pointed directly out of the window, the bird and cage will become silhouettes against an over-light background. Arranging the cage at the side against curtaining, with the viewpoint from the other side is more effective. Direct sunlight will make snapshot speeds quite practicable with fast film.

Shutter speeds of not less than 1/100 second are recommended, although at times the bird may be sharply rendered at 1/50 second. Medium-speed film (panchromatic) is best, allowing for greater enlargement without coarse grain.

Mice. Tame mice are best photographed out of their cage. Outdoor lighting allows faster shutter speeds to be used. The camera must be as close as possible, and a supplementary lens is necessary to get an image of reasonable size.

The mice should be kept off the ground. Posing them on coat sleeves, the shoulder of a child, etc., gives size contrast and scale. Clothing must contrast in colouring and not be strikingly patterned. If the mouse is friendly with another household pet, particularly a cat, placing the two together on a box or table will give effective pictures. One or more helpers are needed for this.

Frontal or overhead lighting should be avoided. Shutter speeds of not less than 1/100 second are advised, and a ready trigger-finger to capture any fleeting action.

Goldfish. Photographing goldfish and similar species presents two major problems: focusing on the fish and coping with movement. The usual dodge is to lower a sheet of glass into the tank, 3 or 4 ins. away from the front, thus confining the fish to a narrow space within a pre-focused depth of field and also restricting rapid movement. Flash is best; the lighting should come from above the water and 45° to the camera axis.

A dark background is recommended and a shutter speed not less than 1/50 second if in sunshine or Photoflood lighting; synchronized flash will give opportunities to increase this speed.

Tortoise. The somewhat drab colouring of the tortoise needs careful choice of lighting and setting to convey its true character. Flat, sunless lighting is unsuitable; low-angle sunshine from the side is best. The camera should be at the same level as the tortoise if possible, but avoid long grass or camouflaging earthy settings. Rough towelling makes a good background and base if out of focus and not creased or folded. If the shell is wetted it will catch highlights and its ribbed markings will be emphasized.

Three-quarters-on or profile are easiest positions for successful pictures. No filter is needed; the shutter speed should be at least 1/50 second.

Although a tortoise moves very slowly, at the close distances necessary for photographing, such movement is equal to quite fast action farther away. P.J.

See also: *Animals; Birds; Cats and kittens; Dogs and puppies; Fish; Insects; Reptiles.*

PETZVAL, JOSEF MAX, 1807–91. Hungarian mathematician. His calculation of a fast achromatic lens in 1840 played a very large part in the rapid development of portrait photography by the daguerreotype process. His portrait lens was of large aperture and achromatic for both optical and chemical rays. Also designed the Orthoscope lens 1857 for landscape and reproduction work. Both these lenses marked a great advance on any previous lenses in use and were widely used for half a century. Biography by P. Ermenyi (Halle 1903).

PETZVAL LENS. In 1840 Josef Petzval, a Hungarian, designed a lens for portraiture, and modern lens manufacturers still use his principle as the basis of many of their designs, especially for projection.

The original Petzval lens was made by combining two colour-corrected combinations—very like an R.R. lens with the negative elements facing inwards with a wide space between the two sets of glasses. It was generally used without a central stop and it gave apertures up to *f* 3·6. To-day modifications of the original form are used in the Dallmeyer portrait lens and in a number of wide aperture lenses for 8 and 16 mm. movie cameras. More complex types have been developed from it with apertures up to *f* 0·85 and are used for cine photography of the image on X-ray screens.

Petzval lenses are used more for projection than for anything else because over the relatively small field covered, the curved field of the Petzval lens is tolerable. In apertures up to *f* 1·5 or larger they are used for projecting lantern slides and all sizes of cine film. They are also used in epidiascopes.

The great drawback of the Petzval lens is its curved field. This restricts good definition to the centre and explains why Petzval lenses are used for projection, where only a narrow angle is wanted anyway, and for portraiture where, although the angle is greater, loss of definition away from the centre does not matter (and in fact may even add to the pictorial effect of the result).

See also: *Lens history.*

PETZVAL SUM. The radius of the curvature of field of a lens depends on the number of glasses making up the lens and their respective powers. This is expressed in terms of a number known as the Petzval sum for that particular lens construction.

The Petzval number varies in magnitude and sign (plus or minus) with the lens construction; the greater the Petzval sum, the smaller the radius of curvature. The normal anastigmat has a positive Petzval sum, indicating that the curvature is concave to the lens. Some telephoto constructions have a negative sum—i.e., their curvature is convex to the lens.

pH. Alkalinity or acidity of an aqueous solution expressed numerically. Figures above 7 represent alkalinity; below, acidity.

See also: *pH value.*

PHENIDONE. 1-phenyl-3-pyrazolidone. Used as a developing agent.

Formula and molecular weight: $C_6H_5N(CH_2)_2NHCO$; 162.

Characteristics: White crystalline powder; acts similarly to metol, but can be used in much lower concentrations and does not tend to cause skin poisoning.

Solubility: Freely soluble in dilute alkalis, slightly soluble in water.

PHENOL. Carbolic acid. Used as a preservative and antiseptic in mountants, also as a developer for diazotype.

Formula and molecular weight: C_6H_5OH; 94.

Characteristics: White crystals with a characteristic antiseptic smell. The solid melts at 42°C (108° F.). Both the crystals and their solution in water attack the skin and must therefore be handled with care. Poisonous. Should be kept well stoppered, as the solid is hygroscopic.

Solubility: Fairly soluble in water.

PHENOSAFRANINE. Used in desensitizers and high speed developers.

Characteristics: Dark red powder (a complex dye).

Solubility: Fairly soluble in water.

PHILATELY. Philatelic photography—i.e., the photography of postage stamps—has three main purposes: for illustrating catalogues or descriptive articles, for investigating flaws and varieties, and for revealing forgeries and fakes.

Illustration. For catalogue or other illustration, stamps are copied at half to one and a half times natural size. Such photographs must convey as accurately as possible the tones and detail of the actual stamp. In black-and-white photography this means using a fine grain panchromatic film with a green correction filter for correct colour reproduction, or a suitable contrast filter to show up differences in tone of multi-coloured stamps.

The lighting set-up for straightforward stamp photography is similar to normal copying lighting, but on a smaller scale.

In some countries (e.g., the U.S.A.) there are legal restrictions on the photography of stamps. Reproduction is sometimes restricted to reduction smaller than $\frac{3}{4}$ natural size, or enlargement larger than $1\frac{1}{2}$ times.

Flaws and Varieties. Close-ups at twice to ten times natural size show up printing or engraving errors, varieties, and other small details which often make a specimen particularly valuable or interesting. To be really informative, such photographs usually show both the normal and abnormal stamp side by side for comparison. Two separate photographs taken under identical conditions may also be used.

Suitable filters may be employed to increase the contrast of the image, especially when photographing pale colours. In this case the filter colour is complementary to the stamp colour. Filters may also be used to eliminate coloured postmarks where these obliterate important parts of the design. The colour of the filter in this case is the same as the postmark. In many instances the postmark is black or so heavy in another colour that filtering can not be done very effectively.

Fakes and Forgeries. Copying on an enlarged scale often reveals mechanical repairs and erasures. For these pictures oblique lighting is used where it is necessary to show up surface irregularities.

Macrophotographs of this type should include a microscale or graticule to permit the exact size of details of the design or of overprints to be checked. Faked overprints are one of the favourite forms of stamp forgery.

Photography by ultra-violet rays is used to identify the paper of suspected counterfeits because most papers have their own characteristic fluorescence under this type of lighting. Chemical cleaning, removed postmarks, and repairs also show under ultra-violet radiation.

Infra-red photography is used to investigate faked overprints, postmarks, and even repainted parts of the original design. Photographs of the genuine pigments and those used in the forged specimen frequently look quite different taken through an infra-red filter on infra-red film. Experiments in photographing stamps by X-rays and grenz rays have also been made.

When examining suspected forgeries, a genuine specimen is usually photographed under the same conditions for comparison.

Watermarks. Photographs of stamp watermarks are taken on a high contrast film or plate to increase the low subject contrast.

Watermarks may be photographed by transmitted or reflected light. The former method gives the better results where the watermark is well defined, and the stamp a mint specimen. By using a filter of the same colour as the stamp, the stamp design can be practically eliminated. But a sheet margin gives the best reproduction of a watermark and this is always used if it is available.

For dealing with cancelled stamps, the most

successful technique is to lay the stamp face down in a black watermark tray and flood it with petrol. The watermark can then be photographed from the back of the stamp. This needs some care, as some inks are soluble in petrol, while on chalk surfaced papers the colours run on contact with water.

Ill-defined watermarks are naturally the most difficult to photograph, but they can usually be made to show up satisfactorily by being photographed by reflected ultra-violet radiation.

Watermark photographs taken from the back of the stamps have to be reversed left-to-right in printing to show the pattern the right way round. L.A.M.

See also: *Coins; Copying; Manuscripts and old documents; Paintings and drawings.*
Books: *Camera Close Up* and *Camera Copying and Reproduction*, by O. R. Croy (London).

PHOSPHORESCENCE. A number of materials possess the property of absorbing light of one wavelength and giving it out as light of another wavelength for some time afterwards. This property is known as phosphorescence. The luminous paint used for the hands and figures of watches and clocks is phosphorescent.

Reproduction processes have been formulated around the phosphorescent effect, using it to store an image which can be used to activate photographic layers placed in contact with it. A lantern slide can be printed on phosphorescent material by prolonged exposure.

It is possible to quench phosphorescence by means of infra-red radiation. This effect has been used to produce positive images from negatives projected by infra-red on to phosphorescent screens.

See also: *Negative-positive viewer; Viewers.*

PHOSPHOROPHOTOGRAPHY. Indirect method of infra-red photography based on the property of infra-red rays of being able to quench or reduce phosphorescence.

The invisible infra-red image—e.g., of a kettle of boiling water—is focused on to a glowing phosphorescent surface; it then creates a visible negative counterpart by reducing the glow in proportion to the intensity of the infra-red rays.

This negative image can then be photographed or transferred by contact on to ordinary sensitized material, giving in both cases a direct positive image.

Using the same principle it is possible to view a negative as a positive. The negative is projected by infra-red radiation on to a phosphorescent surface; where infra-red radiation (from the clear parts of the negative) falls, the phosphorescence is quenched, thus producing a positive image.

See also: *Heat recording; Viewers.*

PHOT. Unit of illumination corresponding to 1 lumen per square centimetre.

PHOTO. (Prefix.) (From Greek, phos, photos = light.) It is used in two senses:

(1) To indicate an association with light, as: photochemistry, the chemistry of the action of light; photogen, a kind of paraffin; photogenic (in physics), emitting light; photology, study of light; photolysis, movement of protoplasm under influence of light; photometer (whence -metric, -metry), instrument for measuring light; photophobia, morbid dread of light; photophone, telephone in which the connexion is made by a modulated light beam; photosphere, the luminous zone surrounding the sun or a star from which the light and heat are emitted; phototherapy, treatment of disease—e.g., lupus—by light rays; photosynthesis, the process by which plants form carbohydrates by the synthesis of water and carbon dioxide under the action of sunlight; phototropism, the bending of plants—e.g., flowers turning to face the sun—under the action of light.

(2) As an abbreviation of photography(ic) as: photo montage, phototelegraphy.

PHOTO-CALL. Full dress turn-out of the actors taking part in a theatrical performance, musical comedy, ice-show or similar entertainment, called by the management so that suitable scenes can be photographed for front-of-the-house and other publicity, for distribution to the press, and for the record.

See also: *Theatre.*

PHOTO-CATHODE. Layer of metallic or semi-conducting material which emits electrons when exposed to light. It may be prepared so that the electrons emerge from the side on which the light is incident (opaque photo-cathode) or from the opposite side (semi-transparent photo-cathode). Photo-cathodes are used in a wide variety of electron-optical systems, including image amplifiers, TV cameras, etc.

The emission process arises from the absorp-

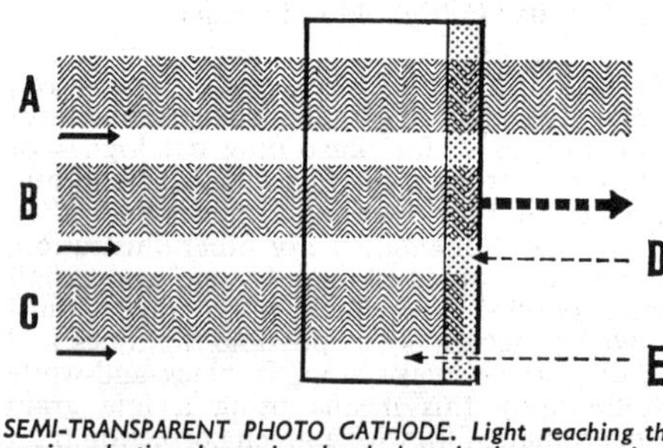

SEMI-TRANSPARENT PHOTO CATHODE. Light reaching the semi-conducting layer is absorbed and electrons emitted instead. This takes place only if the photons (light quanta) have the right energy. A, sufficiently intense photon streams pass right through the photo cathode. B, photons are absorbed at the photo cathode and release electrons. C, if the energy of the photon stream is too low (wavelength too long) the light is absorbed but no electrons emitted. D,-semi-conducting photo cathode. E, transparent (glass or quartz) base.

tion of a quantum of light (photon) by an electron in the material, which thus acquires the energy of the photon. The electron is then able to move through the material, and if it reaches the surface with sufficient energy it can escape. The minimum energy for escape is characteristic of the material, and the photon energy must exceed this if electrons are to be emitted. Consequently a photo-cathode responds only to light of wavelength shorter than a characteristic value, called the photo-electric threshold of the material. Only a few metals and semi-conductors have thresholds in the visible and infra-red.

The sensitivity of photo-cathodes which respond to visible light is often quoted as the current emitted per lumen of white light from a tungsten filament at colour temperature 2870°K. The sensitivity of antimony-caesium cathodes is usually between 30 and 90 micro-amperes per lumen. Even higher sensitivity is obtained from a combination of antimony with sodium, potassium and caesium (multi-alkali photo-cathode). The only known photo-cathode with useful infra-red response is a combination of silver, oxygen and caesium. The threshold lies near 12,000 A, and the sensitivity may be 50 microamperes per lumen. W.L.W.

See also: *Image amplifier.*

PHOTOCHEMISTRY. Study of chemical reactions influenced by the action of light.

The photochemical reaction of main importance in photography is the decomposition of silver halides by the action of light. Other reactions which involve the action of light are:

(1) The reduction of certain ferric salts to ferrous (as in the various iron printing methods such as blue-print process and others).

(2) The tanning of bichromated gelatin or glue (as in the carbon and oil processes and also in a variety of photomechanical methods of reproduction).

(3) The bleaching of dyes (as in diazotype) and a number of further reactions of little photographic application.

Among the principles established as laws of photochemistry, the following are of interest:

The Grotthuss-Draper law states that only those wavelengths of light which are absorbed can cause chemical changes in a system. In the decomposition of silver halide, short wavelengths are the most active. The light rays need not necessarily be absorbed by the reacting substance, but can instead activate certain photo-sensitizers which absorb different wavelengths from those utilized by the photochemical reaction. The energy of the activated photo-sensitizer is then transferred to the non-activated molecules of the reacting system, making the latter sensitive to the wavelengths absorbed by the sensitizer. This is the basis of the sensitization of photographic emulsions to green and red light with help of certain dyes.

The Bunsen-Roscoe law states that the amount of chemical change produced is proportional to the quantity of light absorbed (i.e., its intensity multiplied by the time for which it acts). In practice the change also depends on the intensity of the light. The deviation from this law is the reciprocity failure.

It is now accepted as postulated by Planck, that radiation is emitted and absorbed in definite units or quanta. Einstein's law of photochemical equivalence in its original form stated that the number of molecules reacting should be equal to the number of quanta absorbed. (A quantum of energy is equal to the frequency of the radiation multiplied by a constant known as Planck's constant.) In practice, not every quantum of radiation absorbed necessarily causes a photochemical reaction in a molecule, and the effectiveness of the light absorbed is often expressed in terms of the quantum efficiency of the reaction. This quantum efficiency is the ratio of the number of actual molecules reacting for a given amount of light energy absorbed, to the number which could theoretically react.

In a photographic emulsion each silver halide grain becomes developable when sufficient light has been absorbed to reduce an adequate number of silver halide molecules to produce a sensitivity speck on the silver halide crystal. The primary quantum efficiency of the reaction is nearly 1, and nearly as many silver atoms are liberated as quanta are absorbed. But to produce an effective sensitivity speck, each crystal needs about forty quanta of light and the quantum efficiency of the whole process is in the region of 2½ per cent per crystal. With a quantum efficiency of 100 per cent, i.e., one quantum per crystal, the material would have reached its maximum theoretical speed. This is around 10,000–20,000 ASA for a fairly coarse-grained emulsion. No faster photographic emulsion is therefore ever likely to be produced and it is improbable that even this limit will be reached with silver halide emulsions. L.A.M.

See also: *Development theory; Latent image; Sensitivity.*
Books: *Spectral Studies of the Photographic Process*, by Y. N. Gorokhovskii (London); *The Theory of the Photographic Process*, by C. E. K. Mees (New York); *The Chemistry of Photographic Materials*, by G. F. Duffin; *Processing Chemistry*, by L. F. A. Mason; *Chemical Analysis in Photography*, by G. Russell (all London).

PHOTOCHROMISM. A reversible colour change induced by light energy.

PHOTO CONDUCTOR. Electrical semi-conductor which changes its resistance when exposed to light or certain other types of radiation. Usually the resistance is very high in the dark (so that the conductor becomes a virtual insulator) and drops according to the intensity of the exposing light. Photo-conductors (also known as photo-resistors) are used in various forms in exposure meters, electronic image control devices, etc.

See also: *Electro-luminescence; Photo-electric cell.*

PHOTO COPYING. Fundamentally any document, microfilm or office copying process utilizing light sensitive materials. More specifically the term is generally applied to direct office copying processes.

See also: *Diffusion transfer; Document copying.*

PHOTO-DRAWING. Engineering or architectural plan produced on a photographic background of the object to be detailed. Photo-drawings can present design intentions for modifications to existing installations against a realistic and detailed background. In contrast with conventional plan and elevation drawings, photodrawings provide a clearer visualization of design and working instructions.

Production. The production of photo-drawings involves four main steps:

(1) The taking of photographs to show all relevant details of the existing situation;

(2) Printing on to a film or translucent paper to the desired scale;

(3) The film-print either itself forms the drawing sheet or is taped on to a sheet of drawing material and the draughtsman superimposes a simple perspective view of the engineering work that is to be done;

(4) Reproduction of the drawing, combined with the photograph. This is made on ordinary inexpensive dyeline paper which can be handled in the customary rough manner.

Photo-drawings save drawing time because the photograph shows details that would otherwise have to be measured and drawn as background before new design information can be inserted; they make the job easier to understand and visualize.

Camera angles are selected to permit a properly orientated perspective drawing to be superimposed over the photograph. The scale should be as large as possible to avoid congestion of the drawn work.

The film-print is made by enlarging on to transparent or translucent film or paper. To preserve tonal gradation of the photograph in the dyeline print the contrast of the film-print is kept very low. Contrast is more easily controlled by printing through a half-tone magenta contact screen on to lith-type film. The average density of the film-print must be adjusted to permit good reproduction of the photograph and the drawing simultaneously. When the film-print is made on a base the back of which is treated to provide a drawing surface it can itself be the drawing sheet.

Film-prints can also be made on clear film which can then easily be taped on to the back of tracing paper or film in the desired position.

Application. Photo-drawings are useful to show modification to piping installations, design changes on mechanical and electrical equipment, site clearance planning, structural and architectural modifications, illustration of assembly and maintenance instructions, making perspective drawings for detailing three-dimensional scale models, and a variety of other uses for communication of design and dimensional information.

Non-Perspective Photography. When three-dimensional scale models of manufacturing plant, or prototype models, are used as design tools, photo-drawings can be used to transmit dimensional information in the model to subcontractors, constructors and fabrication teams. In such cases the natural perspective in ordinary photographs can be undesirable. Perspective parallax is reduced by photographing the model with an extremely long focus lens from distances of from 50 to 100 feet. Completely non-perspective presentation is achieved by using a camera with a telecentric mirror-lens system. The image is then formed only by those light rays reflected from the object which are parallel with the optical axis, so that all planes of an object are reproduced at the same scale and the projection if strictly orthogenal. Areas larger than the field of view of the camera, which is restricted to the diameter of the mirror, are photographed in sections. Contact prints on dimensionally stable paper are then assembled to make a mosaic covering the entire object. The technique of photodrawing is used for reproducing the mosaic, when specifications, dimensions and instructions are added prior to duplication by dyeline printing. R.F.

PHOTO-ELASTICITY. Technique of determining, with the aid of light, the stress distribution in bodies under complex systems of loading.

Plane polarized light is passed through a transparent model of the object whose stress distribution is required. The model is subjected to loads proportional to the loads applied to the real object and when viewed through a suitable polariscope a pattern of light and dark lines is observed which can be recorded photographically. In two dimensions the stress condition at any point can be defined by the magnitude and inclination of the two principal stresses and it is possible to determine the stress distribution in the model from these lines.

Two types of lines are observed:

(1) Isochromatic lines, which are coloured when using white light and are contours of stress. The zero stress line is always black.

(2) Isoclinic lines, which are superimposed on the isochromatic lines, are always black, and give the directions of the principal stresses.

Optical Theory. The phenomenon depends upon the fact that an isotropic transparent material becomes double refracting when

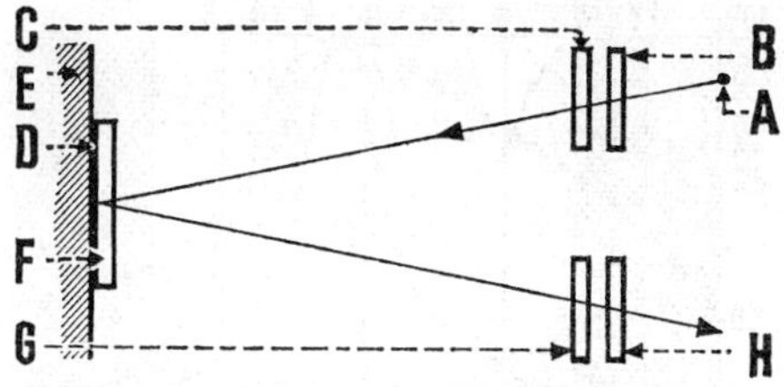

REFLECTING POLARISCOPE FOR PHOTO-ELASTIC COATING TECHNIQUE. The light rays from the source A pass through a polarizer B and a quarter-wave plate C to the metal surface E being investigated. This surface is coated with a reflecting layer D and a photo-elastic coating material F. The latter distorts with the component being tested. The light rays then pass through another quarter-wave plate F and the second polarizing filter (analyser) G.

stressed. Plane-polarized light passing into such a material is split up into two components vibrating in the directions of the principal stresses and travelling through the material at different velocities. If these components are recombined after passing through the model they will interfere if out of phase and produce the isochromatic lines observed.

When the plane of the incident polarized light is coincident with the direction of one of the principal stresses the light will be cut out on recombination in the analyser which is "crossed" (i.e., its plane of polarization at right angles) with respect to the polarizer. A black isoclinic line is thus formed.

The Model. The model is often a section of the object under investigation, the sensitivity of the technique increasing with model thickness. Limitations are set on a model thickness due to difficulties in obtaining constant stress distribution throughout the thickness of the model. A suitable thickness is about ¼ inch.

The material used for the models now is generally one of the epoxy resins such as Araldite CT200. The surface of the model must be polished and areas of the sheet material chosen which are free from initial stresses. Care must be exercised in fabrication of the model that no built-in stresses are produced. If such stresses are present they may be reduced by an annealing process.

Polariscope. A suitable polariscope for investigating and photographing the stress distribution in models consists of a small light source situated at the focal point of a condenser lens. The parallel light emerging from this lens passes through the polarizer, which can be a non-optical quality polarizing sheet, to the model. On emerging from the model the light passes through the analyser which is an optical quality polarizing sheet and then on to a second condenser lens at whose focal point the camera lens is situated.

This second condenser should be corrected for spherical and chromatic aberration although good photographs can be obtained within the central part of the field of view with a simple lens. With the second condenser lens of 12 or 18 ins. focal length and 5 or 6 ins. diameter, and a quarter- or half-plate camera, it is easy to obtain a record of the model which is approximately full size. However, this polariscope may be used with almost any size camera, including miniature cameras using 35 mm. film or 16 mm. cine cameras.

The isoclinic lines can be removed by interposing quarter-wave plates at each side of the model; the light then passing through the model is circularly polarized and the isoclinic lines disappear due to the non-directional character of the light.

With the polarizing sheets and quarter-wave plates in position the condenser lenses need not necessarily be free from residual stresses.

Photographic Materials and Technique. For photographic purposes, a compact source mercury vapour lamp with a green filter is generally preferred and orthochromatic negative material can be used with this light. Alternatively a bank of sodium lamps can be used without any filter or condenser lens but the material must then be panchromatic.

The exposure is adjusted until the light lines of the pattern show a slight density on the negative. The development time of the negative should be increased by 25 or 50 per cent over that recommended for normal subjects in order to obtain a result of sufficiently high contrast. The prints are made on an appro-

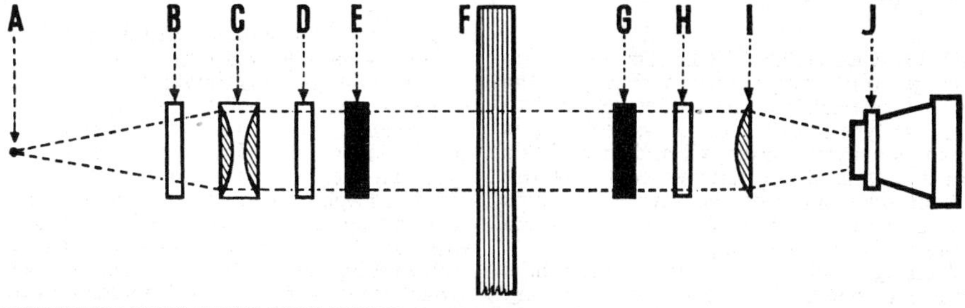

POLARISCOPE FOR PHOTO-ELASTIC STRESS ANALYSIS. The model, made of a suitable resin or plastic, is placed between crossed polarizers, and photographed by a beam of transmitted light. A, light source. B, filter for monochromatic light. C and I, condensers. D, polarizer. E and G, quarter-wave plates to remove isoclinic lines by circularly polarizing the light. F, scale model of object or structure to be tested for stresses. H, analyser (second polarizer). J, camera.

priate high-contrast grade of white, smooth glossy, bromide paper.

Three-Dimensional Stress Analysis by Stress Freezing. Stresses in three dimensions can be analyzed by using certain high polymers (such as Araldite CT200) that retain a stress pattern permanently after they have been subjected to a heating and cooling cycle whilst under load. The method is to construct a fully sealed model of the part to be examined, to subject it to the thermal cycle and, when it is cold, to cut slices from it for examination in the polariscope. The photo-elastic pattern gives the stresses in the plane of the slice.

Photo-elastic Coating Technique. Another method consists of attaching a thin layer of polymer sheet to the surface of a metal part. A reflecting layer is interposed at the metal surface and the effect of loading the part is observed with a reflecting polariscope. This method has the advantage of being applicable to actual hardware. It shows the strains in the metal and can be used for strain above the elastic limit of the metal. G.C.L. (A.F.C.B.)

Books: *Photoelasticity*, by E. G. Coker and L. N. G. Filon (Cambridge); *Photoelasticity* (2 vols.), by M. M. Frocht (London); *Photoelasticity*, by H. T. Jessop and F. C. Harris (London); *Handbook of Experimental Stress Analysis*, by M. Hetényi (New York).

PHOTO-ELECTRET. An electret is an electrically isolating body with persistent internal polarization, i.e. it carries electric charges of opposite sign at two opposite faces. A photo-electret is an electret (formed by subjecting it to an electric field and simultaneous uniform illumination) the polarization of which will disappear imagewise on exposure to radiation of suitable wavelengths. The electrostatic image can be revealed by dusting with a suitable powder. It is also possible to form the photo-electret image by applying an electric field together with the irradiation image, when the surface will be charged more strongly the stronger the exposure.

See also: *Electro-photography*.

Books: *Photoelectrets and the Electrophotographic Process*, by V. M. Fridkin and I. S. Zheludew, transl. by A. Tybulewicz (New York). *Electrophotography*, by R. M. Schaffert (London).

PHOTO-ELECTRIC CELL. Device which either generates, or produces a change in the value of an electric current when exposed to light. The type mostly used in photography is known as a barrier layer cell and in its basic form it consists of an iron disc coated with a layer of selenium over the top of which is applied a film of gold or platinum thin enough to allow light to pass through.

When light falls on the front of the cell, it creates a potential across the boundary between the selenium and the gold. When the photo cell is incorporated in a light meter, this potential is connected to a sensitive micro-

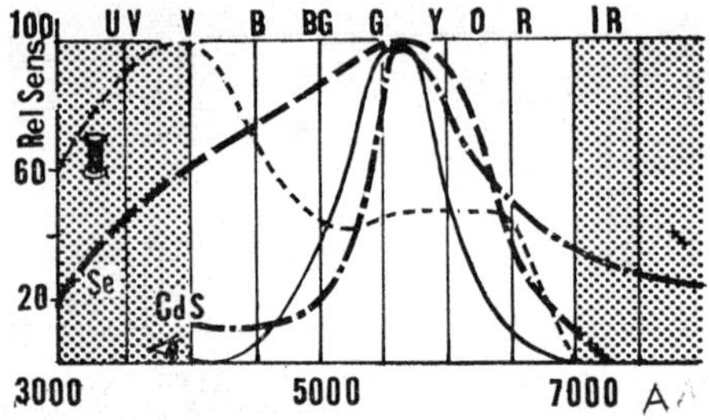

COLOUR RESPONSE OF CELLS. A celeium cell (Se) has its maximum sensitivity in the yellow green region, but responds appreciably also to shorter wavelengths (heavy broken curve). By filters the sensitivity can be restricted to approximately the response of the eye (thin solid curve) but would then appreciably differ from the relative sensitivities of a panchromatic film (thin broken curve). The cadmium sulphide cell Cds follows the eye curve more closely, but also responds to radiation well into the infra-red radiation (heavy dotted curve).

ammeter. The meter needle is thus deflected in proportion to the strength of the light falling on the cell and can therefore be calibrated in terms of the exposure required under that particular intensity of illumination.

Barrier layer cells are self-generating so that there is no need to provide them with an energizing voltage. This enables them to be built into remarkably compact exposure and colour temperature meters, including types incorporated in cameras.

The second type is the photo-conductive cell. This consists of a layer or area of cadmium sulphide between two electrodes, which acts as a semi-conductor. In the absence of light the cadmium sulphide has a high electric resistance, which drops when light falls on the cell. A current flowing through the cell therefore varies according to the light reaching the latter, and a microammeter connected to the circuit can indicate the light strength and can also be calibrated in terms of the exposure required.

A photo-conductive cell needs a battery or other power source to provide the current in the circuit. In modern cadmium sulphide cells—the most popular type of photo resistor used in exposure meters—the usual power source is a mercury miniature cell. This has a life of one to two years.

The cadmium sulphide cell is particularly compact—usually even smaller than a barrier layer cell—and is thus suitable for building into cameras. The sensitivity of the cadmium sulphide cell is appreciably higher than that of a barrier layer cell, and exposure meters of this type can therefore measure very much lower lighting levels. The cadmium sulphide cell is however subject to fatigue effects, and after exposure to strong light does not regain its full sensitivity until after a certain rest period (which may vary from several minutes upwards). Owing to the independent current supply an exposure meter of this type is more versatile for the direct control of camera settings.

A third type of photocell consists of a glass envelope containing a cathode coated with a light-sensitive metal such as caesium or potassium, and an anode which serves to collect the electrons emitted. Sometimes the cathode is formed by a coating on the inside of the bulb. The bulb may be evacuated, or may contain a trace of an inert gas such as argon.

A steady potential of about 90 volts is maintained between the electrodes from an external supply, and under these conditions light falling upon the cathode causes a current to flow to the anode. In the case of a gas-filled bulb this current is intensified by molecular collisions. Over a given range, this current is proportional to the intensity of the light; it can be amplified and used in a number of ways. Photocells of this type are used in the sound head of sound film projectors, for optical triggering of camera shutters and for automatic exposure control in bulk film printing machines.

See also: *Exposure meter.*

PHOTO-ELECTRIC METER. Exposure meter utilising a photo cell which either generates an electric current when light falls on it, or changes in resistance to a current flowing through it. In either case the output current may work a microammeter to give a reading of the lighting level or of the exposure required; or the current from the photo cell can also be used to set—directly or indirectly—the exposure controls of the camera.

Photoelectric systems are also used in semi-automatic printing and enlarging equipment.

See also: *Exposure meter; Photo-electric cell.*

PHOTO-ENGRAVING. Term originally used to mean any process based on photography for the production of a printing surface. It is now generally applied in a more restricted sense to the production of a relief printing surface by a process based on photography.

See also: *Photo-etching; Photomechanical reproduction.*

PHOTO-ETCHING

Process of producing photographically a resist image which protects a surface (e.g. metal) from an etching medium during subsequent treatment.

The etching of patterns or designs into metal by means of acids capable of dissolving the metal is an age-old art. Before the advent of photography the metal was as a rule coated all-over with a suitable acid resistant composition, such as a mixture of beeswax and bitumen and the design was then scraped into the coating by hand or by suitable ruling machinery. The scraped parts were thus bared from etch ground and open to the attack of the acid. After etching, the wax was removed.

Soon after the invention of photography methods were developed to replace the hand or machine engraved etch stencil in wax by a stencil produced by photographic processes, the photo-stencil.

Photo-Resists. The essential requirement for making a photo-stencil is the photo-resist, i.e. a light-sensitive material which can be coated in a thin layer on metal, glass or plastics. The photo-resist layer is exposed to light under a negative, usually but not always produced by photography. The non-exposed areas protected against light, remain soluble in suitable solvents and are removed in such a solvent. The exposed areas become insoluble, forming the photo-stencil which protects the metal image-wise against being attacked by the etching fluid.

Photo-Engraving. Photo-etching by means of photo-stencils is universally used for the production of line or half-tone blocks for letterpress printing and intaglio copper plates or cylinders for gravure. However, even planographic plates, e.g. for offset printing, are quite often etched (deep etch process).

Etching of letterpress blocks in the classical manner raises one problem. The etching, particularly with line blocks, must reach a considerable depth. During the necessary prolonged etching process the acid not only etches in to the depth but it also attacks the metal from the sides of the recesses formed during etching. It gradually widens the etched area, a phenomenon termed under-cutting. Until quite recently this undesirable feature was counteracted by etching in stages. Between each stage the already formed walls of the etching were in some way protected against further attack of the acid during the next etching stage. One of the methods is to powder the etched portions with a suitable resin (e.g. dragon's blood) in such a manner that the resin adheres only to the walls of the etching but does not cover its bottom. This is brought about by a delicate brushing procedure.

The more recent powderless etching methods eliminate the need for the cumbersome powdering or similar procedures. Here etching fluids are used in conjunction with certain organic compounds which form a liquid etching emulsion. Using suitably constructed etching machinery it appears that during etching the organic constituents of the etching fluid are preferentially deposited on the walls of the etched recesses, acting thus to some extent as protection.

By far the largest application of photo-etching is still photo-engraving devoted to the production of printing formes. During and since the end of World War II photo-etching methods have however extended to new fields of increasing importance in mechanical and electrical engineering, such as chemical milling, printed circuit production, etc.

Nameplate Manufacture and Decoration of Metals. Increasing quantities of nameplates are nowadays produced by etching through photo-stencils. On the whole the technique follows the well-established lines of that of block making. However, aluminium and its alloys and anodized aluminium which are not used for block making sometimes require special treatment.

The decoration of metal surfaces by photo-etching is applied to surfaces of all dimensions. So entire panels for lift doors are commercially produced by photo-etching as well as minute dials for miniature watches.

Chemical Milling. Photo-etching can sometimes replace mechanical milling in the manufacture of engineering components. Its object is usually the decrease of weight of metal sheets by removing part of the metal thickness, leaving a structure of ribs of the original thickness intact. This is of special interest for the aircraft industry. The metal in question is usually aluminium and its alloys and the procedure is said to be faster and more economical than mechanical milling. A depth of etching to a quarter of an inch is not unusual.

The term chemical milling is often also applied to the manufacture of perforated metal parts although a distinct term would be desirable.

The important difference between chemical milling in its narrower sense and the manufacture of perforated metal parts by photo-etching lies in the fact that in the latter case the metal in question is etched right through its entire thickness, leaving the perforated metal part free with its desired, frequently quite complicated, contours.

Usually, in order to obtain inner and outer contours of high definition the metal in question is photo-etched from both sides of the sheet. This requires the production of photo-stencils in perfect register on both sides of the metal. So two photographic negatives have to be produced and exactly located relative to each other.

Perforated metal parts, such as springs, electrical contact pieces, sieves, etc., are, of course, produced in huge quantities by stamping them out by means of dies. The newer photo-etching techniques have several advantages.

Dies are very expensive. One does not usually have them cut until the exact shape of an intricate metal part has finally been defined. However, in a time of rapid development it is highly desirable that prototypes of the metal part in question can be produced in small or medium quantities at relatively low cost and without undue delay. It is here that photo-etching scores.

Secondly, die-stamping invariably leaves a burr at the contours of the metal part and the removal of this burr is sometimes a costly additional operation. Also, die-stamping imparts undesirable internal stresses to some metals. Photo-etching, on the other hand, is completely free from these features.

Almost all metals and alloys can be photo-etched, including for instance molybdenum and tantalum, metals which are often a trial for even the hardest dies. Perforated metal parts up to a thickness of 0·04 ins. have successfully been photo-etched and in thinner metal sheets intricate designs have been produced which one can hardly hope to obtain by stamping.

Printed Circuits. There is nowadays hardly an electronic instrument in which the formerly used wiring connections are not replaced by a printed circuit. As the name suggests, printed circuits are to some not inconsiderable extent produced by printing, e.g. by silk screen or offset methods, the ink serving as etch resist. A large and increasing proportion of them is, however produced by photo-etching, as the design of circuits becomes increasingly complicated and the demands on accuracy and definition of the elements of the design grow.

The conventional basis for a printed circuit panel is a copper laminate, i.e. a copper foil, 0·001 ins. thick or more, which is laminated to a non-conductive base, such as a plastic sheet.

The copper surface is coated with a photo-resist. After exposure to light through the negative of the required circuit pattern and processing, the resist forms a photo-stencil. Through this stencil the copper is etched. In contrast to photo-engraving, the metal is etched right through its entire thickness, leaving the design of the conductive circuit on the bared insulating plastic.

Also, instead of a single or a few blocks made from a specific design for the printing press, hundreds or thousands of printed circuits are produced from one and the same pattern. Consequently the whole procedure is mechanized, beginning with the cleaning procedure of the copper surface and ending with the etching stage and the removal of the photo-stencil.

Owing to the thinness of the copper foil there is comparatively little under-cutting but as the demands on quality increase research is on the way to minimize under-cutting by photo-etching.

Transistors and Micro-miniature Circuits. The basis of most transistors is a thin wafer of a semi-conducting substance, such as silicon or germanium, an inch square or even less. From such a wafer a number of transistors, say 100 or more, are produced, i.e. the actual design of each transistor is repeated many times on the wafer.

Frequently the multiple design is produced on the wafer by evaporation of metals through extremely fine masks (evaporation masks), which are laid on the wafer.

Alternatively vacuum deposition is done through a photo-stencil produced on the surface of the wafer.

There are, however, frequent cases where

either a metal, evaporated over the entire surface of the wafer or the material of the wafer itself is etched image-wise through a photo-stencil.

According to the nature of the metal or to the material of the wafer itself etch fluids of various compositions have to be used, ranging right up to mixtures of hydrofluoric acid with other strong acids as for instance nitric acid. Owing to the minute dimensions of the design and to the nature of the acids, photo-stencils of high acid resistance and of very high definition have to be produced on the surface of the wafers.

Similar methods are applied to the production of micro-miniature or miniature circuits. Such circuits might consist merely of an array of conductive lines, when it is a miniaturized version of a printed circuit. However, owing to its small dimensions and by electronic considera-tion and requirements the comparatively crude copper laminate of the printed circuit manufacture cannot any longer be used. Instead, a thin layer of a highly resistant and adherent metal or alloy, is evaporated on glass of suitable properties and the design is photo-etched in this metal.

Other kinds of micro-miniature devices incorporate not only conduction lines but also intrinsic parts of an electronic circuit, such as resistors and capacitors.

The ever increasing demands on smallness of individual devices and on highest definition present problems concerning properties of photo-resists and photo-etching as well as of photo-emulsions and lenses unheard of a few years ago. E.E.L.

See also: *Resists.*

PHOTO FINISH CAMERA. Special camera for producing a photographic record of the order of arrival of competitors at the finish of a track race.

The camera is sighted along the finishing line and the film is moved in the opposite direction to the race at a speed which is approximately equal to the movement of the image across the focal plane. The image is therefore more or less stationary with respect to the emulsion surface. The film is exposed through a fixed vertical slit which coincides with the image of the finishing line formed by the lens.

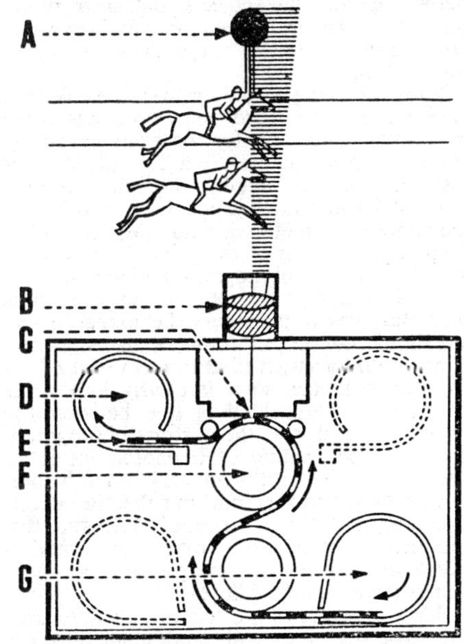

PHOTO-FINISH CAMERA. The film moves past a slit behind the lens, aligned with the finishing line of the race. The film movement is adjusted to the speed of the race. The image of each horse starts to build up the moment the horse passes the finishing post. A, finishing post. B, lens. C, slit. D, take-up cassette. E, film. F, driving drum, G. feed cassette.

Before the leaders arrive, the film simply records a continuous blur left by the stationary image of the finishing line. As soon as the first contestant starts to pass over the line, it begins to record its image on the moving film from the nose backwards along the length of the body. Since the image and the film are moving at the same speed, the record is sharp, and it builds up to form a complete picture as the rest of the field crosses the line.

A time scale in $\frac{1}{100}$s of a second along the film directly shows the winner's time, i.e., the instant at which he crosses the finishing line of the course.

A white cursor line is introduced into the photograph when printed so that the position of the winner in relation to near by runners can be clearly seen. It is common practice now to also have a mirror facing the camera on the far side of the track so that, as the camera points at it from an angle, a second (reflected) image of the runners is also recorded; this helps in distinguishing the positions of the leaders.

Most photofinish installations are now arranged in such a way that the result can be projected as an enlarged negative image directly from the camera room on to an adjustable viewing screen in the Judge's box. The negative can be prepared for projection in 20–25 seconds from the end of the race so saving valuable time by comparison with the earlier procedure of making prints for the judge. The negative is in no way impaired by projection and prints can subsequently be made for display to the public.

PHOTO FINISHING. Work of the commercial establishments which develop and print amateur roll films in quantity for the retail photographic trade. Also known as D. & P. (Developing and Printing).

See also: *Wholesale photo finishing.*

PHOTOFLOOD. Electric filament lamp designed to provide a high intensity source of artificial lighting for a relatively low current consumption. This is done by designing the lamp so that it will be over-run when used on normal household voltage.

Photofloods form a cheap and convenient form of lighting for indoor photography and cinematography.

The brilliance of the light is purchased at the price of its working life; Photofloods last about two hours when switched on for no more than a minute or two at a time. Run continuously, they burn out much sooner. Large photoflood lamps may have a burning life of around 6–8 hours.

The working life can be extended by using a dimmer instead of a switch so that the voltage can be applied gradually.

Another life-saver is to run two Photofloods in series-parallel, switching them on in series for focusing and arranging the subject, and changing to parallel for the exposure.

See also: *Filament lamp; Light sources; Series parallel.*

PHOTOGALVANOGRAPHY. A photo-mechanical printing process patented by Paul Pretsch in 1854. As modified in 1856 it was somewhat akin to photogravure, giving true halftones. A relief picture in bichromated gelatine was used to produce a guttapercha intaglio mould. From the mould, a copper matrix in relief was produced by electrotyping. Finally by a further process of electrotyping, a thick intaglio copper plate was made, from which the proof was printed. The entire process took about six weeks. 400–500 good impressions could be printed from the plate, and a considerable number of plates could be obtained from the matrix.

Pretsch came to England to exploit his invention, and the first number of *Photographic Art Treasures* appeared in November 1856—the first British publication of photographs in printing ink.

PHOTOGENIC. Photogenic drawing was term used by W. H. Fox Talbot to describe his earliest photographic process "by which natural objects may be made to delineate themselves without the aid of the artist's pencil". In modern usage photogenic is an adjective borrowed from the French indicating subjects, more generally people, who are particularly easy to photograph well, possibly because they may combine ease in front of the camera with well-proportioned or expressive features.

PHOTOGRAM. At one time the term photogram was used in a limited way to describe the consciously artistic photograph as opposed to the mere mechanical record. Nowadays, however, the word is generally reserved for pictures of an abstract kind which, though made on photographic material, are created without the use of a camera.

History. In its simplest form the photogram is a direct shadow pattern formed by flat objects placed on the sensitive material during a brief exposure to light, and the image thus produced is a negative silhouette—white on a grey or black ground. Josiah Wedgwood (1730–95), at the end of the eighteenth century, found that he could reproduce the form of a leaf in this way on paper bathed in silver nitrate, but his "sun prints" were not permanent for he knew of no way of desensitizing the paper afterwards. Later Talbot made permanent "photogenic drawings" of plant forms and pieces of lace in the same way, and by using the first print as a paper negative he was able to turn the image into a positive silhouette.

These experiments were, however, merely stages in the early development of photography and it was not until many years later that the aesthetic possibilities of these simple shadow patterns began to be appreciated. Inspired by cubism, Christian Schad in 1918 elaborated the technique by placing strips of torn paper, etc., on the sensitive material to produce overlapping shadow patterns that rather resembled the collages made by the painters. Further developments came in 1922 when Man Ray and L. Moholy-Nagy made their "rayographs" and photograms, using not merely opaque flat objects but also three-dimensional and translucent ones and recording the shadow patterns of varying sharpness produced by them.

Since then, the technique has been further extended and the word photogram has come to be used not merely for these comparatively simple shadow pictures, but for all sorts of abstract pictures made on photographic material by any method at all—including the deliberate (or sometimes, no doubt, accidental) fogging or staining of the print. In fact, anything is fair in photogram making and no method is barred so long as no camera is used. But there would seem to be no reason why one should not project a camera-made negative of a suitably abstract kind to form part of the photogram image, so it is clearly impossible to draw a sharp dividing line between the photogram and, say, the abstract photograph.

Method. The making of a photogram is an uncertain business: it is not easy to tell beforehand how it will turn out, but therein lies the fascination. After a series of dull and disappointing results, the surprisingly beautiful effects that come quite unexpectedly can more than make up for the failures.

Films or plates can, of course, also be used; but large ones are costly and small ones are limiting. But there is the advantage that they can be enlarged to make positive versions (or via contact transparencies, enlarged negative prints also). But photograms made direct on thin paper can also be duplicated by using the original as a paper negative. A certain amount of the texture of the paper will be evident in the photogram, but this will not usually detract from the result and may, indeed, enhance it.

Materials. Literally anything may be used to form the shadow patterns—e.g., paper of varying degrees of translucency, cut or torn to suitable shapes, coiled or crumpled paper or Cellophane, bits of gauze, perforated zinc, bent wire, string, thread, hair, a feather, a watch spring, an electric bulb, or a kitchen implement. Solid objects, if light in tone, will, when not in all-over contact, introduce a partial greying of the white shadow pattern where the light has crept in or is reflected from the underside.

Polished objects will throw out spots and lines of light and, if the exposure is short enough to leave the background fairly light, these will show up in black. Interesting effects of this sort can be got by using strips of tinfoil, curled or twisted in various ways. Somewhat similar effects are produced with glass-ware—e.g., a cut glass stopper, the neck of a broken bottle, fragments of glass, a condenser lens or transparent crystals. These reflect and refract the light in quite unexpected ways, producing lines and splashes of black through both white pattern and background.

The articles may be placed direct on the bromide paper, or they may be arranged on a sheet of glass with a sheet of white paper under it, so that the effects can be to some extent planned and studied. The glass can then be gently lifted and the bromide paper slid under it. This makes it easy to run off several prints, and to alter the arrangement without starting afresh each time, although the objects in question will never record completely sharply by this method. For maximum sharpness the glass used should be as thin as possible, and the exposure should be made under directional illumination, such as is given by an enlarger which is focused onto the bromide paper.

Lighting. Any kind of light source can be used, from a match to a small spotlamp. The character of the shadow shapes will depend on how concentrated the light is, on its distance from the objects and the angle at which it strikes. An electric torch bulb—preferably without reflector—makes a very useful tool because it can be used at close quarters at any angle, either held in the hand or arranged on some improvised stand.

Softer outlines will be created by an ordinary pearl lamp, but this will have to be used farther away or exposures will be too short to control accurately.

For sharp clear-cut outlines the material may be arranged under the enlarger, with the enlarger focused sharply on the bromide paper; a baby spot will produce similar sharp effects and can be applied from any angle, but unless it is very small indeed it will have to be a good way off.

More complex patterns with cross shadows result if two or more lights are used, or the same effects can be got by making multiple exposures with a single lamp in different positions. Curious ghostly effects can be produced by keeping the light moving during exposure.

Endless variations in the shadow shapes can be made by changing the angle of the lighting. This is especially so if the articles stand up well from the glass and are only in contact with it at certain points. With low slanting light the shadows stretch out and widen into fantastic shapes. Bent wire and glass articles are particularly suitable for this treatment. Vaguer patterns are produced if the objects are raised some distance above the paper—e.g., by a second sheet of glass.

Soft all-over patterns can be introduced with netting, or by sprinkling very small articles—say hypo crystals or sugar, or drops of water—over the raised glass.

Variations. The pattern need not, of course, be confined to white shapes on a grey or black ground: by interrupting the exposure and then taking away some of the articles, their shadows will be greyed to a depth of tone depending on the length of the further exposure. When the articles or their shadows overlap this method is extremely effective. The tone of the background also can be varied by holding back parts, as in enlarging. A photogram may be in bold black-and-white or, at the other extreme, in white and pale greys.

And as all is fair in photogram making, the print can be partially fogged or even deliberately stained by exposing it to light before it is completely fixed. Films or plates can also be fogged during development to cause solarization of the image.

There is still another way of making photograms with the enlarger not yet mentioned: it might be called the method of the "hand-made negative".

Cut-outs and various small articles are formed into a design on a small piece of glass which is slid into the position usually occupied by the negative carrier of a vertical enlarger.

Obviously the articles have to be small and with expressive outlines, and many if not all, should be transparent in some degree—e.g., pieces of very thin paper, Cellophane, old film, a mica washer out of an electrical fitting, a small feather, hairs, thread or gauze. Unfortunately, many enlargers are unsuitable for this technique because there is not a big enough gap for such a thick "negative" to be slid into place.

A further extension of this idea which has been applied effectively by some artists is literally to draw the negative by etching on a blackened plate or by applying pigment to a clear plate, and printing the result. H.v.W.

See also: *Abstract photography; Tricks and effects.*

Book: *Design by Photography*, by O. R. Croy (London).

PHOTOGRAMMETRY

Science of making measurements from photographs. Through such measurements taken on aerial photographs it is possible to construct maps and obtain a great deal of information concerning the topography, vegetation and other feature depicted on the photographs. The majority of this type of photography is taken from aircraft using special wide-angle cameras, but photographs taken on the ground are widely used in other applications of this science, namely, in the fields of architecture, deformation surveys, biology and medicine.

By the application of photogrammetric principles to the central perspectives (photographs) obtained, it is possible to carry out the plotting and/or recording of the required data within a great range of scales (e.g. maps are frequently produced from scales of 1 : 100,000 up to 1 : 500) whereas using close range, macro and micro photography scales extending to 100 : 1 can frequently be obtained.

There is little which is basically new in either photographs or their measurement. The photogrammetrists' concern lies largely with the geometrical solution of pairs of perspective views, solid geometry, binocular vision and the accuracies obtainable therefrom; and present techniques are the result of a general advancement in the accuracies obtainable rather than upon new principles.

Despite the development of basic ideas, the practical application of aerial photogrammetry could not develop until the introduction of the aeroplane and its aerial camera, and the plotting apparatus designed for use in analysis and plotting.

As mentioned, photogrammetrical surveys are carried out on various scales, according to the work in hand. There are four main groups used in practice.

PHOTOGRAMMETRICAL SURVEYS

Name	*Use*	*Scale*
Semi-graphical	Unmapped territory	1 : 50,000 to 1 : 25,000
Machine mapping	Development mapping	1 : 25,000 to 1 : 500
Close range	Urban areas	1 : 500 to 1 : 10
Macrophotogrammetry	Hydraulics Machine parts Accidents	1 : 10 to 8 : 1
Microphotogrammetry	Erosion, surface analysis	8 : 1 to 100 : 1

Camera. The basis of any photogrammetrical study is a negative or positive which is dimensionally as near perfect as possible so that reliable measurements may be made from it at any time.

The technique of measuring in this way depends on the fact that if the position of the cameras at the time of exposure are known with respect to any object on the negative, then the size and position of all other objects in the picture space may be determined. Accurate measurements are only possible by using a lens free from distortion, with its axis lying at right angles to the plate.

The design of the lens and camera to be used is basically important; in the case of lenses used in photogrammetry, testing usually concentrates on their resolving power, light distribution and distortion characteristics. Calibration of the position of the "principal point" and the "calibrated" focal length is required so that the most accurate value of the image angles on the object side of the lens may be obtained from image positions recorded on the photograph. The calibration data must remain constant over long periods, consequently the camera is normally rigid. The angular coverage of the lens is usually between 70° and 120°, the lenses used being of normal, wide-angle and super-wide-angle types, i.e. 300 mm. and 150 mm. or 100 mm. nominal focal length when covering a format size of 230 × 230 mm.

Photographic Material. For most aerial surveys film is used with a dimensionally stable base material, although for certain specialized purposes glass plates are occasionally employed. Glass diapositives are prepared from the film negative, and used in connexion with plotting machines for the preparation of maps. Paper prints may also be used for interpretation purposes and for less precise types of survey.

Methods. Although single photographs can be used, the more practical procedures normally require the plotting to be carried out from stereoscopic pairs of photographs. Binocular viewing, the stereoscope, or a similar binocular system, is used extensively for assessing and measuring the three dimensional models seen. The stereo-pair of photographs must satisfy, therefore, the conditions of binoc-

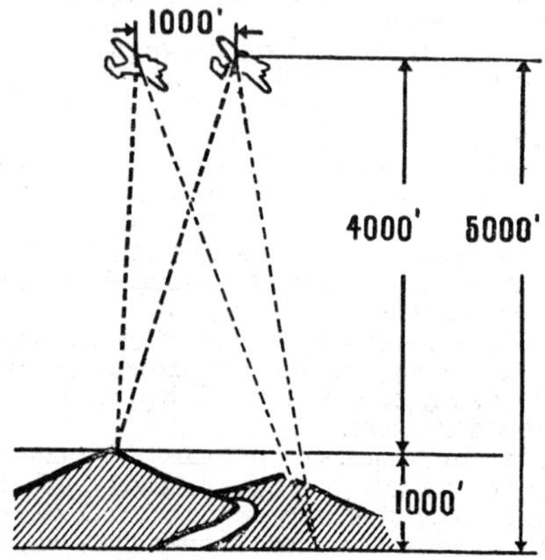

AERIAL PHOTOGRAPHY: Two views taken at a known distance apart from an aircraft provide a hyperstereogram, showing mountains and valleys in stereoscopic relief.

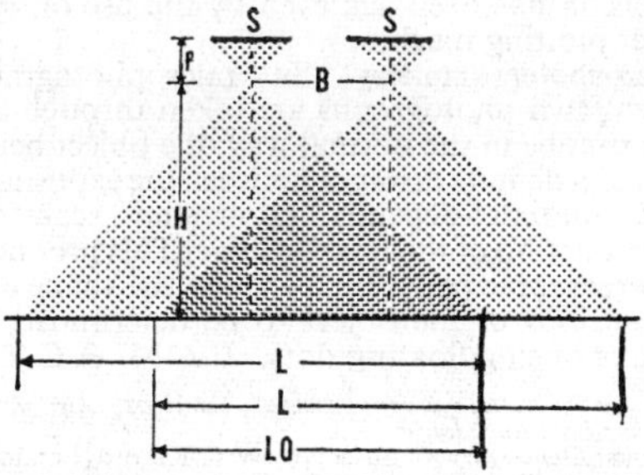

STEREO PHOTOGRAPHIC SET-UP WITH SIGNIFICANT PARAMETERS. S, format side length. B, air base length between successive photographs. f. focal length of camera lens. H, flying height. L, length of ground covered by single photograph. LO, stereo overlap; this should be about 55 per cent of the dimension L. The base to height ratio B/H is important for measurements.

ular vision. In the cases of aerial photography the distance between the reference stations is such that the photographs overlap by at least 55 per cent in the fore and aft direction.

The separation of the perspective centres can therefore be expressed in the form of the base/height ratio $\frac{B}{H}$, where B is the base length (separation of the perspective centres) and H the mean flying height above the ground. In the case of wide-angle photography this ratio, given a 60 per cent overlap, is approximately 3 : 5.

The base to height ratio is of paramount importance, for the accuracy of the measurement of the third dimension depends a great deal upon the parallax measurements which have to be made.

The difference of parallax between selected points is measured, and the actual height of each point above ground level calculated.

Map Making. Today the majority of surveys used for map making are produced by means of photogrammetry, although a certain amount of ground control has to be carried out by the classical ground survey methods. If we take, for example, a pair of overlapping vertical photographs, in which the same area is covered over at least 50 per cent of each photograph, then the angle subtended by the air base will depend upon the distance of the ground point from the air base.

If, for this purpose, we call this angle the parallactic angle and measure its value in the plane containing the air base, we could calculate the distance of the point from some known datum level. It is not usual to measure this angle directly, but to use the image distance measured from the principal point on each photograph.

A stereoscope is frequently used to observe the pair of photographs. When the photographs are rotated until their base lines are colinear, a three dimensional effect is obtained if the separation of the photographs is satisfactory and when the bases are parallel to the eye base of the observer.

With a measuring instrument (which may be a parallax bar), the differing parallaxes can be measured.

With this, a normal procedure is to have a measuring mark for each photograph. Each measuring mark is brought over corresponding points of detail on the steroscopic pair of photographs. These marks are also fused stereoscopically. So when the exact locations of points are obtained, the fused "floating" mark appears to lie on the surface of the terrain at that point in the viewed model.

If the parallax bar is kept parallel to the base line, a measure of the linear parallax can be made at any point with reference to any other point shown on the photograph. This parallax difference is usually measured by a micrometer.

The simple methods outlined can be used on truly vertical photographs, but are subject to many errors if the photographs are tilted from the true vertical. Consequently these methods are not used a great deal and a number of complex plotting machines are employed in practice.

For terrestrial purposes, a phototheodolite is used. As the name implies, this is a combination of a theodolote and a camera. In this case, the two photographs forming the stereoscopic pair are taken at each end of a measured base.

Close Range Applications. Photogrammetric methods are also employed for non-topographic purposes, but they have not yet become as important as the mapping programmes used almost universally throughout the world.

Amongst these non-topographic uses, the following have received much attention: determination of volumes in open cast mines, architectural details, structural movements, medical research and industrial investigations.

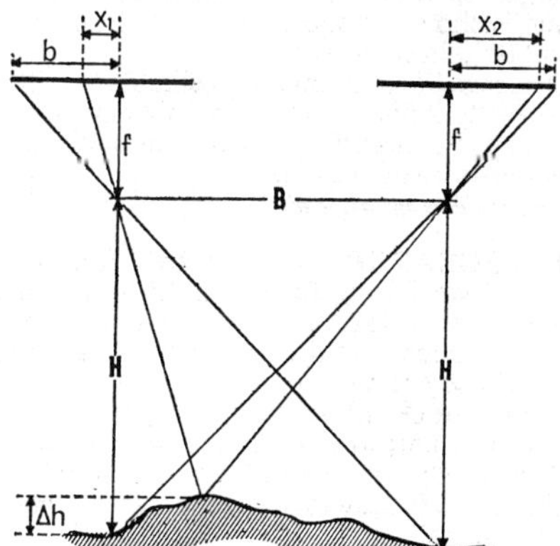

PHOTOGRAMMETRIC HEIGHT MEASUREMENT. The image distance from the principal point in each photograph is measured in the direction of the stereo base. Then $(x_1 + x_2)/f = B/(H - \Delta h)$.

Furthermore, the photogrammetric techniques have been employed for making accurate records of traffic accidents.

Limitations. If the accuracy required of any particular investigation is known, a photogrammetric technique can usually be found which will satisfy the conditions. However, this technique might prove not to be the most beneficial on the score of economy or accuracy.

Undoubtedly, the widespread use of air survey methods for map making has been due to their speed and economy. Certain limitations previously paramount have now been considerably reduced by the introduction of better lenses, better cameras, and better film, whilst precision has been achieved by the use of very exact plotting machines.

Microphotogrammetry. In microphotogrammetry, two photographs are taken through the microscope in the normal way, the object being moved a definite distance between the exposures. The amount depends upon the scale of reproduction of the final positive. This produces a stereoscopic pair which enables the shape and dimensions of the object to be determined by parallax using floating dots. E.A.M. & G.T.S.

See also: *Aerial survey; Aircraft cameras; Air photographs; Phototheodolite.*

Books: *Elementary Air Survey*, by W. K. Kilford (London); *Elementary Photogrammetry*, by C. Crone (London); *An Outline of Photogrammetry*, by K. Schwidefsky (London).

PHOTOGRAPHIC ALLIANCE OF GREAT BRITAIN. Founded in 1930 the Alliance acts as a co-ordinating body for the ten regional federations in Great Britain. By Membership of any Federation the individual clubs and societies, totalling over a thousand, are affiliated to The Photographic Alliance and receive certain services, benefits and privileges.

The Alliance publishes a Year Book in two parts. The Directory Section contains a list of all the affiliated societies, with particulars of meeting place, meeting night, name and address of secretary. The Syllabus Section details the various services provided for affiliated societies, i.e.: Lists of lecturers and lecture titles, lists of judges appointed for exhibitions and competitions; recorded lecture service; filmstrip service; leaflet service; loan collections of prints and slides, etc.

An annual Inter-Federation Competition with trophies and awards in the various sections is organized by the Alliance. A selection of the prints and slides entered for these competitions is retained to form The Photographic Alliance Exhibition of British Photography which is shown at some six or seven centres in Great Britain each year.

Overseas Societies are also eligible for affiliation to the Alliance and a number of services have been instituted for the benefit of clubs and societies in many countries.

See also: *Clubs and associations.*

PHOTOGRAPHIC DEALERS' ASSOCIATION. Association, formed in 1914, as purely a retail organization with headquarters in London. Membership is confined to retail firms whose premises are suitable and who buy, sell and stock photographic apparatus and materials. There are members in all parts of the British Isles; the major proportion of their business is transacted with amateur photographers.

Association affairs are managed by its National Council of 16 elected members (representing areas of the British Isles) and 8 co-opted members, together with a limited number of Past Presidents. In addition, there are a number of standing committees which are concerned with particular aspects of the Association's work.

There are P.D.A. branches in a number of the larger provincial towns.

The P.D.A. policy is to encourage photographic knowledge and efficient service in those behind the retail counter. Considerable attention, therefore, is devoted to education. P.D.A. examinations, held annually, are open to all photographic dealers and assistants. The examinations are of two grades each including written, practical, and oral sections. The syllabus covers a wide range on the theory and practice of photography and the demonstration of all types of photographic apparatus.

The Association publishes a monthly journal, *The Photographic Retailer*, which is free to members. Also free to members and published in collaboration with the Focal Press is the *P.D.A. Reference Book*, with approximately 290 pages of technical, business and legal information.

PHOTOGRAPHIC SOCIETY OF AMERICA. Organization which was incorporated in 1937 as a direct development of the Associated Camera Clubs of America, first formed in 1919. It is a non-profit corporation which aims at "promoting the arts and sciences of photography and furthering public education therein". It is an international organization operating through committees, divisions, and chapters. The sole requirement for membership is a genuine interest in photography.

The Society is governed by its Board of Directors and the National Council consisting of the national officers, the Chairmen of the various Divisions, the District Representatives, the Chairmen of the Standing Committees, the immediate Past-President of the Society, and the Honorary Representatives when in the U.S.A.

The national officers, zone directors, district representatives, and division chairmen are elected by the membership, and the committee chairmen are appointed with the approval of the Board of Directors.

Only the Executive Secretary and his assistants and the Editor and Advertising Manager of the *P.S.A. Journal* receive any remuneration.

The principal classes of society membership are Active, Sustaining, and Organizational. The Active group includes honorary and life membership. Sustaining membership includes business organizations who contribute to the financial support of the Society, and Organizational membership includes camera clubs, libraries, and educational organizations.

Facilities. The society maintains its headquarters in its own building in Philadelphia, where it also houses a permanent collection of outstanding photographs selected by the Committee.

The International Convention is held in a major city each year. Regional conventions are also held in various centres throughout the U.S.A. and Canada. Group and individual activities are conducted by committees with the assistance of district representatives and committeemen.

P.S.A. chapters (formal groups in immediate areas) present important photographic events and promote the use of photography for the benefit of the community—e.g., in the form of town meetings, shows, lectures and so on.

Two of the most important services to camera clubs and other groups are the National Lecture Programme and the Tape Recorded Lectures.

The Camera Club Committee is the chief channel through which P.S.A. serves groups. It makes available, in addition to its regular C.C. Bulletin of news and programme material, all the group services provided by the National Committees and the Divisions.

Other Activities. To provide for the various specialized photographic interests represented in P.S.A., the Society operates through divisions which function independently in their own particular interest under the constitution and bye-laws of the Society. Each has its own elected officers, its executive committee, its regular bulletin, and its own space in the *Journal* and is represented on the Board of Directors and the Committees. The Society dues include membership in any one division, and additional divisions may be joined at any time at a nominal annual fee.

There are separate divisions covering each of the following subjects: colour, journalism, motion pictures, nature, pictorial, stereo and techniques. In January 1955, the Amateur Cinema League of America voted to join the Photographic Society of America and merge with P.S.A.'s Motion Picture Division.

An autonomous Honours Committee awards Honorary Fellowships, Honorary Memberships, Fellowships, and Associateships. There are also Service Awards and Medals for extraordinary service to the Society and a number of special awards given through Society or Divisional Committees.

In addition to the Divisional and Committee Bulletin, P.S.A. publishes the monthly *P.S.A. Journal* and the quarterly *Photographic Science and Technique*.

The P.S.A. membership Directory is issued annually as a supplement to the *P.S.A. Journal*. It lists the full membership alphabetically and geographically. P.S.A. membership extends throughout the world.

PHOTOGRAPHY. Literally "writing with light" (from the Greek, phos, photos—light + suffix graphos—writing). The term is generally accepted as referring to any method of producing a visible image by the action of light—e.g., on light-sensitive silver salts, in the camera. The use of the term was suggested by Sir John Herschel to William Henry Fox Talbot in a letter dated 28th February, 1839. It was also used in the *Vossische Zeitung* of Berlin on 25th February, 1839, in an article over initials which point to the astronomer Johann von Maedler, who was a correspondent of Herschel's.

See also: *Photo.*

PHOTOGRAVURE. Printing process using a recessed image prepared by photo-engraving. The final printed picture derives its tone differences from the amount of printing ink deposited at any one point.

See: *Photomechanical reproduction.*

PHOTO JOURNALISM. The popularity of illustrated magazines and newspapers has created a highly specialized form of journalism in which the author works with a camera and thinks primarily in terms of pictures rather than of text. Hundreds of such examples of camera journalism appear in the press every week, and many of the leading weekly and monthly magazines in particular pay high rates for the illustrated stories they publish. Such stories may range from a series of consecutive shots of some novel subject to a fully documented account of the life and geography of some out of the way part of the world.

Requirements. Since the rewards may be very high indeed for outstanding work, and because few other fields present the same opportunities for travel, adventure and interesting day-to-day experiences, it is not surprising that so many photographers at some stage of their career try their hands at photo-journalism. But it is one of the most difficult branches of the profession, because it requires competence in two fields—photography and journalism—in addition to a high degree of personal organization.

The most common reason for the failure of so many photographers who are certainly good technicians is their ignorance of the rules for creating and maintaining interest; they tell the story exhaustively in every detail, without emphasis or dramatization and the result is almost always boring.

The tyro, on his first assignment, usually produces pictures which are hackneyed, obvious and devoid of surprises, whereas the experienced camera journalist knows that the interest in his pictures often depends not so much on the raw material which confronts him, but on the way he handles it—on story treatment. He mentally plans the page layout as he selects and takes each picture. From experience he has a fair idea of how the editor will want to present the pictures. He knows that most editors look first for a key shot which can be featured—possibly as a full-page illustration—to establish the theme at a glance. Next they want subordinate shots with plenty of variety and contrast. So the old hand deliberately picks his scenes to make it easy for the layout man to arrange them to give the story continuity and development.

With the layout in mind the experienced worker deliberately varies his camera vantage point whenever possible. To establish locale and give the reader a bird's-eye view, he moves back and takes a panoramic shot. (If he cannot move back, he uses a wide-angle lens, to achieve the same effect.) To follow the development of the story, he may choose mainly medium shots, but he will alternate them with close-ups to illustrate or emphasize details. If it is impossible to move the camera close he uses a telephoto lens. This technique of dollying back and forward, or alternating lenses of varying focal lengths, is a standard device of the cine photographer, but it is equally effective for the still man who is working on magazine layouts.

Working Methods. Picture story assignments fall into two general classifications: those in which the photographer can record the events but not control them, e.g., prize fights, or trooping of the colours; and those where he can exercise his skill as a director, e.g., in shooting a rehearsal of a ballet company, or making a picture essay on the life of a country doctor.

Talents and abilities vary as much in this field as in any other; some photographers are good at fast moving events because they have fast reflexes, but they may not be equally successful in an assignment calling for stage-managing ability, or outstanding powers of diplomacy; others may be attracted to stories where they can shape the material. It is a rare man indeed who succeeds in every department. A good picture editor knows the capabilities of his staff photographers and freelances, and gives assignments to one or the other in accordance with their particular talents.

Camera journalism, like writing, has its own overworked and threadbare clichés. A stereotyped, worn-out idea is just as undesirable in an illustration as last year's catch phrase in an article. The old hand is suspicious of the obvious, easy-to-get pictures. He may take them as a matter of record—but for his real pictures he looks for something different. His problem, like the writer's, is often of finding a new way to tell an old story; for while a story may not be identical with one that has already been published, it may be just a repetition of a well-worn theme without a new twist to commend it. To give the theme a new look, the photographer must be inventive and ready to go a little deeper if he hopes to produce story angles which his predecessors have either overlooked or failed to exploit.

The beginner is usually so preoccupied with his technical problems that he is more worried about whether his pictures will be sharp than whether they will be interesting. The old hand does not have to think about his camera; perfect technique with him is automatic, and he is able to turn all his attention to handling his subject. His trained picture sense tells him immediately whether the existing situations will make interesting pictures; if they will not, he tackles the job of reshaping them. This may call for all his skill, personality and tact. Arranging the picture—and the people in it—is more often than not the thing that calls for the time and effort, while the actual camera work may require only a few minutes.

Some photographers object to staging and direction because they feel that it creates pictures that look forced and artificial. But if that is so it simply means that the photographer responsible does not know that part of the job. It takes real mastery to make contrived pictures look candid, and it cannot be done by merely telling people to relax and look natural. Anyone who thinks that people can be commanded to relax and look natural should try taking pictures with himself in front of the camera. He will then see how self-conscious the subject feels.

To get people to look natural in front of the camera lens the photographer must know how to make them stop worrying about themselves. This can be done by keeping their minds occupied with the business at hand. So the approach must be decisive; people are usually co-operative but they do like the photographer to know what he wants. He must be ready not only to explain, but to demonstrate if necessary. This means that he must be something of an actor himself.

Many camera journalists handle the literary as well as the photographic side of their assignments. Theoretically this should result in better integration of text and pictures—but in practice it creates problems. For one thing, it is difficult to establish one kind of relationship with people as a photographer and then set about establishing another kind as a writer. Also, taking photographs is a strenuous business which often exhausts people on both sides of the camera, and during a tiring photographic session no one is in the right mood for discussion; interviews and flash bulbs do not go well together. Photographers who must do their own writing should take the pictures at one session and interview at another.

Rôle of the Caption. Many picture stories need only a hundred or so words of text because everything can be told in the pictures and their captions. In any case there are always captions to be written. It is a fairly safe rule that if the text has any plums, they should be transferred to the captions because there they are in a showcase and their enhanced impact gives a valuable lift to the pictures. This practice also helps to create editorial interest quickly, an important point when it is remembered that editors rarely have time to read the text on the spot and will hold the story only if pictures and captions look interesting.

The art of trenchant caption writing is brevity and condensation—plus research to make sure that every statement is correct and complete. Would-be clever captions that merely skim the surface and convey no real information are usually an admission that the writer does not really know his own story well enough, and should have spent more time looking up the facts. H.B.

See also: *Freelance photography; Magazine photography; Picture series; Press photography.*

Books: *How to Take Photographs that Editors will Buy*, by R. Spillman (London); *Press Photography in Practice*, by N. K. Harrison (London).

PHOTOKINA. International photographic and cine exhibition and trade fair held usually bi-annually in Cologne, Germany. The first was in 1950.

PHOTO LETTERING. Method of producing display headings and lettering layouts for printing by using a photographic enlarger and a master negative of the required lettering characters. In this way headings can be set up in any required layout and the result checked before a printing plate is made.

See also: *Photo type-setting.*

PHOTOLITHOGRAPHY. Normal lithography using an image formed by photographic means instead of by hand drawing or transfer. It is a planographic process—i.e., the printing and non-printing surfaces are at the same level —in which the image is formed of ink-accepting areas and the background of ink-rejecting areas.

In one form of photolithography, the image is made by coating the surface of grained zinc or aluminium with bichromated albumen and exposing it under the photographic negative.

See also: *Photomechanical reproduction.*

PHOTO-LITHO MASTER. Photographically prepared plate from which copies can be taken by offset printing. The term is generally applied to plates made for office-copying processes. Normally, the master material, which can be aluminium, zinc, plastic, plasticized paper or translucent polyester, is sensitized and then exposed in contact with a negative or, in some processes, a positive.

Traditionally, plates are coated with a dichromated colloid on a mechanical whirler. Bichromated albumen is used for the negative process. Exposure hardens the colloid, and after exposure the plate is inked, then developed in water. This washes away the unhardened, i.e. unexposed, areas of the albumen leaving an ink-albumen image from which proofs can be taken. Bichromated gum is used for the positive process in which a negative image is developed on the master surface after exposure, and used as a stencil to produce a positive printing image.

Manufacturers now supply ready-coated plates which eliminate whirling and have good keeping qualities. The simple processing technique associated with these has found special favour in office copying. Prominent is the diazo pre-sensitized plate most usually in negative, but also in positive working form, where diazo compounds take the place of bichromates. Coating solutions which may be wiped on the plate, not whirled, are also available.

There are also many simple methods of master preparation based upon rapid office duplication systems. One such is the diffusion-transfer process using a silver halide material. A negative paper is exposed in contact with the original, then passed through a developer in contact with a master. Solvents in the master surface dissolve silver salts on the negative paper and these are transferred by diffusion on to the master. This is converted to a visible image and, after appropriate treatment, can be used for offset printing.

Transfer processes where dye or carbon is transfered are also used. With these, an exposed and developed sheet is brought into contact with a master on to which the image is transferred and used for printing. F.G.W.

See also: *Document copying; Photomechanical reproduction.*

PHOTOLYSIS. The photochemical reaction of light in decomposing a substance, characteristic of the most common reactions involved in photography.

PHOTOMATON. Proprietary name of automatic portrait camera for producing virtually instant pictures for passports and similar purposes. Usually built into portrait booths incorporating automatic lighting equipment and set up in railway stations, department stores and similar semi-public locations. The exposure cycle is initiated on insertion of the coin by the user, and yields one or more positives within about a minute. The working cycle includes automatic reversal processing of the prints.

See also: *While-you-wait photographhy.*

PHOTOMECHANICAL REPRODUCTION

Photomechanical reproduction, a distinct branch of photography, includes all photographic methods of preparing the surfaces used for printing reproductions of photographs, paintings and drawings in books and in the press. Where the photographer sets out to make a black-and-white or coloured print of his subject, the graphic arts craftsman has to produce a surface which, when treated with printers' ink and pressed into contact with a sheet of paper, will leave the required image behind. His problem is to transfer amounts of printing ink to paper in proportion to the depths of tone of the original picture.

Printing Surfaces. The printing surface must usually be metallic to be strong enough to withstand printing runs of many thousands at high speed. There are three types of printing surface:

(1) The areas which transmit the ink may be raised above the general surface of the metal as in an ordinary metal type character or hand-cut wood block. This is known as cameo (raised), letterpress or typographic printing.

(2) It is possible to print from a perfectly flat surface if parts of it can be made to have an affinity for printing ink while the remainder of the surface repels it. Such processes are known as planographic (printing from a plane) and the most usual method is lithography.

(3) An image may be cut or etched into a flat surface which may then be flooded with ink, the surface wiped or scraped clean, and ink remaining in the incised hollows transferred to paper. Here the image may be recessed into the metal more or less deeply according to the depth of tone of the picture. This principle is called intaglio (meaning "incised" and pronounced "intahlio"). Photogravure is the only photo-engraved intaglio method, although there are many non-photographic techniques such as etching, drypoint and mezzotint which use the principle.

Tone. Only the last of the three basic methods is capable of depositing printing ink in different densities on paper. A photogravure image, for instance, has the quality of a continuous tone picture in which the tones vary in density from point to point as in a bromide print.

The typographic and planographic processes deposit printing ink upon paper at full strength, or not at all—that is, the tone is discontinuous. In these processes the appearance of intermediate tones is brought about by means of an optical illusion. The effect of varying densities is produced by a regular pattern of dots which are small in the light tones and larger in the darker tones. The dots are so small and numerous that the eye sees only the general effect and does not see them separately.

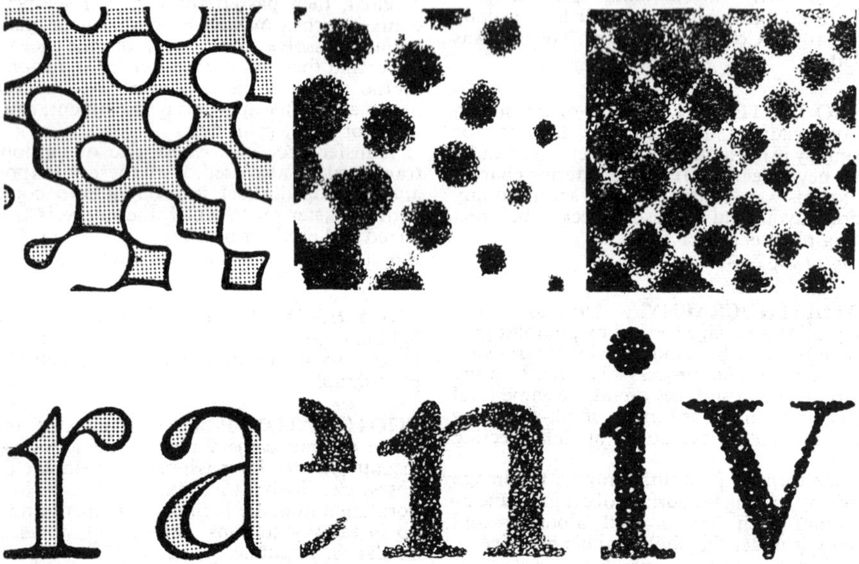

THE PRINTED IMAGE. Differences in printing processes show up under a magnifying glass. ***Left:*** *Letterpress image. Squeezing of ink towards edges of dots and lines produces characteristic ink squash effect. Irregular shaped dots are due to slight differences in image detail.* ***Centre:*** *Lithographic image. There is no ink squash or any sign of an impression mark on the paper, as would be obtained with letterpress printing. The ink deposit is very even; the half-tone dot shape is similar to the letterpress dots.* ***Right:*** *Photogravure image. Half-tone dots are more or less of the same size, but differ widely in density owing to varying ink deposits. Line images also need a screen pattern to provide a bearing surface for the doctor blade after inking.*

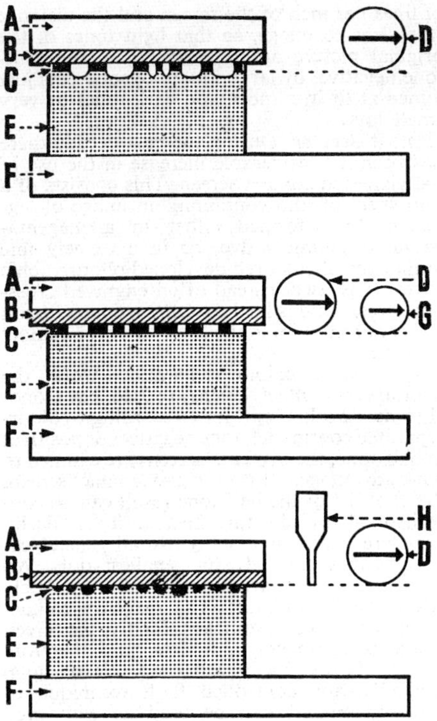

MACHINE PRINTING. Top: *Typography (e.g., letterpress). The printing image is raised and covered with ink by the inking roller, before printing on the paper.* Centre: *Planography (e.g., lithography). The printing image is on the same level as the non-printing surface and is damped before inking.* Bottom: *Intaglio printing (e.g., photogravure). The printing image is etched into the non-printing surface, and filled with ink. A doctor blade cleans off the non-printing surface. In all cases:* A, *platen;* B, *paper;* C, *ink;* D, *inking roller (preceding paper across printing surface in direction of arrow);* E, *printing plate;* F, *machine bed;* G, *damping roller;* H, *doctor blade.*

LETTERPRESS LINE BLOCKS

A line subject is a drawing in black lines on a white background which, for printing purposes, has to be duplicated, line for line, exactly as the original. The copy negative may be enlarged, reduced, or the same size as the original.

Making the Negative. The engraver first makes a photographic negative in which the black lines of the drawing are represented by transparent areas and the white background by dense areas, as nearly opaque as possible. The wet collodion process (invented by Scott Archer in 1851) was commonly used until around 1950. Since the invention of "lith" dry plate emulsions soon after World War II, nearly all photographic work in graphic arts has been done on dry plates and, to an increasing extent, on film. Stripping film and negative paper are also used for ordinary line work.

The process line negative must be laterally reversed, that is, the lettering must be right way round so that when the resulting printing block is inked and pressed on paper the proof will be the right way round. The negative image is reversed by reflection from a mirror placed at 45° in front of the lens. Generally the mirror is a solid glass prism with its long side silvered for convenience and permanence.

Printing on Metal. The powderless etching technique, invented in America in the early 1950s, is rapidly replacing the older methods of making line plates.

For conventional line block-making, a zinc plate about $\frac{1}{16}$ in. thick is coated with a solution of dichromated (and hence light-sensitive) albumen. The metal is spun rapidly in its own plane over a hot plate. This centrifuges off the surplus solution and dries the coating.

The plate is then exposed in close contact with the negative to the light of an arc lamp.

The unexposed dichromated albumen coating is soluble in cold water, but where light has fallen it becomes insoluble. After exposure, the coating is inked very thinly and the plate is put into a bath of water and lightly rubbed with cotton wool. The soluble albumen dissolves away leaving a visible image in the form of light-hardened albumen with a coating of ink. The plate is dried, its image dusted with finely powdered bitumen and finally warmed until the bitumen fuses into an acid resisting layer.

Line Etching. The metal between the lines of the image must next be etched away with nitric acid until it is below printing depth. As the acid begins to etch down between the lines it also tends to etch sideways. This action would very soon eat away the finer detail so the sides of the lines must be protected.

After a very slight etch the plate is treated with a weak gum arabic solution which provides a moist, ink repelling layer over the bare metal. It is then rolled-up with a greasy ink which is accepted by the bitumen image but not by the damp bare metal. The plate is then dried and dusted with a special acid-resisting red resin powder called dragon's blood. This has the property of liquefying into a viscous mass which does not run when it is heated. Upon cooling, it provides an efficient acid resist. The dragon's blood is usually applied and "warmed-in" four successive times between each etch, being brushed across the plate in four opposing directions. In this way the powder piles up against the sides of the lines like a snowdrift, providing an efficient protection against the sideways etch of the acid.

This process of alternately protecting the sides of the lines and etching further is continued until the plate is etched to sufficient depth. The ink and dragon's blood is then washed off, and a layer of hard ink is rolled upon the top of the lines. The plate is treated with a weak etching solution which removes any roughness at the sides of the lines, leaving them with a sturdy pyramidal cross section.

Finally, the plate is cleaned, blackened with a solution of copper sulphate, and the top polished with charcoal.

Mounting. It is possible to etch away all the metal not required to be at type height, but in practice this would waste time and materials, and so all large clear areas are covered with resist before etching and are afterwards cut away on a routing machine before the plate is mounted.

A router is a vertical milling machine with a cutting head that revolves at about 20,000 revolutions per minute. An open line block consists of little more than a skeleton of metal pinned to the wooden base at any convenient points within the work. The finished thin printing plate is mounted to type height (0·9186 inch) usually upon well seasoned oak or mahogany or on synthetic resin-bound chipboard. Plates are frequently mounted on metal bases for use on modern high precision presses.

HALF-TONE BLOCKS

Half-tone blocks can be regarded as line blocks in which the image is made of dense black dots. The illusion of shading or half-tones is caused by variations in the sizes of the dots, not by any difference in the density of the ink deposit. The original photograph is copied on a process camera as for line work with a reversing prism on the lens. In this case, however, a normal copy negative is unsuitable because the image from which the block is etched must consist either of acid resist or clean metal.

The Process Screen. When the photograph or any other type of picture is copied for half-tone reproduction, the image is broken up into a dot formation by a screen placed at a short distance in front of the plate or film in the process camera. The screen consists of two sheets of glass, each ruled with opaque lines equal in width to the spaces between them, and sealed together, face to face, with the sets of lines at right angles to each other and at 45° to the base.

The distance of the screen from the plate is adjusted so that the light passing through each aperture in the screen forms points of full illumination behind the centre of each screen aperture. Around these brilliant points, the intensity of the light falls off until it reaches a very low level behind each line.

If there were plain white paper in front of the lens, a short exposure would produce a pattern of very small dots on the photographic plate corresponding to the central points of full intensity light. These dots would be larger for longer exposures and eventually, with sufficient exposure, they would join together.

When the photo-engraver makes a negative for half-tone reproduction, he adjusts the size of the lens aperture and the distance of the screen from the plate with respect to the number of lines per inch of the screen and the distance from lens to image, so that light tones of the original picture are represented, in the half-tone negative, by large square opaque dots just joined at their corners, and dark tones by very small dots.

Contact Screens. During the past decade there has been a considerable increase in the use of the vignetted contact screen. This consists of a thin sheet of film containing an image of vignetted dots, formed either in a magenta-coloured grainless dye, or in extremely fine grain neutral-coloured developed photographic silver. It is used instead of an engraved cross-line screen, and is held in vacuum contact upon a sheet of lith film in the camera, or for contact printing.

For direct colour separation work the neutral grey contact screens are used. For monochrome work, or for recording colour-separated continuous-tone negatives or positives in half-tone, the magenta screens are claimed to have advantages. With the grey contact screen, the contrast of the half-tone result can be controlled only by adjusting the quantity of "flash" exposure, which is a short overall general exposure affecting only the smallest dots. All other tones form larger dots which completely cover the "flash" dot. The magenta contact screen has three advantages. Firstly, the screen image is completely grainless, and therefore forms thoroughly sharp and clean dots. Secondly, the controlled flash technique to control contrast can be used as with grey screens. Thirdly, the magenta screen yields images of varying contrast according to the colour of the light used for the exposure.

Some manufacturers of grey contact screens claim that they may be used for making both negatives and positives in half-tone. Others consider that the dot formation of the screens should be different for each purpose.

The contrast range of the screen dots can be varied to obtain more, or less, contrast at the top, the bottom, or in the centre of the tone range more effectively than with ruled screens, and this is a most valuable technical advance. In addition, it is almost impossible to obtain a faulty dot structure in a properly exposed contact screen half-tone, as often happened with a ruled screen. Thus the work of the camera operator is considerably eased.

The Photographic Materials. Since the last war the wetplate process has been almost entirely discarded, dryplate emulsions being used almost exclusively, with a rapidly increasing trend towards the use of film, even for colour separation work, now that dimensionally stable film base materials (polyester, etc.) are available. Some monochrome work is done on "ordinary" (blue-sensitive, "colour-blind") emulsions, but "lith" material which is orthochromatic (or panchromatic) and gives extremely dense lines of exceptionally fine grain, or dots in a completely clear ground, is very popular.

Metal Printing. Half-tone negatives are printed on metal about $\frac{1}{16}$ in. thick. Copper is often used for screen rulings finer than about a hundred lines per inch, and zinc for coarser rulings. The metal is coated with a cold-water-soluble fish glue solution containing ammonium dichromate. This coating is dried and then exposed under the negative to arc light in a vacuum pressure printing frame. The glue that is exposed to the light that shines through the clear parts of the negative becomes insoluble in water, but where it is protected by the opaque dots that form the negative image it remains soluble. So, when the metal is washed under a stream of water, the soluble glue dissolves away and leaves a positive of the image in the form of an insoluble glue coating.

In this condition it is difficult to see the image, so it is dyed in a solution of methyl violet. Then, if the appearance of the image is satisfactory, the glue is "burned-in" to an acid resisting carbon enamel by heating the metal very strongly over a gas flame.

Half-tone Etching. Dilute nitric acid is used for etching zinc, and a solution of ferric chloride (iron perchloride) for copper. While the principle of etching is the same for both metals, the technique varies; for instance, copper can be suspended horizontally in an undisturbed solution of ferric chloride (a "still bath") and, as no gases are evolved by the reaction, the plate etches satisfactorily. This method would be useless for zinc because gas is given off during the zinc etching process.

Copper can be etched electrolytically (by the reverse of electro deposition) but the method has never been successful with zinc, although claims have been made for a zinc etching electrolytic machine. Etching of either metal may be carried out either by machines which throw the etching mordant against the plate, or in mechanically rocked baths which keep the solution constantly agitated.

The "underbiting" effect, which has to be prevented in line etching, is allowed for in half-tone etching and is even put to very good use. Before the image is etched, the lightest tones of the picture consist of large squares of resist almost joining at the corners, and the darkest tones consist of a continuous expanse of resist pitted by minute holes. The range of tone gradation is represented by the range of variations between those two extremes.

During etching, the mordant bites sideways as well as downwards, undermining the dots and, as the enamel is brittle and because the effect is helped by brushing, the sides of the dots break off and they become progressively smaller in the light tones. At the same time, the minute pits in the dark tones become enlarged; so that all the tones of the picture become lighter until a stage is reached when the dots representing the lightest tone become so small that they cannot be etched any further without being lost altogether.

Fine Etching. The tones of the final reproduction can be controlled by etching certain selected areas of the block while others are held back by a protective coating. This process is known as fine etching. After a suitable amount of rough etch, sufficient to provide printing depth, the darker tones are painted-in with an acid-resisting ink and the remainder of the image is allowed to lighten during the next etch; then further tones at the lower end of the scale are painted-in and the remainder lightened still further until perhaps all of the picture is covered with resist except for highlights. In a landscape, for instance, light clouds would be lightened in tone just enough to lift them away from the main sky, after which all the "stopping-out" paint is washed off and the plate is ready for mounting.

A certain amount of fine etching is generally necessary to improve the quality of the reproduction, especially when making a block from an over-contrasty bromide print. Photo-engravers can readily increase the contrast of a slightly flat print, but far more often they have to fine-etch for highlight detail or for shadow-detail because the prints are too contrasty and lacking in gradation.

Fine etching is also necessary because of the wide variation possible in printing conditions. The type of ink and paper to be used, whether the plate is to be duplicated, the type of machine and the printing speed, and the actual printing technique all radically affect the type of block that is required.

In colour work the tones of many colours in the different printings have to be altered by fine etching to allow for the faults in printing inks.

Powderless Etching. Since the early 1950s the powderless etching technique, originally used for etching line plates, is used widely throughout the photo-engraving industry. Invented by the Dow Chemical Corporation of America and used for etching magnesium, it has since been adapted also for zinc and copper.

The etching mordant solution (nitric acid for magnesium, or zinc and ferric chloride for copper), contains also an oily substance and a wetting agent, the acid solution and the oily substance being pumped mechanically through the machine and mixed into a semi-emulsion. The mixture of solutions is either splashed by means of paddles or sprayed against the plate to be etched. The oily substance tends to separate out and to be deposited against the sides of the lines or dots of the image, where it resists the action of the acid and automatically builds up a protection at the sides of the image. Thus line or half-tone plates (or indeed, a combination of line and half-tone) can be etched in the machine in one operation.

Obviously this method does not permit tone correction by fine etching half-tone plates, so the correction has to be introduced at the earlier negative stage.

Mounting. The finished plates are mounted to type height in much the same manner as for line work. An ordinary squared up plate is simply bevelled around all four sides and tacked upon the wooden mount with small nails through the bevels. To allow for a caption close to the base of the picture the plate is often trimmed flush at the bottom edge and it can, if necessary, be trimmed on all sides. F.H.S.

Rotary Printing. The traditional method of printing for high quality work is from flat plates or formes containing both types matter and illustrations. For newspaper and other bulk work, this method is too slow and printing cylinders rotating at a high speed are used. Since ordinary metal blocks cannot be bent to fit the curve of a cylinder, curved plates are prepared from the flat blocks via paper or plastic moulds. These are formed by pressing against the flat printing area, they are then bent to the required curvature and the metal (usually a lead-tin alloy) cast into this shape. The resulting curved plates can then be mounted directly on the printing cyclinder.

Electronically engraved blocks and also plastic plates can be made in such a way that they can curve directly round the printing cylinder.

Electronic Engraving. Half-tone blocks are nowadays also made by electronic engravers which bypass the whole process of preparing half-tone screen negatives, printing down on metal, etching, etc. The procedure is briefly as follows.

A fine beam of light in conjunction with a photo cell scans the photograph to be reproduced. That is, the cell and beam move across the picture in successive lines, rather like the lines of a television screen image. At each point the photo cell registers the amount of light reflected by the surface of the picture. A light tone reflects more light than a dark tone.

The current from the photo cell is fed—via appropriate amplifying circuits—to an engraving machine and controls a stylus. This travels across a metal plate in synchronism with the scanning movement of the light beam and photo cell on the picture. The stronger the current from the photo cell, the deeper the stylus digs into the metal while engraving. The stylus is cone-shaped, so that the deeper it digs into the plate, the less plate surface there is left between each scanning line and the neighbouring one.

The final half-tone image may thus be built up of regularly spaced lines of varying thickness. The darker the tone, the broader the lines left standing up on the metal plate; the lighter the tone, the narrower the lines, and hence less ink is transferred by the plate on to the paper.

In the more modern electronic engraving machines the stylus is driven into the engraving material intermittently at the rate of several hundred times per second. This forms image dots similar to the dots of a normal half-tone block prepared photographically, but with a more regular structure.

Electronic engraving permits no etching, but has the advantage of being very quick. A line block can be made in as little as 10–15 minutes. The process is also largely automatic. A considerable measure of tone control and correction is possible by adjustments of the circuit which controls the engraving stylus. It is even possible to engrave a positive block from a negative by reversing the impulses electronically. Also, the transmission from the scanning spot movement to the stylus movement is electric, and can be adjusted to yield blocks of any enlargement or reduction as compared with the original.

Electronically engraved blocks are frequently used in newspaper printing, where the speed is a real asset and fine correction is less important. The electronic engraver can yield both flat printing plates and curves plates for high speed rotary printing.

Plastic Plates. A comparatively recent development is the use of photo-sensitive plastics in photo-engraving. The printing plate is here a layer of plastic coated on a stiff or flexible (according to requirement) metal backing. This is exposed in contact with the line or half-tone screen negative to ultra-violet radiation. The exposure polymerizes the plastic and renders it insoluble; the unexposed areas are then washed away in a suitable solvent, leaving the printing surface in relief. Plastic polymer printing plates are quick to make (the entire operation takes about 15 to 20 minutes). As no etching is involved, the half-tone negatives must be somewhat different in character from those used for conventional blocks. L.A.M.

LITHOGRAPHY

Lithography is the most common planographic printing process. It was discovered and perfected in its entirety by one man, Alois Senefelder of Solnhofen, in Bavaria in 1796. The principle on which it depends is that of the mutual antipathy of grease and water. Senefelder found that if a greasy image is formed on an absorbent limestone which is subsequently damped with water, then the image will "take" a greasy ink and the damp areas will refuse to accept the ink, so that proofs can be taken from the stone.

Senefelder's process has been modified in many minor ways—e.g., by the use of thin grained metal sheets instead of the stone; the production of the image by photographic means, and printing by first laying the image on a stretched "blanket" of rubber and then "offsetting" it to paper. But the basic principles remain unaltered.

For over a hundred years lithographers printed from stones, and their presses had to be enormously strong and heavy to exert the necessary printing pressure. In consequence it was impossible to attain high speeds or to print on very rough paper.

At the beginning of this century—about 1909—a new method of printing came into use. In this method, the printing surface consists of a thin zinc or aluminium plate clamped tightly around a cylinder on the printing machine. As the plate cylinder rotates, the image is first damped and then inked. It then rolls into contact with a rubber-faced "blanket" and the ink image transfers itself on to the rubber surface. This then rolls into contact with the paper and prints the image upon it. The image is said to be "offset" on to the paper and the process is known as "offset" lithography.

Offset machines can be run at high speed and will print satisfactorily even on very rough paper. Colours can be printed in rapid succession by this process.

Photo-litho. To reproduce a continuous-tone image by lithography, it is again broken up into a half-tone image with the aid of a half-tone screen. For photo-litho work contact screens are more popular than the ruled kind, and all contrast and tone control has to take place at the time of the exposure since no etching or fine etching is possible.

The screen negatives are printed on to the grained and coated metal plate. The coating is a dichromated colloid which hardens and becomes insoluble on exposure, while the unexposed portions dissolve away on subsequent treatment. The hardened colloid then is water repellent but accepts greasy ink after the whole plate is dampened.

There are also a number of other processes for preparing photo-litho plates from half-tone or from line originals. An offset photo-litho process is also used for a variety of office copying systems, especially when large numbers of copies are required from one original.

Certain photo-litho plates form a positive image from a positive, and hence require a screen positive as the starting point.

Conversion of Letterpress Blocks. Sometimes an image from a letterpress block, or even typeset matter, has to be used for photo-litho processes. One way is to photograph a printed proof, and use that for preparing a litho plate. It is a roundabout way, and not really practical for half-tone illustrations since the sharpness of the half-tone dot inevitably gets lost on rephotographing.

Various methods have been proposed for photographing direct from metal blocks or metal type. One of the most recent processes uses a pressure-sensitive instead of a light-sensitive film. This is placed in contact with the block or type matter, and bombarded in a special machine with small steel balls. The treatment accurately reproduces the raised letterpress image—including every dot in a half-tone image—on this conversion film, which after suitable treatment yields a dark coloured line or half-tone image. This can then be used directly for printing on to sensitized zinc or aluminium plates to produce litho plates.

PHOTOGRAVURE

Photogravure is a photo-engraved intaglio process. Modern screen rotary photogravure dates from its invention by Carl Klič in 1895, although Fox Talbot's photoglyphic engraving process of 1852 was an intaglio method somewhat akin to photogravure.

By this method the image does not depend upon optical illusion for the effect of tones of the picture as do the letterpress and lithographic processes. In photogravure the image is etched into the surface of the printing cylinder; deeply in the areas representing the dark tones and less deeply in the lighter tones. The image is printed by flooding the cylinder with ink, scraping it off to surface level with a long steel doctor blade and transferring the ink from the etched hollows to the paper.

Printing plates or cylinders for this type of reproduction are given a screen pattern consisting of lines at right angles at surface level to provide a continuous support for the doctor blade. The lines are less than one third of the width of the spaces and of the order of 175 per linear inch, so they are too fine to be seen with the naked eye. The image is etched to its continuously varying depths (according to the tones of the picture) in the minute square pits between the lines.

Partly because of the high printing speed, and also because of the liquid nature of the ink, the screen lines are more or less obscured in the dark tones so that beautifully intense blacks can be obtained and there is a limpid clarity in the lighter tones which cannot be obtained by any other process because of the intrinsic difference between continuous and discontinuous tone.

The Negative. For line work and type a line negative is made of the same kind as for letterpress blocks or photolithographic plates, that is, as opaque as possible in the area representing white paper and transparent in the areas representing the black lines. A positive has to be made from the negative and, if it is to be made by contact printing, then the negative is prism reversed as for letterpress work.

For tone illustrations such as photographs or wash drawings the negative is an ordinary continuous tone copy negative.

The Positive. The image used for preparing the etching resist for the printing cylinder must be a positive transparency. The positives of the illustrations and of the text must lie within specified density ranges because they are planned in page formation and printed as one.

The Resist. The etching resist consists of a negative image of light-hardened gelatin in which the areas representing dark tones of the picture are very thin and the thicknesses increase in proportion to the tones of the subject, the thickest area representing the lightest tone.

Carbon tissue—paper coated with pig-

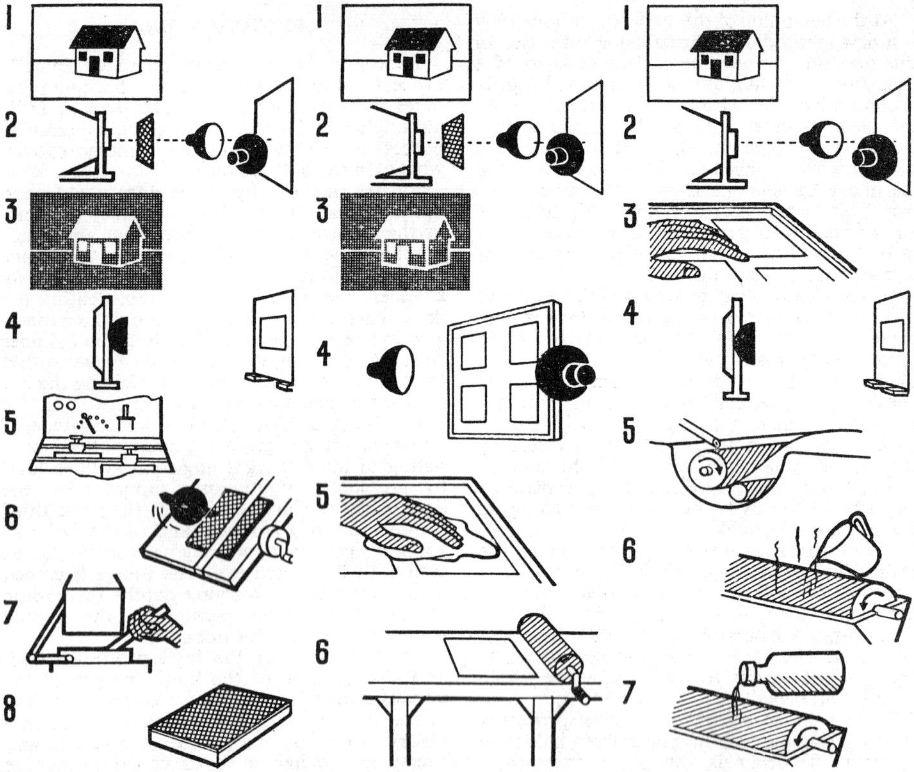

STAGES IN BLOCK MAKING. Left: *Half-tone letterpress block. 1. Original, usually a photograph. 2. Preparation of half-tone negative by copying through a process screen (screen actually inside camera, just in front of plate). 3. Half-tone screen negative. 4. Printing negative on to copper plate carrying bichromated fish glue. 5. Electrolytic etching. 6. Bevel cutting. 7. Hand Proofing. 8. Mounted block.* Centre: *Half-tone photo-litho block. 1. Original. 2. Copying through process screen. 3. Half-tone negative; similar to letterpress negative, but more contrasty. 4. Printing down on grained litho metal coated with bichromated albumen. 5. Developing out hardened albumen image. 6. Pulling proofs on offset press.* Right: *Photogravure block. 1. Original. 2. Making continuous tone negative and positive (no process screen used at this stage). 3. Laying out all positives for printing down. 4. Printing on carbon tissue which has been pre-exposed to a gravure screen to provide the line pattern. 5. Mounting exposed tissue on copper cylinder. 6. Developing resist image in hot water. 7. Etching the copper cylinder to form the intaglio image.*

mented gelatin and similar to that used for carbro printing—is used for the resist. It is rendered light-sensitive by bathing it in a solution of potassium dichromate and then drying in darkness. The sensitized tissue is first exposed to arc light in a vacuum pressure printing frame behind the photogravure screen, and then the screened tissue is exposed to the arc lamp behind the combined positive of the text and illustrations.

The tissue is transferred to a copper cylinder and "developed" in a stream of warm water which washes away the soluble gelatin and leaves only the light-hardened gelatin in the form of an image of varying thickness. The finished "print" is then dried.

Instead of carbon tissue, special photographic films can also be used. After exposure, development hardens the emulsion image-wise. This emulsion is then stripped off the film base and transferred to the copper cylinder, followed by warm water development, etc., as with tissue. Gravure films of this kind have the advantage of being more consistent in their properties, while the visible silver image also makes judging of image density easier.

Etching the Cylinder. The image is etched into the surface of the copper with a solution of ferric chloride (iron perchloride). The result is a surface in which the depth of the etched areas varies according to the tones of the original image. The tone range of the etched image is controlled by suitable variations in the strength of the etching bath. Etching is started with a strong solution.

The particular tone which is to be "brought in" and etched is controlled by the strength of the solution, and the depth of the etching is controlled by the time for which the etching is allowed to continue before the next weaker bath is employed; in that way the tone range of the result can be governed to a surprising degree, but there are definite limitations and any further adjustments have to be taken care of by retouching the original negatives and positives before the carbon printing is done.

The line work and text matter on the cylinder are sometimes etched separately from the illustrations and only two strengths of ferric chloride are used; a slightly weak one to make sure that etching starts on all the work, and a stronger one to continue the etching and to ensure that there is no penetration in the "whites" surrounding the work.

COLOUR

All the standard photomechanical reproduction processes may be adapted to the reproduction of coloured originals.

Principle. It is obvious that if one can print from a letterpress block, a photo-litho plate, or a gravure cylinder upon paper, in black ink, then there is nothing to prevent one from using ink of any colour instead; further, there is nothing to prevent one from printing again in another colour, from another printing surface, in register on the first proof, and so on as many times as required.

So, by taking a set of three-colour separation negatives of any coloured scene, photo-engraving a printing surface from each, and printing them in their appropriate colours in register upon white paper, the result will be a colour print. This is the broad principle of all colour photo-engraving.

Practical Difficulties. The main difficulty is the impossibility of making printing inks which each completely absorb one third of the spectrum of white light while perfectly reflecting the other two-thirds. By reason of the inefficiencies of the printing colours alone, blues, purples, greens and mixed colours such as browns and olive greens are bound to be degraded to some slight extent.

This is one of the two main reasons why colour masking and/or hand correction are necessary in graphic reproduction. The second reason is that the reproduction curve of any of the graphic arts reproduction processes is never a straight line at the top and bottom ends of the scale, even for monochrome reproduction, so it is also necessary to contrive the "tone correction" in one way or another.

Colour and tone correction are usually introduced together, either by "automatic" photographic masking techniques, or by hand correction, often today principally by masking, and the remainder by hand work. For letterpress block making, the hand correction is usually introduced by the colour etcher when the plates are etched. For photo-lithography the initial continuous tone colour separation negatives can be corrected, usually by dye retouching, principally for the light tones (the dense areas of the negatives), and then by "dot-etching" the half-tone positives by ferricyanide-hypo reducing the size of the dots which form the image. For photogravure both the negatives and the positives are in continuous tone, so the negatives are dye retouched, mainly for the light tones, in much the same way as for litho and then the positives, made from them, are also dye retouched for the dark tones (dense areas) of the picture.

Colour Masking. The "automatic" photographic method of colour correction, and often tone correction also, is based upon the general principle of interposing a weak positive, made via one of the colour filters, into the light path when printing an image from a colour separation negative made with another colour filter. Obviously the interposed positive is bound to modify the separation negative selectively, because of the different response of the different colour filters, and if the filters are wisely chosen, then the combined effect can be advantageous. Conversely, it is possible to modify the colour separation from a positive by interposing a weak negative mask made with a different colour filter.

For example, the magenta-red printer (green filter separation) negative from a positive colour transparency is never dense enough in the blue and green colour areas. So, without hand or photographic colour correction, it always prints too much red on the blue colours of the reproduction, degrading them to a purplish colour. It also prints some red on green colours, where it should print none, thus degrading them to brownish tones.

But a red filter negative is, of course, dense in red, orange and yellow colours, and transparent in the areas representing blues and greens (it is a cyan blue-green printer). That negative, at about half strength, may be used as a mask and bound up in register with the colour transparency, the combination being photographed with a tri-colour green filter. Then, because of the mask, more exposure must be given to sufficiently affect the areas representing yellow and orange. As the mask is transparent in the blue and green areas, they are recorded more dense than they would otherwise be in the resulting magenta red printer colour separation negative which is thus "automatically" colour corrected.

This simple form of colour masking may also introduce considerable improvement in the rendering of the darker tones of the transparency. An earlier highlight mask may have helped at the other end of the tone scale. Many colour positive transparencies have a too great contrast range for direct graphic reproduction, but an incidental effect of the negative colour mask is to reduce the overall contrast of the transparency and mask combination to bring the range within the scope of the reproduction process. If the mask is exposed and developed so that only the light end of the tone range is recorded, then the contrast reduction is limited to that end. When the combination is colour separated, the total tone range is shared between the whole of it, thus the lower end of the range is comparatively extended.

Colour Proofs. For all methods of graphic arts colour reproduction it is necessary to take colour proofs so that the progress of the work may be gauged. Many highly skilled craftsmen gain an astonishing intuitive grasp of colour combination quality and they often need to make remarkably slight adjustments to achieve the required results. But that can only be demonstrated by the progressive proofs before the expensive stage of the machine printing.

Progressive proofs are a set of prints from each of the three or four separation printing plates (or cylinders), each in their respective colours and in their combinations. The yellow plate is commonly also printed separately in a darker colour so that its quality can more readily be seen. The proofs not only show that the reproduction work is efficient, but they also provide a valuable guide to the machine printer as to the weight of colour that he needs to obtain in each printing. For this reason the proofs should be on the same paper and in the same inks as will be used for machining, and preferably printed in as nearly as possible the same manner.

The photogravure process is the only one of the three which prints effectively in continuous tone. The image of a tone subject has to be recorded in half-tone dot formation for printing by letterpress or litho, which introduces another problem in colour reproduction. If one lays comparatively even layers of coloured ink at varying intensities, as in a modern integral tripack colour print or a photogravure proof, then the colour combinations are different from the effect obtained by printing various sized dots of the full weight of the colour by the half-tone method. Further, the half-tone colour effect changes according to the position in the tone range because the dots representing tones darker than a mid-tint are likely to fall upon one another, while the dots representing lighter tones are more likely to fall beside one another. This distinctly alters the hue and has also to be allowed for by colour etchers and retouchers.

Screen Angle. The three or four printings of a colour reproduction must not have the dot formations all at the same angle because it is not possible to print with the dots in precise coincidence. If the fit is not exact, the dots fall in and out of "phase" and produce a large blotchy effect (Moiré pattern) which ruins the result. So the screen angle of each printing is set as widely different as possible.

The yellow printing reflects so much light that it is the least likely to "pattern", so it is usually set at the horizontal and vertical position. The magenta-red printing is at 15° the cyan blue-green at 75° and the black 60° farther, at 135° (which is effectively the same as 45°). Thus the three strongest colours are as widely separated in screen angle as possible and the yellow printing lies between the blue and black. For three colour work the blue, being the darkest colour is at 45°, the yellow taking the 75° angle, and the red remaining at 15°.

When more than four colours are used, as for example, in some photo-lithography, then the pink is often run on the red angle and the light blue on the blue angle. Some printers prefer to reverse that order, but one always runs a light colour on a dark colour angle and never two dark colours on the same angle.

The Black Printer. Theoretically it should be possible to reproduce a colour subject in only three printings, but four are generally used for two main reasons. The first is that the accompanying text is usually printed in black, so it is not difficult to include a black printing on the illustrations if there is any advantage.

Secondly it is very difficult indeed to print each of a three colour working with such exact colour balance that the result is perfect. But if the neutral tones and dark colours are produced with the help of a black printing, then some slight lack of balance in the colour printings will be masked and a thoroughly acceptable result is much more readily obtained.

Finally, in the case of "wet-on-wet" printing, where the colours succeed one another on the machine before they have a chance to dry, it is important that no two colours shall overprint at full weight. So the heavy tones must be built up with the help of the black printing.

Electronic Correction and Engraving. Electronic aids are frequently used nowadays to ensure correct balance and effective masking of separation negatives intended for photo-mechanical printing. This correction is introduced at a stage of making positives from a separation negative, and these are then used for producing the screen negatives which lead to the blocks.

Briefly, electronic correction consists of analyzing the three separations simultaneously by scanning them with a flying spot light source and measuring the light intensities coming through the three separation negatives with photo cells. The output of the latter for all three negatives is fed into a unit which controls a further flying spot source which prints each of the separation negatives in turn to a positive. The electronic correction automatically over- or under-exposes areas of the separation negatives corresponding to specific colours, and so introduces the mask effect during a single printing stage. Equipment for this purpose is necessarily complex and expensive.

Three- and four-colour printing blocks can also be produced by electronic engravers which scan a colour transparency (or a colour print) in much the same way as an electronic engraver for a black-and-white picture. Only in this case the colour picture is scanned three (or four) times in succession through the appropriate separation filters. Colour and tone corrections can again be introduced by suitable adjustments of the electrical circuits involved.

OTHER METHODS

During the century since photo-mechanical reproduction came into use, many different processes have been invented; some, like the Woodburytype, were superseded because of economic pressure, others such as collotype have remained in use for special purposes. The few which had thoroughly distinctive features are outlined below.

Typographic Processes. Here the field for originality is limited by the fact that one can only have a raised printing surface and a recessed non-printing surface.

The regular "half-tone" screen dot formation has been varied in many ways for different processes.

(1) By changing the shape of the dots, as with the Schultze screen which makes them rhomboid instead of rectangular.

(2) By changing the size, as with a special Max Levy screen which produces extremely fine dots that coalesce into larger ones in the darker tones.

(3) By changing the pattern as in the Metzograph screen and many others in which the pattern was either broken up or based on straight or wavy lines.

Planographic Processes. There have been more variations on the planographic processes than on any of the other methods. Apart from various offset printing techniques such as litho-deep, dot-etching, bi-metallic plates and photo-composing (which also applies to photogravure), all of which have been assimilated into the parent process, there have also been two complete processes which were radically different: Collotype and Pantone.

Intaglio Processes. The principal variations of this basic reproduction principle are the Bitumen Grain Process, Invert Half Tone and Woodburytype. F.H.S.

See also: *Bitumen grain process; Collotype; Colour separation negatives; Half-tone screen; Lith film, Masking; Pantone; Photo-litho master; Phototelegraphy; Phototypesetting; Process camera; Silk screen printing; Woodburytype.*

Books: *Ilford Manual of Process Work*, by L. P. Clerc (London); *Photographs and the Printer*, by F. H. Smith (London); *Photogravure*, by H. Mills Cartwright (London); *Practical Photo-Lithography*, by C. M. Willey (London).

PHOTOMETER. Instrument for measuring illumination, often used to determine the brightness of a light source. The type of instrument usually called a photometer is strictly a comparison photometer in which the brilliance of a given light is assessed by comparing it with a standard light.

The Bunsen or grease-spot photometer is the simplest type in general use. This consists of a sheet of opaque white material with a central translucent spot or ring formed on it with oil or melted grease. When the spot is illuminated from the front, it looks darker than the surrounding material; when it is illuminated from the back it looks brighter; when the illumination on the front and back is exactly equal, it disappears.

Measurements are made by fixing the standard source at a standard distance from the spot, and adjusting the distance of the other source until the spot disappears. The relative brightness is calculated according to the inverse square law.

Camera Exposures. Until recently photometers were not regularly used in the photographic field although they have been widely used by illuminating engineers. They are also used in photography for measuring subject brightness, i.e., as exposure meters. In one such instrument a telescope is used to sight the scene and a very small mirror occupying about one-quarter degree in the middle of the field is illuminated by a lamp. This lamp is fed from a dry battery which is incorporated in the base of the instrument, and a series resistance is inserted in the circuit of the lamp so that its brightness can be varied. A photo-electric meter, consisting of the usual photocell and micro-ammeter is used to adjust the brightness of the lamp to a standard value, after which it can be used for measuring the brightness of the scene. A pair of neutral density wedges is arranged between the lamp and the illuminated mirror. Adjustment of these wedges to obtain a match actuates a calculator from which the exposure can be read.

Alternatively, the resistance setting required may be taken as a measure of the subject brightness and the resistance control calibrated as an exposure meter.

Practical Advantages. An exposure photometer does not give the answer as quickly as an exposure meter because it often has to be standardized before it is used to measure the actual exposure. Its main advantage is its high sensitivity and small acceptance angle. This is about one-quarter degree and it allows the brightness of a very small area in the scene to be measured. There is another advantage of a small acceptance angle: the brightness of distant objects can be measured from the camera position. In this way the total brightness scale of the scene can be measured, which is often impossible with any other type of meter.

Enlarging. Grease spot photometers are made for use in enlarging to measure the actual brightness of any part of the negative area projected on the easel and hence the exposure and correct grade of paper.

In a typical enlarging photometer there is a translucent disc; the centre is lit by a lamp adjustable for brightness, and the surrounding outer ring by the light from the selected part of the negative.

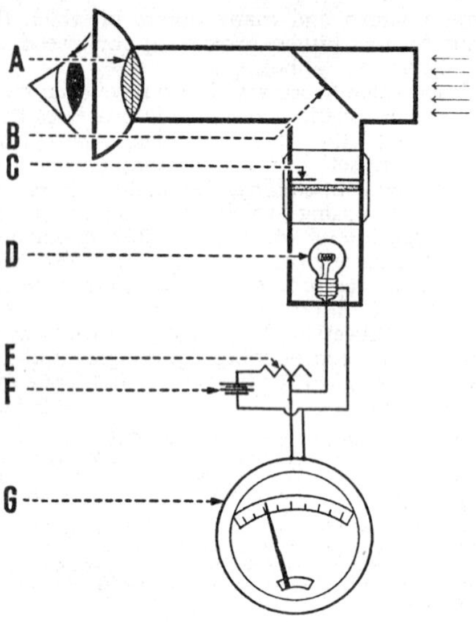

PHOTOMETER LAYOUT. The eye views a grease spot illuminated by the subject from one side, and a small electric bulb (at controlled brightness) at the other. When both sides appear evenly matched (i.e., the grease spot disappears), the position of a diaphragm ring gives exposure readings. A, magnifier. B, grease spot. C, diaphragm controlling amount of light reaching grease spot from bulb. D, light bulb. E, potentiometer. F, battery. G, voltmeter. In some photometers the voltmeter circuit is replaced by a photo-electric cell system.

The brilliance of the two areas is matched by varying the brightness of the lamp in the photometer until there is no difference between the centre and the outer ring. This is done either by varying a resistance in series with the lamp or altering the distance of the lamp from the comparison spot. The latter method is probably more accurate because dimming the lamp by resistance changes the colour of the light and makes comparison difficult.

Once the two parts of the comparison field have been matched, a brightness factor for the selected part of the image is read off either from the resistance calibration or the scale along which the lamp is moved. The factor does not give the brightness in absolute units; it is simply a figure which can be translated into an exposure time for the particular grade and speed of paper in use by referring it to the calculator supplied with the instrument.

Comparison photometers of the type used for estimating camera exposures can be used to arrive at the exposure in enlarging, but they are not as a rule as convenient as those specially designed for the purpose. W.F.B.

See also: *Actinometer; Exposure meter; Exposure tables; Extinction meter; Photo-electric meter.*

PHOTOMETRIC EQUIVALENT Ratio of the mass of silver per unit area to the diffuse density. Thus

$$P = \frac{M}{D}$$

where P = photometric equivalent, M = mass Ag in grams per dm^2 and D = diffuse density. The photometric equivalent varies with the type of emulsion, with the exposure level, with the degree of development and with the conditions of drying.

The latter criterion is the reason why drying marks, or marks of different density from surrounding areas, are sometimes formed in areas which have received uniform exposure and have been developed to uniform density in the wet stages of processing.

See also: *Drying.*

PHOTOMETRY. Measurement of light in terms of the intensity of a standard light source. Measurements are made by comparing the intensity of the given light source with the light given by a standard candle. If the illuminating power of a light source 2 feet away from a surface is the same as that given by a standard candle one foot away, then, by the law of inverse squares the light source has an intensity of $2^2 = 4$ candle power.

Photometry plays an important part in sensitometry—i.e., the study of the effect of varying exposure on the photographic emulsion.

See also: *Illumination; Light units.*

PHOTOMICROGRAPHY

Process by which photographs of minute objects are made with a camera and microscope. It should not be confused with microphotography, the process used for making minute photographs of large objects.

In 1835 Fox Talbot, applying his experimental process of Photogenic Drawing to the solar microscope, obtained photomicrographs of such objects as an insect's wing, magnified 17 times in linear dimensions. The date of 1837 which the Rev. J. B. Reade claimed for his earliest photomicrographs is refuted by his recently discovered letter to his brother, and early 1839 must be taken as the correct date. Friedrich Gerber of Berne obtained enlarged photographic images of microscopic objects on paper sensitized by his independently devised method, before February 1839. Alfred Donné made photomicrographs on daguerreotype plates, of sections of bones and teeth in February 1840, and subsequently he and others took photomicrographs of medical specimens. In July 1840 John Benjamin Dancer photographed a flea magnified by means of a gas microscope large enough to fill a 7 × 5 in. daguerreotype plate. J. Delves, a member of the newly founded

Microscopical Society, successfully took photomicrographs by the wet collodion process in 1852. Thereafter, photomicrography was extensively practised.

BASIC SET-UP

There are always three main units in any set-up for photomicrography: a lamp system for illuminating the specimen, a compound microscope, and a light-tight extension to hold the sensitive material. These three components, fixed rigidly together, make up a photomicrographic optical bench.

This optical bench may be arranged vertically or horizontally; the problem is much the same as in choosing between a horizontal and vertical type of enlarger. The horizontal bench can be made absolutely rigid, and it presents fewer constructional problems. The vertical set-up, however, takes up less space and if it is rigidly constructed it is capable of just as fine work as the horizontal type. In either case absolute rigidity is essential because the slightest tremor at the moment of exposure will ruin the result. It is difficult to make a vertical column rigid unless the top end is anchored to the ceiling by a bracket.

Vertical instruments are useful for photographing liquid preparations. They possess the further advantage that if something of interest is found during visual examination of a specimen, a photograph can be taken without disturbing the set-up. With a horizontal bench, the mere act of turning the microscope to the horizontal position is almost certain to upset the position of the object.

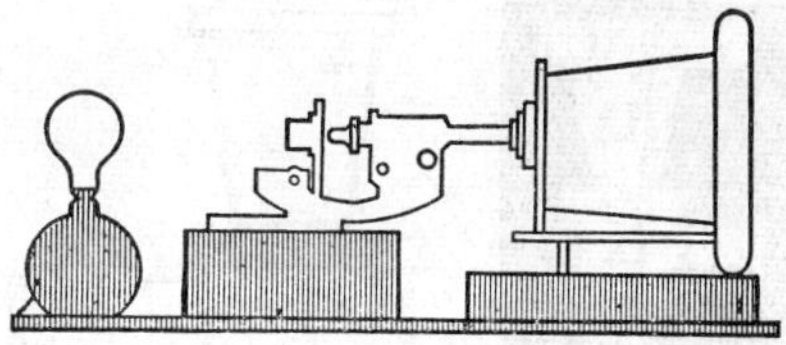

SIMPLE MICRO SET-UP. This uses a normal camera, focused at infinity. The microscope is focused visually, and the exposure found by trial and error. A table lamp serves as the light source for the substage condenser. Such an arrangement is suitable only for low-power work.

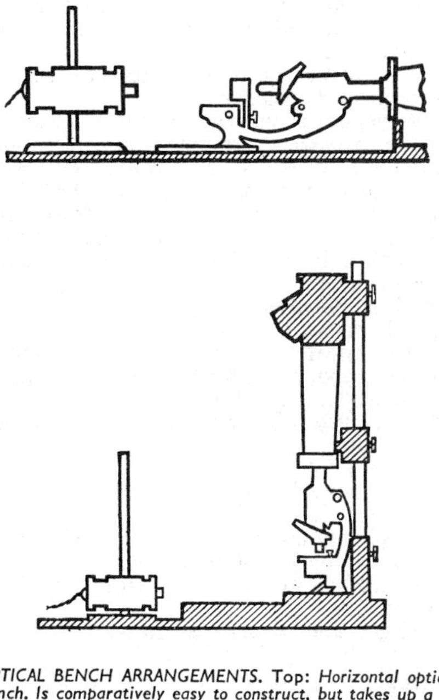

OPTICAL BENCH ARRANGEMENTS. Top: Horizontal optical bench. Is comparatively easy to construct, but takes up a lot of bench room. The microscope lamp directly illuminates the condenser. Bottom: Vertical optical bench; compact, but difficult to construct for the amateur in a sufficiently rigid form. A mirror directs the beam from the lamp into the condenser The camera is usually part of the unit.

The Camera. The camera is no more than a light-tight box fitted with a focusing screen and a plate or film holder, capable of light-tight connexion with the eyepiece of the microscope.

In the simplest form of set-up, any camera may be used if its lens is set to infinity and it is supported close against the eyepiece of the microscope after the specimen has first been focused visually in the ordinary way.

Special eyepiece cameras are made to fit over the draw tube and require no external support. Most makers of miniature cameras supply adaptors so that the camera, minus its lens, can be fixed on to the end of the draw tube of a microscope. A double or triple extension plate camera with the lens removed can also be used in the same way.

The Microscope. Practically any good compound microscope can be used. The main things to look for are rigidity of all parts, freedom from backlash when focusing, and, for horizontal work, absolute steadiness when the microscope is tilted into the horizontal position. Since the excellence of the result depends on the optical equipment, all lenses should be of first quality. Apochromatic objectives with compensating eyepieces are best, but achromatic objectives and Huyghenian eyepieces will give excellent results with monochromatic light, produced by interposing a suitable filter (generally green) between light-source and microscope.

In photomicrography resolving power is more important than mere magnification. The resolving power of a microscope is its ability to separate the images of two points lying very close together. This resolving power depends upon the aperture of the objective given by its numerical aperture (N.A.) which should be as large as possible

The aperture of the substage condenser should be sufficient to provide adequate light for the objective of largest aperture likely to be used.

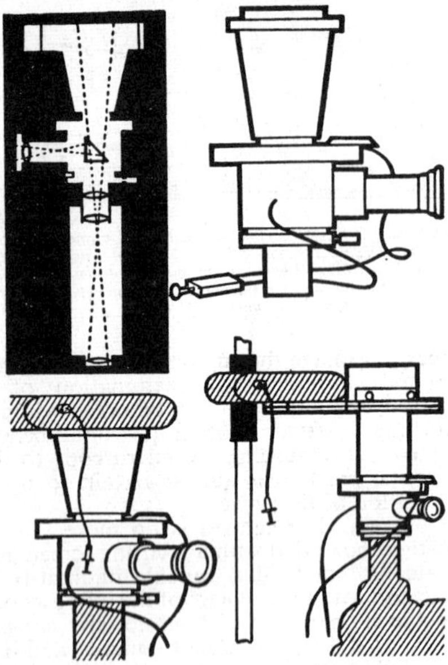

MICRO-ATTACHMENTS. Top left: *Complete optical system of microscope, micro-attachment, and miniature camera, used without the camera lens. A movable prism permits observation of the image.* Top right: *The micro-attachment itself.* Bottom left: *Attachment (with its own shutter) mounted on microscope.* Bottom right: *Attachment used with focusing stage and ground glass screen of miniature camera outfit.*

Illumination. The light on the specimen must be under complete control during focusing and exposure, so artificial light is nearly always used. Most specimens are prepared for microscopic examination in the form of thin sections through which light can pass easily. They can therefore be photographed by transmitted light. The illumination of opaque specimens by reflected light and of very transparent subjects are special techniques (see below).

Special microscope lamps can be purchased, but for many kinds of work other standard types of lamp can be adapted. For very simple subjects, an opal filament lamp from 60 to 100 watts will suffice, but for serious photomicrography a point source of light is required. This can be provided by a motor-car headlamp bulb (6 or 12 volts), or a bicycle dynamo bulb (6 volts), fed from a mains transformer of suitable rating. Better, though more expensive, light sources are the Pointolite lamp, the ribbon filament lamp, and the carbon arc.

The lamp is housed in a light-tight, well-ventilated house like that fitted to an enlarger. For the simplest work the lamp may shine directly on to the specimen, but for anything more critical, some form of condenser is always necessary to concentrate the light into a controlled beam. Most microscopes are equipped with a substage condenser, and a further condenser between the lamp and the microscope is advisable. Each condenser has some means of varying the aperture—generally by an iris diaphragm—and some provision for focusing. A typical lamp condenser consists of a biconvex lens of 3 cm. (1¼ in.) diameter with a power of about 15 diopters (6·5 cm. or 2½ ins. focal length).

The photomicrographic set-up always includes a holder for a filter between lamp and microscope.

Köhler's Illumination. This is the most usual method. It calls for a lamp condenser and a substage condenser, each equipped with an iris diaphragm. First, the lamp condenser is made to project a slightly enlarged image of the light source on the closed diaphragm of the microscope. The substage iris is then opened and a specimen is placed on the stage, and roughly focused. The lamp iris is stopped down, to give an aperture of about 3 mm. diameter, and the substage condenser is used to focus an image of this illuminated aperture on to the specimen. The eyepiece of the microscope is then removed so that the back lens of the objective can be examined.

The substage iris is then stopped down so that its image just appears within the rim of the back lens.

The substage iris is then correctly set for that particular objective.

The eyepiece is replaced, the camera is set up in the same axis as the rest of the apparatus, and the image is focused on the ground glass. This image should be surrounded by a sharply defined black circle—the image of the lamp iris. Adjusting this iris does not affect the brilliance of the image; it varies the size of the illuminated circle and it can be adjusted to give any desired field of illumination.

Bluish marks on the focusing screen show that the apparatus is not correctly centred. Uneven or poor illumination is corrected by moving the lamphouse backwards or forwards, and then correcting the positions of the rest of the apparatus.

TECHNIQUE

The photographic side of micro-technique is comparatively straightforward. Certain points do, however, differ from standard practice.

Focusing. The image is focused on the

NUMERICAL APERTURES

Focal length ...	⅔ inch 16 mm.	½ inch 12 mm.	¼ inch 6 mm.	⅙ inch 4 mm.	1/12 inch oil-imm. 2 mm.
N.A.	0·25	0·35	0·65	0·75	1·25

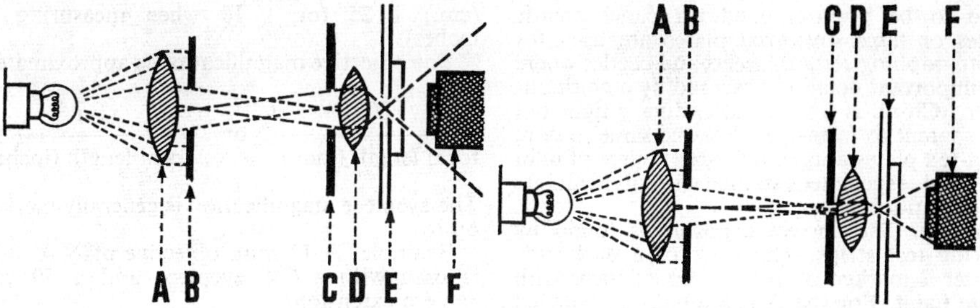

KÖHLER'S ILLUMINATION. Left: *The first condenser focuses an image of the lamp on the second diaphragm. The second condenser focuses an image of the first diaphragm on the specimen. Closing the first diaphragm decreases the illuminated area of the specimen.* Right: *Closing the second diaphragm decreases the cone of rays entering the objective, but not the illuminated area.* A, *first condenser.* B, *first diaphragm.* C, *second diaphragm.* D, *second condenser.* E, *specimen.* F, *objective.*

ground glass. Usually this is too coarse for accurate focusing, but an excellent focusing screen may be made by pencilling a cross in the centre of the ground side and then cementing a cover slip over it with Canada balsam. The result is a clear circle intersected by cross lines. The image is examined by means of a magnifying glass, and should be in focus together with the cross lines. This is best determined by moving the head from side to side while looking through the magnifying glass. If the image is not in focus, it will appear to move in relation to the cross. But when it lies in the same focal plane as the cross, both will move together. When that happens, the image is in sharp focus.

Depth of field in photomicrography is very small. It can only be increased by using an objective fitted with a small iris diaphragm, but this in its turn means poorer resolution.

Filters. Whenever filters are used, focusing is carried out with the filter in position.

Achromatic lenses are generally used with a green filter to hold back the out-of-focus rays. Filters are also used to differentiate between stained areas in prepared botanical and zoological specimens. The choice of the correct filter for any particular specimen relies on the fact that a filter causes its own colour to print light, and its complementary colour to print dark.

Unfortunately, biological stains do not transmit pure spectral colours so that some care is necessary in choosing filters if the final print is to show clear colour contrasts. Gelatin filters are supplied in sets for microscope work; while they give satisfactory service in this form if they are handled carefully, they are better cemented between glass with Canada balsam. This type of filter must be used between lamp and microscope, and not between microscope and camera.

Blue or violet filters are used over the light for photographing fine detail such as the markings on diatoms, and on the chitinous coats of insects.

For work of extremely high resolution, light of a shorter wave-length is used. For diatom photography a blue filter is generally good enough, but for really critical work ultra-violet light must be employed. This necessitates a light source rich in ultra-violet, such as a mercury arc lamp, and an optical system transparent to the rays. So work of this sort calls for quartz lenses and a special photomicrographical bench.

FILTER EFFECTS

Colour of Filter	*Gives Contrast with*	*Gives Detail with*	*Approx. Factor (Tungsten Light)*
Deep magenta or purple	Chitin, diatoms orange, green	Blue	×10–×15
Deep blue (tricolour blue)	Chitin, diatoms, red, orange, yellow	Blue	×15–×35
Blue-green	Diatoms, yellow (compensates achromatic objectives)	Green	×10
Tricolour green	Red, orange (compensates achromatic objectives)	Green	×6–×9
Yellow green	Compensating filter to absorb U.V. and excess red		×1½
Pale yellow	Compensating filter to absorb U.V.		×1½
Deep yellow	Blue, violet	Yellow	×1½–×2
Orange	Blue, violet	Red, orange	×2–×4
Tricolour red	Green, blue violet	Red, orange	×2–×5
Deep red	Green, blue	Red, infra-red	×8–×25
Magenta	Green	Blue, red	×2–×3

Negative Material. Photomicrography calls for negative materials with fine grain and high resolution. Suitable emulsions are now coated on plates, cut films and 35 mm. film.

Process plates or positive film are generally used for normal sized negatives. Photomicrographs on 35 mm. film are generally taken on a document copying type of film. All these emulsions give high resolution and extremely fine grain. Faster emulsions are used only when the subject or the set-up is likely to move.

Most coloured subjects and stained sections

have to be photographed on panchromatic plates or films. Infra-red plates are used for photographing certain insects and beetles where the important detail is obscured by a chitinous layer. (Chitin is opaque to ordinary light but transparent to infra-red.) For the same reason, infra-red plates and an infra-red source of light will resolve specimens stained with silver nitrate which is normally opaque.

Exposure. The correct exposure is found by making test strips. The back of a dark slide shutter is marked off into six equal strips with white paint. For the test, the light is switched off, the shutter is completely withdrawn, and then the light is switched on for 1 second. After 1 second the shutter is pushed in one division, and again after 2 seconds, 4 seconds, 8 seconds, 16 seconds, and so on. In this way each section of the plate receives double the exposure of the one before.

In some cases the exposure may have to be measured in minutes, not seconds. The plate is then developed and the best exposure selected by inspection.

Development. Process materials need no special fine grain development technique; sufficiently fine grain results from development in an M-Q developer or a hydroquinone-caustic where strong contrast is important. Fine grain treatment is only necessary with the faster emulsions.

Naturally, stained sections and similar coloured objects necessitate the use of orthochromatic or panchromatic material, but this is almost always of slow or medium speed. Length of exposure is not usually of great importance, but fine grain is. In addition, the material is processed to give fine grain.

Colour Photography. Photomicrographs in colour are made with Photoflood lamps and Köhler's method of illumination. A suitable filter is used to correct the illumination for the particular colour material. The specimen must be mounted in a colourless medium such as methyl methacrylate or polystyrene, and not in Canada balsam which always has a tint. Achromatic condensers are essential for such work and a compensating filter may be necessary to remove any residual background colour.

Magnification. For the greatest resolving power of the objective, the magnification should be 300 times the numerical aperture. The maximum permissible magnification is about 1,000 times the N.A., so, in practice, a figure of about 600 times the N.A. is aimed at.

The magnification is measured directly by focusing the image of a stage micrometer on the screen. A stage micrometer is a microscope slide marked off in hundredths of a millimetre. The magnification is then given directly by measuring the length of the image of a one millimetre division with a millimetre rule laid on the focusing screen. An approximate figure can be arrived at by using the formula:

Total magnification = objective magnification × eyepiece magnification × camera extension (cm.) ÷ 25 (or ÷ 10 when measuring in inches).

The objective magnification is approximately

$$\frac{160}{\text{focal length (mm.)}} \quad \text{or} \quad \frac{7}{\text{focal length (inches)}}$$

The eyepiece magnification is generally marked on it.

Example: A 12 mm. objective of N.A. 0·35 is used with a 6× eyepiece and a 50 cm. camera extension.

$$\text{Objective magnification} = \frac{160}{12} = 13{\cdot}33\times$$

$$\text{Total magnification} = 13{\cdot}33 \times 6 \times 50 \div 25 = 160\times$$

So to obtain the desired magnification of 600 × N.A. = 600 × 0·35 = 210 ×, the final print will have to be enlarged from the negative a further 210/160 = about $1\frac{1}{3}$ times.

The upper limit of magnification for a 35 mm. camera is usually reached by enlarging to about whole-plate size ($6\frac{1}{2} \times 8\frac{1}{2}$ ins.). Magnification produced by the eyepiece does not increase resolution, it merely increases the size of the image.

Preparing the Specimen. Specimens like pollen, insect eggs, or the wing of a fly are examined dry. Some thick subjects have to be mounted in a cell on the slide before being covered by a cover-glass. But most specimens are sandwiched between slips of glass with a drop of water. The cover-glass is slipped carefully over the specimen so as to exclude air bubbles.

Dilute glycerin is sometimes used instead of water where there is a risk of evaporation during exposure.

Most botanical and zoological specimens must be teased apart with needles, or cut into very thin sections before mounting between glass. Sections may be cut with a flat-ground razor, the specimen being held in the hand and kept wet by constantly dipping it in water. For accurate preparation of slides on any scale, however, a microtome (a form of adjustable mechanical slicer) is more convenient.

Considerable skill is needed to prepare permanent slides, and only the general procedure can be outlined here. First the specimen is killed and its appearance fixed with dilute alcohol or formalin. Sections are cut at this stage. The specimen is left in the fixing fluid until required. As the object is saturated with alcohol it will accept alcoholic stains, such as safranin.

Superfluous stain is washed out with more alcohol, and the alcohol is removed by washing in water.

The section will now accept an aqueous stain, such as Delafield's haematoxylin. Any excess of this is removed by washing in water. The specimen is then immersed in alcohol of various strengths, e.g., 35, 50, 70, 85, 95, and 100 per cent, starting with the weak solution and gradually bringing the specimen up to the strong solution. This removes the water.

The cell contents are opaque at this stage and must be cleared by immersion in clove oil, followed by xylol. The thinnest and best sections are then mounted on a slide and covered with a drop of Canada balsam in xylol. The cover slip is then placed in position, and the slide placed in a dry dust-free place until the balsam has set hard and the slide is safe to handle.

Opaque Specimens. Opaque specimens such as are found in metallurgy may be illuminated by directing a beam of light from the lamp so that it just grazes the surface. In this way every tiny surface irregularity is thrown into sharp relief. It is more usual, however, to employ a vertical illuminator, a special type of objective which receives a ray of light through the side, reflects it vertically down on to the object, and then transmits the rays as they are reflected upwards, towards the eyepiece.

With this type of illumination the loss of light is so great that a high-power light source such as a carbon or mercury arc is usually necessary.

Very Transparent Subjects. Dark ground illumination, used for very transparent objects, can give useful and beautiful results. It is produced by a special substage illuminator or the usual substage condenser fitted with a central stop.

In both systems only the marginal ring of rays is focused on to the specimen. When the rays cross and diverge they do so at such an angle that none of them enters the objective. But where the rays are refracted by the specimen, some of them are directed upwards and enter the objective. The result is a brilliant white image on a black ground.

Exposures are long, even with a powerful light source such as an arc lamp or a Photoflood.

Exposure is controlled by switching the light on and off, or by interposing a card between lamp and microscope to avoid movement or shake. For critical work it is found that practically every shutter introduces vibration.

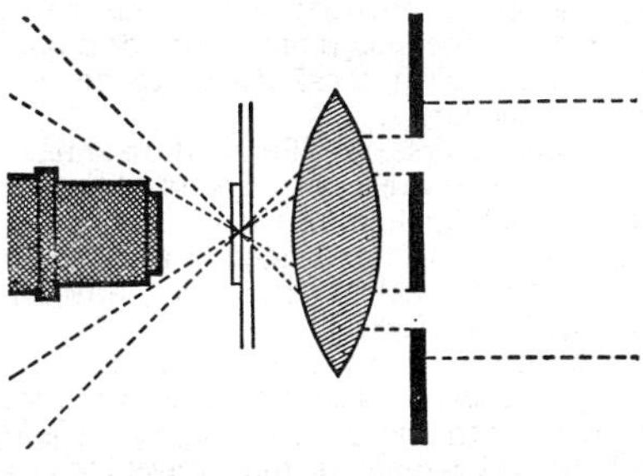

DARK GROUND ILLUMINATION. A central stop between lamp and condenser cuts out the central cone of rays. The marginal rays illuminate the specimen, but do not enter the objective.

SPECIAL EQUIPMENT

Much of the manufactured photomicrographic equipment is intended for such specialized users as industrial research laboratories, hospitals, and universities. But there are three more or less standard types.

Horizontal Equipment. This type of equipment consists of a massive base carrying a lamphouse with condenser and iris, a microscope, and a camera capable of considerable extension. The lamp is usually a high-powered point source or a carbon or mercury arc. All optical units have centring screws by which the whole apparatus is lined up before use. The bed may be mounted on shock absorbers and generally carries remote control rods so that the objective can be focused from behind the focusing screen.

Vertical Equipment. The conventional vertical set-up consists of a camera and microscope running on vertical rails attached to a massive column, with a lamp system on horizontal rails in front of the microscope. The camera has a reflex screen for focusing and the microscope may have a reflex eyepiece so that the image can be observed during focusing. The microscope mirror must always be used.

The Eyepiece Camera. The eyepiece camera fits over the draw tube of an ordinary microscope. It often carries its own shutter and eyepiece so that the image may be observed after the dark slide has been fitted. When the reflex eyepiece is focused accurately on the cross hairs, the instrument is adjusted for the user's sight. The makers incorporate all necessary corrections for the alteration in the tube-length of the microscope. (The optical system of a microscope is corrected for a certain tube-length, usually 160 mm., and if triple nose-pieces or other attachments are added later, the draw tube must be used to correct the tube length.)

The Electron Microscope. The electron microscope is based on an extension of the principle that the shorter the wavelength the higher the resolution. The "light" is an electron stream, and the "lenses" consist of magnets and electrical condensers which focus the beam. The increased resolving power makes it possible to work magnifications up to 100,000 ×. A camera is normally built in to the instrument and used for making permanent records of the image—formed in this case on a fluorescent screen.

Phase-Contrast Microscopes. For dealing with highly refractile specimens there are now on the market phase-contrast microscopes, or attachments for converting an ordinary microscope into a phase-contrast type. These depend on a novel principle—that of slowing down some of the light rays. A.Ja.

See also: *Macrophotography*.

Books: *Amateur Photomicrography*, by A. Jackson (London); *Practical Photomicrography*, by J. Bergner, W. Mehliss and E. Gelbke (London); *The Technique of Photomicrography*, by D. F. Lawson (London); *Photomicrography, Theory and Practice*, by C. P. Shillaber (London).

PHOTOMONTAGE. Composite photograph made by cutting out and pasting together several separate photographs to achieve a particular effect—e.g., to convey an over-all impression in a portrait by combining several aspects of the same face. Alternatively, photo montage may be used in producing photographs covering a very wide angle of view—e.g., in panoramas.

Photomontage was largely practised by pictorial photographers, originally to obtain effects not otherwise possible with the comparatively slow and "colour-blind" negative material. The first to practise photomontage (1858) was H. P. Robinson, most of whose elaborate composition pictures were scissor and pastepot productions. In the 1860s and '70s many amateur artists made photomontages by combining (usually) other people's photographs with their own watercolours. This technique was later forgotten but re-introduced by George Grosz and John Heartfield in 1915. It is nowadays chiefly used in advertising.

See also: *Montage.*

PHOTO MURAL

A photo mural is a photograph that covers an entire wall or section of wall as a decoration. There are two ways of applying a photographic image of the requisite size to the wall surface: the negative is enlarged on to bromide paper in sections and the sections are assembled and stuck on to the wall; the wall may be sprayed with sensitized emulsion and the negative enlarged directly on to it.

Both methods are practised by firms specializing in photo mural work. The second method can only be applied by such specialists, but the first method is within the resources of the normal professional who only wants to use it occasionally. It is easier, however, if a few extra items of equipment—e.g., large dishes and bowls, provision for drying and trimming of the prints—are available.

Uses. One of the most popular applications of photo murals, apart from their use as display pieces, posters, and exhibition features, is as a wall decoration. In hotels, reception halls, and even private houses they are frequently used to give an impression of space. A scene consisting mainly of middle distance and distance is made to cover a whole wall, and enlarged to its natural proportions as viewed from the room.

Sometimes the scene is set in a dummy window frame so that it appears as though seen through an actual window.

Hand colouring may strengthen the impression.

Scenes suitable for this type of photo mural are panoramas of well-varied landscapes: a range of distant mountains seen across a valley, a distant vista of skyscrapers or a fringe of palm trees along the edge of a lagoon. A suitable mural for an air-line booking office would be a photograph taken from an aircraft of a cloudscape in sunlight. In a private house, a blank wall—e.g., of a basement lounge—might be decorated with a view of mountains mirrored in a lake to give a feeling of spaciousness.

The viewing distance must be carefully considered in the production of murals. The effect of the display will be lost if the print is too large and the viewer cannot stand far enough away to obtain a comprehensive view.

The Negative. Because of the great scale of enlargement, the negative must be critically sharp to allow for some inevitable loss of definition in enlarging, and as thin as possible so that printing time is kept within reasonable limits. If it is inclined to be dense, it pays to make a thinner copy negative as big as the enlarger can take, and preferably from an enlarged and retouched print.

It is very much easier to print the whole negative area in one exposure if the space to be covered is not too big for the enlarger used. If it is not possible to find a big enough wall space in the darkroom (or a room that can be darkened) or the enlarger cannot be set up far enough away from the wall to project the whole negative to the required size, then the picture will have to be made in two or more sections. To do this, the negative must be copied in sections on to the required number of plates so that each section occupies as much of the plate area as possible. The plates are all exposed and processed together, or at least under identical conditions.

Paper. In general, glossy paper should be avoided because light reflected from the surface is distracting. White or ivory base papers with a semi-matt or lustre surface have an excellent luminosity and depth of tone and will readily accept retouching and colouring pigments and dyes.

Enlargers. A horizontal enlarger is preferable, though a vertical enlarger with a rotating head can be used. The lens must be of good quality and of a sufficient focal length to cover the negative adequately.

Any unevenness of illumination must be rectified. There are two ways of doing this:

(1) A small circle of fine tissue paper can be stuck to the top of the condenser and the edges of the paper notched to reduce any sharp difference in density, or

(2) A compensating screen can be made by placing an unexposed slow plate in place of the diffusing screen and briefly flashing the exposing lamp. The plate is then developed in a very weak developer to a very low density. The density of the resulting image is greatest

where the light is brightest, and, when it is correctly exposed and developed, it exactly cancels out any unevenness in illumination when inserted in place of the diffusing screen.

Processing Dishes. The normal 18 × 20 ins. dishes are suitable for see-saw processing of lengths of paper up to 3 feet long by 20 ins. wide. Larger and deeper dishes are essential for processing long lengths of paper by the roll method.

A simple method of making temporary dishes of a suitable size is to cover a wooden foundation with waterproof plastic. For the base, a large sheet of wall board or plywood is used, and a raised rim is built around it by fastening lengths of 1 × 2 ins. batten along the edges to form a closed rectangle 2 ins. deep. The plastic sheet material sold by the yard for curtains and table coverings makes a satisfactory waterproof lining. It is simply laid on the wooden base and tacked into position along the outside edges of the side battens.

The size of the dish that can be constructed in this way is limited only by the width of the materials, but both wall board and plastic are made in at least 36 ins. width which is ample for making a dish to handle 20 ins. wide photo mural printing paper. (At least 10 ins. must be allowed on the width of the plastic sheet to include the sides and turnings.) For processing 40 ins. wide paper it will be necessary to join two widths of wall board and cover the dish with 54 ins. wide plastic sheet. Such dishes cost only a fraction of the price of the proper article and the materials can be used again for other purposes.

Printing Surface. If possible the mural should be printed in a single exposure. For this, it is necessary to have a printing surface at least as big as the finished mural with enough distance in front of it to allow the enlarger to be set up to cover that area. There is nothing better than the wall of the room, but if a suitable wall is not available, a surface must be constructed with wall-board on battens.

The finished surface must be free from bumps or unevenness and the enlarger must be set up accurately in position with respect to it. This is best done by removing the negative and measuring up the sides and diagonals of the projected rectangle of light cast by the negative carrier. If only a part of the negative is to be enlarged, then the equivalent rectangle should be marked out in ink on a clear piece of negative material—or scratched on an old dense negative. This rectangle is then projected to the required size and checked for equality of diagonals and opposite sides.

If the photo mural must be printed in sections, when all the sections are assembled it must look as though the whole picture had been enlarged at one time on a single sheet of paper. This means that each section must match the next in regard to its scale, its printing density, and its contrast.

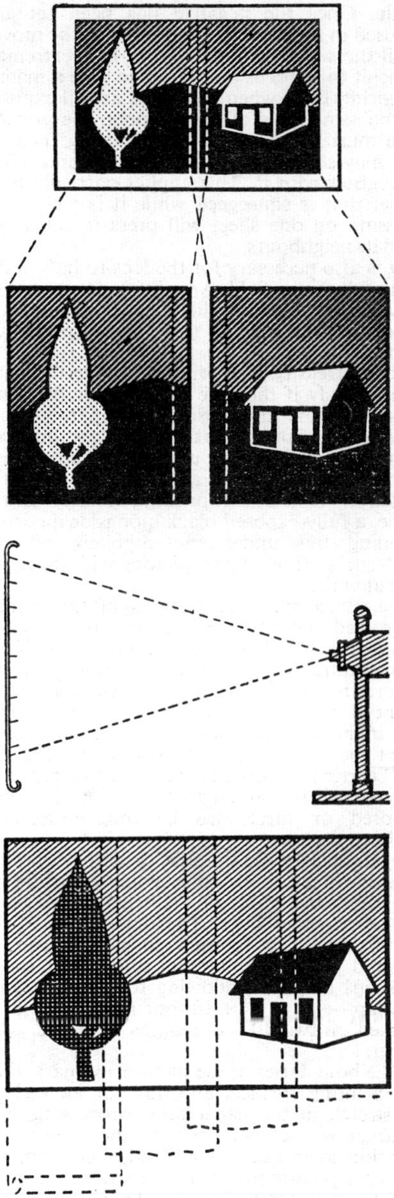

PHOTO MURALS. Top: *The original negative must be critically sharp and reasonably thin.* Upper centre: *Usually it is advisable to prepare a copy negative from the original negative, as large as possible. Where very great wall areas are to be covered, the copy negative may be made in sections.* Lower centre: *The enlargements are made on to strips of paper pinned to the wall. Pins should be placed at an angle away from the lens, to minimize shadows.* Bottom: *The final prints should have sufficient overlap at the edges of each section to permit easy assembly when it is mounted in position.*

Scale. Once the negative has been set and focused in the enlarger, nothing must be moved until the whole job is finished. It is extremely difficult to re-focus a negative to give a matching print. Even when the prints are all exposed at the same scale of enlargement, the greatest care must be exercised in processing them or one may stretch more than another and refuse to register with it. This applies particularly to paper that is squeegeed while it is wet; extra pressure on one sheet will press it out larger than its neighbours.

It is also necessary for the lens to have a flat and undistorted field. Any difference in magnification between one side of the field and the other will show up as soon as the prints are joined up.

Printing Density. The prints will have the same density only if they are exposed for the same length of time under exactly comparable conditions. This means that the enlarger illumination must be absolutely uniform measured over the actual image area, and not simply on the negative. Unless the printing density is the same, a fully-exposed black alongside a corresponding but under-exposed black on the adjacent section of the picture will show the join at once.

Tone. Even when all sections of the picture have had the same exposure, they will not necessarily give blacks of the same tone if the temperature of the developer changes during processing or the solution is worked to the point of exhaustion, or if, by faulty manipulation in the solution, one print gets more development than another. Furthermore, any "dodging" carried out on one print must be repeated exactly when the adjacent section is being exposed or there will be tone differences between the two.

Preparation. In the simplest case, the complete mural will be made in a single exposure by projecting the whole of the negative image on to the paper at once. The dimensions of the mural should be chosen so that it can be covered with the minimum number of sheets or strips of paper, allowing for about a ¾ in. overlap—e.g., a 6 × 10 foot mural would be printed on two 10-foot lengths of 40 ins. wide paper.

The boundaries of the picture are marked on the printing surface, allowing for an inch in 40 stretch in the paper after processing. The enlarger is then set up with the negative in position in the carrier and adjusted until the required picture just fills the marked space.

If the mural must be printed in sections, each section is copied on to the largest negative that the enlarger will take. The size of the first section is marked out on the printing surface and the corresponding negative is adjusted to fill it. Once this has been done, the subsequent sections are printed without adjusting the scale of enlargement again, otherwise the adjacent sections will not match. It should always be possible to print each section on not more than two widths of paper. An allowance of up to 2 ins. should be made down the adjoining edges of the sections to allow for overlapping.

Each negative should be masked down to little more than the exact printing area to cut down the unwanted light as much as possible.

Printing. When the image has been focused and adjusted to fit the area, the lens is stopped down until the definition is satisfactory right to the corners of the picture. Test pieces of paper—about whole plate size—are then exposed in the centre and corners of the picture area to arrive at the right exposure. The test pieces should include representative highlights and shadows and should be cut from the actual roll of paper to be used for making the prints. Each test piece is developed right out, and the exposure (or local control) adjusted until it gives a fully graded image under these conditions.

Once the exposure has been determined, the actual printing can proceed. Whether the mural takes a single exposure, or has to be covered in several, the method is the same. The first length of paper is pinned in position over the lower half of the image area. Then, before pinning up the second half, about an inch is trimmed off along one of the edges to take care of the defective emulsion occasionally found at the extreme edge of this type of paper. The trimmed edge is then lapped about ¾ inch over the adjoining edge of the first length and held in close contact with ordinary pins, placed to cast the smallest possible shadow. (These shadows are subsequently touched out with dye or pigment.) If the paper is left rolled up, emulsion side out, for an hour or two before pinning it up, it will lie flat to the wall.

The exposure is then made by switching the enlarger lamp on and off. It is essential for the whole set-up to be perfectly steady before making the exposure; the slightest vibration at this scale of magnification hopelessly blurs the image.

Processing. One of four methods may be used for processing the exposed strips. For the first three methods the paper should be pre-soaked. All strips must be developed under exactly the same conditions and for exactly the same time as all others for the same mural.

(1) Large dish method. If a large enough tray is available, the strip may be processed flat in the normal manner and should be covered with developer quickly and evenly to avoid uneven development. Another method is to fold the strip in half, emulsion side out, and then fold the doubled strip once more. It is now one quarter of its length and can be processed in a dish of a smaller size by regularly reversing the second fold so as to bring the inside and outside emulsion surfaces into contact with the developer in turn.

(2) See-saw method. This method may be

used by one operator for strips up to 4 feet in length; longer lengths require three operators, one at each end and the third to maintain the loop fully immersed. To facilitate handling, and reduce the danger of tearing, the ends of the paper should be clipped to wooden rods running the full width of the strip.

(3) Deep-dish method. This method is recommended for processing long strips, and requires one operator. The paper is completely immersed during pre-soaking while being continually unrolled and re-rolled from end to end. The unrolling and re-rolling are repeated throughout the whole period of development and fixing; constant movement and complete immersion are essential to even developer action.

(4) Brushing method. A developer with no tendency to stain during prolonged development—e.g., a glycin type—must be used for this method.

The strip should be laid on a non-absorbent surface and swabbed with developer as evenly as possible. Full development of up to 6 minutes may be given to allow areas wetted at different times to even out. Excess developer should then be squeegeed off and an acid stop-bath flooded on.

After a few seconds the stopbath should be drained off, and the fixer swabbed on for the full fixing time, ensuring a continually fresh supply of fixer.

The exposure may be shortened slightly and development time extended should the standard development times prove inconveniently short.

To ensure even density of strips the temperatures of all processing solutions must be maintained at a constant level.

A plain acid fixing bath without hardener should be used to avoid the possibility of the emulsion cracking should the paper be kinked after drying.

Normal washing in running water is preferable; alternatively, the strips should be soaked for 5 minutes in each of twelve changes of water. A hypo test should be made if the prints are for permanent display.

The strips are best dried by clipping them at intervals along the whole length to a stretched cord.

Mounting. If prints are to be mounted direct on to a wall the tendency to contract on drying will not matter. But if mounted on panels for temporary display, the panels should be strongly braced to prevent warping. Alternatively, a sheet of wet kraft paper should be pasted to the back of the mount at the same time as the enlargement is mounted. This will counterbalance the pull of the paper on the other side and leave the mount flat.

If the prints are to be mounted on an absorbent material, the surface must be well sized beforehand. A sheet of muslin may be pasted over the wall to provide a continuous mounting surface.

The adjacent edges of the strips may be trimmed level and butt jointed—in the same way as a paperhanger puts paper on a wall—or they may be overlapped. One method which produces an almost invisible join is to strip the paper support away from the emulsion for about $\frac{3}{4}$ in. along one edge, and to cut away the same amount of emulsion from the corresponding edge of the adjacent strip. (A suitable amount of overlap of the image must be allowed in printing.) The emulsion of one strip is then pasted over the paper base of the strip next to it. The final effect is of an unbroken emulsion surface stretching right over the join. It is doubtful whether such exact methods are justified in photo murals that are to be looked at from a distance, but they may be worth while on the walls of living-rooms, etc.

To apply the paper, it is first soaked in plain water to get it limp and, after draining, it is laid face down and well brushed over with a good water paste, strained if necessary to remove any solid particles. It is then carefully lifted and smoothed into place with a wet flannel rolled around a block of wood and constantly dipped in water to make it slide easily. Any excess paste that squeezes out can be sponged off later. When all strips have been fitted into position, the work is allowed to dry for a day before retouching and finishing.

Finishing. Slight blemishes are retouched with normal bromide spotting dyes, or with water colour. Small areas can be retouched with pencil, provided the surface of the print is subsequently to be protected. Large areas are best worked up with an air-brush.

Oil or water colour can be applied by air-brush, camel-hair brushes, or cotton wool on orange sticks. Colouring work should be frequently viewed from a distance to verify the general effect.

A coat of clear cellulose lacquer provides maximum permanence and weather-proofing, and the lacquer should be applied by means of a paint spray-gun. For outdoor enlargements both print and mount should be sprayed.

In Situ Method. The method of enlarging the negative direct on to a sprayed-on wall-coating of emulsion is a highly specialized techique that is beyond the resources of the ordinary studio.

First of all the surface must be prepared to ensure that no chemicals in the wall or its protective finish can affect the photographic image in the course of time. It may be necessary to spray on a sealing coat before applying the emulsion. Some provision must also be made to collect the solutions that are swabbed on to the surface later.

After the room has been completely blacked out and equipped with a suitable safelight, the emulsion is sprayed on to the wall surface with a special type of paint spray-gun silver plated on all surfaces in contact with the emulsion. So that the operator will be able to tell which parts he has covered, the emulsion incorporates a dye—e.g., a red dye which appears as a dark tone under a light green safelight. The dye is water-soluble and is removed in the course of processing.

The emulsion is supplied as a dry powder which is mixed with warm water immediately before use. It is applied under carefully controlled conditions of room temperature and humidity. When the emulsion dries the negative is enlarged on to it in a single operation, after which it is processed by swabbing or spraying the solutions on to the surface. After washing and drying, the surface is finally sprayed with a protective finishing coat of transparent varnish. A.E.S.

See also: *Giant enlargements; Montage; Murals.*

PHOTON. Least quantity of radiant energy that can be transferred to or from a system. The size of the photon varies according to the frequency of the ray, but it is constant for all rays of one particular wavelength.

See also: *Light units.*

PHOTOPLASTIC RECORDING. A reusable, grainless electrostatic recording process using thermoplastic photoconductive polymeric layer which when molten, deforms in accordance with the pattern of electrostatic charges.

See also: *Thermoplastic recording.*

PHOTO-REPORTAGE. Use of photography to supplement or replace accounts of events and places of current interest written for publication—e.g., in newspapers or magazines.

See also: *Freelance photography; Magazine photography; Photo journalism; Picture series; Press photography.*

PHOTO RESISTOR. Alternative term for photo-conductor, a substance which changes its electrical resistance under the influence of light. With most photo-resistors the resistance decreases the greater the light intensity falling on the resistor. One of the best-known photoresistors for photographic purposes is the cadmium sulphide cell.

See also: *Photo-electric cell.*

PHOTOSCULPTURE. Method of automatic sculpture based on a camera image or images of the subject. Since the middle of the last century workers in various countries have attempted to use the camera for producing either bas reliefs or solid sculptures. Most attempts have succeeded in producing results of a sort, but all were too complicated and costly to survive. Some systems were almost completely automatic; the camera images being translated directly into the corresponding solid form. In others, the images were no more than a guide to the operations of one or more sculptors.

Willème's Method. François Willème (1830–1905), a French painter and sculptor, took out patents for a method of photosculpture in 1860 and received enough support to found La Société Générale de Photosculpture de France (the French General Society of Photosculpture) and erect a pretentious establishment in Paris. He employed a staff of photographers and sculptors whom he trained in the technique.

The studio consisted of a circular room with a glass dome. Twenty-four cameras were disposed regularly around the perimeter and focused on the centre of the room where the sitter occupied a raised dais. A silver ball suspended from the exact centre of the dome immediately over the sitter's head provided a common reference by which all the negatives could later be registered.

All the cameras were operated simultaneously by a common shutter control and an exposure of about 10 seconds gave twenty-four negatives, each showing a profile of the subject from a different angle. These negatives were then projected to the required finished size and a set of templates traced from the projected images, all the templates being registered on the image of the ball and the cord by which it was suspended.

The templates were mounted in turn on a pantograph set up in front of a mass of modelling clay fixed on a rotating platform. The platform could be fixed in any one of twenty-four positions corresponding to the camera angles from which the set of photographs were taken. Each of the templates was set up in turn in such a way that when the operator traced around it with the stylus of the pantograph, he moved a tool which shaped the clay to the same profile. When one profile had been carved out in this way, the clay was rotated into the next position and the corresponding template set up in the pantograph.

The process was repeated until all twenty-four profiles had been cut in the clay, and at that stage the work was continued by hand. A considerable amount of handwork was necessary to fill in the shape between one profile and the next and to add the depressed features —e.g., eye sockets and nostrils—which were bound to be left out by this method. The clay could then be used to make a mould from which copies could be cast in plaster or metal.

In spite of the many practical difficulties and the large amound of skilled workmanship, the method was a craze for a time, and many people patronized the studio, among them such notables as Theophile Gautier, the leading actors of the day, and the entire Spanish royal family. But the enterprise could not pay its way and had to close down in 1867. Various improvements on the method were proposed, notably by Poetschke (1891) and Selke (1892), but were not applied on a commercial scale.

Baes' Method. Carlo Baes, an Italian sculptor, developed a method of photosculpture at the beginning of the century and published an account with a photograph of his studio set-up in 1908. His method was to illuminate the subject by a light source and a ring of mirrors in such a way that the illumination increased uniformly from the front to the back of the subject. One exposure was made in this way and then another was made with the lighting reversed so that its intensity increased from the back to the front.

The two negatives were printed in register on to bichromated gelatin which was processed to make it swell selectively and form a relief image. By combining two complementary negative images in this way all effects of colour and lighting were neutralized and the result was a pattern of tones which varied in density with the relief of the subject.

This method is only suitable for portrait heads of about 3 ins. in height because the relief produced is not sufficiently pronounced to resemble anything like a solid object.

Edmunds' Method. In 1921 H. M. Edmunds published a process in which a spiral light trace was projected on to the subject from a point close to the camera lens. In the resulting picture the spiral was distorted by the irregularities in the surface of the subject. Up to a point, the distortions indicated the shape of the surface and formed the basis for the semi-automatic carving operation that followed.

In the carving process, the distorted spiral image was scanned by a low power microscope linked to a pantograph through which it controlled a power-driven engraving tool. By this means any suitable material could be shaped to a relief image of the subject.

This method possessed the advantage that it reproduced cavities that are inevitably filled in by the various methods in which the illumination falls across the subject from the side.

Givaudan's Method. The inventor of this method, which was described in 1926, called it photostereotomy because in effect it divided up the subject optically into a number of parallel slices. The subject was surrounded by a system of five light sources shining through narrow slits, the total effect of which added up to a thin continuous line of light. This line of light illuminated the edge of an imaginary section of the subject cut parallel to the focal plane to the lens. The light sources were mounted on a sliding carriage so that it was possible to illuminate and photograph a series of sections from the front to the back of the subject. The negative image of each section was printed and the profile cut out. When all the profiles obtained in this way were assembled, the result was a relief image.

By making very small shifts of the lighting between exposures—i.e., making a large number of profiles—the final result was reasonably continuous. In practice, Givaudan combined as many as sixty profiles printed on paper 1/10 mm. thick to form a relief image 6 mm. thick.

It was also possible to insert suitable spacers between the profiles to increase the thickness to its natural proportion, and by combining such reliefs made from both sides of the subject, the result was equivalent to a sculped figure.

Sculptography. A method attributed to Japan and subsequently applied in the United States was described in 1934 under the name of sculptography. In this, the sitter occupies a chair in a dark room on a turntable which can be complete rotated once in four seconds. A thin vertical line of light is projected on to the subject and photographed with a cinematograph camera. The angle between the light projector and camera axes is about 30°, so that the line of light as seen by the camera bends with the contours of the subject. (If both axes lay close together, the camera would see only a vertical line, no matter what the position of the subject.)

The resulting cine record is of a succession of profiles, each taken from a slightly different angle. These profiles are enlarged separately and cut out either mechanically or by etching, and assembled with packing between to form a solid figure. The ridges formed by the edges of the individual profiles are filled with wax and casts are made in the usual way.

Of all the methods that have been proposed or tried, this appears to be the most promising, but like the others, it has never become a widely followed commercial proposition. F.P.

See also: *Statuettes.*

PHOTO-SECESSION. A secession from the New York Camera Club in 1902, led by Alfred Stieglitz with Alvin Langdon Coburn, Frank Eugene, Gertrude Käsebier, Edward Steichen, Clarence H. White and later 41 others. Their aim was to break with conventional pictorialism, to explore fresh photographic subject matter, and to concentrate on rendering the subtleties of light. The Photo-Secession was affiliated to the Linked Ring.

PHOTO-SENSITIVE GLASS. Glass made sensitive to ultra-violet radiation by the addition of certain heavy metal compounds. Exposure forms a latent image which becomes visible on heating the glass to about 600°C.

See also: *Glass (photo-sensitive).*

PHOTO-SENSITIVE PLASTIC. Polymer-type plastics which are made light sensitive and under the influence of light become insoluble or change colour.

See also: *Plastics.*

PHOTOSENSITIVITY. Generic term for the reactions of materials and processes to light energy.

PHOTOSTAT. Proprietary name for a copying camera and process designed to produce quick and accurate copies on a commercial scale. It is mounted on a pedestal with all the necessary controls and its own easel for supporting the document to be copied.

The sensitized material used is a special highly sensitive paper carried on a reel in a magazine in the camera. An automatic processing unit develops and fixes the exposed prints so that they only require washing in daylight before being dried. The Photostat, like the reflex printing and camera processes, produces a negative from which contact copies can be made.

If large true-to-scale positives are required, the first camera picture is usually taken on special film. By using an attachment that converts the Photostat into a projector, enlargements from the film can be made to any convenient scale. Several manufacturers market machines of the Photostat type; their names may vary, but the principle is the same in all cases.

See also: *Document copying*.

PHOTO-STEREO-SYNTHESIS. Method of producing an image which gives an impression of stereoscopic relief when viewed normally. The image consists of a number of positive transparencies made by taking a series of photographs of a still subject, each photograph showing only a single plane sharp. After each exposure the camera is moved nearer to the subject so that the scale of reproduction is the same for each plane. The resulting transparencies are then superimposed so that the sharp planes depicted occur in the same relationship as in the original subject and are proportionately spaced. The impression of relief is appreciable from one observation point only.

PHOTOTELEGRAPHY. Nowadays the transmission of pictures by wire and radio is an everyday occurrence; almost every newspaper possesses at least one transmitter and receiver. Modern equipment produces pictures which are almost indistinguishable from the original, and for this reason many wired pictures are reproduced in our newspapers without any indication of the fact being given in the caption.

History. Picture telegraphy is by no means a recent technique. Attempts at the transmission of pictures by wire were made even before the development of Morse Code telegraphy for normal communication.

The first system was devised and demonstrated in 1842 by Alexander Bain, a Scotsman, using a set-up which satisfied in an elementary way the fundamental requirements for facsimile transmission as it is known to-day. Other early experimenters were Shelford Bidwell (1881), Belin, Korn, Jenkins and Ferre. Even before the introduction of modern electronic valves, practical systems with commercial applications had been developed. These systems used a code for tone values and could transmit reasonably good half-tones.

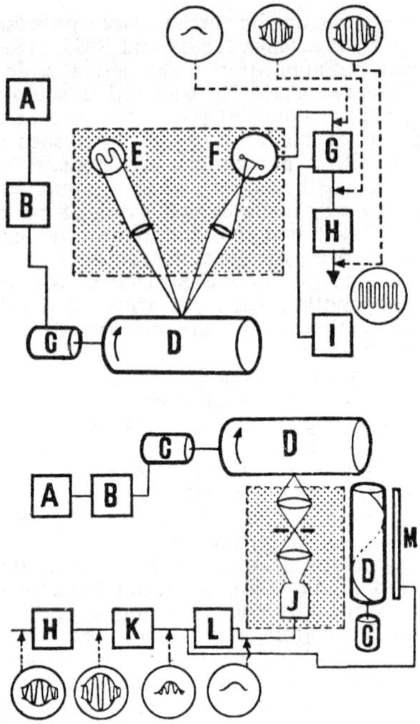

PHOTOTELEGRAPHY. Top: *Transmitter. The picture is wound round a drum revolving at a fixed speed, and scanned by a light source and a photo-cell. The impulses are then superimposed on a carrier wave and fed into the line.* Bottom: *Receiver. The impulses are fed to a controlled light source which, by variations in brightness, records an image on sensitive paper wrapped round a rotating picture drum. The signals can also actuate a writing blade for direct recording. The shaded area is the optical carriage in both cases. A, tuning fork. B, fork amplifier. C, synchronous motor. D, picture drum. E, scanning lamp. F, photo-multiplier cell. G, modulator. H, line amplifier. I, carrier oscillator. J, crater lamp (controlled light source for recording). K, demodulator. L, low-pass filter. M, writing blade bearing on cylinder of direct recording unit.*

With the advent of electronic techniques the transmission of half-tones was made much easier and from the early nineteen-twenties high quality picture transmission became a reality. At the present time the majority of equipment in use is manufactured by Muirhead in England, Belin in France, and Times and Acme in the U.S.A.

Transmission Principles. There is nothing difficult about the basic principles of picture telegraphy transmission. The picture to be transmitted is wrapped round a rotating drum, the surface of which is scanned by a spot of light which traverses it from end to end. Light reflected from the illuminated portion of the picture is focused on to a photo-cell, thus producing a minute current which varies as the drum rotates and the light spot travels over areas of differing tonal density. This current is amplified and transmitted by line or radio to the receiving end. In the receiver it is made to vary the intensity of a spot of light which falls on a sheet of sensitized paper wrapped round a drum similar to, and scanned in synchronism with, the transmitter drum.

The photographic material used may be either paper or film, for by suitably adjusting the recording characteristics of the receiver the incoming signal can be made to record as a negative or a positive image.

Recent Developments. Portable picture transmitters now exist which can be used wherever a telephone line is available, enabling up-to-the-minute photographs to appear in newspapers. The technique of phototelegraphy is also used to transmit complete newspaper pages; the process takes only a few minutes, and allows a newspaper to be printed and published simultaneously in places hundreds of miles apart.

Direct Recording. One of the most notable advances has been the development of direct recording on electrosensitive paper. A recorder of this type uses paper in rolls sufficient for several hours continuous operation. The paper is drawn slowly between a rotating helix and a stainless steel blade, and a fluctuating current controlled by the received signal is passed through the paper at the point of intersection. The chemical action which occurs at the point of contact causes local darkening of the paper. The helix drum is maintained in synchronism with the transmitter drum and the picture is built up in a series of parallel lines.

The paper consists of a resin-bonded sulphite pulp paper of high wet strength and great absorbency, impregnated with an electrolyte and a colour-forming substance. Other chemicals are also added to intensify the colour and limit its spread.

The current decomposes the electrolyte and one of the products of the decomposition reacts with the stainless steel blade to form a ferric salt. This salt reacts with the colour-forming substance and colours the paper. The result is a black-and-white record instantly visible without processing. Tonal gradation is good, and although the definition of photographs received in this way is not as high as those received on photographic paper, it is good enough for newspaper reproduction.

This technique has the advantage that the received copy is instantly usable, no processing of any kind being necessary. Direct-recording machines are used extensively for the accurate dissemination of weather maps and information, there being no possibility of errors occurring during the transmitting/receiving process. Facsimile machines are also used in multi-branch offices for transmitting and receiving graphic material such as documents, drawings, statements, etc. H.S.

Books: *Facsimile*, by C. R. Jones (London); *The Transmission of Pictures by Radio*, by Cole and Small (London).

PHOTO-THEODOLITE. Special type of photogrammetric camera for applying the principles of trigonometrical surveying to record photography. The camera is equipped with means of accurately defining its position and the angle of view of the subject. A grid, giving accurate horizontal and vertical references, is superimposed on the plate at the time of exposure. The field of the camera lens is corrected with respect to the grid lines so that the dimensions of the subject can be accurately deduced. The principle on which the photo-theodolite camera works is that it is possible to determine the position in space of any point object from photographs taken from two different viewpoints when the angle between the two cameras and their respective positions are known.

Similarly, the shape of any solid object can be reconstructed by plotting the intersection of perspective rays from corresponding points on the separate images.

There are two types of photo-theodolite, the simpler in which the photographs are always taken with the plate vertical, and a more complex instrument which may be used with inclined plates.

See also: *Naval photography; Photogrammetry.*

PHOTO-TRANSISTOR. Transistor used as a light-operated switch. All transistors are photosensitive and are normally in an opaque container. In a photo-transistor the active part—a germanium wafer—is enclosed in a glass envelope. A light-diffusing filling of silicone grease ensures that any light falling on the transistor produces the same effect whatever its angle.

There are various ways of connecting a photo-transistor according to the function of the circuit. In the simplest type of circuit for straightforward on-off switching, two of the transistor elements, the emitter and collector, are connected in series with a relay coil and a D.C. supply of 12 to 18 volts. Under these conditions a typical circuit might pass a dark current as low as 300 microamperes, rising to 5 milliamps or more when light falls on the transistor.

Photo-transistors are used in photography as a basis for light-operated shutter release mechanisms and in slave units for multiple flash systems in which one or more separate flash units are triggered off automatically by the flash from the master unit.

See also: *Photo-electric cell.*

PHOTO TYPE-SETTING. (Film Setting.) Photographic method of producing a column or page of text in a single operation from a master photographic negative. Though first proposed at the beginning of the century, it did not become a practical workshop method until the 1950s.

Filmsetter machines are fitted with one or more assemblies of master negatives (photo-matrices), containing all the characters likely to be required. Instead of casting molten type metal to form the printing characters, as in the conventional mechanical typesetting machines, the filmsetter machine projects light through its negative character matrices and forms photographic positive images of them upon film, at the required size and in the appropriate order, thus producing lines, columns and pages of text. The different type sizes are obtained by varying the projection scale.

Photo-typesetting is usually considered in two sections; true filmsetting for the mechanical, keyboard controlled, photographic composition of text matter; and photo-lettering for setting display headings, advertisement layouts, etc. There are a great many photo-lettering machines, varying from quite simple hand positioned devices to comparatively elaborate machines, but their purpose is different from that of the high speed filmsetter machines.

Filmsetter Machines. The standard keyboard operation is used which results in movement of the matrix to bring the selected character into position over a roll of unexposed film. An exposure is then automatically made so that the negative image is recorded on the roll of film. Subsequent operation of keys repeats this procedure, at the same time racking the projected images horizontally to build up each line. The result is lines of text recorded photographically on the roll of film; from this, ordinary line blocks or a negative for photo litho printing can be made.

One electronic machine carries the negative character images on a disc revolving at ten revolutions per second. A high speed flash

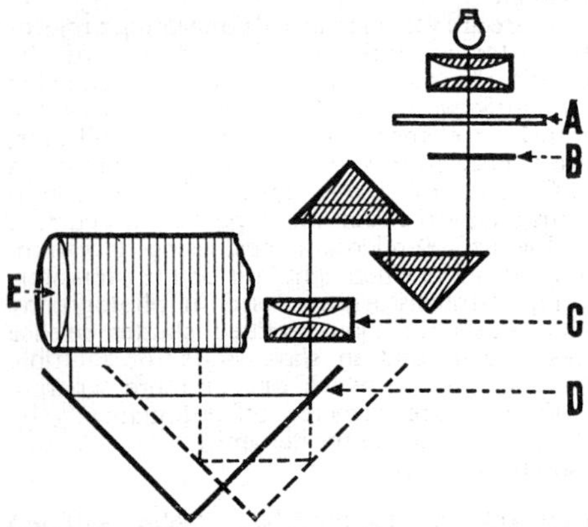

PHOTO-TYPE SETTING. Relative movement of lens and two prisms adjusts letter sizes and focus. Lateral traverse of mirrors moves images along image plane. A, negative matrix. B, shutter. C, lens. D, mirror assembly. E, film drum.

freezes the movement of the selected character at the required instant as it passes the exposure position. The optical system of another machine projects eighty-eight character images (on any one of a number of photo-matrices) into one position at the film plane, a series of louvre shutters covering all but the required character while an electronic flash exposes it.

The lenses used in photo-type setting machines are specially designed and must be accurately positioned as critical definition is essential. Some systems use a turret of lenses to obtain various projection (or type) sizes.

The source of light usually resembles bulbs used in car head-lamps working at 12, 24 or 48 watts, or is an electronic flash tube. The useful light output is concentrated by a condenser lens system to give illumination over the area of the character in the master matrix.

The film is spooled up on to a drum (size of film used is determined by the width of column or page to be set). The film is then fed through the mechanism and held rigidly immobile during exposure. Corrections are made by deleting and inserting full lines. The film is then processed and a positive of the column or page is the final product.

There are six major film setting systems; three American, two British and one Dutch. Of these systems three have been evolved through the adaptation to photography of existing principles in hot metal type setting machines. The others make a completely new approach.

High-speed Filmsetting. Some electronic filmsetter machines can set photographic text images at speeds of the order of 30,000 characters per hour. This is much faster than a keyboard operator can work, so the keyboard output is recorded on punched paper tape, and the tape from a number of keyboards fed into one filmsetter machine. Corrections can be recorded on a separate tape, and fed into the machine by a tape merging system which automatically makes the required corrections at the right positions.

A further system under development is computer control. A computer can be programmed to govern the width of the film set lines, the word spacing and word splitting. Automatic machines are evolving which can even read typed manuscript matter and directly convert this into film set text without the need for a keyboard operator. This makes photo-type setting possible at extremely high speeds.

Photo-Lettering. Photo-lettering equipment for setting display matter and layouts is rather different from a film setter, though the basic principle is the same.

The unit is essentially a photographic enlarger, using a photo matrix containing all the characters of a particular type face in the negative carrier. This is projected on to a sensitive film or paper on a special easel which can be moved horizontally and vertically or angled in any direction.

Operation is manual and consists of locating the required letter on the negative matrix in a suitable central position. A red filter over the enlarger lens permits the placing of the image on the sensitive material in the required place. A white-light exposure records the letter, after which the easel is moved (and also the negative carrier) to bring the next letter in the appropriate position relative to the previous one. In this way a line of a heading is built up, and can be either straight or curved or disposed in any other way.

The sensitive material may be an ordinary photographic film which is then processed and serves as a basis for making a letterpress block or photo-litho plate.

Alternatively special films may be used incorporating a developing agent in the emulsion. Before the setting operation the film is soaked in an alkaline activator, and laid on the easel. After every exposure of a letter the latter then develops up instantly, and is visible as the next letter is projected into place. This simplifies visual alignment and layout. When the display is complete, the film is fixed and used for block or plate making. R.W.D. (F.H.S.)

See also: *Photomechanical reproduction.*

pH VALUE. Number signifying alkalinity on a logarithmic scale. It is the negative logarithm of the hydrogen ion concentration. Pure water, which is neutral, has a hydrogen ion concentration of 10_{-7}, and thus a pH value of 7. Values below 7 are acid and above are alkaline, as indicated in the table.

As the scale of pH values is logarithmic, each unit represents a degree of alkalinity which is ten times as strong as the next below, and one tenth as strong as the next above.

The chief importance of pH values in photography is their effect on the activity of developers. Most normal developing agents work with greatest efficiency in an alkaline solution. The degree of alkalinity and the way in which it is achieved (i.e., by small amounts of strong alkali, or large amounts of weak) have a decisive effect on the characteristics of the developer.

pH VALUES

pH-value	Solution is:
0– 2	strongly acid
3– 4	acid
5– 6	weakly acid
7	neutral (pure water)
8– 9	weakly alkaline
10–11	alkaline
12–14	strongly alkaline

Various ways are used to determine the pH of a solution directly. The main methods measure the potential of a glass, hydrogen, or quinhydrone electrode when immersed in the solution under examination. These are usually incorporated in so-called pH meters, and give a direct reading of pH on a scale of a voltmeter.

In visual pH meters the colour of a universal indicator mixture added to the solution is compared with standard hues of the indicator at known pH values.

See also: *Indicator chemicals.*

PHYSICAL DEVELOPMENT.—In physical development, most of the silver forming the negative image is supplied by the developer and not by the emulsion as in normal development. A physical developer usually consists of a solution of silver nitrate with sodium sulphite, a silver solvent like sodium thiosulphate, and a certain amount of normal developing agent added.

The main reason for the developing agent is that there must be a weak silver image there before physical development can start. The initial silver image is intensified by the addition of silver deposited on it from the solution. The process is therefore a modified form of silver intensification.

The proportions of silver nitrate and developer must be carefully balanced to achieve a suitable rate of silver deposition without fog, and for this reason the chemicals have to be weighed out even more accurately, and clean work is more important, than usual.

Image Characteristics. As very little of the silver halide in the emulsion is reduced, the image grain bears little relationship to the graininess of the emulsion. The grain obtained is, in fact, very fine, and subsequent silver deposition does not materially increase it. And as the image consists of a deposit of silver on the surface of the emulsion, it has very little thickness, so its resolution is also very good.

A certain amount of physical development occurs in the solvent fine grain developers. But its extent, compared with the normal chemical development taking place at the same time, is very much smaller.

Ingredients of Physical Developers. Physical developers can be either acid or alkaline. While the alkaline formulae are more stable, the acid ones act more rapidly.

The silver nitrate is the main constituent of the physical developer. It supplies the silver to build up the image.

The sodium sulphite and the thiosulphate form soluble silver compounds which are not precipitated by the alkali of the developer.

The sodium thiosulphate also dissolves away the silver salts in the emulsion, and puts an early end to the initial chemical development. This, however, reduces the effective photographic speed of the material and since there has to be a fairly high concentration of thiosulphate (about 3–4 per cent), this loss of speed cannot be avoided by reducing the hypo content of the solution.

But the action of the thiosulphate on the silver salts in the emulsion can be slowed down by converting the silver bromide in the emulsion into silver iodide which is much less soluble in thiosulphate solutions. To do this, the film or plate is treated in a forebath of potassium iodide. A thorough rinse between this forebath and the actual developer is essential to prevent precipitation of insoluble silver iodide by potassium iodide carried over.

Almost any developing agent can be used, but metol, or metol and hydroquinone give the cleanest results.

Development Time. The exact amount of developing agent present controls the rate of development. This is greatly affected by even small changes of concentration and it is useful to be able to check the activity before actually using the developer. A check of this sort can be made in the following way.

The actual time of development for any particular negative is first found by experiment, i.e., by developing strips of exposed film for varying times, and comparing them. At the same time, one drop of the same lot of developer is placed on a piece of contact or bromide paper, well fogged by exposure to light. After a short time the paper discolours and turns yellow or pink where the drop of developer was placed.

The time it takes for the paper to change colour is always in a fixed ratio to the development time of the particular film or plate. If, for instance, the spot takes 1 minute to appear, and the best development time for the film in question was found to be 15 minutes, this ratio is 15 : 1. If with another batch of developer the spot takes 1 minute and 20 seconds to change colour, then the same film would have to be developed for 20 minutes to obtain the same result. This relationship only holds of course if the same negative material and the same printing paper are used in both cases. The ratio must be found by trial every time the negative material or test paper is changed.

The development speed for physical development does not follow the same rules as in a normal developer. So it is not possible to estimate the correct development time of various negative materials in a physical developer simply by knowing the relative development times in a normal developer.

But once the ratio of discoloration time to development time has been found for any one combination of negative material and printing paper, it will always be the same for those materials. So it can be used to find the correct time with a new lot of developer.

Formulae. The following formulae are for a physical developer forebath and for solutions A and B of a physical developer.

Forebath				
Sodium sulphite, anhydrous	1	ounce	25	grams
Potassium iodide	175	grains	10	grams
Water to make	40	ounces	1,000	c.cm.
Solution A				
Sodium sulphite, anhydrous	2	ounces	50	grams
Silver nitrate	1¼	ounces	32	grams
Sodium thiosulphate (hypo)	6	ounces	150	grams
Distilled water to make	40	ounces	1,000	c.cm.

First, the silver nitrate is dissolved in about 8 ounces (200 c.cm.) of water. Then the sodium sulphite, dissolved in about 20 ounces (500 c.cm.), is slowly stirred in. When the precipitate which is first formed has redissolved, the thiosulphate is added. Finally, the rest of the water is added to make the correct volume. The solution will keep almost indefinitely.

Solution B				
Metol	35	grains	2	grams
Sodium sulphite, anhydrous	175	grains	10	grams
Hydroquinone	60	grains	3·4	grams
Sodium hydroxide	60	grains	3·4	grams
Distilled water to make	4	ounces	100	c.cm.

Procedure. The film is first immersed in the forebath for 1½ minutes. It is then rinsed well, and transferred to the working developer. This consists of 10 parts A, 1 part B, and 40 parts water; it must be made up immediately before use, and thrown away afterwards. Development takes about 25–30 minutes at 65°F. (18·4°C.).

The film is fixed in an acid fixer containing at least 30 per cent sodium thiosulphate. This may take well over an hour, since the silver iodide dissolves very slowly. Washing also should take at least an hour.

A formalin hardener (1 part formalin solution to 80 parts water) can be used at any stage; the best time is probably between development and fixing.

The film may be milky because of a fine deposit of silver (not to be confused with the milkiness due to incomplete fixation). This can be cleared by a brief immersion in a 0·1 per cent potassium bichromate solution containing 0·2 per cent sulphuric acid. Too long treatment in this bath will, however, affect the image. The same solution, but made 25 times stronger, will remove any silver deposit from the tank, etc.

If the film is found to be under-developed, it can be put into a freshly made solution of the developer again (even after fixing) and the development (which is now pure silver intensification) will continue normally.

Silver solutions stain the skin black so it is wise to wear rubber gloves when handling physical developers. L.A.M.

See also: *Developers; Development after fixing; Intensification.*

Book: *Developing*, by C. I. Jacobson (London).

PHYSICAL REDUCTION. Methods of reducing the density of the silver deposit forming the photographic image by scraping or abrasion.

See also: *Retouching.*

PHYSIOGRAM. Pattern recorded photographically by a point of light attached to a swinging pendulum. The pattern possesses an unusual and delicate beauty and is capable of endless variations.

Making a Physiogram. This is not as complicated as the name might suggest. Beyond a camera, all that is needed is an electric torch without its reflector, and a piece of string to act as a pendulum. The physiogram is made by photographing the track of the torch bulb on the film as it swings on the end of the pendulum which is suspended in a darkened room.

The camera is laid on the floor with the lens pointing upwards so that the torch bulb is directly over the centre of the lens. A lens cap is placed over the lens and the shutter opened on a time exposure. The pendulum is set swinging and when it has steadied itself and the torch has stopped wobbling from the momentum of the swing, the lens cap is removed and the exposure is allowed to continue for a period depending on the effect aimed at.

Varying methods of suspension give rise to different patterns. If the pendulum is made with a single string, the result will be a pattern of ever-diminishing circles or ellipses. If a V shaped pendulum is used, the pendulum proper beginning at the point of the V, about a foot from the ceiling, the pendulum will swing in two movements, one diagonal to the other. Its second movement cuts directly across the first, and the third across the second, and so on until the pendulum comes to a standstill. The result will be either a square or oblong pattern.

If the pendulum is hung by three strings instead of two, the arcs of the pendulum will cut across each other three times instead of two, resulting in another type of physiogram. More strings can be added, each extra string adding variety to the pattern.

Another variation in the suspension of the pendulum can be made by tying the third string from the point of the V to the end of a piece of spring steel attached to the ceiling. This causes the action of the pendulum to last longer and introduces interesting variations to the swing. Other methods of suspension can be developed from the methods described. By giving the pendulum an oval or a circular swing, it will again vary the design of the resulting pattern.

The results of these patterns do not depend on the skill of the photographer. Novices and advanced workers have an equal chance.

With a fast film the aperture should be about *f* 11. The exact exposure can be calculated after several tests.

Colour Physiograms. When making physiograms in colour, the film should be of the artificial light type. The simplest way to produce colour is just to use a coloured torch bulb, but this does not take advantage of the range of colours of which the film is capable. Two or more bulbs of various colours can be suspended together to increase the colour range. Alternatively, multiple exposures can be made with single bulbs of different colours.

A greater range of colours can be had by using filters over the lens instead of coloured bulbs. A selection of filters of equal density can be held alternatively over the lens while the pendulum is swinging. A good idea is to make a graduated colour filter by painting the full spectrum with coloured dyes on a circle of blank sheet film; the colours must, however, be applied evenly. One side of the film is rotated over the lens during the exposure. This causes the pattern to record in continuously changing colours. L.S.

See also: *Abstract photography; Motion study; Tricks and effects.*

PIAZZI SMYTH, CHARLES, 1819–1900. Astronomer Royal of Scotland. Photographed the interior of the Great Pyramid.

See also: *Smyth, Charles Piazzi.*

PICTORIAL HISTORY. The development of pictorial photography has progressed along definable lines. In the earliest stages they were conditioned by technical limitations primarily, but later other factors also became significant.

See also: *Discovery of photography; Painting and photography; Pictorialism.*

PICTORIALISM. Method of photographic representation which aims primarily at aesthetic, emotional and intellectual effects. Design and tone rendering in a pictorial photograph are considered and controlled in such a way as to reinforce or create some appeal of their own which may or may not be inherent in the subject. Pictorialists are less concerned with the subject than with the picture it will yield; they use their cameras not so much to record whatever is there but to stress how they see. Thus pictorialism may be defined as a form of self-expression using the tool of photography with the object of creating images for the sake of enjoying both the process and the product.

In popular controversy the term pictorialism has come to stand for old-fashioned and sentimental conceptions of picture making. Some old-fashioned, sentimental photography may be pictorialism but not all pictorialism is old-fashioned and sentimental. The description pictorialism is too often used arbitrarily; people will never agree about it any more than they will about that other abstraction, "Art".

Past. What is now called pictorialism originated in the late nineteenth century, but the attempt to prove that photography was a medium capable of producing art started in the mid-1850s, and was then called High-Art photography. The term pictorial was chosen as a synonym for artistic photography. In these circumstances it was natural for pictorial photography to follow the subjects and the styles favoured by recognized forms of art as represented in the schools which were then most admired. This is why the most frequently cited pictorial photographs of the 1890's and the early 1900's sometimes resemble paintings, drawings and engravings of the impressionist era. To achieve this resemblance and at the same time broaden the limited tone rendering which the sensitized materials of their day would afford, pictorial photographers initiated and made use of a great variety of controlled printing processes, some of them so ingenious that their products eventually succeeded in approximating to the appearance of hand-made works of art and left little to reveal that they originated in a camera.

The interests and conventions of the early pictorialists remained to some extent preserved through exhibitions and salons which, again following the examples of older arts, were chosen as the chief media for drawing attention to the artistic aspirations of photographers. Their inclination to conserve traditional standards has been reinforced by an increasing isolation from newer trends whose exponents looked for alternative outlets to the public, e.g., in the growing number of illustrated journals. Thus there are frequent and mutual complaints that exhibiting pictorialists will only repeat effects that have degenerated into clichés and those who have anything new to show stubbornly refuse to exhibit.

Pictorialism, started as a controversial gesture, has remained controversial ever since. As it deals in intangible and elusive values it is always equally easy to attack or defend. In addition, its nature and function are often misrepresented by those who identify the term with ideas handed down by and confined to a definite period of photographic history. Yet essentially any photography that puts the picture first and the subject second—that is any photography for the sake of photography—is pictorialism, and that includes the most advanced trends of modern photography as well. Any photograph that neglects or plays down the identity of its subject in order to stress atmosphere, mood, pattern or viewpoint, stands for pictorialism irrespective of its style, period or ideology.

Present. Most present-day pictorialism favours close approach, angled viewpoints, dramatized lighting, emphatic tone rendering, simplified shapes and a variety of associated appeals: it is a blend of the traditions of the exhibition and the salon with modern ideas that have seeped in from illustrative and commercial photography. But trends like these never last long and do not hold the stage by themselves even while they last. What was taboo some years ago and seems only eccentric to-day may become popular next year and yet hopelessly old-fashioned a little later. As with music, the really popular ideas are quickly worked to death while the merit of highbrow experiments takes a long time to sort out and assess.

The dominant qualities of pictorial photography have always been technical perfectionism, deliberate composition and a conscious search for artistic expression. If efforts of this kind are to convey an element of spontaneity and leave a lasting impression they must be based on original vision, subtle craftmanship and sincerity. Lacking creative powers of a high enough order, the pictorialist will be tempted to repeat ideas, patterns and methods originated by others and thus produce clichés. But clichés are not confined to any trend. They simply become more and more frequent as any theme grows older and its scope for variation is gradually exhausted. The pictorialist of to-day would not dream of trying to express anything original in the romantic vein of the school acclaimed a hundred years ago and exemplified by Henry Peach Robinson. Neither is there anything more to add to the atmospheric moods conveyed by the rich prints of sixty years standing produced, for example, by Alexander Keighley, although a few pictorialists may continue to copy his style. The texture-conscious naturalism of thirty-five years ago, represented at its best by Edward Weston, is still modern to many and may occasionally yield some photographic glimpse which is new. What you may seek to pictorialize with your camera and how you do it very much depends upon when you came in to photography, and where.

Whatever trend he may follow the pictorialist is restricted by his self-chosen philosophy of recognizing the supremacy of manner over matter. But although the subject of photography is as endless as the world, its modes of expression are confined to specific and narrow techniques. Paradoxically pictorial progress directly depends on technical progress.

Scope. Photographic pictorialism being a means of self-expression needs to fulfil no other function beyond leaving the photographer contented with his work—particularly if he is an amateur. He is entitled to derive all the satisfaction he may be able to gain by using his camera to capture his own particular day-dreams irrespective of whether these add up to chocolate box prettiness, visions of social discord or surrealistic humour. It is also immaterial whether what he creates is of any interest to others besides himself. Approval or disapproval are of secondary importance to the pictorialist; his work does not depend on a public. The extensive and ecstatic vocabulary that has been used around pictorial photography is but a social game that gives the participants a chance to clothe their personal likes and dislikes in more or less profound statements. These, however much they may analyse the trends of the past or argue about the present, settle nothing for the future. A glance at any dogmatic assertion on pictorialism made twenty, thirty or forty years ago is enough to show how futile specific rules and theories are in this field. Pictorial photography remains free for all.

Photographic pictorialism as a contribution to progress, however, is effective only at the infrequent stages when from time to time new techniques yield novel methods of expression that may fertilize other arts or, more generally, may be borrowed by other photographers who care less about the tool they use and more about the things it can be used for. Sometimes, and by no means in its worst examples, pictorialism appears as sheer sublimated technique. Being so little concerned with subject matter and so much with modes of presentation, pictorialism is free to play virtuoso exercises throughout the whole range of photographic expression.

Pictorialism is photographic art for art's sake; mostly pleasant, sometimes admirable but hardly ever the end in itself it imagines itself to be. The image will always recall its original and any photograph, however successfully pictorial, leaves the onlooker wondering, in some measure, just who the person or where the place or what the thing is at which the lens was looking. No photographic subject will ever wholly submerge in photographic manner. The traditional pictorialist's affectation in suppressing the identity of the subjects behind some wistful, poetic or highbrow caption only interferes with the appreciation of his work and may even irritate the man who looks at it. The natural, popular and effectual photograph will always be the one that has been taken simply to show and to identify something or somebody. Pictorialism **is** a forced growth out of the main body of photography; a hot-house plant perhaps, but no less delightful for that. Being the product of the creative urge it has its own dignity as a human document. Even if it should not record the realities of its time it still does reflect on them by its very evasions. What a man may dream about is often more important than what he does or says when he is very much awake. A.K.-K.

See also: *Painting and photography.*

PICTURE SERIES. Number of pictures built around a single narrative or theme. Picture series have many forms, from such carefully studied and planned sequences as a documentation of a place to the random, spontaneous sequences such as a street riot would provide. Only a relatively few specialists have succeeded in mastering the artistic as well as the technical requirements of the photo-story.

Requirements. Some training in journalism is an asset in this branch of photography. There must be a strong and obvious story; each photograph must convey its own message clearly; captions are used only for information that cannot be expressed in the picture. The most successful picture series is one in which the photographs tell the greatest possible proportion of the story without assistance from the text, always providing that the subject has interest and the treatment is alive and exciting.

Construction. The photographer, with the main idea for a picture series in mind, collects all the relevant facts and looks for the natural sequence either in time or action. Then he looks for possible "highlight" pictures. These are essential to the interest of the series and should in fact be good enough to stand alone. One of these must be an outstandingly good shot which concentrates the whole idea into one picture either for use on the magazine cover, or to dominate the layout of the series.

Presentation. The photographer must also work to produce an interesting page layout, and for this reason he should know something of magazine make-up. A series is apt to look uninteresting if all the shots are taken from the same distance and angle. So close-up and long shots are combined to add variety. Absolute familiarity with the equipment and technical routine is essential so that the photographer is free to give his attention to the story content. And the final series must be more than a mere factual record; it must carry the stamp of the observer's personality in its selection of the critical instants and in its appreciation of the dramatic possibilities of the subject. This essential personal quality comes through more clearly when the picture shows only the significant thing and leaves out everything else.

Selection. A picture series is seldom a record of a single sequence of events made in six or eight consecutive exposures. The action may have to be repeated many times before the photographer gets all the pictures he wants. He takes many more steps in the sequence than he is likely to need, and sometimes photographs each step from a number of different angles and positions.

Most magazine picture stories of about half a dozen pictures are the result of making upwards of a hundred separate exposures and selecting those which express the idea most clearly, and make the most arresting pictures.

Camera. The range of the equipment governs the subjects that can be tackled. A photographer who wishes to specialize in this branch of photography should be able to command a first-class miniature with wide-angle and long-focus lenses, at least one lens with an aperture of $f2$ or larger, and a twin lens reflex. The miniature camera would ideally be a single-lens reflex, and at least one of the cameras should be fully synchronized for flash.

In addition, a 5 × 4 ins. or 9 × 12 cm. camera for colour is almost essential to anyone working for the magazines.

Lighting. Wherever possible the existing lighting is used so that the subject and its surroundings will look natural. If the existing lighting does not allow for the necessary shutter speed and depth of field, then Photoflood lighting is commonly used. This is satisfactory so long as there is no fast action, and it is more economical than the use of flash bulbs where a large number of exposures has to be made. Where the speed of the action or an outdoor night scene rules out Photoflood lighting, then flash must be used. A bulb above and to the side of the camera sight line and a fill-in flash on the camera give a sufficiently natural effect.

Electronic flash is the most satisfactory lighting for very fast action pictures in bad light.

Ideally all technical manipulations and considerations should become almost second nature, allowing the photographer to concentrate solely on his story. M.A.

See also: *Freelance photography; Magazine photography; Photo journalism; Press photography.*

PIEZO-ELECTRIC FLASH. Flash bulb fired by the current derived from a mechanical shock or stress on certain types of crystals. The so-called piezo-electric current produced in this way (similar in principle to the current produced by a crystal microphone) is very small, and can fire only flash bulbs specially designed for this purpose. A piezo-electric system of flash synchronization has, however, the advantage that it is very compact, and does not have to rely on batteries.

See also: *Flash bulbs.*

PIGEON POST. During the Siege of Paris 1870–71, greatly reduced photographs of printed messages on collodion pellicles $1\frac{3}{4} \times 2\frac{3}{8}$ ins. were carried by pigeons from the headquarters of the Government to the besieged capital. There they were projected, copied by clerks and delivered to the addressees. Eventually the messages were printed on photographic paper. This pigeon post was a forerunner of the airgraphs of World War II.

See also: *Dagron.*

PIGMENT PROCESSES. Methods of making positive prints by exploiting the various physical changes produced in a bichromated colloid by the action of light. Certain organic colloids—e.g., albumen, casein, gelatin, gum arabic and shellac—and a few cellulose and polyvinyl esters, sensitized with a bichromate—e.g., of potassium, ammonium or pyridine—change their physical character when exposed to light.

Three principal changes may take place in the colloid: if previously soluble it may become insoluble; it may no longer absorb water and swell up; it may lose its surface tackiness.

The principal processes based on the change in solubility of the colloid are Artigue's process, and the carbon, carbro, Fresson, gelatin-sugar and gum bichromate processes. Here the broad principle is that a pigmented layer of bichromated colloid is exposed under a negative and subsequently developed by washing with hot water which dissolves the unexposed colloid and leaves a positive image consisting of pigment suspended in the exposed and insoluble colloid.

The principal processes based on the swelling of the colloid are bromoil, oil printing and dye imbibition. In these the layer of colloid after exposure under a negative is bathed with water so that the unexposed areas swell and become water-logged. Greasy ink which is then applied with a brush is rejected by the swollen areas and accepted by the exposed unswollen colloid to form an image. Various dye imbibition processes are based on the permeability of the colloid to an aqueous solution. In these, the colloid layer after exposure is bathed in a dye solution which it absorbs in the unexposed areas to give a positive-from-positive image.

The principal process based on loss of tackiness is the dusting-on or powder process. In this a bichromated colloid—e.g., gum arabic—is exposed and "developed" by being lightly brushed over with a finely powdered pigment. Where the light has acted, the surface loses its tackiness and the pigment does not adhere. So an image is formed on the tacky, unexposed areas.

The process gives a positive image from a positive transparency. A positive can be printed from a negative by coating the colloid on a dark support and developing it with a light powder.

See also: *Control processes; Printing processes (obsolete).*

PINACHROME PROCESS. An obsolete colour printing process based upon the use of leuco bases which oxidize on exposure to light, yielding colour images. Three such colour images, derived for instance from separation negatives, can be assembled by superposition.

PINACRYPTOL. Used in desensitizers.

Characteristics: Green (pinacryptol green) and yellow (pinacryptol yellow) dye powders.

Solubility: Slightly soluble in water. Soluble in alcohol.

PINATYPE PROCESS. A method for producing dye imbibition colour photographic prints based on the differential staining action of certain dyes for hard and soft gelatin.

See also: *Printing processes (obsolete).*

PINCUSHION DISTORTION. Particular kind of optical distortion of the image formed by an uncorrected lens. Straight, parallel lines at the edge of the field curve towards the lens axis being closer at their middle than at the ends.

See also: *Aberrations of lenses.*

PINHOLE CAMERA. Photographic camera without a lens but instead with a very small, sharp-edged hole in an opaque diaphragm. Such a hole forms an image at the back of the camera, varying in scale and perspective according to the distance between the diaphragm and the sensitive surface. If the pinhole is close to the plate it gives a wide-angle lens effect; if it is far away it gives a picture equivalent to a long-focus lens.

Image Quality. The quality of the pinhole image depends on the size and sharpness of the hole. Up to a point, the quality improves as the hole is made smaller, but beyond a certain size it deteriorates because diffraction becomes more pronounced.

The definition produced by the pinhole camera is determined by the size of the image patch corresponding to a point object. The diameter of the image patch formed by rays of light from a point source is given by

$$D = \frac{d(u + v)}{u}$$

where D = diameter of the image patch,
d = diameter of the pinhole
where u = distance of the object from the pinhole,
v = distance of the image from the pinhole.

So, if u is large compared with v, even a large pinhole will give a reasonably sharp image. For instance, a keyhole will give a recognizable picture in a darkened room of buildings more than one hundred feet away.

Size of Hole. There is an optimum—though by no means critical—size of hole for any particular hole-to-plate distance. While a lot of theoretical work has been done on this subject and many formulae have been published, for practical purposes it is sufficient to approximate very roughly. A pinhole of about 1/64 in. diameter is satisfactory for normal photography with separations up to 6 ins., and 1/100 in. will give reasonable definition for copying and photographing small objects on a scale of 1 : 1.

Making the Hole. The pinhole itself is worth taking trouble over. It should be a clean, sharp-edged hole with as little depth as possible.

Take a piece of brass about 1/32 in. thick and cut it to form a blank to fit in place of the camera lens. Tap a dimple in the centre of the blank with a blunt-ended centre punch and rub the point of the dimple down with a dead smooth file to form a thin place.

Pierce the thin spot with a sewing needle of suitable diameter, working from the hollow side of the dimple. Carefully rub off the burr with a fine flat carborundum stone and pass the needle through the hole again. Rub off until the hole is seen to be clean and sharp when looked at through a watchmaker's glass. Finally blacken the metal and the pinhole will be ready for use.

Exposures. The exposure depends upon the size of the pinhole and its distance from the plate—in other words, its *f*-value. And the *f*-value can be calculated exactly as for a lens, although it will vary for each hole-to-plate distance. A 1/64 in. pinhole working at 6 ins. from the plate, for instance, will have an *f*-value of

$$\frac{6 \times 64}{1} = f384.$$

Under these conditions, when the light would normally call for an exposure of, say, 1/25 at *f* 11, the exposure time would be 49 seconds.

As there is no need for accuracy with such long exposures, a safe working rule with this pinhole and distance would be to reckon at the rate of one minute of exposure for every 1/25 second of exposure at *f* 8 given by the meter. In the same way a rule of thumb can be worked out for any hole and distance by making one representative calculation.

The Photograph. Pinhole photographs have an all-over softness of definition which is substantially the same for all objects in front of the camera. There is no distortion, and the depth of field extends from the front of the camera to the horizon. Finally the photographer has absolute freedom to vary the perspective of his subject and the scale of reproduction by altering the relative distances of the plate and the subject from the pinhole.

The disadvantages of the pinhole photograph are mainly the lack of crisp definition, the length of the exposure, and the alteration of the "*f*-value" with every change of plate distance. F.P.

PINHOLES. Clear, small spots in the negative caused generally by specks of dust on the emulsion surface at the time of exposure.

See also: *Faults; Spots.*

PIN-UP. (Slang.) Glamorized photograph of a motion picture or stage beauty almost always taken in a bathing suit or lingerie. So called because this type of picture is commonly given away in the form of an insert in a popular magazine so that it can be pinned up as a decoration.

PIPER, C. WELBORNE, 1866–1919. English photographer. Introduced the bromoil process in 1907 as a result of a suggestion by E. J. Wall.

PIX. (Slang.) Abbreviated phonetic for pictures, coined originally—and still mainly employed—in America for its convenience in making up newspaper headlines.

PIZZIGHELLI, GIUSEPPE, 1849–1912. Austrian officer and photographic scientist of Italian origin. From 1878–1884 head of the photographic department of the Army, Vienna. Later president of the Società Fotografica Italiana (Italian Photographic Society). Collaborated with J. M. Eder, notably in the production of silver chloride ammonia emulsions for development papers, 1880–1; this antedated Baekeland's commercial introduction of Velox paper by 12 years. Improved, with von Hubl, Willis' Platinotype process and discovered in 1887 the platinum printing-out process in which no further development is required after printing.

PLANOGRAPHIC PROCESSES. Printing processes in which the printing and non-printing areas of the plate are all at the same level (e.g., in lithography).

See also: *Photomechanical reproduction.*

PLANTS. Photography is used extensively by naturalists and botanists for recording plant growth and characteristics both in the field and in the laboratory.

See also: *Biology; Botany; Flowers; Garden.*

PLASTICS. Organic compounds usually formed by the chemical reaction known as polymerization which combines comparatively simple molecules into long chain structures. The resulting substances are comparatively tough (though not necessarily hard) solids of fairly good strength, chemical resistance and easy workability. The latter property makes them particularly valuable for the manufacture for a wide range of objects and articles of all kinds.

Polymerization can often be induced by light, or a polymer may be cross-linked by a light-catalyzed reaction and these effects are of importance in photography. Further, the physical properties of certain plastics make them suitable for use both as film supports and as material for the manufacture of processing and other equipment.

Photosensitive Plastics. A high polymer is said to be light-sensitive if, after exposure to a suitable light source, the exposed area of a film or coating of the polymer is distinguishable from the unexposed area, either with or without some development step. Among early polymeric materials used in this way was bitumen, which was employed in the original photographic process of Nièpce (1829). Gelatin and albumen sensitized with dichromates have long found application in screen printing and photomechanical work, though they are not strictly speaking plastics.

Synthetic and natural polymers can be made light-sensitive essentially in two ways: (a) the light-sensitive component can be an integral part of the polymer molecule and (b) a substance which may react with a polymer molecule or may photopolymerize itself on exposure to light, in the presence of a light sensitive activator, can be added to the normally non-light-sensitive polymer.

Variations of (b) are known where the addition of only a small amount of a light-sensitive material will bring about a light-induced chemical reaction in the polymer.

Among group (a) systems, polymers containing anthracene or cinnamate residues have been prominent. When a coating containing a polymer having such groups built into the molecule is exposed to light, some of the cinnamate or anthracene groups in one molecule will combine with a similar group in another molecule, producing a change in the physical properties of the molecule. The image can be developed by washing out the unexposed areas with a suitable solvent in which the exposed areas are insoluble. Such an image can be dyed.

The polymers are usually sensitive to ultraviolet rays only, but they can be spectrally sensitized.

Several processes falling into type (b) are known. A doubly ethylenically-unsaturated compound may be added to a large variety of plastic materials ranging from nylon to polyesters and cellulose derivatives. In this case, the further addition of a light-sensitive activator is required. Diesters of acrylic acid and various glycols have been suggested as the unsaturated material and ketones such as biacetyl, benzoin and benzil are the activators. On exposure to light, the unsaturated compounds crosslink the plastic and render it soluble in a given solvent system.

In other processes azides have been used as light-sensitive crosslinking agents. Here also a large variety of high polymers have been used, including proteins, e.g., gelatin and casein, and materials like nylon, rubber and rubber derivatives.

The applications of the various light-sensitive plastic materials are increasing rapidly. They are especially useful in the electronics industry, in the production of printed circuits and various macro- and micro-solid state devices, in the manufacture of name-plates, in chemical milling and etching of engineering components, the production of decorative designs, decorative ceramics, tricolour television phosphor screens and in the engraving of textile-printing rollers. Electrodeposition into a photographic resist image makes use of the high dielectric value of the photosensitive plastic. Letterpress printing plates made of light-sensitive plastic materials have appeared on the market.

Plastic materials can also be used in electrostatic processes similar to xerography. Polymers containing photoconductive groups have been patented. Among the most prominent of these have been polymers containing carbazole and polynuclear hydrocarbons. H.M.W.

Film Bases. Virtually all transparent emulsion supports other than glass are polymer plastics of one kind or another. Before the turn of the century, cellulose nitrate was used as the first flexible transparent film base. This material had many excellent mechanical properties although it was extremely flammable and was not chemically stable.

Nevertheless, it continued to be used as late as the 1950's.

To overcome the poor flammability and stability characteristics of cellulose nitrate, cellulose diacetate was introduced in the early 1920's. However, this material did not have as good mechanical properties as cellulose nitrate and is not widely used today. In the 1930's, film bases composed of mixed cellulose esters were introduced. These cellulose acetate propionate and cellulose acetate buyrate bases had improved physical properties and are currently used for many sheet and amateur films. During this period, cellulose triacetate base was developed and this came into widespread use after World War II. It has excellent strength, toughness, and burning properties and is most desirable for professional motion-picture films. It is currently being used for the majority of photographic films.

Cellulosic film supports do not have the dimensional stability properties desired in some applications, primarily in the graphic arts industry. For this purpose, rigid polyvinyl chloride-acetate base was introduced in the 1940's. While it has low moisture adsorption, it was translucent and had a low softening temperature. It never attained widespread use. In 1954, a biaxially-oriented polystyrene base was marketed and it had wide acceptance because of its excellent moisture resistance. Its mechanical properties were not high and it showed craze marks if kinked. It has been supplanted to a considerable extent by biascially-oriented polyethyene terephthalate base. which is a polyester. This material has excellent dimensional stability combined with mechanical strength which is superior to any other film base. For this reason it has found use not only as a base for graphic arts films, but also for aerial and industrial films. Another widely used polyester type film base is polycarbonate with good moisture resistance. It has been used for graphic arts and for aerial films.

Processing Equipment. The use of plastics in processing equipment is dependent upon both their chemical and mechanical suitability. The former includes both the effects that the processing solutions may have on the plastic such as swelling or plasticizer leaching, as well as any photographic activity of the plastic.

Polyacrylonitrile-butadiene-styrene, cellulose acetate butyrate, polyethylene, polypropylene, and rigid polyvinyl chloride find application as fittings, piping, trays, paddles and tanks. The fluorocarbon polymers are found in pumps, gaskets, packings, bearings and small parts. Nylon has been suitable for gears and bearings, and polystyrene, polyoxymethylene and reinforced phenol-formaldehyde resins for small parts. Epoxys and urethanes are also coated onto metals and employed for film hangers, racks, idler rolls and tanks. Large tanks and other equipment are also constructed from fibreglass reinforced epoxys and polyesters.

This review of plastic usage is a generalization based on trade experience. However, there are considerable variations within any one plastic as to its suitability, as well as wide variations between different photographic processes. P.Z.A.

See also: *Photo-etching; Resist; Supports for emulsions; Tanks.*

PLATE. Photographic material formed by coating a sensitized emulsion on to a sheet of glass. Plates are carried in wood or metal plate holders fitted with a removable draw slide which shields the sensitive surface from the light until the plate is to be exposed in the camera.

See also: *Negative materials; Supports for emulsions; Transparency materials.*

PLATEAU, JOSEPH ANTOINE FERDINAND, 1801–83. Belgian physicist, specializing in optics. Discovered in 1829 the principle of the "wheel of life" which he constructed in 1832 under the name Phenakistiscope; it was discovered simultaneously by Stampfer in Vienna who called his instrument the Stroboscope. Proposed the use of instantaneous photographs (instead of hand drawings of motion series) in 1849; this was attempted in 1851 by Claudet and Duboscq, and later by Uchatius, Muybridge, Marey, Demeny and others.

PLATE CAMERA. Camera designed primarily to take photographs on coated glass plates. The back of a plate camera is equipped to take either the plate in its light-tight holder or sheath, or a focusing screen. The ground surface of the screen and the surface of the plate lie in the same plane when they are in position. The photographer focuses and arranges the subject on the screen and then substitutes the plate in its holder. So long as the subject or camera does not move in the meantime, the image on the plate on exposure will be identical.

Plate cameras are particularly suitable for making single exposures that can be removed and processed at once, unlike roll films which must be all exposed before any can be seen. Plate cameras are generally larger than film cameras and tend to be used mostly by professional photographers.

Although the plate camera is, traditionally, an old type of camera, there are many modern versions designed for specific purposes and the use of sheet film material is today more common than the use of plates in such cameras. It has decided advantages, particularly where great accuracy is important.

Some of the smaller and miniature roll and 35 mm. cameras can be fitted with plate backs to enable them to take single exposures.

See also: *Field camera; Press camera; Technical camera.*

PLATE HOLDER. The sensitized plates used in plate cameras are held in readiness for use in light-tight containers called plate holders. A plate holder is simply a rectangular frame in which the plate is kept in place by some form of clip or slot. The back of the frame is solid and the front is open, but light is normally prevented from reaching the plate by an opaque dark slide. The slide can be withdrawn when the holder is in position on the camera.

The plate holder is designed to fit on the back of the camera in place of the focusing screen so that the sensitized surface lies in the same plane as the ground glass surface of the focusing screen.

There are various ways of attaching the plate holder to the back of the camera. The simplest—and worst—method is for the back of the camera to be grooved to allow either the focusing screen or a plate holder to be slid into place. With this method there is always a risk that the camera will be moved out of adjustment while withdrawing the focusing screen or inserting the plate holder. There is much less risk of movement when the plate holder is designed to clip into position with only a limited sliding action or none at all.

Plate holders are made in single and double form and in either metal or wood.

Single Holders. Single metal holders are supplied mainly with hand cameras and in formats no larger than quarter plate. The draw slide of the single metal holder is a metal plate which is so thin that it can be removed completely to make the exposure without allowing light to leak past the velvet light trap.

Plates are kept in position in the single metal type of holder by a leaf spring along the bottom groove which presses the plate upwards so that the top edge lodges under a pair of metal lugs. To load or unload a plate, it is simply pressed down against the spring pressure until the top edge clears the lugs after which it can be inserted or removed at once.

Double Book Form Holders. This type of holder is mostly made of wood and used to be popular with professional workers, particularly for formats of half-plate and larger. The holder in this case is really two plate holders assembled back to back, hinged together at the bottom and held closed by one or more clips along the edges. Each side of the holder consists of a wooden draw slide and a thin wooden frame with its inner edges rebated to take the plate. A blackened sheet metal separator is held in place by tabs on one side. The separator carries a leaf spring that presses the plates firmly into their frames so that all parts of the plate being exposed are in the focal plane of the lens.

A wooden draw slide for this type of holder has to be about 1/16 in. thick, and if it were completely withdrawn with the holder in place on the back of the camera, the slit left would be too wide for the velvet light trap to close. So the slide is prevented from coming right out by a stop. As it is inconvenient to have a slide sticking out from the camera during the exposure, slides of this type—particularly the larger sizes—usually have a flexible section at the bottom which allows the slide to be folded out of harm's way against the back of the camera. The flexible section consists of strips of wood glued on each side of a width of opaque fabric.

Loading. To load a double book form plate holder, the clips are released and the holder is opened, book fashion. The clips holding the separator are released and a plate is placed in the frame on the same side, emulsion down. The separator is clipped down over the back of the plate.

A second plate is now placed emulsion down in the frame on the other half of the holder. The first half of the holder, with its plate secured by the separator, is then closed down on the plate lying in the open frame, and the two halves are fastened together with the side clips. The holder is now loaded ready for use.

Another widely used type of double dark slide has a cover that fits along the bottom edge of the holder. When this cover is removed the plates can be slipped into position behind the dark slides. A slight variation of this type has at the bottom a separate hinged cover for each plate, which can be opened independently when the appropriate slide is partially withdrawn.

Single metal plate holders are loaded by pulling out the draw slide, pulling back the spring at the top edge, dropping in the plate (emulsion side up) and releasing the spring so as to hold the plate in place.

Plate Identification. Plate holders are usually numbered serially so that the photographer can keep a running record of the exposed plates. A white ivorine panel is sometimes inlaid on wooden slide holders so that brief memoranda may be written on in pencil and wiped off when the plate is changed. Metal plate holders usually have the number embossed on the top of the draw slide.

Safety to Actinic Light. Plate holders are normally kept away from the light as much as possible and are usually kept in their carrying case with the lid closed. Although they are safe to a certain amount of light, there is always a risk of some light leaking through the light trap or being reflected around the edge of a loosely fitting draw slide. When on the camera, the holders should be covered by the focusing cloth, especially when working outdoors in bright light.

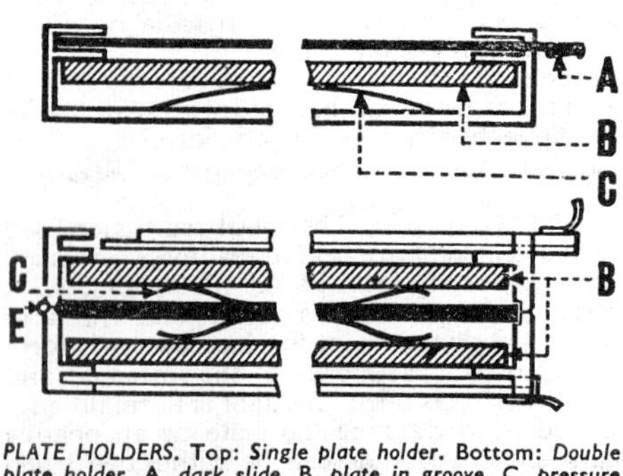

PLATE HOLDERS. Top: Single plate holder. Bottom: Double plate holder. A, dark slide. B, plate in groove. C, pressure spring. E, hinge; opens for loading.

Metal draw slides are safe against most kinds of radiation (except X-rays) but wooden, plastic, and xylonite slides are more or less transparent to infra-red light. For this reason, infra-red sensitized plates must be used only in all-metal holders. There is even some risk from the deep red rays in sunlight if panchromatic plates are left lying about in the open sun in wooden or plastic holders.

Cleanliness. As far as possible, plate holders should be kept away from dust; this traditional source of trouble is always easier to keep out than get out. If the case they are carried in is regularly brushed out and occasionally blown out, and the holders are wrapped in a clean cloth or a square of soft plastic when they are out of the case, they should come to no harm. At the same time, before every loading they should be blown or brushed out with the dark slides withdrawn. The special miniature hand blowers sold for cleaning cameras are useful for this job because they can be used to blow a jet of air right into the grooves and crevices of the holder.

The velvet along the edges of the light trap should be brushed up and the blower used to drive any dust particles out of the pile.

Before loading a plate, a wise precaution is to tap the edge of the plate gently against the work-bench. This removes both dust, and also any static electricity on the plate, which would attract dust.

All these precautions take extra time, but nothing like as much as is wasted in spotting the negative and retouching the print because dust was present during exposure. F.P.

See also: *Film transport; Mackenzie-Wishart envelopes.*

PLATE MARKING. Area embossed on a print or its mount. It imitates the depression of the picture below the border level produced by the pressure of the engraved or etched plate in certain types of art reproduction.

See also: *Embossing prints.*

PLATE RACK. Wooden rack used to hold plates on edge while drying, and to permit complete drainage and free access of air from all sides.

PLATINOTYPE. Contact printing process invented by W. Willis (1873), in which the image was formed of platinum. Due to the high cost of platinum it was succeeded by the palladiotype, which used palladium in place of platinum.

See also: *Printing processes (obsolete).*

PLATINUM CHLORIDE. Platinic chloride; platinum perchloride. Used in certain toners and sensitizers.

Formula and molecular weight: $PtCl_4.5H_2O$; 427.

Characteristics: Red crystals giving a yellow solution.

Solubility: Highly soluble in water.

PLAYERTYPE. Direct copy of a line original made by a now obsolete process invented in 1896 by J. Hart Player.

The original is placed face upwards under a printing light, and a sheet of sensitized paper is placed over it and exposed in contact with it, emulsion side down. A negative image is produced on developing the sensitized paper and this is used to give any number of positive prints by normal contact printing. This process is now used for copying documents (reflex copying).

See also: *Document copying.*

POINT SOURCE LAMP. Special type of arc lamp. In it the arc is struck between tungsten beads sealed into a glass envelope from which the air has been extracted. A small quantity of mercury in the bulb is heated when the lamp is switched on and forms a conducting vapour which carries the arc current. The arc heats up the tungsten until it becomes incandescent and emits a light similar in colour to a tungsten filament lamp but from a concentrated source.

Point source (e.g., Pointolite) lamps are used in projectors, spotlights, and enlargers. The apparatus must incorporate a resistor or transformer to provide a 10 volt supply to the arc.

The efficiency and power of this type of arc can be greatly increased by raising the vapour pressure and operating it at a higher current. The light source is similar to Photofloods in application.

There are various other types of concentrated arc lamps in which metallic electrodes are used instead of carbons—e.g., the zirconium concentrated arc in which the light is emitted by a thin film of molten zirconium. The outstanding characteristic of the zirconium lamp

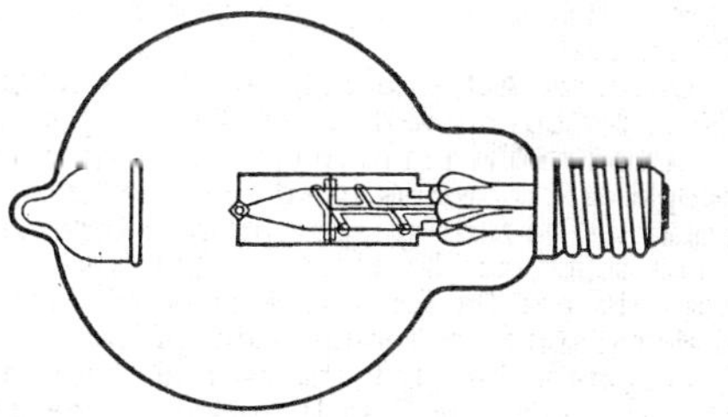

POINT SOURCE LAMP. An arc is struck (either by means of a heated filament, or by a high-voltage pulse) between two beads of tungsten in an atmosphere of mercury vapour. The bulk of the light emanates from the positive electrode, while the mercury contributes little to the colour of the light. As this is produced by incandescent tungsten, it resembles that of a filament lamp in colour. The lamps are made in ratings from 30 to 1,000 candle powers for D.C. operation, and in a more limited range for A.C. use. The voltage drop across the electrodes is only about 10 volts; the optical efficiency due to its near-point source is thus complemented by a high electrical efficiency obtained through the high current density.

is the very small source from as little as 0·003 in. (0·085 mm.) for the 2 watt size to 0·059 in. (1·5 mm.) for the 100 watt size.

Modern low voltage projection lamps frequently have a very concentrated filament which acts virtually like a point source. Lamps of this kind are frequently used with a built-in (proximity) reflector or a special shaped, internally silvered, glass envelope that acts as a reflector. Such a reflector is thus very small, a development only possible with a virtual point source of light.

See also: *Carbon arc lamp; Discharge lamp; Zirconium lamp.*

POISONS

Various chemicals used in photography need special care in handling owing to their poisonous nature. The sale of some of these is governed by law. Substances controlled by the Pharmacy and Poisons Act, 1933, are set out in the Poisons List and are the only poisons for the purpose of the Act. There are other substances used in photographic processes which are either nauseous or which have some element of danger, but which are not officially classed as poisons.

Until 1851 there was no effective legislative control over the sale of poisons in Great Britain, and statistics showed an increasing number of deaths from poisoning. In 1865 a select committee reported that there was need for a bill to provide that only registered persons should sell certain poisons and it was from this beginning that the Pharmacy and Poisons Act, 1933, evolved.

Categories of Poisons. The Poison List is divided into two parts:

Part I—sold only by chemists (authorized seller of poisons).

Part II—sold by authorized seller of poisons and listed sellers of poisons (this list is kept by the local authority).

Poisons are also grouped into schedules, each schedule having specific regulations which are governed by the rules of the Act, but which do not cover photographic use, except in so far as they are referred to below. The first schedule covers the most dangerous poisons in the Poison List.

Under the Act, local authorities must keep a list of persons allowed to sell Part II poisons including particulars of the premises, and these people are known as listed sellers of Part II poisons. The form of application to become a listed seller can be obtained from the local authority and the fee payable is 7s. 6d. for the first year and 5s. per annum subsequently.

Poisons in Part I of the list and those substances containing Part II poisons in the first schedule of the rules can only be sold on registered premises and by or under the supervision of a qualified pharmacist. Other poisons in Part II of the list may be sold by listed sellers.

Part II poisons which are included in the first schedule may be sold by listed sellers only in the form specified against each in the fifth schedule.

The following three groups of photographic chemicals show the position of the photographic dealer in regard to the selling of chemicals under the regulations.

Group I. Poisons which may be sold by the photographic dealer when he has become a listed seller of poisons, are:

Ammonia (liquid).
Hydrochloric acid.
Nitric acid.
Sulphuric acid.
Potassium quadroxalate (Salts of Sorrel, Salts of Lemon).

The following may only be sold in the closed containers in which they are received by the dealer:

Hydrofluoric acid.
Formaldehyde.
Potassium hydroxide (Pot. caustic).
Packed toners containing metallic oxalates.
Sodium hydroxide (soda caustic).

Group II. Chemicals and solutions which can be sold by the photographic dealer without his becoming a listed seller are:

Ammonium oxalate.
Ammonium sulphocyanide.
Paraphenylenediamine and its salts for photographic purposes.
Potassium ferricyanide (red prussiate of potash).
Solutions of ammonia containing less than 5 per cent weight in weight (w/w) of ammonia (NH_3).
Solutions of formaldehyde containing less than 5 per cent w/w formaldehyde. (Photographic glazing and hardening solutions containing formaldehyde have been specially exempted from the provisions of the Act.)
Solutions of hydrochloric acid containing less than 9 per cent w/w of hydrogen chloride (HCl).
Solutions of nitric acid containing less than 9 per cent w/w nitric acid (HNO_3).
Solutions of potassium hydroxide or sodium hydroxide containing less than 12 per cent of potassium or sodium hydroxide respectively. (These cover concentrated developers of the paraminophenol and contrast type.)
Solutions of sulphuric acid containing less than 9 per cent w/w of sulphuric acid (H_2SO_4).

Group III. Poisons which the photographic dealer cannot sell unless he is a registered pharmacist (i.e., authorised seller), include:

Barium sulphide.
Cyanides.
Hydrocyanic acid.

POISONS AND ANTIDOTES

Types of Poison	Description	Symptoms of Poisoning	Antidotes
Alkalis	Potassium hydroxide (Pot. caustic) Sodium hydroxide (Soda caustic) Liq. ammonia	Swelling of tongue and mouth	0·6 per cent solution of glacial acetic acid in water (dilute vinegar will do in an emergency)
*Iodine		Tightness about the throat	Ample emetic, plenty of arrowroot
Lead acetate and nitrate		Constriction in the throat and pit of stomach and stiffness in abdomen	Sodium sulphate or Epsom salts, emetic of zinc sulphate
Mercuric chloride		Acrid metallic taste, constriction and burning in throat and stomach. Nausea and vomiting	White and yolk of raw eggs mixed with milk. In emergency ordinary flour paste may be used
Mineral acids	Hydrochloric, nitric, sulphuric	Corrosion of windpipe and violent inflammation	Sodium bicarbonate or magnesium carbonate, or chalk; even the plaster of the room beaten in water
*Potassium bichromate		Irritant pain in stomach	Emetics and magnesia or chalk
Potassium cyanide		Insensibility, slow gasping respiration, dilated pupils, spasmodic closure of the jaws	No certain remedy, try mixture in water of equal parts of ferrous sulphate and sodium carbonate. Hot and cold douches
Potassium oxalate and oxalic acid		Hot burning sensation in throat and stomach, vomiting, cramps and numbness	Chalk, whiting or magnesia suspended in water. Plaster or mortar may be used in emergency
*Pyrogallic acid		Violent vomiting, collapse, fall in temperature	Emetic, plenty of air or oxygen, hot water on hands and feet

*These three substances are not poisons within the meaning of the Act, but are usually included in lists such as this where antidotes are given.

Lead acetate and nitrate.
Mercuric chloride (mercury bichloride, corrosive sublimate.)
Mercuric iodide.
Oxalic acid.
Para-nitrophenyl acetonitrile.
Potassium oxalate.
Schlippe's Salts (sodium sulphoantimoniate).

Buying Poisons. Poisons of Groups I and II can be bought and sold over the counter in the normal way. Special requirements apply to substances in the first schedule (which includes the Group III poisons listed above).

A substance in the first schedule can only be sold to a purchaser who is known by the seller (or by a pharmacist in the employ of the seller) to be a person to whom that particular poison may properly be sold. A person who wishes to buy a substance in the first schedule from a chemist, and who is not known to the chemist, may be supplied if he presents an appropriate certificate. This must be signed by a householder who is known to the chemist to be a "responsible person of good character". If the purchaser cannot obtain a certificate from a householder known to the chemist, he can obtain one from another householder and get the police of the district to endorse the certificate that the householder is known to the police to be a responsible person of good character.

When a person requires a substance in the first schedule for other purposes than his trade or business, he must also be known to the seller or be certified in the manner described. In addition, he must, before the poison is delivered, affix his signature to an entry made by the seller in the Poison Register.

If a purchaser requires a substance in the first schedule for the purpose of his trade, business or profession only, he may be supplied subject to the following conditions:

(1) The seller must receive before the sale an order in writing giving the name and address and trade, business or profession of the purchaser; the name and quantity of the article and the purpose for which it is required.

(2) The order must be signed by the purchaser, and the seller must be reasonably satisfied that the signature is his and that he carries on the trade, business or profession stated, which is one in which the poison is used.

(3) If the article is sent by post, it must be registered.

(4) The order must be kept by the supplier and numbered, the particulars must be entered in the Poisons Register and in the last column, the words "signed order" must be written, together with the number of the order.

Labelling. The provisions of the Act and rules relating to the labelling of poisons apply with limited exceptions to all transactions involving substances containing poisons.

The label of a poison or of a substance containing a poison must bear the following particulars:

(1) The name of the poison: if the poison is named in the Poisons List, the name in the list must be given. If the poison is described in the list by a group name and is not itself named, the name used must be the name or synonym or abbreviated name used to describe the poison in *The British Pharmacopoeia* or in the *British Pharmaceutical Codex*, or in any supplement to either, followed by B.P. or B.P.C. If the poison is described in neither of these books, the accepted scientific name or name descriptive of the true nature and origin of the poison must be used. If the poison is contained in a preparation in the *Pharmacopoeia* or in the formulary of the *Codex*, the name, synonym or abbreviated name of the preparation followed by the letters B.P. or B.P.C. may be given instead of the name of the poison.

(2) The proportion of the poison to the other ingredients: if the substance containing the poison is a preparation of the *Pharmacopoeia* or of the *Codex*, the proportion of the poison need not be stated if the preparation is named by reference to its *Pharmacopoeia* or *Codex* name. The proportion of the poison present in other substances may be stated in any convenient manner. If it is stated as a percentage, there must be an indication whether the percentage is calculated on a basis of weight-in-weight (w/w), weight-in-volume (w/v), volume-in-volume (v/v), or volume-in-weight (v/w). Where a substance contains among its ingredients a preparation of the *Pharmacopoeia* or of the *Codex* which contains a poison, it is not necessary to state on the label the proportion of the poison to the other ingredients if the proportion of the preparation is declared.

(3) The word "poison" or the prescribed indication of character: poisons and substances containing poisons must be labelled with the word "poison" or with the other prescribed indication of character. It is permissible to label substances not in the Poisons List with the word "poison" but it is undesirable that such a practice be freely adopted.

Substances included in the first schedule must be labelled with the word "poison" or other prescribed indication of character either in red letters or set against a red background. The words must be either on a separate label or surrounded by a line. Within this line there must be no other words except those words (name of poison, proportion of poison, and name and address of seller, or the words "Not to be taken" or "For external use only") with which the container of the poison is required by the rules to be labelled.

(4) The name of the seller and the address of the premises: the container of a poison must be labelled with the name of the seller and the address of the premises on which it was sold. To this general rule there are two exceptions in favour of wholesale distributors.

Firstly, when an article is sold by a wholesaler for the purpose of being sold again in the same container (i.e., when it is sold to a retailer) it is not necessary for the wholesaler's name and address to appear on the label.

Secondly, when a poison is supplied from a warehouse or depot it is not necessary to state the address of the warehouse or depot. The container may be labelled with the address of the supplier's principal place of business, or with the address of the registered office of a company if the supplier is incorporated. When such a poison is subsequently retailed to the public it must be labelled with the address of the premises on which it was sold.

The Isle of Man and the Channel Islands are included in Great Britain and poisons supplied to these places are, therefore, not exported. They are labelled in accordance with the Poisons and Pharmacy Act of Great Britain.

Containers. All containers of poisons must be impervious to the poison and be sufficiently stout to prevent breakage from handling and transport. Glass bottles containing liquids which are not human medicines to be taken internally must be fluted vertically with ridges or grooves recognizable by touch. This does not apply to bottles of a greater capacity than 120 fluid ounces. There is no rule as to the colour of bottles.

Poisons in the U.S.A. There are no existing regulations governing the sale or storage of poisons used in photography, and they can be obtained from chemists without the need for any permit. R.L.T. & J.Br.

POITEVIN, ALPHONSE LOUIS, 1819–82. French engineer, chemist and photographic inventor. Made numerous investigations on galvanography (1847), engraving of daguerreotypes, photography with iron salts, photolithography, photo-ceramics, the reaction of chromates with organic substances (e.g., glue). In 1855 he laid the foundations of both collotype and pigment printing (carbon printing) and also produced direct photolithographs in halftone on grained stone coated with bichromated albumen.

Biography by R. Colson (Paris 1898).

POLACOLOR. Proprietary diffusion-transfer system of colour photography, yielding a positive dye image on development in the camera within about one minute of exposure.

See also: *Diffusion transfer*.

POLAND. There have been photographic enthusiasts in Poland since the introduction of the daguerreotype, and today photography has a wide and active following throughout the nation.

History. Jedrzej Radwanski, a Polish physicist and chemist, is said to have been the first amateur in Poland to take a daguerreotype photograph (1839). The first professional was Moritz Scholz (1842), a German, connected with Volmer's plated-ware factory in Warsaw. It is interesting to note that J. Fraget, another plated-ware manufacturer, also took orders for daguerreotypes in 1839.

In 1845, Karol Beyer opened the first professional photographic studio. His work included daguerreotype photography, photography on china, archaeological photography, numismatic and ethnographic photography, lithography, and later photolithography.

The inventors of certain photographic techniques at this period were Poles, though their nationality is not generally known as having been Polish. O. Anschütz, for instance, a pioneer in the photography of movement, was a Pole from Leszno (originally Anszyc). Warnerke, a pioneer in the use of materials sensitive to light, was also a Pole, originally called Wladyslaw Malachowski.

Many technical improvements were made by Polish photographers in the last decades of the nineteenth century and the first of the twentieth. Proszynski anticipated some of the achievements of the Lumière brothers in cinematography. He also produced the first mechanically operated film camera and a three-blade shutter for film projectors which cut out flicker.

Szczepanik was another inventor who worked on colour photography, using three filters and simultaneous three-fold projection. He used a film embossed with lenticular ridges in colour photography (a system later adopted in the Kodacolor method).

Industry. The Polish photographic industry began to develop at the end of the last century. In 1888 Piotr Lebiedzinski opened a photographic paper factory in Warsaw; he produced the Bromograviura paper which gave pictures with a glossy surface in the light places and matt in the dark. He devised various applications of photography to industry, including a photographic method of checking distortions in railway tracks which was widely used in some countries for many years.

After the first World War a number of factories in Poland began producing photographic equipment: the Foton factory, Ero, Stafra, etc., and one in Bydgoszcz which is still working today. The Fos factory in Warsaw at that time produced very good quality lenses.

It is worth noting that Marie Curie-Sklodowska published one of her first articles on her discovery of polon in the photographic monthly *Swiatlo* (Light) in 1898.

Artistic Aspects. The beginnings of photography as an art can be traced in Poland to the year 1891, when a group of enthusiasts founded the Klub Miłośników Fotograficznych (Club of Photographic Amateurs). A year later the Towarzystwo Fotografików Miłośników (Society of Amateur Photographers) was founded in Cracow. These organizations at first embraced only people of means, who treated photography purely as a hobby and never realized its wider potentialities.

The first Polish journal devoted to the problems of photography appeared in 1898 and was called *Swiatlo* (Light). It was followed by *Fotograf Warszawski* edited by Karolyi, *Kalendarz Warszawski* edited by Uminski, and by *Miesiecznik Fotograficzny* edited by Huber, Mikolasch and Switkowski.

At this time there were a number of genuine enthusiasts who believed that amateur photography offered the true basis for the creation of a photographic art. Even so, after World War I, when cameras became cheap enough to be owned by the man in the street, most people were mere snapshooters and all attempts to raise photography to a higher artistic level by publications or club activities failed for lack of support.

Only a few amateur clubs treated photography as an art. These clubs were of two types. There was the Stowarzyszenie Foto-Amaturów (Photographers' Association), which accepted any amateur, and the Foto-Klub (Photo Club), which accepted as members only photographers with artistic achievements to their credit.

Jan Bulhak has been the greatest personality in the history of Polish pictorial photography so far. He was the founder of the pre-war Polish Photo-Club, for amateurs specializing in pictorial photography. In the post-war Poland he was one of the founders and the first president of the Zwiazek Polskich Artystów Fotografików (Union of Polish Photographic Artists). His death in 1950 was a great blow to Polish photographic art.

Polish pictorial photography of the inter-war period was characterized by a formalistic search after pure lens effects, theatrical compositions, blind addiction to specialized techniques, and a general tendency to make photographs resemble paintings in form and texture.

But a few artists refused to submit to the generally accepted standards of the period and sought to depict the beauty of the homeland. Others developed the artistic portrait, a portrait with psychological meaning.

The war destroyed practically all the achievements of Polish photography but as soon as the war was over the social and political emancipation of the Polish people brought about the emancipation of photography. It now became established as one of the arts.

The whole photographic life in the country has since the war been rebuilt on a completely different basis from that existing before September, 1939.

Organisation. *Zwiazek Polskich Artystow Fotografikow* (The Union of Polish Photographic Artists) was established in 1946. It is a small organization with 290 members, and the conditions of acceptance are exacting. Candidates, before their applications can be considered, must prove a minimum of three years' artistic experience, participation in exhibitions at home and abroad, and publication of works of their own. Graduates of the Cameramen Section of the Higher School of Films, and photo-reporters who are members of the Union of Polish Journalists can also apply.

All candidates for membership of the Z.P.A.F. must also present, with their application, 20 examples of their own work for approval by the Union's Artistic Commission.

The Z.P.A.F. acts as the representative of Polish photographic art at home and abroad (via F.I.A.P.), organizing exhibitions at home and selecting exhibits by Polish photographers for international exhibitions.

The Z.P.A.F. organizes annually an All-Polish Exhibition of Artistic Photography, which is a survey of the Z.P.A.F. members' achievements, and eight regional exhibitions in provincial centres by the branches of the Union. Each year mobile exhibitions visit cities with particularly lively cultural activities and there are approximately 10 individual exhibitions.

Within the framework of cultural interchange with other countries, the Z.P.A.F. organizes annually several exhibitions abroad and, in return, exhibitions by foreign photographers in Poland. Every three years the Z.P.A.F. organizes an International Exhibition of Artistic Photography in Poland. Such events are very popular, and the All-Polish Photography Exhibition in Warsaw is, on the average, seen by some 40,000 visitors.

The Z.P.A.F. published a Yearly Almanac of Polish Photography with approximately 150 examples of the most outstanding works by Polish photographers.

Together with delegates from all the other creative unions, the Z.P.A.F. is represented on the Cultural Council, which is an advisory body to the Minister of Culture and Art. Members of the Z.P.A.F. also serve as artistic consultants to publishing houses and other institutions which concern themselves with artistic and technical photography.

The Z.P.A.F. receives a subsidy from the state which, apart from meeting administrative costs, supplies large sums for the Union's publishing activities, enabling the Union to house its exhibitions in the best available premises. The state in addition grants 10 creative scholarships annually and also finances trips abroad for 2 or 3 of the most outstanding Polish photographers. Another example of patronage by the state is the purchase of work from All-Polish and Individual Exhibitions, which will be included one day in the proposed Museum of Photographic Art. City Councils provide annual awards for the most outstanding artists, including photographers. In all cultural negotiations with other countries, the Z.P.A.F. is the State's representative of Polish photographic art, and the organizer of artistic exhibitions.

Amateur photographic activities, like the amateur activities in other artistic spheres such as painting, sculpture, music, etc., are organized by Cultural Centres, industrial enterprises and institutions, and by schools. All these amateur photography units together constitute the Federation of Photographic Societies whose aim is to popularize photography as widely as possible. At the more advanced amateur centres members of the Z.P.A.F. act as instructors and advisers. A.Jo.

POLAR CURVE. Graphic method of recording the distribution and intensity of light emitted from a source. Points of equal light intensity are plotted along lines radiating from the source over the angle including the emitted rays. When these points are joined up, the result is a polar curve.

Thus, polar curves of a lamp in different types of reflector show at a glance the way in which the light is distributed over the scene in front and indicate the distances from the lamp in all directions where the light intensity—and hence the required exposure—are the same.

Illuminating engineers use polar curves of lamps and fittings when they have to ensure a minimum level of illumination over a given area—e.g., in a schoolroom or a factory machine shop.

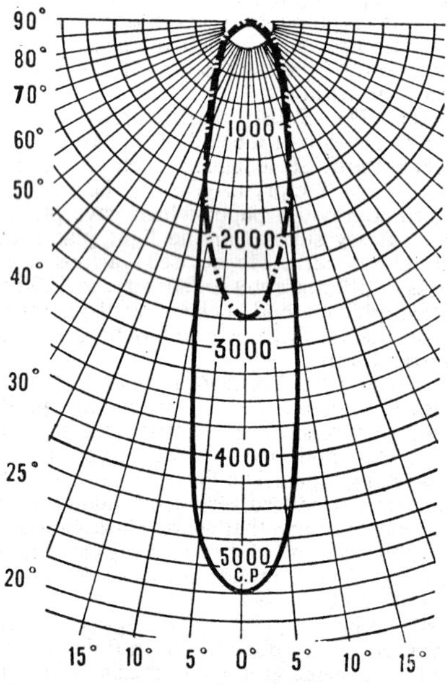

POLAR CURVE. This shows in the form of a graph the light distribution at different angles to the axis of the lamp unit. A bare lamp emits the light evenly all round; its polar curve would thus be a complete circle centred around the pole. Here the solid curve represents the highly directional light of a grid filament spotlight, the dotted curve a ring filament spot.

POLARIZED LIGHT. Ordinary light can be regarded as travelling in waves that vibrate in all directions at right angles to the direction along which the light is travelling. Certain crystals will only allow electro-magnetic waves to pass through them if they are vibrating in one particular plane. So any light that passes through the crystal emerges vibrating in the one plane along which the crystal is transparent. Light which thus has its vibration restricted to a single plane is said to be plane polarized.

It is also a fact that most surfaces, in reflecting light, will polarize it to some extent. Glass for instance, will partially polarize light; in fact, it will effectively polarize all light rays at about 56° to the normal to the surface.

Some organic materials, treated by stretching and orienting the molecular structure, will also polarize light. These materials can be made into films for use in filters over camera lenses.

Filters. Polarizing filters will reduce the brightness of reflections from polished or glossy surfaces providing the angle of reflection is between, say, 70° and 40° to the normal. Much has been written about the usefulness of polarizing filters but it must be said in all fairness that natural conditions very often render such a filter of little value. Under suitable circumstances however, polarizing filters are useful when photographing shop windows through the glass front, copying pictures behind glass, photography of polished furniture (where the reflected glass tends to hide the grain of the wood) and similar purposes.

POLARIZED LIGHT. Top left: Normally light waves vibrate in a large number of planes about the axis of the light beam. Top centre and right. If the waves vibrate in one plane only, the light is plane polarized; e.g., as after passing through a polarizing filter. Centre to bottom: The action of a polarizing filter is analogous to that of a picket fence on a vibrating rope.

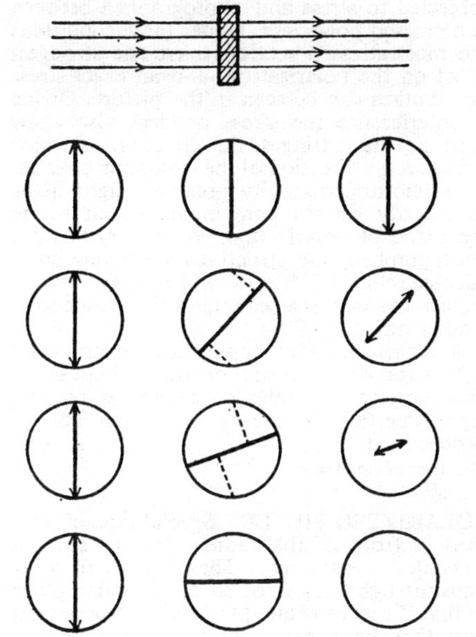

POLARIZING FILTER. The polarizing filter absorbs all light which does not vibrate in its own plane of polarization. As the filter is rotated across a beam of polarized light, the intensity of the beam is progressively reduced as the component vibration in the plane of the filter gets less. With its plane of polarization at right angles, the filter cuts out all light.

Polarizing filters can be used to darken the blue sky, the light from which is to a large extent polarized when the sun is in a specific position in relation to the relevant area of sky and they are, in fact, the only filters that can be used with colour films without affecting the reproduction of other colours.

Many photographic ends and effects may be achieved by the use of polarizing filters over the lens, the light source (artificial), or both.

Photographic Applications. Polarized light is used in a variety of scientific fields to show up and photograph phenomena which would not be visible by normal light.

The technique is the same in most cases. The object to be examined or photographed is placed between two crossed polarizing filters. Normally these would stop all the light from coming through. If, however, the object affects the polarization of light—by rotating the plane of polarization, by depolarizing the light, or by turning it from plane to circularly polarized (i.e., where the plane of polarization is not flat but twisted like a spiral staircase)—it will become visible from behind the filters.

In photo-elastic stress analysis, scale models in glass or transparent plastic of structural engineering units such as girders, etc., are subjected to stress and photographed between two crossed polarizers. Under these conditions the most stressed sections have the strongest effect on the polarization, and an exact stress distribution can be seen in the picture. Owing to interference the stress patterns also show light and dark fringes, the direction of which indicates the direction of the maximum stresses.

In photomicrography polarized light helps to identify certain compounds which rotate the plane of polarization. Where crystals are photographed, the structures will show complicated colour patterns which may have a pictorial as well as a scientific value—especially if taken on colour film.

A rather different application of polarized light is the Kerr-cell shutter used in high speed photography. This relies on changes in polarizing properties of liquids in a high voltage electric field. R.B.M.

See also: *Photo-elasticity*.

POLARIZING FILTER. Special type of filter used in front of the camera lens to cut out unwanted reflections. The filter will only transmit light polarized in a particular plane so that it can be rotated to hold back any light rays that have been polarized by reflection from a polished surface—e.g., a shop window.

See also: *Polarized light*.

POLAROID LAND CAMERA. Trade name of still camera using a diffusion-transfer process to yield finished positive prints within about 10 seconds (black-and-white) or one minute (colour).

The first model of this camera, and also the first camera for general photography using a diffusion transfer system, was introduced in 1948 and produced brown positive prints in one minute. Subsequent developments led to black-and-white prints, to a reduction of the processing time to about 10 seconds, and to special materials yielding black-and-white transparencies, recoverable black-and-white negatives and colour prints. The effective speed of black-and-white materials for this system ranges up to 3,000 A.S.A. (10,000 A.S.A. for special purposes).

The camera takes a negative sensitive material (paper or film) and a non-sensitized positive paper, in two separate rolls. The positive roll also carries pods of jellified developing reagent.

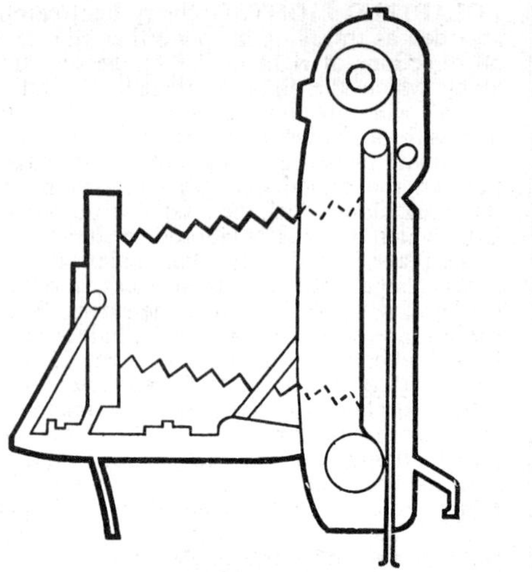

ONE-MINUTE CAMERA. Land camera, utilizing image transfer by diffusion. The lower roll holds the light-sensitive negative paper, the upper roll the non-sensitive positive paper. The image is exposed on the negative paper; after exposure the negative and positive papers pass between rollers which press them into contact and spread a film of jellied processing reagent between them. After one minute, the finished positive can be removed on opening the camera back.

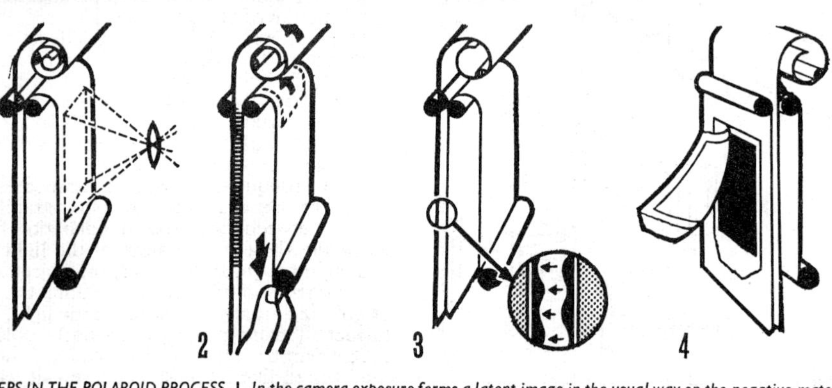

1 2 3 4

THE STEPS IN THE POLAROID PROCESS. 1. In the camera exposure forms a latent image in the usual way on the negative material. 2. On pulling the tab, the processing rollers squeeze the reagent pod between the negative and positive material to start development. 3. The negative image now develops in the negative layer while the unexposed silver salts diffuse into the positive layer to form the positive picture there. 4. At the end of the processing period (usually as little as 10 seconds) the positive print can be peeled away on opening the picture door in the camera back.

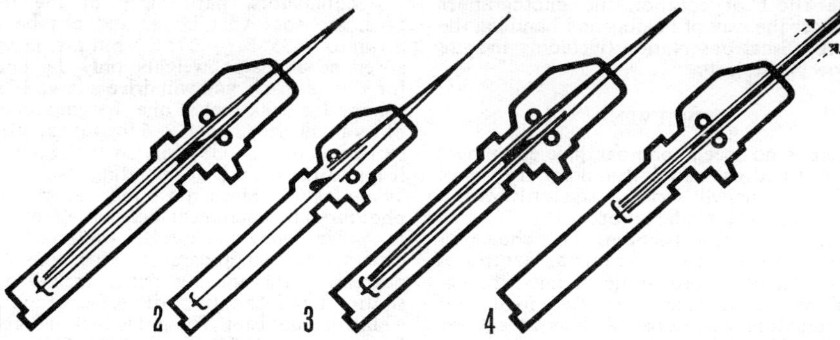

POLAROID SHEET FILM OPERATION. The sheet film, consisting of a negative, a positive and a paper cover plus the reagent pod, is loaded into a special sheet film holder which can be used in most plate cameras. 1. On inserting the film in the holder a clip engages inside and holds the negative film in place. 2. For the exposure the paper cover and the positive paper are withdrawn. 3. After the exposure the paper cover and positive are pushed back again and the processing rollers closed up. 4. On withdrawing the whole sandwich the reagent spreads between the negative and positive layers to start the processing reaction. Processing takes place outside the sheet film holder and the negative and positive are then peeled apart.

The body, shutter, focusing and other camera features follow conventional folding camera lines, but the back is designed to take the two rolls and also to provide for their processing inside the camera. The camera is loaded in daylight and the negative paper or film drawn into position to face the lens for the first exposure.

After the exposure, pulling a strip of paper projecting from the camera brings the exposed negative behind the pressure plate and draws it—in contact with the positive paper—through a pair of steel rollers. Their pressure bursts the pod containing the processing reagent and spread the latter evenly between the two surfaces. This initiaties the diffusion-transfer processing, during which the negative image is developed and the positive formed in the non-sensitized paper layer by diffusion of the unused silver salts into this layer. Processing takes place in a special compartment in the rear of the camera; after the appropriate time a door in the camera back is opened to reveal the back of the positive paper strip. This is perforated around the edges of the picture, and can be lifted out by means of a small tab, by breaking the perforations. The finished print is slightly damp, but completely developed and fixed when peeled off the negative paper.

An alternative camera design takes the negative and positive papers in the form of single cut sheets inside a holder resembling a film pack container. The negative sheets are here stacked on top of each other behind the picture aperture. After each exposure, pulling a tab draws the negative and an associated positive sheet between pressure rollers (which spread the developing reagent between the two surfaces) right out of the camera. Processing here takes place outside the camera and after the appropriate time the negative and positive sheets are peeled apart.

Sheet film adaptors for press and technical cameras take single pairs of negative and positive material. These are packed light-tight, and after exposure in the adaptor are pulled out between pressure rollers which once more spread the activating reagent between the two parts of the sandwich. After the development outside the holder, the negative and positive are stripped apart.

Polaroid roll film backs—without the optical camera elements—also exist for fitting to technical and special cameras for oscillograph recording etc. L.A.M.

See also: *Diffusion transfer*.

POLAR PHOTOGRAPHY

Low temperatures offer special photographic problems, both in the manipulation of apparatus and in taking technique. A distinction is, however, necessary between cold weather conditions prevailing in the winter in temperate climates, and sub-zero temperatures met with in the arctic and antarctic regions.

Temperate winter conditions involve temperatures around or a little below the freezing point of water—ordinary cold weather photography.

Polar photography implies temperatures down to − 40° to − 70° F. (− 40° to − 55°C.). These appreciably affect equipment and materials as well as the comfort and work of the photographer.

The problems relating to equipment include greatly increased viscosity of normal lubricants, leading to sluggishness in the moving parts of the camera; brittleness of shutter blinds and film supports; freezing of condensation on cameras and lenses, etc.

The effect of cold on the photographer influences the way of holding and handling the camera, visual observation (including the use of snow glasses), etc.

EQUIPMENT

There is no special photographic equipment manufactured exclusively for polar use, since most apparatus will serve satisfactorily with a certain amount of adaptation.

Camera Types and Features. The choice of camera is governed by operational features as well as purpose. For arctic or antarctic expeditions, saving weight is often important so a miniature camera has obvious advantages on journeys in the snow and away from the base of an expedition; and its extra lenses, filters, etc., do not greatly add to the total weight. Also the 36-exposure load of 35 mm. film cuts down on the reloading required.

Many people like to use precision twin lens reflex models. Advantages are the absence of bellows, large focusing knobs, viewing without putting the face or eye close to the camera, automatic shutter tensioning, and easy film transport. All the manipulations of picture taking can be carried out with mittens on the hands. However, it is important to avoid breathing on the focusing screen.

Folding roll film cameras are also in use, but the controls are more difficult to set with mittens or gloves.

On polar expeditions it is useful to have at least one medium-size plate camera at the base of the expedition. This serves for high-quality colour shots intended for reproduction, as well as for more advanced survey and similar work.

Bellows cameras should be fitted with first-quality leather bellows not loaded with too much stiffener. Plastic bellows are liable to crack in polar climates. Rigid cameras with extending lens tubes are free from such risks.

Eye-level viewfinders present a risk: a cheek brought in contact with them at temperatures below – 15° F. (– 26° C.) may suffer instant frostbite. If the skin is wet, it will also freeze to the metal. For constant use the eyepiece should be surrounded by fur or cloth. Even rubber eyepieces (as fitted on some cine cameras) can cause freezing of the skin. Frozen rubber becomes brittle, liable to fracture, and spare eyepieces should form part of the kit.

Whatever the equipment of an expedition, it is always advisable to include a simple box-type camera ready for use by anybody at any time.

Cine Cameras. Motors on cine cameras may become unreliable at very low temperatures, and drop in speed even if winterized (see below). Electrically driven equipment suffers from the effects of cold on the batteries. Below 0°F. (about – 18° C.) the current output drops, and at – 10° F. (– 23° C.) becomes insufficient to maintain proper operating speeds for a run of 10 feet.

Accumulators, particularly of the nickel type, are somewhat better and can be used down to – 35° F. (– 37° C.), but are heavy. A silver accumulator weights only 1½ pounds for a 9 volt pack, and will drive a large 16 mm. camera for 7,000 feet at one charging. Because of its small size (2 × 4 × 6 ins.) it can also be carried under the clothing, so that body heat helps to maintain it at operating level.

Winterization. Since most lubricants used in photographic equipment become almost solid in polar climates, special winterization is essential for all apparatus. This consists of degreasing the moving parts, including the shutter, and lubrication with either graphite or a silicone lubricant. The job is best undertaken by an expert, preferably the manufacturer.

Winterization should extend to film transport mechanism, opening and closing movements of the camera, and also to focusing lens mounts. A lens which functions smoothly at normal temperatures may require a pipe wrench to turn it at – 50° F. (or about – 45° C.). This applies particularly to cine lenses because their small diameter and narrow focusing rings make them very difficult to grasp with gloves. A useful device to overcome this difficulty is to thread a small bar radially into the control ring, in much the same manner as is done for zoom lenses, for easy rapid use.

Shutters. Unwinterized shutters cease to be accurate even at moderately low temperatures of 20° F. (– 6 or 7° C.). With suitable low-temperature lubrication diaphragm shutters will work well under all conditions.

Focal plane shutters other than all-metal types are more difficult to lubricate for extremes of temperature, and also suffer from other disadvantages. Thus the opaque material of the blinds easily cracks, and may flake off.

Tensioning of shutters—of any type—should be done slowly to avoid breaking springs. This also applies to other spring-loaded camera parts.

Lenses. Apart from stiffness with normal lubricants in the focusing movement, and condensation, uneven expansion on rapid warming up may cause trouble. With a lens in a heavy focusing mount this may lead to cracking of components. It is more likely to happen with highly complicated lens systems with strongly curved elements. The uneven response of different types of optical glass to temperature changes may separate cemented units.

In all cases the camera should therefore be warmed up very gradually when it is brought indoors from sub-zero temperatures. Certain authorities consider it best to leave the equipment out of doors at all times. If this is done, it will be wise to tension and release the shutter several times before use. Similarly, a short run in a cine camera will help to condition it for the cold. It should, however, be noted that some experts oppose this advice, and maintain that the equipment should be exposed to low temperatures only as long as necessary.

Condensation. Condensation is one of the worst nuisances of polar photography but fortunately does not last very long. Cold instruments brought into warm or moist rooms will promptly collect a sheet of ice. Care should be taken never to blow or breathe on a lens when cold, either inside or out of doors. Hands, if perspiring in heavy wool or fur mittens, should be bared and waved in the air for a few seconds to avoid moisture collection, especially when loading the camera. Otherwise there is a danger not only of the lens icing up, but of the fingers freezing to the metal parts of the camera.

Generally, condensation will evaporate in half-an-hour or so from the front lens element. Condensation or icing on inside surfaces may call for complete dismantling and drying out at room temperature.

Condensation is particularly bad for diaphragm shutters, as the moisture rusts and pits the blades, making smooth working impossible. The iris diaphragm is liable to suffer in the same way. Focal plane shutters are less sensitive in this particular respect.

Dropping a cold camera in the snow at low temperatures is of little consequence. Usually a brief shaking will dislodge any snow particles. With a warm camera matters are worse: thawed snow flakes will immediately form ice which will take days to evaporate (unless of course the camera is warmed and dried).

Accessories. Normally accessories carried in the field tend to be limited to tripods, filters, lens hoods, and exposure meters.

Among tripods the wooden type is less subject to damage by cold or in ice and snow than most metal tripods, and it is easier to handle as there is less risk of condensation and freezing up. Pan and tilt heads will need special attention for degreasing, etc.

A useful aid in snow is to cut a triangle of stout canvas of roughly the same size as the distance the tripod legs are spread apart. The edges are bound for strength, and a brass eyelet fixed in each corner. This should be large enough to pass the spike points of the tripod, but not the tripod legs themselves. In use the triangle may be spread on even soft snow, the tripod points inserted in the eyelets and the camera mounted for operation. This prevents the legs from sinking in the snow.

Filters for polar work should be of solid glass dyed in the mass rather than cemented gelatin (this is unavoidable with polarizing filters).

Lens hoods require no special precaution, but should be easily attachable to the lens to reduce manipulation in the cold.

Exposure meters of the best quality do not vary in reliability at low temperatures, especially if the whole unit is hermetically sealed. It is, however, advisable to take several meters on a polar expedition, as they appear to get out of adjustment more quickly.

Personal Equipment. Apart from the standard warm clothing for sub-zero temperatures, two points require special mention in polar work.

Firstly, it is advisable to wear light finely knitted wool gloves inside the heavy mittens. The thin gloves permit manipulation of camera controls, etc., for a short time when the hands are removed from the heavy outer covering. Some people prefer thin silk or cotton gloves, but once the hands are exposed to the cold—perhaps to the point of numbness—they do not warm again as readily in silk or cotton.

Secondly, snow glasses are usually essential to guard against snow blindness which, while temporary, is always irritating and very uncomfortable. The glasses may be raised for short periods for focusing or other adjustment. Spectacle wearers should have coloured spectacles prepared to their individual prescription.

TECHNIQUE AND SUBJECTS

Polar photographic technique differs from normal procedure in manipulation of the camera, and in exposure, filters, etc.

Carrying and Holding the Camera. The mere handling of intensely cold equipment is a problem, especially with all-metal cameras. Even with wool-lined leather or fur mittens great amounts of body warmth are drawn off by the chilly metal. It is therefore desirable to have sturdy leather cases for equipment if the design permits it. Large cine cameras present even more of a problem in heat dissipation, since they must be grasped firmly if hand-held.

If the camera is small enough, it is best carried under the clothing to keep it reasonably warm and thus in good working order. The photographer should not be perspiring, however, as condensation may form unless the camera is completely encased. With larger equipment the chilling effect of a cold lump of metal may be serious if tucked under a parka or inside a shirt. The use of thin polythene bags for items of equipment is a useful way of reducing the risk of condensation when they are carried in this manner.

Loading and Handling Films. The main effect of intense cold on films is to make them brittle and liable to snap, especially if curved sharply. For roll films, size 120 cameras are therefore better than those taking 620 films because the film leaves, and is rewound on to, a larger core. When curving films to attach them to spools, etc., the film base should be bent only slowly, preferably with bare hands as the body heat will aid in making the film base more pliable. Caution is important with film edges; when cold and hard they are razor-like, and can cause nasty cuts.

To avoid breakage of the film, it should always be advanced slowly in the camera.

Sheet film (in a suitable camera) can be kept flat, but involves the use of comparatively weighty holders. Film packs are unreliable in extremely cold weather, partly because the film easily splits when making the tight turn inside the pack, and partly because it is subject to static marks due to rubbing. The drier and colder the weather, the more likely such marks.

Cine film suffers from similar troubles, especially as it has to be curved and uncurved a lot and moves all the time in the camera. Magazine type cameras generally cannot start at temperatures below −40° F. or C., owing to the set that the film takes in the magazine. Most 8 mm. magazine cameras engaging only one perforation hole during film transport will tear out the film if its stiffness exerts more than 10 ounces of pull.

Exposure. An exposure meter is almost indispensable for light measurement, and the exposure is determined in the usual way.

One complication is the effect of the reciprocity failure and speed loss at very low temperatures. The effective film speed appears to drop to a minimum around − 80 to − 100° F. (− 70° C.), and rises again to its full value around − 180° F. (− 120° C.). This rise is of academic interest, since the lowest temperature at which photographs have been taken in polar regions seems to be about − 70 to − 80° F. (− 60° C.).

Black-and-white films have been found to need one stop more exposure at − 20° F. (− 30° C.), two stops more at − 40° F. or C., and three stops more at − 70° F. (− 56° C.). In practice the latitude of the film takes care of this speed loss down to − 40° F. or C. No actual figures are available for colour film, but the effect is expected to be similar. At the same time the colour balance tends to shift towards cyan. Trial exposures with compensating filters are therefore advisable. Various makes of colour film may behave differently in this respect, and the appropriate manufacturers should be consulted.

Filters. For black-and-white work with clear or partly clouded skies a medium yellow filter enhances pictorial quality. Heavier filtering is useful for special effects. An ultra-violet filter should always be used—for black-and-white as well as colour film—to cut out the high proportion of ultra-violet rays in the daylight.

Polarizing filters are particularly effective in the polar regions and surpass optical absorption filters for clear rendering of distant views. They do, however, seem to call for higher filter factors in very northern or southern latitudes—some 4 to 8 times.

The polarizing filter darkens the north sky most in the arctic, and the southern sky in the antarctic. Clouds, snow-clad mountains, and distant scenery will then stand out sharply against a blue-black background. With the camera pointing east or west, the effect of the polarizing screen is weaker; the sky tone will, however, be uneven, being much darker towards the north (in the arctic) or the south (in the antarctic). The filter factor for eastward or westward views is also somewhat less.

With high speed black-and-white film two partially crossed polarizing screens can be used as a neutral density filter over the camera lens. This helps if the snow scene is so bright that even the smallest lens aperture would still lead to over-exposure (especially in cine work where the shutter speed is often fixed). The method is less satisfactory for colour film.

Processing. Although most photographs are taken in the field, often in very low temperatures, processing is carried out at the base of an expedition or on other semi-permanent premises, and does not differ from ordinary cold weather procedure in more temperate climates. Suitable heating keeps the darkroom and solutions at workable temperatures.

On expeditions the working facilities are somewhat limited, and care is needed to keep irreplaceable equipment in working order.

Water for photography is preferably obtained from glazier ice or snow; pack-ice and icebergs contain salt, but can still be used as a source, if allowance is made during processing for the slight retardation their salt content may cause.

Applications. The rôle of photography on polar expeditions is to study geographical, geological and topographical features of the region under investigation, to record animal and any plant life, and to conduct scientific research. Photography of fossils has also established the existence of a temperate climate in earlier geological eras. Military polar expeditions have, with the aid of photography, investigated the behaviour of clothes and equipment under cold weather conditions.

Aurora Borealis, and other Effects. These sky illuminations of nature can be very impressive and picturesque, but are weak from the point of view of actinic brilliance. The brightest of them still require exposures of the order of ½ minute at *f* 2; the average, especially on colour film, is nearer to 20 minutes.

Because of the quality of the light emitted by the aurora, some lenses appear to be able to produce better images than others.

The phenomenon known as "mother-of-pearl sky" occurs occasionally every few years in the arctic and sub-arctic areas. It can be recorded by normal photographic procedure; slight under-exposure has given the most beautiful results on colour film.

Impressive arctic phenomena are the "sun-dogs". These appear as brilliant balls of fire and are caused by the reflection of the sun from ice crystals floating in the air. They are seen one above and to each side of the sun.

These phenomena are best photographed with a single-lens reflex camera or a plate model with ground glass screen, since exact positioning of the camera relative to the sun is important. The sun must be centred squarely in the lens to eliminate as much flare as possible, as well as the ghost images and reflections of the iris diaphragm and shutter blades. Care is necessary with focal plane shutters, or the sun may burn a hole in the shutter blind.

Polar Night and Day. Beyond the arctic circle the sun does not rise during the winter. That does not mean that it is dark; on a clear day there is a surprising amount of pale blue light. As the sun begins to break the horizon in spring, or descends in the autumn, the bulk of the light received from it is very red. In either case colour film needs appropriate correction filters. By late spring exposure estimation is very misleading without a meter.

The sunsets and sunrises during spring and autumn in the polar and sub-polar regions yield pictures of great beauty, especially on colour film. Again, ground glass screen focusing cameras are preferable for this type of work, as the visual appearance may not correspond with what the film records. S.M. (D.H.O.J.)

See also: *Cold weather.*

Book: *Photography on Expeditions*, by D. H. O. John (London).

POLICE PHOTOGRAPHY. The use of photography in police work dates from autumn 1852, when the Swiss Federal Government authorized the Department of Justice and Police to have photographs taken of all vagrant beggars found in cantons other than their own. In December 1854 the governor of Bristol Gaol, England, issued a circular recommending other prison governors to follow his practice of having a photograph taken of every prisoner in his charge. The same year, a "wanted persons" notice with photograph, made according to the suggestion of Moreau Christophe, was bill-posted publicly in France. This was first done in England in 1861. It was not until 1870 that the photographing of every prisoner in England and Wales was made compulsory. The procedure was justified during the year 1871–72 when 375 arrests were made in consequence of the criminals having been identified by photographs. Today the more advanced photographic techniques are in constant and daily use by all the larger police forces of the world. Well equipped photographic departments are an integral part of police administration and their contribution, especially in crime detection, is an important one.

Great reliance is placed upon photographic evidence by the Courts of this land and hardly a major crime of any importance is solved without photographic evidence of some kind. Evidence of this type is factual, objective and without bias. For recording the scenes of serious crimes its value is inestimable.

Location Work. The photographing of scenes of crime follows well defined practice. The locus is recorded exactly as found and before anything is touched or disturbed. Nothing is moved before complete photographic coverage has been made from several angles to meet all eventualities. If bodies, articles, etc., have been moved they are never replaced in an attempt to approximate their original position. This amounts to reconstruction and reconstructed photographs of scenes of crime are not admissible as evidence in English Courts of Law. Any clues present, such as shoe prints, finger prints, bullet marks, scratches, signs of entry and egress, etc., are recorded with due regard to their relation to each other. Articles bearing on the crime, such as weapons or tools, are photographed in their exact position before removal for examination and further photography.

Wounds on bodies are always photographed as it is possible from a study of the photograph to deduce with a high degree of accuracy the shape and size of the weapon used. Tattoo marks and prominent scars are also photographed in every instance where identification is in doubt.

In recording these scenes of crime the aim is to produce a series of photographs which will tell the story of the crime. All prints must carry a good range of tones and be absolutely sharp from back to front.

Equipment for this type of work can be any good stand camera from quarter to half plate size. It should have a full range of camera movements and be fitted with a double extension. Wide angle lenses are extremely useful on occasions if used with due regard to the risk of distortion. The resultant enlargements up to 10 × 12 ins., a convenient size for presentation in Court, should be indistinguishable from contact prints.

Lighting for interior scenes of crime will vary, but the majority of cases can be covered with expendable flash bulbs. "Painting" the scene with portable half watt lighting is sometimes useful. Electronic flash, owing to the comparative low light output from the portable types, is hardly suitable where large areas have to be covered sharply and well lighted from front to back.

Comparison Records. With many offences a photograph is the only way in which the evidence can be presented to the Courts to supplement the opinion and explanation of the expert. In cases of fingerprint evidence, for instance, the similarity of the prints is quite easily followed by means of the photograph. Ballistic evidence, sometimes of a highly technical nature, is made quite clear through photography and in forgery cases the expert's evidence on handwriting is always supported by means of the photograph.

Comparison photographs, which involve photography of the "trace" found at the scene of a crime, with a known standard are frequently used by the police. Boot and shoe impressions found at scenes of crime can readily be compared with the suspect's footwear, marks made by jemmies and tools can be photographed same-size and the instrument "offered up" to the photograph.

Identity Pictures. Photographs of the criminal aid the investigating officer in his hunt for the criminal. A true likeness is required as opposed to a portrait. Pin sharpness is essential, and under no circumstances must there be any retouching. All facial blemishes, marks and scars must be clearly delineated. Profile and full face photographs are made to ⅛th scale and full length to 1/27th scale. Special apparatus is available where large numbers have to be dealt with, but police photographers usually make use of a special back fitted on to a half plate camera which will accommodate the regulation photograph. Operating indoors under controlled lighting conditions and using time and temperature technique for processing ensures a uniform standard of print.

Questioned Documents. Document reproduction is carried out on a large scale in cases of fraud and processes such as photostat, dyeline reflex printing and Autopos are commonly used.

The examination and photography of questioned documents form a large part of police photographic work and the special properties of the invisible light rays at both ends of the visible spectrum are utilized.

Ultra-violet rays are turned to good account in a number of ways in police investigation. They are used in all cases in the preliminary examination of documents where forgery is suspected. The eradication of original written matter if effected by chemical agency and its substitution by other writing, can generally be detected; any application of chemicals to paper can nearly always be revealed because the area so treated gives off a totally different fluorescence from the remaining part of the document. Usually the original writing is disclosed and when photographed provides irrefutable evidence of forgery.

Special Applications. "Secret" writing with so-called invisible inks or urine, milk, onion juice, etc., all give a characteristic fluorescence and are generally detectable with ultra-violet light. Oil and grease stains give off a particularly brilliant fluorescence and are easily recorded. Blood stains on garments which have been washed to remove the stain can be photographed in U.V. light. Invisible laundry marks which fluoresce strongly under U.V. light are sometimes found on articles of clothing left at scenes of crime and afford valuable evidence when photographed and produced in Court.

The powers of penetration of infra-red rays are used in a variety of ways in police photography. Where written matter is obscured by other writing, scribble, dirt, oil, etc., infra-red photography will give a good recording within certain limits. Partially burnt and charred documents can be photographed with a fair degree of success. Faint pencil writing due to age, wear, or dirt can be "boosted" and rendered legible. Old scars and scarcely visible tattoo marks on bodies can be revealed by infra-red photography. These rays can also be used to determine the distance at which a firearm is discharged from the victim. Powder staining, invisible on dark garments, can be seen quite plainly on the photograph. This evidence is often of great value.

Cine photography is used in police work mainly in recording traffic and crowd control. It has been used on a few occasions to catch the criminal in *flagrante delicto*, but these opportunities, as can be well imagined, are rare. Instructional films for the training of police are made but film strips are to be preferred for this purpose, and are used extensively.

Colour photography, with its well known limitations as regards fidelity in the print, has yet to prove its worth. Colour transparencies, however, are being used on an ever increasing scale and have provided evidence on many occasions. P.G.L.

See also: *Crime photography; Evidence by photographs; Forgeries; Identity photographs; Traffic photography.*

Books: *Photographic Evidence*, by C. C. Scott (Kansas City); *Photography in Crime Detection*, by J. A. Radley (London); *Questioned Evidence*, by A. S. Osborn (New York).

POLYESTER FILM. Plastic film base of high dimensional stability, which has replaced cellulose acetate film bases for certain special applications (notably in graphic arts films, for photogrammetry, etc.).

See also: *Plastics; Supports for emulsions.*

PONTI, CARLO, Dates unknown. Italian architectural photographer in the 1860s specializing in views of Venice, Padua and Verona. Made a pictorial record of Venetian street types in the early sixties.

PONTING, HERBERT G., 1871–1935. English landscape photographer and explorer. Official photographer to Captain Scott's second and last South Pole expedition 1910–1912. Author of *The Great White South* and *In Lotus-Land Japan.*

PONTON, MUNGO, 1801–80. Scottish photographic inventor, secretary to the National Bank of Scotland. Discovered 1839, while experimenting on making Photogenic Drawings with silver chromates, that paper impregnated with potassium bichromate is sensitive to light. This is of fundamental importance for most photomechanical reproduction processes. Ponton used this photographic paper to record thermometer readings, 1845.

PORRO, IGNAZIO, 1801–75. Italian scientist. Described the first tele objective in 1851 (later re-invented and manufactured by Dallmeyer, Duboscq, Miethe, Steinheil) and worked on photogeodetics in 1855.

PORTA, GIOVANNI BATTISTA DELLA, 1538–1615. Italian scientist, usually known as Gianbattista Porta. The invention of the room-type camera obscura without lens is often abscribed to him because he was the first to publish a long and clear description of it in 1558.

PORTRAIT ATTACHMENT. Positive supplementary lens which enables the user of a fixed focus camera to take close-up pictures. Generally, the close-up picture is a portrait (hence the name), but with suitable lenses large-scale pictures of objects as small as coins, postage stamps or flowers can be brought into the scope of the sort of camera that must otherwise work at least seven feet away from its subject.

When a supplementary lens with a focal length of F inches is added to the lens of a fixed focus camera (or any camera with its lens set at infinity) the combined lens is then focused sharply on objects F inches away, and the effective aperture of the lens (and thus the exposure required) remains unchanged.

This is the principle of the portrait attachment. The photographer merely has to fit the attachment over his lens, note the focal length (usually engraved on the edge of the supplementary lens mount) and set up his camera at that distance from the subject.

SUPPLEMENTARY LENSES
WITH FIXED FOCUS CAMERAS

Supplementary Lens			*Range Covered*			
Focal Length		*Power*	*ft. and ins.*		*cm.*	
ins.	*cms.*	*(Diopters)*	*from*	*to*	*from*	*to*
—	—	—	∞	10–0	∞	300
120	300	0·33	10–0	7–6	420	225
80	200	0·5	8–0	5–6	240	165
60	150	0·67	6–0	4–3	180	125
40	100	1·0	3–9	3–0	112	90
32	80	1·25	2–11	2–5	88	73
26	67	1·5	2–5	2–0	73	60

This table assumes that the lens aperture is about *f* 11 to *f* 14.

It is possible to select a series of portrait attachments in such a way that the depth of field on one lens starts more or less where that of its neighbour leaves off. In this way, using not more than about half a dozen lenses, the whole range of distances from a couple of feet to infinity can be brought into focus on a fixed focus camera.

See also: *Close-ups; Supplementary lens.*

PORTRAIT LENS. Lenses intended specially for portraiture do not have to be designed to cover wide angular fields of view. Usually, the focal lengths selected are large in relation to the size of negative in order to avoid the unpleasant distortions of perspective created when the camera viewpoint is too near the subject.

The first lens designed for portraiture was made by the Hungarian, J. Petzval, in 1840. The acceptance of a narrow field of view permitted him to utilize two widely separated doublet components which provided what was, in those days, the remarkably wide relative aperture of about *f* 3·6. This basic lens form has been modified in a variety of ways to suit different optical requirements, but nearly all the derivations are known as Petzval lenses.

The Petzval lens has the characteristic of yielding very sharp definition in the centre of the field with a very rapid deterioration of quality towards the edges. This variation between needle sharp central definition and poor edges is not always desirable or pleasing in portraiture and in more recent times there has been a demand for a more uniform type of definition at a softer or more diffuse level than that produced in the centre of the field of a Petzval lens.

Definition Control. Where the definition of a good modern lens is too hard for portraiture the necessary softness can be produced by devices such as diffusion discs in front of the camera or enlarger lens. There are several methods, both good and bad, of artificially softening definition in this way.

In normal photography every attempt is made to produce a standard of definition as near perfect as possible—i.e., to get the lens to yield a perfect point image of a point in the object. This is not usually the ideal type of image for portraiture; softer definition is obtained if the point image is surrounded by a circular fringe of light of low intensity. By retaining the point image as a hard nucleus in the image patch the lens designer can maintain the ability of the lens to record fine detail and he can vary the degree of softness by controlling the diameter of the fringe and the relative light distribution between it and the hard central nucleus.

This type of image formation can be provided by deliberately introducing a type of optical error in the lens known as spherical aberration. It cannot be provided by merely moving a well corrected lens out of focus, as this gives an image patch of finite size but with a more or less even distribution of light intensity, so that spread of light is too severe and the effect is to spoil definition instead of softening it.

There are two main types of portrait lens constructions which soften the image by introducing spherical aberration: in the first type the amount of aberration is fixed, and in the second, it can be varied.

Fixed Aberration Lenses. In the first type there is a fixed and substantial amount of spherical aberration and at full aperture the image patch corresponding to a point in the object has a pronounced fringe round the central nucleus. When the aperture of the lens is stopped down, the fringe gets smaller until, at about *f* 16, it is negligible and the definition is hard and sharp. In this type, therefore, softness is controlled by choice of relative aperture.

Where a marked degree of softness is not required, some normal lenses not reaching a particularly high standard of performance in other branches of photography can act in a similar manner.

Variable Aberration Lenses. In these types the softness of definition is under control at any aperture. These forms are usually derived from the basic Cooke-Triplet construction with provision for moving one or more of the lens components in relation to the others. This device introduces a variable amount of spherical aberration without seriously impairing the correction of other optical errors. Softness of definition also varies from aperture to aperture, but the control of softness at each aperture is particularly useful when exposure times and illumination levels are dictated by conditions outside the photographer's control.

When using portrait lenses of the second type it is important to realize that the relative movement of the lens components also changes the focal length and the position of focus. For this reason it is essential to focus on a screen in the camera at the aperture and diffusion setting that are to be used in making the exposure.

Focal Length. The focal length of a portrait lens is generally longer than normal in relation to the field covered. Where the normal lens has a focal length approximately equal to the diagonal of the field covered—equivalent to an angular field of about 55°—the portrait lens may have a focal length equal to 1¼ times the diagonal—equivalent to an angular field of 45°.

Portrait lenses are available in focal lengths from 1 in. (25 mm.) for 16 mm. cine cameras to 18 ins. and over for use on 8 × 10 ins. studio cameras. G.H.C.

See also: *Lens history.*

Book: *Photographic Optics*, by A. Cox (London).

PORTRAIT SERIES. In commercial studios these are produced with a semi-automatic camera (Photomaton or Polypose) which enables an unskilled operator to take a series of portraits and deliver a finished strip or sheet of prints. Machines of this type are popular in large department stores.

The sitter is posed under a standard lighting arrangement and the operator simply moves levers to focus, expose, and transport the negative material. The negative material usually consists of a single plate. Only a small section is exposed at a time, and the plate moves to bring successive sections behind the lens until the whole surface has been exposed. One popular system gives 48 exposures measuring 1¼ × 1¼ ins. in 8 rows of 6 on a 9 × 12 ins. plate. The customer is given a contact print of the plate from which he makes his own selection for enlarging.

PORTRAIT STUDIO. Modern portrait studios have developed with the psychology as well as the photography of the sitter in mind. They have grown to be a very different proposition from the early studios where gaslight and daylight were combined in a sort of greenhouse, equipped with clamps to screw the victim's head in position, and prodigiously bulky cameras. In those days it was said to be more pleasant to visit the dentist than the portrait studio.

Throughout the country to-day there are innumerable studios specializing in portraiture. These mostly consist of a small studio with a reception room, and the minimum space for darkrooms. In every case, the studio is tastefully decorated and shows the minimum amount of apparatus. The less ambitious may have walls decorated in pastel shades, bright chintz curtains, and a bowl of flowers on the window-sill. More elaborate establishments may boast of decorative hangings, a period fireplace, and other "props" which give the studio the necessary atmosphere and are also useful as settings.

Equipment. Many studios retain the large studio camera mounted on its heavy table-stand, especially those studios which use contact prints from large negatives to allow for extensive retouching on the negatives. But it is becoming more usual to find a smaller camera of modern design standing unobtrusively in a corner, with lighting equipment that blends with the general scheme of the room.

Portrait studio lighting equipment usually consists of a top-unit where a row of large half-watt bulbs mounted in reflectors provide a bright general light with the minimum amount of shadow. Facial modelling or roundness is achieved with screened floodlights to the right or left of the sitter and the equipment is completed by a good focusing spot for adding local emphasis highlight.

In recent years many of the larger studios have installed high powered speed flash equipment. This allows both brief exposure times and a small lens aperture to be used, to capture fleeting expressions sharply on the film. Such equipment is particularly popular for colour work.

The lights usually also have a low powered half-watt modelling light in each reflector, making it easier to position the lights for the effects desired.

Unlike the commercial studio which has to be prepared to handle a variety of subjects, the portrait studio is concerned with one subject only, and its size as well as the possible viewpoints are limited. So most of the portrait studio equipment can be arranged in advance, leaving only minor adjustments to be made when the sitter arrives. Lights are often fixed on extendible arms, or on easily moved wheeled stands with cables from the ceiling. The atmosphere aimed at is that of a private apartment rather than a technical work-room. In many cases the subject hardly knows when the actual exposure has been made.

Children. Child photography is a very important branch of the portrait studio. Every portraitist has a personal technique; but all seek to create a feeling of freedom and provide something to occupy the young mind. Toys are a necessary part of the studio "props". Complicated apparatus is kept well out of sight, and portable gadgets out of harm's way. The success of the photograph largely depends upon the photographer's ability to handle his subject.

With a young, highly imaginative mind there is no need for expensive and elaborate diversions; ordinary objects and simple ideas are always more successful in establishing the state of mind required.

The psychological reactions of the child must be borne in mind in designing a portrait studio. Apparently trivial points can exercise a profound effect on the mood of the sitter: doors that stick, or that look heavy and as though they would take a lot of effort to open. All such things, and any suggestion of technical mystery tend to destroy the desirable atmosphere of freedom.

Expression. The retouching department is an important part of the portrait studio. The camera, left to itself, does not always give the most attractive rendering of the human face.

Special lighting facilities help to produce more pleasing results, but even so, few people like to see their faces as they appear in an unretouched photograph. This is not vanity; we apply it to our friends as much as to ourselves. The reason is that we are accustomed to see a face in a state of animation. Constantly changing moods and expressions follow one after the other. A still picture freezes one particular expression or one viewpoint to the exclusion of all else; we see in the picture a single phase, instead of the complex sequence we are accustomed to.

Early attempts to soften the too-factual vision of the lens introduced gauze screens and a strange assortment of other objects into the studio. These have their modern equivalents in many studios. On the other hand there are studios that deliberately go in for hard lighting to emphasize character, particularly in photographs of men and elderly people. Such studios have no retouching department in the accepted sense of the word. But in the average studio the camera operator with his lights (and perhaps, a soft focus lens) and the professional retoucher work together as a team to give the public the type of portraits it prefers. And the studio is planned on the basis of this partnership.

Colour. By making use of negative colour film, in conjunction with the services of a commercial colour laboratory, even the small studio can supply colour portraits at a reasonable price. Natural colour photography has almost entirely replaced the hand coloured portraits which used to be popular. B.A.

See also: *Portrait studio camera; Studio; Studio photography.*

Book: *The Business of Photography*, by C. Abel (London).

PORTRAIT STUDIO CAMERA. Type of camera intended for taking portraits, full-length figures, and small groups arranged at one end of the studio and illuminated principally by fixed massed lighting. It is designed to move about on wheels but can be instantly anchored in any position. All controls are large, easy to handle, and silent. The whole design and construction of the camera are aimed at distracting the attention of the sitter as little as possible.

As most studio photographs have to be retouched, the camera is made for large negative sizes—never smaller than half plate, and generally whole plate or 8 × 10 ins.

As it is used exclusively for photographing people in more or less standardized poses under the same sort of lighting conditions, the camera has no use for camera movements like swing back and rising front, and it has no need to be either collapsible or portable.

Camera Body. The body consists of a baseboard with a lens panel and plate back, both of which can be racked backwards or forwards. An extra-long bellows is fitted and the camera has a generous extension to allow the use of relatively long focus portrait lenses. All focusing controls and movements are operated by large hand-wheels or handles. The body is supported on an elaborate stand on wheels which has controls for raising, lowering, and tilting the head, and a device for locking the wheels when once the camera has been pushed into position.

Lens. The studio photographer does practically all his work with lenses of 10 ins. or more focal length and apertures seldom larger than $f 4{\cdot}5$. Most of the lenses used in portrait cameras have a certain softness of definition; this softness can be achieved in several ways, including the exploitation of certain lens aberrations. In some cases the degree of softness can be controlled by stopping down.

Shutter. The favourite shutter for portraiture is the silent-opening, "eyelid" type. This type of shutter is quite noiseless and does not need to be pre-set. It is opened by pneumatic pressure applied by an indiarubber ball held in the operator's hand and connected to the shutter mechanism through a long flexible tube. The shutter stays open until the operator releases the pressure, and the whole action is so silent that the sitter may not even notice.

Negative Carrier. The back of the camera is equipped with a special negative carrier designed, like the shutter, to be smooth-working and noiseless. It consists of a grooved frame carrying a special type of plate holder which is inserted into guides on the camera back.

The guides also carry the focusing screen so that when the plate holder is pushed into position for exposure, the focusing screen automatically slides out of the way. At the same time, a stop on the negative frame engages with the dark slide of the plate holder and uncovers the plate as it is pushed into position. After exposure the dark slide covers the plate once more as the holder is withdrawn.

See also: *Portrait lens.*

PORTRAITURE

Taking portraits can be one of the most rewarding branches of amateur photography and at its highest level it is probably the most difficult work in which a professional can specialize. This is because a successful portrait demands perfect photographic technique, a very definite measure of artistic ability and a flair for handling human beings.

Individual style shows itself here more than in most kinds of photography, but there are certain basic principles that are common to every approach to the subject. These basic principles are concerned principally with the methods of controlling the subject and the lighting.

POSING THE SITTER

The successful portrait photographer is interested in people and can put them at ease, and remain at ease himself while he is taking a portrait. He uses the technical processes purely as a means to arrive at a live and characteristic likeness of the subject. He never forgets that his own mental attitude has an effect on the sitter, and he takes a keen and personal interest in the portrait in hand. His trained observation allows him to see a characteristic pose or expression, even during a session that is sometimes short.

Making a portrait should be a pleasant experience on both sides of the camera, and everything depends on the personality and "managing" ability of the photographer. By keeping up a natural conversation he can break down any reserve or lack of co-operation from the sitter. He should suggest rather than command a pose, and by an occasional word of encouragement and praise, he can aid the confidence of the subject and banish camera-consciousness. A skilled photographer recognizes at once when the sitter looks and feels easy and natural, and he directs his efforts to that end. While there are many individual methods of approach, the object is always to secure a "live", characteristic portrait.

An efficient routine allows the work to proceed smoothly, without fluster or disturbance, and has a favourable influence on the sitter. Operations are usually directed from the camera, where the pose can best be studied. This arrangement allows the sitter to take up the suggested pose without having to be pushed into it by the photographer, a procedure which many people object to. Everything is done to make the sitter feel comfortable and relaxed, but the directing is pleasant and unobtrusive with a complete absence of the professional manner of earlier days. Some subjects are easy to approach, others difficult, and experience in dealing with the various types of sitter helps the photographer to the desired result without wasting time. The most successful portrait photographers are those who have cultivated this understanding of their sitters and the result shows itself in their work.

The Pose. The required natural, easy pose is sometimes a spontaneous action on the part of the sitter, but it is the photographer who decides when it looks right or directs movements to achieve it. For a head and shoulders portrait the subject is rarely taken squarely facing the camera. The body should be turned at an angle away from the camera, and the head slightly back towards the camera, giving a three-quarter or full-face view as desired. The seated model should always lean forward slightly into the picture. In a close-up the head generally fills the frame, and with suitable subjects parts of it may even be trimmed off. A feeling of action can often be imparted by giving the head a tilt either when enlarging or trimming the print.

Three-quarter length poses usually include the whole of the figure to just below the knee when seated. A slight turn of the body is again useful, and the legs can be turned sideways to avoid the appearance of fore-shortening produced when they point directly towards the lens. A more informal look is given in a man's portrait if he crosses his legs, usually by placing the one nearest the camera on top of the other.

This type of portrait includes the arms and hands, and they, too, must be carefully posed to avoid a look of angularity or stiffness. Any suggestion of arrested action is avoided by adopting only the quieter and simpler poses—e.g., a woman with her hands resting in her lap or a man resting his arms and hands on a table. Appropriate accessories that can be held naturally in the hands give the subject something to do with them, and help in achieving an unaffected pose.

A suitable chair or stool of good but not assertive design is a necessary item; it must not be too low or the knees will tend to stick up awkwardly. When the subject is elderly or important the pose should be a simple and dignified pose, but more freedom is permissible with younger people, particularly in dealing with the more sophisticated and decorative types of women. The poise of the sitter—i.e., the manner of holding the body—is a vital part of successful posing: it should be neither stiff nor slouching.

Full length portraits are confined more to the display of ceremonial attire and records of formal occasions. Here, too, great care is taken to secure a well-balanced pose. The subject must not seem to be falling backward by leaning too much on his heels or a support; it is a good rule for the weight to be thrown forward slightly on the ball of the foot. Movement of the subject can be avoided by using a short exposure, but it is always as well to have a support, such as a chair or table, near the subject to prevent the slight swaying that is difficult to control. Much of what has been said about the poses for three-quarter length portraits applies equally to full lengths. The problem of filling the larger area around a full length in order to make a pleasing picture is largely a matter of suitably lighting or furnishing the background.

Expression. To please his sitter, the professional photographer must be able to secure a pleasing or characteristic expression. This requires sensitive observation on his part, and the restraint to avoid extremes. The broad grin or the hearty laugh is out of place in sincere portraiture which is to serve as a vivid and faithful record of the sitter. The expression of the subject is the mirror of the mental state and the penetrating vision of the camera reveals at once any trace of artificiality. The photographer must start by deciding whether he is to portray the transient qualities of the snapshot, or seek to create something more permanent that can be looked at over a period of time without loss of interest. An instantaneous exposure is no guarantee that the best expression will be captured; many a portrait that possesses all the freshness and spontaneity of a snapshot is the result of a time exposure lasting several seconds—it all depends upon the photographer's ability to recognize the right moment to expose.

The features that most directly influence expression are the eyes and the mouth. Apart from any mental or emotional feelings that the eyes may express, it is most important to see that they are looking in the right direction. Turning them to one side, even slightly, can give a furtive, suspicious look, and the whole effect can be ruined if they are looking too high, or too low, or half closed because of the brightness of the light. As a general rule the eyes should look in the direction to which the head is turned. Looking straight into the lens almost inevitably leads to an expression of forcefulness which may or may not be desired. If the subject habitually wears spectacles, he should be taken with them, or he will look and feel strange.

The mouth is a mobile and expressive feature that can convey every emotion from anger to happiness. It is a great revealer of character, and as such must be watched very carefully when timing the instant of exposure. There is often a point before a full smile is reached, or after it has nearly gone, when the lips have a pleasant expression that suits most people. The popular trick of asking the sitter to pronounce some short word which results in a half-opened mouth at the moment of exposure, serves a similar purpose.

With subjects where gravity and dignity are more suitable the photographer must wait for an expression of calm repose. The moods and emotions reflected in the human countenance are many and varied and the eye of the photographer must be trained to recognize them the instant they appear before he can expect the lens to record them.

Features. The type of sitter that the portrait photographer refers to as "photogenic" possesses features of such regularity and proportions that a satisfactory portrait of them can be obtained fairly easily, from any one of a variety of angles. But with most subjects one of the greatest problems in taking a portrait is to choose a viewpoint and pose of the head that will minimize or avoid often considerable irregularities of features and proportions. With a studio portrait which must please the sitter this is imperative, although in a pictorial character study for an exhibition it may not be so necessary. A woman usually has a definite opinion about which is the best side of her face, but the camera does not always agree with her.

If a line could be drawn down the centre of a face, it would be found, in many cases, that the nose was not quite straight, and that one side of the face was fuller than the other, while a line at right angles would show differences in level in the eyes and mouth; classical symmetry is rare.

One of the methods usually employed to overcome the appearance of a mouth that is not level is to turn the head away from the camera to a three quarter or near profile view, keeping the lower corner next to the lens. The same method is adopted when the distance from the outside corner of one eye to the corner of the mouth is greater than the corresponding distance on the other side. The longer side is kept next to the camera. Receding parallel lines appear to converge so in posing a sitter in the manner described, the lack of symmetry in the face is attributed to the normal convergence of such lines, and the face itself is seen as quite symmetrical.

A crooked nose is helped by photographing it from the side towards which it turns, and a broad nose is narrowed by taking it in three-quarter view with careful lighting to put the broader portion in shadow. A long nose or a weak chin is never emphasized by taking them from above or with the head tilted down. With some sitters a sideways tilt of the head can make the angle of the jaw look too prominent, and when the cheeks are seen in outline, they must be compared to make sure that they are symmetrical, as a slight turn can make the face look lopsided.

In portraits of women the neck calls for special attention; too high a viewpoint makes it appear shortened, while the shoulders are made to look higher. The normal height of the camera, at the eye level of the subject, gives a satisfactory rendering and only slight variations of this are permissible, the more extreme angles being reserved for special subjects such as fashion photography.

With men, bald heads are made to appear more prominent by over-lighting, and the light on top of the head should be toned down by a soft-edged shadow from a small head screen, or by a piece of card held in an assistant's hand.

The edge of a beam of light, being softer than its centre, should also be employed whenever possible.

Where the sitter possesses an interesting or beautiful profile, it often makes the best portrait, particularly when certain irregularities can also be played down by taking the picture this way. The choice of the correct viewpoint in all such cases is most important and the movement of the lens an inch or two to one side or the other can have a critical effect on the final portrait.

The lighting effects are chosen both to give the best modelling of the features and to be in keeping with the type of subject. A round, full face, for instance, is made to look even wider and fuller by a broad, frontal lighting, but by lighting it from the side to give the head some bold, firmly-drawn shadows, it can be given a much improved form and structure. Thinner features, on the other hand, can be flattered with the more open frontal type of lighting. Lines of character are brought out in a face by a bold layout of lights and shadows made by the chief source of light.

Hands. The hands in a portrait are next in importance to the head as interesting and expressive features. Like faces, they vary in character and beauty. Some sitters are able to pose their hands easily at the suggestion of the photographer; those who cannot, or who have not good hands, call for a simple and unobtrusive treatment.

Certain views and poses are in any case to be avoided with all types of hands. The broad, flat view of the back or front of the hand to the camera is unpleasing because it looks larger and lacks the slenderness, beauty of line and tapering proportions of the hand viewed from the side. This essential difference can be seen clearly by comparing the effect of placing the hand at the side of the head, and placing it under the chin with the fist closed and the back of it to the camera. Unusual or strange views of the hands should also be avoided—e.g., the acute foreshortened view of a hand pointing directly into the lens, fingers that are amputated, missing, or awkwardly interlaced.

Stiffness, both mental and physical, prevents good poses, and anything that can be done to give the sitter a relaxed, easy feeling helps to produce a better result. A lot of directions make the subject conscious of his hands; it is usually easier to arrive at a satisfactory arrangement by simply giving him something to hold or rest them on. In most cases, and in particular, with men, the simplest treatment with the hands in repose is the best, but where a woman has particularly beautiful hands, they can be given greater prominence by expressive poses that reveal their line and style to advantage. There are many possibilities for elegant and expressive arrangements in posing the wrists, hands and fingers of younger and prettier women in particular.

Children's hands should be left alone: their hands will look well in almost any position so long as it is arrived at naturally. If they are given something to hold, the object should be small, so that their hands can grasp it easily and be seen instead of being hidden by the object.

Backgrounds. The modern studio treatment of the background is generally simple. Painted landscape and rustic bridges belong to a bygone age. Head and shoulder portraits are taken against a plain background of black, dark grey, light cream, grey or white, and the variations in tone are produced by lighting. Even a white background can be made to photograph black if it is far enough behind the subject and all light is kept off it.

When sitters are taken in their own homes or offices, etc., the background is usually confused by a multiplicity of objects, patterns and furniture. As far as possible these should be arranged or removed to leave a simple setting for the subject.

Draperies often make effective backgrounds, especially for portraits of women and children, but the obtrusive patterns must be avoided and the folds should be carefully arranged and not sharply defined. The sheen of silk, or light through a gauzy film of muslin can lend a pleasing decorative effect, but it must always be kept subordinate to the sitter. In three-quarter and full length portraits there is more background in proportion to the subject than in head and shoulder portraits. Here, the judicious use of a piece of furniture, curtains or other suitable props, and variation of the background's lighting can help to create an effective composition and fill in the empty spaces.

Clothes. For the more formal type of portrait the appearance of the clothes, especially of women, needs careful attention. A dress that hangs in badly-arranged folds, or is creased and crumpled, or has a collar disarranged, etc., will not be acceptable to the sitter. So it pays the photographer to see that everything is in order at the sitting, to avoid the labour of correction on negative or print. Even a necklace must be properly arranged or it will mar the appearance of a well-groomed woman, and men's suits must be checked for bulging collars and creases in the wrong places. Service uniforms in particular call for great attention down to the last button. No such care is necessary with the easy, informal type of portrait which may even be improved by a suggestion of untidiness.

Make-up. Modern panchromatic emulsions give a smooth rendering to complexions, but occasionally more pronounced freckles and skin blemishes persist in showing. If these are objectionable to the sitter, they can be masked by the use of the correct make-up. This must be chosen to suit the type of lighting and sensitized material; ordinary make-up is rarely satisfactory. The use of make-up to create character or deliberately alter a face calls for special skill. H.J.Wi.

LIGHTING

Controlled lighting in portraiture, as opposed to mere general illumination, is used for modelling—i.e., to create an impression of roundness and solidity in the two dimensional reproduction of the sitter's head and features—and to give a good likeness. Good lighting is not a matter of using a large number of high-powered lamps. Some of the best portraits are produced by means of very simple lighting, just as many of the greatest masterpieces of portrait painting have been produced by the use of just one simple light source, generally a diffused form of daylight.

Sunlight, diffused daylight, half-watt lighting and flash bulbs are the principal sources of light used in portraiture and there are many types of lighting that are capable of producing a first-class portrait.

Because of its convenience and portability, half-watt lights have become the most popular form of lighting for portrait photography. Such lights are easy to move into any position or any angle, to give front light, side light, and back light as required. Among the most popular light sources for amateurs are Photoflood lamps, which give a very intense light but have a much more limited life than the ordinary half-watt lamp.

These lamps are used in various types of reflectors and on a variety of stands which enable them to be placed in any desired position. The design of the reflector is important. A lamp with a narrow reflector will give a harder image than a lamp in a broader reflector, which tends to diffuse the light more.

Spotlights give a stronger and more concentrated form of light. They are useful principally for back lighting or effect lighting, but can also be used as the main source for strong characterization or realistic skin texture.

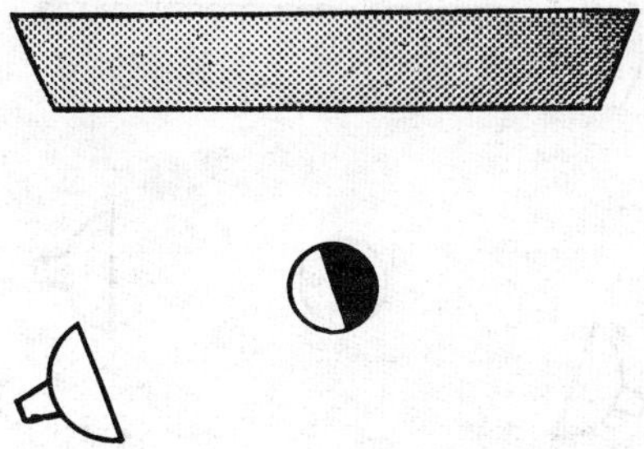

MODELLING LAMP. One main light only produces hard and deep shadows on the side of the subject away from the lamp. The background tone gets darker farther away from the lamp.

Photoflood Lighting. The photographer making a portrait by artificial lighting has to decide how many lamps he will use, how powerful they must be, and where to place them. The commonest mistake is to use too many light sources with the result that one interferes with the other and there is no definite light or shade in the final picture.

A good lighting scheme may include the following basic units: the modelling light or main light, the general light or "fill-in" light, the background light, and the effect light.

Modelling Light. The purpose of this light is to form the actual light and shade areas in the portrait. All other lamps used are subsidiary to this light and are never allowed to compete with it.

Just as there is only one sun in the sky, giving one definite light and shade, so a good portrait produced by artificial light should have one principal source which produces a definite light and shade pattern.

General or Fill-in Light. The purpose of this light is to give an all-round illumination to the subject without throwing any definite shadows. Its object is to illuminate the shadows produced by the modelling light, and so improve the rendering of detail in these areas.

Photographic negative material cannot

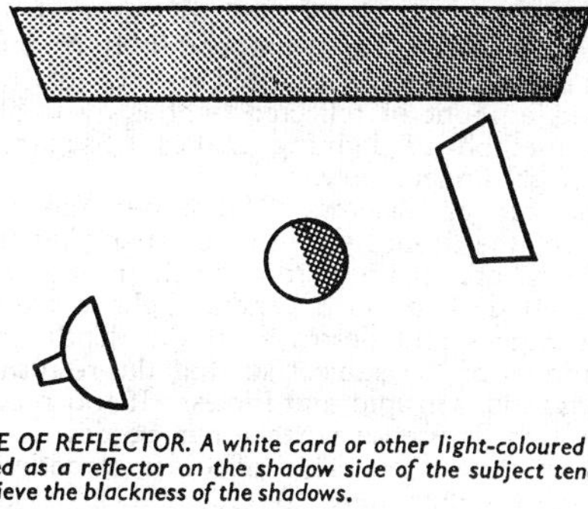

USE OF REFLECTOR. A white card or other light-coloured area used as a reflector on the shadow side of the subject tends to relieve the blackness of the shadows.

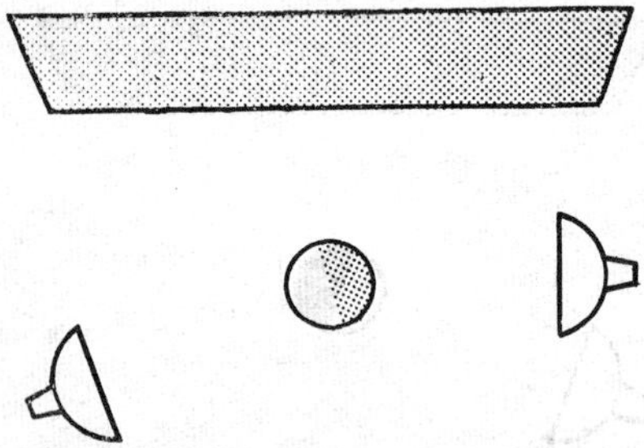

USE OF FILL-IN LIGHT. *This lights up the shadows much more than a reflector can, and produces a more even light balance. The background also becomes more uniform in tone.*

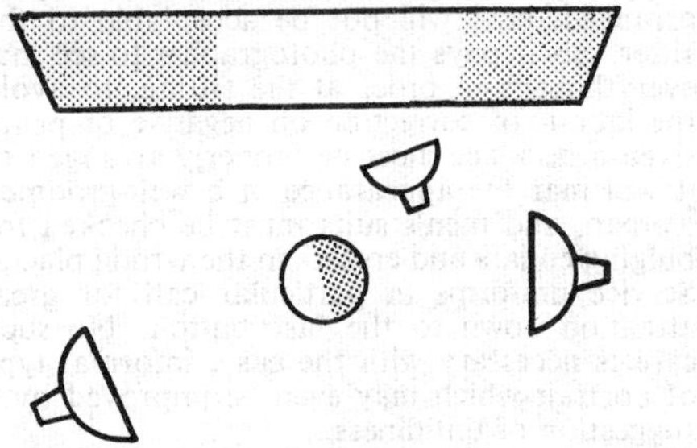

BACKGROUND LIGHT. *A third lamp directed on to the background can give haloes and similar effects. The basic lighting balance on the subject itself is not affected.*

accommodate itself to extreme contrasts in the same way as the human eye and deep shadows in which there is still detail discernible to the eye can frequently appear as blank areas on the negative. So there must always be sufficient general light to give the shadows the necessary detail.

Much of the general lighting can be provided by reflection from the surroundings. If the sitter is placed in a room where there is no daylight and where the surroundings are fairly dark, when the main source of light is switched on the shadows will appear extremely black and the whole lighting effect will be harsh and contrasty. On the other hand, if the walls of the room are white or light the shadows will appear very much softer and more luminous.

The most convenient way of producing the necessary general light is by a second lamp placed at a greater distance from the sitter than the main or modelling light. It may be used direct, or turned to shine on a wall or a white screen to light up the subject indirectly. Great care must be taken that this second source of light does not throw any shadows on to the face. Alternatively, instead of a second lamp, a reflector consisting of a white screen or sheet can be used to give the necessary illumination of the shadow areas.

A room in which there is daylight, particularly if it has light walls, can also be an ideal location for making a portrait, because the diffused daylight reflected from the walls provides an excellent type of general or foundation light. The window blinds or curtains can be opened or closed so that just the desired amount of reflected light is available. This method of lighting makes a separate fill-in light unnecessary.

Importance of Shadows. The power and distance of the fill-in light must be such that the shadows are neither too black nor over-illuminated. Too much general light weakens the shadows and destroys all the depth and roundness of the subject so that the resulting portrait looks insipid and lifeless. If too much fill-in light is used a portrait can be devoid of shadow altogether. This effect is sometimes produced deliberately to create a high key effect. Again, if the fill-in light is incorrectly placed, rival shadows can be produced which spoil the necessary positive character of the lighting.

An excellent way for the beginner to train his eye in the use of artificial light is to practise on a plaster bust, or a display head used for showing off hats in shop windows. In this way he can experiment without tiring his sitter. He should also study the lighting of leading portrait photographers and painters.

Background Light. The background light has a number of possible duties. It may simply be used to brighten the tone of the background where it is much darker than the subject by reason of its greater distance from the main light source. It may also be used to "kill" the shadow of the subject cast on the background by the modelling light. Or it may be used locally on the background to produce intentional differences of light and shade to complete the composition of the picture. It can consist of a half-watt lamp or a spotlight which is projected from behind the sitter or from the side on to the background from a position in which the source does not show in the picture. In such a case the sitter must be a little distance from the background. Illumination of this type helps to detach the model from the background. It may also be made to produce various patterns of light and shade according to the photographer's individual ideas.

It is a fairly common practice to illuminate the part of the background behind the darker side of the model leaving the part behind the lighter side of the model in shadow so that the subject stands out by contrast on each side. By shifting the background light, it is just as easy to allow the darker tones of the subject to merge gently into a dark area of the background. When using a plain neutral background, a background light can add a valuable play of light and shade to relieve an otherwise monotonous grey area.

Effect Light. The effect light is popular for the production of what are known as glamour portraits. In its commonest form it consists of a spotlight or half-watt light directed on to the sitter from the rear, producing a kind of halo, or a rim of highlight. One or more such light sources may be used, and usually the more forceful the light the more glamorized the portrait.

Effect lighting can also be used for emphasizing the contours of the profile, but it must never be allowed to shine on the face generally because it can easily ruin the harmonious lighting of the modelling light. Again, if it shines on existing highlights, they become over-exposed and will print as blank areas in the finished portrait.

Lighting Styles. Broadly speaking, there are two principal styles of lighting in portrait photography: dramatic lighting and classical lighting.

Dramatic lighting is the type of lighting which has been popularized by the movie picture and by the advertising studio. This consists of the use of a great deal of effect light, both on the sitter and also on the background.

Classical lighting is the type of lighting favoured by the great masters of traditional painting. It consists of only one dominating source of light giving a co-ordinated pattern of light and shadow without "effects".

Dramatic lighting is popular today, both in private portraiture and in commercial photography. Classical lighting may be less fashionable, but it is more suited to sympathetic characterization.

Lighting the Sitter. The standard approach in lighting the sitter is to start with one lamp. This is switched on and the effect of the shadows is studied. If these are too heavy for the type of portrait in view, the second lamp or general light is switched on and moved backwards and forwards until the shadows caused by the main light are sufficiently illuminated. The higher the main light, the longer the shadows under the nose and eyes of the sitter. When the main lamp is moved to one side of the sitter it will leave one side of the face more in shadow than the other. As the photographer moves his main light into different positions to achieve the lighting best suited to his sitter, the position of the fill-in light must also be altered so that the shadows are kept illuminated.

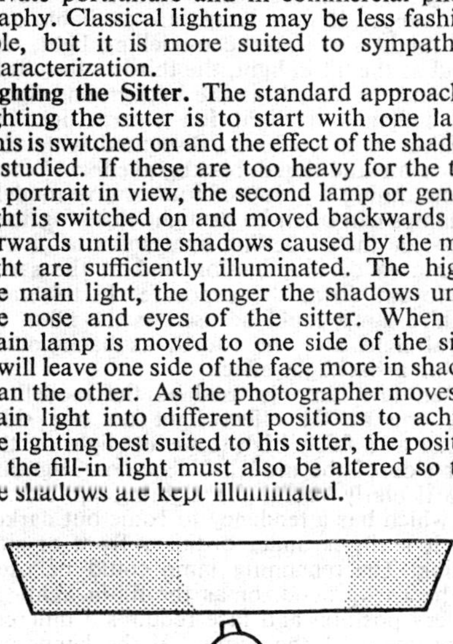

USE OF EFFECT LIGHT. A fourth lamp completes the lighting set-up by adding highlights to the hair, or, from a lower level, producing a rim of light around the subject.

Every sitter calls for a different arrangement of the lamps to suit his or her personal character and this offers tremendous scope for individual expression in manipulating the lighting. Diagrams and examples can provide only a rough guide. The beginner must learn to see photographically: experience alone can develop his feeling for the correct lighting. When he gains this he will instinctively place the lamps in the right position to give the best result for each type of sitter, without referring to diagrams.

The form of lighting which will always give a good likeness is a simple top-front-side-light, but there are many variations to it. It is impossible to lay down any hard and fast rule; the final arrangement is always a matter for the photographer.

Except for special effects—e.g., high key—the main light should never be at the same level as the sitter and shining exactly from the direction of the camera. This arrangement produces shadowless and flat lighting. A main light placed below eye level gives an unnatural type of lighting that is generally not suitable for straight portraiture, although it is sometimes appropriate when taking stage photographs of actors or actresses.

Generally speaking, a good starting point for lighting experiments is with the main light shining down at an angle of roughly 45° on the sitter, and 45° to one side of the camera-subject line.

The character of the subject is brought out by the shadows, so the weaker the fill-in lighting, the stronger will be the impression of character.

A soft general lighting with few or no shadows tends to idealize the subject. When thit effect is desired, well diffused frontal lighting some distance from the camera is employed. The nearer the lamp is brought to the sitter, the stronger the lighting becomes and the more it tends to emphasize heavy, rugged and male characteristics. So when photographing women a softer form of lighting employing diffusing screens is appropriate, whereas with men a stronger type of lighting is more suitable.

Full Lengths and Groups. In a portrait of a full length figure, it is important to have the main light high enough. Lighting that would give a good top-front-side-lighting of a seated portrait would, if used at the same height for a full length standing portrait, give an illumination too near head level. Some lighting stands, particularly those sold for amateur use, will not extend sufficiently high. The remedy is to get an assistant to hold the lamp at the correct height. In such portraits it is also important to remember that the lamp should be far enough away from the sitter to illuminate the whole of the figure. With a head and shoulder photograph this point is not important because the main lamp light may be anything from 4 feet to 10 feet from the sitter and still give excellent lighting. This distance is much too close for a full length portrait.

Artificial light falls off in strength rapidly as the source is moved away from the model. A lamp moved to twice the distance away from the sitter—say from 2 yards to 4—would not require double the exposure but four times. So figure work calls for very much more powerful lighting and when working with full length portraits it is often an advantage to use two or three modelling lights in a vertical row to cut down the exposure while ensuring that the whole of the figure is evenly illuminated. With groups too, it is necessary to move the lamps farther away and allow extra exposure.

One Lamp Only. There is no hard and fast rule about the number of lamps needed for a portrait photograph. Excellent portrait studies can be secured with one lamp only.

When there is sufficient general daylight in the room, then a single lamp can be used. The artificial light will then serve as the modelling light and the daylight as the fill-in light. In rooms with light walls and with windows in more than one side of the room, one lamp can be made to give excellent results. The lighting is best arranged so that the shadow side of the model receives the direct daylight. On the other hand, if there is a strong source of daylight, the model can be brought quite close to the window, and the daylight used as the modelling light, while the Photoflood lamp is used to illuminate the shadows formed by the daylight. But nearly always it is more satisfactory to use artificial light as the modelling light and daylight as fill-in light.

Sometimes the subject can be placed so that the daylight from a window comes from behind and a little to one side. This arrangement creates an attractive backlight effect. The effect can be intensified by bringing the model nearer the window, but if this is overdone it will overpower the main light and the highest effect lights formed by the daylight will be so dense on the negative that they will print white.

A single lamp can also be used with a reflector to light up the shadows where there is no daylight available as the fill-in light. An additional backlight effect can be obtained by using a mirror from the rear of the model to reflect back the main light as an effect light, to illuminate the hair or the contours of the face. Thus, even a single lamp offers a selection of lighting arrangements by which good portraits may be made.

Two Lamps. When there are two lamps available, one is used as the main light, and the other as the fill-in or general light. If there is some daylight available but not enough to light up the shadows, the second lamp may be used to augment it. If the daylight in the room is strong enough to provide all the general lighting, the second lamp may be used as an effect light placed behind the model and a little to one side so that the light comes more or less towards the camera.

The second lamp may also be used for illuminating the background, being placed so that it shines from behind the model on to the wall. It may equally serve as a combined effect and background light. This is done by placing the lamp a little behind the sitter as it would be if it were serving as a background light but turned slightly towards the sitter so that part of its rays outline the model's head and part fall on the background at a slant.

Three Lamps. When three lamps are available it is usual to employ one as the modelling light, one as the fill-in or general light in the absence of daylight, and the third as a background light or to backlight the model. This lamp may also be made to operate as both a background and effect light as already described above.

If there is enough daylight for general lighting, the lighting set-up might be: one lamp as the main light, the second as an effect light, and the third as a background light. Or a double effect lighting might be arranged by having a lamp behind the model on each side. Or the background might be illuminated by having one or both of these effect lamps shining partly on the background at a slant.

Four Lamps. If four lamps are available one lamp might serve as the modelling light, the second as the fill-in light, the third as the effect light, and the fourth as the background light. With sufficient daylight for general lighting, the fill-in lamp could be dispensed with and used as an additional backlight effect. When photographing a group of four or five models two lamps can often be brought close together to provide the modelling light so as to give a bigger area of illumination and at the same time to cut down the exposure. In this case the third lamp could be used as the fill-in or general light and the fourth as background light or a backlight effect.

Often one lamp is used as the modelling light, one as the effect light on one side, another as a fairly high top effect light on the other side of the model. This type of lighting is particularly useful for lighting up blonde hair which has a tendency to come out darker than it really is under ordinary front or side lighting. The remaining lamp could be used for the background, or as the fill-in light.

Every position and face requires a different arrangement of the lamps. If the lamps are in a correct position for a full face or three-quarter face portrait, they must be readjusted if the face is moved to profile position otherwise the profile will be wrongly illuminated.

Lighting the Profile. What has been said above applies to full and three-quarter views of the face. The profile is best illuminated so that the main source of light comes from between the profile and the background either illuminating the profile or the contours of the profile, or shining across the contours so that the near side of the face is illuminated with a patch of light and the remainder of the face is in deep shadow.

Flash. Satisfactory portraits can be made with flash bulbs. The disadvantage of this form of lighting is that unlike half-watt lighting one is unable to see the net result of the lighting before making the exposure. After some actual trials, however, backed by a thorough knowledge of half-watt lighting, the photographer can predict with near certainty the effect of any arrangement of flash bulbs.

Cameras today are mostly sold with an attachment for a flash bulb unit on the top or the side of the body. This by itself does not give satisfactory portrait lighting because it produces no shadows, and the result tends to be flat. In practice a larger size bulb than would be used for flash-on-camera is connected to an extension lead and used to provide the main light. The flash bulbs are, in fact, placed more or less where the ordinary half-watt lighting would be.

A broadly diffused general lighting can be created by "bouncing" a flash off a white wall or ceiling.

Sunlight. Much good portraiture is done with sunlight alone.

Low angle or slightly diffused sunlight gives the best results; high midday sunlight in summer is normally too harsh. Fill in light for the shadows is given by a white card or sheet used as a reflector. A small flash bulb may be used to illuminate the heavy shadows, but the technique must be used with restraint or the shadows become over-illuminated and the whole effect of sunshine is lost.

Flash bulbs are also useful out of doors for illuminating the camera side of the subject in against-the-light effects. On a dull day a flash bulb will give an impression of sunlight out of doors if it is fired from above and to the side of the subject.

Diffused Light. Diffused daylight from a window, with a white card as a reflector, makes a satisfactory lighting set up for portraiture. In this case as the light source cannot be adjusted, the sitter must be moved into the position that gives the most favourable lighting angle. If the window has curtains or blinds these can be opened or closed to control the area and placing of the light falling on the sitter.

When a white card is used as a reflector either indoors or out of doors it is important not to bring it so close that the depth of the shadows is so low that the result will look unbalanced. A.M.

TECHNIQUE

When the subject has been posed and the most suitable lighting arranged, the final important phase is the actual technique of producing the photograph.

Camera. Studio cameras are usually half or whole plate size with a square bellows of long extension, allowing lenses of long focal length to be used, and a ground glass screen on which to compose the picture. The back of the camera should have side and back-swing movements. The lenses can be from 12 to 20 ins. focal length; used on a quarter- or half-plate they give a head and shoulders portrait to fill the frame without going too close to the subject, thus avoiding distortion. The relatively small depth of field of these long focus lenses requires a stop of $f\,5{\cdot}6$ to $f\,11$ to give good definition on important parts of the subject.

A really solid studio stand is necessary to hold the weight of this type of camera and lens. It should be either of the central pillar type with a platform on top, or have a platform mounted between two vertical columns. The stand should be designed to allow the camera to be tilted, raised and lowered, and the whole apparatus should move easily on castors. The shutter is often behind the lens and should be silent. It is usually operated by a bulb and the photographer judges the exposure, although some makes can be pre-set to give the required exposure automatically.

Many successful photographers of children use a studio camera, but for this purpose the miniature or miniature reflex camera with quick focusing and automatic film transport is a very suitable instrument. The mobility and unobtrusiveness of the smaller camera enable the photographer to secure pictures with a spontaneous, unposed look. Excellent portraits of adults can also be taken with a miniature, using a long focus lens for close-ups.

Focusing. In the early days of portraiture lenses of the Petzval type, which gave soft definition except in the centre of the field, made it necessary to focus on the eyes, nose and mouth, leaving the rest of the head softly defined.

This gave a very pleasant effect of roundness, and present-day photographers using an anastigmat do this by selective focusing at a stop that avoids too great a depth of field. The ability of the modern anastigmat lens to give critical definition and show the texture of the skin and every line and wrinkle is exploited by some photographers to achieve a more literal and realistic treatment.

Negative. The miniature camera uses rolls of film, but for the studio camera there is a choice of plates or cut film. The latter has many advantages, being lighter, unbreakable, and less bulky than plates. Some panchromatic emulsions of high speed in artificial light, owing to their great red sensitivity, make the skin look like plaster, the lips pale and blue eyes unnaturally dark. A film of medium speed gives better flesh tones and finer grain.

Over-lighting, over-exposure and over-development are the enemies of good modelling. After a suitable balance of lighting has been secured by a few trials, the production of a good negative should present no difficulties, so long as exposures are kept within fairly narrow limits and the time and temperature of development are standardized.

A good negative is the halfway stage in making a print of fine quality.

Retouching. As some retouching is necessary in professional studio portraits, the image of the head should be at least 1½ to 2 ins. high. The once popular smooth, empty faces with all lines of character removed are fortunately rare now, and many of the best modern portraits are made from unretouched negatives.

Print. The key of the picture, high or low, and the contrast, soft or bold, in relation to the type of sitter and the depth of print are important factors that can be altered to control the final result. The size and position of the image on the enlarging easel and the final trimming of the print can be used to vary the composition. A successful portrait, however, is a work of art and not merely of technique, so that it is not possible to lay down set rules on the subject. The nature of the subject and the individual taste of the photographer should decide the treatment. H.J.Wi.

See also: *Babies; Children; Close-ups; High key; Lighting equipment; Lighting the subject; Low key; Make-up for photography; Portrait attachment; Portrait lens; Portrait series; Portrait studio; Portrait studio camera; Portraiture at home; Portraiture outdoors.*

Books: *The Art and Technique of Taking Portraits*, by H. van Wadenoyen (London); *Erith on Portraiture*, by J. Erith (London); *Faces Before My Camera*, by A. Morath (London); *Group Photography*, by G. Catling (London); *Camera Portraiture*, by H. J. Williams (London); *Lighting for Portraiture*, by W. Nurnberg (London); *Wedding Photography*, by G. Catling (London); *Fame*, ed. by L. F. Gruber (London); *Beauty*, ed. by L. F. Gruber (London).

PORTRAITURE AT HOME

Today, when fast emulsions have made portraits of people by ordinary room lighting a simple matter, the possibilities of this method of approach are not commonly realized. So many photographers nowadays think that a portrait cannot be really effective without elaborate lighting effects in a proper studio. Yet it is in their homes that people and children are to be found in their most characteristic activities and intimate moods. It is there that we get the authentic background that gives the key to their interests and outlook. The comfortable lived-in room, too often looked upon as merely a poor makeshift studio, is an infinitely better place for making true character studies than a studio can ever be. It should be left as far as possible undisturbed; to rearrange it is to destroy its authenticity.

The Best Camera. The camera must be quick in action and have a reasonably fast lens—f 3·5 at least—and a fair selection of shutter speeds, the slower ones being especially desirable. The more foolproof and automatic the camera is the better, so that more attention can be given to the subject. Quick and easy focusing is most important, especially for fairly close-up pictures of moving children; and portability, so that the level and angle of approach can be rapidly changed. Therefore a small reflex or coupled rangefinder camera is indicated.

Coupled rangefinder cameras with interchangeable lenses would seem the ideal, but, for taking moving subjects at fairly close range, many people claim that a focusing screen is quicker. Again, the eye-level position makes the camera more conspicuous and alarming than one held more casually at a lower level. Also, with a camera held to the face, it is difficult to converse with the sitter; with a reflex it is easy and also possible to take unexpected shots.

Camera Stand. The use of some form of support is often necessary when the lighting is too weak for fast exposure. This is unfortunate, because it makes the small camera almost as immobile as the traditional studio outfit.

The perfect camera stand does not exist; the nearest thing is the tripod with adjustable centre pillar, although this does not come down low enough for taking children at their own level on the floor.

Something home-made is generally the answer—e.g., a broomstick firmly held in a heavy base and fitted with a small sliding and tilting platform, held at the desired level by a suitable screw-clamp.

Lights and Reflectors. Though it is always wise to make the best of the existing lighting in order to preserve natural atmosphere, it will often be necessary to amplify or modify it considerably. To be prepared for all emergencies some sort of portable reflector is needed, a few Photofloods for use in ordinary light sockets, and one or two larger lamps on adjustable stands (with plenty of flex).

A synchro-flash outfit that can be used at some distance from the camera is another satisfactory type of light source.

The portable reflector need not be large—if it is it gets in the way; a piece of white oilcloth such as is used for table tops is very suitable. It should be about 3 × 4 feet, attached to two laths so that it can be rolled up when not in use; there should be some sort of collapsible stand to hang it on. It is an advantage to have some arrangement for holding the reflector at an angle.

Reflectors are easy to improvise, e.g., from a folding screen or clothes-horse with a white sheet draped over it.

Mirrors give a more directional lighting and are especially useful for backlighting effects but care must be taken not to reflect any light straight towards the camera lens.

Portable lamps should have **heavy bases** so

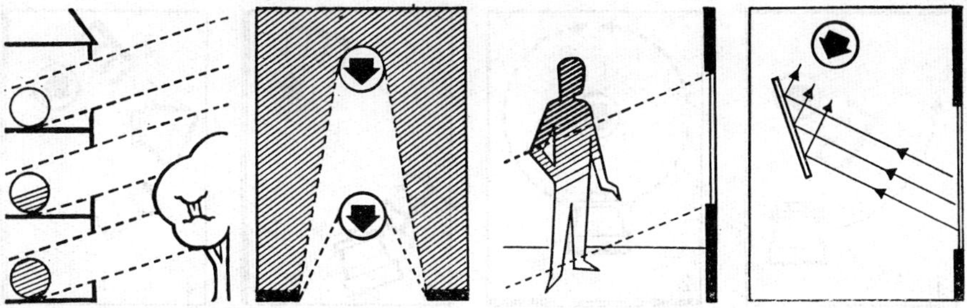

DAYLIGHT INDOORS. Left: *The light is often better in an upstairs room, as there is less obstruction from trees, etc., outside the window.* Centre left: *The farther the subject is from the window, the less light reaches it. The lighting also becomes softer.* Centre right: *The lower part of the room receives more light than the top—an important point when taking full-length portraits.* Right: *A reflector is usually desirable for indoor shots so that the light is well distributed and is not too contrasty.*

that they are not easily knocked over. The flex should either go through a hook or eye at the base, or be suspended overhead and clipped or hooked on to the picture rail or some other suitable anchorage. One lamp should preferably have a fairly large matt reflector, fitted with some form of detachable diffusing screen; the other reflector can be smaller and have a polished surface to provide a sharper, more concentrated light.

Flash has obvious advantages: it is powerful and portable. The disadvantage is that its effect cannot be studied beforehand and the result may show unpleasant surprises in the way of cast shadows and glaring reflections in spectacle lenses and other polished articles. Also, in its most common form—i.e., permanently attached to the camera—it gives a low flat lighting that tends to destroy modelling.

Perhaps the most useful method of employing flash in at-home portraiture is to bounce it from wall or ceiling. This floods the whole scene with soft secondary light which brings out shadow detail that would otherwise be lost.

A similar effect, though hardly as good, can be got by heavily diffusing the light with, say, a couple of handkerchiefs. This device is also useful with flash-on-camera as a method of controlling light intensity to get the right balance with the rest of the lighting.

General Lighting. It is better to look for an appropriate natural setting first and then decide if it is getting enough light for reasonably short exposures rather than to look for a setting where the light is right. There is no such thing as right lighting or wrong lighting in the absolute sense; any kind of lighting may be suitable in that it may be appropriate to the general scene and part of its mood or atmosphere, although it will not necessarily be possible to take photographs by it. However, with a fairly static subject it should be possible to catch the sitter with a characteristic pose and expression quite easily at 1/10 second or perhaps even ½ second.

When exposures must be short, the best thing is to look for light places—i.e., light rooms with the more high windows the better. (If they are in more than one wall, better still, for this will make it possible to get interesting cross-lighting effects.) Upstairs rooms are often better than lower rooms, where the light is apt to be obstructed by trees or buildings. Light-toned walls are to be preferred because they reflect a soft light throughout the room, whereas dark surfaces absorb light, and make for heavy shadows, long exposures and contrasty lighting.

The lightest place is always on the floor near the window and this is the ideal spot for taking children at play. But the light weakens rapidly as you move away in the room and it also becomes softer in character—i.e., the boundary between light and shadow becomes less marked. The upper part of the room always receives less light than the lower and this has to be borne in mind when taking grown-ups at full length, for, unless they are close to the window, the feet will get noticeably more illumination than the head. All this can be easily checked with an exposure meter.

Mixed Lighting. When using electric lamps and daylight together it should be remembered that, though to the eye the artificial lighting may apppear the stronger, the daylight may have the greater effect on the film. The result of any particular combination can only be foretold by practice. Judgement is easier if the excess of red and yellow, which causes the lamp-light to appear so strong to the eye, is held back a little by a pale blue filter in front of the lamp. A circle of pale blue gelatin (described as "daylight blue") such as is fixed on stage lights in the theatre can be used. It should be large enough to filter the direct light from the bulb, but should not fit flush over the lamp reflector or it will over-heat.

The comparative values of artificial light and daylight can be worked out by means of tables and an exposure meter, but in practice past experience will be found the best guide.

Sunlight Indoors. Sunlight effects indoors look most attractive, but the contrast between light and shadow is usually so extreme that the film cannot handle it unless the shadows are flooded with fill-in lighting. It will only very occasion-

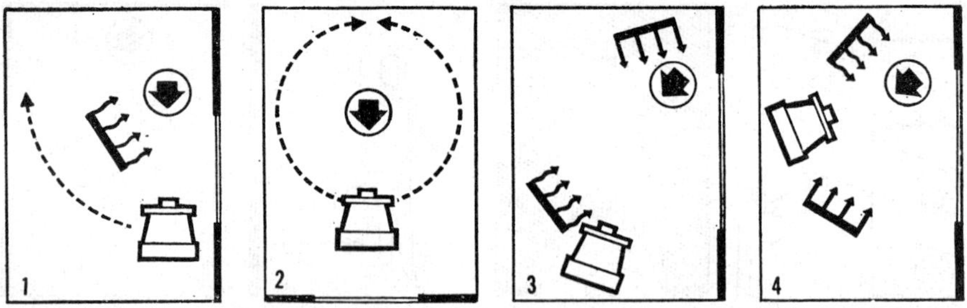

THE LIGHT FROM ONE WINDOW. 1. With the window more or less central in the wall, the sitter should be to one side of it, and a matt reflector on the shadow side. 2. Dead front lighting yields flat results; the more the camera moves round, the more shadow becomes visible, until behind the subject the camera sees only a silhouette outlined by light. 3. A mirror behind the sitter can act like a second window. 4. The mirror can be placed to provide secondary side lighting.

ally be possible to do this with a white reflector placed near enough to the subject for it to be able both to receive direct sunlight, and reflect it in the right direction. Sometimes a mirror will provide the right sort of reflection if it can be kept out of the sun. Direct sunshine reflected from a mirror would be much too fierce and probably also patchy.

The simplest way out is to supplement the fill-in lighting with one or two lamps or, perhaps better still, flash. If there is a light wall not too far away the flash can probably be bounced; if used direct it will have to be a fair distance away or suitably diffused. A photolamp can be used at fairly close range, and preferably diffused to avoid too definite secondary shadows. Two, close together, both diffused and rather farther away are likely to give an even better effect, which will look more like normal reflected light. Usually it is best to let the fill-in light come from the direction of the camera.

Exposure should now be set for the sunlight and will usually turn out to be very much less than might be expected.

Artificial Light Only. When there is no daylight at all, the effect of normal room lighting can be achieved by bouncing a soft flood of light from walls or ceiling with either lamps or flash. If short exposures are required, lamps or flash can be used direct and the general balance of light corrected by putting small Photofloods in the light fittings. This will preserve the natural appearance of a room, when using artificial lighting only, but care must be taken not to make conspicuous shadows and not to take away all the characteristic shadowy corners.

Unless Photofloods are used with restraint the home atmosphere is bound to go, so dramatic or fancy lighting effects should be avoided. For adults one lamp will generally be enough for the main lighting and a second one for shadow illumination. But in a large room it is sometimes wise to light up a distant part of the scene, which would otherwise come out too dark, and so make the subject stand out in perspective.

Most of the best pictures come from taking chances, and accidental effects of lighting often prove much more fascinating than those which are carefully planned. But it is never a good idea to place two lamps at equal distances from the subject, so that the one entirely cancels out the shadows cast by the other; this produces a really ghastly effect.

Light on the Background. When experimenting with the lighting on the sitter and trying various camera angles, the changing background should be taken into account. From one angle it may be the dark corner next to the window, then with a change of camera position a fairly well lighted wall comes into view. These variations

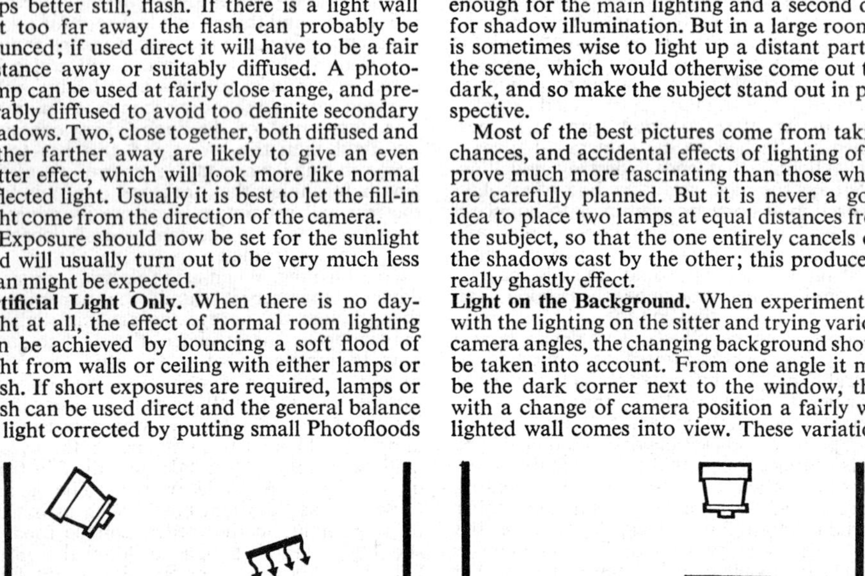

BAY WINDOWS. With the subject actually placed in the bay window, additional lighting effects are possible. 1. If positioned to one end of the bay, the subject receives a combined front and side light against a window background. 2. Placed in the middle of the bay, the subject is lit by double side lighting. The sitter should not be quite in the middle, or the light becomes too uniformly balanced from both sides. The reflector should be strongly directional as the shadows facing the camera are quite heavy.

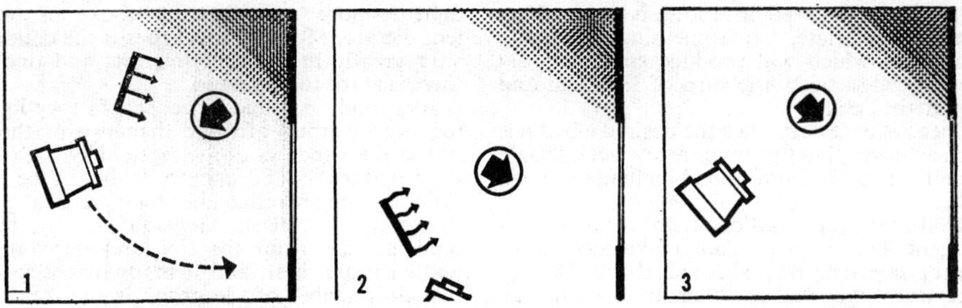

INDOOR BACKGROUNDS. 1. With the subject to one side of the window, taken against the corner of the room, the background gets gradually darker and contrasts against the light side of the face. Movement of camera includes more or less of dark corner. 2. If the subject faces the window, the profile will again stand out against the dark wall. 3. If no reflector is available, moving the subject farther into the room will soften the lighting, but also increase the exposure required.

in tone can be exploited, together with various items of geography such as the position of doors, furniture, pictures, etc., in providing suitable background effects. In the shady parts details are subdued, in the lighter ones they will be more noticeable. Thus the placing of the sitter and choice of camera angle can more or less determine the emphasis given to the features and character of the surroundings.

Special attention should also be given to the effect of each lamp on the background, which in a small room may have to be shielded in some way from direct light to prevent it from becoming conspicuous.

Furniture and pictures can easily be overstressed and there may be awkward heavy masses of shadow or bright reflections from polished edges.

The variations in the tone of the background can be largely chosen to suit the effect of light on the sitter. A dark corner can be used to enhance the effect of strong side lighting, a paler tone will go with a more delicate effect, and so on. Often the background changes gradually as the camera is moved, but sometimes even a very slight shift may bring in view a light patch which may be just what is wanted to show up the shadow side of the model.

There are occasions when the whole setting should be clearly and sharply defined in detail, but where it has no special value or is positively inappropriate it must be played down or completely suppressed. Leaving space behind the sitter helps to put the background out of focus —perhaps the most effective method of suppression. In this way, when taking a close-up portrait, there is no difficulty at all in reducing a fussy background to a series of indefinite shapes and blobs, which can suggest perspective and local atmosphere without really picturing anything in recognizable form. It is helpful if the effect can be studied on a focusing screen so that the abstract shapes may be arranged to the best advantage, and the lighter parts perhaps used to silhouette the contours of the subject on the shadow side. A very slight change in the angle of approach will often make all the difference.

Managing the Sitter. It is only on rare occasions that the subject looks right from the start; generally it is necessary to do a certain amount of staging. This does not mean starting with a rigid preconceived plan, but the subject will often have to be persuaded at least to move to a suitable place.

The methods adopted should be indirect. A definite pose should never be asked for; it will invariably look wrong. A better approach is, "How would it be if you sat there and went on with your knitting?" And, obviously, to ask for a smile merely results in a self-conscious smirk. The only way to get a genuine expres-

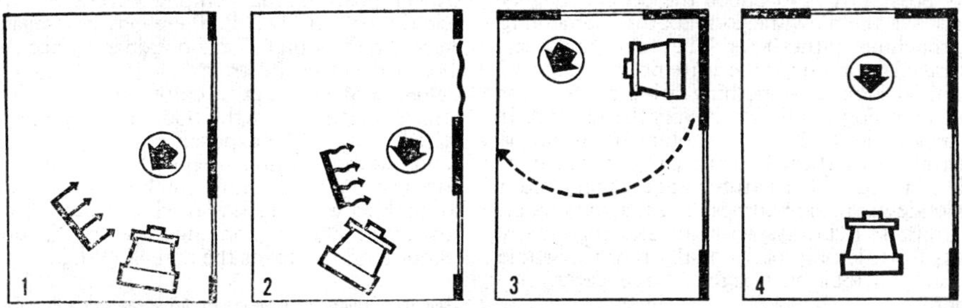

TWO WINDOWS. 1. Two windows in the same wall can be used for side-front and side-rear lighting respectively, with the sitter between them. 2. If one window is covered with net curtains, it can serve as background and at the same time provide a soft back light. 3. Two windows at right angles can serve as main and fill-in light. Varying effects obtainable by camera movement. 4. Movement of the subject adjusts the relative light on it from the windows.

sion, whether amused or serious, is to say something appropriate, do something or create conditions which will produce spontaneously the characteristic subtle turn of the head and look in the eyes.

Speed is necessary once the desired effect has been achieved, but the preparatory work should be unhurried. The sitter must be made to relax first.

Unforced conversation helps and an intelligent friend, who can make occasional appropriate remarks, is the ideal foil. But the spectator who makes comments about the sitter's expression, clothing or hair should be avoided at all cost.

Nearly always the best way of achieving naturalness is to get the sitter to do something—to move rather than just sit. There is no difficulty with children: they may show a little curiosity about the camera at first, but it wears off if they are encouraged to play about. The photographer should be friendly, and treat children as equals, answer their questions honestly and, if necessary, let them look into the camera. He may even enter into their play, although as far as possible he should remain a detached observer. Parents can be useful here, but not those who fuss about clothes and make suggestions at critical moments.

A small infant can be simply put down in a light place; older children can be induced to play near the window if they are told quite honestly why they are wanted to stay there. With a child that has got to the stage where it crawls rapidly from one end of the room to the other there is nothing for it but to follow it round and hope for the best. Patience, an easily moved lamp, and constant adjustment of shutter or stop are the ingredients of success in this branch of at-home portraiture.

The photographer should never allow his sitters to "dress up", or be dressed up: it will make them more than normally self-conscious. Ordinary, everyday things that have had some wear acquire character of their own and look more natural.

Groups. It is not easy to arrange a group in such a way that it does not look posed: the only possibility is to bunch the people roughly and leave them awhile to relax into something approaching naturalness. They should be encouraged to talk or, if there are not too many of them, to look at something, play cards, have tea—anything at all that breaks the tension. It is inevitable that there should be someone with his back turned to the camera, but even this can add to the natural appearance, and a back view can sometimes make a very good portrait. It is always wise to take very many shots for it is only occasionally that the whole group will look just right at the same moment.

Children are on the whole perhaps easier to manage; they can often be induced to play amicably together for a while, and, if later a fight develops, even that may make an excellent picture. Most trouble arises if the children vary greatly in age and interests and one of them is at the roving stage.

It is rarely possible to get satisfactory lighting over a group of more than two or three; unless the windows are exceptionally high the light will always be uneven. This can be put right by augmenting the light source with Photofloods or flash. Generally a soft, even light, slightly from the side and fairly high, will be found best, and it is important to see that no member of the group is too much in shadow.

General diffused shadow illumination is almost essential.

Sharpness. With indoor light it is never possible to stop down very much, but lack of depth at wide apertures has its advantages when dealing with unsuitable backgrounds. But the portrait itself need not necessarily have over-sharpness either,

One part may be emphasized sharply and the rest allowed to go more or less out of focus. The out-of-focus image can still be extraordinarily expressive.

For instance, in photographing a sitter with unusually expressive hands or gestures, the hands might be focused sharply with the face slightly softer; or the focusing might concentrate on the character of the face and let the hands be a mere suggestion to give point to the posture or emphasize the rhythm of a movement.

Again, in a close-up head, there is no reason why a part should not be allowed to go out of focus. This will in fact help to give the portrait a three-dimensional quality. Such treatment is of course a matter of taste, but it should be stressed that over-all sharpness is not necessarily always a virtue.

The other kind of unsharpness due to movement of the subject itself often results in surprisingly live and expressive pictures. Negatives with blur of this kind should never be scrapped without first making a trial print, for they may turn out to be more expressive than the technically perfect exposures. It is always possible that among several blurred negatives of a lively child there may be a masterpiece. The fact that it was obtained by accident should make no difference. H.v.W.

Colour. For shooting in colour, there must be sufficient available light. Colour transparency material should be exposed correctly, so there is no question of under-exposing. Further, it is not possible to use any combination of light sources, as with black and white material, and any fill-in illumination must be of the same colour temperature as the main lighting.

See also: *Babies; Children; Lighting the subject; Portraiture.*

Books: *Pictures by Indoor Lighting*, by H. van Wadenoyen (London); *The Art and Technique of Taking Portraits*, by H. van Wadenoyen (London).

PORTRAITURE OUTDOORS

Nowadays daylight portraiture has by common consent come to mean largely outdoor portraiture. With artificial light, the photographer has at his disposal a variety of lighting effects which he can alter at will; out of doors he has to make the best of the existing lighting conditions. On the other hand, an infinite choice of scene is available in the open and there is no end to the outdoor activities which provide opportunities for portraying people.

The vital qualities of outdoor portraits are naturalness and simplicity; there must be no forced artificial posing. While one may get away with such effects in a studio, outdoors they become conspicuous and insincere. Because of this need for naturalness, outdoor portraiture comes quite near to snapshooting with the difference that it still needs careful planning and technique.

None of the fundamental principles of portraiture may be neglected; one must not go nearer to the subject than the camera will focus, nor take such extreme close-ups with supplementary lenses that the features nearest the camera, for instance the nose or chin, become distorted. Arms and legs must not be stretched out towards the camera.

It is important always to give subjects something to do even if it is only resting on the sand. And their hands should be occupied; they should not just stand somewhere with their arms dangling helplessly.

Backgrounds. One sure way to spoil an outdoor portrait is by selecting an unsuitable background. Unless the background is deliberately made part and parcel of the picture, as for instance in a portrait taken in a boat or car, the subject must be well separated from the background. And the background must not be fussy and distracting, nor should it be ugly—a gasometer is no setting for a glamorous young woman. Trees, bushes and brick walls are popular but unsuitable backgrounds. Exceptions are when the sitter is in the sun and the background is in the shade. If bushes and trees form the background they should be very dense and not show patches of sky through the leaves and branches; light patches produce bright round circles on the print. The subject must be placed so that branches and other objects do not appear to sprout from the head in the finished portrait.

And when portraying one person, passers-by should not be strolling about at random in the background.

The best way to get rid of an unsuitable background is to take the subject against the sky; the outdoor effect is unmistakable and there is nothing to mar its simplicity. Other ways are to make sure that the background is completely out of focus or, with a head and shoulder close-up, to trim off the unwanted background altogether.

Sunlight. Strong sunlight is a much more contrasty light source than anyone would normally employ in a studio. It is unkind to any but very youthful complexions, and emphasizes wrinkles and rough or large-pored skin, simply because even the tiniest irregularity throws a strong though small shadow. The larger shadows thrown by the nose and chin are also strong and very dark, and so is the heavy shadow caused by a large hat brim. This high contrast is greatest when the sun shines from a cloudless blue sky and there is little scattered light to soften the shadows.

White clouds in the sky are very efficient reflectors and throw an appreciable amount of reflected light into the shadow areas, which greatly reduces the contrast and gives a more pleasing portrait. In such conditions there is also less contrast when the subject is placed in a wide open space where reflected light radiates from all sides.

A little light is also reflected from the ground. When the subject is standing on dry sand or snow, the ground reflects enough light to brighten all shadows to an almost ideal degree.

If nature will not help, some kind of artificial reflector must be contrived. It may be possible to place the subject near a brightly painted wall. While it may not be practicable to carry a

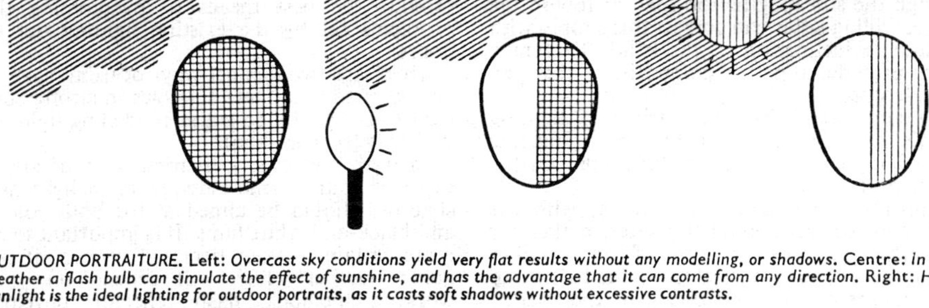

OUTDOOR PORTRAITURE. Left: Overcast sky conditions yield very flat results without any modelling, or shadows. Centre: In dull weather a flash bulb can simulate the effect of sunshine, and has the advantage that it can come from any direction. Right: Hazy sunlight is the ideal lighting for outdoor portraits, as it casts soft shadows without excessive contrasts.

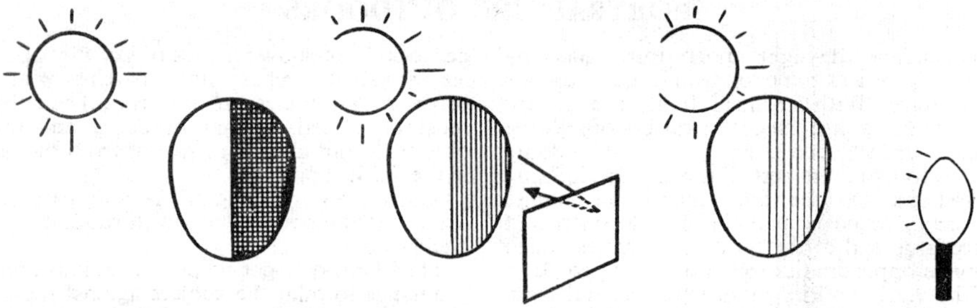

FILL-IN LIGHT OUTDOORS. Left: *Strong sunlight produces strong shadows and brilliant highlights. The lighting contrast may easily become excessive, especially for colour film.* Centre: *A reflector, such as a white card, casts light into too dark shadows.* Right: *A synchronized flash bulb can also act as fill-in light, and has the advantage of being more mobile and permitting wider light control.*

large, light-coloured board outdoors, a sheet of newspaper is perfectly efficient. Its effect can be clearly seen and a few experiments soon show the correct distance. It should not be so near the sitter's face that it cancels the shadows out completely or robs them of their modelling effect.

When the sun is slightly diffused by haze or light cloud, the lighting becomes very soft and ideal for outdoor portraits; contrast is low and shadows become faint and blurred, yet good modelling remains. This is, perhaps, the best lighting for colour and, with correct exposure, natural and pastel tones in the flesh rendering are almost assured—a result which direct sunlight will not yield.

Direction of the Light. The position of the sun in the sky determines the type of modelling of a portrait. Front lighting, that is with the sun behind the camera, can be effective with colour film but it is too flat with black-and-white material. There are no shadows and therefore there is no impression of depth. It is also the most uncomfortable position for the sitter, who will be hard put to keep from squinting or peeping through half-closed eyes; children certainly cannot be expected to keep a natural expression with the sun in their eyes.

Better modelling as well as less discomfort for the sitter is given by the well-known 45° lighting, that is with the sun still behind the camera but very much to one side. It also means that the light must shine down from above the level of the subject and, although the sun may be too high or low in the sky, a willing sitter can correct this simply with a suitable inclination of the head. That may even provide a better composition than an upright pose.

A completely all-round variety of lighting effects can be produced by portraying the sitter lying down—a very natural pose on the beach or in a meadow.

Partial or complete backlighting, with the sun more or less behind the sitter, makes for beautiful portraits. A rim of light, either down the edge of the cheek or in the hair, more than makes up for flat lighting of the shaded part of the face. The exposure must be calculated for the shadows. An efficient lens hood should in any case always be used; but if even that does not prevent the sun from shining into the lens with against-the-light shots, it may be possible to shoot from the shade of a tree or building, or a hat or book can be used to shield the lens if it is held high enough not to intrude into the picture area.

Without Sun. Very pleasing portraits can be taken without sunlight. In fact, when the sun is very strong and there are no means of illuminating the shadows, it is far better to place the sitter in open shade, such as that provided by a large building.

The light, although very diffused, will still give good modelling. Colour portraits taken entirely without sunlight tend to have a blue-green cast, which should be corrected by using a "warm" filter on the camera—e.g. a pink skylight filter.

Under a completely overcast sky the light becomes altogether too flat because it falls on the subject from all sides with almost equal strength, giving at the most only a faint shadow under the chin. In these conditions the sky is not a good background; among other drawbacks, irradiation may occur and blur the outlines. Portraits taken under a dull sky can be given more life and sparkle by printing on a hard grade of paper.

Flash. The compact and in some cases even pocketable flash guns, available for use with the synchronized shutters of modern cameras, provide the best means of overcoming the problems set by the existing lighting out of doors.

There are two distinct ways of using flash in daylight: to illuminate shadows in strong sunlight (fill-in flash), to act as modelling light on dull days (synchro-sun).

In both methods, a lighting ratio of about three or four to one between main light and shadows should be aimed at for both colour and black-and-white films. It is important to be quite certain of your personal version of the guide number which will depend on your working methods, flash reflector, development and the type of negative aimed at. The guide numbers given by the flash manufacturers are usually valid for use indoors in rooms of average size and with light-coloured walls and ceiling. Outdoors it becomes necessary to apply a smaller guide number, the reduction corresponding to one or two stops.

Fill-in Flash. First find the correct exposure for the sunlit subject and set the camera controls accordingly, ignoring the flash. The guide number then determines the distance from the subject at which the flash (not necessarily the camera) has to be used.

Dividing the guide number by the stop gives the distance at which the flash will exactly balance the sunlight, whereas it must contribute only one-third. Therefore the guide number must be divided, not by the stop used, but by the figure corresponding to opening up two stops.

Example: sunlight exposure 1/50 second at *f*11. Set the camera controls accordingly. Correct guide number: 77. Divide guide number by *f*5·6 instead of *f*11 (two stops larger), i.e., 77 ÷ 5·6 = 14. So the flash must be used at 14 feet.

This ratio is correct with black-and-white film, but, with colour film which demands greater care in exposure, the fact that the flash illuminates the sunlit part of the subject as well as the shaded part must be compensated for by closing the aperture by half a stop for absolutely correct exposure.

If it is inconvenient to set up the flash gun at the distance dictated by the guide number, the light output can be reduced by hanging one or more white handkerchiefs over the bulb. Each handkerchief will roughly halve the light.

Synchro-sun. When using flash to produce the effect of sunlight on dull days great care is necessary because the background must necessarily remain dull and darker than the subject.

Over-lighting the subject by using the flash is one of the worst effects of faking in photography, and spoils the picture. A sunlit figure in front of a black open background looks unnatural, but as long as the correct guide number is used the degree of contrast is easily controllable.

Even so, in portraits it is preferable to make the picture a close-up with little background, or take it against the sky. In fact, the slight underexposure of the sky makes it darker (bluer with colour film) and therefore a much better background. Flash, in effect, can then take the place of a filter.

With this method, the daylight provides the fill-in light for the shadows thrown by the flash. So first determine the exposure for the existing light and set the camera controls accordingly, then, to subdue the daylight and achieve the necessary contrast, close the aperture by one stop. Now divide the guide number by the stop in use, and the answer gives the distance of the flash from the subject.

Example: guide number 77. Daylight exposure 1/50 second at *f*8. Set camera to 1/50 second at *f*11. Divide guide number by stop used—i.e., 77 ÷ 11 = 7. The flash must be used at 7 feet.

The ratio of main to fill-in light here appears to be 2 : 1 but this is not so, because the flash is additional to the prevailing light. This method, while satisfactory with black-and-white film, is therefore not so suited to the small latitude of colour material, for the absolutely correct exposure for the existing light plus the flash is difficult to determine. It varies with the quality and intensity of the prevailing light. Stopping down by half a number is a good average allowance to make.

Blue flash bulbs must be used with colour film, but electronic flash is suitable for both colour and black-and-white.

Films and Filters. All films are more or less suitable for daylight portraits. Ortho film is better for children and girls with good complexions and not too much lipstick. It gives good skin rendering but over-darkens lips and blemishes. Pan film is kinder to the not-so-perfect complexion. Slow films with their higher contrast are more suitable for male portraits, giving more vigorous tone rendering.

When taking portraits against a blue sky, it is necessary to darken the sky by means of a filter. A blank, white sky makes a very unpleasant background. With ortho film a light to medium yellow filter should be used; with pan, a yellow or yellow-green. The yellow-green is preferable because it gives skin a beautifully smooth and sun-tanned appearance. With colour film, a polarizing filter will, in certain conditions, darken a blue sky. H.W.

See also: ***Against the light; Daylight; Portraiture.***

Books: *The Art and Technique of Taking Portraits*, by H. van Wadenoyen (London); *Glamour in Your Lens*, by J. McGregor (London); *Outdoor Portraiture*, by W. Mortensen (San Francisco).

PORTUGAL. There is no photographic industry in Portugal, nor have Portuguese inventors or scientists made any contribution to the advancement of photography. A few skilled amateurs have designed and constructed their own cameras and enlargers, but none of their apparatus was intended for sale commercially. In fact, as recently as 1920, Portuguese professional photographers were still preparing their own photographic papers.

Nowadays all photographic apparatus and sensitive materials are imported from the principal firms in France, Germany, Great Britain, Italy, Switzerland and U.S.A. A large variety of cameras, accessories and materials is thus usually available.

Popular Interest. Photography is extremely popular in Portugal; it is a favourite hobby with many thousands of amateurs. Even in the remotest localities roll films are available, and this increasing enthusiasm can be attributed to the artistic temperament of the Portuguese people, combined with the brilliant light and the beauty and variety of subjects which the country offers. It is usual for people of modest means, who do not own a camera, to hire a simple model for such occasions as annual holidays, etc.

Professionals. Portrait photographers are to be found in most towns, and in Lisbon several workers rival the best foreign photographers in this field and have gained international recognition.

Excellent photographic coverage of social, sporting and political events is provided by a number of professional photographers. Lisbon and Oporto are particularly noteworthy for the high quality of photographic work performed in the scientific, industrial and artistic fields. The standard compares favourably with the level of achievement in other countries. Evidence of this fact is provided by some notable publications on the fine arts, also by those issued by government departments and private firms.

In Portugal as in other countries photography is much used in science and industry, particularly in the former. The ministries, departments of state and most of the large laboratories maintain their own photographic departments.

In industry photography is less widely used, though a considerable amount of work is done for documentary, record and publicity purposes.

The availability of high quality apparatus and materials and of the best foreign publications, together with the attractive displays in dealers' windows, enables workers to keep abreast of the latest developments and at the same time educates public taste. Photography can be said to have established itself as an essential element of the national life. The government itself encourages and supports it, notably through its tourist services.

Clubs and Societies. The following is a list of clubs, those with the largest membership being placed first: Foto-Club 6 × 6 (Lisbon); Grupo Câmara (the Camera Group) in Coimbra; Associação Fotográfica do Porto (The Photographic Association) in Oporto; Grémio Associação Fotográfica de Santarém (the Photographic Association) in Santarém. The first four have an average membership of 200. Amateurs throughout the country continue to form groups and clubs which enable members to submit their work collectively to the major international exhibitions abroad. Among such new groups are the Associação Fotográfica do Norte, at Oporto; the Associação Galitos, of Aveiro; the Arte e Sport, of Lisbon; the Sport Lisbon e Benfica, of Lisbon; and the Associação Fotográfica do Sul, of Evora. There are also many sports clubs and industrial and commercial enterprises which include photography among their activities.

Exhibitions. In recent years national and international exhibitions of considerable merit have been held in several Portuguese towns under the auspices of the amateur photographic clubs, which have done much to improve the technical and artistic quality of photographic work in Portugal. The average level of work is as high as in other countries in the forefront of photography, and the Lisbon International Salon is recognized as one of the finest in the world. There are few photographic year-books which do not publish photographs by Portuguese pictorialists of international reputation.

Amateurs and clubs regularly send in both individual and collective entries to the best international exhibitions held in other countries, and high awards have been gained in competition with leading foreign workers.

Every two years, an International Exhibition of Photographic Art, sponsored by the Companhia Nacional de Navegação (National Steamship Company), takes place in Lisbon. This outstanding event has been considered, from the beginning, among the best in the world. The annual International Exhibition of Artistic Photography, sponsored by the Companhia Uniao Fabril of Barreiro and the International Salon organized by the Sport Lisbon e Benfica, have also reached a high standard.

It is interesting to note that, although many amateurs are devotees of the modernist schools at present in favour, the majority prefer to go after typically Portuguese themes, and to treat them in the traditional pictorial manner.

Magazines. Few photographic journals have succeeded in maintaining themselves over a period of years; many have ceased publication after a short period. In spite of the value of such publications, Portuguese photographers generally prefer foreign magazines, which are superior and are on regular sale in Portugal.

Some years ago, during a period of increasing expansion of amateur photography, greater participation in international exhibitions and technical progress, the magazine *Objectiva* appeared. It was an excellent publication and succeeded in maintaining itself for a number of years, performing excellent service for amateur photographers. After its disappearance the magazine *Cinema e Fotografia de Amadores* was published: later came the *Boletim Mensal do Grupo Câmara* of Coimbra and a small magazine entitled *Fotografia*, which was well produced and had interesting articles. At present there are no photographic magazines produced in Portugal. Some years ago the first issue of the Annual *Almanaque Portugues de Fotografia* came out and was well received by the photographic public. A.Pr.

POSING. Technique of arranging people to be photographed in characteristic or pleasing attitudes.

See also: *Banquets; Glamour; Groups; Portraiture.*
Books: *Posing Patterns*, by L. E. Broome (London); *Posing for the Camera*, by H. Shepard and L. Meyer (London).

POSITIVE. In photography, a print or transparency showing an image composed of light (or transparent) and dark (or opaque) areas which correspond to the light and shade of the original image. In a colour positive, the colours of the image correspond to the actual colours of the original subject.

The term positive is also used in electricity to define a contact which has a higher potential of electromotive force than an associated contact (the negative).

POSITIVE LENS. Lens which is thicker in the middle than at the edge, having at least one convex surface. Also known as a converging lens, it is capable of forming a real image, i.e. one that can be focused on a screen.

See also: *Converging lens.*

POSTCARD. Standard size, $3\frac{1}{2} \times 5\frac{1}{2}$ ins., of a picture-card for transmission through the post. Many millions of such cards are printed every year and all makers of sensitized materials supply bromide and contact paper either cut to this size or in continuous rolls of the correct width, and already printed with the standard matter on the back.

See also: *Postcard manufacture; Sizes and packings.*

POSTCARD MANUFACTURE. A few firms specialize in producing photographic postcards; these are sold as picture postcards in all the principal holiday resorts as well as other places of interest. The largest manufacturers can turn out a million photo postcards a week.

The photographs are procured by the firm's own staff photographers working to planned journeys and on continuous tour from May to September. The number of new and replacement photographs taken in a year depends very much on the weather and on features like changes in scenery, etc.

The Negative. The staff photographers work with half-plate size field cameras loaded with cut film. They send their negatives to headquarters at regular intervals for processing. At the factory the negatives are projected for detailed examination, and after the best have been selected, postcard-sized copy negatives are made on film. The original negatives are carefully catalogued and filed after the reduced negatives have been made. The cards are printed from the copy negative.

Printing. Because a single grade of printing paper is used for producing postcards, copy negatives must be standardized to a consistent opacity range. This procedure also prevents the original negatives from becoming damaged.

To make a postcard the following operations have to be carried out: exposure, developing, stop-bath, fixing, washing, backing (i.e. heading spaces for address and correspondence on the back), cutting to single postcards. These operations are generally nowadays performed as a set of sequential steps on bromide paper fed from a roll generally containing 1,000 feet of paper.

From two to four subjects are printed at a time. The copy negatives of the subjects are mounted in a frame which can be slipped into the exposing head of the machine, all the subjects being printed with one exposure. In the exposing head, the web of paper with a latent image is led to tanks containing the necessary processing chemicals for developing, fixing, washing, and bleaching and, if required, toning.

The web then passes over rollers set at the top of each tank and under rollers in the bottom of the tank. Typical times for each process: exposure ·03 sec., developing 2 mins., stop-bath $\frac{1}{2}$ min., fixing 4 mins., wash, at least 20 mins. The web of paper with the subjects fully developed, fixed and washed is led into drying ovens through which it passes by a similar method of top and bottom rollers.

At this stage in some plants the web is reeled up again and later, as a separate operation, the paper may be glazed by being fed into a heated glazing drum whose glazing surface is generally highly polished stainless steel. There then follows the printing, by litho or letterpress, of the backer, and cutting into singles.

Variations that occur from plant to plant are in the number of subjects that are printed at one exposure, the width of the paper used, the speed of the process and the number of operations that are performed without handling or re-reeling the paper. Paper speeds vary from between 7 and 8 feet to over 30 feet a minute.

Picture Requirements. Commercial postcard printers are always willing to consider the work of amateur photographers. Negatives should possess good depth of field. Subjects are best chosen by considering as a picture just what a tourist would readily buy as a souvenir of a visit. The view should be above all readily recognized as local—i.e., it should include some familiar feature, building or landmark associated with the place. G.I.B. (B.K.)

See also: *Processing machines; Roll head printer.*

POSTERIZATION. Extreme tone separation can produce strong poster-like effects in as few as four tones, in which most of the image detail is suppressed.

For pictures in four tones, three separation negatives are required. Excepting the highlight negative, these are processed for extreme contrast to give negatives consisting of an opaque image on a clear ground—like a silhouette.

(Repeated duplication on contrasty films or plates may be necessary to produce the right type of negative.)

The Negatives. The highlight negative image covers only the highlight areas.

The medium tone image covers highlight and medium tone areas.

The shadow negative image masks everything but the deepest shadows.

The highlight, medium tone, and shadow negatives are then printed successively in order on the same sheet of printing paper, and in exact register.

Exposing the Print. The exposure through the highlight negative is the shortest. It should just be long enough to produce a light grey all over the test strip, but leaving the extreme highlights white.

The second exposure through the medium tone negative is longer. The image in this case protects the highlights and medium tones from further exposure, but allows a dark grey tone to print over the shadow area of the picture.

During the final exposure, the image of the shadow negative protects all but the deepest shadows, and these print full black once the whole print has been developed.

The final picture, therefore, has four tones: white highlights, light grey medium tones, dark grey tones, and full black shadows.

Five tones (including three greys: light grey just off white, medium and dark) may be obtained in the same way by using a fourth separation negative.

See also: *Tone separation.*

Books: *Design by Photography*, and *The Complete Art of Printing and Enlarging*, both by O. R. Croy (London).

POTASH ALUM. Hardening agent used in fixing baths. Full chemical name, potassium aluminium sulphate.

See also: *Alum, potash.*

POTASSIUM ALUMINIUM SULPHATE. Full chemical name of hardening agent used in fixing baths. Also known as potassium or potash alum.

See also: *Alum, potash.*

POTASSIUM BICHROMATE. Potassium dichromate. Used in chromium intensifier and carbon, carbro, and gum-bichromate sensitisers.

Formula and molecular weight: $K_2Cr_2O_7$; 294.

Characteristics: Orange red crystals.

Solubility: Fairly soluble in water at room temperature. Freely soluble in hot water.

POTASSIUM BROMIDE. Restrainer in developers, also in bleachers.

Formula and molecular weight: KBr; 119.

Characteristics: Colourless cubical crystals.

Solubility: Freely soluble in water at room temperature.

POTASSIUM CARBONATE. Alkali in developers.

Formula and molecular weight: K_2CO_3; 138.

Characteristics: Very hygroscopic white powder. Keeps in well stoppered bottles.

Solubility: Highly soluble in water at room temperature.

POTASSIUM CHLORIDE. Used in some bleachers and sensitizers.

Formula and molecular weight. KCl; 74·5.

Characteristics: White crystalline powder.

Solubility: Freely soluble in water at room temperature.

POTASSIUM CHLOROPLATINITE. Used in certain toners and intensifiers.

Formula and molecular weight: K_2PtCl_4; 415.

Characteristics: Red crystals.

Solubility: Very slightly soluble in water.

POTASSIUM CHROMATE. Used in certain toners.

Formula and molecular weight: K_2CrO_4; 194.

Characteristics: Yellow crystals.

Solubility: Highly soluble in water at room temperature.

POTASSIUM CHROMIUM SULPHATE. Full chemical name of hardening agent used in fixing baths. Also known as chrome alum by analogy with the chemically related ordinary alum.

See also: *Alum, chrome.*

POTASSIUM CITRATE. Used in many toners.

Formula and molecular weight: $K_3C_6H_5O_7 \cdot H_2O$; 324.

Characteristics: White crystals.

Solubility: Highly soluble in water.

POTASSIUM CYANIDE. Used in reducers and solvents for silver.

Formula and molecular weight: KCN; 65

Characteristics: White lumps. Very poisonous. Must be kept in stoppered bottles to prevent deterioration.

Solubility: Highly soluble in water. Solutions smell of bitter almonds.

POTASSIUM DICHROMATE. Sensitizer used in carbro, gum bichromate and other processes.

See also: *Potassium bichromate.*

POTASSIUM FERRICYANIDE. Red prussiate of potash. Used in Farmer's reducer and some bleachers and toners.

Formula and molecular weight: $K_3Fe(CN)_6$; 329.

Characteristics: Red crystals, giving an intensely yellow solution. Poisonous.

Solubility: Freely soluble in water at room remperature.

POTASSIUM HYDROXIDE. Caustic potash; potassium hydrate. Strong alkali in high energy developers.

Formula and molecular weight: KOH; 56.

Characteristics: White hygroscopic sticks or pellets. Both the solid and the solution are corrosive.

Solubility: Highly soluble in water at room temperature.

POTASSIUM IODIDE. Used in certain bleachers, reducers and intensifiers.

Formula and molecular weight: KI; 166.

Characteristics: White crystals.

Solubility: Highly soluble in water at room temperature.

POTASSIUM METABISULPHITE. Acidifying agent in acid fixers and stop baths, preservative in developers.

Formula and molecular weight: $K_2S_2O_5$; 222.

Characteristics: White crystals.

Solubility: Freely soluble in water at room temperature.

POTASSIUM OXALATE. Used in certain toners.

Formula and molecular weight: $K_2(CO_2)_2.H_2O$; 184.

Characteristics: White crystals. Poisonous.

Solubility: Freely soluble in water at room temperature.

POTASSIUM PERMANGANATE. Used in reducers, bleachers, stain removers.

Formula and molecular weight: $KMnO_4$; 158.

Characteristics: Violet-black crystals, giving an intensely purple solution.

Solubility: Fairly soluble in water at room temperature.

POTASSIUM PERSULPHATE. Used instead of ammonium persulphate in reducers.

Formula and molecular weight: $K_2S_2O_8$; 270.

Characteristics: White powder.

Solubility: Slightly soluble in water at room temperature.

POTASSIUM PYROSULPHITE. Obsolete name for potassium metabisulphite, used as an acidifying agent in stop baths, hardeners, and fixers, and also as a preservative in certain developers.

POTASSIUM QUADROXALATE. Potassium tetroxalate. Used in certain toners.

Formula and molecular weight: $KH(CO_2)_2.H_2(CO_2)_2.2H_2O$; 254.

Characteristics: White powder. Poisonous.

Solubility: Fairly soluble in water.

POTASSIUM SULPHIDE. Liver of sulphur. Used in some sulphide toners.

Formula and molecular weight: K_2S; 110.

Characteristics: When pure, white very hygroscopic powder. The commercial liver of sulphur contains rather more sulphur than suggested by the chemical formula; it is a reddish substance.

Solubility: Highly soluble in water.

POTASSIUM SULPHOCYANIDE. Potassium thiocyanate; potassium rhodanate. Addition to fine grain developers, also may replace ammonium sulphocyanide in gold toners.

Formula and molecular weight: KCNS; 97.

Characteristics: White hygroscopic crystals. Keep in stoppered bottles.

Solubility: Highly soluble in water at room temperature.

POTASSIUM TETROXALATE. Constituent of certain toning baths. It is a white powder which is poisonous.

See also: ***Potassium quadroxalate.***

POTASSIUM THIOCYANATE. Chemical used in fine grain developers and sometimes in gold toners.

See also: ***Potassium sulphocyanide.***

POUNCY, JOHN, *c.* 1820–94. English professional photographer at Dorchester. The first to produce carbon prints (in gum) from photographic negatives, *c.* 1857, which were, however, criticized for lack of half-tone. Published *Dorsetshire Photographically Illustrated* (1857)—the first British book illustrated by photo-lithography, by his own technique. The stones were retouched to some extent.

PRESERVATIVE. Chemical included in a developer mainly to protect the developing agent in the solution from premature oxidation.

Sodium Sulphite. The most common preservative is sodium sulphite. This is available in two forms: powder (anhydrous) and crystals. The two are interchangeable in all formulae; one part (by weight) of anhydrous sodium sulphite is exactly equivalent to two parts of the crystalline salt. The other part is merely water combined in the crystals.

The advantages of using anhydrous sodium sulphite are: it dissolves more quickly and easily, and it is cheaper than the equivalent amount of crystals.

The disadvantage of the anhydrous salt is that any deterioration can only be determined by chemical analysis. Sodium sulphite slowly changes into sodium sulphate which looks much the same as sodium sulphite powder. But on sodium sulphite crystals the sulphate forms a white deposit which can be seen at once. If this deposit is serious, the crystals should be quickly rinsed in cold water and dried before they are weighed out.

The amount of sodium sulphite used in a formula varies with the type of developer.

Normal negative and positive developers contain about 1–4 per cent sodium sulphite in the working solution.

Pyro-metol developers and also special tanning formulae may contain as little as 0·05 per cent, or in some cases none at all.

On the other hand, many concentrated fine-grain developers contain as much as 10 per cent sodium sulphite.

Other Preservatives. The only other preservatives of any importance are the bisulphites and metabisulphites of sodium and potassium. They are freely interchangeable with each other, but not always with sodium sulphite. The chemically equivalent amounts are as below.

For most practical purposes the difference between the bisulphites and metabisulphites can be ignored. The maximum error of about 15 per cent is too small to matter.

EQUIVALENTS OF BISULPHITES

Potassium Metabi-sulphite	*Sodium Metabi-sulphite*	*Sodium Bisulphite*	*Sodium Sulphite (anhyd)*
100 parts	86 parts	94 parts	114 parts
117 parts	100 parts	110 parts	133 parts
106 parts	91 parts	100 parts	121 parts
88 parts	76 parts	83 parts	100 parts

The main difference between sodium sulphite and the bisulphites is that the bisulphites are acid in solution, while the sulphite is weakly alkaline. Bisulphites are particularly useful as preservatives in concentrated developers made up in two solutions, e.g., pyro developers. One solution contains only developing agent and preservative, while the alkali is kept separately as the other solution until required for use.

As most developing agents will keep much better in acid solution, considerably less bisulphite is needed than the equivalent amount of sodium sulphite. The chemical equivalent of the bisulphites is less than for sodium sulphite; the practical equivalents are less still. As the acid bisulphites reduce the alkalinity of the developer, the formula must contain enough excess alkali. L.A.M.

See also: *Developers*.

Book: *Developing*, by C. I. Jacobson (London).

PRESS CAMERA

While press photographers nowadays may use several advanced cameras such as precision miniatures, twin lens and single lens reflex models, etc., there is a traditional camera type specifically associated with press work. This is derived from a medium-size folding plate camera, and notable features are instant readiness for action, higher shutter speeds, and a considerable degree of versatility to cope with any type of subject.

Negative Material. Press cameras mostly use plates or sheet film (nowadays generally the latter) in suitable holders. The usual sizes are from $2\frac{1}{2} \times 3\frac{1}{2}$ ins. (6 × 9 cm.) to 4 × 5 ins. (10 × 12·5 cm.). The 9 × 12 cm. ($3\frac{1}{2} \times 4\frac{3}{4}$ ins.) size is particularly popular on the Continent.

The plates or films are mostly carried in double holders to permit quick changing over from one shot to a second one. The exposures are processed individually—almost a "must" for press photography.

If required, roll film holders can be fitted to several press cameras for use of roll film material. Film packs (in adapters) can also be used as an alternative to plates or sheet film.

A number of press cameras take special magazines for roll film or perforated 70 mm. film. The latter type of magazine holds up to 50 exposures and frequently carries a rapid film transport lever for quick sequence shooting.

Body Construction. The body is usually collapsible, the lens unit being joined to the back by bellows. The standard lens may move on runners on a folding baseboard or alternatively it may be mounted on an extending lazy-tongs system. In either case, positive stops permit rapid opening of the camera. Often several such stops are provided to cater for the different extensions necessary with interchangeable lenses of various focal lengths.

Press cameras with a folding baseboard may incorporate a limited range of camera movements to cope with unusual subjects requiring perspective correction.

A few cameras of this type have a rigid body; this makes for quicker operation but limits the use of interchangeable lenses, as well as the focusing range.

On a number of models the back carrying the film holder rotates through 90° or 360° to permit upright or horizontal shots without changing the grip on the camera.

The film holders slide into retaining grooves in the back, being interchangeable against a ground glass screen. On quite a few modern models the ground glass screen remains permanently in position, being held by spring-loaded arms. The film holder then simply slides underneath the screen, pushing the latter back, for the exposure. This makes insertion and removal of the film holders a specially rapid operation.

A leather strap and sometimes a special handle, provides a steady grip on the camera.

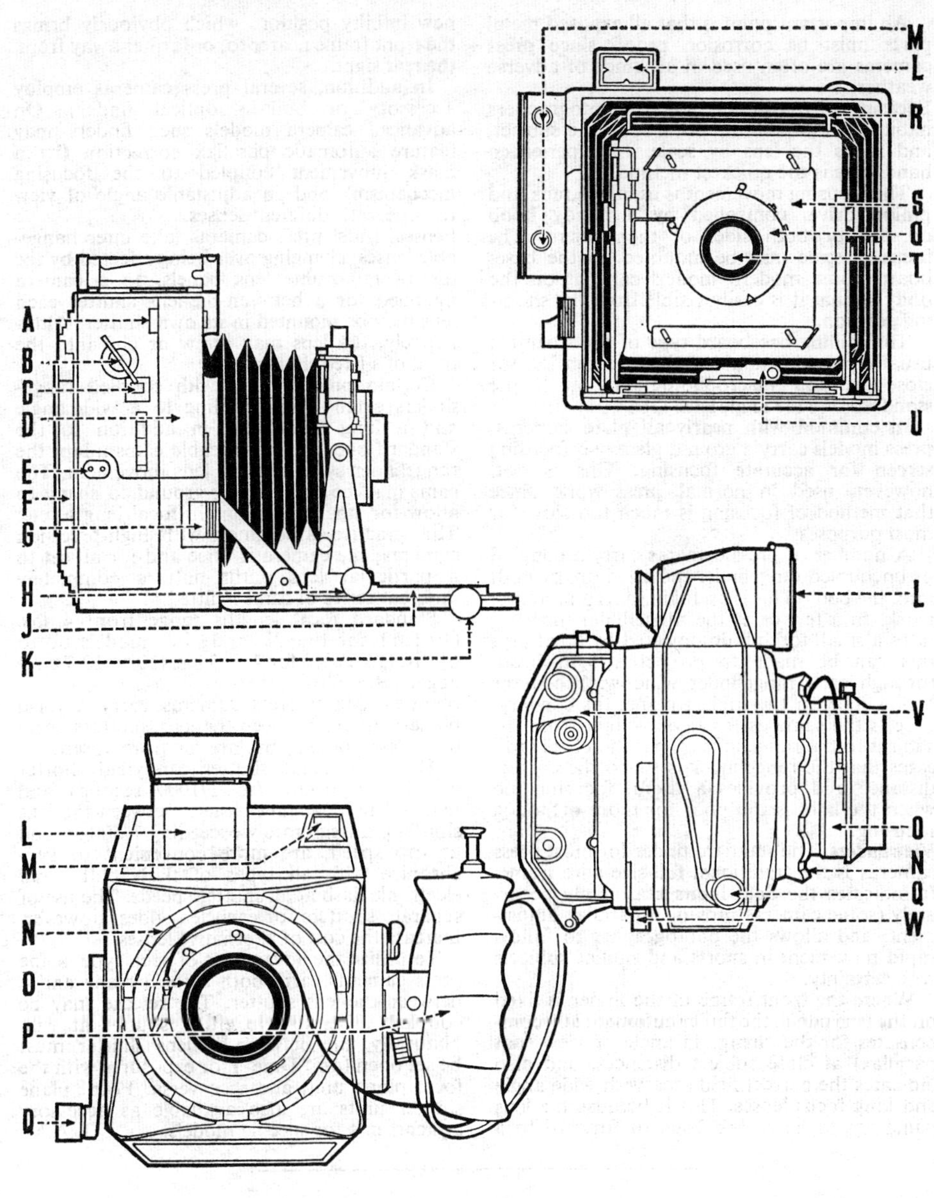

PROFESSIONAL PRESS CAMERA. Although 35 mm. miniature and roll film reflex cameras are widely used for press work, the larger size press camera is still popular among newspaper photographers. This is due to its versatility in the use of different types of material such as sheet film, plates as well as roll film and perforated 70 mm. film in special adaptor backs. Top left: Side view of bellows type press camera with focal plane and diaphragm shutters. Top right: Front view of bellows camera. Bottom left: Front view of rigid press camera. Bottom right: Side view of rigid press camera with roll film back.

A, rangefinder eyepiece. B, winding knob for focal plane shutter. C, focal plane shutter speeds setting. D, coupled rangefinder. E, flash socket. F, spring-loaded ground glass screen for rapid insertion of film holders. G, body release. H, rising front and tilt controls. J, double extension runners on baseboard. K, focusing knob. L, viewfinder window (with multiple finder frames). M, rangefinder window. N, focusing mount. O, interchangeable lens and diaphragm shutter. P, anatomical handgrip with built-in cable release. Q, film transport crank for use with roll film back. R, frame finder. S, interchangeable lens panel. T, release linkage. U, carriage lock. V, roll film back. W, back release.

An important point is that all exposed metal parts must be corrosion proof, since press cameras are often used in all kinds of adverse weather.

Focusing. Professional press photographers habitually estimate the distance of the subject, and focus the lens by scale. In experienced hands this is the quickest method.

The focusing movement is usually a rack and pinion drive, controlled by a handy knob on one or both sides of the camera. The focusing scale may be mounted on the baseboard; some modern models carry it on the body so that it is easily visible from the shooting position.

The folding baseboard type of press camera usually provides double bellows extension for close-ups and macrophotography with the standard or wide angle lens.

In common with nearly all plate cameras, press models carry a ground glass rear focusing screen for accurate focusing. This is not, however, used in normal press work, since that method of focusing is much too slow for most purposes.

A number of press cameras carry a coupled or uncoupled rangefinder mounted on, or built into, the body. This is useful in sport and action work. In a few cases the rangefinder incorporates a small torch bulb powered by a battery; this can be made to project a light beam through each rangefinder window. When the beams meet in the subject plane, the convergence is the same as for rays of light from the subject to the rangefinder. That therefore indicates that the rangefinder is set to the correct distance and provides a useful focusing aid when the light is too poor for more orthodox focusing.

Viewfinders. The standard finder for most press cameras is the traditional full-size wire frame. It indicates the field limits reasonably accurately (some cameras provide a parallax adjustment) and allows the photographer to follow rapid movement in sports and similar subjects with certainty.

Where the front frame of the finder is fixed on the lens panel, the finder automatically compensates for the change in angle of view (not parallax) at close subject distances, and also indicates the correct field area with wide angle and long focus lenses. This is because the lens panel has to be rocked back or forward to a new infinity position, which obviously brings the front frame nearer to, or farther away from, the rear sight.

In addition, several press cameras employ accessory or built-in optical finders. On advanced camera models such finders may feature automatic parallax correction (by a mask movement coupled to the focusing mechanism), and an adjustable angle of view to cope with different lenses.

Lenses. Most press cameras take interchangeable lenses, changing over being effected by the use of removable lens panels. In a camera intended for a between-the-lens shutter each lens may be mounted in its own shutter. Alternatively, the lens may screw or clip into the front of a fixed shutter.

Certain press models with coupled rangefinders extend this coupling to a wide angle and a long focus lens in addition to the standard one. Interchangeable cams adapt the rangefinder system to the lens movement. The cams may be individually ground to shape to allow for small variations in focal length, etc. This practice is common with high-precision cameras; the camera is once and for all set to a particular lens outfit until a completely different series of cams is fitted.

Standard focal lengths range from 4 ins. (10 cm.) for the $2\frac{1}{2} \times 3\frac{1}{4}$ ins. models up to $6\frac{3}{8}$ ins. (16·3 cm.) for 4×5 ins. ($10 \times 12{\cdot}7$ cm.) negatives.

Shutters. Many press cameras carry a focal plane shutter; between-the-lens shutters seem to be less popular, but are far from absent.

The focal plane shutter can yield shorter effective exposures (to 1/1000 second) and makes lens changing easier. Between-the-lens shutters, though rarely exceeding 1/500 second as top speed, are more convenient to synchronize with all types of flash bulbs and electronic flash at all shutter speeds. The use of separate shutters for each lens does, however, increase the cost of alternative lenses.

To utilize the advantages of each type, some press cameras carry both a focal plane and a between-the-lens shutter. The release may be coupled so as to trip either shutter at will. Naturally, the between-the-lens shutter must be set open (on "Time") for exposures with the focal plane unit, and vice versa. Focal plane shutter units are also available as accessory attachments for specific models. L.A.M.

PRESS PHOTOGRAPHY. The function of news photography is to keep the public informed about everyday events, and personalities connected with them, throughout the world. In many newspapers and periodicals photographic illustrations occupy as much space as the text. The press photographer must be able to cope with any situation, nearly always using a hand camera of the press or miniature type. On some occasions he may be expected to supply the accompanying news story.

Sources. News photographs fall into three categories—those taken by newspaper and periodical photographers, those taken by big agencies which have representatives in every quarter of the globe, and (in most countries) those taken by small agencies, free-lances, and amateurs.

Of these the main supply of news photographs of international and general interest comes from the big agencies. They employ a number of photographers, some of whom, like many newspaper correspondents, live in the countries in which they operate; others are continually moving from place to place. Together, these photographers cover every happening of outstanding interest. The more important of their pictures are as a rule sent to the capitals of the world by radio or by wire; they are then "piped" or retransmitted simultaneously to newspaper and agency offices in other cities and towns. This system is very highly developed in the United States. Photographs not of the first importance are sent by air freight or air mail.

Newspapers employ much the same methods, but their operations are usually limited to places where important stories have broken. Today the newspapers tend to employ their own photographers on events comparatively near home, and to rely on the big agencies for international news. The difficulty and expense of complete coverage are so great that no individual newspaper can compete with the agencies in this field. The mass of photographs produced by the small agency, the freelance and the amateur help to fill a considerable amount of space in publications of all kinds. But the pictures from domestic sources, in whatever country they are taken, are usually of no more than local interest. Only amateurs who have the good fortune to be engaged in foreign expeditions, or to be present on some important occasion, or who have similar special opportunities, can hope to secure photographs of any real news value.

Publication of the news photographs from these various sources is shared by the daily and Sunday papers and periodicals. It is probable that the periodical press now devotes more space to illustrations than do the newspapers. The requirements differ: newspapers will give one or two pictures to a subject, but a periodical will frequently spread a subject over several pages, perhaps using as many as twenty pictures.

Equipment. This variation in presentation of a story may call for more than one type of equipment. In general, the larger press camera is more popular for daily newspaper work and the smaller sizes for series or sets of photographs for periodical publications. 4 × 5 ins. and 9 × 12 cm. press cameras are still very popular with newspaper and agency photographers. They are usually loaded with panchromatic plates or cut film, and are fitted with flash bulb or electronic flash equipment (which gives a much faster speed). When series of photographs are in demand for periodical work, 35 mm. or roll film cameras are more suitable. Colour news photography is well established for periodical publications and some Sunday newspapers in the United States.

News subjects cover such a great range that both newspapers and agencies must be equipped with a wide variety of apparatus from copying cameras to special telephoto cameras equipped with lenses up to 60 ins. focal length. Television wire and radio receiving instruments and portable transmitters are also in general use in the bigger offices. The latter instruments are usually operated by trained telegraphists who work in the newspaper or agency offices and accompany the photographers on their assignments. Satisfactory reproductions are made even though the picture may be broken both by the television signal and by the half-tone screen. Many newspapers are equipped to receive pictures by wire from the leading agencies. Often, agencies and newspapers have studios in which fashion and other subjects are photographed for editorial features.

Speed is necessarily the main controlling factor in the production of newspaper illustrations, and though the results of wire and radio transmissions can be excellent, pictures often have to be rushed to such an extent that quality is lost.

Reproduction. Glossy bromide papers are used for newspaper half-tone reproduction. The printer aims at turning out reasonably strong well-balanced prints to reduce time spent in retouching. Even so, all newspaper process departments employ artist retouchers to emphasize or tone down parts of the picture as required by the blockmaker. Retouching is necessary because of the loss caused by the use of relatively coarse half-tone screens of sixty to seventy lines to the inch (which produce the dot formation in the reproduction) and by hurried printing on poor paper. This does not apply in the same degree to periodical printing. Here some retouching is necessary, but the finer screens used and better quality paper ensure a more faithful reproduction of the original photograph. In photogravure printing the retouching is usually done on the positives after the original photographs have been copied. But whatever the process, the better the original print the better the result.

The electronic "block-making" process, in which the printing surface is cut in plastic, or metal, calls for a strong print. In principle this process is similar to that employed for transmitting photographs by wire or radio. The various tones of the picture are scanned by a beam of light transmitted through a photoelectric cell, which operates a stylus. The stylus cuts or burns the plastic or metal and builds up a pattern of dots similar to that formed by the half-tone screen.

Restrictions and Censorship. Freedom to photograph varies in different countries. In Great Britain photography is not allowed in any court of law, or in the Houses of Parliament when they are sitting.

Press photography is restricted at functions

when royalty is present, or wherever lack of space prevents more than a limited number of photographers from working together at the same time. To ensure fairness of representation a rota system of permits, in which newspapers and agencies share, is in operation in Great Britain. In London the system is administered by the Newspaper Proprietors Association, whose secretary, with the advice of representatives of the various concerns interested, allocates permits in rotation from the list of applicants. On the very few occasions when only one photographer is allowed, the pictures are pooled and made available to all publications.

Press photographers have in general a great deal of freedom. In London the majority hold a Press card, carrying a portrait of the photographer, issued by Scotland Yard. In the United States press photographers are still less restricted, and photography is permitted even in courts of law, the Senate, and the House of Representatives. The press photographer is allowed considerable freedom in the countries of Western Europe, though special permits are necessary for important indoor functions. In Great Britain there is little in the way of censorship in peacetime; the exceptions are new aircraft, new weapons, and new naval equipment, but these are rarely visible to the press.

Sometimes during naval exercises the Admiralty ask that all pictures taken by the press shall be submitted to be passed by the Naval Censor, but it is unusual for a photograph to be stopped.

Other photographs which may be classed in the category of news are those supplied to the press generally by the press or public relations' officers of business firms, the big industries, film companies, the entertainment world, transport companies, travel agents, and other bodies. These sources usually offer the pictures of their activities for reproduction without fee in return for the publicity. U.V.B.

See also: *Freelance photography; Magazine photography; Photo journalism; Phototelegraphy; Press camera; Reproduction quality.*

Books: *How to Take Photographs that Editors will Buy*, by R. Spillman (London); *Kemsley Manual of Journalism* (London); *My Way With the Miniature*, by L. Vining (London); *Press Photography in Practice*, by N. K. Harrison (London).

PRETSCH, PAUL. 1808–73. Austrian inventor of photogalvanography, 1854, a photo-electro-engraving process, and manager of the Imperial Printing Establishment in Vienna. Pretsch resided in England for nine years in order to exploit his invention, and started publishing *Photographic Art Treasures*, November 1856—the first publication of photographs in printing-ink, i.e., photo-engraving. Talbot incorrectly alleged that photogalvanography infringed his own photo-engraving patent and, owing to his constant interference, Pretsch's firm failed and he returned to Vienna.

PRICE, WILLIAM LAKE, *c.* 1810–96. English watercolour painter, photographer from 1854. One of the earliest photographers in Britain to compose historical and literary subjects, and to make combination pictures from several negatives (1855). He gave up photography after eight years' activity, during which he specialized in reproductions of paintings. His handbook (1858) was the first to stress the aesthetic aspect of photography.

PRICES OF COMMERCIAL PHOTOGRAPHS. Shrewd and careful pricing of photographic work is essential to the success of a commercial, industrial or portrait studio. The real danger is in fixing charges too low through not making a proper assessment of the costs involved. But the making of photographs is largely a manufacturing process and is subject to detailed costing like any other industry.

Costs can be divided under two headings:

(1) Photographic materials—plates and films, chemicals, printing papers, mounts, mounting tissue, frames, colouring.

(2) Non-photographic expenses—proprietor's salary, staff wages, rent and rates, insurances, services (gas, electricity, water, telephone), depreciation of equipment and repairs, advertising, stationery, packing and postage, professional fees, transport, trade expenses, bank charges.

The percentage of the turnover to be allocated to each of these items varies considerably with the type of business, commercial or portraiture.

Commercial Photography. A large percentage of the over-all expenses is taken up by such items as models' fees, extensive fabricated scenery, travelling, and the transport of heavy equipment. Time is also a big factor; not only the actual time spent operating the camera, but also waiting time. In some industries this is so great that the photographer may find it best to charge a daily rate for his services irrespective of how many photographs are taken; this is preferable to trying to assess charges on a cost per photograph basis.

COMMERCIAL PHOTOGRAPHY CHARGES

Service	*Charge* £	s.	d.
Photographing			
Attending premises within a radius of five miles, taking one photograph and supplying one finished print in black and white ... ... from	3	3	0
Subsequent views up to five taken on same visit, with one finished print of each	2	2	0
Subsequent views after the first five with one finished print of each	1	15	0
Supplying additional prints and slides			
$4\frac{3}{4} \times 6\frac{1}{2}$ ins. or smaller		7	0
$6\frac{1}{2} \times 8\frac{1}{2}$ ins.		8	0
8 × 10 ins.		10	0
10 × 12 ins.		13	0
12 × 15 ins.		17	0
16 × 20 ins.	1	10	0
Over 16 × 20 ins., per square foot		15	0
2 × 2 ins. or $3\frac{1}{4} \times 3\frac{1}{4}$ ins. slides		10	0

In addition to these, the photographer may make the following extra charges:

(1) Models' fees at cost, plus a percentage to cover overheads.

(2) Substantial fee for use of photograph for a poster.

(3) Fee to cover hiring charge of properties, or use of properties if belonging to photographer, plus percentage to cover overheads.

(4) Charge for operator's travelling time on outdoor work beyond a five mile radius of studio, plus 100 per cent to allow for unproductive time.

(5) Operator's hotel and travelling expenses at cost.

(6) Charge for delays caused by customer and not expected at time order accepted, on basis of time lost plus 100 per cent.

Further points are that in cases where additional expense has been incurred owing to work having to be done at abnormal times, a higher rate should be charged to recover the extra costs. Also an original negative should not be lent, but a copy negative made for this purpose, invoiced as for total surrender and credited on its return less any charge which may be agreed. And if prints of stock negatives are supplied to publishers, advertising agencies, art studios, printers, etc., for possible selection, these should be invoiced out and credited on return.

Colour Photographs. Professional photographers who supply colour photographs, either as transparencies or prints, for advertising and reproduction, etc., are under no illusion as to the technical problems encountered arising from the client requiring faithful colour rendering of his subject and from the lack of latitude and other properties peculiar only to colour materials.

As a result of these restrictions and the comparatively high cost of materials, charges for all forms of colour photography should be considerably higher than for black-and-white work. Charges for creative and scientifically accurate colour photography, and with models, should be correspondingly higher, especially to allow for the extra darkroom work involved and it is impossible to lay down specific recommendations for work in these highly specialized fields.

Portraiture. Compared to the commercial photographer, the portrait photographer concentrating on studio work spends relatively little on advertising, travelling, wages and postage, and has an easier task in assessing costs and prices.

Prevailing charges are by no means as stable over the country as those for commercial work and variations can be observed depending on the type of district in which the studio is situated.

The following examples give some idea of the charges made:

(1) A seaside town in Essex, two towns on the south coast and a large town in the south-west: sitting fee, two to four proofs supplied, 10s. 6d. single figures, 15s. groups. Portraits, from 5s. for 3 × 4 ins. to £1. 1s. for 8 × 10 ins.

(2) London, two seaside towns on the south coast, a small town in the south-east, a small seaside town in the south-west and a large town in the Midlands: minimum charge payable at the time of the sitting—3 guineas including sitting, submission of proofs and one finished portrait any size up to 9 × 11 ins., or one 6 × 8 ins., or three 6 × 8 ins., or two 4 × 6 ins.

(3) London outskirts, a fairly large seaside town in the west, a small provincial town in the south-west and a south-west port: minimum charge payable at the time of the sitting—2 guineas including sitting, submission of proofs and three 6 × 8 ins. finished portraits, or four 4 × 6 ins.

(4) A seaside town in the south-west, a town in Kent and Northern Ireland: sitting and one 6 × 8 ins. print, 1 guinea; three copies 2½ guineas. Sitting and one 4 × 6 ins. print, 15s.; three copies 2 guineas. Deposit payable at time of sitting and afterwards deducted from order.

I.I.P. Publications. For further assistance the photographer is referred to the Institute of Incorporated Photographers which can supply information on the methods of charging as follows:

(1) Interior or exterior work away from the studio (monochrome).

(2) Work in studio (monochrome).

(3) Colour photography.

The following publications are available from the I.I.P.: *Conditions of Engagement and Scale of Professional Fees for Commercial and Industrial Photography; Fees and Conditions of Granting Reproduction Rights in Copyright Photographs.*

Prices in the U.S.A. Costing and determination of prices run along simil arlines to British practice, though the expenses, wages, and salaries are of course different.

Typical figures for an average small commercial job are $25 to $35 with one 8 × 10 ins. print; additional prints being charged at $1 each. Portraits run from $25 with one 8 × 10 ins. print, with additional prints from $2.50 each. Average colour jobs cost around $100, and additional 8 × 10 ins. prints from $10 each. J.Bl.

See also: *Costing; Reproduction fees; Selling photographs.*

Book: *Business of Photography*, by C. Abel (London).

PRIMARY COLOURS. Three main colour bands each covering about one-third of the visible spectrum (blue, green, and red) which on additive mixing will produce white. Theoretically, any colour can be obtained by suitable combination of light of the three primary colours only.

See also: *Colour synthesis.*

PRIMULINE PROCESS. Alternative name for the diazotype process, by which an image could be printed on to a variety of supports (e.g., fabrics, paper).

PRIMULINE YELLOW. Bright yellow aniline dye used in diazotype printing. It is used in preparing the sensitized support.

See also: *Fabric printing, Printing processes (obsolete).*

PRINCIPAL AXIS. Straight line about which a lens is symmetrical—i.e., the line passing through the centres of curvature of all the lens elements, and normal to all plane surfaces.

PRINCIPAL LIGHT. Source of subject lighting which creates the dominating highlights and shadows on the subject. This is the light that gives the basic impression of relief, or roundness, to the subject. If there is no principal light the subject is apt to appear flat and lacking in three-dimensional form.

In daylight, the principal light is, of course, the sun.

This is the most usual and natural form of lighting, so when an orthodox photograph is taken with artificial lighting it is usual to place the principal light more or less above the subject to produce the same effect.

See also: *Lighting the subject.*

PRINCIPAL PLANES. Term in optics referring to planes passing through the nodal points of a lens at right angles to its optical axis.

PRINT. In photography, a positive image created on a prepared surface—generally paper—by the agency of a photographic negative.

More than one print can be produced in this way from the same negative, and the positive image may be created by photographic means or it may involve a variety of physical or chemical effects.

See also: *Printing materials.*

PRINTED CIRCUIT. Electric wiring layout usually produced photographically by exposing a negative of the drawn circuit on to a thin copper layer, laminated to an insulating carrier and coated with a sensitized colloid layer. After exposure and hot water treatment the latter forms a resist to protect the areas required for the wiring, while the rest of the copper surface is etched away. After removal of the resist there remains the wiring layout on the laminated surface.

Producing circuits in this way is considerably faster and quicker than the classical method of wiring up small resistors, capacitors, etc. Also, such printed circuits can be made extremely small for specialized applications.

See also: *Photo-etching.*

PRINT FINISHING. Term for all the various non-photographic operations carried out on the print after it has been processed and washed. The finishing operations are intended to improve the appearance of the print and present it in a form which makes the most of its qualities.

The methods in common use are:

(1) Spotting with opaque pigment or dye to obliterate white spots or scratches due to dust, air bubbles, etc.

(2) Knifing away black spots to convert them into white marks which can then be spotted out.

(3) Retouching with artists' brushes or with an airbrush to emphasize or subdue outlines or whole areas of tone.

(4) Building up density by handwork with either oil pigment or graphite powder.

(5) Colouring by hand with oil pigments, water colours or coloured dyes.

(6) Doping or waxing to subdue broken reflections from the surface of the paper and so deepen the dark tones of the picture.

(7) Glazing by which a paper with a glossy surface is given a still higher glaze by squeegeeing it while wet into contact with a highly polished glass or metal surface.

(8) Trimming the edges of the print to remove distracting detail and give better balance to the composition.

(9) Mounting the print on a support to give it importance and to protect its edges from injury.

(10) Framing to isolate the photograph from its surroundings and prevent the attention of the observer from straying away. F.P.

See also: *Air brush; Colouring prints; Doping prints; Flattening prints; Framing photographs; Glazing; Mounting prints; Oil reinforcement; Retouching; Spots; Trimmer; Waxing prints.*

Book: *Retouching*, by O. R. Croy (London).

PRINTING. Process by which an image is made by exposing sensitized material to light passing through a negative or positive transparency. This exposure is made either with the negative in contact with the sensitized material (contact printing) or by projecting an image of the negative on to the material (projection printing). Because it is usually intended to give an enlarged image, projection printing is more commonly known as enlarging.

Printing is mostly done by artificial light, but when daylight is used (as occasionally for contact printing) the process is commonly called "daylight printing".

PRINTING BOX. Improved type of printing frame. A box contains the printing light, and the negative and paper are held down on to a glass window in the top. The exposure is controlled by switching the light in the box on and off.

See also: *Contact printer.*

PRINTING FRAME. Wood, metal or plastic frame in which a piece of sensitized paper is exposed behind a negative to produce a photographic positive print.

See also: *Contact printers.*

PRINTING IN. Selective exposure which is sometimes given, in enlarging, to certain parts of a print. This local increase in exposure is used to bring out detail or increase density in areas of the print that would otherwise appear too light.

The method is to use a sheet of metal or opaque card having a hole of the required shape cut in it. The main exposure for the print is given, and then the mask is held above the printing paper so as further to expose the required area through the hole. The mask is kept in constant motion so that the additional density will blend with the surrounding parts.

To print in large areas, such as a sky, a plain card is used to shade the remainder of the image; its edge may be straight, or cut or roughly torn to the required outline.

Some photographers are skilled in shading with the hands, arranging them to throw the required shadow outline or spot of light.

The term is also used to describe methods of improving a photograph by supplementary exposures through other negatives at the printing stage—e.g., adding clouds to fill in an otherwise blank sky.

See also: *Cloud negatives; Combination printing; Flashing; Shading and spot printing.*

PRINTING MATERIALS. There are two principal types of material used for making positive prints from a negative: printing-out emulsions and development emulsions.

Printing-out Emulsions. These usually are coated on a paper base. When they are exposed to daylight behind a negative they gradually form a visible image. When the image has printed out sufficiently, the unused salts are removed by fixing, leaving a permanent print.

Printing-out papers are suitable for contact printing by daylight only

Development Emulsions. These may be coated on paper for making prints, or on glass to make lantern plates, or positive film for making transparencies. When these materials are exposed behind a negative they form a latent image which must be developed.

Development materials are almost always printed by exposure to artificial light. The slow development papers and transparency materials are used for contact printing, while the faster types are used for printing by projection in an enlarger.

Mostly contact printing and enlarging papers are blue-sensitive only. Mixtures of blue-sensitive and orthochromatic emulsions of different gradations are used in variable contrast papers, and printed through different filters according to the required print contrast. Panchromatic papers are used for making black-and-white prints from colour negatives.

Rapid processing document printing papers and films may incorporate a developing agent in the silver halide emulsion; such papers are suitable for rapid machine processing with roller type processors.

TYPES OF PRINTING MATERIAL

Type	*Characteristics*	*Uses*
Papers		
Chloride	Slow silver chloride emulsion; in up to 6 contrast grades	Contact printing of normal negatives
Slow chloro-bromide	Predominantly silver chloride emulsion with silver bromide; in 1 or 2 contrast grades	Warm-tone contact prints or enlargements
Fast chloro-bromide	Predominantly silver bromide emulsion with silver chloride; in 3–4 contrast grades and wide range of surfaces	Warm-tone enlargements of ordinary negatives
Bromide	Fast silver bromide emulsion; in up to 5 contrast grades and wide range of surfaces	General enlarging work
Waterproof bromide	Bromide paper coated on lacquered paper base which is practically waterproof	High-speed printing, also where dimensional stability is important
Variable contrast	Two bromide emulsions of different colour sensitivity and contrast either mixed or coated on top of each other; contrast of image controlled by varying colour of printing light.	General enlarging work, especially when it is inconvenient to keep several paper grades in stock
Panchromatic bromide	Fully colour sensitized enlarging paper	Black-and-white enlargements from colour negatives
Document	Chloride or bromide emulsions, sometimes orthochromatic, of extreme contrast; may be on extra thin paper	Direct document copying or printing line negatives
Reflex	Chloride emulsion of extreme contrast, for reflex printing processes	Reflex printing of documents and line originals
Rapid processing	Chloride and bromide emulsions of various speeds (especially document papers) incorporating developing agents	Document copying, but also normal prints and enlargements, intended for high speed processing
Direct reversal	Extremely slow pre-fogged emulsion designed for direct reversal	Making positive prints from positive images
P.O.P. (printing-out paper)	Silver chloride emulsion containing free silver nitrate, producing an image by printing out without development	Contact printing of normal negatives by daylight

TYPES OF PRINTING MATERIAL (*continued*)

Type	*Characteristics*	*Uses*
Films		
Positive film	Sheet or 35 mm. miniature film with silver bromide emulsion (sheet film similar to blue-sensitive negative film)	Making positive transparencies and film strips
Contact film	Extremely slow silver chloride emulsion of very high contrast	Contact printing of document and line negatives
Document miniature	35 mm. miniature film with extra-fine grain high contrast emulsion	Printing micro-film negatives
Rapid processing	Chloride and bromide emulsions of high control incorporating developing agents	Making transparent document copies (for use as dyeline masters) by rapid processing systems
Plates		
Chloride lantern	Silver chloride emulsion	Making slides by contact printing only
Warm tone lantern	Silver chloride and bromide emulsion	Making slides by contact or projection, to yield warm-tone images by direct development
Bromide lantern	Silver bromide emulsion	Making slides by contact or projection to yield black tones (blue-sensitive, slow, negative emulsions are sometimes used for same purpose)

Other Materials. Special materials are also manufactured for some modern printing processes. These include matrix film, for making dye transfer colour prints, integral tripack colour printing materials, and stripping emulsions used in certain monochrome and colour processes.

In addition to the commercially available printing materials, there are also numerous formulae for the sensitizing of various supports for printing. Most of these are intended for daylight printing, although it is possible to prepare an emulsion sufficiently sensitive for printing by artificial light. The majority of these formulae originated as early printing processes, but a few still survive for fabric printing, document photography, etc. L.A.M.

See also: *Cine film; Colour materials; Document photography; Fabric printing; Obsolete printing processes; Papers; Printing on special supports; Reversal materials; Slide (transparency); Stripping; Transfer coating; Transparency materials.*

PRINTING ON SPECIAL SUPPORTS. There are four ways of printing photographic images on special supports:

(1) Daylight printing, using sensitive silver salts or dye intermediates. This is mainly suitable for fabrics, but it can also be used for printing on other supports, including metal surfaces. The negative must be the same size as the image required, so it may be necessary to make enlarged negatives for large pictures.

(2) Daylight printing, using bichromated colloids, like gum or gelatin. These methods are mainly suitable for hard, non-porous surfaces like glass, glazed porcelain and enamel. They can also be used on specially prepared metal or even wooden surfaces.

(3) Artificial light printing, using silver salts or emulsions. Under certain circumstances this is suitable for fabrics as well as specially prepared solid surfaces. Such materials can usually be printed by contact or projection, and require development.

(4) Transfer printing methods. These also use silver salt emulsions which are exposed by contact or projection, and have to be developed. They are suitable for almost any prepared hard surface.

The best method for any particular case depends on the nature of the support. Most of the processes can be used equally well for making paper prints.

A number of sensitized fabrics, metals and other special materials are available commercially.

Solid Supports. Photographic images may be printed on to flat surfaces of wood, metal, glass and glazed articles by either daylight or artificial light.

The surface must first of all be prepared to prevent either absorption of the sensitized coating or subsequent chemical action on the image.

Glass and non-porous and chemically-neutral supports like glazed porcelain, china, and enamel, can be coated without further preparation, apart from cleaning the surface.

Wood may be painted with white cellulose enamel, or coated with a 10 per cent gelatin solution containing enough white zinc oxide to hide the grain. Any other surfaces of the wood support that are likely to get wet during processing are waterproofed with wax or varnish.

Metals may be protected by varnishing all the surfaces liable to come into contact with the processing solutions.

Large Supports. Printing and processing have to be modified if the support is part of an object too large for handling in a printing frame or processing dish. In such cases it is generally possible to improvise some form of clamp to keep the negative and sensitized surface in close contact under a sheet of plate-glass during exposure.

Large sensitized surfaces have to be swabbed or brushed over with the various processing solutions. Such surfaces can be most conveniently washed and rinsed by playing a spray or jet of water gently on them.

Daylight Printing. The surface after preparation is brushed over with a 10 per cent solution of potassium bromide, allowed to dry, and then brushed over with a 10 per cent solution of silver nitrate. The final coating must be applied in the darkroom by the light of a bromide paper type of safelight.

The sensitized surface is printed out by daylight behind the negative. Printing is continued until a faint image is visible. This is not the final image; the final image is developed in a 0·1 per cent solution of metol containing 2 per cent glacial acetic acid.

Processing. If the article is too bulky to be immersed in a developing dish, it may be developed by swabbing it over with cotton wool soaked in the developer.

As soon as the image is dense enough, it is washed over with clean water and immersed in or swabbed over with a 10 per cent sodium thiosulphate solution to fix it. A final washing for about thirty minutes completes the process.

Artificial Light Printing. Sensitized silver-halide gelatin emulsions are available for coating on to solid surfaces to be printed photographically. Such emulsions are usually supplied in the form of a powder which is mixed with warm water to give an emulsion thin enough to be sprayed on the prepared surface.

Small objects are coated by whirling as in the wet collodion process, but larger surfaces are coated by spraying. The procedure calls for special equipment and is certainly beyond the scope of the ordinary photographer.

This is the method adopted for coating wall surfaces for printing photo murals *in situ* by projection.

Transfer coating is a much more practical method for the photographer who only needs to make such prints occasionally. This is also used for printing photographic templates on sheet metal to mark it out for machining.

An alternative method is to use the surface as the final support for a print made by the carbon or carbro process.

Fabric Printing. Fabrics like linen, canvas, silk, cotton, etc., can also be used as supports for a photographic image. Usually the fabric is prepared by sizing, followed by treatment in a halide solution and finally in a silver nitrate solution. Alternatively a ferric oxalate and silver nitrate mixture may be used for sensitization. It is then printed out by daylight, washed, and fixed like printing-out paper.

Alternatively, the image may be formed by dyes in the fabric itself, using a process like diazotype. Even a mordant image can be produced by the action of light (e.g., using a bichromate sensitized fabric) which will then take mordant dyes. L.A.M.

See also: ***Carbon process; Carbro process; Dusting on process; Fabric** printing; **Glass pictures; Gum bichromate process;** Stripping; Transfer coating.*

Book: *The Complete Art of Printing and Enlarging*, by O. R. Croy (London).

PRINTING-OUT PAPER (P.O.P.). Sensitive paper which produces a visible image on direct exposure, without the need for development.

The use of printing-out papers derives from Johann Heinrich Schultze's observations in 1727 that silver chloride and other silver salts darken when exposed to light and from Sir John Herschel's discovery in 1819 that the resulting image can be made permanent or "fixed" by immersing in a solution of sodium thiosulphate. The effect of light is to decompose the silver chloride and liberate photolytic silver; the sodium thiosulphate serves to remove the unaffected residual silver chloride.

The use of gelatin or collodion as protective colloids for the silver chloride led to the production of the original print-out emulsions. The presence of excess silver ions was found to increase the efficiency of the print-out process, probably due to absorption of the chlorine liberated during light exposure.

Conventional print-out papers have a very long tonal range. The colour of the photolytic silver is usually reddish-purple, which changes to a yellowish-brown during fixation because the hypo exerts a slight solvent action on the very fine-grained silver image.

Partial transformation of the silver deposit to gold by gold toning prior to fixing strengthens the image and gives it a bluish-purple colour. Toning is effected by bathing the print in a dilute gold chloride-ammonium thiocyanate solution. It was later found feasible to incorporate gold-toning compounds during the manufacture of print-out emulsions, thus originating the so-called self-toning papers which required only fixation after printing.

In general terms the above papers are relatively insensitive and require lengthy exposure to sources rich in ultra-violet, such as sunlight, arc-lamps, mercury vapour lamps, etc. Accordingly, for most photographic purposes, printing-out papers have long since been superseded by the much faster developing-out papers, e.g. chloride, chloro-bromide or bromide types, in which the latent image is rendered visible by development.

Printing-out papers are still employed to some extent for portrait proofing. In recent years there has been a great revival of interest in special print-out papers with unusual properties. These are distinguished by very high sensitivity to ultra-violet light sources, e.g. mercury-quartz lamps. The resulting feeble print-out image is subsequently intensified by a second exposure to light of a longer wavelength such as fluorescent lighting, tungsten lighting or weak daylight. This is effected without appreciable darkening of the background which originally received no ultra-violet exposure. The intensification process can be accelerated by heating the paper during the secondary exposure.

This type of material is extensively used for high-speed galvanometer trace-recording, e.g.

for aircraft and rocket research and for any similar purpose where an immediate dry record is required. The saving of time over wet processing procedures is a great advantage. The combination of high basic sensitivity plus controlled intensification enables very high writing speeds to be recorded.

If required, the print-out image may be made still more stable by the use of so-called permanizing solutions. Alternatively a black image on a white background may be produced by processing in a solvent developer after ultraviolet exposure. H.P.

See also: *Papers.*

PRINTING PROCESSES (OBSOLETE). It would be more correct to describe obsolete printing processes as prototypes of the printing processes in common use today for they relied on the same blackening, bleaching or physical effect of light on certain compounds and used the same sensitizing agents as those required in modern practice; it is only the earlier ways of employing these which are obsolete.

The following are some of those processes which had a spell of popularity during the first fifty years of photography.

Calotype or Talbotype. A good quality letter paper was sponged over with, or floated on, solutions of silver iodide and potassium iodide. When partially dry the excess potassium iodide was removed by bathing in distilled water. Before use, the paper was sensitized in a solution of silver nitrate, acetic acid and gallic acid.

After printing (under a similarly prepared paper negative) in bright sunlight, the feeble image resulting was brought up to the required strength by an application of solution similar to the silver nitrate sensitizer. It produced a rich warm brown image.

Fluorotype. Paper was washed over with a solution containing 2 per cent potassium bromide and ½ per cent sodium fluoride in distilled water. When dry, the paper was made sensitive to light by floating it for two minutes on a 15 per cent solution of silver nitrate. Fluorotype was very sensitive to light and the inventor claimed that an image of a brightly lit scene could be obtained on it in the camera by an exposure of only half a minute. But the image, however obtained, was feeble and had to be intensified by brushing it over with a weak solution of sulphate of iron. It was afterwards fixed with hypo and washed.

Cyanotype. Invented by Sir John Herschel in 1842, it is still in use and known as the ferro-prussiate or blue-print process. Used mostly for copying drawings, it gives a white image on blue ground but can be used for printing continuous tone negatives providing they have plenty of contrast. Suitable well-sized paper is brushed over with a sensitizer which in its simplest form may consist of a solution of potassium ferricyanide and ferric ammonium citrate. Paper so prepared is printed out to the required depth and then washed in several changes of water; the last being slightly acidified with hydrochloric acid.

Pellet Process. This is a variation of cyanotype giving a blue image on a white ground (negative from negative, or positive from positive). It was introduced by H. Pellet in 1878.

Three stock solutions are required: 20 per cent gum arabic, 50 per cent ferric ammonium citrate, and 50 per cent ferric chloride.

The paper is brushed over with:

Gum arabic solution	20 parts
Ferric ammonium citrate solution	8 parts
Ferric chloride solution	5 parts

After printing, the image, which is yellow in colour, is treated with a 20 per cent solution of potassium ferrocyanide which turns it blue.

Ferro-gallic and Ferro-tannic Papers. Introduced by A. Poitevin in 1861, the ferro-gallic process, like cyanotype, relied on the reduction of ferric salts by light. The image was however formed by the combination of the resultant ferrous salts with gallic or tannic acid, yielding a deep purple or brown insoluble organic compound.

A suitable sensitizer was:

Gum arabic	5 ounces	125 grams
Ferric chloride	5 ounces	125 grams
Ferric sulphate	4 ounces	100 grams
Tartaric acid	350 grains	20 grams
Potassium chloride	4 ounces	100 grams
Water to make	40 ounces	1,000 c.cm.

The sensitizer was applied to well-sized paper, and the latter partially printed out under a negative. The image was developed in a 5 per cent solution of gallic or tannic acid containing about 1½ per cent of potash alum, and a trace of hydrochloric acid.

Albumen Prints. Prior to 1850 prints had been made on salted paper (i.e., impregnated with sodium chloride) brushed over with a solution of silver nitrate but these had little vigour. They needed a base that would hold more silver and this need was met when Blanquart-Evrard of Lille suggested coating the paper with a thin film of albumen (egg white) before it was sensitized by floating on a bath of silver nitrate solution. The prints so obtained were toned with gold chloride and the process remained in almost universal use for the next thirty years. It was eventually superseded by printing-out papers using collodion or gelatin as a vehicle for the sensitive silver salts.

Argentotype. Originally invented by Herschel in 1842, this process produced a silver image by utilizing the action of light on ferric salts. Iron salts (principally ferric oxalate), combined with silver nitrate, were mixed with dextrin or starch, coated on to a paper base, and printed out by daylight. The action of light transformed the ferric salts into ferrous salts which, in a suitable solvent developer, liberated a precipitate of metallic silver from the silver nitrate to form the image.

Kallitype. In this process, introduced by W. J.

Nichol, the sensitizer was ferric oxalate, which was reduced by light to ferrous oxalate. When this was dissolved in the presence of silver nitrate (usually incorporated in the paper), the ferrous salt reduced the silver nitrate to metallic silver which formed the image.

In practice a tough cartridge or drawing paper was first sized with starch, then floated on or brushed over with the following (or a similar) sensitizing solution:

Ferric oxalate	4½ ounces	112 grams
Potassium oxalate	500 grains	28 grams
Silver nitrate	500 grains	28 grams
Water to make	20 ounces	500 c.cm.

Printing was done by daylight and continued till only the shadows were visible. The prints developed in:

Borax	1 ounce	25 grams
Sodium potassium tartrate (Rochelle salt)	525 grains	30 grams
Water to make	20 ounces	500 c.cm.

The use of a ferric ammonium citrate sensitizer, followed by development in plain water, yielded warm sepia tones.

The print was finally fixed in a 3 per cent sodium thiosulphate solution rendered slightly alkaline with ammonia.

Palladiotype, Plantinotype. The paper for these processes was purchased ready prepared. The final print consisted of a deposit of platinum or palladium on the paper support.

The negative was printed by contact in daylight and then developed in a solution of potassium oxalate. Development was complete in about thirty seconds. Successive baths of hydrochloric acid were used to clear the print of the yellow stain, and the print was finally washed for about 40 minutes.

Palladium was introduced during World War I when platinum was difficult to get. Both platinum and palladium produced prints with beautiful rich blacks unobtainable with silver. The great advantage of the process was that the image was permanent.

Papers of this type are very sensitive to damp, and special precautions had to be taken to keep them dry after opening the tin in which they were sold. During printing, the paper had to be protected by covering the back with a sheet of rubber.

Owing to the scarcity and high cost of the basic materials, these processes have gone out of use.

Uranium Printing. In this process the action of light reduced uranyl nitrate to a viscous compound. A slightly acidified uranyl nitrate solution (about 15 per cent) was applied to sized paper, printed out, and developed in a 10 per cent potassium ferricyanide solution. A wash in 1 per cent hydrochloric acid completed the process.

Alternatively, silver nitrate might be used for development and deposited silver to produce a grey instead of a brown image.

Powder Processes. Several of these were for a time popular, especially on the Continent. They depended on the discovery that certain substances such as sugar, dextrine, or gum, treated with potassium bichromate, lost their natural tackiness when exposed to light.

A piece of paper or glass so coated after exposure under a transparency or negative and afterwards subjected to a moist air became variously tacky. On dusting with a dark powder this would adhere in proportion to the degree of the action of the light on the hardening of the coating. Plumbago or graphite powder was used and the surplus dusted off. To protect the image the plate or paper was coated with collodion. The same principle, employing certain resins, is still in use in various photomechanical processes.

Pigment Printing. In 1855 Poitevin made photographic prints on paper coated with gelatin mixed with colouring matter and sensitized with potassium bichromate. This was the beginning of the many processes which have led to the carbon printing still practised. The earlier attempts produced only crude contrasty prints. T. W. Swan, in 1864, showed that to retain the more delicate tones the prints had to be developed from the back and, to permit this, devised the single or double transfer system. Since then there has been no basic change in the process, but enormous improvement in carbon tissues with which it is now worked.

A typical pigment process of this type was the ozotype (rendered obsolete by the ozobrome or carbro process) devised by Thomas Manly in 1899. A sheet of bichromate-sensitized single transfer paper was exposed under a negative to give a visible image. It was then applied to a sheet of carbon tissue previously soaked for 1 minute in a solution of 3 c.cm. acetic acid and 1 gram hydroquinone in 1,000 c.cm. water. After drying, the papers were first immersed in cold water for 30 minutes and the pigment image then subsequently developed by dissolving away the untanned gelatin as in the normal carbon process.

In 1905 Manly invented the ozobrome process which eliminated the use of bichromated paper; the pigment image was made direct from a bromide print as in the present-day carbro process (which is an improved version of ozobrome).

Artigue Process. A variation of the carbon process invented by M. Artigue of Paris; it was designed to do away with the necessity of transferring the printed tissue to a temporary support for development. Gum, mixed with any desired colour pigment and sensitized with potassium bichromate, was coated on paper or other support and after exposure under a negative was developed in a soup-like mixture of sawdust and water.

The process was capable of producing very delicate results.

Oil Printing. This process, introduced as a photographic printing method by Rawlins in 1904, was based on lithographic transfer procedures of much earlier date. It depended on the action of light on bichromated but unpigmented gelatin-coated paper. After exposure under a negative, and washing in water, the unaffected areas of the gelatin layer absorbed water and swelled in inverse proportion to their exposure to light. In subsequent application of a greasy ink the swollen areas repelled the ink, while the least swollen ones accepted it most readily.

The pigment image could be transferred to another sheet of paper by pressing the two into contact in a lithographic or a copper-plate printing press. After transfer the original image might be inked up once more and the process repeated for a number of prints. The contrast of each successive print tended to increase up to a point where no more prints of acceptable tonal quality could be made. The type of the original paper and the transfer paper, the character of the ink, and the transfer pressure all had an influence on the quality and brilliance of the transferred image.

As in carbro, the gelatin image could also be produced by chemical action of bichromate on the silver image in the gelatin layer of a bromide print. That process is known as bromoil and is still used by some pictorial photographers.

Chromotype. This was a curious process described by Robert Hunt before a meeting of the British Association in 1843 whereby either a negative or positive image could be obtained depending on the length of exposure, the latter presumably relying on solarization. Describing it, Hunt says: "One drachm of copper sulphate is dissolved in half an ounce of water and to this is added half an ounce of a saturated solution of bichromate of potash. This solution is applied to paper and when dry is exposed under a negative or positive to bright sunshine. The image first changes to a dull brown and, if checked at this stage, we get a negative picture; but if the exposure is continued the brown gives way to a yellow picture on a white ground. If the exposed paper is washed over with a solution of nitrate of silver, a very beautiful picture results." Hunt afterwards added gold chloride to the sensitizing solution and claimed that, according to the degree of solarization, prints could be obtained in crimson, blue, brown or deep black.

Breath Printing. One of Sir John Herschel's interesting discoveries was that latent images could be prepared which could only be seen when breathed on or subjected to a moist atmosphere.

Paper was coated with the dissolved precipitate formed under certain conditions by the addition of silver nitrate to ferro-tartaric acid. Printed under a negative in sunlight the paper so prepared would, by the right exposure, be impressed with a latent image which was not visible unless breathed on or subjected to a light aqueous vapour, when it acquired an extraordinary intensity.

Diazotype. Designed in 1891 for printing photographically on fabrics on which a range of colours could be produced. Prints could be made on paper, but the colours were dull and the whites impure. The term is nowadays applied generally to a number of similar processes.

In practice, the material on which prints were required was first bathed or brushed with a 1 per cent solution of primuline and, when dry, sensitized in a bath of sodium nitrite and oxalic acid. Printing was by daylight (30 seconds to the sun) and the image developed in any one of a number of reagents each of which gave different colour images ranging through red, yellow, orange, purple and brown.

An alternative process used casein instead of primuline for the first solution.

Pinatype. In this process a transparency was made on a soft emulsion of the lantern plate variety and developed in a tanning developer. Alternatively, it could be made on a gelatin-coated plate sensitized with potassium bichromate. The "printing plate" so obtained was bathed for about two minutes in an engraving black or photographic brown pinatype dye; the latter was held in suspension by the gelatin in direct proportion to the extent to which it had been hardened by light action during printing. The plate so charged with dye was then washed free of the surplus dye and finally brought in contact with a sheet of paper coated with plain gelatin. The two were squeegeed together and left for several minutes, when the dye in the printing plate was transferred to the virgin gelatin and, when separated, the latter bore the dye image. Any number of prints could be made by dyeing-up the printing plate and repeating the imbitition process. C.D.

See also: *Ambrotype; Amphitype; Control processes; Discovery of photography; Document photography; Fabric printing; Ivorytype; Opalotype; Pigment processes; Printing on special supports; Sensitized materials history.*

PRINT LIBRARIES. The principal source from which prints can be obtained are the photographic agencies. Names and addressees of these are to be found in directories published for the use of journalists. In addition, the larger newspapers and pictorial magazines usually have a department which supplies prints, not only of pictures which they have published, but also of other photographs in their files. The bigger agencies and periodicals cover a wide range of material; some of the smaller ones specialize in particular subjects.

The fees charged depend to a large extent on the medium in which the print is to be reproduced, the type of reproduction right which is offered, the size of reproduction, and other considerations. Some agencies charge a nominal search fee for pictures which are specifically requested, but which are ultimately not wanted for reproduction.

Following is a list of the principal subjects of which photographs may be wanted with an indication of sources of supply other than photographic agencies.

Aerial views: aeronautical societies: aeronautical journals; commercial air lines; archaeological societies.

Agriculture: agricultural societies; agricultural journals; makers of agricultural machinery; commercial companies growing their own raw materials.

Architecture: architectural societies; architectural journals; housing research organizations; societies for preservation and recording of historic buildings.

Art (paintings, sculpture, ceramics, furniture, etc.): art galleries; museums; art journals; cultural societies; art dealers.

Children: local newspapers and pictorial journals; individual photographers; child welfare societies.

Industry: trade associations; individual companies; large-scale distributors; research institutes; specialized press.

Landscapes: local newspapers and pictorial journals; individual photographers; railway companies; shipping lines; air lines; geographical societies; tourist associations.

Pets: local newspapers and pictorial journals; individual photographers.

Science: scientific societies; scientific journals; professional bodies; research institutes.

Sport: sporting associations and journals.

Personalities: organizations for which the persons concerned are working, including theatrical managements, film companies, broadcasting companies; cultural and scientific societies; direct from individuals.

Zoology (and birds, fish, insects): scientific societies; scientific journals; journals dealing with nature study.

Names and addresses can be found by consulting relevant directories: press directories; directories of museums and art galleries; directories of scientific and learned societies; trade directories; directories of photographers.

If photographs are wanted of anything connected with countries abroad, the press representative at the embassy or legation of that country will usually be willing either to supply prints or to say where they can be obtained. Other sources for photographs of things and places abroad are travel agencies and societies for promoting cultural relations between nations.

Some public libraries have collections of prints which may be borrowed for reference or reproduction and some of them will obtain, on special request, prints of subjects which are not on their own files. Among the more important of such sources are the Central Libraries of Bermondsey (London), Birmingham, Cardiff and Manchester.

Apart from agencies the most important print libraries in the U.S.A. are the George Eastman House of Photography in Rochester, N.Y., the University Library at the Northwestern University in Evanston, Ill., the Library of Congress in Washington, the New York Public Library, the Library of the Standard Oil Co. of New Jersey, and the library in the museum of the City of New York.

To obtain a print of a photograph which has been published in a newspaper or magazine, a letter requesting the print should be addressed to the art editor of the journal. It is advisable to enclose another letter to the photographer who has taken the picture, since the journal may not be able to supply prints and will have to forward requests to him. N.W.

See also: *Agencies; Books on photography; Information and inquiries; Literature on photography; Museums and collections.*

Books: *British Sources of Photographs and Pictures* (London); *Libraries, Museums and Art Galleries Year Book* (London); *Willing's Press Guide* (London); *The Writers' and Artists' Year Book* (London).

PRINT PADDLE. Flat perforated plate fitted to the end of a long rod, and used for moving prints around in the various processing solutions. Print paddles may be made of plastic or stainless steel.

See also: *Forceps.*

PRINT QUALITY. This is a term much used by amateurs—and judges of their work—to indicate technical perfection. It is seldom used in professional and commercial photography. In this field, technical perfection is expected as a matter of course even from the most modest establishment—without it the firm would go out of business.

The commercial photographer can achieve consistently high quality for a number of reasons: he uses appropriate equipment; he very quickly arrives at a rigidly standardized technique; his turnover of sensitized materials and chemicals is rapid enough to ensure that they are always fresh; his printing staff do the same job day after day and they almost always print from negatives of good, standard quality.

All these things work for the professional and against the amateur who generally: has to put up with an improvised darkroom and cheaper equipment; does not do enough routine photography to acquire a flawless technique; does not use materials quickly enough to have them always fresh; is constantly experimenting with new materials and processing techniques, papers and print sizes; and prints from a mixed bag of negatives of a wide range of contrasts and densities but without a correspondingly wide choice of paper grades.

Only the rare amateur possesses the necessary patience and technical knowledge to overcome all these characteristic handicaps and turn out, with much labour and many rejections, a single exhibition print that will stand comparison—at least on technical grounds—with the product of the commercial studio. The result has been the appearance in the amateur vocabulary of the term print quality which has been invested with a false aura of magic and ju-ju. Some prints have it, and others, just as mysteriously lack it; some workers are famous for it, others despair of ever being able to achieve it.

In fact print quality is the result of printing a good negative on the right grade of paper in a safely lighted room with the chemicals recommended by the manufacturer of the paper. The result of following this deceitfully simple specification will be a print which makes full use of the tone range of the paper, reproducing the deepest shadow as a pure, intense black and the brightest highlight as white paper. Between these extreme tones there will be a well-marked variety of shades of grey. All sharp lines will be crisply defined; the print will have no trace of fog, the blacks will be of good colour, and there will be no unintentional blur. As a print it will be technically perfect; it will possess print quality.

The defects that most often rob a print of print quality are: poor blacks, wrong tones, fog, and blur. There is a cause and a cure for each.

Poor blacks can result if the exposure is wrong and an attempt is made to mend matters by giving a longer or shorter development time than the makers of the paper recommend. They are also caused by using stale paper or an incorrect developer or one that is either exhausted or at the wrong temperature. The remedy is to develop according to the maker's instructions using fresh chemicals.

Wrong tones may be the result of using a paper that is either too hard or too soft for the negative. Any of the causes of poor blacks outlined above will also ruin the tone values of the print.

Fog is mostly the result of unsafe light falling on the print—either because white light is leaking into the darkroom, or from the enlarger lamphouse or because the safelight is either the wrong colour or too bright, or too close to the print. Other causes of fog are: stale paper, wrong developer, wrong developer temperature (generally too high) and extended development.

Blur may be caused by incorrect focusing or vibration of the enlarger during exposure. The image may also be diffused by dirt or condensed moisture on the lens. F.P.

See also: *Faults*.

Books: *Enlarging*, by C. I. Jacobson (London); *The Complete Art of Printing and Enlarging*, by O. R. Croy (London).

PRISM. Transparent substance bounded by plane polished surfaces inclined to one another. The simplest form of prism is used to refract light and needs only two polished surfaces, one where the light enters, and the other where it leaves the prism. Reflecting prisms, however, need three or more polished surfaces.

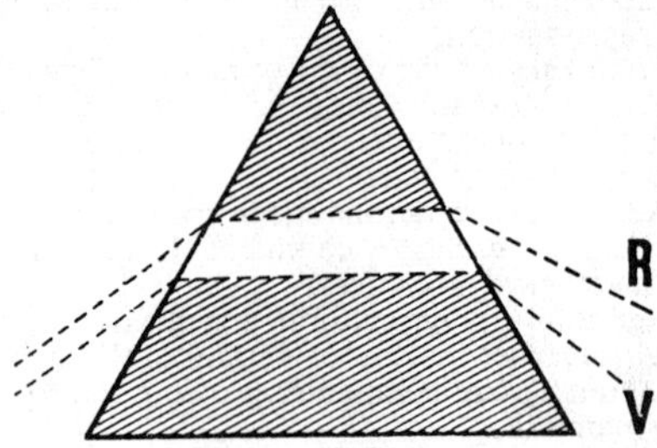

DISPERSION BY PRISM. A prism bends light rays to different degrees, violet rays being bent most and red least. The over-all bending power depends on the refractive index, the difference in refraction for various colours on the dispersive power.

Refracting Prisms. When light rays from a near object point fall on a prism, they meet the surface at different angles, and as the deviation of the ray produced by a prism depends upon the angle between the ray and the first prism face, the various rays of the pencil will be deviated by different amounts. The light leaving the prism no longer proceeds as though it came from a point, so the prism forms an imperfect image.

If the light from the object is white, it is dispersed by the prism into its component colours, the violet light being bent most and the red least. Thus, a small white object looked at through a prism appears as a spectrum with the violet end bent farthest away from the sight line. The colours of the image are the same as the colours of the rainbow, i.e., red, orange, yellow, green, blue, violet.

Refracting prisms are mainly used in spectroscopic instruments.

Reflecting Prisms. Reflecting prisms are often used in place of silvered mirrors because they are free from the disadvantages of back-silvered mirrors which produce a double image due to reflection from the glass as well as the main reflection from the silver. Also a prism is not as delicate and easily tarnished as a mirror made by silvering the front surface of a glass

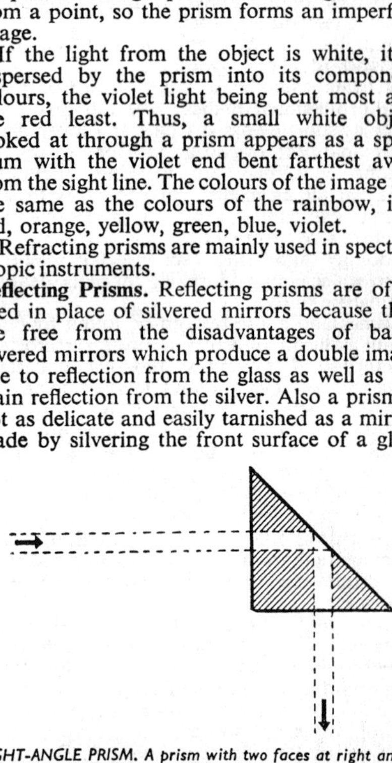

RIGHT-ANGLE PRISM. A prism with two faces at right angles and the third at 45° can act like a plane mirror, and is optically more efficient, losing virtually no light.

plate—a type needing frequent resilvering.

The mirror effect is, if possible, obtained by total internal reflection from the surface of the prism. Even in those cases where total reflection is not possible and the reflecting surfaces must be silvered, the prism can be so arranged that no trouble is caused by the reflections from the unsilvered surfaces. Reflecting prisms are used in the rangefinders which are fitted to some cameras and are also used in front of the lens on certain process cameras for reversing the image from right to left.

Prisms of more complex shape are incorporated in eye-level reflex cameras; there the prism reflects the light several times to turn the ground glass screen image upright and right-way round.

See also: *Pentaprism*.

PROCESS CAMERA

The printing blocks used in the photomechanical reproduction processes are made from negatives which are exposed in a process camera—sometimes alternatively called a copying camera.

The process camera is generally used for making copies to a reduced, or at most, a same-size scale. It is only very occasionally needed to make an enlarged copy.

Construction. The process camera is a permanent piece of equipment which is bolted down to the floor of the room or carried on runners which are bolted to the floor. The camera consists essentially of two vertical frames mounted on a set of horizontal rails. One frame carries the lens panel and the other the plate holder and process screen. The frames are connected by a square bellows fitted with supports to stop it from sagging when extended. The maximum extension is usually a little more than twice the focal length of the normal lens. This allows for same-size reproduction at full extension.

The distance between the copy and the camera is usually varied by moving the copy holder, depending on the design of the apparatus while the lens panel is moved as well, to adjust the focus.

The lens panel and plate back can be moved by some form of mechanical traversing gear—e.g., a rack and pinion, or worm and nut—operated from behind the camera. This enables the operator to focus the image on the screen without changing his position. (When copying on a scale of 1 : 1 with a 20 × 20 ins. camera, the lens panel may be 4 feet or more in front of the focusing screen.)

Mounted in front of the plate back there is a holder for the half-tone screen. The holder can be moved forwards or backwards to bring the ruled screen to the correct position in relation to the surface of the plate. The screen must be moved out of the way when the plate is loaded into the holder.

An indicator or other device at the side enables the operator to bring the screen back to the original position.

Both the lens panel and plate back are strongly constructed and accurately mounted so that they remain rigidly parallel over the whole range of adjustment. Any distortion produces dimensional errors which are intolerable in this kind of photography, particularly when making separation negatives from coloured originals.

With some large cameras it is usual to have the film or plate holder assembly in the darkroom, and the lens panel and copying board in an outer room under normal illumination. This arrangement makes focusing easier, and enables the film or plate to be inserted in the holder without a dark slide and taken directly from the camera to the processing section.

On more advanced cameras the movements of the lens panel and copyboard may be motor driven, requiring only push button operation at the darkroom end. Other remote control operations may include the aperture adjustment, the angling of the half-tone screen (for example when making successive separation screen negatives or positives) and the operation of ancillary gear.

Focusing is usually controlled by percentage focusing charts in connection with graduated scales or tapes on the lens and copyboard movements. For a given scale of reproduction it is then only necessary to set the lens panel and the copy board to the figures derived from the percentage focusing charts. This automatically ensures maximum sharpness of the image.

Some process cameras may have automatic focusing systems, based on a cam coupling between the lens and the copyboard movements (rather like an automatic focusing enlarger). This is however rarer and the image sharpness is subject to deterioration through wear and tear—and hence loss of precision—in the mechanical movements.

Highly accurate focusing systems use vernier scales on the camera adjustments to get the image plane in exactly the right position. Sometimes a cable drive for coarse focusing is combined with a micrometer screw drive for fine focusing.

Overhead and Vertical Cameras. Two variations of the process camera construction are the overhead camera and the vertical camera. The former is adopted for very large models, often taking films or plates up to 48 × 48 ins. The copy board and lens board are here suspended from an overhead gantry which carries the track runners for the movement. The camera back with the image plane is usually fixed and may form one end of the gantry support.

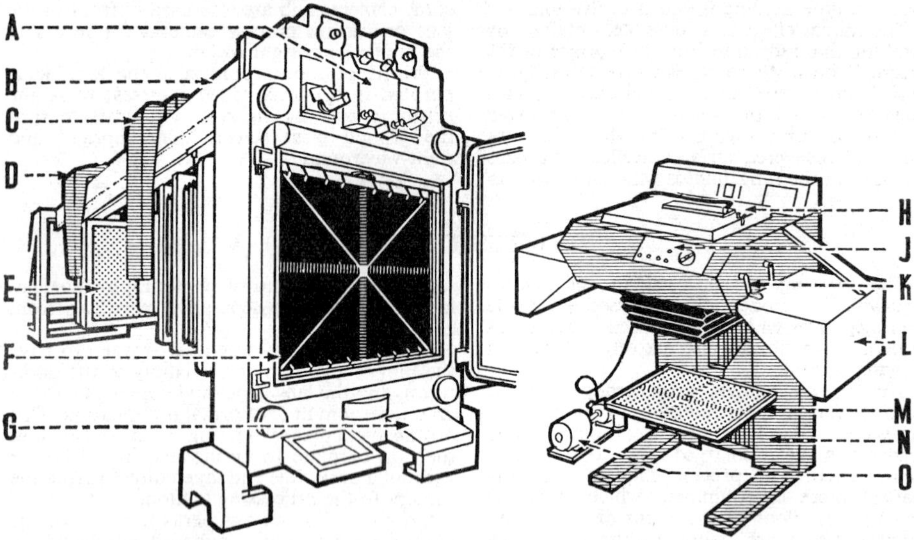

SPECIAL CAMERA SYSTEMS: Left: *Highly automated overhead camera for very large original and film sizes (may go up to* 50 × 60 *and* 48 × 48 *inches respectively).* A, *automatic focusing drives and controls,* B, *overhead rails.* C, *arm holding lens panel.* D, *arm holding copy board* E, F, *focusing screen and film holder.* G, *control panel for exposure timing etc.* Right: *Vertical camera, usually intended for smaller film sizes but particularly compact in set-up.* H, *vacuum back for filmholder.* J, *exposure timing gear.* K, *crank-driven vertical movement of lens panel and copy board.* L, *lighting unit.* M, *copy board.* N, *supports with guide rails between.* O, *vacuum pump.*

The main advantage of this arrangement is that the lens panel and copyboard become more accessible than would be possible with the comparatively bulky supports and runners required with a floor mounted camera.

The vertical process camera is by contrast a comparatively compact unit intended mainly for smaller work up to around 12 × 15 ins. The design is derived from that of a vertical enlarger: a vertical stand carries a copy board (which moves up and down on runners), the lens panel (also movable) and the fixed top with the film or plate holder.

The main advantage of the vertical camera is more convenient operation since the image plane is a horizontal surface making for easy loading of the film, and appreciable saving in space. The main drawback is the reduced versatility, both in available scales of magnification and reduction (typically from about 3 × magnification to about 3 × reduction) and in the type of work that can be done. Thus vertical cameras do not usually take ruled half-tone screens, and any half-tone work must be done with a contact screen. Also, this type of camera is rarely suitable for colour separation work.

Focusing and scale adjustment are usually operated by hand wheels or cranks; vernier focusing scales may be provided.

Certain large size professional enlargers can also be used for process work, especially continuous-tone colour separation and making line and half-tone negatives and positives with a contact screen. The enlarger adaptation consists mainly of fitting appropriate film or plate holders and the use of a vacuum easel (possibly with registration pins) in place of the normal baseboard.

Sensitized Materials. Process cameras take plates or films up to 16 × 20 ins. and larger. Glass plates have the advantage of high dimensional stability and are held between a pair of guides which can be adjusted to take any smaller size.

The use of films has become more common since the availability of polyester base materials of much greater dimensional stability than the older acetate film bases. Polyester film still does not reach the stability of glass plates, but film materials are considerably more convenient to handle.

Films are held in vacuum backs to keep them firmly and flat in the image plane. The back consists of a flat plate carrying a number of channels connected to a vacuum pump. When the film is in position, the pump draws the air out of the channels, so that the film is pressed into complete contact with the back. Different channels are brought into operation for the various film sizes.

Film Registration. Punch register methods are used to ensure the exact position of the images on the film so that their images can subsequently be laid on, for instance, litho plates, in precisely the correct positions so that colour separation images are automatically in register.

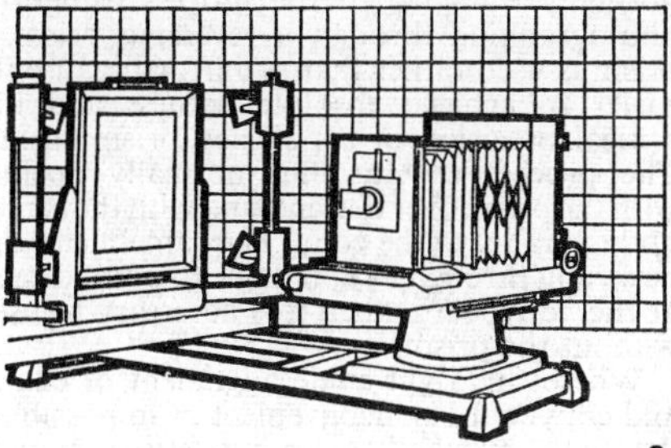

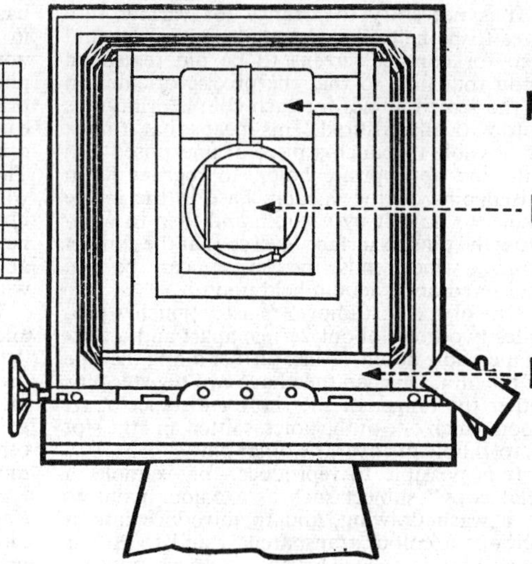

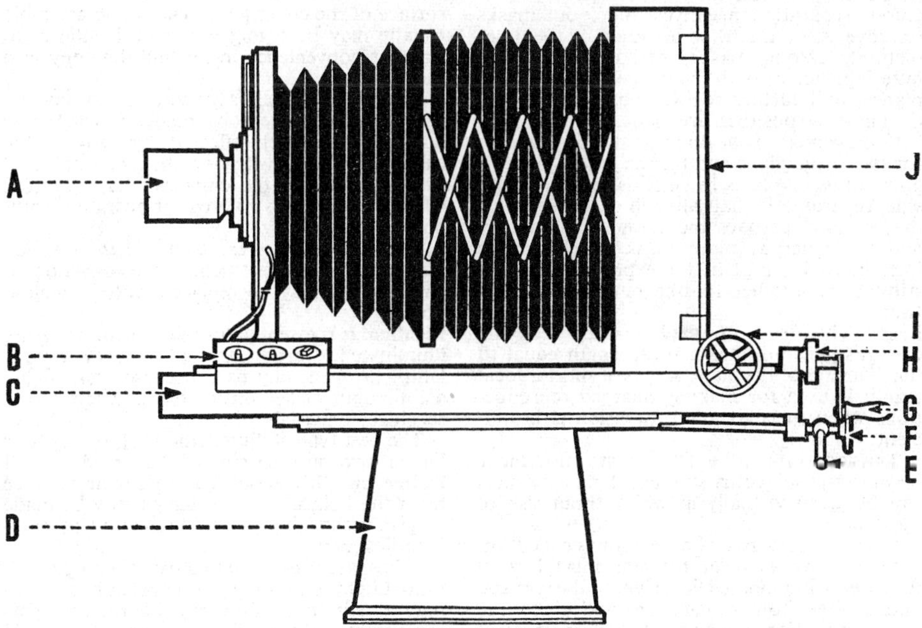

UNIVERSAL PROCESS AND REPRODUCTION CAMERA. These cameras take negatives up to 24 × 30 ins. large, and with the copy holder, stand and carriage may weigh 1 to 1½ tons, occupying a floor space up to 15 × 25 feet. Owing to the large size, all controls are usually arranged at or near the rear of the camera. Top left: Camera set up for working with a process prism, shown against a 1 foot square grid. Top right: Front view (with prism in position). Centre: Side view (without prism over lens). *Bottom left:* Overhead process camera. Bottom right: Vertical camera.

A, lens (may be 12 to 36 ins. in focal length). B, timer control box with seconds timer, minute timer, and selector switch. Connected directly to shutter behind lens. C, camera sub-base. D, stand. E, brake lock for camera carriage. F, fine focusing control to camera carriage. G, focusing control connected to the front body (i.e., lens board, etc.). H, controls for cross front and for rising front movement of lens panel. I, fine focusing handwheels (one on each side of camera). J, stand. K, front body, carrying lens board, prism (when used), and diaphragm control. L, prism to reverse image from left to right when required. M, diaphragm control scale; automatically sets the aperture to the correct value to allow for different camera extensions and scales of reproduction.

It is necessary to arrange for various film sizes to be held in punch and pin register and also for contact screens to be pin registered upon that film so that the precise orientation of the screen angles for each colour printing is automatically ensured. This means that it must be possible to punch a piece of unexposed film with the appropriate holes, to register it on raised pins on the vacuum base in the image plane, to hold it by vacuum and then to withdraw the pins to surface level so that the contact screen, which must be larger, can be pin registered and vacuum held upon it.

One of the best known register punches provides two holes about 2½ ins. apart and a slot-shaped hole about 7½ ins. further along the side of the film. One can then have two register pins either the longer or the shorter distance apart upon which one of the holes is fitted and the slot shaped hole fits over the other pin.

It is possible to reproduce, for example, a "flat copy" subject such as a colour painting or a wash drawing, and to introduce into it perhaps a colour transparency, and to colour mask both images. This entails precise register positioning of intermediate continuous tone colour separation negatives, block-out masks (to cover the parts to be occupied by the other picture), colour masks, perhaps introduced lettering, etc., at both the image and the object planes, and further register positioning when the half-tone positives are made.

Some cameras have double-sided film holders for the image plane and for the object plane to allow films or plates to be registered on either side, separated by half an inch or so. One side may contain perhaps the colour transparency and the other a colour mask which can be introduced for part of the exposure and substituted by another for the remainder of the time.

Lens. The lens in medium-sized copying cameras generally has a focal length equal to the diameter of the plate. A lens of shorter focal length is used for making enlarged reproductions when the maximum extension is insufficient.

The aperture is not of first importance because the subject is still, and long exposures can be given virtually at will without risk of movement.

A very high degree of correction is called for, however; in particular the lens must have a flat, evenly illuminated field right to the corners, and be free from astigmatism and chromatic aberration. For colour reproduction it must be fully corrected to bring all the colours to a focus at the same point.

Reversing Prisms. For direct (i.e. not offset) printing the image on the block or printing plate must be reversed left-to-right to give a right-reading reproduction. As the printing plate is normally produced by contact printing a line or screen negative, the latter must thus also be reversed left-to-right as seen through the base. This is exactly opposite to the requirement in normal photography, and such reversal is normally obtained in the process camera by photographing through a reversing prism. A prism is used rather than a mirror because it is easier to achieve absolute flatness and high optical precision in a polished prism surface. The process camera thus normally operates with the prism over the lens, and with the camera body parallel to the copy. The camera can then be swung through a right angle to point directly at the copyboard when it is necessary to use it without the prism.

Where this right-angle alignment of camera and copyboard is inconvenient or impossible, a "straight-through" reversing prism may be used. This is a double prism arrangement which inverts the image but keeps it in the optical axis of the camera system. The focusing and other optical adjustment scales of the camera must of course be calibrated for use with either type of prism.

Copy Holder. The copy is held vertically in a hinged frame similar to a normal contact printing frame. This has a sprung pressure back which holds the copy flat against the surface of the cover glass. The whole assembly usually may be swung into a horizontal position for convenience in loading the copy into the frame.

If the camera itself is mounted on to runners, the distance of the copy holder is fixed; size adjustment is then effected by moving the camera. Cross movement of the copy holder is provided so that the copy can be re-centred when the camera is rotated for operation without the prism.

For convenience, especially when handling flimsy copy (such as tracings) the conventional copy holder may be replaced with a vacuum holder.

Lighting. It is essential for the copy to be evenly illuminated. Carbon arc or mercury vapour lamps are generally used because their highly actinic light makes up for the slowness of the sensitized materials.

The best type of illumination is given by four lamps, two on each side of the board, spaced from each other about 2·5 times their distance from the board. But two lamps may be made to give reasonably good illumination by careful adjustment.

Transparencies to be photographed by transmitted light are normally supported in a masking frame in front of the copy board. The lighting comes from a white surface illuminated by the normal lamps or by a series of tubes for cold cathode illumination fitted behind the transparency.

Exposure Regulation. Fluctuations in the voltage supply to carbon arc lighting can seriously affect the intensity of light; other factors may also cause variation in the lighting. If this occurs during exposure but after the exposure reading is taken, an incorrectly exposed negative will result. To avoid this, the process camera may be fitted with an integrating exposure meter.

The integrating exposure meter, of the photo-electric type, is fitted on to the copy holder so that it receives the same illumination as the copy. The required exposure is pre-set on the meter which is coupled to an electromagnetic shutter on the camera. The meter then "adds" the total effect of exposure over time, automatically controlling the exposure received by the plate so that fluctuations in the lighting do not affect the final result. F.P.

See also: *Process lens; Registration of images.*

PROCESSING MACHINES

Special machines in which films are automatically developed, rinsed, fixed, washed and dried. They operate in a darkroom and are primarily intended for processing large quantities of exposed film. A typical machine for the processing of short lengths can handle approximately 360 per hour. Machines for continuous processing (as in aerial photography) operate at speeds in excess of 4 feet per minute.

Still Films. Processing machines for roll films and sheet films developed out of the requirement of bulk processing in wholesale photofinishing establishments. Starting from the idea of deep tanks into which films were loaded and transferred in batches, the processing machine in principle features the addition of mechanized transport, timing and temperature control systems.

The machine usually consists of a series of large tanks, containing the appropriate solutions and a suitable system for conveying the exposed films from one to the other. Short film lengths may be suspended from a moving chain driven by suitable gearing so that the films are moved through and lifted from one tank and passed on to the next. Speed of travel of the chain is regulated to give the correct developing time. The time of immersion in any tank can be altered—e.g., by using tanks of differing lengths—to suit the times of processing in the other solutions.

Roll films and 35 mm. films can either hang from bars or be wound round frames. Sheet films are usually supported in special film hangers or frames.

While sheet film processing is normally carried out on a smaller scale in professional darkrooms, not requiring full mechanization, automatic processing machines are used in X-ray departments of large hospitals where the volume of work justifies it.

When continuous lengths of film are used, a slightly different system is employed. The film is fed in and out of each tank over a series of rollers. Times of immersion in each solution are varied by altering the separation between the upper and lower sets of rollers.

The processing of continuous lengths of film is needed for aerial films, microfilm and other material where large volumes of sensitized material of uniform characteristics is processed. There successive lengths of film are spliced together as they are fed into the machine. They then go through the various processing stages, including drying, and are wound up on a final take-up spool. From there the films may be wound off again into their original length, and separated.

In both systems, there is thermostatic control of the processing solutions. The film, after washing, passes through a current of warm air in a drying chamber heated by normal filament heaters or it may pass a series of infra-red lamps.

An alternative system for automatic processing handles the film in batches in a single tank. This is suitable for roll, miniature or sheet film and the batches can vary from half a dozen film lengths up to really big quantities, according to the size of the processing unit installed.

After loading the sensitized material into hangers or other holders for placing in the tank, the appropriate developing and other solutions are pumped into the tank and pumped out again after the required treatment period. The unit contains storage tanks of all the solutions, and one or more pumps plus the necessary pipelines and valve gear to carry the solutions between the individual storage tanks and the processing tank itself. After the processing sequence the film may even be dried in the tank by blowing air through it.

Cine Film Processing. Processing machines for motion picture film are usually larger and more elaborate than for photo-finishing, since they have to handle correspondingly greater lengths of film. For that reason also such machines almost invariably work on the continuous flow principle, with the film passing from tank to tank over rollers. The film is not touched by hand from the point where it is loaded in (straight from the camera spool) until it is wound up on a take-up spool after drying.

Continuous processing machines on this scale frequently also include automatic agitation systems for solutions and automatic replenishment provision. Here replenisher solutions are fed into the individual tanks in previously determined quantities, sometimes with a proportion of the tank stock liquid being drawn off at the same time.

Consistent image quality is particularly important in motion picture work. So frequent quality control tests (short test films exposed under specific conditions) are run through the machine at regular intervals and measured to check the activity of the developer solutions.

In addition to large continuous processing machines for motion picture laboratories, small scale installations also exist to handle short lengths (100 feet or so) of 16 mm. film. In principle such processes are scaled down versions of commercial units, without however automatic replenishment. Some of these outfits are small enough to stand on a medium-size table.

Paper Prints. The only efficient way of handling prints in bulk is in the form of continuous rolls of paper. Hence automatic print processing machines generally work in conjunction with roll head printers for fixed-ratio enlargements. The machine is of the continuous type, similar to continuous running film developing units. The strip of paper again passes through developing, rinsing and fixing tanks over rollers. The treatment time in any one tank depends on the rate with which the paper passes through and on the length of the paper path inside that tank.

At the end of the processing cycle the paper runs over a drying drum similar to a rotary glazer; this in turn may feed the strip to an automatic trimmer which cuts the paper into separate photographs.

Colour. Colour film processing machines are used mostly in the photo-finishing and motion picture laboratories. In principle they work on the same lines as installations for black-and-white film, but are more complex with a greater number of solution tanks to accommodate the extra processing stages required for colour film. Mostly the machines work on the continuous flow principle, with special attention to temperature control, replenishment and quality control. With colour film the slightest variation in processing conditions leads not only to changes in density but also in colour balance.

A special step required for reversal colour films is the re-exposure stage. Here the film passes between banks of lamps to fog the unexposed silver salts prior to the colour development. (Similar re-exposure stages are also used for processing black-and-white reversal film.) With some types of colour film, having no colour couplers incorporated in the emulsion, a number of selective exposure and colour development stages are necessary to treat each emulsion layer individually.

Colour print processing machines often work with continuous rolls of paper in much the same way as black-and-white roll paper processes. Or they may handle colour prints in batches after loading into print baskets or similar containers.

Automated programming. Modern processing machines handling batches of material (black-and-white or colour) frequently have the timing of the individual processing stages and the transfer of the material from one tank to another controlled by a programming unit. Single-tank processers (in which the solutions are pumped in sequence into one processing tank) employ the programming unit to control the valves and pumping stages.

Basically the programmer is fairly simple. The more elementary systems involve locating the switches (electrically controlled) for the various pumps, valves, etc. at a central point and connecting them to one or more time switches. The latter can then be set for the required duration for every individual processing stage. Once this is set up, the processing machine runs through its programme of operations automatically.

More sophisticated programmers have a scanning unit which reads signals recorded on magnetic tape, embossed into plastic panels or punched into cards or perforated paper tape. These signals in turn close relays which control the various programme functions of the machine. The advantage of this arrangements that it becomes very easy to switch over from one process to another (for example with different types of colour film). This means simply changing the programme—feeding a different tape or a different plastic programming panel through the machine—and changing the processing solutions employed. Some machines may have several programmes built in simultaneously. There the machine contains the solution tanks or containers for the two or three processes for which it is programmed and the sensitive material is passed only through those

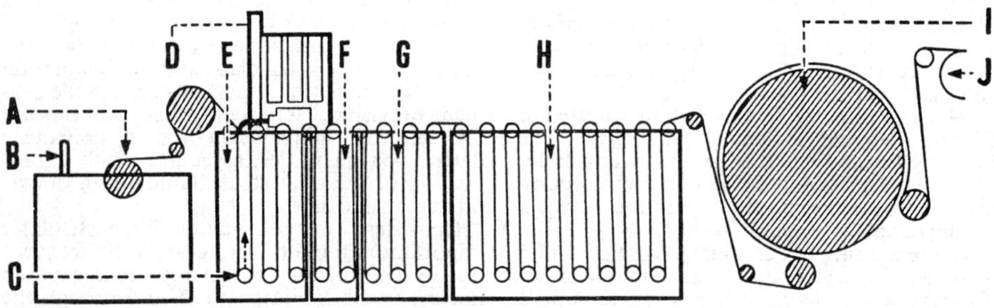

CONTINUOUS PAPER PROCESSOR. This processes long strips of prints, wound up in rolls from a roll-head printer. The unit remains in continuous use, fresh strips being constantly spliced on. A, feed roll. B, splicer unit. C, rack forming bottom of loop. D, solution flow meter. E, developer. F, stop bath. G, fixer. H, washing tank. I, drier and glazer. J, take-up roll. In principle, continuous cine film processors operate similarly, without of course a glazing stage.

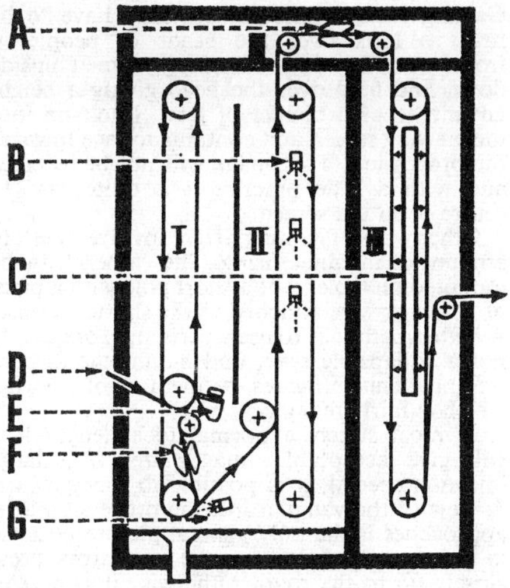

VISCOUS SOLUTION PROCESSOR. A, air squeegee between wash chamber and drying chamber. B, spray nozzles in wash chamber. C, warm air system in drying chamber. D, entry of film. E, coating hopper applying viscous solution to film surface. F, air squeegee. G, spray jet to wash viscous layer off film. I, processing chamber (either a single chamber for mono-bath developers or there may be several chambers for separate development or fixation.) II, wash chamber. III, drying chamber.

solutions belonging to its particular processing programme.

One outfit of this kind consists of a series—up to a dozen—deep tanks set up in a circle. Above the tanks runs a circular rail with a motorized carriage. The latter incorporates a lifting motor to raise or lower the batch of films being processed out of or into the individual tanks. A transport motor drives the carriage along the circular rail.

Contact points along the rail, corresponding to the tank positions, stop the carriage above each tank when the contact point is energized. The programming sequence there controls the lowering and raising of the film batches and the intervals between these actions as well as the points (i.e., the tanks) at which the carriage stops for immersing the film batch. Two or three carriages can run on the top rail simultaneously, thus handling several batches in succession before the first one has gone through the processing sequence.

Rapid Access Processing. Automatic continuous flow systems are also used for certain rapid access processes where the fully developed image is required within seconds of being exposed in the camera. Some of the ideas developed for this purpose have also found use in large-scale normal processing machines.

One example is viscous processing, where the thickened developing solution is spread over the film surface as a very thin layer. This remains in contact with the film long enough to complete the development phase and is then washed off by a water spray. Next a viscous fixing solution layer is applied, followed by further spray washing and drying.

Machines of this kind are also used in regular motion picture film processing on a comparatively small scale. L.A.M. & J.D.C.

See also: *Postcard manufacture; Rapid Access; Wholesale photo finishing.*

PROCESSIONS. There are various kinds of processions and each needs a slightly different approach, but in the main essentials all processions are alike in consisting of a number of people proceeding with colourful pomp along a route lined by a great many spectators.

The best known processions are the state occasions when there are probably a number of horse-drawn landaus forming the main centre of interest but followed and preceded by a troop of cavalry and possibly also a detachment of Foot-guards.

The first thing to do is to find out in advance where the procession is going. If the route is not known well it is advisable to look it over in advance to find a good spot to stand. Try to find a place which is slightly higher than the surrounding area. It will permit a clear view even if with a crowd. Having found the right viewpoint, get there early and stay there.

Work out beforehand from which direction the sun will be shining; don't leave it until the day because once the ceremony has begun no spectators will be allowed to cross the street.

Setting-up. Any kind of camera will serve on these occasions so long as it has a convenient eye level viewfinder. Waist level finders are no good because it is essential to shoot over the heads of the people in front. A temporary eye level finder can be made out of a piece of wire bent into a rectangle the same size as the negative. Mount this on top of the camera, parallel to the lens, and then mount an eyepiece made from another piece of wire bent into a small ring with a stalk on it on the back of the camera, parallel with the film plane. The ring of the eyepiece should be opposite the centre of the frame. A reflex camera can be held upside down above the head to gain height.

Having taken up position, take an exposure reading and set the camera, allowing plenty of depth of field by stopping down well. A fast shutter speed is not likely to be needed: no procession travels at speed. Finally focus on the middle of the road and set the shutter.

Subjects. Several kinds of picture may be wanted, but it will not be possible to get them all. Although the whole of the procession may be wanted in one photograph, it is rarely possible to get it. Most streets are too narrow

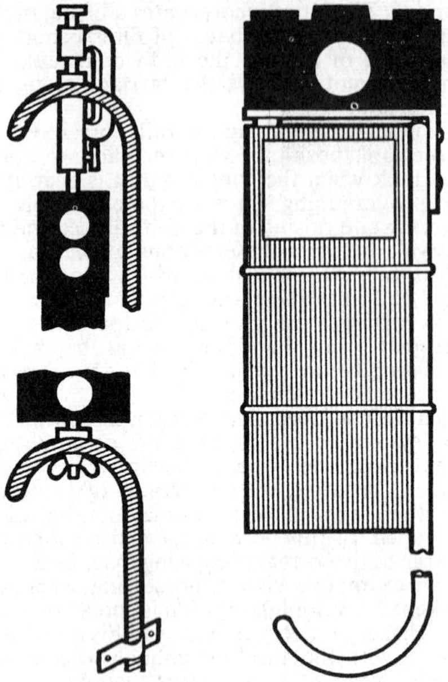

SHOOTING OVER CROWDS. Top left: *A twin-lens reflex camera can be held aloft upside down on a walking stick.* Bottom left: *A tripod screw through the handle of the stick will hold smaller cameras. A bracket on the shaft of the stick will act as sighting device.* Right: *A more elaborate support holds the camera on top of a periscope arrangement.*

to allow inclusion of much more than a section, so under these conditions concentrate on the occupants of the carriages. The closer the better for this kind of picture. But for shots of the mounted cavalry, be either some distance away and include both horse and rider, or as close as possible and shoot against the sky.

Look out for the picturesque costumes of heralds or pikemen in a procession. They make good subjects for close-ups.

If a vantage spot can be found somewhere up in a window overlooking the processional route, it is then easy to include everything in the picture. This may also afford an opportunity to include some of the flags and street decorations which are almost certain to be there. Arrange the picture so that the flags serve to label the scene. For example, a state visit by the ruler of some foreign power might be marked by including his country's flag with that of the nation visited, together with the rulers concerned.

Of course, processions can also include cars instead of open carriages. But there is nothing very interesting in a picture of rows of motor cars, so concentrate on getting as close as possible and then aim the camera through the car window. It will be necessary to open up the aperture by two or three stops to give adequate exposure for the inside of a car.

Gaining Height. The camera may have to be held to look over the heads of people in front. A reflex can be simply turned upside down and held over the head giving a height advantage of a couple of feet. Look up into the viewing screen and continue to face towards the procession. The image will not be inverted and with a little practice it is quite easy to centre it on the screen.

Other ways of gaining height involve a certain amount of manufacturing. The camera can be mounted on the end of a short pole with a piece of string or wire attached to the shutter release. A better method is to use a periscope, preferably not a collapsible type, and mount the camera on top. This provides a method of viewing which is fairly reliable.

In most streets a normal focal length lens will give acceptable image size. If a large image is needed, it is possible to move closer. A view of the vanguard of the procession as it approaches is the only general picture possible in narrow streets where the spectators press closely on to the road. Otherwise it is best to concentrate on close-ups. In wider thoroughfares general views of various units of the procession are easy to get.

If a position is taken some way away, perhaps in a window overlooking the route, use a telephoto lens or attach a telescope in front of the camera lens. This necessitates making a clamp to hold both the camera and the telescope rigidly.

Make some experiments beforehand with this arrangement, to establish the image size produced by the telescope and the minimum aperture for the required depth of field. G.S.D.

PROCESS LENS. The process lens takes its name from the photomechanical processes used in the printing trade. These demand lenses of long focal length in which all the monochromatic aberrations are unusually well corrected relative to the size of the lens. Where colour separation work is required, axial and transverse chromatic aberrations also have to be corrected to a degree beyond anything necessary in ordinary lenses.

Normal achromatic lenses have the ability to bring light of two colours to a common focus, but correction in at least three colours is essential in process work. Lenses corrected in this way are termed apochromatic.

Fortunately, a high lens speed and short exposures are not required and an adequate standard of performance can be provided without undue complexity of lens form at relative apertures in the region of f 11. These small apertures give the designer additional freedom to choose the best types of optical glass for achieving apochromatic correction.

Since lens aberrations are not usually stable with radical change of object distance, process lenses are corrected for use over a small range of copying magnifications and, as a rule, they do not perform very well with distant objects.

Certain photomechanical processes require reversed negatives and this is achieved by placing a mirror or prism in front of, or behind, the lens at an angle of 45° to its axis and pointing the camera at right angles to the object. The mirror or prism has to be inconveniently large if it has to cover a wide angular field from the lens and process lenses are usually designed to cover angular fields varying between 35° in long focal lengths to 50° in the smaller sizes. A coverage of 50° is a wide angle in process lenses.

Large copy and large negative sizes are unavoidable in process work so that focal lengths of these lenses are considerable, reaching a maximum at about 70 ins.

Where the range of conjugate distances only varies slightly from unit magnification, perfectly symmetrical lens forms are adopted. When such lenses are used at unit magnification this symmetry automatically provides perfect correction of coma, distortion and transverse chromatic aberration. Lenses intended for use over a wider range of conjugate distances and some lenses covering wider fields are rarely of the symmetrical form.

Process lenses can also be used for other jobs of a similar nature. G.H.C.

See also: *Lens history; Process camera.*

PROCESS PLATE. Slow plates with a very contrasty emulsion used for copying and photomechanical reproduction. Process plates are usually blue-sensitive only, although both orthochromatic and panchromatic process plates of a variety of speeds and spectral sensitivities are used in colour reproduction of line and tone originals.

See also: *Negative materials.*

PRODUCTION OF PHOTOGRAPHIC GOODS. The production of photographic equipment and materials is undertaken by manufacturers in more than twenty countries of the world, although the greater part of output is concentrated among the more highly developed nations.

In the last decade the expansion of photography in all its aspects coupled with rising economic prosperity, particularly in the United States and Europe, has resulted in an ever-increasing demand for photographic goods, and has served to stimulate established producers and attract new ones to the scene. The increase in both the value and range of equipment and sensitized materials produced has been rapid in recent years (in some cases, notably Japan, spectacularly so), and photographic manufacturing has gained a position among the recognized growth industries.

An objective assessment of the photographic industry is, however, often rendered difficult by the interlinking in many countries of producers with other manufacturing activities. Thus the production of equipment is nearly always to be found in conjunction with optical and scientific instrument making, whilst sensitized materials are frequently manufactured by firms engaged in papermaking and the chemical industry.

A further difficulty, now gradually being remedied, is the general lack of published statistics and other information relating to production and trade in photographic goods, and also the problem of comparability of the available data between countries.

The total production of photographic equipment and materials in Western countries is estimated to have been in the region of £950 million in 1963, and of this figure over 95 per cent was accounted for by manufacturers in seven countries; Belgium, France, West Germany, Italy, Japan, the United Kingdom, and the U.S.A.

The United States is the world's largest producer of photographic goods, some two-thirds in value of whose output is composed of sensitized materials, followed by West Germany and Japan, whose principal interests lie in the equipment field, whilst Belgium, France, Italy, and the United Kingdom are predominantly concerned with sensitized products.

Recent years have seen the growth of links between the chemical industry and photographic sensitized goods producers. Consolidation has been taking place, often by way of mergers and take-overs, and has resulted in the formation of several large international manufacturing groups, each backed by considerable financial and technical resources.

This movement has been particularly marked during 1964, with the linkage of Agfa (Germany) and Gevaert (Belgium), the acquisition of Bauchet (France) and Ferrania (Italy) by the 3 M's Company (U.S.A.), the association of ICI (Britain), Ilford (Britain), and Ciba (Switzerland), and the take-over of Schleussner (Adox) (Germany) by DuPont (U.S.A.). It is partly associated with the desire for manufacturing and marketing facilities within the E.E.C. (European Economic Community), but more particularly with the economies of scale in the production of sensitized film and paper and the high cost of research, especially among colour materials.

Equipment manufacturing on the other hand has continued to be carried on by a large number of independent producers, established principally in West Germany, Japan, and the U.S.A. There are few groupings of any size to be found, apart from the Kodak organization and the Zeiss Foundation, since the production of photographic apparatus can easily be undertaken on a small scale and without necessarily needing highly skilled personnel. Many firms in fact simply purchase individual parts such as lenses, shutters, and lightmeters

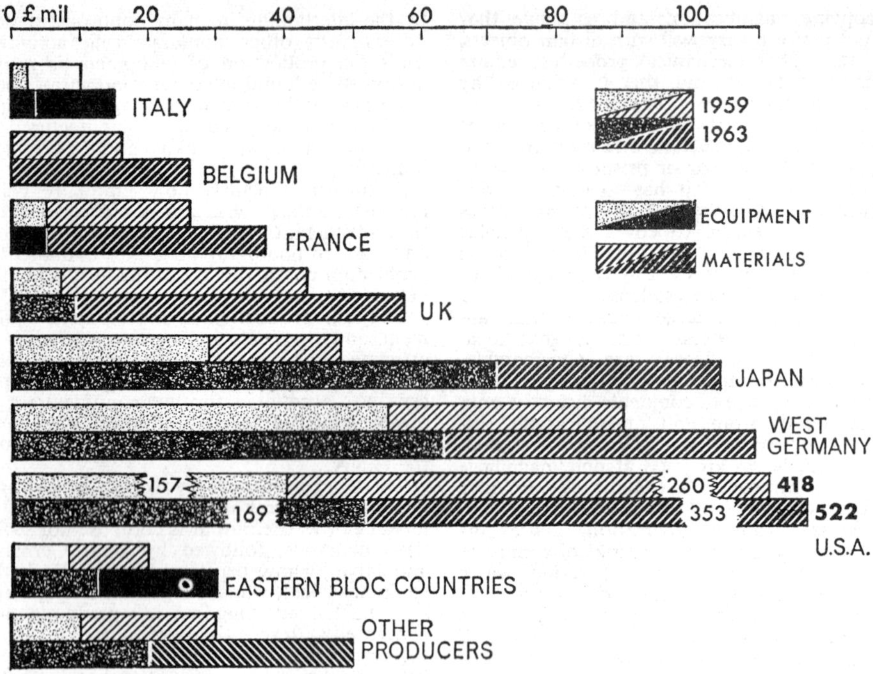

WORLD PHOTO PRODUCTION AND EXPORT. The bars compare the value of the world production of apparatus (stippled) and materials (shaded) for 1959 (white background) and 1963 (black background). While the photo industry has expanded in all countries, Japan showed the fastest rate of growth—particularly in apparatus where production grew nearly 2½-fold during the period covered. After the U.S.A., the biggest producers of sensitive materials are the United Kingdom and West Germany. The Eastern bloc countries in this diagram include East Germany and the U.S.S.R.

from specialist producers and assemble these and other components in their own workshops.

Mention may be made of the Eastern Bloc countries, whose photographic industries are not large but are nevertheless gradually growing in importance.

The two leading producers are East Germany and the U.S.S.R.; the former having had the benefit of gaining the major part of the established (although severely damaged in the war) German photographic industry following the country's partition, whilst the latter has gradually built up production over the last thirty years. Both countries undertake the manufacture of cameras and other equipment, and in addition have a fairly large volume of sensitized materials output.

Other sensitized goods producers are Bulgaria, Czechoslovakia and Hungary, each important manufacturers of paper, whilst Yugoslavia and Poland also make film on a small scale. Additionally, photographic equipment is produced by Czechoslovakia, in particular cine cameras and projectors. Jugoslavia and Poland. An interesting newcomer to photographic manufacturing is China, where the small-scale production of still cameras and black-and-white film and paper is now undertaken. It is estimated that the ouput of cameras is at present in the region of 100,000 units per year, the bulk of which are the simpler types of model. G.R.H. & A.G.B.

See also: *Belgium; Britain; France; Germany; Italy; Japan; United States of America.*

PROFESSIONAL PHOTOGRAPHY. The essential difference between the amateur and the professional photographer is that the amateur undertakes photography for personal pleasure whereas the professional is also concerned with earning his livelihood. The amateur may produce excellent photographs that suit his particular needs, often working without a time limit. The professional photographer, on the other hand, is required to undertake a variety of assignments which must be carried out in accordance with client's requirements.

Scope. The range of professional photography is very wide. One industrial company, for example, may employ a photographer to aid research by such methods as photo-elastic stress analysis, while another company will send their photographer to out of the way parts of the world to make records for public relations or advertising. The photographer may work in a research laboratory, a coal mine, a hospital, a museum, a test farm, in an aeroplane or in the water. He may work amid luxury in a fashion studio or amid dirt and heat in a steel works. He may be occupied in photographing fine buildings, deciphering ancient manuscripts or taking portraits in a studio.

The professional photographer may choose to work for an employer in one of these fields or many others, and with adequate basic training, and continued study he can become a specialist in one branch—e.g., historical research, shipbuilding or portraiture.

While many photographers take up such specialized employment, often in charge of departments, in big industries, public utilities, hospitals etc., many eventually decide to practise on their own account.

Here the field of choice is limited virtually to portraiture, commercial, industrial, and advertising photography, photo-journalism and a few highly specialized applications of photography where the demand is proportionately small.

In most of these the professional photographer must be prepared to serve and study an exacting market. And he will need to know about a bewildering range of regulations and practices—e.g., copyright law, purchase tax, factory act requirements, registration of business names, right to photograph, insurance, liability for clients' goods, and what is done or not done, either on ethical grounds or for good business reasons. In this he can obtain a great deal of guidance and help by joining a professional association.

Before purchasing an established business or starting a new one, he should make sure that he possesses: the right temperament, physical and personal qualities; enough skill, training and experience; the necessary artistic and business ability; and finally sufficient capital to buy, equip, and maintain a business.

In all but the scientific and recording applications of photography he must have some artistic appreciation. It is also essential that he should have general business ability, with a knowledge of simple accountancy and costing. In some branches good salesmanship is indispensable. Whether buying or starting a business, he should engage a good solicitor and a good accountant and he would be wise to budget for no income at all during the first year.

Premises. The choice of the site will depend upon the type of work to be undertaken.

The portrait photographer will consider the type of client he wishes to attract. If he seeks to photograph a large number of people at low fees, he must look for the most prominent site in the busiest street in the town, with a good shop window, and he will have to rely on a big turnover to cover the heavy rent. Very careful study of the neighbourhood by observation and local inquiries is necessary because even a particular street may have a good and a bad side.

If he wishes to serve a more discriminating public and provide high quality work he will need to have premises in quieter and more dignified surroundings.

Position is of less importance to a commercial photographer whose premises simply need to be reasonably accessible to business men and have ample space. The latter generally renders a main road site too expensive.

The industrial photographer only requires an office and workrooms, because most of his work will be done away from the premises. He should at the same time equip a small studio, since he will often be asked to undertake a certain amount of commercial work.

For portraiture the premises must be suitable for conversion into reception room, studio and workrooms. The commercial photographer will need more than one studio, or one big enough to take several sets at once, and there must be large doors and plenty of storage space for articles waiting to be returned. If advertising work is undertaken there must be provision for a model's dressing-room.

In leasing a building it is necessary to bear in mind that in most leases the tenant is responsible for repairs and dilapidations and that heavy expenditure may have to be met when the lease expires. Meanwhile decorations and improvements will be necessary, and it is important that the parts of the premises to which the public will have access should be attractive. This is particularly important in a portrait studio, and the photographer, in choosing his scheme of decoration, must bear in mind the type of client he wishes to attract.

Finally the buyer should take into account that he may in time wish to expand, and it is important to know whether there is room for such expansion.

Security of tenure is of the utmost importance. A short lease is valueless unless it carries an option to renew. An established business with a very short unexpired lease should only be purchased if arrangements can be made with the owner of the property to secure an option for renewal or to take out a new lease.

Equipment. In buying an established business, the photographer is likely to find that some of the equipment may be out of date and capital will have to be available for new equipment.

Some reserve capital is desirable in commercial photography to cover the amount which will always be outstanding in model fees and hire of equipment and apparatus which may not be paid for some time.

Goodwill. When an established business is purchased, goodwill may be worth very little. The goodwill of a popular business in a main thoroughfare may have some value, but an individual business which has been built up on the personality of the outgoing photographer may have little goodwill. There is no standard by which to judge the purchase price. It is currently held that a business may be worth one year's gross takings or three years' net profit, but it is very important to know whether the takings of the preceding years give a true valuation of normal business.

When purchasing a business it is important to include in the contract a clause which will prevent the outgoing photographer from practising as a photographer within a reasonable distance for a specified time and taking the goodwill with him.

Fees. Every business should adopt some system of simple costing, and advice in this respect should be sought from a qualified accountant. The Institute of Incorporated Photographers publishes a guide to professional fees for commercial and industrial photography, and also a schedule for reproduction fees.

Social Contacts. It is not enough just to buy a business. The photographer has to become known and respected in the district where he makes his living. So he should take an active part in the social life of the community.

He must also pursue cultural development. The camera is only a tool, but because the technique of photography is complicated and fascinating, many remain obsessed with it. That is not the way to success. The photographer must constantly broaden his outlook and develop his sensibility. He should be able to converse easily with a variety of people on a variety of subjects. There is, too, an ethical standard to maintain. He must command the respect not only of the public but of his fellow photographers. A.F.B. (J.L.A.H.)

See also: *Agencies; Careers in photography; Commercial photography; Costing; Prices of commercial photographs; Studio photography; Training for photography.*

Books: *The Business of Photography*, by C. Abel (London); *The Professional Photographer in Practice*, by R. P. Hymers (London)

PROGRAMMED SHUTTER. Shutter with a continuous range of exposure settings obtained by interlinking the lens aperture and shutter speed adjustment, yielding one exposure combination at each exposure level. A programmed shutter does not provide alternative aperture speed combinations at a given exposure level. Shutters of this kind are usually fitted in electric eye or fully automatic cameras.

See also: *Shutter.*

PROJECTION LAMP. Special high intensity light source designed to conform to the requirements of projection systems.

See also: *Carbon arc lamp; Filament lamp; Lamp caps and fittings; Light sources; Point source lamp; Projection principles.*

PROJECTION LENS. The principal differences between a projection lens and a photographic lens is the result of their completely different conditions of use. Projection lenses are corrected for colour aberrations to suit the spectral sensitivity of the eye and the character of the light emitted by projector lamps, whereas camera lenses are corrected to suit the panchromatic emulsions used in photography. The projection lens is also used in conjunction with an illuminating system which may focus heat as well as light on the lens, so it must not incorporate any delicate cemented surfaces or unstable components. Again, there is rarely any need for a projection lens to cover a wide angular field, but the demand for high levels of illumination on large viewing screens calls for lenses of wide relative aperture.

A large proportion of the projection lenses now in general use have been developed from the early Petzval lens form originally designed as a portrait lens or from an early form of microscope objective designed by Lister. Both the Petzval and Lister types consist of two widely separated doublet lens components and they are similar in performance. They are both capable of providing relative apertures wider than $f2$ but their field of view is restricted by vignetting and poor definition in the outer parts of the field.

Calculation of Focal Length. The focal length of the lens required for any combination of film size, screen, and length of throw is given by the equation:

$$F = \frac{U}{M + 1}$$

where M is the optical magnification (obtained by dividing the width of the required screen picture by the width of the object or film to be projected) and U is the distance between the projection lens and the screen.

This calculation will give the focal length of the required lens, but in practice it may turn out that none of the available lenses of that focal length has a wide enough covering power. When that happens, the focal length must be increased either by increasing U or decreasing M.

For example, a 16 mm. film (width about 0·4 in.) has to be projected to give a picture width of 5 feet (60 ins.) on a screen 12 feet from the lens. The magnification $M = 60/0{\cdot}4 = 150$. So, from the formula, the focal length of the lens (in inches) $= 144/(150 + 1) = 0{\cdot}96$—i.e., a 1 in. lens. But this value of focal length is only twice the diagonal of the 16 mm. film image to be projected, and in such a lens a high standard of definition at a relative aperture wider than $f2$ is not possible with Petzval or Lister type construction. If only these forms are available the focal length must be increased to at least 1½ ins. by accepting a screen picture width of less than 40 ins. or by increasing the screen distance to more than 19 feet for a greater screen width.

Covering Power. As a very general rule the angular coverage of Petzval and Lister type lenses can only be increased by reducing the relative aperture. A lens with a focal length of approximately twice the diameter of the film image will not usually cover the whole of the area unless it has an effective aperture of f3·5 or smaller.

If the lens must be designed for a larger aperture than this—i.e., so that it gives a more brilliant screen picture—then the focal length must be increased to three times the diagonal of the film image, or more.

Where the lens must have a short focal length in situations where the screen distance is limited, it is necessary to utilize special projection lenses developed from various forms of photographic anastigmat lenses. These alternatives are usually of the Cooke Triplet or Tessar constructions and permit the selected focal lengths to be between one and two times the film diagonal.

Relative apertures of such lenses vary between f4 and f2·8; they sometime replace Petzval type lenses when these medium apertures are acceptable and cost of manufacture has to be kept down.

Alternative Constructions. Whilst fundamental differences between the two constructions prevent projection lenses from being used as photographic lenses, a photographic lens can sometimes be used satisfactorily for projection. But caution must be exercised in using complex lenses in this manner. The damage done by the heat of the projection lamp is not always obvious except by photographic testing and repairs (possibly involving separation and re-cementing) can be costly if the original standard of performance is to be restored after such damage.

Other more complex anastigmatic photographic lens constructions have been considerably modified and developed as projection lenses for special applications requiring a very high standard of definition under critical conditions—e.g., wide-screen cinema projection.

These lenses are more expensive than the normal types, but they have the wide relative apertures necessary to provide adequate illumination and can compete with the good central definition of the Petzval or Lister type. Their ability to maintain such a high standard at the edges of the picture as well as at the centre more than justifies their cost. G.H.C.

Zoom Lenses. Variable focus lenses are sometimes used on projectors. The majority of such projection lenses have been designed for 8 mm. cine projectors, with a comparatively smaller number for 16 mm. projection and for still projectors.

Because of their relative mechanical simplicity most zoom lenses for projection purposes have been of the optically compensated type. This type of zoom lens provides a change of focal length by the equal linear movement of at least two components, these being moved in tandem with respect to fixed members, by a simple slide mechanism. Such an arrangement yields a slight variation in position of the focal plane through the focal length range but this is of little consequence in projection where adequate facilities exist for fine focusing. Mechanically compensated zoom lenses also exist for projection.

The main advantage of a zoom lens for projection is flexibility. A single lens can be used over a varying range of projection distances and screen sizes without requiring exact positioning of the projector. Once the projector is set up the lens may be "zoomed" until the screen is adequately filled. Among lenses of this kind, a zoom ratio of maximum focal length to minimum focal length of about 1 : 3 or 1 : 4 is quite usual. P.A.M.

See also: *Lens history; Projection principles; Projector (still); Zoom lens.*

Books: *Lenses in Photography*, by P. Kingslake (New York); *Photographic Optics*, by A. Cox (London).

PROJECTION PRACTICE

The projection of transparencies and cine films is one of the most important media of entertainment, education and instruction. The applications on the entertainment side range from amateur slide shows presented in a largely casual manner to the commercial cinema. The scope of educational applications extends from individual slides and short cine film lengths presented in the course of a lecture to illustrate specific points, to completely organized slide lectures with recorded commentaries, and to specially planned and made educational motion picture films.

In both cases the main value of projection is that it makes pictures available simultaneously to group audiences. Further, a positive transparency (including in this context a positive cine film) can be the most satisfying form of photographic reproduction because it can present a greater range of tones than any print made on an opaque base. The maximum contrast range of a glossy black-and-white print is limited by the ratio of the brilliance of the whitest parts of the paper to the blackest deposit of silver in the shadows area of the image. This ratio is never greater than about 30:1, while the contrast of a correctly processed transparency may range from areas of completely opaque silver to completely clear film.

Still and cine projection have in common certain requirements of projection conditions, such as the preparation of the (darkening or blackout, etc.), the screen and the seating arrangements for the audience. They may differ appreciably in the organization of a show, depending on the type of picture presented and on the audience and the purpose of the presentation.

PROJECTION CONDITIONS

The projection room and the screen are as important as the projector if the transparencies are to be seen at their best by everyone present. Apart from the obvious requirement that the room should be big enough to seat everyone in reasonable comfort, it must be as dark as possible and well ventilated.

The Blackout. It is possible to show pictures in a normally-lighted room by using a very powerful projector and a screen of high reflecting qualities, but the results are incomparably better if the room is really dark. This can easily be demonstrated by projecting a colour slide in a room with some windows open to the light. As the curtains are drawn, the picture grows steadily brighter and of better quality, but there is an almost magical change as the last rays of extraneous light disappear.

It is not always possible to black-out the projection room completely, especially in halls where exits have to be indicated. But there must be no source of outside light in the line of view of any of the spectators, and no rays of direct light falling upon the screen. Even the mere diffusion of the light by thin curtains can make a good deal of difference so long as the windows are completely covered. Dark venetian or pleated blinds are sometimes used. At the same time, it is generally cheaper to provide an efficient blackout than to purchase and run the high-powered projector needed for daylight projection.

As shows and lectures are often held during the evenings and during the winter season there is no great difficulty in making the room dark, but lights outside in the street and from other buildings must not be overlooked. If a meeting is to be held during the afternoon it is always wise to find out if the hall can be blacked out. This is easily overlooked by the organizers, especially if lectures illustrated by slides are not a frequent occurrence. It is possible to project films and slides with fairly bright room lighting, but only to small audiences.

Ventilation. Unless the show is going to be a very brief one, some extra means of ventilation is necessary to compensate for the closing of all doors and windows. The combined effect of the heat from the spectators' bodies and the projection lamp can spoil the show as decisively as poor quality screen pictures. Some form of forced draught, say by one or more extractor fans, is essential in any room used regularly for projection.

This problem is particularly serious when shows are held in comparatively small halls to large audiences. It sometimes helps to request the audience not to smoke, since a heavily smoke-filled atmosphere is not only unpleasant but also absorbs light and so leads to a dimmer picture on the screen.

In the home ventilation problems can be as serious, especially in small rooms used for showing a film or a slide series. Sometimes it is possible to leave a window open behind the curtains which have been drawn to shut out the light. Alternatively it may be necessary to leave a short break at regular intervals during a show for opening the windows and clearing the air.

The Screen. The kind of screen employed has a marked effect on the quality of the picture and on the viewing conditions. A matt white surface reflects the light from the projector over a wide angle, so that even people at the side of the room get a fairly bright view. This kind of screen surface is the only choice where some of the spectators must watch from the side. Beaded and metallized screens give a more brilliant picture, but have a limited viewing angle.

If the projection room is long enough or if there is another room beyond it, a translucent screen can be used to advantage. The image may be directed upon it straight from a projector fitted with a short focus lens, or indirectly by means of mirrors. One mirror corrects the image laterally, and allows the transparencies to be loaded as for straight projection, instead of having to be reversed. The space behind a translucent screen must be as dark as possible, or the image will be faint.

The angle of view of a translucent screen is

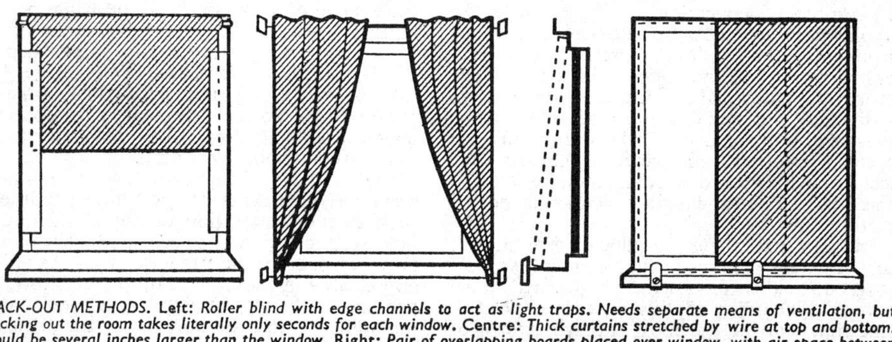

BLACK-OUT METHODS. Left: *Roller blind with edge channels to act as light traps. Needs separate means of ventilation, but blacking out the room takes literally only seconds for each window.* Centre: *Thick curtains stretched by wire at top and bottom. Should be several inches larger than the window.* Right: *Pair of overlapping boards placed over window, with air space between them for ventilation. Is particularly suitable for shows of long duration, though it takes longer to put up.*

narrower than that of a matt white and is almost the same as that of the silver and beaded types.

Such a back projection system needs very careful alignment. Back projection units incorporating their own optical systems are easier to set up, but usually the screen size is severely limited and such units are intended for demonstration purposes and presentation to audiences of two or three people.

For a large-size back projection screen a large sheet of tracing cloth or translucent plastic sold for bathroom curtains and similar purposes are usable. Fine linen can be used, but must be wetted before projection to increase the brilliance of the projected picture. The material is best fixed to a frame to keep it flat.

In rear projection set-ups not using a mirror, the slide must be inserted with the spot pointing towards the front or towards the lens of the projector, to ensure that the audience sees the picture the right way round. With cine projectors this is rarely possible (for example reversal of the film in an 8 mm. projector would bring the perforations on the wrong side) so here a mirror must be used. This must be surface silvered of the highest quality and is placed as close in front of the lens as possible.

This form of projection is best when the room is not easily blacked out. The screen should be in the darkest part of the room and shielded from any extraneous light.

With all opaque screens, most of the light is reflected back towards the projectors. If the projector is set at an angle, the light tends to follow the laws of reflection from a polished surface, and the angle at which it leaves the screen is principally the same as the angle at which it meets it.

The brilliance of the image falls off at each side of the optical axis by an amount that depends upon the type of screen used. Before any particular screen is chosen it should always be tried out under actual working conditions.

The size of the screen in relation to the projector needs to be considered as the illuminating power of the projector governs the screen brilliance. If the audience is limited to twenty or thirty people, a projector fitted with a 300 watt lamp or a 100 watt low-voltage lamp will be quite satisfactory. For a larger number a larger screen and higher power will be required, as well as a lens of longer focal length. If the show is to to be a success the audience must be able to see the pictures in comfort. This is not possible when a large number of people are trying to see a picture on a small screen. If the audience is over fifty in number the screen should not be less than 7 × 5 feet. This in turn will need a projector with not less than 500 watt illumination. For larger numbers greater power will be necessary.

For the audience to see and hear under the most favourable conditions, the picture should be projected over their heads from behind the last row of chairs. This has the advantage that it economizes in seating space, since the best position for viewing in the centre is not taken up by the projector, and it avoids leaving a central gap to keep the screen clear. With the projector low down, heads are apt to appear as a silhouette on the screen, while people behind the projector find their view spoiled by stray light, and possibly by the movements of the operator. The projector should be placed about 5 feet high, and the screen arranged so that the lens is opposite the centre. This will also avoid the necessity for tilting the projector, with the risk of loss of definition and the introduction of a degree of distortion.

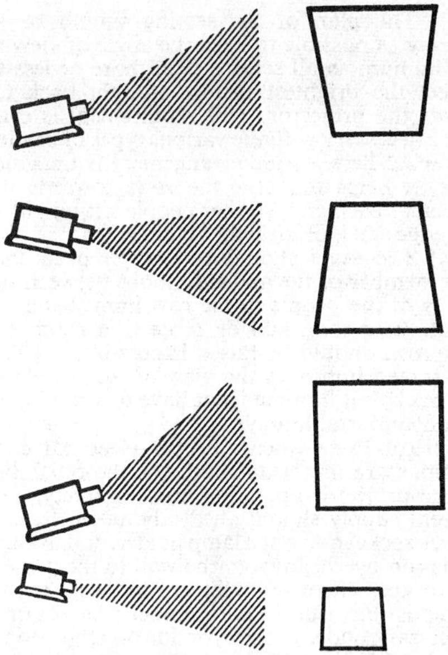

TILTING THE PROJECTOR. Top: *Tilting the projector upwards will distort the image shape.* Upper centre: *Tilting downwards has the same effect but in the opposite direction.* Lower centre: *If a tilt is unavoidable, the screen should be tilted as well, which will square the image again.* Bottom: *A slight tilt on a long throw is permissible, as the distortion is small.*

Seating Arrangements. The seating must be arranged so that the screen can be seen by everybody clearly and without strain. The shape of the room and position of projector and screen do not always lend themselves to a good seating plan, but where possible the following recommendations should be followed:

(1) The front row should be one screen width or more from the screen—never less. This is the shortest distance from which the eye can take in the whole screen at once.

(2) The back row should be eight screen widths or less from the screen—never more. This is the greatest distance from which the normal eye can distinguish detail without strain.

(3) The plan of the seating should be as narrow as possible to keep the angle of view to a minimum as all screens tend more or less to reflect the brightest picture straight back towards the projector. This requirement is even more necessary with the various types of beaded screen. Where a wide arrangement is unavoidable, it helps to curve the rows towards the screen at each end, so that people sitting out at the edge can look straight at the screen.

(4) The seats should be staggered so that any member of the audience looks between the heads of the people in the row immediately in front. In a long hall or projection room the rear rows should be raised if possible. Raising the screen improves the view of the people at the back, but those in front have to strain up at an uncomfortable angle.

Electrical Preparations. Proper electrical connections are important both for the reliability of the projector operation and for safety. The current supply should always be taken from a power socket and not a lamp holder, and flexible leads run overhead from the wall to the projector to keep them safely out of the way. If this is impossible, run the wires under a carpet or if necesssary under a fixed conduit to stop people people from tripping over and—more serious—sending the projector crashing to the floor. Avoid make-shift electrical connections and make sure that the supply is correctly fused for the projector.

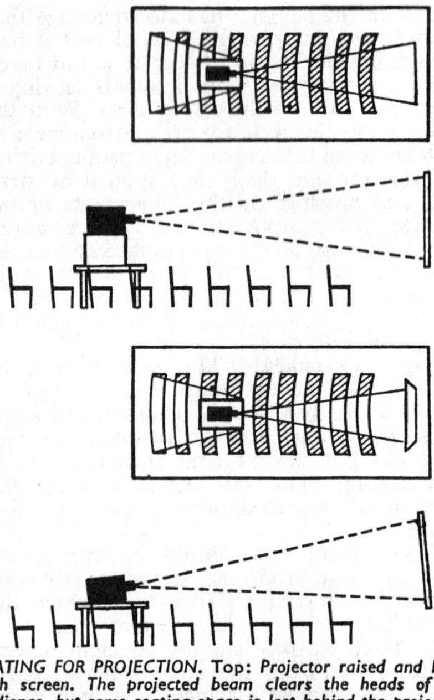

SEATING FOR PROJECTION. **Top:** ***Projector raised and level with screen. The projected beam clears the heads of the audience, but some seating space is lost behind the projector.*** **Bottom:** ***Emergency arrangement with tilted projector, resulting in loss of seating space in front of the projector as well as behind. The screen should be tilted.***

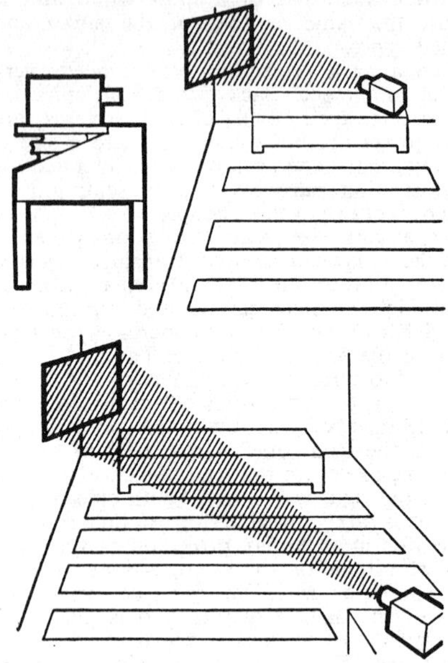

CLASSROOM PROJECTION. **Top left:** ***School desk used as projector support with lid raised by books.*** **Top right:** ***Projector on demonstration bench, permitting operation by teacher. Screen may have to be oblique to projector axis.*** **Bottom:** ***Comer projection in a classroom. Screen may always remain in position without obscuring blackboard.***

A spare lamp and fuses of the correct rating should always be kept handy.

Check also that the current supply is of the same voltage as specified for the projector. This is particularly important with the projection lamp; overrunning a lamp gives brighter light but drastically reduces the lamp life. Underrunning cuts down the image brightness and also affects the colour of colour transparencies or films. Some projectors have a voltage adjustment; this should be set to the available mains voltage. With still projectors voltage variations may affect the speed of the fan and possibly of the slide change; this is not serious. With cine projectors they may also influence the speed at which the projector is running.

SLIDE SHOWS

Owing to the current popularity of reversal colour films, sets of colour slides and a projector are to some extent replacing the photo album as a means of showing one's pictures to friends, etc. At the same time the projector remains an important aid to illustrating lectures and talks. Requirements of the two applications are parallel, though not identical.

Home Projection. The informal slide show needs comparatively little preparation beyond darkening the room and arranging seats for the people watching the pictures. Be sure to prepare the slides, stacked up (or loaded into a magazine in the case of a magazine projector) in the right order and correctly orientated. Where slides are loaded individually into the projector by hand, the identifying spot on the slide must be in the top right hand corner facing the operator behind the projector. With a single-slide manual loading projector every slide must be individually checked for orientation in this way before insertion.

With magazine projectors this preparation can be done beforehand while loading the magazines. Here it is particularly important to check the orientation, since the correction of a wrongly loaded slide in the magazine usually requires removal of the whole magazine from the projectors. On a few models individual slides can be removed while they are in the picture gate for rectification; however this procedure is still obvious to the audience.

With a manual slide projector the projectionist usually has to sit more or less on top of his machine to operate the slide change. Automatic slide changing by remote control is more comfortable for the projectionist and also improves the smoothness of the show. The operator can then sit among the audience and only has to handle the operating button at the end of the remote control cable.

Where card mounted transparencies are being projected the image is liable to "pop" in or out of focus on the screen as the film changes its plane in the card frame. This requires constant refocusing as each new slide is projected. Since that procedure is somewhat cumbersome, it is usually more convenient to focus the first slide after it has "popped" and wait until every successive slide pops into focus rather than readjust the lens setting every time. This takes a few seconds.

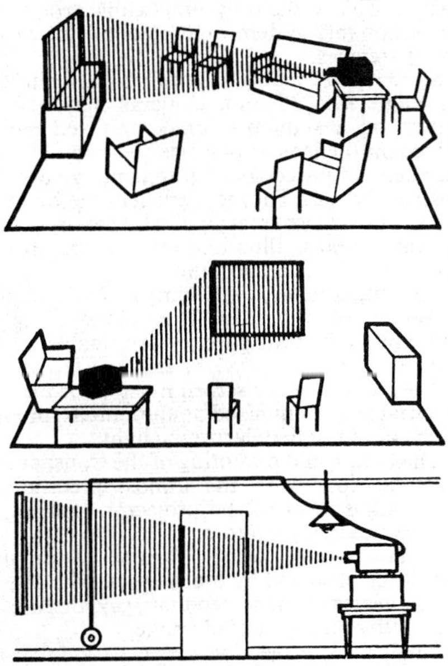

HOME PROJECTION. Top: *Projecting across the long side of a room, with the screen in a recess if available.* Centre: *Where the long side of the room is not suitable for projection arrangements, oblique projection provides some extra throw, though with a certain amount of image distortion.* Bottom: *Projector leads should be run overhead and out of reach.*

Refocusing is simplified with a remote control system that also includes remote focusing, but still remains a job to be done with every new slide.

Transparencies mounted between glass in plastic or other frames are free from this trouble and, provided that all mounts are similar, it is only necessary to focus the first transparency sharply. All the rest will then remain sharp on the screen. Some projectors incorporate automatic means of refocusing the image if the plane of the transparency moves.

A home slide show will not require much in in the way of elaborate commentary. This may be quite informal, though repetitive descriptions like: ". . . this is Jane by the Leaning Tower of Pisa . . . and this is the Colloseum in Rome . . . and this is Vesuvius near Naples . . ." are apt to be monotonous. (The same applies of course to "explanations" accompanying the showing of a photo album.)

The Slides and the Projector. The projector should be suitable for the type of transparency being shown, especially as far as the format is concerned. Most amateur transparencies are of the 24 × 36 mm. size, mounted in 2 × 2 ins. slide frames. Most projectors will also take super slides with an image area of $1\frac{5}{8} \times 1\frac{5}{8}$ ins. (in similar 2 × 2 ins. frames) but some need a change in the condenser. Half-frame 18 × 24 mm. transparencies can also be shown, though special projectors for this format utilize the light output better. A projection lens of shorter focal length is advisable with half-frame transparencies, otherwise the screen image is very small.

The larger $2\frac{1}{4} \times 2\frac{1}{4}$ ins. slides made from size 120 roll film transparencies need a larger projector; some models take both slide sizes interchangably. Big projectors for $3\frac{1}{4} \times 3\frac{1}{4}$ ins. slides are rarely used for home presentations.

It is always worth while to select or process transparencies to be shown in a series so that there are no great differences in brilliance between one picture and the next. This is because the human eye takes some time to accommodate itself to a change in brilliance. So a dense slide following a thin one appears denser than it really is, while a thin one shown after a dark one causes an unpleasant glare.

The first transparencies of the session should be fairly bright because the eyes of the audience will not had have time to become accustomed to the dim light.

Black-and-white slides are at a disadvantage when shown with colour transparencies. They can, however, be improved in this respect by either hand-tinting, toning, staining, or binding them up with a coloured filter.

Transparencies in colour have always been popular, even when hand tinted, and today factory-processed colour films have made it far easier to produce a good colour transparency than to make even a reasonably good black-and-white print.

A good colour transparency will stand as much or more magnification than a mono chrome, but some colour systems suffer from poor definition or an unduly assertive reseau which limits the degree of enlargement.

Colour of Illumination. The colour of the viewing or projecting light is of the first importance in projecting colour transparencies. A projection light of the wrong colour can completely upset the colour values of the transparency. With some types of transparency, the manufacturers specify a certain kind of projection light. But this light is only exactly right when it is used with perfect transparencies.

Where the colour balance of the transparency is defective, it can often be improved by the addition of a tinted filter. In some cases it is better to stain the emulsion, or a filter can be bound up with the transparency. In the same way, a filter can be inserted under a transparency in a viewer to improve its colour balance.

The colour of the filter required must be found by experiment; it is usually very pale in colour. Light blue is the most common, particularly in viewers. Suitable filters can be made from lantern plates, fixed out, bleached if necessary, and stained with dyes. Gelatin or plastic sheet can be dyed and used as a filter, and the palest colours of theatre gelatins are often satisfactory.

In showing colour transparencies, not only should the brightness of succeeding slides be approximately the same, but their colour values should also be similar. The more subdued tints and pastel shades should not follow on very bright colours. (When the eyes look at one colour for a time, and it is then removed, they are left with an impression of the complementary colour. This impression tends to superimpose itself on succeeding colours and alter their values.)

It must be borne in mind that the process of seeing colours is a subjective phenomenon. All that can be done is to aim at producing an effect which will please most of the spectators. It is impossible to overcome all individual differences in colour perception and preference.

Second-rate apparatus and poor viewing conditions will often pass for monochrome work but prove quite inadequate for colour viewing and projection. In particular, colour transparencies can only be seen at their best in complete darkness.

Stereo Transparencies. Stereo transparencies are made in various sizes of which the commonest are:

(1) 18 × 24 mm., made with mirror or twin-lens attachments on normal 35 mm. cameras.

(2) 12 × 13 mm., 22 × 24 mm., 24 × 24 mm. and 24 × 30 mm.; all made on standard 35 mm. film in stereo cameras.

(3) 4·5 × 10·7 mm., 6 × 13 cm., etc.; made on larger films and plates in large stereo cameras.

The transparencies may either be viewed in a stereo viewer, or projected by a suitable projection system.

Since it is not practical to project the two images side by side, chiefly owing to the limitations of human vision, they must be superimposed by projection in twin-lens projectors. The necessary separation of the right-eye and left-eye images can be accomplished in two ways.

For the older method, the anaglyph, the two images are toned red and green respectively (or projected through appropriate filters in front of the projection lenses) and viewed through spectacles of corresponding colours so that each eye sees only its appropriate image. The more modern method is to polarize the two projection beams by two polarizing filters at an angle of 90° to each other. The image is viewed through polarizing spectacles the two glasses of which are orientated in the same sense as the corresponding projected left-eye and right-eye images. This is the only practicable true three-dimensionable system of projection for colour transparencies.

Stereo pairs made with mirror attachments on single lens 35 mm. cameras may also be shown on a standard projector fitted with a device on the lens to polarize the two halves of the beam and accessory equipment to superimpose the images. The more efficient method is to project the transparency pairs by means of twin optical systems. Illumination may be from a single lamp, but on account of the light loss in polarizing systems, twin lamps of the highest power are often needed. The polarizer may be adjacent to the transparency or attached to the lens.

The surface of the screen must be of metallic material (e.g. aluminized or aluminium sprayed) which does not depolarize the light.

The taking and mounting of the transparency must be done with the utmost precision to avoid adjustment of the projector, and inconvenience to the spectators.

A position as nearly frontal as possible should be occupied by the viewers, who should not sit nearer to the screen than, say, three times the width of the projected image.

Small back projection viewers for table use have been evolved. Superimposed polarized images are shown in the usual way, and, of course, polarizing spectacles are required.

Pseudo-stereo Projection. An alternative method of simulating a stereo effect with single transparencies consists of projecting them through a beam splitting optical system. This produces two partially overlapping images on a metallized screen. Polarizing filters in the two projection beams again polarize the two images at right angles to each other. On viewing through corresponding polarizing spectacles a sort of depth effect is obtained. This is because the two images in slightly different positions on the screen induce the eyes to converge less than they would on to a single image in the plane of the screen. A real feature of this depth effect is that the image appears some distance behind the screen (emphasized by having a bold black frame to the latter). The apparent depth effect is most pronounced when objects in the picture decrease in size as they get progressively further away.

SLIDE LECTURES

Where slides are used in connection with a lecture, the lecture and the illustrations must be planned together. A list is needed of the slides available or to be obtained. Each slide must clearly show the point to be made, and it may be necessary to have two or even three. It is not advisable to use colour slides and monochrome together, as one gains at the expense of the other. Preferably colour slides should all be made on the same colour film where possible, to reduce variations of colour quality.

All slides should be of the same size and even of the same transparency format (e.g. all 24 × 36 mm.). Generally 2 × 2 ins. or 2¾ × 2¾ ins. slides are most useful in view of the versatility of projection equipment available for them; the larger 3¼ × 3¼ ins. or similar formats are still used sometimes.

If a magazine projector is used, have the magazines loaded ready beforehand, with all the slides correctly orientated. They should of course be spotted in the standard way.

If the projector is of the older single slide type, the slides are best kept in a transport box with each slide accommodated in a separate groove. This ensures that the slides are kept in a correct sequence and do not slip to the bottom of the box when used. They are also more easily handled by the operator. It is a good plan for such a box to take double the number of slides to be used side by side, so that after use the slide is placed in the opposite empty section. This saves trouble in sorting them next time.

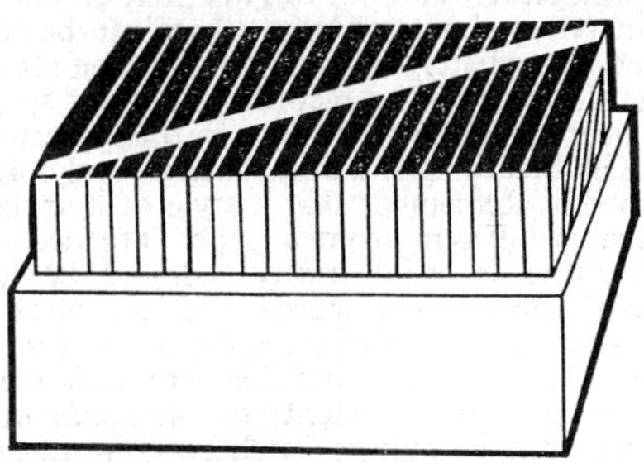

MARKING SLIDE SEQUENCE. A diagonal line across the edges of a series (e.g., a lecture) of stacked slides will show immediately if one slide is out of sequence in a box.

Slide Change and Signalling. Where a projector with manual slide changing (single or magazine type) is used the operator needs some prearranged signal to change the slide. As far as possible the signal should not be seen or heard by the audience. A signal lamp operated by the lecturer at his table is ideal.

Where a magazine projector has remote control, an operator is not always essential. In this case it is necessary to have a projector with a magazine capacity to take all the slides of a lecture. Straightforward remote control cables are usually too short for lecture use where the lecturer may be 20 to 100 feet away from the projector. Direct control devices available for this purpose include supersonic whistles which actuate the automatic slide changing mechanism via an amplifier, and devices that change the slide when the lecturer stops speaking. Radio control is also possible with matchbox size transmitters.

Such remote devices should, however, be able to adjust the picture sharpness as well as change the slide.

Generally it is safer to have an operator in charge of the projector, to deal with breakdowns or other emergencies on the spot. Local government regulations for public halls may make it essential to have a responsible operator for a projector, and incidentally also lay down a variety of other conditions regarding seating, access of the audience to the hall, organization of exits, etc.

Lecture Delivery. Once the illustrations are planned and obtained the lecture should be written out in full.

Next the lecture is rehearsed with the slides, an assistant operating the projector. This will reveal needless duplication or omissions and check the length of time taken for the delivery.

If possible, the lecture should be spoken and not read. Notes are helpful if the lecturer's memory is poor and they lend some confidence to the inexperienced speaker.

For those who are unfamiliar with public speaking here are a few hints:

Make a point of arriving at the hall at least twenty minutes before you are due to start. This will allow you to make contact with the operator of the projector, to show him the slides and their order of projection, and to rehearse the "next slide" signal. It gives you time to meet the Chairman and other officials and all this helps to establish confidence.

When delivering the lecture speak slowly and distinctly. Do not hurry, but on the other hand avoid long pauses. Remember there may be people at the back of the room who have difficulty in hearing. Watch to see if this is so and raise your voice. Keep your head up while speaking.

When it is necessary to indicate some special point on the screen, turn to the screen, point out the subject, and then face the audience. Torch-type pointers that project a bright arrow on the screen where required, are neater than a wooden pointer.

A microphone can be helpful in a large hall and it gives confidence. In using a microphone it is not necessary to raise the voice above the normal conversational level; in fact to do so defeats its own object. A microphone is also useful if written questions are passed up on paper. The speaker is able to repeat the question and reply to the audience.

Film Strips. A film strip consists of a number of illustrations of the same subject arranged in the correct sequence on a single length of 35 mm. film. This is projected in a projector fitted with a film strip carrier of two spools. The film is wound on one and passed through the projector by winding it on to the take-up spool. This may be done by hand or, on modern film strip projectors, by remote control. Many projectors for single slides can be fitted with a film strip carrier.

While a film strip can be prepared for a single lecture, this is usually cumbersome. It involves selecting the illustrations in the form of negatives or making negatives from existing illustrations. These negatives are then printed in the required order on positive 35 mm. film. Film strips are thus more suitable for set lectures to be given in identical form in a number of places and are more widely used by educational bodies and various publicity organizations.

Sources. Film strips and slide sets are supplied by many sources. There are educational film strips for use in schools, and religious bodies lend slides and film strips dealing with their own particular organizations for education or publicity. Film strips are also issued by various departments of horticulture on subjects such as the culture of various plants and the control of plant pests. Travel agencies and railway and steamship companies also supply sets of slides and film strips with the text compiled by the publicity departments concerned. Photographic manufacturers and national photographic organizations also have numerous slide and film strip lectures available; although usually a written commentary is supplied with these, in some cases a lecturer may attend. Many of these sources are free of charge.

Sound Slide Shows. The commentary to a slide show at home or the lecture accompanying a set of slides can also be recorded on tape and played back together with the progress of the images on the screen. This has the advantage that the commentary can be prepared more carefully and the talk delivered with the best effect, rehearsing and re-recording several times if necessary. It tends to make an informal slide show a little more rigid but also more professional looking.

The presentation can become more versatile since other sound effects may be added as well. Such an automatic recorded lecture lacks, however, the personality of a live performance. It is particularly useful for set lectures and demonstrations to be delivered over and over again, for example publicity demonstrations at exhibitions, advertising lectures and shop or shop window demonstration.

With a tape-recorded commentary or lecture the whole presentation can become completely automatic. This involves a projector with remote control facilities and a tape coupler which links the tape recorder with the projector. The tape carries cueing signals which automatically actuate the slide changing mechanism of the projector.

In use the tape runs over a head on the tape coupler (usually attached to the edge of the tape recorder) which picks up the cueing signal. Some tape recorders incorporate special provision for the recording and the reading of such signals without a separate tape coupler.

The production of a tape-recorder lecture or show needs largely the same kind of preparation as an ordinary lecture, but can be done more at leisure. First select the slides to be used and arrange them in the right order. Then prepare the lecture and write it out in full or—if intended for less formal presentations—at least note down the main headings.

For a formal lecture a full scale rehearsal is usually advisable. Set up the projector with the slides and project them one by one as you read the lecture. This rehearsal will show what adjustments are required either to clarify a point or to shorten an overlong description. It is useful to have a trial audience of one or two people to criticize and indicate what the reaction of a normal audience would be. Mark the points of slide changing on the script.

Run through the slides once more, this time recording the lecture on the tape recorder. Set up the microphone in a convenient position in front of you, adjust the tape recorder controls for the best recording level (with a short trial run) and start again from the beginning, delivering the lecture.

Speak clearly and not too fast, but do not be too nervous about mistakes; these can be corrected immediately by running back the tape a short distance and re-recording. Do not speak too near to the microphone, as this otherwise picks up breathing and other background noises of speech. Remember that a style of a spoken lecture is different from a paper intended for publication in print. Speech should be less formal; avoid long words and complicated sentences. On a printed page the reader can go back over a sentence that does not seem completely clear at first sight; the audience in a lecture must be able to catch every word and every idea presented. For that reason more repetition is permissible in a spoken lecture than in print.

It is a good idea to time the delivery of the lecture during the rehearsal. Usually it has to run to a specific time, and the rehearsal allows any necessary adjustments to be made to the length. The beginning and the end of the talk require particular care. The beginning should make immediate contact with the audience while giving the impression that the lecturer just didn't know how to stop.

Finally, when the whole lecture is on the tape, record the cueing signals for the slide change. Usually these are on the second track of the tape and so are completely unnoticed by the audience. Check with a first run-through that everything works satisfactorily; the lecture is then ready for presentation.

With informal slide shows less preparation is necessary. If you have a note of the main headings and items to be mentioned relating to each slide, you can go ahead and record the commentary straightaway. Such a free way of talking often makes the commentary run more easily. However if you are self-conscious in front of a microphone, it may be better to write out the text beforehand, or get someone who is an easy speaker to record it.

Again leave the recording of the cueing signals to the last.

Audio-visual Projection Systems. Complete recorded slide lectures are available for a number of educational purposes as well as for instruction and training in industry, as sales aids, etc. Usually these are played back on combined slide projectors and tape recorders built into a single unit, where the slides are loaded into the projector and the tape threaded into an attached tape deck.

Recorded lectures may also be associated with film strip projectors and may then even be recorded on gramophone records which are played back on a suitable combined machine. This makes the mass production of such lectures easy; since the pictures are permanently combined in the predetermined order, it is more convenient to have the sound in a similar permanent form.

Systems also exist where commentaries are recorded for each picture either on adjacent film frames or on special slide mounts carrying a magnetic track. These require projector systems suitable for such combinations, and incorporating the appropriate amplifying and reproduction components. L.A.M.

CINE PROJECTION

The work that has gone into the making of a film culminates at the moment that the film is presented to an audience. Any standard of projection less than perfection will nullify much of its value.

A hackneyed piece of advice which is often ignored, even by professional 16 mm. projectionists, is to have the projector set up and adjusted before the audience arrives. Make sure the screen is tightened and free from wrinkles, and that no stray light is falling upon it from a partially curtained window. Make arrangements for the control of the lighting.

Check that the projector is properly aligned upon the screen, and that the lens is clean and in focus. When starting the projector, make sure that the film leader is not projected—cover the lens until the opening title arrives in the gate. After every reel, clean the gate and sprockets before rethreading. Make sure that the spools are not bent.

The projector should be set up on a firm table. Books and other objects for raising the projector height or for tilting it are inadvisable, for the image may shake on the screen. The projector table can also hold the extra reels of film (if more than one reel is being shown), the spares and a splicer. The spares, for dealing with unexpected breakdowns, should include a lamp, motor and take-up belts (where the projector has a belt drive) and fuses of the correct rating. A torch and a screwdriver are handy. The splicer is mainly intended for emergency repairs of film breaks. A tape splicer is best, and the faster it can rejoin the film, the better.

Running the Show. Carefully follow the instructions for threading up the projector. To check the correct orientation of the film lift up the film end so that it points upwards from the feed spool and look at it with your back to the screen. The image (for example the title) should then be the right way up and the right way round. Pay special attention to the film loops when threading the film into the projector and (in the case of a sound projector) round the sound head. Attach the film end to the take-up reel and turn the latter through a couple of revolutions to make sure that the film is secure.

Many modern projectors feature automatic threading. There you depress a button or lever, and simply feed the film into a channel above the gate. The projector mechanism then takes the film through (sometimes even through the sound unit) to emerge at the back for attaching to the take-up reel. On some projectors the film even goes to the take-up reel automatically.

When focusing the image on the screen check that the picture is also correctly aligned in the

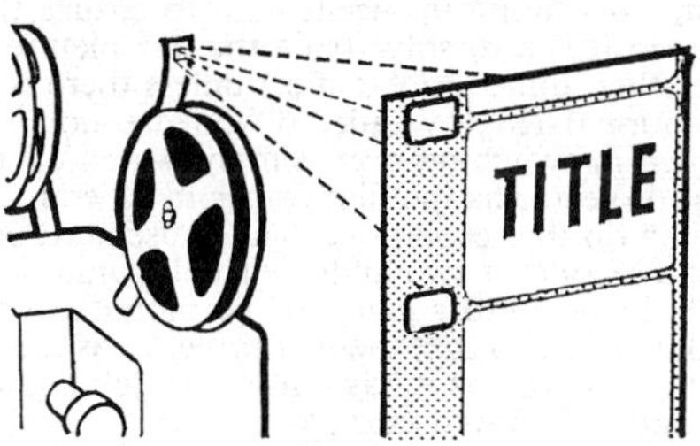

FILM ORIENTATION. To check that an 8 mm. or 16 mm. film spool is correctly mounted on the projector, lift up the film end while standing with your back to the screen, and read the title. If this is upright and the right way round, it will project similarly.

gate; most models have an adjustment to raise or lower the picture so that no frame line is visible on the screen.

Focus while the projector is running rather than when set for still projection; the position of optimum sharpness may not be the same in the two cases.

If the projector has a zoom lens you can at this point adjust the image size to fill the screen; otherwise you have to move the projector nearer or further away from the screen. Needless to say, all this should be done before the start of the show.

When everything is ready, switch off the room light and start the projector running. For a really professional-looking set-up you can use a dimmer. Some projectors have a switch built in for controlling the light; otherwise have just a small table lamp burning near the projector while the rest of the room lights are switched off, and turn off this lamp as the film starts running.

Always start the motor before switching on the lamp. On most projectors the same switch operates the two, and has two positions to start the motor first and the lamp afterwards. If you are projecting several reels, wait with rewinding the individual films until the end of the show. To change over, put the feed spool on the take-up shaft and the new reel on the feed shaft, then thread up as before and continue running. Keep only the pilot lamp near the projector switched on for this purpose, unless you want to make an interval in the performance.

A non-stop show of several reels is possible in two ways. One is to have two projectors and load them with alternate reels. While the first projector is running the second one is loaded up, and switched on as the film runs out on the first machine. This switching over needs a little practice and a knowledge of exactly what happens at the end of the film reel. Always switch off the finishing projector before the trailer goes through the film gate.

If both projectors are lined up on the screen simultaneously you can even switch on the second machine and start it running two or three seconds before the film is finished on the first. This lining up needs care to ensure that the result is a dissolve from the one picture to the other, quite a good effect unless there is an intentional lengthy fade. If you are not very skilled in switching over, simply switch off the first projector as the film comes to its end, and switch on the second one. The pause need only last a second or two and is virtually unnoticed.

Twin projectors can also be mounted on a sliding base, so that the first projector is moved out of the way as it has finished its job and the second one moved into place.

Non-stop multi reel shows with a single projector are a little more involved, since they require the splicing of the film (a temporary splice is adequate) while the projector is running. For this the last 50 or so feet of the film must be pulled off the feed reel while the projector is running, spliced on to the beginning of the next reel, and the latter put in place. The loose loop of the film is best kept in a film bin, and this procedure needs care to avoid twisting the film. It also requires practice to be able to do the job quickly yet safely without panic.

Sound Film Shows. There are three main ways of showing cine films with sound accompaniment: playing music or other effects on gramophone records or a tape recorder, synchronizing sound on a tape recorder with the projector, and projecting an optical or magnetic sound film.

The sound reproducing equipment (amplifier, etc.) is usually built in to the apparatus used. The loudspeaker, where separate, is best positioned above the screen. Two speakers on either side of the screen are better still, as they distribute the sound more evenly, especially in a large hall.

The speaker should not be too low down. Otherwise the sound may be too loud for people sitting in the front seats and too weak for people at the back. If the auditorium (e.g. a large hall) is too reverberant it may be necessary to hang up blankets or carpets to deaden sound reflections. Such reflections are, however, likely to be less prominent when there is an audience than in the empty hall. In the home, sound echoes are less likely, but aim to hold shows in rooms with a carpet.

An almost invariable fault of projection, other than in the cinema, is to have the sound too loud. In a hall it is impossible to judge the quality or volume of the sound when standing by the projector; it is better to have it monitored by a person in the centre of the seating area.

Gramophone Records. Music on gramophone records can be used to accompany silent films. For home projection a simple record player is perfectly adequate, and a 12-inch record at $33\frac{1}{3}$ r.p.m. lasts long enough for a 400-foot length of 8 mm. film or 800 feet of 16 mm. (at 16 f.p.s.).

Silent films of greater length are rare, but you may want to change records even during a film to have a selection of different music pieces. In that case a double turntable is useful with separate volume controls. To change over from one record to the other, lower the second pick up on the record as the first record nears its end or the point where you want to stop. Mix from the first to the second by means of the volume controls. Have the records prepared beforehand; when playing selected portions from a record, a guide for the pick-up to enable it to be lowered at a specific point is useful.

A tape recorder is even more versatile and can be used for playing pre-recorded tapes. Or you can record a tape specially for the film. This may consist of selected items of music combined with a commentary. Keep the latter reasonably general rather than directed at specific scenes requiring accurate timing. Otherwise it becomes difficult to keep the tape and the film synchronized without proper synchronizing equipment. Since tape recorders can run up to an hour even at tape speed of $7\frac{1}{2}$ ins. per second (and twice as long at $3\frac{3}{4}$ i.p.s.) no tape changing need be involved during a film show.

When gramophone records or commercially available pre-recorded tapes are used in public or semi-public film shows, fees may be payable for performance rights (for example to the Performing Rights Society in Great Britain).

Synchronized Sound. When projecting a film carrying an optical or magnetic (stripe) sound track, the only problems involved are keeping the sound level (and, if possible, quality) under control. With optical sound films—i.e. with a photographic sound track printed along side the film edge—the sound and picture are indissolubly wedded together. The sound is got on to the film by printing of a sound track derived from a magnetic tape recording. This usually requires the services of a professional recording studio and laboratory.

When projecting an optical sound film, the projector almost invariably runs at 24 f.p.s., and this is also the standard running speed of the film in the camera.

Careful threading of the sound projector is important to ensure that the length of the film between the gate and the sound head corresponds to exactly 26 frames (on 9·5 and 16 mm.) according to the internationally laid down standards. The sound corresponding to any given frame on the film is thus recorded 26 frames ahead of the picture. This is necessary because the film runs through the projector gate in jerks, but must run smoothly past the sound head. These two opposing requirements cannot be achieved at the same time (and there is in any case no room for the sound head in the projector gate) so the two functions must take place on different parts of the film.

The sound track on a striped film (8 or 16 mm.—also possible with 9·5 mm.) is again permanently wedded to the film. The procedure in projection is much the same as with optical sound, but the projector must of course have the appropriate magnetic heads. The separation between the picture and the sound is here 28 frames for 16 mm. (and also 9·5 mm.) film, and 56 frames for 8 mm. The standard speed of the film through the projector is again 24 frames per second (in all gauges), but 16 f.p.s. can also be used.

A sound track recorded as the slower speed is somewhat inferior in quality (especially in 8 mm.), and cannot be played back on a projector running at any other speed. Sound recording at 16 f.p.s. is useful only if sound has to be added to silent films shot at that speed. When making a film destined for sound projection it is better to stick to the standard sound running speed in the camera. Many magnetic sound projectors can run at either 16 or 24 f.p.s., but a few are limited to 24 f.p.s.

The main feature of striped films is that you can record the sound on the film yourself. With commentary and even music effects this is fairly simple: you record the sound while the edited film is being projected. On most projectors the sound heads can be used both for recording and for playback. The same applies to the various sound adaptors which attach to many 8 mm. projectors to permit the projection of striped film.

Dialogue and live effects are a little more tricky, and these should be available recorded separately on tape. They then have to be mixed into the sound track on the film, a matter which requires accurate timing. However if the recording is not quite right the first time, you can erase and record again on the magnetic stripe.

During projection take special care to see that the controls are set for playback and not for recording, otherwise the record may be erased accidentally. Many sound projectors have special interlocks and warning signals.

Projection with Tape Couplers. There are four main systems of linking a tape recorder with a projector for running the tape and the film in synchronism.

The first uses an attachment fitted to the tape recorder. A loop of tape is run over a capstan and a movable arm connected to a variable resistor. The capstan is geared to the projector motor by a mechanical coupling. The loop arm is spring loaded and moves forward or back according to whether the tape speed increases or decreases relative to the projector motor speed. If the tape runs faster than the film, the loop around the loop arm shortens and the arm reduces the resistance of the variable resistor. This is wired in series with the projector motor and speeds up the latter. If the film runs faster than the tape, the loop lengthens and the resistance is increased, to slow down the projector motor.

In this way the tape recorder and projector automatically remain in step. The actual linear speeds of tape and film need not be the same, as the gearing between the projector motor and the tape capstan can allow for this. (With an 8 mm. projector running at 16 f.p.s. the linear speed of the film is 2·4 inches per second, while the speed of the tape may be $3\frac{3}{4}$ or $7\frac{1}{2}$ ins. per second.) The film and tape can go out of synchronism if the tape stretches or creeps (in other words slips in contact with the capstan). This can be prevented by using a perforated tape running around a sprocket instead of a smooth capstan.

The second method is the mechanical coupling of the tape and the film drives. The tape recorder may be built into the projector, with both driven by the same motor, or the tape driven by a mechanical link from the projector motor.

In an electrical coupling system the tape recorder motor feeds electrical signals into the projector which are compared with similar signals produced by the projector motor. If the two are both running at the correct speed, the signals will be in phase; on going out of phase the speed of the one or the other is automatically adjusted.

The fourth method is a kind of electrical perforation. The tape here carries synchronizing pulses on the second track alongside the recorded sound and the pulses control the projector speed.

When projecting a film and running a tape with the synchronized sound effects, both must start at the right point in the projector and tape recorder respectively. For this purpose starting marks are made on both the film and the tape, and the two threaded up so that these marks are opposite specific index points. Then run the film—before the beginning of the show—until the first title is in the gate and stop both the projector and the tape recorder. To start the show itself simply start both together (usually by a joint switch).

To reduce the risk of synchronization errors it is best to project the film and run the tape on the same machines as were used for synchronizing the film with the tape in the first place (i.e. the same projector and the same tape recorder). Provided the same tape recorder is available for every showing, the loop synchronizing method can even be used with different projectors, where these have similar provision for the mechanical coupling between the tape recorder and the projector.

These methods are cumbersome when multi-reel films are shown, since getting the film and the tape to the correct starting points takes a little more time. However multi-reel films are usually library copies which already carry an optical or magnetic sound track.

While projecting keep monitoring the sound. If the tape and the film go out of synchronism due to the tape stretching or creeping immediate adjustment is needed. Some units have an over-riding control to adjust the projector motor to bring the sound and the picture together again. L.A.M., M.K.K. and R.M.F.

Film Storage. A cool, dry storage place is the best to prevent deterioration of films. Narrow-gauge non-flammable base is the most sensitive to atmospheric conditions. Excessive dryness causes it to become brittle and shrink. Excessive damp causes mildew to attack the emulsion layer.

To counteract the possibility of shrinkage and mildewing there are humidifying solutions—a few drops on a small piece of absorbent material in the can are adequate to keep the roll of film in condition. The humidifier should not be allowed to touch the film itself and in the case of certain colour films is not recommended at all. Fortunately the climate of Great Britain is not usually of the extreme nature that makes the use of humidifiers imperative.

Applications of the Film. Since the early days when animated pictures were still a novelty surrounded by an air of magic, there has been a tremendous development and widening in the scope of the uses of cinematography. Nowadays the existence of film is taken for granted—it is part of our everyday life.

Entertainment is, of course, its main business, made up of such varied artistic offerings as the feature films, cartoons, newsreels, documentaries, and fictional films made specially for children. The commercial world also uses films for advertising; and film sequences are made for the television services where direct television is impossible or inconvenient. But 35 mm. film does not have by any means a monopoly of practicability.

Narrow-gauge films had been introduced solely to cater for amateurs and their personal pleasure. But 16 mm. was soon recognized as having great possibilities in serious work. It has obvious advantages in lightness of equipment and lower cost, and so it has been developed to satisfy the modern needs of many different fields. Good colour is so easily obtained and sound is a routine facility.

Portable equipment made 16 mm. ideal for use in schools. Education films cover a range of subjects that is literally world wide. The inclusion of these visual aids at the right places in the curriculum has proved effective in teaching. The appreciation of film is so great in some schools that the children are encouraged to make their own productions which, although fictional, are considered as part of a constructive policy towards the arts.

Eventually 8 mm. film improved sufficiently in quality to become a serious proposition for the educational and industrial applications originally reserved to 16 mm. film. The current 8 mm. sound film systems with magnetic stripe and even with optical sound tracks have opened a variety of new applications, especially where small and compact projectors make the equipment still more portable than 16 mm. machines. Thus small—and even battery driven—projectors are finding growing use as sales aids, etc. G.H.S. and J.C.

WIDE SCREEN AND THREE-DIMENSIONAL CINEMA PROJECTION

For more than half a century the cinema presented a picture of a format adopted by Lumière and Edison, having an aspect ratio of approximately 1·33 to 1. When the introduction of sound brought the musical film into popularity, it was appreciated that a wider screen would enhance the appeal of many types of scene. But it was not until the 1950's, when the powerful competition of television forced the cinema industry into devising new attractions, that these systems were commercially employed.

As a result, we see pictures on screens of all sizes, in a variety of aspect ratios. The principles adopted can be classified under four headings, as in the next section.

The terms "stereoscopic" and "three-dimensional" have been lavishly misused in reference to these systems. The true stereoscopic film enjoyed a brief period of popularity in the cinema, but did not survive. This was due to three factors; a lack of understanding by producers of the principles of stereoscopy; indifferent projection; and the dislike of audiences to wearing special viewing glasses.

Wide Screen Techniques. It is nevertheless to some extent true that a picture of increased width, enhanced by colour of the high standard that is today possible, helps to create a sensation of depth.

There are four ways of making the picture fill a wider screen: a normal picture can be enlarged to fill the width of the wider screen if some of the foreground and the top of the picture can be sacrified; the camera can be adapted to embrace a wider field; a film wider than normal can be employed; the screen may be filled by running two or more projectors side by side.

(1) Wide screen with standard picture. The first system is the one generally referred to as Wide Screen projection. In this case the term indicates projection of a standard 35 mm. film on to a screen with an aspect ratio which is increased above the standard figure of 1·33 to 1 by reducing the height of the projector aperture, and installing a lens of shorter focal length. The top and bottom of the picture are thus masked off and wasted. Aspect ratios of 1·66, 1·75 and 1·85 are commonly used, i.e., a typical size of screen might be 16 × 28 feet, aspect ratio: 1·75.

Because of the increased magnification of the image, loss of definition inevitably occurs in projection. This is due primarily to the limited resolving power of the negative. A method designed to overcome this fault consists in substantially enlarging the area of the negative frame by running the negative horizontally in the camera and exposing an area greater than two normal frames. This enlarged image is then reduced in printing to the standard frame size so that the release prints can be shown in standard projectors.

(2) Wide screen with wide picture. This system is based on the use of an anamorphic lens on the camera which doubles the angle of view in the horizontal direction only so that the resulting image on the film is "squeezed". In projection, a similar anamorphic optical attachment "unsqueezes" the picture to fill a screen of 2·55 to 1.

In many cinemas this ratio cannot be accommodated without reducing the height of the picture—e.g., a cinema whose normal screen measures 18 × 24 feet may have to reduce the height of the picture to 15 feet in order to install a screen of 38 foot width, to accommodate a picture of this aspect ratio.

In view of the width of the screen, a special sound installation is desirable. The standard 35 mm. film gauge is modified to provide space for a number of magnetic sound tracks carrying records made by microphones spaced across the whole width of the scene. These tracks feed speakers positioned to the right, centre and left of the screen, and an effects track feeds speakers distributed around the auditorium. The audience is thus given the impression of hearing sounds coming from the actual sources on the screen.

(3) Wide screen with wide film. A number of films have been produced on film 70 mm. in width, and with an increased picture height, producing in projection an aspect ratio of 2 : 1. The film also carried six magnetic sound tracks, five feeding speakers behind the screen and the sixth the ambience speakers in the auditorium. Films made in this gauge have shown themselves capable of a standard of projection hitherto unequalled in the cinema.

(4) Wide screen with multiple pictures. In a typical multiple projection system, the taking camera has three lenses, positioned at angles of 48° with their axes crossing. These lenses expose three separate films which, although upon standard 35 mm. stock, have frames six perforations in height instead of four. The sound is recorded upon six magnetic tracks.

In projection, three projectors, two of them at an angle of 48° either side of the axis, cover a large and deeply curved screen, which may measure 50 or 60 feet in chordal width.

This type of projection is possible only in specially equipped halls.

Curved Screen. To secure adequate picture brightness with the increased screen sizes of the new projection systems, highly directional screen surfaces must be used. A more even distribution of reflected light, and hence a more even brightness of the picture as viewed from the side seats, results if the screen is slightly concave to the audience.

The radius of curvature is often made equal to the projection throw.

An audience seated close to a deeply curved picture experiences a sensation of participation referred to as "engulfment". This happens when the angle subtended by the picture to the

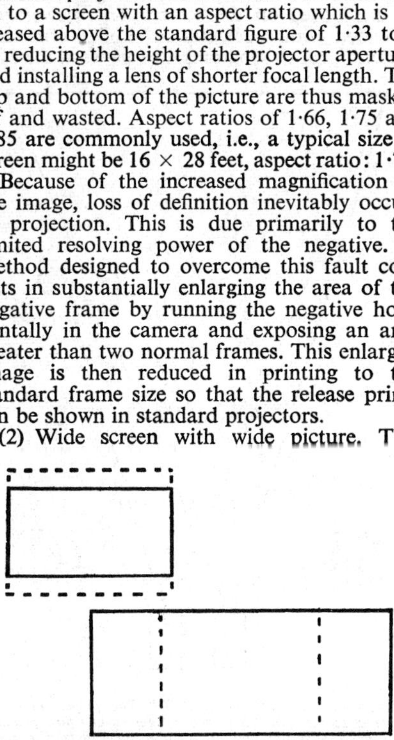

WIDE SCREEN PROPORTIONS. Top: *Projection of a wide screen image on to a normal size screen wastes the top and bottom of the screen area.* Bottom: *Ideally a wider than normal screen should be used.*

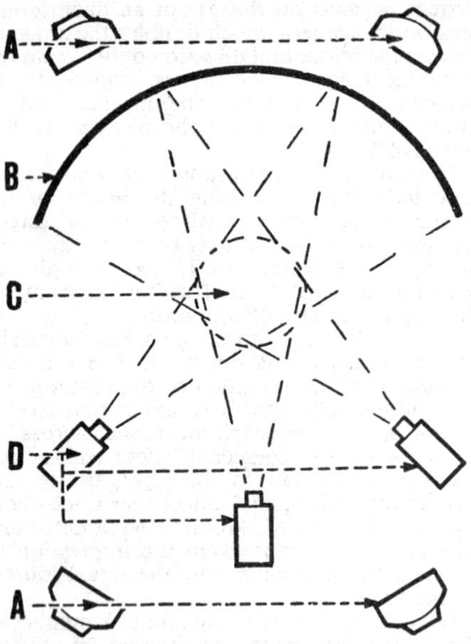

MULTIPLE IMAGE SYSTEM. Three projectors project three images side by side on a curved screen, with slight overlap at the edges. The three films are taken by a corresponding three-camera set-up and must remain accurately synchronized during projection. A, loudspeakers behind screen (actually there are five) and in auditorium for stereophonic sound effects. B, screen. C, optimum viewing area. D, projectors.

viewer approximates to the normal angle of view of the eye. The illusion may be further strengthened by the use of directional sound.

To enable pictures of different aspect ratios to be projected, the masking of the sides of the screen is made adjustable, generally controlled from the projection room.

360° Projection. An ingenious novelty consists in projecting pictures upon a series of screens forming a complete circle, the audience standing inside. In one system demonstrated, 11 projectors are positioned in spaces between the 11 screens, each projector throwing its picture upon a screen opposite it. The films so projected may be of different but related subjects, or may give the effect of a complete 360° angle of vision. For instance one may see a car approaching, turn to see it side view, and turn right round to see it vanishing in the distance.

Stereoscopic Projection. The above systems cannot give more than an approximation to the real sensation of depth created by true binocular viewing in which each eye sees a different picture. This requires that the subject should be photographed from two separate viewpoints, spaced $2\frac{9}{16}$ inches apart, and the resulting films must be projected on to a screen in such a way that each eye sees only the picture made through the corresponding taking camera lens. The two impressions then fuse in the brain to produce the characteristic feeling of depth.

Numerous truly stereoscopic methods have from time to time been worked out. The number of practical projection methods reduces in effect to three, and the underlying principle can be stated as follows.

We can project two images only on the screen, and use special viewing devices to separate them at the eyes; or we can dispense with viewing apparatus altogether if we accept the complexity and expense of providing multiple images and a special screen.

(1) Anaglyph method. Each pair of images is produced in complementary colours. These are superimposed by projection, and spectacles also in complementary colours are used to separate the images. This method can of course only be used for monochrome, and apparatus used to be sold for converting a viewing cabinet into a projector for use with this system. The method was devised by d'Almeida in 1858.

(2) Polarization method. This was the system previously employed in the cinema. The right and left eye images are printed upon separate films, and projected through lenses fitted with polarizing screens. The images are polarized in planes at right angles, and are superimposed on the screen. The viewers are provided with spectacles of similar polarizing material, which separates the images at the eyes,

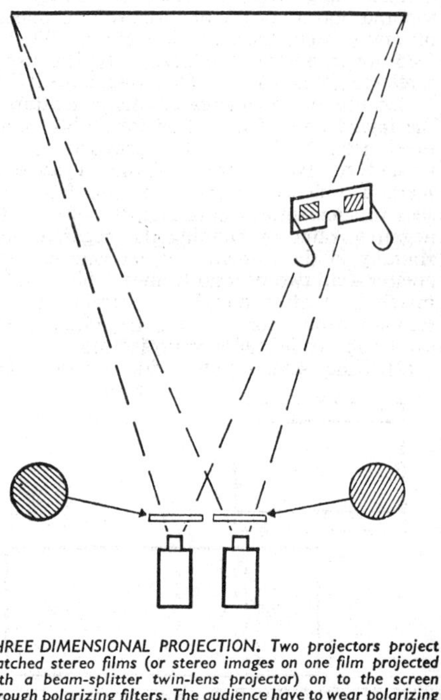

THREE DIMENSIONAL PROJECTION. Two projectors project matched stereo films (or stereo images on one film projected with a beam-splitter twin-lens projector) on to the screen through polarizing filters. The audience have to wear polarizing spectacles with the viewing filters for the left and right eyes oriented with their planes of polarization parallel to the corresponding left and right hand projection filters.

Circular polarization is claimed to possess advantages over the usual plane polarization method. All polarization methods of projection have the disadvantage that considerable loss of light takes place at the polarizing screens.

(3) Special screens. In the special screen systems, the screen is constructed in such a way that the viewers see the separate images of the stereoscopic pair without spectacles.

Raster screens consist of an arrangement of vertical wires interposed between the screen surface and the audience. Two types have been the subject of experiment: one in which the wires from the surface of a truncated cone, within which the projection screen is situated. There are also patented screens in which the projection surface is composed of vertical lenticular prisms, or of spherical lenticular elements graduated in size.

These methods make possible the viewing of stereoscopic films without spectacles or other optical aids, but the apparatus is expensive, and the stereoscopic picture can only be seen from certain positions in the auditorium.

It has been proposed to rotate, or to impart horizontal reciprocating motion to the raster screen to reduce the effect of the vertical wires.

In all systems the projector axes must be kept parallel to avoid geometrical distortion, so the images are superimposed by offset of the lenses. Depolarization of the images in polarized light systems may occur at wide viewing angles, resulting in a double image.

In a more advanced system in development, the polarizing materials are embodied in the actual film coating, one image being carried on either side of a single film.

A different principle necessitates a large convex lens or concave mirror in place of the screen. If at any pair of conjugate points are situated respectively the stereo pair and the eyes of the observer, each eye will see its appropriate image without the need for viewing devices. For obvious reasons the system is limited to a very small number of observers. R.H.C.

Amateur Wide Screen Projection. For amateur use with narrow-gauge film only the wide screen techniques with standard or wide picture are of practical significance. With existing 8 mm. and 16 mm. equipment usually only the squeezing of the image with an anamorphic lens is feasible, followed by "unsqueezing" with a similar attachment on the projector.

Wide-screen projection with a standard picture is also feasible if the film is shot in such a way in the first place, that any loss of image area at the top and bottom of the frame can be allowed for. At least one 8 mm. camera has an adjustable film gate by which the image area may be reduced in height to give a greater aspect ratio. In this case no masking of the projector gate is necessary, provided that the spaces between the frames on the film are opaque. The frame intervals must naturally remain constant.

For projection of such images of reduced height, a wide-angle lens is needed to obtain a screen image of normal height and extended width. The degree of magnification of the image on the screen is then greater than normal, and may therefore show up unsharpness, grain, etc.

A number of ways of modifying the layout of images on narrow-gauge films have been proposed for obtaining pictures of greater aspect ratio. An obvious one would be to modify a normal 8 mm. camera by enlarging the film gate so that the film is exposed across its full width simultaneously instead of down one half only. This therefore would give a 3 : 1 aspect ratio—extreme wide screen—with a frame size of around 10 × 3·3 mm. The camera has to be fitted with a wide-angle lens (i.e. of wide coverage rather than shorter focal length) mounted so as to cover the full film width. The film is projected in a 16 mm. projector, with the gate suitably masked down and the transport system modified to advance the film by an 8 mm. instead of a 16 mm. frame interval at a time. Modifications of this kind have been carried out by enthusiastic amateurs.

The double 8 mm. film used in such a set up must of course not be slit during processing.

A more practical method, sponsored commercially for a short time, is a modified 9·5 mm. film with two perforations between the frames instead of one. The film runs through the camera horizontally and is exposed only along one half of the width (like double 8 mm.). This gives an image of half the area of the normal 9·5 mm. frame, but an aspect ratio of about 1·6. After exposure of the other side of the film and processing, the film is slit in double-8 mm. fashion.

Suggestions have been made that this idea could be adapted to ordinary 8 mm. film, doubling the standard frame area to yield a wider picture through running the film horizontally through the camera. The resulting aspect ratio would be around 1·66. A significant point is that with this system (running the film horizontally through a camera and projector) the wide screen effect is not obtained at the expense of image quality. In a roundabout way this is the nearest equivalent to wide screen projection with a wide film in commercial motion picture theatres. L.A.M.

See also: *Anamorphotic lens; Back projection; Film strips; Projection principles; Projector (cine); Projector (still); Screen for projection; Slide (transparency); Sound synchronization; Splicing; Viewers; Visual aids.*

Books on still projection: *Projecting Slides*, by M. K. Kidd and C. W. Long (London); *Teaching by Projection*, by R. S. Judd (London); *Practical Projection for Teachers*, by N. J. Atkinson (London).

Books on amateur cine projection: *How to Project*, by N. Jenkins; *How to Add Sound*, by D. M. Neale; *How to Produce Magnetic Sound*, by J. D. Chittock (all London).

Books on wide screen and stereo cine projection: *New Screen Techniques*, by Martin Quigley (New York); *3D Kinematography and New Screen Techniques*, by A. Cornwell-Clyne (London). *Introduction to 3D*, by H. Dewhurst (London); *The Theory of Stereoscopic Transmission and its Application to the Motion Picture*, by R. A. and N. Spottiswoode (Californa); *Stereoptics: an Introduction*, by L. P. J. Dudley (London).

PROJECTION PRINCIPLES

When several observers wish to view a positive at the same time, an enlarged image of it may be projected by a suitable optical system on to a viewing screen.

OPTICAL FACTORS

A number of optical factors determine the quality of the image, its apparent brightness and its size.

Image Quality. The quality of the image is affected by aberrations in the optical system. These aberrations are reduced by using optical components specially designed for the purpose.

Then the quantity of light passing through the system must be sufficient to provide an adequate level of illumination on the screen. This is a matter of photometry and it dictates a great part of the design of the optical equipment.

A few of the more important photometric principles and definitions may be summarized before considering any particular projection system.

When a lens system, fulfilling the optical sine condition and having no absorption losses, forms an image of an object on a screen, the illumination is BA/T^2 lumens per square foot (i.e., foot-candles), where B is the intrinsic brightness of the object in candles per square foot, A is the effective area of the lens aperture, and T is the throw or distance from lens to screen.

If the object is a perfectly diffusing surface with 100 per cent reflectivity and is illuminated by one foot-candle, its intrinsic brightness is $1/\pi$ candles per square foot, or one foot-lambert. Similarly, if the viewing screen is a perfect diffuse reflector (or transmitter) and is illuminated by one foot-candle, its apparent brightness is one foot-lambert. Such a screen is said to have a "gain" of unity in all directions.

If the object is a transparency illuminated by a well designed aberration-free condenser system with no absorption losses, the apparent intrinsic brightness of its clear parts, when viewed from the projection lens, is the same as that of the light source. Similarly, the aperture of the projection lens when viewed from the screen appears to have the same brightness and it can in fact be considered as a source of that brightness and of that size.

Image Brightness. The theoretical illumination on the screen, ignoring light lost by absorption and reflection, depends on three factors: it is directly proportional to the intrinsic brightness of the object (and to the brightness of the source in the case of transparencies) and to the effective area of the projection lens aperture, and it is inversely proportional to the square of the throw.

Before we can determine the relative aperture that is required in the optical system, the type of light source, its size and how it is to illuminate the object, we must have some idea of the screen brightness to be aimed at. The following table gives a very rough indication of what is ideal for various applications.

SCREEN BRIGHTNESS

Location	*Screen Brightness Foot-lamberts*
Cinemas	9–15
Schools (partial black-out)	8–12
Schools (complete black-out)	4– 8
Houses (daylight on average curtains) ...	2– 5
Houses (complete black-out)	1– 2

No screen will reflect or transmit all the light incident upon it and if it is perfectly diffusing its gain in all directions will be less than unity. These figures must therefore be divided by the gain of the particular screen to be used before we can decide the necessary illumination in foot-candles.

So, for a school with complete black-out where a screen brightness of 8 foot-lamberts is wanted with a screen of gain $= \frac{3}{4}$, the actual illumination figure would be $(8 \times 4)/3 = 10\frac{2}{3}$ foot-candles.

There are special directional screens which can appear to be much brighter in a particular direction than any diffuse surface.

In cinematography allowance has to be made for the fact that the illumination is intermittent and the projector while running produces only about half the illumination it does while stationary.

Image Size. The size of a screen image depends on three factors: the size of the object being projected, the focal length of the projection lens, and the distance between the projection lens and the screen.

When using a standard projector with a fixed object (i.e., slide or cine frame) size and one projection lens the size of the screen image for a particular screen distance is given by:

$$I = \frac{O\,(U - F)}{F}$$

where U is the distance of the projection lens from the screen, F its focal length, O the width of the object being projected, and I the width of the screen image.

For example, suppose 24 × 36 mm. transparencies are projected by a 4 ins. lens on to a screen 8 feet from the lens. 36 mm. is approximately 1·4 ins. so the screen width is:

$$I = \frac{1{\cdot}4\,(96 - 4)}{4} = \frac{128{\cdot}8}{4} = 32\tfrac{1}{5}\text{ ins.}$$

The screen distance required to yield a given screen dimension from one object size and one projection lens can be obtained by rearranging the above equation into the form

$$U = \frac{F\,(I - O)}{O}$$

With a 4 ins. projection lens a screen image width of 4 feet is obtained when the screen distance

$$U = \frac{4\,(48 - 1{\cdot}4)}{1{\cdot}4} = \frac{186{\cdot}4}{1{\cdot}4} = 133 \text{ ins.}$$

The focal length of a projection lens to suit any combination of object size, screen size and screen distance is obtained from the equation:

$$F = \frac{U\,O}{I + O}$$

In the previous examples the projection lens focal length required to yield a 4 feet wide screen picture 8 feet from the lens is:

$$F = \frac{96 \times 1{\cdot}4}{48 + 1{\cdot}4} = \frac{134{\cdot}4}{49{\cdot}4} = 2\tfrac{3}{4} \text{ ins.}$$

The values of F obtained in this way may correspond to angular covering powers beyond the capabilities of the available projection lenses and the screen width may have to be reduced or the distance increased to reach a more usable focal length. G.H.C.

Zoom Lenses. A projection lens of variable focal length greatly simplifies the problems of setting up the projector to obtain either a given screen image size or a given projection distance. Usually zoom projection lenses have a focal length ratio of 1 : 1½ to 1 : 3. With an average ratio of 1 : 2, the available latitude works out as follows.

For 24 × 36 mm. transparencies, a projection lens with a focal length variable from 3¼ to 6½ ins. and the screen again 8 feet from the lens gives for the shortest focal length:

$$I = \frac{1{\cdot}4\,(96 - 3{\cdot}25)}{3{\cdot}25} = 40 \text{ ins.}$$

and for the longest focal length:

$$I = \frac{1{\cdot}4\,(96 - 6{\cdot}5)}{6{\cdot}5} = 19\tfrac{1}{4} \text{ ins.}$$

So the screen width is variable from 19¼ to 40 ins.

Equally the zoom lens enables a fixed screen image size to be obtained from a range of projection distances. Using the second of the above examples, for a screen image width of 4 feet and a projection lens variable from 3¼ to 6½ ins. the shortest screen distance is

$$U = \frac{3{\cdot}25\,(48 - 1{\cdot}4)}{1{\cdot}4} = \text{approx. } 108 \text{ ins.}$$

while the longest distance is

$$U = \frac{6{\cdot}5\,(48 - 1{\cdot}4)}{1{\cdot}4} = \text{approx. } 217 \text{ ins.}$$

The available range of projection distances is directly proportional to the zooming range of the lens.

The greater the zooming range, the greater obviously the possible combinations of screen size and screen distance that can be used in the equation

$$F = \frac{UO}{I + O}$$

and as long as the resulting value for F falls within the focal length range of the lens, the combination is feasible for a projection set-up with that lens. Nor need the lens be changed when using different combinations of U and I (the object size for a given projector is usually constant).

In practice the even greater value of a zoom projection lens is that the projector can be set up at a required distance without too much concern for the screen image size. This is then adjusted independently by varying the focal length of the lens so that the picture fills the screen.

Zoom projection lenses are predominantly used in the home projection of cine films and still transparencies. Projection lenses of longer focal length are not normally available as zoom lenses. Such long focus lenses are used mainly for projecting in larger halls and auditoriums where both the projection distance and the screen size are more or less fixed and no adaptation to varying conditions is necessary. The only exception is educational cine film projection where a projector may have to be set up in classrooms and other locations with conditions varying from one place to the next.

Image Aspect. Most of the time comparatively little attention is—or indeed can be—paid to the viewing angle of the screen image relative to the audience. It is generally recommended that viewers should not be so near the screen as to get a distorted impression of the picture, nor so far that smaller details become unrecognizable. The usually accepted ratio of nearest and farthest distances is around 1 : 8 and this is the space within which the audience can be accommodated.

For ideal viewing conditions the range is very much reduced. "Ideal" in this context implies a viewing distance where the screen image is seen in the same perspective as the original view. This depends on the focal length of the camera lens used to take the picture or film and on the magnification of the screen image. The product of the two is the ideal viewing distance. Thus for a 24 × 36 mm. transparency taken with a 2-ins. (50 mm.) camera lens and magnified to a screen image width of 4 feet (magnification of approximately 34×) the ideal viewing distance would be about 58 ins. or a little under 6 feet.

As it is rarely practical to concentrate the audience into such a restricted distance range from the screen, the consideration of ideal viewing distance has to be largely ignored. This is helped in practice by the fact that the eye adapts itself in the darkened room to a large number of distortion factors (colour, brightness, etc.) including perspective. Further, as a proportion of still and even more cine views are shot with long-focus lenses the average ideal viewing distance shifts somewhat backwards to take in a deeper audience.

Wide-angle Projection. A special case of the deliberate choice of image aspect relative to the observer is the projection of wide-angle shots (taken with a short focus lens in the camera) on a large screen for viewing at comparatively close distances. As a result the audience sees a much wider view than the eye can take in simultaneously, yielding an impression of increased reality and depth.

This is partly due to the fact that an observer in the audience can let his eyes roam over the screen in much the same way as he would look all round the original scene. The visual viewing angle then becomes more nearly that of the natural angle of observation (rather than optical angle of view) of the eye. Though the latter is around 140°, an image angle of even half this value is already far more impressive than the more usual viewing angle of 20° to 40°.

With a 24 × 36 mm. transparency taken with a 35 mm. (1⅜ ins.) camera lens and the observer 6 feet from the screen, the screen image width would have to be around 6½ feet for this wide-angle effect. For greater viewing distances the screen width must be extended in proportion.

Such a set-up raises two practical issues. First the comparatively restricted ideal viewing distance also limits the size of the audience for such a show. Secondly the image usually also has to be projected by a short focus projection lens which must have an adequate angular covering power. This enables the projector to be set up not too far behind the audience, though that is a practical convenience for projection in smallish rooms. In principle it is the image magnification on the screen, relative to the viewing distance, which counts; the same effect can be achieved by a projection lens of more normal focal length with the projector set up rather further behind the audience.

Such a wide-angle projection arrangement is generally practical only for slides—mainly because the angle of view of wide-angle cine lenses (at any rate for amateur use) corresponds to the angle of a normal still camera lens (rarely exceeding 45° to 50°). Wider angles are then only obtainable with special optical attachments such as anamorphotic lenses. L.A.M.

PROJECTION SYSTEMS

Most classifications of projection systems divide them into two broad groups: projection of opaque originals by reflected light (episcopic projection) and showing transparent originals by transmitted light (diascopic projection). Projectors are accordingly designed for either system; some models provide for both.

Episcopic Projection. The simplest form of projector from the theoretical point of view is the episcope in which the light reflected from a brilliantly illuminated opaque object is focused by a lens system to form an image on a screen.

As the light is reflected by the object in all directions, only a small percentage falls on the lens and passes through to form the image. Because the amount of useful light available is so small, it is not possible to have a large bright screen image.

The screen illumination is given by the equation $E = BA/T^2$ where E is the illumination on the screen, B is the brightness of the perfectly white parts of the object, A is the area of the projection lens aperture and T is the throw.

An illumination on the object of 10,000 foot-candles is possible if it is near a 2,000 candle-power lamp reinforced by mirror surfaces, i.e., B is $10{,}000\pi$. With an episcope giving a magnification of 20 with an 8 ins. f4 lens, the throw

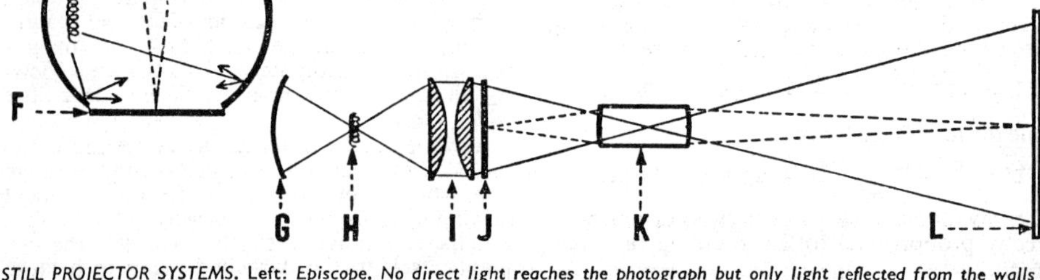

STILL PROJECTOR SYSTEMS. Left: Episcope. No direct light reaches the photograph but only light reflected from the walls of the lamp house. A, screen. B, plane mirror. C, lens. D, lamp house. E, filament lamp. F, photograph. Right: Film strip and slide projector. G, lamp reflector. H, lamp. I, double condenser. J, transparency. K, projector lens. L, screen.

is 168 ins. and the area of the lens is 3·14 square ins. E is therefore

$$\frac{10{,}000 \times 3{\cdot}14}{3{\cdot}14 \times 168 \times 168} = {\cdot}35 \text{ foot-candles.}$$

In practice E would be less than this because the object and the projection lens absorb light and with such a low screen illumination the episcope would have to be used in a completely darkened room. The screen illumination can be improved by using a lens with a wider aperture, being satisfied with a smaller image or increasing the brightness of the object by using a more powerful lamp.

If the projector is on the same side of the screen as the viewer, the image formed by the lens system is, of course, laterally inverted (the right and left sides are interchanged). This is undesirable, particularly if the subject includes writing or print, so the projected beam of light must be reflected from a mirror placed immediately in front of the projection lens to invert it back again to normal. With back projection, no lateral inversion takes place and the mirror is not required.

There is a very similar arrangement for projecting transparencies in which a well illuminated diffusing screen is placed behind the object, making a combination that can be considered as self-luminous. Here again it is difficult to obtain sufficient light on the screen since the amount of illumination diffusely transmitted is not greatly different from that diffusely reflected by a white object under the episcope.

The lateral inversion of a transparency can be overcome by turning it around in the slide holder.

Diascopes. Another form of projection system suitable for transparencies uses a condensing system to form an image of the light source in the pupil of the projection lens whilst the cone of light forming that image passes through the transparency.

If the condensing system is free from aberration, all light rays from the light source passing through the transparency enter the projection lens aperture. So, when looked at from this aperture, all clear parts of the transparency appear equally bright. Provided the size of the image of the light source is slightly greater than the aperture of the projection lens (i.e., the lens is completely "flashed"), the size of the source and its magnification are on the whole immaterial.

Since current consumption and heat emission increase with increasing source size, it is usual to choose the greatest magnification and smallest source that the aberrations of the condenser system permit.

In calculating how much light is projected on to the screen, there is one big difference between episcopic and diascopic projection. Since we can use the light emitted from all the surfaces of the light source in the episcope, we want to know its total candle-power.

With the diascope, however, the condenser system and projection lens will determine how much of the source surface is usefully employed and in this case we want to know its intrinsic brightness in the direction of projection, i.e., how bright the illuminated transparency looks when viewed from the projection lens.

Since the light sources specially designed for this work are relatively small and capable of considerable light output, their brightnesses are more conveniently expressed in candles per square centimetre.

As an example, for an 8 ins. $f4$ lens working at a magnification 20, and a transparency illuminated by a light source with a brightness of 750 candles per square centimetre, the illumination on the screen, at points corresponding to the perfectly clear parts of the object, is (750 × 3 14)/(168 × 168) lumens/sq. cm. or

$$\frac{750 \times 3{\cdot}14 \times 929}{168 \times 168} = 77 \text{ lumens/sq. foot (foot-candles)}$$

A comparison with the very similar episcope example, giving a screen brightness of ·35 foot-candles, shows that the diascope makes a very much better projector.

This intensity of illumination is unnecessarily high for ordinary projection even when an allowance is made for light absorption and scatter in the optical system and for poor screen material. It would be reduced in practice by using a lens of smaller aperture, by increasing the magnification or by using a lamp of lower intrinsic brightness.

The tungsten filament lamps which are almost always used in projection of this sort give off a great deal of heat and if the transparency is in the form of film strip a heat absorbing glass filter will be necessary. This will reduce the amount of light reaching the screen by some 15 per cent.

Epidiascope. Some manufacturers incorporate episcopic and diascopic principles in one instrument which is known as an epidiascope. Its essential features are an enclosed space containing the illuminant and alternative means of illuminating either a transparency mounted at the front or a solid object at the bottom. There must be a lens at the end of a tube in front of the transparency for diascopic projection; and a lens and mirror at the top for episcopic projection.

Back Projection. Another arrangement consists of projecting the image from the rear on to a transulcent screen. While this can be done with any projector, it is usually a feature of compact cine and still projectors used for educational and demonstration purposes. There the screen forms an integral part of the projector housing. This screen is generally fairly small and the image is intended for daylight viewing. The actual brightness in terms of light output utilized is lower than for a reflection screen, but where

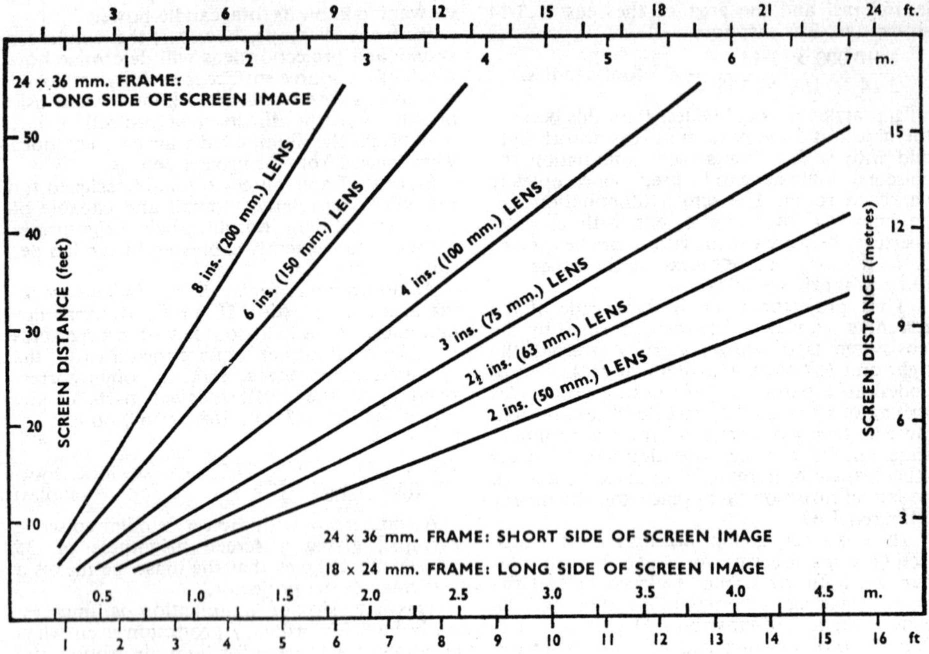

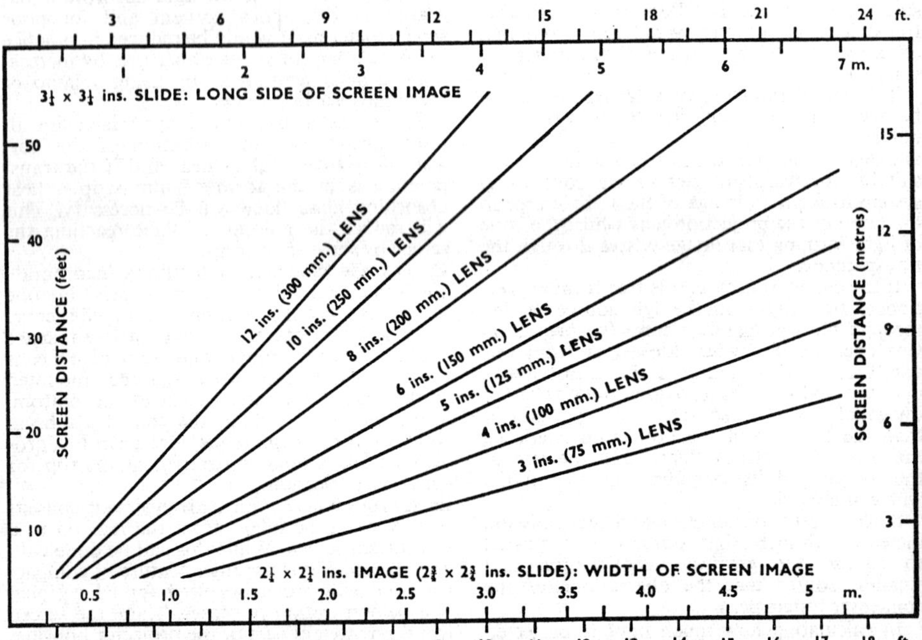

STILL PROJECTION SIZES AND DISTANCES. Top: *Film slides and strips.* Bottom: *Standard and* 2¾ × 2¾ *ins. slides. To find projected size, lay a ruler horizontally across graph, at projection distance to be used. Note distance of point from left edge where ruler cuts line of focal length of projector lens. Same distance on top or bottom edge gives image size. To find projection distance for given screen image, lay ruler vertically on graph, and measure vertical distance from bottom of required focal length line.*

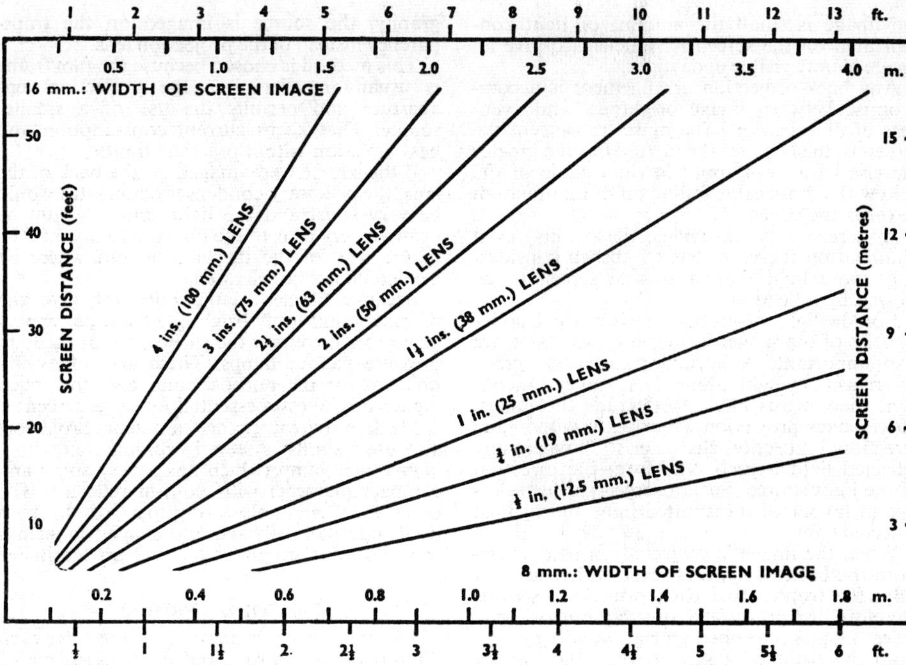

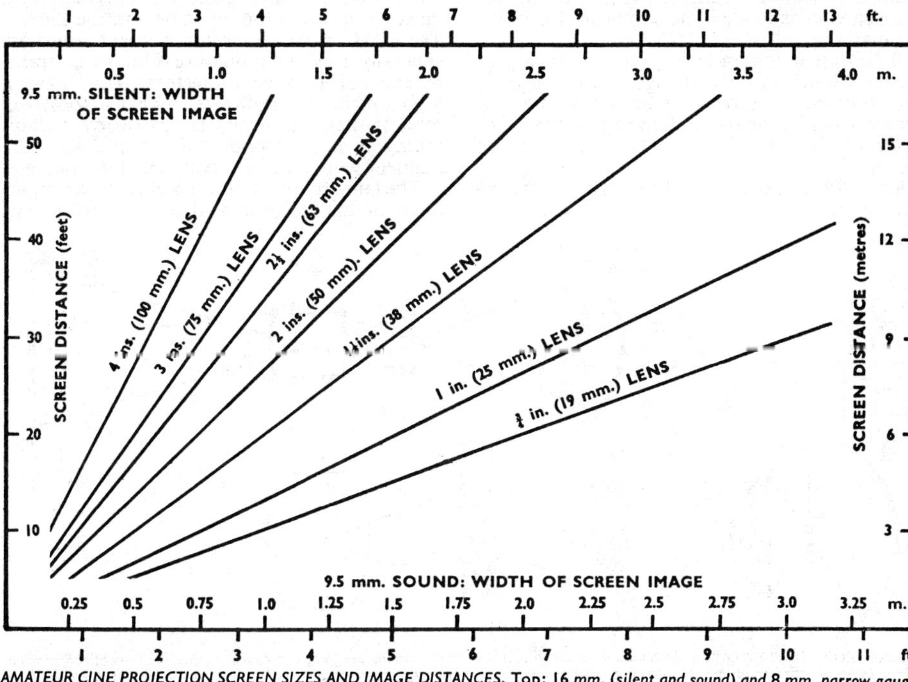

AMATEUR CINE PROJECTION SCREEN SIZES AND IMAGE DISTANCES. Top: 16 mm. (*silent and sound*) and 8 mm. *narrow gauge film. Bottom: 9·5 mm. silent and 9·5 mm. sound film. The height of the screen image in the case of 8 and 16 mm. film as well as 9·5 mm. silent is three-quarters of the width; with 9·5 mm. sound film it is slightly more than three-quarters. All screen image sizes are based on the projector gate aperture, and not on the actual frame size on the film; the latter is slightly larger.*

the image is small the amount of light concentrated on the screen is sufficient to give an image plainly visible in daylight.

Any back-projection arrangement is a compromise between image brightness and evenness of illumination. The more translucent the screen is, the brighter the picture but the greater the risk of a "hot spot" in the middle of the field with a noticeable falling off of illumination towards the edges.

A screen that provides reasonably even illumination therefore has to absorb considerably more light than a surface screen to be viewed by reflection.

For daylight projection the surface characteristics of the screen from the viewing side are also important. A normal translucent screen will reflect any light falling on it, and extraneous light therefore is liable to degrade the image. Special back projection screens exist with highly directional reflection characteristics, where any reflected light is seen only from the direction of the light source. Such a screen is thus selective in favour of transmitted light and against reflected light.

When the image is projected on to a screen from the back, it appears reversed left-to-right from the front. Most back projection set-ups therefore use a surface silvered mirror or a reflecting prism to obtain a right-reading image from the front of a screen when the film or transparency is loaded into the projector in the normal way.

This mirror system also makes the back projection set-up more compact, since normally the projector-to-screen distance behind the screen would have to be the same as when the projector is set up in front with a reflection screen.

Cinema Projection. In the third type of projection system used for large screen cinematography, the source is imaged on the transparency instead of the projection lens.

This method is chosen because the film frame is usually smaller than the projection lens aperture and permits the use of a smaller source. That keeps current consumption and heat emission within practical limits.

If the source were imaged on the back of the lens, the necessary condenser conjugates would be very short and the light source would be brought very near to the film where it would be in the way of the mechanism and where its heat could not be tolerated.

Because filament lamps do not give the necessary uniformly bright light source, cinema projectors use either the carbon arc or the high pressure vapour lamps. These are sufficiently uniform for the purpose, and give the much higher brightnesses essential for large screens.

Modern narrow gauge amateur projectors also use a similar system in conjunction with a low-voltage lamp (which has a very small and compact filament) with built-in reflector. The latter is often a blown-out part of the bulb itself, internally silvered and shaped in such a way that all the light is focused on the image plane.

LIGHTING SYSTEMS

The light system in a projector must give even illumination, it must avoid light losses, and it must not transmit too much heat to the slide.

The Light Source. Tungsten filament lamps are generally used in miniature film strip, transparency or positive projectors. They have a high intrinsic brightness and for practical reasons they are mostly lamps of low voltage taking a large amount of current, with the filament wound in tight coils set close together.

The lamp relies on the high operating temperature of the tungsten and emits more energy

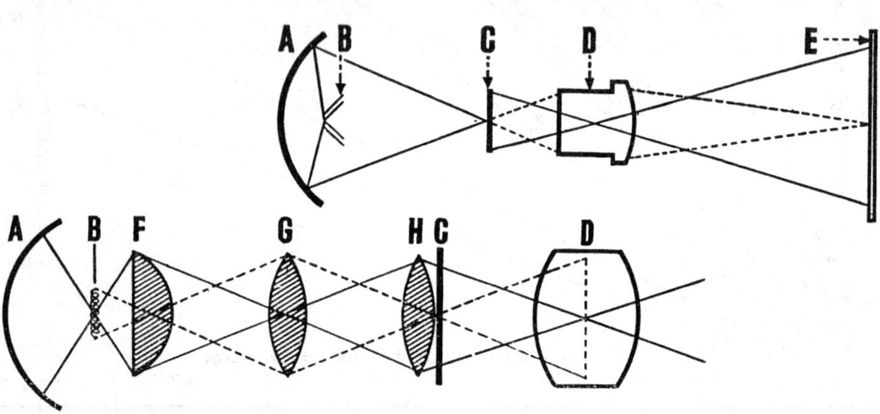

CINE PROJECTOR SYSTEMS. Top: *Commercial cinema projector utilizing an arc lamp and concave reflector. The optical principal of the integral reflector lamp, used without a condenser on amateur cine projectors, is similar.* Bottom: *Filament lamp and relay condenser system. The first condenser forms an image of the source in the plane of the second condenser, while the latter forms an image of the first condenser on the film. The third condenser forms an image of the second one in the plane of the lens. A, reflector. B, light source. C, film. D, lens. E, screen, F, G and H, condensers.*

in the form of heat than light, so the lamp used is always the smallest that the condenser system permits. Smaller lamps are also more economical.

Because of the heat emitted by these lamps, the lamphouse must be well ventilated and often cooled by an electric fan to prevent the glass from softening.

Tungsten melts at 3655° K. and at this temperature has a brightness of about 6,000 candles/sq. cm. The filament in projector type lamps operates between 2,900° K. and 3,400° K. yielding brightnesses of the metal between 1,000 and 3,000 candles/sq. cm. The effective brightness of any useful area of the source includes, of course, non-radiating space due to the turns in the coils and the spacing between them, and an average brightness of about 1,500 candles/sq. cm. is approximately the best that can be hoped for theoretically.

If this type of source is placed at the centre of curvature of a concave mirror, an image of the coils can be formed in the non-radiating spaces, thus increasing the average brightness. The improvement varies with different lamps, but seldom exceeds 50 per cent.

The more recent electric discharge lamp consists of an arc between tungsten electrodes contained in mercury vapour. It is capable of a very high light output without a corresponding increase in heat emission. A high pressure mercury vapour lamp can produce several times the screen illumination given by a tungsten filament lamp of the same wattage with far less heat

Such lamps are more robust and last longer than tungsten filament lamps.

The ordinary mercury vapour discharge lamp is very suitable for monochrome projection but it does not give a perfectly true rendering of colour because its emission of green and blue light is greater than the red emission. Specially made lamps are used for projecting colours. These carry mercury vapour at a higher pressure and also contain other elements such as cadmium so that the light emitted approximates white light.

The xenon arc has the discharge taking place in an atmosphere of xenon gas at high pressure. This type of lamp is progressively replacing carbon arc lamps in professional cinema projectors and is even made in small versions for narrow-gauge projectors. The advantage of the xenon arc is the colour quality of the light (which approaches daylight), and the high light efficiency with considerably simpler operating gear.

The concentrated arc lamp is another light source that is used for very small projectors. It has fixed electrodes sealed in an argon-filled bulb and the light is emitted from a thin film of molten zirconium and the neighbouring vapour. The source itself is quite small—ranging from ·003 in. diameter in a 2 watt lamp to ·10 in. in a 300 watt lamp.

A special discharge lamp developed for cinema projectors is a pulse-arc, a concentrated gas arc with an intermittent light of 72 flashes per second. This is used with shutter-less projectors, the pulsing light taking the place of the continuous light cut off periodically by the projector shutter. This gives greater light efficiency, as no light is wasted during the dark periods.

The lamp itself is only just over 3 ins. long, the brightness is equivalent to a 60-amp. high-intensity arc.

The following table gives an indication of the approximate brightnesses of various types of light source.

PROJECTION LIGHT SOURCES

Light source	*Volts*	*Watts*	*Brightness candles/ sq. cm.*
Tungsten filament	50	250	1,000
Tungsten filament	110	100	600
Tungsten filament	110	500	900
Tungsten filament	110	1,000	1,250
Tungsten filament	230	500	500
Mercury vapour at			
12 atmospheres		100	1,000
10 atmospheres		500	1,000
25 atmospheres		100	12,000
30 atmospheres		500	20,000
Gas arc pulse lamp	420	800	50,000
Zirconium concentrated arc		100	3,900
Zirconium concentrated arc		300	5,200
Low intensity carbon arc ...			17,500
High intensity carbon arc			
70 amperes			70,000
High intensity carbon arc			
150 amperes			86,000
Sun at zenith			165,000

The Condenser System. The optical arrangement of lenses and/or mirrors which illuminates the object cannot affect the definition of the final image to any appreciable extent, but it can and very often does affect the variation of illumination over the field.

A very common form of condenser lens system consists of two plano convex lenses with their vertices almost in contact, and in this case the focal length and *f*-number of each is twice that of the whole. An aperture of about *f*1·3 is the limit for any single element of a system having all surfaces spherical.

Excessive spherical aberration, which varies as the square of the lens diameter, may evidence itself as wide dark annular rings on the screen.

These occur when all the rays from a zone of the transparency do not get through the projection lens, and its aperture will not appear to be filled when viewed from the corresponding image point at the screen.

If the aberrations of a condenser system cannot be accepted, the focal length—and hence the *f*-number—must be increased by using a larger source and a smaller magnification, or alternatively it must be replaced by another system containing more elements with shallower surfaces, or one containing aspheric surfaces.

Another optical system uses what is known as a relay condenser. In this the first component forms an image of the source on a second component which in turn forms an image of the first on the film. At the same time a third component very close to the film forms an image of the second, and therefore of the source, in the aperture of the projection lens.

This type of condenser system provides very even illumination, and is particularly efficient if one or two of the elements have aspheric surfaces.

In the third type of projection a uniform source is imaged on the object. In this case the image-forming cone of light should flush the projection lens aperture, i.e., the condenser and lens apertures have to subtend the same angle at the object.

A very common arrangement uses an ellipsoidal mirror with the source at one focus and its image at the other. A relay condenser for this type of projection consists of only two components (although each component may have more than one element). Both components combine to form an image of the source on the object, whilst the second forms an image of the first in the projection lens aperture. This second image must flash the lens aperture.

In the integral reflector lamp the ellipsoidal mirror is the specially shaped and silvered rear portion of the glass bulb itself. The front of the bulb is also blown out to a spherical shape and silvered, leaving only an opening for the concentrated light beam. The result is a sufficiently intense and focused beam to require no condenser system at all. The film is at the second focus of the ellipsoidal reflector. This arrangement is particularly compact, but is so far available only for 8 mm. and 16 mm. cine projectors.

Light Transmission. The light lost at the various glass-to-air surfaces and absorbed by the thickness of the glass reduces the illumination on the screen. The loss at a glass-to-air surface is 4 or 5 per cent of the light reaching it from preceding surfaces according to the type of glass, and the absorption in the glass itself is, very approximately, 2 per cent per inch. So even with a system of very thin components having fourteen glass-to-air surfaces, only about 55 per cent of the light reaches the screen.

Coating or "blooming" lens surfaces reduces the loss to about 1 per cent at each surface. A thin system of fourteen bloomed surfaces may transmit as much as 85 per cent of the available light.

The coating consists of a film, 4-millionths of an inch thick, of a transparent material having a refractive index as near as possible equal to the square root of the index of the glass. It is usually applied by evaporating magnesium fluoride on to the glass in a vacuum.

This light loss applies equally to the light passing through the condenser system. Hence projectors relying on integral reflector lamps without condensers can, in certain circumstances, utilize more of the light output of the lamp.

Heat Transmission. The heat emitted by very bright tungsten filaments is sufficient to ruin the object placed in an episcope or diascope unless special precautions are taken. Episcopes usually have forced air cooling, and most diascopes include a heat absorbing filter made from special glass. The energy absorbed by the filter is conducted away by air currents.

Although heat filtering glass has a lower coefficient of expansion than usual, it will crack if the illumination on it is not uniform and for this reason the whole surface, including the extreme edges, must be exposed to the radiation.

When it is impossible to obtain uniform heat over a large filter it is divided into several strips about 1 in. wide mounted edge to edge.

A polished sheet of this glass 3 mm. thick absorbs about 91 per cent of the heat energy and about 13 per cent of visible light.

All elements of a condensing system which lie close to a source of great heat, or to an image of it, are made from heat resisting glass or fused quartz.

A further method of reducing the heat reaching the film or transparency being projected involves the use of diathermic lamp reflectors. These are special glass reflectors coated with a dichroic layer which reflects up to 90 per cent of the visible wavelengths of the light, but transmits 80 per cent or more of the long wave infra-red radiation. In this way virtually the only heat reaching the film or slide in the picture gate is the direct radiation from the filament itself.

That is effectively less than half the heat radiated in a conventional lighting system with a fully reflecting reflector.

Diathermic reflectors may be separate and mounted behind the lamp, but a number of modern projection lamps have integral or proximity reflectors of the diathermic type built in. G.H.C. (L.A.M.)

See also: *Anamorphotic lens; Heat filter; Light sources; Magnification; Optical calculations; Projection practice; Projector (cine); Projector (still); Screen for projection; Viewers.*

Books: *Projecting Slides*, by M. K. Kidd and C. W. Long (London); *How to Project*, by N. Jenkins (London); *Photographic Optics*, by A. Cox (London).

PROJECTION PRINTING. Any method of printing in which the image is projected on to the sensitized material by an optical system—e.g., an enlarger. Projection printing is normally employed to produce images of different size from the original, but may also serve for same-size reproduction.

See also: *Enlarging.*

PROJECTOR (CINE)

It is in the cine projector that the motion picture film creates its illusion of movement. The projector has three functions; to hold the long strip of individual images exposed in the camera (the cine film) to feed these images through the projection system at a standard speed, keeping each image in position for an instant; and to project an enlarged picture on to the screen.

The main components of the projector are thus the intermittent film transport; the shutter, the illuminating optics (lamp and means for directing the light on to the film gate), the projection lens, and the film containers (reels) and means for guiding the film past the point where it is projected. In a sound projector there are also the sound unit where the sound recording is picked up, the reproducing amplifier and the loudspeaker.

Sizes. Amateur cine projectors are available for the three main film gauges; 8 mm., 16 mm., and 9·5 mm. Motion picture theatre projectors project 35 mm. or 60 mm. film, and there are semi-professional models—mainly for educational and industrial use—for 16 mm. and occasionally even for 8 mm. film.

Multi-gauge projectors also exist, projecting two or even three narrow-gauge film sizes. Changing over from one gauge to another there involves interchanging the transport elements (claws and sprockets) to adapt the machine to the different perforation intervals, the film channels and the spools and spindles. Multi-gauge models are rather rare; few people make films of more than one gauge and it is easier to achieve maximum precision in a projector designed for one film gauge.

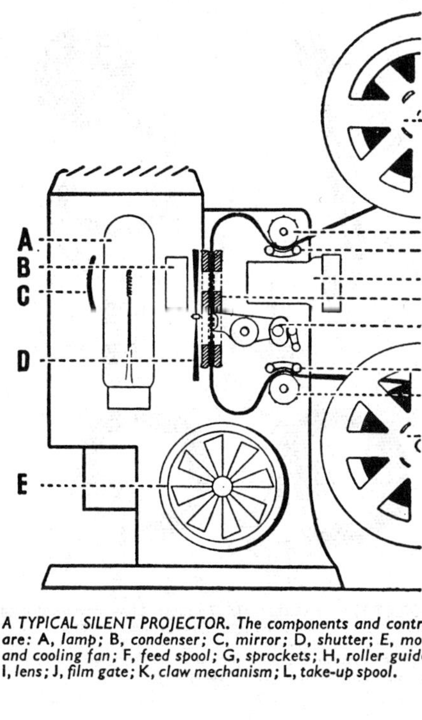

A TYPICAL SILENT PROJECTOR. The components and controls are: A, *lamp;* B, *condenser;* C, *mirror;* D, *shutter;* E, *motor and cooling fan;* F, *feed spool;* G, *sprockets;* H, *roller guides;* I, *lens;* J, *film gate;* K, *claw mechanism;* L, *take-up spool.*

Projector Types. Cine projectors can be grouped according to the film gauge and specifications.

Silent 8 mm. projectors usually have a single projection speed of 16 frames per second, lamp power up to 500 watts (or a low voltage lamp of 50 or 100 watts), and—on modern models—automatic film threading. More advanced projectors may have two running speeds, forward and reverse, still picture device, etc.

8 mm. sound projectors for striped film usually have two speeds of 16 (or 18) and 24 frames per second, and have a sound-head and amplifier unit built on or are provided with a separate unit for direct linking with the projector. Sound projectors using the tape coupling system may have a tape coupler built in or be intercombined with a tape recorder deck.

There are two types of 8 mm. projector (in silent and sound versions) for classical 8 mm. and for Super-8 mm. films. The films differ in the image format and perforations, certain universal 8 mm. machines take both film types.

16 mm. silent projectors are comparatively rare. They usually have a running speed of 16 frames per second, sometimes reverse running and usually a still picture device, and lamps of 500 to 1,000 watts (mains voltage) or low voltage lamps of 150 watts.

16 mm. sound projectors nearly always have speeds of 16 and 24 frames per second, and may have provision for playing back optical or magnetic (striped film) sound tracks or both; sometimes they also have provision for recording a magnetic sound track. On compact models the amplifier and speaker are built into the projector housing. Versions for educational projection in large halls may have separate speakers. Professional 16 mm. sound projectors for cinema use often employ arc lamps.

Mostly 9·5 mm. projectors follow the general specifications for 8 or 16 mm. machines according to complexity. 9·5 mm. sound projectors may screen films with optical or magnetic (stripe) sound tracks.

Special projectors, particularly for 8 mm. film, include compact models with built-in rear projection screen, cassette loading, etc.

FILM TRANSPORT

In the projector the film runs from a take-up spool or reel over a transport sprocket, through the film track and past the film gate (the opening through which the film is illuminated), over another sprocket and—in the case of a silent projector—on to the take-up spool or reel. The

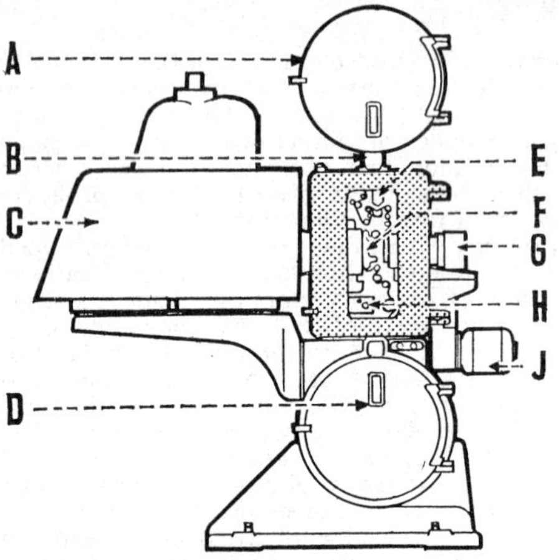

MOTION PICTURE THEATRE PROJECTOR. The most advanced models of this kind can handle 35 and 70 mm. film and sometimes also 16 mm. film with accessory attachments. A, film feed spool in fireproof drum. B, upper fireproof channel C, lamphouse (usually for carbon or xenon arc lamps). D, film indicator window on lower drum containing take-up spool. E, magnetic sound head. F, gate. G, lens. H, optical sound head. J. drive motor.

first sprocket pulls the film smoothly off the feed spool and the second sprocket (sometimes the other side of the same sprocket) passes the film—again at a uniform rate—on to the take-up spool. The latter is usually driven through a friction clutch which enables the take-up spool to reel up just as much film as the second sprocket feeds through.

The Loop. While the sprockets move the film smoothly, the latter passes through the film gate in jerks. This is because the intermittent film movement must keep each individual film frame stationary in the gate for an instant, then quickly pull the film down to hold the next frame for an instant, and so on. To absorb this jerky movement and to prevent it from straining the film, there is a free loop of film above and below the gate. In threading the projector it is important to maintain these loops; if through a fault in the mechanism or damage to the film perforations a loop (usually the bottom one) gets too short, the direct pull on the film through the gate may cause damage.

Intermittent Movement. The commonest means of moving the film intermittently through the gate in a narrow-gauge projector is the film claw. This engages the film perforation, pulls the film down through one frame interval, and disengages to return to its starting point for the next transport cycle. Usually the pull-down time is between one-sixth and one-quarter of the complete transport cycle; for the rest of the time the film is stationary in the gate.

There are many variations of the claw movement, and various means of coupling it to the transport motor to achieve an ideal moving cycle (i.e. where the claw moves as far as possible at right angles to its pull-down direction during engagement and disengagement). The movement may have a double or a triple claw, to pull down a film even with occasional damage to perforation holes.

Heavy duty 16 mm. and practically all 35 mm. projectors use a Maltese cross movement which intermittently advances the film through a special sprocket. This puts less strain and wear on the film perforation, but is a more bulky and more expensive mechanism. Dog or beater movements, acting on the film loop have also been used.

The Shutter. While the film is moved in the gate, a shutter interrupts the light beam to prevent blurring of the picture on the screen. The shutter consists of a rotating blade shaped somewhat like a fan. This not only stops light from reaching the screen while the film moves from one frame to the next, but also serves to eliminate flicker in the picture.

Flicker. It was discovered quite early in cinematography that while a picture change occurring sixteen times a second was sufficient to give an illusion of movement, the audience was aware of light flicker. Tests showed that to obtain an illusion of continuous flicker-free illumination it was necessary to interrupt the light beam forty-eight times per second. For this reason the silent film projection shutter has three blades and three openings. The shutter rotates once for every frame so that it

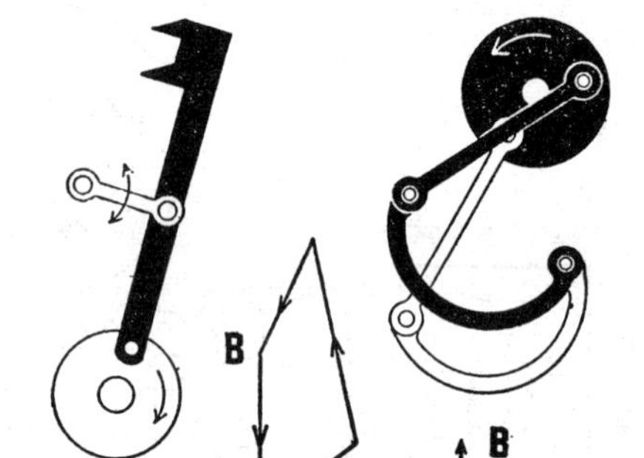

PROJECTOR INTERMITTENTS. Top left: *Claw movement with double claw. The arrowed figure traces the path of the claw; BB is the extent of the film movement.* Top right: *Dog or beater movement. As the beater acts on the film loop,* **BB** *is only half the length.* Bottom: *Maltese cross system.*

produced forty-eight interruptions per second at a projection frequency of sixteen frames per second.

When sound-on-film was introduced, the film had to be run at a faster speed so that the higher sound frequencies could be satisfactorily recorded and reproduced. The picture frequency was increased to twenty-four frames per second and at that speed the projection shutter needs only two blades and two openings to provide the necessary forty-eight interruptions.

The Motor. The sprockets, intermittent movement, shutter and take-up spool are all driven by one projector motor. Usually this runs at a fixed speed to pull the film through at 16 frames per second in a silent projector or 24 f.p.s. in a sound machine. Amateur sound projectors often provide a choice of both speeds to permit the showing of silent and sound films.

A few projectors also incorporate a special slow speed of 5 or 6 frames per second, to create slow-motion effects on the screen with films shot at a normal speed. Owing to the extreme slowing down the image motion tends to be slightly jerky, but this slow-motion projection is valuable for movement analysis, etc. Some specialized projectors for this purpose even have a range of analysis speeds from 1 to 8 frames per second, so that the film can virtually be examined frame by frame.

At such slow motion speed the problem of flicker arises again. Where the shutter is directly geared to the film movement, slow-motion projectors may have a system of splitting up the shutter blades so that the 3-blade shutter for instance becomes a 9-blade one the moment the machine is switched to projection at 5 f.p.s. This gives 45 interruptions per second, near enough to the normally stipulated 48. At still slower projection speed the pull-down time becomes so short in comparison with the period the film is stationary in the gate, that the resulting movement blur on the screen would not be noticeable anyway. So the shutter movement is there disengaged from the film transport, and the shutter continues to rotate at a constant speed to give 48 interruptions per second, irrespective of the film transport rate. (The shutter is still needed for projection at the normal speeds.)

Some amateur projectors have a reversing switch for running the film backwards, to cause amusing effects or enable a scene to be quickly repeated. Frequently a still picture device is also fitted, which holds the frame in the projector indefinitely. In this case an additional heat filter is automatically switched into the optical path to prevent damage to the film. Neither of these features is normally fitted to 35 mm. motion picture projectors.

Spools and Rewinding. The maximum spool size, and hence the film capacity, varies with the different gauges, but not very greatly with different projector models. Usually 8 mm. machines take 400-feet spools, which gives a screening time of just over half an hour at 16 f.p.s., or a little over 20 minutes at 24 f.p.s. Some 8 mm. projectors hold up to 800-feet spools.

Most present-day 16 mm. projectors are sound models and have spool capacity of 1,600 or even 2,000 feet. This provides up to one hour of continuous screening. The capacity of 9·5 mm. projectors also goes up to 1,600 feet. Cinema projectors for 35 mm. usually take 2,000 feet spools. Since this only gives about 20 minutes of screening time at sound-film speed, more frequent change-over is necessary during the cinema performance. This is usually achieved with two projectors and may be controlled automatically or semi-automatically.

Generally amateur projectors incorporate a motorized rewind to wind the film from the take-up spool back to the feed spool for the next performance. For this purpose the take-up and feed spools are interchanged on the spool arms, and the film led directly from the full spool on the feed spindle to the empty one. High speed rewinding however carries some risk of film damage; for that reason it is usually better to rewind the film on a separate rewinding mechanism where the speed is under control; this also permits checking the film for faults during rewinding.

Loading. The instructions with most projectors indicate clearly how the film has to be threaded through the film path and the sound system, where applicable. Usually the projector also carries a threading diagram.

A number of modern projectors have an automatic threading feature. There the film end —which must be trimmed square—is pushed into a slot and automatically picked up by the sprocket. This drives the film through the loops, the film gate, and the various rollers and take-up sprockets. The main constructional feature of such automatic threading devices is a rigid loop guide through which the film must pass and which automatically ensures top and bottom loops of the correct size. Once the film has passed through the transport section, the loops guides are opened so that the film loops are free. Auto threading systems need an accurate construction, otherwise the perforations at the beginning of the film may become damaged. When the film leaves the last sprocket, it has to be attached to the take-up spool by hand; on some advanced auto-threading systems the film end is guided right into the take-up spool and picked up mechanically there.

Automatic film threading is rarer with sound projectors, since the film path round the sound system is more complicated.

A framing control is essential for professional requirements in order to correct any errors in threading the film into the projector. The frames of picture may not be in correct relation to the gate aperture and the effect on the screen will be two incomplete portions of a scene. The framing control enables the projectionist to adjust the picture up or down correctly. The adjustment is of the optical type—that is, it does not change the alignment of the optical axis with the screen.

When a framing control is fitted to narrow gauge machines it is usually of the non-optical type, its disadvantage being that operation of the control moves the projected area bodily up or down off the screen and it becomes necessary to use the tilting adjustment of the projector in order to bring the picture back into place.

ILLUMINATION AND OPTICAL SYSTEM

The optical system of a cine projector consists of the light source, the elements to concentrate the light on to the film image in the gate, and the projection lens which throws the enlarged image on the screen.

Light Sources. Most narrow gauge film projectors use high intensity tungsten filament lamps. Light powers range from 300 to 500 watts in 8 mm. projectors, up to 1,200 or even 1,500 watts in 16 mm. machines. Sometimes the lamps operate from 220 to 250 volt mains supplies; more often 110 volt lamps run via a transformer are used since at the lower voltage the light efficiency is better.

The complete optical arrangement consists of the lamp, a reflector behind it and two or three condensers in front. The latter focus the light beam on to the projector gate to provide a film image of maximum brightness. The lamps are usually designed to have as small a filament as possible, because the concentrated source gives the best results with the optical set-up. The projection lamp and the reflector must be carefully lined up to illuminate the image in the projection gate evenly.

Low voltage lamps with integral or proximity reflectors are also widely used in 8 mm. projectors. Here the lamp has the reflector built in to the bulb, or the rear surface of a bulb may be blown out in to the shape of an ellipsoid reflector. In either case the reflector is so arranged as to concentrate the light from the filament on to the image plane in the projector—often without the need for any condensers at all. This makes for a more compact lighting set-up. Low voltage lamps of this kind usually run at 8 or 12 volts, and come in powers of 50 to 150 watts. The light from the latter is equivalent to the illumination from a 500 watt mains voltage lamp with the more orthodox lighting set-up of a lamp and condensers.

Quartz-iodine lamps are also used in 8 mm. projectors, and have the advantage of being particularly compact as well as providing a light of constant intensity and colour temperature.

Professional motion picture projectors may have tungsten filament lamps, but the more usual light source is a carbon arc or a xenon arc. Carbon arc lamps are the most powerful illuminant, but require bulky lamp houses and automatic feeding gear for the carbons. The high-intensity xenon arc is a more modern light source, nearly as intense as a carbon arc, and easier to handle.

Xenon arc lamps are also used on a small scale in amateur 16 mm. and 8 mm. projectors. Here a 150-watt lamp corresponds in lamp output to about 500 to 750 watts of a mains voltage tungsten lamp.

Pulsed Light. In the usual projector the light source is continuous, and is interrupted by the shutter to produce the dark periods for the film pull-down and also to prevent flicker. The alternative method is to have an intermittent light source which pulses in synchronism with the film transport, and so provides the same effect without the need for a shutter. Pulsed light sources with a frequency up to 72 pulses per second are used in certain cinema projectors specially designed for such operation. The source is a high pressure xenon arc with a power rating of about 800 watts. Since this is the average power, the intensity of the pulses themselves is higher; hence the light output of the projector compares favourably with the light intensity of a conventional projector of twice the power.

Cooling. Some form of cooling device is incorporated in most projectors. Usually a fan, driven by the projector motor (or a separate motor) forces cold air past the film in the gate and past the projection lamp. The maintenance of the correct lamp temperature is important, otherwise the glass envelope may soften and distort.

The lamp and fan motor are generally controlled by separate switches, to enable the fan to be switched on before the lamp. On many projectors the switches are combined in such a way that the lamp can only be switched on once the fan is running.

Ventilation and forced draught can only remove the hot air from the lamp house; they cannot prevent the lamp from radiating heat through the condenser and on to the film. For this reason projectors may incorporate a heat filter in the optical path; more often such a filter comes into action automatically when the projector motor stops for projecting a still picture on the screen. As an alternative to heat absorbing glass, the filter may be a wire mesh or similar material.

35 mm. projectors with their much more intense lighting have a safety shutter which cuts the light off from the gate should the machine stop, otherwise the film would melt. In addition the film reels on a 35 mm. machine are enclosed in fire-proof housings. This is a safety precaution which was particularly necessary in the days when 35 mm. films were still on highly inflammable nitrate base.

Lenses. Normally a projector is fitted with a lens

twice the focal length of the standard camera lens; about 25 mm. for an 8 mm. projector and 50 mm. for a 16 mm. projector. This is suitable for home projection in fairly large rooms and projection in small halls. Modern 8 mm. projectors frequently come with a shorter focus lens about 15 to 20 mm. for bigger screen images in small rooms. Variable focus lenses are also popular; for instance a focal length range of 13 to 25 mm. on an 8 mm. projector covers everything from large images at short projection throws to medium screen sizes with longer throws. Typical projection lens apertures range from $f\,1{\cdot}1$ to $f\,1{\cdot}8$.

The focusing movement is a helical mount; sometimes the lens has a rack and pinion.

SOUND PROJECTORS

The three main ways of projecting sound films are with an optical sound track, a magnetic track (stripe) on the film, and with a separate magnetic tape. The first is largely used for 16 mm. home and educational projectors (though there are also 8 mm. optical sound machines), while the second method is used for all three narrow film gauges. Sound projection with a separate tape is almost entirely confined to 8 mm. film.

With standard 35 mm. motion picture film the optical sound track is the most usual kind; for certain types of motion picture presentations (such as Cinemascope and similar systems) the sound may be recorded on a number of tracks on a separate magnetic film which runs through the projector in synchronism with the picture film.

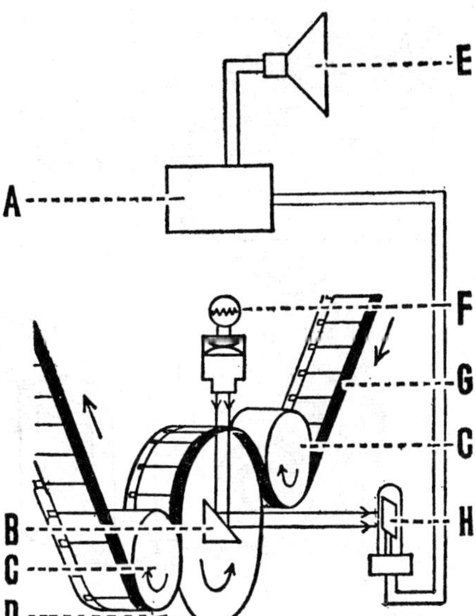

PROJECTOR SOUND HEAD. The principal units of optical sound projection are: A, amplifier; B, reflecting prism; C, rollers to press film against drum; D, sound drum; E, loudspeaker; F, exciter lamp; G, film; H, photo-cell.

Optical Sound. In the optical sound projector the film passes intermittently through the picture gate and then runs over a system of rollers connected to a smoothing device. This ensures that it will move smoothly and continuously through the sound gate. Here light from an exciter lamp shines through the sound track at the edge of the film and falls upon a light-sensitive photo-cell.

The modulations on the sound track vary the intensity of the light falling on the photo-cell. This in turn varies the electrical output of the cell so that it transmits a greater or less current to the amplifier which operates the loudspeaker.

As the sound gate is mounted below, or in advance of, the picture gate, the sound is always printed on the film in advance of the frame of picture to which it refers. In standard practice the sound track starts twenty frames in advance of the picture, on 16 mm. film twenty-six frames in advance and on 9·5 mm. film twenty-five frames in advance.

The reason for this is the impossibility of combining the intermittent movement of the film for the image projection with the smooth movement necessary for the sound. For some special purposes film may also have edge-synchronization—with the sound next to the corresponding image. In this case, however, the sound has to be picked up before the film enters the picture gate in the projector, and then delayed by an interval corresponding to the travel time of the film from the sound pick-up to the picture gate.

Magnetic Sound. The arrangement of the sound head in a magnetic projector is similar, except that there is a magnetic play-back head in contact with the stripe on the film, instead of an exciter lamp and photo-cell. As 16 mm. sound films are available in various forms, with stripes of different widths and in different positions, many 16 mm. sound projectors have interchangeable plug-in magnetic heads for different stripe standards.

The magnetic head may be for playback only, or for recording and playing back. In either case appropriate switching converts the circuit from one mode of operation to the other. Magnetic sound projectors with the playback head only are used for educational and similar purposes where only pre-recorded films are shown.

Many 16 mm. projectors are equipped for showing optical and magnetic sound film. In this case optical and magnetic sound heads are arranged around the sound drum (or there may be two sound drums, one for each system).

Magnetic 8 mm. sound projectors may be of two types. The more advanced design is similar to the 16 mm. sound projector, with the sound drum and sound head built right into the pro-

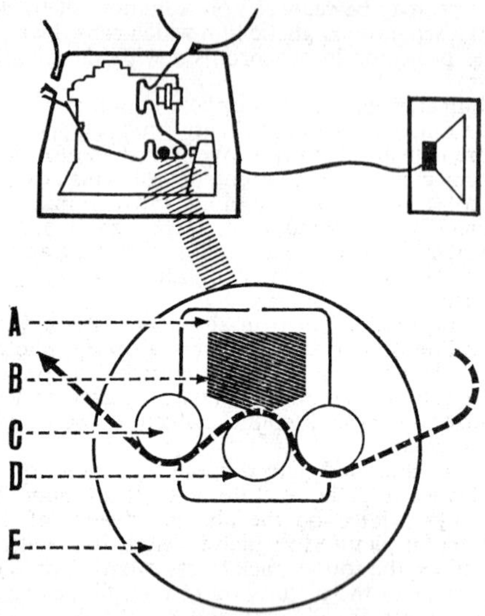

MAGNETIC SOUND PROJECTOR. In addition to the optical soundhead (or instead of it) 16 mm. magnetic projectors carry a magnetic soundhead. A, base plate. B, magnetic head. C, roller. D, sound drum. E, flywheel. The film path is shown dotted.

jector housing. The film then follows the prescribed path around the sound drum and associated rollers.

The second system, utilized with less expensive 8 mm. machines, consists of a straightforward silent projector, plus a sound unit as a separate housing, carrying the amplifier and magnetic sound head. For projecting sound films, the projector is simply placed on top of the sound unit and the film threaded—after the film gate and lower sprocket—over the sound drum of the sound unit. Usually projector and sound unit are designed to clamp together to form a rigid combination.

Sound Adaptors for Magnetic Stripe. 8 mm. projectors designed as two separate units are derived from the idea of the independent stripe adaptor. The latter in fact appeared first and in early versions consisted of magnetic sound head and drum mounted on an arm fixed to an amplifier base. The projector stands on top of this and is positioned so that the film can pass over the sound head after going through the film part of the projector. An alternative system is to have the sound head and drum mounted on a separate panel which can be fitted to existing silent projectors, provided there is room for it.

Tape Synchronization. The second system of providing magnetic sound for silent films is the use of a separate tape. While striped film is suitable both for sound films with a pre-recorded track and for sound recording by the amateur himself, dual film-tape systems are mainly used for amateur-recorded sound tracks.

There are a number of ways of linking the tape and film: loop synchronization, mechanical coupling of the tape recorder and film projector, electrical coupling and pulse synchronization.

Loop Synchronization. This involves a tape coupler which links the tape recorder with the projector. The coupler attaches to the tape recorder and takes the tape between the sound head and take-up spool. The tape runs over a capstan which is linked mechanically to the projector drive. The tape further runs over a loop connected to a variable resistance. If the projector speed and the tape speed are accurately matched, the loop stays constant in length. If, however, the projector runs faster than the tape recorder, the coupler pulls the tape through faster than the recorder is supplying it and the loop shortens. This moves the loop arm, increasing the resistance (which is wired in series with the projector motor), thus slowing down the projector. If the projector runs more slowly, the lengthening of the loop similarly results in increased current being fed to the projector, thus speeding up.

The mechanical connection is usually made to the sprocket of the projector, and can be arranged to suit any combination of tape recorder and projector speed. The tape and film remain in synchronism, unless the tape slips or stretches when running. This can be avoided by using perforated tape with a sprocket capstan.

Some 8 mm. projectors have a tape coupler of

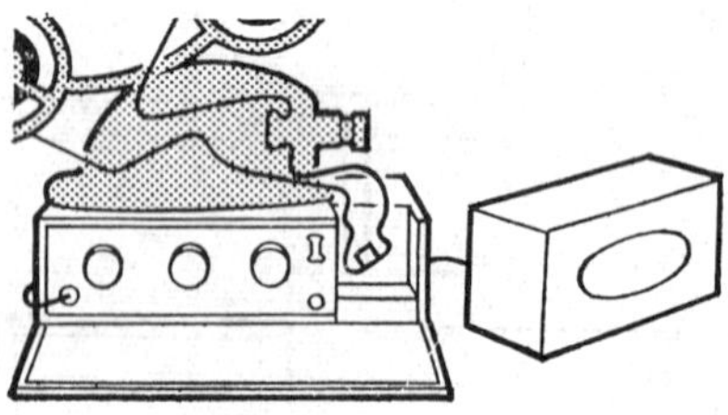

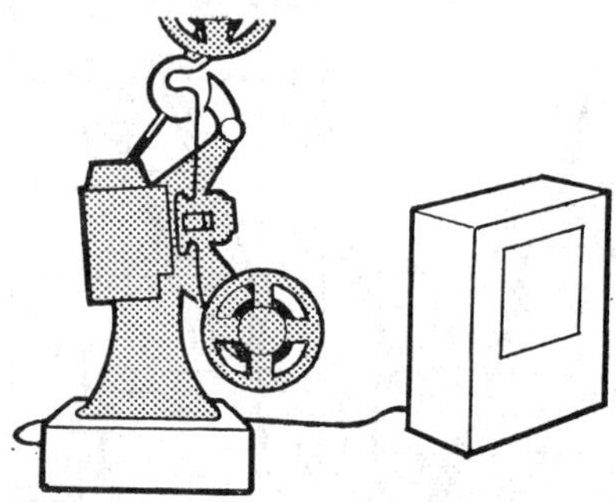

SOUND STRIPE CONVERTER SYSTEMS. Top: Some 8 mm. projectors are designed for conversion to striped film projection through a specially constructed amplifier unit. The projector stands on the latter and the film runs through the sound head built into the amplifier. Bottom: A sound head unit can also be interposed in the film path of some projectors between the supply spool and the picture gate.

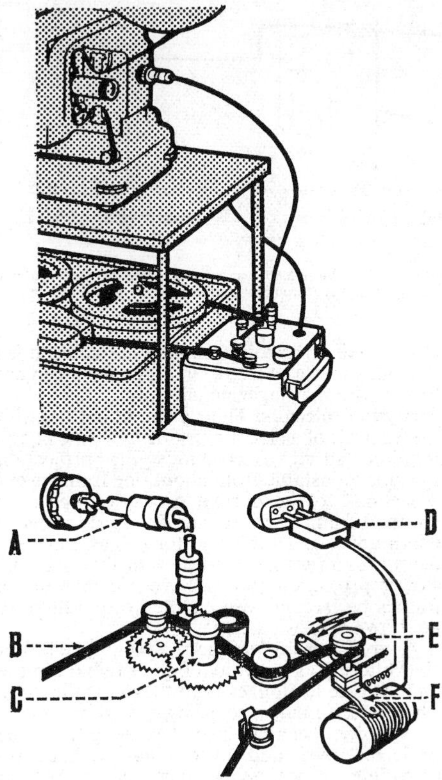

LOOP SYNCHRONIZER FOR TAPE RECORDER AND PROJECTOR. Top: *The synchronizer takes the tape from the recorder and runs it over a tape capstan linked to the projector drive. The system is electrically coupled to the projector motor.* Bottom: *Synchronizer elements.* A, *mechanical interlock with projector.* B, *tape.* C, *tape capstan.* D, *electrical coupling to projector.* E, *loop arm.* F, *variable resistor.*

this kind built into the machine. The tape from the tape recorder then passes over the capstan and other rollers mounted on the projector; the control units are also built into the latter so that setting up is greatly simplified.

Mechanical Coupling. Correct synchronization of the tape and the film—subject again to tape stretch and creep—is also possible by driving the projector and the tape recorder by the same motor. There are several ways of doing this.

First, the projector and tape recorder can be separate units but linked together by a rigid or flexible shaft. Since the one motor drives both units, any slowing down of the film also means a slowing down of the tape and vice versa. The projector in this case is a fairly simple 8 mm. machine, and can be used by itself for silent films or with the associated tape recorder for sound. The two units must be set up appropriately—usually in line with each other—to provide a smooth coupling of the drives.

Secondly, a number of 8 mm. projectors incorporate a tape recorder in the housing. Here again one motor drives both units; the film transport and spools on the one side and the tape reels on the other. This system is widely used also for perforated tape and even for magnetic film (in other words 8 mm. or 16 mm. film coated with a magnetic oxide layer).

Electric Couplers. Completely electric tape couplers can attach to the tape recorder in much the same way as a loop synchronizer. Here, however, the coupler is not linked mechanically to the projector. Instead, the tape—running over a capstan—drives a motor which transmits electric pulses at frequencies depending on the running speed of the tape recorder. These pulses are fed to the projector and automatically adjust the speed of the latter if the tape recorder runs faster or slower than the projector.

Amplifiers. Most 16 mm. sound projectors—optical and magnetic—have the amplifier built into the projector housing. The same applies to 8 mm. stripe projectors. Sound converter systems for striped film have the amplifier in a separate unit which also contains the sound unit through which the film is threaded. Tape coupling systems need no separate amplifier, since this is already built into the tape recorder.

For home projection an amplifier with a 6-watt output is perfectly adequate. Units with 12 watts and even 25 watts output are employed for educational purposes and projection in large halls; commercial and motion picture amplifiers are more powerful still. On modern projectors the whole amplifier system is made fairly compact by the use of transistors.

SPECIAL PROJECTORS

Compact Models. For sale representatives, travellers, etc., there exist specially compact 8 mm. projectors which are easy to carry around and set up at a moment's notice for demonstrations and instruction. They usually have a back-projection screen built into a portable carrying case. The screen is intended for daylight viewing by small audiences and the projector may be mains operated or even battery powered. In the latter case it is completely independent of available current supplies. The screen image is necessarily small, usually less than a square foot.

Projectors of this kind are mostly silent, though a few sound models—for 8 mm. striped film also exist.

Cassette Loading and Loop Systems. For showing short films over and over again, display projectors use a continuous film loop. Here the beginning and the end of the film are spliced together, and the film is wound up on a special spool. This feeds the film from the centre of the spool through the projector, and winds it up again around the outside.

Such loop systems may hold up to 100 feet or

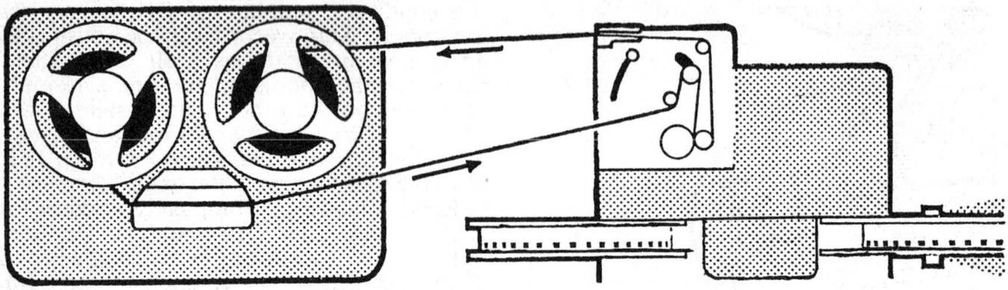

DIRECT TAPE COUPLING SYSTEM. The tape from the tape recorder passes in a large loop over a capstan and tension regulator built on to the projector (right), thus governing the speed of the latter. Precision synchronizing systems of this kind may use perforated magnetic tape.

more of film, giving—in the case of 8 mm. silent film—a projection time of around 8 minutes per cycle. Once the film is threaded up, the projector needs no further attention. Loop systems can be used with silent or sound films.

A special development of this idea is the cassette projector. Here the film, again with its ends spliced together, is contained in a plastic cassette with a built-in film gate and a loop of film carried through this gate. The whole cassette is simply clipped into the projector and engages the film transport. The projector then starts running and can repeat the film indefinitely. The film may be stopped at any point by simply removing the cassette; no other threading or unloading (or rewinding) is necessary.

Cassette systems with automatic loading exist mainly for 8 mm. silent film, and projectors for such cassettes are available both for home projection and for audio-visual use. The idea can also be applied to 8 mm. sound films, but loading in this case becomes a little more complex since the film must be led past a sound head as well. This may be done by having the film run out of the cassette in the form of a loose loop which is then threaded through the film path of the projector. Such a system is adaptable to automatic loading, in which case pressure rollers automatically push the film against the gate and sound head.

Juke Box Projectors. Automatic screening systems are also the basis of coin-operated sound projectors set up in restaurants, coffee bars and similar places in the same way as juke boxes. The equipment consists of a cine projector with the usual optical elements, but a modified layout (to permit automatic placement of the films) and a back projection screen plus the amplifier and loudspeaker. The films are contained in a magazine from which they are brought into position for projection by a push-button selector device. In one system the films are pre-threaded from a feed spool to a take-up spool. The selected spool pair is brought into the projection path and the film then pressed into contact with the gate and sound head by a movable arm carrying sprockets and pressure rollers. At the end of the film-run an automatic rewinding mechanism comes into operation. Other juke box projectors may use continuous film loops in a cassette or other arrangement linked with the magazine selector.

Analysis Projectors. Their purpose is to permit the analysis of a cine film frame by frame at any required speed. Accordingly the projection speed is adjustable from about one frame every 2 seconds to the normal projection speed of 16 or 24 frames per second. The speed range is continuous, and the projector has an independently running shutter to prevent flicker at the slower projection speeds. Speeds of five or six frames per second also enable normal films to be shown in slow motion.

Analysis projectors need specially efficient heat filtering as individual frames remain in the picture gate far longer than they do otherwise. The projector may also have a loop film cassette or spool, to enable short film lengths to be projected over and over again without re-threading—a necessary part of movement analysis.

The film transport usually runs forward as well as in reverse at all speeds.

Wide-Screen Projectors. A few 16 mm. projectors incorporate a variable gate aperture. This enables the film gate to be reduced in height, thus providing an increased aspect ratio similar to that customary with wide-screen motion pictures. To obtain the wide screen effect it is necessary to project on the wide screen with a short-focus lens. Usually the cutting off of the top and the bottom of the image leads to unpredictable losses in the picture itself, unless the film was shot with such a restriction of the format in mind. The greater image enlargement is necessarily associated with a lower image quality.

An anamorphic attachment in front of the projector lens spreads out the normal frame into a wide picture. This of course is only possible when the film was shot in a camera fitted with a similar anamorphic attachment to compress the image sideways on the film.

Apart from anamorphic lenses, two other wide screen projection systems are used in commercial motion picture cinemas. The first utilizes a wider than normal film—65 or 70 mm. instead of the standard 35 mm.—with a long horizontal image format. This gives the increased aspect ratio, but needs of course special projectors for this film size. The second way is to have three projectors linked up to throw three connecting images side by side on an extra-wide curved screen. The films are taken with a cross-coupled triple camera and projected in synchronism. The main technical problem is the hiding of the joins between the three images; various optical devices have been employed for this purpose. L.A.M.

See also: *Cinematography; Projection practice; Projection principles; Screen for projection; Sound recording; Sound synchronization; Visual aids.*

Books: *The Complete Technique of Making Films*, by P. Monier; *How to Project*, by N. Jenkins; *The Bell and Howell Projector Book*, by E. Gilmour (all London).

PROJECTOR (STILL)

From the beginning of photography itself the projection of transparent images with the magic lantern was a popular form of entertainment. Fairly soon still projection became an important adjunct to lectures, both scientific and popular. Today the still projector is an important tool of audio visual presentations as well as being highly popular for private entertainment. Many photographers produce an appreciable proportion of their colour pictures in the form of transparencies for projection.

Still projectors can be divided broadly into slide projectors for individual transparencies in one of several current sizes, film strip projectors for still pictures assembled in the form of a strip of film, and special types such as overhead projectors and episcopes for opaque pictures.

SLIDE PROJECTORS

Projectors for slides can be grouped both according to the slide format they take and according to the degree of automation in loading and slide changing. The most popular slide format is 2 × 2 ins. (5 × 5 cm.) which may accommodate a range of picture formats from the sub miniature 10 × 14 mm. up to the $1\frac{1}{2} \times 1\frac{1}{2}$ ins. (40 × 40 mm.) super slide. However, sub-miniature slides of 3 × 3 cm., as well as larger formats of $2\frac{3}{4} \times 2\frac{3}{4}$ ins. (7 × 7 cm.) and $3\frac{1}{4} \times 3\frac{1}{4}$ ins. (8·5 × 8·5 cm.) to $3\frac{1}{4} \times 4$ ins. (8·5 × 10 cm.) are also used.

Handling differences range from manual loading of individual slides to magazine operation with manual slide change or remote or automatic control.

Optical Layout. The complete optical system of a slide projector consists of a lamp with reflector, a set of condensers, the slide stage and the projection lens. These elements are arranged along the optical axis of the projector, so that the lamp and the condenser system evenly illuminates the slide in the stage, which the lens then projects on to the screen.

For the greatest light efficiency the optical system of the lighting part should be designed for one particular image size. The use of larger slides leads to uneven illumination, while with smaller transparencies part of the light is lost through being masked off by the slide frame.

If a projector is designed for several transparency formats it is usually desirable to have at least one interchangeable condenser in the condenser set to concentrate the light on a larger or smaller image area as required. In practice this is used for multi size projectors, for example when projecting standard 24 × 36 mm. and half-frame 18 × 24 mm. transparencies. Some projectors also have interchangeable condensers for 24 × 36 mm. and 40 × 40 mm. super slide formats. Where,

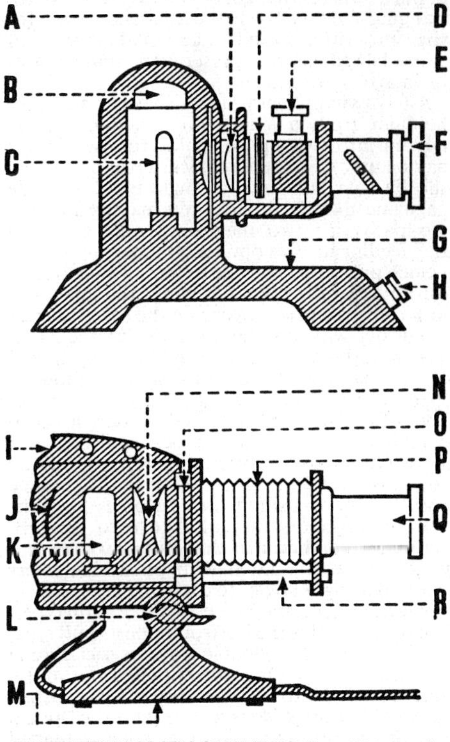

CONSTRUCTION OF STILL PROJECTORS. Top: *Miniature film strip and slide projector of older design. A, condenser system. B, ventilation baffles of lamp house (powerful projectors incorporate a blower unit). C, pre-focus lamp. D, pressure plates for film strips. E, film spools (the whole film strip unit is usually interchangeable against a miniature slide carrier). F, lens in helical mount. G, base. H, tilting adjustment.* Bottom: *Standard $3\frac{1}{4} \times 3\frac{1}{4}$ ins. projection lantern. I, lamp house. J, lamp reflector. K, projection lamp in adjustable lamp holder. L, tilting clamp. M, base. N, condenser system. O, slide carrier. P, square bellows. Q, lens. R, extension rails.*

however mixed slide sets are likely to be projected, it is an acceptable compromise to have a condenser system which illuminates the largest format to be used and put up with the waste of light with the smaller sizes. Most miniature slide projectors therefore cover transparencies up to the super-slide area.

Usually the role of the condenser system is to focus an image of the light source into the back of the projection lens. There is thus a cone of light between the front condenser and the lens. The slide stage is positioned so that the transparency format just fills this cone. With multi-size projectors it is also possible to obtain the maximum light efficiency for different slide sizes, by placing them in different positions in this light cone. Thus at least one dual size projector positions $2\frac{3}{4} \times 3\frac{3}{4}$ ins. slides in a plane comparatively near the front condenser in the widest diameter of the light cone, and 2×2 ins, miniature slides in a second plane further forward where the cone is narrower. When changing over between the two slide sizes (or when projecting mixed slides) the lens must of course be refocused to preserve the sharpness of the image on the screen.

A few twin projector designs also exist. There the light from a single lamp is directed via mirrors into two optical paths (or two lamps may be used). So there really are two projectors side by side built into a single unit. A single magazine may feed slides alternately into the two paths, or two magazines may be fitted each feeding its own optical system.

Such projectors provide a smooth slide transition, since at the point of change the images of both transparencies appear on the screen simultaneously, with one fading away as the other gets more brilliant. This is achieved by linked shutters or iris diaphragms which gradually open in front of one lens while closing in front of the other. Such linked shutters are also available as separate accessories for fitting over the lenses of two projectors arranged side by side to project on to the same screen. The same dissolve-type transition then becomes possible.

Twin projectors may also be used for projecting stereo slides. In that case the two projection lenses are fitted with polarizing filters oriented at right angles to each other.

Single Slide Carriers. There are three main types. The most usual is the horizontal push-pull type, which consists of a double sided carrier sliding between runners. One side is pulled out and a slide is loaded into it. Then the carrier is pushed back so that the slide is in line with the optical axis of the projector. Another slide is loaded into the empty side and when this is being projected the previous one is changed, the carrier being pushed and pulled alternately from side to side.

Another form of slide changer also works horizontally, but instead of the slides being loaded from alternate sides, they are loaded on one side and ejected from the other. They are

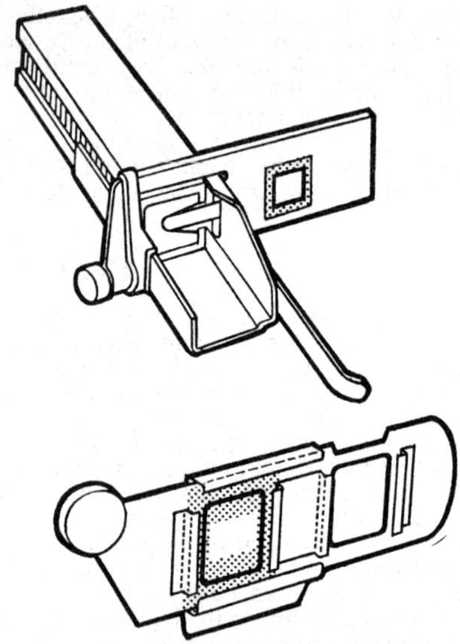

SLIDE CHANGERS. Top: Magazine changer taking slides pre-loaded into magazines. The slides are fed manually from the magazine into the gate and returned again when the magazine advances to the next compartment. This type of attachment can be fitted to most non-magazine projectors; built-in magazine systems work fundamentally in the same way. Bottom: The push-pull slide carrier slides from side to side. It carries two slides which are inserted into and withdrawn from the carrier from alternate sides.

referred to as ejection carriers. The screen is blacked out while the change takes place.

The third type is the vertical slide carrier. The slide is inserted into a slot from the top and pushed down a chute until it is level with the top of the carrier, when it is projected. When the next slide is loaded the previous one is pushed to the bottom of the slot and removed from the side. The slides are held in position by springs which allow slides of varying thickness to be used.

Slide changing with both ejection carriers and vertical chute type carriers can be mechanized and even motorized. For this purpose the slides are loaded into a container in a stack. A motor or a manual control feeds them through the slide stage one by one, restacking them in the original order in a second tray or container at the other side. To work reliably, the slides must not be too thin (e.g. in cardboard frames less than 0·05 ins. thick) otherwise they may be fed in in twos (so that a double image appears on the screen) or jam in the carrier.

Manual Magazine Changers. Although there are considerable differences in design, these carriers all work on the same principle. The slides are loaded into a slotted box, holding from 30 to 70 slides according to make. This box is normally called a magazine in Europe, and a slide tray in the United States.

The magazine is then placed in a channel or tunnel parallel with the optical axis and usually

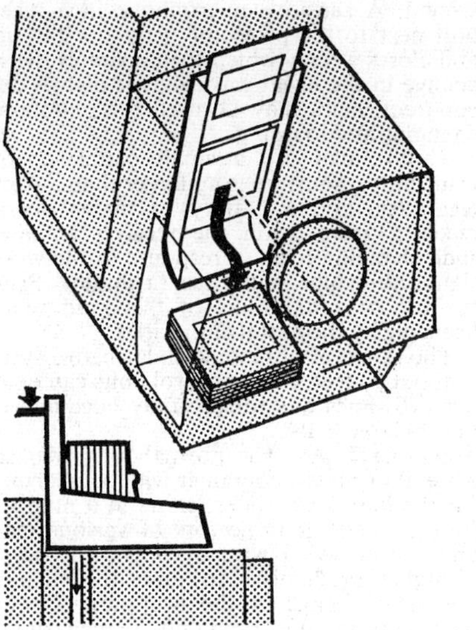

PROJECTOR WITH SLIDE CHUTE. Top: *The slides pass down a vertical chute behind the lens. Each succeeding slide pushes out the previous one which may collect in a stack at the bottom of the projector.* Bottom: *On some machines a slide tray may be fitted on top of the chute and the slides fed through semi-automatically or by a motor drive.*

on the right hand side. Slides are shown by pushing in a lever or sliding bar and withdrawn by pulling the same lever, which returns the slide to its original position in the magazine. Sometimes the magazine jumps forward one slot automatically at this stage, sometimes a small knob at the end of the lever is turned one "click" (the other end has a ratchet which engages on tooth shaped slots on the magazine). A shutter masks off the light from the lamp while the change takes place. It takes only seconds to fit extra magazines, so that it is possible to show any number of slides without any appreciable lapse of time in between.

It is possible to fit separate magazine carriers to some types of projectors supplied originally with single slide changers. These work on the same principle as those built-in.

Editing. In a projector with single slide change the slides are accessible individually. So if one slide happens to be inserted in the carrier the wrong way up or the wrong way round, it can be removed again and turned round. In a magazine projector the slides in the magazine are not accessible unless the whole magazine is removed. Even where the magazine runs along an open track next to the projector, the opening through which the slides are removed faces the slide stage and no removal from above is usually possible.

Some projectors incorporate a so-called editing feature. Here the slide can be removed from the projector when it is in the slide stage. A lever or other control then raises the slide actually being projected through a slot in the top of the projector housing.

Motorized Slide Change. On automatic slide projectors the transport of the slide from the magazine into the slide stage and back again, and also the magazine shift to the next slide are effected by a separate motor. This comes into action on pressing a switch on the projector (or at the end of a remote control cable—see below) and then goes through one slide changing cycle. Normally this advances the magazine step by step; on some projectors the motor can also go into reverse and go through the sequence backwards. This means that any slide just shown can be repeated by pressing the reverse transport button.

A few motorized projectors provide both manual and automatic slide changing, but usually the slides are brought into the optical path of the projector only by the operation of the motor. The magazine may be removed completely or shifted to a different position by

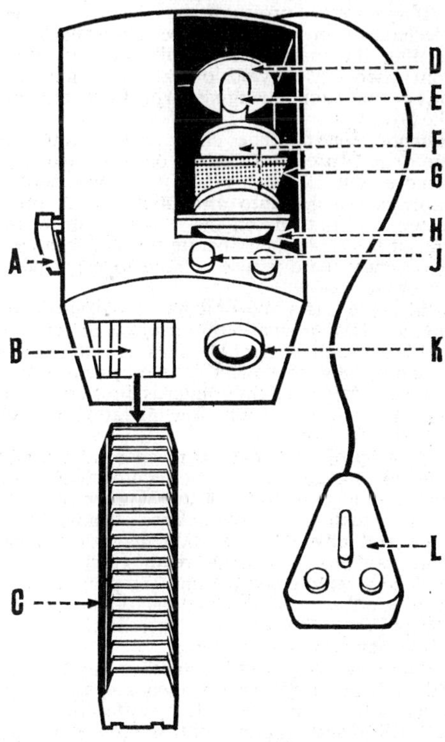

PRINCIPAL ELEMENTS OF MODERN AUTOMATIC SLIDE PROJECTOR. A, *slide change movement, may permit manual slide changing.* B, *magazine track.* C, *slide magazine.* D, *lamp reflector.* E, *lamp (sometimes integral reflector lamps are used).* F, *condenser.* G, *heat filter.* H, *slide stage.* J, *operating buttons for focus and slide change on the projector.* K, *lens.* L, *remote control unit, usually with provision for slide changing forwards and backwards and for remote focusing.*

disengaging the motor drive, provided the changing cycle has stopped with the slide stage empty. Often this is achieved automatically on switching the motor off, when a shutter may close the slide stage to darken the screen.

The motorized slide change usually works with a push-pull arm which slides the transparency from the magazine into the optical path. Some projectors use alternative systems, such as a gripping device that lifts the slide out of its magazine compartment into the projection system, or a magnet that pulls the slide (fitted with a metal edge) into the projector.

Magazine projectors, whether manually operated or motorized, will only work with slides within certain dimensional tolerances. The maximum outside size for 2 × 2 ins. slides is 50·8 mm. Slides bigger than that or unevenly mounted ones with protruding bits of binding tape or not fully closed metal tabs (in the case of metal frames) may get jammed. The range of thicknesses accepted is usually between 0·02 and 0·125 ins.; thinner transparencies (in card mounts) may buckle and jam the projector.

If a motorized projector jams the slide change mechanism may have to be disengaged by turning the motor (or a milled wheel fitted somewhere—usually below the housing) to bring the slide changing arm fully out of the projector.

Remote Control. The slide change can on motorized projectors also be operated by remote control. In this case a remote control cable is plugged into an outlet on the projector, and buttons on the remote unit at the end of the cable then operate the slide change. Frequently there is also remote focusing provision (see below).

In some cases the unit has a built-in light pointer. Here a small lamp projects an arrowhead or similar symbol. This is sufficiently bright to be visible against the projected transparency when the light pointer is directed at the screen, and serves for drawing attention to specific image details.

Some projectors have a mechanical remote control linkage consisting of a Bowden cable with a plunger at the end, Pressing the plunger —like pressing a cable release—changes the slide and advances the slide magazine. This unit may also incorporate remote focusing.

With a remote control unit the projectionist can sit in the midst of the audience and operate the projector from there. The remote control cable may be 6 to 10 feet long.

For slide shows and lectures in large halls this is however not usually adequate, especially where the lecturer himself operates the projector. Cableless remote control units are therefore also available. These can operate either by ultrasonic sound signals or by a radio link.

For ultrasonic operation the slide changing signal is given by a small whistle or other generator producing an inaudible sound impulse of a frequency above 20,000 cycles per second. A sensitive microphone in a receiver unit next to the projector picks up this signal and closes a relay which then operates the slide change in the usual way. Signals of two different frequencies may be used for remote slide changing and focusing.

Radio links use a small battery-operated transmitter, sometimes no larger than a pack of cigarettes, and a battery or mains operated receiver. Here again the receiver actuates a slide changing relay in response to the wireless signal. In some countries (including Britain) the use of radio links of this kind requires licencing by the postal authorities.

The range of ultrasonic slide change systems is about 50 feet; radio control units can operate over 20 times this range—fully adequate even in very large halls.

Magazine Types. The normal long magazine, a metal or plastic container with compartments for the individual slides, exists in a number of forms. These fit projectors of various makes, and are provided with openings for the slide changing mechanism to push the slide into the picture stage of the projector. The compartments are numbered and the magazine also

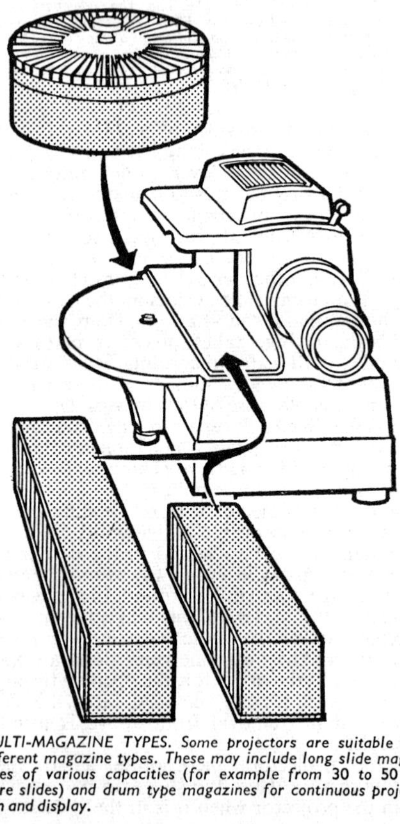

MULTI-MAGAZINE TYPES. Some projectors are suitable for different magazine types. These may include long slide magazines of various capacities (for example from 30 to 50 or more slides) and drum type magazines for continuous projection and display.

carries a ratchet or other device to engage the transport mechanism. Usually the magazines take 30 or 36 slides, though special long magazines for 50 or even 70 slides exist. The orientation of the slides in the magazine depends on the type of changing mechanism.

Since individual projector manufacturers develop their own magazine designs, there is comparatively little standardization. A few groups of manufacturers, especially in Germany, have, however, agreed on the use of a uniform design to fit a number of different projector models.

As an alternative to the usual long magazine, circular slide magazines are also used. These are drum shaped, and may hold 80 to 120 slides in compartments radiating outwards from the centre. There are three types, used with projectors specially designed for them.

Vertically mounted drum magazines are fitted to a horizontal axis alongside the projector, and the slides fed sideways into the projector gate. With this design it is possible to have projectors taking long and drum type magazines interchangeably.

Removable horizontal drum magazines sit on top of the projector, rather like a gramophone disc, and the slides are fed downwards by a gravity system. A special retaining spring inside the magazine prevents the slides from falling out when the magazine is lifted off the projector.

Fixed drum magazines are permanently built into the projector and loaded while on the latter. This type of projector is primarily intended for continuous display and advertising use. The layout is normally horizontal.

Continuous Projection. The projection sequence can be made fully automatic with a motorized slide projector, so that the latter projects one slide after another without attention from an operator. For this purpose some projectors have a built-in time switch which actuates the slide changing mechanism at pre-set intervals—usually from 3 to 60 seconds. Similar timer units are also available separately and can be connected to the projector through the remote control outlet.

If the projector has a drum type magazine, the time switch can repeat a slide series indefinitely, since the drum magazine can go round and round. With projectors having a long magazine some means of reversing the magazine travel is necessary. Special units are available which do this, and may work on two systems.

In one of the projector projects the odd numbered slides in a magazine while the latter goes forward, and then reverses to project the even numbered slides. When the magazine reaches the starting point it reverses once more and repeats the cycle. The magazines must be loaded accordingly.

The other and simpler continuous projection method is to run through the magazine, compartment by compartment, and reverse at the end of the run so that the slide series is then projected in backward order.

Random Selection. On some projectors it is possible to select any given slide in a magazine for projection, irrespective of the order in which the slides are loaded. This method works best with drum type magazines. A multichannel remote control selector chooses any given slide number by means of digital buttons, a telephone type dial or similar device. This sets the transport motor going to rotate the slide magazine and stops it at the selected compartment number to project the slide in it. The magazine only rotates in one direction, and the slide selection may take from 2 to 8 seconds, according to how near the required slide is to the gate before pressing the selector buttons.

Focusing. The basic focusing system of the projector lens usually consists of a helical mount by which the lens is screwed in and out. This serves for both coarse and fine focusing.

During projection of card mounted transparencies re-focusing is, however, often necessary owing to the popping of the transparencies under the warmth of the slide stage. Where slides of different thickness are projected in a set, re-focusing is also necessary because the various slides may not have their image plane in the same position.

Projectors often have some sort of centring device—usually in the form of V guides—to locate all slides in the same plane. This, however, works efficiently only all slides are uniformly mounted with the front and rear halves of a transparency frame of the same thickness.

Slide projectors with remote control systems may incorporate remote focusing of the lens to take care of these variations in image position. The remote focusing may be driven by a separate motor, or the fan motor may be used. There are two basic modes of operation.

In the simpler system, invariably used where the fan motor drives the focusing, an eccentric gearing moves the lens forward and backwards through a short distance. On the screen the image—once reoughly focused with the helical mount—moves through a cycle of sharpness, unsharpness and sharpness as the remote focusing button is depressed. To find the sharpest point takes a little skill, since the motor must be immediately stopped. Otherwise the focusing cycle runs through the position of maximum unsharpness before returning to sharpness again.

The other mode of operation works with a reversible motor controlled by two buttons on the remote unit, These run the lens forward and back, usually through the full range of its travel. It is thus possible to reverse the motor instantly if the point of maximum sharpness has been passed, and the motor may even be used for coarse focusing.

Slide projectors with fully automatic focusing also exist. Here a photo-electric and optical sensing system operates the focusing motor via a relay whenever the image plane of the transparency shifts forward or back. The first slide must be focused accurately by hand; after that the automatic system compensates for any movements or changes in sharpness.

Illumination. Apart from one or two exceptions, the illuminant is a Class A projector lamp with a compact tungsten filament. The lamp usually burns vertically, base down and may vary from 75 to 1,000 watts.

Two or sometimes three condensers are placed in front of the lamp, and they produce a concentrated beam which should just cover the corners of the transparency. Much depends on the design and quality of the condensers as a poor condenser system will lose a great deal of light.

Most projectors employ an asperic rear condenser, placed close to the lamp, with a biconvex front (or field) condenser a short distance in front, just behind the transparency. An aspheric condenser helps to reduce spherical aberration, so that more light can be made to enter the front condenser. The closer the condenser and the greater its diameter the bigger is the cone of light from the lamp filament so adding to the brilliance of the image. In better-class projectors the condensers are coated to reduce the amount of light lost by reflection.

A concave mirror is generally placed behind the lamp to utilize light travelling towards the back of the projector which would otherwise be wasted. The mirror is positioned so that the lamp filament coincides with its centre of curvature, thereby producing a virtual image of the lamp filament in almost the same place as the filament itself. The filament and its reflection do not coincide exactly but come slightly to one side. If the reflected image impinged on the filament of the lamp, this would cause it to overheat and reduce the life of the lamp.

Where the filament consists of several turns spread out in a plane, the ideal positioning of the reflector projects the filament image between the turns of the filament itself. This can be checked by covering the projector lens with a piece of ground glass or translucent paper; the direct image of the filament and its reflected image should just interlace. With compact source projectors the two images of the filament may have to be placed side by side.

Lamp Power and Voltage. The wattage of the lamp is another important factor governing the strength of illumination. Taken on its own, however, it can prove misleading. If everything else were equal, then a high wattage lamp would give a brighter light than one of low wattage, though the difference would not be by any means proportional.

The current flowing through the lamp, i.e. the amperage (watts divided by volts) is, however, equally important. The filament of a high-voltage lamp is thinner and longer than that of a low-voltage lamp of the same wattage. The luminous intensity of the glowing filament increases considerably when the filament consists of a small coil of thicker wire, so that for a given wattage, the light output of a low-voltage lamp is greater.

For this reason low voltage projectors have become popular. Here the filament can become very compact indeed, being sometimes made of a folded ribbon type wire. A 100 watt lamp running at 12 volts can produce light equal in intensity to a 250 watt lamp operating at 200 to 250 volts. In addition the low voltage lamp burns at a higher colour temperature so that it appears to the eye comparable to the light from a 300 watt mains lamp.

As the filament is so small it becomes practical to use a smaller reflector and even to build this into the lamp itself. Such a proximity reflector lamp has the advantage that the mirror is permanently adjusted during manufacture.

Quartz-iodine lamps have also been used in miniature slide projectors. Here the light source is smaller still, and this lamp type has the further advantage of constant light intensity and colour temperature throughout its life. The luminous efficiency of a low voltage quartz-iodine lamp is higher still than of a normal low-voltage tungsten lamp.

The actual lamp power desirable for projection depends on the intended use of the projector. Inexpensive home projectors for showing miniature slides on a small screen at close quarters usually have a 150 watt mains voltage lamp. They are inexpensive mainly because with this lamp power no force ventilation is normally required. That also makes the projector more compact.

Low power projectors need to be operated in complete darkness, especially when projecting colour transparencies. A short distance between projector and screen is advisable, and the transparencies should be kept as thin as possible consistent with good quality. Generally speaking a picture size of 30 × 40 ins. or 40 × 50 ins. is about the largest that projectors of this class will give with reasonably good illumination.

For more comfortable projection a 250 to 300 watt mains voltage lamp (or 100 watt low voltage) is desirable. For large rooms a projector with a 500 watt lamp (or 150 watt low voltage) may be preferable. For really large halls big projectors with 750 watt or more powerful lamps may be needed.

When used with colour slides under good conditions a screen size of 4 × 6 feet is usually well within the limits of a medium power projector and slightly larger picture can often be projected satisfactorily.

With black and white transparencies quite good results can be obtained in classrooms having only partial black-out.

Two other factors influence the image brightness. One is the aperture of the lens. Provided that the optical system of the projector has been carefully designed, so that the image of the lamp filament formed by the condenser just fills the rear element of the lens, more light will come from the lens the wider the aperture. But a wide-aperture lens, without appropriate illumination does not necessarily give more light than a smaller aperture lens.

When projecting mixed slide sets containing very dense and very thin transparencies it is often useful to be able to vary the strength of the illumination. The obvious way, regulating the voltage fed to the lamp, is not suitable for projection of colour pictures because it also changes the colour temperature. Some projectors, however, are, or can be, fitted with an adjustable iris diaphragm over the lens. This can be closed down or opened up to reduce the light for overexposed colour transparencies and to boost it for the denser slides.

In one or two projectors this iris diaphragm is controlled automatically from a photo cell system. This receives the light reflected from the projection screen and responds to the light intensity, adjusting the iris accordingly.

Heat Filter. The second factor is the thickness of the heat filter. This is a special glass which is inserted in the optical system to protect the transparency from the heat of the lamp, at the same time transmitting a high proportion of the light. It is usually placed between the condensers. Most heat filters in 300-watt projectors are 3 mm. thick. This transmits some 85 per cent of the light but only 8·5 per cent of the heat.

In 500 watt and more powerful models, two or even three heat filter glasses may be employed, making a total thickness of perhaps 9 mm. This tends to offset the extra illumination given by the bulb.

Cooling. Unless a projector is designed so that a great proportion of the heat generated by the lamp can escape, there is a danger that the bulb will blister and give an unduly short life. Even more serious, the transparency may be spoilt, and some of the optical components damaged.

The main problem is to provide adequate cooling without excessive light leakage. Too much light escaping from the projector reduces the effectiveness of the black-out and dulls the brilliance of the screen image.

Sometimes light can escape in certain directions and prove disconcerting to people sitting in certain parts of the room.

With low power projectors using lamps up to 150 watts at mains voltage the cooling afforded by the convection airstream is usually adequate. The design of the lamp house, must allow cool air to enter and the hot air to leave freely. More powerful projectors for the larger slide sizes are sometimes also convection cooled, since there the lamp house is big enough to dissipate the heat adequately.

Miniature projectors with lamps of 250 watts or over almost always required forced cooling.

Various types of motor driven fan are employed for this purpose. Some use normal fan blades set at an angle, some are turbine fans shaped like paddle wheels. Other machines use impellers and cylindrical cooling devices with lateral slots in them.

Some of the stream of cool air from the fan is normally used to help cool the transparency. In some cases the air is forced through a duct with a vent just underneath the slide carrier. In an endeavour to prevent popping several magazine type projectors direct the warm air coming from the lamp towards the magazine in order to warm the slides before they are projected.

The warmth of the projector to the touch is not a good guide to the efficiency of cooling. In fact, a cool lamphouse indicates good thermal insulation and nothing more. A hot lamphouse means that the heat at least is getting away somewhere.

As far as the transparency itself is concerned, it is the temperature in the slide stage which counts. This can be measured in various ways, ideally with a thermometer slide incorporating a very thin bi-metal strip with an attached pointer. This is bound up between two thin glass plates to produce a slide of the normal thickness. As the strip inside the slide gets warm, it expands and moves the pointer over a graduated scale indicating temperatures. An efficiently cooled projector should not have a gate temperature exceeding about 40°C.

2 × 2 ins. Projectors. This is the most popular size. Most of them are suitable not only for 24 × 36 mm. transparencies but also for larger ones up to superslide size (4 × 4 cm.) in 2 × 2 ins. mounts. Some models have movable condensers which can be altered to give coverage for all sizes with maximum light concentration. Sometimes extra condensers are available for 4 × 4 cm. slides, and these may add to the price. Users of this size of slide should satisfy themselves that the projector they purchase will give adequate coverage.

The lenses fitted as standard are usually 85 mm. or 100 mm. The shorter of these will give a 3 × 2 foot picture from a 24 × 36 mm. slide at about 8 feet from the screen. With a 10 mm. lens the same image size is obtained at about 9½ feet. These lenses are therefore suitable for home projection in the majority of living-rooms, though longer lenses are sometimes required if the rooms are larger than average.

When used with half-frame (18 × 24 mm.) transparencies—and even more so with sub miniature transparencies mounted in 2 × 2 ins. slide frames—these projectors need a shorter focus lens. The usual focal length for half-frame slides is 50 to 65 mm., and 35 to 40 mm. with sub-miniature transparencies.

Variable focus projection lenses are sometimes used, for example with a focal length range from 85 to 120 mm. These provide a range of screen image sizes from a given projection distance, or permit a variation of the projection distance for a given screen image size. Variable focus lenses are particularly useful for multi size projectors, since they can keep the image size on the screen constant when projecting smaller and larger transparencies in mixed sets.

$2\frac{1}{4} \times 2\frac{1}{4}$ ins. Projectors. These are suitable for $1\frac{5}{8} \times 2\frac{1}{4}$ ins. and $2\frac{1}{4} \times 2\frac{1}{4}$ ins. transparencies made on size 120 roll film. They are also useful for technical lectures where it is necessary to show fine detail which cannot always be obtained on smaller slides. The $3\frac{1}{4} \times 3\frac{1}{4}$ ins. slide size was at one time used for this type of subject but it is less convenient for the production of colour slides.

The standard focal length of lens for this size of projector is 135 mm. to 150 mm. ($5\frac{1}{2}$ to 6 ins.) The picture size is 3 feet square at roughly 8 feet and 10 feet from the screen respecitvely.

In specifications $2\frac{1}{4} \times 2\frac{1}{4}$ ins. projectors are fairly similar to the 2×2 ins. models, being largely scaled-up versions. Thus there are models with manual slide change, magazine operation, remote control, etc. The range of models on the market is however rather more limited.

Lamp power usually ranges from 300 to 750 watts, and virtually no low power projectors of this kind are made. This is because a 150 watt lamp is not adequate to illuminate the full area of the transparency and still give a reasonably bright picture.

Converting to 2×2 ins. It is usually possible to project 2×2 ins. slides in a $2\frac{1}{4} \times 2\frac{1}{4}$ ins. projector. This involves either mounting the 2×2 ins. slides in adaptors inserted in the slide carriers or the fitting of new carriers. Usually it is desirable to change to a shorter focus lens to increase the picture size from the smaller format without having to move the projector farther away from the screen. In certain machines the condenser can also be adjusted to concentrate the light over the smaller area.

Multi-size Projectors. A number of universal slide projectors can be used with both 2×2 ins. and $2\frac{1}{4} \times 2\frac{1}{4}$ ins. slides. Changing over from one slide format to the other is fairly quick, and in at least one outfit both sizes can be projected more or less simultaneously. There are two main types of design: multi-magazine and interchangeable track.

The multi-magazine projector has a $2\frac{1}{4} \times 2\frac{1}{4}$ ins. and a 2×2 ins. magazine running along the two sides of the optical system. There are two slide changing mechanisms, to move the slides into the optical path from alternate sides. Each slide format has its own picture stage, set at different distances from the condenser system to ensure full illumination of the transparency. The lens needs re-focusing when a small slide is projected after a large one, or vice versa. With a projection zoom lens, adjustment of the focal lengths can keep the image size the same on the screen.

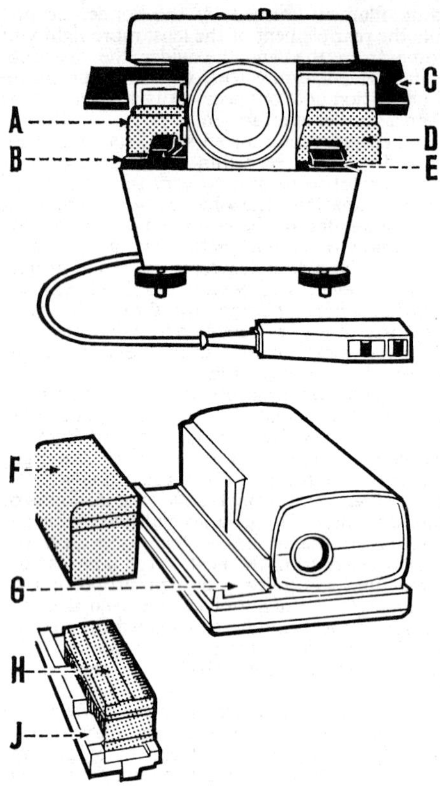

DUAL SLIDE PROJECTORS. Top: Universal projector with two magazine tracks on each side of the optical system. A, miniature slide magazine. B, miniature magazine track. C, slide changer arm. D, large-size magazine with slides. E, large-size magazine track. Bottom: Dual projector with interchangeable magazine track. F, large magazine with slides. G, track for large magazine. H, miniature slide magazine. J, miniature magazine track, fits into larger track.

The interchangeable track system uses a single magazine track for the larger slide magazines. When the smaller magazine is to be used, a second track is inserted in the main one, to locate the alternative size in the correct position. The slide change mechanism is the same for both formats, and special guides are used to locate the larger and the smaller slide accurately in the film gate.

Since only one magazine can be used at a time, there is no possibility of mixing large and small slides.

Large Size Slide Projectors. Projectors for $3\frac{1}{4} \times 3\frac{1}{4}$ and $3\frac{1}{4} \times 4$ ins. slides are very much more bulky and exist practically only with single slide changers. The range of models available is limited, since the demand is comparatively small. Many institutions are, however, compelled to continue using this size of projector as they have large numbers of $3\frac{1}{4} \times 3\frac{1}{4}$ ins.

slides. Colour transparencies in $2\frac{1}{2} \times 3\frac{1}{2}$ ins. and larger formats are also projected with this slide size.

The projector usually has a big lamp house which is often spacious enough not to require forced cooling even with a 500 watt lamp. The focusing movement generally consists of a square folding bellows connecting the slide carrier to the lens panel which moves along two extending rods. The focal length of the lens is normally 8 or 10 ins., so that a fairly long bellows is needed.

Really powerful large size projectors—for example for cinema projection—may use a 1,000 watt tungsten lamp or a carbon or xenon arc.

Converting to Smaller Sizes. It is usually possible to convert a $3\frac{1}{4} \times 3\frac{1}{4}$ ins. projector to show $2\frac{3}{4} \times 2\frac{3}{4}$ ins. and 2×2 ins. slides by fitting alternative slide carriers. Some light is lost when showing $2\frac{3}{4} \times 2\frac{3}{4}$ ins. slides but results are generally satisfactory. For 2×2 ins. slides, however—especially 24×36 mm. or smaller transparencies—the loss of light is more serious, unless the condenser system can be modified. It is also necessary to replace the lens by one of much shorter focal length, otherwise, over a normal distance, the picture size will be too small.

When 2×2 ins. or other small slides are used in a large projector there is of course no possibility of automatic slide changing.

Sub-miniature Projectors. Transparencies produced with sub-miniature cameras in the 8×11 mm., 10×14 mm. or 12×17 mm. format can be mounted in 2×2 ins. slide frames. There are also sub-miniature projectors which takes these transparencies in 3×3 cm. frames. There are not very many on the market, and they are mostly scaled down versions of the normal miniature projector. They include models with manual or magazine slide changing as well as automatic versions. The slide magazine is similarly scaled down to take 30 or more 3×3 cm. slides.

Usually sub-miniature projectors have a low voltage lamp of 100 watts and a 35 or 40 mm. projection lens. Often a small table top screen is included in the outfit for close range projection at low magnifications. This is suitable for daylight projection as well, since the image brilliance on the screen is adequate not to be swamped by the prevailing light.

Stereo Projectors. These consist of two optical systems set up side by side. Usually there are two lamps, but sometimes one lamp only is used, the light being split to illuminate each of the frames. A polarizing filter is placed immediately in front or behind the transparencies, each filter being set at 90° to the other.

The images are projected on to an aluminized or silver screen (normal white or beaded screens are unsuitable) and the lenses adjusted to bring the images into register. Polarizing spectacles are worn by the viewer with the filters arranged at the same angle as the polarizers in the projector, so that the viewer sees only one image with each eye.

A special stereo mounting jig may be used to mount the slides in correct alignment. This reduces the need for adjusting the lenses, and saves time and eye strain.

FILM STRIP PROJECTORS

The continuous film length of a film strip is projected in a film strip projector or a slide projector which can be fitted with a film strip carrier. This is usually the case only in slide projectors with a manual slide changing system, where the whole slide changer is removable and replaced by the film strip carrier.

Film Strip Carriers. The most common and the cheapest form of film strip carrier is the spool type. The film is wound from one spool to another across the film gate. Usually the film

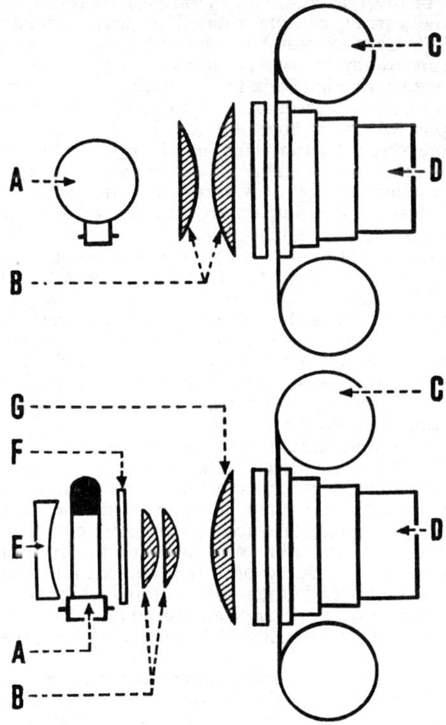

OPTICAL LAYOUT OF FILMSTRIP PROJECTORS. Top: *Low-price film strip projector. This generally uses a low-voltage, compact source, filament lamp of the motor car head-lamp type, and is intended primarily for home use where a high light output is unnecessary.* Bottom: *Precision film strip projector for home and lecture hall use. It utilizes a high-power projection lamp, improved optical system, and incorporates at least a heat filter (often also a fan for forced cooling) to protect the transparencies. In both cases the components are A, lamp B, double condenser; C, film strip; D, projection lens; E, lamp reflector; F, heat filter; G, front condenser (may be aspheric).*

is held flat by two pieces of glass held together by a spring. The projectionist must watch the screen as the film is wound to see when to stop winding. In some cases the glasses must be separated as the film is turned on. In others,the glasses are fixed close enough to keep the film flat, while allowing it to travel freely.

A more elaborate form of carrier transports the film by means of a sprocket wheel which engages the perforations of the film. They normally advance the film four sprocket holes with one turn, that being the number of sprocket holes in a single (half) frame (18 × 24 mm.) Double frame (24 × 36 mm.) transparencies require two turns.

Linked with the sprocket wheel is a cam mechanism which separates the pressure glasses as the film is wound.

Single frame film strips are invariably shown with the film running vertically. Most, but not all, double frame strips are wound horizontally. For this reason, on a general purpose projector the carrier can be rotated at least through 90°. Carriers suitable for double frame strips can usually be masked down for single frames.

Automatic Film Strip Projectors. Some film strip projectors incorporate an automatic film transport which advances the film by one frame on pressing a button. This may be operated via a remote control cable. The film transport may go forward or backwards, but no separate focusing by remote control is usually necessary since the film is flat between the glasses of the carrier. Automatic transport systems are usually designed for film strips where all the images are of the same size and orientation, to avoid the need for manual rotation of the whole carrier.

Film strip projectors intended for display purposes may have a built-in timer (like display slide projectors). For continuous projection loop systems, with the ends of the film strip spliced together, are used.

Random selector devices are also available for film strip projectors. Since a film strip has a very much larger capacity (up to several hundred transparencies) than a slide magazine, the selector devices are accordingly more complex. They run the film strip through the picture stage at high speed until the required image is nearly reached, then they automatically slow down and at the appropriate point a locating pin accurately positions the image in the projector gate.

Like slide projectors, film strip projectors are also available with built-in tape recorders or gramophone turntables for combined sound and picture presentation (see below).

SOUND STILL PROJECTION

A slide show even at home is usually accompanied by some sort of running commentary for the benefit of the audience. Since tape recorders can record such commentary or even lectures, the idea soon arose of using a tape recorder to control the projector. In this way it is possible to organize a combined sound slide show, which virtually runs itself.

Equipment-wise there are three main possibilities: (1) normal automatic magazine slide projectors in conjunction with a tape coupler and tape recorder, (2) combined projectors with a built-in tape recorder; (3) with sound tracks recorded on the slide itself. Corresponding provisions exist for film strip projectors. Such specialized equipment is produced for a variety of audio visual training purposes.

Tape Synchronization. Where a tape recorder operates the slide changing mechanism of a projector, it does so with the aid of signals on the tape. These are inserted in suitable spots within the commentary or other background effects, where a slide has to be changed. There are various systems of doing this, working either with direct electrical contacts or with recorded signals.

Contact Synchronization. This is the older and also more cumbersome method. One procedure is to stick short strips of metal foil to the back of the tape or to splice it into the tape at the appropriate points. During playback the tape runs over a tape synchronizer attached to the recorder. There the tape itself touches a pair of contacts connected to the remote control system of the projector. When the strip of metal foil runs over the contacts, it closes the remote control circuit and the projector automatically changes the slide.

In another system the tape runs between two contacts bearing on opposite sides of the tape. At the required points narrow strips are trimmed out of the centre of the tape, so that there the two metal contacts touch and close the slide changing circuit.

Both these methods have the drawback that the tape is permanently marked, and can thus be used only for the particular slide show for which it has been prepared.

Cueing Signals. The more elegant and modern way is to record slide changing signals electronically on the tape itself. The signal may be an ultra-high frequency one, recorded on the same track as the commentary or lecture, but inaudible during playback through being filtered out. During the projection the tape here runs over a pick up head of the tape synchronizer, and this amplifies the signal, to actuate a relay which closes the slide changing circuit.

Alternatively a cueing signal can be recorded on the second track of a twin track recorder, so that it does not interfere with the commentary at all. Such a signal can be of a normal frequency (for example 1,000 cycles) and it can be erased and re-positioned at will without interfering with the sound effects.

Some tape recorders have a special cueing head built in, so that no separate tape synchronizer is required. In this case a control on the tape recorder records the impulse on the cueing track, and the projector is connected directly to the tape recorder.

Where a film strip projector is fitted with a remote control film transport system, this can also be actuated by cueing impulses from a tape recorder.

Combined Tape Projectors. Certain slide and film strip projectors designed especially for audio visual purposes have a tape recorder built in. In principle this is no different from the combination of a separate tape recorder, tape synchronizer or coupler and projector. When built into one unit it is however more compact and may even include its own projection screen for daylight table-top projection.

Film strip sound projectors sometimes work with gramophone records. Here the signal is an inaudible one, recorded on the disc together with the sound effects and commentary.This arrangement is practical only with film strips, since the sound, once recorded, is permanently keyed to a particular picture series. Since the film strip is also a more or less unchangeable picture sequence, this is a handy combination for mass-produced educational lectures, etc. which are to be widely distributed.

Sound Slides. This is an alternative method of recording a commentary to be played back with a projected picture. It is more flexible than a tape recording, since each slide carries its own associated commentary and can be switched round at will. On the other hand the length of commentary per slide is limited.

The system uses large card or plastic slide frames with a central aperture to take standard 2 × 2 ins. slides. The frame carries a circular magnetic sound track surrounding this aperture. On this the commentary is recorded (during a preliminary projection session) by a rotating sound head inside the projector. Each slide can accommodate about 15 to 20 seconds of speech. During projection the same head picks up the recorded sound and plays it back through an amplifier and loudspeaker system built into the projector. The slides may be changed automatically at the end of the commentary, or manually by remote control or by a time switch. The sound slides themselves are loaded into a magazine—a scaled-up version of the usual miniature slide magazine.

A corresponding system also exists for film strips. Here every other frame on the strip carries an optical sound track. In the projector this is scanned by a travelling sound head which moves back and forth along the track while the picture next to it is being projected.

SPECIAL PROJECTION SYSTEMS

Micro-projectors. Where microscope specimens are to be shown to a class or during a lecture it is sometimes feasible to insert the microscope slide directly in the projector. For some slide projectors special micro-attachments are available, which fit in place of the lens unit. They consist of a microscope slide holder plus a short focus projection lens. The microscope slide is in this case positioned in a narrow part of the projected light beam, so that the lens projects a greatly enlarged and brilliant image of the specimen itself. The complete attachment usually includes a supplementary condenser and may have provision for inserting filters, microcells and similar items. Microprojection with this type of outfit can yield magnifications of 200 to 300 times on the screen.

More elaborate micro attachments may incorporate a complete microscope system in reverse, consisting of an objective and an eyepiece lens with the latter facing the screen. This makes magnifications of 2,000 times and more possible on the screen. An alternative form of micro-attachment uses a mirror housing plus a second mirror, so that the specimen is located horizontally on a microscope stage. Here the projection beam from the projector is deflected upwards through the microscope system, and forward again by the second mirror on to the screen.

Overhead projector. This is a projection unit designed mainly for educational and similar uses, and consists of a light box with an illuminated horizontal top surface. Above this box is a mirror and a projection lens which projects the illuminated area on to a screen behind the projector. The light box takes transparencies up to 10 × 10 ins.

As an alternative to large film transparencies or one-sided translucent documents, a transparent strip of film can also be tensioned across the illuminated area. On this the lecturer can write or draw while talking, and the material then appears on the screen in front of a class. In effect this is therefore an optical blackboard, except that the picture appears in black (or coloured) writing on a white background.

The writing film for this purpose is available as 10 ins. wide acetate strip in lengths up to

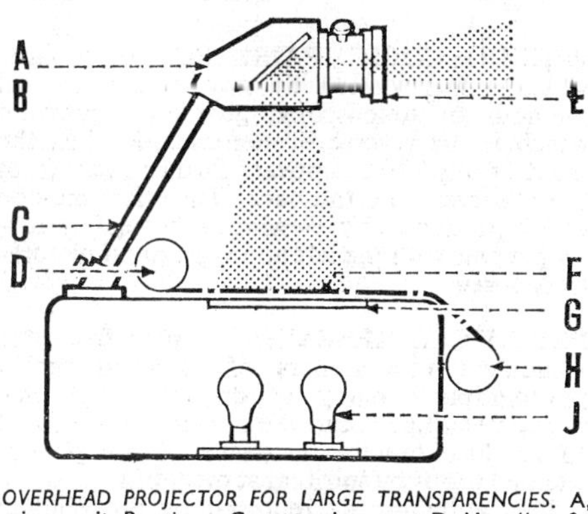

OVERHEAD PROJECTOR FOR LARGE TRANSPARENCIES. A, mirror unit. B, mirror. C, supporting arms. D, H, rollers for continuous strips of transparency film. E, lens. F, film. G, opal glass sheet. J, lamps inside light box.

50 feet. It can be rolled continuously from a feed to a take-up roller to bring new portions over the illuminated area for further writing or drawing.

Overhead projectors usually have a 500 to 1,000 watt lamp, and a projection lens with a focal length of 10 to 15 ins.

With a 14 ins. projection lens a 10 × 10 ins. transparency covers a $6\frac{1}{2} \times 6\frac{1}{2}$ feet area on a screen at a projection distance of 10 feet.

In addition to straightforward drawing and writing, various special effects can be obtained with the aid of polarizing screens, filters, etc.

Episcopes and Epidiascopes. An episcope will project images of opaque objects such as book pages, photographic prints, maps, diagrams and small solid objects. Most episcopes work on the following principle:

The object is placed on a platform which can be raised and clamped under a sheet of toughened glass to hold it flat. The object is illuminated from above by means of a lamp and a number of mirrors, and the light is reflected from the object to a surface silvered mirror (inclined at 45°) at the top of the lamphouse. An episcope lens is placed in front of the mirror, and focused so that an image of the object appears on the screen. Shown on an opaque screen, the image will appear the correct way round. In some episcopes, there is no 45° mirror, in which case the image will be reversed unless back projection is used on a translucent screen. Episcopic projection must normally be done in a darkened room, as reflected light is less intense than light transmitted through a transparency.

Most episcopes will be found to be fitted with a 500, 750 or 1,000 watt lamp, which is cooled either by a motor-driven fan or by convection.

An epidiascope is a combined episcope and diascope so that both opaque objects and lantern slides can be projected. The diascopic attachment can often be obtained separately to convert an episcope into an epidiascope.

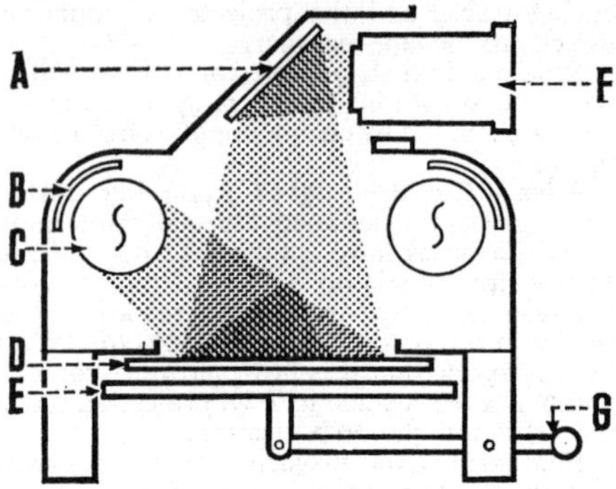

EPISCOPE LAYOUT. The episcope serves for projecting opaque originals which are illuminated by reflected light. A, projection mirror. B, condenser mirror. C, lamp. D, glass plate. E, tray on which items are placed. F, projection lens. G, lever for raising and lowering tray.

The attachment consists of a lens holder, slide carrier and condenser which are placed on the front of the lamphouse, in line with the lamp. The front mirror used to help illuminate the object when the machine is employed as an episcope must be moved clear of the lamp so as to cause no obstruction between lamp and slide.

The diascopic attachment may be for standard slides or miniature slides and film strips, and sometimes a micro-slide adaptor (for projecting microscope specimens) is also available as for miniature projectors.

L.A.M. & C.W.L.

See also: *Filmstrips; Lamp caps and fittings; Light sources; Projection practice; Projection principles; Slide (transparency); Visual aids.*

Books: *Projecting Slides*, by M. K. Kidd and C. W. Long (London); *Photography, Theory and Practice*, by L. P. Clerc (London).

PSEUDOSCOPIC PAIR. Pair of images which, when viewed by any of the normal methods of stereoscopy, give an impression which is the reverse of stereoscopic—i.e., the front of any object appears farther away from the observer than the back. The effect may be easily produced by viewing a normal stereoscopic pair with the left and right hand pictures transposed.

PSYCHICAL RESEARCH. Apart from the fraudulent practitioners of so-called "spirit photography", many investigators of psychic phenomena have used the camera in the effort to produce tangible evidence of the powers claimed for or by spiritualist mediums.

It is authentically reported by Sir William Crookes, that "in the full blaze of the electric light" Katie (a spirit form) and Miss Cook (a medium) were seen together by eight persons in addition to himself. Forty-four photographs were taken and all of these showed marked differences between the medium and the apparition. To the arguments of trickery Sir William replied: "To imagine the Katie King of the last three years to be the result of imposture does more violence to one's reason and commonsense than to believe her to be what she herself affirms."

Some of these photographs are still extant and show Sir William arm in arm with the fair apparition. The sceptics alleged that Katie and Miss Cook were one person and the photographic evidence serves but to confuse the issue.

It is now some eighty years ago since Sir William commenced his investigations into psychic phenomena and the general verdict of experienced investigators is "not proven".

During the years 1908 to 1918, flashlight (magnesium powder) photography was employed in France and Germany to a considerable extent in the study of the mediumship of a young lady who called herself Eva C.

Many experiments were conducted by Mme. Bisson and Baron von Schrenck Notzing, a practising physician in Munich.

The photographic exposures during the sittings were usually carried out by Schrenck Notzing. The medium occupied a curtained corner of the seance room and one or more cameras were trained on the curtain. The room was illuminated by dark ruby light only. The lenses were left uncapped in this light and remained open during the whole of the seance. The magnesium (about a tenth of an ounce) was ignited by mains electricity. It was only necessary to press the switch to make exposures of anything which appeared suitable to the experiment.

The value of the seances, which were recorded in *The Phenomena of Materialization*, rests mainly on the evidence thus obtained. The photographs reveal phantoms in varying stages of development but these tended rather to discount the evidence for their genuineness.

In 1920, Mme. Bisson and Eva C. spent two months in London. Out of forty seances arranged by the Society for Psychical Research more than half were complete blanks and the remainder gave very meagre results. A further fifteen sittings with Eva C. took place at the Sorbonne in 1922.

Thirteen of these sittings were entirely blank and the investigating Committee returned a negative report.

Infra-red illumination would appear to be particularly suitable for psychic investigation. It has long been claimed that white light and psychical phenomena are incompatible. Dark ness or the use of a deep ruby light is considered essential. It would appear, therefore, that infrared radiation would in no way inhibit the production of photographic evidence. On the contrary, it should provide a strong weapon in support of the genuineness of these phenomena. It is unfortunate that infra-red photography has not been applied to any great extent to this end. This is probably due to the fact that for many years there has been little or nothing worthy of investigation. The physical phenomena of spiritualism—if they ever existed—would appear to be a thing of the past. F.B.

See also: *Spirit photography.*

PULFRICH, CARL, 1858–1928. German mathematician and physicist. Produced a stereoscopic rangefinder and devised a stereocomparator in 1901, both for the Zeiss works. Developed stereoscopic measuring methods for photogrammetry. Also devised the well-known refractometer which bears his name.

PULSE LIGHT. Continuous light source which provides an increased light intensity for short times at regular or irregular intervals by applying a high voltage pulse to the circuit feeding the lamp.

Both tungsten and xenon arc lamps can be used for pulse operation. In the former the pulse may be a capacitor discharge through the already burning lamp. This produces a short high intensity flash, but not long enough to damage the filament significantly. Accordingly such a lamp can be used for a large number of such pulse discharges.

Xenon arc lamps can have their light output boosted by short high-power pulses—for example from 500 watts to 5 or 10 kilowatts. The lamp will only stand a pulse duration of a few seconds at this high level, but this is sufficient for high-speed photography or cinematography, where this type of lamp finds its main application.

Pulse xenon lamps have also been used in motion picture projectors. There the lamp, pulsing at 72 cycles per second, can be synchronized with the film transport movement in the projector. This arrangement then needs no rotating shutter as the light itself is already intermittent.

Electronic flash tubes, including stroboscopic flashes are not pulse light sources, but discharges where the current supply is discontinuous.

See also: *Discharge lamp; Light sources; Stroboscopic flash.*

PUYO, E. J. C., 1857–1933. French Army officer until 1902, and amateur photographer from 1887 onward. With Robert Demachy, a leading spirit in the Photo-Club de Paris, cultivated the aesthetic movement in photography by means of gum prints and bromoil in the Impressionist style. A prominent member of the Linked Ring. In 1914, rejoined the service. Wrote a number of books on the aesthetics of photography.

PYRO. Developing agent, once the standard agent in plate developers. Very popular at one time, and still favoured by some photographers.

See also: *Pyrogallic acid.*

PYROCATECHIN. Catechol; ortho-dihydroxybenzene. Developing agent.

Formula and molecular weight: $C_6H_4(OH)_2$; 110.

Characteristics: White powder.

Solubility: Freely soluble in water at room temperature.

PYROGALLIC ACID. Pyro; pyrogallol; 1 : 2 : 3-trihydroxybenzene. Developing agent.

Formula and molecular weight: $C_6H_3(OH)_3$; 126.

Characteristics: White powder or needle crystals. Rapidly oxidized in alkaline or neutral solutions.

Solubility: Freely soluble in water at room temperature.

PYROMETRY Measurement of the temperature of a hot body by studying the nature of its radiation photographically. Usually based on the fact that the brightness of the light emitted by a body hot enough to be self-luminous increases with its temperature.

Body temperatures can also be measured by photography on infra-red film, by photographing the body coated with a heat-sensitive paint, by projecting an image of the body on heat sensitive phosphors, and by electronic scanning techniques.

See also: *Heat recording*.

PYROXYLIN. Nitrocellulose; soluble gun cotton. Used for making celluloid (nitrate) film base and collodion.

Formula and molecular weight: Not constant, depending on origin. A mixture of cellulose nitrates.

Characteristics: Harsh white cottony substance.

Solubility: Insoluble in water or alcohol. Soluble in equal parts of alcohol and ether, and in amyl acetate, acetone, and glacial acetic acid.

Q

QUANTUM. Smallest and basic indivisible unit of radiant energy. It is inversely proportional to the wavelength of the radiation.

QUANTUM EFFICIENCY. Effect of exposure on a photographic emulsion, expressed in terms of the number of light quanta required to render a silver halide grain developable. With present day emulsions the quantum efficiency is about 2½ per cent (1/40), i.e., forty quanta are necessary to produce the slightest trace of developable image.

QUANTUM EFFICIENCY, EQUIVALENT. The scientist's definition of the speed-to-grain ratio. Like photographic granularity, light itself has associated fluctuations due to the quantum nature of radiation. The size of the fluctuations due to granularity can therefore never be less than the limit imposed by the exposing light. This limiting value corresponds to 100 per cent efficiency, and determines the maximum theoretical speed-to-grain ratio.

See also: *Information in the image; Sensitivity.*

QUARTER PLATE. Standard format for negatives and prints, measuring 3¼ × 4¼ ins. A quarter plate is exactly a quarter of a whole plate (6½ × 8½ ins.). Introduced by N. P. Lerebours in 1840.

See also: *Sizes and packings.*

QUARTZ-IODINE LAMP. Tungsten filament enclosed in a narrow quartz (or sometimes glass) tube containing also a trace of iodine. This inhibits the deposition of vapourized tungsten during the burning of the lamp. Quartz-iodine lamps can therefore burn at higher temperatures (and hence higher light efficiency) and have a longer burning life.

Both the colour temperature and the light intensity of these lamps remain constant to the end.

See also: *Filament lamp.*

QUICK-CHANGE BACK. Attachment for plate cameras (particularly those used by the press) which allows the plate holder to be changed quickly. Generally the plate holder is held in place by hinged grooves with a quick-release catch; this allows the plateholder to be lifted straight out instead of being slid out, which takes longer.

Special types of quick-change back are used for making colour separation negatives in the camera by successive exposures. In some designs the three plates are held in a long sliding attachment which moves across the camera back in three stages. Each section may incorporate a tricolour filter, or the back may be used in conjunction with a filter slide or filter turret over the lens. With such a device the three exposures can be made in rapid succession.

The movement may be hand-operated, or worked by a motor. This type of quick-change back is also known as a repeating back.

See also: *Repeating back.*

QUICK-FIRE CAMERA. Camera in which the operations of advancing the film and setting the shutter are coupled to a single control that is either operated by clockwork or by a trigger or lever. Some cameras of this kind incorporate an electric transport motor.

Cameras of the quick-fire type are useful for securing series of pictures of fairly fast movement—e.g., at athletics events—and for press photography where they give the picture editor a selection of photographs instead of one which may or may not have been taken at the best instant.

They are also ideal for taking self-contained picture sequences.

See also: *Film transport; Miniature camera.*

QUINOL. Hydroquinone. One of the most commonly used developing agents, usually combined with metol.

R

RACK AND PINION FOCUSING. Focusing mechanism used on most stand—and the larger hand—cameras. The lens panel is fixed to a slide fitted in guides on the baseboard of the camera.

A focusing wheel or knob at the side of the baseboard is joined to a pinion engaging with a rack on the slide. Turning the focusing wheel slides the lens panel along the baseboard to focus the lens.

See also: *Focusing mechanism.*

RADIATION MEASUREMENT. Applied to the measurement of X-rays, gamma rays and emissions from radio-active substances. Covers two kinds of measurement: intensity (determined by the ionizing effect of such radiations, and detected by an instrument like a Geiger counter); and measurement of dosage to check the exposure of human operators to such radiations. This is done by measuring the density produced on a photographic film carried by the operator while exposed to such radiations.

See also: *Dosimetry.*

RADIATION MONITORING FILM. Film sensitive to X-rays and other types of radiation, packed in a light-tight cassette and worn on the person by workers in radiographic and nuclear establishments. After a specified period the film is given a strictly controlled development and the density of the silver deposit is measured with a densitometer. From this figure it is possible to estimate the radiation dosage received by the film and to switch the wearer to other duties away from the risk of further exposure if the amount should be above an agreed safety level. Where the radiation level varies within wide limits, a double coated film may be used. This has a fast emulsion which responds to low radiation levels, and (on the other side of the support), a slow emulsion for the higher levels. When the radiation dose has been very heavy the fast emulsion is stripped off and the density of the slow emulsion measured. If the slow emulsion is affected only slightly—i.e. when the dose has been light—it is stripped off and the density of the fast emulsion measured. The sensitivities of the two emulsions overlap to give a continuous range.

See also: *Dosimetry.*

RADIOGRAPHY. Branch of photography in which an image is formed on a film or plate by exposure to X-rays. An opaque object—e.g., part of the human body or a metal casting—is placed between the source of the X-rays and the sensitized material; the resulting radiograph shows details of the internal structure which would otherwise be invisible.

Radiographs are widely used throughout industry for the inspection of raw materials, structures and finished products and in medicine and surgery.

See also: *Gamma radiography; Industrial radiography; Medical radiography; Mass miniature radiography; Microradiography; X-ray crystal analysis.*

RAIN. Rain in photographs is nearly always suggested by reflections in wet surfaces, puddles, and open umbrellas, and not by the presence of actual raindrops.

Raindrops in the Picture. Raindrops reproduce best when they are illuminated by the rays of the sun and seen against a background of dark cloud, trees, or buildings in shadow. This means that they must be photographed against the light and that the camera must have a good lens hood. No matter how short the exposure, the drops will register as streaks of light and not as points. And as the subject is being photographed against the light, it will call for a comparatively long exposure, so the streaks will generally be so long that they will tend to obliterate the subject. This is why it is unwise to include the actual raindrops in the picture.

One way, however, in which raindrops not only show up but also materially contribute to the pictorial content, is in shots through rain-washed windows. There are two possibilities: photographing from the outside a subject close to the window pane, and taking the outdoor view from inside. In either case, the window may be only spattered with raindrops, or show the water streaming down on the glass. (Both effects can of course, be produced artificially with a bucket of water or a garden spray.) When shooting out of the window from indoors, the outside view will be blurred by the water; the drops themselves should therefore be brought out sharp by accurate focusing. Often a foreground object will add strength to the picture; neither this nor the window frame must be too dark, and some fill-in lighting is desirable.

When photographing from the outside, a suitable subject behind the window is essential, and it should be as close to the glass as possible. This type of approach has been popular for pictures of children, etc.

Lighting. Showery weather offers best conditions for pictures during or after rain. There is more chance of sunshine and rain occurring simultaneously, thus lighting up the falling raindrops if you wish to record them. Side-lighting or oblique back-lighting shows them to best advantage, particularly when they are striking the ground with force.

Heavy thundery showers during hot weather may produce after-mist and haze which gives plane-separation and "atmosphere", otherwise the air is usually clearer and sunlight harsher after rain, with distant scenes more clearly visible.

There are occasions, however, when the dull monotony of a completely overcast scene expresses the mood that the photographer wants to create. This type of scene calls for a different technique, aimed at avoiding hard contrasts and leaning towards mistiness and soft definition.

Photographs taken on dull rainy days need to be processed so that the contrasts of the subject are retained or even increased. This can be done by extending the development time (up to 50 per cent longer than normal), or by using a suitably contrasty grade of printing paper.

Rain often improves photographs of street scenes taken after dark. When such pictures are taken in dry weather, the foreground is usually a featureless area of black paper. When the streets are wet the foreground becomes filled with reflections and it may even form the principal interest of the picture.

Rainy Reflections. The general impression of a rainy day is strengthened when there are pools and shining wet surfaces in the picture. At such times the reflections themselves may be interesting enough to form the subject. On the other hand, a pool that gives a perfect reflection is bad because it simply duplicates the scene and presents the eye with two identical pictures.

A photograph that looks the same when it is turned upside down is a freak and nothing more.

One way out of the difficulty is to throw a stone into the pool and take the photograph while the surface is still disturbed. Or a low viewpoint can be adopted to reduce the proportion of the picture space occupied by the reflecting surface. In this way the reflection is subordinated and the subject receives its due importance.

If the reflection of the scene is to form the item of principal interest, then the lens must be focused on the reflected image and not on the reflecting surface. The image is as far away from the camera as the subject, no matter how close the reflecting surface may be. If objects on the surface of the water—reeds, waterfowl, floating debris—are to be included, then the lens must be stopped down until the depth of field extends from the reflecting surface to the image.

Sometimes the reflections are not wanted; if they cannot be got rid of by a change of viewpoint, they can be minimized or even cut out altogether by using a polarizing filter. The effect of the filter on reflection can be checked beforehand by visual examination.

Exposure. Exposure for wet objects and scenes should never be greater than that indicated by a meter reading on a medium tone area. Medium or slow films are best for sunless conditions, as fast emulsions are too soft in gradation to suit a subject which is already lacking in contrast. In sunshine, particularly backlighting, the reverse is the case. No filter is needed unless there are blue sky patches present. A lens hood is essential to prevent stray rays from sparkling surfaces striking the lens and causing flare.

For the moody, atmospheric type of rainy landscape, exposure needs to be full enough to register all the shadow detail and flatten the tones of the picture generally. Development should be continued for the full time or—if the flatness is to be exaggerated—a little longer.

The brilliant type of treatment looks best on a glossy or glazed surface, while the atmospheric effect calls for a semi-matt surface.

Protecting the Equipment. No camera should be allowed to get wet. An umbrella held by a companion will prevent this, and the lens hood helps to stop stray drops falling on the lens. If the camera does accidentally get wet it should be dried immediately. The metal parts require particular attention. The lens should be wiped gently with a soft handkerchief or other soft, absorbent material.

Do not hasten drying by heat, especially if the camera is a folding bellows type.

If there is a risk of being caught far away from shelter in stormy weather, it is a good idea to carry a square of plastic material of the kind sold for bathroom curtains. The camera in its case can then be wrapped up and kept perfectly dry in the heaviest downpour. **P.J.**

RANGEFINDER

When a camera has a focusing screen, there is no need for the photographer to know how far he is away from his subject; he simply gets it sharp on the screen and makes his exposure. But as soon as the focusing screen is dispensed with and the camera is focused by a scale, the distance from the camera to the subject must be known. Estimation is usually unreliable, and measurement can be a nuisance; a third alternative is to use a rangefinder.

Principle. A rangefinder enables the photographer to measure the distance to his subject without moving from his position. There are several types of rangefinder, but they all rely on the same basic principle, that when an object is looked at from two separated positions the lines of sight converge. The angle at which the lines converge increases as the object approaches, and decreases as it goes farther away.

Construction. In the type of optical rangefinder used in photography the two viewpoints are provided by windows at opposite ends of a metal tube which is held horizontal and at right angles to the camera-subject line. One of the windows lies on the camera-subject line; the eyepiece of the instrument is opposite this window. When the eye looks through the eyepiece it sees one picture of the subject directly through the window in that end of the rangefinder, and another picture, reflected by two mirrors or prisms, through the window in the other end. Sometimes the windows and mirrors are masked so that the eye sees the top half of one picture and the bottom half of the other. Sometimes, the eye looks through a half silvered mirror and sees the direct and the reflected pictures superimposed.

Included in the optical make-up of the rangefinder is an adjustable device that can be made to deflect the light rays of the reflected picture to bring it into register with the directly-viewed picture. This can be done by altering the angle of one of the mirrors, but it is more satisfactory to interrupt the beam of light by a rotating wedge system or a pair of prisms.

The deflection of the beam is controlled by turning a setting control on the instrument. The amount of rotation that has to be given to the wheel to keep the two pictures of the subject in register is related to the distance of the subject from the rangefinder.

Thus the wheel can be calibrated to show directly the distance of the subject when once the two images have been adjusted to coincide.

Accuracy. The two lines of sight converge more for objects near at hand than they do for those farther away. Because of this, rangefinders of this type are more accurate for objects up to, say, 25 feet than beyond. But as the accuracy required in focusing a normal lens falls off in exactly the same way, the defect is unimportant.

Since however accurate focusing is important with long focus lenses at greater distances, the value of a rangefinder is distinctly limited with lenses of longer focal length. In that case a reflex camera or a reflex attachment is more reliable as a means of getting the image really sharp.

Base Length. The optical accuracy of a rangefinder depends on the base length and on the scale of the image. The longer the base of the rangefinder (i.e., the distance between the two windows) the greater will be the convergence on any given subject, and the greater will be the separation between the two part images for a given focusing error.

The larger the scale of the image, the greater will again be the image separation. In both cases a given focusing error will show up more.

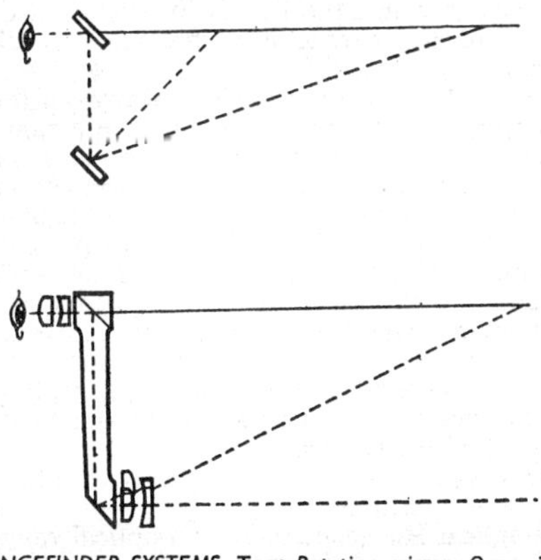

RANGEFINDER SYSTEMS. Top: *Rotating mirror. One mirror is pivoted to alter the angle of deflection and superimpose images at different distances.* Bottom: *Swinging wedge. Sideways movement of the wedge controls the angle of the beams.*

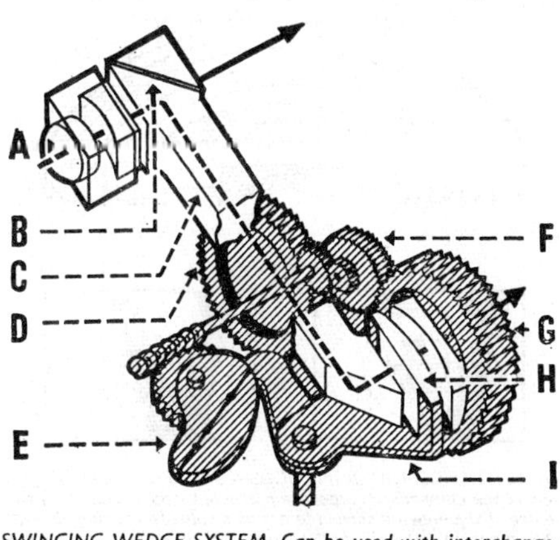

SWINGING WEDGE SYSTEM. Can be used with interchangeable lenses. A, eyepiece. B, semi-reflecting surface. C, prism block. D, lens mount gear. E, cam. F, focusing gear. G, focusing wheel. H, swinging wedge. I, cam lever.

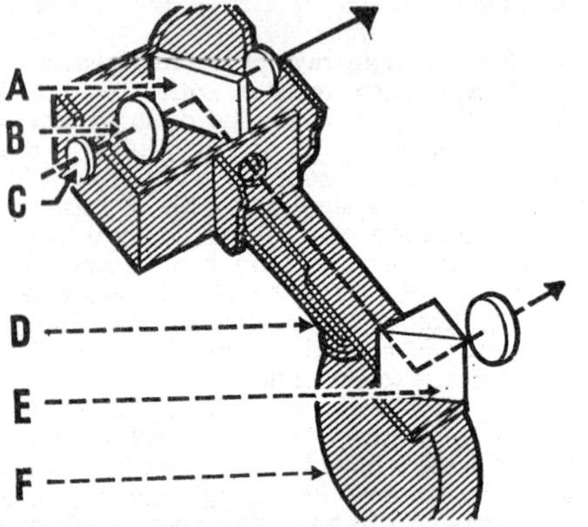

ROTATING MIRROR SYSTEM. *Suitable for interchangeable lens cameras. A, semi-reflecting surface. B, magnifying viewing system. C, eyepiece. D, coupling lever bearing on lens barrel. E, pivoted reflector. F, lens barrel.*

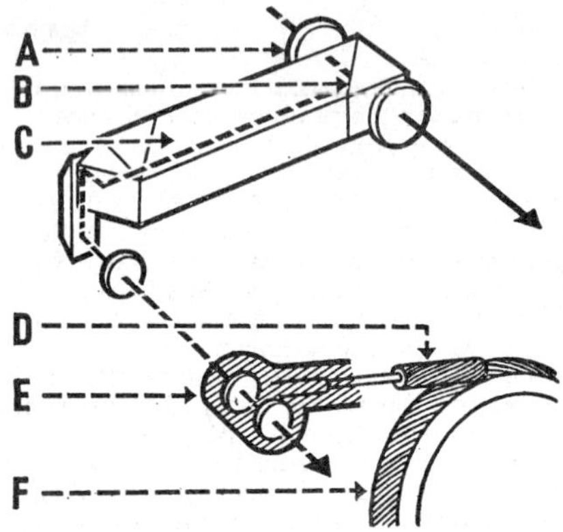

ROTATING WEDGE SYSTEM. *Generally used with non-interchangeable lens. A, viewing system with eyepiece. B, semi-reflecting surface. C, prism block. D, skew gear driving wedges. E, rotating wedges. F, focusing lens mount.*

Commercial camera rangefinders may either use as long a base as the camera body will accommodate, or go for a shorter base, coupled with a magnifier in the eyepiece to increase the image scale and so make up for the loss in accuracy.

A long base has the advantage of minimizing mechanical inaccuracies, but it has the drawback that at extreme settings one of the images may move out of the field of view altogether.

A short base with magnifier makes the instrument more compact but the optical system is more easily influenced by mounting errors.

Coupled Rangefinders. A coupled rangefinder is simply an adjustable optical system of the type described above which has been built into a camera so that the rangefinder and lens focusing work together. A single control operates both systems and when it has been adjusted to make the two rangefinder images of any object coincide, the lens is automatically focused on the object.

The principle can be applied to any type of camera, but it is particularly suited to the miniature where the lens mount is attached direct to the body. This arrangement makes it possible to house the rangefinder mechanism and the lens focusing mount in the same rigid casting. In one popular miniature of this type, the lens focuses in a helical mount and the rear face of the lens tube presses against the end of a spring-loaded lever which carries the tilting mirror on the moving or rotating wedge of the rangefinder system.

With rangefinder and lens focusing systems coupled in this way the lens can be focused without any reference to the actual measured distance from the camera to the subject. At the same time, it is useful to know the exact distance at which the lens is focused, and the system always carries a visible scale of distances.

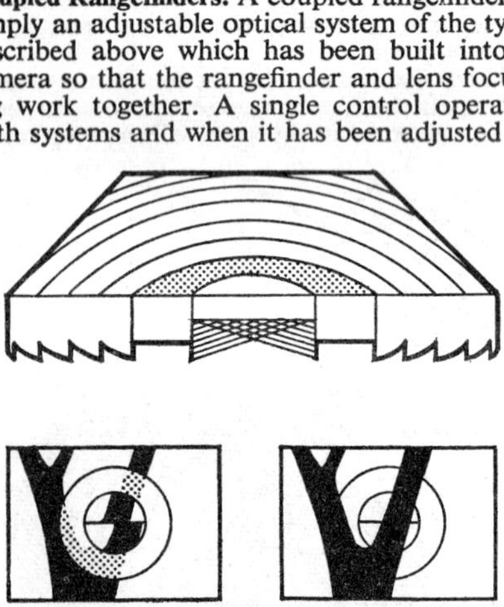

WEDGE RANGEFINDER IN FOCUSING SCREEN. *Many single-lens reflex cameras incorporate a crossed wedge system in the centre of the viewing screen to act as a split-image rangefinder. The image in the split section lines up to become continuous only when the picture is sharply focused. A grained ring surrounding the wedges may also act as a focusing screen for direct observation of the sharpness.*

Focusing Screen Rangefinder. Many current miniature reflex cameras incorporate a rangefinder system in the focusing screen. This consists of a pair of crossed wedges recessed into the screen so that the wedges cross in the plane of the focusing surface. The optical principle is similar to parallax focusing of an aerial image, but the parallax effect appears as a split image which joins up across a dividing line when the lens is set to maximum sharpness. As the rangefinder uses the screen image, there are no moving parts and it is automatically coupled with any lens.

This system works, however, only with fairly large lens apertures.

Non-optical Rangefinder. A non-optical rangefinder can be made from two pointers mounted on a baseboard. The object is sighted along the right hand pointer by the right eye and along the left hand pointer by the left eye. The whole arrangement is kept at a fixed distance from both eyes. The angle between the pointers is calibrated to give the distance of the object directly on a scale marked on the baseboard.

Rangefinders of this type are cumbersome and not very accurate and many people find it impossible to master them. L.A.M.

See also: *Focusing mechanism.*
Book: *Photographic Optics*, by A. Cox (London).

RAPID ACCESS SYSTEM

Photographic camera and processing system that produces a usable picture in the order of a few seconds (in some cases even in a fraction of a second) for immediate evaluation. While normal rapid processing is concerned merely with speeding up conventional processing procedures by concentrated heated solutions, rapid access systems are more integrated set-ups involving special processing techniques—usually applicators for the solution, suitably designed machinery, etc.

Cell and Chamber Processors. These permit the contact of small amounts of processing liquid with the emulsion and the film. A narrow space is provided between the film emulsion and the processing device. In a processor of this type the film, after being located over the sealing frame, is held tightly against the seal by means of a pressure plate. Through the centre bore of the processing platen, metered volumes of processing solutions are brought into the capillary chamber. The liquid could freely overflow the platen but is held within the chamber by

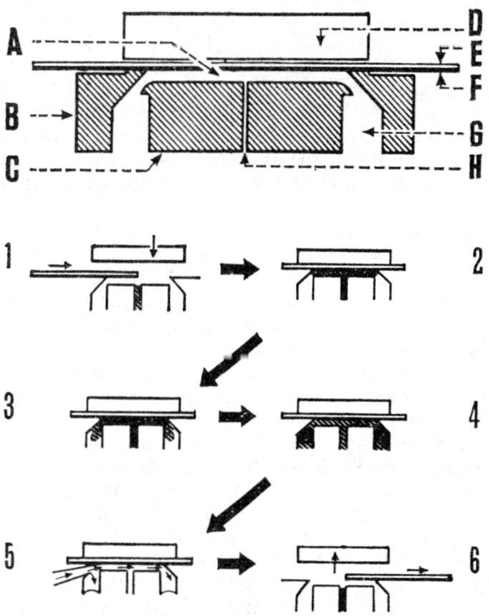

CAPILLARY CHAMBER PROCESSOR. Top: *Capillary chamber. A, chamber. B, sealing frame. C, heater block. D, pressure plate. E, film. F, emulsion. G, overflow. H, orifice.* Bottom: *Processing cycle. 1. Film moves into position at start of cycle. 2. Development. 3, Fixer introduced through the central orifice displaces the developer. 4. Washing water flushes out the fixer. 5. Drying the air blown into the capillary chamber. 6. The film is removed at the end of the cycle.*

capillary forces. The chamber dimensions are critical. Good processing results have been obtained using a chamber depth from 0·010 to 0·030 in. A nearly square frame of 1·16 by 1·06 in. is preferable; however, equal success has been attained with a more rectangular frame of 1·16 by 0·92 in.

The volume of processing liquid required to fill the chamber ranges from 0·3 to 0·5 c.cm., depending on the frame size. The figures given apply to the processing of a 35-mm. double frame. Since the processing solutions contact the emulsion side of the film only, this process is well suited to ultra-rapid processing.

After the development is completed, the fixing solution is introduced into the capillary chamber through the centre bore. This pushes the developer into the overflow and then into the waste container. Any intermixture of the two liquids is counteracted by the strength of the fixing solution, which not only rapidly stops the development but also completes the fixation of the image in the presence of traces of the developing solution. A wash follows the fixer, and a hot, high-velocity air stream which enters the chamber from the side and is directed evenly over the film surface dries the emulsion; drying air, which escapes through the waste spacing into the vented waste container, cleans and dries the chamber as well as the film.

The emulsion of the film and the desired image characteristics determine the selection of processing solutions. The capillary chamber system is flexible, for any wet process can be used. Monobath processing yields acceptable results, although for image requirements of high contrast combined with low fog level development and fixation are best separated. Emulsions especially designed by the manufacturers for rapid processing applications give best results.

A typical processing cycle with a capillary chamber may take 1½ seconds for development, ½ second for fixing, 1 second for washing, and 2 seconds for drying.

The processing platen, which is part of the heater block, contains heating elements and a temperature-sensing probe. The pre-selected processing temperature is automatically controlled and is maintained within ± 0·5°F. The processing temperature can be set for any point from ambient to 130°F. Although high-temperature processing is not essential for ultra-rapid processing, elevated processing temperatures can be maintained more easily in varying ambient conditions.

Another developing chamber with a cavity of

approximately 0·005 in. can deliver recorded photographic data 1/5 second after exposure. The 35 mm. film itself makes a seal with the lips of the chamber during processing. Pre-heated developer at 170 F. is sprayed on the film through a narrow slit and withdrawn through a second slit by a vacuum pump. After a 0·6 second development time hot air saturated with acetic acid vapour dries the film, neutralizes the alkaline developing solution and stabilizes the image. The stabilization, however, is not complete and a yellow light is used for projection to prevent fading of the image. The drying and stabilizing operation takes a little less than 1 second so that the image can be projected approximately 1·5 seconds after exposure.

Viscous Layer Applicators. In selecting a processing method for a machine to be designed for both rapid and simplified processing, the viscous technique has found commercial application on a large scale. A typical machine processes 16 mm. black-and-white positive films at 36 feet/minute, with a dry-to-dry time of 1 minute. The developer and fixing solutions are each coated on the emulsion surface as a viscous layer in an atmosphere saturated with water vapour at 125°F. The viscous processing solutions are packaged in disposable, polythene bags which can be changed without interrupting the processing operation.

The film enters the processing chamber and passes round the coating roller. A coating hopper (a special kind of nozzle) applies a 0·008 in. layer of viscous developer to the emulsion surface. Raising or lowering the roller adjusts the time the developer remains on the film (from 2½ to 7 seconds), to select a specified contrast. The developer coating is removed by a high-velocity water spray jet, and the surface water on the film by an air squeegee. The film then passes round a second coating roller where the viscous fix is applied by another coating hopper, and removed (after 12 seconds) by a second spray jet. The film then passes into the wash compartment for a spray wash of 13 to 17 seconds. After passing through a second air squeegee, the film goes into the impingement air dryer for 15 to 21 seconds.

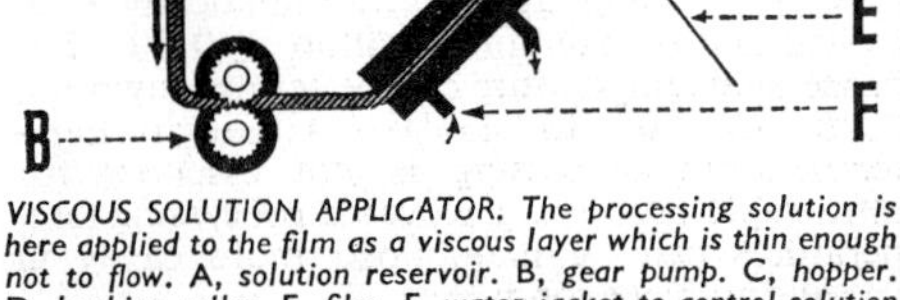

VISCOUS SOLUTION APPLICATOR. The processing solution is here applied to the film as a viscous layer which is thin enough not to flow. A, solution reservoir. B, gear pump. C, hopper. D, backing roller. E, film. F, water jacket to control solution temperature.

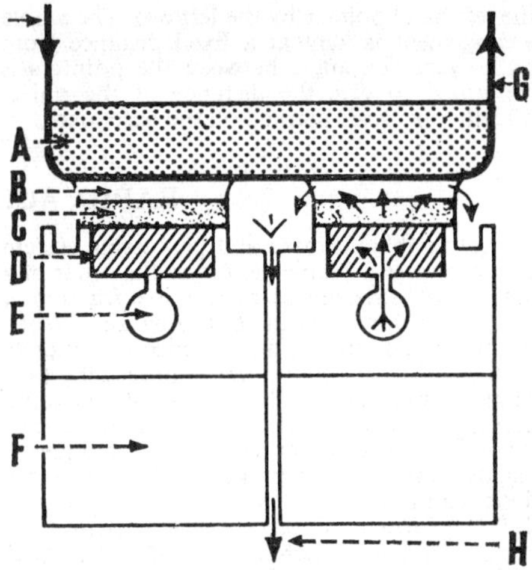

POROUS PLATE APPLICATOR. The processing solutions emerge through a porous plate and form a meniscus in contact with the film passing within a few thousands of an inch of the porous surface. Several applicators may be arranged in series for different processing steps. A, backing plate. B, meniscus of solution. C, porous plate. D, wick. E, solution feed. F, thermostatic heater block. G, film. H, waste.

Porous Plate and Roller Applicators. Rapid access processors based on the application of thin layers of solution to the emulsion surface by means of porous rollers or plates provide another technique. Here thin porous plates, made from sintered stainless steel granules, and of the width of the exposed area on the film are in contact with the film surface. The plates are fastened to a manifold block which contains a feed pipe and a cavity with a wick to disperse the solution flow uniformly through the porous material.

The solution is pumped through each plate as the film moves past it, and different plates in a row can apply different solutions. Cartridge heaters maintain the manifold block—and hence the solutions coming into contact with the film—at the right temperature.

An applicator relying on a meniscus of solution formed between a porous roller and the film emulsion consists of a hollow stainless steel tube plugged at both ends. Small holes are drilled radially through the roller, and die-cut felt washers slipped on the tube and retained by axial pressure of a collar. The processing solution, preferably a monobath, is brought first to the required temperature and then forced under pressure into the applicator tube through small holes in the bearings that coincide with similar holes in the applicator tube. From there it flows out through the felt washers to contact the emulsion.

Jet Spray and Slot Processors. The Kelvin-Hughes system of rapid processing uses a Venturi-type jet spray system operated by compressed air. Atomized droplets of the various solutions are sprayed on to the emulsion at high velocity and at an economical rate. As this is a "total loss system", results are constant and no replenisher system is required.

Jet spray techniques raise condensation problems in rapid access processors carried in aircraft. Slot processors are therefore used in such cases, consisting of a bar of material with three accurately machined slots. Two of these slots carry the processing fluids and, thus have an inlet and outlet feed and return. The centre slot separates the developing slot from the fixing slot by ensuring an area of ambient pressure between these. Each slot may be regarded in itself as an open weir, which, when placed close to the emulsion surface of the film, forms a closed channel through which the processing fluid is drawn in contact with the emulsion surface. Since this fluid is drawn through the channel by negative pressure, it becomes a leak-proof system, which is consistent with airborne requirements.

The film is transported past the slot where it is fully developed in 2 seconds at 40°C.; it then passes a narrow ambient air channel, after which it is fixed via a slot similar to the developer channel in a similar time. Thus, the record is available with an access time of only 5 seconds. This applicator is arranged transversely to the film motion and is so aligned that there is a small air cushion between the emulsion and slot surfaces, thus ensuring turbulent liquid flow in the slot channel and protecting the emulsion.

Roller Applicators. Processing solutions can also be applied to the emulsion surface by non-porous rollers or wheels. Each roller dips into a tank with the appropriate solution to carry the latter into contact with the film as the latter touches the top of the roller. The rollers are driven in a contrary direction to the movement of the film. Such processors permit continuous rapid processing at rates up to 100 feet per minute of cine film.

An alternative arrangement has the lower portion of the rotating drum suspended in the solution, while the drum rotates at 300 r.p.m., in this case, however, in the direction of film travel. A comparatively thick layer of solution is carried round the upper circumference of the drum over which the film is positioned by two small guide rollers. Normally two drums are used for the developer, and fixer respectively, after which the film passes over a spacer roller into a washing tank and finally past a squeegee into a simple drier. The speed of the operation is variable for treatment from 1–10 seconds.

Saturated Web Applicators. Saturated web processing involves the use of a suitable absorbent material acting as the intermediate carrier for the processing solution. Advantages claimed are that the volume of solution used is very small (approximately equal to the volume of the material to be processed), that each area of emulsion is processed with fresh solution, and

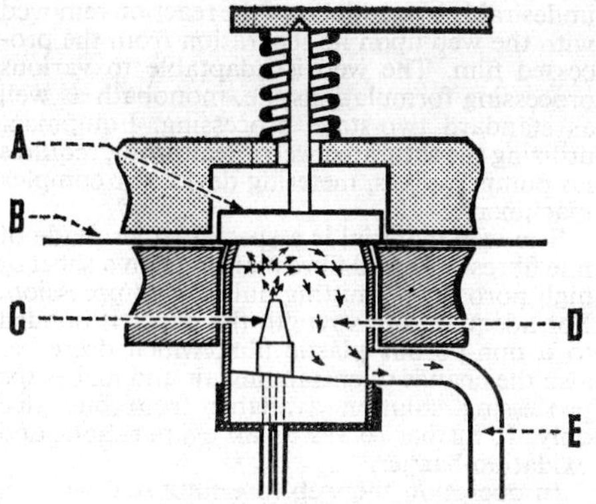

JET SPRAY PROCESSOR. A, pressure pad. B, film. C, jet introducing chemical solutions and air at different stages. D, jet chamber. E, waste flow.

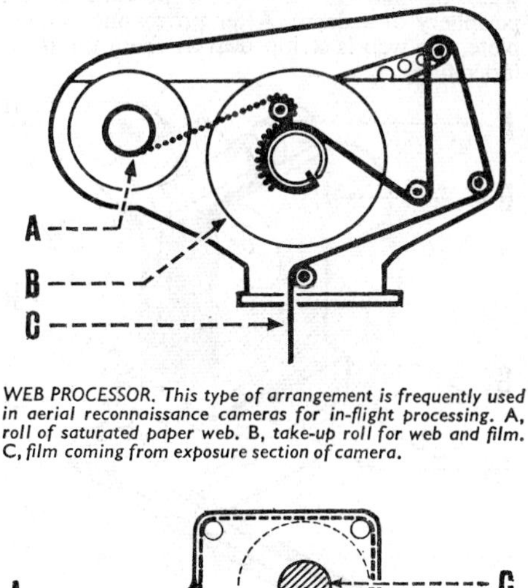

WEB PROCESSOR. This type of arrangement is frequently used in aerial reconnaissance cameras for in-flight processing. A, roll of saturated paper web. B, take-up roll for web and film. C, film coming from exposure section of camera.

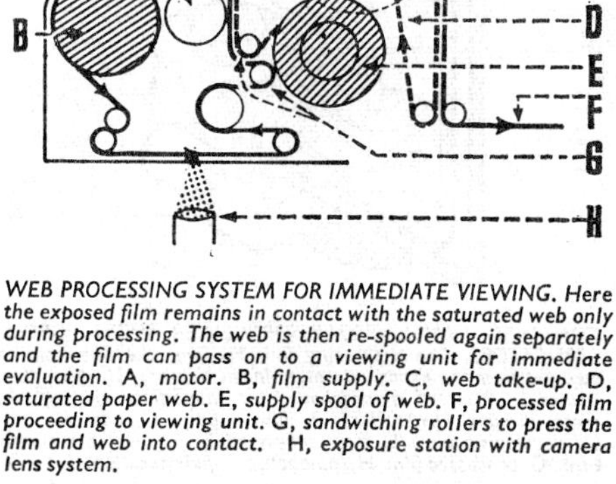

WEB PROCESSING SYSTEM FOR IMMEDIATE VIEWING. Here the exposed film remains in contact with the saturated web only during processing. The web is then re-spooled again separately and the film can pass on to a viewing unit for immediate evaluation. A, motor. B, film supply. C, web take-up. D, saturated paper web. E, supply spool of web. F, processed film proceeding to viewing unit. G, sandwiching rollers to press the film and web into contact. H, exposure station with camera lens system.

undesirable by-products of the reaction removed with the web upon its separation from the processed film. The web is adaptable to various processing formulations, i.e. monobath as well as standard two-stage processing. Equipment utilizing the saturated web is leakproof, requires no pumps, valves, metering devices or complex machinery.

The web material is a special paper made of fine fibres (to avoid fibre markings) in a sheet of high porosity permitting uniform compression. For adequate wet strength the paper is bonded to a non-porous plastic film, which decreases also the chance of entrapping air and makes the processing solution available from one side only. It further serves as an evaporation- and oxidation-barrier.

In operation the web (pre-saturated with an aqueous processing solution such as a monobath) is brought into intimate contact with the emulsion surface of the film by winding the surfaces together in a roll or tensioning the sandwich of material around a portion of the periphery of a drum. After processing is complete, the web is stripped away from the developed film.

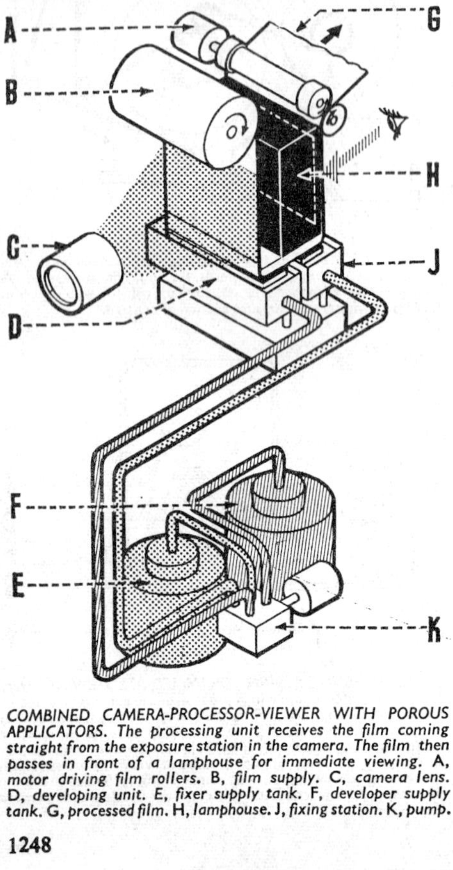

COMBINED CAMERA-PROCESSOR-VIEWER WITH POROUS APPLICATORS. The processing unit receives the film coming straight from the exposure station in the camera. The film then passes in front of a lamphouse for immediate viewing. A, motor driving film rollers. B, film supply. C, camera lens. D, developing unit. E, fixer supply tank. F, developer supply tank. G, processed film. H, lamphouse. J, fixing station. K, pump.

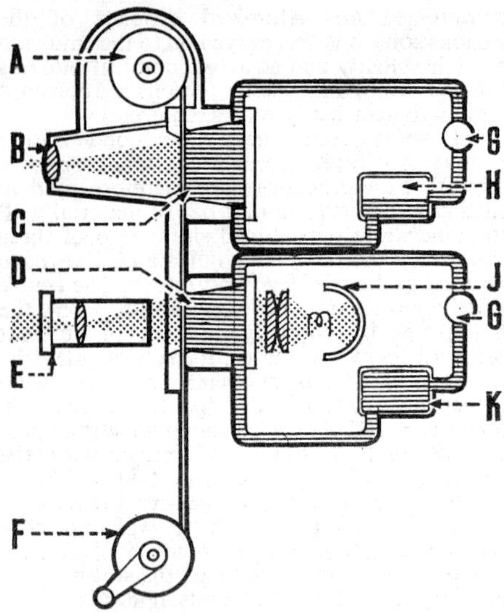

COMBINED CAMERA PROCESSOR AND PROJECTOR. The film is exposed through the base while it is in contact with a developer in a solution-tight chamber. From there it passes to a fixing chamber through which the projector lamp projects the image on to a screen. Pumps circulate the solutions through the two chambers. A, film supply. B, camera lens and system. C, developing chamber. D, fixing chamber. E, projection lens. F, film take-up. G, solution pumps. H, developer reservoir. J, projection optics. K, fixer reservoir.

Diffusion-Transfer. The diffusion-transfer process, as exemplified by the instant-picture camera, is also a rapid access system in the true sense. Here processing takes place inside the camera by spreading an activating reagent—carried in pods on the positive material, between the negative and positive layers of a picture roll or sheet. This yields black-and-white positive prints within as little as 10 seconds, or colour prints in about a minute after exposure. For document copying exposed negative sheets are fed through a roller processing machine in contact with a positive sheet and the sandwich then stripped apart.

Combined Camera-Processor-Viewers. In many cases it is important to shorten the time between photographic recording of information and its presentation for visual inspection either in a permanent or temporary form. This can be achieved with combined camera-processor-viewers or -projectors.

A typical system for projection of the photographic data recording within a fraction of a second after exposure takes a photograph from a cathode-ray presentation, processes it automatically and then projects it immediately. The machine uses two-solution processing to develop the film within 0·2 seconds at 185°F. and to fix it in approximately 0·2 to 0·3 seconds at 205°F. A similar device for radar recording employs a porous applicator which is inserted in a narrow space between a camera portion and the viewing

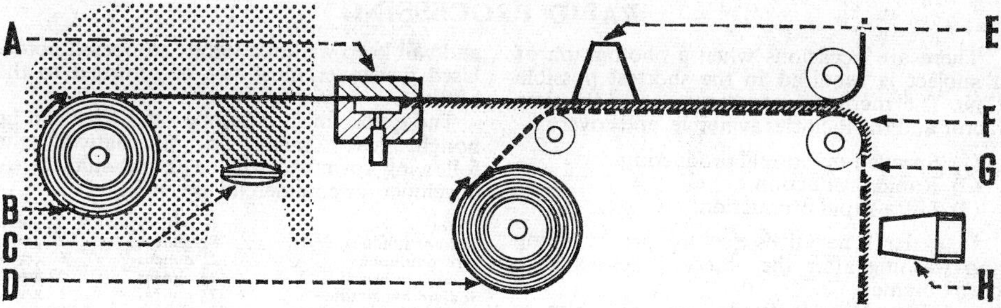

COMBINED CAMERA, PROCESSOR, PRINTER AND VIEWER. A, processing solution applicator. B, negative film roll. C, camera lens of exposure station. D, positive film roll. E, printing light; exposes developed negative image onto positive film. F, processing solution layer. G, positive image. H, viewer system.

station using a U-shaped film path in a relatively small piece of equipment. A machine has also been designed which provides both negative and positive in a rapid processing operation. A sandwich is formed between the exposed negative and a positive film with a monobath processing solution in between. This develops the negative which is immediately printed on the positive material and thus developed by the same monobath.

The processing solution develops and fixes the positive image almost immediately, after which the positive and negative films are separated. Either of them can then be used in the usual way.

Processing Chemistry. For rapid access processing, highly active developers are required making full use of all means of accelerating the developing process, such as high general concentration, pH and temperature. In cases where temperatures considerably exceed normal working temperatures special quick-finish (rapid access) films have to be used which stand temperatures up to say approximately 200°F. (93°C.). The emulsions of these films are specially hardened and also thinly coated. A typical formula for a highly active developer is the following:

Sodium sulphite (anhydrous	6 ounces	150 grams
Potassium carbonate (anhydrous)	4 ounces	100 grams
Hydroquinone	2 ounces	50 grams
Phenidone	20 grains	1·1 grams
Sodium hydroxide	175 grains	10 grams
Potassium bromide	280 grains	16 grams
Benzotriazole	20 grains	1.1 grams
Water to make	40 ounces	1000 c.cm.

This stock solution is normally diluted for use with equal parts of water.

Where machines use monobaths (combined developing and fixing solutions) these are usually based on the accelerating effect of phenidone in hydroquinone developers leading to developers of increased activity compared with the MQ type.

A formula for a suitable monobath as follows:

Sodium sulphite (anhydrous)	2½ ounces	60 grams
Hydroquinone	1⅛ ounces	30 grams
Sodium hydroxide	1 ounce	25 grams
Phenidone	55–90 grains	3·2–5 grams
Sodium thiosulphate (crystals)	3–10 ounces	75–250 grams
Water to make	40 ounces	1000 c.cm.
Formaldehyde (38%)	200 minims	10 c.cm

With two-solution techniques the fixing formula is usually based on ammonium thiosulphate.

Where the permanence of the photographic record is of less importance than utmost speed of processing, stabilization techniques are applied. Here a stabilizer renders the undeveloped silver halide relatively inert to heat, light or moisture.

The main stabilizing agents used for this purpose are ammonium thiocyanate in approximately 20 per cent solution or thiourea in about 3 per cent solution. Prints stabilized in this way are not washed. C.I.J.

See also: *Cine film processing; Diffusion transfer; Processing machines; Rapid processing.*

Books: *Processing Chemistry* by L. F. A. Mason (London); *Developing*, by C. I. Jacobson (London).

RAPID CASSETTE. Name adopted jointly by a number of international camera and film manufacturers for a quick-loading 35 mm. film cassette.

The cassette contains a loose film length sufficient for 12 exposures 24 × 36 mm. (or larger numbers of smaller formats down to 18 × 24 mm.) and is inserted in the camera without any film threading. The film transport of the camera automatically feeds the film into a take-up cassette which is removed when all the frames are exposed. Rapid cassettes must be used in cameras specially designed to take them.

Models of this type have appropriate film chambers and transport systems, and need no film rewind.

See also: *Cassette.*

RAPID PROCESSING

There are occasions when a photograph of a subject is required in the shortest possible time. The methods vary according to the time factor and the facilities available, and cover:

(1) Speeded-up normal production.
(2) Rapid production.
(3) Ultra-rapid production.

In general negatives require proper fixing and washing after the necessary results have been obtained.

Speeded-up Normal Production. The film is developed as laid down by the manufacturers of the developer in use, the negative is fixed in an ammonium thiosulphate rapid fixer until it has cleared, and then given a short rinse in running water or quick changes of water. Agitation of the negative material reduces the time of fixing and washing; a further shortening of the washing operation can be made by a preliminary washing in 0·01 per cent potassium permanganate solutions until the pink colour is no longer discharged, after which only a short wash is required.

For contact printing, any surplus water is removed from the negative by gentle swabbing off with a grit-free piece of chamois leather dipped in water and wrung dry. This is followed by soaking for 2–3 minutes in a solution of industrial methylated spirit with 20 per cent water added, after which the negative is waved gently in the air until thoroughly dry, the negative is then printed in the normal way, the print developed, fixed in a rapid fixer, rinsed for 1 minute in running water, and soaked in methylated spirit. After blotting off with fluffless photographic blotting paper, the print may be waved in the air until dry; or with a little practice a nimble fingered person can set fire to the wet print (without blotting off) thus burning the methylated spirit, being careful not to allow the print itself to burn.

For enlarging with a negative carrier of the glass sandwich type, the method is the same as for contact printing.

With a glassless negative carrier the negative can be enlarged whilst wet. To avoid the formation of droplets of water on the negative (giving unsightly marks on the enlargement) the final rinse of the negative should contain a little wetting agent. This rinse need only be a minute or so, and after draining off the surplus moisture the negative is placed direct into the enlarger, for the normal enlarging procedure.

Rapid Production. In some branches of photography (press work, medical radiography, etc.) it may be necessary to produce photographs in a shorter time than possible by the above methods. For this purpose special developers (in addition to the rapid fixer) are available. Where convenient they may be used for the rapid processing of prints as well as negatives.

The developing solution is reasonably stable and will keep well in a tightly stoppered bottle. Used developer must not be returned to the bottle.

The quick finish developer and fixer can be bought ready made up; alternatively the following formulae will be found suitable. A developer for commercial use is:

Metol	1¼ ounces	32 grams
Sodium sulphite, cryst.	8¾ ounces	220 grams
Hydroquinone	2½ ounces	62 grams
Potassium bromide	2½ ounces	62 grams
Sodium hydroxide	2½ ounces	62 grams
Water (distilled)	80 ounces	2,000 c.cm.

The chemicals are dissolved in warm water. The developer is used undiluted for negatives, development times being 15–30 seconds at 70° F. (21° C.). For papers, dilution of 1 : 1 will give times of 10–15 seconds.

The fixer may be one of the proprietary quick fixers or the following can be used for bulk processing:

Hypo	64 ounces	1,600 grams
Sodium sulphite, cryst.	1½ ounces	38 grams
Potash alum	1 ounce	25 grams
Acetic acid, glacial	1½ ounces	38 c.cm.
Water	80 ounces	2,000 c.cm.

Agitation during the whole process must be vigorous and continuous to avoid development marks and stained prints. Dish development is therefore preferable.

Because of the short time of processing it is advisable to add a few drops of wetting agent to the solutions thus ensuring even development and preventing the formation of clear spots due to air bells.

Negative materials and paper are best transferred without rinsing direct from the developing bath to the fixer hardening bath.

Fixing times at 70° F. (21° C.) are 40–120 seconds, after which the negative is ready for inspection or for printing whilst still wet in the enlarger. If permanence is required then fixing must be increased to twice the time taken for the emulsion to clear, or the negative must be refixed at a later date and thoroughly washed. With ammonium thiosulphate fixers the above times ensure complete fixation.

Bromide paper will fix in 10 seconds at 70° F. (21° C.) in a rapid fixer. After a short wash the prints can be dried by the methods already described.

Ultra Rapid Production. In some specialized jobs it is desirable to increase the speed of production even further. An example of this is the use of photography in race finish recording in which a method has been evolved to produce an 8 × 10 ins. enlargement within 90 seconds of the last competitor crossing the finishing line. To attain this speed the developing and fixing formulae mentioned above can be used, but specialized materials and chemicals have been manufactured for this type of photography.

Quick finish films have an emulsion which is

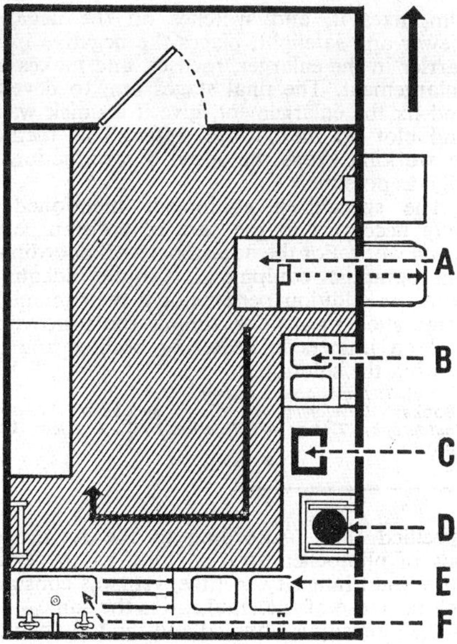

RACE-FINISH LABORATORY. Operating and processing room set up alongside race track for immediate photo-finish prints (short arrow indicates race direction, inside arrow routing of photos). A, camera platform. B, processing unit. C, negative viewer. D, enlarger. E, print processing bench with developer and fixer dishes. F, sink and draining board.

resistant to high temperature processing, and with special processing solutions such films can be developed and fixed in a matter of 30 seconds.

Producing a Photo Finish Print. This procedure is possible only with special pre-hardened film. The film is exposed in a race finish recording camera and varies in length according to the time that elapses between the first and last competitor crossing the finishing line; an average is between 12 and 32 ins. It is developed by hand in a dish of quick finish developer at a temperature of 110° F. (43° C.) for 12 seconds. Without a rinse the negative is then transferred to a dish of proprietary rapid fixer at a temperature of 110° F. (43° C.) for 10 seconds, keeping the negative continually agitated. The fixing will then be complete enough to allow visual inspection and enlargement whilst still wet. Owing to the softness of the emulsion great care must be taken not to touch it, but just allow the moisture to drain off. An 8 × 10 ins. enlargement is then made in the normal way on bromide paper and developed in a quick finish developer (diluted 1 : 1) at 70° F. (21° C.) for about 21 seconds then transferred without rinsing to a dish of rapid fixer. Fixation will take from 10 to 15 seconds. After blotting off any surplus moisture the enlargement is ready for inspection by the judge.

Apart from the special taking and processing materials used, an important factor in achieving speed is the layout of the darkroom and the efficient routing of the work from one piece of equipment to another. To attain the fastest time of production the equipment should be set up with each piece as near as possible to the next and in such a manner that the rotation of work is in one direction only, either left to right or right to left.

A camera operating room designed for photo finish recording uses some specialized equipment to assist still further in speed of production.

In such darkrooms the processing unit is a 12 × 12 × 6 ins. deep metal container, with a stainless steel top cut away to receive two stainless steel dishes 9 × 3 × 2 ins. deep, one for developer and one for the fixer. Inside the metal container is an enclosed heating element, with thermostatic control to enable the operator to keep the chemical solutions at the required temperature. Each dish is fitted with a stainless steel lid; the developing dish has a loose lid to be removed before switching off the light, and the fixing dish is a hinged lid which is kept closed until the film is developed. This prevents any possibility of placing the film in the fixing solution first.

The enlarger is a fixed enlarger, the degree of enlargement depending solely on the area of negative to be enlarged to a 8 × 10 ins. print; focusing is carried out by slight movement of the lens.

The paper is held in a shallow metal box in which there is a sprung platform enabling twelve or more sheets of double weight paper to be loaded at any one time. A white sheath slides over the paper, to make the carrier light-tight and serve as a focusing plate. In use, the negative is focused on the white plate, the light then extinguished, the plate withdrawn and an exposure made. The exposed bromide paper is then removed leaving the next sheet in position for the next exposure. This can be carried on until the paper is exhausted and the carrier reloaded.

An audible electric timer ensures accurate timing for developing and fixing the negatives. This is necessary because of the short duration of these processes. Thermometers are also essential.

A safelight fitted flush with the top of the bench serves as a negative viewer.

All other equipment is of standard design.

The layout of the darkroom is based on the fact that the starting point is the camera. Around the walls from this point is a bench 33 ins. high to take the various items of equipment. As near the camera as convenient comes the processing unit with the dishes one behind the other (the developing solution at the back). Above the unit is the switch that controls the negative viewers and safelight; next comes the negative viewer on which is set the enlarger negative carrier, followed by the enlarger with the paper carrier immediately below and control switch near by, then the print developing and fixing dishes, a sink with a dish for water to rinse the print, and on the opposite side of the sink a small dry bench for blotting off the finished product. Over the print developing and fixing dishes is a 8 × 10 ins. safelight. All switches are placed at the same height above the bench. Finally in a convenient place on the wall at the starting point is fixed the audible timer. The room light switch is positioned as far away as possible from the operating switches to guard against accidental switching on of white light.

The operator proceeds as follows. First he checks all temperatures (for ultra speed work this is very important). He then starts the audible timer, removes the developer dish lid and switches off the light. He then develops the film, fixes it, and switches on the negative viewer and safelight, places the negative in its carrier in the enlarger, focuses, and makes the enlargement. The final stages are, to develop and fix the enlargement, give it a quick wash, and blot off the surface water. This method of working keeps the speed of production as high as possible.

The specialized equipment mentioned is only necessary in the case of constant ultra speed work. For the occasional worker ordinary photographic equipment is quite adequate, with the addition perhaps of a metronome or some such means for timing the processing, and an immersion heater or boiling ring for heating the solution. S.J.H.

See also: *Rapid access system.*
Books: *Developing*, by C. I. Jacobson (London); *Photography, Theory and Practice*, by L. P. Clerc (London).

RAPID RECTILINEAR (R.R.) LENS. Lens system made up of two identical doublet lenses mounted symmetrically about a stop with the negative elements facing out. Dallmeyer in England and Steinheil in Germany hit upon this combination at the same time, 1866. In this way they got rid of all the aberrations except spherical aberration and curvature of field.

When the lens is stopped down to *f*8 it will cover an angle of 40° but the definition near the edges of the field is always softer than at the centre. R.R. lenses went out of general use when genuine anastigmats arrived.

See also: *Lens history.*

RAYOMETER. Special type of density wedge made of graduated thicknesses of aluminium, used in radiography to measure the sensitivity of emulsions to X-rays and the emission of X-ray tubes.

READE, JOSEPH BANCROFT, 1801–70. English clergyman, microscopist and astronomer. Pioneer of photography. Early in 1839 made photomicrographs in the solar microscope on silver chloride paper which was moistened with an infusion of nut-galls (i.e., gallic acid) thereby using tannin as an accelerator. Also the first to use hypo as a fixing salt (1839). Made camera photographs in 1839. Biography by C. H. Oakden (1926).

RÉAUMUR. Name given to a system of thermometry invented by R. A. F. de Réaumur (1683–1757) in 1731 in which the freezing and boiling points of water are 0° and 80°. This scale was mostly used in Russia.

RECIPROCITY LAW. Photographic exposure is the result of allowing light of a certain intensity to act on sensitive emulsion for a specified time. According to the reciprocity law of photochemistry, so long as the exposure (light intensity × time) remains constant, the response of the emulsion is the same—i.e., if the intensity is doubled and the time halved, the blackening produced on development should be unaffected.

In practice, photographic emulsions do not strictly follow this law. A given emulsion is most sensitive to illumination of a particular intensity. When the intensity varies greatly from this ideal, then the reciprocity law no longer applies. Very low intensities of light with correspondingly long times, or very high intensities with very short times, produce less effect.

The resulting reciprocity failure may show itself with the very short exposure of electronic flash, or with very long exposures (e.g., in astronomical photography). In either case more than the calculated exposure (up to three times normal) may be required.

See also: *Latent image.*

RECORD PHOTOGRAPHY. Photography that sets out to state the facts about the subject without any regard for its pictorial qualities. In practice it is not always possible to draw a sharp dividing line between the two, since many photographs taken purely as records often have a strong pictorial appeal—e.g., certain botanical subjects and architectural photographs.

Even in a record photograph, the photographer's personality is bound to show itself in choice of viewpoint, lighting, placing of subject, and in many ways in the photographic treatment of negative and print.

Perhaps the nearest approach to pure record photography is to be found in such operations as aerial survey, document copying, identification photography, and mass radiography.

RECTIFIED SPIRIT. One of the many forms of alcohol, containing about 5 per cent water, but no other additions (unlike methylated spirit). Subject to heavy excise duty.

REDEVELOPMENT. Process of converting the silver salts of a bleached image back into metallic silver by a second development. Bleaching followed by redevelopment is a useful method of improving negatives and prints, particularly for softening over-contrasty negatives and improving the colour of prints.

Redevelopment of Negatives. The contrast, density, and graininess of the negative can be adjusted by first bleaching it in a halogenizing bleacher and then redeveloping it.

A suitable halogenizing bleacher is:

Potassium permanganate	27 grains	1·5 grams
Sodium chloride	2 ounces	50 grams
Sulphuric acid, concentrated	100 minims	5 c.cm.
Water to make	40 ounces	1,000 c.cm.

Instead of the sodium chloride and sulphuric acid, 50 c.cm.—2 ounces—of concentrated hydrochloric acid may be used.

The negative is immersed in this solution until it is completely bleached. It is then transferred to a 5 per cent solution of sodium bisulphite, containing 0·1 per cent sodium hydroxide. The negative is then ready for redevelopment.

An alternative halogenizing bleacher is:

Copper sulphate, crystals	4 ounces	100 grams
Sodium chloride	4 ounces	100 grams
Sulphuric acid, concentrated	1 ounce	25 c.cm.
Water to make	40 ounces	1000 c.cm.

The negative is immersed in the bleaching solution until the whole image is white, after which it must be well washed before redevelopment.

The negative is redeveloped in a soft metol developer by white artificial light, or daylight. When the image is dense enough the negative is fixed and washed in the usual way.

The final image can be made softer and thinner than the original by removing the negative from the solution before it is fully developed.

The image may also be converted into one of finer grain by redeveloping the negative to the required density in a para-phenylene diamine, instead of a metol developer.

Again, the over-all contrast range can be reduced without greatly affecting individual highlight or shadow gradation. The negative is first partially bleached in one of the above bleachers. Bleaching is stopped when the shadows and lower half-tones are completely white, but the highlights are still partially black. The highlights are then reduced in a permanganate reducer which does not affect the bleached shadows. Finally, the whole negative is redeveloped in a normal developer.

Redevelopment of Prints. The object in redeveloping prints is usually to improve the image colour. This may be desirable if the prints were originally processed in stale developer. They may be bleached in the solution given for negatives, above, or in a bath consisting of equal parts of potassium ferricyanide and bromide in about 20 parts of water.

As soon as most of the image has disappeared, the print is thoroughly washed for about half an hour in running water. It is then developed in any fresh print developer by weak white light. Finally the print is fixed and washed in the usual way. L.A.M.

See also: *Contrast control.*

REDUCER. Photographically, a solution which is capable of dissolving silver in a negative or positive image, thus decreasing its density. Chemically, a reagent (such as a developer) which increases the proportion of the metallic or electropositive component of a compound—e.g., by converting a metal salt into the metal. A photographic reducer is generally the opposite of a chemical reducing agent—i.e., it is an oxidizing agent.

See also: *Development theory; Reducing.*

REDUCING. Method of lowering the densities in a negative or print. Although this can be done by physical means (Baskett's reducer) it is more common nowadays to use chemical reducers.

Before starting to reduce a negative or print it should be well soaked in water. This ensures that the solutions used will act evenly.

Negatives. All negative reducers decrease the density of the image. At the same time they may also affect the image contrast. There are thus three kinds of reducer.

(1) Subtractive or cutting reducers which remove equal densities of silver everywhere. These decrease density but do not affect contrast.

(2) Proportional reducers which decrease

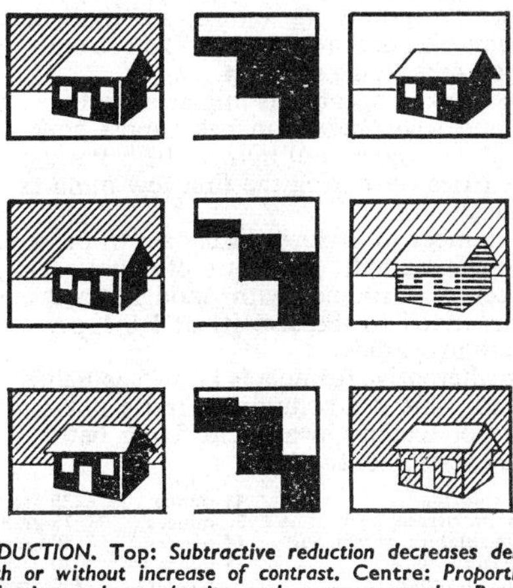

REDUCTION. Top: *Subtractive reduction decreases density with or without increase of contrast.* Centre: *Proportional reduction reduces density and contrast evenly.* Bottom: *Superproportional reduction mainly affects highlights.*

the image density in proportion to the amount of silver already there. These decrease density and reduce the over-all image contrast of the negative image.

(3) Superproportional reducers which hardly attack the shadow densities, but considerably reduce the highlights.

Subtractive Reducers. Potassium permanganate and concentrated potassium ferricyanide (Farmer's reducer) are both subtractive reducers.

The formula for potassium permanganate reducer is:

Stock solution A		
Potassium permanganate	1 ounce	25 grams
Water to make	20 ounces	500 c.cm.
Stock solution B		
Sulphuric acid, concentrated	1 ounce	25 c.cm.
Water to make	20 ounces	500 c.cm.

The concentrated sulphuric acid must be slowly poured into the water and the solution should be stirred all the time.

Water must never be poured into concentrated acid.

For use, 1 part each of A and B are added to 200 parts of water.

The soaked negative is immersed in this solution until it is sufficiently reduced. It is then rinsed in a 2 per cent potassium metabisulphite solution to remove the purple brown stain, and washed.

The formula for Farmer's reducer is:

Stock solution A		
Potassium ferricyanide	2 ounces	50 grams
Water to make	20 ounces	500 c.cm.
Stock solution B		
Sodium thiosulphate	8 ounces	200 grams
Water to make	40 ounces	1,000 c.cm.

For use, 1 part of A and 5 parts of B are made up to 30 parts with water. This mixture does not keep, and becomes useless in about 15 minutes. Its life can be prolonged by adding about 12 grams (½ ounce) of anhydrous sodium carbonate to every 250 c.cm. (10 ounces) of working solution.

The soaked negative is immersed in the solution and when reduction has almost gone far enough it is taken out and washed. Reduction still carries on during the first few minutes of washing.

A number of other reducers act in the same way—for example, a mixture of equal parts of 0·2 per cent iodine (with enough potassium iodide added to dissolve it) and 0·1 per cent potassium cyanide.

An alternative formula is known as Belitzki's reducer. This is a solution of ferric potassium citrate or oxalate in an acid fixing bath. The Kodak version of this is:

Ferric chloride	1 ounce	25 grams
Potassium citrate	3 ounces	75 grams
Sodium sulphite, anhydrous	1½ ounces	40 grams
Citric acid	¾ ounce	20 grams
Sodium thiosulphate	8 ounces	200 grams
Water to make	40 ounces	1,000 c.cm.

The chemicals must be dissolved in the above order.

Proportional Reducers. Dilute Farmer's reducer acts proportionally. The standard formula given above is used, but the working solution is made up with only about one-tenth the amount of potassium ferricyanide.

The Farmer's reducer may also be used in two separate baths. The negative is first immersed in a 1 per cent potassium ferricyanide solution until sufficiently reduced, and then transferred to a 10 per cent sodium thiosulphate bath for 5 minutes.

An alternative proportional reducer can be made up from equal parts of the potassium permanganate reducer (above) and the ammonium persulphate solution (below).

Superproportional Reducers. A common superproportional reducer is ammonium persulphate.

Ammonium persulphate	60 grains	3·3 grams
Sulphuric acid, 10 per cent solution	200 minims	10 c.cm.
Distilled water to make	4 ounces	100 c.cm.

The ammonium persulphate must be pure; the crystals should crackle as they dissolve. The made-up reducer does not keep, and must be used immediately.

The negatives are immersed in the solution until they have almost been reduced enough. They are then immersed in an acid fixer or a 5 per cent sodium sulphite solution.

When the reducer solution begins to go milky, it should be thrown away and a fresh lot made up.

An alternative superproportional reducer consists of a 0·1 per cent solution of benzoquinone containing 0·3 per cent of sulphuric acid.

Prints. The methods and formulae suitable for print reduction are much the same as used for negatives. The solutions are generally used rather more dilute, because the amount of silver deposit that forms the image in a print is much less than that needed to form a negative.

The action of the reducer must be carefully watched and arrested before it bleaches out all the detail in the highlights and leaves only blank white paper.

Usually badly over-exposed or over-developed prints cannot be saved by reduction, nor is it worth the trouble. It is nearly always quicker and more effective to make a fresh print. Farmer's reducer can, however, be used to clear slightly veiled highlights by immersing the print in a dilute solution for about five to ten seconds.

Another clearing bath which acts more slowly consists of three parts ferrous sulphate and one part each of potash alum and citric acid in twenty parts of water.

Local Reduction. Certain reducers, among them Farmer's, may be applied locally with a brush to lighten small areas of the print or to pick out highlights. They may even be used to "paint" clouds into an otherwise empty grey sky.

Chemical Spotting. Local reduction is also useful for removing small black spots, etc., as an alternative to scraping them out with a knife. Chemical removal does not damage the print surface in any way, whereas scraping always leaves a roughened patch. Small spots can be erased in this way with a mixture of equal parts of a 5 per cent solution of iodine in alcohol (with a little potassium iodide added to help it to dissolve) and of 10 per cent thiocarbamide in water.

The print is first swabbed over with a plug of cotton wool soaked in alcohol, to remove surface grease. It is then wiped dry with another piece of cotton wool. The reducing solution is applied with a brush, and immediately wiped off with a piece of cotton wool soaked in alcohol.

The process is repeated until the reduction has gone far enough, and the print is finally fixed in an ordinary hypo fixer.

Bleaching. The print image can be completely bleached out—as in making bromoil prints and also in making pen-and-ink drawings from photographs. L.A.M.

See also: *Bleach-out process; Retouching.*

Books: *The Complete Art of Printing and Enlarging*, by (O. R. Croy (London); *Developing*, by C. I. Jacobson (London).

REDUCTION. Term applied to certain chemical reactions in which the electro-positive (usually metallic) component is increased—e.g., exposed silver bromide in a film is reduced by the developer (a reducing agent) to form metallic silver.

In practical photography the same term is also applied to any operation on a negative or print which lowers the contrast of the image or decreases the density of the silver deposit. The object of reduction is to make a very dense negative easier to print, or to lighten undesirably black areas of a print.

Finally, reduction in printing, in copying and graphic arts work applies to reproduction at a scale of less than 1 : 1 or same size. Thus a three times or 1 : 3 reduction means that the copy is $\frac{1}{3}$ the linear size of the original.

The three uses of the term are unconnected.

See also: *Development theory; Latent image; Reducing; Scale.*

REFLECTANCE. The reflecting power of any surface is measured by comparing it with a perfectly white surface such as magnesium oxide which has been deposited as smoke on a polished silver surface. The ratio of the brightness of a given surface to that of magnesium oxide under the same conditions of illumination is called the reflectance value. In calculations it is treated in the same way as transmission—i.e., it may be quoted as a reflection factor or as a percentage reflection or as a reflection density (the negative logarithm of the reciprocal of the reflectance).

REFLECTION. When rays of light fall on a polished surface like that of a mirror or a pool of still water, they are reflected. The optical laws of reflection are important to the photographer who wants, for example, to take pictures in a mirror, or a scene reflected in water. The same laws apply to the use of a reflector for brightening up shadows.

There are three principal laws of reflection:

(1) The angle of incidence is equal to the angle of reflection.

(2) The incident and reflected rays are in the same plane as the normal at the point of reflection.

(3) The image is as far behind the mirror as the object is in front.

Angle of Reflected Ray. A ray of light striking the reflecting surface bounces off the surface and goes away in a new direction. The angle which it makes with the perpendicular to the surface of the mirror before reflection—angle of incidence—is the same as the angle it makes with the same perpendicular after reflection—angle of reflection.

If the ray of light falls on the mirror at an angle of 45°, it will be reflected at the same angle. It will, in fact, be turned at 45° + 45° = 90° to its original direction. (This is the principle of the reflex viewer in which a 45° mirror is used to swing the light coming through the lens from the subject through 90° to make it form an image on a horizontal focusing and viewing screen.)

This law decides the angle at which a reflector must be placed to throw light from a lamp or the sun into the shadow areas of the subject. Although the matt white surface of reflectors used for this purpose scatters much

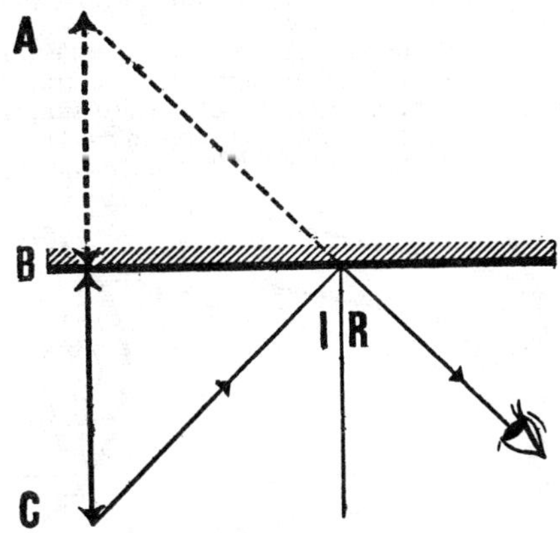

REFLECTION. A, image. B, reflecting surface. C, object R, angle of reflection. I, angle of incidence. The angle of incidence is always the same as the angle of reflection, and the image is as far behind the mirror as the object is in front.

of the light in all directions, most of the rays behave roughly as though they had been reflected from the polished surface of a mirror.

To arrive at the correct angle for a reflector, one imaginary line is visualized as coming from the light source to the reflector and another from the shadow area to the reflector. To throw the maximum amount of light into the shadow area, the reflector must be fixed so that both imaginary lines make the same angle with its surface. In addition, the surface of the reflector must be at right angles to the plane in which both imaginary lines lie.

Position of Reflected Image. When photographing an image reflected in a mirror or a sheet of water, the lens must be focused on the image and not on the reflecting surface. And the image lies as far behind the reflecting surface as the mirrored object lies in front of it. So that the actual distance on which the lens must be focused to reproduce the image sharply is the sum of the distance from the camera to the mirror and the mirror to the object.

A photographer may want to make a self portrait by taking his reflection in a mirror. If he and his camera are 4 feet in front of the mirror, his image in the mirror will be 4 feet behind the mirror, and he will have to focus at 8 feet. F.P.

See also: *Mirror photographs.*

REFLECTIVITY. Term applied to the ability of a surface to reflect light incident on it—the ratio between the intensities of the light before and after reflection. The figure varies with the nature of the surface and also with the direction in which the intensity of the reflected light is measured.

REFLECTOR

A lamp normally gives out light in all directions and unless something is done to control it, only a part of the light falls on the subject while as much, or more, is wasted. In practice the lamp is equipped with a suitably-shaped reflector which intercepts most of the rays of light travelling away from the subject and sends them back and distributes them in the desired direction.

According to the shape and size of the reflector, the light may be spread to give general illumination or concentrated to brighten a small area of the subject.

Surface. Reflectors are usually made of metal to withstand the heat of the lamp.

The surface may be highly polished, in which case it behaves like a mirror, or matted. A polished surface gives a hard light which casts sharp-edged shadows; a matted surface gives a soft diffused light with blurred and more luminous shadows.

Shape. The shape and size of the reflector decide how much of the light is concentrated usefully on the subject. A flat surface simply throws back the light in a spreading beam radiating outwards as though from an imaginary lamp as far behind the reflector as the real lamp is in front. Reflectors of this kind, consisting of a flat, white-painted or silvered board or stretched sheet are used near the subject for lighting up shadow areas; they are not used as lamp reflectors. Flat reflectors of this type are used outdoors in motion film production.

Lamp reflectors are always curved towards the light source. Three principal types of curve are employed—spherical, parabolic, and elliptical. In each case, the name indicates the shape of a cross section through the middle of the reflector.

A spherical reflector is used to produce a broad spreading beam for general illumination of the subject. This is the form chosen for flash and Photoflood lamps used as flood or fill-in lights.

A parabolic reflector is mostly used to give a fairly parallel beam of light for illuminating only a part of the subject or for illuminating the whole of a distant subject. Theoretically, a point source of light placed at the focus of a parabolic reflector gives a parallel beam of light of the diameter of the reflector.

An elliptical reflector gives a converging

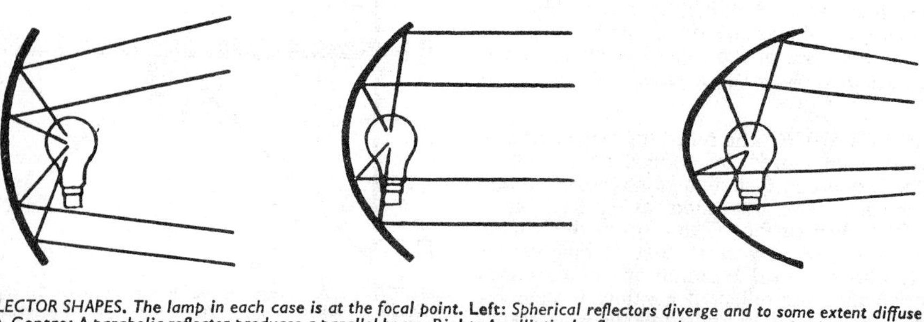

REFLECTOR SHAPES. The lamp in each case is at the focal point. Left: Spherical reflectors diverge and to some extent diffuse the light. Centre: A parabolic reflector produces a parallel beam. Right: An elliptical reflector produces a converging beam. In practice reflectors do not behave in this precise fashion, because of the finite size of the light source.

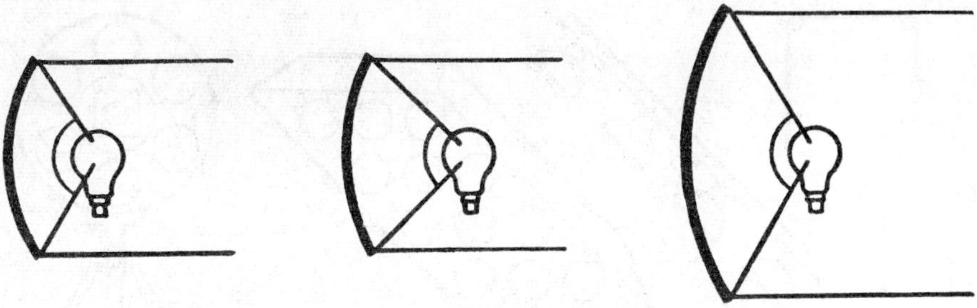

REFLECTOR EFFICIENCY. The efficiency of a reflector depends on the angle subtended at the lamp. As the focus increases, the collection angle drops, and to maintain the same efficiency a larger reflector is required.

beam when the lamp is placed at one of its focal points (the light converging to the other focus). This type of reflector is used for concentrating the light on to a relatively small area, e.g., for illuminating the negative in an enlarger or projector.

But the performance of any one of these reflectors depends on a number of other factors. If the distance of the light source from the reflector is increased, the reflected beam tends to become narrower; if it is shortened, the beam tends to broaden. And even when a beam converges to a point in front of the reflector, it crosses and spreads out after passing through the point, so that although the reflector is of the converging type, the beam it finally produces may be diverging.

Size. The amount of light thrown back by the reflector depends upon how much of the light emitted by the lamp it intercepts. A reflector close to the lamp will intercept more of the light than one farther away.

So the closer the lamp is to the reflector, the smaller will be the reflector needed to throw back a given proportion of the light emitted by the source.

But there are disadvantages in using a small reflector with a short focal length close up to the lamp. A curved reflector will not give evenly distributed illumination unless it is used with a point source of light. The source provided by the normal filament lamp is either an incandescent ring (sometimes a grid) at least 1 in. across (if the bulb is clear) or a luminous opal or frosted bulb of 2 ins. or more in diameter.

Neither of these can be regarded as a point source, but for practical purposes they can be treated as such so long as the focal length of the reflector is large in relation to the size of the source. But as the focal length of the reflector is reduced, the effect of the size of the light source becomes more and more noticeable.

In general, the size of the light source is kept as compact as possible and used in conjunction with the largest convenient reflector. These dimensions are only important for the type of hard lighting given by an approximately point-sized light source and a highly polished curved reflector, e.g., as in projectors, spotlights, and some types of enlarger. Where the light can be more or less diffused however—e.g. for general subject lighting—it is easier to obtain an even flood of light by using frosted or opal lamps in matted reflectors. With this kind of set-up, there is no need for accurate centring or focusing of the lamp, and the exact shape of the reflector is less critical.

In enlarging and projecting apparatus, the reflector is usually a separate (often optically worked) component of silvered glass or metal, but in units for general lighting it is usually shaped to incorporate the lampholder, No adjustment of the relative positions is then possible.

The shape of the reflector as seen from the front is not always a circle. Other shapes are also used—e.g. the square types made to work

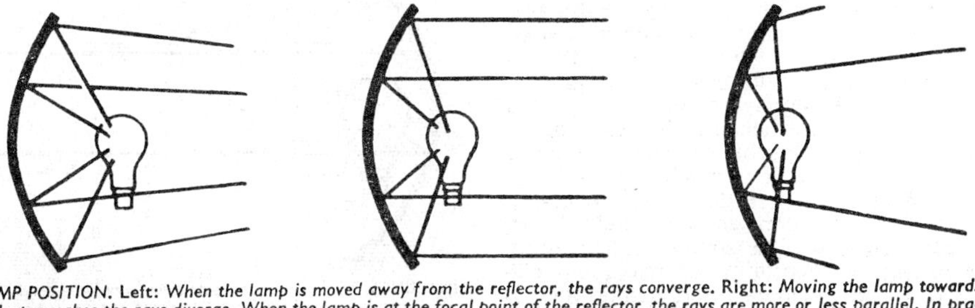

LAMP POSITION. Left: *When the lamp is moved away from the reflector, the rays converge.* Right: *Moving the lamp towards the reflector makes the rays diverge. When the lamp is at the focal point of the reflector, the rays are more or less parallel. In practice the effect is less precise, and converging rays cross and produce a diffused light.*

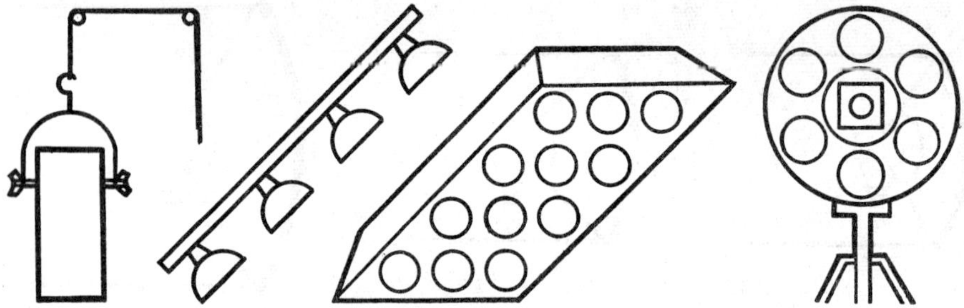

STUDIO REFLECTORS. **Left:** *Masked lamp to cover small area; adjustable for height and angle.* **Centre left:** *Lighting batten, consisting of a number of reflectors and lamps in line.* **Centre right:** *Trough reflector for large number of lamps to provide even flood of light. This and the previous units are usually overhead fittings.* **Right:** *Ring of reflectors surrounding camera lens for shadowless illumination. This is often used in medical record photography.*

side by side, and the long, trough-shaped reflectors made to house tubular lamps.

Folding and Collapsible Types. Reflectors are always a problem when lighting equipment has to be carried about, and several attempts have been made to make them fold or collapse into a smaller space.

One form of collapsible reflector is constructed in the form of an umbrella top without a stick, silvered on the inside. This is commonly used by cinematograph crews on location work; it is too heavy and bulky for normal photographic use.

The normal studio lighting reflector is too awkward for use with small portable flash units so accessory manufacturers have developed various types of collapsible miniature reflector. The simplest of these are the folding reflectors made from segments of stiff card faced with metal foil. These open out to form a shallow cone which reflects a broad flood of light from a lamp mounted at the approximate focus. More accurate reflectors are made from thin blades of polished metal pivoted at one end so that they open up and close like a fan. The blades are shaped and bent in a curve so that they form a fair imitation of a parabolic or spherical reflector.

Large reflectors of this type are expensive to make but they are a practical proposition in the small sizes fitted to the miniature flashguns made for midget flashbulbs. Some cameras with built-in "pop-up" flash guns may use this type of reflector; others house the bulb in a small box-shaped reflector which retracts into the camera body without folding or collapsing.

Quartz-Iodine Lamp Reflectors. Quartz-iodine lamps have a much smaller diameter than a normal filament lamp of the same power and the light source itself is much more concentrated. With a source of this type it is possible to mount a small, deep and accurately shaped reflector close to the source. Accuracy of positioning is important, for maximum efficiency, so some manufacturers market quartz-iodine lamps complete with reflectors in a sealed beam unit. This kind of reflector directs the light with much greater precision so that it spreads evenly over the subject area without spilling a large proportion of it uselessly outside the field of the lens. In addition the whole lamp-reflector assembly is so much more compact that it can even be mounted on the camera accessory shoe. However, the lamp and reflector get very hot and need to be handled with respect after being switched on for even a short time.

Multiple and Special Types. For studio flood lighting whole banks of lamps are often fitted inside a large trough-shaped reflector. This gives even and well diffused over-all illumination.

Where highly localized light is required without the harsh character of a spotlamp, a deep reflector may be employed. That consists in effect of a normal spherical reflector with a long tube added to it (although in fact the whole unit may be made in one piece). The tubular part confines the light to a small area, the size depending on its distance from the source.

A series of separate reflectors arranged in a ring round the camera lens and pointing at the subject will yield completely shadowless illumination. As an alternative the individual

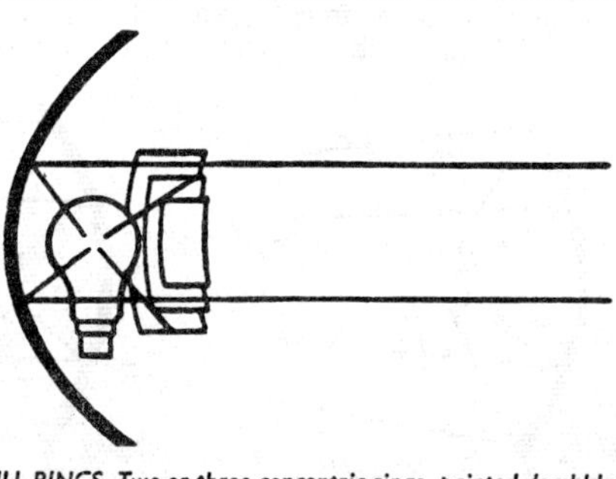

SPILL RINGS. Two or three concentric rings, painted dead black and placed in front of the lamp, intercept and cut out angular rays. This produces a well-defined beam.

lamps can be fitted in a ring-shaped reflector. Similar ring reflectors are available for curved tubular lamps.

Spill Rings. The shape and position of the reflector can be arranged to produce a more or less concentrated area of light on the subject, but the lamp itself sends out direct rays that spread over a much bigger area.

These direct rays are often undesirable but they can be controlled by mounting concentric "spill rings" of blackened metal in front of the lamp.

Spill rings, in proportion to their number and depth, absorb the undesirable wide angle rays emitted by the lamp but allow the useful direct and reflected rays to shine on to the subject. Occasionally a hemispherical metal cap is mounted in front of the lamp. It serves a similar purpose except that it cuts off all the direct rays from the lamp whether they are useful or not, and allows only a flood of reflected light to reach the subject. F.P.

See also: ***Lighting equipment; Projection principles; Reflection; Spotlight.***

REFLEX ATTACHMENT. Focusing device available as an accessory for certain miniature cameras which take interchangeable lenses. It converts the camera into a single lens reflex camera, and thus permits direct focusing of the image on a ground glass screen. This is of great assistance in telephoto work, close-up photography, and photomicrography where accurate focusing is essential.

Separate reflex viewing and focusing units are also available for cameras with non-interchangeable lenses; these are, however, independent viewfinders and not reflex attachments proper, nor do they necessarily show the exact image as seen by the camera lens.

Construction. The reflex attachment is constructed on the lines of a single lens reflex camera but without provision for holding any film. It fits on to the front of the camera and incorporates a movable mirror which deflects the light forming the image on to a ground glass screen in the top of the housing. The distance from the mirror to the screen is exactly equal to the distance from the mirror to the focal plane of the camera. For that reason a reflex attachment has to be specially designed or adapted for use with a particular camera. The front of the attachment carries a bayonet or screw mount for the lens.

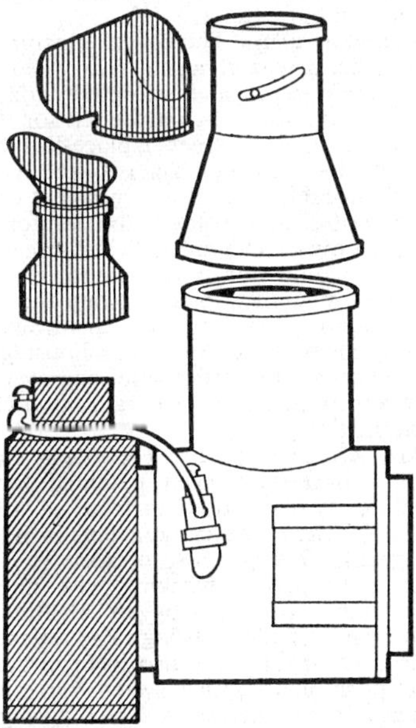

REFLEX VERSATILITY. For maximum adaptability, some reflex attachments provide for interchangeable viewing eyepieces. These permit waist-level or eye-level examination of the image, and may incorporate a pentaprism unit to present the screen picture the right way round. Magnifiers up to 30× are also available for specially exact work.

Viewing. The image on the ground glass screen is observed through a focusing magnifier. Several types are often interchangeable on one and the same reflex housing. The usual magnification is 4× to 5×, but special fine focusing magnifiers are also available with a magnification of 20× to 30×. Other types are:

Vertical magnifiers which serve for observation of the image from above, i.e. for waist-level viewing. These form an upright but laterally reversed image.

Inclined 45° magnifiers for eye-level viewing which incorporate a second mirror or a prism system above the ground glass screen. These form an upright image which may be laterally reversed or (if a pentaprism is used) right-way-round.

Right-angle 90° magnifiers are similar to 45° magnifiers in principle, but suitable for eye-level viewing only. The reflecting system above the screen may be a simple mirror forming a laterally reversed image, or a pentaprism to show a right-way-round image.

Exposure. For exposure, the mirror of the housing is swung out of the way, and the camera shutter released. Some reflex housings use a prism which is moved vertically out of the way.

The movements of the mirror and the camera shutter can be coupled to a common release cable by a special link, or by a cable coupler which takes two separate cable releases and operates them together. The release must be so adjusted that the mirror swings completely clear of the optical path before the shutter starts to open.

On some attachments the mirror swings out

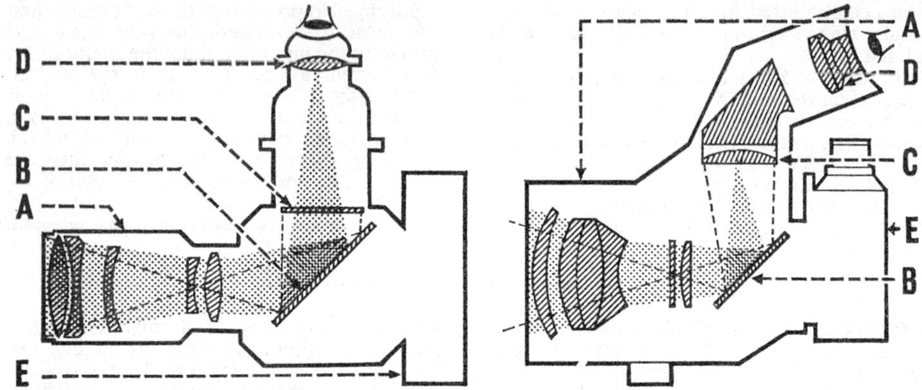

REFLEX VIEWING. Left: *Reflex attachments for miniature cameras convert the camera into a miniature reflex.* Right: *Special reflex attachments designed for long-range sports photography with long-focus lenses may incorporate a magnifying prism system for convenient eye-level viewing of an upright and right-way-round image. A, lens. B, mirror. C, screen. D, eyepiece. E, camera.*

of the way on pressing the release, and moves back again after the exposure. On others the mirror has to be re-tensioned or brought back to the focusing position every time.

Some reflex attachments incorporate special refinements, such as filter turrets and rapid focusing mounts. The former consists of a large rotating disc carrying several filter-apertures which are switched into the optical path between the lens and the reflex housing. This has two advantages. Firstly, it permits the use of one set of filters with a range of lenses. Secondly, changing filters becomes a much quicker matter than attaching them and detaching them in turn from the front of what may be a fairly long lens barrel.

The rapid focusing mount is a device for moving the lens in and out in a sleeve by means of a lever or other system. It provides a quick transition from sharp to unsharp and also enables the lens to be focused much more rapidly than with the usual screw mount. Preset click stops in this focusing movement may be provided to set the lens instantly to any preselected distance.

Close-up Work. The reflex attachment occupies some space between the camera flange and the lens mount, and so acts as an extension tube. It is thus specially suitable for close-up work with the standard, medium long focus, and wide angle lenses of the miniature camera to which it is fitted. The image is focused in the normal way by means of the lens mount, as on the camera.

As the extension is fixed, the range of subject distances is limited. It may however be increased by additional extension tubes, or by adding a bellows extension unit.

The image scale depends on the focal length of the lens and on the depth of the reflex housing.

Telephoto Work. Long-focus and telephoto lenses have to be specially matched to the reflex housing to allow for the depth of the housing and thus permit focusing at infinity. Most telephoto lenses designed for use with such attachments have a shorter barrel than would be necessary when using the lens directly on the camera. Since such telephoto lenses are almost exclusively used only with the reflex attachment, this short barrel is a standard fitting, and a special compensating extension tube is available when the lens is mounted directly on the camera.

Photomicrography. The reflex attachment permits direct screen focusing also when used in conjunction with a microscope adaptor for photomicrography. A refinement for this purpose is a clear glass screen in place of the ground glass for fine focusing. The two screens may be interchangeable, and in one model at least they are mounted in a rotating holder for instant changing over. The ground glass screen is used at low magnifications and the clear glass screen at higher magnifications or when the grain of the ground glass becomes disturbing. The viewing magnifier of the reflex housing is first focused on a set of cross-wires engraved on the clear screen, to permit accurate observation of the aerial image.

Reflex Attachments for Fixed-lens Cameras. Designs have also been patented for reflex attachments to be used in front of a camera lens, e.g. where the camera lens is not interchangeable. The purpose of such an arrangement is to permit working with alternative focal lengths provided by afocal attachments in the front of the housing. The reflex attachment must in this case also incorporate in the finder position an optical system corresponding to that of the camera lens. The focusing screen is placed behind this optical unit. The screen then shows the image in the same way as it would appear on the film in the camera. L.A.M.

See also: *Close-ups; Extension of camera; Focusing stage; Macrophotography; Reflex camera.*

REFLEX CAMERA

Type of camera incorporating a ground glass screen on top of the body which receives an image via a 45° mirror. It has been popular since the beginning of the century. Like the miniature camera, the stereoscopic camera, and the field camera, the reflex camera has its own circle of devotees to whom it is the ideal instrument. There have been two major modifications of this original pattern, so that there are now two distinct types: the single lens reflex (the original type) which includes the modern miniature reflex, and the twin-lens reflex. Each of these has developed along its own lines and possesses special advantages not shared by the others.

SINGLE LENS REFLEX

In the single lens reflex, the image is formed by the taking lens on a ground glass screen on top of the camera. The subject can be viewed and focused right up to the instant of exposure. The image on the screen is seen upright, free from parallax error and in the actual size it will appear on the negative, although laterally everything is reversed from left to right.

Large Size Reflexes. Most of the original large single-lens reflex cameras were of the rigid box form. The film or plate is housed at the back of the box, with usually a focal plane shutter immediately in front of the negative. The front of the box forms the lens panel.

Between the lens and the film is a movable mirror which can either lie flat against the top of the box, or swing down to lie at an angle of 45°. When the mirror is lowered it reflects the rays of light that come through the lens, on to a sheet of ground glass set in the top of the box. When the mirror is raised, it covers the underneath of the ground glass screen and the light from the lens falls on the surface of the plate or film in the back of the camera. The position of the mirror is so arranged that when an image is focused on the screen with the mirror lowered, the same image falls into focus on the plate or film when the mirror is raised out of the way.

Pressure on a lever on the outside of the camera raises the mirror. At first the light falls only on the blind of the focal plane shutter, but as the mirror reaches the limit of its travel it presses against a stop which releases the shutter and makes the exposure.

Shutter. Nearly all reflex cameras have self-capping focal plane shutters. These shutters generally give instantaneous speeds down to 1/10 second with Time and Bulb settings. Some of the smaller models—in particular, the miniature reflexes—have shutters that give a range of slow speeds, and have a built-in delayed action release for self portraiture.

One great advantage of the reflex camera is that it will take any lens whose focal length can be accommodated within the dimensions of the camera. This is because the image is focused visually and not by scale. So it can be used with long or moderately short focus lenses, the correct camera extension being given by sunk mounts to bring the lens nearer to the plate and extension tubes to take it farther away.

A reflex camera is naturally used most for subjects which move about and do not allow any time for slow methods of focusing. For this reason reflex cameras are fitted with fast lenses—generally with an aperture of f 4·5 and never less than f 6·3. The lens is often sunk into the front panel and on old types is covered by a hinged flap. This acts as a sunshade when it is raised to uncover the lens.

Focusing System. In the standard form of reflex camera the lens panel is joined to the body by a short bellows and can be racked out for focusing by turning a knob on the side of the camera. The image appears full size and right-way-up on the focusing screen, and it can be arranged and sharply focused right up to the instant of exposure.

One great disadvantage of reflex focusing shows itself when the lens has to be stopped down. At such times the image on the screen is very difficult to see and often impossible to focus. To get over this difficulty the screen is always fitted with a deep collapsible hood to shield it from the light. Often the hood is extended to reach right up to the eyes of the user.

Formats. Up to about 1930, the most popular format for reflex cameras was undoubtedly the quarter plate size (3¼ × 4¼ ins.). A few were larger than this, but the general trend was towards a less cumbersome size. The first move in the direction of smaller reflexes was the 2¼ × 3¼ ins. roll film camera. This had the disadvantage that the user had either to be satisfied with the normal horizontal format or indulge in acrobatic contortions to hold the camera on its side when he wanted to take vertical pictures. One way of avoiding this was the fitting of a revolving back. The arrival of the 2¼ × 2¼ ins. format solved this problem more easily and reduced the reflex camera to reasonable proportions.

Today, most roll film reflex cameras are made to take a 2¼ ins. square picture. There are also an appreciable number of miniature reflexes that use 35 mm. film.

Focusing Magnifier. The difficulty of focusing the much smaller image was overcome by a built-in focusing magnifier or by substituting a picture-size plano-convex lens for the focusing screen. The lower flat surface of the viewing lens was ground and the image focused on it was automatically magnified.

Eye-level Operation. The bulkier types of reflex are only designed for operation at waist-level with the observer looking down into the screen. For this purpose the camera is generally equipped with a sling so that it can be hung from the user's neck. By mounting it on a tripod and substituting a focusing screen for the plate holder, of course, this type of reflex can be operated and focused like an ordinary stand camera. But the reflex focusing can only be used at or near waist level.

The smaller types of reflex—i.e., those of 2¼ ins. square and smaller formats—are light enough to be held at eye level, and cameras of this class always incorporate some kind of eye-level viewfinder. This may be no more than an open frame finder which folds into the screen hood when not in use. For fast-moving subjects this finder has the advantage that the subject is viewed the right way round and is not dimmed when the lens is stopped down. But it has the disadvantage of dispensing with the reflex screen except for preliminary focusing.

There is another type of eye-level finder which may be fitted instead of or in addition to the open frame type. This consists of a mirror in the screen hood which can be set at 45° to the reflex screen. By holding the camera at eye-level and looking at the mirror, the user sees a reflection of the image on the focusing screen so he can compose and focus right up to the instant of exposure.

The system has the disadvantages associated with the normal reflex focusing screen but in this case the image appears upside down—definitely a further handicap. None of these disadvantages is present in the miniature single lens reflex cameras specially designed for eye-level operation.

Rollfilm Reflexes. The first rollfilm reflex cameras were little more than scaled-down versions of the larger types. Later they were equipped with more accurate shutters and a degree of automatic diaphragm control.

In appearance, current rollfilm single-lens reflexes follow three main patterns, all based on the more or less square box housing the mirror and optical system. The first pattern has the film chambers arranged to each side of this box, thus providing a convenient way of gripping the camera. The film there runs from side to side, and the transport and shutter (if focal plane) controls are ranged on top of the body. The second pattern has the film running vertically between chambers above and below the mirror box. There are thus no protrusions at the side, the camera is slightly more compact, and shutter and transport controls are mounted in the sides of the body. The third arrangement houses the film and transport elements in a separate magazine which attaches to the rear of the camera. This makes the camera body still more compact, and provides interchangeability of film magazines.

Basically the viewing and focusing arrangement is a waist-level screen with folding hood. Frequently this is removable and can be replaced by an eye-level mirror or pentaprism unit. Special focusing attachments with extra powerful magnifiers, etc., are often available.

Most rollfilm single-lens reflex models have a focal plane shutter; some use a diaphragm shutter built into the interchangeable lens.

Miniature Reflexes. The earliest 35 mm. reflex cameras were more or less scaled-down versions of the first design pattern mentioned above for rollfilm reflexes. This shape—with the film chamber and transport systems to each side of the reflex housing—is still the almost exclusive design pattern for 35 mm. single-lens reflex cameras.

The first models of this kind also featured only waist-level ground glass screen focusing. Present day 35 mm. reflexes usually have either an interchangeable finder system to permit waist-level or eye-level focusing, or a permanently built-in eye-level pentaprism unit.

The distinguishing principle of this system is that the image is viewed and focused at eye level through an eye piece which is an integral part of the construction. The eye-piece is a magnifier which gives an enlarged picture of the image formed by the lens.

The ground-glass screen image is reflected by a double mirror system so that it appears upright, though the two mirrors by themselves still show it reversed left-to-right. Most modern eye-level reflexes use a pentaprism with three reflecting surfaces which turn the image upright as well as right-way-round. As the prism is in one piece, it is easily made with a high degree of accuracy, and the reflections do not get out of alignment. Frequently the lower surface of the pentaprism itself acts as the ground-glass screen.

The image is not always formed on a ground glass screen. All or part of the picture may be focused as an aerial image with the aid of a high-power magnifier. This is useful in photomicrography and other specialized applications.

The focusing screen frequently is not ground glass at all, but plastic with a microscopic embossed regular grained texture. This scatters less light than the ground glass, leading to brighter images and greater visual sharpness. The focusing screen may also incorporate a Fresnel field lens (frequently embossed into the same plastic plate) which ensures even visual brightness of the whole image. To increase the brilliance of the screen still further, the greater part of it outside a central area may be clear (without a focusing surface) altogether. Then the centre portion carries either a matted focusing surface or a split-image wedge acting as a rangefinder and surrounded by a focusing screen ring. Within the area of the wedge the unsharp image is split and the two halves displaced relative to each other, as in a split-image rangefinder. When these images line up accurately, the lens is sharply focused on that part of the subject.

An elaboration of this principle is the use of reticular or other patterns in the central area of

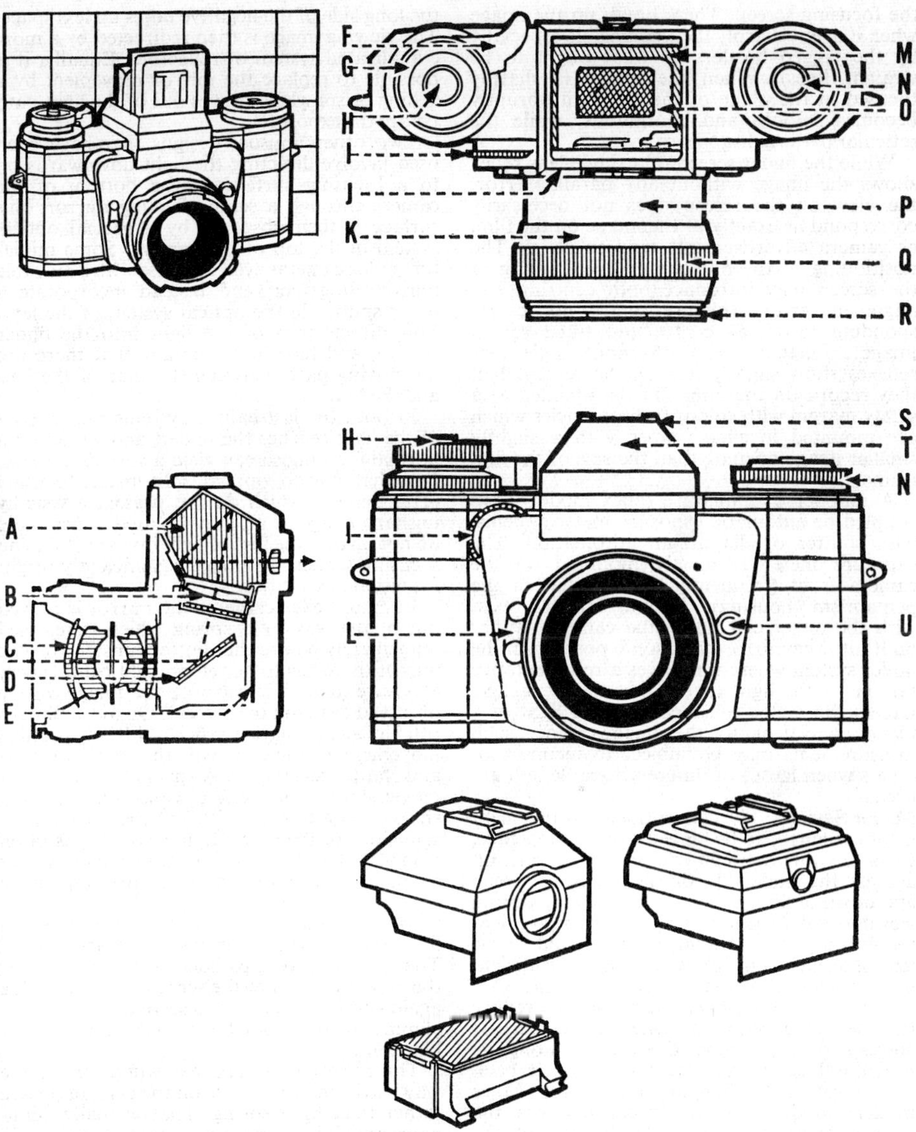

MINIATURE REFLEX CAMERA. The built-in pentaprism system of this 35 mm. model presents an upright and right-way-round image. The pentaprism is however interchangeable against a waist-level finder which shows the image—reversed left-to-right—on a screen which is viewed from above through a magnifier. The screen is usually partly clear, partly ground glass or grained plastic, and incorporates a pair of crossed wedges acting as a split-field rangefinder. On this camera the screen is also interchangeable, permitting the use of a clear screen, a fully ground screen, etc. Most miniature reflexes have interchangeable lenses and may feature either a diaphragm or a focal plane shutter. Usually also the lens aperture is fully open for accurate focusing and viewing, and closes down automatically to a pre-set value on releasing the shutter. Top left: General view of camera with waist-level finder against a 1-in. scale grid. Top right: Top view with waist-level finder. Centre left: Section through optical system of camera. Centre right: Front view of camera with eye-level pentaprism finder. Bottom right: Interchangeable pentaprism and waist-level finder units and interchangeable screen unit. Simpler 35 mm. reflex cameras have a fixed eye-level pentaprism unit, a more restricted range of interchangeable lenses (or none at all) but frequently a built-in exposure meter.

A, pentaprism. B, screen. C, lens. D, mirror. E, film. F, film transport lever. G, film counter. H, release. I, focusing wheel. J, waist-level hood unit. K, distance scale. L, self-timer. M, screen. N, rewind knob with rewind crank and film type indicator. O, focusing magnifier. P, depth of field scale. Q, focusing mount. R, bayonet mounting for front lens accessories. S, accessory shoe. T, pentaprism viewing unit. U, lens changing catch.

the focusing screen. These break up the image when it is not sharply focused (visible especially by the jagged borders of what ought to be straight lines). When the image is sharply focused in the plane of the screen it abruptly becomes smooth and continuous, while the reticular pattern disappears.

While the finder screen of a single-lens reflex shows the image without any parallax error, the view on the screen does not necessarily correspond as exactly to the image on the film, as camera advertisements tend to claim. The positioning of the mirror and the framing of the screen may introduce both centring and alignment errors (picture centre not corresponding to screen centre and tilted screen image). Further, most 35 mm. single lens reflexes show slightly less on the screen than they record on the film. This is intended as a safety margin with colour transparencies which are mounted in slide frames with a slightly smaller image opening than the size of the film frame.

A number of miniature reflex models have coupled or automatic exposure meter systems, with shutter or diaphragm automation. The exposure meter is mostly mounted on the camera front (frequently in the front of the pentaprism housing). Several models exist with the meter cell inside the camera—either built into the mirror or at some position in the finder system where it receives a representative sample of the light that makes up the image. Unless the meter cell measures the light in a plane equivalent to the image plane, such measurements may be subject to inconsistent errors when lenses of different focal length are used.

Mirror Systems. The conventional arrangement is for the mirror to be hinged along its top edge, so as to swing between two fixed positions: up against the underside of the focusing screen and down on to a fixed mechanical stop during viewing and focusing. This mirror movement requires a certain amount of space, and so the lens of a single-lens reflex must have a sufficient back focus to clear the mirror movement.

Various systems have been tried to reduce the clearance required. One of the earliest consists of supporting the mirror along its horizontal centre axis, and swinging it back and up rather than hinging it at the top. More modern designs involve systems where the mirror drops down into the bottom of the camera housing, or folds up along its middle. In both these cases a separate means of obscuring the underside of the focusing screen (e.g. by a roller blind shutter) are required. Alternatively—especially with eye-level reflex models—an auxiliary shutter may close the finder eyepiece.

Another variation, used in certain 18 × 24 mm. miniature reflexes, is a sideways mirror movement. Since the 18 × 24 mm. image is upright in the camera, this movement along the long side of the negative takes up less space. The viewing image is then redirected by a more complicated system of reflectors. Equally, it is possible to replace the mirror movement by a sliding prism which moves out of the light path before the exposure.

Two other unusual designs that have been used involve directing the light downwards on to a focusing surface in the bottom of the camera through a semi-reflecting mirror; this surface is then observed by a special optical system in the top of the camera. Some miniature reflex cameras even dispense with a movable mirror altogether, and instead incorporate a beam splitter in the optical system of the lens. This directs part of the light into the finder system, and has the advantage that there are no moving parts between the rear of the lens and the film.

In both the last-named systems only a part of the light reaches the finder, and would on a normal focusing screen yield a very dim image. For that reason optical split-image focusing arrangements which do not present a visually unsharp image at all may be used; since these do not involve the loss of light due to a diffusing focusing screen, the image is sufficiently bright for visual observation.

In many reflex cameras the mirror is moved out of the way by a spring which is released immediately before the shutter runs down. The mirror there has to be retensioned before every exposure (and in fact before the image can be observed again on the screen). On most modern roll film and miniature reflexes this retensioning movement is coupled with the film transport and shutter tensioning. A spring-loaded mirror movement may sometimes cause some camera shake owing to the "kick" of the mirror as it slams out of the way. High-class designs often balance the mirror movement in such a way as to reduce the momentum of the mirror on impact as far as possible.

Many current miniature (and also some roll film) reflexes feature an instant-return mirror. The mirror there flips back immediately after the exposure, so that the screen image is visible again within a fraction of a second. That almost eliminates the sometimes disturbing image blackout.

The mirror may also be swung out of the way by the mere pressure on the camera release rather than by a spring. That virtually eliminates any "kick", and the mirror returns again into its viewing position once the pressure on the button is released after the exposure. However in this case the travel of the release, and the mechanical pressure required on it, are greater; also if the release is pressed slowly the screen image may disappear a second or two before the actual exposure.

On cameras with a focal plane shutter the protection of the film against light during viewing and focusing raises no special problems: the shutter covers the film at all times except dur-

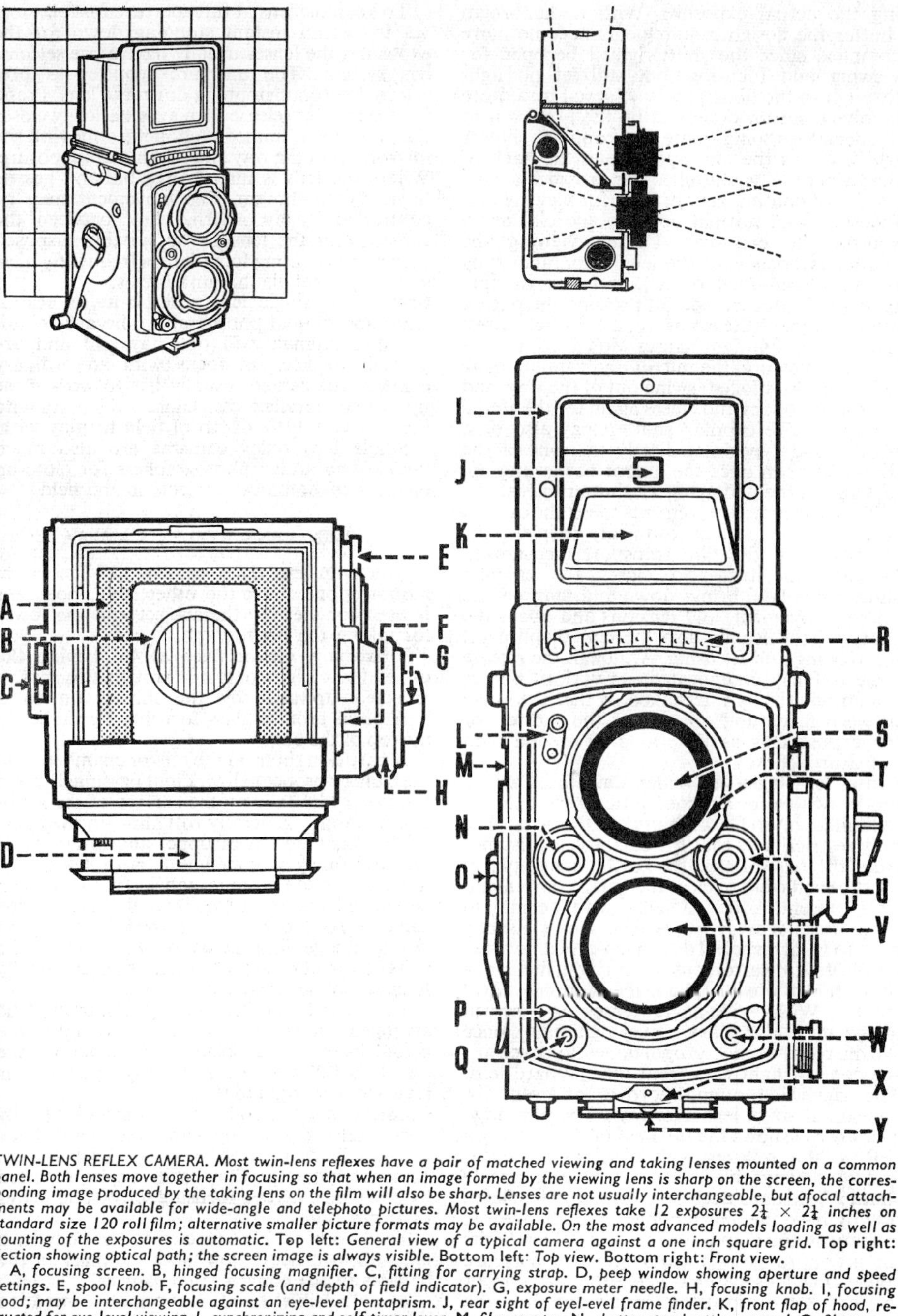

TWIN-LENS REFLEX CAMERA. Most twin-lens reflexes have a pair of matched viewing and taking lenses mounted on a common panel. Both lenses move together in focusing so that when an image formed by the viewing lens is sharp on the screen, the corresponding image produced by the taking lens on the film will also be sharp. Lenses are not usually interchangeable, but afocal attachments may be available for wide-angle and telephoto pictures. Most twin-lens reflexes take 12 exposures 2¼ × 2¼ inches on standard size 120 roll film; alternative smaller picture formats may be available. On the most advanced models loading as well as counting of the exposures is automatic. Top left: General view of a typical camera against a one inch square grid. Top right: Section showing optical path; the screen image is always visible. Bottom left: Top view. Bottom right: Front view.

A, focusing screen. B, hinged focusing magnifier. C, fitting for carrying strap. D, peep window showing aperture and speed settings. E, spool knob. F, focusing scale (and depth of field indicator). G, exposure meter needle. H, focusing knob. I, focusing hood; may be interchangeable against an eye-level pentaprism. J, rear sight of eyel-evel frame finder. K, front flap of hood, retructed for eye-level viewing. L, synchronizing and self-timer lever. M, film counter. N, shutter speed setting control. O, film transport crank. P, time exposure and safety lock. Q, shutter release button. R, exposure meter cell. S, viewing lens. T, bayonet mount for accessories. U, aperture setting control. V, taking lens. W, flash socket. X, back latch. Y, tripod bush.

ing the actual exposure. With a diaphragm shutter the design requirements become more complex, since the shutter must be open for viewing and focusing, but still let no light through to the film. This is achieved by a more involved shutter cycle and by the provision of a special capping plate or capping shutter which closes the film aperture in the back of the camera. The shutter cycle provides a possibility of continuous opening (for viewing and focusing) and normal opening and closing to control the exposure. During viewing the shutter is open (and the lens diaphragm fully open as well—see below), but the capping plate or shutter closed. On pressing the release button to make the exposure, the shutter closes, the lens diaphragm closes down to a preselected aperture, the mirror and capping plate (which is also hinged) swing out of the way, and the shutter opens and closes again for the actual exposure. The capping shutter may also be a roller blind type immediately in front of the film, and open once the mirror has swung up, just before the main camera shutter opens.

The mechanical movements to get the camera ready for the next exposure are then usually coupled with the film transport mechanism. Winding the transport closes the capping shutter or plate, brings down and tensions the mirror movement, and tensions and opens the shutter (and usually also the lens diaphragm) for viewing and focusing. Although the release cycle is fairly complicated, its various phases are automatically interlinked so that the delay between firing and opening of the shutter for the exposure is as little as 1/50 second with miniature reflexes.

Lenses for Single-lens Reflex Cameras. Camera models with a focal plane shutter permit almost unlimited interchangeability of the lens. The only restriction is that the back focus (distance from the rear of the lens to the image plane) must be sufficient to clear the mirror movement. This presents little difficulty with normal to long focus lenses; wide-angle lenses usually have to be of a special design (so-called inverted telephoto or retrofocus types) which have a back focus greater than the nominal focal length. With wide-angle lenses the screen image may also show some vignetting, since the mirror is usually proportioned to cover just the cone of light rays from the standard lens. The vignetting often improves or even disappears at small lens apertures. The film however always shows the full field of the lens.

Roll film reflexes with diaphragm shutter and interchangeable lenses may have a separate shutter built into each lens. Miniature reflexes usually have the lens mounted interchangeably in front of the diaphragm shutter (which limits the usable range of focal lengths) or may use convertible lens systems where the rear unit of the lens is built into the shutter and only the front half is interchangeable to provide alternative focal lengths.

To keep the lens at full aperture for focusing, yet to permit instant stopping down for the exposure, the lenses usually have a pre-selector iris system. The aperture required is preselected without stopping down the lens. Pressing the shutter release then automatically closes the stop down immediately before moving the mirror out of the way and making the exposure. Where the lens is interchangeable and has its own iris diaphragm, this movement may be controlled by an internal link between the camera and the lens or by suitably disposed twin releases. Long focus lenses frequently have simpler pre-set diaphragm systems.

Scope. The single lens reflex is at its best in child and animal photography: because children and animals will not stay still and are difficult to keep in focus with an ordinary camera; the camera usually has to work close up, when parallax can cause difficulties and there is very little depth of field to play with.

Single lens reflex cameras are also much used by naturalist photographers for close-up pictures of botanical subjects in the field.

TWIN LENS REFLEX CAMERA

In a twin lens reflex camera two lenses are mounted one above the other; the upper lens is used for viewing the subject and the lower for taking the photograph.

There is a mirror fixed at 45° behind the upper lens. Light passing through this lens is reflected upwards by the mirror on to a horizontal ground glass focusing screen set in the top of the box.

A partition shuts off the mirror chamber from the camera proper below. Light passing through the lower lens falls on the surface of the sensitive material—generally roll film—in the back.

The focal lengths of upper and lower lenses are accurately matched and both lenses are mounted on a common panel. This panel can be racked out by a knob on the side of the camera. As the knob is turned to bring the visible image into focus on the screen, the image formed by the lower lens is automatically focused on the surface of the film.

So a twin lens reflex allows the subject to be arranged and sharply focused up to and at the actual instant of exposure. The image on the screen is full size and right way up, but it is reversed from right to left.

Construction. The body of the camera is usually a rigid alloy die-casting with a removable back for film loading. Most models take roll film. In some types the film back can be replaced by one which takes a focusing screen and plates or 35 mm. film.

The lens panel is connected to the body by a light-trapped sleeve that takes the place of the bellows in the conventional camera.

A collapsible hood is normally fitted to shield the focusing screen from extraneous light. This folds down and protects the screen

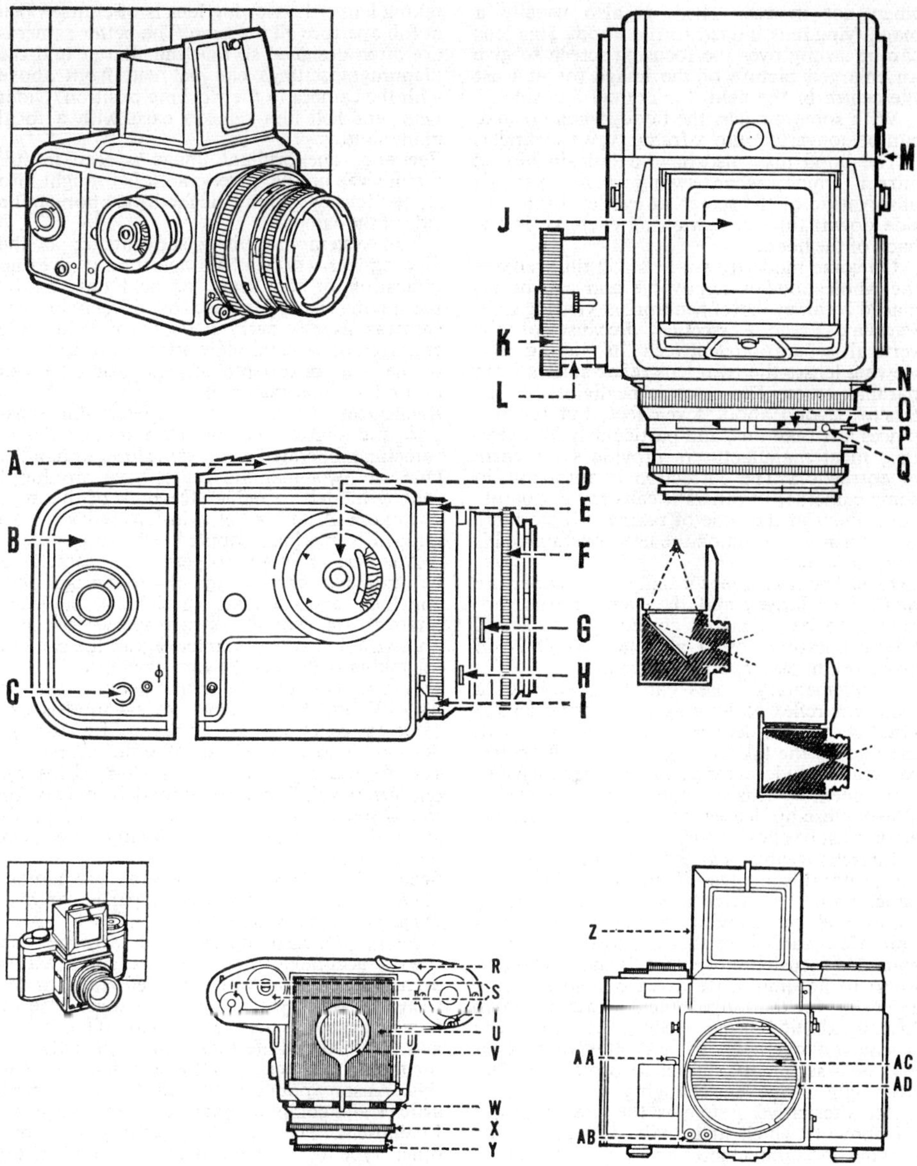

SINGLE-LENS ROLL FILM REFLEX CAMERA. Most cameras of this type are individual designs; typically they take 12 exposures $2\frac{1}{4} \times 2\frac{1}{4}$ ins. on standard size 120 film. The shutter may be a between-lens or a focal-plane type, lenses (and frequently the finder system) are interchangeable and refinements may include interchangeable magazines, plate backs, etc. Top left: *General view of camera against a 1-in. scale grid.* Top right: *Top view.* Centre left: *Side view.* Centre right: *Optical path during viewing and during exposure (with the mirror swung out of the way).* Bottom Left: *General view of an alternative (and older) design against a 1-in. scale grid.* Bottom centre *and* right: *Top and front views (the latter without lens) of alternative design.*

A, *interchangeable focusing hood.* B, *interchangeable roll film magazine.* C, *film counter.* D, *exposure meter calculator.* E, *focusing mount.* F, *shutter speed ring.* G, *exposure value scale and setting.* H, *lens aperture control.* I, *release button, with cable release socket and lock.* J, *focusing magnifier.* K, *film transport knob.* L, *exposure meter cell.* M, *slide of film magazine.* N, *lens changing catch.* O, *depth of field indicator.* P, *synchronizing and self timer lever.* Q, *flash socket.* R, *film transport lever.* S, *shutter speed setting dials.* T, *film counter.* U, *focusing screen.* V, *focusing magnifier.* W, *lens mount.* X, *focusing mount.* Y, *aperture setting.* Z, *interchangeable focusing hood.* AA, *release.* AB, *cable release socket.* AC, *mirror.* AD, *bayonet lens changing mount.*

when not in use. There is also usually a magnifying lens hinged to the hood. This lens can be swung over the focusing screen to give an enlarged picture of the image (or at least the centre of the field) for critical focusing.

With some models, the hood, when erected, can be converted into a frame type viewfinder.

The hood may also have a built-in hinged mirror which, when swung down, permits observation of the screen at eye level (but upside down) through a viewing eyepiece in the back of the hood.

On some modern twin-lens roll film reflexes the whole hood is removable and can be replaced by an eye-level pentaprism viewing unit.

Parallax Error Correction. Because of the vertical separation between its taking and viewing lenses the twin lens reflex suffers from parallax error. The error is negligible at distances beyond about seven feet, but for distances less than this, and particularly for close-ups, most manufacturers provide some form of correction. The correction is automatic in some cameras; in others it calls for a separate adjustment at the time of taking. Where there is no form of correction the user must make his own allowance.

Lenses. The taking lens usually has an aperture of *f*3·5 or larger, and the viewing lens, not having to be so highly corrected, has a still bigger aperture—often as much as *f*1·9—to give a bright picture on the screen.

Supplementary lenses can be used with a twin lens reflex so long as the viewing lens is fitted with a supplementary lens identical with the one on the taking lens. Most of the better cameras include a range of matched pairs of supplementary lenses among their accessories. When close-up lenses are used the parallax error must be allowed for.

Interchangeable lenses are comparatively rare, since this involves making the whole front panel with the matched viewing and taking lenses and the camera shutter a removable unit. Cameras where this is possible have an extended focusing movement (usually with bellows) to accommodate a reasonable focusing extension for the longest focal length available (usually about 180 mm.). Since the taking lens is removed together with its shutter, there must also be a secondary shutter to protect the film from light during lens changing.

An alternative system is the use of afocal attachments which fit in pairs in front of the viewing and taking lenses and convert the latter into wide-angle or telephoto units. The range of focal lengths possible in this way is limited.

The standard focal length for a $2\frac{1}{4} \times 2\frac{1}{4}$ inch twin-lens reflex is about 75 to 80 mm. Special models also exist with a permanently built-in wide angle-lens (around 55 to 60 mm.) or long-focus lens (135 mm. or more). These are generally used for special professional purposes.

Shutter. Most twin lens cameras have a between-lens shutter, and the usual iris diaphragm on the taking lens—the viewing lens is of course open at full aperture all the time. The better cameras are often designed so that the shutter and the diaphragm settings can be read from above with the camera in the working position. Older twin lens roll film reflexes exist with a focal plane shutter.

Focusing. The brilliant image is seen in full picture size on the screen and it is brought into sharp focus by turning the focusing knob on the side of the camera.

The system has one disadvantage; as the viewing lens works at full aperture, it gives no indication of the depth of field covered by the taking lens. Because of this, twin lens cameras usually carry a depth of field table engraved on a metal plate attached to the back of the body or a depth of field scale engraved around the focusing knob.

Refinements. On the better cameras film transport and shutter cocking are interlocked and sometimes operated by the same lever or handle. One or two manufacturers have produced folding models to reduce the bulk of this type of camera. A number of twin-lens reflexes have built-in (sometimes coupled) exposure meters.

Adapters are made to permit the camera to be used for stereoscopy, photomicrography, copying, and similar specialized subjects. Shutters are generally synchronized for flash photography. Other adapters equip the camera for taking plates and 35 mm. cine film.

Formats. The most popular picture size is $2\frac{1}{4} \times 2\frac{1}{4}$ ins. (6 × 6 cm.); a few models use $1\frac{5}{8} \times 1\frac{5}{8}$ ins. (4 × 4 cm.) or even 24 × 36 mm. Several models provide alternative picture sizes from 24 × 36 mm. up to $2\frac{1}{4} \times 2\frac{1}{4}$ ins. on one size of roll film. Special models, mostly for studio portraiture, also exist in formats up to about 4 × 5 inches. These usually take sheet film or plates.

Scope. Twin lens reflex cameras are almost universal in their uses, with a particular bias towards genre, where their ability to focus moving, unposed figures and arrange the subject accurately in the picture space excels even the camera with coupled rangefinder focusing. At the same time, the sheer ease of focusing prompts their use for a lot of landscape and pictorial still-life that would normally call for the field camera. Twin lens reflexes are for this reason popular for illustrating travel articles and compiling picture series sequences.

Limitations. The twin lens reflex, like every other type of camera, has its characteristic limitations. Firstly, there are very few interchangeable lens cameras of this type, as the focusing movement is insufficient to cover wide angle and long focus lenses and also because both viewing and taking lenses would have to be changed.

In addition, the usual waist-level viewpoint tends to make the sitter's chin and lower jaw in portraits more prominent, and the foreground excessive in landscapes.

At the same time, the fact that the viewing and taking lenses are so far apart gives the camera a parallax error that no form of compensation can completely counteract.

Further, the twin lens reflex is both heavy and bulky to carry around because it consists practically of two cameras. It is difficult to make it fold up with the extra complication of the reflex mirror mechanism. There have been successful folding twin-lens reflex cameras, but all modern models are rigid.

In addition it is not possible to equip it with many of the camera movements that are desirable for architectural and other work where it may be necessary to correct perspective distortion and provide special depth-of-field control.

Nevertheless, the twin lens reflex is a popular camera among working photographers and keen amateurs. F.P. & L.A.M.

See also: *Camera history; Diaphragm; Focusing; Pentaprism; Rolleiflex.*

Books: *Cameras*, by W. D. Emanuel and A. Matheson (London); *Keppler on the Eye-Level Reflex*, by H. Keppler (New York and London); *Photography with the Eye-Level Reflex* and *Picture Making with the Reflex*, both by H. S. Newcombe (London).

REFLEX COPYING. Method of contact copying documents, drawings, and other line originals by exposure through the back of the sensitive material. This is placed in contact with the matter to be copied. The dark parts reflect little or no light, while the light background reflects most of the light. This produces on the document paper a reversed negative and the latter yields positive copies by repeating the process. Reflex copying is the only method of copying opaque originals without a camera, and documents with writing on both sides.

See also: *Document copying.*

REFLEX HOUSING. Attachment for fitting between the camera body and lens (usually of a miniature camera) to permit the image formed by the lens to be viewed and focused before exposure.

See also: *Reflex attachment.*

REFLEX PRINTER. Special type of contact printer equipped for making reflex prints up to 8 × 10 ins. or larger. In principle it is no different from normal printing boxes; it is larger than most because it is generally designed for making reflex copies of typescript and drawings.

See also: *Document photography.*

REFRACTION. Change in the direction of a ray of light passing from one transparent medium into another of a different density. Refraction takes place because the velocity of light varies according to the density of the medium it is passing through. The action of both lenses and prisms depends on refraction.

See also: *Refractive index.*

REFRACTIVE INDEX. A ray of light which passes from one transparent medium into another of a different kind is bent either towards or away from the normal—i.e., a line at right angles to the common surface at the point where the ray enters.

The angle between the ray and the normal in the first medium is called the angle of incidence, and the angle it makes with it in the second medium (after being bent) is called the angle of refraction.

No matter what the angle of incidence, its sine always bears a constant ratio to that of the angle of refraction. This ratio is the refractive index for the two substances. It is given by:

$$\frac{\text{sine of angle of incidence}}{\text{sine of angle of refraction}}$$

When the refractive index of a single substance is quoted, it refers to the ratio when the first substance is a vacuum.

So the refractive index of a substance is a measure of its power to bend rays of light. The greater the bending, the higher the refractive index—e.g. for air, the index is 1·00029, and for glass anything from 1·5 to 1·8.

The refractive index also varies according to the wavelength of the light: blue rays are

REFRACTIVE INDEX AND ABBE NUMBER OF GLASSES

Glass Type	*Refractive Index*	*Abbe Number*
Hard Crown	1·518	60·3
Borosilicate Crown	1·509	64·4
Medium Barium Crown	1·576	57·4
Dense Barium Crown	1·613	60·0
Light Flint	1·583	41·8
Dense Flint	1·621	36·1
Extra Dense Flint	1·652	33·6
Double Extra Dense Flint	1·802	25·5

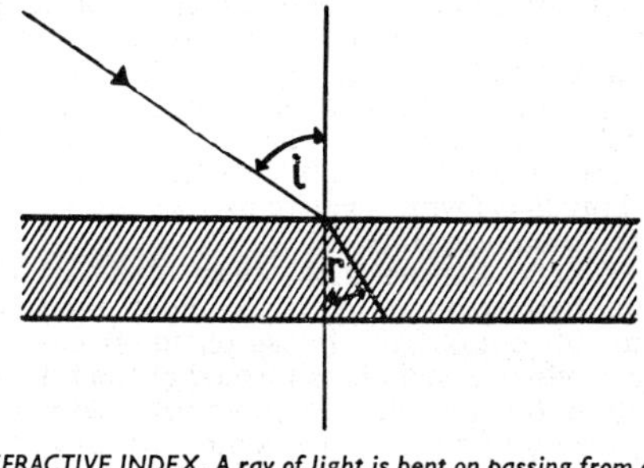

REFRACTIVE INDEX. A ray of light is bent on passing from one transparent medium to another. i, angle of incidence. r, angle of refraction. The refractive index is (sine i)/(sine r).

bent more than red. This causes dispersion of the colours; the extent of this is expressed in Abbe numbers.

The ratio of the velocity of the light waves in vacuum to the velocity in any particular material is also a measure of the refractive index of the material:

$$\text{Refractive index} = \frac{\text{Velocity in vacuum}}{\text{Velocity in transparent material}}$$

See also: *Optical glass.*

REGENERATION OF FIXING BATHS. In use a fixing bath undergoes the following changes:

(1) The concentration of free thiosulphate ions decreases through liquid brought in from previous baths, through loss of fixer carried out of the baths by the material being fixed, and partly owing to the formation of sodium argentothiosulphate.

(2) The concentration of silver increases.

(3) The acidity may decrease as the alkaline developer is carried over into the bath, unless an efficient acid stop bath is used.

(4) The concentration of iodide increases (especially in fixing baths used for negatives) owing to the presence of silver iodide in most negative emulsions.

All these changes interfere with the normal action of the fixing bath and thus lead to loss of activity and eventual exhaustion.

The best course for the amateur or small-scale professional photographer is to discard the fixing bath when it is exhausted. For commercial processing laboratories and other large-scale establishments using large quantities of fixing solution it is often sufficiently economical to regenerate the free thiosulphate, or at least to recover the silver accumulated during fixing in the spent solutions.

In regenerating a fixing bath, all the above-mentioned factors have to be taken into account. Thus it is not enough to add more hypo to make up for the amount used up if nothing is done about the other points, especially the increase in silver. The concentration of iodide is only of importance in negative fixing baths; the iodide converts silver bromide in the unfixed emulsion into silver iodide which takes much longer to dissolve in the hypo. This is also the main reason (apart from possible stains due to backing dyes) why negative fixing baths should not be used for prints.

Electrolytic Regeneration. Electrolysis of the fixing bath removes the silver and at the same time liberates free thiosulphate from the argentothiosulphate ions formed at the time of fixation. It is thus the most complete method of regeneration, but requires comparatively costly apparatus and is economical only for reasonably large-scale use.

Various commercial units are available for the purpose, and some firms even hire them out to processing laboratories in return for part of the value of the silver recovered from the waste fixer.

In electrolysis, the process must not be carried to completion, because silver sulphide may be precipitated as a result of secondary reactions if the silver concentration falls very low, or if the current density in the solution is very high. Generally a current density of about 0·15 amperes per square decimetre can be used for every gram per litre of silver in the bath. Most practical installations use large stainless steel or silver cathodes with a surface area of several square metres. The anode is carbon.

During electrolysis the solution must be well agitated. When the silver concentration falls below 0·5 gram per litre, the bath can be returned to use.

Some silver recovery units work continuously and are simply inserted in the fixing bath during use. They can remain there for several months without further attention beyond a periodic check on the pH of the solution. This system is only practical with large tanks.

Silver Recovery by Precipitation. The silver in a fixing bath can be precipitated by various methods. This does not necessarily regenerate the thiosulphate, but may decompose it and accumulate other by-products in the bath which make it unfit for further fixing. In some cases, however, provided the precipitation is not carried to completion, the fixing bath can be used again. This applies to precipitation by sulphide in acid solution, and by certain proprietary reagents.

With all precipitation methods the silver content of the bath should be determined first in order to decide on the amount of reagent needed for precipitation (especially if precipitation is to be incomplete). This can be done by various analytical methods—e.g., precipitation of a sample to completion, visual comparison of a colloidal silver sulphide solution prepared from the bath with a standard colloidal solution, or testing with indicator papers.

Precipitation by Sulphide. This is the cheapest method, but must take place out of doors and well away from any sensitized materials, as it generates hydrogen sulphide gas.

The exhausted solution is collected in a large barrel with outlets in the bottom and about one-quarter to one-third of the way up. After estimation of the silver content, the requisite amount of 20 per cent sodium sulphide solution is added, the solution well stirred, and left to stand for at least 24 hours. The supernatant liquid can then be drawn off at the upper outlet, and (provided the silver is not completely precipitated) brought up to strength with hypo and acid.

For complete precipitation the bath should first be made alkaline. It then requires less sulphide, but is not suitable for re-use.

The silver sulphide sludge is run off by the bottom outlet of the barrel from time to time, and dried.

Precipitation by Base Metals. For this method the collected volume of exhausted fixer is treated with granulated zinc or zinc dust which precipitates metallic silver. A barrel may be used to accumulate a sludge containing the silver and the general arrangement of equipment is similar to that used in precipitation by sulphide. Other metals—e.g., iron (as steel wool) or copper turnings—can be used.

Other Precipitation Methods. The silver can be precipitated from the fixer with ferrous hydroxide or with sodium hydrosulphite. The latter produces a specially pure deposit of silver, but is comparatively expensive. In the ferrous hydroxide method the solution must be made alkaline before precipitation.

Removal of Iodide. The accumulation of iodide from negative emulsions also limits the life of a fixing bath. It can be removed by adding 5 per cent thallous sulphate solution, which precipitates the iodide as thallous iodide, an orange yellow substance. Some of the bromide will also be precipitated as thallous bromide which is pale yellow. The removal of iodide is reasonably complete when only thallous bromide is precipitated.

The removal of bromide should not be carried to completion, or free thallous sulphate will remain in the bath, and interfere with the fixing action.

Restoration of Thiosulphate Content. After removal of silver by electrolysis, sulphide or hydrosulphite, the fixing bath will need the addition of sodium thiosulphate to bring it up to standard strength. The thiosulphate concentration is best determined either by titration against iodine, or by measuring the specific gravity of the bath.

Restoration of Acidity. The acid neutralized by developer carried over into the bath should be replaced if the acidity drops below the safe level—i.e., if the pH rises above 6·5. The bath may be tested by means of a suitable indicator and re-acidified by sodium bisulphite or sodium or potassium metabisulphite. Hardening fixers containing sulphite and acetic acid can be re-acidified by adding dilute acetic acid. The bath must not become too acid (pH below about 4·5), as that would decompose the thiosulphate. L.A.M.

REGISTERING IMAGES

When making assembly colour prints (e.g., by combining separate, single colour images) some convenient method of accurately registering the images is desirable.

The very nature of some photographs may make it difficult to achieve registration easily, and in photomechanical colour work automatic registration is required.

Photomechanical printing. In sheet-fed printing machines the successive colour printings are registered by the "three-point lay" method. One side of the sheet lies against two points ("lays") fairly wide apart and an adjacent side against a third point. So long as the same sides of the sheet are used every time, it will lie in precisely the same position against the lays, even if the sheet is not exactly square in shape.

In making printing plates in the graphic arts, and in pure colour photography, two register marks are usually used, preferably at the centre of the top and bottom of the subject or at either side. Two points are enough for accurate visual registration although they are not enough for automatic or rapid hand laying of sheets of paper or other material.

Register marks may consist of circles, triangles, segments of a circle, thick and thin lines, etc., but most printers prefer a thin straight line, with three or four lines crossing it at right angles. Originally the main purpose of the crossing lines was for colour proofing, for it was usual to punch a very small hole into the printing plate at one of the intersections of each register line. When one colour had been printed on the paper, a fine needle was passed through the proof from the back in the corresponding positions and the needle points were used to register the proof when printing the next colour. Several intersections were desirable because if there were a number of impressions on each sheet a single hole might become enlarged and give inaccurate registration.

The holes are rarely used today but the marks are, because there is always the risk of a blemish on the intersection of one pair of lines and it is useful to be able to use another.

It is possible to use detail of the subject for registering the successive images in pure colour photography and graphic arts printing processes, but most printers prefer register marks at the margin of the image. It is advisable to place register marks at the edges of any studio set-up even if it is being shot on integral tripack colour film because the marks will be useful if the photograph is subsequently reproduced.

Carbro. Carbro prints are always registered by hand. When the three carbro colour prints are ready on their temporary plastic (usually Perspex) supports, one colour is transferred on to soluble temporary support paper. This is wetted and lightly squeegeed into contact with the next colour image. In this state it can be moved into precise register by hand and squeegeed tightly down for the second transfer. Register marks help but there is sometimes distortion in the images; the register should be checked elsewhere as well as at the register marks.

Dye Transfer. The system from which dye transfer originated was registered in much the same way as carbro. After one dye image had been transferred to the mordanted paper it was covered with a very thin sheet of plastic, laid just to the edge of the image. The next dyed relief was then laid in position and the thin plastic was then slid out from between the two and the second dye image squeegeed down for transfer.

Dye transfer images are registered by the three-point lay system. The three reliefs are first registered by hand and then guillotined along two adjacent sides so that the images will always be in register with respect to the trimmed edges. The special rubber transfer blanket has three raised studs on its surface which provide the lays for registering the squared off sides of the reliefs. The first image is transferred to a sheet of mordanted paper, laid in position on the transfer blanket; the relief is then removed and replaced with the next relief, from which the dye image is transferred into accurate register with the first image. The process is repeated for the third image and perhaps a fourth (black image).

Colour-masking. The three-point lay method is often used in other colour techniques—e.g., of colour-masking, both for graphic arts and for pure photographic colour print making. Process cameras are often fitted with lays consisting of two slightly curved metal strips on which the lower edge of the photographic plate rests, and a third at one side. These strips, being curved, give a point contact and in practice are more efficient than pins. Three-point lays are also fitted in the "screen gear", which holds the process crossline screen, and often at the copyboard end of the camera, particularly in the transparency holder. This system ensures that there are three positions where the subject and images from it may be removed and replaced in accurate register.

The three-point lay system is also used on enlarger easels when making colour separation images by projection from colour transparencies. There are ready made proprietary devices, but the need can also be served by three suitably positioned nails in a board.

Register in Monochrome Printing. In some monochrome printing methods it is necessary to make two or more exposures in register upon bromide paper or on a plate or film; for example, in certain tone separation methods, where the highlight tones of the picture are printed from one negative and the shadow tones from another, because the two halves of the subject tone range can often be recorded better on two different negatives. It is rarely possible to include register marks in the original scene, but it is always possible to introduce the marks into the negatives afterwards.

When the negatives are dry, register marks are scratched at either side, or top and bottom, of the densest negative (the one exposed for the shadow detail) and the position of those marks is reduced out of the other (highlight detail) negative, after which it is re-washed and dried. A contact print is made on a blue-sensitive (colour blind) plate from the negative containing the register marks. A little water is then put on an unexposed ordinary quarter-plate in the darkroom and, after two or three minutes, while the emulsion absorbs the water and swells, the plate is warmed until the emulsion dissolves and some of it can be lifted off on the tip of the finger and transferred to the highlight detail negative in the clear spaces which have been reduced out. (Many modern pre-hardened emulsions do not dissolve easily; dye transfer relief film is ideal for the purpose.)

The applied emulsion is then dried, still in the darkroom, preferably with a fan or a hair dryer, and the register marks can then be printed-in accurately from the positive. The latter is quite easily hand-registered with the negative by laying them together over a safe-light and checking that they obliterate each other, when they are taped together and contact printed. Of course, there will be no image elsewhere because the emulsion was only put in the register mark spaces. After the usual fixing, washing and drying both negatives show the register marks clearly on black.

Other tone separation processes such as may be used in the making of photograms, colour derivations or in posterizing, which start with one negative or one print, lend themselves much more readily to the introduction of register marks; these can be scratched into the initial negative or positive.

Printing into Register. Printing several images with register marks into register on one sheet of bromide paper or other material is comparatively easy. After test exposures to determine the printing times, one negative is exposed, the register marks only developed with a small brush or a piece of cotton wool and stop bathed with 2 per cent acetic acid. After that the other negatives can be exposed upon the same sheet with the register marks as a guide.

Combining a line drawing with a photograph, or double-printing two different negatives into one print can be done by similar means.

The Blue Key Method. A less convenient method for pure photography purposes, but one which is sometimes convenient for certain graphic arts work, is the use of a "blue key". A very weak and soft contact positive is made from one of the negatives to be double-printed (or vice versa from a positive) and it is then toned blue in a suitable toner (e.g., ferro-prussiate, or dye toners). After bleaching out any remaining metallic silver with Farmer's reducer, a clear blue, non-printing image is left. The blue image should be very weak indeed. It is then taped securely into the enlarger so that each negative in turn can be registered against the blue image and exposed. The bromide paper is left in place throughout, and covered while the second, and any other negatives, are placed in position.

Punch and Pin Register. Perhaps the most common method of registration uses location holes and pins on the films and film holders respectively. This is useful for all cases where transparencies have to be registered with masks, or where an image must be projected in exactly the same location on several different films. Punch registration is less reliable with papers, since these are liable to expand unevenly during processing. (Films do, too; but to a much smaller degree.)

The elements of the pin register system are a punch which punches the holes in one side of the film, and register pins fitted in the negative carrier, contact printing frame or the enlarger easel. The two holes punched are usually of slightly different size, one hole being longer than the other. The pins are however the same size, so that one hole serves as accurate lateral location, and the other for alignment. This extra size of one hole prevents buckling of the film if it has slightly shrunk or stretched during processing. For the smaller hole fits exactly on on to its pin; there is no play.

Various manufacturers in the graphic arts field offer punch register systems; most of them have the holes at standard distances apart. Sometimes there are several distances, according to the size of film being used. In any case it is essential to stick to one system throughout.

The register pins themselves may be built into contact printing frames, or into the negative carrier of an enlarger. For use on the enlarger easel, some models have built-in register pins (and a corresponding punch mounted on the easel).

The pins may be partly retractable, for use with vacuum easels. At least one manufacturer offers register pin bars; these can be fitted on any enlarging easel with the aid of double-sided adhesive tape.

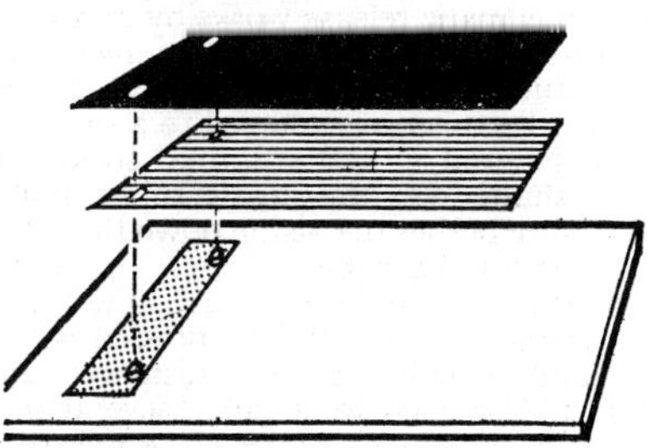

REGISTER PIN SYSTEM. A register bar carrying two standard spaced pins is placed on or built into the enlarger baseboard or the film carrier in a process camera. The negative or colour transparency and masking film are perforated with holes corresponding to the pins. This ensures that the images on the films placed over the bars are always registered in the same way.

Needless to say, the bar must remain in position on the easel during the whole of a sequence of operations.

In making masks by contact printing, the register pins in a special printing frame are used. Both the transparency and the masking film are punched and placed over the register pins of the frame for the exposure. If there is not enough space along the edge of the transparency for the punch holes, a piece of spare film may be taped along one edge for the purpose. The transparency and its mask are—after processing of the latter—placed in register on the pins in the negative carrier of the enlarger. If the separation negatives are also made by contact printing, the transparency, mask and separation film are put together in the printing frame.

If enlarged colour separations are made by projection, the masking film and the separation films are placed in turn on the register pins on the enlarger easel. Alternatively, the mask is made first by enlarging, and placed on top of the separation film for the colour separations. Again the register pins ensure exact superimposition of the processed mask image with the projected image of the transparency (provided that the latter has not been moved in the enlarger).

The punch register method can also be used for monochrome printing, where two or more exposures are made in succession on one bromide paper or film. In this case however it is usually better to make enlarged separation negatives (for example for tone separation processes) and register these visually. The two or more negatives are then taped together, trimmed flush along one edge and punched together. Finally the negatives are contact printed on to the positive material on the enlarger baseboard, the register pins once more serving to keep each negative in register with the positive material. The latter can in this case even be paper, since it is not moved between the exposures and remains on the register pins all the time.

Pin register systems are useful in all cases where images are combined by contact printing, or where two images (for example negative or transparency plus mask) have to be superimposed accurately in a negative carrier. Also it is suitable for all occasions where one image is printed successively on different sheets of printing material (as in masking by projection as described above) where the original is not moved between successive exposures. For projection printing where the negative is moved or different negatives are to be superimposed, visual registration methods with register marks are more suitable.

Pin registration would there require not only a negative carrier with register pins, but also a method of inserting the negative carrier into the enlarger in exactly the same position every time. F.H.S.

Book: ***Colour Prints*, by J. H. Coote (London).**

REGNAULT, HENRY VICTOR, 1810–78. French scientist, professor at the Collège de France. President of the Société Française de Photographie from 1855–68. Proposed (simultaneously with Liebig) pyrogallic acid as developer (1851). Worked on the carbon process.

Described in 1856 an actinometerusing silver chloride paper.

REHALOGENIZATION. Conversion of a silver image back into one of silver halide—e.g., in bleaching for intensification and various toning processes. Generally achieved by treating the image in a solution containing an oxidizing agent (permanganate, bichromate, ferricyanide, etc.) and a halide (potassium chloride or bromide).

See also: *Redevelopment*

REJLANDER, OSCAR GUSTAV, 1813–75. Swedish painter. Worked in England from 1853 on as portrait and genre photographer. A notable exponent of combination printing. His most famous "art photograph" was *The Two Ways of Life*, 1857, made from over 30 negatives. Composed nude and other studies for artists to paint from. Provided a series of photographs of fleeting facial expressions and gestures for Charles Darwin's *Expressions of the Emotions in Man and Animals* (London, 1872).

RELATIVE HUMIDITY. Ratio of the amount of moisture in the air at any temperature to the amount required at that temperature to saturate the air. The relative humidity (RH) of the atmosphere appreciably influences the keeping quality of photographic materials.

See also: *Keeping qualities of materials.*

RELEASE (SHUTTER). Any control or device that releases the shutter mechanism to make an exposure. May take the form of a lever on the shutter itself, or a button on the camera, mechanically linked to the shutter.

See also: *Shutter release.*

RELIEF IMAGE. Three-dimensional image obtained by removing exposed or unexposed areas from the light-sensitive layer, so that the image positions are raised above the surface of the support (e.g., a film base). Used in certain matrix processes such as Dye Transfer. A photomechanical letterpress plate also has a relief image.

See also: *Dye Transfer; Matrix.*

REMBRANDT LIGHTING. Dramatic lighting favoured by the Dutch painter Rembrandt (1606–69). Its effect is to spotlight significant parts of the subject, leaving the remainder in heavy but luminous shadows.

REMOTE CONTROL

Means for operation of equipment at a distance, especially when set up in a position where it is not easily accessible. Remote control systems may be used for still, cine and television cameras as well as for still and cine projectors.

There are various reasons for wanting to operate a camera from a point removed from the actual instrument: self-portraits, pictures of birds or animals from a close-up position while they are being observed from a distance, photographs of inaccessible subjects, etc. In projection, remote control systems enable the operator to sit among the audience or to work his projector while giving a lecture.

In camera operation it may be sufficient to release the shutter only at a distance, visiting the camera before each exposure to reset the shutter and wind on the film. Or a system may have to carry out the whole operation on its own to take one exposure after another without revisiting the camera. Remote control units may also be required to fire flashes during an exposure, or to adjust the focal length and distance setting (and even the exposure) of a cine and television camera. All these requirements have to be met in different ways so that anything more elaborate than a single remote shutter release usually has to be specially made for the job.

Remote Still Camera Releases. There is a certain amount of amateur interest in a remote shutter release for self-portraiture and nature photography. So a number of controls of this type can be bought ready made. There are three kinds: cable, pneumatic and electric.

Long cable releases are suitable for working at camera distances up to about six feet. Beyond that the lost motion caused by spring in the inner cable and its casing makes the release inefficient and other methods have to be used.

The pneumatic release works by transmitting air pressure from a rubber bulb along a narrow plastic tube to a small piston in a closed cylinder. The piston is coupled to a short cable release screwed into the normal socket on a camera shutter. On pressing the rubber bulb the air pressure pushes the piston down the cylinder and operates the cable release. This type of release is reliable and easy to use—the bulb can even lie on the floor and be pressed with the foot—and is effective over distances up to 100 feet. For such long range several plastic tubes can be linked together. The release also works without shock, so that there is no risk of camera shake or damage to the shutter.

An electric release generally involves the use of a solenoid-operated cable release connected to a battery and switch at the end of a remote control cable. The battery may be at either the camera or the control switch end. However, there is a limit to the distance at which this control arrangement will work. If there is too great a length of cable in the circuit, its resistance may prevent the solenoid from drawing enough current from the battery to operate the release. So for long distances it is necessary to locate the battery at the camera end to keep the connecting lead short, and switch it on by a relay connected to the remote control lead. When the remote control switch is pressed, only a few milliamps flow in the control wires connecting it to the relay. This is enough to energize the relay, close the switch in the solenoid circuit and operate the shutter release.

A solenoid used in this way would apply the pressure suddenly and is liable to shake the camera. So it is always linked to a dashpot or gear train or a brake of some sort to transform the jerk into a steady progressive push.

Once the release button has been pressed, the current continues to flow in the solenoid until the shutter actually releases.

Fully Cycling Systems. If the camera has a built-in spring or electric motor, working a remote release in the ways described above also actuates the motor which re-tensions the shutter and winds the film. The camera is then immediately ready for a further exposure.

Cameras not equipped in this way can be remotely controlled by coupling a motor drive to the film transport and adding the necessary electrical and mechanical parts to make it perform a complete cycle. Outfits of this kind are available for specific camera models (usually 35 mm. types, as well as at least one twin-lens reflex).

Completely automatic operation is possible by linking the remote control to a timing device or a system where the subject trips the release. Set-ups of this kind are used for automatic instrument recording, and also in nature photography in conjunction with camera traps. In the latter case an animal may set off the camera by means of a trip wire or by contacts hidden underneath a camouflaged board.

Essential in such systems is, also, a camera of sufficient film capacity to take a reasonably large number of exposures without attention. Cameras for this purpose can often be fitted with special large-capacity magazines. For the same reason only an electric motor drive is practical for automatic operation, owing to the limited capacity of spring motors.

Radio Control. Where the operating distance is very great or where it is inconvenient to run control wires, the shutter can be released by radio. The radio link consists of a transmitter worked by the operator, and a receiver at the camera end. As a rule the transmitter sends out a simple switching pulse when the release button is pressed. The receiver picks up the pulse and closes a pair of contacts. These may simply energize the shutter release solenoid or a motor linked to the shutter release, or they may trip the complete cycle of events involved in releasing the shutter and getting the camera ready for the next exposure.

Flash. Strictly speaking all shutter-synchronized flash guns work by remote control, being fired by an impulse provided by closing the shutter contacts. Normally this involves a lead from the camera to the flash gun. Such a lead can be quite long and several flashes can be connected to take the current from a common battery or charged capacitor.

Where extension leads are in the way, the most convenient way of firing flashes by remote control is by a slave unit. This is a capacitor circuit controlled by a photo cell. One flash is connected direct to the camera, and when this is fired the light triggers off the slave unit and fires the extension bulbs with a negligible time lag. Such slave units are suitable for flash bulbs or electronic flash.

An even more elegant way of firing flashes again involves radio control. Here the transmitter is connected to the shutter contacts of the camera. When the contacts close as the shutter is released, the transmitter sends out a switching pulse. A receiver picks up the pulse, closes the firing circuit of the flash unit and thus fires the flash. Any number of receivers, and hence any number of flashes, can be operated from the single camera transmitter. The range of such a set-up is several hundred yards.

When complete remote-controlled camera set-ups are linked to a flash gun, the flash is of course also fired by the remote release. An electronic flash is usually necessary with a self-restoring system to permit remote-controlled exposure series; with a flash bulb the latter has to be changed after every shot by re-visiting the camera.

Cine Cameras. A spring driven or electric cine camera can be released at a distance by any of the devices available for still cameras (pneumatic release, solenoid, etc.). As a cine camera is in any case self-recycling, no further attention is necessary except for periodic rewinding of the motor in the case of a spring-driven camera.

Cameras with an electric drive often have a socket to take a plug and extension lead with which the camera can be set going from a distance. Sometimes the battery pack with its stop-start switch is detachable and takes a long extension lead between it and the camera. The camera can then be mounted on a tripod and controlled with the switch on the battery pack at the other end of the lead.

Some spring-driven cine cameras can have an electric motor drive attached to a special shaft for complete electric remote operation. Such a remote control is also suitable for use with the single-frame exposure control, for instance for time lapse cinematography. In this case the camera is released automatically by a timing device at pre-set intervals. Usually this arrangement is also coupled with lights which switch on whenever an exposure has to be made. Further elaborations may include automatically controlled screens which obscure the subject from prevailing daylight (for example growing plants) to make exposures at intervals of 1 second to several hours for greatly speeded up films of growth phenomena.

Other functions of the camera, such as the focusing adjustment, zooming (in the case of a zoom lens), aperture control and even the camera mounting may be controlled by relay operated motors in much the same way as with a television camera (see below).

Television Cameras. Television systems are frequently used for observation under conditions where it is impossible or unsafe for a human operator to work the camera (for example observation of events connected with strong radioactivity). So remote-control systems for television cameras are often particularly advanced. Mostly these involve several motors and relay switches. Thus the lens may carry motor drives to change the focal length, to adjust the distance setting or to vary the aperture. A switch at the end of a remote control cable can set each motor in action in either direction—to stop down or open up the lens aperture, zoom to wide angle or long focus settings, etc. In addition built-in motors can provide pan and tilt movements.

Since the control of a number of such functions involves a considerable number of controls wires, various simplified systems are in use. One of these works on the basis of a telephone dial selector; here dialling a number chooses the appropriate camera function, and a single switch then sets the required motor in operation in either direction. If necessary, a number of television cameras can be controlled in this way.

Similar remote control systems also exist for ordinary cine cameras to regulate all functions from a distance. Even still cameras can be controlled by television circuits of this kind. In one case remote-operated motors adjust the lens focus and focal length and swing or tilt the camera. A television camera tube, mounted to receive the image through the viewfinder, shows the operator what the camera is taking in.

Still Projectors. Motorized magazine slide projectors nearly always have provision for remote control. Here the remote unit is simply plugged into the projector and the operator can sit in the midst of the audience and press buttons to change slides, advance the magazine either forwards or backwards, and to operate a motor which focuses the lens. Certain non-motorized slide projectors also permit remote operation by means of a Bowden-type cable which transmits a mechanical impulse to the slide changing mechanism.

Two cable-less remote control arrangements for projectors use ultrasonic signals and radio impulses. In the first case a mechanical or electric whistle emits an inaudible signal (above 20,000 cycles per second) which is picked up by a sensitive microphone attached to the projector. There the amplified signal can close contacts to operate the slide transport. A second separate whistle frequency may also serve for remote focusing.

Radio-controlled slide projectors operate the changing contacts in response to a wireless signal sent out from a battery-powered miniature transmitter and picked up by a receiver attached to the projector.

Impulses recorded on magnetic tape can serve for completely automatic projector control. This involves a tape coupler attached to or built into the tape recorder, and linked with the slide changing contacts of the projector. As the tape with its signal passes over the pick-up head of the tape coupler, the signal impulse closes a relay which starts the slide changing cycle. This system is used for changing slides automatically in a lecture recorded on the same tape.

Cine Projectors. The functions most likely to need remote control are switching the lamp on and off and starting and stopping the projector motor. Reverse running may be sometimes included, as well as single frame projection. The cooling fan normally needs switching on separately; this also avoids the risk of switching on the lamp only by remote control.

The most straightforward remote control system carries wires for each function along a common cable, with appropriate switches at a remote unit at the end. In one system the remote control switch consists of a cube partly filled with mercury and carrying inside the terminals from the various control wires. On turning the cube in different directions, the mercury makes contact across different terminals, so initiating the various functions. Each side of the cube is marked with the control operations that take place when that side is on top.

In motion picture projectors the operations connected with a performance—house lights, interval music, curtains, etc.—may also be coupled to the projector controls. The most advanced systems operate everything by a programming unit which may even take care of the automatic change-over from one projector to another. L.A.M. & F.P.

See also: *Projection practice; Slave flash.*

RENWICK, FRANK FORSTER, 1877–1943. English chemist. Worked from 1898 till 1922, and again as Director of Research from 1925 till his death, with the Britannia Works Company (1898) Ltd. (later Ilford Ltd.). Recently this firm have named a new laboratory after him. From 1922 till 1925 he worked with DuPont at Parlin, N.J., on emulsion manufacture. Specialized in sensitometry, light scatter, photographic chemistry and physics.

During the first World War he developed plates for aerial photography. His name is also connected with the introduction of the Multigrade paper and the Kryptoscreen. He was a President of the Royal Photographic Society from 1927–9. Helped to found the British Photographic Research Association in 1918, the Scientific and Technical Group of the R.P.S. in 1919 and the Photographic Alliance in 1930. Received the Progress Medal of the Royal Photographic Society in 1921.

REPAIRS TO CAMERAS. Whether a photographer should attempt to attend to any fault in his apparatus must depend on his mechanical knowledge in general, his possession of or access to suitable tools and his ability to use them efficiently, as much as on a full knowledge of the equipment and its construction. Even simple repairs are often more difficult than they look because special keys are required to loosen fixtures. Further, not knowing whether a particular screw has a right or left-hand thread may mean damaging the thread and doing more harm than good.

It is generally cheaper, quicker and more reliable to leave repairs to expert mechanics. Many of the larger photographic dealers have their own workshops, staffed by mechanics with specialized experience, who can find the cause of a fault quickly and remedy it. Besides these dealers' repair departments there are photographic repair firms who specialize in repairing photographic apparatus. Most manufacturers' agents maintain mechanics who repair their own products, and only send major work to the factory for attention. This part of a manufacturer's organization also handles repairs within the guarantee period of the instrument.

Today, worn or faulty parts are generally not repaired, but replaced by new original parts. The cost of labour is so high that the expense of, say, repairing an escapement by hand would be out of proportion compared with replacement by a mass-produced, complete item from the factory.

This explains why repairs to obsolete cameras, for which no spare parts are available, are so costly. In these cases no original parts are available to the repairer and he has to make them individually, sometimes at a high cost in time.

It is, however, quite astonishing to find, particularly in countries which have in the past been cut off for years from the country of origin of the camera, the high degree of skill with which workshops have been able to make up gears, screws, levers, etc., to bring instruments anything up to forty years old into perfect working condition.

Common Faults. A survey of some 5,000 repairs to a great variety of cameras shows that about 9 per cent needed no attention at all. The trouble was simply lack of knowledge on the part of the person handling the instrument, particularly in the case of users of the so-called fully automatic (electric eye) cameras. The reasons include refusal to read the instruction manual or a "Camera Guide", misreading a distance scale calibrated in metres for feet, etc., and just plain ignorance of the elementary facts of photography.

Most faults on between lens shutters can be traced to dust, dirt or sand which has found its way into the mechanism, possibly over a period of years. This may slow down the shutter speeds, or block escapement and built-in delayed release action. Gears should generally have a trace of suitable lubricant, but over-lubrication may cause the oil to reach the shutter blades, eventually causing these to stick and seize up.

Inaccuracies in focal plane shutters can frequently be traced to very dry springs or gears. In folding cameras, leaking bellows are high on the list of repairs. The trouble here is mostly caused by gritty particles like sand which have been allowed to remain in the camera and have worked their way through the bellows.

Almost 80 per cent of all folding cameras checked for varying types of complaint have been found with the lens out of parallel to the film plane. This, of course, impairs the over-all definition.

Rangefinders are liable to become inaccurate through shock, a fall or excessive variation in temperature. Viewfinders which show a field different from that of the negative may have become bent or distorted, but it is not unusual, particularly in the lower price range, to find that they have never been accurate, even when new.

Consistently incorrectly exposed results with an electric eye camera indicate a faulty photo-electric meter or shutter. Alternatively, if the meter, lens aperture, and shutter speeds all err in the same direction, the total effect may add up to one, or even two stops over or under exposure. The fault can only be satisfactorily pinpointed with electronic test equipment. The remedy would lie in re-calibration of the meter and/or re-setting of shutter speeds.

Lack of definition can be corrected in many cases simply by resetting an inaccurate distance scale, straightening a distorted camera body or film pressure plate which has lost its spring pressure, and squaring up the lens and film plane. In single lens reflex cameras, out of focus pictures may be the result of inaccurate positioning of a mirror or someone may have replaced the original mirror by one of different thickness or even with a piece of ordinary back-silvered looking glass instead of a surface silvered mirror. Similar complaints in twin lens reflex cameras may be also due to the viewing and taking lenses being out of coincidence.

Scratched negatives, particularly in miniature photography, come fairly high on the list. The cause may be film cassettes with foreign matter, grit, dirt on the light trap through which the film emerges, or a film pressure plate which is not perfectly smooth. (It should be remembered that scratching of the film is caused chiefly when it is kept in a roll. Unrolling, even a few times, is bound to produce some scratching. The remedy is to cut the film immediately after processing into strips of best six frames, which should then be kept flat in a negative wallet.)

Mysterious patches of light in the picture may be traced to internal reflections in the camera. Even a black surface struck by a ray of light will reflect, particularly if it is not a dead matt. Such reflections often reach the sensitized material. To overcome this difficulty, many cameras with large interior surfaces—e.g., twin lens reflexes—are now produced with anti-reflection baffles inside.

Lenses. Repairs to lenses are rarely called for and are only needed if the lenses have been tampered with. The most frequent fault occurs with owners of front cell focusing lenses who remove the lens stop so that they can focus on a subject closer than the normal near limit. In the process the front cell is often fully unscrewed and on re-inserting it the definition is often found to be impaired. This is due to the fact that the large majority of lens threads are of the multi-start type. Without suitable optical and mechanical aids, it is possible to spend days trying to find the correct start for re-insertion.

Another frequent trouble is scratched lens surfaces. Whilst a very few light surface scratches have no noticeable effect on the performance of the lens, a large number will produce light scatter and reduce the definition. As long as they are surface scratches—e.g., of the type caused by incorrect cleaning methods—the lens can be fully restored by repolishing. The problem, however, becomes unsolvable when there are deeper scratches and digs in the lens, or if it is chipped. To remove these by polishing would upset the performance of the lens system and mar the definition. With such lenses the safest thing to do is to fill in the scratch with dead black paint so that it does not create stray reflections. If the coating of a bloomed lens is scratched, this coating can be removed and a new one applied without any ill-effects. W.D.E.

See also: *Bellows; Home-made equipment.*

REPEATING BACK. Special form of plate holder by means of which a plate can be used to take a number of smaller-sized negatives—e.g., for taking four quarter plate negatives on a whole plate, or two $2\frac{1}{2}$ x $3\frac{1}{2}$ ins. negatives on a quarter plate. The holder is constructed to take a special dark slide with a cut-out aperture that can be inserted so that it uncovers each section of the plate in turn. This slide is replaced by the normal slide when the holder is out of the camera.

The term also describes the special holder used in one method of making three-colour separation negatives. In effect it holds three plates, side by side, each plate being brought into position in turn by sliding the holder along a grooved track on the camera back. This allows three quick exposures to be made without changing plateholders and withdrawing dark slide covers. The cover over the three plates is arranged so that it always covers the two not in use, leaving the third uncovered in the camera.

Such repeating backs may have the tricolour filters fitted immediately in front of the plates, so that no manual changing of filters is necessary when taking a set of successive separation negatives.

Sometimes there is a clockwork motor, geared to drive the holder across the camera back and also operate the shutter, so that the three exposures can be made in as little as a second.

REPLENISHER. Chemical that can be added to some types of developer to make good the exhaustion of the active constituents that takes place in the course of development. The practice is common in professional and trade processing where the trouble of draining off developing tanks and making up large volumes of fresh developer is a job to be postponed as long as possible; it is scarcely justified in the case of the amateur who uses only small quantities at a time. In colour processing accurate exhaustion tests (e.g., by processing pre-exposed control strips) are necessary to keep the activity of the working developer constant.

See also: *Developers.*

REPRODUCTION FEES. A reproduction fee is paid by an editor, advertiser or other commercial user for permission to reproduce a copyright photograph. The term also includes television "flashing" of a photograph, the right to exhibit, display or project, and use as a greetings card, calendar or picture postcard.

Editorial Use. An editor who uses a picture for its news value or general interest to his readers pays according to a scale of charges for non-exclusive work, or fixes his own figure in the case of exclusive and valuable pictures.

Non-exclusive pictures, offered to several journals at the same time, are paid for by space rates increasing in proportion to the reproduced size of the picture.

Exclusive pictures, usually of special interest to a certain market, fetch much more. A local paper is seldom, if ever, interested in buying exclusive rights; but national papers are always on the look-out for material of popular interest that is not available to competitors, and pay high fees for exclusive use.

Rights Offered. A photograph of high news value or other outstanding merit should not be submitted without informing the editor or user exactly what rights are offered and on what terms. In particular, when offering a picture for reproduction as a calendar, view card or greeting card, the extent of the licence should be clearly stated. The precise use should be taken into account and, in the case of calendars, the year of use.

Commissioned Work. There is no scale of charges for commissioned work. The copyright is vested in the person—e.g., the editor—who assigns the work. He therefore pays a fee for the work done, not a reproduction fee. Fees may be based on the amount of time taken to complete the commission, but they also take into account the cost of materials and incidental expenses, or the incidental expenses may be billed separately by agreement.

Commissioned work may also be billed according to the amount of space finally used. This is satisfactory if a set of pictures is used over several pages, but not if a quarter-page is allotted to a job that has cost the photographer much time and effort. All magazines, especially, have standard page rates. The leading British weeklies pay a black-and-white rate of about £25–£30 per page.

Other Uses. Fees paid for calendar use depend on the quality of the calendar and its circulation. A calendar intended for sale by a big commercial distributor will obviously fetch more than one to be given away by an engineering firm to its clients. Correct pricing of a picture for calendar use demands experience, and it is sometimes better to leave this to an agent who has specialized knowledge. Payment for calendar pictures is sometimes made in the form of a royalty based on the quantity sold.

Payment for picture postcards and greeting cards is also sometimes made in the form of a royalty based on the quantity sold and the price of the card.

If a photograph is exhibited to the public in any place, the copyright holder is entitled to a fee. In the same way, photographs or negatives to be used wholly or in part as material for lantern slides, film strips, or epidiascope prints, or for advertising, cigarette cards, confectionery lines, jig-saw puzzles, etc., are also subject to reproduction fees.

Minimum Fees. In the United Kingdom leading press groups have adopted certain minimum scales of reproduction fees. Some editors or commercial users pay more. Others follow scales of their own which may be lower. Proportionately higher fees may be asked in the case of pictures of exceptional merit, or whenever the photographer knows that the buyer normally pays a higher rate.

Colour photographs fetch considerably more than black-and-white pictures and there is a moderately good market for first class material, although in the United Kingdom it is difficult to sell 35 mm. transparencies; 4 × 5 ins. material is the ideal minimum size for colour work.

The governing fact is that photographs bought for publication are subject to the elementary economic laws of supply and demand. Because the supply greatly exceeds the demand the level of payment is, on the whole, at the discretion of the consumer—except in cases in which the material offered or its author is particularly sought after. If photographs are submitted without the photographer explicitly demanding a specific fee, publishers will assume that their own standard rate of payment will be acceptable. But it is by no means common and not always politic to stipulate a fee.

Reproduction fees are paid after publication and as a rule on the last day of the month in which it has taken place. Fees for pictures illustrating an article are assessed with the article and paid as one sum.

The not very reasonable attitude of some publishers to refuse reproduction fees to amateurs is becoming rare. There has never been justification for this in any but strictly non-commercial media of publication. R.S.

See also: *Agencies; Copyright and the photographer; Selling photographs.*

Book: *How to Take Photographs that Editors will Buy*, by R. Spillman (London).

REPRODUCTION QUALITY. For satisfactory photomechanical reproduction, a print should be pure black-and-white (neither toned nor tinted), of full tone, and on glossy paper, preferably glazed.

A negative can reproduce correctly tone for tone a fairly long range of subject brightnesses. A bromide print, however, can deal faithfully only with a tone range which is often much shorter than that of the negative. In particular the tones at the light and dark ends of the scale tend to become compressed. Thus if the print is very contrasty, its scale is much too short to accommodate many of the negative tones satisfactorily, and either highlight or shadow tones (often both) get lost.

In a good bromide print for reproduction the whitest tone must not be pure white, nor must the darkest tone disappear into complete darkness. All the tones must be visible and distinct from each other.

Criteria of Quality. When a photograph is to be reproduced in print, a copy negative must first be made. This is used to produce the printing plate. The best test of a photograph for reproduction is therefore whether it will make a perfect copy negative and print. Thus a bright and too contrasty print will, during reproduction, lose its highlight and shadow gradations, while a softer print will give much better results. In fact, prints for reproduction should be soft rather than contrasty.

On the other hand, the tone range of a photographic print is extended (quite apart from the contrast grade of the printing paper) if the darkest possible tone the paper can give is as black as possible, and the paper base as white as possible. This is also important for good reproduction. Therefore a sepia toned print (where the darkest tone is considerably lighter than black) is quite unsuitable. Similarly matt or rough surfaces, when they are as black as possible, appear only as dark grey when compared with a good black tone on a glossy and glazed surface. Shadow gradations which are indistinguishable on a matt surface, can be clearly seen on a glossy one.

For the same reason a tinted paper base will be much darker than a pure white one, again reducing the over-all tone range.

If a toned effect is wanted, the photo-engraver still prefers to work from a black-and-white print, though an example of the toned effect can be supplied separately so that the engraver can modify his methods if necessary, and to help the printer to choose the right coloured printing ink.

Processes. There are three principal methods of reproducing photographs in the printing press, each having its special field of application depending on such factors as the quality of the paper to be printed on, the number of copies to be printed and the standard of reproduction demanded. In each, the method of transferring the ink to the paper is basically different. Although broadly speaking the type of photographic print described above is equally suitable for all three processes, there are points of difference which may profitably be considered when turning out a print specially for reproduction by one particular method.

Typographic Method. The first is the typographic method, where the printing surface consists of raised dots of varying size. These dots carry the ink. They are all equally dark, dark tones and light tones in the finished reproduction being created by what is really an optical illusion. Where the dots are large (and the clear spaces between them correspondingly small) the area seen as a whole looks darker than where the dots are small with larger empty spaces between them.

The typographic or letterpress method can cope with almost any type of subject. It is ideal for crisp subjects; letterpress blocks give good and clean definition. But light subjects and unsharp and diffused prints are difficult to record by this method.

Lithography. The second printing method is lithography. Here again dots are used, but they are on a flat surface, and not raised as in the case of a letterpress printing plate. With this method much softer and more delicate prints are reproduced well, particularly light subjects. Dark subjects are, however, not so well rendered, as the maximum black obtainable is not very deep.

Photogravure. The third printing method is photogravure. Here the tones of the picture are actually formed by the amount of ink deposited on the paper and not by the size of dots into which the picture is broken up. It will produce particularly deep shadow tones, and is therefore ideal for strong and heavy types of subject. It is less efficient for light and delicate subjects, as there may be some loss of tone in the highlights. F.H.S.

See also: *Print quality; Scaling for reproduction.*

Book: ***Photographs and the Printer***, **by F. H. Smith (London).**

REPROGRAPHY. General term applied to photographic techniques of reproducing flat originals (documents, photographs, printed matter, pictures, etc.). Its scope covers copying as well as graphic arts processes.

See also: *Document copying; Photo-mechanical reproduction.*

REPTILES. Reptiles are best photographed in captivity. Trying to locate them in their natural surroundings is a hazardous business and practical experience has proved that unless pictures of a rare specimen are required—and that makes the job even more difficult—it is much simpler and quicker to obtain what will pass for natural photographs by working in a zoo.

Zoos offer very suitable conditions and the collections there cover a great number of species most of which may be photographed in complete safety. Many of the subjects are small, and fascinating pictures may be obtained by close-ups.

Temperature. A frequent source of spoiled pictures is the difference in temperature between the exterior of a reptile house and the interior. The best of the collection are nearly always from tropical climates, and consequently the temperature inside is kept high. This may result in a steamed up lens, not only on the external surface, but on the internal one as well. All condensation must be clear before making the exposure. Fifteen minutes is usually sufficient for the moisture to evaporate as the lens warms up to the temperature of the house.

Lighting. The normal illumination in such places may be adequate for photography, and the fact that the majority of the specimens on most occasions are slow movers provides a great chance to capture shots of many venomous creatures. On the whole, better results will be obtained with the aid of flash. (Permission to use flash should always be sought, as some species may react violently.) The glass panel, which provides safety, sometimes produces reflections; the use of a polarizing filter might help in some instances, although the difficulty can be easily overcome by placing the camera lens very close to the glass of the cage.

Equipment. The following is an ideal set of equipment for reptile photography: a camera capable of quick action both in film winding, focusing and shutter operation—an automatic twin lens reflex covers these requirements ideally—a supplementary lens for close-ups, plus a battery-operated electronic flash of at least 200 joules capacity, preferably split into two flash heads of 100 joules each. The use of two flash heads eliminates the harshness and dense shadows produced by the use of one lamp only, but nevertheless, very good results can be obtained with one lamp.

The Subject. In photographing reptiles, the variety of size is so great that hard and fast rules are out of the question; the subject may be a crocodile, a frog, a snake (such as the Royal Python of some 30 feet in length) or a Pigmy Iguana of a mere few inches. In tackling venomous species—e.g., our native vipers and adders—in the field, where there is no protecting glass, the most important piece of equipment is a pair of stout leg boots. Poisonous snakes are not really dangerous unless provoked by an adversary, but prevention in this case is very much easier than cure.

Field Photographs. Photographing reptiles in their natural surroundings calls first for an intimate knowledge of the subject and then for plenty of patience. Without knowing all about the habits and haunts of the reptile the photographer will not know where to look, and having found a suitable subject he will probably scare it away at once unless he knows how to stalk it.

Whether the reptile is poisonous or not, it must be stalked carefully, from the down-wind side, with the camera at the ready. The exposure should be made from the greatest distance that will give an image of reasonable size. Where possible a long focus or telephoto lens should be used to avoid disturbing the subject.

One of the greatest difficulties in photographing reptiles in their natural surroundings is that the colouring of the subject is often the same as that of the background. Even when the skin carries a bold pattern it becomes lost in the contrasty shadows cast by the bright sunshine sought by most creatures of the species.

One way of making the subject stand out is to photograph it against the light or in the strong low-angle illumination of early or late sunshine in clear weather. Another way is to transfer the creature to a less confusing background—e.g., a patch of sand. This is only possible with a limited number of sluggish snakes—it is out of the question with lizards.

There is no point in going to the trouble of taking pictures of reptiles in the field if the same pictures could be taken more easily in the Reptile House at the Zoo. In other words the subject should, if possible, be photographed doing something characteristic which it would not be able or likely to do in captivity. Behaviour, courtship, and even combat with others of its species are all interesting.

Photographs of the reptile hunting, attacking, and devouring its prey may be difficult to organize, but for that very reason are rare and worth seeking. Pictures of this type are sometimes obtainable by the use of a tethered bait in an open place that the reptile is known to haunt. All field photographs of reptiles should be accompanied by particulars of the date, time of day, temperature and place where taken. This information greatly adds to the value of the photographs as records, even if the subject is being pursued primarily for its pictorial qualities. I.A.

See also: *Animals; Zoo.*

RE-SCREENING. Photography of a half-tone image (e.g. a printed reproduction) through a half-tone screen to produce a new half-tone printing plate or block. This usually requires special precautions to avoid moiré interference patterns between the two screen images.

See also: *Half-tone screen reproduction.*

RESEARCH IN PHOTOGRAPHY. During the nineteenth century photography developed from a crude, slow and impermanent method of making chemical silhouettes, using lace, fern leaves and so on, as stencils, to the point where permanent, minutely-detailed records of nature could be obtained by snapshot exposures in cameras, and reproduced as prints on paper. These developments, largely the efforts of individual amateur workers, were achieved empirically, and the basic question, "How is photography possible at all ?", remained unanswered.

When it was discovered that, by using gelatin as a medium, the photographic material could be stored—and hence sold—the industry, as distinct from the art, of photography became possible, and research into every aspect of the photographic process became essential for its further progress.

At first such researches were one-man efforts by scientists working with limited facilities. The typical photographic manufacturer at that time was a small concern producing a limited number of products usually devised by its founder. During the first quarter of the twentieth century the more research-minded of these firms grew and in some cases amalgamated to form the big manufacturing organizations of today.

Accordingly, although photography still offers scope—if only in the devising of new applications—to the individual scientist and the amateur research worker, most of the research directed to the production of new or improved photograhic materials and equipment is today necessarily carried out either in the research laboratories of the big photographic and optical manufacturers, or by scientists subsidized by these firms while working in university laboratories, e.g. at Bristol and Liège. In addition, research on specialized aspects of photography is carried out in Government Departments and such institutions as the Institut D'Optique (Paris), National Bureau of Standards (Washington), the National Physical Laboratory (London), the Physikalisch-Technische Bundesanstalt (Berlin), the Eidgenössiche Technische Hoch-Schule (Zürich), and the Graphische Versuchs-und Lehranstalt (Vienna).

The work in a typical photographic research laboratory is of three general types: on photographic theory, on materials and processes, and on various projects not directly concerned with photography.

Theory. This covers the physical chemistry of gelatin, the physics and chemistry of the silver halides, the nature of the change which takes place when silver halide is exposed to light, the nature of the development process, the theory of tone reproduction and the effects of the structure of the image and the characteristics of lenses on the resolving power and sharpness of the photographic image.

The use of photographic emulsions to record the sounds and colours as well as the shapes of nature was a consequence of the development of the theory underlying these techniques.

Under the heading "psycho-physics" photography and more particularly, colour and stereo photography, have stimulated many interesting researches on the effective presentation of pictures and on the mental and physical aspects of vision itself.

Materials and Processes. Many hundreds of different forms of photographic emulsion are now available. Whereas the earliest types were capable of recording ultra-violet and blue light only, emulsions are now made for recording the whole or any part of the visible spectrum, infrared and X-radiation. They range in appearance from thin water-clear layers capable of legibly recording the contents of the Bible in an area less than a square inch, to thick opaque layers used to record atomic disintegrations. There is scarcely an industry or a science for which some special form of photographic material has not been specially devised.

Despite the advances made in theoretical studies, the technique of emulsion making is still in part an art in which practice is in advance of the chemical and physical conditions which control the results. The same may be said of the manufacture of gelatin, the medium in which the light-sensitive halides are suspended. The manufacturer of photographic emulsions has to be far more careful in the choice of gelatins than does the manufacturer of table jellies! It is easier to make advances in the applications of photography than in emulsion manufacture itself. Long before there were emulsions fit for the task, for example, the many ways in which colour photography could be achieved, had been fully realized.

Research, however, does not stop when a new form of photographic material has been devised. In addition to the design of conventional cameras and their accessories, such applications as document copying, photofinishing, aerial survey or photomechanical reproduction must also have the equipment devised to use the material, and the research laboratory plays its part in the field studies and statistical investigations necessary to get this into suitable commercial form.

General Research. The physicists, chemists, engineers and mathematicians in a typical photographic research laboratory necessarily uncover in the course of their work fascinating by-paths which are of no apparent importance to photography as such. The large laboratories typical of the photographic industry are always generous with such information they bring to light in this way as their communications to scientific publications frequently bear witness. Occasionally when scientists have explored these by-paths the results have been commercially rewarding. Some methods of manufacturing artificial silk, soluble surgical dressings, coagulating rubber latex—all these and many more benefits derive from the exploration of by-paths opened up during photographic research.

Publication of Results. Much of the successful research carried out by photographic manufacturers is naturally published in the form of patent specifications. Papers on the theory of the photographic process and most of its applications are published as communications to the appropriate scientific journals. The *Journal of Photographic Science* (Great Britain), *Photographic Science and Engineering* (U.S.A.), *Journal of the Society of Motion Picture and Television Engineers* (U.S.A.), *Zhurnal Nauchnoi i Prikladnoi Fotografii i Kinomatografii* (U.S.S.R.), *Uspekhi Nauchnoi Fotografii* (U.S.S.R.), *Tekhnika Kino i Televidenya* (U.S.S.R.), *Zeitschrift für Wissenschaftliche Photographie, Photophysik und Photochemie* (E. Germany), *Journal of the Society of Scientific Photography of Japan*, *Photographische Korrespondenz* (W. Germany), *Sciences et Industries Photographiques* (France), are the principal journals published throughout the world devoted exclusively to photographic subjects. In addition, *Photographic Abstracts* (London), *Abstracts of Photographic Science and Engineering Literature* (New York), and *Referativnyi Zhurnal Fotokinotekhnika* (Moscow) provide a wide coverage of photographic literature.

In the period 1889–1965, over 50 International Congresses and Conferences on the science of photography have been held in different countries, and the proceedings of these congresses provide convenient periodical cross-sections of progress, more particularly in the theory of photography.

In 1950 the first of three volumes of *Progress in Photography* was published as an attempt at reviewing international progress both in theory and in its many applications. After 1959 the quarterly *Perspective* and the volumes of the Focal Library took over the task of providing up-to-date world-wide progress reviews. D.A.S.

See also: *Science and photography.*
Book: *Progress in Photography* (2 vols.), ed. by D. A. Spencer (London).

RESIDUES. Residual processing solutions—e.g., fixing baths—which have removed soluble silver salts from negatives or prints and been treated to recover the silver.

See also: *Regeneration of fixing baths; Silver reclamation.*

RESIST. Image or pattern in the form of a protective layer on a metal or other surface to prevent or control access of chemical solutions (for example etching baths) or physical action (for example sand blasting) on that surface. As a result the reversed pattern of the resist image is etched or otherwise ingrained into the surface. The resist may be applied in various ways to the surface to be treated, including by photography—in which case we are dealing with a photo resist—and is removed from the surface after treatment.

Photo-resists are used in the graphic arts for the production of printing-formes of every description, such as letterpress, line and half-tone blocks, planographic printing plates, gravure plates and cylinders, silk-screen, etc. This classical application dates back more than a century. For several decades photo-resists have also become of major and rapidly increasing importance for engineering and electronic purposes. Examples are "chemical" milling, the production of small two-dimensional metal parts and strain-gauges; furthermore the production of printed circuits and quite recently the production of minute electronic devices, such as transistors and that of micro-miniature circuits. Yet other applications are glass etching (graticules) and etchings in plastics.

Principle. The photo-resist is a solution of a light-sensitive material which is spread as a thin layer on the surface of a substrate (i.e. metal, glass, plastics, semi-conductors, etc.). The layer is usually dried and is then exposed. The exposed areas are rendered insoluble, leaving the unexposed areas soluble in suitable solvents. Alternatively, the reverse takes place, i.e. the exposed areas become soluble whereas the non-exposed areas remain unaffected by the solvent.

In either case a resist stencil is obtained on the surface of the substrate after exposure of the layer under a mask, i.e. usually under a photographic negative. According to the nature of the resist the stencil is either a negative or a positive replica of the pattern of the mask.

The resist stencil thus left on the substrate protects it imagewise against chemical or physical treatment such as etching, electroplating, dying, evaporation and only the areas laid bare by the solvent are affected.

Chemically, photo-resists are solutions of natural or man-made colloids, resins, lacquers or polymers in water or in organic solvents which after drying leave a *coherent* layer on the substrate. The aqueous solutions of colloids, such as gelatin, gum arabic, albumin, polyvinylalcohol, etc. are in themselves not light-sensitive. They become sensitive to light by incorporation of a sensitizer in the solution, usually by that of an alkali bichromate or a diazo compound. The solutions of resins, lacquers or polymers in organic solvents may be light-sensitive in their own right (e.g. natural bitumen or polyvinyl-cinnamate) or they might be rendered sensitive by bichromates, diazo-compounds or azides (examples are shellac solutions, called cold top enamel, and certain rubber derivatives).

Photo-resists in Aqueous Solution. The layers from water-soluble colloids or polymers are usually developed in water or by concentrated salt solutions (e.g. solutions of calcium chloride). Although the exposed areas are strongly tanned by the exposure, the resists still allow aqueous solutions, such as acid solutions or electro-plating baths, to penetrate unless subjected to an appropriate after-treatment. This process, known as burning-in usually consists in heating the resist to a temperature somewhat below the point of charring, i.e. to about 600°F. However, in recent years aqueous solutions of polyvinyl-alcohol in conjuhction with bichromate have become very popular and after development of their layers in water are treated with certain strongly absorbed dyes, such as log-wood dye. Preferably after immersion in a solution of a bichromate or in chromic acid, they are then so strongly tanned that the stencils become impenetrable after burning-in at a much lower temperature.

Owing to the rapidly increasing use of photomechanical procedures for the production of stencils on substrates which are adversely affected by heat (e.g. laminated plastics) or which are very delicate (e.g. thin silicon or germanium wafers) the burning-in process required for water soluble photo-resists renders the latter somewhat obsolete.

Photo-resists in Organic Solvents. Modern photo-resists are therefore almost exclusively based on resins or polymers in organic solvents. These leave on the substrate a resist stencil which is highly resistant to etching or electro-plating solutions without any after-treatment.

The classical representative of this group is the light-sensitive natural bitumen, first used by Nièpce. A thin coating of bitumen becomes somewhat less soluble in organic solvents after exposure to light. However, very long exposure times are required and even then the discrimination between exposed and non-exposed areas is rather low. The bitumen process is therefore obsolete. Recent attempts for a revival do not seem to have met with success.

A group of photo-resists based on either alcoholic shellac solutions or on partly polymerized phenol-formaldehyde and similar resins attained considerable practicability and are used to this day. Exposure of the layers renders the exposed areas insoluble in dilute alkali solution or in alcohol. However, these layers are rather sensitive to temperature and moisture conditions and the chemical resistance of the stencils obtained is relatively poor. They cannot therefore compete with the very stable and reliable photo-resists of modern design.

There are numerous representatives of this group, many of them patented, and the research in this field is very actively pursued. Many, but by no means all, of them are fundamentally based on its first representative, the polyvinyl-cinnamate ester. In itself the resist is not very sensitive to light but its sensitivity is greatly enhanced by addition of certain dyes or other organic compounds to its solutions. Some photo-resists of this description are sufficiently sensitive for allowing camera exposures.

The members of this group are sensitive in their own right. Those of another group are non-sensitive and need sensitization by addition of suitable sensitizers which must be soluble in organic solvents. So certain derivatives of rubber can be sensitized by azides (of high molecular weight). Systems of this description are widely used.

The great advantage of photo-resists in organic solvents is that the coated layers are to all intents and purposes immune against temperature and moisture conditions and that as a coated layer they remain light-sensitive and developable for a long time. In this respect they are very different from bichromated colloid layers which deteriorate within a few days or even hours owing to spontaneous hardening, the so-called dark reaction.

Furthermore, the developed stencils are highly resistant to chemical and physical treatment without the need for any burning-in.

Nevertheless, this type of resists still suffers from two short-comings of lesser importance: firstly, development in organic solvents usually leaves the surface of the substrate in a rather hydrophobic, i.e. water-repellent, state. Consequently the attack of etching solutions is initially not always as uniform as one would wish for. Similarly, image-wise electro-plating requires a perfectly hydrophilic, i.e. water-accepting, surface in order to assure perfect adhesion of the plated metal to the substrate; but it is not always easy to transform the originally hydrophobic, into a satisfactory hydrophylic surface.

Secondly, the very reason that the stencils are almost impenetrable by aqueous solutions and practically insoluble in organic solutions makes it somewhat difficult to remove the stencils without mechanical aids from the surface of delicate substrates after they have served their purpose.

Positive-Positive Resists All photo-resists so far considered yield a positive stencil when exposed under a negative and vice versa. For a long time systems have been known which work in reverse, i.e. they give a positive pattern when exposed under a positive. These systems, such as the ferri-ferro gum system were originally based on water soluble colloids and they have acquired considerable practical significance, though for visual image rather than for stencil formation.

Some resists on the market are coated from solutions of polymers in organic solvents. The coated layer is highly water-, acid-, and to a certain extent alkali-resistant. By exposure the layer is converted in to a state where it is soluble in weak alkali solutions which are in fact used as developer. In this way the exposed areas are removed while the unexposed areas remain on the substrate and form the photo-stencil. Thus a positive stencil is obtained by exposure under a positive mask.

Two-layer Processes These are, strictly speaking, not photo-resists. Here a lacquer or resin layer which is not light-sensitive is coated on the substrate. This first layer is then overcoated by a second photo-resist layer, for instance, by a bichromated polyvinyl alcohol layer. After exposure and development of the top layer the photo-stencil obtained is used for imagewise removal of the bottom layer under the areas laid bare in the top layer. Removal is usually accomplished by means of a viscous solvent for the lacquer or the resin.

Future Development Research and development concerned with photo-resists aims at specimens of highest sensitivity to light combined with high resolution and a maximum of discrimination between exposed and non-exposed areas towards developers. With the present trend of the electronic industry to miniaturization, stencils are required which give safe protection against powerful acids (such as hydrofluoric acid and its mixtures with nitric acid) and against alkaline plating solutions and which are at the same time extremely thin, as only thin layers give the required resolution and edge definition. Furthermore the resist layer ought to adhere firmly to the surface of the substrate, as adhesion is a major factor in successful etching or electroplating. E.E.L.

See also: *Photo etching; Photomechanical reproduction; Template.*

RESOLVING POWER. Ability of an optical or a photographic system to reproduce fine detail. In photography, the image resolution on the final picture depends on the resolving power of the sensitive emulsion, and on that of the lens. The two are not related but the effective resolution is a function of both.

The resolving power in either case is generally specified in terms of the greatest number of lines per millimetre which can be distinctly recorded on the emulsion or separated visually in the image.

The over-all resolving power of the system is always less than the resolving power of either of its components (i.e., lens and emulsion). For reasonably accurate photographic measurements of lens resolution the sensitive material must therefore have a much greater resolving power than the lens; this will reduce the error due to interaction between the image blur of lens and emulsion. Similarly, resolution tests of sensitive materials must be carried out with as perfect a lens as possible and under conditions where it will yield the highest resolution.

Resolving power is by itself only an imperfect measure of sharpness, since other factors also enter into it. A more comprehensive physical measure is acutance.

Emulsions. With films and plates the resolving power depends on the graininess and the contrast of the emulsion, as well as on secondary factors such as thickness, colour of the exposing light, etc. Coarse grained materials naturally tend to resolve less detail than fine-grained ones, but not every fine grain film necessarily has a high resolving power. Double-coated films, as used for amateur snapshooting to ensure great exposure latitude, show very low resolving power owing to the spreading of the image through irradiation and other effects.

Measurement of emulsion resolution must be made under carefully controlled conditions of exposure and development (both of which affect the resolving power) and with a standard test object. This used to be a black-and-white high-contrast line chart. The resolution figure obtained was not, however, always indicative of how the material would behave with medium and low contrast subjects. It is therefore becoming usual to make resolution tests also with charts of reduced contrast.

Lenses. With a lens, the resolving power is a measure of the ability of a lens to form distinct images of fine, close lines. It is stated as the maximum number of lines separated by spaces of line width that the lens can reproduce as separate images. A lens is said to be able to resolve so many lines to the inch or millimetre.

The resolving power of a lens is a maximum on the axis and decreases for off-axis image-points because of the off-axis aberrations. So resolving power must be stated as being measured on the axis or at some specified point off-axis.

Imperfect correction of the coma and astigmatism of a lens produces an image patch of a point-object that is longer in one direction than in another, so the value of the resolving power varies according to the direction of the lines forming the image. For this reason the resolving power is usually specified for lines running radially from the axis, and at right-angles to it, giving radial and transverse resolving powers for any point in the field.

The decrease of resolution for off-axis image-points and the effect of the emulsion on resolution are shown very well by the following example, which is a test on a 100 mm. *f*4·5 anastigmat lens at *f*6·3 when used with commercial medium- and coarse-grain panchromatic films. The table gives the resolving power in lines per millimetre for axial and 25° off-axis image-points.

RESOLVING POWER TEST

	Medium-Grain	*Coarse-Grain*
Resolution on axis	42 lines/mm.	28 lines/mm.
Resolution at 25° off-axis	20 lines/mm.	15 lines/mm.

These results show that the resolution is better on the axis than away from it, and that medium-grain film gives better resolution than coarse.

Effect of Aperture. The effect of certain lens aberrations is reduced at smaller apertures (higher *f*-numbers), so the resolving power improves with stopping down. At small apertures diffraction effects are however liable to lead to a deterioration of image quality.

The best detail resolution of a lens is therefore obtained at a compromise aperture, small enough to lead to good lens correction and large enough to avoid deterioration due to a diffraction.

With some lenses stopping down can reduce the resolving power even before the effect of diffraction becomes noticeable, because the optimum definition may be a matter of balancing the image resolution of the central and of the marginal areas of the lens.

The resolving power depends also on the contrast of the test image used to measure it. The obtainable resolution is always higher with a high contrast image than with a low contrast one. The variation of resolving power with contrast is often expressed in the form of a contrast transfer function curve which gives a more complete indication of lens performance in terms of detail recorded than a single resolving power figure. D.P.C.

See also: *Acutance; Lens performance; Lens testing; Negative materials.*

RESTRAINER. Chemical added to inhibit the fog formation in certain developers where the developer acts both on the exposed and unexposed silver salts in the emulsion. At the same time the restrainer reduces both the development speed (increases the development time) and also the emulsion speed of the sensitive material. Generally, highly active and strongly alkaline developers need a restrainer most, while low energy fine grain developers often do not include a restrainer at all. The amount of restrainer added also depends on the sensitive material. Films and plates can stand more heavily restrained developers than papers.

Potassium bromide is the most frequently used restrainer. Its restraining action depends both on its concentration and on the other ingredients present, especially the developing agent. Generally, contrasty developers are most affected by potassium bromide. Paper developers with high potassium bromide content give warm black or often unpleasant greenish tones.

Sodium bromide can also be used in much the same way. Seven parts of potassium bromide are equivalent to about six parts of anhydrous sodium bromide or eight parts of the hydrated salt.

Ammonium bromide must not be used as restrainer unless specially called for in the formula. An alkaline developer containing ammonium bromide would liberate ammonia and thus be rather uncertain in its activity.

Very rarely, sodium chloride or potassium iodide are used as restrainers.

Anti-fogging agents are also restrainers, and are often more efficient in keeping down fog due to deterioration of the emulsion, oxidation of the developer, and other chemical effects. They do not, however, restrain the activity of the developer to the same extent as potassium bromide.

See also: *Developers.*

RETICULATION. Distortion of the emulsion layer of the negative caused generally by the use of a cold developer followed by a warm fixing bath. The emulsion wrinkles into a regular pattern for which there is no cure. The same thing may occur if a concentrated developer is followed by a plain rinse (especially in colour processing). The primary cause of reticulation—present in the above cases—is excessive swelling of the gelatine.

Reticulation can be avoided by always developing at a lower temperature than 75° F. (24° C.) and keeping all solutions within 3° F. (2° C.) of the same temperature. In colour processing in warm weather, a rinse of 5 per cent magnesium sulphate (instead of plain water) is advisable after the first development to avoid excessive swelling of the emulsion.

Reticulation is sometimes produced deliberately for the special texture it gives to the picture. This breaks up large areas of even tone, and can thus improve the pictorial effect of broad open views. In that it resembles the effect of a texture screen but once the negative is reticulated it cannot be made normal again.

There are two methods of intentional reticulation: the negative is developed at a temperature of 105–115° F. (40–45° C.). This automatically produces reticulation. Negatives that have already been developed and fixed (in an acid hardening fixer) may be bathed in a 10 per cent solution of sodium carbonate at 120–125° F. (49–50° C.). As soon as the gelatin feels wrinkled when touched gently, it is washed in cold water.

Slight accidental reticulation can sometimes be cured if the negative is bathed in alcohol (diluted with 1 part in 5 of water in the case of films) and then treated with an alum hardener.

Reticulation of the gelatin anti-curl backing occurs sometimes on films; in fact the anti-curl backing is more prone to reticulation than the actual emulsion in some cases. Fortunately this fault does not usually show in the final print.

See also: *Faults; Hot weather; Tropical photography.*

RETICULE (RETICLE). American term for graticule, a sighting or measuring aid consisting of a fine line pattern superimposed on the image visible in certain optical instruments.

See also: *Graticule.*

RETOUCHING

Any work carried out on a negative or print by hand with a brush, pencil, knife, etc., is known as retouching. The purpose of retouching is: to remove or disguise blemishes in the negative or print; to do the same thing for the sitter—e.g., to remove wrinkles; to add something to the photograph—e.g., clouds—to make it a better picture; or to work up the tones of a photograph for photomechanical reproduction.

NEGATIVES

Plates, roll and flat films present no great difficulties in retouching, although it is always better to retouch the print if convenient, since a spoilt negative cannot be replaced like a print. 35 mm. and other miniature negatives are best left alone altogether, retouching being carried out either on the print, or by making an enlarged negative and carrying out the work on that.

The plate to be retouched is mounted in a frame on a retouching desk and illuminated from behind. Films to be retouched are held down by the edges on a clear glass plate in the same type of desk.

Those who can manage to use a watchmaker's glass will find it a great advantage in carrying out very fine detail. A 4 ins. glass is about right for most people.

Dark Marks. Small negative areas—e.g., halation rings—can be reduced physically by removing part of the gelatin layer, and with it part of the silver image.

This form of local reduction is carried out by rubbing the negative surface with a fine metal polish paste, or a mixture of finely powdered pumice, vaseline, and a little paraffin wax, or with Baskett's reducer.

Physical reducers are applied on a pad of cotton wool over fairly large areas, or on a suitable stump for small areas that require more local control.

From time to time the paste is washed off

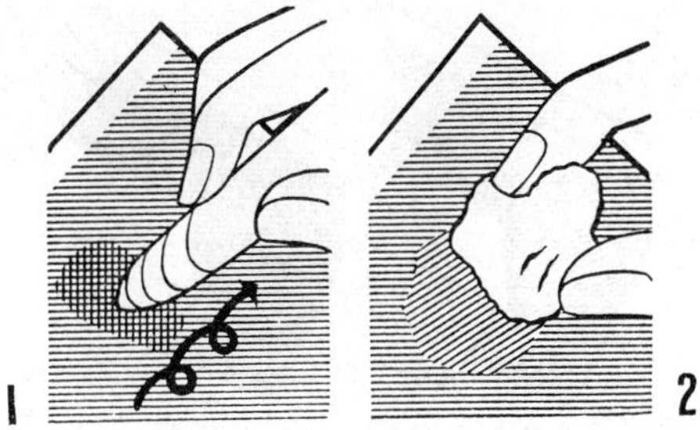

ABRASIVE REDUCTION. Clears larger, less well-defined, areas. 1. Rub stump with reducing paste applied to it over surface until area to be treated is sufficiently reduced. 2. Remove all powder by wiping with cotton wool.

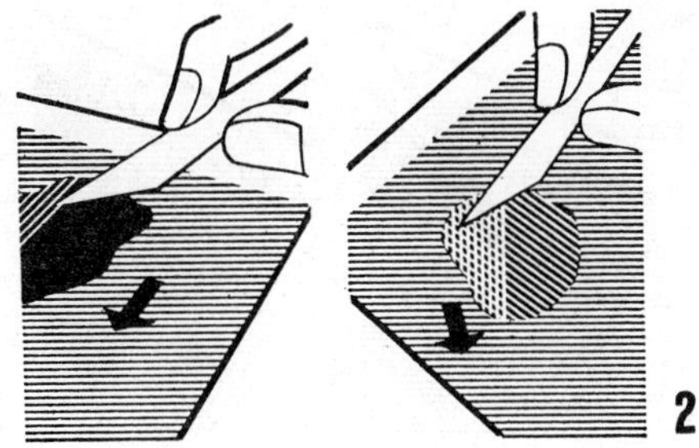

KNIFING. Lightens small areas or fine details. 1. Scrape area in close parallel strokes. 2. Overlay with second set of strokes at an angle to first to hide texture of work.

with alcohol or carbon tetrachloride, to check the progress of the reduction.

With experience, small dark areas can also be reduced by chemical means. The negative is first soaked in water charged with a little wetting agent. Then, using a pad of cotton wool, a weak solution of Farmer's reducer is applied to the area. A supply of fresh water and cotton wool, or better still a dish of clean water, should be kept handy in case any reducer goes in the wrong place: provided that the reducer is not too strong, should this happen a quick rinse in water will prevent any reduction occurring. No attempt should be made at reducing the dark area on the first application of reducer: the reducer should be weak enough to require repeated applications before showing any effect; in this way, patchy, uneven reduction is avoided. When the density has been reduced sufficiently, the negative must be well washed before drying.

Small spots or flaws are more easily removed by chemical reduction. The technique should be the same as for larger areas, except that a brush instead of cotton wool is used to apply the reducer.

Another way of removing small blemishes is by physical reduction with a retouching knife; this is a specially shaped knife, usually with interchangeable blades. The type of blade used depends on whether negatives or prints are to be retouched, but, although this is not critical, once a blade is used for one type of work, it should not be used for any other.

The blemishes are thinned by carefully scraping the gelatin layer with the point of the knife, which must be kept very sharp. The technique is simple, but the beginner should practise on an old negative. The point of the knife barely touches the surface of the negative and each stroke scrapes away only a minute shred of gelatin. Small areas are lightened by scraping first in one direction, and then at an angle until the density of the spot treated is gradually reduced to that of the surrounding image.

It is important to have the knife razor-sharp, as an unsharp edge calls for relatively heavy pressure which digs it into the gelatin and removes too much at once. When knifing is lightly and skilfully carried out the results are practically invisible.

It is even possible for actual areas of tone to be scraped away with a retouching knife. In this way dense areas can be lightened and even have modelling introduced. But the technique is extremely difficult and should only be attempted after extensive experience on waste or unimportant negatives.

Spotting. Small transparent spots and pinholes in the negative are best removed by spotting with dye or water colour.

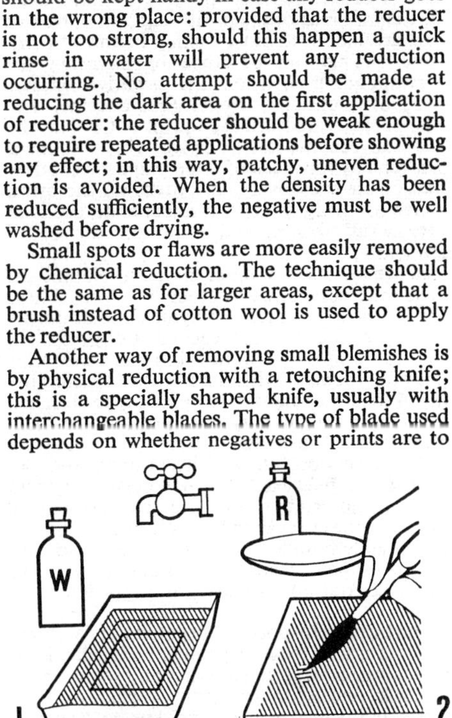

CHEMICAL REDUCTION. Useful for removing dark spots. 1. Soak negative (or print) in water with wetting agent added. 2. Touch out the spot with brush charged with Farmer's reducer.

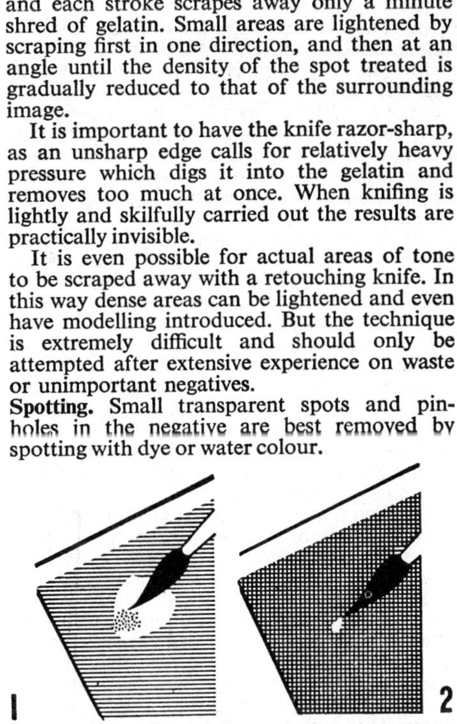

SPOTTING. Darkens small spots or areas on negatives and prints. 1. Stipple larger areas with water colour made up to suitable strength. 2. Fill in small spots in dense areas with single application of strong water colour or indian ink.

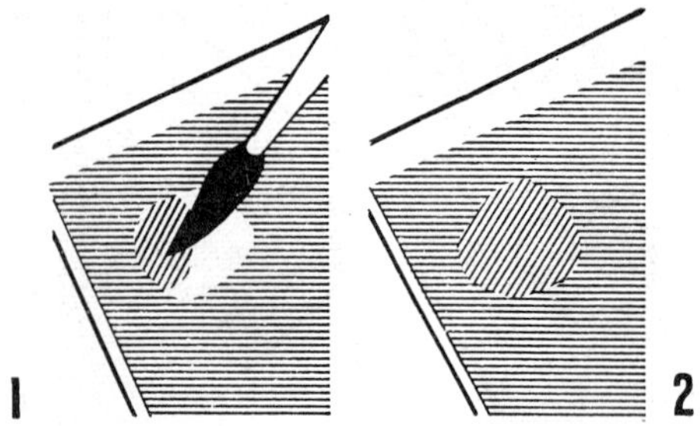

DYE SPOTTING. *Accurately matches small or larger spots. 1. Apply dilute dye to area. 2. Repeat application until surrounding density is matched.*

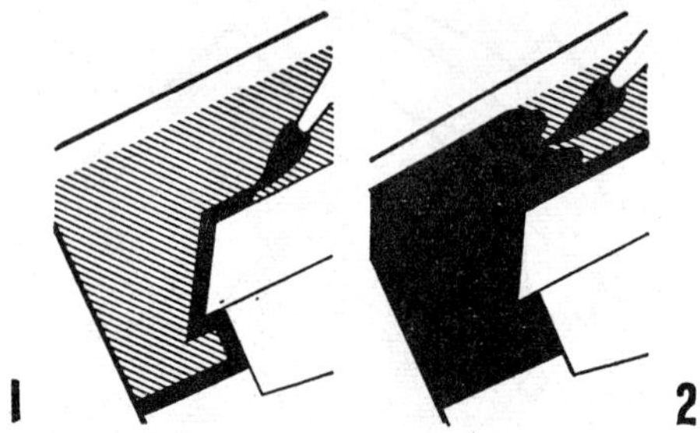

BLOCKING OUT. *Produces completely white backgrounds. 1. Paint band round subject outline on negative with black retouching dye. 2. Fill in rest of area with opaque pigment.*

A fine sable brush with a perfect point is filled with diluted grey retouching dye or water colour pigment, turned to a point on a piece of paper, and applied to the spot. The brush is used fairly dry; a wet "loaded" brush applies the colour in heavy, uncontrollable blobs.

There are distinct differences between pigments and dyes used for spotting:

(1) The dye is a solution which stains the gelatin, while the water colour is a suspension of solid pigment.

(2) The dye sinks into the gelatin, while the pigment settles and dries in a layer on top.

(3) The dye is applied by repeatedly touching the spot with dilute solution until the correct density has been built up. The pigment is applied by touching the spot once only. The strength of the pigment must therefore match the tone of the surrounding area, as repeated application would smudge the pigment already there.

(4) If too much pigment is applied, it is easily wiped off with a piece of moist cotton wool but dye can only be removed by soaking the negative in water for some time.

(5) Dye retouching is grainless, while the pigment application is grainy and may become visible on enlarging.

(6) With dye, large spots are filled in by applying repeated washes of dilute solution. With pigment, larger areas can only be covered by applying many small spots side by side—stippling the solid pigment on to the negative with the point of the brush.

(7) Dye is absorbed by the gelatin emulsion, so it can only be used where the gelatin layer is intact. But pigment, particularly if it is mixed with a little gum, will stick directly to the film or glass support, and so can be used to fill in actual holes or scratches in the emulsion.

Blocking Out. It is often necessary to remove the background of the negative completely—particularly in photographs used for catalogue illustrations, advertising, and composite pictures. This is done by blocking out the image of the subject.

To start with, a band about $\frac{1}{8}$ in. wide is painted around the outline of the subject with black retouching dye or opaque medium sold specially or this purpose. When using dye, a number of fairly weak washes are applied until the band is opaque; this technique gives a soft edge that is much less obvious than that of a layer of pigment put on in a single application.

The area outside the band may then be filled in with opaque pigment painted on the back of the negative, or covered with black paper.

It is often useful to be able to hold back particular areas of the print—e.g., an obtrusive background—without blocking them out completely. This can be done by covering the corresponding area on the negative with a wash of dilute black or red (neo-coccine) dye. The parts treated will then print paler than the rest and become less assertive. This treatment is a remedy for areas of under-exposed shadow that create excessively dark areas on the print. When applying the dye, it is better to use it in very weak washes and build up the depth gradually.

Retouching foils. A more modern method of area tone control uses retouching foils. These are thin cellulose sheets carrying a red or purple dye layer. This layer is bleached out by a special solution supplied with the foil. In use, the foil is placed in contact with the negative (preferably taped together) and the required areas bleached out—over a light box—by applying the bleacher with a brush. In this way fairly intricate outlines can be followed accurately, and desired areas held back (by not bleaching them out) during printing. The negative is of course printed in register with this foil mask. The foil is available in different densities, and two foils can be superimposed with different portions bleached out to provide varying degrees of control.

Portraits. On portrait negatives it is sometimes desirable to remove minor facial blemishes and flaws, or to soften outlines and wrinkles.

Dark marks on the print appear as light areas on the negative and can thus be "touched out" by covering them with pencil. (Light marks on the print which appear as dark marks on the negative are either knifed out on the

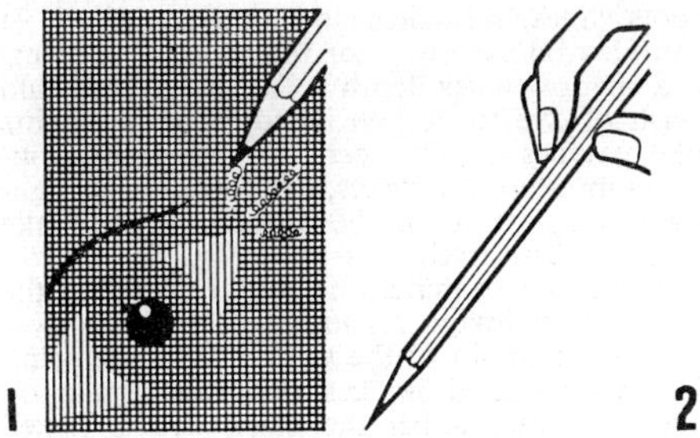

PENCIL RETOUCHING. Removes facial blemishes, etc., on portrait negatives. 1. After treating surface with retouching medium, pencil in light areas. 2. Best hold for pencil.

negative or spotted out on the print.)

As pencil will not take on gelatin, the surface is first rubbed over with a matt varnish. This can be bought or can be made up—e.g., from the following formula:

Gum dammar	2 parts
Canada balsam	1 part
Turpentine	80 parts

Method. A little of this applied to the negative with a piece of flannel, well rubbed in and allowed to dry, will provide a "tooth" for the pencil.

Retouching pencils are made in several degrees of hardness. A hard lead is best for small amounts of darkening and a softer one for adding more density.

During retouching the pencil point is rubbed with a circular motion over the spots requiring treatment. Some workers use short straight strokes, but this style tends to show up more on enlarging.

The density that can be added by pencil retouching is limited. After a few applications over the same place, the pencil point tends to polish the surface so that it will not take any more pencil work. Soft pencils give more blackening than hard ones, but even they refuse to build up density beyond a certain point.

The pressure of the pencil point must be light, or it will polish the surface smooth much sooner. Excessive pressure may push the matt varnish layer aside completely, especially if it has not been given enough time to dry.

The retouching can be removed—together with the varnish layer—by rubbing over the print surface with a plug of cotton wool soaked in petrol or turpentine.

Pencil retouching requires a good knowledge of facial anatomy if the effect is to look natural and blend with the original negative image. Retouching aimed at giving roundness to the face is only permissible where the face muscles could themselves have given roundness; it must produce face shadows only where these could arise naturally from the skin and bone structure.

Retouching Machines. For commercial purposes, where large numbers of negatives have to be retouched, machines for speeding up the operation have been produced. Not only are these fast in use, but they also make it easy for unskilled retouchers to achieve perfect results with a minimum of practice.

The negative is loaded into a carrier which is illuminated from underneath. When the machine is switched on, this carrier oscillates at a very high frequency. Because the actual movement is so slight and fast, the negative appears to be stationary. However, when touched with a pencil, a smooth mark is made which, because of the oscillations, is even and without any sharp edges. In other words, instead of the retoucher moving the pencil, the machine moves the negative. By slowly drawing the pencil across the negative, it is claimed that whole areas of tone can be built up in this way.

PRINTS

Prints to be retouched are laid face up on a flat table or desk in good light.

Most of the spots that have to be touched out on prints could be prevented by greater care at some earlier stage of making the print or negative. It always takes more trouble to touch a spot out than to prevent it.

Spots on the print are either lighter or darker than the surrounding surface. If they are lighter they can be made less noticeable by darkening them on the print. If darker, they can be touched out on the negative or reduced on the print.

Light Spots. There are three ways of darkening out a light spot on the print and they may be used in combination: by filling in the light spots with opaque pigment; by colouring the spots over with suitable dye; by spotting with a soft pencil.

Pigment Retouching. Spotting prints with pigment is a simple operation. The necessary skill can be easily acquired with a little preliminary practice on a waste print.

The expert retoucher uses good quality sable brushes with long tapering points (if the point is bent over by pressing it against a hard surface it should spring out straight as soon as the pressure is released). A Number 0 brush is suitable for normal spotting; Size 2 is used for filling in larger areas.

Good quality artists' water colours are the most suitable pigments. Cheap colours often change after exposure to air and light. Payne's grey and indian ink give the right tints for matching the grey and black tones of bromide prints. Warm tone (chlorobromide) prints may call for a slight addition of burnt umber. Burnt umber is also suitable for sepia toned prints, and if the print is a cold sepia the tone can be matched by adding a little indian ink.

These pigments are suitable for retouching matt or rough surface prints. When a glossy print is spotted with ordinary water colour, the dull brush work stands out on the shiny surface. To counteract this, a very small quantity of liquid gum is added to the water used for mixing the colour, or as a quick dodge, mix a little colour on the back of gummed paper. An alternative method is to cover the pigment with a spot of negative retouching medium.

A small quantity of Payne's grey is squeezed out upon a palette or sheet of clean glass and a small quantity of indian ink is added. A drop or two of water is mixed into the colours with a brush. The colour and depth are checked on a piece of white paper, or on the margin of the print.

It is usually more convenient to start by matching the colour to the deepest shadow so that the pigment can later be diluted with extra water for working on the lighter areas.

If the brush has been used to mix the pigment, it must be well washed, and most of the mixture removed with blotting paper. It is then rolled into the pigment so that it attains a long tapering point and the surplus pigment is wiped off the point on a sheet of clean paper. The brush is then applied lightly and the spot should disappear almost at a touch.

If the defect is too large for it to be treated with a single touch several successive applications are given, each being allowed to dry before the next is applied. The brush is used almost dry and the colour is applied sparingly. No attempt is made to "paint out" large defects; they are covered with a light stipple or a number of small dots of colour.

After the darkest tones have been spotted the pigment is diluted slightly and used to touch out the blemishes on areas that are not quite as dark, and so on up to the highest lights where the faintest trace of very dilute pigment on the brush is sufficient.

When a batch of prints is being dealt with, the darkest tones on all the prints are spotted before going on to the lighter tones.

Light lines on the print are gradually covered by a number of delicate touches or "dots" with the brush.

Those who do not wish to use water colour can purchase print spotting preparations manufactured for the purpose.

This form of retouching, since it is carried out with solid colour, is always visible as a raised mark on the print. For this reason it is unsuitable for glossy surfaces but generally satisfactory on matt or rough papers. (There is, however, nothing against using pigments for retouching glossy prints intended for reproduction.)

Dye Retouching. During recent years another way of finishing prints and modifying tone values has come into popular use: dye retouching.

Dyes are superior to pigments for retouching because they penetrate the emulsion and are quite invisible even upon a glossy surface paper.

Dye retouching can be used to take out white spots, lines, scratches and similar defects. It can also be used to alter the tones of a print; if a shadow lacks depth it can be given a thin wash of dye to deepen its tone, or highlights like patches of sky seen through trees, windows in interior subjects, or a prominent highlight in a portrait can be washed over to make them less assertive.

The aim is to match the colour as well as the depth of the image. As supplied, the black dyes match the colour of the normal bromide print. For warm-toned images or prints made on chlorobromide paper the black dye is mixed with a suitable amount of brown.

The dye is diluted with water and mixed on a porcelain palette or in an odd saucer. After mixing, it must be kept well stirred or the colours may be patchy when they dry. A little dye is taken on the brush and tested on white paper or the edge of the print. If it is too dark, more water is added until it comes to the right depth of tone. The darkest tones are dealt with first and then the dye is progressively diluted to spot correspondingly lighter tones.

The commonest fault is to take too much dye up on the brush. This forms a blob and dries too dark. If too much colour is applied it can be sometimes wiped off with wet cotton wool, but it is generally more satisfactory to soak the prints in cold water to remove the dye, and then make a fresh start.

Glossy prints are first soaked in water for a few minutes. The print is then placed upon a smooth surface, and spotting carried out on the wet print. When the print is dry the spotting will be quite invisible and after re-soaking the print can be glazed as usual. Another way of making spotting invisible on a glazed print is to add a trace of gum to the dye solution.

Pencil Retouching. When light marks on the print are quite small and the adjacent tones not too dark, a pencil may be used for spotting them. Although not so satisfactory as dye retouching, the technique is easier to master and often much more speedy to use.

A pencil with a soft lead is sharpened to a good point. The area to be filled is then lightly stroked with the pencil so as to leave very small, light dashes. The pencil should not be pressed hard against the surface. In this way, by repeatedly covering the area lightly, the tone is built up to match the adjacent tones. If a mistake is made, it can be easily removed by rubbing with a clean piece of bread, or lightened or softened with the finger tip.

Very small spots are filled by the same method—never by pressing the point of the pencil in the spot; always stroke the spot, however small. With practice, it is quite easy to make a stroke in even the smallest of spots, and a surprising speed can be attained.

When the pencil retouching has been completed, the surface of the print must either be sprayed with a fixative or steamed over a boiling kettle; this is essential to prevent the

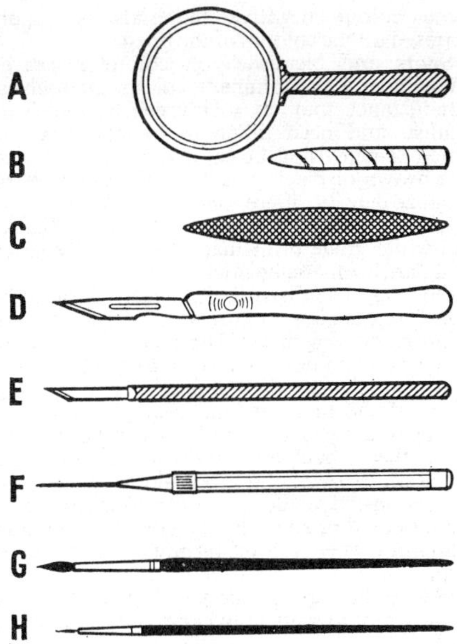

RETOUCHING TOOLS. A, *magnifying glass.* B, *paper stump.* C, *felt stump; both stumps used for applying graphite powder or abrasive reducer to negative.* D, *retouching knife taking interchangeable blades (available with various cutting edges).* E, *retouching knife with fixed blade.* F, *retouching pencil in holder.* G, *dye brush.* H, *spotting brush.*

retouching from coming off. Special fixative—as used by artists—can be sprayed on with a tubular spray which is usually sold with the fixative. If steam is used instead, the print must be held close to the spout of a boiling kettle until the gelatin surface becomes tacky.

Dark Spots. Dye retouching will only cover up light spots on dark areas (because there is no such thing as a white dye). Dark spots on light areas of the print can be avoided by touching out the corresponding clear spots on the negative with dye pigment or pencil. But there are objections to working on the negative because once a negative is spoiled there is no second chance; while miniature negatives are very difficult to retouch successfully.

It is safer, and generally easier, to convert the dark blemish on the print to a white patch by chemical retouching or knifing and then touch it out with dye or pigment.

Chemical Retouching. Chemical retouching is the process of bleaching out dark areas on the print. The process is not confined to the removal of black spots; areas of dark tone can be lightened and highlights made brighter.

The commonest reducer is called Farmer's reducer, formed by mixing potassium ferricyanide and hypo in water. The following is perhaps the easiest method for spotting prints.

The tip of the spotting brush is moistened and applied first to a crystal of ferricyanide and then to the black spot on the print. All the black spots are treated in this way and the print is put into a plain hypo fixing bath.

All black spots on the white parts of the print will disappear in a few seconds; elsewhere they will change to white spots on a darker ground and in that condition they can be easily dealt with by pigment or dye spotting.

The print is removed from the hypo bath, well washed, and dried before any further work is done on it.

If the print contains an area where the tone is too dark, it may be lightened by the same method, but in this case the action must be stopped when it has gone far enough. The action may be controlled by using a mixture of the following solutions:

Stock solution A		
Hypo	4 ounces	100 grams
Water	40 ounces	1,000 c.cm.
Stock solution B		
Potassium ferricyanide	4 ounces	100 grams
Water	40 ounces	1,000 c.cm.

About 100 minims (5 c.cm.) of B are added to 1 ounce (25 c.cm.) of A immediately before use. The mixed solution will not keep; for use it should be pale lemon yellow and should be discarded as soon as its colour disappears or changes to blue green.

The solution is painted over the dark areas of the wet print and as soon as the reducing action has gone far enough, the print is washed vigorously in clean water. To be on the safe side the action should be stopped while the tone is still a little too dark as it goes on reducing for the first second or two in the washing water.

This method of local reduction can be used to give more sparkle to highlights, to add clouds to blank skies, or to improve the tones in a portrait. It can also be used to remove unwanted objects like telegraph poles, printed notices, and advertisements that mar a landscape picture. In this case the unwanted object is completely bleached out and the white area is retouched by successive applications of black or grey dye or pigment to match the surrounding tones.

Knifing. Very small, dark spots can be quickly removed with a retouching knife, using the same technique as for negatives. However, extra care must be taken to leave a smooth surface on the print; some paper surfaces, such as glossy, show up knifing very badly unless the knife is used carefully.

As with chemical retouching, it is better to knife the spot until it is quite clear and then dye it to match the surrounding area.

Reinforcement. Various methods may be used to reinforce the tones in a print by manual application of graphite or pigment. The most popular system uses pigment in an oil medium, and is known as oil reinforcement. This is a special technique, and as such barely comes within the scope of normal retouching.

COLOUR PICTURES

Transparencies. While miniature colour transparencies are too small for retouching of detail, the transparency as a whole may need some overall colour correction to modify colour casts.

Apart from chemical methods of after treatment this can be done by binding up sheets of pale tinted cellophane or plastic with the transparency, or by tinting the whole film. That involves immersing the transparency in a suitable dye bath until the required modification of the colour is obtained.

Larger transparencies permit some retouching with dyes. Suitable dyes for this purpose are similar to those used for tinting and colouring black-and-white prints. Knife retouching and scraping are not possible on transparencies, since this would remove the three layers unevenly.

Colour Prints. Colour print retouching is also possible with water-soluble dyes of the same kind as employed for colouring prints. It is best to use several applications of the dilute dye to build up or modify a colour tone. The more dye is added, the darker the treated detail becomes. Owing to the presence of three emulsion layers, removal of dark spots and other blemishes is impossible also from a colour print. Such faults are best retouched on the colour negative, leaving lighter areas if necessary on the print, which can then be filled in afterwards.

Colour Negatives. As negative colour tones are made up of three dyes, retouching of colour negatives equally needs three colours.

The most suitable medium is a set of coloured retouching pencils in yellow, magenta and cyan plus a dark brown or black pencil to lighten shadow detail where necessary.

Judging the right colour effect needs some experience, not only because of the reversal of the colours but also because the mask tint of some colour negative materials is superimposed on the colours of the image.

Spots and blemishes on colour negatives often have a predominant colour themselves. For instance spots may be greenish, bluish or reddish and need filling in by applying the pencil or pencils of the complimentary colours. The brown or black pencil can then be used to equalize density differences.

It often helps the retoucher to have a black-and-white guide print made from the negative on a panchromatic bromide paper.

One way of assessing the effect of the individual retouching colours separately is to use coloured viewing filters. The effect of the yellow pencil retouching is checked through a blue filter (which greatly subdues the confusing effect of the magenta and cyan dye images). Similarly a green viewing filter can be used to check the retouching work with the magenta pencil on its own, and a red filter for the cyan corrections. The density of retouched spots should equal or smoothly blend with their surroundings when viewed through each filter in turn.

Before the start of the pencil work the film surface requires an application of retouching medium.

For slight corrections this can be on the emulsion side; where a great deal of retouching is necessary both sides can be treated and worked over with the pencils. During the work the negative should be in a retouching desk, lit from underneath either by daylight or by daylight blue tungsten bulbs. It is best to have a rather higher intensity of illumination than for black-and-white negatives, to permit easier judgement of the colour adjustments.

R.M.F.

See also: *Air brush; Brushes; Colouring prints; Doping prints; Intensification; Oil reinforcement; Reducing; Retouching desk; Varnishing negatives; Waxing prints.*

Books: *The Complete Art of Printing and Enlarging*, and *Retouching*, both by O. R. Croy (London).

RETOUCHING DESK. Table or stand to support and illuminate negatives during retouching. It consists of a base supporting a hinged frame which is adjustable to different angles to suit the needs of the worker. In the centre of the frame there is a recessed opening to take the negative. The opening may be fitted with a nest of adapters to hold negatives of different sizes. There may be an adjustable flap over the frame to screen off stray light. The desk is used in a window, over a sheet of white paper which acts as a reflector and illuminates the back of the negative. Instead of daylight, many operators prefer to use the diffused light from a low-powered electric lamp with the rest of the room in semi-darkness.

A retouching desk should be of rigid construction and of fairly large size. It must be capable of supporting the negative as well as the pressure of the operator's hands resting upon it. It should be possible to adjust the angle. The desk should be set upon a rigid table.

In some studios the desk consists of a table with the recessed aperture for the negative or nest of adapters actually cut in the top and illuminated from below by diffused electric lighting. Another type of desk that is popular with amateurs is made in the form of a box with a sloping top cut out to take the negative. Inside the box a small electric lamp bulb—not more than 25 watts—behind a sheet of opal glass provides the illumination. It is important for such boxes to be well ventilated to prevent the heat from the lamp from steaming up the back of the negative or even melting the emulsion.

On professional retouching desks the

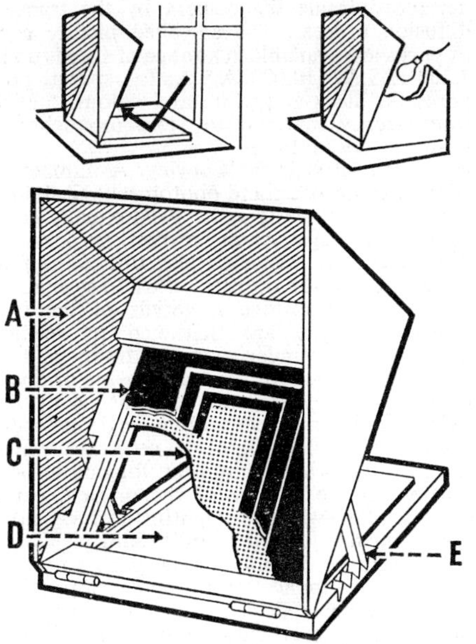

RETOUCHING DESK. Top left: *Used with daylight reflected from white sheet.* Top right: *Desk with built-in lamp.* Bottom: *Construction of the desk: A, hood to keep out stray light; B, nest of masks for different negative sizes; C, opal glass diffuser; D, reflector for even illumination of opal sheet; E, adjustment for most convenient angle of working.*

negative carrier is free to rotate in a circular recess so that the operator can turn it to the easiest position for working on.

For retouching film negatives, the aperture is first fitted with a sheet of glass—opal for preference. The negative is placed in position over the glass and held flat by a masking frame.

Certain mechanized retouching desks support the negative on an electrically-driven vibrating carrier. This simplifies pencil work. The pencil has only to be applied to the areas to be shaded and the vibration of the negative carrier produces the pencil stroke pattern.

On any type of retouching desk the negative carrier and adapters must be a good fit so that light cannot leak around the edges and shine directly into the retoucher's eyes. Very thin negatives that transmit a lot of light and dazzle the retoucher may be roughly masked down with black paper to the actual area being retouched, or the power of the light can be reduced by inserting sheets of paper between the lamp and the negative carrier. R.M.F.

RETROFOCUS SYSTEM. Term frequently applied to lens systems which have a back focal distance appreciably greater than their nominal focal length. Such lenses are used in miniature reflex cameras and in cine cameras as wide-angle lenses, because the back focal distance with a normal lens design is inadequate to accommodate the shutter, reflex mirror or other constructional features of the camera.

Retrofocus lenses are almost invariably of the inverted telephoto type. This construction is predominantly used for wide-angle lenses, but may also be found in lenses of normal focus.

See also: *Wide-angle lens.*

REVERSAL MATERIALS. Most negative materials can be processed by reversal to yield a direct positive. For the best results, however, the negative emulsion should have specific characteristics, and special reversal emulsions are therefore manufactured. Most of the black-and-white materials are narrow gauge amateur cine films, and a few 35 mm. and larger size reversal films are also available for still photography. An appreciable proportion of the colour films marketed for amateur use are designed for reversal processing. In addition, materials using special reversal systems are produced for copying, rapid photography, and similar purposes.

Normal Taking Materials. Reversal films (reversal plates are practically non-existent) for use in the camera have the following characteristics.

(1) A thin emulsion layer. The final image is formed by the silver salts left in the emulsion after the first exposure and development. The total silver halide should not therefore exceed about twice the amount made developable by the normal camera exposure. Otherwise too great camera exposures become necessary to leave only the correct amount of unexposed silver halide, i.e., the effective speed of the emulsion falls, although a perfectly good image is formed by the first exposure.

(2) A continuous and even range of grain sizes from small to large. This is important to permit processing compensation by controlling the second exposure. The silver halide left after dissolving away the negative image is not evenly sensitive everywhere: it is most sensitive where slight first exposure has left comparatively large (and therefore fast) halide grains undeveloped. It is least sensitive where heavy first exposure (corresponding to the highlights of the subject) has left only the smallest (and slowest) halide crystals unaffected. A controlled second exposure, even if not complete, will therefore yield a positive image, corresponding to the negative, on development. The quality of that image will depend on the continuity and range of halide crystal sizes in the emulsion.

(3) Freedom from fog. In the positive image fog degrades the highlights, while in the negative image produced on first development fog reduces the maximum density obtainable in the positive. Reversal films are coated on a clear (non-tinted) base, even in the 35 mm. format.

(4) Reasonably high contrast. The positive image produced on second development has to show adequate brilliance and yield a full range of tones from very thin to nearly black.

Speed of Reversal Materials. The speed criteria on which the usual film speed systems are based do not apply to reversal materials, since the result of the reversal process is a positive and not a negative image. Emulsion speeds, when quoted—even if in a standard system—therefore imply merely that the material is to be exposed like a negative film of the stated speed to yield the best results. Such speed figures are not therefore based on the same sensitometric measurements.

Other Characteristics. Reversal materials are nowadays usually available with panchromatic emulsions, though they could equally well be sensitized only orthochromatically.

As transparencies have to stand up to considerable magnification in projection (especially narrow gauge cine film), most reversal films have fine grain emulsions. The positive image produced by reversal is in any case intrinsically finer in grain than the negative from which it is derived. This is because the latter uses up most of the larger halide grains in the emulsion (see above).

To reduce the risk of halation to a minimum, some reversal films carry an anti-halo layer between the emulsion and the support. The reason for this is that a gelatin layer on the back of the film would very soon get scratched during the normal manipulation of the viewing and projection, while a tinted base would make clear highlights impossible.

Special Taking Materials. Sensitized papers are used as camera material by some street photographers. There the image produced by the camera exposure is partially developed, exposed to daylight, and allowed to develop out fully. The result is a negative with a much stronger positive image superimposed on it, so that the picture appears as a positive. The highlights are of course never white, but the picture is generally acceptable to the average customer of the photographer.

A combination of a sensitized negative paper and a non-sensitive positive paper is employed in the cameras using the Land process. The papers are loaded into the camera in separate rolls, and the negative and positive images developed inside the camera by the transfer diffusion process. The so-called picture rolls or packs are available in a range of speeds up to 3,200 A.S.A. (10,000 A.S.A. for special purposes) in black-and-white and colour, and as transparency material to yield monochrome positive transparencies.

Reversal Materials for Copying. A number of commercial process and photomechanical films and papers are designed to yield equally good results by negative or reversal processing. Special reversal materials also utilize the fog-destruction or the transfer diffusion system.

Several self-positive materials of the fog-destruction type are marketed by leading manufacturers under various trade names. They can be used for contact copying of documents either by transmitted light or by the reflex copying method. Only a single exposure (usually to yellow light) is necessary which destroys the original emulsion fogging in proportion to the density of the original being copied, after which straightforward development and fixation (or stabilization) yields the positive image. In practice the papers may be processed by application of the developer by a brush or similar device while the material is placed on an absorbent porous block, followed by the stabilizer used in the same manner. This permits rapid and certain processing even by unskilled staff in normal office conditions—where this system of copying is frequently used.

Transfer diffusion materials may use two separate supports for the negative and positive images, or carry both emulsions on one support. In the first case the two emulsions are squeegeed together in the processing solution after exposure, and stripped apart when developed; this often takes place in desk-size automatic processing machines for office use. The double emulsion type of material has a tanned positive layer; after exposure and processing the negative layer is dissolved away by warm water. L.A.M.

See also: *Cine film; Colour materials; Document photography; Negative materials; Reversal process; Transfer processes; While-you-wait photography.*

REVERSAL PROCESS

Reversal is a method of developing an image to produce a positive directly on a plate or film that has been exposed in the camera. It can serve equally for making a direct negative from a negative or a direct positive from a positive transparency.

The stages of making first the negative and then the positive are usually carried out in the same emulsion layer instead of on two different sensitive materials. Most amateur cine films and many types of colour film are processed by reversal. The advantage is that no separate printing material is called for. But only one copy can be made at a time. Any further copies have to be made by the relatively lengthy technique of duplication.

Materials. In principle, any sensitive silver halide emulsion may be processed by reversal, but there are materials which carry an emulsion with suitable characteristics for this particular method of processing.

The right characteristics for a reversal emulsion are an even range of large and small silver halide grains in the emulsion; and ability to give reasonably high contrast. Reversal colour films may utilize dyes of different absorption characteristics from those used in negative colour films which have to be printed on a positive colour material.

Special direct reversal materials are also made for document copying and similar purposes. There the positive image is produced at the same time as the negative image by a process of transfer diffusion. An alternative system uses a pre-fogged emulsion which is exposed in such a way that the fog density is destroyed by the image exposure.

Exposure. Reversal materials have a very limited exposure latitude and therefore require accurate exposure. The reason is that the additional control over density provided by the printing process is very largely absent. Over-exposure uses up too much of the sensitive salts in the emulsion for the negative stage, and does not leave enough for the final positive. Similarly, under-exposure leaves too much silver halide for the positive stage.

Exposures with reversal cine film and colour film should be based on the highlights of the subject rather than the shadows.

BLACK-AND-WHITE REVERSAL

The stages in reversal processing are:

(1) First development to form the normal negative image.

(2) Bleaching to remove the negative image. This leaves behind varying amounts of silver bromide. The amount of silver bromide left behind is inversely proportional to the silver salt used to form the first image. Where the negative image was dense, the bleaching process leaves only a little silver bromide behind; where it was thin it leaves a correspondingly greater amount.

(3) Re-exposure to white light to render the remaining silver bromide developable.

(4) Second development to blacken the residual image, producing the positive.

Results vary greatly; they depend both on the sensitive material and on the developer—particularly the first developer—used.

Not all emulsions will yield good reversal images. Very soft materials are not suitable; in fact a fairly contrasty emulsion is preferable. The special reversal materials on the market are best used with the developers recommended for them.

Duplicate negatives can be made by reversal processing of a number of normal negative materials. Usually the manufacturers publish specific recommendations and formulae. Duplicate transparencies from black-and-white transparencies should be made on clear-based emulsions, otherwise the inherent base tint may give the transparency a general and sometimes disturbing tinge.

The First Image. The character of the final image depends to a great extent on the original

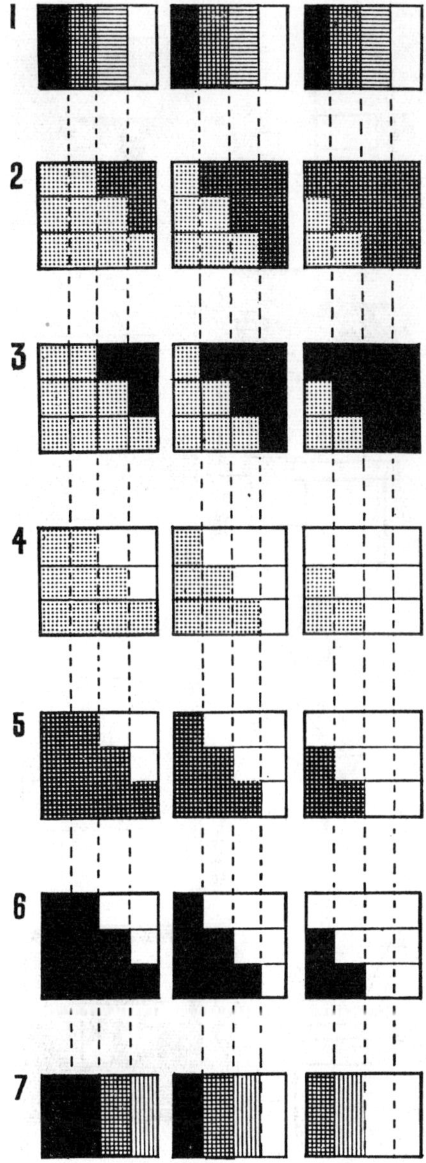

EXPOSURE AND REVERSAL MATERIALS. Left hand column: *Effect of under-exposure.* Centre column: *Correct exposure.* Right hand column: *Effect of over-exposure. In all cases: 1. Subject. 2. Latent image after camera exposure. 3. First development. 4. Silver image bleached out. 5. Remaining silver halide re-exposed. 6. Second development. 7. Corresponding image tones. Under-exposure leaves too much silver halide after first development and destroys shadow gradation of subject since the deeper shadows merge into black in the first image. Over-exposure uses up too much silver halide before second development, leaving too little for the final image, and thus destroys highlight detail.*

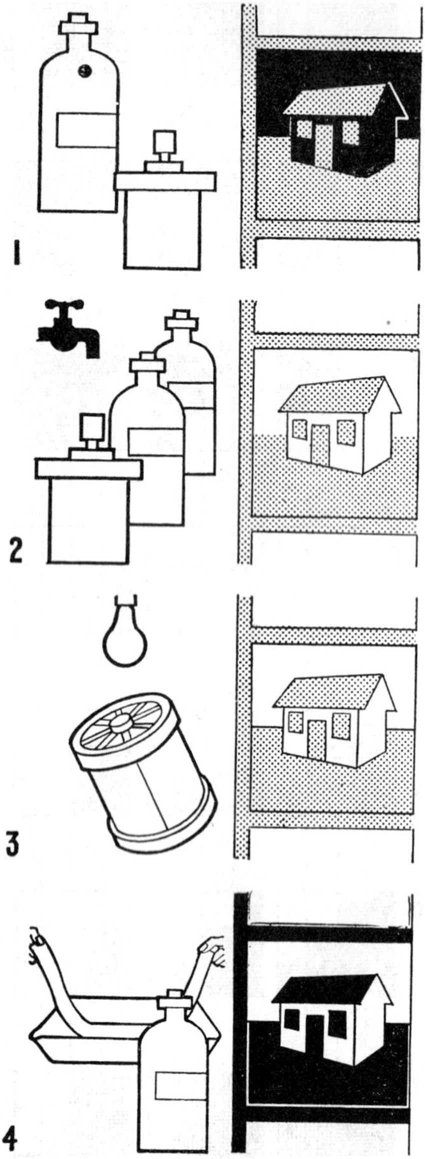

REVERSAL PROCESSING. 1. The exposed film is first developed in a normal negative developer containing a suitable concentration of silver solvent. This produces a negative image. 2. After a thorough wash, a bleacher removes all the silver produced by the first developer, leaving only a positive image of unexposed silver halide. 3. Thorough exposure of the whole film to strong artificial light renders the remaining silver halide developable. All subsequent stages can take place in normal white light (but not daylight unless adequately subdued). 4. Redevelopment in a normal negative developer blackens all the silver halide, and yields the reversed positive image of the original scene on the same film as exposed in the camera.

negative image: a weak first image usually yields a dense, second image; a contrasty and dense first image will yield a contrasty and thin second image.

The best results follow when the silver bromide used up for the first image is about the same as the silver bromide left behind after bleaching. So both the original exposure and the first development are important.

While a certain amount of control is possible in the first development, it is risky because under- or over-development affects the contrast. It is much sounder to standardize the first development so that the original exposure alone controls the density of the negative image. There will thus be very little latitude to spare when making the original exposure.

The Second Image. After bleaching, the simplest procedure is to expose the residual silver bromide fully to white light, and then develop it out completely.

If the original exposure and first development were suitably balanced, no further control is necessary. In that case it does not matter whether the re-exposure takes place before, during, or after, the bleaching process; the white light can in fact be switched on as soon as the film is in the bleacher.

Some control is possible during re-exposure and during the second development, but the latitude is small. A first image that is very thin will not produce a good final picture.

But if the first image is only slightly on the thin side, the re-exposure may be curtailed. The results (tone gradation especially) will now depend upon the stage at which the second exposure is given.

During re-exposure before bleaching, the residual silver bromide is masked by the negative image, while after bleaching the light reaches the whole emulsion evenly. After only partial re-exposure some of the silver bromide will remain undevelopable, and the final image will be less dense.

Automatic Control. In commercial processing of reversal cine film, a photocell, influenced by the amount of light transmitted by the negative image (or alternatively by the residual image) automatically adjusts the strength of the light for the re-exposure. This system is known as automatic exposure compensation. The limits of this method of control are fixed by the shadow detail and the highlight density of the negative image.

If the shadows are so thin that almost all the silver bromide is left in the emulsion after the first development, the shadow gradation in the final image will be negligible, however much the re-exposure is controlled. In other words, the shadows will be solid black.

In such a case a controlled re-exposure through the negative image before bleaching will improve matters. The silver image will itself selectively control the light reaching the shadow areas, and will improve gradation.

If the highlights of the negative image are so dense that very little silver bromide is left behind in the emulsion, even full re-exposure will not bring out much highlight detail, and it may block up the shadow completely. And the fact that the contrast of the negative image tends to be lower at high density (particularly in the shoulder region of the characteristic curve), makes matters still worse.

A certain amount of control is also possible by shortening the duration of the second development. The result is different from partial re-exposure; control during re-exposure affects the density but not the contrast of the image; control during redevelopment lowers both the density and the contrast.

Thin negative images may be controlled to some extent during redevelopment. Reducing the time in the second developer reduces the final image density, but may give far too little contrast. This method is therefore limited to negative images which are fairly thin and at the same time very contrasty. But such images are usually the result of under-exposure in the first place so they will still lack shadow detail.

An incompletely redeveloped image must be fixed to remove all silver bromide still left.

Formulae and Procedure. The first developer is generally a normal negative developer (e.g., metol-hydroquinone, or para-aminophenol). It is used at 3–4 times its normal strength, with 2–3 parts of potassium thiocyanate or sodium thiosulphate added to every 1,000 parts of working solution.

A suitable formula is:

Metol	35 grains	2 grams
Sodium sulphite, anhydrous	3½ ounces	90 grams
Hydroquinone	140 grains	8 grams
Sodium carbonate, anhydrous	2 ounces	50 grams
Potassium thiocyanate	35 grains	2 grams
Water to make	40 ounces	1,000 c.cm.

The material should be developed for the time recommended for it by the manufacturers.

Development is followed by a 2–3 minute rinse in plain water, after which the negative image is removed in the following bleaching solution (also known as the reversing bath):

Potassium bichromate	88 grains	5 grams
Sulphuric acid, concentrated	96 minims	5 c.cm.
Water to make	40 ounces	1,000 c.cm.

The bleacher is best made up in two stock solutions of 5 per cent bichromate and 5 per cent (by volume) of sulphuric acid. These are then mixed, 1 part of each with 8 parts of water, just before use. The used solution must be thrown away afterwards.

When the negative image has completely disappeared, the material is rinsed again for 3–4 minutes, and the yellow stain removed in the following clearing bath:

Sodium sulphite, anhydrous	1 ounce	25 grams
Sodium hydroxide, 10 per cent	96 minims	5 c.cm.
Water to make	40 ounces	1,000 c.cm.

This solution can be used repeatedly until it takes too long to remove the yellow stain.

After yet another wash of 1–2 minutes the emulsion layer is exposed partially (1–2 minutes), or completely (5–7 minutes), to white light (about 12 ins.—30 cm.—from a 100 watt bulb). The exposed image is then developed in a normal negative or positive developer or in the first developer. To give really high contrast a normal hydroquinone developer may be used.

If the second development is curtailed, the material is fixed to remove the undeveloped silver bromide and then washed and dried.

Direct Blackening. The operations of re-exposure and second development may be by-passed by immersing the film in an ordinary sulphide darkener. This has the same effect as complete re-exposure and re-development, but it gives a dense and contrasty sepia image.

A thiocarbamide darkener (below) works just as well, and has the advantage of being odourless.

Thiocarbamide	44 grains	2·5 grams
Sodium hydroxide	265 grains	15 grams
Water to make	40 ounces	1,000 c.cm.

This solution can be made up at 10 times the above strength, and diluted for use.

The direct darkener is useful if the film is wound on a reel in a developing tank, as there is no longer any need for removing the film. It is not, of course, suitable for redevelopment of colour films.

Direct blackening gives no opportunity for control after bleaching.

An alternative method of blackening the image is to immerse the film in a solution of equal parts 0·1 per cent thiosinamine and 0·1 per cent acetic acid. This fogs the silver bromide chemically, and renders it developable without exposure to light. After this treatment, the film need not be removed from the developing tank, but it must still be redeveloped.

IMAGE-ETCH REVERSAL

An excellent formula for reversing line images, especially when on thin film "process" emulsions, such as "lith", is the following:

Solution A		
Water (125–150° F.)	30 ounces	750 c.cm.
Copper sulphate	4¾ ounces	120 grams
Citric acid	6 ounces	150 grams
Potassium bromide	130 grains	7·5 grams
Water to make	40 ounces	1,000 c.cm.
Solution B		
Hydrogen peroxide (10 vol.)		

Before use mix equal parts of A. and B.

The initial negative image should be fully developed, so that a strong image is seen on the base side. After a short wash, and without fixing, the negative is bathed, in white light, in the above-mentioned image-etch bleach solution which dissolves away the developed silver image together with the gelatine containing it. A gentle wipe with wet cotton wool removes the last traces of the initial image, and the film is then thoroughly fogged in bright light and re-developed in the first developer.

SPECIAL METHODS

In addition to normal black-and-white photography, the reversal process has important applications in special processes and is essential in processing certain colour materials.

Colour. Reversal processing of colour materials follows the same general lines as reversal processing of black-and-white materials, except that a colour coupling developer is used for the second development. This develops the positive silver image and at the same time reacts with the colour couplers present in the three sensitive layers of the film, thus producing three-dye images superimposed on the silver images. The final stage is the removal of the silver images, leaving the pure dye images.

Diffusion Transfer. In this process the negative and positive images are developed simultaneously but in different emulsion layers.

The negative emulsion is a silver halide emulsion which is exposed normally. It is developed in contact with a plain gelatin emulsion containing a silver halide solvent and a small quantity of colloidal silver or silver sulphide. During development of the negative image, the silver halide solvent (usually hypo) diffuses from the positive emulsion into the negative. There it dissolves some of the silver halide not used by the negative, and diffuses back into the positive layer where the dissolved silver halide is reduced by the developer and deposited as silver on the silver or silver sulphide nuclei already present. The density of the image produced is proportional to the amount of unused silver halide available in any spot in the negative emulsion. After development the negative and positive emulsions are stripped apart, and washed and dried. Usually the negative requires fixing.

This process is used mainly for copying documents by transmitted or reflected light and special processing machines are available which squeegee the two emulsions together and develop them in close contact.

The Polaroid negative-positive process also uses the same principle for exposures in the camera; there the negative and positive emulsions coated on separate paper supports are loaded into the special camera in rolls. The positive paper carries small pods of jellified processing solutions. Squeegeeing the negative and positive emulsions into intimate contact breaks the pods and spreads the developer jelly evenly between the two layers. After about one minute the negative and positive are stripped apart.

A corresponding process can also yield colour prints. Here the negative material incorporates combinations of developer and dye molecules in special layers which are used to develop the appropriate negative images. The unused dye-developer molecules diffuse into the positive receiving layer and there form a positive colour print corresponding to the negative images developed in the negative film.

Some direct reversal papers using the transfer diffusion system carry both emulsions on one support. The bottom layer is the positive emulsion which is tanned after coating, while the top layer is an untanned silver halide emulsion. After exposure, development, and transfer diffusion the negative image is washed off with warm water which leaves the tanned positive layer unaffected.

Fog Destruction. This is a commercially used method of document copying by contact printing. In this process the material is uniformly fogged during manufacture so as to produce the maximum density required subsequently. When making a direct positive print with such a paper, printing destroys the original fog in proportion to the printing exposure. Development in any rapid working developer then produces the positive image.

The printing exposure generally requires light of a different colour from that used for the fogging exposure, and a yellow filter is therefore employed for printing. The speed of these materials is very low; they can be handled in subdued artificial light. L.A.M.

See also: *Cine film; Cine film processing; Colour film processing; Colour materials; Diffusion Transfer; Polacolor; Polaroid-Land camera; Reversal materials.*

Books: *Developing*, by C. I. Jacobson (London); *How to Process*, by L. J. Wheeler (London); *Processing Chemistry*, by L. F. A. Mason (London).

REVERSED NEGATIVE. Negative that has been deliberately manipulated to record a mirror image—i.e., one that has been reversed, left to right.

Such a negative is needed in certain photomechanical reproduction processes in which the final print is a reversed picture of the negative.

Image reversal is produced by photographing the original through a right-angled (reversing) prism or mirror or by stripping and turning the film.

REVERSE READING. A reproduction which is a mirror image (i.e. reversal left-to-right) of the original when viewed from the image side.

REVERSING CLUTCH. Device inside a 35 mm. camera to disengage the transport sprocket from the transport mechanism. This permits the film to be rewound into its cassette after exposure. Also known as reversing or rewind button (or lever).

See also: *Film transport.*

REVOLVING BACK. Feature of some plate cameras which permits the plate holder to be rotated in its plane. In this way horizontal, upright, or intermediate shots are possible without moving the camera, thus saving a great deal of time in setting up. Revolving backs are often found on large single-lens reflex plate cameras where it is impractical to hold the camera sideways to change the format from horizontal to upright.

REWIND. Mechanism for returning a film in a camera or projector from the take-up spool or reel to the feed spool.

Still cameras using 35 mm. (or 70 mm. or sub-miniature film) in light-tight cassettes without a backing paper progressively wind the film up on a take-up spool as it gets exposed. Since the film is then unprotected (except where a light-tight take-up cassette is also used) it has to be respooled into the feed cassette before unloading from the camera. Miniature cameras usually have a knob or crank to do this; occasionally even powered rewinds are available.

In a cine camera no rewinding of the exposed film is normally involved. However some cameras have provision for winding the film backwards in order to expose a length a second time—for special effects like dissolves and double exposures.

In cine projectors the rewind system spools the film from the take-up reel at high speed back on to the feed reel, ready for the next show.

High speed rewinds are also featured on tape recorders for the same purpose.

See also: *Cine camera; Film transport; Miniature camera; Projector (cine).*

REYNAUD, ÉMILE, 1844–1918. French showman and inventor. Used in 1877 a mirror drum for optical compensation, made in 1889–92 the projection Praxinoscope (a combination of magic lantern and Zoetrope), patented in 1888 perforated flexible strips for the synthesis of motion, and used from 1892 such strips (first drawn, later photographed) in his Théâtre Optique in Paris. After the popularization of the motion picture film he felt frustrated, sold his apparatus for scrap and threw his films into the Seine.

RHEOSTAT. Variable resistance used in electrical circuits to adjust the current flowing in the circuit. In photography rheostats are used principally for controlling the current through over-run and other high intensity filament lamps so that they can be used at a lower level of illumination when focusing the camera and brought up to full brilliance for exposure. This procedure greatly extends the life of such lamps.

It is not advisable to use such dimming resistances for varying the brilliance of lighting for exposure, because the drop in brilliance is always accompanied by a change in colour temperature.

Rheostats are made in the form of coiled lengths of resistance wire connected to the contacts of a rotary or sliding selector switch. The position of the selector arm (and the amount of current flowing in the circuit) is controlled by a knob on the outside of the case. A calibrated scale may be added.

There are two regular methods of varying the resistance in the circuit:

(1) All the resistance is connected in series with the circuit and gradually taken out to increase the current.

(2) Resistance coils are progressively added in parallel to an initial resistance coil connected in series with the circuit. As extra resistances are added, the current through the circuit increases.

Rheostats are also used for controlling the current through arc lamps. The closer the carbons are brought together, the lower the resistance of the arc and the greater the current in the circuit. A variable resistance is one way of keeping the current down to the optimum working value.

See also: *Dimmer.*

RIIS, JACOB, A, 1849–1914. American photo-reporter born in Denmark. Settled in America 1870 and became a police-court reporter for the *New York Tribune.* Began in 1887 taking flashlight photographs in New York slums of social conditions responsible for crime. His pictures influenced Theodore Roosevelt to undertake social reform measures, and today the Jacob A. Riis Neighborhood Settlement stands on the site of once notorious tenements.

RIM LIGHTING. Lighting set-up, used particularly in portraiture, where the light source is more or less behind the subject and thus outlines its contour with a brilliant rim of light.

See also: *Lighting the subject.*

RINSE. Brief wash either with clean water or a special rinsing bath given to a sensitive material in the course of processing. Its object is to remove—or neutralize—chemicals from a previous bath adhering to the negative or print; this stops the image from being discoloured in a subsequent bath and ensures that the chemicals cannot contaminate the subsequent bath and impair its action. The principal rinse is given between development and fixing.

See also: *Stop bath.*

RIPENING OF EMULSION. Essential stage in the manufacture of a photographic emulsion in which it is heated in the presence of a silver halide solvent under controlled conditions. The result is that the average size of the grains —and therefore the sensitivity—increases. Also known as Ostwald ripening.

See also: *Sensitized material manufacture.*

RISING FRONT. Lens panel constructed so that it can slide vertically to allow the lens to be raised above—or dropped below—the normal position. It is used principally to prevent the distortion of verticals in photographs of tall buildings.

See also: *Camera movements.*

ROBERTSON, JAMES. Dates unknown. English photographer. Chief engraver to the Imperial Mint at Constantinople and landscape photographer in Mediterranean countries in the 1850s. His photographs taken in the Crimea immediately following the fall of Sebastopol in September 1855 form a sequel to Roger Fenton's photographic coverage of the siege. In association with A. Beato, Robertson documented the scenes of the Indian Mutiny in 1857.

ROBINSON, HENRY PEACH, 1830–1901. English professional photographer from 1857 onward. Leading pictorialist and exhibitor. Made elaborate compositions by photomontage, the best known and earliest being *Fading Away* 1858. Prolific and influential writer on the art of photography. Founder member of the Linked Ring, 1892.

ROCHELLE SALT. Developing agent used in the kallitype process; also used as a sensitizer.

See also: *Sodium potassium tartrate.*

RODCHENKO, ALEKSANDR MIKHAILOVICH, 1891–1956. Russian artist, photographer and photojournalist. Professor at the Higher Technical Art Institute 1921–31, and theatre art director. First to use photomontage in illustrating printed publications. Considerably influenced the development of Soviet photography by creating new forms of pictorial expression.

ROENTGEN, WILHELM CONRAD, 1845–1923. German professor of physics. Discovered the Roentgen or X-rays and thus radiography (1895), which revolutionized physical theories, medical diagnostics and therapy, and the study of crystal structure. Roentgen received the Nobel Prize in 1901. Biography by F. Dessauer (Olten 1945); O. Glasser (London 1933).

ROHR, MORITZ VON, 1868–1940. German lens designer. Outstanding writer on the history of optical and ophthalmological instruments. A bibliography of his 571 books and articles was published in *Forschungen zur Geschichte der Optik*, 1943.

ROLLEIFLEX. Roll film twin-lens reflex camera, introduced in 1928 by Franke & Heidecke of Brunswick, Germany. This camera and its sister model, the Rolleicord, started the wide popularity of twin-lens cameras which appealed especially to amateurs.

At the same time, the full size image on the ground glass screen made the camera specially useful to pictorial workers and to many professional feature photographers.

The present-day Rolleiflex features automatic crank-operated film transport coupled to the shutter tensioning, automatic loading by means of film feeler rollers, an *f* 3·5 or 2·8 lens, and a Synchro-Compur shutter with exposure value scale and X-M flash synchronization.

On the whole, the Rollei cameras are intended to be self-sufficient and were not visualized as part of a camera system of accessories and special equipment. The lens, for instance, is not interchangeable. The camera thus has definite recognized limitations, and is designed to produce the best possible results within its scope.

Nevertheless the makers have produced certain accessories such as supplementary lenses for close-ups, adaptor backs and kits for using plates or 35 mm. film, and even a supplementary rangefinder.

See also: *Camera history; Reflex camera.*

ROLL FILM. One of the most popular types of negative material, particularly for amateur photography. A roll film consists of a length of celluloid, coated with a sensitive emulsion on one side and a layer of plain gelatin on the other to reduce its tendency to curl.

The film is attached to an opaque paper backing and wound on to a wooden or metal spool. The backing paper is longer than the film and the surplus length wraps around the spool to protect the film from the action of light before and after exposure in the camera. Numbers on the back of the backing paper show through a window in the camera, indicating the position of the successive frames of the film.

The earliest roll film was made in 1875 by L. Warnerke who coated a gelatin emulsion on a paper support and subsequently stripped off the developed negative for printing.

In 1889 the Eastman (later Kodak) Company made the first roll film having a celluloid support with an opaque backing paper to permit daylight loading into the camera and in 1903 they improved it by adding the anti-curl backing layer of gelatin.

See also: *Film transport; Negative materials; Sensitized materials history; Sensitized materials manufacture; Sizes and packings; Spools; Supports for emulsions.*

ROLL FILM ADAPTOR. Attachment for the back of a plate camera to enable it to take photographs on roll films. It consists of a shallow box with cylindrical chambers for the film at each end with a film aperture between, which is kept covered by a dark slide between exposures.

The adaptor can thus be removed from the camera without exposing the film to the light; it is fitted with the usual film transport knob and number window.

ROLL HEAD PRINTER. Mechanized photographic printing equipment in use in photo finishing laboratories for high print outputs. The main characteristic of the printer is that the paper is loaded in a continuous roll of a standard size (e.g. $3\frac{1}{2}$ or 5 inches wide) and fed through the exposure stage of the printing unit. The latter may be a contact printer or (more usually) an automatic enlarger with a limited range of magnifications. All prints are made the same width, the magnification of the enlarger being changed only for different negative formats. Exposures are frequently controlled electronically. After the whole paper roll is exposed, it is processed in a continuous paper processor where it passes through all treatment stages including drying (on a glazing drum) before the pictures are cut off from the roll.

See also: *Contact printer; Wholesale photo finishing.*

ROSCOE, SIR HENRY ENFIELD, 1833–1915. English chemist. Collaborator of Robert Bunsen. Together they carried out investigations on the chemical action of light which resulted in the enunciation of the Bunsen-Roscoe Law (1854/9), on photometry, actinometry and spectrum analysis. Also produced exposure tables based on the position of the sun, and advocated the use of magnesium light (1859). Autobiography: 1906.

ROSS, ANDREW, 1798–1859. English optician; established firm of optical instrument makers in London, 1830. Designed the first photographic lens in Britain for Fox Talbot, 1839, and the first British portrait lens, a double combination similar to Petzval's, for Henry Collen, 1841. Ross mentioned to Talbot the accelerating property of gallic acid, of which he had learned from the Rev. J. B. Reade.

ROYAL PHOTOGRAPHIC SOCIETY. Founded in 1853 to promote the general advancement of photographic science and its applications.

Her Majesty Queen Elizabeth II is the gracious Patron of the Society, an association with the Royal Family which has persisted from earliest days.

There are now over 7,000 members of whom about 35 per cent are resident overseas.

The Associateship (A.R.P.S.) and Fellowship (F.R.P.S.) are open to all members who can produce evidence of outstanding ability in any branch of photography.

Groups of the Society have been formed to study specialized aspects of photography, namely, colour, kinematography, medical, photo-journalism, pictorial, scientific and technical. These groups arrange conferences, lectures, demonstrations, exhibitions, visits, print circulations and criticisms, and publish bulletins, including *Photographic Abstracts*, which summarize current photographic literature. A bibliography of literature of medical photography is also maintained.

The Society publishes the *Photographic Journal* and the *Journal of Photographic Science*. These are free to members.

The Society's exhibitions are world-famous and cover all aspects and applications of photography, contributions being received from almost all nations, irrespective of size.

Over a number of years the Society has built up a collection of masterpieces of photography and of historical apparatus and equipment, which are preserved at the Society's premises and in the Science Museum, South Kensington.

The Society's house has a lounge for the use of members and a Library where books can be consulted or borrowed. Postal facilities are also available to members in the British Isles. Social functions include an annual Dinner Dance in London.

A full and comprehensive programme of lectures and meetings is arranged each year, in London and in the provinces, most of these being reproduced or reported in the Society's publications. These include lectures arranged periodically to perpetuate the memory of distinguished workers in photography.

The Society makes awards for distinguished work, the most outstanding being its Progress Medal, which is awarded in recognition of any invention, research, publication or exhibition which has resulted in any important advance in the science, art or practice of photography. This medal has been awarded on fifty occasions, many to recipients of foreign nationality.

See also: *Clubs and associations.*

RUBBER SOLUTION. Solution of rubber in naptha used in photography as a mountant for prints.

RUDOLPH, PAUL, 1858–1935. German mathematical optician. Collaborated with Abbe and calculated (1889) the characteristics of the first Zeiss anastigmats produced from the new Jena optical glasses in 1890. Rudolph calculated the new Zeiss Tessar lenses (1902) and in 1920 a Double-Plasmat (produced by H. Meyer of Görlitz). He received the Progress Medal of the Royal Photographic Society in 1905.

RUSSELL, MAJOR CHARLES, 1820–87. English amateur photographer. Invented in 1861 the tannin process (dry collodion plates for negatives preserved with tannin), and in 1862 the alkaline pyrogallol-ammonia developer. Worked also on the composition of developers (recommending potassium bromide as a restrainer) and on reversal development (1862). Described in 1866 a yellow anti-halation backing.

RUSSIA AND THE U.S.S.R.

The development of photography in Russia can be divided into two periods: pre-revolutionary and Soviet.

Early Developments. The diplomat A. P. Bestuzhev-Ryumin (1693–1766), when in Denmark, made the important observation in 1725 that light has an influence on iron salts. But photography proper began in Russia in 1839 with the photogenic drawing process. Yu F. Fritsche (1808–1871), a chemist and botanist, reported on Fox Talbot's method at a meeting of the Academy of Sciences on the 23rd May 1839. In the same year I. Kh. Gamel (1788–1862) made the acquaintance of Fox Talbot in London and sent photographs and a description of his method both in May and June 1839 to the Academy of Sciences.

The same year Gamel sent a camera and photographic equipment based on the designs of J. N. Niépce and J. M. Daguerre to the Academy. In 1840 the Moscow photographer A. F. Grekov (at the same time as the French physicist H. L. Fizeau) worked out a method of silvering copper and brass plates to reduce the mirror-like gloss of daguerreotypes. F. Arago reported on this invention by A. F. Grekov to the French Academy of Sciences on 16th November 1840.

Fox Talbot wrote to the Academy of Sciences on the 14th December 1841 enclosing several photographs made by his calotype process. In the proceedings of the Physico-Mathematical Section of the Academy of Sciences of the 7th January 1842, thanks are expressed to Talbot for his communication. Daguerreotype workshops were opened in Petersburg—Davin'on and Fokon'e, Shenfel'd, A. I. Den'er, L. S. Plakhov, Veninger, S. L. Levitskiĭ and others.

I. K. Gamel subsequently received over 160 documents on the history of the invention of photography from Isidore Niépce—letters of Nicéphore Niépce, Daguerre, Isidore Niépce and others. They are in the archives of the Academy, and were published in 1949, under the title *Dokumenti po Istorii Izobreteniya Fotografii* (*Documents on the history of the invention of photography*), edited by T. P. Kravets, corresponding member of the A.N. S.S.S.R. (U.S.S.R. Academy of Sciences).

The wet collodion process developed rapidly in Russia, thanks to the discovery of pyroxylin in 1852 by General K. K. Mann, a Russian chemist, who was in charge of the Glavnaya Laboratoriya Departmenta Vrachebnikh Zagotovlenii (Main Laboratory of the Department of Medical Supplies).

L. V. Warnerke (real name Vladislav Malakhovskiĭ, 1837–1900) was a Polish inventor who lived and worked in Russia up to the end of the 1870s, after which he worked in England and made a sensitive negative material in both his laboratories in Petersburg and London. Dry silver bromide gelatin layers appeared in Russia at the end of the 1870s. In 1878 a paper and plates with dry silver bromide gelatin layers were prepared in Warnerke's laboratory.

In 1881 Boldyrev (born 1850—date of death unknown), a Petersburg photographer, coated a transparent and flexible non-flam film base, first with collodion, and later with a dry gelatin emulsion. In 1882 he successfully demonstrated it at the All-Russian Industrial Exhibition in Moscow, but owing to lack of money was unable to achieve industrial production of his invention.

In pre-revolutionary Russia a number of original cameras and various parts of them were also invented. Of particular interest are: a camera to take spools, paper and film, designed in 1875 by L. V. Warnerke in England and a portable camera, designed in 1879 by D. P. Yezuchevsky (1835–98) and intended for expeditions. This latter camera won a prize in 1882 at the geographical exhibition in Venice, and in the same year it was demonstrated at the Industrial Exhibition in Moscow.

Stereophotography originated in Russia in the 1850s. Aleksandrovsky, in 1854 invented a stereophotographic camera which was capable of producing two photographic images simultaneously for viewing in a stereoscope. He obtained a Russian patent on the 14th December, 1854. D. P. Ezuchevsky invented a stereoscopic camera in 1875 and spent many years perfecting it. The claim for the invention of this stereoscopic camera was lodged in January 1900 after his death and Russian patent No. 5713 dated 14th September 1901 was obtained for his widow by A. Ezuchevskaya.

In 1882 S. A. Yurkovsky, a Vitebsk photographer, invented the first instantaneous roller blind shutter on which many present-day shutters are still based. He published two articles in the journal *Fotograf* giving detailed descriptions of these shutters.

In 1889 K. K. Brandel, a photographer at the Warsaw University, received Russian licence No. 11513 for a hand-operated revolving camera for taking photographs on negative paper.

In 1891 V. A. Dyubyuk (1842–1892), a Moscow photographer, invented a chronophotographic camera for obtaining a continuous sequence of photographs of galloping and racing horses. He demonstrated his photographs at a photographic exhibition in Moscow in 1892, and in the same year in Paris. They were shown on a continuously moving film using a special apparatus.

Colour photography began in Russia in the middle of the 1890s. I. Yanovsky, a photographer in Warsaw, published a description of his Chromograph for three-colour photography in March 1895 in the Journal *Fotograf—Lyubitel* (*The Amateur Photographer*). Z. Kozlovsky, a photographer in Kiev, invented a three-colour photographic camera in 1897 and received Russian patent No. 2661 dated 19th October 1899, on his application dated 1897. S. O. Maksimovich (1876–1942), Petersburg, invented a subtractive-additive method of three colour cinematography in 1909. (German patent No. 229007 of 22nd December 1909.) In 1912 he used a double-sided beam-splitting prism.

In 1899 I. L. Polyakov (born 1877—date of death not established), a student at a Moscow technical college, invented a selenium photometer for the automatic control of exposure and received Russian (No. 10116) and German (No. 117559) patents. I. L. Polyakov's photometer was the first to use two photo-elements connected in opposition, a system widely used nowadays. This patent is the first to mention the use of a photo-element for determining exposure.

In 1902 A. A. Popovitsky, a Petersburg photographer, invented a camera with spherical mirrors instead of a lens which he patented in 1904 in Russia, France and Germany.

Engineer R. Y. Tille designed a multi-lens "Panoramagraph" camera in 1898, for three-dimensional photography from aerial balloons.

During the 1904–5 Russo-Japanese War a long-focus camera designed by the Russian officer S. A. Ulyanin (born 1871; date of death not established) was used.

P. P. Sokolov, afterwards lecturer in the Moscow college of Higher Technology, published in December 1908 in the Journal *Vestnik Fotografii* (*Photographic Herald*) his invention of stereophotography using a lenticular light-sensitive material. In 1911 he published a book entitled *Avtostereoskopiya i Integral'naya Fotografiya* (*Autostereoscopy and Integral Photography*) by Professor Lipmann.

In 1911 the Russian officer V. F. Potte (1866–1918) invented the prototype of a semi-automatic aerial film camera, which was extensively used in the Air Force.

The photographer E. E. Gorin (1881–1951) lodged an application on 27th October 1916 for his invention of electrophotographic apparatus.

Early Photographic Industry. The first photographic manufacturer in Russia was A. N. Sukhachyov of Petersburg who produced papers from 1862 onwards. He was followed in 1877 by Starkov and in 1880 by L. V. Warnerke (who also with A. E. Felish made bromo-gelatin plates). In 1882 N. K. Klyachko of Moscow began the manufacture of cameras and accessories.

Other factories producing bromo-gelatin films were built in 1896—K. I. Freilandt's Vsya Rossiya, Captain Zankovsky's Pobeda and Joseph Pokorny's Iris (first in Libava and later in Moscow).

The first Russian optical firm—FOS—started in Warsaw in 1899, producing anastigmat lenses from optical glass obtained from abroad.

Very few photographic cameras were made in Russia before the revolution. According to the Customs' records, the total number of cameras which were imported into Tsarist Russia during the 25 years prior to World War I was about 500,000. In the year preceding the first World War about 16,000 cameras for amateur and professional use were imported into Russia.

After the Revolution the existing factories and workshops were transferred to the Fotokomiteti v Moskve i Petrograde (Photo-Committees of Moscow and Petrogard) in 1918 and on 27th August, 1919, V. I. Lenin signed a decree of the Soviet Narodnikh Kommisarov (Council of People's Commissars) transferring the photographic industry to the Narodny Komissariat Prosveschenіya (People's Commissariat for Education).

Organizational changes took place in the administration of the photographic industry in 1924. An optico-mechanical industry trust (T.O.M.P.) was formed and in 1926 a photo-cine-chemical trust (F.O.K.K.T.) was set up under the Higher Council for national economy, for the Russkaya Sovietskaya Federativnaya Sotsialisticheskaya Respublika (R.S.F.S.R.—Russian Soviet Federative Socialist Republic). This latter trust marked the beginning of the development in the U.S.S.R. of the photochemical and optico-mechanical industry.

Production of Materials. The chemico-photographic trust started large scale production of photographic papers in 1927. The mass production of black-and-white cinema film stock began in 1931, of multi-layer colour negative and positive photographic materials in 1948. The U.S.S.R. is now second only to the U.S.A. in film production output having risen six times since 1940. Photographic film and paper factories are today distributed in a number of towns in the U.S.S.R. and supply the requirements of the cinema, of industry and of amateurs. The products cover orthochromatic, panchromatic, and infra-red sensitive black-and-white materials for commercial, technical and scientific work, including radiography, spectrography, photomicrography, astronomy, survey and other applications. A wide range of negative materials for special purposes is made, together with cine films of high light sensitivity—up to 350 units GOST (Soviet standard) or 27° DIN.

Colour photography is finding wide application, especially the negative-positive process. Colour films are made for daylight as well as artificial light. There are also films with automatic masking components. Reversal colour film is made mainly for amateur cinematography (narrow gauge film) and application in photography is limited. Paper for colour prints in two grades of contrast and the necessary chemicals for treatment are also on the market.

Photographic papers for black-and-white photography are produced with silver bromide, chloro-bromide, and chloride emulsions in different surfaces and base tints. They are available for amateur use and for technical purposes (document and reflex printing papers, oscillograph and other recording papers, materials for photo-telegraphy, and so on).

The progress in sensitive materials in the Soviet Union may be illustrated by the increasing film speeds available:

GROWTH IN THE LIGHT-SENSITIVITY OF ONE TYPE OF SOVIET FILM

Year of Manufacture	*Spectral Properties*	*GOST Speed*
1928–1929	Orthochromatic	80
1936–1937	Panchromatic	240
1941	Panchromatic	1,000
1945	Panchromatic	1,500
1947	Panchromatic	2,500
1948	Panchromatic	4,000

From 1950 onwards, photographic plates coated with thick layers of emulsion have been prepared for research in nuclear physics and cosmic rays. In 1959 the Yu. Yanonis paper factory in Kaunass began production of electrophotographic papers.

Photographic chemicals are available separately, for mixing to individual requirements, or in sets.

Production of Equipment. Photographic lens manufacture developed in Russia in 1890. The physicist Professor A. L. Gershun (1868–1915) was the founder of the optical machine-tool industry. Academician D.S. Rozhdestvensky (1876–1940) made considerable contributions to the Russian optical and precision mechanical industry. In 1915 I. V. Grebenschikov (1887–1953), academician since 1932, and N. N. Kachalov organized the production of optical glass, the import of which was restricted owing to the First World War.

Two large optical glass factories have been built to supply the motion picture and photographic industries in the U.S.S.R. Their production programme covers a comprehensive range of advanced photographic and optical equipment.

The first mass-produced Soviet cameras using photographic plates were made in 1929 by the co-operative artel Fototrud. They were folding plate cameras of 9 × 12 cm. format fitted with imported lenses and shutters.

In the middle of 1930 T.O.M.P. produced the first models of the Fotokor amateur camera with an *f* 4·5 anastigmat. They went into mass production at the end of 1931 and by 1941 the yearly output reached 200,000. At the beginning of 1934 the Dzerzhinsky labour colony in Kharkov began to produce the FED miniature camera for 35 mm. film. This first appeared on the market in 1935 and modified variations are still being manufactured.

During world war 1941–1945 the production of amateur cameras was stopped in U.S.S.R. It was restarted in 1945, when the optic-mechanical factories began to make new models of cameras and interchangeable optical systems for them.

The number of factories producing cameras grew with the years and their output and variety increased considerably, more than sixty models of different types and classes having been produced. The present-day range covers precision miniature cameras with built-in photo-electric exposure meters, reflex cameras of different types, stereo-cameras, simple cameras for beginners, commercial equipment for copying, process work, printing and cartography, aerial cameras, phototheodolites, medical cameras, precision cine film cameras, and others.

The first experimental Soviet lenses were designed between 1925 and 1928. The production of the Ortogoz anastigmat lens for the Fotokor camera began at the end of 1929. An improved lens—the Industar—followed, and was available in focal lengths from 3·5 to 120 cm., with various maximum apertures. In the post-war period production started of high speed Jupiter lenses (up to *f* 1·5) for miniature cameras. They included normal, wide angle and tele-lenses, with coated elements for increased light transmission.

The coating of lenses was achieved in 1934 by academician I. V. Grebenshchikov. For instance, the Industar—4 lens had a light loss of 37·5 per cent before coating and of only 16 per cent after coating. In 1946, Academicians I. V. Grebenshchikov, A. G. Vlasov and N. V. Suikovskaya published a book entitled *Prosvetlenie Optiki* (*Methods of coating lenses*).

The following figures give some idea of camera production in the U.S.S.R.: there were 355,000 cameras produced in 1940, 1,022,000 in 1955, 1,472,000 in 1958, 1,616,000 in 1959 and 1,764,000 in 1960. The total yearly output of photographic equipment now exceeds 1,700,000 pieces, and continues to rise.

Connected with the change to semi-automatic and full automatic cameras with exposure control, mass production of diaphragm shutters and special objectives started. The making of semi-automatic and automatic single lens reflex cameras began in 1964.

At the same time, output of lenses which satisfy the demands of professional and amateur photographers alike continues to grow. In focal length these range from 20 mm. (for range finder cameras) to 1,000 mm. (for single lens reflex cameras); one is being manufactured with a 37 to 80 mm. variable focal length.

On sale is an adequate quantity of enlargers for all negative sizes. Light filters, auxiliary lenses and many supplementary devices are also available, as well as lighting equipment, including twelve speedlight models.

Some factories are producing photomicrographic equipment, apparatus for the sensitometric testing of photographic materials and for spectral and X-ray analysis, electron microscopes, equipment for micro-filming and other scientific and technical purposes.

Scientific Research. Work on the science of photography began in Russia in the middle of the 1870s. Questions relating to photographic chemistry and optical sensitization were elucidated by K. A. Timiryazev (1843–1920), a botanist and physiologist, afterwards academician, in many of his works: *ob Usvoenii Sveta Rosteniem* (*The Assimiliation of Light by a Plant*) (1875), *Zavisimost' Fotokhimicheskikh Yavlenii of Amplitudy Svetovoi Volny* (*The Relation between Photochemical Phenomena and the Amplitude of the Light Wave*) (1884), *Fotokhimicheskoe Deistvie Krainikh Luchei Vidimogo Spektra* (*The Photochemical Action of the Rays at the Extreme Ends of the Visible Spectrum*) (1893), *Fotografiya pripody c fotografiya v prirode* (*The Photography of Nature, and Photography in Nature*) (1895): and in his Croonian lecture to the Royal Society in London, on the cosmic role of plants (1903).

The electrochemical theory of development of the photographic image originated in 1877. V. V. Lermantov (1945–1919) in his work entitled *O fotograficheskom protsesse* (*The Photographic Process*) (1877), on the basis of his experimental results, put forward the suggestion that during the development of a photographic negative, silver was precipitated electrochemically from the developer onto the silver particles of the latent image. It was the result of the work of a local galvanic element formed by a silver particle which was soluble in the silver salt and in the developer solution.

Russian scientists have made their contribution to the development of dry silver bromide gelatin materials. V. I. Sreznevsky (1849–1937), one of the founders of the photographic section of the Russian Technical Society, and director of its laboratory in the 1880s, was largely responsible for the introduction of dry gelatin materials into Russia.

In 1880 L. V. Warnerke (V. Malakhovsky) designed a sensitometer, the first of its kind in the world, for the quantitative measurement of light sensitivity. It consisted of a glass plate with twenty-five square areas of gradually increasing density. A phosphorescing plate of calcium sulphide, activated by burning a 2·5 cm. length of magnesium ribbon, was used as the standard light source. Exposure was made 60 seconds after igniting the magnesium, and also lasted for 60 seconds.

A. L. Gershun worked on lenses and on the photography of the equatorial solar corona.

N. A. Shilov (1872–1930) explained in his book *K Teorii Fotograficheskogo Proyavitelya* (*Theory of the Photographic Developer*) (1915) the preservative part played by sodium sulphite.

S. W. Maximovich (1876–1942) obtained Russian and German patents in 1909 (No. 229007) for his three-colour method of photography and cinematography.

The work of Soviet scientists was carried on within the framework of a single state plan with constant exchange of information. In the chemical field they devoted themselves to the study of the theory and technology of the production of sensitive materials, the photochemistry of the photographic layer and the problems of the latent image, the processing of sensitive materials, trade and bulk processing methods, etc.

Emulsion Chemistry. The first task in building new factories for sensitive materials in the U.S.S.R. was to work out suitable production methods. In 1929–33 K. V. Chibisov and his colleagues studied the relationship between the conditions of formation of sensitive emulsions and their sensitometric characteristics. The basic factors in the preparation of ammoniacal and non-ammoniacal emulsions were determined as a result of this work.

Developing the theory of emulsion production, K. V. Chibisov and A. A. Mikhailova established the physico-chemical principles of the first and second ripening stages. K. V. Chibisov, together with A. A. Titov and A. A. Mikhailova, investigated the silver-gelatin complex formed during ripening and proved that the sensitivity specks consisted of metallic silver and not of silver sulphide, as had been hitherto presumed.

Worthy of note in latent image research was the work of A. L. Rabinovitch (1893–1942) and K. S. Bagdasaryan on the mechanism of the action of light on silver halide without gelatin. Certain laws of development were observed in the increase of sensitivity of silver halide during the production of photographic emulsions.

Studying the nature of the latent image, T. P. Kravets (1876–1955) and M. V. Savostyanova and their colleagues proved that under the influence of light colloidal particles of silver, which were the centres of the latent image, were formed inside the crystals of silver bromide.

T. P. Kravet's theory is based on an analogy between the photochemical formation of colour centres in alkali halide and silver halide crystals. He pointed out that the colour of coloured silver halide crystals is of the same nature as that occurring in colloidal mixtures of alkali metals, i.e. it is due to the scattering of light by these particles. This theory has been confirmed by the studies of M. V. Savostyanova, P. V. Meĭklyar and A. S. Toporets. In 1930, M. V. Savestyanova, afterwards professor of physico-mathematical sciences, published her study on the nature of the colour centres in rock salt. She made an estimation of the absorption coefficients and scattering produced by different coloured crystals of rock salt with the aim of verifying the colloidal nature of the latter, and established that the position and width of the absorption curve depends on the diameter of the particles. She also demonstrated that the colouration of coloured rock salt is due to its colloidal nature.

The electrical conductivity of the latent photographic image in silver bromide layers was studied by E. A. Kirillov, professor in the Odessa State University, and his students from 1925 onwards. Their studies have established that the spectral distribution of absorption and of the photo-effect change in like manner when a halide compound is used (silver chloride, bromide, iodide) the maximum of the spectral distribution of the photo-effect occurring in the near ultra-violet region.

In 1926 the physicist Ya. I. Frenkel (1894–1952) put forward a quantitative theory of the ionic electro-conductivity of crystals. This was based on the movements of "interstitial" ions and the "hole" which they leave behind them. This "defects" theory is a prerequisite for the theory of the formation of sensitivity centres in an emulsion microcrystal due to "defects" and "dislocations" of the crystalline lattice in the silver halide (cracks, foreign ions, colloidal particles, metal).

K. V. Chibisov (born 1897), corresponding member of the Academy of Sciences of the U.S.S.R. (since 1946), has been investigating the nature of photographic sensitivity since 1928. K. V. Chibisov, A. A. Titov and A. A. Mikhailova from 1946–1948 studied photographic sensitivity by two methods: firstly, through the kinetics of the interaction between silver ions and gelatin in a homogeneous medium, and secondly, through the kinetics of the formation of non-halide silver (metallic silver and silver sulphide) in emulsions prepared in different conditions, and making a parallel observation of the changes in their photographic properties. They came to the conclusion that the formation of the light-sensitive centres is a process of reduction, whereby the kinetics of the changes in the light-sensitivity are not connected with the kinetics of silver sulphide formation but with the formation of metallic silver, since the sulphide compounds are only accelerators of chemical ripening and are not sensitizers. In 1948, Chibisov and Titov stated that the mechanism of latent image formation which was arrived at by Gurney and Mott on the basis of completely general conceptions, agreed most satisfactorily both with the results of an experimental investigation of isolated crystals and with those obtained by a direct investigation of a photographic layer.

Chibisov emphasized that two independent kinds of process, which lead to latent image formation, take place in different regions in the silver halide crystals: the primary process originates inside the monocrystalline sheets, whereas the secondary processes are localized in the structural dislocations in the crystals.

P. D. Dankov established the correspondence between the permanent crystalline lattices in metallic silver and in silver sulphide in diagonal directions.

From 1929 onwards several scientists (V. A. Bekunov, N. V. Makarov, D. I. Virnikov and others) in different research establishments investigated the photographic activity of gelatin and worked out production methods for high activity gelatins as well as establishing a standard of quality for gelatin.

In addition to study on the main questions of emulsion production, intensive work was done on specific technological questions and coating of photographic papers. The processes of emulsion-washing were studied by V. A. Mikhailov (1933) and V. S. Koltsov (1939). Subsequently a number of workers, notably A. A. Titov, worked out various methods for separation of the solid phase from the liquid and thus eliminated washing—the most indeterminate factor in the production of photographic materials—from the industrial process.

The theory advanced by B. V. Deryagin, which established the relationships between the thickness of the emulsion layer, the viscosity of the emulsion introduced, and the speed and temperature of coating, is worthy of note among the work done on the coating of photographic emulsions. This work made it possible to secure the high-quality coating of multi-layer colour photographic materials.

A number of scientists, A. P. Zhdanov, K. S. Bogomolov, N. A. Perfilov and others investigated the preparation of thick layered photographic plates for nuclear physics.

Soviet specialists presented reports at the First and Second International Conferences on nuclear photography in Strasbourg in July, 1957 and in Montreal in 1958. The Third International Conference on nuclear photography was held in Moscow from the 12th–17th July, 1960, in which ninety-five scientists from twenty-two countries took part, and at which Soviet specialists presented thirty-four reports out of a total number of eighty-five. The proceedings of the Third International Conference in Moscow, July, 1960 were published in 1962 under the title *Yadernaya Fotografiya* (*Nuclear Photography*) under the general editorship of K. S. Bogomolov and N. A. Perfilov.

Various new emulsions which have been developed in recent years in a number of laboratories (for instance those of Bogomolov and Perfilov) represent considerable internationally recognized progress.

Scientific studies have been made in recent years on the new photographic processes—electrophotography (xerography) and thermography. Academician A. N. Terenin and E. K. Putseĭko established in 1949 that zinc oxide could be optically sensitized by means of organic dyes. In 1961 I. I. Zhilevich and E. L. Nemirovsky published the book *Elektrofotografiya* (*Electro Photography*). Questions relating to thermography and the use of light-sensitive varnishes and plastics for obtaining images are discussed in the symposium *Sovremennoe Razvitie Fotograficheskikh Protsessov* (*Modern developments in photographic processes*) published in 1960 and edited by Professor N. I. Kirillov.

In the field of optical sensitizers, I. I. Levkoyev, A. I. Kiprianov, N. N. Sveshnikov, and others in the years between 1929 and 1955 prepared a number of extremely effective orthochromatic, panchromatic and infra-red sensitizing dyes. Sensitizers with stable components were also synthesized for multi-layer colour materials. This research provided a basis for the industrial production of the required number of sensitizers.

Alongside applied questions, research was carried out into the physical chemistry of optical sensitization. A. I. Rabinovich and his colleagues (1932–8) studied the adsorption of dyes on silver bromide; Y. I. Bokinin, K. S. Bagdasaryan and others studied the photochemical mechanism of the decomposition of optically sensitized silver halide.

Processing. The development of the motion picture printing industry required considerable research into the photographic processing of film materials. In 1948 N. I. Kirillov put forward a method of continuous flow processing of sensitive emulsions.

At the same time research was in progress on the theory of photographic development and between 1925 and 1928 K. V. Chibisov studied the development process with the aid of photomicrography. A. I. Rabinovich proposed and studied in experimental form the adsorption theory of development, while P. D. Dankov between 1939 and 1949 explained the process of the generation of silver halide as being an electro-chemical reaction. The most important work in this field was done by G. P. Faierman who worked out a general theory of development.

In 1939 D. A. Frank-Kamenetsky stated that the fundamental reaction in development is the electronic oxidation of the developing agent at the surface of a silver centre of development, which, being in contact and equilibrium with silver bromide ions, are dissolved and precipitated practically instantaneously on the surface of the centre. In his opinion, the speed of the resulting process is regulated by the speed of the relatively slow process of diffusion of the developer to the surface of the centre. Kh. S. Bagdasar'yan, in his researches on the kinetics and mechanism of photographic development, proposed a theory of development as a purely electrode process. He considered that the transfer of an electron from a molecule of the developing agent to the surface of the metallic silver is not only the primary, but also the rate-determining step, whereas the secondary reaction, the discharge of silver ions at the silver surface, is very rapid.

The Soviet system of sensitometry GOST—2847—45—2817—50 was worked out theoretically and practically by Professor Yu. N. Gorokhovsky, I. A. Cherny, G. I. Kireev, S. S. Gilev, and F. L. Burmistrov under the direction of T. P. Kravets, corresponding member of the Academy of Sciences of the U.S.S.R. It is a specification for the general sensitometry (determination of the properties of a material in relation to white light) and for the determination of the absolute spectral sensitivity and resolving power of photographic materials.

In 1948 P. V. Kozlov explained the physicochemical basis of ciné film manufacture.

Colour. The production of colour images in a three-layer film by means of colour development was studied experimentally by A. E. Poraĭ-Koshets (1877–1944), a chemist and academician (since 1935). From 1939–1941 G. I. Arbuzov conducted a series of experiments on the formation of colour images in positive multilayer cine film by the silver-dye-bleach process.

Equipment. The main problems of development in the optical and precision engineering fields were computation of photographic lenses, optical apparatus, methods of sound-recording and projection of cinema film, questions of lighting techniques, the design of still and motion picture cameras, cine film processing and printing machinery, and so on.

Of all these problems, lens computation is most directly lined with photography and has been the subject of research by G. G. Slyusarev, A. I. Tudorovsky, D. S. Volosov and M. V. Tsivkin. Original types of wide angle lenses for aerial photographic equipment were calculated by M. P. Rusinov. His Lyar has an angle of 100° and his Russar one of 112°. In 1934 I. V. Grebenshchikov and his colleagues worked out and put into production methods of coating lenses.

A. A. Melnikov designed new types of shutters between 1935 and 1939, while L. P. Riftin and G. Y. Grinevich took the lead in camera design and construction.

Research and Teaching Institutes. At the suggestion of D. S. Rozhdestvensky the Gosudarstvenni Opticheski Institut (G.O.I.—State Optical Institute) was organized in 1918 in Petrograd. A photographic section which aimed to give research service to the photographic industry was established at this institute in 1926.

The Vsesoyuzni Nauchno-issledovatelski Kinofotoinstitut (N.I.K.F.I.—All-Union Scientific Research Institute of Cinematography and

Photography), was set up in Moscow in 1929.

In the same year the Institut Aerosyemki (Institute of Aerial Photography) was established in Leningrad. This latter institute was later reorganized into the Tsentralni Nauchno-issledovatelski Institut Geodezii, Aerofotosyemki i Kartografii (Central Scientific Research Institute of Geodesy, Aerial Photography and Cartography) and it is today located in Moscow.

Problems of photography were also studied by the photo-chemical laboratory of the Nauchno-Khimicheski Institut Imeni Karpova (Karpov Chemical Research Institute) and in the research institute of the printing industry.

In addition to the research institutes listed above, problems of photography are studied in the Department of Photography at Moscow University, by the Institut Metrologii i Standartizatsii (Institute of Meteorology and Standards), in the physics departments of the Universities of Leningrad and Odessa, in the department of organic chemistry of the universities of Kharkov and Kiev, by the Kharkovski Khimiko-Tekhnologicheski Institut (Kharkov Institute of Chemical Technology) and others.

A commission for scientific photography and cinematography was formed within the Akademiya Nauk S.S.S.R. in 1948. Its task is the coordination of research work, information and contacts between research institutes, factories and individuals working in different branches of photography.

The founder of photographic education in Russia was Professor V. I. Kurdyumov (1853–1904) who organized the first photographic research laboratory in 1901 at the Institut Inzhenerov Putei Soobshcheniya (Institute of Transport Communications Engineers).

The Vysshi Institut Fotografii i Fototekhniki (Higher Institute of Photography and Photographic Technique) was organized in 1918 in Petrograd at the suggestion of Petrograd scientists. This institute was later reorganized and is now the Institut Kino-inzhenerov (Leningrad Institute of Film Engineers). The Gosudarstvennaya Shkola Kinematografii (State School of Cinematography) set up in 1919, later became the Vysshi Gosudarstvenni Institut Kinematografii (Higher State Institute of Cinematography) in Moscow.

A photographic institute, opened in Kiev in 1921 was later reorganized into a film institute.

The teaching of photography and the training of qualified specialists in scientific photography and photographic technique is carried on at Moscow, Kiev, Alma-Ata, Kazan and other universities, at the Moskovski Institut Geodezii, Aerofotosyemki i Kartografii (Moscow Institute of Geodesy, Aerial Photography and Cartography), at the Institut Tonkoi Khimicheskoi Tekhnologii (Institute of Technology of Fine Chemicals) and in other teaching organizations.

Scientific Photography. Photography has been used in science since 1874.

Astrophotography and photographic astrometry began in 1874, when the astronomers B. Gessel'berg in the Pulkovsky observatory and V. K. Tserasky in the Moscow observatory obtained photographs which were suitable for making measurements of the transit of Venus across the solar disc. From 1881 to 1888 Gessel'berg systematically photographed sun spots.

Astronomer S. N. Kostinsky (1867–1936) of the Pulkovsky observatory, determined the natural motions of stars from photographs, using a stereoscopic method. From 1896 to 1936 he photographed the satellites of the outer planets. In 1906 he discovered the effect on each other of two neighbouring images on a photographic plate, which is now known as the Kostinsky effect.

Photographic astrospectrography began in Russia in 1879 when B. Gessel'berg started to make a series of spectrographic observations on photographic plates. Academician A. A. Belopol'sky (1854–1934) was one of the founders of photographic astrospectrography. In 1892 he started to photograph the spectra of stars and in 1895 he used photographic plates to determine the velocity of light from the stars by measuring the displacements of the lines in their spectra. He published more than 250 papers on astrospectrography. S. N. Blazhko in 1904 and 1907 obtained photographs of the spectra of meteors.

P. K. Shternberg (1865–1920), an astronomer at the Moscow observatory, obtained 300 photographs of double stars over the years 1902 to 1915.

In 1909 G. A. Tikhov used colour filters to determine the colour of 252 stars in the Pleiades and began to study Mars.

In 1936 A. A. Mikhaĭlov determined the Einstein effect from measurements made on two pairs of plates. In 1951 A. N. Deĭch generalized the photographic observations on double stars in the 61st Cygnus. In 1949 A. B. Severnuĭi and A. B. Gil'varg made quartz interference polarizing filters with transmission bands of 1·8 A in width for studies of the sun, and in 1950 they narrowed the width of the transmission band down to 0·5 A.

Over the period 1949–1952 Academicians G. A. Shaĭn and V. G. Gaze used very narrow-cut red interference filters to photograph the luminous clouds of interstellar hydrogen.

From 1950 onwards, Academician V. G. Tesenkov in collaboration with D. I. Rozhkovsky photographed gaseous nebulae using the red hydrogen line and a D. D. Maksutov type of meniscus telescope.

In 1948 the astronomers A. A. Kalinyak and V. B. Nikonov and the astrophysicist I. I. Krasovsky used an electron-optical infra-red image converter with photographic plates and a suitable colour filter to obtain photographs of the centres of the galaxies in infra-red rays.

On the 7th October 1959 photographs were taken on board the third Soviet cosmic rocket by means of television apparatus equipped with a photographic camera with two lenses and a small programmed installation for the automatic development and fixing of the photographic film. On a signal from the earth, the first photographs of the other side of the moon, at a distance of 65,200–68,400 kilometres from its centre, were obtained and transmitted in the course of 40 minutes. The book *Pervuie Fotografic Obratnoi Stovonui Lunui* (*The First Photographs of the Other Side of the Moon*) was published in 1959.

In recent years phase-contrast and analogous methods have been widely used in photomicrography. In 1955 the microbiologist, Professor M. A. Peshkov perfected a special method of photomicrography for studying cells in vivo.

Electron-micrography, which enables large organic molecules, ultra-microscopic viruses etc. to be studied, is widely used in physics, chemistry and biology.

Radiography of opaque objects is widely used in medicine and in non-destructive testing.

From the beginning of the 1930s fluorography has been widely used for the early detection of lung tuberculosis in mass observations of single groups of people. M. S. Ovoshchnikov, V. N. Ivanov and A. I. Prokofiev designed a tomofluorograph which combined the principles of tomography and fluorography.

From 1938–1940 L. V. Al'tshuller and V. A. Tsukerman developed a method of "impulse radiography" which enabled radiographs to be taken in several millionths of a second.

Electron diffraction photography which is based on photographing the diffractions of high speed electrons by atoms, molecules and crystalline and amorphous bodies, has been in process of development since the 1930s. P. D. Dankov and his collaborators have been making electron-diffraction investigations of the structures of oxide films on the surface of iron, copper and aluminium since 1938.

D. V. Skobel'tsuin obtained photographs of elementary particles in a Wilson chamber in 1926.

The use of photographic plates coated with thick layers of emulsion for studying nuclear reactions was suggested in 1927 by the Leningrad Professor L. V. Muisovsky. A. P. Zhdanov in 1950, in order to make accurate determinations of the nature of fission nuclei, suggested introducing into the emulsions on photographic plates certain pre-selected chemical elements, for studies of the interaction of the latter with cosmic rays. Autoradiography for the detection of invisible radioactive elements is used in biology and medicine. In 1904 Professor E. S. London was the first to use this method for determining the distribution of natural radioactive elements in an anatomical preparation. A. A. Drobkov described experiments in which he used radioactive phosphorus to obtain autoradiographs of tobacco which had been attacked by a virus disease.

Photographic Societies. The Fifth (photographic) section of the Russkoye Tekhnicheskoye Obshchestvo (Russian Technical Society) in Petersburg was formed in 1878. It published the journal *Fotograf* (*The Photographer*) from 1880–1884, also a bulletin *Sborniki Trudov Vth Otdela* (*Collected Papers of the Vth Section*). This Society organized the 1st Congress of Photography in Russia in 1882 and became the main photographic research centre in Russia. Active members of the photographic section included V. I. Sreznevsky, L. V. Warnerke, K. K. Mann, D. I. Mendeleyev, Wagner, the photographers S. L. Levitsky, A. I. Denier, D. P. Yezuchevsky and others.

The Russkoye Fotograficheskoye Obshchestvo (Russian Photographic Society) was founded in Moscow in 1894 and rapidly became an all-Russian photographic organization. It worked for the development of artistic photography in Russia.

The first conference of Russian photographers took place in Moscow in 1896 and more than twenty photographers took part. A second congress was held in 1908 in Kiev.

The Russian Photographic Society discontinued its activities until 1921 and then resumed them until 1930.

At the present time, photographic correspondents are in alliance with the photographic sections of the republican and regional unions of journalists.

Photographic sections exist to serve newspaper photographers. These sections are attached to the unions of journalists.

Photographic Exhibitions. A photographic section was organized in 1882 at the All-Russian Industrial Exhibition in Moscow. The Fifth (photographic) section of the Russkoye Tekhnicheskoye Obshchestvo organized photographic exhibitions in Petersburg in 1888, 1889, 1891, 1894 and 1898.

In 1889 the photographic section of the Society for the Propogation of Technical Knowledge in Moscow, organized a large photographic exhibition to mark the occasion of the jubilee of the discovery of photography in 1839. There were more than 1500 exhibits by professional and amateur photographers, whilst, in the same year, the photographic section of the Russian Technical Society in Petersburg arranged an exhibition, also dedicated to the jubilee of the discovery of photography.

The first international photographic exhibition was arranged in Moscow in 1902 and in Petersburg in 1903. The Akademiya Khudozhestv (Academy of Arts) organized an exhibition of artistic and technical photography in Leningrad in 1924.

A big exhibition, titled *Desyat' let Sovietskoi Fotografii* (*Ten Years of Soviet Photography*), was held in Moscow in 1927 and in 1937 an All-

Union exhibition of photographic art in the Musei Izobrazitelnikh Iskusstv im. A. S. Pushkina (Pushkin Museum of Fine Arts). Exhibitions of colour photography have been organized annually in Moscow in the post-war period. A number of exhibitions were organized also in Leningrad.

From 1960 onwards All-Union exhibitions of pictorial photography have been organized every year. In 1962, 755 photographs by 394 photographers were exhibited.

Exhibitions of Soviet photography have also been held abroad. In 1931 an exhibition of the work of the photojournalists of the magazine *Ogonek* was held in London and other towns in England and Scotland. In 1934 an exhibition of photojournalism in Warsaw met with great success. From 1935–1938 the work of Soviet experts in photojournalism was represented in forty International photographic exhibitions and photo salons.

In 1936 the International Photographic Exhibition in Prague had a large section devoted to Soviet photographic art. In 1937 an exhibition of Soviet Photographic Art was organized in London and other towns in England. The photographic section of the Union of Soviet Friendship Societies participated in a large number of international exhibitions in many countries. This society organizes thematic and personal exhibitions of Soviet photography abroad and exhibitions of foreign photographic art in Moscow.

Documentary Photography. The year 1859 saw the publication of an album *Fotograficheskiye Vidi Ierusalima i Ego Okresnostei* (*Photographic Views of Jerusalem and its Environs*); between 1870 and 1892 five volumes were published in a series *Amur, Vostochnaya i Zapadnaya Sibir* (*The Amur, Eastern and Western Siberia*). A *Turkestanski Albom* (*Turkestan Album*) was published showing the costumes, crafts, homes and architectural monuments of the peoples of Central Asia in those years. An album titled *Tipi Narodnostei Srednei Azii* (*Nationalities of Central Asia*) was specially prepared for the Third International Congress of Orientalists, held in Petersburg.

M. P. Dmitriev (1858–1948), a photographer in Nizhny Novgorod, was responsible for the development of documentary and news photography at the end of the nineteenth and at the beginning of the twentieth centuries. Over a period of years, beginning in 1888, he amassed the Volga Collection of 1,200 documentary photographs of great artistic merit.

P. A. Otzup (1883–1963) was a photographer in both pre-revolutionary and Soviet eras. He photographed the historical events of October 1917, and in 1918 produced the well-known photographs of V. I. Lenin on the platform, in the Kremlin and reading newspapers in his study.

In the Soviet period, photo-reportage work has reached high levels. Leading people in this field include P. K. Novitsky (1887–1942), who was a news photographer from 1911 and later a film camerman, M. M. Kalashnikov (1906–44), who was killed at the front during the War of 1941–5, and D. G. Debabov (1901–49).

The cosmonaut G. C. Titov during his flight in the cosmic ship "Vostok—2" on the 6th–7th August 1961 produced the first cosmic colour photographs.

Artistic Photography. This type of photography began to develop in Russia between 1850 and 1860 and its progress in the latter half of the nineteenth century is linked with the name of S. L. Levitsky (1819–98), who may be justly considered the founder of Russian professional photography. He acquired a daguerreotype camera from A. F. Grekov in 1839 and opened a daguerreotype workshop in Petersburg in 1850. He was a skilled portrait photographer, and his many portraits of Russian writers include N. V. Gogol and A. I. Gertsen. He won a gold medal at the first international photographic exhibition in Paris, 1851. A. J. Denier, a painter and member of the Russkaya Akademiya Khudozhestv (Russian Academy of Arts) opened a daguerreotype studio in Petersburg in 1853 and many leading writers and artists e.g. the Ukranian poet Tarasa Shevchenko, the Russian poets F. I. Tyutchev and N. A. Nekrasov, and the painter I. N. Kramsky were photographed there. In the years between 1840 and 1860 the Russian painters L. S. Plakhov, I. N. Kramskoi, Mikhai Zichi, A. I. Kuindzhi and others took up photography as an art. L. S. Plakhov opened a daguerreotype studio.

The Russakaya Akademiya Khudozhestv in Petersburg included photography in its art exhibition in 1876, and in 1904 and in 1908 organized special exhibitions of artistic photography.

V. A. Karrik (1827–1887) was one of the early masters of photo-journalism in Russia. Born in Edinburgh, he came to Russia with his parents in early childhood, received his primary education in Petersburg and Kronstadt and, at the age of 16, entered the architectural faculty of the Academy of Arts. In 1858–1859 he studied photography in Edinburgh under MacGregor and in 1859 he and MacGregor came to Russia and opened a photographic establishment. His collection included more than 800 photographs of different localities and types of inhabitants of the Russia of that time.

A. O. Karelin (1837–1906), a photographer in Nizhny Novgorod, is one of the leading figures in pre-revolutionary genre and portrait photography. He won a gold medal at the International Photographic Exhibition held in Edinburgh in 1880. Karelin's realistic photographic compositions were much like the painting of the Peredvizhniki (a group of leading Russian realist painters at the end of the nineteenth century).

S. A. Lobovikov (1870–1946) a photographer in Vyatka, specialized in genre photographs of country life in Russia before the revolution. N. P. Andreev (1882–1947) of Serpukhov, caught the spirit of the Russian seasons in his photographs.

The most outstanding craftsmen in artistic photography in the Soviet period include Y. P. Yeremin (1881–1948), who has a considerable collection of Crimean and Caucasian landscapes and country houses near Moscow to his credit. I. A. Bokhonov (1891–1952), instructor in pictorial photography in the Vsesoyuzni Gosudarstvenni Institut Kinematografii (All-Union State Institute of Cinematography) taught many leading photographers and film cameramen and was himself an outstanding craftsman in landscape and portrait photography.

M. S. Nappel'baum (1869–1956) was a noted craftsman in portraiture. He took the well-known portraits of V. I. Lenin in 1918.

A. M. Rodchenko (1891–1956), an artist by profession, was one of the outstanding portraitists of the younger generation. There was also V. D. Shakhovskoi (real name Pogrebshchikov), (1899–1963) photographic correspondent of the magazine *Sovetskii Soyuz* (*The Soviet Union*) who produced a series of fine landscapes, photographic portraits and work scenes, including subjects of a documentary and publicity nature.

N. A. Petrov (1876–1940), Professor at the Kievski Politekhnicheskiinstitut (Kiev Polytechnical Instaitute) is also well known as a photographic artist, a historian and teacher.

Magazines. *Fotograf* of 1864–6 was well up to the level of photographic science of its day. In the 1860's and 70's a number of magazines appeared, the most important of them being *Fotograficheski Vestnik* (*Photographic Herald*) in 1867 and 1887–1910, *Fotograficheskoye Obozreniye* (*Photographic Review*) from 1865 to 1879 and 1895–1902. The best Russian photographic magazine in the second half of the nineteenth century was the journal *Fotograf* (*The Photographer*), the organ of the fifth (photographic) section of the Russian Technical Society (1880–4). Others included *Fotograf-Lyubitel* (*The Amateur Photographer*), 1890–1909, *Russkii Fotograficheskii Zhurnal* (*The Russian Photographic Journal*), 1895–1898, *Zhurnal Russkogo Fotograficheskogo Obshchestava* (*Journal of the Russian Photographic Society*), *Vestnik Fotografii* (*Photographic Herald*), 1908–1918, and *Fotograficheskie Novosti* (*Photographic News*), 1907–1918.

During the Soviet period there have been published *Sovetskoye Foto* (*Soviet Photo*) from 1926, to 1941 and from 1957, and *Fotograf* from 1926–30, being the organ of the *Vserossiiskoye Obshchestvo Fotografov* (*All Russian Photographic Society*); *Optiko-Mekhanicheskaya Promyshlennost* (*Optico-Mechanical Industry*) from 1931–41, 1946–47 and 1956 onwards, *Fotokhimicheskaya Promyshlennost* (*Photo-Chemical Industry*) from 1932 to 1936, *Fotolyubitel* (*Amateur Photographer*), 1936–1937, and *Kinofotokhimicheskaya Promyshlennost* (*Cine-Photo-Chemical Industry*) from 1935 to 1941, being the organ of the Glavnoye Upravleniya Kinopromyshlennosti Vsesoyuznogo Komiteta po Delam Iskusstvo (Central Board of the Cinema Industry of the All-Union Committee of Art Affairs). A *Zhurnal Nauchnoi i Prikladnoi Fotografii* (*Journal of Scientific and Applied Photography*) began publication in 1956. Collections of material titled *Uspekhi Nauchnoi Fotografii* (*Progress in Scientific Photography*) have been appearing since 1951.

Books on Photography. The latest scientific works on photography are: *Osnovui Teorii Fotograficheskikh Protsessov* (*Fundamentals of the Theory of the Photographic Processes*) by Yu. M. Katushev, V. I. Sheberstov (1954). *Teoriya Fotograficheskikh Protsessov* (*Theory of Photographic processes*) by K. S. Lyalikov (Moscow 1960). *Spektral'nuie Issledovaniya Fotograficheskovo Protsessa* (*Spectroscopic Investigations of the Photographic Process*) by Yu. N. Gorokhovsky (Moscow 1960). *Fotokhimiya Krasitelei i Rodstvennuikh Organicheskikh Soedinenii* (*The Photochemistry of Dyes and Related Organic Compounds*) by A. N. Terenin (Moscow and Leningrad 1947). *Fiziko-khimiya Naneseniya Tonkikh Sloev na Dvizhushchuyusya Podlozhku* (*The Physical Chemistry of the Coating of Thin Layers on to a Moving Support*) by B. V. Deryagin and S. M. Levi (Moscow 1959). *Teoriya Nepreruivnuikh Protsessov Obrabotki Svetochuvstvitel'nuikh Materialov* (*The Theory of the Continuous Processing of Light-Sensitive Materials*) by N. I. Kirillov (Moscow 1948). *Osnovui Protessov Obrabotki Svetochuvstvitel'niukh Materialov* (*Fundamentals of the Methods of Processing Light-Sensitive Materials*) by N. I. Kirillov (1954). *Prosvetlenie Optiki* (*The Coating of Lenses*) by I. V. Grebenshchikov, A. G. Vlasov, N. V. Suĭkovskaya (1946). *Metodui Pascheta Slozhnuikh Fotograficheskikh Sistem* (*Methods of Computing Complex Photographic Systems*) by D. S. Volosov (Moscow 1948). *Teoriya i Raschet Svetoopticheskikh Sistem* (*Theory and Calculation of Light-Optical Systems*) by D. S. Volosov and M. V. Tsivkin (Moscow 1960). *Tsvetnoe Proyavlenie Trekhsloinuikh Svetochuvstvitel'nuikh Materialov* (*The Colour Development of Three-Layer Light-Sensitive Materials*) by V. S. Chel'tsov and S. A. Bongardt (1958). *Fotoreproduktsiya Nevidimogo* (*Photographic Reproduction of the Invisible*) by A. I. and G. A. Didebulidze (Tbilici 1948) *Fotograficheskie Metodi Nauchnogo Issledovaniya* (*Photographic Methods of Scientific Research*) by I. V. Sokolov (1953).

Works on the history of photography include: *Dokumentui po Istorii Izobreteniya Fotografii* (*Documents on the History of the Invention of Photography*), edited by T. P. Kravets (1949), *Ocherki po Istorii Fotografii v S.S.S.R.* (*An outline of the History of Photogrhapy in the U.S.S.R.*) by G. M. Boltyansky

(1939), *Sovetskaya Khudozhestvennaya Fotografiya* (*Soviet Artistic Photography*) by S. A. Morozov (1960), *Izobretenie Fotograficheskikh Protsessov, Niepsa, Dagerra i Talbota* (*The Photographic Processes Invented by Niepce, Daguerre, and Talbot*) by I. V. Sokolov (1953).

A number of photographic albums have been published: *Sovetskoe Fotoiskusstvo* (*Soviet Photographic Art*) (Moscow 1939), *Fotoiskusstvo S.S.S.R.* (*Photographic Art U.S.S.R.*) (Moscow 1962) *Dni i Lyudi* (*Days and People*) (1962), *Sovetskoe Fotoiskusstvo* (*Soviet Photographic Art*) (1962), *Sowjetische Meisterfotos* (*Soviet Masterpieces of Photography*), edited by M. Bugaeva (published in the German Democratic Republic, 1962).

Popular books on photography are published in editions of 100,000–350,000 copies. The book by V. P. Mikulin, 25 *Vrokov Fotografii* (*25 Lessons in Photography*) was issued in eleven editions; the number of copies of the last edition reached 1,300,000.

The works of outstanding foreign photographic specialists have been published in the Russian language, among them *The Theory of the Photographic Process* by C. E. K. Mees (New York, 1942) was published in 1949; and *Fundamentals of Photographic Theory* by T. H. James and G. C. Higgins (1948), was published in 1954.

Single articles from the foreign photographic periodicals have also been published, for example those written by T. Evans, J. W. Mitchell, et al. I.V.S. & V.A.Y.-G.

RUTHERFURD, LEWIS MORRIS, 1816–92. American physicist. Photographed the Pleiades in 1864 with wet collodion plates, using a specially corrected lens. In the same year he constructed the first telescope, of 11¼ ins. aperture, corrected for photographic rays. He also produced excellent specimens of diffraction gratings.

Also, in collaboration with Seely in New York, produced a photograph of a solar spectrum two metres long.

RUZICKA, JOSEPH, 1870–1961. Czechoslovakian medical doctor. Settled in the U.S.A. Internationally known amateur pictorialist of the period before World War I.

S

SABATTIER, ARMAND, 1834–1910. French doctor and scientist. Described in 1862 the "pseudo-solarization reversal"; a negative image on a wet collodion plate changed to a positive when daylight fell on the plate during development. This Sabattier effect was exploited for making positives by reversal, a second exposure and development being given after the first exposure and developing.

SABATTIER EFFECT. When a partly developed image is briefly exposed to actinic light, the final developed image becomes reversed. This is because in areas where the negative image had started to form, the emulsion is desensitized by the oxidation products of development. The result is that the extra exposure affects only the shadow areas of the original negative so that the final image is a positive.

This type of reversal is known as the Sabattier effect, after its discoverer.

See also: *Solarization.*

SAFELIGHT. The nature of the working illumination in the darkroom depends on the sensitive material that is being handled. Most of the films, plates, and printing papers normally used in photography are blind (or only slightly sensitive) to certain colours. So it is generally possible to illuminate the darkroom with light of a colour that does not affect the sensitive material in use if it is exposed within reasonable limits of time and distance.

Sensitized Materials. The following safelight screens are recommended for use with the various types of sensitive material. These colours may be accepted as being safe in the absence of any more specific recommendation published by the manufacturer.

Contact printing material—paper, film, lantern plates, document papers, and all very slow, non-colour-sensitive materials: fairly bright yellow.

Enlarging papers—bromide and chlorobromide papers: orange or olive green.

Orthochromatic materials—films and plates: ruby red.

Panchromatic materials—all except the very slowest films and plates: deep green or bluish green.

Very slow panchromatic materials—films and plates: dark green.

X-ray materials—films and paper: light brown or special type of olive green.

Process materials—plates, films, special lantern plates, and all fast non-colour-sensitive materials: dark brown or light red.

Desensitized panchromatic materials—films and plates: bright green.

Colour films and papers—very deep olive green or brown, according to makers' recommendations.

Fast panchromatic materials or colour films are never exposed to the direct safelight for any purpose other than inspection during development, and even then never for more than a second or two. Inspection should not take place until it is thought that development is nearing completion.

Screens. Safelight screens are made of dyed fabric, coloured celluloid or plastic, tinted glass, and coloured gelatin sandwiched between glass plates. The gelatin sandwich type is more reliable as the colours are more accurate.

The screen is fixed in front of the darkroom lamp. There are two kinds of darkroom lamp. In one, the light source shines directly through the screen; in the other, it shines on a reflector behind the screen while the direct rays are cut off by a baffle. Again, some darkroom lamps shine directly down on the work, while others shine on to the ceiling or wall.

The power of the light source depends on the type of lamp and how it is used; the safelight can become unsafe if the light source is too powerful. As a general rule, the lamp is fitted with the lowest power bulb that will give enough working illumination to eyes that have had some minutes in which to adapt themselves. The lamp must have good ventilation because the screen may change colour and become unsafe if it is allowed to get hot.

The following table gives the power of electric light bulbs recommended for various types of darkroom illumination.

POWER OF SAFELIGHT LAMPS

Illumination	*Maximum Watts*
Lamp with bulb shining directly through safe-light screen:	
(a) shining directly on to work not less than 4 feet away	15
(b) with light from lamp reflected from wall or ceiling not less than 8 feet away ...	25
Lamp with indirect illumination from bulb screened by baffle:	
(a) shining directly on to work not less than 4 feet away	25
(b) with light reflected from wall or ceiling not less than 8 feet away	40

Where negatives are likely to be held in front of the lamp for inspection it is an advantage to use a safelight screen bound up with an opal diffusing screen. In the gelatin sandwich type of screen one of the sheets of glass enclosing the gelatin is usually opal.

Coloured Light Sources. The light given out by certain light sources is restricted to a narrow band of frequencies. Such sources can be employed as darkroom safelights for materials which are insensitive to the particular frequencies emitted.

The principal light source of this type is the sodium vapour lamp which is used as a safelight in large printing establishments handling blue-sensitive materials—e.g., document copying and contact papers and negative materials.

Luminescent Panel Lighting. Dark room lighting is one of the principal applications of electro-luminiscent panel lighting. This type of light source consists basically of a layer of electro-luminescent phosphor sandwiched between two layers of conducting material, one of which at least is a conducting layer of transparent glass or plastic. When a voltage is applied to the outer layers, the phosphor glows and emits light through the transparent surface. At present the light output of such sources is low—about 5 watts per square foot—but as it is of a deep green, it is particularly suitable for use in commercial darkrooms where large quantities of parchromatic material are handled.

Self-Powered Light Sources. Luminous paint may be used without risk of fogging the sensitive material for marking the position of objects, controls, clock dials and so on. Another form of self-luminous marker coming into use consists of a sealed glass tube coated internally with a phosphor and containing a small amount of the rare gas tritium. This gas is an isotope of hydrogen which emits low energy beta particles. When these strike the phosphor they cause it to glow. The visible radiation is brighter than luminous paint and has a life of 20 years.

Sizes. Safelight screens are made in a range of standard sizes so that they can be interchanged in the same darkroom lamp according to the sensitized material being handled. The sizes normally available are:

SAFELIGHT SCREEN SIZES

Inches	*Centimetres (approx.)*
5 x 7	13 x 18
8 x 10	20 x 25
10 x 12	25 x 30
5½ circular	14

The darkroom lamps are also of standard sizes.

Testing Safelights. The most satisfactory test for a safelight is to expose a specimen of the relevant sensitized material to it under working conditions.

A piece of the material is laid, sensitized side up, on the darkroom bench with a small opaque object—e.g., a coin, lying on it. The darkroom lamp is then switched on for about twice as long as the material would normally be exposed to it—from the moment of opening the packet to the end of development.

The material is developed for the normal time well away from the safelight and then fixed and examined by white light.

If the light is not safe, the shape of the test

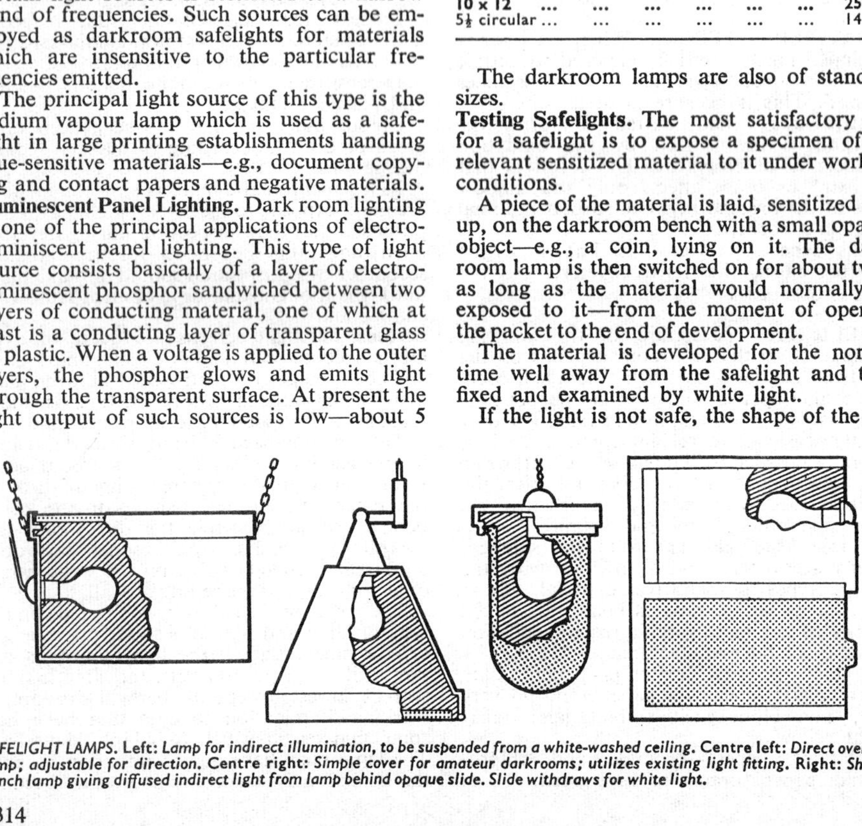

SAFELIGHT LAMPS. Left: *Lamp for indirect illumination, to be suspended from a white-washed ceiling.* Centre left: *Direct overhead lamp; adjustable for direction.* Centre right: *Simple cover for amateur darkrooms; utilizes existing light fitting.* Right: *Shelf or bench lamp giving diffused indirect light from lamp behind opaque slide. Slide withdraws for white light.*

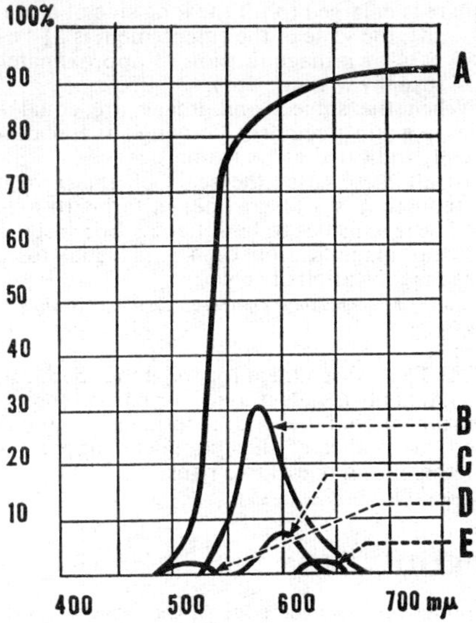

SAFELIGHT FILTERS. Transmission curves (per cent transmission for different wavelengths of light) of safelight filters as used for various types of sensitive materials. A, yellow for chloride contact papers and contact lantern plates. B, orange for bromide and chlorobromide enlarging papers, normal lantern plates, and very slow blue-sensitive materials. C, brown for X-ray films. D, deep green for panchromatic negative materials (for inspection only, not continuous exposure). E, deep red for orthochromatic films and plates.

object will stand out lighter in tone than the surrounding emulsion. In that case the lamp must be fitted with a less powerful bulb, or a deeper screen, or moved farther away from the work bench.

This method of testing does not reproduce the working conditions exactly because the actual print has already been exposed and is therefore more sensitive to further unsafe light than the fresh test piece. So the test tends to be optimistic. This is why the test piece should be left for at least twice the normal period that the printing paper is exposed to the safelight. F.P.

See also: *Filters; Spectral sensitivity.*
Book: *Developing*, by C. I. Jacobson (London).

SAFETY FILM. Negative or positive film support which is not readily inflammable, (by contrast with highly inflammable nitrate films, which were used at one time).

Usually consists of cellulose acetate; sheet film supports often consist of polyester or similar plastics.

See also: *Supports for emulsions.*

SAL-AMMONIAC. Another name for ammonium chloride; constituent of some high speed fixers.

SALOMON, ERICH, 1885–1944. German doctor of law. Became a freelance photoreporter in 1928, after working in the publicity department of a large publishing house in Berlin. With the Ermanox camera, he pioneered modern photo-journalism and is famous for his candid photographs of politicians at international conferences, taken in available light conditions. Aristide Briand called Salomon "le roi des indiscrets". Died in Auschwitz extermination camp, 1944.

SALON. Exhibition which, in photography, is usually of international status. The recommendations made by at least one national photographic organization state that, for an exhibition to be called a salon, it must allow entries from anyone (i.e., is an "open" exhibition, not confined to members of one society) and should be held in a suitable public building such as an art gallery.

See also: *Exhibitions and salons.*

SALT. Term in general use for sodium chloride (common salt).

In chemistry: the name for the class of compound produced by the reaction of an acid with a metal or molecular group of similar characteristics to a metal—e.g., silver nitrate is the salt of silver produced by the action of nitric acid on metallic silver.

SALTED PAPER. Old term for papers sensitized by the user. Their appeal lay in the fact that they allowed the photographer to use papers of a special type and surface, like Whatman paper and other hand-made papers.

The paper was first coated with a solution of common salt and gelatin in water. It was then sensitized in a silver nitrate solution and dried in the dark. The sensitized paper was then exposed under the negative in daylight to give a visible image. This was toned in a bath of potassium chloro-platinate, fixed in hypo, and washed.

See also: *Fabric printing; Printing processes (obsolete).*

SARONY, OLIVER FRANCOIS XAVIER, 1820–79. Canadian born professional photographer. Settled in Scarborough in 1857 and became the most prosperous provincial photographer in England during the carte-de-visite period. A square in the town still bears his name. In the early 1870s Sarony did good business selling licences for his secret enlarging process.

SATELLITE PHOTOGRAPHY. Recording of the flight of space satellites and rockets. Usually requires complex camera installations with telephoto optical systems and precise means of aiming the camera at the missile and keeping it in view.

See also: *Missile photography.*

SATURATED SOLUTION. Any liquid containing the maximum amount of a substance that can be dissolved in it. The amount required to produce saturation depends upon the temperature of the solution.

SAYCE, B. J., 1837–95. English amateur photographer in Liverpool. Introduced with W. B. Bolton the collodio-bromide of silver emulsion—the first workable photographic emulsion—1864.

SCALE. In a photograph, the relation between the size of the subject and the size of the negative image. If a doorway 7 feet high is reproduced as an image ¼ in. high on the negative, the scale of the image is (7 × 12 × 4) : 1 or a reduction of 336 : 1.

In an enlargement, the scale is the relation between the size of the negative image and the size of the enlargement. If a negative 2¼ ins. square is enlarged to fill the long side of a print 6½ × 8½, the scale of the enlargement is 2¼ : 8½, or 9 : 34 or a magnification of approximately 1 : 4 (usually written, 4×).

When the subject and image are equal in size—e.g., in document copying—the scale is usually indicated as s/s (same size).

In all these cases the scale of enlargement or reduction is a linear one; in terms of area, the linear scale has to be squared. For instance a linear magnification of 4× is equal to an area magnification of 16×.

See also: *Magnification; Optical calculations; Scaling for reproduction.*

SCALES. Instrument for weighing chemicals, etc. Actually a scale is a dish or pan on such an instrument, but the plural term is commonly used to indicate the complete apparatus whether it uses one or two pans.

See also: *Balances and scales.*

SCALING FOR REPRODUCTION

In the photo-mechanical reproduction of photographs there is inevitably some loss of detail whatever process is used. To offset this loss as far as possible it is advisable for the process engraver to work from a photograph larger than the size of the proposed reproduction. The size of reproduction is often limited by the width of the printer's column. It is therefore necessary to have some convenient means of scaling down the photograph which will indicate the actual size at which it will reproduce within the limitations of the printing process.

When a photograph is scaled down, all dimensions of the original are reduced in the same proportion, and it often happens that the result cannot be made to fit the printer's requirements without losing some part of the original photograph. Such a situation arises when, for example, a photograph measuring 6 × 8 ins. is required to measure 4 × 6 ins. in the reproduction. This is an impossibility, for one dimension is required at three-quarters original size and the other at two-thirds. Possible dimensions are, amongst others, 4½ × 6 ins. or 4 × 5⅓ ins., so that in either case some portion of the original will have to be trimmed off. In many instances, of course, such trimming can be done without impairing the picture, but in any event it is still necessary to mark the photograph clearly but without damaging it.

Scaling by Diagonal. The best-known way of scaling rectangular illustrations is by the diagonal method. It is only necessary to draw a rectangle the same size as the original photograph and rule in the diagonal. This diagonal is also the diagonal of any other rectangle, larger or smaller, of the same proportions. It is therefore only necessary to take a set square and measure off the width of the required reproduction from the edge of the original to the diagonal. The corresponding depth of the reproduction can then be directly read off at right angles.

By using the diagonal principle in reverse and

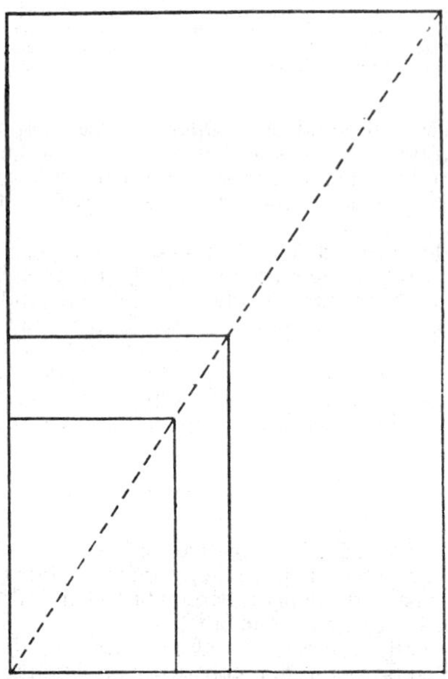

SCALING BY DIAGONAL METHOD. To determine to what size an original will reduce or enlarge, a diagonal is drawn through the required portion. This is then also the diagonal of any other rectangle with the same proportions. On measuring any required width from the edge of the original to the diagonal, the depth from that point is the corresponding depth in proportion to the measured width.

working back from the proportions of the reproduction, one can ascertain how much of an original must be trimmed off so that it can be reproduced in the required size. This is best done by scaling up the reproduction proportions on a piece of tracing paper. When this is laid over the original it will at once show the portion to be trimmed.

If the photograph is not rectangular, or if a portion is to be cut out—e.g., for insertion of a caption—the principle is difficult to apply, and it is better to use one of the alternative methods.

Norman's Rule. This method involves the use of a scale of equal units of length drawn on a strip of elastic and held in a simple frame so that it may be stretched to any desired length and held there. One side of the elastic may conveniently be marked in ¾ in. divisions and the other in 1½ in. divisions by holding the unstretched elastic along a foot rule and making marks on both sides at the required intervals. The short divisions may be used for proportions around the same size and the longer ones for reductions to avoid stretching the elastic unduly.

To use the rule, the elastic is stretched until the original apparently measures the correct size required according to the rule. If the setting of the rule is not altered, all other dimensions of the original can then be measured off against the expanded rule, the resulting readings being in the correct proportion. Thus if a 6 × 8 ins. photograph is required for reproduction at approximately 4 × 6 ins., the rule is stretched so thats the longer dimension apparently measure 6 ins., the shorter dimension, on being measured against the expanded rule, will then be found to be 4½ ins. long.

The great value of this method is that it shows at once just how much of the picture must be trimmed off to obtain the 4 ins. depth, and alternatively how much will be included if it is decided that the reproduction for instance can be 4½ ins. deep.

Another advantage of Norman's rule is that it can be used to discover how many pictures of a batch will "go together"; that is, how many of them it will be possible to reproduce at the same scale so that they can be processed together and afterwards separated into individual printing images. Such a set of subjects, mounted together, is sent to the blockmaker marked "photo together and separate" and can thus be put through at a lower charge than would apply if the pictures were reproduced individually. The pictures which are required at the same scale are found by first scaling one with the rule, then any others which measure their required size without altering the rule can be photographed together with the first.

The different lengths of the units on either side of the elastic rule make it possible to deal with any scale from enlargements to considerable reductions. The longer divisions stretch out to give even bigger units, and so deal with reductions, while the shorter divisions deal with slight enlargements and scales around same size. For scales beyond the scope of the rule it is easy to read two inches as one, or vice versa for enlargements.

The only drawback of Norman's Rule is that it is not absolutely accurate. However one can size up a job to within about $\frac{1}{16}$-inch, which is close enough for most purposes. When the elastic becomes worn, it can easily be replaced.

Slide Rule. A slide rule provides an easy and completely accurate scaling device which in addition indicates the percentage scale of reproduction. On most slide rules reproduction sizes are, however, given in the decimal system but it is possible to obtain special slide rules marked in the more convenient inches, eighths and sixteenths, and in printers' 12 point ems.

Computer Disc. A computer disc, consisting of two concentric scales rotating independently on the common centre, may be used instead of a slide rule. The outer disc is calibrated for the original size, the inner for the reproduction size, and a scale number pointer is provided on the former indicating the percentage magnification. To use the computer disc, the actual width (or depth) of the original is set on the outer scale against the desired width (or depth) on the reproduction scale. All

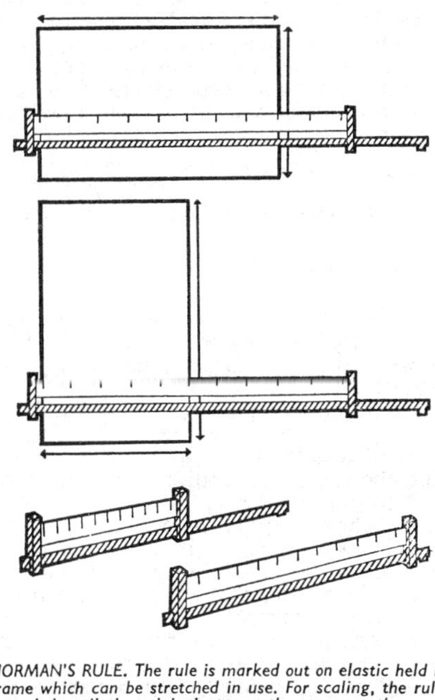

NORMAN'S RULE. The rule is marked out on elastic held in a frame which can be stretched in use. For scaling, the rule is extended until the original apparently measures the correct size in one direction. If the setting of the rule is not altered, all other dimensions of the original will measure the size to which they will reproduce at the set proportion.

other readings are then in proportion, and the percentage magnification is indicated on the scale by the pointer.

The percentage magnification value has two important applications. If several photographs are to be reproduced, all those with the same percentage value can be placed in the process camera at one time, with a corresponding saving in time and cost. Many modern process cameras are scaled directly in percentages and provided with automatic focusing, so that it is only necessary to pre-set the instrument to the required percentage figure and all calculation is eliminated.

Scaling Rule. At least one special scaling rule is available for calculating the relative dimensions of original and reproduction. There is a duralumin rule of the slide rule type with two adjustable pointers. One pointer is set to a dimension of the picture and the other to the size required. This sets the rule for the particular scale of reproduction so that when the first pointer is moved to the length of the other dimension the rule indicates what size it will be when reproduced to that scale. The rule covers any sizes from 1 to 150 (inches or centimetres); it indicates the percentage magnification and it can also be used for calculating the size required to give a certain fraction of the area of the original picture or the dimensions which will produce a stated area.

Marking up Prints. The safest procedure, to avoid damaging the face of the print, is to mark the back of the print with ink and a brush. Alternatively, a piece of tracing paper stuck along the top back edge and folded over the front is useful for indicating the portion to be included and the size. On no account should a ball pen be used for either writing or marking up on either front or back of the print, as the resulting marks can never be erased and will show up in reproduction. Instructions to the process engraver may be lightly written on the back of the print with a soft pencil; this is safer than writing them elsewhere, as they cannot then get lost.

If the whole picture is to be reproduced it is often sufficient to write "reduce to X inches wide" so as to conform to the particular column width, and leave the depth to work out in proportion. Very often the depth of a block does not matter within reasonable limits, as a fraction of an inch one way or another can be taken up by leading the text above or below at the make-up stage.

Marking up Transparencies. On no account should marks be made on the transparency.

The best way with any transparency, monochrome or colour, positive or negative, is to make a simple bromide print from it, preferably to the required size, and to indicate precisely what is required on the print. Whether the print is negative or positive, and whether the tone-values are falsified by the lack of colour sensitivity of the bromide paper, is of no importance, as photo-engravers are accustomed to such effects and they will of course make the reproductions from the transparency itself.

If the whole of the picture is to be included in the reproduction, the dimensions can be stated on the envelope enclosing the transparency, but very often it is necessary to indicate the portion of the picture to be reproduced. This can be done by cutting a paper or card mask and taping the transparency upon it, but it has to be removed for the reproduction work (and probably colour-masking); it is also particularly inconvenient with a miniature transparency. F.H.S.

See also: *Photomechanical reproduction; Reproduction quality.*
Book: *Photographs and the Printer*, by F. H. Smith (London).

SCHEELE, CARL WILHELM, 1742–86. Swedish apothecary and chemist. In 1777 he discovered that the blackening of silver chloride was due to the reducing action of light, and that the black deposit was reduced silver. He also noted that silver chloride blackened more quickly in the violet end of the spectrum than in other colours, but was not influenced by heat rays. He was the first to name the ultra-violet and most refrangible rays of the spectrum the "chemical rays". Also discovered that ammonia was a solvent for silver chloride that had not been affected by light.

SCHEINER, JULIUS, 1858–1913. German astronomer and professor of astrophysics in Berlin. Described a sector-wheel sensitometer using a benzine lamp (1894); an improved version was proposed by Eder in 1898. The Scheiner sensitometer scale, reduced to candle power per second per metre and starting with the threshold value, was largely used in Germany and Austria (and partly in other European countries) for film speed measurement until replaced there by the DIN scale.

SCHEINER SPEED. Logarithmic system of rating the speed of negative materials, introduced by Julius Scheiner.

The Scheiner speed is derived from the formula: speed $= C - (10 \times \log \text{Exposure})$ where C is an arbitrary constant and the exposure is the least that will produce a visible image on development.

This system, in which the speed of the emulsion is expressed in degrees—e.g., 32° Sch.—gives an artificially high rating to the type of emulsion which has a characteristic curve with a very long toe.

See also: *Speed of sensitized materials.*

SCHINZEL, KARL, 1886–1951. Austrian scientist. Inventor of numerous colour photography processes (250 patents). Lived in a small town in Moravia and worked during most of his life without any contacts or assistance from outside. He started in 1905 with work on the three-colour Katachromie process on monopack material and put forward many original and fruitful ideas on subtractive colour photography processes; lost his chance, however, by publishing them too late. Later consultant to the Eastman Kodak Company.

SCHLIEREN PHOTOGRAPHY. Schlieren is a German word meaning streaks or striations. It is applied to a class of optical systems which make it possible to see small local changes of refractive index or thickness in transparent media. Schlieren systems are refined methods for observing those disturbances which in everyday life are seen by virtue of the apparent distortion of the background—e.g., warm air over stoves or imperfections in glass windows.

Many variations of the system are applied to problems such as the flow of air at high speed past aircraft wings and bodies in wind-tunnels and in flight, the flow of heated or cooled gases and liquids, the mixing of different gases and liquids, the movement of sound waves, and the examination of glass. The great value of seeing what is invisible to the naked eye is obvious: that it can be done with simple optical systems is not as well known as it should be considering that Toepler and Mach knew how to apply them almost a century ago.

Typical Arrangement. A typical system for schlieren photography may employ a parallel beam of light formed by a small source at the focus of a lens or spherical mirror. This passes through the transparent object and is brought to a focus again by a similar lens or mirror. Part of this image is obscured, or cut-off, by a knife edge and the light which passes falls on a screen on which the image of the transparent object is focused by an auxiliary lens. If any part of the light is deflected in a direction at right angles to the knife-edge by refractive index gradients or differences in thickness in the object, the image of the source at the knife-edge will be displaced in that direction. More or less light will pass over the knife-edge, and corresponding areas of the image on the screen will lighten or darken in proportion to the displacement (so long as it does not exceed the extent of the image source). Thus the refractive index gradients in the object are represented in light and shade in the picture on the screen, which can then be photographed.

In many cases it is the density variations in the medium which are of interest and, fortunately, the density (d) and the refractive index (n) are simply related:

$$\frac{n^2 - 1}{n^2 + 2} \times \frac{1}{d} = \text{a constant}$$

or, when n is near unity

$$\frac{n - 1}{d} = \text{a constant}$$

The image can thus be easily interpreted as a picture of a density field.

Special Methods. Schlieren systems are used most often for obtaining a qualitative picture, but it is possible to obtain quantitive results in certain cases by relating the change of illumination in the picture with the deflection produced in the object, and thus to the refractive index gradient. Practical difficulties have limited the usefulness of these methods.

By using a multi-coloured filter instead of a knife-edge it is possible to obtain a picture in which the different refractive index gradients are represented by different colours of the spectrum. Similar results can be achieved with a dispersing prism in front of the source and a slit in place of a knife-edge.

It is also possible to achieve special effects by suitable choice of the type of knife-edge or filter arrangement and many such systems have been described.

In recent years a new class of instruments has been developed which are analogous to Schlieren systems in that they record refractive-index gradients but which are really interferometers: they are known as schlieren-interferometers. In one such instrument, the two beams of the interferometer are produced by means of a Wollaston prism. The two beams are of equal intensity, polarized at right angles and laterally separated by a small distance. After passing through the object under observation they are recombined in a similar Wollaston prism and an image is formed on a screen, with fringes representing lines of constant refractive index gradient. With a white light source the fringes are coloured, the scale of colours being those characteristic of Newton's rings and not of the spectrum as in the case of colour-schlieren systems. Sensitive schlieren-interferometer systems are not so much affected by diffraction phenomena as conventional schlieren systems since the sensitivity does not depend on the use of small sources and a knife-edge (or other type of cut-off or filter). However, the image quality is limited by the doubling of the image due to the two beams, and, as the sensitivity of a given system increases with beam separation, the apparent advantage is often illusory.

Photography. There are no special problems in the photography of schlieren images except insofar as spark light sources are concerned. Such sources are often used because they combine the small dimensions necessary in a sensitive system with very high brightness and a short duration otherwise difficult to achieve. The combination of the very short exposure time (10^{-7} second) and the unavoidably low contrast of the schlieren image (about 10 : 1) may give rise to difficulty if an unsuitable emulsion is used. A prolonged development in a high energy developer will give beneficial increases of both speed and contrast. Of two emulsions with the same speed the one with more contrast will be easier to use. Since the output of the spark is concentrated in the blue region of the spectrum there is no advantage in using panchromatic emulsions if blue-sensitive types of equivalent speed are available.

When using colour materials with spark light sources some problems of colour balance may be encountered and compensating filters may be required due to the combined effects of reciprocity-law failure and the spectral quality of the light. With modern colour emulsions and modern spark sources, these effects are small and can usually be ignored. R.J.N.

See also: *Ballistic Photography; Electrophotography; Spark photography.*
Book: *Progress in Photography* (2 vols.), ed. by D. A. Spencer (London); *Schlieren Methods, Note on Applied Science No. 31* (H.M.S.O., London).

SCHLIPPE'S SALT. Constituent of some intensifiers and toners. Another name for sodium thioantimoniate.

SCHMIDT CAMERA. Astronomical camera practically free from chromatic aberrations. It was developed about 1931 by Bernhard Schmidt, maker of reflecting telescope mirrors. The camera, rather like a reflecting telescope, uses a spherical mirror instead of a lens and has a thin aspherical glass corrector plate placed in front of the mirror at its centre of curvature.

The corrector plate has one plane surface while the other is slightly convex around the centre and slightly concave at its rim. Its purpose is to deform the path of incident light rays from an object at infinity—e.g., an astronomical body—to compensate for the spherical aberration of the mirror.

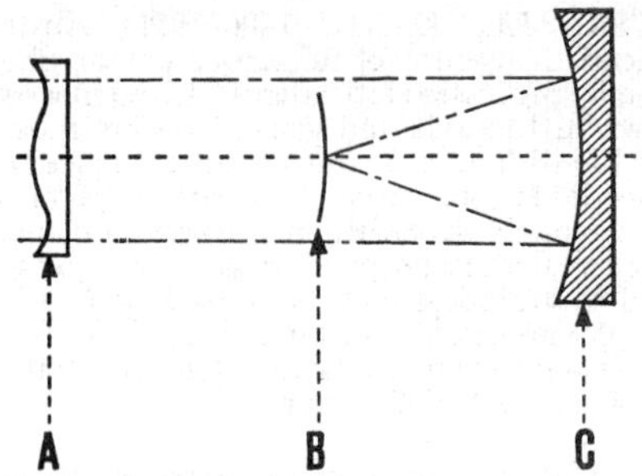

SCHMIDT CAMERA. The essential components are: A, corrector plate; B, film plane; C, mirror. The image is formed on the sensitive film by the spherical mirror. A stop in the centre of curvature restricts the width of the beam and ensures equal reflection at all angles. The only aberration is spherical aberration, which is corrected by deforming the incident rays by means of an aspheric corrector plate. The plate is placed in the centre of the curvature of the mirror.

The image from the mirror is formed on a spherical sensitized surface. This is placed halfway between the mirror and its centre of curvature. Schmidt cameras or cameras of a similar type are now widely used by observatories—including the Mt. Palomar observatory where a 48 ins. aperture Schmidt camera has been used for a complete survey of the sky, comprising about 1,800 exposures.

See also: *Lens history; Mirror lens.*

SCHOOL. Term applied to a group of workers who share the same aims and techniques or whose style and technique exhibits strong national or cultural similarities.

SCHOOL PHOTOGRAPHY

Photography is associated with school life in three principal ways: it can be taught as a creative hobby, it can be used by teachers and pupils for recording school activities, and many commercial photographers make a profitable business or sideline of taking official photographs of school classes, teams and similar groups.

There are many sound reasons for teaching photography to boys and girls at school. It is an excellent introduction to artistic appreciation; it is a practical demonstration of the laws of optics, and the preparation and use of processing solutions gives scope for budding chemists.

By no means the least important reason for encouraging children to take up photography is that a schoolchild who takes the right kind of photograph is laying up for himself a store of memories for the future.

Instruction is best carried out through a school photographic club, the minimum requirements being a knowledgeable and enthusiastic leader, and some funds with which to buy essentials such as a safelight, printing frame, and dishes.

If a darkroom equipped with an enlarger is available the value and interest of the club is considerably enhanced. There will be no lack of premises for meetings.

Suitable Cameras. Beginners should start with a simple, robust camera taking standard roll film and giving either eight or twelve exposures per reel. If possible, the lens should have a focusing movement and an iris diaphragm, and the shutter be provided with variable speeds. A large waist-level viewfinder, such as that fitted to many cameras yielding negatives 2¼ ins. square, is more likely to produce a good proportion of results than the more common eye-level type, which frequently reduces the scale of the image considerably and is rarely fully reliable, especially at close range.

In spite of its obvious attractions, particularly in relation to colour photography, the miniature camera using 35 mm. film is not suitable for children. The negatives need considerable enlarging, general manipulation of the camera occasionally needs special care, especially where precision instruments are concerned, and processing technique throughout is much too critical for young beginners.

Cameras with automatically-coupled exposure meters might yield a higher percentage of satisfactory results, but with such equipment the user does not obtain the valuable experience and knowledge of the basic principles of photography gained by using simple, manually-operated eqiupment.

As the young photographer advances he will need to decide on equipment to suit his aspirations; he may require a technical camera, or a 35 mm., or possibly a single-lens reflex, together with accessories, interchangeable lenses, etc.

Choosing Subjects. Young people need guidance in choice of subjects. Their natural tendency is to snap off the camera on things which seem exciting and novel to them, such as racing cars, aeroplanes and other moving objects. Not only are simple cameras unsuitable for such work even in expert hands; such subjects will lose all interest as the photographer grows up. School children should on the contrary be taught to find their subjects among the incidents of their everyday lives. These may not seem worth recording at the time, but the photographs will appreciate in value as the years go by.

They will find plenty of material inside and outside the classroom and many interesting records to make of the various activities of their classmates and themselves. In particular they should be encouraged to tackle interiors and thus learn how to estimate and use brief time exposures.

Any temptation to use flash as a short cut to correct exposure should be resisted until the basic principles of exposure calculating in daylight are understood.

Young photographers can obtain much instruction and encouragement from studying good photographic work. This should take the form of actual prints, not merely half-tone reproductions, good though the latter may be. They should also be encouraged to follow the progress of their contemporaries studying the other graphic arts. In this way they will learn the essential unity of all art. If they have these opportunities of study, schoolchildren will soon grasp the easier conceptions of good composition and the need to avoid unbalanced arrangements, competing interests and distracting backgrounds.

A pair of L-shaped guides is a most useful tool for instruction in trimming to improve composition.

Darkroom Work. Any small room which can be effectively darkened is adequate for loading tanks and cassettes, but for proper instruction in darkroom technique a good-sized room is needed.

It should have ventilation that is unaffected when doors and windows are sealed, also electricity, water and drainage.

Instruction in processing should begin with contact printing in a box-type printer. After that the pupils can tackle film developing (with ready made-up solutions or powders); and next enlarging to one fixed size, e.g., postcard. Film developing can be carried out in one of the many excellent developing tanks; pupils should practise loading the tank with an unwanted film until they can handle the apparatus in total darkness. It is wise to use a commercially made developer with the appropriate time and temperature tables; those who wish to make up solutions from their own formulae should first experiment with developers for prints or lantern slides, where the developing process can be visually inspected.

Scrupulous cleanliness should be insisted on, and the need for avoiding contamination of one solution by another explained and demonstrated. The children should be taught to use print paddles and tweezers as much as possible and to avoid immersing their fingers in solutions. Failure to do so may result in irritation or poisoning among those with sensitive skins.

In the darkroom special care is needed to guard against electric shocks and short circuits, as these may have serious consequences. All electrical equipment should be earthed and there should be proper fuses and an accessible master switch. On no account should any switch or plug be so placed as to be within reach of anyone standing beside a water pipe. The water supply should have a stop cock. A fire extinguisher is an essential item.

Apparatus. There are many items of photographic equipment which can be made up by the handyman and will give as good service as those in the shops at a fraction of the cost. Every encouragement should be given for their construction. They include: albums, camera cases, darkroom lamps, enlargers, filter holders, flash-guns, lens-hoods, picture frames, tripods and weights and measures.

School children should be taught from the outset to take special care of their equipment. As most things will inevitably get dropped sooner or later, specially fragile items such as cameras, liquid measures and dishes should be carefully chosen with this in mind. The cameras themselves should be of diecast metal rather than thin sheet metal or plastic. They should always be kept in a stout case; this need not be expensive, but can be home-made in wood. Dishes should be unbreakable and liquid measures of transparent plastic.

Colour Photography. There are many applications for colour photography in school, and the filmstrip or slide projectors with which many schools are nowadays equipped provide an admirable means of displaying colour photographs taken by staff or pupils.

The most common projector size is 2 × 2 ins. so transparencies should preferably be on 35 mm. film or be of the superslide size (12 square exposures on size 127 film). Reversal film yields a direct positive and is the most economical means of obtaining a slide for projection; projected slides also usually give the most brilliant and faithful colour rendering. If duplicate colour slides and/or paper prints are required, the exposure is best made on negative film.

Projectors are designed for either 3¼ × 3¼ ins., 2¾ × 2¾ ins. or 2 × 2 ins. slides, or for film strips in the 35 mm. (full or half frame size). Adapters are available to enable the smaller gauge pictures to be projected in the larger projectors.

As exposure in colour photography is much more critical than in monochrome, those about to embark on it for the first time should learn how to use an exposure meter.

Record Photography. The teacher will find the camera an unrivalled tool for recording school activities of all kinds. A roll film camera with focusing lens of *f* 4·5 aperture or larger will give good results for most purposes. The miniature camera for 35 mm. film, with *f* 3·5 or larger aperture lens, is the ideal instrument, provided that the user avoids camera shake and takes special care in handling and processing films. With a miniature it is often possible to take classroom interiors without special lighting at instantaneous shutter speeds and with sufficient depth of field. For good quality interior work it is advisable to use a technical camera on a tripod, and to have a selection of lenses at hand, especially a wide-angle one. Portable tungsten lighting and flash equipment are valuable though not essential.

Examples of school record work are: staff; pupils—new entrants, leavers, houses, prefects; events—sport, excursions, dramatics, exhibitions of pupils' work: specimens of work in art, crafts, needlework, cookery, etc.

Commercial Photography. According to the subject, the equipment needed for commercial photography in schools varies very considerably. Groups of the whole school usually call for a rotating type panoramic camera. This work is highly specialized and is mainly in the hands of a few firms who possess the necessary apparatus and experienced operators.

Smaller groups, such as teams, classes and houses are best taken with a plate camera if large prints are required for mounting and framing. But if enlargements up to 6½ × 8½ ins. are adequate, smaller negatives, including the miniature size, can be used.

The school group photographer must be prepared to work under unfavourable conditions and in limited time. As conditions cannot always be predicted, it is an advantage to have lenses of different focal length available. To keep exposures short, flash equipment is often essential, and at least one extension flash unit will be needed.

Permission to work on school premises can in most cases be granted by principals, who will expect a copy of each group taken. Some teachers allow individual portrait work on their premises.

To obviate the long delays which are common on school orders, and fatal to interest and good business, it is most desirable to deliver finished prints within one week of taking. As an alternative to the usual procedure whereby proofs are first submitted, then orders taken, some workers make an advance estimate of the demand, and aim to produce a number of finished prints in the shortest possible time, supplying them on a sale or return basis.

If a large number of individual portraits are required, and first quality is not essential, it is possible to photograph two or three sitters at a time with a miniature camera against an improvised background.

All the negatives can then be printed and the results supplied on sale or return. The cost of material would in this case be a minor factor.

Interesting film strips can sometimes be prepared by re-photographing groups on 35 mm. film and making positives by reversal or contact printing. M.K.K.

See also: *Groups; Visual aids.*

Books: *Photography at School and College*, by M. K. Kidd (London); *Projecting Slides*, by M. K. Kidd and C. W. Long (London); *Group Photography*, by G. Catling (London); *Making Lantern Slides and Filmstrips*, by C. D. Milner (London).

SCHULZE, JOHANN HEINRICH, 1687–1744. German chemist and natural philosopher. In 1725, discovered that light, though not heat, could darken a solution of chalk moistened with silver nitrate; in this way he produced impermanent contact copies from stencils. Biography by J. M. Eder (Vienna 1907).

SCHUMANN PLATE. Photographic plate containing an emulsion with very little gelatin so that the silver halide grains actually protrude above the emulsion surface. Used by V. Schumann for photography in the far ultraviolet region (below 2,300 A) where the ultraviolet absorption of gelatin would make photography with a normal type of material impossible.

The emulsion of such plates is exceptionally sensitive to mechanical abrasion.

See also: *Negative materials.*

SCHWARZSCHILD, KARL, 1873–1916. German astronomer. Formulated the deviation from the Bunsen-Roscoe reciprocity law in the case of densities obtained by intermittent exposures (Schwarzschild effect, 1899–1900). His assistant E. Kron gave this theory of density a stricter mathematical formulation in 1913.

SCHWARZSCHILD'S LAW. Law formulated by K. Schwarzschild which states that the exposure effect on a photographic emulsion is proportional to $I \times t^p$, where I is the intensity of the exposing light, t, the exposure time, and p, a constant. This is an attempt to formulate reciprocity failure quantitatively, but p is constant only over small ranges.

SCIENCE AND PHOTOGRAPHY

Scientific research is based on observations. Observations lead to the establishment of facts. Facts are grouped together to form theories or hypotheses and hypotheses are tested by further observations.

While observations can be made by means of any of the senses, the use of the sense of sight for this purpose is overwhelming. Whether it be the reading of the scale of a meter, the graduations on a burette or the change of colour of an indicator, the sense of sight alone is employed. It is true that the sense of smell is occasionally used to detect the presence of a chemical substance, the sense of hearing is sometimes useful, and touch plays an important but usually subsidiary part in many kinds of work, but no one would dispute the supreme position of the sense of sight in making scientific observations.

The human eye is a very remarkable instrument consisting as it does of a lens of variable focal length, an iris diaphragm automatically responding to the lighting conditions and a retina capable of presenting the brain with all the information necessary to analyse an image in colour. The retina is also capable of varying its sensitivity over quite a remarkable range.

The photographic camera is an instrument similar to the eye, comprising a lens, an iris diaphragm and a "retina" consisting of the photo-sensitive surface. Sharp focus is achieved in the camera by altering the position of the lens instead of its focal length; the iris is usually manually operated, but in the many automatic cameras now available a photocell determines the iris opening in accordance with the light reflected from the scene; and the photo-sensitive surface can vary in sensitivity also over a very wide range. It is not, however, safe to push the analogy too far; for instance the retina of the eye is responsive to the brightness of the subject viewed, whereas the photo-sensitive surface is responsive to the brightness multiplied by the time of exposure. Thus, if an object is not bright enough to be seen, it is of no help to stare at it for a long time, but an object not bright enough to be recorded on a film in 1 second may be recorded in all its detail by an exposure of 100 seconds. This is a fundamental and sometimes valuable difference as will be seen later.

In view of the use of the sense of sight in making observations it is hardly surprising that a process which is sensitive to light is of great value in research. Furthermore, since the photographic process in conjunction with a lens produces an instrument which has many points of similarity with the human eye, the camera may produce a result of a familiar and therefore easily interpretable form. In fact the camera can be looked upon as an extension of or aid to the eye in much the same way as a telescope can be so regarded.

The photographic process provides an extension of the faculties of the eye, for example by virtue of its sensitivity to ultra-violet light and X-rays; and it helps the eye by providing an image, similar to that produced by the eye, of an event which may be invisible to the eye or transitory.

The camera can never be a substitute for the eye since the eye will always be needed to interpret the result, but certain features of the photographic process make it a most valuable tool in research and an examination of the most important of these features will illustrate why photography is used in such a wide range of fields of work.

Permanent Records. Perhaps the most important feature of photography is its ability to produce a permanent two-dimensional record of an event. Though in most cases the eye can observe the same event it is often not possible for the eye to take in all the important features and it is difficult for it to make rapid and accurate measurements of the relative positions of objects.

Perhaps the commonest example of this is spectroscopy, the measurement of the wavelengths of the light emitted by the excited atoms or molecules of a substance. It is true that the eye can see the spectral lines as produced by a spectroscope, at least in the range of the visible spectrum; and by the use of a suitable graticule in the optical system the relative positions can be measured. But this can become very laborious, and quantitative analysis which depends on the relative amounts of energy of the lines would be well-nigh impossible visually.

In research on the behaviour of aircraft it is frequently necessary to employ a very large number of instruments in the aircraft and determine how their readings vary during certain flight manœuvres. It would in most cases be quite impossible to write down the readings by direct observation at sufficiently frequent intervals to be of use. Photography, however, produces a permanent record for examination at leisure and the interval between photographs can be made as short as is necessary for the purpose in hand.

Straightforward photography from the air has been of great value in archaeological research. Air photographs taken under suitable conditions can reveal features of ancient settlements which are often quite invisible on the ground. In this way not only are new sites discovered, but the ground plans of known sites which can be obtained in detail and to scale are invaluable in subsequent excavation.

Spectral Sensitivity. The permanent photographic record which can be studied at leisure and in detail is of value in nearly all applications of photography and is often used in conjunction with other features. For example, spectrography usually makes use of the extended range of wavelengths to which the photographic process is sensitive. To confine spectrography to the visible spectrum would be a very serious limitation. In fact the ultra-violet region is of much greater general importance.

Photography is commonly used for ultra-violet spectrography, a technique of immense value for the quantitative analysis of metal alloys in which very small quantities of a metal may be present.

The photographic material can be made sensitive to all wavelengths of electro-magnetic radiation from the infra-red to γ-rays. The sensitivity, of course, varies a good deal over this range. In some regions the absorption is so high that only the surface layer of grains is affected, so producing a low density (ultra-violet), and in other regions the absorption is so low that long exposures are usually necessary to produce an adequate density (γ-rays).

In addition to sensitivity to this very wide range the photo-sensitive surface is affected by bombardment with charged particles such as electrons and this fact is made use of in the electron microscope. The electron microscope is similar to the optical microscope except that electrons are used in place of light (this, of course, necessitates the use of magnetic or electro-static lenses instead of glass). It has the advantage that a much greater resolving power is possible owing to the effectively very short wavelength of the electrons. The photographic material is sensitive to high speed electrons and is used to record the image in much the same way as it does in the optical photomicrograph. The electron microscope is a very valuable tool in many branches of science. By its use magnifications of more than 100 times those obtainable with the visual microscope can be achieved. Though the electron image can be made visible by means of a fluorescent screen, photography plays an essential part in recording the image by direct effect of the electrons.

Integration of Light. Apart from colour the eye distinguishes objects by virtue of differences in brightness. The effect on the retina is a function of the brightness of the object and not of the time for which it is observed. The effect on the photo-sensitive surface in the camera is a function of both the brightness of the object and the time during which it is exposed.

If the brightness of an object is gradually reduced there comes a time when the eye can no longer see the object in spite of the enormous increase in sensitivity which takes place in the retina on dark adaptation. With the photo-sensitive material, however, reduction in brightness can be approximately compensated by an increase in exposure time and in this way details can be recorded that are invisible to the eye by reason of their very low brightness.

The adjustment of the exposure to compensate for the brightness of the subject is commonly used in photography throughout nearly all applications from the snapshot upwards. However, in the field of astronomy it has been carried to such an extreme that images of stars and nebulae formed by large telescopes have been recorded photographically by means of extremely long exposures where the amount of light reaching the earth is so small that the stars or nebulae from which it originates cannot be seen by the eye even aided by the same telescope. Furthermore, the light from these bodies can be spread into a spectrum and recorded photographically to yield, for example, information on the velocity with which the distant nebulae appear to be receding from the earth.

At the other extreme, photography can record events in an extremely short time if the intensity of light is sufficient. This may be utilized to record events which are of extremely short duration such as lightning strokes, traces on the face of cathode ray tubes, etc. Similarly, fast moving events may be "frozen" by illuminating them with a pulse of light of short duration or by using a shutter capable of yielding a very short exposure (Kerr shutter). Much valuable work has been done in the sphere of ballistics using this feature. This high speed photography, as it is generally called, could equally well be regarded as making use of the ability of photography to record an event which cannot be properly observed by the eye because of the limited time during which it is available for study.

Inanimate Observation. One of the most valuable features of the photographic process is its ability to produce a permanent record of what the camera sees. This makes the camera superior to a human observer for many purposes: a photograph can record practically all aspects of a subject for study at leisure, whereas we can directly observe only a very limited aspect, especially if the scene is changing at all rapidly. This facility enters into nearly all uses of photography, the camera becoming a sort of inanimate observer with an infallible memory.

The camera can also be sent where it is difficult, dangerous or inconvenient to send an observer, but here too the advantage of a full record to study at leisure is usually of primary importance. Cameras are used in rocket probes and artificial satellites, but only of course where recovery of a capsule containing the camera is intended. If the probe is expendable, such as the first Moon and Mars shots, television cameras relay pictures by a radio link to the Earth. These pictures, however, usually end up as a photograph of a television screen.

Sensitivity to Sub-atomic Particles. The first method used to reveal the actual tracks of particles such as electrons, protons, etc., was the so-called cloud chamber. This consists of a glass-sided chamber saturated with water vapour and capable of being suddenly expanded in volume so that the water vapour becomes super-saturated. If then a charged particle passes through the chamber it ionizes the gas and droplets of water are formed along the track. The track persists for a short time before breaking up and photography plays an essential part in recording the tracks during their brief life.

More convenient for some purposes is the use of the sensitive emulsion itself for recording the tracks direct.

Photo-sensitive materials are sensitive to the passage of charged atomic and nuclear particles as well as to a wide range of electromagnetic radiations. Such charged particles act in the same way as light—they lose energy in passing through the grains and release electrons after which latent image formation follows normal lines.

Any rapidly moving charged particle, an electron, proton, alpha particle or a large ionized atomic nucleus will affect a grain if it passes through a part of it. In this way the path the particle takes is revealed by a succession of developed grains rather like a string of beads. Moreover, the characteristics which affect the appearance of the track depend on the charge, mass and velocity of the particle and it is possible to determine these characteristics of a particle from measurements made on the track it produces.

Special photo-sensitive materials for this purpose are now manufactured and are available in various forms to suit the experiment in hand. They are characterized by the large number of grains per unit volume to minimize the space between the grains in a track, and great thickness to increase the length of track or chance of obtaining a long track.

Combining the characteristics of the photographic material with the ability to record for long periods without fatigue in places not easily accessible to the human observer, much work has been done in the investigation of the nature of cosmic rays.

As alternative methods of recording tracks, such as the cloud chamber, involve heavy equipment, photographic material is being sent up in balloons for recording cosmic ray phenomena at great altitudes. "Stacks" of emulsion weighing up to a hundredweight are used for this purpose. They are composed of many layers of gelatin emulsion with no separate support, clamped together and so marked that they can be taken apart after exposure and developed separately, yet tracks passing across boundaries can be followed and an accurate measurement of the length and angle of the track made. These stacks may be exposed for some hours at great altitudes before returning to earth by parachute. This enables information to be obtained on the nature and energy of the primary rays as they reach the earth's atmosphere and before they suffer the degeneration which takes place by the time they reach sea level and the normal terrestrial laboratory. These primary particles are endowed with more energy than it has been possible so far to confer on any particle in the terrestrial laboratory, though the gap is closing, and so events which are caused by such energetic particles can only be explored at these altitudes.

The technique of autoradiography is another illustration of the use of the photographic sensitive material in its own right. The availability of artificially produced radioactive isotopes has led to their use as tracer elements in all kinds of research. In some techniques the radioisotope of an element, say iodine, is introduced into the body where it is treated in just the same way as normal iodine and some finds its way to the thyroid gland. If then a section of the thyroid is cut and mounted in very close proximity to a photographic layer, the electrons emitted by the radioactive iodine affect the emulsion and produce a latent image. On development and examination with the thyroid section under the microscope the exact location of the iodine is clearly revealed. This technique is comparatively new and is proving a valuable tool, particularly in medical and metallurgical research.

Further Scope and Outlook. In the short space available it is not possible to do more than indicate the reasons why photography is such a useful tool in scientific research.

The applications are so widespread that it would need a volume to do real justice to the subject. There are many researches in which photography has been used for convenience, elegance or just preference. There are many in which photography has been the best available tool for the purpose. There are researches in which there appears to be no alternative to the use of photography, but the ingenuity of man is such that this might be difficult to maintain. However, photography is no longer used solely to record portraits and other pictorial matter. Its value in radiography is quite well known, but it is not always realized that the photographic process has materially helped to bring us to the state of knowledge we have reached to-day, and has helped to develop many techniques which in turn contribute to the higher and still growing standard of living we now enjoy. G.B.H.

See also: *Nuclear physics; Photomicrography; Research in photography.*

Book: *Progress in Photography* (2 vols.), ed. by D. A. Spencer (London).

SCRATCH-PROOFING. Normal hardeners are generally used to prevent an emulsion being damaged while it is still wet—although in many cases the hardening effect will remain.

The emulsion surface, however, can be made specially tough and resistant to injury when dry, by treating it with a tannic acid hardener. This treatment is used to make films more or less scratch-proof. The effect is only apparent when the film has been dried. Films can be hardened in this way straight after washing, or, if they have been dried, after resoaking for 10–20 minutes. A formula for a scratch-proofing hardener is:

Tannic acid	1¼ ounces	30 grams
Formalin solution (40 per cent formaldehyde)	4 ounces	100 c.cm.
Water to make	40 ounces	1,000 c.cm.

The tannic acid is dissolved in almost boiling water, and the formalin is added when the solution is almost cold.

Films to be scratch-proofed are immersed in a mixture of 1 part of the hardener solution and 14 parts water for about 5–10 minutes, and afterwards washed for a further 10 minutes.

See also: *Hardening bath.*

SCREEN (HALF-TONE). Regular mesh consisting of sets of parallel lines normally ruled at right angles, used in blockmaking to break up a photographic or similar continuous tone original when photographing it on to the metal.

See also: *Half-tone screen; Photomechanical reproduction.*

SCREENING (HALF-TONE). Photography of a continuous-tone original through a special screen to convert the lighter and darker tones into areas of smaller or larger dots but of uniform blackness. This is the only way in which tone values can be reproduced in most mechanical printing processes.

See also: *Half-tone screen.*

SCREEN PLATE. Special type of plate used in colour photography consisting of a panchromatic emulsion covered with a "screen" of finely distributed colour filter elements. When processed by reversal and viewed as a transparency, the original colours are seen reproduced in the image by additive colour synthesis.

Some colour films are manufactured on the same principle. The main objections to the system are the serious loss of light and the texture of the colour filter elements which becomes visible on enlargement.

See also: *Colour materials; Colour photography history; Colour synthesis.*

SCREENS FOR PROJECTION

A good screen is essential to enable the best results to be obtained from any cine or still projector. Many different types of screen are made, some suitable for the home, some for the lecture hall or classroom. For front projection an opaque screen is necessary; for rear or oblique projection a translucent one. Opaque screens are available in different surfaces and translucent screens are constructed of different materials. In both cases the best choice will depend on circumstances.

Opaque screens can be considered under two headings, those that are portable and those that are fixtures.

Portable Screens. These are used by the amateur for showing cine films or colour slides in the home. They are also useful to the lecturer who travels from hall to hall, and is obliged to take a screen with him, and are convenient in any classroom or lecture hall where it is not possible to leave a screen in a fixed position.

Most portable screens are made to rest on a suitable table, ledge, or support or stand on the floor by means of a tripod. An essential feature is that they must fold down to a convenient size for storage and portability.

The simplest and cheapest type of screen meeting these requirements has a wooden platform, with feet screwed on underneath. The bottom edge of the fabric is secured to this platform, while the top is fixed to a roller which is supported by side stretchers hinged to each end of the platform. To close the screen, the arms are detached from the roller and folded inwards. The screen is wound round the roller, and the feet turned sideways.

A more elaborate screen is the self-erecting type. It is quicker and easier to erect and close, and less likely to cause wear or damage to the screen fabric. Here the fabric is wound automatically round a spring roller, which is secured at the ends, just above the platform. The screen is erected by lifting the top rail until the fabric is fully extended. Joining this rail to the platform are two arms, hinged and sprung at the centre, so that once fully extended they remain upright. To close the screen, a cord is pulled, or a lever depressed, causing the sidearms to fold inwards and the screen is gently lowered to the platform. Sometimes such screens are built into a box for protection.

Another useful screen also uses a spring-wound roller, which is built into a metal tube. The tube has a narrow slot through which the screen material is pulled. The top bar of the screen is secured to a metal rod which is extended to the required height and locked into position. This rod is hinged to the centre of the tube and folds away when the screen is closed.

Tripod screens, which provide their own support, are a variation of the last type. The tubular body is hinged to an extending bar which has a folding tripod foot. When closed, the tube folds against the extending bar. To erect, the feet are unfolded and the tube placed in a horizontal position. The top of the screen is hooked on the extending bar which is raised up as required.

Non-Portable Screens. A rigid screen is all that is required when it is to be used in a permanent position and does not need to be moved out of the way. In small sizes this may consist of a canvas screen stretched over a wooden framework. In large sizes the fabric is eyeletted and laced to a stout wood or metal framework to keep it perfectly taut and free from wrinkles.

A roller screen can be hung from a fixed position and lifted out of the way when not required. The screen hangs from a top batten, the bottom being attached to a heavy steel or wooden roller which pulls the surface flat. By operating a cord and pulley, the fabric is wound round the roller as it is raised.

Wall type tubular screens consist of a tube and spring-loaded roller fixed to the wall or ceiling while the screen is drawn down into position.

In some makes the bottom of the screen must be held down while the screen is in use, whereas in others a ratchet and pawl are fitted, making this unnecessary.

Screen Surfaces. There are three types of screen surfaces in common use: glass bead, matt white and lenticular.

A glass beaded screen is made by impregnating beads of crystal glass in the surface of the canvas. Such a screen is very directional, giving a brilliant image over an area approximately 15° from the centre. Beyond that angle, the illumination falls off appreciably, though not below that of a matt white screen until the angle is about 30°.

A glass beaded screen is therefore most effective in a long narrow room. For use at close range the beads should be as small as possible, otherwise the image will appear grainy. It is not advisable to allow a beaded screen to remain exposed to strong daylight for long periods, as it may turn yellow and lose efficiency.

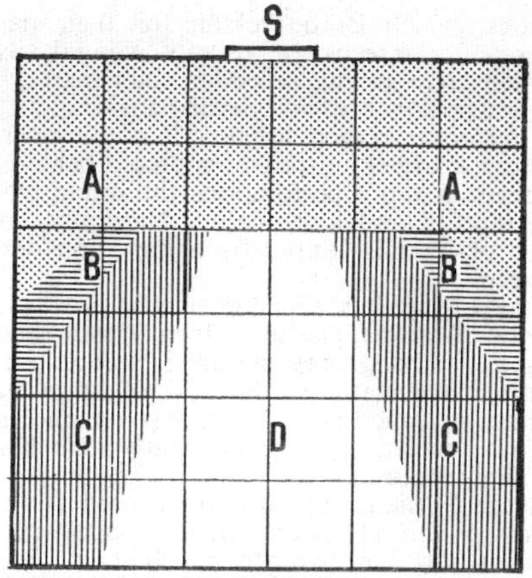

SCREEN SURFACES. The audience should be a minimum distance of two screen widths away from the screen S. The area A is therefore not suitable for viewing with any screen. B, area in which matt white or lenticular screens give best results. C, area in which matt white and beaded screens give equal image brilliance. D, area of maximum brilliance with beaded screen.

Plastic beaded screens are a cheaper version of the glass type. They consist of a layer of plastic applied on a white surface and moulded (by special rollers during manufacture) into a fine bead pattern. These screens are not as efficient as glass beaded screens, and are half-way between the latter and matt white screens in effectiveness.

Silver surface screens used to be popular at one time, but have now largely gone out of fashion. This is because they are less suitable with colour film than a matt white or beaded screen. The silver surface is usually aluminium or other metal paint applied on fabric. The durability of such a screen, especially when rolled and unrolled frequently, is not very high and the surface eventually starts to crack.

The image on a silver surface is less brilliant than on a beaded screen when viewed from a central position, but is fairly even over an angle of approximately 30° from the centre. Beyond that the brilliance rapidly decreases.

Matt white screens are the most diffused form of opaque screen available. They are less costly than beaded screens, and are obtainable seamless in fairly large sizes. They can be viewed from quite a wide angle without any appreciable falling off in illumination, and are therefore favoured for wide rooms and lecture halls.

Lenticular screens are made with a series of vertical ridges and hollows over the surface. Sometimes, there are horizontal ridges as well. Lenticular screens are difficult to mass-produce, as the ridges must be of the right height and curvature, with accurate spacing. They act in much the same way as a number of very small lenses, which is the reason for their name. Lenticular screens are usually painted silver, which gives greater reflection than matt white over a reasonably wide viewing angle.

Perforated screens are used in public cinemas where a large loudspeaker is placed behind the screen. Without perforations the sound would be muffled and lose quality. Naturally, some loss of light is caused by using a perforated screen.

Screen Materials. Opaque screens are usually made of canvas, plastic or linen. Plastic screens are apt to hang unevenly unless secured on all sides, particularly in the case of large sizes. Linen or calico screens are less opaque than canvas screens and consequently give a less brilliant image.

Screen Formats. Screens to be used for cine films or film strips are usually made with an aspect ratio—i.e., width : height—of 3 : 2 and are hung horizontally. In public cinemas the wide screen is now the vogue, but there is no indication yet that screens of this proportion will be used extensively for amateur films although they are available.

A square screen is usually necessary for slides, as pictures may be upright or horizontal.

Daylight Screens. For projection in normally lit rooms, translucent screens probably give the best results. The material may be either white or black plastic, or a special dressed linen or cotton. Plastic screens give a brilliant picture when seen from the centre, but are very directional. Viewed from a wide angle, the image may be almost indiscernible, though quite satisfactory when seen from the centre.

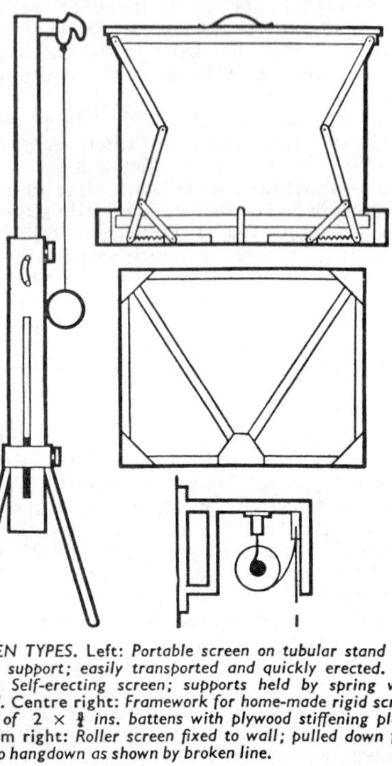

SCREEN TYPES. Left: *Portable screen on tubular stand with tripod support; easily transported and quickly erected.* Top right: *Self-erecting screen; supports held by spring when raised.* Centre right: *Framework for home-made rigid screen, made of* 2 × $\frac{3}{4}$ *ins. battens with plywood stiffening plates.* Bottom right: *Roller screen fixed to wall; pulled down from case to hangdown as shown by broken line.*

The projector can be placed directly behind a translucent screen, in which case the slide or film will need to be loaded the reverse way to the usual. Special screen units are obtainable in which the projector is placed to the side of the screen, the image being projected on the screen via a mirror.

This enables the operator to remain in the view of the audience, and with this method the film or slide is loaded in the same way as for front projection. It is also useful when the projection distance would otherwise be limited.

Translucent screens of dressed linen or cotton absorb more light than the plastic ones, but are more diffuse, enabling them to be viewed from a reasonably wide angle. As plastic screens are only normally available in fairly small sizes, the dressed fabric type is used when a large size is necessary.

Special back-projection screens exist with a high transmission but low reflection of light. In such screens the light falling on the surface from the front is not reflected back to the viewer to any appreciable extent. Hence stray light does not degrade the image and the screen is specially suitable for daylight projection.

An ordinary opaque screen with a silver or beaded surface can be used satisfactorily in a normally lit room if the size of the screen is restricted and the screen hooded and shielded from extraneous light. Some screens are sold complete with hood and side flaps for this purpose.

Ultra Directional Screens. A special screen type is the ultra directional beaded screen, used for producing studio backgrounds by front projection. The screen surface consists of highly refracting glass beads set in an opaque surface. The beads throw back the light only along the direction at which it reaches them. A projected image is thus visible virtually only within one or two degrees of the axis of the projected beam and is there very brilliant. The screen efficiency is around eight to ten times as great as with a normal beaded screen. From any other direction the image is virtually invisible. By adjusting the intensity of the projection beam it is possible to get a picture of normal brightness as seen in the projection axis, and ensure that any objects in front of the screen scatter the light sufficiently to leave no image visible on themselves.

When used for creating backgrounds by front projection methods, a beam splitter system ensures that the camera lens axis and the projection beam are coincident; the camera then sees the background image, but only on the screen surfaces and not on foreground objects.

Stereo Screens. Matt white or beaded screens are unsuitable for stereo projection systems based on polarization, as they depolarize the light. A metallic or aluminized surface is necessary for front stereo projection, and in practice a screen with a silver surface is generally satisfactory. For stereoscopic rear projection a very finely ground glass screen is sometimes employed and a translucent plastic screen is also suitable. Various forms of this exist.

If the projection system employs the split beam method, the screen should be upright to suit the image. C.W.L.

See also: *Back projection; Gain (projection); Projection practice; Projection principles.*

Books: *Projecting Slides*, by M. K. Kidd and C. W. Long (London); *How to Project*, by N. Jenkins (London).

SCRIPT. Shooting plan of a movie film, prepared from the story treatment. The script breaks down the entire action of the film from beginning to end into single shots together with detailed descriptions of the scenes, settings, camera positions, actors and shot durations—as well as other relevant information. Scripting is an essential part of the preparation of the film, since it enables the director to plan the required sets and properties needed, as well as the technical staff and actors for every scene. For the actual shooting the script is usually broken down into groups of shots using the same scenery or setting; these shots are then filmed together to save time and expense. A film is thus very rarely made in the original order of scripted scenes, but put together again afterwards by the editor. Scripts are sometimes also prepared in a similar way for features made up of still photographs, e.g. for a magazine story where the pictures required can be pre-planned.

See also: *Cine technique.*

Books: *How to Script*, by O. Blakeston; *How to Write Film Stories*, by R. Harrison; *The Simple Art of Making Films*, by T. Rose (all London).

SCULPTURE. The photography of sculpture calls for as great, if not greater, skill in the use of lighting and technique as most other branches of photography.

Most photographers have seen portraits with deep shadows in their eye sockets, or one side of the face a glaring white from over-strong backlighting; or distorted features as the result of using a lens of too short focus; or one feature over-emphasized to the detriment of the rest of the face because of a bad viewpoint.

Such mistaken technique and lighting would give as completely false an idea of a piece of sculpture as they do of a human face.

When photographing sculpture the aim of the photographer should be not only to give a correct record of its form, but also to interpret it in such a way as to lift the photograph from the "record" level and place it on the level of pictorial photography. This needs the right lighting, angle of view and background.

Equipment. A suitable camera is the field type with a triple bellows extension, interchangeable lens panel, and a ground glass screen for critical focusing. The single-lens reflex roll film camera which has a focusing screen and interchangeable lenses is also suitable.

The modern miniature camera may also be used very successfully. Its rangefinder and reflex focusing attachments allow for fine examination of focus, and interchangeable lenses and extension tubes permit close-up and long-range photographs. It is especially useful in places where heavier or more cumbersome cameras cannot be taken. A helpful lens to use would be a 13·5 cm. or 20 cm. These prevent the camera being employed too near the subject, which would result in distortion of form.

A firm tripod is necessary to eliminate any possible movement during the long exposures customary in this field. The tripod should also have a pan and tilt head. Reversible legs, spiked one end for outdoor work and rubber-tipped the other end for indoor work, are also useful.

Roll film cameras should have a film with an emulsion of fairly soft gradation—e.g. a fast panchromatic type. For plate cameras, a soft-gradation panchromatic plate is suitable. Miniature negatives require a great degree of enlargement and therefore a thin-coated film of low sensitivity should be used. This has very fine grain and gives a better definition than a double-coated film.

In the Studio. When photographing sculpture in the studio, the photographer has the choice of background, angle and lighting under his direct control. If, when the photograph is printed, any of these is at fault, he can duplicate the same conditions and eliminate the fault.

A spotlight gives hard lighting, casting sharp shadows, and is best for sculpture finished in a rough texture. The beam cast is easier to control than with other types of light. A floodlight, having a larger sized light source than a spotlight, gives a softer effect. Sculpture having a delicate form—children's faces for example—is better rendered with a floodlight. The hard light of the spotlight divides the highlight and shadow areas too sharply, while the soft fringe of the floodlight merges one smoothly into the other. Still softer lighting can be obtained if the floodlight is diffused. Overcast daylight has still greater diffusion.

The secondary light or fill-in calls for a fine degree of judgement. If it is too strong, it overpowers the main light; if it is too weak, the shadow areas lack form. The direction of this light is also important.

Backlighting, if used with restraint, can give sculpture an edge, which can add considerably to the effects of both modelling and texture.

Any shading of tone on the background is best done with lighting. On a plain, neutral background it is possible to get a range of tones from black to white, or to merge gradually from a light tone on one side to a dark tone on the other. This shading can counterbalance the lighting on the sculpture: the light side can be set against the dark side of the background, and the shadowed side against the light part of the background. By arranging the lighting in this way the subject stands well away from the background and the whole outline is thrown in relief without the need for backlighting.

A patterned background is likely to detract from the sculpture, unless it is made part of the composition.

As a very general rule the angle of view should be approximately that from which the sculptor intended his work to be viewed. But very often, by choosing a high or low angle, the pictorial effect can be increased. Having a low angle gives a feeling of strength and height to a subject. A high angle has the opposite effect, which is useful if the height of a large statue needs to be minimized.

As sculpture is three-dimensional, a better idea of its full shape can be conveyed if separate photographs are taken of two or three sides of the subject. This also helps to give a better impression of its dimensions.

Outdoors. Good opportunities for photographing sculpture can be found in the outdoor exhibitions that many local authorities arrange. And even familiar statues, seen by everyone in their home towns, provide an exercise for the photographer's ingenuity in finding an angle that gives them fresh interest.

Unlike studio photography, the factors of background, angle and lighting are not under the photographer's control. When photographing out of doors, the background seen from a particular viewpoint depends on the position of the subject, and the photographer cannot always get the angle he wants. Again the right lighting conditions are generally a matter of luck or patient waiting.

Spring and autumn are probably the best seasons for photographing outdoor sculpture, and during the summer months the mornings and evenings are most suitable. The sunlight then is not too harsh and does not come from too high an angle.

In open places the sky is the simplest background. Some measure of control can be had in the choice of filters used. The deep blue of the sky on a summer's day can be controlled from a light tone (light yellow filter) to a dark tone (orange or red filter) using a panchromatic film. Light statues found near tall buildings can be taken with the dark shadowed side of a building as a background, whereas dark statues are best taken against a light background.

On large statues in particular a good photograph can often be had from one detail, a hand or a foot, or from a small group of details that together make a tight, well knit composition. It is useful to carry a large white folded card to act as a light reflector when photographing details, to balance the lighting of the highlight and shadow areas.

Very large monuments may have an architectural setting which needs a camera with a rising front lens panel so that all vertical lines can be made upright in the photograph. If a roll film camera is used and tilted up, the distortion of the verticals must be corrected on the enlarger. Correcting in this way may introduce some distortion but it is not likely to be very noticeable.

In Churches. Sculpture in churches provides another rich source of subjects. Fonts, effigies, decorations of roof bosses, and bas-reliefs offer an endless choice.

A rigid tripod is essential for taking photographs in a church; the use of a weak tripod may result in camera shake during long exposures. The roll film or miniature camera is quite adequate when taking photographs of subjects which have no verticals to be corrected, but, if there are verticals to correct, then the camera possessing the rise and swing movements is preferable. A telephoto lens is useful in obtaining close-ups of subjects out of reach of the normal focus lens, such as gargoyles and details in vaulted roofs, coats of arms, and the like.

Unless the church is well lit the photographer may have to provide his own lighting with portable equipment. Before attempting any photography the vicar's permission should be sought and, if lights are needed, the voltage of the electricity will have to be checked. Flashlighting is convenient, but the number of bulbs needed for good modelling is apt to be rather expensive. If a single flash is used on the camera, the lighting will be flat and will not show the form of the sculpture; if it is used away from the camera, the lighting will be too hard, unless a fill-in flash is used. Electronic flashes, fired from a succession of points during a time exposure, are an effective way out.

When daylight can be used, which is usually better than flash, the ideal outside lighting conditions are those on a bright day with no strong sunlight. The diffused lighting from such conditions is often preferable to strong sunlight, which would need a bright fill-in light to illuminate the shadow detail. In a church where the light comes from many directions, the lighting is apt to be very flat. A floodlight in this case could be used as the main light source, and the natural light act as a fill-in. In winter months the daylight intensity may change rapidly. An exposure meter should be used constantly to check this.

If a subject has to be photographed in very poor light, it will sometimes be found on developing the negative that, although the correct exposure was given according to the meter reading, it is still badly under-exposed. This is due to reciprocity failure, when the intensity of light reaching the emulsion is too low to respond to the reciprocity law. In such conditions it would be better to use artificial lighting or give a longer exposure.

A photo-electric exposure meter will not usually show a reading at all in a badly lit church and one of the visual extinction type meters is often more useful.

Wood and stone carvings high up or out of reach will need to be photographed with a long focus lens. The long bellows of the field camera are an advantage in this case to give the necessary extension. Wood carvings may need the use of a deep red filter to bring out their details. L.S.

See also: *Lighting the subject; Photosculpture; Statuettes.*
Book: *Focus on Architecture and Sculpture*, by H. Gernsheim (London).

SEASCAPE. Any general view, similar to a landscape, in which the sea, waves, or shore scenery is the dominant feature.

SEASIDE. The seaside is one of the best places to take photographs. The light is generally good, and there is such a diversity of subjects that whatever particular ideas the photographer has, he can always find something to take.

Seascapes. If the photographer is a landscape enthusiast, he will find that very much the same technique is needed to make a successful seascape picture. But it is not so easy to grade the tones of the picture to suggest distance, since the sea is of much the same tone all over, getting only a little lighter towards the horizon. First, to give the impression of depth, some large object is needed in the foreground. A medium or deep yellow filter will help to improve the tone rendering of both the sea and the sky. A lens hood is essential for all beach photography, but, even then, beware of light reflected from the water into the lens. If the hood is not deep enough to keep it out, the only thing to do is to wait until the sun is in a different position or until it is obscured by light cloud.

Those who prefer to take close-ups of still-life subjects will find that they abound on the beach. Pebbles, seaweed, patterns left by the tide on the sand, stranded starfish, are all excellent for close-up studies. Often the best way to tackle these subjects is to take them against the light. This not only gives them greater prominence against the background, but reveals their texture much better than flat lighting.

Here again a tripod is generally useful, although it is not always necessary since the light is usually bright enough for a fast shutter speed. But it is essential to focus accurately on the subject, because the lens has very little depth of field at close ranges. A filter is not necessary except when photographing brightly coloured specimens. To get close to subjects on uneven surfaces and slippery rocks it is best to crouch in the way which gives most steadiness.

People. Those who prefer the very lively scenes of holiday-making must be ready for a little more action. As a rule, people on beaches usually do one of two things: they either sleep or they jump about and play games along the sands and in and out of the water. Pictures of people asleep can often be amusing to look at later, and they make easy subjects: it is simply necessary to take an exposure reading, set the camera, focus and expose. But the viewpoint must be carefully chosen to avoid including confusing objects—e.g., beach-bags and clothing and other people—in the background.

People running about need a fast shutter speed. It is often possible to use about 1/300 second with a yellow filter on the lens since the light on the beach is nearly always very bright and the reflection from the sand helps to fill in the shadows. It is wise to use a low viewpoint to get the sky as a background. The best way to shoot is to pre-focus, follow the subject in the viewfinder and release the shutter as soon as the subject is in the right place. People running towards the camera are less likely to be blurred than if they were moving directly across the line of vision, so, even with a camera which has a shutter speed as slow as 1/125 second, it is still possible to take some action pictures provided the subject is not moving across the camera's view. People jumping must be snapped at the "dead point" of the movement—e.g., when they are at the top of their jump and have stopped going up but have not yet started to come down. At this point they are reasonably still.

People swimming, surf riding and generally enjoying themselves in the sea are good seaside subjects. Stand at the water's edge (or even in the water) and keep the camera rather low down. This cuts out fussy background and shows the subjects against the sky. Use a yellow filter and a fast shutter speed to stop movement of both the subject and the water being splashed about. (Take care that none of the splashes fall on the camera.)

Children. Make a tour of open-air entertainments like concert parties and Punch and Judy shows. Where there are children in the audience, concentrate on their faces; they will be much too interested in the show to take any notice of the camera. If pictures of children are wanted, do not forget to watch for incidents around the ice-cream trolleys on the beach or the promenade; pictures of children buying, eating, or wrangling over ices are always popular.

Children form one of the most worth-while subjects that the beach has to offer; they seem to belong to the beach scene as surely as the sand and the water, whether they are occupied in building castles or covering their sleeping parents with sand. They will generally be much too occupied to notice the photographer. So first take an exposure reading and set the focus at a pre-determined distance. Then just walk up and expose before the subject has time to become aware of the camera.

There are a great many subjects on and around the seashore: interesting people, amusing sights, exciting events, minor tragedies, so it pays to have the camera ready to go into action at a moment's notice.

Boats. Be on the look-out for boats, either drawn up on the shore or riding at their moorings. They are one of the most popular forms of picture material. The best way to take them is to choose a calm day and hire a rowing boat so that a choice of viewpoints can be made. As well as the boats themselves, close-ups of various parts—coils of rope, anchors, blocks—taken with strong side lighting to reveal texture make excellent subjects. Use a shutter speed of about 1/100 second, unless the camera is on a tripod on firm ground to eliminate any risk of shake.

Most seaside resorts have a lifeboat station which is worth visiting. It should be possible to find out when there is to be a practice launch. Take up a position either near the water's edge or close to the ramp and in either case, use a fast shutter speed even though the lifeboat may be coming towards the camera. If the aim is to catch the bow wave as the boat enters the water, there will be no time to alter shutter speed after photographing the boat coming down the ramp. Pre-focus on a spot the same distance away in both pictures. This saves refocusing between exposures.

Care of the Camera. Keep the camera away from the sand and salt water. These two form a most deadly combination for the ruin of a good camera and lens if they come into contact with them.

A small grain of sand can put a shutter out of action and call for an expensive overhaul, while salt water will corrode plating, make screw threads foul and stick, and take the high polish off the surface of a lens.

So take care to avoid splashes falling on the camera and particularly on the lens. A yellow filter placed over the lens will protect it from sand and sea water and will be needed for most shots in any case.

Keep the camera in a case with a lining instead of an ever-ready case and this will help to keep sand out.

If taking photographs in really rough weather, or in a small boat where there is a very real danger of the camera being swamped, obtain a polythene sandwich bag big enough to hold the camera. Keep it in this until it is needed and it will be fully protected. G.S.D.

See also: *Marine photography*.
Book: *Camera Afloat*, by H. S. Newcombe (London)

SECOND-HAND EQUIPMENT

Nowadays most cameras and photographic equipment can be bought second-hand from private owners or photographic dealers and sold to the same people. Such buying and selling is frequently carried on through the classified advertisement columns of photographic magazines. Photographic equipment when intelligently handled suffers less from use than, say, a motor car or a radio set, and it is possible to buy cameras and accessories several years old that are for all practical purposes as good as new. If in good condition, they should also fetch a reasonable price on sale or exchange. At the same time the inexperienced buyer always runs the risk of throwing his money away on a worthless "bargain".

BUYING EQUIPMENT

Some photographic dealers specialize in selling only first rate equipment that they have dismantled and rebuilt, or at least thoroughly inspected, in their own workshops. This equipment carries a guarantee, and the prospective buyer is allowed a week or more in which to try it out.

This is undoubtedly the safest way of buying second-hand photographic equipment, but it naturally costs more—there are no bargains in this market.

The beginner should always protect himself by asking the advice of an expert before buying; the following hints are intended for those who must rely on their own judgement.

Cameras. If you know the make of camera you want to own, get a *Focal Camera Guide* about it if one is published. This will tell you the differences between the various models so that you will be able to recognize an early model from a later one. Remember that one model may be worth twice as much as another of the same make and superficial appearance.

Body. The condition of the plating and covering of the camera body—particularly on sharp edges—will show if the camera has had a lot of use. Burrs on the slots of any of the screw heads are a sure sign that someone lacking in mechanical skill has had the camera to pieces. In that case the camera would probably be dear at any price.

The most important point to watch for in the body of the camera is the rigidity of the lens-mount and erecting mechanism in the working position. The front should open smoothly and click into position positively and without any forceful persuasion. If there is any trace of slackness or wobble anywhere in the erecting or focusing mechanism the camera should be rejected.

Bellows. The bellows of a folding camera should be pulled out to their fullest extent and examined for signs of cracking, wear or repaired places. Even if there are no actual light leaks, the price of the camera should be reduced to cover the cost of new bellows. If everything looks satisfactory, the camera should then be taken into the darkroom and the bellows checked for actual light leaks. Even in apparently sound bellows there may be pinholes that would let enough light through to fog the negative.

To test the bellows, the back of the camera is opened and an electric torch, with the head removed, is inserted. The torch is switched on and the back of the camera covered with a black cloth. With the head of the torch removed, the bulb can be pressed close up to the corners and into the folds of the bellows so as to show up the smallest hole or any thin spots that might give trouble later. It is not a good idea to examine bellows in this way with an ordinary electric light bulb because the heat in such a confined space may scorch the material or turn it brittle.

The leather bellows of large multiple extension cameras can be successfully repaired if they have been accidentally pierced or torn, but if the trouble is due to general wear, patching is never worth while. The bellows of smaller, hand cameras should always be renewed if damaged or worn. New bellows are not expensive, but they should only be fitted by the maker of the camera or by firms specializing in this type of repair.

Viewfinder. Check the accuracy of the viewfinder by comparing the picture seen in the finder with the image on the focusing screen (if there is one) or a piece of tissue paper stretched between the film rollers. It is fairly common for the viewfinder to include less than the image on the screen, but it must never cover more, and the centres of both pictures should coincide when the camera is focused at infinity. This will not hold at closer distances.

Rangefinder. Check the rangefinder calibration by ranging it on a series of objects from the closest distance that the lens will focus to about 30 feet away from the camera. The rangefinder reading should correspond to the distance measured with a tape measure. There must be no slack when the control wheel is moved slightly backwards and forwards.

Focusing Mechanism. Here again, there must be no trace of slack in the focusing mechanism of any camera that focuses by scale, although a certain amount of slack between the rack and pinion of a stand camera is not important. Check the accuracy of the focusing scale if possible by focusing an object with a focusing magnifier on either the focusing screen or a sheet of ground glass held across the negative opening in the back of the camera. Measure the distance of the object focused and compare it with the scale reading, with the lens at full aperture.

The focusing scale of any camera taking a negative smaller than $2\frac{1}{4} \times 2\frac{1}{4}$ ins. cannot be checked with sufficient accuracy this way. The best way for the ordinary photographer to check the focusing scale of a miniature camera is by actual trial as follows. Stand a row of books on a table in front of the camera so that the line makes an angle of 45° with the line of sight. Focus on the lettering on the spine of a book near the centre of the row and make an exposure at full aperture. Do this for a series of camera-subject distances from the closest working distance to 10 feet or so.

When you examine the developed negative through a magnifying glass—or project it on the enlarger—you will see whether the focused book is consistently sharper than its neighbours, or if the book in front or behind is sharper. In this way you can see where the actual focused distance lies and decide either to re-set the scale, or make a suitable allowance when focusing. But the wisest course is to put the onus on the seller and refuse to accept the camera until the scale has been corrected.

Finally check the infinity position by taking a photograph of the horizon line or of objects silhouetted against the sky at least 200 feet away.

Shutter. All that you can do here is to set and release the shutter at each of the marked settings—including T. and B. and delayed action. While you can make your own test of the accuracy of the marked speeds, it is more satisfactory to pay a few shillings to have the shutter tested by any of the photographic dealers who maintain a shutter testing service. These tests are made with the help of equipment beyond the means—and needs—of the ordinary photographer.

Lens. The general condition of the lens can be judged by examining it with the back off the camera and the shutter held open at T.

Both back and front surfaces should have a high polish and be free from dust or grease. A single scratch is not a serious matter. But if the surface looks dull, it is probably covered with very fine scratches caused by rubbing it in the effort to get it clean. Scratches of this type ruin the performance of the lens, and although the surface can be repolished (preferably by the manufacturer) the trouble is a fairly clear indication that the camera has not been looked after and you would be wise to reject it.

Tiny air bubbles in the glass itself are not serious—in fact they are almost always present in high quality optical glass.

If the lens is fairly old and has cemented components it may have a slightly yellow tint. This is simply the effect of ageing on the Canada Balsam joining the glasses. You can ignore it for black and white photography. The tint might affect the rendering of colour, but a lens old enough to have yellowed to such an extent would probably not be fully corrected for colour work anyway.

Cemented lenses should be carefully examined to see if the balsam has "started", that is, if the two glass surfaces are no longer joined by the cement. Starting is usually indicated by small, star-shaped patterns.

Actual lens testing, a special technique, is the only practical way of determining the performance of a lens.

Iris Diaphragm. Test the iris diaphragm with the camera opened up as when examining the lens. The control should work smoothly and in all positions; the leaves of the diaphragm should form a hole of regular shape. Reject the camera if any of the leaves tend to stick and give an irregularly shaped hole.

Bright edges or rubbed areas on the leaves of the diaphragm may cause flare patches or light fog.

Synchronizer Timing. Anyone can check the synchronizer timing for electronic flash quite easily. Simply open the back of the camera, plug in the flash unit and fire the shutter with the diaphragm fully open. Look through the lens at a sheet of paper held in front of the camera and illuminated by the flash and watch the shape of the lens opening.

If the timing is correct, the eye will retain an impression of a circular patch of light the full diameter of the lens aperture. If the impression is of a smaller area with a regular but non-circular outline, the shutter is firing early or late. Remember that the shutter may only be synchronized for electronic flash at one particular speed.

The easiest way to test the timing for flash bulbs is to connect up the flash gun to the camera shutter in the normal way and take a close-up of the bulb itself. The resulting photograph will show whether the shutter has caught the flash before, after, or exactly at its peak.

Plate Holders. See that the draw slides go in and out smoothly and without forcing and that the velvet along the face of the light trap is intact and has not lost its resilience. Reject metal holders that have been bent or buckled—it is practically impossible to repair them so that they will work smoothly and be light-tight again.

Never buy a camera with only two or three plate holders. So few holders are interchangeable that it may prove difficult or impossible to get extra holders to fit the camera. There should be at least six holders to enable you to work without frequent visits to the darkroom to change plates.

Enlargers. First make sure that the enlarger is mechanically sound. The column should be rigidly fixed to the easel and the head should travel up and down smoothly. All controls—focusing, negative tilt, and rise and fall—should work freely but without any slack or backlash.

It is particularly important for the image to remain steady, while focusing, so test this with the enlarging light switched on in the darkroom. Move the focusing adjustment gently to and fro and watch the light on the easel. If the picture moves as you reverse the movement of the control, look for another enlarger; image movement of this type at the critical point of focusing is a perpetual nuisance.

While the enlarger light is switched on, check the evenness of the illumination on the baseboard. To do this, focus a negative sharply and then remove it together with the negative carrier. This will show the performance over the whole working area.

Watch the illumination as you slowly stop down the lens. It may look uniformly bright at first, but presently dark patches and corners may appear. You may be able to get rid of the unevenness by altering the lamp adjustment. Or the lamp may be of the wrong type. But if the unevenness persists after you have checked these points, do not buy the enlarger.

The advice about the general condition of the finish, lens, diaphragm, focusing mechanism, etc., of second-hand cameras applies equally here. In addition, examine the lower face of the condenser for scratches or chips—any defects here tend to affect the image, particularly when the lens is stopped down.

To check the performance of the lens, cover a piece of exposed film with scratches and focus it sharply with a magnifier on the enlarger baseboard with the lens at full aperture. With a good lens, it should be possible to get sharp definition of the scratches all over the baseboard.

Next expose a sheet of bromide paper on the easel, develop it and check that the scratches are equally sharp in the print. This will show whether the lens is free from chromatic aberration.

Flash Equipment. When buying second-hand flash equipment, remember that you may have to renew the battery in a short time. With the ordinary flash gun the cost of a new battery is nothing to worry about, but with a battery capacitor type, and more particularly with a battery operated electronic flash, the cost of a new battery can be quite expensive.

Accessories. The only way to be sure that second-hand photographic accessories are worth buying is to try them. You can accept new equipment on the strength of the manufacturer's guarantee, but when you buy second-hand equipment you rarely have any right of redress if it gives trouble after you have paid for it.

Most dealers will allow you to take equipment away on trial against a deposit, and the classified advertisement departments of most photographic periodicals run a deposit system by which the money is not handed over to the seller unless the buyer is satisfied with the goods.

Price. As a general rule, second-hand equipment in first class order is worth anything from one-half to three-quarters of the list price of the new item. "Shop-soiled" equipment is sometimes sold at a reduction of 10–15 per cent or more.

Accessories made by a camera manufacturer to fit his own particular instrument usually cost more both new and second-hand than the equivalent proprietary items designed to fit any make.

SELLING EQUIPMENT

The photographer with equipment to sell can either take it to a dealer or try to find a private buyer. One way is used as often as the other and each has its special advantages.

To a Dealer. The dealer relies on being able to sell the camera or equipment again at a profit, but in the meantime he must store, maintain, perhaps recondition and often advertise it. So he is not likely to offer as much as a private buyer who wants it for his own use. Against this the dealer offers cash down and an immediate sale which is sometimes the more important consideration.

If the owner of the equipment simply wants to sell it so that he can buy something better, however, the situation is different. His wisest course then is to offer the equipment to a dealer in part exchange for the new outfit. Under these circumstances the dealer may sometimes offer a better price for the old equipment than he would in a straight cash sale and the customer gets what he wants without having to wait.

To a Private Buyer. The private buyer is generally prepared to offer more for any second-hand equipment he wants than the dealer. For the same article, he will expect to pay a price which lies somewhere between the dealer's buying and selling prices.

The private buyer of course has to be found, and there are three principal ways of going about finding him. If the seller is lucky, he may be able to find a buyer among his own circle of acquaintances or at his camera club. He can watch the wanted columns in the classified advertisement sections of the photographic press. He can insert a classified advertisement of his own in the for sale column.

In answering a wanted advertisement, he should first send a detailed and honest description of the equipment, but not, of course, the equipment itself. The deal can proceed from that point to the dispatch of the equipment on approval against the buyer's cheque, or through the deposit scheme of the publication.

If the seller decides to advertise his equipment, the best medium is almost always the classified advertisement columns of the photographic press. Anyone living in a large city or a densely populated area can often do equally well by advertising in the local press because his advertisement is likely to be seen by a fair number of prospective buyers in his own district. If he includes his address and telephone number, he has a good chance of closing the deal by personal contact instead of by laborious and time-wasting correspondence and written bargaining.

An advertisement in the photographic press should give all the essential information about the article briefly: what it is, model or year, condition, any extra items not usually covered by the opening description and price—e.g., Vinto IIIa *f* 2 coated Panchro, equal new (or "mint"), e.r.c., hood, three filters, £75. Name, address and telephone number should only be given if the seller lives in a district where the advertisement would be read by a large number of prospective buyers within calling distance—e.g., Greater London. In most advertisements of this kind a box number is enough. F.P.

See also: *Lens testing; Repairs to cameras; Shutter testing.*

SEEBECK, JOHANN THOMAS, 1770–1831. German physicist. Investigated in 1810 the photochromy of silver chloride: when this is exposed to white light and then to the spectrum colours, the spectrum reproduces itself on the silver chloride in its own natural colours. This interference process is one of the few direct processes of photography in natural colours, but it took 80 years and the work of many inventors before a practical process giving a permanent image was evolved by Lippmann.

SELENIUM. Used in certain toners, either by itself or with sodium sulphide.

Formula and molecular weight: Se; 79.

Characteristics: Reddish powder.

Solubility: Insoluble in water, but soluble in hot solutions of sodium sulphide and sodium sulphite.

Selenium is also the operative component of barrier-layer photo-electric cells, as used in exposure meters.

SELF-PORTRAIT. There are four ways of taking a self-portrait; by using a mirror, a delayed action shutter release, a remote shutter control, and by controlling the lighting.

Mirror Method. The mirror method is popular, but by no means the easiest. It usually means including the camera in the picture and this is not always desirable. The mirror must be fairly large—generally no smaller than 2 feet × 1 foot 6 ins.—and clean and perfectly flat, otherwise the reflection will be distorted. Most ordinary domestic mirrors are unsuitable, as a reflection is obtained from both the mirrored surface and the front glass surface, resulting in double images on the negative. The mirror should be supported in a secure position with plenty of room behind it to allow freedom in arranging artificial lighting (for daylight portraiture it is generally best placed with its back to the window). The lighting must be placed where it does not reflect off the surface of the mirror or shine directly into the lens.

The photographer takes up his position in front of the mirror and arranges his distance from it, his pose, and the lighting until he is satisfied. This procedure calls for some imagination, because the camera will be looking at the effect from a different viewpoint. Even with a reflex camera it is not possible to see the final effect in the viewing screen because if he looks at the screen the photographer cannot see his own face. And unless the portrait is to show the sitter looking straight at the mirror, the effect of looking in any other direction can only be guessed.

If the camera is focused by scale, it is important to set it to the sum of the distance from the lens to the mirror and the distance from the mirror to the subject. (These two distances are not necessarily the same; the camera lens may be six inches or more closer to the mirror than the face of the "sitter", and in a fairly close-up portrait this would be quite enough to put the whole face out of focus.)

When the camera is focused by rangefinder the fact that the camera will be in a different position at the time of exposure must be allowed for.

If possible the photographer should set up the camera on a tripod and sight it on some substitute subject—e.g., a stool standing on the chair that he will be sitting on for his portrait—to make sure that he is not including too much or too little in the finder. It is of course always wise to allow a good working margin in this kind of portraiture.

Finally, the photographer looks in the required direction and releases the shutter. The exposure is the one normally given for the subject and lighting; the presence of the mirror in the set-up makes no difference.

There are two ways of keeping the camera out of the final print. One way is to arrange the picture so that the part where the camera shows can be trimmed off in enlarging. This method is suitable for a picture where the whole head is enlarged and the camera is kept just below shoulder level. The other way is to arrange for the reflection of the camera not to appear in the picture at all—i.e., to keep it well below or to the side. This method inevitably introduces some angular distortion of the image, and the picture that the camera sees is so different from the reflection seen by the photographer that it becomes difficult to visualize and arrange.

When making prints from the negative, this must be done in the enlarger and not by contact printing. This is necessary because the negative must be projected the wrong side up (i.e., emulsion away from the paper) to correct the reversal from left to right caused by the mirror. A negative contact printed in this way will have unsharp definition.

Delayed Action Release Method. The delayed action release offers much more scope and it avoids including the camera in the picture. It is an excellent aid in trying out lighting arrangements, ideas, etc., without having to rely on the co-operation of a model. It also relieves the photographer of worry about whether he is taxing the patience of his stand-in model by playing around with lights and props.

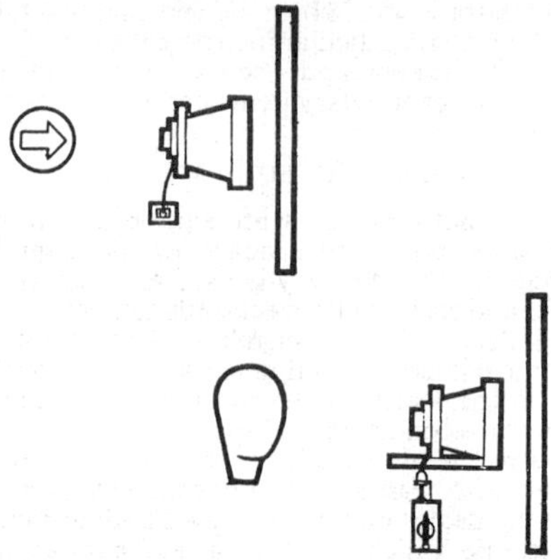

SELF-TIMER PORTRAITS. The camera is set up on a tripod or other support, and the self-timer set in motion. The photographer can then take his place in front of the lens. A mirror behind the camera permits observation of expressions.

Practically any type of camera is suitable; if it is not fitted with a delayed action release, it can generally be fitted with a supplementary one either in conjunction with a cable release or screwed straight into the cable release socket.

The arrangement is exactly the same as for taking a normal portrait except that the photographer, immediately after engaging the delay action and pressing the shutter release, takes up his place in the picture. As the average delay is 15 to 20 seconds, there is always plenty of time to get into position before the shutter clicks.

The safest way to ensure that the picture will include everything is to contrive some form of "stand-in" that the camera can be focused on. A cushion on top of a stool standing on the chair, or some such arrangement, is all that is needed. This substitute is then lighted, arranged and focused, preferably by scale. The lens is then stopped down to give a reasonable margin of depth of field, and the exposure made at leisure. It is better to make sure that the image will be sharp than to shorten the exposure by using a large stop and sacrificing depth of field.

If the portrait is to be focused critically—e.g., to give sharp definition to the front of the head alone, then this "stand-in" method of focusing must be supplemented by something more accurate. The simplest way is to stretch a piece of picture cord (ordinary string is too elastic) from a point behind the camera to the point focused on, marking this position with a knot. The photographer then sits in the chair in front of the camera, pulls the cord taut with the knot against his forehead, and then lets it fall before the shutter clicks.

It is a help to set up a mirror behind the camera so that the photographer can arrange his expression and pose, but it must be remembered that the mirror will give a reversed picture—i.e., transposed from left to right—and will give no indication of the field included by the camera lens. Also, the mirror must be securely fixed so that it cannot fall forward on to the camera—a mishap that would probably prove disastrous for both.

Finally, when taking delayed action portraits, it is always worth while to make one or two "dummy runs" before actually making the exposure. There is a knack in releasing the delay, moving quickly into the picture area, and posing without a flustered air or, what can be just as bad, a look of triumph.

Remote Control. In many ways this is quite the most successful method of self-portraiture because it avoids the last-minute scramble from the camera station to the sitter's chair. The shutter in this case is released by any of the normal types of supplementary remote release—i.e., pneumatic tube, electrical, remote switch-controlled solenoid, extended cable release, or even a length of thread fastened to the release lever. The latter should be used only as a last resort, and then with great care, as camera-shake and even damage to the camera are a potential danger.

The preparations are exactly the same as for normal delayed action release, but when everything is ready, the photographer simply takes up his position and releases the shutter by hand in his own time. If he wants both hands to appear in the picture it is easy enough to arrange to operate the release with his foot.

The remote control connection (tube, wire, etc.) should be suitably hidden—for instance, underneath a ring—so that it does not appear in the picture of a full-figure view.

A mirror is a help also in a remote-controlled self-portrait. For pictures showing the sitter looking into the camera, the mirror may be set up in front of the camera lens with the latter photographing through a cleaned-off window of the silvering.

Lighting Control. This method has certain advantages in special circumstances. It requires the entire subject lighting to be under the control of a single switch which the photographer can operate from the sitter's chair. The ideal form of lighting is remote-controlled flash bulb or electronic flash. The photographer arranges the picture as for a delayed action release and then puts out all the lights. He then leaves the shutter open on Time and takes up his position in the sitter's chair, checking the distance with a knotted cord as described for focusing above. When he is ready to make the exposure he simply switches the lighting on and off, or presses the flash button.

The most notable application of this technique is for making a number of exposures on the same negative with the subject posed against a black background. For this purpose electronic flash is ideal because a succession of exposures controlled by the duration of the flash can be made without revisiting the camera. A typical montage made in this way would be a full face portrait flanked by profiles.

Groups. Although the mirror and lighting control methods are not really suitable for what is probably the most popular of all photographer-in-the-picture work—i.e., groups of the photographer's friends or relatives, the two other methods may be used. In practice the remote control method is not very suitable because the camera must be at a fair distance to include the whole group, and this calls for a control that is inconveniently long and more difficult to exclude from the picture.

The built-in or a supplementary delayed action release is the standard means of taking this type of picture. Where possible, the photographer should pose the group with a space left for himself between two of the members. If he allows space at one end, he may underestimate the width he is going to take up and find part of himself cut off in the picture. And since he will have some distance to cover after pressing the release, he ought to make sure that the way is clear and that the rest of the group know exactly what is going to happen. If they are not fully briefed in advance the picture may show all their eyes directed at the photographer. P.C.P.

See also: *Mirror photographs; Portraiture; Shutter release.*

SELF-TIMER. Another name for a delayed action shutter release. It is more often applied to the supplementary pneumatic or spring-driven devices made for use with shutters not already equipped with a built-in delayed action mechanism. Some models have adjustment for varying the duration of delay.

See also: *Self-portrait; Shutter release.*

SELF-TONING PAPER. Daylight print out papers which need only fixing and no separate toning bath or gold-toning fixer.

See also: *Papers.*

SELLA, VITTORIO, 1859–1943. Italian mountaineer and alpine photographer from 1880. Like his father, Giuseppe Sella, who wrote the first text-book on photography to be published in Italy, 1856, Vittorio combined photography with alpinism. His exploits in the Alps, the Caucasus, Alaska, Central Africa and the Himalayas, some of them as official photographer to the Duke of Abruzzi's expeditions, made Sella's photographs world-famous. Biography: *The Life and Photographs of Vittorio Sella* by Ronald Clark (1948).

SELLING PHOTOGRAPHS

Good photographs are frequently required by editors of the illustrated press and technical journals, book publishers, advertising agents, press agencies, and many other users. Methods of approach vary according to the market, and the fees paid vary even more.

When a photograph is commissioned by an editor or publisher, there is nothing difficult about the transaction; the photographer knows in advance that the prints will be accepted if they fill the agreed specification and he knows what he will be paid for them. (The only point he needs to watch is what rights he is selling—i.e., is he parting with the complete copyright, or simply selling permission to reproduce in one issue of the publication?)

The essentials of a successful sale are: the photograph must be technically perfect (but there are important exceptions to this, e.g., outstanding news pictures), considerably above average interest in its particular class, appropriate to the market, and of the right size.

Technical Perfection. This is taken for granted in prints offered for publication. The only exceptions are photographs of such outstanding news value that technical faults do not matter—e.g., an exclusive but badly focused picture of the assassination of a famous man would be accepted so long as the scene was recognizable. Generally the print should not be smaller than whole-plate and not bigger than 8 × 10 ins. It should be on glossy or glazed white paper with a full range of tones and with no tendency towards strong contrasts. (This holds good even for pictorial subjects; however well they may look on rough surfaced paper with a cream base, the editor is only interested in how well they will reproduce. The blockmaker always prefers to work from a glossy print with a white base.) Finally, the print must be dead sharp.

Interest. Millions of good prints are made every year and most of them are of interest only to the photographer's immediate circle of friends or relations, or to the people who have paid for them to be taken. Only a very small proportion are capable of making a disinterested stranger look twice and, unless a print can be relied on to do at least that, no editor will buy it for publication.

A technically excellent picture of a baby or a kitten or a motor-car, however much it may make the breast of the photographer swell with pride, has no market value. But if the baby is bellowing tearful disapproval just when it is being awarded first prize in a national contest; if the kitten is riding on the back of a tortoise, or if the motor-car is doing a somersault in a road race, then the picture has a good chance of selling in the right market.

The Market. The fact that the market varies according to the subject ought to be obvious, yet thousands of salable photographs waste everybody's time and earn no money simply because they are sent to the wrong type of publication. One of the commonest mistakes is to send a photograph to a national newspaper when its real market is the local press. Before submitting a print, the photographer should always try to forget that he took it and try to imagine himself coming across it for the first time in the publication he has in mind. If it does not take its place naturally alongside the typical contents of that particular publication, then no editorial magic can make it do so. This means that the photographer who hopes to sell a picture must either know the market from personal reading and observation or should study an up-to-date book on the subject. Then he will not make such elementary mistakes as sending a picture series to a magazine that never publishes more than one photograph per page, or sending a red-hot news picture to a week-end paper.

Suitable Presentation. The hard work should all be done by the photographer as far as possible; if he leaves anything that he ought to have looked after himself to the editorial or production staff of the publications he is spoiling his chances of making a sale. So all spotting and retouching must be done as carefully as for an exhibition print; the editor must not be asked to imagine how well the print will look after his art department has removed the blemishes. The print should arrive in perfect shape; if it is packed carelessly and arrives cracked and dog-eared it will call for more work than it is usually worth before it can be reproduced. It should be accompanied by all the relevant information, preferably typed out and stuck to the back; the editor cannot spare the time to hunt for missing facts himself.

When sending trimmed enlargements it is advisable to enclose a contact print as a guide to the editor. On the whole it is, however, better to send untrimmed prints—partly because it gives the art department of the publication more scope to adapt the picture to their layout and partly because the marking up carried out by the art editor will nearly always spoil one's special effect.

Colour. Everything that has been said above applies to both black-and-white and colour photographs, but the editorial requirements for colour are very much more stringent. The cost of preparing the necessary blocks for colour reproduction is so great that only a picture of the highest standard will merit it.

Generally speaking it is rarely worth while submitting miniature colour transparencies as the loss of definition in enlarging them for reproduction is too great. They may be acceptable for sequences (to be reproduced on a small scale) otherwise $2\frac{1}{4} \times 2\frac{1}{4}$ ins. transparencies or colour negatives (accompanied by a colour print) are about the smallest suitable size.

Photographers who specialize in colour for reproduction should submit their work in the form of transparencies of at least quarter-plate size. Carbro or dye transfer colour prints made from three-colour separation negatives are also quite acceptable, but they should be much larger than quarter plate.

Selling Direct. When a print is offered to an editor or other prospective buyer without any conditions, it is understood that payment at the normal rate will be acceptable. The normal rate is the rate that the buyer normally pays for similar pictures from other sources used in a similar way.

Various scales of payment have been proposed in the past, but no definite agreement has ever been—or could ever be—established. The photographer must either be prepared to accept when he receives for the print or he must state how much he wants and what rights he offers. It is useless to protest after the picture has been published that the payment is too low. The only redress open to the photographer is the negative one of not offering further work in that quarter, which is an advisable course of action only when it is obvious that the buyer was trying to exploit the photographer unfairly.

If the reproduction fee asked is too high, the editor will simply return the print—always provided that the necessary packing and stamped and addressed envelope have been included. Where a print is submitted with no stipulation about the reproduction fee, the editor, if he wants to use it, will retain it without further acknowledgement. If he publishes the photograph, payment will normally be made at the beginning or at the end of the month following publication.

Editors sometimes keep prints for an unreasonable length of time without using them. This puts the photographer in a difficult position: if he sends in a bill for a reproduction fee, the editor will almost certainly return the print, and any hope that there might have been of selling it in that particular market will be at an end.

On the other hand, if the sender does nothing about it, the print may lie for months in the editorial office.

When the subject is highly topical, the situation is even more serious, because the print may be worth nothing unless it is published right away. With this type of picture the photographer should either telephone or call on the editor of the publication and negotiate the reproduction fee verbally or put the matter in the hands of an agency.

Selling Through an Agent. An agency will want the negative, not the print. Feature and news agencies will sell either exclusive rights to a single publication, or will submit the material to a large number of editors for reproduction at their standard rates. It is almost certain that the picture will make more money in the hands of an agent than if the photographer handles the sale himself. By knowing his markets exactly, an agent can often sell a picture, even of no great immediate interest, probably to a publication the photographer has never heard of.

The agent's commission will be around 50 per cent of the gross receipts for the picture, but even then the photographer will probably make more out of a topical shot than by his own efforts.

The agent has the advantage of experience in making the most of news pictures, and has the facilities for printing and distributing large numbers of copies of general interest pictures simultaneously to all prospective buyers.

Where the photographer is content with only occasional sales and does not mind how much trouble he takes to get them, there is nothing to prevent him from sending prints off himself to whatever prospective buyer he considers a potential market.

Choosing a Market. A detailed study of various magazines and journals is desirable before submitting any prints. The type of market should, however, first be selected and the following can be taken as a general guide: news pictures will find their best market with local or national daily newspapers; general interest pictures are mostly required by the weekly newspapers and magazines, as well as the monthly popular journals; technical photographs should be sent to the trade publication dealing with any specialized industry connected with the subject of the photograph; pictorial photographs involving human interest are best placed in the hands of an agent, who can circularize them to many customers among advertising agencies, manufacturers of photographic materials or postcard and calendar publishers.

When submitting prints, they should be addressed to the news editor in the case of newspapers, the art editor with general interest and technical magazines, and the art buyer with advertising agencies.

Reproduction Rights. It is always understood, when reproduction rights are offered, that the sender of the photograph either holds the copyright or is authorized to act for the owner. Very broadly, whoever takes the photograph owns the copyright, but he does not own it if he is paid to take it. When he is paid to take the photograph, the copyright belongs to the one who pays him—e.g., his employer or client.

When a print is submitted for publication, the sale of the copyright is not usually involved —only the right to reproduce. If a buyer wishes to purchase the copyright—e.g., to use as a trademark—he must say so, and be prepared to pay much more than he would for the simple right to reproduce. The fee in this case is always a matter for negotiation; there is no standard scale of charges. The purpose of the picture is however some basis for how much to charge; if a shot is suitable for, say, a national advertising campaign it may be worth as much as £50–£100, while a calendar publisher may not be prepared to pay more than £5–£10. Here again the superior experience of an agent helps as he not only knows the most suitable market and the potentialities of the picture, but also the best time for submitting it.

When the photographer sells the copyright, he loses all control over the picture from that date; when he sells reproduction rights he can attach any conditions he wishes. It is usually understood that, when a print is submitted to a newspaper, it may be reproduced in all the editions of the paper on a given date at the normal rate of payment. If in addition the print is reproduced in other newspapers or periodicals of the same company, then each separate use must be paid for, also at the standard rates.

This is common practice, and if the photographer wants his print to be treated in any other way he must make it clear in the covering letter.

A number of prints from the same negative may be sent to a number of different publications and normal reproduction fees will be paid by each user. Unless the covering letter states that the print has not been offered elsewhere, the newspaper will assume that other publications may also be using it. There is no need for the covering letter to mention this, but it should point out if the picture is being offered exclusively.

Exclusive Use. When a print is offered exclusively, it is necessary to define more precisely what is offered than if the offer is simply one of many. The covering letter should say whether the picture may be reproduced on more than one day or in other publications belonging to the same group. It should state whether or not the photographer reserves the right to submit the print overseas, or to offer it for use in other media—e.g., for advertisements, posters, calendars, etc.

Frequently a picture offered exclusively has a very high spot news value, and there is no time for correspondence or haggling. The photographer will in all probability negotiate the sale over the telephone, but, even so, the terms should be discussed at the time and repeated in a covering letter sent with the negative, print, or undeveloped material. It should never be assumed that the copyright automatically goes along with exclusive reproduction rights; copyright sale should always be the subject of separate negotiation and is only effected by written agreement.

Sales Record. All prints offered for sale should be recorded systematically; the details of the system will vary according to the scale of operations. A notebook entry is enough where only a few prints are being sent out, but index cards are essential where more than say a dozen different prints are out at one time. Each card should carry the title and reference number of the negative and give the publication and date of sending of each print—e.g.:

SINKING YACHT AT COWES, p. 634

Publication	*Date sent*	*Result*	
Picture News	5.6.55	Returned	8.6.55
Photo-Recorder	5.6.55	Returned	10.6.55
Yachting Journal	9.6.55	Published	14.6.55
Cowes Review	5.6.55	Published	9.6.55
Sunday Reporter	5.6.55	Published	8.6.55

A check must subsequently be kept to see that all prints published are paid for. It is not normally necessary to send in an account, but if there is any delay beyond the end of the month following publication, an account should be submitted with the date and issue in which the print was reproduced.

The only foolproof way of knowing that a print has been published is for the sender to make a personal search of every issue of the publication in which the picture could appear. This is easy enough if the publication is one that the photographer takes in regularly himself or sees at his work—e.g., trade and technical publications—or at the library or club. But if he sends out numbers of prints to a wide variety of publications the expense and trouble of making such a check are out of all proportion to the rewards.

The photographer has two courses open to him: he can rely on the publications to pay without being prompted (which they usually do) and risk the loss of an occasional reproduction fee; he can place his business in the hands of an agent.

Appointing an Agent. The photographer who normally chooses to sell his own prints direct and only occasionally makes use of an agent when it suits him must be prepared to accept the agent's regular terms of business. If he accompanies an odd picture with a long list of conditions, he will almost certainly get it back. His surest safeguard lies in dealing with an established agency which has a good reputation.

But anyone who is regularly producing marketable pictures and wants to appoint an agent to take all the work of selling them off his hands should only do so on clearly stated terms. It is enough for these to be stated in writing in a letter from one of the parties and acknowledged by the other. The agreement should make clear on the one hand that the agent is appointed by the photographer, and on the other, that the agent agrees to act for him in that capacity. It should further define:

(1) The service expected of the agent.

(2) What prints the photographer is obliged to supply.

(3) An exact definition of the rights to be offered.

(4) Any restrictions on the territory covered.

(5) The period covered.

(6) The amount of commission to be deducted from the gross fees.

(7) Arrangement about credit lines or other acknowledgement.

This type of agreement is particularly necessary for photographers working abroad and sending their prints to an agency at home or for those who want to sell prints in other countries.

Fees. The fees paid vary according to the publication and the size or area of the reproduction. British newspapers have an agreed minimum rate, based on the number of square inches, but the rates for provincial newspapers are only half those paid by the nationals—e.g., for a print offered simultaneously to a number of newspapers, the fee for reproduction up to 30 ins. might be as much as 10 guineas or as little as £1 7s. 6d. On the other hand, for an exclusive topical news photograph, the reproduction fee paid by a national newspaper might be 20 guineas or considerably more.

Weekly and monthly illustrated magazines will pay anything from 8 to 20 guineas per page of photographs and some publications will pay considerably more.

Fees are higher in America—especially when the photographer works on assignment—and very much lower in countries outside Britain and America.

Advertising agents, calendar and greeting card manufacturers and the like will pay from 5 guineas up to about 20 guineas for good, usable prints submitted speculatively, but the photographer should be quite clear about what rights he is disposing of for the fee.

The market for good colour transparencies of high technical quality, and reasonable size, has increased considerably in recent years, and a transparency might sell for one of many purposes for a fee of between 15 and 50 guineas. Colour photographs produced for reproduction should ideally be no smaller than 4 × 5 ins. Expense makes it inadvisable to engage in work of this nature unless it is specifically commissioned.

Return of Prints. When a picture is reproduced, the publication retains the print and may dispose of it in any way except that it may not offer it for publication elsewhere.

In certain cases editors may be persuaded to return a print even after publication—e.g., if it is unique and irreplaceable. This is not, however, standard practice, and requires some preliminary arrangement. Generally, the photographer will have to put up with the fact that a print that has been through the hands of retouchers, blockmakers or engravers, and the various postal personnel between them and the editor, will not be in mint condition any more. It will certainly not be in an ideal state for submission to other publications.

When the print submitted is not accepted for publication, the office is not obliged to return it to the sender. Most offices, however, will return unwanted prints, but the sender should do everything he can to encourage the courtesy—i.e., he should use packing stout enough to stand the double journey and enclose a self-addressed stick-on label.

When submitting prints to a strange editor, it is advisable to include return postage and preferably a self-addressed label. Later, when the photographer becomes known to the editor and his prints are being accepted more or less regularly, this may be unnecessary.

Editors appreciate that the sender has gone to some trouble and expense in submitting the print and in any case most publications rely to a greater or less extent on freelance contributions to keep their picture pages interesting.

And because the print and the covering letter may part company, it is necessary to emphasize once again the value of printing the name and address of the sender on the back of each print with a rubber stamp. F.P.

See also: *Agencies; Copyright and the photographer; Filing negatives and prints; Prices of commercial photographs; Reproduction fees; Reproduction quality; Scaling for reproduction.*

Book: *How to Take Photographs that Editors will Buy*, by R. Spillman (London).

SENEBIER, JEAN, 1742–1809. Swiss librarian in Geneva, and early photo-chemist. Studied the effect of light on plants, and made photometric observations with silver chloride paper, extending Scheele's experiment with the solar spectrum and giving exact times for the darkening of silver chloride under each colour of the spectrum. Senebier also investigated the effect of light on gums and observed the hardening property of some on exposure to light.

SENSITIVITY. A photographic emulsion consists of light-sensitive silver halide grains distributed throughout a layer of gelatin. The grains are extremely small, and a typical emulsion might have grains with an average diameter of around 10^{-4} cm. One frame of a normal 35 mm. film probably contains more than 10^9 of these grains. On exposure the grains react physically with light, and those grains which have received a sufficient amount of light are rendered black during subsequent development. Photographic sensitivity describes the amount of blackening resulting from a given exposure, and is thus closely related to photographic speed.

Originally photographic emulsions were sensitive only to light with wavelength in the violet region of the visible spectrum. Current emulsions are often sensitive throughout the entire visible spectrum and sometimes into the infra-red, and certain types of photographic plates can detect X-rays and nuclear particles.

Quantum Efficiency. When describing the interaction of light with these microscopic silver halide grains, it is instructive to do so in terms of the quantum theory of radiation. According to this, light has a discrete structure, and the smallest unit of light is the photon. Around 10^{11} of these photons are necessary to produce a uniform photographic density of unity over a 35 mm. frame of a medium speed film. The number of photons which are required to make a grain developable is sometimes expressed in terms of the so-called responsive quantum efficiency.

By this representation a 1-photon grain has 100 per cent efficiency, and a 5-photon grain has a 20 per cent efficiency. There is evidence that in modern emulsions the most efficient grains require only 4 photons for developability, but the least efficient grains may require more than a hundred photons. The amount of blackening, and hence the sensitivity, will obviously depend on both the responsive quantum efficiency and the grain size.

If speed were the only criterion of performance for a photographic material, then the best emulsion would be the one containing the most grains with the highest responsive quantum efficiency. But it is not usually sensitivity *per se* that is desirable but, rather, high sensitivity over an extended exposure range, since in most photographic applications the subject covers a large number of brightness levels. The wide spread of responsive quantum efficiencies, along with the spread in grain sizes, plays an important rôle in providing this exposure latitude. If all the grains were of equal size and sensitivity, then all would become developable in a narrow exposure region around the exposure level which provided that on average each grain received the requisite number of photons.

Such an emulsion would therefore have a high contrast and a low latitude, and would probably be unacceptable for most practical purposes, in spite of the fact that its sensitivity might be high.

But in fact there is not only a wide spread of responsive quantum efficiencies, but also a wide spread of grain sizes, and even grains of the same size have an appreciable spread of responsive quantum efficiencies. The probability of a grain receiving the requisite number of quanta is proportional to the size of the grain.

Granularity. Although photographic sensitivity is determined by the grain size and the responsive quantum efficiency, the useful sensitivity depends on a further important factor, which is a direct consequence of the discrete granular structure. For the grains are not distributed evenly throughout the emulsion layer, but are situated at random. Thus one small element of an emulsion may contain twice as many grains as an adjacent element.

This random grain distribution causes the visual phenomenon of graininess, the densitometric representation of which is known as granularity. The consequence is that any form of photographic interpretation is subject to an imposed degree of uncertainty. Generally speaking an emulsion with high speed has large grains and high granularity, and vice versa. It is found that the useful sensitivity depends on the speed-to-grain ratio rather than the speed alone.

Informational Sensitivity. The term informational sensitivity is sometimes used to describe the scientific measure of useful sensitivity. For a high speed-to-grain ratio means that accurate information may be recorded photographically at a high rate. If one emulsion has ten times the speed of another, but both have the same speed-to-grain ratio, then ten exposures of fixed duration using the faster emulsion would yield the same information as one exposure of ten times this duration using the slower one. Thus although the emulsions differ widely in sensitivity, they are rated equal from the point of view of informational sensitivity.

Following these same arguments it is easily seen that the ultimate sensitivity is of less significance than the ultimate informational sensitivity. The former is not a realistic concept, since even a 1-photon grain could be made more sensitive by increasing its size, although its responsive quantum efficiency would remain the same. A 1-photon grain the size of a football would be extremely sensitive, but completely useless for normal photographic purposes.

However the ultimate informational sensitivity is limited by the discrete photon structure of the exposure. The randomness of the photon distribution is similar to the randomness of the grain distribution, and the photographic granularity cannot thus be less than that imposed by the photons in the exposure.

Calculation shows that if the photographic process were capable of the ultimate informational sensitivity, then it would be possible to obtain a good quality photographic image exposure to starlight. At lower exposure levels it is still theoretically possible to obtain a strong photographic image, but the granularity would be very high and in general the picture quality would be unacceptable. R.Sh.

See also: *Information in the image; Quantum efficiency; Sensitized materials history; Sensitizer; Spectral sensitivity; Speed of sensitized materials.*

SENSITIZED MATERIAL. General term for all types of photographic material—plates, films and papers, etc.—that have been rendered light sensitive either by coating with an emulsion containing light sensitive silver salts, or by impregnation with a chemical sensitizer. The types of sensitized material vary considerably, from normal negative materials for use in a camera to special emulsions for scientific work.

See also: *Cine film; Colour materials; Films; Negative materials; Papers; Printing materials; Reversal materials; Supports for emulsions.*

SENSITIZED MATERIALS HISTORY

For about the first forty years of the photographic process the photographer had to prepare his own sensitive material just before making the exposure. The division of labour between photographer and sensitive material manufacturer did not become general until *c.* 1880, although dry collodion plates had been put on the market as early as 1856.

Early Processes. Workable photographic processes were made public in 1839, almost simultaneously by Louis Jacques Mandé Daguerre in France and William Henry Fox Talbot in England. It is correct to honour Fox Talbot as the inventor of photography as the modern processes stem in a direct line from his work. Daguerre's process, although it enjoyed considerable popularity, did not prove capable of further development. Both the calotype, as Talbot called his improved process of 1841, and the daguerreotype are real picture making media and in skilled hands are capable of good results, though the nature of the process limits their applications. Professional photography was founded on the daguerreotype, while the calotype was originally preferred by amateurs who were attracted by the lower cost of the materials.

In the daguerreotype process a highly polished silver plate, or silver faced copper plate, was sensitized by exposing it in the dark to the action of iodine vapour or, in a later method, to bromine and iodine. After exposure in the camera, which might take several minutes in a good light, it was developed by the action of mercury vapour, which was allowed to condense on the surface of the plate and delineated the image. After fixation in a solution of common salt, or later hypo, the image was toned in gold chloride. Fine detail was well reproduced, and the daguerreotype had an attractive, jewel-like quality. The process, however, was expensive and not well adapted for the production of copies—an essential for a popular process.

In Talbot's calotype process, paper was coated with silver iodide by successive immersions in solutions of potassium iodide and silver nitrate, and exposed in the camera while wet. After exposure it was developed in a physical developer containing gallic acid and silver nitrate, and fixed in hypo. The vitally important step of using a developer was due to a suggestion from the Rev. J. B. Reade. The negative so obtained was printed on to similar paper so that multiple copies could be obtained and in this respect it was a great advance on the daguerreotype. The process was better for showing broad effects than for fine detail, but could be used in portraiture, the series of portraits made by D. O. Hill and R. Adamson of Edinburgh in the forties being particularly famous.

Wet Plates. The next important step was the introduction of the wet collodion process by Frederick Scott Archer in 1851. This process consists of coating a sheet of glass with a solution of nitrocellulose containing a soluble iodide, and sensitizing the plate by immersion in a solution of silver nitrate. While wet the plate is exposed in the camera and is physically developed in pyrogallol or a ferrous salt. As the collodion layer must not be allowed to dry it is a most inconvenient process to work, but the results are excellent. Wet collodion has survived to this day for certain applications in the graphic arts field.

With the introduction of the wet plate a sensitive material suitable for topical photographs had arrived, for although the manipulation is exacting, failures are few and for small pictures the speed is adequate for snapshots in a good light. News photography dates from the wet plate, notable early examples being Roger Fenton's photographs of the Crimean War and Brady's pictures of the American Civil War. It was also the medium for a great expansion in portrait photography; compared with the daguerreotype it was cheap and copies were easily made. Dagron's greatly reduced copies, used in the pigeon post which was operated from Paris during the siege in the Franco-Prussian War, were made by means of the wet plate.

Numerous attempts were made to modify the wet plate process so that the sensitized plates would keep. The introduction of collodion emulsion by Sayce & Bolton in 1864 was one of the more successful, but the sensitivity was low, and the method was not popular.

Dry Plates. In 1871 Richard Leach Maddox, an English physician, produced the first workable, though slow, plate using gelatin as the medium to hold the silver bromide. The discovery, after improvement by others, proved outstanding, gelatin dry plates having a great advantage over wet collodion. The plates can be prepared weeks or months before exposure; the sensitivity is far higher; the processing is simple and there is no need to develop the plate immediately after exposure.

It was no longer necessary for the photographer to make his own plates—very soon after Maddox's discovery he could buy them boxed and labelled ready for use. In 1873 gelatin emulsion was being offered for sale by Burgess. In 1867 the Liverpool Dry Plate Company was making plates in England, this being the second factory making sensitive material. By 1878 there were four dry plate manufacturers in England. Johann Sachs of Berlin is reputed to be the earliest manufacturer in Germany, and Carbutt of Philadelphia, 1879, the first maker in the United States. He was soon followed by Eastman, who founded in Rochester the enterprise now known as the Eastman Kodak Company. By the end of the century sensitive material manufacture had been established on a considerable scale in most advanced countries.

Colour Sensitized Plates. In 1873 a discovery of the greatest importance was made by H. W. Vogel in Berlin. While testing some plates which had been dyed to reduce halation he found that they were sensitive to green light. He concluded correctly that this property was due to the dye, which was probably corallin, and set to work to examine a number of dyes. As a result he found several which could act as sensitizers, and predicted that it would in time become possible to photograph even by infra-red rays, with dye-sensitized plates.

In 1884 the first colour-sensitive plates were offered for sale. In the same year Eder discovered the sensitizing properties of erythrosin, which was used in the manufacture of orthochromatic plates for many years. In 1906 the first satisfactory red sensitizing dye was discovered by Homolka working at Hoechst, and was sold as pinacyanol. This dye was used in the first panchromatic plates manufactured in England by Wratten & Wainwright, and continued in use for many years.

The discovery of a great number of new types of cyanine dyes from 1926 onwards led to the introduction of the super-sensitive type of panchromatic materials which of recent years have displaced colour blind emulsions almost completely for ordinary photography and have made possible the production of fine-grained colour-sensitive emulsions, which are essential for miniature camera work.

Plates were used for radiography in the year of Roentgen's discovery of X-rays. In 1920 the plate was replaced by double-coated film, and X-ray films are now made in great quantity and variety.

In the early years of the twentieth century slow process plates were produced for use in the photomechanical production of pictures. A range of special sensitive materials is currently made for the graphic arts industry, an important item being the "lith" type emulsion introduced about 1938, which gives negatives of remarkably high contrast when used with a suitable developer containing formaldehyde.

Films. Fox Talbot used paper as a support for his negative material, but later glass became universal. Warnerke introduced a stripping film for hand cameras in 1875, and in 1889 Eastman started making films on a nitro-cellulose base. The flexible film base was a great advance as it made possible the construction of light daylight-loading cameras and led to the development of cinematography—1895 onwards. Nitrocellulose or celluloid was widely used until about 1930, when the much less inflammable cellulose acetate began to displace it. On account of its good mechanical properties, however, nitrocellulose remained in use for 35 mm. cine film as late as 1947, when it was replaced by cellulose triacetate base.

Printing Papers. The development of sensitive material for making prints proceeded on similar lines to negative materials. The plain paper impregnated with silver chloride introduced by Talbot was replaced about 1850 by albumenized paper which remained in use for over thirty years. It was inconvenient to use as it had to be immersed in a bath of silver nitrate and dried before being printed in daylight. Used with a strong negative, such as is obtained with the wet plate process, it gives good prints when gold toned.

Collodio chloride paper was introduced in 1868 and gelatino chloride printing-out paper was manufactured commercially in 1885. The first bromide paper was introduced by the Liverpool Dry Plate Company as early as 1873 and also by the firms of Morgan & Kidd and Mawson & Swan in 1880. Silver chloride and chlorobromide development papers date from 1883. In the years 1880 to 1914 the platinum process (which depends on the sensitivity of ferric salts, though the image is formed in platinum), and from 1866 on the carbon process (which depends on the sensitivity of chromate) were used to a great extent but are now obsolete. Silver halide development papers, particularly those based on silver chloride or chlorobromide and often dye-sensitized to increase the speed, are now the only important papers in photography.

Sensitized paper coated with diazonium salts, which is produced in great quantities for printing engineers' drawings and similar applications, derives from the discovery of light-sensitive diazo compounds by Green, Cross & Bevan in 1890, although the first practical application was by Koegel in 1923.

In recent years various special papers have been introduced. Among them variable contrast papers dating from 1940, direct positive (Herschel effect) papers dating from 1947, papers for the image transfer process dating from 1950, and papers such as those, introduced in 1952, which yield a direct positive by chemical fogging when used with a special developer.

Colour Materials. The first sensitive material for colour photography to be produced on a commercial scale was the famous Lumière Autochrome plate introduced in France in 1906. This was a screen plate with an irregular mosaic, and yielded colour transparencies by the additive method. A similar screen plate was produced by the Agfa Company some years later and a screen plate with separate screen was manufactured by the Paget Prize Plate Company about 1910. In 1928 Kodak produced 16 mm. colour film using a fine lenticular pattern which images a three component filter on the emulsion. A year or two later Dufaycolor, a film with mechanically printed colour screen which made duplicates possible, was produced.

Kodachrome, the first of the integral tri-pack films yielding colour transparencies by the subtractive process, was introduced in 1935 and proved an immediate success for substandard cine film. In 1936 Agfacolor, a rather similar tripack material, was produced by the Agfa Company in Germany. It differs from Kodachrome in that the colour formers are included in the emulsion, while in the Kodachrome process the colour formers are in the developing solutions. In 1942 Kodacolor, a tripack negative with a corresponding tripack positive material for making colour prints, was produced in the U.S.A. Negative-positive Agfacolor for cine film was made in Germany at about the same time, followed by a paper print material.

Since 1942 a number of tripack films and printing materials have appeared. In 1949 Ektacolor film was introduced. This film contains colour couplers which are themselves coloured and by this means integral colour masking is obtained.

A rather similar form of integral masking is used in Eastmancolor negative cinema film, introduced in 1952.

The demand for colour in all forms continues to grow so that it is probable that a substantial proportion of the output of the sensitive material industry will soon be colour films. The manufacture of modern colour film is one of the most exacting and difficult forms of technology and is far removed from anything the photographer used to undertake for himself. W.H.D.

See also: *Camera history; Development history; Discovery of photography; Obsolete printing processes.*

Books: *A Source Book of Photographic History*, by W. Baier (London); *History of Photography*, by J. M. A. Eder (New York); *The History of the Discovery of Photography*, by G. Potonniée (New York); *The History of Photography*, by H. & A. Gernsheim (London).

SENSITIZED MATERIALS MANUFACTURE

Photographic emulsions as we know them today were prepared by Maddox for the first time in 1871. In those days photographers sensitized their own materials, but soon after (in 1874) photographic plates became available commercially, and from then on the photographic industry developed rapidly.

The present-day photographic materials consist of a light-sensitive emulsion layer proper, containing a suspension of microcrystals of silver halides in gelatin (chloride, bromide, iodide, or mixtures of them), and coated on to a support—e.g., film, glass plates or paper.

Safelighting. From the moment at which silver halides are first formed, the resulting sensitivity to light requires special measures to ensure that the emulsion is not exposed to light which would result in emulsion fog. Any illumination used must be weak and of such a colour that the minimum photographic effect is produced in the emulsion, yet retaining the maximum possible visual brightness. The methods of illumination used are described as safelighting. In general, the same types of safelight are used in manufacture as in processing darkrooms and the same general principles apply.

As long as emulsions are blue-sensitive only, a red safelight will provide the brightest possible safe illumination. For orthochromatic materials a deep red safelight is used. With panchromatic and infra-red sensitive materials the only safelight which can be used at all is a dim green. This colour is chosen not only because many materials have a region of relatively low sensitivity in the green, but also because the dark-adapted eye is most sensitive to blue-green light. Green is therefore the best possible compromise, although for most purposes it is best not to use any light at all with fast panchromatic materials.

In manufacturing rooms light is allowed to fall on the sensitive material only where it is absolutely necessary to control the process. Lights at foot level are frequently used to guide the workers through the rooms; if the sensitive material is carefully shielded these footlights can be relatively bright.

Emulsion Making. The raw materials used in the preparation of the photographic emulsion must be very pure. The silver nitrate and alkali halides used (salts of potassium, sodium, lithium or ammonium) are amongst the purest chemicals available commercially.

The other most important raw material—i.e., the gelatin—is in many respects considerably purer than that used as food. Gelatin is an animal protein obtained from bones or skins of animals such as cows, buffaloes or pigs. The skins and bones are subjected to a liming treatment lasting up to 18 months. An alternative treatment for bones is in sulphuric acid. The material is then cooked in large vats. The resulting brew is boiled down in vacuum until the concentration of gelatin is sufficiently high for the "soup" to set on cooling. The gelatin solution is then evaporated and the resulting dry gelatin is available in the form of powder or flakes containing approximately 15 per cent water.

The next step is to incorporate the silver halide in the gelatin. The nature of the halide varies according to the type of material to be coated. For films and plates the halide is mostly silver bromide, with a few per cent of iodide. Slow papers are coated with a silver chloride emulsion and faster papers with a chloro-bromide, or even 100 per cent bromide or a film-type of emulsion.

The emulsion is prepared in large kettles of chemically inert material such as high-grade stainless steel. There are many ways of preparing the emulsion but the principle is the same in each case. The alkali halides are dissolved in a solution of gelatin (about 1 per cent) and the silver nitrate is added. A chemical reaction takes place, precipitating the silver halide in the form of a very large number of minute crystals (grains) which are prevented by the protective action of the gelatin from sinking to the bottom of the kettle or clumping together.

Ripening. The emulsion crystals or grains are then grown in size to increase their sensitivity. This process of grain growth is described as ripening or cooking of the emulsion. It is carried out by heating the solution in the presence of a solvent for the silver halide. The effect of the solvent is to dissolve the smaller grains and deposit their substance on the larger grains. Different solvents are used for each of the different types of emulsion. Chloride emulsions require a surplus of potassium or sodium chloride, and iodo-bromide emulsions a surplus of potassium bromide or ammonia.

Further gelatin is next added until the concentration is approximately 7 per cent so that the emulsion sets to a jelly pudding on cooling to room temperature. The jelly is then put through a press, which squeezes it out in the form of noodles. In this state it is more easily washed.

Washing is necessary because the process of precipitation which forms the insoluble silver halide also produces soluble sodium or potassium nitrate. This must be removed from the emulsion if it is to be coated afterwards on glass or film base. Any free ammonia must also be removed since these chemicals would form undesirable crystals.

Digestion. The noodles are then melted and the emulsion subjected to further heat treatment known as digestion or after-ripening. No appreciable further grain growth takes place since no silver halide solvents are present, but the sensitivity of the emulsion increases considerably during this treatment for other reasons. Various chemicals are added in very small amounts in order to help increase the sensitivity—e.g., certain sulphur compounds, and more recently gold compounds. During the digestion treatment the chemical fog also tends to increase, so the heat treatment chosen must compromise between increasing sensitivity and keeping fog to a reasonable level.

Further chemicals may now be added to the emulsion to improve its characteristics. Sensitizing dyes extend the colour sensitivity to include, green, yellow, or red, or even the infrared end of the spectrum. Stabilizers ensure good keeping properties. Hardeners give the coated product resistance to surface injury from scratching and abrasion. Plasticizers keep the dry emulsion layer supple, and spreading agents help to make the subsequent coating uniform. These are only a few of the additions that may be necessary to produce the required type of emulsion.

Films. Photographic film base used to consist of cellulose nitrate but in view of its inflammability it has been replaced in recent years by non-inflammable cellulose acetates, in particular tri-acetate, and acetate-butyrate. More recently, polystyrene, polyesters, and polycarbonates have been introduced. The thickness of the base may be between 0·003 and 0·01 in. with an emulsion coat of the order 0·005 to 0·002 in. Very often the emulsion surface is given a separate super-coat to protect it from abrasion.

Many films are multicoated—e.g., for amateur use there may be two emulsion coats, the bottom coat slow, the top coat fast, to give great exposure latitude. Modern colour films possess three emulsion coats often separated by clear or coloured gelatin—all of which involves highly complicated coating operations. X-ray films are often coated on both sides of the base.

To provide protection against irradiation and halation the emulsion may be coated on a grey-tinted film base. This absorbs much of the light that has penetrated the emulsion and prevents it from being reflected back into the emulsion to produce a halo around the true image. Alternatively film is often backed with a layer of dyed gelatin which is dissolved or bleached in processing. This gelatin backing also prevents the film from curling. The backing is usually applied before the film base is coated and there is also a "subbing" layer to ensure good bonding between the base and the emulsion. The subbing layer may consist of a weak solution of gelatin containing a solvent for the film base; it is so thin that it cannot readily be measured.

Film Coating. As a rule the photographic emulsion is applied to its support by moving the support material and keeping the point, or better, line of application of the emulsion stationary. A long web of the film base may be passed over a roller which dips into a pan containing the melted emulsion. According to the viscosity of the emulsion and the rate of movement of the film, a more or less thick uniform layer is picked up. The web of film then passes through a chamber in which cold air is blown on to it causing the emulsion to set, or the film may be passed over a large chilled metal roller. It is then dried, for example by being hung up in festoons from overhead racks which slowly travel through a drying room. Once it is thoroughly dry the film is spooled.

Papers. Photographic paper is produced on machines very like those used for film except that there is less need for chilling because the paper base takes up water and thus concentrates the gelatin solution causing it to set more readily. Paper emulsions are often not washed because the reaction products are absorbed by the porous paper base and do no harm.

The raw material of the base used for photographic papers is carefully selected to be

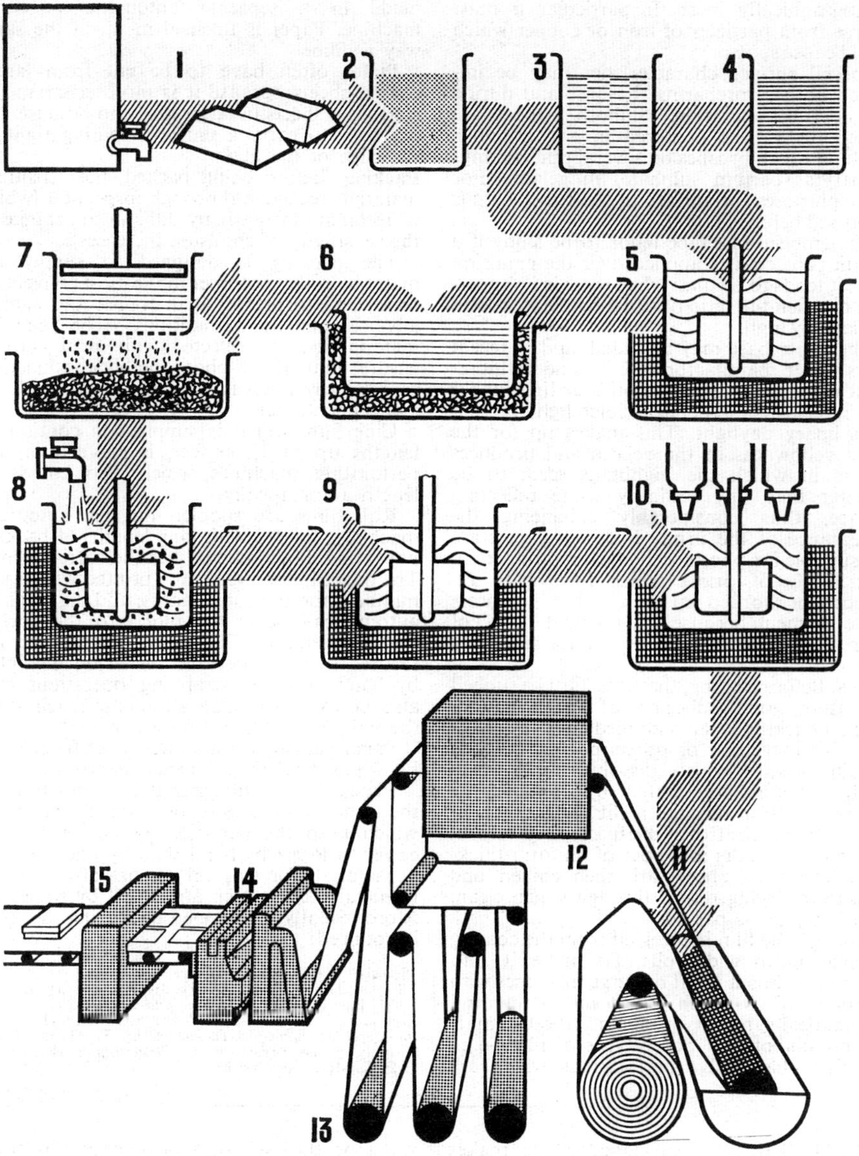

MANUFACTURE OF SENSITIZED MATERIALS. While production details depend of course on the particular manufacturer as well as on the product, this flow chart gives a simplified idea of the stages by which, for instance, sheet film is produced. 1. Essential raw materials: silver and nitric acid. 2. These are used to make silver nitrate. 3 and 4. Further raw materials: gelatin and sodium or potassium bromide and iodide. 5. Mixing of the emulsion under carefully controlled conditions. 6. Cooling to jellify the emulsion. 7. Shredding the jelly. 8. Washing of the shredded jelly to remove the unwanted salts (mainly sodium and potassium nitrate) resulting from the formation of silver halide. 9. Melting of the jelly and digestion (ripening) of the molten emulsion to build up speed and contrast. 10. Addition of colour sensitizers, stabilizers, plasticizers, hardeners, spreading agents, etc. 11. Coating on the film base. 12. Chilling to set the emulsion. 13. Drying. 14. Cutting of the continuous roll of film into required sheet film sizes. 15. Final inspection and packing for dispatch. Various intermediate stages taking place are not shown; these include backing of the film base to provide anti-halo protection and to counteract curling, application of protective supercoat, etc.

photographically inert. In particular it must be free from particles of iron or copper which would cause spots.

Special surface characteristics may be imparted both in preparing the base and during the coating operation. Smooth papers are coated on a base prepared in a continuous machine which brushes on several thin coatings of baryta (barium sulphate) in gelatin. For other purposes the surface of the paper base is embossed before coating.

The emulsion surface tends to be shiny if a gelatin supercoat is applied over the emulsion coat; it tends to be matt when there is no supercoat or when the ratio of silver halide to gelatin is relatively high.

The paper base may be tinted, and in recent years paper manufacturers have added fluorescent materials which give off blue light under the influence of the ultra-violet light present in ordinary daylight. This makes up for the slight yellowness of the gelatin and produces papers in which the highlights seem to be brighter than any ordinary white reflecting surface, thus considerably enhancing the appearance of the print. But such papers are not suitable for photomechanical reproduction, particularly of prints which have been retouched or worked up with ordinary process white pigment because the pigment does not fluoresce and it tends to reproduce as a grey tone.

Plates. Before coating, the glass must be subbed to ensure good adhesion of the emulsion. Often a backing layer is applied; it may consist of a dye in gelatin or other medium such as certain plastics which dissolve during processing. Plates are coated by running the melted emulsion through a slit on to sheets of glass. Careful control of the speed of movement of the glass results in a coat of uniform thickness. The glass sheets are then chilled and stacked in drying rooms through which clean, warm, dry air is passed.

Finishing. The film is received from the coating department in wide rolls. To make it into roll film widths it is slit on large machines with rotating knives which cut it into strips and automatically reel it up again. Sheet film is cut on guillotines. Some types of film—e.g., cine film—have edge perforations which are made in a separate automatic punching machine. Paper is finished in much the same way as film.

Plates often have to be cut from larger coated sheets because it is more economic to coat glass sheets in large sizes and because it is not easy to carry a uniform coating right to the edge of the glass.

Packing. Before being packed, the sensitized materials receive a thorough inspection (which is becoming increasingly difficult to arrange as the sensitivity of emulsion increases).

The packing is designed to supply the material to the customer in the most convenient form, and to protect it in transit and against adverse climatic conditions. The packing material itself is selected with great care to ensure that it is photographically inactive, especially where it is in direct contact with the emulsion surface.

Cine film, which is supplied in continuous lengths up to 1,000 feet, is spooled on the perforating machines which also cut it to length automatically.

Roll films are spooled on semi-automatic machines, loaded with long lengths of backing paper and film already slit to the correct width. The paper is supplied ready printed with frame numbers and indicator marks, and the machine automatically cuts the film into the corresponding lengths. These lengths of film are secured to the paper and sealed on the spool by hand. Further wrapping operations may also be carried out on automatic machines if the volume of production warrants it.

Paper packing can be carried out by brighter light and it raises no particular problems.

Plates are mostly packed by hand because the spacers or sheets of interleaving paper which keep the emulsion surfaces apart are easier to insert by hand than by machine.

Materials for tropical use are often packed in metal boxes which are closed by soldering. More recently heat-sealed packaging has been introduced. W.F.B.

See also: *Emulsion manufacture.*

Books: *Film Coating Theory*, by B. V. Deryagin and S. M. Levi (London); *Making and Coating Photographic Emulsions*, by V. L. Zelikman and S. M. Levi (London); *Photographic Emulsion Technique*, by T. Thorne Baker (London); *The Theory of the Photographic Process*, by C. E. K. Mees (New York).

SENSITIZER. Any substance which makes another react to light, or which extends the range of radiations to which a substance is sensitive, or increases its sensitivity. The two main types used in photography are chemical and optical sensitizers.

Chemical Sensitizers. These act by forming a light-sensitive compound with another substance, e.g., bichromates react with gelatin and other colloids under the action of light to produce a compound which is insoluble in hot water. In the early days of photography silver nitrate was also referred to as a sensitizer, because photographic plates and papers were prepared by coating a support with a solution containing, among other things, a soluble halide. This was then sensitized with silver nitrate, forming light-sensitive silver halide.

In the field of conventional silver halide-gelatin emulsions, "chemical sensitization" refers to ways of producing silver halide grains with a high inherent sensitivity to light in the absorption band characteristic of the particular halide. The usual silver iodo-bromide grain is self-sensitive to ultra-violet and violet light. Chemical sensitization is, in effect, a matter of conditioning the silver halide crystal, or more particularly its surface, so as to render it more fertile for the production of development centres.

Sensitization can be achieved by the controlled production of specks of silver sulphide, of gold, of silver, or of a mixture of these. The gelatin itself plays a most important but not fully understood rôle in sensitization. Without gelatin—or a substitute colloid—the efficiency of the action of light in forming developable centres is very poor. The colloid stabilizes the latent image speck and prevents its spontaneous decay or fading.

Although chemical sensitization can yield emulsions of very high working speed, the sensitivity remains confined to blue light. In normal emulsions the gelatin itself acts as a dense filter against sensitivity to ultra-violet light. If the sensitivity of the emulsion is to be extended to other wavelengths in the direction green-yellow-red-infra-red, it is necessary to resort to "optical sensitization" by means of dyes.

Optical Sensitizers. Vogel found, in 1873, that by the addition of coralline dye the sensitivity of collodion dry-plates could be extended into the green band of the spectrum. Other early sensitizers were chlorophyll (red), eosin (green), and erythrosin (green). An advance in dye sensitization came with the introduction of cyanine dyes of which pinacyanol was notable. This dye was incorporated in the emulsion of the first panchromatic plates of Wratten and Wainwright. In the past 50 years innumerable organic dyes have been specially prepared and tested as sensitizers. In some cases improved results are found when two dissimilar dyes are used together. One dye acts as a supersensitizer towards the other.

In general, a sensitizing dye confers increased sensitivity to the silver halide grain at those wavelengths at which it absorbs light when it is adsorbed to the grain. The colour of the dye, when the latter is present as a close-packed layer of molecules on the surface of a silver halide grain, owes something to the close packing; it shows a marked absorption band at a higher wavelength than the absorption characteristic of the separated dye molecules. Thus, for efficient dye sensitization it is important to achieve a continuous layer, one molecule thick, over the grain surface. This layer of close-packed dye molecules can act as a conductor for light energy, allowing the excitation from an adsorbed quantum to be felt over a large area of grain surface, thus increasing the probability of a fruitful reaction with the silver halide. G.I.P.L.

See also: *Latent image; Sensitized materials manufacture; Spectral sensitivity.*

SENSITOMETER. Instrument used in recording the response of a light-sensitive material to a range of exposures. There are two types of sensitometer; one which subjects the sample of material to a series of lengthening exposures under a light of constant intensity, and the other which exposes adjacent strips of the material for the same length of time to a range of light intensities.

See also: *Sensitometry.*

SENSITOMETRY

The science of sensitometry may be said to be founded upon a paper by Hurter and Driffield entitled *Photochemical Investigations and a New Method of Determination of the Sensitiveness of Photographic Plates*, published in 1890. Today the name is often used to describe the testing of photographic materials.

Sensitometry seeks to establish accurately the relationship between exposure and density for any given photographic material. (In sensitometry, exposure is the product of intensity of light and time of exposure, not just the time of shutter speeds as in common parlance.) This relationship between exposure and density is often very difficult to define in words and such indefinite terms as "high speed", "low contrast" and "poor blacks", etc., have a very limited use. The chief aim of sensitometry is to derive accurate numerical values for the exposure-density relationship of a material to enable guess-work to be eliminated and the results stated in units which can be universally applied and compared.

Many sensitometric standards have now been set up by the British Standards Institution and the American Standards Association as well as standards organizations of other countries, and sensitometry has become almost indispensable both to the scientific and the practical photographer.

Sensitometry is used mainly to measure the effects of the visible spectrum, ultra-violet, infra-red and X-rays, but some work is also carried out with gamma-rays, electrons and nuclear particles.

A sensitometric test on a given sample of light-sensitive material starts with subjecting the material to a series of standard exposures in a sensitometer. It is then processed under controlled standard conditions. Then the sample is examined in a densitometer which measures the densities recorded on the material. Finally a characteristic curve is plotted from the readings obtained in the densitometer. This curve shows the density values on a base of log exposure and provides a complete description of the response of the sensitized material to exposure and processing.

Control of Conditions. For sensitometry to be of maximum value, especially to the practical photographer, it is generally essential for the conditions to be identical with those used in practice. When this is difficult, a correlationship can sometimes be established between the results obtained under the sensitometric conditions and those obtained in the type of work concerned.

The spectral sensitivity of many photographic materials differs from that of the eye. There are four main photographic sensitizings available, blue (ordinary), orthochromatic, panchromatic, and infra-red, and in each of these groups, and sometimes between different manufacturers in the same group, there may be minor differences according to the exact requirements. Because of these factors it is very important that the colour temperature of the light source is suitably chosen when making a sensitometric test.

The exact method of exposure—i.e., whether by an intense light for a short time or a weak light for a long time, and whether steady or intermittent—is extremely important because of the reciprocity failure of photographic materials; the density obtained depends not only upon the product of intensity and time, but upon their individual values. In practice exposures vary widely because of the great differences in light intensity encountered; the astronomer may expose for hours, the studio photographer for seconds, and the sound-recording engineer for 1/50,000 second, while an electronic flash may only last a few millionths of a second.

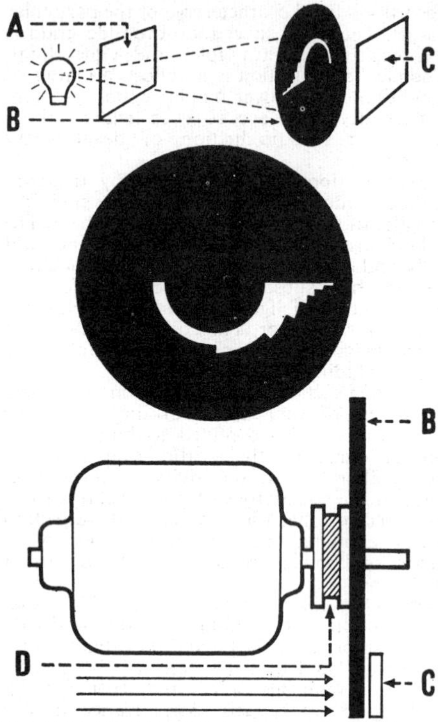

TIME SCALE SENSITOMETER. Top: *General layout. A lamp with a correction filter illuminates the sensitive material behind a sector wheel. A, correction filter.* B, *sector wheel.* C, *sensitive emulsion.* Centre: *The angles of the sector apertures increase in geometric progression, each being twice as large as the preceding one. An automatic shutter exposes the test material during one turn of the wheel.* Bottom: *Arrangement of drive. The wheel is driven through a piece of soft rubber to ensure even speed and to take up any jerkiness in the movement of the motor.* D, *soft rubber coupling disc.*

SENSITOMETERS

The sensitometer is the instrument used to give the sensitized material under test a series of controlled exposures. It consists essentially of a light source, a method of modulating the exposure and a holder for the material being tested.

The Light Source. The unit of luminous intensity adopted within the last few years is the candela. This unit is based upon the light intensity emitted by solidifying platinum and only differs slightly in value from the international candle previously used. It is too inconvenient to use such a light source in a sensitometer and in any case the colour temperature of solidifying platinum is too low for most photographic purposes; correction filters would have the disadvantage of cutting down the light. Secondary standards, in the form of electric tungsten filament lamps, are set up by such bodies as the National Physical Laboratory in England or the Bureau of Standards in the United States of America. In commercial sensitometers, carefully selected and aged Class A projector bulbs are used, being under-run to increase their life. These lamps are usually fed by a D.C. supply because of difficulties arising from the fluctuations in intensity with A.C. supplies. The supply voltage has a marked effect on the light intensity, so it has to be very accurately controlled.

It is very important to control the spectral energy distribution or colour temperature of the light source. Within limits the colour temperature of electric filament lamps may be controlled by varying the voltage across the lamp, but colour temperatures above 3000° K cannot be obtained without considerably reducing the stability and the life of the lamp. Further control of the colour temperature is achieved by light filters known as photometric filters. By their use the light from a stable filament lamp can be adjusted to the required colour temperature. The characteristics of such filters depend on the colour temperature of the source. They may be made of gelatin, glass, or they may be liquid filters. Of the three types, gelatin filters are usually the least accurate. To reproduce daylight, the filter used is the Davis-Gibson liquid filter employed in the BSI and ASA speed systems.

Here a tungsten lamp working at 2360°K in conjunction with the filter gives a colour temperature of 5400°K, referred to as artificial mean noon sunlight.

Modulation of Exposure. The exposure received by any photographic material is dependent upon the product of the time of exposure and the intensity of the light. Consequently, exposure can be modulated either by a time scale or by an intensity scale.

Although time scale modulation is not very common in practical photography, time scale sensitometers are quite frequently used. This is because they are, perhaps, easier to make than intensity scale instruments and within certain limits the reading can be correlated with the intensity scale used in practice. Such instruments are often used for batch control of any one grade of sensitized material where any variations in sensitometric properties are likely to be small.

A common method of time modulation consists firstly of inserting between the light source and the material under test an opaque disc (called a sector wheel) from which has been cut out a series of accurately measured slots of varying length. On rotating this sector wheel at a constant speed for one revolution the material is exposed in strips for a series of times which depend upon the exact contour of the cut-out.

One well-known commercial instrument, widely used in the cine trade, uses a rotating drum instead of a sector wheel. In this case the drum carries a cut-out which gives a series of twenty-one stepped exposure times, each time being $\sqrt{2}$ longer than the previous one. The drum is horizontal and the light enters the inside of the drum and is reflected upwards by a mirror, through the cut-out, on to the material under test.

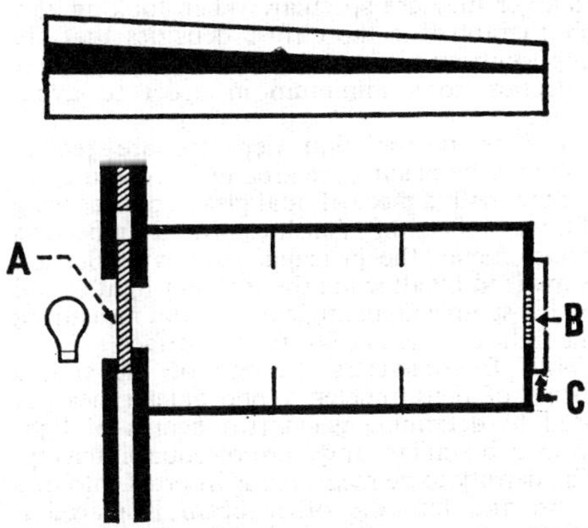

INTENSITY SCALE SENSITOMETERS. Top: Goldberg wedge, of neutral dyed gelatin cast between two inclined glass plates. Bottom: Falling plate sensitometer. A, shutter consisting of plate with slot. B, wedge. C, sensitive material.

An obvious method of setting up an intensity scale is either to expose adjacent strips of the material to light at varying distances or to cover the light with apertures of varying sizes. In both methods the intensity scale is governed by the dimensions of the apparatus. Neither system is very convenient, however, especially as it is desirable to make the whole range of exposures simultaneously on to a single sample of material of reasonable size. Sensitometers based on a range of apertures have been made, but the most common method of obtaining an intensity scale today is to expose the sample under test through a wedge of light-absorbing material.

A typical example is the Goldberg wedge which consists of a wedge of gelatin, containing a neutral colouring material, cast between two glass plates. The greater the density or thickness of the wedge the lower the intensity of the light reaching the material. This is a continuous wedge. Discontinuous or step wedges, where the density increases in discrete steps, are also in common use. The rate of change of density per centimetre in the case of a continuous wedge or the density increase per step in a step wedge is known as the wedge constant. The selection of the wedge constant varies with the job in hand, but for use with negative materials a typical step wedge might have a density range of three, comprising thirty steps, each step having a density 0·1 greater than the previous one. Silver wedges made by the photographic process are frequently used to-day. A section of the wedge may be covered by colour filters so that the spectral sensitivity of the material may also be studied.

Intensity scale sensitometers in addition to the step wedge require an accurate shutter to determine the exposure time that the whole wedge receives. If the exposure time is long, say 30 seconds, then it can be manually controlled fairly accurately, but when the time becomes short—e.g., 1/25 or 1/500 second—an automatic shutter is called for. There are three commonly used shutters: a sector wheel carrying a single wide slot, a falling plate shutter in which a heavy metal plate with a slot in it is allowed to fall under gravity, and a pendulum shutter in which the exposure is timed by allowing a pendulum carrying a disc with a radial cut-out to swing from a fixed distance.

Processing the Material. After exposure in the sensitometer the material must be processed in a uniform and reproducible manner. The development stage in particular has to be carefully controlled for time, temperature and degree of agitation. To reproduce the time and temperature of development used in practice is simple, but to ensure constant and suitable agitation of the developer is less easy. In practice development may be carried out in a deep tank with negligible agitation, in a dish

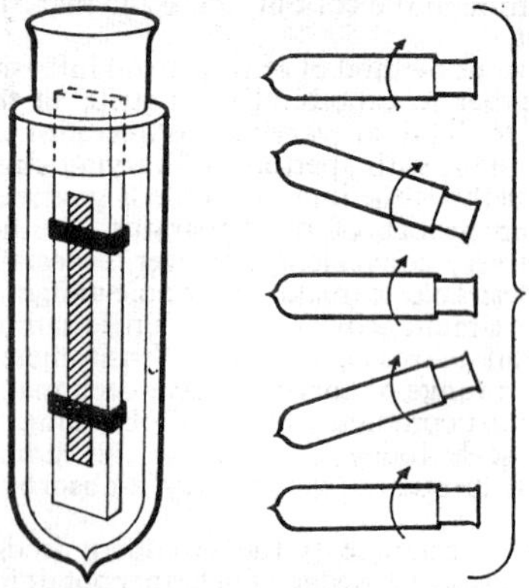

STANDARD DEVELOPMENT. For the B.S. method of sensitometric development the test material, strapped to a glass strip, is constantly and thoroughly agitated inside a vacuum flask containing the developer at standard temperature.

with poor agitation or perhaps in a machine with vigorous agitation. Here it is sometimes better to set up a correlation between laboratory conditions and practical conditions, choosing reproducible laboratory conditions.

In the laboratory, processing is sometimes carried out in a machine in which the sensitometric strips are fastened around a cylindrical drum. This drum is then rotated at a controlled rate in a cylindrical developing bath, the diameter of which is only slightly greater than the diameter of the drum. The agitation is increased by baffles on the inner surface of the developing tank. After development, the drum, still carrying the strips, is transferred first to a stop bath and then to the fixer, both of which are in vessels similar to that used for the developer. All solutions are thermostatically controlled by means of a surrounding water jacket.

Another carefully controlled method of development is that described by the British Standards Institution and the American Standards Association for the determination of speed. Here the development is carried out in a vacuum flask which is agitated systematically.

Fixing and drying conditions should also be constant.

DENSITY MEASUREMENT

After the material has received a series of exposures in the sensitometer and has been processed, it is necessary to measure the degree of blackening, or density, which corresponds to each of the known exposures. Quite a reasonable estimate of density may often be made visually by comparing the processed material with a wedge of known density values, although this method is not very accurate for very low or for high densities. For example, transmission densities between 0·1 and 0·5 can be estimated visually, with a little practice, to within ± ·02.

It is, however, often essential to have a precision instrument capable of measuring a wide range of transmission and reflection densities both accurately and quickly. Such an apparatus is known as a densitometer.

Transmission density (of negatives, etc.) is defined as the logarithm of the ratio

$$\frac{\text{incident light}}{\text{transmitted (emergent) light}}$$

reflection density (from a paper print) is defined as the logarithm of the ratio

$$\frac{\text{incident light}}{\text{reflected light}}$$

From this it will be seen that the densitometer must be able either to measure light intensities or, more often, to be able to adjust two light intensities so that they are equal. When the apparatus is used for such measurements of light in general, it is known as a photometer, but when adapted to the measurement of densities it becomes a densitometer. It is possible to tell when two light intensities are equal by adjusting the instrument so that the two resulting light fields are equally luminous when seen alongside each other in the eyepiece of the photometer.

The transmitted or reflected light must be measured in a special way if the resulting density measurements are to be of value, and diffuse densities are therefore normally employed. In the case of transmission density, the diffuse density based on the total light transmission is measured. This is because diffuse density is well defined optically whereas specular density, based on the portion of the transmitted light that passes through in a straight line, depends upon the optical details of the particular process concerned.

The densities used, for example, in contact printing are diffuse whereas in a condenser enlarger they are specular. When looking at a photograph it is the diffuse densities that are important and it is necessary to keep specular reflection to a minimum in order to avoid glare.

Diffuse transmission densities are readily obtained by placing the area to be measured in contact with a piece of opal glass. In measuring diffuse reflection densities in a reflection densitometer, the principal specular reflection is avoided by allowing the incident light to fall on the sample at an angle of 45° and measuring the light reflected at 90° to the surface.

Visual Densitometers. In one of the simple forms of densitometer a photometer head is used to determine when two beams of light from two similar lamps are of equal intensity. The density to be measured is inserted into one beam and into the other beam is placed a continuous wedge the density of which is known at each point. The position of the wedge is adjusted until the fields balance. At this point the unknown density is equal to the known density of the wedge. The value can then be read from a scale on the instrument. If the density being measured is low, the intensity in the eyepiece will be high, whereas if the density is high the intensity will be low; for this reason such a densitometer is described as a variable intensity type. These instruments are by no means ideal because at low densities the eye is dazzled and at high densities the light is so feeble that the matching of the photometric field becomes difficult.

Another class of visual densitometer works at constant intensity, a well-known example being the Capstaff-Purdy instrument for measuring transmission densities. In this case one lamp is used and its light is split into two beams. One beam, suitably diffused to make it uniform, reaches the photometric head by reflection from mirrors. This is the reference beam and as such remains constant. The second beam of light passes first through a calibrated wedge and then through the unknown density before it reaches the eyepiece. The intensity of the reference beam, when the photometric field is balanced by adjusting the wedge, is equal to the intensity of the other beam after passing both the unknown density and the density of the wedge at the point of balance. It is then simple to calibrate the wedge so that the density value of the unknown sample can be read directly. With this type of instrument continual adaptation of the eye is not necessary and the sensitivity is constant over the whole range. The accuracy of this densitometer depends on the ability of the operator to balance the photometric field and on how carefully the wedge is calibrated.

Photo-electric Densitometers. Visual densitometers are rather slow and tiring to use and sometimes not sufficiently accurate. These drawbacks have been largely overcome by the design of photo-electric densitometers. In these instruments the balancing of the two light intensities can be carried out by means of a photo-electric cell instead of by the eye.

In one commercial densitometer the light transmitted by the test density falls on to a self-generating barrier layer cell. The current generated, which is proportional to the light intensity, is passed to a micro-ammeter calibrated in densities. Another instrument employs a photo-emissive cell coupled to a valve amplifier for the same purpose. E.F.T.

In another type of densitometer the current generated by a photo-electric detector is passed through a resistance, and the voltage across this resistance is balanced against a reference voltage on a potentiometer whose slide is calibrated in density. Photo-electric densitometers can generally be adapted to measure reflection density.

Recording Densitometers. In the manufacture of sensitized materials and in many photographic processes it has become essential to be able to plot numbers of characteristic curves accurately and quickly. Recording densitometers automatically plot such characteristic curves on to a sheet of graph paper.

There are two main types for use in conjunction with step or continuous wedge sensitometers. Step wedge samples are measured on a step densitometer which can consist of a visual or photo-electric densitometer with its manual adjustment control coupled mechanically to a recording pen instead of (or as well as) giving a scale indication. A table or drum holds the graph paper and moves it under the pen. The mechanical gearing from the manual control to the pen is arranged to calibrate the ordinate of the graph paper in density.

The table movement is coupled to the movement of a strip holder in which the sample step wedge is placed. In one action of a control the sample is moved by one step and the table moves a distance (measured on the abscissa of the graph paper) equal to the wedge constant of the exposing wedge, i.e. the abscissa is calibrated in log exposure. For each step of the wedge in turn a balance is obtained on the densitometer by the operator who then lowers the recording pen on to the graph paper to produce a series of points. These can be joined up to form the characteristic curve of the sample material.

Samples exposed in a continuous wedge sensitometer can be measured on a densitometer whose pen automatically plots the complete characteristic curve. A common design of continuous recording densitometer employs a beam chopper. This is a disc of transparent material divided into equal segments which are alternately left clear and silvered. The chopper rotates at high speed in front of a lamp so as to send alternate beams of light along the sample path and a reference path. The two beams are finally focused on the same photo-electric detector whose output is, therefore, a square-

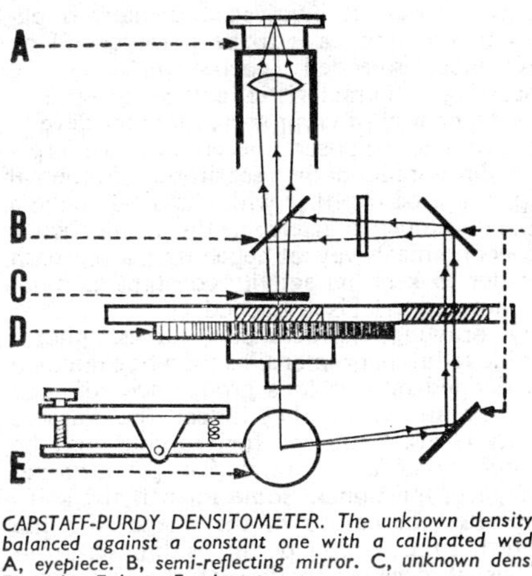

CAPSTAFF-PURDY DENSITOMETER. The unknown density is balanced against a constant one with a calibrated wedge. A, eyepiece. B, semi-reflecting mirror. C, unknown density. D, wedge. E, lamp. F, mirrors.

wave of frequency equal to that of the beam chopper and amplitude proportional to the photometric difference between the reference and sample beams. The signal is amplified and can be used to drive a recording pen or, alternatively, it can be used to drive a servo-operated measuring wedge in either beam to adjust the amplitude to zero. This measuring wedge is coupled to a recording pen which draws out the characteristic curve of the material as the sample wedge and graph paper are moved continuously. The movements of the sample and the graph paper are linked according to the wedge constant of the exposing wedge. R.N.W.

APPLICATION OF RESULTS

Once the characteristic curve has been plotted, various parameters such as speed, gamma and fog can be evaluated.

Particular care has to be taken to ensure that the values derived from the curve are those significant to the process concerned. Sensitometry aims at replacing, at least in part, the skilled observer. The human eye, however, is very sensitive, and the practised photographer, comparing prints for example, can detect differences that are not easily evaluated sensitometrically unless the parameter used is carefully derived from the exact part of the characteristic curve concerned. The H. & D. speed, for example, is of little value to the press photographer when he has to work under very adverse lighting conditions and is only likely to expose on the toe of the characteristic curve; here the DIN speed is far more valuable. On the other hand, a worker in photo-mechanical reproduction is often concerned with the speed at the shoulder of the curve. A similar argument holds for contrast. One process may place more importance on the straight line portion of the curve, whilst others may be governed by the average slope of the toe or the shoulder.

In particular, sensitometry is one of the most powerful tools available to the research worker and is in constant use for investigations into the theory of the latent image, of development and of emulsion manufacture.

In Manufacture. The photographic emulsion on its support of film, paper or glass is an extremely delicate system and considerable vigilance is required to maintain the high degree of reproducibility from batch to batch that is found today. In controlling manufacture and testing the finished product, sensitometry is extensively used for obtaining precise figures for speed, contrast and fog. These figures for the finished product must fall within predetermined limits or the material is rejected.

Determining Speed. Sensitometric measurements form the basis of all methods of exposure determination because to arrive at the exposure required for any particular photograph it is necessary to know the speed of the sensitive material being used. The speed is derived sensitometrically in terms of H. & D., Scheiner, DIN, B.S.I. and A.S.A. units, etc.

Tone Reproduction. A paper print is often the aim of the photographer. In order to obtain the best possible tone reproduction the photographic process has to be well analysed. This has been done very successfully by a method first suggested by L. A. Jones. By considering in turn the luminosity range of the subject, the luminosity range of the image formed by the lens, and the negative and positive characteristic curves, it is possible to build up a tone reproduction diagram of subject luminosity plotted against print density.

Reversal Processes. The latitude of reversal films is increased by either using solvent developers or the method of controlled second exposure. Here the initial image is bleached out and the final positive image is obtained by fogging and developing the remaining silver halide. Without sensitometric control at each stage these processes would become almost unworkable.

Duplicate and Copy Negatives. It is often necessary to prepare duplicate negatives or negatives of a contrast differing from the original. A knowledge of the appropriate sensitometric values of a suitable range of materials enables this to be done without any guesswork and reduces waste to a minimum. This includes deriving the correct exposure time from the known exposure required through a given density.

Cinematography. In a sound film the picture and sound track have each to comply with certain conditions even though they are developed together—i.e., the picture must be uniform and suitable for projection, and the sound track, whether variable area or variable density, has to give the correct pitch, intensity and timbre, and be as free as possible from background noise. In addition a negative-positive process is involved and many duplicates of the film have to be prepared. Here, sensitometry is needed at each stage.

Processing. Characteristic curves provide a convenient way of comparing different developer formulae for speed and contrast. In large-scale developing baths, sensitometric control strips are used together with chemical analysis of the developer to determine the most effective and economical way of replenishing the bath in order to keep its activity constant as more and more material is developed.

By drawing characteristic curves, photographic reducers or intensifiers can be examined and divided into various groups according to the way in which they affect the various density levels. Reducers, for example, may be quickly classified as subtractive, proportional or superproportional. Some intensifying solutions alter the colour of the image with the result that the printing density is different from the visual density because of the difference between the sensitivity of the emulsion and that of the eye. This aspect of photographic photometry, as it is called, can be evaluated sensitometrically.

Colour Photography. Sensitometric control enters into colour processes at many stages. With the most common colour materials, the subtractive integral tripack films, the three sensitive layers must be suitably balanced during manufacture.

The film is prepared so that the speed to blue of the blue-sensitive layer is the same as the green speed of the green-sensitive layer and the red speed of the red-sensitive layer. That condition must hold good over the whole scale of gradations the film can record; in other words, the gammas of the coloured images obtained must be virtually the same for all three layers when developed together for a fixed time.

Separation negatives also need sensitometric balancing to ensure correct colour reproduction. In practice this is best achieved by photographing on each separation negative a step wedge of known and increasing neutral densities at the same time as the subject. If the effective densities of each step (as measured on the coloured positives through a complementary filter) are identical, the component colour densities will balance and the wedge appears grey. For complete balance, the densities must be identical along the whole range. In other words, a curve plotting the density of each step on the positive as a function of the density of the corresponding steps on the original wedge must be a straight line for each positive. In addition, each curve must have the same slope, namely 1·0.

Although the whole cycle of negative—silver positive—colour positive must yield a correct set of wedge curves, the individual steps do not have to correspond. Thus the blue separation negative may have a lower gamma and be compensated in printing, or the positives may have different gammas if the method of dyeing makes this necessary. E.F.T.

See also: *Characteristic curve; Colour sensitometry; Gamma; Speed of sensitized materials.*

Books: ***Progress in Photography*** **(2 vols.), ed. by D. A. Spencer (London);** ***Sensitometry,*** **by L. Lobel and M.** Dubois (London); *General Sensitometry,* by Yu. N. Gorokhovskii and T. M. Levenberg (London).

SEPARATION NEGATIVES

It follows from the principles of colour photography that the complete range of colours can be reproduced by first analysing them into three black-and-white records, one to represent the reds in the subject, one the greens, and the third the blues.

Such analysis records can be made by making three separate exposures of a subject through red, green, and blue colour filters respectively, upon panchromatic emulsions. The three negatives obtained in this way are called colour separation negatives, because in them the spectrum has been separated into three parts.

The three colour records can then be synthesized into a colour print by printing the red record in a minus red colour, blue-green (cyan), the green record in a minus green colour, magenta, and the blue record in a minus blue colour, yellow. The three coloured images when superimposed will give the colour print showing the subject in more or less its original colours.

After printing the colour separation negatives, the three resulting colour records have to be superimposed to reform the picture into colour. The three colour component images must therefore be exactly the same size and shape, otherwise they would not register properly and the picture would show colour fringes.

Separation negatives used to be the basis for specialized assembly colour printing processes such as dye transfer. They are of course still the main method of printing plates or blocks for photo-mechanical printing.

Separation negatives can be made either direct from the subject or from colour transparencies. The former system gives the best results, but obvious practical problems limit its usefulness. The procedure of exposing separation negatives is therefore hardly ever used nowadays. Almost the only exception is the photo mechanical reproduction of paintings—where the original painting can be placed at the disposal of a graphic arts department.

It is still possible to take excellent separation negatives of still live subjects by making three successive exposures in an ordinary camera. There must be no movement either of the subject or the camera, otherwise the negatives will not register properly.

Few photographers use this procedure, since it is easier to take the picture on a colour transparency material. For the same reason one-shot colour cameras and repeating backs have been completely displaced by colour transparency materials as a starting point either for making separation negatives or for printing on multi-layer colour materials.

BASIC PRINCIPLES

The Three Colour Records. Colour prints are made by using three printing colours which are the complementary colours of the filters used when exposing the negatives.

A particularly bad habit which is unfortunately common in the printing trade is to call the magenta and cyan printing colours red and blue respectively. By using the C-M-Y terminology for the printing colours there can be no confusion.

TAKING AND PRINTING COLOURS

Colour Filter for Exposing Separation Negative	Complementary Colour Reflects	Name of Complementary Printing Colour
Red (R)	Blue and green	Cyan (C)
Green (G)	Red and blue	Magenta (M)
Blue (B)	Green and red	Yellow (Y)

Equipment. For successively exposed separation negatives of static subjects the camera must be sufficiently rigid to remain set in exactly the same position while all three negatives are exposed. A very firm tripod is of course essential. The dark slides must be accurately made so that each slide holds the sensitive material in exactly the same plane.

Colour Filters. Separation negatives are almost invariably made on panchromatic material through red, green, and blue colour filters. The hues of the tricolour filters have been fixed for many years, although the exact depth of the filters varies slightly from maker to maker. One maker's filters will work perfectly well with another maker's sensitive material when taking separation negatives direct from the subject, although the filter factors will probably vary.

For colour separation work, it is generally better to use the inexpensive gelatin filters than glass filters of possibly doubtful quality glass.

When making successive exposures, the filters must be held in front of the camera lens by means of a holder which allows the filters to be changed without moving or shaking any part of the camera.

Filter Factors. The manufacturers of plates and films give approximate filter factors for each of the tricolour filters with their various materials. Although quite useful as a guide, these factors can only be accurate if they were determined under exactly the same conditions as will be used for making a particular set of separation negatives.

The reasons for this are as follows:

(1) The colour temperature of the light source may be different from that used by the manufacturer.

(2) The published filter factors are generally based on the same time of development for all three colour records, whereas in most cases the separation negatives need somewhat different times of development.

(3) The exact depth of colour filters varies from batch to batch.

(4) There are variations in colour sensitivity between batches of emulsions.

(5) Reciprocity failure may change the effective sensitivity of the material at different exposure levels.

Filter Ratios. In the past it has been usual to give filter factors, which by definition are relative to the unfiltered (white light) exposure. But now it is becoming increasingly common to speak only of filter ratios—that is, the ratio of the three filtered exposures relative to one of them.

For convenience, one of the three exposures —often the red filter record, being the shortest of the three—is taken as unity. But many workers prefer to consider the green exposure as unity, because the green filter negative looks most like a normal black-and-white negative, and its quality is more easily judged.

With filter factors of 3, 6, and 12 for the R, G and B exposures respectively, the ratios would be 1, 2 and 4, relative to the red, or ½, 1, and 2 when the green is taken as unity.

As reciprocity failure may upset the balance of filter factors when the exposure times are changed, it is preferable to work with the tricolour filters balanced to 1 : 1 : 1 ratios; then all three exposures are always of equal duration.

Tricolour filters can be balanced by adding a suitable neutral density filter to each of the two filters with the lower factors. Each of the three exposures then becomes as long as the longest. Neutral density filters are available in various depths, in increments of approximately 0·05 of density, up to a density of 1·00 (multiplying the exposure 10 times).

For quantity work it is well worth balancing the filters to 1 : 1 : 1 ratios. In some cases, the manufacturers will supply special tricolour filters with the neutral density of the required depth incorporated.

Materials. Plates have always been a good choice for colour separation work, because they cannot shrink or distort and give trouble with register. Nowadays however film is more popular and modern film bases of high dimensional stability (such as polyester) are perfectly satisfactory. Some care is needed to see that each of the three separation negatives receives exactly the same treatment as far as the total time of wet handling and the orientation of the picture when drying is concerned.

A medium speed panchromatic emulsion is the most suitable since it has a long tonal scale. Most manufacturers offer materials specially suitable for making colour separation negatives, sometimes designed to give substantially equal contrast on all three colour records with the same time of development.

Negative Balance. The three separation negatives must naturally bear the correct relationship one with the other, so that they can reproduce the various colours in their correct proportions.

The fundamental requirement of separation negatives is that any neutral present in the original (such as a white, or a grey, or a black) should be reproduced as an identical density on each of the three negatives. This assumes the absence of any colour cast.

A grey scale can be made with pieces of bromide paper, one white (unexposed and fixed out), several greys (lightly fogged and developed), and one black (well fogged and developed). When exposing the separation negatives, the grey scale should be placed so that it is recorded somewhere near the edge of the negatives, but lies just outside the subject matter being photographed. The lighting on the grey scale should as far as possible be the same as on the subject.

The exposure and development of the separation negatives must be adjusted to give as near as possible identical reproduction of the grey scale on each of the three negatives.

Identification. When printing a set of separation negatives, it is of course necessary to know which is which colour record. To identify the negatives include three colour patches, red, green, and blue (marked R, G, B) in the subject, alongside the grey scale. When reproduced on the negatives, one of the three patches will appear dense, this patch indicating which colour filter was used for that negative. For example, in the green filter negative, the green patch will appear heavy, and the other two will be light.

Alternatively use patches of cyan, magenta, and yellow colours. In this case the lightest patch identifies the negative.

The colour patches can be made by using poster colours, and can be painted alongside the edge of the grey scale.

Some makers supply colour separation guides which contain a grey scale and identification patches already made. However, most photographers prefer to make their own in various sizes to suit different types of subject.

Lighting and Exposure. Lighting for colour separation negatives is basically similar to lighting for any other colour work. But where the picture is to be reproduced as a paper print, it is best to avoid harsh contrasts of lighting. All lights on the subject must be of the same colour temperature, or else some of the light will photograph either too blue or too orange.

The filter factors will vary according to the colour temperature of the light. The lower the colour temperature, the redder the light, and so the red filter factor becomes lower and the blue higher. By standardizing on a particular type of light source, the filter factors will always be the same under given conditions.

Colour separation negatives need correct exposure because colour prints show up errors in the negatives that would pass unnoticed in black-and-white work.

Inaccurate filter factors reduce the exposure latitude for the set of negatives. This is because some of the available latitude is used up by the filter factor error on one (or two) of the negatives.

Processing. The three negatives must receive even development to the correct degree of contrast.

Dish development gives the most even results. Three small plates can be developed at the same time, either in a dish with a separating cage in the bottom to prevent them sliding over each other, or in three separate dishes side by side on a rocking platform. The three colour records are always kept in the same order in the dishes, so that each may receive its correct development time. The darkroom temperature should be as near as possible the same as the developer temperature.

First bring the dishes up to the correct temperature with water. At the same time, bring enough developer for all three negatives to the same temperature. Then pour the water from the dishes and divide the developer equally between all three.

The three negatives of a set should always be developed at one operation and receive identical treatment (except of course for times of development) to avoid variations in temperature, agitation, etc. A visible darkroom clock is essential to ensure accurate timing.

Most non-staining developers are suitable, the most common being a metol-hydroquinone developer fairly well diluted to bring the developing times to more than about five minutes. Shorter times are difficult to time accurately.

The developer should be used once only, and then discarded. By using fresh developer for each set of negatives, it is possible to standardize the degree of development quite accurately.

Development Times. Each of the three negatives of a set must be developed to the same degree of contrast. As the colour filters affect the contrast, the three negatives will almost certainly need different development times. Usually the red and green filter records are fairly well matched for contrast and need similar development times. But as the blue filter lowers the contrast, this negative needs increased development to bring the contrast up to the level of the other two.

If no information on processing conditions for a particular material is available, begin by making a suitable green record negative by trial and error. When this is satisfactory, use it as a guide to the others. It is usually safe to give the red record negative the same development time as the green, or a little less, and the blue about 25 per cent more for a first trial.

Checking the Balance. Having made a trial set of negatives, hold them side by side over a light box, and check the balance by comparing the grey scales, one with the other.

A more sensitive method of comparison is to print all three negatives together on one sheet of normal bromide paper. Cut out the prints of the three grey scales and compare side by side for equality of density and contrast. When the negatives are properly balanced, the grey scales will be identical on each.

The best and most scientific method of checking the balance is to measure the densities with a densitometer.

When a densitometer is available, first make gamma-development time curves, and from them find the development time for each colour record to give the desired gamma. For separation negatives from original subjects, a negative gamma of about 0·80 will be required—the exact value depending upon the colour printing process and technique to be used, and on the optical system of the enlarger. A condenser enlarger calls for softer negatives than a cold cathode or diffuser enlarger, or than contact printing. At the gamma value given, the difference between the maximum and the minimum density on the reproduced grey scale will be approximately 1·05.

Having found the correct development times, make a trial set of separation negatives at what are believed to be the correct filter factors. After processing, check the grey scale density readings for equality between the three negatives. If necessary, amend the times, and make another trial set of negatives.

The green record negative, already judged to be correct, provides a standard against which to judge the changes to be made in exposure (to alter density), or in development time (to alter contrast), to improve the balance on future sets of negatives.

For example, if the three negatives are of equal contrast but of different densities, they have received the correct times of development but the filter factors need modification. But if one of the negatives is flatter than the others, it will need more development.

Slight variations in density between the grey scales on the negatives are tolerable and can be corrected in printing. But differences in contrast are much more difficult to correct without upsetting the colour printing technique.

SUBTRACTIVE TRANSPARENCIES

Separation negatives from colour transparencies will not give quite as good prints as negatives exposed directly on the subject. There is, however, the advantage that once the transparency has been prepared, all the work of making the separation negatives can be performed at leisure in the darkroom.

A transparency prepared for subsequent separation should appear of somewhat soft gradation, and slightly heavy, judged by normal transparency standards. The subject should be softly lit, and the transparency should receive about half a stop less exposure than usual.

The coloured image of a subtractive transparency consists of a combination of a yellow, a magenta, and a cyan image, representing respectively the blue, green, and red colour records. When making separation negatives from the transparency, each negative should ideally record only one of the three coloured images. For instance, the red record negative should record the cyan of the transparency, and nothing from the other two colours.

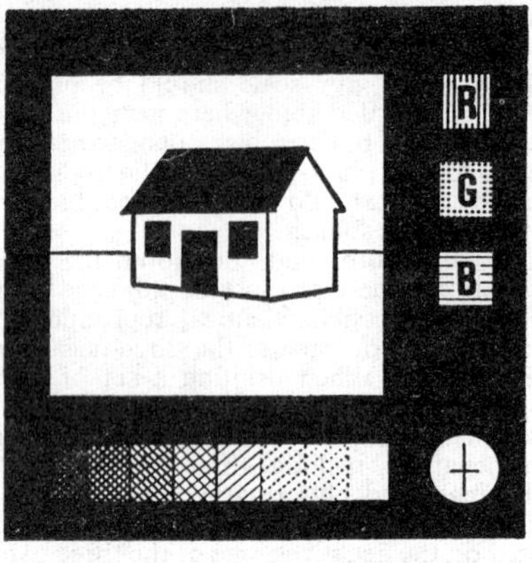

MOUNTING THE TRANSPARENCY. This is taped to the front glass of the printing frame, together with a step wedge, identification colour patches, and a registration mark.

Unfortunately this is never quite possible. All dyes are so impure that the red, green, and blue filters used for exposing the separation negatives cannot sort out one of the coloured images without also recording something of the other two.

So separation negatives made from transparencies tend to show some loss of colour saturation, and also some tone distortion because they are "copies of a copy". They can never be quite as good as negatives made directly from the subject.

Furthermore, all the commercially available transparency materials are made primarily for viewing, and their contrast is rather great for easy reproduction in separation negatives.

All but the softest transparencies therefore need masking to reduce the contrast before making the negatives. Owing to the imperfections of the dyes in the transparency and of the pigments used in most tricolour printing processes, colour corrected masks are generally used to obtain the best colour reproduction. Certain cases may even call for tone correction by means of highlight masks if the original transparency is very light.

Often it is considered more convenient to use unsharp masks because with them registration of mask and transparency is less critical.

Masking Procedure. Masks are almost invariably printed by contact from the transparencies. A robust printing frame should be used and, to ensure good contact, a sheet of firm ⅛ in. thick sponge rubber should be placed between the plate and the back of the frame.

An easily standardized set-up which is convenient for contact printing masks and separation negatives is an ordinary 60 watt pearl bulb in a pyramid type 8 × 10 ins. safelight, which is suspended 36 ins. above the work bench so that the printing frame may be placed immediately below it. (For transparencies larger than about 3¼ × 4¼ ins., this distance should be greater, say 6 feet, and the exposure times lengthened accordingly.) In the front of the safelight is fitted an opaque screen with a 2⅝ ins. square aperture cut centrally in it. A piece of flashed opal glass is fitted over the aperture.

Provision must be made for holding 3 × 3 ins. colour filters over the aperture by fitting a slide into which each filter can be placed in turn. The filters should be fixed in a simple cardboard mount or held between two pieces of glass, with the desired size of aperture.

The lamp should not be left on unnecessarily, since the filters may be damaged by the heat.

Mount up the transparency, emulsion side down, with very thin adhesive tape on the glass of the printing frame and place a step wedge alongside. This wedge should be made by photographing a grey scale on the same type of transparency material. Any other type of wedge is not a reliable guide to colour balance in the final print (although it would of course be a useful guide to contrast). Mount small patches of tricolour filter, each marked with the initial letter R, G or B, alongside the wedge to assist identification of the separation negatives. The clear glass around the transparency, etc., is then masked off with black adhesive paper on the other side of the glass.

Expose the mask on a medium speed panchromatic emulsion. Plates are sometimes preferred to film, because glass plates cannot shrink so they give masks exactly the same size as the transparency, thus assisting registration. However, it is essential to maintain reasonably constant humidity or the transparency itself may change size slightly by shrinking or swelling.

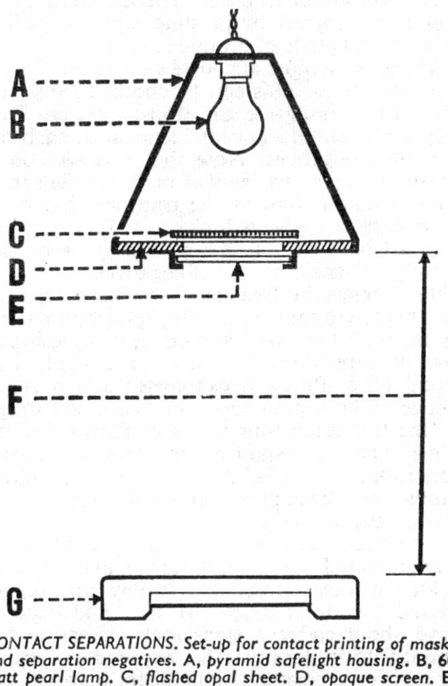

CONTACT SEPARATIONS. Set-up for contact printing of masks and separation negatives. A, pyramid safelight housing. B, 60 watt pearl lamp. C, flashed opal sheet. D, opaque screen. E, colour filter in simple mount. F, printing distance: 36 ins. for small transparencies, 72 ins. for large ones. G, printing frame, with transparency and negative material for separation.

The mask is normally exposed through a red filter. When the two-mask system is used, one mask is exposed through a red filter, and another through a green filter.

Develop the mask to the very low gamma of about 0·35. As a guide: medium to fast emulsions will reach this gamma with about 3 to 4 minutes development at 68° F. in D.76 diluted 1 to 1, developed in a dish with continuous rocking in all directions.

A correctly exposed and developed principal mask will show detail in both highlights and shadows, but it will simply look like a low contrast negative. In general, the correct exposure for a principal mask (i.e., not a highlight mask) is the least that will record the shadow detail of the transparency.

To register the transparency, wedge, and identification patches, with the finished mask, they can best be manipulated over an illuminated ground glass, and a weight used to hold the transparency on the mask while the register is adjusted and checked with a magnifier. When register is correct, fix the transparency to the mask with adhesive tape, and fix the wedge and identification patches similarly in place on the mask. The mask plate is then blocked out, preferably with opaque medium which is physically thin and will not affect contact with the sensitive plate.

Using the normal single mask technique, the same mask is of course used when making all three separation negatives.

When the double mask technique is used, the red and green filter separation negatives are made with the red light mask, and the blue filter separation with the green light mask.

Exposing the Separation Negatives. The three separation negatives are made by printing the masked transparency on panchromatic plates, using red, green, and blue filtered light. The filters must be suited to the type of transparency material. In general, the manufacturer will recommend filters which have relatively narrow spectral transmission bands—so-called narrow-cut separation filters—with peak transmissions to match the spectral characteristics of the transparency material.

The three filters are preferably brought to 1 : 1 : 1 ratio of exposure times by masking down the area on the two filters with the lower factors. By adjusting the filters in this way to give equal exposures, the whole working procedure is greatly simplified. The exposure time can be adjusted to match the density of the transparency.

Expose the separation negatives with the printing frame on the bench under the light source, in exactly the same position and the same way round in each case. This avoids trouble from uneven illumination.

The correct exposure time for an average density transparency is first found by trial, after which the most suitable exposure for other transparencies is estimated by eye, or by measuring the average density of each transparency.

If one transparency passes twice as much light as the standard, it should be given half the standard exposure.

A densitometer may be used to measure the average densities—that is, half of the sum of the minimum and maximum densities—of both the standard and the new transparency, the exposure then being determined from the normal density—transmission relationship—that is, each 0·30 increase in average density requires a 100 per cent (2×) increase in the exposure.

In the case of a transparency with a large brightness range, it is almost always best to expose to suit the highlights, even at the expense of losing the shadows. However, it will generally be found that the mask reduces the contrast enough for the transparency to be properly reproduced on the negative material without undue loss of quality.

The transparency must be graded for density when determining the mask exposure, and again for the separation negative exposures by reading the transparency plus the mask, compared with the standard transparency plus its mask.

Printing Negatives by Projection. It is generally preferred to make enlarged separation negatives from miniature size transparencies, since most workers find the larger negatives easier to handle.

However, it is more difficult to make good separation negatives by projection than by contact printing.

Projection printing inevitably brings in trouble from flare in the lens degrading parts of the wide brightness range of the transparency. This is most noticeable if one colour dominates in the picture. Conventional enlarger optical systems cause some loss of sharpness, particularly with a diffused light source. Condenser enlargers with small source lamps can reproduce all the fine detail in the transparency but will emphasize any scratches as well. And there is the difficulty of heat affecting the transparency.

In an enlarger fitted with the usual tungsten lamp, the density of the sharp-cut tricolour filters makes for long exposures, and a film transparency may heat and distort, or lose register with the mask, during the making of the three negatives. However, a well ventilated enlarger fitted with efficient heat absorbing glasses can be used. An enlarger with a cold cathode light source is even better as regards cool working, although it is less suitable for miniature transparencies as the diffused light source tends to give slightly less sharp results than a small source with a condenser.

A convenient method of projection printing the separation negatives in ordinary room light is to illuminate the transparency by a projection lamp in a horizontal enlarger with the lens removed. A diffuser may be placed immediately in front of the lamp, but the system must be set up to give even illumination first. Photograph the illuminated transparency with a camera accurately lined up in front of it.

The transparency may also be set up in the transparency holder of a process camera, and lit by open arcs shining on to white blotting paper some distance behind the transparency, or by a cold cathode lamp unit. The above remarks on diffuser type light sources of course apply here.

Processing the Negatives. Separation negatives from transparencies are developed in the same way as direct separation negatives. However, the gamma or contrast to which the negatives must be developed will be different. Generally separation negatives from transparencies need to be developed less than direct separations.

As a guide, direct separation negatives are generally developed to a gamma of about 0·80, while separation negatives from masked transparencies need to be developed to a gamma of about 0·75. For separations from unmasked transparencies the gamma need only be about 0·45. It is useful to check that the contrast is reasonably correct by printing a trial negative on a normal grade of bromide paper.

The gamma must of course be the same on all three negatives. This can be checked from the grey scales alongside the transparencies. The grey scales should appear identical in each of the three negatives. Note that this will only apply if the step-wedge was made on the same type of colour film as the transparency. Any other type of wedge will give a useful check on contrast but will not be a reliable guide to the density balance of the set of negatives.

Filter Factors by Sensitometry. Good separation negatives can be made from transparencies by rather tedious trial and error, or by using a manufacturer's specific data as a guide for initial tests. But a densitometer and a step-wedge make testing very simple and accurate.

One test using four plates or films (plus an extra one to establish reasonably correct conditions of working) will give the filter ratios and development times for equal contrast at any desired gamma.

The step-wedge may be a silver or dye type manufactured wedge, or it may be made by the colour worker himself. Preferably the wedge should be about $3\frac{1}{4} \times 4\frac{1}{4}$ ins., and should have about eighteen steps each increasing by about 0·15 in density. Also required are strips of each of the same colour filters that will be used for making the separation negatives. The strips of filter should be at least $\frac{1}{2}$ in. wide,

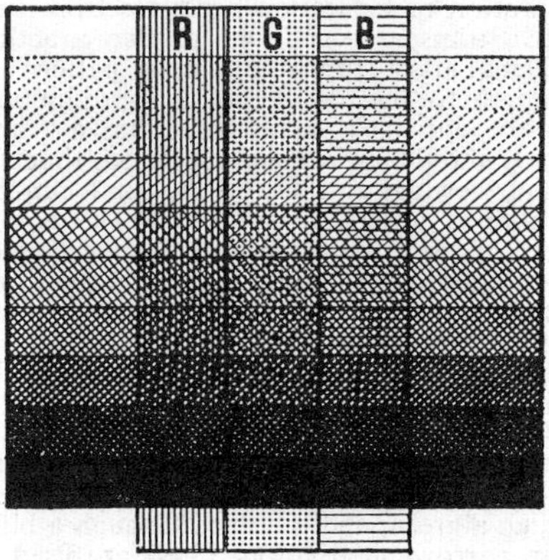

FILTER STEP WEDGE: A grey step wedge, together with strips of the three filters to be used for separation, is used for test exposures to establish correct filter factors and development times. The filter strips each cover the full step range.

and long enough to cover the length of the wedge. If the density increments of the wedge are even steps of, say, 0·15, it will greatly facilitate subsequent plotting of the graphs.

The step-wedge is used as follows. Expose four plates, using the step-wedge covered by the filter strips side by side as the negative in the printing frame. The exposures should be made to the light source described above for masking, but with no filters over the light source, and an aperture of 1½ ins. square. All four plates must receive the same exposure.

Develop the four plates for a sufficient range of times to be sure of covering a suitable range of negative contrasts. For example, if the developing time is thought to be about 5 minutes, develop the plates for 3½, 5, 6½ and 8 minutes. The conditions of development for these tests must be the same as will be used for the actual negatives. Fix, wash, and dry.

Read the densities on each of the three colour filter records on each of the four plates. Tabulate the negative density readings of each step vertically on the graph against the density readings of the respective steps on the original step-wedge, reading from right to left because the highest wedge density gives the least exposure. Both scales must be of the same order.

From these figures, plot the three sets (R, G, and B) of characteristic curves. Use a fresh sheet of graph paper for each set of four characteristic curves representing one colour record at the four different developments.

Having plotted the points on each curve, rule a straight line through the points on the straight line portion of each curve, and extend each line to cut the base line of the graph paper. Ignore the toe and shoulder regions of the curves. Mark off the gamma of each curve.

From the gammas of each record at each of the four developing times, plot the gamma-development time curves, which will give the developing times for each record to give any desired gamma.

Referring back to the three sheets of characteristic curves, choose on each the curve which gave as near as possible the gamma to which the negatives should be developed. Transcribe these three chosen curves of about the right gamma on to a fresh sheet of graph paper marked out just as before. Now the horizontal spacing between these three curves is a measure of the filter ratios which must be used to give balanced negatives.

With tungsten light, the red record will usually have received most exposure, and the blue least, with the green between them. Rule a line horizontally across the graph paper at a level of about 0·4 on the density scale. Now along that line, note the distance (in terms of step-wedge density units) between firstly the green and red record lines, then between the green and blue record lines.

To convert these "horizontal displacement" figures into filter ratios, look up the anti-logarithm of the displacement to the right. The anti-log is the filter ratio (that is, more than 1, in this case), relative to the green—that is, the green filter ratio is 1.

If, however, the displacement is to the left of the green record line, look up the anti-log and then look up the reciprocal of that. The latter figure is the filter ratio (now less than 1).

When performing this test for the first time, it is advisable to expose and process one plate to see that the conditions are about right for the green record image printed from the step-wedge to show a full range of densities from light to dark, with about the right gamma.

To convert the displacement figures to area changes required for the filters over the light source as already described, proceed as follows. For a displacement to the right of the green curve, look up the square root of the anti-log of the displacement figure. For a displacement to the left of the green curve, look up the reciprocal of the square root of the anti-log of the displacement figure. The final result in each case gives the factor by which the length of the sides of the square filter aperture must be changed to give balanced sets of separations with the same exposure time for each of the three colour records.

As the step-wedge used for these tests is not made by the colour transparency process in question, the precise filter ratios may need very slight modification. However, good negatives will be obtained even without this correction.

ADDITIVE TRANSPARENCIES

Since no additive colour films have been commercially available since the early 1950s, the need for making separation negatives from an additive transparency is very rarely likely to arise. The general procedure is similar to the

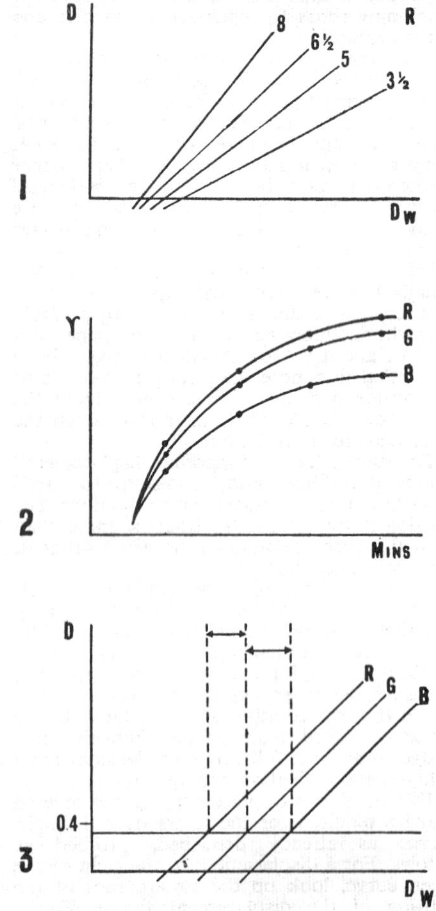

FILTER FACTORS BY SENSITOMETRY. Four negatives are exposed behind the filter step wedge, and are then used for sensitometric determination of the filter factors and development times for ideal working. The four negatives are developed for varying times above and below the normal time for the material. For instance if the normal time is 5 minutes, suitable test times would range from 3½ to 8 minutes. After processing, characteristic curves are plotted for each filter colour to establish the gamma obtained at the different development times. Top: *Typical curves of red filter densities, plotting negative density against wedge density.* Centre: *Gamma-times curves for the three filter negative strips, plotted from the characteristic curves. These give the development times for each record to yield any desired gamma.* Bottom: *Three curves of about equal gamma are finally transcribed from the original set of characteristic curves on to a fresh sheet of graph paper. The horizontal displacement (best measured at a level of about 0·4 on the density scale) is now a measure of the filter ratio required to provide balanced separation negatives. The actual filter ratios are usually based on green = 1. To obtain the blue filter ratio, measure the displacement of the blue filter curve to the right of the green filter curve in terms of step wedge density units. The anti-log of that displacement is the blue-filter ratio relative to green. For the red filter again measure the displacement of the red curve to the left of the green. As that displacement is negative along the exposure axis, the reciprocal of its anti-log is the red filter ratio.*

preparation of separation negatives from subtractive transparencies. One or two practical problems arise from the screen pattern or raster used for the additive colour reproduction.

To be suitable for reproduction, a transparency should appear slightly denser than for normal viewing, and ½ stop less than normal exposure is usually about right. The colour screen of an enlarged additive transparency tends to give a "granular" texture to the print, so large transparencies should be used in the first place where possible.

The additive transparency already carries the three colour records, red, green, and blue; it is only necessary to "sort them out" on to the three separation negatives. This is done by making one negative through a red filter, one through a green, and the third through a blue. Again narrow-cut tricolour filters are used to get good colour separation.

Printing Methods. Additive colour transparencies with an integral screen (the majority of additive transparencies produced) are suitable for printing by contact or projection, but contact printing is usually preferable. This avoids the optical imperfections inseparable from projection. Since the screen pattern prints through, variations of this pattern between the three separation negatives may give rise to interference effects leading to the appearance of a much coarser screen pattern on the final print.

With projection printing special precautions are necessary to avoid overheating the transparency, since the dense narrow-cut filters required lead to relatively long exposure times.

The emulsion side of an additive type transparency must face the lens of the enlarger or copying camera, to give negatives the correct way round.

The pattern of the colour screen of the transparency may be eliminated by using the enlarger or copying lens at a stop small enough to introduce just the right amount of blurring by diffraction. With an aperture small enough to do this the exposure time may be excessive unless a high intensity light source is used.

Processing. The degree of development of the negatives depends upon the contrast of the original and the printing process to be employed. As a guide, negatives from an additive transparency of average contrast should be developed to a gamma of about 0·70 to 0·80.

If no densitometer is available to check the gamma, make a green record negative by trial and error and check the contrast by making a print on a normal grade of bromide paper.

All three negatives must of course be developed to the same gamma. A.P.J.

See also: *Colour camera; Colour print processes; Gamma; Masking; Repeating back.*

Books: *Amateur Carbro Colour Prints*, by Viscount Hanworth (London); *Amateur Dye Transfer Colour Prints*, by Viscount Hanworth (London); *Colour Photography in Practice*, by D. A. Spencer (London); *Colour Prints*, by J. H. Coote (London); *Colour Separation Negatives*, by P. Jenkins (London).

SERIES-PARALLEL. Electrical circuit for connecting up two similar pieces of equipment through a two-way switch so that in one position of the switch the components are in series and in the other in parallel with the power supply. In photography the circuit enables the working life of Photoflood lamps to be extended by running them—in series (at half voltage) for arranging and focusing and in parallel (at full voltage and brilliance) for exposing.

See also: *Dimmer; Switches.*

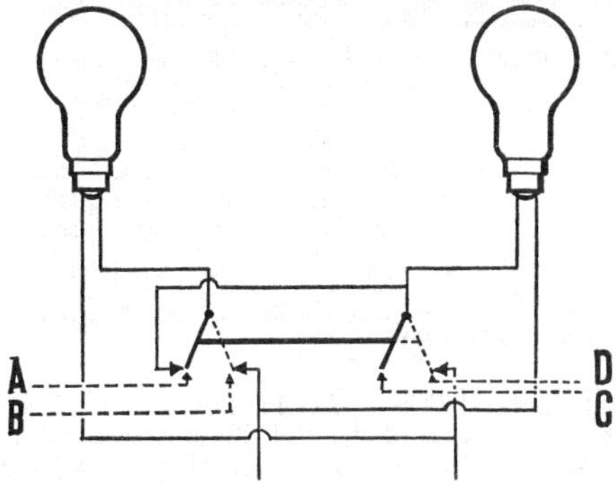

SERIES PARALLEL CIRCUIT. The simplest arrangement for switching over two lamps from series connection into parallel consists of two two-way switches. They are linked together and in the position shown (solid line) the two lamps are in series. On switching over to the dotted positions the lamps are in parallel. The contacts can also be closed separately by means of a rotary four-way switch. In the case setting AC would correspond to series operation, the switches in positions A and D would switch off the first lamp and run the second at full power, positions B and C combined would run the first lamp at full power and switch off the second one, and the setting BD would run both lamps in parallel.

SEYMOUR, DAVID, 1911–56, American reportage photographer born in Poland. Founder member of Magnum Group, 1947 became a free-lance journalist and photographer in Paris, mid-1930s. After the fall of France, 1940, went to New York and joined U.S. Air Force. Returned to Paris, 1944. Made photographic pictures stories in many countries, and was killed while covering the invasion of Egypt during the Suez crisis, 1956.

SHADING AND SPOT PRINTING. Negatives are often reasonably well exposed, except for a few small areas of the image which are either too light or too dark. When exposing the print, the density over such areas can be adjusted by local exposure control called shading, spot printing or "dodging".

This can be done in contact printing or enlarging, but the latter process of course offers far wider scope for control.

Shading consists of holding back a part of the image so as to make it print lighter. Spot printing means over-printing a comparatively small image area to make it darker in the enlargement. The distinction between the two lies mainly in the relative areas in question, since spot printing one part of the picture necessarily involves shading the rest, and vice versa. In practice, the mechanical aids employed for the two techniques are also slightly different.

Sky in Landscapes. One common example of the need for local control is the landscape print in which there is a great disparity between the densities of the foreground and the sky. Both must be given different exposures if they are to print through in the same depth. In this case the whole print is first given the correct exposure for the foreground. Then extra exposure is given to the sky, which is the densest part of the negative and needs the longest exposure. This is done by shading the foreground portion with the hand or a black card and allowing the light to fall on the sky area alone. The hand or the card must be slightly moved during shading, in order to avoid leaving a sharp-edged boundary.

More involved shapes, e.g., a complicated sky line, can be produced by cutting special masks. A duplicate print will serve admirably for that purpose; it should however be slightly smaller so that when held above the enlarging paper at a convenient distance, its shadow is the correct size. Such masks—sometimes with counter masks—are usually essential with most forms of combination printing.

Progressive shading—e.g., when correcting converging verticals—is a similar process, except that shading starts at one edge of the paper and the hand or card is gradually moved across until the whole print is covered.

Local Control. Small areas of the print that would come out too dark are shaded during part of the exposure. The most obvious means of shading are one or both hands which, if held suitably in the path of the light beam from the enlarger lens, will cast almost any shadow outline on the baseboard. (When using both hands, a foot switch or other independent means of working the enlarger light is an advantage.)

For areas in the centre of the picture, suitably-shaped pieces can be cut out of black card and held just above the surface of the paper on the end of a length of stiff wire or a knitting needle.

It is convenient to keep an assortment of such cards of various shapes (squares, circles, triangles, ellipses, etc.), handy to deal with areas of a range of shapes and sizes. Sets of this type can be bought or they can be made from stiff card or thin sheet metal.

The distance of the card above the enlarging paper determines the size of the shadow. The closer the card is to the enlarging lens, the larger and also the more blurred will be the shadow.

The angle of the card can also influence the shape of the shaded area. Thus a circular card held obliquely will throw an ellipse-shaped

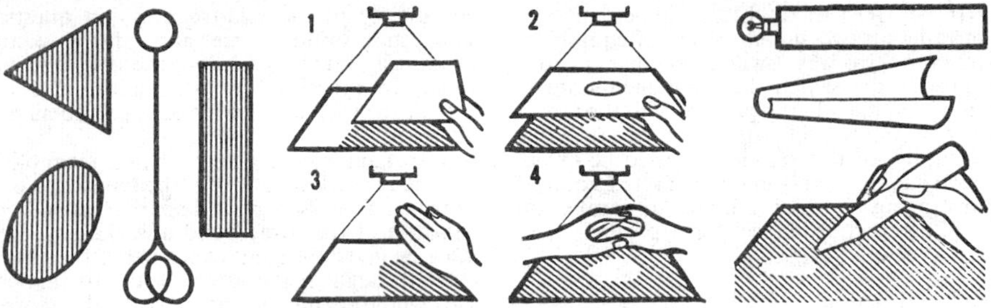

CONTROL TOOLS AND METHODS. Left: *Suitable shading shapes which clip on the end of a thin wire holder.* Centre: *Methods of shading and spot printing.* 1. *Shading large areas with a card.* 2. *Overprinting small area with hole in a card.* 3. *Shading with the hand.* 4. *Spot printing with hands.* Right: *Flashing, i.e., darkening areas of the print by white light, with the aid of a shaded torch.*

shadow, a regular triangle can be compressed in either direction, and so on.

If the card is held in position during the whole of the exposure, the result is a vignette.

Small areas of the print are given extra exposure when necessary by means of a black card with a suitably small hole cut in it. The card is held close enough to the lens to shield the whole of the print, and the spot of light that passes through the hole is moved about the area to be over-printed. This is the technique of spot printing.

Flashing. An alternative method of over-exposing a small image area is to shine white light on it. This is the technique of flashing. The difference between it and spot printing is that during flashing the whole image area concerned is fogged—to complete blackness if desired. In spot printing the tones of the image are darkened in proportion to their density, i.e., the area is simply over-exposed.

A useful way of flashing is to wind a narrow tube of paper round a torch bulb so as to produce a thin pencil beam. The flashing exposure required will vary widely according to the conditions and usually has to be found by trial and error.

Applications of flashing are the production of black edges round a photograph, and vignetting an image to disappear into black near the margins or corners.

Automatic Methods. Electronic enlargers use a flying spot scanner—a light spot that moves over the different parts of the negative—instead of a light source illuminating the whole negative at once. The intensity of this light spot varies according to the density of the image portion being scanned; for this purpose a photo cell system monitors the light coming through the negative. As a result dense negative areas receive more exposure on the print than thin ones. This provides a kind of automatic shading, though it cannot be used selectively for specific negative areas. There are also other methods for controlling the illumination locally, with similar results. Such automatic shading systems are largely used for printing aerial and other photographs where a negative of great brightness range has to be accommodated within the tone range of the printing paper. L.A.M.

See also: *Cloud negatives; Combination printing; Electronic printing.*
Book: *The Complete Art of Prniting and Enlarging*, by O. R. Croy (London).

SHADOWLESS LIGHTING. Diffused, uniform type of frontal illumination which presents the subject completely without shadows when viewed from the camera position. It is arrived at by disposing light sources and reflectors evenly around the camera and directing them towards the subject.

Daylight forms the most convenient source of shadowless lighting. A subject photographed out of doors in light-toned surroundings when the sunlight is diffused by a continuous layer of cloud will be free from shadows.

This type of lighting normally produces a softly-defined high key image.

Shadowless lighting is suitable for high key portraits of babies, children, girls, and young women. It is also useful for playing down lines and wrinkles in portraits of older people.

Catalogue illustrations which must show the maximum amount of detail—e.g., photographs of ceramics, glass and silverware—are best taken by this type of illumination.

When the object is simply to eliminate the shadow cast on the supporting surface, the subject is placed on its side on a sheet of glass mounted some distance above the bench—e.g., bridging two stacks of books—and photographed with the camera looking down on it. The bench is covered with a sheet of white paper and the lighting is arranged so that no shadows are cast on the paper inside the picture area. If necessary, lights can be directed on to the sides of the subject from below the glass support. A sheet of opal glass illuminated from below provides another suitable base for shadowless lighting. This method of lighting is widely used for catalogue illustrations of small parts and "exploded" assemblies because it makes it unnecessary to block out or retouch to get rid of the background.

Ring Illuminator. For special—mostly close-up—subjects, shadowless lighting can be achieved with an electronic flash tube bent in a circle to surround the camera lens. The tube is mounted in a circular trough-section reflector which increases the light falling on the subject and stops it from spilling on to the front of the lens. This type of flash lighting is used largely for medical and dental close-ups and for macrophotography. The lighting gives the best results when the subject detail is marked by differences in colour. This is the nearest practical approach to truly shadowless illumination, i.e. where the light is virtually co-axial with the lens, so that all shadows cast are hidden entirely behind the objects casting them.

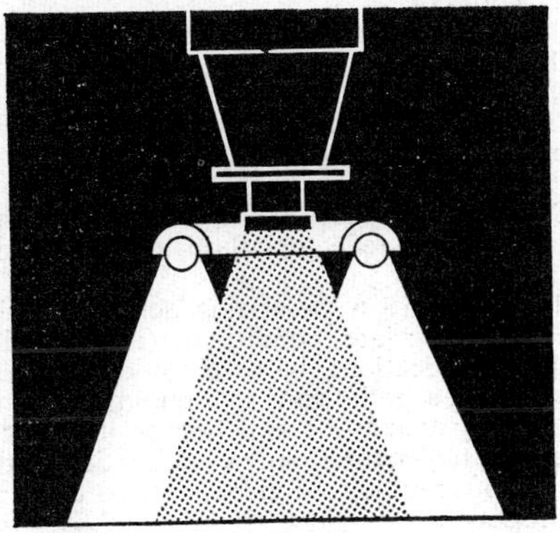

RING ILLUMINATOR. A ring shaped light source in a reflector—either a tungsten lamp or an electronic flash tube—surrounds the camera lens and kills all shadows on the subject side facing the lens.

Lighting Tent. Some subjects like glass, silverware and jewellery that are best photographed by shadowless lighting may be taken in a lighting tent of white muslin or any similar light-diffusing material. The lens is poked through a hole cut in the tent to avoid getting a recognizable portrait of the camera reflected in the polished surface of the subject. Lights are directed from the front and sides on to the tent which diffuses and reflects the rays.

Shadowless lighting can also be provided by a light source consisting of a number of parallel fluorescent tubes mounted behind a diffusing screen of translucent plastic sheet.

See also: *Lighting the subject; Tricks and effects.*
Book: *Camera Close-up*, by O. R. Croy (London).

SHADOWS. The shadows in a photograph are represented by areas of more or less dark grey silver deposit; they correspond to the parts of the subject that received no illumination from the light source and in consequence could not reflect any light into the lens. In most photographs completely black shadow areas are undesirable; they should be noticeably darker in tone than the rest of the picture but it should be possible for the eye to discern the details of the subject even in the apparently unlighted areas. This is only possible if they receive some degree of illumination—either reflected from the principal source or provided by a separate (fill-in) light. It is also necessary for the exposure to be sufficient to allow the shadow detail to record on the sensitized emulsion.

Control of the distribution, shape, and density of the shadows in a picture is an important part of photography for whatever reason it is practised.

See also: *Lighting the subject.*

SHARPNESS METER. Instrument for measuring the sharpness of an image in the image plane of a camera or an enlarger. It utilises the differential response of a photo conductor cell to the different light distributions in a sharp and an unsharp image.

See also: *Focusing (automatic).*

SHAW, GEORGE BERNARD, 1856–1950. Irish playwright, and keen amateur photographer from 1898. A friend of many leading photographers and an influential critic before World War I. He admitted to Helmut Gernsheim that his classic comparison, "The photographer is like the cod, which lays a million eggs in order that one may be hatched", was based on personal experience.

SHAYKHET, ARKADII SAMOILOVICH, 1898–1959. Russian photographic correspondent for *Ogoniok*, a famous illustrated weekly, for many years. Participated in the Ogoniok Exhibition in London, opened by Bernard Shaw in 1930. With others, published *Twenty Four Hours with the Phillippov Family*. Received some Soviet Orders during World War II.

SHEET FILM. Another name for cut or flat film—i.e., a negative material consisting of a sensitized emulsion coated on a stout film base cut in all normal plate sizes. This material has very largely replaced glass plates.

See also: *Films.*

SHELF-LIFE. Period of time before deterioration renders a sensitized material unusable.

See also: *Expiry date.*

SHELLAC. Dark red resin in the form of thin plates used for making varnish—e.g., for varnishing negatives. It is also used for dry mounting.

SHEPPARD, SAMUEL EDWARD, 1882–1948. English photographic chemist, later Assistant Director of Research, Eastman

Kodak Company. Studied with C. E. K. Mees in London. Both worked with Wratten and Wainwright in Croydon and then went to Rochester to found the Research Laboratories of the Eastman Kodak Company (1913). Sheppard's work covered almost every aspect of photographic science, especially latent image theory and chemical sensitization. In 1923 discovered the organic sulphur compounds in photographic gelatin which act as natural sensitizing agents, and thereby founded a new emulsion technique. He received the Progress Medal of the Royal Photographic Society in 1928. Has some 250 publications to his credit.

SHIMOOKA, RENJOH, 1825–1914. Japanese photographer. Was a pioneer in professional photography in Japan, working in Yokohama in 1862.

SHOP WINDOWS. Reflections present the major difficulty when photographing shop windows. During daylight hours the windows reflect people and passing traffic, the opposite side of the street and, of course, the photographer. At night the reflected lights of passing traffic may reproduce as streaks, while the photographer's lighting represents a further problem.

Whenever possible a good quality polarizing screen should be used to reduce reflections and obtain maximum clarity of detail behind the window glass. It should be borne in mind that light reflected at approximately 33° from a glass surface is polarized to the maximum extent, and that the effect is less at other angles, and non-existent at 90°. The angle at which the window must be taken is often dictated by its shape and position relative to pavement width, lamp-posts, adjacent property, etc.

The most attractive window display photographs are those taken by artificial light.

Possibly the best time to photograph a shop window is reasonably late at night, when passers-by and traffic begin to decrease in number and the photographer will have greater freedom of movement.

At such times an assistant should accompany the photographer. Apart from helping with equipment he will be able, for example, to request drivers not to park their vehicles where the lamps will reflect into the lens.

Camera. For commercial work use a technical camera which takes negatives no smaller than 4 × 5 ins. to allow for the degree of enlargement often required and the need for maximum detail.

A firm and rigid tripod is also essential.

Coated lenses are desirable, and should include one of normal focal length, a wide angle and a long focus lens. The latter may be useful when the camera has to be positioned some distance away from the window—e.g., on the opposite side of the road. An extending lens hood adjustable to fit all lenses, or individual hoods for each lens, should always be used.

The shutter should be of the type operated by ball and tube, or by a long flexible cable release. Alternatively, the exposure may be made by means of the lens cap.

A spirit level is desirable for checking the side-to-side and front-to-back levels of the camera.

Photographic Materials. A fast panchromatic anti-halation backed plate or sheet film is best for black-and-white photography, while a reversal colour emulsion is most suitable for colour work where the pictures are to be used for advertising. (Most blockmakers prefer to work from colour transparencies rather than colour prints.)

Lighting. The illumination may be either

(1) Daylight;

(2) Window lighting only;

(3) Window lighting plus high intensity photographic lighting; or

(4) High intensity lighting only.

The photographic lighting may be either Photofloods in reflectors, flash, or a combination of both.

(1) Daylight: a polarizing screen should, under appropriate conditions (i.e.—suitable angles of reflection), be used for daylight exposures to cut reflections to a minimum. Some window lights may be switched on if they are necessary to obtain a more even lighting to the rear, or if they help to give a desired effect to the display.

If the window is accessible, several direct readings should be taken with an exposure meter and a mean selected. If the window is not accessible, similar meter readings should be taken with the meter placed close to the glass and shaded from top light to avoid a false reading from reflected sky.

(2) Window lighting: in the majority of shop windows the lighting tends to give a much brighter illumination to the front of the window. If possible some of the lights should be switched off to obtain a more even balance. The exposure meter reading should be taken close to the window glass to avoid reflected light from street lamps.

(3) Window lighting can be supplemented by high intensity photographic lighting such as Photofloods or flash. Care must be taken to ensure that the lights do not reflect into the lens and that the lighting is even. To avoid sharp shadows, Photofloods may be moved from side to side during exposure, always ensuring that the lights and their reflections remain outside the field of view of the lens. The meter reading should be taken close to the window so as to avoid the reflections of Photoflood lamps.

With flash, invariably more than one flash bulb will be needed to obtain adequate coverage, and the position of each flash firing point must be pre-determined, and marked with chalk. A torch held by an assistant and observed on the focusing screen is a good guide as to whether flash positions are outside the field of view of the lens.

Flash positions should be selected to ensure adequate and even coverage of the area.

As each flash will nominally expose one section of the window, the stop to be used should be calculated in the normal way from the guide number of the flash bulb used.

(4) The procedure for photographic lighting only is the same as for combined light.

Different light sources should not be mixed when taking pictures in colour, as they would produce different colour rendering in various parts of the image. The appropriate colour compensating filter should be used if the film is not balanced for the light source employed.

Exposure and Development. As a rule exposure should be full and a soft working developer used to ensure retention of all the shadow detail without excessive highlight density.

The aperture used should be small to ensure maximum definition and depth of field, and the resultant exposure will frequently be long. In twilight and after dark the lens will have to be shaded from the reflections of the lights of passing vehicles.

People walking between the camera and the window will not register on a long exposure provided they do not pause.

An efficient shield may be made from black photographic paper mounted on a card, or the hand may be used. On no account should the lens be capped or the camera touched.

If the lens cap has to be used for the exposure, it is wise, after removing it, to hold it close in front of the lens for a few seconds, until all vibration has died down, and then remove it completely.

A slight degree of under-development is to be preferred to any degree of over-development. A.E.S.

See also: *Light sources in the picture; Night photography.*

SHORT FOCUS. Description of a lens indicating that its focal length is shorter than the diagonal of the negative it is designed to cover—i.e., shorter than the focal length of a normal lens.

SHUTTER

Mechanical system for controlling the time during which light is allowed to reach the sensitive material in a camera. The shutter consists of some means of covering the lens or the film or plate light-tight, uncovering it for a pre-set time and covering up again.

The early cameras did not need shutters. Plates were so slow that the exposure could always be made by removing the lens cap and replacing it after the required time. As the speed of sensitive materials increased it became possible to use shorter exposures than could be given by hand in this way, so some form of mechanical device became necessary for uncovering the plate for exactly determined periods of less than a second.

DESIGN AND FUNCTION

At first the shutter was manufactured as a separate accessory which could be fitted in front of the lens or, more rarely, between the lens and the plate. In time cameras were manufactured with the shutter included as a part of either the lens assembly or the body.

Nowadays the main types of shutter are the sector and the diaphragm (or leaf) shutter on the one hand and the focal plane shutter on the other.

Most diaphragm and several sector shutters have the iris diaphragm built into the shutter housing to regulate the lens aperture. On cameras with focal plane shutters the iris is of course built into the lens.

Sector Shutters. The sector shutter is simply a circular plate pierced with a hole, mounted in the path of the light entering the camera. When the shutter release is pressed, the plate rotates and the hole uncovers the lens for a period depending on the speed of rotation.

This type of shutter may be mounted in front of the lens, behind it, or even between the components. It is generally fitted to the cheaper cameras with simple lenses.

Tensioning. There are four ways in which the shutter can be re-set for the next exposure:

(1) The shutter lever is arranged to operate in both directions—i.e., downward pressure causes the hole to cross the lens and make the first exposure; the film is then wound on to the next frame and the second exposure is made by pressing the shutter lever upwards.

Used in this way the shutter does not need any arrangement for capping the lens, and a simple mechanism is adequate. Shutters of this type are fitted to the cheapest box and folding cameras.

(2) A second metal plate is arranged to cap the lens while the shutter plate returns to its original position. Before the shutter starts to operate, the capping plate is automatically held out of the way.

This is found in the everset type of shutter. In such shutters the release lever or button is spring loaded and returns to its original position after being depressed. During the return stroke the lens is automatically capped and the shutter plate turned back ready for the next exposure.

The everset shutter is convenient in being always ready for action without first having to be tensioned. It is superior to the simple non-capping sector shutter because the release lever always works in the same direction. But it has the disadvantage of needing a heavy release pressure to compress the return spring. This and the limited range of speeds restrict the everset shutter to cheap cameras.

(3) The shutter may be re-tensioned by a separate tensioning lever. This system makes it possible to operate with a light release pressure. The tensioning lever automatically brings a capping plate over the shutter aperture as it returns to the set position.

(4) The shutter plate can be arranged to rotate always in the same direction—i.e., after uncovering the lens, the plate continues to rotate until it gets back to its starting point. There is no longer any need to cap the lens as the aperture does not uncover it on the return journey. Shutters of this type can be designed as everset or they may be tensioned by a separate lever.

One make of rotating disc shutter is operated by releasing a previously wound spring which provides the power for a number of releases on one winding. As this spring also advances the film after each exposure, it is possible to take a succession of pictures as fast as the release can be pressed.

Diaphragm Shutters. The diaphragm shutter consists of a ring of inter-leaving light metal or vulcanite blades which are pivoted at their outer edges so that they can open outwards and leave a clear hole for the light to pass through. These shutters are designed so that the proportion of the total exposure time occupied by the action of opening and closing is as short as possible—i.e., so that the shutter efficiency will be high.

This principle is widely used in modern cameras, sometimes for shutters of the everset type but chiefly in more complex shutters which require a separate tensioning for the driving spring and operate with very light release pressures. On some cameras the tensioning movement is linked with the film transport, so that the action of advancing the film also tensions the shutter for the used shot.

In most modern diaphragm shutters there are three to five thin metal blades which come together to form a diaphragm between the lens components. On some cameras the blades may be in front of or behind the lens. The latter arrangement permits the use of interchangeable lenses with this type of shutter. On pressing the shutter release the main spring moves a ring which makes the blades swing out to the edge of the aperture and allows the light to pass through the lens. The blades are designed so that the opening starts as a star or a small hole, letting light through the lens more or less uniformly from the centre to the circumference. At the end of the exposure, as

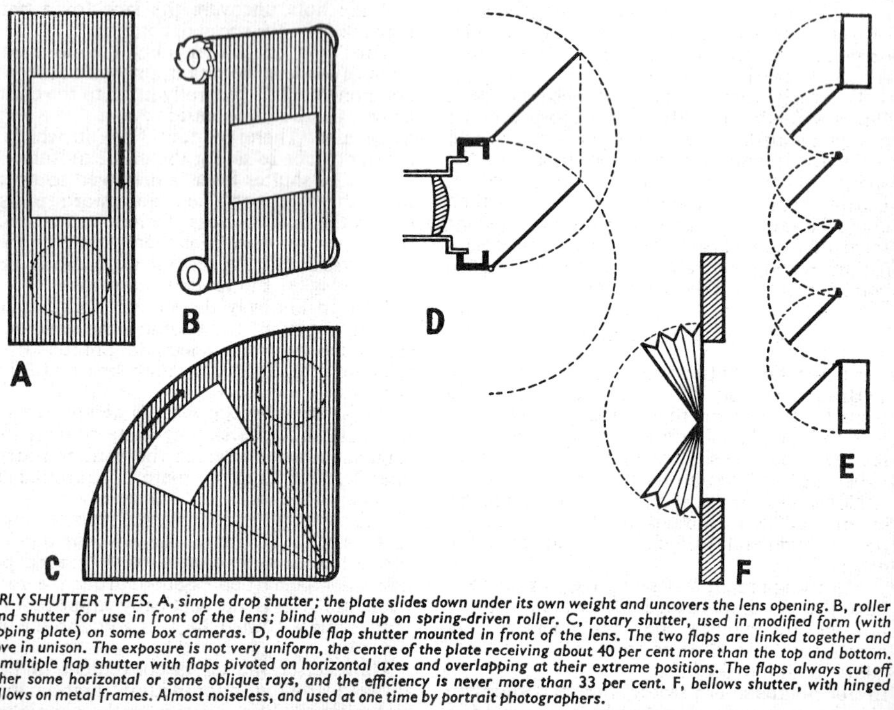

EARLY SHUTTER TYPES. A, simple drop shutter; the plate slides down under its own weight and uncovers the lens opening. B, roller blind shutter for use in front of the lens; blind wound up on spring-driven roller. C, rotary shutter, used in modified form (with capping plate) on some box cameras. D, double flap shutter mounted in front of the lens. The two flaps are linked together and move in unison. The exposure is not very uniform, the centre of the plate receiving about 40 per cent more than the top and bottom. E, multiple flap shutter with flaps pivoted on horizontal axes and overlapping at their extreme positions. The flaps always cut off either some horizontal or some oblique rays, and the efficiency is never more than 33 per cent. F, bellows shutter, with hinged bellows on metal frames. Almost noiseless, and used at one time by portrait photographers.

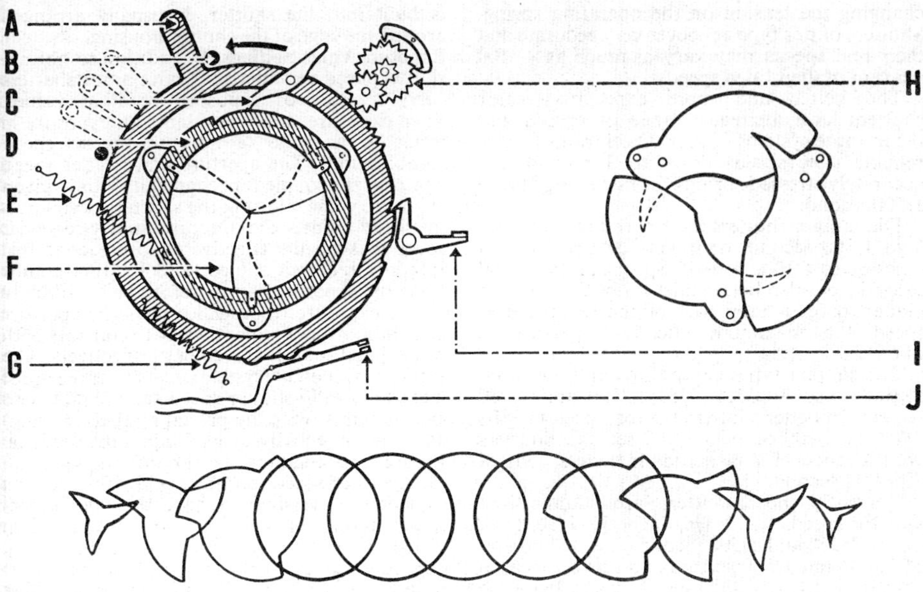

DIAPHRAGM SHUTTER. Usually mounted between lens elements, sometimes also behind or in front. Left: Simplified layout of main components: A, tensioning arm joined to main driving ring, in tensioned position; B, pin to disengage driving pawl; C, driving pawl; D, inner ring opening shutter blades; E, return spring; F, blades; G, main driving spring; H, delayed action escapement, used also to obtain synchronizing delay; I, release catch; J, flash contacts. Right: Shutter blade shapes; blades partially open. Bottom: Shape of the lens opening during the opening and closing of the blades of a three-bladed diaphragm shutter.

the blades close, the hole becomes smaller or changes to a broad-rayed star which becomes thinner and finally disappears, leaving an opaque diaphragm once again.

On a typical diaphragm shutter the time taken by the shutter blades to open fully is about one to two milliseconds.

Shutters of this type are highly complex mechanisms in which all the components are housed in a narrow ring surrounding the lens.

The main spring driving the ring which moves the blades determines the top speed at which the shutter will operate. At the slower speeds an escapement arrests the ring for a given period when the blades are open.

Older shutters of this type use an air-brake (i.e., a piston pushing air out of a cylinder through a small hole) to retard the movement of the blade ring at slow speeds. This system is practically silent, but is less reliable.

On some shutters an additional spring provides extra tension for the fastest speed.

Electronic shutters utilize an electric circuit and a solenoid in place of an escapement to control the shutter time.

Certain diaphragm shutters use a differently shaped set of blades. Each blade is double the normal size, more or less symmetrically shaped and rotates about a central pivot point. During the opening of the shutter one half of each blade moves out of the way to uncover the lens (usually five such blades make up the shutter), while during closing the other half of the blade moves in again to obstruct the light path. Since this design does not involve the reversal of a very fast movement as in conventional diaphragm shutters (where the blades swing out and then back again), inertia problems are greatly reduced and shutters of this kind can yield faster top speeds.

The Release. Most diaphragm shutters carry a special lever to release the shutter, but on some models this is combined with the tensioning lever—movement in one direction tensions the shutter, and movement in the opposite direction fires it.

On the majority of modern cameras a linkage connects the release lever on the shutter with a body release button on the camera body. This makes releasing more convenient and reduces the risk of camera shake. The body release may be linked with a double exposure lock to prevent accidental double exposures.

Usually a threaded socket is also fitted, either on the shutter itself or incorporated in the body release button, to take a flexible cable release. This is used for releasing only; the tensioning must still be carried out separately.

Shutter Speeds. The simpler sector shutters are made to give nominal speeds of 1/25 to 1/75 second. On these cheap shutters there is usually a choice of two speeds obtained by changing the tension on the operating spring. Shutters of this type are never very accurate and their real speeds may vary as much as ± 100 per cent of their rated speeds.

The better—and more expensive—sector shutters have a greater range of speeds and incorporate systems of gear wheel trains to give reliable slow speeds down to 1 second and accurately stressed springs for fast speeds up to 1/500 second.

Diaphragm shutters have speed ranges from 1 to 1/500 second, or in one or two isolated cases up to 1/1000 second. Speeds up to 1/2000 second have also been achieved by allowing the blades to open only part of the way; at this speed the maximum effective aperture is about *f* 8.

The simpler types carry three or four speed settings, usually 1/25, 1/50, 1/100 and 1/200 second. In better shutters the top speed may be 1/250 or 1/300 or even 1/500 second. Shutters with a range of slow speeds include speeds of 1 to 1/15 second.

Currently most shutters are standardized with the speeds 1, 1/2, 1/4, 1/8, 1/15, 1/30, 1/60, 1/125, 1/250 and 1/500 second (or 1/300 instead of 1/250 and 1/500 second). In this series each time is exactly half the preceding and twice the succeeding exposure time.

The speeds are set either by moving a lever along a scale engraved on the rim of the shutter, or by turning a milled ring which forms part of the rim. The speeds are then marked on the ring which is set to an appropriate index mark on the front of the shutter. Alternatively, the speed scale may be marked on the front, and the index mark on the movable ring.

The speed can generally be altered before or after tensioning. The top speed can in some shutters only be set before tensioning.

Intermediate Settings. Where the shutter speeds depend only on the spring tension, as with the simpler sector shutters, intermediate settings between two marked speed figures will yield approximate intermediate speeds.

If the speeds are controlled by an escapement, intermediate speeds may be possible if the selector cam inside the shutter is continuous. In many shutters this cam, which fixes the escapement delay, is stepped and the shutter can only be set to marked speeds.

Even where intermediate speeds are possible, they are not always continuous throughout the range. For instance, many shutters have a definite gap between 1/10 and 1/25 second where either a different cam or a different gearing of the escapement comes into action. Similarly, where the top speed involves an extra spring, no intermediate setting is possible between the top speed and the next slower one.

Certain types of shutters designed for coupling to built-in exposure meters have a completely continuous speed range—from say 1/30 to 1/500 second.

The aperture scale, where the iris diaphragm is built into the shutter, is usually arranged around the edge of the shutter housing.

Exposure Value Settings. In addition to normal shutter speed and aperture scales, some shutters carry a scale of exposure or light values. Each exposure value represents an exposure in terms of a range of aperture-shutter speed combinations. The aperture and shutter speed controls are coupled mechanically; for a given exposure value, altering the shutter speed automatically adjusts the aperture, and vice-versa, so as to keep the effective exposure constant. For any required subject and lighting conditions the exposure can therefore be set directly on the exposure value scale, using the aperture and shutter speed scales merely to control depth of field and subject movement respectively. The exposure value scale system necessitates the use of a constant ratio shutter speed scale (e.g., where each speed is twice the preceding speed setting) as well as equally spaced aperture settings.

Programmed Shutters. To simplify the selection of aperture-speed combinations for correct exposure some shutters have the two adjustments permanently linked together. This then yields a continuous series of settings (e.g. exposure values) without the choice of alternative combinations at each exposure value setting. Thus the shutter may start with a combination of 1/30 second at *f*2·8, and then give—with progressive adjustment of the

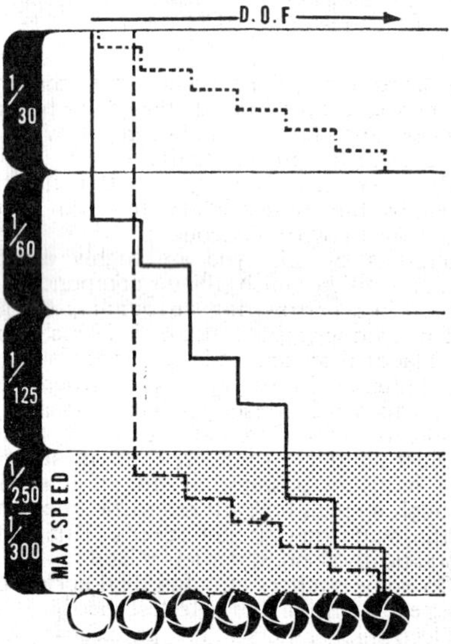

SHUTTER PROGRAMMING. On a programmed shutter—frequently built into cameras with fully automatic exposure control—aperture and shutter speed settings are interlinked and change together from the biggest to the smallest exposure. The usual programme goes in alternate steps of aperture and speed (solid path) some shutters progress first from the slowest to the fastest shutter speed and then from the largest to the smallest aperture (broken path). Apertures can be selected individually at the slowest shutter speed setting for flash shots (dotted path).

setting control—1/30 second at $f4$, 1/60 second at $f4$, 1/60 second at $f5{\cdot}6$, 1/125 second at $f5{\cdot}6$, and so on up to, say, 1/500 second at $f22$. In this arrangement the aperture and shutter speed settings are interlaced, so that each shutter setting yields progressively less exposure on the film. The shutter then carries only an exposure value scale or—when coupled with an exposure meter—no scale at all. This entails a certain loss of versatility, since the choice of slow speeds with a small aperture for great depth of field or of fast speeds with a large aperture for maximum movement stopping ability is absent. Instead each setting is a compromise combining the most generally useful shutter speed with the most frequently used lens aperture for a given exposure level.

The loss of versatility is particularly marked in the middle of the exposure range; at low exposure levels and high ones there would in any case be considerably less choice of alternative combinations. In between the two extremes the actual combinations used as a compromise depends on the programming chosen. With a fully interlaced programme the apertures and the shutter speeds progress in alternate steps. In a fully consecutive programme the shutter speeds may be increased first to the top speed, followed by a reduction of the lens aperture. The latter system therefore always yields the fastest shutter speed usable under the prevailing lighting conditions; this compromise favours movement stopping ability at the expense of depth of field.

Programmed shutters of this kind are mostly associated with cameras featuring exposure automation, where the exposure setting is directly selected by the reading of a built-in exposure meter. They frequently carry a fully continuous adjustment range of apertures and exposure times, yielding actual combinations not provided in orthodox aperture and speed ranges (for example 1/90 second at $f7{\cdot}2$, etc.).

An alternative programming system dispenses with the use of an iris diaphragm and instead employs the shutter blades to regulate the effective lens aperture as well. Here the shutter blades only open partly at the higher exposure value settings, and so close sooner. On varying the shutter adjustment thus both the effective shutter time and the lens opening decrease continuously and simultaneously. The shutter incorporates only a single adjustment which may again be coupled with an exposure meter or automatic exposure control system.

Programmed diaphragm shutters with separate iris and shutter blades usually permit independent adjustment of the aperture at two fixed shutter settings. One of these is a time exposure setting where the shutter opens and stays open until closed again. The other is a fairly slow speed such as 1/30 second and is used for flash exposures where independent control of the aperture is necessary.

Time Exposures. Even the cheapest sector

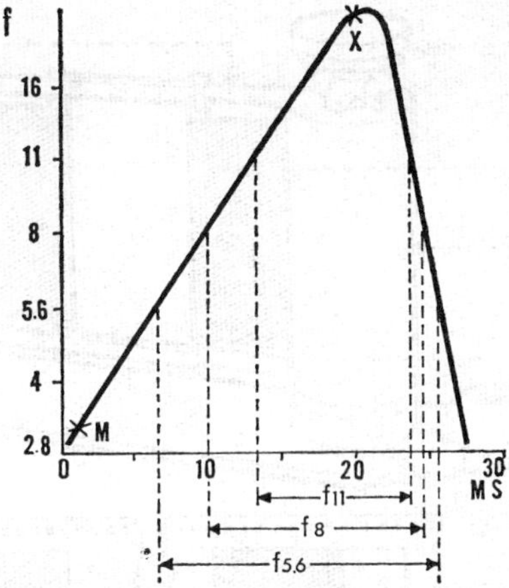

PROGRAMMED SHUTTER SYSTEM WITHOUT IRIS. The aperture and the exposure time are determined simultaneously by the degree to which the shutter opens. For the smaller exposures the shutter curve starts from a point high up and is thus shorter horizontally (representing shorter exposure times) as well as vertically, letting through less light. The effective exposure is given by the area under the curve. Thus at f5.6 the time interval (from 7 to 26 milliseconds) is 19 milliseconds. The effective exposure time—corresponding to the area below the curve at that level—is half this value or about 9.5 milliseconds. Similarly at f11 the total shutter time is about 5 milliseconds, but the effective exposure 2½ milliseconds. The point M and X represent the firing point at M and X synchronization of flash respectively.

shutters have some means of halting the aperture in the open position for making time exposures. The shutter may have to be held open by pressure on the release button, or it may stay open after the first release pressure and close when the button is pressed a second time. The first method is suitable for making brief time exposures and is marked B on the shutter setting control; the second method is more suitable for longer exposures and is indicated by the letter T (Time) on the setting control. Some shutters may be set to give both T and B; most have only a B setting.

Shutters fitted to cameras with a double exposure lock usually have only a B setting for time exposures. The reason is that the double exposure lock blocks the release button once it has been pressed and no second pressure (as required to close the shutter at T) is possible until the lock is freed by advancing the film.

Focal Plane Shutters. A focal plane shutter consists of a blind with a slit in it which travels from one roller to another close to the focal plane of the lens—i.e., just in front of the surface of the sensitive emulsion. The blind is the full width of the negative frame and may move up, down, or across from right or left.

Older types of focal plane shutters have one long continuous blind with slits of varying width at definite intervals. The different speeds are obtained by winding the appropriate slit into position for release. The lens

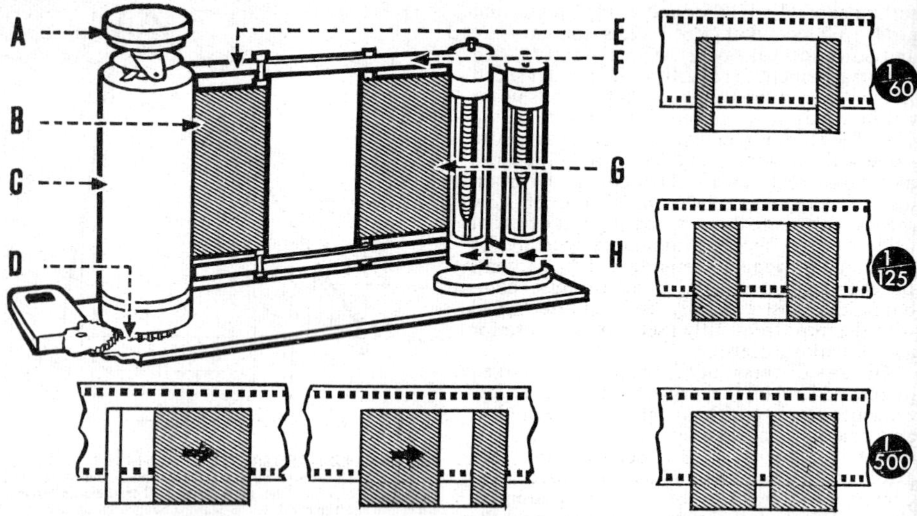

FOCAL PLANE SHUTTER. Two separate blinds move across the picture area just in front of the negative, exposing it from one end to the other. The interval between the blinds (right) determines the exposure time—the wider the effective slit, the longer the time. The main components of a typical miniature camera shutter are: A, shutter speed setting dial; B, second blind; C, main drum, D, gears to transport mechanism; E, tape of first blind; F, tape of second blind; G, first blind; H, rollers on which the two blinds are wound up after the exposure. The blinds move across the film with a slight overlap during tensioning, so that no light reaches the film (self-capping system), and are released one after the other in short succession during the exposure. The interval in the release of the two blinds is controlled automatically by an escapement set according to the selected shutter speed.

has to be capped while winding the shutter.

Most modern focal plane shutters have two blinds, and the slit through which the exposure is made is provided by a gap between the blinds. During rewinding the blinds cross the film or plate closed, i.e., without a gap between them. Such a shutter is self-capping, and the lens does not need capping during rewinding.

As the slit travels across the picture area it exposes each part of the sensitive surface in turn. The amount of exposure that any particular part of the film or plate receives depends upon the width of the slit and the rate at which it travels.

The normal sizes of focal plane shutter usually have some means of adjusting both the width of the slit and the speed of the blind. One knob on the side of the camera gives a choice of slit widths while another alters the tension of the spring which controls the rate of travel. Focal plane shutters on miniature cameras usually operate with a spring of constant tension and the exposure can only be adjusted by altering the width of the slit.

Where both the width of slit and tension can be adjusted, the camera usually carries a printed or engraved table showing the exposure times that can be obtained for any combination of slit and tension. Where only the slit width can be altered, the adjusting knob is calibrated directly in shutter speeds.

Focal plane shutters are also arranged to give T and B settings for which the blinds open up to expose the whole of the negative frame at once, instead of exposing it through a narrow slit. The slower speeds in fact are usually obtained by uncovering the whole frame while an escapement mechanism holds back the second blind for the appropriate time.

The blinds of focal plane shutters are generally made of thin black fabric impregnated with rubber to render it completely opaque, but metal is sometimes used.

Various other designs have also been tried, using folding or hinged multiple metal leaves in place of roller blinds.

Shutters of this type, designed for small roll film and 35 mm. cameras, usually have a range of speeds from 1 second to 1/1000 (1/2000 in special cases) with T, B and delayed action, and incorporate a train of gears, escapement or governor device for slowing down the mechanism for the slower speeds.

Focal plane shutters for the larger formats—e.g., quarter-plate—rarely have speeds below 1/10 second, but the fastest speeds are usually 1/500 or 1/1000 second.

Delayed action may be provided by a separate mechanism which retards the release.

While diaphragm shutters are comparatively independent units which are fitted into the camera or even into an interchangeable lens, focal plane shutters are generally an integral part of the camera body. Thus the same diaphragm shutter, generally made by a specialist manufacturer, may be used in a variety of different camera models. Focal plane shutter designs on the other hand are nearly always tailored to the camera in which they are used. One or two attempts have however been made to produce separate unit focal plane shutters for miniature cameras. Here the whole shutter system is self-contained in an appropriate frame and can be fitted into different camera models. Such shutters generally work with moving metal leaves that swing across the picture aperture in unison, yielding effectively a travelling slit like the roller-blind design.

Acceleration. All focal plane shutters tend to accelerate as the operating spring first overcomes the inertia of the moving parts and then begins to speed them up. The effect of this acceleration is to shorten the effective exposure time progressively as the shutter proceeds across the picture aperture.

There are several ways out of the difficulty. Some shutters are designed so that the width of the slit increases as it traverses the picture aperture, the extra width being calculated to offset the increased speed. Other shutters incorporate some arrangement of a simple governor brake that keeps the speed constant.

Many focal plane shutters—particularly those on the larger size cameras (bigger than $2\frac{1}{4} \times 2\frac{1}{4}$ ins.) have no provision for offsetting the acceleration. In such cameras the slit travels from top to bottom of the plate, so that the lower part of the plate is slightly under-exposed. As this often corresponds to the sky area, under-exposure may be an advantage.

Acceleration effects do not, in any case, tend to be as serious on the larger and less accurately made shutters as on the comparatively delicate and finely adjusted mechanism in, say, a precision miniature camera. On the larger shutters, friction and other imperfections limit the speed that the blind can attain as it crosses the picture area.

Image Distortion. As the slit in the focal plane shutter travels across the sensitized material, the time it takes to expose any particular point on the surface is less than the total time it takes to expose the whole surface—i.e., the total exposure takes longer than the local exposure. When the slit width and shutter tension are set to give an exposure of, say, 1/1,000 second, the slit may easily take 1/100 second or longer to travel across the plate.

The rather long total exposure time of a focal plane shutter is not important with subjects that are still, or moving only slowly in relation to the film or plate. But with rapidly moving subjects, a special type of distortion may occur. If the image is moving across the film at right angles to the travel of the slit, then the image will move appreciably between the exposure of the first and last sections. So the last part to be exposed will be displaced in the direction of motion, and intermediate points will be displaced in proportion.

The result is that the subject appears to be leaning backwards or forwards, depending upon which way the slit moves in relation to the image.

If the slit moves from the top to the bottom of the image, then the bottom of the image will appear to have moved forward during the exposure, and the subject will appear to be leaning backwards.

If the slit moves from the bottom of the image to the top, then the subject will appear to be leaning forward.

This explains why the wheels of racing motor cars, photographed as they pass across the field of the lens, appear oval and lean either forwards or backwards. The shutter in such cases is almost always arranged to expose the bottom of the subject first, because the resulting forward tilt lends an impression of speed.

Focal plane shutter distortion can also be observed in the curved appearance of the air-screw blades of aircraft photographed with their engines running.

If the slit moves in the same direction as the image, it draws it out and makes it look longer in the direction of its motion. If the slit moves in the opposite direction to the subject, it squeezes it up and makes it look shorter in the direction of movement.

Generally, it is better to hold the camera so that the shutter slit moves in the opposite direction to the inverted image on the plate or film—i.e., in the same direction as the subject itself. The effect of shortening the subject in its direction of motion is usually more tolerable than drawing it out.

The distortion produced by a focal plane shutter follows the same rules as the blurring produced by subject movement—i.e., it is worst when the subject is moving across the

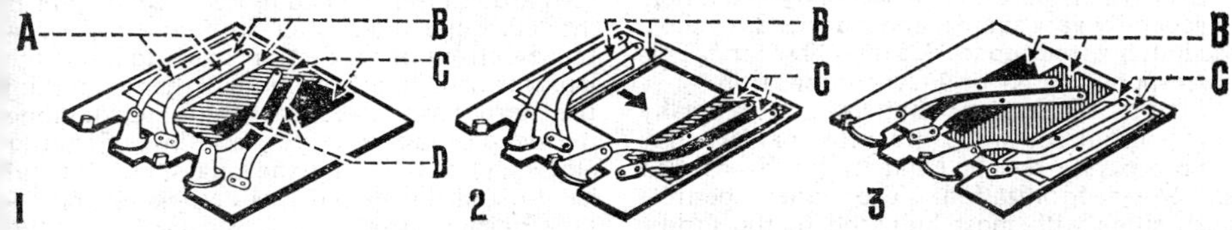

INDEPENDENT FOCAL PLANE SHUTTER. This type of shutter construction is a separate unit for building into cameras. 1. Blade position before exposure. 2. Shutter open. 3. Shutter closed after exposure. A, driving arms for upper blades. B, Upper blades. C, lower blades. D, driving arm for lower blades.

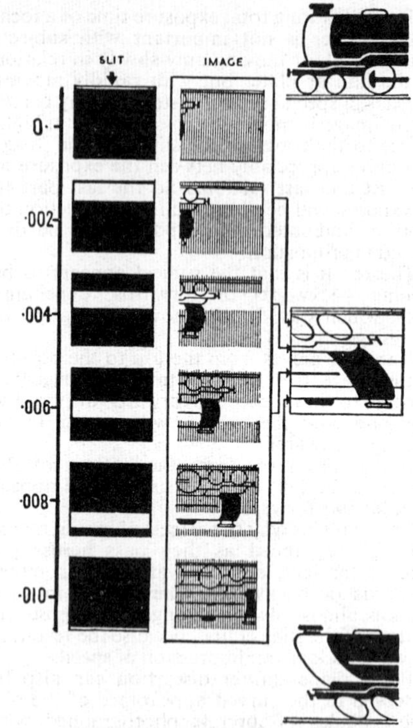

FOCAL PLANE SHUTTER DISTORTION. When the blind of a focal plane shutter and the image of a moving object travel at right angles to each other, a special type of image distortion appears. As the slit of the blind takes longer than the effective duration of the exposure to cross the whole negative area, the image will be in a different position at each instant during the travel of the slit. The assembled effect of all these component images is a picture of the object leaning forward or back. Although undesirable in some cases, it can improve the impression of speed in the picture.

field, when the subject is close to the camera, and when the subject is moving quickly.

Flash Synchronization. Most modern shutters are internally synchronized for flash by means of suitable contacts which close the firing circuit as the shutter runs down. The contacts are generally connected to a suitable flash socket or outlet on the shutter or camera. This socket then takes the flash cable or lead from the flash gun.

The synchronization of sector type shutters is generally very simple, and permits flash shots at shutter speeds up to 1/25 or 1/30 second. This corresponds to X- or F-synchronization.

Simple diaphragm shutters are generally X-synchronized, while precision models may have separate sets of contacts for X- as well as M-synchronization. The latter permits flash shots with most bulbs up to the fastest speed of the shutter.

Focal plane shutters may have one or more sets of flash contacts, often with separate outlets, and corresponding to X- and M synchronization. Owing to the time of trav of the blinds, special long-flash bulbs ar required for synchronization of fast speeds.

Special Shutters. Shutters normally give a audible click when they are released. This nois is undesirable in some branches of photo graphy—e.g., when taking pictures of animals birds, and children. For such work specia silent types of shutter are available. Thes are usually diaphragm shutters with two o more leaves operated by a pneumatic release They are not suitable for high speeds and ar designed for hand-operated exposures of one o two seconds.

Silent shutters are made to be fitted in fron of the lens of the camera to supplement th normal shutter. The camera shutter is set a T and left open while the silent shutter is i use.

Supplementary shutters are also used fo making intentional double exposures o cameras where the normal shutter is inter locked with the film transport mechanism.

Cinematograph cameras have rotary shutter of a special type that work immediately behin the lens at a constant speed. Exposures ar varied in this case by changing the lens aper ture.

The Kerr Cell shutter works on a completel different principle from any of the above. I this a high voltage is used to alter the polariza tion of a liquid mounted between crosse polarizing screens so that it allows light t pass through it. The Kerr Cell shutter can giv exposures as short as 1/100,000,000 second.

Ultra short exposures are also obtained wit various other shutter systems. They depend o obstructing the light path by rendering a optical unit in it suddenly opaque, for instanc by an ultra-fast discharge or miniature ex plosion.

Shutters and Interchangeable Lenses. Since a focal plane shutter operates closely in fron of the image plane of the camera and has n mechanical connection with the lens, it permits virtually unlimited interchangeability of lenses The focal lengths that can be used depend onl on the size and design of the camera mount.

Diaphragm shutters, being built into o fitted just behind the lens raise problems i lens changing. With the lens mounted just i front of the shutter the complete lens can be changed but the range of usable focal lengths is limited. Since the shutter is fitted a fixed distance in front of the film, the back focus of the lens must not be either too small (hence limiting the shortest available focal length) nor too long (to avoid serious vignetting and loss of shutter efficiency), thus limiting the longest usable focal length. Usually special lens designs are necessary for such systems.

Where the shutter is built into the lens itself with the blades between the lens components, two possibilities of interchangeability arise.

The first is to change the focal length of the lens by using alternative front components, with the rear unit and the shutter remaining firmly fixed to the camera. This has been used for certain miniature cameras. The other alternative—employed on large-size professional cameras—is to have a separate shutter with each lens. The complete lens and shutter assembly is then changed. Such an arrangement is very versatile, but also rather expensive.

Obsolete Types. A number of shutter systems, devised in the earlier days of photography, are now of purely historical interest. One or two have still survived for use in, for instance, studio portrait cameras where neither the high speed nor the ultra high precision of modern shutter designs is essential.

One of these is the roller-blind shutter—in its simplest form an opaque fabric blind running between rollers mounted at opposite ends of a wooden frame. When the shutter is released, the blind is drawn from one roller on to the other by a tension spring.

The roller blind shutter is made in various forms to be attached in front of the lens as a separate component. There is either a gap or a circular hole in the blind and as this passes in front of the camera lens it exposes the film. The tension of the spring can be adjusted within limits to make the blind travel faster or slower and thus shorten or lengthen the exposure. The positions of the spring tension adjusting screw are marked with the corresponding exposure values—usually ranging from about 1/10 to 1/50 second. The control mechanism usually includes a means of halting the blind in the open position for time exposures.

The blind may be either continuous, or two blinds may be used as in focal plane shutters. In the latter case the roller blind shutter will be self-capping.

Roller blind shutters are sometimes used in front of the lens on field and studio cameras taking the larger sizes of plate. Where such cameras work with a range of interchangeable lenses, the roller-blind shutter can be switched from one lens to another.

Designs of other earlier shutter types varied widely; a number of types, after a short run of popularity, soon became either obsolete or confined to certain specialized fields.

The drop shutter was simply a plate with a cut-out aperture. It dropped or was moved by a spring across the path of the image-forming rays which passed through the aperture and made the exposure. Various shapes were tried for the opening, but the shutter was never very efficient and the range of speeds was limited and too low for modern materials.

The drop shutter was also known as a guillotine shutter. It was later modified by adding a second plate which "capped" it for setting, and also made instantaneous exposures possible.

The rotary shutter was an extension of the drop shutter principle in which the plate carrying the aperture rotated to produce the same effect. It was the forerunner of the various present day sector shutters.

Flap shutters were all elaborations of a simple hinged flap covering the lens and operated by a pneumatic release. The flap could be outside or inside the camera, and it was simply raised and lowered to make the exposure. Some portrait cameras to-day are fitted with internal flap shutters. A modified type has a number of flaps which open and close simultaneously, like the slats of a Venetian blind.

Bellows, or eyelid shutters, consisted of two concertina-like units which when closed formed a dome-shaped cover over the front of the lens. When the pneumatic release was pressed, the two halves of the dome separated and opened like upper and lower eyelids to make the exposure. F.P.

SHUTTER EFFICIENCY

The efficiency of a between-lens sector or diaphragm shutter is the ratio of the light it actually transmits during the exposure to the amount that it would transmit if it were fully open for the whole duration of the exposure. In practice, no shutter is 100 per cent efficient. The reason for this is that between-lens shutters take some time to reach their maximum opening and some time to close, so they are not fully open all the time. At slow speeds, however, the efficiency of the average between-lens shutter closely approaches 100 per cent.

The efficiency of a diaphragm shutter largely depends upon the time taken to move the shutter blades from one limit of their travel to the other. For this reason shutter blades are always made as light as possible to keep their inertia low.

A small shutter is usually more efficient than a larger one because the blades are smaller, lighter, and have a shorter distance to travel. These advantages, however, may be cancelled out by the reduced power of the mechanism.

Shutters which require separate tensioning can use more powerful driving springs than everset shutters where the same lever tensions and fires the shutter. They are thus more efficient.

The actual times of opening and closing of the blades of a diaphragm shutter do not change much over the range of shutter speeds, and the efficiency is thus higher at low speeds. When the lens is stopped down, the shutter blades do not have to travel as far before uncovering the aperture, and the efficiency is increased.

The efficiency of a focal plane shutter depends on its distance from the film and on the *f*-number of the lens aperture. The optics of the system make it impossible for a focal plane shutter to be 100 per cent efficient.

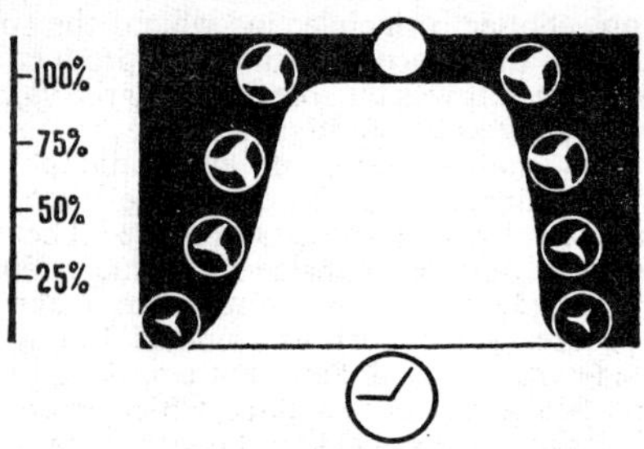

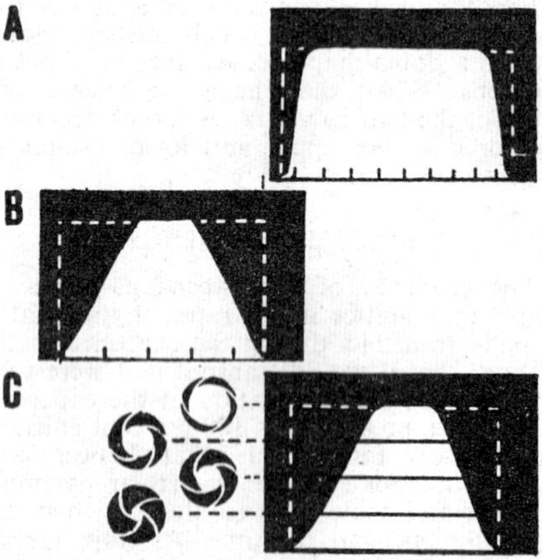

DIAPHRAGM SHUTTER EFFICIENCY. Top: A diaphragm shutter is fully open only during part of the time of its operation; the periods of opening and closing take up some time, and thus result in a certain loss of light. The shutter opening is here plotted against time. Bottom: Comparison of shutter efficiencies. The dotted square represents an ideal shutter opening and closing instantaneously. A, modern shutter at 1/10 second. The time marks represent 1/100 second intervals. B, reduced efficiency of same shutter at 1/250 second. Time marks at 1/1000 second intervals. C, increased efficiency at smaller apertures, as blades clear lens aperture earlier.

The image on the film or plate lies at the tip of a cone of light which converges on it from the lens. When the blind sweeps through this cone of light at a short distance from the film, the light intensity on the film does not become zero instantly, but is reduced to zero as the blind cuts off more and more of the cone of light.

This loss of efficiency is reduced by having the blind as close to the film as possible. As with between-lens shutters, the efficiency increases as the lens is stopped down.

The efficiency of a focal plane shutter can be calculated from the equation:

$$E = \frac{W}{W + d/f}$$

where W = width of shutter slit
f = *f*. number of lens
d = distance of shutter slit from camera focal plane.

The efficiency of the roller blind shutter operating very close to the lens is determined by the slit width and the lens aperture. The greater the slit width compared with the aperture, the higher the efficiency. The efficiency varies with aperture—i.e., it is higher with a small stop which can be uncovered and covered quickly—but for a given slit width, it is the same for all shutter speeds.

Measurement of Efficiency. The efficiency of a shutter is arrived at by comparing the amount of light actually transmitted from the beginning of opening to the end of closing with the amount of light that would have been transmitted if the shutter had been fully open for the whole time.

There are two principal ways of measuring the amount of light actually transmitted by the shutter: by sensitometry, or by using a cathode ray oscillograph.

Sensitometric measurements may be carried out by the following methods:

(1) The shutter can be used to expose film to a standard light source, and the amount of light passed can be calculated from the density of the resultant image. This method has many faults; it is difficult both to calibrate the light source and to control processing with the necessary accuracy. The interpretation of the resultant density is not easy.

(2) A more modern variation of this method is to put a photo-electric cell in place of the film and measure the light falling on it electronically. The interpretation of the result is difficult as the light source must be calibrated and allowance has to be made for the lens aperture. The method is useful, however, for making a quick comparison of shutters against a laboratory standard.

All methods depending on this principle measure the effective shutter speed.

The simplest method of using a cathode ray oscillograph for testing shutters is to photograph a pattern on the tube screen.

A more involved method enables the complete shutter characteristics to be determined almost instantaneously. In this method, light transmitted by the shutter falls on to a photoelectric cell. The signal is amplified and deflects a luminous spot vertically on the screen of the cathode ray tube. The spot may at the same time be deflected horizontally at a known speed to give a time base.

As the shutter opens, the spot is deflected upwards and at the same instant the "time" deflection begins, with the result that the spot draws a graph on the screen of the tube. The "time" deflection is calibrated in fractions of a second either by a grid in front of the screen or by an extra light pattern on the screen.

The whole picture on the screen can be photographed if a permanent record is required. By this means, therefore, the total time of opening, the efficiency, and the effective xposure time may be determined very quickly. Erratic or faulty operation is also obvious.

A graph of the light transmitted by a shutter, ecorded in this way, shows how the amount of ight increases as the shutter opens, remains onstant at full aperture, and then decreases as he shutter closes. The full aperture of the hutter may have been used for 1/100 second luring a total time of opening of 1/80 second.

If the shutter had been ideal, it would have vasted no time in opening and closing, and the graph would rise vertically to fully open, emain at that level for the total time of exposure, and then drop vertically to zero. In practice the graph slopes more or less gradually rom zero to the maximum and then more steeply back again.

The amount of light passed by the shutter is proportional to the area enclosed between the graph and the base line. The efficiency is then simply the area enclosed by the shutter performance graph expressed as a percentage of the area enclosed by the graph of an ideal shutter open for the same total length of time.

Comparison of Shutter Types. The performance of shutters varies very greatly from type to type, from make to make, and from shutter to shutter, and it is difficult to generalize about performance. Calibration, especially of the cheaper everset shutters, is not always reliable. It is only recently that quick and accurate methods of shutter testing have become available, so that modern shutters are likely to be more accurate than older types.

The efficiency of an everset shutter is usually in the 60–70 per cent range for shutter speeds from 1/25 to 1/100 second, which is as far as such shutters usually go.

The efficiency of pre-set shutters is higher, and is generally about 90 per cent at 1/25 second, 80 per cent at 1/100 second, and about 70 per cent for higher speeds, falling to 65 per cent at 1/500 second. These figures are for the smaller shutters, and are usually slightly lower for large ones. The calibration of these shutters is usually more accurate than with the everset type.

Higher effective shutter speeds can be achieved with focal plane shutters, because exposure can be shortened by simply narrowing the slit instead of speeding up the movement of the blind. Although the efficiency falls as the speed is increased, a precision focal plane shutter is usually more efficient than a precision diaphragm shutter at the same speed.

Efficiency and Movement Blur. The time for which a shutter is fully open is less than the total time from the beginning to the end of the exposure. This is of considerable practical importance in photographing moving objects. If the shutter is of low efficiency and takes an appreciable time to open and close, the brighter parts of the moving object will be recorded both before and after the shutter is fully open, and will appear more blurred than darker parts which are only recorded at the maximum shutter opening.

Efficiency and Calibration. As the efficiency of the shutter drops at the highest shutter speeds, this should be allowed for in the calibration of the shutter. Modern precision shutters are in fact calibrated in such a way that the shutter speed represents the exposure that the shutter would give if it were operating at the indicated speed with 100 per cent efficiency, rather than the actual open time of the shutter. Thus the light passed through the shutter at 1/500 second is adjusted to be one-half of the light passed at 1/250 second, one-quarter of the light at 1/125 second, etc. This means that the 1/500 second setting may yield a total exposure time from the beginning of the opening phase to the end of the closing phase of, say, 1/300 second.

Such photometric, as distinct from purely time calibration is important with cameras

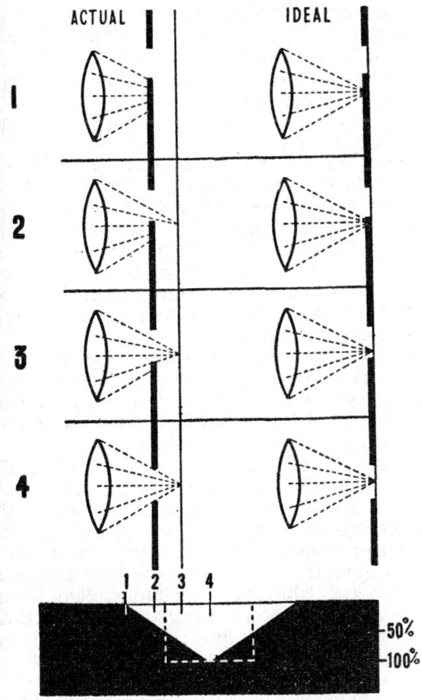

FOCAL PLANE SHUTTER EFFICIENCY. The ideal focal plane shutter operates in the plane of exposure with an efficiency of 100 per cent. Actual shutters operate some distance away from the exposure plane, and thus cut the cone of light coming from the lens plane. This reduces the efficiency; the effect is an increase in the length of the exposure, the total amount of light passed is unchanged. 1. Exposure begins with actual shutter, no exposure with ideal shutter. 2. Actual shutter has already passed appreciable light, ideal shutter none. 3. Ideal shutter just uncovered film, actual shutter nearly uncovers full cone of light. 4. Actual shutter reaches maximum flood of light which is same condition in case of ideal shutter.

using a coupled or automatic exposure meter. If the camera itself selects the shutter speed setting, the photographer may have no indication of the actual speed in use and his exposures may be open to unpredictable errors. Photometric calibration of the shutter largely avoids this. It cannot do so completely where shutter speeds are separately adjustable, because the efficiency (and hence the light passed) depends also on the lens aperture used. In fully automatic cameras with a programmed shutter the effect of the lens aperture can however also be allowed for in calibration. A.S.C

See also: *Exposure automation; Exposure values; Flash synchronization; Shutter manufacture; Shutter release; Shutter testing.*
Books: *Exposure*, by W. F. Berg (London); *Camera Techniques*, by H. J. Walls (London).

SHUTTER MANUFACTURE

The rôle of the camera shutter is to regulate the time during which light is allowed to act on the sensitive material. With simple cameras the action of the shutter merely consists of uncovering and covering the lens opening in an instant, lasting usually approximately 1/20 to 1/40 second. For ordinary snapshooting this is quite adequate, and variations in the effective exposure time from shutter to shutter are usually taken care of by the latitude of the negative material.

Better cameras have more complicated shutters with a range of exposure times which are more accurately determined. To achieve this degree of accuracy, within the small space available for the shutter mechanism in the camera, requires manufacturing methods of appreciable precision. High class shutters are therefore generally made by a small number of specialist firms who supply them to the various camera manufacturers.

There are two fundamental types of shutter: diaphragm shutters which are generally mounted between or just behind the lens elements, and focal plane shutters which work just in front of the sensitive emulsion. The construction and to some extent the materials required for the two types of shutter differ, but the same skill and precision in manufacture are required in each case.

DIAPHRAGM SHUTTERS

The modern diaphragm shutter has to satisfy certain basic requirements relating to the time of action, the control of the speeds, the shape of the sectors and the movement, as well as the subsidiary gear mechanisms which control such items as self-timer and flash synchronization.

The exposure with a diaphragm shutter is spread over three intervals:

(1) The interval during which the blades are opening.

(2) The interval during which the blades remain open.

(3) The period during which the blades are closing.

The first and the third of these intervals must be as short as possible to obtain a maximum effective open interval. The shutter mechanism therefore has to work under much more difficult conditions than, for instance, a watch. The above requirements call for a sudden acceleration, and then a final braking of the shutter blades and the driving elements. The degree of wear and tear is therefore far greater than with a watch where the movement is more or less uniform.

The basic speed of the best diaphragm shutters is 1/500 second; some models may have a basic speed of 1/400 or 1/300 second, and certain special ones up to 1/1000 or even 1/2000 second. Longer exposure times (slower shutter speeds) are obtained by means of a delay mechanism, usually an escapement. This is linked to, but otherwise independent of, the main opening and closing drive.

Other subsidiary mechanisms include the delayed action release or self-timer which is another escapement mechanism to retard the release of the shutter itself. In addition the shutter contains built-in flash contacts coupled in many cases with the delayed action mechanism to time the firing of flash bulbs. The contacts have to be insulated and big enough to carry the firing current.

The shutter blades must exclude all light when closed. To do this the blades have to overlap each other in a fan-like fashion, yet move with the minimum possible friction to permit the short exposure time required.

Owing to the ease of synchronization of electronic flash at all shutter speeds, diaphragm shutters have also been designed for use with interchangeable lenses. This raises special design problems where the shutter is permanently mounted on the camera and the lens units fit into the front of the shutter. (On some cameras the lenses are interchangeable together with the shutter; that only requires an auxiliary shutter in the back of the camera to protect the film against the light during lens changing.) As the interchangeable lenses usually have their own iris diaphragm, the changing system must also incorporate suitable features for coupling and uncoupling the aperture setting.

The design becomes still more complex with diaphragm shutters built into single lens reflex cameras and with automatic exposure control systems. In the first case the shutter tensioning (linked to the film transport) movement must open the shutter and iris diaphragm at the end of the tensioning cycle to permit viewing of the

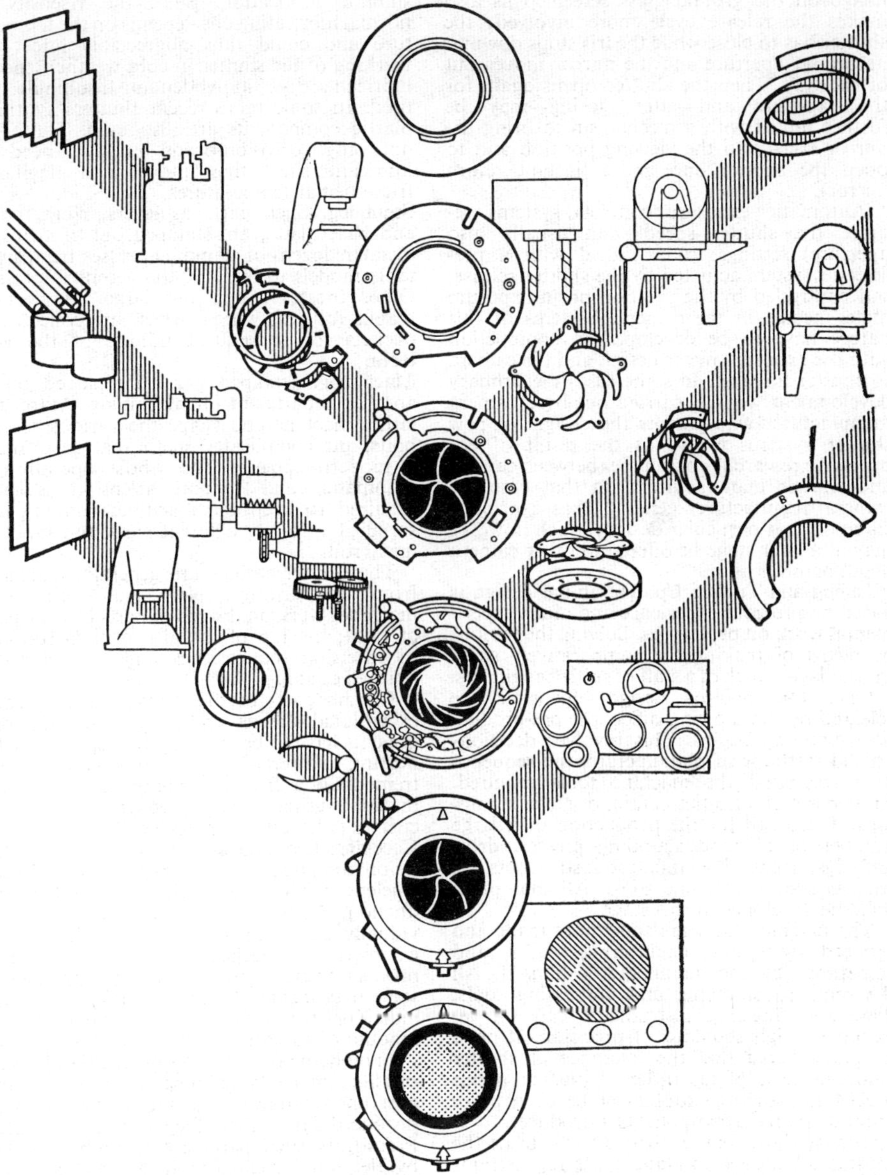

SHUTTER MANUFACTURE. The stages illustrated symbolize rather than attempt to show in detail the production of a between-lens shutter. They are, however, typical in a simplified form for a modern unit. Top: The base plate is stamped by special presses and machined and drilled for the various cut-outs, apertures, and screw holes. Upper left: Most of the levers and rings are cut out from sheet metal, drilled and machined. Upper right: Suitable stamping machines cut the shutter blades from continuous bands of very thin spring steel. Centre left: Screws, shafts, and pins, as well as mechanical components like gear wheels are mass turned on automatic lathes and finished on precision milling machines. Centre right: Iris leaves are stamped from thin steel bands in the same way as the shutter blades. The leaves are assembled inside the main housing together with the rest of the mechanism. Lower left: The cover plate, speed setting ring, etc., are stamped in hydraulic presses and machined. The various assemblies and sub-assemblies are checked at different stages throughout production. Lower right: Speed rings and other scales are engraved after calibration, the synchronizing contacts are set with the aid of a testing circuit. Bottom: Finally the complete shutter undergoes a thorough test for accuracy, consistency, synchronization, etc., before being built with a lens into a camera.

image on the ground glass screen. This also makes the release cycle more involved: the shutter has to close while the iris stops down to its pre-set aperture and the mirror moves out of the way. Then the shutter opens again for the exposure and—after closing—may be further linked with a mechanism to bring the mirror back into the viewing position and to open the shutter once more (instant return mirror).

Automatic exposure controls systems require the shutter speed (and usually also aperture) settings to be linked with a preselector system actuated by the shutter release, but controlled by the position of an exposure meter needle in the camera. Systems of this nature have to be developed in conjunction with the exposure meter design and the camera to shutter linkages. In some cases the primary development work is carried out by a camera manufacturer who patents the design, or new shutter systems evolve as the result of co-operative research programmes between camera and shutter manufacturers. In that case the camera manufacturer generally has a priority claim on that particular design, which however may at a later stage be offered to other camera manufacturers as well.

Planning and Tooling Up. The initial design is based on previous experience and much experimental work on prototypes. Solving the various problems of making the comparatively small shutter is the work of a staff of skilled engineers.

Once the finalized design blueprints are released by the drawing office, the pre-production planning begins. This involves decisions on the methods of manufacture, the sequence of stages, and the machine tools required. Most parts of a shutter consist of metal stampings. These call for the production of a large number of different stamping presses, drills and jigs, measuring and test instruments, as well as special machine tools. All that plant must be developed at this stage.

The machine tools are sometimes made and serviced by special engineering works, and sometimes by the manufacturer himself. All the required stamping and embossing tools, dies, and finishing instruments call for an extremely high standard of precision. It must be remembered that the tolerance of shutter components is of the order of plus or minus 0·0004 in., and the tool must be even more accurate than the components it produces.

Raw Materials. As the various parts inside the shutter often have to stand quite high strains and stresses, raw materials have to be carefully selected. This involves physical and metallurgical tests in a special laboratory which checks incoming deliveries of raw materials.

The materials consist mainly of sheets and rods of various light metal alloys, brass and steel, spring steels for the various springs, and special thin steel plates for the shutter blades.

The moving parts of a shutter also need suitable lubrication. Since the viscosity of normal lubricating oils depends on the temperature and could thus appreciably affect the working of the shutter in cold weather, special lubricants like molybdenum disulphide are used. In some more recent shutters synthetic plastic components are also used. Those are not only corrosion-proof but also need no lubricant since they move with negligible friction at all temperatures.

Stamping. Such parts as levers, rings, cover and base plates, are stamped out of strips of metal in dozens of stamping presses from small table models to large 60-ton eccentric presses. Other presses stamp out large numbers of blanks for gear wheels. These are collected in packets for subsequent milling of the gear teeth.

Machining. Stamped parts are passed on to another department for machining. Automatic lathes machine and shape the various screws, bolts, and complicated shafts that go to make a modern camera. The whole operation is automatic, and human attention is only required to keep the machines running and supplied with the different sizes and types of metal rod.

The shutter casings are generally machined from die castings. Here again automatic machines work on dozens of parts of the casing at once; the job of the attendant is confined to inserting the castings and removing the finished casing.

Certain parts of the shutter have to carry exceptional stresses and strains and the components have to be specially treated by suitable heating and tempering. In some cases the extreme tip of a tiny shaft is hardened in a special machine using a high frequency alternating current to heat up the parts locally.

Checking. Economical mass production of precision components of this type calls for highly efficient checking at each stage in the production. Many of the parts are checked individually by the operator at the machine, and extensively rechecked at the end of each production stage. In some cases a production stage may include as many as fifty to seventy individual operations and involve a considerable amount of careful testing.

In modern production technique the checking of all components by human operators is however uneconomical, since such components are produced faster than they can be examined. Frequently such parts are therefore checked by electronic machines which reject all components falling outside the prescribed tolerances, and at the same time keep a constant tally of the number of items checked as well as the rejection rate.

All measurements boil down to a comparison of the part with an accurate standard. The comparison may be optical or mechanical. Optical checking is carried out by projecting the outlines of the component—e.g., the profile of a tiny screw thread by means of profile projectors. Some of these permit simultaneous measurements of up to ten different components. By comparing the outline of a component with the outline of the appropriate standard projection slide, the tester can see whether the accuracy of machining falls within the permissible tolerances.

Such an instrument—of which there are many in a shutter factory—permits the speedy checking of distances, outlines, angles, etc., with an accuracy of up to 0·0004 in.

Mechanical checking is carried out with gauges linked to suitably calibrated indicators which show instantly if the dimension being measured is correct. Again these gauges operate automatically, controlled by a programming unit, which passes several thousand components through the checking machines.

Finishing. The shutter-speed scales are engraved by hand with the aid of suitable stencils.

To protect the metal surfaces from wear and corrosion and give them an attractive appearance, they are treated by various plating or anodizing processes or by lacquering.

Assembly. The various parts converge first into pre-assembly groups—e.g., escapements and delay mechanisms—and next to the final assembly line.

In modern shutter production even sub-assemblies are put together by automatic machinery. Vibrators and other feeding units supply the individual components in their correct orientation and proper order, while the machine joins them up, rivets pivots etc. Automatic cut-outs stop the machine if any part supplied to the sub-assembly is faulty or if any assembly operation goes wrong.

With a precision mechanism like a shutter, strict cleanliness is essential. The assembly room therefore has to be kept completely free from dust, and at an even temperature and humidity. The most modern assembly departments are air-conditioned and work with closed windows in summer and winter.

The assembly department operates on the principle of the continuous production line. The various parts and various groups are assembled step by step while the shutters move continuously forward towards the finished product. Each worker carries out one operation, often doing no more than adjust a single shutter speed.

Final testing. The shutter speeds are checked by a photocell which measures the light transmitted by the shutter while it is open. An electronic circuit converts the output of the photocell into a direct indication of the shutter speed.

The delay time of the flash contact is measured by a similar device. Built-in calibrating instruments keep a constant check on the speed testing equipment itself.

Final Stages. One of the last production stages is the cutting of the two threads which take the lens components. This operation has to be specially precise to ensure that the axes of the front and rear lens elements exactly coincide. The discrepancy between the two axes must not exceed 0·0006 in., while the deviations of the thread itself must be within a tolerance of the order of 0·0004 in. A special screw-cutting lathe cuts the two threads simultaneously on the finished shutter. W.A.B.

FOCAL PLANE SHUTTERS

The focal plane shutter is not a self-contained unit made separately from the camera. It is built into the camera body and its manufacture is inseparable from that of the rest of the camera. Whereas a diaphragm shutter, perhaps complete with lens, can be obtained ready-made by the camera manufacturer, this is not so with the focal plane shutter.

The main mechanism is built on a scale larger than that of the diaphragm shutter, and the work involved is medium-fine engineering rather than extra-fine. This applies to small camera shutters; the standard can be lower still for larger cameras.

If the shutter has associated mechanisms, such as escapements for slow speeds and delayed action, these are usually similar in quality to the corresponding components of the diaphragm shutter and require a similar standard of workmanship.

Large Cameras. Shutters in current use are of two types—coupled-blind, or independent-blind. The first is used for cameras of large negative size, such as press cameras. Each blind runs between separate pairs of rollers, one at each side of the camera. The two take-up rollers are driven by coil springs, and the winding spools are connected by gears engaged in the position necessary to give the desired slit width. When the shutter is released, the blinds therefore travel across together at constant slit width.

The basic mechanism is simple and presents few manufacturing problems. The design is complicated by the self-capping arrangement, but this does not add any particular difficulties.

Speed calibration is based on slit width. The driving springs are then tensioned until the speeds are correct. Some cameras are provided with a tension control to vary the speeds.

Small cameras. The precision miniature camera shutter is a different proposition. The two blinds operate independently and are not coupled to maintain a constant slit width. Depressing the shutter button releases the first blind, and when this has travelled a certain distance the second blind is automatically released. The exposure time is the time interval between the release of the blinds, and this remains constant providing the travel characteristics of the two blinds are identical—that is, providing the inertia and acceleration are the same. The slit width is not constant, but it becomes wider as the velocities of the blinds increase, which is necessary for even exposure.

Good consistency in performance therefore depends on good design and high accuracy in manufacture; friction losses must be kept to a minimum and the design be such that further friction cannot readily develop. Any differences in the movement of the blinds will result in uneven exposure across the frame; the higher the speed setting the more evident this becomes. The blind fabric in particular should not vary in flexibility over the normal range of temperatures, nor vary greatly with use.

Traditionally focal plane shutters used to be produced to individual designs to suit particular camera models. But such shutter types are also appearing as completely independent assemblies suitable for fitting into appropriate camera models. The assembly and production of such individual focal plane shutters then follows the same principles as with diaphragm shutters, subject to the modifications of design involved.

Usually separate focal plane shutters are all-metal types using sliding blades that move in front of the film, rather than cloth blinds. Some focal plane shutters employ very thin flexible steel blinds covered with black plastic instead of cloth.

Production. In general the production outline for the focal plane shutter of a small precision camera is similar to that for the diaphragm shutter. The methods of making the components depend on the scale of production since this determines the size of the plant available; but the same principles apply as described in detail above.

Assembly and Testing. In assembly the total camera mechanism has to be considered as a whole. The shutter of a small precision class camera is coupled to the film transport; there are the winding, metering and exposure counting arrangements for this, and usually a double-exposure lock. The combined mechanism may therefore be quite complex, and assembly of the shutter becomes part of the general camera assembly. This calls for a team of skilled technicians, each dealing with different stages of the work and each capable of making any adjustments to ensure perfect operation.

The complete assembly is then checked and final adjustments made if necessary to ensure smooth working. The tensions of the driving springs for the blinds are also adjusted at this stage. The usual procedure is to set the shutter to one of the higher speeds, such as 1/200 second, and measure the speed given at each side of the film opening and in the centre. The blind springs are adjusted so that the speed recorded at each point is the same, or as nearly so as it is possible to get it, and so that this speed agrees with the shutter setting. All other speed settings should then be automatically correct, but it is nevertheless usual to check throughout the range. H.H.

See also: *Camera manufacture; Lens manufacture; Shutter testing.*

SHUTTER RELEASE

The normal shutter release is a press button or lever which is ideally mounted in such a position that it can be depressed naturally by the forefinger while the other fingers have a comfortable grip on the camera body.

In practice its position varies; it is usually mounted on the lens-shutter assembly for a between-lens shutter, and on the camera body for a focal plane shutter.

Nowadays many cameras with between-lens shutters have the release button mounted on the body and connected to the shutter by a mechanical link. This is because it is easier to hold the camera steady when pressing a button on the body than one near the lens.

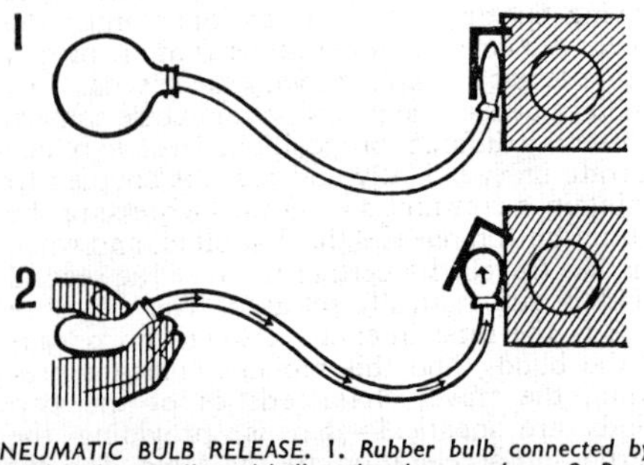

PNEUMATIC BULB RELEASE. 1. Rubber bulb connected by flexible tube to collapsed bulb under shutter release. 2. Pressure on bulb inflates the small bulb and trips shutter.

Some form of supplementary release with a flexible connexion to the camera is always an advantage when making time exposures with the camera mounted on a tripod. At such times direct pressure on the release button on the camera tends to shake it.

Pneumatic Release. The early form of flexible remote release consisted of an india rubber tube which was connected to a cylinder and plunger, or a simple inflatable bag mounted in contact with the shutter release. At the other end of the tube there was a rubber ball which, when squeezed by the operator, caused the plunger or bag to exert pressure on the release.

This type of release is still used on some studio cameras. It has the advantage of being silent, frictionless, and easy to renew.

Flexible Wire Release. The flexible wire (or cable) release is the one in general use for extending the existing shutter control. It consists of a thin stranded steel cable sheathed in a flexible tubular casing made of coiled or braided steel wire.

At one end the outer casing terminates in a

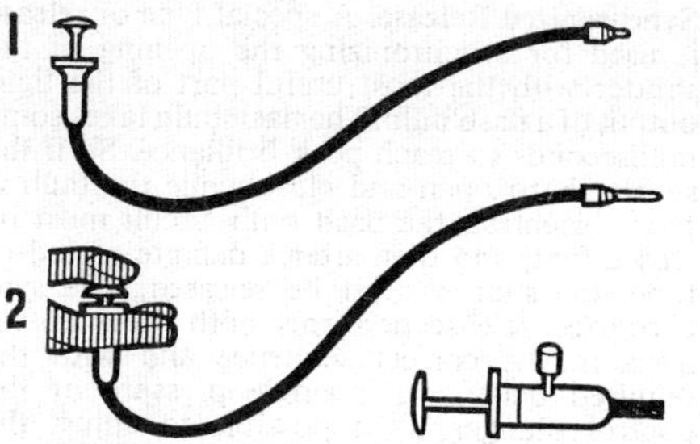

CABLE RELEASE. 1. Stiff cable with plunger at one end and solid pin at other inside a flexible sheath. 2. Pressure on plunger pushes out pin to trip shutter. Bottom: Locking screw for keeping release depressed for long time exposures.

conical screwed nipple through which projects a pin, sweated to the end of the inner cable. This nipple screws into the standard screwed socket which is fitted to most shutters, and the pin engages with the shutter release mechanism.

The other end of the flexible casing is fitted with a collar which acts as a guide for a plunger connected to the inner cable. An internal spring keeps the plunger raised. Pressure on the plunger operates the release.

A short wire release is used when the existing release is not conveniently placed for operating with both hands gripping the camera body.

A flexible release is also used when the camera is mounted on a tripod. In this case it is designed to prevent the operator's hand from shaking the camera when opening the shutter. For this purpose a long release—8 to 12 ins.—must be used; anything shorter than this is likely to transmit movement to the camera and so defeat the whole purpose of the tripod of keeping the camera steady.

This type of remote release is far from being perfect. Its great drawback is the serious amount of friction and lost motion that takes place between the cable and the casing. The result of this is that a short length tends to transmit shakes and vibration while long lengths call for relatively heavy pressure on the plunger and a considerable amount of travel to allow for the large amount of unavoidable slack in the cable.

Where a flexible release is to operate over a distance of more than 3 or 4 feet, this type of cable is unsatisfactory. A more positive action is given by a Bowden type of cable, where the inner cable is under tension and the slack can be taken up by a screw adjustment. This type of release needs a special fitting at the camera end to convert the pull in the cable into the push needed to operate the shutter.

Locking Release. Many modern camera shutters have no T-setting for long time exposures (i.e., where one pressure on the release opens the shutter and a second pressure closes it). This is largely to permit the inclusion of double exposure locks with which two successive pressures on the release—without winding on the film—are impossible. As it is inconvenient to keep the plunger of a cable release pressed down for long periods by hand, many such releases incorporate a lock.

The lock usually takes the form of a small clamping screw fitted in the collar of the plunger. To make a time exposure with such a release, the plunger is depressed to open the shutter (set to B) and the screw tightened to keep the plunger pressed down. Slacking off the screw allows the plunger to return to its original position, thus letting the shutter close.

On some models a second movable collar, arranged concentrically with the plunger, serves as the lock. Depressing the plunger also locks it, depressing the movable collar frees it.

Air-Pressure Release. One type of remote release consists of a long tube with a rubber ball at one end, and a cable release holder at the other. Pressure on the rubber ball actuates the cable release by air pressure. This system can work over appreciable distances (30 feet or more).

Electrical Release. Mechanical remote control of the shutter at distances greater than 10 feet is inconvenient and uncertain. At such distances electrical control is more satisfactory. The shutter is electrically controlled by having some form of electromagnet or solenoid attached to the camera so that it works directly on the shutter release or on the plunger of a cable release screwed into the shutter. A length of twin flex connects the solenoid to a press-button switch at the operating position, and the battery supplying the current may be at either end.

With this method of operation, the full working current required by the solenoid has to flow in the control wires. The control wires have therefore to be fairly thick because a suitable solenoid will require a current of at least 3 amps at 12 volts. So direct operation is only convenient over fairly short distances—not more than 20–30 feet.

A more flexible arrangement is to house the battery close to the camera to keep all the heavy current wires as short as possible. The switching in the solenoid circuit is then done by a relay, operating from the same battery, but consuming a much smaller current. This system makes it unnecessary to run heavy wires between the camera and the shutter release position. Bell-wire or clock flex will carry the small operating current over considerable distances.

Another system uses a spring loaded bar which bears directly on the cable release plunger. The bar is held against the tensioned spring by a catch. A magnetic relay operates the catch so that when the circuit to the relay is closed the bar is released and the cable release plunger operated.

In any system where the camera shutter release is pressed by a remotely controlled mechanical or electrical device, the pressure

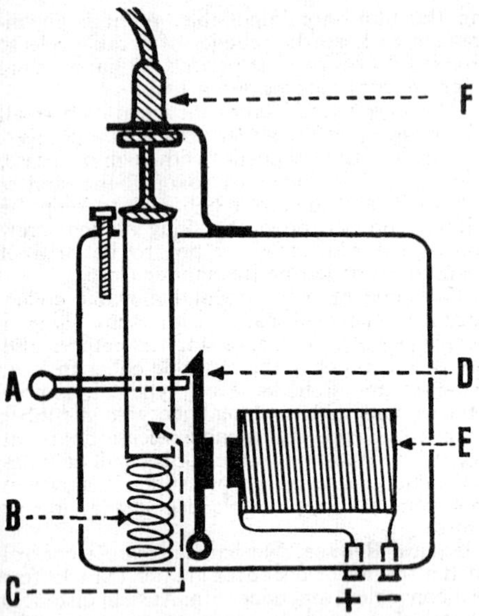

ELECTRIC RELEASE. *Design of an electrically operated remote control release. A spring loaded bar is released by the electromagnetic relay, depressing the cable release plunger. A, re-setting bar (engages relay arm). B, spring. C, plunger. D, relay arm. E, electromagnetic relay operated from suitable point. F, cable release to camera.*

must not be applied with a sudden jerk. Unless the action is similar to the relatively gradual application of pressure by the fingers, the shutter mechanism may be damaged and there will be a risk of camera shake.

The methods described above provide only for the remote release of the shutter, so the camera must be visited after every exposure to re-set the shutter and change the film or plate.

If the entire sequence of operations is to be carried out from a distance then there must be an independent device for re-setting the shutter and changing the negative material. The most satisfactory way of dealing with complete remote operation is to use a clockwork mechanism arranged to carry out the cycle of operations on the receipt of a single signal from the distant end of the circuit.

When the operator presses the remote release, an electric relay at the camera end operates a solenoid; the solenoid lifts a catch on the clockwork unit; the clockwork mechanism then depresses the camera shutter release, winds on the film, and re-sets the shutter, finally coming to rest against the catch ready to repeat the cycle of operations. One winding of the clockwork spring is enough to expose a whole length of film without further attention.

The above release systems are used in nature photography and in scientific work for photographing events in which the operator would be exposed to risk if he remained with the camera.

Synchronized Release. A special type of release is used for synchronizing the opening of the shutter with the most useful part of the light output of a flash bulb. The flash bulb takes some milliseconds to reach peak brilliance. So if the shutter is to open and close while the bulb is at its brightest, the flash bulb circuit must be closed first, and then after a definite period of time, the shutter must be released. The synchronized release performs both these operations in the correct sequence and with the required delay, on a single pressure of the control plunger. It is possible to adjust the delay between the closing of the flash contacts and the operation of the shutter to allow for differences in the burning rate between the various types of bulb. There are several methods of obtaining the delay, mechanical and electromagnetic; some types must be attached to the camera baseboard and others simply screw into the shutter release socket.

Automatic Release. The commonest form of automatic release is the delayed action device by which the photographer can include himself in the picture. Some shutters incorporate a delayed release which can be brought into action when the shutter is set.

With the device in action, when the release is pressed in the usual way, exposure does not take place immediately, but is delayed for 15–20 seconds—long enough for the photographer to take his position in the picture. The delay is usually achieved by allowing a spring to unwind slowly under the retarding action of a train of gears.

Supplementary devices are available for converting ordinary shutters to delayed action. These usually consist of a unit which screws into the ordinary cable release socket and has a plunger which is depressed instead of the normal camera shutter release. A train of gears or a spring-loaded dashpot device holds up the final pressure on the shutter release for the required number of seconds.

Subject-Operated Release. It is often desirable to arrange for the shutter to be operated by the subject—with or without his knowledge. This can be done by making use of one of the

DELAYED ACTION RELEASE. *On winding the lever and releasing the catch, the arm is slowly drawn into the body, and presses the plunger of the cable release until it trips the shutter.*

remote release devices described above.

In big game photography a thread may be stretched across the path of the game and connected to a pin or trigger which when pulled allows a spring to depress the shutter release. The camera is focused on the spot where the game will walk into the thread. For pictures at night, the shutter is synchronized with some form of flash lighting.

Electrical release lends itself to a number of ways of obtaining subject-operated photographs.

One is by pressure operated release. If it is known that the subject will touch or pass over any object, a pair of contacts may be concealed below the object and connected to the remote control wires of the electrical release. This method is particularly suitable for nature photography because it consumes no current when it is waiting to make the exposure and so can be left for long periods after being set, until actuated by the subject.

Another method is by light-operated release. The electrical release may be triggered by a photocell connected to a valve amplifier in the relay circuit of the shutter operating solenoid. The photocell is placed so that when the subject occupies a particular spot (on which the camera is focused) the light is cut off from the cell. This changes its electrical potential and makes an energizing current flow in the relay circuit of the shutter operating solenoid.

The same effect may be produced by directing a beam of invisible infra-red light at the photocell so that when the subject passes through the beam the shutter operating circuit is completed.

This system has many applications in the detection of criminals and in industrial and scientific research. A disadvantage is that the amplifier must be permanently switched on, and a failure of the power supply puts it out of action.

Where absolute reliability is essential, the photo-emissive type of photocell circuit can be replaced by one incorporating a photo-transistor which operates from a 12-volt dry battery and has an extremely low dark current.

Radio Release. The relay of the electrical shutter release can be triggered off by various methods, including a radio signal. This method has been employed in photographing birds and animals from a distance. The quarry is followed through binoculars or a telescope and when it is seen to be within range of a concealed camera, a pulse sent out from a radio transmitter either releases the shutter or switches on a photocell circuit which operates an automatic release mechanism. Radio release is also employed for operating photographic equipment on space rockets or earth satellites.

Acoustic Release. If a microphone is substituted for a photocell, the shutter may be released by sounds. The amplifier circuit can be adjusted to respond to a sound of any given volume or pitch. This system also is used in crime detection and some photographic studios are equipped with it so that the operator can work the shutter release from any part of the room by simply clapping his hands. F.P.

See also: *Flash synchronization.*

SHUTTER TESTING

The purpose of a photographic shutter is to allow the light forming the image in a camera to act upon the sensitive emulsion for a definite time and at a definite moment.

Ideally the shutter should open and close very quickly, but most shutters fall short of this ideal, and there may be wide differences between shutters in this respect. Thus the more detailed tests are concerned not only with determining the period of opening of the shutter, but also with investigating the way in which the shutter opens and closes and how much light it passes during this phase.

Better-quality shutters are surprisingly reliable and consistent in their performance, and individual shutters conform well to type. Consequently, there is little to be gained from attempting to test such a shutter for oneself, unless some fault is suspected, or a precise knowledge of some characteristic is wanted for a special purpose. Some simple tests are possible, however, and examples are given below.

In general, the complete testing of a shutter is rather difficult, and is usually only carried out in the development of new designs or for checking standards in production.

Shutter Characteristics. Important properties of shutters are specified by quantities which now have accepted definitions.

(1) The property of apparently stopping the motion of a moving object is dependent on the time interval during which the shutter is open, or partly open. This is the total open time.

(2) While the shutter is opening and closing it is not passing as much light as when it is fully open. The first and last parts of the exposure are not fully effective. The total amount of light passed is proportional to the effective exposure time, which may be regarded as the period for which the shutter would have had to be fully open to pass the same total amount of light.

(3) The effective exposure time is evidently less than the total open time, but the more effective the opening and closing parts of the cycle can be made, the more nearly equal these quantities become. The ratio of the effective exposure time to the total open time is the efficiency, and is expressed numerically as a percentage.

The time the shutter takes to open is greater, the farther it has to open. With a given individual shutter, the shutter is in effect fully open much sooner when the lens is used at a small aperture. Then part of the opening time no longer leads to an increased amount of light transmission. Accordingly the shutter efficiency at small lens stops is increased and the effective exposure time becomes more nearly equal to the total open time.

By agreement in international standards the shutter speeds marked on a shutter refer to the effective exposure time at the maximum lens opening. From this follows the slightly paradoxical state of affairs that at small lens apertures the effective exposure time of the same shutter becomes longer.

Simple Tests. It is possible to measure the total open time of a shutter quite accurately by photographic methods using the shutter in the camera in the ordinary way; this technique, incidentally, fulfils the requirement, mentioned above, of simulating actual conditions of use.

In the simplest test, all that is required is some small bright object which can be made to move at a constant known speed s, and which can be photographed, using the shutter to be tested, at a known magnification m. The length of the trace of the moving point on the developed film. l, is measured. Then the total open time, t_o, is given by the relation

$$l/ms = t_o.$$

Thus, for example, if a small lamp is carried by a vehicle moving at 30 m.p.h. (528 ins. per second) across the line of sight of the camera, and is photographed from 50 feet away, by a camera with a 2 ins. lens, the calculation is as follows. The magnification is $2/(50 \times 12) = 1/300$. If the streak on the film is 0·10 in. long, the total open time is:

$$t_o = \frac{0{\cdot}10}{1/300 \times 528} \text{ second} = 0{\cdot}057 \text{ second.}$$

If the lamp or brightly illuminated object can be arranged to move in a circle with a known angular velocity, and the angle subtended by the arc on the film is measured, it is not necessary to know the magnification. Thus, if a bright source is mounted near the edge of a gramophone turntable rotating at 78 r.p.m. (468 degrees per second) and is photographed from immediately above the centre of the turntable, the record on the film will appear in the form of a circular arc. The angle subtended at the centre of the turntable by the ends of this arc can be measured with a protractor. It helps if the centre of the turntable is made to show up clearly on the film. A convenient point source can be provided by fixing a small pocket or pen torch on to the turntable, first removing the glass and reflector. Suppose the arc measures 40°; then the total open time is:

$$t_o = \frac{40}{468} \text{ second} = 0{\cdot}087 \text{ second.}$$

Several measurements may be made on one frame if the camera is shifted slightly between exposures. The repeatability of the shutter can be quickly checked in this way, or different shutter speeds can be easily compared on the one single frame.

In the case of focal plane shutters, repeating the measurements at a number of places along the direction of travel of the blind will give a measure of the variation of exposure caused by blind acceleration. The investigation of this point may be made difficult if the performance of the shutter is not repeated exactly, and in any case the trace may be shortened or lengthened if the image of the gyrating source is moving towards or away from the direction of travel of the blind at the time of exposure.

Flashing Light Source. The procedure can be improved if the light which illuminates the moving object can be modulated at a known frequency—e.g., by employing a stroboscopic

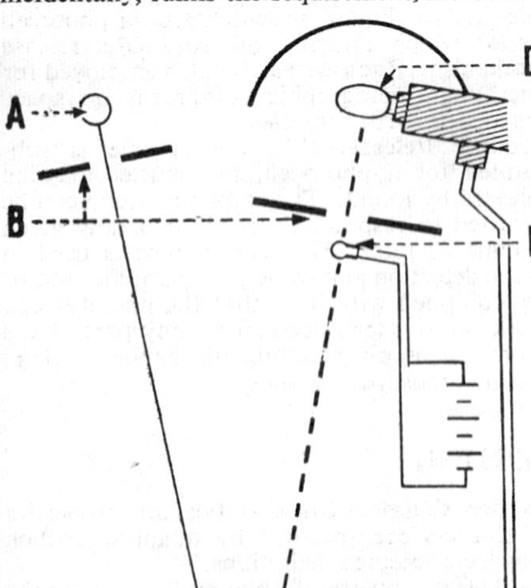

ROTATING MIRROR SHUTTER TEST. A, vertical fluorescent tube. B, screens with slits. C, mirror on motor driven turntable. D, flash bulb (when testing synchronization). E, torch bulb supplying steady light, arranged above or below flash bulb. F, shutter. The film is exposed once to the light from the steady lamp and the flash (if used) at shutter setting to be tested, and then again to the fluorescent tube with the shutter open on time while the light reflected from the mirror sweeps once across the film. When testing a focal plane shutter, the blind must travel parallel to the direction of the fluorescent tube. The flashes from the latter are separated by 10 milliseconds on the film if powered by 50 cycle A.C. mains. Speed of mirror adjusted to give adequate separation of tube images.

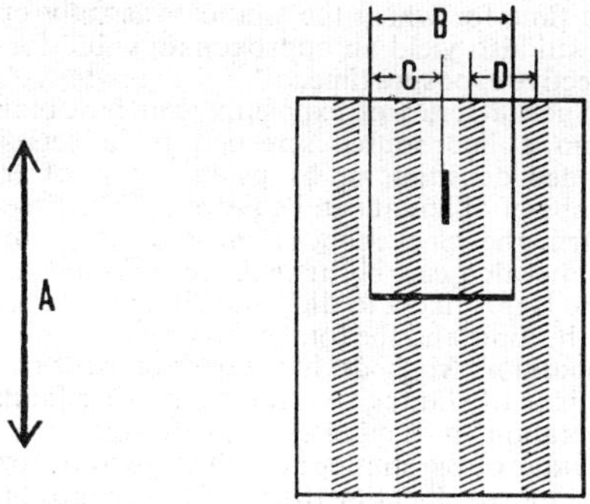

ROTATING MIRROR TEST RESULT. A, direction of focal plane blind (if used). B, streak from steady light. C, flash from bulb. D, interval of tube images (10 milliseconds on 50 cycle mains). Total open time = B/D × 10 milliseconds. Flash occurs C/D × 10 milliseconds after shutter begins to open.

lamp as the source, or by using a discharge lamp run from alternating mains. On 50 cycle supplies, for example, an ordinary fluorescent lighting lamp produces a flash every 1/100 second.

The exposure could then be determined by counting the number of flashes recorded on the film, and it would no longer be necessary to be sure of the constancy of the speed of movement.

The counting method is obviously useless for times less than 1/100 second, but a simple modification enables the whole range of shutter speeds to be tested. A horizontal turntable driven at a constant speed is required, but the speed of rotation need not be known. A speed of about 1 r.p.m. is suitable for times up to 1 second, and about 10 r.p.m. for times less than 1/10 second.

A good quality mirror is mounted vertically along a diameter of the turntable, and the camera is arranged as close as possible, looking into the rotating mirror. The mirror should be large enough to cover the whole field of the camera, if possible.

A fluorescent tube, or other discharge lamp, is set up vertically some distance away, and the width of the tube is masked down to a long narrow slit formed by the edges of two long strips of card, or other opaque material. Near the tube is set up a torch bulb supplied from a battery. The camera is focused on the slit and torch bulb via the rotating mirror, and the shutter and aperture are set to the combination to be tested. Exposures are now made with the steady light from the torch bulb only. Successive exposures on the same frame can be obtained if the lamp is moved vertically between exposures.

Then while the mirror continues to rotate at the same speed, the steady light source is switched off and the discharge lamp is switched on. An exposure is then made to the light from this lamp, using the T or B setting of the shutter and opening the shutter while the mirror causes the image of the illuminated slit to sweep once across the focal plane of the camera.

A series of images of the slit are formed which are somewhat blurred, but whose separation on the film can be measured quite accurately. The separation between adjacent images represents the interval between successive flashes of the lamp.

The exposures made via the shutter to the light from the steady source will give a series of streaks. These represent the movement of the spot of light on the film during the exposure interval, and since the spot was moving at the same rate as the flashing source, a comparison of the length of the trace with the interval between successive images of the flashing lamp gives the total open time, even if this time is less than the interval between successive flashes of the discharge lamp.

Testing Synchronization. This arrangement can be elaborated slightly to test the timing of the synchronizer and the operation of flash equipment.

All that is needed is to set up the flash bulb just behind a slit arranged vertically above the torch bulb in the apparatus just described. The flash is fired by the shutter when the exposure to the steady source is made, and an image of the slit in front of the flash bulb will be recorded in the exact time relation to the opening of the shutter.

This procedure could become rather expensive in flash bulbs if many tests are needed. One possibility is to use another torch type bulb with a thin filament which lights up quickly, operated through the flash contacts, instead of the actual flash bulb used with the camera.

A low wattage, high voltage lamp is suitable, and in cases where the flash contacts make for only a brief time, the lamp can be run with a great deal more than the nominal voltage applied to it. This shortens the delay between the making of the contacts and the instant when the light from the lamp first begins to be recorded on the film. There will, however, still be an appreciable lag, which must be found by a parallel experiment with the proper flash bulb. The method described can then be used to check the operation and timing of the synchronizer.

Cathode-Ray Traces. When a cathode ray oscilloscope fitted with a short-persistence tube is available, the trace on the screen can be used instead of the moving objects described. It is usually possible to modulate the movement in some way to provide a time scale. This can be done by using the mains frequency to deflect the spot vertically, or by arranging a circular trace deflected radially. The scanning raster of interlaced lines as produced on the screen of the domestic television receiver would form an ideal test object for shutter

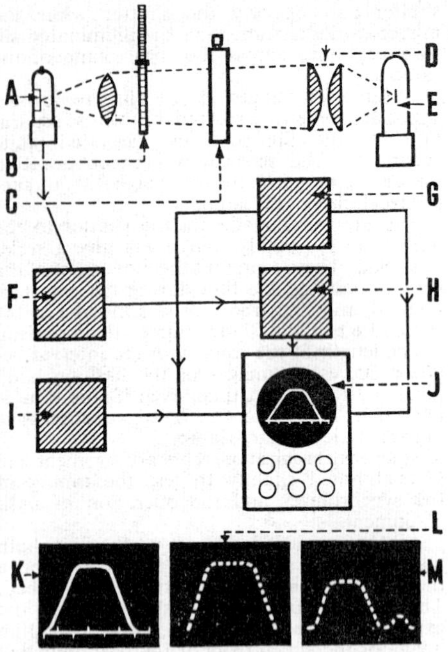

SHUTTER TESTING BY OSCILLOSCOPE. Top: Layout of apparatus. A light beam passes through the shutter to a photocell, and the output is fed to a cathode ray oscilloscope. It is also utilized to trigger off a time base and to control spot brilliance. A timing mark generator provides pips on a base line of a second sweep or interrupts the previous curve at regular intervals. A, photocell. B, compensating grey wedge. C, shutter. D, condenser. E, lamp. F, amplifier. G, single stroke time base. H, automatic brilliancy control. I, timing mark circuit. J, cathode ray oscilloscope. Bottom: Characteristic shutter curves. K, curve with vertical timing marks at pre-set intervals. With a known interval between the marks, the opening, total open, and closing times can therefore be read off directly. L, curve with interrupted trace at pre-set intervals. M, curve indicating shutter bounce; i.e., the blades open again slightly for an instant after the exposure.

testing of times up to 1/25 second, but unfortunately some television tubes have rather a long afterglow period. However, the technique may offer a rapid way of comparing speeds of about 1/50 second.

Effective Exposure and Efficiency. The measurement of the effective exposure time and the efficiency of shutters is rather more difficult than the measurement of total open time. Some idea can be obtained of the time the shutter is taking to open and close, however, by examining the blackening produced on the negative at different points on the streaks. Variations in the blackening are more easily seen if care is taken not to over-expose.

A series of streaks made with the same shutter setting, or, preferably a number made all at the same time, in which there is a progressive increase of density from one streak to the next, gives the best chance of estimating the time for which the shutter was fully open. It will also yield an approximate value for the effective exposure time.

Such a series of exposures can be obtained from a line source screened by a series of graded densities, or by using a row of small lamps of different candle power. If the time for which the blackening is constant at its maximum value can be judged, the effective exposure is not likely to differ much from the mean of this time and the total open time.

Laboratory Methods. In the development of new designs of shutter, and for diagnosing faults in operation, it is necessary to examine the behaviour of the shutter at each stage in the cycle. The more refined methods of testing are therefore aimed at obtaining a record of the light passed by the shutter at each moment in the cycle. Considerably more information is then available for calculating the effective exposure time, but the measurement of the total open time may not be any more accurate than is possible by the simple photographic methods described above.

One technique is, in fact, a refined version of the method described. A light source is modulated mechanically, by a sector disc or by reflection by the mirror of a vibration galvanometer. The light trace is drawn out on the film by some means, and the analysis is carried out in a similar way to that already described.

Another method which is particularly useful for the study of the motion of the shutter blades in inter-lens shutters, makes use of an arrangement which is virtually a high-speed cine camera to photograph the shutter aperture. The transmission of the shutter can be found by measuring the open area in each picture. A graph of shutter opening against time can then

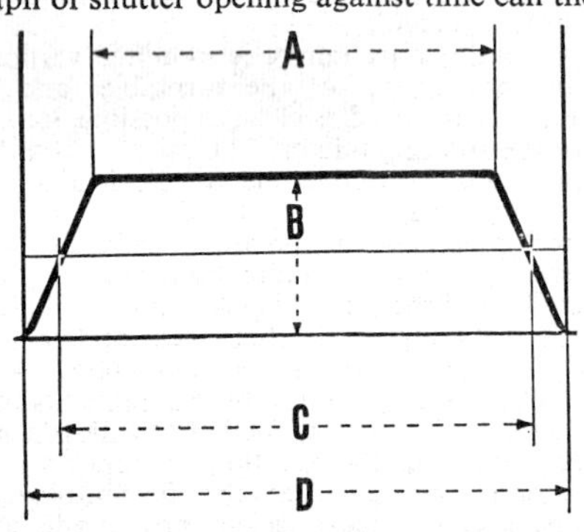

TYPICAL DIAPHRAGM SHUTTER CURVE: A, time of full shutter opening. B, maximum shutter opening. C, effective exposure time set on shutter (corresponds to actual total amount of light passed by an ideal shutter during the effective exposure time). D, total exposure time, taking into account the opening and closing times. If the opening and closing phases of the shutter are equal in length, the effective exposure time can be taken as the distance from the half-open to the half-closed stage of the shutter at full aperture. In practice the effective exposure time is more reliably obtained by integrating the area underneath the curve or by determining the light transmission during the shutter opening photometrically.

be drawn, if the interval between successive pictures is known.

Non-Photographic Methods. For routine testing of shutters in the course of camera manufacture or maintenance, photographic methods are too slow and too expensive. So non-photographic procedures are almost invariably used. Chief of these is the photo-electric method, specified in various forms as a standard test method by standards institutions and associations.

In this kind of test, a broad light source is focused on a photocell by a lens system near the shutter under test. In the case of focal plane shutters, it is necessary in addition to place an opaque screen with a narrow slit near the shutter, in the position normally occupied by the plate. When all precautions have been taken to ensure that the arrangement is optically similar to that of the camera, the illumination on the photocell at any instant is proportional to the transmission of the shutter. The output from the photocell is amplified and applied to a cathode ray tube so as to cause a vertical deflexion of the spot. The horizontal deflexion is obtained from a time base, and a second simultaneous trace is usually provided, with a modulation at a known frequency, for accurate timing.

This method makes visible the transmission-time graph of the shutter, and makes it possible to diagnose instantly such troubles as bounce of the shutter blades after closing. For more detailed examination, and for record purposes, the trace may be photographed.

The method can be made very elegant, but brings additional troubles. For example, tests must be made of the linearity of the amplifier and cathode ray tube. Often it is convenient to have the horizontal deflexion of the spot initiated by the signal from the photocell, but by the time the spot begins to move horizontally the beginning of the upward deflexion may have been lost. This can be overcome by delaying the light signal applied to the vertical deflexion until the trace has begun to move horizontally, but once again this refinement may give rise to non-linearity.

Flash contacts can easily be timed by this method. A simple circuit can be made to brighten the trace when the contacts close, and the relation in time to the opening of the shutter is easily found by measurement.

Digital Readout. Normally the photo-electric testing set-up also uses a diffuser in front of the photo-cell to ensure that the latter takes in the total light passing through the shutter. Alternatively a small aperture of about 0·5 mm. diameter can be placed so that it is uncovered just when the shutter is half open, and covered again as the shutter half closes. With this arrangement the oscilloscope records a more or less square waveform which corresponds in its length to the effective exposure time. If the small opening is placed in the axis of the shutter it measures the total open time.

For focal plane shutters the opening is best a slit placed parallel to the shutter opening slit. The measuring slit can then be positioned in various parts of the picture area and measure varying exposure times over the image width.

Another way of obtaining a square wave form is to connect the photo-cell, reading the total light from the shutter through a diffuser, with a capacitor discharge circuit. With the shutter fully open, a neutral-density filter with a 49 per cent transmission is then placed in the light path. The capacitor circuit is adjusted so that it just fails to trigger the discharge. The neutral density filter is then replaced by one of 51 per cent transmission and the circuit adjusted so that it just triggers the capacitor. In this way the triggering circuit in effect reacts to a light transmission of 50 per cent.

When the shutter is now tested with a diffuser in front of a photo-cell, the triggering of the circuit indicates a light transmission of 50 per cent of the full shutter opening, in other words the moment when the shutter is half open.

During the shutter opening cycle the circuit thus produces two such signals at the half open and half closed stage. These signals again can be displayed on the oscilloscope screen as a square wave, the length of which corresponds to the effective exposure time.

Apart from giving a direct indication of the effective exposure time, such a square wave signal can also be used for digital readout systems. The beginning of the square wave impulse opens a gate circuit which connects an oscillator of for instance 100,000 cycles per second with a counting circuit. This counts the individual impulses from the oscillator during the period that the gate is opened—i.e., the effective exposure time. The counting system then displays the actual exposure time—in multiples of ten microseconds—as a digital readout. (According to the counting circuit used, the initial signal may be a binary number which is then reconverted into the usual decade system.)

Thus when the digital readout shows the figure 800, this corresponds to 8,000 microseconds, or 1/125 second. The precision of this system is plus or minus one step—in this case plus or minus 10 microseconds. For less exacting requirements the counting frequency may be lower, for instance 10,000 cycles per second in which case exposure times are measured to the nearest hundred microseconds (1/10,000 second).

Performance Requirements. Various international standards lay down the required accuracy and performance of camera shutters. The most important of these standards are the British Standard BS 1592: 1958, the German standard DIN 19015 of 1961 and the American standards ASA PH 3·3 and 3·5. They specify the shutter speed calibrations, the tolerances in effective exposure time, the efficiency, the consistency of readings and the delay times and contact times for flash synchronization.

The scale marked on the shutter should be the reciprocal of the effective exposure time in seconds, markings being chosen from the series 1, 2, 4, 8, 15, 30, 60, 125, 250, 500, 1,000, 2,000 etc. This applies to both diaphragm and focal plane shutters.

The tolerance in effective exposure time for diaphragm shutters is ± 20 per cent for exposure times longer than 1/125 second and ± 30 per cent for exposure times shorter than 1/125 second. At full aperture the shutter must have an efficiency of at least 60 per cent.

A focal plane shutter must be accurate within ± 25 per cent at exposure times longer than 1/250 second, or ± 33 per cent at exposure times shorter than 1/250 second. The efficiency with the lens at maximum aperture must be not less than 60 per cent at all exposure settings and on all points of the image plane. With focal plane shutters it is also specified that the maximum effective exposure and minimum effective exposure over the whole frame area should not exceed a ratio of 3·2. In addition errors in measurement of up to 5 per cent are tolerated.

From these limits it is possible to calculate the ratio of the total open time to the marked normal effective exposure time. Even with a shutter which satisfies standard requirements, the total open time may be up to twice as long as the effective exposure time in the worst case. It is easy therefore to over-estimate the power of a high-speed shutter to stop movement.

For flash synchronization with diaphragm shutters the firing contact for X synchronization should close within 1 millisecond of the shutter being 80 per cent open. (The German standard specifies the contact to close between the instants when the shutter is 85 per cent and 100 per cent open.) The firing contact should close for at least 2·5 milliseconds at shutter speeds slower than 1/125 second.

When set for M-synchronization the delay time between the closing of the firing contact and the shutter first being fully opened is laid down as 15 ± 2 milliseconds (16·5 ± 1·7 milliseconds in the German standard). For focal plane shutters the delay from the commencement of firing contact to the moment when the leading edge of the shutter blind begins to uncover the picture area should be 17 ± 2 milliseconds.

For all shutters the firing contact duration should be at least 2·5 milliseconds and the contact sufficiently robust to withstand a current of 2 amps. J.W.G.

See also: *Flash synchronization; Secondhand equipment; Shutter.*

SHUTTER, VARIABLE. Rotating shutter (in a cine camera) with adjustable opening. Normally the shutter has a 180° opening, in other words it lets light through to the film during one half of its revolution and obscures the light during the other half while the film is advanced in the camera. In a variable shutter this opening can be reduced to a smaller angle or even closed down altogether. Smaller shutter openings means shorter exposure of the individual frames (leading to sharper individual images). A variable shutter provides an additional means of exposure control in a cine camera, and can also be used for fading in and out.

In a projector designed for running at slow speeds a variable shutter may increase the number of interruptions of the light beam during projection, to eliminate flicker.

See also: *Cine camera.*

SILHOUETTE. Black-and-white picture consisting of the subject in solid black seen against a white background. Named after a French minister, Étienne de Silhouette, who practised portraiture with paper and scissors as a hobby. The effect is easy to reproduce photographically. Portraits with the subject seen in profile are the most popular subject for silhouette treatment.

There are several ways of making silhouettes photographically, but the same principle is common to all. The subject is posed some distance in front of a brightly illuminated surface and all other light is taken away. This shows the subject as a black shape on a white ground. The farther the subject can be from the surface the better, because all the illumination should come from behind—any rays falling on the edges of the subject weaken the silhouette effect.

For the same reason it is better to use a background just big enough to surround the subject and block out the remainder of the negative rather than to use a background that fills the whole picture area and throws unwanted light on to the sides of the subject.

The camera is focused sharply on the profile of the sitter and the shortest exposure given that will cause the background to print out pure white. There are three ways of providing an illuminated background:

(1) A thin white sheet may be stretched across the doorway of a room and illuminated from behind by one or two Photoflood bulbs. With this type of background an exposure of the order of 1/25 second at *f* 8 on fast panchromatic film would be about normal.

(2) A light wall surface—or a white sheet—may be lighted from the front by Photofloods screened to prevent any light from falling directly on the subject. Exposures with this method would be about four times as long as above.

(3) The Photoflood lighting in either of the

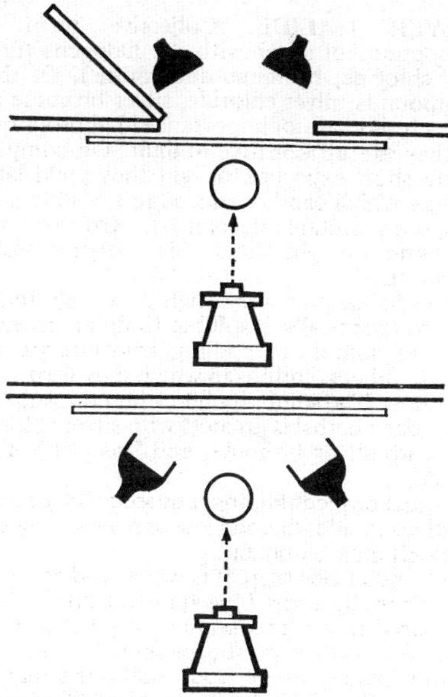

SILHOUETTE LIGHTING. *Top: Subject set up in front of a white sheet or other translucent screen stretched across a doorway. This is illuminated from the rear only, while the camera room remains dark. Bottom: Subject set up in front of a white background which is evenly illuminated by two lamps. The lamp reflectors carry snoots to avoid spilling any light on to the subject which would illuminate shadows.*

above methods can be replaced by flash. In this case the room lights are switched off after focusing the subject, and the exposure is made either by open flash technique or with synchronized flash.

Still life subjects may also be photographed as silhouettes provided that they have a suitable profile. The most usual method is to employ a special baseboard consisting of a deep wooden frame with a sheet of clear glass on top, a sheet of opal or ground glass set halfway down the depth of the frame, and four strip lights fitted at the bottom of the frame underneath the glass diffuser. The subject is placed on the clear glass, and the camera is

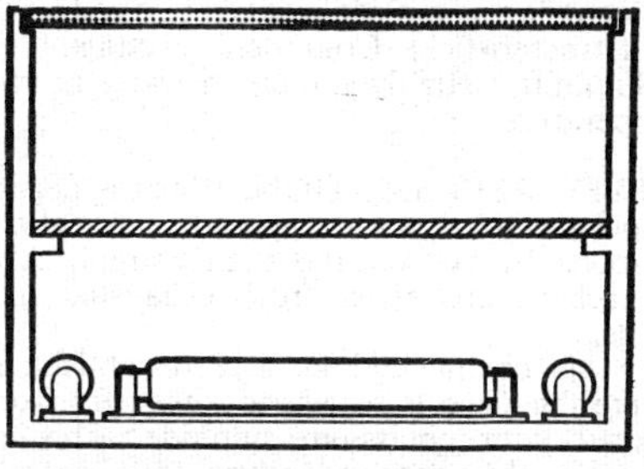

STILL-LIFE SILHOUETTES. *Background is opal glass illuminated by strip lights. Subject is set up in front of opal glass; foreground shapes can be stuck to plain glass front of box.*

mounted pointing vertically downwards above this. When much work has to be done, the set-up can be constructed on the lines of a vertical enlarger with a camera in place of the enlarger head used in filming animated cartoons.

This same equipment can be used to make silhouettes from cut-out figures. By using tissue paper for the figures, in multiple layers, it is possible to vary the tone of some objects in relation to others. For example, with a plain sheet of single tissue for the sky, single sheets cut out for trees and laid over the sky will appear a tone darker; foreground objects of extra thicknesses appear even darker.

An ordinary photograph can also be turned into a silhouette by painting the subject out in indian ink and then bleaching away the rest of the picture as in the bleach-out process for making sketches from photographs. In this case the photographic image should be soft and faint. F.P.

See also: *Abstract photography; Lighting the subject; Photograms; Tricks and effects.*

SILK PRINT. Silk and similar fabrics can be treated so that they can be printed with a photographic image.

The methods available are:

(1) Diazotype printing which leaves a dye image on the material in black or a colour.

(2) Iron-silver printing which gives a black silver image.

(3) Photo-mordant dye printing in which a silver image is first printed on the material. This forms the basis of the final dye image.

See also: *Fabric printing.*

SILK SCREEN PRINTING. Silk screen (retigraphy) is a remarkably versatile printing process by which almost any liquid medium can be printed upon any material with a flat surface or on cylindrical curves, such as bottles. Such media include dye and water colour, paint, printing ink, cellulose, fluorescent pigments, chlorinated rubber, etc., and even semi-conductors of electricity for electronics apparatus. The process is used for the commercial printing of everything from leaflets to posters, for printing on plastics and textiles, and for applying decoration, trade marks, and instructions to all manner of manufactured goods. Printing can be done by hand or by machine. Silk screen is also used extensively for the production of printed circuits for electronics in which lines of electrical conducting material are printed direct on to an insulating surface, or a resist is printed on a copper or other metallic layer and the unprinted areas subsequently etched away to leave only the circuit pattern.

The basic principle of the method is that silk, chiffon, or any other suitable woven material is stretched tightly across a frame and the parts required not to print are filled in with shellac, glue, gum, gelatin, cellulose, paper, or any other suitable substance. The screen is then laid upon whatever material is to receive the image and the "paint" (the general silk screen term for any printing medium) is squeegeed across the screen from the other side, forcing the paint through the open areas of the mesh of the screen on to the material beneath.

Hand-cut Stencils. Stencils can be hand cut out of paper but it is easier to cut them from the special stencil material consisting of a plasticized shellac coating on a translucent paper temporary support. The design is cut with a sharp knife through the shellac film only, and the areas required to print are peeled off. The material is then applied to the screen with a hot flatiron, after which the paper support is removed, leaving the shellac stencil firmly adhering to the screen.

Photo-stencils. The screen itself can be coated with a light-sensitive colloid such as dichromated glue, gum or gelatin, or a gelatino-bromide or chloride photographic emulsion. This is exposed under the positive and leaves a light-hardened stencil after the unaffected areas are washed away on development.

The most common form of photo-stencil is produced by a variation of the carbon process. The carbon tissue may be dichromate sensitized and printed while still wet; or it may be replaced by a special type of pigment paper ready sensitized and mounted on its temporary support.

Photographic stencils may also be made on a special type of permeable membrane which is rendered less permeable by the action of light and so can be used for printing continuous tone images in coloured dyes.

Serigraphy. There is also a fine art process, serigraphy, in which the artist draws directly on the screen using a separate screen for each colour. After filling in the screens with a suitable colloid, the drawn parts are washed-out so that they will print.

Copies of limited editions by this process sell at prices comparable with lithographs, engravings or etchings. F.H.S.

See also: *Photomechanical reproduction.*

SILVER. Used for the production of silver bromide, chloride, and iodide, the halogen salts used in the principal light sensitive photographic emulsions.

Formula and atomic weight: Ag; 107·88.

Characteristics: A lustrous white metal.

Solubility: Insoluble in water and acids, but forms amalgams with many other metals—e.g., mercury.

The photographic image formed in conventional negative and positive black-and-white materials is very finely divided metallic silver in the form of microscopic crystalline ribbons. The usual image colour is black, but may vary through shades of brown according to the fineness of the silver produced.

SILVER HALIDE. Collective term for compounds of silver with the halogens (fluorine, chlorine, bromine, and iodine). Of these compounds silver chloride, silver bromide and silver iodide are of importance in photography, as they are all sensitive to light. On comparatively short exposure to light they yield latent images which can be reduced to a visible silver image by suitable developers. On prolonged exposure to light the halides become visibly discoloured.

The halides are very slightly soluble (chloride) to practically insoluble (iodide) in water, but are soluble in sodium thiosulphate and certain other solutions in which they form complex salts. The solubility in sodium thiosulphate (i.e., fixing baths) is greatest with silver chloride, less with silver bromide, and least with silver iodide.

Emulsions containing appreciable amounts of silver iodide, therefore take a very long time to fix after development.

Silver chloride (AgCl) is white, and comparatively rapidly turns blue on exposure to light. It is used in the preparation of positive emulsions for contact printing (contact papers and lantern plates) where it is usually the main or only sensitive ingredient. Most of these emulsions are rather slow. Silver chloride is also the main sensitive salt in printing-out (daylight) papers.

Silver bromide (AgBr) is yellow and slowly darkens on exposure to light. It is the main light-sensitive salt in negative materials and enlarging papers (to which it gives the name, bromide paper) and yields comparatively fast emulsions.

Warm-tone printing materials contain both silver bromide and chloride.

Silver iodide (AgI) is pale yellow, and darkens only very slowly on exposure to light. It is used in small proportions as an addition to fast negative emulsions and certain specialized materials.

See also: *Sensitizer.*

SILVER NITRATE. Lunar caustic. Used in silver intensifiers, physical developers, and many sensitizers.

Formula and molecular weight: $AgNO_3$; 170.

Characteristics: Colourless crystals.

Solubility: Highly soluble in water at room temperature.

SILVER RECLAMATION. When a negative or print is fixed, the unexposed silver salts are dissolved by the fixer. It is a fairly simple matter to reclaim this silver from exhausted fixing baths.

In big commercial D. & P. establishments where the silver is recovered, the spent fixer is treated with potassium sulphide. This precipitates the dissolved silver salts as silver sulphide which is collected and sold to refiners, who extract the metallic silver by fusing the precipitate in the presence of a reducing agent. In this way it is possible to recover silver to the value of one per cent of the total value of the materials handled.

The silver in a fixing bath may also be recovered electrolytically. This produces very pure silver and is ideal when dealing with large volumes of solution. The principle involved is that when electrodes are immersed in the fixer and a current is passed through it, silver will be deposited. The electrodes are made of special materials such as carbon or silver, the actual choice depending on the composition of the bath.

Similar methods are also employed to regenerate exhausted fixing baths.

In practice, some firms will install silver recovery plant free of any charge (where the quantities used justify it), claiming as profit a proportion of the silver recovered.

See also: *Regeneration of fixing baths.*

SILVERWARE. What applies to photographing silverware applies to other highly polished surfaces like gold, chromium, brass or steel. Articles finished in this way are difficult to photograph.

Squat Subjects. Flat subjects, such as cutlery, trays, coins, etc., are the easiest to photograph. These are best illuminated by one low modelling light to the side or rear and another light directed on to a large sheet of white card behind the subject and leaning forward over it. Light is reflected from this card to illuminate the polished surfaces; the card is adjusted to reflect the maximum amount of light on the subject when viewed from the camera.

For flat objects the camera needs to be fairly high. The tone and texture of the background are chosen to suit the subject. The background is normally darker in tone so that the subject stands out from it.

Tall Subjects. Tall articles are more difficult as the sides reflect everything around them including the background. A kettle or teapot which is smooth and undecorated, for instance, will mirror everything in the room including the camera.

Straight-sided objects, like jugs or tankards, need to be placed in the centre of a large piece of plain white cardboard curved up behind to form a background. The card should extend at least 3 feet each side and to the front of the subject. This arrangement eliminates all reflections provided that the camera viewpoint is not too low. If the sides of the subject taper into the base they reflect a smaller area and call for a smaller surround. This applies to things like bowls or dishes. But if the sides converge towards the top, as might occur with a coffee-pot, the surround often has to be extended to a great distance behind the subject, and be curved upwards towards the edges to avoid reflections.

Round Subjects. With rounded objects it is impossible to avoid all reflections. Even if the object is completely surrounded with white card or sheet, the surface reflects joins or folds and the image of the lens always appears on the front of the subject.

In such cases the reflections can be greatly reduced by using suitable lighting. One almost frontal spotlight directed on to the subject produces a highlight which brings out the sheen of the metal and helps to destroy reflections of the camera. The rest of the illumination is indirect; it is directed on to the background or on to reflectors so that highlights are reflected back on to the subject where required.

Reflections and glare spots which persist even with the most careful lighting can be toned down by dabs from a ball of putty.

The number and position of the light sources vary with different types of subject. A second spotlight is sometimes directed on to the back-

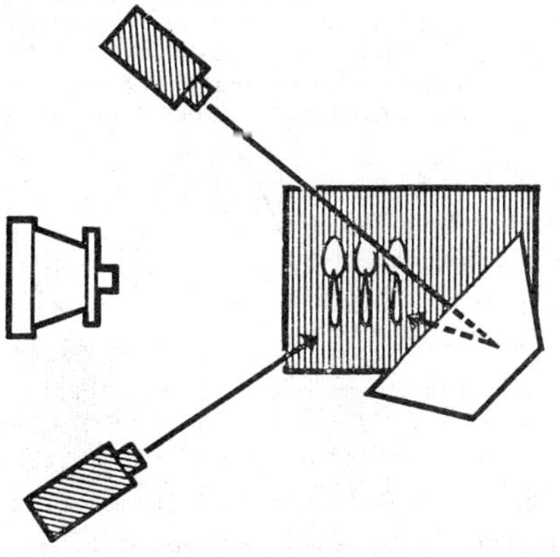

LIGHTING CUTLERY. One spotlamp directly illuminates the subject, while a second spot is reflected by a white card on the far side of the set-up to provide general lighting.

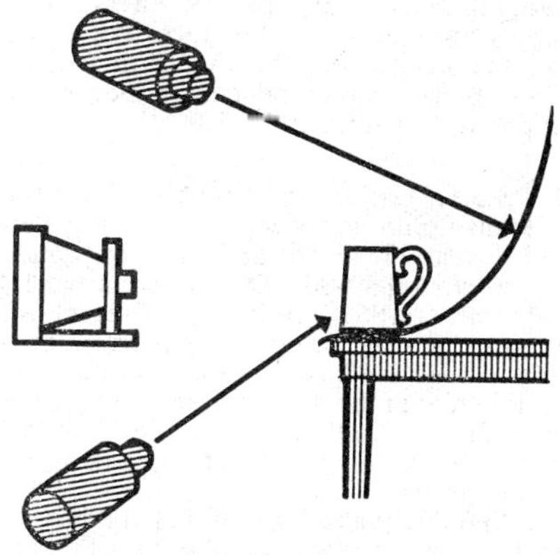

SOLID SILVERWARE. The background is a roll of white paper, lit up by one spot lamp. A second spot, coming from low down illuminates the subject directly.

ground immediately behind the subject to help to outline its shape so that it is not lost in the background.

Equipment. Panchromatic material is not strictly necessary for silver, but it usually gives better quality. It should always be used for coloured metals like gold, brass and copper.

Really high class work for catalogue and "glossy" magazine reproduction is almost always done with a large format plate camera using a long focus lens stopped right down to give the necessary depth of field.

As with most commercial photography, a lens with a long focal length is generally chosen to avoid close-up distortion.

Exposure and Processing. If the subject is being treated purely on its pictorial merits, then a soft, high key treatment will do it most justice. The exposure should be on the long side and the negative given the minimum development time in a soft-working developer.

If the interest is in ornamentation, chasing or engraving on the surface of the silverware, then lighting, exposure and development should tend to produce a fair amount of contrast—e.g., side lighting, minimum exposure and slightly extended development. These expedients need to be employed with great restraint or the subject will lose its characteristic appearance of delicacy and brilliance.

In practice, with this type of subject, the amateur has a chance to arrive at the desired result by trial and error—only a specialist or a professional worker can hope to make a perfect photograph first time. D.L.H.

See also: *Antiques; Ceramics; Coins; Glassware.*

SILVY, CAMILLE. Dates unknown. French diplomat, from 1859 professional portrait photographer in London. His cartes-de-visite of royalty and society earned him an international reputation as "the Winterhalter of photography". About 1868–9 Silvy sold his business and retired to his ancestral chateau. After the Franco-Prussian War, in which he was wounded, Silvy opened another portrait studio in the outskirts of Paris.

SIMPSON, GEORGE WHARTON, 1825–80. English photographer; editor of the *Photographic News* (1860–80) and the *Year Book of Photography* (1860–80). Did valuable work on development (1861), collodion silver chloride emulsions for printing-out papers (1864–5) which formed the basis for the manufacture of Celloidin paper by Obernetter, 1868; on the "photochromy" of collodion silver chloride papers (1866), and on Swan's pigment print (carbon) process.

SINE WAVE. Wave form of a shape based on the variation of the value of the sine of an angle with the angle itself.

The sine of an angle of 0° is 0, and rises to a maximum of 1 for an angle of 90°. It then drops again to 0 at 180° and to — 1 at 270°, returning to 0 at 360°.

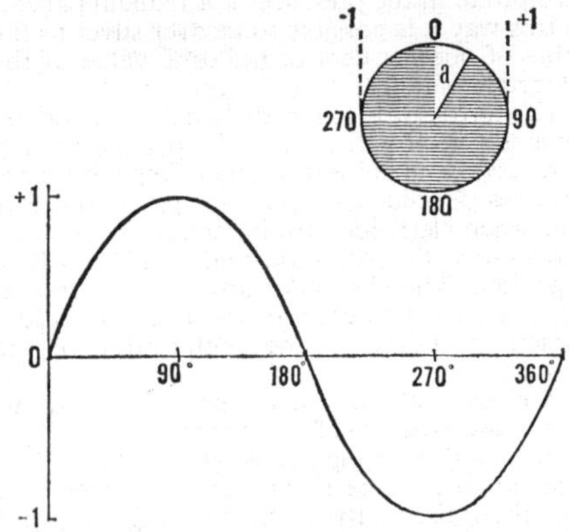

THE SINE WAVE. This is mathematically a plot of the sine of an angle against its size (for example, in the circle top right, the angle a). Since the sine value can never be more than + 1 and never less than — 1, the curve runs regularly up and down between these values.

This wave form is characteristic for many simple oscillating movements, for example, the movement of a simple pendulum, but also including the wave propagation of light and other radiation. It is significant for two reasons.

The first is that even complex wave forms can be analyzed and evaluated mathematically in terms of sine waves (Fourier analysis). This includes mathematical functions used in evaluating optical performance, such as the spread function of an image forming system.

Secondly, following from this, the use of test objects or targets (e.g. as a regular pattern of lines) with a brightness distribution based on a sine wave has been found particularly useful for testing image reproduction systems (including lens and sensitive material) to establish their transfer function. In such a test pattern the brightness varies continuously from 0 to a maximum and back to 0 again; this appears as a series of blurred bars rather than sharply

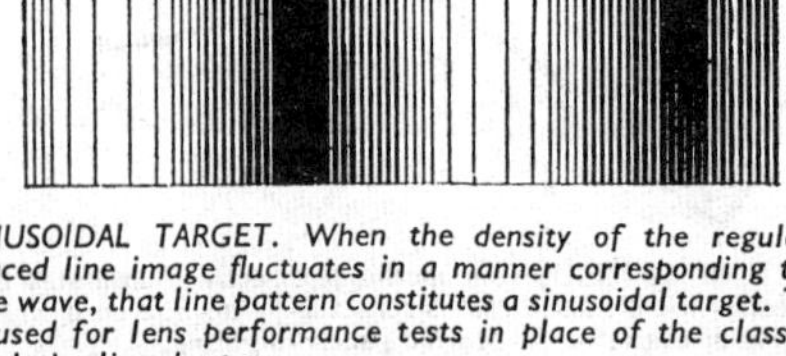

SINUSOIDAL TARGET. When the density of the regularly spaced line image fluctuates in a manner corresponding to a sine wave, that line pattern constitutes a sinusoidal target. This is used for lens performance tests in place of the classical resolution line charts.

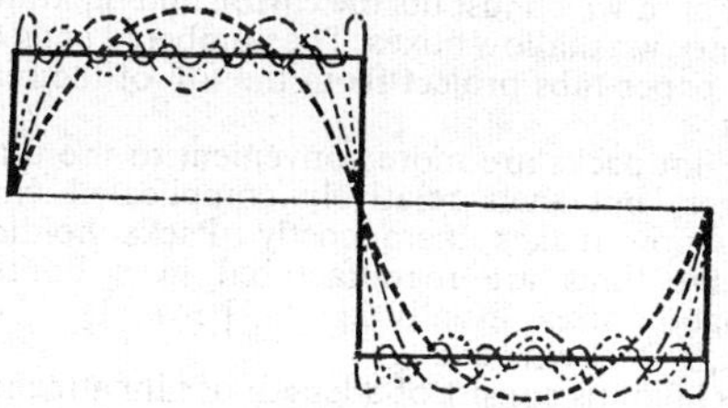

SINE WAVE AND REGULAR WAVE FUNCTIONS. All wave patterns—even a square wave (heavy solid line) can be considered as made up of sine waves of a fundamental frequency (heavy broken line) together with its associated harmonics. This involves a mathematical transformation known as Fourier analysis.

defined lines. Such a brightness distribution is known as sinusoidal.

When a sinusoidal test target is reproduced by a photographic recording, the brightness distribution of the latter remains sinusoidal irrespective of reproduction contrast. This can be measured by a scanning densitometer.

See also: *Information in the image.*

SINGLE-8 mm. Designation of 8 mm. cine film packed in single width for camera use—in contrast to double-8 mm. film. Also a Super-8 mm. type of cassette adopted by a group of manufacturers.

See also: *Cine camera; Cine film; Super-8 mm.*

SIZES AND PACKINGS

Sensitized materials for sale to the public must be packed in such a way that they will be protected from both physical and chemical injury in transit and store, and they must be supplied in certain standard sizes. The method of packing depends to some extent on climatic and other conditions under which the material is to be kept and used—e.g., most materials can be obtained in special tropical packing.

Plates. Plates are usually packed in boxes of twelve in a range of sizes based on either British or metric (for Continental cameras)

BRITISH PLATE SIZES

Ins.	*Nominal cm. Equivalent*	*Ins.*	*Nominal cm. Equivalent*
2½ × 3½	6·5 × 9	10 × 12	25·4 × 30·5
3¼ × 4¼ (¼-plate)	8·2 × 10·8	11 × 14	28 × 35·5
		12 × 15	30·5 × 38·1
3½ × 5½ (post card)	9 × 14	16 × 18	40·6 × 45·7
		16 × 20	40·6 × 50·8
4 × 5	10·2 × 12·7	20 × 24	50·8 × 61
4¾ × 6½ (½-plate)	12 × 16·5	20 × 30	50·8 × 76·2
		24 × 30	61 × 76·2
6½ × 8½ (1/1-plate)	16·5 × 21·6		
		24 × 36	61 × 91·4
8 × 10	20·3 × 25·4	30 × 40	76·2 × 101·6

Usually only photomechanical plates are made in the larger sizes (above 12 × 15 ins.).

units. The plates in each box of twelve are generally split up into three separate wrappings of four plates each, packed emulsion to emulsion. Thin strips of folded card (called separators) are inserted along two edges of each pair of plates to prevent the emulsions from rubbing or sticking.

NOMINAL METRIC PLATE SIZES

Centimetres	*Nominal In. Equivalent*	*Centimetres*	*Nominal In. Equivalent*
4·5 × 6	1¾ × 2⅜	18 × 24	7⅛ × 9½
6 × 9	2⅜ × 3½	24 × 30	9½ × 11¾
9 × 12	3½ × 4¾	30 × 40	11¾ × 15¾
10 × 15	4 × 6	40 × 50	15¾ × 19¾
13 × 18	5⅛ × 7⅛	50 × 60	19¾ × 23½

The actual cutting sizes are laid down in the British Standard No. 1406 of 1947. They are slightly smaller than the nominal sizes:

CUTTING SIZES OF PLATES

British Dimensions		*Metric Dimensions*	
2½ to 5½ ins.	1/32 in. smaller	4·5 to 15 cm.	0·1 cm. smaller
Above 6½ ins.	1/16 in. smaller	Above 15 cm.	0·15 cm. smaller

Film. Film negative material is available in the form of sheet film (cut, or flat film), film packs, roll film, and perforated 35 mm. miniature film. Sheet films are usually packed in boxes of 25 sheets (10 in very large sizes). Nominal sizes are laid down in the British Standard B.S. 1772: 1951:

BRITISH AND AMERICAN SHEET FILM SIZES

Ins.	*Nominal cm. Equivalent*	*Ins.*	*Nominal cm. Equivalent*
2¼ × 2¼	5·9 × 8·2	6½ × 8½	16·5 × 21·6
2½ × 3½	6·5 × 9	8 × 10	20·3 × 25·4
3¼ × 4¼	8·2 × 10·8	10 × 12	25·4 × 30·5
3½ × 5½	9 × 14	11 × 14	27·9 × 35·6
4 × 5	10·2 × 12·7	12 × 15	30·5 × 38·1
4¾ × 6½	12 × 16·5	16 × 20	40·6 × 50·8
5 × 7	12·7 × 17·8		

A number of odd sizes—usually panel-shaped—are also made for stereoscopic photography, spectrography, and other scientific fields.

NOMINAL METRIC SHEET FILM SIZES

Centimetres	*Nominal In. Equivalent*	*Centimetres*	*Nominal In. Equivalent*
4·5 × 6	1¾ × 2⅜	18 × 24	7⅛ × 9½
6·5 × 9	2½ × 3½	24 × 30	9½ × 11¾
9 × 12	3½ × 4¾	30 × 40	11¾ × 15¾
10 × 15	4 × 6	40 × 50	15¾ × 19¾
13 × 18	5⅛ × 7⅛	50 × 60	19¾ × 23½

Certain sensitized materials with specialized uses are made in a number of other sizes—e.g., for document photography, microfilming, stereoscopy, and photomechanical reproduction.

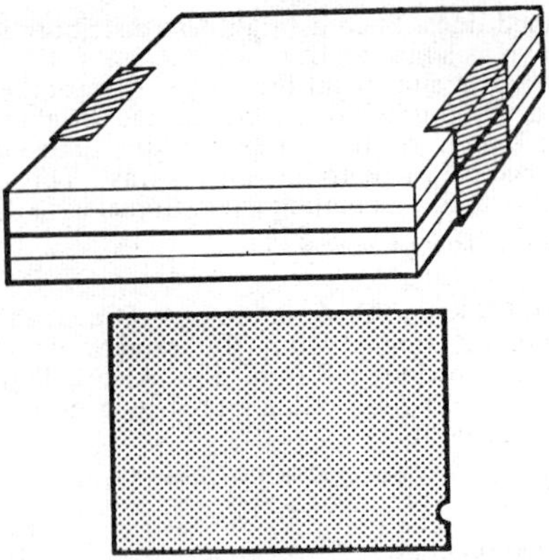

PACKING PLATES AND SHEET FILMS. Top: *Plates are packed back to back in pairs, with paper separators keeping emulsion faces apart.* Bottom: *Notching indicates position of emulsion of sheet films; as shown, emulsion faces observer.*

Sheet films can be loaded into ordinary plate holders by first fitting them into special metal sheaths to hold them flat and compensate for the difference in thickness between the film and the glass plate. Special sheet film holders are also made.

Most sheet films are notched so that the emulsion side can be identified in the dark. When the sheet is held so that the notch is at the top edge in the right-hand corner, the emulsion side is facing the observer. Sizes above $3\frac{1}{4} \times 4\frac{1}{4}$ ins. may also carry extra notches to indicate the type of film.

The notches may be triangular, square, round, etc. Their shape and number indicates the colour sensitivity and speed of the film. The notching codes vary from make to make.

Film packs contain twelve sheets of film in a stiff cardboard carton. Each sheet of film is attached to a backing paper which has a long tab. The films lie one on top of the other in the carton and the outermost film is covered by a blank sheet of paper, also carrying a tab. All the paper tabs are folded under the edge of a stiff card which just fits the carton and separates it into two shallow boxes. The numbered ends of the paper tabs project from the top of the rear box.

Film packs are more convenient to use than plates, but their relatively complicated construction makes them costly. Packs holding twelve films are manufactured in a limited range of sizes, usually $2\frac{1}{4} \times 3\frac{1}{4}$ ins., $3\frac{1}{4} \times 4\frac{1}{4}$ ins., 9×12 cm., etc.

Roll films consist of a length of film attached to a backing paper and wound on to a wood or metal spool.

The backing paper is generally black on the inside, and red, green or even yellow or brown on the outside. The leading end of the film is stuck to the inside (black side) of the backing paper, leaving enough spare paper to reach across the picture aperture to the wind-on spool upon loading in the camera.

The back of the backing paper is printed with numbers which indicate the successive frames of the film as they are brought opposite the picture aperture. The number of frames varies with the format of the picture area and most films carry three sets of numbers—1 to 8 for the normal oblong format; 1 to 16 for cameras taking a half-size picture, and 1 to 12 for cameras taking a square picture.

One row of numbers is printed along each edge of the backing paper and the third is printed down the centre. The numbers are read through a red or green window in the back of the camera, the position of the window being fixed to read the right, left, or centre row of numbers in accordance with the format of the picture area.

When the whole roll has been exposed, it is wound on to the take-up spool, and it can then be removed from the camera in daylight. The backing paper is longer than the film at the trailing end so that the surplus shields the end of the film from the light.

Certain types of roll film are available with only a short length of backing paper overlapping the beginning of the film and another length attached to the film end. In the middle the

ROLL FILM SIZES

B.S. Code	Other Codes	Film Width ins.	Film Width cm.	Exposures per Roll	Nominal Picture Size ins.	Nominal Picture Size cm.
0	888, 88, 828, "Bantam"	$1\frac{3}{8}$	3·5		$1\frac{1}{8} \times 1\frac{5}{8}$	2·8 × 4
				8	$1\frac{5}{8} \times 2\frac{1}{2}$	4 × 6·5
1	127, 27, A8, "V.P."	$1\frac{13}{16}$	4·5	12	$1\frac{5}{8} \times 1\frac{5}{8}$	4 × 4
				16	$1\frac{5}{8} \times 1\frac{1}{4}$	4 × 3
2	120, 20, B2			8	$2\frac{1}{4} \times 3\frac{1}{4}$	6 × 8·2
		$2\frac{7}{16}$		12	$2\frac{1}{4} \times 2\frac{1}{4}$	6 × 6
3	620, Z20, 62			16	$2\frac{1}{4} \times 1\frac{3}{4}$	6 × 4·5
4	116, 16	$2\frac{3}{4}$	7	8	$2\frac{1}{2} \times 4\frac{1}{4}$	6·5 × 11
5	616, Z16			16	$2\frac{1}{2} \times 2\frac{1}{8}$	6·5 × 5·5
*	10			6	$1\frac{1}{4} \times 1\frac{5}{8}$	3 × 4
*	29			6	2×3	5 × 7·5
*	18, 118			6	$3\frac{1}{4} \times 4\frac{1}{4}$	8·2 × 11
*	22, 122			6	$3\frac{1}{2} \times 5\frac{1}{2}$	9 × 14
*	30, 130			6	$2\frac{7}{8} \times 4\frac{7}{8}$	7·3 × 12·5

*Obsolete non-standard sizes; largely discontinued.

film is unprotected. As much less paper is spooled up there is room for more actual film on the spool—usually double the normal film length (e.g. 24 exposures 2¼ × 2¼ ins. instead of 12).

Such double-elength roll films can only be used in cameras with automatic film counting devices and without a red window; this would fog the unprotected film and can only be used to get the first exposure into place (where the film is still backed by paper).

Camerax with automatic counters must be suitable—or adapted—for counting the double number of exposures.

Aerial films are supplied in rolls 5¼, 5½ or 9½ ins. wide and in lengths from 8 to 200 feet. They are usually unperforated or perforated along one edge. The most common standard packings are 5½ ins. × 47 feet and 9½ ins. × 165 feet.

Metric aerial films come in widths of 80 and 190 mm., and length of 5 to 16 meters. Various special lengths and spool types are available to fit specific aerial cameras.

Aerial film is also supplied in perforated 70 mm. widths (see below).

Perforated miniature film is generally packed for use without a backing paper. The film is 35 mm. wide, and is coated on either cellulose nitrate or cellulose acetate base. It is perforated along both edges in the same way as professional motion picture film, with the perforation centres 5 mm. apart. The picture width available between sprocket holes is 24 mm. (approximately one inch).

This type of film is used in miniature cameras for three principal negative sizes: 24 × 36 mm. (just under 1 × 1½ ins.), 24 × 24 mm. (1 × 1 in.), and 24 × 18 mm. (1 × ¾ in.).

35 mm. miniature film is sold in several different packs:

(1) Daylight loading cassettes to give thirty-six exposures, 24 × 36 mm. Some extra film is always allowed for threading. Special daylight loading cassettes holding only twelve or twenty exposures are also made for individual makes of camera. 35 mm. colour film is usually supplied in daylight loading cassettes to give twenty exposures.

(2) Daylight loading spools holding thirty-six (or twelve or twenty) exposures as above, but with a paper leader and trailer for insertion into a daylight cassette.

(3) Darkroom loading refills for reloadable cassettes. These are simply lengths of film wrapped in black paper or rolled in a light-tight tube of black card. The length is usually about 65 ins. or 1·6 metres, which is sufficient for the normal thirty-six exposures in the popular miniature format (24 × 36 mm.) and carries a trimmed leader and trailer.

(4) Unspooled bulk lengths of 25, 50 or 100 feet, or of 5, 10, 17, 25 or 30 metres. For certain purposes bulk lengths are also supplied without perforations.

Perforated 70 mm. film is available for various aerial and also professional cameras in lengths of 30 and 60 meters (100 and 200 feet) and may be loaded into special cassettes for 70 mm. cameras or adaptors.

A variety of packings of 16, 35 and 70 mm. film—both perforated and unperforated—is available for microfilming, recording, fluorographic and other purposes.

Transparency Materials. Lantern plates are made in the following sizes. In each case the first figures give the British or metric units in which the standard is laid down, followed by the equivalent British or metric units in brackets.

LANTERN PLATE SIZES

Lantern Plate	*Sizes*	
Standard British:	3¼ × 3¼ ins.	(8·2 × 8·2 cm.)
Standard U.S.A.:	3¼ × 4 ins.	(8·2 × 10 cm.)
Standard Continental:	8·5 × 10 cm.	(3⅜ × 4 ins.)
Standard Stereo:	6 × 13 cm.	(2⅜ × 5⅛ ins.)
Miniature:	2 × 2 ins.	(5 × 5 cm.)
Miniature Stereo:	4·5 × 10·7 cm.	(1¾ × 4¼ ins.)

Transparency sheet film is made in most paper or sheet film sizes.

35 mm. film strip material is made in continuous lengths (up to 100 feet or more on suitable spools), and may be imperforated or

STANDARD PAPER SIZES AND PACKINGS

Nominal Size *ins.*	*cm.*	**Quantity Packings*
1⅞ × 2¾ (V.P.)	5 × 7	25, 100, 250
2½ × 2½	6·5 × 6·5	25, 100, 250
2½ × 3½ (Size 20)	6·5 × 9	25, 100, 250
2½ × 3¾	6·5 × 9·5†	100, 250
2¾ × 4½ (Size 16)	7 × 11·5	50, 100, 250
3 × 4⅛ (En-print)	7·5 × 10·5†	100, 250
3¼ × 4½ (En-print)	8·2 × 11·5	25, 100, 250
3½ × 4½ (¼-pl.)	9 × 11·5	25, 100, 250
3½ × 4¾	9 × 12†	25, 100, 250
3½ × 5	9 × 12·5	25, 100, 500
3½ × 5½ (P.C.)	9 × 14	25, 100, 250
4 × 5**	10 × 12·5	25, 100, 500
4 × 6	10 × 15	10, 50, 100, 250
4¼ × 5½**	11 × 14	25, 100, 500
4¾ × 6½ (½-pl.)	12 × 16·5	10, 50, 100, 250
5 × 7**	12·7 × 18	25, 100, 500
5⅛ × 7⅛	13 × 18†	10, 50, 100, 250
6 × 8	15·3 × 20·3	10, 50, 100, 250
6½ × 8½ (1/1-pl.)	16·5 × 21·6	10, 50, 100, 250
7⅛ × 9½	18 × 24†	10, 25, 100
8 × 10	20·3 × 25·4	10, 25, 100
8½ × 11**	21·6 × 27·9	10, 50, 250
9½ × 11¾	24 × 30†	10, 25, 100
10 × 12	25·4 × 30·5	10, 25, 100
11 × 14	27·9 × 36	10, 50, 250
11¾ × 15¾	30 × 40†	10, 25, 100
12 × 15	30·5 × 38	10, 25, 100
14 × 17**	36 × 43	10, 50, 125, 250
15 × 18	38 × 46	10, 25, 100
16 × 20	40 × 50	10, 25, 100
20 × 24	50 × 60	10, 25, 100

*Some types of paper (mainly contact paper up to post-card size) are also available in packets of 1,000 sheets. Most American papers are packed in boxes of 500 in place of 250 in Britain and 250 in place of 100. In certain cases, especially contact papers, quantity packets go up to 5,000.

†Standard Continental metric sizes.

**American size, not in use in Britain.

perforated along both edges with standard cine perforations or along one edge with special perforations.

Single perforated film is used for 28 × 40 mm. size (1⅛ × 1⅝ in.) pictures. The standard projector mask cuts this down to 26·5 × 38·1 mm. For camera use, this film is packed in 8 exposure rolls with backing paper.

Double perforated film is used for the following standard picture (frame) sizes in miniature film strips:

Double frame: 24 × 36 mm. (1 × 1½ in.); projector mask 23 × 34 mm.

Single frame: 18 × 24 mm. (¾ × 1 in.); projector mask 17·5 × 23 mm.

Square frame: 24 × 24 mm. (1 × 1 in.); projector mask 23 × 23 mm.

Double frame transparencies are often mounted singly between 2 × 2 ins. cover glasses or in special holders of the same size, and used as miniature slides.

Paper Sizes. Papers are usually cut to sheet sizes which correspond roughly to standard negative sizes. In addition there are several odd sizes available.

HALVABLE PAPER SIZES

16½ × 23½ *ins.* *ins.*	*Range* *cm.*	20 × 28 *ins.* *ins.*	*Range* *cm.*
16½ × 23½	42 × 59·4	20 × 28	50·8 × 71·2
11¾ × 16½	29·7 × 42	14 × 20	35·6 × 50·8
8¼ × 11¾	21 × 29·7	10 × 14	25·4 × 36
5⅞ × 8¼	14·8 × 21	7 × 10	18 × 25·4
4⅛ × 5⅞	10·5 × 14·8	5 × 7	12·7 × 18
3 × 4⅛	7·4 × 10·5	3½ × 5	9 × 12·7
2 × 3	5·2 × 7·4	2½ × 3½	6·5 × 9
		1¾ × 2½	4·5 × 6·5

Printing papers are sold in a standardized range of sizes and packings, which coincide broadly for the most popular sizes in various countries. Some non-standard sizes may be available in certain countries, as well as a few popular sizes not packed elsewhere.

The range of sizes offered nowadays is appreciably smaller than it used to be, when manufacturers produced many little used formats which made cutting and distribution comparatively uneconomical.

Another system proposed for rationalizing

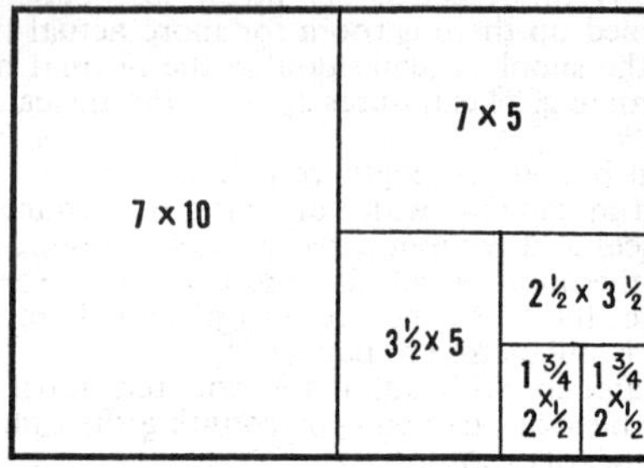

HALVABLE PAPER SIZES: If the sides of a paper format have the ratio of 1 : 1·4, halving the paper produces a smaller size with the same side ratio. This idea is the basis of the principle of halvable paper sizes. There a range of formats can be obtained by cutting down one standard large format—which then is the only one that has to be stocked. A basic format of 10 × 14 ins. as shown here has been proposed as one of the halvable sizes; usually a second basic format is also needed.

paper sizes and reducing the range is based on the principle of the halvable format. Around 1945 two such standard sizes were suggested: 10 × 14 and 12 × 17 ins. (25·4 × 36 cm. and 30·5 × 43 cm.). All other sizes could be derived from these by successive halving, and some were similar to existing standard paper sizes. For a time some manufacturers marketed these halvable formats. There is some more recent interest in this principle, using again the 10 × 14 in. range (based however on four times the area at 20 × 28 ins.) and 16½ × 23½ ins. (42 × 59·4 cm.) as the second range.

The latter series of sizes corresponding to the standard A range of stationery formats used in a number of countries on the Continent of Europe and elsewhere.

Though such a halvable range of sizes does not correspond to negative sizes at all, it has the advantage that the relative proportions, and thus the shape of a sheet of paper, is the same whatever the size.

The main opposition to halvable sizes comes from photographers who regard the sheet as too oblong in shape for pictorial presentation.

L.A.M.

See also: *Cassette; Film transport; Keeping qualities of materials; Negative materials; Papers.*

SKETCH PHOTOGRAPH. Term applied to photographs which have the appearance of a pencil sketch. In practice, this may be achieved in a number of ways. A very light print can be made and the tones built up by careful shading with a pencil; this is only suitable for subjects consisting of a few basic tones—e.g., a portrait taken with almost shadowless lighting. Or it is possible to make sketches by drawing over a faintly printed photographic image and then bleaching it out to leave only the drawn outlines. Another method is to project a negative in the enlarger on to a piece of white paper; the light areas are then shaded in with a pencil until the whole paper assumes an even tone; when the enlarger is switched off, a positive picture built up in pencil is left on the paper.

The term is also sometimes used to describe extreme high key effects in which the subject is softly lighted from the front against a light background. The contours of the subject then appear in very faint tones as though sketched with an artist's pencil. The same remarks apply to bromoil prints which with certain types of subject frequently look like charcoal drawings.

See also: *Bleach-out process; Highkey; Tricks and effects.*

SKIN AFFECTIONS. Photographers come in contact with a number of irritating chemicals in developing, fixing and printing. In many cases these chemicals can produce skin disorders on the hands and face.

Developers. Most developer solutions are alkaline and are the most common cause of dermatitis among photographers. The chief offenders are the bichromates used on blue prints which in either the dry or the dissolved state produce an irritating rash of small ulcers (chronic ulceration) on sensitive skins. Potassium dichromate and ammonium dichromate, which are used for sensitizing gelatin and also in blue printing, cause a greater percentage of dermatitis than any of the organic developers. Next to these, metol is the chief offender.

It is said that chemically pure metol is not a skin irritant and that dermatitis attributed to metol is really caused by traces of a poisonous compound (N.N.-dimethyl paraphenylenediamine) which can be eliminated if the metol is prepared by a special process. Metol poisoning affects some people and not others. It produces cracked, swollen fingertips and painful, spreading sores. Those who are susceptible to it should avoid it altogether. There are developing agents which are less likely to affect the skin and are as good as metol in other respects.

Amidol (diaminophenol) is made by reducing dinitrophenol with hydrochloric acid in the presence of iron filings, and dermatitis has resulted from the process. If the acid vapours are allowed to contact the skin it assumes a yellow appearance. Cyanosis or "blue lip" results from exposure to the vapours of dinitrophenol.

Diaminophenol used in developers can cause dermatitis in hypersensitive individuals. Hydroquinone is another developing agent which has caused dermatitis.

Quinone is a skin irritant to sensitive people and quinone dichloro diamine is said to be even more irritating.

Para-aminophenol is the basis of developers of the Rodinal class. It can be used as the hydrochloride or combined with sodium hydroxide; in both forms it can cause dermatitis.

The phenylenediamines, or diamino benzenes, are used as developers in the form of the hydrochloride, which can prove highly irritant.

Paraformaldehyde used in a hardener or photo developer has been the cause of many cases of dermatitis on the face and hands of photographers.

Other photographic developers which can cause dermatitis are: Adurol or monochlorhydroquinone, Ortol, or ortho-methylamino phenol, and pyrogallol or 1: 2: 3: trihydroxybenzene.

Sensitizers. Certain chemicals are used to increase the sensitivity of photographic plates to various regions of the spectrum. Some of these are said to cause dermatitis by photosensitizing the skin. Others of themselves cause dermatitis. They are used only in minute quantities in the photographic emulsions, but nevertheless, should be handled cautiously by persons with sensitive skins.

The principal sensitizing dyes are the cyanines and isocyanines, which are green sensitizing; the di-cyanines, which are red sensitizing; and the pinacyanols. The dyes used in photography which have caused dermatitis are auramine, aurantia, chrysoidine, crystal violet, methyl violet, malachite green and rhodamine B. The latter is a photosensitizer of the skin.

Other Chemicals. Other skin irritants used in photography are: ammonium bifluoride, used to remove gelatin films from glass negatives; ammonium sulphide, used in toning and in mercurial intensifications; caustic soda, used in alkaline developing solutions; formalin, used in gelatin mixtures and pastes; mercuric chloride and mercuric iodide, used in mercurial intensifications; oxalic acid, used as a preservative of pyro; potassium chloroplatinite, used for plate toning; potassium cyanide, used for bromide prints; potassium sulphide, used as a super toner; sodium carbonate, used in alkaline developing baths; sodium hypochlorite, used as a bleacher and oxidizer, and sodium thiosulpahte as a modern fixing agent.

Solvents, such as turpentine, benzene, etc., can also cause a dermatitis by denuding the skin of its natural lubricants.

Methods of Prevention. The ideal method is to keep the skin out of contact with the chemical solutions by using mechanical handling techniques. This, however, is not always practicable, especially where large numbers of prints have to be made within a limited period (as in press photography). The next best thing is to give the skin some form of protection. This can be provided in two ways: by the use of impermeable gloves or by applying a protective cream which can reinforce the skin to help it resist the action of chemicals.

Gloves have one or two disadvantages. They induce excessive perspiration which can soften and macerate the skin and render it highly vulnerable should the chemical accidentally come into contact with the fingers through penetration or spillage. Gloves also impede dexterity in handling plates and films.

On the other hand, a properly formulated barrier cream can protect and reinforce the skin against the penetration of acids and alkalis, and prevent the development of dermatitis. Creams of this type are made specially for photographers; they are simple to use and very efficient if applied correctly. About half a teaspoonful of the cream is smoothed into the skin of the hands and arms until it disappears. The hands are then held under cold water to set the cream, and then dried. If this technique is practised, there is no danger of contaminating films, plates or paper while processing, and the skin will remain perfectly healthy even if in constant contact with irritant solutions. An application of barrier cream lasts for three hours and the hands can be repeatedly washed without affecting the barrier protection.

These precautions may be elaborate for any one who only occasionally needs to touch processing solutions—e.g., the amateur who may in emergency want to use his fingers for lifting a print out of the developer. Such infrequent immersions can be rendered safe by dipping the fingers into clean water before putting them into the chemical. This ensures that any solution in actual contact with the skin will be too dilute to be dangerous. Immediately afterwards the fingers should again be thoroughly rinsed in clean water so that no traces of chemical will be left to dry on the skin. These precautions are in any event advisable for preventing marks and stains on prints caused by traces of chemical on the fingers. A.S.H.

See also: *Poisons.*

SKLADANOWSKY, MAX, 1863–1939. German showman and pioneer of cinematography. Constructed a crude projection apparatus for short loop films and was the first to show in Germany film projection (1st November 1895) on two projectors running simultaneously at 8 frames per second.

SKY FILTER. Filter in which the depth of tint increases progressively from top to bottom. When placed over the lens with the dense edge on top, it filters the light from the sky but allows the light from the foreground to pass through unaffected. In this way it holds back the highly actinic light from the sky and allows it to print in its natural tone instead of as a white featureless area of paper. The sky is also often very much brighter than the foreground anyway, so the filter also tends to prevent over-exposure of the sky.

The effect of a sky filter becomes less localized—e.g., on the sky—when it is used closer to the camera lens.

Sky filters are no longer as popular as they were some years ago, mainly because most photographers now use panchromatic films which, even without a filter, will give a reasonable rendering of the sky tones. When orthochromatic films were more generally used, they were so disproportionately sensitive to blue that the sky generally reproduced as blank paper unless corrected.

Another reason is that it is necessary to mount a sky filter well in front of the lens to confine the filtering to the sky area. This means that it must also be appreciably larger than an ordinary lens-mount fitting. The problem was not so serious with larger cameras, but the size of the filter and the awkward mounting rule sky filters out for modern cameras.

SKYSCAPE. Scene in which all or the greater part is taken up with the sky and clouds. Many excellent pictorial studies have the sky as their principal feature.

See also: *Clouds; Sunrise and sunset.*

SKY SHADE. Any form of shield attached to the lens mount or camera body for protecting the lens from the direct rays of the sun or strong, reflected light from the sky. Sometimes (particularly in the U.S.A.) used as another name for a lens hood, especially for the scoop-shaped type designed to shield the lens from above only.

The name is also used for the hinged flap fitted over the lens of the older types of camera and sometimes attached to the mount of a lens. As the hinge was at the top, the flap could be set at a suitable angle to shield the lens from sun and sky light.

SLAVE UNIT. Photo-electric relay used in multiple flash photography, which makes use of the light of the main flash to fire one or more auxiliary flashes. The slave unit consists of (1), a light-collecting lens or mirror which is pointed at the main flash and focuses the light on (2) a photo cell or photo-transistor connected to (3), an amplifier which closes the contacts of (4) an electro-magnetic relay.

The triggering lead of the auxiliary flash is connected to the relay contacts by plugging it in to a socket on the slave unit.

A slave unit may work off the mains supply or off a small dry battery and consumes a negligible amount of current. It will fire either bulbs or electronic flash and helps to keep the studio floor clear of trailing leads.

See also: *Flash equipment.*

SLIDE DIAGRAMS. There are three ways of making slides of line subjects—e.g., maps, engineers' drawings and documents: copying with a camera, contact printing, and drawing direct on to the slide.

Copying. Copying is particularly suitable for making slides of large diagrams which are too detailed to re-draw to a smaller scale for direct contact printing. It is the only satisfactory method for making miniature slides.

The subject is first copied to the size of the lantern plate to be used. For a 2 × 2 ins. slide it is necessary to confine the picture area to 1 × 1½ ins. or the area of a 35 mm. negative, if the printing is to be done by contact: 2¾ × 2¾ ins. or 3¼ × 3¼ ins. slides can also be used. The copy may also be made upon a larger size negative and a reduced image printed on the lantern plate by projection. (The enlarger must have sufficient extension for this purpose.) A 35 mm. camera with a copying attachment may be used for making the negatives.

The diagram should be evenly illuminated as in normal copying. The camera must be rigidly supported and focusing exact to ensure the sharpest possible image.

If the subject is black-and-white, the negative is made on a process plate which gives the necessary high contrast. If this type of negative material is not available, a slow fine-grain film is almost as good. For a coloured subject a

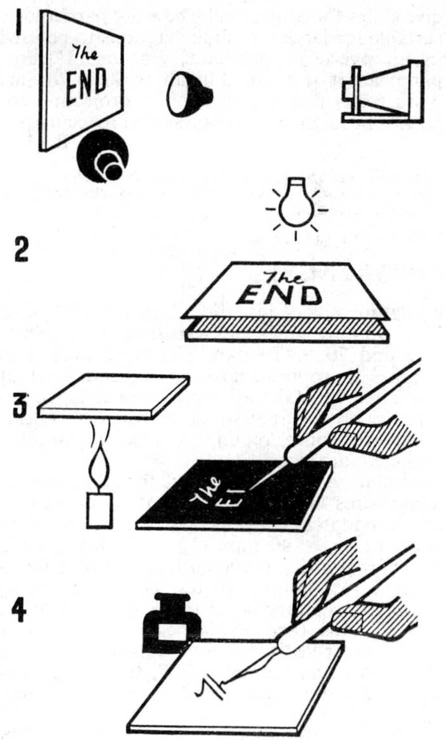

METHODS FOR DIAGRAM SLIDES. 1. Copying a drawing, title, etc., by means of a camera. The negative is then printed by contact or projection on a suitable lantern plate material. 2. Direct contact printing. The diagram or lettering is drawn on a sheet of thin grainless paper, and contact printed directly on to a lantern plate. 3. Drawing on blackened glass. A clear glass plate of the right size is blackened by soot and the diagram scratched on it. 4. Direct drawing. A special opaque ink for glass will yield a black image on a clean glass plate.

slow panchromatic material is used with the appropriate filter.

The exposure is best found by trial. A contrast developer should be used, or the normal time of development increased by 50 per cent. The negative should show all details clearly against an opaque background.

A slow black tone or gaslight lantern plate is best for the print, and a contrast developer is recommended. Plates should be backed to prevent degradation of the image by light scatter. Over-exposure must be avoided and development carried rather further than is necessary for ordinary lantern slides. After fixing, the slide should show the lines very black against a clear background. If the lines are slightly veiled, they can be cleared by a short immersion in a weak ferricyanide-hypo reducer for a few seconds. This will make the slide very clear and transparent.

If desired, the slides may be made on slow or positive 35 mm. film. Care must be taken to expose fully, and to develop for longer than may seem to be necessary if a normal developer is used, otherwise the positives may be thin.

In some cases the negative can be used for projection if there is no objection to showing white lines against a black background. Slight reduction may be necessary to clear the lines of the subject.

Another way is to photograph the original diagram directly on a reversal film to yield a positive transparency (after reversal processing) from the original positive drawing. Suitable reversal films are available as 35 mm. stock; for larger diagram slides many process films can be processed by reversal. Contrastly materials are best to yield a high image density in the black lines against a transparent background. Reversal colour films are suitable for making diagram slides in colour, though here the contrast obtained is more restricted.

Special diffusion transfer films also exist, for use in appropriate cameras, to yield an immediate positive transparency after exposure of the negative material in the camera.

Contact Printing. This method is satisfactory where the diagram is fairly bold and with no fine detail.

The drawing is copied by hand, or with a pantagraph, in black ink, on to a piece of white paper the size of the slide with as little grain or texture as possible. This drawing is then printed by contact on to an ordinary lantern plate which is processed to give plenty of contrast. The diagram when projected will appear as white lines on a black background. If the normal black on white presentation is preferred, the slide is printed again by contact on to a second plate.

This method is not suitable for miniature slides as it is not possible to draw the lines accurately on such a small scale.

Direct Drawing. In this case the diagram is drawn straight on to a specially prepared plain glass plate of the right size. There are several ways of preparing the plate, but the simplest is to blacken it by holding it over the flame of a candle.

The diagram is then scratched in the layer of soot with a needle. This method gives white lines on a black ground. It is the stock device for throwing an urgent announcement on to the screen in a cinema.

Any of the special opaque inks made for writing on glass may be used for drawing on a clear glass plate to give a black on white projected image. A suitable ink can be made by dissolving 1 drachm of borax in 1 ounce of water and stirring it slowly into a solution of 2 drachms of shellac in 1 ounce of methylated spirit. Any precipitate should be dissolved by heating the solution. Enough methylene blue is then added to make the ink reasonably opaque and the solution kept in a tightly stoppered bottle. It may be applied to a clear glass slide with a fine brush or a mapping pen. It is quick-drying and permanent.

Diagrams may also be drawn in indian ink with a pen on a sheet of finely ground glass. When the ink is dry, the surface is painted over with a solution of gum dammar in benzole. This obliterates the grain of the glass and leaves the diagram on a clear ground.

Some of the more modern plastic foils are also suitable for direct drawings of diagrams to give slides for projection. These are particularly suitable for large-size slide diagrams to be used in an overhead projector. For small transparencies it is more difficult to keep the art work really fine and neat, since projection enlarges the slightest roughness and inaccuracy.

R.M.F.

See also: *Inks for photographs; Lettering; Titles.*
Book: *Making Lantern Slides and Filmstrips*, by C. D. Milner (London).

SLIDE (TRANSPARENCY)

Positive transparency on glass support, or on film mounted in a card frame or between glass cover plates. Slides are usually viewed by projection and are invaluable to lecturers and demonstrators because the picture on the screen is visible to a considerable number of people, and can be both large and brilliant. It has distinct advantages over paper prints which can be seen by only a few people at a time: and over the cine picture which is not adapted for the detailed study of static subjects.

For these purposes, any negative that will make a good print will make an even better slide, and even any documents and maps photographed on an appropriate scale may be turned into satisfactory slides.

Many photographers, however, are attracted to slides because they can render a much greater tone range than any paper print. In fact, to reproduce the full beauty of subjects with a long contrast range, a slide may be essential.

Formerly slides were regarded mainly as a means of entertainment by projection in the then comparatively large projection lanterns. For that reason the transparencies projected in this way were commonly known as lantern slides.

The immense popularity of slides and slide projection nowadays is directly connected with the fact that many colour films yield their results in the form of miniature transparencies which are bound in appropriate mounts as slides. In many ways the slide projector has become the modern equivalent of the photo album.

Sizes. There are two commonly used international slide sizes with the outside dimensions of the slide laid down by standards:

Miniature—2 × 2 ins. (50·8 × 50·8 mm.);
Intermediate—70 × 70 mm. ($2\frac{3}{4} \times 2\frac{3}{4}$ ins.).

In addition several other large sizes and a sub-miniature or ultra-miniature size are in limited use:

British large: $3\frac{1}{4} \times 3\frac{1}{4}$ ins. (82·5 × 28·5 mm.);
American large: $3\frac{1}{4} \times 4$ ins. (82·5 × 101·5 mm.);
Continental large—85 × 100 mm. (about $3\frac{3}{8} \times 4$ ins.);
Ultra miniature—30 × 30 mm. (about $1\frac{1}{5} \times 1\frac{1}{5}$ ins.).

The majority of slide projectors take the miniature 2 × 2 ins. slide; larger models take 70 × 70 mm. slides and sometimes both miniature and 70 × 70 mm. The large slide sizes require appropriate large projectors: most of the latter will take all three large sizes, sometimes with alternative slide carriers. Ultra-miniature slides require a special projector designed for them.

Miniature slides, are used for a variety of image sizes. The largest format which can be accommodated conveniently within the slide is about 40 × 40 mm. ($1\frac{5}{8} \times 1\frac{5}{8}$ ins.), made either by printing monochrome negatives on to 2 × 2 ins. lantern plates, or by mounting colour transparencies up to $1\frac{5}{8} \times 1\frac{5}{8}$ ins. large in 2 × 2 ins. card or plastic frames. Slides of this size derived from film transparencies are also known as "super slides", by comparison with the more common 24 × 36 mm. transparency.

The 24 × 36 mm. (strictly speaking 23 × 35 mm. if the masking by the slide frame is taken into account) format is the most widely used for 2 × 2 ins. slides. This is the picture size produced by most 35 mm. miniature cameras, and the colour transparencies for use in such cameras are mounted in slide frames for projection. Other transparency formats, mounted in 2 × 2 ins. slides are the 18 × 24 mm. half-frame, 24 × 24 mm., 26 × 26 mm. and 28 × 40 mm. transparency taken by cameras for these sizes. Ultra-miniature transparencies, usually 10 × 14 or 12 × 17 mm. in size, may also be mounted in 2 × 2 ins. frames.

If a miniature projector is designed to illuminate the largest (i.e. 40 × 40 mm.) image area on a 2 × 2 ins. slide, it can of course also be used for the smaller transparencies—though this entails an appreciable loss of light and a smaller image on the screen. Projectors for transparencies smaller than 24 × 36 mm. are therefore usually available with a modified optical system and a shorter focus projection lens.

The Continental intermediate size of $2\frac{3}{4} \times 2\frac{3}{4}$ ins. is designed to take $2\frac{1}{4} \times 2\frac{1}{4}$ ins. or $1\frac{5}{8} \times 2\frac{1}{4}$ ins. colour transparencies made with roll film cameras. The $1\frac{5}{8} \times 2\frac{1}{4}$ ins. format can, however, also be cut down for mounting as super slides.

The large sizes are falling into disuse.

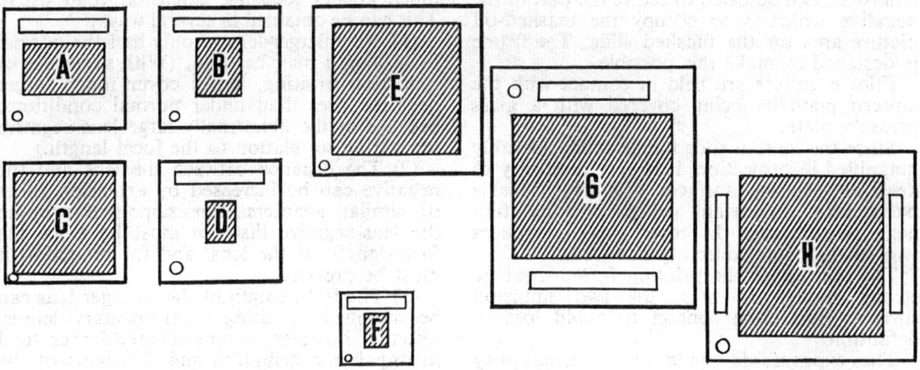

SLIDE SIZES. A, *normal miniature slide with* 24 × 36 *mm.* (*actually* 23 × 35 *mm.*) *transparency.* B, *miniature slide with half-frame transparency.* C, 2 × 2 *ins. super slide.* D, *ultra-miniature transparency in* 2 × 2 *ins. miniature slide.* E, *standard* 2¾ × 2¾ *ins. slide.* F, *ultra-miniature* 3 × 3 *cm. slide.* G, *American and continental* 3¼ × 4 *ins. slide.* H, *British standard* 3¼ × 3¼ *ins. slide*

MAKING MONOCHROME SLIDES

The main ways of producing monochrome slides are by exposure in the camera, printing on 35 mm. (or larger) film for subsequent mounting in transparency frames, and printing on plates.

Exposure in the Camera. Miniature transparencies for mounting as 2 × 2 ins. slides can be made in the camera on reversal black-and-white film. This is exposed in the same way as a normal reversal colour film. Usually the film is processed by the manufacturer, but can also be processed by the photographer.

Negative films can also be processed by reversal, but the resulting transparencies have the inherent base tint of the negative film stock (reversal materials are coated on a clear base). Reversal processing of negative materials is the only way of producing larger transparencies directly in the camera. The film base of roll films is usually clear, and the backing dye virtually disappears during processing.

Special transparency materials are also available for instant picture cameras and produce black-and-white positives ready for mounting in special standard frames, within about a minute of exposure.

Printing on 35 mm. film. Miniature transparencies can be printed on 35 mm. positive film from 35 mm. or larger negatives. Where 35 mm. negatives are used, a film strip printer is the most suitable equipment; larger negatives require projection printing on a reduced scale. Any normal enlarger is suitable for this purpose. It is best used with a short-focus enlarging lens to keep the required lens-to-negative distance within the focusing range of the enlarger. At this increased distance even a short focus lens will adequately cover the larger negative size, and gets over the problem that most enlargers do not provide sufficient lens extension for reduction printing with the normal focal length.

The printing technique is largely the same as for making film strips, though in this case there is no need for pre-planning the orientation or sequence of the images.

Printing on Plates and Sheet Film. Slides can be made by contact or by projection in an enlarger on to glass plates or film. Contact printing gives the best definition, and contact material—i.e., chloride emulsions—give high contrast. Plates for projection printing are either bromide, giving normal black-and-white images, or chloro-bromide, giving warm black and brown tones. Plates for printing slides, are still widely known as lantern plates.

Black-and-white material is well suited to slides for lectures, etc., illustration, especially of documents, line drawings and maps.

Pictorial work is best done on chloro-bromide material. With suitable developers this material gives warm black or brown tones direct, and the addition of potassium or ammonium bromide, etc., increases warmth.

Toning processes include gold chloride, which gives a fine blue, dye-coupled redevelopment, which gives a wide range of colours, and a number of classic metallic toners.

The most suitable sheet films for slide making are slow non-colour sensitive types, such as process, commercial and soft gradation emulsions. The latter are useful for very contrasty negatives, while the other two types give more contrasty images. In general a transparency for projection should be denser and more contrasty than a print for viewing by reflected light, since the projected image can reproduce a much greater brightness range.

Contact Printing. Lantern plates are printed by contact in a special frame designed to hold the plate and a plate negative. The two plates are placed in position, emulsion to emulsion, and adjusted to centre the part of the negative which is to occupy the masked-off picture area on the finished slide. The frame is designed to make this possible.

Film negatives are held in contact with the lantern plate by being covered with a glass pressure plate.

Since the lantern slide image is considerably magnified in projection, it is very necessary to keep all printing surfaces dust-free. This is particularly important when printing film negatives, because the cover glass introduces two further dust-collecting surfaces.

The springs of the printing frame must be strong enough to press the two emulsion surfaces into close contact to avoid loss of definition.

The exposure is made in the usual way either by a printing light or on the baseboard of an enlarger with the negative carrier removed. Some control of the final result can be exercised by shading or overprinting, as in normal contact printing technique.

Projection Printing. Lantern slides can also be made by projection in the enlarger. This method offers much more scope than contact printing and it is, in fact, the way that many people choose to make their slides, even when printing to the same scale. The principal advantage of this method is that the negative, or any selected part of it, can be enlarged or reduced to suit the size of the slide, and the photographer is free to trim and compose the picture to the best advantage, just as in making an enlargement.

Projection printing allows the photographer to make slides of any required size from a variety of types of negative—e.g., from plates, roll films and 35 mm. film. The technique is similar to enlarging except that the resulting "print" is smaller than usual and the enlarger must be equipped for making same-size and even reduced scale prints—e.g., miniature slides and film strips. This may call for some modification to the focusing range.

Focusing Range Adjustment. Enlargers that are to be used for reduction printing need much greater focusing extension than usual. This can be obtained in several ways:

(1) An enlarger lens of only half the normal focal length may be used. (With same-size or reduction printing, lenses cover much larger negative sizes than under normal conditions, because of the abnormally large lens-negative separation in relation to the focal length.)

(2) The distance between the lens and the negative can be increased by extension tubes or similar adapters. For same-size printing the lens-negative distance must be twice the focal length of the lens, and for reduction it must be greater.

(3) The focal length of the enlarger lens can be modified by using supplementary lenses. Usually, however, supplementary lenses tend to impair the definition and sharpness of the projected image, so they are not often used.

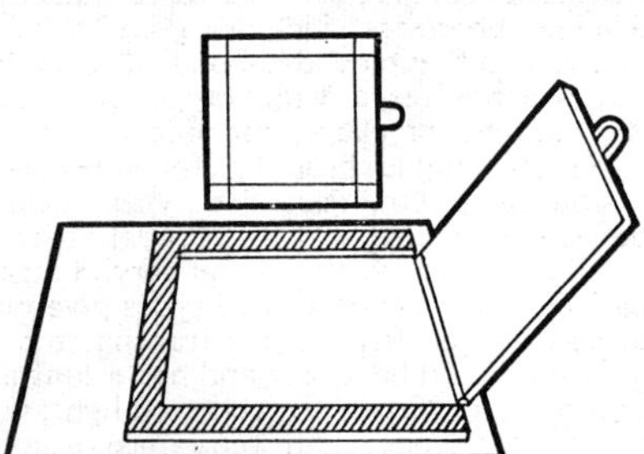

LANTERN PLATE FRAME. A simple jig or holder will locate a lantern plate always in the same place on the enlarger baseboard. A hinged white card (ruled to show loss of image space from binding) serves for focusing.

Plate Holders. Lantern slides are printed in a masking plate holder instead of a paper holder. The mask cuts the picture down to the standard size required for projection.

The lantern plate may also be laid on a sheet of black paper on the baseboard and printed without a mask.

In either case the image is focused on a special focusing plate covered with white paper giving a combined thickness equal to that of the lantern plate. This ensures that when the image is focused on the focusing plate, it will be sharp on the emulsion surface of the lantern plate.

There are various devices and methods for ensuring that the lantern plate occupies exactly the same position as the focusing plate on the enlarger easel. Some means of accurately registering the two is always necessary, because the slide cannot afterwards be trimmed like a paper print to correct errors. The simplest way is to fix the position of the two adjacent edges of the holder with push pins, or to mark it with white ink in the middle of a weighted black card, sliding the card about on the easel until the image occupies the required position.

Exposure. It is not easy to cut up lantern plates into test strips, and it is wasteful to use an entire plate for the purpose. A more practical way is to find a grade of bromide paper which matches the lantern plate in contrast. This is more practical even when the printing material is sheet film.

Testing can then be made with the paper to show the relation between the exposures needed to produce an acceptable bromide print and an acceptable lantern slide.

Once this relationship has been established, all exposure tests can be made on the bromide paper, and the correct exposure adjusted by multiplying or dividing by a fixed factor when printing the lantern plate. The factor must be checked each time a new batch or packet of either lantern plates or bromide paper is opened.

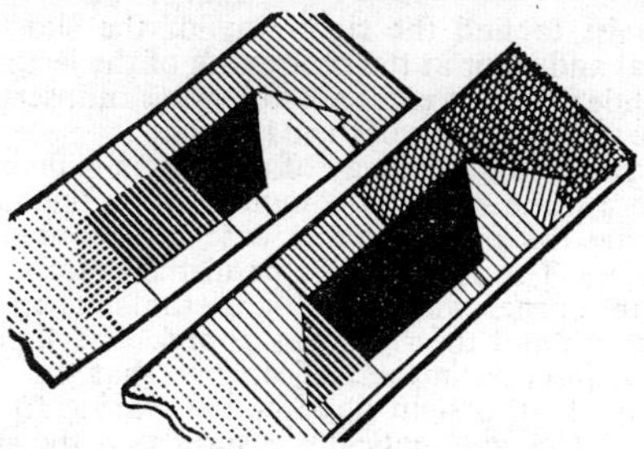

MATCHING TEST STRIPS *Test strips on lantern plates are inconvenient so they are best made on bromide paper. A relative speed factor established by a comparative test between paper and plate will then give the right exposure.*

Judging. There is an appreciable difference between the appearance of a correctly exposed lantern slide and a correctly exposed bromide print when they are seen in the developer or fixer by the darkroom safelight.

During development, while the image still contains a great deal of turbid white silver chloride or silver bromide, its density is best judged by looking at it through the back of the glass support. This gives a fairly good impression of the final visual density to be expected.

When all the white silver salts have been dissolved away in the fixing bath, the slide is best judged by holding it in front of a brightly-lit white surface—e.g., a sheet of white paper. This corresponds most nearly to its appearance when it is projected on to a screen. It is misleading to look at the image directly against the light, because this makes both density and contrast appear to be lower than they really are.

Processing. There is no difference in principle between processing lantern slides and processing prints, but there are certain differences in handling.

The same developer and other processing solutions are used for lantern slides as for prints (either normal or warm-tone).

It is not normally necessary to use a hardener in the acid fixer for prints, but it is advisable for lantern slides, because they are more sensitive to injury than paper prints.

To avoid blistering and reticulation to which lantern plates are more prone than prints, the temperature of the solutions should not vary between developer and rinse, or between rinse and fixer. Developers should not be warmer than 70° F. (21° C.); hot water-bath methods of development are out of the question.

Generally, lantern plates are handled in the same way as plate negatives. Some sort of plate carrier is, however, desirable to immerse the plates in the solutions and lift them out without getting the fingers wet. Washing and drying technique is again similar to that employed for ordinary plates. C.D.M. (L.A.M.)

Transparencies made on sheet film are also handled in much the same way, but using an appropriate film holder or hangar. As with plates, the emulsion is very sensitive to physical injury while wet.

Since sheet films are not normally made as special positive emulsions, they are less suitable for warm tone and other special development methods.

Finishing. Before a slide is made ready for projection, any pinholes caused by dust between the slide and the negative during exposure must be spotted out. If these spots are not dealt with, they will show with distracting brilliance on the screen.

Spots on slides are dealt with in much the same manner as spots on a print, except that the work has to be done even more skilfully or the result will be worse than the original fault. Opaque pigment, water colour, or one of the special spotting preparations, is used. The work is done on a retouching desk on which the slide is illuminated from below and masked so that no light shows around.

A fine-pointed No. 0 brush is used for spotting out pinholes, the pigment being matched to the colour and depth of the surrounding image; just enough is applied to fill in the pinhole. Some photographers use a fine mapping pen instead of a brush because it is easier to manipulate.

Masking. The slide, after retouching, is protected from mechanical damage and from damp by a cover glass, and if necessary it is masked to show only the selected part of the picture and exclude unwanted material. Cover glasses for miniature slides should not exceed ·045 in. (1 mm.) thick, or about 25 to the inch, or the slide may jam in the carrier.

Masking a lantern slide or transparency corresponds to trimming a print. The standard size masks cover only the edges of the slide, leaving the maximum area for projection. There are ready-cut masks of various shapes and sizes, and the holders made for 2 × 2 ins. slides are supplied complete with masks. The mask also cuts off unwanted light from around the slide when it is projected.

Slides may be masked with the same black adhesive binding paper used for binding up the slide, making one lot of paper do two jobs. This is a very effective way provided that the binding paper is stuck down accurately so that all the corners are rectangular. Masks may also be cut from any opaque paper, and it is possible to buy blanks printed with white guide lines for cutting out any required size. Masks of this type can easily be cut out on a sheet of glass using a steel rule and a razor blade in a holder.

Spotting. The slide is clearly marked so that whoever has to project it will be able to see at a glance (even in near-darkness) if he is putting it into the projector the right way around. Few things are more irritating to an audience than having to wait for the projec-

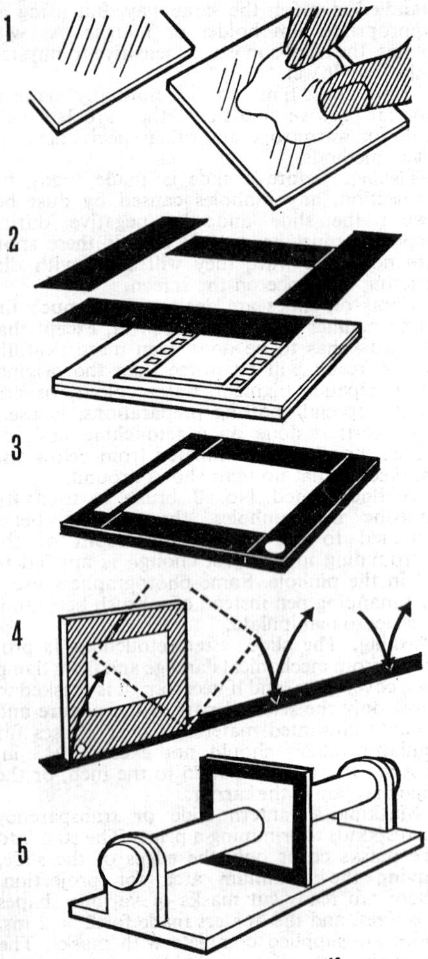

BINDING LANTERN SLIDES. 1. Thoroughly clean cover glasses. 2. Use crossed strips of binding paper to make a mask; (cutting a mask in one piece in 24 × 36 mm. size is not easy). Masks can also be bought ready made. 3. Place spot (in bottom left hand corner) and also title strips on mask. After making sure that all components are completely dry, assemble the cover glasses with the film and the mask. When viewed from the side carrying the spot and title strips the image should be upright and the right way round. With lantern plates no film is used, and only one cover glass becomes necessary. 4. Bind round edges with lantern slide binding tape of correct length. 5. A slide mounting jig makes this easier. 6. Fold tape down on each side, and cut corners for neat finish.

tionist to find the right way of the slide by trial and error at the prompting of the lecturer. While there are eight possible ways of inserting the slide, only one of them is right.

The commonest way of marking a slide is to put a white spot on the mount (or on the mask inside the mount in the case of glass-bound slides). This spot should be in the bottom left hand corner when looking at the slide image upright and the right way round. The slide is then inserted in the projector so that the spot is in the top right hand corner, facing to the rear. This automatically ensures that the slide appears upright and the right way round on the screen.

With slides made by printing on positive materials, the emulsion side of the transparency will face the lamp in the projector. On slides made by exposure in the camera the emulsion side faces the lens.

A titling strip can also be added above or below the mask opening, for noting details of the picture. Usually this is also placed on the same side of the transparency as the spot.

Binding. Finally, the lantern plate with its mask and cover glass are bound in register. **First the slide and the cover glass are warmed** to remove any trace of dampness. The slide must be completely dry before assembly or the heat of the projector will cause condensation. The inner surfaces of the slide and cover glass are then polished clean, wiped free from dust, and brought together ready for binding.

Where films are mounted between glass, slightly uneven contact between the film back and the glass surface may lead to Newton's rings. These are interference fringes and appear as very fine irregularly shaped parallel rings, rather like the coloured fringes on a thin layer of oil. Newton's rings frequently show up during projection, and can be avoided by using cover glasses with a faintly grained surface. The latter should face the film during binding. Such "anti-Newton" glasses are commercially available.

Binding strips are supplied either in 3¼ ins. lengths or in pieces long enough to bind all four edges without cutting. Avoid self-adhesive tape unless specially designed for binding slides. Usually such tapes "creep" especially when they get warm in the projector, causing packs of slides to stick together, while the sticky edge of the tape attracts dust and dirt.

When the 3¼ ins. strips are used, **the slide and its cover glass are held firmly together and bound along opposite edges. A strip is first moistened and then allowed to dry for a few seconds. It is then laid along one edge of the slide so that it forms an even margin, and then turned sharply over to the face of the cover glass and pressed well down. The opposite edge is next treated in the same way, and then the two remaining edges. The moist binding is finally ironed into contact by rubbing it with a knife handle.**

If a single length of binding strip is used, it is attached to one edge of the slide, mitred at the corner, and so on along all the edges, and overlapped at the point where it started. A special mounting jig eases the work of holding the assembly.

Titling. **A slide may be titled or numbered for reference by gumming a printed slip of paper to the mask so that it appears above the binding when the picture is viewed the right way up and from the emulsion side. In this way the title can still be read when the slide is mounted in an exhibition viewing cabinet.**

Filing. **Small numbers of slides may be filed in dovetailed boxes, but collections of any size are more conveniently kept in boxes with compartments taking about twenty slides, without dovetails. Coloured binding tapes, or titles in white ink on the top edges are both** used for identification. In some storage boxes the slides are held at an angle so that the top edges are exposed, allowing titles to be read easily. R.M.F. & C.D.M.

Slide magazines used in many modern still projectors are also suitable for storage. In this case a complete slide set is held ready in the magazine for projection whenever required. The magazines themselves may be stored in suitable cabinets.

COLOUR SLIDES

A number of manufacturers of reversal colour film automatically return processed transparencies ready mounted as 2 × 2 ins. slides. These are generally card frames into which the trimmed pieces of film, corresponding to a single frame, are inserted by a machine.

The card frames grip the edges of the film piece all round. This has the drawback that the film curves when the humidity of the air changes, and is liable to reverse its curl fairly abruptly during projection in the projector. This is known as "popping", and leads to a sudden going out of focus. Some processing laboratories therefore return the film in thin plastic frames which permit a certain amount of movement and so reduce popping. Usually films can also be returned, on request, in uncut strips of 5 or 6 frames for mounting by the user.

Making Colour Slides. While original colour transparencies are almost invariably made by exposure in the camera, duplicates can be produced in various ways. Processing laboratories frequently offer a duplicating service in which the transparencies are printed on to internegative film—a negative colour film specially designed for this purpose with a low image contrast. This negative is then used to produce further positives by contact or projection printing.

Colour negatives can also be printed on to positive film to yield transparencies. The procedure is very much the same as making colour prints, with colour control introduced by filters.

Duplicate colour slides can be made on reversal colour material. Where transparencies are intended for a duplication, they are best made on a special reversal duplicating film of contrast and colour characteristics more suited to duplication. Special printers are available for this purpose, but an ordinary miniature camera with appropriate close-up gear can also be used. Particular care is needed to ensure illumination of the original transparency with light of the correct colour quality.

Mounting Colour Slides. Colour transparencies in film form may be mounted between cover glasses in the same way as black-and-white miniature transparencies. In addition a number of alternative frames for mounting are available. Mostly these utilize glasses of the size of the transparency, let into plastic frames which then can be closed by snapping them together. This makes for fairly rapid mounting, and also permits transparencies to be changed in their frames if necessary.

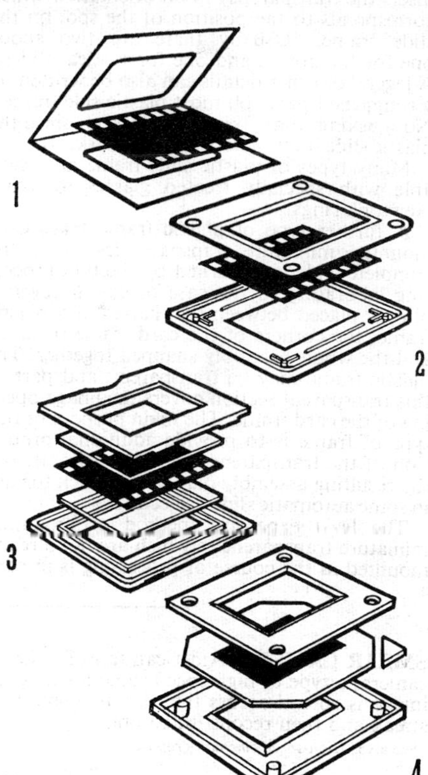

MINIATURE TRANSPARENCY FRAMES. 1. Card frame. 2. Glassless plastic frame. 3. Plastic frame with glasses. 4. Plastic frame for card mounted slides. The centre openings of the upper and lower panels are in clear plastic and are in contact with the film. The corners of the card frame must be trimmed before insertion in this frame.

To mount transparencies in these plastic frames, first thoroughly clean the glasses and make sure that they and the transparency are absolutely dry. Slight warming near a radiator (but not too near) helps.

Next insert the first glass into the plastic frame, place the transparency in position, and cover with the second glass. Finally fit in the second half of the frame and snap shut. The exact method of doing this depends on the design of the frame, and the packings of slide frames usually carry full instructions.

Some plastic slide frames have one side white and the other brown or black. Here the transparency should be inserted to appear the right way round when viewed from the light side of the frame. This side should face the light source in the projector, to reduce the extent to which the slide gets warm.

Plastic slide frames frequently already incorporate a spot for orientation, to make separate spotting unnecessary. In this case insert the transparency in an orientation which corresponds to the position of the spot on the slide frame. (Usually there are two spots, one for horizontal and one for vertical slides). A legend or other details can also be written on a roughened panel on most plastic side frames. No masking of any kind is necessary, since the plastic slide frame acts as its own mask.

Many types of plastic slide frame are available with specially treated glasses to avoid Newton's rings.

A further type of plastic frame takes card mounted miniature transparencies. Here the complete slide, as returned by the film processing laboratory in the case of colour reversal film, is placed between two halves of a plastic frame, the corners of the card frame trimmed and the whole assembly snapped together. The plastic frame may be transparent, and part of this transparent section covers the image opening of the card frame. The main purpose of this type of frame is to provide additional protection of the transparency in its card frame, but the resulting assembly may be too thick for use in some automatic slide projectors.

The least expensive method of mounting miniature transparencies which are not already mounted in the course of processing is in card frames. These consist of a piece of card 4 × 2 ins. in size, scored in the middle for folding and carrying two cut-out apertures of the appropriate size for the transparency. One side of the card is coated with rubber adhesive. The transparency is placed on to one half of the frame, with the adhesive side of the latter facing upwards. Guide lines or other marks show the exact position for the transparency.

As the aperture is slightly smaller than the image area on the transparency, some slight adjustment is possible in placing the piece of film, so as to correct tilts, reduce unwanted foreground areas, etc.

With the transparency in position, the other half of the card is then folded over and pressed down, when the rubber-coated surfaces make contacts and stick together.

The mounted slide then needs spotting and tilting in the usual way.

Suitability for Projectors. Modern miniature slide projectors with automatic slide change working with magazines will only operate satisfactorily if the thickness of a slide falls between certain limits. For this purpose slides should not be thinner than 0·05 in., nor thicker than about 0·125 in. The lower limit is represented by card-mounted slides returned by the processing laboratories of colour film manufacturers; the upper limit corresponds to most plastic frames and glass mounted slides. Thinner slide frames are liable to jam the changing mechanism because the frame may get lodged between the feeding arm and the gate itself. Too thick slides may not be able to move in and out of the magazine compartment freely.

Slides must also be mounted cleanly, with no projections of binding tape at the corners (in the case of glass mounted slides) which may catch in the slide track and foul the mechanism. Plastic slide frames usually have slightly rounded corners and are not likely to give any trouble. However, where the frame is of metal, sharp corners or insufficiently bent down closing tabs of the frame may lead to difficulties.

L.A.M.

See also: *Duplicate colour slide; Film strips; Projection practice; Slide diagram; Transparency.*

Book: *Making Lantern Slides and Filmstrips*, by C. D. Milner (London).

SMEAR CAMERA. American term for streak camera, a type of high-speed camera where the image is spread across the film as a smear or streak and then reconstituted optically.

See also: *High-speed cinematography.*

SMYTH, CHARLES PIAZZI, 1819–1900. Astronomer Royal for Scotland. Photographed by magnesium ribbon light the interior of the Great Pyramid in 1865, using a miniature camera with plates 1 × 3 ins. sensitized and developed in the camera. In 1856 made astronomical investigations from the Peak of Teneriffe and described them in a book published in 1858, the first book illustrated with actual stereo-photographs.

SMYTHE, FRANCIS SYDNEY (1900–1949). English Alpinist and photographer. Chiefly known for his Himalayan explorations on Kanchenjunga, Kamet and Everest, 1933–36–38. Wrote many books about mountains using his own photographs as illustrations.

SNAPSHOOTER. (Slang.) Sometimes applied by non-photographers to photographers in general; otherwise used by "serious" photographers to describe those whose approach is casual and limited to taking instantaneous hand-held exposures without observing the refinements of composition and technique.

SNAPSHOT. Term introduced by Sir John Herschel in 1860 to indicate an instantaneous exposure; nowadays it refers more and more to a casual photograph taken in a box or other simple camera, usually set for fixed focus working.

SNAPSHOT PHOTOGRAPHY. In the early days of time exposures made on plates, only experts could take photographs. Fast films and cheap cameras have changed this state of things and put picture making machines in the hands of millions of people who need to know nothing about photography. The combination of a fast film and a cheap lens working at a small stop makes it possible to turn out a camera that needs no focusing, the simplest shutter, and no stop. All the owner of such a camera needs to know is how to load it with a film, point it at the subject, and press the release. The photographic dealer looks after everything else and will even load and unload the camera. In clear weather out of doors, this simple procedure will produce a print that is reasonably sharp and correctly exposed.

The result of all this has been to create a special type of photograph—the snapshot. The snapshot has no artistic pretensions; it sets out to be no more than a record of a holiday scene, a corner of the back garden, or some place of interest, in which one or more friends or relations of the photographer feature prominently. The idea behind most snapshots seems to be to serve as a souvenir of the people, place, or occasion—or all three. The intention is never to produce a "photographer's photograph".

A snapshot is invariably an instantaneous exposure and it is usually made without a thought for photographic technique. The result is that the snapshot too often falls short of even its own humble standards.

Flash Snaps. Even the cheapest cameras nowadays are equipped for taking flash pictures, and the scope of the snapshooter now includes indoor pictures taken in this way. Cameras in this class are as a rule not synchronized for shutter speeds faster than 1/25 second, and many of them simply have "open flash" contacts. This means that the subject must be still or only moving slowly, so that the choice of subjects narrows down to more or less posed groups, pets, and party festivities.

In many ways flash photographs taken by the average snapshooter have a better chance of being free from faults than those made normally out of doors. This is because there is less risk of camera shake or subject movement, the smaller lens apertures used result in greater depth of field, so focusing becomes much less critical, and the exposure charts supplied by makers of flash bulbs are more accurate than any similar information that can be given about exposure in daylight.

Common Faults. Large numbers of films bought by snapshooters every year used to be wasted, but manufacturers have succeeded in building into their simpler cameras safeguards against most of the common faults. Double exposures and blanks have been eliminated by interlocking film transport and shutter release; wrong exposure (practically confined to under-exposure) is largely prevented by either built-in exposure meters or some system of automatic control; mis-focusing is less likely with the small-format cameras and faster sensitized materials of today, and fog-creating light leaks have disappeared with better standards of design and manufacture.

However, certain traditional mistakes of the beginner still remain and the camera manufacturer is not likely to be able to do much about them. These faults are mainly the result of carelessness—in holding the camera, in using exposure aids, in focusing and in handling the film. The only person who can get rid of these faults is the camera user, and the slight amount of extra trouble involved in avoiding them is fully repaid by the tremendous improvement that follows. F.P.

See also: *Amateur photography; Box camera.*

SNELL'S LAW. Law of optics which states that when light passes from one transparent material to another and suffers refraction, if both rays lie in a plane perpendicular to the common surface of the materials, then whatever the angle of the incident beam, the sine of the angle always bears a constant ratio to the sine of the angle of the refracted beam. This ratio is the index of refraction from the first material to the second.

SNOOT. Shield, often tubular in shape, for lamp reflectors or spotlights. It fits over the lamp and directs the light more or less in one direction. At the same time it also serves to shield light sources in front of the camera position from shining into the camera lens.

SNOW. One of the principal difficulties of photographing a snow-covered landscape is to judge the exposure correctly.

Exposure. The conventional type of photoelectric exposure meter is almost certain to indicate too short an exposure. This is because it is influenced disproportionately by the relatively large expanse of light-toned snow. A snow scene is at its best under a bright blue sky, in direct sunshine, and in such conditions allowance must be made for exposure to the deep shadow areas.

A clear blue sky, especially at high altitudes, is not as brilliant as the snow-covered landscape, so that the procedure of tilting the meter as in landscape work cannot be applied here.

It is necessary to increase the exposure meter reading by at least 50 per cent to make sure of getting a certain amount of detail showing in the shadows. Unless the exposure is enough to bring out this, the resulting photograph will be mostly "soot and whitewash".

Lighting. A snowy landscape without sunshine is apt to make a flat and disappointing photograph because there is very little contrast and the scene tends to reproduce as a more or less uniform shade of grey. The most attractive snow pictures are those in which the sun is shining obliquely across the surface. Lighting of this type casts long, interesting shadows and emphasizes the crisp texture of the snow. Such ideal conditions are rare in the British Isles; they are found most easily in mountainous regions.

Filters. When the sky is included, a pale yellow filter used with either a panchromatic or orthochromatic film will give all the correction necessary.

Even when the sky does not appear in the picture, a yellow filter can still help, in that it darkens the shadows (which will be predominately blue), thus increasing the contrast.

Above about 5,000 feet, even a pale yellow filter tends to render the blue sky tone too dark, and a special colourless U.V. filter is used.

Snowflakes. Snowflakes themselves can be caught by the camera in the same way as raindrops—i.e., if they are in sunshine and the background is in shadow. As it is not possible to focus all the flakes sharply from the camera to infinity, the photograph is best taken from under a shelter so that there are no flakes falling near the camera.

Faking Snowflakes. A passable imitation of snowflakes can be added to a negative by spattering the reverse side with opaque pigment. An old toothbrush is loaded with pigment and the spattering effect is produced by rubbing the finger along the bristles.

The same effect can be applied to the print by covering the paper with a sheet of glass and spattering the top face of the glass. The image of the drops of pigment is blurred by being out of contact with the paper, and the result is a convincing substitute for real snow. P.J.

See also: ***Cold weather; Frost; Polar photography; Winter sports.***
Book: *Photography on Expeditions*, by D. H. O. John (London).

SOCIÉTÉ FRANÇAISE DE PHOTOGRAPHIE. Founded in Paris November 1854 under the presidency of Eugène Durieu and continued its activities until c. 1960, though greatly hampered by lack of premises since the end of World War II. Official organ of the Society was the *Bulletin*, which appeared monthly from January 1855 until 1892 (with certain breaks), fortnightly 1892–1909, and thereafter monthly.

SOCIÉTÉ HÉLIOGRAPHIQUE. The first photographic society in the world, established in Paris in January 1851 under the presidency of Baron Gros. It was discontinued some time prior to the foundation of the *Société Française de Photographie* in November 1854.

SOCIETY OF FILM AND TELEVISION ARTS, LTD. Non-profit-making company registered in 1958 with the object of maintaining the quality and standards of British film-making and television. Incorporates the British Film Academy (founded 1947) and the Guild of Television Producers and Directors (founded 1953).

Members are senior film-makers and production executives, and television producers, directors and senior creative technicians, elected by the Council. There is an Associate membership for other technicians in both media, and affiliateship is open to people interested in the work of the Society, who are not necessarily concerned professionally, but whose work brings them into contact with, film-making or television production. Patrons are companies and organizations which subscribe not less than one hundred pounds a year to the funds.

Weekly meetings are organized for members from September to June. At these meetings members see outstanding films, or engage in technical discussions on film and television subjects with prominent British and foreign technicians.

The Society publishes a quarterly Journal and is continuing the Academy's sponsorship of books concerned with the technique of film-making and television production. There is a library of books and other publications on both media, and a collection of British film scripts. It is hoped shortly to start a selective collection of British television scripts. Two sets of Awards are given annually. The British Film Academy Awards are five (bronzes) for the best films of the year and twelve (plaques) for the best technical contributions and acting performances. These Awards are presented at a Dinner in the Spring each year. The Guild of Television Producers and Directors Merit Awards are fourteen (bronzes) given for overall work during the year in drama, light entertainment, current events, etc., to producers and artists, and for script and design in television. These Awards are presented at the Television Ball held at the end of each year.

SOCIETY OF MOTION PICTURE AND TELEVISION ENGINEERS. American society for promoting the interests of all engaged in cinematography, television instrumentation and high speed photography. It aims to advance the theory and practice of engineering in these areas; to develop standards of equipment and methods; to set and maintain a professional standard among its members, and to publish useful technical information.

The Society was founded in 1916; it now has 6,500 members in 60 different countries. It holds local meetings throughout the year in different parts of the United States and Canada, and holds two annual conventions and equipment exhibitions. In addition to frequent reports on matters of interest, the Society publishes a monthly Journal.

Active membership is open to anyone over 25 years who possesses three years professional experience; Associate membership is open to anyone over 18 years and there are no age or experience requirements for student members. Sponsors are necessary in all cases, and annual dues vary according to the type of membership.

The Society employs a permanent executive, editorial and technical advisory staff.

SOCIETY OF PHOTOGRAPHIC SCIENTISTS AND ENGINEERS. This is a professional association in the United States devoted solely to the advancement of the application of science to photography and of photography to science. Its membership is made up of men who use photography to obtain data in science and industry, or who are engaged in the design, development or production of photographic equipment and materials.

The Society was formed in 1947 as a local group in Washington, D.C. It was reorganized and incorporated in 1948 as the Society of Photographic Engineers to admit to regular membership persons everywhere who are engaged in photographic engineering on a responsible, professional level. The Society adopted its present name in 1957. Associate membership is available to adults with a serious interest in the technical aspects of photography, and full-time university students are eligible for Student membership.

Business organizations who wish to aid the Society are invited to become Sustaining Members.

The Society has members throughout the United States and in countries on every continent. Although its members work in many fields, they are all engaged in pursuits where photography plays a vital part.

The formation of local chapters is encouraged wherever there are sufficient members and the Society now has about a dozen chapters, some of which being in Washington, D.C.; Fort Monmouth, N.J.; New York, N.Y.; Los Angeles, Calif.; Dayton, Ohio; Chicago, Ill.; and Boston, Mass. Chapter meetings are devoted to the presentation of papers and organized discussion. The exchange of experiences affords each member opportunities to learn the techniques of other fields and to utilize them in his own work. Each chapter is self-governing and plans its own activities.

A National Conference on Progress in Photographic Engineering is held annually. Members meet for several days, during which papers are delivered and new equipment displayed.

Each member of the Society receives two bi-monthly publications. The journal, *Photographic Science and Engineering* provides authoritative information about the theory and applications of photographic science, the engineering of photographic materials, optics, and illuminants, and photographic instrumentation and data reduction systems. The *S.P.S.E. News* describes current activities of the Society and its members, the industry, new products and applications.

A Conference on Progress in Photography is held annually. Members meet for several days during which papers on new findings in photographic science and engineering are presented, and new equipment and materials are exhibited by the major companies. The Society also regularly convenes a Symposium on a single topic of suitable importance. These meetings are usually held in the autumn and for several days addresses and discussions study an area of interest, such as rapid processing, photography of electronic displays, or processing of colour materials.

The Society carries on its engineering, business, and professional activities through committees which conduct their work outside the chapter meetings. For example, development of standards is aided by S.P.S.E. representatives serving on each of the several photographic committees of the American Standards Association. Selected S.P.S.E. members are delegates to the Inter-Society Colour Council, while others photographically record the orbits of the man-made satellites for an S.P.S.E. project. Other objectives of the S.P.S.E. include the encouragement of better means for education in photographic engineering; the maintenance of a high level of competence among its members, wider use of scientific thinking and methods; and proper recognition of photographic engineering as a profession.

SODIUM ACETATE. Used in gold toners.

Formula and molecular weight: NaO_2CCH_3; $3H_2O$; 136.

Characteristics: White crystals.

Solubility: Highly soluble in water at room temperature.

SODIUM AURICHLORIDE. Gold sodium chloride. Chemical commonly used in gold toning positives.

SODIUM BIBORATE. Chemical used in certain fine grain developers and for toning. Synonym for sodium borate, or borax.

SODIUM BICHROMATE. Sodium dichromate. Used in intensifiers, bleaching baths, and carbon, carbro, and other gelatin or gum-bichromate processes in place of potassium bichromate. The latter has the advantage of not being deliquescent, but is much less soluble.

Formula and molecular weight: $Na_2Cr_2O_7.2H_2O$; 298.

Characteristics: Deliquescent red crystals.

Solubility: Highly soluble in water.

SODIUM BISULPHATE. Sodium hydrogen sulphate. Used as an acidifying agent in acid fixing baths, especially in dry acid fixing powders where a mixture of sodium bisulphate and acetate acts like acetic acid.

Formula and molecular weight: $NaHSO_4.H_2O$; 138.

Characteristics: Small white crystals.

Solubility: Fairly soluble in water.

SODIUM BISULPHITE. May be used to replace potassium metabisulphite in stop baths and acid fixing baths, or as a preservative in developers. A fairly concentrated solution used to be sold as "bisulphite lye".

Formula and molecular weight; $NaHSO_3$; 104.

Characteristics: White salt, smells of sulphur dioxide.

Solubility: Fairly soluble in water.

SODIUM BISULPHITE LYE. Concentrated solution of sodium bisulphite (approximately 30 per cent), prepared by dissolving sodium sulphite in dilute sulphuric acid. Has the same uses as sodium bisulphite, but is nowadays rarely employed in this form.

SODIUM BORATE. Several borates of sodium are known, and are used as alkalis in developers. The main ones of photographic importance are sodium metaborate and sodium tetraborate (borax).

SODIUM BROMIDE. May be used to replace potassium bromide in bleachers or in developers. Six parts anhydrous sodium bromide are equivalent to 7 parts potassium bromide or 8 parts sodium bromide crystals.

Formula and molecular weight: NaBr; 103, and $NaBr.2H_2O$; 139.

Characteristics: Hygroscopic white crystals or white powder.

Solubility: Highly soluble in water at room temperature.

SODIUM CARBONATE. Soda; soda ash. Alkali in developers. Three parts anhydrous salt are approximately equivalent to 8 parts of crystals.

Formula and molecular weight: Na_2CO_3; 106, or $Na_2CO_3.10H_2O$; 286.

Characteristics: White powder or large colourless crystals. A monohydrate $Na_2CO_3.H_2O$; 124, also exists and is widely used in America.

Solubility: Freely soluble in water.

SODIUM CHLORIDE. Common salt. Used in some bleachers and reducers.

Formula and molecular weight: NaCl; 58·5.

Characteristics: White crystals.

Solubility: Freely soluble in water at room temperature.

SODIUM DIPHOSPHATE. Disodium phosphate. Chemical used as a weak alkali in some developers.

See also: *Sodium phosphate (dibasic).*

SODIUM GOLD CHLORIDE. Sodium aurichloride. Chemical commonly used in gold toning positives.

SODIUM HEXAMETAPHOSPHATE. Calgon. Water softener.

Formula and molecular weight: $(NaPO_3)_6$; 612.

Characteristics: White powder.

Solubility: Highly soluble in water at room temperature.

SODIUM HYDRATE. Caustic soda. Strong alkali. Commonly used in some developers for high activity and rapid processing.

See also: *Sodium hydroxide.*

SODIUM HYDROSULPHITE. Sodium hyposulphite. Used as blackener in reversal processing.

Formula and molecular weight: $Na_2S_2O_4.2H_2O$; 210.

Characteristics: White crystalline powder. Does not keep well; the solution in particular soon deteriorates.

Solubility: Freely soluble in water.

SODIUM HYDROXIDE. Caustic soda; sodium hydrate. Strong alkali in developers.

Formula and molecular weight: NaOH; 40.

Characteristics: Very hygroscopic sticks or pellets. A slightly less pure form is also sold as flakes. Solid and solution are corrosive.

Solubility: Highly soluble in water at room temperature. The solid generates heat as it dissolves.

SODIUM HYPOSULPHITE. Strictly speaking this is sodium hydrosulphite, but in photography the term is frequently applied to sodium thiosulphate.

SODIUM METABISULPHITE. Used in acid fixing baths and stop baths, and as a preservative in developers.

Formula and molecular weight: $Na_2S_2O_5$; 190.
Characteristics: White crystals.
Solubility: Fairly soluble in water.

SODIUM METABORATE. Alkali in developers. Kodalk is a proprietary alkali of this type.
Formula and molecular weight: $NaBO_2.2H_2O$; 102.
Characteristics: White crystalline powder.
Solubility: Freely soluble in water at room temperature.

SODIUM PHOSPHATE (DIBASIC). Disodium phosphate, sodium hydrogen phosphate; sodium diphosphate. Weak alkali in developers.
Formula and molecular weight: $Na_2HPO_4.12H_2O$; 358.
Characteristics: White crystals.
Solubility: Fairly soluble in water at room temperature.

SODIUM PHOSPHATE (TRIBASIC). Trisodium phosphate; sodium orthophosphate. Alkali in developers.
Formula and molecular weight: $Na_3PO_4.12H_2O$; 380.
Characteristics: White crystals.
Solubility: Fairly soluble in water at room temperature.

SODIUM POTASSIUM TARTRATE. Rochelle salt. Used in some toners and sensitizers; also as a developing agent in the kallitype process.
Formula and molecular weight: $NaK(CO_2CHOH)_2.4H_2O$; 282.
Characteristics: White powder.
Solubility: Freely soluble in water.

SODIUM SULPHATE. Glauber's salt. Used in hot weather processing solutions to reduce swelling of the gelatin.
Formula and molecular weight: $Na_2SO_4.10H_2O$; 322.
Characteristics: White crystals. Also exists in an anhydrous powder form, 7 parts of which are equivalent to 16 parts of the crystalline salt.
Solubility: Fairly soluble in water at room temperature.

SODIUM SULPHIDE. Used in sulphide toners.
Formula and molecular weight: $Na_2S.9H_2O$; 240.
Characteristics: Very hygroscopic colourless crystals. Keep in well stoppered bottles. Dilute solutions do not keep.
Solubility: Freely soluble in water at room temperature. The solutions smell of hydrogen sulphide (rotten eggs).

SODIUM SULPHITE. Preservative in developers.
Formula and molecular weight: $Na_2SO_3.7H_2O$; 252.
Characteristics: White hexagonal crystals. Also exists in the form of an anhydrous powder. One part of the latter is equivalent to 2 parts of crystals.
Solubility: Freely soluble in water at room temperature.

SODIUM THIOANTIMONATE. Schlippe's salt. Used in certain intensifiers and sepia (sulphide) toners to obtain a range of comparatively cold tones.
Formula and molecular weight: $Na_3SbS_4.9H_2O$; 481.
Characteristics: Yellowish crystals.
Solubility: Freely soluble in water at room temperature.

SODIUM THIOCYANATE. Sodium sulphocyanide, sodium rhodanate. Used as a solvent in fine-grain developers in place of potassium thiocyanate.
Formula and molecular weight: NaCNS; 83.
Characteristics: White hygroscopic crystals. Poisonous.
Solubility: Highly soluble in water.

SODIUM THIOSULPHATE. Sodium hyposulphite; hypo. Fixing agent; also used in Farmer's reducer.
Formula and molecular weight: $Na_2S_2O_3.5H_2O$; 248.
Characteristics: Colourless hexagonal crystals. Also exists in an anhydrous powdered form. Two parts of the latter are approximately equivalent to 3 parts of the crystalline salt.
Solubility: Highly soluble in water at room temperature.

SODIUM VAPOUR LAMP. Vapour discharge lamp giving an intensely yellow light which has some special advantages for subject lighting in photography—e.g., in scientific work. Also used for darkroom illumination when handling materials which are completely insensitive to yellow.

See also: *Discharge lamp; Monochromatic illumination.*

SOFT FOCUS. Photographers do not always want a critically sharp image. Some subjects look better when the definition is more or less diffused so that the eye is freed from the distracting influence of unimportant detail, surface texture and blemishes—e.g., such as are caused by scratches on the negative.

There are several ways of arriving at a soft focus image: by using a soft focus lens in the camera, by using a diffusion attachment on the camera or enlarger lens, or by printing through a diffusing screen. The image may be softened either in the camera or during printing in the enlarger.

Effect of Diffusion. Diffusion during enlarging differs in one respect from diffusion during exposure in the camera. In the camera diffusion spreads the light from the brighter parts of the image into the darker areas—i.e., the highlights of the negative spread into the shadows. When the diffused negative is printed, the bright parts of the print are surrounded by a kind of luminous halo.

When the diffusion takes place during enlarging, the effect is still to spread the bright parts of the projected image into the dark areas. But in this case the bright areas become the shadows in the developed print. So in the print the shadows spread into the highlights and reduce the brilliance of the picture.

Diffusion during printing also has the effect of subduing small scratches and other flaws, and playing down the grain of the negative when it would be otherwise noticeable.

Soft Focus in the Camera. One of the most effective ways of softening the image is to use a special soft focus lens, but this is only possible when the camera is equipped to take interchangeable lenses. Such a lens is deliberately undercorrected for spherical aberration, with the result that the image formed by the centre of the lens is sharp, while the marginal parts superimpose an unsharp image. The degree of unsharpness is controlled by the lens aperture; stopping the lens down reduces the diffusion.

An alternative method is to use a diffusion attachment. This is a disc made of either glass or clear plastic, engraved with a number of concentric circular lines. The depth and spacing of the lines govern the amount of diffusion.

Most diffusion attachments have a clear area in the centre. This allows a proportion of the light rays to pass without being diffused, and by progressively stopping down the lens from full aperture, the amount of diffusion can be controlled from a maximum to none at all.

Diffusion attachments are fitted with rim mounts and used in the same way as filters. They do not alter the normal exposure.

A diffusion attachment can be improvised from a piece of Cellophane mounted in front of the lens like a filter and having the centre cut away to about half the lens diameter.

While dirt and grease on the surface of the lens will also diffuse the image it would be a mistake to dull the lens deliberately to produce a soft focus effect.

Blur produced by movement or deliberate misfocusing is not a soft focus effect. These methods blur the entire image whereas true soft focus is created by superimposing a certain amount of blur on a perfectly sharp image.

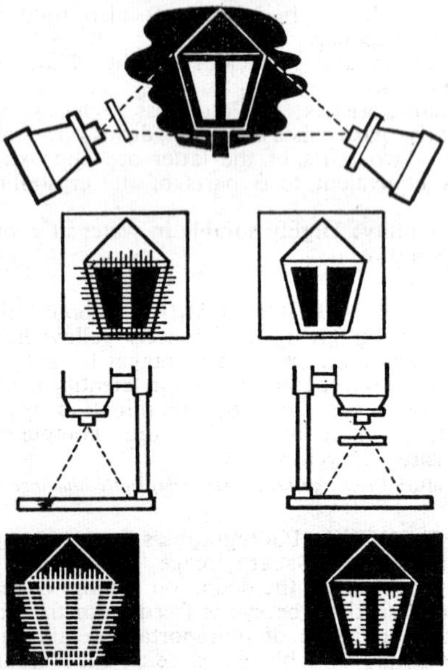

SOFT FOCUS. **Left:** *When the subject is photographed with a soft focus attachment in front of the camera lens (or with a soft focus lens), the light scatter spreads highlights on the negative into the shadows, giving a halo-like glow in the print.* **Right:** *Diffusion of a straight negative during enlarging gives a soft print with shadows spreading into the highlights.*

Soft Focus in the Enlarger. The sharpness of an enlargement is softened for pictorial effect by the use of diffusion attachments mounted in front of the lens or diffusion screens over the paper on the easel.

Any delicate, finely woven fabric—e.g., chiffon, nylon, or silk—can be used to diffuse the image either by being held in front of the lens, or stretched on a frame above the paper. The effect varies according to the weave of the fabric and its distance from the lens or paper.

A coarse fabric superimposes its own texture on the image, giving it the effect of having been printed on—e.g., linen or canvas—but it does not give a soft focus effect.

For a texture effect the fabric or screen should be in contact with, or about $\frac{1}{8}$ in. above, the paper. For diffusion to give a soft focus effect, the screen should be just in front of the lens.

The material used should preferably be black to reduce the risk of fogging the image by reflected rays. Transparent wrapping material, crumpled and then flattened out, can also be used to soften the image in much the same way.

With circular diffusion discs the degree of diffusion can usually be varied by stopping down the enlarger lens. The larger the lens aperture, the greater the diffusion.

Another way of controlling the diffusion is to expose the print for part of the time with, and part without, the diffusion attachment in front of the lens. L.A.M.

See also: *Portrait lens; Texture screens.*

Books: *All the Photo Tricks*, by E. Smith; *The Complete Art of Printing and Enlarging*, by O. R. Croy (both London).

SOFT FOCUS ATTACHMENT. Disc of translucent or transparent material which can be fitted in front of the camera lens like a colour filter to soften the definition of the image.

See also: *Soft focus.*

SOFT FOCUS LENS. Any lens which fails to record crisp pin-point definition can be termed a soft focus lens. Softness of focus is due to the fact that a single bright object point is imaged by the lens as a patch of light having finite size and a complex distribution of light intensity instead of the ideal point image. This softness is brought about by the various forms of lens aberration. In portrait lenses spherical aberration is deliberately introduced to soften the focus.

See also: *Portrait lens.*

SOLARIZATION. Strictly speaking, solarization is the reversal of the image on a film or plate by an extreme amount of over-exposure—about 1,000 times the amount required to give a normal negative image on development.

The term has by this time almost lost its original meaning. Nowadays it is applied—by all practical photographers at least—to the technique for producing a partly reversed image by exposing the negative to unsafe light during development—actually, the phenomenon known as the Sabattier effect. It was first done for artistic effect by Man Ray in 1930.

The effect commonly referred to as solarization is produced when a negative is exposed to too strong or unsafe darkroom illumination during development. The already developed image acts as a negative through which the rest of the silver bromide is exposed. Some reversal of the image occurs and the result is part negative and part positive. If the exposure is heavy enough the resulting positive will, at any rate in the shadows, develop up to a greater density than the original negative image.

Mackie Line. There is a further characteristic effect: a narrow band or rim of low density, called the Mackie Line, is formed at the boundaries between adjacent highlight and shadow areas. This happens partly because there is always an increased concentration of bromide ions in the emulsion at the boundary separating a completely developed area from one that is just developing. The bromide (partly produced from the silver bromide during development, and partly due to the potassium bromide from the developer) along such boundaries greatly retards development, forming a more or less clear rim. In the print the rim appears as a black outline around the principal image contours.

Method. While a solarized negative is often the result of an error, it can be produced intentionally for effect. Negatives with large areas of dark tone and simple outlines are most suitable for the process.

Solarization can equally be applied during printing. Further, the print from a solarized negative can be solarized once more, and there are various degrees to which the process can be carried.

For proper control of the process it is necessary to match the primary exposure of the negative (or print), the initial development, the second exposure that produces the solarization, and the final development.

The primary exposure must be adequate to yield a correctly exposed negative on normal development. If development is interrupted halfway through the process, the image already formed should be adequate to have a reasonable covering effect on the unexposed silver salts in the emulsion, but not so weak as to cause the solarizing exposure to swamp all the initial image formed.

For the same reason the first development must be adequate to form an image which is clearly visible on the surface of the film or paper. Excessive first exposure and first development brings up too much of the image and leaves not enough silver salts to be fogged by the solarizing exposure; hence the latter will not lead to a vigorous continuation of the development process and fail to produce the border outlines.

Once the image is of the right depth the emulsion is brought close to a fairly bright safelight and exposed for a few seconds. The safelight should be definitely unsafe with the material in use. Alternatively, a brief white light exposure can be given.

Development is then continued until the desired appearance is achieved.

The timing of the solarization exposure is important. The thinner half tones gain additional density after the exposure and if the solarization exposure is given too soon, the half tones will be very dense and have poor tone gradation. If the exposure is given too late, the solarization effect will be slight. The solarizing exposure itself must also be sufficient. Too little leads to a flat and fogged positive (on the negative film) or negative on the positive print) or only a partially solarized image with weak border outlines. Excessive solarizing exposure leads to complete blackening of the whole emulsion with little trace of an image.

The duration of the final development is also important. Insufficient development yields a partially solarized image with no sharp outlines. Excessive development after the solarizing exposure kills the border outlines formed in the first place. This is because the retarding effect of the bromide concentration is only temporary. If the final development goes on long enough, the previously underdeveloped outlines catch up with the rest of the image and disappear. It is clear from what has been said that solarization is by no means an exact technique; only the broad principles can be stated; the rest is a matter of individual experiment.

Results similar to solarization can be produced by extreme over-exposure of the negative in the first place followed by a brief exposure to an unsafe light before development.

True solarization often occurs when a very intense light source—e.g., an electric arc—appears in the picture and is grossly over-exposed. Because of solarization, the source reproduces as a black patch on the print instead of a clear white. L.A.M.

See also: *Eberhard effect; Tricks and effects.*
Books: *The Complete Art of Printing and Enlarging*, and *Design by Photography*, both by O. R. Croy (London).

SOLENOID. Electrical device capable of exerting a push or pull when energized with electric current. It consists of a spirally wound coil of insulated wire with a free, soft iron armature. When the armature is displaced from the centre of the coil and a current is passed through the wire the magnetic field created by the passage of the current tends to pull the armature back into the centre of the coil. This force is used to provide remote control of shutter operation, film transport, etc.

See also: *Flash synchronization; Shutter releases.*

SOLUBILITY. Generally understood as the ability of a solid to dissolve in a liquid—e.g., to mix homogeneously with it. Quantitatively the solubility of a substance is the amount of it that will dissolve in a given quantity of the liquid at a specified temperature.

The substance that is dissolved is the solute and the liquid that dissolves it is the solvent.

In photography the solutes are the various chemicals which are used for processing; the solvent is generally water. Some chemicals are very freely soluble (e.g., potassium thiocyanate, sodium hydroxide, and others), some are moderately soluble, and some are practically insoluble (e.g., the silver halides).

A substance may be very slightly soluble or completely insoluble in pure water and yet freely soluble in an aqueous solution of another substance. This is usually due to some chemical interaction between the insoluble substance and the chemical already in solution. For example silver bromide is practically insoluble in water, but dissolves in a solution of sodium thiosulphate because of the formation of complexions or double salts.

Solubilities are generally quoted as the number of grams of solute that will dissolve in 100 grams of solvent. Where the solvent is water the solubility may generally be taken as the number of grams of solute that will dissolve in 100 millilitres. A solution which will not dissolve any more of the solvent is said to be saturated.

Substances with a solubility of 50 grams or more are regarded as highly soluble, 20–50 grams as freely soluble, 5–20 grams as fairly soluble, and below 5 grams as slightly soluble.

Solubility and Temperature. Solubility also depends on the temperature of the solvent. Most substances are more soluble at high temperatures, some do not vary a great deal with temperature, and a few are more soluble at lower temperatures. With a substance that is more soluble at higher temperatures, cooling a saturated or nearly saturated solution causes some of the dissolved solute to reform as solid crystals.

Solubility of Liquids. Liquids which are soluble in each other are termed miscible. Such liquids may be completely miscible in all proportions (e.g., acetic acid and water, or alcohol and water), or they may be partly miscible (e.g., ether and water) or not miscible (e.g., petrol or oil and water). Partly miscible or non-miscible liquids will separate out in two layers if stirred together. Where the liquids are partly miscible each layer is a saturated solution of the other.

Solubility of Gases. Water also dissolves many gases. The solubility of a gas is generally highest at low temperatures, and the gases can be driven off by boiling. Thus tap water contains dissolved air; water for making developer stock solutions should therefore be boiled first to drive off the dissolved air which might otherwise oxidize the solution. A number of aqueous solutions of gases are used in photography, e.g., ammonia (which forms ammonium hydroxide), hydrogen chloride (which forms hydrochloric acid), and formaldehyde (which forms the formalin solutions used for hardening emulsions). L.A.M.

See also: *Chemical calculations; Chemicals; Solutions; Water.*

SOLUTIONS

Most photographic processing operations are carried out by solutions of chemicals in water. That is the only way in which most processing chemicals can act on the light-sensitive compounds in the emulsion of the film, plate, or paper.

Many of the solutions are sold in bottles (usually of concentrated stock solutions which require dilution). Alternatively, the photographer can make up his own solutions from packets or tablets containing pre-weighed amounts of the various ingredients in their correct proportion, or even weigh out the ingredients himself.

Tablets. Tablet chemicals contain the various ingredients ready weighed out and compressed into one or two tablets. They are intended for making working solutions as and when required, as the tablets are very easy to dissolve. The steps in the preparation are:

(1) Measure out the required amount of water in a beaker or dish.

(2) Drop in the tablet or tablets, and crush with a stirring rod. Stir up until all solid has dissolved. (Tablet developers contain the developing agents in one small tablet, and the alkali and sulphite in a larger tablet. The smaller tablet generally has to be dissolved first.)

(3) In the case of paper developers and other solutions, use the bath as it stands. In the case of negative developers it may be advisable to filter the solution.

Packet Developers. Developers sold in packets or tins may make either stock or working solutions. With packets the chemicals are usually contained in two plastic or paper bags, the smaller holding the developing agents and the larger the alkali, preservative, etc. The developing agents in developers packed in tins may be packed separately in glass tubes or contained in a separate compartment of the tin.

The steps in the preparation of the solution are:

(1) Boil enough water for the total amount of solution required, and let it cool down to about 120° F. (50° C.). This drives off most of the dissolved air and thus improves the keeping quality of stock developers.

(2) Pour about three-quarters of the final volume of water required into a glass beaker, or a clean dish (glass, porcelain, glazed earthenware, or enamelled or stainless steel, but not galvanized iron or other metal).

(3) If the water is hard, add the recommended amount of a water softener like sodium hexametaphosphate (Calgon). This is however, not essential.

(4) Pour the contents of the small packet of the developer pack into the water, and stir well until fully dissolved.

(5) Pour in the contents of the main packet, stirring all the time.

(6) Filter the solution into a clean bottle.

(7) Add enough of the boiled water (preferably cooled down) to make up the required final volume. It is best to choose a bottle of the correct capacity so that the solution fills it completely.

The solution can then be used at full strength or diluted as recommended. The bottle should always be correctly labelled; some manufacturers include printed, gummed labels in the pack for the purpose.

Other Packed Chemicals. Proceed in the same way as for developer packets, but note any special instructions regarding the order of mixing, and the temperature of the water—acid hardening fixers, for instance, should not be dissolved in water warmer than 70° F. (21° C.).

Where the chemicals are mixed in one package, it is often convenient to dissolve them directly in the stock solution bottle. In that case fill a bottle of the correct size three-

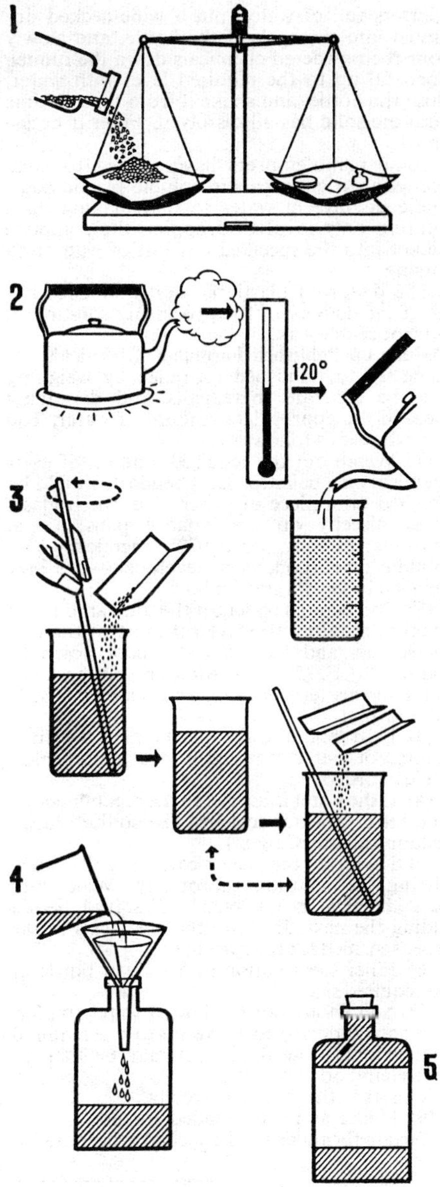

MAKING UP A SOLUTION. **1.** *Weigh out the required amount of each chemical on a balance.* **2.** *Boil enough water for the total amount of the solution to drive off most of dissolved air, then cool to about 120° F. and pour out three-quarters of the final volume into a beaker.* **3.** *Dissolve the chemicals one by one in the order specified in the formula, waiting until each chemical is dissolved before adding the next.* **4.** *Filter the solution into a clean bottle of the right size to take just the final volume of solution.* **5.** *Add enough boiled and cooled water to make up to the final volume specified. Label the bottle.*

quarters full of water, put a wide-necked dry funnel into the neck of the bottle, and slowly pour the powdered chemicals down the funnel. Then fill up to the required level with water, close the bottle, and shake it from time to time until the solid has all dissolved. Filter if necessary.

Some manufacturers have started to pack pre-weighed chemicals in soluble plastic bags. These dissolve in water so that making up a solution only involves dropping the complete packet into the specified amount of water and stirring.

The dissolved plastic is inert and does not affect the activity of the processing solution in any noticeable way.

Making up Published Formulae. Chemicals are made up to published formulae by weighing out the separate ingredients and dissolving them in the appropriate amount of water. The procedure is as follows.

(1) Weigh out the required amount of each chemical on a balance. The chemicals should be weighed on a piece of paper or a watch glass, never directly on the balance pans. Large amounts of any particular chemical—e.g., sulphite or carbonate in developers—are best weighed in several smaller lots.

(2) Boil enough water for the total amount of solution required to drive off most of the dissolved air, and let the water cool down to about 120° F. (50° C.). Alternatively heat the water to this temperature, and dissolve a little sulphite in it.

(3) Pour about three-quarters of the final volume of water required into a glass beaker or clean dish.

(4) If the water is hard, add the recommended amount of a water softener like sodium hexametaphosphate (Calgon).

(5) Dissolve the chemicals one by one, stirring the solution thoroughly. Wait until each chemical is completely dissolved before adding the next. Dissolve the chemicals in the order specified for the formula.

(6) Filter the solution into a clean bottle of the required size.

(7) Add enough of the boiled water (preferably cooled down) to make up to the required final volume. The final level may be marked beforehand on the bottle.

(8) Label the bottle correctly.

(9) Dilute as recommended for use.

Certain formulae need a special procedure for dissolving. Thus some chemicals may have to be dissolved separately and the solutions mixed in a given order.

In some cases cold water is necessary—e.g., with hardening fixers.

Sodium hydroxide should always be dissolved separately in a little cold water.

With solutions incorporating sulphuric acid use either dilute acid or mix the acid by pouring it into the water (never the other way round) with constant stirring.

All the chemicals should be weighed out before starting to dissolve any of them.

Ingredients used in small quantities need accurate weighing—e.g., developing agents must be weighed to the nearest 0·05 gram or to the nearest grain—bulk chemicals may be weighed more roughly. In some cases powders may be measured by volume in suitably calibrated measures. This applies especially to fixing powders and the like where high accuracy is not required since the chemical is used for a reaction that has to go to completion.

Percentage Solutions. Individual chemicals which keep well in solution may conveniently be stored in percentage solutions. This makes measuring out easier, since no weighing is involved apart from making up the original solution. Thus whole formulae—e.g., certain toners—can be made up in a series of percentage stock solutions of the individual ingredients.

Usually 10 per cent solutions are most practical, or 5 per cent solutions if the chemical is not sufficiently soluble at normal temperatures. Substances required in large amounts—e.g., sodium thiosulphate—may be stored in more concentrated solutions provided the solubility is high enough. The same applies to chemicals which deteriorate in dilute solution—e.g., sodium sulphide.

Percentage solutions are also convenient where very small quantities of a chemical are required. It is, for instance, easier to measure out 5 c.cm. of a 10 per cent potassium bromide solution for a developer than to weigh out accurately 0·5 gram. Where the amounts are very small, the solution may have to be more dilute still (e.g., 1 per cent). L.A.M.

See also: *Balances and scales; Bottles; Chemical calculations; Chemicals; Solubility; Stoppers.*

Books: ***Developing*, by C. I. Jacobson (London); *Photographic Chemicals and Solutions*, by J. Crabtree and G. Matthews (New York).**

SOOT-AND-WHITEWASH. (Slang.) Applied to the tones of an excessively contrasty print—e.g., one made from a contrasty negative when printed on a hard grade of paper.

SOUND CAMERA. Cine camera with built-in or connected provision for recording live sound effects or commentary. This may be done on the film actually exposed in the camera (usually carrying a magnetic sound stripe) or on a tape recorder linked with the camera.

Various systems exist for reproducing the sound in synchronism with the film during projection.

See also: *Cine camera; Projection practice; Sound recording; Sound synchronization.*

SOUND RECORDING

Nowadays all commercial motion pictures include sound either recorded at the time of taking the picture, or added afterwards. Even when there is no speech, the film is projected to the accompaniment of background music or the sounds normally associated with the scene being shown. Amateurs too have begun to use sound increasingly since the introduction of wire and tape recorders. A number of commercial recording studios will also add sound records to amateur films at a reasonable cost.

The principal recording systems are disc, optical, magnetic tape, and magnetic stripe. Of these, the magnetic systems are the best that the amateur can normally use most conveniently.

Disc Recording. Recordings on gramophone discs were used for the earliest sound films in both professional and amateur fields. For the amateur, the technical difficulty of sound-on-disc recording used to be an objection, but nowadays these recordings can always be made at a commercial recording studio. The system of using discs, however, is inflexible and little favoured. For accurate synchronization of sound and picture, mechanical or electrical coupling between turntable and projector is necessary to maintain a rigid speed relationship. Any lengths of film that are cut out because of damage produce loss of synchronization since the equivalent part of the sound record cannot be removed at the same time. The limited playing time of discs is less serious at 33⅓ r.p.m.

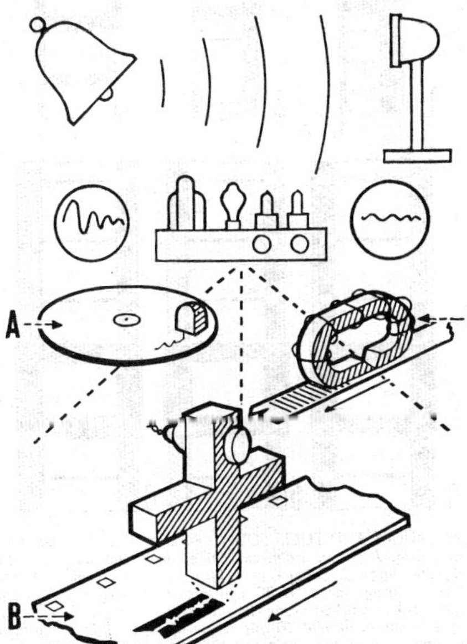

RECORDING SYSTEM. The microphone electrical pulsations, after amplification, may be recorded in various ways. A, mechanical recording on discs; the impulses energize vibrations in a cutting stylus. B, optical recording on film; the impulses modulate the strength of a light source which exposes the sound track. C, magnetic recording on tape; the impulses selectively magnetize an iron oxide coating.

Optical Recording. Optical sound recordings are widely used by both professionals and amateurs. In optical recording the sound is recorded photographically as a variable silver deposit along one edge of the film and so synchronization cannot be lost even when sections of the film are removed. Sound reproduction is effected by scanning the track with a narrow slit of light. Variations in the light transmitted by the silver deposit produce corresponding variations in the current delivered by a photocell on to which the light is directed. The current is subsequently amplified to operate a loudspeaker. Films are projected at a speed of 24 frames per second. Since the film moves intermittently through the picture gate, the sound track is scanned some distance farther on after all irregularities in the film motion have been absorbed.

Optical sound film projectors are available for all three narrow film gauges. The quality of 8 mm. optical sound is rather lower, owing to the reduced linear speed of the film through its projector. Recording sound optically is a highly technical process, but commercial studio facilities are available to users of 16 mm.

Magnetic Tape Recording. Magnetic recorders using wire, or preferably magnetically coated tape, provide the cheapest way of adding high-quality sound to amateur films of all gauges. The sounds in this system are recorded by an electro-magnet as varying degrees of magnetization along a tape wound at uniform speed from one reel to another by a capstan. By operating a switch, the recording machine can be made to play back the sound, thus enabling the record to be checked for volume and quality, or even erased and re-recorded.

The technique of magnetic recording is simple; recordings may be checked immediately without any processing and erased and replaced as often as necessary. As the tape is unperforated, however, it is extremely difficult to maintain a rigid speed relationship between tape and film. The accuracy attainable is good enough for music and commentaries, but not usually for synchronizing speech to the lip movement of characters on the screen.

Recorders operating from A.C. mains run at a speed determined primarily by the mains frequency. Approximate synchronization can therefore be assured by adjusting the projector speed to make the bars of a stroboscopic disc on the flywheel or sprocket appear stationary when illuminated by a neon lamp connected to the same mains supply, or by part of the projector beam itself.

Mechanical coupling between the tape recorder and the projector is a more reliable

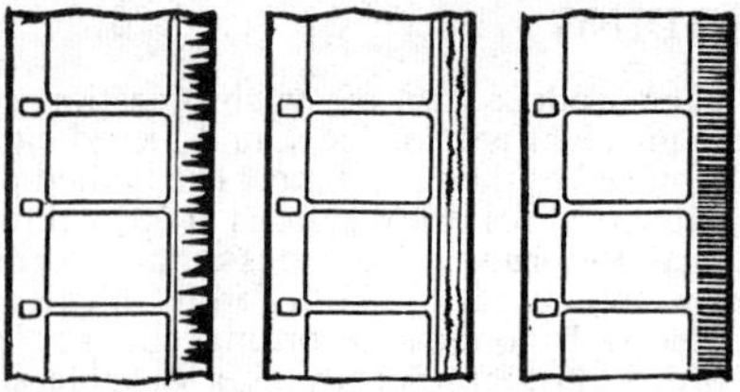

TYPES OF OPTICAL SOUND TRACK. Left: *The variable area track records the sound impulses as a wavy image spreading across the track.* Centre: *Bilateral tracks are more common and are in effect two variable area tracks side by side.* Right: *The variable density track records sound impulses as variations in density along the track.*

method of synchronization. In some cases tape recorder and projector are combined in one, being driven by the same motor.

Slight stretch or shrinkage of the tape and slip between the tape and capstans may still give rise to progressive synchroning errors. To overcome this, some machines use perforated magnetic tape. For constant synchronism magnetic stripe systems are inevitably more reliable.

Magnetic Stripe Recording. Magnetic sound film projectors reproduce sound from a narrow magnetic track applied to one margin of the film. Special attachments with magnetic sound head can also convert existing silent projectors. On 16 mm. films, this magnetic stripe may occupy all or half of the 100 mils (0·100 in.) normally used for the optical sound track. Since the projector is usually designed to reproduce optical sound films as well, either optical or magnetic sound may be reproduced from a film to which a half-track has been applied over an optical track. When striping is used, a balancing stripe is also added to the other margins of the film for even spooling.

To avoid the complications incurred by using single-perforated film in the camera, most amateurs prefer to have 16 mm. edge-striped. In this case a 30-mil track is applied between the perforations and the edge of the film. A similar arrangement is used on 8 mm. films. On 9·5 mm. films the 30-mil stripe is also along the edge of the film and allows the same picture format to be used as for silent projection. Satisfactory recordings of speech can be made on 16 mm. and 9·5 mm. magnetic sound films run at the speed normally used for silent films—i.e., 16 frames per second. But the film must be run at 24 frames per second for quality to compare with that of a good domestic tape recorder. In 8 mm. films, the difference is more pronounced and a speed of 24 frames per second is desirable. With Super-8 mm. film, the greater image spacing and a speed of 18 f.p.s. provide improved sound quality.

As with optical sound films, the sound recording is made some distance ahead of the corresponding picture. This is 28 frames on 16 mm. film, 56 frames on standard 8 mm. film and 18 frames on Super-8 mm. film. The flexibility of editing is thus not as great in the magnetic stripe system as with a separate tape recording. On the other hand the magnetic sound film is simpler to handle and provides infallible synchronization of picture and sound.

Because of the editing difficulties mentioned already, the magnetic stripe system is less convenient for recording the sound while the film is being taken. With careful rehearsal, however, speech can be post-recorded in synchronism with lip movements of the projected, edited picture film. Once recorded correctly, the speech will always be reproduced in synchronism with the action. With magnetic tape this is not generally possible.

Recording on tape can be done during actual filming, and the record later transferred to optical sound film. Small synchronization errors are then corrected as the sound and

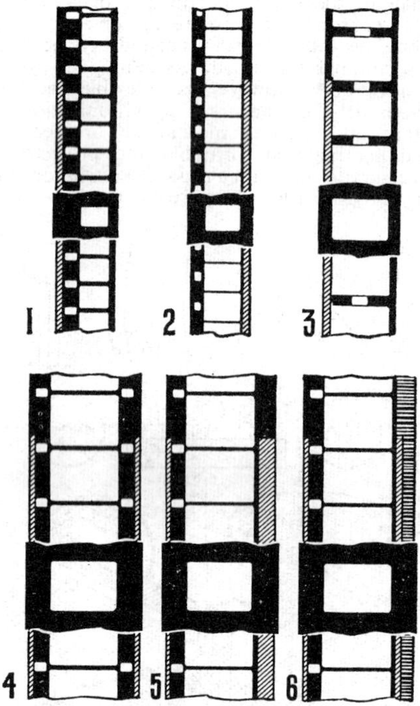

FILM GAUGES AND THEIR SOUND POSSIBILITIES. 1. *Standard 8 mm. film. A 30 mil magnetic stripe can be applied along the perforated edge.* 2. *Super 8 mm. film, with increased pitch and enlarged image area. A 30 mil magnetic stripe or optical sound track is applied along the unperforated edge.* 3. *9.5 mm. magnetic sound film. A 30 mil stripe along the film edge does not detract from the projected picture width even though it encroaches on the exposed picture.* 4. *Double perforated 16 mm. film can have a 30 mil magnetic stripe along one edge outside the perforations. A balancing stripe is always applied along the other edge.* 5. *Single perforated 16 mm. film with 100 mil (full stripe) magnetic track. Again a balancing stripe is applied along the perforated edge.* 6. *A half-stripe of 50 mil width can also be applied over the optical track of 16 mm. optical sound film.*

picture negatives are edited shot by shot. The two negatives are then printed on to a single film to produce a conventional married print, i.e. a positive carrying both optical sound and picture records. D.M.N.

Sound Quality. While the ear remains the best judge of the quality of reproduced sound, there are properties, all of them measurable, by which the performance of different recording and reproducing systems may be compared.

An ideal system would respond equally to sounds in the frequency range 20 to 20,000 cycles per second, which is the region of sensitivity of the human ear. In practice, the very high and very low frequencies tend to be lost, and irregularities often occur in the middle range response. However, variations in level of plus or minus 2 decibels (dB) are unimportant and a response of 30–10,000 c.p.s. ± 2dB would be considered very good for all but the most exacting applications.

The dynamic range of a system also affects the sound quality, that is the range between the faintest and the loudest sound it can handle without distortion. A range of 70dB is very good, and 45dB is adequate for most purposes.

Distortion. Besides failing to reproduce the full frequency and dynamic ranges of sounds, systems can introduce various types of distortion. A non-linear response to changes in signal level has the effect of creating new sound frequencies which interfere with the original sounds to give a harsh, discordant effect.

There are three types:

(1) In harmonic distortion, the spurious tones are direct multiples of the original frequencies, e.g. a note at 1,000 c.p.s. would be accompanied by fainter notes at 2,000, 3,000 c.p.s., etc.

(2) In intermodulation distortion, sum and difference tones are treated by interference between pairs of frequencies, e.g. 1,000 and 1,100 c.p.s. would produce 2,100 or 100 c.p.s., etc.

(3) In transient distortion, sharp changes in signal level are imperfectly followed by some parts of the system. These resonate or "ring" at their sum natural frequency and so blur the sound.

These types of distortion are inherent in valve and transistor amplifiers; disc, tape and film recording; microphones and loudspeakers. They are kept to a minimum by careful design, high quality components and regular maintenance.

Recording Systems Compared. Though disc, optical, magnetic tape and magnetic stripe recording systems are all capable of good quality, the results fall short of the ideal, more especially on amateur grades of equipment.

All these systems involve movement of the recording medium, and they share the disadvantage that a wide frequency response demands a high running speed. Everything else being equal, halving the running speed of a tape machine, for example, reduces the high frequency response by one octave. It is for this reason that the faster speeds of 7½ and 15 ins. per second are preferred for music recording. However, some domestic recorders achieve fairly extended high frequency response at 3¾ and even 1⅞ i.p.s. by incorporating special circuitry and narrow gap record/replay heads.

J.Bo.

See also: *Projection practice; Projector (cine); Sound synchronization; Tape recorder.*

Books: *How to Add Sound,* by I. M. Neale; *How to Produce Magnetic Sound,* by J. D. Chittock; *Sound Facts and Figures,* by John Borwick; *The All-in-One Tape Recorder Book,* by J. M. Lloyd; *Tape Recording and Reproduction,* by A. A. McWilliams; *The Technique of the Sound Studio,* by A. Nisbett (all London).

SOUND SYNCHRONIZATION. The expense and the difficulty of correcting mistakes in the laying of optical sound tracks for cine films have made amateurs and other low budget film-makers turn to magnetic recording for most purposes. Standard ¼-in. tape may be used, or striped film in which the magnetic coating is applied to the processed film, optical prints sometimes being produced finally for exhibition.

When tape is used, it becomes necessary to adopt some method of keeping the tape and film running speeds in precise synchronization. There are various systems.

Stroboscopic Synchronization. The simplest arrangement is to cause the tape to drive a stroboscopic disc and adjust the speed control on the projector so that the disc pattern remains stationary. The disc is illuminated by the flickering light from the projector lens, and the pattern will appear stationary only when the flicker rate is correct. At 16 frames per second, for example, a 3-blade shutter will produce a flicker rate of 48 per second. Should either machine drift from the nominal speed, the disc pattern will appear to rotate. The correction is always applied to the projector speed, of course, since the pitch of the sound would change if the tape speed were altered. In any case, no fine speed adjustment is normally fitted on tape machines.

Alternatively, a stroboscopic scale is sometimes printed direct on the tape. This has the advantage that the possibility of slip between tape and disc is avoided.

Loop Synchronization. Manual control of the projector speed, while watching a stroboscopic pattern, is operationally difficult and replaced by automatic control in a number of systems.

The loop synch system makes the tape travel round a supplementary drive roller which is operated by a flexible coupling cable from the projector motor. At the same time, the tape is looped round a spring-loaded jockey roller in such a way that synchronous running of tape and film maintains an even tension on the jockey arm. However, any discrepancy in the

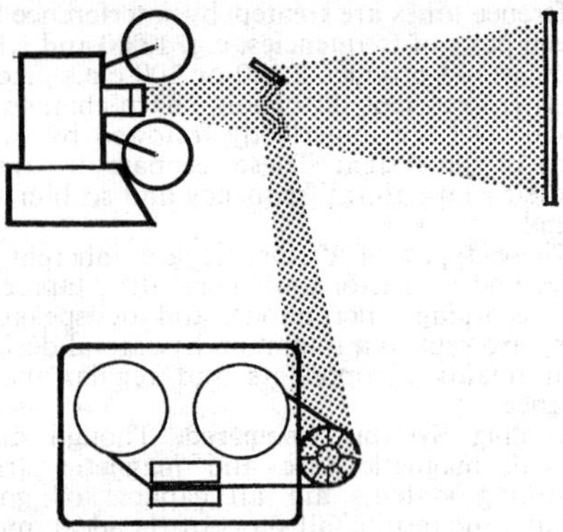

STROBOSCOPIC TAPE AND FILM SYNCHRONIZATION. A mirror reflects part of the projection beam on to a stroboscopic disc driven by the tape of the recorder passing over a pulley. As long as the sectors on the stroboscopic disc remain stationary, the projector and tape recorder are running in step. Adjustment of the projector speed to keep in synchronization must be made manually.

running speed will cause the tape loop to tighten or slacken and the resultant movement of the jockey arm activates a variable resistance in series with the projector. The projector speed is then automatically corrected. This loop synch system can be obtained as a separate unit or built into the projector.

Pulse Synchronization. Another automatic system uses a special synchronizing signal recorded as a separate track on the tape, at the time of filming. In one version, pulses are recorded at twice the frame frequency, and compared with the perforation rate on the film during projection. The control track is inseparable from the sound track and can give perfect synchronization or "lip synch". It is even possible to amplify the control signal and use this to drive the projector. The tape speed then controls the projector speed directly.

Perforated tape has also been used for sound synchronization. The sprocket holes correspond with those on narrow-gauge film and so exact synchronization and dual editing can take place.

Striped Film. The need for more or less complicated synchronizing techniques is eliminated in the other type of magnetic recording referred to, namely striped film. Here the film stock, 16, 9·5 or 8 mm., carries a narrow coating of magnetic oxide and is run on special projectors which possess the necessary erase and record/replay heads and amplifiers. The magnetic stripe is laid along the narrow space between the sprocket holes and the edge of the film on 8 mm. stock, and on the edge opposite the holes on 16 mm. film with perforations on one side only. In the latter case it is usual to lay a thin balancing stripe on the unused edge to prevent the film leaning over as it is spooled up. It is also possible to cut a recess in the film during the striping process, into which the magnetic coating is laid, so that the increase in overall thickness is minimized.

In the usual procedure films are sent to be striped, after processing and editing, at prices around 1½d. per foot. But pre-striped film is used with certain sound cameras which incorporate a recording head to permit simultaneous shooting of picture and sound.

Using ordinary cine cameras, location sound recording must be carried out on separate tape machines. If simultaneous recording is done, the noise of the camera will tend to be recorded on the tape unless the microphone can be taken some distance away or the camera is masked or "blimped" effectively. For certain scenes it is possible to make a separate recording before or after filming, particularly if no speech or action requiring accurate synchronization is included.

When the film is returned from striping and loaded on the projector, it is then possible to record commentary, music and other sounds in the same way as with a standard tape recorder. However, attempts to record the sound track while projecting the film, whether on to the stripe or a separate tape, are spoilt by the continuous background noise of the projector fan and motor. A better practice is to record a series of cue sounds only, at points where specific sounds or scripted phrases are to appear. It is then possible to re-run the tape or film in short sections, as often as may be necessary, placing chinagraph pencil marks on the appropriate spots. Then the recording can be

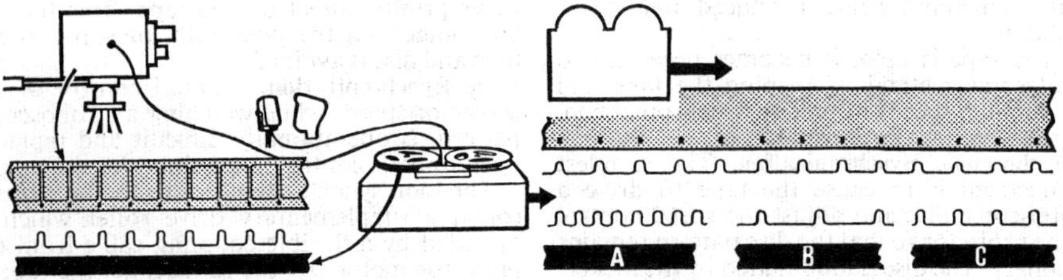

PULSE SYNCHRONIZATION. Left: A contact breaker in the camera closes and opens with the passage of every frame of film. By electrical connection with the tape recorder, this records corresponding pulses on the tape. Right: On transfer to the sound track on the film, the pulses from the tape are "compared" with pulses from the perforations on the sound recording film. This provides a basis for perfect speed matching; if the tape pulses are faster than the film (A) or slower (B), the speed is adjusted until both correspond (C).

made in short takes as each mark reaches the recording head.

Magnetic Film. Yet another method of magnetic recording exists. Magnetic film equipment is available in which a film, carrying a magnetic coating but quite independent from the main film, is used for the sound track or tracks. Greater track widths are possible than with striped film, so that better sound quality is achieved. Widths of up to 0·2 in. are commonly used, which compares very favourably with the 0·029 in. of a 16 mm. edge stripe.

The possibility of laying a number of separate tracks simplifies the building up of complicated sequences of speech, music and sound effects. The British Standard allows for up to six tracks on 35 mm. film and two on 16 mm., but other track arrangements are found, for example to permit stereophonic sound recording.

The independence of magnetic film, besides giving better sound quality, gives greater freedom in editing. But projecting the finished film would involve synchronization problems. The magnetic film is therefore regarded as a means to an end, and the usual practice is to dub the sound eventually to optical film printed alongside the picture. J.Bo.

See also: *Projection practice; Sound recording.*
Books: *How to Add Sound*, by D. M. Neale; *How to Produce Magnetic Sound*, by J. D. Chittock (both London)

SOUP. (Slang.) Processing solution; generally restricted to developer, sometimes one of a home-made type.

SOUTH AFRICA. Although the archives are singularly poor in subject matter relating to the origins and progress of photography in South Africa before Union (1910), old family photographs in the possession of private people help to prove that photography was practised in the Cape Province shortly after its introduction to Europe and England in 1839. Sir John F. W. Herschel, an outstanding pioneer of photography, lived in the Cape in 1834; in all probability he was the first photographer in South Africa.

As the country opened up to the north and around the coastline, photographic studios started to operate in numerous towns and villages. Today every large town in South Africa boasts of numbers of professional photographers, whilst almost every dorp (village) has a studio.

In the years following World War II photography has become enormously popular in South Africa. A high standard of living, together with the excellent climate, help to contribute considerably to this trend.

Due to the clarity of the atmosphere on the Highveld (6,000 feet above sea-level), and also at places like Bloemfontein, Yale University has established observatories with high-powered telescopes to photograph the heavens. As a point of interest, the first President of the Cape Town Photographic Society, Dr. David Gill, was Astronomer Royal at the Cape. Together with a foundation member, A. H. Allis, he took the first known photograph of a star.

Industry and Trade. While the applications of photography in South Africa have kept up with the rest of the world, both amongst amateurs and professionals, little progress has been made towards establishing a photographic industry. No sensitized papers or films are made in the country at all and amongst photographic chemicals, only hypo and sodium sulphite are produced.

In post-war years a certain beginning has been made to manufacture photographic accessories, and this business is likely to expand. Cameras, photographic equipment and sensitized materials are not manufactured in the Republic, but there is limited production of such items as developing dishes, slide mounts, projection stands, screens, floodlamps, dry mounting presses and snap albums.

On the other hand, South Africa is a considerable importer of photographic goods. 78,178 movie and still cameras entered the Republic during 1961, total value at port of entry being £610,899. Also imported were unexposed films valued at £539,338, and other photographic equipment to a value of £385,490. Main sources supplying the cameras were: Japan (24·6 per cent), West Germany (18·9 per cent), United Kingdom (11·2 per cent), Hong Kong (10·3 per cent), and the United States (9·3 per cent). Those made in Hong Kong were very low priced models.

Since World War II, import restrictions have been continually eased or tightened. Since 1962 there has been a striking improvement in the economy, and imports of all products have substantially increased.

Leading overseas film manufacturers have lately been establishing their own processing laboratories in South Africa, or arranging for both cine and still films to be handled by existing local laboratories.

Clubs and Societies. Amateur photography is growing rapidly in the Union of South Africa and there are today over 40 clubs and societies devoted to the pursuit of pictorial photography.

In 1954 the Photographic Society of Southern Africa was formed to act as a liaison body for photographers in all parts of the country. The two largest societies in the Union exist in Johannesburg—the Camera Club of Johannesburg (founded in 1934) and the Johannesburg Photographic and Cine Society (founded in 1930). Cape Town Photographic Society is the oldest society, having been established in 1890. There are professional organizations in three out of four of the Provinces—the Transvaal, the Cape, and Natal.

The amateur bodies organize international salons which are well patronized by world photographers. The Cape International Salon, the Witwatersrand International Salon and the

South African Salon all rate highly amongst photographic competitors everywhere. While the vast majority of 8 mm. movie-camera users are casual photographers who enjoy recording holidays and family activities, there are numerous active cine clubs which encourage their members to create really worthwhile films.
Publications. Two monthly photographic magazines are published in South Africa—*Amateur Photography* was founded in 1959, and in January 1964 *Photography and Travel* began publication.

While such places as the Kruger National Park (the largest wild game sanctuary in the world) have been over-photographed, on film South Africa presents the photographer with a veritable paradise, especially because of the unparalleled climate and interesting native peoples. E.E. (C.H.)

SOUTH AMERICA. Photography as a hobby began to be practised in Latin America towards the end of the nineteenth century, mostly among chemists, curious intellectuals and other knowledgeable people who had heard not only of photography, but also of the work of its creators Daguerre, Niépce, Fox-Talbot and others.
History. At Villa San Carlos, Campina, in the Brazilian State of Sao Paulo, a French chemi s named Hércules Florence, who had arrived in Brazil with an expedition in 1824 and had settled there permanently, continued the experiments he had started years earlier in France with Daguerre himself. Florence had in fact managed to build a "machine for the fixation of light" in 1832, seven years before Arago announced the discovery of photography before the French Academy in Paris. Hércules Florence was not alone, however. He performed his experiments with the co-operation of a Brazilian botanist named Joaquim Correo de Mello whose knowledge of bio-chemistry helped him in the preparation of the necessary formulae to develop and fix exposed images of silver film on paper and glass. Today some essays, a few pictures, and other evidence of the work of this more or less unknown father of photography can still be seen on exhibition at the Museo do Patrimônio Histórico e Artistico Nacional do Brazil (Museum of Historical and Artistic Patrimony in Brazil).

Photography also became fashionable among the Argentine upper classes, and well-to-do families who traditionally travelled in Europe would, on their return home, exhibit their skill with the black-box. In 1894 an Argentinian traveller brought back and exhibited in Buenos Aires a strange apparatus called the kinethoscope, which had been created three years earlier by Thomas Alva Edison, and which was a forerunner of cinematography. In 1896, Edison's vivomatograph, or vitascope, was exhibited in a theatre in Buenos Aires for the first time in South America. In 1901 Argentine photographer Eugenio Py introduced the first movie camera into South America.
Industry. Argentina has a greater photographic industry than the other Latin American countries, and its beginnings date back to 1935. Photographic chemicals, paper, films, enlargers and varied accessories are all produced in Argentina. Two internationally known manufacturers produce photographic paper in Brazil, and these same two companies also have representatives and colour processing laboratories in the principal cities of most Latin American countries.
Societies and Clubs. The Sociedad Fotográfica Argentina de Aficionados (Argentine Amateur Photographic Society), was founded in 1889 in Buenos Aires. It was followed by the Academia de Dilectantes Fotógrafos (Academy of Photographic Amateurs) of Buenos Aires of 1903, the Sociedad Alemana de Aficionados (German Society of Amateurs) in 1917, and the photographic section of the Fine Arts Society of Buenos Aires in 1927.

There are at present more than 150 photo and cine clubs in Latin America. Eighty-two of these are in Argentina, and thirty-two in Brazil. Almost all the Argentinian clubs are affiliated with the Federación Argentina de Fotografía (Argentine Federation of Photography). The Buenos Aires Photo Club has more than 2,000 members. The Argentine Photo Club has over 600 amateur members and a further photographic club, the Peña Fotográfica Rosarina (Rosarina Photographic Group) has 400.

The leading club in Brazil is the Foto Cine Clube Bandeirante (Photo-Cine Club Bandeirante), of Sao Paulo, followed by the Associacão Brasileira de Arte Fotográfica (Brazilian Association of Photographic Art), in Rio de Janeiro. Other prominent clubs in South America are the Foto Club Uruguayo (Uruguay Photo Club), of Montevideo; the Foto Club Chile (Chile Photo Club), of Santiago; the Federación Mexicana de Fotografía (Mexican Federation of Photography); and the Photo Club of Cuba at Havana.

In May 1964 the first Convención Americana de Fotografía (American Convention of Photography) was held in Mexico City. All Federations belonging to the Federation Internationale de l'Art Photographique (F.I.A.P.), and the Photographic Society of America, were represented. The Comité Regional Americano F.I.A.P. (Regional American Committee of Photography) was organized, with the purpose of spreading knowledge of the inter-American world of art and photography.

The Argentine Cine Club was established in Buenos Aires in 1932, and from its ranks have emerged some of Argentina's leading professional movie makers.
Publications. The Argentine Amateur Photographic Society, founded in 1889, produced a periodical magazine, the first of its kind in South America. In 1893 the *Revista Fotografica*

Ilustrada del Rio de la Plata (Illustrated Photographic Review of the River Plate) was founded. In 1920 *Foto Revista* (Photo-Review) was founded, and a year later came *Correo Fotográfico* (The Photographic Mail). These were followed by *Foto Magazin* (Photo-Magazine) in 1927, by *Foto* in 1936 and in 1938 by *Fotocamara-Cinecamara* which, in 1963, became *Fotocamera Con Popular Photography*. Of all the above only the last one still circulates throughout Latin America, being published in Buenos Aires as the official organ of the *Federation Internatiinal de l'Art Photographique* (F.I.A.P.) for all Latin American federations. Also worthy of mention are the *Boletin del Foto Club Uruguayo* (Bulletin of the Uruguayan Photo-Club), of Montevideo, and *Foto-Cine* of the Photo-Cine-Club Bandeirante of Sao Paulo, Brazil.

Exhibitions. Annual Salons with world-wide participation are held in Latin America. The leading exhibitions, in order of importance, are those held by the Peña Fotográfica Rosarina, Rosario, Argentina; the Foto Club Buenos Aires; the Foto Club Uruguayo, Montevideo; the Foto Cine Clube Bandeirante, Sao Paulo, Brazil; and the Foto Club of Valparaiso, Chile. Argentina, Brazil, Uruguay and Mexico are the countries whose photographers have most distinguished themselves in International Salons. Professional Associations hold a River Plate Salon for Argentinc and Uruguayan photographers annually in Montevideo, Uruguay. A national Salon for Brazilian professionals is held in Rio de Janeiro. H.Y.F.

SPAIN. Photography is widely practised in Spain today, and it has been the subject of popular interest since its inception, but, generally speaking, the emphasis is more on artistic output than on the manufacture of cameras and photographic equipment.

History. The first photograph to be taken in Spain by Daguerre's process was exposed at Barcelona on the 10th November, 1839, at one o'clock in the afternoon.

The official test was carried out in what was then known as Plaza de la Constitución, now Plaza de Palacio. A daguerreotype plate was exposed in the presence of the authorities, men of science and artists, besides a large public.

The experiment was carried out under the auspices of the Academia de Ciencias Físicas y Naturales (the Academy of Physical and Natural Sciences) at the instigation of Dr. Pedro Felipe Monlau and on the initiative of Don Ramon Alabern an artist-engraver and disciple of Daguerre., The cost—1946 reals (486 gold pesetas)—of the apparatus used to take this first photograph was borne by fourteen members of the Academy.

The experiment was entirely successful and lots were drawn for the first daguerreotype produced. The camera can now be seen in the Fabra observatory in Barcelona.

The daguerreotype process was popular for a long period, with the result that today many families treasure portraits of their ancestors which were made in this way.

Until the dry plate appeared, photography was practised only by professionals. Leading in the collodion period was Charles Clifford, photographer to the Queen of Spain and an Englishman by birth. When working conditions became easier as a result of this innovation, many amateurs entered the field. After a hesitant beginning, amateurs soon developed a style of photography of considerable merit, for they were not concerned with making profits but rather to give aesthetic enjoyment.

Among practising photographers in Spain at the end of the nineteenth century the most noteworthy was Dr. Ferrán. He carried out many experiments in photographic chemistry, and prepared photographic papers giving results similar to Fresson paper. Some of Dr. Ferrán's most important work is dealt with in his book, *La Instantaneidad en Fotografía*, (Instantaneous Photography), published in 1879 at Tortosa. This book was produced with the collaboration of I. Paulí. This was the period of the pigment processes, which workers of those times used in an effort to get a better rendering than was possible with the materials then available. Pissaca introduced the gum bichromate process into Spain and had a considerable following.

In 1952 the centenary of the birth of Dr. Ferrán was celebrated and a commemorative postage stamp issued bearing his portrait.

Another figure in photography in the late nineteenth century was Don Antonio Cánovas del Castillo.

In 1939, the Agrupación Fotográfica de Cataluña (the Photographic Society of Catalonia) celebrated the centenary of the first photograph taken in Spain and the Academy of Science offered its full collaboration, including facilities for research. The celebration and its associated functions were confined to practising photographers, the population as a whole being engaged in recovering from the difficult times through which the country had just passed.

Clubs and Societies. Throughout the Iberian peninsula there are flourishing groups, societies and clubs of enthusiasts whose aim is the spreading of technical skill and artistic appreciation.

The largest of these bodies is Agrupación Fotográfica de Cataluña (the Photographic Society of Catalonia), established in 1923, which at present numbers about 1,000 members. Next in importance comes the Real Sociedad Fotográfica de Madrid (the Royal Photographic Society of Madrid), which includes photographers of world repute amongst its members; the Sociedad Fotográfica de Zaragoza (the Photographic Society of Saragossa), likewise with a distinguished membership; and the Sociedad Fotográfica Guipuzcoana (the Photographic Society of Guipuzcoa), representing the best Basque photographic talent. All these societies hold an annual international salon. There are also bodies active in Igualada, Alicante, Alcira, Palma (Majorca), Valencia and Bilbao, in addition to others which cater for specialized interests. Many tourist centres have their own photographic clubs and here members tend to specialize in landscapes and mountain studies.

There are today some 94 photographic clubs in the country. The greatest number of active amateur photographers is in Catalonia, where there are 47 clubs; 20 in Barcelona city alone.

Industry and Trade. There are now three important photographic firms engaged in the manufacture of photographic paper, plates and films which they distribute to Spanish dealers.

The manufacture of cameras and accessories is restricted to the simple popular types. There are a few specially built cameras for professional purposes, but even these are fitted with foreign lenses.

The following are typical figures for imports into Spain for the month of December 1962, as published in The Report of Spanish Exports and Imports: Lenses, prisms, and polarizing materials, 1·7 million pesetas; cameras, 12 million pesetas; electronic flash units, 1·4 million; cinematographic equipment, including projection equipment and accessories, 9 million; still projection equipment, and enlarging and reducing apparatus, 1·7 million; apparatus for photomicrography, cine-micrography, and allied work, 2 million; all X-ray equipment, 5·6 million; apparatus and equipment for cine and still processing laboratories, photo-copying equipment, etc., 2·3 million; and films in roll and sheet, black and white and colour, 26·7 million pesetas.

Exports are on a considerably smaller scale, amounting for the entire above-listed materials to only about 900,000 pesetas. The export of cinematographic equipment, including projection equipment and accessories, still projection equipment and enlarging and reducing apparatus account for 800,000 of the above figure, while X-ray equipment of all kinds accounts for the remainder.

The photographic wholesale and retail trade in Spain is centred around Barcelona, and the majority of agents and representatives of foreign manufacturers are also in this area.

Publications. Since the year 1923 there has appeared in Barcelona a monthly bulletin which is the organ of the Agrupación Fotográfica de Cataluña (Photographic Group of Catalonia), an organization with the largest membership in the area. From 1925 to 1930 a Catalan-language magazine entitled *Art de la llum* (Art of Light) was also published. From 1949, for a period of about four years, the magazine *Sombras* (Shadows) appeared in Madrid. At the present time *A.F.* (*Arte fotográfica*), a handsomely produced magazine, regularly publishes contributions from the best-known names in the world of photography. *Afal*, which was originally a bulletin of the Agrupación Fotográfica Almeriense (The Photographic Society of Almeria, a town in southern Spain), and is now a magazine for subscribers only, has in recent years been representative of the latest trends.

Most photographic clubs publish bulletins for their members.

A number of text-books for teaching purposes have been written by the photographer, artist and engraver Francisco Ollé Pinell, of Barcelona. One of the oldest and most complete Spanish manuals is a book by Felipe Picatoste, *Manual de Fotografía*, published at Madrid in 1882. Numerous shorter works on particular items were, however, published some time earlier.

Edicones Omega of Barcelona have produced and published a comprehensive list of technical books on photographic and cinematographic subjects.

Activities. Apart from a few short courses, run either by photographic clubs or by photographic retailers, there is no organized photographic teaching in Spain. Generally speaking, experience is the only way of achieving a mastery of the subject, although some professional workers manage to learn colour processes and techniques abroad, particularly in France and Switzerland.

Active photography is on the increase in Spain, and there are today some 3,000 professionals engaged in one sphere of photography or another on a full-time basis. Furthermore, although Spain does not contribute greatly to the world's photographic trade and scientific development, the country has a large number of workers—both professional and amateur—who are actively engaged in artistic photography.

When portrait photography was first practised, a number of studios were established in the larger Spanish cities, and some of these remain to this day. In 1897 there was in Madrid the Hamer y Menet phototype shop, and at about the same time there were established in Barcelona the Basque engravers Mariezcurrena and Joaristi, who made gravures through the autotype process. From this it may be inferred that at the end of the century the use of photography as a means of graphic communication was well established in Spain. It was, however, mainly the use of the pigment processes by amateurs which made the name of Spain known abroad.

Prominent among those who cultivated these processes, i.e., bromoil, transfer, Fresson, gum bichromate, etc., was José Ortis Exagüe. He made a great impact in international competitions, but his name will be particularly remembered in the history of world photography for his publication of the Series *España, tipes y trajes* (Spain, Types and Costumes), a work in several volumes.

At the present time one of the most comprehensive photographic archives in Spain is that belonging to Francisco Catalá Roca, a photographer who produces work mainly for publication purposes. A notable contemporary photographer is Leopoldo Pomés, who shows very modern trends, and engages mainly in publicity photography. M.C-B. (J-M.C.)

SPARK PHOTOGRAPHY. Method of securing an image of an object moving at high speed by illuminating it with the instantaneous light of an electric spark. The spark is produced by discharging a condenser across a spark gap on one side of the path of the moving object so that it records a silhouette of the object on a plate on the opposite side. There is no need for a lens in the set-up as the spark when placed at a distance behind the subject gives a reasonably close approximation to parallel lighting and forms a sharp edged shadow.

The first demonstration of spark photography was made by F. Talbot in 1851. He photographed a rapidly revolving wheel.

Ernst Mach at Prague, and Dr. Salcher at Fiume, photographed bullets in flight in 1885–86. In 1892 Sir Charles V. Boys published a detailed description of a number of successful experiments on photographing bullets in flight at higher speeds.

The original circuit in modified form is still in use today. It consists of a spark condenser connected on the one hand to a suitable D.C. supply through a current-limiting resistance and on the other to two spark gaps in series. One gap forms the light source while the other is a triggering gap. The combined resistance of both gaps is just too much for the condenser voltage to break down, but, if the small triggering gap is short circuited, the main condenser immediately discharges across the large gap.

Triggering is achieved by discharging a small condenser across the triggering gap, the passage of the current ionizing the air in the gap and lowering its resistance enough to allow the main spark condenser to discharge. The triggering condenser is connected to its spark gap through a pair of triggering leads which are suspended in the path of the projectile. As the projectile strikes the leads it completes the triggering circuit and initiates the photographic spark. (The small spark from the triggering gap is shielded so that it does not affect the sensitized plate.)

The greater part of the actinic light of this type of sparking between open metal electrodes is in the ultra-violet region of the spectrum and of a wavelength to which ordinary optical glass is practically opaque. So the absence of a lens means that all the actinic light is allowed to fall on the plate and what is visually a relatively feeble spark does in fact provide an adequate exposure even during the short period of the flash.

Even in the early days of spark photography exposures as short as one-millionth of a second were achieved, but today exposures at one-tenth of that speed are normal. By modifying the circuit to give a continuously interrupted spark it is possible to produce a succession of profiles of the subject on a moving film up to as many as 200,000 per second.

For many kinds of high speed photography the open spark has been superseded by the electronic flash discharge tube, but it is still used, notably in ballistics. Modern circuits employ main spark condensers of from 0·1 to 1 microfarad, charged from 5,000 to as much as 10,000 volts. The circuit is usually triggered by a mercury tube or thyratron. F.P.

See also: *Ballistic photography; Chronophotography; Flash (electronic); High speed photography; Schlieren photography; Stroboscopic flash.*

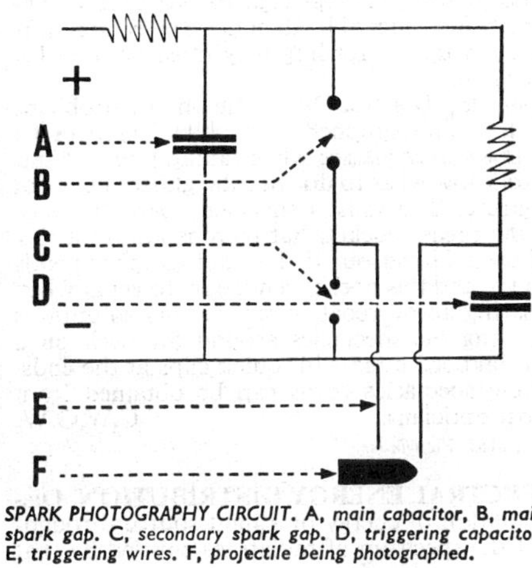

SPARK PHOTOGRAPHY CIRCUIT. A, main capacitor, B, main spark gap. C, secondary spark gap. D, triggering capacitor. E, triggering wires. F, projectile being photographed.

SPECIFIC GRAVITY. Weight of a volume of any substance compared to the weight of the same volume of water. The specific gravity of a liquid (and hence in many cases the concentration of any dissolved substance) can be measured directly with a hydrometer.

SPECTACLE LENS. Single uncorrected lens, unsuitable for photography where good definition is important, but satisfactory for use as a supplementary lens in conjunction with the camera lens. When used in front of the lens to shorten or lengthen the effective focal length, the camera lens should be stopped well down to restore good definition.

SPECTACLES AND EYESIGHT. People who suffer from defective eyesight often wonder whether wearing spectacles will prevent them from taking up photography. Broadly speaking, there is no reason why spectacles should interfere with the use of a camera or any other photographic equipment; there are occasions when they are inconvenient, but there is always a satisfactory answer for those who care to look for it.

Viewfinders. Eye-level finders of either optical or frame types do not affect the clarity of the picture seen by the wearer of glasses. But these finders require the eye to be placed as close as possible to the back sight, and people wearing glasses cannot get close enough. The result is that they see a smaller amount of the scene than the viewfinder—and the negative—actually cover.

There are four solutions to the problem:

(1) If the wearer's sight is good enough to give a reasonable impression without the glasses they should be removed for sighting.

(2) A few experiments will show just how much more will appear on the negative than can be seen in the viewfinder with the spectacles in use. With practice, the wearer is able to make a satisfactory estimate of the actual picture area based on the proportion in the finder.

(3) Any good ophthalmic optician will compute and supply a lens to fit to the viewfinder to take the place of the corresponding spectacle lens. The spectacles can then be discarded while taking the photograph. This course is apt to be too expensive for the casual photographer, but it is the best solution for the professional to apply to his regular camera, at least. The cost of such supplementary lenses varies with the optical prescription. Usually the most expensive part of the arrangement is the mounting and fitting of the lens.

(4) Many busy photographers who must wear glasses habitually for all work find that contact lenses worn on the actual eye itself are the ideal solution.

The large "brilliant" type of finder gives no trouble at all to the spectacle wearer since it is viewed from a distance. Users of this type of finder will see the image clearly if they keep on the glasses that they normally wear for distance. There is nothing to be gained by wearing reading glasses with such finders.

Many spectacled photographers have found that the ordinary frame type of finder can be improved by fastening thin cross wires to mark the centre of the front frame. This helps them to look straight through the centre of the frame at the centre of the subject. Without the cross wires, it is easy for the eye to look through the finder at an angle because of the separation between it and the back sight of the finder.

Focusing Screens. The image given by a focusing screen is a real image, and behaves differently from the virtual image given by optical and "brilliant" types of finder. In this case the image has to be focused like a page of print at the same distance. This applies both to normal and reflex focusing screens. And as these screens are viewed from a distance, there is no difficulty about wearing glasses when using them.

People of middle age who experience difficulty in reading at the normal viewing distance of 10 ins. will have the same difficulty in seeing the image on the focusing screen. If they wear the glasses that they normally use for reading they will not be able to see the subject clearly when they take their eyes off the screen because the glasses are unsuitable for both near and distant sight. There are two answers to this trouble—either to wear glasses with bifocal lenses, or frames fitted with half-lenses which allow the wearer to look over the top for viewing distant objects. Anyone who normally wears bifocal lenses anyway will find them satisfactory for both focusing and direct viewing.

Rangefinders. All types of rangefinder can be used by wearers of spectacles without difficulty. In this case the spectacles should be the ones normally worn for distance. About the only trouble likely to be experienced is from confusing light reflections on the lens of the spectacle, but these can be cut out by simply shielding the viewing eye with the hand or the brim of the hat. Some rangefinders are fitted with a focusing eyepiece which can be adjusted for use with or without glasses.

Where the rangefinder and viewfinder are combined, there may be the difficulty already mentioned of getting the eye close enough when spectacles are worn.

Focusing Aids. All the various types of magnifier used as aids for focusing the image on the camera focusing screen or on the enlarging easel can be used by wearers of glasses without inconvenience. People who normally wear glasses for reading only can dispense with them when using a direct optical magnifier type of aid—i.e., one which does not give the magnified image on a ground glass screen. Where the magnified image is projected on to a ground glass viewing screen—e.g., in one type of enlarger focusing aid—then if glasses are normally worn for reading they must be used for focusing.

Removing Glasses. One of the minor problems of the photographer who either removes or changes over glasses when taking photographs is to know what to do with the glasses he is not wearing. The most convenient parking place is the breast pocket, but there is always a risk of their falling out if the photographer bends down, and this pocket is not easy to get at when wearing an overcoat. A satisfactory solution is to hang the spectacles around the neck on a special cord fitted with quick clips at the ends. These spectacles cords can be obtained from most opticians. C.W.G.W.

See also: *Viewfinder*.

SPECTRAL ENERGY DISTRIBUTION. Distribution of energy of a light source throughout the spectrum of radiations emitted by that source. Usually shown in the form of a curve which plots energy against wavelength. The spectral energy distribution curve completely specifies the colour quality of an illuminant, and is applicable to light sources with a continuous as well as a discontinuous spectrum. It is thus a more reliable specification than the colour temperature. Two light sources of different spectral energy distribution may give the same colour temperature reading when measured with a colour temperature meter.

See also: *Colour temperature meter; Light sources; Spectrum.*

SPECTRAL SENSITIVITY. Response of a photographic emulsion to each of the separate colours of the spectrum.

Ideally, spectral sensitivity should be defined with reference to a spectrum which contains equal amounts of energy in each colour band. From this absolute spectral sensitivity, the sensitivity distribution towards other light sources can be worked out if the distribution of light of different colours in the source is known.

Spectral sensitivities are as a rule determined by means of a wedge spectrograph. This is an instrument which separates white light into its component rainbow or spectrum colours. The light passes through either a glass prism or a diffraction grating consisting of a piece of glass or a mirror on which very many fine lines have been ruled. The ability of a grating to split up light into its component colours can be demonstrated by looking at a lamp through a piece of very finely woven fabric.

Wedge Spectrogram. In the wedge spectrograph a light absorbing wedge of uniformly increasing density is placed across the spectrum formed by the prism or grating, with the density gradient of the wedge at right angles to the extension of the spectrum. The photographic material under test is placed under the wedge. Under any one colour, therefore, the photographic material receives a range of exposures the intensity being relatively high at the thin end of the wedge and low at its dark end. So the degree of blackening on development at each step along the spectrum indicates the relative

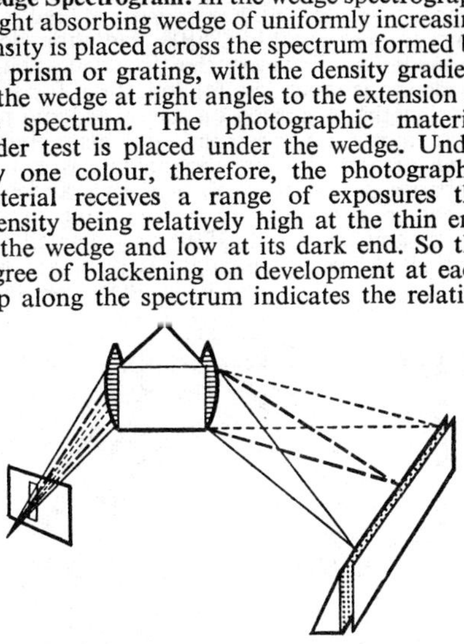

WEDGE SPECTROGRAPH. The light from an illuminated slit is split up into a spectrum and projected through a density wedge on to the material under test, being in contact with the wedge. The result is a wedge spectrogram.

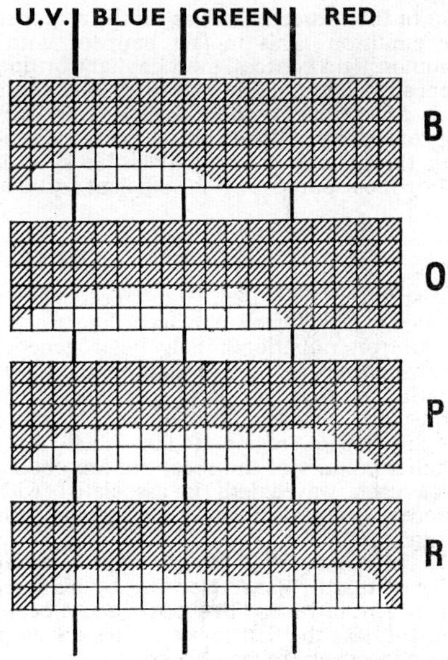

WEDGE SPECTROGRAMS. These show the relative response of the material tested to light of different colours. B, blue-sensitive (e.g., process) materials; slightly sensitive to blue-green as well. O, orthochromatic materials; sensitive to green and yellow as well, but not red. P, panchromatic; sensitivity extended to orange and red. R, panchromatic with increased red sensitivity.

sensitivity of the emulsion to light of that particular colour. If it is completely insensitive, that strip of the material remains clear. If it is sensitive, part at least of the strip under the colour will be blackened.

If now a print is made from the wedge spectrogram on high-contrast paper, all the heavily exposed portions of the spectrogram will appear light in the print, and the weakly exposed parts will appear dark. The separation line between the light and dark portion is a contour line of constant density in the negative. The light portion of the print will be wider at those places where the emulsion is most sensitive to the colour of that part of the spectrum. The wedge spectrogram is thus a very simple method of representing the colour sensitivity of a photographic material.

Correction for Ultra-Violet. The wedge spectrum falls off rapidly towards the ultra-violet because the instruments usually employed hold back ultra-violet rays. So what appears to be a falling off in sensitivity to ultra-violet is not genuine; most materials are evenly sensitive in that region.

Correction for Lighting. Further, the nature of the light used to illuminate the wedge spectrograph must be taken into account.

This should emit equal energies at all wavelengths, or otherwise the contour of the wedge spectrogram will depend on the spectral emission of the source as well as on the response of the emulsion. This in fact happens with all common light sources, even daylight or fluorescent strip lighting. The light from some sources, such as mercury vapour lamps, is not even approximately uniform in its energy distribution throughout the normal visible spectrum.

For most purposes it is sufficient to be able to assess the spectral sensitivity of a material towards average daylight and tungsten light. Wedge spectra for these two types of lighting are frequently published by manufacturers for the convenience of their customers. W.F.B.

See also: ***Negative materials; Sensitizer; Spectrography.*** Book: *Spectral Studies of the Photographic Process*, by Yu. N. Gorokhovskii (London).

SPECTROGRAPHY

A spectrum is a series of light or other electromagnetic rays sorted out according to wavelength from radiation initially heterogeneous—i.e., which is composed of more than one wavelength. The rays can be of any type—X-rays, ultra-violet, visible, infra-red, etc. If they are in the visible region (4,000–7,000 A) the familiar colours of the spectrum are observed, since each wavelength is associated with a different colour—violet, blue, green, yellow, orange, red.

It is commonly, though quite wrongly, believed that Sir Isaac Newton, by his famous experiment in 1666 with a prism and a beam of sunlight, was the first to see an artificially produced spectrum. In fact, his explanation of how the prism separated the colours latent in white light only clarified a conclusion towards which many others had long been groping. Newton did not, of course, associate the colours he saw with wavelengths, since he rejected the hypothesis that light was a wave motion. This was not established until the beginning of the nineteenth century.

After about 1820 rapid progress was made in the production, measurement and classification of spectra. Fraunhofer, shortly after Young's work, produced the first instrument resembling the modern spectroscope, and with it discovered the lines in the solar spectrum still named after him. Spectroscopy was later developed as an analytical tool by many others, particularly Bunsen, Kirchoff and Angstrom. In 1884, Sir W. N. Hartley recorded spectra photographically, and since then valuable contributions have been made by de Gramont, Kayser, Lundegårdh, Gerlach and others.

Spectrographs and Spectroscopes. Spectra are produced and examined today in a spectrograph or spectroscope. Either instrument consists of a slit to admit the light, a prism or grating to disperse it, and a lens system (which must be exceptionally free from chromatic aberration) to focus an image of the slit. In a spectrograph this is received on a photographic emulsion; in a spectroscope it is examined visually through a telescope.

The slit consists of two parallel metal jaws, machined straight to an accuracy of 1/10,000 in. and separated by an adjustable distance, which is usually in practice a few thousandths of an inch. The lenses form two groups—the collimator receiving the radiation from the slit, and the telescope (in the spectroscope) or focusing lenses (in the spectrograph) focusing it after dispersion.

Only the spectrograph uses photography, but much of what is said here applies to either instrument.

The spectrograph usually employs a 60° prism. For the visible region and the nearer (photographic) infra-red (4,000—approx. 12,000 A) it is made of highly dispersive flint glass. Other materials are used for wavelengths to which glass is opaque. Quartz is used for the ultra-violet (wavelengths less than 4,000 A) and rock salt for the far infra-red (wavelengths over 20,000 A).

Instead of a prism, a diffraction grating may be used. This is an optically flat mirror or transparent plate with a large number (up to 20,000 per centimetre) of fine straight parallel lines engraved on the surface. This produces a spectrum by resolving the light through diffraction and interference.

Which method of dispersion is used depends on circumstances. A prism produces the brightest spectra, and is therefore used with faint sources emitting little radiation (e.g., stars). It has, however, the disadvantage that dispersion is much less for long than for short wavelengths. A grating on the other hand produces a normal spectrum, and also gives a much greater absolute dispersion. (A normal spectrum is one in which a given wavelength difference is represented by the same separation at any wavelength.) But only a small proportion of the incident light is dispersed in a grating spectrum, which is therefore faint.

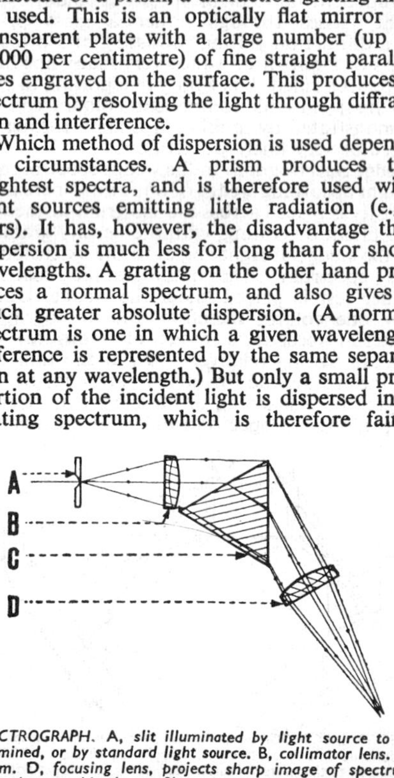

SPECTROGRAPH. A, slit illuminated by light source to be examined, or by standard light source. B, collimator lens. C, prism. D, focusing lens, projects sharp image of spectrum on to photographic plate or film.

TYPES OF SPECTRA. Top: *Emission line spectrum of a luminous source (e.g. discharge tube).* Centre: *Absorption line spectrum (e.g. the line of the solar spectrum).* Bottom: *Emission band spectrum due to molecular groupings in luminous gases.*

Also, a grating produces in fact several spectra, of increasing faintness as they are of higher order, and these, by overlapping, may lead to confusion and difficulty in interpreting the results.

Types of Spectrum. The spectrum of a luminous source may be of one of three kinds.

(1) Continuous. This contains every possible wavelength within the limits of the spectrum. Over the visible region it therefore appears to the eye as a band of continuously varying colour, and is recorded photographically as an unbroken, though not necessarily uniform, exposed strip. Continuous spectra are emitted by incandescent solids and liquids (e.g., an electric lamp filament, molten iron) and by gases under certain conditions. They are in themselves of little scientific utility, but can be used in the examination of absorption spectra.

(2) Line. Gases made luminous, either by heat or by an electrical discharge, normally emit only certain definite wavelengths, which are characteristic of the elements present. Their spectra therefore consist of a series of lines (usually of varying intensity and irregularly spaced), each line being an image of the slit formed by one wavelength.

(3) Band. These are emitted by molecular groupings in a luminous gas, and consist of groups of lines too closely spaced to be resolved except by a very large dispersion. The Swan bands emitted by a carbon arc in air at 3,600–3,900 A are the best known.

These are all emission spectra. Radiation which has passed through a material absorbing certain wavelengths but not others can be dispersed to form an absorption spectrum. This will be a line spectrum if the absorbing substance is gaseous; the Fraunhofer lines in the solar spectrum represent absorption by the sun's atmosphere from the total radiation emitted by the (much hotter) interior. If the absorbing medium is solid or liquid, the spectra consists of series of bands, which are quite diffuse, and cannot be resolved into constituent lines. The position and intensity of these bands is characteristic of the absorbing substance.

Spectra are also classified in practice according to the wavelength region under investigation, different regions needing different techniques.

(1) Infra-red absorption spectra, in the region 2–15μ (i.e., 20,000–150,000 A), cannot be recorded photographically, as they are outside the region to which emulsions can be sensitized.

(2) Visible and near infra-red (4,000 approx.–14,000 A) emission and absorption spectra are usually recorded photographically in a spectrograph with glass components.

(3) Ultra-violet (2,100–4,000 A) emission and absorption spectra are also recorded photographically, but quartz must be used in place of glass.

(4) Far ultra-violet (approx. 100–2,100 A) emission spectra can also be recorded photographically, but very special and difficult techniques are necessary, as both air and gelatin are opaque in this region.

(5) X-ray "spectra": X-rays cannot be dispersed by any mechanical device, as their wavelength is too short (0·05 to, say, 10 A), but they can be diffracted by crystals, which act as natural gratings by reason of the regular arrangement of the atoms in them, spaced perhaps 1–2 A apart. A photographic record after diffraction shows a series of lines representing different angles of reflection. These are not true spectra at all, since "monochromatic" X-rays of a single wavelength are used. True X-ray spectra can be produced, but they are something quite different.

Spectrum Photography. Visible and ultra-violet spectra are recorded on ordinary photographic plates. An emulsion is chosen which has the maximum sensitivity to the particular wavelength region under investigation. Plates are preferred because of the greater accuracy with which line separations can subsequently be measured. Many spectrographs incorporate a device for printing a wavelength scale beside the spectrum.

In some spectrographs the focal plane is slightly curved; the plates must then be coated on extra thin glass, which can be bent slightly.

The sizes of plates used depend on the dispersion. The smallest laboratory spectrograph takes quarter-plates. A much used size disperses the region 2,000–10,000 A over about 9 ins., and takes plates 4 × 10 ins. Still larger prism spectrographs have three times this dispersion, and take three 4 × 10 ins. plates to cover the same region. Still greater dispersion is used for special purposes; spectra over 100 feet long have been photographed on cinema film with a grating of very high dispersion.

In the 2,000–2,300 A region of the ultra-violet, and still more in the far ultra-violet, ordinary emulsions are at a disadvantage owing to absorption by the gelatin. For the region 1,900–2,300 A the emulsion may be coated with a film of fluorescent oil, but this is messy and the oil must be removed before development. Alternatively, special emulsions containing the minimum of gelatin may be used; there are, however, always coarse-grained and inconveniently sensitive to the slightest pressure on the surface. For the far ultra-violet (wavelengths less than about 1,900 A) "emulsions" containing no gelatin at all (Schumann plates) must be used.

X-ray "spectra" are recorded on ordinary X-ray film. Film is usually employed because it can be made more sensitive to X-rays by double coating, and because it may have to be bent round a curve of small radius.

Uses. Spectrography in one form or another is used in practically every field of science and industry. The majority of its uses depend however on two basic facts.

The first of these two is that in line-emission spectra the wavelengths present are characteristic of the elements present in the emitting gas, and can be used as an infallible index of their presence. This gives a now essential tool to modern astromony, providing the only means of ascertaining the composition (and indirectly temperature, mass, etc.) of the stars. The photographically recorded displacement of certain spectrum lines from their normal wavelength is in fact the crucial experimental evidence for the modern conception of the universe expanding like a soap bubble.

The second fact basic to spectrographic analysis is that, in an electric arc or spark, the material of the electrodes is present in the discharge as a luminous gas, at a temperature at which practically all compounds are decomposed into their elements. Hence spectrography of an arc or spark discharge between electrodes composed of, or impregnated with, the material under investigation gives a rapid and convenient method of qualitative elementary analysis of a great variety, of materials. The method is far more sensitive than chemical analysis procedures.

Since most of the analytically useful lines lie in the ultra-violet, photographic recording is essential.

This type of analysis can also be made quantitative. For example, an alloy containing 0·1 per cent of magnesium will show more intense magnesium lines in its spectrum than one containing 0·01 per cent. In general, the intensities of the lines of a minor constituent on a developed plate, as measured by a microdensitometer, relative to the intensities of suitably chosen lines of the major constituent, can be used to determine the proportion of the minor constituent present.

This type of quantitative analysis is widely used for rapid production control of materials whose chemical composition is of critical importance—e.g., alloy in the aircraft and machine tool industries.

There are numerous other special uses of emission and absorption spectra, and of X-ray spectra, as an analytical tool in chemistry, physics, medical research and the applied sciences. H.J.Wa.

See also: *Spectral sensitivity.*
Book: *The Measurement of Optical Radiation*, by G. Bauer (London).

SPECTRUM. A prism disperses rays of light according to their wavelength. Using an illuminated narrow slit as source, a lens and a prism can separate the different wavelengths or colours present in white light, each wavelength forming a separate image of the slit so making a spectrum.

A rainbow is an imperfect spectrum formed by myriads of raindrops.

The visible rays are only a part of the complete spectrum of so-called electro-magnetic radiation; there are others that cannot be seen but which can be detected by other means. The visible part varies in colour from red to orange, yellow, green, blue, indigo and violet. These colours are physical sensations produced by the action of light on a three colour receptor system in the eye, and the resultant interpretation by the mind.

Wavelength Characteristics. The visible spectrum extends from 4,000 to 7,000 A. and the human eye is most sensitive to the wavelength of 5,500 A. corresponding to yellow-green. The near infra-red is usually regarded as extending to about 15,000 A. The near ultra-violet is regarded for all practical purposes as extending to about 3,000 A. Below 3,500 A. all rays are stopped by ordinary optical glass.

For optical systems below 3,500 A. it is necessary to use components of transparent crystalline quartz.

Below 2,100 A. the rays are absorbed by the gelatin of normal photographic emulsions, and special emulsions are used with no or very little gelatin.

Below 1,900 A. it becomes necessary to remove all traces of water vapour from the system.

Between 2,200 and 3,000 A. the absorption of gelatin and of the silver halide emulsion crystals reduces the depth to which the light can penetrate to a mere surface layer. This reduces the density of the silver deposit that forms the image in the developed emulsion.

The silver halide salts in photographic emulsions absorb light of all wavelengths below 5,000 A. This is the region of their normal sensitivity. But certain light-absorbing dyes will extend the sensitivity to include the whole of the visible spectrum and even part of the infra-red. Fast infra-red emulsions are sensitive only as far as 9,000 A., but some slower emulsions with special sensitization go as far as 14,000 A.

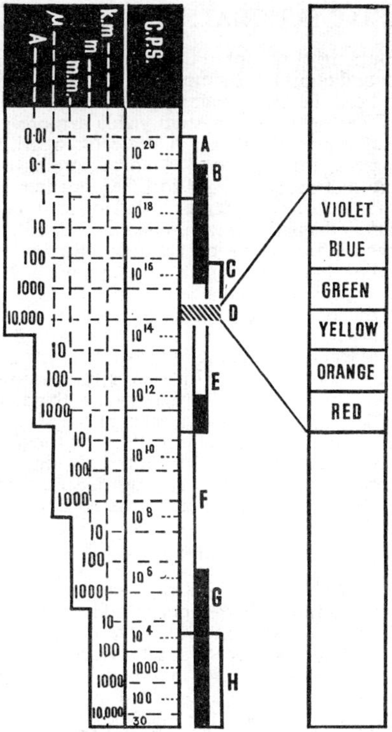

ELECTRO-MAGNETIC WAVES. These waves, travelling through empty space at about 186,000 miles per second, range in wave length from 0·01 Angstrom unit to thousands of kilometres, and in frequency from almost zero to 3 × 10^{20} cycles per second. A, gamma rays, covering 7 octaves. B, X-rays, covering 13 octaves. C, ultra-violet radiations, covering about 4 octaves. D, one octave of visible light and one of near infra-red radiation. All the radiations so far can be recorded photographically. The following ones cannot. E, far infra-red and heat rays. F, Hertzian waves (22 octaves). G, television and radio frequencies. H, audio frequencies.

Equal Energy Spectrum. In the spectrum of white light produced by a prism, the energy is not distributed evenly over the various colours—some colour bands contain more actual light energy than others. This type of spectrum is no use as a basis for studying the way in which the sensitivity of a photographic emulsion or the human eye changes with the colour of the light. What might appear to be high sensitivity to red, for instance, might simply mean that there was more light energy in the red band of the spectrum employed than in the other regions.

The only real basis of comparison is a spectrum in which the energy is distributed evenly over the whole range. Such a spectrum is called an equal energy spectrum; it is the reference for most published colour sensitivity curves for sensitized materials and of similar curves showing the response of the human eye to colour.

Colour sensitivity can be related to an equal energy spectrum without actually producing one. (It would in fact be very difficult to produce at all.) The energy distribution in a given spectrum can be measured accurately in terms of heat energy distribution. Once this is known, it is an easy matter to convert the sensitivity curve it gives into terms of an equal energy spectrum by simple proportion.

For example, if the response ratio of an emulsion to a given spectrum is blue : green : red = 2 : 3 : 5, and the energy is distributed in the test spectrum in the ratio blue : green : red = 1 : 2 : 4, then the sensitivity to an equal energy spectrum would be in the ratio blue : green : red

$$= \frac{2}{1} : \frac{3}{2} : \frac{5}{4} = 2 : 1{\cdot}5 : 1{\cdot}25.$$

These figures show that although the test spectrum gives the sensitivity of the emulsion as being twice as great to red as to green, it is in fact slightly less in terms of an equal energy spectrum. R.B.M.

See also: *Colour; Light; Spectral sensitivity; Spectrography.*

SPEED FLASH. Another name for electronic flash lighting; the duration of the flash is usually extremely short—often 1/1000 to 1/10,000 second or even less—so this type of flash is particularly suitable for freezing the action of very fast-moving objects provided other lighting present is of a relatively low power to the flash.

See also: *Flash (electronic).*

SPEEDGUN. Popular name for a flash unit and mechanical synchronizer designed to give short instantaneous flash exposures by opening and closing the shutter during the period of peak brilliance of the flash bulb.

This type of flash equipment has been almost completely superseded by internally synchronized shutters and by electronic flash lighting.

See also: *Flash equipment; Flash synchronization.*

SPEED LAMP. Alternative name (current mostly in America) for an electronic flash lamp, which is very useful in photographing fast action.

See also: *Flash (electronic).*

SPEED MIDGET (S.M.). Type of now obsolete flash bulb containing an explosive paste (e.g., zirconium hydride) the combustion of which provides the light. These bulbs have no metal wire filling; they are fired in the usual way by an electrically heated metal filament.

Speed Midget bulbs are characterized by a short firing delay and flash duration (both of the order of 5–10 milliseconds). They are specially suitable for use with shutters having simple X or F-synchronization, as fitted to inexpensive amateur cameras.

See also: *Flash synchronization.*

SPEED OF SENSITIZED MATERIALS

The speed of a sensitized material is an expression of its sensitivity to light.

The speed is measured by giving samples of the material a range of exposures under standard conditions of illumination. The amount of blackening produced after carefully controlled development is measured and plotted (usually logarithmically) against the exposure. The speed of the material and much other information can be derived from the resulting graph, which is known as the characteristic curve.

Speed Figures. The data obtained from the characteristic curve have to be translated into a speed figure to state the speed of the material tested. This is done by means of a speed formula, which defines the speed as a function of the exposure at a specific point of the characteristic curve. The choice of such a point has been the subject of a great deal of argument in the past, since various speed systems used to base their criterion on what were to some extent arbitrary considerations.

The aim has been to define for any one photographic material a suitable point on the characteristic curve at which the exposure has a constant relationship to the camera exposure required to produce a satisfactory negative.

Arithmetic and Logarithmic Speeds. Once a point has been chosen on the characteristic curve for speed determination, there are two basic ways of stating the speed. In other words there are two kinds of speed formula.

The simplest type of formula is:

$$\text{Speed} = \text{Constant}/\text{Exposure}$$

The exposure is in this case the sensitometric exposure (time $\times$ light intensity) corresponding to the selected reference point on the characteristic curve. Such a speed figure is known as an arithmetic speed, because it is inversely proportional to the exposure. Thus under a given set of circumstances, a material of twice the speed of another material needs half the exposure of the latter to produce the same image density.

The constant in the speed formula is an arbitrary one, chosen to yield speed figures that are convenient to handle. The most straightforward speed formula uses a constant of 1, with the exposure measured in metre-candle-seconds.

An alternative kind of speed formula yields so-called logarithmic speeds:

$$\text{Speed} = C - K \times \log \text{exposure}$$

The speed figures in this case may be characterized by the sign ° (degrees) to distinguish them from arithmetic speeds.

The constants again serve to make the speed figures easy to handle. Usually K is made equal to 10, to yield suitable intervals of single degrees. A ten-fold increase in speed therefore corresponds to an increase of 10°, or a doubling of the speed to approximately 3°. The second constant C simply is made large enough to keep the speed degrees positive.

The logarithmic system yields a more compact range of figures. A scale of logarithmic degrees from 0 to 30° corresponds to an arithmetic speed from 1 to 1,000. Further, since characteristic curves are plotted logarithmically, curve displacements immediately show the speed difference in logarithmic units.

An alternative logarithmic system is based on a similar formula, but using the logarithm to the base 2 of the exposure and a constant with a value of 1 for K. This yields speed degrees where a doubling of the speed corresponds to an increase of 1°. This is even more compact since an arithmetic speed range from 1 to 1/1,000 corresponds to a scale of only 0 to 10°. This kind of speed scale further has the advantage of giving numerical intervals similar to those of exposure values, and it is suitable also for exposure calculations by the APEX system.

MODERN SPEED SYSTEMS

Black-and-White Negative Materials. After a large number of attempts to define speed in a number of different ways, present-day speed systems use a reference point on a characteristic curve based on the minimum exposure to produce a minimum density above a specified fog level. This is the method employed in current A.S.A., B.S. and D.I.N. standards (A.S.A.: PH2.5–1960; B.S. 1380: Part 1: 1962; D.I.N. 4512, Part 1: 1961).

The speed criterion is an exposure required to produce a density of 0·1 above the base plus fog density. The standards also specify the development with the aid of a second reference point on the characteristic curve. This point is 1·30 log exposure units (measured in metre-

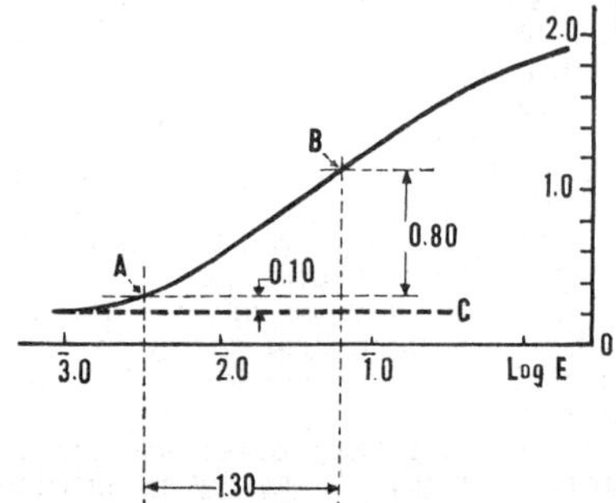

SPEED CRITERION FOR ASA, BS AND DIN SPEED SYSTEMS. The speed point A is located at a density level of 0.1 above fog plus base density. The material is developed so that the density of a second point B, 1.3 log E units farther on than A, has a density of 0.8 above that of the point A. Both arithmetic and logarithmic speeds are derived from the exposure corresponding to the point A on the curve.

candela seconds) to the right of the first reference point. This corresponds to an arithmetic exposure ratio of 200 : 1. When the test sample is developed under the specified processing conditions to the right degree, the density interval of the two points must be exactly 0·8.

From this criterion altogether four speed figures can be derived. The first is the B.S. and A.S.A. arithmetic speed, defined by the formula:

$$\text{Speed} = 0{\cdot}8/E,$$

where E is the exposure corresponding to the point on the characteristic curve at which the density is 0·1 above the base plus fog density under the processing conditions specified above.

The second speed figure is a logarithmic one derived from the equation:

$$\text{Speed (log base 2)} = \log_2 (0{\cdot}24/E)$$

where E is again the same exposure point.

The two speeds are thus related by the equation:

$$\text{Speed (log base 2)} = \log_2 (0{\cdot}3 \times \text{arith. speed}) = 3{\cdot}3 \log_{10}(0{\cdot}3 \times \text{arith. speed})$$

The above arithmetical and logarithmic figures are those used in the B.S. and A.S.A. systems, and the actual figures are rounded off to yield values rising in steps of cube-root—of —2 (1, 1·25, 1·6, 2, 2·5, etc.). The logarithmic figures are rounded off to give intervals of

CURRENT SPEED FIGURES COMPARED

A.S.A. B.S. Arith.	*A.S.A. B.S. log.*	*D.I.N.*	*B.S. log. (Obsolescent)*	*Relative exposure needed*	
3,200	10°	36	46°	1	Special extreme speed
2,500	9·5°	35	45°	1·25	
2,000		34	44°	1·6	
1,600	9°	33	43°	2	Ultra high speed
1,300	8·5°	32	42°	2·5	
1,000		31	41°	3·2	
800	8°	30	40°	4	
640	7·5°	29	39°	5	
500		28	38°	6·4	High speed
400	7°	27	37°	8	
320	6·5°	26	36°	10	
250		25	35°	12·5	
200	6°	24	34°	16	
160	5·5°	23	33°	20	Medium speed B. & W., fast colour
125		22	32°	25	
100	5°	21	31°	32	
80	4·5°	20	30°	40	
64		19	29°	50	
50	4°	18	28°	64	Slow (extra fine grain) B. & W.; med. speed colour
40	3·5°	17	27°	80	
32		16	26°	100	
25	3°	15	25°	125	
20	2·5°	14	24°	160	Extra slow
16		13	23°	200	
12	2°	12	22°	250	
10	1·5°	11	21°	320	
8		10	20°	400	
6	1°	9	19°	500	

0·5°. Arithmetic and logarithmic figures are frequently quoted together: e.g. 100/5°.

The logarithmic base 2 figure is also intended to be the basis of a forthcoming international standard.

The third speed figure is a logarithmic speed to the base 10, derived from the formula:

$$\text{Speed} = 10 \times \log (1/E) = -10 \log_{10}E.$$

This gives intervals where a doubling of the speed corresponds to an increase of 3 steps in the speed numbers. The actual values obtained are again rounded off to give whole numbers. This speed figure is used in the German D.I.N. standard.

A fourth logarithmic figure is similar to the DIN number, but ten steps higher throughout. This is specified in the British Standard No. B.S.1380: Part 1: 1962 as an obsolescent scale, taken over from an earlier British Standard.

The speed figures obtained in these four ways are directly inter-convertible, and are compared in the table above.

Reversal Colour Films. A criterion based on a minimum exposure point is by itself inapplicable to a reversal film, since in a transparency the whole tone range is important. The American Standard PH2.21–1961 and the British Standard B.S. 1380: Part 2: 1963 therefore use as the reference point on the characteristic curve a mean exposure derived from a minimum and a maximum exposure point. The maximum exposure point is one that yields a minimum density of 0·2 above base plus fog density on the colour film; the minimum point is derived by the exposure level at which the shoulder of the curve begins (where the change in the slope reverses its sign) or at a density point 2·0 above the minimum density, whichever is the lower. The mean exposure used as the reference point is the square root of the product of this minimum and maximum exposure level.

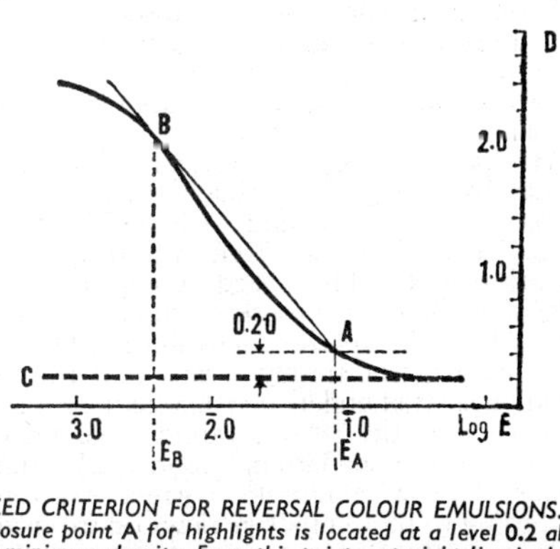

SPEED CRITERION FOR REVERSAL COLOUR EMULSIONS. An exposure point A for highlights is located at a level 0.2 above the minimum density. From this point a straight line is drawn tangent to the D log E curve to establish a shadow point B at a density not more than 2.0 above minimum density. A mean exposure, being the square root of the product $E_A \times E_B$ (corresponding to exposures of points A and B) is the basis of the speed figure.

An arithmetic and a logarithmic speed can be derived from this exposure by the equations:

Arithmetic speed = 8/E

Log (base 2) speed = $\log_2(2{\cdot}4/E)$

In both cases E is the mean exposure point mentioned above. Since this corresponds to a lower exposure (assumed to be on the average by a factor of 10) than the minimum exposure point, the constant in the above speed formulae is ten times larger than for black-and-white negative film speeds.

The speed figures obtained take the same form as for black-and-white negative materials; e.g. 100/5°.

The standards for reversal colour film speeds specify the illumination and exposure procedure, but no processing; the latter is assumed to be carried out in accordance with the film manufacturer's recommendations.

This method of rating colour film speeds is not as arbitrary as it appears at first sight, but was worked out to give a sensitometric method that gave speed figures agreeing closely with empirical speed estimations. The latter were based on the verdicts of large numbers of observers judging colour transparencies exposed to different degrees in the camera.

EARLIER SPEED SYSTEMS

While a reference exposure point based on a minimum density above fog is one of the most obvious systems, early photographic scientists also employed a number of alternative criteria. Some of these were based on minimum exposures to produce a threshold density, on the inertia point, on practical comparative speed tests, on a minimum exposure needed for a negative that would yield a good print, on a point at which the gradient of the characteristic curve reaches a specified value, and so on.

All the earlier systems applied to black-and-white negative materials only. Speeds for reversal and colour materials were expressed as equivalent values, indicating that by setting the quoted speed on an exposure meter or using it in exposure tables, correctly exposed results could normally be expected.

Safety Factors. Some of the earlier film speed ratings, especially the original A.S.A. and B.S. index numbers, were said to include a safety factor. This safety factor is not strictly speaking a feature of the film speed standard, but of exposure meter calibration standards with which these film speed ratings were to be used. The safety factor simply means that the film speed, measured under such a system, and used with an exposure meter, would yield an exposure recommendation about 2½ times higher than a minimum correct exposure. According to more modern views on exposure, such a safety factor is unnecessary and in many cases undesirable. Since it was not practical to revise the standards for exposure meters (which would make all existing meters obsolete) the current A.S.A. and B.S. speed ratings eliminated the safety factor by specifying a higher speed rating for a given exposure criterion. For this reason many films rated according to older speed systems appear to be slower than according to the modern ones.

Scheiner Speed. One of the earliest methods was originated by Professor Julius Scheiner. This criterion is based on the exposure necessary to produce a just visible density above fog.

The speed formula of Scheiner speeds is:

$$°\text{Sch.} = C - 10 \log \text{exposure}$$

Modern high-speed negative materials are reckoned to have European Scheiner ratings between 30° and 38°, colour films between 22° and 34°.

Although this system was used for many years it tends to give misleading exposure recommendations when the characteristic curve of the material has an abnormally extended toe. This raises appreciable difficulties in defining a just visible density. A further drawback is the fact that Professor Scheiner's original specifications for making the sensitometric tests are obsolete, and were based on the use of virtually colour-blind emulsions. The adoption, in the early days of photography, of the Scheiner criterion on the Continent caused what has become known as the "Scheiner inflation". This led to a continuous increase in published speed figures without any real increase in sensitivity.

A number of different Scheiner scales were in existence, differing only in the value of the constant C in the speed formula. The best known was the European Scheiner speed; American Scheiner degrees are about 5° lower. So-called Weston Scheiner degrees are similar numerically to American Scheiner, but are in fact based on Weston speed.

H. & D. Speed. This system was introduced by Hurter and Driffield and utilizes the inertia of the characteristic curve given by the exposure at a point where an extension of the straight-line portion of the characteristic curve cuts the fog level line. The speed formula is:

$$\text{Speed} = 34/\text{exposure}$$

where the exposure is the value in metre-candle-seconds at the inertia point of the material. The speed figures are thus arithmetic.

H. & D. speeds are now virtually obsolete (except in photomechanical applications), partly because modern materials and developers do not behave in the same way with respect to inertia points as the plates used by Hurter and Driffield.

Older DIN Speeds. This criterion for the original German DIN standard was fundamentally the same as for the current one, but specified a development method to bring out the maximum possible film speed. This does not correspond to general photographic practice, and the original DIN speeds do not therefore

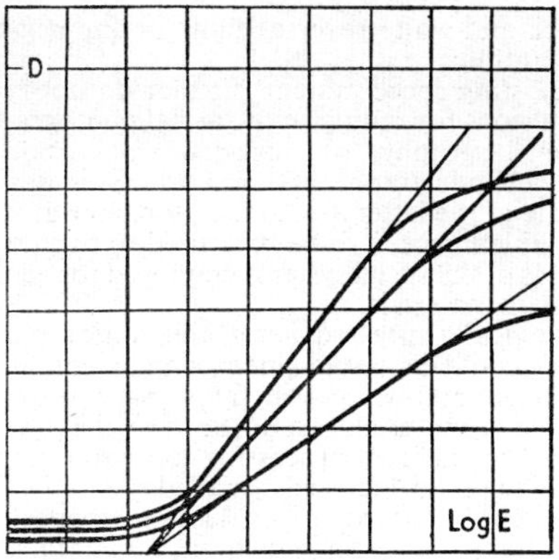

H. & D. SPEED CRITERION. Characteristic curves are drawn for a range of development times in a bromide-free developer, and the inertia point determined where the straight-line portions cut the fog level line.

indicate truly how the materials behaved in normal use.

The speed formula, based on the exposure necessary to produce a density of 0·1 above fog, is logarithmic:

$$\text{Speed} = C - \log \text{exposure}$$

This gave decimal numbers for the speed figures, which were therefore expressed as fractions of 10. A speed of 1·3 was written as 13/10° DIN. The constant C was chosen so as to give a speed number in the numerator of the fraction about 10° lower than the then current Scheiner figures.

The DIN system was the first attempt at large-scale standardization of a speed criterion, and is in its present form widely used in Germany and on the European Continent. The artificial maximum-speed development specification however prevented international adoption. In 1958 the DIN standard was therefore revised to meet these objections, by specifying a more normal development procedure similar to that used today.

The constant C was changed to give numerically similar DIN speeds under the new conditions to the original DIN degrees. At the same time the /10° suffix was dropped, and DIN speeds were expressed as simple degrees, e.g. 15° DIN.

When DIN speed ratings were further revised in 1961 to fall into step with the current A.S.A. and B.S. systems, the degree sign was also eliminated, so that 15° DIN became 15 DIN.

Weston and Other Meter Speeds. To provide reliable speed figures for use with certain makes of exposure meter, some meter manufacturers concerned issued their own film speed systems. These were based not on any sensitometric calculations, but on practical tests and experiments.

The best known of the meter speed systems was the Weston scale which provides arithmetic numbers. Logarithmic Weston speeds (also known as Weston Scheiner) can be obtained for certain purposes by the formula:

$$\text{Log Speed} = 10 \log \text{Weston} + 9$$

Current Weston exposure meters have abandoned the Weston scale and use A.S.A./B.S. arithmetical speed numbers.

A number of other meter speed systems have been published, either numerically similar to Weston speeds, or utilizing lettered speed groups, etc. Practically all these were dropped on the introduction of the B.S. and A.S.A. speed numbers.

GOST Figures. This system is in use in Soviet Russia and a number of East European countries. In principle the criterion is similar (but not identical) with the D.I.N. standard, and several different kinds of development are specified for different types of material. The speed figures are expressed in an arithmetic series, and are fairly similar quantitatively to the A.S.A. and B.S. arithmetic index.

Original A.S.A. and B.S. Index. At a time when existing film speed criteria were based on a variety of not easily reconciled assumptions about speed, attempts were made to establish an empirical system based on judgement of print quality, and to construct a sensitometric method to give a corresponding speed value. With this end in view, Jones and Nelson have correlated various speed criteria with the minimum exposure necessary to produce excellent prints as judged by a large number of observers. They found that the best correlation was given by a criterion which defined the speed by the exposure necessary to reach a point on the characteristic curve where the slope was 0·3 of the average gradient of the characteristic curve between that point and a point where the corresponding exposure was

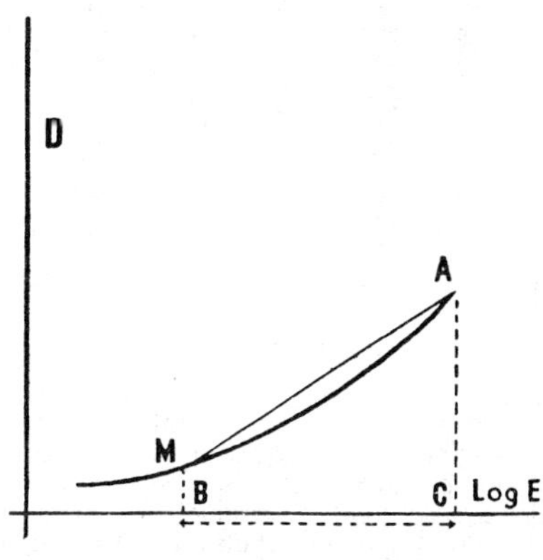

EARLY B.S.I. AND A.S.A. SPEEDS. The slope of the curve at M is 0·3 times the slope MA, where BC is equal to a log exposure range of 1·5. M is then the A.S.A. and B.S.I. speed point.

thirty times as great. The 0·3 criterion where $\bar{G}$ (gee-bar) stands for average gradient, was adopted by both the American and British Standard organizations.

From this criterion several types of speed figure or exposure index were derived.

(1) The speed of a material was defined as 1/E, where E was the exposure measured in metre-candle-seconds (mcs), to reach the speed criterion. Speed figures carried the prefix "0": speed 500 = 0500. The Exposure Index was one-quarter of the speed and was the number recommended for use with exposure aids (125).

(2) The other type of speed figure was the logarithmic exposure index, characterized by the ° sign (degree) given by the formula 5–10 log E. In the above example, the speed would be 32°.

Conversion Tables. The older speed systems are not strictly comparable since they are all based on different properties of the characteristic curve. But the curves of modern photographic materials are sufficiently similar to allow the figures to be compared approximately. The figures in the table below may be regarded as a rough and ready guide.

COMPARISON OF OLDER SPEED SYSTEMS

B.S., A.S.A. Logar.	*B.S., A.S.A. Arithm.*	*Scheiner*	*Weston (Old)*	*DIN*	*H. & D. British*	*GOST (Russian)*
40°	800	41°	640	30°	25,000	720
39°	640	40°	500	29°	20,000	560
38°	500	39°	400	28°	16,000	450
37°	400	38°	320	27°	13,000	360
36°	320	37°	250	26°	10,000	280
35°	250	36°	200	25°	8,000	220
34°	200	35°	160	24°	6,400	180
33°	160	34°	125	23°	5,000	140
32°	125	33°	100	22°	4,000	110
31°	100	32°	80	21°	3,200	90
30°	80	31°	64	20°	2,500	70
29°	64	30°	48	19°	2,000	56
28°	50	29°	40	18°	1,600	45
27°	40	28°	32	17°	1,300	35
26°	32	27°	24	16°	1,000	28
25°	25	26°	20	15°	800	22
24°	20	25°	16	14°	640	18
23°	16	24°	12	13°	500	14
22°	12	23°	10	12°	400	11
21°	10	22°	8	11°	320	9
20°	8	21°	6	10°	250	7
19°	6	20°	5	9°	200	5·5
18°	5	19°	4	8°	160	4·5
17°	4	18°	3	7°	125	3·5
16°	3	17°	2·5	6°	100	2·8
15°	2·5	16°	2	5°	75	2·2
14°	2	15°	1·5	4°	60	1·8

SPECIAL MATERIALS

The speed systems discussed so far apply to continuous-tone black-and-white negative materials—that is, to emulsions intended for ordinary picture taking. Moreover, the speeds refer to daylight. Artificial light speeds—where given by manufacturers—are empirical equivalents.

There is a current standard for reversal colour films (still, not cine) but none for either black-and-white reversal films or for negative colour films.

Existing speed systems further do not cover emulsions for radiography, aerial photography, cinematography, or infra-red work; nor do they include graphic arts and process materials. Some of the latter are, in fact, too slow and too variable in speed to be measured to fit into the scales of the speed systems employed for general camera materials.

Logically infra-red, aerial and motion-picture films could be tested under similar conditions to black-and-white still films. Speeds quoted for such materials would however have to be based on different processing conditions applicable to these materials, and also on possibly different standards of evaluation to meet the requirements under which such materials are used. The Russian GOST system tries to allow for this in part, by specifying different degrees of development and different developers for, for instance, aerial films and motion picture films.

Graphic arts and process films are not amenable to speed measurement by any of these methods, owing to the shape of their characteristic curve. Moreover, since such materials are used under a variety of widely different conditions, any speed figure would have comparatively little meaning.

Printing Materials. There is no standard method of testing or expressing the speed of papers and other printing materials. Generally there is no need for such speed ratings, because printing and enlarging exposures depend on the exposing light and other characteristics of the printing or enlarging set-up and are determined by exposure tests.

Some manufacturers give relative speeds of their different types of printing material. These are neither standardized nor consistent from one manufacturer to another, and do not compare with the speeds of different makes. They do, however, give a basis for exposure tests when changing from one type of paper to another with the same printing equipment. The following is a very approximate comparison.

COMPARATIVE SPEEDS OF PRINTING MATERIALS

Material	*Relative speed*
Contact papers and contact lantern plates ...	1
Slow chlorobromide papers	10–20
Fast chlorobromide papers and warm-tone lantern plates	50–100
Bromide papers and bromide lantern plates, positive films	100–500
(High-speed pan film or plate ... around	100,000)

W.F.B. & L.A.M.

See also: *Characteristic curve; Exposure; Negative materials; Sensitometry.*

Books: *Progress in Photography* (2 vols.), ed. by D. A. Spencer (London); *Sensitometry*, by L. Lobel and M. Dubois (London); *General Sensitometry*, by Yu. N. Gorokhovskii and T. M. Levenberg (London).

SPHERICAL ABERRATION. One of the aberrations of lenses. It causes general loss of sharpness in the image. In some cases it may be introduced intentionally, as is done with soft focus lenses.

See also: *Aberrations of lenses.*

SPILL RINGS. Concentric rings of sheet metal mounted in front of narrow-angle lamp reflectors to cut out the spreading direct rays of light from the light source.

SPIRIT PHOTOGRAPHY. Spirit photography is claimed to have been accidentally discovered by William Mumler of Boston, Mass., in 1861. Mumler, it is said, focused the lens of his camera on a vacant chair with the object of taking a self-portrait. His story was that after uncapping the lens he immediately jumped into position on the chair and sat there for a few seconds. On developing the plate he found an additional image of a little girl sitting on his knee. Mumler claimed that he recognized this photograph as being that of a young relative who had died some twelve years previously. He continued his experiments and stated that the extra figures of his plates became so frequent as to be embarrassing. In this way Mumler became the first professional spirit photographer.

Early Frauds. Mumler claimed that the extra figures were actual objective photographs of spirit forms present in the studio; invisible to the naked eye but capable of being recorded by the camera. For a while, Mumler's theory was accepted until it was found that photographs of living people appeared on his plates. He was accused of fraud and died in poverty in 1884. His results show every evidence of having been produced by double exposure.

Mumler was imitated by many others (usually professional photographers) both in this country and the U.S.A. Almost without exception, these self-styled spirit photographers were convicted of fraud. In many instances they actually confessed to trickery. In 1906, the London *Daily Mail* appointed a committee to investigate the claims in regard to spirit photography, but for lack of the human material with which to experiment the venture produced no results.

The Crewe Circle. It was about this time (1908) that a group of spiritualists, who went by the name of the Crewe Circle, began to attract attention. These people conducted operations from a private house at Crewe. They claimed to be non-professional and had no fixed fees, but the sitters were expected to make a financial contribution towards "expenses". This circle attracted world-wide attention and many investigators made the pilgrimage to the little house at Crewe.

In actual fact, the Crewe circle was founded upon a Mr. William Hope who was undoubtedly a schizophrenic. On one side of his character was an alert, witty and patently honest North-countryman, whilst on the other hand there was the bogus medium who used prayers and psalm-singing as a cloak for his fraudulent operations.

Hope described his work as "psychic photography" and he and his followers maintained that the results produced were not actually photographs of spirits but were spirit-produced photographs.

For the extraneous forms on these so-called psychic photographs they used the name of "extras".

By the simple method of employing several cameras simultaneously, independent investigators easily proved that there was no question of objective spirit forms since the "extra" appeared on one of the plates only—and that, the one under Hope's control.

The usual procedure was for a sitter to take along an unopened box of plates which he loaded in Hope's darkroom. He then sat for Hope to take his portrait after which they both retired to the darkroom again to develop the plates. One of these usually showed an "extra". In every observed instance where there was an "extra" on the plate Hope either handled, or could have handled, the plate before exposure.

Although many of the sitters for whom "extras" appeared were unknown to Hope, it was significant that out of over a thousand cases investigated, where the extra was a recognizable likeness of someone known to the sitter, it was an exact copy of some existing photograph or illustration.

At first Hope appears to have produced his "extras" by substituting his own doctored plates for the sitter's. After he was caught at this he elaborated new methods. One of these appears to have involved the use of luminous paint. A positive transparency backed with calcium sulphide and pressed into contact with an unexposed plate will leave an "extra" negative image in addition to anything else that may be recorded there before or after.

Psychographs. Written messages obtained by the Crewe Circle on photographic plates held between the hands and not exposed in the camera were named "psychographs".

These psychographs were of somewhat rare occurrence and, apart from the actual handwriting, they provided little of evidential value. Many of them were in Hope's handwriting and contained grammatical and spelling mistakes such as he was in the habit of making. Careful investigation proved that in no single instance had Hope produced a message in facsimile handwriting from persons with whom he had not corresponded during their lifetime.

Examination showed that he had built up his psychographs by cutting out words and phrases from these letters and filling in the blanks in his own hand. There were obvious attempts to conceal the cut edges, and Hope's own style was only too evident in the forged writing.

Hope's fraudulent methods were eventually confirmed by one of his greatest friends and supporters, and when he died in 1932 the Crewe Circle died with him.

The late Sir Arthur Conan Doyle was a keen supporter of Hope and his works. It has now been proved that he and others of his time were sadly lacking in the critical faculty desirable for analysing the claims of such people as Hope. The truth is that Sir Arthur was so thoroughly honest himself that he could not believe it possible that trickery was rampant in what, to him, was a very sacred subject.

Investigations. There have been many other variations on the above themes, but in every case impartial investigation has revealed evidence of substituted plates and double exposure. Examination of the wrappings of the plates taken to the seance by the sitter in many cases shows signs of opening and re-sealing. With one medium, the substituted "extra" bearing plates were found to have one roughened edge. This indicated the top of the plate so that it could be inserted in the dark slide the right way for the "extra" to appear upright with respect to the image of the sitter.

Plates that have previously been exposed to give an "extra" can usually be detected by examining the margins; the clean rebate almost always has a doubled edge because the plate never lies in exactly the same position in the holder for the two exposures.

Finally, the Society for Psychical Research have made it clear, in a report covering some hundreds of claims, that so far their investigators have never seen any alleged "spirit" photograph that could not have been produced by natural means.

The publication of this report may explain why spirit photography has now gone out of fashion in Great Britain. Elsewhere, however, claims still continue to be made. F.B.

See also: *Psychical research; Tricks and effects.*

SPLICING

Method of joining length of cine film or magnetic tape during editing or to repair breaks. A good splice should be at least as strong as the original material and it should run through the equipment without any noticeable hitch. A splice can be made by hand or by using a more or less automatic splicing jig. The latter makes a quicker and more accurate job.

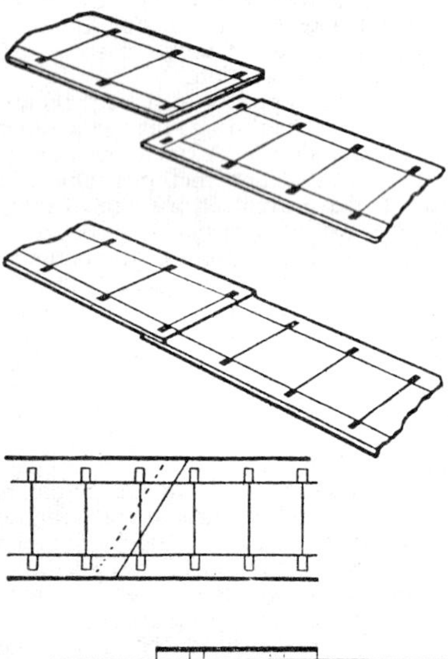

SPLICING CINE FILM. Top: Before joining, the film ends are trimmed together, and the emulsion scraped off one of them. This permits the film cement to weld the two strips together. Upper centre: *Standard square splice.* Lower centre: *Diagonal splice.* Bottom: *Cross-section of two films joined together by a black patch of film. No scraping is necessary as only the shiny sides are in contact.*

Splicing Cine Film. When joining lengths of cine film it is not enough to make a strong and accurately lined-up joint; the spacing of the sprocket holes must be maintained from one side of the splice to the other.

There are three ways of splicing: the usual way is to cut the ends of the film square and join them by lapping one over the other; 16 mm. films are sometimes cut diagonally to give a long splice that does not weaken the film as much as the normal splice; the ends may be cut off square, butted together, and joined by cementing a strip of black film or adhesive tape across the emulsion-free side of the join.

When the splice is made by lapping one end of the film over the other, the emulsion is first scraped away from the coated side of one end. The clear edge is then brushed over with film cement and the other end of the film is quickly brought into register and pressed down on it, uncoated side down. It is essential for the two lengths of film to be in line and for the sprocket holes to coincide. Negative films—subjected to less strain—normally have a narrower splice.

A more modern method of making the splice involves scraping the two ends of the film at an angle. The two bevelled edges are then joined up, so that the join itself is not thicker than the film. This method is particularly suitable for frame-line splicing, where the joint does not encroach on the picture area of the frame on either side of it.

Splicer. The lengths of film to be joined may be held in line with drawing pins. The operation can then be carried out by scraping off the

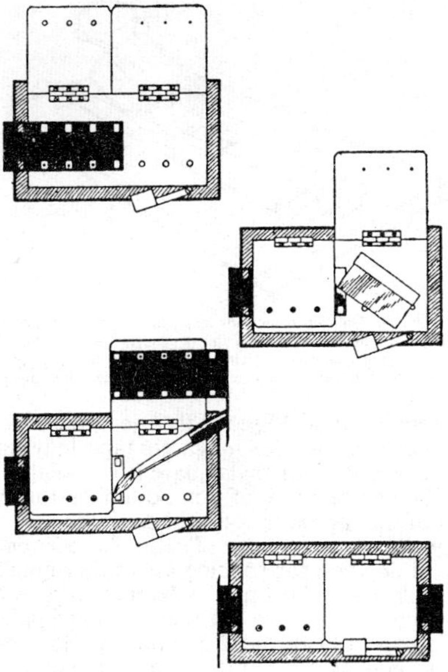

USING A SPLICER. Top left: Trim film end, place on left pins, emulsion up. Top right: Hold down film by cover plate and scrape emulsion off exposed strip. Bottom left: Place end of other film on upper registration pins, emulsion down, and trim flush with pressure plate. Apply cement to scraped portion. Bottom right: Lower the pressure plate, bringing the two films together, and clamp down.

emulsion with a razor blade, applying cement and pressing the ends together by hand.

After some practice it is possible to turn out a good splice in this way in a minute or so, but where there is a lot of splicing to be done there is something to be said for using a special splicer. This consists of a bedplate with pegs for registering the sprocket holes. There are two hinged flaps which can be folded down on to the film when it is in position on the pegs.

One piece of film is trimmed and held in position on the bed by one flap so that a narrow strip the width of the splice projects beyond the flap. The emulsion is scraped away from this strip with a safety razor blade, or a special scraper incorporated in the splicer.

The other half of the film is placed on the open flap and trimmed square with the edge. A smear of cement is wiped over the prepared end of the first piece of film and the second is immediately pressed into contact with it, shiny side down, by folding down the open flap. A spring clip locks the second flap and keeps pressure on the splice until it is set.

Film Cements. There are various proprietary film cements available. In all of them the action is the same; the cement softens the film base as soon as it is applied, and a few seconds later it is absorbed by the base. The two halves of the splice must be pressed together while the base is still in the soft condition. Properly made a splice should be stronger than the film.

The basis of most cements for nitrate film (now obsolete) is amyl acetate and for acetate film, acetic acid.

A suitable formula for both nitrate and acetate films is:—

Acetone	8 parts
Ether	10 parts
Acetic acid, glacial	1 part

To this is added about 1 part of 16 mm. film (with the emulsion removed) per 3–4 ounces (100 c.cm.) of solvent. More may be added to make a thicker cement.

Butt Splicing. A modern method of film splicing uses transparent self-adhesive tape. There the film ends are cut square and abutted end to end A piece of self-adhesive tape is then stuck over the join on both sides. Splices for this purpose have provision for accurate lining up of the film ends and for trimming them; they further perforate the tape once it is in position so that the film runs smoothly through the projector.

The tape should preferably be a polyester tape for maximum strength. Ordinary adhesive cellulose tapes are liable to stretch and to creep, making the join inaccurate.

Where the spliced films carry an optical sound track it is best to paint the join over with black dye (so-called blooping ink) to avoid the otherwise obvious click as the splice goes through the sound head.

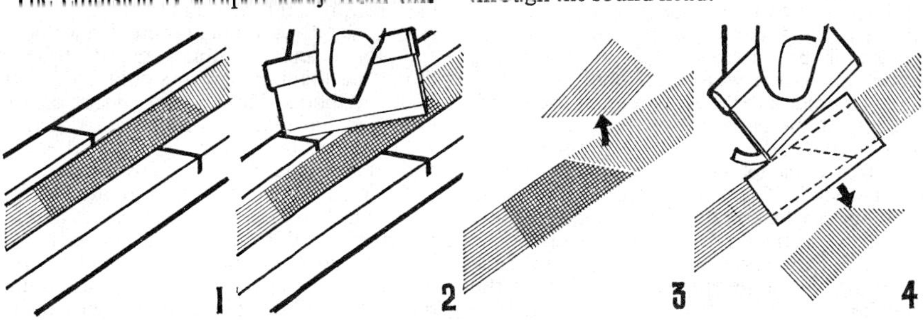

SPLICING TAPE. 1. Place the two tape ends in the channel of the splicing block, overlapping across the cutting line. 2. Cut through both layers of tape with a sharp razor blade at the appropriate angle. 3. Remove the short end cut off at the top. 4. Apply a piece of splicing tape across the join and trim off along the edges.

Splicing Magnetic Tape. There are two ways of splicing magnetic tape: (1) by making a butt joint (where the ends join but do not overlap) with splicing tape, and (2) by making a lap joint with jointing compound. Broadly speaking jointing tape is used for a temporary joint and compound for a permanent joint.

(1) Joining with splicing tape. In this method the ends of the tape are first trimmed across at an angle of 45° with a razor blade or scissors. They are then brought together so that there is no gap between the ends and secured by pressing a short length of special adhesive splicing tape over the join on the polished—not the coated—side of the tape. Ordinary adhesive tape is again unsuitable because the adhesive is too soft and tends to squeeze out of the joint. When it does this it makes a sticky place on the tape. This has two unpleasant effects: it makes the tape jerk as it goes through the sound channel (causing a sudden drop in the pitch of the sound) and it usually transfers itself to the capstan or pressure roller and sets up wow or flutter.

The 45° cut ensures that the splice goes past the sound head with the minimum of noticeable noise.

(2) Joining with jointing compound. This method is basically similar to the one used for splicing cine film. The ends of the tape may be cut across at 45° or 90°. About ⅛ in. of the magnetic oxide coating is removed from one end of the tape by wetting it with tape jointing compound and immediately wiping it off. The clear end of the tape is wet a second time with compound and the other end is pressed on to it so that the polished back of the tape lies over the prepared surface. The joint is kept under slight pressure for 10 to 15 seconds, after which it can be used normally. As magnetic tape may have an acetate, PVC or mylar base it is essential to use the correct type of compound or the two ends of the tape will not weld together.

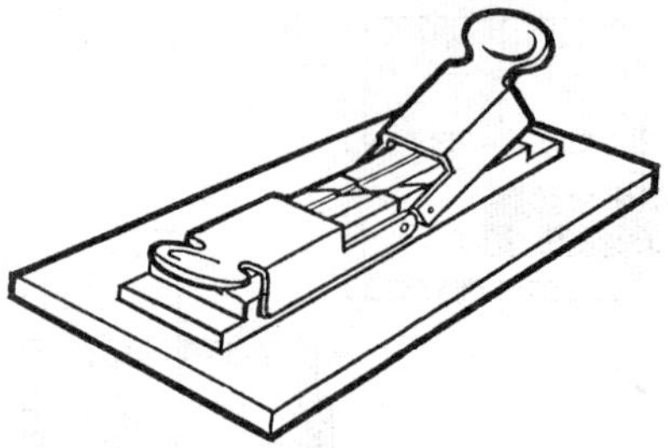

TAPE SLICER. The two tape ends are placed on the channel in the centre of the block and held down by the two clamps. The block carries grooves for cutting the tape at the appropriate angle and for slicing off surplus joining tape from the edges.

Magnetic Tape Splicers. To hold the ends of tape in position for splicing an editing block is generally used. This has a channel to take the two tape ends and may further include a pair of clamps to hold them in position. A 45-degree groove acts as a guide for a razor blade to cut the tape ends at the correct angle before sticking the joining tape on top of them. Further grooves may aid in trimming the tape away at the sides of the join.

The editing block may be plastic or a non-ferrous metal such as bronze (iron might interfere with the magnetization of the tape).

F.P. (L.A.M.)

Books: *How to Edit*, by H. Baddeley (London); *The Tape Editing Guide*, by Ronald Hack (London).

SPOOL. Core or bobbin carrying the film in cameras and film projectors. There is a wide variety of shapes and sizes.

Spools for Still Cameras. Spools for roll film cameras are standardized in size and thickness for the different film widths. Some spools have wooden cores with metal ends, but currently the tendency is to make the spools completely of metal or, in a few cases, plastic.

The spool core carries a slit to take the end of the backing paper, while the spool ends contain keyed holes to engage the peg of the transport shaft in the camera.

Spools of 35 mm. cameras usually form part of the film cassette, being contained inside a light-tight shell. There the spool itself is made of metal or plastic and fitted with a knob at one end which protrudes from the cassette. The other end of the spool has a hole with a suitable fitting inside to engage the shaft of the rewind knob.

The core of a 35 mm. spool carries a slit or similar device to anchor the end of the film.

Most 33 mm. cameras also have a take-up spool on which the film is wound up during exposure. Usually this take-up spool is built into the camera and turned by the film transport through a friction drive. Some cameras, however, have removable take-up spools. The anchoring of the film to the take-up spool must permit easy disengagement after the film has been rewound into the cassette.

Spools for Cine Cameras and Projectors. The film for 8 and 16 mm. cameras (other than magazine loading types) is sold on 25, 50, or 100 feet spools. These again consist of a metal core and flat disc-type metal ends. The latter are considerably larger than the core to hold the greater lengths of film involved.

Projector spools take still greater lengths of film (some up to 2,000 feet) and the metal ends (or cheeks in this case) are usually only frames. They are made of metal (or sometimes plastic) and often carry a gauge for estimating the amount of film wound up by measuring the thickness from the core outwards.

See also: *Cassette; Cine camera; Film transport; Magazine; Sizes and packings.*

SPORT

Taking photographs at sporting and athletic events of any kind usually means taking pictures of action so that movement does not show in the final print. The principal attraction about this branch of photography is that it can present a sharp and permanent picture of a split second of exciting movement which the eye normally misses or sees as a fleeting blur.

Movement can be arrested in this way by the use of a shutter speed so short that the amount of blur caused is too small for the eye to detect.

EQUIPMENT AND TECHNIQUE

The equipment need for sports photography will depend on the type of sport photographed, although simple equipment is satisfactory in many cases. But whatever the equipment, the technique must be studied carefully.

Camera. Any camera—even a simple box—will turn out satisfactory sports pictures so long as it is used intelligently within its own limitations. Pictures of the instant before the start of a sprint race, or long-distance views of a football match, are well within the scope of the box camera, but close shots of the runners in motion or of a scrimmage in the goal mouth call for a camera with a very much better lens and shutter than are found on a cheap camera.

The two cameras most used for outdoor sports pictures are the press and the 35 mm. Newspaper cameramen used to prefer the press camera because they found the relatively large plates or sheet film easier to process and print quickly than the 35 mm. film, but in recent years the 35 mm. camera, especially the single-lens reflex, has gained great popularity among press men.

35 mm. cameras are used more for occasions when the full load of 36 exposures is likely to be needed and where more time can be given to processing the negatives—e.g., when making pictures for weekly or monthly publications.

A camera for serious sports photography must have a shutter with a maximum speed of at least 1/500 second.

Lens. A long-focus lens—or even a telephoto—has distinct advantages for certain sports, but it is generally better to work with a normal angle lens close to the subject wherever this is possible. The telephoto lens gives the subject a flat, unpleasing perspective and tends to dwarf the subject against its background. But for subjects like the Boat Race, batsmen in action, and aquatic sports, a telephoto lens is essential.

No matter what type of lens is used it must have an aperture of $f\,4{\cdot}5$ or larger so that really fast shutter speeds may be used even in poor light.

Viewfinder. An open frame or wire viewfinder is best for sports photography. This type gives a direct view of the subject as a whole with the part included within the picture area clearly marked by the finder frame. The view in the normal direct vision optical finder does not extend beyond the picture area and this makes it difficult to find a fast moving subject.

The Albada type of direct vision optical finder is excellent for sports photography as it combines the compactness of the optical finder with an open-frame view of the subject.

Waist-level reflex viewfinders and focusing screens are unsuitable because they reverse the image from right to left and the camera has to be swung in the opposite direction to the movement of the image—an action that calls for considerable practice.

A number of reflex cameras have eye-level reflex finders with a penta prism or similar optical system. This shows an upright and right-way-round image, and does not suffer from the drawback mentioned.

Filters. Filters are not much used in sports photography because they demand an increase in exposure—i.e., either a slower shutter speed or a wider lens aperture, both undesirable. But a pale yellow filter is sometimes useful for giving some tone to the sky when it forms the background to a low angle shot.

Sensitized Materials. It is almost always advisable to use the fastest available panchromatic material because, in this class of photography, speed is more important than anything else.

Exposure. The exposure is dictated by the strength of the light as in any other kind of photography, but the shutter speed that must be used to arrest the movement of the subject depends upon how far the subject is from the camera and its speed and direction of movement. This in turn limits the smallest lens aperture that can be used, and hence the

IRREGULAR MOTION. The legs and arms of a jumper need a high shutter speed, though his body will come out sharp at a slower speed. The whole of him is practically motionless for an instant at the top of the jump—the dead point.

maximum depth of field at the photographer's disposal.

Lighting. Most sports photography is carried on in daylight. But there are some sporting events—e.g., boxing, fencing, wrestling—which take place indoors. For these sports synchronized flash lighting—either flash bulb or electronic flash—is used where permitted, otherwise the photographer has to make the best of the existing artificial lighting.

NOTES ON TAKING

The technique for photographing sports varies. The following summarizes the points to watch in most sports.

Association Football. The best position is on one of the goal lines, using shutter speeds up to 1/1000 second. Alternatively, shoot from among the spectators, about 20 yards from the corner flag, looking into the goal mouth. Use a long focus lens except in the goal area.

Badminton. The best position is at either end of the net, covering both courts. Use a normal angle lens, synchro flash lighting with shutter speeds of at least 1/500 second.

Billiards and Snooker. The best position is at the top end of the table. A lens of *f* 1·5. or *f* 2·0. will take pictures at 1/15 to 1/30 second by normal table lighting; otherwise use flash and a small aperture to give maximum depth of field. There is no point in exposing during the action; slow shutter speeds at static moments in the game are quite suitable.

Bowls. Needs only normal technique with practically any type of camera. The light is usually good and movement never very fast. Speeds of 1/60 to 1/125 second are fast enough.

Boxing. The best positions are at the ringside, so that the camera can shoot through the ropes, or in the gallery with a long focus lens. A wide aperture lens is needed to keep shutter speeds down to 1/250 to 1/500 second by normal ring lighting (flash is not normally permitted during a bout but may be used before and after).

Cricket. The best position is somewhere above ground level, so that the camera looks down on the players against a background of the pitch. A long focus lens is essential. The shutter speeds for players in action—up to 1/500 second.

Curling. This is very similar to bowls, but played on ice. The movements of players are also similar and the same technique can therefore be used. Shutter speeds of 1/60 to 1/125 second are normally suitable.

Diving. The best position is one where the sun is behind or to the side of the camera and where divers can be caught against the sky. Use a normal angle lens and a yellow filter. A shutter speed of 1/250 to 1/500 second should be used.

Fencing. It is possible to work with slow shutter speeds provided the right moment is chosen for the exposure. Concentrate on one contestant and catch the various stances, or try to get both from an elevated viewpoint. The light is often poor, so flash may be necessary. For dead points in the action, use a shutter speed of 1/50 second, varying up to 1/500 second for fast action.

Field Events. Pictures of jumping, vaulting, throwing the javelin, discus, etc., call more for accurate timing than special equipment. In good weather, most events can be handled with a shutter speed of 1/250 second. Fairly close positions are best where a normal angle lens will give a reasonable size of the image.

Foot Races. The best position, if it can be obtained, is where there is a high oblique or head-on view of the sprinters approaching the tape. The shutter speed must be fast enough to arrest the movement of limbs—e.g., up to 1/1000 second during sprints.

Golf. Players object to the click of the shutter while concentrating on a stroke, so expose after. Otherwise use normal equipment and technique. Shutter speeds for players in action —up to 1/250 second.

Greyhound Racing. The best position is near the track, but as a rule there is no alternative to working from among the spectators with a long focus lens. The lighting of tracks after dark is too dim for high speed photography and flash is not permitted. Action shots of dogs call for speeds up to 1/1000 second.

Hockey. The best position is in range of the "striking circle". Fast movement of sticks and ball need shutter speeds up to 1/500 second.

Horse Racing. The best position is at start and finishing lines, and on the course with a view of a corner where horses will bunch together. Pictures in the paddock are also worth going after. Shutter speeds—up to 1/250 second in the paddock and 1/500 during a race. Use normal and long focus lenses and fastest pan film.

Ice Hockey. The best positions are above ground level—at either end of the gallery, midway between goal and corner. Lighting is best in the smaller rinks where roof and lights are lower. Use normal or long focus lenses. Shutter speeds of 1/150 second can be used at *f* 2 with fastest pan films.

Lawn Tennis. The best positions are at the centre of the side or at the end. Doubles are best photographed from a stand or other elevation. The best method is to set the focusing for one particular zone and shoot when action comes into it. Use a normal focus lens and fast films. Shutter speeds: up to 1/1000 second.

Motor Racing. The best position is a high viewpoint looking down on the width of the track. When shooting from the side, the background should be plain and clear of spectators or somewhat out of focus. Try to include more than one car in each shot. Use a normal angle lens. Shutter speeds—1/250 to 1/500 second, or slower if the camera is carefully "swung" with the moving vehicle. This method, when used with skill, will make the background appear blurred, giving a very realistic impression of speed.

Rowing. The best position is on the river bank as close to the boats as possible, or shooting at an oblique angle from a bridge. The peaks of movement are as the oars enter and leave the water. Use a normal or long focus lens with fast film. Shutter speeds—1/250 to 1/500 second.
Rugby. The best positions are along the side-lines as there is not the same interest in the goal as in Association Football. The greatest interest lies in shots taken with the camera set at medium zone focus; then wait until action takes place in the zone. Close range shots need 1/500 second and distant general views 1/250 second.
Skating. The best position for races out of doors is on the outside of the track where skaters can be taken head-on, the camera being pre-focused on a particular spot and the exposure made as skaters reach it. Indoor shots of figure skating should be taken from a position dictated by placing of the spotlights—preferably after a preview to decide the when and where.

Shutter speeds should be 1/250 second and, for races, up to 1/500 second.
Ski-ing. The best position is often where skiers appear against the light—preferably about 20 feet below a turn, to catch the snow with back lighting. Jumps are best shot from the side and well below the take-off. Shutter speeds—up to 1/500 second.
Speedway Racing. The best position is looking obliquely at the first bend. Beware of flying cinders as riders draw level. Shutter speeds—1/250 to 1/500 second.
Steeplechasing. The best position is in front or at the side of the jump, preferably "open ditch" and water jump. Use a low shooting angle. Shutter speeds—1/250 to 1/500 second, or faster if lighting conditions permit.
Swimming Races. The best position is at the start to catch the swimmers diving into the water; elsewhere there is little interest. Photographs of water sports are often improved by the use of a polarizing screen, which cuts out a lot of the distracting reflections from the surface of the water, though this also destroys atmosphere.

For actual swimming, shutter speeds up to 1/250 seconds should be used; for the start of the race, use 1/500 second.
Table Tennis. The best position for taking doubles is a high viewpoint to prevent one player masking the other. Singles may be taken from a lower viewpoint, concentrating on one player.

Practically the only way of making action shots is to use synchronized speed flash with the lens stopped down to at least *f* 8 to give the necessary depth of field.
Wrestling. As for boxing, except that the action is slower and shutter speeds of 1/125 to 1/250 second at *f* 2 can be used with the existing lighting.
Yachting. Normal technique and equipment are all that is needed. As the light is usually very good, a medium speed film with its finer grain and higher contrast can be used with a filter to make the white sails stand out against the sky. Shutter speeds of 1/125 to 1/250 second are fast enough for most subjects. L.V.

See also: *Flash (electron c); Flash technique; Horses; Ice rink; Marine photography; Motor racing; Movement; Underwater photography; Winter sports.*

Books: *My Way with the Miniature*, by L. Vining, and *Camera Afloat*, by H. S. Newcombe (both London).

SPOTLIGHT. A spotlight combines a compact source lamp, a reflector, and a lens, in a single unit. It is designed to give a uniform light that can be varied at will from a narrow, sharply defined beam to a broad diverging cone.

The concentration of the beam is varied by moving the lamp and reflector together towards or away from the lens. In all such arrangements the lamp is fixed at the focus of the reflector and the two are subsequently adjusted as a single unit.

The duty of the reflector in this combination is to reflect the rays of light back through the plane of the lamp filament. By doing this it enables both the glowing filament and its reflection to be focused at the same time by the lens.

So the arrangement always consists of a spherical mirror with the lamp filament at its focal point—i.e., at the centre of curvature of the mirror.

The focal length of the lens in a spotlight is kept short so that the lamp works close to the lens and throws the maximum amount of direct light on to it. An ordinary type of lens with a suitably short focal length would be unduly

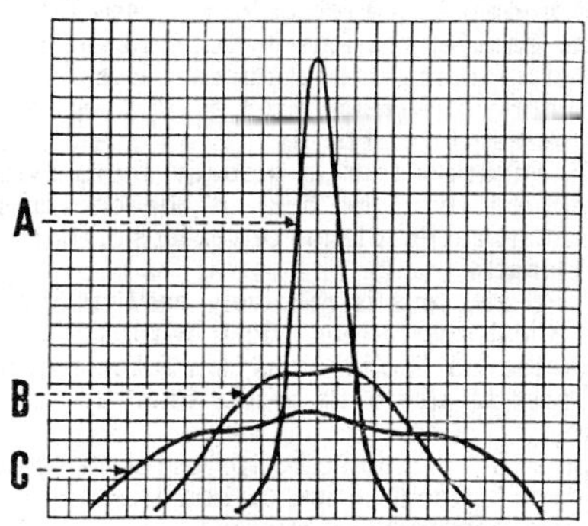

SPOTLIGHT EFFICIENCY. Intensity across illuminated field of a Fresnel spotlight with various lamp positions. A, lamp back from condenser (spot position). B, mid-way position. C, lamp near condenser for full flood effect. The vertical centre line of the graph represents a position directly in front of the lamp.

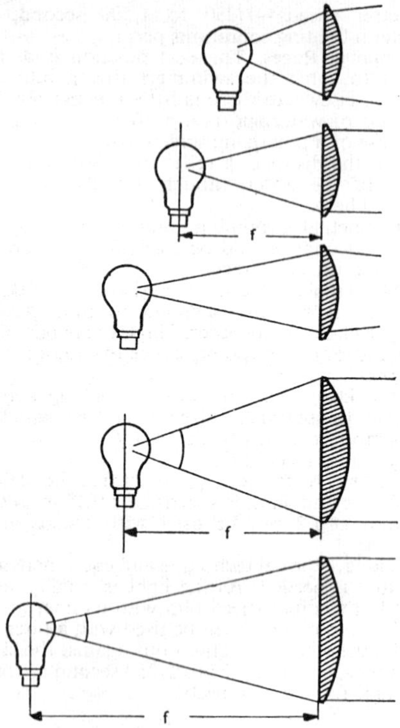

SPOTLIGHT SYSTEMS. Upper group: *When a lamp is placed at a distance from a condenser equal to its focal length, the beam is substantially parallel (actually slightly divergent because of the finite size of the source). When the lamp is brought nearer to the condenser, the beam diverges, and when it is moved away from the condenser, the beam converges.* Lower group: *For a given diameter, a short focus condenser takes in a greater angle of light from the lamp, and therefore is more efficient than one of longer focal length.*

thick and heavy, so a special optical arrangement known as a Fresnel lens is used instead.

The back of the lens is flat, and the front, instead of being a continuous curve, is built up in concentric rings, each being a section of a curved surface. The final effect gives a field of illumination that is actually more evenly distributed than the field of the equivalent, but much heavier, plano-convex lens of the same focal length.

A Fresnel lens is relatively easy to correct for optical aberrations and it withstands heat better than a "solid" lens, since the reduced thickness at all points is less likely to set up strains.

Spotlights are designed for use with projection-type filament lamps, carbon arcs, and high pressure mercury vapour lamps.

The front of the spotlight generally carries a pair of guides to take such accessories as colour and diffusion screens and cut-out metal masks which can be focused by a separate lens mounted in front to project a sharp-edged area of light.

Beam restricting devices—e.g., snoots and barn doors—may also be attached to the front of a spotlight where the spot must be confined to a particular area of the subject.

A slide projector also makes a useful spotlamp, especially for small-scale set-ups.

Integral Reflector Spotlights. Photoflood lamps are also available with an integral reflector which concentrates the light into a narrow beam similar to that of a studio spotlight. While the light cannot be focused like that of the normal studio spot, the cost of the lamp is much lower and it is an acceptable alternative that is attractive to the amateur worker.

Quartz-iodine lamps also exist with permanently built-in reflectors. The whole unit then resembles a sealed-beam motor car headlamp.

Consumption and light output of these lamps are the same as those of the equivalent photoflood—e.g. a typical lamp has a consumption of 500 watts and a life of 25 hours when run off the 240 volt supply.

Another type of integral reflector lamp is manufactured with a narrow beam for illuminating very small areas. These lamps have a concentrated filament run at a relatively high current to approximate as closely as possible to a point source. The inner surface of the glass behind the filament is silvered and shaped to reflect a pencil beam which covers an area of about two inches in diameter at a distance of about 12 ins. The filament runs at a low voltage —12 or 24 volts—and must be connected to the mains through a transformer. Lamps of this type are used in the commercial photography of manufactured goods to provide local highlights. They are also useful sources of illumination in laboratory photographic and photomicrographic work. F.P.

See also: *Lighting equipment; Projection principles.*

SPOTS. Black or white spots due to faulty processing attract attention out of all proportion to their size and should never be allowed to appear on a finished print.

The spots may come from the negative or they may be only on the print. They may be either dark spots on a light area, or light spots on a dark area. They are best avoided by careful technique, but once they are there, they can be removed—or at least rendered unobtrusive—by any of the normal techniques of spotting or retouching.

Spots on the Negative. Dark spots on the negative (which produce light spots on the print), may be caused by particles of dust or solid matter picked up during processing, or they may be specks of developed silver or chemical impurity.

Loose dust particles should always be wiped off the back and front of the negative immediately before printing. One gentle wipe is all that should be given; rubbing builds up static and makes the dust cling tighter (unless an anti-static brush is used).

Specks of developed silver in the emulsion mean that pin-points of light have leaked through to the material either through a punctured blind (in a focal plane shutter) or a defective dark slide in a plate holder.

Dark spots on the negative, of course, leave light patches on the print.

Occasionally, air bubbles or particles of dust cling to the surface of the emulsion in the fixing bath. These may prevent the emulsion from fixing, and subsequently the undissolved silver bromide turns dark or black on exposure to light.

The commonest cause of light spots on the negative is the air bubbles that cling to the surface of the emulsion during development. At this stage the bubbles prevent the developer from acting on the emulsion and when the undeveloped silver bromide is removed during fixing, it leaves a clear patch.

The remedy is either to brush the surface of the emulsion lightly as soon as it is under the developer, or to agitate the solution well. A few drops of wetting agent in the developer will greatly reduce the tendency of bubbles to form.

Light spots on the negative may also be caused by dust on the plate at the time of exposure. The particles shield the sensitive surface from the action of light and leave a clear spot of unexposed emulsion on development. They may also be formed when specks of active chemical fall on the surface of the emulsion when it is left to dry. Certain chemicals—e.g., acid fixer—will bleach out the silver image and leave a clear patch. The remedy is scrupulous care in handling chemicals in the darkroom, and the frequent use of a vacuum cleaner throughout the room.

Light spots can also be the result of physical damage from such causes as particles of grit in the washing water, and contact with the corner of another plate or with the curled-back end of a length of film. The trouble can be avoided by care in handling the material during processing—particularly when developing more than one plate at a time, or when developing a roll film by the see-saw method.

Clear spots on the negative make black spots on the print.

Spots on the Print. Here also the spots may be lighter or darker than the surrounding image. Most of the causes of spots on negatives apply equally to prints.

Those light spots on the print which are not the result of dark spots on the negative, may be caused by dust particles on the glass surfaces of a negative carrier, paper holder, or printing frame. Although the use of glassless negative carriers eliminates several surfaces which could gather dust, particles can still cling to the surface of the negative or paper.

Light spots on the print are fairly easy to deal with by spotting them out with dye or pigment. Dark spots, however, have to be removed by either bleaching or knifing which are more difficult techniques. For this reason it is always better to cover a clear spot on the negative with a spot of dye or pigment even if in the process the spot is enlarged. It will then print out as a light spot which is comparatively easy to get rid of. F.P.

See also: *Dust; Faults.*
Book: *Retouching*, by O. R. Croy (London).

SPOTTING. Technique of filling in pinholes on the negative or white spots on the print with an opaque medium.

The term is also applied to marking lantern slides with white paper spots. This is done to indicate the correct way round for loading slides into a projector.

See also: *Retouching; Spots.*

SPREAD FUNCTION. Density distribution of the actual image of a point of light. Due to the scattering of light within the film layer during exposure, and also to chemical effects during development, the photographic image of such a point of light is spread out along the film layer. This "blurred" density distribution is termed the spread function. A lens system may also be characterized by a spread function, which in this case represents the image degradations due to diffraction and aberrations.

See also: *Information in the image.*

SPROCKET HOLES. Regular perforations made along one or both edges or down the middle of a cinematograph or still camera film so that it can be advanced from frame to frame by the teeth of a sprocket or claw.

See also: *Cine film; Perforations.*

SPY CAMERA. Essentially a small camera disguised to look like something else so that it can be used for taking photographs without attracting attention. Cameras of this type began to appear soon after the invention of the dry plate about 1880—they would have been impracticable with the earlier wet plate process.

For many years designers exercised their imagination to cloak cameras in increasingly improbable disguises. There was one magazine camera, taking up to 24, 4 × 4 cm. plates, which was made to look like a book. The lens looked out of the spine of the book and the shutter was released by pulling a "bookmark".

Other cameras were built into top hats, field glass cases, and even ladies' handbags. A miniature camera, that looked like a watch when closed, had a body built up of concentric telescopic sleeves, like a collapsible drinking cup. Some spy cameras were concealed in articles of dress. There was a cravat camera, a midget that peeped through the button-hole of the jacket, and one that strapped around the waist and took photographs through a lens which was disguised as a waistcoat button.

Most of these cameras used magazines of miniature plates, and incorporated ingenious (but rarely dependable) mechanisms for moving a fresh plate into position after exposure. Some types used miniature films—e.g., the walking stick handle camera, which used 16 mm. film and took 20 exposure 15 × 20 mm.

The vogue for spy cameras was ended by the arrival of the precision miniature and sub-miniature instruments. These are small enough to conceal—often in the palm of the hand—and can be used so unobtrusively that there is no point in trying to disguise them.

The last real "spy" camera was a post-war wrist-watch camera, a sub-miniature taking 8 exposures on a circular flat film through a 12 mm. *f*2·5 lens.

See also: *Camera history; Ultra-miniature camera*

SQUEEGEE. Device for pressing the water out of wet prints and squeezing them into contact with the glazing sheet. The simplest type of squeegee consists of a strip of rubber about ¼ in. thick and 6 to 8 ins. long, mounted in a wooden holder. The most popular type consists of a rubber-faced roller running in plain or ball bearings in a frame fitted with a handle. An efficient squeegee is an essential in the carbro and a number of other control printing processes.

STABILIZATION. Alternative process to normal fixing which may be used when prints are required in a hurry and absolute permanence is not essential. The unused silver halides are converted into more or less stable colourless compounds by the action of suitable agents. After stabilization, there is no need to wash the print. The process thus amounts to one of incomplete fixation.

A 2-3 per cent solution of thiourea acidified with a few drops of acetic acid will stabilize the emulsion in 2–3 minutes. The commonest stabilizing agent, however, is sodium thiosulphate (hypo) itself. A print or negative immersed in the following solution is stabilized in 30 seconds; i.e., the silver thiosulphate complexes formed in that time are reasonably stable, though complete fixation would take longer.

Sodium sulphite, anhydrous	265 grains	15 grams
Sodium bisulphite	2 ounces	50 grams
Hypo	4–24 ounces	100–600 grams
Water to make	40 ounces	1,000 c.cm.

A 30 to 35 per cent solution of potassium or ammonium thiocyanate will also stabilize unused silver halides.

A 5 per cent acid stop bath should be used after development and the stability of the print can be improved by a 10 seconds rinse in water or in a 5 per cent borax solution.

Stabilization techniques are widely used in photo copying systems and also in rapid access photography. In both these fields a ready processed image is usually required in a hurry, but not for long-term filing. If the image is required for archival storage, stabilization can be followed—after the print or negative has served its immediate purposes—by normal fixing in a conventional fixing bath. After that the stability of the image will be equal to one fixed in the normal way.

If conventional fixing is not possible until a long time after stabilization, some interaction between the stabilized image and the emulsion support may take place, affecting the future permanence of the image. Document copying papers with a fine grain image are particularly sensitive to this, especially if the stabilizer solution was strongly acid. The deterioration of a stabilized film image is very much less, and films can be refixed quite a long time after the original stabilization without noticeable disadvantage.

See also: *Rapid processing.*

STAGE PHOTOGRAPHY. Photographs of scenes from shows are regularly used by amateur and professional companies for publicity and front-of-the-house display. Many photographers concentrate on this type of work and take their pictures at special photo-call rehearsals. In addition, modern wide aperture lenses and high speed sensitized materials have made photographs possible from the auditorium during the actual performance.

See also: *Theatre.*

STAINED GLASS WINDOWS. The most suitable time for photographing stained glass windows is on a bright day when the sun is not shining directly into the glass; otherwise the result will be patchy with poor colour rendering. Diffused lighting from a background of pale clouds picks out the pattern very effectively with an even tone range all over.

Equipment. Often a memorial window is so awkwardly situated that it is not easy to include it without tilting the camera, except when the latter has a generous rising front or swing back. For this reason, the most satisfactory apparatus, containing all the requisite movements, is the field or stand camera.

A lens hood must be used to prevent stray light from the other windows reaching the lens, and a rigid tripod is also necessary for time exposures. As an additional precaution it is advisable to fit rubber feet to the tripod legs to provide a reliable grip on the smooth surface of stone church floors. A foundation of coconut matting would be an ideal substitute for rubber feet.

The tripod must also have a strong ball and socket head for tilting the camera or levelling it when on an uneven surface.

Medium speed panchromatic film is best because of its fine grain.

Exposure and Development. A reliable meter is an advantage since the exposure may be anything from a fraction of a second to around one minute. The exact exposure time will depend upon the principal colours in the window. The design containing a greater proportion of dark browns and deep blues, for example, is going to take a much longer time to register on the film than a window consisting mainly of weak colours.

Stained glass windows can generally be well rendered by measuring or calculating the exposure necessary to photograph them from the outside, and giving from 6 to 10 times this exposure inside. This rather generous exposure followed by development curtailed one-third the normal time for general interiors, will give excellent negatives.

Sometimes a subject is a combination of beautiful stonework and design in the glass. This calls for a particularly careful assessment of the exposure in order to secure details of both in one negative: the exposure must be long enough to record the stonework but not so long as to lose detail in the glass.

If the subject is the whole of the church interior, it will be difficult to register detail in the building as well as the windows. An exposure calculated for the shadows will overexpose the windows. This problem may be solved by making two separate exposures on the same plate; between these two exposures the camera must not be moved and the photographer will need the complete co-operation of the custodians of the building. The first exposure is calculated for the window under suitable daylight conditions. The lens is then capped until after darkness when the second exposure is made by flashlight or by the normal illumination with which the building is provided.

Filters. The use of filters requires special care. To render the design of the glass in its correct colour values, a light-green filter will give good results. But occasionally it will be necessary to secure contrast too. If there is much red or green in the design a contrast filter which will darken one colour and lighten another may be helpful (e.g., a red filter would lighten the reds and darken the greens).

Colour. With reversal colour, material balanced for daylight should be used, and the picture should be taken, if possible, against a neutral grey sky—not a blue one. T.P.F.

See also: *Against the light; Architecture; Lighting the subject; Light sources in the picture; Negatives from colour slides; Paintings and drawings.*

STAINING DEVELOPER. Developer in which the oxidation products of the developing agent stain the gelatin. This occurs in particular with developers incorporating pyrogallol or pyrocatechin, but not containing sufficient sodium sulphite to counteract the staining.

The stain is put to practical use in certain maximum energy developers, notably pyrometol, where it helps to provide an adequate printing density in the shadows of the negative. Since the stain also appears strongly in the highlights, the contrast of such negatives is greatly increased. The stain in this case is yellow and thus of comparatively high actinic, even if low visual, density.

The production of a stain during development, which is proportional to the density of the developed silver image, was also the basis of the earliest colour development processes. R. Fischer in 1913 used a pyro developer to produce a yellow stain image, and indoxyl carboxylic acid and thioindoxyl carboxylic respectively for blue and red stain images. The effect of the modern colour coupling developers is similar, though the mechanism is different.

STAINS. Photographic images may show a variety of stains which usually arise during processing and are due to unwanted side reactions in the developer and fixer solutions.

Most stains are some kind of chemical fog caused by deposition of silver from the developer or fixer, oxidation products of the developer, or toning action in exhausted or too warm fixers.

Other causes are insufficient rinsing between development and fixing, prints sticking together in the fixing bath, premature exposure to white light before the process of development has stopped, and not enough attention to cleanliness in working.

Stains may also arise during various toning operations, and are then usually due to incomplete fixing; traces of silver halide remaining in the emulsion become toned in the same way as the image proper.

A number of photographic chemicals—in particular such developers as pyro and amidol—will stain the fingers. But it is never necessary for the fingers to be immersed in such solutions; there is scarcely any process that cannot be carried out with dry hands from start to finish, simply by planning the routine to avoid wetting the fingers and by using tongs to handle the material being processed.

The staining effect of any solution can be minimized by dipping the fingers into clean water before putting them into the solution. This dilutes the chemical contacting the skin.

Pyro and many other developers will not stain the fingers if, after coming into contact with the developer, they are rinsed, dipped for a second or two in an acid fixing bath, and then rinsed again and dried. This treatment will also remove permanganate stains.

Most stains on the fingers disappear if they are rubbed with a crystal of citric acid or a slice of lemon.

An alternative to some of the above methods is to wear rubber gloves. Barrier creams may also be effective.

See also: *Faults; Skin affections.*

STALE MATERIALS. When stored in dry, cool conditions, most sensitized materials will keep for years without suffering, but in a damp or warm place they tend to deteriorate.

If materials have been badly stored, fog and mottle will generally result. However, it is possible to effect some remedy if this happens. The methods of restoring such materials described below must all be carried out in the darkroom by the appropriate safelight for the material concerned.

Plates and Films. Negative materials that have deteriorated through being carelessly stored or which have been accidentally fogged—e.g., by being exposed to an unsafe light in the darkroom—may be treated so that they will once again yield a fog-free image. In the process, however, the sensitivity is reduced up to one-tenth of the original speed.

The materials should be soaked for five minutes in either of the following baths:

(1) Chromic acid	60 grains	3·5 grams
Potassium bromide	120 grains	7 grams
Water to make	20 ounces	500 c.cm.
(2) Potassium bichromate	70 grains	4 grams
Hydrochloric acid	½ ounce	12 c.cm.
Water to make	20 ounces	500 c.cm.

After removal from the bath the material must be thoroughly washed and then left to dry in the dark.

Fast and medium speed plates that have deteriorated or been accidentally fogged can be turned into excellent slow-speed transparency material by soaking for ten minutes or so in a bath prepared as follows:

Potassium bromide	105 grains	6 grams
Potassium iodide	14 grains	·75 grams
Hot water, only sufficient to dissolve		

to this add the following:

Hydrochloric acid	½ ounce	12 grams
Potassium bichromate	105 grains	6 grams
Water to make	20 ounces	500 c.cm.

This converts the latent silver image into undevelopable silver halide again.

After treatment, the plates are thoroughly washed and dried in the dark. The reduction in speed may be twenty times or more depending on the material.

Bromide Paper. If there is any reason to suspect that a package of paper has deteriorated in store, it is always worth trying the effect of one of the proprietary developer improvers before throwing the paper away. These additives delay fog formation, improve the blacks and keep the whites clean. They will often give an image of acceptable quality from old paper that would otherwise give a grey and muddy print.

If the paper has gone beyond this treatment, it will sometimes respond to a bath in a weak acidified solution of potassium permanganate, for example:

Potassium permanganate	4 grains	0·2 grams
Sulphuric acid, 10 per cent	½ ounce	12·5 c.cm.
Water to make	20 ounces	500 c.cm.

After soaking for a minute in this bath the paper is transferred to a clearing bath of 2 per cent (anhydrous) sodium sulphite for a further minute. It may then be rinsed and exposed wet, or dried and used later.

The same treatment will often save paper that has become fogged from any other cause. F.P.

See also: *Faults; Fogging; Keeping qualities of materials.*

STAMPS. Postage stamps are often photographed for record, comparison, catalogues, etc. The operation may call for specialized technical treatment of the stamps.

See also: *Philately.*

STAND. Substantial and adjustable support for the studio type of camera; generally applied also to any type of camera support from an ordinary tripod to a heavy duty stand with full pan and tilt adjustments.

See also: *Camera supports.*

STANDARD CANDLE. Used to be a candle manufactured to a standard specification (burning spermacetti wax at a rate of 120 grains per hour). It has long been obsolete, but a candle power defined in terms of such a candle is used as the basic reference unit of intensity for all light measurements in photometry and sensitometry.

See also: *Light units.*

STANDARDIZATION IN PHOTOGRAPHY

The organized standardization of parts used and manufactured in industry was first introduced by Great Britain in 1901, and since then it has been adopted on an increasing scale all over the world. A great number of photographic products, equipment and techniques has already been standardized, and new items are regularly added to the list.

The principle of standardization, in general, involves the examination of the appropriate aspect of the material or article in question by representatives of the chief manufacturers and users or consumers concerned. In a particular case it may be only one aspect—e.g., dimensions, or purity of the product—that may be considered worthy of standardization. Eventually a range of sizes or limits of composition, as the case may be, will be determined and circulated to, or published for the use of, the appropriate section of the community.

The main advantage of standardization is the economy of labour, plant and materials achieved in manufacture by reducing a wide variety of sizes or compositions of a product to the smallest number sufficient to cover all its applications. Standardization results in a simplification of stocks in the distributive channels and a lower price to the consumer. While manufacturer, designer and customer are all intended to treat such standards as a guide, they are always free to adopt their own individual standards except in special circumstances where standards have statutory effect for reasons of public safety, health, etc.

Authorities. In the United Kingdom the British Standards Institution is the authority on standardization, having been in existence for over sixty years. At the present time there are nearly 5,000 British Standards published, ranging from single- or two-page documents of a few simple requirements to technical specifications with hundreds of pages of clauses, methods of test, formulae, diagrams and tables of data. All standards are kept under constant review so that amendments or revisions may be made to take into account the latest developments in materials, design or manufacturing technique.

International standards are co-ordinated by the International Organization for Standardization (I.S.O.) which was formed in 1946. This body holds meetings of its technical committees from time to time, their members representing national standardizing bodies. The outcome of these meetings determines the issue of I.S.O. recommendations.

The British Standards Institution is not a Government department; it is financed by subscriptions from industry and the Government and from the sale of its publications. Industry recognizes the benefits of standardization by freely providing the services of the thousands of technical experts who serve on the numerous B.S.I. committees. The Institution provides the secretariat for the meetings.

The American counterpart of the B.S.I. is the American Standards Association, Inc. Its constitution is rather different from that of the B.S.I., but it nevertheless fulfils the same functions with comparable effectiveness. Similar bodies exist, and do similar work, in most other countries, notably—from the photographers point of view—Germany and France.

Procedure. The preparation of a British Standard starts with the request from a recognized body representing the manufacturers or the users—e.g., a trade association or a Government department. Even ordinary amateur photographers can have their views considered if presented by someone with standing in the photographic world—e.g. the editor of a photographic journal. If the proposed standard is related to others already published, the matter is dealt with next by the existing standards committee for that particular section of the industry. Otherwise a special committee is constituted for the purpose and the exact scope of the proposed standard is defined for the guidance of the technical committee or panel responsible for the detailed drafting.

Eventually an agreed draft is circulated for the comments of all appropriate industrial organizations, professional institutions, Government departments and kindred bodies in this country, who are not directly represented on the drafting committee (which must obviously be limited to a workable number), as well as to standardizing bodies throughout the Commonwealth. By this system the B.S.I. ensures that the published standards are acceptable generally to the manufacturing and consuming interests concerned, even before publication.

The A.S.A. similarly provides the machinery for creating voluntary standards, although unlike the B.S.I. its constitution enables it to take the initiative and agreement is reached as between members of a federation. Standards so produced are automatically regarded as nationally acceptable, and are designated "American Standard".

There are very many A.S.A. standards covering much the same fields as the B.S.I. Standards. In photography there are one or two instances where nomenclature is standardized, such as the parts of a lens.

Scope. Photographic standards deal largely with sizes of sensitized material and related items such as spools, masks and reels. There are also specifications dealing with qualitative aspects, such as speed, transmission density, performance requirements for components—e.g., resolving power of lenses—equipment for processing and various accessories.

SOME BRITISH PHOTOGRAPHIC STANDARDS

B.S. No.	*Subject*
967	Identity and passport photographs.
1112	Sizes of printing paper.
3368	Identification of exposed colour film.
1019	Photographic lenses; definitions; markings.
1383	Photo-electric exposure meters (under revision).
1496	Safelight screens and housings.
1592	Photographic shutters.
1613	Resolving power of lenses.
1915	35 mm. projectors (still).
1917	Film-strip and lantern slides.
2833	Classification of flash bulbs.
3205	Electronic flash equipment.
1380	Film speed determination
	Part 1. Monochrome negative material.
	Part 2. Reversal colour film.
1437	Determining filter factors.
1384	Determining transmission density.
930	Light output of cine projectors.
1488	Test-films for 16 mm. projectors.
3675	Performance of 16 mm. sound projectors.
CP 412	Optical projection apparatus for education.
2954	Screen luminance for 16 mm. film.
3115	Originals for television.
1153	Storage of microfilm.
1359	Photographic conversion tables.
3216	Stereo still pictures on 35 mm. film.

SOME AMERICAN PHOTOGRAPHIC STANDARDS

A.S.A. No.	*Subject*
PH1.12	Dimensions of printing papers.
PH1.20	70 mm. film for still cameras; dimensions.
PH2.11	Photometric exposure of daylight colour film.
PH2.12	Photo-electric exposure meters.
PH2.13	Testing flash lamps.
PH2.21	Speed of reversal colour film.
PH3.2	Performance of focal-plane shutters.
PH3.3	Performance of front shutters.
PH3.11	Stereo on 35 mm. film; sizes.
PH3.13	Focal length markings of lenses.
PH3.16	Resolving power of projector lenses.
PH3.18	Synchronization of front shutters.
PH3.20	Focusing scales.
PH3.28	Slide-film projectors.
PH3.31	Testing enlargers.
PH3.40	Electronic flash, performance.
PH4.5	Temperature tolerances in processing.
PH4.7	Photographic thermometers.
PH4.14	Definition of a fine-grain developer.
	Cine (excluding 35 mm.)
PH22.27	Transmission density of films.
PH22.41	Photographic sound record (16 mm.).
PH22.42	16 mm. sound-focusing test films.
PH22.43	16 mm. flutter test film.
PH22.44	16 mm. multi-frequency test film.
PH22.53	Resolving power of projector lenses.
PH22.88	Magnetic striping of 8 mm. film.
PH22.89	Scene-change cueing.
PH22.100	16 mm. screen brightness.
PH22.112	Picture-sound separation (16 mm.).
PH22.124	Screen luminance.

SOME GERMAN PHOTOGRAPHIC STANDARDS

D.I.N. No.	*Subject*
4521	Camera lenses; markings.
15741	Projector lenses for 8 and 16 mm. films.
19010	Photo-electric exposure meters.
19011	Guide numbers for artificial light sources.
19012	Flash lamps; technical characteristics.
19013	Flashlight technique.
19015	Measurements on front shutters.
19016	Measurements on focal-plane shutters.
19017	Exposure values;
4512	(1) Sensitivity of monochrome films.
	(2) Blue, green, and red sensitivity of monochrome films.
	(3) Optical density of transparent diffusing layers.
	(4) (draft) Sensitivity of reversal colour film.
108	(1) Still projection (general).
	(2) Preparation of glass slides.
4531	(1) Stereoscopy (general).
	(2) Stereoscopy; sizes, separations, arrangement of pictures.
6170	Anaglyphic stereoscopy.
19021	Light and temperature measurements in still projectors.
15603	16 mm. sound film; relation sound and picture.
15881	8 mm. film; positive with stripe.

SOME FRENCH PHOTOGRAPHIC STANDARDS

N.F.–S. No.	*Subject*
20–001	Speed of fast negative emulsions.
20–002	Sensitometry.
22–001	Exposure-time markings.
22–002	Stop-scale markings.
24–105	Sound-recording (16 mm.).
24–108	Splices in 16 mm. negative film.
28–002	Photometric calibration of camera lenses.
28–003	Measuring optical characteristics of 35 and 16 mm. projectors.
28–006	Testing projector lamps.
28–007	Characteristics of projection screens.

In many fields much of the current technical nomenclature is also standardized, but so far no attempt has been made to standardize photographic terms and descriptions.

Dimensional Standards. Dimensional standardization is of greatest service in fields where small mass-produced items have to be fitted together before use by the general public. This necessary interchangeability, if devised intelligently from the outset, can form an immense incentive to world-wide standardization, with consequent benefits to exporting producers. The problem can, however, lead to difficulties over tolerances which are necessary evils of production.

A designer of a camera which is to use a film having dimensions already standardized must ensure that the camera will accept a fully-loaded standard spool made to the maximum dimensions as well as one made to the minimum dimensions. He must also arrange that the film window is positioned so that the numbers on the backing paper can be seen even when printed at their extreme standard limits. He can then be sure that his camera will accept all films of the size for which it is intended.

Dimensional British Standards are published for photographic paper, recording materials, ordinary sheet film and X-ray film, film strip and lantern slides, amateur (including 35 mm.) and service roll film and spools. Related accessories for which standards are issued include masks, dishes, film clips, safelight screens and many other components dealt with in more comprehensive specifications.

Other Photographic Standards. Other standards of less immediate bearing on photographic products sold to the public apply to routine test and development work. (It is no less important to secure uniformity between laboratory testing methods than to make sure that all 35 mm. film is correctly perforated.) Examples of such standards are those relating to film speed, measurement of transmission density and the determination of filter factors.

Another section of British Standards is concerned with the optical and mechanical performance of lenses and shutters, and equipment—e.g., for processing X-ray film. In processing equipment, of course, a detailed specification is out of the question, but it is of immense help to the user and supplier to know that the standard components incorporated are interchangeable between units and that essential service connexions of standard type are fitted and readily accessible and that suitable requirements for temperature, ventilation and chemical resistance properties of the materials used have been specified.

Other standards concerned with apparatus of a complex nature, like projectors, optical lanterns and microfilm readers, include not only standard dimensions for interchangeability of lenses, etc., but also electrical, optical, and illumination performance requirements to ensure efficient and safe functioning from both the operator's and audience's point of view.

Collaboration Between Standards Organizations. Although full international agreement on standards is achieved through the I.S.O., some degree of collaboration between standards organizations is often desirable in the interests of trade between two countries, as a step towards international standardization, or perhaps to prevent confusion in cases where some standard designation may be used outside its own country. To this end conferences between organizations may be held periodically.

The question of film speeds is an example of collaboration of this kind. The numerous systems that have been current in the past—some sponsored by film-makers, some by makers of exposure meters—led at one time to much confusion. In the last fifteen years or so close co-operation between the standardizing bodies of U.S.A. and Britain has resulted in a method of measuring, and expressing, film-speed common to the entire English-speaking world, and most, if not all, exposure-meter makers have scaled their instruments to suit it. This is the A.S.A. arithmetical system. But Germany, followed by most of Europe, has long used the D.I.N. system, originally based on tests sufficiently different from those used to determine A.S.A. speeds to make it possible for two films having the same speed on one system to have different speeds on the other. By successive concessions on both sides, a compromise system has been approached, and is likely to be internationally standardized very soon; speed will then be determined by the resulting agreed test, and the speed, whether expressed in A.S.A. or D.I.N. figures, will have an unmistakeable meaning.

Many standards deal purely with dimensions; these are of interest to a manufacturer but hardly to a user. A selection of (mainly) non-dimensional standards current in Britain, U.S.A., Germany, and France is listed here; these and other standards can be obtained through the British Standards Institution, 2 Park Street, London, W.1, in whose library, open at that address during office hours to all bona-fide enquirers, classified lists of standards from all countries may be consulted.

F.A.M. (A.L.M.S.)

See also: *Exposure tables; Film transport; Identity photographs; Lens mounts; Perforations; Sizes and packings; Speed of sensitized materials.*

Book: *Progress in Photography* (2 vols.), ed. by D. A. Spencer (London).

STAND CAMERA. Every type of portable camera which is normally mounted on a tripod. Such cameras are usually of large format—half-plate and upwards—and focus by focusing screen.

See also: *Field camera; Technical camera.*

STAND DEVELOPMENT. Development of films and plates for a long time in highly dilute developers, often in special tanks. This method is not generally popular, as the long times required are rarely convenient.

Stand development requires special developers which do not oxidize readily even at low concentration—e.g., glycin—and which do not readily produce aerial fog.

Dilution is in the region of ten to twenty times normal, and development times range from one to four hours.

Agitation is considerably less important than with orthodox development, as the action of the solution is very slow. Consequently the normal molecular movement in the solution is sufficient to keep the concentration of the active ingredients reasonably uniform throughout the liquid. This applies especially if the film or plate is suspended vertically in the solution, as is normal practice in stand development so that the solution is kept moving by gravity as its density changes on becoming exhausted.

Although highly diluted, the developer solution must still contain a sufficient amount of developing agent per film, otherwise the bath would become exhausted before development reached the required stage. Stand development therefore requires large volumes of solution.

See also: *Finality development.*

STARCH PASTE. One of the adhesives commonly used for mounting prints, made by emulsifying starch in boiling water.

See also: *Mountants.*

STATIC. Abbreviation of static electricity; commonly used to describe an electrical charge produced by friction between certain materials—e.g., glass and silk.

In photography static may be produced on a film by the action of unwinding it in very dry weather. It can be recognized by the sharp cracking sound it makes as it discharges. The discharge may also be accompanied by faint blue sparks which can actually produce a developable image on the film. Ordinary roll films are unlikely to discharge enough static to cause visible marks on development, but cinematograph films can be seriously marked. This is because the faster speed of unwinding and the greater length of the film build up much higher charges and produce brighter sparks.

Static also makes it difficult to get rid of dust on glass surfaces. When the surface of a lens or glass negative carrier is polished with a cloth, the friction builds up a static charge on the surface of the glass. This makes the dust cling tenaciously to the surface and fly back to it after being wiped off. Rubbing the surface harder simply makes matters worse. The remedy is to wipe the glass once, very gently, about five minutes after polishing. There are proprietary dopes and impregnated dusters which are claimed to prevent the formation of static and leave the surface dust-free.

Alternatively, glass negative carriers may be brushed with special anti-static brushes. These contain a strip of faintly radio-active material (such as polonium) which discharges the static electricity.

The above remarks apply equally to dust that has settled on the emulsion surface of plates and films. But radio-active anti-static brushes may fog unexposed films.

See also: *Faults.*

STATUETTES. Figures of statuary and of people, children and pets cut out of photographs and mounted on cardboard or plywood. Cut-outs of this type are used for decorative purposes, table top and model photography, and for attaching to calendars. They are of course used on a large scale for advertising and display material in shop windows and at the point of sale, but for this purpose the pictures are printed from half-tone blocks and cut out in one operation on a machine. They may then be folded in such a way as to be self-standing.

The picture chosen should have a clearly-defined and simple outline, and the tones should be bold and with plenty of contrast. There will be no border to isolate the figure from its surroundings and it must be strong enough to assert itself on its own merits. The enlargement should ensure a figure of at least 4 ins. in height and preferably more. Anything smaller is difficult to work on and the finished product looks insignificant.

Mounting the Picture. The figure is cut out of the print by laying it on a flat surface and going around it with the point of a sharp knife, keeping quarter to half an inch away from the outline.

Just enough of the lower part of the print should be left attached to provide a good base for the cut-out.

The roughed-out portion is then stuck down on to a piece of suitably robust support material. Satisfactory supports may be cut from stiff card, but thin plywood is better. The surface must be clean, dry and free from irregularities. It should be rubbed quite flat on both sides with a piece of fine sandpaper wrapped around a block of wood.

Both the back of the print and the face of the support should be thickly coated with a good cold glue. (The popular tube glues are excellent, but not cellulose glue because this type tends to dry before both surfaces are covered.) Rubber solution and ordinary photographic mountants are unsatisfactory because they do not dry hard enough to give a clean edge when the cut-out is sawn down to the outline.

The two surfaces are pressed together, squeegeed into firm contact and then kept under pressure between flat boards for at least twenty-four hours.

The Support. The outline is now sawn out with a fretsaw—if possible of the treadle type since the faster the speed of the saw, the smoother the edges of the cut will be. All edges are rubbed smooth with fine sand-paper. Grey water colour is then painted all round the edge to tone it down to about the average shade of the print.

Wooden blocks are then glued to the back of the cut-out to form feet. With large figures it is a good plan to drill as large a hole as possible at the back of the feet and pour in melted lead so that the weight prevents the cut-out from tipping forward.

Both sides of the cut-out are then given a coat of clear cellulose lacquer so that it will stand up to handling.

If both sides of the statuette will be visible, it is possible to stick a reversed print on to the back. The reversed print is made by reversing the negative in the enlarger, taking care that the second image is exactly the same size as the first.

Another effective method of mounting a cut-out is to sandwich it between two sheets of glass or Perspex held upright in a slotted block of wood. The outline in this case is cut around with the point of a sharp penknife or razor blade. Here again a left-to-right reversed print can be stuck to the back.

Photographers with a liking for this type of handicraft often produce amusing animated versions and children's toys by articulating the joints of cut-out figures. Others turn straight head and shoulder portraits into miniature imitations of busts, with a pedestal base. F.P

See also: *Photosculpture; Sculpture.*

STEINHEIL, CARL AUGUST VON, 1801–70. German astronomer and professor of physics and mathematics at Munich University. Together with Franz von Kobell, Steinheil reported to the Bavarian Academy of Sciences on Talbot's photogenic drawing in April 1839, and at the same time showed their own photographs of views in Munich. These were negatives taken with a cylindrical camera, designed by Steinheil on silver chloride paper fixed with ammonia. Immediately after the publication of Daguerre's process, August 1839, Steinheil began experimenting with the daguerreotype, and in December made the prototype of the first miniature camera, producing pictures 8 × 11 mm. (half the size of a sixpence) which had to be viewed through a magnifying glass. In 1855 Steinheil founded in Munich the optical works which still bears his name. Biography by Rudolf Loher (Munich 1937).

STEINHEIL, HUGO ADOLPH, 1832–93. German designer and manufacturer of lenses, younger son of Carl August von Steinheil. Father and son patented in 1865 the Periskop lens which gave a field of 100° free from distortion. In 1866 purchased his father's optical works and produced his Aplanat, an achromatic rectilinear lens for landscapes, reproduction work and interiors. Later produced wide-angle lenses for cartographic reproductions, an Ahtiplanat (1881) and a telephoto lens (1891).

STELZNER, CARL FERDINAND, *c.* 1806–94. German miniature painter in Hamburg. Became a daguerreotypist in 1842 and for a while worked in collaboration with Hermann Biow, taking the earliest news photographs of the ruins after the Hamburg fire, May 1842. Stelzner opened his own studio, 1843, which was run by an employee after his sight failed in the mid-50s.

STENGER, ERICH, 1878–1957. German professor of photographic chemistry and director of the photo-chemical laboratory at the Technische Hochschule, Berlin, for nearly two decades, in succession to Adolf Miethe. Professor Stenger is best known as an historian of photography. For over fifty years he collected everything appertaining to the history of photography, and published several books, of which *Die Photographie in Kultur und Technik* (1938) is his main work. The Stenger Collection was bought by Agfa, Leverkusen, Germany 1955, with the intention of establishing a museum.

STEP AND REPEAT. Method of photographing or printing a large number of identical images side by side on one film.

Used mainly in photo-mechanical reproduction for preparing printing plates (for example of postage stamps) but also for making printed circuits, etc.

The process is more or less automatic, being carried out by a step-and-repeat camera. This moves the film or plate through a set distance after every exposure, to place the successive images side by side, row after row until the film is full.

See also: *Photo-etching.*

STEP WEDGE. Optical wedge in which the density increases in a series of regular steps from the transparent to opaque or at least to a high density.

The steps are usually chosen so that the light passing through any particular section is twice that passing through its denser neighbour. This requires the density to increase by 0·3 at each step.

Step wedges are used in sensitometry and for determining exposures in enlarging.

See also: *Wedge.*

STEREO-MICROGRAPHY. Except at a very moderate degree of magnification it is normally impossible for the eyes to look through a magnifying glass at a small object and view it stereoscopically. This is because to do so they would have to converge at such an angle that a single magnifying lens would not serve both eyes at the same time, and because the separation between the two viewpoints would be so great that the picture seen by one eye would be quite different from that seen by the other and it would be impossible to fuse them into a single stereoscopic impression.

Enlarged stereoscopic views of small objects can only be obtained by scaling down the separation between the viewpoints of the eyes in proportion to the magnification of the image. This is done in the stereoscopic microscope by a system of prisms which in effect allow the eyes to look through the magnifying system from viewpoints separated by a fraction of an inch instead of normal interpupillary distance.

The same effect can be produced by taking two consecutive photographs of the object through the microscope and either rotating the object, or sliding it sideways between exposures. The extent of the change in position (corresponding to the reduced interpupillary separation) is arrived at by simply dividing the normal separation of the eyes (2½ ins.) by the magnification—e.g., if the object is magnified 50 × life size, then the shift between exposures will be $2{\cdot}5 \div 50 = {\cdot}05$ in. When the resulting negatives are printed and viewed as a stereoscopic pair, the effect will be as though the object had been magnified to fifty times life size and moved fifty times farther away. At that apparent size and distance it would be seen in stereoscopic relief exactly as though it were an enlarged model being viewed normally with two eyes.

See also: *Stereoscopic photography.*

STEREO PROJECTION. Projection of stereo pairs of images to convey an effect of depth by presenting to each eye slightly different images which correspond to the views seen separately by the two eyes. Stereo projection methods may be applied both to still and to cine pictures.

See also: *Projection practice; Stereoscopic photography.*

STEREO-RADIOGRAPHY. Technique of obtaining a three-dimensional impression of the internal structure of an object by X-ray recording. For this purpose two radiographs of the same object are taken in succession at two different positions of the X-ray tube. If measurements in space are to be carried out, the displacement of the tube should be of the order of the interpupillary distance, i.e. approx. 6·5 cm.; otherwise a ratio of tube shift to focus-film distance of 1 : 10 also reveals a good stereoscopic impression.

Change of film cassettes and immobility of

the object between exposures are essential. The latter condition is easily achieved in industrial radiography of castings and similar objects, but is more difficult in medical radiography because of involuntary movements of the patient. In this case the exposures have to be taken preferably by means of rapid cassette changers and the exposures should follow in quick succession, e.g. by using two X-ray tubes.

The two radiographs thus obatined can then be viewed with, for instance, a Wheatstone mirror stereoscope or with a binocular stereoscope. With both instruments a three-dimensional view is obtained. In the Wheatstone type stereoscope distance measurements in space can be carried out. Some people are capable of viewing two such radiographs stereoscopically even without the aid of special instruments.

In stereo-fluoroscopy the object is screened alternatively from two X-ray focal spots at a distance apart and the images on the screen are observed in rapid succession so that the left eye sees the image caused by one focal spot and the right eye that by the other focal spot.

The accurate localization of defects or foreign bodies in the field of industrial radiography can be achieved by taking two exposures (using the tube shift method) on one single film. In another technique two small lead markers are used, one in contact with the upper surface of the specimen, the other in contact with its base. The distance of the tube shift need not be known with accuracy in this case. As the displacement of the image of the defect will lie between that of the images of the upper and lower marker, the position of the defect can be determined from a graph relating the distances of the markers (i.e. thickness of the object) and the image displacement. R.H.H.

See also: *Medical radiography*.

STEREOSCOPE

Special type of viewer designed to accept a stereoscopic pair of pictures so that they can be viewed as a single image in three-dimensional relief.

There are two principal requirements:

(1) Each eye must see in the stereoscope the picture that it would see if it were looking directly at the subject—i.e., the left eye must see the print from the negative taken with the left hand lens of the stereoscopic camera (or with the camera at the left hand end of the stereoscopic slide) and the right eye must see the picture taken from the right-hand side.

(2) The eyes should preferably be focused and converged as though they were looking at the actual object. This condition is not always strictly observed in the design of stereoscopes and any departure from it results in eyestrain after a time.

Unaided Viewing. Some people can manage after a little practice to look at a stereoscopic pair and merge them into one stereoscopic image. To do this, the two pictures are placed side by side with their principal points separated by the normal interpupillary distance of the eyes.

There are then two ways of looking at the pictures:

(1) The observer looks at the pictures from a distance of about 10 in., seeing the left-hand picture with the left eye and the right-hand picture with the right eye. (A piece of black card held between the pictures helps to keep the eyes from wandering.) To see the picture stereoscopically under this condition the eyes must be directed as for an infinitely distant object—i.e., without convergence—and focused as for a near object.

(2) The pictures are transposed from left to right and the observer looks at them with each eye seeing the picture on the opposite side. If the pictures are far enough away, the convergence of the eyes is reasonably near normal, even although the axes are crossed, and the effort called for is not as great as when the axes are parallel.

Both these methods require special practice first and involve considerable eyestrain. Some people are unable to fuse the images at all successfully. They are not suitable for continuous viewing.

Elliot's Stereoscope. In 1839, James Elliot produced a stereoscope based on the second method of unaided viewing described above. It consisted of an oblong box with eye holes spaced at the interpupillary distance at one end and a single hole at the other. The observer, with his eyes to the eyeholes, looked through the hole in the end of the box at a suitably arranged stereoscopic pair of drawings set up at a distance outside the box.

Under these conditions each eye looking at an angle through the end opening could see the picture on the opposite side, but it could not see the picture on its own side. This enabled the

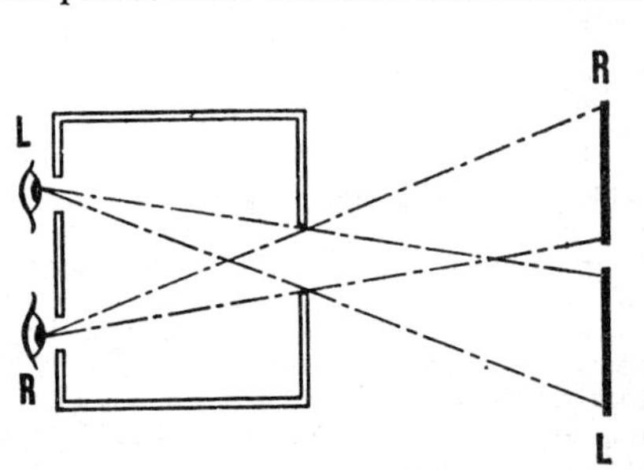

ELLIOT'S STEREOSCOPE. Contains no optical system; the left and right eye rays cross over, so that each eye sees only its own picture and not the other one.

picture for the right eye to be set up where only the right eye could see it, and the left picture where only the left eye could see it; the result was a single stereoscopic image. (In practice the pictures were set up side by side, but with the left picture on the right side and the right on the left; the viewing box was then moved about until the required condition was achieved.)

The Elliot viewer was easier to use than the unaided methods above, but it still called for an unnatural visual combination of focusing and convergence. So only a few specially gifted people could use it and even then its use inevitably produced eyestrain sooner or later.

Wheatstone's Stereoscope. In 1832 C. Wheatstone invented and had prototypes constructed of the reflecting and the refracting stereoscope. In 1838 he fully described to the Royal Society the reflecting type that is still used in modified forms today.

The basic arrangement of this consisted of

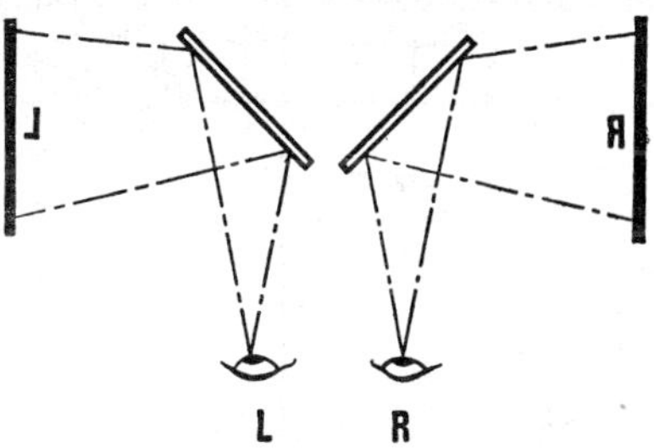

WHEATSTONE'S STEREOSCOPE. Uses two mirrors; each stereo picture must be reversed left-to-right. Suitable for prints of any size, since they can be at any distance from the mirror necessary to include the whole picture.

two mirrors set facing outwards at 45° to the line of sight. The mirrors were arranged to reflect the pictures of a stereoscopic pair of drawings (later photographs) held upright on either side, and facing inwards parallel to the line of sight. Because of the inverting effect of reflection from the mirrors, the pictures had to be printed with the negative reversed if the resulting stereoscopic image was to show left and right in their correct relationship. This was an inherent disadvantage, but even so Wheatstone's stereoscope was the first really practical instrument.

Because the pictures are parallel, there is no limit to their size as there is in side-by-side arrangements. So the Wheatstone principle is still employed for viewing large images like those of X-ray film stereoscopic pairs.

Cazes's Stereoscope. A similar type of viewer to the Wheatstone was made by L. Cazes in 1895. In this, two extra reflecting mirrors are introduced so that pairs of any size may be observed side-by-side in one plane. Converging or diverging lenses can be combined with the

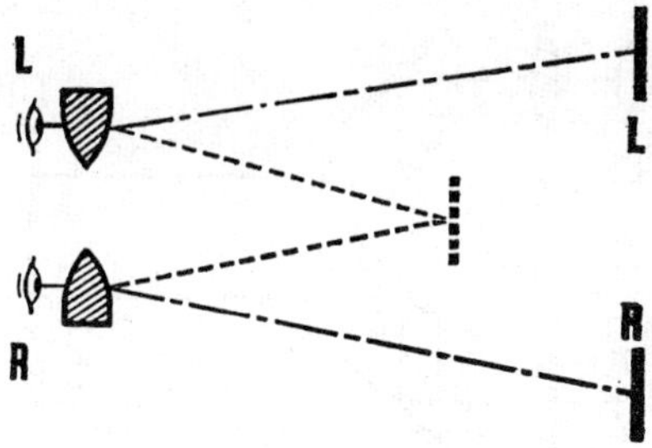

BREWSTER'S STEREOSCOPE. Two prisms or half-lenses allow the eyes to converge naturally, though the pictures are appreciably separated. Principle of many modern stereoscopes.

eye pieces to alter the scale and apparent distance of the image. Such viewers have adjustments for arranging the separation of the pairs and the viewing distance to give natural convergence of the eyes no matter what the magnification or reduction of the optical system. Instruments of this type are used for giving magnified stereoscopic images of aerial reconnaissance photographs.

Brewster's Stereoscope. In 1849 Sir David Brewster introduced a refracting stereoscope based on the principle of Wheatstone's second stereoscope of 1832. Originally it consisted of a side-by-side viewer in which each eye viewed its own picture through a plane prism. The two prisms were placed, base out, across the line of sight; after refraction, the bundles of rays coming from each picture were bent away from each other, thus appearing to come from a point. So the respective right and left image rays, instead of appearing to come from two spaced pictures, seemed to come from a single stereoscopic picture at the meeting point of the two converging bundles of rays. This arrangement allowed the eyes to converge naturally on the single virtual image. It went into production in 1851 for viewing daguerreotypes.

An immediate improvement on this system was to substitute lenticular prisms for the plane prisms, thus giving a magnified image. By making the prisms to a suitable angle (in

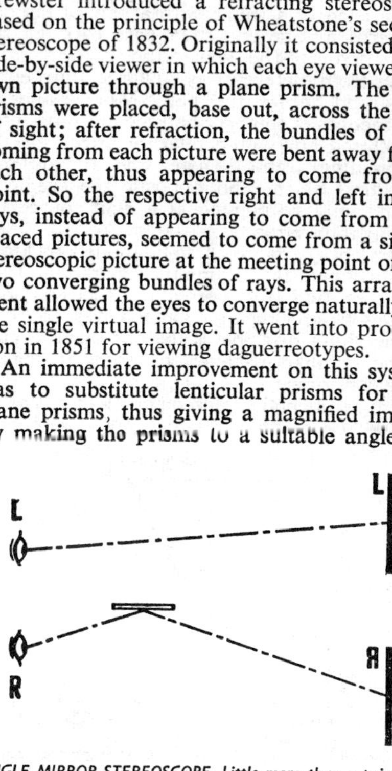

SINGLE MIRROR STEREOSCOPE. Little more than a primitive viewing aid, it permits the eyes to converge naturally. One of the prints must be reversed left-to-right

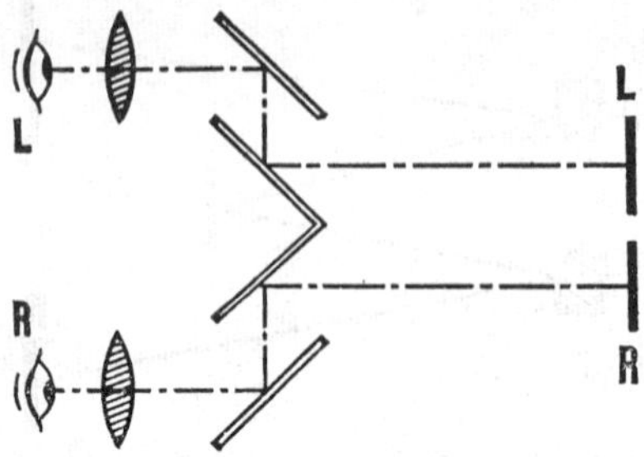

STRAIGHT 35 mm. STEREO VIEWER. Designed for viewing untransposed transparency pairs taken side by side on one 24 × 36 mm. frame with the aid of beam splitter lens attachments working on a similar principle.

practice they are formed by a pair of half lenses) it is possible to view pairs separated by as much as 3 ins. without discomfort. This enables larger pictures to be viewed than are possible with the normal interpupillary separation of about $2\frac{1}{2}$ ins.

This type of stereoscope is the one commonly used for amateur stereo photography and for all portable and compact viewers. In its most popular form (at least up to about 1940) it consisted of a frame with a pair of hooded eyepieces at one end of a base carrying the print holder. The separation of the print holder from the eyepieces could be adjusted to focus and merge the pair. More elaborate stereoscopes of this type had the units mounted in a closed box with a reflector at the top for illuminating prints. There was an optional opal glass at the end opposite the eyepieces which could be pointed at a light source for viewing transparencies.

Some of the more complex models included an adjustment for varying the separation of the eyepieces and for correcting focusing disparities between the right and left eyes.

The above principle is the one incorporated in the familiar machines on piers at coastal holiday resorts.

A fundamental requirement of this type of stereoscope is that for natural stereoscopic presentation of the subject, the effective focal length of the viewing lenses should be identical with that of the taking lenses. This requirement is difficult to meet in practice, particularly with stereoscopic pairs made on 35 mm. film in miniature cameras. The cost of short focus prismatic lenses, suitably corrected, is out of proportion to the cost of the rest of the equipment, so that it is customary to use lenses of longer than the ideal focal length. This results in over-emphasis of the foreground, but the worst of the distortion can be removed by incorporating additional meniscus lenses in the eyepieces.

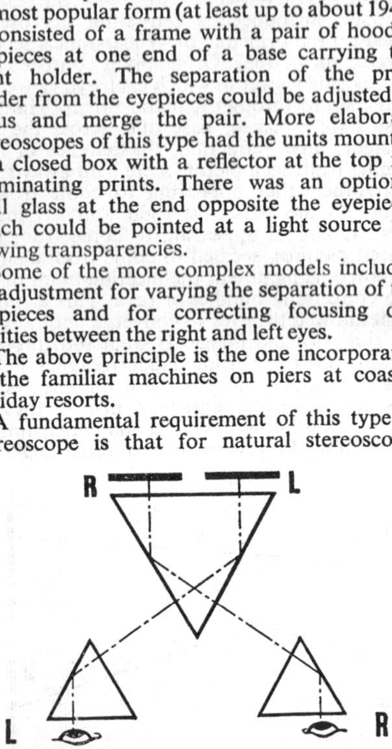

TRANSPOSING 35 mm. STEREO VIEWER. Designed for viewing transposed transparency pairs taken with twin-lens stereo unit in interchangeable-lens 35 mm. cameras. The viewing rays cross over in centre prism.

Highly developed stereoscopes on this principle, with their own illuminating lamp and battery, are now widely used for viewing stereoscopic colour transparencies taken on miniature cameras with stereo attachments. One of the commonest applications of such instruments is as a sales aid carried by commercial travellers to display goods—e.g., machinery, furniture, motor cars—too large to take around to show to the buyer. But this

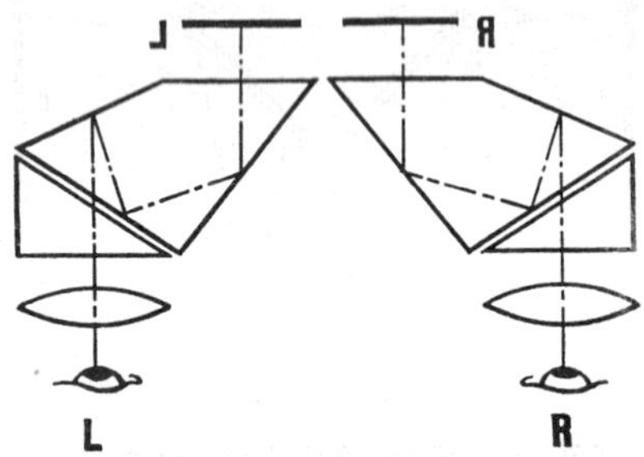

TRANSPOSING 35 mm. STEREO VIEWER. The transposed transparency pair is viewed from the back; the prism system for each eye reflects the image three times, resulting in a right-way-round picture and correct stereo view.

kind of visual demonstration has become virtually obsolete, in view of the more impressive procedures of demonstration cine films.

Other Systems. Several variations are based on the method of viewing in which, say, the left eye looks directly at the left picture while the right eye converges on to an image of the right-hand picture which is displaced towards the centre by either a pair of mirrors or a refracting prism of suitable shape.

A variation on this method requires the pair to be set at an angle like the sides of an almost fully opened book. One eye looks directly at one page, while the other, converging naturally on the line of sight of the first, sees the second image reflected in a mirror. This method lends itself to the examination of stereo pairs bound in book form.

In this case the side which is viewed in the mirror must be printed in reverse since the image is reversed on reflection. F.P.

See also: *Beam splitter; Stereoscopic photography.*

Books: *Stereo Photography*, by K. C. M. Symons; *Stereoscopy*, by N. A. Valyus (both London).

STEREOSCOPIC CAMERA

A stereoscopic camera is normally designed to take two simultaneous photographs of the scene in front of it. The photographs are taken from viewpoints separated by the same distance as a pair of normal human eyes. When the subsequent positive prints are looked at through a suitable stereoscopic viewer they combine to give a three-dimensional reproduction of the original scene. The first binocular stereoscopic camera was put on the market in Paris by A. Quinet in 1853.

Some stereoscopic cameras are made to fold, but the most popular type is simply a rectangular box in which the essential parts of two box-type cameras are mounted side by side. The axes of the two cameras are 2½ ins. apart—2½ ins. being the normal separation of the human eyes. The camera may be designed to take pictures on either pairs of plates, pairs of films, separately wound, or on alternately spaced frames of a single film.

The earliest roll film stereo cameras produced 2¼ × 2¼ inch images; hence a stereo pair could consist of neighbouring film frames which were automatically separated by the required 2½ ins. Roll film cameras for smaller film widths—such as size 127 roll film—were at one time made, yielding interlaced stereo pairs.

A more modern attempt to produce stereo pictures on roll film consists of a camera with the two lenses arranged side by side across the width of the film. The film there runs vertically up or down. This automatically reproduces two adjoining stereo images, though the separation of the optical axes is not the normally required 65 mm. The images on the film are transposed when viewed upright and right way round; prints made from negatives therefore need re-transposing, while transparencies have to be viewed in an appropriate transposing viewer.

Miniature Stereo Cameras. Stereoscopic photography gained a new lease of popularity with the introduction of camera models taking perforated 35 mm. film. This also brought colour within the range of the stereo photographer and made it an economic proposition.

Stereo cameras using 35 mm. film take pairs of either 23 × 24 mm. vertical pictures or 24 × 30 mm. horizontal ones. In the first case the two corresponding images are separated by two frames, and in the second case by one frame. This leads to two distinct film transport systems.

In cameras which produce 23 × 24 mm. pictures the film is advanced by two frames for every exposure. The arrangement of images on the strip of film is therefore:

1a. 0. 2a. 1b. 3a. 2b. 4a. 3b. 5a. 4b. 6a., etc., the space between 1a and 2a being blank (1a, 2a, 3a are the right-hand images, 1b, 2b, 3b the left-hand ones). The stereo pairs are thus interlinked throughout the whole length of the film which cannot be cut without splitting at least one stereo pair. A complete 36-exposure load of film yields 27 or 28 stereo pairs.

Where the camera takes 24 × 30 mm. pictures, the film is advanced alternatively by one and three frames for every pair. The images are arranged in a different fashion:

1a. 2a. 1b. 2b. 3a. 4a. 3b. 4b. 5a. 6a. 5b., etc.

Each set of four frames constitutes two complete stereo pairs. A full 36-exposure film yields 21 to 22 stereo pairs. Usually the single and triple frame advance are controlled automatically with the transport knob geared to advance four frames at one revolution. Two film locks, spaced one-quarter revolution apart, then yield the interval of one and three frames alternately.

Many miniature stereo cameras can be used for single shots if desired. One of the two film apertures is then blanked off by means of a built-in masking plate (or one of the lenses covered) and the film transport locks reset to advance the film frame by frame.

Optical System. The lenses must be accurately matched because any discrepancy would mar the stereoscopic effect. Matching wide aperture lenses is an expensive business, so very few stereoscopic cameras have lenses of aperture greater than f6·3. In the miniature formats

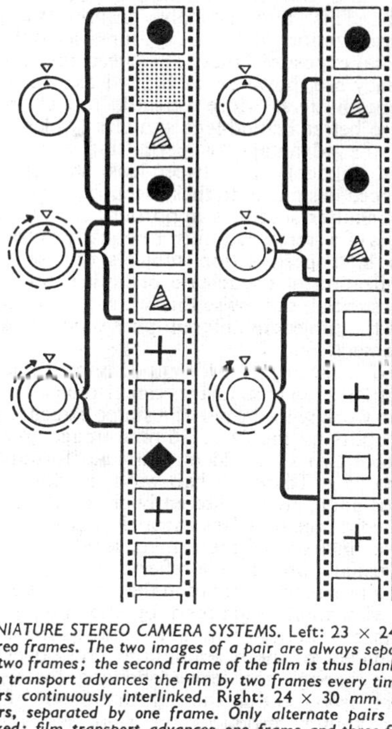

MINIATURE STEREO CAMERA SYSTEMS. Left: 23 × 24 *mm. stereo frames. The two images of a pair are always separated by two frames; the second frame of the film is thus blank. The film transport advances the film by two frames every time. All pairs continuously interlinked.* Right: 24 × 30 *mm. stereo pairs, separated by one frame. Only alternate pairs interlinked; film transport advances one frame and three frames alternately by use of special transport lock.*

however, some cameras are made with lenses of aperture $f3{\cdot}5$ and even $f2{\cdot}8$. The iris diaphragms of both lenses are coupled to a common control.

A brilliant type of viewfinder is mounted on the top of the box mid-way between the lenses. One advanced type of stereoscopic camera is built on the twin-lens reflex principle and has a viewing lens and focusing screen in place of the viewfinder.

If the finder is mounted between the lenses, parallax correction is simple; the finder area is masked vertically to include only the field covered by both lenses. The masks may be fixed (for all subjects beyond 3–5 feet) or movable (for close-up work).

Some cameras have their lenses fitted in screw focusing mounts, coupled to a single lever which indicates the distance of sharp focus on a calibrated scale. In others, the lenses are mounted on a focusing panel racked forward by a thumb wheel on the side of the camera.

A number of miniature stereo cameras incorporate a coupled rangefinder. As the camera itself has a comparatively long body, the rangefinder can have a base length of up to 6 ins. or thereabouts without difficulty.

Shutter Units, The simplest stereo camera shutter consists of a long metal plate with two apertures, spaced the appropriate distance apart, and moving horizontally behind the lenses. This automatically ensures perfect synchronization of both exposures as well as identical exposure times. The speed range and accuracy are, however, as limited as with box camera shutters. More versatile models may feature better behind-lens shutters (which still have the advantage of permitting a reliable direct linkage) or between-lens shutters of advanced design. With these accurate matching at all shutter speeds is important, for the two images of a stereo pair must be equally exposed. With high-speed action subjects precise synchronization of the release of both shutters is also essential, otherwise the two stereo pictures show the subject in different phases or stages of movement.

Stereo Attachments. Normal 35 mm. cameras giving 24 × 36 mm. images can also be used for stereo work with the aid of a stereo attachment. This splits up the 24 × 36 mm. image area to produce two 18 × 24 mm. images forming a stereo pair. There are two ways in which this can be done, giving differently orientated pairs.

The first method uses a straightforward beam splitter in front of the camera lens. This consists usually of a double reflector unit, possibly with a prism, which takes in the same view from two viewpoints separated by the interocular distance of about 65 mm. The beams from the two halves of the beam splitter cross over on going through the lens (following the usual optical laws) so that the right hand side of the beam splitter produces the right eye image upside down on the left-hand half of the film frame (as seen from the back of the camera). Similarly, the left-hand beam produces the left-eye image on the right-hand half of the frame. If the complete frame is turned upside down (for example after processing) and viewed so that the images are right-reading, the left-eye half of the stereo pair is on the left- and the right-eye half on the right. The miniature stereo pair, if taken on reversal black-and-white or colour film, can be mounted in standard slide frames and viewed in a straight (non-transposing) viewer.

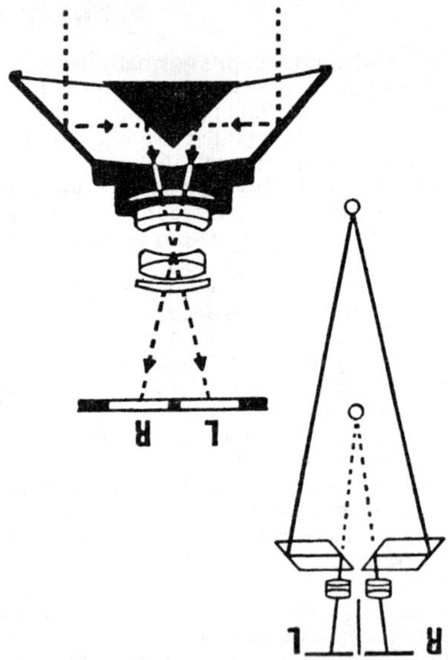

STEREO ATTACHMENTS. Left: Beam splitter to fit in front of normal camera lens, produces two untransposed images side by side on the normal negative area. The prism unit ensures correct separation of viewpoints. Right: Twin-lens attachment for interchangeable lens cameras. The images are inverted (as with beam splitter) and also transposed. Can be used with a prism unit, or, for very near subjects, without it.

Beam splitter attachments of this kind have been produced at various times for a number of cameras; usually they are specially tailored to a camera model, but do not have to be. As long as the beam splitter is correctly centred in front of the lens and close to it, it will work with any miniature camera.

The second way is only suitable for cameras with completely interchangeable lenses. Here the normal camera lens is replaced by a pair of short focus lenses mounted side by sidef forming their images on the left- and right-hal, respectively of a standard 24 × 36 mm. negative. Since the separation between the lens axes is usually of the order of 15 to 17 mm., a prism stereo attachment is also used. This is not a beam splitter, but a pair of prisms which shifts the effective viewpoint of each lens of the pair outwards to obtain the correct interocular distance. For close-up photography the

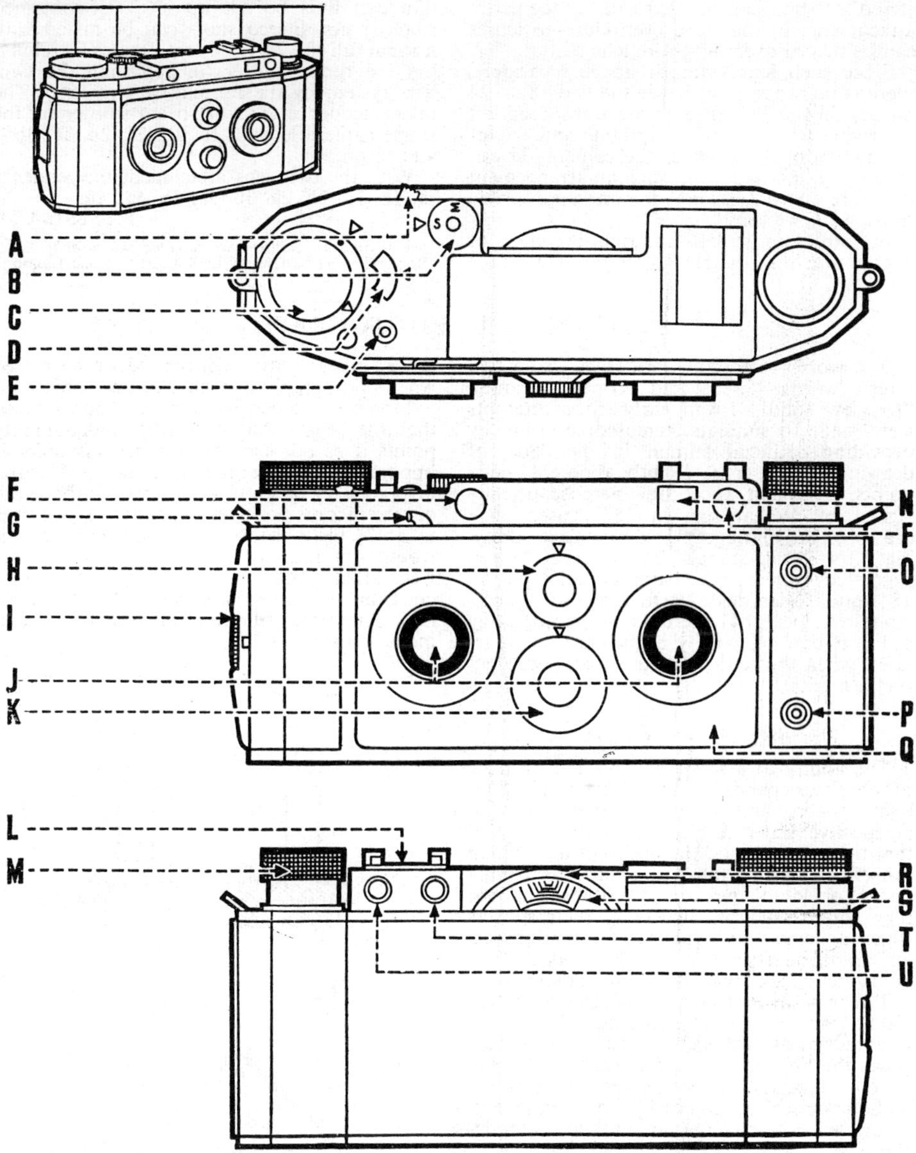

STEREO PRECISION MINIATURE CAMERA. This model takes up to 21 pairs of 24 × 30 mm. stereo frames on a standard 36 exposure load of 35 mm. film. The matched lenses are mounted on a common lens panel for simultaneous focusing, while aperture and shutter controls are coupled for both. The individual images of a stereo pair are separated by one frame; the film transport therefore advances the film alternately by one and three frames, with a complete turn of the knob corresponding to a cycle of two stereo pairs. The camera will also take about 42 single frames. Top left: General view of camera against 1 in. square grid. Top right: Top view. Centre right: Front view. Bottom right: Rear view.
A, reversing lever. B, stereo and single frame control; when set to single frames, covers up one of the film apertures with a blanking plate, and gears the transport knob to advance one frame every time. C, film transport knob. D, film counter. E, release button. F, rangefinder window. G, lever for independent shutter tensioning. H, aperture setting knob. I, back lock. J, lens. K, shutter speed setting knob. L, accessory shoe. M, rewind knob. N, viewfinder window. O and P, flash sockets. Q, lens panel. R, focusing wheel and scale. S, depth of field indicator. T, viewfinder eyepiece. U, rangefinder eyepiece.

stereo lens pair can be used without the prism attachment; in that case even close-up lenses can be placed over the entire lens pair.

Since each lens forms it image separately, there is no cross-over; hence the two 18 × 24 images on the 35 mm. film are transposed left to right when viewed upright and right-reading from the back of the camera. Transparency slides made with such an arrangement therefore need viewing in an appropriate transposing viewer.

In both cases the left- and right-eye views of a pair are immediately adjoining; the whole film can thus be cut up into self-contained stereo pairs. Stereo shots can be mixed with normal full size 24 × 36 mm. frames by removing the beam splitter (or replacing the twin lens system by the normal camera lens). The image sequence with the beam splitter on the single camera lens is: 1a. 1b. 2a. 2b. 3a. 3b—and so on.

With the twin-lens attachment the sequence is: 1b. 1a. 2b. 2a. 3b. 3a. 4b. 4a., etc.

F.P. & L.A.M.

See also: *Beam splitter; Stereoscopic photography.*
Book: *Stereo Photography*, by K. C. M. Symons (London).

STEREOSCOPIC PHOTOGRAPHY

The word stereoscopic is derived from two Greek words stereos and scopeo, meaning "to view solid". From early times attempts were made to simulate stereoscopic vision by providing artificial stimuli in the form of drawings made from slightly different viewpoints. The first successful stereoscope, designed by Wheatstone in 1832, used such drawings, but photography soon replaced drawings, for Wheatstone's friend, Fox Talbot, suggested stereoscopic pairs of calotypes in 1841 and Collen made the first ones for Wheatstone in August, 1841. Stereoscopic photography did not, however, really begin until ten years later, when the refracting stereoscope was put on the market.

BINOCULAR VISION

The ability to assess the mass and distance of objects depends to some extent on their light, shade, and colour contrast, and on perspective, but it results mainly from the fact that the two eyes and the brain form a human rangefinder. When the eyes are directed from a distant object to a near one, two things happen: they rotate inwards slightly in their sockets (this movement is known as convergence); an adjustment of focus takes place (this adjustment is known as accommodation).

The most distant plane in which the human vision can distinguish relief is around two hundred yards, although for all practical purposes the limit is closer. The nearest plane in which objects can be viewed comfortably without optical assistance is about ten inches, below which convergence and accommodation are difficult, and may cause discomfort.

The approximate separation of the human eyes when fixed on infinity is assumed to be $2\frac{9}{16}$ ins. (65 mm.). This figure, known as the interpupillary distance, is important in the design of stereo cameras and viewers.

The Stereoscopic Slide. A stereoscopic slide consists of a pair of positive prints or diapositives made from negatives taken from two slightly different viewpoints. Normally the separation is equal to the interpupillary distance, but it may be increased or decreased when photographing distant or near objects.

Modern practice is to mount positives so that the separation of corresponding infinity points does not exceed 65 mm.: this point is important because the eyes do not easily diverge. This fact imposes a limit on the width of stereoscopic positives. At one time, positives up to $3\frac{1}{2}$ ins. wide were used. So that the eyes would not have to diverge, prismatic lenses were fitted to old-time stereoscopes; they were cut from the transposed halves of a bi-convex lens. There are still many of these old instruments in use.

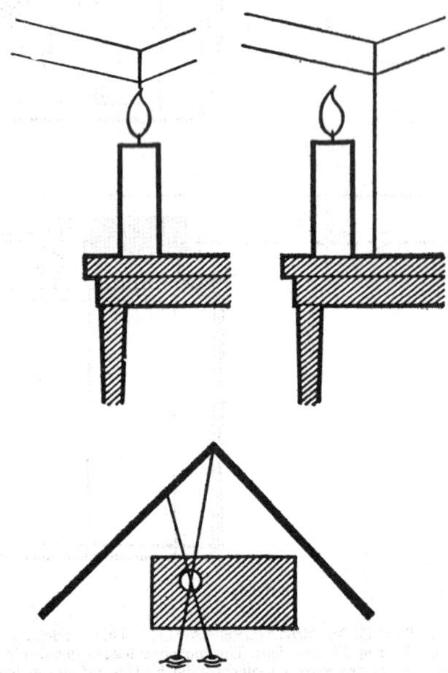

BINOCULAR VISION. As the two eyes occupy distinctly separate positions when viewing a scene, they also see slightly different views. The difference is particularly apparent in a relative displacement (parallax) of near objects against their background (e.g., the candle against the corner of the room). This gives the scene its appearance of depth.

Distortions. Stereoscopes are fitted with lenses so that the slide can be brought close to the eyes, and fill the field of view more or less completely. The focal lengths of the camera and stereoscope lenses should theoretically be equal, but in practice viewing lenses of increased focal length are commonly used: these cover the field more easily and with less marginal distortion, and some people prefer the effect they give. In practice, the error introduced by an excess focal length of as much as 30 per cent may escape detection.

If the viewing lenses are of longer focal length than the taking lenses, the subject will appear more distant, and elongated in a direction away from the observer. The picture will in addition appear rather small. Conversely, if the taking lenses are of longer focal length than the viewing lenses, the subject will appear to be nearer, and compressed. The picture will be over-magnified, and the corners may not be covered.

If the stereo negatives are taken from viewpoints spaced wider apart than the normal interpupillary separation, the subject appears nearer, and relief is exaggerated. Conversely, if the taking separation is reduced, the subject appears more distant, and the impression of relief is diminished. Excessive separation should be avoided: it may cause difficulty in viewing and is unnecessary—in fact many stereoscopic cameras have been made with less than normal separation.

STEREOSCOPIC EXPOSURES

Any ordinary camera can be used for making stereoscopic negatives—even a simple box camera. The size of negative should preferably be less than $2\frac{1}{4} \times 3\frac{1}{4}$ ins. If the negatives are made with a smaller camera, and the prints are intended for use in a large viewer, the viewing conditions can be improved by making small enlargements up to 65 mm. wide.

Stereoscopic negatives may be made in two different ways: by sequential exposure or by simultaneous exposure.

Sequential Methods. In sequential methods, the negatives are exposed one after the other in the same camera, from viewpoints separated by the normal interpupillary distance. With this method, several seconds may elapse between one exposure and the next, so it is only suitable for still subjects. (In simultaneous methods, both negatives are exposed a the same time, so the subject may be moving as quickly as for normal photography.)

(1) Hand held exposures. It is possible to make a stereoscopic picture with an ordinary camera held in the hand in this way:

Stand with the feet apart, and the weight of the body on one foot. Make an exposure, wind the film, transfer the weight to the other foot, and make a second exposure. With luck and good judgement the result will be a stereoscopic pair of negatives. By moving the body in the same direction as the film is moved for the next exposure, the pictures will be taken in the correct relative positions for viewing, but probably at an incorrect separation. The camera may be tilted up or down, but it must be kept level from side to side, and the same area should as far as possible be included on both negatives.

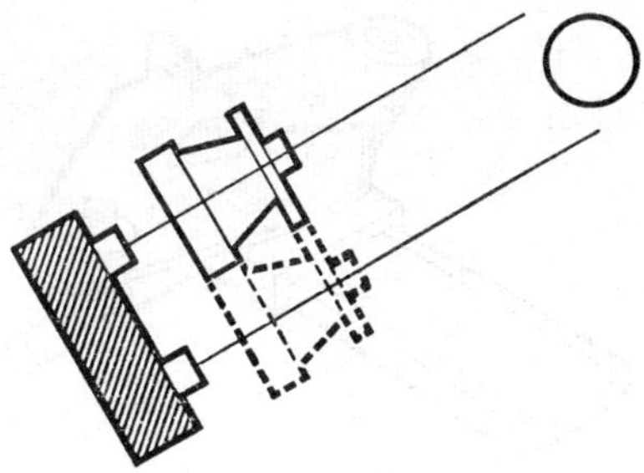

STEREO-PHOTOGRAPHIC METHODS. Two photographs of a subject, separated by the interocular distance, can be made either with a twin-lens stereo camera, or by successive exposures with a normal camera.

(2) Tripod attachments. A suitable stereo slide can be made from a shallow wooden tray, 65 mm. wider than the camera. It should have a bush on the under side for attaching it to a tripod. This enables two exposures to be made with certainty from the correct distance apart. Exposures are made with the camera first at one end and then at the other. The camera may be lifted off the tray when setting the shutter and winding the film. Commercial attachments are available for many of the better makes of roll film and miniature cameras.

(3) Moving the object. This method is useful for small objects and has the same effect as moving the camera. The subject must be taken against a plain background; the movement between exposures should be $\frac{1}{4}$ in. per foot distance from the lens (2 cm. per metre). The lighting must be diffused, or must move with the subject.

(4) Rotating the object. A small object can be stood on a turntable and rotated about four degrees between exposures. The lighting must be diffused, or rotate with the subject.

(5) Rotating the camera. This method is suitable for use when taking small objects with a plate camera. A piece of black card is arranged inside the camera back so as to obscure half the plate. After the first exposure, the card is changed to the other side, and the camera is rotated on its support so as to bring the image to the other side of the plate. This method gives a pair of exposures on one plate.

(6) Displacing the lens. The body of a plate camera can be divided by a central partition, and the lens, mounted on a long panel, is moved from one side to the other between exposures. Alternatively, it can be fitted to one side of its panel, and the panel inverted between

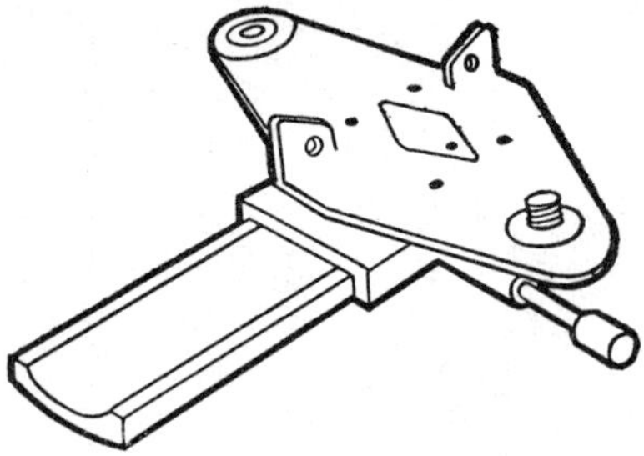

STEREO SLIDE. *For stereo photography by successive exposures in a normal camera, the camera displacement and positioning have to be accurate. The stereo slide moves the camera by the exact distance required between exposures, while a spirit level helps to check that it is truly level. In use the whole slide is mounted on a tripod, and the camera on the sliding part.*

exposures. The slide must be closed during the change unless the camera has a focal plane shutter.

Simultaneous Methods. These methods may give two separate negatives or a pair of negatives on opposite halves of a single plate.

(1) Using two cameras. Two identical cameras can be fixed together at the required separation, the shutters being released together. Commercial attachments for this method are made for some miniature cameras, and many workers find an economical way of making stereos is to mount two cheap box cameras side by side.

(2) Using a stereoscopic camera. A stereoscopic camera resembles two cameras in one body. The lenses are matched, and the diaphragms, shutters, and focusing movements are coupled, so as to ensure that the negatives are identical except for the separation of the viewpoints.

(3) Using beam splitters. Two pairs of surface silvered mirrors, or a similar arrangement of surface treated prisms, are arranged in front of the lens in such a way as to impress two images on opposite halves of one frame of film. The resulting negatives are of rather a narrow angle; and exposures must be increased from 50 to 100 per cent because light losses take place.

In another pattern, now obsolete, a pair of mirrors fixed at an obtuse angle was placed at 45° in front of the lens, and the picture was taken with the camera at right angles to the subject. Commercial beam splitters are available for most of the better miniature cameras. One such instrument incorporates a pair of short focal length lenses and includes a normal angle of view.

Subject and Exposure. Subjects should generally be selected for their depth, but a distant background should not be included with a close-up subject. Apart from difficulty with depth of field, eyestrain is likely to result when such pictures are viewed.

The lighting should preferably emphasize roundness although *contre jour* subjects make attractive slides. Since stereoscopic photography is essentially the photography of foregrounds, a foreground of some kind should always be included. Even a fence will often make the difference between success and failure.

Good definition is necessary over the whole picture area, so the depth of field must be adequate. Generally the apertures should be small and a tripod should be used if possible. The exposure must be generous enough to give printable detail in dark foreground objects: at least double the exposure necessary for a normal photograph should be given. Areas of solid black that can be tolerated in a normal photograph will ruin a stereo picture.

Special Techniques. Objects more than about 200 feet away do not possess stereoscopic relief, but the normal separation between the exposing positions can be increased by moving the camera to provide relief in objects at any distance. The separation of the camera stations can be calculated on the basis of one-fiftieth of the distance to the nearest plane to be included, or it can be estimated by changing the camera position until a noticeable displacement occurs between the foreground and background on moving from one camera station to the other. This technique is known as hyperstereoscopy.

Separation is sometimes increased to provide extra relief in such subjects as worn coins or inscriptions.

Objects under the microscope may be photographed in stereoscopic relief either by taking two photomicrographs from slightly separated viewpoints or by tilting the stage.

A stereoscopic impression of the X-ray image can be created by taking two radiographs from separated viewpoints. This technique is used principally in medicine, to determine the depth of organs or foreign objects in the human body.

When taking close-ups with a single lens camera, the separation can be adjusted to the subject, and if necessary the lens can be converged to keep the subject central on the film. With a stereoscopic camera, the separation can be varied from the normal by capping each lens in turn, and making sequential exposures, moving the camera the required distance.

NEGATIVE MATERIAL AND PROCESSING

In the long history of stereoscopic photography, many formats for taking or viewing have been used. They vary from frames on 16 mm. film up to 5 × 7 ins., and the total number so far tried is in excess of twenty. Current formats for negatives include the 35 mm. film sizes, 45 × 107 mm., and 6 × 13 cm., and viewing formats include the above, 41 × 101 mm., and the standard card mount for prints, which measures 3½ × 7 ins. The complete list of formats includes a numbe of obsolete sizes, and apparatus for some of these is still in circulation. For example, 7 × 13 cm. and 3¼ × 6¾ ins. cameras are still used, but supplies of plates or film are difficult to obtain. These sizes, however, are equally effective in viewers of correct design.

Plates, sheet film, roll film, and 35 mm. film are all currently used, and at one time film packs were popular. The positives are not only viewed under magnification, but they receive closer scrutiny than any enlargement of equal magnification, and for this reason grain must be avoided. Fine grain material should be used whenever possible: certainly for small sizes.

Monochrome reversal material can be used with success; it is becoming increasingly popular, and some workers use reversal processing methods with normal negative material.

A normal developer can be used for sizes above 45 × 107 mm., but for smaller sizes fine grain processing is best. Below full size 35 mm. frames, fine grain technique is essential, and processing must be carried out with great care. Over-development, leading to blocked highlights, must be avoided; there should be printable detail in all parts of the negative.

In stereo photography it is always easier to prevent blemishes than to cure them; even the spotting of paper prints is difficult, and it is practically impossible to retouch a blemished transparency. For this reason, cleanliness is important. Dust must be avoided at all stages, all processing solutions must be filtered, and sensitive material should be handled as little as possible, so as to reduce the risk of mechanical damage. The camera itself should be carried with the lenses pointing downwards, so that dust tends to fall away from the plate or film. Airbells should be avoided by using a wetting agent in the developer.

Making Positives. The aim should be to produce a print or transparency with a full range of tones, and with detail in every part. Areas which are devoid of detail appear either as meaningless dark patches, or as holes in the picture through which the grain of the diffuser in the viewer is visible.

On the other hand, a slide which appears dull and flat when held in the hand, and which would be quite unsuitable for use as a normal photograph, may look quite satisfactory when viewed in the stereoscope. This is because points which are scarcely to be distinguished from others by contrast occupy different positions in space. It is in fact common for a dull slide to brighten up in the viewer.

Prints on Paper. Daylight paper (P.O.P.), contact, bromide, and chlorobromide papers are all suitable for making stereoscopic prints. Many workers still favour daylight printing for its long scale of tones and for the beauty of the toned image.

It is essential to use a glossy surface, but the prints need not be glazed: some people consider that the reflections from a glazed print are distracting. Opinion generally favours a warm tone, as black tones appear unnaturally cold.

Transparencies. Transparency plates are obtainable in the 45 × 107 mm. and 6 × 13 cm. formats coated with emulsions which correspond to contact, bromide, and chlorobromide papers. Sheet film coated with a fine grain ordinary or a process emulsion can also be used. These materials provide a range of contrasts which enables a reasonably good slide to be made from most negatives, with warm or cold tones, which can in turn be toned to provide a further range of image colours.

Colour Material. Colour film may be used in the form of sheet film, roll film or 35 mm. film. Some of the additive processes are satisfactory in sizes above 45 × 107 mm., but the reseau on which these processes depend becomes annoyingly visible under high magnification. However, 45 × 107 mm. slides made by the additive processes appear satisfactory if a viewer with lenses of abnormally long focal length is used.

Slides of exceptional beauty can be made in all sizes including 35 mm. on subtractive colour film because of the absence of reseau or grain.

Transposition. If a negative made in a stereoscopic camera is held up to the light with the emulsion facing away, it will appear the right way up and the right way round, but the left hand image will be on the right hand side, and vice versa. The reason for this is that the image

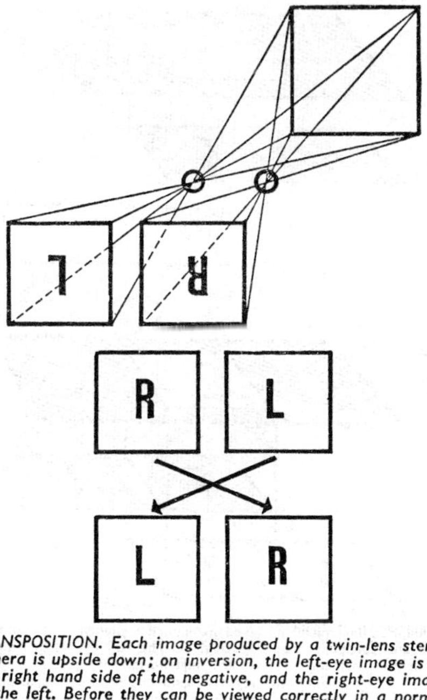

TRANSPOSITION. Each image produced by a twin-lens stereo camera is upside down; on inversion, the left-eye image is on the right hand side of the negative, and the right-eye image on the left. Before they can be viewed correctly in a normal stereoscope, these images must be transposed so that each is on its own side for viewing by the correct eye.

was upside down in the camera: inverting it has brought the frames to the wrong side of the plate. As it is essential that each eye should view the picture made from its own corresponding viewpoint, it is necessary to transpose the pictures. Negatives on film can be cut and transposed. and at one time apparatus for cutting and transposing glass plates could be obtained. (There was a risk of damaging the plate in cutting it.) Similarly, if the positive is on paper or sheet film, the two frames can be cut apart, and mounted in their correct positions.

For making positives on glass, it is usual to use a special type of printing frame called a transposing frame. Several types are made, but the most usual has a central opening of the correct size, surrounded by an opaque mask. The positive and negative plates overlap at the central aperture, and extend at each side, their positions being governed by a rebate. After the first exposure, their positions are reversed, and a second exposure of the same duration is made.

The length of the rebate is calculated to ensure that the separation of corresponding points on the positives does not exceed 65 mm. in the case of 6 × 13 cm., and 63 mm. in the case of 45 × 107 mm.; but as a small tolerance is allowed in cutting the plates, it may be necessary to adjust the length by gluing thin cards in the end rebates.

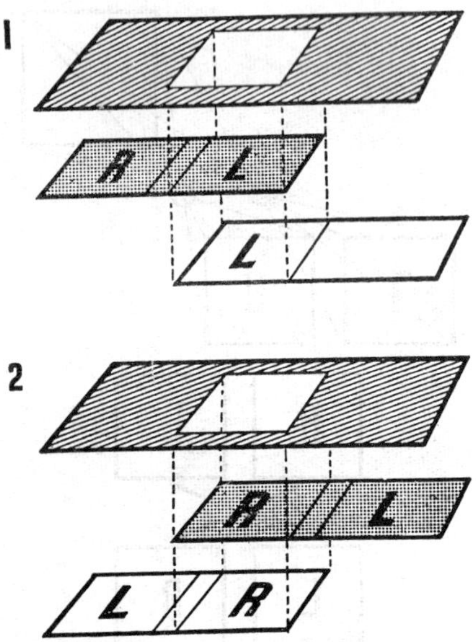

TRANSPOSING FRAME. Used for contact printing the images of a transposed stereo negative. 1. The left-eye (right-hand) negative is printed first on the left-hand side of the printing paper. 2. By sliding the negative and paper in opposite directions, the right-eye (left-hand) negative is then printed on the right-hand side of the paper, yielding a correctly oriented stereo pair for normal stereo viewing.

VIEWING METHODS

Many experienced workers are able to view slides stereoscopically without apparatus, having trained themselves to dissociate convergence from accommodation. The eyes are fixed on infinity, and the slide, held about eighteen inches from the eyes, is raised into the line of sight. If all goes well, the pair of pictures will fuse, and will be seen in full stereoscopic relief. This technique calls for special ability.

Viewers. Normally correct reconstruction can only be realized with specially designed viewing apparatus. The following are the main types of viewer.

(1) Brewster type viewers. These are box type viewers used for transparencies. They have a pair of lenses at one end, and a slot for the slide at the other. Daylight types have a diffuser behind the slot, but modern miniature types are usually self-illuminated. In some viewers an opaque slide can be illuminated through a window in the top.

(2) Holmes viewers. These are open type viewers, having at one end a pair of well-hooded lenses, which are frequently prismatic. The carrier that holds the slides is adjustable for focusing along a horizontal rod. Viewers of this type are used almost exclusively for opaque slides mounted on standard 3½ × 7 ins. cards.

(3) Cazes viewers. This type of stereoscope, designed by L. Cazes, closely resembles a beam splitter of the four-mirror type, and is used for the examination of stereoscopic pairs taken from the air. The same principle, on a smaller scale, is used in viewers designed for the examination of stereoscopic pairs made on 35 mm. film with a beam splitter, and mounted as 5 × 5 cm. slides. In this case, the mirrors produce images with the required interpupillary separation from positives printed side by side on a single frame of 35 mm. film.

(4) Cabinets. The need for changing the slide by hand is avoided by using a cabinet type viewer. The early types carried a number of slides (usually thirty or more, but models were made which held 300) attached to a continuous chain or band. This was operated by turning a handle, which caused the pairs of prints to appear one after the other in the viewer.

Cabinets of an improved type carry the slides in groups of twenty-five in grooved plastic trays. By operating a handle, each slide in turn is raised vertically into the viewing position, and the eyepieces are obscured during the change.

(5) Special viewers. A number of special viewers exist. Among these is one for using uncut filmstrip. This has an internal prismatic system, which renders transposition unnecessary. Another interesting viewer is made for use with discs which carry seven pairs of positives. These are changed by operating a lever,

the eyepieces being obscured during the change.

Other Methods of Viewing. Auto-stereoscopic methods are becoming popular for advertising and publicity. As many as six or more exposures are made with special apparatus, and from the resulting negatives a positive is made which consists of a number of narrow vertical picture elements. This is displayed behind a screen which may either consist of vertical grid elements, or of moulded vertical lenticular prisms. The grid or lenticular elements enable only one aspect to be seen at a time by each eye, and the person viewing can obtain a satisfactory impression of relief at near or distant points within a wide arc.

Photographs made by these systems are sometimes known as parallax stereograms.

Anaglyphs, originally popular in the first quarter of this century, have recently appeared again in magazines. In making an anaglyph, a pair of images printed in ink of complementary colours (usually red and green), is superimposed. When these are viewed through spectacles of similar colours one image becomes invisible to each eye while the other appears in dark grey tones. In this way the correct images for the right and left eyes can be presented as they would appear when viewed normally, and the result appears in monochrome. Carbro, dye imbibition processes, photogravure, and half-tone are all suitable processes for the production of photographic or printed anaglyphs. This method dates from 1891.

A vectograph consists of a pair of stereoscopic transparent images located on each side of a plastic polarizing support, which is backed by a reflecting metallic surface. When viewed by reflected light, the images are polarized in two separate planes at right angles to each other. A pair of polarizing spectacles is worn when viewing, the eyepieces being so arranged as to select only the image appropriate to each eye.

TRIMMING, MOUNTING AND MASKING

Positives on paper are usually mounted on dark toned card of size 3¼ × 7 ins. Prints made from a stereoscopic negative must be cut apart, trimmed, and transposed, but before doing so they are first marked. This is best done by reversing the print, and drawing a light pencil line from the centre of one print to the centre of the other. When separated and correctly transposed, the line cuts the outer edges of the prints.

Care must be taken to trim the prints square, and to mount them at the correct separation. It is important also to see that the prints do not slope at an angle. Inaccuracy will be evident in what is called a "floating edge"—the sign of careless work.

In order to obtain the correct separation, a millimetre rule is used to set a pair of distinct homologous points in the background to the normal distance apart. A centre line and a base line are ruled on the mount as a guide, using a white pencil, which is easily erased.

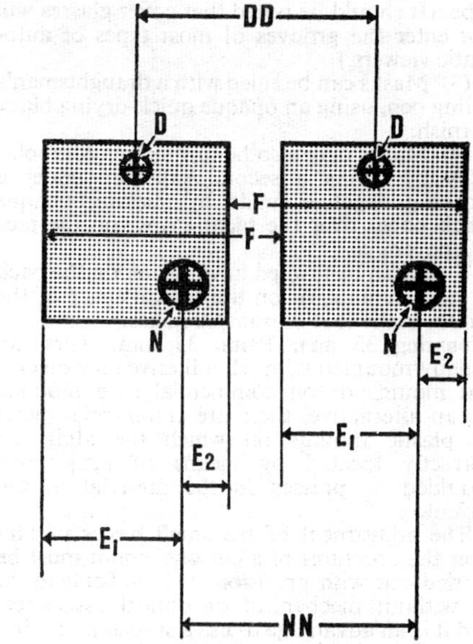

WINDOW EFFECT. Making the separation of the picture edges the same as the separation of two near point images produces an effect of a picture window located at the same apparent distance as the near points. D, distant image points. F, frame separation. N, near image points. NN must be the same as F, and E_1 as well as E_2 must be the same for each image. Making F equal to DD locates the stereo window in the plane of the distant subject points, behind the foreground.

Window Effect. It is the customary, but not universal, practice to mount prints or to make transparencies so as to produce the effect of looking at the subject through a thin opaque window. This is done by making the separation of the edges of the mask or prints equal to the separation of a point at the distance at which it is desired to locate the imaginary window. The distance chosen is usually 3, 6 or 9 feet, and the separation can be calculated.

In practice, it suffices when trimming prints to do it by a method of trial and error. The rule is as follows:

Trim a little more from the right-hand edge of the right-hand print than you trim from the right-hand edge of the left-hand print, then place the prints together and trim them to the same size. Experience soon shows how much to trim off to give the effect of the distance required.

Masking Transparencies. The following are the main methods of masking:

(1) The mask may be cut from thin black paper with a steel straight edge and a razor blade.

(2) Or a mask may be built up on the slide or cover glass with adhesive black binding

tape. (It should be noted that cover glasses will not enter the grooves of most types of automatic viewers.)

(3) Masks can be ruled with a draughtsman's ruling pen, using an opaque quick-drying black varnish.

(4) Masks may also be printed photographically by giving a second exposure under a masking negative made by sticking opaque black paper over the picture area on the face of a cover glass.

Titles can be printed in white ink on the back of card mounts, or on the central space of the mount in the case of transparencies.

Mounting 35 mm. Pairs. 35 mm. pairs are usually mounted with self adhesive tape on cut-out mounts or on commercial card mounts. As an alternative, there are commercial metal or plastic mounts on which the films are correctly located by means of projections moulded or pressed in the material of the mount.

The adjustment of the small pieces of film over the apertures of a cut-out mount must be carried out with precision: it is difficult to do so without mechanical or optical assistance, and it is an advantage to use a special mounting jig sold for the purpose.

This holds the films in the correct position during mounting.

STEREOSCOPIC ACTIVITIES

The interests of the stereoscopic photographer are rarely catered for by the average photographic society, but there exist in Britain, France, Canada, Australia, and America, societies devoted to the subject.

British. The oldest is The Stereoscopic Society, with headquarters in London. Founded in 1893, it circulates print and transparency portfolios in Britain, America, Australia, New Zealand, and Canada, and has also a local transparency folio circulating only in Britain. The Society also holds meetings, and organizes reunions and outings.

Foreign. In America, there is an active branch of The Stereoscopic Society, and also The Stereo Guild, which circulates 35 mm. transparencies. A number of local societies either specialize in or cater for stereoscopic interests.

In France, the Stereo Club Francais holds monthly meetings in Paris, and circulates an illustrated monthly bulletin.

Two major exhibitions in Great Britain, and several in America, have classes for stereoscopic slides. K.C.M.S.

See also: *Hyperstereoscopy; Lenticular system; Parallax stereogram; Stereo-micrography; Stereo-radiography; Stereoscope; Stereoscopic camera.*

Book: *Stereo Photography*, by K. C. M. Symons; *Stereoscopy*, by N. A. Valyus (both London).

STERRY PROCESS. Forebath for producing a soft print from a contrasty negative. The exposed print is immersed in a 1 per cent solution of potassium bichromate or ferricyanide. After about a minute, it is drained and transferred to the developer for normal processing.

While the process softens the gradation of the print it tends to change the image colour from pure black to warm black with a greenish tinge.

Potassium ferricyanide or bichromate both tend to make the developer inactive and may cause stains, so the developer must be changed frequently.

See also: *Contrast control; Thorium nitrate forebath.*

STIEGLITZ, ALFRED, 1864–1946. American photographer, editor of *Camera Work*, 1903–17. Studied engineering and photography in Berlin 1882–90, then worked for five years in the family photo-engraving business in New York. Stieglitz condemned artificial picture making and advocated straight photography of life, opened up by the hand camera. Founded the Photo Secession, 1902. Exhibited pictures of *avant-garde* artists as well as photographers in his New York gallery "291". A prominent exhibitor at international photographic exhibitions and influential leader of American photography for forty years. Received Progress Medal of the R.P.S. 1924. Biography by Waldo Frank (New York, 1936).

STILB. Photometric unit of brightness (luminance) of a light source or illuminated object in terms of intensity per unit area, equal to one candle (candela) per square cm.

See also: *Light units.*

STILL LIFE. To the photographer, still life subjects offer opportunities for experiments in composition, lighting, colour arrangements, studies in texture reproduction and so on. They have the advantage of requiring little space, and a wide variety of effects can be achieved with simple lighting equipment.

The basic essential of all still life work is a satisfying, self-contained arrangement of shapes. From this point of departure the photographer can branch off into a number of directions. He can use shadowless lighting technique so that all the emphasis falls on the tones of the subject, or he can employ strong lighting to present his subject in terms of shadow patterns or contrasted surface texture. He can project the shadows of his subject on to the background or foreground so that the shadows themselves become a part of the subject, or he can light the background to kill the projected shadows so that the whole interest is once more framed within the subject.

The best type of camera for still life work is undoubtedly a technical camera of at least quarter-plate size, with a focusing screen and a selection of lenses.

A reflex camera is excellent, but whatever the camera, it must be possible to examine the image full size on a focusing screen.

A long focus lens is essential so that the subject can be photographed from a fair distance, to ensure good perspective. A wide-angle lens is useful for special effects, such as deliberately distorted perspective. The maximum aperture of the lens is unimportant since the exposure can be of any duration without fear of subject movement. It is frequently desirable to stop the lens well down, in order to achieve overall sharpness.

The only props of any importance are one or two large sheets of thick paper of various tints to use as background material. As a rule the background should be unobtrusive, and for that purpose it is always safe to stand the subject on a large sheet of paper and curve the paper up behind to form the background.

See also: *Antiques; Ceramics; Close-ups; Commercial photography; Flowers; Food; Glassware; Lighting the subject; Models (scale); Sculpture; Silverware; Table top.*
Books: *Camera Close-up*, by O. R. Croy (London); *Creative Table Top Photography*, by E. Heimann (London).

STIRRING ROD. Glass or plastic rod used to stir solutions while dissolving chemicals. Some stirring rods have a thick disc at one end for crushing tablets, crystals, etc., to make them dissolve quicker. Some photographic thermometers also have a thickened end for use as stirring and crushing rods.

STOCK SOLUTION. Processing solution which needs only to be diluted with another chemical or to be mixed with another stock solution to be ready for use. Developers which will not keep in normal solution are often stored as two separate stock solutions which will keep indefinitely apart and be always ready for mixing when required.

See also: *Solutions.*

STOP. Originally, the term referred to a metal plate with a hole in it which was inserted through a slot in the mount of early lenses to act as a fixed diaphragm. Stops with holes of different sizes could be interchanged to alter the working aperture of the lens. Each metal plate was associated with the corresponding *f*-number, and nowadays it is customary to refer to the diaphragm settings as stops whether they are formed by a perforated plate or a variable iris diaphragm.

See also: *Diaphragm.*

STOP BATH. After a negative or print has been developed, it is usual to rinse it in clear water for a minute or so before transferring it to the fixer. In these circumstances development does not stop completely for some seconds.

Normally the slight amount of extra development that takes place is not important but there are times when it is desirable to end development decisively with a special stop bath. A solution of 2–5 per cent acetic or citric acid, or potassium metabisulphite is commonly used for this purpose. It immediately neutralizes the alkalinity, and thus the activity, of the developer.

Commercial stop baths sometimes incorporate an indicator dye to show at a glance whether the acidity is still sufficiently high. When the stop bath is exhausted, it changes colour.

An acid stop bath should always be used after a high speed developer and for hot weather processing.

An acid hardening bath acts as a stop bath and also hardens the gelatin.

See also: *Hardening bath.*

STOPPERS. Ordinary corks are unsuitable for bottles of photographic processing solutions because they rapidly become soggy and disintegrate. They can be made to last longer by being soaked in hot paraffin wax, but even then they are never completely reliable for more than short-term use.

Glass stoppers are not affected by processing liquids, but they are by no means ideal. Their greatest drawback is that they tend to stick stubbornly if the bottle has been left for any length of time and the liquid has crystallized around the neck.

A smear of vaseline applied evenly around the stopper will prevent it from sticking, but it is apt to make the fingers greasy and cause finger marks on any sensitized material handled afterwards.

A glass stopper that has become jammed in the neck of the bottle can often be freed by tapping it smartly on the side with the handle of a table knife.

Another popular method is to wrap a cloth soaked in hot water around the neck of the bottle to expand it away from the stopper.

Rubber is by far the best material for stoppers for photographic solutions. It is not affected by any of the normal chemicals, and it cannot become jammed in the neck of the bottle. Encrusted solution can easily be washed off the surface, and the resilience of the material means that the stopper gives a completely air- and solution-tight seal. Rubber stoppers can be bought in all sizes, and many photographic chemicals are sold in rubber-stoppered bottles which, when empty, can be used again for stock solutions.

For some purposes, plastic screw caps may be better than rubber stoppers; they can be screwed on tightly without any fear of them coming undone again.

Plastic stoppers are now also being made and possess the same advantages as rubber stoppers.

See also: *Bottles.*

STOPPING DOWN. Decreasing the size of the lens aperture—i.e., using a smaller stop. This may be done either to increase the depth of field covered sharply by the lens or to control exposure.

With field cameras, single lens reflex cameras, and all other types in which the image formed by the taking lens is focused visually on a ground glass screen, it is always easier to focus on the principal plane with the lens at its widest aperture. Focusing done at wide aperture is the most accurate, since the acceptable depth of field at a wide aperture is comparatively small, and hence the transition from sharpness to unsharpness is most noticeable. Afterwards the lens aperture is reduced to a value (found either by actual inspection or by referring to a depth of field table) which gives the required depth of field.

If the lens has to be stopped down considerably, the image becomes difficult or impossible to see. In numerous reflex cameras this problem has been solved by a pre-selecting mechanism. This enables the user to focus at full aperture and then press the shutter release normally. Before the shutter actually operates, the diaphragm is automatically stopped down to a pre-selected value.

Stopping down the lens also reduces the amount of light passing through, and together with the shutter speed controls the effective exposure.

When taking flash photographs with open flash technique the usual way of controlling the exposure at a particular flash-to-subject distance is by stopping down the lens. In fact the flash exposure is expressed in terms of a guide number which is constant for any one film speed, and any one type of flash bulb or flash unit, and which is the product of the distance and aperture number.

Stopping down may also be necessary in enlarging, after focusing the image visually at the full aperture, to increase the exposure to a manageable length. If the exposure is made at full aperture, it may be too short to be estimated accurately, and leaves no time for "dodging" and other forms of control during printing. Stopping down also assures better overall sharpness of a print in the case of an enlarger which may not be aligned quite accurately. It becomes essential when the planes of the paper and negative are deliberately tilted with respect to each other to correct converging verticles.

See also: *Aberrations of lenses; Diaphragm; Focusing.*

STORAGE OF SENSITIZED MATERIALS. All kinds of light-sensitive material tend to lose speed and deteriorate in the course of time. There are many reasons for this. Some are beyond the control of the purchaser—e.g., the effect of impurities in the emulsion support—but the conditions under which the materials are stored have a great bearing on how long they will keep in good condition. In general, if films, plates and papers are stored in a cool, dry, dark cupboard, away from the fumes of gas or open fires, or—most deadly of all—sulphide solutions, they will lose none of their original speed or quality at least up to the expiry date given by the makers, and generally for very much longer.

See also: *Filing negatives and prints; Film storage; Keeping qualities of materials; Polar photography; Tropical photography.*

STREAK CAMERA. Special type of high-speed cine camera which records an image as a continuous spread-out picture rather than as a series of separate frames. This image needs special viewing equipment to analyze it and reconstitute individual pictures.

See also: *High-speed cinematography.*

STRESS MARKS. Streaks and other areas of density on a developed photographic image caused by pressure or friction on the emulsion before development.

This form of pressure can quite easily render parts of an emulsion developable just as if they had been exposed to light.

See also: *Faults.*

STRIKE CAMERA. Special cine camera mounted in aircraft to record "strikes" or hits during aerial gunnery practice. There are two main types: (a) 16 mm. cameras with a long focus lens mounted alongside the guns of an aircraft to record the target during bursts of firing; and (b) 35 mm. cameras with wide-angle lenses mounted in radio-controlled target aircraft to photograph projectiles or missiles approaching the target. The latter type of strike camera often works at high framing rates and may need special protection to permit recovery of the film if the target aircraft crashes.

See also: *Aircraft camera; Air force photography; Gun camera.*

STRIPING. Process of applying a narrow strip of a magnetic oxide layer to a cine film for subsequent sound recording. This narrow strip, commonly known as the stripe, may be coated direct on the emulsion surface of the film during manufacture (pre-striped film) or after processing (post-striped films). Alternatively, the stripe may be coated on a very thin plastic support and the latter applied to the film.

Dimensions and positions of magnetic stripes on different types of film stock are laid down by various standards.

See also: *Sound recording; Sound synchronization.*

STRIPPING. There are times when it is useful to be able to remove the film from a glass plate or celluloid support and retain the negative image intact. If the glass of a plate negative or lantern slide is cracked, the emulsion can be stripped off and transferred to another plate.

When large numbers of plate negatives or slides are to be stored for reference, it is more convenient to store the emulsion film without its glass support.

Stripping film, which could be peeled off the glass plate, was mentioned by Frederick Scott Archer in his communication of the collodion process in March 1851. Other methods of stripping the negative from its glass or paper support were put forward by Alexander Parkes (1856), C. M. Ferrier (1857), Leon Warnerke (1875), Thiébault (1883), Georges Balagny (1883). George Eastman introduced stripping film for use in the Eastman-Walker roller-slide in 1886.

Some printing processes reverse the image, left to right. In such cases it may be more convenient to strip the film off the plate and put it back reversed so that the print appears the right way round.

There are several ways of removing the film. The quickest methods rely on hydrofluoric acid which is dangerous to handle because it attacks glass and calls for special containers and processing vessels.

It is generally wiser to avoid using this acid. The methods described below may be slower, but they are completely safe and can be carried out with the normal darkroom equipment.

Methods. The simplest method is as follows:

(1) Prepare a solution of:

Potassium carbonate	90 grains	5 grams
Glycerine	100 minims	5 c.cm.
Formalin	100 minims	5 c.cm.
Water to make	10 ounces	250 c.cm.

(2) Filter off the precipitate.

(3) Soak the plate for 30 minutes in the solution.

(4) Drain, wipe, and allow to dry slowly for at least 12 hours.

(5) Cut through the emulsion all round with a sharp knife about ⅛ in. from the edge.

(6) Raise one corner of the film with the point of the knife and peel the whole film layer off the glass.

The following method is suitable for either plates or films but it has one disadvantage: it increases the dimensions of the film by about 3 per cent.

(1) Soak the negative in this solution:

Caustic soda	100 grains	5 grams
Formalin	¼ ounce	6 c.cm.
Water to make	10 ounces	250 c.cm.

(2) When the emulsion begins to leave the support, transfer the plate or film to the following solution:

Hydrochloric acid	½ ounce	12·5 c.cm.
Glycerin	½ ounce	12·5 c.cm.
Water to make	10 ounces	250 c.cm.

(3) Float off the emulsion and squeegee lightly down on to the new support while still wet.

A suitable adhesive for sticking the film down on to the new support is made as follows: soak 0·25 grams (4 grains) of gelatin in 50 c.cm. (2 ounces) water, dissolve by heating, and add about 10 drops of formalin. Coat the glass with this solution immediately before applying the wet film.

The following method allows the dried emulsion to be stripped off a glass plate:

(1) Soak the negative in water for at least an hour.

(2) Transfer it to a saturated solution of potassium carbonate in water and leave for 5 minutes.

(3) Wipe off the surplus solution and cut through the film ⅛ in. from one edge.

(4) When dry, insert the point of a knife in the cut and the film will peel away.

Stripping Materials. There are also several photographic processes which rely on stripping the emulsion from the support. For this purpose special transfer or stripping materials are available.

These sensitive materials usually have a specially hardened emulsion, coated on top of a layer of unhardened gelatin. When the material is soaked in warm water, the soft gelatin dissolves and allows the emulsion layer to be stripped away from the film or plate support with ease.

The surface of the final support may have to be prepared to receive the emulsion by being coated with a layer of gelatin or other suitable adhesive.

In the process of being transferred from one support to the other, the image is reversed from left to right. If it is to appear the right way round, it is transferred twice: first to a temporary support of non-adhesive plastic or waxed celluloid, and then to the final support.

Negative transfer emulsions are mostly used in photomechanical work. In the form of positive materials they are used in three-colour photography.

With processes like Carbro—where the three colour images have to be brought together on one support the stripping (by transfer to a temporary support) takes place during the development procedure.

Silver halide stripping films have also been used for producing the etching resist of photogravure cylinders as one alternative to the traditional carbon tissue.

Transferred positives are also used for decorating walls, ceramics, and other surfaces with photographic images. The emulsion may in certain cases even be transferred to the final support first and then exposed and processed in position.

This method is used in making templates and similar industrial work. L.A.M.

See also: *Colour print processes; Emulsion removal; Printing on special supports; Transfer coating.*

STROBOSCOPIC FLASH

Electronic flashes fired repeatedly at high frequencies of hundreds to thousands of flashes per second. This method has various applications in speed photography and motion study.

Technically a stroboscopic flash unit is based on an electronic flash discharge circuit, with certain modification to permit a high flashing rate.

Historically, electronic flash lighting is almost as old as the negative-positive photographic process, since W. H. Fox Talbot used both about one hundred years ago. True, he covered only a small subject with flash, but his writings prophetically describe the modern electrical flash equipment, as used in photography today.

Theory of the Single Flash. Light is produced when electrical energy stored in a charged condenser is discharged into the flash tube. During the discharge, the instantaneous power is very large (it may be as much as several million watts). However, the practical criterion is the light energy or exposure—i.e., the integral of the light power in candle power during the flash time in seconds.

Possibly the most important component in a flash unit, other than the flash lamp, is the condenser (or capacitor). This accounts for most of the weight, volume and cost, of the usual equipment. The capacitor is the energy storage component of the circuit. It can take in energy at a slow rate over a period of seconds or even minutes and then discharge it into the lamp in a fraction of a second at the required megawatt rate. Many of the advances in modern electronic flash equipment design have been concerned with improved capacitors and further improvements are expected.

Electronic flash tubes are made in many sizes and forms, such as straight, U, or spiral shapes, as required for different applications. The gas pressure and electrode spacing are arranged so that the tube will not flash by itself in most applications. But when a triggering pulse is applied to an external electrode, internal ionization results, and the tube flashes.

Most of the flash tubes used today are filled with xenon gas at a pressure of a fraction of an atmosphere. A series of 1 microsecond exposures of the arc in a typical tube made with a magneto-optic shutter shows that the arc starts as a narrow filament adjacent to the flash-tube wall on the side of the external triggering electrode. Soon, if the energy is sufficiently large, the arc filament enlarges to fill the entire flash tube.

Flash duration for a specific flash tube and circuit is best defined by plotting the light as a function of time. This shows the initial delay, the initial rise, the peak, and the decay of the light. From a practical standpoint, the actual duration is negligible, if the motion of the subject is effectively stopped. One practical way to indicate duration is to define it as the time between the initial and final instants when the light is 1/3 of the peak.

The output of a flash tube is the integral of the light against time. An approximate value can be obtained by taking the product of duration by the peak light. If the peak light is measured in candle power, and time in seconds, then the output will be in units of candle-power-seconds. The lumen-second output is greater by a factor of 4π, or about 2·5, when the angular asymmetries of the flash tube are considered.

The electronic flash tube has a most unusual volt-ampere characteristic, At the start of the discharge, the resistance is infinite, since no current is flowing. Then, as current starts to flow, the resistance drops rapidly to a value of a few ohms for the main portion of the discharge. Finally, the current again becomes zero, and the resistance infinite. During most of the flash, the volt-ampere curve is similar to that of a resistor and the circuit transients can be calculated approximately. Extrapolations to other lengths and diameters of flash tubes can be made as though the tube were a resistor of uniform conductivity. Then flash duration can be estimated from the equation:

$$\text{Flash duration} = \frac{RC}{2} \text{ seconds.}$$

where C = capacitance in farads.
R = resistance.

There is a large variety of electronic flash tubes available for the equipment designer to consider. For any specific case important factors are:

(1) The shape of the tube—linear, spiral, U-shaped, etc. A concentrated lamp is more effective in a reflector, where control of the light is required.

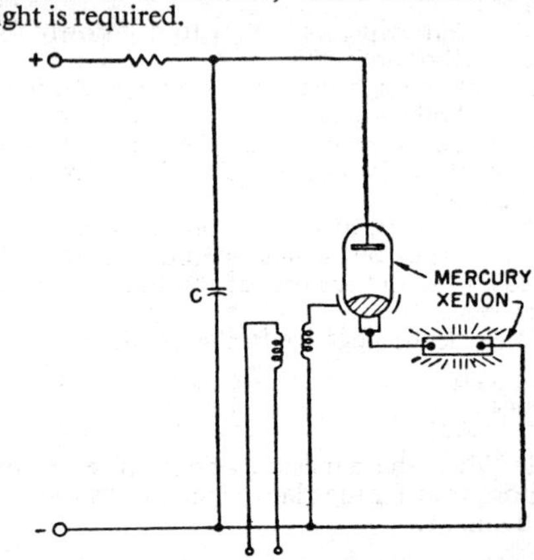

MERCURY POOL CONTROL CIRCUIT. The mercury tube in series with the main flash tube sharply starts and interrupts the current for the flash, and handles high power rates. Permits several thousand flashes per second. C, main capacitor.

(2) The efficiency at the required energy input. Each lamp has its maximum efficiency for a particular input. The designer should attempt to use the lamp at or near this maximum efficiency point, although in practice it is generally necessary to work at a lower value of efficiency to prevent the tube from overheating and to prolong its life.

(3) The flash duration is a function of the tube and the circuit. If a specific duration is required, the design is usually fixed.

(4) The voltage at which the flash is desired. Item (2) above needs to be reconsidered in terms of voltage, since the efficiency usually drops when the tube is used at low voltage.

There are many times when the most efficient flash tube cannot be used—particularly in stroboscopic work, at high frequency and high power—because of heat storage and heat conductivity problems.

Stroboscopic Photography. The name "stroboscopic" photography has come to mean multiple-flash exposed photographs. Some of the earliest exploiters of this general system are Muybridge, Marey, Cranz, Bull, etc., whose excellent pictures of horses, people, and bullets are still used as examples today. The first multi-exposure photographs were made on a single, moving plate using a slotted disc as a shutter or on separate plates with a series of cameras. The modern method is to use a succession of electronically produced flashes of light separated by accurately controlled intervals of time.

Two practical problems arise when an electronic flash tube is required to run as a stroboscopic source:

(1) The flash tube becomes so hot that it does not function properly—e.g., it may miss occasional flashes, due to not starting properly.

(2) The flash tube fails to de-ionize, thus preventing the capacitor from building up a charge for the subsequent flash. A low value continuous current flows in the flash tube. This condition is called "hold-over".

A hot tube may fail for several reasons: such as, puncture of the glass by the external sparking circuit, short circuiting of the triggering spark by the conduction of the hot glass, or actual collapse of the glass wall of the tube. Failure of a tube to de-ionize results in the continuous arc hold-over condition, where the capacitor charging current flows continuously into the lamp. A further difficulty may result when the tube self-flashes, as the capacitor recharges, due to the lowering of the hold-off voltage by residual ionization or temperature.

Any or several of the difficulties mentioned above are soon experienced when a flash tube is operated at a fast rate of flashing with high energy per flash.

Tubes of quartz are better than glass tubes at high power rates, since quartz has the higher melting temperature.

Several special circuits have been used to operate flash tubes at high rates. Examples are the series mercury-arc rectifier of the pool type, and the hydrogen thyratron, as used in radar modulators.

Mercury-pool Control Tube Circuit. The only new element added to the conventional electronic flash operating circuit is the mercury tube. This mercury tube is connected directly in series with the main discharge current path of the capacitor to the flash tube. Thus the mercury tube must be designed to handle adequately these high-valued peak currents. Immediately after a discharge, the mercury tube de-ionizes quickly, due to its low-pressure, and thereby prevents the previously mentioned hold-over current from flowing in the flash tube.

The mercury tube connected in series is also beneficial in starting a flash tube, because the igniting spark circuit from the mercury cathode goes directly between the two main electrodes. This circuit makes it possible to start flash tubes at very low voltages on the flash capacitor.

Operation at several thousand flashes per second with an input of 1 kilowatt are practical with the mercury-connectron tube.

Hydrogen-thyratron Tube. The thyratron serves as a switch to discharge the capacitor energy into the flash tube. This thyratron has a very quick de-ionization time, thus enabling the starting of the flash tube to be controlled at high frequency, far above rates where the frequency is limited by the slowly de-ionized high-pressure gas in the flash tube. High-pressure (10 to 70 cm.) of xenon gas is required for efficient light production in the xenon flash tube.

The circuit with the hydrogen thyratron is similar to that used in radar transmitters, except that the flash tube is replaced by a magnetron. In either application, the function

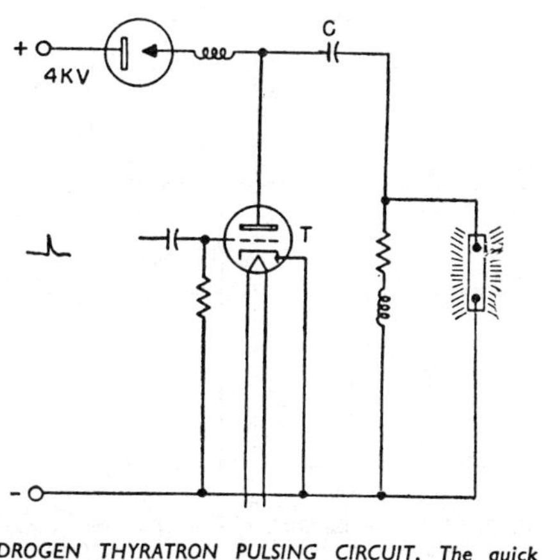

HYDROGEN THYRATRON PULSING CIRCUIT. The quick deionization time of the thyratron tube permits higher flashing frequencies. C, main capacitor. T, thyratron tube.

is the same: namely, to pulse the lamp or magnetron with high-voltage and high-power energy. One of the properties of the hydrogen thyratron is the ability to supply large peak currents without change to the thyratron cathode.

A typical commercial high speed stroboscope using a hydrogen thyratron control tube operates at the following electrical conditions: lamp voltage, 8,000 volts; capacity, 0·01, 0·02, or 0·04 microfarads; frequency, 6,000 max. per second (0·01 microfarads); flash duration, 1 to 3 microseconds; energy per burst, 1,500 watt seconds (limit set by lamp heating).

The flash tube in such equipment must be designed with high resistance to reduce the peak-current requirement of the hydrogen thyratron. This is accomplished by the use of a small diameter (of the order of 1 mm.) and a long arc length (up to 4 ins.). A small percentage of hydrogen gas is mixed with the xenon to reduce the afterglow in the discharge.

The stroboscopic light can also be synchronized with the motion of the film to produce framed pictures for projection. This is commonly done with 16 mm. high-speed cameras. Magnetic pickups are used to trigger the flash lamp at the correct instants of time.

Multi-capacitor Circuits. There is another general type of multiflash equipment that is capable of great speed and flexibility, and which has utility when a limited number of photographs can be used. Each flash for this method is powered by a separate capacitor into either separate lamps or a common lamp. This circuit becomes bulky when a large number of flashes are desired, since each flash requires a separate storage system, as contrasted with a single storage capacitor discussed in the previous method. There is a frequency limit set by trigger-tube de-ionization for these last two circuits, even if one operation is used. Control air-gaps with three electrodes triggered by a time-delay element can be used.

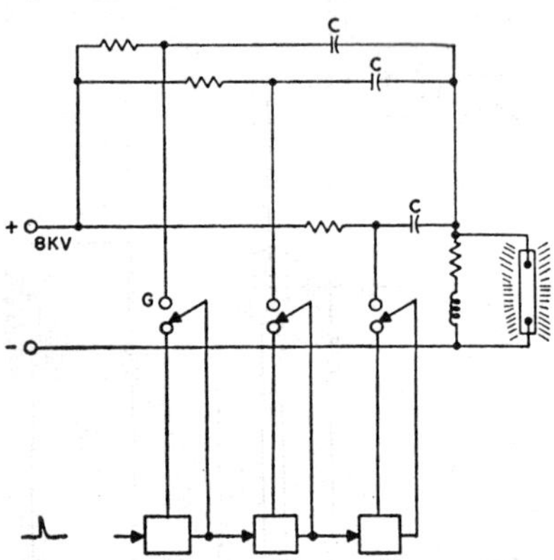

MULTI-CAPACITOR CIRCUIT. A series of separate capacitors are discharged in turn into the same flash tube. Unit is flexible but bulky. C, power capacitors. G, controls gaps. T, time delay units for triggering the successive flashes.

Applications. Stroboscopic flash has two main applications: analysing a rapid movement by photographing successive phases on one film or plate, and arresting a rapid cyclic or other periodic movement for visual observation and photography.

Examples of the first kind are multiple photographs of dancers, hurdlers, etc.; the classical one was a shot of a golfer driving off. For this two flash tubes were used, each driven by a mercury pool tube. Each flash tube was operated from a 10 watt-second charge at 100 flashes per second. The power into each lamp was about one kilowatt. Normally, such a lamp will soon overheat unless it is artificially cooled, but this tube did not overheat, since the operation time was only half a second.

A bullet in flight was taken by multiflash photography with a xenon tube operated at 6,000 flashes per second, with about half watt second per flash for a brief period. The bullet was photographed against a background of reflecting material to reflect the light of the lamp which was mounted directly above the camera lens. The film was a strip of 35 mm. film. The flashing frequency was controlled by an electronic oscillator. There was no blur on the continuously moving film, since the exposure was only about one microsecond in which time neither the bullet nor the film had moved appreciably. In this method there is no need for synchronization between the film motion and the frequency of flash, it is only necessary to move the film so fast that the photographs do not pile up on top of each other.

For the second type of application the flashing frequency is matched to the frequency of the cycle of the movement. As a simple instance, a machine component rotating at 5000 revolutions per minute may be illuminated by one or more stroboscopic tubes flashing at 5000 flashes per minute. If both rates are accurately matched the flash will illuminate the same phase of the movement every time and the component will appear stationary. It can then be observed or photographed in that way —a useful procedure in industrial photography where it is not practicable to stop a machine for picture taking.

By slightly shifting the phase of the flash between successive exposures a whole picture series of complicated movement cycles is obtained permitting the study of motion that would be impossible to observe in any other way. H.E.E.

See also: *Chronophotography; Flash (electronic); Flash equipment; Flash technique; High speed photography; Spark photography.*

Books: *Flash! Seeing the Unseen by Ultra-high Speed Photography*, by H. E. Edgerton and J. R. Killian (Boston); *The Photographic Study of Rapid Events*, by W. D. Chesterman (Oxford).

STUDIO

Room equipped and set aside for photography—in particular, for portraiture. By extension the term is also applied to any business in which photographs are taken in a studio on the premises—e.g., portrait studio, commercial studio.

Size. There is no upper limit to the size of a studio. Commercial studios may be big enough to accommodate large objects like motor cars and well spaced groups of people. At the other end of the scale it is possible to be more specific. It is desirable for the studio to be long enough to permit full-length portraits to be taken with a normal angle lens. So assuming a maximum subject height of 6 feet and allowing another 1 foot clearance above and below—i.e., 8 feet altogether—the distance of the lens from the subject can be obtained by simple proportion. If the lens is, say, of 6 ins. focal length working on a quarter-plate ($3\frac{1}{4} \times 4\frac{1}{4}$ ins.) negative and D is the distance from lens to subject, then

$$6 : 4\tfrac{1}{4} = D : 8$$

and $D = 6/4\frac{1}{4} \times 8 = 11{\cdot}3$ feet approximately.

This is the bare distance between the lens and the subject. To this must be added the focal length of the lens (in this case, 6 ins.), at least 3 feet space between the subject and background, and another 2 feet behind the camera for the photographer. So that, no matter what the size of the studio camera, if it is equipped with a lens of normal focal length the studio must be at least 17 feet long to give full length portraits with a reasonable amount of elbow room for the photographer and subject.

For normal operations, the studio should be half as wide as it is long and the ceiling, overhead cables or roof girders should all be over 10 feet from the floor. All the above dimensions are the least that should normally be accepted in choosing or planning a studio. It is of course possible to work in a smaller space with smaller subjects or with a greater amount of inconvenience.

Daylight Studio. In the early days of photography there were no suitable forms of artificial lighting, and the photographer who wanted to work indoors had to use a specially constructed or adapted room which admitted the maximum amount of daylight (but not direct sunlight) and was equipped with some means of controlling it. Daylight studios are still in use today in out-of-the-way places without an electricity supply.

The majority of daylight studios have a glass roof facing north (in the northern hemisphere) and sloping at an angle of not less than 60°. This minimum angle ensures that the direct rays of the sun will not shine into the studio. It applies only to latitudes around 50°; towards the equator the angle must be increased and eventually it becomes impossible to keep out the direct rays and they must be diffused with a curtain of fabric or by white-washing the glass. In addition to the roof, the wall away from the sun should be glazed down to a height of about 3 feet.

Roller blinds are fitted directly under the roof and over the glazed part of the wall to allow the light to be controlled. There should be two sets of blinds—one of opaque fabric to cut off the rays completely, and the other of light diffusing muslin or similar translucent cloth. Each set of blinds should consist of a separate roller and blind at the top and bottom of the glazed area so that the light can be cut off at either end.

One type of daylight studio was glazed at one end only—immediately over the sitter—while the other consisted of a low unlighted annexe in which the photographer could focus the image without the need for a focusing cloth.

A favourite method of constructing a daylight studio extension to another building is to make it in the form of a glass lean-to against the side of a wall facing away from the sun.

The great objection to daylight studios is the unreliability of the daylight and its variation from hour to hour and season to season. For this reason daylight studios have now been almost entirely superseded by those entirely lighted by electricity.

Professional Studios. In professional studios there is a wide variation in design, size and equipment, but the basic furnishing conforms to a regular pattern. There is a camera with interchangeable lenses (or a range of such cameras) mounted on a substantial camera stand which can generally be moved about or locked in position at will.

Lighting consists of banks of overhead and side floodlights for general illumination of the subject area; one or more high power modelling lights; spot lights for background lighting and effects, and various white reflectors and mobile screens. The background may be provided by roll-up fabric back-cloths, by portable "flats", or there may also be a translucent screen with space behind for projected background scenes. In addition to normal lighting, many modern studios are now equipped for multiple electronic flash lighting.

If the studio has any ordinary windows, these can be completely blacked out by roller blinds or shutters so that the lighting is limited to the controlled electric sources.

The studio is usually lavishly wired, with socket outlets at many points in the floor or suspended from the ceiling. The object of this is to reduce the number and length of trailing wires to a minimum.

"Props" vary according to the nature of the work undertaken, from comfortable chairs for a portrait studio to comprehensive kitchen fittings for a studio specializing in food photography, recipe illustration and the like.

Amateur Studios. Few amateurs are fortunate enough to be able to set aside a room exclusively to serve as a studio and even camera club studios have to revert to normal use after the club night. So the problem is to equip the studio with lighting and other props that can be quickly brought out and arranged and just as quickly dismantled and packed away.

Clubs usually have no difficulty in finding a room of the right size, and they can nearly always count on having a cupboard where the equipment can be stored from one club night to the next. The worst problem is apt to be the electricity supply. Too often the room has only a single power plug which is neither conveniently placed nor of a high enough rating to take the lighting load. It is of course possible for the club to have the necessary extra points installed, but the expense can be considerable.

Where this is out of the question, the best plan is to connect a heavy cable from the nearest power point to a portable box fitted with a number of outlets up to the maximum load on that particular circuit—e.g., a 15 amp outlet on a 230-volt supply would safely provide for a load of 3,000 watts consisting of four Photofloods (1,200 watts), two 500 watt spotlights (1,000 watts) and a number of 150 watt tungsten lamps for effect and focusing lights. The cable and portable distribution box enable all the lights to be kept on short leads and avoid numbers of trailing cables between the power plug and the subject position.

Backgrounds and light reflectors in such circumstances must be either sheets of fabric tied on to collapsible frames, or arranged like roller blinds and hung from the picture rail. Other properties must be limited to the furniture normally in the room supplemented or disguised by draped fabric. An effective method of providing unusual backgrounds is always available in the club projector; if this is quite powerful, it can be used for projecting made-up transparencies of a suitable nature from either in front of or behind a suspended sheet of white or translucent material.

Home Studio. The amateur who works indoors at home must be prepared to improvise and compromise to a much greater degree. His studio is usually the living-room or a bedroom which can only be spared from its normal functions for an hour or two at most. The room itself is usually not long enough and there is insufficient space to accommodate lighting equipment on safe, broad bases.

The best way of solving the problem of the length of the room is to shoot through the open doorway and have the subject on the far side of the room. This adds 3 or 4 feet at least to the camera-to-subject distance and gives a much better perspective.

Everything should be cleared away from the studio end of the room, leaving it completely bare. If the carpet has an assertive pattern, it should be covered over with a plain, neutral-tinted sheet and a similar sheet should be suspended from the picture rail—preferably on a batten—to form the background. It is sometimes convenient to use a piece of background material long enough to cover the floor under the subject as well. This gets rid of the horizontal line at the floor level, but it also makes the background bulkier to handle and store.

If there is a fireplace at the only convenient end of the room, it is better to arrange to let the fire go out for the evening and heat the room with a portable heater. The background can then be hung so as to cover the fireplace, leaving the whole of the end of the room clear for the subject and lamps.

The portable distribution point described earlier is the best method of supplying the electricity for the lighting, and it should always be connected to a 15 amp outlet if more than one Photoflood bulb is being used. It is certain that the room lighting will be wrongly placed to provide any useful light on the subject, so if possible it should be switched off and the subject arranged by the photographic lighting only. One of the best aids to this is a series-parallel switch enabling two Photofloods to be connected in series for continuous lighting and switched on to full brilliance in parallel for the duration of the exposure only.

In the average living-room there is not enough floor space for more than two lighting stands in addition to the subject and the normal furniture. It is generally best to use these stands for the main (modelling) light and the back light, as these generally are used at a reasonable height. For the remaining lights, the type of lamp reflector that stands on a table or a chair or on the floor or that can be clamped on to any handy piece of furniture or on to the edge of a door at any height may be used. This type of support is much more rigid than the folding telescopic stand, and it calls for no extra space.

A typical amateur studio set-up would consist of a background sheet suspended from the picture rail along one end of the room and lighted by a Photoflood lamp in a reflector standing on the floor or on a chair. Main lighting would be provided by a Photoflood lamp in a reflector and on a stand. Another Photoflood lamp in a reflector for fill-in lighting would be clamped on to the back of a chair or other suitable item of furniture. Back lighting, or rim lighting, to highlight the hair, etc., is generally given by a spotlight, but a flood in a reflector, and on a stand, can do the same job provided the lamp is placed so that it does not shine straight into the camera lens.

This basic set-up would be satisfactory for head and shoulder portraits, but would have to be considerably boosted for full-length figures. The most desirable additions would be a spotlight and an adjustable light reflector—e.g., a square of plywood painted white on one side and mounted on a universal clamp-on tripod head.

A set-up of this kind can be stored in an ordinary cupboard; it will transform a normal living-room into a studio in about a quarter of an hour and take no longer to clear away.

The amateur who is lucky enough to have a room that he can reserve exclusively for photography has no conversion problems. He can afford to have his lamps mounted on adjustable stands, and permanent wiring run to conveniently-sited points not only along the floor but along the picture rail. In such circumstances it is a great convenience to have one perfectly plain wall to serve as a background. If the skirting board can be removed and the wall plastered and distempered or papered right down to floor level it will get rid of a perpetual distraction. In addition to the wall surface (which gives a background free from folds and creases) there should be a roll-up background mounted on brackets near the ceiling and long enough to curve forward under the subject. All wall and ceiling surfaces should be light cream or off-white.

The electric wiring should allow for room heating as well as photographic lighting: it should not be necessary to use the heating point for any of the lamps.

Finally, the floor should be covered with plain unpolished lino or drugget.

Power Requirements. The total amount of electric power consumed by the studio will depend on its size, the type of work undertaken, and whether heating as well as lighting is provided by electricity. The average amateur equipment can usually be connected to the normal lighting and power outlets of the room without overloading the distribution system of the whole house (in case of overloading the fuses of the individual lighting or power circuits will blow first).

On the other hand, the maximum demand of a professional studio is apt to exceed the capacity of the average distribution system and may call for special provision in the form of heavier supply cables, metering and wiring. In any case it is advisable to add up the individual consumption of all the apparatus likely to be switched on at one time and make sure that the supply arrangements are adequate.

A typical commercial studio undertaking general still photography will require about 10 kilowatts (i.e., about 45 amperes at 200–250 volts) for lighting alone and double that figure if sub-standard cinematography is included. A purely portrait studio can operate on 5 to 10 kilowatts, while the amateur will have to be satisfied with a maximum of about 2½–3 kilowatts (i.e., 10–15 amperes at 200–250 volts) if he wishes to run all his lighting off the normal room wiring (e.g., 15 ampere power plugs). The consumption of any heaters, motors, dry-mounting presses and other equipment must be included in the above totals.

Wiring. Modern studios are wired on the ring main system whereby a number of 13-amp. socket outlets are supplied by a closed loop of power cable. Current to any socket can travel around the loop in either direction, thus doubling the current-carrying capacity of the system. In Great Britain, the main fuse supplying the ring is rated at 30 amps. so that it is only likely to blow if a short-circuit occurs on the distribution cable itself. Each separate piece of equipment carries its own correctly rated cartridge fuse in a holder, inside the plug connecting it to the mains. The result is that a short circuit in the equipment simply blows the fuse in the plug and leaves the rest of the system unaffected.

Ring mains outlets are designed to take standard 13-amp. fused plugs which have flat pins so that there is no danger of inserting an ordinary round pin plug. This type of circuit is particularly advantageous in a professional studio since it allows further power socket outlets to be connected straight on to the ring instead of having to be wired all the way back to the main distribution panel.

A large studio may require more than one ring to carry the total load, but a single ring should be adequate for an amateur set-up.

To be on the safe side any commercial or professional set-up should be installed—or at least checked over—by a competent electrical contractor. The amateur can always get the representative of his local electricity authority to tell him if his lighting arrangements are likely to overload the supply equipment. F.P.

See also: *Commercial photography; Dimmer; Electricity; Lighting equipment; Portraiture at home; Studio photography; Wiring.*
Book: *The Business of Photography*, by C. Abel (London).

STUDIO PHOTOGRAPHY

Even the smallest town nowadays has at least one professional photographic studio which caters for the photographic needs of the general community. It is an essential part of the make-up of a typical township, and although businesses of this type vary in size and scope, they conform to a fairly regular pattern. There is usually a retail shop with a photographic studio attached, and facilities for taking portraits, weddings, etc., and a certain amount of industrial and commercial photography both on and away from the premises.

Premises. In the days before electricity became available for lighting photographers had to depend upon daylight and the photographic studio of the past was constructed largely of glass. The roof faced north to secure even, shadowless light and it was usual for one or more of the walls to be in the form of a large window. Because of these vast expanses of

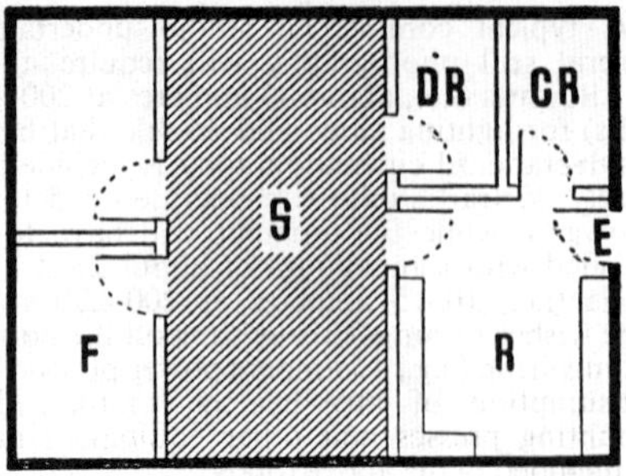

SIMPLE SMALL-TOWN STUDIO. Scale about 1 in. = 18 feet. O, office. F, finishing room. S, studio space. DR, darkroom. CR, changing room (situated near darkroom for easy plumbing). R, reception room. E, entrance. The arrangement is rather crowded, especially as far as processing is concerned (a single darkroom serves for developing, printing, chemical mixing, and chemical storage), but it frequently is dictated by the facilities and space available in small premises. Display shelves are fitted in the reception room which sometimes also serves as a shop, at least for selling films and materials and for accepting developing and printing orders of customers' films.

glass it was difficult to keep the place warm in winter and cool in summer; it was also difficult to make it waterproof. The essential equipment included an enormous camera with its black focusing cloth, a rose-garden background and an assortment of decorative furniture. This was the photographer's workroom.

The studio of today is a very different proposition. Daylight can be dispensed with altogether, backgrounds are plain and easily interchangeable, furniture is comfortable, lighting units are efficient and portable and the camera for many assignments can be small and compact. Modern heating and ventilating appliances help to make clients comfortable and the style of decoration is chosen to impart a feeling of restfulness and security.

Although a studio may be intended primarily for portraiture it is usual (particularly in the country and small-town businesses) for all kinds of photographs to be taken in it. These include copying, photographing paintings and taking photographs for catalogue illustrations, as well as photographing many other small objects and collectors' pieces. Only a few exclusive specialists can afford to keep a studio solely for portraiture.

Planning. The planning, organization and running of a studio and workrooms are essentially matters of economics. It is of little use building up a fine connexion if by mis-management and lack of organizing ability the money earned is wasted by general inefficiency of the supporting departments. Whatever the reason for which photographs are taken, the basic functions of making a photograph are always the same: negatives have to be processed, prints and enlargements have to be made and subsequently washed, dried, mounted and finished. Even a small studio will be severely handicapped if all these jobs are carried out in one or even two rooms. It is much wiser to plan or rearrange the premises so that each section of the work can be conducted in a separate room, and not interfere with the next.

In addition to the studio, office and reception rooms, there should be a small room or cubicle where negative materials are unpacked and loaded into slides, a negative processing room and a printing and enlarging room with separate accommodation for washing and drying.

There should be a separate room set aside for mounting, finishing and retouching. In the darkrooms, provision should be made for wet and dry areas and they should be kept solely for these purposes. This avoids trouble from spotty negatives and prints. It is also worth while, whatever the size of the studio, to set aside a small room solely for mixing chemicals —a job which should never be done in darkrooms or in draughty places.

The layout of darkrooms and benches in all departments should be on the continuous chain principle so that as materials go from one dish or tank to the next they do so without passing over solutions or benches previously used. Every studio layout should be checked by a little time/motion study to analyse the separate movements involved in doing specific jobs and to see that wasteful walking about and reaching up and down are reduced to a minimum. No studio is complete without one or two toilet rooms and cloakrooms a pleasant and comfortable waiting room and a lock-up stockroom where a careful check can be kept of all materials consumed.

A business, artistic or otherwise, can never make healthy progress by neglecting matters such as costing, fixing prices and assessing profits. If the photographer cannot attend to them himself then he should employ someone who can.

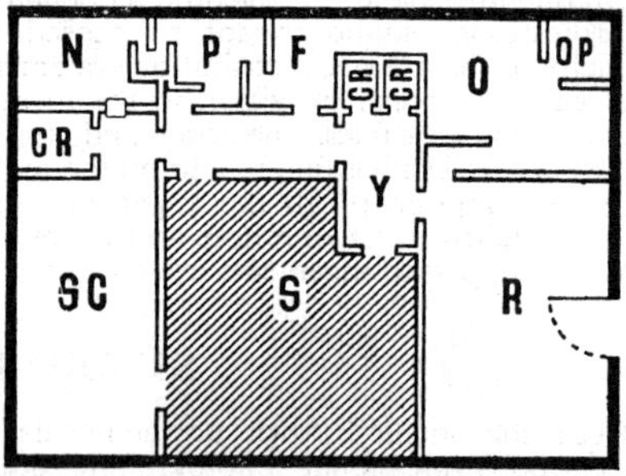

EXCLUSIVE PORTRAIT STUDIO. Typical layout for a larger city location. Scale about 1 in. = 32 feet. N, negative processing room. P, printing room. F, finishing room. O, office. OP, private office. CR, changing rooms. SC, special children's studio. S, main studio. Y, waiting room. R, reception room. The finishing and printing rooms are easily reached by the receptionist. Separate negative and positive processing rooms increase the flow of work and the capacity of the business. This type of studio can handle advertising orders (using hired models), as well as normal portrait photography.

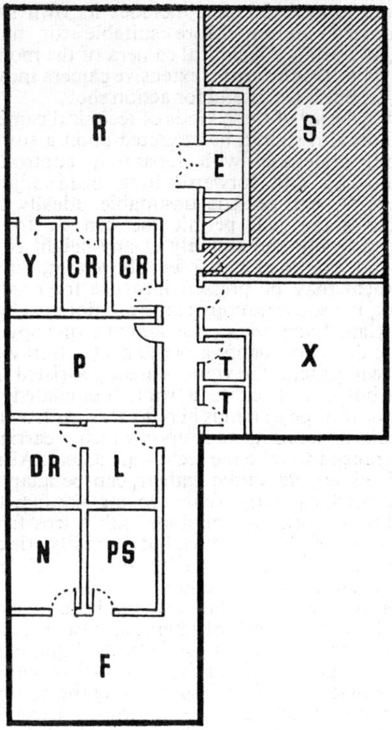

UNIVERSAL STUDIO IN SMALL TOWN. Upstairs location. Scale about 1 in. = 22 feet. R, reception room. E, entrance. S, studio. Y, storage room. CR, changing rooms. P, printing room. DR, general darkroom for loading, film storage, etc. L, laboratory for mixing and storing chemicals, etc. N, negative processing room. PS, photo-finishing printing room. F, finishing room. T, toilets. X, framing room. The studio handles portraiture, commercial work, photo-finishing, framing, etc., for the town as well as for a wide surrounding district. The reception room can also be used as a retail shop.

Photography is now largely a manufacturing process and as such the prices charged for the finished product must bear a fair relationship to the raw materials used, the processes involved and overhead expenses including salaries, depreciation and rents.

The husband-and-wife studio is a combination which has brought success to many enthusiastic workers. In practice, the wife is receptionist/secretary/retoucher/colourist and finisher while the husband takes on the operating, processing and getting business. With one general assistant to handle routine jobs of washing, mounting, making up solutions, etc., this type of studio can profitably undertake work of all kinds.

Studio Layout. The layout and equipment of a photographer's studio should be dictated as far as possible by the kind of work he mainly practices. Whatever type of work is undertaken, it is essential to interpose a small room between the studio itself and the outside world, which can act as a reception-place. Privacy is essential to portraiture, and careful studio work of any kind demands a degree of concentration which necessitates insulation against the voices and eyes of chance visitors.

Except for the most simple passport studio, the portrait studio should be at least 15 × 12 feet in size, with a 9 foot ceiling. Full-length shots require at least 20 × 15 feet of floor-space and a ceiling of 10 to 12 feet is the minimum needed for overhead lighting of standing groups. It is difficult, however, to establish a personal atmosphere for portrait work in too large a studio and a degree of intimacy might be introduced by partially cutting-off a small area decorated in a more domestic fashion. In setting up a studio for the first time, the owner may be unsure of the channels through which his future work will come, so it is wise to install equipment which is adaptable to several different kinds of photography and to avoid apparatus

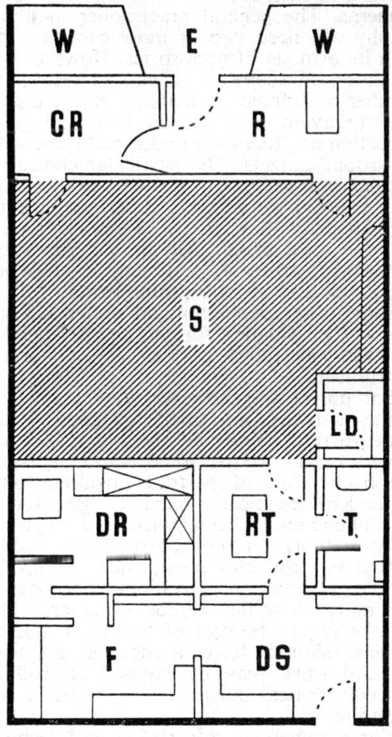

SUBURBAN PORTRAIT STUDIO. Scale about 1 in. = 11 feet. W, display windows. E, entrance. CR, changing room. R, reception room. S, main studio space. LD, special loading darkroom for reloading plate holders and storing films and plates. DR, general darkroom; used at night and in the early morning for developing, during the day for printing. RT, retouching room, also negative filing room. T, toilet. F, finishing room for washing and drying of prints as well as trimming, mounting, etc. DS, dispatch and mailing room. The studio uses studio and reflex cameras, as well as a roll film twin-lens reflex for child photography

which is firmly attached to walls or ceiling until the real need for these is proved.

Backgrounds. As part of the decor, a number of differently textured wall panels may be installed, using genuine materials or some of the excellent replicas offered by display specialists. Brickwork, rough and textured plasterwork, timber weatherboarding and stone-walling are available in simulated form. A panel 8 feet high by 6 feet wide in each is a minimum for general background use, though smaller pieces can be used for head and shoulder work. A realistically painted sky may be used like a theatre cyclorama, when exterior settings have to be built. Rolls of background paper, 9 feet wide, may be suspended from the farthest studio wall, to be pulled down when required. Black, white and middle grey are satisfactory for black-and-white, but many colours are available for colour work. The paper is cheap enough to be readily discarded by the yard, as it becomes dirty or buckled at each use.

Cameras. The general practitioner in photography will need two or more cameras, each with its own set of accessories. However wide the intended scope, it is wise to restrict the number of different formats which are used, if only to avoid need for an impossibly large collection of processing tanks, racks and other darkroom gadgets. The particular choice will depend on the nature of the work itself. Action and ballet photography, reportage, identity portraits, etc., may require 35 mm. equipment, and the ideal choice will be a really sound precision instrument, preferably a single lens reflex, taking inter-changeable lenses.

Some photographers might prefer 120 roll-film as their smallest format, and a single-lens or twin-lens reflex as their most portable instrument.

For portraiture, however, the $2\frac{1}{4} \times 2\frac{1}{4}$ ins. frame is really too small for good quality negative retouching and the old type studio camera on its massive stand is too clumsy for the lively kind of portrait favoured today. A modern technical camera, as used for the bulk of commercial and industrial work, is also quite suitable for portraiture. The 4×5 ins. format is most convenient, and if a roll film adaptor is available the equipment becomes all the more versatile. Lenses of several focal lengths should be part of the outfit, and the camera should have a double extension. General work may be exposed on rollfilm, specially critical portraits may be made with sheetfilm.

For commercial, advertising and technical photography a technical camera with front and rear focusing, all movements, and a triple extension is required. This may be of the rangefinder type, rapidly convertible to use as a press-type camera, but with a focusing back into which may be inserted sheet film holders, film magazines, Polaroid adapters and rollfilm holders. Each lens on its own panel is coupled to the rangefinder and includes its own synchronized shutter. More suitable for most studio work is a technical camera of the monorail type. This has more extensive camera movements, but is less useful for action shots.

Whichever of these kinds of technical camera is chosen, it should be mounted upon a sturdy stand, preferably with separately controlled side-tilting and forward-tilting head adjustments. A ball-head is unsuitable. Ideally the studio stand will permit the camera to be placed firmly and securely at any height from ground-level to at least 7 feet above that. Either of them may be pressed into use for coping work, if a separate copy-camera is not installed.

Copying. Even where the amount of copying work does not justify a permanent set-up with its own camera, the work is greatly assisted if a copyboard is fixed to a wall, illuminated by two or four pearl lamps in reflectors, with a pair of rails squared up to it upon which a carriage is arranged to take the technical camera. Alternatively an old studio camera can be adapted and used as permanent copying equipment. Ordinary work may be done with a lens from one of the studio cameras, but for really critical reproductions it is worth obtaining a lens especially made for the job.

Lighting. The small studio should install adaptable lighting equipment which may also be taken to the homes or factories of clients. Equipment must be good and sturdy, and quality should come first, even if it means limiting the number of units which can be afforded. Excellent portrait and simple still-life photography is possible with only a couple of 500 watt spotlights, two diffused floodlamps and one or two 500 watt photopearl reflectors. All should be on well-balanced wheeled stands which can be lowered to within 24 ins. of the ground and raised to 6 feet or more high. If these are fed from loose distribution-blocks with 3-pin sockets, properly earthed, drawing their current from heavier 3-way cables from one or more wall-points, any part of the studio may be rapidly used when working. The cabling serves as feeders on the outside jobs as well, with one or two adapter-plugs to match the local installation.

A single 13-amp ring socket or 15 amp power point will safely feed the above collection of lighting equipment, but it is more convenient and safer still to split the load into two. For more complex projects one or two more spotlights, a couple of the smallest 100 watt "inkie-dinkies" for catch-lights and perhaps one or more "broadsides" of banked 500 watt photopearl lamps, in a trough holding 4, 6 or 8 such lamps together, might be added. A further higher-powered spotlight to simulate sunlight, ceiling battens and troughed rows of photopearl lamps at floor level would make fully competent fashion photography a possibility. Installation should be carefully executed to the highest safety standards.

In addition to the above tungsten lighting, some kind of flash equipment will sometimes be required. A portable electronic unit, with one extension lamp can deal satisfactorily with child portraiture; but if fashion in colour, food-products and fast-moving subjects are to be photographed, a heavy-duty studio electronic unit with several lamps containing their own modelling-lights, will be needed. Synchronized flash bulb illumination from two or three flash heads is also useful for many varied types of work.

Commercial Photography. The small-town studio will have to undertake a certain amount of commercial work and, if space is limited, some careful planning is necessary to allow both types of work to proceed at the same time. That does not mean that two operators will ever be using the studio at the same time but that it should be possible for sitters to be taken without disturbing a complicated set-up of a subject and without their being distracted by its presence.

For instance, a local engineering firm may require a number of photographs to be taken of a small mechanism or perhaps a layout of tools for a catalogue and because such work cannot be hurried it may occupy the studio for several days. For normal portraiture to be done at the same time, therefore, it must be possible to divide the studio by a temporary partition. In addition, as the lighting of these subjects is usually straightforward, the work can often be done with portable units at the opposite end of the studio.

If it is at all possible, however, commercial operating should be carried out in a separate room because it is essential that sitters should be allowed complete privacy while being photographed. In addition, this work often involves the use of numerous wooden blocks, supporting stands, etc., which it is impossible to keep tidy.

In addition to commercial work on his own premises, the photographer must be prepared to undertake work outside the studio. This will include architectural photographs for estate agents and surveyors, etc., shop windows, local views and factory installations. A progressive worker who is the only photographer in the district will also be expected to attend local functions, photograph people and children in their own homes and, in fact, to record any event or scene for all manner of private, business or official clients.

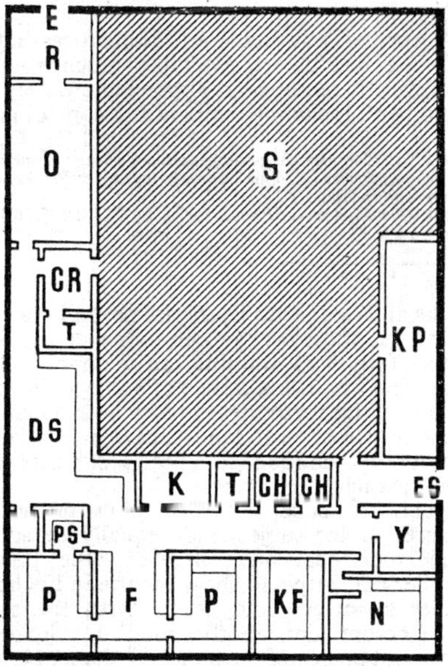

COMMERCIAL AND ILLUSTRATIVE STUDIO. Designed by owner. Scale about 1 in. = 19 feet. E, entrance. R, reception room for props (used largely in advertising and illustrative work). T, toilet. DS, dispatch room. K, storage room. CH, chemicals rooms; one is intended for storage, the other for mixing solutions. ES, service entrance. Y, film loading room. N, negative processing room. KF, filing room for negatives and prints. P, positive processing rooms. F, finishing room. PS, special printing room for contact prints. The two main processing rooms feed prints from the fixing bath direct to the finishing room by means of service hatches. The frontage of the studio is taken up by display windows.

Wedding Photography. A professional studio undertaking wedding photography should formulate a policy regarding this work and stick to it. The soundest principle is to attend a wedding only when specifically commissioned to do so, with a deposit or understood charge for the service. It is not uncommon for a certain type of freelance photographer to appear uninvited at the ceremony, to take pictures which are rapid-processed and brought back to the reception as proofs by young ladies who hawk them among the guests. What is first seen is often purchased in such circumstances, and unless one meets this method with a similar approach or is covered by agreement with the client, it is doubtful whether wedding photography can be profitable.

Side-lines. It is usual for all but the most exclusive portraitists to sell a range of films and photographic goods and to stock a limited number of side-lines. These include frames of all kinds, coloured miniatures and often pottery and artists' materials. In addition, the small-town photographer will accept amateur films for developing and printing. It is a mistake, however, for the processing to be attempted by a one-man business. Developing and printing business is a specialized branch which, to be conducted economically, must be operated on a large scale by a highly skilled fast working team of operators in workrooms equipped specially for the purpose. If a studio proprietor wishes to handle developing and printing because it brings people into his shop, then the processing should be put out to one of the many firms specializing in the work. The same thing applies to retouching, colouring, framing and perhaps copying. These all represent routine work which, for a start, it is better to pass on to specialists.

Colour Photography. So long as daylight can be excluded, there is little difficulty in taking colour photographs with normal studio lighting, particularly if the nature of the work requires the use of negative colour film. Although, ideally, all kinds of colour film should be exposed only by illumination which matches that for which the film is balanced, negative colour film is more adaptable to different kinds of lighting. But all the lamps used must match each other in colour temperature.

The C.P. range of projector and floodlamps covers all reasonable needs from a modified Photoflood of 275 watts, which costs only a few pence more than the common Photoflood lamp, to giant projector lamps used in film and television studios. All are designed to produce light at a nominal colour temperature of 3,200° K. for which the present-day studio colour films are mostly balanced. As they are equally suitable for black-and-white photography, there is no reason why this kind of lamp should not be fitted by a studio which anticipates colour work as part of its business. If the demand will be small, then a set of C.P. lamps can be stocked for use only when required, for the cost of ordinary lamps is slightly lower.

Reflectors should be white or neutral silvered and kept clean. When C.P. lamps become blackened by use, they should be replaced for colour photography, though they can be kept in use for some time for monochrome work.

Reversal colour films which produce transparencies must be exactly matched to the illumination. It is advisable to use either films specifically intended for 3200°K studio lamps, or possibly the type which is balanced for Photofloods (3400°K) with a suitable filter on the camera.

Alternatively, daylight colour film may be used with electronic flash.

Except for deliberately unusual effects all colour subjects must be illuminated with lamps of uniform colour quality. The normal room lighting of the studio should be extinguished, especially fluorescent tubes. Blinds or curtains should be drawn tightly and any areas of brightly coloured walls which are not a feature of the picture itself are best covered with white or neutral cloth or paper.

Colour Processing. When the volume of colour work is great, processing and printing of colour negatives might be carried out on the premises, but good colour prints require a skilled hand and it is advisable for a small studio to send out this work to a trade laboratory. Transparency films in roll and miniature sizes are processed by the manufacturer with the processing cost included in the price of the film. Sheet films are charged separately for processing and this is again worthwhile carrying out on the premises where the volume of work is sufficient.

A further advantage of user-processing is that the results are available much sooner. However, it requires an appreciable outlay on equipment, including temperature control gear.

J. Bl. (J.V.)

See also: *Commercial photography; Costing: Insurance; Darkroom; Prices of commercial photographs; Professional photography; Studio.*

Book: ***The Business of Photography*, by C. Abel (London).**

STYLE. Any medium of artistic expression must allow a certain amount of control. The peculiar combination of methods favoured, and the variation of degree in applying them, produce a particular style. The more flexible the medium, the more pronounced the style possible, because the artist is allowed greater scope in selection and emphasis.

When the style of an artist is well developed it shows consistent tendencies—his choice of subjects, the particular range and hue of colours he favours, his special manner of applying the paint, the effects of lighting he chooses—and so on. Any one of these things on its own would not give him style. But the very particular combination of effects he favours and the emphasis he gives to each, add up to a unique and singular arrangement. Alter or vary one of the effects and the sum no longer gives the same answer. That is style.

The medium of photography does not allow the same amount of individual control that is possible, for instance, in painting. But some control is possible; so the photographer can achieve a certain degree of style.

Personal Treatment. There are several ways in which the photographer can influence the final result of his work and thus create an individual style. The first is his choice of subject. He may have a liking for portraiture or he may find a peculiar attraction to landscapes. He may even narrow his outlook to special subjects like trees, or yachts, or flowers.

Next, a tendency to emphasize certain features of the subject may equally be something individual and singular. One photographer may strive to show Nature in his landscapes as rugged and fierce. With the same scene before him, another may—by choice of viewpoint and careful emphasis—depict Nature as delicate and beautiful. This tendency of the photographer can be evident no matter what subject he chooses.

He can regulate exposure to concentrate interest on the highlights in a picture; or on the shadows.

He can vary or at least select the effect of lighting to give the picture depth and shape. Or he can go to the other extreme and display the subject flatly with no differentiation of planes. (Interest then would be concentrated on outline rather than form, pattern rather than shape.) In the same way he can emphasize or play down texture.

By suitably adjusting both exposure and lighting he can introduce predominantly high or low key tones with their attendant effects.

He may arrange the composition to make a picture appear graceful and passive. Or he can vary it to suggest vigour and movement.

By selective focusing he can concentrate interest on only one plane, or give an over-all sharpness to the picture. Or he can use diffusion to give the whole picture softness.

Even blur caused by camera shake or subject movement can be exploited to give a picture spontaneity or a feeling of action.

The photographer who has matured and developed a style will use devices such as these. But it is his actual selection and combination of methods, and the consistent adoption of this pattern, that creates his style. It may not be the result of a conscious effort indeed it seldom is.

Style is dependent on the photographer's background, training and outlook. Generally, the more distinctly individual the photographer's character is, the more consistent and easily recognizable will be his style.

But while style itself is the result of the combination of a mass of details, the final product can only be general. This special pattern of method can only be seen by a broad, over-all examination of a photographer's work. In fact, it is only its constant repetition that makes it style at all.

Style however should be individual. It is an intricate pattern, peculiar to the photographer. Copy it and it becomes unconvincing and lacks the vitality of sincere art. Its development must be spontaneous and not forced, its growth natural and unhurried.

Schools of Style. Just as the individual can have style, so can nations or communities. The style then is no less individual, but rather are the people who contribute to it less individual in their outlook. Such a unity of style among people is known as a school.

Photographs from certain countries will often show this tendency to a common style. It may be the result of national characteristics. Or it may be a changing outlook reflecting economic and social conditions.

When just a group of people share a common style—although often still with personal differences—it is usually the result of a conscious and organized effort. This kind of common purpose inspired many of the great schools of painting. More often than not it begins as a revolt against convention or tradition in art. The unified outlook of the group then favours the natural development of a style.

Natural development—that is the essential condition of style, whether of the group or the individual. And style can only come naturally by forgetting about it.

In a photograph style makes the work original, belonging to one photographer and not any photographer. It is the stamp of the artist that renders a signature unnecessary. Indeed, style is frequently the very means by which unsigned work is identified. J.D.C.

See also: *Painting and photography; Vision.*

SUBJECTIVE PHOTOGRAPHY. Self-chosen name of a contemporary school of creative photographers with distinct leaning towards non-representational patterns. Initiated by Prof. Otto Steinert in Germany after the second World War and allied in outlook to the Bauhaus after the first World War. The earlier trend arose among amateurs and was more original; the later among young professionals and is more mature in its products.

See also: *Abstract photography.*

SUB-MINIATURE. Alternative name for ultra miniature camera, i.e. still cameras taking pictures 12 × 17 mm. or smaller on 16 mm. or narrower film.

See also: *Ultra-miniature.*

SUB-STANDARD CINE FILM. Obsolescent term for narrow gauge film (8, 9·5 or 16 mm. wide) for amateur cinematography, as distinct from the standard 35 mm. cine film used for professional motion picture work.

See also: *Cine film.*

SUBSTRATUM. Binding layer which helps the emulsion to adhere to the glass or film base.

See also: *Sensitized material manufacture.*

SUBTRACTIVE SYNTHESIS. Method of reproducing colours in which white light passes successively through, two or more coloured filters each of which absorbs (subtracts) a region of the visible spectrum which is transmitted by the others. The colour produced depends upon the quantity of light absorbed by each filter. Usually filters of the colours cyan, magenta and yellow are used, and these are often termed minus red, minus green, and minus blue since they absorb the colours red, green and blue respectively. When those filters are in the form of dye images superimposed in register on a transparent support, the result is a colour transparency.

See also: *Colour synthesis; Additive synthesis.*

SULPHIDE TONING. Method of changing the colour of the normal bromide print to give brown or sepia tones instead of black.

See also: *Toners.*

SULPHURIC ACID. Oil of vitriol. Used in reducers and clearing baths.

Formula and molecular weight: H_2SO_4; 98.

Characteristics: Corrosive oily liquid. Evolves a great deal of heat when mixed with water. The acid should always be poured into the water while stirring the mixture, never the water into the acid.

Solubility: Mixes with water in all proportions.

SUNK MOUNT. Type of lens mount in which the major part of the lens lies behind the lens panel instead of projecting in front. Sometimes also applied to print mounts with plate markings.

See also: *Lens mount; Mounts.*

SUNRISE AND SUNSET. Light at sunrise and sunset is richer in red and orange rays than in the middle of the day.

Photographs of the sky at sunrise and sunset are apt to be disappointing when seen in black-and-white as the main attraction about such a scene is its colour. At the same time much of the beauty of the sky can be reproduced by careful technique. Generally the exposure should be kept to the minimum value; anything like a full exposure gives a dense negative and a print that lacks the necessary feeling of luminosity. For the same reason the negative should be processed in a soft-working developer to preserve the delicacy of the tone rendering in the highlights.

Panchromatic films are always best for black-and-white photographs taken early or late in the day because they are sensitive to red light.

A sunrise or sunset is best photographed in colour. Exposures should be kept to a minimum, otherwise the scene will be colourless and have a washed-out appearance. A daylight colour film without filter should generally be used, as such a scene would lose considerably if its characteristic red quality were removed. When photographing scenes illuminated by a low sun, and not the sunrise or sunset itself, however, it will often become necessary to use a correction filter to render the scene less red.

At such times of day the light changes quickly from minute to minute and any special effects need to be anticipated and photographed without delay.

See also: *Daylight; Sunshine.*

SUNSHADE. Alternative name for lens hood, current in the U.S.A. Its use is not limited to sunshine, though it is of course most effective then.

SUNSHINE. Sunshine is nature's light source Being a point source of light, it requires reflecting surfaces which will illuminate shadow areas; this applies to all photography, and to colour work in particular. Reflectors may be clouds, light coloured walls, a sandy beach etc.

Tone Range. The correct use of reflecting surfaces ensures that the ratio of the highlight intensity to that of the shadows is within the acceptable latitude of the photographic emulsion, the resulting photograph having detail in both the highlights and shadows.

Instead of reflectors, flash may be used as a fill-in for shadows. Where it is not practicable to employ any of the above means, it is wise to use a fast emulsion negative material, which yields softer results. Further reduction of contrast may be obtained by over-exposing the negative and under-developing it. Unless one is aiming for special effects, exposures should be based on the more important shadow areas, and not the highlights, in black and white photography.

It is not always necessary to reduce the extreme contrast of a scene in direct sunshine. The photographer sometimes aims deliberately at making a print where the shadows are black shapes. He may do this to emphasize the brilliance of the sunshine, or simply to create a black and white pattern in which intermediate tones would spoil the effect.

Generally speaking, the direct light of the sun is bad for portraiture, and should be particularly avoided when high in the sky and shining down almost vertically.

Variations in Quality. Sunshine varies in quality and strength according to time of year and day, and the locality. In Britain it seldom reaches the brilliance and high colour temperature that it does in many other countries. These inconsistencies produce great differences in contrast, exposure and, in the case of colour films, of colour.

Early and late in the year and/or day, sunshine is weaker in actinic value. Its red-content is apparently greater in the evening and in the morning, although in fact it is the blue content which is lower at those times, and this, too, influences colour film to a marked degree.

Between 11 a.m. and 3 p.m. in June, July, and August, sunny scenes should be avoided. The sun is then too overhead, giving downward, short shadows, devoid of modelling or pictorial interest. Sidelighting at angles of 45°–80° to the line of vision is admirable for depicting texture on sides of buildings, scenes, and objects of all kinds. Oblique back-lighting (coming from slightly behind the subject at angles of 100°–160° to the line of vision) is excellent for isolating the subject from its background and "glamorizing" mundane objects. It is sometimes difficult to prevent it striking the lens from this position, and a lens hood is therefore necessary.

Artificial Sun Effects. Sunshine may be simulated outdoors by means of flash held to one side of and well above the camera. The flash and daylight must be carefully balanced, so that the background does not reproduce unnaturally dark. The same can be achieved indoors by judicious placing of lights, with a plain medium-grey backcloth or painted cloud-scene if expertly executed and rendered out-of-focus. No shadows must be allowed to fall on this and so reveal the artifice.

A sunny atmosphere can be strengthened by using a diffusion disc on the camera lens, thereby spreading the highlights into the shadows and softening definition with its halo effect. P.J. (V.G.)

See also: *Daylight; Sunrise and sunset.*

SUPERCOAT. Many sensitized materials have a coating of unsensitized gelatin on top of the sensitive emulsion. This supercoat is added to protect the sensitive layer from physical injury, in particular, from abrasion and stress marks (visible marks produced on development due to pressure on the sensitive emulsion).

With special materials, like transfer stripping films, where the image is to be transferred to another support, the supercoat may also serve as adhesive layer upon transfer to attach the emulsion to the new support. Stripping materials are not always manufactured with such an adhesive layer.

High resolution materials and films and plates intended for photographing ultra-violet radiation never have a supercoat; in the latter case it absorbs ultra-violet.

See also: *Negative materials.*

SUPER-8 mm. Type of 8 mm. film with a 50 per cent larger image area as compared with traditional 8 mm. film. The Super-8 mm. film also has an increased frame height and perforation pitch, so that there are only about 72 frames per foot (instead of 80). The perforation size itself is smaller. The film can only be used in appropriately designed cameras and projectors.

See also: *Cine film; Perforation.*

SUPPLEMENTARY LENS

Extra lens used to change the focal length of the camera lens. Usually fitted in a mount to slip, clip, or screw on to the front of the camera lens like a filter.

Positive Supplementary Lenses. A popular way of photographing close-up subjects is to shorten the focal length of the camera lens by using a positive supplementary lens in front of it. When the focal length of the lens is shortened in this way, the camera must be brought closer to the subject to focus it sharply, and this automatically gives a bigger image of the subject.

There are two methods of focusing with positive supplementary lenses:

(1) If the camera lens is set at infinity, adding a supplementary lens focuses it on objects at a distance equal to the focal length of the supplementary lens. A 50 cm. supplementary lens added to any camera lens focused on infinity will focus the camera on objects 50 cm. in front of it.

This method of focusing does not require a focusing screen, so it is a popular method of extending the range of fixed focus cameras and those which focus by scale alone. It is a simple device for adapting fixed focus cameras for portraiture, and supplementary lenses sold for this purpose are called portrait attachments.

So long as the subject is at a distance exactly equal to the focal length of the supplementary lens, the image will be sharp. The disadvantage of working close to the subject in this way, however, is that it tends to distort the proportions of the subject.

(2) With screen focusing cameras (including single lens reflex types) the supplementary lens may be used over a range of subject distances. In this case the camera is focused by examining the image on the screen in the usual way, the only difference being that it now works closer to the subject and gives a bigger image than with the normal taking lens alone.

Used in this way the greatest subject distance is given with the camera lens set at infinity, when the subject distance is equal to the focal length of the supplementary. The normal focusing extension then enables the camera to be focused for distances less than this.

Even if the camera is not equipped with a focusing screen, its normal focusing extension can still be utilized for close-ups at distances less than the focal length of the supplementary. Under these circumstances the distance focused sharply for any point on the focusing scale is given by the expression:

$$d = \frac{us}{u + s}$$

where d = the actual distance focused on
s = focal length of supplementary lens
u = scale setting of camera lens.

Example: if the camera lens focusing scale is set at 3 feet (36 ins.) and the supplementary lens has a focal length of 24 ins., adding the supplementary lens will enable the camera to focus on (36 × 24)/(36 + 24) ins.—i.e., 864/60 = 14.4 ins.

A positive supplementary lens can also be built up of several converging and diverging units. If the separation between these units is adjustable, the complete system can act as a variable focus supplementary lens. In other words, the power of the combination in dioptres depends on the actual separation between the elements, and the supplementary lens can therefore cover an appreciable range of continuous close-up distances. This has obvious advantages over the use of a number of simple supplementary lenses, each of which covers only a narrow range of subject distances.

Variable power supplementary lenses usually have their adjustment calibrated to show both magnification and required subject distance at different settings; they are, however, best used

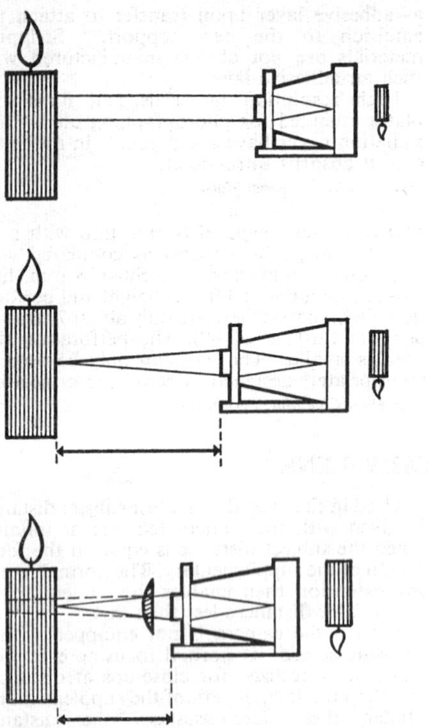

USE OF POSITIVE SUPPLEMENTARY LENS. Top: At medium subject distances, the image of a smaller subject becomes very small on the film. Centre: The closest focusing distance of most cameras is still in the region of 3–4 feet. At this range the subject may still be too small in the picture. Bottom: A positive supplementary lens enables the camera to approach the subject closer for the same focusing movement of the camera lens. This focusing movement now covers a continuous, but limited range of near subject distances, the farthest of which (with the camera lens set to infinity) is equal to the focal length of the supplementary lens.

with reflex or other cameras that permit visual focusing on a screen.

Wide-angle Conversion. The fact that a positive supplementary lens shortens the focal length of the camera lens makes it suitable as a wide-angle attachment.

Such a combination can of course only be used with a camera where the bellows movement permits the lens to be racked back nearer to the negative. The effective focal length of the combination is (s × f)/(s + f), s and f being the focal lengths of the supplementary and camera lens respectively.

In practice the optical quality of such a wide-angle combination is not very good, and the image is likely to show appreciable distortion near the edges of the field.

Negative Supplementary Lenses. A negative (diverging) supplementary lens increases the focal length of the camera lens. The new focal length of the combination is given by the expression (s × f)/(s − f) where s is the focal length of the supplementary lens and f, the focal length of the camera lens.

This means that it is not possible to focus an image sharply unless the lens is moved farther away from the plate. When the lens-plate separation is equal to the new focal length—i.e., (s × f)/(s − f)—objects at infinity will be in sharp focus. For nearer objects, the lens must of course be moved farther out still.

Because the negative lens increases the focal length of the camera lens it automatically reduces the field covered—actually, in the ratio s : (s–f). In other words it sees less of the subject in front of the camera. And since less of the subject now occupies the same picture size, it is presented on a magnified scale.

So, without moving the camera position, a negative supplementary lens can be used to magnify the scale of the image. The increase in image size is proportional to the increase in focal length—i.e., it increases in the ratio s : (s − f).

Example: if the negative supplementary lens has a focal length of s = 12 ins. and the camera lens a focal length of f = 6 ins., the focal length of the combination will be (12 × 6)/(12 − 6) = 12 ins. And the increase in scale of the image will be 12/(12 − 6) = 12/6 = 2—i.e., the supplementary lens will double the size of the image obtained with the normal lens alone.

But a negative supplementary lens can only be used to give a bigger image in this way if the camera will extend far enough to allow for the increased focal length. Furthermore, the aberrations introduced by adding an uncorrected lens become serious at magnification over about 2×—i.e., where the focal length of the supplementary lens is less than twice the focal length of the camera lens.

Afocal Attachments. A lens system acting like a telescope presents to the eye, on looking through, a visual image at infinity. Such a

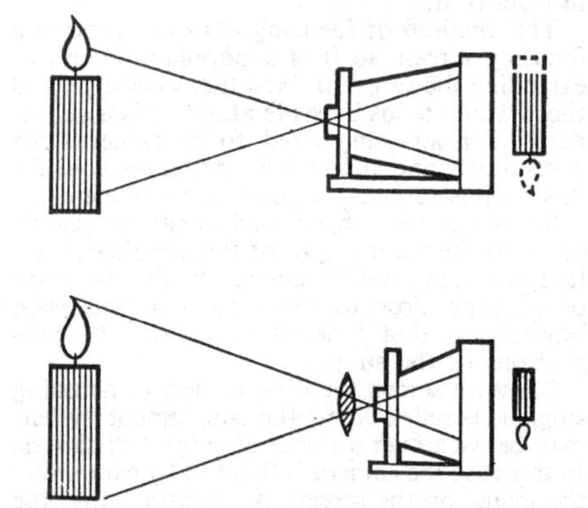

WIDE-ANGLE SUPPLEMENTARY LENS. A positive supplementary lens shortens the focal length of the camera lens. The combination covers a larger angle of view, yields smaller image scale, and needs reduced bellows extension.

APPROXIMATE FOCUSING DISTANCES WITH POSITIVE SUPPLEMENTARY LENSES (INCHES)

Setting of Camera Lens (feet)	*Focusing Distance for Focal Length of Positive Supplementary Lens* 80 / 200 / 0·5	60 / 150 / 0·67	53 / 133 / 0·75	40 / 100 / 1	32 / 80 / 1·25	26¾ / 67 / 1·5	20 / 50 / 2	16 / 40 / 2·5	13⅜ / 33 / 3	11⅜ / 28·5 / 3·5	10 inches / 25 cm. / 4 diopters
∞	80	60	53	40	32	26¾	20	16	13⅜	11⅜	10
100	75	57	51	39	31	26	19¾	15¾	13¼	11¼	9⅞
50	70	55	49	37½	30½	25½	19¼	15½	13⅛	11⅛	9¾
25	63	50	45½	35	29	24½	18¾	15¼	12⅞	11	9⅝
20	60	48	43½	34	28¼	24	18½	15	12¾	10⅞	9⅝
15	55	45	41	32	27	23¼	18	14¾	12½	10¾	9½
12	51	42½	39½	31	26	22½	17½	14⅛	12¼	10½	9⅜
10	48	40	37	30	24½	22	17	14	12	10⅜	9¼
8	44	37	34	28¼	24	21	16½	13¾	11¾	10⅛	9
7	41	35	32½	27	23½	20¼	16	13½	11½	10	8⅞
6	38	33	30½	25¾	22	19½	15½	13	11¼	9⅞	8¾
5	34	30	28¼	24	21	18½	15	12⅛	10⅞	9½	8⅝
4	30	26¾	25¼	21¾	19¼	17	14	12	10½	9¼	8¼
3	25	22½	21½	19	17	15¼	12¾	11	9⅝	8⅝	7⅞

The figures are rounded off to the nearest ⅛-in. up to 15 ins., to the nearest ¼-in. up to 30 ins., to the nearest ½-in. up to 50 ins., and to the nearest inch above 50 ins.

APPROXIMATE FOCUSING DISTANCES WITH POSITIVE SUPPLEMENTARY LENSES (CENTIMETRES)

Setting of Camera Lens (metres)	*Focusing Distance for Focal Length of Positive Supplementary Lens* 200 / 0·5	150 / 0·67	133 / 0·75	100 / 1	80 / 1·25	67 / 1·5	50 / 2	40 / 2·5	33 / 3	28·5 / 3·5	25 cm. / 4 diopters
∞	200	150	133	100	80	67	50	40	33	28·5	25
25	185	141	126	96	77	65	49	39·5	32·5	28	24·7
15	177	136	122	94	76	64	48·5	39	32·5	28	24·6
12	172	133	120	92	75	63·5	48	38·5	32	28	24·5
10	167	130	118	91	74	63	47·5	38·5	32	27·5	24·4
7	155	124	112	88	72	61	46·5	38	31·5	27·5	24·2
5	143	115	105	83	69	59	45·5	37	31	27	23·8
4	133	109	100	80	67	57	44·5	36·5	30·5	26·5	23·5
3	120	100	92	75	63	55	43	35	30	26	23·1
2	100	86	80	67	57	50	40	33·5	28·5	25	22·2
1·6	89	77	73	62	54	47	38	32	27·5	24	21·6
1·2	75	67	63	55	48	43	35	30	26	23	20·7
1	67	60	57	50	44·5	40	33·5	28·5	25	22	20
0·8	57	55	50	42	40	36·5	31	26·5	23·5	21	19

The figures are rounded off to the nearest 0·1 centimetre up to 25 centimetres, to the nearest 0·5 centimetre up to 50 centimetres, and to the nearest centimetre above 50 centimetres.

APPROXIMATE SCALES OF REPRODUCTION WITH SUPPLEMENTARY LENSES

Focal Length of Camera Lens inches	cm.	*Scale of Reproduction with Supplementary Lens of Focal Length* 80 / 200 / 0·5	60 / 150 / 0·67	53 / 133 / 0·75	40 / 100 / 1	32 / 80 / 1·25	26¾ / 67 / 1·5	20 / 50 / 2	16 / 40 / 2·5	13⅜ / 33 / 3	11⅜ / 28·5 / 3·5	10 inches / 25 cm. / 4 diopters
2	5	0·025	0·033	0·038	0·050	0·063	0·075	0·10	0·13	0·15	0·18	0·20
2½	6·25	0·031	0·042	0·047	0·063	0·078	0·094	0·13	0·16	0·19	0·22	0·25
3	7·5	0·038	0·050	0·057	0·075	0·094	0·11	0·15	0·19	0·22	0·27	0·30
3½	9	0·044	0·058	0·066	0·088	0·110	0·13	0·18	0·22	0·26	0·31	0·35
4	10	0·050	0·067	0·075	0·100	0·125	0·15	0·20	0·25	0·30	0·35	0·40
4½	11·2	0·057	0·075	0·085	0·112	0·140	0·17	0·23	0·28	0·34	0·39	0·45
5	12·5	0·062	0·083	0·094	0·125	0·156	0·19	0·25	0·32	0·38	0·43	0·50
5½	13·7	0·069	0·090	0·104	0·138	0·172	0·21	0·28	0·35	0·42	0·47	0·55
6	15	0·075	0·100	0·113	0·150	0·188	0·23	0·30	0·38	0·45	0·53	0·60
7	17·5	0·088	0·117	0·132	0·175	0·219	0·26	0·35	0·44	0·51	0·61	0·70
8	20	0·100	0·133	0·151	0·200	0·250	0·30	0·40	0·50	0·60	0·70	0·80
10	25	0·124	0·166	0·189	0·250	0·313	0·38	0·50	0·63	0·75	0·88	1·00
12	30	0·150	0·200	0·216	0·300	0·375	0·46	0·60	0·75	0·90	1·05	1·20

The scale of reproduction is the ratio of image size/object size. Thus values smaller than 1 are reductions, larger than 1 are magnifications.

The table holds good for a camera lens set at infinity.

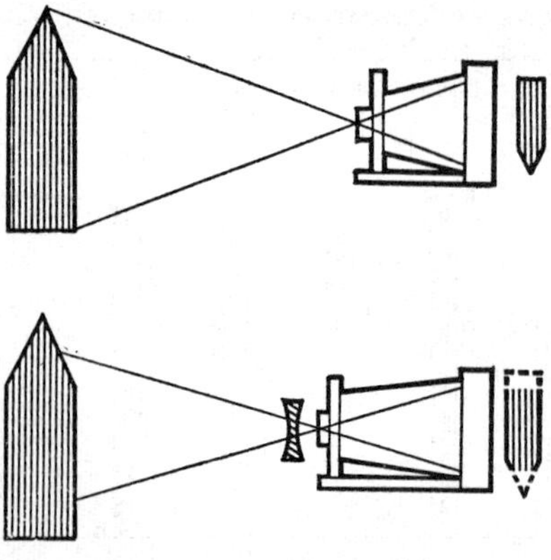

NEGATIVE SUPPLEMENTARY LENS. Increases focal length of camera lens. Combination yields larger scale images at given distance, covers smaller angle, and needs increased bellows extension on camera to retain sharp focus.

system therefore has no focal length, but still magnifies the scale of the visual image. On placing such an afocal unit in front of a camera lens focused at infinity, the camera records the image on an enlarged scale, as through a telephoto lens. In the same way an afocal system can reduce the image scale, and so increase the angle of view.

The first type of system is used as a supplementary tele-attachment, and the second type as a wide angle attachment, on cameras with fixed (not interchangeable) lenses. This can provide a range of focal lengths, especially for telephoto purposes. The use of suitably designed binocular or monocular units can achieve image magnifications of 8, 10, or more times. Afocal attachments of variable magnification also exist and produce an effect similar to that of a zoom lens.

Effect on *f*-number. For most practical purposes it can be assumed that a positive supplementary lens when used for close-up work with the camera lens set at infinity does not affect the *f*-number of the camera lens. This must be so because the size of the stop remains the same and the lens-plate distance—i.e., the effective focal length—is unchanged. (Neither of these assumptions is quite true, but the error is not normally appreciable.)

But if the combined lens is focused on distant objects, the *f*-number of the camera lens must be divided by $(s + f)/s$ to give the true relative aperture. A camera lens of $f = 6$ ins. used with a supplementary of 20 ins. would thus have its effective aperture divided by $(20 + 6)/20 = 26/20 = 1{\cdot}5$—i.e., at *f* 8, the true relative aperture would be $(8 \times 2)/3 = 16/3 =$ *f* 5·3. In practice, this theoretical increase in aperture is offset by light losses and aberrations.

A negative supplementary changes the *f*-number of the camera lens in proportion to the increased lens-plate separation. So the *f*-number of the camera lens must be multiplied by $s/(s - f)$—i.e., in the example for a negative lens the *f*-number would be multiplied by $12/(12 - 6) = 2$. So that the supplementary lens in this case doubles the *f*-number. Doubling the *f*-number e.g., from *f* 8 to *f* 16—calls for a 4× increase in exposure which is another limitation of the method.

Definition. Supplementary lenses are generally simple uncorrected elements of the spectacle lens type—i.e., with both surfaces curved. Such additions upset the corrections of the camera lens and introduce most of the aberrations in some degree. So the camera lens must always be stopped down to give a sharp image right to the corners of the plate. Even when stopped down, the definition is never as good as with the camera lens alone.

Meniscus lenses—i.e., with one convex and one concave surface—give distinctly better definition than the spectacle (biconvex or biconcave) type and are generally meant to be fitted with the concave side to the subject.

Since a camera lens is always designed to give its best performance on its own, the use of any form of supplementary lens tends to impair maximum definition and evenness of illumination. In high-quality supplementary lenses and afocal attachments, designed specifically for a given camera lens, the loss of quality can be negligible.

Attachments of the monocular or binocular type used for tele-photography must also be designed with a larger exit pupil to avoid vignetting of the image.

Corrected Supplementary Lenses. Fully corrected supplementary lens assemblies are supplied by some manufacturers for providing a range of focal lengths in conjunction with a permanent rear lens element and shutter system. These give the equivalent of a set of interchangeable lenses of short, normal, and long focal length for the cost of what is virtually a set of three corrected supplementary lenses and a fixed camera lens. To change from one focal length to another, the front component is simply unscrewed and interchanged, leaving the back component, shutter and diaphragm undisturbed.

With some cameras such an interchangeable front component can yield a lens system that is already focused at a close or very close distance. With an interchangeable macro lens of this kind, up to same size copying may be possible at a considerably higher level of lens performance than is normally achieved by adding supplementary close-up lenses to a complete camera lens.

A corrected supplementary lens is also used in an accessory which converts an ordinary camera into a coupled rangefinder focusing type.

The device consists of a supplementary lens assembly with front-cell focusing coupled to a split-image rangefinder. The camera lens is set at infinity and the supplementary assembly is fitted in front of it. Under these conditions, operation of the coupled rangefinder automatically focuses the camera on the object ranged upon.

Supplementary lens assemblies of the types described above are highly corrected and capable of giving definition equal to that of a normal camera lens.

Use With Twin-Lens Reflex Cameras. Supplementary lenses can be used with twin-lens reflex cameras provided that both viewing and taking lenses are fitted with identical supplementaries so that the image on the viewing screen is exactly the same as the image on the film. Some manufacturers of twin-lens reflex models supply special sets of matched positive and negative lenses for their own cameras.

Scale of Reproduction. The scale of reproduction with supplementary lenses is given by:

$$M = \frac{F}{S}$$

where M = Scale or reproduction (image size-object size)
F = Focal length of camera lens
S = Focal length of supplementary

provided the camera lens is focused on infinity. In this case the focal length of the camera lens must be known; in all other supplementary lens calculations, it is unnecessary.

Both focal lengths must be measured in the same units. Where the power of the supplementary lens is stated in diopters the reciprocal of the number of diopters is equal to the focal length in metres. L.A.M.

See also: *Close-ups; Extension of camera; Macrophotography; Optical calculations.*

Books: *Camera Close-up*, and *Camera Copying and Reproduction*, both by O. R. Croy (all London).

SUPPORTS FOR EMULSIONS

Two main types of support are used for negative emulsions: glass plates and celluloid film. Both are transparent and allow the negative to be printed by contact or projection.

Positive emulsions are coated on paper, film and glass plates.

Paper is occasionally also used for making the sensitive material known as negative card because of its cheapness.

Various plastics, transparent and opaque, are also used as support for special sensitive materials.

Glass Plates. Photographic plates consist of a support of flat glass plate coated with the sensitive emulsion. As a rule, the larger the size of the plate, the thicker the glass.

The advantages of glass as an emulsion support are dimensional stability and chemical inertness.

Glass plates do not stretch or curl during processing. They are therefore used in preference to films where high dimensional accuracy is required, as in photogrammetry, aerial survey, and certain types of astronomical photography.

Plates are also used when making colour separation negatives where the images are later to be combined in exact register.

The glass support is chemically neutral and unaffected by all solutions used in photographic processing. As glass does not decompose or deteriorate in storage, emulsions coated on it keep better than on any other support.

The disadvantages of glass are its thickness, weight, low adherence of emulsion and the fact that it is easily broken.

Because of the thickness of the glass, it is not possible to reverse the plate in contact printing to give a transposed positive, since the positive image would no longer be sharp. This is a disadvantage in certain transfer processes. When a plate negative is reversed in the enlarger, the plane of the emulsion moves out of focus by the thickness of the glass. This shift must be allowed for if the plate is being enlarged in an automatic focusing enlarger.

Plate negatives are more likely to show halation than films, because the greater thickness of the glass support allows the light reflected from the back surface to scatter over a wider area.

Gelatin emulsions do not adhere as well to glass as they do to celluloid. When the emulsion on a glass plate has been wetted, it is more likely to frill and leave the support near the edges, so glass plates call for more care in handling and processing.

Plates are fragile and relatively heavy so they are more troublesome to transport and store than films.

Because of these disadvantages sheet film is steadily taking the place of glass for general

PLATE SIZES AND THICKNESSES

Nominal Sizes ins.	Nominal Sizes cm.	Average Thickness ins.	Average Thickness mm.
Smaller than			
2½ × 3½	6 × 9	0·04	1·0
2½ × 3½	6 × 9	0·05	
to	to		1·3
4 × 6	10 × 15	0·05	
4¾ × 6½	12 × 16·5	0·06	
to	to		1·5
8 × 10	20 × 25	0·06	
9½ × 11¾	24 × 30	0·08	
to	to		2·0
12 × 15	30 × 38	0·08	
16 × 18	40 × 45	0·11	
to	to		2·8
20 × 30	50 × 76	0·11	
Larger than			
24 × 30	61 × 76	0·13	3·3

purpose materials, particularly in the U.S.A. The present tendency is to use plates only where high dimensional stability is of first importance.

Film Materials. Film negative material consists of a sensitive emulsion coated on sheets or rolls of thin, flexible, transparent material.

Advantages of film supports are flexibility, thinness and low weight.

Films can be handled and processed without fear of breakage. They can be bent and rolled, and they lend themselves to film-changing systems which are far more reliable and convenient than any of the methods possible for changing glass plates. Film negatives will bend easily to fit the curved focal plane used on some cheap cameras with the idea of producing reasonably sharp negatives from lenses uncorrected for spherical aberration. They are used in certain special scientific cameras and in panorama cameras where curved focal planes are sometimes necessary for other reasons.

Flexible film is the only practicable material for cinematography.

Film supports are thin enough to be used either way round in the camera or in enlarging or projection apparatus without calling for an appreciable change in focusing adjustment. This is important in such work as colour photography where the negative may have to be exposed through the support.

Film negatives are less affected by halation than plates because light passing through the emulsion and film support travels only a short distance before it is reflected by the back of the support.

Films are so much lighter than the equivalent number of glass plates that they are very much easier to store and transport. This is one of the principal reasons for the greater popularity of films.

The disadvantages of film are low dimensional stability and liability to decomposition.

The commonly used types of film base stretch when they get wet during processing. They do not stretch evenly in all directions, and do not always go back to their original shape when dry. Certain special types of film base with high dimensional stability are, however, made for use in photogrammetry and similar special fields.

Types of Film Base. For many years the main type of film base in general use was cellulose nitrate film. This had the advantages of great strength and comparatively high dimensional stability, but the serious drawback of being highly inflammable and even explosive. Because of its dangerous nature certain legal requirements covered its storage and use in quantity. With the advent of acetate and other chemically more stable film supports, cellulose nitrate gradually went out of use, and is today completely obsolete.

The most important current film bases are the range of cellulose acetate and polyacetate compounds on the one hand, and various polymerized plastics on the other. While these still burn under suitable conditions, they will not do so easily; the fire risk is usually less than for an equivalent weight of paper.

Cellulose tri-acetate is the commonest film base for general use today, forming the support for most roll, sheet, and miniature films, as well as standard and narrow-gauge cine film. While this is not quite as strong nor quite as stable dimensionally as the old nitrate film, it does not lag far behind. Cellulose tri-acetate is the successor of the earlier types of acetate film—originally referred to as safety film to distinguish it from cellulose nitrate—which was inferior in dimensional stability and strength.

Tri-acetate film is made in thicknesses from 0·003 to 0·009 ins. (0·08 to 0·22 mm.).

Graphic arts, photogrammetry, and other special fields require still more stable film bases, and a wide variety of polymerized compounds have been tried.

At present the main supports to achieve success are polystyrene, polycarbonate, and polyester films. The latter in particular surpass nitrate film in toughness, and in certain cases approach glass in dimensional stability. Further advantages of some polymer plastics are increased resistance to creasing, reduced absorption of moisture, and consequently more rapid drying after processing.

Paper. The great advantages of paper as an emulsion support are that it is cheap, flexible and light in weight.

Its principal disadvantages are that it is easily torn and damaged, is apt to change its size and shape when it is processed, and is difficult to free from chemical impurities.

The last objection is largely overcome by the baryta coating which most photographic paper receives in manufacture before being coated with the sensitized emulsion.

Paper has one inconvenient characteristic from the point of view of processing: it is porous. Consequently paper prints require long washing times after processing (since the paper fibres retain harmful soluble chemicals more tenaciously) and also longer drying times. Special waterproof papers are therefore made where the paper surface is sealed by varnishing or similar treatment and becomes impervious to the processing solutions. Such papers dry much more quickly.

Paper bases are available in a variety of thicknesses, according to purpose. Normal enlarging and contact printing papers come as single weight or double weight (card) base. Single weight paper is handier for mounting; the increased thickness of double weight paper gives it greater protection where unmounted prints are handled a lot. Document papers (and others also) are frequently available in an extra light weight base which take up less room in filing. In addition document papers also exist on a translucent paper base which permits contact printing through the paper. The translucent base is more uniform in its structure, and so does not show up the paper grain so much during printing. Such papers are largely used as masters for mass printing processes such as dyeline.

Other Supports. Although cellulose acetate and polymerized plastics are the commonest bases, other materials, both flexible and rigid, are occasionally used.

Opaque supports, which are printed either by reflex methods or projection, include steel, aluminium—which must have a protective coating to prevent interaction between the emulsion layer and the support—resin-impregnated glass cloth and white plastics used for reversal colour print materials.

For high dimensional stability laminated papers may be used.

These papers contain a piece of metal foil sandwiched between layers of paper which thus cannot stretch or shrink easily during processing and drying respectively.

Laminated papers with an opaque layer between the normal paper layers are sometimes used for special purposes where no light must reach an emulsion layer through the back. This might be the case in transfer-diffusion processing, where the negative and positive sandwich are exposed to ordinary lighting during processing.

Transparent supports include nylon-type plastics for plates. L.A.M.

See also: *Negative materials; Plastics; Sizes and packings.*

SURFACE DEVELOPMENT. During the process of development the image may either form first at the surface of the emulsion and gradually penetrate deeper, or it may come up simultaneously throughout the depth of the emulsion layer. The first is surface development, the second, depth development.

The mechanism of the process depends on the composition of the developer.

With rapid acting developing agents (e.g., metol, para-aminophenol, etc.) the developer acts as soon as the solution reaches the surface of the emulsion layer. It therefore builds up a surface image of appreciable density before penetrating deeper.

With slow-acting developing agents (e.g., hydroquinone) the developer begins to reduce the silver halide only after a certain induction period, by which time the solution has penetrated fully into the emulsion layer. The image therefore builds up gradually throughout the depth of the layer.

Because of irradiation, the image tends to be more blurred below the emulsion surface. Provided development is stopped early enough, surface developers largely ignore the blurred deeper image, and therefore yield images of higher resolution than depth developers. In a similar way, surface developers also help to minimize halation (which first affects the deepest areas of the emulsion). On the other hand, the contrast and maximum density obtainable at the surface are limited, so the depth developers yield negatives of the greater density range.

See also: *Physical development.*
Book: *Developing*, by C. I. Jacobson (London).

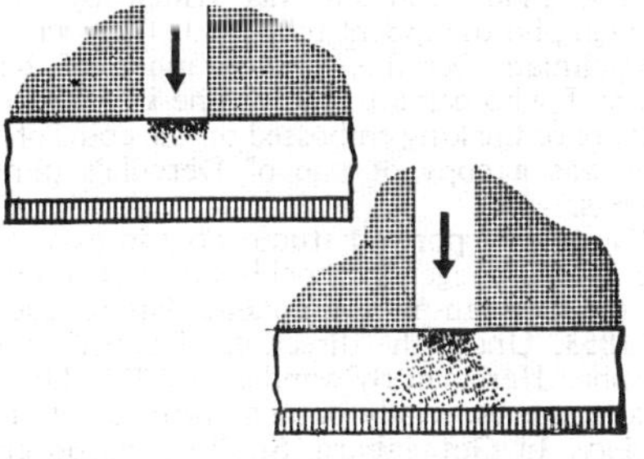

SURFACE AND DEPTH DEVELOPMENT. Left: *Surface developers confine their action largely to the upper part of the emulsion layer.* Right: *Depth developers work in the whole emulsion and may bring out irradiation.*

SURGICAL PHOTOGRAPHY. Photography is used in hospitals for "before-and-after" records of surgical cases, and both still and moving pictures are taken in the actual operating theatre.

See also: *Dental photography; Medical photography; Medical radiography; Ophthalmic photography.*

SURVEYING PROPERTY. Apart from specialists engaged in map making or mining, the bulk of surveyors are concerned with land in its broadest sense. "Land" in its accepted legal meaning includes not only the ground but all buildings or erections thereon. In general the professional surveyor is concerned with land and what is on it, its location, value, design, construction, condition and use.

Photographic Records. Since he is invariably occupied with on-the-spot recording of facts, it is not unnatural that the surveyor finds a great deal of use for the camera as a means of making records to supplement his written notes. With few exceptions he uses either a folding roll film camera or a 35 mm. miniature, and there are times when the Polaroid type of camera, yielding a finished print within seconds of the exposure being made, is invaluable.

On arrival at the building to be surveyed he will probably walk round the outside to get a general idea of the property and take photographs of the exterior elevations. He will not trouble much about the lighting; his concern is simply to make records in accordance with a set routine. The survey invariably starts with the roof and works down. The surveyor writes down a description of each part of the building he visits and notes dimensions, condition, features and defects, and uses the camera to augment his notes—e.g., he will probably photograph the roof, taking one or two general views, and will take close-up pictures of any defects such as fractures to the chimney stacks. In this way he then goes through the building from the attic to the cellar recording data.

On all interior pictures he uses a tripod and gives a time exposure or, alternatively he might use flash, because auxiliary lighting is bulky to carry (and, in any case, many buildings will be vacant and have their electricity disconnected).

When the survey is finished it is usual to take an exterior "prestige" picture to show the building to best advantage.

Presenting the Information. At the office the film is developed. Roll films are generally contact printed, and 35 mm. films enlarged to quarter-plate. Then, from his written notes and his photographic prints, the surveyor compiles his report. This report, a typed, bound and signed document, is handed to the client requesting the survey. It is supplemented with drawings and plans, but generally includes only one "prestige" photograph. The other prints are indexed and dated and filed with the written notes, and are not shown to the client unless for some specific reason. This is because the report deals with all the factors affecting a property and an odd photograph of one part might mislead rather than help the client. The "prestige" picture, however, precedes the written matter in the report folder and helps the client to understand the written part.

Shop premises often depend for their value more on their location than any other type of property, so it is usual to include in the survey photographs of both sides of the street. With a miniature camera using a short focal length lens it is possible to include about four shops in each oblique shot and a thirty-six exposure cassette usually holds enough to cover most thoroughfares. The film can then be printed as a positive film strip by contact or better still, by projection. The leader strip can be marked with the date and the name of the road.

This film strip can be run through a projector to show the whole of the main shopping thoroughfare to a prospective buyer of the property offered. M.H.T.

See also: *Architecture; Interiors; Photogrammetry.*

SUTCLIFFE, FRANK M., 1859–1940. English professional portrait photographer in Whitby, England, from 1875. Mainly known for his landscape and genre subjects. Prominent in the Naturalistic school of photography and founder member of the Linked Ring, 1892.

SUTTON, THOMAS, 1819–75. English writer on photography. Encouraged by Prince Albert and partnered by Blanquart-Everard, he established a photographic printing works in Jersey (1855). Founder and editor of the periodical *Photographic Notes*, 1856–67. Invented a fluid globe lens, 1859, used in his panoramic camera 1861; patented the first reflex camera in 1861. Published the first Dictionary of Photography in English (1858).

SWAN, SIR JOSEPH WILSON, 1828–1914. English inventor, chemist and photographic manufacturer. Perfected the carbon process for printing photographs in permanent pigment, 1864 (introduced 1866) and applied it also to a mechanical form of carbon printing from an electrotyped copper mould-photomezzotint, 1865. Discovered the ripening process by heat in making gelatin emulsion, 1877, when his firm Mawson & Swan in Newcastle began producing commercial dry plates. Patented the first automatic plate-coating machine in Britain, 1879. At the same time started the manufacture of gelatin bromide paper. Greatly advanced half-tone printing with his line screen turned halfway through the exposure, 1879. Swan took out over 60 patents, his best known invention being the carbon filament electric light bulb, 1881. Biography by M. E. and K. R. Swan, London 1929.

See also: *Carbon process.*

SWEDEN. It is only in recent years that photography has really made its presence felt in Sweden. A great interest in photography had always existed and, as in other lands, it entered into all spheres of human activity. But not until after World War II was photography acknowledged as an essential cultural factor.

The Past. Photography arrived in Sweden soon after its discovery. In the autumn of 1839 the description of Daguerre's invention was published in a Swedish translation. The first portrait studio in Sweden was opened in August, 1841 by J. A. Seven. At the beginning of the 1840's Sweden was visited by many foreign photographers and among them was the Frenchman Derville, photographer of King Oscar I who came to the throne in 1844. The picture of the king embossed on the coins of the day was a copy of one of Derville's photographs.

The oldest portrait studio now in existence, the Studio Jaeger in Stockholm, was founded by the German photographer Johannes Jaeger in 1853. Under the direction of court photographer Herrman Sylwander (1883–1948), the studio was especially popular with royalty and society. In Gothenburg, Studio Jonason celebrated their 100 years jubilee in 1964.

Another internationally known photographer, Ferd. Flodin (1865–1935) was for a large number of years one of the foremost champions of artistic portrait photography. One of the greatest fighters for the cause of the professional photographer was court photographer Ernest Florman (1862–1952), a founder of the Swedish Photographers' Association in 1895, and its president for thirty-six years.

The best-known names in Swedish photography were Henry B. Goodwin (1878–1931) and Jan de Meyere (1880–1950). Henry B. Goodwin was born in Munich and for several years was lecturer on German at Uppsala University. He first became interested in photography as an amateur but he later opened a portrait studio in Stockholm and attracted a lot of attention through the excellence of his portraits. Dr. Goodwin published several books and often gave lectures; he spoke before the Royal Photographic Society in London and was a member of the London Salon of Photography.

Jan de Meyere was a Dutchman and came to Sweden in 1914. At first he was concerned only with painting, sculpture and handicrafts, but eventually photography took his whole interest and during the 1930's he was the best-known portrait photographer in Stockholm. He was the first in Sweden to work in high key and his pictures soon became known throughout the world. He took part in nearly all the international exhibitions, and he too was a member of the London Salon of Photography.

Present-day Activities. Today photography is accepted as a means of artistic expression, a view supported by the big daily newspapers, where mention is made of exhibitions and newly published books—by foreign as well as Swedish photographers—and by the principal museums of art, where permanent collections of photographs are being built up.

Since 1945 a leading newspaper, *Svenska Dagbladet*—Sweden's *Times*—has presented a gold medal and a large money prize for the year's best photography. The authorities have also taken notice of photography and photographers were included among those entitled to art and cultural scholarships presented by State and Municipal authorities. Previously only artists, composers, actors and authors received such scholarships. Numerous publishing houses and the Swedish Photographes' Association also award valuable scholarships each year.

In 1954 scholarships worth £500 were given to two photographers, and the Sweden-America Foundation awarded a photographer $1,500 to study in America. Other institutions and associations and the Swedish Photographers' Association give scholarships to professional photographers.

There are several municipal training schools for photographers and the Swedish Photographers' Association arranges courses at the National Institute for Handicrafts in Stockholm.

Many of the Swedish news-photographers cover the whole world and their work is published internationally. The production and appreciation of good photographs is also helped by the large number of books on photography published every year.

Exhibitions and Collections. The Swedish photographers often arrange their own exhibitions and participate in the various international photographic exhibitions with great success. Special attention has been paid to the work of the younger photographers, and when Edward Steichen gave his exhibition of European post-war photography at the Museum of Modern Art in New York in 1953 Swedish photographers were well represented.

The Modern Museum in Stockholm, the Art Museum in Gothenburg, and the Värmlands Museum in Karlstad all have fine collections of modern photography.

The Royal Institute of Technology in Stockholm has a large and modern institution for scientific photography. This scientific institution was previously under the direction of John Hertzberg (1871–1935), who became world famous in 1930 when he developed the negatives from S. A. Andrée's polar expedition of 1898. All the members of the expedition perished and for thirty-two years the films lay in the expedition's last camp. Thanks to John Hertzberg's achievement a great and unique collection of photographs was made available to the world.

At the Museum of Technology in Stockholm there is a large historical collection, both of photographic apparatus and pictures, among which are included several photographs by Hill and Talbot. In Stockholm City Museum and in several others too there are large photographic collections.

Industry and Professional Photography. Sweden imports photographic material and apparatus to the value of about £12 million yearly. The country's only factory for the manufacture of photographic material is Ceaverken, in Strängnäs. The factory, which was started in 1939 manufactures chiefly X-ray film, but also produces a certain amount of photographic paper. A world-renowned $2\frac{1}{4}$ in. sq., single lens reflex camera, the Hasselblad, is manufactured at Gothenburg.

There are a large number of photographic organizations in Sweden. The professionals' organization is the Swedish Photographers' Association; the country's 173 photographic clubs come under the National Association of Swedish Photography, which has 8,000 members; there is a Press Photographers' Club, founded in 1930; and there is also a special organization for the picture agencies that sell photographs to the press. The photo-dealers have an association, and the importers have their Photo-Contractors' Union. The Scandinavian professional photographers' organizations are united in the Nordic Photographers' Association.

There are three photographic magazines, *Foto*, *Popular Photography* (Swedish edition), and *Nordisk Tidskrift för Fotografi*, which mainly cover amateur and pictorial photography, but also include technical news and reviews. *Fotonyheterna* is a journal on professional, technical and industrial photography and *Svensk Fotografisk Tidskrift* is published by the Swedish Photographers' Association. K.S.

SWING BACK. Back of a screen-focusing camera hinged so that it can be turned at an angle to the vertical—and often the horizontal—axis of the camera to improve the perspective of the image and extend the depth of field covered sharply by the lens.

See also: *Camera movements.*

SWING FRONT. Lens panel of a camera hinged about its vertical—and sometimes horizontal—axis so that it can be used, in conjunction with the swing back, to control perspective and, independently, to control depth of field.

See also: *Camera movements.*

SWINGING THE CAMERA. Technique for photographing objects moving more or less across the line of sight. The camera is moved to keep the subject in the same position in the viewfinder as the shutter is released. The resulting image of the subject is sharp while the background is blurred, thereby increasing the impression of motion.

The term "panning" is sometimes used to indicate swinging of the camera. Panning, however, is a motion picture expression that refers to a slow horizontal swing of the camera to take in the whole sweep of a stationary scene.

See also: *Movement.*

SWITCHES. There are switches of two main types; single- and double-pole. A single-pole switch disconnects one wire only; the double-pole disconnects both. Single-pole switches are always connected in the live wire so that when they are in the off position, the live wire goes no farther than the switch. If a switch is connected in the neutral wire, even when it is off, the live wire is still left connected to the equipment. Double-pole switches cannot be incorrectly wired up in this way; they are generally stronger than the single-pole types and are used for more responsible work.

Switches are normally of the "single way" type; they complete a circuit in the "On" position and break it in the "Off" position. However, there are also two- and three-way switches which will connect, say, a live wire to any one of two or three separate circuits according to the position of the switch dolly. A two-way, double pole switch with a centre "Off" position can be used for a number of complicated switching duties—

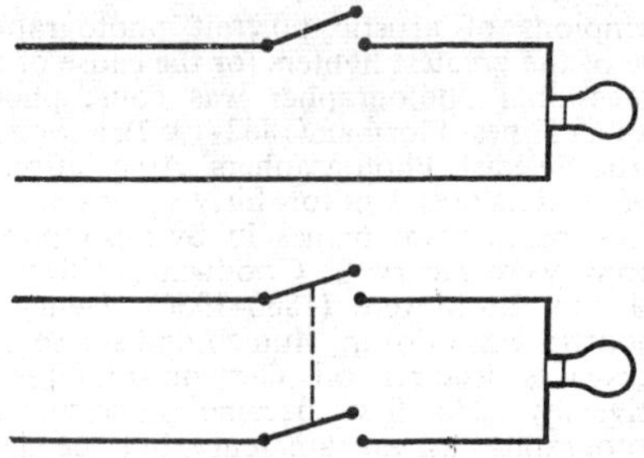

ON-OFF SWITCHES. Top: Single-pole switch. Bottom: Double-pole switch, completely isolates circuit from electricity supply by breaking both sides of the connexion

e.g. to switch a pair of Photofloods (or groups of Photofloods) from series connexion (in which they run at half voltage) to parallel (in which they run at their full voltage and brilliance at the expense of shorter burning time). This arrangement allows the lamps to be run economically while the subject is being arranged and only switched to full brilliance for the actual exposure.

There used to be only one type of switch for both A.C. and D.C., but now there are special switches that are made for A.C. only. These switches are simpler, cheaper, and quieter than the traditional tumbler switch, but they cannot be used in D.C. circuits. Switches of this type are always marked "A.C. only".

Foot Switches. A foot-operated switch is often the most convenient form of control for electrical equipment in the darkroom and studio; it leaves both hands free and allows the power to be switched on and off near the equipment instead of at the outlet on the wall or skirting board.

All the various types of foot-operated switch conform to the same general pattern. A heavy metal box houses the switch and is connected to the supply socket outlet by a length of cable and a plug. On the side of the box is a socket outlet where the lead from the equipment is plugged in. The equipment is switched on or off by pressure on a pedal mounted on the box. Some foot switches are made with more than one socket outlet so that, for example, alternative lighting may be switched on by foot pressure on a second pedal.

Time Switches. Time switches automatically switch off the electricity supply to any piece of equipment at the end of any chosen interval of time. The switch has a length of flex and a plug for connecting it to the mains, and the equipment is plugged into a socket outlet on the switch.

In principle, the switch resembles an ordinary alarm clock. There is a dial, marked in hours and minutes, or minutes and seconds, a hand or hands that show how long the mechanism has been running, and a movable hand that can be set to the time when the switch is to open.

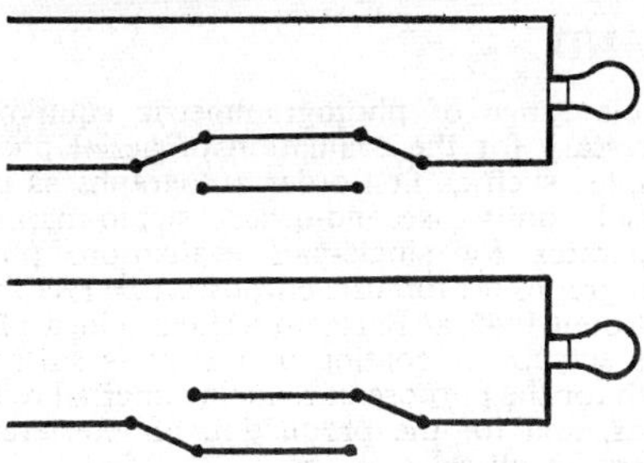

TWO-WAY SWITCHES. Appliance can be switched on and off from either switch. Top: Switched on, either switch breaks circuit. Bottom: Switched off, either switch closes circuit.

Switches that cover periods of more than a minute are ordinary pieces of clockwork mechanism, but the type of switch used for controlling enlarging and printing lamps is a more elaborate affair. Printing times may be as short as half a second, and it is costly to design clockwork for such intervals. Instead of using clockwork, switches of this type are operated by an electronic device and a relay. The electronic circuit measures the passing of time in terms of the time taken to charge a condenser.

Thermal Switches. Thermal switches are designed to open or close when the temperature rises or falls to a chosen level. In photography they are used to switch off current to room, tank, or dish heaters when the temperature reaches the working temperature.

Switches of this type are generally built into the equipment they control, but some are manufactured as separate accessories which can be connected to a warning light or bell, or individual pieces of processing apparatus.

Most thermal switches utilize the unequal expansion of a bimetal strip or spiral to open or close a pair of contacts. Where the switch is required to break a heavy current, the first pair of contacts merely breaks or makes the circuit of a relay. It is the relay that does the real work of switching the main circuit.

Photo-electric Switches. A photo-electric switch is operated by a change on the level of the illumination falling on it.

A photo-electric cell is connected in the grid circuit of a thyratron valve which is biased to pass no current under normal conditions. When the strength of the light falling on the cell changes, it changes the potential applied to the grid of the valve, and current flows in the plate circuit. This current is made to operate a relay which makes or breaks the main circuit.

Photo-electric switching is used for certain kinds of photography in which the subject automatically releases the shutter—e.g., nature photography and crime detection. For such work the photocell is illuminated by a beam of invisible infra-red light, and interruption of the beam releases the camera shutter and fires a synchronized flash.

One of the most interesting applications of the photoelectric switch in photography is in the operation of electronic flash "slave units." These are electronic flash units consisting of a portable power supply, tube, and reflector, with a photocell circuit operated solely by light from the master flash.

Any number of slave units can be set up to illuminate the subject, and when the master flash is fired, it operates the slave flashes with a delay of no more than a few microseconds. This system cuts out any wires between the master and slave units.

Two-way Switching. Two-way switches are designed to give control of a piece of equipment —generally a lamp—from more than one point. A room with a door at each end may have a switch by each door to control the same room light, or a bank of floodlights may be controlled at will by switches at each side of the studio. In each case, either switch will put the light on or off without reference to the other.

Ceiling Switches. Lamps over darkroom sinks are best operated by ceiling switches. These are simply ordinary make and break switches incorporating a toggle action worked by pulling on a hanging cord. The switch is fixed to the ceiling and all contact with wet hands is avoided —giving safety to the operator and longer life to the switch.

Miscellaneous Types. A number of other switches are used in photographic lighting and power circuits.

Three-position tumbler-type switches may be employed to connect either of two circuits to the supply—e.g., a yellow safelight (left), enlarger lamp (right), and both off (centre).

Two tumbler switches coupled by a bar may be used for series-parallel switching of lighting circuits—e.g., two Photofloods: in series (at half brilliance for arranging the subject); in parallel (at full brilliance for making the exposure).

Switch fuses are heavy duty multiple outlet switches for controlling a number of circuits in the permanent wiring of the building. They are generally fitted in iron boxes attached to the wall near the incoming supply and contain a set of fuses, one of which is associated with each out-going set of cables. The switch is connected directly to the meter and is operated by an external handle that can only be placed in the "On" position if the door of the fuse box is closed.

Ratings. The rating of a switch is the current in amperes that it will safely carry and break without overheating or arcing. This safe current is clearly marked on every switch designed to be connected to a public electricity supply. The total current that the switch will have to carry when all the equipment connected to it is switched on should never exceed the rating. Overloading a switch is dangerous and in time renders it useless. F.P.

See also: *Electricity; Series-parallel; Wiring.*

SWITZERLAND

In the past Swiss scientists have made notable contributions in the optical and chemical fields of photography, and today there is a thriving Swiss photographic industry.

Historical. Even before the invention of photography, Swiss experimenters had shown an interest in photo phenomena. Jean Senebier (1742–1809), a clergyman turned librarian, studied the visible change caused by the action of light on silver chloride. He established that it was the blue and violet rays that had the strongest effect (five years earlier the Swedish chemist, Scheele, working independently had discovered the same thing).

Pierre Louis Guinard (1748–1824) improved the manufacture of optical glass; his methods are still used today.

Shortly after the publication of Daguerre's invention by the Académie Française in 1839, the painter Johann Baptist Isenring (1796–1860) introduced the process into Switzerland, and held a large exhibition of his own daguerreotypes (including portraits) at St. Gall in the following year (August 1840): he improved the method by the introduction of retouching and colouring with fast and indelible pigments.

Half a century later the photographic process, which had in the meantime made considerable progress, was put on a firm scientific basis by the introduction of sensitometry—an advance in which Swiss workers can claim considerable credit. The Swiss Ferdinand Hurter (1844–98), together with his friend Vero Charles Driffield (1848–1915), laid the foundations by formulating the conceptions of transparency, opacity, and density, and by defining the characteristic curve (1891).

At Zürich, in 1907, the chemists J. H. Smith and W. Merkens developed a process for obtaining coloured paper prints from colour-transparency originals, under the name of Utocolor Paper. This had a black layer which was bleached by irradiation and assumed the colour of the incident light. This very elegant method of colour reproduction, which others had tried before without success, pursued the same end, although by different means, as the modern colour reversal processes.

In 1912, W. R. Hess, 1950 Nobel Prize Winner in medicine, published an ingenious process for the production and viewing of stereoscopic pictures by means of a lenticular screen film. The results could be viewed without any optical accessories and were so pleasing that for a time the process was very popular.

Industry. Switzerland is the home of a considerable industry for the manufacture of photographic products and accessories.

Heerbrugg, in the extreme north-east of the country, possesses a large factory producing high-quality aerial cameras equipped with lenses which combine high speed and excellent resolving power. The same plant makes the whole range of photogrammetric equipment necessary for the evaluation of aerial photographs, such as first-order autographs as universal units, second-order stereo-mapping apparatus for single-pair evaluation, police autographs for forensic purposes, and rectifiers.

About 1945, at Balzers-Liechtenstein, a plant was set up for coating optical glass surfaces, both for the purpose of reducing internal reflections, and for the production of interference filters for all wavelengths.

High-vacuum coating machines to carry out these processes are also mass-produced and sold at Balzers.

A further project has been developed at Mauren, Liechtenstein, for the manufacture of 8 mm. cine-cameras. A plant in Schaffhausen, with a subsidiary branch in Murten, produces a technical camera which is widely exported.

Zurich is famous for the manufacture of cameras, and as early as 1880 the first miniature camera, made of wood and taking 40 mm. roll film, was made there. Today Zurich is the home of two manufacturers of note. One produces automatic reproducing machines and photographic registration equipment and the other makes large format mirror reflex cameras.

The town of Aarau has an efficient optical precision engineering industry manufacturing microscopes, photographic lenses and various optical instruments for medical and other uses.

St. Croix is a centre of the precision engineering industry. Here, in addition to the famous Swiss watches, is made a range of world-famous, narrow gauge cine gear.

In the centre of Switzerland, at Schwyz, there is a factory for film supports and magnetic recording tapes.

A well-known miniature reflex camera is made in Ballaigues, and in Lugano, in the south, there is a comparatively new company producing sub-miniature cameras.

At Burgdorf, not far from Berne, the federal capital, sensitized photographic materials for various types of technical photography are manufactured, including X-ray films and materials for photocopying and typographical and photomechanical reproduction processes. The factory is unique in having been the first in the world to specialize in this field.

Another plant, at Fribourg, produces photographic contact and enlarging papers, and has finally taken up the manufacture of colour materials. Recently this firm was amalgamated with a very large chemical concern in Basle, which is beginning to build up a world-wide photographic industrial organization.

At Baden, an industry is engaged in a patented process in which silver salts are deposited in an oxide layer electrolytically produced on aluminium plates, on which mechanically and chemically resistant pictures can be reproduced.

Lastly, there is a well known company at Winterthur making photographic gelatin.

In addition there are in Switzerland many young companies making high quality apparatus for varied photographic purposes, not to mention the numerous Swiss firms which act as representatives for foreign products.

Professional Organizations. There is no institute of photographers—i.e., an organization representing photography on a national level—in Switzerland. The commercial interests in the photographic field, however, are organized in the Verein der Photoindustrie (Federation of the Photographic Industry and its representatives) in Switzerland, with offices at Zürich. The federation covers both Swiss manufacturers and the numerous representatives of foreign interests.

The Verein der Photo-Industrie und deren Vertretungen in der Schweiz (Union of Photographic Manufacturers and Representatives in Switzerland) provides an organization for the phographic retail trade, and the Schweizerische Photographenverband (Swiss Photographers' Union), represents the intrests of professional photographers and also takes care of professional training. The Schweizerische Verband für Photohandel und -gewerbe (Swiss Photo-dealers' Association), acts in a similar manner on behalf of the dealers. The Schweizerische Fotografen Bund (Swiss Photographic Federation) represents photographic employees. The amateur photographers are united in the Schweizerische Amateur-Photographen-Verband (Swiss Amateur Photographers' Association).

Education. Photographers receive their training either by serving a normal apprenticeship, which may be followed by gaining a master's diploma, or by taking a course at the photographic section of the technical college (Ecole des Arts et Métiers) at Vevey, Lake of Geneva.

The Kunstgewerbeschule (Art School), in Zurich also has a photographic department.

Scientific photographers with a special final certificate are taught at the Photographic Institute of the Eidgenössische Technische Hochschule (Swiss Federal Institute of Technology). This institute, founded in 1886, is remarkable in being one of the very few establishments in Europe offering photography on an academic level within the framework of the college teaching programme. It covers theory and practice as well as conducting scientific research. There are regular lectures and photographic symposia by well-known scientists and Swiss and foreign experts. These lectures, in which the various experts report on the nature of their work, enable the Photographic Institute of the E.T.H. to maintain permanent contact, on a technical as well as a personal basis, between the circles interested in the scientific and technological aspects of the profession.

The first world exhibition of photography, covering practically every field of pictorial work and photographic applications, was held in Lucerne in the summer of 1952.

Photographers. Mention has already been made of the daguerreotypist, Isenring. Johann Baptist Taeschler (1805–66) was another pioneer in this field. He made a name for himself in the early stages of photography, opening a photographic studio at St. Fiden near St. Gall in 1846, where he produced work of high artistic merit; he gained awards at the first international photographic exhibitions. His four sons took over the business under the name of Taeschler Bros.

Fred Boissonas (1858–1946) was one of the first and most outstanding Swiss landscape photographers. At Paris, in 1891, he was awarded a Gold Medal for the first telephotograph of Mont Blanc ever produced; it was a masterpiece, exposed through a telephoto lens on a home-made orthochromatic plate. He derived his main fame from his pictures of Greece. The mountain and landscape photographs by Emil Meerkämper (1877–1948) and Jean Gaberell (1887–1949) are also world-famous.

In those days photographers would frequently also be active in a technical capacity, as for example Carl Koch (1845–97) who was a mountain photographer and also manufactured dry plates for the Company of Tronel & Koch, Schaffhausen. Other technicians to be mentioned in this connection are Emil Suter (1850–1936) who made cameras and lenses as well as other optical apparatus, Gottileb Zulauf (1871–1959) and Carl Pfann (1869–1929),, both well known producers of cameras from Zurich.

As an aerial photographer, the pilot Walter Mittelholzer (died 1937) not only produced captivating alpine landscapes of his own native country, his aerial photographs of other continents are to this day acclaimed as some of the best pictures of their kind. A number of outstanding specialists in this field are still to be found in Switzerland.

The pictorial journalists Paul Senn (died 1953) and Werner Bischof, who met with a fatal accident in the Andes in 1954, were of the same outstanding calibre in their own special branch of pictorial journalism. Examples of their work have appeared in all important illustrated journals the world over.

Camill Ruf (1872–1939) was perhaps the greatest Swiss portrait photographer. He was the author of excellent group pictures and many exquisite examples of bromoil and gum-printing techniques.

The work of other great portraitists of that time, such as Francis de Jongh (1864–1928) and Boissonnas deserve mention here.

Switzerland has also produced some eminent advertising photographers—a branch of photography which gives scope to the special abilities of Switzerland's professional workers.

The photomechanical industry in Switzerland has reached a standard of high quality. Striking examples of this are the excellent ordnance maps of the country, produced by the Federal Ordnance Survey by a special photomechanical process. Swiss posters offer the most impressive evidence of the fruitful application of photography and graphic art in the field of advertising. International poster competitions, such as those held for the Olympic Games, have repeatedly been won by Swiss designers.

Magazines. The main Swiss publications for professional photographers are the *Schweizerische Photo-Rundschau* (Swiss Photo Review) which is the official publication of the Schweizerische Photographen-Verband (Swiss Association of Professional Photographers), and the *SFB Mitteilungen* (Reports of the Swiss Photographic Union). There is also a dealer's magazine, *Der Schweizer Photo-Händler* (Swiss Photo Dealer), and a semi-professional periodical *Camera.*

The main professional cine periodical is *Schweizer Film* (Swiss Film) and *Schweizer Schmalfilm* is a monthly publication for cine-amateurs. A number of religious organizations also issue their own cine magazines.

Amateur photographers are served by a number of club magazines, the main one being *Photo-Amateur* (of the Schweiz. Amateur-Photographen-Verband), and *Foto-Rundschau* (Photo Review), the official publication of the Arbeiter Fotovereine (Working Mens's Photo Clubs). There are also a few amateur cinematographers' magazines sponsored by amateur cine associations and clubs, as well as one or two minor independent publications.

J.E. & K.P.

SYMMETRICAL LENS. First of the true anastigmats, made possible by the discovery of the new Jena optical glasses at the end of the nineteenth century.

A symmetrical lens consists of two identical, or at any rate closely similar, sets of glasses with a stop between. Up to five glasses may be used in each set. The distortions introduced by one half of the combination tend to be cancelled out by the same distortions of an opposite nature introduced by the other half.

But this obliging characteristic depends on the object-lens-image distances. Where they are roughly the same, as in copying, an exactly symmetrical lens will serve, but where there is a big difference, as in camera lenses, the designer has to make changes in one half of the combination to balance matters. For this reason a symmetrical lens designed for copying or small amounts of enlargement is of no use in an ordinary camera.

Symmetrical designs nowadays are used mainly for wide angle and process lenses although a few modern camera lenses are derived from the symmetrical type.

Two distinct types of symmetrical lens are commonly used in cameras.

One is intended as a long focus lens for 35 mm. cameras. In this type, the centre of the field is highly corrected for the small area it has to cover. The second type is for use in normal cameras and is corrected to cover a bigger area. So it is possible to have two lenses of the same focal length and aperture, one for use as a long focus lens on a miniature camera, and the other a normal lens on a larger camera.

See also: *Lens history.*
Book: *Photographic Optics,* by A. Cox (London).

SYMMETRY. Effect of similar or identical arrangement of subject matter on each side of an imaginary central dividing line or point. Aesthetically symmetry is rarely pleasing because it looks too formal and studied. It is occasionally effective where the intention is to draw the observer's attention specially to the perfect symmetry of a particular subject.

See also: *Composition.*

SYNCHRO-FLASH. Technique of flash photography in which the firing of a flash bulb is initiated by contacts which are generally in the shutter mechanism, but at other times externally connected to it, so that the useful part of the flash illuminates the subject at the instant the shutter is open. For very short exposures the timing of the flash in relation to the shutter opening is arranged so that the exposure starts and finishes during the peak brilliance of the flash. With electronic flash, which fires almost instantaneously upon contact being made, and which is of very short duration, the flash circuit is closed at the instant the shutter is wide open.

See also: *Flash synchronization.*

SYNCHRONIZER. Mechanical device for firing a flash bulb in synchronism with the opening of the camera shutter. A synchronizer usually incorporates the following parts—a release button, a pair of electrical contacts which can be connected to the flash head and flash battery through a flexible lead, a plunger for operating the shutter. Finally there must be an adjustable coupling between the flash contacts and the shutter release plunger so that the contacts can be timed to allow the flash to rise to peak brilliance before the shutter opens.

Synchronizers of this type have mostly been rendered obsolete by internally synchronized camera shutters.

See also: *Flash synchronization.*

SYNCHRO-SUNLIGHT. Balanced combination of sunlight and synchronized flash used for taking photographs.

See also: *Flash technique.*

T

TABLE TOP

There are three different branches of photography open to the photographer who chooses to work within the limits of a table top studio: still life studies, photography of small scale models and creative composition. The last is the true "table top photography", and it offers the greatest scope to the photographer with imagination and artistic ability.

Still Life. Still life arrangements are best if kept simple. Whether the subject is a flower study, a group of porcelain, or a pile of nuts and bolts, the effect is strengthened by straightforward presentation.

The still life photographer is fortunate in having complete control over the background of his picture. Generally a plain background is to be preferred as it helps to concentrate interest on the picture material.

A plain background is easily contrived from a large sheet of off-white or grey paper which covers the top of the table, hangs over the edge at the front, and curves gently upwards and out of the picture at the back.

Shadows cast on the background by the lighting can be exploited to make an important contribution to the composition. At the same time, unless they are carefully watched, they can ruin the whole effect.

Back lighting can be attractive in throwing useful shadows that help to fill the foreground and unify the composition. Other shadows may be introduced from objects outside the picture.

In straightforward still life studies of this type there is no attempt at deception; the aim is simply to arrive at a pleasing piece of decoration, using recognizable objects suitably arranged and lighted. In flower studies, however, the lighting and background may be deliberately arranged to give the effect of a natural outdoor setting by using shaped masks and cut-outs in front of the background light to cast shadows, suggesting sky and clouds.

Scale Models. In dealing with a scale model, the aim is generally to make the finished print look like a photograph of the real thing. This is done by using a suitable lens and viewpoint, and by careful attention to the details of the model and its setting.

The height of the viewpoint is best chosen to correspond with eye-level on the actual full-sized object. A doll's house, for instance, would be photographed with the lens at the eye-level of an imaginary figure standing in the doorway. And the natural viewing distance is reduced in the same proportion—i.e., if the actual object would be viewed from, say, ten feet away, and the model is 1/10 scale, then ideally the model should be photographed from 1 foot away and with a 1 in. lens.

In practice these conditions cannot always be achieved, but it is possible to get a more natural perspective by using a short-focus lens.

Whereas a 6 ins. lens on a quarter-plate is satisfactory for the average still-life and table-top shot, a lens of 3 or 4 ins. focal length will be found more suitable for scale models. If such a lens is not available, the camera should be brought nearer to the subject. The illusion can be strengthened by using a viewpoint below table level, with the aid of the rising front of the camera.

The effect of perspective can be further increased in the building of the set. If the model is to be shown in a landscape, any foreground trees or buildings are made large, and the others are arranged to decrease in size as they recede into the distance. The foreground may be built of small stones, the middle distance of fine gravel and the far distance of sand.

Creative Effects. True table top photography is not concerned with merely reproducing facts.

It can be a source of amusement or a useful medium for book illustration and advertising· and at its best it is a branch of creative art.

Much of the interest of this branch of photography lies in building up scenes with all manner of odds and ends. Pieces of paper, Cellophane, cotton wool, glass shapes and other materials, when seen by the camera out of focus, can be made to assume the appearance of mountains, water, icebergs and so on. For instance, it may be more effective to represent a forest by a few simple alternating lines of light and shade, than by using actual twigs.

Interior Sets. Interiors are built like stage sets, with three sides and open at the top. Thin strawboard covered with paper with a small pattern (or the back of a piece of wall paper whose pattern is not too obtrusive) serves admirably for walls. Furniture may be made to scale from cardboard or plywood, although it is generally easier to buy suitable ready-made toy furniture. The fewer props and pieces of furniture there are the better; suggestion is more effective than actual portrayal.

Exterior Sets. Sand and assorted stones can be used to form the basic shapes for most landscape scenes. Pieces of velvet will do for valleys and hills, large stones or coal for mountains, and moss and felt with a rough nap for bushes and grass. Snow can be made from cotton wool, salt or flour; Cellophane over a light or dark base can suggest water, and walls can be made from modelling clay or children's building bricks.

Small trees, carefully pruned branches, or roots of heather, can be used as trees. Trees in leaf are difficult to suggest, but small pieces of ornamental shrubs, ferns, or asparagus plants will often give the desired effect.

Figures. Those with sufficient skill can make their own solid figures from plaster, plasticene or modelling clay. Adjustable figures can be made from bent wire padded with cotton wool and dressed. Often, suitable figures can be bought from toy or fancy goods shops, and blown glass figures are especially attractive. But whatever form they take, they must be well finished because at such close distances any faults or blemishes stand out and destroy the illusion, particularly if the final enlargement shows the figures larger than actual size. Dolls of foam rubber or neoprene, having wire skeletons enabling them to be bent into any pose, are very useful for table-top work.

Models made of glazed china or finished in glossy paint may have to be dabbed with putty to deaden surface reflections. Average figures should measure 4 to 6 ins. in height, and though smaller ones can be used, any extra accessories—like umbrellas or sports equipment—will be correspondingly smaller and more difficult to make.

Treatment. A table top photo can be made to express an idea, tell a story, illustrate a quotation, or put over a topical skit. It offers the photographer a completely flexible medium of expression for humour, fantasy, drama, or pure decoration. Space and gravity can be defied in the world of the table top, and the photographer is free to portray the supernatural and the impossible with equal ease.

In any set-up the subject matter should be simple and all fussy detail avoided. Most table-top pictures fail either because they include too many items or because they try to represent the subject accurately and in detail instead of simply suggesting it in a broad effect.

As an example of correct technique, a desert scene might be constructed on the following lines:

A pool of water is suggested by a piece of creased Cellophane and some sand. (A piece of mirror would give unbroken reflections.) Two or three toy camels and a cardboard palm are arranged around the pool. The foreground can be a piece of draped cloth or creased paper, and the background a sheet of translucent paper.

The camera is focused sharply on the pool at a wide aperture. This throws everything in front and behind partly out of focus, and emphasizes the principal item. The lighting is arranged to show only the pool and its immediate surroundings. Shadows can be cast on the background from the back and be made to suggest clouds and receding distant hills.

This method of treatment does not attempt to disguise the nature of the subject material, but softens it and presents it in the pleasing light of fantasy.

Glass-top Technique. A technique which might be called glass-top photography offers even wider scope. This calls for some means of holding sheets of glass horizontally at different levels. The objects to be photographed are laid on the glass, each glass sheet representing a different plane in the picture. The camera looks down vertically on the glass, and the subject is lighted principally from the sides.

This method presents figures and objects as though they were suspended in mid-air without any means of support.

The Camera. No matter what camera is used, it should have a strong, steady tripod, with a tilting top or a universal head.

All forms of table top photography call for a camera with a focusing screen. If possible, it should also have double extension, a drop

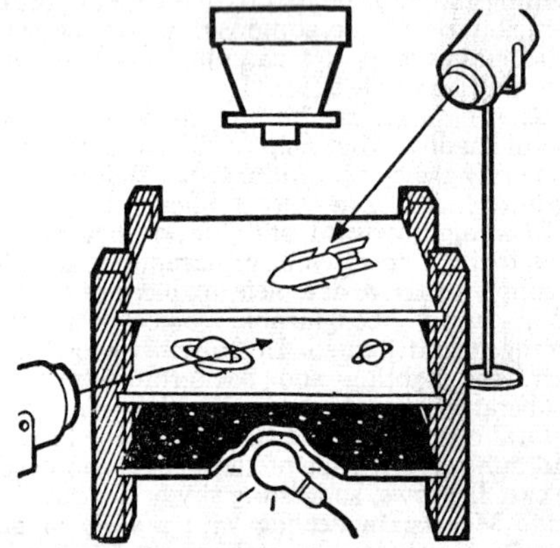

GLASS-TOP PHOTOGRAPHY. Some subjects are best built up on layers of glass against a suitably illuminated background. This permits wide excursions into realms of fantasy.

TABLE TOP SETS. *Near, medium distant, and distant objects in a table top set-up have to be built to different scales. This is to give an effect of reality to the subject and to simulate the perspective recession of a life-size scene within the confines of the set-up. For the same reason, the actual proportions of model buildings must also be distorted to exaggerate perspective.*

and rising front, and a swing back. Front side swing is useful for subjects (e.g., a toy train) which recede at an angle. For photographing scale models the camera lens should be capable of being stopped down to at least *f*64.

The creative branch of table top photography demands a wide-aperture lens. All-over sharpness is not always required because many effects depend upon differential focusing.

Lighting. Table top photography does not call for elaborate lighting equipment. Exposure times should, however, be kept within those limits where the reciprocity law applies, as reciprocity failure will lead not only to incorrect exposure, but also to incorrect colour rendering. The instruction leaflet supplied with each film should be studied carefully.

Light sources of varying colour temperatures should not be mixed when using colour, unless special effects are wanted.

A great deal can be done with ordinary table lamps fitted with ½-watt pearl or opal bulbs. Small spotlights are invaluable, especially in glass-top photography, and there should be at least one focusing spot in which the aperture can be adjusted to throw a narrow beam.

In the simplest form of lighting the principal light is placed to one side of the camera at an angle of about 45° to the subject. The fill-in light to give shadow detail should be diffused, or it can be indirect light produced by shining the lamp on to a white wall, a piece of card, or even the ceiling. This light should not throw any noticeable shadows, but if it does, they should be weak enough to be killed by the principal light.

The beginner should start with one light only, placed slightly to one side, with white cardboard reflectors adjusted to throw light into the shadows. Pieces of mirror or shiny silver paper can be employed to provide local spots of reflected light, but they need careful adjustment to avoid throwing patches of light of their own shape. Lamps must be screened off so that no stray light reaches the lens. The direct light from the lamps should be prevented from falling on the camera, and the lens should, in any case, always be protected from stray reflections with an efficient hood.

Exposure. If a photo-electric meter is used readings should be taken of the darkest and lightest areas to be rendered in detail. This will give the contrast range of the subject. If the range is found to be excessive it should be reduced by using a more powerful fill-in light or extra reflectors. An exposure should then be selected which will not leave the shadow areas under-exposed nor the brightest lights over-exposed. This procedure is very necessary with colour film.

In any case, a few test exposures should be made. They will always provide a useful basis for future work because, on the whole, conditions do not vary greatly with artificial lighting. It is always worth while to make a note of all exposure details, together with rough diagrams of the set-up.

Where extension tubes or bellows, or extended camera extensions, are used, allowance must be made for the change in the effective aperture of the lens, and an accordingly longer exposure must be given.

Sensitive Material. Almost any type of sensitive material can be used for table top work because speed is unimportant, but a medium speed panchromatic emulsion may be preferred for its more correct rendering of colour values. It is a wise precaution not to dismantle the set before the film is processed, if this is at all practicable.

E.H.

See also: *Animation; Models (scale); Silhouette; Tricks and effects.*

Books: *Camera Close-up*, by O. R. Croy (London); *Creative Table Top Photography*, by E. Heimann (London).

TAKING PHOTOGRAPHS

Certain basic steps are common to every field of photography outside the laboratory.

The minimum equipment required is: a camera, sensitized material (film or plate), a support for the camera (the hand, or a tripod) some means of measuring the distance from the camera to the subject (rangefinder or tape), some means of estimating or measuring exposure (calculator or exposure meter), some form of lighting (daylight, electric light, or flash).

Basic Steps.

(1) Load the camera with the film or plate.

(2) Sight the subject in the viewfinder and choose the best distance and viewpoint.

(3) Determine the distance from the camera to the subject and focus the lens.

(4) Determine the exposure and choose the stop and shutter speed that will give it.

(5) Tension and release the shutter.

(6) Change the film or plate.

(7) Process the film or plate to a negative.

(8) Print the negative.

Choice of Equipment and Materials. Many normal cameras cover a wide range of subjects. Intelligent photography, particularly when dealing with specific subjects, goes still further and includes selecting the most useful apparatus and the most suitable sensitive materials.

The following tables summarize the requirements from this point of view.

In the case of equipment the specifications are the ideal ones. With some subjects special apparatus is virtually essential. With many others, the ideal camera type will do the job best—and is the obvious choice for photographers specializing in that subject. But most average or better outfits will usefully tackle such subjects occasionally, often with special accessories, though perhaps not as conveniently as the apparatus made for the purpose.

Where emulsions are concerned, it is much easier to choose the best film or plate for the job Again the materials listed are the ideal ones.

Different aspects of a subject may call for different ranges of properties and the choice of emulsions is intended to cover these, rather than to indicate substitute requirements.

SUBJECTS AND SUITABLE EQUIPMENT

Subject	*Ideal camera*	*Accessories and Special Features*	*Remarks*
Action	Miniature or press	Rangefinder	High shutter speeds desirable
Aerial photography	Special aerial camera	Haze cutting filters	
Animals	Twin-lens reflex or miniature	Close-up lenses for small animals	Other types suitable, subject to limitations
Architecture	Technical or field	Filters, tele lenses, wide angle lenses	Camera movements desirable
Astronomy	Plate camera, special type		Invariably used with telescope
Candid shots	Miniature or twin-lens reflex	Short focus lens useful for depth of field	
Close-ups	Technical, single lens reflex or miniatures	Extension tubes, special gear for miniatures	Fixed lens cameras usable with close-up lenses
Commercial	Technical or field	Filters, various lenses	Camera movements desirable
Copying	Special copying cameras, also single-lens reflex	Extension tubes or close-up lenses, filters	Many simple cameras and miniatures suitable for limited work
Feature	Twin-lens reflex, miniature or press	Filters, interchangeable lenses, flash gear	Camera should be versatile to cover variety of subjects
Flash	Almost any, provided shutter is synchronized	Flash gun and equipment	XM-synchronized shutter permits high speed shots
Groups (large, formal)	Field or technical, or special types	Wide angle lens for confined locations	Large negative and ground glass screen desirable
Holiday	Any roll or miniature	Yellow filter	Simple operation useful
Industrial	Technical or reflex, interchangeable lenses necessary	Flash and other portable lighting equipment	Camera movements desirable
Landscapes	Field or reflex, but very simple cameras suitable	Filters, interchangeable lenses (if available)	Ground glass screen helps composition
Mountaineering	Small folding or miniature	Filters	Light weight deciding factor
News and Press	Press (taking sheet film and plates) with fast lens and shutter	Rangefinder, flash equipment, large frame finder	Miniature and $2\frac{1}{4}$ in. sq. also used. Rapid action important
Night	Any suitable for time exposures	Tripod, cable release with time lock	Lens speed not important with static subjects
Photomicrography	Single-lens reflex or miniature	Microscope and adaptor	Miniatures need special equipment
Portraits and studio groups	Studio or reflex	Long focus lens for head and shoulder portraits	Ground glass screen preferable
Snapshots	Any roll or miniature	Yellow filter	Simple operation useful
Sports	Press or miniature with fast shutter	Long focus lens, frame finder	Instant readiness for action important
Theatres and shows:			
from auditorium	Miniature with fast lens	Long focus lens	Preview of performance before shooting session desirable
photo-calls	Press or reflex	Flash, tripod	Subject carefully posed

SUBJECTS AND SUITABLE NEGATIVE MATERIALS

Subject	*Speed ASA*	*Colour Sensitivity*	*Gradation*	*Grain*	*Other Properties*
Aerial photography and photogrammetry	50 or faster	R, IR	h.	mf. or f.	Glass or low-shrink support, thin emulsion, high resolving power
Architecture	Medium or slow	O P	m.	mf. or f.	Glass or low-shrink support for accurate surveying work
Artificial light	160 or faster	R, P	m.	As fine as available	
Astronomy	As fast as available	O. R, IR	h.	As fine as available	Special colour sensitization may be needed
Colour separation	50–100	P	m. or s.	As fine as available	Equal contrast of all three separation negatives for the same development time. Plates preferable.
Commercial and technical	50 or faster	O, P, R	m. or h.	As fine as available	Colour sensitivity depends on colours of subject. Plates and sheet film preferred, as they allow single development
Copying (line)	Slow	B, O, P	eh. or uh.	f. or ef.	Colour sensitivity required depends on colours of subject
Daylight and outdoors (incl. amateur)	50 or faster	O, P	m.	As fine as available	
Copying (half tone)	Medium or slow	B, O, P	m. or h.	f. or ef.	Colour sensitivity required depends on colours of subject
Copying (micro)	Medium or slow	B, O, P	h. or eh.	ef.	Colour sensitivity required depends on colours of subject. Usually carried out on 35 mm. or even 16 mm. cine film.
Flash photography	80 or faster	R, P, O	m. or s.	As fine as available	Low reciprocity failure materials of higher contrast advisable for electronic speed flash
Fluorography	40 or faster	O, B	h.	As fine as available	
Forensic photography	25 or faster	R, P, IR, B	m. or h.	f.	Infra-red materials needed for special purposes, blue-sensitive emulsions for ultra-violet photography
Graticules	Very slow	B	uh.	uf.	Extreme fine grain and highest resolving power essential
Instrument recording	16 or faster	B, O	h. or eh.	f.	Material often supplied and used in continuous rolls
Industrial	Medium or fast	O, P	m. or h.	As fine as available	Artificial light work may need red pan materials. Plates and sheet film preferred, as they allow single development and individual control
Photogrammetry	As fast as available	R, IR	m. or h.	f. or mf.	Low-shrink support desirable. Infra-red materials often used in aerial photogrammetry
Photomechanical work	Very slow	B, O, P	h. or eh.	f.	Panchromatic materials needed for coloured originals and for colour separation work. Line copying needs the greatest contrast
Photomicrography	25 or slower	O, P	h. or eh.	f.or ef.	Blue-sensitive materials suitable for colourless specimens, panchromatic for stained specimens requiring filters
Portraiture	Medium	P, O	m. or s.	mf.	Different types of materials may call for different make-up
Spectrography	Varied	As relevant	eh.	mf. or f.	Colour sensitivity required depends on spectrum range to be recorded. Special U.V. sensitization may be necessary. Sensitivity should be uniform over the spectral range used

Abbreviations. Colour sensitivity: B—blue; O—orthochromatic; P—panchromatic; R—red panchromatic; IR—infra-red. Gradation: s—soft; m—medium; h—hard; eh—extra hard; uh—ultra hard. Grain: mf—medium fine; f—fine; —ef extra fine; uf—ultra fine.

See also: *Against the light; Angle shot; Artificial light; Background; Camera; Camera movements; Camera shake; Camera support; Cases; Cassette; Changing bag; Close-ups; Colour materials; Colour technique; Composition; Contrast; Daylight; Depth of field; Depth of focus; Diaphragms; Distance estimation; Double exposure; Exposure; Faults; Films; Filters; Fine grain technique; Flash equipment; Flash technique; Focusing; Focusing mechanism; Foreground; Holding the camera; Key; Lens hood; Lighting the subject; Light sources in the picture; Miniature camera technique; Movement; Over-exposure; Negative materials; Parallax; Permits to photograph; Perspective; Pictorialism; Rangefinder; Shutters; Spectacle and eyesight; Style; Sunshine; Supplementary lens; Swinging the camera; Travel photography; Tricking the subject; Tricks and effects; Under-exposure; Viewfinder; Vision; Visual appeal; Zone focusing.*

Books: *Successful 35 mm. Photography*, by Andrew Matheson (London); *The All-in-One-Camera-Book*, by W. D. Emanuel (London); *Camera Techniques*, by H. J. Walls (London); *Basic Photography*, by M. Langford (London).

TALBOT, WILLIAM HENRY FOX, 1800–77. English archaeologist, chemist, linguist, mathematician. Inventor of the negative-positive process of photography. Educated at Harrow and Trinity College, Cambridge. Used the camera obscura on a trip to Italy in 1823 and the camera lucida in 1833; this led him to experiment on the fixing of the images by chemical means. Investigated the action of light on paper treated at first with nitrate of silver and later with chloride of silver. He experimented to make his paper more sensitive, and in 1835 produced a picture of his home. Lacock Abbey, near Chippenham, Wiltshire. He used silver chloride paper to copy botanical specimens, lace, engravings and manuscripts. The announcement of Daguerre's invention in January 1839, led Talbot to describe his "Photographic Drawing" to the Royal Society in London on the 31st January, 1839. The manipulation was made known to the Royal Society three weeks later.

Soon afterwards he also discovered the great light-sensitivity of silver bromide paper. He next investigated the sensitivity of silver iodide and discovered that a latent image was produced which could be developed with gallic acid. The exposure needed for this calotype process, patented 1841, was much shorter than that for the old ("Photogenic Drawing") process. The paper negatives were made transparent by waxing, and positive prints were obtained by contact copying on to silver chloride papers. In 1844 to 1846 published *The Pencil of Nature*, the first book illustrated with photographs (showing many applications of photography, e.g., document copying). In 1852 patented Photoglyphy, having independently discovered the light-sensitivity of bichromated gelatin; this was a method of heliographic etching on steel as well as on copper. In 1858 he patented an improved process giving fairly good half-tone rendering.

Talbot patented most of his inventions, and demanded licence fees from those using them. In 1852 he publicly permitted the non-commercial use of his patented inventions.

See also: *Discovery of photography.*

TALBOTYPE (Calotype). Original paper negative process patented by Fox Talbot in 1841. It soon became obsolete after the introduction of the collodion process in 1851.

See also: *Calotype; Discovery of photography.*

TANKS

Many different types of tank have been produced for processing films and plates. Some are intended for use entirely in the darkroom, others must be loaded in the darkroom but may be used in daylight. A third type is intended for full daylight loading and operation without the need for any darkroom.

ROLL AND 35 MM. FILM TANKS

Tanks used for roll and 35 mm. film differ in design according to various requirements—e.g., daylight loading, wet loading. Some may not even allow processing in daylight at all.

Darkroom Tanks. The simplest, intended for use entirely in the darkroom, are straight, flat trays, sometimes provided with a lid, in which the film is held stretched out flat along the bottom. These are just slightly modified forms of a developing dish.

For processing roll film by the see-saw method, special dishes are available which have a roller or a pair of "dimples" under which the film passes to keep the loop of film under the processing solutions.

Even more elementary is a trough-like dish in which the film is developed by rolling and unrolling the strip of film upon itself.

The earliest type of tank for 35 mm. film was simply a long tubular container holding a straight length of film. Then came the developing drum consisting of a glass cylinder supported on a pair of "spiders" and turned by a handle. The film was wrapped around the drum and the ends secured to it with clips. The drum was supported over a developing dish and rotated so that the lower part of the drum dipped into the processing solution. This equipment is still used for developing cine film and long lengths of positive film for filmstrips.

Daylight Processing Tanks. Daylight processing tanks are of two types: apron and spiral loading. Probably the earliest of the apron-type tanks was that made by Kodak. It consisted of two main parts: a loading box, and a developing tank. In the loading box, the film and its backing paper were transferred from the camera spool to a special developing spool.

The film was eventually wound on to this

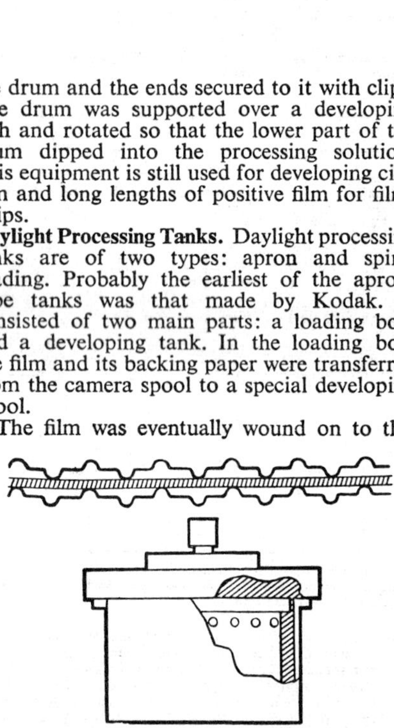

APRON REEL. The film is wound up on the tank reel together with a celluloid or plastic band (the apron) carrying embossed projections in both directions along the edges. These keep the turns of film apart and permit access of solutions.

developing spool along with an apron which had corrugated edges. The apron itself was opaque to light but was permeable to processing solutions; so, when the film and apron had been wound on to the developing spool, the whole thing could be removed from the loading box and placed in the developing tank for processing in daylight.

Since the back of the film was in contact with the apron during the whole time of processing, it was only possible to develop films which had no coloured backing. Such tanks are useless for present-day films.

The Correx type tank is much used for processing 35 mm. film. The Correx system, originally used for processing 35 mm. professional motion picture negatives in lengths of several hundred feet, uses a celluloid band with dimples along its edges.

The film and celluloid band are together wrapped around a spool and loaded into a light-trapped tank into which processing solutions can be poured without danger of light leaks. The dimples along the edge of the apron keep it away from the film and allow the solutions to circulate freely between the layers.

Two lengths of 35 mm. film can be loaded on to a spool, back to back, thus doubling the effective developing capacity of the tank.

A more recent tank uses a special apron with very shallow dimples to save space. This will take a full 36-exposure length of film yet its capacity is little more than 4 ounces.

Correx type tanks are also made in sizes suitable for developing films of V.P. size and $2\frac{1}{4} \times 3\frac{1}{4}$ ins. size. The capacity of these tanks is about 25 ounces.

The spiral loading tank has a spool with continuous spiral grooves moulded on the inside faces of the cheeks. The spool is loaded by first removing the film from its backing paper (or cassette) and then threading it into the spiral track, the leading end passing round the spiral until it comes to the end of the track near the core of the spool. The loaded spool is then placed in the tank and covered with the light-tight lid, after which processing can be carried out in daylight.

In some models of the spiral loading tank the width between the two cheeks may be adjusted to suit any of the standard sizes of film between 16 mm. and $4\frac{1}{4}$ ins. wide.

Since the length of a 36-exposure 35 mm. film is appreciably greater than that of a size 127 or 120 roll film, adjustable tanks may incorporate a spacer to surround the centre core when using the shorter roll film lengths. This reduces the solution capacity of the tank, providing some saving in processing solutions.

The contents of both apron-type and spiral loading tanks are agitated by a separate spindle which is inserted through a light-trapped hole in the lid of the tank and engages with the core of the spool. This makes for even development. A thermometer may be inserted in place of the turning spindle to check the temperature of the tank contents.

The more modern way of agitating the solution inside the tank is by inverting the whole tank periodically. Many tanks therefore have a more or less leak-proof lid with a sealing cap which fits over the central opening. This arrangement allows little or no liquid to escape while the tank is inverted for a few seconds. The tank body is in this case rather larger than usual, to allow most of the liquid to drain away from the film reel during inversion.

One spiral loading tank is constructed entirely of stainless steel. The spool is made up from a spider which supports a spiral made of heavy gauge wire. In the centre of the spool there is a wire clip which anchors the end of the film. A special guide which curves the film laterally feeds the film into the spiral from the centre outwards. The final result is the same as with the other spiral tanks except that loading proceeds from the centre outwards instead of from the outside to the centre.

Multiple and Bulk-length Tanks. For processing several films at the same time some tank reels have an additional cheek in the middle between the two outside ones. That gives in effect two reels in one, permitting the two films to be loaded in the upper and lower spiral. Usually this is intended for 35 mm. film, and the same

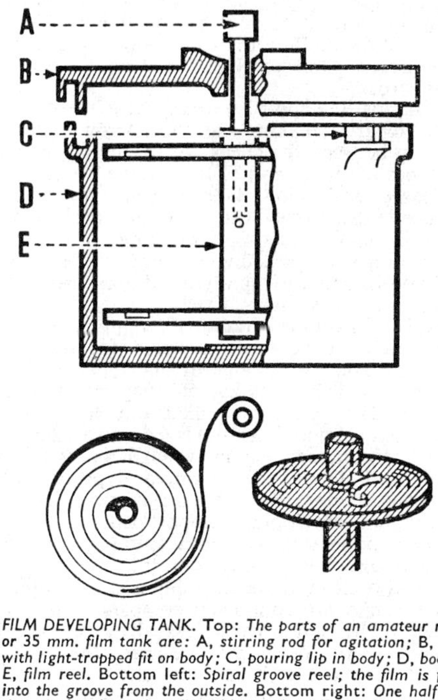

FILM DEVELOPING TANK. Top: *The parts of an amateur roll or 35 mm. film tank are:* A, *stirring rod for agitation;* B, *lid with light-trapped fit on body;* C, *pouring lip in body;* D, *body;* E, *film reel.* Bottom left: *Spiral groove reel; the film is fed into the groove from the outside.* Bottom right: *One half of the spiral reel may be adjustable to take different film widths.*

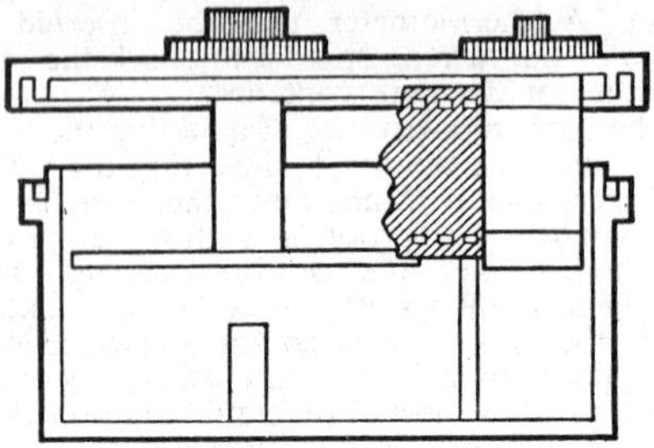

DAYLIGHT LOADING 35 mm. TANK. The film cassette is loaded into a special chamber, and the film transferred to the reel while inside the light-tight tank.

reel can be used without the centre cheek for wider roll film sizes.

Some developing tanks take two or three (or even more) normal reels, again for simultaneous processing of several films. Here the tank body is of the appropriate extra height. When fewer reels than the full capacity are used, a spacer unit keeps the reels in position, avoiding excessive movement if the tank is inverted for agitation.

Spiral-loading tanks working on the same principle are used for processing 16, 35 or 70 mm. film of lengths up to 100 feet. The extra large spiral reel is loaded on a special stand. The film is led from its container through a long spout which reaches right into the centre of the spool. By rotating this spool—which may be done by a motor—the full length of 100 feet of film automatically winds into position between the spiral grooves from the centre outwards.

The loading spout squeezes the edges of the film inwards, making the film narrow enough to slip between the cheeks of the spool until it reaches the inner end of the groove where it is released and allowed to spring into place. Processing is carried out in circular darkroom-type dishes or scaled-up tank bodies of suitable size and, when washing is complete, the film still on its spool is rapidly dried in a current of warm air, on a rotary dryer.

Daylight Loading Tanks. A number of roll film tanks have been designed for daylight loading and processing. In one type the film is separated from its backing paper inside the tank. The paper is pulled out through a light-tight slot and the film is led into a separate vertical compartment in which it is stretched taut and developed. A similar type of tank of much more simple construction had a horizontal developing compartment.

One popular tank has a spiral spool into which the film is fed from the centre by a special built-in guide. Loading and developing of the film are conducted completely in daylight. The tank needs less than the usual quantity of developer because the spool rotates about a horizontal axis and is only half-immersed in the processing solution. In this type of tank the spool must be kept rotating continuously throughout processing.

Several manufacturers have produced 35 mm. daylight-loading developing tanks. In these the film is placed in the tank in its cassette and the loading end of the film is connected to the centre of the spool. The film is drawn into the spiral and then cut off at the mouth of the cassette. The cassette is removed and processing is completed as in a darkroom-loading type of tank.

The difficulty of persuading a length of film to traverse the spiral groove of the ordinary reel has been overcome in a number of tanks by a self-feeding device. This is produced by allowing one cheek of the reel to turn slightly on the centre spindle. The grooves of the spiral are moulded to give the effect of a ratchet, so that when the end of the film is started in the spiral and the cheeks of the reel are rotated backwards and forwards in opposite directions, the film is automatically drawn into the grooves.

This type of loading is particularly useful for developing colour film which must be removed during processing and fed back on to the wet reel.

Colour Film Tanks. For processing colour film there are elaborate stainless steel tanks. These may consist of five or six individual tanks built into a thermostatically controlled water-bath, which keeps the processing solutions at the recommended temperature. Stainless steel is not recommended for colour-processing solutions like certain bleach baths, for which a lead-lined tank is safer. Glazed porcelain or hard rubber tanks are very much cheaper than stainless steel and are suitable for all colour processing solutions.

Nearly all current spiral reel tanks have transparent perspex cheeks. These are essential for reversal processing of colour films and

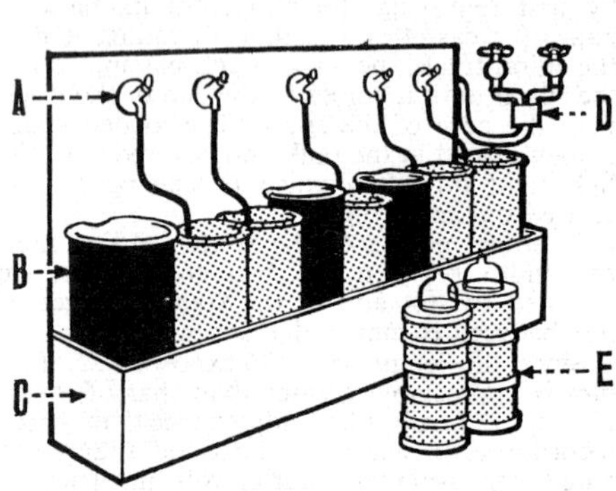

SMALL PROFESSIONAL TANK SET FOR COLOUR DEVELOPMENT. The films are loaded into deep tanks in hangers and transferred from tank to tank. The tanks themselves stand in a water bath kept at a constant temperature. A, water for rinsing tanks (shown shaded) fed from temperature-controlled supply. B, processing tanks (shown solid). C, water bath. D, mixing valve. E, spiral reel sets for loading roll and miniature films.

permit the necessary exposure to white light to be made without removing the film from the reel.

Agitation in commercial colour film tanks usually takes place by nitrogen burst equipment which passes bursts of nitrogen gas from cylinders through the solution at regular intervals. This stirs up the bath to bring fresh solution into contact with the film surface. The timing and length of the bursts is usually controlled by an automatic programming unit.

Aero Film Tank. Aero film is developed in daylight in a tank having two spools with winding handles. The film is wound backwards and forwards, from spool to spool, through the developer. Development proceeds reasonably evenly and, if the film has been thoroughly wetted, adjacent turns of film do not stick to one another when wound on the spools. Some models of this tank were provided with an automatic electrical drive since to wind 200 feet of film backwards and forwards for half an hour is tedious work.

Commercial Tanks. Wholesale photo finishing (D. and P.) establishments use large earthenware or plastic tanks of 12, 24, or 48 gallon capacity, deep enough to hold complete lengths of roll film.

Films are unrolled, weighted at the bottom, and supported at the top on a special double clip, which carries an identifying tag. These top clips are suspended from rods which rest across the top of the tank.

The tanks have two bung-holes at the bottom. One for a tap for draining the tank and the other to provide a lead-in for the immersion heater cable.

These establishments have to handle films in large numbers and they usually employ six tanks.

(1) Developing (10 minutes).
(2) Rinsing.
(3) First fix (10 minutes).
(4) Second fix (10 minutes).
(5) First wash (10 minutes).
(6) Final wash (10 minutes).

Since fixing and washing are the slowest parts of the process, taking twice the time of development, twice the number of tanks is used and the films spend the same time in each. Washing the films is usually carried out by leading water into the tank through the lower bung-hole and allowing it to flow to waste over the top of the tank. All tanks stand on a duck-board over a drain.

The conveyor-band principle is used for even higher production. After films have been loaded, together with their control tickets, on to double clips, they are unrolled, stripped of their backing paper, weighted and hung on a travelling band, which conveys them, in batches of ten, through the developer. They remain there for a predetermined time, and then they pass through a spray rinse and into the fixing bath. When fixed, they pass through the washing tanks, get a rinse in wetting agent to make them dry evenly, then go through a current of warm air in a drying tunnel.

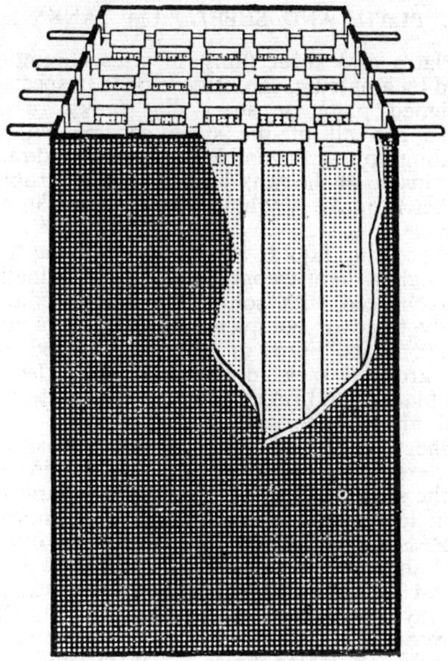

PHOTO-FINISHER'S TANK. Developing tanks for large-scale processing at D. & P. establishments are usually tall containers taking the films suspended by clips from hangers or rods. A tank may have a capacity of anything up to 150 films suspended in rows on their hangers. Separate containers are used for developer, stop bath, fixer, etc., the films being moved from one tank to the next in batches on their hangers.

An alternative conveyor band method passes the films through the various tanks on hangers or racks. In more mechanized establishments the movement of the racks from one tank to the next (and the time they spend in each solution) is controlled by an automatic programming unit. This can be set to different programme sequences for various types of film.

If sufficient tanks are used, a programme can also be set up for colour film processing.

Large-scale processing machines, especially for 35 mm. colour and document copying film, may use a continuous film transport system. There the films are spliced together in a long band and passed through the processing solutions in the individual tanks in big loops. The size of the loop in each tank and the speed with which the film moves continuously determines the immersion time at each stage. The processing stages include the final wash and often even the drying process. After that the film is wound up on large reels, from which it is later unwound, separated in the individual strips, and where necessary cut up into single frames of transparencies and mounted in cardboard mounts.

PLATE AND SHEET FILM TANKS

Plates and sheet films, in sizes commonly used by amateurs, can be developed in specially designed plate or sheet film tanks. These tanks are all of the same general pattern, holding up to a dozen plates in a grooved rack. The inside of the tank lid usually has a rubber gasket to make it solution-tight when clamped down.

Processing liquids are poured into the tank through a light-trapped spout. When loaded and charged with solution, the tank can be safely tilted or completely inverted to agitate the contents. Films or plates are loaded into the grooved rack in the darkroom. Usually the plates stand on edge, but in one pattern of tank they lie horizontally.

Sheet films are held by the edges in special carriers or frames which slip into the grooves in the same way as a plate. After the tank has been loaded in the darkroom, all the normal processing operations can be carried out in daylight.

Metal Tanks. These plate tanks are, nowadays, usually made of stainless steel; formerly they were of nickel. Both these materials are vastly superior to plated brass which is sometimes employed in the cheaper types of tank. The plating soon wears or chips off the surface, leaving the brass exposed. Copper in the brass is readily dissolved by the developer and leads to chemical fog during development. Only a few parts per million of copper in the developer will cause severe fogging of any photographic material during development. Hence brass or other copper-containing materials are never used in photographic processing equipment. Porcelain and glass are probably the best materials, though they are readily broken.

Stainless steel has the advantage of being unbreakable and if correctly treated during the manufacture of the equipment, can give excellent service. It cannot be used in conjunction with certain solutions (e.g., bleaching baths) used in colour processing. All stainless steel equipment must be thoroughly washed after use to prevent chemicals (particularly hypo) from crystallizing out on the surface, otherwise it soon corrodes. Stainless steel also corrodes if it comes into contact with steel wool.

Ebonite Tanks. Hard rubber (ebonite) is often used in the construction of processing equipment for professional use and gives very satisfactory service in temperate climates. But prolonged exposure to heat softens it and may distort its shape. Hard rubber tanks cannot be manufactured with the same precision or finish as the moulded plastic types.

Plastic Tanks. One type of moulded plastic tank holds four plates in closely-spaced grooves formed in the walls of the tank. Tanks of this sort have a remarkably small solution capacity but the spacing between plates is so narrow that efficient agitation is difficult.

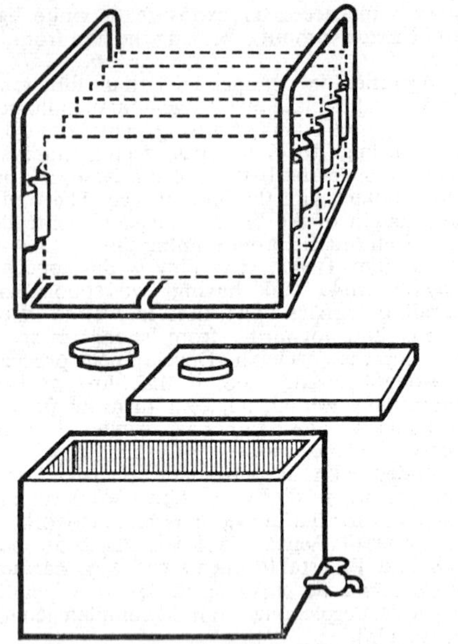

PLATE TANK. The plates are carried in a stainless steel rack which in turn fits inside the tank. The latter has provision for filling and emptying. It closes hermetically, and is turned upside down for agitation. Sheet film may be accommodated in special stiff frames of plate size. The racks also serve for washing, and even drying. Alternative plate tanks may have grooves in the tank body to take the plates.

Film Pack Tanks. Film packs are usually developed in special tanks because the film base on which they are coated is thinner and less rigid than that of sheet film.

One of the earliest film pack tanks made use of a partitioned cage into which the individual films were loaded in the darkroom. The cage was divided into twelve sectors by radial partitions and the individual film pack sheets were folded lengthways, emulsion inwards, to fit into the compartments. The loaded cage was placed in the tank, which had a water-tight lid, and processing took place in the darkroom. Darkness was only necessary when pouring solution into or out of the tank.

A more recent tank for film packs contains a spool with rectangular ends which have curved grooves moulded on their inner faces. Film pack sheets are loaded into the spool by sliding them along the grooves from the open end so that they lie on edge and slightly curved, facing the core of the spool. Up to twelve films can be loaded and developed at one time. The tank is fully light-trapped and a darkroom is needed only for the actual loading.

Commercial Plate Tanks. Tanks for development of plates and sheet film of sizes used in commercial photography are a good deal larger than those commonly used by amateurs.

They are intended entirely for darkroom use and are not adapted for daylight loading or manipulation. They are mostly made of hard rubber, although stainless steel is becoming steadily more popular, in spite of its higher cost.

These tanks are usually employed in sets of two or three; one each for developer, fixer and, possibly water rinse or acid stop-bath. They are provided with close-fitting covers and a floating lid to preserve the developer from oxidation whilst the tank is not in use.

Since these tanks are used to develop films or plates up to 8 × 10 ins. or larger—and several plates may have to be developed simultaneously—they are usually of several gallons capacity.

Special stainless steel hangers support the plates or films in the solutions in the tanks. Films are usually held taut in a frame provided with spring-loaded clips which grip the film firmly at each corner. Plates slide into a frame of channel-section.

For colour processing commercial tank sets incorporate some temperature control system, usually in the form of a thermostatically controlled water bath. Such tank sets for the professional darkroom are designed to handle not only sheet films in hangers or racks, but also roll films and colour papers.

Modern commercial tank sets are used in conjunction with semi-automatic or even automatic programming systems that ensure absolutely standardized processing conditions. The solutions in colour processing tanks are usually agitated by nitrogen burst equipment.

X-ray Film Tanks. Still larger tanks are used for X-ray film. These tanks must be able to take films up to 14 × 17 ins. They are sold in sets of four tanks, one for developer, one rinse tank, one fixing tank, and one washing tank. All four tanks are supported on a strong wooden stand, and the developing tank has a lid to protect the film from the darkroom illumination during development.

For processing X-ray films in quantity (up to 60 per hour in the largest sizes), as in the radiographic department of a hospital, special processing units are available. These have a stout tank, divided into two compartments; the larger holds the solution tanks and the smaller serves as a washing chamber. The processing tanks stand in water which is maintained at a constant temperature.

Three processing tanks are provided, one for developing and two for fixing; between the developing and first fixing tanks there is a spray-rinse compartment where a water spray removes surplus developer from the film and prevents it being carried over into the fixer. Two fixing tanks are provided because fixing takes about twice as long as developing and to prevent congestion in the fixing tank each film spends the same length of time in each tank.

Washing takes place in a strong current of water in the washing compartment, and a small tank in the washing compartment holds wetting agent into which the washed films are briefly dipped to assist even drying. G.B.M.

See also: *Developing negatives; Processing machines.*
Book: *Developing*, by C. I. Jacobson (London).

TANNIC ACID. Tannin. Hardening agent in scratch-proofing solutions.

Characteristics: Brown powder.

Solubility: Very slightly soluble in water at room temperature. It is best dissolved in boiling water, cooling the solution afterwards.

TANNING DEVELOPER. Type of developer which tans or hardens the gelatin in a sensitized emulsion. This tanning action is proportional to the silver forming the image in the emulsion.

Tanning developers are necessary for certain processes in which a relief image is required—e.g., dye transfer. The effect can also be employed to prevent excessive highlight densities in a negative.

See also: *Developers; Dye transfer colour prints.*

TANNIN PROCESS. One of the many methods of preserving wet collodion plates in a "dry" (or rather tacky) condition. Introduced in 1861 by Major C. Russell. The plates could be kept for several weeks after having a tannin solution poured over them, but were only suitable for inanimate subjects since the exposure was six to eight times longer than for wet collodion.

TAPE RECORDER

Apparatus for recording and reproducing sound on magnetic tape. Tape recorders have two main uses in photography: they provide background music or sound effects while cine films or still transparencies are being projected and they provide a synchronized sound track for a cine film.

The recording medium is a magnetic tape ¼ in. wide, consisting of a thin cellulose acetate or plastic base coated with magnetic iron oxide.

The Tape Deck. The tape deck is a flat platform carrying two spool holders, one each side of a slot through which the tape passes from one spool to the other. Magnetic heads mounted in the slot (sound channel) either record a signal on the tape or play it back. On some machines one head does both jobs; on others there are separate heads. In addition there is an erase head which clears any existing magnetic record from the tape before it passes over the

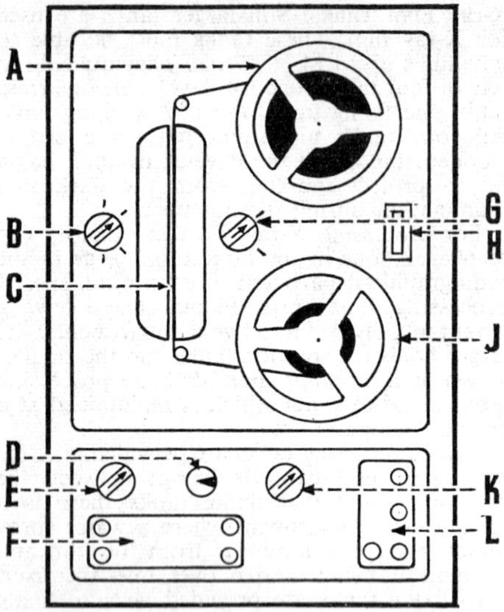

THE BASIC TAPE RECORDER FEATURES. A, *feed spool of tape.* B, *operation control switch (record, playback, fast wind forward and reverse).* C, *sound channel.* D, *recording level indicator.* E, *recording level volume control.* F, *output socket panel.* G, *speed selector.* H, *footage counter (tape position indicator).* J, *take-up spool.* K, *on/off switch of amplifier.* L, *input socket panel. The actual controls on commercial tape recorders may of course be differently shaped and differently positioned.*

recording head. The erase head is put out of action automatically during playback.

As the tape passes through the sound channel it is gripped between two rollers, one of which, the capstan roller, is power-driven at a constant speed. This may be varied to give more than one of the standard tape recording speeds: 15/16, $1\frac{7}{8}$, $3\frac{3}{4}$, $7\frac{1}{2}$ or 15 ins. per second. The take-up spool which stores the tape coming out of the sound channel is driven through a slipping clutch which maintains just enough tension on the tape to wind it neatly on to the spool. While recording and playing back, a light brake prevents the supply spool from over-running and tangling the tape.

The tape transport mechanism also includes provision for winding the tape rapidly in either direction, a digital tape position indicator to facilitate finding any item on the tape, and a "pause" control which allows the tape to be stopped and re-started instantly.

The Amplifying Unit. This serves for either recording or playing back or both.

In recording, a section of the amplifying unit varies the strength of the incoming signal to the correct recording level actually at the head and indicates it on a magic eye or a meter dial. Most recorders have separate inputs for high and low level sources—e.g. one for a microphone which may deliver only a few microvolts and another for a radio or gramophone pickup which may deliver anything up to 200 or 300 millivolts. The playback section of the amplifier boosts the feeble signal from the tape to make it powerful enough to drive the loudspeaker. This section includes circuits for tone control.

In addition to the built-in speaker there are usually outlet sockets which enable the tape to be played back through an external speaker or an external amplifier and speaker.

Recording. To make the record, the tape is drawn at a constant speed over the polished face of the recording head. This face carries a narrow gap where the magnetic signal is concentrated. The width of this gap and its position in relation to the tape determine the number of tracks that can be recorded on the tape.

For the highest possible quality the head is designed to make one record over the whole width of the tape. This system, known as whole track recording, being uneconomical, is used mostly by broadcasting and commercial recording companies. Normally, tape recorders use half-track recording in which a single head records two tracks on the tape. In this method, the gap in the head is slightly less than half the width of the tape and is placed so that it records along the top half of the tape with the tape running from the supply spool on the left to the take-up spool on the right. When all the tape has passed over the head, the full take-up spool and the empty supply spool are transposed. The tape then runs through the sound channel a second time, when a record is made on the blank half of the tape.

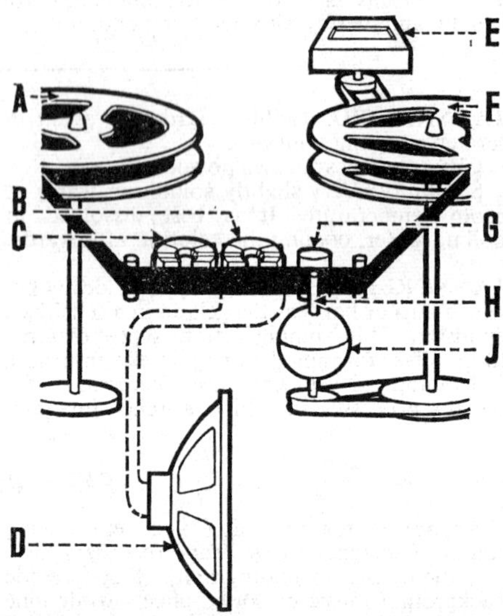

MAIN COMPONENTS OF TAPE DECK. A, *feed spool.* B, *recording and playback head.* C, *erasing head.* D, *monitoring or play-back loudspeaker (earphones may also be used; in either case the connection goes through the amplifier).* E, *footage counter, driven from take-up spool.* F, *take-up spool.* G, *pressure roller.* H, *capstan.* J, *motor driving the capstan and the take-up spool.*

The quarter track method uses two heads, one above the other, each covering rather less than one quarter of the width of the tape. The upper head records along the top quarter of the tape and the lower head along the next-but-one track down, leaving a space equal to one quarter of the width of the tape between the two heads. A track selector switch enables either head to be switched in or out of circuit.

The four tracks are recorded this way: with the upper head switched on, the tape is run through from left to right, making a record on the upper track. The tape is then wound back on to the left hand spool and run through a second time with the lower head switched on, making a second record on the next-but-one track below. The spools are now changed over and the above process is repeated, filling in the two remaining blank tracks.

To play back the record the tape is run over the head at exactly the same speed and in the same direction as for recording but with the erase head switched off and the record/play head connected for playback.

Normally only one track can be played at a time, but some multi-track recorders have facilities for playing back two tracks at once.

Mixing. It is often an advantage to be able to mix sounds from two independent sources—e.g. a continuous musical background and a spoken commentary. Most tape recorders have separate inputs for connecting to signal sources according to type—e.g. microphone, radio and gramophone pickup. These inputs are usually all connected to a common input control so it is not possible to mix them or record two at a time. However some machines have separate controls for two or more inputs and these allow two separate signals to be mixed—i.e. adjusted independently.

A similar result can be achieved on a multi-track tape recorder by recording the signals to be mixed on separate tracks and then playing them back together. It is usually possible with these machines to listen to the first recording on headphones while making the second recording in synchronism with the other head on the associated track. The second recording can be faded in and out and varied in volume as required.

Stereo Tape Recorders. A stereo tape recorder has two sets of heads and two recording amplifiers. The heads are one above the other and record the right and left channel signals simultaneously on the tape. There may be two separate playback amplifiers and speakers, or there may be simply two output sockets for connecting to an external stereo amplifier and its speakers.

On half-track recorders, the single stereo record takes up the whole width of the tape, the left hand stereo channel being recorded on the upper half of the tape and the right hand on the lower half. On quarter-track recorders the heads are arranged as on a 4-track monaural recorder, so that each complete stereo record occupies only two of the quarter tracks.

A 4-track recorder can be used for playing back pre-recorded stereo tapes if it has facilities for playing back one track through its built-in speaker and the other through an external amplifier and speaker (or both tracks through an external stereo reproducer).

Types. From the photographer's viewpoint there are two principal types of tape recorder; mains operated and battery operated. In these two categories prices cover a very wide range. Broadly speaking, mains machines are heavier and bigger than battery machines and have more volume and better quality. Some machines can use either battery or mains.

Most mains operated recorders are designed to run off the normal alternating current supply, the usually incorporate an amplifier capable of delivering anything from 3 to 10 watts and will take tape spools up to 7 ins. diameter. While there is always a built-in speaker, the amplifier is generally powerful enough to give even better volume and quality when connected to a larger external speaker. Mains operated recorders are primarily for recording, dubbing (transferring a record or parts of one from a tape on a second machine) and playing back indoors. As they are powered by current alternating at a fixed frequency, the speed of the motor is constant within extremely narrow limits and does not vary either with time or the length of tape on the spool. They are mostly too heavy (20 to 60 lb.), and bulky to be carried far.

Battery operated recorders are designed for recording in the field so they are usually small, light and easy to carry.

The reproduction quality is never as high as from mains machines but it is good enough for recording sounds to go with projected films and transparencies. The exceptions are the cheapest models with a performance little better than a pocket radio and the expensive machines costing ten times as much which are used by roving reporters working for broadcasting and commercial studios.

On the better types of battery operated tape recorder, the capstan is driven by a governed motor and the tape runs at one of the standard speeds. On the cheaper machines there is no capstan and the motor drives the take-up spool. This means that the tape speed varies with the diameter of the tape wound on the spool, starting slowly and increasing as the tape builds up. Tapes recorded in this way must be played back on the same machine.

Some cine camera manufacturers produce miniature recorders of this type for recording a sound track as the film is shot. The recorder is carried on a shoulder strap and is connected to the camera by a plug-in lead which couples the pause control on the tape recorder to the stop-start button on the camera.

Tape and Transparencies. A tape recorder can provide a background of sound to a transparency projection session without any special equipment. The machine may be simply set up in the normal way to reproduce suitable music while the transparencies are being shown. However it is better for the tape recorder to be placed alongside the transparency projector and connected to an extension speaker either behind or just to one side of the screen. The built-in speaker of the tape recorder is switched off so that all the sound comes from the screen end of the room. This arrangement allows the slide projector operator to control the tape recorder as well and results in a more finished presentation.

Synchronizing devices used in conjunction with the tape recorder can operate the slide changing mechanism of an automatic projector. Here a signal recorded on the tape (e.g. on the second track) actuates a relay when the tape runs past a special playback head (often on a separate tape coupler) to change the slide in the projector. The first track carries the recorded sound commentary.

Tapes and Cine Films. A tape recorder can be used for playing background music and sound effects to accompany a cine film. If the sound and action are to be synchronized, special coupling systems are necessary to keep the speed of the tape recorder in step with that of the cine projector.

Most such systems control the speed of the projector motor from the tape and do so mechanically or electrically.

The sound may be recorded as the film is being shot, or it can be recorded afterwards while the finished film is being projected. When the tape and film are to be run through together, the tape recorder is started, run through to the cue sound and held stationary on the pause button.

The film is then started; as soon as the marked cue frame on the leader flashes on the screen, the tape is released and from then on the sound and picture run in synchronism. Even when the sound track is recorded on a magnetic stripe on the actual film it is generally preferable to record and edit the sound on tape separately and then dub it from the tape on to the stripe with the aid of a synchronizing attachment. F.P.

See also: *Magnetic Tape; Projection practice; Projector (Still); Projector (Cine); Sound recording; Sound synchronization.*

Books: *Tape Recording*, by A. A. McWilliams; *The All-in-One Tape Recorder Book*, by J. M. Lloyd.

TAUPENOT, J. M., 1824–56. French professor of chemistry and physics. Invented in 1855 the Taupenot Dry Plates. These were glass plates coated first with silver iodide collodion and then with albumen; they were prepared in advance and required only supplementary sensitization in a silver bath before use. They produced good images and were convenient for outdoor work, but required longer exposures than wet collodion plates.

TAYLOR, HAROLD DENNIS, 1862–1943. English lens designer and manufacturer of optical instruments. In 1893 designed and patented the Cooke triplet lens which, owing to its simple construction and outstanding performance, became very popular. The Cooke design provided a foundation on which many subsequent advances in photographic lens design have been based. In 1896 suggested the use of thin coatings on lens components to diminish reflections. Patented in 1885 an exposure meter, the first of many patents for which he was responsible. Also wrote on optics; his book *A System of Applied Optics* (1906) was the first important English book on applied optics to have been published for nearly 100 years. Another of his books, *The Adjustment and Testing of Telescope Objectives*, is still a current reference work. H. D. Taylor was not related to the firm of Taylor, Taylor and Hobson Ltd., although they manufactured his triplet lens. Received the Progress Medal of the Royal Photographic Society in 1935.

TAYLOR, JOHN TRAILL, 1827–98. English writer on photography. Editor (from 1864 to 1898) of the *British Journal of Photography* and the *British Journal Photographic Almanac*. The Royal Photographic Society instituted in 1898 the Traill Taylor Memorial Lectures in his memory.

TECHNICAL CAMERA

The modern technical camera is a photographic instrument designed for maximum control of image scale and perspective. It therefore provides a high degree of versatility by camera movements, adjustable extension, interchangeable lenses and—in some cases—different negative formats and negative materials.

The earliest technical cameras closely resemble the large size studio camera from which they were derived. A movable plate holder and lens panel provide independent front and back adjustments, comprising movement along and at right angles to the optical axis, as well as swings and tilts about a vertical and a horizontal axes. Extra long bellows produce double and triple extension systems. The lens standard moves along a flat bed or baseboard which pulls up to close the camera. Cameras of this type are also known as field, stand or view

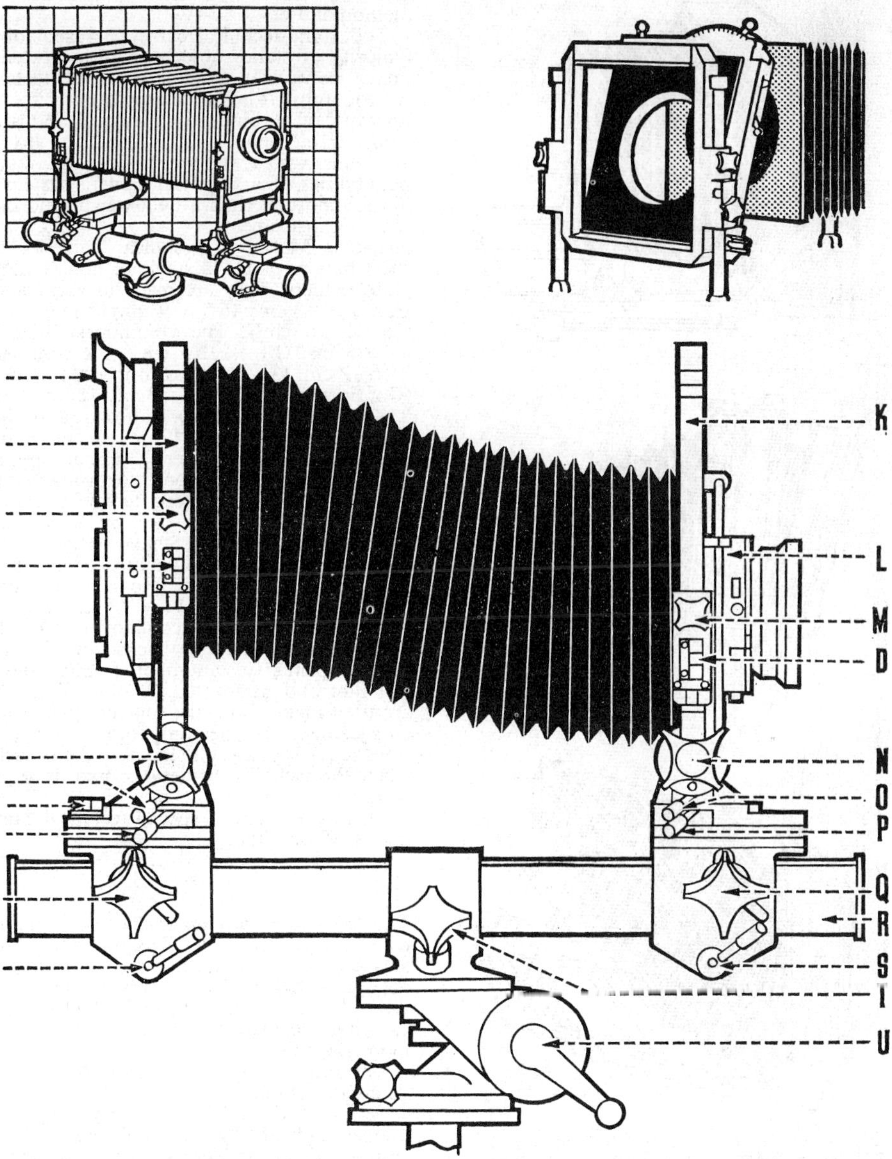

MONORAIL TECHNICAL CAMERA. Derived from the older view camera type, the present-day technical camera is built on the lines of an optical bench based on the monorail. All units are individually adjustable in every direction to give the maximum operational flexibility. Lenses and shutters are interchangeable and sheet films or plates are invariably used. Negative sizes range from $2\frac{1}{2}$ to $3\frac{1}{2}$ ins. to 8 × 10 ins. Top left: *General view.* Top right: *Fitting of behind-the-lens shutter between the front standard and the bellows.* Bottom: *Side view.*

A, ground glass screen holder and fitting for film or plate holder. B, rear standard. C, knob to control vertical rise of rear unit. D, spirit level. E, knob and scale to control vertical swing (about horizontal axis) of rear standard. F, lock for cross movement of rear. G, lock for rear swing about vertical axis. H, rear focusing knob. J, rear focusing lock. K, front standard. L, interchangeable shutter and lens unit. M, knob to control rising front movement. N, vertical swing (about horizontal axis) of front standard. O, lock of cross-front movement. P, lock of front swing about vertical axis. Q, front focusing knob. R, monorail. S, front focusing lock. T, centre support. U, pan-and-tilt tripod head.

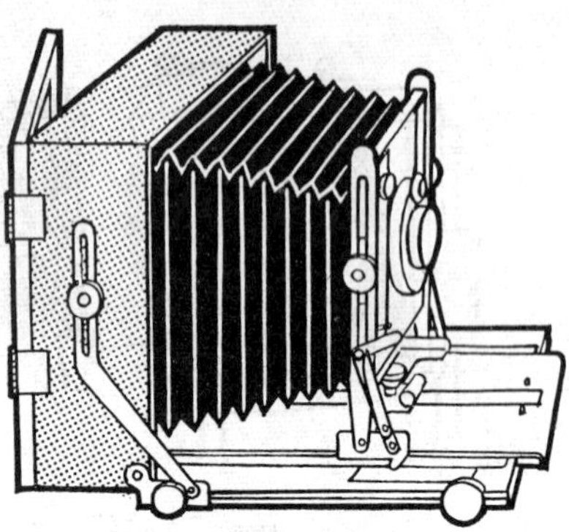

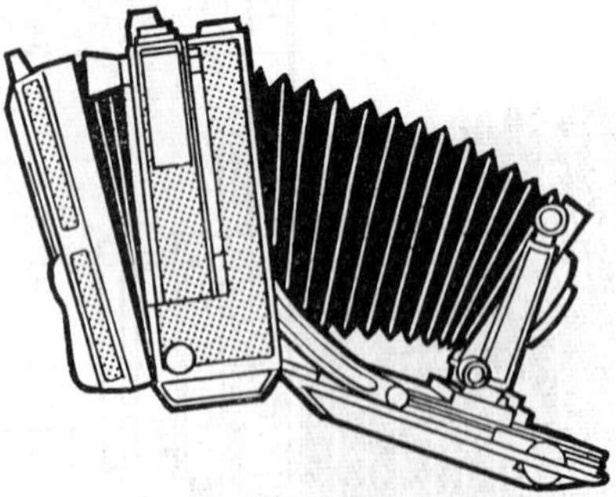

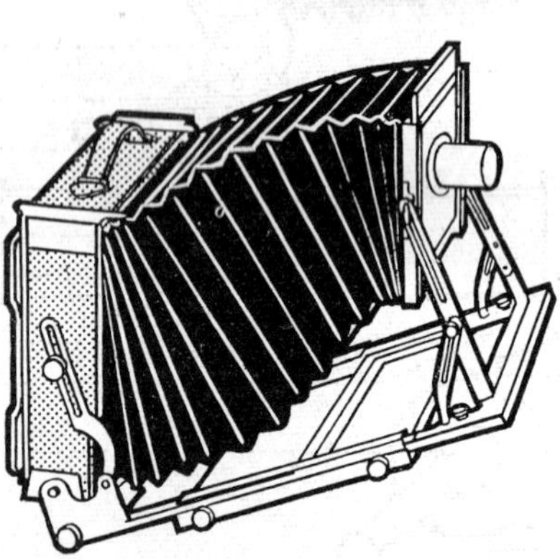

ALTERNATIVE TECHNICAL CAMERA DESIGNS: Top: *Old type field camera. The hinged-on baseboard permits a swing of the wooden camera back about a horizontal axis only. The lens panel has rising front and tilt movements.* Centre: *Modern technical hand camera. This is usually of all-metal construction. The hinged-on baseboard in conjunction with a ball-jointed extension of the film holder permits a limited back swing in all directions. The lens panel has a limited swing-and-tilt movement as well as rising and cross front.* Bottom: *Flat bed camera. The back, baseboard and lens panel are of wood. The baseboard provides up to triple extension. Both the back and front carry swing-and-tilt movements; the front also has a reasonable degree of vertical rise. This camera type is nearly as versatile as a monorail design.*

cameras, reflecting their applications and method of use.

Today the technical camera is used for the majority of static subjects requiring extensive image control in commercial and technical photography. It has wide applications in close-up work of all kinds, architectural, industrial, medical, etc. photography—in fact virtually all fields where the versatility and control possibilities of a technical camera system are more important than a light-weight and quick working instrument.

Negative Material. Technical cameras take sheet film or plates in single or double film or plate holders. They are made in various sizes from $2\frac{1}{2} \times 3\frac{1}{2}$ ins. (6.5×9 cm.) to 11×14 ins. (28×36 cm.). The commonest sizes are 4×5 ins. (10×12.5 cm.), the continental 9×12 cm. ($3\frac{1}{2} \times 4\frac{3}{4}$ ins.) format, 5×7 ins. (Continental equivalent 13×18 cm.), and 8×10 ins. (20×25 cm.). Half-plate and whole-plate ($4\frac{3}{4} \times 6\frac{1}{2}$ ins. or 12×16 cm. and $6\frac{1}{2} \times 8\frac{1}{2}$ ins. or 16×21 cm.) formats are also still in use especially among older camera types.

A number of advanced technical cameras take several negative sizes by using interchangeable rear units. Smaller negative sizes can be used with virtually all technical cameras by inserting reducing adaptors in the camera back.

Special materials, accommodated by suitable adaptors, include backs for diffusion-transfer prints (instant photography) and for roll film and perforated 70 mm. film. These are used mainly with the smaller sizes of technical camera, and then only for special applications requiring large film loads, or for occasions when the technical camera is used as a press camera.

There exists also a special technical camera taking 35 mm. film.

TYPES

The Monorail Camera. The contemporary technical camera design has no camera body as such, but is constructed on the principle of an optical bench. The basis is a single (sometimes double) rail unit with independent frames carrying the plate holder and the lens panel along the rails.

The rail itself may be square, triangular or round in section.

As in an optical bench, every part can be aligned individually. Apart from the camera movements themselves, the carriers run to and fro along the rail, usually by separate rack and pinion movements. On better camera types these can also be locked. This provides front and rear focusing and permits very precise control of sharpness as well as image scale.

The whole rail unit is usually mounted on a tripod or stand by means of a tilting head, providing a full range of inclinations and tilts in all directions.

Flat-bed cameras. A number of technical cameras still are based on the older design of a baseboard hinged to the camera back. The baseboard contains sets of extending runners which in turn carry the lens standard. Movement along one set of runners provides a normal focusing range; the extension of the runners gives extra long lens-to-film distances for close-ups or with long-focus lenses. Usually the runners extend to a total of twice or three times the lens-to-film distance with the normal lens set to infinity (double and triple extension cameras), permitting magnifications up to twice natural size. Macrophotographs on still larger scales of reproduction are possible with short-focus lenses.

The movement of the bottom unit of runners is generally linked to a rack and pinion drive or a pivoted lever movement for fine focusing. The additional extension is adjustable by hand. This controls the overall extension and coarse focusing.

Normally the baseboard, when open, is fixed at right angles to the camera back, being held rigid by means of struts. The baseboard can also drop down at a greater angle so as to be out of the way when using wide angle lenses; otherwise the front end of the baseboard might intrude in the field of view of the lens.

Flat-bed technical cameras usually have a more limited range of camera movements, especially for the rear frame, and back focusing is rarer. Sometimes the back is a shallow box hinged to the baseboard, and accommodates the bellows and lens standard when the camera is closed. Cameras of this design are sometimes also used as press cameras in the smaller sizes (up to 4 × 5 ins.) and are then fitted with a rangefinder coupled to the focusing movement, as well as an independent finder (since in press photography there is rarely time to focus an image on the ground glass screen at the back).

Technical Camera Systems. A number of modern technical cameras are designed on a construction set principle, with interchangeable components which can be put together in a variety of ways. The components are the rail, the front standard for the lens, the bellows, the

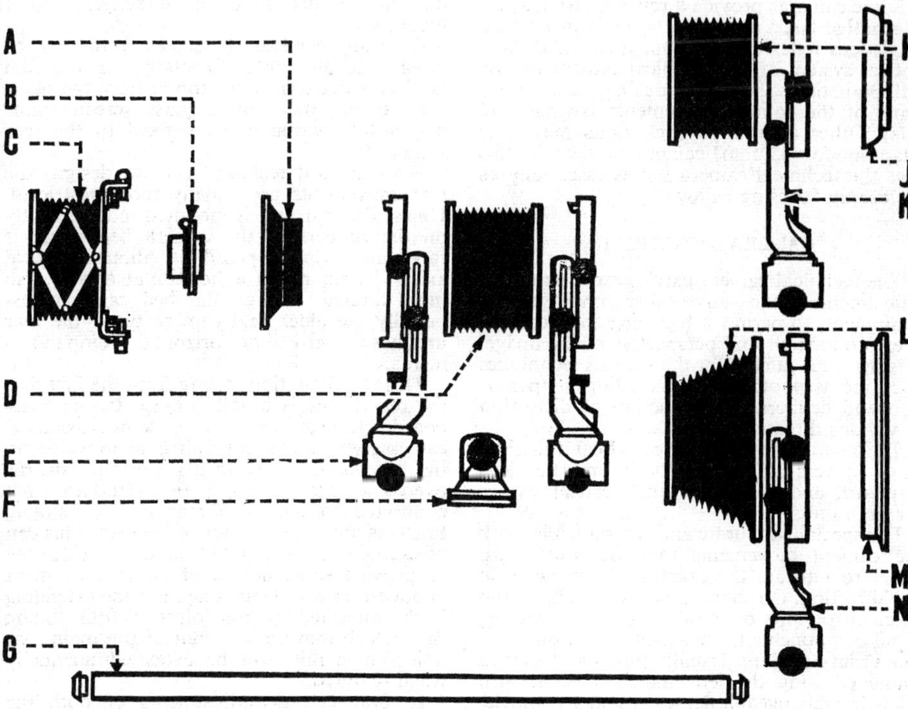

UNIVERSAL MONORAIL CAMERA SYSTEM. On some modern monorail cameras all components are interchangeable, providing amongst others a series of different camera backs for various film or plate formats. A, lens panel. B, interchangeable lens (usually with shutter). C, bellows-type lens hood. D, first bellows. E, front frame and carrier. F, central rail support. G, rail (may be extendable). H, second bellows for smaller format. J, ground glass screen frame and film holder for smaller size. K, rear frame and support for smaller size. L, second bellows for larger format. M, ground glass screen frame for larger format. N, rear frame and support for larger format.

rear standard for the ground glass screen and film holders, etc. The rail will carry not only the front and rear units, but can if necessary take a centre frame as well, to provide support for extra long bellows extension with additional bellows units. The latter clip into the frames and are available in different forms and lengths. Frequently base rails of different length are also provided (or the standard rail can be extended by additional sections). The focusing range is thus very great, permitting close-ups down to a few inches and macro-photographs with a continuously adjustable magnification up to several times natural size. Alternatively, such an arrangement can accommodate a wide variety of focal lengths of lens from extreme wide-angle to very long telephoto. Special bellows are often provided for wide-angle photography to enable the lens and screen frames to be brought really close together.

Usually the rear unit is also interchangeable, providing ground glass screen frames of different sizes for various negative formats. One outfit of this kind might thus be built up into a 4 × 5 ins., 5 × 7 ins. or 8 × 10 ins. camera.

Some cameras provide a reflex housing as an alternative back unit. The latter is in effect a single-lens reflex camera, joined to the normal optical system of the technical camera on the rail. Sometimes the reflex housing incorporates some of the camera movements (swings and tilts). Other alternative back units may take adaptors for 35 mm. camera bodies; in this case the technical camera serves essentially as a close-up focusing bellows.

CAMERA MOVEMENTS

The technical camera carries more extensive adjustments than any other photographic equipment, to permit a high degree of control over proportions and perspective of the image. It is the ideal camera for the various techniques of corrective photography, including compensation and deliberate introduction of distortion in various directions.

The main adjustments are vertical and sideways movements of the back and the lens standard, and horizontal and vertical swings for both units.

For specially accurate and reproducible work the camera movements may be fitted with scales to indicate the exact angle of swing in any direction. On some cameras a positive stop or catch permits quick aligning and centering of all components in their basic position.

The Front Standard. Usually this is a U-shaped frame carrying the lens panel. This in turn takes interchangeable lenses (usually complete with shutter) on lens boards. The latter clip into position on the lens panel.

The lens panel moves up and down in its guides (rising front) and this movement may be controlled by a rack and pinion gear. Sometimes the panel will drop down below its normal (centre) position. In less expensive cameras the rising front movement is combined with a tilt about the horizontal axis, enabling the lens to point upwards or downwards as well. On better cameras the tilt is separate from the rising front movement, and may be either a centre tilt (about a horizontal axis running through the centre of the lens) or a base tilt (about the horizontal axis at the bottom of the U shaped carrier). The two types have various advantages; for example the centre tilt is quicker to adjust since it requires no refocusing of the image, while the base tilt permits a greater movement range.

The whole U-carrier frequently also slides sideways on its mounting (cross-front movement) and swings about a vertical axis.

The Rear Frame. On the modern monorail camera the rear unit usually has identical adjustments to the front one. This unit takes the ground glass screen frame and the film or plate holders. On older cameras the ground glass screen has to be removed before the film or plate holder can be inserted; on modern designs the screen frame is spring-loaded and the film holder slides in underneath after focusing.

On many cameras the screen frame can be fitted vertically or horizontally (or the film holders inserted from the top or from the side); alternatively the ground glass screen frame may rotate within a panel fixed to the rear standard.

With flat-bed technical camera designs the rear movements are usually more restricted. Unless the rear unit is mounted independently on the runners of the camera bed, there is rarely any swinging possibility about a vertical axis. The tilt about a horizontal axis is also more limited, and on flat bed cameras—especially the older field camera type—the rear unit has no vertical or horizontal sliding movements.

This is a limitation arising from the fact that the rear frame is usually hinged to the baseboard (though there are some technical cameras where the rear unit is as movable on the baseboard as the front). Certain cameras overcome this by using an extending back connected to the main rear unit by sliding brackets and a second set of bellows. This can be extended at the top, bottom or one side only to provide some degree of swing movement about either axis. One design has the extending back connected by ball joints to four sliding rails which move in and out of the main rear unit. These rails can be extended unequally when required.

In practical use adjustments of both the front and rear units are frequently required simultaneously. That is where the monorail type of technical camera scores over the older designs. Thus by tilting the rail upwards and tilting both the rear and the front units forward parallel to each other, a degree of front rise results which is far in excess of that provided by the direct rising front movement. The same applies to the swings about the vertical exis (which can be used as a substitute for a cross-front movement), while opposite tilts of the front and rear standard provide a greater degree of depth of field control and simultaneous focus and perspective control.

EQUIPMENT

Lenses. The choice of lenses for a technical camera is governed by much the same considerations as with any other interchangeable lens camera. Thus wide-angle lenses are required to cover large subject areas at short range, and long-focus lenses provide large-scale images at greater object distances. (This applies to distant as well as close-up views.)

Some early view cameras had convertible lenses consisting of two main components which could be used either together or separately, providing three different focal lengths. Modern demands on image quality are better met by having completely interchangeable separate lens systems.

Lenses for use in technical cameras require a greater covering power to permit utilization of the various camera movements. While the normal lens on a rigid camera rarely has to cover an image angle beyond 55°, a technical camera may require a "standard" lens covering up to 70°. Otherwise part of the image becomes vignetted or cut off as soon as the rising and cross front movements or the swing or tilt of the lens panel are brought into action. Similarly, wide-angle lenses need a useful image angle of 90° or over to retain the possibility of using the camera adjustments.

One way of obtaining extra versatility is to use as the standard lens a wide-angle lens system intended for a larger format (for example a wide-angle lens for a 5 × 7 ins. camera as the standard lens—the focal length should be the same—for a 4 × 5 ins. format).

Shutters. The subjects taken with a technical camera rarely call for fast shutter speeds, and comparatively long exposure times used to be quite common. Accordingly many earlier technical cameras have a simple shutter or none at all (some photographers still make their exposures by removing and replacing the lens cap).

The shutter on modern technical cameras may either be built into the lens or separate. In the former case every lens has its own shutter and is changed together with it on the lens panel.

Separate shutters may fit in front of the lens or may attach to the lens standard behind the lens board. One shutter, provided it has a large enough clear opening, can in this case serve a number of lenses.

Technical camera shutters usually have a separate opening and closing control, in addition to the normal shutter speed settings, to permit focusing on the ground glass screen. A few shutter designs incorporate an automatic diaphragm control that also closes down the lens aperture (from full opening for focusing and viewing) to a pre-set stop just before the exposure.

Tripods. The camera is always used on a tripod or other support. Sometimes the tripod uses the baseboard of a flat bed technical camera as a head on which the legs are assembled. Since technical cameras are comparatively heavy, tripods are nearly always bulky; they also have to be really rigid. An adjustable tripod head to permit tilting or inclining the camera in any direction is nearly always part of the camera support.

As the picture is arranged and focused on the ground glass screen, operation is slow; but the subjects photographed with a technical camera are nearly always static. L.A.M.

See also: *Camera movements.*

TECHNICOLOR. Colour process in which positive prints in colour of cine films are made by dye transfer or imbibition.

The basic principle of the process lies in the use of separate matrices which superimpose yellow, cyan, and magenta dye images in register on the final support.

The matrices may be produced in a number of ways. For many years there was a special Technicolor three-strip camera which exposed three separate films at once to provide three colour separation negatives. Coloured cartoons were photographed on a single strip of film, the three separations for each frame being initially photographed on to three frames, each through the appropriate tri-colour filter. Both these processes of photography are obsolete and it is now the general practice for photography to be undertaken on integral tri-pack materials which can be used directly, or indirectly, to produce the separate matrices.

The outstanding feature of the system is its flexibility; several negative-making methods may be used in shooting one picture and all are subsequently printed in Technicolor.

A great advantage of Technicolor is that large numbers of identical prints can be made at a relatively low cost because the raw stock used in the final prints is very similar to ordinary black-and-white stock, whereas most processes call for highly specialized materials.

See also: *Dye transfer colour prints.*

TELECINE. Systems and equipment for transmitting motion picture films by television. Telecine transmission involves projecting the film with a projector into a television camera and requires a number of provisions to adapt the characteristic manner and the time in which the images change in a cine projector to the properties of a television transmission system.

See also: *Television recording.*

TELECINEMA. Live or recorded television transmissions projected on a large screen in a motion picture theatre. The television programme itself is produced in the normal way (it may be recorded on tape) and transmitted to a special television receiver in the cinema. This receiver is used for projecting the image on the usual large cinema screen.

To obtain enough light for projection, ultra-intense cathode ray tubes operating at up to 50,000 volts are used. The image displayed on a tube can then be projected by a conventional projection lens system.

An alternative method is to use a secondary projection system which does not derive its light from the power of the television tube. For instance the electron beam can scan an automatically and perpetually renewing oil film while a powerful light beam is focused on this film. The electric charge of the electron beam causes distortions in the oil film which the illuminating beam can project on a screen in the form of a visible image. Such secondary telecinema systems have the advantage of permitting projected images of much greater brightness.

TELEPHOTO ATTACHMENT. Afocal lens system to fit in front of the normal camera lens, converting the latter into a telephoto lens. The optical system is usually similar to that of a terrestrial telescope, often in the form of a prism monocular.

See also: *Afocal lens; Supplementary lens; Telephotography; Telephoto lens.*

TELEPHOTOGRAPHY. Long-distance photography depends on having the right apparatus and on knowing how to exploit it.

We can employ a long focus lens to produce a larger image on the negative than a normal lens would, or we can enlarge a small portion of a negative when printing, or we can combine both methods. Enlargement from a small portion of a negative may be accompanied by an unacceptably large grain size in the finished print. Also, sharpness decreases the greater the enlargement. For these reasons a good long focus lens on the camera is the better method. The effect of perspective must also be considered because it will appear correct only when viewed from the right distance and this is not always possible.

Rigidity. When using long focus or telephoto lenses on any camera, absolute rigidity is essential. When highly magnified images are recorded, the slightest camera movement will also be magnified and blur the image.

A reflex camera is particularly suitable for use with long focus lenses, firstly because most reflex cameras have provision for considerable camera extension, and secondly because the magnified image can be examined on the top focusing screen up to the actual moment of making the exposure. However, heavy lenses used with a long camera extension upset the point of balance when the camera is attached to a tripod. For this reason it is often helpful to mount the camera on a long baseboard which offers a direct support for the lens and can be mounted on the tripod at the point of balance. Some very long focus lenses have a tripod bush on the mount, and the camera (as the smallest part of the assembly) is attached at the end of the lens barrel.

With miniature cameras, telephoto lenses are often set in very light alloy mounts to reduce the tendency to overbalance.

Prism monoculars (one half of a prism binocular) can also be mounted in front of the camera lens for telephotography. They can increase the focal length of the camera lens 8 or more times, depending on the magnifying power of the monocular; certain models are marketed specially for this purpose with provision for rigid mounting.

The whole question of rigidity becomes very acute when long-distance photographs are taken from high viewpoints and exposed positions subject to strong winds.

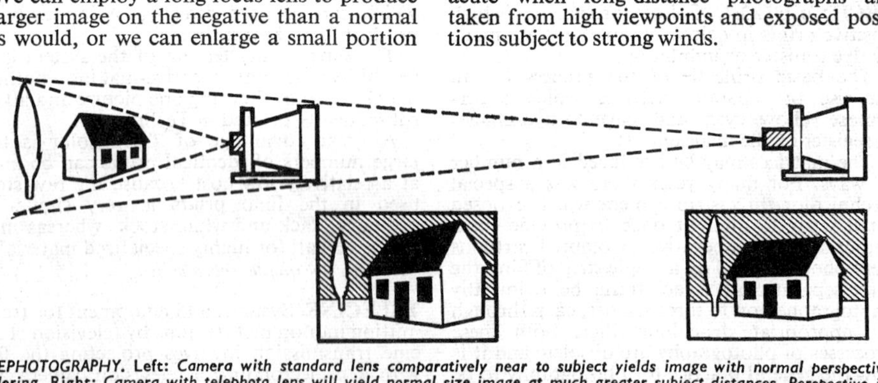

TELEPHOTOGRAPHY. **Left:** *Camera with standard lens comparatively near to subject yields image with normal perspective rendering.* **Right:** *Camera with telephoto lens will yield normal size image at much greater subject distances. Perspective is, however, flattened, and the subject appears to have little depth in the picture.*

Lens Hood. When a very long focus lens is adapted to a small camera, only a comparatively small centre portion of the image is recorded on the negative. The lens hood will have to be exceptionally long and narrow-angled to be really effective.

In expensive miniature outfits, stray light is not so serious because the lenses are so designed as to give no internal reflections.

With such lenses good negatives can usually be obtained with no lens hood at all.

Filters. Light coming from distant objects is apt to be "scattered" on the way. The maximum scatter is at the blue end of the spectrum and, in particular, in the ultra-violet. The human eye is especially sensitive to green and it is not nearly as sensitive to ultra-violet light as a photographic film. The scattered blue and ultra-violet light, unless filtered out, records as a haze on the negative. This is why distant objects seen clearly by the eye so often appear unsatisfactory when recorded in a photograph. To add to this effect, the intense brilliance of many distant views in comparison with the dullness of the immediate foreground may lead to over-exposure.

With panchromatic film, a pale pink or yellow filter designed to eliminate the ultra-violet band of the spectrum may be helpful without affecting the general colour balance of the subject. A green filter—i.e., a so-called full correction filter—will do the same thing to a greater degree and also improve the general colour balance. Much greater distant detail can be recorded by the use of a red filter, but only at the expense of the natural colour rendering. The maximum haze-penetrating effect (and greatest falsification of tones) is obtained on infra-red film used in conjunction with infra-red filters. By this method it is often possible to record distant detail not detectable by the human eye.

At the same time, light scatter in the shorter wave bands is produced by the presence of haze in the atmosphere. This can be penetrated by infra-red rays. But where dirt or carbon particles (smoke and dirty fog) are present it is impossible to obtain a clear result even by the use of infra-red technique. Therefore long-distance views in manufacturing areas are often disappointing. As a rule the best effects are obtained after the air has been washed by heavy rain, when clear visibility follows for a short period.

Perspective. Because of the abnormal focal length used in a telephoto picture, the natural perspective appears falsified if the picture is viewed from the usual distance of 10 ins. The nearest object in the picture may be a building about a quarter of a mile away, whilst the most distant object is some ten or fifteen miles beyond. In the resulting picture the two objects will appear comparatively close. This effect is further emphasized where deep filters or infra-red technique has removed even the natural haziness of the distance. Falsified tone values are then added to the flattened perspective to exaggerate the unfamiliar effect.

The great variation in the techniques, available when photographing long-distance views makes it possible to exercise considerable personal taste. Unless the photographer is deliberately sacrificing everything else to record the maximum detail, the various alternatives at his disposal can be adapted to produce creative effects far beyond the scope of mere topographical records. Many distant views are enhanced by the pictorial subject matter either in the foreground or middle distance. To narrow the view by means of a long focus lens may be to eliminate much of the charm. But if some nearer objects are included, the problem is usually how to compress the tone scale to obtain sufficient exposure in the foreground whilst not over-exposing the distance. The solution is usually to be found in using suitable colour filters and taking care not to over-expose the negative. B.A.

Book: *Photographic Optics*, by A. Cox (London).

TELEPHOTO LENS. Type of lens introduced in 1891, giving a bigger image than a normal focus lens without moving the camera closer to the subject and without using more than the usual normal camera extension.

This is especially useful with miniature and cine cameras taking interchangeable lenses, where long focus constructions are apt to be disproportionately heavy and bulky compared with the size of the camera. A telephoto lens under those circumstances yields images of equal magnification for only a fraction of the size of the unit.

The optical construction of a telephoto lens comprises two groups of lenses. The front group acts as a converging lens system, while the rear group—often separated by an appreciable distance from the front one—acts as a diverging lens unit.

Each group may be quite complex, comprising up to three elements per group to correct aberrations.

The front unit converges the rays from the subject, while the rear unit partly cancels this convergence. As a result the cone of light reaching the film appears to converge from a point some distance in front of the front unit. In other words the rear nodal point of the complete lens system is not inside the lens at all, but in front of it.

The distance from the rear glass of the lens to the film or plate is the back focus and is important when mounting the lens in the camera. In this instance the back focus is much shorter than the equivalent focal length (i.e., the distance from the rear nodal point to the focal plane)—a characteristic feature of all telephoto lenses.

Power of Telephoto Lens. The equivalent focal length divided by the real distance of the back

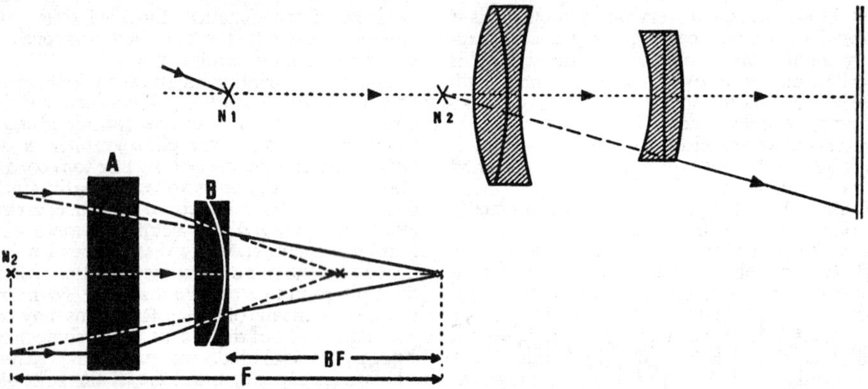

TELEPHOTO LENS. Top: *The lens consists of a front converging, and a rear diverging element. Both nodal points are in effect in front of the lens system.* Bottom: *The back focus of a telephoto lens is shorter than its focal length.* A, *front element.* B, *rear element.* N_1 *and* N_2, *nodal points.* F, *equivalent focal length of whole system.* BF, *back focus.*

of the lens from the film (the "back focus") gives the power of the telephoto combination. If the equivalent focal length is say, 12 ins. and the back of the lens is 4 ins. away from the film when focused on an object at infinity, the power of the lens is 12/4 = 3, usually written 3×. The power is approximately equal to the magnification produced by the lens.

Early telephoto lenses were made to give magnifications up to 10×, and the power of some types could be varied by adjusting the distance between the back and front components. A few lens makers supplied an accessory that turned the normal camera lens into a telephoto. This was simply a diverging lens group mounted at the back of a tube. The tube was fitted in place of the normal lens, and the normal lens was then mounted at the front of the tube, giving a conventional telephoto combination.

Modern telephoto lenses for still cameras have a fixed magnifying power, generally of around 4× to 8× and are made in apertures up to *f*3·5 and greater.

Telephoto lenses of greater magnification and aperture than this are made for motion-picture photography and press work. Variable focus lenses are used to-day on movie cameras to give distance-to-close-up transition effects without moving the camera.

Increased focal lengths of telephoto lenses are obtained with mirror lens designs. Here the saving in overall length as compared with the nominal focal length is greater still, for the light path is folded back on itself two or three times inside the lens system. Mirror telephoto lenses are mostly in focal lengths of 500 and 1,000 mm. but extreme tele units exist with a focus of several metres. They are used with miniature or larger reflex cameras.

Inverted Telephoto Lenses. The telephoto principle is used in reverse to give longer instead of shorter back focus distances. This extra distance between lens and film is necessary in cameras like those used for Technicolor work where prisms and other mechanism have to be accommodated between the lens and the film. A diverging component in front of the normal converging camera lens in this case gives a short-focus combination with a long back focus.

Variable Focus Types. Variable focus telephoto lenses have some means of altering the distance between the front converging and rear diverging elements.

Lenses of this type can be made to give magnification from 2× to 10× and over, but they are not easy to correct for aberrations over the whole range.

Increasing the focus of the lens by moving the back component automatically lowers the relative aperture: if the magnification is doubled, the light falling on the film has only a quarter of its original intensity and the relative aperture of the lens falls two stop values.

This type of lens was popular in the days when all cameras had to be focused on a focusing screen, and they are still sought by owners of old type single lens reflex cameras. But they are generally unsuitable for use with scale focusing cameras. This and the need for correcting for change of *f*-number at every change of magnification, has resulted in their becoming almost obsolete.

Film studios use a special type of variable focus lens to give a quick transition from a distant to a close-up shot. In this lens the change of relative aperture and the effect of aberrations is automatically compensated for as the handle is turned to alter the focus. F.P.

See also: *Lens history; Mirror Lens; Telephotography; Wide angle lens; Zoom lens.*
Books: *Optics* and *A System of Optical Design*, both by A. Cox (London).

TELEVISION

Continuous transmission of a visible picture by means of electrical signals. The picture is, converted into the signal in a special camera, and reconstituted as our image in a receiver. The link between the two may be a cable connection (closed circuit television) or wireless waves (broadcast television).

Scanning. For sound radio communication it is necessary to transmit an electrical signal corresponding to only one variable, the instantaneous amplitude of the sound wave impinging upon the microphone. In television the equivalent variable is scene brightness, but information is required simultaneously from all parts of the area of the scene. It would theoretically be possible to use a large number of parallel communication channels each one carrying information on the brightness of one picture point, but in practice a single channel must be used. An image of the scene is therefore scanned in the television camera in exactly the same way as a reader's eyes scan the words and lines of a page of print, and information on the brightness of successive picture points is transmitted in time sequence.

The picture is reconstituted at the receiver by a scanning process syncronized with that occurring in the camera, and persistence of vision is relied upon to give the effect of simultaneous presentation of all picture points.

In European systems the picture is scanned with either 405 or 625 lines, the entire picture scanning being repeated 25 times a second. The corresponding figures for American and some other systems are 525 lines and 30 times a second.

In practice interlaced scanning is used; each frame is scanned twice, first covering all odd-numbered lines, and then all even-numbered lines.

Television Camera. The camera used for taking the television picture is similar in principle to an ordinary photographic camera—i.e., it consists of a light-tight box with a lens in a focusing mount at one end, and a light sensitive surface at the other. The light reflected from the scene to be televised is collected by the lens and focused on to the light sensitive surface. Here the similarity between the two types of camera ends.

In the photographic camera, the optical image is transformed into a negative consisting after it has been processed of metallic silver. The density of the silver at any point is roughly proportional to the brilliance of the optical image; the brighter the image, the denser the deposit.

In the television camera, the optical image is transformed into an electrical signal. It is the electrical potential of the signal in this case that is proportional to the brilliance of the optical image.

A lens system, often a turret or zoom, forms an optical image on an electro-photo-sensitive surface inside the evacuated glass envelope of the camera tube. Two types of tubes are in common use, the image orthicon and the vidicon.

In the image orthicon, the photo-sensitive surface (the photo-cathode) emits electrons proportionally to scene brightness and an electron image is formed. This electron image is focussed by a magnetic lens on to a second surface where the impinging electrons form an electric potential image, i.e. the voltage to which a given point of the surface is charged is proportional to the scene brightness.

In the vidicon, the optical image is formed on a photo-conductive layer, the electrical conductivity of which increases with incident illumination. The increased conductivity of those parts of the layer which are brightly illuminated allows an electric charge to be conducted through to the back of the layer, where an electric potential image is thus formed.

In both types of tube, the electric potential image is scanned by an electron beam from an electron gun which discharges the potentials. The discharge currents form the output signal from the tube.

The Electronic Chain. In a television signal, the amplitude (voltage) corresponds to brightness, and non-linear electrical circuits are used to

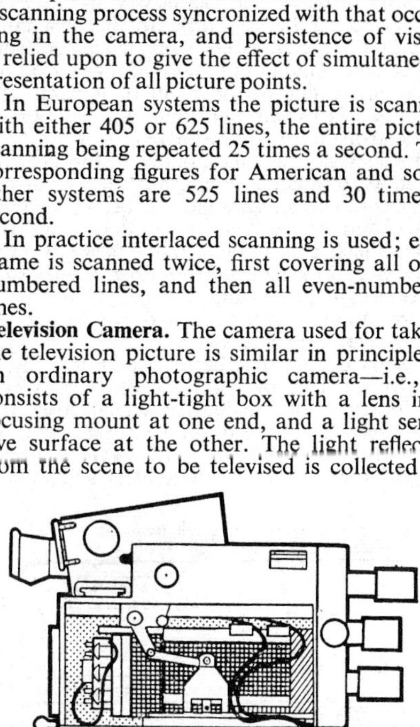

TELEVISION CAMERA. Any one of the three lenses in the lens turret can be brought to focus the image on the photo-sensitive cathode of the image tube. Alternatively a zoom lens is used. The viewfinder is mounted on top. The whole unit is on a sturdy tripod or camera crane.

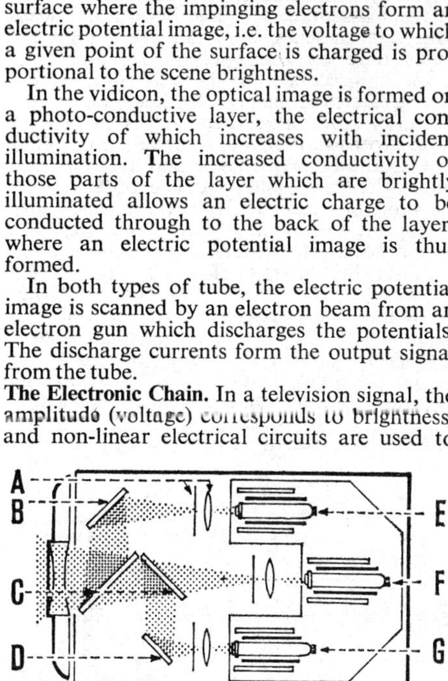

COLOUR TELEVISION CAMERA SYSTEM. A beam splitting system focuses the image on to the photo-sensitive surfaces of the three ridicon tubes. A, trimming filter and objective lenses. B, and D, front surface mirrors. C, dichroic mirrors. E, F, G, blue, green, and red camera tubes.

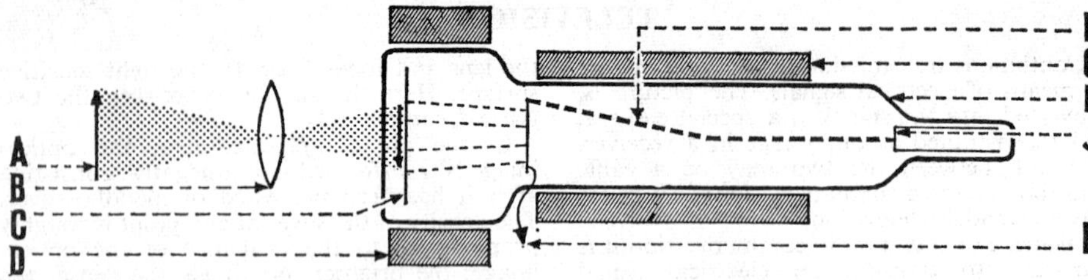

IMAGE ORTHICON CAMERA TUBE. The lens projects its image on a photo-cathode—usually about 5 ins. in diameter. The photo-cathode emits electrons in proportion to the amount of light falling on it. This electron stream is focused by magnetic coils on to the target connected to the output circuit. A scanning beam from an electron gun scans this image and produces the output signal. A, object. B, lens system. C, photo-cathode. D, magnetic lens coil. E, output signal. F, scanning beam. G, beam focusing and scanning coils. H, glass envelope. J, electron gun. (The electron stream is shown dotted.)

correct or modify the grey-scale (gamma) characteristics of the picture.

The resolution of a television system is, in general, different in the vertical and horizontal directions. Assuming that the camera optical system is of adequate resolution and that the camera and receiver scanning beams are of adequately small diameter, the vertical resolution depends only on the number of scanning lines employed. Thus a 625 line system is superior in vertical resolution to a 405 line system by some 50 per cent.

Again assuming adequate optical and scanning arrangements, horizontal resolution depends only on the bandwidth (number of megacycles/second) of the electronic chain. The greater the bandwidth, the greater the resolution. But large bandwidths necessitate more expensive design and greater transmitter power; allow fewer channels in a given communication band; and introduce more "noise", the random electrical fluctuations which give an appearance similar to film granularity. The chosen bandwidth must therefore be a compromise: that selected for 625 lines is 5.5 Mc/second. The transmitted channel bandwidth must be greater than this in order to accommodate the accompanying sound signal.

Synchronizing. A certain level, often 30 per cent, of the electrical picture signal is arbitrarily defined as black level. Any signal greater than 30 per cent will give increasingly lighter greys. Any signal less than 30 per cent will be black and therefore invisible on a correctly adjusted receiver.

The zero-to-30-per cent range ("blacker than black") is used to transmit synchronizing pulses, which serve to lock the receiver scanning circuits so that camera and receiver scanning proceed exactly in synchronism. Two types of pulses are transmitted, one to synchronize the horizontal or line scan, the other to synchronize the vertical or field scan.

Television Receivers. In the receiver, the picture is finally displayed on the electro-luminescent screen of a cathode-ray tube. In this evacuated glass tube, a beam of electrons is ejected from an electron gun and impinges on the phosphor, which emits light at the point of impact. The brightness of the emitted light is a function of the electron density of the beam.

The television signal is applied to the electron gun so as to modulate the beam density: as the signal voltage increases, so the beam becomes denser and more light is emitted from the screen. Electrical circuits scan the beam over the screen area in synchronism with the television camera scanning, so that the bright and dark picture points appear on the screen in the correct relative positions.

Colour Television. By the use of optical filters (usually dichroic mirrors) the red, green and blue portions of the light reflected from the scene may be directed on to three separate camera tubes mounted in a single camera body. The camera then yields three electrical signals

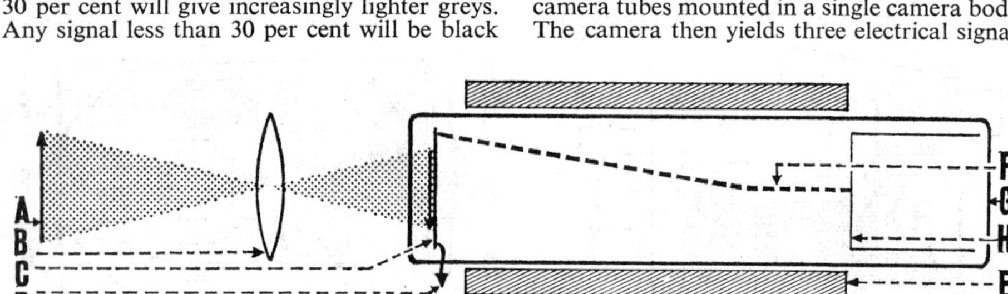

VIDICON CAMERA TUBE. This is considerably smaller than the image orthicon and hence more suited to compact television cameras. The image is projected on to a photo-conductive layer which is scanned by a scanning beam from the electron gun. The gun discharges the charge formed by the optical image, and thus produces the output signal. A, object. B, lens system. C, image formed on photo-conductive layer. D, output signal. E, beam focusing and scanning coils. F, scanning beam. G, glass envelope. H, electron gun.

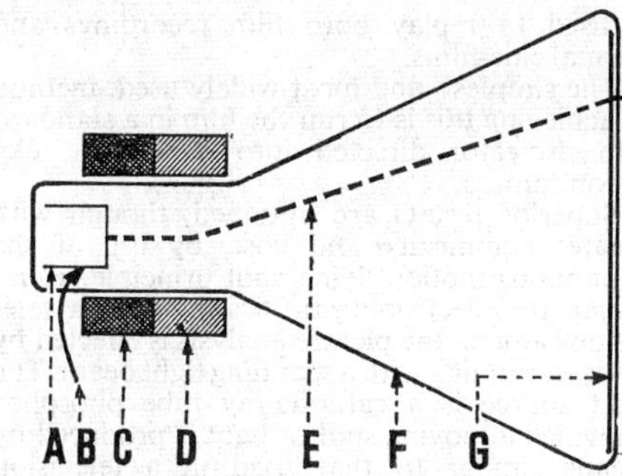

RECEIVER CATHODE RAY TUBE. The electron beam modulated by the transmitted picture signal scans over a luminescent phosphor layer inside the tube to build up an image. The receiver tube may have a front face diameter of up to two feet or so. A, electron gun. B, signal input. C, focusing magnet. D, scanning coil. E, scanning beam. F, glass envelope. G, phosphor layer.

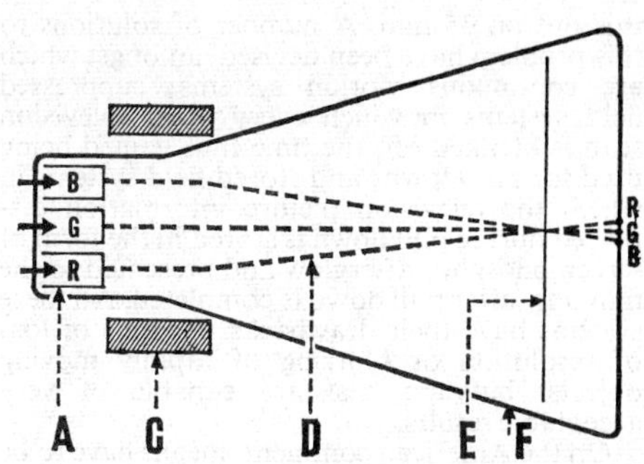

SHADOWMASK COLOUR RECEIVER TUBE. Three electron beams pass through a masking plate to a phosphor mosaic of dots emitting blue, green and red light respectively to build up an additive colour image by simultaneous scanning. A, electron guns. C, focusing and scanning coils. D, electron beam. E, masking plate. F, phosphor mosaic. R, G, B, red, green and blue electron guns or phosphor dots.

corresponding to the three colour components of the scene.

These three signals could be simultaneously transmitted on three separate communication channels and the final colour picture formed by optical superimposition of the three separate primary colour pictures. This would, however, be wasteful of channel space and would require a cumbersome receiver.

To conserve channel space, use is made of the fact that normal black-and-white television makes very inefficient use of the bandwidth occupied. By a system of special electrical coding, signals carrying colour information are fitted into parts of the band left unoccupied by the black-and-white signals. At the receiver, decoding restores the original primary colour signals. Some interference occurs between the chrominance and luminance signals when coded as described, and this gives rise to a pattern of small dots on the final picture. One well-known coding system is that of the N.T.S.C. (National Television Systems Committee) and was originated in the United States.

To avoid the use in the receiver of three separate tubes with coloured screens, the shadowmask tube was developed. The screen of this tube consists of a large number of triplets of phosphor dots. A dot of red emitting phosphor, a dot of green and a dot of blue are laid down close together, these triplets being repeated over the whole screen area. Three separate electrons guns are employed, and the electron beams pass through a plate containing many small holes. The guns are so arranged that any one gun on "looking" through the perforated plate can "see" phosphor dots of only one colour: thus one gun activates only red dots, one only green and the third only blue. The three primary colour electrical signals can thus each be applied to modulate an electron beam activating only the required colour. The phosphor dots are sufficiently closely spaced to give the effect, at a normal viewing distance, of a continuous colour picture. D.P.L.

See also: *Television recording.*
Book: *The Technique of Television Production*, by G. Millerson (London).

TELEVISION RECORDING. Recording of electrical television signals, both picture and sound, in such form that subsequent replaying will reproduce the original electrical signals as faithfully as possible. Two recording media are in general use, photographic film and magnetic tape.

Recording finds many uses in television operations. Delayed transmission, repeats, export, archive and editing are the most important.

Film Recording. Basically this is achieved by displaying the television picture on a cathode ray tube screen and photographing the display with a motion picture camera. The end product is a film which is almost identical to one produced by normal motion picture techniques. The line structure of the original television picture is as far as possible eliminated in the recording process, so that a film recording made on one television standard (e.g. 405 lines) may be replayed on a different standard (e.g. 625 lines): this greatly eases the problem of exchanging programmes between countries using different standards.

A fundamental design problem in film recording concerns pull-down time, since the time interval between successive television pictures is very short, of the order of $1\frac{1}{2}$ thousands of a second. It is found just possible to pull down 16 mm. film in this short interval, but larger dimensions and greater mass rule this out on 35 mm. A number of solutions to this problem have been devised, amongst which are continuous motion systems; suppressed field systems in which every other television scan is blanked off, the time thus gained being used for pull-down; and stored field systems in which the television picture information displayed during pull down is stored in the form of screen phosphor afterglow and presented to the film only after pull down is completed. All these systems have their drawbacks in terms of loss of resolution or blurring of rapidly moving objects, but the best are capable of very acceptable results.

On the American continent, means have to be found in terms of non-regular pull-down cycles to convert the television picture repetition rate of 30 per second to the film rate of 24 per second. On European standards the television repetition rate is 25 per second, and film is accordingly run at this rate since this differs from the normal 24 per second by only 4 per cent.

Telecine. Telecine equipment converts a film image into an electrical television signal, and

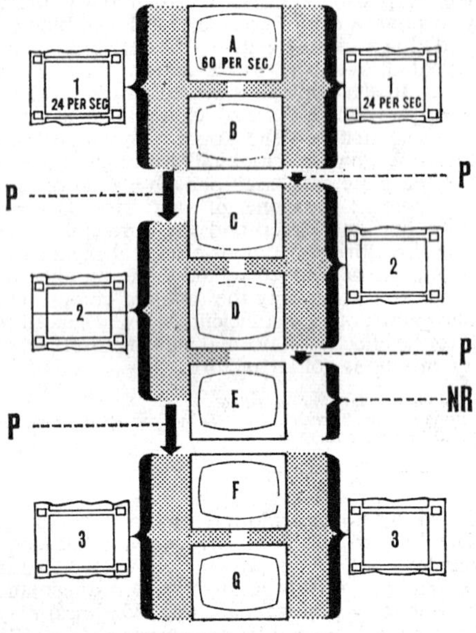

SYNCHRONIZING WITH 60 CYCLE SYSTEM. When recording television pictures at the picture frequence of 30 per second from film at 24 frames per second, one television field in every five must be lost. The American method (left) uses a half field pull down time of 1/120 second (i.e. the film is advanced from one frame to the next in the cine camera while the television receiver displays half a field). Thus the first film frame records two fields, the second frame one full and two split half fields, the third frame again two fields and so on. This requires careful adjustment. The British system (right) uses a fast inter-field pull down at a non-regular pull down frequency. A shutter eliminates every fifth field. The television fields A, B, C, etc., are displayed at 60 per second, corresponding to 30 complete pictures per second.

is used to replay both film recordings and normal cine-films.

The simplest, and most widely used, method of achieving this is to run the film in a standard film projector directed into a vidicon television camera.

Superior results are obtained, though with greater complexity and cost, by use of the continuous motion flying spot principle. Here, unlike the electron beam scanning of a television camera, the picture analysis is effected by optical scanning with a scanning light beam. The light source is a cathode-ray tube phosphor on which a moving spot of light is produced by means similar to that used in a television receiver. The scanning light beam, after passing through the film, is directed on to a photoelectric cell which produces the required electrical signal. The film is moving continuously at 25 frames per second, no intermittent mechanism being required. This system is suitable only for European standards where the picture repetition rate is 25 per second.

Video Tape Recording. The recording medium here is unperforated magnetic oxide tape very similar to that used for sound recording, although video tape is 2 ins. wide.

The electrical television signal is recorded on the tape in the form of magnetic variations and there is no visible image. A basic problem is to compress the television signal information into a frequency band which the tape can accommodate, and this is solved by causing the television signal to modulate a higher frequency carrier (between 5 and 10M c/s). Although this results in higher frequencies, the number of octaves occupied by the signal is reduced. Frequency modulation is employed to minimize the effect of variations of tape sensitivity, etc.

To enable such high frequencies to be recorded and reproduced, a very high head-to-tape velocity is required. The tape is moved longitudinally at only 15 ins. per second, but rotating heads are used which sweep laterally across the tape at 1600 ins. per second., laying down a series of closely spaced lateral tracks. To provide for longitudinal positioning of the tape during replay, a control track is recorded in addition to the signal recording: this is the video tape counterpart to film sprocket holes. Elaborate electro-mechanical servo systems are required for correct tape positioning. Lower tape tracking speeds can be used, but lead to lower image quality. This may be acceptable for certain purposes—e.g. industrial closed circuit systems.

A video tape recording can be replayed only on the television standards on which it was recorded: this renders video tape particularly inconvenient for programme exchange between countries using different field rates, since the unsatisfactory process of optical standards conversion must then be employed.

Sound is recorded longitudinally along one edge of the tape.

Film versus Video Tape. At its best, 35 mm. film recording can rival video tape in picture quality and, with separate magnetic sound film, surpass it in sound quality. On 16 mm. the picture is less satisfactory, and even on 35 mm. first class results are difficult to obtain consistently. However, the end-product of film recording is in essence a standard motion-picture film and can be handled, edited, viewed and copied by standard methods, equipment and personnel. It is the medium to use if much editing is required, if many copies are needed, or if copies are to be sent to an organization using different television standards.

Video tape offers the advantage of first-class quality, immediate play-back without processing and comparatively low running costs since tape can be erased and re-used. It is subject to the disadvantages of combined sound and difficult editing, restriction to one television standard, and the difficulty of making further copies. Video tape recording should be used for time delay purposes of any sort but is inconvenient when much modification of the original programme is contemplated.

It is also possible to record still pictures magnetically. In this case a magnetic disc may be used, rotating at high speed. Each revolution of the disc corresponds to a frame scan. The scans are presented to the receiver in continuous repetition. D.P.L.

See also: *Television.*
Book: *Video Tape Recording*, by J. L. Bernstein (London)

TELEVISION SCREEN PHOTOGRAPHY. It is a fairly easy matter to take reasonably well defined photographs of pictures on the screen of a television set. Within limits any camera can be used, providing that it is possible to give the suitable exposure time.

The Screen Picture. The image on the television screen is produced by a very narrow beam of electrons generated in the cathode ray tube. The screen consists of a coating of fluorescent material on the end of the tube.

When the electron stream (the cathode ray) strikes the screen, the fluorescent material glows visibly producing a bright spot on the end of the tube. This spot glows more or less brightly according to the strength of the cathode ray which in turn is controlled by the signal received by the set. When the set is working, the spot is made to "scan" the whole area of the screen—i.e., it is made to travel very rapidly in a pattern of horizontal parallel lines. As the spot traverses the picture space it is "modulated" by the incoming signal and generates an image.

Minimum Shutter Speed. In the television system in general use in Europe, the receiver picture is the sum of two sets of lines, one set occupying the spaces between the other. Each scans the picture space in 1/50 second, so it takes 1/25 second to produce a complete picture. It follows that an exposure of less than 1/25 second will show only part of the screen image. This limits the shutter speed to not less than 1/25 second. American systems with a scanning time of 1/60 second call for a shutter speed of 1/30 second.

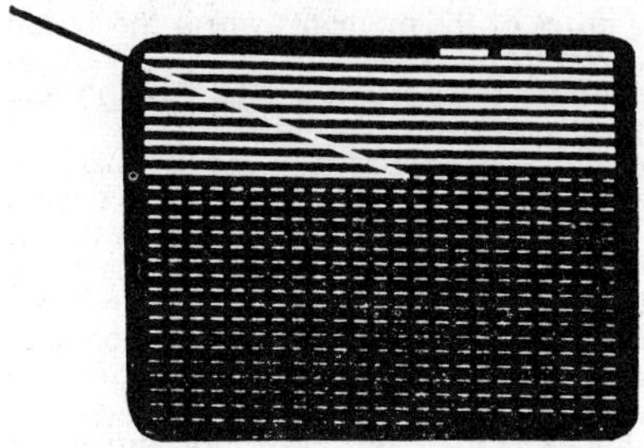

TELEVISION SCREEN IMAGE: The picture is built up by a scanning beam which in one cycle covers the screen with a set of scanning lines, followed by a second interlaced set (starting top right). The camera exposure must cover at least one, or a whole number of, complete image cycles.

Definition. There are on European systems 625 or 405 lines across the width of the picture, and this number is the same no matter what the size of the screen. This figure governs the definition of the picture.

If the screen is 10 ins. high, then the image will have—on a 625-line system—62·5 lines per in. A photograph of the image 5 ins. high would therefore have 125 lines per in. So the lines are always more obvious than the grain of the fastest film or the shortcomings of the cheapest lenses. The figure of 62·5 lines per in. limits the rendering of detail. On American screens the number of lines is 525.

Taking the Photograph. The camera should be set up on a steady tripod with the lens opposite the centre of the screen and the back parallel to its surface. For a 12 ins. screen, the camera should be about 3 feet away—if it is closer than this, the depth of field of the lens will not be sufficient to allow for the curvature of the end of the tube. When measuring the distance it is necessary to add an inch or so for the space between the end of the tube and its protective glass window.

As the screen image tends to be over-contrasty, the receiver controls should be set to give a soft picture. Tube characteristics vary so the best exposure must be found by trial and error (an exposure meter is useless). With the lens stopped to *f*4, test exposures of 1/5, 1/10, 1/25 second should give one satisfactory negative. A shutter speed less than 1/25 second will not give a complete picture.

Shutter speeds longer than 1/25 second may lead to image movement, caused by the registering on the negative of more than one scan of the T.V. screen, and these longer exposure times can only be used when there is little or no movement on the screen. So it is advisable to choose a programme that is going to be repeated and to use the first transmission for making notes of the moments where the action is suitably restrained.

As grain is unimportant it is as well to use a high speed panchromatic film.

Care should be exercised in the use made of pictures taken from the television screen. While generally the television companies would have no objection to photographs being taken for personal use, it should be remembered that copyright may exist in their productions and no commercial use should be made of the photographs without consent.

Television Monitor for Press Recording. A television unit is made by at least one firm specially for photography of the televised picture. It is intended principally for use in the news departments of the daily press.

The unit has a camera with electrically operated shutter release. Operation of the release automatically blacks out the picture tube, opens the camera shutter, brightens the picture tube for one or two complete frames, blacks out the picture tube, closes the shutter and then restores the television picture. This operation takes approximately $\frac{1}{8}$ second.

The advantage of this system is that there is no possibility of recording a fraction of a complete frame on the screen—as might occur with slight shutter speed variation. In other words the television screen and not the camera shutter controls the exposure. F.W.G.

See also: *Cathode ray tube traces.*

TEMPERATURE COEFFICIENT (DEVELOPER). Factor by which the development time of a developer must be multiplied to compensate for a fall in temperature of 10° C. (18° F.). The figure varies according to the developer—usually from 1·25 to 2·8.

See also: *Developing negatives.*

TEMPERATURE OF SOLUTIONS. The temperature at which a processing solution is used is important because it affects the rate of action of the chemicals involved and the gelatin of the emulsion.

Effect on Rate of Action. The warmer a solution, the quicker it works, but it always works at the same rate at any given temperature.

Development times are therefore always quoted at a standard temperature (usually about 65–68° F., or 18–20° C.). So long as the temperature of the developer is kept constant its rate of action will remain constant (other things being equal). For temperatures above and below the normal, the rate of action can be found by multiplying the standard time of development by the temperature coefficient.

At low temperatures special rapid acting developers must be used, as otherwise processing times are inconveniently long. At very low temperatures anti-freeze agents may be necessary.

Certain developing agents are almost inactive below 55° F. (13° C.).

Fixing baths, intensifiers, reducers, etc., all work faster at high temperatures than at low.

Effect on Gelatin Layer. The higher the temperature of a processing solution, the more water the gelatin layer absorbs, and the more it swells. At the same time it becomes increasingly soft and sensitive to scratches and other injury.

In its swollen state the emulsion must therefore be handled very carefully. The surface should never be touched with the fingers, or they may leave permanent marks. The extra warmth of the fingers on the swollen emulsion may even cause it to melt.

At temperatures above 75° F. (24° C.) special hardeners must be used to counteract this softening and all solutions should contain neutral chemicals to prevent excessive swelling.

Sudden changes in temperature from one solution to the next—specifically going from cold to warm—cause the gelatin layer to ex-expand irregularly. It then forms tiny wrinkles all over the surface, which look—and print—like very coarse grain. This effect is known as reticulation. The emulsion may also frill or form blisters by locally leaving the support as a whole.

So all processing solutions must be at the same temperature, particularly in hot weather. There should be no more than one degree difference between developer and rinse or between rinse and hardening fixer, and no more than two degrees between the hardening fixer and the washing water.

Temperature control systems. Methods for ensuring the correct processing temperature can vary from primitive water bath set-ups for occasional amateur purposes to complex thermostatically-controlled installations for professional and trade processing.

The various solutions can be kept at a uniform temperature throughout by standing all bottles, tanks, and dishes in a large dish of water. This should be a few degrees warmer (or, in hot weather, colder) than the processing temperature. The solutions should be left there for about half an hour before processing. Hot or cold water may be added from time to time to keep the temperature of this water bath constant.

The film is therefore inserted in the developing solution in the tank, rather than the developer poured into the tank. This is the only reliable way of accurate temperature control in processing (see below). Even there the reel holding the film should also be at approximately the correct temperature to avoid appreciable cooling or warming of the developer. Having a large solution volume helps, since the cooling effect of a film on a reel immersed in the bath is comparatively smaller.

Water baths—with an appropriately large volume of water—are also the most reliable method for controlling the temperature in large-scale installations. There a pump keeps the water circulating, and a thermostatically-controlled heater keeps the temperature at a preset level.

If the consumption of water is no economic problem, a temperature control system can use a water bath fed by continuously-running water from a thermostatically-controlled mixing tap. This adjusts the proportion of hot and cold water to the required temperature level. Mixing taps of this kind are frequently used in any case for the rinsing and washing stages in commercial processing.

For specially accurate temperature control the developing tank can have its own smaller water bath with an extra-sensitive thermostat and heater unit.

Where absolute accuracy of temperature levels is less important, direct heating of the solutions is also a suitable method. In large tank installations it provides the same accuracy as a water-bath; but direct heating of small volumes of solution may lead to unpredictable temporary temperature variations within the solution. Available methods include the use of heating (and also refrigerating) coils immersed in the liquid in the tank. The heating or cooling unit is then brought into action by a thermostat system.

Temperature control by thermo-elements has also been proposed. There the cold end of a set of thermo-couples cools down the processing solution (or a water bath surrounding a processing tank) while a heating coil warms it up. The two are controlled thermostatically.

In amateur darkroom practice a simple dish heater, consisting essentially of a light-tight metal box containing one or more 60 watt light bulbs, keeps developing dishes warm in cold weather. Such a unit can easily be made home-made, but requires appropriate precautions regarding efficient earthing and wiring. The degree of control is only approximate, and temperatures have to be checked with a thermometer in the bath.

Optimum Processing Temperatures. The best temperature for developing, fixing, and after-treatment is the highest at which the material can be handled without risk of damage to the emulsion.

The normally useful range of processing temperatures for films and plates is about 60–75° F. (15–24° C.), with 65–68° F. (18–20° C.) as the optimum. Above this range there is a real danger of excessive softening of the gelatin, below it the development and other times become very long.

In regions where extreme climatic conditions make it impossible to observe these temperatures (e.g., tropics, arctic, etc.) special processing solutions and procedure may be required.

For papers, the safe range of temperatures extends up to about 85° F. (30° C.), and even higher. It is quite usual to process printing papers at 70–75° F. (21–24° C.).

Emulsions coated on paper are less sensitive to temperature changes than films or plates, as the gelatin layer is much more intimately bonded to the paper grain.

Colour processing. Accurate control of solution temperatures is specially important when processing reversal colour materials. For the first development and colour development stages the manufacturer's recommended temperatures should be adhered to within ½°F. or ¼°C. This is because the rate of formation of the different colour images may be affected in slightly different ways by changes of temperature; a higher or lower processing temperature may thus lead—even with corresponding decrease or increase of the development time—to a shift in colour balance.

With negative colour films the developing temperature is specified with similar precision, though here some colour correction is possible in making the print. The other solutions (stop baths, hardeners, bleaches, etc.) offer somewhat more latitude, and the temperature should be exact within about 2°F. or 1°C. either way.

Special Processes. Certain processes, especially colour development and print toning, have very sharply defined optimum temperatures.

Processes which are based on dissolving away part of the emulsion layer, such as carbon, carbro, bromoil, gum-bichromate, etc., need temperatures round about 110–120° F. (45–50° C.). Most of the printing methods involving the reduction of chromates or bichromates in contact with gelatin or other organic materials fall into this category.

A few special toning methods such as hypo-alum are worked at about 140° F. (60° C.). At such temperatures even paper emulsions must be hardened. L.A.M.

See also: *Cold weather; Hot weather processing; Polar photography; Tropical photography.*
Book: *Developing*, by C. I. Jacobson (London).

TEMPLATE. Exact, full-scale pattern used as guide in cutting out or machining some part of a manufactured article or mechanical structure in quantity.

Templates (which have been in use for over 150 years) are usually flat, made of metal, wood or plastic, and carry constructional details to aid quick assembly. Although photographic production of such templates was suggested as early as 1910, this method was not applied on any scale until 1940, when it was adopted for use in the aircraft industry. A modern aircraft contains thousands of parts cut from templates, and if templates laid out manually from data derived from blueprints are replaced by templates marked out photographically thousands of pounds per prototype aircraft can be saved and the time between design and test flight shortened by months.

In a typical photographic system, the working drawings are laid out on white painted metal in the drawing office (in the case of small parts), or on lofting tables in the case of the larger sheets (up to 5 × 18 feet) used in aircraft, boat and automobile drafting. The drawings are made on a dimensionally stable material, so it is practical to transfer the finished layout photographically to the metal from which the part is to be made.

For small parts the original drawing may be made on ground glass or copied by reflex printing from metal on to polyester direct-positive matt film—the reflex negative being then used for printing on to metal sheets coated with photographic emulsion. These polyester materials give good line positives directly, and have remarkably good dimensional stability. For larger parts the original drawing is photographed on to glass plates and the negatives are subsequently projected at the required magnification on to metal sensitized by spraying with a photographic emulsion.

The metal prints are then cut out and become the templates or patterns. This projection process not only facilitates the production of scale models, but permits the correct allowances to be made for the shrinkage of castings when dies in other metals are required.

Dye-line sensitizing solutions (diazo compounds) can also be applied to suitably treated metal sheets. Drawings made on low-shrink translucent materials can be contact-printed, using UV-rich light sources. D.A.S.

TESSAR LENS. Anastigmat lens designed by Paul Rudolph and introduced by Zeiss in 1902. It was one of the early forms to break away from the symmetrical type of lens which suffered from zonal aberrations and which was therefore limited to medium relative apertures of *f*5 or *f*6.

The Tessar utilizes collective and dispersive lens elements separated by air spaces to correct curvature of field and achieves wider relative apertures and useful angular fields. The change from a single rear collective element in the triplet to the cemented doublet at the rear of the Tessar was a logical step forward which provided further improvements of zonal aberrations.

The presence of a cemented contact surface in the rear component provides a useful control over higher order aberrations and, since the expiry of the original Zeiss patents of 1902 and later, this form has been modified for various purposes by nearly every lens manufacturer. In the shorter focal lengths covering medium angular fields the relative aperture has been extended up to about *f*2·8. The Tessar form has been used extensively in medium price and expensive cameras for all focal lengths at apertures of *f*3·5 and *f*4·5 covering angular fields between 50° and 60°.

The Tessar form has been found particularly useful for front-cell focusing since it can be designed so that variation of the front separation between lens elements provides the required change of focal length whilst the correction of aberration suffers only small departures from the ideal balance. Inevitably there is a certain loss of definition and, generally, the lens is adjusted on the camera in such a way that the separation yielding best definition will focus objects at a medium distance. This ensures that focusing for infinity or short distances does not demand considerable variations from the ideal separation.

With the introduction of new types of optical glass in recent years the Tessar type of lens has undergone further development and, whilst its general shape and construction remains unchanged, the modern lens attains a standard of performance superior to that of the earlier forms. G.H.C.

See also: *Lens history*.

TEST PAPERS. Strips of absorbent paper impregnated with indicators for testing the acidity, silver content, etc., of solutions.

See also: *Indicator chemical*.

TEST STRIP. Narrow strip of sensitized material which is used to find by trial the most suitable exposure—generally for a contact print or enlargement, but also occasionally for a negative.

The strip is exposed to light section by section in such a way that each section receives a constant multiple of the exposure given to the one preceding. The right exposure is determined by examining the developed image.

It is quicker and cheaper to find the correct exposure with test strips than by trial and error. The disadvantage of the method is that each section of the strip shows a different part of the picture—and not always a characteristic part.

There are special test-strip holders designed to overcome this disadvantage. The holder usually takes a strip of paper about 1 in. wide and 6 ins. long in a flat metal sheath. Only a small rectangle of the paper—about 1 in. by ¾ in.—is exposed, the remainder being

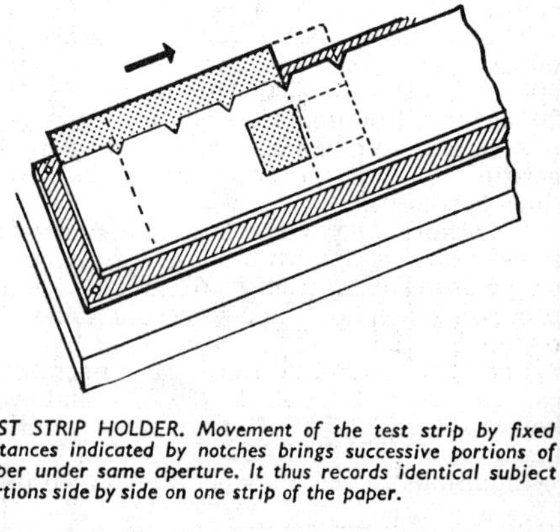

TEST STRIP HOLDER. Movement of the test strip by fixed distances indicated by notches brings successive portions of paper under same aperture. It thus records identical subject portions side by side on one strip of the paper.

covered by the metal mask that forms the top of the holder. The edge of the holder is marked to indicate how much the paper must be moved after each exposure step to give a continuously exposed strip.

The holder is laid on the enlarging easel so that the open section lies under a typical part of the negative. It is then pinned or weighted in position, and the paper strip is given the desired series of exposure. In this way the same part of the negative image is used for each step along the test strip.

See also: *Enlarging.*

TEXTURE. Strictly, the arrangement of the constituent parts of a material—e.g., the weave of a fabric, the fibrous composition of wood, or the granularity of sandstone.

In art—and by extension, photography—the term refers to the structure of the material in so far as it is apparent on the surface. The characteristic surface texture—e.g., of cloth, skin, brickwork—can be made more obvious by the use of hard lighting from the side.

See also: *Lighting the subject.*

TEXTURE SCREEN. A special pictorial purpose is sometimes served in breaking up the image by printing the enlargement through a texture screen.

Such screens usually consist of a transparent film or plate, printed or engraved with a fine background pattern. This pattern is projected with the negative or printed by contact on the picture.

The texture relieves the monotony of large areas of tone in much the same way as rough or textured paper surfaces. The effect is mainly suitable for subjects consisting of broad tone masses rather than a lot of intricate and distributed detail.

The range of patterns obtainable in texture screens is very much wider than that provided by paper surfaces.

Screened Print Method. One way of giving a texture to the image is to lay a suitable screen on the printing paper. Screens for use in this way may consist of an open fabric like silk, muslin, etc. There are also special screens sold which imitate the effect of etchings, etc.

The screen is laid on the enlarging paper and pressed down with a glass plate to keep it in good contact. The print is then exposed in the normal way with possibly some adjustment in exposure.

The screen is not often left in position during the whole of the exposure. With large patterns in particular, only a partial exposure is given, so that the real image can be seen even in the texture pattern.

Screened Negative Method. The other way is to place the screen in contact with the negative in the negative carrier of the enlarger.

The screen used may again be thin fabric like silk or muslin, or it may be a special texture negative made by photographing any desired pattern (e.g., textures of wood, stone, cloth, leather, etc.) as background to the picture.

Where the screen is in contact with the negative, the texture effect may be diffused by the normal soft-focus methods. When very fine texture patterns are diffused in this way, they disappear in the shadow areas being overlaid by the spreading of light on the paper, and show through only in the highlights. L.A.M.

See also: *Soft focus; Tricks and effects.*

Books: *The Complete Art of Printing and Enlarging*, and *Design by Photography*, both by O. R. Croy (London).

THEATRE

There are two main approaches to theatre photography—the professional and the amateur.

The professional is chiefly concerned with taking commissioned pictures of the production on the stage or in the studio. Such pictures may be used for display in the theatre foyer and for various publicity purposes. In practice the field is restricted to a handful of specialists who possess an intimate acquaintance with the needs of both the theatre managements and the artists, and who are experienced in working within the peculiar limitations of the stage itself.

The amateur is mostly interested in taking photographs of the actual performance from the auditorium. This branch of photography has been made possible by the 35 mm. camera with its wide aperture lens equipment and the extremely fast panchromatic films which are available today.

ON STAGE

Stage photography involves taking photographs on or around the stage itself, usually of a complete production. The management of the show, or their press representative, commissions the photographer, who is usually given the fullest co-operation by the actors and technicians. A period of one to four hours is set aside for this photo-call, usually on a day just before or shortly after the show opens. The prints are used for display in frames outside the theatre, and for issue to the newspapers and magazines from whom editorial publicity is sought. (Editorial use of pictures means their reproduction in the parts of a publication under the direct control of the editor and his contributors. This type of space cannot be purchased and, whilst it is given free of charge, there is always the possibility that the comments of the critic will be unfavourable. Photographs supplied by a theatre for editorial use are usually expected to be made available free of reproduction fee.)

To be of value the whole procedure must be rapid from submitting proofs to completing the press and display prints. In addition, as the photographer does not know which press copies may be sent to newspapers with coarse-screen blocks and which to weeklies with the highest standards of reproduction, every print must be very carefully finished. In close-ups, the stage make-up and wig or false hair lines require a lot of discreet retouching if they are to be acceptable. For this reason much of the best stage photography has been done on quarter-plate or larger negative sizes.

Cameras. Theatre time is expensive, and some forty different pictures may have to be made in about three hours. So some stage photographers prefer a quarter-plate single-lens reflex because it allows them to see the image right up to the moment of exposure, which makes for rapid working. Others however, work almost as fast with the half-plate technical camera. Their familiarity with their instrument enables them to judge visually the effect of minor variations which may occur—or have to be made—between closing the shutter and drawing the dark slide ready for the actual exposure. In either case, at least two interchangeable lenses are carried: one of about 45°–50° angle of view, for groups and full-stage shots; and one of about 30° angle of view for close-ups and small group shots from a box or the circle of the auditorium.

A plentiful supply of film holders is also necessary, as there is no time to spare for reloading.

Oddly enough, the ruthless time limits, far from being a handicap, actually benefit the photographer who does not let himself get flustered by the pace and difficulties of theatre life. These conditions tend to produce interesting and lively pictures, because the actors can build up and retain the character and tempo of the piece much more effectively when they are not kept waiting about with little to do.

This is especially so in very light farces or revues, and for such shows a 6 × 6 cm. twin- or single-lens reflex may well be used, as occasional minor defects in make-up and décor are of secondary importance to the capturing of some of the sparkle and animation of the show. It is also possible to use this type of camera for character plays or unglamorized dramas in small theatres, as such pictures also require little or no retouching.

Emulsions. High-speed (but not super-speed) panchromatic films or plates are the usual negative material.

Lighting. Most experienced stage photographers carry lighting equipment of their own. This simplifies exposure estimation and reduces loss of time in handling unfamiliar and often heavy stage lamps. The portable equipment usually consists of two or more 500 watt spotlights on telescopic stands with rubber wheels, and at least one 1,500–2,000 watt floodlamp, similarly mounted. One or two hand-floods and ample cables and distributors feeding from, and interchangeable with the standard theatre plugboards, complete the outfit. Every advantage is taken of the apparatus of the theatre itself, and, in general, basic lighting is arranged to retain the atmosphere of the production. It is usual, however, to employ the footlights or a portable floodlamp to give more fill-in illumination than would be desirable at a public performance.

For ballet and acrobatic performances, but seldom for other work, two or more flash-heads, mounted on stands and used as if they were spotlights, are synchronized with the shutter. In view of the modest fees current in stage work, flash bulbs are less frequently used than electronic flash.

Exposure. The stage photographer must see earlier rehearsals of the play so that he can prepare notes of the points in its progress which will make good pictures or which are essential to his collection of record shots (not always the same thing). Where possible, he notes down the actual spoken words to identify particular moments in the action. From these notes he prepares a photo-plot giving, scene by scene, the moments selected, with a note of the characters involved and other essentials. This is given at the photo-call to the stage director, whose job it is to call and check costumes of the actors and to supervise the changing of settings and props as the photographer works through the play.

Each picture is roughly grouped and lit, the actors do a brief rehearsal of that particular passage of the play, then the photographer decides at which precise point the movement is to be arrested for the exposure. Then the grouping is improved by rearrangement of the characters, the lighting finally adjusted and the camera positioned exactly and made ready. Now, the actors go through the scene—acting as if in a public performance—until the determined moment arrives. At this point the photographer tells them to "hold it", and while they "freeze" the action for an instant, he makes his exposure. Then he goes right on to the next picture; there is no time for retakes, unless the need is recognized immediately.

Properly handled, more feeling of movement can be compressed into such a picture in this way than in any taken during a performance. Exposures vary widely; but average conditions will permit ¼ second at *f* 11 on medium-speed panchromatic material. With high-speed panchromatic material and development for maximum speed, the 2¼ × 2¼ ins. roll film camera may use 1/50 second at *f* 11 under equivalent conditions.

Processing. The larger cameras produce negatives of suitable quality for enlargement if they are developed in a soft working formula.

Development by time and temperature at tank strength is standard practice if only because of the large number of negatives made at each session. $2\frac{1}{4} \times 2\frac{1}{4}$ ins. roll film users will need a fine grain formula giving full speed rating.

Printing. The whole-plate or 8×10 ins. glossy prints for issue to the press must be made rapidly and in quantities, but the quality must be good enough to permit any of them to be reproduced full-size in good quality magazines. It is impossible to know in advance which may be selected for this purpose from the dozens made. The display publicity prints, 12×15 ins. in size and frequently larger, must be equal to any salon prints in quality, yet some thirty or forty of these must be produced within two or three days of taking. These prints will hang outside the theatre if the play is a success for a year or more, in all weathers, flooded with sunshine for hours each summer day, becoming humid in wet weather. Prints exposed to such conditions rapidly deteriorate unless they are given really thorough fixing and washing. For the same reason, only artists' quality water-colour may be used for print finishing. Dye is apt to fade or diffuse into the surrounding image areas.

Portraiture. Although at first sight a stage portrait is not essentially different from any other portrait, it is, in fact, a highly specialized type with its own very strict set of requirements. Portraits of stage personalities are needed for three main purposes: publicity and fan-mail, theatre display frames and job-getting. One portrait can seldom fulfil the requirements of all three purposes.

Deceptive retouching or "rejuvenation" has now more or less passed out as a fashion in the theatre. The average actor has found that interviews based on such pictures waste the time of casting directors and his own time. Nevertheless, no one wishes to draw attention to facial defects which make-up can successfully disguise. Rather large features look well on the stage, at audience distance. They may not look so well, or so co-ordinated, in a close-up, unless the photographer is very clever in his choice of viewpoint and lighting.

The aim in a job-getting portrait is to obtain a true likeness with some hint of the range of character types and "style" which are the special talent of the actor. For publicity, there must be some pictures which do this and which are also suitable for reproduction on a small scale in coarse-screen newspaper columns. Fan-mail pictures in postcard size usually show the star in more casual and informal poses and surroundings than would be employed for normal display and publicity. For theatre display, in the actor's private frames, pictures may be somewhat more imaginative, and are certainly more successful if they have some artistic merit. For display exciting photographs can be used, even if unsuitable for newspaper reproduction or casting purposes.

The retouching of all theatre portraits must be impeccable and the quality and permanence of the prints themselves must conform to the same standard. Many actors and actresses work exclusively for the theatre and radio without appearing before film or television cameras in their customary stage rôles. It is legitimate in such cases to use retouching to make the close-up portraits of such artists appear as they would do to a theatre audience, but to go further is undesirable, and discretion is more important than skill in applying artificial glamour.

COLOUR IN THE THEATRE

Theatre managements cannot usually afford to photograph an entire show in colour and purchase large colour prints for display. Nevertheless, they are frequently asked for colour shots by editors of illustrated magazines and the best way of doing this economically is to use reversal transparency film. Transparencies should not be less than $2\frac{1}{4}$ ins. square, preferably larger, owing to difficulties of cropping and photo-engraving from very small images.

The time-factor may well necessitate the use of film which can be processed by the photographer himself or by a professional laboratory which offers a rapid service. Very slight under-exposure of reversal film is far preferable to over-exposure, even though such transparencies may seem a little heavy in viewing.

For colour prints, negative type colourfilm should be used as colour prints from transparencies can be disappointing unless using highly expensive masking-methods. From colour negatives paper prints may be made and also enlarged transparencies for rear-illumination. In either case, it should be remembered that the dyes of the positives are not completely permanent. Display within the building, where the sun does not penetrate, will give a reasonably long life to the pictures, but exposed to daylight they will fade in a few weeks, even if protected by a final bath in an ultra-violet conversion substance before drying, and later coating them with a UV-absorbing lacquer. The client should be advised of these facts and encouraged to order duplicates when the first set of display copies are made. Their price is thereby lower than if reprinted at a later date. Greater permanence is obtainable in carbro and dye-transfer colour processes, but the high cost makes their use rare.

Historically important productions will justify the taking of negative colour pictures because the photographer may preserve the negative safely over many years and yet provide from them colour-slides and prints without parting with his original. J.V.

FROM THE AUDITORIUM

Modern high-speed emulsions in conjunction with lenses with an aperture of $f3{\cdot}5$ or wider make it possible to take fully exposed pictures with a hand camera from the auditorium by normal stage lighting.

Lens. The minimum speed for a hand-held exposure is 1/25 second, and under favourable stage lighting and with the fastest panchromatic films this speed calls for an aperture of about $f4$. So this is the smallest aperture that can be considered for this branch of photography. But a lens of this class leaves no margin for action or for shooting scenes with a low level of illumination, and anyone who wishes to take stage photography seriously should be prepared to work at $f2$.

The focal length of the lens is as important as the aperture, because it decides how much of the stage will be included from a given position in the auditorium. It is obviously wise to arrange for the width of the scene to fill the negative so as to keep the amount of enlargement down to a minimum.

There are three ways of tackling the problem: to choose a seat from which the camera with a particular lens will just include the width of the stage and cover the depth of the stage sharply at its maximum aperture; to accept whatever seat is available (as one must in a heavily booked show) and use a lens of a suitable focal length to give the necessary cover; or to choose the position that gives the best view of the stage and let it dictate the focal length of the lens.

The answer, in any case, will depend on the dimensions of the auditorium and stage. In a typical theatre with a stage 30 feet wide, 20 feet deep and 50 feet away from the front of the dress circle, the approach in each case would be along these lines:

(1) A camera equipped with a normal angle lens would cover the stage from a distance of 30–35 feet—i.e., from about the 6th–8th row of the orchestra stalls (allowing for the depth of the orchestra pit and a sparing of about 3 feet per row). A camera with no provision for interchanging lenses will get the biggest picture from this position. From a photographic point of view, however, this may be a poor choice—i.e., the feet of the performers may not be visible and the camera may have to be tilted unduly.

Under these conditions most hand cameras with the lens at $f3{\cdot}5$ will cover the whole depth of the stage sharply with the focusing scale set at 50 feet.

(2) If the only seat obtainable is closer than about the 6th row of the orchestra stalls, there would be no point in using a wide angle lens to get the whole of the stage into the picture because the perspective would be unpleasant and the viewing angle too steep. But if the seat is very much farther away, then a long focus lens can be used to put matters right—e.g., from the back of the stalls or the front of the dress circle, a lens of about double the normal focal length will be satisfactory. And even at $f3{\cdot}5$ the lens will cover the depth of the stage comfortably.

(3) In most theatres the best view of the stage is to be had from the front of the dress circle. From this position a normal angle lens will use only about one half of the width of the negative. To keep the scale of enlargement down, the camera should have a lens of one and a half to two times the normal focal length.

A long focus lens is also useful at closer range for taking individual pictures or small groups. If the photographer is at liberty to move about—e.g., during an amateur performance or at a photo call—a long focus lens will be most useful. But when working from a seat in the stalls, the camera should also have a normal angle lens for shots of the whole stage.

Lighting and Exposure. The stage lighting—and therefore the exposure—varies with the scene, the size of the stage, and the nature of the scenery. It is impossible to guess the combined effect of these things in advance. Once the curtain has gone up, the experienced photographer can make a reasonable estimation of the light value. An exposure meter is difficult to read in the darkness of the auditorium and even then it will probably give either an unreliable reading or none at all.

While the lighting may range from dim to dazzling and from white through all the colours of the spectrum—all in the course of a single performance—extreme lighting effects are generally reserved for special scenes. The average set is lighted to give reasonable visibility all over the stage with extra lighting on the principals from spots high up at the sides and front of the house—e.g., from the sides of the gallery, and the front of the circle or upper circle.

Tests have shown that in a London theatre with a very small stage, the exposure under full stage lighting might be 1/50 second at $f2$ for a fast film developed in a non-solvent developer. Under the same conditions, the exposure for another theatre with a very large stage is about 1/250 second at $f2$.

For average stages with full lighting the beginner should start with a trial exposure of 1/50 to 1/100 second at $f2$ on fast pan film.

Predominantly blue or pink lighting will call for 1/25 second; and green, yellow, mauve or red, 1/10 second.

Amateur companies cannot command the banks of high powered lights available on the professional stage, but on the other hand the stage is generally smaller and the lights have a proportionately greater effect. So for a well-lighted scene, there will not be a great deal of difference between the exposures for an amateur or a professional show.

The important thing is to expose for the highlights and let the less strongly lighted areas of the stage look after themselves. Any attempt to produce good shadow detail will leave the highlights as staring white areas, right in the most important part of the picture. This technique may give a picture of which two-thirds of the area are under-exposed, but this is infinitely better than one in which the principal feature is a patch of blank paper.

Viewpoint. The best position for the amateur is a seat on a level with the stage or slightly above it—i.e., in the dress circle or balcony—and about a quarter of the way across; not exactly in the centre. A viewpoint away from the centre conveys a better impression of the depth of the stage and avoids the formal appearance of a fully frontal picture. Photographs can be taken from a seat in the front row without interfering with the view of the people behind, whereas a photographer sitting in the body of the audience may be considered a nuisance. However, in most theatres the angle of view from the front row is too low.

Types of Show. The actual technique varies with the type of show; only the broad differences can be indicated here:

(1) Plays. The lighting is usually moderately good and not coloured. Normally an exposure of 1/25 second at *f*3·5 or *f*2 will give a reasonable print. It is as well to study the action of the play beforehand and choose a dramatic moment. Photographs taken during the ordinary course of the play are uninteresting. The normal angle lens will include the whole of the stage, so the exposure should be made when the action is spread out.

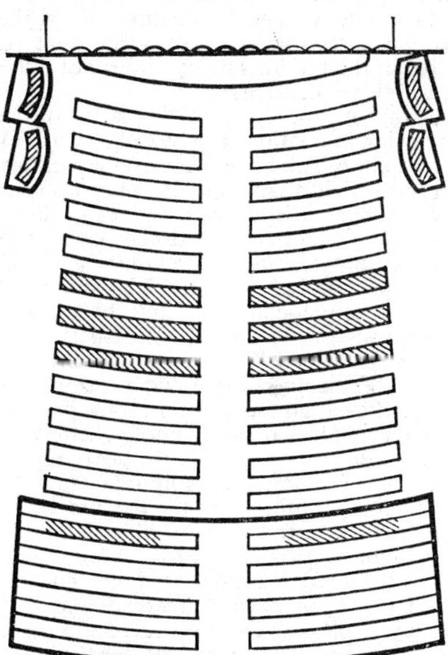

THEATRE PHOTOGRAPHY. Suitable camera positions for taking pictures during a performance are shown shaded. They are the centre stalls (about sixth row onwards), and the front row of the dress circle. Avoid seats dead centre of the stage. Seats in the boxes will give good angle shots, but may take in the wings at the far end of the stage.

(2) Musicals. The lighting here ranges from excellent to impossible and to make matters worse it is generally changing colour. The principals, however, are usually spotlighted and can be taken with a relatively short exposure. One good time to shoot is on the entrance of the star. There is usually plenty of warning and when she or he makes the traditional entrance and pauses for effect there is plenty of time to give the exposure required even with a modest lens. When taking sketches, the grouping should be watched and the exposure made when the players are mutually occupied—not when action is disintegrated.

(3) **Pantomime.** The conditions here are similar to the ordinary music hall but the changes of scene, performer and lighting are apt to be more abrupt. On the whole it is better to use the fastest lens and film and take pictures of everything that comes along in the hope that some at least will be worth printing. The fun is so fast and furious that planning in advance is impossible. Any really worthwhile pictures that fail to come up to expectations at the first attempt can always be made the subject of a second visit.

(4) Opera. On the whole this is an easier and more rewarding subject than the straight play. So far as the opera is a play set to music, the two techniques are the same, but the expression, gestures and grouping generally are more dramatic and sustained. Anyone with the most elementary musical sense can anticipate the instants of climax and crisis. But the singers should never be taken when they are singing open-mouthed right at the camera. The time to photograph a singer is when he is singing to another member of the cast.

Permission. Photography in the theatre is not as a rule permitted, and there is usually a notice printed on the programme to this effect. And while provincial theatre managers will often allow a local photographer to take pictures, this permission can be overruled by the manager of a touring company.

Amateur and repertory companies, on the other hand, are generally very pleased to allow a certain amount of photography at their dress rehearsals provided they are supplied with some prints for the outside of the theatre.

At a Performance. When the photographer has permission to take photographs during a performance, he should take care not to let his actions inconvenience other members of the audience. He will be less likely to cause annoyance if he can arrange to have a seat at the end of a row. He must on no account stand up, in spite of the strong temptation to do so.

It is a mistake to try for pictures at the first visit to a show. Wherever possible two visits should be made. The first should be given up entirely to taking notes of the shots to be made later.

There should be a note of the point in the action that is going to be taken together with an indication of any suitable cue that heralds its arrival. There should also be a note of the lighting of the scene, the proposed exposure and the type of lens, normal or long focus. This information must then be memorized since it will not be practical to try to refer to notes during the performance. When the photographer makes his second visit, this time with his camera and a plan of action, he is much more likely to get correctly exposed and well-timed pictures.

Amateur Theatricals. There is always more freedom for the photographer at an amateur show than in a professional theatre. Most producers are eager to co-operate with anyone who wants to take photographs and will usually afford special facilities at dress rehearsals for the purpose. Even during an actual performance, the photographer will be permitted to roam around and shoot from several viewpoints —even in the centre of the gangway, so long as he does not obstruct anyone's view of the stage.

Under these conditions he can place himself at the right distance to fill the field of view of his lens from whatever angle he is shooting. This means less enlarging and better picture quality.

At a rehearsal, things are easier still. The photographer can have the action of the play arrested or repeated and he can have the lighting modified to give the best photograph without reference to the dramatic aspect. (This means calling for plenty of general white lighting from above and cutting down side and footlights to a safe, fill-in light level.) This avoids the characteristic harsh contrasts of photographs made by stage lighting. L.V.

See also: *Circus; Dancing; Ice rink.*
Book: *My Way With the Miniature*, by L. Vining (London).

THERMOFAX. Proprietary document copying process using a heat-sensitive paper. Exposure takes place to infra-red radiation by the reflex method. The black portions (lettering or lines) reflect the heat radiation strongly and darken the corresponding portions of the paper in contact with the original. The paper requires no further processing, but does remain heat sensitive. The process is only suitable for originals printed, written or drawn in opaque pigment inks, but not suitable for many dye images which do not reflect infra-red.

See also: *Document copying.*

THERMOGRAPHY. Processes of image formation by heat. These are not new—Niépce de St. Victor and Robert Hunt were experimenting along these lines over 100 years ago and Hunt coined the name thermography in 1840. But it was not until 1952 that the first commercial heat process was introduced.

All heat processes are based on the fundamental fact that when a document having an infra-red absorbing image is subjected to an intense source of heat for a brief period, the image will become hotter than the background: the temperature difference can be as high as 30°. This heat image is converted into a visible image in an adjacent copy material by utilizing a suitable physical or chemical change that takes place over a small temperature range. The most common change is melting, but boiling and decomposition are also used.

The melting or fusion processes can be further subdivided according to whether the change is restricted to a physical one, or initiates a chemical change, or the softened image is transferred to another support.

Physical Processes. Basically, the image is formed from a layer of dispersed solid that melts when heated image-wise and remains as a super-cooled liquid when cooled. The liquid provides a window in the translucent layer of dispersed solid and an underlying dark layer is rendered visible. The liquid can also be made visible by dusting the surface after exposure with dye particles that will dissolve in the liquid. The dye does not adhere to the non-melted areas.

In a variant, the dye particles can form part of the original coating. It is well known that a given weight of solid dye is a much less efficient absorber of light than when in solution. Hence in the image areas the dye dissolves to give an image of high density on a tinted background. A similar process uses a cellular layer of a thermoplastic polymer. In the heated areas the cellular structure collapses and the dispersed dye dissolves in the softened polymer.

Chemical Processes. This group comprises the majority of the thermographic processes described in the patent literature. In some processes the image is formed by a simple chemical decomposition that occurs with heat, e.g. the formation of black lead sulphide from lead dithioformate. But most of the patents describe image formation from chemical reactants that are brought together by fusion or softening to form a coloured reaction product. The reactants may be solids that are dispersed in the same layer, or they can be coated in adjacent layers. Examples of chemical reactions that have been described are those between iron stearate and gallic acid, a hydroquinone and a quinone, a leucotriphenylmethane dye and an oxidizing agent, and an aromatic triazene and an azo dye coupler.

Transfer Processes. These utilize two surfaces to provide the final copy. The heat image softens the layer in the copy material immediately adjacent to the original and the softened substance is transferred image-wise to a receptor support. The transferred image, which is laterally reversed, can be used for making many copies by spirit duplication or lithography. Alternatively if the receptor material is translucent, the image can be read in the correct sense by turning the sheet over and viewing from the back.

Distillation. In one distillation process a thin film of oil is applied to the original. The higher temperature of the image (i.e. printed text) when it absorbs infra-red radiation causes the oil to distil selectively from the image and it condenses image-wise on an adjacent sheet of plain paper. The oil image is rendered visible by dusting with dye particles, such as the toner powder used for electrophotographic development. These particles adhere only to the oil and can be fixed by heat treatment as in electrophotography.

The copy is laterally reversed but if the receptor sheet is translucent the copy can again be read from the reverse side.

Another process involving distillation uses reactants contained in layers on two separate supports. One reactant is volatile and distils from a heat image to condense on the adjacent material where it combines with the second reactant to form a visible image.

Thermography in Practice. Since the image in the original must absorb infra-red radiation, the thermographic process will not reproduce every printed document. Carbon black images (pencil, typescript, letterpress printing) and metallic images such as silver are all copied well. Aniline dyes, as are used in ball-point inks, do not normally absorb infra-red so they are not copied.

The resolution of the process is poor but good enough to give readable copies of quite small print. The best definition is obtained if the original and heat-sensitive layer are in direct contact, but adequate copies are still obtained if the copy material is exposed to heat with the uncoated side in contact with the original, since the heat image passes by conduction through the support. This ability to obtain copies by both methods is of advantage with transfer processes.

Exposure can be made either by reflex, when the copy material is between the original and the source of heat, or, for single sided originals, with the copy material on the side away from the infra-red source. The best-known heat-copying paper is exposed by reflex with the heat-sensitive layer in contact with the original. This layer is coated on to thin (40–50 grams per sq. metre) tracing paper and the reproduction is viewed through the support. In one type of machine, the original and copy sheet pass close to a 1200 watt infra-red tungsten filament lamp, so that the original is scanned by infra-red in a total time of about 4 seconds. Each section is heated for not more than half a second. In order to avoid overheating, controls are provided so that the lamp lights up only when the paper passes through. The non-image areas are still heat-sensitive and the paper will blacken uniformly if heated to about 100° C. In practice, the copies are sufficiently stable in a file but the background will darken on exposure to sunlight.

The thermographic process has enjoyed wide popularity for office convenience copying because it will reproduce both single- and double-sided originals quickly by a completely dry process.

Special products are available that, by a combination of light and heat exposures, will allow of the reproduction of both dye and carbon black images. D.J.F.

Infra-Red Scanning. Heat radiations emitted by objects at normal or increased temperatures can be accosted by optical and mechanical scanning techniques with an infra-red detector. The output of the latter modulates the intensity of a light beam which scans over a photosensitive material. The resulting image shows colder and warmer portions of a subject as darker and lighter areas. Very accurate evaluation of the temperature of any point of the object is possible in this way, and such techniques have been used for military surveys, industrial testing and medical and biological investigations.

See also: *Document copying; Heat recording; Infra-red photography; Pyrometry.*

THERMOMETER. Instrument for measuring temperature.

The commonest form consists of a liquid filling a bulb connected to a long and very narrow glass tube which is sealed off at the other end. As the temperature around the glass bulb rises or falls, the liquid expands or contracts, and fills more or less of the tube The position of the liquid level is therefore a measure of the temperature, and the tube may be calibrated in degrees of the scale used.

The liquid is most commonly alcohol (with a blue dye added to make it more easily visible) or, in more accurate thermometers, mercury. The latter, owing to its high boiling point, can also be used at much higher temperatures. Such mercury or spirit thermometers are available in various sizes for darkroom use.

Amateur developing tank thermometers have to be sufficiently thin to go down the central opening of the film reel. In addition the useful part of the scale, i.e., about 50–80° F. must protrude well outside the tank for easy legibility.

Some developing tanks have a hollow and partly transparent agitating rod which takes a thermometer. The latter is thus permanently immersed in the solution, yet protected.

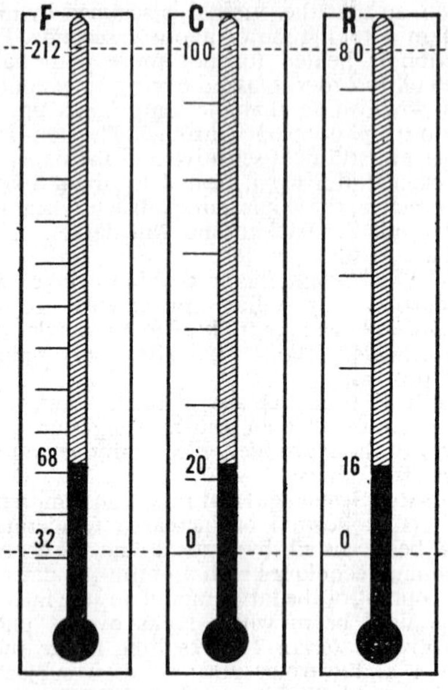

THERMOMETER SCALES. Left: *Fahrenheit.* Centre: *Centigrade (Celsius).* Right: *Réaumur. The boiling points of water are 212°, 100°, and 80° respectively, the freezing points 32°, 0°, and 0°. Usual processing temperature is 68° F., 20° C., or 16° R.*

Dish thermometers often incorporate a clip or similar fitting to hang or fix the thermometer on the edge of a developing dish out of the way, but easily seen. Certain types of dish thermometer may have a bent stem to ensure that the bulb is always fully immersed in the solution.

Large-scale professional processing tanks may have built-in thermometers. Alternatively a normal glass thermometer can be fitted in a floating holder, or suspended in a cage near the top of the tank.

Other forms of thermometer may consist of a strip consisting of two layers of metal with different coefficients of expansion; the strip therefore changes its shape at different temperatures, and is connected to a pointer moving over a scale. Many dial thermometers are of this type.

There are three common scales of thermometry, called after their inventors: Fahrenheit, Celsius and Réaumur. The Celsius scale is better known in Britain as the Centigrade. The freezing and boiling points of water on the various scales are given below.

COMPARISON OF SCALES

Scale	*Freezing point*	*Boiling point*
Fahrenheit	32	212
Centigrade	0	100
Réaumur	0	80

To convert Centigrade degrees to Fahrenheit degrees, multiply by 9/5, and add 32 to the result. Thus 20° C. = (20 × 9/5) + 32 = 68° F. Conversely, to convert Fahrenheit degrees to Centigrade, subtract 32 and then multiply the remainder by 5/9. For instance 50° F. = (50 − 32) × 5/9 = 18 × 5/9 = 10° C.

To convert to and from Réaumur degrees, proceed in the same way, but multiply by 9/4 or 4/9 instead of 9/5 or 5/9 (as for centigrade) respectively.

To convert Centigrade degrees to Réaumur degrees multiply by 4/5 (without adding or subtracting anything); to turn Réaumur degrees into Centigrade multiply by 5/4.

The Fahrenheit scale was invented in 1714 and has been used exclusively in Britain until the present century when Centigrade temperatures have been increasingly used.

The Centigrade scale was invented in 1740 and has been used on the Continent and in many other parts of the world since then.

The Réaumur scale, dating from 1731, has been popular only in Russia.

See also: *Weights and measures.*

THERMOPLASTIC RECORDING. System of image formation relying on the physical stress induced in thin plastic layers under the influence of an electrostatic field. The field is produced by an electrostatic image, after which the application of heat followed by immediate cooling converts the stresses in the plastic layer into a physical pattern. The latter can be made visible under suitable conditions of illumination. The image is suitable for playback immediately after recording, and can also be erased by re-warming the plastic layer.

The recording material is a thin layer of heat-sensitive plastic on top of a transparent con-

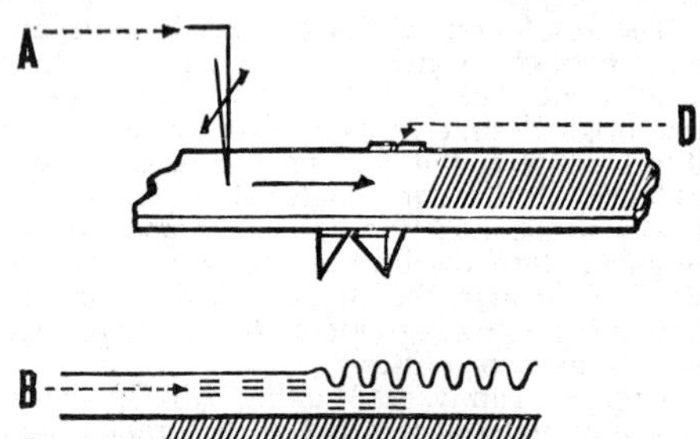

THERMOPLASTIC RECORDING. Top: *An electron beam scans the moving tape and records a charge pattern on it. As the tape passes over a heating stage which softens the plastic surface, the charge pattern distorts this surface into "wrinkles".* Bottom: *Section through thermoplastic tape. The charge pattern is discharged at the heating stage, but as the tape immediately cools the wrinkle pattern remains. A, electron beam. B, thermoplastic layer. C, film base with transparent conducting layer between it and the thermoplastic layer. D, heater.*

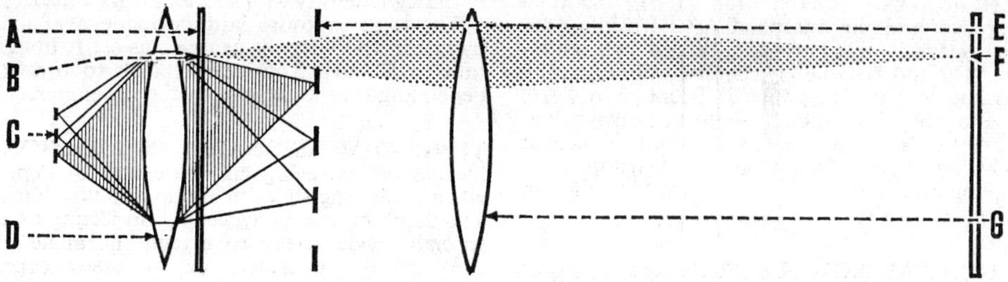

PROJECTING THE THERMOPLASTIC IMAGE. A series or slit light sources is so arranged, that a condenser system projects their light on to a grid of opaque bars between the thermoplastic film and the projection lens. Hence no light reaches the latter, except where the wrinkle pattern B in the film scatters light between the bars of the grid so that the lens records the wrinkle as a point of light at F on the projection screen. A, thermoplastic film with wrinkle images at various points such as B. C, slit light source. D, condenser lens. E, parallel grid of opaque bars. F, screen. G, projection lens.

ducting layer which may be coated on a plastic or film support. There are two possibilities of image formation. In the first an electron beam scans the surface of the plastic film and charges it in a pattern corresponding to the input signals. This produces an electrostatic imagewise distribution of charges. Alternatively, the plastic layer can be pre-charged by a coronary discharge and an image exposed on this afterwards by light.

Immediately after the exposure the plastic film passes over a heating unit which partially melts the thermoplastic layer (usually by induced currents in the transparent conducting coating). The conducting coating attracts the charges in the plastic layer and locally depresses the surface of the latter. This produces a ripple pattern which is immediately frozen into the film surface as the plastic cools again.

The image in the plastic film can be viewed directly if illuminated by light at the correct angle. For projection a system based on a Schlieren-optical set-up is used. This consists of a series of line light sources ranged in a plane behind a condenser lens, a grid some distance in front of the condenser lens, and a normal projection lens. The light is focused in such a way that the image of each line source falls on one of the bars of the grid. Hence no direct light reaches the screen at all. When the film with the thermoplastic image is introduced in front of the condenser lens, the surface ripples of the plastic layer deflects the light rays, so that they pass between the bars of the grid, forming an optical image on the screen.

Colour images can also be recorded on thermoplastic material by superimposing a diffraction grating on each picture element. In an electronic recording system this can be produced by an extra input signal to another electrode of the electron gun. The grating is an additional physical pattern on the processed thermoplastic layer and during projection produces a spectrum to each side of the central beam. The slots of the grid in front of the condenser permit only one colour to pass through the projection lens. The colour of a projected image portion is thus produced by the diffraction grating on the thermoplastic film; the colour depends on the spacing of this grating.

Applications. Originally thermoplastic recording was developed for direct recording of images produced in a manner analogous to a television screen image. In a recording unit the film can there run through at a standard speed (about 10 ins. per second) while the picture is scanned in a sequential frame fashion.

The thermoplastic film can also be used for information storage from a computer in the form of a binary digital system. The storage capacity of thermoplastic film is about 100 times as great as of magnetic tape. Thermoplastic film can record and store the contents of complete libraries in equipment the size of an office desk. Its potential applications also include optical and non-optical data recording in satellites and space ships, where the more favourable ratio of storage capacity to weight gives it advantages over straight-forward television and photographic systems. L.A.M.

See also: *Electro-photography.*

THIOCARBAMIDE. Thiourea. Used in certain warm tone developers, stabilizers, clearing baths and toners.

Formula and molecular weight. $CS(NH_2)_2$; 76.

Characteristics: White crystalline powder.

Solubility: Fairly soluble in water at room temperature.

THIOUREA. Chemical used in some warm tone developers and toning baths. It may also be used in clearing baths.

See also: *Thiocarbamide.*

THIRTY-FIVE MM. CAMERA. Miniature camera using perforated cine film 35 mm. wide. Generally 35 mm. cameras are regarded as the true miniature cameras, and are associated with the special techniques developed for making normal photographs from very small negatives with an area of 24 × 36 mm. or less.

THOMSON, JOHN, 1837–1921. Scottish explorer and photographer. After travelling for ten years in the Far East, settled in London *c.* 1870 and became instructor in photography to the Royal Geographical Society. In 1876, made the first social photo-documentation, *Street Life in London* (1877). 1880, pioneered another field—"At Home" portraiture of celebrities. Also had a portrait studio in Grosvenor Street, London.

THORIUM NITRATE FOREBATH. Forebath which produces a soft print from a harsh contrasty negative. It consists of a 10 per cent solution of thorium nitrate, with enough nitric acid added (if necessary) to clear the solution of any turbidity.

Prints to be treated in this forebath must be given two to three times the normal exposure. They are then immersed in the solution for two to three minutes, rinsed, and developed in the normal way.

The thorium nitrate forebath does not decrease the over-all contrast in the same way as the Sterry process, but mainly effects the shadows, extending their tone range. It does not change the image colour.

See also: *Contrast control; Sterry process.*

THREE-BATH DEVELOPMENT. In warm-tone development of chlorobromide papers, it is possible to make satisfactory pictures from very much over-exposed prints as long as the exact tone colour is not important. This type of paper has a wide exposure latitude when developed in diluted and restrained developer, but the tone varies with the exposure.

This latitude can be fully exploited by three-dish development.

Three dishes of developer are required:

Dish No. 1 contains full strength developer with no added potassium bromide.

Dish No. 2 contains quarter-strength developer with one part 10 per cent potassium bromide in every 50–100 parts of developer.

Dish No. 3 contains even more restrained developer: diluted eight times, with one part 10 per cent potassium bromide added to every 20–30 parts of working developer.

The approximately correct exposure is first found for an average negative to produce prints of the right density when developed in Dish No. 2. This development time is noted.

The exposures for other negatives are then guessed and the development of each exposed print is started off in Dish No. 2. If the image appears much sooner than it ought to for a correctly exposed print, the print is immediately transferred to Dish No. 3 and developed there to the right depth.

If the image appears much later than it would do with a correctly exposed print, the print is rinsed and transferred to Dish No. 1, which contains no restrainer and can produce images of the right density with shorter print exposures.

The total exposure latitude made available by this method may be as much as **4 : 1.** Image tones will vary from warm black to brown, depending on the paper and the developer.

THREE-DIMENSIONAL PROJECTION. System of conveying an impression of depth when projecting slides or motion picture films. The depth may be conveyed by projecting on a specially wide screen to extend the angle of view of the audience, or by stereoscopic methods.

See also: *Projection practice.*

THRESHOLD EXPOSURE. Smallest exposure given to a sensitized material that will just produce a visible density above the fog level on development.

See also: *Characteristic curve; Speed of sensitized materials.*

TIME BASE. Time scale photographed along the edge of cinematograph films recording high-speed phenomena. The scale—generally a wave form or a succession of dots—is produced by impulses from an electrically excited tuning fork or an oscillatory circuit of known constant frequency (in electronics, also called a time base).

By referring the action in the film to the time base, the period taken up by any particular incident can be accurately measured, irrespective of any speed variations in the camera motor.

See also: *High speed cinematography; Timer.*

TIME EXPOSURE (T). Exposure made with a device fitted to some shutters whereby an initial pressure on the release opens the shutter and a second pressure closes it. This method of exposure is used when the shutter is to be left open for more than a second or two. The shutter setting is indicated by the letter T, or, on Continental shutters, Z (Zeit).

For short-time exposures—up to two or three seconds—the B (Brief or Bulb) setting is used in which the shutter stays open only as long as pressure is maintained on the release. Cameras with double exposure lock have only a B setting; there a long time exposure must be made with a locking cable release.

TIME GAMMA CURVE. Graph of change of gamma with time of development. The slope of the curve varies with the sensitized material and the nature of the developer. The usefulness of such curves lies in the guidance they give on the length of time a sensitized material should be developed in a given formula under certain conditions to produce a desired gamma value in the negative.

See also: *Developing negatives; Gamma.*

TIME LAG. With most photographic electric light sources there is a time lag between the initial switching on and the arrival at full working brilliance.

In filament lamps the lag may be as long as a second. This is not serious in normal photography, but in automatic recording camera systems—e.g., for time lapse work or for instrument dial recording—it must be taken into account. The mechanism that operates the shutter in such cases must be arranged to switch on the lamp in advance so that when the shutter opens the illumination is at its normal brilliance. (Under these conditions it is advisable to keep a small residual current flowing in the filament to protect it from the shock of repeated switching from cold.)

With flash bulbs, the time lag from closing the contacts to the peak brilliance of the flash varies from two or three to twenty or more milli-seconds. Where the whole light output of the bulb is being used—i.e., in open flash working—the time lag is unimportant, but if the exposure is so short that it occupies only a fraction of the burning time of the bulb, it becomes important to know the exact length of the delay. The shutter synchronizing mechanism must be adjusted to allow for the time lag or it will open before the bulb has reached its peak brilliance.

Certain types of compact source lamps as used for projection and in film studios have an appreciable time lag on first switching on, and once they are switched off they must be given time to cool before the arc can be started again.

See also: *Flash synchronization.*

TIME LAPSE PHOTOGRAPHY

Technique of taking a series of photographs at set intervals, so that the changes in an object that slowly alters with time are recorded for future study.

For its simplest form, a still camera is used and exposures are made manually by an operator. The resultant photographs will form a series from which comparisons or measurements can subsequently be made.

More usually, however, a motion picture camera is employed so that the separate exposures form a motion picture film. On projection at a normal speed, this will synthesize the changes in the object which will show as movement at a greatly accelerated rate.

For work of this type, involving hundreds or perhaps thousands of exposures, manual operation is impossible and a mechanism must be provided to operate the camera automatically.

Time lapse photography is a technique that can be applied to an enormous variety of subjects, but some fundamental requirements are common to all, and the principles can be applied to any particular circumstances.

In its simpler applications, the technique presents no serious difficulty, particularly if the user has some degree of mechanical ability, though in the more specialized fields of work, highly complicated equipment is necessary.

Time lapse cinematography is usually carried out with botanical subjects such as germination of seeds, growth of plants and development of flowers and fruit. The beginner should start on an easy subject such as the opening of a well-developed flower bud, cut and kept in water. Work can be carried out in a darkened room and exposures made with artificial light. The flower will open in a reasonably short time, usually without any difficulty.

Precautions. As a general rule, slow growth is much more difficult to record than rapid changes, such as flowering, because the plant must be kept growing under favourable conditions and this may be difficult over long periods of time.

All plants require warmth, moisture, air and light for active growth, and if the arrangements for photography interfere with their proper requirements, either by reducing or increasing them, serious trouble may be encountered.

Incandescent lighting is usually not suitable for long periods owing to the large proportion of heat present and the complete absence of ultra-violet radiation. The plant will probably wilt sooner or later. Fluorescent lighting can, on the other hand, be used successfully over quite long periods but several fluorescent tubes will be needed in order to get sufficient illumination, particularly with colour film.

For the maintenance of good growing conditions over periods in excess of, say, one month, daylight is probably essential and this necessitates the provision of some arrangement of shutters to exclude light at the times of exposure. The shutters must be arranged to open and close automatically and with complete reliability, which introduces further mechanical complications.

It is not generally appreciated that many growing plants require better ventilation than human beings. For example, a box of germinating seeds may fail to grow in a closed room. The damp soil tends to go "sour" and moulds may develop. An open door and window will prove beneficial. Care, however, must be taken that the plant is not disturbed by draughts as even slight movements are distressingly apparent on the screen.

Watering must be sufficient and not overdone. Water may darken the colour of the soil, which may be undesirable. The watering of seeds in soil is difficult as the water must not disturb either the seeds or the soil. A very fine spray, frequently applied, is often the best method.

The most important point in this field is to

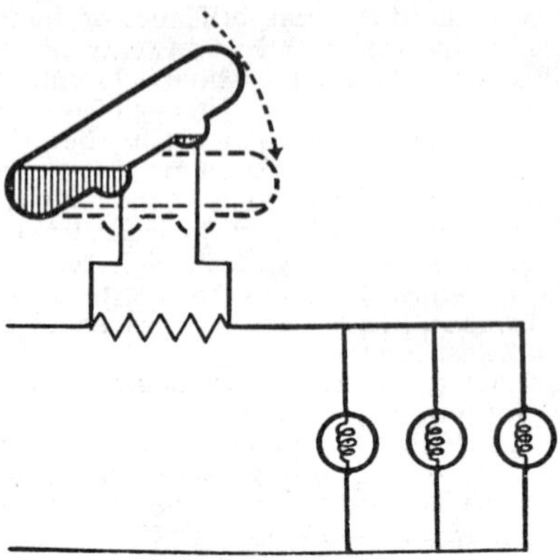

MERCURY SWITCH. Switch short-circuits resistance in series with continuously underrun Photofloods, and brings them up to full brilliance whenever required for exposing one frame.

provide suitable conditions for growth and development of the plant and to avoid any interference with these conditions. The balance is not always easy to maintain but this provides at least some of the interest in the work.

Still Camera: Manual Operation. Simple types of time lapse photography can be undertaken with almost any type of camera provided an operator can carry out the required sequence.

The basic operations will be as follows: load the camera or transport film, set the shutter, turn on lights (if used), make the exposure and turn off the lights.

Still Camera: Automatic Operation. If conditions make manual operation of the equipment impracticable, some mechanism has to replace the operator.

The automatic mechanism of any equipment for time lapse photography can be divided into two main parts: the timing unit and the operating unit.

The timing unit can be of any type, mechanical clock, electric clock, electronic timer or any other device whereby at the conclusion of any suitable interval of time, an event occurs that causes the next photograph to be taken. The operating unit replaces the human operator and must perform the sequence of operations for taking a photograph.

In some simple types of equipment both functions may be carried out by a single mechanical unit but, even so, this unit consists of these two essential parts, the timer and the operating unit.

Certain types of miniature still camera with built-in or accessory spring motor drive lend themselves to simple time lapse work. All they need is some elementary form of timing mechanism which will trigger the camera at ascertained intervals. Cameras of this type are quite often used to record the readings of a bank of instrument dials; they are known as dial recording cameras.

Cine Camera: Manual Operation. Many motion picture cameras have provision for taking single exposures either when spring driven or by means of the single turn handle. The scope of such simple equipment is limited mainly by the patience and endurance of the operator. The method has the advantages that the equipment is portable and that electric supplies are not necessary.

Many films have been made manually in circumstances when automatic operation would be impossible.

Cine Camera: Automatic Operation. For work over long periods, automatic operation is clearly essential. At least one camera for this type of work is available commercially. It is a 16 mm. motion-picture camera which is specially adapted and normally holds 50 feet of film in a daylight loading cassette, other magazines being available to hold up to 200 feet of film.

The camera is electrically driven (no spring to run down at the critical moment), and a variety of lenses is available. There are two models which can be run from either 12 or 24 volts D.C., and a mains unit is also obtainable. For normal operation, speeds of 8, 16 and 24 frames per second are possible, but for time-lapse work an interval timer, driven by a synchronous electric clock motor, is supplied which allows intervals up to 60 minutes. For situations where there is not much light, extended exposures are also possible.

For work with the microscope, a special form of observation eyepiece is required, whereby the operator can see the preparation and maintain focus whether or not the camera is running. Such eyepieces are available from a number of firms.

For use with the Bolex camera a complete outfit with an electronic interval-timer (maximum time interval three hours) can be obtained commercially. This apparatus also embodies

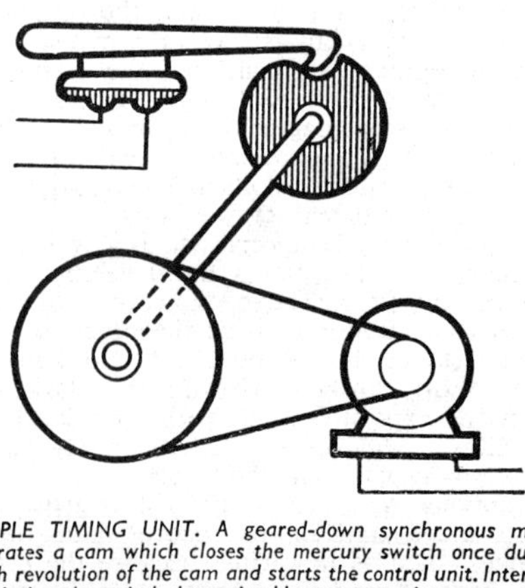

SIMPLE TIMING UNIT. A geared-down synchronous motor operates a cam which closes the mercury switch once during each revolution of the cam and starts the control unit. Intervals of closing the switch determined by motor gearing.

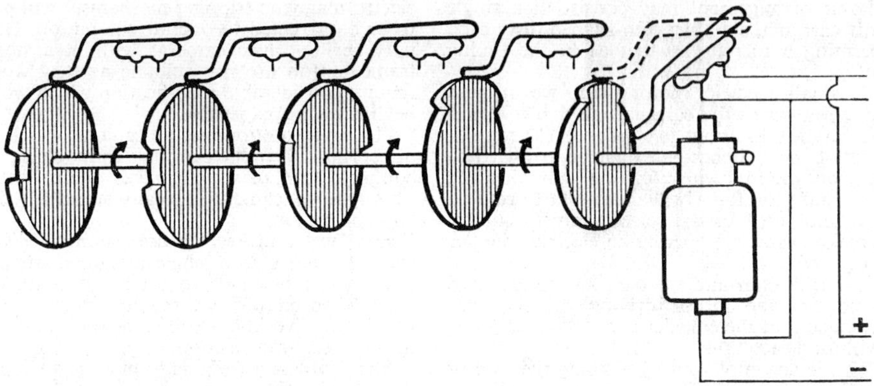

COMPLETE OPERATING UNIT. A geared down motor drives five cams, each operating a mercury switch. Four switches in turn switch on lights, start camera drive motor, release shutter, and operate an electric counter. The fifth switch controls the motor driving the sequence switch, so that the camshaft makes one revolution only at a time.

an electro-magnetic shutter whereby the preparation on the microscope stage is protected from possible harmful effects of the illumination, except at the actual moment of exposure. It also has a special electric motor drive for the Bolex camera.

Construction of Equipment. The user can also design and construct his own equipment to suit his requirements.

Complete reliability of operation is of paramount importance. If the apparatus should fail to function during a run, not only is film stock wasted, but also time. On long runs, waste of time may be much more important than waste of film stock and, in any case, breakdowns are always annoying.

In designing any equipment for time lapse work, every unit should as far as possible be the most reliable of its kind. For example, mercury switches control electrical circuits more reliably over longer periods than ordinary switches and contacts. Springs should be avoided and gravity control used instead. Electric motors should be of generous proportions and rated for continuous operation.

When considering reliability, the frequent switching of lamps will often lead to early failure. This can be largely overcome by using a considerable number of lamps and providing supervision. If one of a number of lamps should fail the result may not be very noticeable, whereas with only a few lamps the result will be spoilt.

The better solution, however, is to use a normal lighting arrangement and to arrange for the lamps to be on continuously but very much under-run, and only brought up to full brightness when an exposure is to be made. This is very easily arranged by means of a resistance (which itself may be a lamp of suitable rating) wired in series with the lamps so that the lamps are normally fairly dim. Closing a mercury switch across the resistance then applies full voltage and brings the lamps up to full brilliance for the exposure. Since the lamp filaments are never allowed to cool down and are only run at full brightness for a few seconds per exposure, their life is very much prolonged.

The Timing Unit. It is not a difficult matter to construct a simple and reliable timing unit using the movement of an electric clock. Such movements are available commercially and consist of a small synchronous motor and a train of gears, the final shaft of which makes one revolution in one minute. By means of suitable gears or pulleys a cam-shaped disc can be made to turn, say, once every 5 minutes and tilt a small mercury switch. Closing the switch initiates the automatic cycle of operations leading up to exposure.

The Operating Unit. This again can be constructed in simple form to do whatever is required by anyone with mechanical aptitude. A

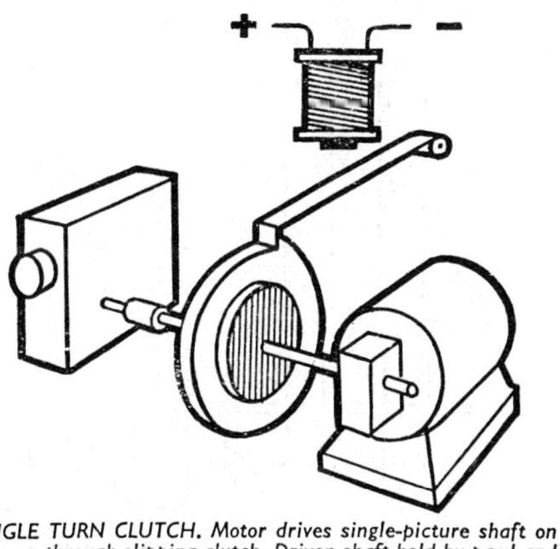

SINGLE TURN CLUTCH. Motor drives single-picture shaft on camera through slipping clutch. Driven shaft held by pawl and notched disc. To make an exposure, the electromagnet raises the pawl to release the driven shaft for one revolution.

suitable arrangement may consist of a single shaft carrying, say, five cam-shaped discs each operating a mercury switch as in the timing unit.

A small electric motor with worm gear reduction drives the camshaft so that it makes one rotation in about five seconds (12 r.p.m.). Four of the switches can then be arranged to carry out the following: turn on lights or bring them up to full brilliance; start camera drive motor (if used); release camera shutter to make exposure; operate an electric counter, if required.

The fifth cam and mercury switch are used to control the motor driving the sequence switch so that the camshaft can make only one revolution each time.

This is accomplished by shaping the cam of the timing unit so that the mercury switch is only closed for a short time. The fifth cam of the sequence switch is shaped so that it closes the mercury switch immediately the motor starts and holds it closed for almost the whole revolution. Both mercury switches are wired in parallel and are in series with the supply to the motor of the sequence switch.

If the timer motor is running, both switches will be in the off position and the sequence switch will be stationary. Immediately the timer switch closes, the sequence switch motor will start. The timer switch opens quickly, but the other switch is still closed and the sequence switch motor continues to run. When the sequence switch has made, say, three-quarters of a turn, the switch opens and the sequence switch motor stops in a position to operate again as soon as the shaft starts to revolve.

The Camera Release. The motion picture camera must be arranged to take a single exposure when required. The details will depend upon the type of camera to be used.

If the camera is spring-driven it must have a single-frame release button and some form of electro-magnetic tripping mechanism will have to be constructed to operate the button. Ordinary spring-driven cameras without a single-frame button are not suitable as they would require considerable modification which would be difficult and expensive.

Nearly all motion picture cameras which are not spring-driven have a single frame shaft, one complete turn of which makes one exposure. This type of camera is ideally suited for time lapse operation.

Single Turn Clutches. To use a motion picture camera fitted with a single picture shaft, provision must be made to drive this shaft at a suitable speed for one revolution only. This usually involves the use of some form of clutch in the drive transmission.

The simplest method is to provide a slipping disc clutch. The driving shaft is driven at low speed, say 120 r.p.m. or one turn in half a second, and the driven shaft is held back by means of a pawl and a notched disc. To make the exposure, the pawl is raised by an electromagnet to release the driven shaft.

This type of friction clutch is fairly satisfactory but remains reliable only if carefully supervised and maintained. The chief difficulties are that with insufficient friction between the two discs the clutch will slip when it should be operating the camera. On the other hand, when the friction between the two discs is high, considerable power will be required from the motor which may slow it down. On releasing the pawl, the motor speed will then increase and the time of exposure may be somewhat variable.

In addition, the faces of the discs become either polished or scored with use, which may lead to further unreliability in the operation of the clutch.

More involved designs exist for a clutch that runs quite freely but is capable of transmitting the full power of the motor to the camera. A

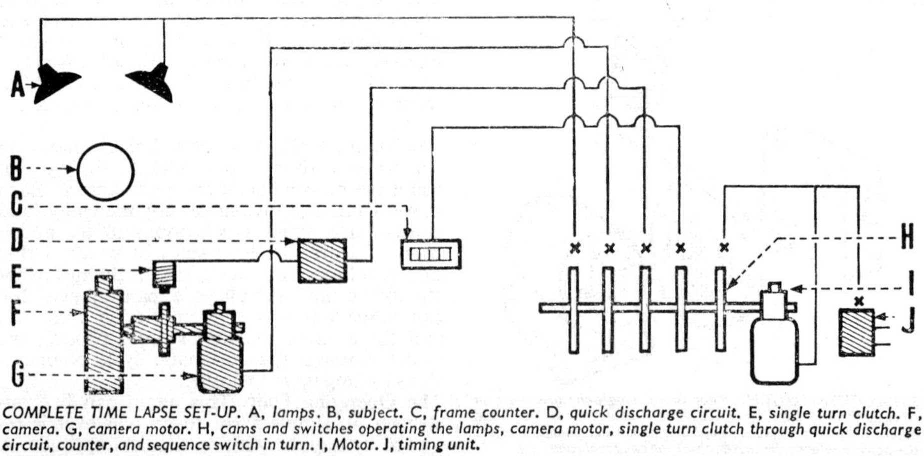

COMPLETE TIME LAPSE SET-UP. A, lamps. B, subject. C, frame counter. D, quick discharge circuit. E, single turn clutch. F, camera. G, camera motor. H, cams and switches operating the lamps, camera motor, single turn clutch through quick discharge circuit, counter, and sequence switch in turn. I, Motor. J, timing unit.

clutch of this type can be made without much difficulty but must be more of a precision job if it is to operate satisfactorily.

Quick Discharge Circuits. If the single turn clutch is driven by a motor so as to make one revolution in a quarter of a second or less, the electrical impulse to operate the clutch must be brief enough to avoid taking more than one picture.

An impulse of short duration cannot be expected from the tilting mercury switch of the sequence switch and a quick discharge circuit will be necessary.

This may consist of a simple resistance-capacitor circuit supplied with direct current so that the capacitor charges up and is then discharged through the magnet coil of the single turn clutch.

Selecting the Time-Interval. Selecting a suitable time interval should not present any difficulty when all the factors are known.

Supposing it is desired to photograph the opening of a flower and it is known that this takes ten hours. If this is to be shown on the screen in 20 seconds, the total number of exposures required will be 16 (for a silent film) or 24 (for a sound film) multiplied by 20, say 480 frames. Ten hours is 600 minutes, in which time 480 pictures have to be taken. The time interval is thus 600/480 minutes or 1¼ minutes.

R.McV.W.

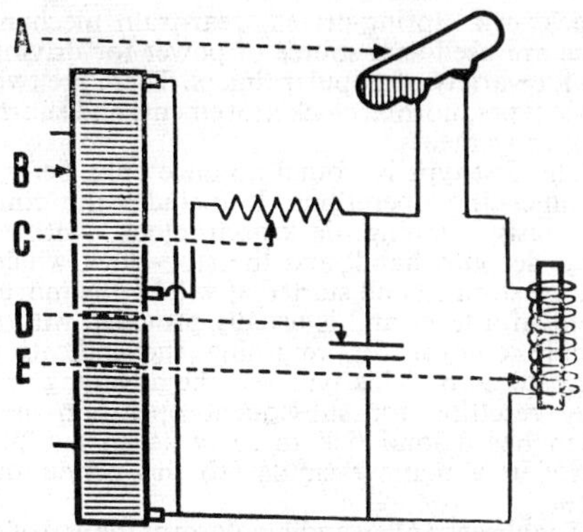

QUICK DISCHARGE CIRCUIT. A, *mercury switch.* B, *mains metal rectifier.* C, *high resistance.* D, *capacitor.* E, *magnet coil of single picture clutch. Mains voltage through rectifier charges capacitor. On closing switch, capacitor momentarily energizes coil, and camera takes one picture.*

See also: *Chronophotography; High speed cinematography; Picture series; Quick-fire camera.*

Book: *How to Do Tricks in Amateur Films,* by J. Caunter (London).

TIMER

Some form of timing device is always needed in the photographic darkroom to time the processing of negative materials and to measure exposure times in printing. While an ordinary clock with a seconds hand, or a wrist-watch with a sweep-second hand, will suffice, they are often inconvenient. A specially designed timer is more easily observed and places less strain on the photographer's attention.

The design of a timer will vary according to the precise requirements of the instrument—i.e., it may function solely as a clock or stop-watch, it may include an audible alarm, or it may automatically operate a printing light. The type of drive or timing mechanism involved —pendulum, clockwork, electric motor, or electronic circuit—will also, of course, affect the design.

According to their intended application, timers are designed to function primarily within a range of minutes or of seconds. A timer calibrated in minutes is usually intended for negative processing and one calibrated in seconds is designed for timing printing exposures.

The main features required in a timer are that it should give an audible alarm at the end of a variable pre-set period and should have a bold, easily observed dial. This latter feature is of particular importance in seconds timers which are normally operated in near darkness.

Automatic interval timers provide a convenient way of controlling enlarging exposures in particular. They consist of some form of timing device incorporating a switch. An indicator, which is usually a rotary pointer (or sometimes a telephone-type dial), is first set to the length of exposure required. Starting the timer running (generally by means of a push-button) switches on the exposing light and allows the pointer or dial to rotate back to its initial position. When the required time has elapsed, the switch is automatically opened and the light extinguished.

The chief characteristics of the various types of timers are listed below

Pendulum. This type of timer is based on the fact that the time of swing of any given pendulum depends only upon its effective length and is always the same, no matter how great the swing. The commonest example of this principle is a musician's ordinary metronome. A simpler form comprising a hanging pendulum has been marketed for photographic use. This gives audible signals at 1 second intervals for 5 minutes or so after being set in motion by the touch of a finger. Both forms are especially useful in "dodging" since the audible ticks can be counted mentally leaving the photographer free to keep his eyes on the work while shading, or dodging intricate shapes. The main disadvantage is the inconvenience of counting long exposure times.

Clockwork. Spring-driven, gear-train mechanisms are used as a source of power for driving a wide variety of popular timers. There are two basic types, normal clock movements and short-time movements.

The first type is wound up once only, before commencing operations. It includes the continuously running darkroom clock, with a large seconds hand, and the stop-clock which can be stopped and started at will by the movement of a lever and is usually provided with a simple control for returning the indicating hand to zero. This type has the advantage of easy resetting for subsequent operation and often has a loud tick to serve as an audible timer in a manner similar to the pendulum type.

The absence of an audible alarm at the end of a desired period is, however, a disadvantage as the operator has to keep one eye on the dial.

The second type of clockwork timer is similar to that found in many kitchens, in which the action of setting the indicating pointer to the required time interval also winds the spring. On pressing a release button the pointer rotates steadily back to zero, whereupon a bell or similar alarm is sounded. This audible alarm is an advantage but resetting is a two-handed job which can become tiresome, particularly in repetition work.

Automatic clockwork timers are usually based on the second type of movement above, the alarm being replaced by a switch which closes when the release button is pressed and opens at the end of the set period.

To obtain a high degree of accuracy over timed periods from a second, or less, to several minutes, some automatic clockwork timers are made double range clocks. The whole of the setting dial is used for both ranges and two separate escapements are used.

Electric. These are basically clockwork timers in which the spring is replaced by a synchronous electric A.C. motor. The continuous running form is no more than a standard electric clock with an extra large seconds hand, while more powerful types operating an alarm or automatic switch are available. The advantage of the really accurate time-keeping provided by the synchronous motor is somewhat offset by the absence of a tick for audible timing.

Electronic. Electronic timers work on the principle that an electrical condenser takes a definite time to charge to a definite voltage (or, alternatively to discharge) through a fixed resistance. The actual time depends upon the capacity of the condenser, the value of the resistance, and the applied voltage. The time may be controlled by varying any of these factors but usually it is the resistance that is made variable in relation to a scale calibrated directly in seconds. The condenser capacity or the working voltage is sometimes also changed to alter the range covered by the variable resistance.

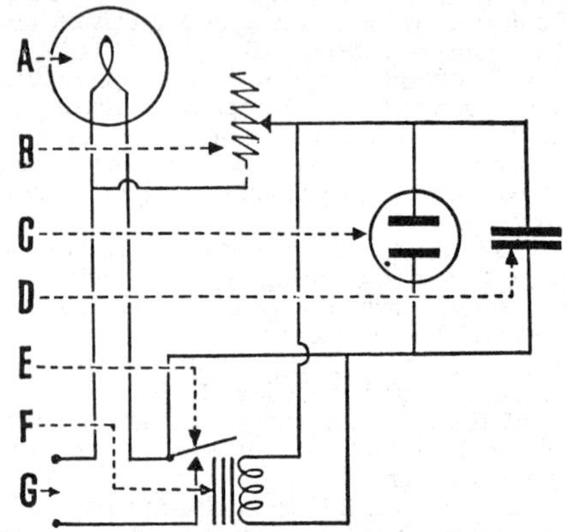

ELECTRONIC TIMING CIRCUIT. A, *lamp.* B, *variable resistance.* C, *neon tube.* D, *condenser.* E, *switch.* F, *magnet holding switch closed during exposure, released on flashing of neon tube.* G, *mains supply (must be D.C.).*

The most elementary form of electronic timer is simply a relaxation oscillator. This is arranged by connecting a neon lamp across a condenser which is being charged from a D.C. supply through a high resistance. The voltage across the condenser terminals builds up until it reaches the breakdown potential of the neon lamp. At this point, the lamp flashes and discharges the condenser. This sequence repeats automatically as long as the voltage is applied, causing the neon lamp to flash at regular intervals. By suitable adjustment of the condenser or resistance the neon may be made to flash at any desired intervals—e.g., 1 second. Since the light emitted by a neon lamp of this type does not fog printing paper, the flashes may be counted as a means of timing exposures in enlarging. The flashes are readily observed without the visual concentration required by a clock dial and the device has advantages similar to the pendulum timer.

Such a simple arrangement, however, is rather unreliable where great accuracy is essential. In practice, more complex circuits are employed, with a thermionic valve or a transistor system which trips a relay at a particular point on the charge or discharge of the condenser. This relay may be made to operate a signal lamp, an audible alarm, or a switch, so electronic timers are especially suited for automatic operation. These timers have the advantage that they operate by pressure of a button, and repeat the timing, without resetting, as often as the button is pressed.

Electronic timers may be designed for A.C. or D.C. supplies but most makes are available for A.C. operation only. The main disadvantages of electronic timers are their relatively high cost, large bulk and limited time range. Few models provide for intervals greater than 1½ minutes, although at least one expensive make will time up to 35 minutes.

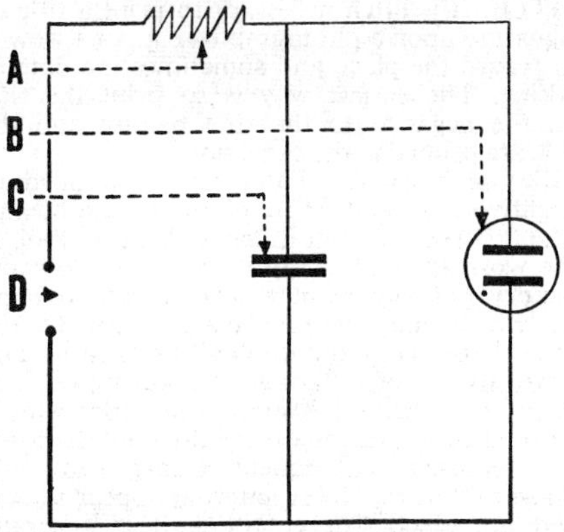

NEON FLASHER. A, resistance. B, neon tube. C, condenser. D, D.C. supply. The supply charges the condenser which discharges when the striking voltage of the neon tube is reached This cycle repeats itself, making the tube flash at regular intervals as long as the supply is connected.

Tape Recordings. For timing a complex sequence of operations, e.g., in colour, processing, a tape recording is useful. The instructions are recorded at the appropriate instant on a continuously running tape. On playing back, the operator receives these instructions without having to observe or separate times.

Special Timing Devices. Highly specialized timers are used in the various departments of scientific, industrial and commercial photography. These are usually complex equipment beyond the scope of this article, but the following brief comments cover a few of their uses.

In instrument recording by cinematography it is often essential to have a time record in addition to the readings of the various scientific meters; for this reason a clock is often included in the field of view. This takes up valuable picture space, however, and it is not always convenient to keep the clock in focus. To overcome both these disadvantages, edge-mark time impressions may be made on the film by a small neon lamp. This lamp shines on the edge of the film through a slit inside the camera and is flashed by an accurate pulse generator.

A time scale is also included in the photo-finish pictures of racing events. There are two systems in use. Both rely on the fact that the horizontal dimension of a photo-finish picture is a function of time and not distance along the race track. One system reproduces a series of short white vertical lines, looking something like a fence in the background of the picture. The white lines mark off fiftieths of a second and each represents a flash from a beacon on the finishing post opposite the camera. They enable the time difference between any two runners to be accurately estimated.

The other system prints small numbers along the base of the picture which indicate the time elapsed from the start of the race in minutes, seconds and hundredths of a second. The timer in this case is built into the photo-finish camera and has a separate optical system through which a revolving drum counter is photographed on to the moving film. The speed of rotation of the counter drums is electrically controlled by an extremely accurate quartz oscillator, and a synchronized shutter is used to produce a record every fiftieth of a second.

Electronic timers, in particular, lend themselves to adaptation for special purposes. Two timers may be linked to repeat a chosen exposure continuously and automatically, with any desired interval between operations. This is of use in some forms of D. & P. work. Another modification of electronic timer for printers provides automatic compensation for changes in developer temperature by using a resistance thermometer in the basic condenser charging circuit. An automatic timer for mass-produced enlargements has been designed in which the condenser charges through a photo-electric cell. The cell detects the brightness of the image on the enlarger baseboard, so the exposure time is automatically compensated for degree of enlargement, negative density, lens aperture and lamp brightness. This timer needs resetting only on change of paper speed, and this can be readily done by switching in different capacitors or changing the scanning aperture of the photo-electric cell.

For commercial contact printing one form of timer has been devised to speed up output by avoiding long exposure times. A synchronous electric motor rotates a make-and-break cam giving a fixed exposure time of only five-eighths of a second. To compensate for the short exposure, a relatively powerful lamp is used and its brightness is controlled by a variable resistance. Printers of this type can turn out 750 to 1,200 prints an hour.

For blue-sensitive materials, such as bromide paper, the effective intensity of a tungsten filament is proportional to the fifth power of the lamp voltage. In order to make printers that are independent of supply voltage fluctuations, electronic timing circuits have been developed which effect changes in exposure intervals which are inversely proportional to the fifth power of the changes in the supply voltage. The result is that the exposure (intensity × time) given will be substantially constant irrespective of any alterations in the mains voltage—provided that the timer and lamp are run off the same supply.

Time lapse photography is performed by a special timer which switches on the lamps to illuminate the subject, operates the camera shutter and switches off the lamps in the correct sequence at the required (pre-set) intervals.

This method reduces the power consumption of the lights and makes constant attention to the apparatus unnecessary. R.E.

TIME-TEMPERATURE DEVELOPMENT. Method of developing exposed sensitized materials. The material is immersed in a developer of a particular activity, maintained at a constant temperature and given a definite amount of agitation. Under these conditions the same material will always develop to the same gamma in the same length of time. The makers of the film or plate publish a table of recommended development times for different temperatures (and, sometimes, concentrations) of the developer for each type of material. In practice, the photographer soon learns to make any allowances necessary to produce the particular type of negative he prefers and to modify the times to suit special subject conditions—e.g., flat or contrasty lighting. Where the negative material is a length of film or a tank of plates covering a range of subjects, it is not possible to make allowances for individual subjects in development, but any discrepancies can always be put right at the printing stage—e.g., by choosing a paper of the correct grade to suit the contrast of any individual negative.

See also: *Developing negatives.*

TINTING. Term for colouring prints and lantern slides by applying suitably coloured dyes with a brush. Tinting cannot change the greys and blacks of the silver image and is mostly effective only over the highlights; nevertheless the results often succeed in being pleasing.

TINTYPE. Another name for a ferrotype—a photograph made on a wet collodion emulsion coated on a japanned metal plate and developed in an iron developer. Introduced in 1852, it was still popular in the earlier part of this century with beach and street photographers.

See also: *Ferrotype.*

TISSUE. Very thin and loosely textured soft paper. One type, known as Japanese tissue, is used as a texture screen to give enlargements a bromoil print appearance.

The term is also applied to:

(1) Dry-mounting tissue. This is a shellac impregnated tissue paper used for mounting prints. The shellac in the tissue, laid between print and mount, melts on application of heat, bonding the print to the mount.

(2) Carbon and carbro tissues. These are somewhat thicker paper with a pigmented gelatin coating which is rendered partly insoluble on sensitization with bichromate and exposure to light or chemical interaction with a silver image. Used for producing pigment images by the carbon or carbro process.

(3) Photogravure tissue. A carbon tissue which provides the etching resist on gravure cylinders.

TITLE. It is often necessary to print a title or signature upon a photograph—e.g., of a view—to record the place and sometimes the date of taking. The easiest way is to print the title on the negative so that the picture and the title are printed at the same time.

Titles on Negatives. Titles may be applied to negatives by hand or by photographic means. The first is quick, but it demands some skill. If the work is done neatly it can be quite satisfactory but anyone not accustomed to hand lettering is unlikely to make a good job. On the other hand the photographic title is easier and generally neater, although it takes more time. Whichever method is employed, titles should be used on negatives when only contact prints are required; subsequent enlargement will make all but the finest lettering appear ragged and imperfect. For enlarging it is generally better to apply titles directly on to the print—either by hand or photographic methods.

If the photographer is capable of producing good lettering, the easiest method of titling negatives will be by hand. Photographic methods, although suitable for the inexperienced artist, generally take much more time to prepare.

Hand Titles. If the negative is on film the title can be lettered on the celluloid side in block letters or script. The lettering is best done with a very fine mapping pen.

Ordinary writing ink is not suited to this purpose; the opaque medium used for blocking out negatives and indian ink used by draughtsmen are both satisfactory. If either medium is too thick, or it is thought the lettering will print too white, it can be thinned with water.

The dyes used for retouching—red, grey, or black—are also suitable. These also may be diluted and one of their advantages is that they can be removed at any time by soaking in cold water.

When the title is put on the celluloid side of the film there is some slight diffusion in printing owing to the thickness of the film, but this is not objectionable.

Those who are not skilled in the art of lettering can use printed typescript as a guide for tracing the title in this way:

First cut out a sufficient supply of letters or words from some suitable source—e.g., old magazines or Christmas cards. Then paste them on a sheet of thick paper or card, using the negative as a guide for spacing and size, and starting at the left hand side of the picture. The title is then fixed to the negative with transparent adhesive tape to keep it in the same place during tracing. It is now easy to copy the outline of each letter on the celluloid side of the film.

There must not be too much of the medium on the pen. It is best to work with a pen nearly dry, and to make two or three applications of the medium to build up the letters.

If the title is printed on the emulsion side of

the negative, the lettering must be reversed. This can be done by first tracing the letters on to a sheet of transparent material such as an old film negative bleached in Farmer's reducer. This is placed in position on the glass side of the plate and the title can then be traced on the film side. Alternatively the title may be traced on a strip of clear celluloid and attached to the negative with cellulose tape.

Black lettering is often preferred as being less glaring than white while being just as easy to read. The lettering is scratched on the film side of the negative with a sharp point—e.g., of a darning needle—mounted in a cork and pressed into a pen holder. The lettering in this case must be drawn reversed as for titles on plate negatives. The clear lines left on the negative come out as black lines on the print.

Another way of introducing black lettering is to write on the negative with a small brush dipped in Farmer's reducer. This is not an easy method as the solution is not easy to control, and there is some risk of spoiling the negative. It is not suited to small negatives.

Photographic Methods. Although these involve rather more work they give more pleasing results especially in the hands of beginners and they avoid risk of damage to the negative. Moreover if the title is no longer needed it can easily be removed.

For black lettering the title should be lettered in indian ink on stiff white paper or card. The characters should be $1\frac{1}{2}$ to 2 times the final size as this makes the work easier and helps to minimize any small imperfections. It is possible to build up a title using typescript cut out and pasted on paper or card or to have it set up by a printer, and the "pull" used.

Another alternative is to build the title with the card or plastic letters used for titling cine films. The title is set up in front of the camera and photographed. The negative should be used as a guide to see that the title is in the correct position and of the correct size.

High contrast film of the process type and a contrast developer should be used. This combination will ensure that the lettering is sharp and clear against an opaque background. Any slight veil can be removed by treatment (for a few moments only) with Farmer's reducer.

The title is printed as follows: expose the picture negative under the enlarger or in the printing frame; remove it, and replace with the title negative, taking care that the title comes in the right place; expose through the title negative and develop the print.

If white letters are preferred it is only necessary to make a contact positive from the first negative and print it simultaneously with the picture negative. Title strip negatives may be made in this way and attached to the film or plate in the correct position with transparent adhesive tape.

Titles on Prints. If the print is on matt or rough paper, the title may be written on the surface in

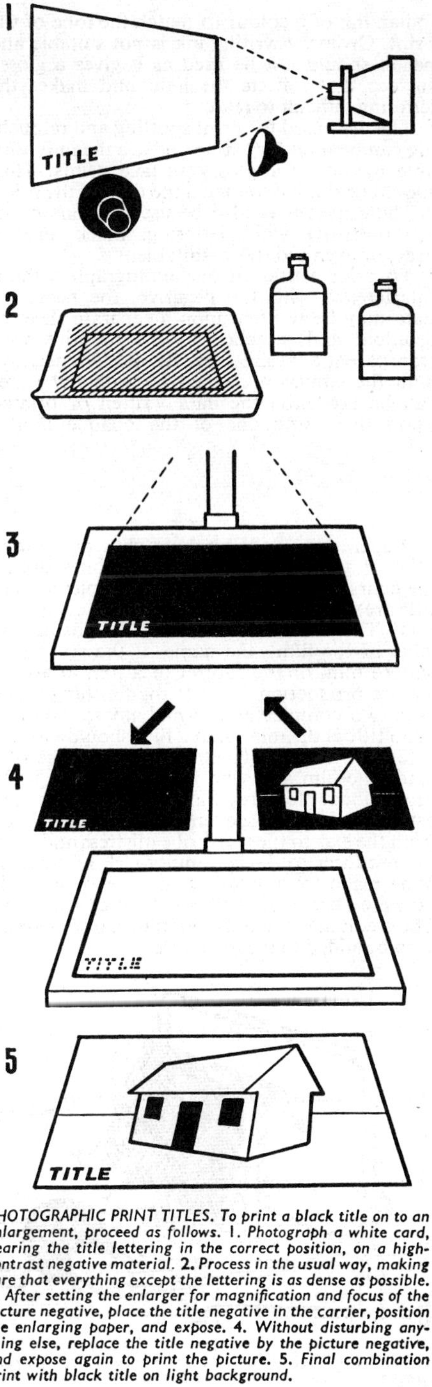

PHOTOGRAPHIC PRINT TITLES. To print a black title on to an enlargement, proceed as follows. 1. Photograph a white card, bearing the title lettering in the correct position, on a high-contrast negative material. ***2.*** *Process in the usual way, making sure that everything except the lettering is as dense as possible. 3. After setting the enlarger for magnification and focus of the picture negative, place the title negative in the carrier, position the enlarging paper, and expose. 4. Without disturbing anything else, replace the title negative by the picture negative, and expose again to print the picture. 5. Final combination print with black title on light background.*

indian ink of a colour to match the tone of the print. Ordinary writing ink is not suitable and pencil should not be used as it gives a glossy surface that reflects the light and makes the lettering difficult to read.

The dyes used for print spotting and retouching can be used for tilting and are the only suitable medium for glossy surface prints. They should be diluted to match the tone of the print.

These dyes may also be used to tone down too obtrusive white lettering. Black, brown, grey, or even red dye is suitable.

To print a title on the photograph without interference with the negative, the necessary data may be written upon a separate sheet of celluloid and attached to the negative with transparent adhesive tape. An old film negative with the emulsion cleaned off with hot water can be used and the data written or printed upon this using one of the opaque liquids employed for blocking out. These will take easily upon celluloid by using a hard pen nib with a fair amount of pressure. A small trace of liquid gum added to the medium helps. Alternatively the celluloid may be given a matt surface by rubbing with valve grinding paste or fine emery paper. The data may then be written on the surface with ordinary pencil.

For black lettering, fog a sheet of film, place it in a developer until opaque, and fix, wash and dry. The title or reference can be scratched on the film side with a darning needle. The title is then mounted over a hole in a sheet of opaque paper and placed over the exposed print. A white light is then shone on the title for a second or two and, when the print is developed, the title appears in black lettering. R.M.F.

See also: *Caption; Greeting cards; Lettering.*

Book: *The Complete Art of Printing and Enlarging*, by O. R. Croy (London).

TITLING

Preparation and photography of written titles at the beginning and sometimes during the course of a cine film. The principle types of title are:

(1) The main title. This is the identifying label of the film—the name of the story in a feature film, or the subject of a film in a non-feature production. It adds the finishing touch even to a completely informal amateur film. A main title is the minimum a film should have.

(2) Credit titles. These tell the audience who made the film and who helped to make it. In professional feature films credit titles are usually a lengthy sequence, listing everybody from the star to the wardrobe mistress and from the producer to the continuity girl. In amateur films made by a small team or even a single movie maker one credit title is often sufficient. The credit line of a one-man film can sometimes even be added to the main title.

(3) Sub-titles. These tie together groups of sequences, establish the location and where necessary the time. In silent films they may also indicate the dialogue.

(4) The end title.

Some processing laboratories prepare titles for cinematographers according to texts supplied and following certain styles of lettering and presentation. Home-made titles give, however, a greater scope as well as individuality.

Titling involves preparing the title itself and filming it with the cine camera. The latter is pointed squarely at the title, illuminated by suitable lamps. The title is filmed either at the normal camera speed or—for special effects—by single-frame exposures. To be suitable for titling, the camera must also be able to focus on near subject distances. A variety of commercial titling units are available which accurately line up the camera opposite the title and may

HORIZONTAL TITLER. A, *title board (here with flap-over feature).* B, *title frame.* C, *titler base.* D, *adjustable lamp.* E, *close-up attachment.* F, *camera.* G, *twin rail base.*

incorporate the lighting units and provision for other titling accessories.

With fixed focus cameras supplementary lens provide appropriate near focusing distances. If the title size is constant, a single supplementary lens is sufficient.

Titlers. Basically a titler consists of an easel or other arrangement to hold the title, a support for the camera, and the lighting unit or units to illuminate the title. The lay-out of the titler may be horizontal or vertical. A horizontal titler carries the camera on carriage or support moving along the single or double rail to keep it accurately centred opposite the easel. It also permits it to move nearer or farther away while remaining centred.

In a vertical titler the easel is horizontal, and the camera moves up and down on the vertical column.

Horizontal titlers with a vertical easel are usually more rigid, and adequate for printed or drawn titles or stick-on title letters. Vertical titlers permit more scope in arranging and animating title solid letters on the horizontal easel.

Ideally, titles should be made at least 8 or 10 ins. large; the easel dimensions will then be correspondingly larger. Smaller titles can be produced. But they need a closer camera-to-title distance for filming, and any roughness in the lettering shows up far more obviously on the screen.

The distance the camera has to be from the title depends also on the focal length of the lens. With an 8 mm. camera an 8 × 10 ins. title can be filmed at about 2½ feet from the camera fitted with a standard 12·5 mm. lens. For a 9·5 mm. camera the titling distance is just over 2 feet with a 20 mm. lens. A 16 mm. camera with a standard 25 mm. lens would cover an 8 × 10 ins. title area again from about 2½ feet.

In practice the exact title area covered by the camera at various lens-to-easel distances is best determined by test exposures, or by visual

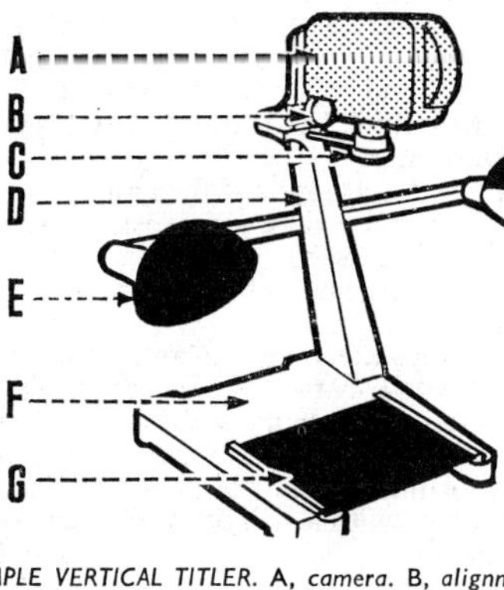

SIMPLE VERTICAL TITLER. A, camera. B, alignment control. C, supplementary lens. D, column. E, lamp. F, baseboard. G, title in guides.

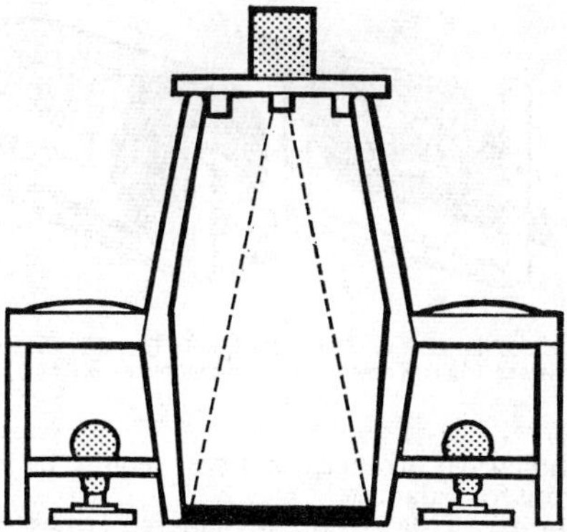

EMERGENCY TITLING SET-UP. The camera is supported above two chairs pushed together. The title board lies between them and is illuminated by two lamps on each side. The seats of the chairs shield the camera lens from direct light.

observation—if the camera has a through-the-lens reflex finder.

Lettered Titles. Titles should in most cases be as short as possible, to aid readability and also to reduce the amount of lettering required.

At one time virtually all titles were white letters on a black background. This is now considered unnecessarily dull, and black or grey letters on a tinted or light background are preferable.

Very short titles can be drawn or painted free-hand, but should still be neatly set out. Stencils are a useful way of producing letters of various kinds. So are transparent plastic sheets of self-adhesive letters which only need cutting out and pressing into position on a suitable background.

Plastic, cork and felt letters are easy to handle on a horizontal titling board, and are obtainable in a wide range of sizes, styles and colours.

Backgrounds. The easel of the titler can be covered with drawn, painted or other background patterns. Textured backgrounds are also effective, especially in conjunction with solid title letters when illuminated by a low glancing light to emphasize the depth relief.

Titles can also be set up against a still photograph or a live scene. In professional motion picture work the latter type is produced by combination printing; the amateur method is to film the live scene and then wind that section of film back in the camera for re-exposure a second time to the title. The latter thus appears in super-imposition on the scene.

Alternatively, the title may be set up on a transparent sheet and arranged in front of the camera against a live background. In this case the title should be rather larger, and small lens apertures are needed to make the background at least recognizable.

Titles should always be carefully centred within the title frame. Deliberately non-

ACCESSORIES FOR EFFECT TITLES. Left: *Flap-over title board.* Centre left: *Drum for rotating title.* Centre right: *Shutters for title wipes.* Right: *Travelling title on continuous rollers.*

symmetrical arrangements still require careful layout for balance.

Informal Titles. Formal lettered titles are always obvious in a film, and when used to link scenes or sequences, may break up the smooth flow of the film. The information to be given in the title can however also be conveyed in a variety of other less obvious ways. Thus sign-posts establish locations in holiday films. A close-up of a calendar on the wall indicates the date. Linking shots of a map, posters advertising a sports event (being filmed), and even newspaper headlines are examples of informative shots of this nature which can obtrusively replace orthodox set titles. Even dialogue titles can be largely cut down in number by linking scenes which make it obvious to the audience what is being said.

Shooting. The exposure for a title depends of course—apart from the film speed—on the intensity of illumination provided by the lamps. These should be set up in such a way that they evenly illuminate the titling area, yet do not reflect any glare (especially from shiny-surfaced title cards) into the camera lens.

Where titles are shot at the normal camera speed of 16 or 24 frames per second, fairly intense lighting is necessary, comparable to that recommended for indoor shooting with the film employed. With colour film the appropriate artificial light type emulsion should be used, unless the titles are superimposed on an outdoor shot. With colour film, titles can, however, be lit by coloured lights for special effects.

A meter measuring the light reflected from the title give a reliable indication of the exposure. The meter should take in just the title area and the title should have a fairly balanced tone range. Alternatively, the light reflected from a white card in the position of the title area may be read and the camera lens set to two stops larger than the indicated value.

The footage devoted to a title depends largely on its length. Credit titles and sub-titles should be held long enough to be read comfortably by an average audience. Approximately two words per second is suitable for descriptive titles. Main titles need a longer time.

For trick titles (e.g. letters unscrambling themselves) the time to count is the period for which the fully readable title stays on the screen. Running titles should be held long enough for reading about two-thirds of the text before the title starts to move. Hold it again after it has come to rest.

Special Effects. Titles may incorporate the usual camera tricks and effects such as feeds, dissolves, wipes, etc. which are largely made in the same way as when shooting live scenes. In addition there are moving titles obtained in several ways.

A long descriptive title—or even a series of credit titles—can be made to move past the field of view by being mounted on a drum or on a long strip of paper wound from one roller to another.

Moving titles to be created in the titler also include the flap-over where a pivoted title board carries successive titles on its two sides and swings over for the change. Another classic way of changing from one title to the next is to have title sheets arranged in a book, and to turn over the pages one by one.

Letters that appear in jumbled fashion and then sort themselves out to the titles are usually filmed by reverse action. The title is set up with movable plastic or cork letters, filmed with the camera upside down, and the letters then scrambled up by suitable means. On splicing this sequence into the film upside down in the desired unscrambling effect. (In 8 mm. this effect requires a camera with reverse drive since splicing in a film length upside down brings the perforation to the wrong side of the film. Reversing the film left-to-right would also reverse the lettering.

Advancing titles are easy if the camera has a zoom lens; otherwise they involve moving the camera smoothly towards the titling easel during the shot. The title must be held and sharp in its final position.

Animated titles are made in much the same way as animated cartoons, by drawing or arranging lettering step by step while photographing them with the single frame release of the camera. Animated or inanimate titles can also be combined with a still or moving cartoon background. L.A.M.

See also: *Animation; Cine technique.*

Book: *How to Title*, by L. F. Minter (London).

TOMOGRAPHY. Special method used largely in medical radiography to reveal details of internal structure normally shielded by other parts. During exposure the X-ray tube is swung in an arc of a circle which is centred on the point of interest. Meanwhile the film cassette, which is connected to the tube by a lever, is slid in the opposite direction. In the resultant radiograph, all structures above and below the pivot plane will be blurred, leaving a relatively clear impression of the selected object.

See also: *Medical radiography.*

TONE. Area of uniform density in a positive or negative image which can be distinguished from darker or lighter parts. A picture may consist of two tones—e.g., a black and white woodcut; or a number of tones—e.g., a photograph.

TONE-LINE PROCESS. Photographic method of converting a continuous-tone image into a line drawing, while preserving full detail of all outlines and supressing all tone values. The resulting line image may be negative or positive. Such line images can also be reproduced photo-mechanically for printing by line blocks or plates.

Principle. The process uses a negative and an exactly matching positive transparency of the same contrast and density. The two are superimposed with a thin spacing layer between them (the film supports of the negative and positive material are enough for this purpose) and contact printed on to another film by an oblique moving light source. By going through the negative and positive images at an angle, the light can spill past outlines of the superimposed images, recording these outlines on the printing material. The greater the separation between the negative and positive, and the more oblique the angle of the light, the thicker the lines are reproduced. The resulting image shows black lines on a white background; this can be printed directly on a printing paper to yield white outlines on black, or indirectly via another transparency to yield a print with black outlines on white.

Method. The image on the original negative should be as sharp as possible and with a density range of about 1·5. If the original image is on miniature film it is usually preferable to make a large-size duplicate negative.

The negative is printed by contact on to a continuous tone film and developed in the normal way. Exposure and development of this positive transparency should yield an image that matches the negative as closely as possible in contrast and density.

The negative and positive images are then registered back to back and placed in contact with a high contrast line or process film. This combination is now printed with a lamp placed at an angle of 45° to the surface of the sandwich. During the exposure the printing frame

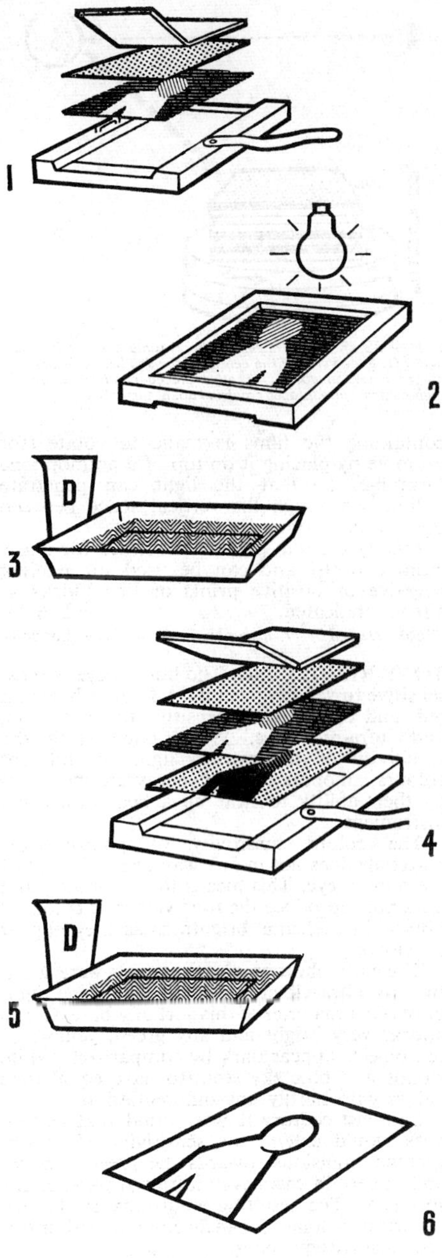

THE TONE-LINE PROCESS. 1. Loading the selected negative into a printing frame with a positive film. 2. Exposure. 3. Development. 4. Combining the negative and the positive in a printing frame with a further unexposed film. 5. Exposure and development. Exposure takes place with relative movement between the printing frame and the light source. 6. Resulting line image.

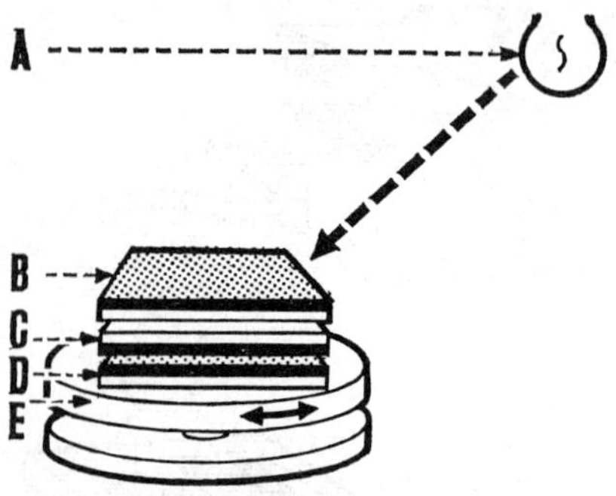

EXPOSURE ARRANGEMENT. A, *pearl lamp about 3 ft. above and 3 ft. to one side of the centre of the sandwich.* B, *positive, placed emulsion up.* C, *negative placed emulsion down.* D, *high-contrast film, placed emulsion up.* E, *turntable.*

containing the films is made to rotate (for example by placing it on top of a gramophone turntable) so that the light can penetrate obliquely through the vertical space between the outlines from all sides.

Finally the film print is processed for maximum contrast and can be used for making negative or positive prints or line blocks as already indicated. L.A.M.

Book: *Design by Photography*, by O. R. Croy (London).

TONE RENDERING. The human eye is most sensitive to yellow-green light, least to blue and red, and completely insensitive to ultra-violet and infra-red rays. When considering the sensitivity of the eye to light of different colours, amounts of light energy are measured by their ability to heat up a heat measuring instrument.

The colour sensitivity of photographic materials does not in any way resemble that of the human eye. This means that a photograph does not reproduce the tone values of coloured objects in the same brightness as they appear to the eye.

The early photographic materials were sensitive to ultra-violet radiation and blue light only. To a material of this sort any blue objects appear very bright and any green, yellow, or red objects appear dark by comparison. White clouds and blue sky seem to have equal tone values with hardly any differentiation.

Late last century it was found that certain dyes would extend the sensitivity of photographic emulsions towards the green and red and in certain cases even infra-red parts of the spectrum. The resulting emulsions are known as orthochromatic, panchromatic, and infra-red sensitive emulsions.

Orthochromatism. Orthochromatic emulsions are sensitive to blue and green light and give a much better tone rendering than the earlier blue-sensitive emulsions. But they are insensitive to orange and red and tend to make them too dark in the print. Even so, orthochromatic emulsions are widely used for general photography and their performance is much improved if they are used with a yellow filter which cuts off some of the excessive sensitivity to blue.

Panchromatism. The colour sensitivity of panchromatic emulsions is nearest to that of the human eye. Nevertheless they are still excessively sensitive to blue and somewhat too sensitive to red light. So correction filters are used to give correct colour rendering.

Fairly accurate rendering of colour in terms of different tones of grey can be obtained with a panchromatic emulsion and a yellow filter which cuts out the excessive sensitivity to blue and ultra-violet light.

Even more exact colour rendering is obtained by using a greenish filter which corrects the excessive sensitivity to red light as well as cutting down the blue.

Contrast Filters. The photographer does not necessarily want to reproduce every colour according to its brightness to the eye. The human eye distinguishes between different objects in a scene not only by their different brightness, but by the contrast in colour.

Unfortunately, the colour contrast cannot be captured by the normal photographic material and it is quite possible for the effective contrast between two objects to be lost entirely in a black-and-white photograph. For instance red poppies in a green meadow may become quite inconspicuous. In such cases the photographer often resorts to contrast filters to suppress one colour and increase the contrast between it and another colour. Filters can be used in this way to give striking, though unnatural effects.

Orange and red filters, for instance, turn a blue sky almost black in suitable lighting conditions, and are often used to add drama to sky and cloud scenes.

Infra-Red Emulsion. Infra-red sensitive materials give completely unnatural tones but they have a number of specialized applications in which truthful tone rendering is not important. Thus to an infra-red sensitive material the green of meadows and trees appears a brilliant white since the chlorophyll responsible for their green colour reflects the infra-red light. So an aerial reconnaissance photograph, if taken on infra-red material, will clearly show up otherwise perfect camouflage.

Infra-red material is also used for photographs of distant objects. This is because atmospheric haze tends to allow infra-red rays to pass while it scatters blue light. Therefore if a photograph is taken with an infra-red sensitive material and a filter which cuts out blue light, haze is much less noticeable than it would be with ordinary photographic materials. W.F.B.

See also: *Filters; Monochromatic illumination; Negative materials; Spectral sensitivity.*

Books: *Filter Practice*, by H. Clauss and H. Mensel; *How to Choose and Use Film and Filter*, by L. A. Mannheim (both London).

TONERS

Toners change the colour of the black silver image of a print. They are commonly used for giving a black-and-white print a colour more suited to the subject of the photograph—e.g., for making a sepia-coloured print of a portrait or a subject in sunshine, or for making a blue or green coloured print of a sea study. Most toners can be used equally well on lantern slides or other positive transparencies.

There are four general types of toner:

(1) Sulphide and selenium toners, which convert the silver image into silver sulphide and/or selenide. These usually give a range of reddish brown tones.

(2) Metal toners which convert the black silver image into a compound of the desired colour. This compound may be an insoluble coloured silver salt or a double salt with another metal.

(3) Dye toners, which act by depositing a dye image on top of the silver image. The silver image can, if necessary, be removed completely in a suitable reducer.

(4) Colour developers, which form a dye image as well as a silver image. The dye is formed during development by oxidation products.

SULPHIDE AND SELENIUM TONERS

In indirect sulphide toning, the image is bleached out in a halogenizing bleacher, washed, and then darkened in a sulphiding solution.

The Bleacher. The standard bleacher is a solution of 1–5 per cent each of potassium ferricyanide and potassium bromide. A permanganate or copper sulphate halogenizing bleacher containing sodium chloride will work equally well.

Variations in tone can be obtained by using an alkaline potassium ferricyanide and potassium iodide bleacher. There are several alternative bleachers which give slightly different tones from those obtained by the standard process. One of these is a solution of 0·5 per cent potassium bichromate containing 4 per cent hydrochloric acid.

The Darkener. In sulphide toning the standard darkener is a 1–5 per cent solution of sodium sulphide—usually made by diluting a 20 per cent stock solution, since dilute sodium sulphide solution does not keep well. Those who object to the smell of sulphide toning with sodium sulphide solutions may use a darkener containing 0·25 per cent each of thiocarbamide and caustic soda, with or without up to 2 per cent potassium bromide. It is possible to control the tone to some extent by varying the proportions of thiocarbamide and caustic soda.

For selenium toning the darkener may be a 1 per cent solution of sodium sulphide containing 0·05–0·1 per cent selenium—dissolved in the hot solution. Those who object to the smell of the sulphide may use the thiocarbamide darkener (above) containing 0·1 per cent of selenium.

With most bromide papers, sulphide toners give a warm sepia image colour. There are some papers, however, which cannot be made to yield anything better than an unpleasant yellow-brown. Such papers are unsuitable for sulphide toning.

Selenium toners give a colder purplish tone.

Controlled Sulphide Toner. The image tone obtained by sulphide toning can be varied to colder brown or even warm-black by adding up to equal parts of 2·5 per cent mercuric chloride solution to the standard ferricyanide and bromide bleacher. The higher the proportion of mercuric chloride, the colder the tone obtained. At the same time, the more mercuric chloride there is in the solution, the more the print will be intensified.

After any bleacher containing mercury, the prints should be rinsed in several baths of 1·5 per cent hydrochloric acid, before being washed and toned.

Warmer tones (to brownish-red) follow the use of the standard bleacher with a darkener containing a proportion of 1 per cent sodium thioantimonate (Schlippe's salt) solution. This adds orange-yellow antimony sulphide to the image. The warmest tones of all are given by the thioantimoniate solution only as a darkener.

Direct Toning. For direct sulphide toning the print or transparency is immersed in the toner without previous bleaching. The bath converts the image into silver sulphide or selenide.

The solutions take rather longer to act than in indirect toning—some 10–20 minutes instead of 3–5 minutes. The prints are taken out of the solution just before they reach the required tone, because the action continues slightly during washing.

Most direct sulphide toners are of the polysulphide type. A suitable direct toner can be made by adding powdered sulphur to boiling 50 per cent sodium sulphide solution until no more will dissolve. When cold, the solution is diluted ten times for use. A 1 per cent solution of liver of sulphur (potassium polysulphide) is equally satisfactory.

On the same principle, a direct selenium toner can be made by dissolving powdered selenium in hot 10 per cent sodium sulphide solution. This toning solution tends to stain the prints, but it can be improved by mixing 3 parts of it with 20 parts of a 15 per cent sodium thiosulphate (hypo) solution containing 5 per cent of potassium metabisulphite.

An odourless selenium toner can be prepared by dissolving 1–2 parts of selenium powder in 150–200 parts of 15–20 per cent sodium sulphite, and adding 30–40 parts solid ammonium chloride. This solution may be diluted with an equal volume of water.

Hypo-Alum Toners. Hypo-alum toning is another method of direct sulphide toning. The prints are first hardened in a formalin or similar hardener and then immersed in a hot solution of 20 per cent sodium thiosulphate (hypo) containing 2·5–5 per cent potash alum. The tone obtained depends on the temperature of the bath: the hotter the solution, the colder the tone. About 120–130° F. (50–55° C.) is the optimum temperature. Toning usually takes 10–15 minutes.

A fresh hypo-alum toning bath has a slight reducing action on the image. This action can be exhausted by immersing a few waste prints in it or by adding a small amount of 10 per cent ammoniacal silver nitrate solution (about 1 part in 200).

Because of the high temperature of the solutions, hypo-alum toning is not suitable for lantern slides or other transparencies.

METAL TONERS

With most metal toners the silver image is first converted into an insoluble silver salt in a bleaching bath. The silver salt is then acted on by a darkener which converts it into a coloured compound consisting of a double salt of silver and the metal contained in the darkener. The salts most commonly used in this double salt formation are ferrocyanides.

Some toners contain both the bleaching and the darkening agent in one solution, so that the whole process takes place in one bath.

In other toners—e.g., iron toners—the silver itself may even be removed altogether, leaving only the coloured metal salt in its place.

A few metal toners deposit another metal directly on the silver image. The deposited metal has to be a metal nobler—i.e., of lower chemical activity—than silver, e.g., gold.

Uranium Toning, In uranium toning the silver image is converted into silver and uranium ferrocyanide which is reddish-brown in colour. The image is not, however, as permanent as the original and tends to be destroyed by alkaline solutions.

For this reason, all rinses and the final wash should take place in slightly acidified water—e.g., 0·1 per cent acetic acid.

For indirect uranium toning the image is bleached in a 3 per cent potassium ferricyanide solution containing a trace of ammonia. It is washed and then toned in a 2·5 per cent uranyl nitrate solution containing 1 per cent of acetic acid and 0·5 per cent of potassium bromide.

For direct uranium toning the print or transparency is treated in a toner containing uranyl nitrate and potassium ferricyanide in a suitably acid buffer solution—e.g., the Kodak T-9 formula:

Uranyl nitrate	44 grains	2·5 grams
Potassium quadroxalate	44 grains	2·5 grams
Ammonium alum	105 grains	6 grams
Potassium ferricyanide	18 grains	1 gram
Hydrochloric acid, conc.	10 minims	0·5 c.cm.
Water to make	40 ounces	1,000 c.cm.

The prints are toned to the desired colour, and washed in changes of 0·1 per cent acetic acid until all stain is removed.

Uranyl nitrate solutions are slightly sensitive to light, and should be stored in the dark.

Gold Toning. There are two methods:

(1) A plain black-and-white print or transparency can be made to turn a rich bluish-black by immersing it in the gold toner. In the resulting image, each individual silver particle is gold-plated. This type of image is more permanent than the original silver image. The process can also be used to improve greenish-black tones caused by exhausted or excessively restrained developers.

(2) A sulphide-toned print or slide can also be toned subsequently with gold. The effect of this is to turn the image chalk-red.

The usual gold toner contains 0·1 per cent of gold chloride and 1 per cent of potassium or ammonium thiocyanate. This can be made up by mixing the components as 1 per cent solutions in the appropriate proportion of 1 : 10.

The image is toned to the colour required, and the print or transparency is then washed and dried in the usual way.

Nickel Toners. In normal nickel toning the print is first immersed in the following bleaching bath, in which the toner converts the silver image into silver and nickel ferrocyanide.

Nickel nitrate	2 ounces	50 grams
Potassium citrate	12 ounces	300 grams
Citric acid	1 ounce	25 grams
Formalin, 40% solution	4 ounces	100 c.cm.
Water to make	40 ounces	1,000 c.cm.

Just before use, 5 parts of this solution are mixed with 1 part 15 per cent potassium ferricyanide.

After bleaching, the print is immersed in a 5 per cent sodium thiosulphate solution for 5–10 minutes.

For bright red tones the print is then toned with a dimethyl glyoxime toner:

Dimethyl glyoxime, saturated solution (7–8 per per cent) in alcohol	10 parts
Ammonia solution, 0·880	1 part
Sodium hydroxide, 5% solution	1 part
Water to make	100 parts

Final washing in changes of 0·1 per cent acetic acid solution completes the process.

For violet tones the nickel-toned print may be immersed in a 25 per cent solution of ferric alum containing 0·5 per cent of sulphuric acid and 10 per cent of potassium bromide.

For bluish-green tones the original black-and-white image is bleached in the nickel bleacher, washed, and immersed for five minutes in 5 per cent ferric chloride, followed by hydrochloric acid diluted 1 : 6. The print or transparency is then washed in several changes of 1 per cent citric acid, fixed in 5 per cent sodium thiosulphate solution, and washed again in eight to ten changes of 1 per cent citric acid.

Lead Toning. In lead toning the silver image is converted into yellow lead chromate.

The print or transparency is first bleached in a 2 per cent potassium bichromate solution containing 0·5 per cent of thiocarbamide. It is then washed and toned in 5 per cent lead acetate containing 0·5 per cent of acetic acid. After toning it is fixed in 5 per cent sodium thiosulphate and washed again.

Cobalt Toning. Cobalt toning gives green tones. The image is bleached in a 3 per cent potassium ferricyanide solution containing 0·5 per cent of potassium bichromate, cleared in 3 per cent potassium metabisulphite solution, and toned in a solution containing 2·5 per cent each of crystalline cobalt chloride and acetic acid, and 0·5 per cent of ferrous sulphate. Washing, fixing, and a final wash complete the process.

Vanadium Toning. Vanadium toners have also been used to obtain green image tones. They are, however, uncertain in their action, and can now be considered obsolete.

Iron Toning. Iron toners convert the silver image into a brilliant prussian blue ferro-ferricyanide compound. These blue toners also intensify the image, so prints or slides should be made slightly lighter if they are to be toned blue afterwards.

All iron toners consist essentially of a mixture of a ferric salt—usually a more stable double ferric ammonium salt like ferric alum or the double oxalate or citrate—and potassium ferricyanide, in acid solution. The silver is converted into silver ferrocyanide which immediately forms a ferro-ferricyanide. Various other substances may be added to stabilize the toner and improve its consistency and keeping qualities.

A suitable formula is:

Ammonium persulphate, 10% solution	1 part
Ferric alum, 10% solution	3 parts
Oxalic acid, 5% solution	12 parts
Potassium ferricyanide, 10% solution	2 parts
Ammonium alum, 10% solution	10 parts
Hydrochloric acid, diluted 1 : 9	2 parts
Water to make	200 parts

The solutions are mixed in the above order, and filtered if necessary.

The prints or transparencies are immersed in the toning solution until they are completely blue, and washed in changes of slightly acidified water.

For green tones the blue-toned image may be treated in a solution containing 0·5 per cent each of sodium sulphide and concentrated hydrochloric acid. It should afterwards be washed in slightly acidified water.

DYE TONERS

Mordant dye toning depends on the fact that certain dyes will not stain ordinary gelatin, but they will dye gelatin which has been impregnated with certain insoluble compounds, known as mordants, on which the dye can be adsorbed. Suitable mordants are the insoluble thiocyanates, ferrocyanides, and halides of some metals.

For mordant dye toning the silver image is converted into a mordanting compound of this type and then dyed in a dye bath.

The formula for copper thiocyanate mordanting bleacher is:

Stock solution A		
Copper sulphate (blue vitriol)	1 ounce	25 grams
Potassium citrate	4 ounces	100 grams
Glacial acetic acid	390 minims	20 c.cm.
Water to make	40 ounces	1,000 c.cm.
Stock solution B		
Ammonium (or potassium) thiocyanate	1 ounce	25 grams
Water to make	10 ounces	250 c.cm.

For use 1 part of B is slowly stirred into 4 parts of A. The solution must be clear, otherwise it is useless.

The image is bleached until it is a dirty grey, soaked in several changes of still water, and dyed in a dye bath.

The formula for a silver iodide mordanting bleacher is:

Potassium iodide	2 ounces	50 grams
Iodine	265 grains	15 grams
Glacial acetic acid	1 ounce	25 c.cm.
Water to make	40 ounces	1,000 c.cm.

The image is bleached in this solution, washed, and dyed. This mordant gives clearer colours than the copper thiocyanate because the silver iodide is practically colourless.

Other mordanting bleachers can be prepared with uranium nitrate and potassium ferricyanide, or with potassium iodide, thiocyanate, and thiocarbamide.

The following dyes are suitable for the dye bath:

TONING COLOURS

Colour	*Dye*
Violet	Methyl violet, benzyl violet
Blue	Methylene blue, thionine blue
Green	Victoria green, malachite green, methylene green
Yellow	Auramine, thioflavine T.
Orange	Chrysoidine 3R
Red	Rhodamine, basic fuchsine, safranine A, neophosphine
Brown	Chrysoidine brown

Most of these dyes can be mixed in any proportions to give intermediate shades.

A suitable dye bath consists of 1 part of the dye dissolved in 500 parts of 1 per cent acetic acid. Higher dye concentrations dye the image more rapidly, but produce less contrast; weaker dye baths act more slowly, but produce a contrastier image.

If the highlights tend to become stained, they can be cleared by adding more acetic acid to the dye bath.

When the dye image has reached its required density, the print is washed. This operation must be carefully watched, because prolonged washing may remove some of the dye from the image.

COLOUR DEVELOPERS

For colour development, the image is bleached in a normal halogenizing bleacher and re-developed in a chromogenic (colour generating) colour developer. The oxidation products of the developing agent combine with the colour-formers in the solution to form a stain image at the same time as the developer develops the silver image. The silver image may be removed, leaving only the stain image.

Suitable developing agents usually consist of derivatives of paraphenylene diamine or para-aminophenol. Proprietary compounds of this type are commercially available under various trade names.

The colour obtained with each colour-former varies slightly according to the developing agent used.

A suitable colour-coupling developer formula is:

Diethyl para-phenylene diamine hydrochloride or bisulphite	44 grains	2·5 grams
Sodium sulphite, anhydrous	44 grains	2·5 grams
Sodium carbonate, anhydrous	265 grains	15 grams
Potassium thiocyanate, 10% solution	190 minims	10 c.cm.
Colour former solution	95 minims	5 c.cm.
Water to make	40 ounces	1,000 c.cm.

The developer should be made up immediately before use.

The colour formers can be stored as solutions in alcohol (colourless industrial methylated alcohol will do), acetone, or a mixture of equal parts of alcohol and acetone.

COLOUR FORMERS

Colour	*Colour Former*	*Solvent*	*Strength of Solution*
Blue	Alpha-naphthol	Alcohol	1–2%
Blue	5-nitro-alpha-naphthol	Alcohol	1–2%
Blue-green	2 - 4-dichloro-alpha-naphthol	Alcohol	1–2%
Green	Pentabromo-alpha-naphthol	Alcohol	3%
Pale-green	Dichloro-ortho-cresol	Alcohol	2–3%
Yellow	2 - 5-dichloro-aceto-acetanilide	Alcohol or acetone	5%
Yellow	Ortho-chloro-aceto-acetanilide	Alcohol	5%
Yellow-brown	Cyanacetanilide	Alcohol and acetone	1%
Orange	Benzoylacetanilide	Alcohol or acetone	5%
Red	1 : 3 : 5-phenyl-methyl pyrazolone	Acetone	3–5%
Magenta	Para-nitrophenyl aceto-nitrile	Alcohol or acetone	5–8%

The above colour formers may be mixed in any proportion to produce intermediate tints.

The silver image can be fully removed with Farmer's reducer, or, better still, it can be bleached in a ferricyanide and potassium bromide bleacher, followed by a plain 10 per cent sodium thiosulphate fixing bath.

If the bleaching process is arrested before it is complete, leaving the silver image only partially removed, the colour tones will be darkened by a certain amount of black, silver deposit. If the bleaching bath is followed by a sulphiding solution, the colours will be combined with brown. L.A.M.

See also: *Chromogenic development; Colouring prints; Flexichrome process.*

Books: *Enlarging*, by C. I. Jacobson (London); *The Complete Art of Printing and Enlarging*, by O. R. Croy (London).

TONE SEPARATION. Since a paper print hardly ever shows a tone range of over 30:1, it can rarely record the true tones of the negative. On a soft grade of paper the tone gradation from high lights to shadows may be compressed to as little as one-tenth of the range of the negative—i.e., if the negative has a long scale of tones ranging from deep shadows to bright highlights, or if the negative includes tones at both extremes of the density scale. While the darkest part of a negative may be 300 times as opaque as the thinnest section, the blackest shadow of the print cannot be more than about 30 times as dark as whitest paper tone. This means that the print will look flat, even if it includes a complete range of tones from black to white, simply because the separation between the individual tones is not great enough.

Selective Printing. One way out of the difficulty is to print only the medium tones and part of either the shadow or the highlight detail on a contrasty paper—i.e., some of the tones are deliberately sacrificed, so that the whole of the available tone scale of the paper can be exploited fully in reproducing a relatively narrow section of the negative tone scale. This method is satisfactory if the tones that are sacrificed are unimportant or even completely absent in any case.

Separation Printing. The other way out is to print the highlights and the shadows separately, and to allow the tone scale of each to overlap. This method is called tone separation; it is particularly useful for subjects consisting of extremes of light and shade, with nothing much in between them.

While tone separation falsifies the gradation of the subject, it is capable of giving a print that looks more natural than one made by normal printing procedure with its resulting compressed tone scale. The reason for this lies in the differing psychological reactions that are brought into play when we look at a real object and a two-dimensioned representation of it in tones of grey.

There are two steps in making tone separation prints; making the highlight and shadow negatives from the original film or plate; printing the two negatives, one after the other, in exact register on to the same sheet of printing paper.

The separation negatives can be made by direct printing and reversal processing, or via an intermediate transparency. In either case, they may be the same size as the original or enlarged to exactly the same degree.

The highlight negative should be thin and soft, with sufficient detail in the highlights, but none in the shadows. The shadow negative should be dense and contrasty with the highlights almost opaque, but full of detail in the shadows.

The correct type of negative in each case is arrived at by suitably adjusting the exposure, development, or both. If the original negative is dense and contrasty, it may be used as the shadow negative. In this case it is only necessary to make the highlight negative.

The original negative is marked with a couple of crosses on opposite margins. These crosses are reproduced on the separation negatives and are used to bring the images into exact register on the printing paper.

The shadow negative is printed first. When the negative has been focused, and just before it is exposed, the registration marks are marked on the paper with pin-pricks. The highlight negative is then printed on the same sheet of paper, after the registration marks have been made to coincide exactly with the position of the pin-pricks.

The relative exposures are best found by test strips. The exposure for the shadow negative should be long enough to render the shadows satisfactorily. The highlight negative will need about one-twentieth to one-tenth as long; the highlights should just print through.

The character of the result depends on the length of each exposure, as well as on the density and gradation of the separation negatives.

Certain methods of processing—e.g., water-bath development—or aftertreatment of the original negative may also be employed to adjust the highlight and shadow gradation independently to give a measure of tone separation on printing. The control possible in this way is, of course, much more limited.

Person Process. Another variation of the principle of tone separation was described in 1930 by A. Person. His method was to expose and process the negative so that it embraced a range of densities much greater than the softest paper could reproduce. From this negative he made a positive transparency in which the highlights were printed in full detail and the shadows were allowed to block up completely. He then made a contrasty negative of the high-

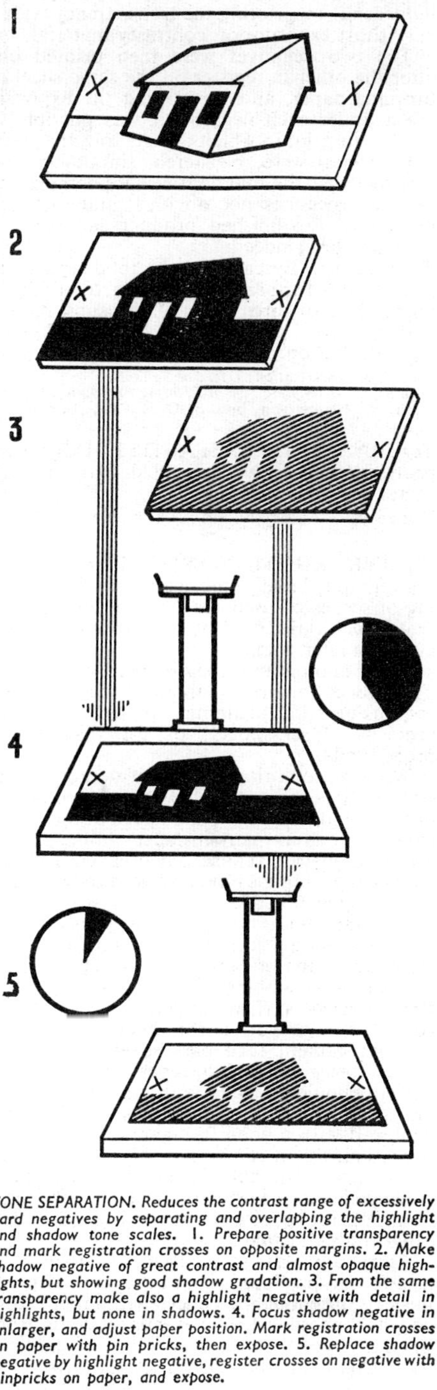

TONE SEPARATION. Reduces the contrast range of excessively hard negatives by separating and overlapping the highlight and shadow tone scales. 1. Prepare positive transparency and mark registration crosses on opposite margins. 2. Make shadow negative of great contrast and almost opaque highlights, but showing good shadow gradation. 3. From the same transparency make also a highlight negative with detail in highlights, but none in shadows. 4. Focus shadow negative in enlarger, and adjust paper position. Mark registration crosses on paper with pin pricks, then expose. 5. Replace shadow negative by highlight negative, register crosses on negative with pinpricks on paper, and expose.

light detail by printing the transparency with a very short exposure on contrasty material.

The two negatives were then printed one after the other in register on the same sheet of bromide paper, and by varying the exposure given under each negative it was possible to produce a print in which the shadow and highlight tones were registered simultaneously. The fact that the middle tones are compressed by the process is not obvious, and the tone quality of the finished product can reach a very high level indeed.

Posterization. By carrying the tone separation process still further it is possible to eliminate all but two or three tones of the subject, producing a broad poster-like effect. This is known as posterization. L.A.M.

See also: *Posterization; Tricks and effects.*

Book: *The Complete Art of Printing and Enlarging* and *Design by Photography*, both by O. R. Croy (London).

TOURNACHON, GASPARD FELIX, (known professionally as Nadar), **1820–1910.** French writer and photographer.

See also: *Nadar, Gaspard Felix Tournachon.*

TRAFFIC PHOTOGRAPHY. Part of police photography largely connected with the recording of scenes of road accidents of all kinds so that a pictorial record of the incident is available at a later date.

These photographs may be used as evidence in a prosecution or at an inquest; to assist those responsible in planning improvements at accident "black-spots"; or for road safety propaganda.

Working conditions vary considerably as the operator has to be available at any time of day or night and in all weathers. Speed of operation is essential as in many instances traffic is held up by damaged vehicles, debris, etc., which may be blocking the road and need to be moved as quickly as possible.

A certain amount of survey work is done by photography and a limited amount of surveillance by means of cameras attached to patrol cars and in helicopters.

Road Accidents. The camera in general use for this type of work is the modern 4 × 5 in. technical camera. It enables the operator to deal satisfactorily with both types of work. The camera should be fitted with a coupled rangefinder for quick and accurate focusing and equipped with a good flashgun for night time operation. Stand cameras of $4\frac{3}{4} \times 6\frac{1}{2}$ ins. are still in limited use but are slower to use and not so convenient in busy streets or at night-time. Some operators also work with 35 mm. and other roll film cameras.

Advantages of the larger negative if it has to be produced in Court are that the image is big enough to be seen without difficulty and it is possible to make enlargements from small areas of the negative. A lens of standard focal length for the camera should always be used as there is a tendency for the wide-angle lens to present an elongated view, i.e. with the planes separated. A long focus lens has the opposite effect—it tends to flatten the planes. Where it will assist the enquiry on the spot prints are produced by an instant-picture camera.

A fast panchromatic film is generally used to permit the use of a small aperture. Process film may be used on occasions when dealing with such things as scratch or fabric marks on vehicles.

Fairly large flash bulbs having a maker's flash factor of about 720 with the film in use are generally favoured, though electronic flash may be employed for close up work.

The number and nature of the photographs may vary with the type of accident being dealt with. It is better to take too many photographs than too few. Everything likely to assist in the investigation or to apportion the blame, if any, should be recorded. No attempt must be made to favour one side or the other, the photographs should be a true and accurate record of the scene. The operator should find out before starting if anything has been moved from its position immediately following the incident. No attempt should be made to put things back into position or to reconstruct the scene after things have been moved.

"General view" photographs looking in the direction of travel of the vehicles involved are necessary to begin with. Maximum depth of field is required here as everything on the photograph should be reasonably sharp. Therefore as slow a shutter speed as possible is chosen in order to allow the operator to stop down to a fairly small aperture. When taking these photographs the operator should stand roughly in the position that an approaching driver should have been in were he in his correct position on the road. This will show the view available to the driver as he approached the point of impact. Care is necessary when showing bends as a wrong viewpoint can give a false impression of the rate of curve, etc. Where possible, road warning signs should be included as should buildings or overhanging trees which may have obscured vision.

These photographs are followed by close-up shots to show the final positions of vehicles, skid and tyre marks and damage to vehicles to establish if possible point of impact, point of contact and speed of travel. Invariably some distance is travelled following the collision before the vehicles come to rest. When photographing damage to vehicles it is as well to include the registration plate if at all possible for identification purposes. Any broken part which was allegedly the cause of the accident should be photographed before the vehicle is moved as further damage may ensue during removal.

Skid and tyre marks may have to be taken from a variety of angles in order to record them properly as some can only be seen from certain positions.

At night the photography is relatively short range work, the aim being to record as much as possible and produce that which is recorded in a clear and sharp form. Close-up shots present no difficulty but with general views the light from the flash-bulb falls off rapidly towards the rear of the field of view. Large apertures are no advantage as the depth of field is small, therefore the operator is confined to apertures of *f* 11 or possibly *f* 8 when using a single bulb in a flash gun attached to the camera. Greater areas can be covered by mounting the camera on a tripod and, by the open flash method, firing a series of flash bulbs at various distances along the field of view. Time exposures where street lighting is of assistance coupled with fill-in flash also produces good results. Flood-lighting and motor car headlights can also be utilized but these latter methods are not easy on busy roads or where time does not permit their setting up. The scene is usually re-visited in daylight and further pictures of the road etc. taken.

Photographs are occasionally taken to corroborate the evidence of an eye witness. In these cases an eye level shot from the position the witness was in and looking in the direction he was looking in will show what he was able to see.

Survey. In road traffic survey work photographs taken at ground level are often supplemented by overhead shots. They are taken at road junctions, black spots and of traffic conditions at peak times, etc., to assist those concerned with road construction or alteration. They also assist in the planning of bus routes, diversions, etc. The plan-type view of the overhead photograph is superior to the ground level one in these cases. Such pictures are taken by means of a roll film camera with a wide-angle lens on a 60 foot extending pole mounted at an angle on a vehicle, or from a helicopter or light aircraft. A fixed length marker can be placed on the ground and photographed with the scene; this will allow measurements to be taken from the photograph as if it were a scale plan. The helicopter is more suitable for this type of work due to its manœuvrability and range of heights. High shutter speeds are necessary to overcome unsharpness due to a certain amount of vibration in the machine.

Surveillance. Cameras mounted on brackets in patrol cars are used to record incidents of dangerous or careless driving when witnessed by the car crew. Examples of these are vehicles crossing the double white line, overtaking three abreast or overtaking on the brow of a hill or other dangerous place.

Generally 35 mm. cameras are used and they are mounted on a bracket inside the car horizontally and facing square to the front when the car is level. The shutter is set at 1/100 sec. and all the operator needs to do is to select the aperture suitable to the prevailing light conditions and fire the shutter by means of a long cable release when the occasion arises. The photograph should always show the registration number of the vehicle and its position to the road. These photographs may be used in evidence at Court or for road safety propaganda.

Special automatic cameras with motorized film transport have been produced for this work; some types include devices that simultaneously record one or two instrument dials on one part of each frame (e.g. clock time and speed of the vehicle from which the exposures are made).

Photographs for the same purposes are also taken from the helicopter flying overhead or slightly on one side and travelling parallel to the road. H.Sc.

See also: *Crime photography; Police photography.*

TRAINING AND EDUCATION IN PHOTOGRAPHY

Instruction in photography can be along two lines: pure training in the practical aspects of the craft, and more elaborate schooling covering both theory and practice. The first may involve some sort of apprenticeship under the supervision of a professional photographer. The second is usually connected with study at a special school or course, following a syllabus under one or more instructors.

In considering either methods of approach different requirements must be taken into account. Teaching courses vary accordingly.

The craftsman needs a very practical training in a specific aspect of photography; he gets it during his hours of employment. If he wishes to progress and take examinations he can also attend evening classes, or take a correspondence course in photography.

The technician will require more theoretical background and may be released by his employer to attend college one day a week, or be encouraged to attend evening classes. Alternatively, he may wish to pass examinations before seeking employment in which case he will attend college as a full time student.

The student professional photographer must receive a thorough and comprehensive education in photography rather than be limited to specialization at an early stage in his career. For this reason alone a full-time course of studies at a recognized school of photography is the quickest and surest route. Part-time courses are available to those who need a job to live.

The needs of the photographic technologist are best met by the "sandwich" course, which in other industries has become a recognized method of education and training combined.

Practical training, carefully programmed, is received during the periods spent in the industry and the basic discipline, principles and applications of the subject are studied by the student during College terms.

The dealer in photographic equipment needs training in salesmanship and business. He must still know enough about cameras, equipment and photographic materials to advise customers with assurance.

BRITAIN

Photography is considered to be a subject for further education in Britain. It is practically never taught at secondary school level; a student is expected to have finished his general education before embarking on a vocational course of studies. Students applying to join courses of photography at colleges that offer them, must satisfy certain requirements.

Courses in photography are conducted throughout the country in regional, area and local technical colleges, schools of art and evening institutes. All of these are financed from public funds and come under the jurisdiction of the local education authorities. The latter finance the grades of work up to intermediate level, the higher levels being paid for by the Ministry of Education. Students can apply for financial assistance from their local educational authority. This system makes it possible to offer courses in photography at a comparatively low fee, which would be economically impossible in an independent organization.

Private schools of photography existed before the last war, but conditions are no longer favourable for them. Some private ventures are known to offer specialist courses, e.g. in medical photography and in film technique.

On the whole the British photographic industry does not offer financial assistance to colleges, or present items of equipment on the lavish scale as is done in European countries. However, considerable assistance is given in a less obvious manner, e.g. by granting scholarships on an advanced level and in specialized fields.

Professional Courses. For professional examinations a full-time course is most satisfactory. It requires five days a week attendance at college for two to three years according to the level aimed at. The courses offered vary in entry requirements, course content and standards.

The intention of a full-time vocational course is that the student should enter the course soon after he has finished his secondary education, having passed the necessary examinations. He receives a thorough education in the principles and practice of photography. Usually the courses consist of lectures, demonstrations, tutorials and practical work, on a general level for a period of two years, followed by an intermediate examination.

The third year is usually devoted to concentrated study of a defined area of photography: commercial and industrial, advertising and fashion, portraiture, photography for publications, scientific applications of photography, cinematography. Selected schools of photography, which are equipped for the purpose, prepare their students for the final examination in photography during the third year of the course, the examination being sat at the end of the third year.

Some schools include ancillary subjects in their curricula, as is the case with professional courses in other subjects, e.g. mathematics, physics, design, art appreciation, communication, business studies. These ancillaries are spread over the three years of the courses.

The usual requirements for admittance to professional courses in photography are a minimum age (between sixteen and eighteen), and a minimum number of passes in the General Certificate of Education examinations at Ordinary or Ordinary and Advanced levels. Some schools specify certain subjects for admittance to a full-time day course. English is an essential requirement for all courses which prepare for the professional examinations; mathematics is often required, sometimes also are physics, chemistry and art. Examination equivalents from other countries are usually taken into account, and, if a foreign applicant can furnish evidence of proficiency in the English language, he may be accepted into the course.

Study for the professional examinations can also be conducted on a part-time basis. Students employed by photographers or a photographic department in Government service or industry, may attend college one day a week provided the employer is willing to sponsor them, and facilities are available. In addition, some colleges offer evening courses for the professional examinations. Part-time courses take four to five years to cover the same ground (in abbreviated form) as a full-time course in three. Preparation for the final examination by attendance at night school can only succeed if the student has adequate scope for practical experience and experimental work.

Full-time courses recognized by the Ministry of Education are held at the following institutes in England and Wales:—

Berkshire College of Art, Reading
Birmingham College of Arts and Crafts
Blackpool Technical College
Bournemouth College of Art
Derby & District College of Art
Ealing Technical College & College of Art
Gloucester College of Art
Guildford College of Art
Harrow Technical College
Hull College of Arts & Crafts
Leeds College of Technology
London College of Printing
Manchester College of Science & Technology
Nottingham College of Arts and Crafts
The Polytechnic, Regent Street, London

Salisbury College of Art
Staffordshire Technical College, Wednesbury
Stow College of Printing, Glasgow.

Part-time courses are available at:—
Newcastle-upon-Tyne College of Art
Coventry College of Art
S.E. Essex Technical College, Dagenham.

Craft and Technician Courses. These are usually organized on a part-time attendance basis and supplement the training the student receives during his employment. They prepare students for examinations in photography and cover tuition in theoretical aspects as well as a variety of practical procedures. The applicant must be of a certain minimum age (sixteen), employed in photographic work and, for some courses, must also have the General Certificate of Education in a specified number of subjects. The majority of part-time students attend evening classes, but day release courses are becoming popular with employers and employed alike. Duration of courses varies according to the examinations for which students are being prepared. On the average the intermediate level takes two years' study, and the final level a further two years.

Technological Courses. At the present time one course of this character is organized for the employees of the photographic manufacturing companies. Another course, to be introduced at a higher level, will provide suitable photographic technologists, for the photographic manufacturing industry, the wholesale photographic finishing industry and the research and development sections of major industries and Government Departments.

Short Courses. In the colleges these consist of "refresher", specialist, or post-graduate courses. The large manufacturing companies of photographic materials organize short courses. In the main these are intended for professional photographers and instruct in the characteristic features of the manufacturer's own materials, and also how to achieve the best results from these materials.

Non-Vocational Courses. Many such courses exist, especially in schools of art and evening institutes, and can take one of two forms. First, and particularly in the London area, evening classes in photography thrive as a cultural pursuit or as instruction in the hobby. Second, photography is gaining favour as a suitable ancillary subject for inclusion in curricula of art and design courses. Most departments of photography within art colleges are expected to provide such a service. Photography is sometimes also considered a suitable subject for inclusion as an ancillary in science courses in technical colleges.

Cinematography. This subject is incorporated in certain three-year courses; it is now primarily concerned with documentary and industrial film-making, and includes the scientific applications of cinematography. There are also part-time courses in cinematography.

Examinations. In Britain examinations in photography are organized by three examining bodies: The Institute of Incorporated Photographers, The City and Guilds of London Institute, and The Photographic Dealers Association. A few colleges also award their own diplomas to successful students who satisfy the requirements of the courses.

The Institute of Incorporated Photographers (formerly Institute of British Photographers) offers membership only to professional photographers. Candidates for the examinations at intermediate and final levels must be registered student members of the Institute. The Institute holds examinations at three levels: (1) Preliminary, (2) Intermediate, (3) Final. The first two are concerned with the general principles and practice of photography. The Final examination consists of one written paper taken by all candidates, the subject of the other two papers (one written and the other a searching practical test of fifteen hours duration over a period of two and a half days) being selected by the candidate in advance. The areas in which the candidate may choose to be examined are: (*a*) commercial and industrial photography; (*b*) advertising photography; (*c*) medical photography; (*d*) scientific applications of photography; (*e*) portraiture.

These examinations are recognized by the Ministry of Education, and are steadily gaining public recognition and confidence. For the examination at all three levels candidates have to be sponsored by the college or school where they are studying, or by the employer who is training them. For the intermediate and final examinations an additional entry requirement is the General Certificate of Education with Ordinary level passes in English Language or Literature and three other subjects.

The City and Guilds of London Institute is solely an examining body covering a wide range of craft and trade subjects. The examinations in photography are predominantly suited to the needs of the photographic technician, but are fairly closely aligned with the I.I.C. examinations. The C.G.L.I. holds their examinations in photography at intermediate and final levels. There are no educational entry qualifications but students attending courses must be sponsored by their college.

The Photographic Dealers' Association offers examinations solely concerned with aspects of photographic dealing, including however some theory and practice of photography. The examinations are for the members of the Association, i.e. people engaged in the wholesale distribution or retailing of photographic equipment and materials.

Qualifications in Photography. The Institute of Incorporated Photographers awards Associateships and Fellowships. The Associateship is granted to members of the Institute who have passed the Final examinations and have spent two years in photographic practice. Licentiate members of the Institute can also apply for Associateship. The fellowship is a higher qualification. Applications must be accompanied by photographs or a thesis indicating distinction and outstanding ability in photography in a specified field.

The City and Guilds of London Institute grants an Insignia Award which is very sparingly given in recognition of technological advances or developments in the subject in which application is made. Only two awards have been made in photography.

Information on Educational Facilities. In most areas the local education authorities can give information whether there are facilities for studying photography. The Institute of Incorporated Photographers publishes a list of schools recognized by it as providing entry qualifications or exemption for the I.I.P. examinations. Colleges of further education provide information and prospectuses on direct application. This information details entry conditions, application dates, curricula, etc.

UNITED STATES

Photography is recognized at university level in the U.S.A. and degrees are granted in the subject. A number of institutes also have their photographic departments, such as the Rochester Institute of Technology. Such organizations may be members of the American Council on Education and other educational associations. The institutions mentioned in this article are privately-endowed non-government institutions. Some vocational secondary schools have courses in photography.

In addition there are many private schools of photography offering generalized or specialized courses and sometimes sponsored by professional associations. There are also correspondence courses, some of which cover specialized aspects such as camera repairs.

Each institution which offers a course in photography conducts its own examination and awards its own degrees, diplomas or certificates.

The courses range from full-time photographic training to evening classes and refresher courses.

In addition to the degrees, diplomas and certificates conferred by universities, colleges and schools offering courses in photography in the U.S.A., a number of organizations award qualifications to their members who satisfy the required conditions. The largest trade association is The Professional Photographers' Association of America Inc. Members qualify to the Master of Photography certificate.

Syllabus. The first year in a full-time course of photography usually covers photographic theory, technology, practical photography, mathematics and physics. Succeeding stages may offer specialization in photographic science, professional or illustrative photography.

Photographic science courses are concerned with the materials and processes of photography, with special reference to product development, research, and manufacture of photographic material, as well as the application of photography in science and engineering.

The School of Photography at the Rochester Institute of Technology is the only one with both graduate and undergraduate courses in photographic science and instrumentation.

Professional courses embrace the various branches of practical photography as a career, including also photographic technology and business management.

Illustrative photography covers the more creative aspects including design, graphics as well as communication. Press and feature photography for magazines and other publications and also advertising photography come into this category.

Entry requirements stress academic ability evidenced by graduation results in high school with a special accent on science subjects.

There is a great deal of interest on the college level in photographic illustration (expressive photography). There are courses in the Fine Arts or Art in design departments of fifteen to twenty colleges and universities in the U.S.A.

The major institutions where such courses are organized are the University of Ohio, the University of California at Los Angeles, the Institute of Design of the Illinois Institute of Technology, and the School of Art of the Philadelphia Museum. The Department of Cinema of the University of Southern California offers motion picture courses at the same level, while the School of Engineering of Columbia University has special lecture courses in photographic technology.

EUROPE

Many European countries restrict all professional and commercial activities to duly trained and certificated practitioners, often organized in trades and crafts guilds. In the German Federal Republic, for instance, the guild statutes for photographers specify a three year apprenticeship, followed by an intermediate examination, a five-year term as an assistant, and a final qualifying examination. This entitles the newly fledged photographer to registration in the guild and to conduct his own business. Hence apprenticeship training for photographers is still very much in force, often supplemented by evening classes.

Entry into Professional Training. The main possibilities are (1) apprenticeship combined with evening classes; (2) technical school training concurrent with apprenticeship; (3) full-time courses at schools of photography.

The first course is for young people with a full secondary school education who start their apprenticeship at 16 years with a qualified photographer. They may also take evening classes at a technical college. The apprenticeship is followed by a year of practical experience after which the student can enter a school of advanced photography on a full-time basis for two years.

The second alternative is for students who leave school at 14 and then receive a supplementary general education and technical training in a technical school. The apprenticeship training follows on almost the same lines as in the first case but the apprentice attends evening classes to bring him to the secondary school certificate level.

Thirdly, full-time courses in elementary photography at technical colleges take students straight from secondary school at 16 years. These courses are regarded as the equivalent of apprenticeship and are often considered to give a more liberal training in photography.

There are variations in different countries on these three basic systems which are firmly established as methods of providing education and training in photography in the German Federal Republic (West Germany).

Photographic Trade. Germany has courses for photographic dealers and for photographic chemists at certain special schools. Dealers serve a three-year apprenticeship in a photo-retailer's establishment and attend a day-release school during this time. An examination at the end of the apprenticeship leads to a diploma as a commercial assistant, specializing in photography.

A similar three-year apprenticeship for the photographic chemist specializes far more on pharmacy. A number of special schools of pharmacy provide photographic training as part of the general three-year course for photographic chemists. Further specialized training is available at certain advanced schools.

Technical colleges in Germany also have classes for photographic darkroom assistants, but are not listed as special photographic schools.

A wide range of private schools also exists, but are not recognized as professional training for photographers and students cannot therefore set up businesses of their own.

College Courses. Because of the limitations of apprenticeship and the variations in the capacity of individuals to impart knowledge, there is a growing demand for, and facilities of specialized photographic courses in schools and colleges, either as a supplement to apprenticeship or in place of it. Some of the better-known colleges are:

- Graphische Lehr-und Versuchsanstalt, Vienna, Austria
- Ecole Belge de Photographie, Brussels, Belgium
- Ecole Technique de Photographie et de Cinematrographie, Paris, France
- Fachschule für Optik und Fototechnik, Berlin, Germany
- "Lette-Verein", Technische Berufsfachschule, Berlin, Germany
- Staatliche höhere Fachschule für Photographie, Cologne, Germany
- Fachschule für Photographie, Hamburg, Germany
- Fachschule des Deutschen Photo-und Kinohandels, Kiel, Germany
- Bayerische Staatslehranstalt fur Lichtbildwesen, Munich, Germany
- Institut for Bildjournalismus, Munich, Germany
- School voor Fotografie & Fototechniek, The Hague, Holland
- Istituto Professionale Statale "Ceasare Correnti", Milan, Italy
- Istituto Statale d'Arte, Rome, Italy
- Civica Scuola Avviamento Fotografico "G. Pacchiotti", Turin, Italy
- National Institute for Handcrafts, Stockholm, Sweden
- Ecole des Arts et Metiers, Vevey, Lake Geneva, Switzerland.

Many of the European colleges which offer courses in photography are financed and sponsored in respect of those courses by the State, the photographic industry and the photographic trade.

Qualifications. The qualifying examination in photography in many European countries comes under the authority of the photographers' guilds, and is recognized as the state final examination. In other countries there is no qualifying examination. In Italy, for instance, it is not necessary to furnish proof of qualifications or training before setting up in business as a photographer.

There is no comparable organization in Europe to Britain's Institute of Incorporated Photographers. Probably the nearest associations of professional photographers are the Swedish photographers' association and the Scandinavian photographers' association. However neither of these organizations hold examinations or grant qualifications.

Higher Photographic Education. Education in photography is being extended in Britain, Europe and the U.S.A. into the higher levels of education by certain major colleges.

At the three best-known colleges of this kind, The Rochester Institute of Technology, New York, the Polytechnic of Regent Street, London; and the Staatliche Höhere Fachshule für Photographie in Cologne, Germany, the subject is covered comprehensively at the professional level. Both science and art aspects are offered either as separate courses or as separate areas of study at the final level. In at least two of these three colleges new courses are being planned with an aim to degrees in photographic science and technology. The schools have the advantage of being independent of any other discipline so that their work can develop on a broad front to a high level. M.F.H.

See also: *Careers in photography; Examinations; Professional photography; Training of photographic technologists and scientists.*

TRAINING OF PHOTOGRAPHIC TECHNOLOGISTS AND SCIENTISTS

A photographic technologist or scientist (PTS), for our purposes, is an academically trained person. In other words, he received his education at an institution which confers a Ph.D degree or its equivalent as its highest award for successful formal education. He is therefore defined as a person with a research degree. He may have taken this in the form of a diploma, where this entails several semesters of work (such as is now general, for example, in the German and Czech Universities) or of a Doctorate. Those trained in technical colleges or institutions of similar standing will therefore not be considered in any detail.

Approach to Training. It is a peculiar fact that facilities for formal training in photographic technology and science (PTS) are limited to a few geographical areas. Notably, the Anglo-Saxon countries have never gone in for this kind of training. There are obvious pros and cons for the various approaches.

Photography, like any other industrial activity, implies, and makes use of, all modern scientific tools. In order to advance any one aspect, chemists, physicists and mathematicians of high calibre are required. Indeed we are at a point, when even in a field as narrow as photographic technology and science, no single person can hope to master the whole subject. People tend to become specialists in one small area, and the most they can do is to have a somewhat superficial knowledge of the rest.

This being so, industry is inclined to engage men with the best possible education in basic sciences and train them up in photography within their own laboratories. Industry has not been particularly interested in men who specialized in photography during their formal courses of study, on the grounds that such specialization was bound to be at the expense of their basic training.

These reasons are behind the fact that (with the possible exception of the U.S.S.R.) no academic institution exists that offers a full-time undergraduate course in photographic science and technology. Such courses are available exclusively at technical colleges. Our photographic technologist and scientist is therefore a man who qualified in one of the basic sciences and who did research on a photographic subject.

Types of Academic Training. Training in PTS can be had in two types of institution: the photographic institutes proper, and in departments engaged in one of the basic sciences, where the head or a member of the staff happens to take an interest in photographic matters. A whole spectrum of attitudes and outlooks exists, resulting in different facilities for training and study. At one extreme is the scientist who will attempt to boil down a photographic problem to a basic one in physics or chemistry. At the other is usually the head of an ostensibly photographic institute, who uses photographic systems and means for his research.

An institute specifically devoted to photography ought to be able to combine these attitudes. Whilst the individual research student could work on a problem—which is basic, although the questions to be answered stem from the photographic field—he should gain some insight into the whole of photography from seminars, colloquia and from observation of the activities of the other students and members of the staff.

It is perhaps remarkable that so few photographic institutes are characterized by an element of purely photographic activity, which is also missing from the usual science department: the taking of photographic records on a routine or service basis, and perhaps even the training of apprentices. Where this sort of work is done, it provides a background, not dissimilar from that of an industrial laboratory, which can be decidedly helpful in training a useful PTS.

Consultation and Service. Photographic institutes, technical universities and engineering schools which enjoy the advantages of close contact with industry, are often asked for expert advice, as they are free of manufacturing interests. By participating in consultation and service work, the PTS under training shares the advantages of industrial contact and gets training additional to his research work proper. This develops his feel for industrial work and helps him decide what type of activity suits him best.

Location of Institutes. The considerations above at once provide a clue for the peculiar geographical distribution of training facilities. Where large industrial laboratories exist, such specific training is often regarded as unnecessary or even—from the viewpoint of the potential employer—undesirable. Where the manufacturing units are small, so that they cannot give adequate training themselves, a supply of trained PTS's is desirable. Hence we find photographic institutes at technical universities (Technische Hochschulen) in countries as small as Sweden and Switzerland. There are no less than three in Germany, two of which were established in the lst century, when manufacturing units in that country were still small.

A case which illustrates this point is that of the British Photographic Manufacturers' Research Association which, although not a teaching institution, effectively gave postgraduate training to some of the men now in leading positions in the industry in England. This association was dissolved at a time when industry began to build up its own research facilities and thus became capable of training its own men.

These remarks are strictly historical: it cannot be gainsaid that some of the institutes flourish to this day and fulfil a useful function.

Syllabi. In very few instances are photographic courses compulsory at undergraduate level; some of the exceptions are students engaged on photogrammetry, cartography, etc., or those wishing to take a research degree on a photographic subject. A typical set of courses the aspiring photographic technologist and scientist could take at undergraduate level might be as follows:

Introductory course of lectures: 2 hours/week for two semesters, or three terms giving a general survey of the process and its applications.

Advanced lectures: 1 hour/week for one semester or 1–2 terms, may be arranged from time to time, on colour photography, radiography, photographic optics, image transfer theory, and emulsion technology.

Practicals: beginners, advanced, special, one semester or term each, 2 to 4 hours per week.

The beginners' practical covers the elementary aspects of the process; the advanced practical may take the student as far as colour printing including masking processes. The special will include the study of unusual processes and advanced sensitometry.

There are, as a rule, no formal post-graduate courses in the European technical universities, at which photographic institutes are situated, a weakness which in time will have to be remedied by following the American example. Attendance of special lecture courses in the field of research of the PTS under training should be encouraged; specifically photographic lectures, seminars and colloquia, in which the students will take over the preparation and presentation of at least some of the lectures, are useful training grounds. Similarly beneficial is attendance at conferences where the young scientist can rub shoulders with more experienced men in his field, take part in the discussions, or even present the results of his own work.

Research projects vary widely according to the interests of the department head, from emulsion technology on the one hand to the mathematics of information theory on the other.

Jobs for the Trained PTS. All the trained photographic technologists and scientists land up in industry, most of them in salaried jobs, a few in independent positions in manufacturing or consultation work, and none, directly, in academic posts.

It is perhaps natural that academic positions in this field should tend to be filled by men with industrial experience, since without this the professor's feet may not be firmly on the ground.

The applicant will, as a rule, have to sacrifice financial prospects; it is nevertheless surprising that at least one chair of photography (Stockholm) could not be filled to the satisfaction of the authorities for several years.

The employment of the photographic technologist and scientist naturally varies with the trends of industrial development. No generalizations are possible beyond the obvious one that the tendency towards advanced instrumentation —*vide* space research—has attracted many into this field.

Here, as in any other branch of technology, there is a great shortage of trained men.

Borderline Cases. A number of departments do not strictly come under the heading of training institutes in this subject, since they do not award a research degree. The outstanding one is the School of Photography at the Rochester Institute of Technology (RIT), which accepts the students as undergraduates and which runs specific courses on photographic technology. Plans for similar courses exist at the Polytechnic in London. A photographic section exists at the physical-chemistry department of Basle University; here again no research work is done. A series of courses on photographic science, engineering and technology is being built up at Columbia University at New York. So far, this does not include practical work or research.

Institutions. Many of the points above result from a questionnaire that set out to obtain answers on various aspects of training. The accompanying tabulation is the result of this enquiry; not all questions were, or could be, answered, and some institutions gave incomplete answers. Hence the detailed figures shown may not be strictly comparable, but they do provide an overall picture. The preponderance of Continental European teaching of photography at academic levels is one of the most remarkable features of the table.

A country which seems to do outstanding work in this field is Russia. The approach is so different, that the information cannot all be fitted into the table.

The Leningrad Institute of Motion-Picture Engineers provides a five-year course in photography. For three years chemical technology is taught as for all students in the college; thereafter, photographic specialization begins. *Year* 3, second half: General photography, 2 hours/week, plus a "short" practical. *Year* 4; Theory of the photographic process, *4* hours/week, two semesters. Practical on sensitometry, 10 hours total/semester. *Year 4*, second half/ *Year 5*, first half: Manufacturing technology, 3 hours/week for two semesters; practical, 75 hours/semester. Processing technology, 3 hours/week for two semesters; practical, 75 hours/semester. *Year 4*: Technology of photographic supports, 2 hours/week; practical, 40 hours total/semester.

No fewer than five professors and a lecturer run this course. There is no other formal training to be had anywhere which can compare with this.

Some eleven people took the course in 1960, thirteen in 1961; the aim is to take twenty-five per annum who will be designated as Photographic Engineers. W.F.B.

See also: *Careers in Photography; Conferences; Examinations.*

Photographic Institutes	Year Established	Character of Institute	Percentage of photographic activity	Practicals No. of stages	Practicals Hours per week (No. of participants)	Lecture Courses Intr. (No. of participants) [Level in semesters]	Lecture Courses Adv.
Berlin: Institut für angewandte Photochemie und Filmtechnik (Freie Universität)	1863	photographic	100	3 1 sem. each	3–6 (50)	4h/2 sem. (30) [all]	4h/1 sem. [higher]
Dresden: Wissenschaftlich-photographisches Institut der T.H.	1865	photographic	90	3 1 sem. each	2 (10) 4 (10) 10 (10)	4h/2 sem. (15) [5–6] 2h/1 sem. (60) [5–6]	2h/1 sem. [higher]
Moscow	1949	photographic	100			2h/2 sem.	Seminar total 50h
München: Institut für wiss. Photographie der T.H.	1958	photographic	100	2 1 sem. each	4 (60)	2h/1 sem. (15) [all]	1h/1 sem. (10–15)
Stockholm; Kuengl. Tekniska Hoegskolan Fotografiska Institutionen	1930	photographic	100	1 1 sem.	3 (42)	3h/1 sem. 2½h/1 sem. [all]	1h/1 sem. [7–8]
Wroclaw: Katedra Fototechniki, Politechniki Wroclawskiej	1946	photographic-technological	100	3 2 sem. 1 sem.	12 thesis	7h/2 sem. (12) [9 and 10]	
Tokyo: Institute of Industrial Science of the University (Chemistry Section)		electro-chemistry and applied photography		1	4 afternoons/ week	2h/2 sem. (40)	2h/1 sem. every 2 years (40) [post-graduate]
Zürich: Photographisches Institut der E.T.H.	1888	photographic	100	3 1 sem. each (up to 72, 12, 12)	2, 4 and 4 (12) (12)	2h/2 sem. (150) [1–5]	1h/1 sem. (12)
Photographic Sections at Academic Institutions	*Phot. work since*						
Darmstadt Physikalisches Institut der T.H.	1963	physics	10			2h/1 sem. (35) [all]	2h/1 sem. (15) [higher]
Erlangen: Institut für physikalische Chemie der Universität	1950	phys. chem.	10	1 sem. (10–15)	2	2h/1 sem. (20) [6–8]	
Frankfurt: Institut für angewandte Physik der Universität	last few years	appl. physics	50	2 1 sem. each (48)	3	1–2h/2 sem. (30–60) [1–5]	1–2h/1 sem. (30)
Kyoto: Dept. of Industrial Chemistry, University	1897	industrial chemistry	10	1 sem. (50)	3	1h/1 sem. (200)	2h/1 sem. (20)
Liège: Laboratoire de Physique General	1938	physics					
Prague: Physikalisches Institut der math.-phys. Fakultät der Universität	1950	physics		1 sem.	3	2h/2 sem. (8–10) [6, 7]	2h/2 sem. (4–6)
Prague: Institut für physikal. Chemie der naturwiss. Fakultät der Universität	1956	phys.-chem.	50	1 sem. (6–12)	4	2h/1 sem. (4–10) [7]	1–2h/1 sem. (3–6)
Stockholm: Physics II and Institute of Optical Research of the T.H.	1950	physics and optics	10	1 sem.	2 (15)		
Wien: Abt. Angewandte Photochemie am Institut für physikalische Chemie der T.H.	1946	phys.-chem.	20	1 sem. (15)	5	2h/1 sem. (15–20) [5]	none

PHOTOGRAPHIC TECHNOLOGY AND SCIENCE

Colloquia	Research Students			Staff; general remarks
	No. in last ten years	Type of qualification	Time taken for research degree	Note: *figures in parenthesis followed by the letter r denotes regular staff where known*
yes	38	Dipl. Phys. Dipl.Chem. Dr.	1 year 2 to 3 years	Prof; Akad. Rat; 7 assts (5r); (4 lab. assts (2r)); 1 secr. Emulsion work, dosimetry for penetrating radiation solid-state dosimetry, colorimetry of colour pictures; Free subject
1/mo.		Dipl. Phys. Dipl. Chem. Dr.	1 year max. 4 years	Prof; lecturer; 6 assts; 4 lab. assts; 5 mech.; 1 secr. Latent image, development, phot. optics, colour photography, cine. Obligatory for students of Materials testing and Measurement and for Cartographers
				Prof; 6 lecturers and assts. Theory of photographic process and technology. 1,500 students take the basic course; no information on research students and their degrees
2/mo.	6 (since 1958)	Dipl. Chem. Dipl. Phys. Dr.	9 mo. 2 years	Prof; 6 assts (2r); 4 lab. assts (1r); 1 mech.; 1 secr. Nuclear track recording; chemical and spectral sensitization; detail rendering; electro-photography
1/week		Tekn. Lic. Dr.	2 years	Prof; 3 assts (r) plus demonstrators; 1 lab. asst; 1 mech.; 1/2 secr. Obligatory for technical physicists and surveyors. Some photographic service
1/week	96 2	Mgr. Ing. Dr.	 3 to 4 years	Prof; asst Prof; senior asst; 2 assts; 1 research asst; 2 lab. assts; mechanic. Latent-image and emulsion theory, tone rendering, granularity; some photographic service
1/week	5 5 3	Graduates Master Dr.	 2 years 3 years after master	2 Profs; 2 assts; 2 research assts; 4 lab. assts; 1 secr. Potentiometry in photography; mechanism of development; mechanism of sensitization; electro-photography; photo-sensitive resins. Photographic service and apprentices
2/mo.	19	Dr.	min. 2 years	Prof; 5 assts (3r); 3 lab. assts; 1 mech.; 1 secr. Latent-image theory; physical properties of developed image; graphic arts. Obligatory for cartographers. Photographic service, especially high-speed still and cine. Photographic apprentices
2/sem.		Dipl. Dr.	1 year 2 years	Prof; 4 assts; 1 lab. asst. Radiation damage in solids (change of lattice parameter in irradiated crystals), optical, absorption, X-ray small angle scattering, photo-conductivity, imperfections in crystals, spectral sensitization
3/sem.	7 4	Dipl. Dr.	1 year 2 years	Prof; 1 asst. Institute facilities. Electro-photography, development.
seminar 2–3h/week coll. 2/mo.	5	Dipl. Phys. Dr. rer. nat.	2 years 2–3 years	Prof; 4 assts (2r); Institute facilities. Nuclear particle recording; latent-image; electro-photography, all from point of view of solid-state physics
2/week	7 2	Dipl. Dr.	2 years over 5 years	Prof; 2 assts (r); 1 secr. Photographic sensitivity, sensitization and desensitization by dyes, photochemistry of solid state. Course obligatory for junior students.
	since 1938 3 4 6	 agr. Dr. lic.		Prof; Latent-image and emulsion theory, granularity, detail rendering. Photographic activity confined to research.
none	17	Dipl. Phys.	1½ years	Prof; asst; 2 lab. assts. Institute facilities
none	4	Dipl. Chem.	1½ years	2 assts; 1 lab. asst; Institute facilities. Development, covering power; electrochemical methods, esp. polarography
yes	20 altogether 2 in phot'y	M. Eng. Dr.	varies	Prof; 7 assts; technicians. Detail rendering and transfer function.
none	3	Dipl. Dr.	3–6/mo. 2 years	Prof; 1 asst (r); Institute facilities. Methods of increasing sensitivity of commercial materials for spectroscopy.

TRAINS. People photograph locomotives and trains either because they are interested in recording their technical features or because they want to make pictures of them.

The pictorial photographer is usually content to work from a distance so that he can include the setting and exploit the appearance of the smoke and steam from the locomotive. He may be able to get all his pictures without being on railway property. But the technical enthusiast must be able to work close up to his subject. This means getting permission to take photographs from line sides and in and around the locomotive sheds. As a rule, an application for permission addressed to the Public Relations Officer of the Region is willingly granted. When the photographer receives his permit he should regard himself as a privileged guest of the authority and behave accordingly.

No permit is required for taking photographs from the railway platform or on railway property to which the public is normally admitted.

The technical photographer must work to a plan if his prints are to be of any value as records. Generally the characteristic features of a particular class of locomotive or rolling stock call for at least two views (three-quarter front, and side) of each side and one or more views of engines of that type at work.

The steam locomotive, in virtue of its shape, is an awkward problem for the photographer, as the tone range of the subject extends from brilliant reflections off the shiny boiler top to the dull black of dirt-grimed wheels and frames, usually deep in the shadows of the running plate. The smoke deflectors standing out at each side of the smoke box, like huge blinkers, produce deep shadows in which it is easy to lose the details of the black painted smokebox, especially if there is back lighting. Under these conditions the only thing to do is to expose for the shadows or fill in the shadows with a synchronized flash bulb.

It is important to photograph the engine or train from an angle that will show as many important features as possible but which avoids awkward perspectives or depth of field problems. Where it is only possible to get one record picture of a locomotive it is best taken with the subject at an angle of between 45° and 30° to the focal plane. This shows most of the side of the engine quite clearly and also most of the front detail. If more than one view can be taken then there should be at least one broadside view, a view showing the front, and another showing the cab and the arrangement of injector piping and so on, below the footplate.

Moving Trains. The photographer should aim at taking pictures of moving trains with as little blurring of the subject as possible. Three things control the blur—the speed of the train (the faster the train, the greater the blur will be); the direction of movement in relation to the camera position (movement towards or away from the camera gives least blur; movement broadside on to the camera gives most blur); for any given direction of movement, the faster the shutter speed, the less the blur.

As a head-on view is very seldom required (and not very practicable) and a broadside view is almost sure to be blurred (unless the camera is panned), moving trains are best taken with the fastest possible shutter speed when they are approaching at between 30° and 60° to the sight line of the camera. This way records the maximum amount of detail and shows the subject in good perspective.

Photographing a train from the outside of a curve exaggerates the relative movement (and hence the blur) and also causes distortion, as the train is tilted away from the camera by the super-elevation of the outside rail. Pictures taken from the inside of a curve are much more satisfactory because the relative movement is less and the train leans towards the camera.

The height of the camera in relation to the train also affects the appearance of the subject. Normal pictures are taken with the lens at the level of the footplate or a little higher and with the camera held level. The camera should not normally be tilted to include the top of the subject. If necessary, a more distant viewpoint must be chosen, or the photograph taken with a camera equipped with a rising front. When shots are made from above—e.g., from an over-bridge—the viewpoint should be far enough to the side to show the side of the engine and train. Low angle shots are best made with the camera on a level with the running rail.

Background. When photographing steam locomotives the pictorial worker relies mainly upon the clouds of steam and smoke to make his picture. Such pictures are, however, increasingly difficult to obtain, and pictorial photography of diesel and electric locomotives depends more upon the beauty and effectiveness of the background setting than upon the train itself. With all forms of motive power, the background can make or mar the picture, and poles, wires, and lamp-posts sprouting from the top of a locomotive or train are highly distracting, and may confuse important technical detail. The photographer should watch carefully for such things and avoid their inclusion in his picture whenever possible. If it is impossible to avoid them they should be manœuvred into the least offensive position by moving the camera one way or another. The unsightly objects may then be removed by painting them out on the negative. Subsequently the sky is reduced on the print until the white outline of the blanked-out object disappears. Iodine cyanide or iodine carbamide reducers are the best to use for this purpose.

Camera. For photographing stationary locomotives, almost any type of camera may be used, so long as the lens will give good definition over the whole field at the fairly wide apertures which are often needed in the murk and grime of a locomotive yard, especially in poor weather. It is also possible to photograph quite fast-moving trains with relatively slow shutter speeds if the subject is kept well away from the camera and fairly head-on to it. But this technique will not create outstanding pictures.

The photographer who takes this subject seriously will not be satisfied with this kind of compromise. If he wants to get close enough to take impressively large pictures of moving trains he will need a more specialized type of camera. A shutter speed of 1/250 second is necessary for even slow-moving trains at normal angles of view, and speeds up to 1/1000 second are necessary for real action. A press camera with a focal plane shutter and an open wire frame finder has no equal for the work, although this type of shutter tends to distort fast-moving images.

In the hands of the expert, the high grade miniature can be used to give nearly comparable results but it is not an ideal instrument for train photography. Generally, the simpler the controls the better, but a rising front and swing back are very useful; the latter enables the whole of a long train to be brought into focus when a wide lens aperture is used with the train oblique to the camera.

The lens should have an aperture of at least $f4{\cdot}5$ to permit the high shutter speeds necessary even in good light and be of normal focal length for the plate size. Some photographers prefer to use a slightly longer focal length from a more distant viewpoint in order to obtain a slight "bunching up" of the perspective of the train. This is the familiar telephoto effect and it avoids the drawn-out appearance of a long train tapering away to nothing across the whole width of the print. Special care in focusing is necessary when using such a lens as the depth of field is reduced.

Materials. It is safest to use fast films and to stick to one make. In railway work it is often necessary to act quickly, so the photographer must be thoroughly familiar with his camera and film; he cannot afford to experiment.

A soft working developer giving maximum shadow detail is generally the most suitable, and prints should normally be made on glossy bromide paper, properly glazed unless intended as exhibition prints, when a smooth lustre surface may be preferred. Owing to the great contrast between the dark tones of the subject and the bright sky, it is usually necessary to "dodge in" the sky and often also the boiler top and other highlights when printing, but care must be taken to avoid any suggestion of a white "ghost" round the engine.

Colour. Steam, diesel and electric locomotives all make fine transparencies, the first because of the clouds of steam and the latter two on account of their brilliant colours.

Steam locomotives are frequently very grimy, and in colour appear only as an ugly black mass unless extra exposure, which may be as much as four times that required for the rest of the train, is given to show their detail. Usually, a compromise exposure which gives acceptable colour rendering of both locomotive and train must be determined from experience.

Brightly-painted diesel and electric locomotives are less of a problem though even with these some increase of exposure over that given by a general meter reading is often necessary to show the black running gear.

A good lens capable of critical definition at wide aperture is necessary to obtain good colour transparencies of moving trains, and medium-speed colour films in larger sizes have been found to give the truest rendering. P. R-W.

See also: *Movement.*

TRANSFER COATING. Transfer coating is suitable for sensitizing any flat and hard surface like metal, wood, and plastic.

The working material consists of a sensitive silver bromide gelatin emulsion layer, coated on a stripping paper. This is transferred on to the surface to be sensitized, exposed and processed. This process is used in various industries for making templates, patterns, and for copying working drawings on to the material to be shaped or machined.

The surface of the support is first lacquered by brushing or spraying on a special white or clear lacquer. White lacquer is used for most surfaces, clear where the surface is light already. Metal articles which might react with the processing solutions must be sealed off with a coat of lacquer on the surface, the back and sides.

When the lacquer coat is dry, it is softened again by applying a special softener, and the sensitive transfer coating paper is immediately squeegeed on to it. After a minute or two the lacquer sets and the paper backing can be stripped off. This leaves the sensitized layer ready for exposure and processing.

When the layer has been exposed, it may be treated like a normal development paper with suitable modifications in processing technique where the sensitized article cannot be immersed in solutions.

See also: *Stripping; Template; Transfer processes.*

TRANSFER FUNCTION. When a waveform is analyzed into frequency components, the effects of any linear transmitting or recording medium may at any frequency be expressed by a simple scale factor. The variation of this factor with frequency is known as the transfer function. The transfer function of an optical or photographic system describes how the system's response to fluctuations in exposure depends on the spatial frequency. It is a much more complete measure of image quality than such concepts as resolution and sharpness.

See also: *Information in the image; Lens performance; Lens testing.*

TRANSFER PROCESSES. Processes involving the transfer of an image from one emulsion or support to another. The transfer often involves physical reversal (left—right) and may involve tone reversal (negative—positive). There is a variety of such processes working on fundamentally different principles and used for a wide range of purposes. They fall into the following general groups: chemical image transfer, physical image transfer and emulsion transfer.

Chemical image transfer: the image is transferred by chemical diffusion from one emulsion layer to another placed in intimate contact with it. The image is in fact generated in the second emulsion; the only physical movement taking place is that of the chemicals generating the second image. The latter may be silver (as in various reversal transfer processes) or hardened gelatin (as in bromoil, carbro, and similar processes).

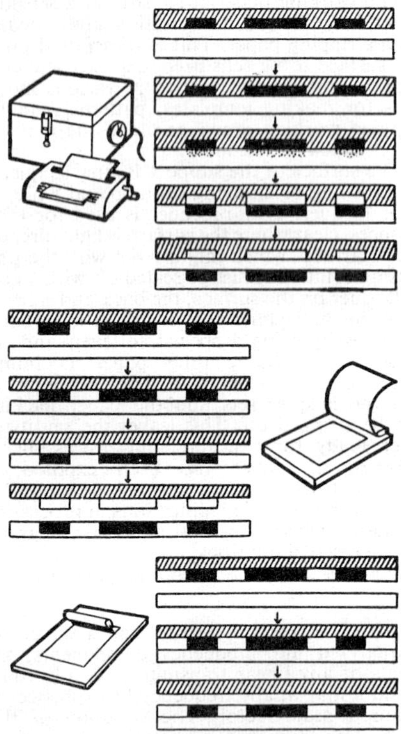

TRANSFER PROCESSES. Top: *Chemical transfer, as used in certain document copying processes for direct positives, in the carbro process, etc. Chemical action in the original images causes image forming reactions in the transfer layer.* Centre: *Physical transfer, as used in dye imbibition and similar processes. A matrix image, charged with dye, is brought into contact with a gelatin layer, and transfers a dye image on to the latter.* Bottom: *Emulsion transfer, as used in a number of colour print processes, etc. The image in its emulsion layer is transferred as a whole from one support to another.*

Physical image transfer: most of these processes involve the transfer of a pigment or dye image absorbed or adsorbed on a gelatin matrix to a plain or gelatin coated paper or similar support. Examples are bromoil transfer (pigment ink), dye transfer (dyes), and various photomechanical and document copying processes.

Emulsion transfer: here the whole emulsion or a gelatin matrix is stripped from one support and transferred to another. This may be necessary with many matrix processes in order to obtain a laterally correct image—i.e., one that is not reversed left to right—or to ensure that the matrix has a firm support—e.g., if the gelatin of the matrix is hardened from the top, the gelatin next to the support would be dissolved on development so the whole layer must be transferred to a temporary or permanent support first.

Another application of emulsion transfer is the preparation of photographic templates. Here the unexposed emulsion is transferred on to the material to be worked, and exposed and developed there. L.A.M.

See also: *Diffusion transfer; Stripping; Transfer coating.*

TRANSFORMER. Electric circuit for changing the voltage of an alternating current. It consists essentially of two coils of insulated wire wound around a closed iron core. If an alternating voltage is applied to the ends of one of the coils it sets up a voltage in the other. The ratio of the voltages in the two coils is the inverse of the ratio of the number of turns in each coil. So if there are ten times as many turns in the first coil (the primary winding) as in the second (the secondary winding), the voltage set up in the second coil will be one tenth of that applied to the first.

A transformer may have more than one primary winding, to provide for input voltages of different values—e.g. 230 and 110 volt—and more than one secondary winding, to provide a range of output voltages—e.g. 6, 12, and 110 volts. These windings may consist of completely separate coils with both ends led out to terminals for connecting to external circuits, or they may be connected as one continuous winding with tappings taken off at suitable points. For example, if a winding consisting of 10,000 turns is tapped at the 9,000th turn, the voltage across the 9,000 turns will be 9/10 of that across the complete winding, and the voltage across the odd 1,000 turns will be 1/10 of that figure. If the input is connected across the 9,000 turns then the voltage available at the ends of the winding will be 10/9 = 1 1/9 times the input voltage. A transformer of this type—i.e. with the same winding for the primary and secondary coils—is known as an autotransformer. The autotransformer principle is used in mains voltage regulating devices and in constant voltage transformers—e.g. to compensate for the small variations in

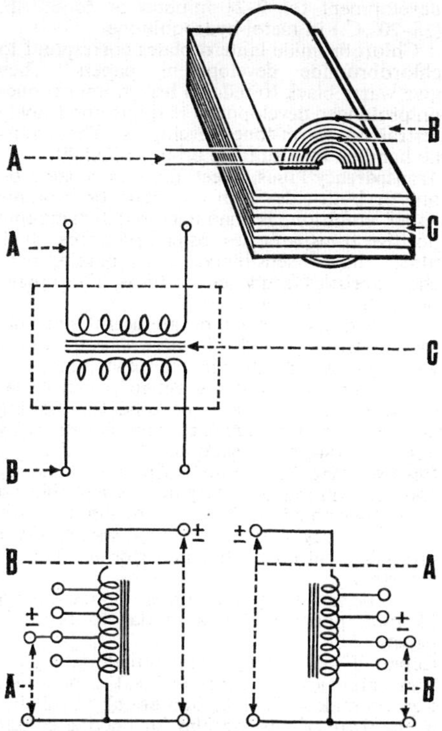

TRANSFORMER TYPES. Top: Basically a transformer consists of two coils wound around a laminated iron core. A, primary winding. B, secondary winding. C, core. Centre: Conventional circuit representation of a transformer, with the same components as before. Bottom: Auto-transformer used as step-up (left) or step-down (right) system. In both cases the lower voltage fed in or drawn out respectively is adjustable by connecting one side of the circuit to one of a series of tappings.

mains voltage that occur between one district and another or even from one period of the day to another in the same district.

Transformers are used in photography for stepping down mains voltages of 230 and 110 volts to provide power for low voltage equipment such as projector lamps, spot lights, rechargeable batteries and control circuits.

TRANSMISSION. Passage of light through a transparent or translucent medium. The amount passed may be quoted as a percentage transmission; the more scientific term, especially when applied to silver deposits in a photographic image is transparency. A transmission of 100 per cent corresponds to a transparency of 1.

See also: *Transmission efficiency.*

TRANSMISSION EFFICIENCY. When light from an object falls on a lens, it does not all pass through to form an image on the plate. Some of it is lost by absorption and some by reflection. If the lens transmitted all the light—which no lens ever does—it would be 100 per cent efficient.

The absorption of light is reduced to a minimum by using the purest materials for the manufacture of optical glass; most of the losses occur because of reflection at the various glass-air surfaces.

Usually at each glass-air surface only about 96 per cent of the incident light is transmitted, and if the surface is only slightly dirty, the transmission is reduced to about 90 per cent. At first sight, this reduction in the amount of light transmitted may not seem very serious, but a modern photographic lens may contain eight glass-air surfaces, all of which actually reduce the transmission to 70 per cent, even when the surfaces are perfectly clean.

See also: *Coated lens.*

TRANSPARENCY. Positive image on a transparent or translucent support—e.g., glass or film base—intended to be viewed by transmitted light or projected on to a screen. The image may be in monochrome or natural colour.

Transparencies are popular in spite of the extra trouble involved in looking at them, for they can reproduce a greater range of tones than any method of printing on an opaque base.

This is because the tone range of a print is limited at one end of the scale by the whiteness of the paper base and at the other by light reflections from the surface which prevent the attainment of a pure black tone. The tone range of a transparency on the other hand extends from a true black where a completely opaque area of the image blocks out the light altogether, to a highlight which can be as brilliant as the light source itself—i.e., where the emulsion is quite clear. In practice this means that while the contrast range of a bromide paper print can never exceed about 1 : 40, a transparency can reproduce the actual contrast range of the subject which may be considerably more than 1 : 200.

The result is that a lantern slide or film transparency in black-and-white or colour can give a more faithful rendering of the tones and modelling of the subject, and this is the favourite method of presentation among serious workers. Slides and transparencies are popular in camera clubs and societies because, in addition to their superior tones, they are more convenient for viewing by a number of people. Most competitions include classes for slides.

The term transparency is also a measure of the light passed by a transparent or translucent medium. When light rays enter a material and are partly absorbed, the proportion of the original light that passes through unaffected is known as the transparency of the material. So transparency is a figure that may be anything between 0 and 1·0. The transparency factor is the reciprocal of the transparency. Often the transparency is quoted as a percentage value (from 0 to 100 per cent) and called transmission.

If a material absorbs some of the white light but the light that gets through is still white, then the material is said to be non-selective. But if only certain wavelengths are absorbed so that what passes through is coloured light, then the material is said to be selective. The extent to which the material absorbs the visible wavelengths can be shown by an absorption curve, which is a graph of percentage transmission in relation to wavelength.

The absorption curve is typical of the material, and is in fact an accurate means of identifying it. The power of the absorption depends on the concentration of absorbing material and the thickness of the layer; as these increase, the colour deepens, becoming more saturated.

In photography, the density of the negative—i.e., transparency of the developed silver deposit—is expressed in terms of the logarithm of the transparency:

Density = − log transparency = log 1/transparency where transparency lies between 0 and 1·0. R.B.M.

See also: *Contrast; Projection practice; Projection principles; Slide (transparency); Transparency materials; Transparency viewing and projection.*

TRANSPARENCY MATERIALS. The sensitive emulsions used for transparency materials closely resemble those used for printing papers but they are coated on glass or film instead of paper.

There are three principal kinds of transparency materials:

(1) Lantern plates, which are positive emulsions on glass and used to make lantern slides.

(2) Transparency film, which is a positive emulsion coated on a film base. (The positives must be bound between glass to make slides.)

(3) Positive cine film, which is perforated 35 mm. film used for making film strips.

Lantern Plates. There are three main types of lantern plate: chloride, bromide, and chlorobromide.

Chloride lantern plates are used for making ordinary contact prints from negatives. Chloride lantern plates are coated with the same type of emulsion as contact paper. They give black to blue-black images on normal development for ¾–1 minute in a metol-hydroquinone developer at 65–68° F. (18–20° C.). Like contact paper, they can be handled in subdued artificial light or by a yellow safelight.

Bromide lantern plates correspond to bromide development and are used for projection printing in the enlarger. Their sensitivity is much higher than that of chloride contact plates, and they must be handled by an orange safelight. They give black tones on normal development for 2–2½ minutes at 65–68° F. (18–20° C.) in metol-hydroquinone.

Chlorobromide lantern plates correspond to chlorobromide development papers. They give warm black to reddish brown image tones on prolonged development in dilute and heavily restrained warm-tone developers. They must be handled by a suitable orange safelight.

Transparency Films. Sheet film base, coated on one or both sides with a contact or bromide paper emulsion, is manufactured for making positive transparencies instead of prints. Such films have sensitivity and development characteristics equivalent to those of ordinary bromide or contact paper.

These transparency films are useful, not only for projection, but also in large sizes for direct viewing in suitable illuminated viewing cabinets.

Ordinary non-colour-sensitive negative flat or sheet films may also be used for making positive film transparencies from negatives by either contact or projection.

Positive Cine Film. Film strips are made by contact printing or enlarging the negative on to positive cine film. Positive cine film used for this is a fairly slow non-colour-sensitive silver bromide emulsion coated on perforated 35 mm. film stock.

In its sensitivity and other characteristics 35 mm. positive film is similar to bromide paper.

Colour Films. Many colour materials yield transparencies by direct reversal processing. Certain makes of colour film are also available in the form of positive film for making colour transparencies from colour negatives.

Contrast Grades. Most lantern plates and some positive films are made in several contrast grades to provide for negatives of varying contrast. The range is, however, much more limited than with printing papers, because the contrast of a transparency can be controlled during development in a way that is not possible with printing papers. Where there is only one grade manufactured it is usually contrasty.

Physical Characteristics. The surface of a transparency is always perfectly smooth and flat. Transparency materials are hardly ever made with a base tint. The finished transparency can, however, be tinted afterwards.

The thickness of the glass used for lantern plates depends on their size. Miniature (2 × 2 ins.) lantern plates are usually 0·03 ins. (0·75 mm.) thick and standard (3¼ × 3¼ ins.) plates about 0·043 ins. (1·1 mm.). Stereo lantern plates are available in both thicknesses.

Transparency or positive cine film is coated either on nitrate base 0·003 in. (0·08 mm.) thick, or on acetate safety base 0·005 in. (0·12 mm.) thick.

Manipulation. Lantern slides and film strip material need greater care in handling and processing than paper prints, as the emulsion is more easily damaged, especially when it is wet. In particular, the temperature of all solutions must be the same or the emulsion may frill and leave the support in places (mostly with plates) or it may reticulate. L.A.M.

See also: *Cine film; Colour materials; Film strips; Papers; Reversal materials; Sizes and packings; Slide (transparency); Supports for emulsions; Transparency.*
Books: *Colour Book of Photography*, by L. Lorelle (London); *Colour Transparencies*, by C. L. Thomson (London); *Making Lantern Slides and Filmstrips*, by C. D. Milner (London).

TRANSPARENCY VIEWER. Various types of viewer are sold for looking at small transparencies without using a projector. There is usually some form of frame for holding the film and a simple optical magnifier through which the observer sees an enlarged image of the transparency.

Some viewers are intended to be held up to the light, others have their own electric light source mounted behind the transparency and supplied from a battery or the mains.

See also: *Viewers.*

TRANSPOSING FRAME. Frame for printing stereoscopic negative pairs. If the negatives made in a twin-lens stereo camera are printed normally and the prints are looked at in a viewer each eye sees the picture that ought to be seen by the other, i.e., the effect is a pseudoscopic one, with the space relationships apparently inverted.

This is because the negative images are both reversed and laterally transposed in the camera. The image reversal is restored by the normal printing process, but the lateral transposition remains. For this reason, the printing frame is made so that the negative can be moved to print the right hand image on the left hand print and vice versa.

See also: *Stereoscopic photography.*

TRAUBE, ARTHUR, 1878–1948. German photographic chemist and inventor. Worked in Berlin with Miethe on colour photography and found in 1902 that ethyl red, the first of the isocyanine series of sensitizing dyestuffs, was an orange-red sensitizer (used in 1904 for the panchromatic Perutz Perchromo plate). Also worked on mordanting dye images (Diachrome, patented 1906), invented in 1916 the Uvachrome three-colour transparency process, and in 1925 the Uvatype colour paper print process. Also developed in 1905 (with A. Tellkampf) the Fotoldruck, a blue-print hectograph process.

See also: *Uvachrom.*

TRAVEL PHOTOGRAPHY

Any kind of pleasure or holiday trip, whether at home or abroad, presents excellent opportunities for photography.

The sort of holiday, the method of travel, and the photographer's interests all affect the photographic results, and some advance planning is always advisable whether for the nearest seaside resort or the next continent.

Equipment. Amateur photographers, at least, are apt to carry large quantities of equipment which they never find time or the need to use.

Basically, only a camera and film are necessary for taking photographs. Since, however, the object will be to produce the best possible photographs, it is as well to include an exposure meter, a lens hood and, for protection, an ever-ready camera case. A lens cap adds nothing to weight or bulk and keeps the lens cleaner even inside the ever-ready case which, if velvet-lined, seems to attract dust, sand, fluff and other unwanted foreign matter.

For hikers, cyclists, campers, climbers, and all who have to keep baggage to a minimum—this list represents the photographic minimum.

For motorists, coach tourists, hotel and guest-house residents, more equipment may be comfortably included but even they need not go beyond a couple of filters and a light-weight tripod. The latter is better than a camera-clamp because there is not always a suitable feature situated where the camera might be placed. It is better to be occasionally inconvenienced by an accessory that is usable when needed, than have one that is easy to carry but may turn out to be useless. A light yellow and a deep yellow or green filter should be enough, although in mountain and seaside territories it is advisable to take an ultra-violet-absorbing (UV) filter.

What camera to take is no problem. The camera to be taken on any holiday trip is the one that is completely familiar. The make, type and size matter little in comparison with ease and certainty of operation. Every traveller should carry his own favourite instrument, whatever the size—so long as he can be sure that negative material will be readily obtainable for it.

Holidays and journeys are no occasions for trying out a new camera. Nor is it wise to use a borrowed camera; unless it is one that has been borrowed so often before that all its idiosyncrasies are known.

Much the same applies to accessories. Nothing new or untried should be taken. An incorrect lens hood can cut corners off irreplaceable negatives; unfamiliar filters can produce unexpected effects.

Whilst the majority of people have only one camera and use the two types of material turn and turn about, others carry two cameras, one for colour and the other for black-and-white pictures, which is the ideal method. Accessories and film for both instruments add to the baggage but the gain in scope is well worth it.

On a holiday in Britain it is rarely necessary to carry films for the entire trip, as few places are so remote and off-the-track to be ever out of stock of the popular makes, at least. For a foreign trip, however, the only way to be really sure is to take a stock of negative material. Films are undoubtedly as plentiful in continental shops to-day as they are in British shops but it is usually unfamiliar makes that are most freely obtainable, and these are the ones to be avoided. Developing unknown makes of film which are not named in any time-and-temperature tables is an undesirable end to a holiday.

There is a further objection to buying films away from home. Some manufacturers have factories in different parts of the world, each manufacturing the materials under the same brand name. While in theory these materials may have identical characteristics, in practice they may vary widely. The photographer who buys his favourite brand of film in a foreign country may find that it is anything but the same when he comes to use it.

Customs. Foreign holidays involve Customs formalities; and photographic equipment and materials are generally of special interest to Customs authorities everywhere. Unfortunately for the holidaying photographer no two countries appear to see eye to eye on what is and what is not permissible, and Customs departments seldom provide detailed statements of their policy to help the wandering cameraman.

Very broadly speaking, no trouble need be expected if equipment has been honestly come by and if holiday intentions are equally honest. The British Customs are concerned only to see that nothing is illegally imported and that all necessary dues are paid on incoming goods, if they have not already been paid. Many holiday-makers pass unquestioned through the British Customs on leaving the country and on re-entry; others may be examined once; and some twice. The safest and easiest thing is to carry the receipted dealer's invoice as proof of purchase of each camera carried, and to include with this any Customs slips available.

Most countries have rules about the number of cameras allowed per individual, and many countries restrict the quantity of unexposed films. Two cameras, sometimes three, may be taken into and brought out of the majority of European countries. The rules about unexposed films are not rigidly enforced and it is possible to carry a good supply even though the regulations may permit only two or three.

Customs rules and regulations vary from one country to another so it is useful to know that European countries often have consulates in the larger cities in Britain where guidance on the subject can be had from the consular staff of the country to be visited. Travel agents can sometimes supply reliable information, and can in any case provide the addresses of the English offices of all national tourist bureaux operated by foreign countries for the convenience of tourists.

It is always wise to find out in advance the regulations of the country to be visited because the British Customs authorities are only concerned with what may be taken into Britain and not with what happens at the other side.

Technique. Continental holidays in summer-time frequently present the British photographer with conditions that he never encounters at home. Intense sunshine, clear atmosphere, white and light-coloured buildings, all present a brilliant picture to the eye. But in the dazzling scene, in spite of the high reading of the exposure meter, are sure to be found some very dark shadows. Such high-contrast subject-matter calls for care both in exposure and development.

Since meter-readings often appear extravagantly high, by English standards, there is a strong temptation to use slow-speed films, yet these have more inherent contrast than the faster emulsions. The latter, being softer, are more suitable for the conditions, and when handled correctly can be expected to give a more workable negative.

It is helpful to develop a film early in the holiday so that the result can be used as a guide to later work. If this plan is to be a success the test film must be processed with the same developer and under the same conditions normally used at home. This means that developing materials and equipment—chemicals, tank and thermometer—must be carried, thus increasing the amount of luggage. And unless a changing-bag is also taken the tank must be loaded under the bedclothes or in the hotel wardrobe or in some such uncertain manner. If really serious work is intended it is certainly wise to pack a changing bag.

Another scheme is to have an early film trade processed, but unless the developing data can be ascertained this course only indicates in a general way how the exposures are faring. If possible, the photographer should try to find a dealer who operates an individual processing service.

Subjects. A holiday trip, especially one abroad, is a source of considerable joy and every chance of recording it for re-enjoyment afterwards should be taken without hesitation. Places are being seen which will probably never be visited again. Any picture—even if it may not be a salon print—is better than none.

It is as well to remember however that in all countries, including Great Britain, there are certain subjects that may not be photographed without permission—e.g., inside many museums, public and private buildings, and churches. Some authorities forbid photography; others allow it by special permission only; still others permit it on payment of a fee, usually small, which goes towards upkeep.

The places for photography are limitless: landscapes, seascapes, street scenes, zoo pictures, beach scenes, all these and countless others by day and by night will offer themselves to the travelling camera enthusiast.

Foreign holidays especially offer vast scope for human-interest photographs. Strange people in exotic settings are irresistible to all photographers, from the casual snapshotter to the ardent exhibition worker.

There are many ways of taking characteristic photographs of people abroad. If a good candid picture cannot be obtained unobtrusively the subject will often agree to pose; this assumes that the photographer has enough knowledge of the language to explain what he is about and to arrange an appropriate pose which gives the subject something to do and prevents self-consciousness and embarrassment. Unposed shots are very much easier to take with the co-operation of a friend or two, to act as a shield. (In taking quick shots of this type the background should be watched carefully because such pictures are often ruined by unsightly objects beyond the subject but in the field of view of the lens.)

People are waiting to be photographed everywhere: in streets and cafés, on river banks and quay sides, at markets and shopping-centres, in parks and gardens, at bus stops and railway stations. But many of them are photographically uninteresting. It is a waste of time to search for the interesting types; the best plan is to be alert and ready to shoot quickly when the right characters appear—as they generally do when they are least expected. The very young and the very old make especially good unposed studies.

Pictures en Route. Rewarding photographs can be taken en route to and from the destination, as well as on excursions made during the holiday.

It is possible to get good pictures from almost every form of transport. Normally airlines do not object and air-to-ground shots are fairly straightforward providing the camera is held just away from the window surface and a fast shutter speed (say 1/250 second) is given. If the traveller has the choice he should select a seat ahead of the wings, where the heat trail from the engine exhausts would not interfere with the negative sharpness. In aircraft having engines mounted at the rear, any seat is satisfactory.

Pictures from fast-moving trains are best taken through an open window whilst standing in the corridor away from the wheels—i.e., midway between the bogies. A shutter speed of 1/250 or 1/500 second is sufficient to ensure a sharp negative when shooting diagonally at an approaching or receding scene. Only the feet should be in contact with the train; the body should be held away from the window-frame and allowed to sway with the motion, arms in to the side and camera at eye-level. Pictures taken from moving trains cannot be composed. They are not seen until a split second before exposure, and a split second later they are gone for ever. The only satisfactory method is to look ahead for likely villages, churches, tree groupings, waterfalls, which will form a centre of interest and to have the camera ready to shoot in an instant. The photographer must simply hope that telegraph poles, embankments, walls, trees, signal boxes, and other obstructions will not spring up to spoil the view just as the shutter is being released.

Shooting from a stationary train is simple and scenes taken in foreign stations can be extremely interesting and sometimes even picturesque. Railway stations abroad are usually bright and airy places, often decorated with flowers; and teeming with character studies ready to be taken easily from the unseen vantage point of the train window.

Taking pictures on board ship, even during a none too smooth crossing, is again a question of using a fairly fast shutter speed and letting the body go with the ship's movement. Shipboard scenes, during embarkation and disembarkation particularly, are a satisfying—and essential—part of any overseas holiday record.

Coaches and motor buses have the drawbacks of more pronounced vibration and less room to manœuvre, but open windows and sliding roofs help; and at stopping places the raised viewpoint is of much use in picturing village scenes, wayside inns and bystanders.

Perfect negatives are easy to take during rides on the funicular railways, and also cable-cars, which take tourists up mountains. From this type of transport, cameras without fast shutters are just as capable of producing good results, as the cars operate in a very steady and slow manner.

Stowing the Camera. If the photographer is a passenger—e.g., by boat, train or motor car—and there are likely to be pictures about on the journey, he can carry his camera at the ready in the normal way. But the driver of a car or motor-cycle and the cyclist will have to stow the instrument somewhere on the vehicle.

In a car, the ideal place for the camera is in a hold-all case on the back seat. The worst place is in the cubby hole at the side of the dashboard where it gets all the vibration and heat from the engine, and is always in danger of being pulled out on to the floor when someone is hunting for a map or a packet of cigarettes. Some cars have a net for small packages stretched across under the roof. This is an excellent place for a small camera, but it should not be left there while the car is parked under a direct hot sun for any length of time.

Cyclists and motor-cyclists are best advised to carry their cameras slung over one shoulder under the jacket or overcoat. There is no place on the machine where it will be as well protected from rain and vibration. The cyclist can always strap a light tripod along the cross-bar of the machine but the motor-cyclist will have to sling it across his back or invest in one of the small pocket-size folding types.

When the camera is packed in a suitcase, the best place is in the centre or near the handle. It should be packed last, and on top of soft clothing. Finally it should be covered with a book or map to protect it from damage through the case.

Precautions. Costly equipment can be safe-guarded by insurance for a comparatively small sum, and the surprising thing is that more photographers do not take out a policy. At holiday-time the camera is in use or being carried about for many hours every day, and the risk of loss or damage is obviously greater. The risk can be minimized by a number of commonsense precautions. If cameras have to be packed away they should be placed in a well-protected position in the suitcase or ruck-sack, where they are still reasonably accessible. On the other hand, if the camera will be carried constantly it will be safer if the strap is kept fairly short so that the camera cannot swing far away from the body.

A foolproof way of carrying the camera is to sling it over one shoulder before putting on the jacket. Once the jacket is on, it is physically impossible for the strap to slip off the shoulder, or for the camera to swing about. At the same time the arrangement offers no hindrance to quick and efficient operation as the case is easily pulled up and forward to the taking position. This idea also makes it impossible for the camera to be left behind any-where or for it to be left unattended.

When all is said and done most of the information about travel photography boils down to applying common sense, standard routine with slight adaptation, and good manners.

There are certain countries abroad where a stranger carrying a camera is regarded with suspicion, and others where a foreigner is not allowed to take photographs without a special permit. In such places it is foolhardy to attempt to take pictures by stealth—the photographer who is caught at it runs the risk of losing his equipment permanently and his freedom at least for a time. To a camera journalist after a scoop, the risk might be justified; for the ordinary tourist it can never be worth while. The same thing applies to attempts to photograph people—or personalities—who are known to be hostile to photographers. Here the photographer runs the two-fold risk of damage to his equipment and his person. A.H.

See also: *Air photographs; Customs; Insurance; Marine photography; Permits to photograph.*

TREES. Trees offer unlimited scope to the pictorialist, not only as part of a landscape but as studies in themselves. To be successful, tree photography calls for the ability to make a suitable choice of type or species of tree, lighting, setting, and treatment.

Apart from their intrinsic beauty trees are useful in landscape composition. The fore-ground use of branches (with or without the trunk) will often mask a bald sky, and will also assist in preserving the impression of space by throwing back a distant hill outline, a valley seen from a ridge, or other open landscape.

A single tree, or group of trees, stunted or weird in shape, will often convey the character or mood of a particular place, as, for example, along the Atlantic coast of Britain, where the south-west wind bends the trees inland at an acute angle, causing the branches all to grow in the one direction.

Type of Tree. Because the composition relies chiefly on balancing masses, the shape of the tree and the distribution of its branches are important. Spreading trees like cedar, oak and chestnut lend themselves to horizontal picture shapes, whilst tall elms, poplars, etc., need the upright format.

The base of the tree is always included in the picture, otherwise the result looks unnatural and top-heavy. Any branches that project towards the camera must be in focus as well as the main trunk, so the taking distance and aperture must be chosen to provide the necessary extra depth of field. Ugly limbs and foliage-shapes are avoided as far as possible by experimenting with different viewpoints.

Lighting. Frontal or flat lighting is rarely used because a tree is a round object and must have modelling. This point is not as serious in mist or fog, where the separation of the picture into planes is more important. Low-angle lighting with its long shadows gives a much more pleasing effect than overhead lighting, and for this reason the best tree pictures are unlikely to be made during June, July and August, between the hours of 11 a.m. and 3 p.m., when the sun is high in the sky.

Backlighting is seldom successful with sparse or small-leafed trees like ash and hawthorn, but it can be very effective with large-leafed trees like chestnut, sycamore, lime, etc. There is little to be said for photographing "open" or bare trees in backlighting conditions because the results are almost bound to be hard and spoiled by halation, caused by spreading of the light into the image of the branches.

Setting. The setting of the subject includes the sky behind it, the other trees around it and the physical features of the landscape. The greatest difficulty usually lies in isolating the one chosen tree or group of trees from its neighbours.

The photographer should always look beyond his subject to ensure that the background is

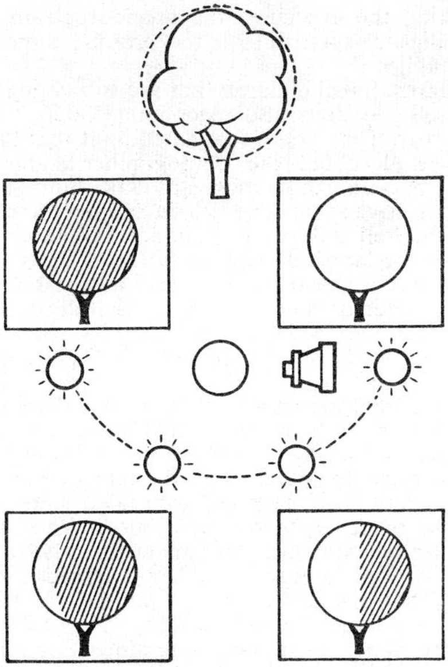

LIGHTING ON TREES. *A tree can often be regarded as a sphere, and follows the same rules for the effect of light on modelling.* Top left: *Back lighting tends to yield silhouettes.* Bottom left: *Side-rear light produces pronounced rim of light.* Bottom right: *Side lighting gives good modelling, and picks out the pattern of leaves.* Top right: *Front lighting is usually dull.*

free from unwanted objects like buildings, telegraph poles, etc., and he should move his viewpoint if necessary to get rid of them. He should also avoid mixing trees of various kinds; firs do not blend well with oaks, limes, etc., and tall trees should not be surrounded by low, scrubby specimens. The sky generally makes the best background, provided that the area is interesting and not simply blank white paper.

Treatment. Unless the camera is fitted with a rising front, trees will present the same sort of perspective problems as high buildings. If the photographer tries to include the whole of the subject from a close viewpoint, he is forced to tilt the camera up. But if he tilts the camera, the subject will look as though it is toppling backwards because its vertical lines will lean in towards the top.

The answer is to move away to a more distant viewpoint where the camera can be held level and still include the whole subject. This inevitably gives a top-heavy picture in which the lower half consists of empty foreground and all the subject matter is crowded into the top half. But the problem is easily solved by leaving out the unnecessary foreground when the negative is enlarged, or trimming it off the print.

The selected branch or group of branches should be focused sharply at a wide aperture to keep the background out of focus. A sky background of cloudless blue is often better than one covered with patchy cloud. A 3× yellow, or green filter with panchromatic film prevents the green foliage from being reproduced in unnaturally dark tones and gives depth to the blue sky.

A long focus lens is a great help in capturing detail in blossom, and avoids the distortion produced by using a short focus lens at close range.

Trees in Winter. Trees in winter generally call for semi-close or close-up viewpoints which stress the characteristic signs of the season—i.e., snow, rain or mist. Sunshine, no matter how feeble, is necessary for snow and rain effects and the feeling of mist is heightened when there are trees at varying distances from the camera to divide the picture into close, medium and distant planes.

Gaunt, angular trees against an angry sky convey a better impression of the mood of winter than trees with graceful flowing branches, even when they are laden with snow.

Trees in Spring. Spring studies call for close-up viewpoints to show the opening leaves, young buds and blossom. As in so many other branches of photography, the part is generally more effective than the whole.

Often a single branch or spray photographed with care so as to show plenty of crisp detail makes the best picture of all. This demands careful arrangement and viewpoint to make the branch fill the picture area in a pleasing fashion. Fairly strong, oblique daylight improves the rendering of detail and helps to keep exposures short. Shutter-speeds of 1/100 second or faster are usually required to counteract the movement of the branch in the wind.

Trees in Summer. During the summer months, trees are better handled in terms of their masses and shadow effects than as detailed studies. The full, dusty foliage and harsher lighting, characteristic of this season, make it the least interesting to the photographer.

Trees in Autumn. Autumn, on the other hand, with its varied tints and the sparser, lighter-toned foliage of the fall, offers splendid opportunities for photography in both colour and monochrome. Back lighting turns close and semi-close golden leaves into delightful picture material. An orange filter can be used to good effect in this type of picture as it lightens the golden tones.

Large-leafed trees generally provide the best pictures when they are photographed with the sun behind them in this way.

Woodlands. Woodland studies are more difficult, because of their extreme lighting contrasts. The very wide range of tones demands a fast, panchromatic film to give sufficient exposure in the deep shadows without "burning out" the highlights. Slower, more contrasty emulsions are quite unsuitable.

The photographer who is not prepared to

take extra trouble in processing to deal with the extreme contrasts of this type of scene should avoid the interior of the wood and make his pictures in clearings or on the outskirts of the wood.

Avenues of limes, elms, horse-chestnut, etc., make natural subjects at almost any time of the year.

Light patches between branch formations should be avoided as far as possible because they tend to irritate the eye.

Camera. Almost any camera can be used for making pictures of trees, but the photographer who wishes to specialize in it should invest in a half- or whole-plate field camera with all the usual camera movements and a selection of lenses of various focal lengths.

For this class of work, while the great depth of field of the miniature camera is an advantage, it lacks facilities for perspective correction and cannot give the crisp detail obtainable with the larger camera. P.J.

See also: *Landscapes*.

TRICHROME CARBRO PROCESS. Method of producing colour prints by the assembly, in register, of a set of three colour carbro images made from separation negatives.

See also: *Carbro colour prints*.

TRICKING THE SUBJECT. One of the difficulties of photographing living creatures, from human beings to insects, is to get the subject to pose for the camera in a natural manner. With human beings—and some animals—there is often the added difficulty of getting them into position and photographing them unawares. The photographer needs to know something about stage management, misdirection and psychology.

Animals. There are many standard tricks for shooting animals, birds and fish. Seagulls in flight, for example, can be enticed into position if a companion stands close to the photographer and throws some morsels of food at the right distance. Fish in an aquarium may be kept in the plane of focus with a sheet of plain glass slipped into the water behind them.

Bait is a great help when photographing birds, animals and insects at liberty. A few crumbs will soon entice all sorts of birds to the desired spot; a single flower that has no other of the same species nearby will attract butterflies, rotting meat similarly attracts flies and other insects. Birds of prey can be attracted by imitating an owl-call, and monkeys can be kept interested with a child's toy.

With dogs and cats there is usually some suitable line of approach suggested by the animal's habits or training. It is often easy to make a placid sleepy dog look bright and alert by rattling his lead or getting dressed as though off for a walk. A cat will generally respond to the clinking of a milk jug. All such stratagems shoutd be introduced immediately before making the exposure; the precious gleam of intelligence soon dulls if the promise remains unfulfilled.

Children. Small children that are too young to be self-conscious also react immediately to a new toy. They are so fascinated by it that they forget all about the photographer standing near by—though he must still make himself as unobtrusive as possible. When making pictures of children indoors it is important never to shine the lamps directly on them at the start. It is best to turn the reflectors away at first, then switch on and swing them carefully round towards the field of view. In this way the children will not be scared by the sudden glare.

Candid Shots. In candid photography there are several tricks that the photographer can employ to avoid drawing attention to himself or his camera. In cold weather the camera may be slung round the neck and concealed under an ample overcoat, worn unbuttoned. When the right moment comes the camera is ready to go into action. In warm weather it is better for the photographer to stroll around casually with his hands behind his back. No one will notice the camera he is holding. For this, however, he must learn to set the camera by feel alone. All adjustments must be made before the camera comes into action.

A right-angle viewfinder deceives the victim by leading him to believe that the photographer is aiming the camera at someone else.

The "swing-round" technique is an excellent method of securing candid shots where the camera cannot be used unobtrusively. Here the idea is to focus on an object at about the same distance from the camera as the intended victim but in a different direction—usually rather more than a right angle. The real subject is observed out of the corner of the eye and, at the last instant, the camera is swung round on to the target and the exposure made right away. Any start of surprise on the part of the subject usually occurs after the shutter has closed.

Portraiture. The following trick saves sensitive material often wasted when the photographic session has been pre-arranged. It is common experience that some time is needed for the model to overcome the initial self-consciousness.

The last exposures of a series are often the best, because by then the model is sufficiently relaxed and at ease and the feeling of stiffness has disappeared.

Why then waste film on the initial exposures? The trick here is not to make them at all, but simply to go through the motions without actually exposing any film. With a plate camera the dark slide can be left in position, and the sitter is led to think by the noise of the shutter that an exposure has been made.

Most photographers know that the sitter appreciates a certain amount of retouching on the portrait. But apart from normal retouching, one useful trick is worth mentioning: deliberate elongation of the image. The procedure is the same as that used in correcting the distortion of converging verticals, but in this case the object is to make the subject look slimmer. The image is elongated in enlarging by tilting the baseboard. So long as the trick is not overdone, the result will please sitters who have begun to put on weight. O.R.C.

See also: *Candid photography.*

TRICKS AND EFFECTS

Tricks in photography may be executed for their own sake, or as a means to some more serious end. Even when making a photograph by normal procedure, there are occasions when trick methods save time and give a better result.

There are hundreds of situations and subjects where tricks can be used to advantage—e.g., to achieve some special effect or to overcome limitations and shortcomings inherent in the photographic process. In most cases the tricks are simply a matter of applying an established process or technique in an unorthodox manner.

There is scope for tricks at each stage of the photographic process.

OPTICAL TRICKS

Many kinds of trick photography rely on purely optical methods. Generally this is merely a question of using normal equipment in an unorthodox fashion. Once the photographer understands the basic principles the number of tricks he can perform will depend more upon his imagination than his equipment.

Tricks with Lenses. The use of a lens of longer or shorter focal length than normal is not a trick, because these are regular accessories. It is a trick only when some unorthodox optical equipment is used.

A telescope is an example of unorthodox equipment; fitted in front of the camera lens it brings distant objects nearer. There are only two things to watch: the telescope must be rigidly supported, and the coupling between it and the camera lens must be made light-tight—e.g., with electrician's insulating tape. The discrepancy between the small-diameter eyepiece of the telescope and the camera lens is overcome by stopping the camera lens down to *f* 8 or *f* 11.

The resulting photographs resemble genuine long-distance photographs. The slight unsharpness at the edges is in many cases hardly noticeable. One element of a pair of binoculars may be used in the same way as a telescope.

Telescopes and binoculars increase the focal length; the simplest method of shortening it is with a single spectacle lens as used for long-sight. (All supplementary front lenses are fundamentally spectacle lenses of different diopter value: the higher the number, the shorter the focal length.) As all such glasses reduce the optical precision of the camera lens, the latter should be stopped down to *f* 8.

Spectacle lenses may also be used in place of the camera lens. When used in this way as a principal lens they produce a special effect. Being uncorrected, they suffer from all the aberrations. The faults combine to give soft focus.

Ordinary spectacle glasses of from +3 to +6 diopters may be used (concave side out) either to obtain soft outlines, for example to conceal facial wrinkles in a portrait, or to handle an extreme contrast range when absolute sharpness is not essential. The softened focus and lowered contrast are, however, only produced if the lens is not stopped down.

Instead of substituting a spectacle lens for the camera objective, a diffusing disc may be placed in front of the camera lens. This is a piece of glass engraved with a fine texture pattern and used like a filter. It produces very much the same effect as a simple spectacle lens, provided the camera lens is used at or near full aperture.

If the texture pattern engraved in the glass is coarse instead of fine, there is no soft focus effect but instead a multiplication of the image. Thus glasses ground in pyramid form with three, four or more "steps" produce a similar number of symmetrical, overlapping images. (This trick is often used in motion pictures.)

Tricks with Mirrors. A mirror, too, can be made to give multiple images. If an object is placed on a mirror, a double image is formed, the upright image being laterally correct, while the reflected image will appear upside down and laterally reversed. The images form a symmetrical pattern.

If the edge of a mirror is placed against a page of a newspaper, the text appears printed backwards in the mirror. If two mirrors are stood vertically on a newspaper with their reflecting surfaces at an angle of 60°, the newspaper text will be seen multiplied six times.

Every change in the angle between the mirrors will produce a further range of image multiplication.

Curved mirrors make a variety of other effects possible. A cylindrical mirror shortens the image when placed flat and lengthens it when stood up vertically. But it is necessary to focus on the reflection, not the subject. This applies equally whether curved, cylindrical or spherical mirrors are used. If a new silver spoon is held in front of the camera lens and the reflection in it of the surroundings photographed, this will at once show how it is done.

With this type of photograph, the photographer and his camera will appear in the picture unless special precautions are taken. One way is to turn the reflecting surface towards a plain wall, and set the camera up to the side. The resulting image is then no longer rectangular, but the reflecting surface is blank and the distortion of the mirror frame can easily be corrected when enlarging the negative.

Tricks with Distorting Glasses. All tricks where uneven-surfaced glass plates are interposed between subject and camera can also be classified as optical.

Certain types of ornamental and window glass are designed to give privacy by breaking up images seen through them by a pattern embossed in the glass. Distorting glass of this kind can be used to produce a number of novel effects by taking photographs through it. The effect on the image varies according to the distance between the glass, the camera and the subject.

Tricks with the Enlarger. Some control methods used in enlarging can also be classified as optical tricks.

First there is the soft focus effect, designed to reduce the contrast of an excessively hard negative and conceal obtrusive detail. Instead of placing a soft focus attachment in front of the lens, all that is needed is a sheet of glass rubbed with a smear of vaseline. This, of course, spreads the shadows into the highlights of the picture whereas a soft focus disc or lens used in the camera spreads the highlights into the shadows.

The technique used for correcting converging verticals in the enlarger may be used to bring about deliberate distortion. Distorted lines in the negative are squared up by inclining the baseboard. In the same way all kinds of distortion can be introduced by inclining the baseboard in various directions. A similar class of distortion can be produced by bending the bromide paper into concave or convex shapes and holding it there with pins or tape during the enlarging exposure.

In this way the image may be elongated so as to give an outsize lady a sylph-like figure; or to make a sitter's face look broad or narrow.

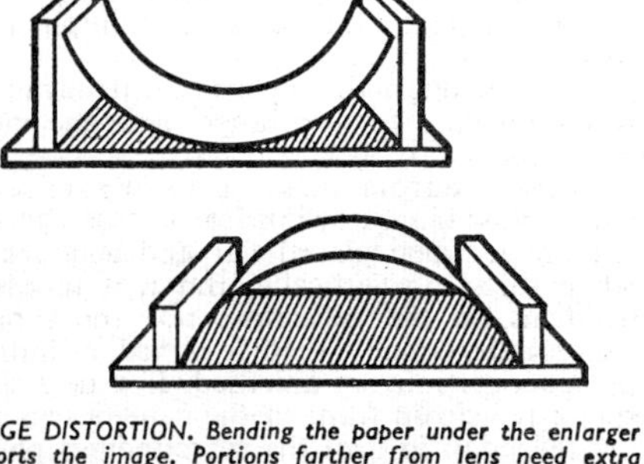

IMAGE DISTORTION. Bending the paper under the enlarger distorts the image. Portions farther from lens need extra exposure, also small stop to ensure sufficient depth of focus.

CHEMICAL TRICKS

Trick photographic effects produced by chemical methods usually involve a departure from normal working procedure, often the deliberate production of well-known faults. As chemical manufacturers are at pains to render their products foolproof, such faults are today often hard to induce.

Tricks During Development. If a white or unsafe light is switched on during development, the upper layer of silver already darkened by the developer acts like a negative at the printing stage and masks the light off from the layer underneath which has remained unaffected. At the same time, the clear, unexposed emulsion is exposed, and is darkened by the developer. So a negative and a positive image are formed in the same emulsion, one beside the other. The potassium bromide that is normally formed during development collects along the borders of the image where chemical activity is smallest. Being a restrainer it prevents further darkening there and creates a transparent line in the negative which shows in the print as a dark line around the edge of the positive image. This effect, which resembles a line drawing, is much used as a stunt in modern photography. It is known as the Sabattier effect.

Deliberate alteration of development time can also produce a trick effect. If development is unduly prolonged, it produces an excessively hard negative. From this negative both soft and hard transparencies may be printed on positive film. When separate negatives made from these positives are enlarged in register, the result is a positive with a particularly rich scale of tones, giving a distinctly plastic effect.

Tricks in the Fixer. If the temperature of the fixing bath is raised to 86°–96° F. (30°–35° C.), the emulsion curls up and breaks the picture up into an irregular network of lines like those of a miniature jigsaw puzzle. A wrinkled pattern is thus superimposed on the normal rendering of the subject, giving it a novel texture that can be most effective under certain conditions—e.g., to imitate the brush work of an oil painting.

While the emulsion is still wet and swollen from the final rinse, it will soften and melt if exposed to heat. Here is the basis of a number of photographic tricks. A variety of grotesque distortions may be produced by holding wet plates over a spirit lamp flame (films should be held over an electric cooker or radiator) until the emulsion melts. By tilting the plate one way or another the emulsion can be made to run in any direction. When it cools, it sets permanently in the desired shape.

The normal method of reducing a dense negative is to place it in Farmer's reducer. But the action of the reducer may also be confined to selected parts of the negative by applying it with a brush or a wad of cotton wool. If

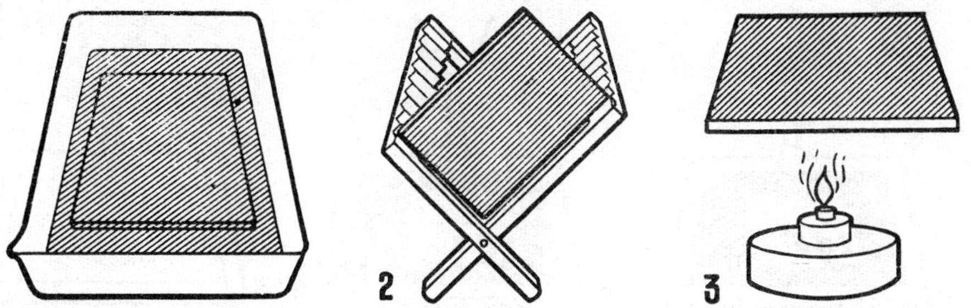

MELTING EMULSION. This produces distorted images of irregular shape. 1. Soak the negative (must be a plate, not film) in water for 10 minutes. 2. Leave to drain thoroughly. 3. Carefully heat glass side over a spirit burner; when the emulsion starts to melt, tilt the plate to and fro to get the molten emulsion to run in any desired direction.

necessary, the whole of the background or foreground may be removed. This method is one way of making a montage. Or the background may be removed from one negative and the foreground from another, the two negatives being then superimposed and a composite print or enlargement made.

MECHANICAL TRICKS

Mechanical trick effects on the negative are limited to physical alteration of the emulsion surface. Dense areas of the emulsion may either be lightened by rubbing with silver polishing powder (containing oil) or scraped away altogether with a retouching knife. The latter method is used for putting a black border around the picture. To do this the knife is drawn along the edge of a steel ruler until a clear even line has been scraped in the emulsion.

Mechanical treatment of positives covers both cut and paste montage and all methods involving pencil and brushwork.

Cut and Paste Montage. Normal cut and paste montage is the simplest of these methods. Two prints of similar gradation and surface are cut up with the scissors along a particular outline, then pasted together to form a new picture. After copying, any number of prints may be made from the negative. To prevent the cut edges from showing, all prints should be on single weight paper and cut with sharp scissors or a safety razor blade. The cut edges should be dyed or painted black, grey, or left white to match the adjacent print area. O.R.C.

Picture Patterns. Making a pattern from a mosaic of photographs is a good variation of normal cut and paste montage. Very simply the trick is to combine sixteen prints—eight printed the right way round, and eight reversed—to produce a repeating pattern.

The most suitable type of photograph is one with strong geometrical shapes. If the original print consists mainly of flat tones then the finished mosaic will lack interest.

The prints should be about quarter-plate size, otherwise the finished mosaic may be too large. When selecting the composition, an L-shaped piece of card as normally used for composing is required; also a pair of hand-mirrors. Two sides of the print are selected by the piece of card, and the other two sides selected by the mirrors. In this way a combination of four prints can be visualized, two the right way round and two in reverse. In this way it is possible to get an idea of the final effect.

All sixteen prints must be in good register. To ensure this, when making the viewing print, a reversed print of exactly the same size is made. The negative in its carrier must also lie in exactly the same plane for both prints. When the portion of the print has been marked up, the two prints should be placed back to back, registered by viewing them against a strong light, bound together with adhesive tape and trimmed to the marked up area. The reversed print will then have the same area.

Before making the first run of prints, the area of the projected negative is masked according to the viewing print. The exposures should be accurately timed by a clock, and developed to strict time and temperature. A guide print is first made to get the exposure right and then the first batch of prints (ten prints, eight for use and two spares) are given the same exposure. The whole batch are processed by placing them in the developer at 5-second intervals. When the first print has been developed all the subsequent prints are removed after the same intervals. All prints will then be of exactly the same tone. Having made the first batch of prints, the negative is reversed in the enlarger and masked according to the reversed guide print to make the second batch.

The prints can then be trimmed and mounted accurately on card with rubber solution and the joins touched out. L.S.

Drawings from Prints. Most paper surfaces can be worked over with pen and waterproof indian ink, so a light toned enlargement may be transformed into a line drawing by inking it in with the pen, and then placing the print in a strong solution of Farmer's reducer. This bleaches out all photographic detail and leaves only the hand work.

DRAWING IN THE ENLARGER. This requires no sensitive materials. 1. Project the negative on to a sheet of suitable drawing paper. 2. Shade in the light parts of the projected image with a pencil, gradually working over the picture until the paper is uniform in tone while the image is superimposed on it. 3. Switching off the enlarger reveals a positive pencil image.

Direct Pencil Drawings. Pencil drawings, unlike pen drawings, contain middle tones, but they may be made from photographs by the following method. The negative is placed in an enlarger and focused on a sheet of drawing paper. The bright parts of the enlarged image are then shaded in with the pencil until the image projected on the paper looks a uniform grey all over. When the paper is examined in ordinary light a positive "print" will be found on it. It will be rough or delicate according to the draughtsman's skill, and will display all the characteristics of a genuine pencil sketch.

Endless photographic tricks can be evolved by the imaginative use of such mechanical treatment as retouching and titling.

EXPOSURE TRICKS

Many photographic trick effects are possible at the exposure stage by deliberately disregarding the normal rules.

Long Exposures. Instead of taking an action shot at a shutter speed short enough to arrest the motion, a long exposure may be given. When this is done, highlights become comet-tails that follow the path of the moving object. The more the exposure exceeds the normal minimum the longer the tails will be. Such pictures, though themselves static, convey an unusual impression of movement.

For example: the permissible shutter speed for a man doing a knees-bend would be about 1/50 second. If (after stopping down the lens accordingly) an exposure of 1 second is given, the negative will suggest the whole of the action even though the man performing it will only be seen as an unrecognizable blur. And the work gains a new significance. Instead of being, say, "Father doing a knees-bend", it achieves the status of a study, "Knees-bend by an unknown performer".

Movement offers unlimited scope for trick effects. If, for instance, a small pocket torch is fastened to one hand of an orchestral conductor and a sufficiently long exposure given in a dimly lighted music room, all the conductor's movements will be visible on the negative as a sharp line of light on the dark background and the ghostly orchestra. The photograph no longer shows a particular instant but all the rhythm of a complete cycle.

Records of inanimate objects in continuous motion open up another field of possibilities.

A pendulum, for instance, may be made to swing in an elliptical or circular path, or have its motion interrupted or broken by some obstacle. Its movement will gradually become less and less until it is stationary. An interesting photographic trace of its movements can be made in some such way as the following. A pocket torch is placed in a tin with a small hole in it so that the beam can be seen through the hole. The tin is then suspended from the ceiling by a cord about five feet long. This pendulum is started on any desired pattern of motion in the darkroom and the camera shutter is left open to record the ever decreasing track of the light. A print from the negative will have the appearance of a delicately drawn pattern executed with absolute geometrical precision.

Stars also are moving points of light in the dark sky. An exposure of several hours will record the movement in the same way. After dark the camera is placed on a tripod and aimed northwards towards the pole star. According to the length of the exposure (½ to 5 hours) the stars will trace their own orbits around the pole as a pattern of arcs of light.

The trick of giving a too-long exposure comes in useful when, for instance, a realistic rendering of running water is required. If a very short exposure is used to photograph a waterfall, the water looks dead and frozen. Longer exposures of the order of 1/5 and 1/10 second produce blur and impart a sense of movement to the whole mass of water.

Photographing a lightning flash calls for more exposure trickery. The camera is directed towards the approaching storm front at night and the shutter is opened after the first flash, as experience shows that a second flash is certain to follow a few seconds afterwards. A succession of flashes may be recorded one after another on the negative if the shutter is left open sufficiently long.

Multiple Exposure. But very long exposures can also be used to eliminate moving objects. The trick can be exploited, for instance, to photograph a busy street as though it were empty. An instantaneous exposure of about 1/100 second would show the traffic clearly. But if, instead of a single exposure of 1/100 second, ten exposures of 1/1,000 second each are made on the same plate, allowing some time to elapse between exposures, the moving objects are recorded so indistinctly at each exposure that the resulting image is overlaid and eliminated by the other nine exposures. The result is a street that appears to be entirely empty of all traffic.

There is another way of using multiple exposure to eliminate moving objects. In night photography out of doors, motor car headlamps appear as lines of light running across the picture. The remedy is to hold a hat in front of the lens whenever a car or other illuminated vehicle passes across the field of vision.

Multiple exposure technique is another way of assembling a photo-montage. Several different subjects may be exposed on the same plate with or without a mask to fill the picture space with either normal images or "ghosts".

Double Exposure Tricks. A number of double exposure tricks are possible with any camera not fitted with a double exposure lock.

The simplest trick of this type is to show a person and his "double" on one plate, against a black background. The same person is placed first on the left of the picture area and one exposure made, then on the right for the second exposure without moving the camera. Trick photographs of this kind (which may even consist of more than two exposures) can show the same person playing cards with himself, drinking his own health, etc.

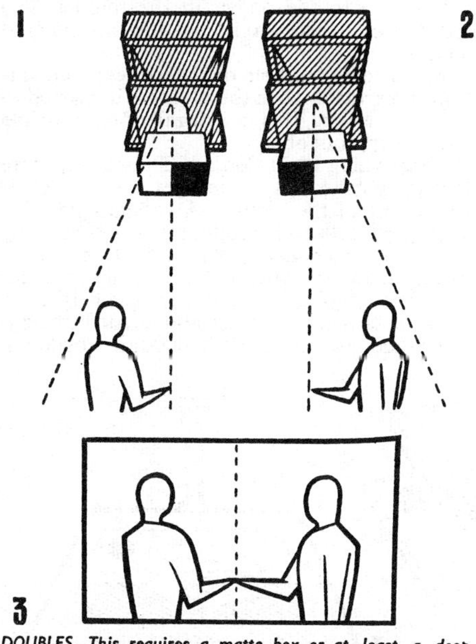

DOUBLES. This requires a matte box or at least a deep lens hood capable of covering up either half of the field of view with masks. 1. Make first exposure of subject with mask covering half of the view. 2. Change mask over to cover other half of view, and also switch round subject. 3. Result shows subject twice, e.g., conversing with himself.

Ghost photographs can also be made by double exposure. The ghostly outline of the "apparition" is left by the image of a sitter who is present in the scene for the first half of the exposure but absent when the shutter is opened on the same scene the second time.

If persons are to be photographed with their "doubles", the unoccupied half of the set must be masked off each time with a piece of black paper held in front of the lens hood. Otherwise the images will appear as transparent ghosts against the background.

Double exposure is frequently used in advertising and instructional pictures—e.g., a portable typewriter may be photographed so that both the machine with the keyboard, and the cover are visible. Double exposure (once with and once without the cover) produces a "phantom" picture in which the details show through the lid as though it were transparent.

LIGHTING TRICKS

The subject lighting is perhaps the most versatile tool available for trick effects in photography.

Special Light Sources. A simple but powerful concentrated light source for trick effects consists of a hollow glass globe filled with water. When this is placed in a beam of light from an ordinary lamp or spotlight it throws a bright patch of light over a short distance and objects in its path cast a strong shadow. The same effect is produced if a large hand magnifier is held in the path of the rays. The magnifier may be adjusted to any desired position by pushing the handle into a pot filled with sand. A concave shaving mirror may also be used to reflect light on to a small area. The light from the mirror is then brighter, and provides the main illumination, while the lamp itself provides a weaker beam that can serve as a fill-in light. Light may also be concentrated on a small area with a large sheet of tinfoil rolled into the shape of a cone and placed over the reflector. Beyond a certain point, concentrating the light on a smaller area by this means produces no appreciable increase in brightness.

Similar effects accompanied by still more marked shadow formations can be obtained from the light of a miniature slide projector. This type of projector can be used to provide additional lighting effects—e.g., if a piece of fine-mesh net is placed in the slide carrier, it will cast its pattern over the whole subject and make it look as if lit by sunlight shining into the room through net curtains. Other slides may be inserted to cast a pattern of rings, lines, stars or any other design over the subject at will.

Projected Backgrounds. Suitable transparencies may be projected on to a white surface behind the subject, to give the appearance of a natural background. But the relatively weak light from the projector calls for long exposures, and precludes the use of powerful foreground lighting, which would "kill" the background.

Foregrounds. The converse of the above method is foreground projection. A translucent white material such as tulle is stretched over a large frame. The model sits behind it, and behind the model is placed the lamp. The lamp is directed on to the tulle which shines brightly wherever the light strikes it, but remains dark and transparent in the shadow cast by the sitter. All background detail in the bright surround is thus lost, the only visible detail being within the shadow cast by the sitter.

Silhouettes. If opaque white material like linen is used, the same arrangement of material, subject and lighting produces a silhouette. If the sitter is placed immediately behind the screen with the light directly behind him the outlines will be sharp even with ordinary amateur lighting equipment.

Shadowless Lighting. The different lighting tricks described above are mostly intended to provide strong shadows and enhance the sense of depth. But there are occasions that call for flat, shadowless lighting. This can be arranged in several different ways. One way is to put a Photoflood bulb in a lamp socket in the ceiling without a reflector. If the walls are light coloured, its light will be diffused and while powerful enough for short exposures, it will cast no shadows. Photographs taken by such lighting have all the appearances of daylight exposures because the typical contrasts of artificial light are absent.

Such soft lighting is particularly suitable for family group photographs. For still life or other static subjects, a moving lamp is preferable. The lamp is held in the hand throughout the exposure and moved backwards and forwards in an arc, as if to fill in the details in the subject with the light. When photographing china and glass, this trick avoids the bright dots on the subject produced by the reflection of stationary lighting, and the highlights are suitably diffused.

This method of lighting will not serve for bright metal objects, as such surfaces reflect not only light and shade, but the surrounding objects, including the photographer and his camera. The surroundings must therefore be masked off altogether with a "tent". A suitable tent may be made from a cone of tracing paper placed over the subject like a roof. This diffuses the light falling on the subject. The lens is pointed through a small slit and apart from the neutral walls of the tent there is nothing to create reflections.

Shadowless lighting of small objects can be achieved as follows. The seat is removed from a chair and replaced by a sheet of glass. A piece of white cardboard is placed below it, resting across the legs at an angle of 45°, and turned so that it reflects light from the window on to the glass. The object to be photographed is placed on the glass with an empty lamp reflector inverted over it facing downwards. The camera lens is pointed down through the hole in the reflector to make the exposure. The light rays pass from the evenly illuminated cardboard to the inside of the reflector, which in its turn throws them back in a ring-shaped pattern on to the object in the middle. The result is that there are no shadows in the final picture.

An extreme example of shadowless illumination is provided when the object is immersed in a dish of water. This is naturally only possible with waterproof objects.

Photographing Light Sources. Double exposure is usually called for if the light source itself is to be part of the picture. A candle flame, for example, is relatively bright, but its illuminating power is very small, so a still life study by candle light, showing the candle flame itself, is a photographic impossibility. The brightness contrast between the flame and its surroundings is beyond the scope of the sensitive material.

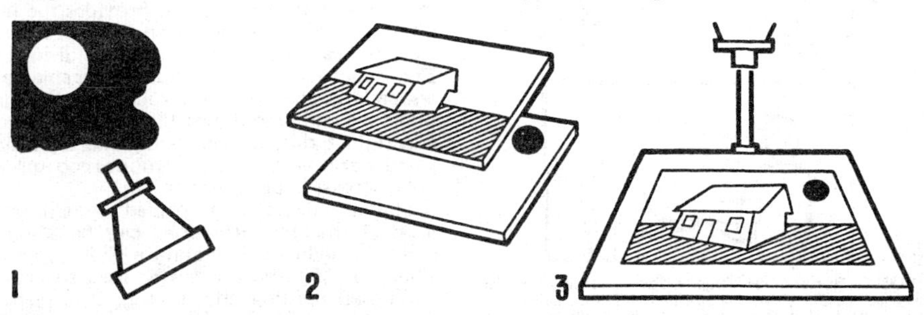

COMBINATION PRINTING. Adding a moon to a night picture. 1. Photograph the moon only with a long focus lens, to obtain a moon negative of the right size and scale. 2. Superimpose the landscape and the moon negatives. 3. Place the sandwich in the negative carrier of the enlarger, and enlarge as one negative in the normal way.

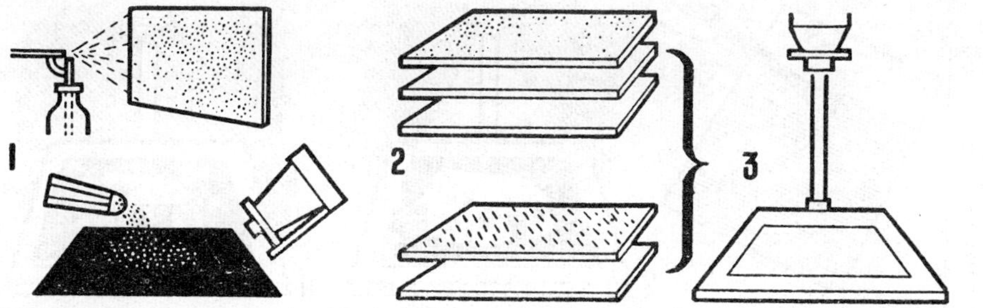

FAKING SNOW. 1. Prepare a snow negative, either by spraying a glass plate with red ink, or by photographing salt sprinkled on to black velvet while moving the velvet slightly during exposure. 2. Combine the snow negative with a winter landscape (use a spacer between them in the case of the red ink plate). 3. Enlarge the two together like one negative.

Some form of general illumination must be provided, though it must not be bright enough to destroy the atmosphere. The following method gives a natural effect. A preliminary exposure is made with the candle burning in a normally lit room. The candle is then extinguished and a second exposure made, three to five times as long as the first. In the resulting picture, the candle flame retains its typical appearance and the surrounding objects are adequately lighted.

The same trick of giving a short preliminary exposure for the light source followed by a second, longer exposure with the object extinguished, may also be used for photographing electric bulbs, gas flames and open fires.

TRICKS IN THE DARKROOM

There is more scope for photographic trickery in the darkroom than at any other stage of picture making. As with most other branches of trick photography, the effects are produced generally by applying orthodox techniques in an unorthodox manner.

Negative-Positive Combinations. If a particularly plastic rendering of an object is required, two exposures are made on separate negatives, the light being held to the left of the camera for the one and to the right for the other. A positive transparency is then prepared from one of the negatives. The positive transparency is then held in exact register with the other negative, emulsion to emulsion, the two are placed in the enlarger and printed. The dense highlights combine with the empty shadows to give an exaggerated impression of relief and modelling to the positive. (This process sounds very much more complicated than it really is.)

Two further tricks are possible when a positive transparency is made and bound up with its own negative. If the transparency is printed very slightly out of register with the negative, a photographic bas-relief is produced. When the two are in exact register, they combine to form a more or less neutral grey, but slight movement out of register brings out the light and dark edges that are characteristic of a bas-relief.

A number of fantastic and unreal effects can be produced by combinations of two identical negatives, or of one negative and a positive transparency. The effect varies according to the relative position of the two images.

Simultaneous enlargement from two different negatives produces two further useful effects:

Snowstorm Effects. A snowstorm is difficult to photograph; it is easy to print in afterwards. The snow negative is made by sprinkling some granulated sugar on a piece of black velvet. The camera is mounted above the velvet, lens pointing downwards, and a time exposure

PICTURES ON VASES AND TUMBLERS. **Left:** *For images etched or painted on to glass tumblers, wrap a sheet of bromide paper round the glass. Print the image directly on to the paper by means of a small torch bulb in the centre of the tumbler.* **Right:** *In the case of a pattern on an opaque vase, wrap the paper round, and illuminate from the back (i.e., a kind of reflex printing).*

LUMINOUS PICTURES. 1. Coat a sheet of plain paper with luminous paint. 2. Expose for a few minutes under a negative in the enlarger. 3. Switch off the enlarger light, and a luminous image will be visible in the dark.

made. During the exposure the velvet is drawn diagonally across and out of the field of view. A print from this negative shows a mass of bright and slightly blurred lines looking exactly like snowflakes. When printed together with a negative of a landscape, the result looks as if taken in a snowstorm.

Alternatively, a small sheet of glass can be finely sprayed with red or black ink and when dry, bound up with the subject negative with a spacer in between to throw the snow slightly out of focus.

Rainbow Effects. A rainbow can be faked in much the same way. It is drawn in with a pair of compasses in dilute red ink on an unexposed, fixed-out film. To soften the outline of the rainbow, a sheet of clear glass must be sandwiched between it and the principal negative (which is accurately focused in the enlarger to print sharply).

Moon Effects. Moonlight landscapes which include the moon itself are also obtained by combination printing. The moon in a straight print always appears too small. A separate exposure of the moon in a clear sky is therefore made with a lens of two to three times the normal focal length. This negative is then printed in over any night scene, and the result corresponds to the actual visual impression.

Texture Effects. Negatives or transparencies of artificial patterns may also be printed in—e.g., dotted or lined screens, coarse woven cloth, nets, or surfaces with wood-grain or crackle finish. The material itself may even be used. When suitable patterns are chosen, they can add a useful touch of character to the print.

Objects which cannot be inserted in the carrier of the enlarger together with the principal negative may be placed on the bromide paper itself. By laying a piece of netting, grains of rice or sand, or small nails on the paper before exposure, the surface of the enlargement may be broken up according to the nature of the objects used.

Photograms. By laying objects directly on the paper it is even possible to dispense with a negative in the enlarger entirely. This kind of print is called a photogram. A direct print can easily be made from the leaf of a tree, a feather or even a cut-glass dish, provided only that the latter is sufficiently flat. Taller objects such as wine glasses, vases and statuettes lend themselves to more imaginative effects.

The enlarger lamphouse may shine down on the object or be inclined so that the beam of light casts an oblique shadow. The image will be foreshortened and distorted according to the positions of light and paper.

Using the same principle, it is fairly simple to print flat pictures from a curved surface. Where the surface is transparent, such as an etched glass tumbler, a sheet of bromide paper is

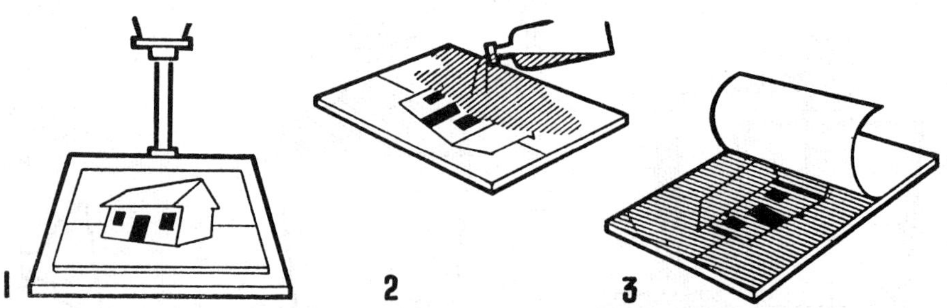

SILVER OR GOLD BACKGROUNDS. A metallic surface to the image, in place of a white paper one, enhances the detail and brilliance. 1. Make a thin transparency in the normal way by enlarging the negative on to a slow ordinary plate. 2. Coat the emulsion side with linseed oil, or a suitable adhesive. 3. Stick down a sheet of gold or silver paper foil, and trim to shape.

wrapped round the outside with the sensitized side facing the pattern; the exposure is then made by a small torch bulb inside the tumbler. With an opaque object, such as a vase, the paper is wrapped round in the same way, but the exposing light is placed outside the vase so that it shines through the back of the paper; this produces an image by reflection from the pattern on the vase (reflex printing).

Luminous Pictures. An effective way of making pictures that will glow in the dark is to coat a sheet of plain paper with luminous paint. This is left to dry in the dark until it has lost its luminosity. It is then re-activated by long exposure under a positive transparency, which will produce a positive and luminous picture when viewed in the dark. This image is not permanent and will be "fogged" if exposed over-all to light.

A method of producing permanent pictures which have brilliance is to back positive transparencies with gold or silver foil. The transparency must be weaker in density than normal and have no heavy shadows. The emulsion side of the transparency is coated with linseed oil or Canada balsam diluted with a little xylene; one side of the metal foil is treated similarly. The two coated surfaces are then pressed together into firm contact, avoiding air bubbles between them. O.R.C.

See also: *Abstract photography; Anamorphoscope; Back projection; Bas-reliefs; Cinematography; Combination printing; Converging verticals; Faults; Flash (electronic); Giant enlargement; Greeting cards, High key; High speed photography; Hill cloud lens; Lighting the subject; Light sources in the picture; Low key; Masks; Mirror; Mirror photographs; Montage; Moonlight; Night photography; Photograms; Photo murals; Photosculpture; Physiograms; Posterization; Reticulation; Retouching; Sabattier effect; Shading and spot printing; Silhouette; Sketch photograph; Soft focus; Solarization; Spirit photography; Statuettes; Stroboscopic flash; Table top; Texture screen; Tone separation; Vignetting.*

Books: *All the Photo Tricks,* by E. Smith (London); *How to do Tricks in Amateur Films,* by J. Caunter (London); *The Complete Art of Printing and Enlarging,* and *Design by Photography,* both by O. R. Croy (London).

TRICOLOUR FILTERS. Set of three filters—red green, blue—used for making the separation negatives in the three-colour process of colour photography.

TRIHYDROXYBENZENE 1 : 2 : 3. Chemical name for pyrogallol or pyrogallic acid. One time very popular developing agent.

TRIMMER. A trimmer is a tool for making a straight cut through a print or mount and leaving a clean edge. It is used principally for removing distracting objects lying near the edges of the print, or for giving the picture more pleasing proportions. It is, in fact, a useful means of improving the composition.

Scissors make poor trimmers because they cannot cover the whole edge in a single cut, and each shift of the blades jags the edge of the paper—unless the scissors are unusually long, or the print is very small.

Prints can be trimmed quite well with a sharp pen-knife or a safety razor blade in a holder. The blade must be steadied against the edge of a steel ruler and enough pressure used to cut through the paper at a single stroke. The print should be on a piece of stiff board; cutting on to a sheet of glass soon blunts the edge of the blade. A strip of medium sand-paper glued to the lower face of the steel rule prevents it from slipping about.

But for those who can afford it, the commercial print trimmer is the best tool for the job. There are two types of trimmer—the desk trimmer, and the guillotine.

Desk Trimmer. This type of trimmer consists of a board, hinged to a base and kept raised an inch or so at the free end by a spring. A cutter blade runs the length of the free edge of the board and is fixed, edge upwards, on the base. A steel strip is fixed along the raised edge of the board, leaving a slit just wide enough to allow the edge of the print to be pushed under.

In use, the print is laid on the board so that the piece to be trimmed off projects under the steel strip. The board is then pressed down against the spring pressure and the fixed knife slices off the surplus paper against the edge of the steel strip. When the board is released, it automatically springs up into position ready for the next cut. Desk trimmers are not usually made to cut prints longer than twenty-four inches.

Some models also have a small ruler fixed along one edge of the board (at a right angle to the blade). This serves as a guide edge for cutting the print squarely; it may also be calibrated in inches or centimetres (or both) so that it indicates the length the print will be after the cut is made.

Guillotine Trimmers. The guillotine type of trimmer can be more easily made to cut prints up to any size. It is a board with a metal edge

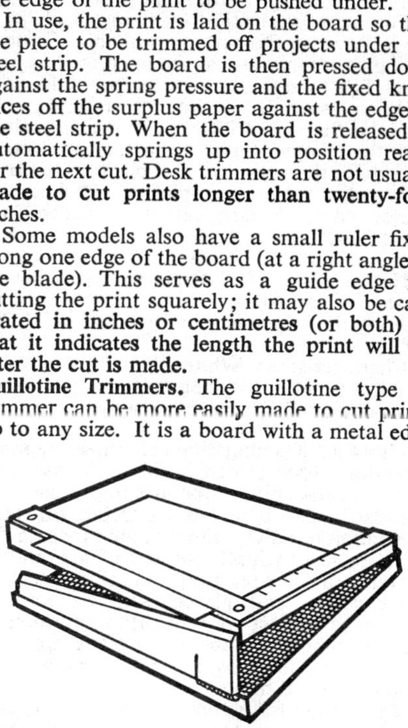

DESK TRIMMER. The top board is hinged to the base at the far end. The paper protrudes over the near edge against which the blade bears. Downward pressure on the board cuts the paper.

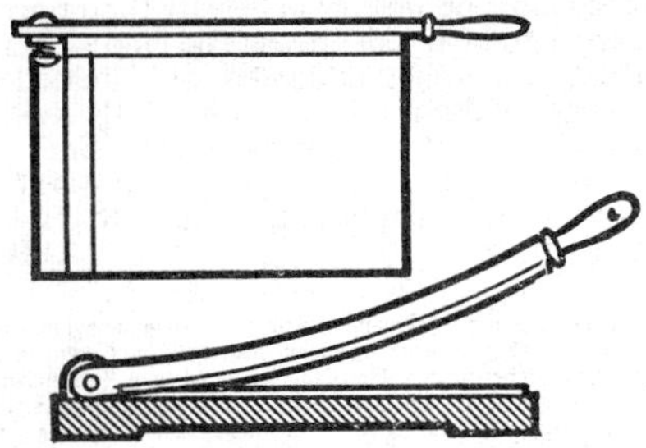

GUILLOTINE TRIMMER. The trimming blade is hinged at one end, and is pushed down to cut off the paper protruding over the edge of the board. The blade may have a return spring.

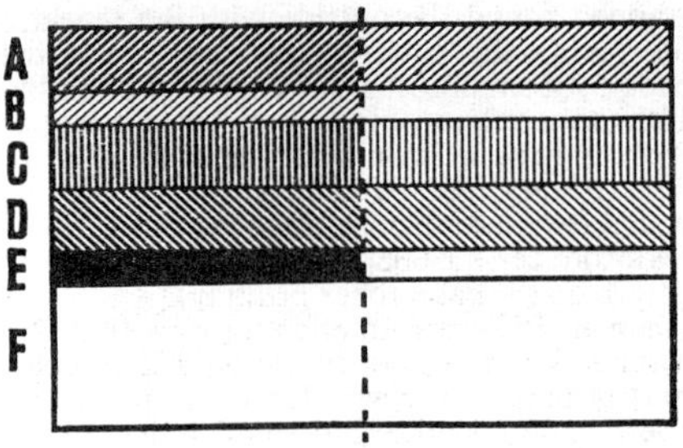

INTEGRAL TRIPACK FILM. A, blue-sensitive layer. B, yellow filter layer. C and D, green and red sensitive layers. E, anti-halo layer containing an even silver density. F, film base.

which has a knife blade fitted with a handle hinged at one end. The paper is held on the board overlapping the metal edge where it is to be cut. When the knife is pulled down it makes a sharp, clean cut. The advantage of a guillotine is that it will easily cut several thicknesses of paper, and can also be used for trimming cardboard mounts.

Guillotine trimmers are also made to cut with a deckle edge. Whilst a deckle edge is out of place on large prints, some people like it for prints of about postcard size or less.

For mass production trimming, such as in D. & P. work, the cutting mechanism is operated by a treadle so that the operator has both hands free to move the print as required.

When using either type of trimmer, it is essential that the print is held in firm contact with the trimmer board when the cut is made Otherwise the action of the blade is liable to pull against the print, displacing it and causing a crooked cut or else tearing the paper if the blade is not dead sharp. F.P.

TRIOXYMETHYLENE. Hardening agent used instead of formalin.

Formula and molecular weight: $(CH_2O)_3$; 90.

Characteristics: White crystalline powder. Unlike formalin, it is odourless.

Solubility: Freely soluble in water.

TRIPACK. Combination of three separate emulsions, each coated on its own base, used in some early colour processes under the name "tri-pac". The object is to secure the three separation negatives with a single exposure. The usual form of tripack consists of a front film carrying a blue-sensitive emulsion which is dyed yellow and placed with its sensitive surface in contact with an ortho emulsion on another film. The latter has a transparent red backing in contact with a red-sensitive emulsion on a third film.

The exposure is made through the blue sensitive emulsion layer. The layers are then separated and developed normally to give separation negatives. Unfortunately, although two of the three emulsions can be placed in contact, the remaining emulsion is slightly out of focus. The process is no longer used.

The modern development of the tripack is to coat all three emulsions and the intervening filter layer on a glass or celluloid base. This composite material, called an integral tripack, is developed to form the three colour separation negatives in the respective layers. It is then further processed to leave only three corresponding dyed positive images which combine to produce a natural colour transparency. Alternatively, it may be processed to yield a colour negative for printing on to another tripack coated on a paper base.

See also: *Bi-pack; Colour photography history; Colour materials; Colour printing.*

TRIPLE EXTENSION. Amount by which the front of a field, technical or similar camera can be extended—e.g., to three times the focal length of the normal lens. It is primarily required for taking close-ups. A triple extension enables a camera to be used for reproducing to a scale twice as large as life—i.e., a subject 1 in. high can be reproduced 2 ins. high.

See also: *Extension of camera.*

TRIPLET LENS. Popular form of lens assembly consisting of three single glasses: a converging front crown glass lens, a diverging flint glass lens and a converging crown glass lens. The elements are air-spaced.

The first lens of this type, the Cooke Triplet, was developed by H. D. Taylor in 1893 and it was subsequently widely adopted.

See also: *Lens history.*

TRIPOD. Support for camera, consisting of three legs usually hinged together at one end to a component which carries a tripod screw.

See also: *Camera supports.*

TRISODIUM PHOSPHATE. Alkali, used mostly in fine-grain developers as it is to some extent self-buffering.

See also: *Sodium phosphate (tribasic).*

TROPICAL PHOTOGRAPHY

Heat, moisture, insects, air-borne sand and dust are the main factors which complicate photography in the tropics. There are two characteristic combinations of heat and moisture: a high and widely varying temperature accompanied by a low relative humidity such as is found in desert regions, and a very high humidity with comparatively low maximum temperature, i.e., the conditions that may be expected in the rainy season or on tropical islands. Owing to these great variations in both temperature and humidity, general photographic data may not apply in the tropics. The fullest use possible should be made of any available local information.

Materials and Equipment. The construction of tropical equipment is important. Aluminium and unprotected iron are liable to corrosion. Where wood is used, teak or mahogany are to be preferred, with all joints fastened by rustless screws. Glued joints should be avoided.

Modern plastic or light metal non-folding cameras, which minimize the entry of moisture into the air-space of the body, are particularly suitable for tropical use. Collapsible lenses are best left permanently extended.

Between-lens shutters are normally more trouble-free than the focal-plane shutter.

Sensitive material should be ordered in tropical packings. The packages of sensitized materials should be kept as cool and dry as possible. A cold store should be used if available, but as the humidity in a cold store or a refrigerator is usually high, all packages so stored must be made completely waterproof. Materials required for use must be removed from the cold store about twenty-four hours before use, and allowed to warm up gradually to air temperature.

Colour films should be purchased in hermetically sealed tropical packings. The seals should not be broken until the films are required for use, and all colour films should be sent for processing as soon as possible after exposure, and should on no account be stored.

Should no cold storage be available, unventilated buildings and the top floors of uninsulated buildings should be avoided, and materials stored well away from outside walls and steam pipes. Storage pits are the next best to a cold store.

Photographic chemicals should always be purchased in tropical packing. In general, anhydrous forms are preferable to crystalline. Chemicals which deliquesce (i.e., absorb so much water that they dissolve in it), such as sodium sulphite, sodium hydroxide and potassium carbonate should be kept in corked bottles and wax-sealed. Efflorescent chemicals, such as sodium carbonate and sodium sulphate, must be similarly treated. After mixing, solutions should be stored in dark bottles filled to the neck, and kept in a cool, dark place.

All apparatus should be kept as cool, dry, and clean as possible. In humid regions, all metal parts should be greased regularly, and leather parts treated with a wax polish. In dusty or sandy regions, lenses should be protected, when not in use, by a lens cap, and during use by a clear glass "filter" or a haze filter. In such places no grease or oil should be applied to metal parts.

Processing Films and Plates. Films and plates should be processed as soon as practicable after exposure because of the possible fading of the latent image. Should delay in processing be inevitable, the materials should be dried before repacking in their original containers.

To dry materials, a quantity of indicator silica gel should be placed with the material in a sealed air-tight container. The silica gel should be replaced every 24 hours until there is no change in colour from blue to red. The drying properties of silica gel may be revived by heating to 400° F. (205° C.) and cooling in an air-tight container.

The hands must be kept as dry and free from perspiration as possible by soaking them in iced water or spirit and thoroughly drying them afterwards. To avoid fingerprints, film clips, film hangers, plate racks and print forceps should always be used.

High-temperature processing affects the photographic characteristics of emulsions, giving rise to an increase in fog and a decrease in gamma. Loss of gamma is especially noticeable with process materials normally developed in caustic developers which are too alkaline to use above 75° F. (24° C.). Use of an anti-fogging agent tends to lessen the fog. With high-speed emulsions there is also an increase in graininess together with a decrease in resolving power.

A wetting agent may be added to developers, but must not be added to fixing baths. If fresh water is restricted, chlorinated or sea water may be used for mixing developers. Any sludge formed should be allowed to settle, after which the clear decanted liquid may be used. Acid hardening-fixing baths are quite satisfactory if made with sea water. Washing in sea water must be followed by two or three short rinses in fresh water to remove residual salts. This may be done at a later date.

If all solutions can be maintained below 75° F. (24° C.), standard formulae and processing technique may be used for processing plates and films. Should the washing water be above 75° F., an acid hardening-fixing bath must be used, and the washing time should not be excessive.

With processing solutions and wash water between 75° F. (24° C.) and 90° F. (32° C.) any normal developer loaded with sodium sulphate may be used. Up to four ounces of sodium sulphate may be added to 20 ounces of solution (up to 200 grams per litre), the maximum quantity being required at the higher temperature. To compensate for the resultant slowing down of development, 20 per cent should be added to the normal tropical development times. Highly alkaline M.Q. formulae should be avoided because they tend to form fog with prolonged development times.

A special tropical developer of the following type may also be used:

Metol	100	grains	5·7	grams
Sodium sulphite, anhydrous	3½	ounces	90	grams
Kodalk	33	grains	1·9	grams
Sodium sulphate crystals	4¼	ounces	105	grams
Water to make	40	ounces	1,000	c.cm.

The average development times are:
6–8 minutes at 75° F. (24° C.).
3½–4½ minutes at 85° F. (29·5° C.).
2½–3 minutes at 90° F. (32° C.).

Development should be followed by a 3-minute immersion in a chrome-alum stop bath. Fixing is then carried out in a tropical acid hardening-fixing bath for 10 minutes.

Tropical acid-hardening fixing baths may be mixed from packed proprietary brands, or made up to the following formula:

Sodium thiosulphate, anhydrous	6	ounces	150	grams
Sodium sulphite, anhydrous	265	grains	15	grams
Acetic acid, glacial	300	minims	17	c.cm.
Boric acid	130	grains	7·5	grams
Potassium alum	265	grains	15	grams
Water to make	40	ounces	1,000	c.cm.

If the processing solutions exceed 85° F. (29° C.), and a very humid atmosphere has caused the emulsion to become excessively swollen the materials should be dried, as described above, before development is attempted.

When all solutions, including the wash water, are at a temperature higher than 90° F. (32° C.), a tropical pre-hardening bath in conjunction with a sodium sulphate loaded developer should be used. In this case the chrome-alum stop bath should be omitted.

Pre-hardening baths may be mixed from packed proprietary brands, or made up to the following formula:

Sodium sulphate	8	ounces	200	grams
Formalin (40 per cent solution)	1	ounce	25	c.cm.
Water to make	40	ounces	1,000	c.cm.

The materials should be fixed in a tropical acid hardening-fixing bath for 10 minutes.

Where running water is available, washing time should not exceed 30 minutes for temperatures of 75–90° F. (24–32° C.), and 20 minutes for temperatures of 90–105° F. (32–41° C.). Alternatively, the material may be immersed for 5 minutes in each of three changes of still water, allowing 16 ounces (450 c.cm.) of water per square foot of material in each bath. If the emulsion is very soft, a 2–3 minute wash in still water should be given, and the material refixed and rewashed as above within three days. Washing times can be reduced by the use of proprietary hypo-eliminators.

The final rinse, or a proprietary fungicide bath if used, should contain a wetting agent.

Before drying, the emulsion may be carefully wiped down with a viscose sponge or a chamois leather, provided the emulsion is not very soft, and there is no appreciable frilling or stripping. Drying may be assisted by the use of a spirit bath diluted with 30 per cent water. Fungicide baths may be precipitated by alcohol, and the two treatments should not be given at the same time. Drying can be accelerated by placing the material in a current of air—e.g., from a fan. Whatever method of drying is used, negatives must be protected from insects and dust by the use of gauze or mosquito netting. Suspending films from a wire gives protection from cockroaches which might eat the wet gelatin off the film or plate.

If maximum permanence is desired, films and plates should be given the following treatment within twenty-one days of processing, in solutions preferably at about 75° F. (24° C.).

(1) Soak in water containing a wetting agent for 5 minutes.

(2) Fix in a freshly prepared tropical acid hardening-fixing bath for 10 minutes.

(3) Wash in running water for 30 minutes.

(4) Soak in a 0·03 per cent (by volume) solution of ·088 ammonia for 5 minutes.

(5) Wash for 5 minutes.

(6) Immerse in a fungicide solution for 5 minutes before drying.

Processing Paper. For processing paper at temperatures between 75–95° F. (24–35° C.), a standard print M.Q. developer may be used without pre-hardening. The development time will become appreciably shorter as the temperature increases. When development is complete, the prints should be rinsed for a few seconds in an acid stop bath before transferring to an acid hardening-fixing bath. Two fixing baths should be used, the prints remaining in each bath for 5–10 minutes. A maximum of sixty-five whole-plate prints, or the equivalent area, per gallon may be fixed in the first bath, which should then be replaced by the second bath, and a fresh second bath made up.

Single-weight prints should be washed for 40 minutes and double-weight prints for 60 minutes. If prints are required quickly, or it is necessary to conserve water, a hypo eliminator may be used, provided the solution temperature is brought below 75°F.

Prints may be dried by heat, or laid out to dry in a dust-free atmosphere.

For maximum permanence, prints should be toned with either a selenium or a hypo-alum toner.

For selenium toning the prints are fixed normally, using two baths in succession, washed (single-weight prints for 40 and double-weight prints for 60 minutes) in running water, toned in a selenium toner, and re-washed.

For hypo-alum toning the prints are fixed normally, using two baths in succession, washed in running water for 10 minutes, and toned in a hypo-alum toning bath. This should be followed by sponging in luke-warm water to remove any sediment, and washing (single-weight prints for 40 and double-weight prints for 60 minutes) in running water.

Storage of Processed Material. Negatives and prints destined for storage in the tropics should be specially treated during processing, as detailed above.

All processed materials should be stored in conditions with a relative humidity below 60 per cent and as cool as possible.

An air conditioning system, if available, helps to ensure favourable and consistent storage conditions.

Films to be stored may be given a coating of film lacquer which will prevent fungus growths in the emulsion and will also protect the surface of the film from abrasion in handling. Most proprietary makes of film lacquer contain a fungicide.

Prints and negatives in fairly regular use may be stored in fungicide envelopes. Single kraft paper envelopes impregnated with a fungicide (a 2 per cent solution of sodium pentachlorphenate is most effective), are satisfactory for this purpose. The envelopes should have narrow edge seams, rather than a centre seam, and carry a folding flap. Certain pastes are very hygroscopic and envelopes secured by these should be avoided. All material stored in this way should be examined regularly and any signs of deterioration dealt with immediately.

Colour transparencies must be kept as cool and dry as possible, and preferably stored in a dark place.

Materials for long term or archival storage should be dried out thoroughly and then placed in laminated foil envelopes, which can be hermetically sealed, and stored in as cool a place as possible. Care should be taken to ensure that packages to be re-opened are allowed to condition to normal room temperature before the seal is broken.

Cine films to be stored in the tropics should be specially washed after processing to remove any residual traces of processing chemicals. A lacquer coating is recommended for all valuable negative material and prints intended for archival storage. If air-conditioned storage space is not available, the films should be carefully dried out and then stored in cans which should be tape-sealed to prevent the entry of humid air. A drying agent may be included in the can, but care must be taken that the film does not become so dry as to make it brittle. The cans should always be laid horizontally so that the weight of the film rests on the edge of the roll. A.E.S. (L.G.F.)

See also: *Frilling; Hardening baths; Hot weather processing; Reticulation.*
Books: *Camera in the Tropics*; by G. C. Dodwell; *Photography on Expeditions*, by D. H. O. John (both London).

T-STOPS. Aperture values measured on a scale based on the actual transmission of light at different stop diameters instead of on the relation of the diameter to the focal length of the lens (*f*-stops).

See also: *Diaphragm; Transmission efficiency.*

TUNGSTEN FILAMENT LAMP. Electric lamp consisting of a filament of tungsten in a glass envelope containing an inert gas. The filament is rendered incandescent by the passage of an electric current. Lamps of this type are among the most useful light sources.

See also: *Filament lamp; Light sources.*

TUNGSTEN IODINE LAMP. Term preferred by some lamp manufacturers for the quartz-iodine lamp because this type of lamp is also made with an envelope of heat resisting glass.

TURRET HEAD. Rotary multiple lens housing on cine cameras and some experimental miniature still cameras. The head holds several lenses of different focal lengths and permits them to be rapidly interchanged. The required lens is brought into the taking position by simply rotating the turret.

See also: *Cinematography.*

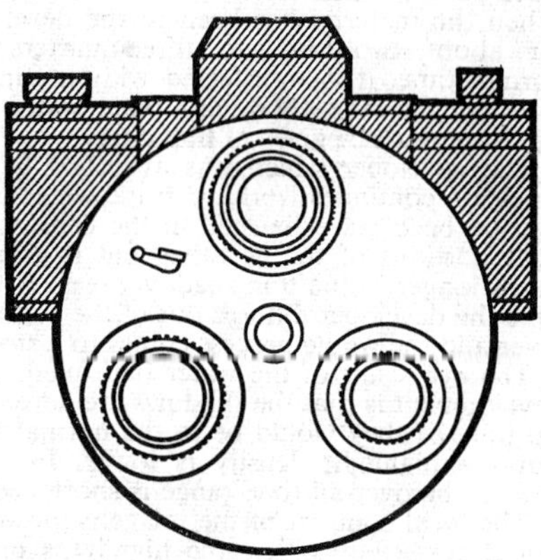

TURRET HEAD. Special type fitted on 35 mm. still camera. Any of the three lenses (wide angle, standard, long focus) can be brought into use by the turret.

TWIN LENS CAMERA. Camera consisting of two compartments, each with its own lens; the top compartment carries a focusing or viewing screen and the other the sensitized material. The focusing movement is usually common to both lenses so that when the image appears sharp on the screen it is also focused sharply on the sensitized material. The system has the advantage that the lens of the viewing camera remains fully open while that of the taking camera can be stopped down as required. So the screen image is seen at full brilliance all the time.

A number of early workers had the idea of coupling two identical cameras in this way. In 1864 Disderi constructed a "double appareil" on this plan and there was a camera of this type manufactured by Marion in 1888.

There have been twin lens reflex cameras since 1891, the upper, viewing half of the instrument being equipped with a reflex mirror and a focusing screen in register with the focal plane of the taking lens. In particular, Dr. Heidecke built a twin lens plate reflex in 1908. But such cameras were extremely bulky and the idea did not become popular until it was applied to the small-format roll film camera in 1928 by the manufacturers of the Rolleiflex, and subsequently by a number of other firms.

See also: *Camera history; Reflex camera.*

Book: *The Twin Lens Reflex Camera Companion*, by **H. S. Newcombe (London).**

TWO-BATH DEVELOPMENT

Two-bath development, as its name implies, is carried out in two solutions, and two separate stages.

In the first solution, the emulsion layer is saturated with developer. At this stage, some development may or may not take place.

For the second stage the material is transferred to another bath which may be a dilute alkali solution or even plain water. Here the image starts to develop (or continues developing) until the developer carried over in the emulsion becomes exhausted or too diluted to have any further effect.

Water Bath Development. The simplest form of two-bath development uses a normal developer for the first stage, and plain water for the second.

Generally, most of the development takes place during immersion in the developer. When the material has been in the developer for about two-thirds to three-quarters the normal time, it is transferred without rinsing to a dish or tank containing plain water. No agitation must be given at this stage.

The developer which was absorbed by the emulsion continues working in the water bath. It soon becomes exhausted in the dense highlight portions of the negative, but it goes on much longer in the thin shadow areas. After a time the developer diffuses out of the emulsion layer and the whole process comes to a stop.

The net result of the water bath method of development is that the shadows are developed up fully as they would be in the normal way, but the highlight density is lower. In other words, the over-all tone range is shortened.

The local contrast of the image is, however, not decreased in either the highlights or the shadows. Since the developer still goes on working for a while, the contrast will, if anything, become higher. This is particularly so in the shadows where the developer diffuses out of the emulsion and is diluted long before it is exhausted.

On the other hand, the developer is used up so quickly on the highlight areas that they stop developing before they have a chance to block up. Flattening of the tone gradation occurs mainly in the middle tones and produces an effect akin to tone separation.

This result cannot be achieved by shortening the development time, since that would reduce the contrast all over, particularly in the thin shadow tones.

The method is most useful for negatives which have a well-defined range of highlight tones, an equally well-defined range of shadow tones, but nothing much in between, like shots against the light and indoor photographs which include parts of a view through the windows. If carried to excess, staining will result.

Although the main use of water bath development is in processing negatives, it can also be applied to prints. There it reduces the contrast and so enables an ultra-hard negative to be printed on a paper that would not otherwise accommodate the negative tone range.

To obtain a noticeable effect in practice, it is necessary to use a hot water bath for prints, transferring the print to the water as soon as the image begins to appear.

Intermittent Development. The effect of water bath development depends to a large extent on the amount of development allowed to take place in the developer. The sooner the negative is transferred to the water bath, the greater the compression of the extreme tones, but the softer the final image.

But the effect of the water bath can also be controlled by intermittent development. This procedure is mainly suitable for development by inspection, as it depends to some degree on judging the partially developed image.

The material is immersed in the developer until the first highlights begin to appear. It is then transferred to the water bath, and left there for 2 minutes. After this time development will have stopped almost completely. The negative is placed again in the developer for about two to three minutes, followed by another two minutes in the water bath. The procedure is repeated until the negative has built up sufficient density. The contrast can be increased by extending the time in the developer.

A fairly energetic developer—e.g., a concentrated contrast metol-hydroquinone formula gives the best results. Easily oxidized developers, like pyro or caustic hydroquinone, are less suitable as they are apt to stain.

Alkaline Two-bath Development. Two-bath development is similar to water bath development in principle and results, but the water bath is replaced by an alkali bath, e.g., 1 per cent borax or sodium hydroxide.

The first bath—i.e., the developer itself—is often made up with little or no alkali so that hardly any development takes place in it. The gelatin emulsion merely becomes soaked with developer solution which is not activated until the negative reaches the alkali bath. Here development soon stops in the dense portions while it goes on in the thin shadow areas for much longer until the solution diffuses out of the emulsion.

One of the advantages of alkaline two-bath development is that neither time nor temperature are critical. Both baths must be kept at the same temperature to avoid reticulation and similar troubles.

Immersion in the first bath should last about three to four minutes. After that there is practically no increase in the amount of solution absorbed by the gelatin. But the time is not critical.

The immersion time in the second bath should again be three to four minutes. By that time most of the developer solution has diffused out of the gelatin and further immersion only softens the emulsion layer. Since development stops altogether, the negative cannot be overdeveloped and the time of immersion in the second bath also is not critical.

There must always be enough alkali solution in the second bath to "swamp" the developer carried over and diffused out by the film, or development will continue instead of stopping after about three to four minutes. At least ten to twenty ounces (250–500 c.cm.) of alkali bath should be used and for the same reason it should not be used for more than one session.

At higher temperatures the developer acts more rapidly, but it also diffuses out of the emulsion sooner. These two effects more or less neutralize each other, so the exact development temperature is not important.

The concentration of the first bath is usually varied to compensate for the increased or decreased development speeds of different materials. Slow developing materials are given a concentrated first bath so that the developer will continue working for a long enough time in the second bath. Rapid developing materials are given a more diluted first bath.

The concentration of the alkali bath also influences the rate and thus the amount, of development, but in practice it is usually kept constant.

A normal metol-hydroquinone negative developer, made up at about two to four times its normal strength, makes a suitable first bath. Since practically nothing happens in this

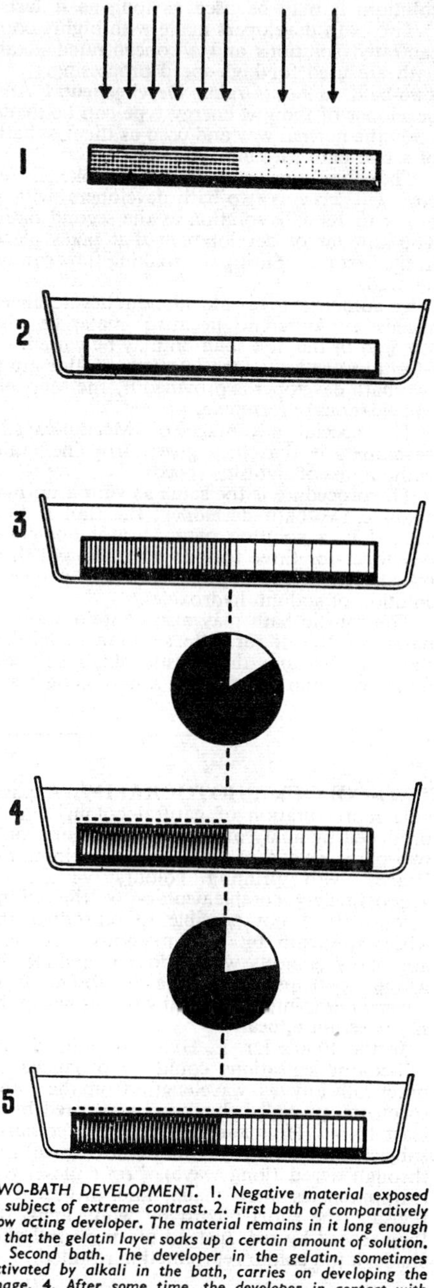

TWO-BATH DEVELOPMENT. 1. Negative material exposed to subject of extreme contrast. 2. First bath of comparatively slow acting developer. The material remains in it long enough so that the gelatin layer soaks up a certain amount of solution. 3. Second bath. The developer in the gelatin, sometimes activated by alkali in the bath, carries on developing the image. 4. After some time, the developer in contact with heavily exposed areas becomes exhausted, while the developer in the lightly exposed parts still carries on working. 5. Finally, development stops when all developer has diffused out of the emulsion layer. The tone difference is, however, less than it would have been with straightforward development.

solution, it may be used as long as it lasts.

Two-bath developers made with highly concentrated solutions and a concentrated alkali bath are used for high-speed processing.

Two-bath Fine Grain Development. Any developer of the low energy type can be made up in the normal way and used as the first bath of a two-bath formula.

Therefore fine-grain developers like D-76, etc., will serve as two-bath developers, with a borax or Kodalk solution as the second bath. The amount of development that takes place in the first bath during the soaking time can be ignored.

To compensate for the different development speeds of different negative materials the strength of the first solution may be varied.

Meritol-Caustic. A special type of fine-grain two-bath developer is provided by the Meritol-caustic series of formulae.

The special advantage of Meritol-caustic developers is that they give fairly fine grain without loss of emulsion speed.

The procedure is the same as with a normal alkaline two-bath developer: the film is first soaked in a solution of a Meritol fine-grain developer for three to four minutes, and then transferred to the caustic bath (a 1 per cent solution of sodium hydroxide).

The caustic bath may also contain about 5 parts per 100 of formalin solution to harden the film. Because the caustic alkali solution tends to soften the gelatin, a hardening fixer, preceded by a thorough rinse, must be used after development.

The strength of the first bath again determines the degree of contrast obtained. For average films, the Meritol developer (e.g., Meritol Super Fine Grain, or M.C.M. 100 without alkali) is diluted with an equal volume of water. Slow developing films need a stronger first bath, rapid developing films a weaker one.

There is a fixed relationship between the strengths of the two baths for any particular degree of contrast. Strong developer and a weak alkali bath (say 0·5 per cent) give the finest grain, but this is offset by a certain amount of fog, and loss of shadow detail. A weak Meritol bath with a stronger alkali solution (over 1 per cent) gives better shadow detail, but slightly coarser grain.

The backing dyes on roll films may interfere with the action of the Meritol. They must be removed before development by soaking the film in a 1 per cent solution of sodium sulphite for about one to two minutes. After this fore-bath the Meritol has to displace the sulphite solution already absorbed in the gelatin. The film must therefore be left longer—up to five minutes—in the Meritol bath.

Developer improvers can also be added to the Meritol bath in the proportion of about 1 part in 200. These give slightly lower emulsion speeds, but considerably less fog. L.A.M.

See also: *Contrast control; Three-bath development.*
Book: *Developing*, by C. I. Jacobson (London).

TWO-COLOUR PHOTOGRAPHY. A passable representation of natural colour can be produced by analysing the spectrum into only two parts instead of three, and combining the images with printing colours which are approximately complementary to the taking colours. It is not possible to reproduce the whole spectrum by such methods, but there are many subjects which do not include the whole spectrum—e.g., blue is absent from normal flesh tints and in many scenes where the sky does not appear.

In the 1950's Dr. E. H. Land claimed that full-colour sensations could be produced by practically any two wavelengths from the visible spectrum, provided they were separated by at least 10 mμ. He made negatives of a coloured subject through a green filter (short wave) and through a red (long wave). From these, two transparencies were made, and projected in register, the short record by white light, the long record by red. Instead of an overall pink shade, a full colour reproduction resulted.

It was shown that colour sensations were not stimulated in the eye by the overall relative intensities of the two records, but by *differences* in the relative intensities from point to point in the picture. Moreover the pattern of intensity ratios had to be haphazard rather than orderly.

Although certain wavelength combinations produced an incomplete colour range, this could be rectified by broadening the wavebands. Wavelengths over much of the spectrum could act as both long and short records, when matched with shorter or longer wavelengths respectively. However, when the short (green) record was projected by red light and the long (red) record by white, colour reversal occurred.

Land showed that the eye could experience the same colour sensations simultaneously from several sets of wavelengths. He concluded that, in sunlight, the eye sees many "colour worlds", separating the wavelengths into long- or shortwave stimuli about a balancing wavelength of 588 mμ in the yellow region. P.B.M.

See also: *Colour photography history; Colour synthesis.*

U

ULTRA-MINIATURE CAMERA. Type of camera which takes films narrower than 35 mm., such as perforated or unperforated 9·5 or 16 mm. cine film.

Some types take a number of exposures (usually 8) on round discs of film about 1½ ins. in diameter.

An ultra-miniature camera is usually much smaller than any other type, varying from matchbox size to about 4 × 2 × 1 ins. Many ultra-miniatures are derived from spy cameras which were popular at the beginning of the century.

Scope. There are two distinct groups of ultra-miniatures: very simple cameras, and precision models. The former are comparable with box cameras in performance and standards of construction. The precision models have the refinements of a normal precision miniature.

The negative sizes of the ultra-miniature cameras range from about ¼ in. square to around ½ × ¾ ins. so if the negatives are to be enlarged to any degree, the optical and mechanical requirements have to be very strict. With the best instruments of this kind, the largest print size for acceptable quality is around whole-plate or 6 × 8 ins., but in most cases postcard size enlargements seem to mark the limit. For this reason the simple ultra-miniatures are little better than toys, and very few models of even the intermediate specifications can compare with a typical medium priced camera of the normal roll film type.

The commonest shape is that of an oblong rectangular box, similar in appearance and dimensions to a cigarette packet. Ultra-miniatures have also been designed in other forms, occasionally based on the layout of an amateur cine camera. The latter type sometimes takes normal 16 mm. cine film in standard 25 foot spools, yielding several hundred exposures at one loading.

Features. High-class ultra-miniatures are, despite their small size, fully equipped as serious photographic instruments, and are built with the precision of a watch.

Most cameras use film in special cassettes, holding up to 50 exposures. These cassettes are often made to be reloaded by the photographer in his own darkroom.

There does not exist a standardized frame and film size or packings for ultra-miniatures as there are for 35 mm. miniature cameras. Attempts at standardization have, however, been made both in Germany and in Japan; the German standard suggests a 12 × 17 mm. frame on unperforated 16 mm. film, the Japanese a 10 × 14 mm. frame on single perforated 16 mm. film. Neither of these standards is universally adopted, and other sizes are also in use.

Accordingly, films are often made in packs for specific camera models (sometimes marketed by the makers of the camera) and only a limited range of films exists for any one camera.

The lens generally has a focal length of ⅓ to 1 in., with a maximum aperture in most cases around *f* 3·5. Even at full aperture such a lens will give a depth of field extending from infinity down to about 3 feet.

The focusing range generally starts as close as 20 ins.

Interchangeable lenses are very rare in ultra-miniatures, mainly probably because the mechanical tolerances in positioning different lenses in a minute lens changing mount are liable to be too great to preserve the utmost precision necessary for maximum image quality. (Making more substantial lens mounts would increase the bulk of the camera.) Some ultra-miniatures take afocal wide-angle and tele lens attachments to provide effective alternative focal lengths.

Shutters are mostly of the sector type working either in front of or behind the lens, and speeded to 1/1000 second.

Interlocked shutter tensioning and film transport, and a built-in rangefinder are common features on these cameras. Other features sometimes included are built-in coupled exposure meters, built-in filters and automatic film transport.

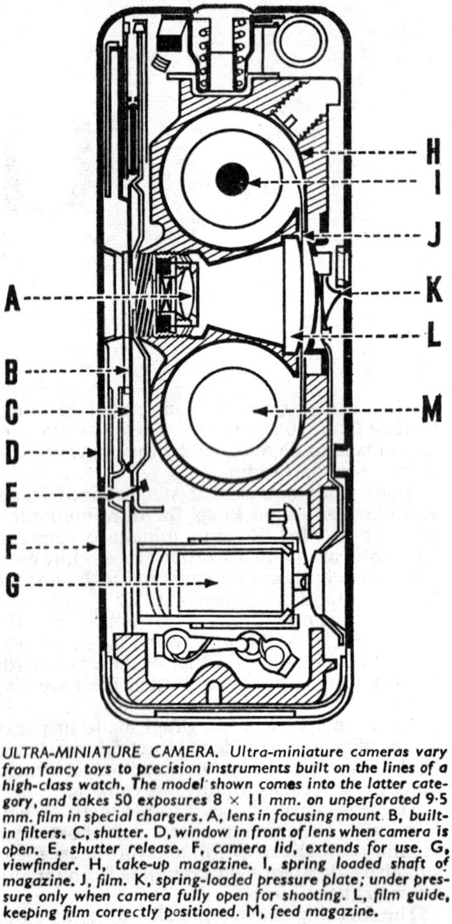

ULTRA-MINIATURE CAMERA. *Ultra-miniature cameras vary from fancy toys to precision instruments built on the lines of a high-class watch. The model shown comes into the latter category, and takes 50 exposures 8 × 11 mm. on unperforated 9·5 mm. film in special chargers. A, lens in focusing mount. B, built-in filters. C, shutter. D, window in front of lens when camera is open. E, shutter release. F, camera lid, extends for use. G, viewfinder. H, take-up magazine. I, spring loaded shaft of magazine. J, film. K, spring-loaded pressure plate; under pressure only when camera fully open for shooting. L, film guide, keeping film correctly positioned. M, feed magazine.*

Systems. Since originally standard developing tanks and enlargers for 35 mm. film did not accept narrow-gauge ultra-miniature films, most such cameras were brought out with a complete system of accessories including their own developing tank, film loading equipment, enlarger, and so on. These, together with the comprehensive range of accessories offered by manufacturers of the leading types of ultra-miniature form a self-contained camera and equipment system akin to that of the classical miniature cameras.

More recent developing tanks are frequently adjustable to take 16 mm. film as well, and there are special attachments and accessories to adapt normal enlargers to ultra-miniature films. Complete processing outfits for specific makes of ultra-miniature camera are therefore nowadays less frequent.

See also: ***Miniature camera; Spy camera.***

ULTRASONIC IMAGE RECORDING. Production of visible images formed by high-frequency sound waves in the ultrasonic region (20,000 to 1,000,000 cycles per second). Such sound waves resemble light waves in their ability to undergo reflecton, refraction and diffraction, and can form acoustic images and shadow images analogous to optical ones.

Since no photographic materials are directly sensitive to ultrasonic vibrations, the effect of the latter is made visible by their action on the process of photographic development. Sound waves can agitate processing solutions; if such agitation is sufficiently localized, it can lead to differential action on different areas of the surface of a film or plate. The resulting image is thus based on local under and over development.

Method. One set-up for producing ultrasonic images consists of a water tank containing an ultrasonic generator (a quartz crystal oscillator). An evenly fogged but undeveloped photographic plate is immersed in dilute developer solution contained in a perspex tank. The latter is placed in the water tank. A suitable object (for example a plastic saucer with a prominent relief pattern) is placed close in front of the photographic plate, so that the sound waves produce an acoustic image in the plane of the plate. This image causes differential agitation of the developer on the emulsion surface, yielding a reproduction of the pattern of the saucer (due to diffraction) in the form of a developed image.

Similar physico-chemical sound images can be obtained by immersing a uniformly fogged and developed plate in a silver solvent solution (for example acid potassium bichromate). This selectively dissolves away the silver image according to the ultrasonic agitation of the solution.

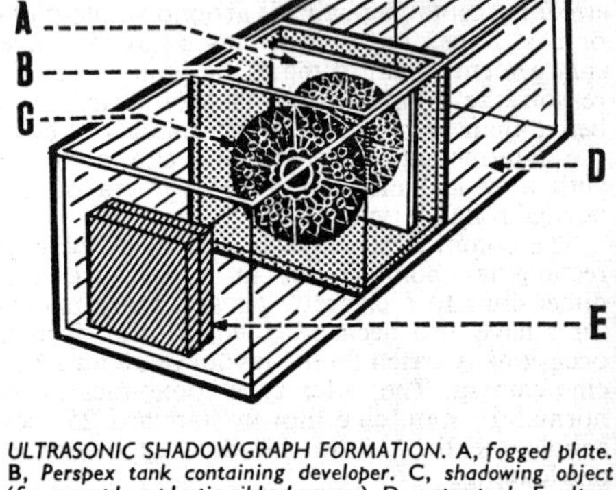

ULTRASONIC SHADOWGRAPH FORMATION. *A, fogged plate. B, Perspex tank containing developer. C, shadowing object (for example a plastic ribbed saucer). D, water tank. E, ultrasonic generator. The sound waves from the latter are held back to a greater or lesser extent by the object and so produce local variations in the agitation of the developer, leading to the development of a shadow image of the object on the fogged plate.*

Images and image patterns can also be obtained by using ultrasonic acoustic lenses and by placing the photographic plate in the path of standing waves produced in the tank.

The Ultrasonic Image. High sound frequencies (usually above one million cycles) are necessary to yield any sort of resolution. The resolving power is limited by the wavelength; at one million cycles this is 0·3 mm., and the resolution at such frequencies is of the order of one line per mm.

Ultrasonic Projector Control. Ultrasonic sound signals have been used for remote control of slide projectors and other equipment. This requires a transmitter (in effect an ultrasonic whistle) and a receiver built into the projector which responds to these signals and closes a relay to change the slide, etc.

ULTRA-VIOLET (U.V.). Beyond the violet end of the spectrum there is a band of invisible radiation to which the term ultra-violet or U.V. is applied. The region includes rays with a wavelength from 4,000 down to a few hundred Angstrom units, but below about 2,000 A. both gelatin and air cease to be transparent to the rays. Sunlight is rich in ultra-violet rays of the 3000–4000 A. region, although the greater proportion and all radiation below 3000 A. is filtered out by the upper layers of the atmosphere before the sun's rays reach the earth.

Carbon arc and mercury vapour lamps are the commonest artificial sources of ultra-violet light; and if necessary, the visible rays can be removed completely by a special filter of deep violet glass.

Filtered ultra-violet light is used in photomicrography because its short wavelength gives better definition of fine detail. The microscope lenses must be made of quartz for this work; ordinary glass does not transmit ultra-violet.

Ultra-violet light is highly photo-actinic and it is the U.V. content of sunlight which causes sunburn.

Some inorganic substances (e.g., uranium salts), many synthetic organic ones (e.g., eosin—as used in red inks) and most animal and plant materials (except blood) fluoresce strongly under ultra-violet radiation. Most materials fluoresce to a greater or less degree according to their composition. This fact is exploited in various ways in science, medicine, forensic science when dealing with forged documents, etc., examination of precious stones, industry, etc.

See also: *Spectrography; Spectrum; Ultra-violet and fluorescence photography.*

ULTRA-VIOLET AND FLUORESCENCE PHOTOGRAPHY

The limits of the colours visible to the human eye lie between violet at one end of the spectrum and red at the other. Beyond these limits, however, there are radiations that the eye cannot see—i.e., ultra-violet and infra-red rays, both of which are made use of in photography. Ultra-violet (or U.V.) radiations are often loosely and incorrectly referred to as U.V. "light" and sometimes "black light".

Ultra-Violet Radiations. The U.V. region of the electro-magnetic spectrum adjoins the short wavelength end of the visible spectrum at about 4,000 A. and extends to about 140 A. Only the "near" U.V., i.e., between 2,000 to 4,000 A. is accessible without special apparatus and techniques, for beyond 2,000 A. the gelatin of the photographic emulsion absorbs all the radiation and beyond 1,800 A. the air begins to absorb the rays.

Conventional optical glass absorbs all U.V. of wavelengths shorter than about 3,500 A.; quartz lenses must be used for the shorter wavelengths. Therefore, in the absence of special equipment, photographs are normally restricted to the 3,500 to 4,000 A. band.

Fluorescence. Fluorescence may be described as the property of some substances of absorbing radiation of one wavelength, and in its place emitting light of another wavelength or colour. The phenomenon ceases immediately the source of radiation is cut off. As used here the term fluorescence refers to the visible "glow" which results when invisible U.V. rays impinge on certain substances. Fluorescence may be of any colour from violet to red and this feature may be of analytical importance. Since all materials also reflect some U.V. rays, fluorescence will be masked in a photograph unless unfiltered U.V. radiations are prevented from reaching the emulsion.

Some substances have the property of continuing to glow for a time after the exciting rays (light, U.V. or X-rays) are shut off; this glow is known as phosphorescence and is often mistaken for fluorescence. In some instances it may be difficult to detect as the afterglow may last for only a fraction of a second. The luminous clock dial and the coating of fluorescent light tubes offer two examples of phosphorescence. The phenomenon is of little photographic importance.

There are two ways of making photographs by the agency of U.V. radiations: U.V. photography of the reflected radiations and fluorescence photography.

Sources. Sunlight provides only a 5 per cent U.V. emission as compared with 40 per cent visible light and 55 per cent infra-red. Carbon arcs have a strong emission at about 3,900 A. but are inconstant in operation. This is especially true of the enclosed variety. The most practical, convenient and constant source is some form of mercury vapour arc-lamp because mercury vapour emits a strong and

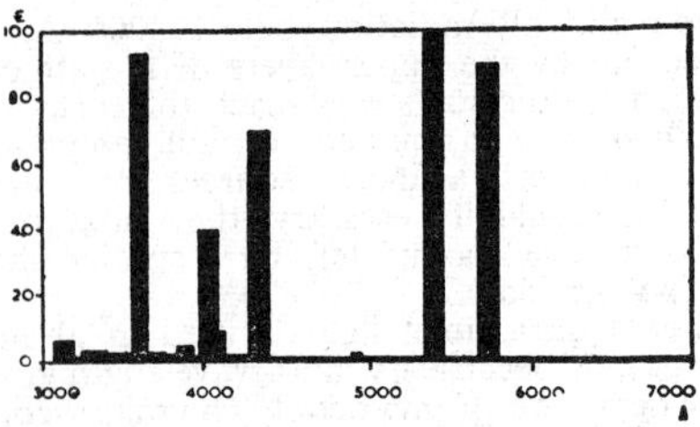

MERCURY LAMP SPECTRUM. The spectrum of the high pressure mercury vapour lamps used for ultra-violet photography is concentrated in one line in the yellow and yellow-green regions, two in the violet, and one in the near ultra-violet. In practice, all the visible light is filtered out, so that only invisible ultra-violet radiation reaches the subject.

photographically useful radiation at 3,650 A. Visible radiations can be subsequently filtered out by a suitable screen.

A convenient lamp is made by at least one firm. Essentially it is a mercury-vapour discharge lamp enclosed in a glass envelope, which restricts its U.V. emission, but having the advantage of a built-in 60° reflector.

One very suitable make of lamp is run in series with a choke and is designed to operate continuously; it takes about ten minutes before it reaches full brilliance and after it is once switched off it will not "strike" again as long as it is hot. It is convenient to mount this lamp in a light-trapped, ventilated box with slots in front to accept the various screens as required. Two sources may be used to advantage, but this necessitates greater capital outlay on equipment which is not merited for occasional use.

Low-pressure sources similar in form to domestic fluorescent tubes are available with emissions at 3,650 A., 3,100 A., and 2,537 A.

Electronic flash offers a useful and readily obtainable source of U.V. radiation especially in the 3,650 A. band. If living or moving objects are involved, this may be the most appropriate source. Ultra-violet flash may be applied effectively to both ultra-violet and fluorescence photography, although in the latter instance a continuously burning source will also be required, for visual inspection.

U.V. lamps must be used with great care. Wavelengths around 3,000 A. have a stimulating effect on the skin but too much exposure causes a painful "sunburn".

As U.V. lamps will probably be left burning for a considerable time during photography, the eyes of anyone engaged in this work should be protected by goggles of U.V.-absorbing glass as a matter of routine, more particularly if there is appreciable transmission at 3,130 A. U.V. may cause severe inflammation of the eyes (conjunctivitis) which only becomes noticeable some hours after exposure and which is both dangerous and unpleasant.

Screens. The U.V. transmitting filters which are used to screen off unwanted visible light are called U.V. screens or Wood's glass screens. If they are used over the radiation source instead of the lens, both reflection and fluorescence work may be undertaken with the same basic set-up. Some lamps possess an integral screen: with lamps that do not, a special U.V. passing screen should be used. These screens transmit a very narrow band of U.V. from roughly 3,000–4,000 A. (with a maximum at 3,600 A.). They are thus well matched to the 3,650 A. mercury line. A home-made liquid filter of a 1 cm. layer of saturated cobalt chloride (in a suitable glass cell) can also be used.

Filters. The term U.V. filter is applied to the U.V. absorbing filters used in the photography of the visible fluorescence emitted by some substances when exposed to U.V. radiation; they are always used over the camera lens. These filters vary according to make and type, but are generally designed to begin maximum cut at anything between 3,800–4,800 A. A home-made liquid filter can be a 1 cm. layer of 10 per cent sodium nitrite solution.

General U.V. Arrangement. In the practice of U.V. photography there may be some difficulty or confusion in the exact interpretation of results, so a control photograph is made by half-watt light wherever possible. To make the examination complete, it may also be necessary to make an additional record by the fluorescence method and, until the user is thoroughly familiar with U.V. techniques and subsequent interpretation, these three steps should be taken as a precaution.

An inexpensive set-up for both reflected and fluorescence photographs consists of the following items:

(1) A mercury vapour lamp as the source of U.V. radiations.

(2) A U.V. screen that can be placed in front of the source for cutting out visible light when taking fluorescence photographs.

(3) A U.V. filter for use in front of the camera lens for cutting out all but U.V. radiations when taking reflected U.V. photographs.

(4) A U.V. absorbing filter for use in front of the camera lens to absorb U.V. radiations when taking fluorescence photographs.

It should be remembered that if a glass-fronted lamp and a glass lens are used the recording range is restricted to radiations of 3,500–4,000 A.

Reflected U.V. Photography. This method requires U.V. radiations falling on the subject, and only U.V. radiations entering the camera. This is achieved by using a screened or unscreened source, and a filter over the camera lens which passes only U.V.

Use of an unscreened source has the advantage that focusing can be done by the visible emission of the source without the camera filter in position. The latter is then placed over the lens for the exposure.

Any normal camera may be used. After

initial setting-up the normal room lights should be extinguished.

As most of the normal sensitized materials are sensitive to U.V. radiations, there is no particular point in using panchromatic materials for this reflected U.V. photography; although they may be faster, speed is not usually of any great importance.

Fluorescence Photography. This method requires U.V. radiations only (and no visible light) falling on the subject, and visible light (but no U.V.) entering the camera. The source must be screened, and an U.V. absorbing camera filter is essential.

The same set-up described above for U.V. photography can be used with the addition of a U.V. absorbing filter over the camera lens. Thus, theoretically, any U.V. radiations reflected by the subject are absorbed by the filter so that, in the absence of a fluorescing substance, no photographic image should be formed. Polished metal will reflect U.V. radiations without fluorescing as well and can be used to test the efficiency of the screen/filter combination. Any image obtained is probably due to a violet component passing through the filter and it is useful to know the extent of this "background" exposure. It is sometimes produced deliberately to obtain a faint orientating image for reference in studying the finished record.

As in U.V. photography, the actual exposure should be made in the absence of room lighting.

Where the fluorescence is gross and of one colour, a black-and-white record may suffice but a colour photograph is more satisfactory in every way. Even when the brightness range extends beyond that acceptable by colour materials a better idea of relative brightness and colour is obtained. If 35 mm. transparencies are adequate, there is little difference in cost.

For black-and-white records a panchromatic emulsion may be necessary as the fluorescence is often reddish in colour. A contrast filter of the same colour as the fluorescence will help to separate a dimly fluorescing subject from its background.

General Applications. The uses to which reflected U.V. and fluorescence photography have been put are numerous. Many of these applications will not even concern the practical photographer in his daily work. The appearance of an object photographed by U.V. and/or fluorescence may be quite unrelated to its normal visible appearance. There is no general rule as to which is the particular method to use for any particular purpose: if the detail which is to be shown can be seen (by presence or absence of fluorescence) then the fluorescence method may be used, but otherwise, only experiment will show whether the required information can be revealed by U.V. photography.

Some of the commoner applications of U.V. and fluorescence photography are:

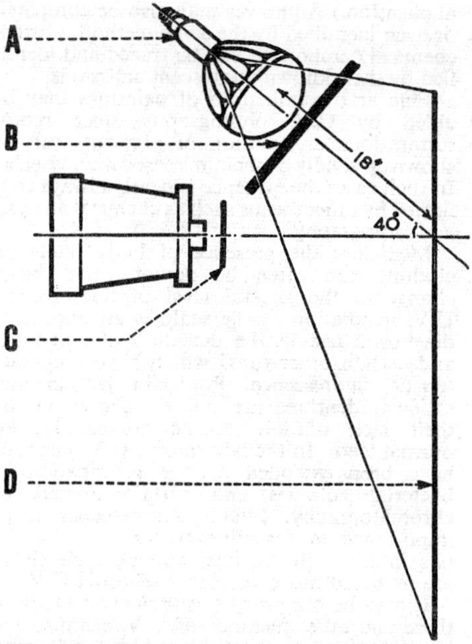

ULTRA-VIOLET PHOTOGRAPHY SET-UP. The main difference between reflected ultra-violet and fluorescence photography is that an ultra-violet absorbing filter is essential over the camera lens for fluorescence photographs. A, mercury vapour lamp. B, visually opaque, ultra-violet transmitting filter. C, ultra-violet absorbing filter (for fluorescence). D, object plane.

Forensic science: for the examination of forged or altered documents; reflection may reveal the character of erased writing and fluorescence may show up the defaced paper surface or indicate the presence of invisible inks, chemical erasers, etc. Fingerprints on confusing backgrounds may be made to stand out clearly by fluorescence if they are first dusted with a fluorescing powder like anthracene. Sealing waxes differ in fluorescent properties and any tampering with seals may be detected by fluorescence photographs. Some precious and semi-precious stones can be differentiated from paste counterfeits and natural pearls can be distinguished from artificial by fluorescence. Seminal and other body fluid stains fluoresce brightly. Traces of oil and greases can be characterized by their fluorescence, especially after partial fractionation by absorption in filter paper.

Philately: removal of cancellation marks and authenticity of watermarks may sometimes be established by comparison under U.V. irradiation. Watermarks are often forged with oils which may fluoresce, whereas the true watermark would photograph normally.

Industry: textiles, dyes, stains and markings may be compared by fluorescence. (The "invisible" laundry mark which becomes visible under U.V. is a familiar example of such an application.) Adhesives may also be compared or even identified by the same method. Certain chemical compounds can be traced and identified by their known fluorescent emissions.

Fine art: examination of paintings may be aided by U.V. photography since recent restorations or addition of pigments can be shown up vividly by their increased fluorescence. In such cases the evidence is nearly always confirmed by other means such as chemical analysis and radiography.

Medicine: the presence of body fluids on clothing can often be demonstrated by a change in the general level of fluorescence. U.V. irradiation of the scalp is an important diagnostic test in the detection of ringworm and certain other fungi which have a characteristic fluorescence. Porcelain jackets and artificial dentures are quickly shown up by their lack of fluorescence compared with normal teeth. In the laboratory, U.V. methods have been extended to the examination of bacterial colonies and certain aspects of chromatography. Finally, fluorescence is of importance in revealing traces of industrial oils in the skin crevices, and skin blemishes are rendered more visible by reflected U.V.

It may be necessary to supplement many of these and other methods of U.V. examination by radiography, chemical analysis and even infra-red photography, and in any case interpretation of the results calls for special experience because they present differing visible translations of invisible states.

Surface texture: perhaps the least obvious use of reflected U.V. photography is the rendering of any surface texture. In general, within the spectral range from infra-red to U.V., the shorter the incident wavelength, the more it tends to become reflected from a surface without penetration. Thus infra-red radiation is mostly used to reveal detail beneath surfaces which it can penetrate and U.V. for recording the fine detail of a surface itself. Fluorescence, however, can be excited within some structures by penetration.

Special Applications. The foregoing techniques and applications require normal equipment and make use of only a narrow spectral band. This type of work is within the reach of any competent photographer. A few more precise techniques requiring specialized apparatus involve microscopic methods:

U.V. microscopy: very high magnifications and resolving power can be attained by microscopes designed to utilize U.V. radiation in place of visible light. Because of the selective absorption of the rays, structures can be differentiated without staining, and negligible depth of field makes "optical sectioning" possible—i.e., the photography of thin sections selected simply by focusing and not by actual cutting.

Since glass absorbs all radiations shorter than about 3,500 A. every optical component, including slides and coverslips, must be made of quartz for this type of work. And because it is difficult to achromatize objectives for more than a single waveband, a cadmium arc is commonly used as the source of U.V. radiations. This source gives strong emission lines at 2,750 A. and 2,573 A. either of which can be selected to illuminate the specimen by a monochromator.

The optical system is first centred and focused visually using a mercury-vapour arc as the illuminant. After changing over to the cadmium high-tension spark system the image is no longer visible and final adjustments are made by means of an "artificial eye" consisting of a fluorescing uranium glass screen over which is mounted a strong field lens. This separate focusing system presents a small visible image and bears a definite relation to the true focal plane at which the actual exposures are made. Optical sections of the preparation are photographed by making a series of exposures between which the microscope focusing control is advanced by a known small amount.

In this way it is possible to make extremely detailed studies of planes $\frac{1}{4}\mu$ thick at magnifications ranging from 1,000 to 3,000 diameters.

Colour translating U.V. microscope: an ingenious elaboration of normal U.V. microscopy has been developed in recent years. The system was originally intended to provide coloured images of unstained biological specimens, the colour differences being related to the chemical constitution of the material under investigation.

Three U.V. photomicrographs are prepared of the same field using wavelengths of 2,630, 2,580 and 2,370 A. respectively. Differing selective absorption in these three bands produces corresponding differences in the three negatives. These three photographs are then projected through red, green and blue filters, the individually coloured images being superimposed for final examination. In the original system the negatives obtained by the shortest U.V. band were projected through a red filter and the longest through a blue filter, thus inverting the sequence of wavelengths, but other combinations are possible.

Fluorescence microscopy: the general principles of fluorescence photography may be applied with equal effect to microscopy.

Normal microscope equipment may be used, although any mirror should be made of polished aluminium as silver is a poor reflector of U.V. For best visual results a heavy yellow (minus blue) filter is placed within or over the viewing eyepiece; but for photography it may be preferable to allow a certain amount of visible blue and violet light to be transmitted to give a direct indication of the relationship between the specimen and the fluorescence. This technique is now in general use, and is applied to the examination of unstained material (primary fluorescence), or to preparations stained with specific fluorochrome dyes (secondary fluorescence).

In some instances fluorescence can be excited merely by placing a deep blue filter over the normal microscope lamp, in which case a green or yellow absorption filter is used at the eyepiece. All such techniques should be regarded as a supplement to normal photomicrography. P.H.

See also: *Fluorescence; Negative materials; Spectrography.*
Books: *Infra-red and Ultra-violet Photography*, Eastman Kodak (Rochester, U.S.A.).

UNDER-DEVELOPMENT. Fault indicating that the developing solution did not act long enough on the sensitized material to produce an acceptable result.

Negatives. An under-developed negative looks thin and lacks contrast. All the detail is clearly visible, but the highlights lack density. It is easy to distinguish from an under-exposed negative which has been correctly developed. The under-exposed negative is thin also, but in this case there is no shadow detail, and the brightest highlights may be somewhere near maximum density.

A normal print from an under-developed negative is flat, lacking in contrast and with no real black tones.

Assuming that the developer is correctly mixed and fresh, under-development may be caused by a mistake in working out the time of development for the sensitized material in the particular developer used or it may be simply due to faulty timing or to insufficient agitation. Other possible causes are that the developer concentration was too low, or exhausted, or that the time of development was not adjusted to allow for a drop in temperature or for the number of films already developed in the same solution.

When developing by inspection, under-development may be due to the photographer's inability to judge the appearance of a fully developed negative.

An under-developed negative may be bleached and redeveloped as in intensifying. This treatment can result in a negative indistinguishable from one correctly developed in the first place. If the amount of under-development is slight, the remedy may simply be to print it on a harder grade of paper to restore the contrast.

Prints. An under-developed print looks pale and washed out. If it was correctly exposed, it will show detail in both highlights and shadows, but there will be no black tones even in the deepest shadow areas. If anything, the tones will be mottled, tinged with a trace of reddish brown. The image will tend to show up more strongly in some parts than in others.

An under-exposed print, that has received normal development, looks quite different. In this case the image is weak, but there is no highlight detail and the shadows are light in tone with only a few, if any, real black areas.

Since prints are developed by inspection, the only reason for failing to leave the print in the solution long enough is sheer inability to recognize a fully developed print. Most beginners tend to lift the print out too soon: the tones always look darker when the print is in the developer than when it is finally dried and looked at in daylight.

Contact paper requires a more concentrated developer than bromide paper, and under-development will result if contact prints are developed in a solution mixed in error to bromide paper strength.

If the temperature of a MQ developer is allowed to fall much below 60° F. (15·5° C.) it will produce under-development.

See also: *Developing negatives; Developing prints; Faults; Under-exposure.*

UNDER-EXPOSURE. Fault indicating that the image formed on the sensitized material was either not bright enough or was not allowed to act long enough during exposure to produce an acceptable image on development.

Negatives. The appearance of an under-exposed negative is characteristic: the shadow areas are clear emulsion without any trace of detail or gradation and only the highlights show a black silver deposit. It is easy to tell the difference between an under-exposed negative and one that has been correctly exposed, but under-developed. The under-developed negative will bear a distinct but thin image of shadow detail, but the highlight areas will lack density and appear greyish. An under-exposed negative looks contrasty but generally gives flat contrast in the shadows. An under-developed negative looks flat and gives a soft, flat print with no strong contrast at either end of the scale.

The commonest cause of under-exposure is simply failure to estimate or measure the light intensity accurately, and to adjust the shutter speed and lens aperture to give an adequate exposure to the sensitized material in use.

In flash photography it may be the result of faulty synchronization or of wrong calculation of the stop and subject distance.

In the first place, the method of estimation may be faulty, or the exposure meter may be faulty, or it may not have been set for the right speed of sensitized material. It is also possible that the meter has been used incorrectly—e.g., pointed too much towards the sky—or that the correct allowance has not been made for such additional factors as subject brightness and contrast, the filter in use, the direction or colour temperature of the light, and the developer which will be used—some fine-grain developers reduce the effective emulsion speed and call for a substantial increase in exposure time.

Finally, the shutter itself may be faulty.

There is very little that can be done about an under-exposed negative. If the exposure has been insufficient to record detail in the shadows, no subsequent processing can put it there. Continued development simply increases the density of the highlights, making them print out whiter and whiter, while the shadows remain a featureless black.

Intensification is no more likely to succeed than over-development, and the results are similar. The only aftertreatment likely to yield a better picture is optical intensification—i.e., the negative is set up against a white background and copied by reflected light. Even this method is useless unless there is some slight suggestion of detail in the shadows.

Prints. An under-exposed print looks pale and washed-out. The shadows are grey and the highlights are featureless areas of blank paper. Such a print cannot be improved by being left in the developer.

An under-exposed print can be distinguished from one that has been correctly exposed but under-developed by the fact that in the under-developed print all the detail in both highlights and shadows is visible but there are no good black tones anywhere. If anything, the tones in an under-developed print tend to be of a warm or muddy colour and in extreme cases may be mottled. In an under-exposed print, most of the detail is absent and all visible tones are a light grey except in the extreme shadows.

The print has not been left long enough under the printing light. If a test strip was used, then the section selected as giving the best reproduction may have been too pale. It is important to remember that the image tones always look darker under the safelight, especially a deep orange one, than they do in the finished print in daylight.

In enlarging, the trouble may be that the scale of enlargment has been increased without making a corresponding change in the exposure. Or the lens may have been opened up a stop and the exposure halved when, owing to the concentration of light rays at the centre of the lens, very little reduction should have been made.

If an automatic timer is used for the exposure, it may not be giving times that correspond with the times of the test strip.

Nothing can be done to improve the tones of an under-exposed print.

The remedy for under-exposure is to make one or more tests on strips of paper from the same packet and to choose the section that looks just too dark under the safelight. If a big change is made in either the magnification or the lens aperture, a fresh test should be made.

Both the test strip and the finished print should be developed for the full time recommended by the maker of the paper to obtain the best image quality.

See also: *Exposure; Faults; Under-development.*

UNDERWATER PHOTOGRAPHY

Underwater photography remains a somewhat esoteric art despite the fact that most people are now familiar with underwater photographs. The reason for this is probably that it is a branch of photography that must be embarked upon wholeheartedly or not at all. The would-be underwater photographer must obviously be able to dive, and if using a standard camera, he must be prepared to expend some effort in preparing it for a sub-aquatic environment.

History. Underwater photography was first attempted in 1856 by William Thompson at Weymouth, England. The camera, in a watertight box, was lowered from a boat to the seabed 18 feet below, and the drop shutter held open for 10 minutes by pulling a long string from the boat. A passable picture of seaweed and sand was obtained. Ten years later a French photographer named Bazin went down in a diving bell with watertight port-holes through which he photographed by electric light from a Bunsen battery. His purpose was to make the reclamation of sunken ships easier. In 1893 Louis Boutan photographed in a diving suit in the Mediterranean. He used a magazine plate camera sealed in a metal box with a glazed opening for the lens. Exposures of ten to thirty minutes were necessary, depending on the depth. Below 15 feet, artificial light was needed. After experimenting with two forms of magnesium light, Boutan in 1899 used with greater success submarine searchlights charged from a dynamo on board a boat, and a camera with an electro-magnetic shutter operated from the boat, with which instantaneous pictures could be taken and greater depths explored. Boutan wrote the first book on underwater photography *La Photographie Sousmarine* (1900) in which he foresaw that one day underwater photography would be practised by amateurs as a new sport.

EQUIPMENT

Underwater Cameras. As the result of the increasing popularity of underwater photography, cameras designed specifically for this purpose are available.

The bodies of such cameras are constructed of pressure- and corrosion-resistant materials, with the controls operating through water-

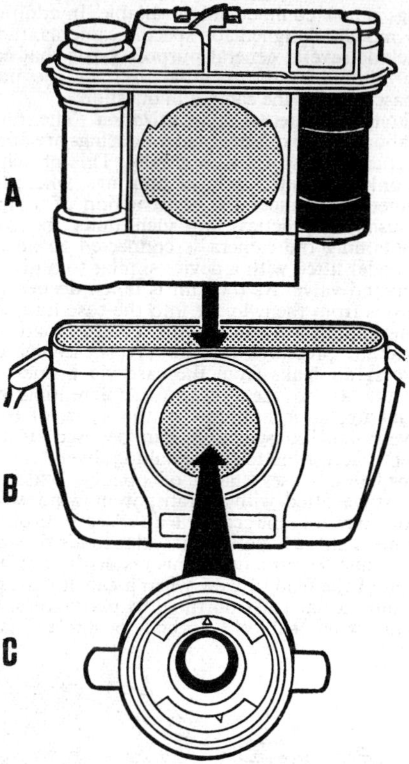

SELF-CONTAINED UNDERWATER CAMERA. This type of design needs no separate pressure proof-housing. The main parts of the camera—the fim holder and transport unit A, the outer housing B, and the lens unit, C—fit together with pressure-tight seals. The interchangeable lens unit incorporates the aperture and distance controls which connect through water-tight glands. A single lever, also working through a gland, operates the fim transport and shutter release.

tight glands. Standard lenses are usually hermetically sealed behind flat glass portholes. This enables the cameras to be used above and below water without modification.

Standard Cameras. Because of the limited range of underwater cameras, the majority of divers use a standard camera enclosed in some form of waterproof housing. As the cost of such cases is linked with the number of controls, and the ease with which they can be remotely manipulated, the choice of camera is of considerable importance. Rigid housings have to withstand considerable pressures underwater and are kept as compact as possible.

The number of controls necessary to operate the camera should be kept low as possible. These have to operate through watertight glands in the case and are therefore possible sources of leakage. Also, they are susceptible to mechanical malfunction. Thus a small camera with a coupled exposure meter and a mechanism that advances the film and cocks the shutter is preferable to a larger model that requires four separate manipulations to operate the same number of controls. These parameters are satisfied by a wide range of 35 mm. cameras. So 35 mm. models are the most popular size of still camera used underwater. The same criteria can be used to judge the suitability of cine cameras for underwater use.

The range of waterproof housings used to accommodate standard cameras for sub-aquatic use can be divided into three basic categories: pressure-compensating, pressure-resistant, and internally pressurized.

Pressure-Compensating Housings. The simplest type of case employs the pressure compensating principle used by Boutan. His first camera housing had a small air-filled balloon attached to a short length of tubing that connected with the inside of the case. As the diver descended the increasing water pressure compressed the air inside the balloon, which in turn transmitted the pressure to the inside of the camera case. The build up of considerable pressure differences between the inside and outside of the case was thereby prevented, and the possibility of water leaking into housing reduced.

Nowadays, however, the pressure compensating device has a dual purpose. It usually takes the form of a rubber glove through which the diver manually operates the camera controls. The advantage of this type of housing is that it does not have to be tailored to a specific camera. As the controls are operated by hand, a range of cameras can be used in a single case.

There are two main disadvantages: one is the relatively fragile and perishable nature of the thin rubber components. The second, and perhaps more important, reason is that the limiting depth to which such cases can safely be taken is usually in the region of 30 feet. The reason for this is that the air enclosed inside the housing decreases in volume according to Boyles Law. At a depth of 33 feet the pressure is double that on the surface. At this depth, therefore, the volume of air inside the case is reduced by half when the pressures equalize.

Pressure-resistant Housings. The most widely used type of case consists basically of a rigid box capable of withstanding the pressures encountered under water. The forces to which such housings are subjected are quite considerable, as the pressure increases by 14·7 pounds per square inch for every 33 feet the diver descends. An 8-in. cube, for instance, is subjected to a force of over 3 tons at a depth of 60 feet.

A sphere is the ideal shape to withstand such pressures, but the inherent difficulties of construction usually preclude its use. Instead the housing takes the form of a rectangular box with a removable side or lid. The camera is fixed inside the box and "looks" through a glass or Perspex window. The camera is operated via

UNDERWATER HOUSING. Usually underwater housings have to be designed for individual cameras. This model for a twin-lens reflex has outside controls for all the camera operations, and will also take a flash unit. The controls are linked to a suitable coupling mechanism inside the case, and pass through water-tight glands in the body.

controls that pass through watertight glands in the body of the housing. A variety of methods are employed to provide these watertight seals. One popular method utilises a greased string packing compressed round the shaft of the control rod. A more sophisticated method is to use "O" ring seals upon which the compression can be varied. The seal between the box and its gasket is achieved with the aid of a continuous flat rubber gasket, or an "O" seal.

A considerable range of materials is used for the construction of such housings. Perspex is popular because it is transparent and has the advantage that the camera settings and controls can be inspected visually. Wood, metal and fibre glass are also used.

Rigid housings are often fitted with valves which enable them to be internally pressurized with a hand pump. Increasing the internal pressure serves two purposes. Firstly it helps to balance the external pressure when the camera is taken deep. Secondly, it acts as a leak detector. For if the camera is immersed in water a tell-tale stream of bubbles issues from any leaking seals.

The weight of a housing is usually adjusted so that it is slightly negatively buoyant when it contains the camera. This enables it to be comfortably carried by the diver with the aid of a neck strap.

A fairly wide selection of rigid camera housings is now commercially available. In addition to models designed for specific cameras, these include several general purpose types that can be tailored by the manufacturer, or the purchaser, to suit the camera in question.

Automatic Internal Pressurization. The most elaborate type of underwater housings are automatically pressurized internally. This principle is only warranted in large housings, where the dimensions render the construction of a rigid housing impractical. In such units the case containing the camera is connected to an air cylinder fitted with a device similar to a diver's demand valve. As the unit is taken deeper, air passes from the cylinder into the case until the internal pressure is slightly greater than the external pressure. As the diver ascends the excess air leaks from the case via a one-way valve. As no great pressure differentials are built up, large housings of this type, of relatively light construction, can be used to any depth accessible to the aqualung diver.

Portholes. Cases made of opaque materials must be fitted with a transparent window or porthole for the camera to "see" through. When light rays pass from water to air they are bent due to refraction. This results in a reduction of the field of view when a camera is used behind a flat glass porthole to photograph an underwater subject. Special portholes have

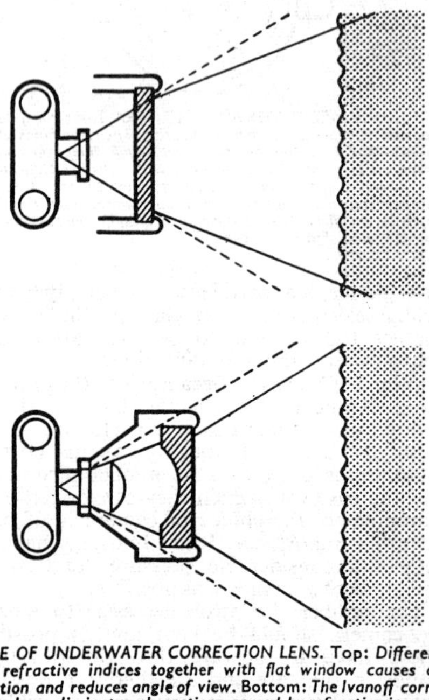

USE OF UNDERWATER CORRECTION LENS. Top: Difference of refractive indices together with flat window causes distortion and reduces angle of view. Bottom: The Ivanoff correction lens eliminates aberrations caused by refraction in water, and compensates for reduced angle of view and for apparent reduction in subject distances. Beyond the wavy line (the "blue wall") water turbidity makes sharp images impossible.

been designed to overcome this effect. They are, however, expensive, and have to be accurately aligned with the camera. They are seldom used. Most housings are fitted with optically flat portholes made of glass or Perspex.

Condensation. As the temperature of the water is generally lower than that of the air at the surface, there is always a tendency for moisture to condense on the inside of the viewing and taking windows in plastic containers. This is particularly so with the flexible pressure-equalizing types when the air becomes compressed. One way of counteracting this is the inclusion of a dehydrating agent—e.g. capsules of silica gel—in the housing. In a metal container any condensation takes places on the metal which is colder than the glass.

Viewfinders. If the diver can see through the viewfinder when the camera is encased in its underwater housing there is no reason why he should not use it underwater, as it will indicate the true field of view accepted by the camera. Housings for reflex cameras may include prism and other devices for direct observation of the screen image. Usually, however, it is impracticable to use the viewfinder in this way. Most underwater camera housings are therefore equipped with an external direct vision frame finder. The separation of the front and back sights is increased to compensate for the reduced field of view. The need to correct such viewfinders for parallax is also important because of the increased separation of the viewfinder from the camera. This is usually achieved by making the back sight adjustable. When located above the camera the rear sight is raised for close-up shots.

Rangefinders. As with viewfinders, it is usually impossible to use a rangefinder attached directly to the camera when it is enclosed in its watertight case.

The refractive index of water is 1·33. This has the effect of making underwater objects appear closer to the diver. As the lens of a camera behaves in much the same way as the lens in the eye it must be focused at the apparent, and not the real distance of the subject, to obtain a clear picture. If a rangefinder can be used, the camera lens should be set at the distance indicated by the rangefinder as it will automatically correct for the foreshortening effect of the water.

Distances greater than 3 feet can usually be guessed by the photographer with sufficient accuracy. Distances of less than 3 feet can be measured with the aid of a focusing stick. This is a stick 4 feet long, marked to represent 3 underwater feet.

Underwater Lighting Equipment. When light strikes a water surface some of it passes into the water, and a fraction of it is reflected. The percentage of the light reflected varies with the angle of the source, and is a minimum when the light rays are perpendicular to the interface. The intensity of light just below the surface is therefore always less than it is above water. The light that passes into the water can be attenuated in a number of ways; by bubbles, suspended particles, and absorption by the water itself, which absorbs various colours to different extents. The overall result of these effects is that the light intensity underwater is lowered, its intensity decreasing with increasing depth. In addition, there is a pronounced change in its colour balance. As a result it is often necessary to supplement the natural lighting in some way to obtain adequate exposure for black-and-white films. To take colour films with a full colour range it is essential to use additional lighting below a depth of 20 feet.

For the still photographer, additional lighting can be achieved cheaply and simply by the use of expendable flash bulbs. The bulbs can be used in direct contact with water. They can be placed in an external reflector and changed by hand. The reflector should be positioned as far away from the camera as possible to reduce the particle illumination effect to a minimum. The electrical components of the flash gun can be located inside the camera case. Some underwater housings have special compartments for this purpose.

The low voltages used to fire flash bulbs will do no more than make the photographer jump should he accidentally get a shock. The same cannot be said of the circuits used to energize electronic flash tubes which operate at much higher potentials. The voltage from these could be lethal. So when electronic, or strobe, flash is used under water, the entire unit must be enclosed in a reliable watertight housing.

The use of electronic flash underwater was pioneered in the 1950s by Dimitri Rebikoff who housed the components in a long metal tube. The flash tube was placed in the nose, and the unit named the "torpille", or "torpedo". As a result of the current trend in miniaturization, it is now possible to construct, or buy a much more compact unit. It is a practical proposition to incorporate an electronic flash unit into a small separate housing.

The cine photographer must have a continuous source of lighting, which has to be of considerable power if it is to be of any use under water. For lighting large areas, waterproof lamps connected via heavy cables to generators on the surface are useful. For close up work, self-contained, battery-powered units can be used. Rebikoff utilized a torpedo, containing silver-zinc batteries, with a projection lamp located in the nose. The batteries were also used to power a propeller at the rear end of the torpedo. With the camera, enclosed in its case, mounted on the side of the torpedo, he had a mobile, self-contained, filming unit. Underwater lighting units containing standard batteries and highly efficient, cool running, tungsten iodide projection lamps are now commercially available.

Diving Equipment. An obvious prerequisite to

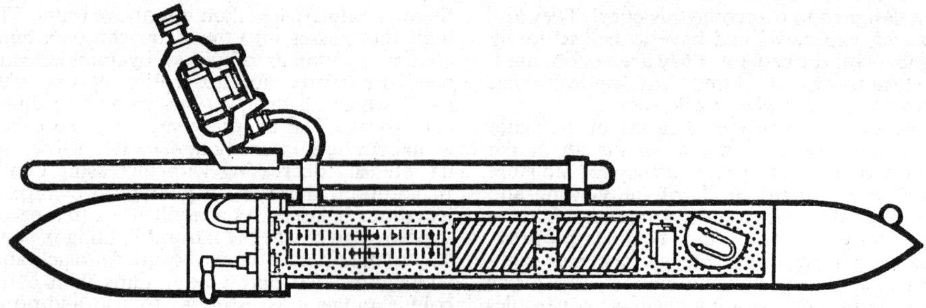

UNDERWATER TORPEDO. The complete casing incorporates an electronic flash unit, and on some models an electric driving motor which even pulls the swimmer along. The front of the torpedo carries the flash tube, the centre the capacitors and accumulators, and the rear portion holds the controls as well as the camera in its own pressure-proof casing.

taking underwater pictures is that the photographer should be able to dive. The use of standard diving equipment—consisting of a rubber suit, and a copper helmet into which air is pumped from the surface—is becoming obsolete, and is not suitable for underwater photography. The alternatives are snorkel diving, and free diving.

When snorkel diving, the underwater swimmer uses only basic equipment, i.e., facemask, fins, and snorkel tube. The facemask enables the diver to see clearly underwater, the fins increase his swimming power and manœverability, leaving his hands free to operate the camera, and the snorkel tube enables him to breath with his facemask submerged—permitting him an uninterrupted view of the underwater scene. Some divers can attain depths of 100 feet using such equipment. The duration of their stay underwater is governed by the length of time they can hold their breath.

The free diver can use either an oxygen rebreathing set, or an aqualung. Oxygen sets, which were developed for military purposes, can be used to a maximum depth of about 40 feet. They can be hazardous and are not therefore recommended.

The most suitable diving equipment for the underwater photographer is an aqualung. This consists basically of a cylinder of compressed air fitted with a valve that supplies the diver with air at the pressure of his surroundings. It is simple to use, and perfectly safe, providing the diver follows a well-defined code of rules. Before commencing with his photography, the diver should master the use of the aqualung so that he can deal confidently, and automatically, with any situations that he is likely to meet under water. Once this is achieved he can give all his attention to photography.

MATERIALS AND TECHNIQUES

Only in very foggy conditions does the visibility above water fall below 40 yards. Underwater a visibility in this region would be regarded as very good. It is usually much less. This, coupled with the change in colour balance and considerable variations in light intensity, present the photographer with a challenging and rather unique set of conditions which require a special approach.

Films. The underwater world is one of soft diffuse lighting and low contrast subjects. The use of the slow films which usually produce a negative of high contrast is therefore indicated for black-and-white photography. Unfortunately, the low light intensity usually precludes the use of the slowest films. A compromise choice is the slowest film commensurate with the depth and lighting intensity of the underwater scene.

The equivalent difference in colour brilliance between slow and fast colour films is not so apparent. Thus a fast colour film is recommended. The speeds of some colour films can be considerably up-rated by modifying the development procedure. This factor can sometimes be used to the photographer's advantage by extending the maximum depth at which colour photographs can be taken.

Filters. Minute suspended particles reflect and scatter light underwater, thereby reducing the visibility and producing a hazy effect. The scattered light lowers the contrast and definition of monochrome pictures, tending to give them a flat and lifeless look. As much of the scattered light is in the blue end of the spectrum it can be partially eliminated by the use of yellow filters. Although the use of yellow filters enables brighter pictures to be taken when conditions are poor, they cannot be regarded as a substitute for good visibility. Collectively, suspended particles present a physical barrier through which light cannot penetrate, and filters cannot increase transmission. As a general rule, clear pictures can be taken when the distance between the subject and the camera is less than one-third of the underwater visibility.

The use of filters in colour photography is a controversial subject. With increasing depth underwater there is a change in colour balance in addition to the overall lowering of light intensity. As no red light penetrates more than 30 feet underwater, photographs taken below this depth have a monochrome blue-green appearance. The photographer can therefore either reproduce the scene as he sees it (by not using filters), or he can try and reverse the change in colour balance and give his pictures an "on the surface" colour rendering. The loss in red light can be corrected for by the use of CC-R filters which have a high red transmission. The strength of filter to give full correction is determined by the total underwater path length of the light, approximately four filter units being added for every foot. A CC 50R which should be used when the path length is about 12 feet requires an increase in exposure of one stop. It is impossible fully to correct for the loss of red light at depths below 20 feet. Only supplementary lighting can yield pictures with a full colour range below this depth.

Exposure. The percentage of light penetrating to a given depth depends upon many variables. The use of an exposure meter is essential unless the diver is very familiar with conditions. The meter is used in the usual manner. It can either be located in the camera case, or in its own separate housing. It is sometimes advisable deliberately to underexpose monochrome films and prolong the development in order to obtain a higher contrast negative.

Flash. When using flash above water the exposure is derived from the flash factor, which assumes the flash is the sole source of illumination. The attenuation of light under water, and the general lack of reflecting surfaces, considerably reduces the illuminating power of a flash bulb. The loss depends upon the prevailing conditions, but as a general rule dividing the normal flash factor by five gives a reasonable correction.

When the flash reflector is located near to the camera, small particles suspended in the water close to the camera are relatively intensely illuminated. As they are usually out of focus they appear as unsightly circular white blotches on the final print. The flash reflector should therefore be positioned as close to the subject as possible—but out of the field of view of the camera. In this position the illuminating power of the flash is increased, and the particle-illuminating effect is reduced to a minimum. The haze produced by scatter of the flash light can be reduced by the use of yellow filters. With colour film, however, such filters give the picture an unnatural yellow cast.

Flash bulbs for use with colour daylight films are normally coated with a blue filter to remove some of the excess red light. However, water itself preferentially absorbs red colours tending to balance the spectral distribution of the light. Colourless bulbs are therefore preferable for colour pictures under water.

Subjects. The underwater world abounds with subjects. An obvious one for the diver is his diving companions. The weightlessness of the diver under water can be used to advantage by the photographer as it enables the subject to assume a great variety of positions. If the model is wearing an aqualung, wait until a plume of bubbles is rising from the demand valve before releasing the shutter. The play of light on the bubbles not only has an intrinsic beauty, but it does gives a picture a truly underwater atmosphere. The use of wide angle lenses enables the camera to get closer to the subject. This greatly enhances distortion due to perspective. Divers should therefore move parallel to the camera and keep their hands by their sides.

A flat sandy seabed is usually relatively lifeless, and is not very photogenic. A rocky area contains more abundant life and a greater variety of natural backgrounds. Caves, gulleys and arches can all be exploited to provide a framework in which the major subject can be posed. Wrecks are ideal objects for exploration and can provide the cameraman with a wealth of exciting material.

With increasing experience the diver subconsciously acquires the lore of the underwater world. He can find and photograph the habits and haunts of underwater animals that pass unnoticed by the novice. A cardinal rule when photographing fish is to swim as quietly as possible. Most fish can swim much faster than a diver and they will quickly take flight if the diver makes vigorous erratic movements. Many fish are naturally inquisitive, and will swim right up to the photographer if he sits quietly and patiently on the bottom. Fish can

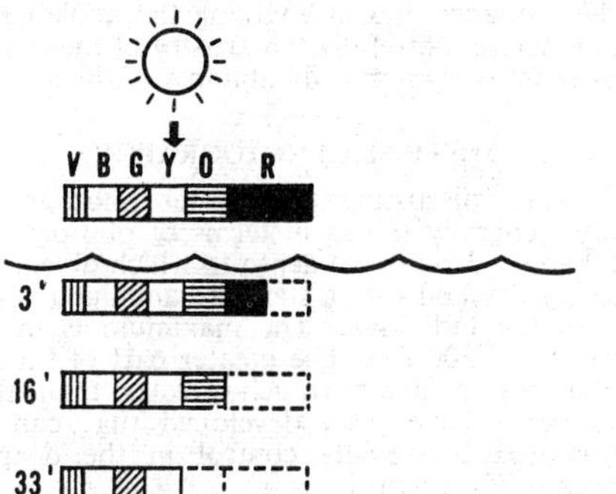

COLOURS UNDER WATER. Water acts as a strong filter and cuts out some of the colours of white sunlight. At 3 feet below the surface, half the red is filtered away; at 16 feet all red, at 33 feet all orange, at 66 feet all yellow and violet. At 100 feet everything looks blue-grey. Electronic flash or other artificial light will of course show a full range of colours.

often be attracted to a specific site by scattering pieces of broken-up shell fish in the area. This practice, known colloquially as chumming, was used by Boutan who could not readily move his cumbersome camera housings.

The problem of the small size of the subject can be solved only by working close to it and producing a large image in the first place. Working with a long-focus lens from a distance is not satisfactory because of the murkiness of the water; for the same reason the quality of a small image taken from a normal viewing distance is too poor to give a good enlargement.

When selecting a subject for underwater photography it is the result that counts, not the risk to life and limb that goes into procuring it. The diver will find that his best results with natural lighting are obtained in shallow water where the light is brighter and the subjects have a much greater contrast than when photographed at depth.

Darkroom Techniques. Increasing the development time will give negatives of greater contrast. If the photographs were taken in dull conditions the film should be developed to a gamma of 0·8 to 1·0. The development times necessary for this can be found in the data sheets available from the film manufacturers.

Underwater negatives are often too soft to produce a bright print on normal printing paper, and very hard papers may be necessary to obtain the desired result. As such papers have a limited exposure latitude, differential shading of various parts of the print may be called for.

CINE

For underwater filming, 8 mm. cameras with fast wide-angle or zoom lenses, and coupled exposure meters are ideal, and underwater cases for many such cameras are available.

The shorter focal length of lenses used on 8 mm. cameras yields a considerable depth of field, even at the widest apertures. This gives well-defined pictures over a wide range of camera-to-subject distances. Fixed focus lenses are therefore quite adequate for most purposes.

An 8 mm. camera, with a fixed focus lens and a coupled exposure meter, is very simple to use above and below water. It is the ideal medium for recording people and the natural world that lies beneath the waves. Weightlessness gives the underwater movements of man and fish an intrinsic grace and beauty that can never be captured in a still photograph. 16 mm. film should be used when the product is intended for showing to large audiences, or for television.

It is sometimes advantageous to fit horizontal fins to the camera case. They help stabilize the camera, and enable the diver to take smooth tracking shots.

It is often difficult to keep the camera steady in shallow water where there is a strong tidal swell. One solution is to attach the camera to a tripod. An alternative is for the diver greatly to overweight himself so that he can sit, or kneel, on the bottom with some stability. A somewhat safer alternative is for the diver to provide himself with a rigid support of some kind about which he can entwine a leg or arm.

As in normal cine filming, underwater sequences should be planned beforehand. As it is impossible to communicate verbally underwater, the photographer must give the participants in the film clear directives on the surface. A prearranged code of simple hand signals can be used underwater.

The photographer should bear in mind that a large portion of his audience is likely to be unfamiliar with the underwater world. He should therefore hold his underwater shots for much longer than above water sequences. The audience can be prepared for the sub-aquatic sequences by being shown the divers getting ready to enter the water. Underwater, the scene can be set by showing a group of divers leaving the surface and heading towards the sea bed. This can be followed by a sequence showing a single diver reaching the sea bed. The photographer can then show the natural inhabitants of the underwater world in close-up.

Fast-moving fish are best photographed at a faster camera speed. When the film is projected at normal speed the movements are slowed down.

This has the double advantage of prolonging the sequence and of enabling the audience to appreciate more fully the artistry of movement of those ever-mobile inhabitants of the sea.

DEEP SEA PHOTOGRAPHY

When discussing underwater photography one generally thinks in terms of photographs taken by divers. The depth to which divers can safely descend with a standard aqualung varies with the individual. The maximum is in the region of 300 feet. The greater part of the sea bed lies at depths much greater than this. Cameras have been developed that can be operated by remote control in the deepest oceans of the world.

Foremost in the field of deep sea photography is Professor Harold E. Edgerton, of the Massachusetts Institute of Technology. He has developed a photographic unit capable of withstanding pressures in the region of 17,000 pounds per square inch. Working in conjunction with a pioneer of underwater exploration, Commander Jacques Cousteau, Professor Edgerton has provided conclusive proof that life does exist in the depths of the ocean.

One of the cameras developed by Professor Edgerton can take up to 2,000 pictures on 35 mm. film stock at signalled time intervals. For use at great depths, the camera is mounted on a frame and lowered by cable from the research ship. The frame also carries an electronic flash unit that is synchronized with the camera. The height of the camera above the sea floor is determined by a sonar (radar) pinger on the camera. The pinger simultaneously sends a pulse, or signal, downwards, and another upwards. The camera to bottom distance can be determined by the difference in time it takes the pulse bounced from the bottom, and the direct signal, to reach a receiver under the ship. In this way it is possible accurately to position the camera a few feet above the sea bed despite the fact that it may be more than 15,000 feet from the surface. H.E.D.

See also: *Marine photography.*
Book: *Camera Under Water*, by H. Dobbs (London).

UNIFORM SYSTEM (U.S.) System of stop numbers proposed in 1881 by the Royal Photographic Society. Based on *f*4 as the unit, a stop requiring twice the exposure was marked 2; three times the exposure, 3; half the exposure, ½, and so on. The idea never became popular and has not been used for at least twenty years.

However, the practice of marking enlarging lenses to indicate relative exposures is on the increase. These lenses are marked with numbers in the series 1, 2, 4, 8, 16 and so on, taking the larger stop as equal to 1.

See also: *Diaphragm.*

UNION OF SOVIET SOCIALIST REPUBLICS. Photography in the U.S.S.R. had its beginnings in the pre-revolutionary period. The development of photography in the Soviet Union is therefore considered together with this earlier phase.

See also: *Russia and the U.S.S.R.*

UNITED STATES OF AMERICA

Towards the end of September, 1839, less than a month after the public disclosure by the French Government of Daguerre's invention, daguerreotypes were being taken in the United States. Of the several scientifically-minded men who re-created Daguerre's technique from the detailed specifications contained in his instruction manual, the most prominent was Samuel F. B. Morse (1791–1872), best known for his work in the development of the electric telegraph.

Daguerreotypes. Morse showed his first successful daguerreotype, a view of a church in New York taken with an exposure of 15 minutes, on 28th September, 1839. Most early daguerreotypes, like Morse's, were views of buildings, to judge from the accounts of them. Only one has survived; a small (1¼ × 2 ins.) and imperfect view of the Philadelphia Mint, taken by Joseph Saxton in October, 1839. Not until François Gouraud (died about 1848) arrived from Paris in November, to teach and demonstrate the process, did daguerreotyping become firmly established in America. His first step on arriving in New York was to arrange an exhibition of daguerreotypes taken by Daguerre himself. The display was a revelation, for Americans had no idea of the potentials of the process, and the pictures on view were vastly superior to anything which had been done in America. Gouraud gave private instruction in daguerreotype; S. F. B. Morse was one of his first pupils.

Within a matter of months, Americans began to improve the process, particularly with a view to making portraits. In New York, Alexander S. Wolcott (1804–44) and John Johnson (active 1839) patented a new kind of camera, in which a mirror was used instead of a lens, and with their invention they were able to take miniature portraits. It is hardly possible, owing to lack of documentation, to single out any one of the pioneer daguerreotypists as the first to make a portrait. There is evidence that Wolcott, Morse, John W. Draper (1811–82), and the Philadelphian Robert Cornelius (1809–1903) all took portraits before the year 1839 had come to a close. The question is academic, for portraiture was not practical. The sitters had to suffer the direct sun for minutes on end and the results were very small.

The world's first public portrait studio was opened by A. Wolcott in New York, March, 1840. By 1841 portraitists opened galleries in all leading American cities. Yankee ingenuity overcame many of the problems of preparing plates and making them more light-sensitive, and soon American daguerreotypes received critical acclaim abroad; in the London Crystal Palace Exhibition of 1851, for example, more prizes were awarded for photographic displays to Americans than to any others. Daguerreotyping was not limited to portraiture, and magnificent daguerreotypes of city views, landscapes, buildings, locomotives, ships, steamers and even news events exist.

The daguerreotype lingered longer in America than in England or on the Continent, and was not entirely displaced until the late 1860's.

The paper negative technique published by William Henry Fox Talbot in London just before the disclosure of the daguerreotype was hardly used at all in America. The only professional use of Talbot's calotype was by the Philadelphia firm of William Langenheim (1807–74) and Frederick Langenheim (1809–79), who bought the American rights for the patent. They had, however, small success in promoting the process, perhaps because Americans preferred the brilliant detail of the daguerreotype, and perhaps because photographers resented having to pay a licence fee, when the daguerreotype could be practised at no cost.

Two glass plate techniques were patented in 1850, both apparently identical with the albumen plate process of the Frenchman Niépce de Saint-Victor. The Langenheim brothers showed, at the London Crystal Palace, positives on glass which they called hyalotypes, and in Boston John Adams Whipple (1823–91) launched his crystallotypes. Although neither process was suited to portraiture, owing to extreme lack of sensitivity of the albumen emulsion, they proved useful for lantern projection and making copy negatives of daguerreotypes from which an unlimited number of paper prints could be made.

Collodion Process. To the nineteenth century American daguerreotypist, photography meant but one thing: the collodion process of Frederick Scott Archer, invented in England in 1851. One by one, and not without real regrets, daguerreotypists gave up the silvered copper plate, buff stick, coating box and mercury bath for the glass plate, collodion vial, and silver bath of the wet plate worker.

The new collodion process was first used in America to make imitations of daguerreotvpes. The ambrotype was simply a direct positive, made on a collodion plate, backed with black. The technique, unsuccessfully patented in 1854, was in use through the Civil War. The tintype, called the ferrotype or the melainotype when it was patented in 1856, was identical with the ambrotype, except that a sheet of iron, japanned black, was substituted for glass. Because the materials were inexpensive, the technique simple and the processing rapid, tintypes became extremely popular for cheap portraiture.

The first widespread use of the collodion process for making negatives was by portraitists who supplied cartes-de-visite and by travelling photographers who made stereoscopic views.

Stereoscopy. If stereoscopic photography was slow in starting in America, it quickly grew into a major part of the photographic industry. Following the publicity of the stereoscopic daguerreotypes shown at the Crystal Palace in London in 1851, a few experiments were tried in America. Southworth and Hawes in Boston built a mammoth Wheatstone viewer to take whole plate ($6\frac{1}{2} \times 8\frac{1}{2}$ ins.) daguerreotypes, and charged admission. A daguerreotype case, with viewing lenses set in a flap inside the cover, was patented in 1853. But the daguerreotype was not suited to stereoscopic work, because it was difficult to view the mirror-like images.

At the time the Civil War broke out, stereo was part of the American scene. Oliver Wendell Holmes (1809–94), physician, philosopher and writer, devised a simple skeleton-type viewer and wrote on the pleasures of collecting in the *Atlantic Monthly*. The stereoscopic camera became an important tool for showing America to Americans.

Documentary Photography. The value of photography as a means of preserving for the future the very look of a place and a time was early recognized in America. Portraits of famous Americans were collected to form National Galleries of Daguerreotype Miniatures. One of the first to form such a collection was Mathew B. Brady (1823–96), a maker of leather cases for instruments and miniatures, who became interested in daguerreotypes when he began to make cases to hold them. In 1844, he opened a daguerreotype gallery in New York, which became the fashionable place to have one's likeness taken. Soon he had a gallery in Washington. He published a folio volume of lithographs made from daguerreotypes which he had collected. It was a failure, but Brady was established as a photographic historian.

When the Civil War broke out, Brady determined to make an historical record by means of photography. Collodion plates had replaced the daguerreotype, and Brady sent out operators into the field with a travelling darkroom. The result of this undertaking was the creation of a vast record of the war. Thousands of pictures were taken, which Brady published. His cameramen were equipped with both view cameras and stereoscopic cameras, so that the record exists in three-dimensional form as well as in large single prints. The scheme was a bold one, but it was the financial ruin of Brady. After the war people did not want to be reminded of what the country had been through and did not buy prints. One part of his collection was sold for non-payment of storage and acquired by the War Department; it is now in the National Archives. Another part eventually found its way into the Library of Congress.

Some of the photographers who had created Brady's project went West with government surveys to photograph the unknown parts of the country and the building of the transcontinental railroad; from their photographs we have a vivid picture of the opening of the American West.

Eadweard Muybridge (1830–1904), an Englishman who had emigrated to California, was the first successfully to photograph a horse in full gallop. He was asked to undertake this seemingly impossible assignment in 1872 by Governor Leland Stanford, who owned extensive stables in California and was a great student of hippology as well as a racing man. Muybridge succeeded in the summer of 1878 by using a battery of cameras, a white-washed background, and a track sprinkled with powdered lime.

Although he was using the slow wet collodion process, he secured remarkably detailed photographs showing clearly the position of a horse's feet in the gallop, the trot, the walk and the pace. In 1885, using dry plates and perfected cameras, he produced nearly 800 sequence photographs of the human figure and all kinds of animals. They were published in folio plates as *Animal Locomotion.* His analytical photographs of action were at first disbelieved. In 1880, to prove their validity, he reconstructed the original action by projecting each photograph of the sequence on a screen in rapid succession. The result, though crude, was the first moving picture.

Dry Plates. American photographers shared with their overseas brothers impatience with the wet-plate process, which chained the photographer to a dark tent, in which to prepare the fugitive sensitive material. Attempts to increase the sensitive life of collodion by the addition of hygroscopic materials to keep the emulsion moist were not successful. Carey Lea of Philadelphia (the first American photographic scientist to win international recognition) devised a dry collodion technique, but the plates were very slow.

Gelatin dry plates, which eliminated all drawbacks of the collodion process, were invented by the Englishman, Richard Leach Maddox, in 1871 and soon improved by others. The first American known to have used this process was a Rochester, New York, bank clerk named George Eastman, who in 1877 imported gelatino-bromide emulsion in shredded form (known as pellicles) from England. He had taken up photography as an amateur, and quickly became dissatisfied with the collodion process. Reading in English periodicals of the new dry plate experiments, he at once began to investigate the process. Within a year he was making his own emulsion, following Charles Bennett's formula. His friends found his plates so good that he decided to go into the business of making them. He visited England in 1879 with drawings for a plate-coating machine which he patented, and on his return began to build a factory. In 1880 he was in business for himself.

Roll Film. Before many years, Eastman had devised a substitute for glass plates: rollable sensitive material to fit into a roll holder that could be put on any standard camera. He called this new material American Film. It was paper, coated with, first, a layer of soluble gelatin, and second, light-sensitive gelatin. After development and fixing, he stripped the layer of gelatin which bore the camera image from the paper support, dried it in contact with a piece of clear gelatin, and thus produced a transparent negative. It was an awkward technique so far as the processing was concerned. But it enabled him to design in 1888 a small, hand-held camera, loaded with material for one hundred pictures, which was simple to operate. After the last picture had been taken, the entire camera was returned to the Eastman factory, where the film was processed by experts, and a hundred mounted prints were returned. Eastman named this new camera of his the Kodak, a word invented by him.

The Kodak was typical of the many hand cameras which made possible a new kind of photography. No skill was required, and anybody could take pictures of a sort simply by pressing a button. Within a year, Eastman introduced a much superior sensitive material, a true film of transparent cellulose nitrate, coated with gelatin emulsion. It was this film which Edison used in his motion picture system. The first inventor of celluloid roll film was, however, the Rev. Hannibal Goodwin.

The Amateur. The dry plate and film developments led to an enormous increase in amateur photography, which had begun in America with the introduction of the collodion process. A lively Amateur Photographic Exchange Club flourished just before the outbreak of the Civil War in 1861. But the new dry processes of the 1880's so facilitated photography that more and more people were attracted to it as an avocation.

It was the amateur who established pictorial photography in America. One man was responsible for focusing the attention of American photographers on the aesthetic: Alfred Stieglitz (1864–1946). He first became interested in photography in Berlin, where he became a student of Hermann Wilhelm Vogel, the great photographic scientist who discovered spectral sensitizing of photographic materials by the use of dyes. Under Vogel's instruction, Stieglitz gained a thorough technical knowledge and put this to use in making photographs which instantly won critical praise. England was the home of artistic photography in the 1880's, and Stieglitz's photographs appeared in every exhibition. P. H. Emerson, the founder of the "naturalistic photography" movement, awarded him the first prize. Critics were so enthusiastic over his work that they regretted he was not an Englishman.

On his return to America in 1890, Stieglitz was disappointed to discover that artistic photography hardly existed in his native land. He urged the formation of an American salon, became the leader of the New York Camera Club, made its periodical one of the finest in the country and through his energy brought into being what was dubbed the New American School. In 1902 the Photo Secession—a society based on the Linked Ring of England—was formed, and Stieglitz began the publication of the famous magazine, *Camera Work*, which won for him the Progress Medal of the Royal Photographic Society.

The Modern Style. Around the time of the outbreak of World War I, photography all over the world reached a turning point. If the established styles did not imitate painting, they certainly emulated that medium. Photographers now became aware that the camera had an aesthetic of its own. Except for occasional experimental departures, Stieglitz had always used the camera direct, without distorting the image. Edward Steichen (1879–), his close associate in the Photo Secession and a leader of the New School of 1900, changed his style completely at the end of World War I, preferring the sharp, unaltered image of the lens to the soft-focus effects on which he had built his fame. Charles Sheeler (1883–) painter and photographer, began to use his camera to document American forms with a direct simplicity that had almost been forgotten. Paul Strand (1890–) exhibited, at Stieglitz's gallery in 1916, photographs in which abstract form was emphasized by choice of viewpoint and the use of extreme close-ups. Alvin Langdon Coburn (1882–) showed that the camera could produce abstractions as readily as the painter's brush.

In the late 1920's this new spirit found another champion in Edward Weston (1886–1958), a Californian with an international fame for his soft focus, atmospheric portraits and figure studies. Weston abandoned this style for the image of a well-stopped anastigmat printed by contact on glossy paper. An organization was formed in San Francisco in 1932 of photographers who shared this point of view. They chose the name "Group *f*64", held exhibitions and fostered the new approach.

A distinct contribution made by America has been the use of photography for sociological purposes. In 1888 Jacob Riis (1849–1914), the author of *How the Other Half Lives*, took a series of photographs of slum areas to supplement his verbal descriptions of the plight of the under-privileged. Lewis W. Hine (1874–1940), a trained sociologist, produced in the first decade of the twentieth century photographs of child labour which led to the adoption of legislation. The term "documentary photography" was applied to photographs taken from this standpoint. Later, in the 1930's, the Farm Security Administration of the Government, under the direction of Roy E. Stryker, used photography to document agricultural conditions. The project, as it developed, became a vast documentation of rural America and some of America's leading photographers were employed: Walker Evans (1903–), Dorothea Lange (1895–), and Arthur Rothstein (1915–).

Two technical innovations from Germany exerted a powerful effect upon American photography: the 35 mm. miniature camera (1925) and the electric photoflash bulb (1929). The frontiers of photography were pushed farther than ever before, especially when, in 1938, sensitizers were discovered which greatly increased the speed of film. News photography became increasingly important, and magazines began to develop a new use of pictures. *Life*, for example, founded in 1936, demanded photographs which were unposed and realistic, and established definite styles of picture taking.

Colour. In 1935 the Eastman Kodak Company announced Kodachrome film. The invention of Leopold Mannes and Leopold Godowsky, this film differed radically from other colour material. On a single support were three coatings of light sensitive emulsion, which recorded the red, blue, and green components of the scene. By the technique of dye coupling development, each coating was converted from metallic silver to the complementary dye (cyan, magenta, yellow). This film, which was quickly followed by Ansco Color, could be used in any camera. Sheet film monopack material was introduced and commercial photographers adopted the new material in preference to the older forms of colour photography which involved the taking of three separate negatives, or the use of material with ruled or mosaic screens.

A further development was the production of colour negative material which yields, upon processing, a negative in which the brightnesses and also the colours are inverted. From these colour negatives, prints can be made on transparent material or on paper.

Equipment Trends. Enormous strides have been taken in increasing the sensitivity of both black and white and colour films. In 1947 the fastest film available of domestic manufacture was 200 A.S.A. Today black and white film is available at A.S.A. 1250 and colour film at A.S.A. 160. Although this is partly accounted for by a change in the method of determining speed rating by the American Standards Association, the increase is still significant.

Chester F. Carlson's Electro-photography, a dry, non-chemical process, by which an image on an electrostatic field is produced and then dusted with powder, was introduced commercially for document copying as *Xerox* in 1950 by the Haloid Co. (now Xerox Corp.) of Rochester, N.Y. A similar system, *Electrofax* was announced by the Radio Corporation of America in 1956. While perhaps not a branch of photography, the tape-recording of electronically transmitted images (video tape), first demonstrated in 1951, became commercially practical with the *Ampex* system in 1954.

In 1947 Edwin H. Land announced at a meeting of the Optical Society of America his invention of a one-step photographic system, and by the end of 1948 cameras and sensitized material were put on sale. With this Polaroid Land camera, processing took place within the camera. Originally processing was complete in one minute but in 1960 the time was reduced to 10 seconds. Sensitized material in sheet form, to yield pictures 4 × 5 inches in size and to be used in a special holder fitting most 4 × 5 press or view cameras, was introduced in 1958, and a further improvement followed in 1961 when a negative-positive film packet was announced, enabling the photographer to obtain both a film negative and a positive in one operation. The speed of the Polaroid-Land film was increased to 3,000 A.S.A. in 1959. In 1963 a new colour material, based on the diffusion through triple emulsions, of dye developer molecules, was introduced. Named *Polacolor*, this system enables colour prints to be made in one stage within the camera, the whole process being complete within 50 to 90 seconds of the exposure.

Cameras equipped with aperture and/or speed setting controlled automatically for light conditions by a photo-electric cell became popular in the late 1950's. In 1963 Kodak introduced its *Instamatic* magazine-loading camera taking 12 black-and-white negatives, or 20 colour transparencies or negatives per loading in the format of 28 × 28 mm.

American Photography Today. The two largest societies are the Photographic Society of America (membership largely amateur) and the Professional Photographers' Association (membership strictly professional). Specialists have their own organizations: the National Press Photographers Association, the American Society of Magazine Photographers, the Biological Photographic Association, the Society of Motion Picture and Television Engineers, and the Society of Photographic Scientists and Engineers. Three societies are active on behalf of photographic manufacturers, dealers, and photofinishers: N.A.P.M. (National Association of Photographic Manufacturers), P.M.D.A. (Photographic Manufacturers and Distributors' Association, Inc.), and M.F.D.F.A. (Master Photo Dealers' and Finishers' Association). Every city and town has its camera club, ranging from large organizations with well equipped darkrooms and studios, down to small groups of enthusiasts who meet at each other's homes. Four magazines, with a total circulation of well over a million copies per month, are directed primarily at the amateur: *Popular Photography*, *Modern Photography*, *U.S. Camera* and *Camera 35*.

In 1961 sales in the United States of photographic equipment and supplies totalled $1·18 billion. Of an estimated 40 million still cameras in use, 28·7 million were of the simple box type. Some 52 million Americans were taking photographs, including 2·5 million advanced amateurs, and 92,000 professionals. Photographers had a choice of 440 different cameras, of which 53 were fully automatic. $355 million was spent on photo-finishing, and 624·7 million flash bulbs were shipped by manufacturers. Of the 2·3 billion photographs taken during 1961, 41 per cent were in colour.

If the bulk of consumers of photographic goods are amateurs, the most significant progress is being made by professionals, especially in the field of what has been called "photo journalism". Both magazines and newspapers have made a greater use than ever of photography since World War II, and notable picture stories have been produced for *Life* magazine and others by W. Eugene Smith, Henri Cartier-Bresson, Alfred Eisenstaedt, etc. The recent trend in news as well as feature photography has been away from the 4 × 5 in. hand camera to 35 mm. and 2¼ in. square formats, the single-lens reflex camera having become particularly popular. The many picture magazines absorb a great quantity of photographs, and demand that these pictures have vitality and style.

The Museum of Modern Art, the George Eastman House, and the Art Institute of Chicago hold frequent exhibitions. Under the direction of Edward Steichen, the Museum of Modern Art held in 1955 a mammoth photographic exhibition, *The Family of Man*, of 500 photographs, mostly of unusually large format; the exhibition was later shown throughout the world. The United States Information Agency displayed an equally large exhibition *A Nation of Nations*, organized by Nancy Newhall, designed by Herbert Bayer. *Photography in the Fine Arts*, a movement to encourage the exhibiting and collecting of photographs, sponsored by the *Saturday Review* and directed by Ivan Dmitri has circulated four exhibitions since 1959 through leading art museums.

Another important field of work is photographic illustration for advertising purposes. Studios are maintained which rival the motion picture studios of Hollywood. The natural scene is reproduced indoors, and artificially lighted—for advertisers cannot wait for the seasons and for good weather. A great deal of ingenuity goes into the making of the sets, and the success of a photo illustrator is judged more by his skill as a stage manager than as a master of the photographic medium.

In contrast to these commercial uses, the direct simplicity, aesthetic integrity and masterful technique of such creative photographers as Edward Weston, Ansel Adams, Brett Weston and Paul Strand have set standards which are recognized internationally. There are few photographers in any land whose work is consistently of equally high artistic value. B.N.

Industry. The United States possesses the world's largest photographic industry, both in terms of production value and the number of manufacturers, which in conjunction with the general expansion of the national economy has shown continued strong growth over the past five years. The scope and volume of output is sufficient to fulfil most of the needs of the domestic market, and consequently the exports and imports of photographic goods, although considerable by European standards, are of relatively minor importance.

The Census of Manufactures held in 1958 recorded that there were then in existence 158 photographic manufacturing establishments with twenty or more employees each, which together provided employment for 61,200 persons, of whom 39,300 were classed as production workers. About two-thirds of output is accounted for by producers located in New York State, whilst other important centres of production are Illinois and New Jersey.

The largest photographic manufacturer in

the United States is the Eastman Kodak Company of Rochester, New York, which produces still and cine equipment and all types of sensitized materials, and also has interests in the manufacture of chemicals, plastics, and synthetic fibres, being the parent of a chain of subsidiary companies and distributors throughout the world. In 1963, the company employed 46,000 persons in the U.S.A., and the total value of sales reached $1,106·7 million, of which amateur photo products accounted for about 25 per cent, and commercial and professional photo goods for some 32 per cent.

Among other leading manufacturers are Bell and Howell, producers of high quality still and cine cameras and other equipment, employing approximately 8,600 persons and having a sales figure of $150·0 million in 1963, and the Polaroid Corporation who invented and market cameras and film using the unique Land on-the-spot developing process, with 3,600 employees and 1963 sales of $110·0 million.

Mansfield Industries is one of the largest photographic manufacturing groups, with plants in both the U.S.A. and Japan, undertaking the production of 8 mm. cine equipment. This group acquired the Argus Cameras Division of Sylvania Electric Products, making still and cine apparatus and accessories. The combined annual sales of these exceeded $40 million in 1963, whilst the number of employees was about 1,100.

Important companies in the field of sensitized materials include the Dynacolor Corporation, which derives the major part of its income from the processing of Kodachrome film but also manufactures its own film and sensitized papers, with a sales figure of $15·0 million in 1963, and the General Aniline and Film Corporation, a Division of which is concerned with the production of film, papers, and chemicals and photographic equipment marketed under the trade name of Ansco, having an overall 1963 sales total of $190·0 million.

The industry is served by its own trade body, the National Association of Photographic Manufacturers.

Production. The value of photographic goods produced in the United States has risen by over one-fifth since 1959 to reach a total of $1,460 million in 1963. Approximately one-third of this is now accounted for by apparatus which, in contrast to the considerable growth shown by the output of sensitized materials, has followed a gradually declining trend during the period.

The largest group among apparatus consists of still picture equipment, within which the major product appears to be still cameras, whilst cine cameras, projectors and accessories are a close second in importance.

No breakdown of the totals is available beyond the 1958 Census, but the figures for that year showed the output of roll film cameras of 35 mm. or smaller format as 239,600 units valued at $9.2 million, and that of other types larger than 35 mm. as 535,900 cameras worth $34·6 million. At the same time, the production of slide and film-strip projectors totalled 654,900 units with a value of $24·1 million, whilst among cine equipment the output of 8 mm. cameras reached 1,093,700 units valued at $45·6 million, and that of 8 mm. projectors attained 684,600 units worth $36.1 million.

The total of $1,040 million reached by sensitized materials in 1963 continued the rising trend of output over recent years, the leading item being film and plates, followed by sensitized paper and cloth. Again, the only breakdown available for individual products is that of the 1958 Census, which showed the production in that year to be as follows:—

	Million sq. feet	*$ million*
X-ray film	218·3	78·4
Plates, Sheet and Graphic Arts Film	171·1	58·5
Other Still and Cine Film	514·1	257·9
	869·5	394·8

Imports. Imported photographic equipment and materials totalled $88·6 million in 1963. The greater part of this continued to be attributable to apparatus, among which still and cine cameras are the chief products.

Still cameras imported in 1963 reached 1,287,800 units with a value of $19·9 million, of which 875,100 cameras worth $10·6 miillon came from Japan and 144,200 units valued at $7·9 million from West Germany. These two countries supply nearly all of the models with a declared value of $10 or more each, and are also important sources together with Hong Kong of the other types.

Imports of cine cameras have shown rapid growth in recent years, due very largely to the activities of Japanese producers, to reach 311,100 units with a value of $10·9 million in 1963.

Of these, Japan supplied 272,000 cameras worth $7·6 million, the bulk of which were 8 mm. models, whilst Switzerland supplied 16,000 cameras with a value of $1·8 million.

Lenses form another important item among imports, with a total of 508,100 units worth $5·8 million in 1963. Japan and West Germany are the chief sources of these, the former supplying 393,700 lenses worth $2·5 million and the latter 94,100 units valued at $2·0 million during the year.

Imports of sensitized materials have risen one and a half times in value since 1959 to a total of $40·1 million in 1963. The leading product is photographic paper, the silver sensitized type accounting for $10·1 million of the $15·7 million total in the year, the principal suppliers of which are Belgium ($6·9 million's worth) and West Germany ($2·9 million's worth).

X-ray film is another important item, and

imports have registered a noticeable increase in the value during the period 1959 to 1963 to reach $11·6 million in 1963, of which $5·8 million's worth came from Belgium and $4·1 million's worth from the U.K. Cine film imports also rose substantially during the year, resulting from an increase in stock of less than one inch in width (to 154·8 million linear feet worth $1·5 million) whilst the broader gauge type expanded slightly (to 210·5 million linear feet valued at $2·3 million). The main sources of both types are Belgium and Italy, with Japan in third place.

Exports. Photographic goods exported by the United States have risen by 65 per cent in value over the last five years to a total of $148·2 million in 1963.

Two-fifths of this amount was accounted for by apparatus, among which still cameras are an important product. Exports of these reached 1,166,900 units with a value of $9·2 million during 1963, of which the hand-type fixed focus models amounted to 986,000 cameras worth $3·7 million. The largest single market for all types of still cameras is W. Germany, which received 246,600 units valued at $1·4 million in the year, whilst France took 232,900 cameras worth $1·3.

Exports of cine equipment have shown little change during the period from 1959 to 1963. In the latter year those of cameras totalled 74,800 units with a value of $4·1 million, whilst cine projectors reached 58,400 units valued at $4·9 million, the greater part of both being represented by 8 mm. models. The leading market for 8 mm. cameras is France (22,800 units worth $0·7 million), whilst the U.K. is the major purchaser of this gauge projectors (16,400 units with a value of $0·6 million).

Among other products exported an important place is held by microfilm and photocopy equipment and also processing and finishing

ESTIMATED UNITED STATES PRODUCTION OF PHOTOGRAPHIC GOODS
$ Million
(Manufacturers' Net Selling Price ex-works)

	1959	1960	1961	1962	1963*
Still cameras, projectors and equipment	n/a	160·0	159·0	146·1	n/a
Cine cameras, projectors and equipment	145·3	151·4	136·4	127·8	n/a
Lenses	16·0	15·5	14·5	7·0	n/a
Microfilm and photocopy equipment and parts	75·7	86·4	90·0	127·7	n/a
Other equipment and parts	n/a	n/a	n/a	n/a	n/a
Total apparatus	440·9	427·7	416·5	429·1	n/a
Sensitized paper and cloth	244·8	268·7	286·2	313·9	n/a
Film and plates	440·6	478·6	511·5	546·4	n/a
Chemicals	43·0	46·4	51·3	69·8	n/a
Total sensitized materials and chemicals	728·4	793·7	849·0	930·1	1,040·0
Total production	1,169·3	1,221·5	1,288·2	1,359·2	1,460·0
(£ million)	417·6	436·2	460·0	485·4	521·4

*Breakdown of 1963 figures not available

UNITED STATES IMPORTS OF PHOTOGRAPHIC GOODS
$ Million c.i.f.

	1959	1960	1961	1962	1963
Still cameras	21·9	18·0	18·8	23·2	19·9
Cine cameras	3·1	3·8	5·7	9·9	10·9
Lenses	5·3	4·9	5·1	5·2	5·8
Still and cine camera parts	3·0	4·4	4·0	3·8	3·5
Other equipment and parts	0·7	7·1	6·7	8·4	8·4
Total apparatus	34·0	38·2	40·3	50·5	48·5
Photographic paper	13·6	14·0	16·1	20·4	15·7
Roll film	1·8	1·9	2·4	2·9	3·8
X-ray film...	5·4	6·1	6·8	10·4	11·6
Other still film and plates	3·6	4·7	4·7	6·4	5·3
Cine film	2·8	2·3	2·7	3·9	3·7
Total sensitized material	27·2	29·0	32·7	44·0	40·1
Total imports	61·2	67·2	73·0	94·5	88·6
(£ million)	21·9	24·0	26·1	33·8	31·5

UNITED STATES EXPORTS OF PHOTOGRAPHIC GOODS

$ Million f.o.b.

	1959	1960	1961	1962	1963
Still cameras	3·8	5·5	5·5	5·4	9·2
Cine cameras	3·1	4·5	4·9	3·8	4·1
Still projectors	2·1	2·0	2·7	2·8	3·3
Cine projectors	4·1	4·4	4·2	3·9	4·9
Cine projection equipment	3·1	3·1	2·9	3·1	3·6
Microfilm and photocopy equipment and parts	4·5	6·1	7·7	7·6	10·5
Processing and finishing equipment ...	5·1	6·1	8·0	8·7	10·6
Other parts and accessories	9·3	9·6	11·0	10·7	12·4
Total apparatus	35·1	41·3	46·9	46·0	58·6
Photographic paper	8·3	10·4	12·9	15·8	20·8
X-ray film	10·4	10·9	10·7	9·8	9·8
Roll film	9.5	10·3	12·1	13·6	17·7
Other still film	7·5	9·0	11·0	13·4	16·1
Cine film	17·7	20·3	20·5	22·7	22·6
Chemicals	1·3	1·6	2·0	2·0	2·6
Total sensitized material and chemicals	54·7	62·5	69·2	77·3	89·6
Total exports	89·8	103·8	116·1	123·3	146·2
(£ million)	32·1	37·1	41·5	44·0	52·9

UNITED STATES EXPORT MARKETS FOR SENSITIZED MATERIALS 1963

$ Thousand f.o.b.

Country	*Sensitized Paper*	*Roll Film*	*X-Ray Film*	*Cine Film*
European Common Market:				
Belgium	450·7	455·2	302·9	410·3
France	1,179·0	483·2	62·5	111·4
W. Germany	1,391·0	1,030·4	134·9	1,243·7
Italy	909·0	690·4	7·4	495·5
Netherlands	277·2	499·6	362·6	504·1
E.C.M. total	4,206·9	3,158·8	870·3	2,765·0
European Free Trade Area:				
Austria	146·7	682·5	442·7	436·6
Denmark	334·2	253·9	163·4	190·3
Finland	68·8	32·5	6·5	85·3
Norway	237·0	282·9	56·5	16·2
Portugal	32·5	84·7	44·3	69·7
Sweden	903·5	1,223·7	111·5	575·3
Switzerland	1,105·1	1,522·4	398·3	957·5
U.K.	1,055·0	649·9	29·8	992·0
E.F.T.A. total	3,792·8	4,732·5	1,253·0	3,322·9
Australia	601·3	667·9	245·1	1,010·9
Canada	3,420·9	1,766·1	1,889·9	2,116·5
Japan	986·8	738·0	338·6	4,503·9
Spain	104·8	549·5	29·3	2,049·6
Union of S. Africa	574·0	543·9	172·1	601·2
All other countries	7,159·5	5,541·1	5,004·8	6,184·1
Total exports	20,847·0	17,697·8	9,803·1	22,554·1

equipment. The principal market for both of these in 1963 was Canada, which received $2·5 million's worth of the former and $2·0 million's worth of the latter item.

Sensitized materials have shown a substantial growth in value since 1959, to reach a total of $89·6 million in 1963. The principal product is cine film, which rose to 783·6 million linear feet with a value of $22·6 million during the year, of which 35 mm. gauge stock accounted for 287·2 million linear feet worth $11·3 million. Leading markets for all types of stock are Japan and Canada, whilst Australia, the U.K. and West Germany are also large purchasers.

Other major items which have recorded considerable growth in the period are photographic paper and roll film, reaching respective totals in 1963 of 430·0 million square feet worth $20·8 million and 24·6 million square feet valued at $17·7 million. The principal destinations of these products are shown in the accompanying table.

In contrast to these, exports of X-ray film have shown relatively little increase since 1959, totalling 32·0 million square feet with a value of $9·8 million during 1963. Important markets are Canada (6·0 million square feet worth $1·9 million) and Brazil (3·4 million square feet valued at $1·0 million). G.R.H. & A.G.B.

UNIT VALUE. Single number assigned to lens aperture or shutter speed settings of a camera in accordance with a logarithmic exposure function. Each increase of 1 in the unit value halves the light transmitted through the lens opening or the shutter. Unit values are used in certain exposure calculation systems.

See also: *Apex system.*

UNIVERSAL DEVELOPER. Name given to the many proprietary—usually M.Q.—developers intended for use with all types of black-and-white negative and positive materials.

URANIUM NITRATE. Uranyl nitrate. Used in toners and intensifiers.

Formula and molecular weight: $UO_2(NO_3)_2.6H_2O$; 502·2.

Characteristics: Greenish yellow salt.

Solubility: Highly soluble in water at room temperature.

UVACHROM COLOUR PROCESS. A three-colour dye transfer process invented by Dr. Arthur Traube in collaboration with Prof. Adolf Miethe in 1904. Named after Dr. Traube (Latin *uva* = grape = Traube (German)). Commercialized by Uvachrom A.G., Munich. From the three separation negatives of a coloured object, three diapositives were made on a silver bromide gelatine film. In a bleaching solution the metallic silver was converted into silver iodide. After rinsing, they were put in three different dye solutions: the one taken from the blue filter negative, in a yellow dye; that taken from the green filter negative in a red dye; and that from the red filter negative in a blue dye. After further rinsing the diapositives were fixed, and then superimposed to form a colour transparency. There was a great deal of control to obtain perfect colour balance by locally adding or removing the dye according to need.

V

VACUUM BACK. Holder for sheet films, especially in process cameras, or for flat originals in copying set-ups. It depends on the use of a vacuum to hold the sheet in position and flat.

The vacuum back consists of a flat plate carrying recessed grooves which are joined to a suction pump. When the pump is started it draws the air out of the grooves and so pulls the film or other sheet perfectly flat against the back.

This is also the ideal way of ensuring intimate contact between two sheets of material such as a negative film and a contact screen used in photo-mechanical reproduction.

See also: *Process camera.*

VADAS, ERNÖ, 1900–62. Hungarian photographer. Founded Magyar Fotomüvészek Szövetsége (Association of Hungarian Artistic Photographers); had leading influence on pictorial photography in Central Europe. Won Emerson Medal and was honoured by Hungarian Government.

VALENTA, EDUARD, 1857–1937. Austrian chemist, professor and later director of the Graphische Lehr und Versuchs-Anstalt (Graphic Arts Institute), Vienna. Used stereo radiography as early as 1896. Discovered the red sensitization of ethyl violet and the green-yellow-red sensitization of glycine red for silver bromide emulsions. Worked on isocyanine sensitizers (1903), silver chloride printing-out papers, toning, platinum printing, silver phosphate emulsions, tracing papers, printing materials, photochemistry and spectroscopy. Wrote numerous books.

VALENTINE, JAMES, 1815–80. Scottish landscape photographer from mid-1850s. Founder of firm in Dundee still in existence as one of the largest producers of photographic postcards in Britain.

VANISHING POINT. Point to which parallel lines converge when they recede from the observer. When the parallel lines are horizontal, the vanishing point lies on the horizon. All lines parallel to a given line converge towards the same vanishing point—e.g., the roof, eaves, window-sills, etc., of a rectangular building seen in perspective have a common vanishing point. There is a vanishing point for each of the visible sides of such a building.

Once the vanishing point of a set of parallel lines has been determined—graphically or by observation—it gives the inclination of any associated parallel lines. The conception of a vanishing point, therefore, is perhaps more important to graphic artists than to photographers.

See also: *Perspective.*

VAPOUR DISCHARGE LAMP. Type of electrical lamp which derives its illumination from an electrical discharge through a vapour or gas—e.g., mercury vapour—instead of through a wire filament.

See also: *Discharge lamp.*

VARIABLE CONTRAST PAPER. Certain bromide papers are made in a single variable contrast grade.

These papers carry two sensitized emulsion layers, one coated on top of the other. One emulsion is very contrasty and sensitive to yellow-green light. The other emulsion is soft and is sensitive only to violet and blue. (The sensitivities can equally well be reversed: soft yellow-sensitive and hard blue-sensitive.) An alternative method is to mix two emulsions, so that the sensitive layer contains silver halide grains of the different sensitivities side by side. The image is then all in one plane, but this makes no noticeable difference.

With such an arrangement, the contrast of the resultant print depends on the colour of the printing light. Exposure to blue light—e.g., through a blue filter held in front of the enlarging lens—affects mainly the soft emulsion and yields a print of low contrast. Used in this way the paper is suitable for hard negatives.

Similarly, exposure to yellow light, with a yellow filter in front of the enlarging lens, affects only the yellow-sensitive, contrasty layer and gives a print of high contrast. Used in this way the paper is suitable for printing from soft negatives.

The print can be given any intermediate degree of contrast by exposing it for part of the time through the blue, and the rest through the yellow filter. The final contrast is decided by the relative exposure times through each filter.

The filters themselves may be held in a filter slide or turret which brings the selected filter in front of the enlarger lens. Annular filters are also used, in which a yellow filter has a central clear aperture. The contrast there depends on the lens aperture employed; at the largest lens opening most of the light passes through the yellow filter, yielding high positive contrast. At small lens apertures the light through the yellow filter is largely cut out, leaving only light passing through the clear centre of the filter and leading to lower print contrast.

Some advanced enlargers incorporate variable density filters of the two colours in the lamp house. Setting an appropriate control there selects the proportion of light passing through each filter, mixing the light of the two colours thoroughly before it reaches the negative.

Where continuous variation of contrast is not essential, filter sets for variable contrast papers may include three or more intermediate filter densities between the maximum blue and maximum yellow. In that case the filter corresponding to the actual paper grade required is fitted in front of the lens.

An alternative method of obtaining a range of print contrasts from a single paper grade involves pre-exposure or flashing. This method is popular in automatic photo finishing printers where a single high-contrast paper grade is used. By varying degrees of pre-exposure to a fogging light, the effective print contrast is reduced. The actual pre-exposure must be balanced with the image-forming exposure for the print. This is frequently done semi-automatically.

See also: *Flashing; Papers.*

VARIABLE FOCUS LENS. Lens system in which the focal length can be changed continuously by mechanical displacement of one or more lens elements. Such a lens can thus replace a range of interchangeable lenses of fixed focal length.

See also: *Zoom lens.*

VARNISHING NEGATIVES. Particularly valuable or heavily retouched negatives can be given extra protection against scratches or other injury by varnishing.

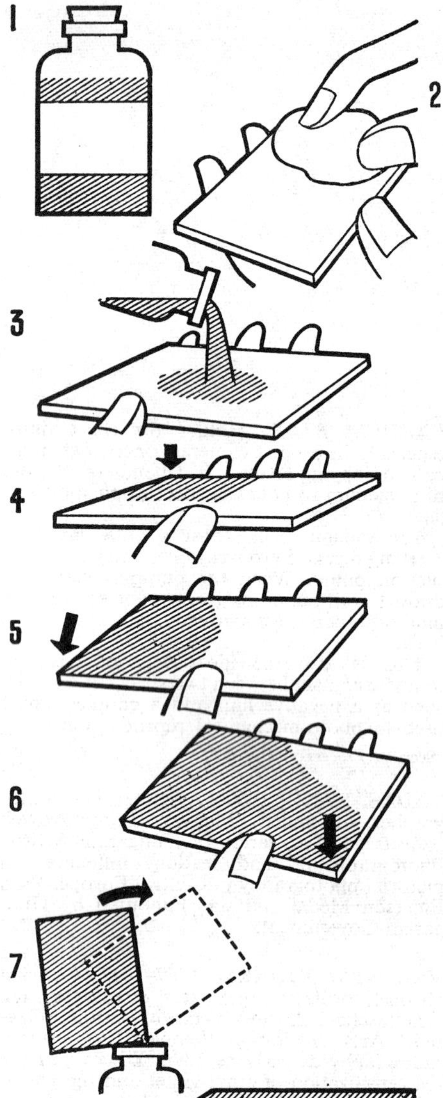

APPLYING NEGATIVE VARNISH. Varnishing protects the emulsion layer against damage, and also provides a retouching surface for pencil work. 1. Choose a suitable varnish, either hot or cold. 2. Clean and thoroughly dry the negative. 3. Pour a pool of varnish on to the centre of the plate. 4. Carefully tilt the negative so that the varnish flows towards one corner. 5. Tilt in the opposite direction to cover the second corner. 6. Cover the third corner in the same way by further tilting of the negative. 7. Drain off the surplus liquid from the last corner of the negative into the bottle. 8. Lay down the negative to let it dry and to give the varnish a chance to settle in an even layer and harden completely.

There are two types of negative varnish, hot and cold.

Hot Varnish. Hot varnishes are suitable for plates only, since they must be applied to a rigid surface, and the hot solvent may attack the film base.

They usually consist of shellac or other resin dissolved in alcohol. They are coated on to the negatives while hot. These varnishes dry very quickly, and are usually hard as soon as they are cold, by which time virtually all the solvent has disappeared. After a further drying out time of a day or two, the varnished negatives will stand a higher degree of rough treatment. A formula for a hot varnish is:

White shellac	5 parts
Sandarac...	5 parts
Gum mastic	1 part
Oil of turpentine	1 part
Methylated alcohol to make	40 parts

The methylated alcohol should preferably be the colourless (industrial) variety, or the denatured but clear type sold by some chemists as surgical spirit.

For hot varnishing the plate negative should be quite dry, and slightly warm to the touch. The plate, emulsion side up, is held level by supporting it on the finger tips of one hand, and a pool of hot varnish (heated up on a water bath) is poured on to the centre.

The plate is then tilted each way in turn, so that the varnish flows over the whole negative surface.

When the whole of the surface has been wetted with the varnish, as much as possible is drained off at one corner, and the plate set level again to dry.

Cold Varnish. Cold varnishes are usually suitable for films and plates. The finish they produce is as hard as that of a hot varnish but they may take several hours to dry. They are thus more likely to pick up dust particles than the quicker-drying varnishes. Film varnishes never contain alcohol, since it tends to soften acetate film bases.

A formula for a cold film varnish which is perfectly safe is:

Celluloid	1 part
Amyl acetate	50 parts

The celluloid can be obtained by boiling old film negatives in 5 per cent sodium carbonate solution to remove all gelatin. The films are then cut up into small pieces and dissolved in the amyl acetate. The solution is very inflammable and should therefore not be handled near a naked flame.

The negatives are immersed in the varnish, immediately drained, and hung up to dry in a dust-free place.

A 5–10 per cent solution of gum dammar in benzene or petroleum spirit (lighter petrol) can be used in the same way. A formula for a cold plate varnish is:

Canada Balsam	1 part
Bleached shellac	15 parts
Sandarac...	15 parts
Methylated alcohol, to make	100 parts

This mixture is brushed on to the plates and allowed to dry in the usual way. L.A.M.

VECTOGRAPH. Composite stereoscopic positive in which the pairs are bound up with a polarizing filter and a metal reflector. The light reflected from the pairs is differentially polarized and when seen through spectacles fitted with suitable polarizing lenses, the pictures fuse to a three dimensional impression.

See also: *Projection Practice.*

VEIL. Alternative name for fog; a uniformly distributed silver deposit or density in a negative, transparency, or print not forming part of the image itself.

See also: *Fogging.*

VERIFAX. Proprietary transfer method of document copying, utilizing a paper which produces a dyed gelatine matrix on exposure to the original and processing. The original is usually copied by the reflex method, and the processed matrix pressed into contact with a specially prepared but non-sensitized paper to which the pigment image is transferred.

See also: *Document copying.*

VERTICAL ENLARGER. Type of enlarger in which the head assembly—lamphouse, negative carrier, and lens—is attached to a vertical column so that the negative image is projected downwards on to a horizontal easel.

VESICULAR IMAGE. Image with density variations derived from the differential scattering power of microscopic bubbles in a transparent layer. Usually obtained by a diazo process.

See also: *Diazo processes.*

VIDAL, LÉON, 1834–1906. French photographer, professor of photography at the Conservatoire National des Arts et Métiers (National School of Arts and Crafts), Paris. Described in 1861 one of the first plate-changing cameras, the Autopolygraph. Worked in the 1870's on three-colour pigment pictures, chromolithography and later on three-colour additive projection. Founded the Association du Musée des Photographies Documentaires (Museum of Documentary Photographs Association), Paris.

VIDEO TAPE. Magnetic tape used to record electric impulses of a television scanning system and to reproduce these, through suitable equipment, as a television image. The term Videotape (written as one word) is a trade mark.

See also: *Television recording.*

VIEW CAMERA. American equivalent of the British field camera; in each case, usually a camera of quarter-plate size or larger, and designed to give the main camera movements. Originally intended for outdoor "views", these cameras are also used by many commercial photographers for subjects like formal groups, school photographs, architecture and general technical work.

See also: *Field camera; Technical camera.*

VIEWERS

Very small photographic positives cannot be viewed with comfort unless some apparatus is used to increase their apparent size. Optical projection gives the most pleasing result but, when the positives are only to be viewed by one person for short periods of time, something simpler is needed.

For this purpose, various types of viewers are made.

Mostly they are intended for viewing transparencies—partly because it is this form of photograph which is too small for viewing in other ways and partly because the transparency needs to be viewed by a transmitted light.

Holding the transparency up to the window or the sky or another light source is not normally satisfactory: it neither provides the even illumination necessary nor is the light quality usually the best to show the colours in their correct tones.

Viewing under the right conditions not only brings out the true beauty of a transparency, but shows up an even greater brightness range than is seen on projection. This is because some light scatter inevitably takes place on projection, so that the darker parts of the image on the screen receive some light from the brightest areas and so get slightly degraded. On viewing by transmitted light in a properly designed viewer no such scattering occurs. In a viewer with a directional light source (lamp and condenser) the brightness range may even be enhanced by the Kallier effect (the preferential scattering—and hence loss—of light in the dense areas of a transparency). This is particularly pronounced with black-and-white silver images.

Magnifying viewers for prints are rarer. On the one hand when an enlarged view of a part of a print is required, it is quite sufficient to look at it through a normal magnifying glass. A positive print, if enlarged to the right degree, can only be taken in as a whole without any viewing aid, anyway. Apart from that, magnification of a positive print brings out detail of the image structure (grain, the paper grain, etc.) rather than additional image detail. The latter is best obtained by enlarging the print to a higher degree.

OPTICAL PRINCIPLES

The apparent size of an object depends upon its distance from the eye; the closer the eye comes to the object, the bigger it looks. But there is a limit; when the eye is very close, it can no longer focus the object. By using a magnifying lens, however, objects can be examined at distances that would be too close for the unaided eye.

Ten ins. is the normal comfortable unaided viewing distance, so, by using a lens of 5 ins. focal length, we can see an object distinctly when it is only 5 ins. away. (Such a lens would have a magnification of 2 × and, in the same way, the magnification given by any lens can be found by dividing 10 by its focal length in inches. When the power of a lens is given in diopters, its magnification can be found by dividing the figure by 4, i.e., a lens with a power of 12 diopters has a magnifying power of approximately 3.)

Effect of Aberrations. The human eye can accept considerable amounts of aberration in magnifying lenses but, after a certain point, astigmatism, field curvature and distortion set a limit to magnification and field of view. The best shape for a single magnifying lens is a meniscoid with its concave surface towards the eye. Weak lenses of this form are used in spectacles. If plano-convex lenses only are available, the plane side should be towards the eye.

Double convex lenses suffer badly from astigmatism and are only usable over a narrow angular field.

Limitations. A magnification of 3 is about the limit for a single lens covering a reasonable field, and 5 where only a small area has to be examined. Higher magnifications call for a more complicated design which must be specially computed. The corrections in such lenses have to suit an imaginary stop positioned in the centre of rotation of the eye. Camera lenses are not always suitable as magnifiers because the eye is not close enough to the internal stop and the head must be moved about to see different parts of the field.

One objection to the use of monocular viewers is that both eyes are not relaxed. This causes a form of eye-strain because it is impossible to relax with one eye held shut. Viewing is much more comfortable if both eyes are open and if the amount of light falling on them is about the same. This may be a little difficult at first but the knack is easy to acquire by practice.

Binocular Viewers. Binocular viewers can only be used with comfort if they allow the eyes to converge naturally in accordance with the apparent distance of the object. If both eyes are to focus comfortably on an object which is apparently 10 ins. distant, their lines of sight must converge to a point at that distance.

A simple form of binocular magnifier consists of a low power lens (giving magnifications up to about 2) made wide enough for both eyes to look through at once. In this case the lens itself provides the correct relationship between apparent viewing distance and convergence.

Binocular magnifiers using separate lenses can incorporate deviating prisms to provide the necessary convergence, but their use, especially when they are mounted in the form of spectacles, is not recommended. The reason is partly psychological and partly due to optical aberrations and the difficulty of fusing two images with opposite forms of key-stone distortion.

Binocular viewers of a special kind are used for looking at stereoscopic pairs. In this case there is no problem of convergence because the eyes are looking almost straight at two objects, not turned to look in at one.

Magnification. The magnification required from the viewing lenses is related to the focal length of the camera lens. When the eye views the positive through a magnifier with the same focal length as the camera lens, it sees a scene exactly as the camera saw it.

If the separation between the two camera viewpoints is the same as the distance between the human eyes, and if the two positives are mounted so that each eye sees the appropriate image, they will fuse to give a stereoscopic effect corresponding exactly to the effect created by the original scene.

If the viewing lenses are of shorter focal length than the camera lenses, the stereoscopic effect will be increased. At the same time, the scene will appear larger in proportion so that the net effect is that of looking at the original scene from a closer viewing distance. The perspective will not be true, but otherwise there is no unnatural representation.

Exaggerated stereoscopic effects are obtained by varying the separation of the two camera viewpoints. They are not caused by the design of the viewer or by enlarging the positives. G.H.C. (L.A.M.)

The Transparency. The correct density for a transparency for viewing depends upon the type of viewer to be used. Some viewers call for a relatively thin positive, others may need greater density. In general it is better for an image to err on the side of weakness than for it to be very dense. The power of the light source can be reduced to deal with a thin image where a very dense positive simply cannot be seen at all.

Transparencies are usually looked at in a magnifying viewer which inevitably magnifies any defects present. So it is more important than usual to avoid, or conceal, any blemishes, however small. Pinholes and dust are particularly noticeable faults.

In the same way irrelevant detail, which may not be noticed on a small print, is apt to attract undue attention in a transparency, and has to be toned down or, better still, avoided.

A transparency has no mount to help it out; as a rule there is just a large black area round the picture. This is not pictorially suitable for all subjects, but there is no alternative. In a positive transparency made by contact or reversal, masking takes the place of trimming, so if the original includes too much of the subject, the unwanted portions are cut out by the mask, even though this means a smaller picture.

Transparencies are not considered as ready for viewing until they are properly mounted, titled and spotted (i.e., have a white spot placed on the mount to show proper orientation). The standard system of spotting is a single spot in the left-hand lower corner, when the transparency is viewed upright and the right way round. Transparencies for projection are always mounted between sheets of glass of standard thickness to avoid variations in focusing.

Conditions for Viewing. The only light allowed to reach the viewer of a transparency should be the light used for illuminating the image. All other light either distracts the eye or degrades the picture.

The power and colour of the illuminant is chosen to suit the transparency. A viewer designed for mixed types of transparency, which vary greatly in density from one to another, should incorporate a control for increasing or decreasing the brilliance of the light source.

With any method of viewing there is a limit to the number of people who can see a picture of good size and quality; the method adopted is decided by the size and distribution of the audience. In general, if the width of the picture is about a quarter of the distance from the spectator the result will be satisfactory, although much smaller images are often used with success.

Transparencies are viewed directly to the best advantage when they are mounted in the centre of a large black mask—far larger than the size usually supplied for the purpose. A diffusing screen between the light and the transparency and a few inches away from it greatly improves the character of the illumination. M.K.K.

CONSTRUCTION

Most viewers are designed for 2 × 2 ins. mounts to accommodate 35 mm. transparencies $1\frac{5}{8} \times 1\frac{5}{8}$ ins. (4 × 4 cm.) superslides, and size 828 (28 × 40 mm.) transparencies. There are also viewers for $2\frac{3}{4} \times 2\frac{3}{4}$ ins. slides. Special models are made for viewing narrow gauge film transparencies of 10 × 14 mm. and 8 × 11 mm. formats, in 2 × 2 ins. or smaller mounts, certain models can be adapted for viewing slides of smaller dimensions than those for which they are designed, and some can be adopted for viewing uncut lengths of film.

Most viewers are much cheaper than projectors, but they give only a limited magnification. There are occasions however, when a viewer is more convenient than a projector. It can be more easily carried about and can be used at a moment's notice without any of the preparation associated with the projection of transparencies on a screen.

There are many different kinds of viewers ranging from folding types that fit unobtrusively into the pocket to the larger and more elaborate viewer which is used on the desk or table. Special viewers are made for showing stereo slides and films.

Illumination. Daylight is the illuminant for inexpensive viewers, but has a number of limitations. One is the variable strength and colour. Most colour films for miniature transparencies are designed to yield the best colour compromise when the transparency is viewed or projected by a light source of a colour temperature of around 3,200°K. On viewing by daylight such transparencies therefore tend to appear too bluish. This is particularly pronounced if the picture is held up against the blue sky. Colour films intended for larger transparencies—especially sheet films—are, however, usually balanced for viewing by a daylight-quality illuminant. This is because they are mostly intended for display in light boxes lit by fluorescent or similar tubes.

Most slide viewers therefore have some sort of artificial light source built in. A miniature bulb fed from a torch battery, as used in cheap viewers, also suffers from variable colour quality. As the battery runs down, the light becomes progressively weaker and more reddish. In any case, the light of a torch bulb is considerably more reddish (lower in colour temperature) than the ideal illumination.

With mains-operated viewers the light source is at least constant in quality, though even there often too red. The use of high-power lamps for a whiter light is rarely practical, because then the intensity becomes too great for comfortable viewing. One possible correction is to interpose a very pale bluish filter (best found by experiment) in the light path of the viewer. Behind the diffusing screen is usually the best place.

Even illumination is also important. The only way of achieving this in a daylight viewer is by a diffusing screen placed behind the transparency. Flashed opal glass or opal plastic are best since they have virtually no grain structure of their own. Where no magnifying viewer at all is used, the transparency should be placed against a suitable diffusing material before being held up to the light. Some manufacturers of colour films return their transparencies after processing in plastic sleeves with one side clear and the other translucent to provide a certain degree of diffusion.

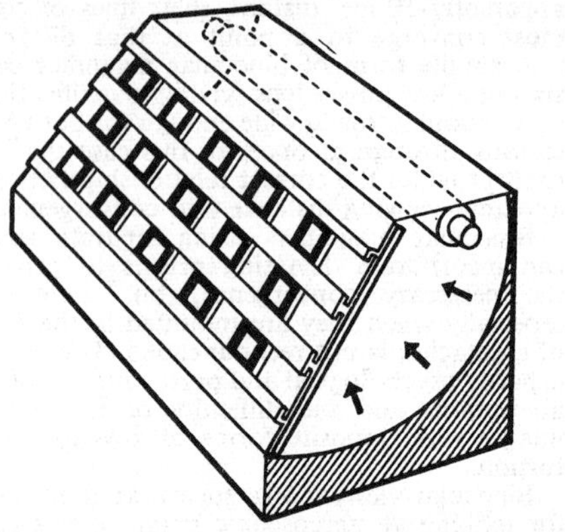

SLIDE DISPLAY CABINET. The slides are supported in grooves in front of a sheet of glass. A tubular filament or fluorescent lamp illuminates a white reflector facing the front panel. Careful choice of illuminant is desirable for colour slides.

Diffusing screens are also used in the less expensive viewers with incorporated light source. The screen should in this case be not too near the lamp (otherwise the illumination is again uneven) nor too near the transparency if the screen has a grain texture.

Condenser illumination in small viewers is comparatively rare. However it provides a much more brilliant light and better utilization of the output of the lamp. Condenser illumination is, however, also liable to lead to unevenness in lighting, unless the condenser system is of high optical quality and correctly adjusted.

Hand Viewers. These consist of a plastic or metal housing with a magnifying lens at one end and a slot to take the transparency behind it. A diffusion screen is fitted at the opposite end to the magnifier and the transparency is viewed

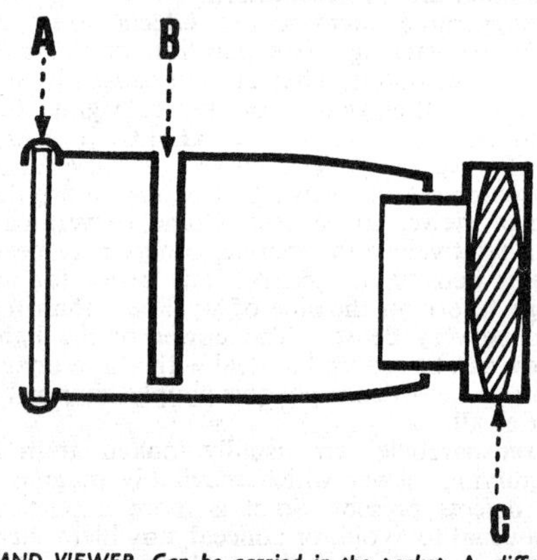

HAND VIEWER. Can be carried in the pocket. A, diffusing screen. B, slot to take slide. C, magnifier eyepiece. In use the viewer is held up against a lamp or window; the diffusing screen then provides an evenly bright background for the slide.

by placing the lens to one eye and looking through the viewer at a suitable light source. The lens is usually a simple meniscus, and may be either in a fixed position or in a sliding tube to enable it to be focused on the transparency. The diffusion screen may be of etched or flashed opal glass or translucent plastic. The grain in the diffuser is rendered less noticeable by placing it some distance away from the transparency.

In folding viewers, the slot for the transparency and the diffuser are placed close together for compactness. They are made to fold down on to one end of a rectangular platform and the magnifiying lens folds down on the other. Sometimes the centre platform incorporates a slide which can be extended as required to enable the transparency to be brought into focus. When folded the viewer, which is often sold complete with a leather purse, can easily be slipped into a pocket. The transparency is not boxed in with this type of viewer, and the image may appear less brilliant than it would if it were entirely enclosed and all the light came directly through it.

It is obviously an advantage for the viewer to provide its own illumination, and some hand viewers are made to do this while retaining the attraction of portability.

The main part of the viewer is a battery case, holding two, or sometimes three, torch batteries, which light one or two torch bulbs. The bulbs are covered with a diffuser just large enough to cover the transparency. Viewers with a magnifying system of one lens will yield only limited magnification without aberration and distortion. To increase the image two or three lens units are employed. Certain models have a switch to turn on the light, but it is more common to make the necessary electrical contact simply by means of depressing the slide into the viewer. In some cases the slide can be locked into position, with the light remaining on, by pressing the slide both downwards and also backwards into a small recess.

Table Viewers. These are bigger and more rigid than the illuminated hand viewers and intended to rest on a table or desk when in use. The component parts are the same as in the illuminated hand viewers but are not designed to fold into a small space. In some cases the base is adjustable to allow the viewer to be tilted to a convenient angle for viewing. Their larger size permits the use of bigger lamps, which may be operated from the electric mains. Certain battery models are fitted with small transformers, which enable them also to be used off the mains. Some viewers are designed to hold a magazine, which does away with the need to load each individual slide separately for viewing.

The viewers mentioned so far give only a relatively low magnification, and can be used by one or two persons only at a time. A more elaborate kind of table viewer shows a bigger

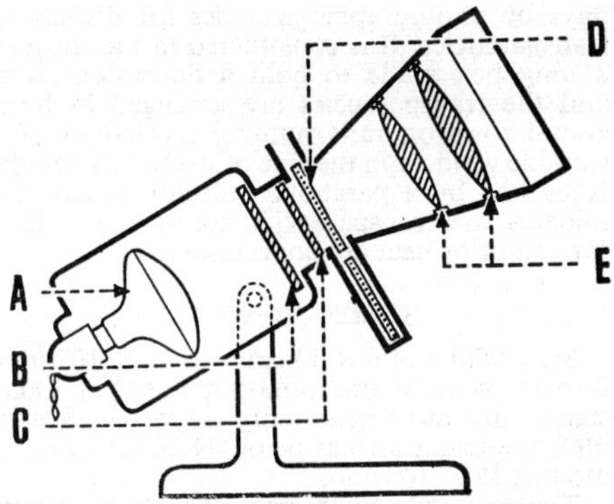

TABLE VIEWER. Gives maximum enlargement of small transparencies, incorporates own lamp. A, reflector lamp. B, diffusing screen. C, heat filter. D, slide in chute. E, magnifying system deeply hooded from stray light.

picture which can be seen by several people at once. This viewer uses a low power projector lamp and condenser system which project an image of the transparency via a system of surface-silvered mirrors on to a translucent screen measuring up to 7 × 7 ins. Since this type of viewer is really a self-contained projector and screen, it is much more expensive than the ordinary directly-viewed magnified image type.

Viewers for Exhibition Stands. For exhibition and advertising, sets of transparencies are often mounted in cabinets and illuminated from behind. The cabinets are placed in as dark a situation as possible, and the illuminated area is surrounded by the largest possible area of black. Display cabinets of this sort incorporate diffusing screens to spread the illumination evenly over the transparency, and filters both for correcting the colour values of the transparency and for filtering out the heat rays.

A better type of cabinet is being used nowa-

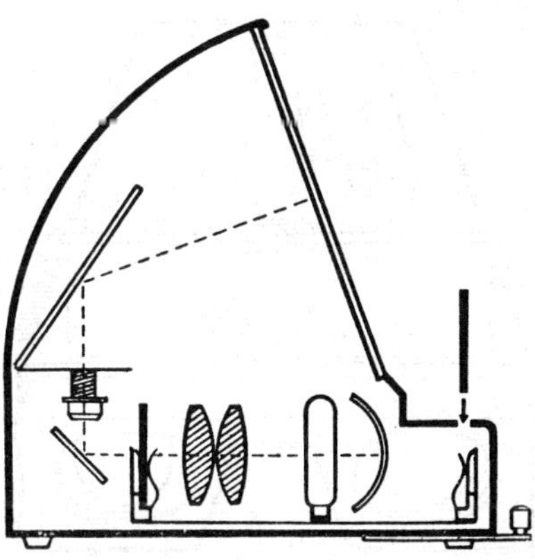

TABLE PROJECTOR. Although shaped like a large viewer, the instrument incorporates a projection system. This throws an enlarged image on to a screen by back projection.

days by photographic societies for displaying transparencies. It is constructed in the form of a long box made to hold a fluorescent tube and the transparencies are arranged in horizontal rows on rails running the length of a suitable window in the side of the box. The tube is backed by a parabolic reflector which illuminates the transparencies evenly, and filters for colour or heat are unnecessary.

STEREO VIEWERS

Stereoslides of 6 × 13 cm. and 45 × 107 mm. formats were at one time popular, but today stereos are most commonly taken on 35 mm. film, the usual format being 24 × 23 mm., in mounts 45 × 100 mm.

The essential requirement of a stereo viewer is to present the two transparencies recorded by the camera so that the left eye sees only the image taken by the left camera lens, and the right eye the image taken by the right camera lens. This is achieved by using two viewing lenses placed approximately 2½ ins. apart, and by mounting the transparencies with a similar separation distance between them. As with monoscopic viewers, stereo viewers may be of the illuminated kind with a lamp or of the more simple type to be used with an external light or daylight.

The positives are mounted in positions where they give the correct convergence and where they can be fused with ease. This condition is met when lines joining corresponding points in the two prints are parallel to one another and to the lines joining the centres of the viewing lenses, and when the positives are spaced so that beams from these points diverge very slightly towards the magnifying lenses and the eyes.

It is impossible to fuse the positives if they are spaced wider than this and discomfort is soon felt when they are closer.

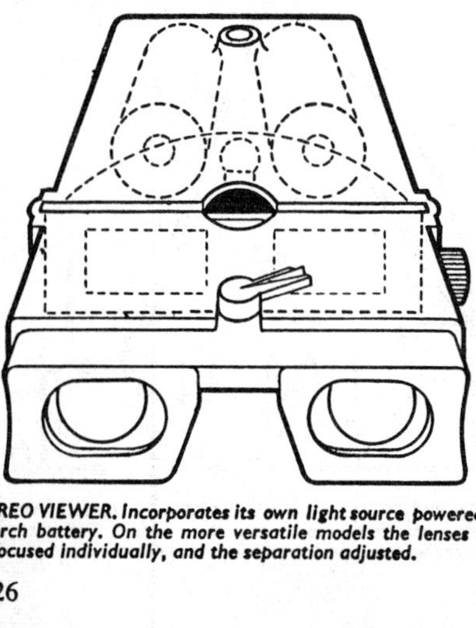

STEREO VIEWER. Incorporates its own light source powered by a torch battery. On the more versatile models the lenses can be focused individually, and the separation adjusted.

If the pictures themselves are narrower than the correct spacing, no difficulty arises, but if they are so wide that they would overlap, some prism or mirror system must be employed so that they can be mounted clear of each other, while at the same time appearing to have the normal separation. A number of stereoscopes have been designed for viewing large prints in this way, using a suitable beam splitter.

Non-illuminated stereo viewers bear a close resemblance to the hand viewers used for viewing flat or two dimensional transparencies, apart from the fact that there are two lenses instead of one and the viewer is correspondingly wider.

Inexpensive viewers in metal or plastic are obtainable with or without focusing adjustment and they may be of the enclosed box type or open folding type.

Illuminated stereo viewers are usually made of plastic and the illumination provided by two torch batteries operating a single bulb. The better models are fitted with achromatic lenses, and in addition to a focusing adjustment, there may also be some means of varying the interocular distance of the lenses.

Split Frame Viewer. Special viewers exist for slides taken with beam-splitting devices used with 35 mm. cameras. No additional cutting, transposing or mounting of the transparencies is required. The transparency is inserted in the viewer in its cardboard mount, just as it is returned from processing. Two pairs of surface silvered mirrors or prims present each image to the appropriate lens with the correct separation.

The simple viewers of this type have a plastic or pressed metal case, have no built-in illumination, the lenses are of fixed focus and the interocular distance cannot be varied.

Prismatic Film Viewers. Another better viewer for this type of beam-splitter stereo film employs a system of prisms in place of the mirrors. The prisms enable the film to be viewed without transposing. Each eyepiece focuses independently, and the interocular distance can be varied. This type of viewer is held in the hand and pointed at the light. Both split frame attachments for use on 35 mm. cameras and the necessary viewers have now gone off the market.

A similar type of viewer, which is no longer in production, makes it possible for films taken in twin-lens 35 mm. stereo cameras to be viewed without cutting. The viewer uses the same system of reversing prisms which makes it unnecessary to transpose frames. The film is transported by sprockets, the transport being similar to that of the camera so that the frames come into view by turning the knob as far as it will go. The lenses can be focused and the distance between them varied. This viewer can be fitted to a stand and can either be used in conjunction with external light or it can be fitted with a special lighting attachment working from the mains. C.W.L.

Stereo Print Viewers. Since stereo pictures in the

form of paper prints still require special viewing to see the depth effect, various stereo viewers for such prints also used to be produced. In their simplest form, such viewers consist of a holder for the stereo pair of prints, and a pair of eye-pieces (or simple magnifying lenses) set an inch or two in front of the stereo prints. Frequently such viewers are designed to fold for easy carrying around in the pocket.

Usually print viewers are intended to be used by daylight or other prevailing illumination, and only require holding in such a way that both stereo prints are equally strongly illuminated. More elaborate types may include their own light source, adjustable eye-pieces, etc. Stereo prints usually come in standard sizes mounted on 6 × 13 cm. or 45 × 107 mm. mounts for direct insertion in the viewer. Stereo positives in 35 mm. size are rarer.

Stereosopic Aerial Photographs. Photographs taken for aerial survey and reconnaissance work are frequently exposed at intervals—as the aircraft flies along—which make each picture overlap its neighbour by about 60 per cent. Such adjacent frames are usually viewed stereoscopically in print form with the aid of special viewers. This permits much more detailed examination and recognition of ground features in the photograph. Prints—or, more usually, transparencies—may also be viewed with the aid of stereographic plotting instruments for accurate evaluation and survey, for example for map making. In these instruments different points of the image, apparently seen in depth, are matched up with floating marks. These indicate by their movement not only the lateral disposition of picture details but also their apparent height above the ground. Such stereoscopic surveying requires elaborate viewing equipment which must not only be highly precise but may also be linked with stereo-autographic plotting and mapping gear.

SPECIAL VIEWERS

Negative-positive Viewers. Sometimes it is useful to be able to see a negative image as a positive (or occasionally vice-versa). Several viewers exist for this purpose, and depend on one of two principles: (1) the quenching effect of infra-red radiation on the emission of fluorescent light by fluorescent screens excited by ultra violet rays; or (2) electronic image reversal in a television playback system.

The quenching type of viewer is the simpler of the two. It consists of a projection set-up, with a lamp, condenser and lens. The lamp is specially rich in infra-red radiation, and is usually filtered to eliminate most or all visible light. The projector projects the negative on to a fluorescent screen excited by ultra-violet rays.

Where the infra-red radiation reaches the screen, it quenches or subdues the fluorescence in proportion to the intensity of the infra-red rays falling on it. Thus where little light comes through a negative (the dense portions) the image on the fluorescent screen is brightest. Where a lot of light comes through the thin areas (shadow portions) the fluorescence is subdued more and the image appears darker. The result is a tonally reversed picture, in other words a positive from a negative.

The apparatus for this purpose is comparatively bulky and complex, since it must include both the projection system and infra-red source as well as an ultra-violet emitting lamp. The latter is also filtered to allow only ultra-violet rays to reach the fluorescent screen.

Usually such viewers yield a slightly enlarged image of the negative; they can also be designed for adjustable magnification. Their main use is in the quick evaluation of negatives in a number of industrial fields where a positive is not required at the time or where making a positive would be too lengthy. They are also of potential use to photo dealers, to show customers what their negatives look like as prints.

When an image is reproduced by a closed-circuit television chain (camera and receiver) it is a comparatively simple matter to adjust the circuit so that the reproduced image is negative with respect to the original. This is one of the aspects of the contrast control; hence a negative cannot only be shown as a positive, but also as one of any selected contrast.

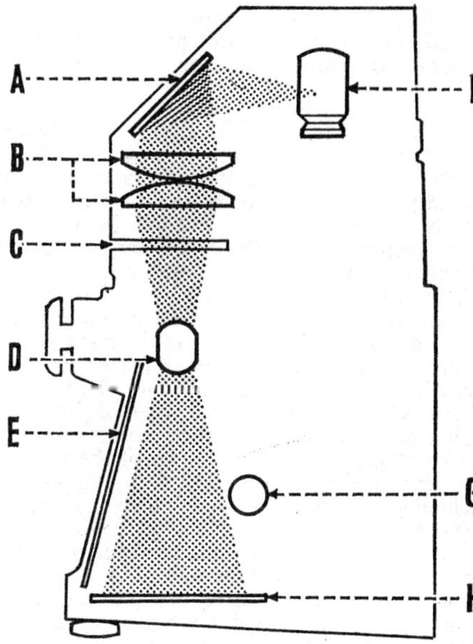

NEGATIVE-POSITIVE VIEWER SYSTEM. An infra-red beam is projected through the negative on to a flourescent plate excited by an ultra-violet lamp. The infra-red radiation coming through the negative partially extinguishes the fluorescence, so showing a positive on the plate. A, mirror. B, condensers. C, slot for negative. D, projection lens. E, green screen for viewing the image on the fluorescent plate. F, infra-red projection lamp. G, ultra-violet light source. H, fluorescent plate.

Television systems for this purpose are far more expensive than their utility warrants, but they have some potential use for determining the paper grade required for printing a given negative.

Of more interest is the application of this idea to colour printing. With a three-colour television system it is not only possible to reverse a negative to a positive, but also to transform the negative colour effects to positive ones. Television systems for a negative-positive colour viewing have therefore been used for colour grading—i.e. determining the type of filtering necessary to produce a correct positive colour print from a negative.

Basically this involves setting up specific control positions corresponding to the filtering required for a perfect colour print (determined by independent tests) and adjusting the controls until the positive colour image on the television screen corresponds to that obtained in the print. For other colour negatives the required printing correction can then be obtained from the necessary readjustment of the controls on the television receiver to achieve a good colour reproduction in positive from the new negative. Systems of this kind are in use particularly for the colour printing of motion picture positives and provide a speedier method of negative grading than orthodox test procedures.

Microfilm Viewers. Usually microfilms are viewed and read with the aid of microfilm readers—table-top projectors which project the microfilm image on to either an easel or a back projection screen at a magnification of 10 to 15 times. Most of these readers will take 16 mm. or 35 mm. microfilm in continuous rolls, or will project individual unitized microfilms (film cards, aperture cards, etc.). Some microfilm readers also project positive prints (microcard images) since that is the only way of viewing these positives which require magnifications up to 25 times to restore them to original size.

Small microfilm hand viewers also exist, mainly for quick checking of microfilm frames rather than detailed reading. They are generally compact instruments, often with their own light source (battery powered) and yield a 5 to 10 times magnification of the microfilm frame.

Cine Viewers. 8 and 16 mm. cine film is almost invariably projected for observation, since the projector must also incorporate the means of moving the film at the required rate through the optical system. A few hand viewers also exist, however, and consist of a simple hand-driven transport system, a viewing magnifier and either a built-in battery-powered lamp or a diffusing screen for holding the viewer up to a suitable light source. For viewing, the film is cranked through the viewer while the eye looks through the eyepiece at the rear. The image is still very small, and mostly such hand viewers are to be regarded as toys. L.A.M.

See also: *Back projection; Micro-copying; Projection practice; Projector (still); Stereoscope.*

Book: *Colour Transparencies,* by C. L. Thomson (London).

VIEWFINDER

When taking a picture it is necessary to know not only in what direction to aim the camera but also how much of the subject will be included in the picture. For this reason practically every camera has a viewfinder of some kind.

The most straightforward way of checking the image recorded on the film or plate is to observe it on a screen placed in the back of the camera. The screen is then replaced by the sensitive material for the exposure. This is the oldest viewing system and is still used in technical and certain other cameras. The screen shows exactly how much of the subject the lens projects on to the film, and—if a ground glass screen is used—it also indicates the sharpness of the image.

The use of a rear screen is however practical only if the camera is mounted on a tripod, to photograph comparatively static subjects. When the camera is held in the hand or used to photograph moving objects, a back focusing screen is out of the question.

There are two ways of getting over this problem. One is to have a focusing screen not at the back where the film is positioned, but on top. A movable mirror inside the camera then directs the image on to the top screen for viewing and focusing, and moves out of the way for the exposure so that the lens can project the picture on the sensitive material. This arrangement is a single-lens reflex camera and permits viewing and focusing right up to the instant of exposure.

The other way is to build a separate sighting system on to or into the camera, and align it sufficiently accurately to indicate the field of view taken in by the camera lens (or lenses if interchangeable).

A hybrid solution is the twin lens reflex where the viewing device is almost a camera in its own right, with a mirror and a top focusing screen. Only in this case the mirror is not movable but the so-called viewing camera is attached to the camera that takes the picture.

INDEPENDENT FINDERS

Viewfinders which are built on to or into the camera may be simple framing devices, rather like the sights of a rifle or they may employ optical systems derived more or less indirectly from a Galilean telescope.

Frame Finder. The frame finder, which is one of the simplest and best of all viewfinders,

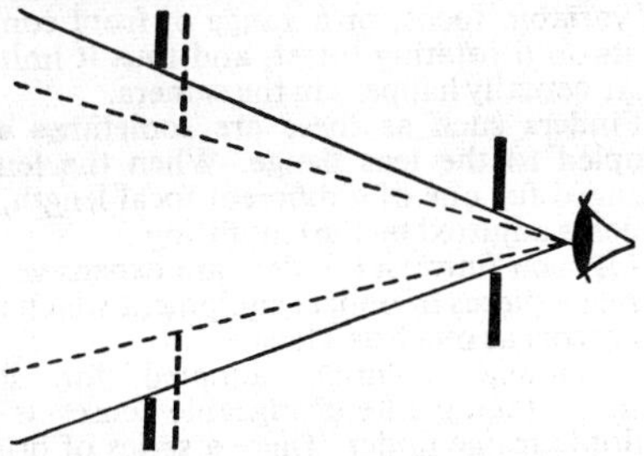

FRAME FINDER. *The eye looks through a rear sight centred behind a front frame which indicates the limits of the field of view. Alternative or interchangeable front frames may be provided for lenses of various focal lengths. The surroundings of the subject are visible outside the frame.*

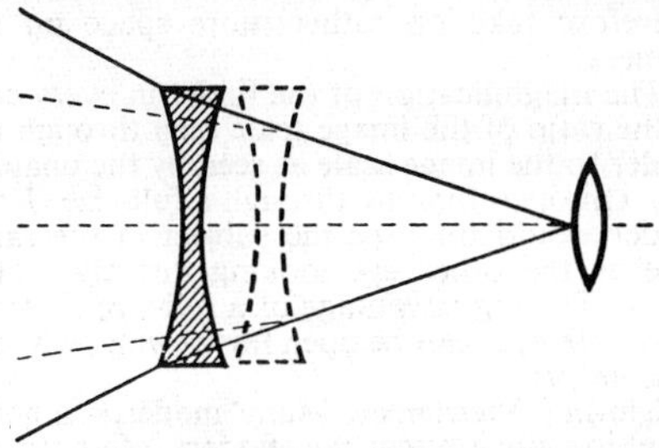

OPTICAL FINDER. *Uses the principle of an inverted Galilean telescope to present a reduced image of the subject. The separation between the lenses controls the angle of view (and hence the size of the subject field taken in) and scale of the finder image as visible to the eye.*

consists of a metal—usually stiff wire—frame corresponding to the shape and size of the negative, and a small peep-sight. The frame is mounted above the lens and in the same plane, and the peep-sight is mounted on the camera body in the same plane as the sensitive material. The axis of the frame and peep-sight is parallel to the optical axis of the lens.

When the eye is brought close to the aperture of the peepsight, the view bounded by the wire frame is almost exactly the same as the view seen by the lens. The only error is that of parallax, which is more or less negligible for subjects beyond about six feet. Apart from this the frames of the two pictures are identical. More than this; they are the same for any lens of normal construction—wide angle or long focus—and all camera movements provided that the front section of the finder always moves with the lens as one unit.

The frame finder unfortunately makes an awkward projection that in effect almost doubles the bulk of the camera but it can also be designed to be collapsible.

It is possible to make a small frame finder by mounting a proportionately reduced frame closer to the peep-sight. But as the frame comes closer to the peep-sight it becomes more and more blurred until it gives only a hazy idea of the picture boundaries.

Direct Optical Viewfinders. Optical viewfinders, consisting of a negative lens with a small positive lens mounted behind it, are probably more used than any other type. They are also sometimes known as Newton finders. The limits of the field are masked on the negative lens, and the scene is looked at through it with the camera at eye level.

In older models the front and back components of this type of finder are simply mounted in frames which fold flat against the side of the camera; they are mounted in a fixed tube, or built into the camera body.

This type of finder is reasonably accurate, and it gives a bright picture that is neither inverted nor reversed. When it is mounted in a separate tubular mount, it may even carry some form of automatic or adjustable tilting device to compensate for parallax.

Albada Viewfinder. The Albada viewfinder (named after its inventor) is a special type of frame finder in which a white line, framing the negative area, is seen apparently suspended in the plane of the subject. This image is the reflection of a miniature frame painted on the inside of the rear plain glass window of the finder. The reflection is seen in a half-silvered mirror of curved glass which forms the front window of the finder.

This finder has a number of advantages: the image is undistorted; the frame is sharply focused by the eye; part of the subject is visible outside the frame—a great help when following moving subjects.

Magnification. The inverted Galilean telescope type of optical system generally shows the view on a greatly reduced scale. The optical components of such a finder can however be very small and so take up little space in the camera. By increasing the separation between the front and the rear lenses, suitably matching the power of the two lenses used and making the front lens large enough, it is possible to increase the image scale so that the finder shows the subject in nearly natural size. The current preference is for such full-size finders, which

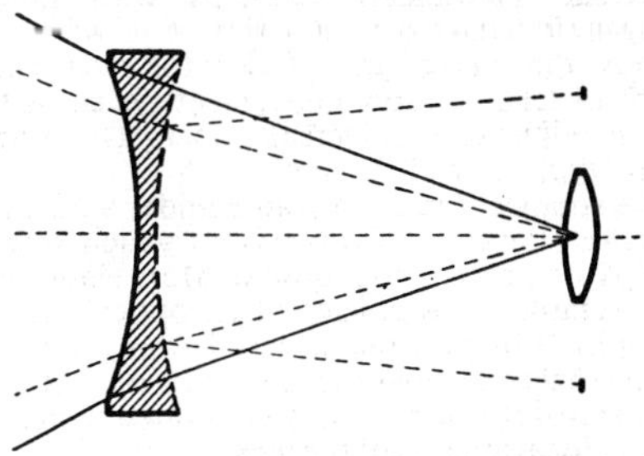

ALBADA FINDER. *This is an optical finder, with a partly reflecting rear surface to the front lens. This surface reflects a white frame which surrounds the inside of the rear lens. The frame thus appears suspended in mid-air visible at the same time as the subject, outlining the field area.*

therefore take up rather more space on the camera.

The magnification of the finder in every case is the ratio of the image scale seen through the finder to the image scale as seen by the unaided eye. One eyes looking through a full-size (1 : 1) finder would thus see the subject in the same size as the other eye looking directly at the scene. The big advantage of a full-size finder is that both eyes can be open for viewing, without discomfort.

Bright-line Viewfinder. Many modern cameras combine the optical viewfinder—often in an enlarged form to give nearly full visual image size—with the Albada principle to produce the bright-line finder. This again shows the subject area outlined by a brilliant reflected frame. It is also known as a brilliant-frame, reflected-frame, or crystal-frame viewfinder.

Modern brilliant frame finders frequently achieve the same effect by reflecting a white frame from a second finder window into the optical system. This second window is covered with a diffusing screen, and the brilliant frame is cut into a masking plate situated behind the screen. A prism or mirror behind this reflects the frame into a semi-reflecting mirror in the optical path of the finder, and from there into the eye. This arrangement permits a brighter frame to be combined with the finder image, and is also more suitable for full-size brilliant frame finders. In addition it provides a way of incorporating automatic parallax compensation, by linking the masking plate movably with the focusing adjustment of the lens (see below).

Another type of brilliant-frame finder uses only a diffusing window with a masking plate behind it. Here one eye looks through the system to see the bright frame, while the other eye observes the subject directly. By a process of visual fusion the frame then appears superimposed on the scene. Using this kind of finder needs a little practice, and it does not work with people whose binocular co-ordination is in any way restricted (such people would not be able to see a stereo pair of images as a three-dimensional picture).

Universal Viewfinder. Cameras with interchangeable lenses need a series of viewfinders to show the correct field of view for each focal length of lens. A more convenient system is to use an adjustable universal finder. There are several types of such finders.

The simplest kind is a frame finder with two or more alternative front frames which swing or clip into position as required. More elaborate and versatile types consist of an optical finder with a built-in masking device. Turning a knob or ring makes the mask frame larger or smaller. The control ring is generally calibrated with the various focal lengths of the lens.

In another type of universal finder, the frame remains the same size and the image increases or decreases according to the focal length of the lens. This type of finder incorporates a lens of variable focus, or a range of front components on a rotating turret, and thus it imitates what actually happens in the camera.

Finders such as these are sometimes even coupled to the lens flange. When the lens is changed for one of a different focal length, the finder is adjusted by the lens fitting.

All such universal finders are expensive and intricate pieces of optical equipment which may easily cost as much as a lens.

A simpler solution adopted for some cameras taking interchangeable lenses is the multiple frame finder. There a series of bright-line frames are marked in the standard camera finder, showing also the field of view with two or more alternative lenses. The frames may be reflected from a separate glass plate; by incorporating a movable mask it is possible to arrange for only one frame to be visible at a time. The frame changing control is then sometimes linked with a pin or other fitting on the lens for automatic changing over when fitting a different lens in the camera.

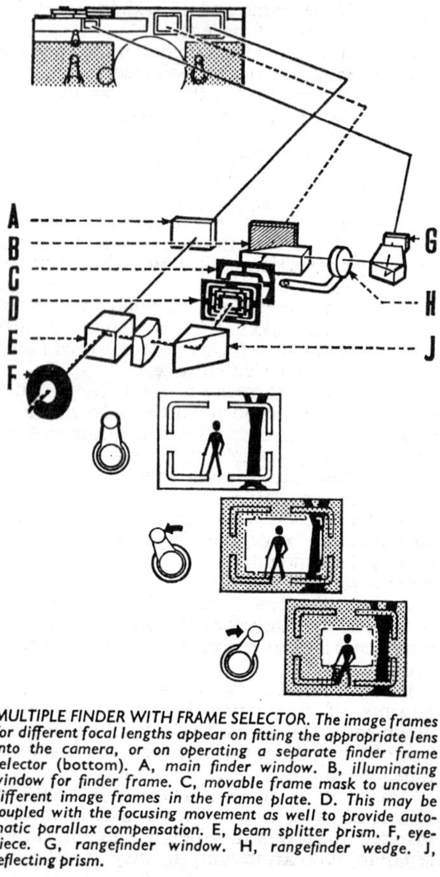

MULTIPLE FINDER WITH FRAME SELECTOR. The image frames for different focal lengths appear on fitting the appropriate lens into the camera, or on operating a separate finder frame selector (bottom). A, main finder window. B, illuminating window for finder frame. C, movable frame mask to uncover different image frames in the frame plate. D. This may be coupled with the focusing movement as well to provide automatic parallax compensation. E, beam splitter prism. F, eyepiece. G, rangefinder window. H, rangefinder wedge. J, reflecting prism.

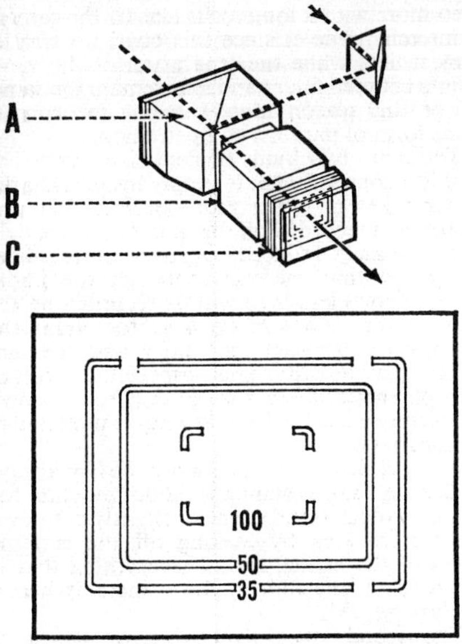

STRAIGHTFORWARD MULTIPLE FRAME OPTICAL FINDER. *Several image frames are visible in the finder field, indicating the view taken in by different focal lengths—for example, 35, 50 and 100 mm. A, beam splitter prism (to combine the viewfinder with the rangefinder beam). B, field lens. C, graticule plates with finder frames.*

Reflecting Viewfinder. The elementary reflecting type of finder has nothing but its cheapness to explain a long run of popularity in low and medium-priced cameras. It consists of a short-focus positive lens which forms the image, and a 45° mirror which reflects the image on to a horizontal ground glass or a second positive lens.

The image seen in the top of the finder was never very bright, it was reversed, and its boundaries were uncertain. Finally, because the finder had to be held below eye-level, it imposed an unnaturally low viewpoint.

In spite of all these shortcomings, however, makers continued to fit such finders to their cameras until the early 1940's. Since that date, the small reflecting finder has practically disappeared in favour of the optical direct vision finder, and the large brilliant finder.

Brilliant Viewfinder. The brilliant type of finder is simply an enlarged and improved reflecting finder. It has a positive image-forming lens, a 45° mirror or prism and a large, brilliant viewing lens which takes up practically the whole of the top of the camera. The size of the picture shown in the viewing lens is often as big as the actual negative frame.

Because this type of finder is always built into the camera body and is too large to leave room for a second similar finder, the camera can only be used in one position. This rules out the rectangular formats which demand two positions—"portrait" and "landscape"—so the large brilliant viewfinder is generally fitted only to cameras that take a square picture.

Special reflecting finders are generally included in the accessories available for expensive miniature cameras. One of the main reasons for wanting to use such a finder is that the camera attracts less attention when it is held at waist level than at eye-level. It can also be used with the camera pointing at subjects to the right or left of the photographer. So it is often a help for taking candid pictures. It can also be used for taking photographs over the heads of the people in a crowd if the camera is held upside down.

Accessory Finders. In addition to the finder built into the camera, additional viewfinders are frequently available for precision 35 mm. cameras. These usually cover the fields of view taken in by the various interchangeable lenses, and follow one of the designs described above.

Accessory finders are also produced for certain special purposes: right-angle optical finders for viewing in one direction while the camera points in another; close-up finders which are specially corrected for the field of view at specific close ranges and may incorporate a rangefinder; and various finder attachments to adapt the field of view to lenses of alternative focal length or to specially close subject distances.

Mostly such supplementary finders fit into the accessory shoe on top of the camera.

Finder accuracy. The value of a viewfinder depends largely on how accurate it is. It can be subject to three main errors: directional, framing and field size.

If the finder does not point in the right direction, it will not take in the view that the lens projects on the sensitive material. Ideally the finder axis should be parallel with the lens axis, though for near subjects it may swing in towards the axis. Directional errors are mostly found on inexpensive cameras, and may arise if for instance a reflecting viewfinder is attached to the camera by a thin and easily bendable bracket.

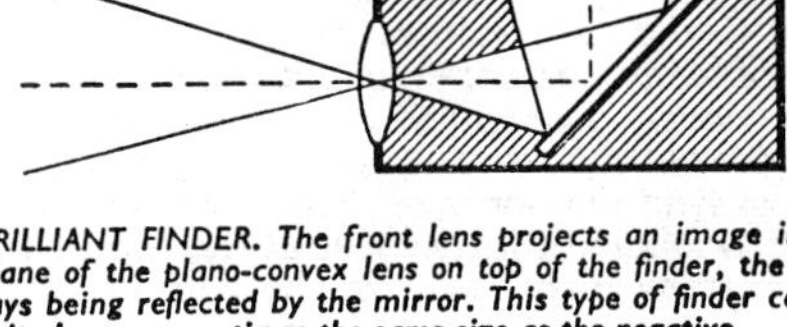

BRILLIANT FINDER. *The front lens projects an image in the plane of the plano-convex lens on top of the finder, the light rays being reflected by the mirror. This type of finder can be quite large—sometimes the same size as the negative.*

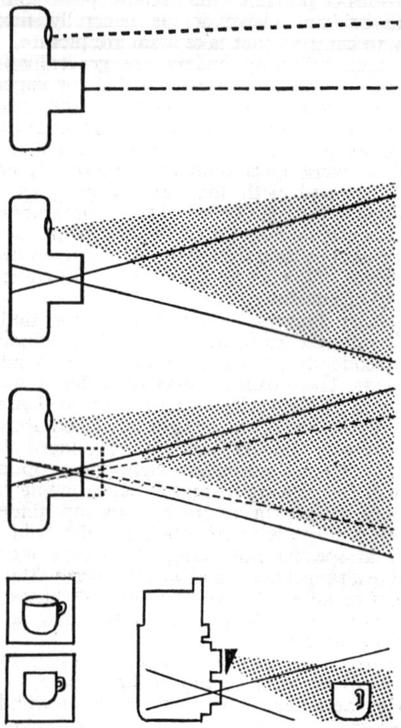

VIEWFINDERS ERRORS. Top: *Sighting error. The viewfinder does not point in the same direction as the camera lens.* Upper centre: *Error of covering angle. The finder field takes in more than the lens, so parts of the picture visible in the finder may not appear on the film.* Lower centre: *Close-up loss of field angle. The increased lens extension at close distances reduces the angle of view of the lens so that the film then takes in less.* Bottom: *Genuine parallax error. Even with full correction for the different fields taken in by the viewfinder and the camera lens, there remains the perspective difference between the view of the finder and the camera lens. This becomes particularly noticeable at close distances.*

When the viewfinder, even though pointing in the same direction as the lens, takes in a field of view that is displaced relative to the view projected on the sensitive material, it suffers from a framing error. This is commonly referred to as the parallax error. It arises from the difference in viewpoint between the finder and the camera lens—which may be one to two inches apart. The finder, then includes an inch or two more at the top and perhaps also at the side of the view, with a corresponding reduction at the bottom and at the other side.

With a distant subject which may cover a field of many yards or even miles an inch or two is obviously negligible. At close range, for instance with the camera focused on 3 feet, the subject field may only be 2 to 3 feet wide and there a couple of inches lost at the top of the image frame may result in the cutting off of important subject matter. The error will be even more with a long focus lens (if the camera is interchangeable) since this covers a smaller subject field while the separation of the viewpoints between finder and lens remain the same.

For this reason most cameras incorporate some form of parallax compensation.

There are two kinds of field size error. The first is a constant one; it simply means that the finder takes in either more or less than the image field recorded on the film. Often the field is deliberately reduced to cover other finder errors, so that the view through the finder always shows less than will be recorded on the film. If the field size error is too great the discrepancy between the finder and the film images may be disturbing—especially in colour transparencies where it is not easy to cut down the picture when the transparency is mounted in a slide frame.

A small field size reduction is nearly always deliberate, since standard slide mounts for 35 mm. colour transparencies slightly cut down the picture area by making off the extreme edges of the image. So not everything that is recorded on the film in the camera is necessarily projected.

The second type of field size error is variable and arises with cameras where the lens is racked forward to focus on near distances. As the lens-to-film distance increases, the angle of view of the lens decreases. The viewing angle of the finder however remains constant. At close range the finder may therefore include a greater subject field—quite apart from any parallax error—than the film. Some cameras also incorporate a compensation for this field size error.

Parallax Correction. There are various ways of compensating for the framing error due to the finder parallax (the difference in viewpoint between the viewfinder and the camera lens). The simplest is to cut down the finder field so that it always takes in less than the film. With distant subjects the image will then include more at the top than the finder, while at near distances there will be more recorded at the bottom of the picture.

Some cameras have a viewfinder which tilts to point at the centre of the subject field at different distances. This tilt may be mechanically linked with the focusing movement for automatic parallax compensation. Or it may have a manual adjustment which is set to the distance at which the lens is focused.

Bright-line finders usually have one of two types of correction. The simplest is an outline of a second frame superimposed on the picture, and used to frame subjects nearer than say 6 feet. Frequently the normal bright-line frame carries two small marks near the top. The upper limit of the picture are at close distances is an imaginary line connecting these two marks. Similar correction marks may also be incorporated in a normal optical finder.

On other cameras the view is framed by a

sliding mask which is moved by a cam coupled to the focusing mechanism. As the lens focuses on nearer objects the mask slides down to take in more of the bottom and less of the top of the picture. (Where the finder is off-set to one side of the lens as well, the mask similarly moves both downwards and sideways.) Bright-line finders, where the brilliant frame is provided by the light coming through a masking plate reflected into the finder system, may have this plate movable and coupled with the focusing mechanism in the same way.

The field size error may be compensated by designing a sliding mask which reduces the picture field as it moves downwards (and sideways, where applicable). Usually the fact that the parallax correction marks in a brilliant frame include a smaller picture area than a full frame approximately compensates for the field size error as well.

On some cameras the finder is permanently set to take in the field corresponding to a subject distance of 6 feet. In that case only the field size error at still closer distances has to be taken into account, but for distant subjects the film again includes more than the finder shows.

None of these devices for dealing with parallax error can compensate for the perspective error also associated with parallax. The closer the subject comes to the camera, even with all framing errors fully corrected, the more the viewfinder tends to look down on what the camera sees straight in front. For this reason only a single-lens reflex camera system, which uses the camera lens to produce the finder image, is really reliable at very close distances. For most subjects however the compromise of normal parallax compensation methods is good enough.

Viewfinders that Include Other Information. A number of manufacturers use the viewfinder to display other essential information to simplify the operation of the camera. On rangefinder cameras the viewfinder and rangefinder windows are frequently combined so that a small central area of the viewfinder frame carries the double image formed by the rangefinder. This system allows the photographer to focus on any subject while arranging it in the finder, and to keep it in focus even if it is moving about.

Viewfinder windows are also used to display exposure data. Sometimes the needle and scale of the meter appear superimposed on the scene. Sometimes there is a second pointer which is coupled to the exposure setting ring. The correct exposure can then be set for any subject by pointing the camera at the scene and turning the exposure setting ring until the adjustable pointer coincides with the exposure meter needle.

Cameras with automatic exposure control frequently incorporate a colour signal, e.g. red and green in the viewfinder. This warns the user when the prevailing light is inadequate for a correct exposure with the automatic control

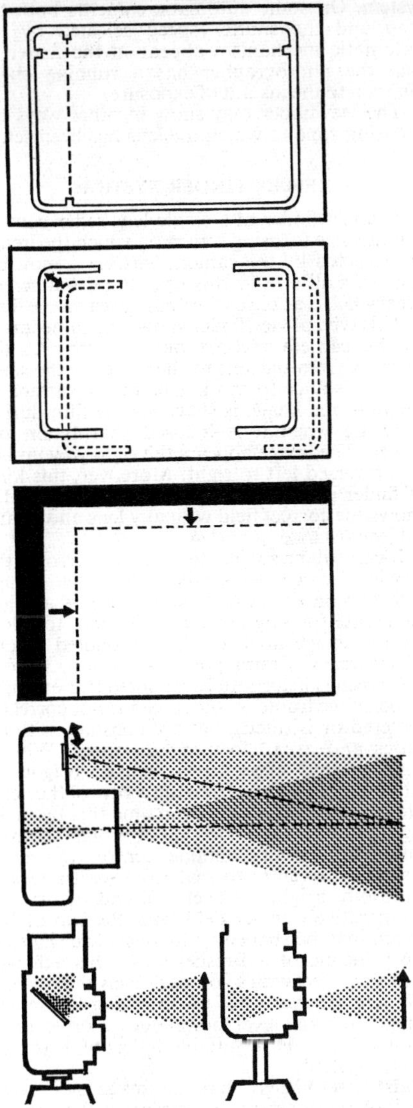

SYSTEMS OF PARALLAX CORRECTION. Top: *Brilliant frame finders on simple cameras often carry correction marks which indicate the limits of the field of view at closer distances.* Upper centre: *More advanced cameras may have the brilliant frame linked with the focusing movement, to shift in the finder field.* Centre: *On some cameras a mask in the finder moves to cover up the portions of the finder field not taken in by the film at closer distances.* Lower centre: *Sometimes the finder axis swings towards the lens axis at nearer distances.* Bottom: *The only absolute method of parallax correction with twin lens reflex cameras is to raise the camera between viewing and exposure so that the camera lens is in the position previously occupied by the finder lens.*

system. On some automatic cameras both the stop and the shutter speed selected by the automatic mechanism appear in the finder so that the photographer has a running check right up to the instant of exposure.

The viewfinder may show in other ways the focusing zone to which the lens has been set.

REFLEX FINDER SYSTEMS

The viewfinder of a single-lens reflex camera is a full-size focusing screen on which the image is projected by the camera lens via a movable mirror. Full size in this case implies a screen area which is as big as the image area on the film.

This type of viewfinder shows the same image as the camera records on the film. As the distance from the lens to the screen is the same as the distance from the lens to the surface of the film, the image is sharp on the film during exposure when it is focused sharply on the screen. The image appears the right way up but it is reversed left to right. Moreover, this kind of finder is free from any parallax error and it shows the correct field with any lens and at any subject distance.

It can still be subject to a framing error if the viewing screen is smaller than the image recorded on the film. Usually there is a slight deliberate framing error of this kind, to allow for the image area lost in a mounted colour transparency. Framing errors can also arise if the screen is inaccurately mounted (for example if the image frame on the screen is not correctly centered or is tilted), but are constant with all lenses and at all focusing distances. With a plain ground glass screen, image brightness falls off rapidly away from the centre of the field. Modern reflex cameras overcome this by making the screen the underside of a plano-convex lens. This reflects oblique centre rays back towards the centre to make the screen appear uniformly bright all over. Instead of using a comparatively bulky field lens, the top of the screen may be shaped with concentric rings on the principle of a Fresnel lens. This achieves the same effect with a comparatively thin screen.

There are two ways of observing the screen image: at waist level looking down on it, or at eye-level through a suitable reflecting prism or mirror system.

Waist Level Finder. Here the image is observed by looking down on the screen in the camera top. The viewing screen is surrounded by a fairly deep hood to keep side light and reflections off the screen. The hood nearly always folds up when not in use. It incorporates a magnifier since the image is usually too small for really precise estimation of sharpness. Frequently the hood also has a built-in frame finder or Albada type of optical finder for eye-level viewing (though not focusing).

Waist level reflex finders of this kind are found on single-lens roll film reflex cameras and sometimes on 35 mm. models.

Eye-level Reflex Finder. The viewing and focusing screen itself can be observed at eye-level by placing a pentaprism on top of it. This reflects the screen image into an eyepiece where it appears upright and the right way round. This is the arrangement employed in the majority of modern 35 mm. single-lens reflexes. The image is viewed with the camera held at eye level, looking straight ahead. On a few models the viewing system is so designed that the photographer looks into the eyepiece at an angle of 45°.

Some eye-level reflex cameras use two mirrors instead of a pentaprism. This is a cheaper system but the image is either right way up and reversed laterally (when the camera is held horizontally) or upside down and the right way round (when the camera is held vertical). A more complex mirror system (Porro mirrors) is sometimes used and has the advantage of being lighter in weight than a pentaprism (especially in the case of larger roll film reflexes); on the other hand the Porro mirror is often more bulky.

On a number of precision single-lens reflexes (both roll film and 35 mm. film types) the finder system is interchangeable. Either a waist-level hood with magnifier can be fitted or an eye-level pentaprism unit.

In some cases the screen itself can also be changed to provide a variety of alternative screen types (plain ground glass, ground glass with central split-image rangefinder, clear screen for aerial focusing, etc.).

Image Magnification. When a focusing screen image is observed at waist level with the unaided eye, the visual scale is not constant; the image appears larger the closer the eye is to it, just as any normal object looks bigger from a closer distance. The term "full-size" is frequently applied to the focusing screen image, to imply that the image scale is the same, and the image size approximately the same, as the picture recorded on the film. This is somewhat confusing when a screen image is compared with the visual image of other kinds of viewfinder.

When the screen is observed through an eyepiece magnifier, as it is with all eye-level

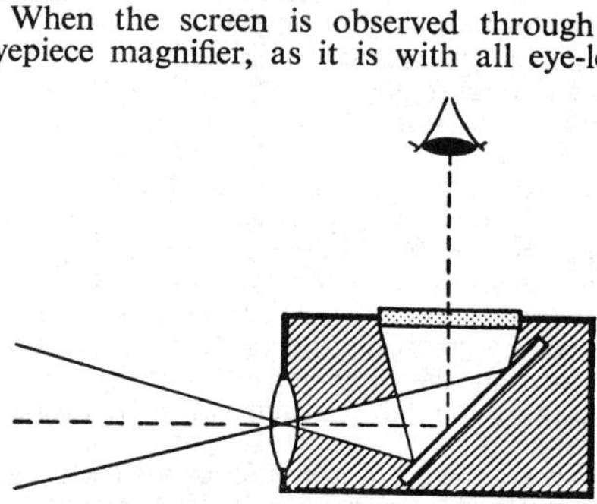

REFLEX FINDER. A mirror reflects the rays from the lens on to the ground glass screen on top of the finder. Large precision reflex finders, as used on the twin-lens reflex camera, also show the image sharpness.

reflexes, the visual image scale can again be compared with the scene as observed directly in front of the camera. The scale of the finder image then depends on the focal length of the camera lens used and on the power of the eyepiece magnifier. Normally the system is arranged to yield a 1 : 1 visual image (in the same sense as with an optical finder, as described above) with a lens of standard focal length (50 to 55 mm. on a 35 mm. camera). The image scale changes slightly as the camera is focused on different distances, but this is unimportant.

The eye-level reflex image as observed through the eyepiece appears on a smaller scale when a wide angle lens is fitted to the camera; it appears enlarged with a long focus or telephoto lens.

Twin-lens reflex. A twin lens reflex camera has a viewfinder constructed on the principle of a single-lens reflex, but mounted independently on top of an ordinary box camera. The viewing and taking lenses have the same focal length and are mounted on the same panel. This ensures that when a viewing lens is focused on the viewfinder screen, the taking lens will be focused on the exact plane of the film. On a few of the cheaper cameras of this kind the viewing lens is fixed and the taking lens focused by scale. This reflex system is just a viewfinder.

Twin-lens reflexes are usually roll film cameras with a picture format of $2\frac{1}{4} \times 2\frac{1}{4}$ ins. (6×6 cm.), but may be convertible to take smaller pictures. A few 35 mm. twin-lens cameras have also appeared from time to time. However with such a small format it is not easy to see when the image is sharply focused and some magnifying viewer is essential.

Twin-lens reflex systems are usually limited to one focal length. Where the lens is interchangeable, both the finder and the taking lens must be changed together and are mounted in pairs on a common interchangeable panel.

Usually the finder of a twin lens camera is of the waist level type with a folding hood; sometimes this is interchangeable against an eye-level pentaprism.

Since the viewing lens is some distance above the taking lens, the twin lens reflex is again subject to parallax error. This may be corrected in several ways. The simplest is a permanent masking down of the screen area so that it never takes in more than the taking lens records on the film. On better twin-lens reflexes the finder system incorporates a moving mask coupled to the focusing mechanism. The mask then uncovers the correct field of view corresponding to the picture on the film. On a few cameras the finder lens tilts downwards during focusing to point at the centre of the subject area at all distances. For extreme close-ups with supplementary lenses—which require matched lenses fitted in front of both the finder and the taking lens—the finder attachment may incorporate an optical wedge which deflects the finder beam downwards.

Since the camera and finder lenses move forward together during focusing, the twin lens reflex system is free from any variable field size error. As the angle of view taken in on the film decreases, so does the same angle of view on the finder screen.

The methods of parallax compensation with a twin lens reflex again cannot correct the perspective or viewpoint error associated with the different positions of the finder and taking lens. For really accurate viewing at very close distances the only reliable method is to mount the camera on a tripod, determine the view through the viewing lens, and then raise the whole camera through a distance corresponding to the separation of the two lenses, to record exactly the same view on the film.

Reflex Attachment. The optical viewfinder built into normal cameras is unable to provide accurate viewing and focusing at very close subject distances (macrophotography, etc.). Reflex attachments or housings are frequently available for 35 mm. precision miniatures with interchangeable lenses. There the lens extension tube, necessary for close work, is replaced by a tube of the same length equipped with a reflex attachment. This is simply a hinged 45° mirror and viewing screen mounted inside an accurately made housing designed for the camera, and transforms it into a single-lens reflex.

The mirror-raising lever is coupled to the shutter and the whole set-up works exactly like a reflex camera. The axial distance from the focusing screen to the lens is the same as the distance from the lens to the film surface.

The same type of attachment is necessary for use with very powerful telephoto lenses. In this case the angle of the lens is so narrow that the camera must be sighted very accurately to frame any particular area of the subject. It would be difficult to make a separate viewfinder to cover exactly the same view as the lens, and it would be impossible to check the accuracy of the set-up just before exposure. So a reflex attachment is built into the lens mount behind the rear component of the lens. F.P. & L.A.M.

See also: *Cinematography; Pentaprism; Reflex attachment; Reflex camera.*

Book: ***Photographic Optics*, by A. Cox (London).**

VIEWING FILTER. Another name for a panchromatic vision (P.V.) filter which shows the subject in its approximate brightness values, as it would be recorded in monochrome by a panchromatic emulsion.

VIEWPOINT. Position of the camera in relation to the subject at the time of exposure. Viewpoint has a considerable influence on both the practical and aesthetic value of the photograph.

There are two ways in which the viewpoint may vary: in distance and in height. Both factors may be varied independently and together they form one of the most important methods of control available to the photographer.

Distance. The closer the camera comes to the subject, the bigger the image. A telegraph pole 50 yards away produces an image twice as big as one 100 yards away; at 25 yards the image is four times as high, and so on.

But bringing the camera close does more than this; it increases the size of the subject in relation to whatever is behind it. Suppose the camera is looking straight down a road lined by telegraph poles 50 yards apart. Then, a pole 50 yards away will give an image twice the height of the pole behind it (100 yards away). But a pole 500 yards away will give an image almost the same height as the pole behind it.

This means that a particular subject appears bigger in proportion to its background as the viewpoint comes closer. So the apparent size of an object in relation to things farther away from the camera can be increased by bringing the camera closer to it.

Height. The normal viewpoint is at eye-level—i.e., between 5 and 6 feet from the ground. Most photographs are taken from this level when the intention is simply to produce a straightforward reproduction of the subject. Viewpoints above and below this level are used either to show details not visible from eye-level, or to present the subject in an unaccustomed aspect for special interest.

A high viewpoint shows more of the tops of objects than is usually seen. It also places the subject against a background of the ground and includes the other objects around it. So a high viewpoint tends to produce a "busy" photograph in which the subject has to compete with a mass of surrounding detail, most of it equally sharp since it is about the same distance from the camera.

There is very little novelty or shock value in a viewpoint above eye-level because it is fairly common for people to look at things this way—e.g., from upper windows.

A low viewpoint shows the unfamiliar underside of things but, what is more important, it tends to isolate the particular thing looked at from the surrounding objects. The background in this case becomes a plain wall, or the sky, or distant hills. To the photographer seeking to simplify his picture by eliminating all but his subject, this is the great advantage of the low viewpoint.

It explains the familiar kneeling stance of the press photographer.

Distortion. Whether the camera looks down or up, it exaggerates the proportions of the things nearest to it. So a high viewpoint makes the tops of things look big in proportion to the lower parts, giving an unpleasing top-heavy appearance to the subject. This effect, however, is often exploited in portraiture to increase the apparent size of a small forehead.

A low viewpoint has the opposite effect; it exaggerates the base of the subject, so giving it an air of strength and stability which is most appropriate with certain subjects—e.g., "strong man" studies.

For the same reasons lines which are parallel when viewed from eye-level converge towards the base when seen from a high viewpoint and towards the top when viewed from below. The most familiar example of this is the result of tilting the camera up so as to include the top of a building.

Sometimes this type of viewpoint distortion is used deliberately for effect, but generally it is undesirable and must be corrected either by the use of camera movements at the time of exposure, or by manipulation in the enlarging process.

The most familiar type of distortion is that caused by a too close viewpoint. As mentioned above, the closer the viewpoint to the subject, the bigger the near parts of it appear in relation to the more distant. So in pictures of people sitting with their legs stretched out towards the camera, their feet appear unnaturally large and in portraits a prominent nose or chin pointed towards the camera appears larger than in real life. For this reason a close viewpoint should be avoided—even 12 feet is not too far away for a natural portrait—and the limbs and attitude of the sitter should be disposed across rather than stretched towards the camera. The same procedure should be adopted in photographing any other subject if the proportions are not to appear exaggerated.

Perspective. The perspective of the picture—i.e., the relationship between the apparent sizes of near and far objects—is controlled by varying the distance of the viewpoint from the subject.

This is clear from the example of the telegraph poles discussed above. It does not depend on whether the subject is photographed with a telephoto, normal or wide angle lens. From the same viewpoint, the relative sizes of near and far objects will always be the same.

Scale. The scale of the subject—i.e , the relationship between the size of the subject and the size of the image formed in the camera—is decided by both the distance of the subject and the focal length of the lens. A closer viewpoint increases the size of near objects more than that of objects farther away; the scale of the image of an object can, however, be kept the same by simultaneous change of viewpoint and focal length. A lens of longer focal length increases the size of all objects in the picture area in the same proportion, whether they are near or far away, so it increases the scale of the picture. But it does not affect the perspective of the image.

So if the aim is to make the subject look larger in relation to the background, the viewpoint must be moved closer. But if the aim is to get a bigger picture of the subject without moving closer to it, the answer is to use a lens of longer focal length. F.P.

See also: *Perspective.*

VIGNETTING. Technique, in exposing prints, in which the picture is faded away more or less gradually towards the edges of the paper leaving a soft-edged, white border.

Vignetting is often used to eliminate unwanted background or other secondary matter —e.g., in portraits.

A card with a shaped hole slightly smaller than the area to be vignetted is held or supported a little above the surface of the paper.

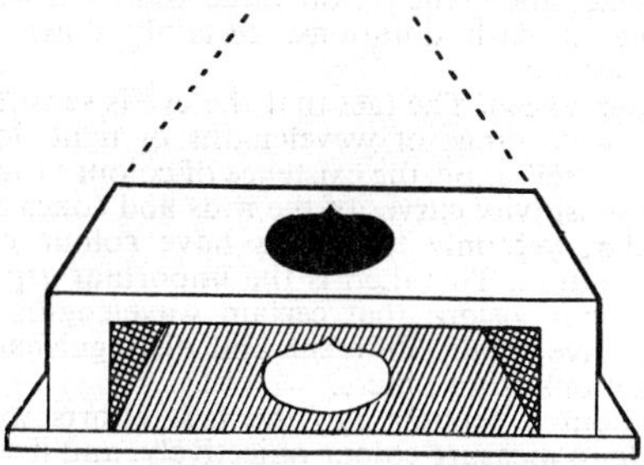

MAKING VIGNETTES. The area to be vignetted is cut out of a stiff card, and supported above the enlarger baseboard. The printing paper is placed underneath the cut-out, so that the image is projected on to the paper through the vignette mask.

The negative and paper are arranged (with the orange cap over the lens) so that the image occupies the desired position inside the masked area. The print is then exposed normally.

The photograph can be vignetted into a dark surround by "flashing" the edges of the paper.

The term is also used to describe a similar effect resulting from the inability of a lens to cover the negative so that the outer edges are under-exposed and print out darker than the rest of the picture.

See also: *Covering power; Masks; Shading and spot printing; Tricks and effects.*

VINCI, LEONARDO DA, 1452–1519. Italian painter, sculptor, man of letters, engineer, optician, scientist, anatomist and architect. Described the camera obscura in his secret notebooks, made numerous discoveries in optics and developed a theory of vision, including binocular vision. Also described "nature printing".

VINYL FILM. Sensitive emulsions coated on a plastic base consisting mainly of polyvinyl chloride acetate. Its advantage is low shrinkage as compared with cellulose nitrate or acetate, but it is not dimensionally stable above 40° C. It has been used for certain process materials and cartographic materials where accuracy of image size and scale are essential and no distortion must take place during processing.

Processing must take place at sufficiently low temperatures to avoid softening the film.

See also: *Supports for emulsions.*

VISCOSE SPONGE. Synthetic type of sponge used to wipe surplus water off roll films or miniature films before they are hung up to dry after processing. Films wiped in this way will dry more quickly.

Viscose sponges used for this purpose must be kept absolutely clean, as the slightest particle of grit in the sponge may scratch the film from end to end. Many photographers prefer not to wipe their films at all.

Some viscose film wipers consist of two viscose sponge pads mounted on the inside of a U-shaped plastic or metal strip. The film is drawn through between the two pads, which thus wipe both sides at the same time.

VISCOUS PROCESSING. Method of processing films—especially cine film—by spreading the developer, etc. on to the film surface in the form of a viscous paste or jelly. Used also in in-flight processing of aerial films, since the viscous solution is not as liable to flow under conditions of low atmospheric pressure and with the movement of the aircraft.

See also: *Rapid access.*

VISION

Vision is more than just sensitivity to light. Many of the lower animals and most plants respond to light but cannot "see" in the normal sense of the word. Only when the light sensitive organ can offer considerable information about the quality, dimensions and especially the movement of the outside world can the term "vision" be used.

The process of vision involves both the registering of light impressions and their interpretation. The first function is largely physiological, the second partly physiological and partly psychological.

PHYSIOLOGY

The process of seeing involves the reception of light impressions by the sensitive cells of the eye (the rods—highly sensitive but without colour discrimination—and cones—less sensitive but able to react to colour differences) and the transmission of signals corresponding to those impressions to the brain via the optic nerve.

The signals contain information about the intensity of the visual stimuli, their colour, and disposition. The brain interprets these signals as a picture of the scene perceived. In doing so it relies in part on comparison with previously stored sense impressions (in the memory—not necessarily derived only from visual stimuli). The interpretation is affected by a variety of physiological factors as well as by the physical signal itself. The latter can, however, be modified by the circumstances of observation as distinct from the observed object itself.

Apart from straightforward deception the eye is therefore subject to various phenomena of adaptation which affect our perception of detail, brightness, contrast, and colour. Further, our sense of vision is also able to extract information from simultaneous signals presented to the two eyes, and derive from that an impression of depth and distance.

Discrimination of Detail. Apart from colour effects which will be dealt with later, detail is produced by a variation in luminance from one point to another in a scene, and results in a corresponding variation of illumination across the retina. The ability to detect such detail is a basic property of vision. It has often been described in two parts: the detectability of small luminance differences and the visibility of different sizes of pattern. The first part has been used to describe the tone-scale requirements in photographic materials, the second to investigate sharpness and graininess requirements. More recent treatments have emphasized the inter-dependence of the two parts of the description. Thus visual threshold contrast (i.e. minimum detectable luminance difference) is found to vary with the size of the detail and the luminance level.

Flicker and Fusion Phenomena. If the luminance of a luminous area varies rhythmically, the sensation of flicker is often experienced. Whether or not flicker is seen depends on three main factors: the rate at which the luminance oscillates

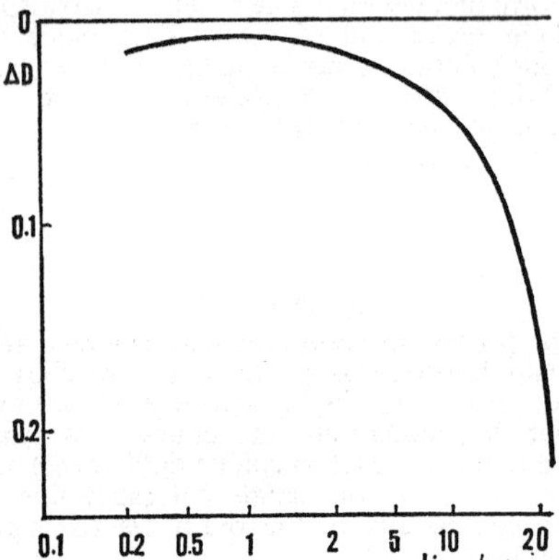

VISUAL SENSITIVITY TO DETAIL. To distinguish fine detail to be distinguished. The detail here is assumed to be viewed at 10 ins. from the eye, with an illumination level greater than 10 foot-lamberts.

(cycles/sec), the average luminance level, and the degree of variation of the luminance.

This is an extremely important aspect of ciné projection. The data may be reduced to a convenient form which for average cine scenes, expresses the critical flicker frequency for different average luminances on the screen. Below the critical flicker frequency the scene appears to flicker while above that frequency it appears steady. In this steady state fusion of the successive flashes of light occurs due to the property of persistence of vision. Above the fusion frequency it has been shown that the apparent luminance is given by the value of the time average of the actual luminances. This relation is accurate for a large range of frequencies above the fusion value and for a wide range of flash durations, certainly down to 10^{-8} second.

Colour Vision. The fact that the eye is sensitive to a wide range of wavelengths of light does not of itself imply the existence of colour vision. The sensitivity curves of the rods and cones are similar, yet only the cones have colour discrimination. This then is the important aspect of colour vision: that certain wavelengths of light give totally different and distinguishable sensations from others.

About a century ago the first efforts were made to measure colour objectively, and it was found that mixtures of blue, green and red light (called primary lights) could match nearly all known colours, including "white". Over the years these experiments have been embellished but they still form the basis of our ideas about colour vision.

Grassmann in 1853 propounded these basic ideas in the form of laws. In these he stated that the eye appreciates only three variables in the sensation of colour. How these variables are chosen depends on the context in which they are to be used, but there is never more nor less than three needed. In dealing with colour mixture he stated that once a colour match has been established between two lights, they can be treated as identical in any further colour mixture or colour matching measurements. The surprising thing here is that the statement is still true even if the two matching lights have quite different spectral compositions. Such a match is called metameric.

Mechanisms of Colour Vision. The association of colour vision with the cones in the retina is well established. Colour discrimination is at its best in the luminance range 1 to 10^4 ft.-lamberts where rod response is negligible and yet the cones are not over-loaded by the retinal illumination level.

Although a large number of theories have been propounded and examined, no completely satisfactory explanation of all the facts of colour vision is yet available. Nevertheless, several of the proposed theories fit the facts of colour mixture sufficiently well to form good working hypotheses.

The trivariant nature of colour vision implies that at least three different cone responses are possible either by each cone having three modes of response or by there being three types of cone. Histological examination of the cones has shown no differences between them. The presence of a pigment in the rods which appears to take part in the visual processes of those cells suggests a similar mechanism might produce different cone responses. Three differently coloured pigments either present together in each cone, or distributed separately throughout the cones, could result in the three required spectral responses, somewhat in the same way that colour filters may be used to produce colour separation negatives. Unfortunately, no sign of any pigments has yet been found in the cones. This may be simply an indication of the difficulty of detecting pigments which if they were like the rod pigment, would bleach when exposed to light. Nevertheless, many authorities still consider that the trichromatic theory assuming three pigments in the cones is the most reasonable explanation of colour vision.

The other, usually more complex, theories of colour vision have been devised mainly to explain more correctly the facts of defective colour vision. Hering's theory is perhaps remarkable in stressing the psychological aspects of colour, in particular the presence of pairs of opposite sensations—white-black, yellow-blue, red-green.

The relation of light to colour still remains very complex; for instance, many different spectral compositions of light are incapable of evoking different colours. On the other hand, with an identical light stimulating the eyes in different circumstances, very different colours may be produced. Indeed, there are many circumstances, apart from the colour stimulus itself, which are capable of affecting markedly the perceived colour; these include the state of adaptation of the observer's eyes, the focus of his attention, the retinal location where the stimulus impinges, size of stimulated area, duration of stimulation, and contrast of the immediate surroundings with the colour stimulus.

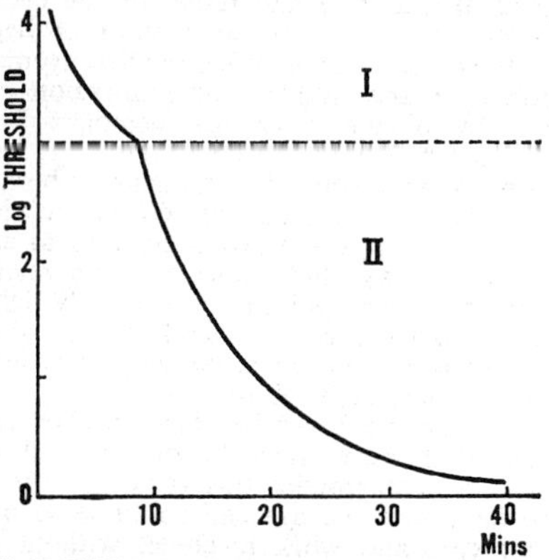

DARK ADAPTATION OF THE EYE. After about 8 minutes the cones in the eye have reached their lower limit of response (photoptic phase—I) and thereafter the rods take over the adaptation process (scotoptic phase—II).

Adaptation. If there is a change in luminance, the retina changes its response characteristics to give the best performance possible at the new luminance level. This process is luminance adaptation, and allows us to maintain much the same picture of the outside world under a wide range of luminance levels.

For sudden rises in luminance adaptation is rapid but it is generally much slower for sudden falls in luminance. In particular, adaptation from day-vision (photopic) to night-vision (scotopic) may take as long as 30 minutes.

This type of adaptation is obvious to any user of photographic darkrooms as well as to any traveller on a dark winter's night. It is probably less obvious that it plays a part in the correct appreciation of projected pictures.

Perhaps of even more importance in the viewing of colour pictures (especially projected transparencies) is the related phenomenon of chromatic adaptation. When we move from daylight to an indoor room lighted with incandescent lamps, we are at first aware of a big change in the quality of the colours around us—everything looks more yellow. Very rapidly, however, this sensation disappears as under chromatic adaptation the colours revert to their "normal" appearance. There is often a residual change in certain colours even after adaptation. This is due to true colorimetric changes in the relationship of different colours when we change the quality of the light—one of the reasons why we insist on daylight for critical matching when buying dress and furnishing fabrics.

This overall chromatic adaptation can also occur for much more pronounced changes, although there comes a stage at which the eye can no longer compensate for the changes. In the projection of colour transparencies chromatic adaptation can make us much less critical of a slight defect in colour balance if it is consistent through a series of slides. When we insert a slide of slightly different balance, however, the difference is easily apparent, and in a magnified form, until the eye is adapted to the new conditions.

Colour contrast is a related effect of considerable importance. If two different colours are viewed side by side, each tends to enhance the colour of the other. In other words, the difference between the colours is magnified by viewing them together. A different effect occurs if the two coloured areas are quite small and especially if they form a repeating pattern. The two colours then tend to blend and the difference between them is reduced. If the individual coloured areas are smaller still the pattern will look of one uniform colour derived from a mixture like the two actual colours. This effect is found frequently in fabrics.

After-Images. A further effect related to adaptation is that of after-images. These occur if we stare hard at a small brightly coloured object, let us say a red box, and after 20 or 30 seconds we suddenly look away at a piece of grey card. The after-image of the red box is a blue-green colour rather than red. The red receptors of the area of the retina where the image of the red box fell were fatigued (adapted). On looking at the grey surface the red receptors therefore responded much less than the blue and green. Thus a blue-green colour was seen. The red receptors soon recover so that the after-image is seen only for a second or so. Similar effects occur with other combinations of colours.

Defective Colour Vision. As with other physiological properties there are small differences between the colour vision of one person and another even when they are normal. Some people however (about 8 per cent of men and about ½ per cent of women) have defective colour vision, which in most cases is of an hereditary nature. In such cases the discrimination between one colour and another is reduced.

The most common form of colour vision defect results in a confusion of reddish and greenish colours. There are two forms of this defect: one in which the relative luminous efficiency curve is much the same as in normal vision, the other in which this curve shows much lower values in the yellow and red regions of the spectrum. When the defects are present in their full form these are termed deuteranopia and protanopia respectively. In a less pronounced form they are called deuteranomaly and protanomaly.

We know from rare cases where only one eye is defective that in these types of defect all colours look like mixtures of yellow, blue and grey.

Much more rare forms of defect involve an absence of blue from the perceived colours (tritanopia) and a complete absence of colour perception of any sort (monochromatism).

These defects tell us a great deal about the mechanisms of normal colour vision, and at the same time constitute the major difficulties in the theories of colour vision.

Methods of detecting colour vision defects fall into two classes. The first is the use of coloured patterns in which the subject is asked to distinguish figures or shapes (e.g. the Ishihara Tests). The second is the use of simple colour matching instruments (anomaloscopes) where the subject has to adjust a mixture of lights to match a standard colour. F.J.B.W.

Colour Association. Despite all the factors affecting the colour appearance, colours are normally perceived as belonging permanently to physical objects or situations out in the external environment. By a casual look round, one may perceive multitudes of colours as belonging to sky, clouds, soil, grass, trees, flowers, animals, people, clothing, conveyances, signs, lamps, decorations, buildings, and photographs. This effect of "belonging" is a consequence of perceptual development.

Colours seem to belong to objects partly because objects tend to continue to look the same even when the light illuminating them is changed and partly because they regularly reflect, transmit and emit light in ways which are characteristic of them. Another reason why colours seem to belong to objects is that the orientation, size, shape, and distance away of an object determine the size, shape, and orientation of the colour stimulus pattern from the object to the eyes; therefore, the colour may seem to coincide with the object whether it is moving or motionless. Such realationships assure close association of particular colours with particular objects as a phase of the general perceptual development of any infant. The colours of many familiar objects come to be well-remembered along with other characteristics of the objects. Thus, with a rose regularly looking red, and even so recalled, while other flowers regularly look other colours, it seems inevitable that red should be perceived as belonging to the rose. S.M.N.

Monocular Vision. An observer looking at a scene with one eye sees it as a picture in two dimensions in the same way as the single lens of the camera sees it. That being so, he should receive the same impression as he would from a photograph in true natural colours seen from the same viewpoint. Such a picture would convey the outlines and colours of objects, but it would give no idea of how far one object lay behind another. There would be no real depth in the scene although all the objects would appear in perspective.

In fact the impression on the observer's brain would be better than this. The brain adds something of its own knowledge and experience to the purely geometrical reproduction of the scene on the retina of the eye. It continues to create an impression of depth from the presence of other factors that normally supplement the stereoscopic impression produced by binocular vision: memory of the true relative sizes of objects in the scene helps the observer to estimate their distance; because of parallax, any slight movement of the viewpoint makes objects move in relation to each other in a way that indicates their relative positions; distant objects are generally lighter in tone than nearer ones; and the eye of the observer is constantly refocusing as it shifts between near and far objects.

So that, while the optics of monocular vision suggest that a scene viewed with one eye should appear flat, in practice this is not so. People who have lost one eye can continue to drive motor cars and work machines without any difficulty. At the same time, the impression of depth is by no means as complete as that given by two eyes.

Binocular Vision. When two eyes look at an object, they converge more or less—according to the distance from the object—so as to form the separate images on corresponding parts of the retina. This automatic act of converging is coupled to the focusing mechanism of the lenses of the eyes. The effort of making the two images coincide focuses the lenses for the particular distance which corresponds to the angle at which the eyes converge (the principle is the same as that of the optical rangefinder). This ability to converge and focus the eyes at the same time is acquired during the first few months of life.

At the same time, because two eyes view the object from two separate points of view, each sees a slightly different picture. The brain fuses these two pictures and thus helps to convey the solidity of the object and indicate its place in the depth of the scene. When the eyes look at an object more than about 200 feet away, the two lines of sight are practically parallel, and the two pictures are substantially the same. So that beyond 200 feet away binocular vision no longer helps to give the impression of depth, though special techniques in stereophotography can bring out depth.

Seeing in Perspective. The scene in front of the eye is reproduced as an image in perspective on the retina. A camera, built to the same specification, would produce the same image. But the human eye is more than a camera, it is part of a brain, and the brain interprets the picture in its own way, which is not always the camera's way.

The eye accepts a photograph which shows horizontal railway lines converging as they recede. But it quarrels with a photograph which shows the same effect taking place with the walls of a vertical tower. In fact, both pictures are right according to the laws of perspective, but while the eye is familiar with the horizontal picture, it finds the vertical picture strange.

The eye sees the picture of converging railway lines in one piece, as the camera takes it. But when the eye looks up at a tall tower, it scans it from bottom to top and, in fact, looks at it in sections. The broad base is never consciously compared with the apparently narrow top until the two are seen in a photograph taken by pointing the camera upwards. Seen in the same picture, the proportions appear completely false.

Sometimes the apparent falsity is worth seeking deliberately for its novelty or shock value. Many photographers make use of it for such reasons. But it is generally objectionable in normal photography where it is wiser to satisfy the eye by showing all vertical lines as parallel to each other and to the sides of the picture. This can be done on the negative or, where the lines converge on the negative, on the print.

Animal Vision. The eyes of animals are similar in most respects to those of human beings, but there are probably differences in the way they see things and in the colour sensitivity of the retina. Experiments to find out more about the physiology of animal vision have been carried out, using the actual lenses from the eyes of dead animals.

Successful photographs have been taken with sheeps' eyes in the course of these experiments. The eye of a slaughtered sheep has a window cut in the back of it and a ¼ in. disc of sensitized material is inserted. The tiny negative is exposed by flash lighting; developed and enlarged it yields a recognizable picture.

This method could bring useful information about the physiology of human vision. R.B.M.

PSYCHOLOGY

Vision is one of the tools the mind uses to find out something about the world around it. The carpenter uses it to make sure that two screws are the same size; the toddler uses it to watch his step; the photographer needs it to find out what kind of picture he can make from a scene of cows and trees. In each case, light rays are being focused on the sensitive surfaces of the retinae, and the images thus produced in the eyes are transmitted to the appropriate parts of the brain by means of the fibres of the optic nerves.

Physiologists have some information on what happens to the optical stimuli in the brain. But the facts they know will help us little when we wish to find out why and according to what rules the carpenter, the toddler, the photographer see what they see. At the present time the best answers to these questions are not found by studying the nervous system but by looking at what is accessible to all of us, namely experience.

Here, then, the psychologist takes over. His task is complicated by the fact that vision is not an isolated operation. It is constantly influenced by other activities of the mind, which looks at things for a purpose, reacts to them with desire or refusal, remembers things observed and learned in the past, thinks about them, and applies standards as to what they should or should not be like. The strivings, past experiences, thoughts, and judgements of values that are particular to the photographer make his visual experiences different from those of the carpenter or the toddler.

Even so, there are certain basic rules that seem to hold for all seeing. Some of them will be taken up first.

Organization. The visual field presents well segregated units, such as buildings, trees, human figures. This is surprising if we remember that the retina of the eye consists of millions of relatively self-contained receptors or groups of receptors, each of which transmits but one point of the image.

How are the point-sized stimulations grouped

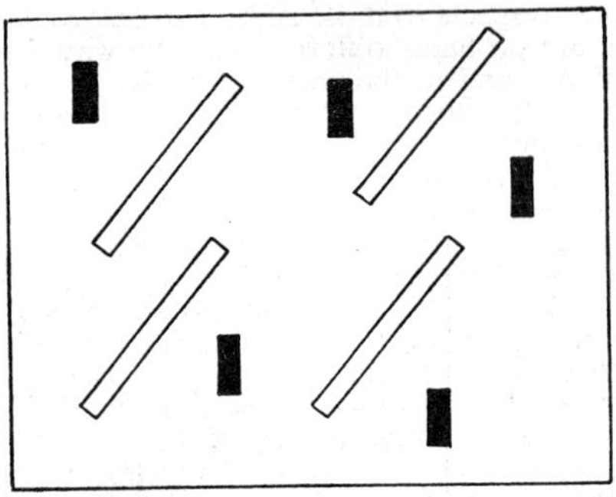

ORGANIZATION. Visual elements tend to combine spontaneously into groups which share common perceptual properties, such as shape, size, colour, brightness, spatial orientation, etc. Here the black, short, vertical bars form one group, whereas the white, long, obliquely oriented ones form another. If the elements of a group show repetitive characteristics of orientation, a certain rhythm presents itself.

together in such a way as to form images of objects?

The rule which controls visual organization can be stated in two ways: "from above", that is, with regard to the whole, and "from below", that is, with regard to the parts.

From above: "The elements of a visual field will spontaneously group themselves in such a way that the simplest possible organization of the whole results." Example: if the colour, shape, texture, and position of an insect are adapted by nature to those of a twig, the insect will be seen as a part of the twig, our better knowledge notwithstanding. By unifying twig and insect a simpler over-all pattern is obtained than by separating them.

From below: "The elements of a visual field will combine spontaneously to the extent to which they share perceptual properties." That is, objects that resemble each other in shape, size, colour, brightness, spatial orientation, etc., are related to each other visually.

The photographer does not often concern himself with mere geometrical patterns; however, it is essential that he learn to look at the more complex things of nature as though they were meaningless forms. Only in that way will he observe the purely visual properties that determine what is seen. Our eyes profit surprisingly little from what we know intellectually about the objects represented in a picture. What is only known but not seen makes a poor effect photographically. We may understand very well that a certain area in the picture represents a tree and another a house; but if the two areas have visual properties that make them fuse rather than separate, knowledge will not remedy the defect. In the same way, if a duck looks like a patch of mud, it will not be transformed into a duck by the camera.

The photographer should look at things as though he had never seen them or heard about them. For what he photographs is, strictly speaking, never the object "itself" but certain visual properties of the object. If you set out to photograph your wife you are likely to produce a non-committal, uncharacteristic, vaguely female effigy unless you ask yourself, consciously or unconsciously: what particular properties of this person do I want to bring out? What visual factors express these properties? The erectness of the body, the curve of the jaw line, the light reflections on her hair? If so, it is these features you must strengthen and sharpen by the choice of the right moment, the control of camera angle, distance, lighting, etc. Nothing in this world is anything but the totality of its properties.

Identification. To identify objects effectively in a photograph requires more than enabling the beholder to guess what is in the picture. In practical life, all the cognitive capacities of the mind, namely, sensory perception, memory, and reasoning, co-operate in steering the person through the business of the day, and in Western culture the share of vision in the work of this team has been severely reduced. The average person has become accustomed to relying on his or her eyes for little more than a minimum of cues, sufficient to distinguish red light from green light, avoid running into lamp posts and passers-by, tell a customer from a salesman, and so on. Therefore, such a person seems half-blind when called upon to look at a painting or to take a photograph; because these activities require the observation of a maximum rather than a minimum of visual features and make it necessary to understand by means of the eyes what otherwise is taken care of by routine identification based on learning in the past.

Photographic identification requires that the pattern in the picture be relatable visually to the image the observer has formed of it, or is expected to form of it, in his mind. This does not necessarily mean that a great deal of detail has to be reproduced. In fact, the visual identity of an object is most effectively conveyed by the "structural skeleton" of its main axes and proportions, easily rendered by a simple line drawing, but often hidden by a photograph, so that animals, young children, primitives, who immediately understand an elementary drawing, may see nothing but a jumble of spots in a snapshot.

It is relatively simple to recognize the picture of a woman executed in the most extravagant modern style so long as it respects the basic structural skeleton of the object. It is more difficult to recognize a natural form when viewed from an unusual angle, which foreshortens and all but destroys the familiar structural skeleton. No matter what angle or lighting the photographer may choose, he must see to it that in his picture the main visual characteristics of the object clearly dominate all the detail of shape, colour, and texture.

IDENTIFICATION. **Left:** *It is comparatively easy to recognize an unusual stylization of a familiar object*—**e.g.,** *a Picasso-like figure of a woman—as long as it respects the basic structural skeleton. In other words, the layout of the figure must correspond in its sense—even if not in its proportions—to a natural object.* **Right:** *It is more difficult to recognize a natural figure when it is presented from an unusual angle that does not immediately reveal the familiar structural shape.*

Emphasis. What main characteristics of the object should we emphasize? This depends on the purpose of the photograph. A documentary picture taken to serve technological or scientific ends must bring out the particular features which are relevant for the competent handling of the physical object. It must visually define the local colours and textures, the typical shapes and proportions, the decisive spatial relationships. The artistic photograph, on the other hand, is meant to convey the expression rather than the physical properties of the object, that is, it must stress the features that carry the feeling to be transmitted by means of the object, the ruggedness of a mountain, the sensitivity of a hand, the gracefulness of a vine.

Expression. Expression results from the "forces" transmitted by visual qualities. No knowledge of the mood inherent in the subject matter of the picture will replace appropriate photographic expression. A beach scene rendered by sagging contours, heavy masses, and gloomy contrasts is no beach scene but a funeral.

What makes for expression? The main rule indicates that nothing is expressive unless it fulfils two conditions: it must be visually clear-cut, and it must be dynamic.

First, an unsure contour, an undefinable colour, a disorganized arrangement are incomprehensible to the eye and therefore lost for the message the picture is supposed to transmit. Secondly, a visual pattern which does not unmistakably produce the fitting measure of tension and relaxation, stability and instability, crescendo and diminuendo, is not expressive. Oblique position, overlapping, distortion, asymmetry, wedge shape, make for tension and instability. Vertical and horizontal orien-

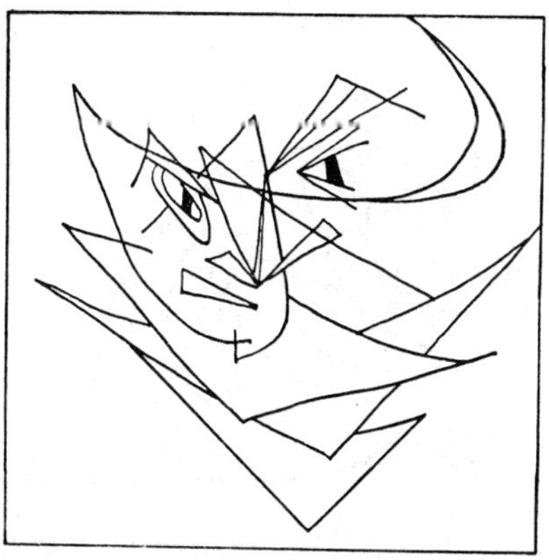

EXPRESSION OF TENSION. Oblique positioning, overlapping of lines, distortion of shapes, asymmetry, wedge forms, and pointed shapes make for tension and convey instability.

EXPRESSION OF RELAXATION. Vertical and horizontal orientation, completeness, normal aspect, symmetry of shapes, and regular outlines give a relaxed impression and convey stability.

tation, completeness, normal aspect, symmetry, make for relaxed stability.

Knowledge. We have said that knowledge must not be relied upon to replace interpretation by visual means. Nevertheless, it is highly desirable that the photographer should have some information about the things he is portraying and that he should have done his own thinking on them. Nobody can be expected to make a scientifically useful photograph of a medical preparation, a machine, or the behaviour of a group of animals unless he is familiar with the meaning of what he sees and unless he knows what the essentials of the object or happening are. Many of the best artistic photographs show that their makers have not only been intrigued by the play of pattern and texture, but have understood through their eyes what lies beyond the reach of vision and, in turn, have succeeded in expressing by the visual surface some of the things they have learned and thought about their subjects.

Translation. A photographic picture differs in many respects from what we see in the world around us. Also our attitude toward what we see is not the same in the two cases. Therefore if you want something in the picture to look as it does in everyday life, you cannot simply make a mechanical copy of it. You must translate it into a new medium, that is, adapt the object of the "real" world to the conditions of the picture world.

Selection *v.* Contemplation. In everyday life we use our eyes mainly for the purpose of practical orientation: we pick out the things we need to know about and ignore the rest. This attitude is enhanced by the fact that the visual field is largely accidental, crowded with irrelevant things, and lacking in any definite pattern and colour scheme. Under these conditions, the best we can do is to assume the "selective attitude". When people look at pictures, a different attitude is appropriate. It may be called "contemplative". Contemplation means receiving what is offered rather than searching for what is desired. It means dealing with the visual matter as an organized whole, in which every element is a necessary and significant part. An accidental "slice of reality" will leave the contemplative eye confused. The main reason why the amateur so often produces poor results is that he persists in the selective attitude while handling his camera. Later, when he looks at his prints, he finds himself partially transformed from a practical into a contemplative man, and discovers to his bewilderment that his picture is populated with chopped-off feet, stray boxes, and alien background figures, which "did not count" when he looked at his subject.

Space. The world we see around us is practically boundless. The picture cuts a piece out of this continuum. The frame changes almost all visual qualities because they depend on relations within a larger context. A kitten may cease to be small when the large things around it are not shown. The striking quietness of a tree may have come about by contrast with its restless environment, and the active gesture of a mother may be incomprehensible without the child to which it refers. On the other hand, objects reveal hidden virtues when relieved from an overwhelming context. The selected portion of the visual field must be self-contained and self-explanatory. Furthermore, the frame does not only remove an environment but also adds one. It provides a severe rectangle, to whose verticals and horizontals all shapes in the picture must clearly relate themselves.

In everyday life, any object we look at has a well defined place in the total space which contains all other objects, notably the observer himself. The framed space of the picture has no such connexion with its surroundings. Distant mountains may cease to look distant

FRAME AND CONTENT. The frame of the picture can alter the visual qualities and apparent content of the subject. Left: The active gesture of the subject is incomprehensible by itself. Right: It makes sense when the child is included in the frame to complete the picture. Such a change of framing also alters the dynamic equilibrium of the picture from an unstable to a stable one.

because distance is a relation between near and far. The surface of a lake, taken from a bridge, may look upright in the picture, and a tilted camera will produce a tilted building. Which may or may not be what the photographer wished to obtain.

Time. The camera frames the subject in time as well as in space. It picks a momentary phase from a time continuum. Psychologically, however, a snapshot is not experienced as a part of a time context but as a translation of action into a timeless picture, in which movement can be expressed only by the pictorial means of oblique orientation, outspread legs, blurred shape, etc. The photographer is always in search of the "pregnant moment", which epitomizes the features of a given situation.

Depth. The ability of a photograph to convey depth and volume is limited. Unless the stereoscopic technique is used, both eyes of the beholder receive the same image and thereby reveal the flatness of the picture. This does not rule out depth entirely since effective means of creating the third dimension are available within the visual pattern itself. Perspective, overlapping, shading, light contrasts, etc., must be stressed by the photographer if he wishes to compensate for the lack of "binocular parallax".

However, even at best the range of depth is limited in a picture. The weaker the depth effect, the less there is of what psychologists call the constancy of size and shape, that is, the tendency of everyday vision to eliminate perspective distortions and to see objects, and parts of objects, in their physically correct size relationships. Hence the gigantic hands of portrait sitters and the desperately small background landscapes.

This effect is reinforced when a photograph is viewed from too far away, that is, when the angle at which it is seen is smaller than the one at which it was taken.

Colour. Black-and-white photography reduces the colours of everyday vision to a brightness scale, whose particular nature depends on the sensitivity of the film to the various wavelengths of light. It reduces the three-dimensional world of hue, brightness, and saturation to the one-dimensional world of the grey scale. Thereby it simplifies the relations between the elements of the picture.

This is like translating a symphony into a tune that is played on a flute. Things that looked different in reality may look alike in the picture. This may make them harder to identify, but at the same time the smaller number of visual values can enhance the order and clarity of the composition. Also by controlling the lighting the photographer is free to assign to any particular object almost any value of the brightness scale from black to white.

In colour photography this freedom is severely curtailed, and the bewildering variety of visual values makes it difficult to unify the composition. Here the photographer can help himself best by choosing subjects which contain a relatively small number of hues. The complexity of the richer medium will reward the colour photographer for his greater effort. It offers him a life-likeness and an emotional appeal in comparison with which a black-and-white picture looks cold and abstract.

Photography *v.* Painting. While it is true that photography draws well shaped visual patterns from the largely accidental raw material of reality, the very nature of the photographic medium indicates that it offers its best in a realm somewhere in between everyday vision and the highly controlled creations of the painter.

Not by accident was photography discovered in a century which had found in the momentary aspects of surface, shape and action a new instrument for the interpretation of life. The human mind always roams between the poles of abstracted order and the bewildering freshness of direct experience. Vision reflects both order and complexity and is constantly in need of either one. At a time when painters are taking full advantage of a medium that permits them to render the basic melodies of human existence by means of remote patterns, it is the function of the mechanical technique of picture making to remind us that reality remains an inexhaustible source of challenging disorder. R.A.

See also: *Binocular vision; Colour; Colour measurement; Colour vision; Composition; Eye; Perspective; Visual appeal.*

Books: *Light, Colour and Vision,* by Yves Le Grand (London), *Color in Business, Science and Industry,* by Deane B. Judd and G. Wyszecki (New York), *Art and Visual Perception,* by R. Arnheim (London).

VISUAL AIDS

On a wide definition visual aids would include anything that helps a teacher to teach, or a learner to learn, by means of an appeal through the eye—e.g. the blackboard, charts, models, etc.

More specifically the term, however, denotes visual material developed or made available for schools in the form of projection images such as cine film, film strips or slides. Such material is comparatively inexpensive, since it is produced by mass copying, and is easy to store.

Cine Films. Educational films are supplied to schools in 16 mm. size on safety film; they may be in colour or black-and-white, in sound, silent or mute versions. In content and style of production they are distinct both from entertainment films and from general documentary films.

Each educational film is planned to serve a

definite purpose in the teaching of a specific subject to a given age-range of pupils. As the film is an aid to the teacher and not a lesson in itself, it can deal with the part or aspect of the subject which is most suitable for visual presentation. For satisfactory incorporation in a lesson the film is kept short; 10–12 minutes is the common length, though some classroom films run for 20 or 25 minutes. A wide variety of photographic techniques is employed: direct photography, animated diagrams, photomicrography, speeded-up and time lapse photography, and other devices; these techniques are not used for their stunt or surprise value, but for clarifying and instructing.

In sound films the sound track consists usually of a commentary, which is carefully planned to suit the subject and the age-range of the children, and is exactly synchronized with the visuals. Sound tracks also include natural sounds, for example in films of animals; or natural dialogue, for example in French teaching films. Music is used only when it is relevant to the subject, for example in a film on ballet or on the symphony orchestra; the inclusion of background music, as so commonly used in feature and documentary films, is generally condemned by teachers as distracting and unsuitable. It is usual to supply with an educational film a brief booklet of notes for the teacher, explaining the aim of the film and describing its contents.

Usually the sound track on such films is an optical one, which is cheaper to produce on a mass basis and incapable of being erased or otherwise interfered with. Magnetic sound tracks are easier to produce for films in small numbers, and also offer the facility of providing single copies of the film with sound tracks in different languages.

Narrow-gauge 8 mm. film is to some extent replacing 16 mm., mainly because it is cheaper. Magnetic 8 mm. sound films have the same applications as similar films in the 16 mm. gauge. Optical 8 mm. sound offers the advantage of still less expensive mass production of the films and protection against accidental erasing of the sound track. The quality of magnetic sound on 8 mm. is however, still superior to that of optical sound. The latter is adequate for speech and commentary.

Another application of 8 mm. film developed only in this gauge is the concept film. This is a short film length, running for about one to two minutes, illustrating a single idea or concept in dynamic terms. This may be a simple movement cycle, for example the action of the heart for a biology class, filmed either from a live subject or in the form of an animated diagram film. Such concept films are interpolated in the normal lecture or class whenever required, and do not take up an appreciable time proportion of a lesson. For easy presentation such concept films may be shown on special daylight projectors taking cassettes of the film in a continuous loop. The movement cycle can thus be shown without virtually any preparation.

Film Strips and Slides. The 35 mm. film strip is very widely used in schools. Single frame size (18 × 24 mm.) is generally adopted as the standard. Most teaching film strips are in black-and-white form, though colour is considered desirable in subjects such as nature study and art, and also for all material planned to be shown to young children. The number of frames varies; but the commonest practice is to supply about thirty frames, often divided into sub-sections according to subject matter. As with cine films, the visual material is carefully planned for a specific purpose, and is accompanied by a booklet of teachers' notes. Film strips are generally prepared from master-cards, which may be photographs, diagrams, or drawings in a wide variety of styles; elaborate montages are not often successful.

The double frame (24 × 36 mm.) finds favour with some users.

Film slides 2 × 2 ins., which some teachers prefer, are an alternative to the film strip. A few such slides are sold in commercially prepared sets; but it is more common for teachers to make their own slides by cutting up film strips and inserting individual frames in special glass or paper mounts which are marketed for this purpose. The old-fashioned glass lantern slide $3\frac{1}{4} \times 3\frac{1}{4}$ ins. is no longer widely used, but $2\frac{3}{4} \times 2\frac{3}{4}$ ins. slides are.

Sound and Slides. Current systems of coupling between automatic slide projectors and tape recorders permit complete lessons to be assembled in slide form, with the teacher's explanations recorded on tape. The tape can moreover carry cueing signals to actuate the slide change by remote control at the appropriate points of the lecture. This system is flexible as the teacher can prepare his lesson whichever way he wants to. He can include as many or as few slides as he wants, subject to the capacity of the projector. In the last resort the presence of the lecturer or teacher is only necessary for direct contact with the pupils, such as answering questions. This is indeed the way in which audio-visual systems are frequently used for briefing and instruction in industry, the military forces, etc.

To some extent the same considerations apply to sound in the form of recorded lectures in conjunction with film strips. As, however, the sequence of images on the strip is unchangeable, film strips with sound accompaniment are more set in the character of their lesson. The visual and sound records have to be specially planned and prepared with a view to production in larger numbers.

With film strip there is also the possibility of recording the sound on gramophone discs. The fact that such a sound record does not permit changing or adaptation is in this case no handicap. The combination also has the advantage that it is easier to prepare for projection. A number of film strip projectors combined with gramophone record players exist for this purpose.

Sound On Its Own. Sound recordings by themselves, without picture, are also a fairly widespread teaching aid, especially for learning foreign languages. The oldest and best known system is the gramophone record, and language lessons on records have been issued by certain firms for several decades. The tape recorder is a more modern and much more flexible means for this purpose, since it enables the student to record his own voice alongside a teaching track (with a twin track recorder) for comparison of pronunciation, etc.

An elaboration of this idea is the language laboratory where a number of pupils are gathered in one class, each equipped with a tape recorder inside a more or less sound-proof booth. The tape machines are connected to a master machine for the instructor who can communicate with all pupils simultaneously or with any selected one individually. Pupils record their exercises on their own tapes, can play them back and have them corrected by the teacher.

Language laboratory instruction is of course combined with visual work to cover the written aspect of a language. This may be achieved with a remote-controlled projector which projects sentences and other written material on to a screen in front of the whole class. Or the teacher may fall back on the blackboard or an overhead projector.

Overhead Projection. Cine, film strip and slide projectors can show material reduced photographically to the image size which such a projector can handle. This therefore implies preparation.

The overhead projector is suitable for impromptu visual demonstration, and is the optical equivalent of the blackboard. The teacher there can write or draw material on a transparent foil over an illuminated light box, and the image is simultaneously projected on to a screen behind the instructor via a mirror system. The writing surface is large enough for convenient operation, usually about 10 × 10 ins. The overhead projector can also be used to project large transparencies on the screen. It usually needs at least partial darkening of the room. Though this reduces the degree of teacher control over the class in the case of younger children, it increases the concentration of the students on the material presented.

Other Aids. Other items of teaching equipment generally included as school visual aids are: wall charts presenting information by means of selected photographs or pictures with some verbal explanation; microslides; geographical maps; displays of samples and specimens arranged to convey information; and models of various types.

Uses. The basic principle in the use of such visual aids in the classroom is that they in no way supersede the teacher, but help teacher and learner by giving a vivid impression through the eye. Visual aids are used as part of a lesson; they are often preceded or accompanied by explanations from the teacher and followed by questions, discussion and other classroom activities by the pupils.

The particular value of the visual medium in school teaching lies in the appeal which pictures have to all children; the emotional response of pleasure and interest is exploited by the teacher and linked with the intellectual effort of learning facts and understanding cause and effect. There are obviously many classroom topics that can be understood better when the teachers' words or the textbook explanation can be illuminated by pictorial illustration. Some children who have difficulty in following a verbal presentation can understand a well-planned pictorial explanation.

Visual aids (in particular films, film strips and wall charts) are used also in different teaching situations outside the ordinary school: for apprentice training in factories; for religious teaching in Sunday schools and churches; for mass-education; for adult education and technical instruction; and for advertising. Content and production vary according to the special circumstances for which the material is required.

Equipment. The use of films and film strips in schools involves the provision of a certain amount of apparatus. Many firms specialize in the production of school 16 mm. cine projectors and classroom film strip projectors; the emphasis is on robust construction, portability, easy maintenance and low price.

Equally, the machine should be easy to set up ready for projection and not require too much expert handling by the teacher who also acts as the projectionist. Hence some modern projectors incorporate automatic film threading and certain other refinements. For school room use a projector with the loudspeaker built into the same case is preferable to one with a separate speaker—which needs extra wiring up, the running of cables, etc.

Since 16 mm. films come with optical or magnetic sound tracks, a combined optical and magnetic projector is preferable to one suitable for optical tracks only. Where a school has several projectors, only one need have provision for recording magnetic tracks; the rest can be (and usually should be) pure playback projectors. The one projector with recording facilities can be used for adding sound tracks to films produced specially for the school.

A zoom projection lens is a help, especially if one machine has to be used in several classrooms of different size, but not essential.

An 8 mm. projector for school use must again be robust, easy to handle (with automatic threading) and capable of projecting striped 8 mm. sound films.

Projectors with an integral back projection screen have certain important advantages. One is that a film can be presented without darkening the room. Further, the projector is set up in front of the class instead of at a point behind it, so that the teacher retains direct contact with his audience. Special projectors of this kind are also available for the screening of concept films in the form of pre-loaded film cassettes. These are simply clipped into the projector to set the machine going immediately, and require no threading whatever. With sound film projectors for cassette loaded films some threading through the sound head is usually necessary, but even this is minimal. As the film is arranged as a continuous loop in the cassette, the same movement cycle can be shown repeatedly to facilitate understanding by the class.

Back projection screens are also useful with film strip and even slide projectors. For projection on a large screen it is generally best to have the screen itself permanently in front of the class, either behind the blackboard (many modern classrooms are designed to make this possible) or in the form of a roller type screen suspended above the blackboard and pulled down when necessary.

If the projector is set up behind the class, some form of remote control is desirable, so that the teacher can remain in front. The remote control units normally supplied for still projectors come with rather short leads (intended mainly for home projection). Alternative remote control systems, suitable for classroom use, employ either ultrasonic sound signals or wireless signals—in either case eliminating the need for a cable connection between the projectionist and the projector altogether.

Like a complete 16 mm. or 8 mm. sound film, a slide or film strip series with a sound commentary on a pre-recorded tape is more or less a self-contained lesson or part lesson. The teacher then has no other function during the show than to supervise the class watching the film or slides. Any questions and answers must come afterwards.

There are two other systems for providing a more flexible method of instruction, still using pre-recorded sound accompaniment. The first is the sound slide, consisting of miniature slides mounted in special large frames carrying a magnetic track surrounding the slide itself. A brief explanation or commentary can be recorded on that track (usually lasting about 20 seconds per slide) and the slides changed by the teacher by remote control. The teacher thus determines the rate at which the pictures are shown, and can interrupt the presentation at any point for questions and answers. A similar system also exists for film strips, with an optical or a magnetic sound track recorded on every alternate frame next to an image frame.

The other type of projector is much more a teaching machine, consisting of a slide or film strip projector with built-in tape recorder. The complete lesson is pre-recorded in image and sound, but the pupil controls the rate at which the pictures (and their corresponding sound commentary) are shown. He advances the image once he has taken in the information presented, and can also back track to the previous illustration and sound when required. Such machines require no teacher at all (except for the preparation of the material), and have been used also for instruction of apprentices, mechanics on a factory bench, maintenance engineers following a complicated repair or maintenance routine, and so on.

All these adaptations require special equipment. They have therefore not yet spread very far into general student instruction.

Other projection aids for educational use include episcope and overhead projectors. Episcopes are used for the showing of opaque pictures, but are largely falling into disuse owing to the comparative dimness of the projected images. Usually it is more practical to copy an illustration required from a book on to a reversal film and use the transparency for projection in a normal slide projector.

Micro projectors are sometimes supplied to school science departments. These enable microscope slides to be shown to a whole class simultaneously, where it would be impractical to let every student look in turn through a microscope with the specimen set up.

There are numerous commercial firms producing classroom film strips and wall charts; school film production is severely restricted by economic difficulties. There is also a certain amount of amateur production of school visual aids, particularly of film strips; this material is often of limited or localized use, but some of it is used on a national scale.

Schools obtain wall charts and other pictures in much the same way as they obtain text-books. Films and film strips are obtained from various sources. Film strips are generally purchased and kept in the schools. Films are usually hired, either from a library organized by the local education authority, or from a commercial library, or from the library of the Educational Foundation. There are some free sources of films and film strips, but the material they supply is not always appropriate for classroom use.

Concept films are also available through the library of the Educational Foundation. These are mostly sold outright, since they are required repeatedly in basic lessons. It is therefore more worthwhile for a school to build up its own library of such concept films. In Britain these titles are sponsored by the makers one of the 8 mm. cassette projectors. Short films of other single concepts are also available for educational use as part of the Scientific Film Encyclopaedia. This is a collection of basic movement records produced by a scientific film institute in Germany, but available internationally under the auspices of the International Scientific Film Association.

Television. Unlike the other visual aids which are directly under the control of the teacher for any one class, television is a more centralized affair. Educational programmes are available over the national broadcast television networks. But they are tied to specific broadcast times, and thus oblige the teacher to organize his schedule to fit in with such times. Experiments have been carried out in Britain and elsewhere to produce special programmes for a school or group of schools, relayed by a closed circuit system from a central point. This, however, still ties a class to a fixed time for the transmission, especially where the latter is largely made up of live material.

There are two possible developments in the use of television, especially in closed-circuit systems.

The first is the use of tele-cine transmissions, playing a film through a closed-circuit system to a class or a number of classes. Where several classes are involved, this requires co-ordination. But once a school is equipped with closed-circuit television, this is permanently installed in all classrooms, requires no preparation or setting up, and is available for daylight observation. Also a single film—usually obtained from a library—can be shown to a larger number of pupils.

The other possible development is video tape. Here the film is entirely replaced by a tape recording which may be played back either on a central video tape machine or even on individual tape playback units in specific classrooms. Such a set-up, however, requires expensive equipment.

National Organization. Many difficulties have been encountered and overcome by educationists during the recent development of visual education in this country. To promote and assist the wider use of visual methods in the classroom is the function of the National Committee for Audio-Visual Aids in Education, set up in 1946. Composed partly of representatives of teachers' organizations and partly of representatives of local education authorities, it is responsible for evolving general policy and watching the interests of teachers. Its parallel organization, the Educational Foundation for Visual Aids, was set up in 1948 as a non-profit-making educational trust to arrange for the production of new visual aids, for the supply of equipment to schools, and for the distribution of visual material for hire and for sale. These two national bodies work in close association to solve the many problems involved. They have a free advisory service on technical matters and on visual material; organize courses and conferences on visual education; and run a film library with a sales service of films and film strips with free preview facilities. Their information service, called VENISS (Visual Education National Information Service for Schools) provides combined national catalogues of available films and film strips, a monthly magazine called *Visual Education*, a year-book of visual aids, and occasional pamphlets and information lists.

Teachers are directly represented on the National Committee by means of a special sub-committee called the Central Committee of Teachers' Visual Aids Groups. Throughout the country groups of teachers who are specially interested in visual education combine in visual aids groups, for the discussion of visual aids topics, the review of new visual material, the production of their own films and film strips, or for training in the use of equipment, and other activities. These groups elect representatives to the Central Committee and hence are constantly in touch with national developments.

The production of new visual material is an especial concern of the national bodies. Contact with teachers through the groups supplies information on what material is required. Efforts are made to secure production of selected topics through negotiations with industrial and other sponsors; the Educational Foundation itself sponsors production as far as possible out of its trading profits. In every case a practising teacher is appointed as educational adviser to co-operate in the production of each visual aid so that the producer may know what is acceptable to teachers. The national bodies are always prepared to co-operate with commercial producers, to give free advice to amateur producers, or to provide information for sponsors on the production and distribution of visual aids.

Standards. Most of the dimensions and other specifications of audio visual materials are subject to national and international standards. These define slide and film strip sizes, 8 mm. and 16 mm. film, screen brightnesses, image definition, and sound tracks and sound systems. J.A.H. & L.A.M.

See also: *Cine film; Film strips; Projection practice; Projector (cine); Projector (still); Slide (transparency); Standardization in Photography, Viewers.*

Books: *Projecting Slides*, by M. K. Kidd and C. W. Long; *Making Lantern Slides and Film Strips*, by C. D. Milner; *Teaching by Projection*, by R. S. Judd; *Teaching with Tape*, by J. G. Jones.

VISUAL APPEAL. Visual appeal denotes the ability to catch the eye and hold the attention. It transcends the subject matter and artistic merits of a picture and includes the potential response to their combined attraction. That is to say that visual appeal depends on both what is shown and who is looking at it.

Any snapshot of a baby would provide ample visual appeal for its family. Comparatively dull photographs of a new piece of machinery are still very likely to interest its potential users. But to produce or find pictures of sufficiently wide appeal that will draw the eyes of hundreds of thousands to some magazine cover or advertising display is not quite so simple.

Pictures that meet such conditions are sometimes caught by photographers who either just follow their instincts or are lucky enough to stumble on the right opportunities. More often, however, they are supplied by skilled professionals who repeat and vary effects based on certain fixed formulae—even when they are not conscious of doing so.

Methodical analyses of visual appeal are largely speculative. Although controlled investigations of the subject in terms of experimental psychology are conceivable, the data we possess are neither complete enough nor sufficiently documented. It is generally agreed, however, that visual appeal is a complex phenomenon and whilst the terms applied to its components vary a great deal and may be grouped in many ways they usually cover three categories: novelty impact, emotional response and signalling force.

Novelty Impact. Novelty impact is governed by news value, personality content and action.

(1) The value of any news clearly depends on just how new, how big and how close it is to the man whom it seeks to inform.

(2) The more a personality is known the greater his or her pictorial impact. Yet the face of any person still has a stronger impact than a scene lacking human content.

(3) Anything dynamic will attract attention. Action, movement, expression, visually score over repose.

Emotional Response. Emotional response is evoked primarily by the human element in the picture. This can be broken down under the more specific facets of survival, sex, success, experience and escape. Most of these appeals are elementary and people are hardly ever indifferent to them. They remain effective irrespective of whether the reaction to them is positive or negative: approving or disapproving and whether it involves pleasure or revulsion, temptation or fear.

(1) Death being the ultimate problem of everybody alive, any pictorial demonstration of survival is universally fascinating. This explains the unfailing attraction of all young creatures—but also of catastrophe, murder and any form of violent fate from which the reader may feel reasonably sheltered.

(2) There is an effective appeal of sex, not only in the obvious display of it but also in the associations that the subject may evoke. The world of entertainment, for example, is an inexhaustible source of such associations as it draws mass attention to its stars and blends their personalities with the parts they play.

(3) Other people's success stimulates curiosity, challenges ambition and evokes envy. All these varying responses look for the identical reassurance that the person behind some achievement is as human as ourselves.

(4) It is human to cherish any confirmation of one's own experiences. Familiar faces, places and situations are likely to meet with warm response. Everyday aspects of everyday subjects may have little appeal, but there is a great deal of it in unusual aspects.

(5) There is a promise of thrill in change, distance and adventure; pictures that feed people's dreams are popular.

Signalling Force. Signalling force is the power in a picture to make itself noticed by its sheer formal qualities; these are clarity, pattern and emphasis.

(1) The primary task of any picture is to concentrate on its subject, lift it from its surroundings and show off exactly as much detail as its story needs.

(2) A picture should be obviously self-contained; its unity may be visibly reinforced by distinctive framework, balancing elements or graphic space management.

(3) Emphatic angles of view and perspectives, a strong suggestion of direction, speed and tension as well as contrasts of size, tone, light or colour possess strong pictorial momentum.

Application. Any successful picture will have something of all the three basic categories that add up to visual appeal. But only a few of the many elements on which any of these categories rest are likely to be present at any one time.

It is useful training discipline for photographers, designers and art editors to apply some such method of classification as this to analysing existing publications and photographs.

This is done, as a rule, by using a points system based on credits that represent these several appeals.

Although both the classification used and the evaluations gained in an exercise of this type are bound to be arbitrary they help to standardize habits of pictorial judgement provided the experimenter will force himself clearly to account for his motives in crediting points rather than worry too much about exactly how much credit this or that subordinate element would deserve. At any rate such an approach is a great deal more effective than just following the time honoured but vague advice to visit art galleries and look at the work of the masters. A.K.-K.

See also: *Composition; Vision.*

VOGEL, HERMANN WILHELM, 1834–98. German chemist. Became the first professor of photography at the forerunner of the Technische Hochschule (Institute of Technology), Berlin (1864). Founded several photographic societies and an important periodical, *Photographische Mitteilungen* (1864). In 1873 discovered the dye-sensitization of collodion plates and laid the basis for ortho and panchromatism and for three-colour photography. In 1884 used

Azaline for sensitizing in green, yellow and orange; this was a mixture of quinoline red and cyanine, but was later superseded by the isocyanine dyestuffs. In 1884 developed with Obernetter and Perutz the Vogel-Obernetter eosin silver plate which was the first orthochromatic dry plate. Also worked on sensitometry and described widely used printing paper photometers and tube photometers. His *Handbuch der Photographie* (from 1867 onwards) was one of the most important textbooks of the late nineteenth century.

VOIGTLÄNDER, PETER WILHELM FRIEDRICH, 1812–78. Austrian optical manufacturer. Established in 1849 a new branch of the family firm (founded 1756 in Vienna) at Brunswick. Manufactured from 1841 the famous portrait and landscape lens calculated by Petzval in 1840. Produced in 1841 the first metal daguerreotype camera. Biography by H. Harting (Brunswick 1925).

VOLT. Unit of pressure or potential in electricity. When using photographic equipment that requires an electrical supply, it is important to ensure the supply voltage is the same as that specified on the equipment. This can be particularly important in colour photography with electric lighting, as variations in supply voltage affect the colour temperature, as well as the intensity, of the light. Voltages above those recommended for a lamp also radically shorten lamp life.

VOLTAGE STABILIZER. Electric circuit, usually a variable transformer, designed to deliver a constant output voltage despite fluctuations of input voltage. The main use of voltage stabilizers is in colour photography and colour printing, where voltage fluctuations in the mains supply would affect the colour temperature of the exposing or printing lights, and thus might upset the colour balance. Voltage stabilizers are also essential in sensitometry to keep the light of lamps constant.

See also: *Transformer.*

VORTOGRAPH. Name given to any abstract picture made with the kaleidoscopic apparatus first used by Alvin Langdon Coburn in 1917. Three mirrors were fixed together, as in a kaleidoscope, so that the subject could be photographed through the middle of them to produce repeating patterns.

See also: *Abstract photography.*

VULCANITE. Black plastic material made from indiarubber and used for a number of camera parts (including the outer body covering in place of leather) and darkroom utensils (e.g., print paddles, forceps, etc.)

W

WALL, EDWARD JOHN, 1860–1928. English chemist, photographer, editor and instructor. Worked first as chemist for B. J Edwards & Co., London, who introduced the first orthochromatic plates in England; later with European Blair Camera Co. and several American motion picture film companies. Wrote a widely used *Dictionary of Photography* (from 1889 onwards in many editions), the *History of Three-colour Photography* (1925), and other books. In 1907 he described the principle of bromoil printing.

WARM TONE DEVELOPER. Certain positive developers which can produce image colours ranging from warm black to reddish-brown—depending on the developer or development time. Warm tones are often more effective than the normal blue-black, particularly for sunny subjects and portraits printed on a cream-tinted paper.

WARNERKE, LEON, (Vladislav Malakhovskii) 1837–1900. Russian civil engineer. Established in London (1870) a private photographic laboratory. Lived from 1880 partly in London and partly in St. Petersburg and established a photographic factory in Russia. Invented the Warnerke sensitometer which was the first practically serviceable device for measuring speed (1880); this he used in his investigations on the silver bromide collodion and gelatin dry plates. Also discovered the tanning action of pyrogallic acid and invented (1875) a roll-holder for silver bromide collodion stripping paper. Manufactured silver chloride gelatin papers from 1889. Received the Progress Medal of the Royal Photographic Society in 1882.

WASHERS. Both negative and positive materials must be thoroughly washed after processing to remove any remaining chemicals which might attack the silver image—or destroy or stain the base.

In each case the efficiency and speed of washing are governed by the rate with which the fresh water can replace the contaminated water in contact with the emulsion.

The processes of washing may be carried out in special washers in which the materials may be left unattended until washing is complete.

Negative Washers. Roll films processed one at a time call for no special apparatus. Bulk films are transferred from the developing tank and suspended from rods in a similar tank in which jets of fresh water spray on to the films and

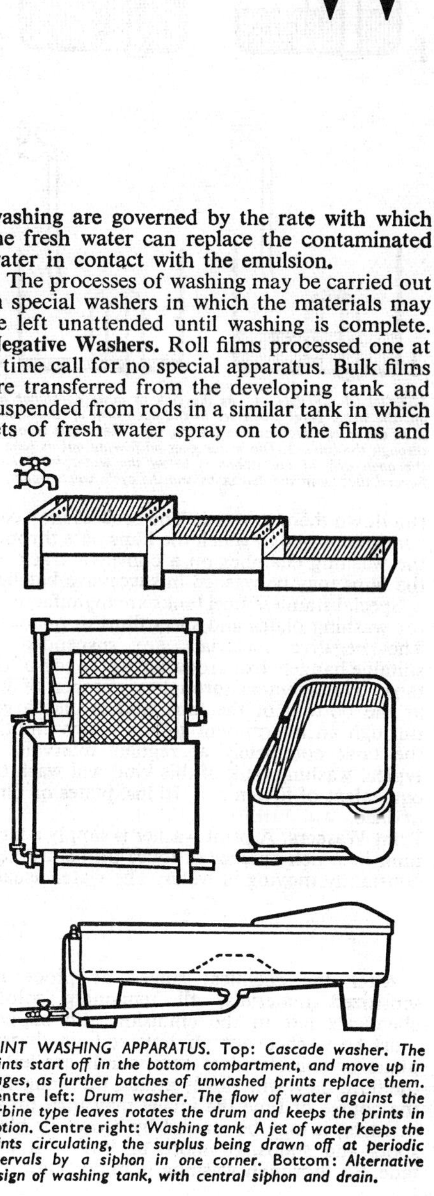

PRINT WASHING APPARATUS. Top: *Cascade washer. The prints start off in the bottom compartment, and move up in stages, as further batches of unwashed prints replace them.* Centre left: *Drum washer. The flow of water against the turbine type leaves rotates the drum and keeps the prints in motion.* Centre right: *Washing tank A jet of water keeps the prints circulating, the surplus being drawn off at periodic intervals by a siphon in one corner.* Bottom: *Alternative design of washing tank, with central siphon and drain.*

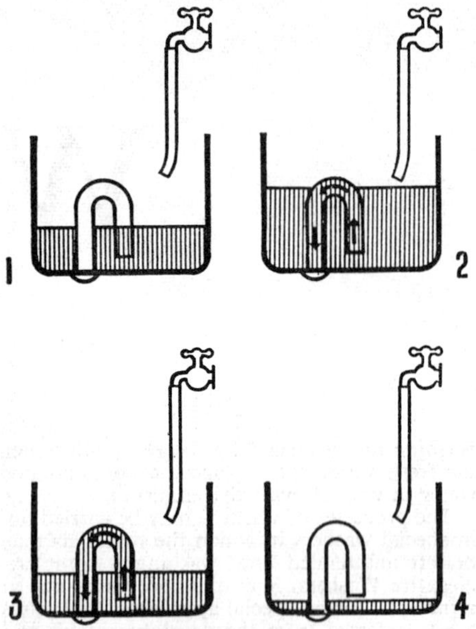

ACTION OF SIPHON. 1. As the sink or other container fills up, the water level also rises in the open arm of the siphon. 2. As the water level reaches the top, the water starts emptying through the tube. 3. The water goes on flowing out as long as the open arm of the siphon is below the water surface. 4. Beyond that point the flow stops and the cycle starts again.

run down them to waste. The process may be a continuous one in which the films pass through the washing chamber on a conveyor track, or the films may be washed in successive batches.

Special stainless steel tanks are manufactured for washing plates and sheet films in quantity. The negative materials are suspended in suitable hangers from rods across the top of the tank. Water enters through a number of jets at the bottom of the tank and is discharged through an intermittent siphon which empties the tank completely at regular intervals. A typical washing tank of this kind will wash the equivalent of fifteen 8 × 10 ins. plates or films at once.

Print Washers. A print washer is simply a large tank in which batches of loose prints are kept constantly moving in water. The water usually enters in a series of jets around the bottom of the tank. The jets prevent the prints from sticking together and ensure a constant flow of water over the surface of the paper. An intermittent automatic siphon completely drains the tank every few minutes every time the water reaches a certain level.

Washers of this type are used principally by professionals and wholesale photo finishers who keep them in constant use. They are not popular with amateurs because of the space they waste when idle.

Those who only occasionally need an automatic print washer can use a sink conversion unit. This is simply a short vertical length of piping pushed through a plug in the drain hole of the ordinary domestic sink and bent back so that the free end is close to the bottom of the sink.

As the water fills the sink, it rises inside the pipe until it reaches the bend. At that point it spills over and runs out of the drain. This starts a siphoning action that quickly empties the sink. As soon as the water level falls below the open end of the pipe again, air enters, stops the siphon, and the sink fills once more.

For this arrangement to work efficiently, the water should enter the sink with a fair amount of force and at an oblique angle to give a swirling motion to the water in the sink.

Self-Agitating Tanks. These rely on the velocity of the incoming stream of water to keep the prints in a state of constant movement. There are many designs which work on this principle but they all consist basically of (1) a large cylindrical tank, usually of moulded plastic, incorporating (2) a pipe for connecting the tank to the water supply and (3) an overflow which discharges into the sink. The inlet nozzle is set at an angle so that it directs a jet of water around the edge of the tank keeping the mass of water constantly rotating while the prints turn over and over. As the water is moving all the time there is no need to siphon it from the bottom of the tank and a simple overflow is all that is needed. This type of tank has the advantage that it is always full to the brim and there is no danger of the prints being crumpled by the sudden siphoning off of the entire contents. F.P.

See also: *Washing.*

WASHING

After the chemical stages of processing sensitized materials, all remaining soluble chemicals left in the emulsion and support must be washed out. If any sodium thiosulphate or soluble silver compounds are left in the film, they will slowly attack the image. In the course of time the image may turn brown in parts; it may be marred by brown or black spots and stains, or more or less of it may fade. So thorough washing is important.

WASHING NEGATIVES

Films and plates can be washed either in running water or in successive baths of fresh water.

Running Water. Running water is the most convenient means of washing negatives.

During washing the water must renew itself continually to carry away the chemicals diffusing out of the emulsion. So there must be no stationary pockets by-passed by the main

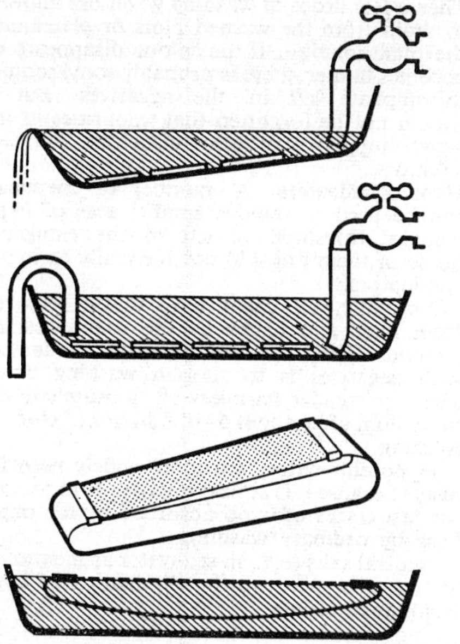

WASHING NEGATIVES. Top: *Plates washed in a tilted dish, letting the water flow in at the top, and out at the bottom.* Upper centre: *Alternatively, a siphon may be used to empty the dish periodically, provided it is deep enough.* Lower centre: *One way of washing roll films is to clip the ends to a board.* Bottom: *This board then floats in the washing sink or dish.*

stream. The most efficient washing systems are those where the water enters at one end and has to go through the whole system before it can flow out. Any arrangement is unsatisfactory where the water flows out as it flows in, leaving the bulk of the liquid undisturbed.

Sheet films can be washed lying face up in a dish placed in the sink. The water should flow in at one end of the dish from a rubber tube connected to the tap, and the dish should be slightly inclined so that the water overflows at the opposite end.

Another method is to place the plates or filmo at tho bottom of o doop dioh ond ollow the water to flow in direct from the tap. There should be an outlet near the bottom of the dish, or the water may be drawn away from the bottom through a siphon. Washing tanks of this type are available in a variety of forms.

The stream of water should never flow very rapidly or fall directly on to the emulsion surface, or it may peel the gelatin layer off its support.

Small particles of grit or other solid matter present in the water supply should be caught in a wire gauze filter or a commercial filter unit fitted over the tap.

Continuous batches of negatives are best washed in a cascade washer.

There is no satisfactory way of leaving loose roll films to wash in a dish. The ends of the film may be clipped together to prevent them from curling up and damaging the film surface, but inevitably the rest of the film curls up and gets into a dangerous tangle. Roll films should be either see-sawed through changes of water or left on the spool in the developing tank and washed by passing a continuous stream of water through the tank.

With most types of developing tank, the films or plates can be washed in the tank itself. The water should be directed into the centre of the reel of the ordinary roll film tank so that it has to pass through the turns of the film before it can emerge.

Washing in plate tanks is only satisfactory if the inflow is at the top, and the outflow near the bottom. If there is no outlet at the bottom, a rubber tube from the tap can be made to lead the water in at the bottom of the tank and let it flow out over the top. This will ensure very satisfactory mixing of the water.

With some small plate tanks neither method is possible. In that case the plates are best washed in changes of water, or removed and washed in a dish.

So long as the water is renewed completely at least once every 3 minutes, a total washing time of 30 minutes will be sufficient whatever method is used.

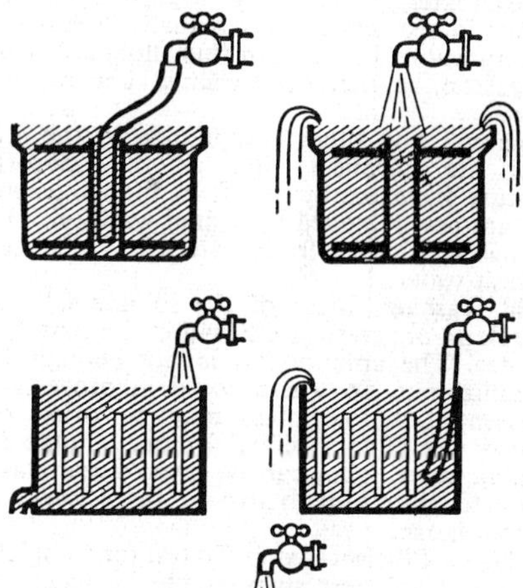

WASHING IN DEVELOPING TANKS. Top left: *Ideally the water should flow into the bottom of the tank and displace the hypo-laden water upwards.* Top right: *Placing the tank under the tap does not renew the lower levels of water.* Centre: *Plate and sheet film tanks often have a bottom drain plug for washing. Otherwise a rubber tube can be used.* Bottom: *Incomplete renewal of washing water.*

Changes of Water. Where no running water is available or when water is scarce, negatives can be washed satisfactorily in changes of water.

The dish or tank with the negatives in it is filled with clean water and left for about five minutes. It is then emptied, drained, and refilled with fresh water. The process is repeated every five minutes.

In this way the salts left in the emulsion are diluted up to 50 times at each stage. After about 4 to 5 such changes the amount of soluble matter left in the film or plate is so small that it is harmless.

Miniature tanks holding less than 10 ounces (250–300 c.cm.) of solution need about 8–10 changes.

Temperature Control. Often the washing water is considerably colder than the fixing bath and, as sudden changes of temperature are undesirable, the temperature must be stepped down gradually.

The procedure is the same as for washing in changes of water, but the temperature at each successive change is reduced by about 3° F. (2° C.). When the temperature finally reaches the level of the tap water, the films or plates may be washed in running water, or in changes of water, in the normal way.

The step-down changes count as part of the total washing period.

In Sea Water. If necessary, negatives can be washed in sea water. This actually speeds up the removal of the sodium thiosulphate left in the gelatin, so that the washing time can be halved.

The salt and other compounds in the sea water are harmless. But as they absorb moisture from the air if left in the emulsion, they are best removed by a final wash of 5–10 minutes in running fresh water, or 2 changes of fresh water.

Rapid Washing. Most of the soluble salts in the gelatin are washed out within the first 2–3 minutes. The amount left is not enough to crystallize out. So where speed in processing is essential the negatives may be dried or printed wet after washing for one minute in running water. They can be given a real wash at a later date—up to two or three weeks—without harm.

Testing for Efficient Fixing. To test for complete fixing, a blank test strip should be fixed together with other prints. A drop of a 1 per cent sodium sulphide solution is placed on the strip; if a brownish stain forms, fixing is incomplete. This usually means that the fixing bath is nearly exhausted.

Test for Efficient Washing. There is a simple test for complete removal of sodium thiosulphate. This requires a 0·1 per cent solution of potassium permanganate containing about 0·2 per cent of sodium hydroxide.

First, enough of this solution is poured into a glass of clean water to turn it faintly pink. Then a few drops of washing water are allowed to drain from the washed films or plates into the pink solution. If the colour disappears or becomes fainter, there is probably some sodium thiosulphate left in the negatives. But it should not be forgotten that impurities in the water supply may also discolour the test solution.

Hypo Eliminators. A number of chemicals can be used to remove small traces of hypo (sodium thiosulphate) left in the emulsion. Some of them are sold commercially as hypo-eliminators.

Normally, all chemicals can be removed from films and plates by the usual washing methods. The main use of hypo eliminators with negatives is to shorten washing times. They will render harmless all thiosulphate left in the film after about 5–10 minutes of efficient washing.

Hypo eliminators are more widely used for prints because it is almost impossible to remove the last traces of hypo absorbed by the paper base by ordinary washing.

Neutral salts (e.g., in sea water as mentioned above) also speed up the elimination of hypo from the emulsion.

WASHING PRINTS

Prints can be washed in running water in a dish or bowl; in special print washers or in changes of clean water.

Running Water. The simplest method is to put the prints in a large dish or bowl, placed in the sink with one side slightly raised. A rubber tube is then connected to the tap and arranged so that the water flows in at the raised edge of the dish. When the water fills the dish, it overflows at the lower edge. The rubber tube may be dispensed with and the water allowed to run directly into the dish, but in that case it will probably splash and be a nuisance.

The prints should be washed in this way for at least 45 minutes. Thorough washing demands continual renewal of the washing water, so that it can carry away the chemicals diffusing out of the prints.

This system takes up a comparatively large amount of room, and for large batches of prints a cascade washer, a rotary washer, or a special washing tank with siphon are more convenient.

Changes of Water. This method is more economical in water and is useful where no running water is available, or when water is scarce.

The prints are placed in a large bowl or dish of clean water, and left for about five minutes to allow most of the soluble salts to diffuse out of the prints into the water. The water is then poured out and replaced by fresh water. This dilutes still further the very small amount of soluble salts still left in the paper. After another five minutes, the water is changed again; after eight to nine such changes, the prints will be practically free of harmful salts.

Test for Complete Washing. The washing water can be tested for complete removal of sodium thiosulphate in much the same way as for washing negatives.

Hypo Elimination. The most thorough washing still leaves minute traces of silver salts and sodium thiosulphate (hypo) in the print. These are mainly absorbed in the fibres of the paper.

Tests have shown that the slight traces of harmful chemicals still remaining in the print, even after prolonged washing, may be enough to endanger the permanence of the silver image, particularly the fine-grain images of chloride and chlorobromide papers. The amount of chemicals left behind in a single-weight print is not enough to harm the coarser-grained bromide paper images, but a double-weight paper may absorb enough to affect even these images in time.

Washing is improved by bathing the prints for a few minutes in a 1 per cent borax solution after fixing. This neutralizes the acid of the acid fixing bath—which is one of the harmful chemicals absorbed in the fibres of the printing paper base, and at the same time speeds up the removal of hypo.

A more efficient method is to use a hypo eliminator which renders all chemicals absorbed in the paper harmless by converting them into neutral salts. L.A.M.

See also: *Faults; Hypo eliminator; Washers.*

Books: *Developing* and *Enlarging*, both by C. J. Jacobson (London).

WASHING SODA. Domestic form of sodium carbonate. It is often too impure for photographic purposes.

WATER. The chief rôle of water in photography is as a solvent for developers and other processing solutions. In its pure state it freezes at 0° C. and boils at 100° C. at 760 mm. barometric pressure, and has a pH of 7·0, which is taken as the reference point of neutrality.

Impurities of Water. Natural and tap water are never pure, and some of the impurities may interfere with one or other of the ingredients of photographic processing baths, in particular:

(1) Calcium and magnesium salts: these may be bicarbonates, or the sulphates, nitrates, etc. Water containing calcium and magnesium salts is generally referred to as hard water; very hard water may contain up to 0·05 per cent.

Hardness caused by the bicarbonates is said to be temporary because the carbonate can be precipitated by simply boiling; with sulphates the hardness is permanent, as it can only be removed by chemical means—e.g., by using sodium carbonate to precipitate the calcium or magnesium carbonates, or adding complex phosphates (sodium hexametaphosphate or Calgon) to prevent their precipitation altogether.

(2) Chlorides of calcium and magnesium: in addition to causing hardness the chloride affects certain silver salt solutions, restrains developers, and may interfere with bleaches.

(3) Dissolved gases: of these oxygen is the most important. Dissolved gases can largely be removed by boiling, though they may return when the water cools.

(4) Sulphides fog emulsions. Copper salts cause development fog, and reduce the keeping quality of colour developers. Iron, manganese, lead, tin, and zinc salts may cause stains.

(5) Other impurities: organic and mineral matter other than mentioned above is normally removed in the purification of the water supply and present in too small quantities to matter.

All mineral dissolved solids and most of the gases are removed from water by distillation, so distilled water is suitable for all, and essential for certain, photographic purposes.

Water equal to distilled water in purity can also be prepared by percolation through certain ion-exchange resins. Suitable units are supplied by laboratory furnishers and simply fit over the water tap. Small purifiers in the form of plastic bottles containing the resin are also sold by photographic dealers. The bottle is filled with tap water and left to stand for some minutes after which the pure water may be poured out. The resin is re-usable several times until a colour indicator shows exhaustion.

Water for Processing. Water for developers: the main impurities that interfere with developers are the calcium and magnesium salts which cause hardness. These precipitate their carbonates or phosphates from developers incorporating sodium carbonate or phosphate or an alkali. The precipitate is chemically harmless, but may be deposited as a scum on films or plates. Developers should therefore be filtered before use. The loss of alkalinity due to this precipitation is usually negligible.

Dissolved oxygen impairs keeping qualities of developers and should be removed by boiling before use. Dissolved carbon dioxide can be ignored, except with developers using caustic alkalies; there it will slightly reduce the activity.

Water for fixers: the normal impurities of tap water do not interfere with fixing solutions.

Water for aftertreatment and toners: in certain cases hardness or chlorides will cause precipitates, especially with solutions containing salts of lead, silver and other heavy metals. Such solutions must be made up with distilled water. Distilled water is also essential for all sensitizers, etc., where impurities may cause unpredictable variations in the characteristics of the sensitive material.

Water for washing: hard water—even sea water—will not appreciably interfere with washing, but may leave scum deposits on drying negatives or prints. The trouble can be avoided by using a final rinse of distilled water, or by adding a trace of acetic or hydrochloric acid to the final rinse.

Negatives and prints intensified or toned by certain processes—e.g., uranium—should be washed in slightly acid water, as traces of alkali will tend to dissolve the image.

Sea Water. Sea water contains appreciable quantities of dissolved sodium chloride and other salts. It is not suitable for making up photographic processing solutions, but may successfully be used for washing negatives and prints. The salt content of the sea water actually aids the removal of sodium thiosulphate from the emulsion and washing times can therefore be reduced. A final wash of 5–10 minutes in fresh or distilled water is, however, necessary.

See also: *Solutions.*

WATER BATH DEVELOPMENT. Method of improving the rendering of shadow detail in a negative or of softening the contrast of a print by transferring the material to a bath of still water after a short immersion in the developer. The result is that the heavily exposed areas quickly exhaust the developer they have soaked up, while the lightly exposed areas go on developing and building up detail. The water merely helps to keep the action of development even.

In practice, the negative or print is developed until a faint image appears. It is then transferred to a dish of water where the action is watched until no further increase in density occurs. If the over-all density of the material is insufficient, it may then be returned to the developer for a short immersion; or, if the contrast is still rather hard, the process can be repeated until the correct contrast is obtained. At no time should the negative or print be agitated when it is in the water; that would help the carried-over developer to diffuse out of the emulsion and so stop its intended local action considerably before the light tones have built up sufficient density.

Although water bath development provides a very useful form of contrast control, if carried to excess it will cause staining on the material. Liberal use of developer improvers and, if possible, a developer that is resistant to oxidation (e.g., glycin) is therefore recommended for this method.

See also: *Two-bath development.*

WATERHOUSE, JAMES, 1842–1922. English officer and scientist. Writer on the history of optics and of photography, and on photographic chemistry. Discovered in 1875 that eosin was a green sensitizer for silver bromide collodion dry plates. As chief of the Cartographic Service, Calcutta, developed photographic transfers for photo-zincographic halftone printing and produced in 1894 the first three-colour photogravure prints.

WATERHOUSE STOP. Early form of aperture adjustment consisting of fixed stops in a range of sizes, cut in flat plates so that they could be inserted in a suitable slot machined in the lens mount. The stops were numbered arbitrarily; the figures had no direct relation to the relative aperture.

John Waterhouse of Halifax invented the system in 1858; it is now obsolete, having been replaced by the iris diaphragm calibrated in *f*-numbers.

See also: *Diaphragm.*

WATERPROOF PAPER. Photographic paper coated on stock which has been treated, usually by lacquering both surfaces, so the processing solutions penetrate only negligibly into the support. Hence a print on such a paper dries much more rapidly after processing than a normal paper print.

See also: *Supports for emulsion.*

WATKINS FACTOR. With every developer used for development by inspection, there is a fixed relation between the time of first appearance of the image, and the total time required to give a fully developed image. The Watkins factor is the figure by which the first time must be multiplied to give the second. Its use is now virtually obsolete.

See also: *Developing negatives.*

WATT. Unit of power in electricity. It is the power generated by a current of one ampere acting over a potential difference of one volt. It is used to express the power consumption of electrical equipment such as lamps, etc. The wattage is there obtained by multiplying the voltage at which the equipment is run by the current in amperes taken. (Conversely, the current loading is obtained by dividing the wattage by the voltage.)

For some practical purposes such as high-power lamps, heaters, etc., the watt is too small and the kilowatt is then used. One kilowatt is equal to 1,000 watts.

WATT-SECOND. Unit of energy, numerically equal to the joule. Used to express exposure with artificial light sources (watt-seconds = watts × seconds) and also to measure the energy discharged through the flash tube of electronic flash units. With electronic flash, 1 watt-second corresponds to a light output of approximately 40 lumen-seconds.

WATZEK, HANS, 1848–1903. Viennese amateur photographer from 1890 and prominent Salon exhibitor during the Impressionistic period of photography. Professor Watzek favoured the gum print for his portraits, landscapes and still-lifes.

WAVELENGTH. Distance between corresponding points on two successive waves, applied particularly to electromagnetic radiations. These radiations range in wavelength from miles (wireless waves) down to fractions of a millionth of an inch (X and gamma rays). The wavelengths of visible light are measured in Angstrom units (1 Angstrom unit = one ten-millionth of a millimetre) and range from 4,000 A (violet band limit) to 7,000 A (red band limit).

See also: *Frequency; Light units; Spectrum.*

WAXED PAPER PROCESS. Photography on paper reached its culmination in the 1850s with this variant of the Calotype invented by Gustave Le Gray in 1851. The paper was waxed before iodizing, instead of—as in the Calotype process—merely after exposure to facilitate printing. The wax made the paper so transparent that as fine detail as a glass negative was obtained. The waxed paper process was popular with travelling photographers because the paper could be prepared ten to fourteen days beforehand (instead of the day before as with the Calotype) and did not need to be developed until several days after the picture had been taken (whereas the Calotype had to be developed the same day). Exposures were about the same, but development took one to three hours.

See also: *Calotype.*

WAXING PRINTS. It is often noticed that a print upon a matt or semi-rough surface paper has lost some of the depth and brilliancy it had when wet. The shadow parts of the picture especially seem to suffer.

A thin coating of wax applied to the print will restore some of the wet print quality. All that is necessary is to rub a good quality white wax furniture polish or cream well into the print, taking care to cover the whole surface. The wax must be applied sparingly with a piece of soft fluffless rag or the ball of the finger, and the print must be laid on a smooth surface and be quite dry.

The print is put aside for a few minutes and then polished with a soft dry cloth. It is important to polish off all the surplus wax, as if any remains dust will settle upon, and tend to adhere to, the surface.

As an alternative to the wax the print may be treated by doping the surface, but this method takes longer.

Waxing will not only restore the wet print finish, but it will also help to make any spotting or other hand work in the finishing of the print less obvious. R.M.F.

See also: *Doping prints.*

WEATHER PHOTOGRAPHY. Recording on film the state of the atmosphere. This can range from the size of raindrops and size and form of ice crystals and snow from various altitudes to the detection of hot and cold layers or turbulence over the hot desert. The size and shape of clouds, particularly as they change shape and direction in short periods of time, can best be recorded on film and the results used in forecasting heat or cold, wetness or dryness, calm or storm, clearness or cloudiness and many other meteorological phenomena. The rate of acquisition of data has increased exponentially in recent years and the modes of atmospheric circulation changes are clearly illustrated in the reversal of flow through the seasons. Weather photography also extends to cover photography of icebergs for detection of breakup by darkening, and rocket photography of airglow at 200 km. for atmospheric studies in the ultra-violet.

Observation of the stratosphere, that region of the earth's atmosphere bounded on the lower side by the first temperature minimum (the tropopause) and on the upper side by the first temperature maximum (the stratopause) has revealed that it is not, as first believed, an uninteresting, isothermal, quiescent region. The troposphere jet streams accomplish complete circuits of the northern hemisphere as they circle the cold low pressure region near the North Pole. These have been observed by cameras in the meteorological rockets which have revealed a steady northward flow at 50 km. altitude of appproximately 10 miles per second.

Filters are used for cloud photography and lightning photography and other electrical discharges. The most usual ones are yellow filters to darken a blue sky for cloud effects, and red filters which render red and yellow objects lighter and likewise penetrate aerial haze. High speed aerial films are in use for weather photography because of high contrast and sensitivity.

Until recently meteorologists relied almost entirely on observations or photographs of the atmosphere from balloons, in rockets or on the ground at favourable locations. In regions of sparse observations such as the oceans, tropics or polar areas, the lack of data was a serious obstacle. The use of TIROS (acronym for Television and Infra Red Observation Satellite) was highly successful. Since 1960, satellite observation of cloud coverage has been made by small vidicons which take thirty-two consecutive pictures at intervals of 10 to 30 seconds, transmit these pictures to ground, or store them on magnetic tape. The pictures are then regenerated at ground stations, where they are played out on a kinescope and reproduced on 35 mm. film to produce the positive copies for operational use. The pictures are often rectified by projection through a pinhole on a convex surface representing the curvature of the earth.

These photos are compared with aerial photographs of clouds or radar observations, covering a circular area with a diameter of 90 km. Brightness, texture, structure, pattern shape and dimensions are observed. Water clouds always appear brighter than ice clouds of the same thickness, under similar conditions of illumination. Brightness increases with thickness up to 15,000 feet. Ice reconnaissance was performed by Tiros IV, and also provided some 32,500 photographs for the first U.S. manned orbital flight.

As a typical example of a weather satellite, Tiros contains wide and narrow angle lenses of 5·3 mm. and 40 mm. focal length and short exposures are used because the satellite travels at 7 km. per second. The satellite is space-oriented so that its spin axis changes only slowly in space. The cameras point toward earth a maximum time during daylight hours.

The ½-in. vidicon camera takes 2 seconds to scan at a 1-millisecond exposure at *f* 1·5 from the vehicle that spins once per 5 or 6 seconds, and stores it on a video tape recorder. This stores a sequence of thirty-two pictures, at a resolution of 400 lines per in. Various controls permit the operator to enhance the contrast of the picture on playback. Areas of greatest brightness in clouds are enhanced to identify regions of uniform thickness, moisture content or energy levels. These are of particular value in studying the origin and growth of destructive hurricanes, and several storms have been forecast by this satellite.

Satellite weather photography is occasionally correlated with the photographs of the hemisphere of the sky using Condit's design of a convex mirror which forms a vertical image of the entire sky, reflected on a plane mirror to a 16 mm. cinecamera at 6 frames per minute or more. L.H.N.

See also: *Missile photography.*

WEB PROCESSING. Method of processing continuous lengths of film, for example motion picture film or aerial film, by winding it up in contact with a paper web saturated with a monobath developer-fixer. This makes the finished result available within a few seconds, and also has the advantage of not involving solution flows—useful for processing on board aircraft or missiles.

See also: *Rapid access.*

WEDDINGS

Wedding photography demands a high degree of technical skill, the ability to handle people, and a feeling for the "human angle".

Faultless technique is necessary because the ceremony cannot be repeated. If by mishandling of apparatus or by carelessness in processing the pictures are spoilt there is no second chance to make good the loss.

The photographer's ability to handle people can easily play a decisive part in the success or failure of the ceremony. This is no exaggeration. In a modern wedding the photographer is present at every stage of the proceedings, and his manner and working methods influence everyone around him. His ability to introduce the very desirable human touch into the finished result depends upon his own character and his own personal regard for the occasion.

To many hardened cameramen, a wedding is just another job to be done, often as quickly as possible and with complete lack of sentiment. The established professional, on the other hand, regards it as a special assignment: an intimate glimpse of the lives of two people embarking on a serious undertaking, the uniting of two families. He must not only go to endless trouble to secure the best possible pictures, but he must do all he can to make the event a successful social occasion.

Preparations. On being booked to attend a wedding the photographer first notes the date and exact time the ceremony will take place. Also, in his diary will go the name of the church, where the reception is to be held (with, if necessary, road directions), the bride's future name and address and the date of return from the honeymoon. Then he records the style of photograph or wedding album selected and the sum deposited.

If the wedding is not to be a white one the photographer must find out what the bride will be wearing so that he will be able to identify her quickly. (There have been occasions when the wrong person has been photographed!) He will also want to know how long the reception will last and whether the couple will be going away by road, rail, sea or air.

Then the bridal pair need to be briefed on what is expected of them photographically: arriving at the church, outside the church after the service, (the "official" groups will be taken either here or at the reception), at the cake cutting ceremony and as they leave.

Equipment. Over the last ten years or so the technique of wedding photography has undergone many changes, the major difference being that the posed groups at the studio have been replaced by more informal groups taken either at the church or at the reception. It is also quite common these days to make a complete illustrated story of the occasion.

"Official" group negatives should not be smaller than half plate and should be taken with a stand camera to allow full attention to be given to the arranging and posing. Large negatives are also preferable because they can be retouched. For the remainder of the photographs, the action pictures and the cake cutting ceremony, a coupled rangefinder hand camera or twin lens reflex should be used. Flash will be required for all interior shots, so the shutter should be flash synchronized.

Exposure. With white weddings, exposure can be extremely critical. The brightness range between the bride's white gown and the bridegroom's dark suit can be considerable. So when estimating the correct exposure, preferably with a photo-electric exposure meter, particular care should be taken. If the brightness range is more than the film can handle, exposure should favour the highlights; this is desirable in order to show detail in the bride's gown—more important than the bridegroom's suit.

Arrivals at the Church. The first photographs to be taken are of the arrivals of relatives and friends at the church. As it is impossible to include everyone, the photographer must use his own judgement. Two early arrivals will be the bridegroom and best man and the photographer must be there in time to catch them.

After all the guests have arrived there will be a short lull followed by the arrival of the bridesmaids. They should be given time to get together after the ride and then photographed either actually walking up the path or halted specially for the purpose. There will then be another longer pause.

The next arrival will be the bride who will be escorted into church on the arm of the man who

AT THE CHURCH. Camera positions for various stages and effects. 1. Arrival of bridegroom, best man, bridesmaids. 2 and 3. Arrival of bride. 4. The ceremony, taken from behind the choir stalls. 5 and 6. Groups outside church (facing away from sun). 7. Bride and bridegroom walking to car. 8. Confetti throwing and crowds. 9, Bridal couple driving off.

will give her away, usually her father. Here again, they should be allowed a moment or two to compose themselves after getting out of the car before being photographed. Often the photographer will get a better picture by asking them to pause on the church steps.

It is very rare for photographs to be taken inside the church during the service, but if such photographs are requested it is essential that permission from the vicar is obtained first. Time exposures are quite suitable as little movement occurs during most of the ceremony. Care must be taken to avoid making any noise; this applies particularly when using plate cameras with focal plane shutters, which on some models tend to be noisy.

After the Service. While the service is in progress the photographer can be making up his mind on his best place for taking the formal groups. If they are to be taken at the reception then he will only take action shots of the party leaving the church. The light in the porch may not be ideal for photography, but this is undoubtedly the best setting to convey the atmosphere of the ceremony. With a little tact the photographer can usually manage to get rid of any untidy notices together with any unsightly scraps of paper that may be on the ground. If the sun happens to be shining directly into the porch, it will mean screwed-up faces, but the difficulty can generally be overcome by posing the group in one of the other porches where the lighting is more favourable.

Taking the Groups. Whether at the church or the reception the most satisfactory method of working is for the photographer to operate the camera while an assistant—preferably his regular receptionist—sees to the posing by his instructions. This is much quicker and less confusing than when one person keeps running between camera and subjects.

The first photograph to be taken is of the bride and bridegroom, particular care being taken to see that the bride holds her bouquet at a becoming angle and that the bridegroom is upright without being wooden, and that his clothes are free from creases. Spectators should not be visible in the background. Two pictures are normally taken, one with the couple looking at the camera and one looking at each other. Their expressions should be pleasant, but broad laughter is best avoided.

The next photograph is of the bride alone, usually half-turned towards the camera with her face looking straight at the lens. She should hold her bouquet gracefully, smile pleasantly, and have her dress and train well arranged around her feet.

The third photograph is the bridal group—i.e., bride and bridegroom, best man and bridesmaids. Again particular care should be given to bouquets, hats and trouser creases, as a single fault can make the whole group look foolish.

The last picture consists of the previous

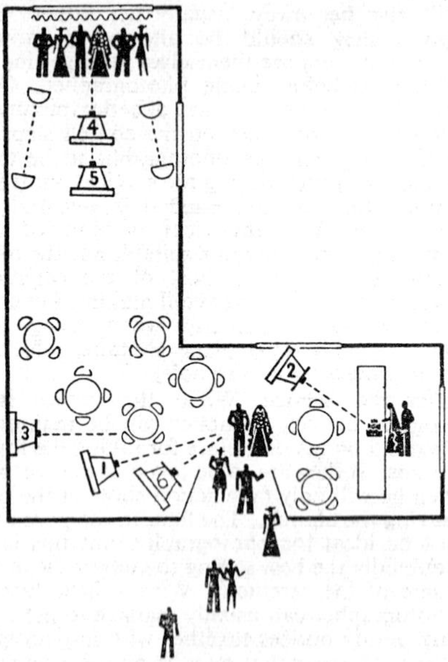

INDOOR RECEPTION. Camera positions for different groups and activities. Photographed mainly with press camera and synchronized flash. 1. Bride and bridegroom receiving guests (taken from high viewpoint, standing on a chair). 2. Cutting the cake. 3. Making the speeches; should include some of audience. 4 and 5. Formal groups, taken with stand camera and Photoflood lamps. 6. Final departure.

group with the addition of the two families on the respective sides of the bridal couple. Cramping should be avoided and the spectators (and guests not to be included in the group) should be kept tactfully but firmly well out of the way.

Tradition has for so long associated the throwing of confetti with weddings that at least one picture must be secured of the bridal couple undergoing this happy ordeal.

As soon as the last family group has been taken, the photographer should move quickly away ahead of the crowd. He should take up a position from which he can photograph the procession walking down the path to the waiting cars.

For these shots he will of course use a hand camera. Here he will be able to catch the bride and bridegroom covered in confetti entering their car. The photographer should shoot from the "open" side of the car door so that the latter does not obscure the couple. Getting this picture is always a scramble, but it is well worth the effort.

After a final general view for record purposes of the crowd, the church and the cars, the cameraman can follow the bridge and bridegroom and go on to the reception.

At the Reception. There are so many opportunities for securing valuable record pictures at the reception that it pays to carry an electronic flash outfit to save the inconvenience of carrying and disposing of flash bulbs. This should be a powerful portable outfit (100–200 watt-seconds) to provide adequate light even for medium to large size groups.

First the bride and bridegroom are taken receiving their guests; it is as well to take several shots to be sure of getting at least one in which everybody comes out sufficiently clearly and looks happy.

Next, two or three general views of the room should be taken with the guests in their places at the tables.

After the speeches comes the cutting of the cake. This is one of the most popular pictures and the photographer who can be sure of securing one or two happy pictures of this part of the ceremony will be well rewarded. The couple should be carefully posed with knife held by the bride in full view and her hand held by the bridegroom's. It is usual, but not always possible, for the bridesmaid also to appear in this picture. If the table looks untidy a few bouquets placed around the base of the cake will prevent the débris appearing in the pictures.

One or two photographs of the speech-making may next be taken. If group photographs have not been possible outdoors, these should also be taken at the reception; for

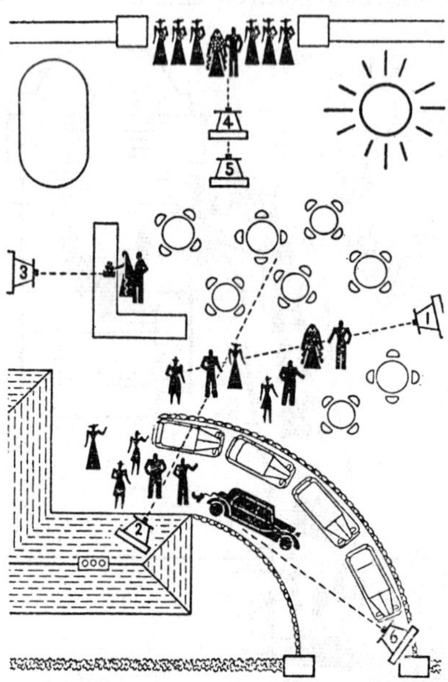

OUTDOOR RECEPTION. Camera positions for groups and activities. Photographed mainly with press camera. 1. Bride and bridegroom receiving their guests. 2. General view of assembly and speech making, taken from an upstairs window. 3. Cutting the cake; guests in background. 4 and 5. Formal groups. 6. Final departure, with wing of house as background. Coverage also includes incidental shots of guests at tables.

these, extra lighting will certainly be needed.

The last indoor photograph to be taken is of the bride and bridegroom leaving the reception. Immediately afterwards, the alert cameraman can often capture an unexpected scoop by making his way outside ahead of the couple to photograph the "decoration" of the car.

Finally comes the last photograph of the series as the couple drive away to the accompaniment of waving and cheers.

The Wedding Album. The experienced professional photographer avoids sending out rough proofs; he prepares a complete set of finished photographs and sends them out in a specimen wedding album.

This in its best form is a well presented album in which the photographs are mounted in the order in which they were taken. The first two or three pages are usually reserved and suitably titled to allow details of the wedding to be entered.

With the album goes a price list of both complete albums and individual copies, all prints being plainly numbered to simplify ordering.

A good set of wedding photographs calls for a surprising amount of planning and organization, but it is too serious a task to justify slipshod or careless work. The sincere photographer never loses sight of the fact that he is recording an occasion that his subjects will treasure as a precious memory. J.Bl.

See also: *Banquets; Groups.*
Book: *Wedding Photography*, by G. Catling (London).

WEDGE. In photography, a strip of material—such as glass—covered with a layer which is clear at one end and becomes more and more opaque towards the other end.

The layer itself may be dyed gelatin, or a suspension of pigment (such as carbon black) in gelatin, or a developed silver halide emulsion. The increase in density may be smooth and continuous or in regular sections. The latter type is known as a step wedge.

The Goldberg wedge consists of a wedge-shaped edge of dyed gelatin cast between two inclined sheets of glass.

The density range of a wedge is usually from 0 to 5 or 6. On a uniform continuous wedge the increase in density is specified by the wedge constant which is the density increase per cm. of wedge length. Thus a wedge with a constant of 0·3 increases in density by 0·3 every cm.

Step wedges increase by a constant density from step to step, but for really accurate work the wedge is calibrated individually.

Wedges are used mainly for sensitometric tests of emulsions. For this purpose the wedge is laid on the surface of the sensitized material and a single exposure is made through it. The result is as though the sensitized material had been given a number of separate exposures increasing gradually from one end of the wedge to the other. After development, the test piece can be examined with a densitometer, and the characteristic curve of the material plotted.

If instead of using white light for the exposure, a spectrum of light of different colours is allowed to fall on the wedge so that the whole range of colours crosses each step, the result will indicate the spectral sensitivity of the material because each colour will have registered a range of exposures from one end of the wedge to the other. The developed test piece in this case is called a wedge spectrogram.

See also: *Sensitometry.*

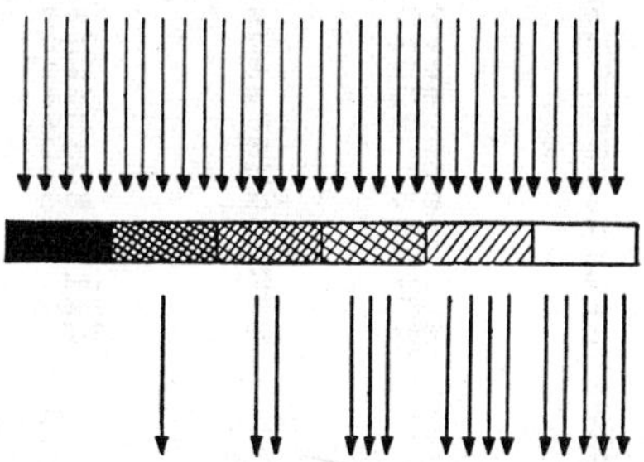

OPTICAL WEDGE. Steps of increasing density hold back progressively more of the incident light. Sensitized material exposed behind wedge thus receives series of graduated exposures at a standard exposure time.

WEDGE SPECTROGRAM. Record of the spectral sensitivity of a sensitized material made by giving it a series of differing exposures to a spectrum projected on to it. In practice the exposure scale is produced by placing a neutral density wedge over the sensitized material in such a way that each step of the wedge lies across the whole projected spectrum. This ensures that the material is subjected to the full range of exposures over every colour.

See also: *Negative materials; Spectral sensitivity Spectrography.*

WEDGWOOD, THOMAS, 1771–1805. English amateur scientist, son of the famous potter, Josiah Wedgwood. Published in 1802, with Humphry Davy, an *Account of a method of copying paintings upon glass and of making profiles by the agency of light upon nitrate of silver*. Was the first to attempt photography and to produce images (e.g., silhouettes and botanical specimens) on paper or leather moistened with silver nitrate solution, and in the solar microscope. But he and Davy were not able to fix their images. Biography by R. B. Litchfield (London 1903).

WEIGHTS AND MEASURES

There are three current systems of weights and measures in the English language countries: American, British Imperial, and Metric.

The metric system is consistent throughout the world whereas the British pint, quart and gallon differ from the American measures of the same name. For instance, the British pint is 20 fluid ounces while the American pint is 16 fluid ounces. However, as the latter are American fluid ounces the corresponding British equivalent is something like 16⅔ fluid ounces.

This gives some idea of the confusion encountered when using weights and measures in the British Imperial and American systems. Many people still prefer to use them, however, so conversion tables are necessary.

The metric unit of volume is the cubic centimetre, but the scientifically correct liquid unit of capacity is the millilitre. As the difference between the two (0·0028 per cent) is too slight to matter in photographic formulae, the cubic centimetre is taken here as the equivalent practical unit.

Units of Capacity

IMPERIAL LIQUID UNITS

60 minims	= 1 fluid drachm
8 fluid drachms	= 1 fluid ounce
20 fluid ounces	= 1 pint
2 pints	= 1 quart
4 quarts	= 1 Imperial gallon

METRIC LIQUID UNITS

1 litre	= 1,000 millilitres
1000·028 cubic centimetres	= 1 litre

UNITED STATES LIQUID UNITS

60 U.S. minims	= 1 U.S. fluid dram
8 U.S. fluid drams	= 1 U.S. fluid ounce
16 U.S. fluid ounces	= 1 U.S. pint
32 U.S. fluid ounces	= 1 U.S. quart
4 U.S. quarts	= 1 U.S. gallon

Weights

BRITISH UNITS (AVOIRDUPOIS)

437·5 grains	= 1 av. ounce
7,000 grains = 16 av. ounces	= 1 av. pound
14 pounds	= 1 stone
28 pounds (or 2 stone)	= 1 quarter
4 quarters	= 1 hundredweight
20 hundredweights	= 1 ton

BRITISH UNITS (APOTHECARY)

20 grains	= 1 scruple
3 scruples	= 1 drachm
8 drachms	= 1 ap. ounce
12 ounces	= 1 ap. pound

METRIC UNITS

1,000 milligrams	= 1 gram
1,000 grams	= 1 kilogram

UNITED STATES UNITS

Use avoirdupois weights up to	1 pound then
2,000 pounds (av.)	= 1 short ton
2,240 pounds (av.)	= 1 long ton

Lengths

BRITISH UNITS

12 inches	= 1 foot
3 feet	= 1 yard

METRIC UNITS

10 millimetres	= 1 centimetre
100 centimetres	= 1 metre

Linear Conversions

ENGLISH (OR AMERICAN) TO METRIC

Inches	*Millimetres*	*Inches*	*Millimetres*
1/64	0·397	7/32	5·56
1/32	0·794	¼	6·35
3/64	1·19	9/32	7·16
1/16	1·59	5/16	7·94
5/64	1·97	11/32	8·75
3/32	2·38	3/8	9·52
7/64	2·78	7/16	11·1
⅛	3·18	½	12·7
9/64	3·57	9/16	14·3
5/32	3·97	⅝	15·9
11/64	4·37	11/16	17·5
3/16	4·76	¾	19·2
13/64	5·17	⅞	22·2

Inches	*Centimetres*	*Feet*	*Metres*
1	2·54	1	0·305
1½	3·81	2	0·610
2	5·08	3	0·914
2½	6·35	4	1·22
3	7·62	5	1·52
3½	8·89	6	1·83
4	10·2	7	2·13
4½	11·4	8	2·44
5	12·7	9	2·74
5½	14·0	10	3·05
6	15·2	12	3·66
6½	16·5	15	4·57
7	17·8	20	6·10
7½	19·1	25	7·62
8	20·3	30	9·14
8½	21·6	40	12·2
9	22·9	50	15·2
9½	24·1	100	30·5
10	25·4	200	61·0
10½	26·7	300	91·4
11	27·9		
11½	29·2		
12	30·5		

METRIC TO ENGLISH (OR AMERICAN)

Centimetres	*Inches*	*Metres*	*Feet*
1	0·394	¼	0·820
2	0·787	½	1·64
3	1·18	¾	2·46
4	1·57	1	3·28
5	1·97	2	6·56
6	2·36	3	9·84
7	2·76	4	13·1
8	3·15	5	16·4
9	3·54	6	19·7
10	3·94	7	23·0
11	4·33	8	26·2
12	4·72	9	29·5
13	5·12	10	32·8
14	5·51	15	49·2
15	5·91	20	65·6
16	6·30	25	82·0
17	6·69	30	98·4
18	7·09	40	131
19	7·48	50	164
20	7·87	75	246
25	9·84	100	328
30	11·8		
40	15·7		
50	19·7		
60	23·6		
70	27·6		
80	31·5		
90	35·4		
100	39·4		

Area Conversions

ENGLISH (OR AMERICAN) TO METRIC

Sq. ins.	Sq. cm.	Sq. ins.	Sq. cm.
1	6·45	60	387
2	12·9	70	452
3	19·4	80	516
4	25·8	90	581
5	32·3	100	645
6	38·7	200	1290
7	45·2	300	1935
8	51·6	400	2581
9	58·1	500	3226
10	64·5	600	3871
20	129	700	4516
30	194	800	5161
40	258	900	5806
50	323	1000	6452

METRIC TO ENGLISH (OR AMERICAN)

Sq. cm.	Sq. ins.	Sq. cm.	Sq. ins.
1	0·155	80	12·4
2	0·310	90	14·0
3	0·465	100	15·5
4	0·620	200	31·0
5	0·775	300	46·5
6	0·930	400	62·0
7	1·09	500	77·5
8	1·24	600	93·0
9	1·40	700	109
10	1·55	800	124
20	3·10	900	140
30	4·65	1000	155
40	6·20	2000	310
50	7·75	3000	465
60	9·30	4000	620
70	10·9	5000	775

Lens Power

DIOPTERS TO INCHES AND CENTIMETRES

Power in Diopters	Focus in Centimetres	Focus in Inches
¼	400	157½
½	200	78¾
¾	133·3	52½
1	100	39⅜
1¼	80·0	31½
1½	66·67	26¼
1¾	57·14	22½
2	50·0	19 11/16
2¼	44·45	17½
2½	40·0	15¾
2¾	36·99	14 5/16
3	33·33	13⅛
3½	28·58	11¼
4	25·0	9 27/32
4½	22·22	8¾
5	20·0	7⅞
5½	18·20	7 5/32
6	16·67	6 9/16
6½	15·38	6 1/16
7	14·29	5⅝
7½	13·34	5¼
8	12·50	4 59/64
8½	11·75	4⅝
9	11·11	4⅜
10	10·0	3 15/16
11	9·09	3 37/64
12	8·33	3 9/32
12½	8·00	3 5/32
13	7·69	3 1/32
14	7·14	2 13/16
15	6·67	2⅝
16	6·25	2 29/64
17½	5·72	2¼
20	5·00	1 31/32

Temperature

DEGREES FAHRENHEIT TO DEGREES CELSIUS (CENTIGRADE)

Degrees F.	0	1	2	3	4	5	6	7	8	9
20	−6·7	−6·1	−5·6	−5·0	−4·4	−3·9	−3·3	−2·8	−2·2	−1·7
30	−1·1	−0·6	0·0	0·6	1·1	1·7	2·2	2·8	3·3	3·9
40	4·4	5·0	5·6	6·1	6·7	7·2	7·8	8·3	8·9	9·4
50	10·0	10·6	11·1	11·7	12·2	12·8	13·3	13·9	14·4	15·0
60	15·6	16·1	16·7	17·2	17·8	18·3	18·9	19·4	20·0	20·6
70	21·1	21·7	22·2	22·8	23·3	23·9	24·4	25·0	25·6	26·1
80	26·7	27·2	27·8	28·3	28·9	29·4	30·0	30·6	31·1	31·7
90	32·2	32·8	33·3	33·9	34·4	35·0	35·6	36·1	36·7	37·2
100	37·8	38·3	38·9	39·4	40·0	40·6	41·1	41·7	42·2	42·8
110	43·3	43·9	44·4	45·0	45·6	46·1	46·7	47·2	47·8	48·3
120	48·9	49·4	50·0	50·6	51·1	51·7	52·2	52·8	53·3	53·9
130	54·4	55·0	55·6	56·1	56·7	57·2	57·8	58·3	58·9	59·4
140	60·0	60·6	61·1	61·7	62·2	62·8	63·3	63·9	64·4	65·0
150	65·6	66·1	66·7	67·2	67·8	68·3	68·9	69·4	70·0	70·6
160	71·1	71·7	72·2	72·8	73·3	73·9	74·4	75·0	75·6	76·1
170	76·7	77·2	77·8	78·3	78·9	79·4	80·0	80·6	81·1	81·7
180	82·2	82·8	83·3	83·9	84·4	85·0	85·6	86·1	86·7	87·2
190	87·8	88·3	88·9	89·4	90·0	90·6	91·1	91·7	92·2	92·8
200	93·3	93·9	94·4	95·0	95·6	96·1	96·7	97·2	97·8	98·3
210	98·9	99·4	100·0	100·6	101·1	101·7	102·2	102·8	103·3	103·9

DEGREES CELSIUS (CENTIGRADE) TO DEGREES FAHRENHEIT

Degrees C.	0	1	2	3	4	5	6	7	8	9
0	32·0	33·8	35·6	37·4	39·2	41·0	42·8	44·6	46·4	48·2
10	50·0	51·8	53·6	55·4	57·2	59·0	60·8	62·6	64·4	66·2
20	68·0	69·8	71·6	73·4	75·2	77·0	78·8	80·6	82·4	84·2
30	86·0	87·8	89·6	91·4	93·2	95·0	96·8	98·6	100·4	102·2
40	104·0	105·8	107·6	109·4	111·2	113·0	114·8	116·6	118·4	120·2
50	122·0	123·8	125·6	127·4	129·2	131·0	132·8	134·6	136·4	138·2
60	140·0	141·8	143·6	145·4	147·2	149·0	150·8	152·6	154·4	156·2
70	158·0	159·8	161·6	163·4	165·2	167·0	168·8	170·6	172·4	174·2
80	176·0	177·8	179·6	181·4	183·2	185·0	186·8	188·6	190·4	192·2
90	194·0	195·8	197·6	199·4	201·2	203·0	204·8	206·6	208·4	210·2
100	212·0	213·8	215·6	217·4	219·2	221·0	222·8	224·6	226·4	228·2

Weight Conversions

AVOIRDUPOIS (OR AMERICAN) TO METRIC

Grains	Grams	Grains	Grams
1	0·065	35	2·28
2	0·130	40	2·60
3	0·194	45	2·92
4	0·259	50	3·24
5	0·324	60	3·89
6	0·389	70	4·55
7	0·454	80	5·20
8	0·518	90	5·83
9	0·583	100	6·50
10	0·648	110	7·13
11	0·713	120	7·78
12	0·778	150	9·72
13	0·842	180	11·7
14	0·907	200	13·0
15	0·972	220	14·3
16	1·04	240	15·6
17	1·10	300	19·4
18	1·17	360	23·3
19	1·23	400	25·9
20	1·30	420	27·2
25	1·62	440	28·5
30	1·94	480	31·1

Ounces	Grams	Ounces	Grams
¼	7·09	6	170
½	14·2	7	198
¾	21·3	8	227
1	28·3	9	255
1½	42·5	10	284
2	56·7	11	312
2½	70·8	12	341
3	85·0	13	369
3½	99·1	14	397
4	113	15	426
5	142	16	454

METRIC TO AVOIRDUPOIS (OR AMERICAN)

Grams	Grains	Grams	Grains
0·1	1·54	5	77·2
0·2	3·09	6	92·6
0·3	4·63	7	108
0·4	6·17	8	123
0·5	7·72	9	139
0·6	9·26	10	154
0·7	10·8	11	170
0·8	12·3	12	185
0·9	13·9	13	201
1	15·4	14	216
1·5	23·1	15	231
2	30·9	16	247
2·5	38·6	17	262
3	46·3	18	278
3·5	54·0	19	293
4	61·7	20	309
4·5	69·4	25	386

Grams	Ounces and Grains	Grams	Ounces and Grains
30	1- 26	175	6- 76
40	1-180	200	7- 24
50	1-334	250	8-358
60	2- 51	300	10-253
70	2-205	350	12-152
80	2-360	400	14- 47
90	3- 76	450	15-380
100	3-231	500	17-279
125	4-179	750	26-200
150	5-127	1000	35-120

APOTHECARY TO AVOIRDUPOIS (OR AMERICAN) AND METRIC

Scruples	Grains	Grams
¼	5	0·324
½	10	0·648
¾	15	0·972
1	20	1·30
1½	30	1·94
2	40	2·59
2½	50	3·24
3	60	3·89
4	80	5·18
5	100	6·48
6	120	7·78

Drachms	Grains	Grams
⅛	7·5	0·486
¼	15	0·972
½	30	1·94
¾	45	2·92
1	60	3·89
2	120	7·78
3	180	11·7
4	240	15·6
5	300	19·4
6	360	23·3
7	420	27·2
8	480	31·1

Ounces (Apoth.)	Ounces (Av.) and Grains	Grams
¼	120	7·78
½	240	15·6
¾	360	23·3
1	1- 43	31·1
1¼	1-163	38·9
1½	1-283	46·7
1¾	1-403	54·4
2	2- 85	62·2
2¼	2-205	70·0
2½	2-325	77·8
2¾	3- 8	85·5
3	3-128	93·3
3¼	3-248	101
3½	3-368	109
3¾	4- 50	117
4	4-170	124
4¼	4-290	132
4½	4-410	140
4¾	5- 93	148
5	5-213	156
5¼	5-333	163
5½	6- 15	171
5¾	6-135	179
6	6-255	187
6¼	6-375	194
6½	7- 58	202
6¾	7-178	210
7	7-298	218
7¼	7-418	226
7½	8-100	233
7¾	8-220	241
8	8-340	249
8¼	9- 23	257
8½	9-143	264
8¾	9-263	272
9	9-383	280
9¼	10- 65	288
9½	10-185	295
9¾	10-305	303
10	10-425	311

Conversion Factors

WEIGHT

Grains	*Grams*	*Drachms (ap.)*	*Ounces (av.)*	*Ounces (ap.)*	*Pounds (av.)*	*Kilograms*
1·00	0·065	0·017	0·002	0·002	0·00014	0·000065
15·4	1·00	0·257	0·035	0·032	0·0020	0·001
60·0	3·89	1·00	0·137	0·125	0·0086	0·004
437·5	28·3	7·29	1·00	0·911	0·0625	0·028
480	31·1	8·00	1·10	1·00	0·0686	0·031
7000	454	117	16·0	14·6	1·00	0·454
15432	1000	257	35·3	32·2	2·205	1·00

LENGTH

Millimetre	*Centimetre*	*Inch*	*Foot*	*Yard*	*Metre*
1·00	0·10	0·039	0·0033	0·0011	0.001
10·0	1·00	0·394	0·033	0·011	0·010
25·4	2·54	1·00	0·083	0·028	0·025
305	30·5	12·00	1·00	0·333	0·305
914	91·4	36·00	3·00	1·00	0·914
1000	100	39·4	3·28	1·09	1·00

LIQUID AND CAPACITY

Imperial Minims	*U.S. Minims*	**Millilitres*	*Imperial Fluid Drachms*	*U.S. Fluid Drams*	*Cubic Inches*	*Imperial Fluid Ounces*	*U.S. Fluid Ounces*
1·00	0·961	0·059	0·017	0·016	0·004	0·002	0·002
1·04	1·00	0·062	0·017	0·017	0·004	0·002	0·002
16·9	16·2	1·00	0·282	0·271	0·061	0·035	0·034
60·0	57·6	3·55	1·00	0·961	0·217	0·125	0·120
62·5	60·0	3·70	1·04	1·00	0·226	0·130	0·125
277	266	16·4	4·61	4·43	1·00	0·577	0·554
480	461	28·4	8·00	7·69	1·73	1·00	0·961
500	480	29·6	8·33	8·00	1·80	1·04	1·00

***Millilitres are here taken as equal to cubic centimetres.**

U.S. Pints	*Imperial Pints*	*Litres*	*U.S. Quarts*	*Imperial Quarts*	*U.S. Gallons*	*Imperial Gallons*	*Cubic Feet*
1·00	0·833	0·473	0·500	0·416	0·125	0·104	0·017
1·20	1·00	0·568	0·600	0·500	0·150	0·125	0·020
2·00	1·67	0·946	1·00	0·833	0·250	0·208	0·033
2·11	1·76	1·00	1·06	0·880	0·264	0·220	0·035
2·40	2·00	1·14	1·20	1·00	0·300	0·250	0·040
8·00	6·66	3·79	4·00	3·33	1·00	0·833	0·134
9·61	8·00	4·55	4·80	4·00	1·20	1·00	0·161
59·8	49·8	28·3	29·9	24·9	7·48	6·23	1·00

WEIGHT PER VOLUME OR VOLUME PER VOLUME TO METRIC

Units A	*Units B*							
	Imperial Fluid Ounce	*U.S. Fluid Ounce*	*U.S. Pint*	*Imperial Pint*	*U.S. Quart*	*Imperial Quart*	*U.S. Gallon*	*Imperial Gallon*
Grains	2·3	2·2	0·14	0·11	0·07	0·057	0·017	0·014
Imperial minims	2·1	—	—	0·1	—	0·05	—	0·013
U.S. minims	—	2·1	0·13	—	0·067	—	0·016	—
Imperial fluid drachms ...	125	—	—	6·3	—	3·1	—	0·8
U.S. fluid drams	—	125	8	—	4	—	1	—
Avoirdupois ounce ...	—	—	60	50	30	25	7·5	6·25
Imperial fluid ounce ...	—	—	—	50	—	25	—	6·25
U.S. fluid ounce	—	—	62·5	—	31	—	8	—

To convert Units A per Unit B to grams (or ml. or c.cm.) per litre, trace from the Units A in use on the left to the vertical column corresponding to the Unit B and multiply Units A by the factor. To convert Units A per Unit B to percentages (grams per 100 ml. or volume per volume) convert to grams or millilitres per litre, and divide by 10.

Example—To convert 60 grains per Imperial pint to grams per litre, trace from "grains" on the left to "Imperial pint". The factor is 0·11. Therefore the answer is 60 × 0·11 = 6·6 grams per litre.

METRIC TO WEIGHT PER VOLUME OR VOLUME PER VOLUME

Units A	Units B							
	Imperial Fluid Ounce	*U.S. Fluid Ounce*	*U.S. Pint*	*Imperial Pint*	*U.S. Quart*	*Imperial Quart*	*U.S. Gallon*	*Imperial Gallon*
Grains	0·44	0·46	7·3	8·75	14·5	17·5	60	70
Imperial minims	0·48	—	—	10	—	20	—	80
U.S. minims	—	0·48	7·7	—	15	—	60	—
Imperial fluid drachms ...	0·008	—	—	0·16	—	0·32	—	1·25
U.S. fluid drams	—	0·008	0·125	—	0·25	—	1	—
Avoirdupois ounce ...	—	—	0·017	0·02	0·033	0·04	0·13	0·16
Imperial fluid ounce ...	—	—	—	0·02	—	0·04	—	0·16
U.S. fluid ounce	—	—	0·016	—	0·032	—	0·125	—

To convert grams (or ml. or c.cm.) per litre into Units A per Unit B, trace from the Units A in use on the left to the vertical column corresponding to the Unit B and multiply the number of grams by the factor. To convert percentages (grams per 100 ml. or volume per volume) to Units A per Unit B, multiply by 10, and treat as grams or millilitres per litre.

Example—To convert 15 ml. or c.cm. per litre to U.S. minims per U.S. fluid ounce, trace from "U.S. minims" on the left to "U.S. fluid ounce". The factor is 0·48. Therefore the answer is 15 × 0·48 = 7·2 or about 7 U.S. minims per U.S. fluid ounce.

The figures in the above two tables are rounded off, but are accurate to within 4 per cent (which is sufficient for all photographic purposes).

Capacity Conversions

ENGLISH TO METRIC AND AMERICAN

English Minims	c.cm.	*American Minims*
1	0·059	0·961
2	0·118	1·92
3	0·178	2·88
4	0·237	3·84
5	0·296	4·80
6	0·355	5·76
7	0·414	6·73
8	0·474	7·69
9	0·533	8·65
10	0·592	9·61
15	0·888	14·4
20	1·18	19·2
30	1·77	28·8
40	2·37	38·4
50	2·96	48·0
60	3·55	57·6

English Drachms	c.cm.	*American Drams and Minims*
1	3·55	57·6
2	7·10	1-55
3	10·7	2-53
4	14·2	3-50
5	17·8	4-48
6	21·3	5-46
7	24·9	6-44
8	28·4	7-41

English Fluid Ounces	c.cm.	*American Fluid Ounces and Minims*
¼	7·1	115
½	14·2	231
¾	21·3	346
1	28·4	461
2	56·8	1-442
3	85·2	2-424
4	114	3-405
5	142	4-386
6	170	5-367
7	199	6-348
8	227	7-329
9	256	8-310
10	284	9-292
20	568	19-103
40	1137	38-207

METRIC TO ENGLISH AND AMERICAN

c.cm.	*English Fluid Ounces and Minims*	*American Fluid Ounces and Minims*
0·1	1·69	1·62
0·2	3·38	3·25
0·3	5·07	4·87
0·4	6·76	6·49
0·5	8·45	8·12
0·6	10·1	9·74
0·7	11·8	11·4
0·8	13·5	13·0
0·9	15·2	14·6
1	16·9	16·2
2	33·8	32·5
3	50·7	48·7
4	67·6	64·9
5	84·5	81·2
6	101	97·4
7	118	114
8	135	130
9	152	146
10	169	162
15	253	243
20	338	325
25	422	406
30	1- 27	1- 7
35	1-111	1- 88
40	1-196	1-169
45	1-280	1-250
50	1-365	1-332
55	1-449	1-413
60	2- 54	2- 14
65	2-138	2- 95
70	2-223	2-176
75	2-307	2-257
80	2-391	2-338
85	2-476	2-420
90	3- 80	3- 21
95	3-165	3-102
100	3-249	3-183
120	4-108	4- 24
150	5-134	5- 32
180	6-162	o- 38
200	7- 19	6-366
250	8-384	8-212
300	10-268	10- 69
350	12-235	11-390
400	14- 37	13-252
500	17-287	16-435
600	21- 55	20-125
700	24-470	23-300
800	28- 75	27- 6
900	31-324	30-190
1,000	35- 95	33-365

AMERICAN TO ENGLISH AND METRIC

American Minims	*English Minims*	*c.cm.*
1	1·04	0·062
2	2·08	0·123
3	3·12	0·185
4	4·16	0·246
5	5·20	0·308
6	6·25	0·370
7	7·29	0·431
8	8·33	0·493
9	9·37	0·555
10	10·4	0·616
15	15·6	0·924
20	20·8	1·23
30	31·2	1·85
40	41·6	2·46
50	52·0	3·08
60	62·5	3·70

American Drams	*English Drachms and Minims*	*c.cm.*
1	1- 2	3·70
2	2- 5	7·39
3	3- 7	11·1
4	4-10	14·8
5	5-12	18·5
6	6-15	22·2
7	7-17	25·9
8	8-20	29·6

American Fluid Ounces	*English Fluid Ounces and minims*	*c.cm.*
¼	125	7·39
½	249	14·7
¾	374	12·2
1	1- 20	29·6
1½	1-269	44·4
2	2- 39	59·2
3	3- 59	88·7
4	4- 79	118
5	5- 98	148
6	6-118	177
7	7-138	207
8	8-157	237
9	9-177	266
10	10-197	296
20	20-393	591
40	41-307	1183

Weight per Volume Conversions

ENGLISH TO METRIC AND AMERICAN

Grains per English Pint	*Grams per Litre*	*Grains per American Pint*
1	0·114	0·833
2	0·228	1·67
3	0·342	2·50
4	0·456	3·33
5	0·570	4·16
6	0·684	5·00
7	0·798	5·83
8	0·912	6·66
9	1·03	7·49
10	1·14	8·33
15	1·71	12·5
20	2·28	16·7
25	2·85	20·8
30	3·42	25·0
45	5·13	37·5
50	5·70	41·6
60	6·84	50·0
90	10·3	74·9
110	12·5	91·6
120	13·7	99·9

ENGLISH TO METRIC AND AMERICAN

Grains per English Pint	*Grams per Litre*	*Grains per American Pint*
150	17·1	125
180	20·5	150
220	25·1	183
240	27·4	200
300	34·2	250
360	41·1	300
400	45·6	333
420	47·9	350
440	50·2	366
480	54·7	400

Ounces per English Pint	*Grams per Litre*	*Ounces and Grains per American Pint*
¼	12·5	91
½	24·9	182
¾	37·4	273
1	49·9	364
1½	74·8	1-109
2	99·8	1-291
3	150	2-218
4	200	3-145
5	249	4- 71
6	299	4-436
7	349	5-363
8	399	6-289
9	449	7-216
10	499	8-143
12	599	9-434

METRIC TO ENGLISH AND AMERICAN

Grams per Litre	*Ounces and Grains per English Pint*	*Ounces and Grains per American Pint*
0·1	0·877	0·730
0·2	1·75	1·46
0·3	2·63	2·19
0·4	3·51	2·92
0·5	4·38	3·65
0·6	5·26	4·38
0·7	6·14	5·11
0·8	7·02	5·84
0·9	7·89	6·57
1	8·77	7·30
1·25	11·0	9·13
1·5	13·2	11·0
1·75	15·3	12·8
2	17·5	14·6
2·25	19·7	16·4
2·5	21·9	18·3
2·75	24·1	20·1
3	26·3	21·9
3·5	30·7	25·6
4	35·1	29·2
4·5	39·4	32·9
5	43·8	36·5
6	52·6	43·8
7	61·4	51·1
8	70·2	58·4
9	78·9	65·7
10	87·7	73·0
15	132	110
20	175	146
25	219	183
30	263	219
40	351	292
50	1- 1	365
60	1- 89	1- 1
70	1-176	1- 74
80	1-264	1-147
90	1-352	1-220
100	2- 2	1-293
125	2-221	2- 38
150	3- 3	2-220
175	3-222	2-403
200	4- 4	3-148
250	5- 5	4- 76
300	6- 6	5- 3
400	8- 8	6-296
500	10- 10	8-151

AMERICAN TO ENGLISH AND METRIC

Grains per American Pint	*Ounces and Grains per English Pint*	*Grams per Litre*
1	1·20	0·137
2	2·40	0·274
3	3·60	0·411
4	4·80	0·548
5	6·01	0·685
6	7·21	0·822
7	8·41	0·959
8	9·6	1·10
9	10·8	1·23
10	12·0	1·37
15	18·0	2·05
20	24·0	2·74
25	30·0	3·42
30	36·0	4·11
45	54·0	6·16
50	60·1	6·85
60	72·1	8·22
90	108	12·3
110	132	15·1
120	144	16·4
150	180	20·5
180	216	24·7
220	264	30·1
240	288	32·9
300	360	41·1
360	432	49·2
400	1- 43	54·8
420	1- 67	57·5
440	1- 91	60·3
480	1-139	65·7

Ounces per American Pint	*Ounces and Grains per English Pint*	*Grams per Litre*
¼	131	15·0
½	263	30·0
¾	394	44·0
1	1- 88	59·9
1½	1-351	89·9
2	2-176	120
2½	3- 0	150
3	3-264	180
3½	4- 88	209
4	4-352	240
4½	5-175	270
5	6- 2	300
6	7- 90	359
7	8-178	419
8	9-266	479
9	10-354	539
10	12- 4	599
12	14-180	719

APOTHECARY OUNCES TO AVOIRDUPOIS, METRIC AND AMERICAN

Ounces (Ap.) per English Pint	*Ounces (Av.) and Grains per English Pint*	*Grams per Litre*	*Ounces and Grains per American Pint*
¼	120	13·7	100
½	240	27·4	200
¾	360	41·1	300
1	1- 43	54·7	400
1½	1-283	82·1	1-162
2	2- 85	109	1-362
3	3-128	164	2-324
4	4-170	219	3-286
5	5-213	274	4-248
6	6-255	328	5-211
7	7-298	383	6-173
8	8-340	438	7-135
9	9-383	493	8- 97
10	10-425	547	9- 59
12	13- 73	657	10-421

Makeshift Weights. If proper weights are not available certain coins can be used as a substitute. These weights are accurate enough for most practical purposes, but it is important that new or at least reasonably unworn coins should be used.

It is useful to remember that with English and American silver coins weight is directly proportional to monetary value.

EQUIVALENT WEIGHTS

English Coins	*Grains*	*Grams*
Sixpence	43·6	2·83
Shilling	87·2	5·65
Two shillings	175·4	11·38
Half-crown	218	14·11
Farthing	43·75	2·84
Halfpenny	87·5	5·66
Penny	145·8	9·44

American Coins	*Grains (approx.)*	*Grams (approx.)*
Dime	40	2·5
Quarter-dollar	100	6·5
Half-dollar	200	13
Silver dollar	400	26
Cent	50	3·25
Nickel	80	5

Carefully measured lengths of copper wire, in any standard gauge, will provide very accurate makeshift weights. It is, however, important to ensure that the wire is free from corrosion and has not been stretched.

The following tables show the weight of lengths of copper wire in various gauges:

METRIC WEIGHTS

Weight in Grams	*30 s.w.g. (cms.)*	*24 s.w.g. (cms.)*	*18 s.w.g. (cms.)*	*12 s.w.g. (cms.)*
0·01	1·45	0·46	—	—
0·05	7·23	2·30	0·48	—
0·1	14·46	4·60	0·96	—
0·5	72·3	22·98	4·82	1·03
1	144·6	45·95	9·65	2·06
5	—	229·75	48·2	10·3
10	—	—	96·5	20·6
50	—	—	—	103
100	—	—	—	206

ENGLISH WEIGHTS

Weight in Grains	*30 s.w.g. (ins.)*	*24 s.w.g. (ins.)*	*18 s.w.g. (ins.)*	*12 s.w.g. (ins.)*
¼	0·92	—	—	—
1	3·68	1·17	—	—
5	18·4	5·86	1·23	—
10	36·8	11·71	2·46	0·52
50	184	58·55	12·3	2·63
100	—	117·1	24·6	5·25
109⅜ (¼ oz. Av.)	—	128·1	26·9	5·75
200	—	—	49·2	10·5
218¾ (½ oz. Av.)	—	—	53·8	11·5
437½ (1 oz. Av.)	—	—	—	23·0

For easy manipulation thin wire is best coiled up neatly and tied in a packet. J.D.C.

See also: *Balances and scales; Chemical calculations Chemicals; Solutions; Thermometer.*

WELLINGTON INTENSIFIER. Formula for intensifying negatives first published by J. B. B. Wellington in 1889 and then in an improved form in 1911.

See also: *Intensification.*

WESTON, EDWARD, 1886–1958. American photographer. Started as an itinerant portrait photographer, opened a studio 1911, and until mid-1920s made conventional Salon pictures for exhibitions. Changed his style completely during a three-year stay in Mexico City, under the influence of the artist Diego de Rivera. Weston made portraits for a living, but his reputation rests on his remarkable nature photographs taken from 1927 onward. Founder member of the F.64 Group, 1932. *The Day Books of Edward Weston* edited by Nancy Newhall (1962).

WESTON SPEED. Arithmetical system of emulsion speed classification based on practical tests with reference to the photo-electric exposure meter of that name and now no longer generally used.

See also: *Speed of sensitized materials.*

WET COLLODION PROCESS. Method of making negatives on supports coated with iodized collodion and immersed in a silver nitrate sensitizing bath immediately before exposure.

See also: *Collodion process.*

WET PLATE. Early type of photographic plates which were coated with a collodion emulsion and exposed wet in the camera. They are still used in process work, but even there are being increasingly displaced.

See also: *Collodion process.*

WETTING AGENT. Chemical which reduces the surface tension of water or a watery solution when added in extremely small amounts—about one part in 1,000 or even less.

By lowering the surface tension of a processing solution in this way, the solution is made to spread evenly and quickly over the surface of the sensitized material.

A wetting agent reduces the risk of air bells and makes for uniform development—particularly in daylight roll film, plate, or sheet film tanks.

A trace of wetting agent in the final washing water of films and plates makes the water drain off the surface faster and stops it from drying in tears and leaving marks.

Wetting agents are sold in liquid form, and sometimes even in a solid powder, under a number of proprietary names.

WHEATSTONE, SIR CHARLES, 1802–75. English scientist. Inventor of stereoscopy and of electric telegraphy. Made experiments on stereoscopy (with geometrical drawings) 1832–8, and invented the mirror stereoscope and the refracting stereoscope in 1832. Published his results in 1838.

WHILE-YOU-WAIT PHOTOGRAPHY. The term "while-you-wait photography" applies, strictly speaking, to the work of the itinerant camera operator who plies his trade at country fairs, at the seaside and in places of recreation. He makes his money by supplying finished prints, of a sort, to the customer within a few minutes of taking his portrait.

There are other departments in which prints are produced in minutes—e.g., in newspaper offices and in photo-finish work—but these are not really branches of while-you-wait photography.

There is also a special type of folding hand camera which delivers a positive paper print within seconds of taking the photograph and which lends itself, at least, to use in this particular field.

A fully automatic sitter-operated machine is also made for while-you-wait photography.

Ferrotype Method. A great deal of what is generally accepted as while-you-wait photography of this kind is done by means of the ferrotype or "tintype" method. In this a black enamelled piece of thin sheet iron of the required size of the finished photograph has coated on its surface a sensitive emulsion fast enough to give brief exposure in average daylight. The camera and lens are of the type used in other forms of photography except that the back of the camera is designed to allow the exposed plate to drop directly into a container of developing solution attached underneath. The developer can be poured off and replaced by the fixer. As soon as the plate is fixed, it is brought into the open, rinsed in a convenient bucket of water, waved in the air to dry and clamped into a simple frame. It is usually not quite dry when handed to the customer, but dries completely a very few minutes later. The appearance of the photograph is quite normal because the particular emulsion used forms a milky image and the background is black enamel. This process was invented by a Frenchman and introduced to the French photographic world in 1852. In various forms it has since been very largely used by itinerant photographers and is still practised to-day.

Paper Negative Method. Another method occasionally used is to make the exposure on negative card, which is nothing more than ordinary negative emulsion coated on a cardboard base. This is developed and fixed in the normal way but inside a special compartment in the camera with light-tight sleeves for the operator's arms. It is then re-photographed on to similar material so as to produce a positive. There is no need to dry the negative card before re-exposure, as it does not come into contact with the positive print. The quality of print produced in this way is not equal to that obtainable from a transparent negative, but it is still quite acceptable for inexpensive portraiture. Quick drying can be facilitated by coating the emulsion on a waterproof based card.

Diffusion Transfer Cameras. A further system of instant photography uses the Polaroid Land camera in which a positive print is produced within about 10 to 15 seconds. The camera is designed mainly for owner use, but is also employed to an increasing degree by while-you-wait photographers and instant picture services both in photographic studios and at fair grounds and similar places. The sitter here receives a positive print which is taken straight out of the camera. With a special colour film, colour prints can also be obtained.

In the camera itself the exposure produces a negative—usually on paper—which is processed in contact with a non-sensitized positive material by spreading a layer of jellified activating solution between the two. This develops the negative and at the same time transfers the unused silver salts to the positive layer where they are reduced to form metallic silver. The colour material works in a similar way, though using a more complex system of dye transfer.

Sitter-Operated Camera. In suitable sites—e.g., the big stores—there are special automatic while-you-wait portrait booths. The user sits in a chair inside a cubicle about the size of a telephone box and adjusts his position to bring his eyes level with a mark on a mirror in the booth. Other mirrors in the cubicle help him to assume the desired expression. When he is ready to be photographed he simply inserts a coin in a slot and the machine goes into action. The photographic lighting switches on, an automatic camera makes the exposure, or series of exposures, and within one minute the machine delivers the finished portrait in the form of a strip of brownish or black-and-white paper prints. Sometimes the prints come singly in a metal frame.

Behind the scenes there is an automatic camera fitted with a wide-aperture short focus lens, and loaded with reversal paper, in rolls. After exposure, the strip of paper with the 4 to 6 pictures on it is cut off from the roll and passes in turn through the various processing solutions.

Alternatively, the paper may be mounted in single pieces in a metal frame. The frame then drops into a horizontal position after exposure and a measured quantity of each of the various processing solutions is squirted in turn on to its sensitized surface. The metal frame acts like the sides of a dish and prevents the processing solution from running off.

The processing routine follows the normal reversal procedure—i.e., develop, rinse, bleach, clear, second exposure, and final development. No fixing is of course required and the print is reasonably permanent without further treatment.

The manipulation and processing are all carried out automatically and the temperature inside the mechanism is kept at 90° F. (32·2° C.) by a thermostat. All processing tanks are fitted with floating lids and the solutions are renewed when necessary. The times of the various operations in the automatic cycle can be adjusted to suit local conditions. P.W.H.

See also: *Diffusion-transfer; Polaroid Land camera; Reversal process.*

WHITE, CLARENCE HUDSON. 1871–1925. American photographic exhibitor, advocate of straight photography, and founder member of the Photo-Secession, 1902. Lecturer on photography at Columbia University, New York, from 1907, and at his own summer school of photography at Sequinland, Maine (later Canaan, Conn.). White was also instructor at the Brooklyn Institute of Arts and Sciences, 1908–21.

WHITE, HENRY, 1819–1903. English solicitor and amateur photographer, London. One of the earliest artistic landscape photographers, White's close-ups of plants and walls in the 1850s and '60s were forerunners of the non-romantic approach to the subject.

WHOLE PLATE. Standard format for negatives and prints measuring 6½ × 8½ ins. (about 16·5 × 21·6 cm.).

See also: *Sizes and packings.*

WHOLESALE PHOTO FINISHING

The photographic dealer who operates a developing and printing service for amateur roll films very rarely carries out the work himself. He is much more likely to send the films to a wholesale photo finishing establishment specially organized and equipped to deal with films in large quantities. Such establishments are also known as D. & P. (i.e., Developing and Printing) works. They vary in size from the so-called "one-man" business to huge concerns capable of processing many thousands of films per day.

With the introduction of more automatic processing equipment, small firms—and even some of the larger businesses—have tended to follow the modern trend to merge several small units into one large concern. The main reason for this is that the electronic and mechanical equipment which is now almost essential in providing a profitable output is extremely costly and the small firms usually cannot find sufficient capital, or they cannot in any case justify the outlay. This applies even more to

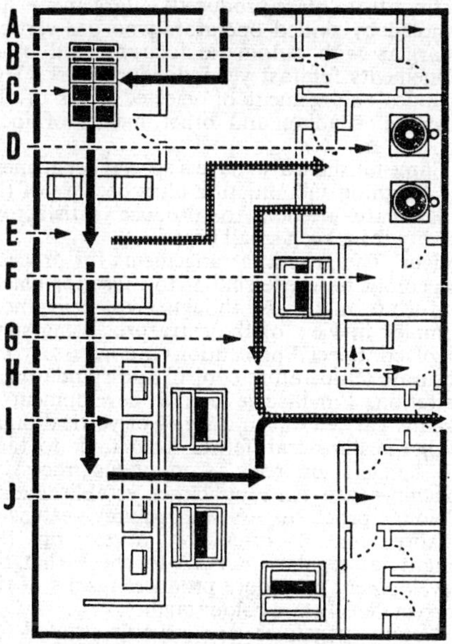

PHOTO-FINISHING LABORATORY. Solid arrows show flow of work for standard developing and contact printing orders, dotted arrows for enlargements. A, stores. B, office. C, film developing. D, enlarging. E, film drying and sorting. F, chemicals and mixing. G, print finishing (washing, drying, glazing) and packing. H, men's cloakroom and toilet. I, contact printing. J, women's cloakroom and toilet.

finishers who handle colour material. Not only each type (negative, reversal) but also every make of material calls for its own processing "line", that is to say a line of processing tanks and replenisher units, together with a temperature control unit, so that each material receives its correct processing treatment. Only the very large concern can provide such facilities and operate them on an economical basis.

The dealer therefore sends colour films to his local finisher; but the latter, instead of carrying out the processing himself, forwards it to one of the large colour processing houses, for which service he receives a small percentage of the retail price.

Organization. When the customer's film is handed to a retail photographic dealer or chemist for developing and printing, it is passed on to the photo finisher who processes and prints it. It is then returned to the dealer with the prints and finally handed back to the customer in as little as twenty-four hours from the time he took it in, or in three or four days in the case of colour films. For acting as the agent of the photo finisher, the dealer receives a percentage of the retail price.

While all work is handled in bulk, each film is recorded separately. The dealer attaches a numbered order form to the roll and this form stays with the film all through the processing and until the finished negatives and prints are ready for handing to the customer. The form is stamped with a dealer's number which tells the finisher to which account the order should be charged. The finisher retains half of the form to enable him to make up his accounts.

Finishers supply their dealers with order pads, price lists and window display showcards and as a rule they run a motor van collection and delivery service to the dealers' shops.

The D. & P. works consists generally of six departments: office, tank-room, drying-room, contact and enprint printing section, enlarging department, and finishing-room. Many works now use automatic continuous processing machines and dispense with the film drying-room, the processed films being conveyed straight into a special drying section at the end of the processing line. Similarly, contact printing is often done on electronic machines. These are designed to make projection prints, one-to-one (or same size as the negatives), as well as the popular enprints (or enlarged prints), so that all this work is done in the one department. There are often several other sections, for instance, those which deal with checking of work, packing and dispatch, miniature film processing, chemical mixing, maintenance, etc.

In the smaller establishments the work is usually handled in ten-minute units, films being divided into batches which will take ten minutes to develop, ten minutes to print, multiples of ten minutes to wash and so on. This standardizes the flow of work through the various departments and maintains an easily controlled and supervised production. In the larger works, where continuous processors are used for films and prints an even flow of production is maintained by the nature of the equipment.

Developing. Films are developed in vertical tanks made of glazed stoneware, stainless steel or hard rubber. These vary in size from 12 to 48 gallons capacity in small works, whereas the larger plants use special developing machines.

In either case the films are stripped of their protective backing-paper and are placed in spring clips by which they are suspended in the tanks from stainless steel rods with the order forms clear of the solution. Solutions are warmed by immersion heaters and the temperature in each tank is maintained at a constant level by a thermostat. Replenishing solution is added to the developing tanks, often automatically, to maintain a constant level and strength so that all processing times can be kept standard. In developing machines the rods of films are raised from one tank and lowered into the next quite automatically, times being controlled by the setting of a time-clock. Once the films have been loaded on to their rods, the operator does not touch them until they are dry.

Printing. After the final wash, the films are dried in heated cabinets at controlled temperatures and passed on to the printers.

There are various types of printing machine —e.g., the old type "judgement printer" with which a skilled operator examines each negative and relies on experience in judging the correct type of sensitive paper and the right exposure time for each print, but these are now little used; the later automatic machine which "reads" the negative through a photo-electric cell.

Finally there is the "roll head" printer on which the exposure is either "read" on a dial and set manually or which may set itself automatically as it "reads". With such machines the exposures are made on sensitized paper in rolls of 250 feet or 500 feet., and these rolls are subsequently placed on continuous strip processors which develop, rinse, fix, wash and glaze the prints without any work from the operator other than a watchful eye. The strips of prints are even cut up into single prints by an electronically operated trimmer. The roll-head printer may also be used by the smaller works without a continuous processor, the prints being then cut off into strips and processed by hand in long trays of solution, and washed in tanks with jets of water. The most advanced machines of this type control the exposure using a single grade of printing paper. The operator's job is then confined to selecting the number of prints per negative, the print size (where adjustable) and making overriding corrections for negatives of special tone distribution which do not yield good results with automatic measurement.

Finishing. After prints have been glazed and trimmed in either of the above processes, they are stacked into piles, one to each order and are placed in wallets after having been identified by "printing numbers" which are stamped on both order forms and prints in the printing-room.

Meanwhile the rolls of film have been cut into separate negatives and placed into wallets to await the prints. Completed orders are passed to checkers who examine the work for good quality and correctness of execution. Faulty prints are returned to the printing department to be reprinted.

Special Services. Orders for high-quality (as distinct from mass-produced) enlargements are handled by skilled operators who use vertical enlargers with automatic focusing and other refinements for fast yet individual work. This includes enlargement of selected parts of the negative, shading and other means of local control, etc.

Some finishers also have a special department for handling subminiature films because of the great care needed to produce satisfactory quality from such small negatives.

Colour. The general arrangements for processing colour films are similar to those for black-and-white materials, though necessarily more complex in view of the extra processing steps involved. Special precautions are necessary for accurate temperature control and standardized agitation. Small-scale colour development is usually carried out in a series of vertical tanks with the films transferred from tank to tank on hangers or reels. Continuous-processing machines are used by larger establishments. Colour processing also requires extensive quality control—often by developing test lengths at regular intervals to check that the development procedure produces results of the correct density and colour quality.

Colour negatives are usually printed on semi-automatic enlargers taking continuous rolls of paper and producing fixed-size enlargements. Operators have to be specially trained, as they must also be able to grade the colour negatives, i.e. select the right filtering for correct colour reproduction. On some printers even this is automatic, but reliable in only about 80 per cent of the negatives printed. A final quality check pinpoints prints of unacceptable colour rendering; the negatives are then sent back to operators for individual reprinting.

High-quality colour enlargements require greater skill and are carried out by specialist operators.

The continuous rolls of colour paper pass through automatic processing machines similar to the rolls of paper from roll-head printers for black-and-white enprints. Individual enlargements on single sheets of colour paper are processed in tank sets which take the prints in special processing baskets, or in small rapid colour processors. L.G.S.

See also: *Darkroom; Processing machines.*

WIDE-ANGLE LENS

Any lens which covers, across the diagonal of the negative, an angular field of view of 60° or more can be classified as a wide-angle lens. This field corresponds to a negative diagonal slightly in excess of the focal length of the lens.

There are many types of wide-angle lens ranging from very simple forms which can only operate at very small relative apertures to complex constructions either operating at large apertures or at extreme image angles.

Design. With the exception of the inverted telephoto system, wide-angle lenses tend to be more symmetrical about a central diaphragm than other photographic lenses and their lens components are usually strongly meniscus in shape with the concave surfaces facing inwards.

Simple wide-angle lenses can be made to cover fields of view of 90° or 100° but at these angles useful results are only obtainable at apertures of *f* 11 or *f* 16. At such small apertures poorly lighted subjects cannot be focused and composed satisfactorily on a screen in the camera. Some manufacturers choose a lens form which gives, for this purpose, an acceptable visual image at about $f\,6{\cdot}5$. These forms must, however, be stopped down to about $f\,16$ before they reach the necessary photographic standard of definition.

There are of course many applications which demand relative apertures more in keeping with modern photographic practice. These needs have been met by modifications to more complex lens constructions to cover fields of view of about 75° at relative apertures up to about $f\,4$.

Illumination of the Field. The increased angular field of wide-angle lenses draws attention to the difficulties of providing sufficiently uniform illumination throughout the image.

Even a perfect lens provides a variation of illumination across the negative which varies as the fourth power of the cosine of the semi-angular field measured from the lens axis.

For example, at a point in the image corresponding to 30° of the semi-angular field such a lens can only provide a level of illumination 56 per cent of that in the centre of the image. At an angular field position of 45° the illumination falls to 25 per cent of that in the centre. These percentage reductions in illumination are not affected by change of relative aperture.

Vignetting. Well corrected wide-angle lenses, however, have appreciable axial length and can be considered as hollow cylinders through which axial and oblique beams of light have to pass. Such a cylinder obviously cannot pass an oblique beam as wide as the axial beam because the front and rear apertures of the cylinder vignette (i.e., cut off part of the diaphragm opening as seen from the edge of the field).

This type of mechanical vignetting aggravates to a considerable extent the optical cosine vignetting referred to above. In many cases wide-angle lenses are designed so that their front and rear apertures can be increased beyond those required for the axial

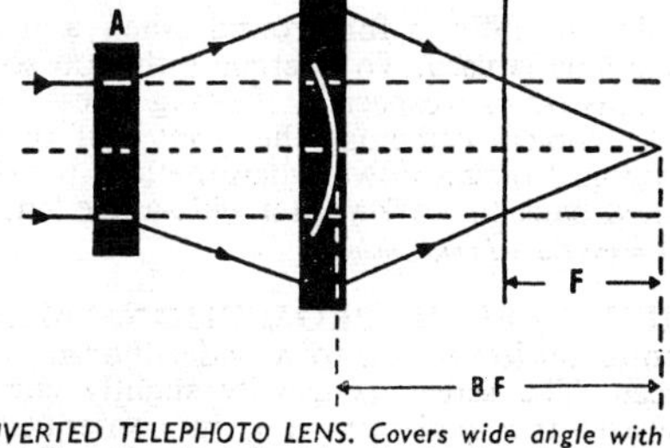

INVERTED TELEPHOTO LENS. Covers wide angle with long back focus for short focal length. A, diverging lens group. B, converging group F, equivalent focal length. BF, back focal distance.

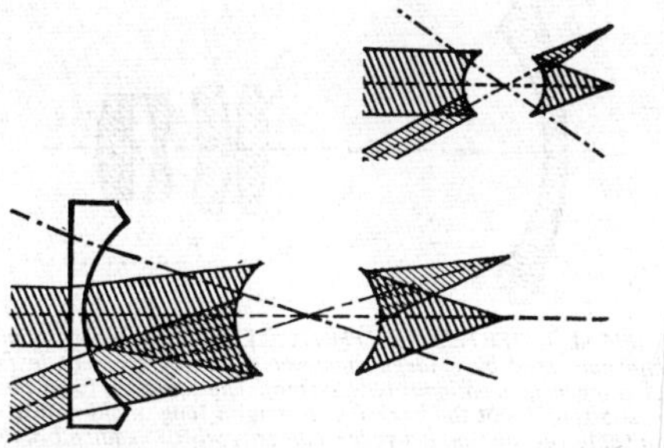

INVERTED TELEPHOTO SYSTEM. Compared with a normal lens (top) an inverted telephoto system of the same focal length, aperture and angle of view can provide an increased oblique aperture, thus reducing vignetting.

beam alone thereby reducing this mechanical vignetting as much as possible.

Stopping down a lens of finite length by means of a central iris reduces the diameter of the axial beam and since front and rear apertures are not affected the mechanical type of vignetting will be reduced. It is usually completely removed at intermediate apertures leaving a lens which approximately obeys the cosine law.

Inverted Telephoto System. Entirely different forms of wide-angle lenses are based on what is known as the inverted telephoto system. In these the lens comprises a front dispersive component of relatively large diameter widely spaced from a rear collective component. The arrangement results in a distance between the rear surface and the focal plane which can exceed the focal length. This is particularly advantageous in small cameras in which the focal length has to be small and shutters or other mechanisms are situated between the lens and the film.

Thus wide-angle lenses for single-lens reflex cameras are almost invariably of the inverted telephoto design. This permits the lens to be mounted sufficiently far forward of the film plane to accommodate the movable reflex mirror. At present inverted telephoto systems of this kind are available with the back focal distance up to twice the nominal focal length.

Owing to this property, inverted telephoto lenses are also known as retrofocus systems.

The inverted telephoto wide-angle lens is largely restricted to these small cine and miniature cameras since it is a very bulky form of lens relative to its focal length. Distances between the front lens surface and focal plane of up to five times the focal length are quite common and diameters, especially at the front, are correspondingly large. This large bulk and obvious departure from symmetry make inverted telephoto lenses readily recognizable.

The inverted telephoto lens has also gained appreciably in importance with the development of interchangeable lens systems for cameras with a diaphragm shutter. Since the

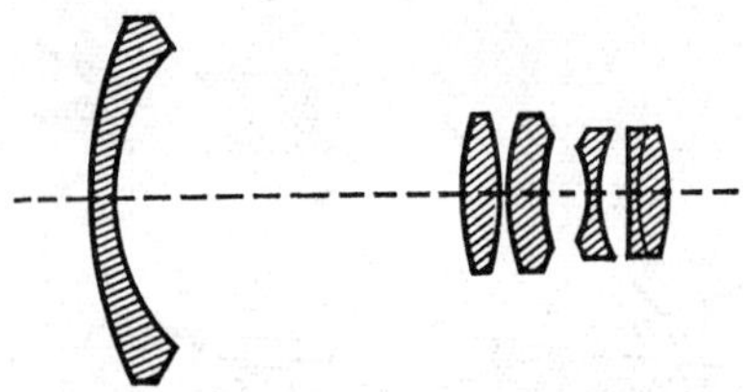

TYPICAL INVERTED TELEPHOTO LENS. This construction is characterized by a large diameter diverging element in front of a more conventional lens system. The effective focal length is shortened, but the back focus remains long, making this lens suitable for single-lens reflex cameras which require plenty of space between the lens unit and the film for the mirror system.

shutter is permanently built into the camera, the lens has to be mounted in front of it. Alternative lens systems must therefore all have the same back focal distance within fairly narrow limits.

An alternative method of obtaining a range of focal lengths on a camera with built-in shutter is to split the lens itself and have the rear component also built-in behind the shutter. Alternative focal lengths are obtained by interchangeable front components—a construction which permits short focal lengths to be obtained only by a use of inverted telephoto designs.

This form of lens has other very desirable characteristics. It can be designed with greatly improved vignetting and achieve, simultaneously, wide angle and large aperture. Angular fields of about 75° can be covered by this form at apertures greater than *f* 2—a performance not yet equalled by more symmetrical lens forms.

The better vignetting characteristics result from both the greater back focal distance and the large front component. The increased focal distance reduces the obliquity of beams of light forming off-axis images. The front dispersive component has to be large enough in diameter to accommodate the increased width of oblique beams that can be transmitted by the rear component. This leads to an increased oblique aperture for this lens design, and hence less vignetting.

Distortion. Provided the object lies solely in a single flat plane at right angles to the lens axis and also that the sensitive emulsion is at right angles to the lens axis, a wide-angle lens corrected for distortion will record, throughout its field of view, a correct geometrical reproduction of the object.

Although distortion in the aberrational sense may be absent from the lens itself, there are independent geometrical factors affecting perspective which, under certain viewing conditions, can be distorted to an objectionable extent.

Perspective. When wide-angle lenses are employed it is often impossible to maintain the ideal conditions for rendering perspective. In wide-angle photography it is likely that the camera viewpoint will be too near the object and also that the focal length of the camera lens (multiplied by the enlarging factor, if any) will be less than the viewing distance of eye to photograph. Both these factors introduce an unusual perspective which appears as a disproportionately large variation in the scale of foreground objects to background objects.

This violent variation in size between near and far objects apparently increases the depth dimension in the photograph to an unnatural degree. Since the wide-angle lens has a good view of the sides of objects at the edges of the field, this increase in depth is more apparent there than at the centre.

Thus the distortion in three-dimensional objects at the sides of the photograph is such that they appear to be too large in the direction parallel to the lens axis.

Perspective can be restored to some extent by viewing the photograph from such a distance that it subtends the same angle at the eye as the original scene subtended at the camera. It should be realized, however, that this can only be achieved for monocular vision and that, although the eye then sees exactly what the lens saw, different mental processes are employed in viewing a real scene and in viewing a single plane reproduction. G.H.C.

See also: *Covering power; Lens history; Telephoto lens.*
Books: *Photographic Optics* and *A System of Optical Design*, both by A. Cox (London).

WIDE-ANGLE RACK. Supplementary focusing rack on the baseboard of a stand camera which allows the lens panel to be racked back from the normal infinity stop so that a wide-angle lens may be focused. This is necessary because the short focal length of the wide-angle lens brings it closer to the plate than a normal lens.

Sometimes the sliding member on the baseboard of a stand camera can be racked right out and reversed to bring the lens closer to the plate.

Occasionally the dimensions of the camera are such that the front edge of the baseboard would cut off the foreground when a wide-angle lens is used. To overcome this on some cameras—more especially folding Press type and technical cameras—the front half of the baseboard hinges down when the lens panel is racked back to work with a wide-angle lens.

See also: *Camera movements.*

WIDE SCREEN PROJECTION. Motion picture projection on to a wider-than-normal screen. The screen is usually slightly curved towards the audience who, in consequence, experience a sensation of participation in the action of the film.

There are various systems for achieving more or less similar effects.

See also: *Projection practice.*

WIENER SPECTRUM. Alternative term for the noise power spectrum, derived from a Fourier analysis of the density distribution due to granularity in a photographic image.

See also: *Information in the image.*

WILLIS, WILLIAM, 1841–1923. English engineer, bank employee and photographer. Patented the Platinotype process in 1873 and founded the Platinotype Company which manufactured platinum development paper from 1878 onwards. Later worked out the Satista (silver-platinum) and the Palladiotype papers. Received the Progress Medal of the Royal Photographic Society in 1881.

WILSON, GEORGE WASHINGTON, 1823–93. Scottish professional portrait and landscape photographer in Aberdeen from 1853. Was among the earliest to take instantaneous street scenes, 1859. Appointed photographer to Queen Victoria in Scotland, *c.* 1860. Built up an extensive business in publishing photographs of all countries. The firm was also the largest producer of lantern slides in Britain.

WINTER SPORTS. Snow scenes can be taken with any camera, but for winter sports a small camera is best. Any size, from 35 mm. to $2\frac{1}{4} \times 3\frac{1}{4}$ ins. is suitable, but the miniature is preferable for long ski tours and similar winter sports; it reduces the problem of carrying film and does not weigh down the photographer, who is already loaded with his gear and sufficiently tired by the exertion of ski-ing. The lighter the camera and the handier its controls, the better the results will be when fingers are stiff from the cold.

For sports photographs a long focus lens will be found useful, to fill the negative with figures which would otherwise be disappointingly small—e.g., a ski-jumper coming down a steep slope. Experts usually prefer a frame finder as it shows everything in its natural size and makes it easier to follow fast movement.

Miniature cameras are also to be preferred for another reason: they can be carried under the outer clothing and so kept warm in colder weather. Old shutters give slower exposure times in lower temperatures, and the 1/100 second setting may easily become a 1/50 second or more. With a camera that cannot be protected from the cold, the slowing down may have to be corrected by selecting a shorter exposure.

If the camera is to be carried on the chest whilst ski-ing or climbing, a length of elastic around the camera and chest will prevent it from swinging or bouncing.

In dazzling snowscapes a lens hood is an essential accessory and also, above about 3,500 feet, a light or medium yellow filter and a U.V. filter.

Film and Filter. A glance at any winter resort brochure shows how effectively colour film can record winter sports. The faster and less contrasty films are best for snow scenes, and the excess blue in the shadows must be filtered out. At altitudes up to about 3,500 feet, a colourless U.V. filter will be sufficient on sunny days, but on sunless days and at higher altitudes a skylight filter is necessary.

For black-and-white pictures, too, the medium to fast films are best able to deal with the high contrast of scenes with people and/or objects in the snow. When using black-and-white material, in addition to the above-mentioned filters, filters of various colours may also be used to achieve desired effects,—e.g., yellow, orange or even red.

In the lower reaches, a medium yellow filter will generally improve the sky tones and snow texture; at medium heights, a light yellow filter is all that is necessary. At altitudes from 3,500 feet up a U.V. filter is essential. The U.V. filter is practically colourless glass, which removes a large amount of the ultra-violet rays always present in the light at great heights. For pictures in the mountains, filters must be used with caution. If the filter is too deep, the photographs will lack the characteristic atmosphere of the mountain scene.

Deeply tanned faces produce a filter problem of their own. Correct filters show the tanned faces and the deep blue sky in nearly the same grey tones. Stronger filters make the faces lose their deep tan and render the sky too dark. Light filters make the faces too dark against a very light sky. The final choice must be a compromise depending on the conditions and the photographer's taste.

Light and Exposure. Snow photographs are only effective when taken in sunshine. Without sun, snow scenes always look flat. Early morning and the afternoon, when shadows are strong and clear, are the best times for photographing.

Good photographs can be obtained with side lighting; but the best are almost always taken against the sun, when it is high in the sky. Only strong sidelighting or clear light shining towards the lens can give the snow its true sparkle, and this sparkle is essential for conveying the optical effect of snow.

The exposure times must be correct; there is little or no margin one way or the other. Fully exposed photographs have a tendency to lack detail in the highlights with normal development. It is preferable to expose fully and curtail development to about two-thirds normal.

With the abnormal amount of general reflection coming off the snow, the exposure meter can only be relied on up to a point. The readings are correct for the snowscape by itself, but as soon as the foreground includes shadow detail the exposure must be doubled. For near shots close-up readings are necessary.

The Subject. A picture taken on a slope lying diagonally across the light or illuminated from the side by the sun, so that it sparkles with reflected light, should generally have one interesting item sharply defined in the foreground to provide a base for the composition.

Ski photographs come out best when there is freshly fallen powder snow about. This sweeps up on both sides with the movement of the skis, and the cloud of powdered snow, taken when it is a little windy and against the sun, can make a most impressive picture.

When taking photographs of skiers or ski-jumpers the choice of the correct positioning is everything. The shadow side of a hill, with the sun shining across the crest will always make a good vantage point for photography. As the skier moves downhill, the clouds of snow he raises are brilliantly illuminated against the sun. Under these conditions a long focus lens and frame finder are particularly useful accessories.

Pictures of skaters are generally easier to take because they can if necessary be rehearsed and repeated. The most interesting subjects are undoubtedly figure skaters, and the light is generally bright enough to allow exposures of 1/200 to 1/500 second which is short enough to arrest the movement. Pictures of racers are only interesting if they are taken close enough to catch the tense expressions of the competitors and at an instant when the grouping is dynamic. Mere speed as such does not register convincingly.

The last rule holds good also for pictures of toboggan and bobsleigh races. These should be taken on the corners where the racers mount high up the banking and thus convey an impression of their speed. At such points there is also a chance to photograph a spill. Zone focusing and a shutter speed of 1/500 second (less if the camera is swung) are the best combination for such subjects.

Curling is a game that moves more slowly and is usually played by individuals of a more assorted and characterful type. Their gestures and antics aimed at guiding the "cheese" make amusing pictures. No great shutter speeds are called for; near or medium zone focusing at about 1/125 second should give good results.

In ice hockey the most profitable area for the photographer is the area in front of the goal. This is where nearly all the shots for goal are made and where the most exciting incidents happen. As the puck can be a dangerous missile, it is wiser to work at a distance with a long focus lens. Action shots will require exposures of about 1/500 second and shots of players taking a corner or free hit, no more than 1/100 second.

Making the Print. The negative must be well graduated but not too thin, or the print will lack vigour. On the other hand, if the negative is too hard, the result will look more like white plaster than snow. It is true, however, that no print on paper can truly reproduce the sparkling brilliance of a snow scene. A transparency is much more likely to produce the desired effect.

For the best results from paper prints all snow pictures should be printed on highly glossy paper with plenty of contrast and clean highlights.

Chloride papers generally give better results, as their tonal range is greater than that of bromide paper; the blacks are deeper and the half-tones richer. R.K.

See also: *Ice rink; Movement; Snow; Sports.*

WIRING. The electrical wiring of a building carries the mains supply to the various lamps, heaters, and other pieces of equipment that make up the permanent installation. It also supplies convenient points to which items of portable or movable equipment can be connected by flexible leads.

Permanent Wiring. Permanent wiring consists of copper conductors insulated with vulcanized indiarubber (V.I.R.), or nowadays more often in plastic. These insulated wires may be run in metal conduit, sheathed with lead, sheathed with tough rubber, or simply supported away from the wall surface on porcelain cleats.

The thickness and number of the copper wires is always arranged to be more than sufficient to carry the heaviest load that the circuit is likely to impose, and the insulation is always tested to withstand many times the supply voltage.

The installation or extension of the permanent wiring of a building calls for a qualified contractor; amateurs and handymen who insist on treading on such dangerous ground should at least consult the I.E.E. regulations on the subject.

Flexible Wiring. The wiring to standard lamps may be made of twin twisted insulated flexible wire; all other portable equipment must, by regulations, be connected to the supply through three-core circular braided flexible cable. This cable consists of three insulated conductors, coloured red, black and green, laid up with hemp worming into a circular section and covered with either indiarubber, braid, or both.

The colours indicate how the flex is connected to its plug and to the equipment: red, to the live pin of the plug and the switch on the equipment; black to the neutral pin of the plug; and green to the earth pin of the plug and all exposed metal parts of the equipment. (At one time, the earth wire was coloured brown, but green is now the approved colour.) This colour scheme applies to Britain only.

The current rating of flex should never be less than the maximum working current of the equipment to which it is connected, to avoid overheating and danger of fire.

Socket Outlets. The supply points to which equipment is plugged in are known as socket outlets. There are three standard sizes: 2 amp, 5 amp and 15 amp. Nowadays two-pin sockets are no longer fitted; all sizes are of the three-pin type. The socket for the large pin is connected to earth and when the socket is looked at with this hole at the top, the hole on the left (marked N) is connected to the neutral (black) wire, and that on the right (marked L), to the live (red) wire.

When the outlet is controlled by a switch, the assembly is known as a switch socket outlet. The better types of socket outlet have an automatic shutter that covers the live and neutral sockets when the plug is withdrawn. The shutter is released by the earth pin on the plug so that the arrangement prevents two-pin or non-standard plugs from being used.

Plugs. The plugs by which the equipment is connected to the socket outlet are made in 2, 5 and 15 amp sizes. Most plugs nowadays incorporate some type of clamp which firmly anchors the flex to the body of the plug and prevents any strain from being taken by the wire at the terminals.

Three-pin plugs are wired in this way: looking on the back of the plug, with the earth pin at the top—brown or green wire to the earth pin; black wire to the left pin (marked N); red wire to the right pin (marked L).

Distribution. The main supply cable is connected to a pair of fuses as close as possible to its point of entry into the building. These fuses fix the total amount of current that can be drawn from the mains by the equipment in the building. The main fuses are there to protect the supply cable; they are sealed in their own fuse box and if they "blow" they must be replaced by the supply authority.

The cables that emerge from the main fuse box pass via the meter (which measures the number of units used in the building) to one or more distribution boxes. These distribution boxes hold sets of fuses which are connected to the various sub-circuits of the permanent

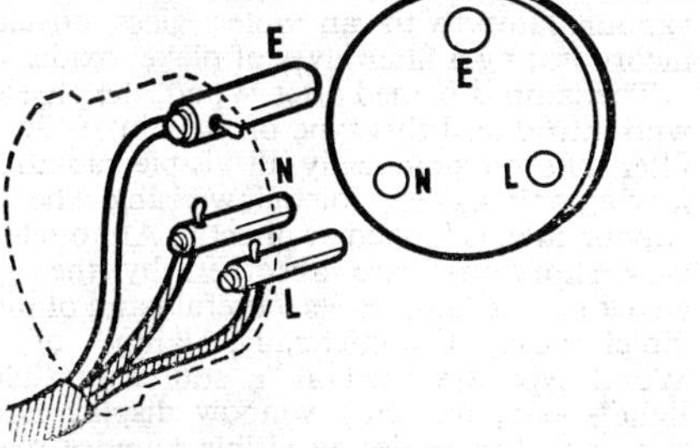

THREE-PIN PLUG AND SOCKET. E, earth pin, connected to green wire of cable or flex. N, neutral pin, connected to black wire. L, live pin, connected to red wire.

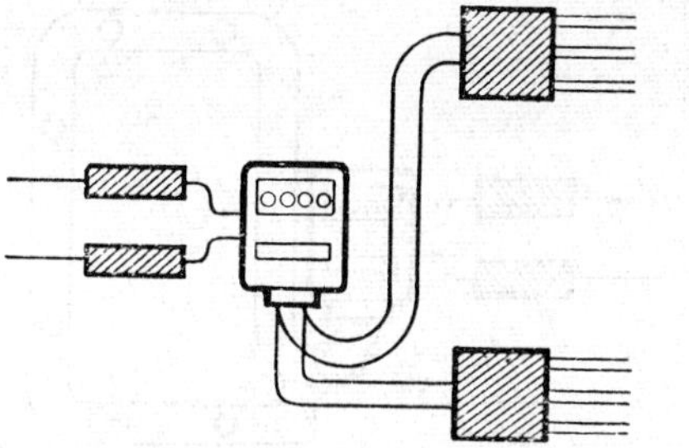

STANDARD DISTRIBUTION NETWORK. The current supply reaches the meter through the service fuses, and is then distributed to subsidiary fuse boxes and hence to permanent lighting and power circuits

wiring. Each sub-circuit supplies its own group of points—lamps, heaters, etc.

The maximum safe load that the sub-circuit wires will carry fixes the type and amount of equipment that can be connected to it. It also fixes the value of the fuses in that outlet.

In planning the distribution it is assumed that only a proportion of the equipment on any sub-circuit will be switched on at one time. So if the maximum safe current that can be carried by the cable in a sub-circuit is 15 amps, the capacity of the fuse must also be 15 amps. But the circuit could include enough equipment to impose a load that, if it were all switched on at the same time, would be more than 15 amps.

The idea of dividing the distribution network into sub-circuits is to make the system safe. In a correctly planned system, as soon as any distribution cable, flex, or piece of equipment carries substantially more than its rated current, a fuse blows and that part of the system becomes dead.

This is why it is not good practice to connect a piece of equipment that only takes a small current—a 25 watt safelight, for instance—to a power point. A short circuit in the thin flex of the safelight could carry enough current to heat it up and start a fire without blowing the 15 amp fuse of the power circuit. If the safelight is connected as it ought to be, to a circuit reserved for lightly-rated equipment, the current need rise no higher than two or three amps before blowing a fuse.

Ring Main Distribution. Until recent years there was no alternative to the sub-circuit system of distribution. Nowadays, however, there is a new system—Ring Main Distribution—which may in time become universal. This method employs a single heavy-duty cable in place of the multitude of sub-circuits. The only fuse in the wiring is the main fuse, rated to carry the entire current of all the equipment in the building.

All the heating, and power connexions are provided by 13 amp socket outlets connected directly to the main cable. These 13 amp socket

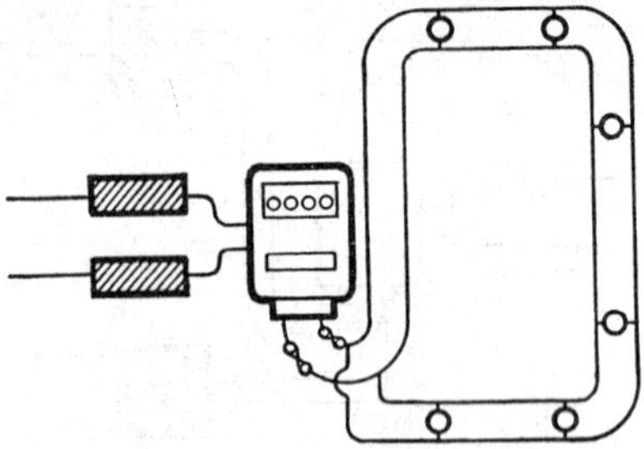

RING MAINS. The mains supply reaches the meter through the service fuses, and thence goes along the ring main all round the building. Outlets wired as required, used with fused plugs.

outlets will accept only the special flat-pin fused plugs designed for them, and each piece of equipment carries its own fuse in the plug that connects it to the ring main.

The size of the fuse depends upon the current rating of the equipment, but the plug is always the same size, whether it belongs to a 3 kW. heater or a 25 watt safelight. All fuses are of the cartridge type and the plugs are designed to allow the fuse to be changed easily. F.P.

See also: *Electricity; Fuse; Lamp caps and fittings; Switches; Transformers.*

WOLCOTT, ALEXANDER, 1804–44. American instrument maker and daguerreotypist. Designed a camera without a lens which the light entered to be reflected by a concave mirror on to the plate (which was turned away from the subject). On the 7th October 1839 probably took the first successful portrait ever made. In March 1840 opened the world's first photographic portrait studio in New York.

WOLLASTON, WILLIAM HYDE, 1766–1828. English physician and scientist. Made many improvements to optical instruments, especially the microscope and the camera obscura, for which he invented the meniscus lens in 1812. Invented in 1807 the camera lucida. In 1802 observed the blackening of silver chloride in the ultra-violet part of the spectrum. Also observed the dark lines (better known as the Fraunhofer lines) in the solar spectrum.

WOODBURY. WALTER BENTLEY, 1834–85. English professional photographer. Invented in 1864 the Woodburytype process of printing with pigmented gelatin from lead moulds which had been obtained from chromated gelatin reliefs. The process was widely used as it produced prints with all half-tones, high definition, and without any grain or screen. It was superseded in the late 1880's by Collotype and other processes whose prints did not need trimming and mounting. Invented stannotype 1879. Woodbury also worked on balloon photography, on rotary printing and on stereo-projection. Received the Progress Medal of the Royal Photographic Society in 1883 for his stannotype process.

WOODBURYTYPE. This obsolete photo-mechanical printing process produced magnificent reproductions in monochrome and was probably the only photo-engraved method which had absolutely no grain or half-tone dot formation. It was named after its inventor and was used from about 1864 until the turn of the century.

A high gelatin relief was made on special carbon tissue by printing from a continuous tone negative. The relief was dried and pressed into lead in an hydraulic press. This lead cast provided the "printing plate".

A warm solution of pigmented gelatin was poured into the lead mould, allowed to set and then transferred on to a sheet of paper pressed into contact with it. Woodburytype was a true continuous tone intaglio process, but, although its results were beautiful, it was superseded by other more rapid processes.

Woodburytype prints can always be identified by their complete lack of any grain and by the fact that they are always trimmed flush and mounted separately into the page.

WOOD PRINT. Wood and similar materials can be treated so that they can be printed with a photographic image. The printing may be carried out by contact printing by daylight or artificial light, or by enlarging.

After the surface of the wood has been sealed and, if necessary, coated with a white base for the image, it may be printed on by any of the following methods:

(1) The surface may be brushed over with a sensitized coating or emulsion, exposed under the negative by contact or in the enlarger, and then developed.

(2) It may have the image transferred to it by the method of transfer coating.

(3) It may take the place of the final support in the carbon or carbro processes.

See also: *Printing on special supports.*

WOOD'S LAMP. Source of ultra-violet radiations consisting of a high pressure mercury vapour lamp with an outer glass envelope incorporating a filter layer of nickel oxide.

The lamp is named after Wood, the physicist who introduced this type of filter in 1913. The filter cuts off practically all visible radiations, leaving only a deep violet glow visible when the vapour tube is looked at directly. Although the very short rays are absorbed by the glass envelope, the lamp emits a useful band of ultra-violet rays. Commercially, lamps of the Wood type are used as a source of "black light"—e.g., for shop window display novelties—and for rendering visible laundry marks printed in colourless fluorescing ink. For photography by reflected ultra-violet rays, it is more convenient to filter out the unwanted rays by placing a suitable filter in front of the lens instead of over the source as in a Wood's lamp.

Lamps of this type must be used in conjunction with a choke since the internal resistance is not constant. For this reason the lamp cap has three pins to prevent its being accidentally inserted in an ordinary two-pin bayonet holder. As the lamp does not give its full efficiency until the temperature has reached a certain value (which varies according to the lamp's construction and characteristics) it must be switched on five or ten minutes before it is wanted. Once switched off, the lamp cannot be switched on again until cold enough to allow the arc to strike again.

See also: *Discharge lamp; Ultra-violet and fluorescence photography.*

WORKING APERTURE. Largest aperture at which a lens will give a satisfactory negative, as opposed to the largest aperture to which the diaphragm can be opened. A wide-angle lens may have a large aperture to give a bright enough image on the focusing screen to allow the picture to be arranged, but its standard of definition at this opening may be unsatisfactory. Such a lens needs stopping down to its known working aperture before makingt he exposure.

See also: *Diaphragm.*

WOTHLYTYPE. A printing-out paper invented by J. Wothly of Aachen, 1864. The paper coated with collodion containing uranium and silver nitrate, was considerably more sensitive than the albumen paper then in general use, but the claim for permanence, due to the uranium, was not fulfilled, and after a couple of years the Wothlytype was abandoned.

WRATTEN, FREDERICK CHARLES LUTHER, 1840–1926. English photographic manufacturer, founder (1878) of the firm of Wratten and Wainwright, Croydon. Invented in 1878 the "noodling" of silver bromide gelatin emulsions before washing, and manufactured silver bromide gelatin dry plates in 1878. Produced the first panchromatic plates in England in 1906 and became a famous manufacturer of photographic filters.

X

XENON. Rare gas used in a variety of discharge lamps and in most present-day electronic flash tubes. Xenon discharge lighting has an energy distribution sufficiently similar to daylight, so that xenon discharge lamps can be used for colour photography with daylight type colour films.

See also: *Discharge lamp*.

XEROGRAPHY. Electro-photographic system involving the image-wise discharge under the influence of light of an electrostatically charged selenium plate. The areas that remain charged attract oppositely charged pigment particles during subsequent "development". This image of pigment particles is then transferred on to paper and fused into the paper support. The system has wide applications in photo-copying.

See also: *Electrophotography*.

X-RAY CRYSTAL ANALYSIS. Method of deriving information about the structure of a crystalline substance by means of diffraction patterns produced when a beam of X-rays is passed through a single crystal or powdered crystals of the substance.

X-ray Diffraction. When X-rays traverse a crystalline substance, the regular network of atoms of which the substance is composed acts as a three-dimensional diffraction grating and the X-rays are scattered according to certain rules. The diffraction effects are most simply interpreted in terms of reflection by planes in the atomic network and this leads to the well-known Bragg relationship

$$2d \sin \theta = n\lambda$$

where d is the interplanar spacing, θ the angle of incidence of the X-rays on the "reflecting" planes, λ the X-ray wavelength and n an integer.

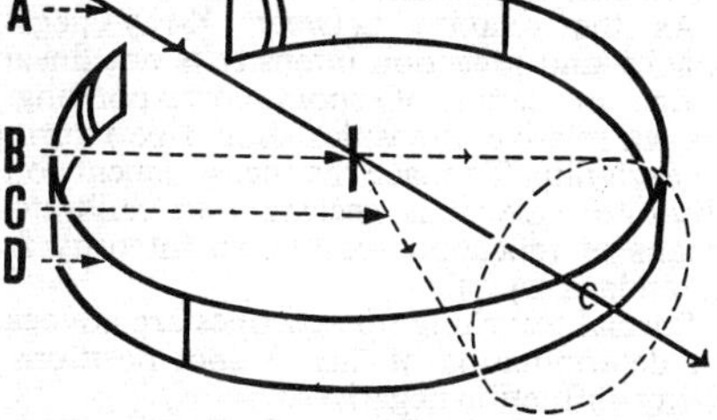

DIFFRACTION CAMERA SET-UP. Production of X-ray powder diffraction pattern. A, beam of incident X-rays. B, powder specimen to be examined. C, cone of diffracted X-rays. D, cylindrical film surrounding specimen.

The diffracted X-rays cannot, however, be focused by lenses to form a direct image of the crystal structure. All that can be done is to make a record of the diffraction pattern. The problem of X-ray crystal analysis then lies in postulating or calculating the positions and distribution of atoms in the structure cell that have given rise to the intensities of spots or lines found in the diffraction pattern.

The discovery of X-ray diffraction provided an enormously powerful method of advancing our knowledge of the structure of matter. Besides the solution of crystal structures there are many other industrially important applications of X-ray analysis. The precise identification of a substance, the study of alloy formation, and the investigation of the causes of strain in the crystal lattice are just three examples of such applications.

Photographic Recording. X-ray diffraction patterns are commonly registered photographically. It should be remembered, however, that one of the original methods by which crystal structure determinations were made was the Bragg spectrometer in which an ionization chamber was used to detect and measure the intensities of X-ray reflections. Today the basic principles of the Bragg spectrometer are incorporated in the so-called counter diffractometer instruments. In such instruments the ionization chamber has been replaced by one of the three well-known types of counter, Geiger-Müller, proportional, or scintillation.

Many different types of camera have been designed for the photographic recording of X-

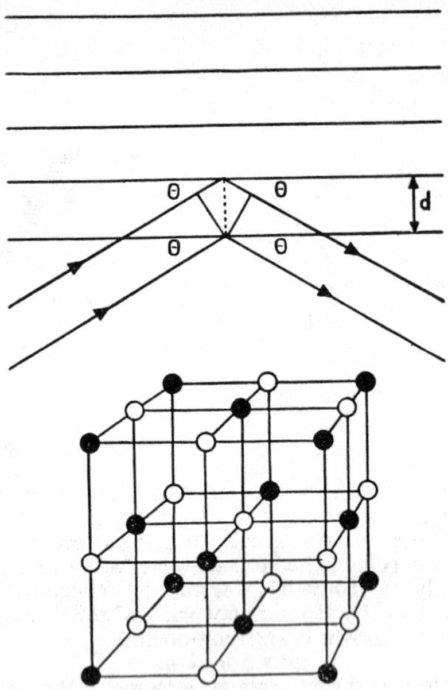

X-RAY DIFFRACTION. Different planes of crystal lattice reflect X-rays. Diffraction takes place when incident and reflected angles are equal, and 2d sin θ is a multiple of the X-ray wavelength, and produces patterns on the film.

ray diffraction patterns. In the simplest arrangement the crystal is mounted on a goniometer device which allows easy choice of different orientations towards the incident collimated X-ray beam. The diffracted effects are intercepted by a flat film held in a cassette mounted behind the crystal. When the specimen consists of a multitude of small crystals instead of one crystal, and the incident X-rays are monochromatic, the diffracted beams travel along the surfaces of a set of coaxial cones and the diffraction pattern registered on the film comprises a series of concentric rings. For some purposes it is sufficient to record only small parts of these rings, and then a strip of film is held on a narrow cylindrical surface with the powder specimen at its centre. The resultant line diffraction patterns are commonly described as "powder photographs". The powder photograph is widely used for identification, crystal size measurement, and other applications of practical importance.

Single crystal photographs are generally required for the complete solution of crystal structures. Interpretation is assisted by recording the patterns with specially designed cameras in which during the exposure the photographic film is moved in synchronism with rotation of the crystal. Movement of the film effects a separation of reflections which would otherwise overlap.

The X-ray film normally employed is similar to the non-screen double-coated film used in industrial radiography. In the majority of applications a large part of the incident X-ray energy is absorbed in the front emulsion, but the second coating adds something to the over-all density obtained.

Exposure times range from a few minutes to many hours in some single crystal work.

A special fine-grain high-contrast film is advocated when fine detail is of primary importance, but there is the accompanying disadvantage of a three- or four-fold increase in exposure time. In certain techniques, involving thirty times enlargements of a single X-ray reflection, Lippmann type plates are necessary.

If the X-ray reflections are likely to be incident obliquely on a flat film, it is occasionally desirable to use single-coated X-ray film. This avoids ambiguities arising because of lack of register between the front and back images on a double-coated film.

Counter Diffractometers. The counter diffractometers which were referred to above are commonly coupled to pen recorders, which enable for example an X-ray powder diffraction pattern to be recorded on a chart as a continuous trace. Peaks on this trace register clearly both the positions and intensities of the "reflections" which comprise the pattern. The instruments also incorporate means for studying selected "reflections" in detail by time counts. Such equipment lends itself to automation. Thus applications of the X-ray diffraction method to the monitoring of production processes in which phase identification and concentration must be controlled become possible. Other types of diffractometers have been developed for recording automatically the intensities of the multitude of "reflections" from single crystals, thus enormously increasing the speed of collection of data from which atomic arrangements have to be computed.

Interpretation. To evaluate the information on an X-ray diffraction pattern, both the intensity and the position of the X-ray reflections have to be determined.

The intensities of the reflections are obtained by comparison of the densities of lines or spots comprising the pattern. Visual estimation may sometimes give acceptable results, but photometric measurement is often required.

As the relation between X-ray negative opacity and reflection intensity is not linear, a wedge or series of spots corresponding to known relative intensities should be imprinted on each film. To place the measurements on an absolute scale, it is necessary to calibrate by means of reflections of known intensity from a standard crystal.

Special travelling microscopes are advocated for determination of line or spot positions on X-ray diffraction negatives.

Many of the problems of reflection-intensity measurement may be minimized by use of counter diffractometers.

Much of the laborious computation involved in complete structure determination may be reduced by an experimental approach involving optical diffraction methods. One optical arrangement that has been suggested is known as the "fly's eye". A plane mask or pattern is constructed to correspond to the proposed crystal structure viewed from some chosen direction. The optical diffraction pattern, obtained when monochromatic light from a point source is passed through the mask, is compared with the X-ray diffraction pattern to determine whether the proposed atomic arrangement is correct. The masks are prepared by a photographic reduction method, since the element of a pattern must be reproduced perhaps two or three hundred times in any given mask.

Attempts have also been made to produce directly an optical image of the crystal structure or key projections of it. The masks then have to imitate particular features of the X-ray diffraction pattern, and a section of what is termed the reciprocal lattice, which can be readily derived from it, is constructed. The problems of adjusting the intensities of the X-ray beams and of determining their phases are both dealt with by taking advantage of the polarizing properties of mica. By covering the holes in the mask with sheet mica in various orientations the transmitted light beams can be modified in appropriate ways, and different possibilities examined by trial and error. The development of this so-called diffraction spectrometer has provided a powerful new tool for the structure analyst. H.P.R.

Books: *The Crystalline State*, Vol. I, by Sir Lawrence Bragg; Vol. II by R. W. James; Vol. III by H. Lipson (London); *The Interpretation of X-ray Diffraction Photographs*, by N. F. M. Henry, H. Lipson and W. A. Wooster (London); *The Powder Method in X-ray Crystallography*, by L. V. Azaroff and M. J. Buerger (New York); *X-ray Diffraction by Polyscystalline Materials*, ed. by H. S. Peiser, H. P. Rooksby and A. J. C. Wilson (London); *X-ray Powder Photography*, by R. W. M. D'Eye and E. Wait (London); *Crystals and X-rays*, by K. Lonsdale (London).

X-RAY DIFFRACTION. Coherent scattering of X rays by the atoms of a chemical substance. Where the arrangement of the atoms is regular, e.g., in a crystal lattice, a beam of X-rays produces a diffraction pattern which is recorded photographically and this pattern yields substantial information about the crystal structure of the substance.

See also: *X-ray crystal analysis*.

X-RAY FILM. Sensitive material designed specifically for radiography. X-ray films consist of a safety base carrying an emulsion on both sides, thus utilizing the penetrating power and absorption of X-rays in both emulsion layers. Double coating gives increased speed and contrast over a single coating.

Direct and Screen X-ray Films. X-ray films are exposed either to X-rays directly or in conjunction with a pair of intensifying screens. The purpose of the latter is to reduce to exposure time which would be required without screens. There are essentially two types of screen: salt and lead screens. X-ray films differ in characteristics according to whether they are determined for direct exposure to X-rays and in combination with lead screen or for salt screen exposure.

Salt screens consist of a support of cardboard or other flexible layer coated with calcium tungstate or barium lead sulphate. Both substances emit bluish fluorescent light when exposed to X-rays. An X-ray film is sandwiched between a pair of such screens. Although excited by X-rays the image obtained is essentially due to the action of bluish light.

Lead screens consist of a flexible support provided with a thin lead foil which emits electrons when struck by X-rays. The X-ray film is sandwiched between a pair of lead screens and the intensification achieved is due to the action of electrons on the photographic emulsion.

Direct X-ray films have an emulsion highly sensitive to X-rays and to the electron emission from lead screens. The screen type emulsion is highly sensitive to the blue light emission from salt screens and may have a relatively low X-ray sensitivity. The spectral emission of the calcium tungstate screens with a peak of about 4,300 A. almost coincides with the natural spectral sensitivity of an X-ray silver bromide emulsion later. The emission of barium lead sulphate screens is shifted slightly more towards the U.V. radiation, to which the emulsion is still very sensitive.

Exposure Characteristics. In direct X-ray exposures it has been found that one X-ray quantum absorbed per silver halide grain renders this developable, whereas this is not so for screen light exposures. Hence for direct X-ray exposures there is no reciprocity failure and results are independent of the rate of exposure. In the initial part of the density-exposure curve the number of grains rendered developable is proportional to the number of quanta absorbed, hence there exists proportionality between exposure and density which holds until more and more grains receive several X-ray quanta hits. Then the proportionality ceases and the density-exposure curve gradually leads to saturation at higher densities. In some non-screen X-ray films this proportionality holds up to densities of 2 and 3 and may be utilized in scientific applications, such as X-ray crystallography.

In screen exposures where the image is produced essentially by light, reciprocity failure may occur, particularly in long exposures using low X-ray intensities. The intensify-factor is defined as

$$\frac{\text{exposure required without screens}}{\text{the exposure required with screens.}}$$

in order to obtain the same density on the same type of film. The intensifying factor may reach values of up to 100 and even more depending on the quality of radiation and the rate of exposure applied.

Medical Radiography. In medical use most exposures are made with screen films. Most intimate contact between screens and film is required to obtain optimum resolution. As the individual crystals of screens emit light in all directions the resolution obtainable with screens is limited. Hence exposures of extremities, where high definition of bone structure is expected are usually taken with direct X-ray films. Lead screens are not used in medical applications, as the screens would not permit any reduction of exposure time within the range of kilovoltage applied in medical radiography.

Industrial Radiography. Most exposures for industrial radiography are made on non-screen X-ray film, usually in conjunction with lead intensifying screens. The latter become useful only at kilovoltages above 140 kV. Resolution is not impaired when using lead screens if these are in perfect contact with the film. In contrast to medical radiography, where exposure times range from fractions of a second to seconds, industrial radiographic exposures run to minutes and even hours. Hence the use of screen film with screens is very limited owing to the reciprocity failure of the film.

For both medical and industrial radiography a range of X-ray films is manufactured covering different requirements of speed, contrast and definition. Dental films which are of the non-screen type are individually packed, often with a thin lead foil behind the film to reduce scattered radiation originating from the gum. Screen and non-screen films are usually supplied folder wrapped, but non-screen films are also obtainable in individual envelopes. These permit the handling of films in daylight without the use of special cassettes.

X-ray films are preferably stored in lead lined containers to protect them from accidental exposure to X-rays.

Processing. X-ray films are normally processed in tanks using strict time and temperature control. During processing the films are supported by stainless steel hangers which also permit inspection after processing whilst the films are still wet. In hospitals and industrial laboratories where a great number of radiographs are taken per day, the use of automatic processing units has become an essential feature of a national operation R.H.H.

See also: *Antoradiography; Industrial radiography; Medical radiography.*

X-RAY MICROGRAPHY. Technique for examining the internal structure of thin opaque specimens not normally revealed by photomicrography. A specimen of the material—e.g., a metallurgical section—1 mm. thick, or less, is placed in contact with an extremely fine-grained sensitized emulsion and a radiograph is made.

The radiograph is then examined or photographed under the microscope. By this means the internal structure of specimens may be seen at magnifications of up to 300×.

See also: *Microradiography.*

X-RAYS. Rays discovered and so named by Röntgen in 1895. They are able to penetrate solid matter and to produce a developable image in the photographic emulsion. X-rays were shown to be electro-magnetic waves by von Laue in 1912. They occupy the region of the spectrum between gamma rays and the ultra-violet, having wavelengths from about 0·05 A ("hard" rays) to over 100 A ("soft" rays). Hard rays are more penetrating; rays of more than a few A wavelength have no practical application.

X-rays are generated by the impact of high-velocity electrons on matter. In the original tubes an electric discharge in a rarefied gas (cathode rays) was used. Modern tubes (Coolidge tubes) are evacuated as completely as possible; the electrons are emitted by a hot tungsten filament, and are accelerated by a potential of several thousand volts to strike the anticathode or target, which emits the X-rays. The tube also contains parts designed to focus the electrons on to a small area of the target. The material used for the target depends upon the type of radiation desired, but it must be very infusible and/or a good conductor of heat, as the impact of the electron beam has an intense heating effect; for the same reason the target is always water-cooled. The target is frequently made of tungsten, but copper, molybdenum, iron, cobalt, chromium, silver or rhodium are also used for special purposes.

Very hard X-rays, in the production of which potentials of up to 1 million volts are employed, are used therapeutically, in the treatment of cancer, etc. Photography is not concerned at all with this use.

The other uses of X-rays employ their penetrative powers. These depend upon the density of the material under examination, bone being more opaque than the surrounding soft tissues, blowholes and other faults in castings more transparent than the surrounding sound metal, etc.

The records of X-ray examinations of these types are normally made photographically on radiographic films.

The photographic recording of X-rays is used in: medical diagnosis, for which comparatively soft rays are used; industrial radiography. Very hard and penetrating rays (wavelengths 0·05–1 A) are used to reveal the internal structure of castings, etc. A new development is micro-radiography, which uses soft rays to reveal the structural details of comparatively transparent materials such as paper, plastics, etc.; a fine-grain maximum-resolution emulsion is exposed behind the sample, and the developed image examined under a microscope. X-ray crystallography, which uses rays of 0·5–2 A wavelength, is the most important scientific, as distinct from technical, application of X-rays. It depends upon the diffraction of the rays by the regularly-spaced atoms in crystals (analogous to the manner in which visible light is diffracted by a grating), and is a most powerful tool in the exploration of molecular structure. The chemical constitution of penicillin was elucidated in this way, for example.

Although X-rays act on photographic emulsions, they are less effective in this respect than light, since their penetrative power allows them to a large extent to pass through the emulsion without affecting it. X-ray film is therefore always coated on both sides, in order to increase the total emulsion thickness. Coarse-grained emulsions are also used wherever possible, since these are more sensitive to X-rays than fine-grained ones. H.J.Wa.

See also: *Industrial radiography; Mass miniature radiography; Medical radiography; X-ray crystal analysis.*

Y

YARD (YD.). Unit of lineal measure equal to 3 feet or 36 ins. Originally the yard was based on the ell (from Lat. ulna, the forearm) which, in England was standardized as the length of the arm of King Henry I. The present standard is given by the distance between two gold inserts in a bar of platinum in the Exchequer Offices, London.

See also: *Weights and measures.*

YELLOW. The red-green colour reflected or transmitted by a surface which absorbs blue, also known for this reason as minus blue. One of the three primary colours used to form the image in subtractive colour processes, the other two being minus red (cyan), and minus green (magenta).

YLLA (Camilla Koffler) ?–1955. Hungarian-born photographer, settled in Paris in 1920s and trained as photographer. Her talent for photographing animals manifested itself during her student days. She emigrated to U.S.A., 1940, and after the War her picture stories of animals won international reputation. She had a fatal fall from her truck while photographing big game in Africa.

YUGOSLAVIA. At the time photography was invented, the people of present-day Yugoslavia were either under foreign rule as Croatians and Slovenes, or were just creating their state as Serbians, after five centuries of Turkish occupation. In spite of this, besides showing keen interest in photography the Yugoslavs also contributed to its development.

History. Daguerreotypes were being made by numerous workers between 1840 and 1850, among them being Novaković and Boš in Zagreb and J. Kapiseri and A. Dajć in Belgrade. It was the Slovenian Catholic clergyman Janez Puhar and the Serbian Anastas Jovanović, however, whose work made an impact beyond the boundaries of Yugoslavia.

Puhar (1814–64) found it difficult to procure the necessary material for daguerreotypes, particularly the silver plates which would have been somewhat expensive for a young clergyman of limited means, in the small Austro-Hungarian town of Kranj. He began using glass plates as early as 1842—five years before Niépce de St Victor, who is officially considered the inventor of photographs on glass plates—and made transparent pictures by means of sulphurous, iodine and chromium vapours. Puhar's invention was registered by the Viennese Academy of Science but was forgotten, in spite of his being elected a member of the French Academie Nationale, Agricole, Manufacturiere et Commerciale in 1852, and being awarded a diploma in which he is referred to as inventeur de photographie sur verre (Inventor of photography on glass plates). Puhar exhibited his photographs at the World Fair in London in 1851, and also in New York in 1852. Anastas Jovanović (1817–99) was first a printer in Belgrade, and later the first Serbian lithographer. He went to Vienna on a state scholarship to study art at the Art Academy of St. Anna. There he saw the first daguerreotypes, and became so enthusiastic about them that he immediately bought a camera with a Petzval objective. He first made talbotypes and was the first to take stereophotographs of Vienna which he exhibited at a show organized by the Viennese optician Prokesch in 1854. Jovanović made a complete portrait gallery of eminent Serbian personalities in the first half of the nineteenth century, using the talbotype method.

Clubs and Associations. The photographic section of the Croatian Art Society, founded in Zagreb in 1890, is doubtless the oldest amateur photographers' organization in the country. The Slovene amateur photographers had two clubs at the beginning of 1900. The Belgrade photo club was founded in 1901, and was the first in the Kingdom of Serbia. The Balkan War and the First World War interrupted photographic activities over a long period, and they were continued only after the unification and creation of a common state in 1918. The *Fotoklub Zagreb* (Zagreb Photo Club) was established in 1923, Belgrade founded its club in 1928 and the club in Ljubljana also resumed its activities at that time. Amateur photographers were extremely active within the photo-sections of mountaineering societies, and these often developed into independent clubs later.

The first amateur photographers' organizations after World War II were founded in Belgrade, Zagreb and Ljubljana. Interest in photography having become far keener in the post war years, the membership of these clubs grew rapidly.

The Federation of Yugoslav Amateur Photographers was founded in 1949 with the aim of raising the standard of technical training, and of promoting photography generally. Since 1950 the Federation has been a member of the Federation Internationale de l'art Photographique (F.I.A.P.) Photo clubs have been established throughout the country with the help of the Federation, photography courses have been organized, and instruction- and work-schedules drawn up for newly established photographic organizations. By 1950 there were 63 clubs, with a total of 9,835 members. The Federation also directed its attention to amateur cinematographers, who were operating under great financial difficulties, and by 1956 seven cine-clubs were in operation. News reporters and press photographers are organized within the Federation of Journalists' Associations of Yugoslavia.

The Photographers and Cinematographers Federation of Yugoslavia (FKSJ) consists of six separate associations, each of which operates independently on its territory. Each association co-ordinates the activities of the clubs on its territory. Both amateurs and professionals are eligible for membership. The Federation has introduced the ranks of Master of Photography and Candidate Master, these titles being conferred by the Arts Council.

In order to operate as a professional photographer it is necessary to pass specified examinations. The general standard of work is not as yet as high as in most Western countries. At present there are no photographic schools in Yugoslavia although there is a photographic section within the School of Applied Art, this being the only centre of photographic learning in the Balkans.

Industry. The Yugoslav photographic industry is still young, and the first films and plates were manufactured as recently as 1947. A Zagreb factory, the only one of its kind in the Balkans, is producing a wide assortment of photographic materials including 35 mm. films, papers, X-ray materials, positive 16 and 35 mm. cine films, photostat materials for copying and registration, photographic chemicals, optical gelatin filters and lantern plates. The manufacture of materials for colour photography is currently in preparation. Photographic materials are exported to several countries such as the U.S.A., Holland, Austria and the Scandinavian countries, as well as to India, Egypt, Lebanon and Turkey where demands are very great. Twenty-five per cent of the roll films produced in 1962 were sold to the U.S.A. Brazil is beginning to show an interest in Yugoslavia's photographic products. High quality motion picture projectors for 16 mm. film are being manufactured in Kranj, as well as slide projectors and other equipment and accessories for photo-laboratories. The optical glasses for these are manufactured in Ljubljana. Cameras and accessories, motion picture cameras, colour films and optical glasses are all imported into the country. High quality cameras come mostly from West Germany, while the less expensive models for the average amateur are supplied by East Germany, Czechoslovakia and the Soviet Union. During the past three years cameras and other photographic materials and accessories have also been imported from Japan on a fairly large scale.

Exhibitions. The country's first photographic exhibition was held at the Arts and Crafts fair in Zagreb in 1891. The Serbian amateur photographers held their first exhibition in Belgrade in 1901, and the Slovenians had their first exhibition in Ljubljana in 1911. Foreign exhibitors took part in exhibitions held in Zagreb in 1910 and 1913.

After 1935, when the International Exhibition of Slavic Photography was held, the Zagreb photographers club organized five international exhibitions before the beginning of World War II. The Ljubljana photo club also organized international exhibitions in 1934, 1936 and 1938. The Zagreb amateur photographers club exhibited at the Royal Photographic Society in London in 1938, and also in Argentina and Holland. The newly formed Association in 1949 resumed the activities cut short by the war with still greater enthusiasm, and so far twelve all-Yugoslav exhibitions, with about 4,000 photographs, have been held. Today the all-Yugoslav shows are always accompanied by lectures and discussions at which the individual problems and aspects of Yugoslav photography are examined in detail.

International exhibitions have been held since 1951 and apart from this, small-scale exhibitions are being organized by the individual clubs. Yugoslav amateur and professional photographers are taking part regularly in shows and competitions abroad and some artists have held one-man shows in various European countries. One-man shows within Yugoslavia received a vigorous boost at the beginning of 1962 through the inauguration of the exclusive Photograph Gallery in Belgrade, whose aim is to acquaint the public with the work of individual Yugoslav photographers and distinguished foreign amateurs and professionals.

Publications. Although the first publication, the review *Fotografski pregled* (Photographic Survey), appeared in Belgrade in 1911, Yugoslav photographic literature is limited. There were ten photographic reviews published in Yugoslavia from 1911 to 1940, and of these the international review *Galerija*, (*die Gallerie*) published in 1933, was perhaps the most noteworthy. *Fotozevija*, which specialized mostly in technical problems of photography, appeared from 1931 to 1940. The review *Savremena fotografija* (Contemporary Photography), was founded in 1941, but was forced out of existence by the war. In 1948 the Yugoslav Federation of Amateur Photographers founded *Fotografija*, a universal review of photography which is now published under the title of *Fotokino revija* (Photographers and Cinematographers Review). Apart from this, professional photographers published their own review *Naša Fotografija* (Our Photography), from 1947 to 1951, and the Belgrade amateur photographers published *Beogradski objektiv* from 1952 to 1958 as well as the young photo-amateurs' review *Oko* (Eye). Foreign books and publications on photography are widely read, and two years ago a Zagreb publishing house began issuing books devoted to the art of photography.

Artistic Trends. At the beginning of the century, photography in Yugoslavia, as in the rest of the world, tended to take its inspiration from the field of painting and thus portrayed such subjects as idyllic village scenes and artificial genre. There was a lack of any definite objective criteria. In the 1930s, the Neue Sachlichkeit (New Reality) movement and the Bauhaus experiments emanating from Germany, had a brief influence, but served in the long run, only to confirm belief in the silver bromide process.

The most active groups between 1930 and 1940 were the Fotoklub Ljubljana (Ljubljana Amateur Photographers Club) and the Fotoklub Zagreb (Zagreb Amateur Photographers Club). The Ljubljana school of photography was based on pictorialism and emphasized fine technique and craftsmanship, and much use was made of the soft focus lens to gain an impressionistic effect. The Zagreb school, on the other hand, favoured portrait and genre. They aimed for realism and simplicity and regarded the negative as inviolable, retouching only to remove blemishes.

Towards the end of the 1930s the Klub Fotoamatera Beograd (Belgrade Amateur Photographers Club) practised the same style but with greater originality and freshness owing to their comparative remoteness from international exhibitions. Their activities were stopped by World War II.

After the liberation of the country, life in all its forms and manifestations was the basic theme of the new Yugoslav photography. International exhibitions are inspired by a new spirit and new concept of photography with which the Yugoslav amateur has become gradually acquainted through foreign publications. Subjective photography and Italian neo-realism may also be considered to have had a noticeable influence on the development of Yugoslav photography. The *Family of Man* exhibition made a particularly great impact, as Yugoslav photographers have for a long time had a strong affinity for this kind of work.

Photography plays a prominent part in business publications, prospectuses, and catalogues; in publications on tourism and on art: in the numerous lavish presentations of the Yugoslav publishing house; and in the prominent monthly magazine, *Review*.

After World War II, and under the new social conditions, photography acquired a status equal with the other arts, and photographers were able to join the Association of Visual Artists alongside painters and sculptors. As the impact of photography on the country grows efforts are being made both to spread interest in this field to those more remote regions where it is as yet little known, and also to introduce photography and photographic facilities into the schools, as an optional subject. B.D.

Z

ZEISS, CARL, 1816–88. German optical manufacturer. Founder of an optical firm in Jena. With collaborators and successors (especially Abbe and Rudolph) designed or improved microscopes, field-glasses, optical instruments, photographic lenses (anastigmat, Tessar) and cameras. Biography in *Zur Geschichte der Zeissischen Werkstätte* by M. von Rohr (Jena 1936).

ZINCOGRAPHY. Photomechanical printing process in which the negative image is printed on zinc to produce a litho plate from which reproductions can be taken in printing ink.

ZIRCONIUM LAMP. Concentrated arc lamp used in projectors and enlargers.

The electrodes are a ring-shaped metal anode surrounding a tiny tube of zirconium oxide, and the arc occurs in the narrow annular gap (a few thousandths of an inch) separating them. At the high temperature of the arc, the oxide becomes coated with molten incandescent zirconium which gives a brightness comparable with that of a carbon arc at a colour temperature of about 3,000° K. All the light is emitted through the hole in the anode in a forward direction, so it is a highly efficient source.

A 2 watt lamp (37 volts) has a light source of 0·003 in. diameter and a 100 watt type (15·4 volts) about 0·059 in.

The electrodes are enclosed in a glass bulb.

See also: *Point source lamp.*

ZOËTROPE. Device invented by W. G. Horner of Bristol in 1833 and named Zoëtrope in 1867. It consisted of a light sheet metal cylinder about a foot in diameter, mounted on a central pivot so that it could be spun around by hand. The top half of the cylinder was pierced with a series of slits, and around the inside of the lower half was fastened a paper band. This band carried a series of pictures on the lines of an animated cartoon, creating the illusion of moving pictures when rotated.

The number of pictures around the band was equal to the number of slits in the cylinder, and the whole series covered a complete action sequence—e.g., a boy leap-frogging over another's back, or turning a cart wheel.

When the drum was set turning, an observer looking through the moving slits at the pictures on the opposite side of the drum caught a glimpse of each picture at the instant when it reached the opposite side of the drum. Before and after that instant, his view of the picture sequence was cut off by the side of the drum.

So the observer's eye received a series of impressions of still pictures, each successive picture showing another step in the progressive development of the action. Because of persistence of vision, all the separate impressions added up to a moving picture, repeating the action at each revolution of the cylinder.

The popularity of this form of entertainment fell off abruptly with the introduction of cinematography.

Other types of apparatus on the same principle in one form or another were variously called a wheel of life, a phenakistoscope, stroboscope, a phantascope, and a kaleidorama.

The earliest and simplest form was invented simultaneously by Plateau of Ghent and Stampfer of Vienna in 1832.

See also: *Cine history.*

ZONE FOCUSING. For fast-moving subjects, sports events, children at play, and animals, there is no time to focus and set the aperture before each exposure. Under these circumstances the experienced photographer sets his lens to cover the entire depth of the zone in which the subject is likely to be moving. Then, so long as the subject keeps inside the zone, there is no further need for focusing, and the photographer can concentrate on making exposures.

Once he has decided on the near and far limits of the field he has to cover, he con-

SETTINGS FOR ZONE FOCUSING

Lens Focus	Aperture	Close-up			Focusing Distance Group			Landscape		
1⅜ in. (3·5 cm.)	f 8	3–1½	**4–0**	5–6½	4–3	**6–0**	10–4	7–0	**14–0**	∞
1⅝ in. (4·25 cm.)	f 8	3–11	**5–0**	7–1	5–5	**8–0**	15–2	8–6	**17–0**	∞
2 in. (5 cm.)	f 8	4–8	**6–0**	8–5	6–3	**9–0**	15–10	10–6	**21–0**	∞
2½ in. (6·25 cm.)	f 8	5–6	**7–0**	9–7	8–9	**11–0**	19–0	13–0	**26–0**	∞
3 in. (7·5 cm.)	f 8	6–4	**8–0**	10–9	9–2	**13–0**	22–3	15–6	**31–0**	∞
3½ in. (8·75 cm.)	f 8	7–3	**9–0**	12–0	10–7	**15–0**	25–5	18–0	**36–0**	∞
4 in. (10·0 cm.)	f 8	8–8	**11–0**	15–0	12–10	**18–0**	33–6	21–0	**42–0**	∞
4½ in. (11·25 cm.	f 8	9–6	**12–0**	16–2	14–0	**20–0**	34–10	23–6	**47–0**	∞
5 in. (12·5 cm.)	f 11	7–11	**10–0**	13–7	11–3	**16–0**	28–0	19–0	**38–0**	∞
5½ in. (13·7 cm.)	f 11	8–8	**11–0**	15–0	12–0	**17–0**	28–9	21–0	**42–0**	∞
6 in. (15·0 cm.)	f 11	10–1	**13–0**	18–3	14–0	**20–0**	35–6	23–0	**46–0**	∞
7 in. (17·5 cm.)	f 11	10–5	**13–0**	17–3	14–6	**20–0**	32–0	26–6	**53–0**	∞

The heavy figure in each group gives the distance on which the camera should be focused. The lighter figures give the limits of the depth of field. The zones assume that the lens is in each case the normal focal length (i.e., neither wide angle nor telephoto or long focus) for the negative size it is used with.

sults the depth of field tables for the lens he is using to find what focused distance and lens aperture will give him the required depth. This will then cover the particular subject or a series of subjects (e.g., a picture sequence) within the selected zone.

Next he finds what shutter speed he must use at this lens aperture to give him the correct exposure.

If this shutter speed is fast enough to arrest movement of the subject, he can make the necessary adjustments of focused distance, aperture, and shutter speed, and fire away with that fixed setting.

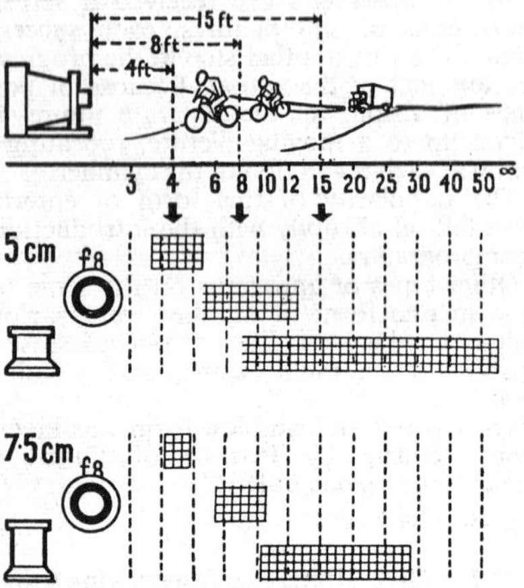

ZONE FOCUSING. Focusing zones for miniature camera with interchangeable lenses. **Top:** *Near, medium, and distant zones covered by 5 cm. lens focused on 4, 8, and 15 feet respectively, and stopped down to f8.* **Centre:** *Corresponding zones with 7·5 cm. semi-long focus lens.* **Bottom:** *Corresponding zones with 10 cm. long focus lens at same settings.*

But sometimes the shutter speed will not be fast enough to arrest movement of the subject at the range of distances included in the focused zone.

There are two ways out of the difficulty, according to whether sharpness to arrest movement or the subject depth covered is more important:

(1) He can move farther away from the subject. At a greater focused distance, the depth of the zone of sharpness is also greater. As the photographer will then have more depth than he wants, he can sacrifice some of it to allow him to open up the lens aperture and speed up the shutter.

Unfortunately when he moves farther away from the subject, he is forced to make do with a smaller image.

(2) He can use a bigger aperture which will in turn let him give a faster shutter speed. This has the disadvantage of narrowing the zone of sharp focus.

Both expedients carry penalties; the final choice is bound to be a compromise that will vary with the circumstances and the photographer.

Zone Indicators. Many cameras show the settings for one or two focusing zones on the distance scale.

This may take the form of special symbols to which the scale is set; such a setting, combined with an appropriate aperture (sometimes marked in red on the aperture scale) then yields one or two useful focusing zones.

A number of modern simple cameras dispense with the focusing scale or else use in addition a scale of symbols indicating the most suitable zone for portraits, groups and more distant views. (This is in fact substantially the same thing as the two point or three-point focus settings on older simple cameras.)

Occasionally the lens mount clicks into place at specific focusing zone settings. With some cameras the viewfinder may even indicate a selected focusing zone by a suitable signal (for example 1, 2, or 3 coloured dots or some symbols and a movable pointer).

See also: *Depth of field.*

ZOO. Photography at the zoo calls for patience above everything else. The photographer has no control over the animal and so he must be prepared to wait for it to take up a suitable position. It is sometimes possible to exercise a certain amount of control by tempting the subject with food, and a slight noise will often make it look more alert. But beyond this, there is little that the photographer can do except wait—for hours, if necessary. It is better to wait for one really good picture than to hurry away with a batch of second-rate snaps.

The beginner will usually find it easier to start with the larger and tamer animals. Deer, camels, zebras and the like are easier to get at and offer fewest technical difficulties. With wild animals there are the bars to contend with from the outside of the cage, and the risk of attack when taking photographs inside the cage. And smaller animals are very much more difficult to photograph; it is considerably easier to make a good picture of an elephant than a mouse.

Camera. Almost any type of camera can be used to make zoo pictures, so long as the photographer works within its limits. A box camera will take satisfactory photographs of any animal from an elephant down to a goat. But it is quite unsuitable for the smaller animals and birds.

Close-ups of small subjects call for a camera that can be fitted with a long focus or telephoto lens; this will allow the photographer to work at a comfortable distance and still get a big enough image. Long focus and telephoto lenses are useful in open-air zoos where the animals are generally seen a long way off. Under such conditions a camera with a normal lens gives a picture in which the animal comes out as a tiny speck in a wide expanse of uninteresting landscape.

Two very suitable cameras for zoo photography are the 35 mm. miniature and the single-lens reflex, both taking interchangeable lenses.

Tripod. A tripod is often more of a handicap than a help. Most zoos discourage the use of cameras on tripods because they obstruct the public pathways at the busiest points. Then a tripod is useless when the camera lens has to be pushed between the bars of a cage. And finally, most animals are frightened by a tripod; to them it is a stick, and they know that sticks can hurt.

Flash. A flash outfit can be useful for taking pictures in dark surroundings and for lighting up strong shadows, particularly on dark fur or feathers. But it is always necessary to ask permission of the keeper first. Some antelopes, for instance, are so timid that they would dash themselves against the fence and break their limbs if they were startled with a flash.

Backgrounds. One of the greatest difficulties in zoo photography is to avoid an ugly or uncharacteristic background. Too often the photographer has to make the best of existing and unalterable surroundings. When this happens—as when the animal is in a cage or other confined space—it is best to coax the animal away from the background, or wait until it moves away from it, and then focus sharply on the subject at a wide aperture. This technique throws the background out of focus so that it is blurred and very much less assertive in the final photograph.

Another useful way of separating the subject from its background is to choose a time of day when the background is in shadow and the animal can be caught in sunshine.

Dealing with Bars. Zoos without bars are to be preferred for animal photography, but even in the conventional type of cage there are plenty of opportunities for making good pictures.

It is often stated that if the camera lens is pushed right up against fine wire mesh, the wire will not show in the photograph. This is not true, and wherever possible the lens should be poked through the mesh or between the bars so as to be quite clear of them. Before going as close as this, however, the photographer should always be sure that it is safe. When the animal is dangerous, there is always a notice on the cage saying so, and in that case there is nothing to be gained by experimenting.

When a photograph is taken with the lens pushed through the bars in this way, there is a danger that the animal will be too close. In these circumstances a friend—or the keeper for preference—should throw pieces of food to tempt the animal away from the bars. The photographer himself should never let the animal see him with food or it will probably come right up and lick his lens.

On the whole, better pictures result from simply waiting until the animal moves naturally over to the right spot; food should only be used as an incentive when time is short.

The shadow of the bars is always a potential nuisance as it tends to turn all animals into zebras. If the cage is so situated that the sun always lays the shadows of the bars across the animal and its background, the only solution is to wait for a bright sunless day.

Finally, if the bars cannot be kept out of the picture by any of the methods described, it is important that they should be shown absolutely vertical—i.e., parallel to the sides of the print.

Behind the Bars. The keeper can often be persuaded to allow the photographer to take photographs inside the cages of the tamer animals. Photography at such close quarters and with no bars or wires in front of the subject offers much greater scope for individual work.

The photographer who is lucky enough to get permission to enter a cage must learn to move slowly—this is necessary not only when approaching the animal, but for all movements. Often even the act of drawing out a dark slide quickly is enough to scare the animal just when everything was set for a really good photograph.

Animal Portraits. It is very tempting to concentrate on portraiture inside the cage, and close-ups that show every hair are very satisfying. But many animals have long noses and at such short distances they must be taken in profile or their faces will be distorted.

In making portraits of animals, it is not always possible to get the whole of the head in focus, but the eyes should always be perfectly sharp.

The expression on the features of the animal can make or mar the portrait. Even a photograph that is technically perfect is unattractive if the subject has half-closed eyes and a bored expression. The keeper can co-operate by making a suitable noise, or perhaps rattling a food bucket at the right instant. This is generally sufficient to make the animal prick its ears and look alert and interesting.

Small animals call for a special photographic technique and are much more difficult to deal with successfully. They have to be photographed at relatively short distances, even with long focus lenses, and the depth of field available is very limited; it is narrowed further by the need for using big lens apertures and fast shutter speeds because of the speed of movement of the subject. Even when a mouse stands still, for instance, its whiskers vibrate rapidly. W.S.

See also: *Animals; Big game; Biology; Birds; Botany; Fish; Flowers; Insects; Pets; Reptiles.*

ZOOM LENS

Lens of variable focal length which can be adjusted continuously by the movement of one or more groups of elements in the lens system. Also known as variable focus lens.

Zoom lenses were originally developed for motion picture and television cameras, where the ability to vary the focus while the camera is running makes it possible to simulate tracking shots without moving the camera itself forward or back. In motion picture parlance the effect of the image on the screen rapidly getting larger is also known as zooming in, from which the term zoom lens is derived. Zoom lenses are also available for still cameras and projectors and in all cases have the advantage of providing a continuous range of focal lengths without the need for changing the lens on the camera.

Optical Principles. The basis of all zoom lenses is the fact that in a lens system consisting of positive and negative elements a change in the separation of the elements also changes the focal length. This however also alters the lens to image distance for maximum sharpness. (A front cell focusing movement of a normal lens

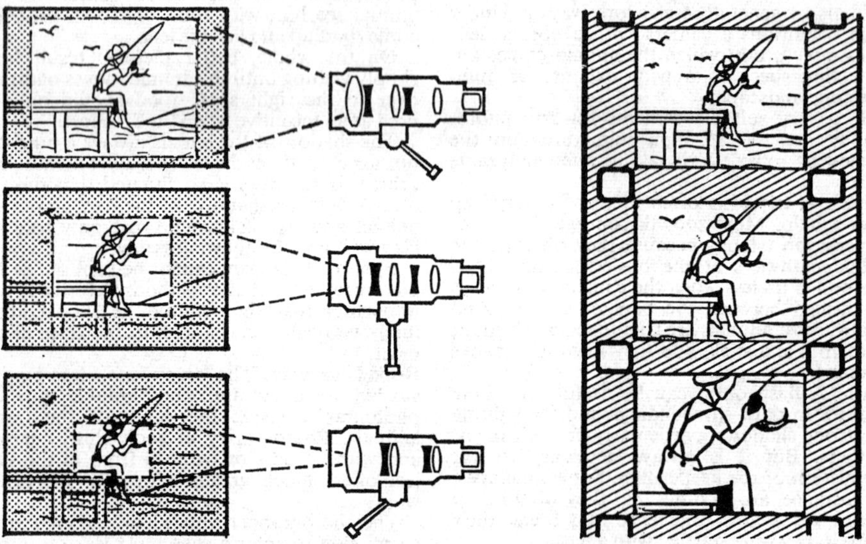

ZOOMING IN CINEMATOGRAPHY. As the focal length of the zoom lens is increased, the angle of view taken in narrows (left). The effect on the film (right) is that of a progressive enlargement of the image scale, resembling a tracking shot—but from a fixed viewpoint.

is a way in which this is adapted to focus the lens on different distances by keeping the lens to image distance more or less constant.)

With a zoom lens, especially in cine work, it is usually regarded essential to vary the focal length without changing the image sharpness. To achieve this it is necessary to move one or more of the other components as well. There are two main ways of doing this.

The movement of the lens element which changes the focal length may be linked with the movement of another lens element which compensates for the change of back focus and so keeps the image sharp. These two movements are unequal, and are therefore linked by a cam system. This is known as mechanical compensation. The constancy of focus at different focal lengths depends largely on the precision of the cam.

Alternatively two (or more) lens components may be connected together in such a way that they both move through the same distance yet keep the image sharp at the different focal length settings. This is known as optical compensation, since no variable separation between the moving elements is involved.

In this case the movable components are mounted in a travelling sleeve shifted by a lever or ring on the lens mount. The advantage of this arrangement is that image definition does not deteriorate if mechanical wear takes place in the system.

The variation of focal length also affects the

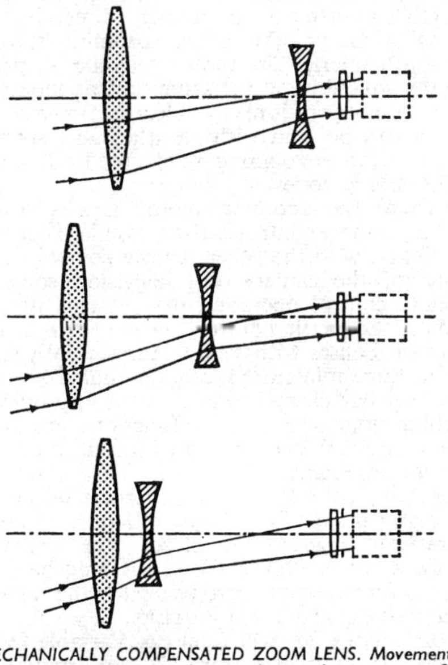

MECHANICALLY COMPENSATED ZOOM LENS. Movement of the negative component, changes the angular coverage of the lens. To maintain the system in an afocal adjustment, the front positive component must also move separately. The two are therefore linked by a cam (mechanical) movement.

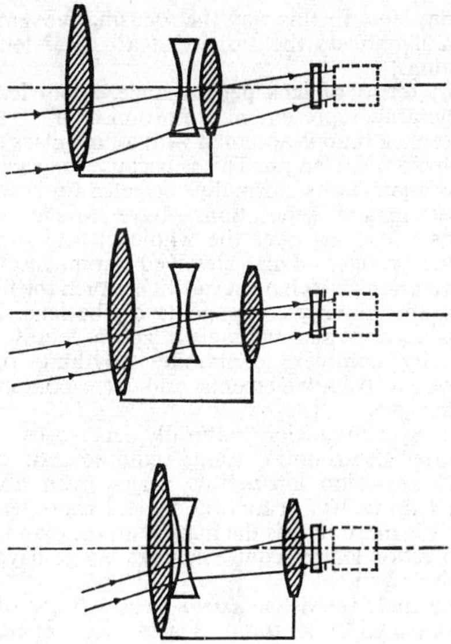

OPTICAL COMPENSATION. The movable parts of the zoom lens are linked together and shift forward (for the telephoto setting—top) or back (for the wide angle setting—bottom) in unison.

relative aperture. Again it is desirable that the lens should not need resetting of the aperture at different focal lengths to get the same exposure, and there are two ways of ensuring this.

The first is a mechanical linkage between the focal length adjustment and the iris diaphragm. As the focal length increases the iris opens, and as it decreases the iris closes so that the relative aperture remains the same.

The second method involves a different approach to the optical construction. Here the lens system essentially consists of two parts: a basic imaging lens and an afocal attachment. The lens diaphragm is built into the imaging lens, while the variable focus system only changes the magnification. Since the optical diameter of the iris and its distance from the image plane remain constant, the relative aperture also remains constant. This is the commonest zoom lens system today.

Afocal attachments of variable magnification also exist on their own and are available for some movie cameras to fit in front of the normal (fixed) lens.

Focusing. The straightforward way of focusing a lens, namely changing the lens to film distance, is not normally practical with a zoom lens. This is because the required lens extension varies with the focal length of the system. Hence the image could stay in focus only at one distance setting when the focal length is varied.

The focusing movement is therefore obtained by introducing an additional front component which precedes the afocal section of the whole zoom lens. In this way the focusing movement equally affects the system at all focal length settings.

In terms of lens performance, zoom lenses inevitably represent a compromise of greater tolerance than is accepted with a high class lens of fixed focal length. This is because the various lens aberrations, normally corrected for specific inter-element separations, have to be kept constantly low over the whole optical movement range. This involved compensating aberrations which must cancel between the fixed and the moving components of the lens. For this reason also high-class zoom lenses are usually complex, containing anything from eleven to twenty elements, and correspondingly expensive.

The complexity naturally increases the greater the zooming range required. For cine and television lenses this varies from about 3 : 1 up to 10 : 1; for still camera zoom lenses 4 : 1 is more or less the maximum ratio and for projection lenses requirements rarely go beyond 2 : 1 or 3 : 1.

Cine and Television Zoom Lenses. Low price (comparatively) zoom lenses for amateur narrow gauge cine cameras start with a zooming range of 3 : 1 and maximum apertures of *f* 1·8 or *f* 1·9. For an 8 mm. camera the range might thus cover from 9 to 27 mm. or 10 to 30 mm., going from a slight wide angle lens to a 2½× telephoto. Frequently the zooming range can be changed by a fixed afocal attachment, which might bring the lower limit of the focal length range down to about 6·5 or 7 mm., correspondingly reducing also the tele end of the zooming movement.

For 16 mm. cameras a corresponding zooming range might be 20 to 60 mm.

High class cine zoom lenses come with a focal length range of 1 : 4 to 1 : 6. For 8 mm. cameras this might cover focal lengths from 7·5 to 35 mm. or 8 to 48 mm., the latter extreme being nearly a 4× telephoto focal length and the longest ever likely to be required for amateur purposes. For 16 mm. cine and television purposes (with vidicon cameras) a typical zooming range might be from 17 mm. to 70 or 85 mm., in other words from a normal wide angle to a 3½× telephoto position. Lenses of extra long zooming range are however available, covering focal lengths from 12 to 120 mm. and are principally used in professional cinematography, newsreel work, etc. In addition there are also zoom lenses of specially long focal length, covering a telephoto range from 1½ to 2× up to 6×. Some types of this kind go up to very extreme focal lengths.

Zoom lenses for 35 mm. cameras may have a focal length range from 35 to 40 mm. up to 150 mm. or more. For image orthicon television cameras the focal length range usually starts at 60 or 70 mm. and may extend to 200 mm. or more. Again telephoto zoom lenses, mainly for newsreel work, are also available.

Viewfinder Systems. Early cine cameras with zoom lenses used to have (and less expensive ones still do) a separate zoom finder coupled with the zooming movement of the lens. This shows the variable field of view at different focal lengths.

More modern cameras usually have a through-the-lens reflex finder. There a beam splitter within the lens system diverts part of the light into a second optical system which is the viewfinder. It may incorporate a ground glass screen or a clear image plane for aerial focusing, or some sort of split-image rangefinder system. Sometimes part of the finder beam is further split with a proportion of the light diverted to a photo cell for exposure automation.

Alternative beam splitting systems utilize a reflex shutter or an oscillating mirror which also acts as the shutter. In all these cases the finder image corresponds to the view seen through the lens.

Professional cine and television cameras nearly always have reflex finders of this kind.

Automatic and Remote Control. On amateur cine cameras the zooming movement is usually controlled by a lever or ring around the circumference of the lens, or by a fore-and-aft lever connected with a sliding lens barrel through a gear. On some cameras a knob in the camera side changes the focal lengths.

On a number of amateur cameras, and most professional ones, the zooming movement is motorized. A small motor drives the focal length adjustment in either direction on pressing one or other of two zooming buttons on the camera. The motor may be a spring motor wound up at the same time as the main camera motor or it may be electric. Some zoom lenses can be used with an attachable supplementary power zooming motor, fed by separate miniature batteries.

Where the zooming motor is electrically driven, remote control is also possible. This may be linked with the other remote control functions of the camera. On television cameras remote control operation may extend also to the distance setting of the lens and the aperture control. Lenses with these features usually have up to three miniature motors mounted around the lens barrel, and geared to the appropriate setting ring. The various functions are then operated from a control panel at the end of an appropriate cable.

Remote control is useful wherever television or cine cameras have to be mounted in comparatively inaccessible places (for example traffic control, etc.) or where it would be dangerous for the operator to be next to the camera —as in the case of radiation risks.

Zoom Lenses for Still Cameras. Variable focus lenses are available for 35 mm. miniature cameras from a number of manufacturers. They are primarily intended for reflex cameras, where the finder permits a direct and con-

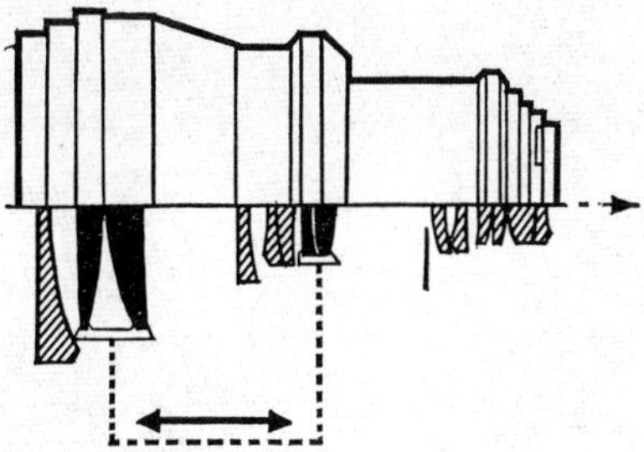

ZOOM LENS FOR MINIATURE REFLEX CAMERA. The front element serves for focusing. The two elements shown solid move in unison for changing the focal length.

tinuous check on both the subject area and the picture sharpness at the different focal length settings. Constancy of focus for zoom lenses of this kind is still regarded as desirable, though there are one or two types where the lens needs refocusing whenever the focal length is changed.

Where the zooming range includes the standard focal length, it usually extends from about 35 to 45 mm. up to 80 to 100 mm. Zoom lenses of this kind are only moderately popular, mainly because of their bulk as compared with that of a normal camera lens or even a couple of interchangeable lenses.

Tele-zoom lenses with a focal length range from about 80 up to 200 to 300 mm. are also available. Since a tele lens of 135 mm. or over is mostly used with the camera on a tripod, the extra bulk of the zoom lens is less of a handicap and the availability of several focal lengths in one lens is here a real advantage.

Projection Zoom Lenses. Variable focus lenses for still and cine projection—especially the latter—are becoming a standard feature of many better-class projectors. Here the availability of a variable focus lens permits the projector to be used with a range of screen sizes at a given projection distance, or adapted to a range of projection distances for a given screen size. It thus eliminates the need for moving the projector about to get a required image size.

With multi-size slide projectors (for example half-frame 18 × 24 mm. and 24 × 36 mm., or 24 × 36 mm. and $2\frac{1}{4} \times 2\frac{1}{4}$ ins. transparencies) a zoom lens provides a neat way of quickly changing over from one transparency size to the other while ensuring a constant screen image size.

The most usual zooming range for 8 mm. cine projectors is 15 to 25 mm., and for 16 mm. projectors about 40 to 60 mm. For still projectors the zooming range frequently extends from 80 or 90 mm. up to 110 or 120 mm.; projection zoom lenses with a 3 : 1 range however also exist.

See also: *Cine camera; Lens history; Projection practice; Tele-photography; Telephoto lens; Wide-angle lens.*
Book: *Optics*, by A. Cox (London).

ZOOPRAXISCOPE. An instrument devised by Eadweard Muybridge in 1880, based on the phenakisticope of Joseph Plateau and the stroboscope of Simon Stampfer, both invented in 1832. The zoopraxiscope was the first instrument in which drawings based on photographs were projected on a screen in quick succession to demonstrate, by means of persistence of vision, the phases of movements which had been instantaneously photographed. It was a forerunner of cinematography.